Endothelial Biomedicine

The endothelium, the cell layer that forms the inner lining of blood vessels, is a spatially distributed system that extends to all reaches of the human body. Today, clinical and basic research demonstrates that the endothelium plays a crucial role in mediating homeostasis and is involved in virtually every disease, either as a primary determinant of pathophysiology or as a victim of collateral damage. Indeed, the endothelium has remarkable, though largely untapped, diagnostic and therapeutic potential. This volume endeavors to bridge the bench-to-bedside gap in endothelial biomedicine, with the goal of advancing research and development and improving human health. The book is the first to systematically integrate knowledge about the endothelium from different organ-specific disciplines, including neurology, pulmonary, cardiology, gastroenterology, rheumatology, infectious disease, hematology-oncology, nephrology, and dermatology. Moreover, it is unique in its interdisciplinary approach, drawing on expertise from such diverse fields as evolutionary biology, comparative biology, molecular and cell biology, mathematical modeling and complexity theory, translational research, and clinical medicine.

Dr. William C. Aird received his medical degree from the University of Western Ontario in 1985. After completing his internal medicine and chief medical residency at the University of Toronto, he undertook a Hematology fellowship at the Brigham and Women's Hospital, Harvard Medical School. Dr. Aird received his postdoctoral training in the Department of Biology at the Massachusetts Institute of Technology. In 1996, he established an independent research program in the Division of Molecular Medicine at the Beth Israel Deaconess Medical Center. Dr. Aird is currently Associate Professor of Medicine at Harvard Medical School and Chief of the Division of Molecular and Vascular Medicine, at Beth Israel Deaconess Medical Center, Boston, Massachusetts.

Endothelial Biomedicine

Edited by

William C. Aird, M.D.
Chief, Division of Molecular and Vascular Medicine
Beth Israel Deaconess Medical Center
Associate Professor of Medicine
Harvard Medical School

CAMBRIDGE
UNIVERSITY PRESS

CAMBRIDGE UNIVERSITY PRESS
Cambridge, New York, Melbourne, Madrid, Cape Town, Singapore, São Paulo, Delhi

Cambridge University Press
32 Avenue of the Americas, New York, NY 10013-2473, USA

www.cambridge.org
Information on this title: www.cambridge.org/9780521853767

First published 2007

Printed in the United States of America

A catalog record for this publication is available from the British Library.

Library of Congress Cataloging in Publication Data

Endothelial biomedicine / edited by William C. Aird.
p.; cm.
Includes bibliographical references and index.
ISBN 978-0-521-85376-7 (hardcover)
1. Endothelium. I. Aird, William C.
[DNLM: 1. Endothelium – physiology. 2. Endothelial
Cells – physiology. 3. Endothelium – physiopathology. QS 532.5.E7 E561 2007] I. Title.

QP88.45.E5387 2007
611′.0187–dc22

2007007717

To my wife Renee, and children Xan, Ali, and Jamie

Contents

Metaphors

PART II: ENDOTHELIAL CELL AS INPUT-OUTPUT DEVICE

Input

Coupling

Output

Editor, Associate Editors, Artistic Consultant, and Contributors

EDITOR

William C. Aird, M.D.
Associate Professor of Medicine
Harvard Medical School
Chief, Division of Molecular and Vascular Medicine
Center for Vascular Biology Research
Beth Israel Deaconess Medical Center
Boston, Massachusetts

ASSOCIATE EDITORS

Helmut G. Augustin, D.V.M., Ph.D.
Department of Vascular Oncology and Metastasis
German Cancer Research Center Heidelberg (DKFZ) and Center for Biomedicine and Medical Technology Mannheim (CBTM), Germany

Ary L. Goldberger, M.D.
Professor of Medicine, Harvard Medical School
Director, Margret and H.A. Rey Institute for Nonlinear Dynamics in Medicine and Associate Director
Division of Interdisciplinary Medicine and Biotechnology
Beth Israel Deaconess Medical Center
Boston, Massachusetts

Mansoor Husain, M.D.
Director, Heart & Stroke Richard Lewar Centre of Excellence in Cardiovascular Research
Associate Professor of Medicine
University of Toronto, Toronto
Ontario, Canada

Manfred D. Laubichler, Ph.D.
Professor of Theoretical Biology and History of Biology
Affiliated Professor of Philosophy School of Life Sciences and Centers for Biology and Society and Social Dynamics and Complexity
Arizona State University Tempe
Tempe, Arizona

Jane Maienschein, Ph.D.
Regents' Professor, President's Professor, and Parents Association Professor
Director, Center for Biology and Society
Arizona State University Tempe
Tempe, Arizona

Jan E. Schnitzer, M.D.
Scientific Director
Professor of Cellular & Molecular Biology
Director of Vascular Biology & Angiogenesis Program
Sidney Kimmel Cancer Center
San Diego, California

Michael Simons, M.D.
A.G. Huber Professor of Medicine and Pharmacology and Toxicology
Chief, Section of Cardiology
Director, Angiogenesis Research Center
Dartmouth Medical School, Dartmouth Hitchcock Medical Center
Hanover, New Hampshire

ARTISTIC CONSULTANT

Steven Moskowitz
Advanced Medical Graphics
Boston, Massachusetts

CONTRIBUTORS

Md. Ruhul Abid, M.D., Ph.D.
Assistant Professor of Medicine
Harvard Medical School
Division of Molecular and Vascular Medicine and Center for Vascular Biology Research
Beth Israel Deaconess Medical Center
Boston, Massachusetts

José A. Adams, M.D.
Director, Division of Neonatology
Mount Sinai Medical Center
Miami Beach, Florida
Tatiana A. Afanasieva, Ph.D.
Institute of Physiology
Technical University
Dresden, Germany
Eugene R. Ahn, M.D.
Fellow, Wallace H. Coulter Laboratory
Division of Hematology/Oncology
Department of Medicine
University of Miami School of Medicine
Miami, Florida
Yeon S. Ahn, M.D.
Professor of Medicine
University of Miami School of Medicine
Wallace H. Coulter Laboratory
Division of Hematology/Oncology
Department of Medicine
Miami, Florida
William C. Aird, M.D.
Associate Professor of Medicine
Harvard Medical School
Chief, Division of Molecular and Vascular Medicine
Center for Vascular Biology Research
Beth Israel Deaconess Medical Center
Boston, Massachusetts
Eivind Almaas, Ph.D.
Center for Network Research and Department of Physics
University of Notre Dame
Notre Dame, Indiana
John A. Ambrose, M.D.
Professor of Medicine
University of California at San Francisco
Chief of Cardiology
University of California at San Francisco
Fresno, California
Gary An, M.D.
Assistant Professor of Surgery
Division of Trauma/Critical Care
Northwestern University Feinberg School of Medicine
Chicago, Illinois
Wadih Arap, M.D.
M.D. Anderson Cancer Center
University of Texas
Houston, Texas
Stephen R. Archacki, M.D.
Center for Cardiovascular Genetics
Department of Molecular Cardiology
Lerner Research Institute
Cleveland Clinic and Cleveland State University
Cleveland, Ohio
Chaitanya A. Athale, Ph.D.
Complex Biosystems Modeling Laboratory
Harvard-MIT (HST) Athinoula A. Martinos Center for Biomedical Imaging
Massachusetts General Hospital-East
Charlestown, Massachusetts
Helmut G. Augustin, D.V.M., Ph.D.
Department of Vascular Oncology and Metastasis
German Cancer Research Center Heidelberg (DKFZ) and Center for Biomedicine and Medical Technology Mannheim (CBTM), Germany
Sandra Austin, Ph.D.
Department of Molecular Medicine
Conway Institute, University College Dublin, Belfield
Dublin, Ireland
Barbara J. Ballermann, M.D.
Professor of Medicine
CRC Chair in Endothelial Cell Biology
Director, Division of Nephrology and Immunology
University of Alberta, Edmonton
Alberta, Canada
Péter Balogh M.D., Ph.D.
Department of Immunology and Biotechnology
Faculty of Medicine, University of Pécs
Pécs, Hungary
Albert-Laszlo Barabasi, Ph.D.
Emil T. Hofman Professor of Physics
Department of Physics, Center for Network Research and Department of Physics
University of Notre Dame
Notre Dame, Indiana
Rajat S. Barua, M.D., Ph.D.
Cardiology Fellow
Division of Cardiology
Department of Medicine
University of California at San Francisco
Fresno, California
Howard D. Beall, Ph.D.
Department of Biomedical and Pharmaceutical Sciences
The University of Montana
Missoula, Montana
Bryan Belikoff, Ph.D.
New England Inflammation and Tissue Protection Institute
Northeastern University
Boston, Massachusetts
Orina Belton, Ph.D.
Department of Molecular Medicine, Conway Institute
University College Dublin
Belfield, Dublin, Ireland
Robert Benezra, Ph.D.
Member, Memorial Sloan-Kettering Cancer Center
New York, New York

Laura Benjamin, Ph.D.
Associate Professor of Pathology
Harvard Medical School
Department of Pathology
Beth Israel Deaconess Medical Center
Boston, Massachusetts

Resham Bhattacharya, Ph.D.
Department of Biochemistry and Molecular Biology
Mayo Clinic College of Medicine
Mayo Clinic Cancer Center
Rochester, Minnesota

Diane R. Bielenberg, Ph.D.
Assistant Professor, Harvard Medical School
Vascular Biology Program
Department of Surgical Research, Children's Hospital
Boston, Massachusetts

Angelika Bierhaus, Ph.D.
University of Heidelberg
Department of Medicine I and Clinical Chemistry
University of Heidelberg
Heidelberg, Germany

Kaiser M. Bijli, Ph.D.
Research Associate, Division of Neonatology
Department of Pediatrics, School of Medicine and Dentistry
University of Rochester
Rochester, New York

David G. Binion, M.D.
Associate Professor of Medicine
Medical College of Wisconsin
Director, IBD Center, Froedtert Memorial Lutheran Hospital
Milwaukee, Wisconsin

Miri Blank, M.D.
Research Unit of Autoimmune Disease and Department of Medicine B, Sheba Medical Center, Tel-Hashomer and the Sackler Faculty for Medicine
Tel-Aviv University, Israel

Andrew D. Blann, Ph.D.
Hemostasis Thrombosis and Vascular Biology Unit
University Department of Medicine, City Hospital
Birmingham, United Kingdom

Natalia V. Bogatcheva, Ph.D.
Vascular Biology Center
Medical College of Georgia
Augusta, Georgia

Roberto Bonasio, Ph.D.
The Immune Disease Institute
Department of Pathology, Harvard Medical School
Boston, Massachusetts

Rene M. Botnar, Ph.D.
Professor of Biomedical Imaging
Department of Nuclear Medicine
Technische Universität München
Munich, Germany

Jonathan D. Brown, M.D.
Cardiovascular Division, Brigham and Women's Hospital
Harvard Medical School
Boston, Massachusetts

Alf O. Brubakk, M.D., Ph.D.
Professor of Applied Physiology
Department of Circulation and Medical Imaging
Norwegian University of Science and Technology
Trondheim, Norway

Jon A. Buras, M.D.
Research Associate Professor
New England Inflammation and Tissue Protection Institute
Northeastern University
Boston, Massachusetts

Warren W. Burggren, Ph.D.
Dean, College of Arts and Sciences, Professor, Biological Sciences
University of North Texas
University of North Texas
Denton, Texas

Jane C. Burns, M.D.
Professor and Chief, Division of Allergy, Immunology, and Rheumatology
Department of Pediatrics
Rady Children's Hospital San Diego and UCSD School of Medicine
La Jolla, California

Christopher V. Carman, Ph.D.
Assistant Professor of Medicine
Harvard Medical School
Division of Molecular and Vascular Medicine and Center for Vascular Biology Research
Beth Israel Deaconess Medical Center
Boston, Massachusetts

Rob Cartotto, M.D. FRCS(C)
Ross Tilley Burn Centre, Sunnybrook Health Sciences Centre
Department of Surgery
University of Toronto
Toronto, Ontario, Canada.

Lucy A. Carver, Ph.D.
Vascular Biology & Angiogenesis Program
Sidney Kimmel Cancer Center
San Diego, California

Anna Cattelino, Ph.D.
IFOM-Fondazione Instituto FIRC di Oncologia Molecolare
Milano, Italy

Shampa Chatterjee, Ph.D.
Research Assistant Professor
Physiology
University of Pennsylvania
Philadelphia, Pennsylvania

Irshad H. Chaudry, Ph.D.
Professor of Surgery, Microbiology, Physiology & Biophysics
Vice Chairman, Department of Surgery
Director, Center for Surgical Research
University of Alabama at Birmingham
Birmingham, Alabama

Qing Chen, M.D., Ph.D.
Department of Immunology
Roswell Park Cancer Institute
Buffalo, New York

Pavan K. Cheruvu
Medical Student, Harvard Medical School
Beth Israel Deaconess Medical Center
Boston, Massachusetts

Augustine M.K. Choi, M.D.
Division of Pulmonary
Allergy, and Critical Care Medicine
The University of Pittsburgh School of Medicine
Pittsburgh, Pennsylvania

Sameer S. Chopra, Ph.D.
Departments of Medicine, Pharmacology, and Cell & Developmental Biology
Vanderbilt University School of Medicine
Nashville, Tennessee

Aldo Ciau-Uitz, Ph.D., M.R.C.
Molecular Haematology Unit
The Weatherall Institute of Molecular Medicine
University of Oxford, John Radcliffe Hospital, Headington
Oxford, United Kingdom

Kristen Clancy
Department of Immunology
Roswell Park Cancer Institute
Buffalo, New York

Gilles Clermont, M.D., M.Sc.
Department of Critical Care Medicine
University of Pittsburgh Center for Inflammation and Regenerative Modeling
University of Pittsburgh
Pittsburgh, Pennsylvania

J. Douglas Coffin, Ph.D.
Associate Professor for Molecular Genetics
The University of Montana
Department of Biomedical & Pharmaceutical Sciences
Missoula, Montana

Robert Colman, M.D.
Sol Sherry Professor of Medicine
Temple University School of Medicine
Philadelphia, Pennsylvania

Valéry Combes, Ph.D.
Department of Pathology
Faculty of Medicine
The University of Sydney
Camperdown, Australia

Edward M. Conway, M.D., M.B.A., Ph.D.
Group Leader, Professor of Medicine
VIB Department of Transgene Technology and Gene Therapy, K.U. Leven
Leuven, Belgium

Judith Coppinger, Ph.D.
Department of Molecular Medicine
Conway Institute, University College Dublin
Dublin, Ireland

Loren Cordain, Ph.D.
Department of Health and Exercise Science
Colorado State University
Fort Collins, Colorado

Nicholas L.M. Cruden, B.Sc. (Hons), M.B., Ch.B., M.R.C.P.
Specialist Registrar in Cardiology University of Edinburgh
Edinburgh, United Kingdom

Xavier Cullere, Ph.D.
Department of Pathology
Center for Excellence in Vascular Biology
Brigham and Women's Hospital
Boston, Massachusetts

David T. Curiel, M.D., Ph.D.
Division of Human Gene Therapy
Departments of Medicine, Obstetrics and Gynecology, Pathology, Surgery, and the Gene Therapy Center
University of Alabama at Birmingham
Birmingham, Alabama

Myron I. Cybulsky, M.D.
Professor of Laboratory Medicine and Pathobiology, University of Toronto
Senior Scientist
Toronto General Research Institute, University Health Network, Toronto
Ontario, Canada

Björn Dahlback, M.D., Ph.D.
Professor of Blood Coagulation Research
Lund University Department of Laboratory Medicine
Clinical Chemistry, Wallenberg Laboratory, University Hospital
Malmö, Sweden

Peter F. Davies, Ph.D.
Robinette Foundation Professor of Cardiovascular Medicine, Pathology and Laboratory Medicine, and Bioengineering
Director, Institute for Medicine & Engineering, University of Pennsylvania
Philadelphia, Pennsylvania

Silvia Deaglio, M.D.
Visiting Assistant Professor
Harvard Medical School
Beth Israel Deaconess Medical Center
Boston, Massachusetts

Thomas S. Deisboeck, M.D.
Assistant Professor of Radiology
Harvard Medical School
Director, Complex Biosystems Modeling Laboratory
Harvard-MIT (HST) Athinoula A. Martinos Center for Biomedical Imaging
Massachusetts General Hospital-East
Charlestown, Massachusetts

Laurie D. DeLeve, M.D., Ph.D.
Professor of Medicine
Division of Gastrointestinal and Liver Diseases
University of Southern California Keck School of Medicine
Los Angeles, California

H. William Detrich III, Ph.D.
Professor of Biochemistry and Marine Biology
Department of Biology, Northeastern University
Boston, Massachusetts

Thomas G. Diacovo, M.D.
Department of Pediatrics and Pathology
Columbia University, New York
New York

Daniel Dumont, Ph.D.
Professor, Department of Medical Biophysics
University of Toronto
Director, Molecular and Cellular Biology
Sunnybrook Research Institute, Toronto
Ontario, Canada

Winston Dunn, M.D.
GI Research Unit
Division of Gastroenterology and Hepatology, Mayo Clinic
Rochester, Minnesota

Ann M. Dvorak, M.D.
Professor of Pathology, Harvard Medical School
Department of Pathology, Beth Israel Deaconess Medical Center
Boston, Massachusetts

Harold F. Dvorak, M.D.
Professor of Pathology, Harvard Medical School
Department of Pathology, Beth Israel Deaconess Medical Center
Boston, Massachusetts

S. Boyd Eaton, M.D.
Departments of Anthropology and Radiology
Emory University Atlanta, Georgia

Jay M. Edelberg, M.D., Ph.D.
GlaxoSmithKline, King of Prussia, Pennsylvania
Department of Medicine, Weill Medical College of Cornell University, New York
New York

Olav S. Eftedal, M.Sc. (Eng.), M.D., Ph.D.
Department of Circulation and Medical Imaging
Norwegian University of Science and Technology
Trondheim, Norway

Arye Elfenbein, Ph.D.
Angiogenesis Research Center
Dartmouth Medical School, Dartmouth Hitchcock Medical Center
Hanover, New Hampshire

Sharon S. Evans, Ph.D.
Professor of Oncology
Department of Immunology, Roswell Park Cancer Institute
Buffalo, New York

Heinz Feldmann, Ph.D.
Special Pathogens Program
National Microbiology Laboratory
Public Health Agency of Canada
Department of Medical Microbiology and Infectious Diseases
University of Manitoba, Winnipeg
Manitoba, Canada

Claire Fernandez, Ph.D.
M.R.C. Molecular Haematology Unit
The Weatherall Institute of Molecular Medicine, University of Oxford
John Radcliffe Hospital, Headington
Oxford, United Kingdom

Aron B. Fisher, M.D.
Professor, Physiology and Medicine, Director, Institute for Environmental Medicine
University of Pennsylvania Medical Center
University of Pennsylvania
Philadelphia, Pennsylvania

Danilo Fliser, M.D.
Assistant Professor of Medicine
Department of Internal Medicine
Division of Nephrology
Medical School Hannover
Hannover, Germany

Kimberly E. Foreman, Ph.D.
Associate Professor
Department of Pathology
Breast Cancer Research Program
Cardinal Bernardin Cancer Center, Loyola University Chicago
Maywood, Illinois

William Foster, Ph.D.
Hemostasis Thrombosis and Vascular Biology Unit
University Department of Medicine
City Hospital, Birmingham
United Kingdom

Jay W. Fox, Ph.D.
Professor and Assistant Dean of Research Support, Microbiology
University of Virginia School of Medicine
Charlottesville, Virginia

Roy Freeman, M.D.
Professor of Neurology
Harvard Medical School
Director, Center for Autonomic and Peripheral Nerve Disorders
Beth Israel Deaconess Medical Center
Boston, Massachusetts

Yasuhiro Funahashi, Ph.D.
Department of Obstetrics and Gynecology
Institute of Cancer Genetics
Columbia University Medical Center, New York
New York

Daniel Gale
Harvard University
Undergraduate Student, Beth Israel Deaconess Medical Center
Harvard Medical School
Boston, Massachusetts

Joe G.N. Garcia M.D.
Lowell T. Coggeshall Professor of Medicine
Chairman, Department of Medicine, Pritzker School of Medicine
University of Chicago
Chicago, Illinois

Jean-Marc Gauguet, M.D., Ph.D.
The Immune Disease Institute
Department of Pathology
Harvard Medical School
Boston, Massachusetts

Jean-Philippe Girard, Ph.D.
Head, Laboratory of Vascular Biology
Deputy-Director, Institute of Pharmacology and Structural Biology, CNRS UMR 5089
Toulouse, France

Mark T. Gladwin, M.D.
Chief, Vascular Medicine Branch, NHLBI
Critical Care Medicine Department, Clinical Center
National Institutes of Health, Vascular Medicine Branch
National Heart Lung and Blood Institute
Bethesda, Maryland

Ary L. Goldberger, M.D.
Professor of Medicine, Harvard Medical School
Director, Margret and H.A. Rey Institute for Nonlinear Dynamics in Medicine and Associate Director Division of Interdisciplinary Medicine and Biotechnology
Beth Israel Deaconess Medical Center
Boston, Massachusetts

James A. Gordon, M.D., M.P.A.
Director, Gilbert Program in Medical Simulation
Harvard Medical School
Assistant Professor of Medicine
Department of Emergency Medicine
Massachusetts General Hospital
Boston, Massachusetts

Tommaso Gori, M.D., Ph.D.
Department of Internal
Cardiovascular and Geriatric Medicine
University of Siena
Siena, Italy

Georges E.R. Grau, M.D.
Chair of Vascular Immunology
Department of Pathology
Faculty of Medicine
The University of Sydney
Camperdown, Australia

Andreas Greinacher, M.D.
Professor and Head, Department of Transfusion Medicine
Institute of Immunology and Transfusion Medicine
Ernst-Moritz-Arndt-University-Greifswald
Greifswald, Germany

Marion Gröger, Ph.D.
Department of Dermatology
Division of General Dermatology Medical University Vienna
Vienna, Austria

Peter L. Gross, M.D.
St. Michael's Hospital, University of Toronto, Toronto
Ontario, Canada

Rajiv Gulati M.B., Ch.B.
Assistant Professor of Medicine
Division of Cardiovascular Diseases, Mayo Clinic
Rochester, Minnesota

Kalpna Gupta, Ph.D.
Assistant Professor
Department of Medicine
Division of Hematology
Oncology and Transplantation
University of Minnesota Medical School
Minneapolis, Minnesota

David Haig, Ph.D.
Professor of Biology
Department of Organismic and Evolutionary Biology
Harvard University, Cambridge, Massachusetts

Charles A. Hales, M.D.
Professor of Medicine, Harvard Medical School
Pulmonary and Critical Care Unit, Department of Medicine
Massachusetts General Hospital
Boston, Massachusetts

Mitchell L. Halperin, M.D.
Emeritus Professor of Medicine
University of Toronto
Division of Nephrology, St. Michael's Hospital, Toronto
Ontario, Canada

Hans-Peter Hammes, M.D., Ph.D.
Department of Medicine V, University of Mannheim
Germany

Shivalika Handa, Ph.D.
Candidate, Department of Laboratory Medicine & Pathobiology
University of Toronto, Toronto
Ontario, Canada.

Per M. Haram, M.D., Ph.D.
Department of Circulation and Medical Imaging
Norwegian University of Science and Technology
Trondheim, Norway

Alan R. Hargens, Ph.D.
Professor of Orthopaedic Surgery
University of California, San Diego
UCSD Medical Center
San Diego, California

Elizabeth O. Harrington, Ph.D.
Pulmonary Vascular Research Laboratory, Providence VA Medical Center
Associate Professor of Medicine
The Warren Alpert Medical School of Brown University
Providence, Rhode Island

Ossama A. Hatoum, M.D.
Froedtert Memorial Lutheran Hospital
Medical College of Wisconsin
Milwaukee, Wisconsin

Gary J. Hausman
Research Physiologist, USDA – ARS and University of Georgia
Athens, Georgia

Robert P. Hebbel, M.D.
Regents Professor, George Clark Professor
Vice-Chairman for Research
Department of Medicine
Director, Vascular Biology Center, University of Minnesota
Minneapolis, Minnesota

Wayne J.G. Hellstrom, M.D.
Professor of Urology
Chief, Section of Andrology Tulane University Health Sciences Center
Department of Urology
New Orleans, Louisiana

Erik Henke, Ph.D.
Memorial Sloan-Kettering Cancer Center, New York
New York

Timothy Hla, Ph.D.
Professor of Cell Biology
Director, Center for Vascular Biology, University of Connecticut Health Center
Farmington, Connecticut

Laszlo M. Hoesel, M.D.
University of Michigan Health Systems
Department of Pathology
Ann Arbor, Michigan

John B. Holcomb, M.D., F.A.C.S., C.O.L., M.C.
U.S. Army, Clinical Professor of Surgery, University of Texas Health Science Center, San Antonio
Texas Trauma Consultant for The Surgeon General, Commander
Commander, U.S. Army Institute of Surgical Research
Fort Sam Houston, Texas

Anthony Hollenberg, M.D.
Associate Professor of Medicine
Harvard Medical School
Chief, Thyroid Unit, Beth Israel Deaconess Medical Center
Boston, Massachusetts

Young-Kwon Hong, Ph.D.
Assistant Professor, Department of Surgery and Biochemistry & Molecular Biology
Norris Comprehensive Cancer Center
University of Southern California
Los Angeles, California

Lawrence Horstman, BS
Research Associate, Wallace H. Coulter Laboratory
Division of Hematology/Oncology
Department of Medicine
University of Miami School of Medicine
Miami, Florida

James B. Hoying, Ph.D.
Associate Professor, Department of Surgery
University of Louisville
Chief, Division of Cardiovascular Therapeutics
Cardiovascular Innovation Institute
Louisville, Kentucky

Grace Huang, M.D.
Instructor in Medicine, Harvard Medical School
Director, Office of Educational Technology, Shapiro Institute for Education and Research
Beth Israel Deaconess Medical Center
Boston, Massachusetts

Sui Huang, M.D., Ph.D.
Assistant Professor of Surgery, Vascular Biology Program
Department of Surgery
Children's Hospital, Harvard Medical School, Boston and Harvard Stem Cell Institute
Cambridge, Massachusetts

Deborah A. Hughes, Ph.D.
Department of Biochemistry and Molecular Biology
Mayo Clinic College of Medicine
Mayo Clinic Cancer Center
Rochester, Minnesota

Mansoor Husain, M.D.
Director, Heart & Stroke Richard Lewar Centre of Excellence in Cardiovascular Research
Associate Professor of Medicine
University of Toronto, Toronto
Ontario, Canada

Richard O. Hynes, Ph.D.
Center for Cancer Research and Howard Hughes Medical Institute
Massachusetts Institute of Technology
Cambridge, Massachusetts

José M. Icardo
Department of Anatomy and Cell Biology
University of Cantabria
Santander, Spain

Donald E. Ingber, M.D., Ph.D.
Judah Folkman Professor of Vascular Biology
Department of Pathology, Harvard Medical School
Vascular Biology Program, Departments of Surgery and Pathology
Children's Hospital, Boston, Massachusetts

David A. Ingram, M.D.
Department of Pediatrics, Herman B. Wells Center for Pediatric Research
Indiana University School of Medicine
Indianapolis, Indiana

Kaikobad Irani, M.D.
Associate Professor of Medicine, Cardiovascular Institute University of Pittsburgh Medical Center
Pittsburgh, Pennsylvania

Mukesh K. Jain, M.D.
Case Western Reserve University
Cleveland, Ohio

Joaquin J. Jimenez, M.D.
Assistant Professor of Medicine, Wallace H. Coulter Laboratory Division of Hematology/Oncology
Department of Medicine
University of Miami School of Medicine
Miami, Florida

Peter Lloyd Jones, Ph.D.
Associate Professor of Pathology & Laboratory Medicine
Director, Penn-CMREF Center for Pulmonary Hypertension Research
Institute for Medicine & Engineering
University of Pennsylvania, Philadelphia

Klaus D. Jürgens, Ph.D.
Professor of Physiology, Zentrum Physiologie, Medizinische Hochschule
Hannover, Germany

Wenche Jy, Ph.D.
Assistant Professor of Medicine
Wallace H. Coulter Laboratory Division of Hematology/Oncology
Department of Medicine, University of Miami School of Medicine
Miami, Florida

Darren Kafka, Ph.D.
Department of Surgery and Norris Comprehensive Cancer Center
Keck School of Medicine, University of Southern California
Los Angeles, California

Kamel S. Kamel, M.D.
Professor of Medicine, University of Toronto, St. Michael's Hospital
Toronto, Ontario, Canada

Kenneth L. Kaplan
Associate Director, Collaborative Initiatives at MIT
Associate Director, Urban Design Lab, Earth Institute
Columbia University

Kai Kappert, M.D.
Cancer Centrum Karolinska
Department of Oncology-Pathology, Karolinska Institutet
Stockholm, Sweden

S. Ananth Karumanchi, M.D.
Associate Professor of Medicine
Harvard Medical School, Beth Israel Deaconess Medical Center
Boston, Massachusetts

Jaswinder Kaur, Ph.D.
Candidate, Department of Physiology and Biophysics
University of Calgary, Calgary
Alberta, Canada

Muammer Kendirci, M.D.
Department of Urology, Sisli Etfal Training and Research Hospital
Istanbul, Turkey

Levon M. Khachigian, Ph.D., D.Sc.
Professor and Head Transcription and Gene Targeting Laboratory Centre for Vascular Research
Prince of Wales Hospital
University of New South Wales
Sydney, Australia

Jan T. Kielstein
Assistant Professor of Medicine
Stanford University Medical School
Falk Cardiovascular Research Center, Stanford, California; and Department of Internal Medicine, Division of Nephrology
Medical School Hannover, Germany

Hong Pyo Kim, Ph.D.
Division of Pulmonary, Allergy, and Critical Care Medicine
The University of Pittsburgh School of Medicine
Pittsburgh, Pennsylvania

Rolf Kinne, M.D., Ph.D.
Max Planck Institut für Molekulare Physiologie
Dortmund, Germany

Jan Kitajewski, Ph.D.
Departments of Pathology and Obstetrics and Gynecology
Institute of Cancer Genetics, Columbia University Medical Center, New York
New York

Michael Klagsbrun, Ph.D.
Patricia K. Donahoe Professor of Surgery
Department of Surgery, Harvard Medical School
Vascular Biology Program, Children's Hospital
Boston, Massachusetts

Karolina Kolodziejska, Ph.D.
Candidate, Department of Laboratory Medicine & Pathobiology
University of Toronto, Toronto
Ontario, Canada

Peter Koopman, Ph.D.
Institute for Molecular Bioscience
The University of Queensland
Brisbane, Australia

Paul Kubes, Ph.D.
Director, Institute of Infection, Immunity and Inflammation
Department of Physiology and Biophysics
University of Calgary, Calgary
Alberta, Canada

Frank Kuhnert, Ph.D.
Postdoctoral Fellow, Department of Medicine
Division of Hematology
Stanford University School of Medicine
Stanford, California

Rainer Kunstfeld, M.D.
Department of Dermatology
Division of General Dermatology, Medical University Vienna
Vienna, Austria

Peter Kurschat, Ph.D.
Harvard Medical School
Vascular Biology Program, Department of Surgical Research, Children's Hospital
Boston, Massachusetts

Michael J.B. Kutryk, M.D., Ph.D.
The Terrence Donnelly Vascular Biology Laboratories
St. Michael's Hospital, and The McLaughlin Centre for Molecular Medicine and the Department of Medicine
University of Toronto, Toronto
Ontario, Canada

Eckhard Lammert, Ph.D.
Max Planck Institute of Molecular Cell Biology and Genetics
Dresden, Germany

Manfred D. Laubichler, Ph.D.
Professor of Theoretical Biology and History of Biology
Affiliated Professor of Philosophy School of Life Sciences and Centers for Biology and Society and Social Dynamics and Complexity Arizona State University Tempe
Tempe, Arizona

Jeffrey Laurence, M.D.
Professor of Medicine, Weill Medical College of Cornell University
Attending Physician, New York Presbyterian Hospital
New York, New York

Jack Lawler, Ph.D.
Professor of Pathology, Harvard Medical School
Department of Pathology, Beth Israel Deaconess Medical Center
Boston, Massachusetts

Michelle Letarte, Ph.D.
Molecular Structure and Function, Hospital for Sick Children, University of Toronto, Toronto
Ontario, Canada

Marcel Levi, M.D., Ph.D.
Professor of Medicine, University of Amsterdam
Chairman, Department of Medicine
Academic Medical Center, Amsterdam
The Netherlands

Dean Y. Li, M.D.
Department of Oncological Sciences
Program in Human Molecular Biology and Genetics
Division of Cardiology, School of Medicine, University of Utah
Salt Lake City, Utah

Jian Li, M.D., Ph.D.
Assistant Professor of Medicine
Harvard Medical School
Division of Cardiology, Beth Israel Deaconess Medical Center
Boston, Massachusetts

James K. Liao M.D.
Associate Professor of Medicine
Harvard Medical School
Director, Vascular Medicine Research, Brigham and Women's Hospital
Boston, Massachusetts

Andrew H. Lichtman, M.D., Ph.D.
Associate Professor of Pathology, Harvard Medical School
Department of Pathology, Brigham and Women's Hospital
Boston, Massachusetts

Stefan Liebner, Ph.D.
Institute of Neurology, University Frankfurt
Frankfurt, Germany

Zhiyong Lin, Ph.D.
Case Western Reserve University
Cleveland, Ohio

Gregory Y.H. Lip, M.D.
Consultant Cardiologist and Professor of Cardiovascular Medicine
Director – Haemostasis Thrombosis & Vascular Biology Unit, University Department of Medicine
City Hospital, Birmingham, United Kingdom

Robert Loewe, M.D.
Department of Dermatology
Division of General Dermatology Medical University Vienna
Vienna, Austria
Frank W. LoGerfo, M.D.
William V. McDermott Professor of Surgery
Harvard Medical School
Chief, Division of Vascular Surgery
Beth Israel Deaconess Medical Center
Boston, Massachusetts
Jin Ning Lou, Ph.D.
Institute of Clinical Medical Sciences
China-Japan Friendship Hospital
Beijing, China
Charles J. Lowenstein, M.D.
Professor, Departments of Medicine and Pathology
The Johns Hopkins University School of Medicine
Baltimore, Maryland
Qing Lu, Ph.D.
Pulmonary Vascular Research Laboratory
Providence VA Medical Center
Assistant Professor of Medicine
The Warren Alpert Medical School of Brown University
Providence, Rhode Island
F. William Luscinskas, Ph.D.
Associate Professor of Medicine
Harvard Medical School
Brigham and Women's Hospital
Boston, Massachusetts
Nigel Mackman, Ph.D.
John Parker Professor of Medicine
University of North Carolina at Chapel Hill
Chapel Hill, North Carolina.
Patricia B. Maguire, Ph.D.
Molecular Medicine Group, Conway Institute
School of Medicine and Medical Science
University College Dublin
Dublin, Ireland
Jane Maienschein, Ph.D.
Regents' Professor, President's Professor, and Parents Association Professor
Director, Center for Biology and Society, Arizona State University Tempe
Tempe, Arizona
Mark W. Majesky, Ph.D.
Professor, Departments of Medicine & Genetics, Associate Director
Carolina Cardiovascular Biology Center
University of North Carolina at Chapel Hill
Chapel Hill, North Carolina
Warren J. Manning, M.D.
Professor of Medicine and Radiology
Harvard Medical School
Section Chief, Non-invasive Cardiac Imaging, Beth Israel Deaconess Medical Center
Boston, Massachusetts
Mark Martinez, M.D.
Instructor in Medicine, University of Utah School of Medicine
Salt Lake City, Utah
Alan E. Mast M.D., Ph.D.
Investigator, Blood Research Institute
The Blood Center of Wisconsin
Milwaukee, Wisconsin
Tanya Mayadas, Ph.D.
Associate Professor of Pathology, Harvard Medical School
Department of Pathology
Center for Excellence in Vascular Biology, Brigham and Women's Hospital
Boston, Massachusetts
Joseph H. McCarty, Ph.D.
Department of Cancer Biology, Unit 173,
University of Texas – M. D. Anderson Cancer Center, 1515 Holcombe Boulevard,
Houston, Texas
Rodger P. McEver, M.D.
Eli Lilly Distinguished Chair in Biomedical Research, Member and Head
Cardiovascular Biology Research Program
Oklahoma Medical Research Foundation
Oklahoma City, Oklahoma
John H. McVey, Ph.D.
Group Leader Haemostasis and Thrombosis, Reader in Haemostasis and Thrombosis
MRC Clinical Sciences Centre, Imperial College London, Hammersmith Hospital Campus
London, United Kingdom
Janice V. Meck, Ph.D.
NASA/Johnson Space Center
Houston, Texas
George A. Mensah, M.D., F.A.C.P., F.A.C.C., F.C.P. (SA) Hon
Distinguished Scientist, Heart Disease and Stroke Prevention, National Center for Chronic Disease Prevention and Health Promotion, Centers for Disease Control and Prevention
Atlanta, Georgia
Christine N. Metz, Ph.D.
Associate Investigator, Director, Laboratory of Medicinal Biochemistry
The Feinstein Institute for Medical Research, North Shore-LIJ Health System
Manhasset, New York
Valerie C. Midgley, Ph.D.
Center for Vascular Research, Prince of Wales Hospital
University of New South Wales
Sydney, Australia

Takashi Mikawa, Ph.D.
Department of Cell and Developmental Biology
Cornell University Medical College
New York, New York
Ronald W. Millard, Ph.D.
Professor of Pharmacology & Cell Biophysics
University of Cincinnati College of Medicine
Cincinnati, Ohio
David Milstone, Ph.D.
Department of Pathology, Brigham and Women's Hospital, Harvard Medical School
Boston, Massachusetts
Takashi Minami, Ph.D.
Associate Professor, University of Tokyo
Tokyo, Japan
Jamie Mitchell, M.D.
Beth Israel Deaconess Medical Center
Harvard Medical School
Boston, Massachusetts
Aigul Moldobaeva, Ph.D.
Research Associate, Division of Pulmonary and Critical Care Medicine
Johns Hopkins Asthma and Allergy Center
Johns Hopkins University
Baltimore, Maryland
Bruce Molitoris, M.D.
Division of Nephrology
Indiana Center for Biological Microscopy
Indiana University School of Medicine
Indianapolis, Indiana
Thomas S. Monahan, M.D.
Division of Vascular Surgery
Beth Israel Deaconess Medical Center
Boston, Massachusetts
Craig N. Morrell, Ph.D.
Departments of Pathology and Comparative Medicine
The Johns Hopkins University School of Medicine
Baltimore, Maryland
Steven Moskowitz
Advanced Medical Graphics
Boston, Massachusetts
Debabrata Mukhopadhyay, Ph.D.
Professor, Department of Biochemistry and Molecular Biology
Mayo Clinic College of Medicine
Mayo Clinic Cancer Center
Rochester, Minnesota
Masahiro Murakami, M.D., Ph.D.
Angiogenesis Research Center & Section of Cardiology
Dartmouth Medical School
Dartmouth Hitchcock Medical Center
Hanover, New Hampshire
Silvia Muro, Ph.D.
Research Assistant Professor of Pharmacology
Institute for Translational Medicine and Therapeutics
Institute for Environmental Medicine
University of Pennsylvania Medical School
Philadelphia, Pennsylvania
Allan G. Murray, M.D.
Associate Professor of Medicine
University of Alberta, Edmonton
Alberta, Canada
Vladimir R. Muzykantov M.D., Ph.D.
Associate Professor of Pharmacology and Medicine
Director, Targeted Therapeutics Program, Institute of Translational Medicine and Therapeutics (ITMAT)
Sr. Investigator, Institute for Environmental Medicine (IFEM)
University of Pennsylvania School of Medicine
Philadelphia, Pennsylvania
Peter P. Nawroth, M.D.
University of Heidelberg
Department of Medicine I and Clinical Chemistry
University of Heidelberg
Heidelberg, Germany
Randolph M. Nesse, M.D.
Professor of Psychiatry
University of Michigan
Ann Arbor, Michigan
Debra K. Newman, Ph.D.
Associate Investigator, Blood Research Institute
The Blood Center of Wisconsin
Milwaukee, Wisconsin
Peter J. Newman, Ph.D.
Vice President for Research
Walter A. Schroeder Associate Director, Blood Research Institute
The Blood Center of Wisconsin
Milwaukee, Wisconsin
Mark R. Nicolls, M.D.
Associate Professor of Medicine
Pulmonary and Critical Care Medicine Division
University of Colorado at Denver and Health Sciences Center
Denver, Colorado
Anne Nicholson-Weller, M.D.
Professor of Medicine
Harvard Medical School
Divisions of Allergy/Inflammation and Infectious Diseases
Beth Israel Deaconess Medical Center
Boston, Massachusetts
Max Nieuwdorp, Ph.D.
Department of Vascular Medicine
Cardiovascular Research Institute Amsterdam
Academic Medical Center and University of Amsterdam
Amsterdam, The Netherlands

Niaobh O'Donoghue, Ph.D.
Department of Molecular Medicine
Conway Institute, University College Dublin
Dublin, Ireland
Dessy Nikova, Ph.D.
Institute of Physiology
Westfalia Wilhelms University
Muenster, Germany
Peter Oettgen, M.D.
Associate Professor of Medicine
Harvard Medical School
Division of Cardiology
Division of Molecular and Vascular Medicine, and Center for Vascular Biology Research
Beth Israel Deaconess Medical Center, Boston
Kenneth R. Olson, Ph.D.
Professor of Physiology, Indiana University School of Medicine – South Bend, South Bend, Indiana
Arne Östman, Ph.D.
Professor of Molecular Oncology, Cancer Centrum Karolinska
Department of Oncology-Pathology, Karolinska Institutet
Stockholm, Sweden
Sareh Parangi, M.D.
Assistant Professor of Surgery
Harvard Medical School
Department of Surgery, Beth Israel Deaconess Medical Center
Boston, Massachusetts
Kye Won Park, Ph.D.
Department of Oncological Sciences
Program in Human Molecular Biology and Genetics
University of Utah
Salt Lake City, Utah
John D. Parker, M.D., F.R.C.P.(C)
Professor of Medicine and Pharmacology, University of Toronto
Head, Division of Cardiology, Department of Medicine
University Health Network and Mount Sinai Hospital
Toronto, Ontario, Canada
Michael J. Parker, M.D.
Assistant Professor of Medicine, Harvard Medical School
Division of Pulmonary and Critical Care Medicine
Beth Israel Deaconess Medical Center
Boston, Massachusetts
Renata Pasqualini, Ph.D., M.D.
Anderson Cancer Center, University of Texas
Houston, Texas
Michael J. Passineau, Ph.D.
Division of Human Gene Therapy
Departments of Medicine, Obstetrics and Gynecology, Pathology, Surgery, and the Gene Therapy Center
University of Alabama at Birmingham
Birmingham, Alabama
Roger Patient, Ph.D.
M.R.C. Molecular Haematology Unit
The Weatherall Institute of Molecular Medicine
University of Oxford
John Radcliffe Hospital
Headington, United Kingdom
Cam Patterson, M.D., F.A.C.C.
Professor of Medicine (Cardiology), Pharmacology, and Cell and Developmental Biology
The University of North Carolina at Chapel Hill
Chief, Division of Cardiology
Director, Carolina Cardiovascular Biology Center
North Carolina
Knut Pettersson, Ph.D.
Athera Biotechnologies, Fogdevreten 2b, SE-171 77 Stockholm
Sweden
Peter Petzelbauer, M.D.
Department of Dermatology
Division of General Dermatology Medical University Vienna
Vienna, Austria
Jorge Plutzky, M.D.
Cardiovascular Division
Associate Professor of Medicine
Harvard Medical School
Brigham and Women's Hospital
Boston, Massachusetts
Jordan S. Pober, M.D., Ph.D.
Professor of Immunobiology, Pathology and Dermatology
Vice-Chair of Immunobiology for Human and Translational Immunology
Yale University School of Medicine
New Haven, Connecticut
Thomas J. Poole, Ph.D.
Department of Cell and Developmental Biology
SUNY Upstate Medical University
Syracuse, New York
Sonja Praprotnik, Ph.D.
University Clinical Center
Ljubljana, Slovenia
Dexter Pratt
VP Knowledge Modeling, Genstruct Inc.
Cambridge, Massachusetts
Ralph E. Purdy, Ph.D.
Department of Pharmacology University of California, Irvine
Irvine, California
Anthony E. Pusateri, Ph.D.
Associate Director, Novo Nordisk Research U.S.
North Brunswick, New Jersey
Susan E. Quaggin, M.D.
Associate Professor of Medicine, University of Toronto
Mt. Sinai Hospital, University of Toronto, Toronto
Ontario, Canada

Deborah A. Quinn, M.D.
Assistant Professor of Medicine
Harvard Medical School
Pulmonary and Critical Care Unit
Department of Medicine, Massachusetts General Hospital
Boston, Massachusetts

Jacob H. Rand, M.D.
Hematology Laboratory, Montefiore Medical Center
Bronx, New York

Arshad Rahman, Ph.D.
Associate Professor of Pediatrics, Division of Neonatology
Department of Pediatrics, School of Medicine and Dentistry
University of Rochester
Rochester, New York

May J. Reed, M.D.
Associate Professor, Department of Medicine
Division of Gerontology and Geriatric Medicine
University of Washington
Seattle, Washington

Wende R. Reenstra, M.D.
Research Associate Professor, New England Inflammation and Tissue Protection Institute
Northeastern University
Boston, Massachusetts

Carl L. Reiber, Ph.D.
Associate Dean, College of Sciences
Professor, School of Life Sciences
University of Nevada
Las Vegas, Nevada

John M. Rhodes, Ph.D.
Angiogenesis Research Center & Section of Cardiology
Dartmouth Medical School, Dartmouth Hitchcock Medical Center
Hanover, New Hampshire

Simon C. Robson, M.D.
Professor of Medicine
Harvard Medical School
Beth Israel Deaconess Medical Center
Boston, Massachusetts

Sharon Rounds, M.D.
Pulmonary Vascular Research Laboratory, Providence VA Medical Center
Professor of Medicine and of Pathology and Laboratory Medicine
The Warren Alpert Medical School of Brown University
Providence, Rhode Island

Jonathan Rubin, Ph.D.
Department of Mathematics, University of Pittsburgh, Pittsburgh
Pennsylvania Center for Inflammation and Regenerative Modeling
University of Pittsburgh
Pittsburgh, Pennsylvania

John E. Rush, D.V.M., M.S.
Department of Clinical Sciences, Cummings School of Veterinary Medicine at Tufts University
North Grafton, Massachusetts

Stefan W. Ryter, Ph.D.
Division of Pulmonary, Allergy, and Critical Care Medicine
The University of Pittsburgh School of Medicine
Pittsburgh, Pennsylvania

Paul W. Sanders, M.D.
Nephrology Research and Training Center, Comprehensive Cancer Center, and Cell Adhesion and Matrix Research Center
Division of Nephrology, Professor of Medicine and Physiology & Biophysics and AM Chief
VA Renal Section, Department of Medicine, and Department of Physiology & Biophysics
University of Alabama at Birmingham
Birmingham, and Department of Veterans Affairs Medical Center
Birmingham, Alabama

Gernot Schabbauer, Ph.D.
Departments of Immunology and Cell Biology
The Scripps Research Institute
La Jolla, California

Herrmann Schillers, Ph.D.
Institute of Physiology, Westfalia Wilhelms University
Muenster, Germany

Ann Marie Schmidt, M.D.
College of Physicians and Surgeons, Columbia University, New York
New York

Hans-Joachim Schnittler, M.D.
Institut für Physiologie
Technische Universität, Medizinische; Fakultät; Fetcherstraβe 74
Dresden, Germany

Jan E. Schnitzer, M.D.
Scientific Director, Professor of Cellular & Molecular Biology Director of Vascular Biology & Angiogenesis Program
Sidney Kimmel Cancer Center
San Diego, California

Daniel L. Schodek, Ph.D.
Professor of Architectural Technology
Department of Architecture, Harvard Design School
Cambridge, Massachusetts

Anthony Sebastian, M.D.
Department of Medicine, Division of Nephrology
Clinical and Translational Science Institute
University of California – San Francisco
San Francisco, California

Jochen Seebach, Ph.D.
Institute of Physiology, Technical University
Dresden, Germany

Chantal Sguin, M.D., C.M., B.Sc. (N), F.R.C.P.C.
Assistant Professor of Medicine, Montreal General Hospital, Division of Hematology
McGill University Health Centre, Montreal
Quebec, Canada

Frank W. Sellke, M.D.
Johnson & Johnson Professor of Surgery
Harvard Medical School; Division of Cardiothoracic Surgery
Beth Israel Deaconess Medical Center
Boston, Massachusetts

David Semela, Ph.D.
GI Research Unit, Division of Gastroenterology and Hepatology, Mayo Clinic
Rochester, Minnesota

Gregg L. Semenza, M.D., Ph.D.
Professor, Departments of Pediatrics
Medicine, Oncology, Radiation Oncology, and the McKusick-Nathans Institute of Genetic Medicine
Director, Vascular Biology Program, Institute for Cell Engineering
The Johns Hopkins University School of Medicine
Baltimore, Maryland

Haruki Senoo, Ph.D.
Department of Cell Biology and Histology
Akita University School of Medicine
Akita, Japan

Solange M.T. Serrano, Ph.D.
Laboratório Especial de Toxinologia Aplicada/CAT-CEPID
Instituto Butantan
São Paulo, Brazil

William C. Sessa, Ph.D.
Professor of Pharmacology
Director, Vascular Cell Signaling and Therapeutics Program
Yale University School of Medicine, Boyer Center for Molecular Medicine
New Haven, Connecticut

Vijay Shah, M.D.
GI Research Unit, Division of Gastroenterology and Hepatology, Mayo Clinic
Rochester, Minnesota

Carrie J. Shawber, Ph.D.
Department of Obstetrics and Gynecology; Institute of Cancer Genetics
Columbia University Medical Center
New York, New York

Ichiro Shiojima, M.D., Ph.D.
Molecular Cardiology/Whitaker Cardiovascular Institute, Boston University School of Medicine
Boston, Massachusetts

Sruti Shiva, Ph.D.
Vascular Medicine Branch, National Heart Lung and Blood Institute
Critical Care Medicine Department, Clinical Center, National Institutes of Health
Bethesda, Maryland

Yehuda Shoenfeld, M.D.
Head, Department of Medicine B and Center for Autoimmune Diseases, Sheba Medical Center Tel-Hashomer
Tel-Aviv University, Israel

Eric V. Shusta, Ph.D.
Assistant Professor, Department of Chemical and Biological Engineering
University of Wisconsin-Madison
Madison, Wisconsin

Nicholas W. Shworak, M.D., Ph.D.
Associate Professor of Medicine
Director, Ultrasound Biomicroscopy Core
Section of Cardiology
Angiogenesis Research Center; Dartmouth Medical School
Hanover, New Hampshire

Christopher C. Silliman, M.D., Ph.D.
Professor of Pediatrics and Surgery
University of Colorado at Denver School of Medicine
Associate Medical Director, Bonfils Blood Center
Denver, Colorado

Robert D. Simari, M.D.
Professor of Medicine, Chair of Cardiovascular Research, Mayo Clinic
Rochester, Minnesota

Michael Simons, M.D.
A.G. Huber Professor of Medicine and Pharmacology and Toxicology
Chief, Section of Cardiology; Director, Angiogenesis Research Center, Dartmouth Medical School
Dartmouth Hitchcock Medical Center
Hanover, New Hampshire

Rashmi Sood, Ph.D.
Blood Center of Wisconsin
Blood Research Institute
Milwaukee, Wisconsin

Elmar Spuentrup, M.D.
Associate Professor of Radiology
Technical University of Aachen, Department of Diagnostic Radiology
Aachen, Germany

Radu V. Stan, M.D.
Assistant Professor, Departments of Pathology and of Microbiology and Immunology
Angiogenesis Research Center, Dartmouth Medical School
Hanover, New Hampshire

Elliot J. Stephenson, Ph.D.
Department of Medicine, Division of Hematology, Oncology and Transplantation
University of Minnesota Medical School
Minneapolis, Minnesota

Troy Stevens, Ph.D.
Professor, Department of Molecular and Cellular Pharmacology Director
Center for Lung Biology
University of South Alabama College of Medicine
Mobile, Alabama

Duncan J. Stewart, M.D.
Dexter Hung-Cho Man Chair and Director
Division of Cardiology, University of Toronto
St. Michael's Hospital, Toronto
Ontario, Canada

Erik Stroes, Ph.D.
Vascular Medicine, Cardiovascular Research Institute Amsterdam
Academic Medical Center and University of Amsterdam
Amsterdam, The Netherlands

Ute Stroher, Ph.D.
Special Pathogens Program, National Microbiology Laboratory, Public Health Agency of Canada
Department of Medical Microbiology and Infectious Diseases
University of Manitoba, Winnipeg
Manitoba, Canada

Heidi Stuhlmann, Ph.D.
Professor, Department of Cell and Developmental Biology Weill Medical College of Cornell University, New York
New York

Laimute Taraseviciene-Stewart, Ph.D.
Assistant Professor of Medicine, Pulmonary and Critical Care Medicine Division
University of Colorado at Denver and Health Sciences Center
Denver, Colorado

Glenn J. Tattersall, Ph.D.
Department of Biological Sciences, Brock University
St. Catharines
Ontario, Canada

Eugene Tkachenko, Ph.D.
Angiogenesis Research Center; Dartmouth Medical School
Dartmouth Hitchcock Medical Center
Hanover, New Hampshire

Alex Toker, Ph.D.
Associate Professor of Pathology, Harvard Medical School
Department of Pathology, Beth Israel Deaconess Medical Center
Boston, Massachusetts

Mourad Toporsian, Ph.D.
Molecular Structure and Function, Hospital for Sick Children
University of Toronto, Toronto
Ontario, Canada

Kevin J. Tracey, M.D.
Director and Chief Executive, Feinstein Institute for Medical Research
North Shore-LIJ Health System
Manhasset, New York

Lisa D. Urness, Ph.D.
University of Utah, Division of Cardiology, School of Medicine, University of Utah
Salt Lake City, Utah

Bernard M. van den Berg, Ph.D.
Division of Molecular and Vascular Medicine and Center for Vascular Biology Research
Beth Israel Deaconess Medical Center
Boston, Massachusetts

Marlies Van de Wouwer, Ph.D.
VIB Department of Transgene Technology and Gene Therapy, K.U. Leuven
Leuven, Belgium

Alexander D. Verin, Ph.D.
Professor, Vascular Biology Center and Department of Medicine, Medical College of Georgia
Augusta, Georgia

Alan. S. Verkman, M.D., Ph.D.
Departments of Medicine and Physiology
Cardiovascular Research Institute, University of California
San Francisco, California

Aristides Veves, M.D., D.Sc.
Research Director, Microcirculation Lab and Joslin-Beth Israel Deaconess Foot Center
Associate Professor of Surgery, Harvard Medical School
Boston, Massachusetts

Volker Vielhauer, Ph.D.
Department of Pathology, Center for Excellence in Vascular Biology
Brigham and Women's Hospital
Boston, Massachusetts

Hans Vink, Ph.D.
Department of Medical Physics, Cardiovascular Research Institute Amsterdam
Academic Medical Center and University of Amsterdam
Amsterdam, The Netherlands

Yoram Vodovotz, Ph.D.
Director, Center for Inflammation and Regenerative Modeling
Associate Professor of Surgery, Immunology, and Communication Science and Disorders
University of Pittsburgh
Pittsburgh, Pennsylvania

Norbert F. Voelkel, M.D.
The Hart Professor of Emphysema Research, Pulmonary and Critical Care Medicine Division
University of Colorado at Denver and Health Sciences Center
Denver, Colorado

Ulrich H. von Andrian, M.D.
Mallinckrodt Professor of Immunopathology, Department of Pathology
Harvard Medical School, Senior Investigator, The Immune Disease Institute
Boston, Massachusetts

Elizabeth M. Wagner, Ph.D.
Professor of Medicine, Division of Pulmonary and Critical Care Medicine, Johns Hopkins Asthma and Allergy Center
Johns Hopkins University
Baltimore, Maryland

Victoria Wahl-Jensen, Ph.D.
Special Pathogens Program, National Microbiology Laboratory, Public Health Agency of Canada
Department of Medical Microbiology and Infectious Diseases
University of Manitoba, Winnipeg
Manitoba, Canada

Kenneth Walsh, Ph.D.
Molecular Cardiology/Whitaker Cardiovascular Institute
Boston University School of Medicine
Boston, Massachusetts

Qing K. Wang, Ph.D., M.B.A.
Director Center for Cardiovascular Genetics
Department of Molecular Cardiology
Lerner Research Institute, Cleveland Clinic
Cleveland, Ohio

Wan-Chao Wang, M.D.
Department of Immunology, Roswell Park Cancer Institute
Buffalo, New York

Xue Wang, Ph.D.
Division of Pulmonary, Allergy, and Critical Care Medicine
The University of Pittsburgh School of Medicine
Pittsburgh, Pennsylvania

Michael R. Ward
The Terrence Donnelly Vascular Biology Laboratories
St. Michael's Hospital, and The McLaughlin Centre for Molecular Medicine and the Department of Medicine
University of Toronto, Toronto
Ontario, Canada

Peter A. Ward, M.D.
Stobbe Professor of Pathology
University of Michigan Health Systems, Department of Pathology
Ann Arbor, Michigan

Theodore E. Warkentin, M.D.
Professor, Department of Pathology and Molecular Medicine, and Department of Medicine, McMaster University, Hamilton
Ontario, Canada

Jean-Luc Wautier, M.D., Ph.D.
University Lariboisiere-Saint Louis and Institut National de la Transfusion Sanguine
Paris, France

David J. Webb, M.D., D.Sc., F.R.C.P., F.R.S.E., F.Med.Sci.
Christison Professor of Clinical Therapeutics and Pharmacology, Centre for Cardiovascular Science, Queen's Medical Research Institute
University of Edinburgh, Edinburgh
United Kingdom

Alan B. Weder, M.D.
Professor, Department of Internal Medicine
Division of Cardiovascular Medicine
University of Michigan
Ann Arbor, Michigan

Christian Weidenfeller, Ph.D.
Department of Neurology
University of Wuerzburg
Weurzburg, Germany

Hartmut Weiler, Ph.D.
Investigator, Blood Center of Wisconsin
Blood Research Institute, Milwaukee, Wisconsin

Andrea J. Wiethoff, Ph.D.
Philips Medical Systems, Best
The Netherlands

Stuart K. Williams, Ph.D.
Professor, Department of Surgery, University of Louisville
Scientific Director, Cardiovascular Innovation Institute
Louisville, Kentucky

Ulrik Wisløff, Ph.D.
Department of Circulation and Medical Imaging
Norwegian University of Science and Technology
Trondheim, Norway

Kenneth K. Wu, M.D., Ph.D.
Professor and Director, Center for Vascular Biology
Associate Director, The Brown Foundation Institute of Molecular Medicine
Division of Hematology
The University of Texas Health Science Center at Houston, Houston, Texas
and Distinguished Investigator and President
National Health Research Institutes
Zhunan, Taiwan

Xiao-Xuan Wu, M.D.
Hematology Laboratory, Montefiore Medical Center
Bronx, New York

Quansheng Xi, Ph.D.
Center for Cardiovascular Genetics
Department of Molecular Cardiology, Lerner Research Institute, Cleveland Clinic
Cleveland, Ohio

Munekazu Yamakuchi, Ph.D.
Departments of Medicine and Pathology
The Johns Hopkins University School of Medicine
Baltimore, Maryland

Xiaoqiang Yao, Ph.D.
Professor, Li Ka Shing Institute of Health Science and Department of Physiology
Chinese University of Hong Kong

Susan Bok Yeon, M.D., J.D.
Assistant Professor of Medicine, Harvard Medical School
Associate Director, Noninvasive Cardiac Imaging, Cardiovascular Division
Beth Israel Deaconess Medical Center
Boston, Massachusetts

Mervin C. Yoder M.D.
Department of Pediatrics, Herman B. Wells Center for Pediatric Research
Indiana University School of Medicine
Indianapolis, Indiana

Yukihiro Yokoyama, M.D., Ph.D.
Assistant Professor, Division of Surgical Oncology
Department of Surgery
Nagoya University Graduate School of Medicine
Nagoya, Japan

Sun-Ah You, Ph.D.
Center for Cardiovascular Genetics, Department of Molecular Cardiology
Lerner Research Institute, Cleveland Clinic
Cleveland, Ohio

Neville Young, Ph.D.
Institute of Cancer Therapeutics
University of Bradford, Bradford
United Kingdom

Hai-Tao Yuan, M.D., Ph.D.
Beth Israel Deaconess Medical Center
Harvard Medical School
Boston, Massachusetts

Tao P. Zhong, Ph.D.
Assistant Professor, Departments of Medicine, Pharmacology, and Cell & Developmental Biology
Vanderbilt University School of Medicine
Nashville, Tennessee

Guy A. Zimmerman, M.D.
Professor of Internal Medicine
Director, Program in Human Molecular Biology & Genetics
University of Utah School of Medicine
Salt Lake City, Utah

Amado J. Zurita, Ph.D., M.D.
Anderson Cancer Center
University of Texas
Houston, Texas

Preface

This volume was born out of a yearning to develop a synthesis of the field of endothelial biology from bench to bedside. The book has several important – and complementary – goals. These are discussed in turn:

CELEBRATE THE HISTORY OF THE FIELD

William Harvey's discovery of the circulation of blood in 1628 is considered to be one of the greatest achievements in the annals of science and medicine. Another 200 years would pass before Wilhelm His, a Swiss anatomist, "discovered" the endothelium. In the late 19th century and the first half of the 20th century, most investigators employed physiological tools to study the role of the endothelium in inflammation and permeability. In the 1960s, electron microscopy opened a new window into the endothelium, providing a fascinating glimpse into the structure (and, by inference, the function) of this cell layer. The successful in vitro culture of endothelial cells (ECs) in the 1970s marked the beginning of a new era in endothelial biology, spawning some 100,000 articles in peer-reviewed journals over the next 40 years. More recently, there has been an increased awareness of the need to study the endothelium in the context of its native environment. To that end, investigators have made remarkable progress in integrating in vitro and in vivo model systems. An important goal of this book is to highlight the rich historical foundation upon which today's field is built. The first chapter charts the history of vascular and endothelial biology. Subsequent chapters integrate – wherever possible – relevant historical considerations.

APPROACH THE ENDOTHELIAL CELL AS AN INPUT-OUTPUT DEVICE

Part II of the book is centered on the analogy of the EC as a miniature adaptive input-output device. Input includes biochemical and biomechanical stimuli. The various input signals are covered in separate chapters. Output (or cell phenotype) includes changes in gene and protein expression, leukocyte trafficking, permeability, hemostasis, and antigen presentation, to name just a few. Again, these topics are covered in individual chapters. Input is coupled to output through a large array of nonlinear signaling pathways that typically begin at the cell surface and end at the level of post-transcriptional modification or gene expression. Thus, other chapters focus on important signaling pathways and transcription factors involved in transducing signal input. One drawback in emphasizing the theme of the EC as an input-output device is that the system (input, output, and coupling) is reduced to a series of component parts, and the principal unit of investigation is the cell. Thus, there is a risk of overlooking *interactions* between the parts, and ignoring higher levels of organization. To minimize this risk, we have emphasized, wherever possible, relevant in vivo data. In addition, we have included selected overview chapters (for example, on hemostasis, barrier function and leukocyte trafficking). Finally, each section begins with an introductory essay with the goal of weaving together and providing context to the ensuing chapters.

ENGAGE DIVERSE FIELDS AND CREATE NEW KNOWLEDGE

Although the current volume is anchored on a solid foundation of cellular and molecular biology, it integrates a number of areas not typically associated with endothelial biology. Examples include history of medicine, evolutionary biology, comparative biology, metaphors, complexity theory, education, and public health. Some of the authors have little background in endothelial research. In writing their chapters, they have, for the first time, applied their unique perspective (and research tools) to the endothelium and, in doing so, offer new and exciting insights that are not otherwise available in the existing literature.

INCORPORATE BROAD TIME AND SPACE SCALES

All too often, biomedical sciences are approached on limited time and space scales. Little consideration is given to the rich historical foundations upon which current discoveries are based. Moreover, there is a tendency to focus on a single level of investigation. In this volume, we consider multiple time scales – evolution, history of medicine, the present and future – and a broad spectrum of space scales, from molecules to cells, blood vessels, organs, organisms, and populations.

CREATE A VIRTUAL COMMUNITY

The endothelium traverses every organ and is involved in virtually every disease state, either as a primary determinant of pathophysiology or as a victim of collateral damage. Many organ-specific disciplines have initiated efforts to study ECs in their respective fields. In some cases, such as cardiology (coronary arteries), pulmonary medicine (pulmonary and bronchial circulations), and neurology (blood–brain barrier), there is a long tradition of vascular biology, whereas in other cases (e.g., nephrology), the field of endothelial biology is in its infancy. However, owing to the current infrastructure of biomedicine, EC biologists are far more likely to interact with their departmental colleagues than they are with like-minded endothelial biologists from other organ-specific disciplines. At present, few venues exist for cross-fertilization between diverse investigators. This book represents an early effort to establish a virtual community of EC biologists.

BRIDGE THE BENCH-TO-BEDSIDE GAP

A large gap exists between the bench and bedside in endothelial biomedicine. Although no shortage exists of published articles pertaining to ECs and the endothelium, there is a virtual absence of appreciation for this cell layer at the bedside. The endothelium has enormous, as yet untapped, diagnostic and therapeutic potential. An important goal of this book is to bridge the gap and promote an increased awareness of the endothelium in human health and disease.

William C. Aird, M.D.

ACKNOWLEDGMENTS

This book would not have been possible without the help of many, many individuals. I am particularly indebted to the authors responsible for writing the chapters. I thank them for their efforts, timeliness, and patience with the editor, and most importantly for their excellent contributions. I had invaluable input from a group of Associate Editors (Helmut Augustin, Ary Goldberger, Mansoor Husain, Manfred Laubichler, Jane Maienschein, Jan Schnitzer, and Michael Simons) who helped with the design and editing of the book and who contributed thoughtful, integrative essays to help weave together the various chapters. In particular, I wish to acknowledge Manfred Laubichler and Jane Maienschein for their courageous foray into a field in which they had no prior experience. It took some coaxing to get them involved, but once they were on board, they provided me with endless inspiration and wisdom. The book is so much better for their involvement. I have worked for years with Steve Moskowitz, a medical artist. Although he has no formal training in endothelial or vascular biology, he has an uncanny ability to take words and concepts and bring them to life with his drawings. Steve played an important role in designing the layout of the book. In addition, he has spent countless hours editing and advising on the artwork, not an easy task with so many individualized chapters. I also thank Susan Glueck for her assistance in the early stages of the editing process, and Dena Groutsis for her administrative support. I would like to thank Joann (Annie) Woy for her immense efforts in copyediting, and Navdeep Singh for his work in typesetting. I am indebted to many individuals who provided support, guidance and encouragement in one way or another during the preparation of this volume, including (but by no means limited to) Robert Rosenberg, Jeffrey Flier, Peter Weller, Robert Moellering, Mark Zeidel, Alan Rigby, Kate Spokes, David Beeler, and members of my immediate and extended family. Finally, I would like to acknowledge the support of the National Institutes of Health (HL076540), the American Heart Association (Established Investigator Award), and the Mount Desert Island Biological Laboratory.

We hope that this edition is the first of many. I invite readers to e-mail me (waird@bidmc.harvard.edu) with their comments. What aspects of the book did you like? What didn't you like? How would you organize the book differently? What suggestions do you have for new topics? The value of this book – not only as a resource to, but also as a product of, the endothelial community – will depend on your continued critical feedback.

PART I

CONTEXT

History of Medicine

1

The Endothelium in History

Manfred D. Laubichler*, William C. Aird†, and Jane Maienschein*

*Center for Biology and Society, Arizona State University, Tempe; †Beth Israel Deaconess Medical Center, Harvard Medical School, Boston, Massachusetts

The endothelium is only now beginning to gain acceptance as a physiologically relevant organ with potential clinical significance. Yet the cell layer called the endothelium was first identified well over a century ago. In this chapter, we explore the circumstances leading to the slow recognition of the endothelium as a system with untapped diagnostic and therapeutic potential. We trace historically important steps toward increased interest in the endothelium, beginning with ancient discussions of the heart and blood vessels, and the conviction that blood derives from nutrition and is continually used up by the body. We see that, in Western medicine, the dominant culture of the Catholic Church impeded new discovery and instead emphasized reliance on accepted ideas for nearly 1,500 years. Only in the context of the Scientific Revolution of the seventeenth century could anatomists such as William Harvey challenge prevailing dogma and reach the conclusion that blood circulates and that it does so through a system of connected vascular vessels.

In this chapter, we examine those contributions and the developments that followed, slowly and gradually, the rise of new technologies for observation and the framing of new questions. We ask what caused researchers to focus on cells and tissues, and then, during the last part of the nineteenth century, to identify endothelial cells (ECs) as a unique structure, distinguishable structurally, physiologically, and developmentally from the epithelium that researchers initially had seen as closely connected to it. Then, we explore what implications the identification and naming of this particular set of cells had for the biomedical sciences.

The potential for applying principles of endothelial biology to the clinic has been much less well developed, and one goal of this book is to help change that. We return to the current research situation and to the medical potential of EC research in the final chapter (see Chapter 196).

PRE-ENDOTHELIAL HISTORY

Medical researchers often blame the second-century physician Galen for holding back progress in understanding the vascular system. These same researchers point to seventeenth-century physician William Harvey as the heroic founder of modern medical research. Galen certainly did maintain a theory-driven interpretation of arteries and veins as conduits for all manner of things. Inhaled air, expended air, nutriment, and blood all flow through the same blood vessels, according to Galen, responding to the needs of the body as a whole (Figure 1.1, *left side*).

Almost inevitably, medical researchers and textbooks refer to Galen as "in error." Of course, from our twenty-first century perspective of accumulated knowledge, he *was* wrong. However, such a clear-cut judgment ignores the context of the times, Galen's reasoning, and his potentially positive contributions. As surgeon and historian Sherwin B. Nuland explains, what really matters to historical judgment is whether Galen should have known better (1). Given that he did not dissect humans, but relied on animal studies alone, we can excuse some of his descriptions, which deviate from what he would have seen had he been able to look as carefully at humans as we can today. And, given that he could not see the microscopic capillaries and that what he *could* see showed differently colored and textured arteries and veins, we can understand his descriptions of the arterial and venous networks as two largely separate vascular systems. After all, the two systems do look different. Arteries are thick, pulsate, lie deep within tissues and carry red-colored (as we now know oxygenated) blood; veins have thin walls, do not pulsate, are often superficial (such as those on the back of the hand), and contain bluish (deoxygenated) blood.

But later studies, often praised as exemplary (notably William Harvey's), did not differ significantly from Galen's in the physical observations that they were based on; these studies also relied on animal models and naked-eye observation. The difference lay in the questions asked, the assumptions made, and in the nature of the search for additional new information. Harvey drew on a diverse mix of experimentation, observation, and calculation in a way that Galen only argued that researchers should do. When Nuland calls Galen "The Paradox of Pergamon," he emphasizes the irony that, during

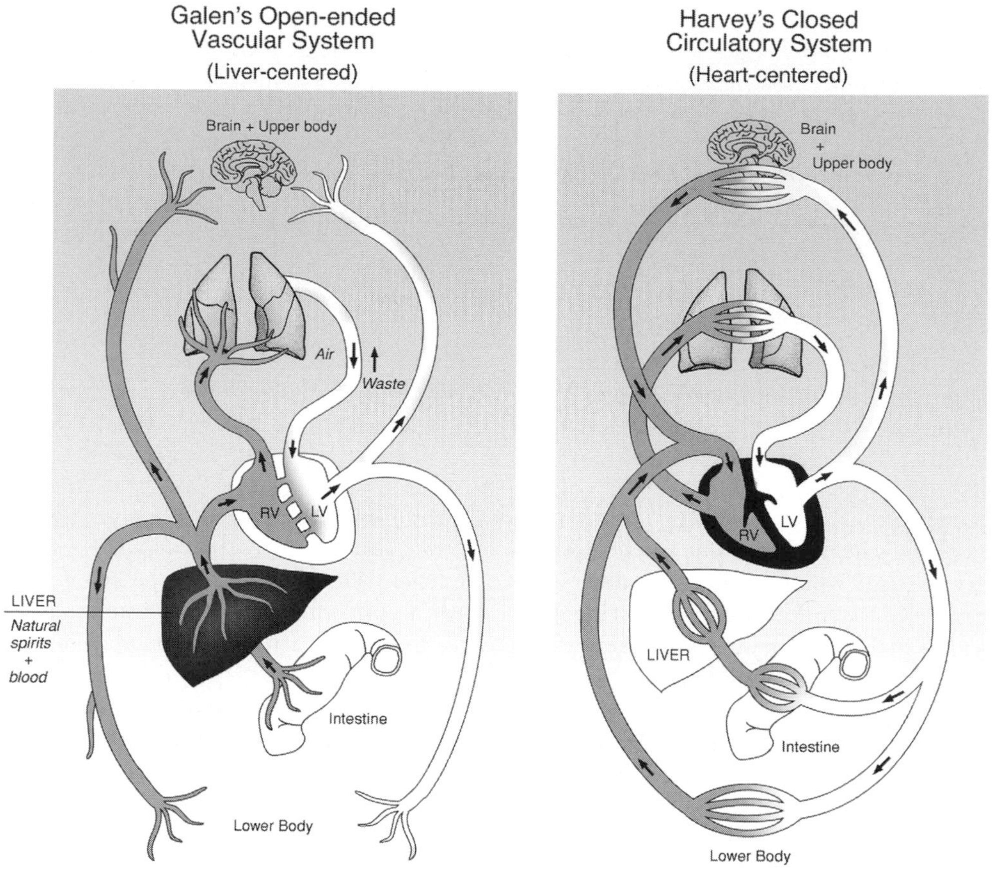

Figure 1.1. Schematic of the vasculature system as viewed by Galen (*left*) and Harvey (*right*). Galen did not recognize blood as circulating. He believed that arteries and veins functioned as distinct, open-ended systems, with veins carrying blood (synthesized in the liver), and arteries carrying both blood (derived from venous blood through invisible interventricular holes) and pneuma (derived from the lungs). Harvey employed simple yet elegant physiological experiments, including ligating arteries and veins, to prove his hypothesis that blood circulates.

his lifetime, this physician, noted for his progressive demand for evidence based on experience and for his questioning the authority of others, did not allow similar questioning of his own authority and did not question his own experiences and interpretations further. Thus, Galen was "wrong," both at the time and in retrospect, in his inconsistent application of his own evidence-based epistemology.

These faults had a lasting effect. It is no exaggeration to say that Galen's ideas, his insistence on adhering to them, and their unquestioning acceptance and promulgation by Catholic Church–run medieval universities effectively held back Western biomedical discovery for about 1,500 years. The universities adopted the ancient learning of Galen, Aristotle, and others ex cathedra, to be taught through rote lecture and memorization and without question. Medical students did not carry out their own dissections, nor did they question existing knowledge or add new discoveries. Although Galen did not create this climate of uncritical acceptance of dogmatic ideas, his own attitudes and writings did not discourage such blind acceptance – as long as it was acceptance of his own ideas.

What, then, was the impact of Galen's interpretation? We can ask whether his "mistakes" actually captured something worth noting, and whether they were reasonable in the context in which he worked. In his insistence that the arteries and veins allowed blood, air, and nutriment all to flow in the same vessels and in both directions, as needed by the body for nutritive reasons, he actually assigned the blood vessels an active role in helping to determine which direction and at what rate the flow would occur. In this, he saw something that those "moderns" missed who viewed the system as passive plumbing that merely allowed fluids to pass through the body.

For Galen, as for the already legendary Hippocrates of the fifth century B.C., the arteries and veins both play important regulatory roles in maintaining function in a balanced, healthy body. Although we know little about Hippocrates the individual, or even about Hippocratic ideas about blood and vessels, we do know that the Hippocratic ideal retained its attraction well into the twentieth century. With its system of interacting humors and responses to the environment, the Hippocratic body was active, with an observable structure, a function that responded to environmental conditions, and developed over the life of the individual as the baby grew into the adult. Structure, function, and developmental responses to environment are all parts of the Hippocratic body, and Galen largely adopted that set of assumptions. This ancient model, which dominated medicine for nearly 2,000 years, was internally active and reactive within its environment in ways often ignored in later times.

CIRCULATION

Galen insisted that the heart has invisible pores that allow the movement of blood through the thick walls of the septum (see Figure 1.1, *left*). This must be the case, he surmised, because he could not see how fluids could travel from the arterial to the venous system otherwise, as they surely do. Generations of medical students absorbed this lesson as their professors read from the Galenic texts. When they looked at bodies, it was to read off the lessons of the texts: "See, here we observe exactly what the great Galen tells us that we must see."

Only in the early sixteenth century did Andreas Vesalius join a small number of anatomists who were beginning, especially in Padua, to actually look at the body with their own eyes and to ask questions that went beyond Galen's doctrines. At first, these questions focused mainly on filling in details and correcting small errors. Vesalius began by asking how it is that the blood can pass through the presumably small pores that Galen had described in the heart's septum. In 1541, Vesalius contributed to a new edition of Galen. Two years later, in 1543, *De Fabrica* appeared under his name. There he wrote that:

> The septum is formed from the very densest substance of the heart. It abounds on both sides with pits. Of these none, so far as the senses can perceived, penetrate from the right to the left ventricle. We wonder at the art of the Creator which causes blood to pass from right to left ventricle through invisible pores (2).

Although Vesalius had made many new observations that disagreed with Galen, he did not challenge Galen's interpretation of the blood's movement. If Galen said that the blood passes through pores in the heart's septum, even if those pores are invisible, then it must be so.

Vesalius continued looking and continued thinking, however. By the second edition of his book, he concluded that he had not seen what Galen said he should see and that, therefore, the pores through the septum are simply not there. Galen was simply wrong about this. As Vesalius wrote in his 1555 edition:

> Not long ago I would not have dared to turn even a hair's breadth from Galen. But it seems to me that the septum of the heart is as thick, dense and compact as the rest of the heart. I do not see, therefore, how even the smallest particle can be transferred from the right to the left ventricle through the septum (2).

This was a tremendous breakthrough. Despite the attacks he received for the impertinence and even perceived sacrilege in challenging Galenic authority, Vesalius and his contemporaries had opened the door for further questioning of anatomical and physiological details. They also laid the groundwork for the basic methods of biomedical science: Start with one's own observations rather than blindly accepting established doctrine. In particular, Vesalius opened the way for the study of the blood system of heart and vessels, and this focused attention on the anatomical structures that seemed important for physiological function. Medicine moved away from the idea of Hippocratic humors that run throughout the body and serve as a unifying holistic tie. Instead, a new emphasis on blood began a trend toward breaking the body into smaller and smaller units, looking for localization of function within defined structures and, eventually, localization of disease within specific structures and functions.

William Harvey carried the investigation further. Building on Vesalius's work (and his questioning the existence of pores in the septum) and on the observations of Hieronymous Fabricius of Aquapendente (who had discovered valves in the veins but not the arteries and had asked why), Harvey found the Galenic interpretation of the movement of blood through the heart and vessels unsupportable. As he noted in the opening section of *De motu cordis* in 1628:

> When they say that the left ventricle draws material, namely air and blood, from the lungs and the right sinus of the heart for the formation of spirits, and likewise distributes spirituous blood into the aorta; that sooty vapours are sent back to the lungs through the vein-like artery and spirit forwards into the aorta; what is it that keeps the two streams apart? And how do the spirits and the sooty vapours pass in opposite directions without mixing or getting into disorder (3)?

And so on to the point that they "would have it that the mitral valves should hinder its return. Good God! How do the mitral valves hinder the return of air, and not of blood?" (3). The fact that, in the same introduction, Harvey also apologized for having to challenge Galen's authority almost 1,500 years later shows just how long Galen's grip on medical theory lasted during the Middle Ages and the Early Modern Period. But that grip was loosening as Vesalius, Harvey, and others opened their own eyes and trusted their own senses.

Harvey famously went on to outline his arguments that blood must circulate through the body, moving out through the arteries and back through the veins after having passed through the tiny anastomoses that connected the two systems. Even though these connections and the passage of blood through them was not yet visible, for Harvey, the overwhelming accumulation of evidence compelled him to the conclusion that blood must move from one system to the other and that, therefore, the connections must exist (Figure 1.1, *right*).

The overpowering logic, the diversity of converging types of empirical evidence relating to blood's quantity and movement, and the accumulated anatomical evidence eventually carried the day in favor of Harvey's interpretation, although not without a fight. Gradually, after 1628, blood was accepted as circulating through an essentially closed system of blood vessels. In connection with interpretations from mechanical philosophers such as René Descartes in 1632, the heart came to be seen as a pump or a furnace, pushing blood out into the arteries by the action of contraction (4,5). For the mechanists, blood flows along its constrained path until it finally reaches the heart again, and it flows in from the veins to fill the void left by yet another contraction that has sent out yet more blood into the arteries.

The weight of argument in favor of the circulation model was overwhelming, even though Harvey himself could not actually see the connections between the arterial and venous systems. They must be there, but it would take new technology to see them. Sure enough, when Italian anatomist Marcello Malpighi used the newly available compound microscope to look at blood flow in the lungs of frogs in 1661, he directly observed the connecting capillaries (6). His reports drew on the direct, meticulous observation of diverse tissues and experimental manipulations to enhance observation—for example, injecting colored fluids into the vessels to observe their paths. Malpighi's capillaries were so small and so important in allowing the blood to circulate that they naturally became a focal point for understanding how the transmission of blood from arteries to veins works. With Antoni van Leeuwenhoek's confirmation using his higher-powered single-lens microscopes in the 1670s, the circulation of blood was largely accepted.

By the mid-seventeenth century, then, a very neat anatomical picture formed that was clearly "right" in the sense of accurately describing the physical phenomena of blood flow. But it largely missed the physiological action and life of the system, and it also lacked any sense of how the system develops or whether it simply exists, already connected from the very earliest stages of any individual. The focus remained primarily on structure: Harvey's followers had turned Galen's active and reactive system into a machine, with arteries and veins serving as mere passive plumbing. That Harvey himself did not hold such a mechanistic view is evident from his vision of the blood as the body's revitalizing agent. For Harvey, the circulation of blood brings renewal, similar to the cycle of evaporation and spring rains that renew the soil, or like the heavenly bodies orbiting and returning every year. Circulation brings life, and the parallels between circulation in the macrocosm of the heavens and in the microcosm of man stamped a sort of confirmation on the circulation hypothesis.

Yet Harvey's vitalistic picture had given way to a largely mechanistic world view in medicine. The mechanistic conceptions of the body also resonated with the emergence and increasing popularity and importance of sophisticated mechanical contraptions, such as clocks or pumps (7). Indeed, it can be argued that the prevailing images and metaphors of the organism during the seventeenth and eighteenth century were all derived from the technologies of these times, which provided both the instruments for studying biological phenomena as well as the interpretative framework for its understanding. The best known and one of the more far-reaching analyses is Julian Offray de la Mettrie's "Man a Machine." Although he did not focus on anatomy as such, La Mettrie (8) saw a close connection between the fluids circulating in the vessels and the maintenance of the "elasticity of the blood vessels on which their own circulation depends."

The emerging new world order of early globalization and increased trade also contributed to the prevailing view of the importance of circulation and well-defined channels of transport. Here, as in most instances of the development of scientific ideas, the exchange of metaphors went both ways: On the one hand, the existing social and economic order shaped ideas about the organism (including concepts of pathology), while on the other hand, the biological conception of the organism also became a model for ideas about the organization of the state and the economy (9,10).

The mechanistic conception of the organism, together with the increased understanding of anatomy, also contributed to the development of a new conception of disease as a localized deficiency in a particular part of the body. Not unlike a broken machine, a sick body was considered to have a broken part. Pathology emerged in the nineteenth century as a scientific discipline that investigated both the symptoms and causes of disease within this framework of machine-like organisms (11).

SPECIALIZATION IN BIOMEDICAL RESEARCH

The nineteenth century brought a new view of the circulatory system and its blood vessels, in terms of tissues and then cells. Rather than seeing vessels as long, essentially unstructured pipes through which the blood passively flows, researchers began to see the vessels as structured and constructed of parts. In particular, cells came to be seen as making function possible and developing over time.

The new view arose partially because of increased knowledge. Improved achromatic microscopes and microscopic techniques made it possible to observe smaller and smaller parts of the organism. Technology and inquiry reinforced each other: The desire to see more stimulated the push to develop new technologies and, simultaneously, new technologies stimulated new questions. At the same time, biology was emerging as a field of study, with an emphasis on examining structure (through anatomy and cytology), function (through physiology), and development (focused on cells and organisms). Although "biology" as a field by that name only emerged in the nineteenth century, and only fully developed in the early twentieth century, already the study of life was beginning to be differentiated into specialized subfields of study, localized in different specialties within medical schools and research institutes.

Development

Early in the nineteenth century, Karl Ernst von Baer and others had carefully examined eggs and discovered that mammals have eggs too (first seen in a family dog sacrificed for the cause of science) (12). Observing the processes of development, they saw an emergence of form from what appeared to be unformed matter. That is, they saw form coming into being only gradually (or epigenetically), with the egg developing layers and only then differentiating into organs and systems.

Von Baer joined fellow embryologist Christian Heinrich Pander in noticing that the process of development forms "germ layers" (13). These connected but distinct layers of matter then become the various parts of the differentiating organism. Perhaps the embryo was always divided into the outer ectoderm, inner endoderm, and middle mesoderm layers, or perhaps the layers arise epigenetically through the developmental process? That remained to be determined, and some researchers held each position. (It was not until the late nineteenth century that researchers understood that these layers arise only at the gastrulation stage.) In addition, the biological significance of the layers remained to be determined: Did they provide the start of differentiated body parts, and therefore have embryological significance? Did they represent tissues that would give rise to different functions, and therefore have physiological significance? Or were they just structurally different, and changing with time? These were central questions for early nineteenth-century biologists.

Cells

Early in the nineteenth century, Matthias Schleiden and Theodor Schwann focused on cells (14,15). They saw cells as the vital units that make up organisms, and they offered a theory of cell development whereby accumulating cells make up a growing and differentiating organism. The history of ideas about the formation of new cells during the mid-nineteenth century shows how contemporary philosophical and theoretical conceptions can shape the interpretation of observations. Schwann, who was committed to a unified theory of nature, first conceived of the formation of new cells as analogous to crystallization, which was an established mode for the emergence of new forms. He thought that existing cells secrete material, and new cells emerge through a process analogous to crystallization. It took several decades of painstaking and detailed observation to establish the mechanisms of nuclear and cell division.

By mid-century, with advancing microscopic techniques, a growing community of biological researchers had generated a picture of the embryo as a fertilized egg cell that undergoes cell divisions, develops germ layers, and then differentiates into specialized types of cells and tissues (16–20).

Cell Pathology

Cells also assumed the central role in understanding disease, with Rudolph Virchow presenting the case for *cellular pathology* (21,22). Although the "morbid anatomists," as the early pathologists were called in Paris (led by Pierre Louis, Xavier Bichat, and others), had emphasized localization of disease in organs, Virchow localized disease in the cells. Medical science needed to understand which cellular changes were associated with which diseases, he urged, and also how cells contributed to causing disease. Cells work together at times to form membranes, Virchow asserted, including that lining the capillary (21): "A capillary vessel is a simple tube, in which we have, with the aid of our present appliances, hitherto only been able to discover a simple membrane, best at intervals with flattened nuclei . . ." This is "a membrane as simple as any that is ever met with in the body." Although he did not call this membrane the endothelium, it was, in effect, what he was describing. And, as in later contributions, he argued that the "simple membrane" results from the cells working together. For Virchow, medicine should focus on cells and how they work together to make up functional tissues and organs. Pathology should examine the failures that occur at each level, down to the cellular.

Pathologists also began to distinguish even more finely among different types of cells and tissues. For example, diseased linings of organs and parts called for identification; Viennese surgeon Theodor Billroth used the prefix endo- to describe as an "endothelioma" those tumors occurring in what came to be known as *endothelial cells* (23).

Connecting the Pieces

In the dissecting rooms and in pathology labs, researchers were looking at ever finer distinctions in their search to link disease with localized material. Physiologists sought to link functions to the localized parts of organs and cells, asking how the parts cause the observed responses. Embryologists wondered how the parts and their functions arise, although they had no way to make much progress in studying human development as yet. Structure, function, and development began to hold their specialized places in medical education. Meanwhile, the clinical ideal remained largely Hippocratic, focused on the whole organism and its interactions.

William Osler exemplified the clinician's perspective on and wish for – if not the reality of – holism and integration. He did not look inside vessels for an endothelial lining, but instead emphasized the whole system and its actions and failures. As he wrote in *Diseases of the Arteries*, the arteries reflect the whole of life, with its "wear and tear." For "Among organs, the bloodvessels (sic) alone enjoy no rest . . . like other organs, they live under three great laws – use maintains and in a measure sustains structure; overuse leads to degeneration; in time they grow old, in threescore or in fourscore years the limit of their endurance is reached and they wear out (24)." Osler's remained largely a structural view, but one that saw the organism as an organic whole:

> The stability of tubing of any sort depends on the structure and on the sort of material used; and so it is with the human tubing. With a poor variety of elastic and

> muscular fibers in the bloodvessels, some are unable to resist the wear and tear of everyday life, and have at forty years of age arteries as old as those of others at sixty . . . not only are there individuals, but whole families with "shoddy" bloodvessels. Hence the truth to the old saying attributed to Cazalis, "a man is as old as his arteries." In the building of the human body, as of chaises, there is, as Autocrat says, "always somewhere a weakest spot," and too frequently this is in the circulatory system. The conditions of modern life favor arteriosclerosis, as a man is apt to work his body machine at high pressure . . . Living quieter lives and with less stress and strain, women are not so frequently the subject of arterial changes, and in consequence they last longer (24).

THE ENDOTHELIUM

The Swiss anatomist Wilhelm His introduced the term *endothelium* in 1865, in a programmatic essay titled "Die Häute und Höhlen des Körpers (The membranes and cavities of the body)" (Figure 1.2). Halfway into his tenure as professor of anatomy and physiology in Basel (from 1857 to 1872) His introduced an academic research program that became the foundation for his work in developmental mechanics. It was based on the conviction of a "tight connection between histological embryology and the most fundamental problems of general physiology" (see Ref. 25, p. 33). Programmatically, His continued the work of Xavier Bichat, who began his short but extremely productive career with a monograph on the membranes of the human body (26). Following Bichat, His's program was to identify the embryological origins and further developmental differentiation of tissues that have structural and functional meaning for the organism.

> Nobody doubts, as was first recognized by Bichat, that all the capacities of the living body can, in the end, be explained by the coordinated interactions of the capacities of its tissues. These capacities of the tissues are, however, a direct consequence of their organization... A cell, even though it is endowed with rich internal capacities, only develops in closest dependency of its external conditions, it even responds promptly to the most fleeting external cause, either through changes in its vegetative state, or through other changes in its vital functions. . . . These phenomena will be revealed by means of pathological–anatomical and experimental as well as embryological analysis (25, p. 34).

During the following decades, His constantly refined his initial program, always ready to adopt new technologies. After the mid-1880s, these included advanced apochromatic microscopes and microtomes (that His helped to refine) that allowed meticulous serial sectioning as well as new methods for the three-dimensional representation of anatomical structures. As a result, anatomical details became observable both in adult and in embryonic specimens. His's program sought theoretical generality, but was based on observed particulars in both human and vertebrate (mostly chick) specimens.

His triggered an immediate and at times rather heated debate about the appropriateness of this new concept of endothelium. His's specific focus was on the cavities and membranes of the third germ layer, the mesoderm, which include the vascular system, pleural spaces, and the pericardium and peritoneum. He focused especially on the importance of developmental history (*Entwicklungsgeschichte* or descriptive embryology) in understanding histology and anatomy. During that time, the respective contributions of different germ layers to various organs systems were still debated, as were the actual mechanisms of organogenesis. His's own program emphasized the movements and foldings of germ layers as a strictly mechanical and material cause for differentiation, development, and function. His's focus remained on early developmental stages, rather than on the later anatomical results and their biological and medical implications. In the context of increasing specialization, this mattered, because many medical researchers did not yet hold the early developmental stages as important. Researchers questioned his claims about the developmental process, about observations based on manipulative techniques that necessarily destroyed the organism being studied, and about the claims that these cells and tissues were really distinct and deserving of special consideration.

One of the peculiar features of the mesoderm, which His and other embryologists clearly recognized, is the formation of inner cavities within the differentiating mesodermal tissues (e.g., the vascular system, the lymphatic system, or the pleural spaces) and the histological differentiations associated with these structures. Among the differentiated structures connected to these cavities were so-called inner membranes, which show a remarkable diversity and thus proved to be a serious challenge for microscopic anatomists and histologists.

One problem was conceptual. How should one refer to those cell layers that line these inner cavities of the mesoderm? Common practice at that time was to refer to them as an epithelium, in strict analogy to the epithelia covering the outer surfaces of organs (e.g., the keratinocytes that cover the skin or epithelial cells that form the inner lining of the digestive system) and protecting these organ systems from their environment. In this case, the generic term *epithelium* simply meant a layer of cells serving as a lining. But, as His pointed out, the cells that line the cavities of the inner germ layer (mesoderm) exhibit certain characteristics that differentiate them from those epithelial cells that originate from the two outer germ layers (endoderm and ectoderm). Therefore, these structures should be identified by their own designation.

One alternative was to call them "false epithelia." His found that unsatisfactory and instead introduced a new term, *endothelia*:

DIE

HÄUTE UND HÖHLEN DES KÖRPERS.

ACADEMISCHES PROGRAMM

VON

WILHELM HIS

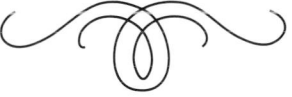

BASEIL,
SCILWEIGHAUSERISCHE UNIVERSITYTH-BUCHDRUCKEREL,
1865.

Figure 1.2. Frontispiece of Wilhelm His: *Die Häute und Höhlen des Körpers*, published in 1865. In this publication, an outline of His's research program, he first defined the endothelium as the lining of the vasculature and the lymphatic system.

It is customary to refer to the cell layer that lines the vascular and inner cavities as an epithelium. The same designation is also used for the inner cellular linings of the joint cavities and those on the back of the cornea. However, all these cellular layers that line the cavities of the inner germ layer [mesoderm] display such a large number of similarities and, from their first appearance during development, they differ from those cellular layers that have their origin in one of the outer layers [endoderm and ectoderm] to such a degree that it is well justified, especially with respect to understanding their physiological functions, to identify those by means of a special designation, either referring to them as "false" epithelia in opposition to the "true" epithelia, or by calling them endothelia [sic], thus reflecting linguistically their relationship to inner membranes (25, p. 18).

His went on to describe the differences between endothelial and epithelial cells:

> Beginning with early development, the contrast between serous and vascular endothelia on the one hand and true epithelia on the other hand is already visible. The former develop, as we have seen, from lymphoid cells, the least differentiated cell type that the inner germ layer [mesoderm] can produce and which are also the precursor (Mutterform) for all others. Soon they take on their characteristic flattened shape and become transparent and after reaching this stage they barely change anymore nor do they participate in any significant way in growth processes within the body (25, p. 18).

From these statements, it is clear that, by 1865, His did not recognize the participation of ECs in blood vessel formation. The perceived passivity of ECs in development is also in stark contrast to the activity of epithelial cells, which were already recognized to continue to grow and participate in changes during development.

A second, physiological difference between endothelial and epithelial cells was recognized by His. Whereas epithelial cells produce all those substances that form the secretions of the various glands in the body, in contrast, ECs were not seen to produce any form of secretions. As His emphasized, "we have no reason to ascribe to endothelia any secretatory functions" (25, p. 18). The final difference between endothelial and epithelial cells that His mentioned relates to their function as barriers: Although blood serum can pass freely through ECs, which therefore do not provide a clear separation between the blood in the vessel and the surrounding intercellular substrate, epithelial cells act as a much stronger barrier, especially with regard to larger molecules:

> There is another aspect in which true epithelia and endothelia are in stark contrast to each other; serum can freely pass through the latter at any place; sometimes serum filters through the endothelia and leaves the blood vessels in order to nourish the surrounding tissues; sometimes it passes from the tissues into the lymphatic system or the serous cavities, following a simple pressure gradient. This implies that endothelia do not provide a strict boundary between cavities and intercellular substances of the inner germ layer [mesoderm]; therefore physiologically these have to be seen as a whole, as they equally contribute to the function of containing the general nutritional fluids. The situation is different with true epithelia (25, p. 19).

Summarizing His's arguments, which we present here at some length because of their historical significance, we see that the concept of the endothelium as a separate and clearly distinguishable part of the body arose as a consequence of three different considerations. First, the endothelium can be distinguished because of its embryological origin from the mesoderm, becoming a layer of cells that covers the cavities of the inner germ layer (mesoderm). Second, ECs have a clearly recognizable structure, with the endothelial layer clearly identified as a connected layer of flattened cells. And third, ECs were not considered to be active in physiological secretion. Instead of having an active role that would have been considered physiological at the time, the endothelial layer was seen as providing a somewhat porous lining for the vascular system and related mesodermal cavities. Endothelium was more a matter of providing structure to support the vascular plumbing system, rather than as anything more active.

The New-Found Endothelium

In the years following His's introduction of the term, not everyone immediately adopted his proposal to identify the endothelium as a separate entity. Arguments continued about the usefulness of separating the endothelium from epithelium. Was there really something different here and, if so, did it deserve its own name? A survey of textbooks and published articles from the later nineteenth century suggests that leading anatomists such as Joseph Hyrtl, Carl Gegenbaur, and Philipp Stöhr – who all argued against the separateness of the endothelium – seemed to have the upper hand. For them, the epithelium and presumed endothelium had fundamental similarities in function and in morphology. If it was important to make distinctions of type, they preferred using additional descriptive terms to specify the origin of these "thelia," such as mesenchymal epithelium. This interpretation was codified in some histology textbooks, which typically defined an epithelium as a connected layer of cells covering the surface of the body, an organ, or an inner cavity. Under this definition, endothelium was simply a specific form of epithelium consisting of flattened cells (*Plattenepithelium*) that lined the blood vessels.

Narrowing the Endothelium to Blood and Lymphatic Cells

Increasingly, however, others did take seriously the differences, because the term *endothelium* had its uses. Increasing acceptance that something specialized called the endothelium existed was reinforced after the 1880s and 1890s because of the advanced microscopic and histological techniques and improved equipment that made possible a much more detailed and wider range of observations. Specifically, researchers began to reliably distinguish the endothelium as a layer of cells that together serves as a membrane lining blood vessels, the lymphatic system, and (for some) parts of the nervous or other systems. The influential Heinrich Wilhelm Gottfried von Waldeyer, for example, suggested restricting the term to those cells that make up the innermost layer of blood and lymph vessels and the posterior lining of the cornea. He thus excluded some of the other "thelia" also derived from the mesoderm and that His had included in his definition of the endothelium

(27). Waldeyer's approach, based on detailed observations, provided considerable clarification about what should be included as endothelium.

The types of observations on which Waldeyer drew illuminated fine anatomical structure, and discussion of function followed. If this layer of cells is truly differentiated from others, then why? What function does it serve? Also, questions arose about how ECs develop and whether that developmental history might provide clues to their identity and use.

By 1908, knowledge about the endothelium and its use had already begun to achieve standardization and become accepted within the body of established scientific knowledge. Even such a general standard source as *Gray's Anatomy* reflected the new understanding about an endothelial lining for the vascular system. What had been called the epithelial layer was now the endothelium, and the capillaries, in particular, were seen as "very small endothelial tubes which connect the venous system with the arterial system." "The nucleated endothelial cells which constitute the wall of a capillary are flat, irregular in outline, and are united by a cement material." This passage reflects the state of knowledge: Endothelium is connection, the cells together make up the vessel walls, and they are united by some unknown material. Furthermore, they make up vast connected networks or systems of "endothelial tubes throughout the entire blood-vascular system. The heart is a great muscular thickening around a portion of the system of endothelial tubes" (28, pp. 586–587).

In 1910, A. A. Böhm and colleagues' textbook was one of many specialized texts to appear on the study of histology, which had become a standard field in medical education. For these authors, the message was clear about what types of cells constituted ECs and the significance of their embryonic origins from the mesodermal germ layer: "*Endothelial* cells are differentiated mesenchymal cells. They line the blood- and lymph-vessels and lymph-spaces." It was also important to describe them: "Endothelial cells are in structure like those of the mesothelium. In blood- and lymph-vessels they are of irregular, oblong shape, with serrated borders" (29). By the beginning of the twentieth century, although debate might continue about the existence or importance of the serrations or about the ways in which the cells connected into tissue, histologists largely followed His's initial characterization. They saw endothelial tissue as lining vessels and providing more nuance to the structural plumbing system for blood circulation.

Thus, the beginning of the twentieth century brought a growing awareness of the complexity of what had once looked like such simple tissues and cell structure. Now, it was seen that capillaries consisted of a complex coating, including ECs that themselves seemed to be connected. The understanding of structure had changed dramatically, and it remained for others to interpret what this meant for function. It now seemed that vessels do more than passively carry blood or lymphatic fluid around the body: They play some more active functional role as well. But how? Most importantly, are the vessels made of permeable cells, or do they have porous passages to allow fluids to pass from the vessels to the body? If so, how? To answer such questions, researchers studied both the fine structure and the related function of the ECs under both normal and pathological conditions.

Physiology and Pathology: The Endothelium in Action

With accepted structure came ascribed function. Pathologists and physiologists each worked to understand endothelial function. Already in 1884, pathologist G. Hare Philipson discussed, in his Bradshawe Lecture "On the Pathological Relations of the Absorbent System," the endothelium, which he defined as the lymphatic and circulatory vessels. As he put it, "Alterations in the amount of transudation may thus be referable, not to disturbance of the circulation, but to changes in the vessel wall, and especially in their endothelial lining" (30). The vessel wall is not a passive or essentially dead membrane, but is living and active. It helps regulate the flow of fluids either along the vessels or perhaps even through the vascular walls. But how?

Physiological studies undertaken during the first half of the twentieth century supported the concept that, in addition to the visible flow of blood through capillaries, there existed an "invisible flow of water and dissolved materials back and forth through the capillary walls," and that this "invisible component of the circulation takes place at a rate which is many times greater than that of the entire cardiac output" (31). In 1927, Eugene M. Landis demonstrated that flow through the capillary wall is directly proportional to the difference between hydrostatic and osmotic forces acting across the membrane (32). Other investigators compared the flow of graded-size molecules through artificial porous membranes and capillaries and used mathematical formulas to calculate capillary filtration coefficients and pore dimensions.

These studies gave rise to a pore theory of capillary permeability, which predicted the presence of two distinct pathways across capillaries: a small pore system for the passage of small hydrophilic solutes through water-filled channels, and a large pore system for the passage of macromolecules. Proof for the existence of these structures would have to wait for the development of higher-resolution microscopy; the advent of new microscopic techniques clarified the structure of individual cells and, especially, how cells interact.

More than a Layer Lining Blood Vessels: Form and Function Together

The discussions about the very existence and then the nature and role of the endothelium during the first decades of the twentieth century reflect epistemological and methodological discussions within medicine and biology more generally. A developmental approach reflected the belief that biological understanding required knowledge of the origin and genesis of structures. A morphological/histological approach argued for the comparative evaluation of the internal configuration and external appearances of structures, whereas a functional/pathological approach emphasized the role that certain structures have in the orderly operations of the body.

In the endothelium, we see all three at work, as researchers sought to discover whether, what, and why endothelia exist. Even though the picture became much clearer as research moved into the early twentieth century and the endothelium was accepted as a specific entity, continuing debates about its structure and role reveal the competing underlying assumptions of the researchers involved. These debates also reveal the tendency of researchers in different specialty areas to work at different levels of analysis.

Ramon y Cajal's histology textbook in Spanish, revised and translated in 1933, provides an example. Ramon y Cajal emphasized the structural flatness of the endothelial layer of cells. Those flat cells that form a thin, pavement-like layer over a surface are endothelial, he maintained, wherever we find them and whatever their origin. The metaphor of the pavement not only conveyed that this was a flat covering, but also that it contained a sort of cement that held the individual cells together to do their job of providing a solid lining (33). This raised the question of how the cells adhere. Histologists, especially in the 1920s and 1930s, used silver nitrate preparations to discover detailed cell appearance. In the blood vessels, ECs were seen as being connected by a "resistant cement," but in the corneal surface they seemed to have connecting filaments holding the individual cells together. Details of these phenomena were laid out in textbook after textbook, each recognizing that the meaning and role of these cells remained largely unknown, and each emphasizing different aspects of what was known.

Edmund Cowdry's widely used textbook of the next year (1934) sought to bring the structure and function closer together, looking at the "functional significance of cells and intercellular substances." Cowdry pointed to a debate about the importance of ECs with respect to red blood cells. Florence Sabin and others believed that red blood cells must be produced intravascularly, because the endothelium forms a tight barrier to control the movement of cells. Alternatively, if blood cells form outside vessels, then this requires movement across the endothelial layer into the vessels. Cowdry felt that the issue was not resolved but that, in either case, the endothelium played a central role (34).

In his discussion, Cowdry recognized that it was crucial to understand the nature and extent of the endothelial layer's permeability as well as what regulates movement across that layer. Are the individual cells penetrable, or are there pores between cells that either allow, enable, or restrict movement? And how? As details about the structure became clearer, these functional questions came more sharply into focus as well.

A few years later, in Cowdry's 1938 edition, he noted that the endothelium of capillaries in the brain "holds back protein quite effectively; for the cerebrospinal fluid – a special tissue fluid – is remarkably free from protein. Permeability naturally increases with the inevitable thinning of the endothelium in capillary dilatation or stretching. This happens when the liver swells on receipt of absorbed substances from the digestive tract after meals; but the brain cannot expand to anything like the same degree because it is limited by a bony case" (35, pp. 125–126). By the end of the 1930s, questions about the extent, nature, and effects of endothelial permeability assumed increasing importance.

Leading histologist Alfred Kohn and others argued that, especially for the vascular and lymphatic system, the notion of the endothelium as a cover and barrier facing a hollow space is misguided. Rather, because these vessels are always filled with blood and other internal substances, the vascular system is actually a complex unit composed of moving mesenchymal elements (blood and lymphatic fluids), a layer of cells (endothelium), and the vascular wall. Functionally, these cells all interact; furthermore, the ECs mediate important exchanges between the blood and other tissues (36).

According to Kohn, the functional role of the endothelium to mediate the important material exchanges within the body also accounts for the differentiation among ECs. "Fatter" ECs will be found in those areas of the body where a more intensive exchange occurs, whereas more flattened cells will occur in areas of less functional importance. Some authors referred to this as the reticulo–endothelial system, whereas others rejected the idea and continued to look at individual cells as the mechanism of control. The seeming clarity of the twentieth-century's early decades was once again giving way to debates and alternative interpretations, and evolving and expanding questions. Different levels of analysis and new techniques would bring new information, new insights, new questions, and new debates, as new technologies allowed new observations.

TECHNOLOGY AND ITS IMPACTS AND IMPLICATIONS

Modern science is characterized by a close connection between the development of scientific concepts and theories and the available technologies and methods. Indeed, we find this connection at the birth of modern science during the so-called Scientific Revolution, which was greatly aided by telescopes, microscopes, and shortly thereafter by clocks and the pendulum. These instruments and their associated new methodologies not only allowed for increased precision in measurement, they also carried with them (or in a sense embodied, as the modern historiography of science would have it) a specific conceptual structure. Instruments and research methods are therefore not just a more precise extension of our senses. Rather, through their interactions and intervention with "nature," they constitute – or in a very real sense create – scientific objects. And these scientific objects are not just material objects "out there in nature" with specific physical properties; they are simultaneously conceptual abstractions that also have theoretical significance.

The history of endothelial biomedicine is a good illustration of these patterns. We have already seen how early ideas of the circulatory system and the introduction of the concept of the endothelium reflected larger scientific trends of the times.

The innovations included contributions of then-prevalent research methodologies and technologies. His's recognition of the endothelium as a separate and distinct entity of the body drew both on his theoretical ideas about the role of germ layers during embryogenesis and also on the newly available microscopes, microtomes, and histological techniques that revealed the morphological characteristics of ECs and their mesodermal origins. The interactions of ideas, questions, and technologies accelerated during the course of the twentieth century (37–39). As we discuss below, physiological approaches gave way to ultrastructural studies; electron microscopy, in turn, was largely supplanted by cell culture.

New Technologies, New Discoveries: Electron Microscopy

Electron microscopy brought great advances in understanding structure. In doing so, however, it also moved away from an appreciation for development and dynamic interactions of the cells and their surroundings. Much of the initial excitement generated by electron microscopy was due to its ability to reveal details about the internal structures of cells and resolve some of the long-enduring arguments over whether these were artifacts. Many of the crucial techniques for employing electron microscopy in the study of cells were developed to advance the understanding of cytoplasmic structures. But electron microscopy also helped reveal some of the distinctive features of ECs.

Although a number of new procedures for improving the resolution of the light microscope were developed during the 1940s (e.g., phase contrast and ultraviolet microscopy [40,41]), the electron microscope offered the greatest promise for generating images at finer resolutions that could reveal structures in cell cytoplasm. The electron microscope depended on the idea that electrons have wave properties, with wavelength inversely proportional to electron velocity, advanced by Louis de Broglie in 1924. Several investigators began to develop electron microscopes in the 1930s, and the first commercial microscopes were introduced by Siemens in 1939. Over the ensuing years, advances in microtome sectioning, embedding protocols, and fixation techniques set the stage for a new generation of investigators who employed electron microscopy to generate new information about the structure of cells. Discoveries included George Palade's description of mitochondrial cristae and endoplasmic reticulum–associated ribosomes (42,43).

At the meetings of the Electron Microscope Society of America in 1953, Palade presented a paper entitled "Fine Structure of Blood Capillaries," in which he presented micrographs of capillaries in different organs such as skeletal muscle, heart, intestine, and pancreas. He reported that, in these organs, "the endothelial cells form a continuous lining," and that the cells contained a "large number of vesicles concentrated immediately under the cell membranes facing both the capillary lumen and the pericapillary spaces." Surprisingly, electron microscopy studies failed to reveal pores of the dimensions postulated by physiologists. Palade proposed that the vesicles "may represent a system for transporting fluids across the capillary wall and may account for the high permeability rate of the capillaries." In a subsequent study employing ferritin as a molecular tracer, Palade concluded that "endothelial cell vesicles are the structural equivalent of the large pore system postulated in the pore theory of capillary permeability" (42,43).

A major result of the early electron micrographic studies of ECs was the discovery, in 1964, of Weibel-Palade bodies, which are unique to ECs. These organelles, now known to store and secrete von Willebrand factor and P-selectin, were described by George Palade and Ewald Weibel as "hitherto unknown rod-shaped cytoplasmic component which consists of a bundle of fine tubules, enveloped by a tightly fitted membrane." They concluded that, due to the regularity of its appearance in ECs, the rod-shaped bodies must have "functional significance which for the moment remains obscure" but suggested that they figure in vascular or blood physiology (44).

Another important consequence of electron microscopic studies during the 1950s was the finding of structural differences between capillaries in different organs. As noted by Hibbs and colleagues in 1958, "some variation in the structure of capillaries and arterioles normally occurs from one organ to another, and even among vessels of the same organ" (45). In the 1960s, electron microscopy was used to confirm the functional relevance of structural heterogeneity, demonstrating, for example, that the tight junctions between ECs in the brain form a highly functional blood–brain barrier, whereas the loose, somewhat disorganized junctions of postcapillary venular endothelium correspond to their predilection for solute and leukocyte trafficking during inflammation.

Cell Culture

It may be argued that physiological and morphological studies – in and of themselves – were beginning to yield diminishing returns. That would change in the early 1970s, when Eric Jaffe and Michael Gimbrone independently reported the first successful isolation and primary culture of human ECs from the umbilical vein (46,47). The cells, which were obtained by collagenase digestion, could be maintained in culture for weeks to months, and were identified as endothelium by the presence of Weibel-Palade bodies and von Willebrand factor (VIII-associated antigen). The capacity to culture ECs provided researchers with a new and powerful tool to dissect cell structure and function under controlled (albeit artificial) conditions. These studies have addressed every conceivable in vitro property of ECs including – but certainly not limited to – subcellular organelles (e.g., Weibel-Palade bodies, mitochondria, Golgi apparatus, and endoplasmic reticulum); membrane microdomains such as caveolae or lipid rafts; cell signaling (from cell surface receptor to the level of gene transcription and posttranscriptional regulation); and the cytoskeleton.

While the cultured EC became a focal point for research in vascular biology, increasing evidence pointed to the highly complex topology of the intact endothelium. In the 1980s, several groups carried out systematic immunohistochemical analyses of the endothelium in various organs (48–50). These studies, which built on the earlier results of electron microscopy, revealed differential expression of lectins and antigens in the intact endothelium. In other words, the endothelium displayed *molecular* heterogeneity. Implicit in these observations is a critical – though largely overlooked – message which was articulated by Robert Auerbach in 1985:

> The concept that vascular endothelial cells are not all alike is not a new one to either morphologists or physiologists. Yet laboratory experiments almost always employ endothelial cells from large vessels such as the human umbilical vein or the bovine dorsal aorta, since these are easy to obtain and can be readily isolated and grown in culture. The tacit assumption has been that the basic properties of all endothelial cells are similar enough to warrant the use of the cells as in vitro correlates of endothelial cell activities in vivo (51).

According to Auerbach, a key to understanding structural and functional heterogeneity was to isolate and study microvascular ECs from different organs. This approach makes the most sense if site-specific phenotypes are retained in culture. However, as we will see throughout this book, this assumption is only partially correct. For example, many site-specific properties of ECs depend on extracellular signals. Thus, when ECs are removed from their native microenvironment and cultured in vitro, they undergo phenotypic drift and lose many of their original properties.

Beyond Electron Microscopy and Cell Culture

Although electron microscopy and cell cultures largely contributed to the structural and functional characterization of the endothelium, other techniques have provided new insights into the endothelium. For example, during the past few years, investigators have employed novel genomic and proteomic techniques to uncover an enormous array of site-specific properties (so-called vascular addresses or zip codes) of the endothelium. Together, these studies suggest that "far from being a giant monopoly of homogenous cells, the endothelium represents a consortium of smaller enterprises of cells located within blood vessels of different tissues . . . while united in certain common functions, each enterprise is uniquely adapted to meet the demands of the underlying tissues" (52). The modern techniques of fate mapping emphasize their developmental perspective. Again, differences in methods matter, enabling researchers to highlight different aspects of development. Similarly, the availability of new model organisms for research in endothelial biology, such as genetically modified mice and zebrafish, have had a profound impact on theoretical conceptions of the endothelium.

One consequence of science's dependence on methods and technologies is that knowledge accumulation is rarely continuous. We see this in the gap existing between the availability of a large number of empirical details about ECs and our limited understanding of the endothelium as a functional whole with clinical importance. We lack relevant theoretical models in systems biology that could help illuminate this bigger and more complex system in which ECs reside. As a result of the specialized thinking within separated lines of study, the dominant view of the endothelium is still rooted more in the conceptual structure of isolated cell cultures, which has implications for the clinical applications of endothelial biology.

ENDOTHELIUM IN DISEASE

The first evidence of a potential role for the endothelium in disease (as victim, if not perpetrator) is found in published reports from late nineteenth century, which describe the abnormal morphology of ECs in a number of disease states, including tuberculosis and malaria. In the 1870s, Julius Cohnheim studied the frog tongue to demonstrate that leukocytes adhere to the blood vessel wall of venules (so-called pavementing of leukocytes), many of which passed through the wall into the extravascular tissues (leukocyte emigration). These observations were later confirmed in mammalian species. Studies in an ear chamber model demonstrated that leukocytes adhere to the damaged side of a blood vessel, suggesting that the blood vessel wall – as distinct from the leukocyte – is primarily responsible for mediating adhesion (53). However, the mechanisms underlying inflammation-induced leukocyte adhesion remained elusive for decades. According to one theory, the endothelium secreted a gelatinous substance that trapped leukocytes (54). Others claimed that electrostatic forces were responsible for mediating the endothelial–leukocyte interactions (55).

As with so many other aspects of EC biology, the elucidation of the role for endothelium in inflammation and the molecular basis of leukocyte trafficking would wait for the successful culture of ECs. In the 1980s, several investigators demonstrated that treatment of cultured ECs with inflammatory mediators resulted in phenotypic changes that included increased expression of cell adhesion molecules, leukocyte adhesion, antigen presentation, and procoagulant activities. These changes, many of which were subsequently shown to occur in vivo, were termed *endothelial activation.*

Another term that came into favor in the late twentieth century was *endothelial dysfunction.* In the 1970s, subsequent to Russell Ross's response-to-injury hypothesis to explain the mechanisms of atherosclerosis, there was a growing appreciation that the *intact* endothelium may actively contribute to disease initiation and/or progression (56). Although the term *endothelial cell dysfunction* was first coined in 1980 to describe the hyperadhesiveness of the endothelium to platelets, it was quickly adopted by the field of cardiology to describe abnormal endothelial vasodilator function (57). Indeed, it is not uncommon today to find full-length publications on endothelial dysfunction that refer exclusively to altered vasomotor tone in coronary arteries. Of course, the endothelium has many functions beyond the control of vasomotor tone, and it is distributed widely throughout the body. Thus, endothelial dysfunction is not restricted anatomically to the heart, nor is it limited in disease scope to atherosclerosis.

CONCLUSION

Since Galen, we have determined that a layer called the endothelium exists, and that it is made up of cells called *endothelial cells*. The structure is derived from the mesodermal germ layer, is widely distributed throughout the body, is highly active with multiple functions, and is remarkably heterogeneous in structure and function. New techniques have illuminated more detail while simultaneously raising questions about the earlier simple assumptions.

Despite breathtaking advances in EC biology and a growing appreciation for its role in disease states, endothelial biomedicine has made little progress as a field. There are several possible reasons for this. First, overspecialization in medicine hampers cross-disciplinary approaches to the endothelium and keeps researchers in one area from learning about the ideas and approaches of another. Second, because the endothelium displays emergent properties and is so highly adapted to its microenvironment, it must be approached in the context of the whole organism. As long as cell culture studies are sufficient for publication and funding, there is little incentive to study endothelial biomedicine as a whole, or to look at the diverse and dispersed roles of endothelium in the body. Finally, it may be argued that by figuratively stripping the endothelium from the blood vessel and employing it as a frame of reference (i.e., "endothelial biomedicine") we are "inventing" a field with little clinical value.

Some might argue that the endothelium should not be considered outside the context of the blood vessel. They might point out, for example, that endothelial–smooth muscle cell interactions are essential and should constitute the minimal unit of investigation and inquiry. Admittedly, this argument stands for conduit and resistance blood vessels. Alternatively, it may be argued that the functional unit of capillaries (which comprise the vast majority of the surface area of the vasculature) is the endothelial–pericyte–extracellular matrix interaction. An investigator in an organ-specific discipline may make a case for studying the endothelium in the context of the whole tissue – for example, the nephrologist, who wishes to understand the role that the glomerular endothelium plays in urinary excretion.

We are beginning to appreciate previously hidden levels of communication between ECs and other cell types and systems. For example, increasing evidence suggests a tight developmental, structural, and functional link between the endothelium and the nervous system. Perhaps, one day, these endothelial–neural connections will define the term *vascular*.

So, it is really a matter of perspective. In choosing the endothelium as our frame of reference, we acknowledge the importance of the endothelial–mural cell interaction and the unique value of each and every vascular bed/organ (including the heart), but treat them as one of many configurations of the endothelium.

This book provides collective knowledge about the endothelium. It strives to answer the following questions:

- What do we know about the biology of the endothelium – for humans as well as other animals?
- What do we know about its structure, function, and development, including evolutionary development?
- What do we know about the clinical implications of endothelial function and the foundations on which those rest?
- How do we go about interpreting what we know, through metaphors, models, and methods?

This volume examines all these issues. In the final chapter, we return to reflect on the knowledge provided by the whole and on what it means for endothelial biomedicine.

ACKNOWLEDGMENTS

Special thanks to William Bechtel, who added tremendously to the chapter with his advice and knowledge about electron microscopy and its impacts. We recommend his *Discovering Cell Mechanisms: The Creation of Modern Cell Biology* (Cambridge University Press, 2005) for further information on that topic.

REFERENCES

1 Nuland SB. *Doctors. The Biography of Medicine*. New York: Alfred A. Knopf, 1989.
2 Vesalius A. *De Fabrica* (1543; 1555). In: Debus AG, trans. *Man and Nature in the Renaissance*. Cambridge: Cambridge University Press, 1978.
3 Harvey W. *Movement of the Heart and Blood in Animals* (1628). In: Franklin KJ, trans. *The Circulation of the Blood and Other Writings*. London: Everyman's Library, 1963.
4 Descartes R. *The Passions of the Soul*. In: Haldane ES, Ross GRT, trans. *The Philosophical Works of Descartes*. 2 Vols. (Especially articles VII–IX.) Cambridge: Cambridge University Press, 1970.
5 Descartes R. *Description of the Human Body*. In: Cottingham J, Stoothoff R, Murdoch D, trans. *The Philosophical Writings of Descartes*. 3 Vols. Cambridge: Cambridge University Press, 1991.
6 Adelman H. *Marcello Malpighi and the Evolution of Embryology*. Ithaca: Cornell University Press, 1966.

7 Westfall RS. *The Construction of Modern Science: Mechanisms and Mechanics*. Cambridge: Cambridge University Press, 1978.

8 de La Mettrie JO. *Man a Machine* (1748). La Salle, IL: Open Court, 1991.

9 Otis L. *Networking Communicating with Bodies and Machines in the 19th Century*. Ann Arbor: University of Michigan Press, 2001.

10 Weindling PJ. *Darwinism and Social Darwinism in Imperial Germany: The Contribution of the Cell Biologist Oscar Hertwig (1849–1922)*. Stuttgart: Gustav Fischer, 1991.

11 Ackerknecht E. *Medicine at the Paris Hospital, 1974–1948*. Baltimore: The Johns Hopkins University Press, 1967.

12 von Baer KE. *De ovi mammalium et hominis genesi. Epistola ad Academiam Imperialem Scientiarum Petropolitanam*. Lipsiae: L. Vossius, 1827.

13 Pander CH. *Dissertatio sistens historiam metamorphoseos, quam ovum incubatum prioribus quinque diebus subit*. Wirceburgi: T. E. Nitribitt, 1817.

14 Schleiden MJ. *Beiträge zur Phytogenesis*. In: Syndenham Society, trans. *Müller's Archive Anat Physiol Wissenschaftliche Med*. Vol. 12, 1838.

15 Schwann T. *Mikroskopische Untersuchungen uber die Übereinstimmung in der Struktur und dem Wachsthum der Tiere und Pflanzen*. Berlin: Verlag der Sander'schen Buchhandlung, 1839.

16 Bracegirdle B. *A History of Microtechnique*. Ithaca: Cornell University Press, 1978.

17 Bradbury S. *The Microscope Past and Present*. Oxford: Pergamon Press, 1968.

18 Harris H. *The Birth of the Cell*. New Haven: Yale University Press, 1999.

19 Hughes A. *A History of Cytology*. London: Abelard-Schuman, 1959.

20 Baker JR. *The Cell Theory: A Restatement, History, and Critique* (1948–1955). New York: Garland Publishers, 1988.

21 Virchow R. *Die Cellularpathologie in ihrer Begründung auf physiologische und pathologische Gewebelehre*. Berlin: A. Hirschwald, 1858; English translation, New York: R. M. Dewitt. Frank Chance, trans., 1860.

22 Rather LJ. *Disease, Life, and Man. Selected Essays by Rudolf Virchow*. Palo Alto: Stanford University Press, 1958.

23 Billroth T. Ein Fall von Meningocele spuria cum fistula ventriculi cerebri. *Archiv für klinische Chirurgie*. 1862;3:398–412.

24 Osler W. *Diseases of the Arteries*. Philadelphia: Lea & Febiger, 1908.

25 His W. *Die Häute und Höhlen des Körpers*. Basel: Schwighauser, 1865.

26 Bichat X. *Traite des membranes en general, et de diverses membranes en particulier*. Paris: Chez Mme. Veuve Richard, Chez le C. Maequignon, 1802.

27 Waldeyer W. Kittsubsatnz und Grundsubstanz, Epithel und Endothel. *Archiv für mikroskopische Anatomie und Entwicklungsmechanik*. 1901;57:1–8.

28 DaCosta JC, Spitzka EA. *Anatomy Descriptive and Surgical by Henry Gray*. Philadelphia and New York: Lea and Febiger, 1908.

29 Böhm AA, Davidoff M. von. *A Text-book of Histology Including Microscopic Technique*. Huber GC, ed. Philadelphia: W. B. Saunders, 1910.

30 Philipson GH. Pathological relations of the absorbent system. *Lancet*. 1884;24:309–311.

31 Pappenheimer JR. Passage of molecules through capillary walls. *Physiol Rev*. 1953;33(3):387–423.

32 Landis E. Micro-injection studies of capillary permeability. II. The relation between capillary pressure and the rate at which fluid passes through the walls of single capillaries. *Am J Physiol*. 1927;82:217–238.

33 Ramon Y Cajal S. *Histology* (1928). Tello-Muñoz JF, ed. Fernan-Nuñez M, trans. Baltimore: William Wood and Company; 1933.

34 Cowdry EV. *A Textbook of Histology. Functional Significance of Cells and Intercellular Substances*. Philadelphia: Lea and Febiger, 1934.

35 Cowdry EV. *A Textbook of Histology. Functional Significance of Cells and Intercellular Substances*. Philadelphia: Lea and Febiger, 1938.

36 Kohn A. Endothel und Epithel. *Anatomischer Anzeiger*. 1935;81:106–116.

37 Nachman RL, Jaffe EA. Endothelial cell culture: Beginnings of modern vascular biology. *J Clin Invest*. 2004;114(8):1037–1040.

38 Aird WC. Spatial and temporal dynamics of the endothelium. *J Thromb Haemost*. 2005;3(7):1392–1406.

39 SoRelle R. Nobel prize awarded to scientists for nitric oxide discoveries. *Circulation*. 1998;98(22):2365–2366.

40 Baker JR. *Cytological Technique*. London: Muthuen & Co., 1951.

41 Wyckoff RWG. *Optical Methods in Cytology*. New York: Academic Press, 1959.

42 Palade GE. An electron microscope study of the mitochondrial structure. *J Histochem Cytochem*. 1953;1(4):188–211.

43 Palade GE. A small particulate component of the cytoplasm. *J Biophys Biochem Cytol*. 1955;1(1):59–68.

44 Weibel ER, Palade GE. New cytoplasmic components in arterial endothelia. *J Cell Biol*. 1964;23:101–112.

45 Hibbs RG, Burch GE, Phillips JH. The fine structure of the small blood vessels of normal human dermis and subcutis. *Am Heart J*. 1958;56(5):662–670.

46 Gimbrone MA Jr., Cotran RS, Folkman J. Human vascular endothelial cells in culture. Growth and DNA synthesis. *J Cell Biol*. 1974;60(3):673–684.

47 Jaffe EA, Hoyer LW, Nachman RL. Synthesis of antihemophilic factor antigen by cultured human endothelial cells. *J Clin Invest*. 1973;52(11):2757–2764.

48 Kumar S, West DC, Ager A. Heterogeneity in endothelial cells from large vessels and microvessels. *Differentiation*. 1987;36(1):57–70.

49 Page C, Rose M, Yacoub M, et al. Antigenic heterogeneity of vascular endothelium. *Am J Pathol*. 1992;141(3):673–683.

50 Turner RR, Beckstead JH, Warnke RA, et al. Endothelial cell phenotypic diversity. In situ demonstration of immunologic and enzymatic heterogeneity that correlates with specific morphologic subtypes. *Am J Clin Pathol*. 1987;87(5):569–575.

51 Auerbach R, Alby L, Morrissey LW, et al. Expression of organ-specific antigens on capillary endothelial cells. *Microvasc Res*. 1985;29(3):401–411.

52 Aird WC, Rosenberg RD. *Vascular Endothelium: Physiology, Pathology, and Therapeutic Opportunities*. Born GVR, Schwartz CJ, eds. New York: Schattauer, 1997.

53 Allison F, Jr., Smith MR, Wood WB Jr. Studies on the pathogenesis of acute inflammation. I. The inflammatory reaction to thermal injury as observed in the rabbit ear chamber. *J Exp Med*. 1955;102(6):655–668.

54 McGovern VJ. Reactions to injury of vascular endothelium with special reference to the problem of thrombosis. *J Pathol Bacteriol*. 1955;69:283–293.

55 Bangham AD. The adhesiveness of leukocytes with special references to zeta potential. *Ann NY Acad Sci*. 1964;116:945–949.

56 Schwartz SM, Gajdusek CM, Reidy MA, et al. Maintenance of integrity in aortic endothelium. *Fed Proc*. 1980;39(9):2618–2625.

57 Ludmer P, et al. Paradoxical vasoconstriction induced by acetylcholine in atherosclerotic coronary arteries. *N Engl J Med*. 1986; 315:1046–1051.

Evolutionary, Comparative Biology, and Development

2

Introductory Essay

Evolution, Comparative Biology, and Development

William C. Aird* and Manfred D. Laubichler†

*Beth Israel Deaconess Medical Center, Harvard Medical School, Boston, Massachusetts;
†Center for Biology and Society, Arizona State University, Tempe

WHY STUDY EVOLUTION?

An understanding of the evolutionary origins of the endothelium has important implications for human health and disease.[1] As discussed by Nesse and Weder (Chapter 15), when approaching mechanisms of structure and function, evolutionary explanations ("why") complement proximate considerations ("how"). They provide important insights into the design constraints, path dependence, trade-offs, and selective pressures that underlie endothelial function and vulnerability to disease. Another critical and often underappreciated consideration is that, in the course of human evolution, our body (including the endothelium) has evolved to maximize fitness in a far earlier era, perhaps tens of thousands of years ago, which is the time it takes for the gene pool to be filtered by natural selection. In Chapter 16, Eaton and his colleagues write about the environment and lifestyle of the hunter-gatherer, and remind us about the importance of cultural (as distinct from genetic) evolution in mediating the modern-day predisposition to atherosclerosis. As another example of how evolutionary theory may be applied to an understanding of human disease, Haig discusses how evolutionary conflicts between maternal and fetal genes may underlie the pathophysiology of preeclampsia (Chapter 17).

Piecing together the evolutionary origins of the endothelium is challenging and necessarily speculative. Because the human endothelium does not fossilize, interpretations rely on a combination of molecular phylogeny and comparative biology/physiology, with the assumption that what works for extant organisms may have worked for ancestral species. In Chapter 14, McVey demonstrates how molecular phylogenetic approaches may be used to gain insights into the evolutionary history of the endothelium. In Chapter 3, Burggren and Reiber provide a comprehensive overview of the comparative biology of cardiovascular systems. Based on these data, they offer possible explanations for the evolutionary origins of a vascular lining. However, as they readily acknowledge, the selective pressures underlying the transition from an epithelial-like cell layer in invertebrates to a true endothelium in vertebrates remain enigmatic.

Endothelial cells (ECs) are absent in invertebrates, cephalochordates, and tunicates, but present in the three major groups of extant vertebrates: hagfish (myxinoids), lampreys, and jawed vertebrates (gnathostomes). The last common ancestor of hagfish and gnathostomes was also the last common ancestor of all extant vertebrates, which lived at some time more than 500 million years ago. As discussed by Aird in Chapter 6, the fact that the endothelium is shared by jawless and jawed vertebrates is evidence that the endothelium was present in this ancestral vertebrate. The absence of endothelium in amphioxus indicates that this structure was not present in the common ancestor of cephalochordates and vertebrates, and must have evolved following the divergence of amphioxus from vertebrates, between 550 and 510 million years ago (Figure 2.1).

WHY STUDY COMPARATIVE ENDOTHELIAL BIOLOGY?

In addition to providing insights into the evolutionary origin of the endothelium, comparative studies also provide powerful conceptual frameworks for approaching the physiology and pathophysiology of the human vascular system. From the standpoint of oxygen (O_2) transport, the body plan of vertebrates (and many invertebrates) may be simplified according to the scheme shown in Figure 2.2. According to this model, O_2 delivery (hence consumption) is governed by two principal forces: diffusion and convection. Diffusion takes

1 The application of evolutionary principles to an understanding of human health and disease represents the foundation of a fledgling field termed Darwinian medicine, as reviewed in Nesse RM, Williams GC. Evolution and the origins of disease. *Sci Am.* 1998;279:86–93. See also Nesse RM, Williams GC. *Why We Get Sick. The New Science of Darwinian Medicine.* New York: Crown, 1995.

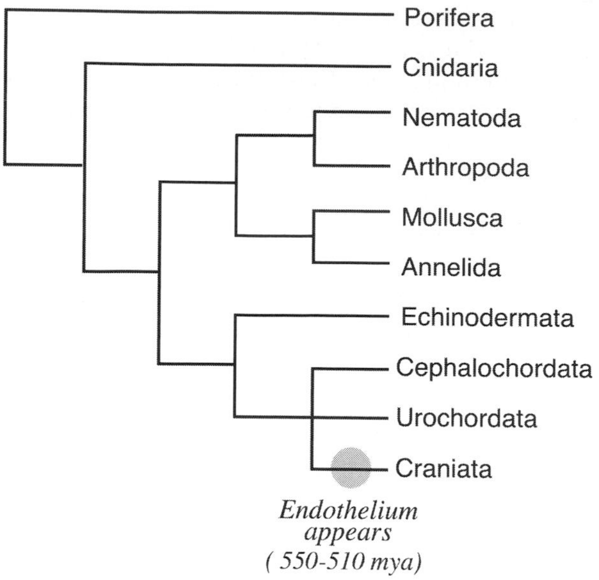

Figure 2.1. Phylogenetic tree.

place between the external environment and a highly vascularized surface and internally between blood and underlying tissue. Convection provides bulk flow delivery of O_2 from the external environment and bulk flow delivery of oxygenated blood throughout the body.

As evidenced by the examples presented in the following chapters, many ways exist to "solve" the fundamental challenge of O_2 delivery. Importantly, each variation conforms to the laws of physics. In fish, diffusion of O_2 occurs in the gills; in terrestrial vertebrates, diffusion of O_2 occurs at the level of alveoli. In both cases, a dedicated pump is present to deliver O_2 by bulk flow. In Chapter 9, Tattersall discusses an interesting exception to the gill/lung paradigm, namely the obligate skin-breathing amphibian. As predicted by Fick's law of diffusion, the skin of these animals is extremely thin, displays high surface area (by way of skin folds), and comprises a dense capillary network within the epidermis. To compensate for the lack of a dedicated ventilatory pump, these animals have evolved voluntarily movements (e.g., bobbing) as a means of replenishing O_2 at their skin surface.

Another fascinating exception to the general body plan is the Antarctic icefish, which is discussed by Detrich in Chapter 7. These fish are unique among vertebrates in that they completely lack circulating red blood cells and hemoglobin. As a result, their O_2-carrying capacity is extraordinarily low. To meet their O_2 requirements, these icefish display several adaptations, including increased cardiac output, mitochondrial density, and capillary density. Detrich offers an interesting hypothesis to explain how the evolutionary loss of hemoglobin (and myoglobin) was physiologically coupled to adaptive changes in capillary and mitochondrial density.

The giraffe, the tallest living land animal on the planet, experiences unparalleled gravitational forces and requires special adaptations to prevent gravity-dependent edema in the legs. As discussed by Hargens and his colleagues (Chapter 12), a consideration of these adaptations illuminates the importance of the Starling-Landis law, and provides potential insight into mechanisms underlying autonomic dysfunction and postural hypotension in humans, as occurs in the elderly.

Icardo's chapter on fish endocardium demonstrates not only that endothelial heterogeneity exists in early vertebrates (teleosts), but also that certain homeostatic mechanisms (e.g., secretory function, scavenger function) may be carried out by distinct organs and/or vascular beds across species, again pointing to different solutions to similar problems and challenges (Chapter 8).

The shrew is the smallest mammal on the planet. The physiology of the shrew, which is reviewed by Jürgens in Chapter 13, emphasizes the importance of scaling. Scaling is the structural and function consequences of a change in size (or in scale) among similarly organized animals. When considering mammals, certain properties scale in direct proportion (exponent 1.0) to body size. For example, although the heart size is larger in an elephant than a mouse, the heart represents 0.6% of the total body mass in each animal. Other properties

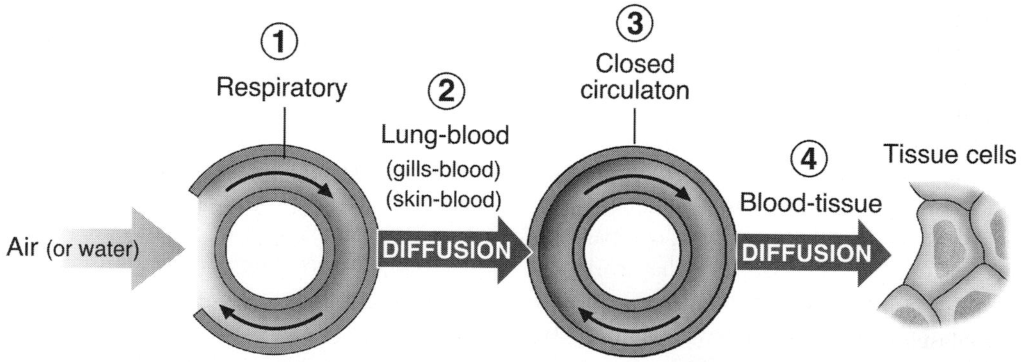

Figure 2.2. Body plan of vertebrates. This schematic simplifies oxygen transport according to four steps: (1) convection of oxygenated air or water from environment to a highly vascularized surface (skin, gills, or lungs); (2) diffusion of O_2 across the gas exchanger into the blood; (3) convection of oxygenated blood around to the various tissues of the body; and (4) diffusion of oxygen to the individual cells of the tissues. (Adapted with permission from Weibel ER. *The Pathway for Oxygen: Structure and Function in the Mammalian Respiratory System.* Cambridge: Harvard University Press, 1984.)

that scale in simple proportion to body weight are lung volume, tidal volume, and blood volume. In contrast, some features exist that do not scale proportionately with weight. The most widely studied of these is basal metabolic rate or O_2 consumption (hence, O_2 delivery). Consider for example an elephant and a mouse. Although the elephant weighs 1 million times the mouse, each gram of mouse tissue consumes more than 100 times the amount of O_2 compared with each gram of elephant tissue. In other words, the rate of O_2 consumption per gram decreases with increasing body size. It has been hypothesized that the high surface area:volume ratios of smaller animals result in greater evaporation, thus necessitating higher rates of O_2 consumption (metabolic heat) to maintain body temperature. Arguing against this supposition (as the sole explanation) is the observation that "cold-blooded" invertebrates also demonstrate regression lines with similar slopes.

WHY STUDY DEVELOPMENT?

The circulatory system is the first functional organ to develop in the embryo. This is intuitive, because all other organs depend on bulk flow delivery of O_2 and nutrients. Moreover, as mentioned below and covered in more detail by Lammert (Chapter 22), ECs provide important cues for the development of other tissues and organs. Many human diseases involve abnormalities in blood vessels. Signaling mechanisms are likely to be conserved between adult and developmental blood vessel formation. An important goal in regenerative medicine is to induce the formation of new blood vessels in certain ischemic diseases. Blood vessel growth is an important therapeutic target in diseases associated with excessive angiogenesis, including tumors, retinopathy, and rheumatoid arthritis.

Vasculogenesis, the process that describes the in situ differentiation of endothelial precursor cells (angioblasts) from embryonic mesoderm, results in the formation of the earliest vascular plexus (also termed *primary capillary plexus*) in the embryo proper. In a process common to all vertebrates, these angioblasts, once differentiated, migrate to the trunk midline and coalesce to form the primary axial vessels – the dorsal aorta and cardinal vein. In some cases, angioblasts differentiate directly from mesodermal cells, whereas in other cases they differentiate from an intermediate cell type, the hemangioblast. As discussed in several of the chapters in this section, the molecular basis of mesoderm-to-angioblast and mesoderm-to-hemangioblast-to-angioblast differentiation is an area of intense research interest. Among the mediators implicated are fibroblast growth factor, Indian hedgehog, and vascular endothelial growth factor (VEGF)-A.

Later development of the mature vessel system involves angiogenesis, with proliferation and sprouting of new vessels from existing ones. Sprouting angiogenesis[2] is controlled by gradients of extracellular matrix–bound growth factors interacting with specialized cells at the growing edge of the vessel, termed tip cells. These latter cells express growth factor receptors and possess filopodia for "exploring" the microenvironment. Programmed branching, or stereotypic patterning, of new blood vessels is governed by a delicate balance of attractive and repellent guidance cues. Increasing evidence suggests that these cues (e.g., ephrins, semaphorins-neuropilins, netrins, and Robo-Slit) play a common role in mediating the patterning of both blood vessels and nerves (these mechanisms are covered in Part II of this volume).

Arterial-venous (A-V) identity is determined before the onset of blood flow. Several markers of arterial versus venous endothelium exist. Based on studies in zebrafish and mice, some of these markers have been shown to play a role in determining venous-arterial identity, including notochord-derived sonic hedgehog, somite-derived VEGF-A, and endothelial-derived Notch, gridlock, EphB4, and ephrinB2. The prevailing view is that Notch induces arterial fate by suppressing venous phenotype. Stabilization or maturation involves the recruitment of mural cells, a process that involves transforming growth factor (TGF)-β, TGF-β RI and RII receptors, platelet-derived growth factor (PDGF)-B, PDGF-Rβ, and angiopoeitin-Tie2 interactions (these mechanisms are covered in Part II of this volume).

Advances in the field of vascular development are based on complementary approaches. Some investigators focus on mice, others on fish (zebrafish), amphibians (*Xenopus*), or birds. Mice have the advantage in that they are closer to humans and are amenable to knockout and overexpression studies (see Chapter 20). As discussed by Chopra and Zhong in Chapter 19, fish develop externally from the mother, are optically clear, receive their O_2 via passive diffusion (and can thus survive in absence of blood vessels), and are amenable to unbiased forward genetic analysis. In Chapter 18, Patient and his colleagues review the benefits of studying vascular and endothelial ontogeny in *Xenopus*. Like zebrafish, aquatic frogs are easy to maintain and develop externally. Moreover, frog embryos lend themselves to microinjection and microdissection. An advantage of the bird as a research tool is that it is amenable to transplantation studies (reviewed by Poole in Chapter 10). Certain labs focus on a signaling pathway, for example VEGF/FLK-1, throughout development and even into the postnatal and adult periods. Others place their investigative lens on a certain developmental process, such as A-V identity, which requires an understanding of multiple ligand–receptor pairs.

Two important caveats must be stated. First, while much attention is being paid to genetic predeterminants of arterial and venous fate, increasing evidence suggests that many of these properties are plastic and are in fact critically dependent on the microenvironment (particularly blood flow). Second, vasculogenesis is not confined to early stages of embryogenesis, but continues to play an important role throughout development and into adulthood.

2 Another form of angiogenesis is termed *intussusceptive angiogenesis.*

THE ENDOTHELIUM SEEN FROM THE PERSPECTIVE OF EVOLUTIONARY DEVELOPMENTAL BIOLOGY

As we have said before, establishing the exact evolutionary history of the endothelium involves a lot of conjectures. We can readily establish the molecular phylogeny of certain crucial genes, such as VEGF (Chapter 14), or provide an argument as to why an endothelial lining of the vascular system is a beneficial, or even crucial, adaptation of larger organisms, and why we would expect that natural selection favored these structures (Chapter 3). All these studies and associated ideas contribute to our understanding of the evolution of the endothelium. But they do not yet provide us with a detailed mechanistic understanding of the origin and evolutionary history of the endothelium. For this, we must realize that at one point the endothelium had to originate within an ancestral condition; in other words, at one point in the course of evolution the endothelium emerged as an evolutionary novelty. In all likelihood, the emergence of the endothelium was a gradual process – we do not expect such a complex structure to come into being all at once – but nevertheless, the specific developmental changes and differentiations leading to the formation of what His called the "membranes of the inner cavities" (see Chapter 1) began at some point in evolutionary history. From what we have said so far, it is clear that any explanation of evolutionary novelties necessarily involves an account of how these structures develop ontogenetically in the first place, as well as how they were subsequently modified in the course of evolution. Such explanations – involving both developmental (i.e., proximate) and evolutionary mechanisms – are the purview of the relatively new discipline of evolutionary developmental biology (also termed *Evo Devo*).[3] The endothelium as a crucial innovation in the evolution of higher organisms and a rather diverse structure, both ontogenetically and phylogenetically, is thus a prime candidate for an Evo Devo analysis (Figure 2.3).

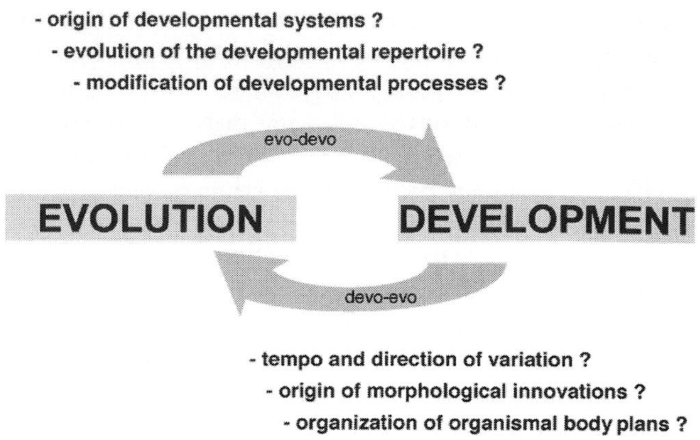

Figure 2.3. The dualistic structure of Evo Devo questions. The two subagendas, evo-devo and devo-evo, address different kinds of questions. For the study of the endothelium these involve: (a) the developmental origins of the endothelium; (b) the evolution of the developmental trajectories of the endothelium; and (c) their modification; (d) the phylogenetic differentiations of vascular systems and their lining; (e) the endothelium as an evolutionary novelty; and (f) the organization of a generalized vascular system (body plan).

Methodologically, evolutionary developmental biology combines approaches from developmental and evolutionary biology to explain the patterns and processes of phenotypic evolution. Besides the notion of evolutionary novelties, one of the central concepts in Evo Devo is the idea of constraints as a limiting factor on phenotypic variation. All the chapters in this section, especially Burggren and Reiber's overview of the comparative and evolutionary biology of vascular systems (Chapter 3), point out that the circulation of gases and nutrients is one of the central problems of animal physiology. In many organisms, especially those of a particular size and activity level, a circulatory system is an essential component of organismal design and, given the limitations set by the physics of gas exchange and diffusion, a closed system with a vascular lining provides an optimal solution to these challenges. This engineering perspective on organismal design can be combined with evidence from the evolutionary history and developmental mechanisms to arrive at a better understanding of the origin of vascular systems. In the course of evolution, several different types of circulatory systems emerged in different lineages, in part as independent evolutionary events or novelties, in part as modifications of existing ancestral designs. All these circulatory systems represent a way to overcome the functional limitations imposed by the basic physical laws of diffusion.

Changes in organismal design are limited by certain functional and physical constraints. Moreover, the patterns, mechanisms, and historical contingencies of individual development also constrain what kind of phenotypic variations are possible (so-called developmental constraints).[4] The structure of the vascular system and the endothelium is no exception. Comparative studies of the structure and formation of the

3 All evolutionary transformations, although they manifest themselves on a population level, must materialize within the constraints and possibilities of a functioning developmental system of individual organisms. Although developmental biology focuses on individual developmental sequences describing the transformation from simple (a fertilized egg) to complex (an adult organism) through the differentiation of cells and the emergence of anatomical and histological structures, Evo Devo tries to understand how in the course of evolution something new can emerge within these regulated developmental sequences.

4 The concept of constraint has received a lot of attention in evolutionary biology, and a variety of different types of constraints have been proposed. Constraints are seen to have two different effects on the patterns of phenotypic evolution: (a) *limiting*, in the sense that certain phenotypes are not possible for a variety of different reasons, such as architectural constraints – when certain shell shapes in bivalves are geometrically not possible; functional constraints – in cases where the diameter of a vessel is too large, so that the pressure is too low to assure blood flow; or developmental constraints – when certain phenotypes cannot be realized within the existing developmental systems of certain lineages; and (b) *enabling*, in the sense that preexisting aspects of developmental systems facilitate certain types of novel phenotypic variants. See Maynard Smith J, et al. Developmental Constraints and Evolution. *Q Rev Biol* 1985;60:265–287 for a discussion of developmental constraints and Kirschner M, Gerhart J. *The Plausibility of Life.* New Haven: Yale University Press, 2005, for a discussion of the idea of facilitated variation.

vasculature, and more specifically the endothelium, reveal these trends and illustrate exactly what kind of phenotypes are possible based on certain invariant properties of vascular architecture and vascular development. An example of an architectural constraint is the closed-loop geometry and branching pattern of the vascular tree, which of necessity engender microdomains of disturbed flow at curvatures and bifurcations, respectively. Previous studies have established that ECs in these regions are primed for activation and vulnerable to disease (see Chapter 28). There is simply no way around this constraint. As another example, the differentiation of mesoderm-derived angioblasts into ECs, which is common to all vertebrates, must involve certain core processes whose disruption would compromise the very development of this cell layer. These core mechanisms, then, place a boundary on the nature and degree of phenotypic variation. A less theoretical and more practical consequence of these phenomena can be seen in the maternal–fetal conflict, described by David Haig (Chapter 17). Haig's analysis also makes it clear that many solutions to functional challenges represent compromises between different functional demands and are therefore not optimal for any one given task.

Developmental constraints, however, do not just limit the phenotypic possibilities of organismal design. Quite the contrary, they also enable phenotypic evolution in the first place. Indeed, the very regulatory programs that mediate core biological processes also serve as an important substrate for modification and natural selection. Françoise Jacob once described phenotypes as the result of *bricolage*, an opportunistic assembly of contingent properties of biological systems.[5] But this process is far from being random, as the term *bricolage* might suggest. Rather, the historical and developmental constraints of biological systems, while limiting the range of possible phenotypes in one sense, are also the foundation for the future evolution (evolvability) of these systems in the first place. Marc Kirshner and John Gerhart have recently called this property of biological systems *facilitated variation*.[6] Facilitated variation is a consequence of the constraints of developing systems, and mediates between random mutations and the more structured expressions of observable phenotypic variation. We can see how the evolutionary history of the vascular system and the endothelium (as described in the chapters of this section) is a case of facilitated variation that enables certain evolutionary modifications, while at the same time ruling out alternative solutions (see Figure 2.3).

The last topic of interest in the evolution of the vascular system and the endothelium that we want to discuss here are the mechanisms that largely make evolution of the endothelium possible. Recent studies in comparative genomics have shown that only a small number of often highly conserved genes are responsible for the observed diversity in phenotypes. It has thus become clear that phenotypic evolution is mostly a consequence of changes in regulatory gene networks and/or epigenetic control mechanisms, rather than downstream effector genes. This emphasis on regulatory evolution provides an important interpretative framework for the evolution of the endothelium. For example, it is likely that the mutation and selection of a small number of pre-existing regulatory genes (e.g., transcriptional control circuits) permitted the evolution of the endothelium. It is far less likely that the appearance of a completely novel gene, or the simultaneous mutation of multiple functional genes (e.g., those involved in mediating vasomotor tone, hemostasis, and leukocyte trafficking), was responsible for this event.

Regulatory changes, both ontogenetically and phylogenetically, operate through a complex web of signaling, gene activation, and cell-to-cell communication events based on complex signaling pathways and regulatory networks. As Mikawa shows in Chapter 21, fate mapping studies have revealed some of the details of the timing and location of vascular cell commitment and differentiation. Detailed studies of ECs throughout the vascular system of individual organisms have shown high degrees of localized specialization, as is evidenced by the variations in their functional properties and their molecular signatures and gene expression profiles (reviewed in Part II). Although some site-specific phenotypes are epigenetically programmed (i.e., rendered mitotically heritable), others are maintained through spatial and temporal differences in the extracellular microenvironment.

Kirschner and Gerhart argue that phenotypic diversification in multicellular organisms has evolved in the context of relatively fixed, basic body plans.[7] They propose a series of mechanisms by which the inherent constraints of core biological processes are reduced and allow for evolvability. Among these mechanisms for deconstraining evolutionary change is exploratory behavior, as exhibited by the genesis of neuronal connections and angiogenesis. Exploratory mechanisms are flexible and robust, and these are the very properties that are highly conserved.

> Angiogenesis is a well established example of functional selection. Although large vessels are probably placed by instructive processes, moderate and small vessels are not, as indicated by bilateral differences in pattern. In the vascular system (and in the analogous tracheal system in insects), local oxygen deprivation may play a role in controlling the growth and branching of moderate and small vessels.[8]

These principles may be more broadly applied to the endothelium. The endothelium evolved in the common ancestor of vertebrates, presumably through one or a series of mutations in regulatory genes involved in mesoderm function.

5 F. Jacob. Evolution and tinkering. *Science.* 1977;196:1161–1166.

6 Kirschner M, Gerhart J. Evolvability. *PNAS.* 1998;95:8420–8427, and M. Kirschner M, Gerhart J. *The Plausibility of Life.* New Haven: Yale University Press, 2005.

7 Viewed from this perspective, the endothelium was an "add-on" structure to the basic body plan of chordates, the so-called *pharyngula*.

8 Kirschner M, Gerhart J. Evolvability. *PNAS.* 1998;95:8420–8427.

Such changes did not compromise the basic body plan, but rather built upon that plan. The endothelium evolved as an exploratory system, meaning that, although its requirements for effective function are minimal, it has a remarkable capacity to sense and respond to extracellular signals and is thus able to adapt to the diverse needs of the local microenvironment.

The actual state or differentiation of individual cells is thus a consequence of rather complex regulatory contexts that often involve genetic, epigenetic, and environmental factors. In some cases, these decisions can be reversed, at least to some degree, whereas in others differentiation is irreversible. The diversity of ECs is thus a consequence of context-specific effects. Understanding these properties of ECs and their differentiation has enormous therapeutic implication, because clinical interventions often must be tailored to rather specific locations.

3

Evolution of Cardiovascular Systems and Their Endothelial Linings

Warren W. Burggren* and Carl L. Reiber†

*University of North Texas, Denton; †University of Nevada, Las Vegas

The evolution of large, multicellular, complex metazoans from their simple unicellular animal ancestors comprises a rich history covering many hundreds of millions of years. Pieced together from fossil remains, DNA analysis, and comparative morphological studies of extant animals, biologists have described a staggering array of adaptations that have arisen (and in some cases been lost and arisen again) through the course of animal evolution. Describing the specific evolutionary pathways from unicellular organisms to human is, of course, well beyond the scope of this chapter. Rather, in this chapter, we focus specifically on the general evolutionary pathway leading to the cardiovascular systems of extant animals, especially highlighting the general principles that continue to shape cardiovascular form and function to this day. In doing so, we intend to create a backdrop and context for other chapters in this book that present more specific information on the vascular endothelium in specific groups of animals (e.g., fish, amphibians, reptiles, birds, and mammals).

Before delving into cardiovascular systems and their evolution, we first must address a semantic issue regarding our use of the term *endothelium*. The invertebrate literature is replete with discussions of "endothelium" – whether it exists, whether it is discontinuous or continuous, what functions it serves, and the like. Much less often, the vascular lining of blood spaces (where it occurs) in invertebrates is described more generically as a "vascular cellular lining." Of note, the position taken in this book is that "true" endothelium is found only in vertebrates, and it has distinguishing characteristics, such as being derived from a lateral plate mesoderm, possessing caveolae and Weibel-Palade bodies, and more. Although the vascular cellular linings in some invertebrates are analogous to the endothelium of vertebrates, these linings are not necessarily homologous. Consequently, in this chapter we carefully distinguish between the two tissues, although in fact their functions may overlap extensively. Future research directed specifically at a comparison and contrast of the cells that line vascular spaces is much anticipated.

Now, let us consider why animals have cardiovascular systems – lined or unlined.

WHY DID CARDIOVASCULAR SYSTEMS EVOLVE?

A key feature of almost all complex metazoans is that they have an internal system for the convection of body fluids – a cardiovascular system. Although enormous species diversity exists in cardiovascular form and function, fundamental components include a fluid pump (or pumps) that propel often specialized body fluids into a series of distribution, exchange, and collection vessels. Numerous variations occur on this basic circulatory plan of heart and vessels, ranging from the relatively low pressure, "open" circulation of many invertebrates to the typically higher pressure, closed system of vertebrates. Yet, the almost ubiquitous presence of a circulatory pumping of fluid through vessels begs the following questions: "*Why did cardiovascular systems evolve?*" The answer, as is often the case in biology, is to be found in chemistry and physics.

A feature common to both the earliest single-celled animals and today's 100-metric ton blue whale is that they both ultimately depend on diffusion to supply oxygen (O_2) and nutrients to, and remove carbon dioxide (CO_2) and wastes from, their intracellular sites for oxidative metabolism. Perhaps because physiologists often study diffusion over the ultrathin boundaries between blood and gas in the alveolus, or blood and tubular fluid in the nephron, we tend to attribute near-magical properties to the process of diffusion. In reality, diffusion is a highly imperfect physico-chemical process for the bulk movement of materiel. Although energetically inexpensive, diffusion is also a very slow process; a diffusing molecule takes a very long time to move by diffusion alone. Indeed, it is precisely because diffusion is so slow that the ultrathin boundaries of the alveolus, nephron, and other structures have evolved to reduce transmembranous diffusion times and thus maximize the bulk diffusion of molecules, ions, and other particles.

The transmembranous diffusion of molecules depends on several factors including the pathway length for diffusion, surface area across which diffusion occurs, molecular weight of the diffusion molecule, solubility of the substrate

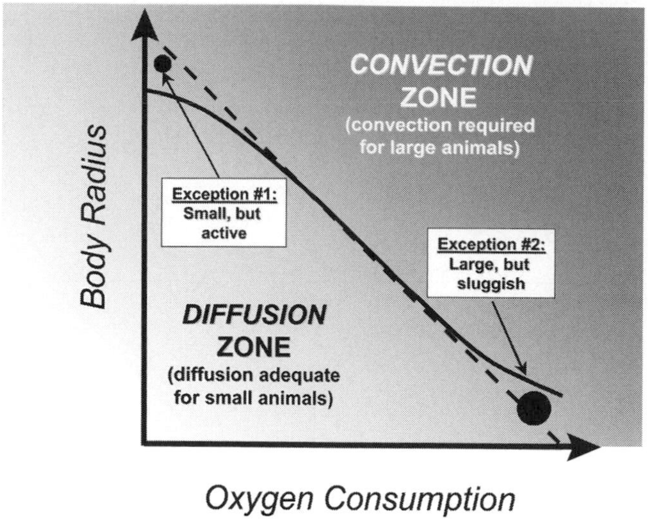

Figure 3.1. The role of diffusion and convection in the O_2 consumption of animals of different body size. Generally, small animals can thrive on diffusion alone, because the diffusion distances are small. The larger an animal's body radius, the more it depends on convection to support its O_2 consumption. However, very small yet active animals may require internal convection for delivery of O_2, whereas large but inactive animals may exist on diffusion alone.

for the diffusion molecule and, of course, the partial pressure or concentration gradient for the diffusion molecule. As a general rule of thumb, a hypothetical animal with a cylindrical shape, expressing a modest metabolic rate of about 100 μmol O_2/sec in a fully oxygenated environment, could survive on diffusion alone for its transport until it grew beyond a radius of about 2 mm (Figure 3.1) (1). Beyond this radius, diffusion alone cannot provide for substrate delivery and byproduct removal.[1] Some animals without internal circulations considerably exceed this size by having a very low metabolic rate, especially when combined with a large surface area-to-volume ratio–for example, the large, pancake-like marine Polyclads (Platyhelminthes) – as well as a modest internal convection system created by the muscular contractions of the animal's body wall during locomotion. Conversely, some animals with high metabolic rates, although near microscopic and thus possessing short diffusion distances, have evolved circulations for the internal transport of nutrients, wastes, and respiratory gases (e.g., small crustaceans) (see Figure 3.1).

Thus, the physico-chemical limitations of diffusive O_2 supply (together with the attendant limitations on the diffusion of CO_2, metabolic substrates, and nitrogenous and other wastes) have likely served as a strong selection pressure for the evolution of internal cardiovascular systems. Let us now consider the diversity of cardiovascular systems, beginning with relatively simple invertebrate systems and moving to the high-pressure systems used to perfuse large, complex vertebrate homeotherms.

CONVECTION OF INTERNAL BODY FLUIDS: DIVERSITY OF EVOLUTIONARY ADAPTATIONS

The mechanisms by which fluids are moved through a circulatory system evolved in loose association with both the complexity of the conduit system and particular species' metabolic demands. The advent of a "true" centralized muscular pump – that is, a heart that generates the pressure gradients used to move fluids (hemolymph or blood) through the circulatory system – is common in a variety of forms and to a broad array of animal taxa. Importantly, the evolutionary lineages of these pumps and their associated circulatory systems have multiple phylogenetic origins. Consequently, identifying homologous structures that cross phyla is difficult at best. It is, however, possible to identify analogous components that have arisen independently due to common selective pressures. Thus, distributed throughout the body and evident in a number of phyla are a variety of less familiar systems used to generate fluid pressures to facilitate blood movement and thus gas exchange.

INVERTEBRATE CIRCULATIONS

The evolution of the invertebrate cardiovascular system is not a continuum of homologous structures seen from taxa to taxa, but rather appears to have evolved independently in several phyla in response to limitations in diffusional gas exchange and high metabolic rates (for a thorough review of invertebrate cardiovascular systems see reference 2). Most invertebrates that have evolved a circulatory system do not show "vertebrate-like" circulatory complexity and lack sophisticated regulatory mechanisms. They also typically generate relatively low flows and pressure (Figure 3.2). Nonetheless, several examples occur in which invertebrates have evolved highly efficient, high-pressure, high-flow systems. These high-performance invertebrate cardiovascular systems show a number of convergent evolutionary[2] traits that are comparable to vertebrate taxa: muscular pumps capable of developing appropriate driving pressures and flows, cardiac regulatory mechanisms, a complex branching circulatory system and, of particular significance, a cell-lined vasculature, to list a few. A current misperception holds that the invertebrate cardiovascular system is sluggish, poorly regulated, and "open," wherein blood bathes the tissues directly as it moves through a system of ill-defined sinuses and/or lacunae, without an endothelial boundary (thus, in part explaining the use of the term *hemolymph* or a blood–lymph mixture versus "true" blood, which is

1 Interestingly, these same principles also apply to the sizes that tumors can achieve without additional circulation.

2 *Convergent evolution* is the evolution of similar characteristics by taxonomically different organisms.

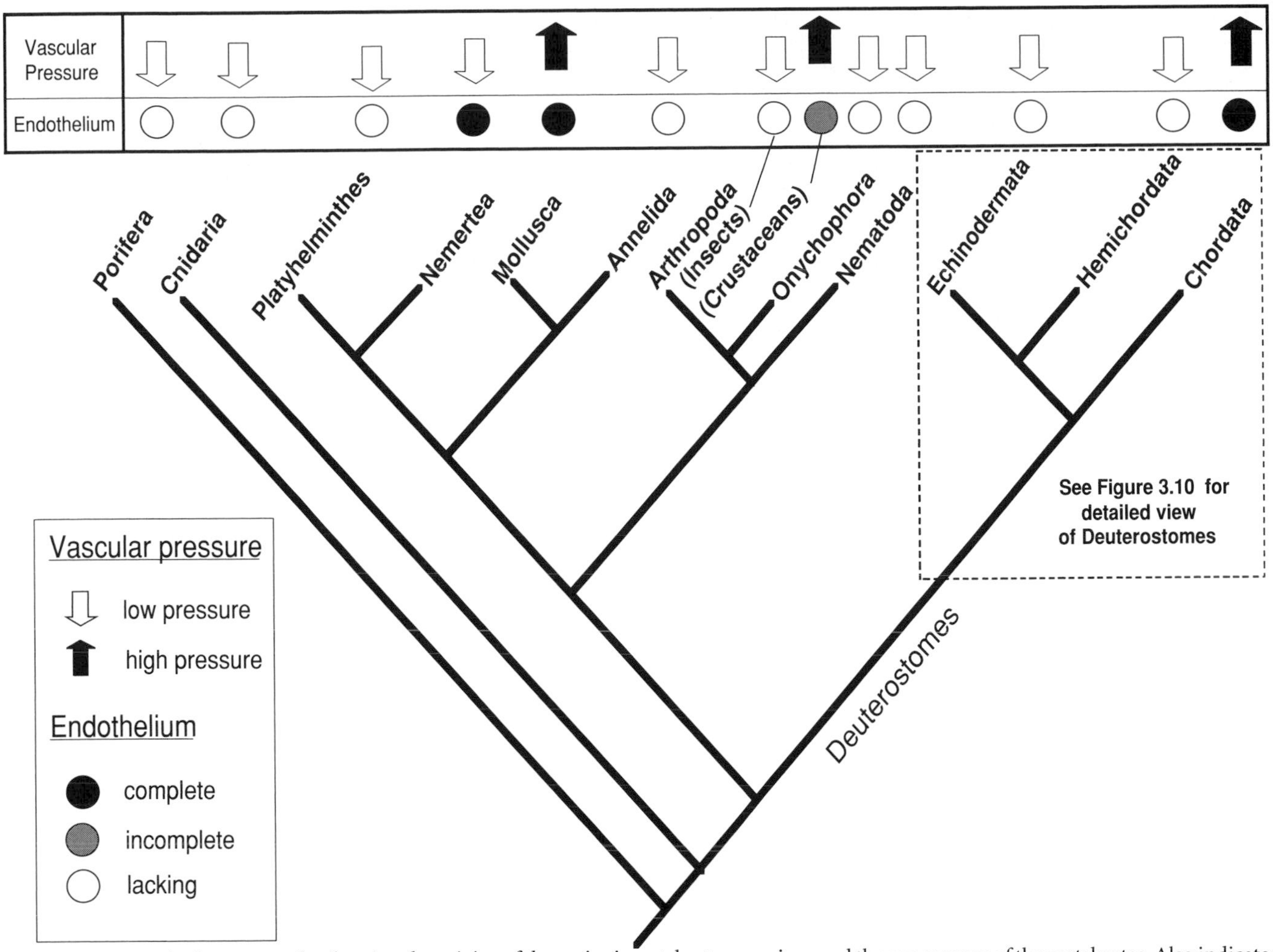

Figure 3.2. An evolutionary tree for showing the origins of the major invertebrate groupings and the appearance of the vertebrates. Also indicated is the presence of a complete or incomplete endothelial vascular lining. The arrows (*down* for low blood pressure, *up* for high pressure) indicate the taxon's blood pressure. (For comparison with vertebrates, see Figure 3.10).

typically contained within defined vessels and capillaries) (3). To the contrary, many invertebrates have evolved partially complete or even continuously cell-lined vessels (as revealed below). Why have both invertebrates and vertebrates coevolved these structures? What are the selective pressures behind the development of such systems? Several corollaries must be investigated, the most interesting being that animals having highly active lifestyles require an efficient pump and delivery system to maintain O_2 delivery (metabolic demand) and thus also require high-pressure, high-flow cardiovascular systems. In most examples of a high-pressure system, the vessels in that system are lined with an "endothelium." Why? The high vasculature pressures observed are the result of a well-developed muscular pump, with fluid contained within a robust arterial system followed by a capillary-like vascular bed to support gas exchange. Does a high-pressure cardiovascular system require the development of a cell-lined vasculature? A review of the literature in the area of invertebrate cardiovascular physiology, followed by a survey of the vertebrates, follows with a specific view toward supporting this hypothesis.

The Sponges: Internal Circulation of Seawater

One of the simplest yet more interesting pressure-generating systems is that of the phylum Porifera.[3] Sponges have circulatory systems for distributing seawater internally, the architecture of which incorporates a diffuse yet highly efficient pumping and circulatory system (4–5). A brief description of this circulatory system begins where seawater enters the sponge through in-current dermal pores (ostia) spaced throughout the exterior covering of epithelial-like flattened cells (the pinacoderm) (Figure 3.3). Internally, seawater then moves through the in-current canals and into a series of interconnected flagellated or radial chambers (the spongocoel or atrium), then

3 Porifera, a phylum within the kingdom Animalia, consists of approximately 5,000 species of sponges. Sponges are among the simplest animals. It is believed that they evolved from the first multicellular organisms. Sponges are filter feeders and have no true digestive tract; instead, they digest food within cells. They lack the gastrula stage during development, and the three cell layers present are not homologous to the body layers of most animals.

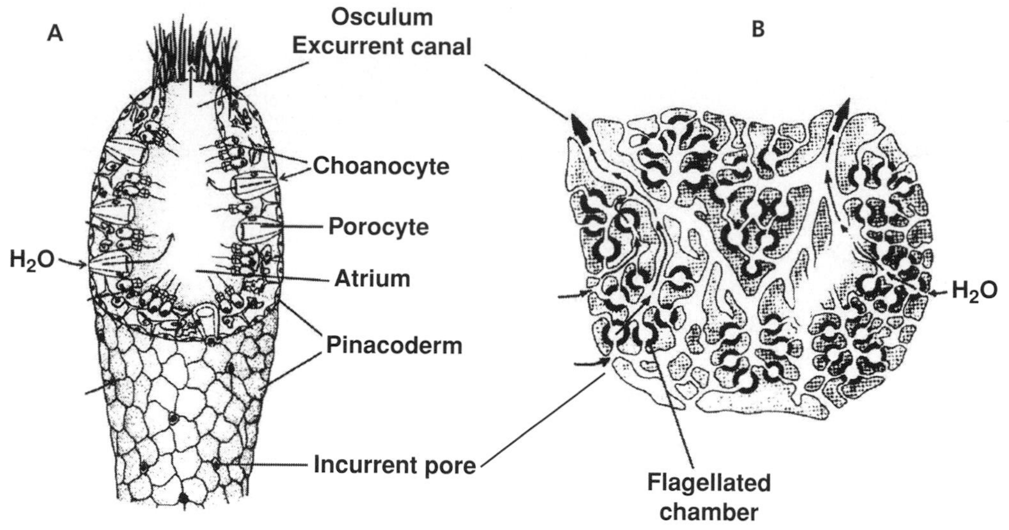

Figure 3.3. Circulatory system in sponges. The sponges are one of the simplest of multicellular animals, yet they have developed an internal circulatory system for the distribution of seawater. (**A**) The general architecture and cell types are illustrated in a simple Asconoid sponge, shown with arrows indicating the flow pattern of water into and out of the sponge. (**B**) The more complex pattern of water flow through a Leuconoid-type sponge. (Reproduced with permission from Ruppert EE, Barnes RD. *Invertebrate Zoology*, 6th ed. Fort Worth: Saunders, 1991.)

into major ex-current canals, and ultimately out of the sponge through the osculum (6–7).

Convective water movement through the sponge is generated by the beating action of the flagellated choanocytes that line the spongocoel or flagellated chambers. This noncentralized pump generates sufficient pressure differentials to move a substantial volume of water through the sponge, which facilitates not only filter feeding but also the more "typical" cardiovascular functions of O_2 uptake and waste removal (8–9). Although the historical view of sponge circulation was thought to be both simple and unregulated, it now appears that its pressure–flow dynamics are regulated through simple yet effective mechanisms. Pressures generated by flagellar beating may be started and stopped on a diurnal basis in some species (e.g., Demospongiae, Halichondria) and, in many sponges, changes in ambient water conditions may influence choanocyte flagellar activity. Moreover, flows and pressures can be regulated throughout the various chambers of the sponge by either opening or closing the in-current ostia or by increasing or decreasing the size of the osculum. This regulatory activity is accomplished through the action of specialized mesohyl cells (myocytes), which surround the osculum and act much like vascular smooth muscle cells with regard to contractility. Modulation of osculum radius has a tremendous effect on flow and pressure (10–12), just as modulation of arteriole diameter has profound effects in vertebrate cardiovascular systems.

The Cnidarian Gastrovascular Cavity

In many invertebrate phyla, "circulatory systems" consisting of muscular pumps generate pressure for the primary purpose of digestion, yet these invertebrates can and do use these systems secondarily for gas transport and waste removal (13–14). The internal pressure-generating systems of other invertebrates are used primarily for locomotion and are only secondarily adapted to generate pressures for the movement of internal body fluids or ingested food materials. Examples include the radial canals and gastrovascular cavity of the cnidarians[4] (Figure 3.4) and the highly convoluted gut of the Platyhelminthes (discussed in the next section). The primary function of these structures is to provide a large surface area for food digestion and absorption, as well as provide the structural elements for the animal's hydraulic skeleton. However, the disruption of internal boundary layers to maintain internal diffusion gradients could be an important positive functional consequence of internal convective fluid movement. In the cnidarians, the complex network of gastric pouches and blind-ending cavities of the gastrovascular system allows nutrient uptake by cells located some distance from the pharynx and "stomach" (14). Pressures in the gastrovascular cavity are generated to move food and fluids into and out of the gastrovascular system by the contraction of the locomotory muscles of the velum in more primitive species, by coronal muscles of the subumbrella in jellyfish and, in the more complex sea anemones, via the

4 Cnidaria, a phylum within kingdom Animalia, consists of hydras, jellyfish, sea anemones, and corals. This group has two key characteristics: The first is that all Cnidaria possess stinging or adhesive structures called *cnidae*, and each cnida resides in a cell called a *cnidocyte*; the most common example of which belongs to the nematocysts. The second characteristic most directly related to this chapter is that the mouth leads directly into the gastrovascular cavity (a digestive cavity). Because no anus is present, excess food and waste products must exit through the mouth; thus, the digestive system is considered incomplete. This gastrovascular cavity is filled with fluid that services internal cells, provides body support, and helps maintain structure.

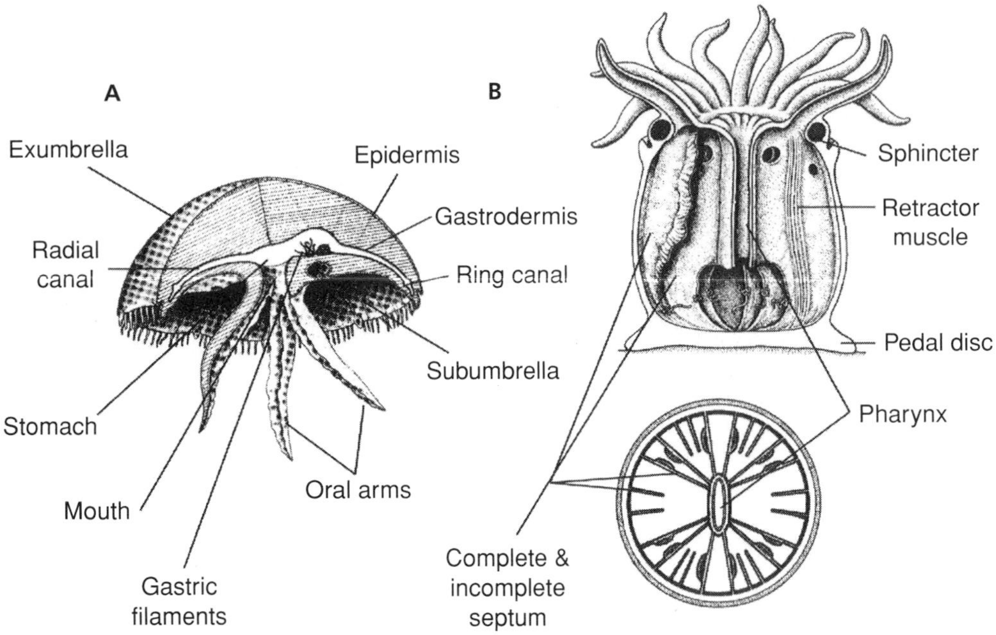

Figure 3.4. Circulatory system in cnidaria. The cnidaria are able to make dual use of their mouth, stomach, and gastrointestinal track for both ingestion and digestion but also to generate internal pressures and circulate fluids. (**A**) This panel illustrates an Aurelia jellyfish showing the mouth, stomach, gastrodermis, and radial canal. (**B**) The internal structure of a sea anemone shows a more complex series of internal compartments that are used to both increase the internal surface area for nutrient absorption as well as an internal pressure-generating system. (Reproduced with permission from Ruppert EE, Barnes RD. *Invertebrate Zoology*, 6th ed. Fort Worth: Saunders, 1991.)

digestive septal retractor muscles and pharynx (15–16). Given the size of some of the cnidarians and the complexity of the gastrovascular cavity, gas exchange is likely also facilitated by the pressure-driven movement of fluid throughout the body of these animals.

The Platyhelminth Gut

The phylum Platyhelminthes (flatworms),[5] and specifically the marine polyclads of the class Turbellaria, exhibit similarities to the cnidarians with the potential for their highly convoluted gut to facilitate both digestion and nutrient/gas exchange (Figure 3.5). Platyhelminths can be active predators that can grow to several centimeters in width and length. The larger size and active nature of these animals requires not only a well-developed gastrovascular system (gut) but also some form of internal fluid circulation to support nutrient, waste, and gas exchange, which may be seen in the pseudocoel of some turbellarians (17–20). It would appear that both the larger size of these animals and their more active behavior are in part possible because of their highly convoluted gut, which provides a very large surface area for the exchange of nutrients and other fluids.

5 The phylum Platyhelminthes includes all flatworms in the kingdom Animalia. Platyhelminthes are bilaterally symmetrical, having an anterior, posterior, dorsal, and ventral end and two seemingly identical lateral surfaces. However, flatworms lack any type of body cavity or fluid-filled space located between the digestive tract and the body wall. In flatworms, this area is filled with tissue. Flatworms have primitive brains, comprised of clusters of nervous tissue and two long nerve cords that run the length of the body. These cords branch to form small nerves running throughout the body. Another hallmark of flatworms is the lack of a complete digestive tract. Flatworms take in food and excrete waste from one single opening.

Nemerteans: The First Cell-Lined Circulatory System

The first true cell-lined "circulatory system" is seen in the nemerteans or ribbon worms (21–22). The relatively large body size and robust muscular body wall of ribbon worms presents a significant barrier to diffusion and thus requires internal convective fluid flow through the action of a blood circulatory system that facilitates gas exchange, nutrient cycling, and waste removal (23). Additionally, a second, rhynchocoelan circulatory system is used to perfuse the head and proboscis and also functions as a hydrostat to evert the proboscis for feeding (24–26). In its simplest form, the nemertean circulation consists of two lateral vessels that parallel the gut and anastomose at the anterior and posterior ends of the animal. More complex nemertean circulatory systems (e.g., as in *Tubulanus* and *Amphiporus*) have an additional dorsal vessel with multiple connecting transverse vessels (Figure 3.6) (27).

The vessels of this system are surprisingly complex, with a complete vascular cell lining that, in some species, consists of myoepithelial cells with cilia facing into the vessel lumen (26–27). The vessels are surrounded by both circular and longitudinal muscle to facilitate vascular contractions that, in combination

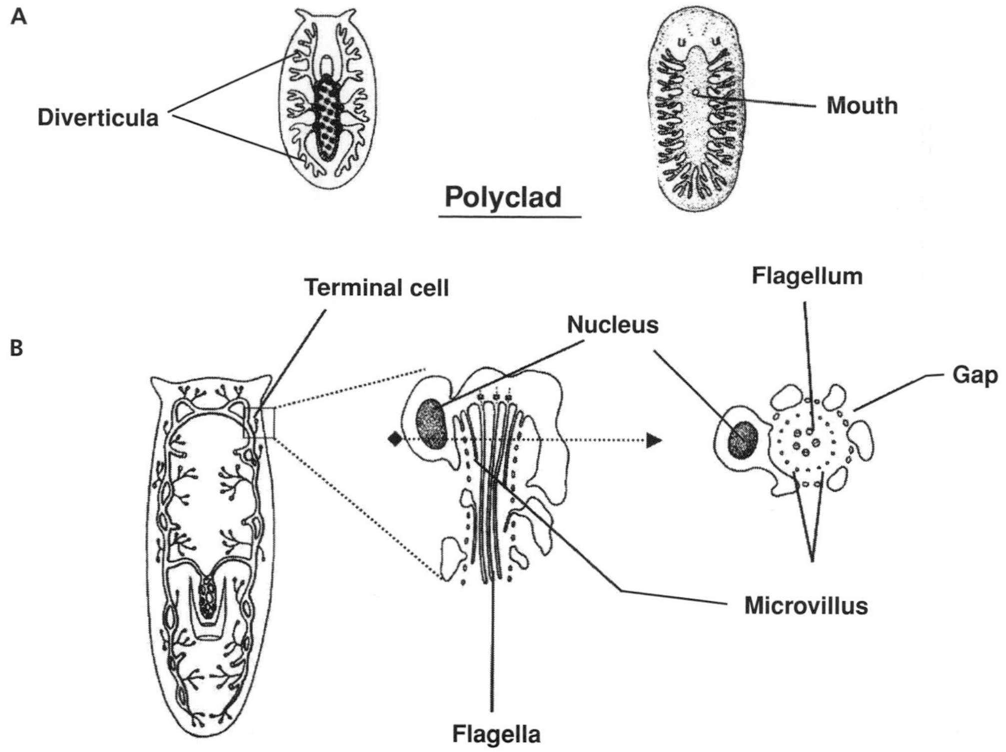

Figure 3.5. Circulatory system in platyhelminthes. The gut of the platyhelminths can show extensive diverticula, as is the case in the polyclads. These diverticula, which are small and vessel-like, serve to increase the absorptive surface area and distribute nutrients throughout the animal (**A**). An extensive network of complex anastomosing ducts interconnect the protonephridia (primitive renal structures) of the Turbellaria. The terminal cells of the protonephridia end within the tissues, where nitrogenous waste as well as other fluids are accumulated and then transported to exterior pores via ciliary action (**B**). (Reproduced with permission from Ruppert EE, Barnes RD. *Invertebrate Zoology*, 6th ed. Fort Worth: Saunders, 1991.)

with contractions of the body wall, provide the pressures to move blood within the circulatory system. The presence of a vascular cell lining is unique at this phylogenetic level and complexity within invertebrates, and it closely resembles the coelomic lining of other coelomate invertebrates. Thus, the nemertean circulatory system is proposed to have its evolutionary origins as specialized coelomic channels that have become secondarily adapted to perform circulatory functions. The myoepithelial differentiation of these coelomic cells has given rise to the endothelial lining (23,28–29). Debate exists as to the functional role of this "endothelium" or, more correctly stated, "vascular cell lining" with regard to its permeability, diffusional characteristics, and functional impact on hemodynamics, none of which has been investigated (29). Equally interesting is the tight integration between the nemertean circulatory system and the protonephridia of their excretory system. It would appear that, in some species, the protonephridia take advantage of significant circulatory pressures to aid in nephridial filtration for purposes of osmoregulation, and potentially, nitrogen excretion (29). An analogous situation exists in some platyhelminths, in which a proteonephridial system is integrated with the fluid-filled spaces of the pseudocoel. In the nemerteans, however, these two systems appear to be more functionally linked, with circulatory pressures driving filtration.

The Molluscan Cardiovascular System: High Efficiency Muscular Pumps

With the exception of a fluid circulation within the pseudocoel of a variety of coelomate worms (four minor phyla comprising fewer than 600 species – the Sipuncula, Echiura, Pogonophora, and Vestimentifera) and the lacunar system found in the epidermis of the acanthocephalans, which are not cell-lined systems and do not appear to take on a significant circulatory role in terms of pressure development or flow, few examples of circulatory systems and no examples of centralized pumps ("hearts") are seen until the advent of the molluscs[6] (23,30). Members of this phylum have developed extensive circulatory networks, with highly efficient centralized pumps (true hearts,

6 The phylum Mollusca is made up of over 150,000 diverse species. All molluscs have a muscular foot used for locomotion, as well as a mantle, an outgrowth that covers the animal. Molluscs have a coelom, which is developed from solid cell masses. Thus, all species in this phylum are protosomes. All organs are suspended in mesentery tissue within this mesodermic coelom, between the outer covering (ectoderm) and the digestive tube (endoderm) of the animal. The phylum Mollusca includes the following classes: Aplacophora (solenogasters), Monoplacophora, Polyplacophora (chitons), Gastropoda (snails and slugs), Bivalvia (clams and oysters), Scaphopoda (tusk shells), and Cephalopoda (nautilus, squids, cuttlefish, and octopi).

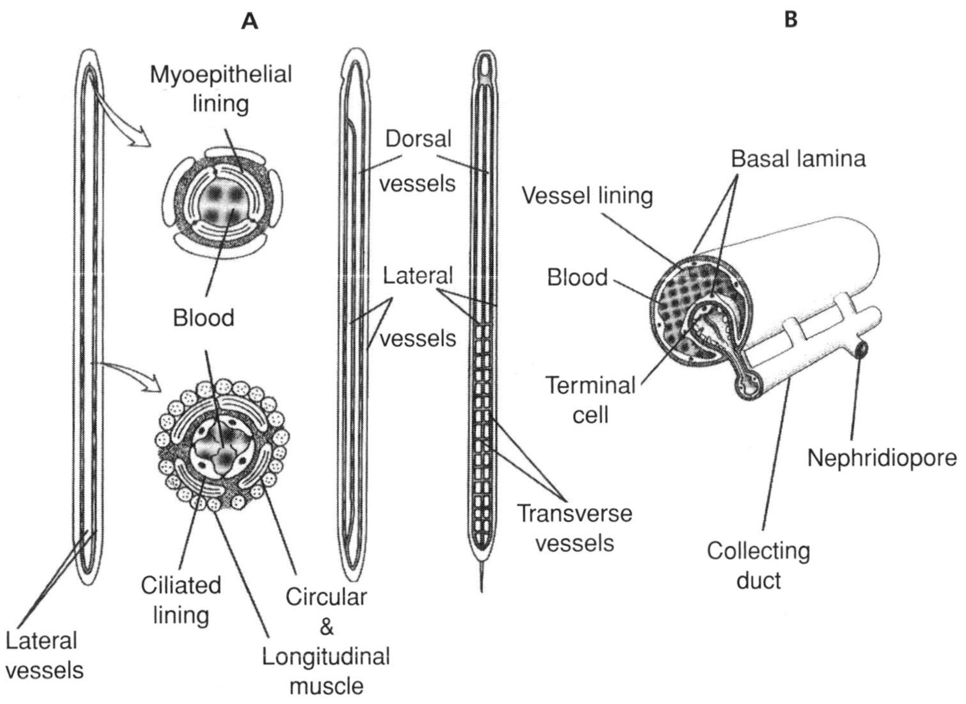

Figure 3.6. Circulatory system in nemerteans. The nemerteans show the first endothelial-lined circulatory system (**A**), which not only facilitates the movement of blood to promote gas exchange but also is integrated tightly with the excretory systems (**B**), which are pressure-dependent. (Reproduced with permission from Ruppert EE, Barnes RD. *Invertebrate Zoology*, 6th ed. Fort Worth: Saunders, 1991; and Turbeville JM. Nemertinea. In: Harrison FW, Bogitsh BJ, eds. *Microscopic Anatomy of Invertebrates*. New York: Wiley-Liss, 1991.)

composed of cardiomyocytes and capable of generating pressures and flows comparable to some lower vertebrates); these hearts function in an integrated fashion with a variety of other physiological systems (Figure 3.7). Moreover, the most complex molluscs (the highly active cephalopods) have evolved a cell-lined vascular system able to generate pressures and generally perform at a level comparable to some of the simpler vertebrates (3,22,29).

The generalized molluscan body plan has been extensively modified in both form and function to allow this group of animals great adaptive radiation (23,31–33). With specific regard to the heart and circulatory system, the typical mollusc (monoplacophorans and cephalopoda) has evolved a robust muscular heart consisting of a ventricle supplied by two atria or auricles, which drain the gills and sit within the coelomic cavity (the pericardial chamber) (34). The circulatory system is functionally integrated with both the respiratory structures (gills) and the metanephridial systems (kidneys) through both direct vascular connections and blood-pressure–dependent filtration, respectively. Arterial O_2-rich hemolymph (blood) is pumped anteriorly through a major aortic vessel by the contractile action of the muscular ventricle. This aortic vessel then branches into smaller vessels to supply defined tissues through hemolymph sinuses (32). Deoxygenated venous hemolymph then moves from the tissue sinuses into the gills via afferent vessels and the hemocoel. Hemolymph flow through the gills (branchial circulation) is countercurrent to mantle water flow, which, as in fish gills, maximizes gas exchange. Hemolymph then exits the gills through an efferent vessel and enters the atria or auricles for recirculation (3,35). Metanephridial, or renal, function depends on vascular pressures generated by the heart, with hemolymph filtration occurring through the walls of the atria (auricles) and their associated podocytes, which serve a similar function as podocytes of the visceral layer of the mammalian glomerular capsule. Pressure-driven filtrate moves through the walls of the atria (auricles) and into the pericardial cavity (renopericardial cavity), where it moves into the nephrostome and on into the kidney tubules. Excretion ultimately occurs via the nephridopore in the mantle cavity (36).

The fundamental design of the molluscan cardiovascular system has been functionally modified to meet a variety of demands, as is seen in the diversity of molluscan classes. The cephalopods show the most extensive evolution and specialization of the cardiovascular system, with blood driven at high pressures by the heart through an endothelial-lined (i.e., more closed than open) circulatory system that contributes to the maintenance of metabolic rates almost equivalent to some vertebrates (3,37). To sustain such high O_2 uptake rates, paired branchial hearts have evolved to pump blood through the gills, after which the now-oxygenated blood flows to the ventricle, where it is pumped into the systemic circuit. Functionally, the

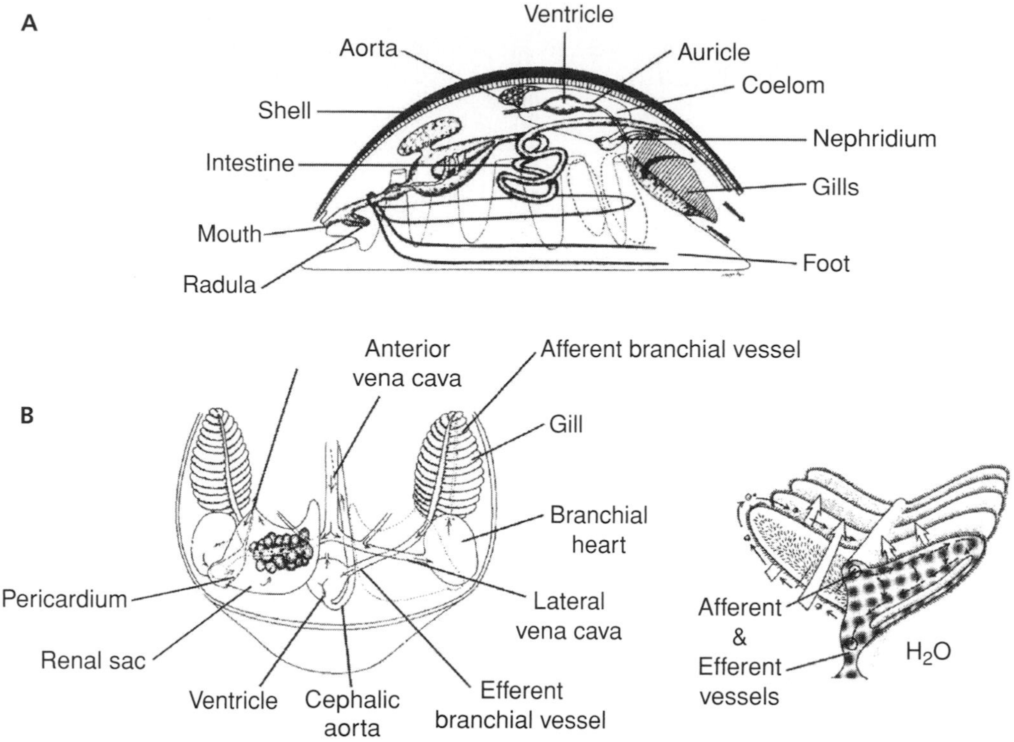

Figure 3.7. Circulatory system in molluscs. The generalized molluscs (**A**) have a well developed heart and circulatory system. The cephalopods have evolved highly efficient hearts, pumping blood at high pressure through both a branchial and systemic circuit (**B**). The inset illustrates the complex branchial circulation. The vessels of the cephalopods are lined with endothelium. (Panel **A** reproduced with permission from Ruppert EE, Barnes RD. *Invertebrate Zoology*, 6th ed. Fort Worth: Saunders Press, 1991; panel **B** reproduced with permission from Packard A. Cephalopods and fish: The limits of convergence. *Biol Rev.* 1972;47:241–307.)

cephalopods have evolved a multichambered heart capable of maintaining separation between venous and arterial blood and regulating branchial and systemic circulations. Additionally, this group of animals has developed the cardiorespiratory regulatory mechanisms needed to integrate cardiovascular and ventilatory performance with metabolic demands (38). The anatomical complexity of the cardiovascular system, along with the development of cell-lined, thin-walled, capillary-like exchange vessels and the appropriate regulatory mechanisms, appears to have been selected for in this group by increased activity patterns associated with predatory behavior, swimming, and jet propulsion (37).

The larger and more active cephalopods, such as the nautilus and the squid (the giant squid being the largest of any invertebrates), swim using jet propulsion. The use of this system does have some significant cardiorespiratory implications, in that a large volume of water must be moved through the funnel at some minimal acceleration to propel the animal. This results in excess O_2-carrying capacity in the ventilatory stream or more water flowing over the gills than is necessary for purely ventilatory purposes. O_2 uptake must be independent of jet volume to prevent postexercise mismatches in ventilatory volume and the repayment of an O_2 debt (39). The nature of the jet propulsion system requires that blood delivery to both the gills and tissues be at relatively high rates of flow and pressure. Ultrastructural examination of the vasculature of these very active cephalopods reveals a complete endothelial vascular lining, with highly branched "capillary-like" vessels that perfuse metabolically active tissues (3,40,41). Octopi are less active than squid, because they have reverted secondarily to a more benthic (i.e., bottom dwelling) existence. Octopi are restricted in the use of jet propulsion because the mantle edges have fused dorsally and laterally to the head, which limits water movement into the mantle cavity. Thus, their primary means of locomotion is crawling, through the use of their arms; they use their jet propulsion system only for shorter distances and when startled (23). The cardiovascular systems of these two cephalopods with different activity levels are similar, but discrepancies arise in reports as to the nature of the endothelia lining the vessels and the degree of tissue perfusion.

The inner cell lining of the cephalopod vasculature (like that of the nemerteans) appears quite permeable. It may not serve as a selectively permeable barrier, as it does in the vertebrates, which would make the cephalopod vascular system more "open" than "closed" (3,30,40,41). This leads to the question: Why have cephalopods evolved a cell-lined vascular system? It can be hypothesized that the invertebrate endothelia has evolved for reasons more to do with hemodynamics and the maintenance of laminar flow than with the array of functions ascribed to the vertebrate vascular endothelia. It appears the embryonic origin of the invertebrate vascular system and the associated vascular cell lining found in some taxa is derived from the coelom. Those taxa that have evolved such a vascular cell lining appear to be more active, with higher metabolic demands, and also have well-developed centralized pumps, an extensively branched vasculature with "capillary-like" vessels, and relatively high blood pressures (see the section on crustacean circulatory system for further details).

Laminar flow is required to minimize the energy needed to move blood through these complex vascular systems. Laminar flow through a cylindrical tube can be predicted based on vessel diameter, mean blood velocity, and blood density and viscosity (Reynold's number).[7] However, turbulent flow can result if sudden spatial variations occur in vessel diameter or irregularities appear along the vessels walls. In turbulent flow, a significantly greater pressure is required to move a fluid through the vessels, as compared to laminar flow. This is best exemplified by the fact that, in turbulent flow, the pressure drop is approximately proportional to the square of the flow rate, as opposed to laminar flow, in which the pressure drop is proportional to the first power of the flow rate (42–43). It would require a more robust heart and would be energetically inefficient to move blood in a turbulent pattern through a vasculature that changes shape abruptly and where the interiors of the vessels are not smooth, as seen in many lacunar systems. Thus, a vascular cell lining may have evolved to facilitate laminar flow to minimize the energy required to move blood through the cephalopod circulatory system.

The Annelid Blood–Vascular System

Members of the phylum Annelida[8] contain some of the largest examples of wormlike invertebrates. Thus, in many ways, these animals should be comparable to the molluscs with regard to cardiovascular development. However, because annelids tend to be sluggish, with relatively low metabolic rates, the high rates of flow and pressure characteristic of molluscans, for example, generally are not seen in annelids (Oligochaeta are an exception, as discussed in later paragraphs) (44–46). The segmented annelids have evolved several mechanisms by which to enhance internal transport between compartments, the most primitive being the development of a coelom and coelomic circulation. They also have evolved intracellular, iron-based O_2-binding pigment (hemoglobins). The most advanced annelid system is a fairly well-developed blood–vascular system (47–48). In the smaller annelids, few cardiorespiratory adaptations exist, however; in the larger more active worms (Oligochaeta), such as the giant Australian earthworm, the heart and vasculature are well developed (49).

The general pattern of circulation through an annelid is best seen in the polychaete worms, and starts with a dorsal vessel that runs just above the digestive tract. Blood moves anteriorly at the point where the dorsal vessel anastomoses with a ventral vessel either directly or by several parallel connecting vessels. The ventral vessel runs under the digestive tract and moves blood posteriorly. Each segment of the animal receives a pair of parapodial blood vessels that arise from the ventral vessel. The segmental parapodial vessels supply the parapodia, the body wall (integument), and the nephridia, and give rise to intestinal vessels that supply the gut. Blood moves from the ventral vessel through the parapodial system, and it returns to the dorsal vessel through a corresponding segmental pair of dorsal parapodial vessels (Figure 3.8). When gills are present and integrated with the blood vascular system (as apposed to being perfused with coelomic fluid), they contain both afferent and efferent vessels (23). Pressures are generated by contractile peristaltic waves through the dorsal vessels. These blood vessels and their associated blood sinuses lack an endothelium and are lined by only the basal lamina of overlying cells (33).

Although many of the anatomical variations observed in the cardiovascular system of annelids appear to have evolved due to activity patterns, feeding behaviors, and environment, the most complex systems are seen in the class Oligochaeta. The basic anatomical pattern seen in polychaetes is maintained: segmental vessels providing blood flow to well-developed integumental capillary beds, to support gas exchange across the skin. The pressure-generating system of oligochaetes is better developed than that in the polychaetes, with primary pressure generation coming from the contractile dorsal vessel. The "hearts" of the oligochaetes are actually robust contractile vessels that connect the major dorsal and ventral blood vessels and act as accessory organs to aid in blood movement. These hearts, along with the other major blood vessels, contain folds in their walls that act as one-way rectifier valves for blood flow. The number of accessory hearts varies among Oligochaeta species, with *Lumbricus* having five pairs of hearts and *Tubifex* only one (23).

The blood vascular system of the annelids is complex in its architecture, yet has not evolved to the level seen in the cephalopod molluscs. The cardiovascular physiology of annelids is one of low pressures and flows designed to support the relatively low metabolic rates of slow-moving or benthic animals. With the exception of the capillary bed in the integument, oligochaetes lack the high degree of tissue perfusion seen in more active animals. The vasculature lacks any cell lining, which may reflect the generally low flows and pressures observed in these systems and the animals' relatively low metabolic rates and activity patterns.

7 $N_R = 7745.8\,u\,d_h/v$, where, N_R = Reynolds number (nondimensional) u = velocity (ft/s); d_h = hydraulic diameter (in); v = kinematic viscosity (cSt) (1 cSt = 10^{-6} m^2/s). The Reynold's number (N_R) is used to predict the occurrence of turbulent versus laminar flow based on the ration of inertial to viscous forces.

8 The phylum Annelida is made up of approximately 15,000 species of segmented worms. Body segmentation, a key characteristic of annelids, was a major step in the evolution of animals. Classes include Polychaeta (sand-, tube-, and clamworms), Oligochaeta (earth- and freshwater worms), Myzostomida (flattened, oval, and/or aberrant annelids), and Hirudinida (leeches).

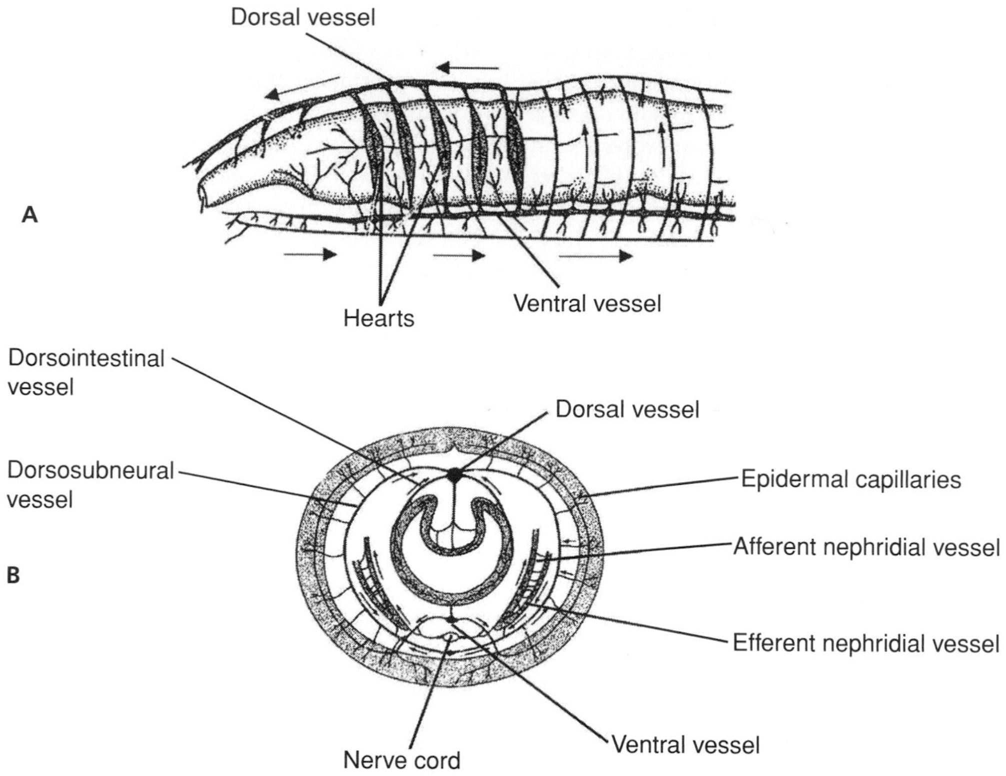

Figure 3.8. Circulatory system in annelids. The annelid circulatory system, although extensive, is a low-pressure system with contractile vessels for pumps. It lacks an endothelial lining. (**A**) Longitudinal section and (**B**) cross-section of the earthworm (*Lumbricus*). Reproduced with permission from Brusca, RC, Brusca GJ. *Invertebrates.* Sunderland, MA: Sinauer, 1990.)

The Arthropod Cardiovascular System

The arthropods[9] are by far the largest animal phylum, and they exhibit tremendous adaptive diversity. Of the major arthropod subphyla – Chelicerata, Crustacea, and Uniramia – only the crustaceans have evolved a complex cardiovascular system. The chelicerates have a relatively undifferentiated cardiovascular system when compared with the general arthropod model. The class Merostomata is best exemplified by the horseshoe crab (*Limulus*), which has a tubal heart, segmentally arranged ostia, and segmental vessels that are blind ending and supply large sinuses. The Arachnids are also part of this subphylum, and they have a similar cardiovascular system, with a tubal heart that supplies major vessels to the anterior and posterior of the animal. A unique structure found in this class is the *book lung*, which is a modified gill open to air via spiracles. This structure is perfused with hemolymph through the ventral sinus, which then returns to the heart by way of the pericardial sinus.

The insects (Uniramia) are an incredibly diverse group of arthropods, yet this diversity has apparently come without major evolutionary advances in the heart or circulatory system, which shows very little change from the Chelicerata. The success of the insects in part can be attributed to the development of the tracheal system. The trachea are a series of tubes that open to the environment through spiracles located on the lateral edges of the abdomen. The trachea then branch repeatedly, allowing for the diffusion of O_2 and CO_2 between the environment and cells that can be augmented by convective mechanisms using muscular abdominal compressions. Thus, instead of a circulatory system carrying O_2 to the cells, a tracheal system replaces capillaries to meet gas exchange demands (23,33,50).

The arthropods exhibit a segmental structure similar to that of the annelids and, in the Crustacea, vestiges of segmentation can be seen in the anatomy of the heart and circulatory system. The heart (primary pressure pump) of the crustacean in its primitive form is an elongated tube and, in the more highly evolved examples, is a globular, box-like structure.

9 The phylum Arthropoda is by far the largest, with estimates of over 50 million species in existence. Arthropods range in size from less than 1 mm in length to over 4 m (the Japanese spider crabs), and show tremendous diversity in body form. Primary characteristics include bilateral symmetry; a tough, regularly shed chitinous exoskeleton; a coelom reduced to portions of the reproductive and excretory systems and the main body cavity an open hemocoel; a muscular dorsal heart with lateral ostia for blood return; and a complete gut with a well-developed nervous system. Major extant subphyla include Cheliceriformes, with classes including Merostomata (horseshoe crabs) and Arachnida (spiders, scorpions, mites, and ticks); Uriramia, with classes including the Myriapoda (centipedes, millipedes, insects) and Insecta (insects); Crustacea, with classes including the Branchiopoda (fairy, brine, and tadpole shrimp and cladocerans) and Malacostraca (true crabs: crabs, shrimp, lobster, crayfish, and pillbugs).

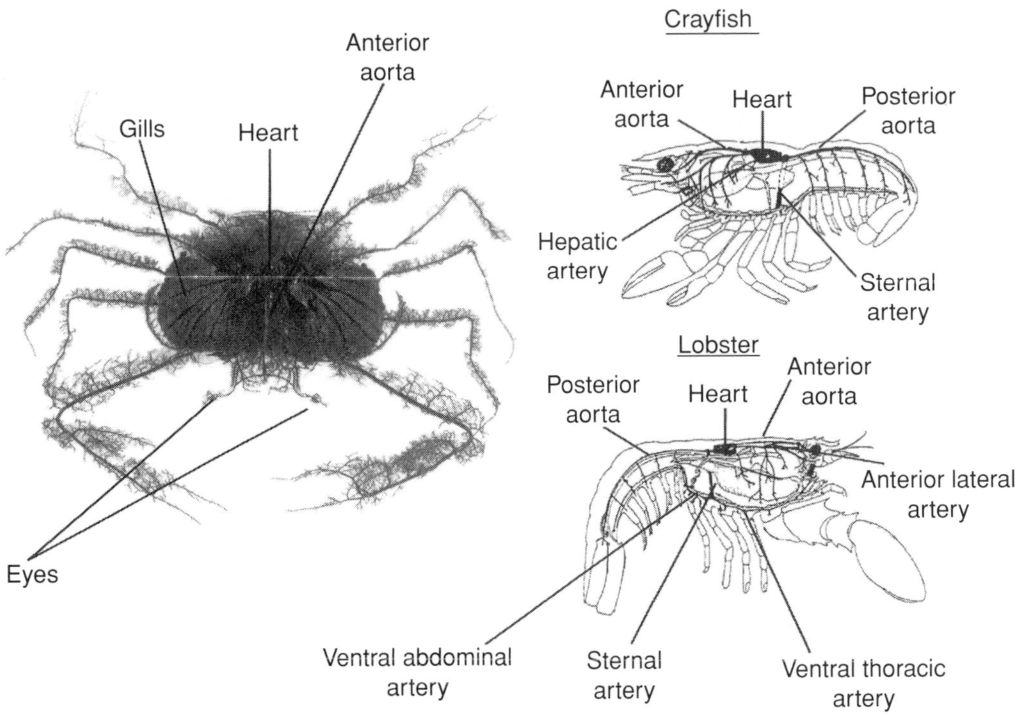

Figure 3.9. Circulatory system in crustaceans. The cardiovascular system of decapod crustaceans is highly developed, with a globular heart capable of delivering hemolymph at relatively high pressures and capillary-like vessel flows supplying metabolically active tissues. This distribution system is dramatically illustrated (*left*) in a corrosion cast of the blue crab's circulatory system and in the schematics of the crayfish and lobster. (Blue crab circulatory system reproduced with permission from McGaw IJ, Reiber CL. Cardiovascular system of the blue crab (*Callinectes sapidus*). *J Morph.* 2002;251:1–21; crayfish and lobster schematic reproduced with permission from Reiber CL, McMahon BR, Burggren WW. Cardiovascular functions in two macruran decapod crustaceans (*Procambarus clarkii* and *Homarus americanus*) during periods of inactivity, tail flexion and cardiorespiratory pauses. *J Exp Biol.* 1997;200:1103–1113.)

The ventricle is located within a pericardial sinus, where it receives hemolymph directly from the hemocoel through three paired ostia (51). In the more primitive tubal heart, the ostia appear to be segmentally arranged; in the globular hearts, however, these ostia have migrated, as the cardiac muscle has folded upon itself to form a box-like structure (52). Additionally, in the more primitive crustaceans, segmental arteries are seen branching from the anterior and posterior vessels. In the more advanced decapods, the development of the cephalothorax hides external evidence of segmentation, yet the vasculature specific to the posterior aorta in the abdomen has collateral arteries that branch off the main artery at each segment (Figure 3.9).

Typically, seven arteries leave the heart of a decapod crustacean. Arterial flow into each vessel is controlled by a single cardioarterial valve. The seven arteries consist of an anterior aorta, paired lateral arteries, paired hepatic arteries, a sternal artery, and a posterior aorta, each supplying a defined region or tissue. These vessels branch up to three times before ending in either a hemolymph sinus or in what appears to be true capillary beds (53) (see Fig. 3.9A). In the specific case of the anterior aorta, which supplies the esophageal ganglia, mouth parts, antenna, and eyes, an accessory hemolymph pump is present, the cor frontale. This structure is composed of muscle surrounding the anterior aorta, and aids in maintaining pressure and flow through the complex series of capillary-like blood vessels in the anterior regions of the animal. Hemolymph passes through the sinuses and capillaries, where gas exchange occurs, then follows well-defined venous pathways to major inferior sinuses that supply the gills or the branchiostegal tissue (used by terrestrial crabs as a lung) (54). Once hemolymph has moved through the gills or the branchiostegal sinus, it flows into the pericardial sinus and back into the heart for recirculation (53).

Hemolymph flows and pressures are relatively low in many of the lower crustaceans (55). However, in the highly active shrimps, crabs, lobster, and crayfish, metabolic demands are met through high hemolymph flow rates and relatively high driving pressures (55–57). Intracardiac pressure has been measured previously in a number of decapod crustaceans, with mean pressures ranging from 36.3 to 6.1 mm Hg during systole to 16.2 to 1.2 mm Hg during diastole (57,60,61). The hearts of these animals are able to generate considerable pressures and flows during periods of both brief and long-term exercise (swimming). In many ways, the decapods are analogous to the cephalopods in their development of an extensive vasculature and heart. The decapod cardiovascular system has been shown to be highly regulated through both intrinsic and extrinsic mechanisms, and it responds to a variety of internal and external stresses (56–59). The similarities with cephalopods continue at the level of the fine vessels. Although these vessels are larger than those of their vertebrate counterparts, they appear to serve the same function (53). Many examinations have been performed of the fine structure of decapod vessels, but few conclusions have been reached as to the nature of the cells that line the vasculature. Several authors have observed an endothelial-like lining of both capillaries and hemolymph sinuses. Clearly, the walls of decapod vessels do not contain muscle; thus, vasodilatation and contraction cannot be used to regulate hemolymph flow. A systematic examination of the vasculature for the presence of a cellular lining in sedentary versus very active decapod crustaceans would be interesting, and could support the hypothesis that active high-pressure systems will develop a vascular endothelia. Additionally, the nature of this "endothelium" must be examined to determine the permeability of the vasculature. In other words, is this system "open" as previously thought, or is it more appropriately viewed as "closed"?

VERTEBRATE CIRCULATIONS

Vertebrate circulations generally are characterized by having:

- A single muscular pump for propelling body fluids at relatively high arterial pressures
- A series of closed vessels leading from arteries to arterioles to capillaries to venules to veins
- Vasculature with a complete endothelial lining and walls invested with both smooth muscle and sympathetic and parasympathetic innervation
- Blood with both a fluid matrix and a cellular component consisting of several different cell types

A few fascinating discrepancies exist, either because they represent extremely primitive lineages (e.g., hagfish with multiple hearts) or they represent extreme adaptations in more advance groups (e.g., lack of red blood cells in the Antarctic icefish). Despite some common design features, great variation occurs in vertebrate cardiovascular systems. Consequently, it would be inaccurate to leave the reader with the sense that a heart structure or vascular layout is common to all vertebrates. With that in mind, let us now consider the taxonomic variation in vertebrate cardiovascular systems.

Protochordates

The Protochordates (Hemichordates, Urochordates, Cephalochordates) are an informal grouping of marine animals that are pelagic (i.e., free swimming) as larvae and largely benthic as adults. They are best described as occupying an evolutionary position between the ancestral invertebrates leading to the vertebrates and the true vertebrates (see Figure 3.2) (62–64).

The phylum Hemichordata (acorn worms) are worm-like filter feeders. Their blood contains amebocytes (leukocyte-like cells showing amoeboid motion) and hemoglobin in solution, which is synthesized from epithelial tissue lining the coelomic cavity (64). Blood is pumped by the peristaltic actions of a dorsal and ventral blood vessel through a combination of distinct vessels and large, open sinuses (65). The dorsal vessel, having a dilation often referred to as the "heart," pumps blood into the glomerulus, presumably for the excretion of wastes through filtration. Blood draining the glomerulus flows into the ventral vessel and on to the digestive tract and other organs. A cell lining is lacking in both vessels and sinuses.

The phylum Urochordata (tunicates) comprises three classes best recognized by the class Ascidiacea or sea squirts. The cardiovascular system of ascidians has been examined fairly carefully because of the pivotal position held by the Urochordates on the proposed evolutionary pathway from echinoderms through hemichordates to urochordates and cephalochordates to vertebrates (Figure 3.10). Most recent research has taken a cellular or molecular approach toward the differentiation of cardiac mesoderm, with a view to identifying homologies with vertebrate hearts (see, for example, 66–67). Previous anatomical and physiological studies reveal a defined heart, formed of a single layer of muscle cells, that lies within a fluid-filled pericardium (68). Major aortas convey blood from both cephalic and caudal ends of the heart into a series of interconnected sinuses. Like the hemichordates, no recognizable endothelial linings are present in the heart, vessels, or sinuses of urochordates (69–72).

The phylum Cephalochordata or lancelets more closely resembles the primitive living fishes, in part because they are free-living and have a very clear bilateral symmetry, with distinct head and tail regions. Their developmental biology has been investigated extensively (see 70). Much less is know about the cardiovascular morphology and physiology of cephalochordates, but several general accounts of their cardiovascular layout have been offered (62,68,70,72). Briefly, one of the main distinguishing features of *Amphioxus* (*Branchiostoma*) is that it lacks a heart. Instead, blood is propelled through the branchial and systemic circulations by three contractile vessels. Superficially, the circulation resembles that of fishes, in that a distinct branchial circulation serves some 50 gill bars used in filter feeding, and a systemic circulation receives blood from the efferent branchial arteries and distributes it via vascular spaces and sinuses to the body tissues. However, the circulation of cephalochordates lacks a complete cellular lining (22,69,73), although an incomplete (discontinuous) lining has been described recently (74). Moreover, the gill bars comprise only 4% of the animal's total gas-exchange surface, compared to 84% for the lining of the coelomic cavity in which the gills reside. Thus, the function of the gills appears to be almost entirely as mechanical structures for feeding, and the circulation of blood has little gas-exchange function (74–76). Circulation times are approximately 1 minute, as is the interval

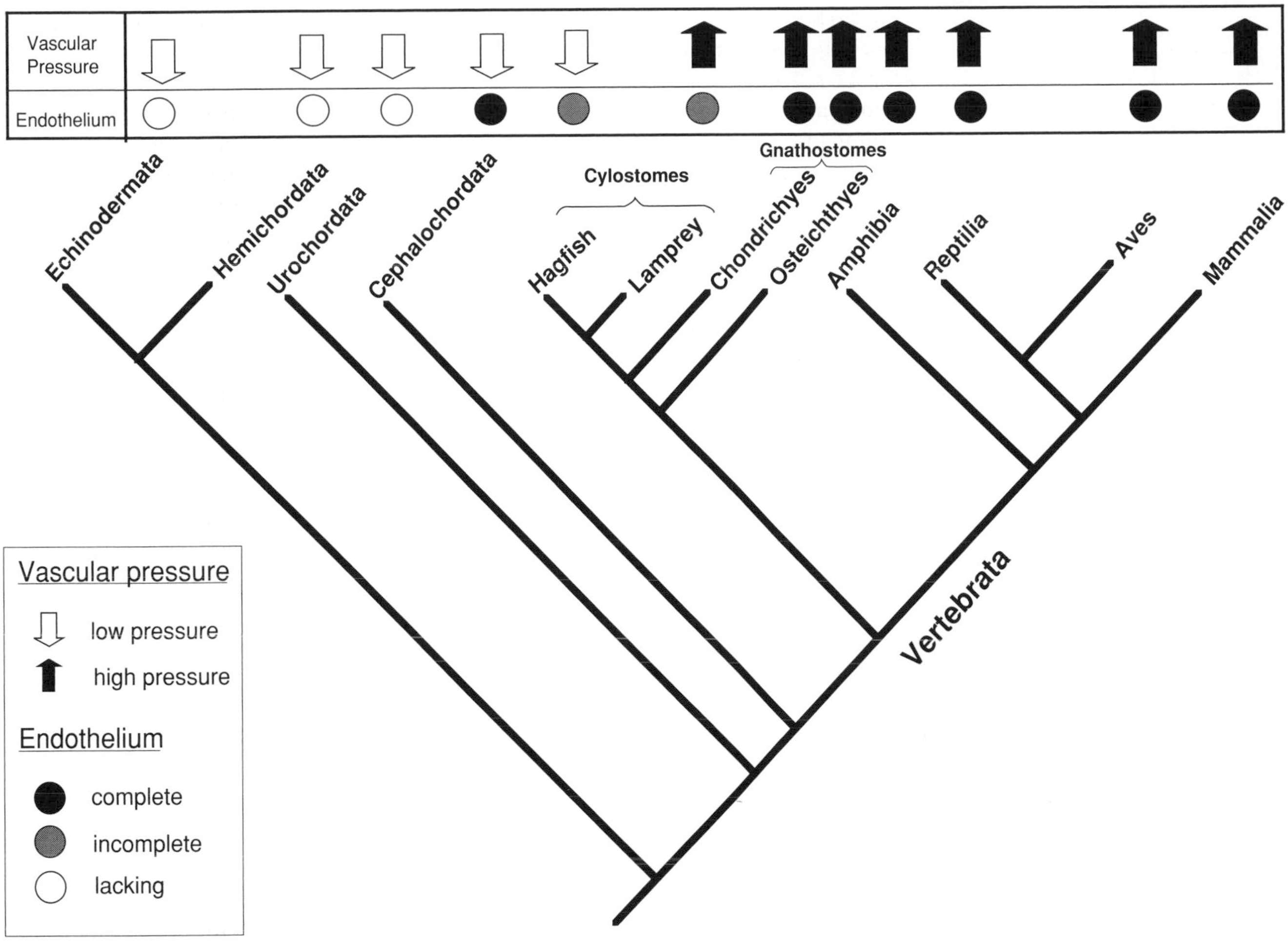

Figure 3.10. An evolutionary scheme for the protovertebrates and vertebrates. Also indicated is the presence of a complete or incomplete endothelial vascular lining. The arrows (*down* for low blood pressure, *up* for high pressure) indicate the taxon's blood pressure. The hatched perpendicular bar indicates the ancestral onset of endothelium, whereas the solid perpendicular bar indicates the ancestral onset of higher blood pressures.

between pulsatile vessel contractions (68). Intravascular pressures are thus presumed to be very low, although they have yet to be measured.

Cyclostomes (The Agnathan Fishes, Hagfish, and Lampreys)

The most primitive fishes are the Cyclostomes, represented today by the suborders Myxinoidea (hagfishes) and Petromyzontia (lampreys). The hagfish (*Myxine*) exhibits a cardiovascular system that, in many respects, resembles that of elasmobranch and teleost fishes (see Chapter 6). A heart homologous with that of more highly evolved fishes sits centrally and ventrally, and propels blood into a ventral aorta that carries blood to the gills via the branchial vasculature. From there, a dorsal aorta carries blood to the systemic tissues. However, this heart has several unusual features, including a lack of autonomic innervation and a relative insensitivity to extracellular calcium (Ca^{2+}) (see 77–78). Supplementing the "systemic heart" is a series of "accessory hearts" serving the posterior of the body (caudal heart), the abdominal cavity (portal heart), and the anterior region of the body (cardinal heart). Entrance into each of these hearts, which pump venous blood, is guarded by one-way valves. Although the peripheral circulation of the hagfish superficially resembles that of gnathostome fishes, it has large open sinuses devoid of a continuous endothelial lining (77–79). Mean blood pressures in the hagfishes *Myxine* and *Eptatretus* are on the order of 5 to 10 mm Hg, placing them among the lowest recorded for any vertebrate (see 80 for references).

The lampreys appear to occupy a morphological and functional position intermediate between hagfishes and the gnathostome fishes (78,83). The lamprey cardiorespiratory physiology has been extensively investigated (see 81–85). Unlike the hagfishes, lampreys have only a single ventrally located heart. Although the heart is innervated, vagal stimulation and acetylcholine enhance rather than inhibit cardiac performance (86). Mean arterial blood pressure is about 25 to 35 mm Hg, well within the range of many gnathostome fishes. Endothelium lines their capillaries (22,87–88).

Gnathostome Fishes (Cartilaginous and Bony Fishes)

The cartilaginous fishes (Chondrichthians) and bony fishes (Osteicthyians) generally show a common piscine cardiovascular layout, characterized by a single heart and a series of completely endothelial-lined blood vessels comprising the typical vertebrate vascular bed: arteries, arterioles, capillaries, venules, veins (see reviews 78,89–92) (also see Chapter 5). Deoxygenated blood returning from the systemic tissues enters into a large atrial chamber. Deoxygenated blood pumped from the atrium enters the single ventricle. Subsequent ventricular contraction creates mean arterial pressures typically in the 15 to 30 mm Hg range, although tunas and other similarly highly active fishes may exhibit mean pressures as high as 50 to 100 mm Hg and, remarkably, may generate peak systolic pressures in excess of 150 mm Hg (for references, see 78,93). Deoxygenated blood ejected from the heart traverses a short ventral aorta, then enters the branchial circulation and perfuses the lamellae of the gill arches. After gaining O_2 and losing CO_2 and ammonia, oxygenated blood enters the dorsal aorta for distribution to the systemic tissues. The branchial capillary network represents a significant resistance to blood flow; consequently, blood pressure in the dorsal aorta may be reduced by 20% to 50% compared with that in the ventral aorta.

Air breathing has independently evolved several times in fishes (for review see 62,81,94–95). Because the circulation plays a fundamental role in gas exchange, whether aquatic or aerial, it is thus not surprising that the introduction of an additional gas-exchange organ in the form of a gas bladder (e.g., *Lepisosteus*, the gar pike) or a true lung (the lungfishes, *Lepidosiren*, *Protopterus*, and *Neoceratodus*) has resulted in major modifications to the basic piscine circulatory pattern. To operate at highest efficiency with maximal air-to-blood O_2 diffusion gradients, an air-breathing organ must preferentially receive deoxygenated blood from the tissues via a newly evolved, dedicated arterial pathway, while oxygenated blood draining the air-breathing organ must be preferentially directed towards the tissues. Because most air-breathing fishes retain reasonably well developed gills (even if just for osmoregulation), the dilemma exists that O_2 picked up during air breathing might actually be lost from the gill surfaces into hypoxic water surrounding the gills. Thus, in many air breathing fishes, some or all of the gill arches are modified to present a much reduced surface area to the water flowing over them. This is particularly true in the lungfishes, in which oxygenated blood leaving the air-breathing organ enters the left atrium by a single pathway – the pulmonary artery (78). Concurrently, oxygenated blood from the systemic veins enters a right atrium. Although the left and right atria empty into a single, muscular ventricle, the trabecular nature of the ventricular lining, as well as the hemodynamic blood flow patterns through the ventricle, results in partially maintained separation of these two streams of arterial blood as they are ejected into the ventral aorta. A spiral valve that runs longitudinally down the internal length of the ventral aorta rotates the streams of blood such that deoxygenated blood is primarily directed into the last two pairs of gill arches (from which are derived the pulmonary arteries), while oxygenated blood is preferentially directed into the first pair of gill arches (from which arises the dorsal aorta leading to the main systemic circulation). The anterior-most arches tend to be small and poorly developed, with lower surface area. These arches act more as shunt vessels that allow oxygenated blood to transit the branchial circulation without diffusing back out into the potentially hypoxic water irrigating the gills.

Although the lungfish has evolved a relatively specialized circulation, fishes using nonpulmonary bladders, labyrinth organs, or modified guts for air breathing tend to be plumbed into the circulation in ways broadly similar to those described for lungfishes (62,81,89,96). However, only the lungfishes have evolved a pulmonary artery leading to a distinct left atrium (62,81). Consequently, oxygenated blood from a nonpulmonary air-breathing organ typically re-enters the general systemic venous drainage. Thus, the presence of an air-breathing organ that is not homologous to a lung leads to a general elevation of blood O_2 levels in the fish, but does not lead to the peak efficiencies of gas exchange to be found in the lungfishes.

It is tempting to view the three living genera of lungfishes as on a direct evolutionary pathway to more terrestrial amphibians, simply because they have functional cardiorespiratory features intermediate between fish and tetrapods. However, the modern-day tetrapods, beginning with the amphibians, emerged from a vertebrate ancestor, which the lungfishes share with tetrapods. Not surprisingly, then, the amphibian circulation has its own evolutionary innovations – and limitations – as we will now explore.

Amphibians

Amphibians typically have two distinct atria but only a single, muscular ventricle generating mean arterial pressures generally in the range of 10 to 30 mm Hg (see 78,97–98 for reviews). Like the lungfishes, amphibians have a well-developed spiral valve in the single larger arterial outflow tract leaving from the heart. This valve maintains partial separation of oxygenated and deoxygenated blood streams derived from left and right atrium, respectively. In anuran amphibians (frogs and toads), oxygenated blood preferentially enters the systemic arches, while deoxygenated blood is directed primarily into paired "pulmocutaneous" arches. Each pulmocutaneous arch immediately splits into a pulmonary artery serving the simple, nonalveolar lungs, and a cutaneous artery, which perfuses the thin, highly vascularized skin (Figure 3.11). Amphibians are skin as well as lung breathers, often deriving more than 50% of their O_2 from cutaneous O_2 uptake under resting, normoxic conditions (99–100) (see Chapter 9). Noteworthy, however, is that the skin also receives a blood supply from the normal systemic circulation. Thus, any given patch of skin is likely to receive both deoxygenated blood via the cutaneous artery and oxygenated blood via the systemic circulation. Oxygenated

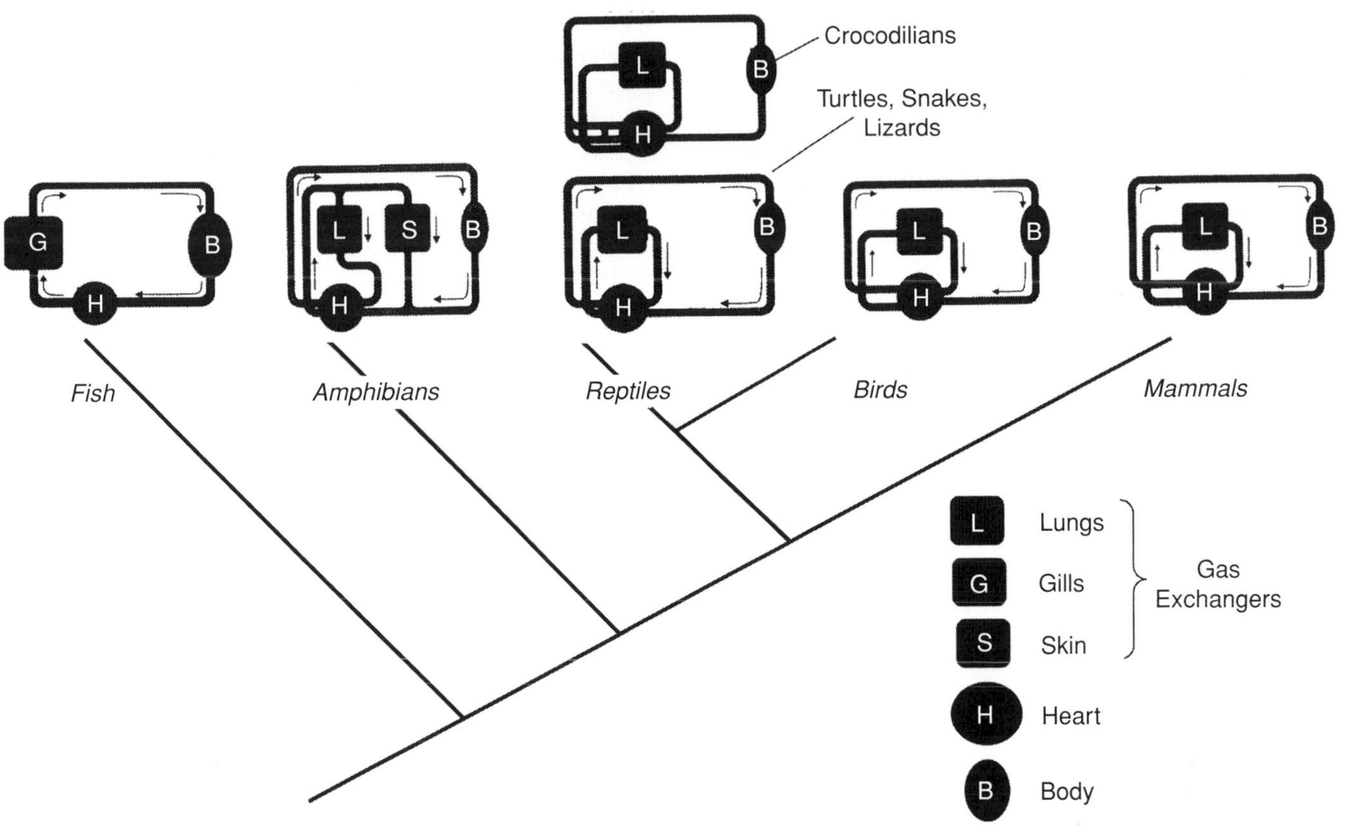

Figure 3.11. A schematic of vertebrate phylogenetic relationships and the basic cardiovascular layout of the vertebrate classes. Note that no smooth continuum of evolutionary improvement exists. Rather, each cardiovascular system is well designed to provide gas exchange at rates and in patterns appropriate for each group's environment and lifestyle. See text for further description. (Modified with permission from Hicks JW. The physiological and evolutionary significance of cardiovascular shunting patterns in reptiles. *News Physiol Sci.* 2002;17:241–245.)

blood draining the lungs returns to the distinct left atrium via pulmonary arteries. Oxygenated blood draining the skin, however, enters the general systemic venous circulation and eventually enters the right atrium.

Salamanders (Urodeles) differ from anurans in that they lack the distinct cutaneous artery of anurans, with the skin being served only by the systemic circulation typical of other vertebrates. Nonetheless, the skin still provides a significant boost to O_2 uptake. An interesting evolutionary oddity in the urodeles are the terrestrial Plethodontid salamanders, which have secondarily lost their lungs and subsist entirely on cutaneous gas exchange (100). These animals also have lost pulmonary arteries and veins, and retain only a vestigial left atrium. Some salamander groups have evolved a partially divided ventricle (e.g., *Siren*, 101). Although the physiological performance of these unusual amphibians has not been evaluated, improved separation of deoxygenated and deoxygenated blood would be expected on a purely morphological basis.

Reptiles

Reptiles are a highly polyphyletic group (that is, the group includes various distantly related taxa) and so, not surprisingly, as a vertebrate class, they show three highly distinctive patterns, each characterized by profound differences in heart structure and vascular layout (62). The reptiles show some of the most complex cardiovascular patterns of all vertebrates. Indeed, the notion that the "reptile heart" is a three-chambered heart somehow awaiting evolutionary repair to strive for the heights of the superior avian or mammalian heart is now a historical view that has been replaced by a considerable appreciation of the sophistication of the various patterns that evolved in the reptiles.

Chelonians (turtles and tortoises) and squamates (snakes and lizards) have two distinct atria that pump blood into a complex ventricle that – rather than being viewed as a single ventricle, as in amphibians – is best regarded as containing three distinct chambers (see 102–103 for reviews). The left atrium pumps oxygenated blood into a cavum arteriosum, which has no direct arterial outlet. The right ventricle empties deoxygenated blood into a large cavum venosum, separating the cava arteriosum and pulmonale. The major systemic arches arise from the cavum venosum. Clearly, the function of this heart depends on highly complex patterns of flow through this partially but complexly divided ventricle. The hemodynamics and, especially, the patterns of blood flow of the chelonian heart are highly complex and involve both spatial and temporal separation. Their description is beyond the scope of this chapter. Suffice it to say that the anatomically distinct chambers are functionally interconnected and, with each systole, blood traverses prominent intraventricular muscular ridges separating the cava. Mean arterial pressures are generally in the range of 25 to 40 mm Hg in chelonians (78), with a single pressure measured throughout the ventricle during systole despite the anatomical complexity of chambers within the ventricle.

One of the key features of chelonians is that they are typically intermittent breathers (breathing is of course suspended during diving in aquatic turtles, but breath-holding is also characteristic of terrestrial tortoises). Gas Po_2 and Pco_2 levels in the modestly alveolar and subdivided turtle lungs vary with the breathing–diving cycle; so too does the gas exchange performance of the lungs (see numerous authors in 98,104–105). To match ventilation to perfusion, chelonian reptiles are able to control potentially large intraventricular blood shunts by creating large changes in relative resistance between the systemic and pulmonary circulation (see 78,102–103,106 for reviews). Thus, during breath-holding, pulmonary resistance increases greatly due to constriction of a sphincter at the base of the pulmonary artery and vasoconstriction in the pulmonary vasculature. Because blood leaving the heart ultimately follows literally the path of least resistance, pulmonary vasoconstriction creates a large net right-to-left shunt leading to reduced pulmonary blood flow that, in extreme situations, is nearly eliminated, even as systemic blood flow continues. At the termination of the dive, the right-to-left shunt is reversed, and a modest left-to-right shunt appears to facilitate not only rapid reoxygenation of arterial blood, but also an elimination of CO_2 sequestered in tissues during the breath-holding event (102–103). Thus, turtles and tortoises have evolved a highly flexible circulation that is admirably adapted to a lifestyle characterized by breath-holding and low metabolic rates.

Squamate reptiles (lizards and snakes) tend to show cardiovascular morphology and performance similar to chelonians. However, fascinating exceptions exist that show that pressure separation within the complex ventricle of reptiles has independently evolved several times within the reptile group. The varanid lizards (probably best known by the Komodo dragon, *Varanus komodoensis*, although numerous smaller and less frightening species inhabit this genus), are capable of high levels of metabolism and brief periods of great activity, such as predation (107). Supporting this atypical reptilian performance is a modified ventricular structure (108). Although the three ventricular chambers – cava arteriosum, venosum, and pulmonale – are still in evidence, their relative sizes and interrelationships have been modified to allow pressure separation during ventricular systole. The net effect is that the pulmonary circulation of varanids is perfused at a mean pressure of about 10 to 15 mm Hg, while the systemic circulation is perfused at mean pressures that can exceed 60 to 70 mm Hg. This decidedly mammalian-appearing separation of the circulation into a high-pressure systemic and low-pressure pulmonary circulation represents an intermediary position on a functional continuum between the heart of lower vertebrates and mammals (109). The snake, *Python molurus*, also is capable of producing a functional separation of ventricular chambers during systole, leading to markedly different systemic and pulmonary perfusion pressures (110). However, it bears emphasis that the varanid lizards and pythons are, in an evolutionary sense, cardiovascular oddities that are in no way on the main pathway to the endothermic vertebrates (birds and mammals). However, that such circulations *can* evolve serves as evidence of an "evolutionary feasibility study" (111). The existence of these circulatory patterns has potentially profound implications as we ponder cardiovascular performance, metabolic rates, and attendant behaviors of those reptiles that do represent a more mainstream evolutionary pathway – the dinosaurs (112–113).

The Crocodilia – crocodiles and reptiles – exhibit a third pattern, representing a highly adapted cardiovascular performance that permits great efficiency (for reviews see 78,102–103,106,114,115). With respect to their heart structure, crocodilians have the same completely divided, highly muscular left ventricle and somewhat less muscular right ventricle as do birds and mammals. The fascinating adaptation in crocodilians comes from the numbers and positions of arterial arches. Crocodilians have a left arch that arises from the left ventricle and carries blood to the descending aorta, as in the homeotherms. However, crocodilians have retained a right aortic arch that derives from the right ventricle, as does the pulmonary arch. The bases of these two aortic arches are connected by a very short foramen of Panizza. How does this apparent oddity, which at first glance conjures up images of congenital pathologies in human patients, actually function? During active air breathing, high systolic pressures on the left side of the heart not only provide high driving pressures for systemic perfusion, but also transmit high pressure through the foramen of Panizza into the base of the right aorta, keeping shut the valve at the base of the right aortic arch. During air breathing, the low pressures generated by the right ventricle are enough to open the valve at the base of the pulmonary artery and perfuse the lungs, but not high enough to allow blood from the right ventricle to enter the base of the right aorta. Thus, during active air breathing, the crocodilian circulation is functionally indistinguishable from the circulation of homeotherms, sending equivalent amounts of blood under low pressure to the lungs and, under high pressure, to the systemic circulation. During what can be extended periods of breath-holding during diving, however, the same physiological degradation of pulmonary gas-exchange efficiency that occurs in diving turtles also occurs in crocodilians. Pressures generated by the left ventricle (which had been in the range of 60 to 80 mm Hg or higher), decrease greatly due to an as-yet poorly understood mechanism involving a decrease in left ventricular contractility. No longer held closed by high left aortic arch pressures, the valve at the base of the right aortic arch opens with each systole, and a proportion of right ventricular blood potentially bound for the lungs actually is diverted back into the systemic circulation via the right aortic arch. By virtue of this extraventricular right-to-left shunt via the right aortic arch, crocodilians, like turtles, can bypass the lungs and save the considerable energy that would have been expended on their continuing perfusion during a dive. Diving physiologists, in many respects, view the crocodilian circulation as the pinnacle of efficiency for an intermittent breather.

Birds and Mammals

The circulation of birds and mammals has been extensively investigated for literally centuries, and it is not our intent to repeat the efforts of the many authors who have recently reviewed the circulation of homeothermic vertebrates. Suffice it to say that, although we often lump the avian and mammalian circulation together, actually fairly major quantitative differences are present in their structure (e.g., relative size and wall thickness of the ventricles show considerable species diversity) (see Chapter 10). Another difference involves the developmental physiology of avian and mammalian cardiovascular systems. Unlike the placental arrangement supporting the mammalian fetus, the avian embryo develops its own extraembryonic chorioallantoic membrane (CAM). These anatomical distinctions in early development lead to both qualitative and quantitative differences in cardiovascular performance at all developmental stages, and especially during the transition from pulmonary breathing associated with hatching or birth. (For reviews of avian embryonic circulations, see 116.)

EVOLUTION OF CELLULAR LININGS IN INVERTEBRATE AND VERTEBRATE HIGH-PRESSURE CIRCULATIONS

As evident from the earlier discussions, cardiovascular systems have followed a series of complex and diverse pathways during their evolution. Against this backdrop of cardiovascular evolution, it is interesting to speculate on *why* endothelial linings evolved in cardiovascular systems. These speculations can be formed as a series of hypotheses.

The "Endothelium as Barrier" Hypothesis

Although our observations are simply correlative rather than causative at this point, it is interesting to note that, in the invertebrates, most of the cardiovascular systems that generate and experience relatively high pressures have evolved a continuous cellular vascular lining, as have all the high-pressure systems in vertebrates (see Figure 3.10). Of course, relatively high pressures generated by a heart or hearts in a circulation provide a hydrostatic gradient for moving blood longitudinally through the vessels. However, the generation of pressures within a vessel also generates a hydrostatic pressure gradient *across* the vascular walls. Whereas in some structures (nephridia of invertebrates, nephrons of vertebrates), this is a useful – indeed, necessary – phenomenon, another by-product of high intravascular pressure is the ultrafiltration of fluid from vascular into nonvascular fluid compartments in nonexcretory body tissues. Thus, this particular hypothesis holds, a cellular vascular lining may have evolved to provide some degree of control over the loss of fluids from across the vascular walls, presumably by decreasing the filtration coefficient, Kf, in the Starling equation for transcapillary fluid flux. Once endothelium as a physical barrier had evolved, secondary functions of endothelium could have evolved subsequently.

Morphological support for this thesis comes from an examination of the few but very significant appearances of cellular vascular lining in invertebrates. Shigei and colleagues (22) propose that a cellular vascular lining (endothelium or otherwise) is essentially a vertebrate adaptation. In citing numerous examples of studies that have failed to find a cellular vascular lining, these authors acknowledge but pay little attention to the fact that a cellular lining for the vasculature has indeed evolved independently from vertebrates in the Nemertea, cephalopods, and crustaceans. That these three groups of invertebrates also have a high blood pressure lends credence to the notion that the appearance of a vascular lining is in some way associated with higher sustained blood pressures. Additional evidence comes from the lack of continuous endothelium in *Amphioxus*. Although physiological examinations of this species are largely lacking, at least the morphological arrangement of the cardiovascular systems suggests a low-pressure circulation – certainly in comparison to even the most primitive of fishes. Then, in the hagfishes (with atypically low blood pressures for a vertebrate), the endothelial lining of the circulation is incomplete, being interrupted by unlined sinuses. However, speaking against the notion that endothelium serves as some form of barrier in high-pressure systems, Chapter 5 of this volume describes how the endothelium in fishes is generally a poor barrier against ultrafiltration (except in the gills and brain, where vascular permeability is much reduced).

Alternative Hypotheses for Evolution of Cellular Vascular Linings

Other equally valid hypotheses for the evolution of a cellular vascular lining, if not outright endothelium, relate to the many demonstrated functions of endothelium, including hormone secretion, leukocyte trafficking, control of blood clotting, and control of vascular tone. It is beyond the scope of this chapter on cardiovascular evolution to delve into these hypotheses. However, one bears consideration as an example. Muñoz-Chápuli and colleagues (29) present a "nonhemodynamic" hypothesis for the evolution of endothelium. They suggest that the epithelium may have first evolved as a system for immunological cooperation of "protoendiothelial cells" and, in a good example of an "evo–devo" approach, these authors present both ontogenetic and evolutionary evidence.

Although presenting a useful summary of the cell lineages lining the coelom of various invertebrates, these researchers (29) do not recognize the existence of cellular vascular lining in some cephalopods. Thus, their hypothesis focuses more narrowly on the invertebrate–vertebrate transition of the cardiovascular system. In so doing, they suggest a more hypothetical linear transition from invertebrate to higher vertebrate evolution than the widely accepted messy, nonlinear transition with numerous false-starts. Indeed, the presence of a cellular vascular lining in the cephalopods and other invertebrates remains a stumbling block to attributing the evolution of such linings strictly to the invertebrate–vertebrate transition.

The search for *the* hypothesis to explain all data for the evolution of vascular cellular linings is likely to be frustrated by the fact that such linings have so many critical functions, and these functions are highly likely to have evolved concurrently rather than in sequence.

Future Directions

Although the general anatomical layout of the cardiovascular system of most extant phyla is well documented, we know very little about the cardiovascular performance of the vast majority of invertebrates. In fact, we estimate that not even a single species in more than half of the phyla possessing internal circulatory systems has been assessed for blood pressure, blood flow, or even whether the circulation functions in respiration. We are discovering more and more evidence for sophisticated physiological regulation in "open" circulations (formerly viewed as inferior by physiologists, incidentally) and in lower vertebrates. Thus, we also should outline experiments that look at these same higher vertebrate-like functions that occur in invertebrate and protochordate vasculature. We also should explore how endothelial linings – wherever they exist – contribute to the regulation of blood or hemolymph flow at the tissue level, and whether this involves autoregulatory mechanisms, nitric oxide, adenosine, adenosine diphosphate, adenosine monophosphate, and the like.

We also know very little about even basic processes in the cardiovascular development of most invertebrates (e.g., the relative roles of vasculogenesis and angiogenesis in formation of the circulation), and even a surprising number of chordates. This leaves the currently popular and effective evo–devo approach to understanding physiology (117–118) waiting on the sidelines for want of even basic data for analysis. Not until we have a more comprehensive understanding of how cardiovascular systems develop and perform across development and taxa can we begin to piece together the selection pressures leading to particular suites of cardiovascular characteristics (e.g., regulated distribution of blood to selective vascular beds, efficient matching of blood or hemolymph perfusion to gas-exchanger ventilation).

In summary, taking a broad evolutionary perspective, we have learned a great deal about the cardiovascular systems of only a very small number of extant animals. A more sophisticated understanding of internal body-fluid circulation system evolution will require a more synthetic, integrative, "systems biology" approach that spans the fascinating diversity of living animals.

KEY POINTS

- Cardiovascular systems evolved when the combination of increasing animal size and increasing metabolic rate necessitated the internal convection of body fluids for gas exchange, metabolic substrate delivery, and waste removal.
- Many invertebrates have relatively low-pressure systems that slowly direct either ambient water or hemolymph through vascular spaces rather than through the endothelium-lined vessels characteristic of vertebrates. Contrary to historical perspectives, however, certain invertebrates (e.g., cephalopods and some crustaceans) have relatively high-pressure cardiovascular systems capable of the regulated preferential direction of hemolymph through cell-lined vessels.
- Evolution has led to a wide variety of cardiovascular patterns in nonmammalian vertebrates.
  - The relatively simple system of aquatic fishes becomes more complicated when the capability for air-breathing evolves, whether by appearance of a bladder-like, air-breathing organ or a true lung.
  - Amphibians have developed a dual atrial system and, in some taxa, possess specialized circulation for cutaneous respiration.
  - Reptiles show three distinct cardiovascular patterns. Chelonian circulations are well adapted to intermittent breathing by creating pulmonary bypass shunts during diving, when pulmonary respiration is less effective. Varanid lizards, with their functionally separated ventricle and high-system/low-pulmonary pressures, represent an evolutionary side track but an interesting indication that an ancestral intermediate step between typical reptiles and birds could occur. Crocodilian reptiles represent the most highly evolved reptilian system, allowing mammalian-like performance during air breathing, but the energy-saving performance of chelonians during diving.
- A correlation exists between the evolution of relatively high pressures and the presence of a cellular vascular lining. Importantly, a cellular vascular lining associated with elevated vascular pressures appears to have evolved independently multiple times (e.g., cephalopod molluscs, crustaceans, vertebrates). One (of many) possible hypotheses is that ECs might have

evolved as a barrier to the ultrafiltration of plasma in systems with high transmural vascular pressures. However, no single hypothesis is likely to account for the evolution of cellular vascular lining.

REFERENCES

1 Burggren, W. W. What is the purpose of the embryonic heart beat? or How facts can ultimately prevail over physiological dogma. *Physiol Biochem Zool.* 2004;77:333–345.

2 McMahon BR, Wilkens JL, Smith PJ. Invertebrate circulatory systems. In: Dantzler WH. *Handbook of Physiology*, Vol. II. Section 13: Comparative Physiology. Oxford: Oxford University Press; 1997:931–1008.

3 Bourne GB, Redmond JR, Jorgensen DD. Dynamics of the molluscan circulatory system: open versus closed. *Physiol Zool.* 1990;63(1):140–166.

4 Pavans de Ceccatty M. Coordination in sponges: the foundations of integration. *Am Zool.* 1974;14:895–903.

5 Willenz P, Hartman WD. Micromorphology and ultrastructure of Caribbean sclerospones. I. *Ceratoporella and Nicholsoni and Stromatospongia norae. Mar Biol.* 1989;103:387–401.

6 Lawn ID, Mackie GO, Silver G. Conduction system in a sponge. *Sciences.* 1981;211:1169–1171.

7 Langenbruch PF, Weissenfels N. Canal systems and choanocyte chambers in freshwater sponges. *Zoomorphology.* 1987;107:11–16.

8 Brauer EB. Osmoregulation in the freshwater sponge, *Spongilla lacustris. J Exp Biol.* 1975;192(2):181–192.

9 Reiswing HM. Particle feeding in natural populations of three marine demospones. *Biol Bull.* 1971a;14(3):568–591.

10 Reiswing HM. In situ pumping activities of tropical Demospongiae. *Mar Biol.* 1971b;9(1):38–50.

11 Vogel S. Current induced flow through the sponge, *Halichondria. Biol Bull.* 1974;147(2):443–456.

12 Palumbi SR. How body plans limit acclimation: Responses of a demosponge to wave force. *Ecology.* 1986;67(1):208–214.

13 Gladfelter EH. Circulation of fluids in the gastrovascular system of the reef coral *Acropora cervicornia. Biol Bull.* 1983;165:619–636.

14 Sandrini LR, Avian M. Feeding mechanism of *Pelagia moctiluca*, laboratory and open sea observations. *Mar Biol.* 1989;102:49–55.

15 Lewis JB, Price WW. Patterns of ciliary currents in Atlantic reef corals and their functional significance. *J Zool.* 1976;178: 77–89.

16 Lewis JB, Price WS. Feeding mechanisms and feeding strategies of Atlantic reef corals. *J Zool.* 1975;176:527–544.

17 Jennings JB. Studies on the feeding, digestion, and food storage in free-living flatworms. *Biol Bull.* 1957;112:63–80.

18 Prusch RD. Osmotic and ionic relationship in the freshwater flatworm *Dugesia dorotocephala. Comp Biochem Physiol.* 1976; 54A:287–290.

19 Tempel D, Westheide W. Uptake and incorporation of dissolved amino acids by interstitial *Turbellaria* and *Polycheata* and their dependence on temperature and salinity. *Mar Ecol Prog Ser.* 1980;3:41–50.

20 Reiger RM. Morphology of the *Turbellaria* at the ultrastructural level. *Hydrobiologia.* 1981;84:213–229.

21 Morre J, Gibson R. The evolution and comparative physiology of terrestrial and freshwater nemerteans. *Biol Rev.* 1985;60:267–312.

22 Shigei T, H Tsuru, N Ishikawa, K Yoshioka. Absence of endothelium in invertebrate blood vessels: significance of endothelium and sympathetic nerve/medial smooth muscle in the vertebrate vascular system. *Jpn J Pharmacol.* 2001;87:253–260.

23 Ruppert EE, Barnes RD. *Invertebrate Zoology*, 6th ed. Fort Worth: Saunders Press, 1991.

24 Clark RB, Cowey JB. Factors controlling the change of shape of certain nemertean and turbellarian worms. *J Exp Biol.* 1958; 35:731–748.

25 McDermott JJ, Roe P. Food, feeding behavior and feeding ecology of nemerteans. *Am Zool.* 1985;25:113–126.

26 Tureville JM, Ruppert EE. Comparative ultrastructure and the evolution of nemertines. *Am Zool.* 1985;25:53–72.

27 Turbeville JM. Nemertinea. In: Harrison FW, Bogitsh BJ, eds. *Microscopic Anatomy of Invertebrates.* New York: Wiley-Liss, 1991.

28 Turbeville JM. An ultrastructural analysis of coelomogenesis in the hoplonemertine *Prosorhochmus americanus* and the polychete *Magelona sp. J Morphol.* 1986;187:51–60.

29 Muñoz-Chapuli R, Carmona R, Guadix JA, Macias D, Perez-Pomares JM. The origin of the endothelial cells: an evo-devo approach for the invertebrate/vertebrate transition of the circulatory system. *Evol Develop.* 2005;7(4):351–358.

30 Browning J. Octopus microvasculature: Permeability to ferritin and carbon. *Tissue Cell.* 1979;11(2):371–383.

31 Jespersen A, Lutzen J. Ultrastructure of the nephridio-circulatory connections in *Tubulanus annulutus* (Nemertine, Anopla). *Zoomorphology.* 1987;107:181–189.

32 Harrison FW, ed. *Microscopic Anatomy of Invertebrates*, Vols. 5 and 6: Molluscs. New York: Wiley-Liss, 1992.

33 Brusca, RC, Brusca GJ. *Invertebrates.* Sunderland, MA: Sinauer, 1990.

34 Vagvolgyi J. On the origin of mollusks, the coelom and coelomic segmentation. *Syst Zool.* 1967;16:153–168.

35 Mangum CP. Oxygen transport in invertebrates. *Am J Physiol.* 1985;248:505–514.

36 Potts WTW. Excretion in molluscs. *Biol Rev.* 1967;42(1):1–41.

37 Packard A. Cephalopods and fish: the limits of convergence. *Biol Rev.* 1972;47:241–307.

38 Smith PJS. Integrated cardiovascular control in the Mollusca. *Physiol Zool.* 1990;63(1):12–34.

39 Wells MJ. Oxygen extraction and jet propulsion in cephalopods. *Can J Zool.* 1990;68:815–824.

40 Barber VC, Graziadei P. The fine structure of cephalopod blood vessels II. The vessels of the nervous system. *Z Zellforschung Midrosk Anat.* 1967;77(2):147–161.

41 Barber VC, Graziadei P. The fine structure of cephalopod blood vessels III. Vessel innervation. *Z Zellforschung Midrosk Anat.* 1967;77(2):162–174.

42 Nichols WW, O'Rourke MF. *McDonald's Blood Flow in Arteries: Theoretic, Experimental and Clinical Principles*, 3rd. Philadelphia: Lea and Febiger, 1990.

43 Berne RM, Levy MN, eds. *The Arterial System in Cardiovascular Physiology*, 3rd ed. Saint Louis: CV Mosby, 1991.

44 Toulmond A. Adaptations to extreme environmental hypoxia in water breathers. In: Dejours P, ed. *Comparative Physiology of Environmental Adaptations*. Vol. 2. 8th ESCP Conference, Strasbourg. Basel: Karger, 1986.

45 Mangum CP, Colacino JM, Vandergon TL. Oxygen binding of single red blood cells of the annelid bloodworm *Glycera dibranchiata*. *J Exp Zool*. 1989;249:144–149.

46 Fritzsche D, von Oertzen JA. Metabolic responses to changing environmental conditions in the brackish water polycheates *Marenzelleria viridis* and *Hediste diversicolor*. *Marine Biol*. 1995; 121:693–699.

47 Vinson CR, Bonaventura J. Structure and oxygen equilibrium of the three coelomic cell hemoglobins of the *Echiuran* worm *Thalassema mellita* (Conn). *Comp Biochem Physiol B*. 1987;87(2): 361–366.

48 DeJours P, Toulmond A. Ventilatory reactions of the lugworm *Arenicola marina* (L.) to ambient water oxygenation changes: A possible mechanism. *Physiol Zool*. 1987;61(5):407–414.

49 Pörtner HO. Anaerobic metabolism and changes in acid-base status: quantitative interrelationships and pH regulation in the marine worm *Sipunculus nudus*. *J Exp Biol*. 1987;131: 89–105.

50 Nikam TB, Khole VV. *Insect Spiracular Systems*. New York: Wiley, 1989.

51 Maynard DM. Circulation and heart function. In: Waterman TH, ed. *The Physiology of Crustacea*. Vol. 1: *Metabolism and Growth*. New York: Academic Press, 1960.

52 Wilkens JL. Evolution of the cardiovascular system in Crustacea. *Amer Zool*. 1999;39(2):199–214.

53 McGaw IJ, Reiber CL. Cardiovascular system of the blue crab (*Callinectes sapidus*). *J Morph*. 2002;251:1–21.

54 Greenaway P, Farrelly C. Vasculature of the gas exchange organs in air breathing brachyurans. *Physiol Zool*. 1990;63:117–139.

55 McMahon BR, Wilkens JL. Ventilation, perfusion and oxygen uptake. In: Mantel L. *Internal Anatomy and Physiological Regulation*. Vol. 5 of Bliss DE, ed. *Biology of the Crustacea*. New York: Academic Press, 1983.

56 Blatchford JG. Haemodynamics of *Carcinus maenas* (L.). *Comp Biochem Physiol*. 1971;39A:193–202.

57 Reiber CL. Hemodynamics of the crayfish *Procambarus clarkii*. *Physiol Zool*. 1994;67:449–467.

58 Wilkens JL, Wilkens LA, McMahon BR. Central control of cardiac and scaphognathite pacemakers in the crab, *Cancer magister*. *J Comp Physiol B*. 1974;90:89–104.

59 McGaw IJ, Airriess CN, McMahon BR. Peptidergic modulation of cardiovascular dynamics in the Dungeness crab, *Cancer magister*. *J Comp Physiol B*. 1994;64:1–9.

60 Airriess CN, McMahon BR. Cardiovascular adaptations enhance tolerance of environmental hypoxia in the crab *Cancer magister*. *J Exp Biol*. 1994;190:23–41.

61 Reiber CL, McMahon BR, Burggren WW. Cardiovascular functions in two macruran decapod crustaceans (*Procambarus clarkii* and *Homarus americanus*) during periods of inactivity, tail flexion and cardiorespiratory pauses. *J Exp Biol*. 1997;200:1103–1113.

62 Kardong KV. *Vertebrates: Comparative Anatomy, Function, Evolution*. New York: McGraw-Hill, 2002.

63 Graham A. Evolution and development: rise of the little squirts. *Curr Biol*. 2004;14(22):R956–958.

64 Benito J, Pardos F. Hemichordata. In: FW Harrision, EE Ruppert, eds. *Microscopic Anatomy of Invertebrates*. Vol. 15: Hemichordata, Chaetognatha, and the Invertebrate Chordates. New York: Wiley-Liss, 1997.

65 Kozloff EN. *Invertebrates*. Philadelphia: Saunders, 1990.

66 Davidson B, Levine M. Evolutionary origins of the vertebrate heart: specification of the cardiac lineage in *Ciona intestinalis*. *Proc Natl Acad Sci U S A*. 2003;100(20):11469–11473.

67 Kusakabe T. Decoding cis-regulatory systems in ascidians. *Zoolog Sci*. 2005;22(2):129–46.

68 Randall DJ, Davie PS. The hearts of uro- and cephalochordates. In: Bourne GH, ed. *Hearts and Heart-like Organs*. Vol. 1. Comparative Anatomy and Development. New York: Academic Press; 1980: 41–59.

69 Ichikawa A. The Fine structure of the tunicate heart. In: Uyeda R, ed. *Electron Microscopy*. Vol 2. Tokyo: Maruzen Co.; 1966:695–696.

70 Holland LZ, Laudet V, Schubert M. The chordate amphioxus: an emerging model organism for developmental biology. *Cell Mol Life Sci*. 2004;61(18):2290–308.

71 Kriebel ME. Wave front analyses of impulses in tunicate heart. *Am J Physiol*. 1970;218:1194–1200.

72 Moller PC, Philpott CW. The circulatory system of Amphioxus (*Branchiostoma floridae*). Morphology of the major vessels of the pharyngeal area. *J Morphol*. 1973;139:389–406.

73 Rähr H. The circulatory system of Amphioxus (*Branchiostoma lanceolatum* Palas). A light-microscopic investigation based on intravascular injection technique. *Acta Zool*. 1981;60:1–18.

74 Casley-Smith JR. The fine structure of the vascular system of *Amphioxus*: implications in the development of lymphatics and fenestrated blood capillaries. *Lymphology*. 1971;4(3): 79–94.

75 Moller PC, Philpott CW. The circulatory system of *Amphioxus* (*Branchiostoma floridae*) I. Morphology of the major vessels of the pharyngeal area. *J Morph*. 1973;139:(4);389–406.

76 Schmitz A, Gemmel M, Perry SF. Morphometric partitioning of respiratory surfaces in amphioxus (*Branchiostoma lanceolatum* Pallas). *J Exp Biol*. 2000;203(22):3381–3390.

77 Johnsson M, Axelsson M. Control of the systemic heart and the portal heart of *Myxine glutinosa*. *J Exp Biol*. 1996;199(6):1429–1434.

78 Burggren WW, Farrell AP, Lillywhite HB. Vertebrate cardiovascular systems. In: Dantzler W, ed. *Handbook of Comparative Physiology*. Oxford: Oxford University Press; 1997:215–308.

79 Johansen K. The Cardiovascular system of *Myxine glutinosa* L. In: Brodal A, Fange R, eds. *The Biology of Myxinge*. Oslo: Universitetsforlaget, 1963.

80 Johanson K, Burggren WW. Cardiovascular function in lower vertebrates. In: Bourne G, ed. *Hearts and Heart-like Organs*. New York: Academic Press, 1980;61–117.

81 Burggren WW, Johansen K, McMahon BR. Respiration in primitive fishes. In: Foreman RE, Gorbman A, Dodd JM, Olsson R, eds. *The Biology of Primitive Fishes*. New York: Plenum; 1986:217–252.

82 Forster M E, Davison W, Axelsson M, Farrell AP. Cardiovascular responses to hypoxia in the hagfish, *Eptatretus cirrhatus*. *Respir Physiol*. 1992;88(3):373–386.

83 Johansen K, Lenfant C, Hanson D. Gas exchange in the lamprey, *Entosphenus tridentatus*. *Comp Biochem Physiol A*. 1973;44(1): L107–119.

84 Tufts BL. Acid-base regulation and blood gas transport following exhaustive exercise in an agnathan, the sea lamprey *Petromyzon marinus*. *J Exp Biol*. 1991;159:371–385.

85 Rovainen CM. Feeding and breathing in lampreys. *Brain Behav Evol*.1996;48(5):297–305.

86 Nilsson S. *Autonomic Nerve Function in the Vertebrates*. New York Springer-Verlag, 1983.

87 Tsujii T, Naito I, Ukita S, Ono T, Seno S. The anionic barrier system in the mesonephric renal glomerulus of the arctic lamprey, *Entosphenus japonicus* (Martens) (Cyclostomata). *Cell Tissue Res*. 1984;235(3):491–496.

88 Ellis LC, Youson JH. The anionic charge barrier in the renal corpuscle of the pronephros in the lamprey, *Petromyzon marinus L*. *Anat Rec*. 1991;231(2):178–184.

89 Randall DJ. The circulatory system. In: Hoar WS, Randall DJ, eds. *Fish Physiology*. Vol IV. New York: Academic Press; 1970: 133–172.

90 Satchell, GH. *Physiology and Form of Fish Circulation*. Cambridge: Cambridge University Press, 1991.

91 Bushnell PG, Jones DR Farrell AP. The arterial system. In: Hoar WW, Randall DJ, Farrell AP, eds. *Fish Physiology*. Vol. XIIA. San Diego: Academic Press; 1992:89–139.

92 Farrell AP, Jones DR. The heart. In: Hoar WW, Randall DJ, Farrell AP, eds. *Fish Physiology*. Vol. XIIA. San Diego: Academic Press; 1992:1–88.

93 Braun MH, Brill RW, Gosline JM Jones DR. Form and function of the bulbus arteriosus in yellowfin tuna (*Thunnus albacares*): dynamic properties. *J Exp Biol*. 2003;206(19):3327–3335.

94 Randall DJ, Burggren WW, Farrell AP, Haswell, S. *Evolution of Air Breathing*. Cambridge: Cambridge University Press, 1981.

95 Little, C. *The Colonisation of Land: Origins and Adaptations of Terrestrial Animals*. Cambridge: Cambridge University Press, 1983.

96 Brauner CJ, Matey V, Wilson JM, Bernier NJ, Val AL. Transition in organ function during the evolution of air-breathing; insights from *Arapaima gigas*, an obligate air-breathing teleost from the Amazon. *J Exp Biol*. 2004;207(9):1433–1438.

97 Feder ME, Burggren WW, eds. *Environmental Physiology of the Amphibia*. Chicago: University of Chicago Press, 1992.

98 Burggren WW. Central cardiovascular function in amphibians: qualitative influences of phylogeny, ontogeny and seasonality. In: Heisler N, ed. *Mechanisms of Systemic Regulation*. Vol. 1: Respiration and Circulation. Berlin: Springer-Verlag; 1995:175–197.

99 Feder ME, Burggren WW. Cutaneous gas exchange in vertebrates: design, patterns, control and implications. *Biol Rev*. 1985a; 60:1–45.

100 Feder ME, Burggren WW. Skin breathing in vertebrates. *Sci Amer*. 1985b;253(5):126–143.

101 Putnam JL. Septation in the ventricle of the heart of *Siren intermedia*. *Copeia*. 1975;1975:773–774.

102 Burggren WW. Hemodynamics and regulation of cardiovascular shunts in reptiles. In: Johansen K, Burggren W, eds. *Cardiovascular Shunts: Phylogenetic, Ontogenetic and Clinical Aspects*. Copenhagen: Munksgaard; 1985:121–142.

103 Hicks JW. The physiological and evolutionary significance of cardiovascular shunting patterns in reptiles. *News Physiol Sci*. 2002;17:241–245.

104 Hopkins SR, Wang T, Hicks JW. The effect of altering pulmonary blood flow on pulmonary gas exchange in the turtle *Trachemys (Pseudemys) scripta*. *J Exp Biol*. 1996;199(10):2207–2214.

105 Jackson DC. Acid-base balance during hypoxic hypometabolism: selected vertebrate strategies. *Respir Physiol Neuro Biol*. 2004;141(3):273–283.

106 White FN. Comparative aspects of vertebrate cardiorespiratory physiology. *Annu Rev Physiol*. 1978;40:471–499.

107 Clark TD, Wang T, Butler PJ, Frappell PB. Factorial scopes of cardio-metabolic variables remain constant with changes in body temperature in the varanid lizard, *Varanus rosenbergi*. *Am J Physiol Regul Integr Comp Physiol*. 2005;288(4):R992–997.

108 Burggren WW, Johansen K. Ventricular hemodynamics in the monitor lizard, *Varanus exanthematicus*: pulmonary and systemic pressure separation. *J Exp Biol*. 1982;96:343–354.

109 Burggren WW. And the beat goes on. *Nat History*. 2000; April:62–65.

110 Wang T, Altimiras J, Klein W, Axelsson M. Ventricular haemodynamics in *Python molurus*: separation of pulmonary and systemic pressures. *J Exp Biol*. 2003;206(23):4241–4245.

111 Burggren WW, Bemis WE. Studying physiological evolution: Paradigms and pitfalls. In: Nitecki MH, ed. *Evolutionary Innovations: Patterns and Processes*. Oxford: University Press, Oxford; 1990:191–228.

112 Seymour RS, Lillywhite HB. Hearts, neck posture and metabolic intensity of sauropod dinosaurs. *Proc Biol Sci*. 2000;267(1455): 1883–1887.

113 Rogers SW. Reconstructing the behaviors of extinct species: an excursion into comparative paleoneurology. *Am J Med Genet A*. 2005;134(4):349–356.

114 Nilsson S. The crocodilian heart and central hemodynamics. *Cardioscience*. 1994;5(3):163–166.

115 Axelsson M. The crocodilian heart: more controlled than we thought? *Exp Physiol*. 2001;86(6):785–789.

116 Burggren WW, Keller BB. *Development of Cardiovascular Systems*. Cambridge: Cambridge University Press, 1997.

117 Gilbert SF. The morphogenesis of evolutionary developmental biology. *Int J Dev Biol*. 2003;47(7–8):467–477.

118 Arthur W. The effect of development on the direction of evolution: toward a twenty-first century consensus. *Evol Dev*. 2004;6(4):282–288.

4

The Evolution and Comparative Biology of Vascular Development and the Endothelium

J. Douglas Coffin

University of Montana, Missoula

One of the least considered aspects of vascular biology is vascular evolution and the clues it may hold to understanding endothelial function and pathology. Comparative problems between species are encountered on a regular basis when using laboratory animals and cultured endothelial cells (ECs) to model human conditions. Further complicating these problems is the endothelial heterogeneity found within those animal models and within humans. The evolutionary process underlies the marvelous diversities among vascular systems both within an organism and across the animal kingdom. But that same evolutionary process has generated the abundance of technical problems and complexities that are necessary to resolve as practical matters in research and medicine. The resolution of these problems lies in understanding the fundamental physiological and biochemical nature of comparative vascular biology and the subsequent application of that knowledge to human pathologies. Thus, understanding the origins of the vascular system is a vital component for the resolution of human vasculopathies.

Evolution is a biological fact; much the same as gravity or electromagnetism are physical facts. Fundamental evolutionary forces, such as mutagenesis and natural selection, are scientifically, qualitatively observed and quantitatively measured on a regular basis. The application of those evolutionary forces to the endothelium and vascular development provides a model (Figure 4.1) that correlates vascular specialization and complexity with larger or more complex organisms. Further consideration of these evolutionary trends in a developmental context leads to "Evo Devo" constructions. These reveal tissue perfusion as a basic need leading to the adaptations in vascular evolution. The need for tissue perfusion is evident for both the evolution of a heterogeneous, specialized vascular system in large, complex organisms and for the precocious neovascularization that occurs during the rapid growth and differentiation phases in embryogenesis. This need for perfusion supplies the impetus for the evolution of a specialized EC that is capable of rapid growth, and the specialized physiological needs required to line and regulate the vascular system.

A comparison of the endothelium between evolution and development may appear to invoke "recapitulationist" arguments. Farrell (1) provides a good, short critique of Haekel's recapitulation theory, relative to the cardiovascular system, and how current perspectives on developmental evolution suggest that developmental processes, similar to most other evolutionary trends, are conserved or adapt without teleological or recapitulation implications. The model presented here simply suggests that tissue hypoxia can occur during two biological circumstances: Evolution of large, complex organisms and in utero development of complex, specialized embryos. The biological response in both cases is evolution, or the development of an endothelial-lined vascular system concomitant with the size and complexity of either the organism or the embryo. No teleological or recapitulationist interpretation for the events in this model are necessary – it is simply based on a common occurrence (hypoxia) with a similar response (neovascularization) in both circumstances.

CARDIOVASCULAR AND DEVELOPMENTAL EVOLUTION

Chapter 3 provides an elegant discussion of cardiovascular evolution. The descriptions in that chapter provide an essential conclusion for the arguments in this chapter: The evolution of cardiovascular systems correlates with increasing animal size. It also provides an abundance of phylogenic examples for the evolving endothelium in a variety of species. This chapter relates vascular morphogenesis and the developmental adaptations that have led to the more complex vascular systems in an evolution of development or "Evo Devo" context similar to other discussions on vascular evolution (2). The adaptations apparent during the evolution of the human/mammalian endothelium include gastrulation and mesodermal

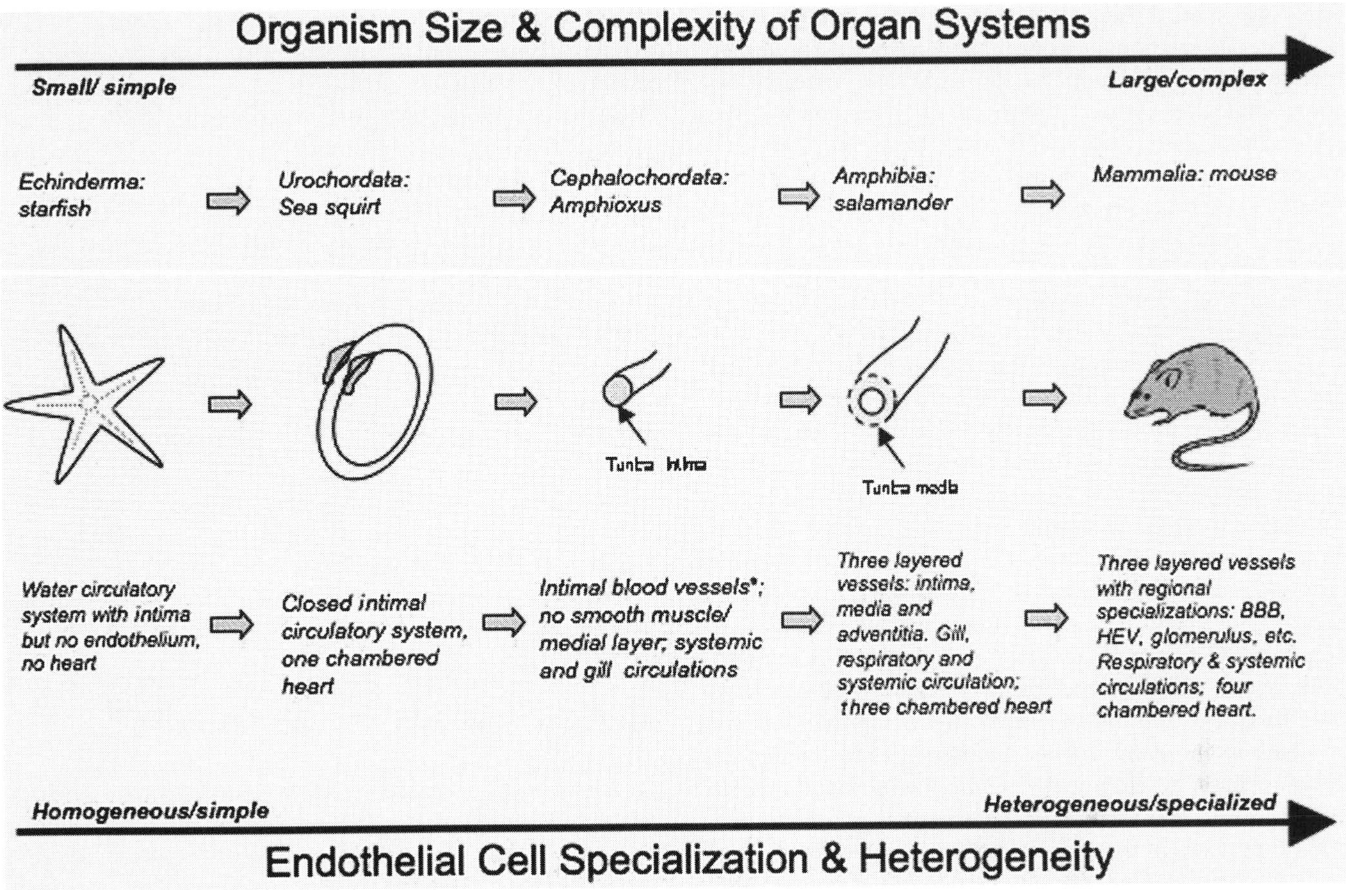

Figure 4.1. Proposed model for evolution of the simplest to the most specialized vascular systems. The upper axis represents adaptations (*left to right*) from smaller, simple organisms to larger, more complex organisms with more organ systems. The lower axis represents a homogeneous, simple endothelium evolving (*left to right*) to a more heterogeneous, specialized endothelium. In this model, the more specialized vessels serve the more complex organs and organ systems present in larger animals. Examples of increasingly specialized animals are listed from left to right. Below each animal is a short description of its respective circulatory system, which also increases in complexity. The asterisk for Amphioxus represents some debate over whether they have a true endothelial layer, and this analysis assumes the affirmative. Note that intima (tunica intima) and endothelium are often interchangeable, but not necessarily in this case. An organism may have an inner lining layer (intima) that is not a true endothelium. The terms *smooth muscle* and *media* (tunica media) are considered interchangeable here; note that adventitial layers are not represented. The first major adaptation is acquisition of an intima, between echinoderms and urochordates. The next adaptation, from a single-chambered to a multichambered heart, represents another major endothelial specialization distinguishing arteries and veins. In Ciona, the endothelium already has begun a transition, completed by Amphioxus, to the microvascular form. Amphibians then have medial layers for smooth muscle control of blood pressure, which is regulated through the endothelium by nitric oxide (NO). Further specializations that are present in some amphibia continue in mammals. BBB, blood–brain barrier; HEV, high endothelial venule.

differentiation, the "hemangioblast," the heart, an open circulatory system, and then a closed circulatory system. The selective forces that integrate these adaptations to form a closed vascular system correlate with the specializations found in terrestrial vertebrates.

Phylogenetic trends show that the smaller, less complex or the least-specialized organisms have fewer organs or organ systems and, therefore, they have the least need for perfusion and subsequently the least-developed circulatory systems (3). The least-developed circulatory systems have epithelial-, or perhaps endothelial-lined circulations without the medial or adventitial layers found in mammals. Essentially, in the simplest vascular systems, the intimal layer is the vasculature. It is unclear whether the intimal layer is always an "endothelium."

The most specialized organisms in evolutionary terms have ten different organ systems. These systems have evolved to serve specialized, interdependent functions including digestion, mobility (muscular), support (skeletal), sensory and control (nervous), gas exchange (respiratory), circulation, plasma salt and protein balance (urinary), reproduction, protection and thermoregulation (integumentary), and endocrine regulation of physiological functions (4). These specializations, typically occurring in larger terrestrial mammals, are associated with more the complex vascular systems concomitant with EC heterogeneity.

The circulatory system is among the first functioning organ systems to form during development (5). In chordates, the first vessels form in midgastrulation, and the heart begins pumping blood through these vessels during the transition from late gastrulation to early organogenesis (6). The key feature requiring circulation for the embryo is increased size and complexity. As the embryo grows larger and becomes regionalized, with a noticeable head, trunk, and hind, the increased size and complexity demand perfusion of those regions for the provision of nutrients and oxygen (O_2) and also the removal of respiratory waste. Based on size alone, simple diffusion is no longer adequate to provide all cells and tissues with O_2, and a vascular system expands to meet the demand.

Developmental trends from midgastrulation through organogenesis also reveal that the endothelium forms the morphological base for the vascular system both in utero (7) and in ovo (8). The evolutionary and developmental demands for a pump/heart, contractility, immunology, and other functions come at later stages. The evolutionary conservation of developmental systems is similar to the conservation of anatomical structures and physiological systems: They are conserved because no advantage accrues from change. A consideration of developmental stages from conception through midgastrulation, when the endothelium begins to develop, suggests that the evolution of developmental mechanisms and processes in chordates has been conservative by building on the most basic developmental requirements of the simplest organisms (9). Preservation of the branchial arches and pharyngeal pouches in the embryo is likely the best example of that conservation. They are conserved in earlier, aquatic adult organisms (i.e., fishes and amphibians), but remodeled in adult terrapods. Perhaps this adaptation reflects the demand for gas exchange between the atmosphere and a liquid circulatory system on land.

The evolution of vascular development is quite similar. It appears that the most basic elements of a simple circulatory system are conserved, then adaptations occur at later stages. For example, evolution of cardinal veins and a dorsal aorta precede the heart in some aquatic cephalochordates such as Amphioxus (10). The evolution of lungs in terrapods results in a reduction in the size of the dorsal body space commensurate with an increased ventral body space. Relative to the vasculature, these changes are commensurate with a reduction of the posterior cardinal veins and evolution of the caval system for anterior venous drainage (11). Yet, the posterior cardinal veins still are evident in the early stages of the vascular development of more specialized chordates, including *Homo sapiens*. However, like the gill system, they are remodeled during organogenesis, relatively late in development. In essence, the evolutionary trend for vascular biology is to conserve the most basic processes early in development and allow remodeling during late gastrulation and organogenesis for the development of specializations.

Comparisons of vascular development among chordates show the same evolutionary conservation, in which vasculogenesis and angiogenic processes develop very similar vascular patterns. Most chordates have dorsal and ventral aortae, a cardinal system, and a primitive single-chambered heart. These structures are conserved in simpler chordate adults such as lampreys, but remodeled in larger more specialized mammalian terrapods such as mice (see Figure 4.1) (10). Most of that remodeling occurs in organogenesis, during those stages of development when most of the specialization occurs; the earlier developmental stages are conserved. The basic, primitive vasculature develops during gastrulation; later, during organogenesis, remodeling of the basic, primitive system occurs, thereby generating specializations for each individual organ and subsequent endothelial heterogeneity.

Overall, for tissues to survive in increasingly larger, complex organisms, they must circulate fluids. Thus, as the organism becomes larger during evolution or the embryo becomes larger during development, circulation becomes a primary requirement. First, the perfusion demand is met by a simple vasculature composed mostly of a homogeneous endothelium. Later, as the organism or embryo becomes more specialized, a more specialized vasculature develops with the addition of a heart, smooth muscle, and a regionalized heterogeneous endothelium.

BIRTH, DEATH, AND GASTRULATION

"It is not birth, marriage, or death, but gastrulation, which is truly the most important time in your life" according to a classical description provided by Louis Wolpert (5). Perhaps an exaggeration, the emphasis must not be lost regarding gastrulation that entails the formation of germ layers with specific fates and, regarding mesoderm, in reference to the circulatory system. Gastrulation is the major developmental specialization in evolution. It sets the stage for the physiological specializations occurring when organisms increase in size and complexity. Only with the evolution of gastrulation came evolution of the circulatory system, commensurate with the perfusion of tissues with O_2 and nutrients and the disposal of carbon dioxide and nitrogenous wastes. Concurrent with these specializations came enhanced locomotion through extended use of the musculature and advanced neurological functions for sensory and motor control. It is not long into gastrulation that the first pluripotent mesoderm begins differentiation into angioblasts, at the start of vascular development (12,13). In this developmental scheme, hemangioblasts, first observed nearly a century ago (14–16) are the pluripotent precursors for hematopoietic cells and ECs.

THE HEMANGIOBLAST

Hemangioblasts were purposed as the most primitive or undifferentiated prevascular cell because they were first observed on blood islands in the yolk sac and the periphery of the embryo before the embryonic vasculature developed. The early models for vascular development held that the hemangioblast (17) gave rise to blood cells and angioblasts from yolk sac precursors that migrate into the embryo proper to form the vascular anlagen.[1] These are later remodeled into the embryonic circulatory system and hematopoietically formed elements (15,16). However, for many years, the hemangioblast was a theoretical concept because it had never been characterized, isolated, and proven pluripotent for both the blood and endothelium. Recent advances using differentiation markers such as FLT-1, TAL1/SLC, and platelet-endothelial cell adhesion molecule (PECAM)-1 (18–20) applied to in vitro stem-cell technology have characterized the hemangioblast a pluripotent cell upstream from both the angioblast and hematoblast (21,22). The hemangioblast offspring are freely differentiating, proliferating, and migrating angioblasts that form the basis for the "vasculogenesis" mode of vascular development (23). These new data have not only confirmed the existence of the hemangioblast, but they also have provided insights into differentiation pathways that regulate the mesoderm to hemangioblast differentiation and the subsequent hemangioblast to angioblast differentiation. The models include familiar endothelial regulatory genes such as TAL1/SLC and bone morphogenetic protein (BMP)-4 (20), hypoxia-inducible factor (HIF)-1α (21), and HoxB5 (22). These are required for mesodermal differentiation into hemangioblasts; then, vascular endothelial growth factor (VEGF) and its receptor, FLK-1, are required for hemangioblast-to-angioblast differentiation (24–26).

In evolutionary terms, hemangioblasts have been described in invertebrates, aves, amphibians, and mammals (7,27). Across species, they represent a common, pluripotent stem cell capable of differentiating to either formed blood elements or the endothelium. However, a consideration of echinoderms, which have circulatory channels but no endothelium, and cephalochordates, with a primitive endothelium but no formed elements, suggests that the angioblast and endothelium may have evolved independently or before the hematopoietic precursors. Several models show that angioblasts differentiate directly from mesoderm without hematopoietic potential (12,28–30). This revives an enduring question regarding the pluripotency of mesodermal derivatives: If an angioblast arises from lateral plate mesoderm and forms a blood vessel, does it have hematopoietic potential in the interim? This problem, represented in Figure 4.2, does reveal a scenario in which redundancy for the sake of evolutionary fortitude may allow both processes to occur simultaneously in the same embryo. This type of diversity in vascular development would certainly confer the greatest evolutionary advantage.

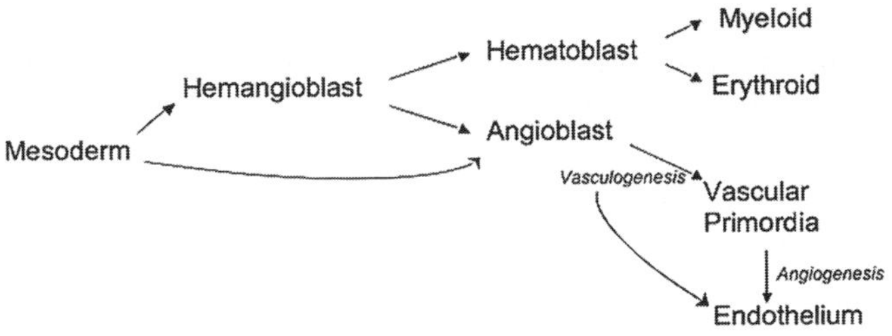

Figure 4.2. In evolutionary and phylogenic terms, vasculogenesis is the precursor to angiogenesis, and the same holds true for embryonic vascular development. Current models hold that hemangioblasts and angioblasts both can arise from mesoderm. No current data suggest that either cell can be derived from endoderm or ectoderm. The central question is whether all angioblasts arise from hemangioblasts, or whether angioblasts that have no hematopoietic potential arise independently, directly from mesoderm. Perhaps both are true. Angioblasts then form vascular primordia de novo through vasculogenesis. Further neovascularization then occurs by angiogenesis or vascular sprouting from existing vessels. Several models show that circulating or free angioblasts can contribute to vascular sprouts as an alternative form of vasculogenesis, thus adding more diversity to the developmental system.

COMPARATIVE EMBRYOLOGY OF VASCULAR DEVELOPMENT

The first and best descriptions of vascular development published in peer-reviewed scientific journals came in the early twentieth century by Evans (14) and Sabin (15,16). Sabin, in particular, described two fundamentally different modes of blood vessel development in de novo and branching terms (16). Subsequent descriptions of vascular and lymphatic development (31,32) predominantly described the branching morphogenesis aspects vascular development, because that represented the technical limits for observation through compound microscopes of ink-injected embryos.

Increasing interest in tumor angiogenesis during the 1970s, led by Folkman (33,34), generated advancements in understanding vascular development during the 1980s, when use of the QH-1 monoclonal antibody for immunofluorescence microscopy and immunoelectron microscopy led to more detailed descriptions of the de novo process for vascular development and its relationship to branching vascular morphogenesis or angiogenesis (12,13,35). The commonly used nomenclature, "vasculogenesis" for de novo vascular development and "angiogenesis" for branching vascular morphogenesis were then applied to in vitro (36) and in vivo (37,38) developmental systems.

A consideration of how these two modes of vascular development evolved provides insight into how the endothelium may have evolved. Primitive vascular systems in arthropods and insects are "open" circulatory systems, in which hemolymph is pumped in a directed, consistent manner to provide circulation of molecules and cells. In these organisms, body fluids that are separate from the feeding process are circulated within a body cavity enclosed by an exoskeleton that provides increased perfusion of organs and immune protection. The addition of a muscular pump or heart is a specialization that increases perfusion. Subsequent advancement to the closed system observed in invertebrate annelids and chordate tunicates provides even more protection, better perfusion, and thermoregulation.

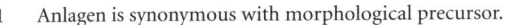

1 Anlagen is synonymous with morphological precursor.

The initial evolution of an open circulatory system appears to occur through a vasculogenesis mechanism. Condensations of differentiating mesodermal cells in the embryos form the anlagen of the initial, open channels that later acquire musculature (heart or smooth muscle) in organisms such as *Drosophila* (39). This is also true in developing birds and mammals, in which the first vessels form through vasculogenesis (40–42). Evolutionary conservation of the vasculogenic mechanism as the first means to develop the embryonic circulation is noteworthy. First, it demonstrates the importance of gastrulation and mesoderm as adaptations that allow formation of a circulatory system. Second, it shows a consistent, conserved function for blood vessel growth and development in ontogeny (i.e., the perfusion of rapidly growing, enlarging embryonic tissues that become ischemic without circulating blood).

Angiogenesis, in terms of sprouting or branching vascular morphogenesis from the existing vasculature, is a logical adaptation from vasculogenesis as the organism enlarges in size during development, and hypoxic regions require perfusion. Intuitively, it seems to be a more efficient process than vasculogenesis. As shown in Figure 4.2, the existing endothelium generally is considered the source of cells for this extension of the existing vessel. In angiogenesis, the new vessels are simply produced by mitotic expansion, in contrast to vasculogenesis, which requires recruitment, differentiation, adhesion, and maturation of surrounding, pluripotent mesoderm that may not be readily available. However, several studies have shown that circulating angioblasts are capable of homing to a sprouting vessel and undergoing incorporation into the endothelium (43,44). Thus, the source of ECs for embryonic angiogenesis is not necessarily limited to the existing endothelium. Blocking experiments in embryos conducted during fate mapping experiments (37), and the relative difficulty encountered in effectively blocking angiogenesis during tumor neovascularization, suggest that the evolution of neovascularization has provided a robust, opportunistic angiogenic process.

Perhaps the best means to distinguish the two modes of vascular development relative to function and evolutionary significance is to survey where and when either vasculogenesis or angiogenesis occur. Overall, vasculogenesis is mostly a developmental process that initially establishes the embryonic vascular pattern during gastrulation. Then, during organogenesis and in the adult, the angiogenic process dominates to perfuse organs and limbs. Finer resolution of the processes at work during vascular development reveals that both modes are at work in the earliest stages. Based on avian and mammalian data (12,13,42), the first vascular tissues form through vasculogenesis as capillary strands on the hypoblast or splanchnopleure during midgastrulation, when the first somites appear. Within a few hours, these structures, including the primitive dorsal aortae, have expanded to the length of the embryo and begin to sprout between the somites as intersomitic arteries. Later, the cardinal veins on the dorsal surface form likewise, and the same combination of vasculogenesis and angiogenesis forms the pattern for the endocardium that connects the aortic arches. Once the aortic and cardiac circulations fuse, the embryo begins to circulate blood; then the transition from vasculogenesis to angiogenesis begins. Angiogenesis becomes the favored mode of vascular development and predominates into the adult stages to maintain vascular homeostasis in processes such as wound healing and in pathological processes such as tumor angiogenesis.

A comparison of the events occurring in vascular development during ontogeny with phylogenic adaptations reveals a similar basic function for expanded, specialized vascular systems: meeting the perfusion demands of larger, more complex body forms. The earliest circulatory systems in marine invertebrates and insects appear to develop through initial condensations of mesoderm. This may be followed by some sprouting, but the latter is very limited, if occurring at all. The earliest vascular systems circulate water through marine invertebrates such as echinoderms for feeding and ambulation (ambulacral systems). These organisms have vascular channels with an intimal layer, but this is not considered an endothelium proper (3). However, echinoderms are deuterostomes and, therefore, they are distantly related to vertebrates and humans. This suggests that the echinoderm circulatory systems may be the evolutionary precursors for the closed systems of chordates.

A comparison of vascular systems in the chordates is revealing. Within the urochordates some species lack a circulatory system; others have a closed fluid, noncellular system; and then sea squirts and tunicates, for example, have both a closed circulatory system with an "endothelium" and a cellularized circulatory fluid (3,10). A comprehensive comparison suggests that several independent evolutionary pathways result in the most specialized form of a circulatory system. The simplest forms manifest an open circulation of noncellular fluids within a cellular tube with a nonseptate muscular band that serves a pump. The most specialized forms contain an endothelium, smooth muscle, a four-chambered heart that serves serial pulmonary and systemic circulations, and a urogenital system (see Figure 4.1).

The common element in the independent, convergent evolution of the endothelium and advanced circulatory systems is gastrulation and the subsequent formation of mesodermal tissue. No phylogenic or ontogenic evidence suggests that endothelium arises from any embryonic tissue except mesoderm. Mesoderm is a product of gastrulation; consequently, the principal means for the initial formation of blood vessels in all developmental systems across all species is vasculogenesis—mesodermal differentiation and the subsequent aggregation of the angioblast products.

EVOLUTION OF ENDOTHELIAL HETEROGENEITY

Once vasculogenesis has occurred, however, the angiogenic process takes over and drives the vascularization of many organ rudiments and organs (40,41,45). The fact that the greatest

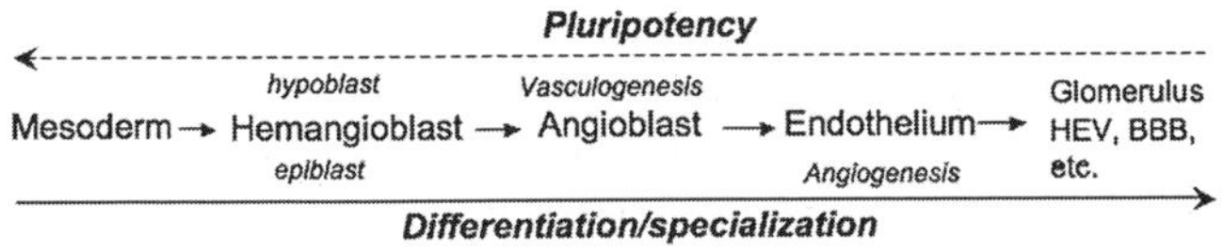

Figure 4.3. The evolution of developmental processes is represented by the application of fundamental processes, such as induction and progressive determination, that are utilized in cleavage and gastrulation of simpler organisms. In more complex, specialized organisms, induction and progressive determination are again found in vascular morphogenesis, exemplifying a more specialized developmental and physiological system. Note the key element of progressive determination, in which pluripotency decreases as differentiation and specialization progress. Application to the hemangioblast model suggests that a point is reached in differentiation at which an angioblast is fated toward a specific endothelium and no longer capable of forming a hematopoietic cell. HEV, high endothelial venules; BBB, blood–brain barrier.

bulk organ perfusion occurs in organogenesis, the developmental phase following gastrulation, once again reveals the necessity of regional, vascular specialization within larger, more complex organisms (Figure 4.3). This also leads the specialization of the endothelium and the reiteration of endothelial heterogeneity in larger vertebrates. In short, the question arises: If we are discussing the evolution of the endothelium, and the endothelium in advanced organisms is heterogeneous, which endothelium are we discussing during "evolution of the endothelium"? A comparison of the various types of endothelium is discussed in other chapters, but in evolutionary terms the discussion of heterogeneity itself and how it arises is important. At what point in development, and at what point in evolution, does the heterogeneity arise? Transplants and fate mapping in embryos suggest that the endothelium undergoes induction and progressive determination (30,37,38,46,47).

The most important developmental concepts applying to evolution and the development of endothelial heterogeneity are embryonic induction and progressive determination. First conceived by Spemann and Mangold (48), an application of the induction principle to vascular development suggests that the surrounding tissues provide cues for local angioblast differentiation. The progressive determination of the endothelium implies that the mesoderm, in the earliest stages of vascular development, is pluripotent. Pluripotency suggests that earlier angioblasts, which have experienced little differentiation, have the ability to adapt to different regions early in development (i.e., the transplanted angioblasts incorporate into the existing vessels and assume the host phenotype or specialization). The transplant data in avian systems support this principle, in which heterogeneous transplants conducted during early gastrulation show no phenotypic differences (30,43,49). According to progressive determination, however, at later stages of development, the angioblasts become more differentiated, losing their pluripotency and their ability to adapt new phenotypes as they develop into more specialized endothelium. The avian transplant data also support this concept, exemplified by results showing that trunk angioblasts transplanted into the head generally fail to incorporate into brain endothelium (38). Overall, the application of progressive determination appears to be valid based on the results from fate mapping and transplants in experimental embryology and in clinical human organ transplants, in which vascularization of the transplanted organs is often one of the major problems in rejection (50,51). However, relative to other adult tissues, the vasculature still retains considerable pluripotency (witness the "arterialization" of saphenous vein grafts in coronary artery bypass grafts).

An application of the developmental concepts to evolution and phylogeny suggest that smaller, simpler organisms without complex organ systems will have relatively more endothelial homogeneity. It follows then, that larger organisms with complex organ systems will have less endothelial homogeneity and a more heterogeneous endothelium organ-to-organ or region-to-region. (However, as discussed in Chapter 6, the endothelium of hagfish is heterogeneous, suggesting that EC heterogeneity arose as an early feature of this cell layer in the ancestral vertebrate.) In evolutionary terms, the simplest form of vascular heterogeneity is arterial versus venous endothelium. This was acquired as the simplest heart evolved atria and ventricular regions.

Lymphoid endothelium and vascular endothelium have an evolutionary branching point separate from venous and arterial vessels. It is unclear exactly when the lymphoid system evolved and whether it arose separately from the vascular endothelium. Did the lymphoid system exist previous to the vascular system, with the latter evolving from the primitive lymphatics to circulate blood? Or perhaps the lymphatics arose separately following the paradigm of Figure 4.1, when increasing size necessitated drainage of the interstitium to supplement the existing circulatory system. Gene expression profiles suggest that lymphatics represent a specialized endothelium derived from the vascular system early in the evolutionary process (52–56).

Examples for the comparison of regionally different vascular endothelia include the blood–brain barrier (BBB) in the brain, the high endothelial venules of the lymph nodes, microvascular and large vessel endothelium, and the glomerular endothelium. Differences in microvascular and large-vessel endothelium are discussed elsewhere in this volume and are described extensively in the literature. Based on the earlier discussion, the evolution of these differences is likely a function of size and specialization, working from a model in which the phylogenically simplest endothelium in smaller organisms resembles the microvasculature of larger, more complex organisms (see Figure 4.3). In smaller organisms, such as the sea squirt and annelids, no associated smooth muscle cells, internal or external elastic laminae are present. The endothelium in the phylogenically simplest forms resembles an open capillary in complex mammalian organs, such as the spleen. An extension of this discussion for deuterostome vascular evolution suggests that closure of the circulatory system and commensurate increases in size and complexity of the organism correlates with the acquisition of the smooth muscle cells and elastic fibers in the medial layer that are key morphological features of large-animal endothelium. Whether the pericytes of the microvasculature are the precursors for smooth muscle, as some in the smooth muscle field suspect, is an argument for another venue. For the purpose of discussing EC evolution, changes in gene expression, EC kinetics, and cell adhesion by the large vessels during progressive determination occur as the vessel becomes functional during organogenesis. Similar changes occur phylogenically as the organisms become highly specialized. This suggests that, aside from recapitulation arguments, vascular specialization in the embryo is an adaptation resulting from increased specialization that is conserved during developmental evolution.

The correlation between increasing size and complexity noted for the evolution of specialized large-vessel endothelium can be said for the acquisition of a BBB. This feature is found in large, specialized organisms that have both a closed circulatory system and a closed central nervous system containing cerebrospinal fluid, with the two closed fluid systems in anatomical apposition. In simpler organisms, such as annelids, the smaller, less-complex central nervous system does not require a specialized vasculature for cerebral perfusion. However, as the brain becomes large and more complex with increased size and folding of the cortex, increased cerebral perfusion in required. The acquisition and evolutionary advantage of the cerebrospinal system, although still requiring perfusion from the systemic circulatory system, suggests a need for the evolution of a specialized endothelium at the interdigitation of the two tissues.

The kidney and acquisition of the highly specialized glomerular endothelium is yet another, very well studied system in vascular development. Elegant experiments using explants from the chick metanephros onto chorioallantoic membranes reported that the developing kidney became vascularized through an angiogenic mechanism (41). As the metanephron became larger and more differentiated, the ECs migrated in from the exterior to first provide perfusion, and then subsequently differentiated into glomerular endothelium. Biochemical signals from morphogenic growth factors such as fibroblast growth factor (FGF) and transforming growth factor (TGF)-β are associated with both the initial neovascularization and the subsequent differentiation into glomerular endothelium. Later studies reported that relatively undifferentiated kidney mesoderm associated with VEGF regulation is capable of differentiating into glomerular endothelium (57). These data provide some doubt that the neovascularization of the developing kidney is strictly an angiogenic mechanism with a sole, external source of endothelial precursors. The current model (57) suggests that both angiogenic and vasculogenic modes of development are at work, with the early and predominant neovascularization occurring through angiogenesis. The perfusion of more specialized structures, such as the glomerulus, occurs through a vasculogenic mode that requires tissue induction at the site of the developing glomerulus. In evolutionary terms, this scenario demonstrates considerable developmental specialization. The evolutionary need to acquire a renal organ and urinary system during the transition from aquatic to land animals once again correlates with a demand for perfusion of those specialized organs from the systemic circulation.

The development of a complex, specialized immune system also provides an excellent example of endothelial specialization in both evolution and development. Two striking examples exist of specialized endothelium associated with the interaction of the systemic circulatory system and the immune system: lymphatic endothelium and the high endothelial venules in lymph nodes. The question of whether the lymphatic circulation or the systemic circulation evolved first has been deferred. The simplest deuterostomes, echinoderms, have a closed circulatory system with a specialized fluid that bears a resemblance to lymph. Lymphatic endothelium has unique characteristics that suggest it appeared early in evolutionary terms. The evolutionary need for separate lymph and circulatory systems follows the common theme of increased size and specialization of the organism. The primitive, open circulatory systems of insects contain "hemolymph," containing cells in a lymphatic fluid that circulates under the exoskeleton. Insects and deuterostomes are not on the same phylogenic branch, so it is difficult to determine whether convergent or divergent evolution allows any association in the evolution of the lymphatics. Some lower deuterostomes have a systemic circulation but lack lymphatics, suggesting they either had lymphatics that were later lost in favor of a circulatory system, or never had lymphatics at all. In organisms with lymphatics, the need to transfer lymphoid cells between the lymphatic and systemic circulations has been met through the evolution of the very specialized high endothelial venules (HEVs), which not only allow the transfer of cells but also participate in the differentiation of T cells. Much like the BBB, evolution of HEVs is another example showing the concomitant evolution of specialized body systems in larger, complex organisms requiring the protection of an immune system and perfusion through a circulatory system.

CONCLUSION

Two common themes emerge when considering the evolution of the vasculature and the evolution of vascular development. First, terrestrial adaptation and/or the increasing size and complexity of organisms is commensurate with the evolution of specialized organ systems. Thereafter, specialized organs are consistent with vascular specializations for improved perfusion of organs, distribution of nutrients and O_2, and disposal of nitrogenous wastes. Likewise, a larger, more complex embryo is commensurate with a more specialized embryonic vascular system. Second, the embryonic vascular system develops through both vasculogenic and angiogenic modes of vascular development, with vasculogenesis occurring first and angiogenesis second. These processes also are found in the evolution of more complex organisms, with vasculogenesis evolving first, then angiogenesis second. As the embryos of organisms become larger and more complex, vasculogenesis and then angiogenesis are integrated into vascular development. This suggests that these developmental processes are adaptations to fulfill the need for increased perfusion, and that they are conserved in evolutionary terms. Furthermore, the perfusion of specialized organs during embryonic organogenesis results in a specialized, heterogeneous vasculature within those organs in adults. The evolution of ECs that are capable of both precocious growth to meet perfusion demands and physiological specializations for immunity, renal physiology, and the like are the focal point of vascular evolution.

KEY POINTS

- The evolution of a specialized vasculature and the evolution of vascular development are both adaptations based on the demand for perfusion of the more specialized organs found in increasingly larger, complex organisms and embryos.
- The EC has evolved as an adaptation to perfuse tissues.
- Adaptations such as gastrulation and the hemangioblast preceded the EC.
- Two modes of embryonic vascular development—vasculogenesis and angiogenesis—have evolved to perfuse embryos. During embryonic development, vasculogenesis occurs first, then angiogenesis expands and remodels the embryonic vasculature.
- Vasculogenesis precedes angiogenesis in endothelial evolution. These two evolved, developmental processes are conserved in embryogenesis.
- The specialization of the vascular system in organisms is consistent with the size, complexity, and the need for perfusion, of the organs within.
- The increased size and complexity of an organism is concomitant with increased endothelial heterogeneity.

Future Goals

- To define the molecular basis for evolution of the EC, vasculogenesis, angiogenesis, and endothelial heterogeneity
- To characterize the evolutionary processes that led to the developmental, cellular and molecular heterogeneity of the vascular system, and the endothelium
- To transfer knowledge about endothelial evolution to biomedical applications for newer, advanced treatment of cardiovascular diseases

REFERENCES

1 Farrell, A. Evolution of the cardiovascular systems: insights into ontogeny. In: Burggren W, Keller BB, eds. *Development of the Cardiovascular Systems*. Cambridge: Cambridge University Press; 1997:101–103.

2 Munoz-Chapuli R, Carmona R, Guadix JA, et al. The origin of the endothelial cells: an evo-devo approach for the invertebrate/vertebrate transition of the circulatory system. *Evol Dev*. 2005;7:351–358.

3 McMahon B, Bourne GB, Chu KH. Invertebrate cardiovascular development. In: Burggren W, Keller BB, eds. *Development of the Cardiovascular Systems*. Cambridge: Cambridge University Press; 1997:127–144.

4 Liem K, Bemis WE, Walker WF, et al. *Functional Anatomy of the Vertebrates*. Belmont, CA: Brooks/Cole-Thompson, 2001.

5 Gilbert S. *Developmental Biology*. Sunderland, MA: Sinauer Associates, Inc., 1988.

6 Hamilton H. *Lillie's Development of the Chick*. New York: Henry Holt & Co., 1952.

7 Arey L. *Developmental Anatomy*. Philadelphia: Saunders & Co., 1966.

8 Patten B. *Early Embryology of the Chick*. Philadelphia: Blakiston Co., 1951.

9 Oppenheimer J. Problems, concepts and their history. In: Willier B, Weiss P, Hamburger V, eds. *Analysis of Development*. Philadelphia: Saunders & Co.; 1955:1–24.

10 Wischnitzer S. *Atlas & Dissection Guide for Comparative Anatomy*. New York: W. H. Freeman & Co., 1993.

11 Liem K, Bemis WE, Walker WF, et al. The circulatory system. In: *Functional Anatomy of the Vertebrates*. Belmont: Brooks/Cole-Thompson; 2001:603–631.

12 Coffin JD, Poole TJ. Embryonic vascular development: immunohistochemical identification of the origin and subsequent morphogenesis of the major vessel primordia in quail embryos. *Development*. 1988;102:735–748.

13 Pardanaud L, Altmann C, Kitos P, et al. Vasculogenesis in the early quail blastodisc as studied with a monoclonal antibody recognizing endothelial cells. *Development*. 1987;100:339–349.

14 Evans HM. Development of the dorsal aorta, cardinal and umbilical veins and other vessels of the vertibrate embryos from capillaries. *Anat Rec*. 1909;3:498–518.

15 Sabin FR. Preliminary note on the differentiation of angioblasts and the method by which they produce blood-vessels, blood-plasma, and red blood-cells as seen in the living chick. *Anat Rec*. 1917;13:199–204.

16 Sabin FR. Studies on the origin of blood-vessels and of red blood-corpuscles as seen in the living blastoderm of chicks during the second day of incubation. *Carn Cont Emb*. 1920;36:213–262.

17 Murray P. The development in vitro of the blood of the early chick embryo. *Proc R Soc Lond Biol Sci*. 1932;111:497.

18 Yamashita JI, Ogawa M. Medroxyprogesterone acetate and cancer cachexia: interleukin-6 involvement. *Breast Cancer*. 2000;7:130–135.

19 Wang L, Li L, Shojaei F, et al. Endothelial and hematopoietic cell fate of human embryonic stem cells originates from primitive endothelium with hemangioblastic properties. *Immunity*. 2004;21:31–41.

20 Park C, Afrikanova I, Chung YS, et al. A hierarchical order of factors in the generation of FLK1- and SCL-expressing hematopoietic and endothelial progenitors from embryonic stem cells. *Development*. 2004;131:2749–2762.
21 Ramirez-Bergeron DL, Runge A, Dahl KD, et al. Hypoxia affects mesoderm and enhances hemangioblast specification during early development. *Development*. 2004;131:4623–4634.
22 Wu Y, Moser M, Bautch VL, et al. HoxB5 is an upstream transcriptional switch for differentiation of the vascular endothelium from precursor cells. *Mol Cell Biol*. 2003;23:5680–5691.
23 Cogle CR, Scott EW. The hemangioblast: cradle to clinic. *Exp Hematol*. 2004;32:885–890.
24 Munoz-Chapuli R, Perez-Pomares JM, Macias D, et al. Differentiation of hemangioblasts from embryonic mesothelial cells? A model on the origin of the vertebrate cardiovascular system. *Differentiation*. 1999;64:133–141.
25 Saha MS, Cox EA, Sipe CW. Mechanisms regulating the origins of the vertebrate vascular system. *J Cell Biochem*. 2004;93:46–56.
26 Schatteman GC, Awad O. Hemangioblasts, angioblasts, and adult endothelial cell progenitors. *Anat Rec A Discov Mol Cell Evol Biol*. 2004;276:13–21.
27 Mandal L, Banerjee U, Hartenstein V. Evidence for a fruit fly hemangioblast and similarities between lymph-gland hematopoiesis in fruit fly and mammal aorta-gonadal-mesonephros mesoderm. *Nat Genet*. 2004;36:1019–1023.
28 Cox CM, Poole TJ. Angioblast differentiation is influenced by the local environment: FGF-2 induces angioblasts and patterns vessel formation in the quail embryo. *Dev Dyn*. 2000;218: 371–382.
29 Flamme I, Frolich T, Risau W. Molecular mechanisms of vasculogenesis and embryonic angiogenesis. *J Cell Physiol*. 1997;173: 206–210.
30 Poole TJ, Finkelstein EB, Cox CM. The role of FGF and VEGF in angioblast induction and migration during vascular development. *Dev Dyn*. 2001;220:1–17.
31 Clark E, Clark EL. Microscopic observations on the growth of blood capillaries in living mammals. *Am J Anat*. 1939;64:251–301.
32 Clark E, Clark EL. Microscopic observations on the extra-endothelial cells of living mammalian blood vessels. *Am J Anat*. 1940;66:1–49.
33 Folkman J. Tumor angiogenesis: therapeutic implications. *N Engl J Med*. 1971;285:1182–1186.
34 Folkman J, Merler E, Abernathy C, et al. Isolation of a tumor factor responsible for angiogenesis. *J Exp Med*. 1971;133:275–288.
35 Poole TJ, Coffin JD. Developmental angiogenesis: quail embryonic vasculature. *Scanning Microsc*. 1988;2:443–448.
36 Risau W, Sariola H, Zerwes HG, et al. Vasculogenesis and angiogenesis in embryonic-stem-cell-derived embryoid bodies. *Development*. 1988;102:471–478.
37 Coffin JD, Poole TJ. Endothelial cell origin and migration in embryonic heart and cranial blood vessel development. *Anat Rec*. 1991;231:383–395.
38 Poole TJ, Coffin JD. Vasculogenesis and angiogenesis: two distinct morphogenetic mechanisms establish embryonic vascular pattern. *J Exp Zool*. 1989;251:224–231.
39 Baylies MK, Michelson AM. Invertebrate myogenesis: looking back to the future of muscle development. *Curr Opin Genet Dev*. 2001;11:431–439.
40 Ekblom P, Sariola H, Karkinen-Jaaskelainen M, et al. The origin of the glomerular endothelium. *Cell Differ*. 1982;11:35–39.
41 Sariola H, Ekblom P, Lehtonen E, et al. Differentiation and vascularization of the metanephric kidney grafted on the chorioallantoic membrane. *Dev Biol*. 1983;96:427–435.
42 Coffin JD, Harrison J, Schwartz S, et al. Angioblast differentiation and morphogenesis of the vascular endothelium in the mouse embryo. *Dev Biol*. 1991;148:51–62.
43 LaRue AC, Lansford R, Drake CJ. Circulating blood island-derived cells contribute to vasculogenesis in the embryo proper. *Dev Biol*. 2003;262:162–172.
44 Lin Y, Weisdorf DJ, Solovey A, et al. Origins of circulating endothelial cells and endothelial outgrowth from blood. *J Clin Invest*. 2000;105:71–77.
45 Thisse C, Zon LI. Organogenesis – heart and blood formation from the zebrafish point of view. *Science*. 2002;295:457–462.
46 Arashavskii Iu I, Kashin SM, Litvinova NM, et al. [Turning over in the ophiuroid *Amphipholis kochii*]. *Zh Evol Biokhim Fiziol*. 1977;13:39–43.
47 Ambler CA, Nowicki JL, Burke AC, et al. Assembly of trunk and limb blood vessels involves extensive migration and vasculogenesis of somite-derived angioblasts. *Dev Biol*. 2001;234:352–364.
48 Spemann H, Mangold H. Induction of embryonic primordia by implantation of organizers from a different species (1923). *Int J Dev Biol*. 2001;45:13–38.
49 Noden DM. Origins and assembly of avian embryonic blood vessels. *Ann NY Acad Sci*. 1990;588:236–249.
50 Denton MD, Magee C, Melter M, et al. TNP-470, an angiogenesis inhibitor, attenuates the development of allograft vasculopathy. *Transplantation*. 2004;78:1218–1221.
51 Webster KA. Therapeutic angiogenesis: a complex problem requiring a sophisticated approach. *Cardiovasc Toxicol*. 2003;3: 283–298.
52 Hirakawa S, Hong YK, Harvey N, et al. Identification of vascular lineage-specific genes by transcriptional profiling of isolated blood vascular and lymphatic endothelial cells. *Am J Pathol*. 2003;162:575–586.
53 Maillard I, Fang T, Pear WS. Regulation of lymphoid development, differentiation, and function by the notch pathway. *Annu Rev Immunol*. 2005;23:945–974.
54 Makinen T, Jussila L, Veikkola T, et al. Inhibition of lymph angiogenesis with resulting lymphedema in transgenic mice expressing soluble VEGF receptor-3. *Nat Med*. 2001;7:199–205.
55 Makinen T, Veikkola T, Mustjoki S, et al. Isolated lymphatic endothelial cells transduce growth, survival and migratory signals via the VEGF-C/D receptor VEGFR-3. *EMBO J*. 2001;20: 4762–4773.
56 Partanen TA, Paavonen K. Lymphatic versus blood vascular endothelial growth factors and receptors in humans. *Microsc Res Tech*. 2001;55:108–121.
57 Freeburg PB, Abrahamson DR. Hypoxia-inducible factors and kidney vascular development. *J Am Soc Nephrol*. 2003;14:2723–2730.

5

Fish Endothelium

Kenneth R. Olson

Indiana University School of Medicine, South Bend

Fishes are the most phylogenetically ancient vertebrates, and their cardiovascular system presumably resembles the ancestral prototype. Nearly half of all vertebrate species are fishes, and they are also the most physiologically diverse. For example, hagfish have the highest plasma sodium and chloride concentration of any vertebrate (>400 mM), elasmobranchs (sharks skates and rays) adjust plasma osmolarity with 300 mM urea, and bony fishes maintain plasma osmolarity ~300 mOsm in either saltwater (1,000 mOsm) or freshwater (<1 mOsm). Body temperature varies with the environment and can range from supercooled −1.8°C in Antarctic icefish to 40°C for desert pupfish. Ambient oxygen (O_2) varies diurnally, seasonally, and with strata from supersaturation to anoxia, and various mechanisms have evolved to deal with hypoxia; the crucian carp is unaffected by up to 5 days in anoxic water, and many fishes breathe air, either using modified gills or as lungfish with a primitive lung. Hagfish have the lowest blood pressure yet measured in any vertebrate (7 to 10 mm Hg), whereas ventral aortic pressure approaches 90 mm Hg in tuna, similar to mammals. The gill and systemic circulations of fishes are arranged in series thus, unlike most other vertebrates, delicate respiratory tissues are exposed to the highest blood pressures. Also unique to fishes is the apparent lack of a lymphatic system; the impact of this on transcapillary fluid balance is largely unknown.

Despite this variability, the general features of cardiovascular systems in all extant vertebrates are surprising similar (1). Thus, it seems likely that the piscine cardiovascular system can provide clues into the factors that shaped the evolution of the vertebrate cardiovascular system and defined its plasticity.

The endothelium is the least understood component of the fish cardiovascular system. The intent of this chapter is to highlight some of its unique physiological aspects. Evolution of the endothelium, the role of the endothelium in vasculogenesis, and the specialized attributes of the piscine endocardial endothelium are described elsewhere in this volume.

ANATOMY

Relatively few attempts have been made to systematically characterize endothelial cell (EC) anatomy in fishes, either along the vascular tree, among organ systems, or between species. In general, ECs forming systemic capillaries are fairly typical, 4 to 10 μm diameter and 500 to 1,000 μm long. They are surrounded by a basement membrane (2). Although most endothelia are continuous, they are discontinuous in the liver and spleen and fenestrated in renal glomeruli (2). Endothelia in the swim bladder, gill, and secondary circulations are the most extensively studied and appear to be the most specialized.

SWIM BLADDER

The swim bladder, or gas bladder, is a gas-filled sack present in most fishes. It is used to regulate buoyancy and occasionally to sense vibrations in the water. Many fishes actively secrete O_2 into the swim bladder with aid of a gas gland that consists of a countercurrent system of long arterial and venous capillaries called a *rete mirabile* (3–7). This countercurrent system short-circuits hydrogen (H^+), which unloads O_2 from incoming blood and sustains an O_2 gradient into the bladder. Retial ECs contain large vacuoles, and they have a tubular network that, in some instances, has been shown to be continuous with both plasma (lumenal) and interstitial (ablumenal) spaces. Water and O_2 traverse the entire cellular surface, whereas intercellular junctions are the primary pathway for small solutes, and plasma proteins appear to traverse the capillary via the tubules. Arterial capillary ECs are thicker and have more intracellular organelles than do venous capillaries. Stimuli that increase intracellular calcium, such as cyclic adenosine monophosphate (cAMP), increase the permeability of the paracellular (and perhaps tubular) route, and decrease water and O_2 permeability (7).

GILL VESSELS

The fish gill is the primary organ of respiration, osmoregulation, nitrogen excretion, acid–base balance, and a major factor in regulating many blood-borne biomolecules. Three vascular pathways are present in the gill: respiratory, nutrient,

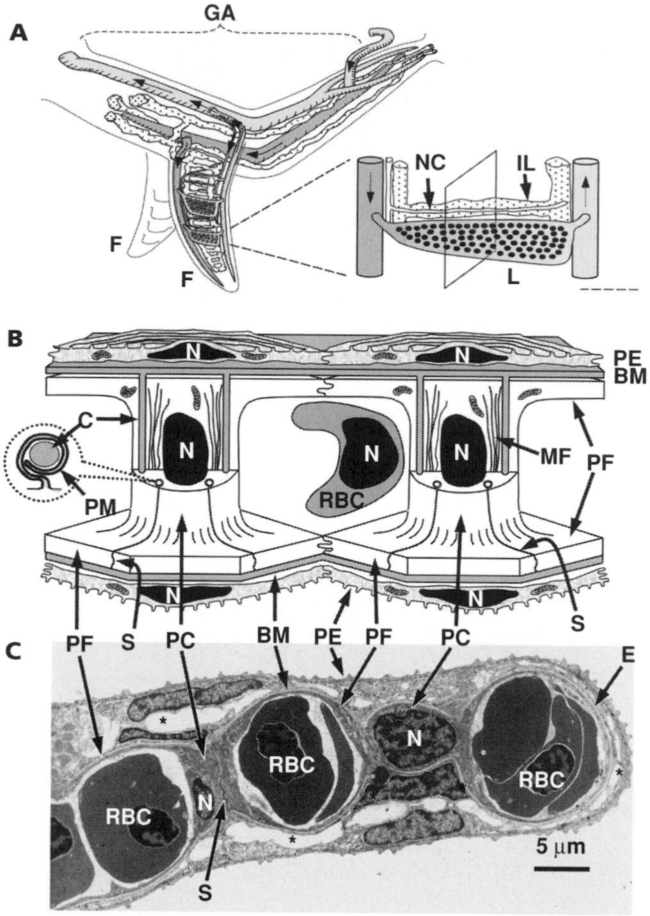

Figure 5.1. Fish gill vasculature. Schematic of a single gill arch (GA) and associated vessels is shown in (**A**). A 300-g rainbow trout (*Oncorhynchus mykiss*) has eight gill arches, each arch has two rows of ~200 paired filaments (*F*), and each filament has ~200 respiratory lamellae (*L*) on each side (~400/filament). Nutrient capillaries (*NC*) and interlamellar vessels (*IL*) are present between lamellae on each side of filament (enlarged diagram on right). The plane depicts a cross-section through *L*, *IL*, and *NC* as described in (**B**) and (**C**), and in Figure 5.2. The respiratory lamella is shown schematically in (**B**) and the transmission electron micrograph in (**C**). Pillar cells (*PC*) are sandwiched between two sheets of pavement epithelial cells (*PE*) and underlying basement membranes (*BM*). Flanges (*PF*) of adjacent PC interdigitate and delimit the vascular space. Collagen columns (*C*) connect the two BM and are enveloped by the PC membrane (*PM*), forming a seam (*S*) on the PC. Axial microfilaments (*MF*) in PC appear to be contractile. The only true EC (*E*) is found in the outer margin of the lamellar sinus. N, cell nucleus; RBC, red blood cell; *, interstitial space. *Dotted inset* shows cross-section of PM enveloping C. (**A, B** are modified with permission from Olson KR. Vasculature of the fish gill: anatomical correlates of physiological function. *J Electron Microsc Technique.* 1991;19:389–405 and Olson KR. Vascular anatomy of the fish gill. *J Exp Zool.* 2002;293:214–241. Micrograph is courtesy of J. Mallatt and R. McCall.)

and interlamellar (Figure 5.1). These form the most complex circulatory network in vertebrates (8,9). The respiratory lamellae are thin, parallel, plate-like sinusoids. Each lamella consists of two parallel sheets of pavement epithelium, with underlying basement membranes, sandwiched over spool-like pillar cells (see Figure 5.1). Typical ECs are found only in the outer lamellar margin, and these are the only cells to contain classical Weibel-Palade bodies (WPBs) (10). Pillar cells line over 95% of the lamellar vascular space but lack WPBs. Cytoplasmic flanges at each end of the pillar cell flare out to attach to neighboring pillar cells, thereby delimiting the vascular space. Each pillar cell envelops four to ten columns of collagen strands, the latter connecting the two apposed sheets of basement membrane (11). The collagen columns presumably provide structural support against the distending intravascular pressure, and their thrombogenicity is minimized by the surrounding pillar cell membrane (11). It is intuitively easy to visualize how the pillar cell plasma membrane could wrap around the collagen strands in the middle of the pillar cell (see Figure 5.1B); however, this also implies that the pillar cell flange is not continuous but rather consists of a number of broad spokes, each interrupted by a central column of collagen. Axially oriented immunoreactive filaments of actin, myosin, and 5.5 LIM protein 5 also have been observed in the pillar cell body surrounding the collagenous columns (12,13). These may provide additional structural support, and they appear to have contractile activity (14). Pillar cells are not innervated (15). Communicating (gap) junctions have been reported in pillar cells from the hagfish (16), but they have not yet been reported in teleosts. Pillar cells have been shown to actively participate in regulating plasma hormones and other blood-borne biomolecules (see below). Perhaps one of the most significant questions that remains unanswered is the degree of pillar cell involvement in actively regulating lamellar perfusion.

Interlamellar and nutrient circulations originate from postlamellar vessels in the body of the gill filament and are often apposed closely to each other (9). Although both are technically systemic circulations, their anatomy is strikingly distinct (Figures 5.2). Interlamellar capillaries contain few or no red cells, and the plasma appears to contain relatively little protein. Interlamellar ECs are thin, the cytoplasm contains few organelles, and the junctions between adjacent cells often overlap. Interlamellar vessels appear very distensible. When deflated, their EC nuclei bulge into the lumen (see Figure 5.2); when distended, they form a voluminous network that may nearly completely envelop an adjacent nutrient vessel (see Figure 5.2). Although fishes lack a lymphatic system (17), the interlamellar vessels may be its antecedent, because many of the anatomical features of interlamellar vessels are reminiscent of mammalian lymphatic capillaries.

Compared to interlamellar vessels, nutrient capillaries have a more regular circumference, frequently contain red cells, and contain more protein in the plasma. Nutrient ECs are also thicker, contain an abundant tubular system and vacuoles, and cell–cell junctions do not overlap. In some fishes, the entrance to either interlamellar or nutrient circulations is lined with ECs bearing numerous, long microvilli that perhaps restrict or regulate blood cell access to these pathways. Interestingly, corneal ECs in the blowfish contain a central cilium and numerous button-like microvilli (18).

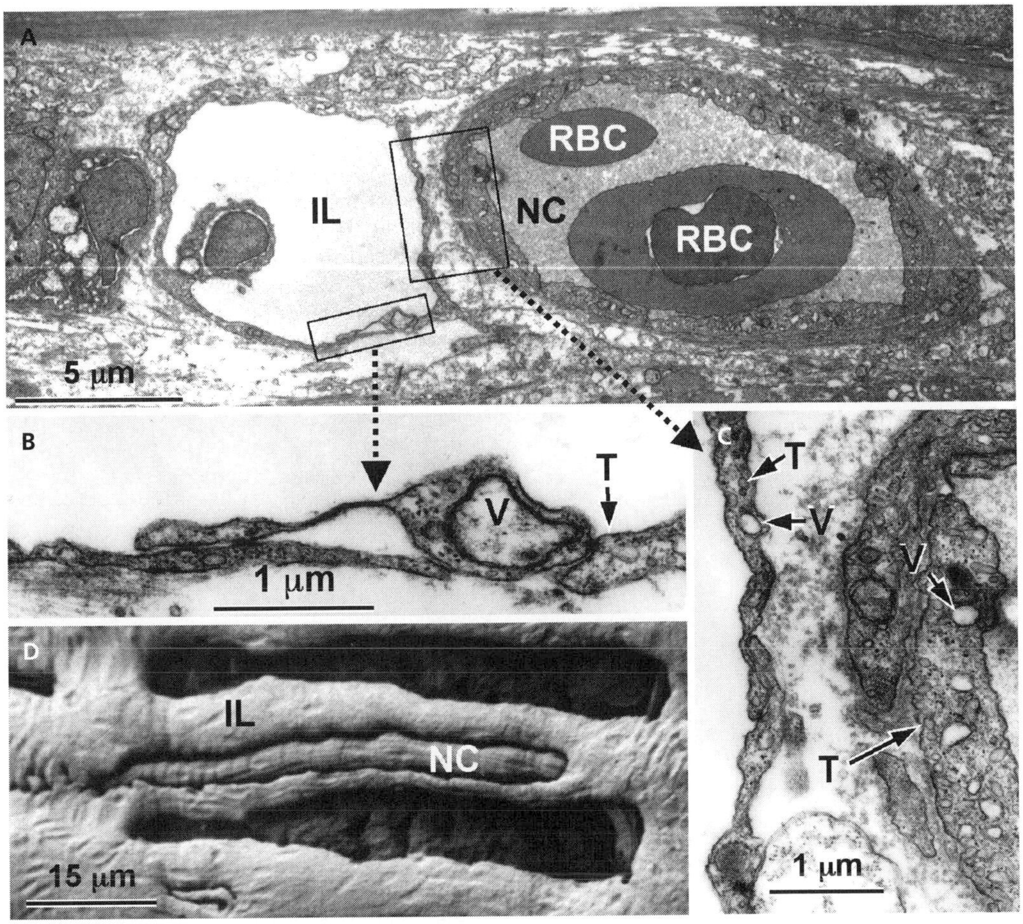

Figure 5.2. Vascular architecture of fish gills. **(A–C)** Transmission electron micrographs of interlamellar (*IL*) and nutrient capillaries (*NC*) in the non-respiratory gill filament of rainbow trout, *Oncorhynchus mykiss.* **(A)** Although IL and NC are closely associated, NC are thicker, have numerous red blood cells (*RBC*), and more plasma protein, vacuoles (*V*) and tubules (*T*). Rectangles are enlarged **(B)** and **(C)** showing thin, overlapping valve-like endothelia in the IL **(B)** and thicker endothelium in NC **(C)**. (Adapted with permission from Olson and Kingsley, unpublished and Olson KR. Secondary circulation in fish: anatomical organization and physiological significance. *J Exp Zool.* 1996;275:172–185.) **(D)** Scanning electron micrograph of a methylmethacrylate vascular corrosion replica of the dogfish shark (*Squalus acanthias*) gill showing a distended interlamellar vessel (*IL*) enveloping a nutrient capillary (*NC*). (Adapted with permission from Olson KR, Kent B. The microvasculature of the elasmobranch gill. *Cell Tiss Res.* 1980;209:49–63.)

SECONDARY CIRCULATION

The secondary circulation of bony fishes is unique among vertebrates in that it is an additional vascular network arranged in parallel with the primary system (19–21). It is formed from numerous narrow-bore arterioles (from several hundred to thousands per mm length of primary vessel) (22) that originate from larger systemic arteries. These arterioles then repeatedly anastomose to form progressively larger arterioles and arteries that ultimately supply a secondary capillary circulation in the skin, fins, some skeletal muscle, and buccal and opercular epithelia. Entrance to the secondary system often is guarded by ECs studded with 3- to 7-μm long microvilli, suggesting that it is similar to the interlamellar system in the gill. It has been suggested that the secondary system is involved in osmoregulation because it is usually associated with epithelial tissues; it also has been proposed to be the progenitor of lymphatics and a volume reservoir. Little is known about the structure of secondary capillaries or their physiology.

PHYSIOLOGICAL PROPERTIES

Permeability

Fish capillaries in general appear far more permeable to fluid and macromolecules than do the capillaries of other vertebrates (23,24), and bulk fluid movement between intravascular and interstitial compartments can be rapid and substantial. The half-time for transcapillary fluid transfer, either following plasma injection or loss due to hemorrhage, is only 15 minutes (24). In fact, following hemorrhage, fishes can replace nearly their entire plasma volume with interstitial fluid within a half hour (25). Because O_2 solubility increases as temperature decreases, many cold-water fishes, such as trout, can survive extended periods with a hematocrit of less than 5.

The whole-body transcapillary filtration coefficient has been measured recently in trout (24). At 5 mL per mm Hg filtration pressure/kg body weight/min, it is five times greater than that reported for mammals. Both capillary hydraulic and oncotic pressures in fishes are between 5 and 15 mm Hg (26). These values are considerably lower than the corresponding 20 to 30 mm Hg pressures found in mammals. Most fish capillaries are freely permeable to protein, and the whole-body reflection coefficient is essentially 0 (0 = protein permeable; 1 = protein impermeable). Permeability measurements of perfused eel rete mirabile support the whole-animal studies: albumin permeability is only slightly less than half that of the extracellular marker, inulin (6), and this difference could be attributable to size exclusion effects (26). Notable (and necessary) exceptions to this are the gill and brain, in which reflection coefficients appear very close to 1 (23). The gill is located between the heart and systemic circulations, and intravascular pressures in the respiratory lamellae can exceed 50 mm Hg. Were it not protein-impermeable, any edema would jeopardize gas exchange. The brain, which is enclosed in a rigid vault, is also edema sensitive.

The high protein permeability of fish capillaries calls into question the relevance of Starling forces (that is, the ability of protein retained in the plasma to counter capillary hydraulic pressure). If extravasated fluid could return to the capillaries with equal ease, the Starling forces would be moot, and a lymphatic system would be unnecessary. A neutrally buoyant environment (i.e., zero hydrostatic pressure) and the relatively low metabolic rate typical of ectotherms, both of which were undoubtedly present in the primordial cardiovascular system, would allow arterial blood pressure to remain relatively low. This would help keep filtration low and thereby require less reabsorption. The addition of hydrostatic pressure, attendant with a terrestrial environment, necessitated an extravascular system to return extravasated protein, a feature common in all other vertebrates. However, the prototypical lymphatic system also may be present in the gill, appearing as the interlamellar circulation, in which higher reflection coefficients and intravascular pressures necessitate an alternative route for removing extravasated protein.

Metabolic Functions of the Endothelium

Fish ECs have the capacity to modify plasma constituents through metabolism and transmembrane transfer. To date, most information has been obtained from the heart and gill.

The gill, like the mammalian lung, is ideally suited for regulating plasma-borne molecules (27,28). The gill is in series with the systemic circulation and, because no other organ receives more than 20% of the cardiac output, the gill will have five times more capacity than any other organ with the same metabolic activity. The gill also has an extensive endothelial surface area. Calculations based on pillar cell shape and lamellar geometry show that the vascular surface area is 80% greater than the respiratory surface area, approximately 900 cm^2 for a 300 g trout. Much of the gill's metabolic capacity is likely due to pillar cells, because their surface area greatly exceeds that of other ECs.

Table 5-1: Percent Removal (or Activation) of Biomolecules from Circulation by Perfused Gills (p) or Gills in Vivo (i)

Biomolecule	*% Removal (or Activation)*
Peptides	
Angiotensin II	(65)
Bradykinin (p,i)	25, 20
Atrial natriuretic peptide (p,i)	60, 60
Arginine vasotocin (i)	20–35
Endothelin-1 (i)	65
Amines	
Epinephrine (p)	5
Norepinephrine (p)	16
Serotonin (p)	80
Prostanoids	
Arachidonic acid (p)	99
PGE_2 (p)	4
PGI_2 (p)	13
Purines	
Adenosine (p)	40

(Reproduced with permission from Olson KR. Gill circulation: regulation of perfusion distribution and metabolism of regulatory molecules. *J Exp Zool.* 2002;293:320–335.)

Biomolecules can be inactivated following their physical removal from the circulation or by endothelial enzymes in contact with plasma. Table 5-1 illustrates the combined efficiency of these processes on a number of physiologically important hormones. Angiotensin converting enzyme (ACE, or kininase II) has been identified in the vascular beds of many fishes. In the gill, it activates 65% angiotensin I to angiotensin II (Ang-II) or inactivates 25% of bradykinin in a single pass. In systemic vessels, Ang-II activation is 100% (29). Gill nutrient and interlamellar pathways play a minor role in Ang-II catabolism, most of which occurs in the systemic circulation. The gill is the predominant site at which natriuretic peptides (NPs) and endothelin (ET)-1 are physically removed from the circulation. NPs are extracted by C-type clearance receptors (NPR-C), whereas the mechanism for ET-1 removal is unknown. ACE and C-type clearance receptors have been identified in gills of the most primitive fishes, hagfish (30,31), and are indicative of the long lineage of endothelial metabolism. Inactivation of biogenic amines can occur via a number of uptake mechanisms and metabolic pathways, and these have not been fully resolved in fishes. Serotonin is removed avidly by the gill and does not circulate. Norepinephrine (NE) inactivation exceeds that of epinephrine; the latter is the predominant circulating amine. Pillar cells accumulate radiolabeled NE, but most inactivation appears to be via systemic vessels. Gill extraction of PGE_2 and PGI_2 is relatively low.

Endothelial enzymes also are involved in the metabolism of nonvasoactive biomolecules. Carbonic anhydrase, present in endothelium of the swim bladder (32), speeds CO_2 hydration in the process of acid-driven O_2 secretion. Scavenger ECs in the gills of cyclostomes and elasmobranchs and in the heart or kidney of bony fishes also remove a variety of physiological and foreign waste molecules from the circulation (33). The endothelial clearance of xenobiotics by gill pillar cells may be an important adjunct to hepatic metabolism in many fishes, and this provides a first line of defense against environmental contaminants (28,34). Cytochrome P4501A (CYP4501A), which metabolizes polycyclic and polyhalogenated aromatic hydrocarbons, has been found in gill pillar cells (35) and vascular endothelium from most organs (35,36). CYP4501A levels in many endothelia are upregulated by these hydrocarbons as a concerted response to these toxicants. Pillar cells also phagocytose particulates such as colloidal carbon and latex spheres. As an additional defense mechanism, P-glycoprotein and multidrug-resistance protein 2 in brain capillaries of elasmobranchs and teleosts secrete xenobiotics in reverse, driving them from the central nervous system into the blood (37).

Paracrine and Endocrine Functions of the Endothelium

A definitive identification of endocrine and paracrine signaling molecules produced by fish ECs is very limited, and most evidence is rather indirect. For instance, ET-1 and C-type natriuretic peptide (CNP), both known to be synthesized by mammalian ECs, also are produced by fishes (38,39); however, the tissue of origin is unknown. Evidence for prostanoid-type endothelium-derived relaxing factors (EDRFs) has been based on the similar ability of indomethacin and de-endothelialization to prevent the vasodilatory response of elasmobranch and teleost vessels to a number of stimuli (39–44). Prostaglandin production by fish endothelium has not been measured directly, although a bradykinin-stimulated increase in plasma PGE_2 in intact trout (45) suggests this is a likely candidate. Whether or not nitric oxide (NO) is an EDRF in fishes is controversial. Evidence supporting NO as a vascular EDRF has been obtained only from intact or perfused preparations (46–51). Little evidence supports NO production from isolated vessels; the inhibitors of NO synthase (NOS), or the addition of the NO precursor, L-arginine, are ineffective in isolated vessel preparations, and antibodies to NOS do not react with the endothelium of either sharks or eels (41,52,53). Fritsche and colleagues (54) observed NOS immunoreactivity in the dorsal vein of 3- to 6-day-old zebrafish larva; however, it is not known if this remains in the adult. Compounds known to inhibit cytochrome P450-mediated production of endothelium-derived hyperpolarization factor in mammals generally are ineffective in isolated trout vessels (D.G. Smith and K.R. Olsen, unpublished observations). Thus, the evidence to date suggests that either nonprostanoid EDRFs first evolved in terrestrial vertebrates or, if present in fishes, occur only in early development. The former scenario seems more likely.

TECHNICAL ASPECTS

Detailed methods to isolate and culture ECs from capillaries of the eel swim bladder rete mirabile, swim bladder artery and vein, bulbus arteriosus, and ventricle have been described by Garrick (55). Cells can be grown to confluence in 3 to 6 weeks, and subcultures grown in 2 weeks. These methods should be valuable for detailed studies of the physiology of fish vascular endothelium.

NOTE ADDED IN PROOF

Recent evidence obtained from developing zebrafish embryos has shown that a thin-walled vessel, termed the *thoracic duct* (56) appears between the dorsal aorta and posterior cardinal vein and possesses numerous attributes of a true lymphatic vessel – its development requires Prox1, neuropilin 2, and VEGF-C signaling; the vessels drain the interstitium; they are devoid of erythrocytes; and they are structurally similar to lymphatics of higher vertebrates. Although these studies support the hypothesis of a piscine lymphatic system, it should be noted that similar vessels have been described in a variety of adult fish, in which both arterial and venous anastomoses, characteristic of the secondary circulation, have been identified (57).

KEY POINTS

Research on the piscine endothelium has lagged far behind mammalian efforts, yet a number of similarities and differences are becoming evident:

- Systemic ECs in fishes are anatomically similar to those of mammals, whereas gill pillar cells are unique.
- The fish endothelium is, in general, more permeable to macromolecules.
- As in their mammalian counterparts, fish ECs actively regulate circulating biomolecules via a variety of metabolic pathways and transport mechanisms.
- Fish ECs produce vasodilatory prostanoid(s) but, at this time, little evidence supports endothelial production of either NO or a hyperpolarization factor.

The physiological and ecological diversity of fishes has undoubtedly pushed the adaptability of the cardiovascular system to limits not encountered by mammals. Although the zebrafish has been invaluable in developmental studies, the aspect of the piscine endothelium as a model system of vascular adaptability is only beginning to be explored. These studies should provide considerable insight into the forces that shaped the evolution of the vertebrate cardiovascular system and the plasticity of the endothelium.

ACKNOWLEDGMENTS

The author's research has been supported in part by NSF Grant No. IBN 8616028, 9004245, 9105247, 9723306, and 0235223.

REFERENCES

1 Olson KR, Farrell AP. The cardiovascular system. In: Evans DH, Claiborne JB, eds. *The Physiology of Fishes*. Boca Raton: CRC Press, 2005.
2 Satchell GH. *Physiology and Form of Fish Circulation*. New York: Cambridge University Press, 1991.
3 Bendayan M, Rasio EA. Transport of insulin and albumin by the microvascular endothelium of the rete mirabile. *J Cell Sci.* 1996;109:1857–1864.
4 Bendayan M, Rasio EA. Evidence of a tubular system for transendothelial transport in arterial capillaries of the rete mirabile. *J Histochem Cytochem.* 1997;45:1365–1378.
5 Chen S-H, Liu K-M, Wagner RC. Three-dimensional analysis of vacuoles and surface invaginations of capillary endothelia in the eel rete mirabile. *Anat Rec.* 1998;252:546–553.
6 Rasio EA, Bendayan M. Sequential morphological and permeability changes in the rete capillaries during hyperglycaemia. *Micros Res Tech.* 2002;57:408–417.
7 Rasio EA, Bendayan M, Goresky CA. Effects of second messengers on the permeability and morphology of eel rete capillaries. *Circ Res.* 1995;76:566–574.
8 Olson KR. Vasculature of the fish gill: anatomical correlates of physiological function. *J Electron Microsc Technique.* 1991;19: 389–405.
9 Olson KR. Vascular anatomy of the fish gill. *J Exp Zool.* 2002; 293:214–241.
10 Boyd RB, DeVries AL, Eastman JT, Pietra GG. The secondary lamellae of the gills of cold water (high latitude) teleosts. A comparative light and electron microscopic study. *Cell Tissue Res.* 1980;213:361–367.
11 Hughes GM, Weibel ER. Similarity of supporting tissue in fish gills and the mammalian reticuloendothelial system. *J Ultrastruct Res.* 1972;39:106–114.
12 Mistry AC, Kato A, Tran YH, et al. FHL5, a novel actin-binding protein, is highly expressed in eel gill pillar cells and responds to wall tension. *Am J Physiol Regul Integr Comp Physiol.* 2004;287: R1141–1154.
13 Smith DG, Chamley-Campbell J. Localization of smooth-muscle myosin in branchial pillar cells of snapper (*Chrysophys auratus*) by immunofluorescence histochemistry. *J Exp Zool.* 1981;215: 121–124.
14 Stensløkken K-O, Sundin L, Nilsson GE. Cardiovascular and gill microcirculatory effects of endothelin-1 in Atlantic cod: evidence for pillar cell contraction. *J Exp Biol.* 1999; 202:1151–1157.
15 Hughes GM, Wright DE. A comparative study of the ultrastructure of the water-blood pathway in the secondary lamellae of teleost and elasmobranch fishes – benthic forms. *Z Zellforsch.* 1970;104:478–493.
16 Bartels H, Decker B. Communicating junctions between pillar cells in the gills of the Atlantic hagfish, *Myxine glutinosa*. *Experientia.* 1985;41:1039–1040.
17 Vogel WOP, Claviez M. Vascular specialization in fish but no evidence for lymphatics. *Z Naturforsch.* 1981;36:490–492.
18 Collin SP, Collin HB. The corneal endothelium in the blowfish (*Torquigener pleurogramma*). *Cornea.* 2000;19:231–235.
19 Olson KR. Secondary circulation in fish: anatomical organization and physiological significance. *J Exp Zool.* 1996;275:172–185.
20 Vogel WOP. Systemic vascular anastomoses, primary and secondary vessels in fish, and the phylogeny of lymphatics. In: Johansen K, Burggren W, eds. *Cardiovascular Shunts. Phylogenetic, Ontogenetic, and Clinical Aspects.* Copenhagen: Munksgaard; 1985:143–159.
21 Steffensen JF, Lomholt JP. The secondary vascular system. In: Hoar WS, Randall DJ, Farrell AP, eds. *Fish Physiology*. Vol. XII, Part A. The Cardiovascular System. San Diego: Academic Press; 1992:185–213.
22 Skov PV, Bennett MB. The secondary vascular system of *Actinopterygii*: Interspecific variation in origins and investment. *Zoomorphology.* 2004;123:55–64.
23 Bushnell PG, Conklin DJ, Duff DW, Olson KR. Tissue and whole-body extracellular, red blood cell, and albumin spaces in the rainbow trout as a function of time: A reappraisal of the volume of the secondary circulation. *J Exp Biol.* 1998;201:1381–1391.
24 Olson KR, Kinney DW, Dombkowski RA, Duff DW. Transvascular and intravascular fluid transport in the rainbow trout: Revisiting Starlings forces, the secondary circulation and interstitial compliance. *J Exp Biol.* 2003;206:457–467.
25 Duff DW, Olson KR. Response of rainbow trout to constant-pressure and constant-volume hemorrhage. *Am J Physiol.* 1989; 257:R1307–1314.
26 Olson KR. Blood and extracellular fluid volume regulation: role of the renin-angiotensin system, kallikrein-kinin system, and atrial natriuretic peptides. In: Hoar WS, Randall DJ, Farrell AP, eds. *Fish Physiology*. Vol. XII, Part B: The Cardiovascular System. San Diego: Academic Press; 1992:136–232.
27 Olson KR. Hormone metabolism by the fish gill. *Comp Biochem Physiol.* 1998;119A:55–65.
28 Olson KR. Gill circulation: regulation of perfusion distribution and metabolism of regulatory molecules. *J Exp Zool.* 2002;293: 320–335.
29 Olson KR, Chavez A, Conklin DJ, et al. Localization of angiotensin II responses in the trout cardiovascular system. *J Exp Biol.* 1994;194:117–138.
30 Cobb CS, Frankling SC, Thorndyke MC, Jensen FB, Rankin JC, Brown JA. Angiotensin I-converting enzyme-like activity in tissues from the Atlantic hagfish (*Myxine glutinosa*) and detection of immunoreactive plasma angiotensins. *Comp Biochem Physiol B.* 2004;138:357–364.

31 Toop T, Donald JA, Evans DH. Localisation and characteristics of natriuretic peptide receptors in the gills of the Atlantic hagfish *Myxine glutinosa* (Agnatha). *J Exp Biol.* 1995;198:117–126.

32 Würtz J, Salvenmoser W, Pelster B. Localization of carbonic anhydrase in swim bladder of European eel (*Anguilla anguilla*) and perch (*Perca fluviatilis*). *Acta Physiol Scand.* 1999;165:219–224.

33 Seternes T, Sørensen K, Smedsrød B. Scavenger endothelial cells of vertebrates: a nonperipheral leukocyte system for high-capacity elimination of waste macromolecules. *Proc Natl Acad Sci.* 2002;99:7594–7597.

34 Miller MR, Hinton DE, Stegeman JJ. Cytochrome P-450E induction and localization in gill pillar (endothelial) cells of scup and rainbow trout. *Aquat Toxicol.* 1989;14:307–322.

35 Grinwis GCM, Besselink HT, van den Brandhof EJ, et al. Toxicity of TCDD in European flounder (*Platichthys flesus*) with emphasis on histopathology and cytochrome P450 1A induction in several organ systems. *Aquat Toxicol.* 2000;50:387–401.

36 Malmström CM, Koponen K, Lindström-Seppä P, Bylund G. Induction and localization of hepatic CYP4501A in flounder and rainbow trout exposed to benzo[α]pyrene. *Ecotoxicol Environ Saf.* 2004;58:365–372.

37 Miller DS, Graeff C, Droulle L, Fricker S, Fricker G. Xenobiotic efflux pumps in isolated fish brain capillaries. *Am J Physiol Regul Integr Comp Physiol.* 2002;282:R191–198.

38 Inoue K, Naruse K, Yamagami S, Mitani H, Suzuki N, Takei Y. Four functionally distinct C-type natriuretic peptides found in fish reveal evolutionary history of the natriuretic peptide system. *Proc Natl Acad Sci.* 2003;100:10079–10084.

39 Wang Y, Olson KR, Smith MP, Russell MJ, Conlon JM. Purification, structural characterization, and myotropic activity of endothelin from the trout, *Oncorhynchus mykiss. Am J Physiol Regul Integr Comp Physiol.* 1999;277:R1605–1611.

40 Donald JA, Broughton BRS, Bennett MB. Vasodilator mechanisms in the dorsal aorta of the giant shovelnose ray, *Rhinobatus typus* (Rajiformes; Rhinobatidae). *Comp Biochem Physiol A Mol Integr Physiol.* 2004;137:21–31.

41 Evans DH, Gunderson MP. A prostaglandin, not NO, mediates endothelium-dependent dilation in ventral aorta of shark (*Squalus acanthias*). *Am J Physiol Regul Integr Comp Physiol.* 1998;274:R1050–1057.

42 Farrell AP, Johansen JA. Vasoactivity of the coronary artery of rainbow trout, steelhead trout, and dogfish: lack of support for non-prostanoid endothelium-derived relaxation factors. *Can J Zool.* 1995;73:1899–1911.

43 Kagstrom J, Holmgren S. Vip-induced relaxation of small arteries of the rainbow trout, *Oncorhynchus mykiss*, involves prostaglandin synthesis but not nitric oxide. *J Auton Nerv Syst.* 1997;63:68–76.

44 Olson KR, Villa J. Evidence against non-prostanoid endothelial-derived relaxing factor(s) in trout vessels. *Am J Physiol Regul Integr Comp Physiol.* 1991;260:R925–933.

45 Olson KR, Conklin DJ, Weaver L Jr., Herman CA, Wang X, Conlon JM. Cardiovascular effects of homologous bradykinin in rainbow trout. *Am J Physiol.* 1997;272:R1112–1120.

46 Eddy FB, Tibbs P. Effects of nitric oxide synthase inhibitors and a substrate, L-arginine, on the cardiac function of juvenile salmonid fish. *Comp Biochem Physiol C, Pharmacol Toxicol Endocrinol.* 2003;135:137–144.

47 Haraldsen L, Söderström-Lauritzsen V, Nilsson GE. Oxytocin stimulates cerebral blood flow in rainbow trout (*Oncorhynchus mykiss*) through a nitric oxide dependent mechanism. *Brain Res.* 2002;929:10–14.

48 Hylland P, Nilsson GE. Evidence that acetylcholine mediates increased cerebral blood flow velocity in crucian carp through a nitric oxide-dependent mechanism. *J Cereb Blood Flow Metab.* 1995;15:519–524.

49 Mustafa T, Agnisola C. Vasoactivity of adenosine in the trout (*Oncorhynchus mykiss*) coronary system: involvement of nitric oxide and interaction with adrenaline. *J Exp Biol.* 1998;201:3075–3083.

50 Mustafa T, Agnisola C, Hansen JK. Evidence for NO-dependent vasodilation in the trout (*Oncorhynchus mykiss*) coronary system. *J Comp Physiol B.* 1997;167:98–104.

51 Söderström V, Hylland P, Nilsson GE. Nitric oxide synthase inhibitor blocks acetylcholine induced increase in brain blood flow in rainbow trout. *Neurosci Lett.* 1995;197:191–194.

52 Jennings BL, Broughton BR, Donald JA. Nitric oxide control of the dorsal aorta and the intestinal vein of the Australian short-finned eel *Anguilla australis. J Exp Biol.* 2004;207:1295–1303.

53 Park KH, Kim K-H, Choi M-S, Choi S-H, Yoon J-M, Kim YG. Cyclooxygenase-derived products, rather than nitric oxide, are endothelium-derived relaxing factor(s) in the ventral aorta of carp (*Cyprinus carpio*). *Comp Biochem Physiol A Mol Integr Physiol.* 2000;127:89–98.

54 Fritsche R, Schwerte T, Pelster B. Nitric oxide and vascular reactivity in developing zebrafish, *Danio rerio. Am J Physiol Regul Integr Comp Physiol.* 2000;279:R2200–2207.

55 Garrick RA. Isolation and culture of capillary endothelial cells from the eel, *Anguilla rostrata. Microvasc Res.* 2000;59:377–385.

56 Yaniv K, Isogai S, Costranova D, et al. Live imaging of lymphatic development in the zebrafish. *Nature Med.* 2006;12:711–716.

57 Kuchler Am, Gjini E, Peterson-Maduro J, et al. Development of the zebrafish lymphatic system requires VEGF-C signaling. *Curr Biol.* 2006;16:1244–1248.

6

Hagfish

A Model for Early Endothelium

Pavan K. Cheruvu*, Daniel Gale*, Ann M. Dvorak*, David Haig†, and William C. Aird*,‡

*Beth Israel Deaconess Medical Center, Harvard Medical School, Boston, Massachusetts; †Harvard University, Cambridge, Massachusetts; ‡Mount Desert Island Biological Laboratory, Salisbury Cove, Maine

Evolution can be viewed as a gradual accumulation of adaptive mutations against a background of neutral change. Intraspecific diversity may be the substrate of evolution, but intraorganismal diversity – the emergence of new proteins, new cell types, and new tissue functionalities – is its exquisite product. Each organism, indeed each organ, is a complex, interlocking "bundle of adaptations."[1] The endothelium is no exception. Over the last three decades, the notion that the endothelium represents an uncomplicated permeability-selective barrier between blood and interstitium has been revised considerably. Endothelial cells (ECs) are now known to regulate inflammation, hemostasis, vasomotor tone, growth and proliferation of other cells, antigen presentation, extravasation of immune cells, and metabolism of tissue- or blood-derived hormones. These functions are differentially regulated in space and time (reviewed in 1). Endothelial cell heterogeneity reflects the capacity of the endothelium to adapt and respond to the unique demands of the underlying tissue (reviewed in 2). An important goal is to dissect the complex molecular mechanisms underlying endothelial cell function and dysfunction.

THE CALL FOR AN EVOLUTIONARY APPROACH

Why study comparative and evolutionary vascular biology? First, evolution is at heart an historical science, and curiosity drives us to understand the narrative of biological change as well as patterns that may underlie this change. Second, the study of simple models has for centuries been an indispensable tool for sharpening our analysis of more complex biological systems. One is reminded of the contributions of Mendel on *Pisum sativum*, Morgan on *Drosophila melanogaster*, and Kandel on *Aplysia californica*. Third, genomic comparisons between phylogenetically distant species facilitate the identification of essential coding and noncoding regions, because a greater interval of divergence allows neutrally evolving sequences to be discriminated from functionally constrained ones. Fourth, evolutionary thinking allows us to broaden our focus from proximate functions to include evolutionary explanations of EC function. In particular, early adaptations borne of trade-offs that are no longer applicable may contribute to design flaws in a later system. The extreme path-dependence of evolutionary change – imagine trying to build a jet engine from a propeller through slight modifications of preexisting parts – might demarcate the limits of modern endothelium and account for its propensity to disease.

HAGFISH AS A MODEL ORGANISM

Three major groups of extant vertebrates exist: hagfish (myxinoids), lampreys, and jawed vertebrates (gnathostomes).[2] All these groups possess ECs. Therefore, it is probable that ECs were already present in their most recent common ancestor. Hagfish and gnathostomes are the most distantly related

1 Here we use *adaptation* in the general sense to include any character (or even loss of character), which equips an organism to better survive in its environment and reproduce, without an implied axis of progression or growth in complexity.

2 The relationships among these groups is currently controversial. Morphological data tend to favor a sister-group relationship between gnathostomes and lampreys, with hagfish more distantly related. Some molecular data favor a sister-group relationship between hagfish and lampreys, with gnathostomes more distantly related. In either case, the last common ancestor of hagfish and gnathostomes was also the last common ancestor of all extant vertebrates. Features of ECs that are shared between hagfish and gnathostomes can be inferred to have already been present in this ancestral vertebrate.

modern animals to possess ECs.[3] This common ancestor lived more than 500 million years ago. Modern hagfish are represented by about 43 species referred to six genera, including *Myxine*, which is discussed in this chapter. These eel-like animals burrow into sea bottoms, from which they emerge on occasion to scavenge dead or morbid offal and prey on benthic invertebrates. They inhabit the nontropical waters of the Northern and Southern Hemispheres, preferring a lifestyle of low temperatures (6°C to 10°C) and moderate depths (30 to 1,500 m).

Hagfish are well-placed in the evolutionary tree to call attention to certain heuristic difficulties that accompany the invertebrate–vertebrate divide. No modern organism, of course, is a true intermediate form. Yet several features of the hagfish body plan bridge (and blur) the morphologic transition from invertebrate to vertebrate. It shares several features common to vertebrates, including segmental architecture with repeating myomeres, closed circulation, cellular blood, an endoskeleton, and ECs. Unlike jawed vertebrates, however, hagfish lack a biting apparatus, a third semicircular canal in the inner ear, spleen, and paired appendages; conversely, they possess a large notochord that is retained into adulthood, horny teeth, and internal gills. The uniqueness of hagfish is further evidenced by the existence of an aneural systemic heart, disseminated pancreatic tissue (within the submucosa of the gut), monomeric hemoglobin, slime-gland pores, and degenerate eyes. Research in hagfish biology may spark insights regarding the most basic features required for an endothelium in the vertebrate body plan.

In this chapter we examine the gross vascular anatomy, histology of local vascular beds, and ultrastructure of ECs in the Atlantic hagfish (*Myxine glutinosa*). We conclude by speculating on evolutionary pressures that may have led the early ancestors of craniates to develop an endothelium, and posit open questions for future research in comparative endothelial biomedicine.

CARDIOVASCULAR STRUCTURE AND FUNCTION

Overview

Hagfish have a closed circulation (Figure 6.1). They maintain the lowest arterial blood pressures (between 3 and 8 mm Hg) and highest relative blood volumes (18%) of any vertebrate (Table 6-1). Cardiac output is a "respectable" 8 to 9 mL min^{-1} kg^{-1} (reviewed in 3), which approaches the values seen in some teleosts. According to Ohm's law, cardiac output (Q) is a function of pressure gradient (ΔP) over total peripheral resistance (TPR) (Equation 1).

$$Q = \Delta P / TPR \qquad \text{(Equation 1)}$$

Because the pressure gradient, but not the cardiac output, is extraordinarily low (approximately 6 mm Hg), TPR must be negligible. Indeed, TPR has been estimated to be the lowest of any vertebrate.

Like other fishes, hagfish have a double-chambered heart containing a single atrium and a single ventricle. Unlike other vertebrates, the hagfish heart lacks cardioregulatory nerves.[4] The heart rate ranges from 18 to 26 beats per minute. The slow heart rate is attributed to the relatively long conduction times of the cardiac action potential – in fact, hagfish seem to be the only animals for which atrial repolarization (P_t) normally occurs prior to the onset of ventricular depolarization (QRS complex). Hagfish have a number of accessory hearts: the portal heart, two caudal hearts, and two cardinal hearts. Most notably, the portal heart – a unique adaptation not shared by teleosts, elasmobranches, or lampreys – serves to increase blood pressure in the common portal vein. The single chamber of the portal heart beats asynchronously with the systemic heart, as it propels blood from the gut into the liver sinusoids.

Similar to the case in other fishes, the branchial (gill) circulation is found in series with the systemic circulation. The gills are unique among vertebrates in that they are internalized and organized as pouches, typically six on each side in *M. glutinosa*.

The hagfish circulation also possesses a series of blood sinuses that are in direct communication with systemic vessels, leading some to characterize the circulation as "semi-closed."[5] The most prominent of these is the large subcutaneous vascular sinus located between skeletal muscles and the skin, stretching from the tentacles of the snout to the caudal fin fold. Hagfish can hold up to 30% of their blood volume within the sinus system. It is believed that the caudal and cardinal hearts function to reintroduce sinus blood into the systemic circulation.[6]

Oxygen Delivery

Oxygen (O_2) delivery is a function of cardiac output (CO) and O_2 content of the blood (Equation 2). Cardiac output is a

3 Invertebrates either have no cardiovascular system, or an open circulation with a heart and aorta, in which "blood" (hemolymph) is ejected into an open body cavity (hematocoele). The lobster and crab aorta is lined by an innermost acellular internal lamina (consisting of fibrous matrix of unbranched fibrils and fine branched filamentous material), a thick middle lamina containing fibroblasts and striated muscle cells, and an external lamina of dense connective tissue, containing granulated cells. In contrast, the whelk (snail) aorta is lined by irregularly/randomly positioned smooth muscle cells, with fibroblasts and granulated cells scattered between the smooth muscle cells.

4 The heart is said to be *aneural*, meaning that the brain lacks direct control of the heart rate. Accumulating evidence suggests that subendocardial chromaffin cells might release catecholamines constitutively to maintain heart rate in vivo.

5 It is interesting to speculate that the sinus system represents a vestige of the open circulatory system or the beginnings of a lymphatic system. The cellular lining of the sinuses has yet to be characterized.

6 Each caudal heart contains two skeletal muscle chambers separated by a continuous cartilaginous plate. The cardinal hearts are not hearts in the true sense of the word, because they operate by the contraction of extrinsic muscles.

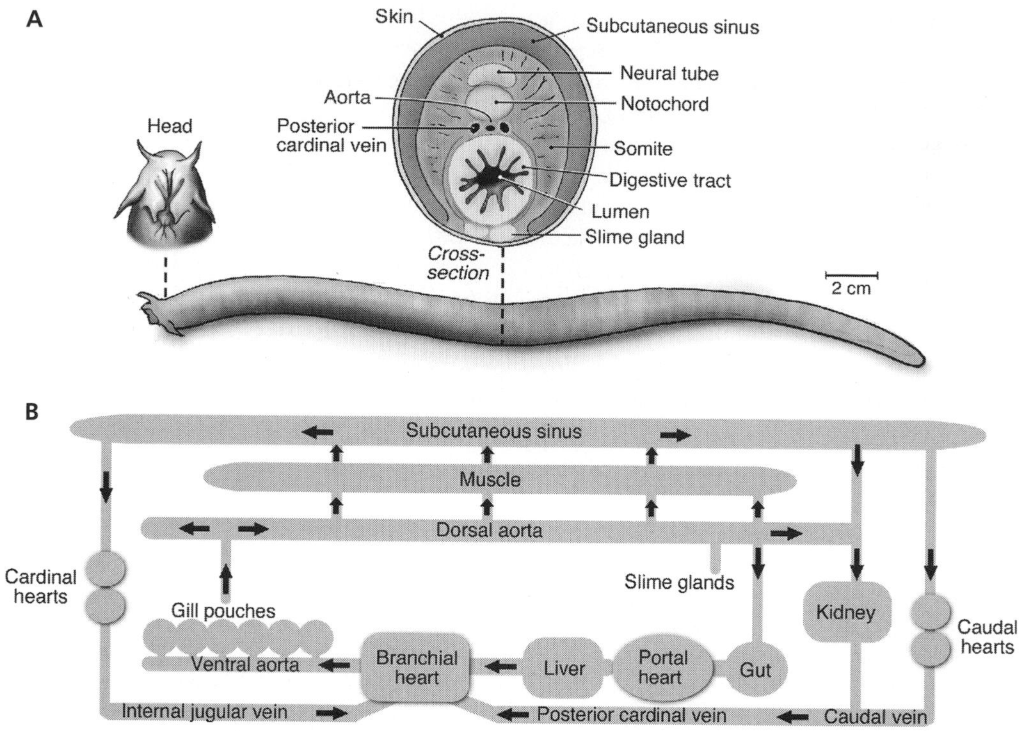

Figure 6.1. Organization of the hagfish circulation. (**A**) Schematic representation of adult hagfish cross-section in mid-region. Several features are typical of other vertebrates, including the arrangement of myomeres, neural tube, aorta, and digestive tract. Features that are unique to hagfish include the large subcutaneous sinus between skin and skeletal muscle, the retention of the notochord in the adult, and the presence of slime glands on the ventrolateral surface. (**B**) Schematic of hagfish circulation. Like other fish the gills and systemic circulation are in series. Unlike other fish the gills are internalized (there are 2 rows of 6 gills in *Myxine*; only one row is shown). Blood enters the subcutaneous sinus via skeletal muscle capillaries and reenters the systemic circulation via accessory hearts (the caudal and cardinal hearts). The portal heart pumps blood from the intestinal vasculature into the systemic heart via the common portal vein.

function of stroke volume (SV) and heart rate (HR), whereas O_2 content is determined primarily by hemoglobin (Hb) levels and O_2 saturation (Equation 3).

$$O_2 \text{ delivery} = \text{CO} \times O_2 \text{ content} \qquad \text{(Equation 2)}$$

$$O_2 \text{ delivery } \alpha \ (SV \times HR) \times (Hb \times O_2 \text{ Saturation}) \qquad \text{(Equation 3)}$$

Although hagfish are capable of fully saturating their blood in the gills, they have unusually low Hb levels and thus low O_2 content. Moreover, because the heart lacks coronary arteries, the myocardial cells derive their O_2 from poorly oxygenated mixed venous blood (reviewed in 3). As a result, the myocardium depends heavily on anaerobic metabolism. The heart contains extensive glycogen stores and is believed to generate adenosine triphosphate (ATP) from glycolysis.

Hagfish spend most of their time burrowed in the floor of the ocean. However, they exhibit short bursts of activity, for example when they feed. During exertion, cardiac output is increased. Heart rate demonstrates minimal changes. Thus, increased stroke volume must account for the change in cardiac output. Indeed, the systemic heart has been shown to be responsive to filling pressures. Body movements help to force blood from the sinus system into the circulation, resulting in elevated preload.

Large Vessels

Similar to large arteries of other vertebrates, the ventral aorta (and presumably other large arteries) of the hagfish contains three layers: the endothelium, media, and adventitia (4,5). Unlike other vertebrates, the medial smooth muscle cells are not concentrically aligned. The adventitia contains an unusually high number of ganglion cells and nerve fibers. An important property of vertebrate arteries is that they function as elastic reservoirs. During systole, arteries inflate and store energy, which is then released in diastole. Their elastic properties provide for continuing blood flow during diastole and protect against rupture at high pressures. Vertebrate arteries typically contain a combination of elastin and collagen, and display nonlinearly elastic behavior, in that they become stiffer with

Table 6-1: Hagfish Properties

	Properties	*Functional Significance/Implications*	*References*
Variables involved in oxygen delivery			
Hematocrit	13.5–15% (systemic) (vs. 30% in trout); 4.3% in sinuses	Low oxygen carrying capacity	14
Oxygen content	2.3% volume (vs. 14.6% volume in trout)	High anaerobic capacity	14
Blood pressure	Lowest among vertebrates (1–8 mm Hg)	Optimizes gas counter-current exchange in gills	15
Afterload	Lowest among vertebrates	Facilitates cardiac output	15
Blood volume	Highest among vertebrates (180 ml/kg) (18%)	High filling pressures	14
Cardiac output	Comparable to some teleosts (resting 8–9 ml/kg/min; peak 24 ml/kg/min)		15
Hearts			
Branchial (systemic)	Large (relative to teleosts); cardiac muscle; lacks rigid pericardium; aneural; no coronary circulation; demonstrates Frank-Starling curve	Myocardium bathed by poorly oxygenated venous blood; thus heart relies on anaerobic metabolism; anoxic tolerance based on high glycogen content (glycolysis-mediated generation of ATP)	15, 16
Portal	Cardiac muscle	Believed to "boost" blood flow from mesentery to systemic heart	15
Caudal	Skeletal muscle	Provides propulsion of blood from sinus system to systemic circulation	15
Cardinal	Skeletal muscle	Provides propulsion of blood from sinus system to systemic circulation	15
Blood vessels			
Aorta	Thin (compared with other vertebrates) wall consisting of three layers (endothelium, media, and adventitia), but no elastin	Lack of elastin may account for low (even nonexistent) diastolic pressure; thick wall and elastin may not be needed because of low blood pressures	4, 5
Blood–brain barrier	Capillaries 9–14 μm diameter		7
Skin	Well-developed dense capillary network in dermis	Together with lack of scales may facilitate cutaneous respiration; also serves to deliver precursors and nutrients to mucous-producing cells in epidermis	17
Endothelial cells			
Kidneys	Rare fenestrae; many vesicles; thick basal lamina	Filtration believed to occur via pinocytosis	12
Aorta	Simple squamous layer, membrane bound spherical granules; surface vesicles on basal side; long tubules		4,5
Skin	Lectin positive (HPA, PNA, UEA I); tubules; prominent basement membrane, coated pits		9
Brain	Functionally tight BBB; multiple vesicles; EC contain tubular invaginations	Tubular invaginations believed to be involved in calcium regulation, not in transport	10, 11

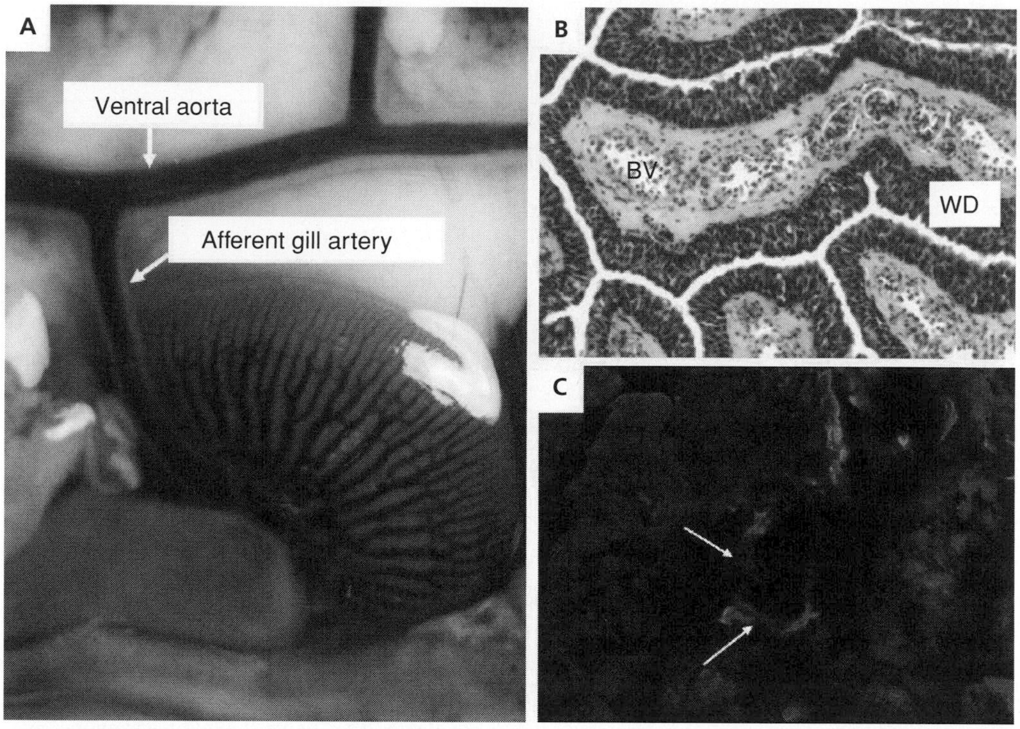

Figure 6.2. Hagfish gills. (**A**) Evans blue was injected into the posterior cardinal vein and appears in the ventral aorta, afferent gill artery and radial arteries of gill pouch. (**B**) H&E staining of a 10 μm frozen section from the gill of hagfish, BV indicates one of several blood vessels cut in cross section, WD indicates water duct or channel. (**C**) Immunofluorescent staining of 10-μm frozen section from the gill of hagfish using FITC-labeled UEA-1. Arrows point to UEA-1-positive ECs.

increased stretch. Although the aorta of hagfish (and lamprey) displays nonlinearity as evidenced by in vitro inflation experiments, these vessels do not contain elastin (4). Rather, they possess compliant elastin-like microfibrils, which are likely responsible for their capacity to function as elastic reservoirs.

In aortic ring preparations, acetylcholine (Ach), nitroprusside, and nitric oxide were shown to cause vasoconstriction, whereas prostaglandins (PGE_1, PGE_2, and PGI_2) resulted in vasodilation (6). The paradoxical effect of nitroprusside and nitric oxide was observed in hagfish, lamprey, and cartilaginous fishes, but not in teleosts. These findings argue against an evolutionarily conserved role for the nitric oxide system in mediating vasodilation.

Capillaries

The capillary bed has been described in the brain, gills, muscle, and dermis of hagfish. In microvascular corrosion casts of the brain examined by scanning electron microscopy, capillaries have large diameters (9.15 to 14.35 μm) and are poorly demarcated from arterioles and venules (7). These data are consistent with the notion that blood–brain barrier diameter decreased during vertebrate phylogeny. A clue to the selective advantage of reduced capillary diameter may be found in Fick's law of diffusion: Smaller diameter allows for increased vascular density, hence increased surface area (S), and shorter distance for O_2 to travel (t) (Equation 4).[7]

$$O_2 \text{ diffusion } \alpha \; S/t \; (\Delta P) \qquad \text{(Equation 4)}$$

The ventral aorta splits into one pair of afferent brachial arteries for each pair of gills. Each artery then joins an afferent ring artery at the base of the gill folds, which gives rise to afferent radial arteries (Figure 6.2). These radial arteries anastomose and fill out a *corpus cavernosum* in the lateral aspect of the gills. Both are lined with phagocytic high endothelium and smooth muscle cells. The afferent ring artery lassoes the efferent branchial duct as it leaves the gill pouch, so that gas exchange takes place against countercurrent flow with seawater. Gills are irrigated through the pumping of the velum, which draws ocean water in the nasopharyngeal duct and into the afferent branchial duct of each gill. The velum is a relatively weak muscle compared to the buccal and opercular apparatuses used by teleost fishes for tidal ventilation. Correspondingly, ventilatory inflow in *M. glutinosa* is thought to be only 15% to 25% of that observed for teleosts, although O_2 extraction is just as efficient.

7 It is also possible that smaller-diameter capillaries result in increased deformation of red blood cells, resulting in increased red blood cell–EC contact, thus increased surface area for gas exchange.

In skeletal muscle, 97% of the capillaries are located next to red fibers, whereas 53% of white fibers lack adjacent capillaries (8). These findings, which are typical for most vertebrates, are in keeping with the relative O_2 requirements of each fiber type.

The dermis contains an extensive and elaborate network of capillaries, which are located mainly on the outer one-third of the layer (9). These capillaries are lined by ECs, overlying a prominent basement membrane that is shared by occasional pericytes. The presence of an organized capillary network in the dermis may facilitate cutaneous respiration when the hagfish is burrowed and thus less able to draw in water over the gills. Some have argued that the diffusion distance from the water to the capillary beds is too large for skin breathing to be feasible. An alternative, although not mutually exclusive, explanation is that high capillary density is required to meet the high metabolic demands of the slime-producing mucus cells, which lie in the epidermis.

Endothelium

Electron microscopic studies of the ventral aorta demonstrate an endothelial lining whose fine structure is "similar to those described lining the walls of blood vessels in most vertebrates" (5). These cells are rounded or flattened, have irregularly shaped nuclei, and are linked to neighboring cells via junctional complexes. The cytoplasm contains endoplasmic reticulum (ER), Golgi apparatus, few mitochondria, and "variable numbers of membrane-bound spherical granules." Surface vesicles were found predominantly on the basal side of the cell, where they formed "long tubules." ECs in the dermal capillaries of hagfish have been shown to bind lectins, including helix pomatia agglutinin (HPA), peanut agglutinin (PNA), and Ulex europaeus agglutinin I (UEA I) (9). However, it is not clear whether lectins bind to the sugars of the cytoplasm, as in mammals, or to subendothelial structures (9). In electron microscopic studies, dermal ECs display highly variable thickness, flattened nuclei, and cytoplasmic smooth tubules (also termed tubular invaginations), coated pits, coated vesicles, Golgi apparatuses, mitochondria, ER, and "dense bodies" (9). These ECs were noted to display "flange-like" lateral extensions that form desmosome-like interdigitations between adjacent cells. In our own studies, we have demonstrated that dermal capillaries contain continuous endothelium, with junctional complexes typical of the adherens and tight junctions found in the endothelium of other vertebrates (Figure 6.3).

Although the blood–brain barrier of the hagfish is tight (10), the endothelial lining of the blood–brain barrier contains many more vesicles when compared with other vertebrates. Cerebral endothelium is characterized by tubular invaginations, or infoldings of the EC membrane (11). The function of these invaginations is unknown. In the kidney, endothelial cells display rare fenestrae, but contain numerous vesicles (12). Finally, recent studies have demonstrated that ECs in gill arteries function as scavengers, much like the ECs in the endocardium of teleosts or the liver sinusoidal ECs of higher vertebrates (13).

EVOLUTIONARY CONSIDERATIONS

Hagfish are the most distant human relatives to possess an endothelium. Thus, the study of its cardiovascular system may provide insight into the evolutionary origins of the endothelium. ECs are found in hagfish (and all other vertebrates), but are absent in invertebrates, cephalochordates (e.g., amphioxus), and tunicates. The fact that the endothelium is shared by jawless and jawed vertebrates is evidence that the endothelium was present in the ancestor of these animals. The absence of endothelium in cephalochordates and tunicates suggests that this structure evolved after the divergence of these groups from the vertebrate lineage, between 550 and 510 million years ago.

Many invertebrates, cephalochordates, and tunicates possess muscular hearts that pump hemolymph through tubes into an open cavity. Although this latter design – a so-called open circulation – is not lined by endothelium, its mere presence indicates that contractile muscle as a means of providing bulk flow to various cells of the body evolved far earlier than did the endothelium.

The hagfish cardiovascular system has several unique features, including low blood pressures, the presence of a portal heart, and a large sinus system. An interesting question is whether these properties are derived or ancestral. Consider, for example, the portal heart. All vertebrates (and many invertebrate animals) have a single heart. If the accessory portal heart of the hagfish is indeed ancestral, then it must have been lost during evolution of the jawed vertebrates. Alternatively, the portal heart may have derived in the hagfish lineage, perhaps as an adaptation to maintain adequate venous return in the face of increased O_2 demands. One way to address the question of derivation versus ancestry would be to compare the cardiovascular system of hagfish to lamprey. However, such an approach is rendered dubious by the ongoing controversy regarding the relationship of these two groups.

Further analysis of the hagfish endothelium may provide important information about the evolutionary history of endothelial structure and function. The endothelium of higher vertebrates is involved in many aspects of homeostasis, including regulation of vasomotor tone, leukocyte trafficking, hemostasis, antigen presentation, barrier function. For each of the above properties, one may pose the following question: Is that feature present in hagfish and thus present in a common ancestor, or is it absent in hagfish and thus derived during vertebrate evolution? For example, does endothelium in hagfish exhibit all or some properties of vasomotor tone, leukocyte trafficking, hemostasis, antigen presentation, and barrier function? The findings in the aortic ring experiments suggest that hagfish (and perhaps cartilaginous fishes) employ unique mechanisms to control vasomotor tone. If hagfish endothelium contributes to hemostatic balance, does it

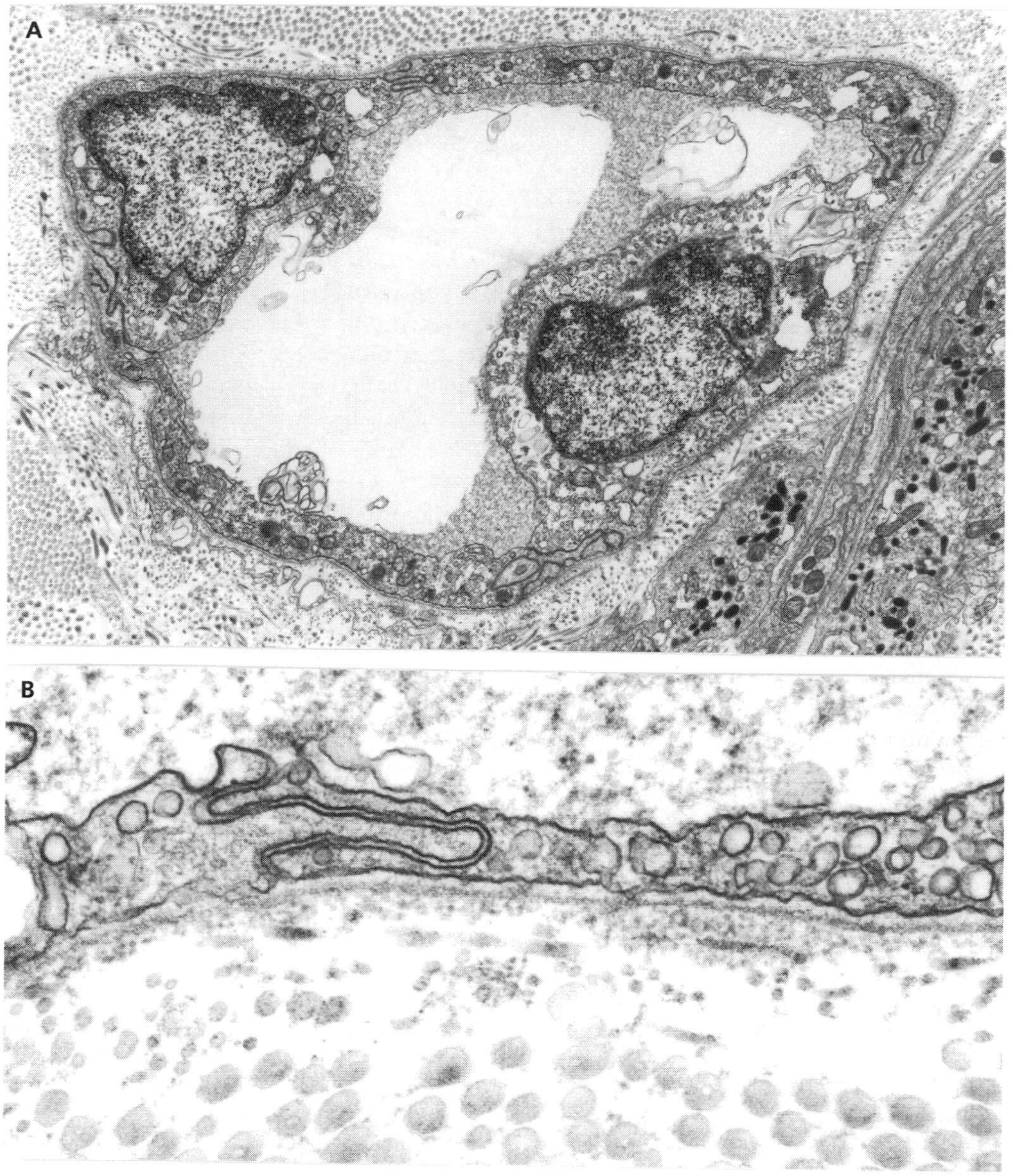

Figure 6.3. Dermal microvascular endothelium. (**A**) EM of dermis reveals a microvessel in cross-section. Cells containing melanin granules are seen in the extravascular space on the right hand side of the microvessel. (**B**) Higher magnification of a similar vessel shows well-developed lateral borders of two adjacent ECs, caveolae, and basal lamina.

express the same complement of procoagulants and anticoagulants as does mammalian endothelium? Using transmission electron microscopy, we have recently observed classic Weibel-Palade bodies in ECs from hagfish (18). These observations strongly suggest that hagfish express von Willebrand factor, a highly complex molecule that is absent in invertebrates.

The endothelium is a mosaic of phenotypes, exhibiting site-specific differences in morphology and activity. Such complexity is likely to reflect the cell's ability to adapt to the varying needs of underlying tissues. We have recently demonstrated that the endothelium of hagfish displays vascular bed–specific heterogeneity in ultrastructure, lectin binding and leukocyte adhesion. These findings suggest that structural, molecular, and functional heterogeneity evolved as an early, core feature of this cell type.

NOTE ADDED IN PROOF

We have recently published a study demonstrating the structural, molecular, and functional heterogeneity of hagfish endothelium (18).

KEY POINTS

- Comparative physiology and morphology provide important insight into the evolutionary origins of endothelial structure and function.
- The fact that the endothelium is shared by hagfish and jawed vertebrates is evidence that the endothelium was present in the ancestor of these animals. The absence of endothelium in amphioxus indicates that this structure was not present in the common ancestor of cephalochordates and vertebrates, and must have evolved following the divergence of amphioxus from vertebrates, between 550 and 510 million years ago.
- Phenotypic heterogeneity is an evolutionarily conserved feature of the endothelium.

Future Goals

- To determine extent of EC heterogeneity in hagfish and to determine whether hagfish endothelium exhibits all or some properties of vasomotor tone, leukocyte trafficking, hemostasis, antigen presentation, and barrier function

ACKNOWLEDGMENTS

This work was supported by an MDIBL New Investigator Award and an Established Investigator Award from the American Heart Association (to WCA).

REFERENCES

1 Aird WC. Spatial and temporal dynamics of the endothelium. *J Thromb Haemost*. 2005;3(7):1392–1406.
2 Aird WC. Mechanisms of endothelial cell heterogeneity in health and disease. *Circ Res*. 2006;98(2):159–162.
3 Forster ME, Axelsson, M, Farrell AP, Nilsson S. Cardiac function and circulation in hagfishes. *Can J Zool*. 1991;69:1985–1992.
4 Davison IG, Wright GM, DeMont ME. The structure and physical properties of invertebrate and primitive vertebrate arteries. *J Exp Biol*. 1995;198(Pt 10):2185–2196.
5 Wright GM. Structure of the conus arteriosus and ventral aorta in the sea lamprey, *Petromyzon marinus*, and the Atlantic hagfish, *Myxine glutinosa*: microfibrils, a major component. *Can J Zool*. 1984;62:2445–2456.
6 Evans DH, Harrie AC. Vasoactivity of the ventral aorta of the American eel (*Anguilla rostrata*), Atlantic hagfish (*Myxine glutinosa*), and sea lamprey (*Petromyzon marinus*). *J Exp Zool*. 2001;289(5):273–284.
7 Cecon S, Minnich B, Lametschwandtner A. Vascularization of the brains of the Atlantic and Pacific hagfishes, *Myxine glutinosa* and *Eptatretus stouti*: a scanning electron microscope study of vascular corrosion casts. *J Morphol*. 2002;253(1):51–63.
8 Flood PR. The vascular supply of three fibre types in the parietal trunk muscle of the Atlantic hagfish (*Myxine glutinosa*, L). A light microscopic quantitative analysis and an evaluation of various methods to express capillary density relative to fibre types. *Microvasc Res*. 1979;17(1):55–70.
9 Potter IC, Welsch U, Wright GM, Honma Y, Chiba A. Light and electron microscope studies of the dermal capillaries in three species of hagfishes and three species of lamprey. *J Zool Lond*. 1995;235:677–688.
10 Bundgaard M, Cserr HF. Impermeability of hagfish cerebral capillaries to radio-labelled polyethylene glycols and to microperoxidase. *Brain Res*. 1981;206(1):71–81.
11 Bundgaard M. Tubular invaginations in cerebral endothelium and their relation to smooth-surface cisternae in hagfish. *Cell Tissue Res*. 1987;249:359–365.
12 Heath-Eves MJ, McMillan DB. The morphology of the kidney of the Atlantic hagfish, *Myxine glutinosa*. *Am J Anat*. 1974;139:309–334.
13 Seternes T, Sorensen K, Smedsrod B. Scavenger endothelial cells of vertebrates: a nonperipheral leukocyte system for high-capacity elimination of waste macromolecules. *Proc Natl Acad Sci U S A*. 2002;99(11):7594–7597.
14 Bernier N, Fuentes J, Randall D. Adenosine receptor blockade and hypoxia-tolerance in rainbow trout and Pacific hagfish. II. Effects on plasma catecholamines and erythrocytes. *J Exp Biol*. 1996;199(Pt 2):497–507.
15 Forster ME. The blood sinus system of hagfish: its significance in a low-pressure circulation. *Comp Biochem Physiol*. 1997;116A: 239–244.
16 Johnsson M, Axelsson M. Control of the systemic heart and the portal heart of Myxine glutinosa. *J Exp Biol*. 1996;199(Pt 6):1429–1434.
17 Lametschwandtner A, Weiger T, Lametschwandtner U, et al. The vascularization of the skin of the Atlantic hagfish, *Myxine glutinosa* L. as revealed by scanning electron microscopy of vascular corrosion casts. *Scanning Microsc*. 1989 1989;3(1):305–314.
18 Yano K, Gale D, Massberg S, et al. Phenotypic heterogeneity is an evolutionarily conserved feature of the endothelium. *Blood*. 2007;109:613–615.

7

The Unusual Cardiovascular System of the Hemoglobinless Antarctic Icefish

H. William Detrich III

Northeastern University, Boston, Massachusetts

The possession of erythrocytes containing hemoglobin has long been regarded as a sine qua non of the vertebrate condition. Imagine the surprise of zoologists when they read J. T. Ruud's *Nature* article on icefishes, "Vertebrates without erythrocytes and blood pigment" (1). Ruud himself was skeptical when he heard about the *blodlaus-fisk* (i.e., lacking hemoglobin-bearing erythrocytes) that whalers reported to inhabit the waters of South Georgia in the Antarctic. He explained:

> I first heard about these "bloodless fish" on a visit to South Georgia in 1929; but no specimens were forthcoming, and I did not take them seriously. I was reminded about their existence, however, when Mr. D. Runstad, biologist of the *Norvegia* Expedition (1927–28), presented me with some photographs of a "white crocodile fish" caught by him at Bouvet Island, mentioning the fact that its blood was colourless (1).

When, years later (1953), he captured several specimens of the "white crocodile fish" *Chaenocephalus aceratus* (family Channichthyidae, suborder Notothenioidei; Figure 7.1) at South Georgia, Ruud measured the hematological parameters of its blood (1). He reported that fresh blood is nearly transparent, contains leukocytes at <1% by volume, is iron poor, and lacks erythrocytes and hemoglobin. Ruud also determined that the oxygen (O_2)-carrying capacity of *C. aceratus* blood is ~10% that of two red-blooded Antarctic fishes that belong to a closely related family (Nototheniidae, Notothenioidei). In all other respects, hematopoiesis in icefishes appears normal and yields the full complement of teleost nonerythroid blood cell lineages: heterophils, granulocytes, lymphocytes, and thrombocytes (2). Ruud surmised that the evolutionary loss of hemoglobin and red blood cells, which are derived characteristics (i.e., synapomorphies) shared by all 16 members of the Channichthyidae, could only occur in well-aerated and very cold waters,[1] which typify the Southern Ocean surrounding Antarctica (O_2 near saturation, temperature range −1.9°C to +2°C):

> Since these fish presumably descend from ancestors with hemoglobin in their blood, one imagines that only in the cold water of the polar regions could a fish survive which had lost its blood pigment (1).

ANTARCTIC ICEFISHES: AN EVOLUTIONARY PERSPECTIVE

Implicit in Ruud's hypothesis is the recognition that a phenotype having the absence of erythrocytes and hemoglobin, which would be deleterious or disaptive[2] for fishes living at high temperature, may be selectively neutral or perhaps disadvantageous but nonlethal at low temperature. Although the icefishes may have benefited from loss of red cells due to a concomitant reduction in blood viscosity (3–5) (this possibility remains unresolved), the development of compensatory readaptations that enhance O_2 delivery, which include large increases in cardiac output, vascular density, and blood volume; substantial cutaneous uptake through a scaleless skin; and a modest decrease in O_2 consumption (6,7), indicates that the phenotype was probably maladaptive under conditions of stress.[3] How, then, does one explain the successful

1 Cold water dissolves gases, such as oxygen, to a much greater extent than does warm water.

2 Per Montgomery and Clements (3), I define an adaptation as an organismal trait that evolved by natural selection for a particular biological function. Disaptations are characteristics inferior to phylogenetically antecedent characters, and readaptations are characteristics produced by natural selection that reverse, at least partially, prior disaptations (3).

3 Evolution of these cardiovascular changes in the ancestral channichthyid may have occurred before, or concomitant with, loss of hemoglobin expression, but the factor(s) that would drive this alternative path in icefishes, but not other notothenioids, are not apparent.

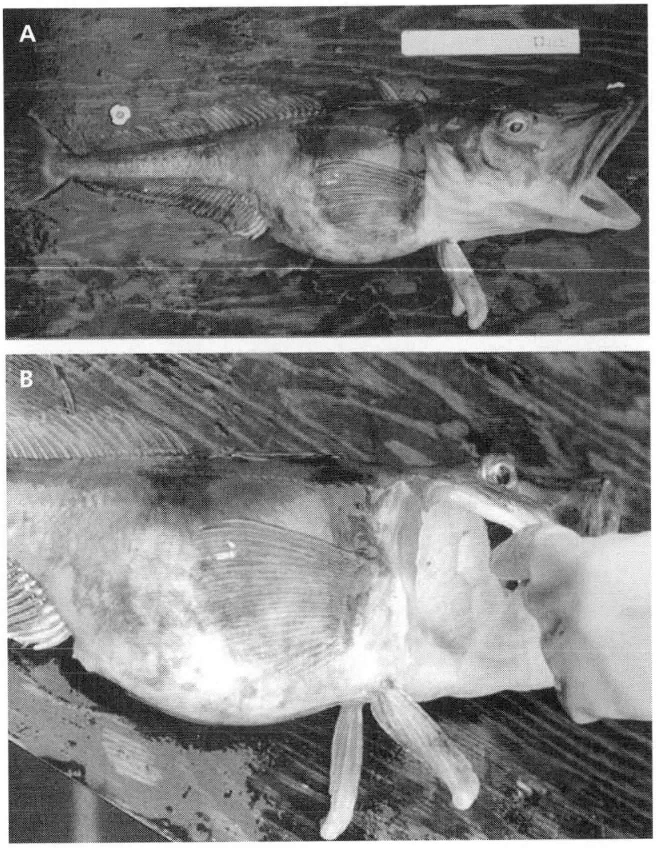

Figure 7.1. The icefish *Chaenocephalus aceratus*. **(A)** A living, gravid female. The ruler measures 15 cm. **(B)** Lifting the gill operculum reveals the white complexion of the gills.

diversification of the notothenioid fishes over the past 10 to 15 million years and the icefishes over the past 5 to 8 million years (8,9)? The answer most likely lies in the well-documented collapse in species diversity in the Southern Ocean that occurred during the mid-Tertiary (~40 million years ago) (5,10).

Whether or not the Antarctic notothenioid fishes possessed the "right stuff," they came to dominate the fish fauna of the Southern Ocean because tectonic and oceanographic events (e.g., the opening of the Drake Passage ~25 million years ago, and formation of the Antarctic Circumpolar Current and of the Antarctic Polar Front) led to continental glaciation, ice-shelf scouring of the continental shelf, and the cooling of the Southern Ocean to freezing temperatures by 10 to 14 million years ago (10,11). The subsequent loss of habitat and changes in trophic structure caused the extinction of more metabolically active groups. In the absence of significant niche competition, the notothenioids radiated rapidly, such that this suborder makes up over 50% of the species diversity and 90% of the biomass of the Antarctic fish fauna (10). The formation of ice-free marine embayments during the periodic recession of the continental glacial ice cover provided the cold refugia that facilitated notothenioid diversification (8). Thus, icefishes evolved and diversified as selection pressure for a cellular O_2 transport system was relaxed in a cold, stable, and O_2-rich marine environment. Indeed, virtually all icefish genomes have lost the globin gene complexes of the red-blooded relatives from which they evolved (12–16).

THE MYOGLOBIN AND MITOCHONDRIAL CONNECTIONS

The loss of O_2-binding proteins by icefishes is not restricted to hemoglobin; six of the 16 species do not possess myoglobin in the heart (17,18), the only tissue to express this hemoprotein in the group. Failure to produce cardiac myoglobin by the six species has occurred by four independent mutations and at least three distinct molecular mechanisms (18–20). Furthermore, the myoglobin-expression status of icefishes is associated with quantitative changes in mitochondrial density and morphology. The mitochondrial density of *C. aceratus*, a myoglobin nonexpresser, is dramatically expanded (36% of cardiomyocyte cell volume) compared to those of the icefish expresser *Chionodraco rastrospinosus* (20%) and the red-blooded, myoglobin-expressing notothenioid *Gobionotothen gibberifrons* (16%) (21). Moreover, the packing density of cristae within *C. aceratus* mitochondria is low compared to those found in the mitochondria of myoglobin expressers (21). Assuming myoglobin loss and mitochondrial expansion are physiologically related, what role does the latter play in O_2 delivery to cells?

The expansion of mitochondrial density observed so strikingly in the myoglobinless icefish *C. aceratus* is likely to enhance O_2 flux in the heart by two mechanisms (22,23). First, high densities of mitochondria reduce the mean diffusional path length for O_2 transfer from capillaries to these organelles. Second, the elaboration of additional mitochondrial membrane, a necessary consequence of the organelle expansion, may provide a "lipid highway" that enhances the diffusion of O_2 through the cell because O_2 is more soluble in lipid than in aqueous media. These features should compensate for the absence of myoglobin-mediated intracellular O_2 diffusion (24).

THE ICEFISH CARDIOVASCULAR SYSTEM

Having examined the molecular and cellular consequences of hemoglobin and myoglobin loss, I now turn to tissue- and organ-level properties of the icefish cardiovascular system. The icefish heart is a spongy myocardium (lacking coronary arteries) (Figure 7.2) that functions as a high-volume, low-pressure pump (ventral aortic pressures are ~2 kPa [26]). Compared to red-blooded notothenioids of the same size, icefishes have very large hearts that produce a weight-normalized cardiac output that is four- to fivefold larger, blood volumes that are two- to fourfold greater, and dense beds of capillaries throughout the body (6,17). The latter point is illustrated dramatically by Figure 7.3, which shows the retinal vasculature of a red-blooded notothenioid *Pagothenia borchgrevinki* and of the icefish *C. aceratus*. Furthermore, the mean capillary diameter in icefishes

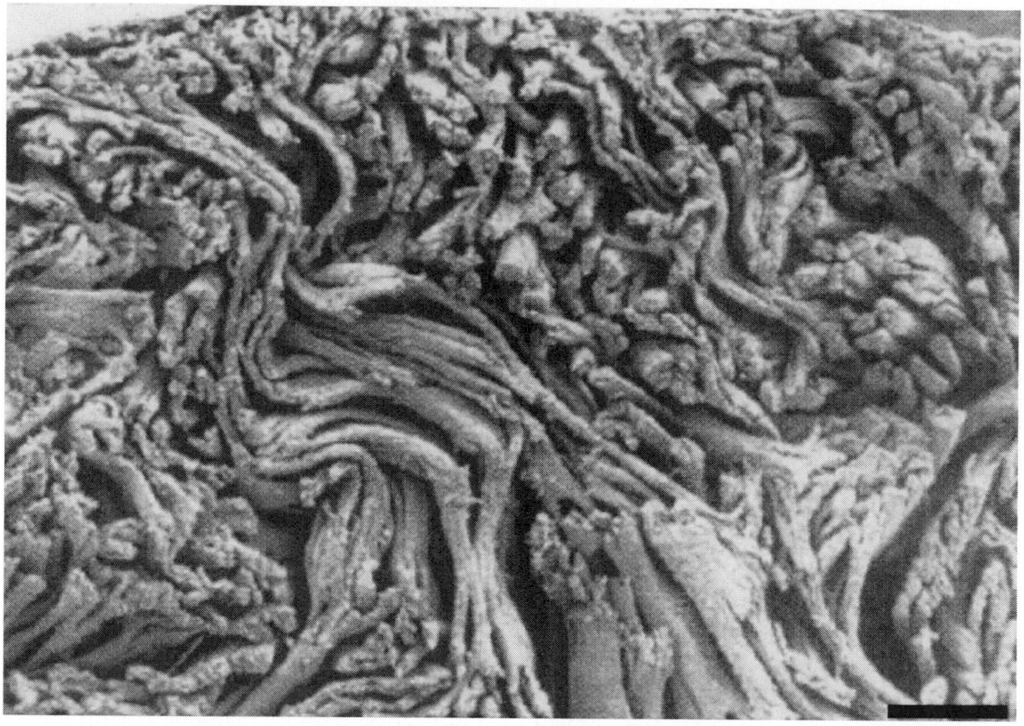

Figure 7.2. Myoarchitecture of the spongiform ventricular wall of *C. hamatus*. Scanning electron micrograph showing the myocardial trabeculae and the intratrabecular lacunae of this volume pump. Bar, 100 μm (25). (Reprinted with permission from Pelligrino D, Palmerini CA, Tota B. No hemoglobin but NO: the icefish (*Chionodraco hamatus*) heart as a paradigm. *J Exp Biol.* 2004;207:3855–3864.)

is 1.5 times greater than in red-blooded notothenioids (29). Because the resistance for laminar flow in a tube varies inversely with the fourth power of the radius and directly with viscosity in a Newtonian flow regime, relatively small changes in capillary diameter are expected to contribute significantly to efficient cardiovascular performance.

Although flow resistance and arterial blood pressure are low, the cardiac work of *C. aceratus* and other icefishes is estimated to consume more than 22% of resting metabolic energy production (26). Nevertheless, icefishes have a substantial reserve capacity for the uptake and transport of O_2 and are certainly able to perform energetic, nonresting activities such as vertical migration in the O_2-saturated water column in search of prey (7).

JUST SAY NO: THE SIDELL HYPOTHESIS

The roles of nitric oxide (NO), which is produced by several isoforms of NO synthase (NOS) as a modulator of vasodilation (30) and as a signal that stimulates angiogenesis (reviewed by Conway et al. [31]), are well known. Recently, NO also has been shown to stimulate mitochondrial biogenesis (32,33). Should an organism experience long-term systemic elevation of steady-state levels of NO, one might predict that two outcomes would be (a) the expansion of tissue capillary density through angiogenesis and (b) the increase of mitochondrial densities in heart muscle and other aerobic tissues. These predicted physiological responses correspond to two of the critical adaptations found in the cardiovascular systems of Antarctic icefishes.

Relatively little is known about NO production and levels in Antarctic notothenioids, but two recent reports suggest that this molecule may play an important role in the cardiovascular adaptation in the icefishes. First, Morlà and colleagues (34) reported five icefish species that express neuronal NOS (nNOS or NOS I) constitutively in skeletal muscle at levels that are dramatically higher than those for six red-blooded notothenioids. Second, NO regulates several cardiovascular activities in the heart of the icefish *Chionodraco hamatus*, and NOS activity is present in the endocardial endothelium and cardiomyocytes (25). The NOS isoform of the endocardial endothelium is probably endothelial NOS (eNOS or NOS III), whereas the myocardial isoform is demonstrably inducible NOS (iNOS or NOS II).

We now come full circle to the hemoprotein status of the icefishes. The oxygenated forms of hemoglobin and myoglobin are recognized as the major proteins responsible for the degradation of NO to nitrate (35,36). Thus, I can paraphrase the hypothesis originally developed by Sidell and O'Brien (28): Loss of the expression of hemoproteins during evolution of the Antarctic icefishes results in increased steady-state NO levels that mediate modification of their vascular systems and expansion of mitochondrial populations in oxidative tissues.

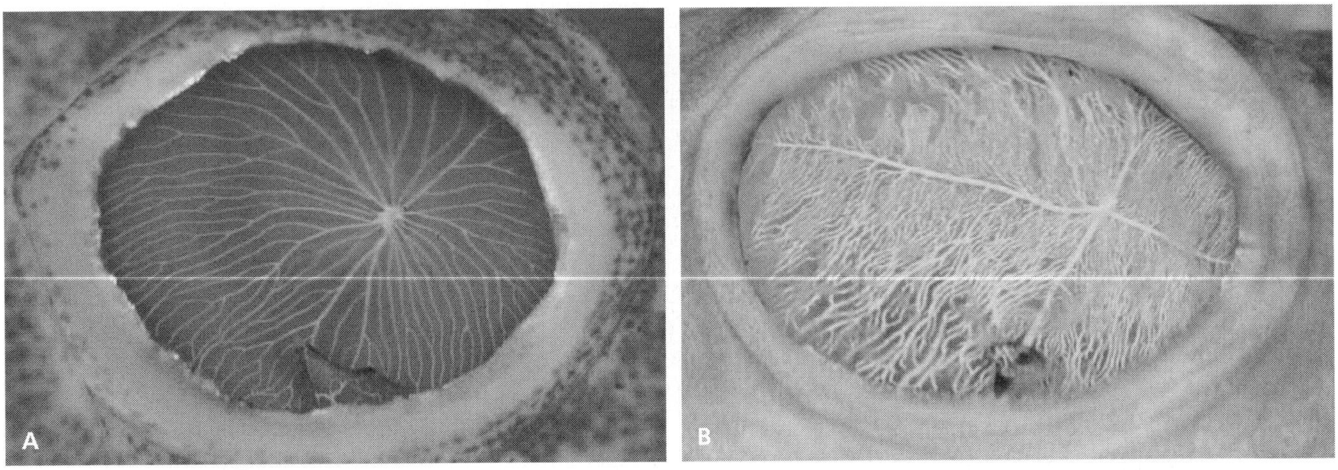

Figure 7.3. Retinal vasculature of red- and white-blooded notothenioids. (**A**) *Pagothenia borchgrivinki*, a hemoglobin-expressing notothenioid (27). (Reprinted with permission from Eastman JT, Lannoo MJ. Brain and sense organ anatomy and histology in hemoglobinless Antarctic icefishes (Perciformes: Notothenioidei: Channichthyidae). *J Morphol.* 2004;260:117–140.). (**B**) *C. aceratus*, a hemoglobin- and erythrocyte-null icefish (28). (Reprinted with permission from Sidell BD, O'Brien KM. When bad things happen to good fish: The loss of hemoglobin and myoglobin expression in Antarctic icefishes. *J Exp Biol.* 2006;209:1791–1802.). The vasculature has been filled with Microfil latex compound. Lenses and vitreous humors have been removed.

In the cold, stable, and O_2-rich Southern Ocean, the disaptive loss of hemoglobin by the icefishes (and cardiac myoglobin in a subset of the family) may have been compensated evolutionarily by a self-rectifying readaptation. Although this hypothesis may be a simplification and the causality perhaps indirect, it is nonetheless clear that the icefishes possessed the physiological tools to correct a deleterious, but nonlethal O_2-transport phenotype.

KEY POINTS

Antarctic icefishes are a uniquely valuable system for understanding the physiology and biochemistry of cardiovascular adaptation, and the results of such study may suggest novel therapeutic approaches to endothelial diseases:

- Evolution of the erythrocyte-null condition of the Antarctic icefishes could have occurred, almost certainly, only in the Southern Ocean, which is characterized by extremely cold temperatures, high O_2 availability, and low niche competition.
- Loss of hemoprotein expression in icefishes has been compensated by numerous cardiovascular adaptations.
- Recent evidence suggests that at least two of these adaptations, vascular expansion and increased mitochondrial biogenesis, result from the physiological function of the NO/NOS system.
- Key to evaluating the Sidell hypothesis will be a quantitative comparison of the signaling pathways of white- and red-blooded notothenioid fishes that function downstream of NO.
- If causality is direct, the Sidell hypothesis predicts that inhibition of NOS in icefishes, particularly those species that lack both hemoglobin and myoglobin, would lead to regression of vascularity and reduced densities of mitochondria in aerobic tissues.
- The last two proposals are currently under investigation.

ACKNOWLEDGMENT

This work was supported by National Science Foundation grants OPP-0089451 and OPP-0336932 (to H.W.D.).

REFERENCES

1 Ruud JT. Vertebrates without erythrocytes and blood pigment. *Nature.* 1954;173:848–850.
2 Rowley AF, Hunt TC, Page M, Mainwaring G. Fish. In: Rowley AF, Ratcliffe NA, eds. *Vertebrate Blood Cells.* Cambridge: Cambridge University Press; 1988:19–127.
3 Montgomery J, Clements K. Disaptation and recovery in the evolution of Antarctic fishes. *Trends Ecol Evol.*2000;15:267–271.
4 Macdonald JA, Montgomery JC, Wells RMG. Comparative physiology of Antarctic fishes. *Adv Mar Biol.* 1987;24:321–388.
5 Eastman JT. *Antarctic Fish Biology: Evolution in a Unique Environment.* San Diego: Academic Press, 1993.

6 Hemmingsen EA, Douglas EL, Johansen K, Millard RW. Aortic blood flow and cardiac output in the hemoglobin-free fish *Chaenocephalus aceratus*. *Comp Biochem Physiol A*. 1972;43:1045–1051.

7 Hemmingsen EA. Respiratory and cardiovascular adaptations in hemoglobin-free fish: resolved and unresolved problems. In: di Prisco G, Maresca B, Tota B, eds. *Biology of Antarctic Fish*. Berlin: Springer-Verlag; 1991:191–203.

8 Bargelloni L, Marcato S, Zane L, Patarnello T. Mitochondrial phylogeny of notothenioids: a molecular approach to Antarctic fish evolution and biogeography. *Syst Biol*. 2000;49:114–129.

9 Near TJ. Estimating divergence times of notothenioid fishes using a fossil-calibrated molecular clock. *Antarctic Sci*. 2004;16:37–44.

10 Eastman JT. The nature of the diversity of Antarctic fishes. *Polar Biol*. 2005;28:93–107.

11 Kennett JP. Cenozoic evolution of Antarctic glaciation, the circum-Antarctic ocean, and their impact on global paleoceanography. *J Geophys Res*. 1977;82:3843–3860.

12 Cocca E, Ratnayake-Lecamwasam M, Parker SK, et al. Genomic remnants of alpha-globin genes in the hemoglobinless antarctic icefishes. *Proc Natl Acad Sci U S A*. 1995;92:1817–1821.

13 Zhao Y, Ratnayake-Lecamwasam M, Parker SK, et al. The major adult alpha-globin gene of Antarctic teleosts and its remnants in the hemoglobinless icefishes: calibration of the mutational clock for nuclear genes. *J Biol Chem*. 1998;273:14745–14752.

14 Cocca E, Detrich HW, III, Parker SK, di Prisco G. A cluster of four globin genes from the Antarctic fish *Notothenia coriiceps*. *J Fish Biol*. 2000;57:33–50.

15 di Prisco G, Cocca E, Parker S, Detrich HW, III. Tracking the evolutionary loss of hemoglobin expression by the white-blooded Antarctic icefishes. *Gene* 2002;295:185–191.

16 Lau DT, Saeed-Kothe A, Parker SK, Detrich HW, III. Adaptive evolution of gene expression in Antarctic fishes: Divergent transcription of the 5′-to-5′ linked adult alpha1- and beta-globin genes of the Antarctic teleost *Notothenia coriiceps* is controlled by dual promoters and intergenic enhancers. *Am Zool*. 2001;41:113–132.

17 Moylan TJ, Sidell BD. Concentrations of myoglobin and myoglobin mRNA in heart ventricles from Antarctic fishes. *J Exp Biol*. 2000;203:1277–1286.

18 Grove TJ, Sidell BD. Fatty acyl CoA synthetase from Antarctic notothenioid fishes may influence substrate specificity of fat oxidation. *Comp Biochem Physiol B Biochem Mol Biol*. 2004;139:53–63.

19 Small DJ, Vayda ME, Sidell BD. A novel vertebrate myoglobin gene containing three A+T-rich introns is conserved among Antarctic teleost species which differ in myoglobin expression. *J Mol Evol*. 1998;47:156–166.

20 Small DJ, Moylan T, Vayda ME, Sidell BD. The myoglobin gene of the Antarctic icefish, *Chaenocephalus aceratus*, contains a duplicated TATAAAA sequence that interferes with transcription. *J Exp Biol*. 2003;206:131–139.

21 O'Brien KM, Sidell BD. The interplay among cardiac ultrastructure, metabolism and the expression of oxygen-binding proteins in Antarctic fishes. *J Exp Biol*. 2000;203:1287–1297.

22 O'Brien KM, Xue H, Sidell BD. Quantification of diffusion distance within the spongy myocardium of hearts from antarctic fishes. *Respir Physiol*. 2000;122:71–80.

23 Sidell BD. Intracellular oxygen diffusion: the roles of myoglobin and lipid at cold body temperature. *J Exp Biol*. 1998;201:1118–1127.

24 Wittenberg JB, Wittenberg BA. Myoglobin function reassessed. *J Exp Biol*. 2003;206:2011–2020.

25 Pelligrino D, Palmerini CA, Tota B. No hemoglobin but NO: the icefish (*Chionodraco hamatus*) heart as a paradigm. *J Exp Biol*. 2004;207:3855–3864.

26 Hemmingsen EA, Douglas EL. Respiratory and circulatory adaptations to the absence of hemoglobin in chaenichthyid fishes. In: Llano GA, ed. *Adaptations within Antarctic Ecosystems*. Washington, DC: Smithsonian Institution; 1977:479–487.

27 Eastman JT, Lannoo MJ. Brain and sense organ anatomy and histology in hemoglobinless Antarctic icefishes (Perciformes: Notothenioidei: Channichthyidae). *J Morphol*. 2004;260:117–140.

28 Sidell BD, O'Brien KM. When bad things happen to good fish: the loss of hemoglobin and myoglobin expression in Antarctic icefishes. *J Exp Biol*. 2006;209:1791–1802.

29 Egginton S, Skilbeck C, Hoofd L, Calvo J, Johnston IA. Peripheral oxygen transport in skeletal muscle of Antarctic and sub-Antarctic notothenioid fish. *J Exp Biol*. 2002;205:769–779.

30 Palmer RM, Ferrige AG, Moncada S. Nitric oxide release accounts for the biological activity of endothelium-derived relaxing factor. *Nature*. 1987;327:524–526.

31 Conway EM, Collen D, Carmeliet P. Molecular mechanisms of blood vessel growth. *Cardiovasc Res*. 2001;49:507–521.

32 Nisoli E, Clementi E, Paolucci C, et al. Mitochondrial biogenesis in mammals: the role of endogenous nitric oxide. *Science*. 2003;299:896–899.

33 Nisoli E, Falcone S, Tonello C, et al. Mitochondrial biogenesis by NO yields functionally active mitochondria in mammals. *Proc Natl Acad Sci U S A*. 2004;101:16507–16512.

34 Morlà M, Agusti AGN, Rahman I, et al. Nitric oxide synthase type I (nNOS), vascular endothelial growth factor (VEGF) and myoglobin-like expression in skeletal muscle of Antarctic fishes (Notothenioidei: Channichthyidae). *Polar Biol*. 2003;26:458–462.

35 Kerwin JF, Jr., Lancaster JR, Jr., Feldman PL. Nitric oxide: a new paradigm for second messengers. *J Med Chem*. 1995;38:4343–4362.

36 Flögel U, Merx MW, Gödecke A, Decking UK, Schrader J. Myoglobin: a scavenger of bioactive NO. *Proc Natl Acad Sci U S A*. 2001;98:735–740.

8

The Fish Endocardium

A Review on the Teleost Heart

José M. Icardo

University of Cantabria, Santander, Spain

Research interest in the fish heart has been focused classically on physiological activities and on gross morphological aspects (1–2). Thus, most surveys of the fish heart define the endocardium as a continuous lining formed by cells that may be squamous, cubic, or high. This simple definition summarizes the interest aroused by the endocardium in the field of fish research. However, increasing evidence suggests that the endocardium plays a crucial role in heart physiology. Moreover, the fish endocardium displays functions that are relevant for the entire organism. Most of these data are derived from studies of the teleost heart.

STRUCTURE OF THE ENDOCARDIUM

It is true that the endocardial cells lining the different chambers of the teleost heart may be squamous, cubic (with a more or less convex surface), or high (3–6). Less recognized is the fact that the histological appearance of endocardial cells may differ in the various heart chambers. In some species, endocardial cells bulge into the lumen of the bulbus arteriosus but are extremely flattened in the ventricle (Figure 8.1). In other species, the converse pattern is found. Histological differences in the endocardium also may occur in different areas of the same heart chamber (4). The extent to which these morphological differences reflect underlying functional heterogeneity remains unknown.

Under the transmission electron microscope, endocardial cells may show rough and smooth endoplasmic reticulum, Golgi apparatus, small mitochondria, microfilaments, and surface microvilli. The cells form a continuous endothelium, and are joined primarily through tight junctions and desmosome-like plaques (6). The interdigitations are complex and may result in cells with overlapping cytoplasm. The abluminal basement membrane is poorly organized.

Some, but not all of these features are unique to the teleost endocardium. For example, extensive interdigitations with overlapping cytoplasm occur across species, from fishes to mammals (7). Desmosome-like junctions also have been described in other classes of fishes such as in elasmobranches (7). In mammals, desmosomes do not normally occur between endocardial cells, but their presence can be induced by the injection of proterenol (8). Furthermore, the presence of zonula adherens junctions is well documented in mammals (7).

Perhaps the most striking structural characteristic of the teleost endocardium is the presence of membrane-bounded moderately dense bodies (MDBs) (Figure 8.1). It has been suggested that the histologic appearance of the endocardial cells (high versus flattened) correlates with the presence of these granules (9). However, a more important factor contributing to the degree of endocardial cell bulging is cell density relative to the area occupied by the basal endocardial surface.

MDBs demonstrate heterogeneity in density, structural appearance, and content. In some endocardial cells, they are highly abundant and closely packed, occupying most of the cell cytoplasm. In other cases, only a small number of MDBs can be observed (see Figure 8.1). Most MDBs are rounded and regular, and are structurally similar to the atypical secretory granules, the so-termed Weibel-Palade bodies, described in the endothelium of mammalian endothelial cells (ECs) (10). However, other MDBs are oval or irregular. Depending on the species, MDBs may contain homogeneous material of different electron density, fibrillar material, or tubules (3–6). MDBs may be present in only one or in all of the heart chambers. The presence and distribution of MDBs appear to be unrelated to fish size, swimming capabilities, lifestyle, or ecology (9).

MDBs are not unique to teleosts. Endocardial-specific granules also appear in primitive fishes (hagfish and lamprey) and in elasmobranches (11). Hence, the presence of MDBs seems unrelated to the phylogenetic position of the fish species.

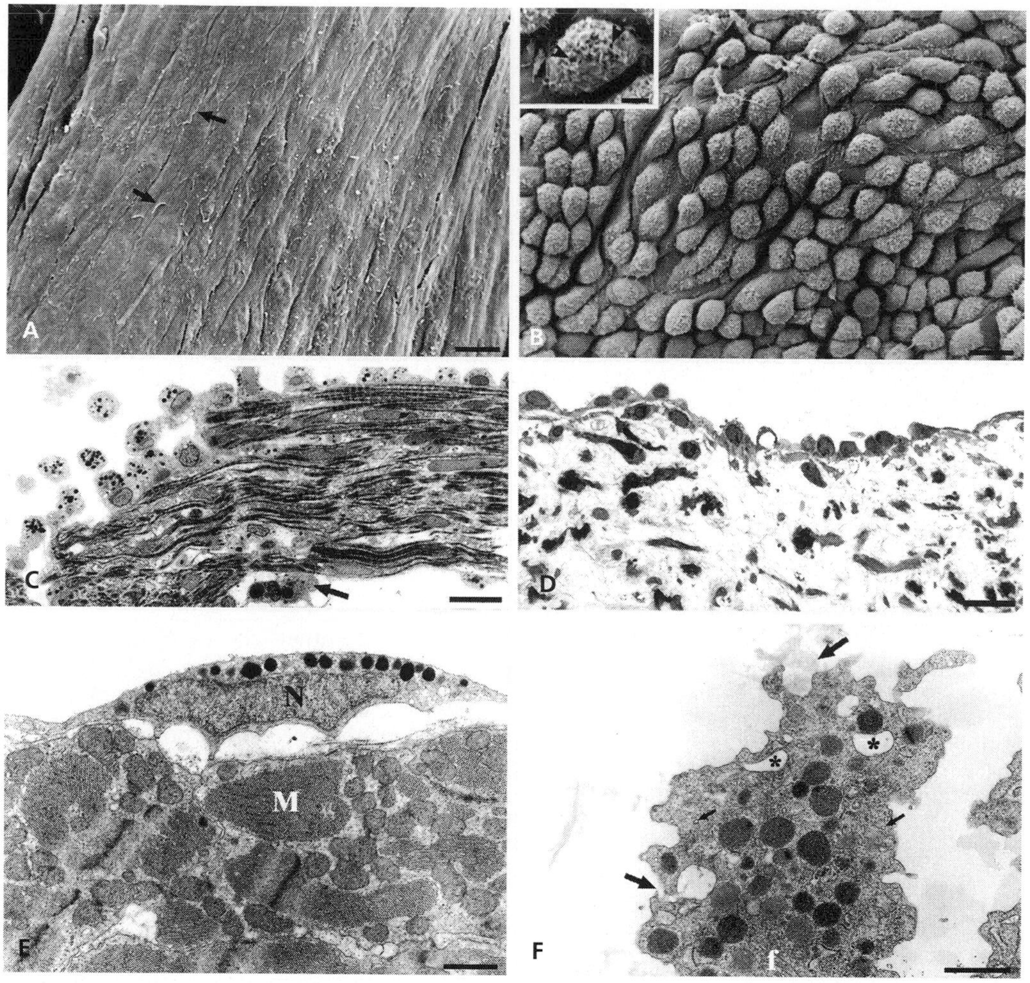

Figure 8.1. Differences in endocardial cell morphology across the heart chambers and across species. **(A, B)** Scanning electron micrographs from the Antarctic teleost *Trematomus bernacchii*. Endocardial cells in the ventricle (**A**) are flattened cells (*arrows*, surface microplicae). In the bulbus arteriosus (**B**), cells are prominent and show convex surfaces, bulging nucleus, and surface microvilli. *Inset*: Detail of a nuclear bulging. *Arrowheads* indicate cell surface pits. (**C, D**) Light microscope micrographs from the polar teleost *Arctogadus glacialis*. Endocardial cells in the ventricle (**C**) are tall, prominent cells that contain a limited number of rounded, dark MDBs. *Arrow* indicates a subendocardial macrophage. In the bulbus (**D**), flattening of cells is more noticeable. MDBs are not evident. (**E, F**) TEM micrographs. (**E**) Ventricular endocardial cells of the European eel are flattened and contain numerous MDBs of small size. Endocardial cell extensions contact the subjacent myocardium (M). N, endocardial nucleus. (**F**) Bulbus endocardial cells of the Antarctic teleost *Notothenia coriiceps* are tall cells showing numerous MDBs with contents of different electron density. Some MDBs are releasing flocculent material into the bloodstream (*thick arrows*); others are empty (*asterisks*). Note the presence of small coated vesicles (*thin arrows*), cisternae of rough endoplasmic reticulum, and discrete microfilament bundles (F). Bars: **A–D**: 0.5 μm; **E–F**, 1 μm. Inset: 1 μm.

THE ENDOCARDIUM AS A SECRETORY ORGAN

MDBs represent the morphologic expression of the secretory capabilities of endocardial cells. The MDBs of many species appear to contain neutral mucopolysaccharides (3–4,9). In other species, MDBs react with histochemical probes specific for catecholamines (12). MDBs appear to be the storage site of material prior to secretion. The precise mode of MDB formation is unknown. It is also unclear how the MDB content is released. Direct discharge into the heart lumen has been observed (3) (see Figure 8.1). It presently is unknown whether the content of the MDBs contributes to the formation of the subendocardial extracellular matrix, whether it contains peptides that may regulate local activities or have systemic effects, or all of the above. Because of the great morphologic MDB variability across species, any general statement would be an oversimplification.

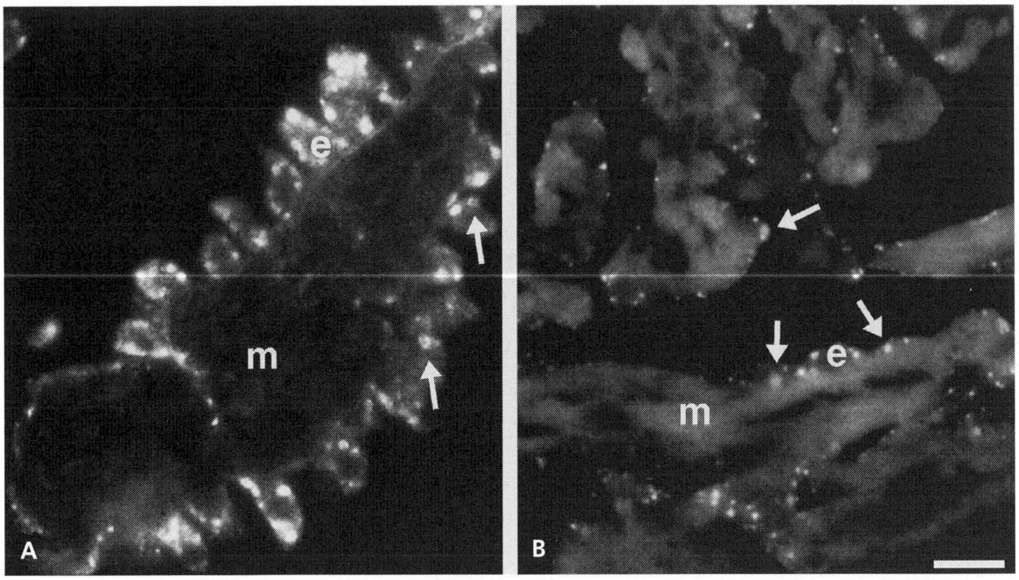

Figure 8.2. Fluorescence micrographs from the atrium (**A**) and ventricle (**B**) of the Atlantic cod. One hour after intravenous injection of FITC-hyaluronan, labelled hyaluronan (*arrows*) appears in endocardial (e) cells. Note the different morphology of the endocardium. Endocardial cells are tall in the atrium and flat in the ventricle. m, myocardium. Bar: 10 μm. (Reprinted with permission from Sorensen K, Dahl LB, Smedsrod B. Role of endocardial ECs in the turnover of hyaluronan in Atlantic cod (*Gadus morhua*). *Cell Tissue Res.* 1997;290:101–109.)

The secretory capabilities of the teleost endocardium appear to be enhanced greatly in some species. An examination of the bulbus arteriosus of the icefish *Chionodraco hamatus* (3) revealed that the endocardium invaginates into the subjacent tissue and forms solid cell cords that constitute complex secretory structures. Endocardial cells in these cords contain numerous MDBs, which appear to discharge their content directly into the bloodstream. On the basis of the structural appearance, it was suggested that these MDBs may contain mucins or mucopolysaccharides that may protect the endocardium against physical stress (such as cold), thus performing an antifreeze function.

THE ENDOCARDIUM AS A SCAVENGER ORGAN

Circulating blood contains a significant amount of soluble waste macromolecules resulting from the various cellular activities and connective-tissue metabolism. In terrestrial vertebrates, these macromolecules are removed from the blood by the ECs of the liver and by macrophages resident in different tissues, such as those located in the proximity of the sinusoids in liver, spleen, and bone marrow. In a broad sense, the two cell types constitute the backbone of the reticuloendothelial system.

Experimental studies in teleosts have shown that this scavenger function is performed primarily by the ECs of the kidney or by the heart endocardium (13). (Clearance and degradation of biomolecules by the gill endothelium is discussed in Chapter 5.) In recent years, atrial and ventricular endocardial cells of several teleost species have been shown to endocytose hyaluronan (Figure 8.2) (14), collagen (15), carbon particles (16), and ferritin[1] (12). As a general rule, the endocardium takes up soluble material, a process mediated by receptor-mediated endocytosis (17). However, in some cases (e.g., the medaka), endocardial cells also are able to take up particulate material such as India ink (18), ferritin (12), and latex beads (14). It is formally possible that the latter findings represent an experimental artefact associated with high concentrations of injected macromolecules. This caveat notwithstanding, endocardial cells appear to be the main site for the clearance of molecules such as hyaluronan and collagen (14).[2] Indeed, some MDBs could correspond to secondary lysosomes, involved in the degradation of phagocytosed material.

1 The location of ECs with scavenger capabilities varies along the phyletic scale. In primitive fishes (hagfish and lampreys) and in elasmobranches, uptake of soluble macromolecules occurs in the gill endothelium. In teleosts, the location of the scavenger endothelium appears to be species-specific. For example, in the Atlantic cod, medaka, and plaice, it occurs in the heart endocardium; in salmonids, in the endothelium of the head and trunk kidney; in higher vertebrates, scavenging activity is carried out primarily by liver ECs. Scavenger cells also occur in the so-termed pericardial cells of invertebrates. In addition, the kidney of the lamprey, the kidney and spleen of the medaka, and the kidney, spleen, and liver of the cod contribute in variable degree to the clearance of both soluble and nonsoluble macromolecules (for references, see 13, 16).

2 In the Atlantic cod, hyaluronan also is cleared by cells in the kidney and in the spleen. In higher vertebrates, this occurs in liver ECs (for references, see 14).

The heart of all teleosts studied contains subendocardial resident phagocytes (16) (see Figure 8.1C).[3] These cells are able to ingest carbon particles (16), and have been shown to contain degenerated cardiac parasites (19). Endocardial cells and subendocardial macrophages appear to have established complementary and/or cooperative functions. In the teleost medaka, macrophages with phagocytosed material aggregate to form nodules located under the endocardium (18). Direct contact often is observed between the endocardial cells and the macrophages. These aggregations are reminiscent of the melano-macrophage centers, which appear to be the precursors of the germinal centers in higher vertebrates. However, the immune function of the aggregates remains unclear.

The role of the teleost endocardium as a defence barrier may well be even more complex. In the Atlantic cod (*Gadus morhua*), the endocardium appears to be the main site of bacterial lipopolysaccharide (LPS) uptake (17). One hour following injection of radiolabeled LPS, atrial and ventricular endocardial cells are shown to contain nearly half the injected dose. LPS is taken up by endocytosis, via scavenger receptors (17). Atrial (but not ventricular) endocardial cells collaborate with the spleen and the kidney in the clearance of bacterial antigen (20). The exposure of fish to different carcinogens induces the expression of the catalytic cytochrome P450 (CYP450) in the heart endocardium (21). CYP450 localizes in microsomes and appears to be involved in the degradation of endogenous and exogenous compounds. It should be stressed that most research on the scavenger function of the endocardium has focused on the atrium and ventricle. It appears that the bulbus endocardium does not have a scavenger capacity (12–13), although subendocardial macrophages in the bulbus are able to phagocytose ferritin particles and store them in lysosomes (12).

OTHER ENDOCARDIAL FUNCTIONS

The endocardium has been shown to bind different molecular forms of the natriuretic hormones. Experimental studies carried out on Antarctic teleosts (22) have shown that ventricular endocardial cells bind the two types of natriuretic peptides described in fishes. Understanding of the functional significance of natriuretic hormones in fishes is far from complete. However, it is likely that endocardial cells are able, through the presence of surface receptors for these hormones, to control the physiologic activity of the subjacent myocardium and to participate in the control of body homeostasis.

Nitric oxide (NO) modulates cardiac performance in the teleost *Anguilla anguilla* through the activation of a protein kinase cascade. Furthermore, NO signalling requires the presence of the endocardium (23). Although information on fishes is still very scarce, the teleost endocardium appears to be an important source of endogenous NO. Endocardial NO, probably acting through different activation pathways, appears to be involved in the regulation of fish heart function through autocrine-paracrine mechanisms (23).

ENDOCARDIAL HETEROGENEITY AND THE SUBENDOCARDIUM

The endocardium is not a uniform cell layer, but rather demonstrates structural heterogeneity both across and within species. Most probably, the phenotypic differences are associated with functional heterogeneity (e.g., the secretory activity determined by the presence of MDBs, or the restriction of natriuretic peptide binding activity to the ventricular endocardium). These differences cannot easily be explained from an embryological point of view. Fate mapping studies have shown that endocardial progenitors originate in a very limited area of the early blastula and are incorporated into the developing heart with no evidence of chamber restriction (24). Thus, it seems likely that the structural and functional differences are determined – to a large extent – by microenvironmental cues from subjacent tissue (e.g., the myocardium in the atrium and ventricle, and smooth muscle in the bulbus arteriosus). Any such communication is certain to be bi-directional, with endocardial-derived signals influencing the form and function of underlying muscle.

ENDOCARDIAL ADAPTATION TO THE ENVIRONMENT

The endocardium is able to perform complex secretory functions at below-zero temperatures. The case of the icefishes (3) is paradigmatic in this regard. However, it could be postulated that this is only an example of the long evolutionary adaptation to extreme cold. Experimental studies indicate that the scavenger function of the medaka endocardium is maintained over a wide range of temperatures (4°C to 23°C). Nevertheless, the ability to take up carbon particles is markedly decreased at 4°C (16), reflecting a decrease in general metabolism. Much less is known of the ability of the fish endocardium to respond to other environmental challenges. Our own unpublished studies on the endocardium of the African lungfish indicate that the endocardial cells undergo structural changes during the aestivation period,[4] losing both central surface microvilli and ruffling activity. This is accompanied by changes in the

3 Subendocardial cells are common in the vertebrate heart. In primitive fishes, such as hagfish and lamprey, subendocardial interstitial cells contain catecholamines, show the characteristics of chromaffin cells, and appear to represent the adrenal medulla of these animals. Macrophages probably do occur in these primitive fishes, but their presence has not been reported. Fibroblasts occur in the subendocardium of elasmobranches, and occasional macrophages and fibroblasts also have been reported in the subendocardium of terrestrial vertebrates (for references, see 7, 11).

4 The aestivation is a state of dormancy or torpor, similar in some ways to mammalian hibernation. During the drought season, lungfish bury in the mud, secreting mucus that envelops the entire body and forms a cocoon, thus allowing them to survive for months without ingesting food or water. During the aestivation period an overall depression occurs of both the general metabolism and of the cardiovascular and renal functions.

nuclear surface and in chromatin distribution, probably reflecting modifications of the cell synthesis activity. Although the evidence indicates that the fish endocardium undergoes functional and structural adaptations during temperature and climate changes, this remains a largely unexplored field.

Endocardium and Phylogeny

In teleosts, the presence of trabeculae and lacunary spaces in the ventricular chamber, as well as the presence of intricate folds in the bulbus arteriosus, accounts for a relatively greater endocardial mass than in that of the mammalian heart (25). Thus, the fish endocardium represents a large cellular compartment whose importance may be greater than in mammals, from both structural and functional points of view. Indeed, the teleost endocardium appears to perform activities that become restricted to other endothelia (or to other cell types) along the phyletic scale.

KEY POINTS

- The teleost endocardium is a complex organ with various capabilities. Most functions appear to be lost through evolution.
- The endocardium is a secretory organ of endocrine and paracrine importance.
- The endocardium acts as a scavenger organ.
- The endocardium regulates the functional activity of the subjacent myocardium.

Future Goals

- To identify the secretory products and functional analysis of local and systemic effects

ACKNOWLEDGMENTS

Supported by CGL2004–06306-C02–01/BOS from the Ministerio de Educación y Ciencia.

REFERENCES

1 Satchell GH. *Physiology and Form of Fish Circulation.* Cambridge, England: Cambridge University Press, 1991.
2 Farrell AP, Jones DR. The heart. In: Hoar WS, Randall DJ, Farrell AP, eds. *Fish Physiology.* Vol. 12. The Cardiovascular System. Part A. San Diego: Academic Press; 1992:1–87.
3 Icardo JM, Colvee E, Cerra MC, Tota B. Bulbus arteriosus of the Antarctic teleosts. I. The white-blooded *Chionodraco hamatus. Anat Rec.* 1999a;254:396–407.
4 Icardo JM, Colvee E, Cerra MC, Tota B. Bulbus arteriosus of the Antarctic teleosts. II. The red-blooded *Trematomus bernacchii. Anat Rec.* 1999b;256:116–126.
5 Icardo JM, Colvee E, Cerra MC, Tota B. Light and electron microscopy of the bulbus arteriosus of the European eel (*Anguilla anguilla*). *Cells Tissues Organs.* 2000a;167:184–198.
6 Icardo JM, Colvee E, Cerra MC, Tota B. The bulbus arteriosus of stenothermal and temperate teleosts: a morphological approach. *J Fish Biol.* 2000b;57(SupplA):121–135.
7 Andries L. *Endocardial Endothelium: Functional Morphology.* Austin TX: RG Landes Company, 1994.
8 Turcotte H, Bazin M, Boutet M. Junctional complexes in regenerating endocardium. *J Ultrastruct Res.* 1982;79:133–141.
9 Benjamin M, Norman D, Scarborough D, Santer RM. Carbohydrate-containing endothelial cells lining the bulbus arteriosus of teleosts and the conus arteriosus of elasmobranches (Pisces). *J Zool London.* 1984;202:383–392.
10 Weibel ER, Palade GE. New cytoplasmic components in arterial endothelia. *J Cell Biol.* 1964;23:101–112.
11 Yamauchi A. Fine structure of the fish heart. In: GH Bourne, ed. *Hearts and Heart-Like Organs.* Vol. 1. Comparative Anatomy and Development. New York: Academic Press; 1980:119–48.
12 Leknes IL. Fine structure and endocytic properties of subendothelial macrophages in the bulbus arteriosus of two bony fishes. *Acta Histochem.* 1986;79:155–160.
13 Seternes T, Sorensen K, Smedsrod B. Scavenger endothelial cells of vertebrates: a nonperipheral leukocyte system for high-capacity elimination of waste macromolecules. *Proc Natl Acad Sci USA.* 2002;99:7594–7597.
14 Sorensen K, Dahl LB, Smedsrod B. Role of endocardial endothelial cells in the turnover of hyaluronan in Atlantic cod (*Gadus morhua*). *Cell Tissue Res.* 1997;290:101–109.
15 Smedsrod B, Olsen R, Sveinbjornsson B. Circulating collagen is catabolized by endocytosis mainly by endothelial cells of endocardium in cod (*Gadus morhua*). *Cell Tissue Res.* 1995;280:39–48.
16 Nakamura H, Shimozawa A. Phagocytotic cells in the fish heart. *Arch Histol Cytol.* 1994;575:415–425.
17 Seternes T, Dalmo RA, Hoffman J, Bogwald J, Zykova S, Smedsrod B. Scavenger-receptor-mediated endocytosis of lipopolysaccharide in Atlantic cod (*Gadus morhua* L.). *J Exp Biol.* 2001;204:4055–4064.
18 Nakamura H, Shimozawa A, Kikuchi S-I. Melano-macrophage centre-like structure in the heart of the medaka, *Oryzias latipes. Ann Anat.* 1993;175:59–63.
19 Overstreet RM, Thulin J. Response by *Plectropomus leopardus* and other serranid fishes to *Pearsonellum corventum* (Digenea: Sanguinicolidae), including melano-macrophage centres in the heart. *Aust J Zool.* 1989;37:129–142.
20 Arnesen SM, Schroder MB, Dalmo RA, Bogwald J. Antigen uptake and immunoglobulin production in Atlantic cod (*Gadus morhua* L.) after intraperitoneal injection of *Vibrio anguillarum. Fish Shellfish Immunol.* 2002;13:159–170.
21 Stegeman JJ, Miller MR, Hinton DE. Cytochrome P450IA1 induction and localization in endothelium of vertebrate (teleost) heart. *Mol Pharmacol.* 1989;36:723–729.
22 Cerra MC, Canonaco M, Acierno R, Tota B. Different binding activities of A- and B-type natriuretic hormones in the heart of two Antarctic teleosts, the red-blooded *Trematomus bernacchii* and the hemoglobinless *Chionodraco hamatus. Comp Biochem Physiol.* 1997;118A:993–999.

23 Imbrogno S, Cerra MC, Tota B. Angiotensin II-induced inotropism requires an endocardial endothelium-nitric oxide mechanism in the in-vitro heart of *Anguilla anguilla*. *J Exp Biol.* 2003:206:2675–2684.

24 Lee RKK, Stainier DYR, Weinstein BM, Fishman MC. Cardiovascular development in the zebrafish. II. Endocardial progenitors are sequestered within the heart field. *Development.* 1994;120:3361–3366.

25 Sys SU, Pellegrino D, Mazza R, Gattuso A, Andries L, Tota B. Endocardial endothelium in the avascular heart of the frog: morphology and role of nitric oxide. *J Exp Biol.* 1997;200:3109–3118.

9

Skin Breathing in Amphibians

Glenn J. Tattersall

Brock University, St. Catherines, Ontario, Canada

Skin breathing, or cutaneous, gas exchange is an important route of respiration in many aquatic or semiaquatic vertebrates, and is particularly well developed in the amphibians. The skin of amphibians contains a unique vasculature that facilitates oxygen (O_2) uptake and carbon dioxide (CO_2) excretion. Cutaneous gas exchange can fulfill routinely 0% to 100% of O_2 uptake and 20% to 100% of CO_2 excretion (1). Amphibians pay a price for this: They require a relatively thin epidermis and, as a result, suffer from high rates of water loss. Thus, amphibians are, for the most part, tied to an aquatic or semiaquatic life. Furthermore, unlike lungs or gills, the skin lacks a dedicated ventilatory pump and, as such, has been thought to be a poorly regulated respiratory organ, with little scope for change. Research over the past couple of decades has revealed that amphibians may exhibit partial control over the cutaneous vasculature, and that such control is under both neural and possibly hormonal control. For the purposes of this chapter, most information will be taken from literature on ranid frogs (order Anura), because the majority of research has been done on this group.

AMPHIBIAN CUTANEOUS CIRCULATION

Most amphibians possess a double circulation – the pulmonary and systemic – consisting of a right and left atrium and an undivided ventricle. A similar blood flow distribution pattern exists in reptiles and certain air-breathing fishes (2) (see Chapter 5). Primarily deoxygenated (deoxygenated systemic blood plus oxygenated blood from the skin are mixed) blood returns to the heart via the right atrium, whereas oxygenated blood from the lung only returns to the heart via the left atrium. Both atria connect to the single, undivided ventricle, although considerable flow separation of blood is achieved through the extensive system of trabeculae within the ventricle (3,4). Upon ventricular contraction, deoxygenated blood primarily travels through the spiral valve to the pulmocutaneous artery, which subsequently divides into the pulmonary and cutaneous artery. Oxygenated blood, on the other hand, travels via the aorta to supply blood to the tissues and other parts of the skin. Thus, the skin generally receives blood from both the systemic and pulmocutaneous circuits (Figure 9.1). Specifically, at least in *Rana catesbeiana* (the American bullfrog), the dorsal skin receives blood primarily from the cutaneous artery, whereas skin surrounding the legs and, to some extent, the ventral surface receives mainly systemic blood (5). The skin circulation is drained by a pair of large cutaneous veins that feed into the subclavian veins and back to the heart. Thus, the skin receives both oxygenated and deoxygenated blood, but returns oxygenated blood to the heart, which is subsequently mixed with deoxygenated blood from the systemic circulation before returning to the right atrium.

A further aspect of import of the amphibian circulation concerns the innervation of the primary circulations to the lung and skin. The cutaneous artery is innervated by adrenergic fibers of sympathetic origin, which cause constriction. The pulmonary arteries themselves are innervated by cholinergic fibers of parasympathetic origin (i.e., the vagus) which cause constriction of the pulmonary artery and dilation of the cutaneous artery (6,7). This reciprocal autonomic innervation of the pulmonary and cutaneous arteries leads to situations in which blood can be directed to the respiratory organ that has the greatest potential for gas exchange. In most cases, this will be the lung, although in cases when the lung is hypoxic, hypercapnic, or poorly ventilated (e.g., during diving), blood flow to the lung will be decreased and blood flow to the skin will at least remain constant, if not increase (8,9).

CUTANEOUS VASCULATURE

In many amphibians, 20% to 95% of the entire body's respiratory capillaries are in the skin (Figure 9.2) (10). Indeed, in the smaller frogs, skin capillaries can be more abundant than pulmonary capillaries (10). In Czopek's examination of amphibian skin, he observed that the epidermis is only four to seven cell layers thick. Overall, the dorsal epidermis is 10% to 20% thinner (18–24 μm) than the ventral epidermis (32 μm). Unlike in most vertebrates, the skin capillaries are located inside the epidermis rather than the dermis, making

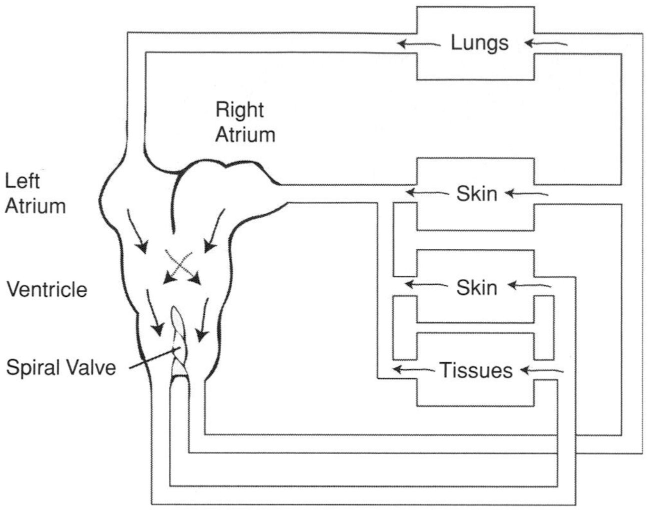

Figure 9.1. Schematic of the generalized circulation in amphibians. *Arrows* indicate the direction of blood flow. *Dotted arrows* indicate that a portion of blood can be shunted from one side of the heart to the other side within the undivided ventricle.

the barrier to gas exchange much shorter (see Figure 9.2). The values quoted above, however, were from a rather unique amphibian. The epidermis of most frogs is two to three times thicker, although the respiratory capillaries still lie within the epidermis. This barrier for gas exchange (18–60 μm) exceeds that of lung epithelium by one to two orders of magnitude, making the diffusive transfer of gases a relatively slow process.

The respiratory capillaries form a vast network of anastomosing vessels (Figures 9.3 and 9.4) lying within the epidermis and forming a virtual sheet for gas exchange (11). These respiratory capillaries are connected to the skin arteries via branching arterioles, and they drain into subcutaneous veins via venules that penetrate the dermis. Very few true (i.e. systemic) capillaries connect the arterioles directly to venules or the venular network. Flow within the respiratory capillaries themselves exhibits a high degree of heterogeneity (12). Furthermore, the number of capillaries perfused can significantly affect gas exchange (13), although the nature of the regulation of these flows is unknown (i.e., whether it occurs at the arteriolar level or within the capillary network itself). Some clue to this may come from studies examining microvascular permeability (14). Although the respiratory capillaries themselves lack smooth muscle cells, microfilaments within endothelial cells (ECs) have been proposed to be part of a contractile machinery within the frog skin capillaries. These ECs within the capillaries alter their shape and appearance under electrical stimulation (14). Whether they operate in this manner in vivo, respond to neural or hormonal stimulation, or respond to changes in respiratory gases is unknown.

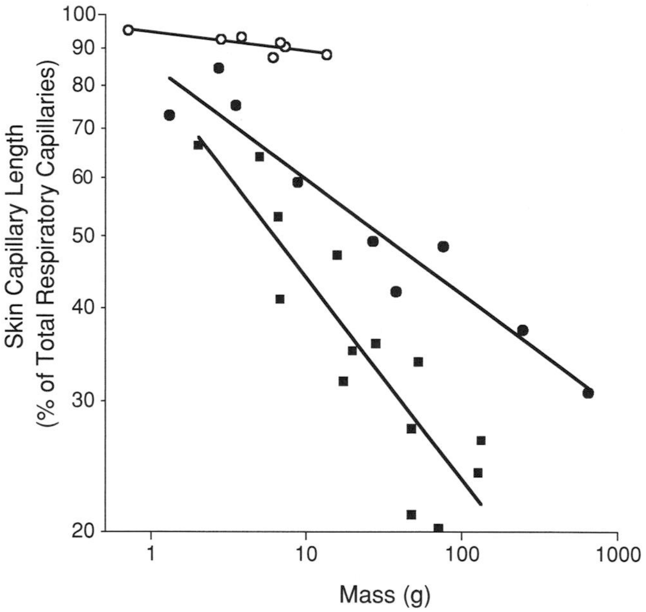

Figure 9.2. The percentage of respiratory capillaries that are cutaneous in origin differs among groups of amphibians. Lungless salamanders (*open circles*) have greater than 90% of respiratory capillaries in the skin, with the remainder found in the buccal cavity. Lunged salamanders (*filled circles*) and frogs (*squares*), on the other hand, have substantial respiratory capillaries in the lungs, buccal cavity, and skin. Note the trend with increasing body size for the skin to play a decreasing role in providing a vascular network for gas exchange. (Values adopted from Szarski H. The structure of respiratory organs in relation to body size in amphibia. *Evolution.* 1964;18:118–126.)

SKIN AS A GAS EXCHANGER: ROLE OF DIFFUSION

As in any respiratory organ, the diffusion of gases occurs across a layer of tissue and into the nearby bloodstream. Across large distances, the molecular movement of respiratory gases, which is governed by Fick's law of diffusion is not a very efficient process[1]; therefore the larger the barrier between environmental concentrations of respiratory gases and the internal concentrations, the longer it takes for diffusion to occur. Some amphibians have anatomical modifications that capitalize on Fick's law of diffusion by increasing the total surface area available for gas exchange. The salamander, *Cryptobranchus* (15), and the Lake Titicaca frog, *Telmatobius* (16), have marked folds in their skin that allow for greater diffusive exchange area, and these species are thought to make very little use of their lungs for gas exchange (15). In addition, the male hairy frog, *Astylosternus*, develops epidermal "hairs" during its breeding season that may supplement the total gas exchange surface (17).

Gas exchange across the skin of amphibians is believed to occur between the skin and an "infinite pool" of gases from the direct, surrounding environment (18). However, theoretical considerations and empirical evidence show that an external

1 Fick's law of diffusion: $= \frac{dQ_g}{dt} = D_g A \frac{dP_g}{dx}$ where $\frac{dQ_g}{dt}$ is the rate of diffusion of gas (g), D_g is the Krogh diffusion coefficient, A is the surface area, dP_g is the partial pressure gradient, and dx is the diffusion distance.

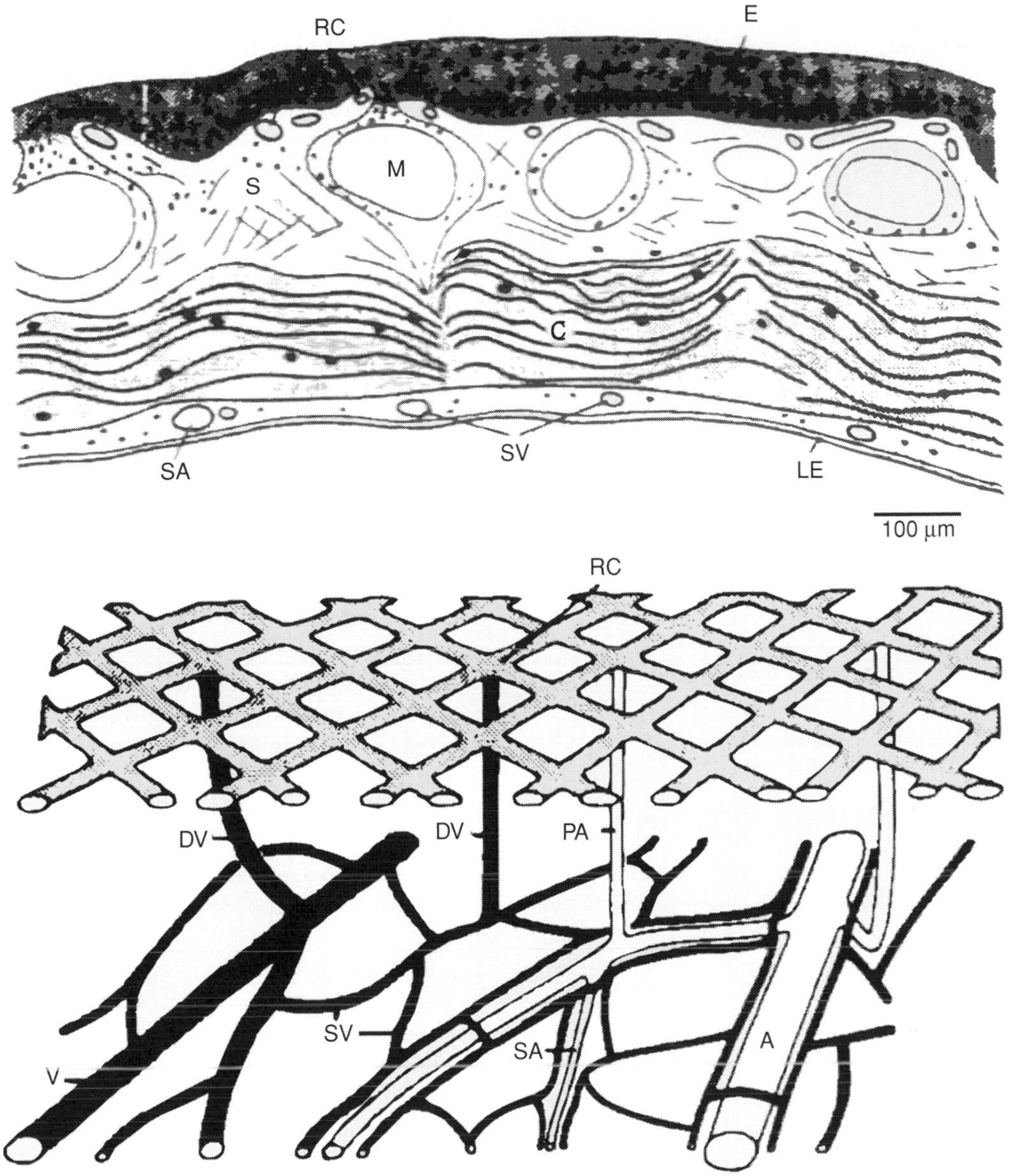

Figure 9.3. Organization of skin microcirculation in amphibians. *Top*: Cross-sectional schematic of frog skin, showing the relative placement of blood vessels within and beneath the epidermis. *Bottom*: Schematic of the network of blood vessels within the skin. Mucous gland (M), stratum compactum (C), stratum spongiosum (S), skin arteries (A), branching arterioles (SA), subepidermal or respiratory capillary network (RC), arteriolar branches (PA), venules (DV), subcutaneous veins (V), subcutaneous venular plexus (SV), epidermis (E), lymph endothelium (LE). (Reproduced with permission from Olesen SP, De Saint-Aubain ML, Bundgaard M. Permeabilities of single arterioles and venules in the frog-skin – a functional and morphological-study. *Microvasc Res.* 1984;28:1–22.)

boundary layer of poorly mixed gases contributes to limiting cutaneous gas exchange (19,20). These diffusion-boundary layers result from decreased fluid velocity close to solid surfaces. The decreased fluid velocity prevents bulk flow from occurring close to the solid surface, resulting in a stagnant layer of fluid, through which diffusion is the only means of gas movement. The resulting diffusion-boundary layer poses a limitation to gas exchange (especially O_2) primarily because the partial pressure of O_2 at the skin–fluid medium interface is low, and thus presents a relatively small partial pressure gradient (i.e., driving pressure) across the skin (20). Because it is the partial pressure gradient that ultimately determines the rate of diffusion of a gas across any surface, low values at the skin–water interface result in decreased rates of uptake and subsequent limitations to cutaneous gas exchange.

Amphibians have had to develop methods for coping with these diffusion boundary layers (namely, the hypoxic boundary layers). Increased water velocity acts to minimize or disrupt

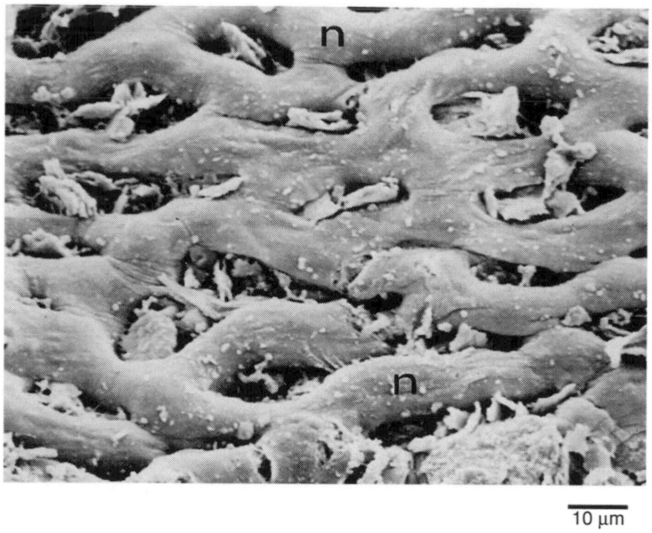

Figure 9.4. Scanning electron micrograph of the respiratory capillary network of the tree frog, *Chiromantis petersi*. Note the relatively flat anastomosing capillaries that form the respiratory exchange network. The black bar denotes 10 μm. (Reproduced with permission from Maina JN. Is the sheet-flow design a 'frozen core' (a Bauplan) of the gas exchangers? Comparative functional morphology of the respiratory microvascular systems: illustration of the geometry and rationalization of the fractal properties. *Comp Biochem Physiol A Mol Integr Physiol.* 2000;126:491–515.)

the boundary layer surrounding the amphibian skin; thus, increased activity or movement could be one method of alleviating boundary layers (21,22). The primarily skin-breathing Lake Titicaca frog, the exclusively aquatic salamander *Cryptobranchus*, and cold-submerged (and thus skin-breathing) bullfrogs all show behaviors described as "bobbing" or "rocking" when confronted with ambient hypoxia (15,16,23), suggesting an integrated detection of O_2. These behaviors act to disrupt the hypoxic boundary layers at the skin (18) and promote increased O_2 diffusion gradients and therefore O_2 uptake across the respiratory surface (24). Increased locomotory activity has been observed in response to hypoxia in frogs (23,25), and has been interpreted as a means of disrupting and minimizing hypoxic boundary layers by increasing skin ventilation (9,23). This detection of low O_2 has been suggested by behavioral studies of overwintering amphibians that demonstrate a moderate ability to detect and avoid hypoxic environments (26). This is further supported by data showing that these same submerged, skin-breathing frogs exhibit a decline in preferred body temperature while in hypoxic environments, suggesting a sensory mechanism that modifies behavior (27).

Experimentally, Pinder and Feder (23) demonstrated that increased fluid velocity decreased the actual size of the hypoxic boundary layer around the skin. At high velocities ($5.2\ cm{\cdot}s^{-1}$), hypoxic boundary layers were approximately 200 μm, whereas at low velocities ($0.2\ cm \cdot s^{-1}$) they were nearly four times as large, and increased to several millimeters when no water movement was present (23). Increasing the flow of water over the skin also results in a reduced hind limb capillary recruitment, suggesting that local external hypoxia at the skin surface can serve to regulate the open probability of many capillaries. How this autoregulation of blood occurs in amphibian skin is unknown. Furthermore, when boundary layers are disturbed, the overall cutaneous O_2 uptake rises in frogs (21), suggesting either a limitation of surface PO_2 to gas exchange, a reflex increase in localized capillary blood flow or recruitment, or both.

SKIN AS A GAS EXCHANGER: REGULATION OF CUTANEOUS RESPIRATION

Because the skin lacks a dedicated ventilatory pump characteristic of gills (buccal and opercular muscles) and lungs (intercostal and diaphragmatic muscles), it often has been considered a poorly regulated respiratory organ (21), incapable of matching external medium flow with internal blood perfusion. Ample evidence suggests, however, that cutaneous blood flow can be altered in physiologically meaningful ways (28) and is most strongly affected when amphibians are removed from water; in other words, cutaneous blood flow decreases when the skin is exposed to air, presumably to minimize water loss (13,29). It is not normally the overall rate of blood flow that changes in individual capillaries, but rather the total number of capillaries through which blood flows appears to be altered (13).

The simplest and most compelling evidence that amphibians can, to a large extent, regulate cutaneous gas exchange comes from evidence of obligately skin-breathing amphibians. Lungless salamanders are capable of maintaining a constant rate of O_2 uptake in hypoxic environments as low as 8% O_2 (~60 mm Hg) (30). Furthermore, numerous species of frogs that overwinter under the ice are functionally and exclusively skin breathers for up to 6 months of the year. These species show a capacity to maintain constant O_2 uptake down to 5% O_2 (~40 mm Hg) (24,26). Some obligately skin-breathing amphibians are capable of elevating O_2 uptake three to five times if made to exercise (31), suggesting that the skin is not 100% limited by its diffusing capacity, and also that there exists a degree of scope and regulation available that is presumably derived from vasculature adjustments and changes in the partial pressure gradients.

Recent work on frog skin suggests that some similarities may exist between the cutaneous artery circulation and the circulation to the lung. In mammals, a well-established relationship exists between alveolar O_2 levels and blood flow, whereby poorly ventilated alveoli receive little blood (referred to as hypoxic pulmonary vasoconstriction). Based on experiments in which local hypoxia is created over certain regions of the skin, changes in cutaneous perfusion have been observed to occur (28). Within 1 to 1.5 minutes of regional and localized hypoxia, red cell flux velocity decreases by 50%, and the temporal heterogeneity of blood flow within capillaries increases by 30% (32). This response to hypoxia is not likely centrally driven, because the areas of skin exposed were too small to elicit changes in arterial Po_2, but is rather an autoregulatory response brought about by the cutaneous vasculature. This suggests that the O_2 sensor is located within the skin, although where and how it regulates capillary flow is unknown.

Simple and definitive evidence exists, however, for some neural involvement in skin blood flow, at least in the regulation of global changes in skin blood flow. Descending influences from the brain are required for the vasodilatory response of the frog hind web, as evidenced by the fact that hind web capillaries of pithed frogs do not vasodilate under normal stimuli (33). Furthermore, α-adrenergic blockade in bullfrogs increases the number of perfused capillaries (29) or, at the very least, prevents the normal "closure" of capillaries upon air exposure. Moreover, nonthermal vasodilation in frog hind web capillaries is blocked by guanylate cyclase inhibitors, suggesting that nitric oxide (NO) and calcium (Ca^{2+}) signaling also are implicated in the responses of smooth muscles (presumably only in the arterioles or precapillary junctions) (34). However, no definitive studies have attempted to integrate and compare the various roles of circulating versus neural influences.

When examining cutaneous respiration, it also is useful to consider the association between perfusion limitation and diffusion limitation. At low blood perfusion rates, O_2 uptake is limited by the blood flow through the capillaries and should be theoretically directly proportional to the rate of blood, or hemoglobin, flow. At higher perfusion rates, the epidermal barrier limits the exchange of O_2, due to a relatively stable diffusive conductance. In general, O_2 is mainly diffusion-limited across amphibian skin (35). Increasing blood flow does little to alter cutaneous respiration; plus, most of the parameters determining diffusive conductance are usually difficult to adjust in response to acute challenges (35). Such parameters include cutaneous surface area, capillary radius, distance between capillaries, thickness of the skin, and the diffusion constant of O_2. Therefore, it seems reasonable to suggest that, for significant blood flow regulation of O_2 uptake to be possible, the overall O_2 requirement of the animal must be relatively low, allowing for regulation to occur within the range of blood flows that accommodate the metabolic demand.

CUTANEOUS BREATHING: SPECIAL CIRCUMSTANCES IN NATURE

Many temperate amphibians become obligate skin-breathers for up to 6 months of the year when they overwinter under the ice of frozen lakes. Not only are they capable of movement and behavioral thermoregulation (27), but they also enter into a state of reversible metabolic suppression (36,37). Because submerged frogs are capable of regulating constant levels of O_2 uptake down to extremely hypoxic levels, it is possible that they are utilizing alterations in skin perfusion or arterial gases to achieve these results. When exposed to ambient hypoxia or the hypoxemia associated with diving, O_2 delivery to certain tissues is drastically altered. Although perfusion of muscle or other tissues can decrease, the skin must continue to be perfused at an elevated rate to maintain O_2 uptake (38,39), because pulmonary O_2 uptake is no longer possible. Studies on lungless salamanders suggest that these skin-breathing amphibians are primarily diffusion-limited for gas exchange (13,40,41). As a result, increasing cutaneous blood flow should be an ineffective means of increasing O_2 uptake when required. However, most of these experiments were conducted at higher temperatures, at which skin perfusion is likely already at its maximum; therefore, extrapolating to much lower temperatures may lead to erroneous conclusions about the effectiveness of cutaneous gas exchange in overwintering frogs (24,29). For example, cold-submerged bullfrogs can regulate cutaneous O_2 uptake either through rocking or small movements (i.e., spontaneously ventilating the skin) that disrupt hypoxic boundary layers or by the recruitment of otherwise underperfused capillaries in the skin (24).

Early work also suggested that cutaneous CO_2 conductance does not change with temperature in amphibians (42, 43). These results, however, came from frogs at higher temperatures in air. It appears that the constant cutaneous CO_2 conductance prediction does not fit the scenario at these low temperatures in submerged frogs. Tattersall and Boutilier (44) showed that arterial P_{CO_2} remains constant in submerged frogs between 0°C and 7°C, despite an incredible increase in whole animal metabolism (and thus CO_2 excretion), suggesting that CO_2 conductance increases at higher temperatures in submerged frogs, at least up to a critical point. This is quite opposite to the conclusion reached by Mackenzie and Jackson (42) when observing air-breathing frogs at higher temperatures. This is likely to be achieved by the recruitment of blood flow through otherwise underperfused capillaries, thus presenting a greater functional surface area for the exchange of CO_2 to the environment (13). The probable reasons why cutaneous CO_2 conductance changes with temperature in cold-submerged frogs are (a) because their total metabolic requirements are so much lower, and (b) these metabolic demands must be met solely through the skin. Air-breathing frogs at higher temperatures would have sufficient control of O_2 uptake and CO_2 excretion through the combined contributions of the lungs and the skin, but the skin itself will be governed by other external influences, namely that of water conservation. Because any increase in cutaneous conductance for respiratory gases is bound to accelerate the evaporation of water from the skin, air-breathing amphibians likely keep cutaneous conductance at a minimum across a wide temperature range through reductions in cutaneous blood flow, thus minimizing their water loss (45).

KEY POINTS

- The amphibian skin is a relatively simple, yet not entirely understood respiratory organ, particularly with reference to the regulation of its circulation and the role of the endothelium.

- The amphibian skin serves numerous functions, including respiration, osmoregulation, thermoregulation, protection, reproduction, and communication; as such, it is little wonder that compromises in the efficiency of any one particular system are made to accommodate the function of another. Thus, the amphibian's skin is adaptive in the sense of its flexibility in serving multiple needs (e.g., thermal, hydric, and respiratory), even though it is not optimal for any one particular function.
- The regulation of cutaneous respiration is primarily determined by the diffusive properties of the skin and, ultimately, the boundary layer around it. Capillary recruitment, coupled with voluntary ventilation of the skin, which disrupts the boundary layer, combine to provide amphibians with a unique way of regulating respiration.

Future Goals

- To determine what aspects of the cutaneous vascular bed are responsible for regulating gas exchange under conditions related to potential ventilation:perfusion heterogeneities and in conditions relevant to what amphibians experience in nature (i.e., diving, overwintering)
- To determine how the skin appears to detect both external Po_2 and internal Po_2 and to respond with appropriate physiological adjustments; in light of recent research into putative cellular O_2 sensors, the amphibian skin may become a tractable and interesting model for future exploration

REFERENCES

1 Feder ME, Burggren WW. Skin breathing in vertebrates. *Sci Am.* 1985;253:126–142.

2 Withers PC. *Comparative Animal Physiology.* Fort Worth: Saunders College; 1992: xxii, 949, [111].

3 Tazawa H, Mochizuki M, Piiper J. Respiratory gas transport by the incompletely separated double circulation in the bullfrog, *Rana-catesbeiana. Respir Physiol.* 1979;36:77–95.

4 Shelton G, Boutilier RG. Apnoea in amphibians and reptiles. *J Exp Biol.* 1982;100:245–273.

5 Olesen SP, De Saint-Aubain ML, Bundgaard M. Permeabilities of single arterioles and venules in the frog-skin – a functional and morphological-study. *Microvasc Res.* 1984;28:1–22.

6 Smith DG. The innervation of the cutaneous artery of the toad, *Bufo marinus. Gen Pharmacol.* 1976;7:405–409.

7 de Saint-Aubain ML. The morphology of amphibian skin vascularization before and after metamorphosis. *Zoomorphology.* 1982;100:55–63.

8 West NH, Burggren WW. Control of pulmonary and cutaneous blood flow in the toad, *Bufo marinus. Am J Physiol.* 1984; 247:R884–R894.

9 Pinder AW, Storey KB, Ultsch GR. Estivation and hibernation. In: Feder ME, Burggren WW, eds. *Environmental Physiology of the Amphibians.* Chicago: The University of Chicago Press; 1992: 250–274.

10 Czopek J. Distribution of capillaries in the respiratory surfaces in two species of *Batrachophrynus* (Amphibian, Anura, Leptodactylidae). *Zoologica Poloniae.* 1983;30:211–226.

11 Maina JN. Is the sheet-flow design a 'frozen core' (a Bauplan) of the gas exchangers? Comparative functional morphology of the respiratory microvascular systems: illustration of the geometry and rationalization of the fractal properties. *Comp Biochem Physiol A Mol Integr Physiol.* 2000;126:491–515.

12 Malvin G. Microcirculatory effects of hypoxic and hypercapnic vasoconstriction in frog skin. *Am J Physiol.* 1993;264:R435–R439.

13 Feder ME, Burggren WW. Cutaneous gas exchange in vertebrates: design, patterns, control and implications. *Biol Rev.* 1985;60:1–45.

14 Tyml K, Weigelt H, Schafer D. Are endothelial-cells rich in filaments involved in the phenomenon of electrically induced stoppages of flow in frog capillaries. *Int J Microcirc Clin Exp.* 1985; 4:121–130.

15 Boutilier RG, Toews DP. Respiratory, circulatory and acid-base adjustments to hypercapnia in a strictly aquatic and predominantly skin-breathing urodele, *Cryptobranchus alleganiensis. Respir Physiol.* 1981;46:177–192.

16 Hutchison VH, Haines HB, Engbretson G. Aquatic life at high altitude: respiratory adaptations in the Lake Titicaca frog, *Telmatobius culeus. Respir Physiol.* 1976;27:115–129.

17 Noble GK. The integumentary, pulmonary, and cardiac modifications correlated with increased cutaneous respiration in the amphibia: a solution of the 'hairy' frog problem. *J Morph Physiol.* 1925;40:341–416.

18 Feder ME, Pinder AW. Ventilation and its effect on "infinite pool" exchangers. *Am Zoologist.* 1988;28:973–983.

19 Vogel S. *Life in Moving Fluids: The Physical Biology of Flow.* Princeton NJ: Princeton University Press; 1994:xiii,467.

20 Booth DT, Feder ME. Formation of hypoxic boundary layers and their biological implications in a skin-breathing aquatic salamander, *Desmognathus quadramaculatus. Physiol Zool.* 1991; 64:1307–1321.

21 Burggren WW, Feder ME. Effect of experimental ventilation of the skin on cutaneous gas exchange in the bullfrog. *J Exp Biol.* 1986;121:445–449.

22 Pinder AW, Burggren WW. Ventilation and partitioning of oxygen uptake in the frog *Rana pipiens:* effects of hypoxia and activity. *J Exp Biol.* 1986;126:453–468.

23 Pinder AW, Feder ME. Effect of boundary layers on cutaneous gas exchange. *J Exp Biol.* 1990;143:67–80.

24 Pinder AW. Cutaneous diffusing capacity increases during hypoxia in cold submerged bullfrogs (*Rana catesbeiana*). *Respir Physiol.* 1987;70:85–95.

25 Bradford DF. Winterkill, oxygen relations and energy metabolism of a submerged dormant amphibian, *Rana muscosa. Ecology.* 1983;64:1171–1183.

26 Tattersall GJ, Boutilier RG. Behavioural oxy-regulation by cold-submerged frogs in heterogeneous oxygen environments. *Can J Zool.* 1999;77:843–850.

27 Tattersall GJ, Boutilier RG. Balancing hypoxia and hypothermia in cold-submerged frogs. *J Exp Biol.* 1997;200:1031–1038.

28 Malvin GM. Cardiovascular Regulation of Cutaneous Gas Exchange. In: Heatwole H, ed. *Amphibian Biology*. Chipping Norton, NSW: Surrey Beatty & Sons; 1994:147–167.
29 Burggren WW, Moalli R. 'Active' regulation of cutaneous gas exchange by capillary recruitment in amphibians: experimental evidence and a revised model for skin respiration. *Respir Physiol.* 1984;55:379–392.
30 Sheafor EA, Wood SC, Tattersall GJ. The effect of graded hypoxia on the metabolic rate and buccal activity of a lungless salamander (*Desmognathus fuscus*). *J Exp Biol.* 2000;203:3785–3793.
31 Tattersall GJ, Boutilier RG. Does behavioural hypothermia promote post-exercise recovery in cold- submerged frogs? *J Exp Biol.* 1999;202:609–622.
32 Malvin GM, Macias S, Sanchez M, Parkrapid A. Hypoxia-induced hemoconcentration in the toad *Bufo woodhousei*: role of spleen and lymph heart. *FASEB J.* 1992;6:1529A.
33 Poczopko P. Further investigations on the cutaneous vasomotor reflexes in the edible frog in connexion with the problem of regulation of the cutaneous respiration in frogs. *Zoologica Poloniae.* 1957;8:161–175.
34 Miura M, Okada J. Non-thermal vasodilatation by radio frequency burst-type electromagnetic field radiation in the frog. *J Physiol.* 1991;435:257–73.
35 Pinder AW. Gas exchange in isolated perfused frog skin as a function of perfusion rate. *Respir Physiol.* 1991;85:1–14.
36 Donohoe PH. Factors effecting metabolic rate reduction during hibernation in the frog, *Rana temporaria*. Department of Zoology. Cambridge: University of Cambridge; 1997:113.
37 Donohoe PH, Boutilier RG. The protective effects of metabolic rate depression in hypoxic cold submerged frogs. *Respir Physiol.* 1998;111:325–336.
38 Poczopko P. Respiratory exchange in *Rana esculenta* L. in different respiratory media. *Zoologica Poloniae.* 1959;10:45–55.
39 Armentrout D, Rose FL. Some physiological responses to anoxia in the great plains toad, *Bufo cognatus. Comp Biochem Physiol.* 1971;39:447–455.
40 Gatz RN, Crawford EC, Piiper J. Kinetics of inert gas equilibration in an exclusively skin-breathing salamander, *Desmognathus fuscus. Respir Physiol.* 1975;47:151–164.
41 Piiper J, Gatz RN, Crawford EC. Gas transport characteristics in an exclusively skin-breathing salamander, *Desmognathus fuscus* (Plethodontidae). In: Hughes GM, ed. *Respiration of the Amphibious Vertebrates.* London: Academic Press; 1976:339–356.
42 Mackenzie JA, Jackson DC. The effect of temperature on cutaneous CO_2 loss and conductance in the bullfrog. *Respir. Physiol.* 1978;32:313–323.
43 Moalli R, Meyers RS, Ultsch GR, Jackson DC. Acid-base balance and temperature in a predominantly skin-breathing salamander, *Cryptobranchus alleganiensis. Respir Physiol.* 1981;43:1–11.
44 Tattersall GJ, Boutilier RG. Constant set points for pH and P(CO2) in cold-submerged skin-breathing frogs. *Respir Physiol.* 1999;118:49–59.
45 Szarski H. The structure of respiratory organs in relation to body size in amphibia. *Evolution.* 1964;18:118–126.

10

Avian Endothelium

Thomas J. Poole

SUNY Upstate Medical University, Syracuse, New York

THE AVIAN EMBRYO

Chick and quail (*Coturnix coturnix japonica*) embryos have advantages for studying embryological events because they can be accessed through the eggshell or in whole embryo culture for microsurgery, microinjection, or electroporation (1). The monoclonal antibody QH-1 that labels quail angioblasts and endothelial cells (ECs) has been used extensively to study the origin of ECs and their assembly into the initial vascular pattern (2–5). This also allowed observation of quail angioblast migration in quail–chick chimeras (4). Although no equivalent monoclonal antibody exists for the chick embryo, fluorescent lectins have been used that bind to the endothelium when injected into the vasculature (6). Mouse tissue also has been successfully grafted into developing quail embryos. The mammalian angioblasts migrate extensively in the avian embryo and respond to host signals that pattern the vasculature (7). The accessibility in whole embryo culture, even into the later developmental stages, has made the quail embryo the ideal organism to follow nerve and blood vessel development in the limb, where antibodies and growth factors have been delivered by bead implantation to study the molecular basis of neurovascular congruence (8). In summary, the ability to manipulate the avian embryo has led to important discoveries of the relative roles of vasculogenesis (vessel formation from angioblast assembly) and angiogenesis (sprouting of new vessels from preexisting vessels) in embryonic vascular development.

MOLECULAR MARKERS AND BLOOD FLOW

Arterial (ephrinB2, neuropilin 1) and venous (EphB4, neuropilin 2) markers are present on the endothelium of chick, mouse, and zebrafish embryos before blood flow begins (9). The chick embryo has been used to demonstrate that changes in these markers can be experimentally induced by changes in the pattern of blood flow (10). Thus, flow manipulations in the chick embryo before 11 days of incubation can transform arteries into veins and veins into arteries. This ability of arterial ECs to transform into venous ECs (and the reverse) also is used during normal development as the yolk sac vasculature develops. Ephrins and Eph receptors have been shown to be important in establishing boundaries during both neural and vascular development. Increasingly, a number of growth factors and other signaling molecules have been shown to be important for the formation both of nerves and blood vessels. These signals appear to be the same for birds and mammals.

THE AVIAN LUNG

The avian lung does not have the typical branching tree structure (bronchioalveolar) of the mammalian lung. Instead, it has what is termed a *parabronchial lung* that is said to be even more efficient physiologically (11,12). The structure of the avian lung, with its parallel parabronchi, is said to resemble a sponge. The extra efficiency of the bird lung is probably due both to the parallel and serial arrangement of the parabronchi and the exceptionally thin capillaries at the sites of gas exchange (11). The origins of the lung vasculature for both the chick and the mouse appear to be primarily by vasculogenesis (13,14). One difference between the chick and the mouse is the evidence for hematopoiesis in the developing chick lung. A recent study of the chick embryo combining vascular filling, serial sections, and immunohistochemistry of endothelial markers and vascular endothelial growth factor (VEGF) demonstrated that the pulmonary vasculature forms by vasculogenesis (14).

SPECIAL ADAPTATIONS OF THE AVIAN ENDOTHELIUM

The chorioallantoic membrane (CAM) is one aspect of the avian endothelium not found in mammals. The CAM allows for efficient gas exchange through the eggshell and has been used both in the shell and in culture dishes as a model angiogenic assay for screening substances for angiogenic or anti-angiogenic activity (15). The CAM is an interesting vascular bed that begins to shut down as soon as it is fully formed. Another concept that emerged from the study of the chick

CAM and the postnatal rat lung is intussusceptive microvascular growth or nonsprouting angiogenesis. This describes the process in which a capillary bed increases its complexity and vascular surface through the splitting of existing vessels by insertion of transcapillary pillars.

Birds also experience special hypoxic conditions when flying at high altitude or diving. Very little is known about any special endothelial adaptations for this purpose. Some evidence suggests changes in physiology (16,17). Avian eggs incubated in hypoxic conditions similar to high altitude are hypervascular (16). Embryos from certain species can hatch successfully at altitudes of up to 6,500 meters. Although no endothelial modifications are known for high-flying species, such as barheaded geese, they do have amino acid substitutions in their hemoglobin that cause a small increase in oxygen affinity (16).

Another interesting question is whether the avian nucleated red blood cell affects the efficiency of gas exchange across the capillary wall or otherwise makes capillary passage different from that of the mammalian red blood cell that lacks a nucleus and has the shape of a biconcave disc.

KEY POINTS

- The avian embryo is a useful experimental model for the study of vascular development. EC molecular markers have been well studied in birds, but little is known about special endothelial adaptations for hypoxic conditions.
- The quail embryo has been used extensively to understand the contributions of vasculogenesis and angiogenesis to the vascular pattern.
- The avian lung has a morphology that makes it more efficient than the mammalian lung.
- Future experiments might address how birds can function at high altitude or during diving, and how the nucleated red blood cell interacts with the endothelium.
- Now that the chicken genome has been sequenced, avian embryos will continue to be key model organisms for understanding the molecular mechanisms of vasculogenesis, angiogenesis, and neurovascular congruence.

REFERENCES

1 Stern CD. The chick; a great model system becomes even greater. *Dev Cell.* 2005; 8(1):9–17.

2 Poole TJ, Finkelstein EB, Cox CM. The role of FGF and VEGF in angioblast induction and migration during vascular development. *Dev Dyn.* 2001;220(1):1–17.

3 Pardanaud L, Altmann C, Kitos P, Dieterlen-Lievre F, Buck CA. Vasculogenesis in the early quail blastodisc as studied with a monoclonal antibody recognizing endothelial cells. *Development.* 1987;100(2):339–349.

4 Poole TJ, Coffin JD. Vasculogenesis and angiogenesis: two distinct morphogenetic mechanisms establish embryonic vascular pattern. *J Exp Zool.* 1989;251(2):224–231.

5 Coffin JD, Poole TJ. Embryonic vascular development: immunohistochemical identification of the origin and subsequent morphogenesis of the major vessel primordia in quail embryos. *Development.* 1988;102(4):735–748.

6 Jilani SM, Murphy TJ, Thai SN, Eichmann A, Alva JA, Iruela-Arispe ML. Selective binding of lectins to embryonic chicken vasculature. *J Histochem Cytochem.* 2003;51(5):597–604.

7 Ambler CA, Nowicki JL, Burke AC, Bautch VL. Assembly of trunk and limb blood vessels involves extensive migration and vasculogenesis of somite-derived angioblasts. *Dev Biol.* 2001; 234(2):352–364.

8 Bates D, Taylor GI, Minichiello J, et al. Neurovascular congruence results from a shared patterning mechanism that utilizes Semaphorin3A and Neuropilin-1. *Dev Biol.* 2003;255(1): 77–98.

9 Eichmann A, Yuan L, Moyon D, Lenoble F, Pardanaud L, Breant C. Vascular development: from precursor cells to branched arterial and venous networks. *Int J Dev Biol.* 2005;49(2–3):259–267.

10 le Noble F, Moyon D, Pardanaud L, et al. Flow regulates arterial-venous differentiation in the chick embryo yolk sac. *Development.* 2004;131(2):361–375.

11 Maina JN. A systematic study of the development of the airway (bronchial) system of the avian lung from days 3 to 26 of embryogenesis: a transmission electron microscopic study on the domestic fowl, *Gallus gallus* variant *domesticus. Tissue Cell.* 2003;35(5):375–391.

12 Maina JN. Developmental dynamics of the bronchial (airway) and air sac systems of the avian respiratory system from day 3 to day 26 of life: a scanning electron microscopic study of the domestic fowl, *Gallus gallus* variant *domesticus. Anat Embryol (Berl).* 2003;207(2):119–134.

13 Maina JN. Systematic analysis of hematopoietic, vasculogenetic, and angiogenetic phases in the developing embryonic avian lung, *Gallus gallus* variant *domesticus. Tissue Cell.* 2004;36(5):307–322.

14 Anderson-Berry A, O'Brien EA, Bleyl SB, et al. Vasculogenesis drives pulmonary vascular growth in the developing chick embryo. *Dev Dyn.* 2005;233(1):145–153.

15 Ribatti D, Nico B, Vacca A, Roncali L, Burri PH, Djonov V. Chorioallantoic membrane capillary bed: a useful target for studying angiogenesis and anti-angiogenesis in vivo. *Anat Rec.* 2001;264(4):317–324.

16 Leon-Velarde F, Monge CC. Avian embryos in hypoxic environments. *Respir Physiol Neurobiol.* 2004;141(3):331–343.

17 Butler PJ. Metabolic regulation in diving birds and mammals. *Respir Physiol Neurobiol.* 2004;141(3):297–315.

11

Spontaneous Cardiovascular and Endothelial Disorders in Dogs and Cats

John E. Rush

Tufts University School of Veterinary Medicine, North Grafton, Massachusetts

Cardiovascular disease is common in both dogs and cats, although few efforts have been made to identify the roles that endothelial dysfunction plays in these diseases. Although cardiovascular diseases such as myocardial, pericardial, and valvular heart diseases are a frequent cause of disability in companion animals, atherosclerotic coronary artery disease is almost nonexistent in companion animal species.[1] Tests commonly used to characterize cardiovascular disease in clinical veterinary medicine include history, physical examination including auscultation, direct and indirect blood pressure measurement, thoracic radiography, electrocardiography, and echocardiography. Echocardiographic techniques include two-dimensional, M-mode, spectral Doppler, and color-flow Doppler echocardiography. Specific clinical tests to evaluate endothelial function, such as flow-mediated vasodilation and venous plethysmography, commonly are not used, and normal values have yet to be published for clinical application in dogs and cats. Preliminary investigations into the biochemical markers that may reflect endothelial dysfunction in certain diseases have been reported for certain cardiovascular disorders; however, these tests are not used in clinical practice. Although the study of endothelial function in veterinary patients is still in its infancy, many investigators believe that endothelial dysfunction is associated with the common cardiovascular disorders in dogs and cats. This chapter approaches the topic from a brief review of the common cardiovascular diseases in dogs and cats with an accompanying summary of information related to endothelial function.

1 Atherosclerotic heart disease is virtually nonexistent in any species that veterinary cardiologists deal with. The frequency of this disease in humans is truly the anomaly in the animal kingdom, excepting certain purpose-bred animal models (rabbits, etc.).

COMMON SPONTANEOUS CARDIOVASCULAR AND ENDOTHELIAL DISEASE IN DOGS

Chronic Valvular Heart Disease

Chronic valvular heart disease is the most common cardiovascular disease in dogs and is the most common cause of congestive heart failure (CHF) (1). Chronic valvular heart disease is most analogous to the floppy mitral valve syndrome in humans, and it is a progressive, degenerative disorder of the valve. Progressive thickening of the valve results from an accumulation of glycosaminoglycans and the accompanying loss of collagen and elastin from the valve and chordae tendineae. In most cases, the mitral valve is affected most severely; however, the tricuspid valve can be affected concurrently in up to one-third of dogs. Small- to medium-sized dog breeds are predisposed, and male dogs are predisposed compared to female dogs (by an approximate ratio of 1.5 to 1). This slowly progressive disease can be first apparent by 2 to 5 years of age as valvular thickening on echocardiography, sometimes with accompanying mitral valve prolapse. As valve thickening progresses, the mitral valve becomes incompetent and a cardiac murmur is readily identifiable. Disease in the chordae tendineae and progressive enlargement of the heart causes worsening valve coaptation, progressive mitral regurgitation and, eventually, CHF.

The etiology of chronic valvular heart disease is unknown; however, a genetic predisposition to develop the disease has been proven for Cavalier King Charles Spaniels and is suspected in Dachshunds (1,2). The disease is primarily a noninflammatory myxomatous degeneration of the atrioventricular valves and is commonly referred to as endocardiosis, acquired valvular heart disease, or simply mitral regurgitation.

Dogs may be identified with mild to moderate disease based on the presence of a cardiac murmur over the left apex of the heart, with progressive disease associated with a murmur of increasing intensity. The earliest clinical signs usually are due to early CHF or progressive left atrial enlargement, and include intermittent cough, nocturnal dyspnea, exercise intolerance, altered sleep habits, or abdominal distention. In most dogs, left-sided CHF predominates with the development of cardiogenic pulmonary edema. Syncope also may be the initial sign of chronic valvular disease and can occur secondary to vasovagal (neurocardiogenic) syncope, supraventricular arrhythmias, low cardiac output, or as a result of paroxysmal coughing (e.g., tussive syncope). CHF can often be managed through a combination of exercise restriction, dietary sodium restriction, furosemide or other diuretics, angiotensin-converting enzyme (ACE) inhibitors, digoxin, β-blockers, and antiarrhythmics as required. Common clinical endpoints include severe and refractory CHF, sudden death due to arrhythmia or left atrial rupture with cardiac tamponade, and euthanasia due to poor quality of life, side effect of cardiac medications, anorexia, or worsening clinical signs.

In recent years, significant enthusiasm has been generated for the role that endothelial dysfunction may play in dogs with chronic valvular heart disease. Peripheral arterial disease has been reported in Cavalier King Charles Spaniels, a breed recognized to develop chronic valvular heart disease at a very early age. The peripheral arterial disease is identified first by either a lack of femoral arterial pulse or weakening of femoral arterial pulse.[2] One study noted a unilateral or bilateral weakened arterial pulse in 26% of dogs with compensated chronic valvular heart disease, and the pulse was absent in 2% of dogs (3). A weak arterial pulse was found to be caused by reduced femoral artery diameter, with histopathological findings including intimal thickening with breaks in the internal elastic lamina in occluded segments and organized thrombus (3,4). Because of excellent collateral circulation, these femoral arterial findings rarely result in overt clinical signs; however, they are believed to be a marker of more widespread endothelial dysfunction. In addition, several investigators have reported that small- to medium-sized coronary arteries have evidence of intramural (nonatherogenic) arteriosclerosis in dogs with chronic valvular disease, and that these lesions can lead to reduced left ventricular systolic function (5,6).[3] These arteriosclerotic lesions can result in focal myocardial necrosis or infarction, with the latter lesions (seen in severely affected dogs) having been referred to as *multifocal intramural myocardial infarction*. It is presumed that the mural coronary arteriosclerotic lesions in these cases result from endothelial disease and contribute to reduced systolic function and eventually CHF.

2 This is not a result of emboli from the valves, and it is not seen in any of the other countless breeds that get this form of heart disease. The peripheral arterial disease seems to be fairly unique to the CKCS breed, and the relationship, if any, to valvular heart disease is not well understood.

3 Although arteriosclerosis also has been documented as a histopathological finding in dogs with dilated cardiomyopathy and in cats with various forms of cardiomyopathy, these lesions have only been clearly documented to cause small vessel occlusion and microscopic infarction in dogs with chronic valvular disease.

A limited number of studies have examined biomarkers that may indirectly support a role for endothelial dysfunction in dogs with chronic valvular disease. Two studies have investigated the role that activation of nitric oxide (NO) systems may play in dogs with chronic valvular heart disease. One study identified an increase in serum nitrite/nitrate concentrations in dogs with chronic valvular disease (7), whereas another study failed to identify any increase in serum nitrite/nitrate in affected Cavalier King Charles Spaniels (8). Another study looked at nicotinamide adenine dinucleotide phosphate-oxidase diaphorase (NADPH-d) activity in canine myxomatous mitral valves and identified the role that nitric oxide formation may play in the pathogenesis of chronic valvular heart disease based on increased NADPH-d activity in the valve (9). Although the canine mitral valve does not appear to stain heavily for renin-angiotensin system activity in either normal dogs or affected dogs, endothelin (ET)-1 receptor density was found to be increased in parallel with the severity of disease in dogs with chronic valvular heart disease (10). Finally, scanning electron microscopy has identified a variety of endothelial alterations on the surface of the mitral valve from dogs with chronic valvular disease, including denudation of endothelial cells (ECs) with exposure of the basement membrane and subendothelial collagen matrix, as well as loss of orderly arrangement of ECs and increased EC plasmalemmal microappendages compared to cells on unaffected valves (1,11). These findings suggest that endothelial disease is present in dogs with chronic valvular disease and that alterations of the endothelium may contribute to the pathogenesis of chronic valvular heart disease or the accompanying peripheral or coronary arterial disorders.

Dilated Cardiomyopathy

Dilated cardiomyopathy is the second most common heart disease in dogs. Large and giant breed dogs are predisposed, and the disease is particularly common in certain breeds of dogs such as Doberman Pinschers, Great Danes, Boxers, Newfoundlands, and Irish Wolfhounds. It has been estimated that up to 50% to 60% of all Doberman Pinschers will develop dilated cardiomyopathy. A genetic origin for the disease has been described in certain breeds based on pedigree analysis; however, a specific molecular cause has not yet been determined for any breed. With the exception of Portuguese Water Dogs, dilated cardiomyopathy is an adult-onset disease with clinical signs developing in middle-aged or older dogs. Each breed appears to have specific features that are characteristic of cardiomyopathy in that breed. For example, serious ventricular arrhythmias leading to syncope or sudden death are common in both Doberman Pinschers and Boxers with dilated cardiomyopathy. The histologic finding of attenuated wavy fibers is typical for Newfoundlands and other giant-breed dogs, whereas cardiomyopathy in the Boxer appears to mimic arrhythmogenic right ventricular cardiomyopathy in humans, with fatty infiltration and loss of myocytes in the right ventricular wall (12). In most cases, canine dilated cardiomyopathy

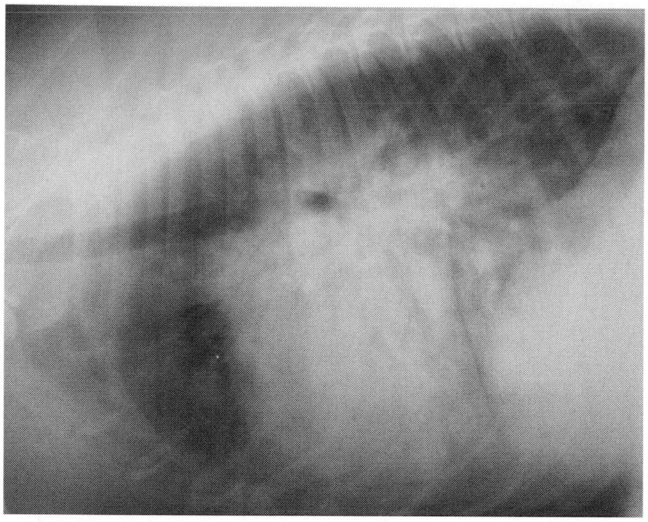

Figure 11.1. Lateral thoracic radiograph from a Doberman Pinscher with dilated cardiomyopathy and CHF. Note the pulmonary edema in the perihilar region.

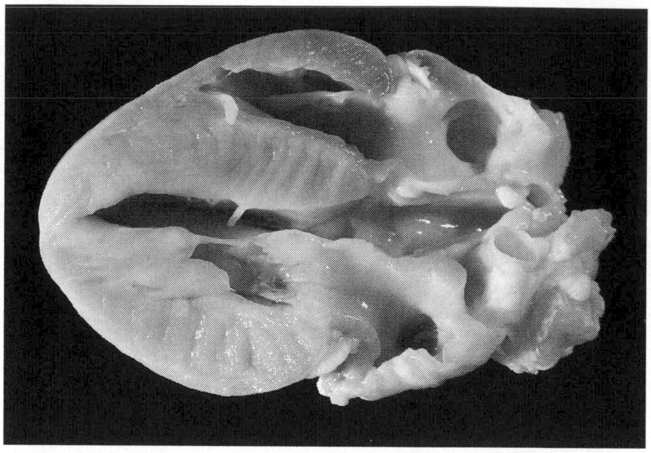

Figure 11.2. Gross pathology specimen from a cat with hypertrophic cardiomyopathy. The heart has been sectioned to demonstrate the right atrium and right ventricle (*top*), aorta (*center and extending to the right*), left ventricle, and left atrium. Note the hypertrophy of the interventricular septum and left ventricular free wall.

eventually results in CHF (Figure 11.1), sudden death, or euthanasia due to poor quality of life, progressive signs of heart failure, drug intolerance, or anorexia.

Less is known relative to endothelial dysfunction in dogs with dilated cardiomyopathy. Serum nitrite/nitrate concentrations are reported to be elevated in dogs with dilated cardiomyopathy (7), and plasma ET-1 immunoreactivity was reported to be elevated in dogs with dilated cardiomyopathy and CHF (13).

Hemangiosarcoma and Pericardial Effusion in the Dog

The third most common cardiac disease in dogs is pericardial effusion, leading to cardiac tamponade. Pericardial effusion in dogs can result from a number of neoplastic disorders as well as idiopathic causes. Neoplasia is the most common cause and, of the neoplasms that cause pericardial effusion, hemangiosarcoma of the right atrium and right auricular appendage is the most common. The cause of hemangiosarcoma in the dog is unclear; however, it is known to originate from ECs.

Hemangiosarcoma often is associated with disseminated intravascular coagulation, and canine hemangiosarcoma cells express vascular endothelial growth factor (VEGF) receptors 1 and 2, CD31, and CD146, and also produce growth factors and cytokines such as VEGF, basic fibroblast growth factor, and interleukin-8 that are stimulatory to EC growth (14).

SPONTANEOUS FELINE CARDIOVASCULAR AND ENDOTHELIAL DISEASE

The most common cause of cardiovascular disease in the cat is cardiomyopathy, and hypertrophic cardiomyopathy is the most commonly described form (Figure 11.2) (15). The cause of hypertrophic cardiomyopathy in the cat is undocumented for most breeds; however, a genetic basis is suspected in the majority of cases, and a genetic mutation of myosin binding protein C has recently been documented in Maine Coon cats (16). Maine Coon cats, Himalayans, American Short Hairs, Rag Doll cats, and Norwegian Forest cats are predisposed to hypertrophic cardiomyopathy. Most surveys identified that male cats and cats of larger size also are predisposed to hypertrophic cardiomyopathy. Dilated cardiomyopathy, once a common cause of cardiovascular disease in the cat, is now rarely seen due to the discovery that dietary taurine deficiency caused this reversible form of cardiomyopathy. Various other forms of cardiomyopathy have been described, including restrictive cardiomyopathy, arrhythmogenic right ventricular cardiomyopathy, and various unclassified forms of cardiomyopathy.

Hypertrophic cardiomyopathy in the cat can result in syncope or sudden death; however, the most common clinical presentations are CHF or arterial thromboembolism. CHF is thought to result from diastolic dysfunction, with progressive left atrial enlargement and eventually pulmonary edema or biventricular heart failure, leading to pulmonary edema and plural effusion. CHF often is managed with a combination of dietary sodium restriction, diuretics such as furosemide, ACE inhibitors, and either β-blockers (atenolol) or calcium channel blockers (diltiazem) to help reduce heart rate and improve diastolic function. Narrowing of intramural coronary arteries is identified on histopathology; however, the most impressive changes are myocardial hypertrophy with increased fibrous connective tissue and myofiber disarray in the left ventricle and interventricular septum. The role of endothelial function in the development or progression of cardiomyopathy in the cat has not been well studied. (A new study describes a test for endothelial function in the dog [17].)

Arterial Thromboembolism

Arterial thromboembolism is a common complication of virtually all forms of feline heart disease. In most cases, the thrombus originates in the left atrium, dislodges, and is carried downstream, where embolism of the terminal aorta,[4] and/or iliac arteries creates severe clinical signs. In 90% of affected cats, the thrombus lodges at the aortic trifurcation, although thromboembolism to the front limb, central nervous system, kidneys, or other sites also is reported. Aortic thromboembolism results in an acute onset of severe posterior paresis and pain, with loss of femoral arterial impulses, cyanotic nail beds, and the development of cool limbs and firm gastrocnemius muscles. Aortic thromboembolism is a catastrophic complication of cardiomyopathy in the cat: only 40% to 50% of cats ever regain function of the rear limbs. Of those cats who do regain function of the rear limbs, many experience repeated episodes of arterial thromboembolism. Vascular stasis in the dilated left atrium, exposure of subendothelial collagen due to left atrial stretch, and altered coagulation are all proposed to play a role in the development of intracardiac thrombus. It appears that vasoactive factors mediate a significant component of the clinical signs, because simple ligation of the terminal aorta of the cat results in no clinical signs; however, double ligation with addition of thrombus leads to severe vasoconstriction of collateral vessels and classic clinical manifestations of arterial thromboembolism. Earlier reports indicated that serotonin may play a role in the development of this scenario; however, the exact interplay of endothelial factors that lead to the vasoconstriction is unclear at this time. One recent study identified that plasma arginine concentrations are reduced in cats with arterial thromboembolism (18), leading to the hypothesis that the lack of collateral circulation could result from reduced NO production. Manual surgical removal of the thrombus requires anesthesia and is associated with a high rate of complications. Thrombolytic drugs also have not been uniformly successful, because they can lead to bleeding complications or severe reperfusion injury with life-threatening hyperkalemia and metabolic acidosis. Most therapy is aimed at prevention of thrombus formation with use of aspirin, low-molecular-weight heparin, or Coumadin.

4 The term *saddle thrombus* often is used to describe arterial thromboembolism (ATE) of the terminal aorta. On necropsy, the thrombus itself is located in the distal aorta and extends a short distance into the iliac artery, but the thrombus typically does not extend down into the femoral artery. Simple ligation of the terminal aorta in normal cats fails to produce the clinical signs of ATE. A substance(s) is suspected to be elaborated by the thrombus that causes vasospasm and "shuts down" the collateral circulation, because a thrombus seems to be necessary to cause clinical signs of ATE. Some cats have thrombus in the terminal aorta large enough to only occlude one iliac artery, and they have single rear-limb signs, but this is less common. Clinically "silent" thromboembolism to the kidneys commonly is noted on abdominal ultrasound or necropsy of these cats, and ATE to the coronary vasculature is suspected based on focal left ventricular wall thinning and dysfunction in selected cats – this is thought to account for at least some cases of sudden death.

KEY POINTS

- Valvular, myocardial, and pericardial cardiovascular diseases are common in dogs and cats, although atherosclerotic coronary artery disease is extremely rare in these species.
- Common cardiovascular diseases in dogs include degenerative mitral valvular disease, dilated cardiomyopathy, and pericardial effusion due to hemangiosarcoma. Hypertrophic cardiomyopathy and resulting arterial thromboembolism are the most common cardiovascular problems in cats.
- Preliminary research indicates that endothelial dysfunction contributes to the development or progression of cardiovascular disease in dogs and cats.

Future Goals

- To determine whether therapeutic interventions to improve endothelial function can alter the development or progression of cardiovascular disease in companion animals
- To develop a reliable, minimally invasive, and clinically applicable test for endothelial function, such as flow-mediated vasodilation, to aid in the characterization of endothelial and cardiovascular disorders in dogs and cats

REFERENCES

1 Häggström J, Kvart C, Pedersen HD. Acquired valvular heart disease. In: Ettinger SJ, Feldman EC, eds. *Textbook of Veterinary Internal Medicine*, 6th ed. St. Louis: Elsevier Saunders; 2005: 1022–1039.
2 Pedersen HD, Kristensen BO, Lorentzen KA, Koch J, Jensen AL, Flagstad A. Mitral valve prolapse in 3-year-old healthy Cavalier King Charles Spaniels. An echocardiographic study. *Can J Vet Res.* 1995;59:294–298.
3 Tarnow I, Olsen LH, Jensen MB, Pedersen KM, Pedersen HD. Determinants of weak femoral artery pulse in dogs with mitral valve prolapse. *Res Vet Sci.* 2004;76:113–120.
4 Buchanan JW, Beardow AW, Sammarco CD. Femoral artery occlusion in Cavalier King Charles Spaniels. *J Am Vet Med Assoc.* 1997;211:872–874.
5 Falk T, Jonsson L. Ischaemic heart disease in the dog: a review of 65 cases. *J Small Anim Prac.* 2000;41:97–103.
6 Jönsson L. Coronary arterial lesions and myocardial infarcts in the dog. A pathologic and microangiographic study. *Acta Vet Scand Suppl.* 1972;38:1–80.
7 deLaforcade AM, Freeman LM, Rush JE. Serum nitrate and nitrite in dogs with spontaneous cardiac disease. *J Vet Intern Med.* 2003;17:315–318.
8 Pedersen HD, Schutt T, Sondergaard R, Qvortrup K, Olsen LH, Kristensen AT. Decreased plasma concentration of nitric oxide metabolites in dogs with untreated mitral regurgitation. *J Vet Intern Med.* 2003;17:178–184.

9 Olsen LH, Mortensen K, Martinussen T, Larsson LI, Baandrup U, Pedersen HD. Increased NADPH-diaphorase activity in canine myxomatous mitral valve leaflets. *J Comp Pathol.* 2003;129:120–130.

10 Mow T, Pedersen HD. Increased endothelin-receptor density in myxomatous canine mitral valve leaflets. *J Cardiovasc Pharmacol.* 1999;34:254–260.

11 Corcoran BM, Black A, Anderson H, et al. Identification of surface morphologic changes in the mitral valve leaflets and chordae tendineae of dogs with myxomatous degeneration. *Am J Vet Res.* 2004;65:198–206.

12 Tidholm A, Jonsson L. Histologic characterization of canine dilated cardiomyopathy. *Vet Pathol.* 2005;42:1–8.

13 Prosek R, Sisson DD, Oyama MA, Biondo AW, Solter PF. Plasma endothelin-1 immunoreactivity in normal dogs and dogs with acquired heart disease. *J Vet Intern Med.* 2004;18:840–844.

14 Akhtar N, Padilla ML, Dickerson EB, et al. Interleukin-12 inhibits tumor growth in a novel angiogenesis canine hemangiosarcoma xenograft model. *Neoplasia.* 2004;6:106–116.

15 Rush JE, Freeman LM, Fenollosa NK, Brown DJ. Population and survival characteristics of cats with hypertrophic cardiomyopathy: 260 cases (1990–1999). *J Am Vet Med Assoc.* 2002;220:202–207.

16 Meurs KM, Sanchez X, David RM, et al. A cardiac myosin binding protein C mutation in the Maine coon cat with familial hypertrophic cardiomyopathy. *Hum Mol Genet.* 2005;14:3587–3593.

17 Puglia GD, Freeman LM, Rush JE, et al. Use of a flow-mediated vasodilation technique to assess endothelial function in dogs. *Am J Vet Res.* 2006;67(9):1533–1540.

18 McMichael MA, Freeman LM, Selhub J, et al. Plasma homocysteine, B vitamins, and amino acid concentrations in cats with cardiomyopathy and arterial thromboembolism. *J Vet Intern Med.* 2000;14:507–512.

12

Giraffe Cardiovascular Adaptations to Gravity

Alan R. Hargens*, Knut Pettersson†, and Ronald W. Millard‡

* *University of California, San Diego;* † *Kungsladugårdsgatan 110, Göteborg, Sweden;*
‡ *University of Cincinnati College of Medicine, Ohio*

GRAVITY AND THE CARDIOVASCULAR SYSTEM

The physiological systems of animals have adapted to Earth's gravity over the past hundreds of millions of years. In general, gravitational adaptations of the cardiovascular system are more pronounced in terrestrial species with greater height and thus greater gravity-dependent gradients of blood pressure from head to feet. For example, dinosaurs (1), tree-climbing snakes (2), giraffes (3), and other tall animals have evolved mechanisms to provide adequate blood flow and nutrition to their brains while restricting blood flow and tissue swelling in their legs. Terrestrial animals of short stature and marine animals probably require much less sophisticated cardiovascular adaptive mechanisms. At the other extreme, aquatic snakes have little ability to withstand gravity out of water and rapidly "faint" when placed head above tail (2). Moreover, when gravity is absent even over short periods of time, astronauts experience orthostatic intolerance upon readaptation to gravity (see Chapter 58). Because humans are relatively tall compared to other species of animals, they too have developed extensive and sophisticated regulatory mechanisms to maintain cerebral perfusion and prevent lower extremity edema while in an upright posture. In fact, most understanding of gravitational mechanisms to date relates to observations in humans. However, taller terrestrial animals, such as the giraffe, may allow better understanding of the physiological adaptations to gravity. For example, blood pressure in giraffes is high to pump blood to their brain, but high blood pressures in their feet would theoretically cause severe dependent edema.

Cardiovascular systems generate and regulate blood pressure to provide flow to tissues. This blood flow nourishes tissues by supplying oxygen (O_2) and other nutrients, and by removing carbon dioxide (CO_2) and other waste products. Transmural pressures (*Px*: pressure gradients from inside to outside the circulation) dictate the wall stress (*S*) and degree of openness or closure (vessel radius *r*) experienced by circulatory structures according to the Law of Laplace: $Px = S/r$ (4). Normally, five components determine transmural pressure in the circulation: (a) the dynamic pressure resulting from cardiac pumping against peripheral resistance to blood flow; (b) dynamic pressures due to inertial forces of blood during activity (e.g., the footward acceleration of blood in leg vessels at heel strike during locomotion); (c) pressure due to the finite compliance (volume/pressure characteristics) of the systemic vasculature, especially the veins (this pressure is commonly called the mean circulatory filling pressure); (d) extravascular pressure of tissues and interstitial fluid (e.g., total tissue pressure); and (e) an intravascular hydrostatic pressure due to gravity (gravitational pressure). Although extravascular tissues and fluid have mass, gravity affects extravascular hydrostatic pressures to a minor extent because tissue fluids are bound within cells or are discontinuous in most extravascular spaces (except for cerebrospinal fluid). Structural and contractile tissues respond dynamically, within physiological limits and over a long period of time, as adaptive structural and functional remodeling, to increased (or decreased) levels of physical stress by increasing (or decreasing) their functional capability and/or mass (5). Although physiological responses to stress occur within seconds to minutes, longer-term functional and structural compensatory adjustments to the initiation of stress are detectable within hours (6).

TRANSCAPILLARY FLUID BALANCE

Transcapillary fluid shifts between the microcirculation and the surrounding tissues (Figure 12.1) are dictated by capillary surface area and the ability of capillary basement membrane and endothelium to retain plasma proteins. These fluid shifts are determined by hydrostatic and colloid osmotic pressures according to the Starling-Landis equation (7):

$$J_c = L_p A[(P_c - P_t) - \sigma_p(\pi_c - \pi_t)] \qquad \text{(Equation 1)}$$

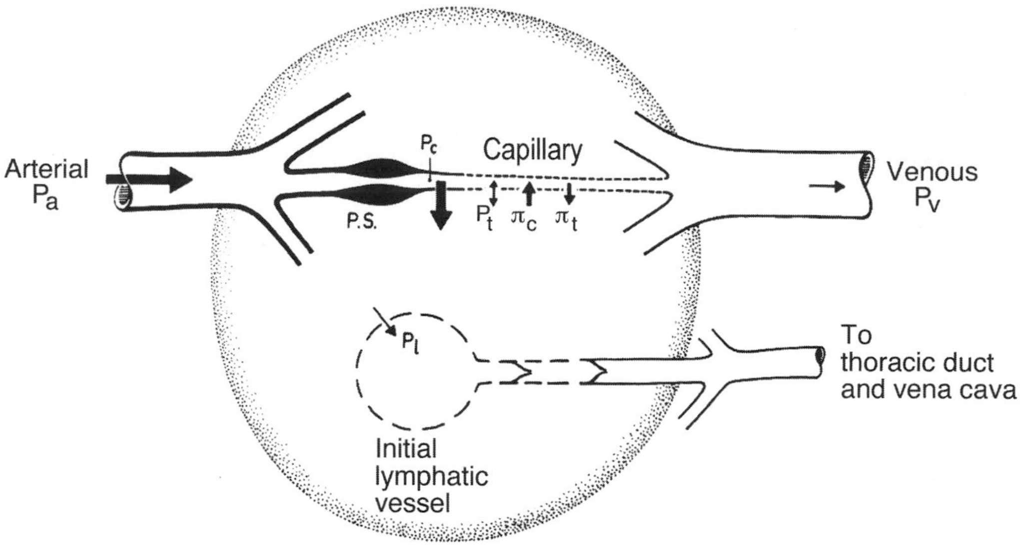

Figure 12.1. Starling pressures, which regulate transcapillary fluid balance. Pressure parameters that determine directions and magnitude of transcapillary exchange include capillary blood pressure P_c, tissue fluid pressure P_t (directed into capillary when positive or directed into tissue when negative), plasma and capillary colloid osmotic pressure π_c, and tissue fluid colloid osmotic pressure π_t. Precapillary sphincters (P.S.) regulate P_c and capillary flow. It is generally agreed that a hydrostatic pressure gradient ($P_t >$ lymph pressure P_l) drains off excess tissue fluid under conditions of net filtration. Relative magnitudes of pressure in resting tissues are depicted by the size of the arrows. (Reproduced with permission from Hargens AR. Interstitial Fluid Pressure and Lymph Flow. In: Skalak R, Chien S, eds. *Handbook of Bioengineering.* New York: McGraw-Hill; 1986:1–35.)

where J_c is net transcapillary fluid shift, L_p is hydraulic conductivity of capillary wall, A is capillary membrane filtration area, P_c is capillary blood pressure, P_t is tissue fluid pressure, σ_p is reflection coefficient for plasma proteins, π_c is capillary blood colloid osmotic pressure, and π_t is tissue fluid colloid osmotic pressure. Again, because blood has continuous columns within the body, local P_c is greatly affected by gravity. On the other hand, because tissue fluids are discontinuous in the body, P_t is relatively unaffected by gravity.

In most tissues, a net flux of fluid and plasma proteins occurs across the capillary membrane (endothelial cells [ECs] and basement membrane), through the interstitial spaces, and into the lymphatic system (see Figure 12.1). The hydraulic conductivity of capillary wall L_p is determined by endothelial function as well as capillary basement membrane thickness. Generally, the reflection coefficient for plasma proteins, σ_p, is determined by the permeability properties of the capillary wall. Some tissues, such as muscle and skin, have relatively high σ_p ($\approx$0.9) such that protein transport rates are low. In other tissues that have lower σ_p, such as lung and liver, π_t is almost equivalent to π_c. In these tissues, L_p and filtration rates are relatively high and lymphatic drainage is critical for tissue fluid balance. Recent evidence also indicates that lymphatic vessels have a two-valve system, with primary valves at the level of the initial lymphatic endothelium and a secondary valve system in the lumen of the transporting lymphatics that facilitates unidirectional transport towards the lymphatic nodes and thoracic duct (8).

MICROCIRCULATORY ADAPTATION TO GRAVITY

Transcapillary ultrafiltration, J_c, in the human calf is significantly lower than that in the forearm in supine subjects (9). Although postural changes are associated with increases in precapillary resistance, altering pressure and flow characteristics within exchange vessels of the dependent limb, no changes occur in the capillary surface area available for exchange (10). The autonomic regulation of capillary hemodynamics in dependent limbs appears to involve a veno-arteriolar reflex (11). Exploration of dependent limb capillary filtration has revealed that the α-adrenergic receptor-mediated vasoconstrictor response in the circulatory system of the human leg is enhanced upon standing and may represent an adaptive mechanism that limits blood pooling and transcapillary ultrafiltration in the leg during upright position (12). However, should autonomic nervous system dysfunction (i.e., dysautonomia) exist, profound hypotension will occur without the typical reflex tachycardia when standing abruptly. In such individuals, dependent limb vasoconstriction control is decreased significantly (13). In this regard, orthostatic hypotension is a common consequence of such dysfunction in elderly patients.

Because the human feet and head experience the greatest change in blood pressure with altered vectors of gravity, it is worthwhile to understand their structure and functional responses to gravitational stimuli. Previously, we found that elevation of blood pressure to the head produced by a transition from 60 degrees head-up tilt (HUT) to 6 degrees head-down tilt (HDT) increases microcirculatory blood flow in skin of the head (14). This pressure and flow increase may be expected in a tissue not commonly exposed to increased blood pressure and therefore, not able to regulate blood flow closely to prevent edema. On the other hand, moving the body from 6-degree HDT to 60-degree HUT actually reduces microcirculatory flow in the foot. This response suggests that the microcirculation of the lower body, normally adapted to upright posture on Earth, is able to regulate capillary pressure and flow adequately to prevent dependent edema. Combined with vasoconstriction and a thicker capillary basement membrane in dependent tissues to reduce hydraulic conductivity (15), other mechanisms for preventing dependent edema include lymphatic drainage (7) and skeletal-muscle venous pumps in the leg (16).

GRAVITATIONAL PHYSIOLOGY IN THE GIRAFFE

The giraffe, *Giraffa camelopardalis*, represents a unique mammalian model for developmental studies of tissue adaptation to increasing load bearing. Whereas fetal giraffes develop in a quasi-weightless milieu, postpartum giraffes must contend with increasing load bearing in their dependent tissues as they grow to heights of over 5 meters. Previous studies of the adult giraffe cardiovascular system (17–19) indicate that this system is certainly a unique model for investigating the physiology of being tall and adaptations to large and variable gradients of blood pressure. These early studies of blood pressures document that arterial pressure near the giraffe heart is about twice that in humans to provide more normal blood pressure and perfusion to the brain. In the context of higher systemic blood pressure and longer hydrostatic columns of blood from heart to feet, the giraffe circulatory system exhibits significant increases in vascular wall muscle thickness, capillary basement membrane thickness, and other vascular control systems in dependent blood vessels that, in concert, participate in capillary fluid dynamics and edema prevention.

During the 1985 Giraffe Physiology Expedition to Africa, our studies focused on hemodynamic adaptations and edema prevention in the legs of adult giraffes (3,20). Mean values for arterial and venous pressures for giraffes and humans qualitatively match the expected gravitation pressure gradients, using the heart as a reference for fluid discontinuity between the head and foot (Figure 12.2). Although P_c was not measured directly in the giraffe, it is probably near local venous pressure (P_v) in the feet (150 mm Hg) and about 10 to 20 mm Hg at the top of the giraffe's neck (3). Importantly, π_c was identical in the giraffe and human and, therefore, contrary to an early, unpublished theory by August Krogh. Thus, π_c in the giraffe foot offers no unusual resorptive pressure for preventing dependent edema. Because of rapid circulation and mixing of blood within the cardiovascular system, it is difficult to imagine how π_c could be significantly higher in the dependent tissues of the giraffe. Although some P_t values in the neck were negative, the average bodily P_t ranged between 1 and 6 mm Hg, except under the tight skin and fascia of the extremities, where mean P_t was 44 mm Hg. Interestingly, π_t in legs was very low (1 mm Hg), except in foot samples that were contaminated by blood. This finding provided evidence that giraffe capillaries are highly impermeable to plasma proteins and that σ_p approximates unity in lower limbs. This conclusion was supported by studies of peripheral lymph that indicated only trace amounts of protein were present and π_l equaled zero.

Of importance, we found that the blood and tissue fluid pressures that determine transcapillary ultrafiltration are highly variable with exercise (Figure 12.3). These pressures, combined with a tight skin and fascial "antigravity suit," move venous blood and tissue fluid upward against gravity, thus preventing pooling of blood and edema in dependent tissues. More proximally, an active skeletal-muscle pump aids venous return. The nonhydrostatic pressure gradient down the giraffe's jugular vein indicates that blood cascades down from the head, and that circulation above heart level does not depend on a siphon-like principle, as earlier proposed (21). Other edema-preventing mechanisms include dependent precapillary vasoconstriction, abundant lymphatic drainage, and very low capillary permeability to plasma proteins.

Developmental alterations are present in the load-bearing tissues of newborn compared with adult giraffes (22). For example, vascular wall thickness closely relates to local blood pressure in the developing giraffe. Tissue samples were collected from four 5- to 6-year-old, 3.5- to 4-meter, male and female giraffes during the 1985 Giraffe Physiology Expedition; and from three newborns and two 25- and 35-year-old, 5-meter adult giraffes from the Cincinnati and Cheyenne Mountain Zoos. Arterioles and other microcirculatory vessels were sampled from skin and muscle of the head, neck, thorax, legs, and feet. Although we never had the opportunity to measure blood pressures in newborn giraffes, based on comparable studies of other species, the arterial pressures of baby giraffes are probably significantly lower at heart level and in dependent tissues than are those in adult giraffes. It is apparent that dependent arteries in adults had much thicker walls than did those in newborn giraffes (Figure 12.4). In adult animals, arterial wall hypertrophy correlates directly with the degree of tissue dependency (Table 12-1). Thus, the wall thickness-to-lumen ratios of large arteries increase from head to foot (except for one "ankle" artery that didn't follow this pattern).

We believe the developing giraffe provides an excellent model for investigations of adaptive mechanisms to natural hypertension found in the adult giraffe. Arteries of the feet

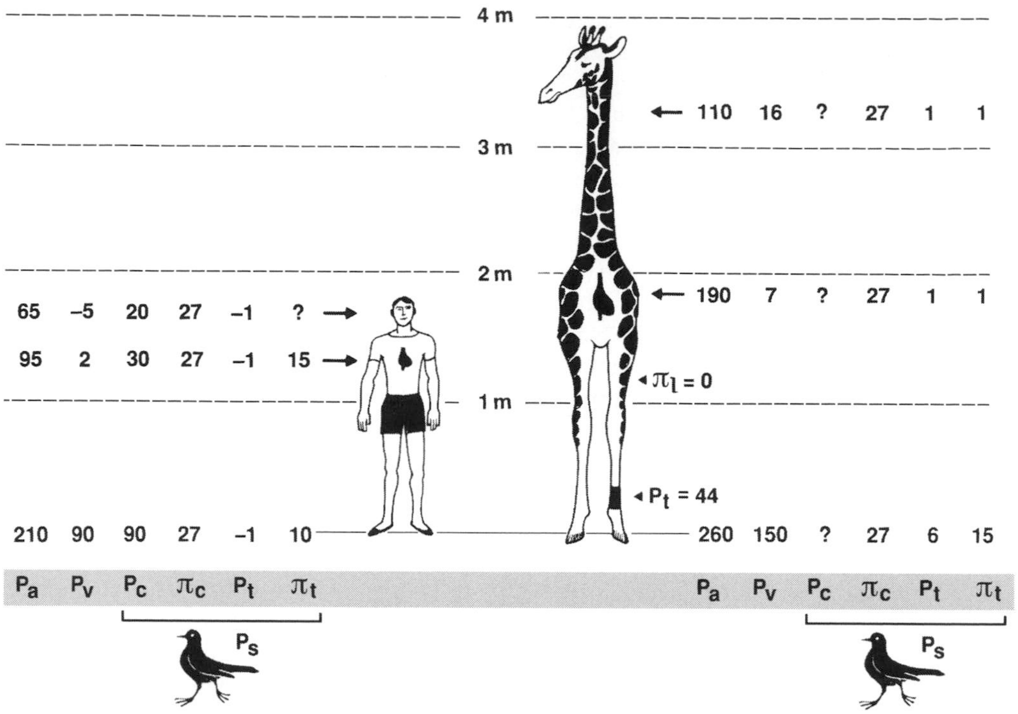

Figure 12.2. Mean arterial P_a, venous P_v, and Starling pressures (P_c, π_c, P_t, and π_t) in giraffe (*right*) as compared to human (*left*) at hydrostatic levels between the head and feet. In the giraffe, lymph samples obtained from the leg had only trace amounts of protein and $\pi_l = 0$. Foot samples for π_t often were contaminated by blood and therefore were less reliable. P_t beneath the tight skin and fascia of the legs ranged between 40 and 50 mm Hg and indicated the presence of a "natural anti-G suit" in the giraffe. Data for pressures in humans are obtained from several sources in addition to our previous studies. (Reproduced with permission from Hargens AR, Millard RW, Pettersson K, van Hoven W, Gershuni DH, Johansen K. Transcapillary Fluid Balance in the Giraffe. In: Staub ND, Hogg JC, Hargens AR eds. *Interstitial-Lymphatic Liquid and Solute Movement, Advances in Microcirculation.* Basel, Switzerland: S. Karger; 1987:195–202.)

are exposed to blood pressures greater than 400 mm Hg in adult giraffes (3). These vessels have developed pronounced smooth-muscle hypertrophy and narrow lumens to accommodate these extraordinarily high blood pressures. Our previous report of a less-than-normal arterial pressure gradient from heart to foot in the upright, stationary giraffe (3) suggests that the reduction in lumen cross-sectional area plays some role in blood pressure reduction to dependent tissues. It is interesting that the arterial wall hypertrophy apparently is confined to dependent vessels with diameters over 400 μm microns in adults and is not observed in newborn giraffes or in vessels near the head of adult giraffes.

CAPILLARY BASEMENT MEMBRANE OF THE GIRAFFE VASCULAR SYSTEM

In 1971, Williamson and co-workers (15) measured the capillary basement membrane width (CBMW) in muscle obtained from human infants (n = 19, 9F/1M, 1-day to 6-month-olds), children (n = 4, 1F/3M, 3- to 5-year-olds), and adults (n = 17, 6F/11M, 36- to 75-year-olds) obtained within 6 hours postmortem. They also evaluated muscle from one adult giraffe for capillary basement membrane thickness. The measurements in human samples illustrate a significant age-dependent effect, with the thickest basement membrane measured in the adult capillaries. The results also illustrate increased CBMW in the lower legs of adults. CBMW in human infants ranges from 538 to 563 Å and does not vary significantly with anatomic location. In children, CBMW is significantly thicker (806 to 1,076 Å) in all anatomical sites sampled (pectoral, abdominal, thigh, calf) than that in the same sample sites of infants. As in infants, no difference in the CBMW is noted among anatomical locations in children. In contrast, the CBMW of adults is significantly greater than that in children and infants, and shows progressive thickening in dependent tissues moving down toward the feet (pectoral = 964 Å, calf = 1,894 Å). In giraffe, the CBMW increases from 780 Å at the neck to 1,437 Å in the front and hind legs. These data are consistent with the hypothesis that CBMW is strongly influenced by intravascular hydrostatic pressure, which is most extreme in the long hydrostatic column of the giraffe circulatory system. This increase in CBMW in the leg of giraffes may be an important contributing factor to microcirculatory fluid homeostasis in dependent tissues with high intravascular pressure.

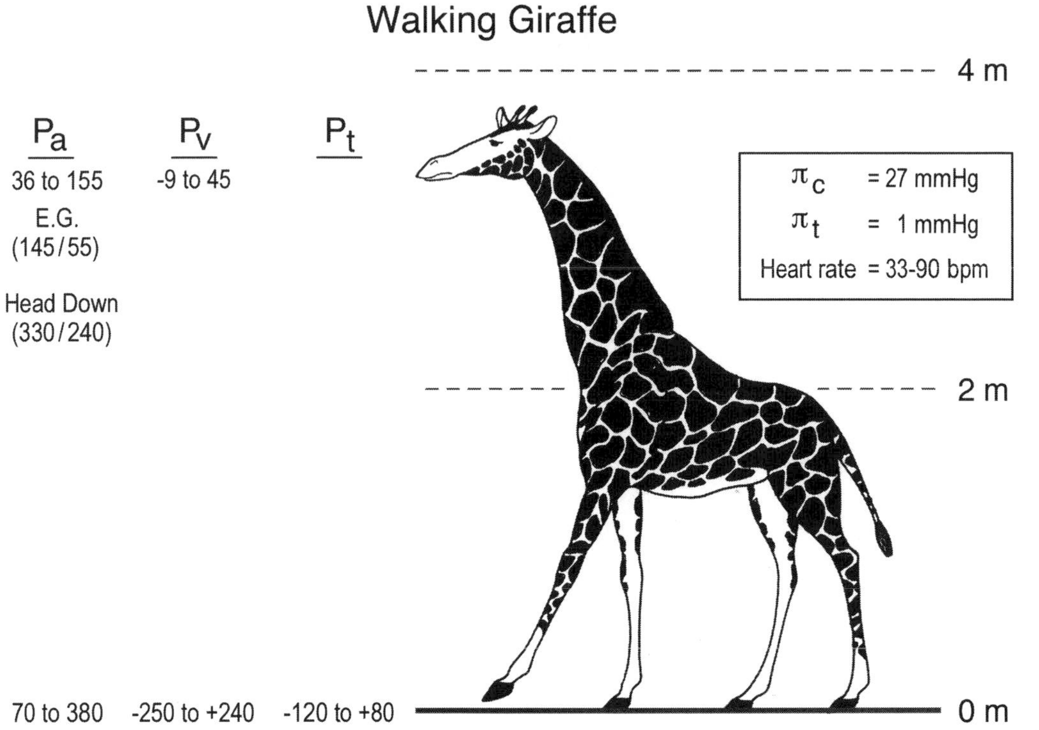

Figure 12.3. Range of averaged mean blood and tissue fluid pressures in neck and foot of walking giraffes. Negative venous and subcutaneous pressures during exercise help prevent dependent edema. With head upright, mean arterial pressure below the jaw ranged between 36 and 155 mm Hg (e.g., 145/55 mm Hg for systolic/diastolic pressures). Drinking water in head-down posture raised carotid pressures to 330/240 mm Hg. (Reproduced with permission from Hargens AR, Millard RW, Pettersson K, van Hoven W, Gershuni DH, Johansen K. Transcapillary Fluid Balance in the Giraffe. In: Staub ND, Hogg JC, Hargens AR eds. *Interstitial-Lymphatic Liquid and Solute Movement, Advances in Microcirculation.* Basel, Switzerland: S. Karger; 1987:195–202.)

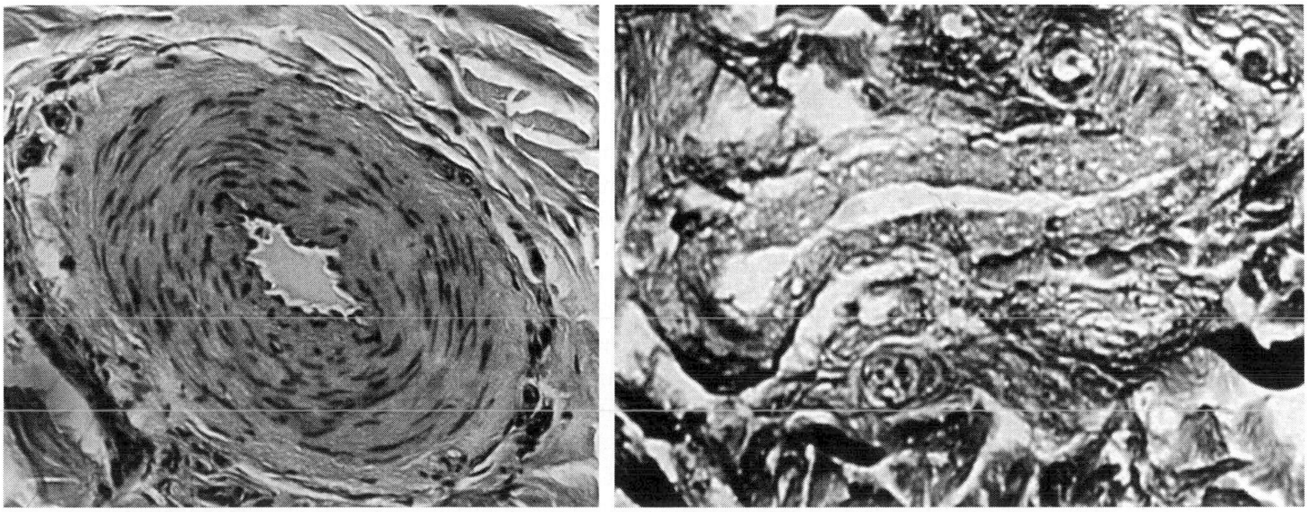

Figure 12.4. Four-hundred-micron diameter artery from foot skin of adult giraffe has thick wall (*left*) compared with a similar artery from foot skin of newborn giraffe (*right*).

Table 12-1: Ratios of Smooth Muscle Wall Thickness/ Lumen Radius (w/r) for Arteries from the Neck to the Feet of Adult Giraffes

Artery	*w/r*	*Outer Diameter (mm)*
Carotid	0.15–0.20	9.5–1.0
Brachial	0.33–0.43	5.5–7.5
Femoral	0.65–0.68	4.0–4.2
Ulnar	0.70	7.2
Radial	0.70	5.4
"Ankle"	0.33	4.6
Metatarsal	0.51–0.81	3.1–4.8
Digital	0.56	4.0

(Reproduced with permission from Hargens AR, Gershuni DH, Danzig LA, Millard RW, Pettersson K. Tissue adaptations to gravitational stress: newborn versus adult giraffes. *Physiologist.* 1988;31:S110–S113.)

Table 12-2: Innervation of Giraffe Vasculature

Blood Vessels		*Adventitial Innervation*	*Media Innervation*
Load-bearing vessels	Tibial artery	+	−
	Popliteal artery	++	−
	Metatarsal artery	+	−
	Tarsal artery	++	+
	Saphenous vein	−	−
	Metatarsal vein	−	−
Central vessels	Ascending aorta	++	++
	Pulmonary artery	+	−
	Superior caval vein	−	−
	Inferior caval vein	−	−
Head and neck	Carotid artery	+++	+++
	Jugular vein	+	−

The density of nerve terminal (axons) near smooth muscle cells ranges from − (no nerves observed) to +++ (a dense network present). (Reproduced with permission from Nilsson O, Booj S, Dahlstrom A, et al. Sympathetic innervation of the cardiovascular system in the giraffe. *Blood Vessels.* 1988;25:299–307.)

INNERVATION OF THE GIRAFFE VASCULAR SYSTEM

Giraffes are tall animals, and changing posture will therefore cause major circulatory changes, including an increase of hydrostatic pressure in the head and neck during head-down drinking. However, the structure of the cardiovascular system in the giraffe is adapted gravitationally to the normal upright posture of the animal. Because short-term regulation of blood pressure and flow is determined by activity in the sympathetic nervous system, we examined and compared the innervation of different segments of the cardiovascular system (23). In general, efferent, sympathetic nerves control cardiac function, arterial resistance, and venous capacitance.

Tissue specimens were obtained from two young male giraffes killed as a part of a conservation-culling program in Africa. We used antisera against dopamine-β-hydroxylase (DβH, marker of adrenergic nerves), neuropeptide Y (NPY, a peptide transmitter in sympathetic nerves), as well as general markers for nerve terminals (synapsin 1) and nerve fibers/axon bundles (neurofilament, NF). With these antibodies, a network of nerve terminals and fibers was demonstrated. The distribution of nerve elements was similar using different antisera. However, although all antibodies visualized varicose nerve terminals indicative of adrenergic fibers, the NF antibody also marked nonautonomic axon bundles (Table 12-2).

In all arterial blood vessels, a network of nerve terminals with DβH-, NPY-, and Syn1-positive fibers was found. NF-positive fibers were sparsely located at the medioadventitial border, and thick paravascular bundles were found frequently. In the medial layer of the popliteal artery, NF-positive nerve fibers also were observed; but the other antibodies used gave no labeling, indicating that the nerves were not adrenergic. In general, arteries from the extremities had a relatively sparse innervation and were usually restricted to the medioadventitial border, although some evidence for a sparse sympathetic innervation of the outermost layer of the media was found. The ascending aorta had a denser innervation, with sympathetic fibers frequently penetrating into the outer medial layers. The carotid arteries had the densest innervation of all arteries. The nerves penetrated deep into the media, with fibers reaching almost to the intima layer.

In veins, the sympathetic innervation was sparse. In fact, veins from the lower part of the body and caval veins were devoid of adrenergic innervation. However, NF-positive, paravascular axon bundles were found, and the expected adrenergic fibers at the medioadventitial border were absent. In the jugular veins, however, we observed such adrenergic fibers, although their density still appeared sparse.

These observations show that a sympathetic innervation of the vascular system exists in giraffes, but the innervation pattern is nonuniformly distributed. It is surprising that the caval and load-bearing veins apparently lack sympathetic innervation. This is in sharp contrast to man, wherein veins in the extremities, and especially capacitance veins, have a rich supply of adrenergic fibers, the activity of which is important for venous return from the legs (24). We can only speculate on why giraffes lack this type of innervation, or alternatively, use other transmitters than those for which we stained. However, the larger veins in the legs of giraffe run beneath both the tight skin and a stiff fascia. Thus, the capacitance of these veins may be limited despite their high venous pressures reported (see above), so little capacitance is present to regulate.

Sympathetic nerves were, however, found in the jugular vein. During upright posture, these veins do not need active regulation, but this could be important during head-down drinking, when the pressure gradient is reversed and jugular pressure increases. Valves also are present in the jugular veins (25), and these could—together with sympathetic nerves—contribute to prevent blood flow in the retrograde direction and pooling of blood in the head during those short periods when giraffes lower their head below heart level to drink.

The innervation of limb arteries is similar to that found in most mammalian species, and innervation is restricted largely to the medioadventitial border. However, as the thickness of the medial layer can be extreme in these vessels, the efficiency in the spreading of excitation from the nerves must be very efficient if these nerves are to control the contractions of the entire medial sheet. The density of the innervation of large arteries increases from foot to head. In the aorta, we found fibers in the outer medial layer. In the carotid arteries, sympathetic fibers are found penetrating deep into the media, almost reaching the intima layer. The physiological significance of this dense innervation is not known. However, the pattern we observed is consistent with an adaptive pattern, with dense innervation in vessels subjected to the largest transient fluctuations during head-down drinking, and a less dense sympathetic innervation occurring in the vasculature of load-bearing tissues.

In summary, our results concerning tissue adaptations to increased load bearing in the developing giraffe, along with those of other investigators, provide intriguing findings that deserve further investigation, especially from an EC standpoint. More studies of the developmental biology of giraffes are needed to elucidate those mechanisms by which tissues adapt to increased blood pressure in the feet. Such knowledge may provide useful information for understanding structural and functional adaptations of blood vessels in human health and disease.

KEY POINTS

- Despite very high local blood pressures, dependent edema in the giraffe is prevented by a tight "antigravity suit," muscle pumping of blood and lymph toward the heart, and vascular adaptations.
- Vascular walls and capillary basement membranes increase thickness in dependent tissues.
- Some vascular adaptations in the giraffe are developmental in nature.
- Surprisingly, arteries and veins of the lower body have sparse sympathetic innervation.

Future Goals

- To understand better EC structure and function and their role in the local regulation of giraffe circulations of blood and lymph
- To determine which adaptations are developmental in nature
- To investigate cerebral circulations, in order to understand adaptive mechanisms during head-down drinking behavior in giraffes

ACKNOWLEDGMENTS

These studies were supported by grants from NSF (DCB-8409253), NIH (HL-32703), National Geographic Society (3072–85), Department of Veterans Affairs, and NASA. We thank Professor Wouter van Hoven and the Department of Zoology, University of Pretoria, South Africa, for providing giraffes and facilities for our 1985 Giraffe Physiology Expedition. We thank Dr. D.G.A. Meltzer and Roodeplaat Research Labs, Dr. Richard Burroughs and the National Zoological Gardens in Pretoria, and Dr. Frik J. Stegmann and the University of Pretoria Veterinary School at Onderstepoort. We thank Dr. Paul Calle, the Cheyenne Mountain Zoo, and the Cincinnati Zoo for providing specimens of baby and aged giraffes.

REFERENCES

1 Millard RW, Lillywhite HB, Hargens AR. Cardiovascular system design and barosaurus. Comments on the article by D. S. J. Choy, P. Altman. The cardiovascular system of barosaurus: an educated guess. *Lancet*. 1992;340:543–536. Letter published in *Lancet*. 1992;340:914.

2 Lillywhite HB. Circulatory adaptations of snakes to gravity. *Am Zool*. 1987;27:81–95.

3 Hargens AR, Millard RW, Petterson K, Johansen K. Gravitational haeomodynamics and edema prevention in the giraffe. *Nature*. 1987;329:59–60.

4 Fung YC. *Biomechanics: Mechanical Properties of Living Tissues*. New York: Springer-Verlag; 1981:16–19.

5 Hargens AR, Akeson WH. Stress Effects on Tissue Nutrition and Viability. In: Hargens AR, ed. *Tissue Nutrition and Viability*. New York: Springer-Verlag; 1986:1–24.

6 Fung YC, Liu SQ. Changes of zero-stress state of rat pulmonary arteries in hypoxic hypertension. *J Appl Physiol*. 1991; 70:2455–2470.

7 Hargens AR. Interstitial Fluid Pressure and Lymph Flow. In: Skalak R, Chien S, eds. *Handbook of Bioengineering*. New York: McGraw-Hill; 1986:1–35.

8 Schmid-Schönbein GW, Bronzino JD, Hargens AR, eds. Mechanics of Tissue/Lymphatic Transport. In: *The Biomedical Engineering Handbook*. Boca Raton FL: CRC Press; 2006:1–16.

9 Mahy IR, Lewis DM, Tooke JE. Limb capillary filtration coefficient in human subjects: the importance of the site of measurement. *Physiol Meas*. 1998;19:339–343.

10 Gamble J, Christ F, Gartside IB. The effect of passive tilting on microvascular parameters in the human calf: a strain gauge plethysmography study. *J Physiol*.1997;498:541–552.

11 Sejrsen P, Henriksen O, Paaske WP. Effect of orthostatic blood pressure changes upon capillary filtration-absorption rate in the human calf. *Acta Physiol Scand*. 1981;111:287–291.

12 Pawelczyk JA and Levine BD. Heterogeneous responses of human limbs infused adrenergic agonist: a gravitational effect? *J Appl Physiol.* 2002;92:2105–2113.

13 Brown CM, Stemper B, Welsch G, et al. Orthostatic challenge reveals impaired vascular resistance control, but normal venous pooling and capillary filtration in familial dysautonomia. *Clin Sci.* 2003;104:163–169.

14 Aratow M, Hargens AR, Meyer JU, Arnaud SB. Postural responses of head and foot cutaneous microvascular flow and their sensitivity to bed rest. *Aviat Space Environ Med.* 1991;62:246–251.

15 Williamson JR, Vogler NJ, Kilo C. Regional variations in the width of the basement membrane of muscle capillaries in man and giraffe. *Am J Pathol.* 1971;63:359–370.

16 Rowell LB. *Human Cardiovascular Control.* New York: Oxford University Press, 1993.

17 Goetz RH, Warren JV, Gauer OH, et al. Circulation of the giraffe. *Circ Res.* 1960;8:1049–1058.

18 Patterson JL, Jr., Goetz RH, Doyle JT, et al. Cardiorespiratory dynamics in the ox and giraffe with comparative observations on man and other mammals. *Ann NY Acad Sci.* 1975;127:393–413.

19 Van Citters RL, Kemper S, Franklin DL. Blood flow and pressure in the giraffe carotid artery. *Comp Biochem Physiol.* 1968; 24:1035–1042.

20 Hargens AR, Millard RW, Pettersson K, et al. Transcapillary Fluid Balance in the Giraffe. In: Staub ND, Hogg JC, Hargens AR eds. *Interstitial-Lymphatic Liquid and Solute Movement, Advances in Microcirculation.* Basel, Switzerland: S. Karger; 1987:195–202.

21 Badeer HS. Does gravitational pressure of blood hinder flow to the brain of the giraffe? *Comp Biochem Physiol.* 1986;83A:207–211.

22 Hargens AR, Gershuni DH, Danzig LA, Millard RW, Pettersson K. Tissue adaptations to gravitational stress: newborn versus adult giraffes. *Physiologist.* 1988;31:S110–S113.

23 Nilsson O, Booj S, Dahlstrom A, et al. Sympathetic innervation of the cardiovascular system in the giraffe. *Blood Vessels.* 1988; 25:299–307.

24 Folkow B and Neil E. *Circulation.* New York: Oxford University Press, 1971.

25 Millard RW, Hargens AR, Pettersson K, Johansen K. Blood pressure and intervalve distances in giraffe veins. *Fed Proc.* 1987; 46:793.

13

Energy Turnover and Oxygen Transport in the Smallest Mammal

The Etruscan Shrew

Klaus D. Jürgens

Zentrum Physiologie, Medizinische Hochschule, Hannover, Germany

Life on Earth started with the formation of small organisms and, in general, larger ones evolved from smaller ones. Today, the size range of adult mammals spans eight orders of magnitude, from 1.5 to 150,000,000 g. Although all mammals have a common design principle, they are by no means geometrically similar, either in form or in function. If an elephant had the shape of a blown up shrew, its legs would not be able to carry the weight of its body. If the energy turnover rates of the elephant and shrew were directly proportional to differences in body mass, the elephant's body temperature would be at boiling point. The field of biology dealing with the scaling of the structural and functional properties of organisms is called *allometry*. Allometric tools are used not only to phenomenologically describe the body size dependence of parameters but also to investigate the underlying scaling laws. One of the most important parameters governing life is energy turnover or metabolic rate, usually measured as oxygen (O_2) consumption. For terrestrial mammals, the best fit of the relationship between measured basal metabolic rate (BMR) and body mass (M) is the power function BMR $= 3\ M^{0.7}$ (BMR in watts, M in kg) (1).[1] On a double logarithmic scale, this function is a straight line with a slope of 0.7 (Figure 13.1). In addition to studying the biological basis for scaling laws, it is also of interest to investigate why some species deviate significantly from the allometric mean. When considering extremely small endotherms,[2] the following questions arise: What are the structural and functional properties that constitute the lower end of the mammalian size range, and why do no smaller mammals exist?

The smallest extant mammalian species is the Etruscan shrew (*Suncus etruscus, S.e.)* (Figure 13.2). The average body mass of this species is 1.8 g (range 1.3 to 2.4 g), which is close to that of the bumblebee bat (*Craseonycteris thonglongyai*), for which a mean body mass of about 2 g has been reported (see 2). Small shrew-like animals are thought to be among the earliest species in mammalian evolution, appearing about 200 million years ago. A fossil record from the early Eocene, about 50 million years ago, of an extinct shrew weighing only 1.3 g has been described (2).

An important parameter that appears to limit the smallness of mammals is energy expenditure. Depending on geometric conditions, the relation of body surface to body mass increases with decreasing size. Therefore, heat loss through the body surface to the environment compared to heat production in the body core is larger in smaller species. Thus, the smaller the mammal, the higher is the mass specific energy turnover necessary to maintain a constant body temperature. High energy turnover requires high rates of food intake as well as high O_2 consumption rates; hence the need for appropriate adaptation of the O_2 and substrate transport systems.

Because muscles play a major role in the capture and chewing of food, in convective O_2 transport, and in the generation of heat, these organs must be adapted to provide the enormous mass-specific rate of energy consumption to the body. Exceptionally rapid muscle contractions, together with an enormous store of adipose tissue, serve to optimize thermogenic potential.

The O_2 transport properties of the Etruscan shrew, especially its cardiovascular system, have been studied extensively during the last several years. The shrew's mass specific performance is higher than that of any other studied mammal. Among other parameters, the capillary densities of heart and

1 Different mass exponents are calculated for different parameters. If the exponent is close to 0, the parameter is independent of size or mass; if it is 1, the parameter is size dependent, and it scales linearly with mass (M). Many parameters fall between these values, such as metabolism (0.7 to 0.75) or some rates (respiratory and heart approx. −0.25). For example, the stroke volume is proportional to M (i.e., exponent 1), and heart rate has a mass exponent of −0.25; thus, the product stroke volume times heart rate or cardiac output scales with 0.75, which is in accordance with metabolism. This is as it should be, because all the O_2 must be delivered by the circulation.

2 Warm-blooded mammal.

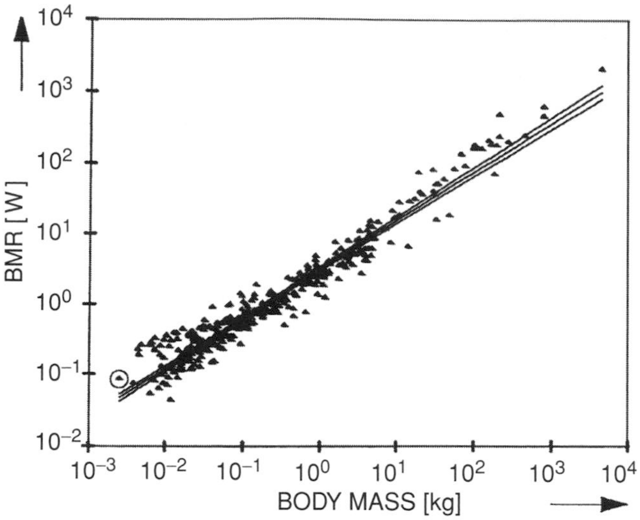

Figure 13.1. Allometric plot of BMR of terrestrial mammals as function of body mass. *Circled triangle* depicts Etruscan shrew.

skeletal muscle (3,4) and of lung are higher than in any other mammal, indicating very large diffusion capacities per unit body mass (Table 13-1).

A large vascular bed means a large endothelial surface. As yet, no studies have been carried out to determine whether the endothelia of species living under such extreme metabolic conditions show special adaptations with respect to the enormous fluxes of respiratory gases and nutrients that are required.

OXYGEN CONSUMPTION AND METABOLISM

The shrew demonstrates the highest O_2 consumptions per unit body mass (mass specific VO_{2max}) of any endotherm studied to date. This holds for basal as well as for maximal metabolic rates. Taylor (5) measured a maximal O_2 consumption of 870 mL O_2/(kg min) in the red-toothed shrew *Sorex minutus*

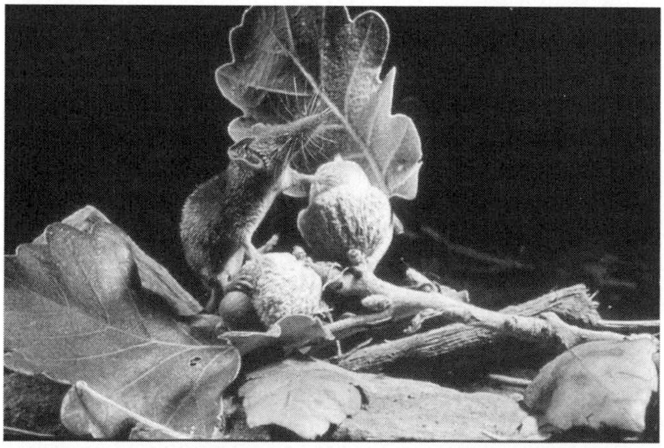

Figure 13.2. Etruscan shrew (*Suncus etruscus*) standing on acorns. (Photo courtesy of Roger Fons.)

Table 13-1: Comparison of Energy Turnover–Related Data of Etruscan Shrew and Man

Parameter	*Suncus etruscus*	*Man*
Mean adult body mass (g)	1.8	70, 000
Oxygen consumption (mL O_2/(kg min))		
Basal	100	4
Maximal	1,000	80
Respiratory		
Maximal tidal volume (mL/kg)	16–20	40
Maximal respiratory rate (min^{-1})	894	70
Estimated maximal ventilatory oxygen transport rate (mL O_2/(kg min)	970	150
Alveolar surface (m^2/kg)	8.5	1.9
Lung capillary surface (m^2/kg)	6.5	1.7
Estimated maximal diffusional oxygen transport rate (mL O_2/(kg min))	1,050	180
Oxygen-carrying capacity		
Red cell volume (μm^3)	28	88
Red cell count (10^6 cells/μm^3)	18	5
Hemoglobin concentration (g/L)	175	150
Oxygen capacity of blood (mL O_2/L blood)	240	200
Blood oxygen half saturation pressure (mm Hg)	35	27
Oxygen extraction (mL O_2/mL blood) (maximal work)	0.19	0.15
Cardiovascular		
Maximal heart rate (min^{-1})	1,511	220
Stroke volume (mL/kg)	3.5	2
Estimated maximal circulatory transport rate (mL O_2/kg/min)	1,000	80
Tissue		
Capillary density (per mm^2)		
Heart	7,300	2,700
M. soleus	2,800	290
Content of mitochondria in muscle (Vol%)		
Heart	36	25
Diaphragm	35	12

(3 g body mass) exposed to cold in a helium/O_2 atmosphere (6). This value is almost the same as that measured in a small hummingbird during hovering flight (7). For the white-toothed shrew *Suncus etruscus* (2 g body mass), values close to 1,000 mL O_2/(kg min) have been recorded during exercise (8). This value is 250 times the mass specific resting O_2 consumption of humans and about 12 times their mass specific maximal O_2 consumption. To make such an outstanding metabolic rate possible requires many special adaptations: (a) the food intake per day equals the animal's body mass (9); (b) the intestinal passage time is less than an hour, so an enormous transport rate of sugars, amino acids, and fatty acids takes place through

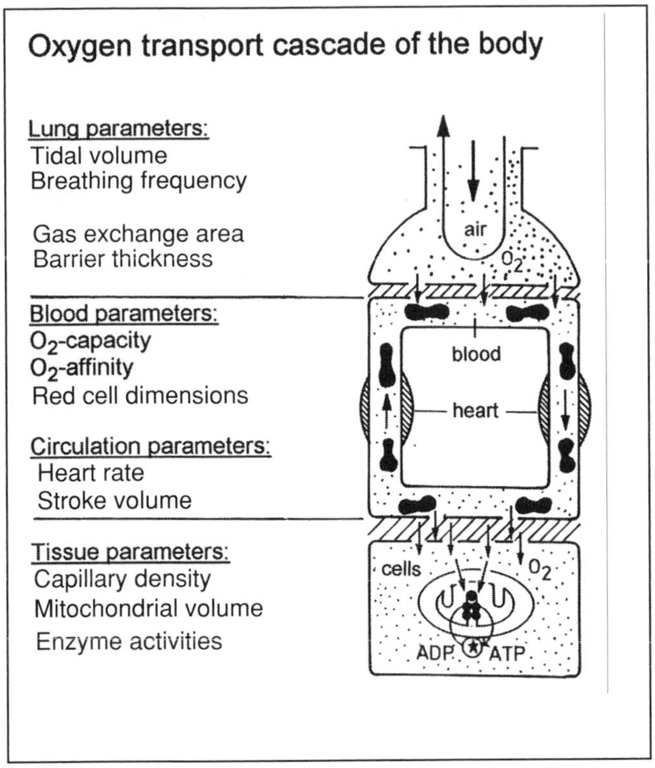

Figure 13.3. Oxygen transport cascade and transport parameters in mammals.

intestinal epithelia, blood, and lymph vessels; (c) the activities of enzymes needed for aerobic metabolism, such as citrate synthase (CS), exceed those of any other mammal; and (d) the activities of enzymes needed for anaerobic metabolism, such as lactate dehydrogenase (LDH), are lower than in any other mammal, leading to a CS:LDH activity ratio in *S. etruscus* gastrocnemius muscle that is 100 times higher than that in the muscle of a 70 kg human (10). The shrew's skeletal muscle cells consist of 35% mitochondria (human muscle cells have 10% to 20%) (11,12), which have richly folded inner membranes and, therefore, an extremely large inner respiratory surface. Moreover, remarkable adaptations of the O_2 transport system are seen at all levels of the O_2 transport cascade (Figure 13.3).

RESPIRATORY SYSTEM

The mean maximal respiratory rate measured at 22°C was 758/min in *S. etruscus*, and the highest single value amounted to 894/min or 15/s. The shrew has a total lung volume of 80 to 130 μl (8,13). Assuming a maximal tidal volume of 40% of the lung volume, these values are consistent with a mass-specific tidal volume of 16 to 20 mL air/kg. O_2 extraction per breath, which is independent of body size in mammals, is approximately 50 mL O_2/L air (8). Thus, the Etruscan shrew has a calculated maximal ventilatory O_2 uptake rate of 720 to 970 mL O_2/(kg min), which is consistent with the measured maximal O_2 consumption in this species. Because relative tidal volumes and O_2 extraction are not significantly size-dependent,[3] the high ventilatory O_2 transport rate in *S. etruscus* is achieved exclusively by its high respiratory rate (8).

How well do the diffusional transport properties of the lung match this high ventilatory O_2 uptake rate? Gehr and colleagues (13) determined morphometrically a mean specific O_2 diffusion capacity of 11 mL O_2/(kg min mm Hg) and a maximal value of 15 mL O_2/(kg min mm Hg) for *S. etruscus* (compared with an average of 6 mL O_2/(kg min mm Hg) in humans). The high diffusion capacity of the shrew lung is due to a very large alveolar surface (8.5 m^2/kg body mass) and capillary surface (6.5 m^2/kg body mass), compared to 1.9 and 1.7 m^2/kg in man, respectively, as well as to an extremely thin diffusion barrier, 0.27 μm, compared to 0.61 μm in man, as determined morphometrically. In addition, the diffusion distance within red cells is relatively short in the shrew; the mean cell diameter is only 5.5 μm (mean cell volume 28 μm^3) compared to 7.5 μm (88 μm^3) in humans (14).

Assuming a rather large mean maximal Po_2 difference between alveolar gas and lung capillaries of 50 to 70 mm Hg, which is appropriate for conditions of heavy work in mammals, one predicts from the maximal diffusion capacity a maximal diffusional O_2 flow of 750 to 1,050 mL O_2/(kg min). This means that the maximal O_2 consumption in Etruscan shrews is limited by the structurally determined diffusional O_2 transport capacity as well as by the ventilatory O_2 transport capacity (8). This situation is in contrast to larger mammals and humans, in which the morphometrically determined diffusion capacity of the lung generally exceeds the functional diffusion capacity required for maximal aerobic performance (13).

CARDIOVASCULAR SYSTEM

Measurements of the maximal heart rate in *S. etruscus* have yielded an average of almost 1,100/min. The maximum single value recorded, 1,511/min (25/s), is the highest heart rate ever reported in an endotherm. It exceeds the highest values reported so far for small red-toothed shrews like *Sorex minutus* (3 g, 1,091/min), small bats such as *Myotis lucifugus* (5 g, 1,370/min), and small hummingbirds like *Calliphlox amethystina* (2.4 g, 1,280/min) (8,15). Nevertheless, small mammals such as shrews have considerably lower resting and maximal heart rates than predicted from their body mass using allometric relationships. For resting conditions, 1,140 beats/min are predicted, but only 840/min were measured. In larger mammals, the resting heart rate increases up to threefold due to exercise; in *S. etruscus*, the maximal increase is only 1.8 times the resting and 1.3 times the allometrically predicted basal heart rate. The maximal heart rate of the Etruscan shrew is only six to eight times that of man, whereas

3 Examples of parameters that are allometrically independent of body mass include blood pressure, hematocrit, or capillary diameter. Examples of mass-dependent properties include organ sizes and functional parameters, such as metabolic rate and heart rate, or life history parameters, such as maximal lifespan and gestation period.

the maximal respiratory rates such as the VO_{2max} are 12 times that of humans (see Table 13-1). This means that, unlike in the respiratory system, O_2 transport by the circulatory system requires adaptation not only of heart frequency but also of stroke volume and blood O_2 extraction. Indeed, the relative heart mass and the relative heart volume of *S. etruscus* is higher than that of man (14), providing a higher relative stroke volume for the shrew (estimated 3.5 mL/kg body mass compared to 2 mL/kg in trained humans). The shrew's blood is characterized by a large O_2 capacity (0.24 mL O_2/mL blood), based on an increased hemoglobin concentration and a low O_2 affinity or large O_2 half-saturation pressure (P_{50}), respectively (P_{50} at 37°C = 35.2 mm Hg) (humans 0.20 mL O_2/mL blood, 27 mm Hg). A mixed venous Po_2 of 20 mm Hg, as may occur under conditions of maximal work in mammals, would result in an 80% O_2 desaturation of the shrew's blood, thus allowing an increased O_2 extraction (0.19 mL O_2/mL blood) at maximal work when compared to humans (0.15 mL O_2/mL blood). These values lead to an estimate of the maximal circulatory O_2 transport rate of 760 to 1,000 mL O_2/(kg min) in the Etruscan shrew (8), thus matching the measured VO_{2max}.

Studies reveal that all three maximal O_2 transport rates – ventilatory, diffusional, and circulatory – correspond fairly well with the experimentally obtained maximal O_2 consumption rate. Moreover, all these transport capacities seem to match at this lowest end of the mammalian body size range; that is, the concept of symmorphosis (16) is ideally realized in the smallest mammal (see Table 13-1). The principle of symmorphosis suggests that morphogenesis is regulated in a way so that the formation of structural elements in an organism satisfies but does not exceed the functional requirements of the system.

MUSCLE TISSUE

The heart and skeletal muscles of *S. etruscus* are functionally and structurally adapted to the requirements of an enormously high energy turnover. The tissue diffusion capacities are mainly determined by capillary surface area and intercapillary distances and, hence, capillary densities. Heart muscle of *S. etruscus* contains 7,300 capillaries/mm^2 (papillary muscle), compared with 2,700/mm^2 in human heart (3). The capillary density of M. soleus is 2,800/mm^2, compared with 290/mm^2 in man (4). These data lead to mean maximal diffusion distances of 6.6 μm in shrew heart muscle (compared with 10.8 μm in humans), and 10.7 μm in shrew soleus muscle (versus 33.1 μm in humans).

The high metabolic rate in shrews is associated with rapid muscle contractions. At 37°C, a contraction cycle is completed within 13 msec in the extensor digitorum longus and within 18 msec in the soleus muscle, allowing these muscles to contract at extremely high contraction frequencies. Skeletal muscles of *S. etruscus* contract up to 900/min in respiratory muscles (diaphragm), up to 780/min during running, and up to 3,500/min during shivering (10). In contrast to larger mammalian species, in which muscles used for locomotion (fast twitch muscles) have different fiber types and different myosin heavy chain patterns compared with muscles predominantly used for supporting the weight of the body (slow twitch muscles), all skeletal muscles of the Etruscan shrew completely lack slow-twitch type I fibers and consist of only fast-twitch type IID fibers. IID fibers seem to be most appropriate when a combination of a high shortening velocity and a high aerobic capacity is required. IIB fibers show even higher shortening velocities than do IID fibers, but rely predominantly on glycolytic metabolism. IID fibers are considerably thinner in diameter than IIB fibers; they exhibit smaller distances for O_2 diffusion. Therefore, IID fibers seem to be the best compromise between a high adenosine triphosphate (ATP) turnover, which is required not only for physical performance but also for the heat production needed to cope with the great heat loss of a small mammal, and a small fiber diameter which, together with a high capillary density, makes a high O_2 flux into the fiber and high ATP production possible (10).

THERMOGENESIS AND BROWN ADIPOSE TISSUE

During food shortage or at low ambient temperatures (T_a), the Etruscan shrew can lower its body temperature (T_b) and enter a torpid state to reduce its resting energy expenditure. During torpor, the T_b of the shrew has been shown to drop on average to 12°C at a T_a of 4°C (with levels falling to as low as 6°C in some animals) (17). Heart frequency decreased to 100/min and respiratory rate to 50/min at 12°C. A mean warm-up rate of nearly 1°C/min in the body temperature range of 12°C to 37°C was measured in *S. etruscus*, which is among the highest values measured in heterothermic mammals (17). Because active rewarming from torpor, like exercise, requires a high metabolic turnover, the mass-specific O_2 transport rate is also very high under these conditions. Estimation of the mixed venous O_2 partial pressure showed that, at the tissue level, the rewarming process corresponds to heavy work conditions.

Below a T_b of 17°C, shivering does not significantly contribute to rewarming, as shown by electrocardiograph/electromyograph (ECG/EMG) recordings. In this temperature range, nonshivering thermogenesis by brown adipose tissue (BAT) is responsible for heat production and rewarming. It is useful for this purpose that most of the fat mass of *S. etruscus* consists of BAT. On average, it amounts to nearly 10% of total body mass, the highest value reported for a mammal. After the onset of shivering, the rate of rewarming increases, and EMG recordings show up to 3,500 bursts per min, probably caused by asynchronous firing of several motor units. The ventilatory system is adapted to such an extent that, during rewarming from torpor, in addition to the appropriate O_2 transport capacity, a capacity also exists for hyperventilation. This is necessary because, as known from other heterothermic mammals, animals in the phase of rewarming from torpor have to compensate for acidosis caused by CO_2 accumulation during transit into torpor (17).

LIFESPAN AND REPRODUCTION

Allometric investigations show that, within many mammalian orders, maximal lifespan (LS) increases with body mass (M) proportional to $M^{0.2}$ (1). Metabolic rates (MR) of mammals (basal [BMR], mean active, or field as well as maximal MR), scale with body mass according to $M^{0.7}$ to $M^{0.8}$. Thus, the product LS × MR, which is the amount of energy consumed in a mammal during maximal lifespan, is roughly proportional to M, or, the mass-specific lifetime energy expenditure LEE = LS × MR/M is independent of body mass. For the LEE of mammals, a mean value of 5,000 MJ/kg is calculated (1). A possible mechanism underlying this relation is the production rate of reactive oxygen species (ROS), which increases with metabolic rate or O_2 consumption, respectively. ROS damage membranous and cytosolic cell components and, although antioxidants and repair mechanisms continuously fight against the processes impairing cell function, a steady accumulation of cell aging events seems to occur. Due to the shrew's extreme O_2 consumption rate, its LEE is used up within a short time. Despite being equipped with an outstanding antioxidant capacity, the maximal lifespan of *S. etruscus* occurring in wild-living animals is only 19 months, whereas under ideal laboratory conditions 26 months have been observed (9).

In terms of survival of the species, this short lifespan is compensated for through a very high reproduction rate. On average, 3.5 (maximal six) litters per year with a mean litter size of four newborns have been recorded. The altricial newborns are suckled until they have reached their adult body mass 3 weeks after birth (9). On average, a female shrew produces offspring equivalent to 20 times its own body mass. Thus, in addition to the energy expenditure required for maintaining its own homeostasis, a considerable amount of lifetime energy expenditure is needed for gestation and lactation – for the growth and homeostasis of offspring.

KEY POINTS

- The smallest mammal, the shrew *S. etruscus*, exceeds all other animals with respect to its maximal O_2 consumption and the performance of its O_2 transport cascade.
- Maximal ventilatory, diffusional, and circulatory O_2 transport capacities are matched in a perfect way, supporting the concept of symmorphosis.
- Entrance into torpor is used to save energy. Rewarming times are very short.
- The extreme metabolic rate leads to a large production rate of ROS and a short maximal lifespan, according to the concept of a constant maximal lifetime energy expenditure.

Future Goals

- To investigate whether the large transport rates require special properties of the endothelia

REFERENCES

1 Jürgens KD, Prothero JW. Lifetime energy budgets in mammals and birds. *Comp Physiol Biochem A.* 1991;100:703–709.

2 Jürgens KD. The consequences of being small: Etruscan shrew muscle. *J Exp Biol.* 2002;205:2161–2166.

3 Pietschmann M, Bartels H, Fons R. Capillary supply of heart and skeletal muscle of small bats and non-flying mammals. *Respir Physiol.* 1982;50:267–282.

4 Jürgens KD, Pietschmann M, Yamaguchi K, Kleinschmidt T. Oxygen binding properties, capillary densities, and heart weights in high altitude camelids. *J Comp Physiol B.* 1988;158:469–477.

5 Taylor JRE. Personal communication, publication in progress. 2006.

6 Ochocinska D, Taylor JRE. Living at the physiological limits: field and maximum metabolic rates of the Common shrew (*Sorex araneus*). *Physiol Biochem Zool.* 2005;78:808–818.

7 Epting RJ. Functional dependence of the power for hovering on wing disc loading in hummingbirds. *Physiol Zool.* 1980;53:347–357.

8 Jürgens KD, Fons R, Sender S, Peters T. Heart and respiratory rates and their significance for convective oxygen transport rates in the smallest mammal, the shrew *Suncus etruscus*. *J Exp Biol.* 1996;199:2579–2584.

9 Fons R. Premières données sur l'écologie de la pachyre étrusque *Suncus etruscus* (Savi, 1822) et comparaison avec deux autres Crocidurinae: *Crocidura russula* (Hermann, 1780) et *Crocidura suaveolens* (Pallas, 1811) (Insectivora, Soricidae). *Vie Milieu.* 1975;25:315–360.

10 Peters T, Kubis HP, Wetzel P, et al. Contraction parameters, myosin composition and metabolic enzymes of skeletal muscles of the Etruscan shrew *Suncus etruscus* and the Common European white toothed shrew *Crocidura russula*. *J Exp Biol.* 1999;202:2461–2473.

11 Mathieu O, Krauer R, Hoppeler H., et al. Design of the mammalian respiratory system. VII. Scaling mitochondrial volume in skeletal muscle to body mass. *Respir Physiol.* 1981; 44:113–128.

12 Hoppeler H, Lindstedt SL, Claassen H, Taylor CR, Mathieu O, Weibel ER. Scaling mitochondrial volume in heart to body mass. *Respir Physiol.* 1984;55:131–137.

13 Gehr P, Mwangi DK, Amman A, Maloiy GMO, Taylor CR, Weibel ER. Design of the mammalian respiratory system. V. Scaling morphometric pulmonary diffusion capacity to body mass: wild and domestic mammals. *Respir Physiol.* 1981;44:61–86.

14 Bartels H, Bartels R, Baumann, R, Fons R, Jürgens KD, Wright PG. Blood oxygen transport and organ weights of two shrew

species (*S. etruscus* and *C. russula*). *Am J Physiol.* 1979;236:R221–R224
15 Jürgens KD, Bartels H, Bartels R. Blood oxygen transport and organ weights of small bats and small non-flying mammals. *Respir Physiol.* 1981;45:243–260.
16 Taylor CR, Weibel ER. Design of the mammalian respiratory system. I. Problem and strategy. *Respir Physiol.* 1981;44:1–10.
17 Fons R, Sender S, Peters T, Jürgens KD. Rates of rewarming, heart and respiratory rates and their significance for oxygen transport during arousal from torpor in the smallest mammal, the Etruscan shrew *Suncus etruscus*. *J Exp Biol.* 1997;200:1451–1458.

14

Molecular Phylogeny

John H. McVey

MRC Clinical Sciences Centre, Imperial College, London, United Kingdom

An important challenge in studying evolutionary biology is to reconstruct the past based primarily on evidence from the present by studying the comparative biology and molecular phylogeny of present-day organisms. Studying the fossil record may help fill some of the gaps in our understanding. Unfortunately, the cardiovascular system does not fossilize; therefore, any rendition of the evolutionary history of the endothelium is at best speculative. In constructing a hierarchical pattern of evolution based on the comparative analysis of structure and function, it is assumed that two structures that look similar are closely related, however, structures also may look similar because they have evolved similar adaptations, so called *convergent evolution*. For example, the arthropod *Tachypleus tridentatus* (Japanese horseshoe crab) contains a clotting system consisting of a protease cascade of three serine protease zymogens and a clottable protein (coagulogen). The components of this cascade are stored in granules in hemocytes and released in response to foreign substances such as lipopolysaccharide (LPS). Hemolymph coagulation in the horseshoe crab shares many common features with the vertebrate-clotting system (cascade of serine protease zymogens, clottable protein, and LPS response), but the proteins involved are quite different and have arisen independently of each other. Conversely, structures that no longer look similar nor appear to share a common function may have arisen from a common ancestor through *divergent evolution*. Similar pitfalls befall the analysis of molecular sequences: Because genomes have evolved through duplication events, many coding sequences may share a degree of sequence identity with each other; however, this does not mean that they share a common function. For example, it is widely believed that large regional or whole genome duplications have contributed to the structure of vertebrate genomes (1,2). The mechanisms by which these have arisen are controversial, but it has been proposed that the evolution of the vertebrate genome involved two whole-genome duplication events that are thought to have occurred early in the vertebrate lineage, some 500 million years ago (3). However, local duplication and translocation also could account for the evolution of multigene families on separate chromosomal regions, whereas other genetic events, such as gene conversion and gene loss, can further complicate our understanding of duplicate gene evolution. (To clarify certain terms used in evolutionary comparative analysis, a glossary is provided at the end of this chapter.)

Central to a discussion of molecular phylogeny is the distinction between orthologous genes (*orthologues*) and paralogous genes (*paralogues*). The degree of sequence identity may make it difficult to differentiate orthologous genes in different species, which have evolved directly from a common ancestral gene, from paralogous genes, which have evolved from a common ancestral gene through duplication and therefore share common structural features, but have since diverged from the parental copy through mutation and selection or drift and have completely diverged functions. The sequence composition of the genome of a particular organism leading to *codon usage bias* may further complicate analysis. For example, the genome of the lamprey (*Lamptera fluviatilis*) is very GC-rich: Three members of the vitamin K-dependent blood coagulation factor family, sharing the domain structure Gla-EGF1-EGF2-SP, were recently identified (4,5). Phylogenetic sequence comparisons, however, all suggested that these sequences were more related to each other (because of sequence composition bias) rather than other known mammalian coagulation factors, making it impossible to assign them to appropriate orthologous groups and thus specify function.

Evolution, while providing a solution to one problem, may also lead to a new set of challenges. For example, the development of a high-pressure cardiovascular system, although providing a means for overcoming the oxygen (O_2) transport problem and thereby paving the way for the evolution of larger animals, introduced the potential life-threatening consequence of rapid exsanguination following vascular damage and the potential entry and dissemination within the organism of life-threatening pathogens. The organisms we see today reflect multiple, parallel evolutionary events that may have arisen independently but have coevolved because of their reliance on each other. Thus, all extant vertebrates have a similar vascular network, specialized blood cells that transport O_2 and fight infection, and a coagulation network that provides a rapid localized response at the site of vascular damage. The endothelium in man has coevolved to provide a complex series of physiological functions that include the control of vasomotor tone, trafficking of cells and nutrients, maintenance of blood fluidity, regulation of permeability, and formation of new blood vessels, among other functions. The endothelium expresses appropriate receptors or molecules that interact with blood cells (lymphocytes, monocytes, platelets) and complex molecular systems (for example, the blood coagulation network) to maintain normal blood flow and exclude blood cells from peripheral tissues. Upon vascular injury or inflammation, this same system responds by promoting cell adhesion, extravasation, and/or controlled fibrin deposition to prevent blood loss.

The endothelium cannot, however, be regarded as a simple homogeneous cell type. Endothelial cell (EC) phenotypes are differentially regulated at any given point in time, at any given location. Endothelial heterogeneity occurs between different organs, within the vascular loop of a given organ, and even between neighboring ECs of a single vessel. Indeed, it has been suggested that it is more realistic to consider an EC as an adaptive input–output device responding to the local environment in which it is located (6). If it is so difficult even to directly compare ECs within a single organism, it is likely that comparisons across different species may be even more difficult.

At the gross anatomical level, it is clear that all extant jawed vertebrates possess a similar vascular system, which at least in part fulfils a similar role to that of man. At the molecular level, it also is clear that all jawed vertebrates share at least some of the differentiated features found in man. This chapter reviews approaches for studying molecular phylogeny and illustrates how, using molecular phylogenetic analysis of the vascular endothelial growth factor (VEGF) family, one can study the evolution of the endothelium.

PHYLOGENETIC ANALYSIS

The starting point for the construction of a multiple-sequence alignment that optimally defines or represents a protein family is an analysis of current sequence databases (for example, NCBI's nonredundant [nr] database) using sequence analysis tools such as TBLASTN or Position-Specific-Iterated BLAST (PSI-BLAST) (http://www.ncbi.nlm.nih.gov/BLAST/) (7–9). Translated DNA-protein searches are capable of finding more homologues than searching by DNA-DNA comparison because of the redundancy of the genetic code. These two methods work in very different ways: TBLASTN searches by comparing the query protein sequence against a nucleotide database translated dynamically into all six reading frames. In contrast, PSI-BLAST takes a query protein sequence and compares it to a protein database; it generates a profile from the results of the search, which is then applied to subsequent searches in an iterative manner. As new sequences are added, the profile is refined. The process is repeated until no new relationships are identified. The PSI-BLAST search is an extremely powerful tool that can identify distantly related sequences with less than 20% amino-acid sequence identity. To refine the approach, it may be useful to select a subset of the query sequences encoding a known conserved functional domain to maximize the number of significant scores. Web-based tools, such as SMART (http://smart.embl-heidelberg.de/) (10,11), can aid the identification of conserved functional domains within the sequence of interest. The aim is to identify distant relatives of a given sequence while minimizing the number of false positives. All BLAST-based programs assess the statistical relevance of sequence similarities through an E-value, which is defined as the number of sequences expected to have a similar or better score by chance alone. The current default threshold for the NCBI Web-based PSI-BLAST analysis is 0.005, which is a 1 in 200 chance of obtaining a false positive. Those sequences that share significant sequence identity and are thought to represent orthologous sequences or genes may then be aligned using tools for multiple sequence alignment (for example, the ClustalW program [12]) and then optimized either manually or using other tools (for example, Gblocks http://molevol.ibmb.csic.es/Gblocks/Gblocks.html [13]) if necessary. These multiple alignments then can be used to infer phylogenetic trees.

Two crucial steps must be taken in phylogenetic inference: the identification of *homologous characters* and tree construction. In phylogenies based on morphological data, the reliability of a phylogenetic tree depends on the quality of the assignment of homologous characters; similarly, in phylogenetic tree building, the reliability depends on the quality of the multiple sequence alignment. A high-quality alignment containing accurate and carefully chosen sequences is the single most important prerequisite for secure phylogenetic analysis.

Three reconstruction methods commonly are used for inferring phylogenies: distance-based methods, maximum parsimony, and likelihood methods. *Distance-based methods* first convert the sequence alignment into a distance matrix that represents the evolutionary distances between all pairs of sequences in the data set. The phylogenetic tree then is inferred from this distance matrix using algorithms such as neighbor joining (NJ). Pairs of sequences are compared, and those pairs with the lowest distance score (i.e., having the shortest distance) are linked to form a *node*. This pair is then treated as a single unit; the process is repeated to identify the sequence with the least distance score from this node until all sequences have been placed in the tree. In contrast, the *maximum parsimony methods* consider all possible evolutionary trees that could explain the observed sequence relationships and then select the tree that requires the minimum number of character changes. Finally, *likelihood methods* create all possible trees and then use statistical methods to evaluate the probability that a given tree could have produced the observed data. This is possible for small data sets; however, for large numbers of sequences the number of trees generated becomes so large that computational analysis becomes limiting.

No ideal method exists at present, and the best tree from any program may not represent the "true" tree. Once a tree has been generated, the reliability of the tree can be assessed using statistical indices, such as bootstrap percentages obtained using bootstrap analysis and Bayesian posterior probabilities. The *bootstrap* procedure randomly samples the original data set to generate new pseudoreplicate data sets, which then are analyzed. Bootstrapping typically involves resampling subsets of data 1,000 times; the bootstrap value is the frequency with which a given branch occurred exactly within that topology. Branches having bootstrap values of 95% to 100% indicate a high level of confidence, but values as low as 70% often are cited as reliable.

VASCULAR ENDOTHELIAL GROWTH FACTOR FAMILY

In the vertebrate embryo, the development of a vascular system involves two fundamental processes: vasculogenesis and angiogenesis. VEGFs are members of a family of growth factors known as the cystine-knot super family (14). These factors are characterized by the presence of eight conserved cysteine residues essential for embryonic vasculogenesis and angiogenesis and for neovascularization in cancer and other diseases (reviewed in [15]). In mammals, VEGF-A is the prototypical member of a family of related growth factors that includes placental growth factor (PlGF), VEGF-B, VEGF-C, and VEGF-D. The biological functions of these growth factors are mediated by a family of cognate protein tyrosine kinase receptors (VEGFRs): VEGF-A binds to VEGFR1 (also known as FLT-1) and VEGFR2 (also known as FLK-1); VEGF-B and PlGF bind only to VEGFR2; and VEGF-C and VEGF-D both bind to VEGFR2 and VEGFR3 (also known as FLT-4). VEGF-A has a critical role in vascular development as evidenced by the observation that targeted disruption of the murine gene results in impairment of angiogenesis resulting in death between embryonic day (E)11 and E12. In contrast, a biological role for VEGF-B has not been clearly established. VEGF-C and VEGF-D are strongly implicated in lymphangiogenesis. Overexpression of VEGF-C and VEGF-D in transgenic mice induces the formation of hyperplastic lymphatic vessels. Conversely, inhibition of VEGF-C and/or VEGF-D by overexpression of a soluble (nonsignaling) form of VEGFR3 leads to inhibition of lymphatic vessel growth. PlGF appears to be weakly angiogenic; mice lacking PlGF are viable and develop normally but show reduced angiogenesis in pathophysiological situations such as ischemia.

A simple text search of the Entrez database (http://www.ncbi.nlm.nih.gov/gquery/gquery.fcgi) identified the human VEGF amino acid reference sequences for VEGF-A, -B, -C, -D, and PlGF (GI: 76781481, 4507887, 4885653, 4758378, and 20149543, respectively). These paralogous sequences were then aligned using ClustalW. This sequence comparison identifies conserved sequence motifs essential for the structure-function of the VEGF family. The highly conserved cystine-knot domain is clearly identified (Figure 14.1). The consensus sequence for the cystine-knot motif is C_1-$(X)_n$-C_2-X-G-X-C_3-$(X)_n$-C_4-$(X)_n$-C_5-X-C_6, in which cysteines 2, 3, 5, and 6 form a ring by disulfide bonding between C_2–C_5 and C_3–C_6. The third disulfide bond formed by C_1–C_4 penetrates the ring, forming a knot (Figure 14.2). In the VEGF family, two additional highly conserved cysteine residues exist. Each of the reference sequences (VEGF-A, -B, -C, -D, and PlGF) was then

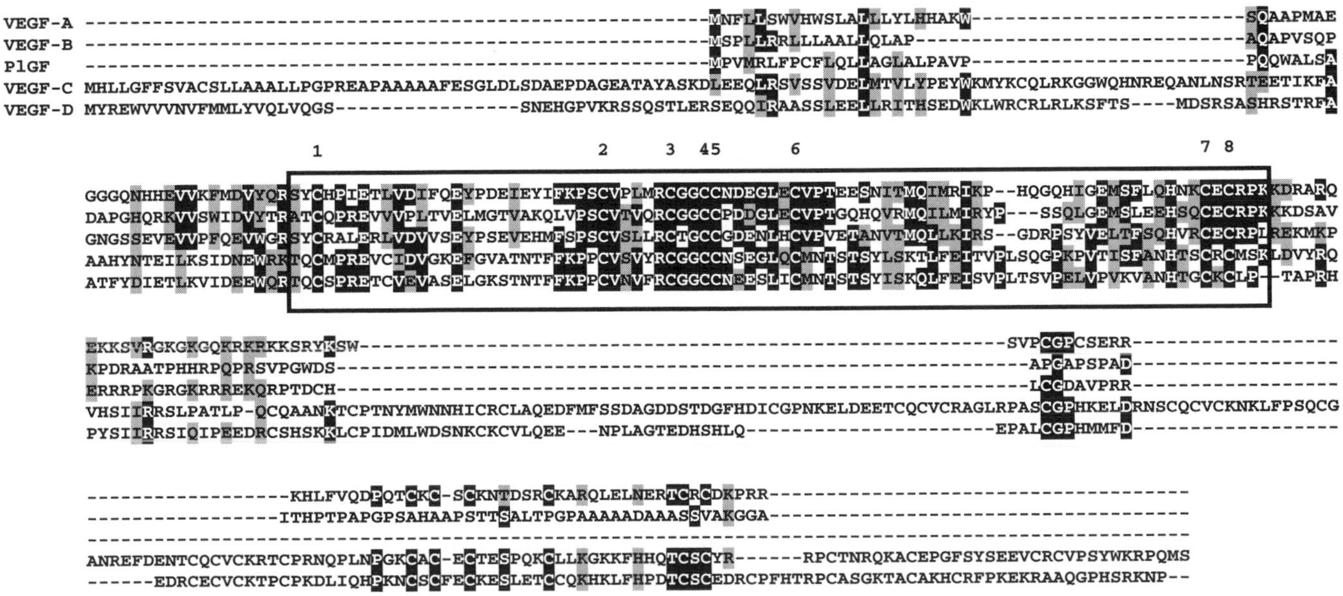

Figure 14.1. Amino acid alignment of the human VEGF family sequences. The sequences were aligned using ClustalW multiple sequence alignment program and displayed using Bioedit (http://www.mbio.ncsu.edu/BioEdit/bioedit.html). *White on black* indicates identity with the consensus; *black on gray* indicates similarity with the consensus. The cystine-knot motif is boxed, and the eight conserved cysteine residues are numbered.

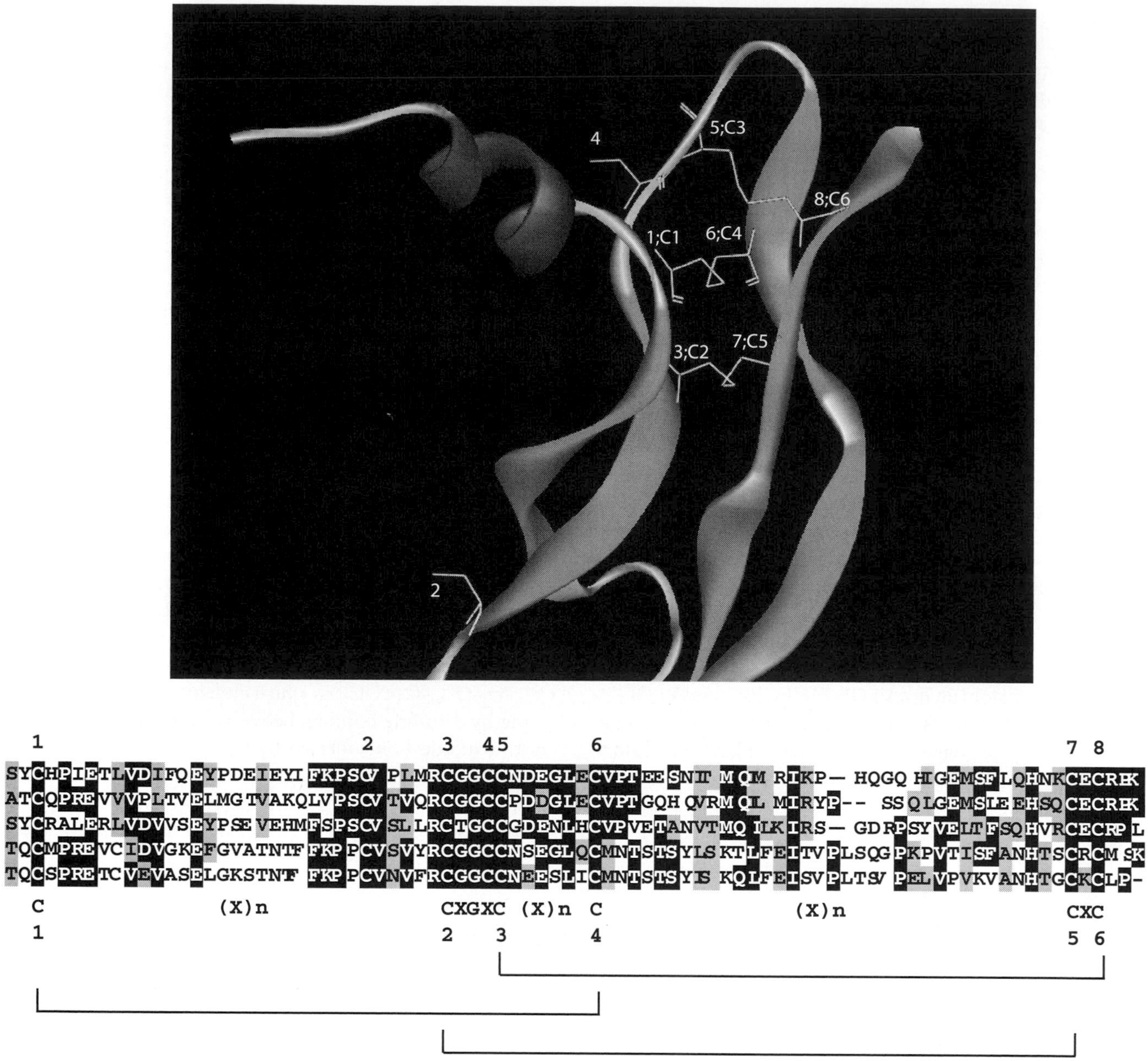

Figure 14.2. Molecular model of VEGF-A generated from the structure 1VPP (http://www.wwpdb.org/index.html) using WebLab Viewer Lite (http://www.accelrys.com/). The disulfide bonds Cys57–Cys102 (C_2–C_5), Cys26–Cys68 (C_1–C_4), Cys61–Cys104 (C_3–C_6), and the cysteine residues Cys51 and Cys60 are shown in stick representation. The alignment of the conserved cystine-knot of the human VEGF-family members VEGF-A, -B, PlGF, -C and -D generated by ClustalW is shown with the conserved cysteines numbered 1–8. The consensus sequence for the cystine-knot is shown below the alignment, and the disulfide bonds are indicated by connecting lines.

used in a PSI-BLAST of the NCBI nr database and, after five or six iterations, no new sequences with significant alignment were obtained. Limiting the search to the conserved domain retrieved fewer sequences by reducing the number of non-relevant sequences. The conserved cystine-knot domain of all five human reference sequences generated the same output sequence data set. The reference sequences identified were then aligned using the complete sequence rather than limiting the alignment to just the cystine-knot sequence, and an unrooted neighbor-joining tree was generated by ClustalW (Figure 14.3).

This tree demonstrates the relationship of the cystine-knot family members across a range of species. Each sequence is assigned to a distinct *clade* or family based on the multiple sequence alignment. All the sequences annotated as platelet-derived growth factors (PDGFs) from the various species are grouped together in their respective families and form a distinct group separate from the VEGF family members. Within the VEGF family, five distinct groups are obtained with all the annotated sequences apparently assigned to their appropriate clade. However, the reader should be aware that the annotation of most sequences in the available databases is

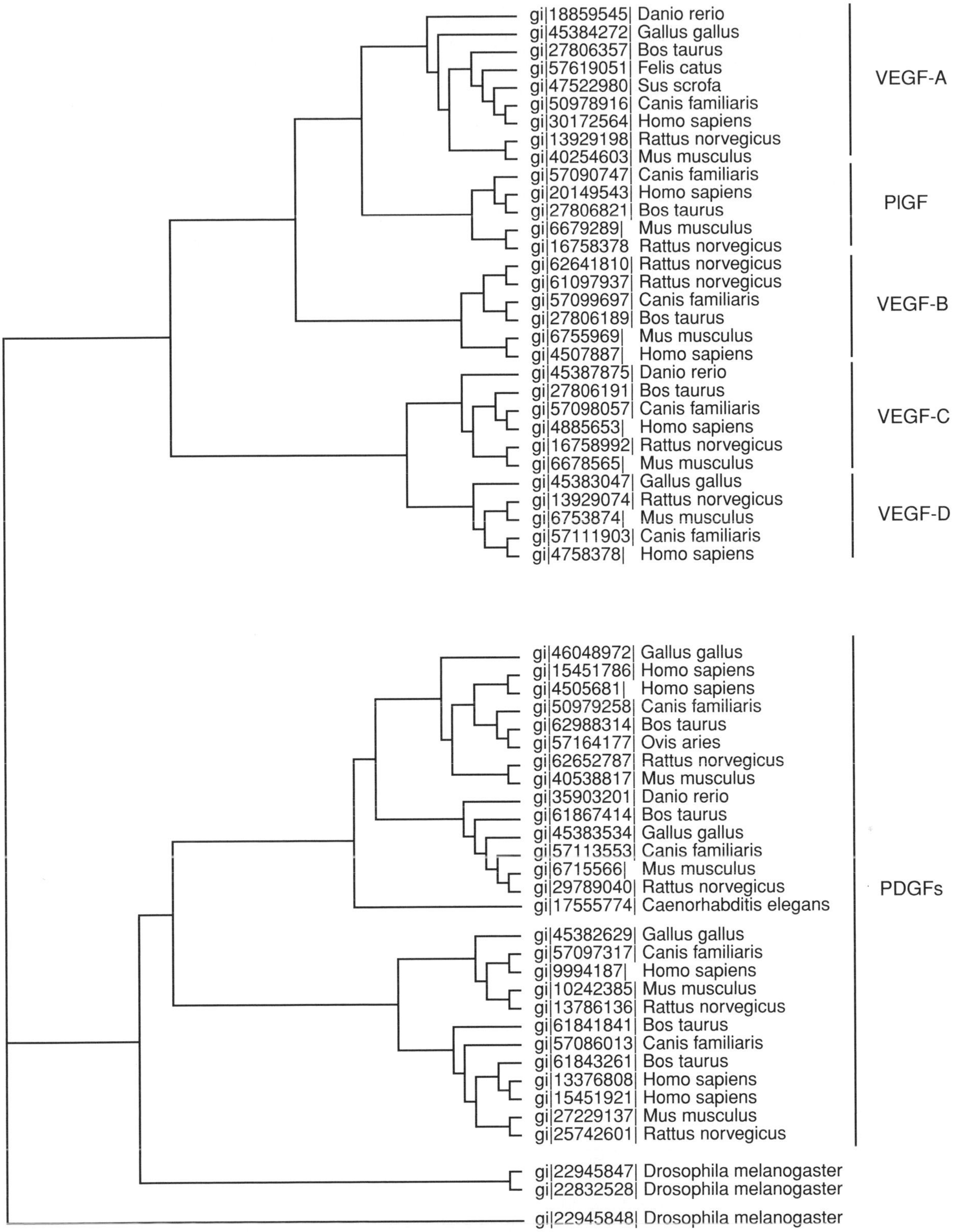

Figure 14.3. Neighbor-joining tree inferred from the alignment of sequences with sequence similarity to the cystine-knot of the human VEGF family, illustrating the phylogenetic relationship of sequences. The sequences were aligned using ClustalW, following a PSI-BLAST search with the boxed sequences from Figure 14.1. The figure was generated by TREEVIEW (http://taxonomy.zoology.gla.ac.uk/rod/treeview.html; [21]). The GI accession number for each sequence and the species of origin is given. The VEGF clades are assigned based on the annotations of the human amino acid sequences.

based on exactly this sort of automated analysis, rather than on any kind of functional analysis. Therefore, this concordance of *clade* with annotation is, to a certain extent, a self-fulfilling exercise! To establish ancestor–descendant relationships in phylogenetic analysis it is necessary to root a tree with an *out-group* sequence—that is, a distantly related sequence that still shares some sequence identity. For example, the *Danio rerio* (zebrafish) PDGF sequence could be used to root a VEGF tree; however, this will limit the sequence alignment to the cystine-knot domain only, because that is the sequence information the out-group sequence shares with the sequences to be analyzed. In this particular case, however, the rooted tree had the same topology as the unrooted tree. Both trees suggest that, at the amino acid sequence level, VEGF-A and -B are most closely related to each other, and VEGF-C and -D are most closely related to each other, consistent with their known functional relationship. PlGF appears to be most closely related to VEGF-A and may have been a more recent evolutionary acquisition, because only mammalian species have VEGF sequences assigned to this clade. However, without the complete genome sequence of other vertebrates, this is only a hypothesis. Indeed, soon after running these analyses, the NCBI database was updated and additional sequences with sequence identity to the cystine-knot of human VEGF were identified in *D. rerio*, thus illustrating the need to continually reanalyze available sequence databases. Two of these new sequences were annotated as "predicted: similar to PlGF" (GI: 68388133 and 68436095). Sequence alignment and phylogenetic analysis confirm that these sequences are most similar to the known PlGF sequences; however, close inspection of pairwise sequence comparisons reveals that the sequence identity is limited to the cystine-knot (Figure 14.4); thus, these sequences may not represent *D. rerio* orthologues of PlGF.

Further analysis of the intron–exon structure of a gene is a useful approach to provide additional supporting evidence, because orthologous genes would be expected to share similar gene structures. For example, the genomic organization of the mammalian *VEGF-A* genes is highly conserved. A *D. rerio* "*VEGF-A* gene" has been characterized and shown to have a similar organization to the human gene, thus supporting the assignment based on sequence comparisons (16). Ultimately, functional characterization will be required to

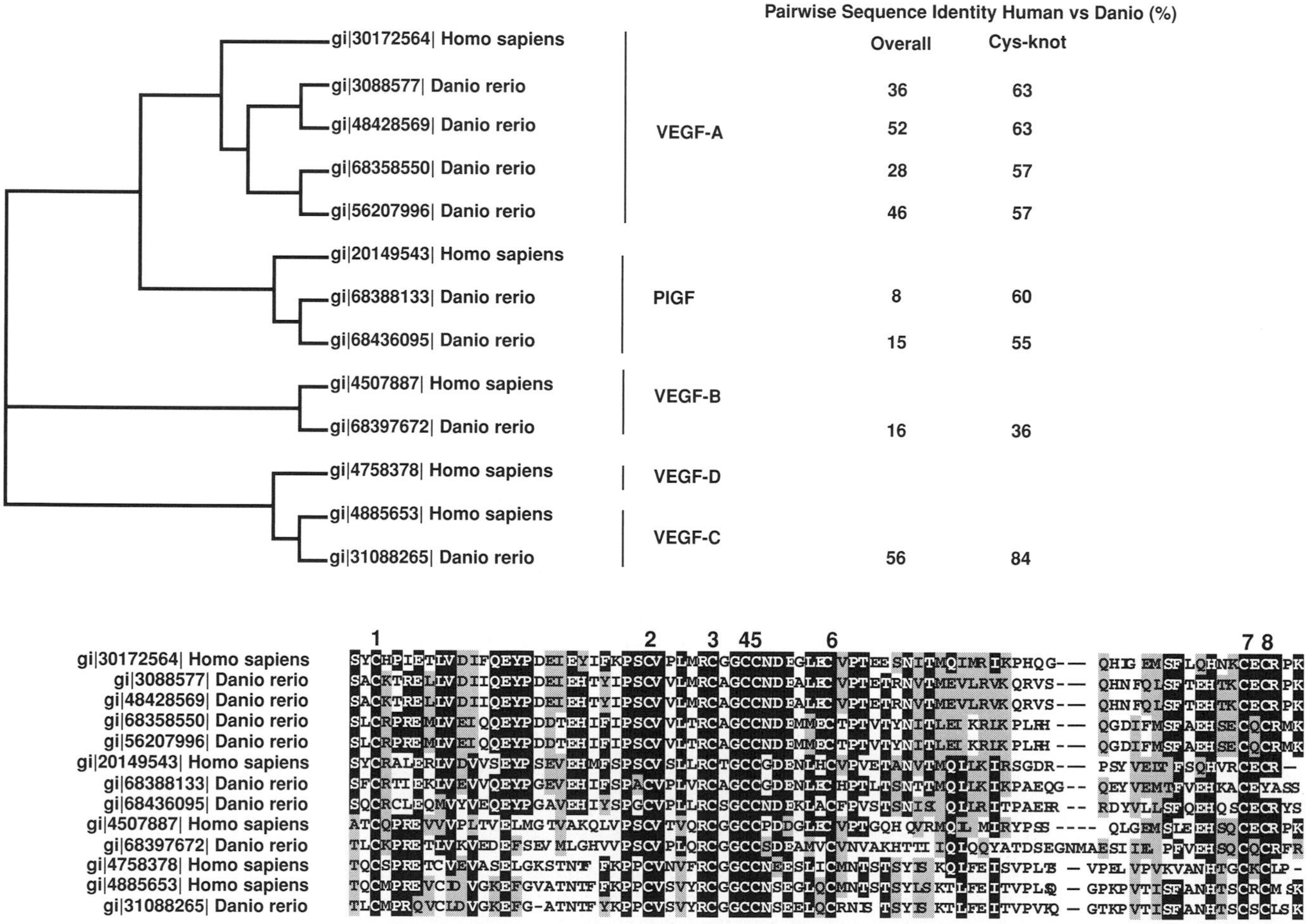

Figure 14.4. Comparative sequence analysis of human and *Danio rerio* VEGF-related sequences. The sequences were aligned using ClustalW, following a PSI-BLAST search. The GI accession number for each sequence and the species of origin is given. The clades are named according to the human amino acid sequences. The amino acid alignment of the cystine-knot is shown with the conserved cysteines numbered 1–8. The sequences are displayed in the same order as presented in the tree. The pairwise sequence identity was calculated using a PAM250 scoring matrix using BioEdit.

unambiguously assign these sequences. It also must be noted that these sequences are predicted to encode expressed proteins using automated gene prediction methods, and they may not encode a functional protein as the process of genome evolution proceeds through a process of gene duplication and gene loss.

The analysis of the additional *D. rerio* sequences demonstrates the presence in the *D. rerio* genome of at least four genes predicted to encode proteins with significant sequence identity to mammalian VEGF sequences (see Figure 14.4), with all eight cysteine residues and their spacing within the cystine-knot conserved, with the exception of 68388133, which lacks cysteine 8. Because cysteine 8 (C_6 in the cystine-knot consensus) forms one of the essential disulfide bonds (C_3–C_6; see Figure 14.2), it is highly likely that this sequence does not represent a functional VEGF molecule. The presence of multiple VEGF-like sequences in *D. rerio* suggests this gene family evolved prior to the divergence of bony fish from tetrapods over 430 million years ago. This is perhaps not surprising given the highly conserved vascular system found in all vertebrates and the importance of the VEGF family in vascular development. However, further analysis will be required to demonstrate that these molecules have identical functions in mammals and teleosts.

Proteins with sequence identity to mammalian VEGFs also have been identified in several invertebrate species, *Drosophila melanogaster* (fruit fly; Pvf1, GI: 22832528; Pvf2, GI: 22945847; Pvf3, GI: 22945848), *Caenorhabditis elegans* (nematode; GI: 17555774), and *Podocoryne carnea* (jellyfish; GI: 45934767) (17). *D. melanogaster* has three genes predicted to encode proteins with sequence similarity to human VEGF sequences (Figure 14.5). The sequence similarity is primarily confined to the cystine-knot and is relatively low; however, all eight cysteine residues are conserved, with the exception of Pvf2 (GI: 22945847), which lacks cysteine 2. This cysteine does not contribute to the cystine-knot (see Figure 14.2) and therefore this may still represent a functional VEGF-like molecule. Molecular phylogenetic analysis fails to assign the *D. melanogaster*

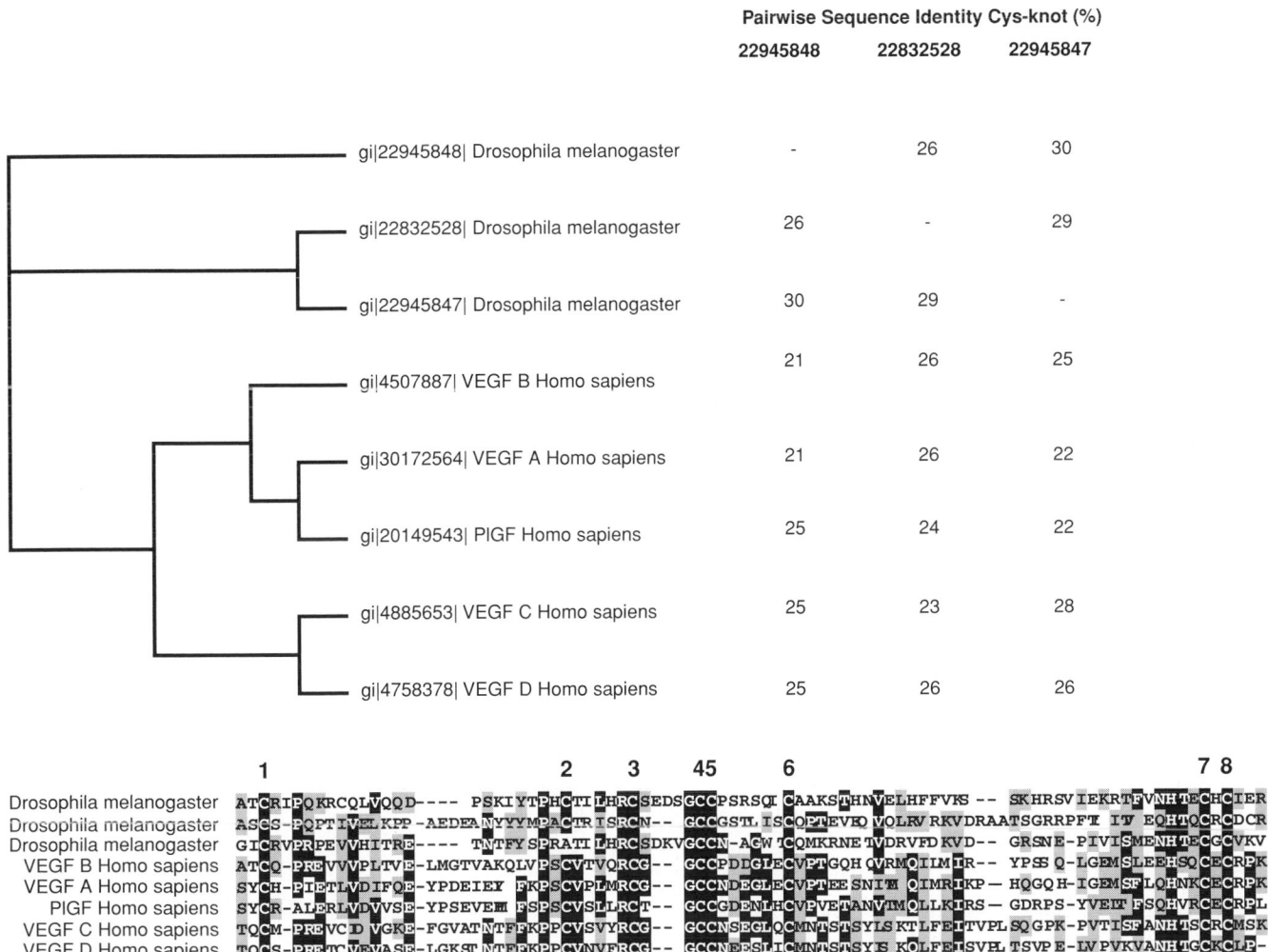

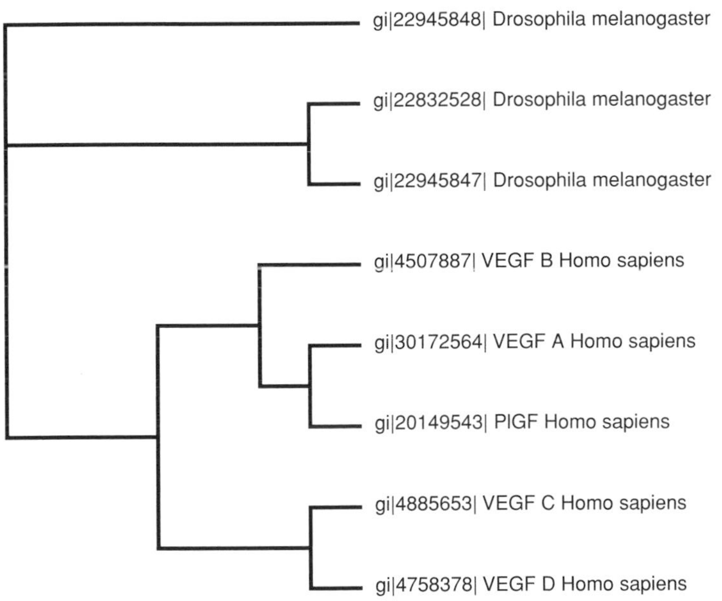

Pairwise Sequence Identity Cys-knot (%)

Sequence	22945848	22832528	22945847
gi|22945848| Drosophila melanogaster	-	26	30
gi|22832528| Drosophila melanogaster	26	-	29
gi|22945847| Drosophila melanogaster	30	29	-
gi|4507887| VEGF B Homo sapiens	21	26	25
gi|30172564| VEGF A Homo sapiens	21	26	22
gi|20149543| PlGF Homo sapiens	25	24	22
gi|4885653| VEGF C Homo sapiens	25	23	28
gi|4758378| VEGF D Homo sapiens	25	26	26

```
                             1                              2   3    45      6                                              7 8
Drosophila melanogaster  ATCRIPQKRCQLVQQD---- PSKIYTPHCTILHRCSEDSGCCPSRSQI CAAKSTHNVELHFFVKS -- SKHRSV IEKRTFVNHTECHCIER
Drosophila melanogaster  ASCS-PQPTIVELKPP-AEDEANYYYMPACTRISRCN-- GCCGSTL ISCQPTEVEQ VQLRV RKVDRAATSGRRPFTI ITV EQHTQCRCDCR
Drosophila melanogaster  GICRVPRPEVVHITRE---- TNTFYSPRATILHRCSDKVGCCN-AGW TCQMKRNETVDRVFDKVD-- GRSNE-PIVISMENHTECGCVKV
VEGF B Homo sapiens      ATCQ-PREVVVPLTVE-LMGTVAKQLVPSCVTVQRCG-- GCCPDDGLECVPTGQH QVRMQIIMIR-- YPSS Q-LGEMSLEEHSQCECRPK
VEGF A Homo sapiens      SYCH-PIETLVDIFQE-YPDEIEY FKPSCVPLMRCG-- GCCNDEGLECVPTEE SNITM QIMRIKP— HQGQH-IGEMSFLQHNKCECRPK
PlGF Homo sapiens        SYCR-ALERLVDVVSE-YPSEVEH FSPSCVSLLRCT-- GCCGDENLHCVPVETANVTMQLLKIRS— GDRPS-YVEIT FSQHVRCECRPL
VEGF C Homo sapiens      TQCM-PREVCID VGKE-FGVATNTFFKPPCVSVYRCG-- GCCNSEGLQCMNTSTSYLS KTLFEITVPLSQGPK-PVTISFANHTSCRCMSK
VEGF D Homo sapiens      TQCS-PRETCVEVASE-LGKST NTFFKPPCVNVFRCG-- GCCNEESLICMNTSTSYIS KQLFEISVPL TSVPE-LVPVKVANHTGCKCLP-
```

Figure 14.5. Comparative sequence analysis of human and *Drosophila melanogaster* VEGF-related sequences. The sequences were aligned using ClustalW, following a PSI-BLAST search. The GI accession number for each sequence and the species of origin is given. The amino acid alignment of the cystine-knot is shown with the conserved cysteines numbered 1–8. The sequences are displayed in the same order as presented in the tree. The pairwise sequence identity was calculated using a PAM250 scoring matrix using BioEdit.

sequences to any particular clade relative to the characterized mammalian sequences (see Figures 14.3 and 14.5), and comparison of the exon-intron structure of the genes does not provide any additional insight into any possible relationship between the *D. melanogaster* genes and the mammalian VEGF genes. Is it possible to infer a functional role for these molecules in *D. melanogaster* based on our knowledge of their function in mammals? Conversely, is it possible to infer functional roles for VEGF molecules in mammals based on experimental data obtained from *D. melanogaster*? To infer a functional role for a mammalian gene from studies of a gene in *D. melanogaster*, one must be confident about their orthologous relationships. *D. melanogaster* does not have ECs or blood vessels, but has an open circulatory system, with an aorta and heart represented by a contractile tube lined by a layer of myoepithelial vascular cells called *cardioblasts*. It is therefore intriguing that, in *D. melanogaster*, VEGF appears to play a role in blood cell migration and differentiation; recent analysis of *D. melanogaster* lymph gland development has provided evidence for a "hemangioblast"-like cell that can differentiate into heart, aorta, or blood cells (18,19). Disruption of the *Vegfr2* gene in mouse suggests a close association between endothelial and hematopoietic cells in the mouse embryo. Furthermore, human hematopoietic cells purified from cord-blood or bone marrow on the basis of CD34 and *VEGFR2* expression generate both haematopoietic and ECs in vitro.

In contrast, in *C. elegans*, an organism completely devoid of a vascular circulatory system, receptors resembling the VEGFR of vertebrates were expressed in cells of neural origin (20). Expression of VEGFR molecules in neural cells occurs in mammals but not in *D. melanogaster*. Whereas in *P. carnea*, a member of the most basal phylum of the animal kingdom having tissue organization and a complex nerve net but no blood cells or blood vessels, the VEGF and VEGFR-like genes are expressed in the tentacles and developing radial and ring canals. These structures share a common tubular formation, and the genes appear to play a role in tube formation (17). It is tempting to suggest that during evolution these molecules may have been recruited for the purposes of cell migration and cell adhesion in similar processes (for example, neurogenesis, angiogenesis, hematopoiesis) in different species. However, at present it is not possible to conclude whether these processes used genes derived from the same ancestors or different genes and, consequently, whether the sequence similarity is due to homology by descent or to convergence.

KEY POINTS

- Molecular phylogenetic analysis and comparative sequence analysis are powerful tools to identify and characterise molecules that may share similar function(s) in diverse organisms.
- If an orthologous relationship can be defined securely between genes in different species, then important insights may be obtained about the function of these molecules; however, it is often the case that orthologues and paralogues cannot be distinguished.
- Comparative sequence analysis can identify amino acid residues critical for the structure and function of these proteins.
- Sequence similarity shared by two proteins does not necessarily imply shared function.
- The addition of complete genome sequences of many more organisms will increase the confidence of phylogenetic analysis, in particular the discrimination between orthologues and paralogues.

REFERENCES

1 Abi-Rached L, Gilles A, Shiina T, Pontarotti P, Inoko H. Evidence of en bloc duplication in vertebrate genomes. *Nat Genet*. 2002;31:100–105.

2 McLysaght A, Hokamp K, Wolfe KH. Extensive genomic duplication during early chordate evolution. *Nat Genet*. 2002;31:200–204.

3 Ohno S. Gene duplication and the uniqueness of vertebrate genomes circa 1970–1999. *Semin Cell Dev Biol*. 1999;10:517–522.

4 Davidson CJ, Tuddenham EG, McVey JH. 450 million years of hemostasis. *J Thromb Haemost*. 2003;1:1487–1494.

5 Davidson CJ, Hirt RP, Lal K, et al. Molecular evolution of the vertebrate blood coagulation network. *Thromb Haemost*. 2003; 89:420–428.

6 Aird WC. Endothelium as an organ system. *Crit Care Med*. 2004; 32(5Suppl):S271–S279.

7 Jones DT, Swindells MB. Getting the most from PSI-BLAST. *Trends Biochem Sci*. 2002;27:161–164.

8 Altschul SF, Koonin EV. Iterated profile searches with PSI-BLAST – a tool for discovery in protein databases. *Trends Biochem Sci*. 1998;23:444–447.

9 Altschul SF, Madden TL, Schaffer AA, et al. Gapped BLAST and PSI-BLAST: a new generation of protein database search programs. *Nucleic Acids Res*. 1997;25:3389–3402.

10 Letunic I, Copley RR, Schmidt S, et al. SMART 4.0: towards genomic data integration. *Nucleic Acids Res*. 2004;32(Database issue):D142–D144.

11 Schultz J, Copley RR, Doerks T, Ponting CP, Bork P. SMART: a web-based tool for the study of genetically mobile domains. *Nucleic Acids Res*. 2000;28:231–234.

12 Thompson JD, Higgins DG, Gibson TJ. CLUSTAL W: improving the sensitivity of progressive multiple sequence alignment through sequence weighting, position-specific gap penalties and weight matrix choice. *Nucleic Acids Res*. 1994;22:4673–4680.

13 Castresana J. Selection of conserved blocks from multiple alignments for their use in phylogenetic analysis. *Mol Biol Evol*. 2000; 17:540–552.

14 Vitt UA, Hsu SY, Hsueh AJ. Evolution and classification of cystine knot-containing hormones and related extracellular signaling molecules. *Mol Endocrinol.* 2001;15:681–694.
15 Ferrara N, Gerber HP, LeCouter J. The biology of VEGF and its receptors. *Nat Med.* 2003;9:669–676.
16 Gong B, Liang D, Chew TG, Ge R. Characterization of the zebrafish vascular endothelial growth factor A gene: comparison with vegf-A genes in mammals and Fugu. *Biochim Biophys Acta.* 2004;1676:33–40.
17 Seipel K, Eberhardt M, Muller P, Pescia E, Yanze N, Schmid V. Homologs of vascular endothelial growth factor and receptor, VEGF and VEGFR, in the jellyfish *Podocoryne carnea. Dev Dyn.* 2004;231:303–312.
18 Mandal L, Banerjee U, Hartenstein V. Evidence for a fruit fly hemangioblast and similarities between lymph-gland hematopoiesis in fruit fly and mammal aorta-gonadal-mesonephros mesoderm. *Nat Genet*. 2004;36:1019–1023.
19 Ward EJ, Skeath JB. Characterization of a novel subset of cardiac cells and their progenitors in the *Drosophila* embryo. *Development* 2000;127:4959–4969.
20 Popovici C, Isnardon D, Birnbaum D, Roubin R. *Caenorhabditis elegans* receptors related to mammalian vascular endothelial growth factor receptors are expressed in neural cells. *Neurosci Lett*. 2002;329:116–120.
21 Page RD. TreeView: an application to display phylogenetic trees on personal computers. *Comput Appl Biosci.* 1996;12:357–358.

GLOSSARY

Convergent evolution Convergent evolution results when unrelated species are acted on by the same selection pressure, and adaptive traits tend to resemble traits in other species without a shared evolutionary lineage. The development of flight in birds and certain insects is an example of convergent evolution.

Divergent evolution Selection pressure may also result in divergent evolution, where related lineages evolve different traits in response to different selection pressures.

Orthologue Orthologues are equivalent genes in different species that evolved from a common ancestor by speciation.

Paralogue Homologous sequences (that is, sequences that share a common evolutionary ancestor) that diverged by gene duplication (within one organism).

Codon usage bias Most amino acids are specified by more than one codon. For example, the codons GCU, GCC, GCA, and GCG all code for the amino acid Alanine and are therefore described as synonymous. However, genes of many organisms use these codons with different frequencies. The bias in synonymous codon usage may be determined by the sequence composition of the organism, in particular the GC content of the genome.

Homologue Homologous proteins have evolved from a common ancestor, and their evolutionary relationship is evident from similarities in sequence, structure, and/or function.

Distance-based methods Method that takes the input data and derives from them some measure of similarity/difference and from this constructs a tree that tries to match this data.

Parsimony methods Method based on the principle that the simplest explanation, the one that requires the fewest hypotheses, is the one most likely to be correct and therefore the one that requires the fewest number of character transformations.

Likelihood methods A statistical method that calculates the probability of the observed data under varying hypotheses, in order to estimate model parameters that best explain the observed data and determine the relative strengths of alternative hypotheses.

Bootstrap analysis A type of statistical analysis used to test the reliability of specific branches in an evolutionary tree.

Clade A lineage (or family) made up of its ancestor and all its descendants (derived from *klados*, Greek for branch).

Out-group A closely related species that is used for comparison; for example, to infer the ancestral versus the derived state of a polymorphism.

15

Darwinian Medicine

What Evolutionary Medicine Offers to Endothelium Researchers

Randolph M. Nesse and Alan Weder

The University of Michigan, Ann Arbor

Almost all our detailed knowledge about the endothelium is proximate knowledge about its structure, development, and functioning, and how its dysfunctions lead to disease. The other, evolutionary half of endothelial biology has been neglected. Some researchers are unfamiliar with the fundamental distinction between proximate and evolutionary questions, and methods for formulating and testing evolutionary hypotheses about the endothelium remain underappreciated. This chapter offers a brief overview of the distinction between evolutionary and proximate explanations, followed by an introduction to evolutionary medicine and its applications to the endothelium.

Evolutionary, or Darwinian, medicine simply brings the power of evolutionary biology to bear on the problems of medicine. Some areas, such as pathogen evolution and population genetics, are well developed. Another aspect, emphasized in Darwinian medicine, is the enterprise of formulating and testing hypotheses about why natural selection has not shaped bodies that are more resistant to disease. The six key reasons for traits that leave bodies vulnerable to disease are reviewed below. Each is illustrated with examples from general medicine, and then with examples from a major disease of the endothelium, atherosclerosis.

TWO SEPARATE BIOLOGICAL QUESTIONS

Ernst Mayr passed away early in 2005 at age 100, shortly after publishing his twentieth book and the last of his 700+ scientific articles. One of his enduring contributions was his dogged emphasis on the importance of distinguishing proximate from evolutionary (sometimes called *ultimate*) biological explanations. His treatise, *The Growth of Biological Thought*, traces these two only-occasionally intersecting threads in biology: one that studies how things work (proximate), the other that studies why organisms are the way they are (evolutionary) (1). His final book, *What Makes Biology Unique*, continues and updates the theme (2). Over and over again, he makes the point that every biological trait needs both a proximate explanation of its structure and operation, and an evolutionary explanation of why it is the way it is. For example, an explanation of the structure of chlorophyll and how it captures photons and traps their energy in chemical bonds is essential. Equally important, however, is an evolutionary explanation of the biological history of chlorophyll and how its particular form gives a selective advantage. As many will know, the story is intriguing, with the machinery for photosynthesis evolving in one-celled organisms about 3.5 billion years ago, probably in the precursors to cyanobacteria. It has been only about 1 billion years, however, since those organisms were incorporated into the eukaryotic cell as chloroplasts. A complete biological explanation of chlorophyll needs both a proximate description of how chloroplasts work, and an evolutionary explanation of the selection forces and phylogenetic pathways that gave them their current characteristics, instead of others.

The distinction between "how" and "why" questions was developed further by Tinbergen, in a seminal article describing his famous "Four Questions" (3). Here, he divides proximate questions into two subtypes: those about the structure of a trait and those about its ontogeny. He also divides evolutionary questions into those about how a trait gives a selective advantage and those about the phylogeny of a trait. Implicit in this classification is the need to address four separate questions when offering a full biological explanation for any trait. All too often, the four questions are inadequately distinguished or incorrectly taken to be alternatives. Sometimes the evolutionary questions are not taken seriously or are ignored completely.

Much of this situation is understandable because education about these aspects of evolutionary biology often is limited, even for some biologists. It remains lamentable nonetheless. Biology would progress faster if evolutionary questions about every trait were addressed systematically. In medicine, this is of particular importance. Many physicians have naïve notions about the body, such as the belief that a normal genome exists, or that the body is designed like a machine, or that its flaws result mainly from the limited powers of natural selection. Asking evolutionary questions reveals that the body is a bundle of trade-offs, with no trait perfect, and substantial vulnerabilities persisting not just because selection is too weak, but also for several other reasons.

DARWINIAN MEDICINE

In the past decade, rapid progress has been made in using evolutionary principles to understand why bodies are so vulnerable to disease (4–8). The obvious explanation, the limits of natural selection in the face of mutation and drift, is certainly important. However, five additional reasons must be considered in any assessment of why a trait appears suboptimal. Each of these reasons deserves separate consideration for medical conditions in general, and for aspects of the endothelium in particular.

Novel Environments

Much, even most, of modern chronic disease arises because we live in environments far different from those our bodies were selected for (see Chapter 16). Paradoxically, our greatest threats now come from our grand success at providing ourselves with all manner of tasty food available with little effort at any time. A third of adults are overweight in many developed countries, and rates of obesity are rising quickly everywhere food is plentiful, giving rise to Syndrome X and all its complications (9). The preferences that shaped our current food supply shape our personal dietary habits. On the African savannah, tendencies to eat fat, sugar, and salt and minimize exercise were generally useful; today, they are fatal (10).

Infection and Vulnerabilities from "Arms Races"

A second major reason why natural selection cannot make bodies that defend themselves better from disease is because natural selection is simultaneously making other organisms better able to penetrate defenses (11). Worse yet, the rate of evolution in bacteria and viruses is orders of magnitude faster than is the rate for slowly reproducing organisms such as humans.

Limits on What Selection Can Do

This brings us to what selection can and cannot do. Some scientists attribute flaws in the organism mainly to the inevitability of mutations and the slow rate at which selection can purge them from the gene pool. Such mutations and the limited speed of selection are indeed significant factors explaining the body's imperfections. However, they are by no means the only or the most important factors. Furthermore, their importance often is overrated by those who hold fundamentally mistaken views, such as that the body is a well-designed machine, or that there exists a single normal genotype/phenotype. The body does work like a machine, but its origins were far different from any design shaped with the forethought of an engineer. It is cluttered with remnants of previous features that limit optimality. Consider the path of the recurrent laryngeal nerve, from the brainstem all the way down to just above the lung, where it wraps around the subclavian artery before ascending again to pass behind the thyroid gland on the way to the larynx. The vas deferens makes an equally circuitous path (12). Such "design flaws" exist because the body is profoundly path-dependent: Just like the arrangement of keys on a typewriter keyboard, once structures are in a particular conformation, changing them can be very expensive or impossible. An automobile designer can move the gasoline tank if it proves vulnerable to rupture, but natural selection can never alter the awkward and dangerous path of childbirth through the narrow ring of the pelvis.

Trade-Offs

Trade-offs are another source of traits that seem less than perfect. Here the problem is the same as that faced by any human designer. No trait can ever be completely perfect because at some point making one trait better will make others worse. Bones could be thicker, but they would then be heavier and less mobile. Joints could be more flexible, but that would risk damage to muscles. The immune system could be more aggressive but only at the cost of constant tissue damage.

Selection Maximizes Reproduction, Not Health

It is a common mistake to think that the body is designed for health, and all disease arises from design flaws. Actually, bodies are products of evolution, not design. They are merely phenotypes that maximize the transmission of genes (13,14). If a gene increases reproductive success, it will spread, even if it causes disease or shortens life. Aging itself may result, in part, from pleiotropic genes that cause aging but that are nonetheless selected for because they give a benefit in youth when selection is strong (15,16). Likewise, much of the increased mortality in males compared to females comes from the reproductive benefits of increased allocation of effort to competition at the expense of tissue repair (17). Increased reproduction at the expense of health does not seem to be an obvious major factor in shaping the endothelium, but men develop atherosclerotic disease more rapidly than do females, almost certainly because sexual selection has resulted in systems that allocate more effort to competition and less to tissue protection and repair in males as compared with females (17,18). In particular, traits that benefit a fetus may be selected for, even if they damage the mother.

Defenses Are Useful Despite Their Costs

Finally, there exist the body's many protective responses such as pain, fever, vomiting, cough, and inflammation. Although they cause suffering and often bodily damage, they are useful responses, not defects. Their costs are trade-offs for the benefits they offer. It is all too easy to imagine that defenses are problems. For instance, the fever, pain, fatigue, and cough that accompany a cold do not seem to be obviously useful. This "clinician's illusion" that they are problems is bolstered because blocking them with drugs seems to cause few complications. However, this is only because the body has so many redundant defense systems, and because the systems that regulate defense expression are shaped to express defenses whenever that is worthwhile. If there is any uncertainty in when to express them, this means a normal system will have many false alarms. Many defenses, such as vomiting or panicked flight, are quite inexpensive compared to the huge cost of not responding to a real danger such as intestinal infection or an approaching tiger. The defense should be expressed whenever the average benefit is greater than the average cost, so many false alarms are normal and to be expected. This has been called "The Smoke Detector Principle," because we tolerate false alarms from burnt toast to ensure that the alarm sounds every single time even a small fire occurs (19). Disrupting defenses is not, however, completely benign; using drugs to block fever, cough, or diarrhea can lead to serious consequences (6).

EVOLUTIONARY ANALYSIS OF ENDOTHELIAL DYSFUNCTION AND ATHEROSCLEROSIS

The healthy endothelium contributes to several processes of obvious adaptive advantage: control of hemostasis, protection of subendothelial structures from circulating blood, local modulation of hemodynamics, and defense against infection. Endothelial dysfunction, broadly defined, can be lethal, as evidenced by von Willebrand disease and endotoxic shock. Subtler endothelial dysfunction is characteristic of early atherosclerosis, an important disease of the developed world. Although we cannot know for certain, it is unlikely that atherosclerosis caused significant morbidity or mortality in the pre-agricultural era, and indeed modern hunter-gatherers are little affected by atherosclerosis (see Chapter 16). However, atherosclerosis is now epidemic in developed societies worldwide, suggesting a near universal predisposition for the disease, with some of the variation in vulnerability arising from genetic differences. Although some common vulnerability factors, such as low levels of high-density lipoprotein (HDL) cholesterol, may result from the accumulation of multiple rare alleles in a specific population, most probably they result from common allelic variants found in all populations (20). The apolipoprotein E story is particularly germane in this regard (21). Common and widely dispersed genetic patterns are of ancient origin, and thus presumably predispose all humans to disease susceptibility when they interact with modern conditions. Indeed, even in recent historical times, transitions from relatively low to high rates of atherosclerosis have been observed to accompany cultural transitions.

At this early juncture, it is important to acknowledge explicitly that we do not propose that "genes for atherosclerosis" exist. Rather, we believe that those suites of genes, shared by most of us, that now exact a net cost were preserved because they have yielded a net benefit over evolutionary time. This may be because dietary changes interact with these genes to impose costs throughout life, or because the benefits during early life are larger for some genes than are their adverse consequences during adulthood (*antagonistic pleiotropy*). The massive amount of work constituting the bulk of the current volume provides insights about the proximate mechanisms by which atherosclerosis arises and progresses, and we draw heavily on that mechanistic knowledge. Our task is to suggest reasons why natural selection over the past hundred thousand years conserved an essentially universal vulnerability to such an important disease process.

Although early atherosclerosis usually is characterized as "endothelial dysfunction," and the progression of atherosclerosis as a "response to injury," close examination of early events shows the endothelium to be functioning in a largely normal manner, at least as regards one of its primary functions – immunity. For most of hominid history, infections have represented grave threats, whether arising from injury, peripartum infection, or endemic pathogens. The nature, although not the risk, of infectious disease changed as settled communities developed (11). Increased population density permitted human-to-human disease transmission, leading to the rise of epidemic disease, and animal domestication led to the emergence of novel zoonoses, especially evident now as avian influenza threatens. Innate and acquired immunological functions that eliminate or suppress infections are highly conserved as powerful and often apparently redundant defenses. The evolution of such systems cannot reasonably be attributed to anything other than incremental beneficial genetic changes that improved the reproductive success of those individuals who had them. The endothelium contributes very importantly to immunity, and one of its functions – the controlled transport of intravascular substances and cells to the subendothelial space – seems to predispose us to atherosclerosis. Although this could be interpreted as a simple defect, it appears much more like an adaptation that has benefits that are equal to or greater than its costs in a classic trade-off.

Atherosclerotic lesions begin as low-density lipoprotein (LDL) particles accumulate in the subintimal space (22). The concentration of LDL in the blood is an important determinant of the rate of accumulation, as is the concentration of HDL available for reverse cholesterol transport out of the tissue. However, the trafficking of lipoproteins across the endothelium is not passive. Rather, LDL and HDL cross endothelial cells (ECs) by means of *pinocytosis*, which allows the delivery of triglycerides and cholesterol for metabolic support, cell growth, and steroidogenesis.

The macrophage is another key player in atherosclerosis (22,23). Although macrophages normally resident in the subintima can handle some foreign antigens, the local response to tissue infection triggers an elaborate signaling system marshalling a defense reaction that includes the transendothelial migration of large numbers of monocytes and subsequent differentiation into macrophages to engulf bacteria and expose them to lysosomal enzymes. Macrophage phagocytosis and the intracellular lysosomal degradation of bacteria are efficient, but via coevolution, some organisms have derived strategies to frustrate macrophage capabilities. Thus, although bacteria that can be opsonized are cleared efficiently, many bacteria (e.g., *Listeria* and tuberculosis) have evolved resistance to the constitutive killing mechanisms used by macrophages such that they can survive, even intracellularly. *Chlamydia pneumoniae* may be of particular importance (24).

The process of the development of bacterial resistance is reminiscent of how LDL particles take up residence in the subendothelium, escape macrophage metabolism, and cause disease. Native LDL particles attach to macrophage LDL receptors and are efficiently engulfed and metabolized, just as bacteria are. As part of this normal process, macrophage LDL receptors are down regulated and, in consequence, individual macrophages do not become overloaded with LDL particles. However, under oxidizing conditions, LDL particles are transformed to oxidized LDL particles that are taken up by macrophages via alternative high-capacity scavenger receptors (CD36 and type A scavenger receptor [SR-A]) normally employed for the ingestion of bacteria. These receptors are not down-regulated by the accumulation of oxidized LDL; macrophages continue to accumulate lipoprotein moieties in a relatively unregulated manner, leading to the cellular congestion evident in foam cells (22). Note, however, that during the early stages, both ECs and macrophages are doing what they are well adapted to do: that is, prevent invasion by microscopic threats. It is probably only in our modern environment that these processes are corrupted by excess lipoprotein levels and oxidative stress that leads to disease.

While macrophage functions are being subverted, the endothelium overlying the early atherosclerotic lesion is intact and functional. Indeed, it appears that the normal vessel is able to heal the early effects of subintimal foam cell accumulation. Fetuses show foam cell lesions, termed "fatty streaks," during early development, particularly in hyperlipidemic mothers in which the fetal blood cholesterol level correlates closely with that of the mother (25). These early fatty streaks appear to regress later in pregnancy either as fetal cholesterol demand decreases or fetal production increases; fetal cholesterol requirements become less dependent on maternal sources, and fetal blood cholesterol concentration is no longer correlated with that of the mother. The initial accumulation probably reflects an example of fetal–maternal conflict: The rapidly developing fetus requires more cholesterol than it can synthesize and is dependent on transplacental delivery, thus creating a potential conflict with the mother who could use these caloric resources for herself or a subsequent offspring. David Haig has pioneered this line of thought (26), as exemplified in his chapter in this volume on such conflicts and their role in preeclampsia (Chapter 17). The consequences of poor fetal lipoprotein access are profound: Cholesterol deficiencies in the mother can cause severe fetal developmental defects, particularly in the lipid-rich brain. Not all the vascular changes in the fetus may be reversible, however, because postnatal atherogenesis may be accelerated in the offspring of hypercholesterolemic mothers who develop hypercholesterolemia themselves (27).

In part, the healing process itself may induce atherosclerosis. Lipoprotein(a) is an LDL-like particle with putative roles in tissue repair and inhibition of fibrinolysis that may also act as a surrogate for vitamin C in animals that have lost the ability to synthesize the vitamin de novo. Lipoprotein(a) is limited to a subgroup of primates (and interestingly appears to have evolved independently in the hedgehog) in which it may have evolved when injuries resulted from trauma and infection. When expressed in a modern milieu, in which endothelial dysfunction is widespread (27), the actions of lipoprotein(a) promote atherosclerosis. In fact, just as for macrophages and the endothelium, lipoprotein(a) is doing what it does best – trying to heal injury.

The aspects of the modern environment predisposing to atherosclerosis in the developed world are well known: dietary excess, sedentary lifestyle, cigarette smoking, air pollution, and diabetes, in addition to the acquired lipoprotein disorders of high LDL and low HDL concentrations. Although it is clear that the nonlipid risk factors can result in endothelial dysfunction, as evidenced by diminution of flow-mediated dilatation or responses to infused acetylcholine, it appears that atherosclerosis develops only in the face of the lipid disorders. The key player is almost certainly LDL – isolated low HDL-cholesterol or high blood pressure in knockout mice without LDL are incapable of causing atherosclerosis; high LDL cholesterol seems to be the *sine qua non* of atherosclerosis. However, native LDL-cholesterol itself does not appear to be sufficient for triggering atherosclerosis; there appears to be a requirement for oxidation. As noted above, LDL oxidation seems to be a key feature of the macrophage's inability to regulate LDL particle uptake, and oxidation likewise plays an important role in the progression of lesions by stimulating subintimal lipoprotein retention, cytokine elaboration, and the chemoattraction of participating inflammatory cells by macrophages. Oxidized LDL uptake is one of the factors promoting the release of inflammatory mediators. The inflammatory response decreases nitric oxide (NO) production and contributes to "endothelial dysfunction" (by which is meant the vasodilator response to NO-mediated agonists) (28).

Macrophage responses may not be confined to vessels, as recently suggested by observations of macrophage infiltration into adipose tissue, where elaboration of cytokines also occurs. Lerke and Lazar (29) have recently suggested the possibility that fat cells are recognized as invaders by macrophages, triggering responses similar to those causing atherosclerosis. If so, the sequence would add credence to the idea that macrophage activity and its attendant ill effects on the endothelium are conserved evolutionary responses of broad utility that are rendered harmful by modern environments. Although many aspects of the environment contribute to atherosclerosis, all seem to operate through LDL oxidation which, in a sense, serves to protect LDL particles from intracellular metabolism and, in that way, mimics one of the strategies followed by bacteria.

Returning to the question of why this most useful process has come to cause atherosclerosis, the most obvious factor is the level and kinds of lipids characteristic in contemporary human diets. Hunter-gather and traditional rural societies have serum cholesterol levels far below those of Westernized populations. Indeed, less than 5% of adults in the United States have a total cholesterol of less than 150 mg/dL, the average value in the more biologically "normal" populations (30). Thus, in essence, almost everyone in the United States, and in many other Westernized countries, is hypercholesterolemic, and the situation for triglycerides is similar. Undoubtedly, our societal hyperlipidemias are overwhelmingly due to diet, both its macronutrient composition and its caloric excess, particularly when such a diet is coupled with a sedentary lifestyle. The effect of modern lifestyle is Syndrome X, which in addition to promoting hyperlipidemia also increases oxidative stress, resulting in increased levels of atherogenic oxidized LDL (31).

One way of thinking about atherosclerosis is to view it as the result of an evolved adaptive response that protects against infection but that results in endothelial injury in modern environments. Because the consequences of the injury occur late in life, natural selection preserves those systems promoting atherosclerosis in preference to those suppressing them, a classic case of antagonistic pleiotropy. This is a trade-off, but not quite a classic one. Because the costs were probably minimal until the past century, and because they caused no harm until modern times, the genetic variations that increase vulnerability to atherosclerosis are not really "defects," but are instead excellent examples of "genetic quirks" that give rise to untoward effects only when they interact with factors encountered in modern environments. These speculations about the adaptive roots of atherosclerosis give rise to a specific prediction that individuals who have a genetic predisposition to atherosclerosis may be less vulnerable to infection and more susceptible to other inflammatory diseases. In regards to the question of autoimmune diseases, data demonstrate a strong association between rheumatoid arthritis and accelerated atherosclerosis (32), although the authors of that article suggest that the chronic low-grade inflammation of arthritis activates the atherosclerotic process rather than vice versa. In truth, the relationship may be bidirectional, with each process reinforcing the other.

The adaptive significance of the systems that leave us vulnerable to atherosclerosis is only one half of a full evolutionary explanation. We also need to know the phylogeny of those systems, and particularly if path dependence has constrained them so that natural selection cannot shape a less vulnerable mechanism. Although it would conceivably be possible to develop alternative ways of preserving the functions of lipoproteins necessary for blood-borne triglyceride and cholesterol delivery while preventing excessive pinocytotic delivery to subendothelial spaces, such an adaptation would probably seriously compromise cholesterol-dependent cellular and immune processes. Because immune function has such obvious selective value – whereas atherosclerotic disease, primarily manifest late in life, presumably is not a major selection force – there would seem to be little tendency to shape such an adaptation. However, too much of a good thing is dangerous, as illustrated by chronic inflammatory diseases such as rheumatoid arthritis accelerating atherosclerosis. Conversely, in situations in which innate immunity is compromised (e.g., by inactivating mutations of the Toll-like receptor 4 (TLR4) receptor, a pattern recognition receptor involved in lipopolysaccharide clearance), atherosclerosis is decreased (33). Thus, in the trade-off of immune function versus atherosclerosis, the benefits of immunity provide greater advantages early in life when selection is stronger, thus leaving us with a strong predisposition to atherosclerosis. A signal detection analysis of the costs and benefits of various regulatory settings could further illuminate the situation (19).

Are other evolutionary factors at work? A direct infectious hypothesis has been championed by Ewald, who contends that atherosclerosis is triggered by infections that have coevolved with humans (24). Of particular interest has been the relationship of *Chlamydia pneumoniae*, an organism capable of living inside inflammatory cells in blood vessel plaques. Ewald suggests that the bacterial infection, rather than being a bystander colonizing a disrupted vascular wall, is pathogenetic. He further notes that in the case of inflammatory arthritis, a higher incidence of joint-fluid *C. pneumoniae* infections are associated with a higher prevalence of the apolipoprotein $\varepsilon 4$ genotype. Because this is a risk factor for coronary disease, it is possible the mechanism mediating risk is an increased susceptibility to *C. pneumoniae* infection and vascular inflammation. Clearly, if true, the early lesions could attract macrophages, increase local oxidative activity and, in the presence of LDL particles, trigger atherosclerosis. An attempt at testing this hypothesis by treating patients with established atherosclerosis with antibiotics was negative (34), but it is still plausible to assume that the role of the bacterium is prominent in the early disease but not after atherosclerosis has progressed. *C. pneumoniae* is not the only infection thought to accelerate atherosclerosis: Many chronic subclinical infections, including periodontitis, sinusitis, bronchitis, and diverticulitis – in fact any source of chronic endotoxemia – may provoke an inflammatory response and increase the risk of atherosclerosis (35). The other major player in the atherosclerotic picture, HDL, may be a defense against atherosclerosis-promoting infectious agents by acting as a scavenger of lipopolysaccharide (36).

CONCLUSION

The implications of an evolutionary view of the endothelium are several. First and most obviously, in the absence of profound changes in the environment or human nature, it is unlikely that atherosclerosis can be prevented in the general population through the public health measures known to be effective – diet and exercise – although they can be effective for those individuals capable of modifying their behavior (37). The burgeoning epidemic of obesity suggests that the lifestyle causing hyperlipidemia is triumphing to the extent that reducing cholesterol values to those of hunter-gatherers by lifestyle interventions is simply unrealistic. As a society, we seem to be much more willing, even eager, to embrace antiatherosclerotic strategies that depend on the administration of costly drugs to already-ill individuals nearing the end of the natural history of atherosclerotic diseases. Given current costs, inequalities in health care delivery, and the poor public health infrastructure, such treatments, unfortunately, cannot reasonably be provided for population-wide treatment and, in any case, they are not that effective. More promising preventive approaches include modifications like statins and antioxidants, although one line of argument developed above suggests the need to be on guard for the possibility that antioxidants may increase other vulnerabilities. Epidemiological evidence supporting the benefits of dietary antioxidants is encouraging, although the effectiveness of antioxidants in intervention trials has been disappointing (23). Trials in established atherosclerosis may, however, be a case of too little, too late, as atherogenesis is already well advanced, perhaps beyond the period in which antioxidants could be effective. What is clear is that only with the development of effective preventive approaches that can be widely applied will any progress be made in preventing the complications of atherosclerosis.

One potentially beneficial feature of modern society is found in Barker's hypothesis, which implies that improved maternal nutrition may decrease the development of features described as the "thrifty phenotype" – a set of metabolic adjustments programmed to promote calorie retention but which increase the risk of late-life atherosclerosis (38,39). Unfortunately, to realize the benefits of improved maternal nutrition, the environment of the offspring throughout life would have to promote avoidance of calories and fat excess, which again seems impracticable.

Genetic engineering has been held out as a possible cure for atherosclerosis. We believe the evidence is overwhelming that only a small min the focus on "gene hunting" for atherosclerosis should therefore be on finding common allelic variants or haplotypes controlling "normal" homeostatic functions. This is not to say that elucidation of monogenic diseases will not lead to a better understanding of the vulnerabilities of atherosclerotic disease progression, just as a molecular understanding of the LDL receptor in familial hypercholesterolemia contributed so importantly to the development of drugs useful in preventing further progression of atherosclerosis. The identification of the Milano Apo A1 mutation (40), which exerts a powerful protective effect against atherosclerosis by markedly enhancing reverse cholesterol transport and presumably acting to heal the early atherosclerotic lesions before the sequence of reinforcing mechanisms is recruited, may be another example of how a monogenic disease can inform the development of prevention strategies.

Just as genetic determinants are likely to be "normal" alleles, normal immune processes predispose to atherogenesis. Could atherosclerosis be prevented by the selective modulation of immune function, specifically the physiological functions of macrophages involved in atherogenesis? Similarly, perhaps endothelial transport of LDL molecules can be targeted. The endothelium has its own protective mechanisms against the transport of larger lipoproteins such as very-low density lipoprotein (VLDL). Perhaps a "vaccination" that slowed LDL-cholesterol transport could have promise for slowing atherosclerosis. Ceruloplasmin is a potential source of oxidized LDL (41); perhaps the copper-depleting drug developed for the treatment of Wilson's disease of the liver, trientine, could possibly have an antiatherosclerotic effect. Such speculations demonstrate the heuristic value of an evolutionary perspective on the endothelium. We anticipate that readers of this volume will bring a mass of sophisticated knowledge about the proximate mechanism involved in endothelial diseases. We hope that this chapter will encourage some of them to ask new evolutionary questions about why the endothelium is the way it is, and why it leaves us so vulnerable to such common and devastating diseases. If they do, progress in endothelial biomedicine will occur even faster.

KEY POINTS

- Posing and testing evolutionary hypotheses about characteristics of the endothelium will speed scientific progress.
- Every aspect of the endothelium needs both a proximate and an evolutionary explanation.
- The same design features that lead to vulnerability to atherosclerosis offer other substantial benefits.
- Few genes that predispose to endothelial disease are defects; most are quirks that pose risks only in the modern environment.

Future Goals

- To describe every aspect of the endothelium in relation to an evolutionary explanation
- To describe every disease of the endothelium in relation to an evolutionary explanation that encompasses the six reasons for vulnerability

REFERENCES

1 Mayr E. *The Growth of Biological Thought: Diversity, Evolution, and Inheritance.* Cambridge: The Belknap Press of Harvard University Press, 1982.

2 Mayr E. *What Makes Biology Unique?: Considerations on the Autonomy of a Scientific Discipline.* New York: Cambridge University Press, 2004.
3 Tinbergen N. On aims and methods in ethology. *Zeitschrift für Tierpsychologie* 1963;20:410–433.
4 Stearns SC, Ebert D. Evolution in health and disease: a work in progress. *Q Rev Biol.* 2001;76: 417.
5 Nesse RM, Williams GC. Evolution and the origins of disease. *Sci Am.* 1998;279(5):86–93.
6 Nesse RM, Williams GC. *Why We Get Sick.* New York: Times Books, 1994.
7 Stearns S, ed. *Evolution in Health and Disease.* Oxford: Oxford University Press, 1998.
8 Williams GC, Nesse RM. The dawn of Darwinian medicine. *Q Rev Biol.* 1991;66:1–22.
9 Neel JV, Julius S, Weder A, Yamada M, Kardia SL, Haviland MB. Syndrome X: is it for real? *Gen Epidemiol.* 1998;15:19.
10 Eaton SB, Strassman BI, Nesse RM, et al. Evolutionary health promotion. *Prev Med.* 2002;34(2):109–118.
11 Ewald P. *Evolution of Infectious Disease.* New York: Oxford University Press, 1994.
12 Williams GC. *The Pony Fish's Glow: And Other Clues to Plan and Purpose in Nature.* New York: BasicBooks, 1997.
13 Williams GC. *Adaptation and Natural Selection: A Critique of Some Current Evolutionary Thought.* Princeton NJ: Princeton University Press, 1966.
14 Dawkins R. *The Selfish Gene.* Oxford: Oxford University Press, 1976.
15 G. C. Williams. Pleiotropy, natural selection, and the evolution of senescence. *Evolution.* 1957;11:398–411.
16 Austad SN. *Why We Age: What Science Is Discovering about the Body's Journey Throughout Life.* New York: J. Wiley & Sons, 1997.
17 Kruger D, Nesse R. Sexual selection and the male:female mortality ratio. *Evolutionary Psychology.* 2004;2:66–85.
18 Kruger DJ, Nesse RM. An evolutionary life-history framework for understanding sex differences in human mortality rates. *Human Nat.* 2006;17:74–97.
19 Nesse RM, Natural selection and the regulation of defenses: a signal detection analysis of the smoke detector principle. *Evolution and Human Behavior.* 2005;26:88–105.
20 Cohen JC, Kiss RS, Persemlids A, Marcel YL, McPherson R, Hobbs HH. Multiple rare alleles contribute to low plasma levels of HDL cholesterol. *Science.* 2004;305:869–872.
21 Finch CE, Sapolsky RM. The evolution of Alzheimer disease, the reproductive schedule, and apoE isoforms. *Neurobiol Aging.* 1999;20(4):407–428.
22 Steinberg D. Atherogenesis in perspective: hypercholesterolemia and inflammation as partners in crime. *Nat Med.* 2002;8:1211–1217.
23 Chisolm GM, Steinberg D. The oxidative modification hypothesis of atherogenesis: an overview. *Free Radic Biol Med.* 2000;28 (12):1815–1826.
24 Ewald PW, Cochran GM. Chlamydia pneumoniae and cardiovascular disease: an evolutionary perspective on infectious causation and antibiotic treatment. *J Infect Dis.* 2000;181(Suppl 3): S394–401.
25 Napoli C, D'Armiento FP, Mancini FP, et al. Fatty streak formation occurs in human fetal aortas and is greatly enhanced by maternal hypercholesterolemia. Intimal accumulation of low density lipoprotein and its oxidation precede monocyte recruitment into early atherosclerotic lesions. *J Clin Invest.* 1997;100: 2680–2690.
26 Haig D. Genetic conflicts in human pregnancy. *Q Rev Biol.* 1993; 68:495–532.
27 Lippi G, Guidi G. Lipoprotein(a): from ancestral benefit to modern pathogen? *QJM* 2000;93:75–84.
28 Murohara T, Kugiyama K, Ohgushi M, Sugiyama S, Ohta Y, Yasue H. LPC in oxidized LDL elicits vasocontraction and inhibits endothelium-dependent relaxation. *Am J Physiol.* 1994;267(6 Pt 2):H2441–H2449.
29 Lehrke M, Lazar MA. Inflamed about obesity. *Nat Med.* 2004; 10:126–127.
30 Eaton SB. Humans, lipids and evolution. *Lipids.* 1992;27:814–820.
31 Holvoet P, Kritchevsky SB, Tracy RP, et al. The metabolic syndrome, circulating oxidized LDL, and risk of myocardial infarction in well-functioning elderly people in the health, aging, and body composition cohort. *Diabetes.* 2004;53:1068–1073.
32 Sattar N, McCarey DW, Capell H, McInnes IB. Explaining how "high-grade" systemic inflammation accelerates vascular risk in rheumatoid arthritis. *Circulation.* 2003;108(24):2957–2963.
33 Tobias P, Curtiss LK. Thematic review series: The immune system and atherogenesis. Paying the price for pathogen protection: toll receptors in atherogenesis. *J Lipid Res.* 2005;46:404–411.
34 Cannon CP, Braunwald E, McCabe CH, et al.; Pravastatin or Atorvastatin evaluation and infection therapy-thrombolysis in myocardial infarction 22 investigators. Antibiotic treatment of *Chlamydia pneumoniae* after acute coronary syndrome. *N Engl J Med.* 2005;352(16):1646–1654.
35 Stoll LL, Denning GM, Weintraub NL. Potential role of endotoxin as a proinflammatory mediator of atherosclerosis. *Arterioscler Thromb Vasc Biol.* 2004;24:2227–2236.
36 Wu A, Hinds CJ, Thiemermann C. High-density lipoproteins in sepsis and septic shock: metabolism, actions, and therapeutic applications. *Shock.* 2004;21:210–221.
37 Jenkins DJ, Kendall CW, Marchie A, et al. The Garden of Eden – plant based diets, the genetic drive to conserve cholesterol and its implications for heart disease in the 21st century. *Comp Biochem Physiol A Mol Integr Physiol.* 2003;136:141–151.
38 Barker DJ. Fetal origins of cardiovascular disease. *Ann Med.* 1999;31(Suppl 1):3–6.
39 Hales CN, Barker DJ. The thrifty phenotype hypothesis. *Br Med Bull* 2001;60:5–20.
40 Roma P, Gregg RE, Meng MS, et al. In vivo metabolism of a mutant form of apolipoprotein A-I, apo A-IMilano, associated with familial hypoalphalipoproteinemia. *J Clin Invest.* 1993;91:1445–1452.
41 Fox PL, Mazumder B, Ehrenwald E, Mukhopadhyay CK. Ceruloplasmin and cardiovascular disease. *Free Radic Biol Med.* 2000; 28:1735–1744.

16

The Ancestral Biomedical Environment

S. Boyd Eaton*, Loren Cordain†, and Anthony Sebastian‡

*Emory University, Atlanta, Georgia; †Colorado State University, Fort Collins;
‡University of California, San Francisco

Behaviorally modern humans evolved in Africa perhaps as early as 100 thousand years ago (kya) (1) and, by 50 kya, they began spreading throughout Eurasia and Australia. Since that evolutionary watershed, the human genome has changed little. For millions of years, human ancestors, like all other organisms, had responded to altered environmental circumstances solely through biological evolution (Table 16-1). However, during the past 50 millennia, humans have increasingly been able to adapt to new ambient circumstances through cultural innovation, in addition to the underlying (and slower) genetic change. Since agriculture and animal husbandry first appeared, perhaps 10,000 years ago, hemoglobinopathies and adult lactose tolerance are almost the only generally acknowledged genetic modifications. On the other hand, our lifestyle has changed radically: Nutrition, physical activity, reproductive experience, psychosocial relations, microbial interactions, and toxin/allergen exposure are all vastly different now from what they were for ancestral humans and prehumans during the period when our primary genetic makeup, including those factors relevant to endothelial health and disease, was selected.

The resulting discordance or mismatch between our genes and our modern lives is a likely contributor to many common chronic diseases and probably to various forms of endothelial dysfunction, so an appreciation of its potentially pertinent elements may further our understanding of pathophysiology. Also, awareness of the ancestral human lifestyle may suggest new avenues of prevention research applicable to the endothelium.

Reconstructing the biomedical circumstances of Stone Age humans is a fascinating, if sometimes frustrating endeavor. Several categories of data exist. Human skeletal remains are amenable to gross anatomical, microscopic, and biomechanical evaluation, as well as to radioisotopic analysis. Archeological finds, mainly at living sites, include the bones of animals consumed as food, botanical remains, artifacts (such as hearths, tools, weapons), and cave or rock wall paintings. Hunter-gatherer (forager)[1] groups studied in the last century (no true hunter-gatherers have survived into the twenty-first century) were the best, if imperfect, surrogates for Paleolithic humans. Their subsistence patterns, obligatory physical activity, reproductive experience, biomarkers (serum cholesterol levels, blood pressure, glucose responsiveness, etc.), and microbial interactions can be considered representative of patterns for the humans of 50 to 25 thousand years ago. Proximate analyses of the game animals and uncultivated plant foods consumed by recent hunter-gatherers provide further information regarding our ancestors' nutrient intake.

Most available evidence is thus indirect. We cannot take blood samples from, biopsy, observe, or interrogate our behaviorally modern ancestors of 25 kya. We will never have electron micrographs of their endothelium. Nevertheless, by dint of ongoing multidisciplinary efforts, a defensible and increasingly detailed picture of their life and health is emerging. It affords a view of the lifestyle for which our genes were originally selected, and it provides further understanding of the epigenetic milieu[2] that influenced their function.

ANCESTRAL HUMAN BIOLOGY

Blood Pressure

Numerous studies in varied geographic settings have established that the blood pressure of hunter-gather populations averages 100 to 110 systolic/70 to 75 diastolic (2), somewhat below the long-accepted "normal" values for Americans

1 Hunter-gatherers (synonymous with foragers) subsist on hunted wild animals, gathered wild plant foods, and aquatic resources. They have no domesticated plants or animals except the dog.

2 The epigenome consists of proteins and chemical molecules that surround and adhere to protein-encoding DNA. By amplifying or muting genetic expression in response to bioenvironmental circumstances (e.g., nutritional changes), the epigenome provides rapid response capability – genetic flexibility independent of DNA evolution as classically understood.

Table 16-1: The Time Scale of Human Evolution

Million years ago (Paleolithic)	
8–6	Human and chimpanzee ancestral lines split. (*Orrorin, Sahelanthropus*). Beginnings of upright posture and bipedalism
6–4	*Ardipithecines*
4.5–2	*Australopithecines* (including "Lucy")
~2.5	*H. habilis* (or *A. habilis* – genus disputed); initial appearance of stone tools
1.75	*H. erectus/H. ergaster*; achievement of modern human height and body proportions.
1.75–1.5	*H. erectus* migrates to southern Asia; present until 30 kya or later
	Presumably ancestral to *H. florensis*
1.75–0.2	Brain enlargement with little technological innovation
Thousand years ago (Paleolithic)	
750–500	*H. heidelbergensis*; ancestral (in Europe) to *H. neanderthalensis* and (in Africa) to *H. sapiens*
200–150	Anatomically modern *H. sapiens* appears in Africa with little or no technological advance
100–50	Behaviorally modern *H. sapiens* appears in Africa (improved linguistic capability is thought to be the key evolutionary change)
	Rapid technological innovation begins
60–5	*H. sapiens* spreads worldwide
20–10	*Epipaleolithic*; incipient agriculture
Thousand years ago (Neolithic)	
10–2	Ice Age ends, agriculture becomes widespread
	Sedentary living with towns and states; "civilization"

and other Westerners. Furthermore, the same studies have invariably found that forager blood pressure remains low as they age. Conversely, for inhabitants of developed nations, blood pressure typically rises as the population grows older. For hunter-gatherers, hypertension is almost nonexistent (2), whereas over 25% of American adults have elevated blood pressure (3).

Several factors appear to underlie these repeatedly confirmed observations. An obvious contributor is the availability of commercial salt. Hunter-gatherers are not the only low blood pressure societies. Traditional horticulturists, agriculturists, and pastoralists[3] whose economies lack commercial salt are also normotensive throughout their lives (2). As for all other free-living terrestrial mammals (both carnivores and herbivores), forager diets provide far more potassium than sodium (five to ten times as much) (4).[4] In stark contrast, the foods consumed by Americans yield more sodium than potassium, a nutritional electrolytic inversion with manifold pathophysiological implications (4). Of these, acid–base balance may be most important. As for recent hunter-gatherers, a high Stone-Ager intake of fruits and vegetables (twice the present consumption) tended to drive systemic pH toward alkalinity, whereas modern consumption of cereal grain products and dairy foods is net acid-producing (5). Over a lifespan, the corrective metabolic measures required to maintain homeostasis produce urinary calcium loss and accelerated skeletal mineral depletion while increasing the risk of urolithiasis (4). Effects on endothelial function have been little investigated.

Another significant influence is body size; greater body weight increases the risk of hypertension. The body mass indices (BMI) of recently studied hunter-gatherers average 21.5 (6), whereas the American mean is 26.5 (7). It is not clear that body mass independently influences blood pressure. Probably BMI is a marker for body composition, and it may be that excess fat relative to lean tissue is the pertinent biological variable. Forager skinfold thickness measurements, which are typically half or less those of age-matched North Americans, show that their fat content, as a percentage of total body weight, is typically below 15% for men and 25% for women (2). Many Westerners with "normal" BMIs are nonetheless overfat and/or undermuscled (8). When a convenient, inexpensive, and accurate method of determining body composition becomes available, this biomarker is likely to supplant BMI as a health status indicator.

Recently studied foragers have been aerobically fit with VO_2 max values placing them in the good to superior range, well above the poor to average values more characteristic of age-matched Americans (9,10). This difference clearly reflects the circumstances of hunter-gatherer life: travel, subsistence activities, and recreation all entail physical work.

Walking, running, carrying, digging, and dancing are prominent features of a lifestyle without motorized equipment or draft animals. Increasing one's aerobic power through a program of endurance exercise usually results in lower blood pressure (and a lower resting heart rate) (see Chapter 56) (11). For Stone Agers, such a program was obligatory, not elective.

Carbohydrate Metabolism

The insulin responsiveness of five different hunter-gatherer groups, on three continents, was evaluated in the last century (12). All showed remarkable insulin sensitivity, and it seems reasonable to assume that preagricultural human ancestors shared this desirable metabolic characteristic. Insulin resistance, an increasingly common finding in Americans and other Westerners, seems rare to nonexistent among foragers (12). However, migrant studies and serial determinations among societies undergoing acculturation to Western ways in their own homelands indicate that a population with initially favorable insulin responsiveness may become transformed within a generation or two into one with highly prevalent

3 Pastoralists practice animal herding as their primary economic activity (e.g., the Samburu and Maasai of East Africa).

4 The homeostatic response to chronic diet-derived net acid load includes release from bone of alkaline calcium salts (phosphates, carbonates) to buffer hydrogen (H^+). The liberated calcium and phosphorus are lost in the urine without compensatory gastrointestinal absorption; thus bone mineral content gradually declines.

insulin resistance (13,14). This shows that genetic makeup, while important, is not the main driving force regarding carbohydrate metabolism.

The chief behavioral changes associated with secular increases in insulin resistance have to do with diet and body habitus. The common nutritional elements that characterize populations with desirable insulin responsiveness include a limited intake of simple carbohydrate such as sugars and goods made from refined flour. Carbohydrate consumption itself may be substantial for insulin-sensitive groups, but is derived from whole grain (for many agriculturists) or from uncultivated fruits and vegetables (for hunter-gatherers). Such foods typically exhibit low glycemic indices and, especially, low glycemic loads (15). In addition, dietary fiber intake for normoglycemic populations tends to be considerable (16). The fiber may be predominantly insoluble (nonfermentable), as from wheat and rice, or it may be largely soluble (fermentable), as from wild fruits and vegetables. Uncultivated plant foods tend to be much more fibrous (three to four times as much fiber/100 g) than are the fruits and vegetables popular in Western nations (16). Groups tending to have a low prevalence of insulin resistance generally consume much more fiber (of all sorts) than do Americans, sometimes 70 to 100 g/day (16,17). Added sugar (i.e., extraneous to basic food items themselves) contributes up to 25% of American food energy each day (18). Foragers loved honey, an equivalent foodstuff, but it was difficult to obtain and usually available only on a seasonal basis. The best current estimate is that, over the course of a year, honey generally contributed about 2% to 3% of dietary energy for hunter-gatherers (4).

As discussed earlier, recently studied foragers have been lean, with BMIs in the low-normal range; corpulence is essentially unknown. On the other hand, Venus statuettes dating from 25,000 years ago prove that at least some ancestral humans were obese. However, the mere fact that such statuettes were made suggests they represented special, high-status individuals such as religious figures (shamans) or political leaders (19). Given the exertional requirements of hunter-gatherer existence, obesity would have been impossible for rank-and-file group members. Skeletal remains from the Late Paleolithic (50 to 10 kya) indicate average muscularity similar to that of today's superior athletes (10). This finding tends to confirm that strenuous physical activity was a hallmark of the ancestral lifestyle. Although depictions of humans are uncommon elements of cave and rock wall paintings from the Paleolithic, those available show lean and muscular individuals. No painted or etched figures matching the Venus statuettes have been found.

Higher BMIs are strongly correlated with a risk for insulin resistance, and individuals whose BMI is within normal limits also have increased risk if their body composition includes excessive adipose tissue. Both these circumstances were most uncommon among Stone Agers, and it seems likely that the metabolic milieu within which the human endothelium was designed to function was one in which hyperinsulinemia was rare.

Lipid Metabolism

During the twentieth century, the serum cholesterol values of six different hunter-gatherer groups living on four continents were determined. The mean was 123 mg/dL (3.2 mmol/L), and none of the groups had an average value exceeding 150 mg/dL (3.9 mmol/L) (2). (We have no data on the HDL:LDL partition.) The hunter-gatherer mean falls within the range observed for serum cholesterol levels in free-living higher primates –90 to 135 mg/dL (2.3 to 3.5 mmol/L) – much below those of average Americans (~200 mg/dL [5.2 mmol/L]) (2,20).

The cholesterol levels of hunter-gatherers (and other traditional peoples with originally low serum values) rise strikingly when they adopt a more Western lifestyle, whether by migration or by the introduction of new ways within their own region (21,22). This again indicates that, although genes are important, nutrition and other modifiable bioenvironmental factors are even more so for determining a population's blood lipid values.

Although foragers, horticulturists, traditional rural agriculturists, and pastoralists all have low serum cholesterol levels, their intake of dietary lipids varies substantially. Hunter-gatherers and most pastoralists (e.g., Maasai) consume a considerable amount of fat and cholesterol, whereas the diets of horticulturists and rural agriculturists typically provide relatively little of either. The common nutritional features are a low level of saturated fat and almost no harmful trans fatty acids (4), while dietary intake of cholesterol and total fat appears to be of limited importance regarding serum cholesterol levels (23).

During the Late Paleolithic (50,000 to 15,000 B.P.) our ancestors were consummate hunters, obtaining about 50% of their dietary energy from animal sources (24). Although game animals are much leaner than the commercial animals from which our supermarket meat is obtained, their cholesterol content is quite similar, so cholesterol consumption for Stone Agers averaged about 500 mg/day (versus <300 mg/day for contemporary Americans) (25). In Northeast Africa, the region where behaviorally modern humans are thought to have evolved (1), fats probably contributed about 35% of daily energy intake (26). However saturated fat consumption was only about half that of Americans (~7% versus 12% to 15%) because, during most of the year, animal fat from game is predominantly monosaturated with a far lower proportion of saturated fat than is found in commercial meat (27). Harmful trans fat intake was essentially nil (versus ~2% of total energy in the United States). Wild ruminant flesh contains 3% to 5% trans-vaccenic acid, but this is converted, after absorption, into conjugated linoleic acid (CLA) isomers for which limited evidence indicates anticarcinogenic and antiatherosclerotic activity (7). Polyunsaturated fat consumption for Paleolithic humans was nearly double that of Americans, and the $\omega 6:\omega 3$ partition was much different, approximately 2:1 ($\omega 6:\omega 3$) for ancestral humans (28); at least 10:1 currently (29).

Micronutrient Intake

Typical Stone Agers were more physically active than are average Westerners: estimates of 5.2 MJ (1,240 kcal)/day versus 2.3 MJ (555 Kcal)/day as energy expenditure during physical activity have been proposed (9,10). Contrary to popular misconception, our remote ancestors (after the appearance of *Homo erectus* 1.7 million years ago) were about as tall as are average Americans (30). Consequently, preagricultural humans required considerable dietary energy each day – perhaps 12.1 MJ (2,900 kcal) for adult males (9,10). Uncultivated plants and wild game tend to provide high levels of micronutrients relative to their energy content, unlike today's often calorically concentrated foods. Hence, retrojected ancestral intake of vitamins, minerals (including antioxidants), and – probably – phytochemicals was high, ranging from 1.5 to 8 times current intake depending on the specific nutrient (25,31). For example, mean daily vitamin C intake for American adults is about 90 mg, whereas retrojected ancestral consumption is estimated at just over 500 mg (25).

The exception, previously noted, is sodium. Preagricultural intake is thought to have been less than 1 gram per day (25), whereas contemporary Westerners typically consume over 4 g/d because of salt added during food processing, meal preparation, and at the table (4). The key factor is commercially available salt – not latitude. Tropical Venezuelan Yanamamo consume extremely little sodium, less than 500 mg/d. Conversely, the pattern for potassium follows the general rule – much less intake now than for Stone Agers (less than 3 g/d currently versus up to 10 g/d in the past) (4).

Tobacco Abuse

Tobacco abuse was an impossibility for the earliest behaviorally modern humans of 60 to 50 kya because they were Africans, whereas the genus *Nicotiana* is indigenous to the Americas, Australia, and certain Pacific Islands (2). By about 50 kya (or even slightly before) humans had spread as far as Australia, while Paleoindians, the ancestors of Native Americans, were in the New World by 12,000 B.P (and perhaps earlier; this dating is contentious). However, contemporary human ancestors who came from Europe, Asia, or Africa had no experience with tobacco until the voyages of Columbus 500 years ago. This relatively brief exposure has been insufficient to develop any genetically based resistance to its adverse health effects. Indeed, Australian Aborigines and Native Americans appear similarly prone to tobacco-linked disorders, so even 50 millennia have failed to provide significant immunity.

Microbial Interactions

In his book, *Plague Time*, biologist Paul Ewald argues that microorganisms, especially *Chlamydia pneumoniae*, but also *Porphyromonas gingivalis*, and perhaps others are the underlying, basal cause for atherosclerosis (32). He and numerous other investigators suggest that chronic vascular wall infection by such intracellular organisms produces the initial mural damage, either directly or by triggering an immunological inflammatory response. In either event, these authors maintain, such primary injury triggers the complex pathological process leading to atherosclerosis. Absent this chronic infection, atherosclerotic vascular disease fails to develop, no matter what commonly accepted risk factors are present.

Epidemic infectious diseases are thought to have been rare to nonexistent among Stone Agers because of the small group size, because their nomadic existence minimized the sanitation problems that became critical after fixed settlements appeared about 10,000 years ago, and because several important microbial illnesses are thought to have spread from animals to humans after domestication, again beginning about 10,000 years ago (33).

Conversely, endemic infectious disorders are thought to have been major killers of Paleolithic humans (34), and both *C. pneumoniae* and *P. gingivalis* are endemic conditions, at least in the contemporary Western world. If they were similarly prevalent during the Stone Age (and among those current populations largely free from atherosclerosis), their role in vascular pathogenesis would have to be considered necessary, but not sufficient. The modern epidemic of coronary, cerebral, and peripheral vascular disease may have resulted from the superimposition of tobacco abuse, hypercholesterolemia, hypertension, and other factors on an underlying infectious or inflammatory predisposition that has existed perhaps as long as there have been humans.

POTENTIAL RESEARCH INITIATIVES

A recapitulation of selected biomedical differences between ancestral and contemporary humans, potentially pertinent to endothelial health, is shown in Table 16-2.

Their collective or individual influence on endothelial microanatomy and physiology provide many possibly rewarding opportunities, for example:

- Other factors being equal, how does the endothelium respond when total blood cholesterol is varied between 3.2 mmol/L (125 mg/dL) and 5.2 mmol/L (200 mg/dL) (35)?
- Is the endothelium affected when a typical American sodium–potassium intake pattern is changed to one in which sodium is restricted to <1,000 mg/day and potassium increased to >5,000 mg/day (36,37)?
- What is the endothelial impact of prolonged magnesium supplementation (to match retrojected Stone Age intake) (38)?
- Does a diet rich in long-chain (C20 and above) polyunsaturated fatty acids with an $\omega 6{:}\omega 3$ ratio of about 2:1 protect the endothelium against some or all known harmful influences (39,40)?
- How does an elevated intake of antioxidants (to ancestral levels) affect the endothelium (41–43)?

Table 16-2: Ancestral versus Modern Humans

Differences	*Contributing Factors*	*Paleolithic*	*Contemporary*
Blood Pressure			
Paleolithic:			
Normotensive through life cycle	Dietary intake of electrolytes	$Na^+ << K^+$	$Na^+ > K^+$
	BMI	Lower	Higher
Contemporary:			
Tends to rise with age; many hypertensives	Aerobic power	Higher	Lower
Insulin Responsiveness			
Paleolithic:			
Nearly all insulin sensitive	BMI	Lower	Higher
	Body composition	More muscle Less fat	Less muscle More fat
Contemporary:			
Many insulin resistant	Glycemic load	Lower	Higher
	Fiber	High intake	Low intake
Lipid Metabolism			
Paleolithic:			
Lower serum cholesterol	Saturated fat intake	Lower	Higher
	Trans fat	Almost nil	2% of energy intake
Contemporary:			
Higher serum cholesterol	PUFA	More	Less
	$\omega 6{:}\omega 3$	~2:1	10:1 (or more)
	Simple CHO	Lower	Higher
	Dietary fiber	Higher	Lower

- Over time, does a consistently low glycemic load diet affect the endothelium beneficially (44,45)?
- Does the endothelium respond differently to an alkalinizing dietary pattern compared with one that is acid-producing (46)?

More important than any single investigative lead is the integrative, theoretical framework provided by the concept that the human genome was selected through evolutionary experience for the biobehavioral circumstances of ancestral life and that contemporary chronic diseases arise, in large measure, from the discordance that has been created between our genes and our lives. Deviation from and reversion toward the original human pattern may underlie, respectively, both disease causation and prevention. Perhaps a program of research based on the evolutionary discordance paradigm could make a vital contribution to the understanding and ultimate conquest of chronic illnesses, including those involving the endothelium.

KEY POINTS

- We are genetic Stone Agers living in a Space Age biomedical milieu.
- Discordance between our Stone Age protein-encoding DNA and our contemporary (especially affluent) lifestyles promotes development of chronic degenerative diseases, including those affecting the endothelium.

REFERENCES

1 Klein RG. *The Human Career. Human Biological and Cultural Origins.* Chicago: University of Chicago Press; 1999:494.

2 Eaton SB, Konner M, Shostak M. Stone agers in the fast lane: chronic degenerative diseases in evolutionary perspective. *Am J Med.* 1988;84:739–749.

3 American Heart Association. *Heart and Stroke Statistics – 2004 Update.* Dallas: American Heart Association, 2003.

4 Cordain L, Eaton SB, Sebastian A, Mann N, Lindeborg S, Watkins BA, O'Keefe JH, Brand-Miller J. Origins and evolution of the Western diet: heath implications for the 21st century. *Am J Clin Nutr.* 2005;81:341–354.

5 Sebastian A, Frassetto LA, Sellmeyer DE, Merriam RL, Morris RC Jr. Estimation of the net acid load of the diet of ancestral preagricltural *Homo sapiens* and their hominid ancestors. *Am J Clin Nutr.* 2002;76:1308–1316.

6 Jenike MR. Nutritional ecology: diet, physical activity and body size. In: Panter-Brick C, Layton RH, Rowley-Cowy P, eds. *Hunter-Gatherers. An Interdisciplinary Perspective.* Cambridge, UK: Cambridge University Press; 2001:205–238.

7 Institute of Medicine. *Dietary Reference Intakes. Energy, Carbohydrate, Fiber, Fat, Fatty Acids, Cholesterol, Protein, and Amino Acids.* Washington DC: National Academies Press, 2002.

8 Prentice AM, Jebb SA. Beyond body mass index. *Obesity Reviews.* 2001;2:141–147.

9 Cordain L, Gotshall RW, Eaton SB, Eaton SB III. Physical activity, energy expenditure and fitness: an evolutionary perspective. *Int J Sports Med.* 1998;19:328–335.

10 Eaton SB, Eaton SB III. An evolutionary perspective on human physical activity: implications for health. *Comp Biochem Physiol Part A.* 2003;136:153–159.

11 Kelley G, Tran ZV. Aerobic exercise and normotensive adults: a meta-analysis. *Med Sci Sports Exerc.* 1995;10:1371–1377.

12 Eaton SB, Strassman BI, Nesse RM, et al. Evolutionary health promotion. *Prev Med.* 2002;34:109–118.

13 Murphy NJ. Diabetes mellitus in Alaskan Yup'ik Eskimos and Athabascan Indians after 25 years. *Diabetes Care.* 1992;15:1390–1392.

14 Cohen AM, Marom L. Diabetes and accompanying obesity, hypertension and ECG abnormalities in Yemenite Jews 40 year after. *Diabetes Res.* 1993;23:65074.

15 Foster-Powell K, Holt SH, Brand-Miller JC. International table of glycemic index and glycemic load values:2002. *Am J Clin Nutr.* 2002;76:5–56.

16 Eaton SB. Fibre Intake in prehistoric times. In: Leeds AR, ed. *Dietary Fibre Perspectives. Reviews and Bibliography*, 2nd ed. London: John Libby; 1990:27–40.
17 Brand-Miller JC, Holt SHA. Australian Aboriginal plant foods: a consideration of their nutritional composition and health implications. *Nut Res Rev*. 1998;11:5–25.
18 Guthrie JF, Morton JF. Food sources of added sweeteners in the diets of Americans. *J Am Diet Assoc*. 2000;100:43–48, 51.
19 Duhard J-P. Upper Paleolithic figures as a reflection of human morphology and social organization. *Antiquity*. 1993;67:83–90.
20 Eaton SB. Humans, lipids, and evolution. *Lipids*. 1992;27:814–821.
21 Marnot MG, Syme L. Acculturation and coronary heart disease in Japanese Americans. *Am J Epidemiol*. 1975;102:481–490.
22 Stanhope JM, Sampson VM, Prior IA. The Tokelau Island migrant study: serum lipid concentrations in two environments. *J Chronic Dis*. 1981;34:45–55.
23 Howell WH, McNamara DJ, Tosca MA, Smith BT, Gaines JA. Plasma lipid and lipoprotein responses to dietary fat and cholesterol: a meta-analysis. *Am J Clin Nutr*. 1997;65:1747–1764.
24 Richards MP, Hedges RM. Focus: Gough's Cave and Sun Hole Cave human stable isotope values indicate a high animal protein in the British Upper Paleolithic. *J Archeol Sci*. 2000;27:1–3.
25 Eaton SB, Eaton SB III, Konner MJ. Paleolithic nutrition revisited: a twelve-year retrospective on its nature and implications. *Eur J Clin Nutr*. 1997;51:207–216.
26 Cordain L, Brand-Miller J, Eaton SB, Mann N, Holt SHA, Speth JD. Plant-animal subsistence ratios and macro-nutrient energy estimations in worldwide hunter-gatherer diets. *Am J Clin Nutr*. 2000;71:682–692.
27 Cordain L, Watkins BA, Florant GL, Kebler M, Rogers L, Li Y. Fatty acid analysis of wild ruminant tissues: evolutionary implications for reducing diet-related chronic disease. *Eur J Clin Nutr*. 2002;56:181–191.
28 Eaton, SB, Eaton SB II, Sinclair AJ, Cordain L, Mann NJ. Dietary intake of long-chain polyunsaturated fatty acids during the Paleolithic. *World Rev Nutr Diet*. 1998;83:12–23.
29 Kris-Etherton PM, Taylor DS, Yu-Poth S, et al. Polyunsaturated fatty acids in the food chain in the United States. *Am J Clin Nutr*. 2000;71:(Suppl 1):1795–1885.
30 Walker A, Leakey R. Perspectives on the Nariokotome *Homo erectus* skeleton. In: Walker A, Leakey R, eds. *The Nariokotome Homo Erectus Skeleton*. Cambridge, Massachusetts: Harvard University Press; 1993:411–430.
31 Eaton SB III, Eaton SB. Consumption of trace elements and minerals by preagricultural humans. In: Bogdon JD, Klevay LM, eds. *Clinical Nutrition of the Essential Trace Elements and Minerals*. Totowa, NJ: Humana Press; 2000:37–47.
32 Ewald PW. *Plague Time: How Stealth Infections Cause Cancer, Heart Disease, and Other Deadly Ailments*. New York: Free Press; 2000:107–122.
33 Cohen MN. *Health and the Rise of Civilization*. New Haven: Yale University Press; 1989:32–54.
34 Hill K, Hurtado AM. *Ache Life History. The Ecology and Demography of a Foraging People*. New York: Aldine De Gruyter; 1996:45–66.
35 Steinberg HO, Bayazeed B, Hook G, Johnson A, Cronin J, Baron AD. Endothelial dysfunction is associated with cholesterol levels in the high normal range for humans. *Circulation*. 1997;96:3287–3293.
36 He J, Ogden LG, Vupputuri S, et al. Dietary sodium intake and subsequent risk of cardiovascular disease in overweight adults. *JAMA*. 1999;282:2027–2034.
37 Young DB, Ma G. Vascular protective effects of potassium. *Seminar Nephrol*. 1999;19:477–86.
38 Maier JAM. Low magnesium and atherosclerosis: an evidence-based link. *Mol Aspects Med*. 2003;24:137–146.
39 DeCaterina R, Liao JK, Libby P. Fatty acid modulation of endothelial activation. *Am J Clin Nutr*. 2000;71:213S–233S.
40 Kris-Etherton PM, Harris WS, Appel LJ. Fish consumption, fish oil, omega-3 fatty acids, and cardiovascular disease. *Arterioscler Thromb Vasc Biol*. 2003;23:20–30.
41 Engler MM, Engler MB, Malloy MJ, et al. Antioxidant vitamins C and E improve endothelial dysfunction in children with hyperlipidemia. *Circulation*. 2003;108:1059.
42 Esposito K, Nappo F, Giugliano F, Gugliano G, Marfella R, Giugliano D. Effect of dietary antioxidants on postprandial endothelial dysfunction induced by a high-fat diet in healthy subjects. *Am J Clin Nutr*. 2003;77:139–143.
43 Stangl V, Lorenz M, Ludwig A, et al. The flavonoid phloretin suppresses stimulated expression of endothelial adhesion molecules and reduces activation of human platelets. *J Nutr*. 2005;135, 172–178.
44 Liu S. Willett WC, Stampfer MJ, et al. A prospective study of dietary glycemic load, carbohydrate index, and risk of coronary heart disease in US women. *Am J Clin Nutr*. 2000;71:1455–1461.
45 Dickinson S, Brand-Miller J. Glycemic index, postprandial glycemia and cardiovascular disease. *Curr Opin Lipidol*. 2005;16:69–75.
46 Frasetto L, Morris RC, Sellmeyer DE, Todd K, Sebastian A. Diet, evolution and aging. The pathological effects of the post-agricultural inversion of the potassium-to-sodium and base-to-chloride ratios in the human diet. *Eur J Clin Nutr*. 2001;40:200–213.

17

Putting Up Resistance

Maternal–Fetal Conflict over the Control of Uteroplacental Blood Flow

David Haig

Harvard University, Cambridge, Massachussetts

All expenditures involve an opportunity cost. This is true in economics: Money spent on one activity is unavailable for other activities. But it is also true in evolutionary biology: Time, resources, or energy expended on one fitness-enhancing activity is unavailable for other fitness-enhancing activities. Organisms, like consumers, are faced by trade-offs. Beyond a certain level of reproductive expenditure on any particular offspring, a parent's resources are better allocated to other uses, say to fighting disease or laying down fat to survive the next winter. And these other uses, either directly or indirectly, translate into less investment in other offspring. An organism usually maximizes its expected number of surviving offspring, not by investing everything in a single offspring, but by spreading its reproductive effort across multiple offspring (1).

Trivers (2) recognized that the reproductive trade-off between continued investment in one particular offspring and investment in other offspring implied the existence of an evolutionary conflict between parents and offspring. A parent is equally related to all its offspring (a gene in the parent has a 50% chance of being transmitted to each offspring), but an offspring is more closely related to itself than to its sibs (a gene in an offspring is definitely present in that offspring but has only a probability of being present in the other offspring that compete for parental resources). Therefore, offspring will have evolved to extract more investment from parents than parents have evolved to supply. It should be emphasized that this is not a conflict in which every gain to the offspring is a loss to the parent, but one in which conflict arises over how much the parent supplies. Human parents will be familiar with the problem of "setting limits" to offspring demands. As former children, we are all familiar with how unreasonable these limits can seem.

Parturition does not mark any fundamental change in the genetic relationship between mother and offspring. Therefore, the logic of intergenerational conflict should apply to interactions before, as well as after, birth. Haig (3–7) has argued that the physiological details of human pregnancy do not make sense without an appreciation of the element of conflict in maternal–fetal interactions. In this chapter, I focus on conflict over the level of maternal blood flow through the intervillous space of the placenta and how this helps to illuminate the causes of preeclampsia. The maternal endothelium emerges as a major player in fetal attempts to direct a greater proportion of maternal blood to the placenta.

HEMODYNAMICS OF PREGNANCY

A simple model of the maternal circulation during pregnancy is presented in Figure 17.1. Cardiac output from the left side of the heart is shared between two subcirculations arranged in parallel. The uteroplacental subcirculation (with resistance, R_p) represents all maternal blood diverted through the intervillous space of the placenta, whereas the nonplacental subcirculation (with resistance, R_m) represents the systemic blood supply to maternal tissues. The fetal share of maternal cardiac output is determined by the ratio of nonplacental resistance to the sum of the uteroplacental and nonplacental resistances, namely:

$$\textit{Fetal share} = \frac{R_m}{R_m + R_p}.$$

The theory of parent–offspring conflict predicts that fetal genes have been selected to appropriate a greater share of maternal blood than maternal genes have been selected to supply. Therefore, placental factors are predicted to increase R_m and decrease R_p, whereas maternal factors are predicted to decrease R_m and increase R_p (7). If R_m were increased without a change in R_p, the arteriovenous pressure difference would

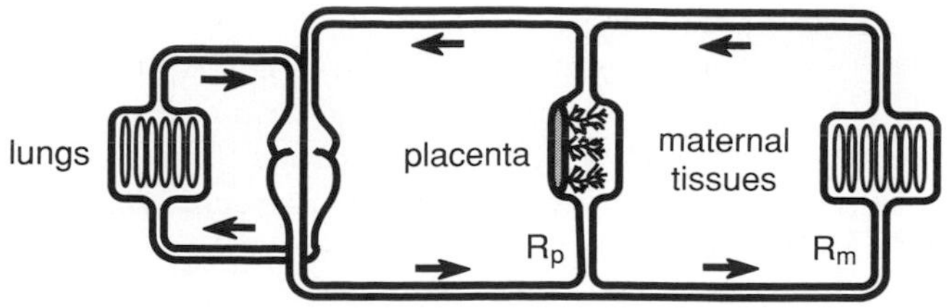

Figure 17.1. Schematic diagram of the maternal circulation during pregnancy. A proportion of maternal systemic blood flow is diverted through the intervillous space of the placenta. This proportion is determined by the relative values of the uteroplacental resistance (R_p) and the nonplacental resistance (R_m).

have to rise for nonplacental blood flow to be maintained. Uteroplacental blood flow would then increase in proportion to the rise in pressure head.

Figure 17.1 illustrates another important consequence of hemochorial placentation: The placenta can release substances directly into maternal blood. Therefore, all aspects of maternal physiology and metabolism that can be influenced by blood-borne factors come under contested control during pregnancy. The placenta is expected to release a cocktail of substances into the maternal circulation that can act at remote sites to benefit fetal fitness. From the perspective of natural selection, the mother's body becomes a somewhat unruly extension of the fetal soma. The mother lacks a comparable ability to influence fetal physiology, because maternal substances must first cross a layer of trophoblast and then the fetal endothelium before the substances gain access to the fetal circulation. These layers allow the fetus to exclude, or metabolize, unwanted maternal factors.

UTEROPLACENTAL CIRCULATION

The endometrium is supplied with blood via spiral arteries that undergo cyclic regeneration and degeneration over the course of the menstrual cycle. A pregnancy is established when an implanted embryo releases human chorionic gonadotropin (hCG) into the maternal circulation to prevent menstruation. Extravillous trophoblast cells invade the endometrium and remodel the spiral arteries, destroying the vessels' muscular lining and greatly expanding their radius. In the process, the endothelium is, at least partially, replaced by trophoblast. The remodeled arteries open into the intervillous space of the placenta, where maternal blood bathes the large surface area of the chorionic villi. The villi and the intervillous space are lined with syncytiotrophoblast and constitute the principal site for fetal uptake of nutrients and release of wastes. Maternal blood then drains from the intervillous space via uterine veins (8–10).

Maternal blood flows through the intervillous space from at least as early as 8 weeks of gestation, although flow initially is torpid because the lumens of the remodeled arteries are partially clogged by trophoblast (11,12). The oxygen tension of the placenta increases steeply from 8 to 12 weeks of gestation, as these trophoblastic obstructions are cleared to allow a more vigorous intervillous circulation (13). The trophoblast-mediated remodeling of maternal arteries continues into the second trimester of pregnancy and extends into the myometrium (14,15). This process of remodeling may continue into the third trimester in some pregnancies (16).

The uterus receives a small proportion of cardiac output in nonpregnant women, probably less than 1% (9). This proportion increases progressively during pregnancy as the fetal demand for resources escalates. The uterine share of maternal cardiac output is estimated as 9% at 20 weeks gestation, 12% at 24 weeks, 14% at 28 weeks, 15% at 32 weeks, and 16% at 36 to 38 weeks (17). This increase in fetal share must be associated with a progressive increase in the ratio of R_m to R_p. The continued invasion of trophoblasts during second trimester could contribute to this change by reducing R_p, either by extending remodeling deeper into the myometrium (for already modified arteries) or by the modification and recruitment of additional arteries as the placenta spreads laterally.

The theory of maternal–fetal conflict predicts that placental factors will act to decrease, and maternal factors act to increase, R_p. Therefore, the theory would be supported if trophoblast were principally responsible for the remodeling of spiral arteries into large-diameter, low-resistance vessels. The situation appears somewhat more complex than this simple prediction. The earliest signs of remodeling (vascular dilation) occur in the absence of trophoblasts, but the full expression of the "physiological changes" depends on the presence of trophoblasts (8,10,18). The latter changes include the loss of muscular and elastic elements from the vessel walls, thus abrogating the mother's ability to control the flow of blood to the intervillous space by vasoconstriction.

What is the evidence for maternal counteradaptations to increase R_p? The arterioles that supply the endometrium increase rapidly in length during the first weeks after ovulation, becoming highly convoluted as they outpace the growth in thickness of the endometrium (hence their designation as spiral arteries) (19–21). The vessels' increase in length and tortuosity would, by itself, increase R_p. However, another factor also may be at work. The rapid growth of the arterioles greatly increases the size of the "target" for modification by the initially tiny embryo, potentially slowing the early invasion of trophoblasts and allowing the mother to prepare defenses in depth. The peri-implantation period also is accompanied by decidualization of the endometrial stroma. This process is initiated around spiral arteries but then spreads to the rest of the endometrium (22). Decidualization involves the deposition of a basement membrane-like pericellular matrix (23) and has been conjectured to create a barrier that limits the extent of trophoblast invasion (3,24). It is worth emphasizing that the evolutionary challenge facing mothers is to "set limits" to trophoblast invasion, not to prevent it altogether.

NONPLACENTAL CIRCULATION

Maternal cardiovascular changes during pregnancy (increased heart rate, increased cardiac output, increased plasma volume, decreased arterial pressure, peripheral vasodilation) have been compared to the adaptive response to an arteriovenous fistula (25,26), with the low-resistance connection between spiral arteries and uterine veins playing the role of the fistula. This analogy has fallen into disfavor because the cardiovascular changes are initiated before substantial blood flow occurs through the intervillous space (27,28). The opening of the intervillous "fistula" is a predictable event, however, and thus can be *anticipated*, whereas fistulas caused by traumatic injuries are unpredictable events. In Aristotelian terms, the analogy is false from the perspective of efficient causes (physiological mechanism) but perhaps not from that of final causes (adaptive function).

The theory of maternal–fetal conflict predicts that placental factors will act to increase nonplacental resistance (R_m), but maternal factors will act to decrease R_m. Conflict over R_m is predicted to intensify as pregnancy progresses, because the nutritional requirements of the fetus steadily increase and because growth of the placenta confers greater power on the conceptus to influence maternal physiology. Early in pregnancy, a decrease in R_m and a concomitant decrease in arterial pressure occurs (29,30). The theory interprets these changes as a maternal adaptation to reduce the fetal share of cardiac output. Arterial pressure reaches a nadir in the second trimester, before a progressive rise toward term (29). Some studies report little increase in cardiac output after 24 weeks (31). Others report a decline toward term (32,33) or find that cardiac output in the third trimester is widely divergent among subjects (34). An increase in arterial pressure without an increase in cardiac output implies that total systemic vascular resistance (a combination of R_m and R_p) has increased. That is, the increase in arterial pressure during third trimester appears to be at least partly explained by an increase in vascular resistance. The theory of maternal–fetal conflict interprets this increase as a fetal adaptation to direct more blood to the intervillous space.

A corollary of these interpretations is that the factors responsible for the early decline in peripheral resistance should be maternal in origin (28), whereas the factors responsible for the late-pregnancy increase in resistance should be placental in origin. Placental factors that increased R_m would not increase fetal share if these factors caused a proportional rise in R_p. Therefore, placental factors are predicted to preferentially target R_m. Specificity of effects for resistance vessels of the nonplacental circulation is probably facilitated by the physiological changes to uteroplacental vessels (large diameter, lack of smooth muscle, partial replacement of endothelium by trophoblast).

PREGNANCY-INDUCED HYPERTENSION AND PREECLAMPSIA

For a subset of pregnant women, the rise in blood pressure during the second half of gestation is sufficient for a diagnosis of pregnancy-induced hypertension (PIH). A subset of this subset develops a far more serious condition known as preeclampsia, in which high blood pressure is accompanied by proteinuria (see Chapter 161). Bosio and colleagues (35) reported a longitudinal study of hemodynamic measures in 378 primigravidas recruited between 10 and 14 weeks of gestation. Of this cohort, 44 women (12%) developed hypertension, including 20 (5%) who developed preeclampsia. Women who subsequently developed hypertension had higher cardiac output at 10 to 14 weeks than did women who did not develop hypertension, and this difference became more pronounced as pregnancy progressed. For the women who developed hypertension but did not develop proteinuria (gestational hypertension; $n = 24$), high cardiac output was maintained until term. For the women who developed hypertension with proteinuria (preeclampsia; $n = 20$), a dramatic fall in cardiac output occurred with the onset of clinical disease concomitant with a dramatic increase in peripheral resistance. Thus, the progression to preeclampsia involved a crossover from a state of high cardiac output with low peripheral resistance to a state of low cardiac output with high peripheral resistance.

Preeclampsia can be associated with damage to most maternal organ systems. Its maternal pathophysiology exhibits necrosis and hemorrhage in multiple tissues, best explained by reduced perfusion secondary to vasospasm (36). Roberts and colleagues proposed that endothelial dysfunction was the common factor that united the diverse manifestations of preeclampsia and that this dysfunction was caused by factors released into the maternal circulation by the placenta (37). This model of the immediate cause of preeclamptic symptoms has attained wide acceptance, although debate continues on which circulating factors are principally responsible for endothelial dysfunction and on the causal antecedents for the release of these factors.

Fetal nutrition is compromised in many cases of preeclampsia. Numerous studies have reported that fewer spiral arteries are remodeled by trophoblasts in preeclamptic pregnancies (14). Significantly, increased uteroplacental resistance and reduced placental perfusion precedes the onset of preeclampsia. Therefore, birth weight is decreased and the incidence of preeclampsia increased for first pregnancies compared with subsequent pregnancies (38,39), for twin pregnancies compared with singleton pregnancies (39,40), and for women giving birth at high altitude compared to low altitude (41,42). Uteroplacental blood supply may improve in second and subsequent pregnancies because of permanent changes to uterine vessels that occur during the first pregnancy and may be impaired at high altitude because fewer spiral arteries are remodeled (43–45). Average birth weight is reduced in twin pregnancies, presumably because two fetuses compete for limited maternal resources.

Haig (3,7) interpreted preeclampsia as an expression of a fetal adaptation to increase uteroplacental blood flow by causing vasoconstriction of the mother's nonplacental circulation. High peripheral resistance does not occur in all pregnancies.

Therefore, the induction of this state was proposed to be a conditional strategy of the fetus to increase blood flow to the uterus in pregnancies in which the supply of one or more nutrients is inadequate for fetal needs. The strategy is risky because of the possibility of substantial impairment of the health of the fetus's principal postnatal caregiver (its mother). For such a risk to be justified, the increased flow of nutrients to the placenta must have been associated with major increases in the probability of infant survival. From this perspective, the induction of preeclampsia may be a stereotyped response to increase placental perfusion whenever a fetus stands to gain a relatively large benefit from an increased blood supply, whereas the reasons for activating this response may be diverse. As a corollary, the release of factors that cause maternal vasoconstriction should be interpreted as evidence of normal, rather than aberrant, placental function. In this view, preeclampsia has two kinds of "causes" that should be clearly distinguished: First are the "predisposing" factors (inadequate trophoblast invasion, twin pregnancy, etc.) that cause activation of the adaptive fetal response; second are the "effectors" of the adaptive response – the placental factors responsible for maternal symptoms.

Uteroplacental blood flow can be increased either by increasing cardiac output (for given R_m and R_p) or by increasing the ratio of R_m to R_p (for a given cardiac output). Bosio and colleagues reported that preeclampsia preferentially occurred in pregnancies with higher than average cardiac output before the onset of disease. They proposed that high cardiac output could contribute to endothelial damage and the onset of preeclampsia (35). An alternative interpretation is that high cardiac output is the fetus's "preferred" mechanism of enhancing uteroplacental blood flow, with peripheral vasoconstriction employed as a last resort if the preferred mechanism proves inadequate.

I have neither the space nor competence to adequately review the many circulating factors that are increased or decreased in preeclampsia. Instead, I will pick three factors that have attracted recent attention (sFLT-1, AT_1R autoantibodies, and leptin), and use these to illustrate the kinds of interpretation that are suggested by viewing preeclampsia as a fetal adaptation rather than as physiological malfunction. My discussion is prefaced by two general observations. First, adaptations may be complex. If preeclampsia is an expression of a fetal adaptation to increase nutrient supply, then this adaptation may involve multiple placental factors that target multiple maternal systems. These may include factors that increase the nutrient content of maternal blood as well as those that increase the volume of blood flowing through the intervillous space. Second, mothers are expected to have evolved countermeasures to limit fetal manipulation of maternal physiology and to limit the costs to maternal health. Therefore, fetal actions must be clearly distinguished from maternal responses. Both may be hyperactivated in preeclampsia: The former are predicted to exacerbate symptoms, whereas the latter are predicted to ameliorate symptoms.

Maynard and colleagues (46) identified soluble *fms-like tyrosine kinase 1* (sFLT-1) as a prime candidate to be the placental factor responsible for the endothelial dysfunction of preeclampsia. Levels of sFLT-1 were elevated fivefold in the sera of preeclamptic patients compared with normotensive controls, but these high levels declined rapidly after delivery. Moreover, sFLT-1 was shown to be responsible for the antiangiogenic properties of preeclamptic sera and to induce preeclampsia-like symptoms (hypertension, proteinuria, glomeruloendotheliosis) when administered to rats. sFLT-1 is believed to antagonize the actions of vascular endothelial growth factor (VEGF) by preventing VEGF binding to membrane-bound receptors. Several independent groups have comfirmed the association of preeclampsia with elevated sFLT-1 (47–51).

The maternal–fetal conflict hypothesis sees endothelial dysfunction and vasoconstriction as the *function*, rather than as a side effect, of elevated sFLT-1. There is no need to invoke some other function for placental release of sFLT-1 in preeclampsia. The conflict hypothesis also provides a ready interpretation of why sFLT-1 and VEGF should be simultaneously elevated in preeclamptic sera (46): The former is a fetal adaptation to reduce angiogenic repair; the latter, a maternal adaptation to enhance repair.

Wallukat and colleagues (52,53) have reported that the sera of most women with preeclampsia contain antibodies that stimulate *angiotensin* AT_1 *receptors* (AT_1Rs), but that such antibodies are absent from the sera of most normotensive pregnant women (54). In one woman, anti-AT_1R antibodies were undetectable in serum at 21 weeks of gestation but were present at 24 weeks, after the onset of preeclampsia. Anti-AT_1R antibodies disappear from the serum of preeclamptic women after delivery (52). The theory of maternal–fetal conflict suggests that anti-AT_1R antibodies may be part of a fetal adaptation to increase uteroplacental blood flow, in this case by targeting maternal AT_1Rs. This interpretation would be strengthened if it could be shown that anti-AT_1R antibodies are a placental rather than a maternal product (or are induced by exposure of the maternal immune system to placental AT_1Rs).

The human placenta is a significant source of *leptin*, most of which is released into the maternal circulation (55). Maternal plasma leptin levels are higher in preeclamptic pregnancies than in normotensive pregnancies, and this difference precedes the onset of symptoms (56). Levels of leptin mRNA are higher in placentas from preeclamptic, than from normotensive, pregnancies (57,58). A plausible function of placental leptin is to promote maternal lipolysis, thus enhancing the nutrient supply to the fetus (59). If preeclampsia results from the activation of an integrated fetal response to increased nutrient supply, then increased production of leptin may be one component of this response that specifically targets maternal adipocytes.

The hypothesis that preeclampsia is an outcome of a fetal adaptation to increase uteroplacental blood flow has a strong and a weak version. In the strong version, preeclampsia itself has been associated during human evolution with benefits to fetuses, relative to otherwise-equivalent fetuses that did not cause preeclampsia. In the weak version, preeclampsia is the occasional nonadaptive outcome of fetal adaptations that operate in nonpreeclamptic pregnancies to increase uteroplacental blood flow.

The key prediction of the strong version of the hypothesis – that the peripheral vasoconstriction of preeclampsia increases blood flow to the placenta – has yet to be demonstrated. The ideal test would be a longitudinal study that measured uteroplacental blood flow before and after the onset of preeclampsia. The strong version predicts that the onset of preeclampsia should be associated with an *increase* in placental perfusion. Such a test would be technically challenging. Another prediction is that preeclampsia should increase fetal fitness (strictly, that preeclampsia should have increased fetal fitness in the recent evolutionary past). However, this prediction is confounded with the subsidiary prediction that preeclampsia should be preferentially induced by fetuses with lower than average expected fitness. Preeclamptic pregnancies are more likely than normotensive pregnancies to result in small-for-gestational-age babies, for both preterm and term deliveries (60). Can this be used as evidence against the prediction that preeclampsia improves fetal nutrition? Not necessarily, if the fetuses were already small-for-gestational-age before the onset of preeclampsia. It is not obvious what should be the appropriate comparison group. Preeclampsia is not always associated with impaired fetal growth, however, because the incidence of large-for-gestational-age babies is increased when preeclampsia develops near term (60,61).

Twin pregnancies provide an interesting comparison. The incidence of preeclampsia is increased presumably because two fetuses are competing for maternal resources, and the uterus contains two placentas, not because remodeling of spiral arteries is impaired in twin gestations. Thus, twins from preeclamptic pregnancies are less likely to have viability compromised for other reasons that predispose to preeclampsia in singleton pregnancies. Two studies have reported that average birth weight and gestation length are increased, and perinatal mortality is decreased, in comparisons of preeclamptic twin pregnancies to normotensive twin pregnancies (62,63).

In the weak version of the hypothesis, preeclampsia is viewed as a pathological malfunction that benefits neither mother nor fetus. This is similar to conventional views. Even so, I hope that an appreciation of the element of conflict could contribute to understanding the causes of preeclampsia in two ways. First, it could provide a set of hypotheses about the physiological functions in normotensive pregnancies of the factors that overwhelm the mother's homeostatic mechanisms in preeclamptic pregnancies. Second, it could provide an understanding of why homeostasis should be particularly unstable in pregnancy. Communication between parties with conflicting interests is more difficult than communication between parties with identical interests. If conflicts of interest are possible, messages cannot always be trusted, and untrustworthy messages are often best ignored. One consequence is that maternal physiology during pregnancy will lack many of the checks, balances, and feedback controls that are present in the nonpregnant state. Suppose, for example, that a signaling system existed whereby a mother informed the fetus whenever she found herself in situations that required a temporary reduction in fetal demands. If the fetus responded accordingly, those maternal genes would then be favored that caused mothers to send the same signal when their need was less dire. This would, in turn, favor fetal genes that gave a muted response to maternal signals. Such a signaling system is inherently unstable. In this view, preeclampsia may result from the inability of a physiologically threatened mother to credibly communicate her perilous state to a demanding fetus (5,6).

Finally, evolutionary considerations may help to explain why preeclampsia is a relatively common condition of human pregnancies, but similar conditions are rare or absent in other mammals. An important constraint on the physiological demands that a fetus can make on its mother is the offspring's dependence on a healthy mother for postnatal care. This constraint is relaxed in humans relative to other mammals because mothers who experienced difficult pregnancies would often have received support from other members of their social group. Thus, temporary incapacitation of a mother need not have been a death sentence for her newborn (indeed, human infants sometimes survive maternal death). A consequence of human social evolution may therefore have been a shift in the relative costs and benefits toward favoring greater fetal demands during pregnancy.

KEY POINTS

- An evolutionary conflict exists between maternal and fetal genes over the amount of maternal blood that flows through the intervillous space of the placenta. In preeclampsia, placental factors cause damage to the maternal endothelium, resulting in systemic vasoconstriction and increased blood flow to the placenta.
- Placental factors are responsible for the endothelial dysfunction observed in preeclampsia.
- sFLT-1 appears to be one of these placental factors.
- Multiple maternal systems may be targeted by multiple placental factors.

Future Goals

- To determine whether the vasoconstriction of preeclampsia causes an increase or decrease of maternal blood flow to the placenta
- To identify the different placental factors (and their maternal targets) responsible for the complex symptoms of preeclampsia

REFERENCES

1 Stearns SC. *The Evolution of Life Histories*. Oxford: Oxford University Press, 1992.

2 Trivers RL. Parent-offspring conflict. *Am Zool.* 1974;14:249–264.

3 Haig D. Genetic conflicts in human pregnancy. *Q Rev Biol.* 1993; 68:495–532.

4 Haig D. Gestational drive and the green-bearded placenta. *Proc Natl Acad Sci USA.* 1996;93:6547–6551.

5 Haig D. Placental hormones, genomic imprinting, and maternal – fetal communication. *J Evol Biol.* 1996;9:357–380.

6 Haig D. Altercation of generations: genetic conflicts of pregnancy. *Am J Reprod Immunol.* 1996;35:226–232.

7 Haig D. Genetic Conflicts of Pregnancy and Childhood. In: Stearns SC, ed. *Evolution in Health and Disease*. Oxford: Oxford University Press, 1999:77–90.

8 Kam EPY, Gardner L, Loke YW, King A. The role of trophoblast in the physiological change in the decidual spiral arteries. *Hum Reprod.* 1999;14:2131–2138.

9 Kliman HJ. Uteroplacental blood flow. The story of decidualization, menstruation, and trophoblast invasion. *Am J Pathol.* 2000; 157:1759–1768.

10 Kaufmann P, Black S, Huppertz B. Endovascular trophoblast invasion: implications for the pathogenesis of intrauterine growth retardation and preeclampsia. *Biol Reprod.* 2003;69:1–7.

11 Carbillon L, Challier JC, Alouini S, Uzan M, Uzan S. Uteroplacental circulation development: Doppler assessment and clinical importance. *Placenta.* 2001;22:795–799.

12 Jauniaux E, Hempstock J, Greenwold N, Burton GJ. Trophoblast oxidative stress in relation to temporal and regional differences in maternal placental blood flow in normal and abnormal early pregnancies. *Am J Pathol.* 2003;162:115–125.

13 Jauniaux E, Watson AL, Hempstock J, Bao YP, Skepper JN, Burton GJ. Onset of maternal arterial blood flow and placental oxidative stress. A possible factor in human early pregnancy failure. *Am J Pathol.* 2000;157:2111–2122.

14 Brosens JJ, Pijnenborg R, Brosens IA. The myometrial junctional zone spiral arteries in normal and abnormal pregnancies. A review of the literature. *Am J Obstet Gynecol.* 2002;187:1416–1423.

15 Lyall F. Development of the utero-placental circulation: the role of carbon monoxide and nitric oxide in trophoblast invasion and spiral artery transformation. *Microsc Res Tech.* 2003;60:402–411.

16 Khong TY, Sawyer IH, Heryet AR. An immunohistologic study of endothelialization of uteroplacental vessels in human pregnancy – evidence that endothelium is focally disrupted by trophoblast in preeclampsia. *Am J Obstet Gynecol.* 1992;167:751–756.

17 Konje JC, Kaufmann P, Bell SC, Taylor DJ. A longitudinal study of quantitative uterine blood flow with the use of color power angiography in appropriate for gestational age pregnancies. *Am J Obstet Gynecol.* 2001;185:608–613.

18 Craven CM, Morgan T, Ward K. Decidual spiral artery remodelling begins before cellular interaction with cytotrophoblasts. *Placenta.* 1998;19:241–252.

19 Daron GH. The arterial pattern of the tunica mucosa of the uterus in *Macacus rhesus. Am J Anat.* 1936;58:349–419.

20 Markee JE. Menstruation in intraocular endometrial transplants in the rhesus monkey. *Contributions to Embryology*, No. 177. Carnegie Institution of Washington Publication 518 1940;28: 219–308.

21 Maas JWM, Groothuis PG, Dunselman GAJ, et al. Endometrial angiogenesis throughout the human menstrual cycle. *Hum Reprod.* 2001;16:1557–1561.

22 Frank HG, Kaufmann P. Nonvillous parts and trophoblast invasion. In: Benirschke K, Kaufmann P, eds. *Pathology of the Human Placenta.* 4th ed. New York: Springer-Verlag, 2000:171–272.

23 Wewer U, Faber M, Liotta LA, Albrechtsen R. Immunochemical and ultrastructural assessment of the nature of the pericellular basement membrane of human decidual cells. *Lab Invest.* 1985;53:624–633.

24 Fothergill WE. The function of the decidual cell. *Edinburgh Med J.* 1899;5:265–273.

25 Burwell CS. The placenta as a modified arteriovenous fistula, considered in relation to the circulatory adjustments of pregnancy. *Am J Med Sci.* 1938;195:1–7.

26 Burwell CS, Strayhorn WD, Flickinger D, Corlette MB, Bowerman EP, Kennedy JA. Circulation during pregnancy. *Arch Intern Med.* 1938;62:978–1003.

27 Schrier, RW, Briner VA. Peripheral arterial vasodilation hypothesis of sodium and water retention in pregnancy: implications for pathogenesis of preeclampsia-eclampsia. *Obstet Gynecol.* 1991; 77:632–639.

28 Chapman AB, Zamudio S, Woodmansee W, et al. Systemic and renal hemodynamic changes in the luteal phase of the menstrual cycle mimic early pregnancy. *Am J Physiol.* 1997;273:F777–F782.

29 Redman C. Hypertension in pregnancy. In: de Swiet M, ed. *Medical Disorders in Obstetric Practice.* 2nd ed. Oxford: Blackwell Scientific, 1989:249–305.

30 Duvekot JJ, Cheriex EC, Pieters FAA, Menheere PPCA, Peeters LLH. Early pregnancy changes in hemodynamics and volume homeostasis are consecutive adjustments triggered by a primary fall in systemic vascular tone. *Am J Obstet Gynecol.* 1993;169: 1382–1392.

31 Clapp JF, Capeless E. Cardiovascular function before, during, and after the first and subsequent pregnancies. *Am J Cardiol.* 1997;80: 1469–1473.

32 Adams JQ. Cardiovascular physiology in normal pregnancy: studies with the dye dilution technique. *Am J Obstet Gynecol.* 1954;67:741–759.

33 Hennessy TG, MacDonald D, Hennessy MS, et al. Serial changes in cardiac output during normal pregnancy: a Doppler ultrasound study. *Eur J Obstet Gynecol Reprod Biol.* 1996;70:117–122.

34 van Oppen ACC, Stigter RH, Bruinse HW. Cardiac output in normal pregnancy: a critical review. *Obstet Gynecol.* 1996;87:310–318.

35 Bosio PM, McKenna PJ, Conroy R, O'Herlihy C. Maternal central hemodynamics in hypertensive disorders of pregnancy. *Obstet Gynecol.* 1999;94:978–984.

36 Roberts JM. Pregnancy-Related Hypertension. In: Creasy RK, Resnik R, Iams JD, eds. *Maternal-Fetal Medicine.* 5th ed. Philadelphia: Saunders, 2004:859–899.

37 Roberts JM, Taylor RN, Musci TJ, Rodgers DM, Hubel CA, McLaughlin MK. Preeclampsia: an endothelial cell disorder. *Am J Obstet Gynecol.* 1989;161:1200–1204.

38 Karn MN, Penrose LS. Birth weight and gestation time in relation to maternal age, parity and infant survival. *Ann Eugen.* 1951;16: 147–164.

39 MacGillivray I. Some observations on the incidence of pre-eclampsia. *J Obstet Gynaecol Br Empire.* 1958;65:536–539.

40 McKeown T, Record RG. Observations on foetal growth in multiple pregnancies in man. *J Endocrinol.* 1953;8:386–401.

41 Palmer SK, Moore LG, Young DA, Cregger B, Berman JC, Zamudio S. Altered blood pressure course during normal pregnancy and increased preeclampsia at high altitude (3100 meters) in Colorado. *Am J Obstet Gynecol.* 1999;180:1161–1168.
42 Zamudio S, Palmer SK, Regensteiner JG, Moore LG. High altitude and hypertension during pregnancy. *Am J Hum Biol.* 1995;7: 183–193.
43 Pankow Dr. Graviditäts-, Menstruations- und Ovulations-sklerose der Uterus- und Ovarialgefässe. *Archiv Gynäkol.* 1906;80: 271–282.
44 Khong TY, Adema ED, Erwich JJHM. On an anatomical basis for the increase in birth weight in second and subsequent born children. *Placenta.* 2003;24:348–353.
45 Tissot van Patot M, Grilli A, Chapman P, et al. Remodelling of uteroplacental arteries is decreased in high altitude pregnancies. *Placenta.* 2003;24:326–335.
46 Maynard SE, Min JY, Merchan J, et al. Excess placental soluble fms-like tyrosine kinase 1 (sFLT1) may contribute to endothelial dysfunction, hypertension, and proteinuria in preeclampsia. *J Clin Invest.* 2003;111:649–658.
47 Ahmad S, Ahmed A. Elevated placental soluble vascular endothelial growth factor receptor-1 inhibits angiogenesis in preeclampsia. *Circ Res.* 2004;95:884–891.
48 Chaiworapongsa T, Romero R, Espinoza J, et al. Evidence supporting a role for blockade of the vascular endothelial growth factor system in the pathophysiology of preeclampsia. *Am J Obstet Gynecol.* 2004;190:1541–1550.
49 Koga K, Osuga Y, Yoshino O, et al. Elevated serum soluble vascular endothelial growth factor receptor 1 (sVEGFR-1) levels in women with preeclampsia. *J Clin Endocrinol Metab.* 2003;88: 2348–2351.
50 McKeeman GC, Joy A, Caldwell CM, Hunter AJ, McClure N. Soluble vascular endothelial growth factor receptor-1 (sFlt-1) is increased throughout gestation in patients who have preeclampsia develop. *Am J Obstet Gynecol.* 2004;191:1240–1246.
51 Tsatsaris V, Goffin F, Munaut C, et al. Overexpression of the soluble vascular endothelial growth factor receptor in preeclamptic patients: pathophysiological consequences. *J Clin Endocrinol Metab.* 2003;88:5555–5563.
52 Wallukat G, Homuth V, Fischer T, et al. Patients with preeclampsia develop agonistic autoantibodies against the angiotensin AT_1 receptor. *J Clin Invest.* 1999;103:945–952.
53 Wallukat, G, Neichel D, Nissen E, Homuth V, Luft FC. Agonistic autoantibodies directed against the angiotensin II AT_1 receptor in patients with preeclampsia. *Can J Physiol Pharmacol.* 2003;81: 79–83.
54 Xia Y, Wen H, Bobst S, Day MC, Kellems RE. Maternal autoantibodies from preeclamptic patients activate angiotensin receptors on human trophoblast cells. *J Soc Gynecol Invest.* 2003;10:82–93.
55 Linnemann K, Malek A, Sager R, Blum WF, Schneider H, Fusch C. Leptin production and release in the dually *in vitro* perfused human placenta. *J Clin Endocrinol Metab.* 2000;85:4298–4301.
56 Anim-Nyame N, Sooranna SR, Steer PJ, Johnson MR. Longitudinal analysis of maternal plasma leptin concentrations during normal pregnancy and pre-eclampsia. *Hum Reprod.* 2000;15:2033–2036.
57 Dötsch J, Nüsken KD, Knerr I, Kirschbaum M, Repp R, Rascher W. Leptin and neuropeptide Y gene expression in human placenta: ontogeny and evidence for similarities to hypothalamic regulation. *J Clin Endocrinol Metab.* 1999;84:2755–2758.
58 Reimer T, Koczan D, Gerber B, Richter D, Thiesen HJ, Friese K. Microarray analysis of differentially expressed genes in placental tissue of pre-eclampsia: up-regulation of obesity-related genes. *Hum Mol Reprod.* 2002;8:674–680.
59 Roberts JM, Lain KY. Recent insights into the pathogenesis of preeclampsia. *Placenta.* 2002;23:359–372.
60 Vatten LJ, Skjærven R. Is pre-eclampsia more than one disease? *Br J Obstet Gynaecol.* 2004;111:298–302.
61 Rasmussen S, Irgens LM. Fetal growth and body proportion in preeclampsia. *Obstet Gynecol.* 2003;101:575–583.
62 Campbell DM, MacGillivray I. Preeclampsia in twin pregnancies: incidence and outcome. *Hypertension Preg.* 1999;18:197–207.
63 Sibai B, Hauth J, Caritis S, et al. Hypertensive disorders in twin versus singleton gestations. *Am J Obstet Gynecol.* 2000;182:938–942.

18

Xenopus as a Model to Study Endothelial Development and Modulation

Aldo Ciau-Uitz, Claire Fernandez, and Roger Patient

Weatherall Institute of Molecular Medicine, University of Oxford, United Kingdom

The endothelial cells (ECs) of the primary vascular plexuses are among the first mesodermal derivatives to differentiate and become functional in the vertebrate embryo. Anatomically, ECs can be seen developing long before the heart initiates beating and, in gene expression studies, cells expressing endothelium-specific genes can be detected just after the end of gastrulation. In mammalian models, in embryo studies of the early events involved in the induction of ECs from their mesodermal precursors, a process termed *vasculogenesis*, are hampered by the technical difficulties inherent to the in utero development of the embryo. In addition, gene perturbations that result in impaired vasculogenesis are usually fatal and, as a consequence, the further remodeling and maturation of the vascular system, angiogenesis, cannot be investigated. Although embryonic stem cells and EC lines represent useful systems for the investigation of the signaling mechanisms involved during vasculogenesis and/or angiogenesis, due to their cell homogeneity and low anatomical complexity, these signaling mechanisms may well not be fully applicable to the embryo environment. Thus, model systems such as *Xenopus* and zebrafish, in which embryos develop externally and in a less circulation-dependent manner, offer an invaluable opportunity for the study of the early development of the vascular system. However, because the zebrafish is less amenable for transplantation and lineage-labeling experiments and, critically, because it lacks smooth muscle cells and a lymphatic system, *Xenopus* appears to be more suited for the study of some aspects of blood vessel development. Therefore, fate maps of the different blood vessels can be generated in *Xenopus*, and the importance of the interaction between ECs and smooth muscle cells during blood vessel formation can be studied.

During the last 50 years, the African clawed frog, *Xenopus laevis*, has been one of the most useful models to study early vertebrate development. After its initial worldwide introduction to the laboratory as a pregnancy bioassay,[1] *Xenopus* became the preferred model for developmental biologists due to its easy cultivation and capacity to produce large numbers of homogeneous, nonseasonal embryos (1). Together with its external development, the large size of the *Xenopus* embryo makes it ideal for experiments involving microsurgery, tissue transplantation and, very importantly, lineage labeling and the microinjection of genetic material. The application of modern molecular techniques, combined with traditional embryological methods, has led to the discovery of numerous developmental mechanisms that later have been proven to be relevant to all vertebrates. Here, we summarize the current knowledge on the early development of vascular ECs in *Xenopus* and discuss the prospects for using the *Xenopus* embryo as a model for drug discovery and therapeutics.

ORIGINS AND DEVELOPMENT OF VASCULAR ENDOTHELIAL CELLS

Knowledge of the origins, developmental history, and tissue interactions of the mesodermal precursor cells giving rise to ECs is essential for a better understanding of the hierarchy of signaling pathways governing vasculogenesis. Few systems are as suited as *Xenopus* to the investigation of these fundamental issues: The embryos are large and, in a small percentage of them, the cleavage planes are perfectly regular and symmetric, which allows reliable and reproducible cell labeling and fate mapping experiments (2). In fact, the contribution of single blastomeres of the 32-cell stage embryo to tissues of the

1 The South African clawed frog (*Xenopus laevis*) was formerly widely used in a pregnancy test, in which female frogs were injected with an extract of urine from the women to be tested. The presence of a characteristic pregnancy hormone (gonadotropin) induces ovulation in female frogs.

tailbud-stage embryo has been determined (3–4), and some attempts have been made to investigate the origins of the major blood vessels of the tadpole (5–6). Surprisingly, one of the fate mapping efforts concluded that all blastomeres of the 32-cell stage embryo contribute to some extent to the endothelium of the major blood vessels (6). This is in clear contradiction to previous fate maps, which showed that the animal-most[2] blastomeres of the 32-cell stage embryo only give rise to ectodermal derivatives (3–4); therefore, the endothelium, which is a mesodermal derivative, would not derive from these blastomeres. This discrepancy may be a consequence of the use of irregularly cleaving embryos,[3] which have an altered fate map (2); their animal-most blastomeres do give rise to mesodermal derivatives (7). In a 32-cell stage regularly cleaving embryo, the blastomeres giving rise to the mesodermal layer are well defined (Figure 18.1A for color reproduction, see Color Plate 18.1.), and only those blastomeres would be expected to give rise to the endothelium of the blood vessels. Alternatively, the contribution of endothelium from the animal-most blastomeres may reflect a neural crest contribution to this cell lineage (8).

To familiarize the reader with the blood vessels of the *Xenopus* tadpole, we will describe the development of the embryonic vasculature before giving an account of their origins. The first anatomically distinguishable ECs in the embryo are seen a couple of hours before the initiation of heart beating, whereas the first cells expressing endothelium-specific genes are detected one day earlier (Figure 18.1B, *arrow*). A comprehensive anatomical description of embryonic vascularization can be found elsewhere (9 and references therein); here we focus on the development of the vasculature up to stage 40, just after the primary circulatory circuit is established, as visualized with molecular markers.

Using the endothelium-specific gene *Tie2* as a marker, the first ECs are observed ventral to the heart field just after the end of neurulation[4] (see Figure 18.1B, *arrow*). As development progresses, the ventral region of the embryo is rapidly populated by ECs that form a network of vitelline vessels (Figure 18.1C–K, *VitV*). Initially, the vitelline vessels surround the developing ventral blood island (VBI)(10), but later they extend both dorsally and ventrally. Dorsally, they extend to cover the whole of the lateral plate and join the developing posterior cardinal veins (PCVs, see Figure 18.1C–F); ventrally, they reach the ventral midline as the red blood cells of the VBI mature and enter circulation (Figure 18.1G–K). Interestingly, a large yolky area at the posterior of the embryo is never vascularized (see Figure 18.1E–I, *star*); this yolk mass is continuously reduced through development until it disappears at around stage 40. After stage 40, the vitelline vessels further remodel to form a single major vessel in the ventral midline, the median abdominal vein, which is connected to the caudal vein by ventral commissures (9).

By stage 32 a number of vessels can be seen developing anterior to the vitelline vessels (see Figure 18.1F). The anterior end of the vitelline vessels connects to the sinus venosus (SV) via the omphalomesenteric vein (OMV). Further anterior the SV drains into an already lumenized endocardium (End). Anterior to and from the endocardium extends the developing arterial system; the ventral aorta (VA) extends toward the head, where it branches to form a pair of external carotid arteries (ECAs), one on each side of the embryo. Almost immediately, the ECAs extend dorsally and internalize to connect with the internal carotid arteries, which in turn extend posterior to the head as the paired dorsal aortae (pDA). The eyes are one of the first organs to be irrigated. To fit this demand, the hyaloid artery (HyA) develops from the anterior end of each internal carotid artery. Also at this stage of development, a complex vascular network appears to be developing around and posterior to the eyes, and later a number of cranial vessels will develop from this vascular plexus (Figure 18.1G–J). The VA is connected to the aortic arches (see Figure 18.1G, *AoArch*), which extend dorsally to join the pDA. The aortic arches develop in series from anterior to posterior and irrigate the gills of the tadpole. Six pairs of aortic arches develop in the early embryo, but they remodel constantly until only three of them remain at stage 50: the first, known as the ECAs; the third, which are the largest and known as the larval aortae; and the fourth (9). In the head, the venous system appears to develop later than the arteries, because at stage 32 only the primordia of the anterior cardinal veins (ACV) can be seen developing just anterior to the pronephric sinus (see Figure 18.1F).

The PCVs are the first of the major axial vessels to develop (see Figure 18.1D). Although anatomically they are first seen at stage 32, their precursors are seen as early as stage 26 (see Figure 18.1D), in the area dorsal and posterior to the pronephric anlagen (the dorsal lateral plate, DLP), where they are already associated with the precursors of the pronephric sinus (see Figure 18.1D, *PS*). Later, the PCV precursors extend posteriorly all along the trunk and finally, at around stage 34, they reach the tail. At the level of the tail, the two PCVs fuse in the midline to form the caudal vein (see Figure 18.1J, *CauV*). Meanwhile, the pronephric sinus grows ventrally forming the duct of Cuvier (DCuv) and later connects to the SV. As the nonvascularized yolk mass disappears, the two PCVs fuse to form the interrenal vein. Fusion initiates in the posterior at stage 35 to 36 and extends slowly anteriorly until,

2 *Xenopus* mature oocytes exhibit an animal–vegetal polarity, in which the animal hemisphere is characterized by being strongly pigmented, whereas the vegetal hemisphere is lightly or not pigmented at all. After fertilization, the animal–vegetal polarity is conserved so that, at the 32-cell stage, eight blastomeres (Tier A, Figure 18.1A) form the animal pole of the embryo; these are the animal-most blastomeres.

3 After fertilization, *Xenopus* embryos undergo a series of synchronous cell divisions that can occur both symmetrically or asymmetrically. In symmetrically dividing embryos, the animal–vegetal division planes meet at both the animal and vegetal poles (Figure 18.1A) and produce blastomeres of equal size. On the other hand, in asymmetrically dividing embryos, the animal–vegetal division planes do not meet at the vegetal pole; as a result, blastomeres produced are of different sizes (2). The former embryos are called *regular cleavers*, whereas the latter are *irregular cleavers*. Both are found in the wild and are able to develop normally.

4 The developmental stage in which the rudiment of the central nervous system is laid out in the embryo is known as *neurulation*. During this process, neural cell precursors undergo morphogenetic movements to form the brain and neural tube.

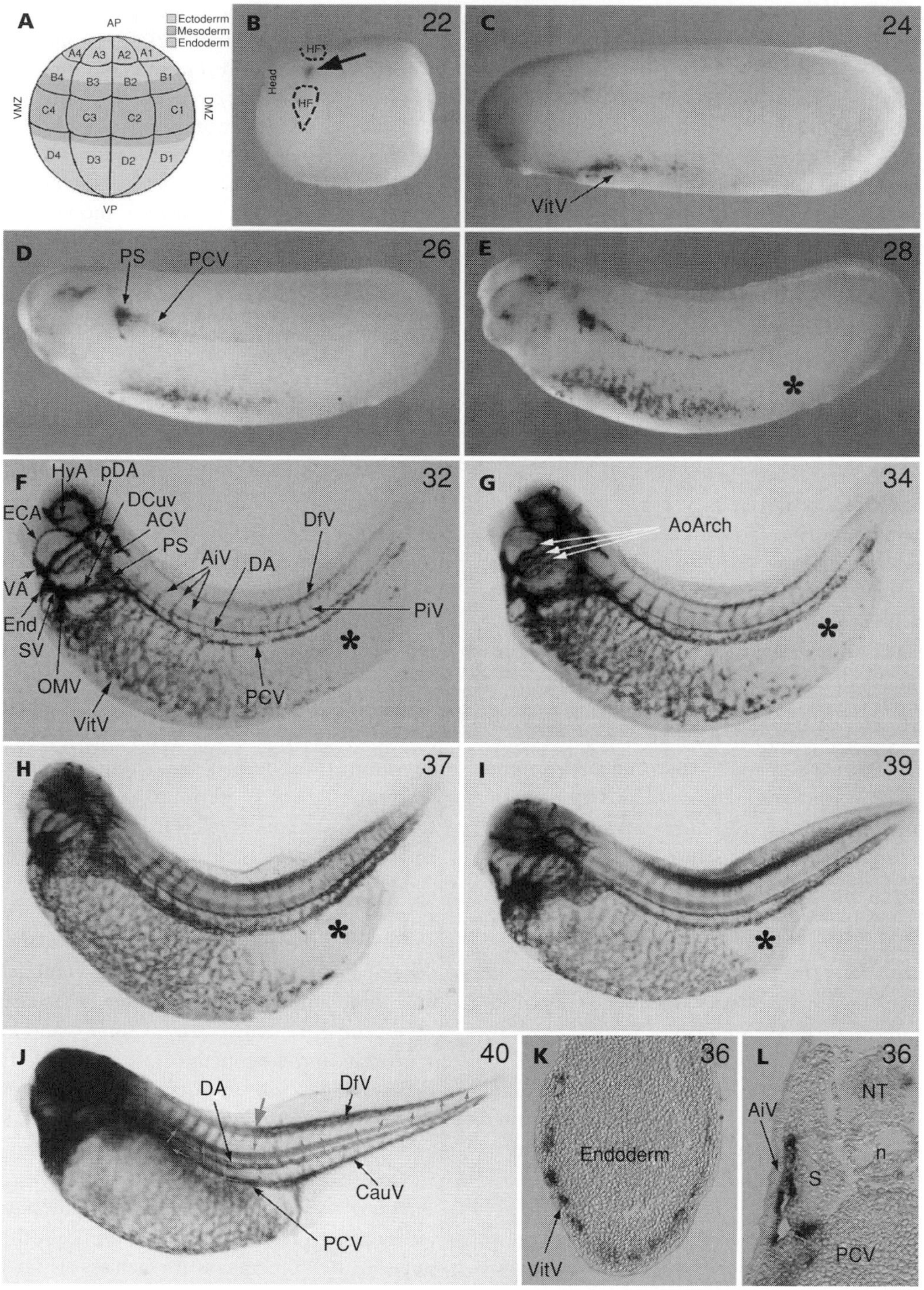

Figure 18.1. Early development of the vascular system in *Xenopus* embryos. (**A**) Regions of the 32-cell stage embryo contributing to the three primary germ layers. (**B–L**) Development of the vascular system as illustrated by the expression pattern of the angiopoietin tyrosine kinase receptor, Tie2; numbers at top right corner indicate developmental stage. **B**, Ventral view; **C–J**, lateral view, anterior to the left; **K–L**, transverse sections through the trunk. Asterisk (*) indicates nonvascularized tissue in the posterior. Small arrows in **J** indicate direction of circulation, whereas the large arrow indicates the anterior limit of circulation through the DfV. ACV, anterior cardinal vein; AiV, anterior intersomitic vein; AoArch, aortic arches; AP, animal pole; CauV, caudal vein; DA, dorsal aorta; DCuv, duct of Cuvier; DfV, dorsal fin vein; DMZ, dorsal marginal zone; ECA, external carotid artery; End, endocardium; HF, heart field; HyA, hyaloid artery; n, notochord; NT, neural tube; OMV, omphalomesenteric vein; PCV, posterior cardinal vein; pDA, paired dorsal aortae; PiV, posterior intersomitic veins; PS, pronephric sinus; S, somite; SV, sinus venosus; VA, ventral aorta; VitV, vitelline vessels; VMZ, ventral marginal zone; VP, vegetal pole. For color reproduction, see Color Plate 18.1.

at stage 48, it reaches and joins the posterior vena cava (9). The largest artery of the embryo, the dorsal aorta (see Figure 18.1F, *DA*) extends posteriorly from the fused pDA. Anatomically, the pDAs are first seen at stage 31, at the level of the first branchial arch; by stage 33 to 34, they have reached the level of the pronephros and, by stage 35 to 36, they unite in the midline to form a median DA. The DA extends posteriorly, so that, at stage 37 to 38, it penetrates the tail as the caudal artery (9). At the molecular level, DA precursors already can be seen all along the trunk and tail by stage 32 (see Figure 18.1F, *DA*).

The development of the DA is very interesting for several reasons. First, the origins of the DA differ from that of the pDAs: Although most of the pDAs differentiate from head mesoderm, the DA and the most posterior portion of the pDA develop from the DLP mesoderm (10–11). Second, at stage 26, the precursors of the PCVs and the DA are both located in the DLP; however, from stages 28 to 30, a group of cells migrate from the DLP toward the midline; later, these migrating cells differentiate into the DA (11). Migration of the DA precursors appears to be under the control of vascular endothelial growth factor (VEGF) signaling, because the migrating cells express the VEGF receptor, Flk-1, whereas the hypochord, a structure located in the midline, expresses a diffusible form of VEGF (11–13).

As early as stage 31, intersomitic vessels can be seen developing from the axial vessels. Interestingly, in the posterior of the embryo, dorsal to the nonvascularized yolk mass, both the veins and DA produce intersomitic vessels (5,14) (see Figure 18.1F, *PiV*) whereas, more anteriorly, in the trunk, intersomitic vessels only sprout from the PCVs (Figure 18.1L, *AiV*), the significance of which is currently unknown. Also at stage 32, the dorsal ends of the PiV appear to coalesce to form a large dorsal fin vein (DfV), which only exists in the posterior of the embryo and drains into the PCVs at the level of the cloaca, through the most anterior PiV. Circulation initiates at stage 35, when the first red blood cells are seen in the heart; however, not until stage 39 to 40 can cells be seen in circulation throughout the embryo. After stage 40, the vasculature continues to remodel, and new vessels develop to meet the demands of the developing thoracic organs. A description of these events can be found elsewhere (9). Finally, in parallel, the rudiments of the lymphatic system are being laid out in the pronephric region, where the lymphatic heart[5] (Figure 18.2B, *LH*; for color reproduction, see Color Plate 18.2.) can be morphologically distinguished from stage 33 to 34 (9). Interestingly, the lymphatic heart endothelium appears to have a gene expression profile similar to that of the endothelium of the arteries and veins (A. Ciau-Uitz and R. Patient, unpublished observations). Later, after stage 40, the lymphatic heart endothelium extends both anteriorly and posteriorly to form the main lymph vessel of the larva.

5 In embryos with a well-developed lymphatic system, the lymphatics drain into the most anterior end of the posterior cardinal vein. At this junction, the lymphatic vessel attached to the posterior cardinal vein appears to rhythmically pump lymph into the venous system, for which reason it has been called the *lymphatic heart.*

Our lineage-labeling of regularly cleaving embryos indicates that three areas of the early blastula give rise to the majority of the ECs of the tadpole:

- The anterior vitelline vessels, omphalomesenteric veins, endocardium, ventral aorta, and the most anterior portion of the external carotid arteries derive from blastomeres C1 and D1 of the 32-cell stage embryo (10–11).
- The posterior vitelline vessels derive from blastomere D4 (11).
- The major vessels of the trunk and tail derive from blastomere C3 (10–11) (Figure 18.2A–D).

Interestingly, blastomere C3 also gives rise to the lymphatic heart (see Figure 18.2B), indicating that the PCVs, the DA, and the lymphatic hearts all derive from the early DLP precursors. In mouse embryos, the lymphatic vessels sprout from the PCVs; therefore, it is well possible that the *Xenopus* lymphatic hearts also sprout from the PCVs. Because all arteries irrigating the future thoracic organs sprout from the DA, and most of the veins from these organs drain into the PCVs, it appears that blastomere C3 gives rise to most of the vasculature of the adult. The origins of the head vasculature and the aortic arches have not been investigated in regularly cleaving embryos; however, in irregularly cleaving embryos, the origins of the aortic arches have been assigned mainly to blastomere D2 (6).

HEMANGIOBLASTS, HEMOGENIC ENDOTHELIUM, AND HEMATOPOIETIC STEM CELLS

For over 80 years it has been known that the blood and ECs of the chick and mouse yolk sac differentiate from homogenous cell clusters, with the peripheral cells becoming ECs while the internal cells give rise to blood. Since then, it has been postulated that both lineages derive from a common precursor cell called the hemangioblast (15). The existence of such a cell is supported by the common origins of blood and ECs during development, as well as by their common gene expression profiles. In addition, zebrafish mutants and mouse knockout and knockin mutants further support the existence of the hemangioblast, both in embryos and adults. However, only recently has direct evidence been presented supporting the existence of hemangioblasts in differentiated mouse embryonic stem cells (16) and in the mouse yolk sac (17). In *Xenopus* embryos, we have described the existence of two hemangioblast populations (10), the first located ventral and anterior in the neurula-stage embryo (Figure 18.2E–H), and the second in the DLP of tailbud embryos (10–11) (Figure 18.2I–K). We also have demonstrated that the ventral anterior hemangioblast gives rise to vitelline vessels, OMVs, endocardium, VA, and the most anterior portion of the ECAs as well as to the red blood cells and myeloid cells of the embryonic lineage (10, A. Ciau-Uitz and R. Patient, unpublished observations). In its turn, the DLP

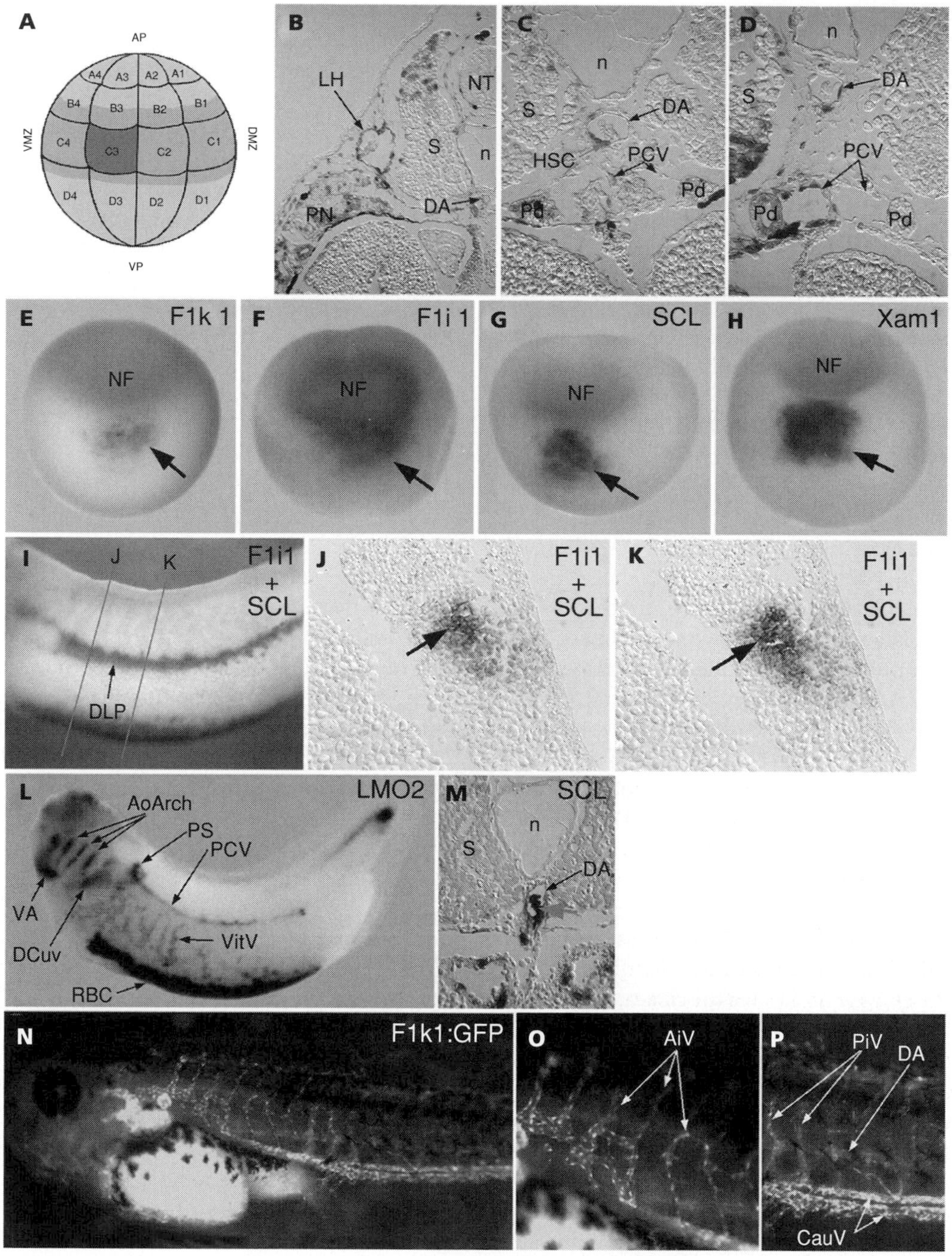

Figure 18.2. Development of the hematopoietic system occurs in close association with ECs. **(A–D)** Contribution of blastomere C3 to the vasculature. **(B–D)** Ten-μm transverse sections of a stage-43 embryo at the levels of the pronephros, midtrunk, and posterior trunk respectively. **(E–H)** Neurula-stage embryos stained for blood and endothelial genes; arrows indicate the population of hemangioblasts giving rise to embryonic hematopoietic cells, vitelline vessels, endocardium, and ventral aorta. **(I–K)** Tailbud-stage embryo double-stained for blood and endothelial genes showing a population of hemangioblasts located in the dorsal lateral plate (DLP, *arrows*); later, this population gives rise to the axial vessels and adult hematopoiesis. **(L)** Expression of LMO2 in blood vessels. **(M)** Hematopoietic stem cells (*arrow*) emerging in association with the ventral wall of the dorsal aorta and expressing the blood gene, SCL. **(N–P)** Transgenic embryos showing expression of GFP in blood vessels. GFP expression was under control of the *Xenopus FLK-1* promoter as indicated in reference 25. LH, lymphatic heart; NF, neural fold; Pd, pronephric duct; PN, pronephros; RBC, red blood cells. For color reproduction, see Color Plate 18.2.

hemangioblast gives rise to the pronephric sinus, PCV, and DA as well as to hematopoietic stem cells (HSCs) (10–11) (Figure 18.2C and M, *arrows*). It appears that the generation of hematopoietic cells in the *Xenopus* embryo always occurs in close association with ECs; however, the opposite is not true, because the head vasculature is never seen associated with developing blood cells. Thus, according to their ontogeny, two types of ECs exist in the embryo: those deriving from hemangioblasts, and those not deriving from hemangioblasts. Interestingly, some blood genes such as *CBFβ*, *MafB*, and *LMO2* are seen associated with different maturing and/or differentiated blood vessels. Between stages 32 and 40, LMO2 is expressed in a large number of vessels (Figure 18.2L), whereas in the same period, CBFβ and MafB are mainly expressed by the vitelline vessels (data not shown). This may suggest that different hematopoietic genes have different roles during the specification and/or maturation of the vasculature.

One feature strikingly conserved through evolution is the formation of hematopoietic clusters in the ventral wall of the mature DA and other large arteries of the embryo (10,18–20) (Figure 18.2C and M). These intraaortic clusters express a number of genes later expressed in mature endothelium or blood, including CD34, CD31, GATA2, LMO2, AML1/Runx1, and CD45. Furthermore, transplantation experiments have demonstrated that the first HSCs localize to the DA and the vitelline and umbilical arteries of the developing mouse embryo (21). Taken together, it is widely accepted that the intra-aortic hematopoietic clusters represent the first emerging HSCs in the embryo. However, how these clusters are generated is still controversial (22). Several mechanisms are proposed, including: (a) transdifferentiation, in which the mature endothelium of the DA gives rise directly to the HSC; (b) dedifferentiation-differentiation, in which the mature endothelium dedifferentiates into a less mature mesodermal cell that subsequently differentiates into the HCS; (c) mesodermal differentiation, in which a mesodermal precursor localized within the endothelial lining of the DA differentiates into the HSC; and finally (d) colonization, in which a mesodermal precursor located in the surroundings of the DA differentiates into HSCs that migrate and colonize the ventral wall of the DA. By whichever mechanism, the ECs of the DA undoubtedly play a significant role during the emergence of the HSC.

FUTURE DIRECTIONS

A body of evidence suggests that the genes and signaling mechanisms involved during the development of the vasculature are well conserved between *Xenopus* and the mouse. Key regulators of hemangioblast development, vasculogenesis and angiogenesis–such as VEGF-A and its tyrosine kinase receptor, Flk-1; the angiopoietin tyrosine kinase receptor, Tie2; and the hematopoietic transcription factors GATA-2, SCL, LMO2, and AML1/Runx1 – have been described for *Xenopus*. Also reported are genes shown to be specifically expressed by ECs and/or involved in EC physiology and/or migration; among them are Xmsr, members of the Eph-ephrin system, neuropilins, and the Ets transcription factors Fli1, Ets-1, Ets-2, and Erg. However, the isolation and characterization of key genetic components such as the VEGF receptor tyrosine kinases FLT-1 and FLT-4 is yet to be accomplished in *Xenopus*. Fortunately, several libraries of expressed sequence tags (ESTs) have been generated in the past few years and, undoubtedly, these libraries will greatly facilitate the isolation of the factors currently missing in the endothelial genetic machinery of *Xenopus*.

Collections of thousands of *Xenopus* ESTs are available commercially in the form of microarrays. The employment of such microarrays in gain- and loss-of-function experiments can be used not only to unravel the genetic cascades involved during the embryo's vascularization, but also as a powerful tool for the discovery of unknown components of the endothelial genetic machinery. Until recently, gain-of-function experiments in *Xenopus* have been achieved by mRNA injection into early embryos. Similarly, loss-of-function has been achieved by injecting the mRNA of dominant negative forms of the gene of interest. Undoubtedly, future researchers will make use of transgenic technologies (23) to achieve tissue- and temporal-specific gene perturbation. New loss-of-function technologies also are being developed, and so far the employment of modified antisense oligonucleotides, in particular morpholinos, has greatly benefited the *Xenopus* community. Thus, morpholinos can be used to perform rapid, low-cost loss-of-function studies of endothelial genes and, because *Xenopus* embryos are less dependent than their mammalian counterparts on circulation for oxygen transport during early stages of development, loss-of-function analyses can be performed at later stages of development.

The *Xenopus* animal cap induction assay also has been an informative tool for developmental biologists (24). The *Xenopus* blastula contains a large blastocele, the roof of which is called the *animal cap* and which consists of a few layers of ectodermal cells. If the animal cap is excised and cultured for a few days in saline medium, it forms atypical epidermis with no evident structural differentiation. When the animal cap is cultured in media containing growth factors however, differentiation is induced, and tissue differentiation can be identified by morphology and/or molecular markers. Tissue-specific differentiation can be achieved by using particular growth factors at particular concentrations, and a wide range of mesodermal and endodermal derivatives, including blood, has been produced (24). The animal cap assay also can be used to investigate the roles played by individual genes in these processes. Fertilized eggs are injected in their animal poles before the first cleavage with mRNA encoding the factor of interest, the animal caps are excised at the blastula stage and cultured for a few days, and then morphological and molecular analyses are performed. The combination of single-gene animal cap assays and EST microarrays could be used as a high throughput protocol to identify downstream genetic targets and new components of the endothelial genetic machinery.

Xenopus laevis is a pseudotetraploid species, and this fact has hampered genetic studies in this organism. In recent years, a close diploid relative of *X. laevis*, *X. tropicalis*, has been proposed as an alternative model for genetic studies. The developmental biology of *X. tropicalis* appears to be very similar to that of *X. laevis*, and all the technologies employed in *X. laevis* also can be applied in *X. tropicalis*. This has prompted the genomic characterization of *X. tropicalis*, and its genome is currently being sequenced. The genome sequence will facilitate not only the identification of conserved motifs in the regulatory sequences of endothelial genes, but also will provide much-needed promoters and enhancers for the generation of endothelium-specific transgenic lines. Furthermore, several large-scale mutagenesis screens currently are being carried out using *X. tropicalis* (25). The generation of mutations affecting the development of the vascular system may lead to a better understanding of the roles of currently known factors and to the discovery of new genes.

In vivo studies of gene regulation using both *X. laevis* and *X. tropicalis* can now be performed thanks to the development of a simple and reliable transgenesis protocol (23). Transgenic lines can be generated using promoters isolated from different organisms but, as its genome is becoming available in the form of clones and sequence information, more and more lines will be established using *X. tropicalis* sequences. By fusing the regulatory sequences of the gene of interest with the coding region of the green fluorescent protein (GFP), gene expression can be monitored in real time as the embryo develops (Figure 18.2N–P). Furthermore, transgenic lines have been generated using binary gene expression systems such as GAL4-UAS, tetracycline-controlled gene expression, or Cre/FLP-mediated recombination; these can be used to produce uniform and tissue-specific transgene expression. At the moment, only the promoters of *Xenopus Flk-1* (25) (see Figure18.2N–P) and mouse *Tie2* (25) have been used to drive GFP expression in ECs in *Xenopus*. In the future, the use of hemangioblast-, vein-, artery- or lymphatic-specific promoters will greatly facilitate the identification of the mechanisms involved in the differentiation of these cell lineages. Transgenic lines also could be used as models for the identification and validation of novel drugs targeted to the vascular system. Although both zebrafish and *Xenopus* embryos are suitable for the screening of chemical libraries with the aim of discovering novel chemical modulators of angiogenesis, the absence of smooth muscle cells and lymphatic vessels in the zebrafish leaves *Xenopus* as more suited for these studies (26). In the longer term, this could assist with the identification of pharmacological agents for the treatment of numerous pathological disorders in which ECs play an important role – for example, tumor development. Importantly, screenings also will provide new tools to study the development and physiology of ECs under normal conditions.

KEY POINTS

- The *Xenopus* system offers an exciting opportunity to study the development and genetics of ECs and their modulation. In particular, *Xenopus* should make valuable contributions to our understanding of how ECs are specified in the early embryo and the importance of their interaction with other cell lineages.
- The endothelial genetic machinery appears to be conserved between *Xenopus* and mammals, including the close association between hematopoietic and EC programming during development.
- In *Xenopus*, well-developed embryological techniques are available, such as fate mapping, tissue transplantation, and explant culture, which facilitate the study of EC development.
- The generation of transgenic lines, together with the external development of the *Xenopus* embryo, greatly facilitates the in vivo study of the embryo's vascularization.

Future Goals

- To sequence the genome of *X. tropicalis*, together with large scale mutagenesis screens, to provide new sequences for the generation of endothelium-specific transgenic lines and new mutant lines carrying endothelium-specific mutations
- To develop EST microarrays in conjunction with animal-cap induction to facilitate the investigation of the EC genetic cascade and the discovery of new components of the genetic machinery
- To establish endothelium-specific transgenic lines for in vivo study of the development of the vascular system, as well as provide models for assessing the toxicity, effectiveness, and specificity of new drugs modulating angiogenesis, before clinical trials are implemented
- To use *Xenopus* as an ideal system for the low-cost, high-throughput screening of chemical libraries with the aim of discovering new drugs modulating particular aspects of angiogenesis

ACKNOWLEDGMENTS

We thank Paul Krieg for providing the Flk1 promoter used for transgenesis. We apologize to those investigators whose work could not be cited because of space constraints. This work was supported by the MRC.

REFERENCES

1 Gurdon JB, Hopwood N. The introduction of *Xenopus laevis* into developmental biology: of empire, pregnancy testing and ribosomal genes. *Int J Dev Biol.* 2000;44:43–50.

2 Walmsley M, Ciau-Uitz A, Patient R. Tracking and programming early hematopoietic cells in *Xenopus* embryos. *Meth Mol Med.* 2005;105:123–136.

3 Dale L, Slack J. Fate map for the 32-cell stage of *Xenopus laevis*. *Development*. 1987;99:527–551.

4 Moody S. Fates of the blastomeres of the 32-cell stage *Xenopus* embryo. *Dev Biol.* 1987;122:300–319.

5 Rovainen CM. Labeling of developing vascular endothelium after injections of rhodamine-dextran into blastomeres of *Xenopus laevis. J Exp Zoo.* 1991;259:209–221.

6 Mills KR, Kruep D, Saha MS. Elucidating the origins of the vascular system: a fate map of the vascular endothelial and red blood cell lineages in *Xenopus laevis. Dev Biol.* 1999;209:352–368.

7 Lane MC, Smith WC. The origins of primitive blood in *Xenopus*: implications for axial patterning. *Development*. 1999;126:423–434.

8 Collazo A, Bronner-Fraser M, Fraser SE. Vital dye labelling of *Xenopus laevis* trunk neural crest reveals multipotency and novel pathways of migration. *Development*. 1993;118:363–376.

9 Nieuwkoop PD, Faber J. *Normal Table of Xenopus laevis (Daudin)*. Amsterdam: North Holland Publishing, 1967.

10 Walmsley M, Ciau-Uitz A, Patient R. Adult and embryonic blood and endothelium derive from distinct precursor populations which are differentially programmed by BMP in *Xenopus. Development*. 2002;129:5683–5695.

11 Ciau-Uitz A, Walmsley M, Patient R. Distinct origins of adult and embryonic blood in *Xenopus. Cell.* 2000;102:787–796.

12 Cleaver O, Tonissen KF, Saha MS, Krieg PA. Neovascularization of the *Xenopus* embryo. *Dev Dyn.* 1997;210:66–77.

13 Cleaver O, Krieg PA. VEGF mediates angioblast migration during development of the dorsal aorta in *Xenopus. Development.* 1998;125:3905–3914.

14 Levine AJ, Munoz-Sanjuan I, Bell E, North AJ, Brivanlou AH. Fluorescent labeling of endothelial cells allows *in vivo*, continuous characterization of the vascular development of *Xenopus laevis. Dev Biol.* 2003;254:50–67.

15 Murray PDF. The development *in vitro* of the blood of the early chick embryo. *Proc Roy Soc B.* 1932;111:497–521.

16 Choi K, Kennedy M, Kazarov A, Papadimitriou JC, Keller G. A common precursor for hematopoietic and endothelial cells. *Development*. 1998;125:725–732.

17 Huber TL, Kouskoff V, Fehling HJ, Palis J, Keller G. Haemangioblast commitment is initiated in the primitive streak of the mouse embryo. *Nature*. 2004;432:625–630.

18 Dieterlen-Lievre F, Martin C. Diffuse intraembryonic hemopoiesis in normal and chimeric avian development. *Dev Biol.* 1981; 88:180–191.

19 Ottersbach K, Dzierzak E. The Endothelium – The Cradle of Definitive Hematopoiesis? In: Godin I, Cumano A, eds. *Hematopoietic Stem Cells.* Austin TX: Landes Bioscience, 2004. (http://www.eurekah.com).

20 Peault B, Tavian M. Hematopoietic stem cell emergence in the human embryo and fetus. *Ann NY Acad Sci.* 2003;996:132–140.

21 de Bruijn MF, Speck NA, Peeters MC, Dzierzak E. Definitive hematopoietic stem cells first develop within the major arterial regions of the mouse embryo. *Embo J.* 2000;19:2465–2474.

22 Godin I, Cumano A. The hare and the tortoise: an embryonic haematopoietic race. *Nat Rev Immunol.* 2002;2:593–604.

23 Kroll KL, Amaya E. Transgenic *Xenopus* embryos from sperm nuclear transplantations reveal FGF signalling requirements during gastrulation. *Development*. 1996;122:3173–3183.

24 Okabayashi K, Asashima M. Tissue generation from amphibian animal caps. *Curr Opin Genet Dev.* 2003;13:502–507.

25 Warkman AS, Meadows SM, Small EM, Cox CM, Krieg PA. Cardiovascular genomics: the promise of *Xenopus. Drug Discov Today: Disease Models.* 2004;1:249–255.

26 Ny A, Koch M, Schneider M, et al. A genetic *Xenopus laevis* tadpole model to study lymphangiogenesis. *Nat Med.* 2005;11:998–1004.

19

Vascular Development in Zebrafish

Sameer S. Chopra and Tao P. Zhong

Vanderbilt University School of Medicine, Nashville, Tennessee

The use of zebrafish (*Danio rerio*) as a vertebrate model has yielded tremendous insight into the complex cellular and molecular events that underlie embryonic vascular development. Compared to more traditional model organisms such as chicks, mice, and frogs, zebrafish offer distinct advantages and have made unique contributions toward our understanding of vascular development. Recent studies in zebrafish have led to the identification of mutations and molecules that are responsible for the specification of endothelial progenitor cells (or angioblasts), differentiation of arterial and venous cells, and patterning of the dorsal aorta (DA) and intersegmental vessels. Zebrafish embryos develop externally and are optically clear, affording noninvasive and high-resolution access to nearly the entire vascular system. The ability of zebrafish embryos to survive temporarily without blood circulation (oxygen is obtained via diffusion) permits the study of defects in vascular development that would otherwise be embryonic lethal in other organisms, and without the confounding effects of hypoxia. Given their fecundity, small size, and brief generation time, zebrafish are highly amenable to genetic manipulation, including large-scale mutagenesis screens. Importantly, the zebrafish genome has been mapped and sequenced by the Sanger Center in Britain. Genomic and positional cloning reagents, such as genetic maps and libraries, are available. Forward genetic approaches,[1] in combination with gene mapping and positional cloning, already have been employed successfully to identify genes that disrupt the formation and patterning of embryonic vasculature and other organs (1,2). The discovery of *gridlock* (*grl*) – a *hairy*-related basic helix-loop-helix (bHLH) transcription factor involved in arterial endothelial development – demonstrates the power of mutagenesis screens to uncover novel genetic pathways and developmental mechanisms with no presupposition about the role of genes in biological processes (discussed below) (3,4).

To date, zebrafish studies have helped to clarify the cellular and molecular mechanisms underlying critical phases of embryonic vascular development including: (a) specification of angioblasts and differentiation of arterial and venous endothelial cells (ECs); (b) organization of ECs into a primary circulatory network comprised of the DA and the cardinal vein; and (c) growth and remodeling of this primary network into intersegmental vessels. This chapter discusses some of those important events that have been analyzed mostly in zebrafish, and also compares some of these events to those in other organisms.

HISTORICAL OVERVIEW

The earliest published studies using zebrafish date back to the 1950s, when several researchers took advantage of the transparency of zebrafish embryos to examine the effect of toxic compounds on development. George Streisinger, Charles Kimmel, and colleagues at the University of Oregon later championed the use of zebrafish in the 1980s, through a series of important publications describing its genetics and embryonic development. The zebrafish reached widespread popularity in biomedical research a decade ago, with reports of the successful conduct of two large-scale zebrafish forward genetic screens.[2] The power of forward genetics to uncover new pathways that govern early development and organogenesis previously only had been demonstrated in invertebrate models such as *Drosophila*. Interestingly, a number of zebrafish mutants exhibited features resembling human disease phenotypes, and the positional cloning of specific mutations often identified genes orthologous to human disease genes. The heart and fin defects in the *heartstring* mutant, for example, result from a nonsense mutation in the zebrafish orthologue to *TBX5*, a T-box transcription factor whose loss of function underlies the heart and limb abnormalities of human patients suffering from Holt-Oram syndrome (5). The otic vesicle defects in the *mariner* zebrafish mutant are caused by mutations in the myosin VIIA gene, which is disrupted in human Usher 1B deafness syndrome (6). The poor cardiac contractility and

1 Mutagenesis screens are conducted to isolate mutants with phenotypes, then genes are cloned from the mutants.

2 An entire issue of the journal *Development* was devoted to these reports (*Development.* 1996;123(1):1–461.).

perturbed sarcomeric assembly in the *pickwick* mutant result from a mutation in the gene encoding the zebrafish homolog of *titan*, a gene whose aberrant function results in human dilated cardiomyopathy (7). Zebrafish *sau* proved to be due to a defect in the erythroid aminolevulinate synthase (*ALAS-2*) gene, a critical enzyme that regulates the first step in heme biosynthesis in embryonic red blood cell (8). Mutations in *ALAS-2* gene cause human X-linked congenital sideroblastic anemia. In addition to modeling diseases with known causes, zebrafish mutants may also hold promise for identifying new disease genes in humans. Such was the case with *weissherbst*, a zebrafish mutant displaying hypochromic anemia caused by a mutation in the gene encoding the iron transporter ferroportin 1 (9,10). Later, it was discovered that patients with type 4 hemochromatosis have defects in the same transporter. In the *gridlock* mutant, the aortic bifurcation is blocked, in a manner similar to coarctation of the aorta, a human congenital cardiovascular disease of unknown etiology (3). There also exist zebrafish mutants with cystic kidneys, such as *doublebubble*, which may represent polycystic kidney disease of human (11). Both *gridlock* and *doublebubble* also one day may demonstrate that genes discovered through zebrafish forward genetics can contribute to the discovery of novel human disease genes.

THE ZEBRAFISH EMBRYONIC VASCULATURE

The anatomy and development of the major embryonic vascular network, the circulatory loop in the trunk, is highly conserved between zebrafish and other vertebrates. The zebrafish axial vessels, the DA and the posterior cardinal vein (PCV), form a simple circulatory loop in the trunk prior to the first heartbeat (Figure 19.1 for color reproduction, see Color Plate 19.1) (12). The rudiments of these vessels assemble through vasculogenesis, in which endothelial progenitors (angioblasts) migrate from the lateral mesoderm and coalesce first into cords and subsequently into vessels with lumens (13). Elaboration of the simple circulatory loop occurs through angiogenesis, in which new vessels sprout, grow, and remodel to form an anastomotic arterial and venous network. One key feature differentiating zebrafish from other vertebrates is the vasculature of the yolk sac. Whereas other vertebrate

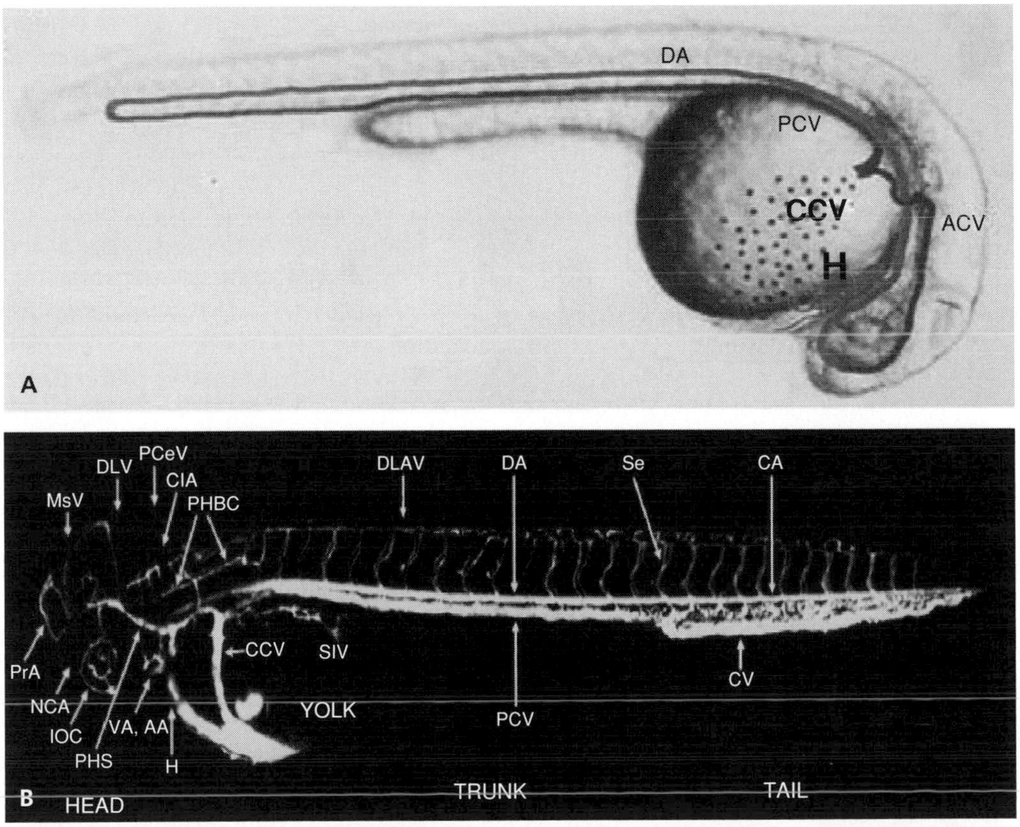

Figure 19.1. Anatomy of the zebrafish embryonic vasculature. (**A**) Diagram of a circulatory loop at the zebrafish embryo at 30 hpf. (**B**) Fluorescent confocal microangiogram showing the zebrafish embryonic vasculature at 60 hpf. H, Heart; VA, ventral aorta; AA, aortic arch; PHS, primary head sinus; NCA, nasal ciliary artery; PrA, prosencephalic artery; MsV, mesencephalic vein; DLV, dorsal longitudinal vein; PCeV, posterior cerebral vein; CctA, cerebellar central artery; PHBC, primary hindbrain channel; DLAV, dorsal longitudinal anastomotic vessel; DA, dorsal aorta; Se, intersegmental vessel; CA, caudal artery; CV, caudal vein; ACV, anterior cardinal vein; PCV, posterior cardinal vein; SIV, subintestinal vein; CCV, common cardinal vein (duct of Cuvier). For color reproduction, see Color Plate 19.1.

embryos form a complex capillary plexus network that undergoes pruning and remodeling to form the vitelline artery and vitelline vein, the zebrafish yolk sac does not develop these extraembryonic vascular structures. Instead, part of the cardinal venous system extends into the yolk sac (see Figure 19.1).

THE ORIGIN AND SPECIFICATION OF ANGIOBLASTS

Angioblasts are endothelial precursors that have certain characteristics of ECs but have not yet organized into a cord (13). In zebrafish embryos, angioblasts arise in the lateral plate mesoderm at the 5-somite stage and can be detected by their expression of vascular endothelial growth factor receptor (*VEGFR2*; *flk1*) (14–16). Cell lineage analysis indicates that angioblasts migrate to the midline and assemble into the DA and the cardinal vein (Figure 19.2; for color reproduction, see Color Plate 19.2) (4). Similarly, in avian embryos, angioblasts, which are detected by QH1 antibodies, originate from the lateral plate mesoderm (17). Angioblasts from the splanchnopleural mesoderm, a ventral compartment of the lateral plate mesoderm in contact with endoderm, populate the floor of the DA and endothelial lumens of vessels in visceral organs (18). Hematopoietic stem cells that eventually reside in the ventral side of aortic lumen also are derived from the splanchnopleural mesoderm. In contrast, angioblasts derived from the somitic paraxial mesoderm contribute to vessels in the neural tube, limb buds, and the roof of the aorta and never invade the visceral organs. The somatopleural mesoderm, a dorsal layer of the lateral plate mesoderm in contact with ectoderm, produces few angioblasts (18). Thus, angioblasts in the embryo proper originate primarily from splanchnopleural mesoderm and somitic paraxial mesoderm.

In the yolk sac of avian embryos, the earliest sign of angioblast formation is the appearance of blood islands that emerge from the extra-embryonic mesoderm in association

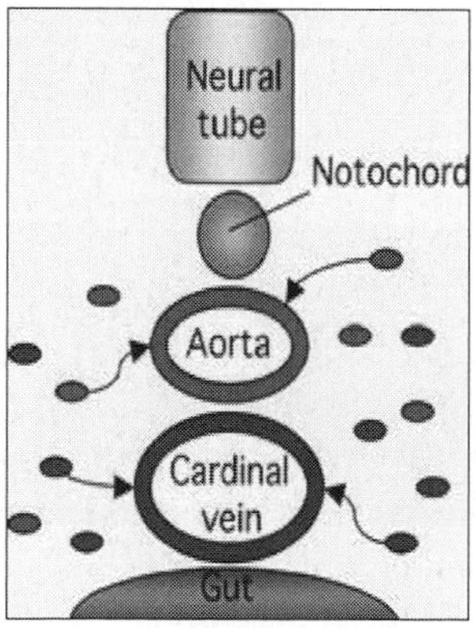

Figure 19.2. Assembly of the dorsal aortae and the cardinal veins in zebrafish embryos. Schematic drawing indicates the crossover section of zebrafish embryos at the trunk region, showing the neural tube, notochord, aorta, cardinal vein, and gut. Angioblasts appear to migrate and assemble into the dorsal aorta and the cardinal vein. For color reproduction, see Color Plate 19.2.

with visceral endoderm (19). The external cells of blood islands differentiate into ECs, and the internal round cells develop into hematopoietic stem cells. Based on the simultaneous emergence of angioblasts and hematopoietic cells in blood islands, it is postulated that angioblasts and hematopoietic cells originate from a common precursor termed the "hemangioblast" (20,21). Multiple lines of evidence support the existence of such progenitors with dual potential. Both angioblasts and hematopoietic progenitors express similar sets of transcription factors and surface receptors. In addition, single-gene mutations have been found to perturb the development of both angioblast and hematopoietic lineages (22).

Although numerous studies suggest that direct interactions between endoderm and mesoderm might be required for angioblast specification[3] in the embryo proper (23,24), more recent studies in *Xenopus* suggest the endoderm is not required for this process (25). These findings imply that signals for angioblast specification may be contained within the mesoderm, or that this process occurs by an intrinsic cellular mechanism. This suggests that interactions of mesoderm and endoderm might not be required for angioblasts specification. Critical molecular insight into the specification of angioblasts has come from studies of VEGFR2/FLK-1 and its ligand VEGF. Targeted inactivation of FLK-1 in mouse embryos causes embryonic lethality and results in the absence of both endothelial and hematopoietic cells, indicating that FLK-1 is essential for angioblast specification (26). In zebrafish, the *schwentine* mutation encodes a truncated *Flk-1* gene (27). Surprisingly, *schwentine* mutant embryos do not display defects in angioblast specification, although formation of both intersomitic and subintestinal vessels is disrupted. This milder phenotype in zebrafish *schwentine* mutants, compared to mice with inactivated *Flk-1*, might be explained by partial retention of VEGF signaling by the truncated FLK-1 receptor. Alternatively, a second copy of *flk1* in the zebrafish genome may compensate for the defective *flk1* in *schwentine* mutant embryos (Len Zon, personal communication).

VEGF, the ligand for FLK-1, is expressed in the paraxial mesoderm and in the surrounding endoderm adjacent to angioblasts (28). In mice, inactivation of a single *VEGF* allele results in impaired yolk sac vasculogenesis and a defective DA, suggesting that a critical level of VEGF is required to sufficiently activate the FLK-1 receptor during vasculogenesis (29). Unexpectedly, complete inactivation of VEGF activity in both zebrafish and mouse embryos does not inhibit angioblast specification. These results suggest that additional VEGF ligands (i.e., VEG-F-B and VEG-F-C) may interact with the FLK-1 receptor during vasculogenesis to induce angioblasts (30). Alternatively, yet unidentified mesodermal-derived factors may promote angioblast specification via the FLK-1 receptor.

Unlike *schwentine*, a second zebrafish mutant line termed *cloche* is characterized by a defect in angioblast specification (31). Similar to mouse embryos with inactivated *Flk-1*, *cloche*

3 Mesodermal cells differentiate into angioblasts.

mutants also fail to develop hematopoietic lineages (32). Transcripts of both endothelial and hematopoietic markers including *flk1*, *fli*, *scl*, *gata1*, *tie1*, and *tie2* are absent in *cloche* embryos at very early stages; at later stages, their expression is minimal and restricted to caudal regions of the embryo. Cell transplantation studies reveal that the *cloche* mutation is cell-autonomous[4] for the generation of endothelial lineages but non–cell-autonomous for hematopoietic lineages (32). These findings suggest that *cloche* might act as an intracellular factor for inducing angioblast formation. Improved understanding of the molecular mechanism of *cloche* in vasculogenesis will depend on the success of ongoing efforts to identify the *cloche* gene via a positional cloning approach.

Current studies using a gain-of-function approach have placed *cloche* in a putative molecular pathway. Forced expression of *scl*, a basic helix-loop-helix (bHLH) transcription factor, in *cloche* embryos restores expression of the endothelial marker *flk1* and the hematopoietic marker *gata1*, and increases the number of both endothelial and blood cells (32). These findings suggest that *cloche* acts upstream of Scl for hemangioblast specification, which in turn controls differentiation of angioblastic and hematopoietic lineages. Additional zebrafish studies suggest that Scl acts synergistically with Lmo2, a LIM domain transcription factor, in converting paraxial mesoderm into angioblasts (33). Forced expression of *scl* mRNA in zebrafish embryos induces ectopic expression of endothelial markers *flk*, *fli*, and *flt4* in somitic paraxial mesoderm. The restriction of angioblast induction to somitic paraxial mesoderm correlates with Scl-induced *lmo2* expression in this region. Consistent with these findings, misexpression of *lmo2* and *scl* mRNA induces mesodermal cells in the region of head, heart, and pronephros into ECs at the expense of myocardial and pronephros fates (33). Thus, expression of both *scl* and *lmo2* can induce angioblast specification and endothelial differentiation in paraxial mesodermal cells. Together, these data suggest that Scl and Lmo2 synergistically act downstream of *cloche* in controlling angioblast formation. Although these gain-of-function studies in zebrafish indicate an important role for Scl and Lmo2 in angioblast specification, loss-of-function studies of these genes in mice and zebrafish suggest that neither is essential for this process. Targeted inactivation of mouse Scl or Lmo2, for example, causes defects in both hematopoiesis and angiogenesis but does not affect angioblast specification (34–36). Similarly and consistent with the loss-of-function phenotype in mice, knockdown of zebrafish *scl* using antisense morpholinos results in absence of hematopoietic cells but not a loss of angioblasts or ECs (37,38).

Taken together, these results suggest that Scl and Lmo2 are required for hematopoiesis and angiogenesis but are not essential for angioblast specification. In mammals, members of fibroblast growth factor (FGF) are suggested to induce angioblast specification. However, in zebrafish, mutation in *fgf8* does not affect the formation and differentiation of angioblasts and ECs, but causes deficiency of cardiac precursors (39), suggesting that Fgf signaling is not required for angioblast specification but is essential for induction of cardiomyocytes in zebrafish.[5]

In summary, angioblasts arise from the lateral plate mesoderm that associates with the endoderm. The induction and differentiation of angioblasts and ECs require molecular mechanisms that are critically dependent on *flk1* and *cloche*.

FORMATION OF THE DORSAL AORTA AND THE CARDINAL VEIN

The assembly of angioblasts into endothelial cords and subsequently into lumens of the DA and the cardinal vein represents the next major milestone in vascular development. Although little is known about cardinal vein assembly, a number of zebrafish studies have clarified the structural and molecular requirements for formation of the DA. The zebrafish *floating head (flh)* mutation occurs in *Xnot*, a homeobox gene, and inhibits development not only of the notochord and hypochord, but also of the DA. *Flk*-1-expressing ECs, however, still appear to form a venous luminal structure in the region of the cardinal vein (14). The hypochord is an endodermally derived, transient structure in fish and amphibian embryos that lies ventral to the notochord and immediately dorsal to the aorta. Flh acts cell-autonomously with respect to the development of the notochord, floor plate,[6] and hypochord because transplanted wild-type cells give rise to all three structures in *flh* host embryos (40). In addition, the aorta can be rescued only in the region immediately beneath the cell-transplant notochord and possibly the hypochord (14). Similar to zebrafish *flh*, *no tail (ntl)* embryos that have a mutation in the *brachyury* gene lack a differentiated notochord and hypochord but develop one axial vessel, which appears to be the cardinal vein (14). Collectively, these observations suggest a tight correlation between presence of the notochord (and possibly hypochord) and the DA. It is not clear, however, whether signals from the notochord or hypochord directly induce formation of the aorta or of adjacent tissues, which then subsequently provide the requisite signals for aorta assembly. Nonetheless, these observations provide a novel avenue for investigating the molecular basis of DA formation.

Studies conducted in *Xenopus* suggest a putative model for DA assembly that involves both the hypochord and VEGF signaling (41). VEGF has been found to play a critical role in controlling DA development (29,42). Although *VEGF* is expressed in the hypochord, somites, and endoderm during frog and fish embryonic development, *VEGF-121* appears to be preferentially expressed in the hypochord relative to *VEGF-165*

4 Genes that encode proteins with roles in the intracellular rather than extracellular environment.

5 It is possible that redundancy in the *fgf* family may compensate for the loss of individual fgf members for angioblast formation.

6 The ventral part of the neural tube.

and *VEGF-188*. Furthermore, ectopic *VEGF-121*-expressing implants are sufficient to promote angioblast migration. Based on these findings, it may be hypothesized that a gradient of VEGF established between the hypochord and the lateral plate mesoderm directs the migration of angioblasts to the midline to form the aorta. This hypothesis would be supported by evidence that loss of *VEGF-121* in the hypochord impairs arterial angioblast migration and leads to abolishment of aorta formation in *Xenopus* embryos. It also remains to be resolved how the loss of VEGF activity in mice (which, like other mammals, lack a hypochord) results in the complete absence of DA formation. It is possible that other embryonic structures, such as somites, express VEGF and thus serve a role analogous to the hypochord in inducing angioblast migration and aorta assembly.

Recent studies on vascular tubulogenesis in zebrafish have uncovered cellular events required for reshaping endothelial cords into tubes (43). At the 22-somite stage, arterial and venous angioblasts coalesce into single cords at the midline. Thereafter, arterial and venous angioblasts gradually segregate from each other in order to occupy distinct domains and then align in the form of rudimentary tubes. A factor EGF-like domain 7 (Egfl7) secreted from ECs was identified and thought to regulate the cord-to-tube transition. Inactivation of *Egfl7* in zebrafish embryos results in failure of the separation of arterial angioblasts from venous angioblasts, resulting in lack of lumen formation of the aorta and cardinal vein (43). Because the secreted Egfl7 protein remains in the vicinity of ECs and associates with the extracellular matrix, it is proposed that Egfl7 provides a permissive substrate around angioblasts to facilitate the local movement and segregation of arterial angioblasts from venous angioblasts.

PATTERNING OF THE INTERSEGMENTAL VESSELS

In the vertebrate vascular development, intersegmental vessels sprout from the DA and cardinal vein and grow to form the vascular system of the trunk. Taking advantage of the optical transparency of zebrafish embryos, studies have determined that this vascular network actually forms via a two-step mechanism (44). Primary sprouts emerge bilaterally from the DA and elongate dorsally, branching cranially and caudally along the dorsolateral roof of the neural tube to form paired dorsal longitudinal anastomotic vessels (DLAV) (Figure 19.3; for color reproduction, see Color Plate 19.3). Secondary sprouts grow from the PCV, with some connecting to the base of primary segments. The primary segments with patent connections to secondary segments become intersegmental veins, while the primary segments that remain connected to the DA become intersegmental arteries (44,45).[7] These intersegmen-

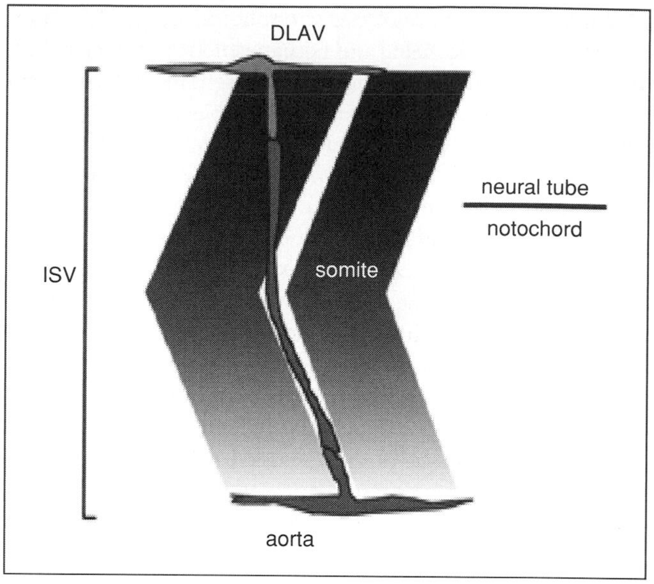

Figure 19.3. Model of the construction of an intersegmental vessel in zebrafish embryos. An intersegmental vessel is composed of three types of ECs: the dorsal connection to the DLAV is a T-shaped EC; the ventral connection to the aorta is an inverted T-shaped EC; the middle connection is an elongated EC. ISV, intersegmental vessel; DLAV, dorsal longitudinal anastomotic vessel. For color reproduction, see Color Plate 19.3.

tal vessels develop via migration and growth directly from axial vessels. Each of the intersegmental vessels is comprised of three types of ECs (see Figure 19.3) (44,45). The first one has its cell body in the DA and branches dorsally, the second one connects the first one and the third one, and the third one has its cell body in the DLAV and branches ventrally. These ECs derive from the lateral plate mesoderm, incorporate into the DA and move up to become part of the intersegmental vessels.

An analysis of the molecular events underlying this phase of vascular development has been facilitated by several zebrafish mutant lines. For example, *schwentine*, *y10*, and *out of bounds* (*obd*) mutations have linked FLK-1, phospholipase Cγ-1 (Plcg1), and plexinD1 (PlxnD1), respectively, to angiogenic growth, sprouting, and patterning of intersomitic vessels (27,45,46) (MGH screen, unpublished observations). Because Flk1 and Plcg1 are implicated in the intracellular transduction of VEGF signaling, the phenotypes resulting from mutations in these genes are consistent with the known function of VEGF for promoting blood vessel growth. In *obd* mutants, intersegmental vessels grow but are mispatterned, sprout precociously at irregular sites, and neither elongate along the intersomitic boundaries nor form interconnected branches along the trunk (47). Reduction of *plxnD1*, a semaphorin ligand receptor, using antisense morpholinos, causes defects indistinguishable from zebrafish *obd* mutants and similar to those displayed by mice with targeted inactivation of *PlxnD1* (47). These data implicate *PlxnD1* as a gene required for proper blood vessel path-finding during angiogenesis. It is notable that, in the developing nervous system, axonal path-finding is

7 The normal circulatory loop in the intersomitic vascular system is as follows: (a) blood flow into the dorsal longitudinal anastomotic vessel (DLAC) via intersomitic arteries from the dorsal aorta, (b) flow into intersomitic veins from DLAV, and (c) flow from intersomitic veins into the cardinal veins.

guided by repulsive cues from semaphorin ligands that bind to plexin receptors in migrating neuronal growth cones (48). Semaphorins possibly could serve a similar role in the embryonic vasculature by providing repulsive signals that restrict the navigation of intersegmental vessels to the intersomitic boundaries. Further support for this model comes from studies of the netrin receptor Unc5B and its ligand Netrin-1a, which appear to control endothelial tip-cell filopodia and branching (49). *Unc5B* is expressed selectively in endothelial tip cells, and the reduced function of this gene in mice causes excessive vessel branching in the internal carotid artery, intersegmental vessels, and vessels of the nervous system. In zebrafish embryos, reduced Unc5B function causes intersegmental vessels to exhibit ectopic filopodial extensions and form extra vessel branches at variable positions, resulting in aberrant anastomosis. These data thus suggest that, during angiogenesis, blood vessels are guided by repulsive mechanisms that employ the same cues and receptors that guide axons.

In summary, the optical clarity of embryonic zebrafish and their genetic and experimental accessibility have enabled the rapid imaging and analysis of the vascular "wiring" and patterning of intersegmental vessels in developing transgenic embryos, in which the expression of green fluorescent protein is driven by the *fli1* promoter. These studies represent the first comprehensive investigation into the development of angiogenic networks.

MOLECULAR SIGNALS IN ARTERIAL AND VENOUS DIFFERENTIATION

The mechanism by which ECs assume an arterial or a venous identity has been elusive for many years. It has been hypothesized that arterial and venous EC fate is established by anatomical location and local physiological conditions in the primary vascular network. Mounting evidence suggests, however, that arterial and venous fates actually are determined *prior* to blood circulation (3,4,50–52). Since the discovery that *ephrinB2* (*efnb2*) (a transmembrane ligand) and *EphB4* (a cognate receptor for *efnb2*) expression are restricted to the arterial and venous endothelium, respectively, multiple molecules including Notch, Gridlock(grl)/hey2, and VEGF have been demonstrated to regulate arterial and venous EC differentiation (Table 19-1). Inactivation of either *ephrinB2* or *EphB4* in mice results in defects of vascular remodeling and maturation in angiogenesis. However, these mutations do not exhibit defects in arterial and venous differentiation (50). Nevertheless, this pair of transmembrane-anchored signaling molecules defines the identity of arterial and venous endothelium prior to the onset of circulation.

DELTA-NOTCH SIGNALING

The expression of Notch receptors and ligands is observed in arterial but not venous endothelium. Furthermore, perturbed

Table 19-1: Markers of Arteries and Veins

Gene	*Species*	*References*
Artery specific		
EphrinB2	Chick, zebrafish, mouse	50,51
deltaC	Zebrafish	51
Dll4	Mouse	54
Jagged1	Mouse	55
Jagged2	Mouse	55
Hey1	Mouse	60,61
grl/Hey2	Zebrafish, mouse	3,60
HeyL	Mouse	60,61
Neuropilin-1	Chick, mouse	74,66
Notch1	Mouse	55
notch1b	Zebrafish	T.P. Zhong, unpublished data
Notch3	Mouse	55
Notch4	Mouse	55
notch5	Zebrafish	51
Connexin37	Mouse	59
CD44	Mouse	66
Vein Specific		
EphB4	Zebrafish, mouse	50,51
flt4	Zebrafish	16

Delta-Notch signaling results in defective arterial differentiation in mice and zebrafish. In zebrafish, *deltaC*, *notch5*, and *grl* (a component of the Notch pathway) are expressed in arterial ECs (3,51). This observation also was made for *notch1b* (T.P. Zhong, unpublished observations). Because *notch1a* and *notch1b* are duplicate copies of the *notch1* gene in the zebrafish genome (53), *notch1b* could be the counterpart of the mammalian *Notch1* gene. In mice, Notch1, Notch3, and Notch4 receptors and Jagged1, Jagged2, and Delta-like4(dll4) ligands are all expressed in arteries but not in veins (54,55). Among those, Notch1, Notch 4, Delta-like4 (dll4), Jagged1, and Jagged2 are expressed in arterial ECs and Notch3 and Jagged1 are expressed in arterial smooth muscle cells (55). Although zebrafish homologues of *jagged1* and *jagged2* have been isolated, whether they are expressed in arterial ECs is not known. Future in situ hybridization experiments are required to test their expression patterns.

Functional studies of Notch signaling in both zebrafish and mice suggest a critical role for this pathway in arterial–venous differentiation. The zebrafish *mind bomb* (*mib*) mutation occurs in a gene encoding a ubiquitin ligase that is essential for Delta-expressing cells to activate Notch receptors in neighboring cells (56). During vascular development, *mib* mutants display disorganized axial vessels and cranial hemorrhage. Some mutant embryos also have reduced DA size, arterial–venous shunts, and ectopic intersomitic sprouts (51). Characterization of *mib* demonstrates that the expression of certain arterial gene markers are absent (*efnb2*, *notch5*) or reduced (*deltaC*), whereas others are unaltered (*grl*, *tbx20*)

(51). This suggests that ECs in *mib* mutants have established arterial fate but are defective in subsequent arterial differentiation and maintenance. *Mib* mutant embryos also display ectopic expression of *flt4* – a gene whose expression is normally restricted to venous endothelium 24 hours post fertilization (hpf) – in arterial endothelium. Furthermore, perturbation of Notch signaling by misexpression of a dominant negative *Su(H)*, a transcriptional mediator of Notch signaling, results in the absence of arterial markers and, in some cases, partial loss of the DA with simultaneous expansion of a contiguous region of the cardinal vein (4,51). These data support the idea that arterial fate is specified and maintained via repressing venous fate. Because the milder phenotype of *mib* mutants potentially may be explained by the expression of the *mib*-related gene, it will be important to test whether inactivation of both genes causes more severe defects in arterial endothelial specification. Consistent with these findings in zebrafish, mice with targeted deletions in *Dll4*, *Jagged1*, *Notch1*, or *Notch1/Notch4* display marked abnormalities in vascular development, including yolk sac vascular remodeling abnormalities, a reduced DA, massive hemorrhages, and embryonic lethality (57–59).

The Role of *grl/hey2*

The zebrafish *grl* gene is related to the recently described *Hey2/Hrt2/ Hesr1/Chf1* in mouse, belonging to a novel group of the *hairy* and *E(spl)*-related bHLH family (3,60,61). The *hey* gene family consists of three members, *hey1*, *grl/hey2*, and *heyL*, in human, mouse, and zebrafish, with only a single counterpart in *Drosophila* (3,62,63). In mice, members of the *Hey* gene family show both partially overlapping and complementary expression patterns. All three members of this family, for example, are expressed in the presomitic mesoderm, somites, and aorta, but *Hey1* and *Grl/Hey2* expression are restricted to the atrium and ventricle of the heart, respectively (60,61). Zebrafish *grl/hey2*, in contrast, is the only *hey* gene expressed in the aorta and the heart. In addition, *grl/hey2* is expressed in angioblasts at the lateral plate mesoderm prior to vessel formation and subsequently becomes restricted to arterial ECs (3,63). *Grl/hey2*, therefore, is expressed at the correct time and in an appropriate region for regulating arterial endothelial differentiation in zebrafish. Furthermore, two conserved Su(H)/rbp-1-binding sites (CGTGGGAA) have been identified in the upstream regulatory region of zebrafish *grl* and its mouse homologue *Hey2*, suggesting that *grl/hey2* acts as a transcriptional target of Su(H) (4,64). Indeed, transcriptional activity of *Hey2* is regulated by Notch signaling in cultured cells, and expression of *grl/hey2* is downregulated in zebrafish and mouse embryos lacking Notch activity (4,64–66).

The functional importance of Grl/Hey2 has been analyzed in mutant zebrafish embryos whose hypomorphic *grl* allele causes artery-specific defects, including a disruption of the aortic bifurcation (67). In addition, *EGFP* expression under control of the *flk1* promoter reveals that the aorta in *grl* mutants has a decreased lumen size. Reduction of Grl function using a high dose of morpholino-modified antisense oligonucleotides specifically inhibits the formation of the DA, whereas a low dose of antisense morpholinos phenocopies the defects in *grl* mutant embryos (4). This suggests that, in zebrafish, Grl is required for arterial endothelial differentiation in a dosage-sensitive manner. Furthermore, *grl* mutant embryos express the early arterial marker *notch1b* but not the late arterial markers *deltaC*, *notch5*, *ephrinB2*, and *tbx20*, indicating that *grl* mutants (like *mib*) have acquired an initial arterial fate but are defective in subsequent arterial differentiation, maintenance, and maturation.

In mice, *Hey* genes function redundantly in vascular development because of their overlapping expression in the aorta. Targeted inactivation of *Hey2* causes fatal cardiomyopathy, ventricular septation defects, and high lethality, but the arterial endothelial system is not affected (68). However, the combined loss of *Hey1* and *Hey2* prevents arterial endothelial differentiation, reduces the size of the DA, inhibits both vasculogenesis in the placental labyrinth and vascular remodeling, and is embryonic lethal (66). Thus, the phenotype of *Hey1/Hey2* double–knockout mouse embryos is similar to that of zebrafish embryos with reduced Grl activity. These defects also are similar to the phenotype of mouse embryos lacking either *Notch1* or *Dll4*, suggesting that Hey1/Hey2 function in the Notch pathway to regulate arterial endothelial differentiation and development. In humans, expression profiling has identified *grl* as part of a group of signaling genes that includes Notch receptors and is selectively expressed in human arterial ECs (HUAECs). In addition, expression of Grl in human umbilical venous ECs (HUVECs) induces the expression of a group of artery-specific genes (69). The apparent conservation of Grl signaling from zebrafish to humans underscores the potential importance of this pathway in arterial EC differentiation.

VEGF Signaling

Among its other roles, recent studies suggest that VEGF is also important for promoting differentiation of the arterial endothelium (52,70). In cultured murine embryonic ECs, the addition of VEGF-A_{120} or VEGF-A_{164} induces *ephrinB2* expression in up to 50% of cells, whereas the addition of bFGF, BMP2, PDGF, IGF-1, NGF, NT-3, BDNF, or Sonic Hedgehog can only promote *EphrinB2* expression in approximately 10% (70). Importantly, both VEGF isoforms induce arterial differentiation without promoting endothelial proliferation at the low concentration of 2.5 pM (70). In zebrafish, reducing VEGF activity using antisense morpholinos causes loss of expression of the arterial markers *ephrinB2* and *notch5* and leads to ectopic expression of venous marker *flt4* (52). Conversely, misexpression of *Vegf-121* mRNA or *Vegf-165* mRNA causes ectopic *ephrinB2* expression in the cardinal vein (52). Taken together, these results suggest that VEGF is sufficient to promote arterial endothelial differentiation in vitro and in vivo. What remains unclear, however, is the manner in which VEGF influences this process. One proposed model suggests a differential effect of Vegf on arterial ECs in zebrafish, in which some angioblasts that migrate medially first preferentially receive Vegf that is secreted from the somites and hypochord. These angioblasts therefore become specified as arterial angioblasts and subsequently differentiate into arterial ECs (71). Alternatively, angioblasts with prespecified arterial fate determined by an intrinsic cellular mechanism respond to Vegf induction, which triggers overt arterial differentiation. Indeed, the proportion of *ephrinB2*-expressed embryonic ECs induced by Vegf never exceeds 50%, suggesting that isolated *ephrinB2*-negative embryonic ECs contain prespecified arterial and venous angioblasts in a 1:1 ratio, respectively. Thus, Vegf only promotes prespecified arterial angioblasts to subsequently differentiate into arterial ECs.

Gain- and loss-of-function studies using VEGF as an entry point in zebrafish have led to the placement of several genes in a putative pathway that regulates arterial endothelial development. The hedgehog (Hh) signaling pathway has been shown to be important for arterial differentiation and development in zebrafish (52). The zebrafish *sonic-you* (*syu*) and *you-too* (*yot*) genes encode the orthologues of shh and *gli2*, a downstream effector of Hh signaling (72,73). Mutations in either *syu* or *yot* result in loss of *vegf* expression in somites, which in turn downregulate *ephrinB2* expression in the aorta and suggest that the somatic source of Vegf is important for arterial differentiation. Forced expression of *vegf-121* or *vegf-165* in *yot* mutant embryos rescues *ephrinB2* expression. Furthermore, constitutively active Notch signaling rescues expression of *ephrinB2* and *notch5* in embryos with reduced Vegf activity, whereas misexpression of *Vegf-121* cannot restore *notch5* expression in *mib* mutant embryos. These data suggest that Vegf acts downstream of Hedgehog signaling but upstream of the Notch pathway for regulating *ephrinB2* and *notch5* expression during arterial endothelial differentiation.

Plasticity of Arterial and Venous Fate

Although multiple signaling molecules appear to regulate the identity of arterial and venous ECs prior to blood flow, studies using quail-chick transplantation have revealed that arterial and venous EC fate remains plastic until embryonic day 7 (74). Before this time, the embryonic DA or cardinal vein can colonize both arteries and veins in chick host embryos when transplanted from quail donors. These transplanted arterial and venous ECs express the avian arterial markers *ephrinB2*, *neuropilin-1* (*nrp-1*), and *tie2* when they integrate into the lining of arteries. Cells that integrate into venous endothelium, however, do not express these markers. After embryonic day 7, arterial and venous ECs gradually lose their plasticity and become committed. It is possible that the smooth muscle cell layer in the maturing vascular wall plays a role in maintaining the arterial endothelial fate. Thus, arterial and venous EC fates are prespecified but not fully committed at early embryonic stages and can be altered by the local vascular environment. Furthermore, genes such as *ephrinB2*, *nrp-1*, and *tie2* may not be involved in determining arterial and venous endothelial fates, but rather serve as markers of arterial differentiation and/or maintenance. Indeed, *ephrinB2* is not expressed in angioblasts at the lateral plate mesoderm in zebrafish embryos.

It would be interesting to determine whether constitutive expression of Notch receptors and downstream Grl/Hey2 transcription factors leads to commitment of arterial endothelial fate. Likewise, it will be important to examine whether *notch* and *grl* are ectopically induced in venous ECs when they convert into arterial ECs after transplantation. Venous arterialization occurs when a vein segment is transposed into an arterial region in therapeutic procedures (75). The most recent evidence suggests that venous ECs can adapt to a novel vascular environment. High levels of Notch and Grl/Hey2, for example, are ectopically induced in human femoral veins when exposed to arterial flow conditions (G. Garcia-Cardena and M. Gimbrone, personal communication). This suggests that venous arterialization in humans is coincident with a molecular program similar to that required for arterial EC differentiation during embryonic development. It remains to be seen, however, whether molecular evidence of arterial transformation actually leads to structural and functional adaptation of the vascular wall.

KEY POINTS

- Despite their relatively recent arrival as a vertebrate model system, zebrafish have already contributed much to our understanding of embryonic vascular biology. Uniquely amenable to both forward and reverse genetic approaches as well as in vivo gain-of-function analysis, the zebrafish model is primed to uncover key signaling pathways conserved among vertebrates and essential for vascular development.
- Zebrafish studies have provided novel insight into angioblast specification, differentiation of arterial and venous endothelium, assembly of the DA through vasculogenesis, and the sprouting, growth, patterning, and remodeling of vessels during through angiogenesis.

Future Goals

- Future comparative genetic studies in zebrafish, mice, chicks, and other organisms will be necessary to decipher the tremendous complexity underlying the emergence of a vascular network and circulation during vertebrate embryogenesis.
- Benefits of this formidable effort are numerous and may include important insights for regenerative biology and novel therapeutic approaches for patients suffering from a variety of diseases involving the vascular endothelium such as atherosclerosis and cancer.

ACKNOWLEDGMENTS

We thank all members of the Zhong laboratory, at Vanderbilt University, who have contributed to some of the ideas and work summarized here. Work in the Zhong laboratory has been supported by grants from the National Institutes of Health (NIH), the American Heart Association (AHA), March of Dimes Foundation, and the Vanderbilt Academic Venture Capital Fund.

REFERENCES

1 Stainier DY, Fouquet B, Chen JN, et al. Mutations affecting the formation and function of the cardiovascular system in the zebrafish embryo. *Development*. 1996;123:285–292.
2 Weinstein BM, Schier AF, Abdelilah S, et al. Hematopoietic mutations in the zebrafish. *Development*. 1996;123:303–309.
3 Zhong TP, Rosenberg M, Mohideen MA, Weinstein B, Fishman MC. Gridlock, an HLH gene required for assembly of the aorta in zebrafish. *Science*. 2000;287 (5459):1820–1824.
4 Zhong TP, Childs S, Leu JP, Fishman MC. Gridlock signaling pathway fashions the first embryonic artery. *Nature*. 2001;414(6860):216–220.
5 Garrity DM, Childs S, Fishman M. The heartstrings mutation in zebrafish causes heart/fin Tbx5 deficiency syndrome. *Development*. 2002;129 (19):4635–4645.
6 Ernest S, Rauch GJ, Haffter P, Geisler R, Petit C, Nicolson T. Mariner is defective in myosin VIIA: A zebrafish model for hereditary deafness. *Hum Mol Gen*. 2000;9(14):2189–2196.
7 Xu X, Meiler SE, Zhong TP, et al. Cardiomyopathy in zebrafish due to a mutation in an alternatively-spliced exon of titin. *Nat Genet*. 2002;30(2):205–209.
8 Brownlie A, Donovan A, Pratt SJ, et al. Positional cloning of the zebrafish sauternes gene: A model for congenital sideroblastic anaemia. *Nat Genet*. 1998;20(3):244–250.
9 Donovan A, Brownlie A, Zhou Y, et al. Positional cloning of zebrafish ferroportin1 identifies a conserved vertebrate iron exporter. *Nature*. 2000;403(6771):776–781.
10 Fraenkel PG, Traver D, Donovan A, Zahrieh D, Zon LI. Ferroportin1 is required for normal iron cycling in zebrafish. *J Clin Invest*. 2005;115(6):1532–1541.
11 Drummond IA, Majumdar A, Hentschel H, et al. Early development of the zebrafish pronephros and analysis of mutations affecting pronephric function. *Development*. 1998;125(23): 4655–4667.
12 Isogai S, Horiguchi M, Weinstein BM. The vascular anatomy of the developing zebrafish: an atlas of embryonic and early larval development. *Dev Biol*. 2001;230(2):278–301.
13 Risau W, Flamme I. Vasculogenesis. *Ann Rev Cell Dev Biol*. 1995;11:73–91.
14 Fouquet B, Weinstein BM, Serluca FC, Fishman MC. Vessel patterning in the embryo of the zebrafish: guidance by notochord. *Dev Biol*. 1997;183:37.
15 Liao W, Bisgrove BW, Sawyer H, et al. The zebrafish gene cloche acts upstream of a flk-1 homologue to regulate endothelial cell differentiation. *Development*. 1997;124:381–389.
16 Thompson MA, Ransom DG, Pratt SJ, et al. The cloche and spadetail genes differentially affect hematopoiesis and vasculogenesis. *Dev Biol*. 1998;197:248–269.
17 Coffin DJ, Poole TJ. Embryonic vascular development: immunohistochemical identification of the origin and subsequent morphogenesis of the major vessel primordia in quail embryos. *Development*. 1988;102:735–748.
18 Pardanaud L, Luton D, Prigent M, Bourcheix LM, Catala M, Dieterlen-Lievre F. Two distinct endothelial lineages in ontogeny, one of them related to hemopoiesis. *Development*. 1996;122: 1363–1371.
19 Haar JL, Ackerman GA. A phase and electron microscopic study of vasculogenesis and erythropoiesis in the yolk sac of the mouse. *Anat Rec*. 1971;170:199–223.
20 His W. Lecithoblast und angioblast der wirbeltiere. *Abhandl Math-Pys Ges Wiss*. 1900;26:171–328.
21 Murray PDF. The development in vitro of the blood of the early chick embryo. *Proc Roy Soc*. 1932;11:497–521.
22 Orkin SH, Zon LI. Genetics of erythropoiesis: induced mutations in mice and zebrafish. *Annu Rev Genet*. 1997;31:33–60.
23 Wilt F. Erythropoiesis in the chick embryo: the role of endoderm. *Science*. 1965;147:1588–1590.
24 Pardanaud L, Yassine F, Dieterlen-Lievre F. Relationship between vasculogenesis, angiogenesis and haemopoiesis during avian ontogeny. *Development*. 1989;105:473–485.
25 Vokes SA, Krieg PA. Endoderm is required for vascular endothelial tube formation, but not for angioblast specification. *Development*. 2002;129:775–785.
26 Shalaby F, Rossant J, Yamaguchi TP, et al. Failure of blood-island formation and vasculogenesis in Flk-1-deficient mice. *Nature*. 1995;376:62–66.
27 Habeck H, Odenthal J, Walderich B, Maischein H, Schulte-Merker S; Tubingen 2000 screen consortium. Analysis of a zebrafish VEGF receptor mutant reveals specific disruption of angiogenesis. *Curr Biol*. 2002;12:1405–1412.
28 Liang D, Chang JR, Chin AJ, et al. The role of vascular endothelial growth factor (VEGF) in vasculogenesis, angiogenesis, and hematopoiesis in zebrafish development. *Mech Dev*. 2001;108:29–43.
29 Carmeliet P, Ferreira V, Breier G, et al. Abnormal blood vessel development and lethality in embryos lacking a single VEGF allele. *Nature*. 1996;380:435.
30 Conway EM, Collen D, Carmeliet P. Molecular mechanisms of blood vessel growth. *Cardiovasc Res*. 2001;49:507–521.
31 Stainier DY, Weinstein BM, Detrich HW 3rd, Zon LI, Fishman MC. Cloche, an early acting zebrafish gene, is required by both the endothelial and hematopoietic lineages. *Development*. 1995;121:3141–3150.
32 Liao EC, Paw BH, Oates AC, Pratt SJ, Postlethwait JH, Zon LI. SCL/Tal-1 transcription factor acts downstream of cloche to specify hematopoietic and vascular progenitors in zebrafish. *Genes Dev*. 1998;12:621–626.
33 Gering M, Yamada Y, Rabbitts TH, Patient RK. Lmo2 and Scl/Tal1 convert non-axial mesoderm into haemangioblasts which differentiate into ECs in the absence of Gata1. *Development*. 2003; 130:6187–6199.
34 Porcher C, Swat W, Rockwell K, Fujiwara Y, Alt FW, Orkin SH. The T cell leukemia oncoprotein SCL/tal-1 is essential for development of all hematopoietic lineages. *Cell*. 1996;86:47–57.
35 Visvader JE, Fujiwara Y, Orkin SH. Unsuspected role for the T-cell leukemia protein SCL/tal-1 in vascular development. *Genes Dev*. 1998;12:473–479.

36 Yamada Y, Pannell R, Forster A, Rabbitts TH. The oncogenic LIM-only transcription factor Lmo2 regulates angiogenesis but not vasculogenesis in mice. *Proc Natl Acad Sci USA*. 2000;97:320–324.
37 Dooley KA, Davidson AJ, Zon LI. Zebrafish scl functions independently in hematopoietic and endothelial development. *Dev Biol*. 2005;277:522–536.
38 Patterson LJ, Gering M, Patient R. Scl is required for dorsal aorta as well as blood formation in zebrafish embryos. *Blood*. 2005;105:3502–3511.
39 Reifers F, Walsh EC, Leger S, Stainier DY, Brand M. Induction and differentiation of the zebrafish heart requires fibroblast growth factor 8 (fgf8/acerebellar). *Development*. 2000;127(2):225–235.
40 Halpern ME, c, Walker C, Kimmel CB. Induction of muscle pioneers and floor plate is distinguished by the zebrafish no tail mutation. *Cell*. 1993;75:99–111.
41 Ondine C, Krieg P. VEGF mediates angioblast migration during development of the dorsal aorta in *Xenopus*. *Development*. 1998;125:3905.
42 Ferrara N, Carver-Moore K, Chen H, Does M, Hillan KJ, Moore MW. Heterozygous embryonic lethality induced by targeted inactivation of the VEGF gene. *Nature*. 1996;380:439.
43 Parker LH, Schmidt M, Jin SW, et al. The endothelial-cell-derived secreted factor Egfl7 regulates vascular tube formation. *Nature*. 2004;428:754–758.
44 Isogai S, Lawson ND, Torrealday S, Horiguchi M, Weinstein BM. Angiogenic network formation in the developing vertebrate trunk. *Development*. 2003;130:5281–5290.
45 Childs S, Chen JN, Garrity DM, Fishman, M. C. Patterning of angiogenesis in the zebrafish embryo. *Development*. 2002;129:973–982.
46 Lawson ND, Mugford JW, Diamond BA, Weinstein BM. Phospholipase C gamma-1 is required downstream of vascular endothelial growth factor during arterial development. *Genes Dev*. 2003;17:1346–1351.
47 Torres-Vazquez J, Gitler AD, Fraser SD, et al. Semaphorin-plexin signaling guides patterning of the developing vasculature. *Dev Cell*. 2004;7:117–123.
48 Raper JA. Semaphorins and their receptors in vertebrates and invertebrates. *Curr Opin Neurobiol*. 2000;10:88–94.
49 Lu X, Le Noble F, Yuan L, et al. The netrin receptor UNC5B mediates guidance events controlling morphogenesis of the vascular system. *Nature*. 2004;432:179–186.
50 Wang HU, Chen ZF, Anderson DJ. Molecular distinction and angiogenic interaction between embryonic arteries and veins revealed by ephrin-B2 and its receptor Eph-B4. *Cell*. 1998;93:741–753.
51 Lawson ND, Scheer N, Pham VN, et al. Notch signaling is required for arterial-venous differentiation during embryonic vascular development. *Development*. 2001;128:3675–3683.
52 Lawson ND, Vogel AM, Weinstein BM. Sonic hedgehog and vascular endothelial growth factor act upstream of the Notch pathway during arterial endothelial differentiation. *Dev Cell*. 2002;3:127–136.
53 Westin J, Lardelli M. Three novel Notch genes in zebrafish: implications for vertebrate Notch gene evolution and function. *Dev Genes Evol*. 1997;207:51–63.
54 Shutter JR, Scully S, Fan W, et al. Dll4, a novel Notch ligand expressed in arterial endothelium. *Genes Dev*. 2000;14:1313–1318.
55 Villa N, Walker L, Lindsell CE, Gasson J, Iruela-Arispe ML, Weinmaster G. Vascular expression of Notch pathway receptors and ligands is restricted to arterial vessels. *Mech Dev*. 2001;108:161–164.
56 Itoh M, Kim CH, Palardy G, et al. Mind bomb is a ubiquitin ligase that is essential for efficient activation of Notch signaling by Delta. *Dev Cell*. 2003;4:67–82.
57 Xue Y, Gao X, Lindsell C, et al. Embryonic lethality and vascular defects in mice lacking the Notch ligand Jagged1. *Hum Mol Genet*. 1999;8:723.
58 Krebs LT, Xue Y, Norton CR, et al. Notch signaling is essential for vascular morphogenesis in mice. *Genes Dev*. 2000;14:1343–1352.
59 Duarte A, Hirashima M, Benedito R, et al. Dosage-sensitive requirement for mouse Dll4 in artery development. *Genes Dev*. 2004;18:2474–2478.
60 Leimeister C, Externbrink A, Klamt B, Gessler M. Hey genes: A novel subfamily of hairy- and Enhancer of split related genes specifically expressed during mouse embryogenesis. *Mech Dev*. 1999;85:173–177.
61 Nakagawa O, Nakagawa M, Richardson JA, Olson EN, Srivastava D. HRT1, HRT2, and HRT3: a new subclass of bHLH transcription factors marking specific cardiac, somitic, and pharyngeal arch segments. *Dev Biol*. 1999;216:72–84.
62 Kokubo H, Lun Y, Johnson RL. Identification and expression of a novel family of bHLH cDNAs related to *Drosophila* hairy and enhancer of split. *Biochem Biophys Res Commun*. 1999;260:459–465.
63 Winkler C, Elmasri H, Klamt B, Volff JN, Gessler M. Characterization of hey bHLH genes in teleost fish. *Dev Genes Evol*. 2003;213:541–553.
64 Nakagawa O, McFadden DG, Nakagawa M, et al. Members of the HRT family of basic helix-loop-helix proteins act as transcriptional repressors downstream of Notch signaling. *Proc Natl Acad Sci USA*. 2000;97:13655–13660.
65 Iso T, Chung G, Hamamori Y, Kedes L. HERP1 is a cell type-specific primary target of Notch. *J Biol Chem*. 2002;277:6598–6607.
66 Fischer A, Schumacher N, Maier M, Sendtner M, Gessler M. The Notch target genes Hey1 and Hey2 are required for embryonic vascular development. *Genes Dev*. 2004;18:901–911.
67 Weinstein BM, Stemple DL, Driever W, Fishman MC. Gridlock, a localized heritable vascular patterning defect in the zebrafish. *Nat Med*. 1995;1:1143–1147.
68 Donovan J, Kordylewska A, Jan YN, Utset MF. Tetralogy of Fallot and other congenital heart defects in Hey2 mutant mice. *Curr Biol*. 2002;12:1605–1610.
69 Chi JT, Chang HY, Haraldsen G, et al. Endothelial cell diversity revealed by global expression profiling. *Proc Natl Acad Sci USA*. 2003;100:10623–10628.
70 Mukouyama YS, Shin D, Britsch S, Taniguchi M, Anderson DJ. Sensory nerves determine the pattern of arterial differentiation and blood vessel branching in the skin. *Cell*. 2002;109:693–705.
71 Torres-Vazquez J, Kamei M, Weinstein BM. Molecular distinction between arteries and veins. *Cell Tissue Res*. 2003;314:43–59.
72 Schauerte HE, van Eeden FJ, Fricke C, Odenthal J, Strahle U, Haffter P. Sonic hedgehog is not required for the induction of

medial floor plate cells in the zebrafish. *Development*. 1998;125: 2983–2993.

73 Karlstrom RO, Talbot WS, Schier AF. Comparative synteny cloning of zebrafish you-too: mutations in the Hedgehog target gli2 affect ventral forebrain patterning. *Genes Dev*. 1999;13:388–393.

74 Moyon D, Pardanaud L, Yuan L, Breant C, Eichmann A. Plasticity of ECs during arterial-venous differentiation in the avian embryo. *Development*. 2001;128:3359–3370.

75 Henderson VJ, Mitchell RS, Kosek JC, Cohen RG, Miller DC. Biochemical (functional) adaptation of "arterialized" vein grafts. *Ann Surg*. 1986;203:339–345.

20

Endothelial Cell Differentiation and Vascular Development in Mammals

Cam Patterson

Carolina Cardiovascular Biology Center, University of North Carolina, Chapel Hill

As the developing mammalian embryo expands beyond a small mass of cells, it acquires a need to establish circulation in order to facilitate communication among different parts of the embryo. It also must establish a mechanism for nutrition and oxygenation once simple diffusion becomes a limiting mechanism. These requirements provide a rationale for why the cardiovascular system is the first to function during mammalian development. An analysis of vascular development therefore provides a paradigm for understanding the general principles of organ development in mammals. Practical reasons also exist to elucidate the mechanisms by which blood vessels form. It is clear that defective vascular development contributes to a number of congenital conditions in humans, from hereditary angiomas to more complicated developmental syndromes. In addition, we now know that many pathological events in adults follow rules that resemble developmental events, such that understanding the latter can inform how we approach the former. In this chapter, we briefly review the field of vascular development from an historical perspective, after which the key steps in mammalian endothelial differentiation will be considered, with a focus on the molecular and cellular events that underlie the principles of developmental blood vessel formation.

HISTORICAL OVERVIEW

Consistent with its central importance in embryonic development, the history of research into vascular development is populated with seminal histologists such as Wilhelm His and Florence Sabin (see Chapter 1). Ideas about endothelial origin followed shortly after the development of histology and cell biology as fields of study. Beyond characterization of the fundamental anatomy of vascular development (in many timeless analyses that bear rediscovery by present-day investigators), the first investigators in this field participated in one of the classic debates in all of developmental biology: Where and when do endothelial cells (ECs) – and hence blood vessels – arise in the developing embryo? Because blood vessels are first observed in the yolk sac in avian and mammalian embryos, initially it was assumed that all blood vessels arise from extraembryonic tissues. However, careful histologic analysis subsequently indicated that isolated foci of ECs also can be observed in the embryo proper, which suggested that blood vessels arise from an intraembryonic source (specifically, the mesoderm) rather than via colonization (1,2). The latter model has been successively reconfirmed by increasingly sophisticated approaches to isolate and track blood vessel development in mammalian systems.

In addition to taking advantage of advances in cell biology, and subsequently in the rich tools of the molecular biology revolution, several fundamental approaches have greatly accelerated the understanding of endothelial differentiation and vascular development. Techniques to culture embryos in vitro and to perform fate mapping with viruses and dyes have markedly simplified the discovery of antecedent–decedent relationships within vascular cell lineages. The ability to manipulate the mouse genome via homologous recombination has facilitated reverse genetic approaches for understanding the role of specific genes and pathways in vascular development. Simultaneously, high-throughput phenotype-driven genetic approaches in mammalian systems have permitted the identification of unanticipated molecular participants in vascular developmental pathways. Finally, the ability to manipulate embryonic stem cells in culture and to differentiate these cells down the endothelial lineage has served as a powerful tool for recapitulating the developmental process in a refined and controlled situation. These and many other approaches are the tools for discovery that have led to the understanding of endothelial development described in the remaining sections of this chapter.

GASTRULATION AND POSTGASTRULATION EVENTS

Although considerable debate has arisen about what is a surprisingly elusive question, it is now generally accepted that the

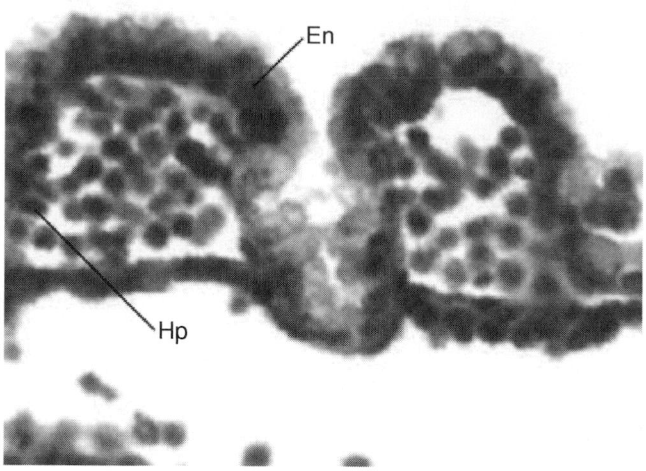

Figure 20.1. Blood islands. Blood islands are foci of loose hematopoietic (Hp) and adjacent endothelial (En) precursors forming a thin lumen. These structures represent the initial morphological characteristic of vascular development in extraembryonic tissues in mammals.

mesoderm is the exclusive source of EC progenitors. Although efforts have been made to identify specific loci within the mesoderm that preferentially specify endothelial fates, it seems clear that precursors to endothelium arise within all intraembryonic mesoderm, with the exception of the notochord and prechordal plate (3), as loose aggregations of cells bearing early markers of the endothelial lineage and with the potential to coalesce into vascular cords.

The signals for angioblast induction are not entirely clear, but data support roles for members of the fibroblast growth factor (FGF) and bone morphogenetic protein (BMP) families in this inductive process. The initiation of vascular differentiation within the embryo occurs in opposition to endoderm, which suggests that endoderm-derived signals are required for angioblast commitment within mesoderm. Indian hedgehog is appropriately expressed in a spatial and temporal pattern within the endoderm to fulfill this role, and is required for normal vascular development (4). However, recent evidence suggests that angioblast commitment is not totally dependent on endoderm-derived signals, although later morphogenetic steps in vascular development may require endoderm–mesoderm interactions (5).

The pluripotency of vascular precursors also has been a vexing issue within the field. The close colocalization of endothelial and hematopoietic precursors (so-called "blood islands," Figure 20.1) within the embryo has been recognized since early in the twentieth century (6). This raised the possibility that both lineages arise from a common precursor, the hemangioblast. Subsequent studies have indicated that both lineages bear certain common molecular markers, and cells with the potential to produce either lineage have been isolated from differentiating mouse embryoid bodies. The pluripotency of these putative hemangioblasts has not been completely characterized, although some data support the possibility that these multipotent cells also may assume characteristics of the smooth muscle lineage under appropriate inductive conditioning (7). It also does not seem likely that hemangioblast identity is a required step for all endothelial development and, in fact, the direct differentiation of angioblasts from mesoderm is also a well-supported phenomenon. The teleological rationale for direct endothelial differentiation from mesoderm under some circumstances, rather than passage through a hemangioblast intermediary stage, is not necessarily clear, but the direct differentiation of ECs appears to occur primarily within somites (8).

In spite of data supporting the existence of the hemangioblast, definitive isolation of these cells and localization within the developing embryo has been an extraordinary challenge. However, recent evidence indicates that cells with hemangioblast properties are present transiently in the posterior segment of the primitive streak during gastrulation and are defined by the coexpression of the vascular endothelial growth factor (VEGF) receptor FLK-1 and the mesodermal T-box gene brachyury (9). These data also suggest that the endothelial and hematopoietic fates of embryonic cells are established prior to the appearance of blood islands during development. In fact, some data suggest that the hemangioblast lineage may be committed prior to gastrulation within the epiblast (9).

EXTRAEMBRYONIC VASCULOGENESIS

Blood vessel formation classically is divided into two categories. *Vasculogenesis* refers to the in situ differentiation of ECs to form blood vessels, with or without associated angioblast migration. In contrast, *angiogenesis* refers to the formation of new blood vessels via extension or remodeling of existing blood vessels. Angiogenesis occurs throughout development and in adulthood, whereas vasculogenesis generally is thought to occur during a limited period early in embryonic development. (The term *vasculogenesis* is occasionally used to refer to the development of blood vessels during adulthood, especially when associated with circulating vascular progenitor cells. This is an inappropriate use of this term based on its natural definition.) Vasculogenesis is further subdivided depending on whether it occurs within the extraembryonic or intraembryonic compartments. The best available evidence suggests that these two waves of vasculogenesis are temporally and spatially distinct, and molecular studies indicate that they are also partially distinct at the mechanistic level.

Extraembryonic vasculogenesis precedes intraembryonic vascular development, and, in most mammals, is first apparent as blood islands assembling within the mesodermal layer of the yolk sac. (It is important to note that not all species have yolk sac blood islands; for example, teleosts have similar structures that appear in the intermediate cell mass that arises during gastrulation. This suggests that extraembryonic vasculogenesis is an evolutionarily late adaptation in blood vessel development.) Blood islands are foci of hemangioblasts that differentiate in situ, forming a loose inner mass of embryonic hematopoietic precursors and an outer luminal layer of angioblasts. Blood islands eventually coalesce into a functional vascular network that constitutes the vitelline circulation, which is adapted to transfer nutrients from the yolk sac to the embryo proper. Recent evidence indicates that extraembryonic blood vessels also may arise independently of blood islands via the direct differentiation of angioblasts from mesoderm (10). Vessels arising via yolk sac vasculogenesis communicate with the fetal circulation via the vitelline vein, but otherwise do not contribute to intraembryonic vasculature. Likewise, embryonic hematopoiesis arising within the yolk sac is eventually replaced by blood cell populations derived from definitive intraembryonic hematopoietic differentiation within the spleen, liver, and, ultimately, the bone marrow. Establishment of the extraembryonic circulation is particularly dependent on transforming growth factor (TGF)-β signaling, which is required for matrix assembly, vascular remodeling, and coordinated hematopoietic differentiation within the yolk sac (11).

INTRAEMBRYONIC VASCULOGENESIS

Intraembryonic vasculogenesis initiates within the para-aortic mesoderm (the aorta-gonad-mesonephros or AGM region) (Figure 20.2) and the allantois, and gives rise to the dorsal aortae and the cardinal veins. (The para-aortic mesoderm is analogous to the lateral plate mesoderm in *Xenopus* and zebrafish systems.) The endocardium originates in mammals from clusters of migrating angioblasts derived from presomitic cranial mesoderm that enter the pericardial area to form a vascular plexus adjacent to the developing myocardium. This plexus undergoes remodeling to form the endocardial tube, which is the first intraembryonic vasculogenic event in the mammalian embryo (10). The development of the aorta follows the endocardium from a distinct subset of angioblasts derived from the AGM region that assemble into a primary vascular network. This network remodels in a bidirectional fashion to generate the bilateral embryonic aortae (10). The AGM also seems to be the major site for the initiation of intraembryonic hematopoiesis (12). Convincing evidence indicates that the allantois also contains dispersed precursors that differentiate and assemble in situ to form the allantoic vasculature shortly after the aortae are assembled (13).

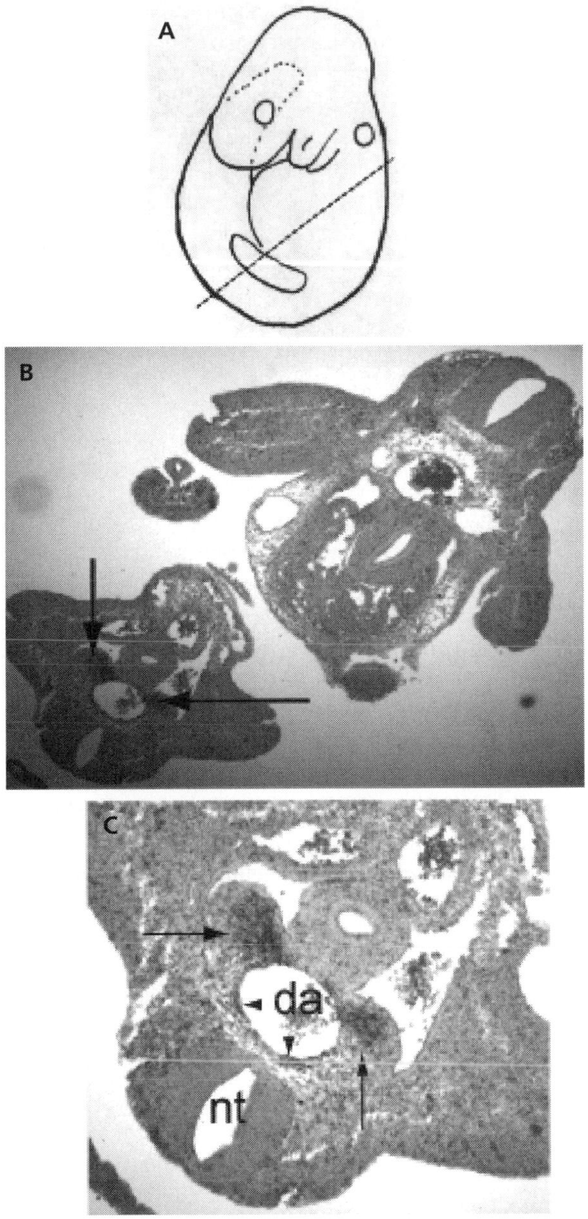

Figure 20.2. The AGM region. **(A)** Diagram of an E9.5 mouse embryo with a line indicating the location of subsequent sections. **(B)** A section at the indicated level with arrows indicating the AGM regions. **(C)** A high-magnification view showing the AGM (*arrows*) in relationship to the dorsal aorta (*da*) and neural tube (*nt*).

Subsequent waves of intraembryonic vascular development occur primarily via angiogenesis. However, some endoderm-derived organs are capable of supporting vasculogenesis. This appears to be due in part to signals derived from undifferentiated endoderm that support angioblast differentiation from adjacent mesoderm. Remarkably, the developing vasculature in turn provides signals that are necessary for endodermal organogenesis (14) (see Chapter 22), which indicates that a rich interplay exists between endoderm-derived organs and their vasculature during the developmental process. As a corollary of this general mechanism, ectodermally derived organs such as the brain are strictly dependent on angiogenesis for vascularization.

DEVELOPMENTAL ANGIOGENESIS

Beyond the vascular beds described above, the majority of vascular development occurs via the process of angiogenesis – the growth of new blood vessels from existing vessels. Anatomically, angiogenesis occurs via at least two mechanisms. *Sprouting angiogenesis* is the classical mechanism, and involves the invasion of new capillaries into unvascularized tissue from an existing mature vasculature. Sprouting requires coordinated degradation of matrix proteins via the secretion and activation of matrix metalloproteinases, detachment and migration of ECs from the mature vessel into the avascular tissue, and proliferation of ECs at the leading edge of the extending vessel. Eventually, communications must be established between ECs in adjacent vessels for flow to be established, although the mechanisms by which this occurs are incompletely understood. Sprouting angiogenesis generally is activated in response to a gradient of growth factors. This method is the predominant mechanism for the vascularization of organs that lack the ability to develop vessels via vasculogenesis.

A second, less intuitive mechanism for vessel development is referred to as *intussusceptive* (or nonsprouting) *angiogenesis.* Intussusception involves the remodeling of existing vessels in such a way that a vessel enlarges, pinches inward, and eventually splits into two vessels that are invaded by pericytes and matrix to add stability. This awkward process is actually a complicated ritual of cellular rearrangement that requires flow-dependent signaling, alterations of cell–cell interactions, and rearrangements of blood flow. Intussusceptive angiogenesis is a versatile mechanism for new vessel development in organs that undergo extensive branching morphogenesis, presumably because rearrangements of vascular structure are obligatory. Intussusceptive angiogenesis plays a role in both early capillarization and in network remodeling and the formation of larger vessels in the embryo (15).

MOLECULAR MECHANISMS

The anatomic basis for vascular development in the embryo has been clarified through studies that have been performed over the past century. In contrast, the molecular underpinnings for EC differentiation have become clear only in the last decade. In this section, we review briefly the interrelationships among some of the key steps in endothelial differentiation. Many of the signaling and transcriptional events that regulate endothelial differentiation are characterized in detail in other chapters of this anthology, and the reader will find more detail about individual pathways in these chapters.

Key to any inductive process in the embryo is the elaboration of signals that are integrated to mark the initiation of a developmental cascade. In the case of early events in endothelial differentiation, these cues must determine the spatial and temporal appearance of hemangioblasts from undifferentiated mesoderm, and also the further maturation and assembly of these precursors into nascent blood vessels. Of the well-defined signaling pathways, the best data indicate that crucial roles exist for members of the FGF and BMP families as proximal inductive cues for the hematovascular lineage. Remarkably, less is known about the roles for these families in the vascular development in zebrafish (Chapter 19), which raises the possibility that they play a greater role in mammalian vascular development than they do in teleosts, although no data are available to exclude a role for these proteins in zebrafish. FGFs appear to be key components required for hematovascular differentiation in differentiating embryonic stem cells (16). BMPs and their receptors within the BMP2/4 series have critical roles in early vascular development, as indicated by gene deletion experiments and by the potent effect of the endogenous endothelial BMP inhibitor BMPER in suppressing endothelial differentiation (17). In addition, Indian hedgehog is implicated as an endoderm-derived signal for vascular competence (4). However, remarkably little is known about how these upstream factors interact to specify vascular identity of precursors. In addition, none of these signals has specificity for the vascular lineage.

The first secreted molecule with specificity for the endothelium during the developmental cascade is VEGF, and the role of VEGF family members in developmental and adult angiogenesis is well-described. Cells that respond to VEGF must first express its receptors, FLK-1 and FLT-1 (VEGFR2 and VEGFR1), which are exquisitely restricted to the endothelial and early hematopoietic lineages. This indicates that VEGF cannot itself be the most proximal signal for endothelial differentiation, and it is still not clear precisely how early VEGF participates in vascular specification. Nevertheless, the impressive phenotype of mice lacking a single copy of the *VEGF* gene – which die in utero soon after vascular development initiates – indicates the critically necessary role of this signaling pathway (18).

The appearance of FLK-1 expression is presently recognized as the earliest marker available for the endothelial lineage during development; FLK-1 is expressed in earliest endothelial and hematopoietic cells and remains expressed in mature endothelium. It is, however, downregulated within the hematopoietic lineages as they mature. This pattern suggests that Flk1 may be a marker for hemangioblasts during early development, and this is borne out by studies that indicate that the coexpression of FLK-1 with mesoderm-derived transcription factors brachyury or Scl/Tal1 denotes hemangioblasts (9,10). It is also important to note that, because FLK-1 marks angioblasts and mature ECs, it cannot be used solely to distinguish endothelial precursors or hemangioblasts at any stage of development. Nevertheless, upstream cues that initiate FLK-1 expression would be likely participants in early specification of the vascular lineage. Evidence supports BMP signaling as a proximal stimulus for FLK-1 expression and, at the transcriptional level, there appears to be necessary roles for GATA family proteins as well as the homeodomain protein HoxB5 in upregulation of FLK-1 during development (19,20). Nevertheless, an entire transcriptional program of the endothelial lineage has not been established, and this represents a major limitation in our understanding of how cell-autonomous mechanisms coordinate the development of vascular cell lineages.

Downstream of the early events, the cues required for blood vessels to form become better known, if also more complex. Among paracrine factors, the angiopoietin family members bind to receptors structurally similar to the VEGF receptors and play crucial roles in EC survival and remodeling of capillary plexi, platelet-derived growth factors are required for the recruitment of pericytes and smooth muscle cells to invest developing arteries and establish vasomotor tone, and ephrins and their counter-receptors (Ephs) are required for arteriovenous communications. The specific molecular roles of each of these signaling pathways is discussed in detail in chapters devoted to these topics.

Likewise, much is known about downstream transcriptional events that participate in vascular assembly. These transcription factors include vascular endothelial zinc finger, GATA-2, several members of the Krüppel-like family of zinc finger proteins, and Ets proteins, all of which have all been linked to steps in angiogenesis by virtue of their cellular phenotypes of defined transcriptional targets. Homeodomain proteins such as HoxB3 and HoxD3 participate in morphogenetic events responsible for vascular tube formation. The transcription factor hypoxia-inducible factor 1α and its family members play crucial roles in sensing changes in tissue oxygen (O_2) tension and stimulating gene expression changes that enhance blood vessel growth into hypoxic tissues during postgastrulation development (when the passive diffusion of O_2 and nutrients becomes limiting due to the size of the embryo) as well as in pathological angiogenesis. The specific roles of each of these transcriptional events also are described in respective chapters. Remarkably, we have a much less well-informed understanding of the transcriptional cascades that determine the programs of development and plasticity of ECs and other vascular lineages than we do of, for example, cardiomyocytes or skeletal muscle cells. This may be due to the fact that greater heterogeneity and plasticity exists among vascular cells, making impractical the relatively simple mechanisms that are responsible for determining other lineages. Nevertheless, future effort will be needed to clarify those transcriptional hierarchies that culminate in blood vessel formation.

CONCLUSION

The experimental database that we use to understand endothelial differentiation and vascular development is exceptionally rich and draws on approaches that range from the imaginatively observational to the rigorously inductive. A number of key controversies in developmental biology have concerned the origin of ECs and the means by which blood vessels are assembled. It has become obvious during the past decade that many of the principles of blood vessel development also apply to the assembly and disassembly of blood vessels in the adult, particularly in pathological circumstances. A notable example of this is the well-defined role for VEGF and its receptors in both developmental and pathological angiogenesis (21,22). Activators and inhibitors of developmental pathways have been tested for their ability to modulate angiogenesis in early-phase clinical trials and, in the case of anti-Flk1 antibodies, clinical utility has been demonstrated for antitumor strategies.

Likewise, the explosion of interest in stem cell biology and the potential for regenerative medicine have caused many to reconsider the utility of understanding vascular developmental events, with the notion that many of the pathways identified may be recapitulated in adult stem cells as they are coaxed toward the vascular lineage. Analyses of circulating endothelial progenitor cells, which have angiogenic potential, do indeed suggest that similarities exist in the biology of these cells compared with developmental endothelial precursors. Stem cell therapeutics therefore represents another potential arena for the translation of insights from vascular development to clinical practice.

Even though our understanding of endothelial development is much richer than it was even a few years ago, and despite the potential applications of this knowledge in clinical medicine, a number of key issues on this topic still remain to be resolved. Precisely how early are endothelial precursors specified during development, and what is the nature of this progenitor cell pool? What are the relationships among signaling pathways that specify endothelial fates in a coordinated fashion? Is there a transcriptional hierarchy that regulates vascular development? The answers to these and other questions about endothelial development are likely to be forthcoming in the near future as experimental methods continue to evolve.

KEY POINTS

- Endothelial differentiation during embryonic development is tightly coordinated via spatial considerations, and is regulated through the interplay of diverse signaling and transcriptional cascades to effect the appropriate formation of the vasculature.
- The vascular endothelium arises from mesodermal precursors.
- Endothelial differentiation is tightly linked to the development of the hematopoietic system and to the differentiation of other vascular lineages.
- Developmental pathways are recapitulated in angiogenic processes in adulthood.

Future Goals

- To elucidate a transcriptional hierarchy of endothelial differentiation
- To apply knowledge of vascular development to understand congenital disorders of the vasculature and to modulate angiogenic processes that recur during adulthood

ACKNOWLEDGMENTS

Work in the author's laboratory is supported by NIH grants GM61728, HL65619, AG02482, and HL61656. C.P. is an Established

Investigator of the American Heart Association and a Burroughs Wellcome Fund Clinical Scientist in Translational Research. Thanks to Holly McDonough for critical reading of this chapter.

REFERENCES

1 de Bruijn MF, Speck NA, Peeters MC, Dzierzak E. Definitive hematopoietic stem cells first develop within the major arterial regions of the mouse embryo. *EMBO J*. 2000;19:2465–2474.

2 Muller AM, Medvinsky A, Strouboulis J, Grosveld F, Dzierzak E. Development of hematopoietic stem cell activity in the mouse embryo. *Immunity*. 1994;1:291–301.

3 Noden DM. Embryonic origins and assembly of blood vessels. *Am Rev Respir Dis*. 1989;140:1097–1103.

4 Dyer MA, Farrington SM, Mohn D, Munday JR, Baron MH. Indian hedgehog activates hematopoiesis and vasculogenesis and can respecify prospective neurectodermal cell fate in the mouse embryo. *Development*. 2001;128:1717–1730.

5 Vokes SA, Krieg PA. Endoderm is required for vascular endothelial tube formation, but not for angioblast specification. *Development*. 2002;129:775–785.

6 Sabin FR. Preliminary note on the differentiation of angioblasts and the method by which they produce blood-vessels, blood-plasma and red blood-cells as seen in the living chick. *Anat Rec*. 1917;13:199–204.

7 Choi K, Kennedy M, Kazarov A, Papadimitriou JC, Keller G. A common precursor for hematopoietic and endothelial cells. *Development*. 1998;125:725–732.

8 Pardanaud L, Dieterlen-Lievre F. Emergence of endothelial and hemopoietic cells in the avian embryo. *Anat Embryol*. 1993;187: 107–114.

9 Huber TL, Kouskoff V, Fehling HJ, Palis J, Keller G. Haemangioblast commitment is initiated in the primitive streak of the mouse embryo. *Nature*. 2004;432:625–630.

10 Drake CJ, Fleming PA. Vasculogenesis in the day 6.5 to 9.5 mouse embryo. *Blood*. 2000;95:1671–1679.

11 Goumans MJ, Zwijsen A, van Rooijen MA, Huylebroeck D, Roelen BA, Mummery CL. Transforming growth factor-beta signalling in extraembryonic mesoderm is required for yolk sac vasculogenesis in mice. *Development*. 1999;126:3473–3483.

12 Nishikawa SI, Nishikawa S, Kawamoto H, et al. In vitro generation of lymphohematopoietic cells from endothelial cells purified from murine embryos. *Immunity*. 1998;8:761–769.

13 Downs KM, Gifford S, Blahnik M, Gardner RL. Vascularization in the murine allantois occurs by vasculogenesis without accompanying erythropoiesis. *Development*. 1998;125:4507–4520.

14 Lammert E, Cleaver O, Melton D. Induction of pancreatic differentiation by signals from blood vessels. *Science*. 2001;294:564–567.

15 Djonov V, Schmid M, Tschanz SA, Burri PH. Intussusceptive angiogenesis: its role in embryonic vascular network formation. *Circ Res*. 2000;86:286–292.

16 Flamme I, Risau W. Induction of vasculogenesis and hematopoiesis in vitro. *Development*. 1992;116:435–439.

17 Moser M, Binder O, Wu Y, et al. BMPER, a novel endothelial cell precursor-derived protein, antagonizes BMP signaling and endothelial differentiation. *Mol Cell Biol*. 2003;23:5664–5676.

18 Carmeliet P, Ferriera V, Breier G, et al. Abnormal blood vessel development and lethality in embryos lacking a single VEGF allele. *Nature*. 1996;380:435–439.

19 Minami T, Rosenberg RD, Aird WC. Transforming growth factor-beta 1-mediated inhibition of the flk-1/KDR gene is mediated by a 5′-untranslated region palindromic GATA site. *J Biol Chem*. 2001;276(7):5395–5402.

20 Wu Y, Moser M, Bautch VL, Patterson C. HoxB5 is an upstream transcriptional switch for differentiation of vascular endothelium from precursor cells. *Mol Cell Biol*. 2003;23:5680–5691.

21 Millauer B, Shawver LK, Plate KH, Risau W, Ullrich A. Glioblastoma growth inhibited in vivo by a dominant-negative Flk-1 mutant. *Nature*. 1994;367:576–579.

22 Aiello LP, Pierce EA, Foley ED, et al. Suppression of retinal neovascularization in vivo by inhibition of vascular endothelial growth factor using soluble VEGF-receptor chimeric proteins. *Proc Natl Acad Sci USA*. 1995;92:10457–10461.

21

Fate Mapping

Takashi Mikawa

Cornell University Medical College, New York, New York

Endothelial cell (EC) precursors, termed *angioblasts*, form primitive vascular plexuses de novo, a process termed *vasculogenesis* (1–3). Cells of the primitive vascular plexus further differentiate into arterial, venous, endocardial, and lymphatic ECs. New vessels also sprout out from the existing vessels via *angiogenesis*. Angioblasts are highly migratory, often originating far from their destination (1,3). By tracking the migratory patterns and lineage specification of EC precursors, the mechanisms involved in EC commitment and diversity can be defined more precisely. The fate maps of groups of cells provide clues regarding the tissue–tissue interactions involved in EC differentiation and vessel development. Single-cell labeling and tracing can determine the specific timing and site of lineage segregation. This chapter describes the ontogeny of cardiac and noncardiac ECs as revealed by genetic and nongenetic fate mapping and lineage analysis.

CELL TAGGING METHODS

Successful fate mapping or lineage analysis depends on a reliable cell tagging method that allows researchers to label a defined group of cells and to restrict the labeling to its progeny without dilution or horizontal spread. Several methods of genetic and nongenetic cell labeling have been developed.

Nongenetic Approaches

Labeling with a lipophilic fluorescent dye, such as DiI (1,1′-dioctadecyl-3,3,3′,3′-tetramethyl-indocarbocyanine · perchlorate), is a widely used nongenetic cell labeling method for fate mapping studies (4,5). DiI stably integrates into cell membranes and generates a high fluorescent signal, but exhibits little diffusion to adjacent cells and insignificant cell toxicity. Because of their stable integration into the cell membrane, lipophilic dyes are passed mainly to the progeny of the labeled parental cells. The injection of fluoroprobe-conjugated dextran particles into a cell is another nongenetic cell tagging method best suited for tagging a large cell such as a frog blastomere (6). Laser-assisted activation of a caged fluorescent dye in a single cell or a group of cells is another nongenetic option (7). This approach is suitable for fate mapping and cell lineage studies within a few cell layers of the embryonic surface, and is particularly powerful in zebrafish (*Danio rerio*) and *Caenorhabditis elegans* embryos. These methods allow us to trace migration patterns and/or morphogenetic movements of fluorescently labeled cells and to determine the cell types generated by the differentiation of a tagged cell population during a given developmental window. Obvious limitations of these approaches include: (a) physical accessibility of cells of interest; (b) the potential uptake of a dye by nonlabeled cells via cell fusion or phagocytosis; and (c) the exponential drop in the concentration of dye per cell as cells proliferate.

Genetic Cell Labeling Methods

The long-term tracing of cell fate requires a stable tag in all daughter cells of the originally labeled parental cell(s) without dilution by cell proliferation. Cell transplantation from a donor tissue carrying a recognizable trait to a host is often used in amphibian and avian embryos (8,9). Host cells are replaced with donor cells from an identical region of stage-matched embryos; the implanted donor cells are expected to follow the same developmental program in the host embryo. *Xenopus* chimeras are suitable for fate mapping of individual blastomeric cells (9). The chick–quail chimera approach often is used for fate mapping studies on a group of cells during and after gastrulation. Mouse chimeras also can be generated by mixing blastodermal cells from two strains carrying distinct traits (10). A broad range of mosaicism can be achieved simply by altering the ratio of cells from each strain. It is currently difficult, however, to control the position of transplanted cells either in the mouse blastocyst or inner cell mass, which is critical for fate-mapping studies. The use of chimeras requires skill in microsurgical techniques, and thus is not available to all researchers.

A number of methods also have been developed to introduce a genetic tag to a target cell population. These include the introduction of an exogenous tracer gene via microinjection, electroporation, or viral-mediated gene transfer, and the activation of tracer gene expression via Cre-lox knockin or random mutation. The direct introduction of an expression vector plasmid DNA encoding a tracer such as bacterial β-galactosidase (β-gal) or green fluorescence protein (GFP) frequently is used in *Xenopus laevis*, in which individual cells are accessible and large in size (11). This method is suited for short-term cell-fate tracing, because most injected plasmid DNAs remain as an episome without integrating into the host genome. This results in an uneven distribution between daughter cells, and the tracer gene often is lost in some daughter cells as cells proliferate.

A Cre-lox approach is based on a recombination induced by Cre recombinase through which constitutive expression of a tracer gene, such as β-gal or GFP, is turned on and remains on in all daughter cells (12). The success of the Cre-lox strategies for fate mapping depends on the availability of a promoter that is active within a specific cell population only briefly and at a specific time during embryogenesis. Intragenic recombination is another approach developed for retrospective clonal analysis in the mouse embryo (13,14). This approach requires a mouse line that carries a tandem duplication of a reporter gene driven by a tissue-specific promoter. The reporter protein is produced only when an intragenic recombination takes place. Because intragenic recombination is a rare event, cells positive for reporter expression are likely to be clonally related. Because it relies on a tissue-specific promoter, this method is not suited for the study of cell lineage diversification. The growth rate of a clonal population cannot be precisely studied by this method, because an intragenic recombination can occur long before cells become permissive for the promoter function.

Replication-defective retroviruses and adenoviruses have been developed to trace cell fate and lineage (15–18). In both cases, the virus is engineered to encode a reporter gene that replaces the genes necessary for viral replication; as a result, the reporter gene is inherited only by progeny of the originally infected cells without horizontal spread. Adenoviral vectors can infect cells regardless of their proliferative state, but the copy number of the introduced gene cannot be controlled and may provide drastically different levels of marker expression. Because the adenoviral DNA remains as a linear episome in a host cell, the gene often is not stable and is exponentially diluted over time by subsequent cell divisions. Thus, the adenoviral approach is suited for a short-term cell tracing but not for long-term lineage studies. In contrast, a retrovirus-mediated gene transfer assures the stable integration of a single copy of the reporter gene into the host DNA. However, because the integration process requires a host-cell division, the retroviral approach is applicable to dividing cells, but not postmitotic cells. The stable integration of retrovirally introduced genes makes retroviral gene delivery ideal for long-term lineage studies, because marker expression does not decline over time. Retroviruses have been the major tool used in the study of EC lineage in birds and mammals.

ORIGIN OF CARDIAC ENDOTHELIAL CELLS

Four EC types are present in the heart: endocardial, coronary arterial, venous, and capillary. Endocardial ECs exhibit phenotypes distinct from coronary ECs. They form an epithelial tube with a large diameter but never undergo angiogenic sprouting into the myocardium, whereas coronary ECs establish a complex but typical vessel network throughout the myocardium. Despite close proximity, subendocardial coronary vessels do not fuse with the endocardium or open to the heart chamber. Although the exact mechanism that confers these distinct phenotypes to endocardial and coronary ECs is not known, a series of retroviral cell lineage studies in the chick embryos have identified distinct ontogenies for these two cardiac EC populations.

Origin of Endocardial Endothelial Cells

The heart begins to function before any other organ system is established during embryogenesis (Figure 21.1). Fate

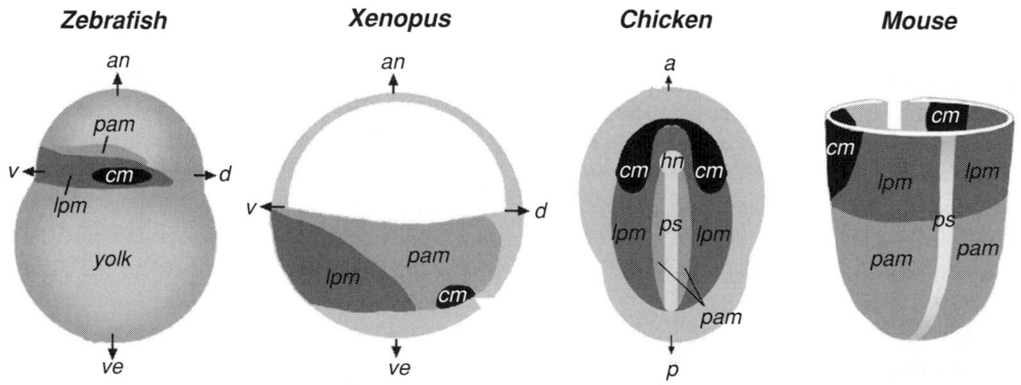

Figure 21.1. Schematic illustration of fate maps of progenitor cells for cardiogenic mesoderm (cm, *black*) in various vertebrate species. The relative position of the cardiogenic mesoderm progenitor to paraxial mesoderm (pam, *light gray*) and lateral plate mesoderm (lpm, *dark gray*) is illustrated. an-ve, the animal–vegital axis; v-d, the ventral–dorsal axis; a-p, the anterior–posterior axis; ps, the primitive streak.

map studies in chick embryos during early gastrulation have revealed that endocardial and cardiomyocyte precursors both arise from a rostral portion of the primitive streak (19). As gastrulation begins, they migrate rostrolaterally, forming bilateral heart fields called cardiogenic mesoderm (20–21). During the lateral body fold, the bilateral cardiac fields fuse at the ventral midline of neurula-stage embryos and form a single tubular heart that soon begins to beat (22). During this fusion process, the majority of cardiogenic cells differentiate into the presumptive myocardium, while a minor population segregates from the original epithelioid myocardium and forms the presumptive endocardium via a vasculogenic mechanism (22, 23). Thus, both cardiomyocyte and endocardial EC lineages arise from the cardiogenic mesoderm.

The induction of these two cardiac cell types in the heart field mesoderm requires paracrine factors secreted by the underlying endoderm (24). Cardiomyocyte-specific gene expression can be induced by endoderm-derived paracrine factors such as fibroblast growth factors (FGFs), bone morphogenic proteins, insulin-like growth factor, and activin (25–28), whereas EC differentiation is promoted by angiogenic peptides, such as vascular endothelial growth factor (VEGF) and FGFs (29,30). It was once postulated that cells of the cardiogenic field are bipotent progenitor cells that differentiate into both cardiomyocytes and endocardial cells, based on expression patterns of myocardial and endocardial cell markers in cardiogenic mesoderm (23) and an immortalized myogenic cell line (31). However, because these signaling molecules are expressed broadly in the underlying endoderm, the bipotent progenitor model does not explain how they instruct individual cells of the heart-field mesoderm to enter either the cardiomyocyte lineage or the endocardial cell lineage. It is also unclear how more myocardial cells than endocardial cells are generated from the cardiogenic mesoderm.

To clarify these issues, the fate of individual cells within the primitive streak and the heart-field mesoderm has been studied in the chicken embryo using the retroviral single-cell tagging and tracing method. The retroviral cell lineage studies have demonstrated that individual cells in the rostral half of the primitive streak and those in the heart field give rise to a clone consisting only of one cell type, either endocardial or myocardial (32–34). These data support a model in which these two cell lineages of the heart are segregated already within the primitive streak, significantly prior to their migration to the heart field. It is likely that the precardiomyocytes and pre-endocardial cells already may be receptive to inductive signal(s) from underlying endoderm. Consistent with this model, it has been suggested that EC commitment may occur before and independent of gastrulation, whereas myocyte commitment has not yet occurred at the prestreak stage (35). Cell fate studies in zebrafish also have identified a blastomere population that produces only endocardial or myocardial cells (36). These studies in both fish and avian embryos indicate that the separation of these two lineages can occur at the blastula stage, prior to formation of mesoderm. In the mouse embryo, it is known that endocardial and myocardial cells also arise from bilateral clusters of distal epiblast cells (37), but it remains unknown when these two cardiac cell lineages are established.

Origin of Coronary Vascular Endothelial Cells

Coronary vessels are originally absent from the presumptive heart tube but develop as the myocardium becomes thickened (38). Although the coronary vessel network was once thought to be established by angiogenic sprouting from the aorta (39–42), this model did not explain the presence of discontinuous EC channels in the myocardium prior to the establishment of a closed coronary vessel network (43,44). A clear understanding of the embryonic origin of coronary vasculature is critical to address the mechanism of coronary vessel formation. Chick–quail chimera studies have shown that smooth muscle cells of the great vessel arise from a subset of ectoderm-derived neural crest cells termed *cardiac neural crest.* This finding raised the possibility that the coronary arteries originated from neural crest derivatives. Another possibility was that they arose from the heart-field mesoderm, as do the myocardium and endocardium. However, retroviral labeling of cardiac neural crest or the heart tube never gave rise to tagged cells within the coronary system (32,45,46).

In addition to cardiac neural crest, another cell population migrates to the heart. The mesoderm-derived proepicardium (47,48) is a transient embryonic tissue located at the septum transversum. Proepicardial cells form multiple epithelial cell villi that protrude toward the myocardium of the looping-stage heart and that, once attached to the heart, begin to envelop the myocardium, forming the epicardium. The retroviral tagging of proepicardial cells produces either virally tagged cardiac fibroblasts or coronary vessels containing tagged endothelial and smooth muscle cells (45,46). These studies have provided direct proof for the proepicardial origin of coronary vasculature, including its ECs. This conclusion has been further confirmed using both adenoviral tagging (18) and chick–quail chimera analysis of the proepicardium (49,50). Importantly, individual virally labeled clones contain only one cell type, either ECs, smooth muscle, or fibroblasts, suggesting these lineages are already determined within the proepicardium. Regardless of the cell type of the clone, each clone occupies only a segment of a coronary vessel, and in no case does it populate the entire length of a vessel (45,46). The data have provided the experimental proof that the coronary vasculature arises from the proepicardium and forms via a vasculogenic mechanism (51).

ORIGIN OF NONCARDIAC ENDOTHELIAL CELLS

The embryonic vasculature was originally thought to form by the invasion of the embryo by extraembryonic blood vessels (52,53). This model was challenged by the finding that vascular development occurs in embryos from which the extraembryonic tissue has been ablated prior to blood vessel

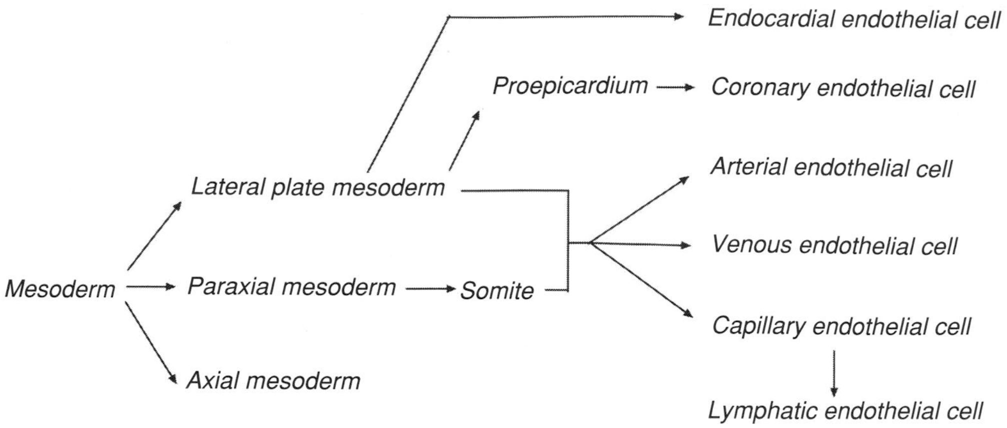

Figure 21.2. Lineage relationships between multiple EC types generated from mesoderm. As three germ layers form via gastrulation, endocardial ECs differentiate from the cardiogenic mesoderm, while noncardiac arterial, venous, and capillary ECs are generated from the lateral plate and somatic mesoderm. Coronary vasculature, including ECs, differentiates from a distinct mesodermal tissue, the proepicardium. Lymphatic ECs appear to differentiate from capillary or venous ECs during budding off.

formation (54). Thus, the embryonic germ layers are sufficient for producing embryonic vasculature (Figure 21.2). Further studies have established that, of three germ layers, only the mesoderm gives rise to ECs (1). In the avian embryo, it has been demonstrated that all mesoderm, with the exception of prechordal mesoderm, has the potential to produce angioblasts (55,56).

Fate mapping studies in zebrafish suggest that EC populations may be committed to a vessel-type fate within the lateral plate mesoderm before vascular assembly (57): Laser-assisted labeling of single angioblasts with a caged dye in zebrafish embryos has shown that individually labeled ECs were found in either arterial or venous vascular beds. On the contrary, data from the avian embryo suggest that the arterial or venous identity is determined only on recruitment into the vascular network by local environmental cue(s). For example, donor quail venous ECs can incorporate into the arteries of the host chick embryo and express an arterial marker, ephrinB2 (58,59). Conversely, donor arterial ECs can incorporate into the veins of host embryos and downregulate ephrinB2.

In contrast to extensive studies of vascular EC origins, the ontogeny of the lymphatic EC has been examined only recently. It was once thought that the lymphatic system of the forelimbs develops through the sprouting of the jugulo-axillary lymph sacs into the limb buds (60). However, recent studies using markers for the lymphatic endothelia have detected lymphangioblasts in the wing buds 1 day prior to the formation of the jugulo-axillary lymph sacs (61). Currently, it is thought that a local budding from existing capillary or venous network is the key event in initiating lymphatic vessel formation. It has further been shown that dorsal somatic cells at the wing bud level migrate into the limb and differentiate into lymphatic endothelia (62) in a process similar to how the vascular ECs of wing blood vessels arise from somatic mesoderm (63).

KEY POINTS

- Fate mapping and single cell lineage studies have identified both the timing and location of vascular cell commitment and differentiation.
- The cell fate mapping and lineage data have provided clues for defining cell–cell and molecular interactions that regulate EC differentiation and diversity.
- Cell lineage studies of cardiac ECs have led to new questions regarding the induction and patterning of endocardial and coronary EC populations
- Understanding the mechanisms that govern these processes will substantially further our understanding of vascular system development and patterning during embryogenesis.

Future Goals

- To understand how are endocardial precursors specified within the cardiogenic mesoderm, while exposed to the same signals as cardiomyocyte precursors
- To understand how is the proepicardium, the source of coronary vasculature, induced at a specific embryonic site
- To understand what mechanisms govern the oriented migration of coronary progenitors from the proepicardium to the looping-stage heart
- To understand what cell–cell and molecular interactions are responsible for patterning of the coronary endothelia and the establishment of a connection with the proximal aorta

- To understand the degree of EC plasticity to their arterial, venous, capillary, or lymphatic identities in the embryonic heart

REFERENCES

1 Noden DM. Origins and assembly of avian embryonic blood vessels. *Ann NY Acad Sci.* 1990;588:236–249.

2 Poole TJ, Coffin JD. Developmental angiogenesis: quail embryonic vasculature. *Scan Microsc.* 1988;2(1):443–448.

3 Risau W, Flamme I. Vasculogenesis. *Ann Rev Cell Dev Biol.* 1998; 11:73–91.

4 Honig MG, Hume RI. Fluorescent carbocyanine dyes allow living neurons of identified origin to be studied in long-term cultures. *J Cell Biol.* 1986;103(1):171–187.

5 Fraser SE. *Methods in Avian Embryology.* San Diego, CA: Academic Press, 1996.

6 Dale L, Slack JM. Fate map for the 32-cell stage of *Xenopus laevis. Development.* 1987;99(4):527–551.

7 Serbedzija GN, Chen JN, Fishman MC. Regulation of the heart field in zebrafish. *Development.* 1998;125:1096–1101.

8 Le Lievre CS, Le Douarin NM. Mesenchymal derivatives of the neural crest: analysis of chimaeric quail and chick embryos. *J Embryol Exp Morphol.* 1975;34(1):125–154.

9 Krotoski DM, Fraser SE, Bronner-Fraser M. Mapping of neural crest pathways in *Xenopus laevis* using inter- and intra-specific cell markers. *Dev Biol.* 1988;27(1):119–132.

10 Papaioannou VE. Lineage analysis of inner cell mass and trophectoderm using microsurgically reconstituted mouse blastocysts. *J Embryol Exp Morphol.* 1982;68:199–209.

11 Lane MC, Sheets MD. Primitive and definitive blood share a common origin in *Xenopus:* a comparison of lineage techniques used to construct fate maps. *Dev Biol.* 2002;248(1):52–67.

12 Soriano P. Generalized lacZ expression with the ROSA26 Cre reporter strain. *Nat Gen.* 1999;21(1):70–71.

13 Nicolas JF, Mathis L, Bonnerot C, Saurin W. Evidence in the mouse for self-renewing stem cells in the formation of a segmented longitudinal structure, the myotome. *Development.* 1996; 122(9):2933–2946.

14 Mathis L, Bonnerot C, Puelles L, Nicolas JF. Retrospective clonal analysis of the cerebellum using genetic laacZ/lacZ mouse mosaics. *Development.* 1997;124(20):4089–4104.

15 Cepko C. Retrovirus vectors and their applications in neurobiology. *Neuron.* 1988;1(5):345–353.

16 Sanes DH, Poo MM. In vitro analysis of position- and lineage-dependent selectivity in the formation of neuromuscular synapses. *Neuron.* 1989;2(3):1237–1244.

17 DA Fischman, T Mikawa. The use of replication-defective retroviruses for cell lineage studies of myogenic cells. *Meth Cell Biol.* 1997;52:215–227.

18 Dettman RW, Denetclaw W, Jr., Ordahl CP, Jbristow J. Common epicardial origin of coronary vascular smooth muscle, perivascular fibroblasts, and intermyocardial fibroblasts in the avian heart. *Dev Biol.* 1998;193(2):169–181.

19 Garcia-Martinez V, Schoenwolf GC. Primitive-streak origin of the cardiovascular system in avian embryos. *Dev Biol.* 1993;159(2): 706–719.

20 Rawles ME. The heart forming regions of the chick blastoderm. *Physiol Zool.* 1943;16:22–42.

21 Rosenquist RL, DeHaan GC. Migration of precardiac cells in the chick embryo: a radioautographic study. *Carnegie Institute of Washington Publication.* 1966;625:111–121.

22 Stalsberg H, DeHaan, RL The precardiac areas and formation of the tubular heart in the chick embryo. *Dev Biol.* 1969;19(2):128–159.

23 Linask KK, Lash JW. Early heart development: dynamics of endocardial cell sorting suggests a common origin with cardiomyocytes. *Dev Dynam.* 1993;196(1):62–69.

24 Sater AK, Jacobson AG. The specification of heart mesoderm occurs during gastrulation in *Xenopus laevis. Development.* 1989; 105(4):821–830.

25 Lough J, Barron M, Brogley M, Sugi Y, Bolender DL, Zhu X. Combined BMP-2 and FGF-4, but neither factor alone, induces cardiogenesis in non-precardiac embryonic mesoderm. *Dev Biol.* 1996;178(1):198–202.

26 Schultheiss TM, Lassar AB. Induction of chick cardiac myogenesis by bone morphogenetic proteins. *Cold Spring Harbor Symposia on Quantitative Biology* 1997;62:413–419.

27 Antin PB, Yatskievych T, Dominguez JL, Chieffi P. Regulation of avian precardiac mesoderm development by insulin and insulin-like growth factors. *J Cell Physiol.* 1996;168(1):42–50.

28 Yatskievych TA, Ladd AN, Antin PB. Induction of cardiac myogenesis in avian pregastrula epiblast: the role of the hypoblast and activin. *Development.* 1997;124(13):2561–2570.

29 Folkman J, Klagsbrun M. Angiogenic factors. *Science.* 1987; 235(4787):442–447.

30 Flamme I, Risau W. Induction of vasculogenesis and hematopoiesis in vitro. *Development.* 1992;116(2):435–439.

31 Eisenberg CA, Bader D. QCE-6: a clonal cell line with cardiac myogenic and EC potentials. *Dev Biol.* 1995;167(2):469–481.

32 Mikawa T, Cohen-Gould L, Fischman DA. Clonal analysis of cardiac morphogenesis in the chicken embryo using a replication-defective retrovirus. III: Polyclonal origin of adjacent ventricular myocytes. *Dev Dynam.* 1992;195(2):133–141.

33 Cohen-Gould L, Mikawa T. The fate diversity of mesodermal cells within the heart field during chicken early embryogenesis. *Dev Biol.* 1996;177(1):265–273.

34 Wei Y, Mikawa T. Fate diversity of primitive streak cells during heart field formation in ovo. *Dev Dynam.* 2000;219(4):505–513.

35 von Kirschhoffer K, Grim M, Christ B, Wachtler F. Emergence of myogenic and EC lineages in avian embryos. *Dev Biol.* 1994; 163:270–278.

36 Lee RK, Stainier DY, Weinstein BM, Fishman MC. Cardiovascular development in the zebrafish. II. Endocardial progenitors are sequestered within the heart field. *Development.* 1994; 120(12):3361–3366.

37 Kaufman MH, Navaratnam V. Early differentiation of the heart in mouse embryos. *J Anat.* 1981;133(Pt 2):235–246.

38 Rychter Z, Ostadal B. Mechanism of the development of coronary arteries in chick embryo. *Folia Morphologica.* 1971;19(2):113–124.

39 Aikawa E, Kawano J. Formation of coronary arteries sprouting from the primitive aortic sinus wall of the chick embryo. *Experientia.* 1982;38(7):816–818.

40 Hirakow R. Development of the cardiac blood vessels in staged human embryos. *Acta Anatomica.* 1983;115(3):220–230.

41 Conte G, Pellegrini A. On the development of the coronary arteries in human embryos, stages 14–19. *Anat Embryol.* 1984; 169(2):209–218.

42 Hutchins GM, Kessler-Hanna A, Moore GW. Development of the coronary arteries in the embryonic human heart. *Circulation.* 1988;77(6):1250–1257.

43 Bogers AJ, Gittenberger-de Groot AC, Poelmann RE, Peault BM, Huysmans HA. Development of the origin of the coronary arteries, a matter of ingrowth or outgrowth? *Anat Embryol.* 1989;180(5):437–441.

44 Waldo KL, Willner W, Kirby ML. Origin of the proximal coronary artery stems and a review of ventricular vascularization in the chick embryo. *Am J Anat.* 1990;188(2):109–120.

45 Mikawa T, Fischman DA. Retroviral analysis of cardiac morphogenesis: discontinuous formation of coronary vessels. *Proc Nat Acad Sci USA.* 1992;89(20):9504–9508.

46 Mikawa T, Gourdie RG. Pericardial mesoderm generates a population of coronary smooth muscle cells migrating into the heart along with ingrowth of the epicardial organ. *Dev Biol.* 1996; 174(2):221–232.

47 Ho E, Shimada Y. Formation of the epicardium studied with the scanning electron microscope. *Dev Biol.* 1978;66(2):579–585.

48 Hiruma T, Hirakow R. Epicardial formation in embryonic chick heart: computer-aided reconstruction, scanning, and transmission electron microscopic studies. *Am J Anat.* 1989;184(2):129–138.

49 Vrancken Peeters MP, Gittenberger-de Groot AC, Mentink MM, Poelmann RE. Smooth muscle cells and fibroblasts of the coronary arteries derive from epithelial-mesenchymal transformation of the epicardium. *Anat Embryol.* 1999;199(4):367–378.

50 Poelmann RE, Gittenberger-de Groot AC, Mentink MM, Bokenkamp R, Hogers B. Development of the cardiac coronary vascular endothelium, studied with antiendothelial antibodies, in chicken-quail chimeras. *Circ Res.* 1993;73(3):559–568.

51 Mikawa T. Cardiac lineages. In: RP Harvey, N Rosenthal, eds. *Heart Development.* New York: Academic Press; 1999:19–33.

52 W His. *Untersuchungen über die erste Anlage des Wirbeltierleibes.* Lepzig, Germany: Vogel, 1868.

53 His W. Lecithoblast und Angioblast der Wirbeltiere. *Abh Math-Phys Kl Ges.* 1900;26:171–328.

54 Reagan FP. Vascularization phenomena in fragments of embryonic bodies completely isolated from yolk sac endoderm. *Anat Rec.* 1915;9:329–341.

55 Coffin JD, Poole TJ. Embryonic vascular development: immunohistochemical identification of the origin and subsequent morphogenesis of the major vessel primordia in quail embryos. *Development.* 1988;102(4):735–748.

56 Pardanaud L, Dieterlen-Lievre F. Manipulation of the angiogenic/hemangiopoietic commitment in the avian embryo. *Development.* 1999;126:617–627.

57 Zhong TP, Childs S, Leu JP, Fishman MC. Gridlock signalling pathway fashions the first embryonic artery. *Nature.* 2001;414 (6860):216–220.

58 Moyon D, Pardanaud L, Yuan L, Breant C, Eichmann A. Plasticity of ECs during arterial-venous differentiation in the avian embryo. *Development.* 2001;128(17):3359–3370.

59 Othman-Hassan K, Patel K, Papoutsi M, Rodriguez-Niedenfuh M, Christ B, Wilting J. Arterial identity of ECs is controlled by local cues. *Dev Biol.* 2001;237(2):398–409.

60 Sabin FR. The lymphatic system in human embryos, with a consideration of the system as a whole. *Am J Anat.* 1909;9:43–91.

61 Schneider M, Othman-Hassan K, Christ B, Wilting J. Lymphangioblasts in the avian wing bud. *Dev Dynam.* 1999;216(4–5):311–319.

62 Wilting J, Schneider M, Papoutski M, Alitalo K, Christ B. An avian model for studies of embryonic lymphangiogenesis. *Lymphology.* 2000;33(3):81–94.

63 Wilting J, Brand-Saberi B, Huang R, Zhi Q, Kontges G, Ordahl CP, Christ B. Angiogenic potential of the avian somite. *Dev Dynam.* 1995;202(2):165–171.

22

Pancreas and Liver

Mutual Signaling during Vascularized Tissue Formation

Eckhard Lammert

Max Planck Institute of Molecular Cell Biology and Genetics, Dresden, Germany

The cardiovascular system is the first functional organ system to develop in the mammalian embryo. Early during development, it consists mainly of vascular endothelial cells (ECs), which form tubes connected with the heart. Later in development, these vascular tubes develop and branch into a more complex tubular system with a variety of tissue-specific properties. Some of these vascular properties develop when ECs receive signals such as growth factors from surrounding non-vascular tissue cells.

Because blood vessels are present in the embryo at the onset of organogenesis, it is reasonable to expect that they shape the development of organ tissues. Indeed, the pancreas and liver consist of tissues whose features are shaped by signals derived from vascular ECs.

The fact that tissue cells are modulated by signals from ECs and that, in turn, EC phenotypes are influenced by tissue-derived signals, suggests that the development of vascularized tissues is critically dependent on mutual signaling. This mutual signaling results in structural and functional coupling between tissues and their respective vascular beds.

THE PRINCIPLE OF MUTUAL SIGNALING

The dorsal aorta is the first intraembryonic artery to form in vertebrates. Despite some differences between vertebrate species, the aorta normally develops through migration of angioblasts from the lateral plate mesoderm towards the midline of the embryo and subsequent formation of a vascular tube connected with the heart. The venous blood vessels, such as the cardinal veins, develop at around the same time and, together with heart and aorta, form the first circulatory system within the embryo. In mammals, additional blood vessels are present, such as those needed to connect embryonic and maternal circulations.

After formation of the primitive circulatory system, blood vessels begin to sprout, thus increasing the complexity of the vasculature. During this time, inner organs start to develop (1). The liver bud forms at the onset of blood circulation, and the pancreas starts to develop shortly thereafter. Other inner organs such as the lung and kidney similarly form during the formation of the primitive vascular plexus.

In mammalian models, a developmental blockade of the circulatory system leads to embryonic lethality (2,3). In general, the stronger the defect in the circulatory system, the earlier the embryo is aborted. In contrast to mammals, fish embryos may meet their oxygen (O_2) needs through simple diffusion and can develop much further in the absence of a circulatory system (4). Thus, the prevailing view is that blood circulation is important during development insofar as it is required to supply the bulk flow delivery of O_2 and nutrients. However, in mouse and frog, defective tissue development can be observed, when ECs are absent, even before blood flow takes place (5,6). These observations suggest that the vascular system is critical not only for delivering O_2 and nutrients to tissues, but also for providing EC-derived cues necessary for tissue development (Table 22-1). It is noteworthy that this EC requirement for early tissue development might not exist in fish (7).

It has long been recognized that mutual signaling between blood vessels and tissues plays a critical role in optimizing O_2 supply and demand. In brief, a reduction in tissue oxygenation induces the production of hypoxia-responsive angiogenic signals, such as the vascular endothelial growth factor (VEGF)-A, in tissues. VEGF-A binds to its signaling receptor VEGF receptor 2 (VEGFR2) on vascular ECs and thereby initiates a variety of different responses, the outcome of which is the growth of blood vessels (hence, O_2 delivery) into the hypoxic tissue. Once the tissue is adequately oxygenated, VEGF-A levels return to baseline, and new blood vessel formation ceases (8).

This model of mutual signaling is oversimplified, because some tissues express high basal levels of VEGF-A despite adequate oxygenation and in the absence of neoangiogenesis (9,10). These limitations notwithstanding, the above example demonstrates the importance of bidirectional crosstalk and provides a rationale for therapeutic intervention. For example, the finding that tumor growth depends on a vascular supply

Table 22-1: Endothelial Signals and Responding Tissue Cells

Endothelial Signals	*Responding Tissue Cells*	*Developmental Change in the Tissue Cells*
BMPs	Neural crest cells	Autonomic neuronal differentiation
BDNF	Neurons	Neuronal cell survival
TGF-β	Mesenchymal cells	Mural cell specification
PDGF-β	Mural cells	Chemoattraction
FGFs	Pericytes	Cell proliferation
HGF	Hepatocytes	Hepatocyte proliferation
MMPs	Podocytes	Glomerular assembly
Jagged-1	Podocytes	Glomerular epithelium formation

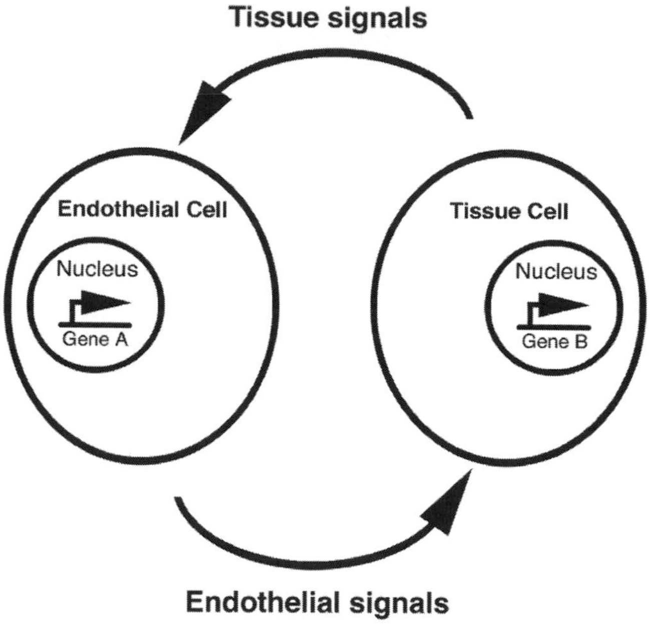

Figure 22.1. Development of a vascularized tissue. The suggested principle behind the development of a vascularized tissue is the mutual signaling between vascular ECs and nonvascular tissue cells. The result of these signaling events is a tissue with specific cell types (e.g., pancreatic β-cells) and ECs optimized for the function of these cell types (e.g., secretion of insulin). Tissue cells secrete signals or harbor signals on their cell surface that permit and change the development of adjacent ECs, for example, by turning on the transcription of Gene A. In turn, ECs secrete signals or harbor signals on their cell surface that permit and change the development of adjacent tissue cells, for example by turning on the transcription of Gene B. The tissue signals and endothelial signals exchanged during development assist the formation of a functional vascularized tissue.

represents the foundation for antiangiogenic therapy in cancer (11). An important obstacle in the field of tissue engineering is the limited growth potential of grafts that lack a vascular supply. Indeed, an urgent goal is to develop strategies that promote the vascularization of tissue implants.

In addition to matching O_2 supply and demand, it is clear that mutual signaling is involved in many other aspects of tissue development (Figure 22.1). In one direction, vascular ECs provide the developmental signals required for morphogenesis, differentiation, and growth of embryonic tissues (Figure 22.2). In the other direction, cues from the microenvironment play a critical role in mediating EC phenotypes. This bidirectional exchange of signals results in a match between the needs of tissue cells and vascular bed–specific ECs, allowing organs to fulfill their physiologic function in the context of the entire organism (Tables 22-1 and 22-2) (1).

TISSUE CELL–DERIVED SIGNALS

A variety of tissue-derived paracrine signals promote vascular development (Table 22-2), in part by altering EC gene expression (see Figure 22.1). When gene expression patterns are compared between cultured ECs derived from different tissues, tissue-specific differences in gene expression become apparent (12). In addition to these mitotically heritable differences, many other environmentally dependent vascular bed–specific transcriptional profiles are likely to exist in vivo, which are otherwise lost upon in vitro culture. In disease states, alterations in tissue-derived signals may cause ECs to undergo changes in gene expression (13).

ECs are capable of sensing and responding to a multitude of extracellular signals, including growth factors, cytokines, chemokines, and serine proteases. At any given site in the body, ECs are exposed to a unique mix of tissue-derived signals. For example, endocrine cells express a number of different VEGF molecules (9), including the tissue-specific angiogenic factor, endocrine gland-VEGF (EG-VEGF) (14). These growth factors are likely to play a role in regulating endothelial permeability and hence secretion of hormones. In contrast to endocrine glands, paracrine factors derived from astroglial cells in the brain are important in mediating the tight junctions of the blood–brain barrier.

The site-specific properties of ECs render the vasculature vulnerable to focal pathophysiology (15). For example, in diabetes mellitus, systemic hyperglycemia is associated with early

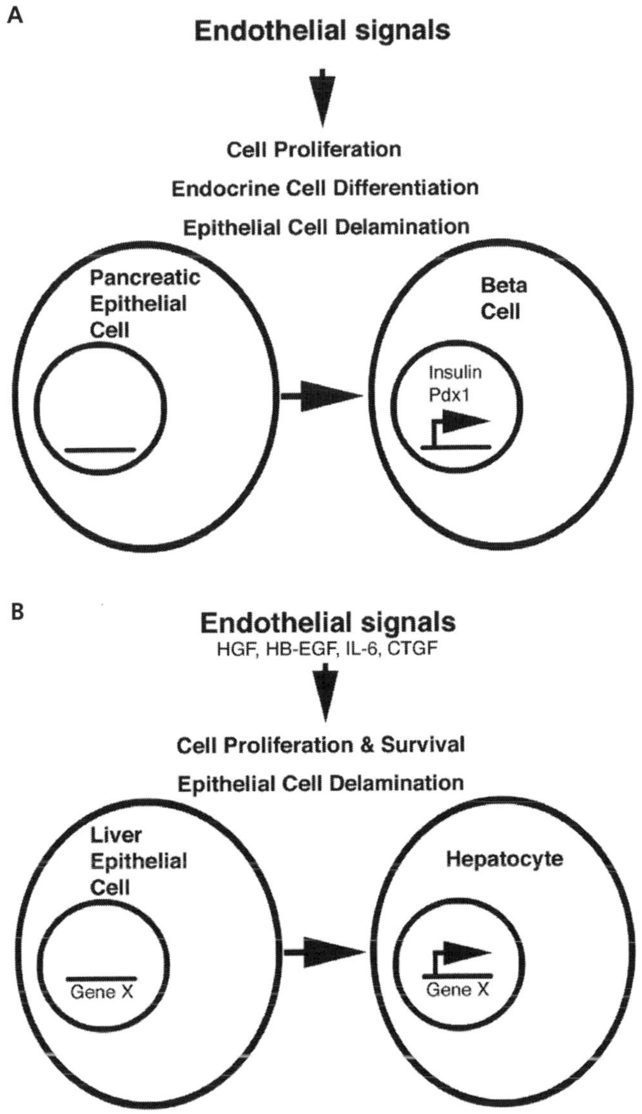

Figure 22.2. Endothelial signals in pancreas and liver development. ECs provide developing tissues with signals that promote developmental processes as diverse as cell proliferation, cell differentiation, and delamination of cells from an epithelium. **(A)** During pancreatic development, ECs induce the outgrowth of the dorsal pancreatic epithelium. Moreover, ECs are capable of inducing endocrine pancreatic differentiation. In particular, the expression of the transcription factor *Pancreas-duodenum homeobox (Pdx)-1* and the p48 subunit of the *Pancreas transcription factor 1* (*Ptf1*) as well as the β-cell hormone insulin are turned on when pancreatic epithelial cells interact with ECs. The inductive signals for endocrine cell differentiation are not yet known. **(B)** During liver development, ECs are required for the delamination of liver cells, which is an essential part of liver morphogenesis. In addition, ECs stimulate the proliferation of hepatocytes by secreting growth factors such as hepatocyte growth factor (HGF), heparin-binding epidermal growth factor (HB-EGF), interleukin (IL)-6, and connective tissue growth factor (CTGF). The genes turned on in developing hepatocytes in response to endothelial signals are not yet known. The comparison of the EC-induced processes during pancreas and liver development reveals both similar and diverse outcomes of endothelial signals during tissue formation.

Table 22-2: Tissue Signals and Developmental Changes in ECs

Tissue Signals	*Developmental Change in ECs*
VEGFs	Angiogenesis, cell proliferation, cell survival, migration, fenestration
EG-VEGF	Steroidogenic gland angiogenesis, cell proliferation, cell survival, migration, fenestration
FGFs	Angiogenesis, cell proliferation, migration
Angiopoietin 1	Blood vessel stabilization, migration, sprouting
Angiopoietin 2	Cell proliferation, migration, sprouting
EGF, TGF-α	Angiogenesis, cell proliferation
Colony stimulating factors (CSFs)	Cell proliferation, migration
Insulin-like growth factor (IGF)-1	Cell proliferation
TGF-β	Angiogenesis, tube morphogenesis, blood vessel stabilization
Tumor necrosis factor (TNF)-α	Angiogenesis, migration, tube morphogenesis
IL-8	Angiogenesis

vascular complications in the arterioles of the retina and kidney, suggesting that these vascular beds are preferentially susceptible to the elevated blood glucose levels. An understanding of the unique EC–tissue-cell interactions underlying focal vasculopathy might provide important new insights into therapy.

The remarkable heterogeneity of the endothelium has been exploited for purposes of drug targeting (16). Based on studies of in vivo phage display, Pasqualini and her colleagues have advanced the concept of "vascular zip codes," in which each vascular bed harbors specific molecules to which peptides or chemical compounds can bind specifically (17) (see Chapter 100). Linking drugs to such peptides or chemical compounds would help to more specifically deliver drugs to the tissue of interest. For example, linking cytotoxic compounds to peptides, which specifically home to a tumor tissue, has been shown to enable a more specific elimination of the tumor (18). Discovering compounds specific for vascular beds would enable a much better treatment of diseases in which a specific tissue is affected. Rather than affecting all tissues, homing drugs to the diseased tissue would prevent side effects in nontarget tissues and allow lower doses of drugs to be used.

ENDOTHELIAL CELL–DERIVED SIGNALS

As shown in Table 22-1, a number of EC-derived molecules that signal to tissues both during development and in the adult have been identified. In general, these molecules are expressed by some, but not all ECs and include members of the fibroblast growth factor (FGF), transforming growth factor (TGF), Wingless (Wnt), and platelet-derived growth factor (PDGF) families; Notch ligands; neurotrophins; cytokines; and matrix metalloproteinases (MMPs).

EC-derived signals act in a paracrine manner to influence the phenotype of nonvascular tissue cells, including changes in gene expression (Figure 22.1). Studies on embryonic neuronal development have provided examples of such signals. For example, the development of sympathetic neurons from neural crest cells was observed to occur adjacent to the dorsal aorta. Bone morphogenetic proteins (BMPs) expressed by aortic endothelial and mesenchymal cells were found to be responsible for inducing sympathetic neuronal differentiation (19,20).

Because endothelial signals induce neuronal differentiation in the embryo, the question was raised as to whether such signals could be used to regenerate neurons in humans (e.g., in spinal cord injuries). For a long while, it was believed that neurons residing in the central nervous system could not be renewed. However, in mammals, adult neuronal proliferation and differentiation was found to take place in the hippocampus, a brain region responsible for learning and memory, among few other brain regions. Proliferating precursor cells reside in a region of the adult hippocampus particularly rich in capillaries, suggesting that capillaries could promote the proliferation of neurons (21). Direct evidence for endothelial signals in adult neuronal development was provided by studies on male songbirds, which sing at high rates when testosterone levels are high (22). It was shown that testosterone induces VEGF-A in the high vocal center of the songbird brain, where it leads to local EC proliferation. The proliferating ECs secrete the brain-derived neurotrophic factor (BDNF), which stimulates precursor cell migration as well as neuronal maturation and survival. This results in an expansion of the entire high vocal center, including the neurons of the songbird brain. Moreover, ECs stimulate self-renewal and expand neurogenesis of neural stem cells in vitro (23). These observations point to the possibility of treating neuronal injuries, such as spine injuries, with ECs or endothelial signals. In addition, VEGF-A directly promotes the proliferation, survival, and differentiation of VEGFR2-expressing neurons and neuronal stem cells (24). Thus, VEGF-A might have both direct and EC-mediated positive effects on neuronal development.

ENDOTHELIAL CELLS IN PANCREATIC DEVELOPMENT

The inner organs, including the liver, lung, stomach, and pancreas, develop by budding from the embryonic gut epithelium (the endoderm). The buds then start to grow into morphologically distinct organs with organ-specific cell types. Several lines of evidence support the notion that vascular ECs induce pancreatic budding as well as the development of endocrine pancreatic cells in the dorsal gut epithelium (summarized in Figure 22.2A) (6).

First, when the precursor cells of the aortic ECs are removed in frog embryos, the aorta does not form, and endocrine pancreatic genes (NeuroD/BETA2, Pax6, and insulin) are not expressed in the gut epithelium of these embryos (6). This experiment showed that the aorta is required for endocrine pancreatic development, but it did not discriminate between the effect of the blood circulation and the vascular endothelium. To determine whether the aorta ECs are responsible for inducing endocrine gene expression in the embryonic gut epithelium, mouse embryonic tissue co-cultures were performed using isolated gut epithelial cells and aortic ECs (6). In these co-culture studies, embryonic gut epithelial cells were shown to express endocrine hormones in the presence, but not absence, of aortic ECs. Moreover, the epithelium grows and forms buds. These experiments provided evidence for aortic ECs as a source of signals inducing endocrine pancreatic development.

A second line of evidence derives from studies of transgenic mouse embryos, in which VEGF-A was ectopically expressed in the embryonic gut epithelium fated to become duodenum and stomach (6). As a result of VEGF-A expression, ECs were attracted to this nonpancreatic gut epithelium. It was shown that these ECs were able to induce the ectopic development of insulin-producing β-cells in the epithelium of the posterior part of the stomach and anterior part of the duodenum, thus showing that ECs can instruct some nonpancreatic epithelial cells to adopt a pancreatic β-cell fate.

Finally, in a converse experiment, endocrine pancreatic development and dorsal pancreatic budding did not take place in VEGFR2-deficient mouse embryos (25) that lacked ECs (26). In these mice, the pancreatic transcription factor Ptf1a was not turned on, and the expression of the pancreas–duodenum homeobox transcription factor Pdx1 was not maintained in the dorsal pancreatic epithelium. These data provided further evidence that ECs induce endocrine cell differentiation and dorsal pancreatic budding in the embryonic gut epithelium (see Figure 22.2A). Because the pancreas develops from ventral and dorsal buds, it is noteworthy that the ventral buds do develop in the VEGFR2-deficient mouse embryos. This suggests that ECs are not the only source for signals capable of inducing organ buds.

Endothelial signals act on a so-called prepatterned embryonic gut epithelium. The term prepatterned implies that the distinct regions of the gut epithelium can only develop into a limited number of cell types (27). It was observed that only cells within the stomach and duodenum can be induced to become pancreatic β-cells if they express the pancreas–duodenum–homeobox gene Pdx1 (E. Lammert, unpublished observation). In contrast, Pdx1-negative gut epithelium, such as the anterior part of the stomach, would not start to express insulin in response to endothelial signals. It is possible that these gut regions have a different *pre-pattern* and might respond to endothelial signals by turning on their own tissue-specific genes. Common responses to endothelial signals also exist, because inner organs have some morphogenetic events in common. For example, budding morphogenesis is impaired in both the developing pancreas and liver when ECs are missing (5,6,25).

Many research laboratories are currently aiming toward producing pancreatic β-cells for the treatment of diabetes mellitus, for example, by using human embryonic stem cells (28). Thus, it is useful to identify the endothelial signals involved in pancreatic β-cell development. In the juvenile, β-cells seem to proliferate to a certain extent (29), and the discovery of signals (possibly derived from ECs) that support β-cell proliferation would help to expand the β-cell mass in diabetic patients. A detailed molecular analyses of factors secreted by ECs would help to make progress in generating more β-cells for the treatment of insulin-dependent diabetes mellitus.

ENDOTHELIAL CELLS IN HEPATIC DEVELOPMENT

During liver development, hepatocytes undergo an epithelial–mesenchymal transition and grow into the surrounding mesoderm. This happens at a time when ECs surround the liver bud. In VEGFR2-deficient mouse embryos, which lack ECs (26), hepatocytes fail to detach from the epithelium, suggesting that ECs are required for liver morphogenesis (5). Tissue culture experiments showed that liver buds did not grow when an angiogenesis inhibitor prevented EC proliferation, thus showing that liver morphogenesis depends on EC-derived signals (see Figure 22.2B).

In the adult liver, endothelial signals play an important role in liver cell growth and survival (30). It was shown that the systemic administration of VEGF-A in mice increased liver size and induced cell division in both liver sinusoidal ECs and hepatocytes. In vitro, VEGF-A did not stimulate the growth of hepatocytes unless liver sinusoidal ECs were present, suggesting that EC-derived signals are necessary for hepatocyte mitogenesis.

Interestingly, depending on which VEGF-A receptors were activated, different sets of endothelial signals were produced. Activation of VEGFR1, but not VEGFR2, led to secretion of hepatocyte growth factor (HGF) by ECs, which promotes liver cell division. In contrast, activation of both VEGFR1 and VEGFR2 receptors led to responses in the gene expression of other endothelial signaling molecules including interleukin (IL)-6, heparin-binding epidermal growth factor (HB-EGF), and connective tissue growth factor (CTGF). During toxic injury, activation of both VEGF receptors led to improved liver-cell survival. From a medical point of view, it is interesting that selective activation of VEGFR1 stimulated hepatocyte, but not EC growth. Moreover, VEGFR1 agonists reduced liver damage in mice exposed to a hepatotoxin. These experiments point to the possibility that VEGFR1 agonists may have therapeutic potential in certain liver disorders (30).

The exchange of signals between hepatocytes and ECs illustrates once again the principle of mutual signaling between tissue cells and ECs. Because angiogenic factors activate distinct receptors, the composition of such factors might help to determine the distinct properties of the adjacent vascular beds. The finding of tissue-restricted angiogenic factors (such as EG-VEGF or Bv8) supports this idea (31).

SPECIAL FEATURES OF ENDOTHELIAL CELLS IN PANCREATIC ISLETS

Adult pancreatic islets are oval-shaped cell aggregates (Figure 22.3A; for color reproduction, see Color Plate 22.3A) that consist of five endocrine cell types: the insulin-secreting β-cells that form the islet core, and the non–β-cells (α-, γ-, PP-, and ϵ-cells) that form the islet mantle. Pancreatic islets usually are between 50 and 500 μm in diameter and express high VEGF-A mRNA and protein levels. The mRNAs of the other members of the VEGF family (VEGF-B, -C, -D, and placenta growth factor PlGF) also are expressed, in contrast to EG-VEGF, which is not present in islets (9). The islet vasculature is characterized by one to three arterioles that enter the islet core and branch into a dense network of capillaries (see Figure 22.3) (32). These capillaries drain into venules at the periphery of the islet. About 12% of the islet mass is made up by EC bodies, and every β-cell is in contact with at least one vascular EC.

Islet capillaries are characterized by thin, perforated or fenestrated EC bodies separated from the islet cells by a basement membrane (see Figure 22.3B and Table 22-3). The fenestrae of islet capillaries have a diaphragm, which is a mesh composed of fibrils. The function of the diaphragm is unknown. Because of its structure, it might prevent certain macromolecules or particles from passing through the EC. The observation that horseradish peroxidase quickly passes through these fenestrae, when injected into the pancreatic bloodstream, shows that these fenestrae are permeable to at least some proteins (33).

In contrast to the pancreatic islets, the exocrine pancreas does not express high levels of VEGF-A, and the capillary network is about five times less dense than is the capillary network of islets. Moreover, ten times less fenestrae are present in the capillaries of the exocrine tissue, when compared with islet capillaries.

To understand the function of VEGF-A expression in pancreatic islets, VEGF-A was specifically deleted in the pancreatic epithelium of mice (9,10). It was shown that without VEGF-A, fewer capillaries form in islets, and the few remaining capillaries have fewer fenestrae. These results show that the VEGF-A expressed by endocrine islet cells is required for the density and morphology of the islet capillary network. In addition, mice with VEGF-A deficient pancreatic islets reveal defects in steady-state blood glucose levels and glucose tolerance, demonstrating that VEGF-A is required for islets to

Table 22-3: Pancreatic Islet and Liver Sinusoidal ECs

Property	*Pancreatic Islet ECs*	*Liver Sinusoidal ECs*
EC cell body	100-nm thick at many locations	100–300 nm thick at many locations
Fenestrae	Abundant, ~100 nm in diameter (mouse), contains a diaphragm	Abundant, ~100 nm in diameter (mouse), contains no diaphragm
Extracellular space	Continuous basement membrane	No continuous basement membrane
Microvilli	No microvilli from pancreatic β-cell	Microvilli from hepatocyte

optimally fulfill their physiologic function: insulin secretion in response to blood glucose (10).

It is noteworthy that VEGFR2 is not expressed on pancreatic β-cells, but only on the ECs within islets. Similarly, VEGFR1 is not expressed by β-cells. Thus, the diabetic phenotype of mice with VEGF-A–deficient islets is not due to the missing autocrine VEGF-A signaling of β-cells, but results from the missing paracrine VEGF-A signaling from β-cells to ECs. How ECs are involved in the ability of β-cells to better secrete insulin is currently under investigation. The working hypothesis is that: (a) β-cells require endothelial signals to improve their insulin secretory function (mutual signaling) and, (b) β-cells require ECs to pass insulin into the blood stream (transendothelial permeability) (32).

SPECIAL FEATURES OF LIVER SINUSOIDAL ENDOTHELIAL CELLS

Liver sinusoidal ECs (LSECs) account for up to one-fifth of the total number of liver cells (34). These ECs are fenestrated, and the fenestrae lack a diaphragm. During liver development, ECs produce a continuous basement membrane between LSECs and hepatocytes, which disappears once the liver tissue is fully developed (35). Although the adult liver lacks an organized basement membrane, liver injury may lead to the abnormal formation of a continuous basement membrane.

The space between hepatocytes and LSECs is called the space of Disse (SD). Lipid particles occasionally are found in this space, suggesting a role for this site in lipid exchange. The morphology of LSECs might be suited to facilitate the bidirectional exchange of macromolecules and particles between blood plasma and hepatocytes. In the case of the endocrine pancreas, ECs are next to a continuous basement membrane, and they have fenestrae with a diaphragm (see Figure 22.3B), whereas, in the case of the liver, ECs do not have any continuous basement membrane, and their fenestrae are without any diaphragm. The endocrine pancreas primarily

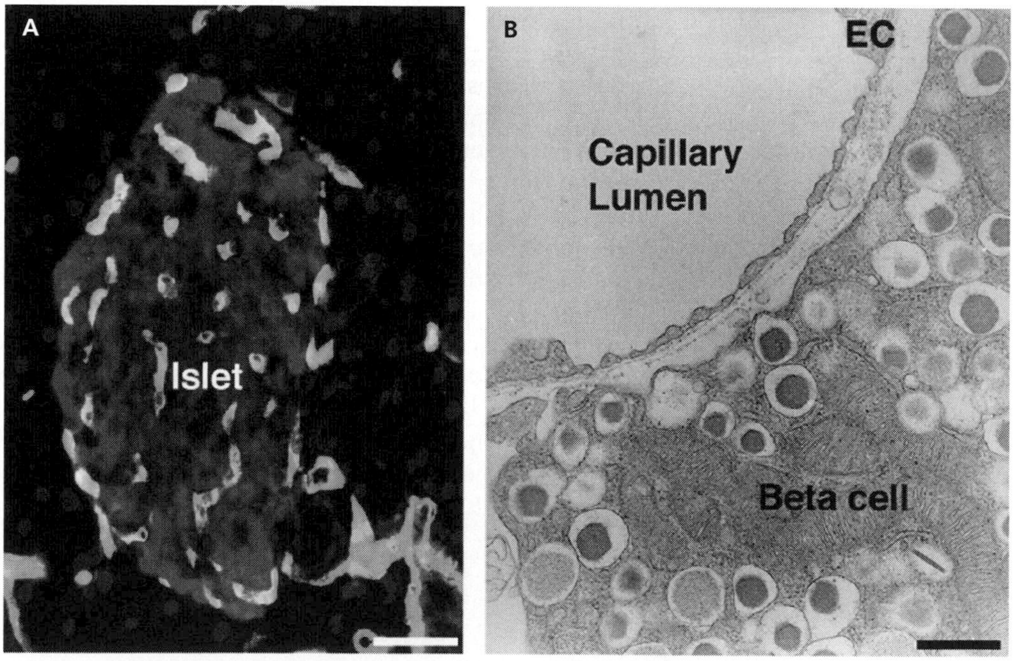

Figure 22.3. The vascular system of pancreatic islets. Islets are vascularized by a dense network of fenestrated capillaries. **(A)** A confocal image of a section through a mouse pancreas is shown. The β-cells are shown in gray (insulin), the ECs in white (PECAM-1), and the cell nuclei in dark gray (DAPI). Scale bar: 50 μm. **(B)** An electron microscopic image of a mouse pancreatic islet is shown. A β-cell is adjacent to a thin and fenestrated capillary EC, which surrounds a capillary lumen. Scale bar: 500 nm. For color reproduction, see Color Plate 22.3A.

secretes polypeptide hormones such as insulin, whereas the liver must be capable of exchanging a large variety of heterogeneous substances with the blood, including lipid particles, serum proteins, blood toxins, and bilirubin. Thus, it is possible that the fenestrated liver sinusoidal capillary endothelium with no basement membrane and no diaphragm is optimally suited to the exchange of these diverse substances and particles.

CONCLUSION

As illustrated for pancreas and liver, tissue development is accompanied by mutual signaling events taking place between nonvascular tissue cells and vascular cells, in particular ECs. Endothelial signals have been shown for other cellular processes, including kidney morphogenesis (36), neural stem cell proliferation and differentiation (23), as well as for neural stem cell transdifferentiation (37). The result of mutual signals is a tissue in which the nonvascular tissue cells and their vascular beds are matched to enable the physiologic function of the tissue. Liver and pancreas are organs that develop from the embryonic gut epithelium. Vascular ECs are required for the early morphological development of these organs. Later during development, ECs obtain tissue-specific properties to enable the specific physiologic function of the tissue.

KEY POINTS

- ECs provide tissues with developmental signals and, in turn, receive diverse angiogenic signals from the tissues.
- ECs are required for hepatic and pancreatic development; pancreatic islet ECs and LSECs differ in their morphology.

Future Goals

- To identify endothelial signals and their application

REFERENCES

1 Nikolova G, Lammert E. Interdependent development of blood vessels and organs. *Cell Tissue Res.* 2003;314(1):33–42.
2 Ferrara N, Carver-Moore K, Chen H, et al. Heterozygous embryonic lethality induced by targeted inactivation of the VEGF gene. *Nature.* 1996;380(6573):439–442.
3 Carmeliet P, Ferreira V, Breier G, et al. Abnormal blood vessel development and lethality in embryos lacking a single VEGF allele. *Nature.* 1996;380(6573):435–439.
4 Stainier DY, Weinstein BM, Detrich HW, 3rd, Zon LI, Fishman MC. Cloche, an early acting zebrafish gene, is required by both the endothelial and hematopoietic lineages. *Development.* 1995;121(10):3141–3150.
5 Matsumoto K, Yoshitomi H, Rossant J, Zaret KS. Liver organogenesis promoted by endothelial cells prior to vascular function. *Science.* 2001;294(5542):559–563.
6 Lammert E, Cleaver O, Melton D. Induction of pancreatic differentiation by signals from blood vessels. *Science.* 2001; 294(5542):564–567.
7 Field HA, Dong PD, Beis D, Stainier DY. Formation of the digestive system in zebrafish. II. Pancreas morphogenesis. *Dev Biol.* 2003;261(1):197–208.
8 Maxwell PH, Ratcliffe PJ. Oxygen sensors and angiogenesis. *Semin Cell Dev Biol.* 2002;13(1):29–37.
9 Inoue M, Hager JH, Ferrara N, Gerber HP, Hanahan D. VEGF-A has a critical, nonredundant role in angiogenic switching and pancreatic beta cell carcinogenesis. *Cancer Cell.* 2002;1(2): 193–202.
10 Lammert E, Gu G, McLaughlin M, et al. Role of VEGF-A in vascularization of pancreatic islets. *Curr Biol.* 2003;13(12):1070–1074.
11 Kerbel R, Folkman J. Clinical translation of angiogenesis inhibitors. *Nat Rev Cancer.* 2002;2(10):727–739.
12 Chi JT, Chang HY, Haraldsen G, et al. Endothelial cell diversity revealed by global expression profiling. *Proc Natl Acad Sci USA.* 2003;100(19):10623–10628.
13 Whitfield ML, Finlay DR, Murray JI, et al. Systemic and cell type-specific gene expression patterns in scleroderma skin. *Proc Natl Acad Sci USA.* 2003;100(21):12319–12324.
14 LeCouter J, Kowalski J, Foster J, et al. Identification of an angiogenic mitogen selective for endocrine gland endothelium. *Nature.* 2001;412(6850):877–884.
15 Aird WC. Endothelial cell heterogeneity. *Crit Care Med.* 2003; 31(4Suppl):S221–S230.
16 Brown DM, Pellecchia M, Ruoslahti E. Drug identification through in vivo screening of chemical libraries. *Chembiochem.* 2004;5(6):871–875.
17 Pasqualini R, Ruoslahti E. Organ targeting in vivo using phage display peptide libraries. *Nature.* 1996;380(6572):364–366.
18 Arap W, Pasqualini R, Ruoslahti E. Cancer treatment by targeted drug delivery to tumor vasculature in a mouse model. *Science.* 1998;279(5349):377–380.
19 Shah NM, Groves AK, Anderson DJ. Alternative neural crest cell fates are instructively promoted by TGF-beta superfamily members. *Cell.* 1996;85(3):331–343.
20 Reissmann E, Ernsberger U, Francis-West PH, Rueger D, Brickell PM, Rohrer H. Involvement of bone morphogenetic protein-4 and bone morphogenetic protein-7 in the differentiation of the adrenergic phenotype in developing sympathetic neurons. *Development.* 1996;122(7):2079–2088.
21 Palmer TD, Willhoite AR, Gage FH. Vascular niche for adult hippocampal neurogenesis. *J Comp Neurol.* 2000;425(4):479–494.
22 Louissaint A, Jr., Rao S, Leventhal C, Goldman SA. Coordinated interaction of neurogenesis and angiogenesis in the adult songbird brain. *Neuron.* 2002;34(6):945–960.
23 Shen Q, Goderie SK, Jin L, et al. Endothelial cells stimulate self-renewal and expand neurogenesis of neural stem cells. *Science.* 2004;304(5675):1338–1340.
24 Storkebaum E, Lambrechts D, Carmeliet P. VEGF: once regarded as a specific angiogenic factor, now implicated in neuroprotection. *Bioessays.* 2004;26(9):943–954.

25 Yoshitomi H, Zaret KS. Endothelial cell interactions initiate dorsal pancreas development by selectively inducing the transcription factor Ptf1a. *Development*. 2004;131(4):807–817.

26 Shalaby F, Rossant J, Yamaguchi TP, et al. Failure of blood-island formation and vasculogenesis in Flk-1-deficient mice. *Nature*. 1995;376(6535):62–66.

27 Wells JM, Melton DA. Early mouse endoderm is patterned by soluble factors from adjacent germ layers. *Development*. 2000; 127(8):1563–1572.

28 Cowan CA, Klimanskaya I, McMahon J, et al. Derivation of embryonic stem-cell lines from human blastocysts. *N Engl J Med*. 2004;350(13):1353–1356.

29 Dor Y, Brown J, Martinez OI, Melton DA. Adult pancreatic beta-cells are formed by self-duplication rather than stem-cell differentiation. *Nature*. 2004;429(6987):41–46.

30 LeCouter J, Moritz DR, Li B, et al. Angiogenesis-independent endothelial protection of liver: role of VEGFR-1. *Science*. 2003; 299(5608):890–893.

31 Ferrara N, LeCouter J, Lin R, Peale F. EG-VEGF and Bv8: a novel family of tissue-restricted angiogenic factors. *Biochem Biophys Acta*. 2004;1654(1):69–78.

32 Konstantinova I, Lammert E. Microvascular development: learning from pancreatic islets. *Bioessays*. 2004;26(10):1069–1075.

33 Like AA. The uptake of exogenous peroxidase by the beta cells of the islets of Langerhans. *Am J Pathol*. 1970;59(2):225–246.

34 Braet F, Wisse E. Structural and functional aspects of liver sinusoidal endothelial cell fenestrae: a review. *Comp Hepatol*. 2002; 1(1):1.

35 Martinez-Hernandez A, Amenta PS. The hepatic extracellular matrix. II. Ontogenesis, regeneration and cirrhosis. *Virchows Arch A Pathol Anat Histopathol*. 1993;423(2):77–84.

36 Serluca FC, Drummond IA, Fishman MC. Endothelial signaling in kidney morphogenesis: a role for hemodynamic forces. *Curr Biol*. 2002;12(6):492–497.

37 Wurmser AE, Nakashima K, Summers RG, et al. Cell fusion-independent differentiation of neural stem cells to the endothelial lineage. *Nature*. 2004;430(6997):350–356.

23

Pulmonary Vascular Development

Peter Lloyd Jones

Institute for Medicine & Engineering, University of Pennsylvania, Philadelphia

"The subtle blood is urged forward by a long course through the lungs . . . and is poured from the pulmonary artery into the pulmonary vein, where the blood is mixed with inspired air." So wrote Michael Servitus (1511–1553), a Spanish physician and theologian who was burnt at the stake for his beliefs. Although the contemporary pulmonary vascular biologist is unlikely to face such a fate, the field of lung vascular developmental biology is nevertheless fraught with historic and technical obstacles that have impaired evolution of a unified theory of lung vascular development. On the other hand, tremendous advances have been made in comprehending morphogenesis of the lung at the level of the epithelium. When coupled with the knowledge that a dynamic and reciprocal relationship exists between the developing lung epithelium and their endothelial cell (EC) neighbors, the time appears ripe to re-evaluate the mechanisms controlling genesis of the pulmonary circulation. What is more, the recognition that a cache of postnatal lung diseases that perturb or destroy respiratory function may arise, in part, as a result of pulmonary vascular defects underscores the need to re-examine the "lesser circulation" (known as such because the flow path is short and lower pressure is required for it to operate efficiently), not only within its unique historical context, but also in the setting of modern biology. In fact, recent discoveries in fields as diverse as developmental biology, in silico modeling, and stem cell and gene therapy suggest that this time already has arrived. In this chapter, I begin by outlining the existing state and outstanding questions in the pulmonary vascular field and then detail the molecular mediators believed to be critical to pulmonary vascular development, with particular emphasis on homeobox gene transcription factors. Finally, I argue for the need to re-evaluate and potentially treat certain chronic lung disease from an endothelial perspective, using information derived from an understanding of development.

LEARNING MORE ABOUT THE LESSER CIRCULATION: UNIQUE FEATURES AND CHALLENGES

The primary role of the postnatal pulmonary endothelium – the largest vascular bed in the human body, which eventually reaches the size of a tennis court in surface area – is to maintain an efficient and functional interface with the alveolar epithelium in which ventilation and perfusion are precisely matched. To achieve this, lung capillaries that bridge the pulmonary arterial and venous beds must develop and remain in close apposition with the terminal gas-exchanging units of the lung, the alveoli. Pulmonary arteries deliver deoxygenated blood to the alveoli, where gas exchange occurs. Pulmonary veins collect the oxygenated blood, delivering it to the left atrium of the heart (Figure 23.1). Extrapulmonary bronchial arteries branch off to the esophagus and perihilar region, whereas intrapulmonary bronchial arteries feed the airways, visceral pleura, and walls of the pulmonary artery (vasa vasorum). At these sites, bronchial arteries function to supply oxygen (O_2) and nutrients before draining into the venous network (see Chapter 127). Thus, a minimum of six distinct lung EC subtypes (i.e., those from pulmonary arteries, alveolar-associated capillaries, pulmonary veins, bronchial arteries, bronchial veins, and bronchial capillaries) coexist within the lung (still other EC subtypes exist within arterioles, venules, and lymphatics). Moreover, even neighboring ECs within a given segment of the vasculature may display phenotypic heterogeneity (see Chapter 126). Thus, the pulmonary endothelium cannot and should not be viewed as a single entity.

Another distinctive feature of lung ECs is that their biophysical and biochemical niche in utero is radically different from their postnatal microenvironment: Although the pulmonary vascular bed receives approximately 8% to 10% of the cardiac output volume prior to birth, this hemodynamic

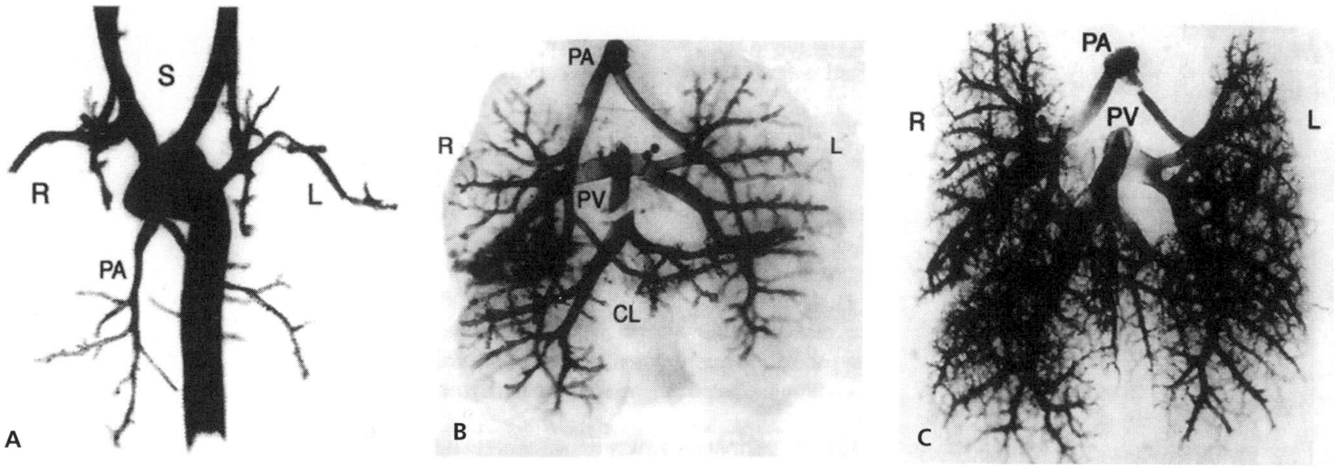

Figure 23.1. Development of the pulmonary arterial tree. Mercox casts of the pulmonary circulation show development of peripheral vessels over time in a (**A**) 12-day, (**B**) 14-day, and (**C**) 15-day mouse fetus. S, systemic; R, right; L, left; PA, pulmonary artery; PV, pulmonary vein; CL, cardiac lobe. (Reproduced with permission from deMello DE, Sawyer D, Galvin N, Reid LM. Early fetal development of lung vasculature. *Am J Respir Cell Mol Biol.* 1997;16(5):568–581.)

state changes at birth, when pulmonary blood flow that was previously diverted away from the lungs via the ductus arteriosus (prior to its programmed closure at birth) increases eight- to tenfold at birth. Accordingly, the pulmonary endothelium must develop in a manner that fully prepares it for the switch at birth from a low-flow, high-pressure, high-resistance setting, to a high-flow, low-pressure, low-resistance environment that is suddenly in contact with oxygenated blood at the levels of the alveolar capillaries and veins. Altered hemodynamics and O_2 tension, together with other environmental factors (including changes in the underlying extracellular matrix [ECM]), contribute to structural and functional alterations in the vessel wall, such as flattening of the endothelium, alterations in vessel lumen diameter, and a concomitant fall in pulmonary artery pressure.

Like other cell types, ECs from the lung are studied most often as monolayers in a noncompliant, biomechanically inert microenvironment under conditions of normoxia (i.e., 21% O_2). Note, however, that depending upon their location, cells within tissues most probably experience only between 6% to 16% O_2, with O_2 tension in the thymus being registered as low as approximately 1% (1). Although these conventional systems provide a valuable first approximation, standard tissue culture conditions fail to fully recapitulate many of the important features that define normal pulmonary EC architecture and behavior in vivo. Similar limitations apply to studies of lung epithelium. Indeed, modern-day assays for epithelial branching morphogenesis frequently incorporate relevant variables, including tissue biomechanics, fetal O_2 tension, and the composition and compliance of the ECM (2,3). Similarly, researchers studying lung vascular development are beginning to pay heed to the notion of organ recapitulation (4). For example, several groups have employed static organ cultures to assay for vascular development, in which whole fetal lung explants are isolated and used. Even in the absence of blood flow, these explants develop a discernible vasculature. These studies have shown that fetal O_2 tension, which is designated as 1% to 3% O_2 in culture studies, is essential for fetal lung vascular development, whereas increasing O_2 to 21% limits epithelial branching morphogenesis (4). Furthermore, it already has been established in multiple tissue systems that shear stress- and adhesion-dependent alterations in cell shape control cell behavior, including EC growth, differentiation, tubulogenesis, and survival (5–8). These types of considerations lead to additional outstanding questions, such as: How do pulmonary ECs sense and interpret physical forces, O_2 tension differentials, and changes in cell adhesion and shape? How are these factors interrelated? And, do all ECs within the lung respond to these and other stimuli in the same manner as ECs at other sites?

The complex interactions and functional codependence that exist between pulmonary arterial and capillary ECs and adjacent airway epithelium represents another unique feature of lung vascular development, and a breakdown in communication between these distinct compartments can result in loss of normal respiratory structure and function. However, the intimate juxtaposition and functional interrelationship between the epithelium and pulmonary vessels does not exist for all vessels within the entire lung. For example, apart from sites of coalescence with the capillary bed at the alveolar level, pulmonary veins are distributed in a more peripheral pattern within the lung. Although pulmonary veins and arteries branch to the same extent, pulmonary veins differ in that they do not follow the airway architecture (9). Understanding the origins, intrinsic behavior, and external cues that combine to generate these distinct lung vascular networks may help to explain their unique spatial relationships to the epithelium, as well as to one another. It is likely that the spacing patterns between different pulmonary vascular networks also rely on intricate and overlapping gradients of specific microenvironmental cues emanating from the adjacent epithelium, the vessels themselves, or from other tissue elements within the lung that have been largely ignored, such as fibroblasts, pericytes, and nerve cells.

Once all the principal players that dictate lung vascular patterns have been identified, it should be possible to model in silico spatially distributed properties that exist for multiple cell types within the lung, including epithelial cells and ECs. Toward this end, Tawhai and colleagues have developed a combined morphometric and computational space-filling three-dimensional Voronoi diagram,[1] to generate the alveolar honeycomb-like geometric mesh, which is then superimposed and covered with a two-dimensional pulmonary capillary network in silico (10,11). Furthermore, blood flow through the capillary network from arteriole to venule has been modeled implementing a solution procedure, which takes into account the hematocrit and capillary dimensional changes in each segment (12). This solution procedure determines the pressure at each junction and flow and red blood cell concentration in each segment, and red blood cell and neutrophil transit times across the network. Such mathematical tools, based on the integration of anatomical, biophysical, cellular, and molecular information, also have been implemented to generally model capillary morphogenesis (13,14). For example, Merks and colleagues have developed a cell-based model that assumes that ECs chemotactically attract one another. This cell model allows the user to assign different properties (i.e., intercellular adhesion, cell shape, and chemoattractant saturation) to each cell and its surroundings, and allows the user to study how these properties affect vascular patterning, based on in vitro capillary formation on Matrigel (Figure 23.2). Ultimately, it is hoped that these types of systems also will be adopted by pulmonary vascular biologists as standard tools in order to pretest hypotheses and to suggest new experiments. Of course, continued progress in this area will rely on collaborations between investigators in diverse fields, as well as retraining of lung EC biologists to embrace and unleash the power of information technology, engineering, and mathematics.

In the meantime, cellular and molecular biologists alike have continued to generate important information that has provided new details regarding EC differentiation and patterning in the developing lung. For example, systematic transcript profiling of genes expressed within the whole lung has been potentially useful in identifying new molecules involved in specification, patterning, and homeostatic control of the pulmonary endothelium. This technology not only has confirmed the expression of markers already known to be expressed in pulmonary ECs in vivo, such as vascular endothelial (VE)-cadherin, CD31, vascular endothelial growth factor receptor 1 (VEGFR1), VEGFR2, Tie1, Tie2, and angiopoietin 2 (Ang2), but also has led to the identification of other mediators that were not previously known to be expressed by lung

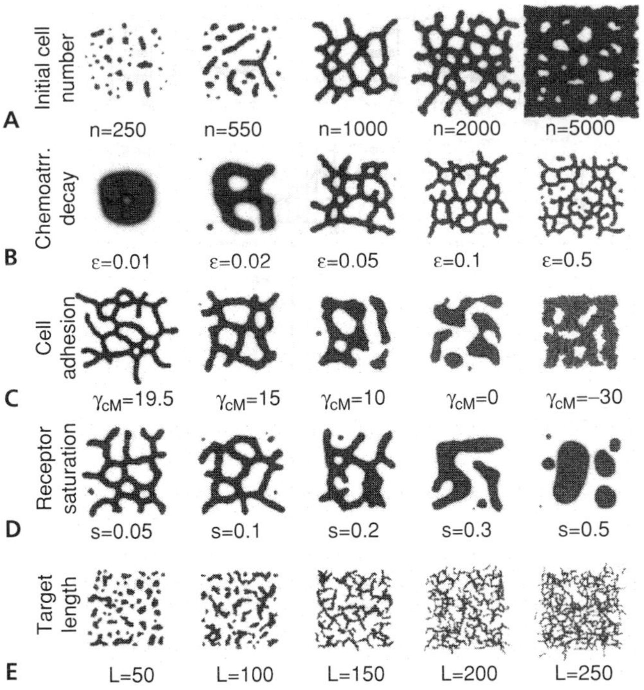

Figure 23.2. *In silico* modeling of vascular development. Overview of computational parameter sweeps showing development of a cell-oriented approach to model in vitro vascular network formation, based on the Gambini-Serini chemotaxis model. Computational models of individual human microvascular ECs were constructed that mimic experimentally observed phenomenology, including response to external chemical signals, cell elongation, cell adhesion, chemotaxis, and haptotaxis. These simulated cells are then released into a virtual tissue culture dish and their macroscopic, collective behavior is quantitatively compared. In conducting these studies, the researchers attempt to recover minimal sets of physiochemical properties and behavioral rules for the cells that reproduce the experimentally observed tissue-level pattern. (**A**) When cell density is high, $n > 1{,}000$, all cells interconnect. (**B**) Size of lacunae depends on the rate of chemoattractant decay. In **C–E**, the Gambini-Serini model is systematically modified by changing the biophysical properties of the individual cells. In this model, cell adhesion is essential (**C**). If adhesivity is reduced, a network no longer forms, and the cells aggregate as islands. Figure **D** shows the effects of changes in *s*, the threshold for chemotaxis saturation (sensitivity). Up to about $s = 0.01$, vessel-like patterns emerge, whereas for larger values more amorphous patterns with larger clusters form. The effect of cell elongation at low cell densities was also tested (**E**), and this indicates that cell elongation facilitates the interconnection of isolated parts of the vascular network pattern. (Reproduced with permission from Merks RM, Brodsky SV, Goligorksy MS, et al. Cell elongation is key to in silico replication of in vitro vasculogenesis and subsequent remodeling. *Dev Biol.* 2006;289(1):44–54.)

ECs, such as semaphorin 3C and its receptor neuropilin 1 (NRP1) (15). Yet major intrinsic problems also are associated with this type of approach. For example, one recent microarray study using isolated tissue showed that mRNA transcripts encoding the ECM protein tenascin (TN)-C are enriched in nonbranching regions of the developing lung epithelium when compared with distal branching regions, which harbors the

1 Also known as a Voronoi tessellation or decomposition; named after Georgy Voronoi, this type of modeling is a representation of metric space determined by distances to a specified discrete set of objects in the space.

capillary bed (16). At face value, these data suggest that TN-C may serve as a negative regulator of lung branching and capillary morphogenesis, yet TN-C is known to be a potent inducer of both developmental and tumor angiogenesis (3,17,18). In fact, TN-C is subject to posttranscriptional control in the fetal lung, and protein levels are enriched around the distal branching airways and capillaries (3,19). Thus, the powerful method of gene profiling, when applied to the lung (or any other tissue for that matter), must be utilized with a full understanding that newly identified genes exist within a highly dynamic environment, and that such genes are controlled at multiple levels that can modify their expression and function. In this regard, ongoing proteomics studies represent a necessary and complimentary approach.

Although numerous molecular features are shared between the pulmonary and systemic circulations, a misconception is that all information gleaned from the systemic circulation can be automatically extrapolated to the pulmonary circulation and vice versa. For example, in developing systemic vessels, ephrinB2 (a transmembrane ligand) and EphB4 (a receptor for ephrinB2) are expressed on embryonic arterial and venous cells respectively, where they are believed to specify boundaries between developing arteries and veins (20). In human fetal lungs, however, EphB4 is expressed on both arteries and veins (9), indicating that the specification of arterial and venous identity in the developing pulmonary circulation relies on other factors. Answers to these and other intriguing lines of inquiry will be revealed over time, but they are eclipsed by a major unresolved question: How do pulmonary ECs arise within the lung?

Table 23-1: Lung Development by Vascular Phenotype

Stage of Lung Development	*Vascular Phenotype*
Embryonic	Vasculogenesis within immature mesenchyme, pulmonary arteries branch from 6th aortic arch, veins as outgrowths from left atrium
Pseudoglandular	Parallel branching of large pulmonary arteries with central airways, lymphatics appear
Canalicular	Increased vessel proliferation and organization into capillary network around airspaces
Saccular	Marked vascular expansion with thinning and condensation of mesenchyme, thin air–blood barrier, double capillary network in septae
Alveolar	Accelerated vascular growth, fusion of the double capillary network with thinning of septae
Postnatal	Marked vessel growth + remodeling, as surface area increases >20-fold

TOWARD A UNIFIED THEORY OF PULMONARY VASCULATURE DEVELOPMENT

Formation of the Primary Lung Vascular Plexus

Pulmonary vessels originate both outside and within the lung in utero, and they continue to develop at all stages of lung development through early adulthood (i.e., 16 to 18 years in humans) via repeated rounds of growth and regression until the final template is established (Table 23-1), in which each generation of airway is accompanied by a pulmonary artery, as well as numerous supernumerary arteries, defined as vessels that branch off from pulmonary arteries, away from the primary vascular tree (21). During fetal life, the pulmonary trunk, which is derived from the truncus arteriosus, represents the most proximal part of the developing pulmonary circulation, and it connects to the pulmonary arch arteries, which are derived from the sixth branchial arch arteries by 7 weeks of gestation in humans, or approximately 9 to 9.5 days of gestation in mice. These proximal vessels supply the upper poles of the right and left lung and infiltrate these tissue elements through one or more mechanisms, including sprouting and intussusceptive and expansive angiogenesis, the latter being defined as a slow expansion of vascular networks via an increase in volume. Similarly, angiogenesis from a pair of intersegmental arteries, which arise from the dorsal aorta, supplies the lower poles of the lung via penetration up through the diaphragm.

Sprouting angiogenesis is initiated by the loss of cell–cell adhesion and proteolytic degradation of the underlying ECM, which is composed predominantly of basement membrane proteins. The extension of EC microspikes into the resulting gaps signals the onset of migration of sheets of cells into the surrounding space. A proliferation of cells then ensues in the lagging cells, resulting in an increase in sprout length. Sprouts continue to branch and fuse, and eventually they form a slit-like lumen, before reassimilating a basement membrane and recruiting surrounding mural cells. Intussusceptive angiogenesis occurs when opposing ECs form a capillary interendothelial bridge, which is then sealed and organized into an interendothelial junction. The endothelium then thins centrally and gives way to invading components of the interstitium that divide the capillary channel (22).

As for the distal intrapulmonary vessels, it was originally proposed that these vessels form exclusively through vasculogenesis (23), defined in this instance as a differentiation of lung mesodermal cells in situ into hemangioblasts, angioblasts, and ultimately ECs. These mesodermal cells develop into isolated and discrete nests within the distal lung mesenchyme and ultimately form primitive networks of angioblasts that envelope pools of blood cells.

Based on morphological and vascular casting methods using polymers injected into the pulmonary circulation, it was thought that the onset of pulmonary vascular development was initiated at embryonic day (E)13 to E14 in the mouse (23), that angiogenesis was restricted to proximal areas of the lung, and that vasculogenesis occurred only in distal regions. Recent studies using developmental biology approaches, however, have radically changed this overall picture. For example, by using a β-galactosidase tagged *Flk-1* locus, Schachtner and colleagues demonstrated that connections already exist between the lung mesenchymal vascular network and the aortic sac by as early as E10.5 (24). This study not only supports the idea that central vessels may arise via vasculogenesis, rather than exclusively by angiogenesis, but it also supports the idea that blood vessel formation occurs concurrently with development of the lung epithelium, which starts to bud from the ventral foregut into the neighboring mesenchyme by E9.5 in the mouse. Furthermore, a recent study suggests that distal vessels may arise via angiogenesis (25). In this instance, lungs of *Tie2–LacZ* transgenic mice were examined from the onset of lung development (E9.5) until the pseudoglandular stage (E13.5). Using this approach, these investigators reported that from the first morphological sign of lung development, a clear vascular network was evident that was in contact with the proximal vasculature arising from the great vessels of the heart. Overall, these models indicate that capillary networks surround the terminal buds and expand by formation of new capillaries from pre-existing vessels as the lung bud grows. Figure 23.3 summarizes these findings, as well as other previous models of pulmonary vascular development. Clearly, alternative approaches still are needed, such as live tracking of the developing lung vasculature in utero, or in the very least, organ culture or animal models that will allow testing of these models ex vivo or ex utero. In this regard, murine lung allograft models still represent the most sophisticated means for recapitulating and controlling lung epithelial and vascular development (26,27), whereas isolated fetal lung mesodermal cell cultures have proven to be useful in identifying factors involved in both vasculogenesis and angiogenesis (19,28).

ASSEMBLING THE PULMONARY VASCULAR WALL

Once the primary lung vascular plexus has formed, prior to birth, the pulmonary vascular wall begins to assemble and mature. Different cell layers become apparent, and this is accompanied by the deposition of specialized extracellular matrices between distinct cell layers, as well as the recruitment, proliferation, and differentiation of surrounding mural cells (i.e., pericytes, fibroblasts, and vascular smooth muscle cells [VSMCs]) that are recruited from the surrounding mesenchyme, depending on their position within the pulmonary vascular tree. For example, in the fetus, preacinar arteries and arteries at the level of the terminal bronchiolus (i.e., vessels that are not in proximity to the alveolar duct) are muscular, whereas the intra-acinar arteries (which accompany the respiratory bronchioli) are partially muscular (i.e., α-smooth muscle actin positive) or nonmuscular. Arteries at the level of the alveolar duct and alveolar wall are not invested with smooth muscle, although they are surrounded by pericytes, and this state remains throughout infancy. In childhood, however, the preacinar arteries become partially muscularized, whereas those in the alveolar wall remain largely nonmuscular, even in the adult. How ECs recruit or repel these different mural cells to the developing pulmonary vascular wall is not known, yet studies in pulmonary hypertensive vessels, where this process is recapitulated (albeit at exaggerated levels) may provide some clues (see the section Molecular Control of Pulmonary Vascular Development).

It is important to note that most of the aforementioned studies refer to the interconnected and contiguous pulmonary arterial, capillary, and venous beds, and not to the lymphatic or bronchial circulations. Until such time as lung lymphatic and bronchial ECs can be accurately distinguished at the cellular and molecular levels it will not be possible to state with certainty how ECs arise within these other vascular beds. For example, the hyaluronan receptor lymphatic endothelium–specific hyaluronan receptor (LYVE)-1 is widely accepted as a lymphatic endothelial marker, but this receptor also is expressed by pulmonary capillary ECs (15). To further complicate matters, lung vasculogenesis and angiogenesis most often are described as separate processes, yet they likely share overlapping cell differentiation and maturation makers, and both networks rely on common cellular processes, including cell proliferation, migration, and the ability to recruit surrounding mural cells. Further, the fact remains that the currently available methods used to assess the connectivity and spatial positioning of emerging lung ECs and networks derived thereof (e.g., casting of vascular channels using high-viscosity polymers, immunohistochemistry on serial sections, and electron microscopy) are unable to provide a complete and accurate temporal–spatial map of lung vascularization by pulmonary, bronchial, and lymphatic networks. This process will rely ultimately on the use of new technologies, such as real-time imaging of lung EC network formation using fluorescently labeled early EC precursor markers and the identification of lung EC-specific markers that can be used to generate conditional transgenic animals. In the meantime, headway has been made in identifying some of the molecular factors involved in generating and maintaining the pulmonary vasculature.

MOLECULAR CONTROL OF PULMONARY VASCULAR DEVELOPMENT

Numerous genetic and epigenetic factors collaborate (in both positive and negative ways) to generate, remodel, restrict, and maintain the pulmonary vasculature, including O_2 tension, hemodynamics, peptide growth and antigrowth factors, transcription factors, and components of the ECM. It should be stated, however, that despite the demonstration of the importance of many of these factors in cardiovascular development, the systematic functional evaluation of the majority of

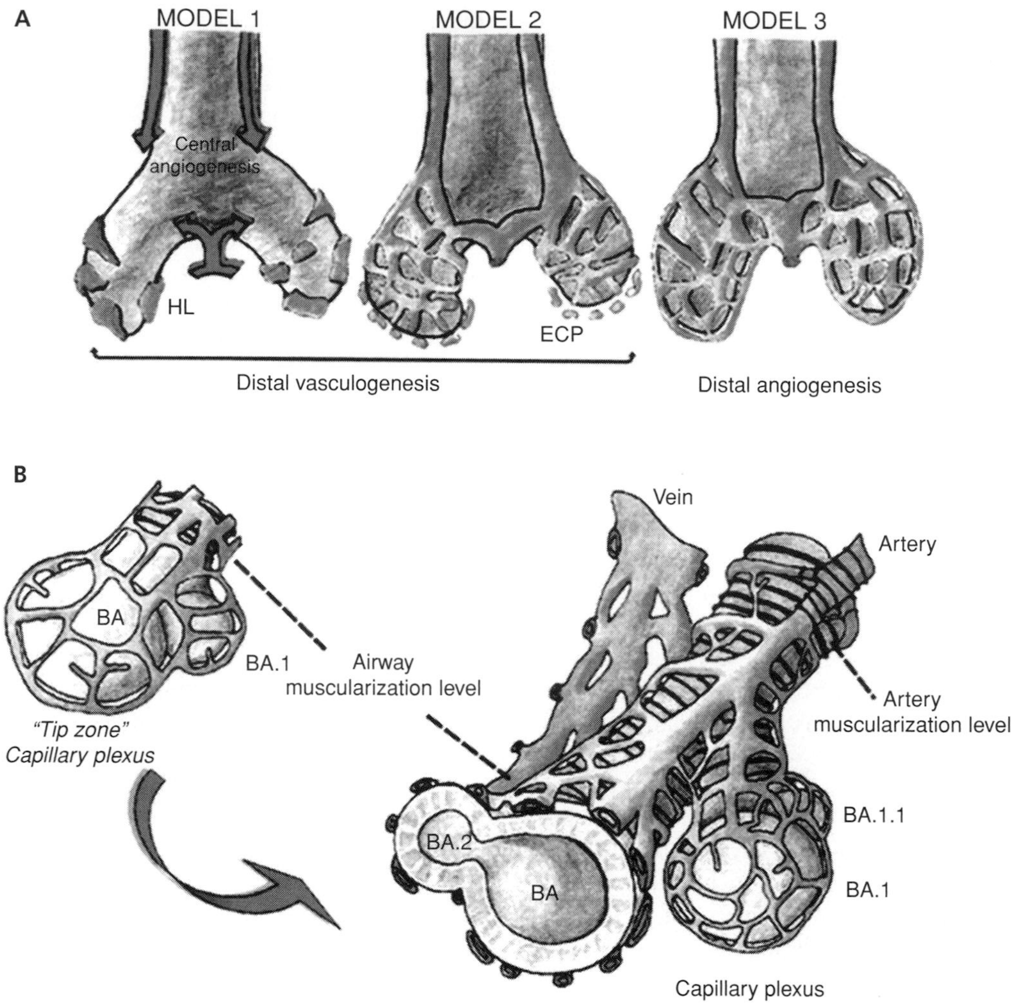

Figure 23.3. Models of lung vascular morphogenesis. (**A**) Model 1, proposed by deMello describes two mechanisms of lung vascular morphogenesis, central angiogenesis (sprouting of arteries and veins from central vascular trunks), and distal vasculogenesis (development of hematopoietic lakes in the mesenchyme). Connection between the two vascular beds would occur at E13–14. Model 2, proposed by Hall, shows distal vasculogenesis (development of new vessels from EC precursors) as the mechanism of the lung vascularization. Model 3, proposed by Parera, shows that distal angiogenesis also occurs. HL, hematopoietic lakes; ECP, EC precursor. (**B**) Detail of the distal angiogenesis model proposed by Parera. Formation of new capillaries from pre-existing vessels takes place in the "tip zone," where airway and capillary networks expand in a coordinated way through epithelial–endothelial interactions. Newly formed vessels remodel dynamically as they form part of the afferent or efferent component. BA, airway branch A; BA.1, and BA.2, daughter branches from branch A. (Reproduced with permission Parera MC, van Dooren M, van Kempen M, et al. Distal angiogenesis: a new concept for lung vascular morphogenesis. *Am J Physiol Lung Cell Mol Physiol.* 2005;288(1):L141–L149.)

these factors in the lung has mostly been restricted to descriptive studies that do not prove directly their role in lung vascular development. Such conclusions will rely on the introduction of inducible systems to control gene expression specifically in the pulmonary vasculature; currently, no such markers exist. Nevertheless, Table 23-2 summarizes the nature of a number of molecules that have been examined and described in detail in the lung and/or systemic circulation, and that are known or believed to contribute to vascular development. As can be seen from the table, the initial template and final fractal patterns that define the developing pulmonary vasculature are likely to be controlled by numerous sets of genes and signal transduction networks that collectively dictate positional identity and morphogenesis. The question remains, however, as to which genes exert ultimate control over these multiple factors. In this regard, homeobox (Hox) gene transcription factors represent ideal candidates, because these evolutionarily conserved transcription factors are able to simultaneously regulate and coordinate the expression of collections of genes already implicated in vasculogenesis and angiogenesis, and they represent

Table 23-2: Examples of Homeobox Genes Expressed in the Lung Vasculature and Cells Controlled by the Lung Vasculature and Vice Versa

Gene	*Expression Pattern, Regulation, and Function*	*References*
Hoxa-1	Null mice are anoxic and die at birth	32
	Co-expressed with *Hoxa-3, Hoxa-5, Hoxb-3, Hoxb-4, Hoxb-6, Hoxb-7,* and *Hoxb-8* in branching region of E11.5 mouse lung	42
	Hoxa-1/b-1 double knockouts possess smaller lungs	78
	Hoxa-1–regulated-62 (HA1R-62) gene represents putative target in adult lung	79
	Transcripts are expressed in normal adult lung	48
Hoxa-2	Expression pattern suggests role in differentiation of proximal mesenchyme derivatives and in vasculogenesis	39
Hoxa-3	Null mice die shortly after birth, conceivably of pulmonary failure	33
Hoxa-4	Expressed between E8.5–12.5 in lung mesenchyme	80
	Expression in lung relies on a conserved retinoic acid response element (RARE) in the 5′ flanking region	81
	By E14.5, *Hoxa-4* restricted to proximal mesenchyme, smooth muscle cells, subepithelial fibroblasts, and alveolar cells; expression may be determined by opposing effects of RA and TGF-β_1	82
Hoxa-5	Expressed in E8–E13 lung, but not by E18	83
	Null mice display improper lung morphogenesis, respiratory distress, and perinatal lethality; controls tissue interactions, because loss of mesenchymal Hoxa-5 leads to surfactant deficiency and altered expression of Nkx2.1, HNF-3, and N-myc in pulmonary epithelium	34
	Induced by RA in human bronchial fibroblasts	84
	Induction by RA relies on *Hoxa-4*	82
Hoxa-10	Expressed in normal human lung and primary pulmonary hypertension	48
Hoxb-1	Expression in foregut endoderm, which gives rise to lungs, depends on a 3′ RARE	85
Hoxb-2	Coexpressed with *Hoxb-5* in distal lung mesenchyme at E10.5–E14.5; *Hoxb-3* and *b-4* detected in mesenchyme of trachea, mainstem bronchi, and distal lung; indicates that specific combinations of *Hoxb* genes specify differences between proximal and distal mesenchyme	86
	Present in human fetal and adult lung, as well as in emphysema and primary pulmonary hypertension	48
Hoxb-3	Co-expressed with *Hoxb-4* and *b-5* in E9.5 foregut	86
	Expressed in prenatal lung, involved in retinol-induced gene expression of Clara-specific secretory protein	45
	Reduces expression of surfactant protein C	
Hoxb-4	Expressed in embryonic lung mesenchyme	87
Hoxb-5	In fetal mouse lung, mesodermal cells surrounding branching epithelial cell layer accumulate high levels of Hoxb-5	88
	In E13.5 mouse to postnatal day 2, expressed in conducting airways and surrounding mesenchyme; at E14.5, levels decrease in mesenchyme distal from airways, but persists in fibroblasts underlying conducting airways	89
	Expressed with *Hoxb-6* in developing chick respiratory tract; prior to branching, expressed around ventral–distal tip of lung bud, eventually demarcating trachea, bronchial tree, and air sacs; may specify positional identity, because *Hoxb-6* through *Hoxb-9* expression corresponds to morphologic subdivisions of air sacs along proximo distal axis	90
	Required for branching morphogenesis in cultured fetal lungs; controls morphogenesis of the first airway divisions from the mainstem bronchi; RA-induced alterations in branching are mediated in part by Hoxb-5	91
	At E14, primarily in mesenchymal cells; by E19, appears mainly in epithelial cells of prealveolar structures and adjacent subepithelium; potential role in lung maturation, because expression persists in hypoplastic lung mesenchyme throughout development and in postnatal lung	92
Hoxb-6	*Hoxb-6* expression reduced during lung development; potential role in distal airway branching	39
Divergent homeobox genes		
Barx2	Expressed in developing mouse lung buds	93
Gax	Expressed in adult rat cardiovascular tissues, including lung	94
	Inhibits vascular SMC proliferation and migration	56
		58

(*continued*)

Table 23-2: ***(continued)***

Gene	*Expression Pattern, Regulation, and Function*	*References*
Hex	Expressed in newborn mouse lung	35
	Detected in E12.5 and E15.5 lung	95
	Expressed at E11.5, especially in distal lung mesenchyme; suppressed by E16.5, and upregulated in adult lung; indicates role in later stages of development and in control of gene expression in adult lung	96
Hlx	Expressed in E10.5 mouse foregut lung mesenchyme; suggested role in controlling tissue interactions	97
Irx1 and *Irx2*	Expressed in E8.5 foregut region; during glandular development, present in lung epithelium; expression declines at end of cannilicular stage; suggested control epithelial–mesenchymal tissue interactions in concert with Gli1, -2, and -3	98
Nkx2.1	Controls surfactant protein A and B gene promoters	99 100
	Antisense oligonucleotides inhibit epithelial branching morphogenesis in embryonic mouse lungs explants	101
	Activates transcription of Clara cell-specific protein	102
	Null mouse lungs consist of dilated sacs arrested at E11–E15	103
	Reduction in *Bmp-4* expression may account for this branching defect; controls VEGF and pulmonary epithelial surfactant expression	
	Transactivation of *Nkx2.1* gene promoter in pulmonary epithelial cells relies on interactions with Sp1 and Sp3 transcription factors	104
	Detailed examination of null mice indicates that distal lung morphogenesis is Nkx2.1-dependent	105
	Siblings with heterozygous deletion of *Nkx2.1* display congenital thyroid dysfunction and recurrent acute respiratory distress	106
	Controls RA stimulation of surfactant protein B promoter activity	107
	Appears to cooperate with C/EBPα in determining high level, lung epithelial-specific expression of Clara cell secretory protein during later stages of lung development and in adult lung	108
	SMAD3 interactions with Nkx2.1 and HNF-1 underlie basis for TGF-β-dependent repression of surfactant protein B	109
Pitx2	Null mice develop right pulmonary isomerism	110
Prx1 and *Prx2*	Expressed in walls of developing muscularized chick pulmonary arteries; *Prx1* especially is prominent at adventitial–medial boundary; *Prx2* also expressed in media; indicates roles in assembly of the vascular wall by promoting recruitment and segregation of smooth muscle and fibroblasts, as well as matrix synthesis and deposition	50
	Prx1-null mouse exhibits cardiovascular defects and die of respiratory distress soon after birth	111
	Activate transcription of TN-C, an ECM protein that promotes pulmonary artery SMC proliferation and survival	62
	Required for in vitro and in vivo lung vasculogenesis	19
Six4 and *Six5*	*Six4* is present in embryonic lung, whereas *Six5* is expressed in adult lung	112

major determinants of pattern formation and morphogenesis throughout development.

Homeobox Genes as Tools to Comprehend Lung Vasculogenesis and Angiogenesis

All proteins encoded by Hox genes contain a 60-amino acid DNA-binding motif – designated the homeodomain – which folds into three α helices. The base sequence for the homeodomain is virtually identical in all the genes that contain it, and it has been found in many organisms, from fruit flies to man. This motif arises from an 180-nucleotide sequence designated the homeobox, so called because mutations in some of these genes result in homeotic transformations in *Drosophila*, in which one body structure replaces another (29). Since the discovery of Hox genes in *Drosophila*, homologous genes have been identified in vertebrates. In mice and humans, these Hox genes are arranged as four unlinked complexes, designated *Hoxa, Hoxb, Hoxc,* and *Hoxd*, which contain a total of 39 genes that are believed to have arisen through the gross duplication of an ancestral cluster. Thus, corresponding genes (or paralogues) located in different clusters on separate chromosomes resemble one another closely, and members of the same paralagous group often collaborate with one another during development (30). Another feature of Hox genes is that those located at the 3′ end of each cluster are expressed prior to those at the more 5′ end, and this is also reflected in their spatial pattern of expression – that is, 3′ genes are expressed in the anterior to thoracic positions, whereas 5′ genes are more posterior (31). This colinearity likely explains why the majority of Hox proteins found in the developing lung are encoded by genes located at the more 3′ end. In addition to Hox genes, at least 160 other divergent homeobox genes have been identified that lie outside Hox clusters, all of which possess a homeodomain.

A review of the prevailing literature indicates that at least 30 homeobox genes are expressed in the developing lung. Although the precise functions of many of these genes remain to be defined, knockout experiments in mice reveal that a number of them may control airway and vascular development, or both. For example, *Hoxa-1*–knockout mice fail to initiate breathing (32), and *Hoxa-3*–null mice die shortly after birth, possibly as a result of pulmonary failure (33). Hoxa-5–deficient mice display respiratory distress at birth due to profound dysmorphogenesis of the developing lung and decreased expression of surfactant proteins (34). Furthermore, expression of Nkx2.1 and hepatic nuclear factor (HNF)-2α (genes that regulate the production of surfactant-associated proteins in the endoderm) is altered in *Hoxa-5*–null mice (34). Because Hoxa-5 expression is confined to the respiratory tract mesenchyme (i.e., embryonic connective tissue developing mainly from the embryonic mesoderm), which interacts with the overlying endoderm during lung morphogenesis, this study exemplifies how homeobox genes may regulate lung tissue interactions between the epithelium and mesenchyme during development.

Although it is still not clear how cells within the lung mesenchyme are selected to undergo vasculogenesis, early evidence indicates that the divergent homeobox gene, *Hex*, may be involved (35,36). This gene is expressed in a range of multipotent progenitor lines and is generally downregulated during terminal differentiation (37). In the mouse embryo, Hex first materializes in the primitive endoderm of the implanting blastocyst. Subsequently, it appears in the foregut endoderm and within the mesoderm, where it is transiently expressed in the nascent blood islands of the yolk sac and later in the embryonic angioblasts and endocardium (36). In the embryonic mouse lung, the highest levels of Hex are observed in the distal lung mesenchyme (35). Taken together with the aforementioned studies, Hex appears to be well poised, both temporally and spatially, to drive vasculogenesis. In fact, a comparison with FLK-1 expression indicates that Hex is an earlier marker of endothelial precursors (36) that is required for the initial stages of EC selection within the mesoderm, rather than differentiation, an idea substantiated by the finding that expression of the *Hex* homologue *hHex* in zebrafish leads to early and ectopic expression of FLK-1 (38). In addition to Hex, in situ hybridization studies suggest that Hoxa-2 may play a role in lung vasculogenesis (39), yet ablation of this gene has no effect on the normal lung phenotype (40). Other approaches, including analysis of compound knockouts involving one or more *Hoxa-2* paralogues will no doubt reveal whether this or other Hox genes truly contribute to pulmonary vascular development.

As well as regulating vasculogenesis, Hox genes also have been implicated in angiogenesis. Once again, little is known about how these genes control this process in the lung, but studies in other tissues indicate that they are likely to be involved because a number of their targets have been shown to be important for lung development and homeostasis (41). For example, Hoxb-7, which is expressed in the fetal lung (42), upregulates a variety of angiogenic stimuli, including fibroblast growth factor 2 (FGF2), vascular endothelial growth factor (VEGF), and matrix metalloproteinase (MMP)-9 (43). Similarly, HOXD3 (i.e., the human homologue of mouse Hoxd-3) colocalizes with integrin $\alpha v\beta 3$ in angiogenic vessels formed in response to FGF2 on a chick chorioallantoic membrane assay, and it also promotes the invasive phenotype of human umbilical vein and dermal microvascular ECs by stimulating the expression of urokinase-type plasminogen activator (u-PA) and $\alpha v\beta 3$ (44), recognized components of the angiogenic cascade. Although HOXD3 and HOXB7 participate in the invasive phase of angiogenesis, HOXB3 (a HOXD3 paralogue expressed in prenatal lung tissue [45]) promotes capillary morphogenesis on basement membrane ECM by controlling the expression of ephrinA1 and subsequent activation of the EphA2 receptor (46).

Following capillary morphogenesis, angiogenic ECs become quiescent. Appreciating that, during development, Hox genes at the 3′ end of each Hox cluster are expressed prior to those at the more 5′ end, Myers and colleagues hypothesized that subsequent to the initial stages of angiogenesis, maturing capillaries would begin to express more 5′ Hox genes (47). Indeed, HOXD10 expression is greater in quiescent ECs when compared with their angiogenic counterparts. Consistent with this, the expression of HOXD10 in vivo reduces the angiogenic response of ECs to FGF2 (47). Mechanistically, HOXD10 suppresses the expression of u-PA receptors and blocks FGF2–dependent migration of ECs into fibrin gels, a process that relies on u-PA receptors. Because HOXD10 appears to be expressed in adult but not fetal lung tissue (48), it is conceivable that this gene actively maintains the fully differentiated state of ECs within the mature lung.

Prx1 represents another homeobox gene that may support both vasculogenesis and angiogenesis within the lung (19). This Hox gene (which is also known as *Mhox*) encodes a divergent, paired-related homeobox gene. In the developing chick cardiovascular system, Prx1 is first evident in the endocardial cushions and valves, the epicardium, and in the walls of the great arteries and veins (49). In the embryonic chick and mouse pulmonary vasculature, Prx1 appears in the endothelium from early stages onward, and later in the adventitial and medial cell layers (19,50), indicating that this homeobox gene may promote differentiation and patterning of pulmonary vessels. In support of this, *Prx1*-null mice exhibit a variety of pulmonary and systemic vascular anomalies, including abnormal positioning and awkward curvature of the aortic arch and an elongated ductus arteriosus (51). As well, *Prx1*$^{-/-}$ mice are cyanotic and die soon after birth from respiratory distress, which had been suggested to arise from palatal defects. Alternatively, it is possible that Prx1 affects respiratory function, because it is required for one or more stages of lung vascular development. Indeed, our group has shown that Prx1 can not only promote differentiation of fetal lung mesodermal precursors to an EC phenotype, but that this gene product also promotes capillary morphogenesis in vitro by inducing transcription of the *TN-C* gene, a proangiogenic ECM protein (19). In keeping with these data, lungs from *Prx1*$^{-/-}$ mice express reduced levels of TN-C, they fail to form capillary networks, and their epithelium is hypoplastic (19). Thus, ablation of lung mesenchymal Prx1 not only affects vascular cell behavior, but also results in epithelial defects and early demise following birth due to respiratory distress.

The lymphatic vasculature in the lung collects extravasated immune cells and keeps the airways dry by collecting fluid leaked from blood vessels. Identifying factors that regulate lymphangiogenesis is a crucial, yet understudied, aspect of lung vascular developmental biology. However, a relatively recent study reveals that the *Prox1* homeobox gene is essential for this process (52). Prox1 is expressed in a subpopulation of ECs that, after budding from veins, give rise to the mammalian lymphatic system. In *Prox1*-null embryos, budding is arrested at E11.5, resulting in embryos that are capable of undergoing vasculogenesis and angiogenesis, but not lymphangiogenesis. Furthermore, ECs that ordinarily express lymphatic markers – including VEGFR3 and LYVE-1 – are absent in these mice, being replaced by markers that characterize the blood vascular phenotype, Thus, Prox1 appears to be required for the maintenance of budding of venous ECs and differentiation toward the lymphatic phenotype. In light of this work, it will be important to determine when and where Prox1 is expressed in the developing lung lymphatics, and whether this gene is dysregulated in lung lymphatic diseases, such as familial pulmonary lymphangiectasia.

Homeobox Genes and Assembly of the Pulmonary Vascular Wall

Homeobox genes also have been implicated in a variety of processes required for assembly and elaboration of the vascular wall, including the promotion and inhibition of VSMC proliferation, differentiation, and changes in cell adhesion to the ECM. For example, Hoxb-7 – which appears in the developing lung (42) – is expressed more highly in fetal than in adult VSMCs (53), indicating that this gene may play a role in the early expansion and differentiation of VSMCs. Consistent with this, Hoxb-7 promotes proliferation, as well as inducing the expression of early VSMC markers, such as SM22-α (54). In contrast to Hoxb-7, the divergent homeobox gene product Gax (which is expressed in adult cardiovascular tissues including heart, lung, and blood vessels) inhibits VSMC proliferation (55). Within the vasculature, Gax is expressed in quiescent VSMCs and is downregulated when these cells are stimulated to proliferate (56,57). In keeping with this growth-suppressive role, overexpression of Gax inhibits VSMC proliferation by activating the cell cycle inhibitor protein p21(57). Gax also attenuates VSMC motility by downregulating integrins $\alpha v\beta 3$ and $\alpha v\beta 5$ in a manner that is dependent on the cell cycle (58).

Prx1 and another homeobox gene product designated Barx2 are believed to regulate VSMC differentiation (59,60), and both of these genes are expressed in the developing lung. Functionally, Prx1 and Barx2 have been shown to control smooth muscle-specific gene expression in vitro by enhancing the binding of serum response factor (SRF) to CarG elements, a DNA motif present in many smooth muscle-specific gene promoters (60). Furthermore, the *Barx2* gene promoter contains an SRF-binding serum response element, which is essential for its activity (61), indicating that this gene may be autoregulated. However, whereas Barx2 is expressed in the differentiated adult aorta, Prx1 only appears in the walls of developing and injured blood vessels, when VSMC differentiation is being actively suppressed in favor of proliferation (50,62). It is possible, nonetheless, that Prx1 and SRF cofunction in other cell types within the blood vessel wall (i.e., activated adventitial fibroblasts) that express VSMC markers during development and in pathologic remodeling.

In addition to arteries, homeobox genes appear to play a role in the assembly of pulmonary veins. *Hoxa-3*–null mice die shortly after birth and frequently have heart and blood vessel defects, including hypertrophy of the atria and enlargement of the pulmonary veins (33). This finding is intriguing, given that the wall of pulmonary veins not only possesses EC lining and smooth muscle tunica media, but also an outer layer of cardiac troponin I-positive cells, indicating that these may arise from the developing heart (63). Considering the phenotype of *Hoxa-3*–null mice, it is plausible that this gene functions to limit cardiac cell migration during pulmonary venous development.

Studies with homeobox genes in the setting of occlusive pulmonary vascular disease associated with pulmonary arterial hypertension also have provided potentially important clues regarding the role of these transcription factors in the assembly and disassembly of the pulmonary vascular wall. As in development, the formation of occlusive vascular lesions in the lung also depends on the migration and proliferation of ECs, VSMCs, and fibroblasts, as well as inappropriate expression of VSMC markers and excessive catabolism and synthesis of specific ECM proteins. It is therefore not surprising that many of the homeobox genes involved in pulmonary vascular development are re-expressed in pulmonary vascular disease. In keeping with this, a comparison of Hox gene expression in human fetal lungs versus tissue derived from patients with primary pulmonary hypertension and emphysema indicates that this is the case for certain Hox genes, including HOXA5 and HOXB2 (48). On the other hand, the tissue distribution of some of these Hox genes appears to differ between fetal and remodeled adult lung. For example, whereas HOXA5 expression is confined to the mesenchyme of the respiratory tract in the developing lung (34), this gene has been reported to be expressed by ECs in hyperproliferative, occlusive plexiform lesions in adult patients with primary pulmonary hypertension (48). It is possible, however, that HOXA5 also is expressed in those muscle cells that are known to contribute to these lesions.

Recent work in our laboratory indicates that both Prx1 and Prx2 (a gene arising from duplication of *Prx1*), contribute to pulmonary vascular disease by promoting the expression of TN-C, an ECM glycoprotein that supports VSMC proliferation and fibroblast migration (62,64). TN-C is expressed in the developing vasculature, but it is suppressed in normal, quiescent pulmonary vessels. In hypertensive vessels, however, TN-C reappears at the adventitial–medial boundary, and later throughout the vascular wall, where it surrounds proliferating and migrating VSMCs and fibroblasts (65–67). At a functional level, TN-C promotes the proliferation and survival of pulmonary artery VSMCs via its ability to cross-modulate the activity of receptor tyrosine kinases. In terms of its effects on migration, TN-C destablizes focal adhesions, thereby facilitating cell detachment from the underlying ECM (64). Given the importance of these functions in pulmonary vascular development and disease, it is of interest to determine how this ECM protein is regulated.

Multiple factors induce TN-C, including ECM-degrading MMPs. For example, inhibition of MMP activity suppresses TN-C expression and pulmonary artery VSMC proliferation, and reduces the severity of vascular lesions (68). These studies indicate that MMPs lie upstream in an adhesion-dependent signaling pathway that controls TN-C. Consistent with this, the *TN-C* gene promoter contains an ECM-responsive element that is silenced on native type I collagen, an $\alpha 2\beta 1$ integrin ligand, but is activated on the MMP-proteolyzed or denatured type I collagen, which represents an $\alpha v\beta 3$ integrin ligand (69). This ECM/$\beta 3$ integrin-responsive element in the *TN-C* gene promoter was subsequently shown to harbor a homeodomain binding site, suggesting that the induction of TN-C in response to changes in VSMC adhesion might depend on homeobox proteins and, in this regard, Prx transcription factors represent ideal candidates. For example, like TN-C, Prx1 and Prx2 are expressed during embryogenesis, predominantly in mesenchyme-specific patterns. In the developing cardiovascular system, Prx1 and Prx2 expression overlaps with TN-C in the endocardial cushions and valves, the epicardium, and the wall of the great arteries and veins. Further, *Prx1* and *Prx2* gene knockouts exhibit skeletal defects and vascular anomalies, including a twisted aorta and elongated ductus arteriosus (50), defects that conceivably arise due to dysregulated TN-C synthesis. In further support of a link between Prx proteins and TN-C expression, examination of hypertensive adult rats lungs showed that Prx1 and Prx2 are re-expressed at the adventitial–medial boundary of remodeling pulmonary arteries, and later within the arterial wall, colocalizing with TN-C protein and proliferating VSMCs (62). Moreover, identical to TN-C, Prx1 and Prx2 expression also is suppressed and induced by native and denatured type I collagen respectively, and overexpression of Prx1 promotes *TN-C* gene transcription and VSMC proliferation. Whether Prx1 promotes VSMC proliferation by inducing TN-C remains to be determined. Because expression of Prx1 and TN-C both depend on changes in cell adhesion to the ECM via integrins, we hypothesized that the activation of focal adhesion kinase (FAK), an integrin-activated nonreceptor tyrosine kinase that is required for VSMC and fibroblast migration as well as EC differentiation and capillary assembly (70), supports the Prx1-dependent induction of TN-C. In keeping with this, migration-defective fibroblasts and E8.5 embryos devoid of FAK express reduced levels of Prx1 and TN-C when compared to their wild-type counterparts, whereas overexpression of Prx1 in *FAK*-null cells restored TN-C expression and cell migration (64). Taken together with previous studies, a tenable working hypothesis is that, during vascular development and following injury, MMP-dependent activation of FAK stimulates Prx1-dependent induction of TN-C, which subsequently contributes to the formation of an ECM that supports VSMC and fibroblast proliferation and migration, as well as EC differentiation and network formation. The consequences of perturbing one or more of these processes during lung development are discussed in the next section.

SOCIETAL COLLAPSE: VASCULAR DEVELOPMENTAL DEFECTS ASSOCIATED WITH LUNG DISEASE

Numerous experimental and clinical data indicate that reciprocal tissue interactions between the developing pulmonary vasculature and the adjacent epithelium are critical for normal lung development and postnatal function. For example, when fetal rat lung mesenchyme is cultured alone with fetal lung epithelium, differentiated ECs emerge from within the distal mesenchyme. However, when fetal lung mesenchyme is cultured alone, EC differentiation is suppressed (71). A major factor that appears to be critical for lung-tissue interactions is VEGF, which is expressed at high levels by the epithelium. For example, mice that do not express the VEGF-A_{120} isoform fail to develop a normal peripheral lung vasculature (72). Similarly, the genetic ablation of hypoxia-inducible transcription factor (HIF)-2α results in diminished VEGF levels in alveolar cells and a defect in vascularization of alveolar septa (73). These studies indicate that compromising the pulmonary vasculature might have major effects on respiratory function. Indeed, numerous lung diseases are defined by vascular developmental defects, including persistent pulmonary hypertension of the newborn, bronchopulmonary dysplasia, congenital diaphragmatic hernia, lung hypoplasia, alveolar capillary dysplasia, arteriovenous malformations, pulmonary hemangiomatosis, pulmonary lymphangiectasis, pulmonary venoocclusive disease, pulmonary vein stenosis, and many forms of congenital heart disease.

Bronchopulmonary dysplasia (BPD) represents a good example of how knowledge of pulmonary EC developmental biology eventually may be applied to the clinical treatment of a human disease. Historically, pulmonary hypertension was regarded as a major determinant in the pathophysiology of BPD, with the assumption that pulmonary vascular abnormalities were a secondary consequence of primary injury to the airspace. Recent studies, however, indicate that alterations in the growth and structure of the pulmonary vasculature may contribute directly to the abnormal and hypoplastic alveolarization that characterizes this disease. Although treating many premature infants via assisted ventilation and O_2 therapy is a necessary step for their survival, this standard therapy is believed to contribute to the lung vascular injury associated with BPD. Most research into this disease has been focused on alterations in vascular tone and reactivity in infants and children with BPD, with less emphasis being placed on disordered vascular growth as an underlying feature of this disease. However, studies with neonatal rats demonstrate that pharmacologic inhibition of VEGF receptor activity and angiogenesis attenuates pulmonary vascular growth, leading to a BPD-like phenotype characterized by reduced alveolarization (74). Work in our laboratory has shown that ablation of pulmonary vessels via inactivation of the *Prx1* homeobox gene also results in a BPD-like hypoplastic epithelial phenotype (19). Also, premature baboons placed on mechanical ventilation exhibit decreased lung capillary density, and this is associated with a significant reduction in VEGF levels in the lung (75). This finding is interesting, because VEGF expression levels are decreased in human infants with BPD who die from this disease (76). Although more basic studies are required to understand the relationship between the developing lung vasculature and the airspace, it is hoped that implementation of therapies that will result in the restoration of the pulmonary endothelium will eventually become standard practice for the treatment of BPD and other chronic or fatal lung diseases, including asthma and pulmonary hypertension. Along these lines, Stewart and colleagues have demonstrated that bone marrow–derived endothelial-like progenitor cells, re-engineered to express human endothelial nitric oxide synthase, can restore microvascular structure and function in a rodent model of pulmonary arterial hypertension (77). This compelling experimental study argues strongly that stem cell–based therapies represent a viable option for the treatment of lung diseases underscored by vascular changes, and that restoring the pulmonary endothelium may represent a novel therapeutic option whose time has come.

KEY POINTS

- Recent studies have led to major insights into the mechanisms underlying the development of the normal pulmonary circulation and the critical relationship between the vasculature and its surrounding tissue elements, including the epithelium.
- Nevertheless, much still is to be learned and many scientific and political challenges lie ahead. However, it is hoped that continued study of the lesser circulation will ultimately herald a new era in respiratory medicine, especially one that reconsiders the importance of the pulmonary endothelium as a central regulator of normal lung structure and function.

Future Goals

- To resolve the origin of the pulmonary vasculature
- To develop computational modeling to predict pulmonary vascular cell behavior in development and disease
- To integrate biology, medicine, physical, and mathematical sciences to generate new hypotheses in pulmonary vascular development and disease
- To restore the pulmonary vasculature and respiratory function via gene- and stem cell–based therapies

REFERENCES

1 Hale LP, Braun RD, Gwinn WM, et al. Hypoxia in the thymus: role of oxygen tension in thymocyte survival. *Am J Physiol Heart Circ Physiol.* 2002;282(4):H1467–H1477.

2 Moore KA, Polte T, Huang S, et al. Control of basement membrane remodeling and epithelial branching morphogenesis in embryonic lung by Rho and cytoskeletal tension. *Dev Dyn.* 2005; 232(2):268–281.

3 Gebb SA, Jones PL. Hypoxia and lung branching morphogenesis. *Adv Exp Med Biol.* 2003;543:117–125.

4 van Tuyl M, Liu J, Wang J, et al. Role of oxygen and vascular development in epithelial branching morphogenesis of the developing mouse lung. *Am J Physiol Lung Cell Mol Physiol.* 2005; 288(1):L167–L178.

5 Jones PL, Chapados R, Baldwin HS, et al. Altered hemodynamics controls matrix metalloproteinase activity and tenascin-C expression in neonatal pig lung. *Am J Physiol Lung Cell Mol Physiol.* 2002;282(1):L26–L35.

6 Boudreau NJ, Jones PL. Extracellular matrix and integrin signalling: the shape of things to come. *Biochem J.* 1999;339 (Pt3): 481–488.

7 Chen CS, Mrksich M, Huang S, et al. Geometric control of cell life and death. *Science.* 1997;276(5317):1425–1428.

8 Connolly JO, Simpson N, Hewlett L, Hall A. Rac regulates endothelial morphogenesis and capillary assembly. *Mol Biol Cell.* 2002;13(7):2474–2485.

9 Hall SM, Hislop AA, Haworth SG. Origin, differentiation, and maturation of human pulmonary veins. *Am J Respir Cell Mol Biol.* 2002;26(3):333–340.

10 Burrowes KS, Hunter PJ, Tawhai MH. Anatomically based finite element models of the human pulmonary arterial and

venous trees including supernumerary vessels. *J Appl Physiol.* 2005;99(2):731–738.

11 Tawhai MH, Burrowes KS. Developing integrative computational models of pulmonary structure. *Anat Rec B New Anat.* 2003;275(1):207–218.

12 Huang Y, Doerschuk CM, Kamm RD. Computational modeling of RBC and neutrophil transit through the pulmonary capillaries. *J Appl Physiol.* 2001;90(2):545–564.

13 Merks RM, Brodsky SV, Goligorksy MS, et al. Cell elongation is key to in silico replication of in vitro vasculogenesis and subsequent remodeling. *Dev Biol.* 2006;289(1):44–54.

14 Peirce SM, Skalak TC. Microvascular remodeling: a complex continuum spanning angiogenesis to arteriogenesis. *Microcirculation.* 2003;10(1):99–111.

15 Favre CJ, Mancuso M, Maas K, et al. Expression of genes involved in vascular development and angiogenesis in endothelial cells of adult lung. *Am J Physiol Heart Circ Physiol.* 2003; 285(5):H1917–H1938.

16 Lu J, Qian J, Izvolsky KI, Cardoso WV. Global analysis of genes differentially expressed in branching and non-branching regions of the mouse embryonic lung. *Dev Biol.* 2004;273(2): 418–435.

17 Jones FS, Jones PL. The tenascin family of ECM glycoproteins: structure, function, and regulation during embryonic development and tissue remodeling. *Dev Dyn.* 2000;218(2):235–259.

18 Tanaka K, Hiraiwa N, Hashimoto H, et al. Tenascin-C regulates angiogenesis in tumor through the regulation of vascular endothelial growth factor expression. *Int J Cancer.* 2004;108(1): 31–40.

19 Ihida-Stansbury K, McKean DM, Gebb SA, et al. Paired-related homeobox gene Prx1 is required for pulmonary vascular development. *Circ Res.* 2004;94(11):1507–1514.

20 Wang HU, Chen ZF, Anderson DJ. Molecular distinction and angiogenic interaction between embryonic arteries and veins revealed by ephrin-B2 and its receptor Eph-B4. *Cell.* 1998;93(5): 741–753.

21 Hislop A, Reid L. Intra-pulmonary arterial development during fetal life-branching pattern and structure. *J Anat.* 1972;113(1): 35–48.

22 Burri PH, Tarek MR. A novel mechanism of capillary growth in the rat pulmonary microcirculation. *Anat Rec.* 1990;228(1):35–45.

23 deMello DE, Sawyer D, Galvin N, Reid LM. Early fetal development of lung vasculature. *Am J Respir Cell Mol Biol.* 1997;16(5): 568–581.

24 Schachtner SK, Wang Y, Scott Baldwin H. Qualitative and quantitative analysis of embryonic pulmonary vessel formation. *Am J Respir Cell Mol Biol.* 2000;22(2):157–165.

25 Parera MC, van Dooren M, van Kempen M, et al. Distal angiogenesis: a new concept for lung vascular morphogenesis. *Am J Physiol Lung Cell Mol Physiol.* 2005;288(1):L141–L149.

26 Schwarz MA, Zhang F, Lane JE, et al. Angiogenesis and morphogenesis of murine fetal distal lung in an allograft model. *Am J Physiol Lung Cell Mol Physiol.* 2000;278(5):L1000–L1007.

27 Vu TH, Alemayehu Y, Werb Z. New insights into saccular development and vascular formation in lung allografts under the renal capsule. *Mech Dev.* 2003;120(3):305–313.

28 Akeson AL, Brooks SK, Thompson FY, Greenberg JM. In vitro model for developmental progression from vasculogenesis to angiogenesis with a murine endothelial precursor cell line, MFLM-4. *Microvasc Res.* 2001;61(1):75–86.

29 McGinnis W, Krumlauf R. Homeobox genes and axial patterning. *Cell.* 1992;68(2):283–302.

30 Chen F, Capecchi MR. Paralogous mouse Hox genes, Hoxa9, Hoxb9, and Hoxd9, function together to control development of the mammary gland in response to pregnancy. *Proc Natl Acad Sci USA.* 1999;96(2):541–546.

31 Duboule D. Vertebrate hox gene regulation: clustering and/or colinearity? *Curr Opin Genet Dev.* 1998;8(5):514–518.

32 Lufkin T, Dierich A, LeMeur M, et al. Disruption of the Hox-1.6 homeobox gene results in defects in a region corresponding to its rostral domain of expression. *Cell.* 1991;66(6):1105–1119.

33 Chisaka O, Capecchi MR. Regionally restricted developmental defects resulting from targeted disruption of the mouse homeobox gene hox-1.5. *Nature.* 1991;350(6318):473–479.

34 Aubin J, Lemieux M, Tremblay M, et al. Early postnatal lethality in Hoxa-5 mutant mice is attributable to respiratory tract defects. *Dev Biol.* 1997;192(2):432–445.

35 Bogue CW, Gross I, Vasavada H, et al. Identification of Hox genes in newborn lung and effects of gestational age and retinoic acid on their expression. *Am J Physiol.* 1994;266(4 Pt 1):L448–L454.

36 Thomas PQ, Brown A, Beddington RS. Hex: a homeobox gene revealing peri-implantation asymmetry in the mouse embryo and an early transient marker of endothelial cell precursors. *Development.* 1998;125(1):85–94.

37 Pellizzari L, D'Elia A, Rustighi A, et al. Expression and function of the homeodomain-containing protein Hex in thyroid cells. *Nucleic Acids Res.* 2000;28(13):2503–2511.

38 Liao W, Ho CY, Yan YL, et al. Hhex and scl function in parallel to regulate early endothelial and blood differentiation in zebrafish. *Development.* 2000;127(20):4303–4313.

39 Cardoso WV, Mitsialis SA, Brody JS, Williams MC. Retinoic acid alters the expression of pattern-related genes in the developing rat lung. *Dev Dyn.* 1996;207(1):47–59.

40 Gendron-Maguire M, Mallo M, Zhang M, Gridley T. Hoxa-2 mutant mice exhibit homeotic transformation of skeletal elements derived from cranial neural crest. *Cell.* 1993;75(7):1317–1331.

41 Jones PL. Homeobox genes in pulmonary vascular development and disease. *Trends Cardiovasc Med.* 2003;13(8):336–345.

42 Mollard R, Dziadek M. Homeobox genes from clusters A and B demonstrate characteristics of temporal colinearity and differential restrictions in spatial expression domains in the branching mouse lung. *Int J Dev Biol.* 1997;41(5):655–666.

43 Care A, Felicetti F, Meccia E, et al. HOXB7: a key factor for tumor-associated angiogenic switch. *Cancer Res.* 2001;61(17): 6532–6539.

44 Boudreau N, Andrews C, Srebrow A, et al. Induction of the angiogenic phenotype by Hox D3. *J Cell Biol.* 1997;139(1):257–264.

45 Nakamura N, Yoshimi T, Miura T. Increased gene expression of lung marker proteins in the homeobox B3-overexpressed fetal lung cell line M3E3/C3. *Cell Growth Differ.* 2002;13(4):195–203.

46 Myers C, Charboneau A, Boudreau N. Homeobox B3 promotes capillary morphogenesis and angiogenesis. *J Cell Biol.* 2000;148 (2):343–351.

47 Myers C, Charboneau A, Cheung I, et al. Sustained expression of homeobox D10 inhibits angiogenesis. *Am J Pathol.* 2002; 161(6):2099–2109.

48 Golpon HA, Geraci MW, Moore MD, et al. HOX genes in human lung: altered expression in primary pulmonary

hypertension and emphysema. *Am J Pathol.* 2001;158(3):955–966.

49 Leussink B, Brouwer A, el Khattabi M, et al. Expression patterns of the paired-related homeobox genes MHox/Prx1 and S8/Prx2 suggest roles in development of the heart and the forebrain. *Mech Dev.* 1995;52(1):51–64.

50 Bergwerff M, Gittenberger-de Groot AC, DeRuiter MC, et al. Patterns of paired-related homeobox genes PRX1 and PRX2 suggest involvement in matrix modulation in the developing chick vascular system. *Dev Dyn.* 1998;213(1):59–70.

51 Bergwerff M, Gittenberger-de Groot AC, Wisse LJ, et al. Loss of function of the Prx1 and Prx2 homeobox genes alters architecture of the great elastic arteries and ductus arteriosus. *Virchows Arch.* 2000;436(1):12–19.

52 Wigle JT, Harvey N, Detmar M, et al. An essential role for Prox1 in the induction of the lymphatic endothelial cell phenotype. *EMBO J.* 2002;21(7):1505–1513.

53 Miano JM, Firulli AB, Olson EN, et al. Restricted expression of homeobox genes distinguishes fetal from adult human smooth muscle cells. *Proc Natl Acad Sci USA.* 1996;93(2):900–905.

54 Bostrom K, Tintut Y, Kao SC, et al. HOXB7 overexpression promotes differentiation of C3H10T1/2 cells to smooth muscle cells. *J Cell Biochem.* 2000;78(2):210–221.

55 Gorski DH, Walsh K. The role of homeobox genes in vascular remodeling and angiogenesis. *Circ Res.* 2000;87(10):865–872.

56 Perlman H, Luo Z, Krasinski K, et al. Adenovirus-mediated delivery of the Gax transcription factor to rat carotid arteries inhibits smooth muscle proliferation and induces apoptosis. *Gene Ther.* 1999;6(5):758–763.

57 Smith RC, Branellec D, Gorski DH, et al. p21CIP1-mediated inhibition of cell proliferation by overexpression of the gax homeodomain gene. *Genes Dev.* 1997;11(13):1674–1689.

58 Witzenbichler B, Kureishi Y, Luo Z, et al. Regulation of smooth muscle cell migration and integrin expression by the Gax transcription factor. *J Clin Invest.* 1999;104(10):1469–1480.

59 Hautmann MB, Thompson MM, Swartz EA, et al. Angiotensin II-induced stimulation of smooth muscle alpha-actin expression by serum response factor and the homeodomain transcription factor MHox. *Circ Res.* 1997;81(4):600–610.

60 Herring BP, Kriegel AM, Hoggatt AM. Identification of Barx2b, a serum response factor-associated homeodomain protein. *J Biol Chem.* 2001;276(17):14482–14489.

61 Meech R, Makarenkova H, Edelman DB, Jones FS. The homeodomain protein Barx2 promotes myogenic differentiation and is regulated by myogenic regulatory factors. *J Biol Chem.* 2003; 278(10):8269–8278.

62 Jones FS, Meech R, Edelman DB, et al. Prx1 controls vascular smooth muscle cell proliferation and tenascin-C expression and is upregulated with Prx2 in pulmonary vascular disease. *Circ Res.* 2001;89(2):131–138.

63 Millino C, Sarinella F, Tiveron C, et al. Cardiac and smooth muscle cell contribution to the formation of the murine pulmonary veins. *Dev Dyn.* 2000;218(3):414–425.

64 McKean DM, Sisbarro L, Ilic D, et al. FAK induces expression of Prx1 to promote tenascin-C-dependent fibroblast migration. *J Cell Biol.* 2003;161(2):393–402.

65 Jones PL, Cowan KN, Rabinovitch M. Tenascin-C, proliferation and subendothelial fibronectin in progressive pulmonary vascular disease. *Am J Pathol.* 1997;150(4):1349–1360.

66 Jones PL, Crack J, Rabinovitch M. Regulation of tenascin-C, a vascular smooth muscle cell survival factor that interacts with the alpha v beta 3 integrin to promote epidermal growth factor receptor phosphorylation and growth. *J Cell Biol.* 1997;139(1): 279–293.

67 Jones PL, Rabinovitch M. Tenascin-C is induced with progressive pulmonary vascular disease in rats and is functionally related to increased smooth muscle cell proliferation. *Circ Res.* 1996;79(6):1131–1142.

68 Cowan KN, Jones PL, Rabinovitch M. Elastase and matrix metalloproteinase inhibitors induce regression, and tenascin-C antisense prevents progression, of vascular disease. *J Clin Invest.* 2000;105(1):21–34.

69 Jones PL, Jones FS, Zhou B, Rabinovitch M. Induction of vascular smooth muscle cell tenascin-C gene expression by denatured type I collagen is dependent upon a beta3 integrin-mediated mitogen-activated protein kinase pathway and a 122-base pair promoter element. *J Cell Sci.* 1999;112 (Pt 4):435–445.

70 Ilic D, Kovacic B, McDonagh S, et al. Focal adhesion kinase is required for blood vessel morphogenesis. *Circ Res.* 2003;92(3): 300–307.

71 Gebb SA, Shannon JM. Tissue interactions mediate early events in pulmonary vasculogenesis. *Dev Dyn.* 2000;217(2):159–169.

72 Galambos C, Ng YS, Ali A, et al. Defective pulmonary development in the absence of heparin-binding vascular endothelial growth factor isoforms. *Am J Respir Cell Mol Biol.* 2002;27(2): 194–203.

73 Compernolle V, Brusselmans K, Acker T, et al. Loss of HIF-2alpha and inhibition of VEGF impair fetal lung maturation, whereas treatment with VEGF prevents fatal respiratory distress in premature mice. *Nat Med.* 2002;8(7):702–710.

74 Jakkula M, Le Cras TD, Gebb S, et al. Inhibition of angiogenesis decreases alveolarization in the developing rat lung. *Am J Physiol Lung Cell Mol Physiol.* 2000;279(3):L600–L607.

75 Maniscalco WM, Watkins RH, Pryhuber GS, et al. Angiogenic factors and alveolar vasculature: development and alterations by injury in very premature baboons. *Am J Physiol Lung Cell Mol Physiol.* 2002;282(4):L811–L823.

76 Bhatt AJ, Pryhuber GS, Huyck H, et al. Disrupted pulmonary vasculature and decreased vascular endothelial growth factor, Flt-1, and TIE-2 in human infants dying with bronchopulmonary dysplasia. *Am J Respir Crit Care Med.* 2001;164(10 Pt1): 1971–1980.

77 Zhao YD, Courtman DW, Deng Y, et al. Rescue of monocrotaline-induced pulmonary arterial hypertension using bone marrow-derived endothelial-like progenitor cells: efficacy of combined cell and eNOS gene therapy in established disease. *Circ Res.* 2005;96(4):442–450.

78 Rossel M, Capecchi MR. Mice mutant for both Hoxa1 and Hoxb1 show extensive remodeling of the hindbrain and defects in craniofacial development. *Development.* 1999;126(22):5027–5040.

79 Shen J, Gudas LJ. Molecular cloning of a novel retinoic acid-responsive gene, HA1R-62, which is also up-regulated in Hoxa-1-overexpressing cells. *Cell Growth Differ.* 2000;11(1):11–17.

80 Behringer RR, Crotty DA, Tennyson VM, et al. Sequences 5′ of the homeobox of the Hox-1.4 gene direct tissue-specific expression of lacZ during mouse development. *Development.* 1993; 117(3):823–833.

81 Packer AI, Crotty DA, Elwell VA, Wolgemuth DJ. Expression of the murine Hoxa4 gene requires both autoregulation and a conserved retinoic acid response element. *Development.* 1998;125(11):1991–1998.

82 Packer AI, Mailutha KG, Ambrozewicz LA, Wolgemuth DJ. Regulation of the Hoxa4 and Hoxa5 genes in the embryonic mouse lung by retinoic acid and TGFbeta1: implications for lung development and patterning. *Dev Dyn.* 2000;217(1):62–74.
83 Dony C, Gruss P. Specific expression of the Hox 1.3 homeo box gene in murine embryonic structures originating from or induced by the mesoderm. *EMBO J.* 1987;6(10):2965–2975.
84 Bernacki SH, Nervi C, Vollberg TM, Jetten AM. Homeobox 1.3 expression: induction by retinoic acid in human bronchial fibroblasts. *Am J Respir Cell Mol Biol.* 1992;7(1):3–9.
85 Huang D, Chen SW, Langston AW, Gudas LJ. A conserved retinoic acid responsive element in the murine Hoxb-1 gene is required for expression in the developing gut. *Development.* 1998;125(16):3235–3246.
86 Bogue CW, Lou LJ, Vasavada H, et al. Expression of Hoxb genes in the developing mouse foregut and lung. *Am J Respir Cell Mol Biol.* 1996;15(2):163–171.
87 Graham A, Papalopulu N, Lorimer J, et al. Characterization of a murine homeo box gene, Hox-2.6, related to the Drosophila Deformed gene. *Genes Dev.* 1988;2(11):1424–1438.
88 Krumlauf R, Holland PW, McVey JH, Hogan BL. Developmental and spatial patterns of expression of the mouse homeobox gene, Hox 2.1. *Development.* 1987;99(4):603–617.
89 Volpe MV, Martin A, Vosatka RJ, et al. Hoxb-5 expression in the developing mouse lung suggests a role in branching morphogenesis and epithelial cell fate. *Histochem Cell Biol.* 1997; 108(6):495–504.
90 Sakiyama J, Yokouchi Y, Kuroiwa A. Coordinated expression of Hoxb genes and signaling molecules during development of the chick respiratory tract. *Dev Biol.* 2000;227(1):12–27.
91 Volpe MV, Vosatka RJ, Nielsen HC. Hoxb-5 control of early airway formation during branching morphogenesis in the developing mouse lung. *Biochem Biophys Acta.* 2000;1475(3): 337–345.
92 Chinoy MR, Nielsen HC, Volpe MV. Mesenchymal nuclear transcription factors in nitrofen-induced hypoplastic lung. *J Surg Res.* 2002;108(2):203–211.
93 Jones FS, Kioussi C, Copertino DW, et al. Barx2, a new homeobox gene of the Bar class, is expressed in neural and craniofacial structures during development. *Proc Natl Acad Sci USA.* 1997;94(6):2632–2637.
94 Gorski DH, LePage DF, Patel CV, et al. Molecular cloning of a diverged homeobox gene that is rapidly down-regulated during the G0/G1 transition in vascular smooth muscle cells. *Mol Cell Biol.* 1993;13(6):3722–3733.
95 Keng VW, Yagi H, Ikawa M, et al. Homeobox gene Hex is essential for onset of mouse embryonic liver development and differentiation of the monocyte lineage. *Biochem Biophys Res Commun.* 2000;276(3):1155–1161.
96 Bogue CW, Ganea GR, Sturm E, et al. Hex expression suggests a role in the development and function of organs derived from foregut endoderm. *Dev Dyn.* 2000;219(1):84–89.
97 Lints TJ, Hartley L, Parsons LM, Harvey RP. Mesoderm-specific expression of the divergent homeobox gene Hlx during murine embryogenesis. *Dev Dyn.* 1996;205(4):457–470.
98 Becker MB, Zulch A, Bosse A, Gruss P. Irx1 and Irx2 expression in early lung development. *Mech Dev.* 2001;106(1–2):155–158.
99 Bohinski RJ, Di Lauro R, Whitsett JA. The lung-specific surfactant protein B gene promoter is a target for thyroid transcription factor 1 and hepatocyte nuclear factor 3, indicating common factors for organ-specific gene expression along the foregut axis. *Mol Cell Biol.* 1994;14(9):5671–5681.
100 Bruno MD, Bohinski RJ, Huelsman KM, et al. Lung cell-specific expression of the murine surfactant protein A (SP-A) gene is mediated by interactions between the SP-A promoter and thyroid transcription factor-1. *J Biol Chem.* 1995;270(12):6531–6536.
101 Minoo P, Hamdan H, Bu D, Warburton D, et al. TTF-1 regulates lung epithelial morphogenesis. *Dev Biol.* 1995;172(2):694–698.
102 Zhang L, Whitsett JA, Stripp BR. Regulation of Clara cell secretory protein gene transcription by thyroid transcription factor-1. *Biochim Biophys Acta.* 1997;1350(3):359–367.
103 Minoo P, Su G, Drum H, et al. Defects in tracheoesophageal and lung morphogenesis in Nkx2.1$^{(-/-)}$ mouse embryos. *Dev Biol.* 1999;209(1):60–71.
104 Li C, Ling X, Yuan B, Minoo P. A novel DNA element mediates transcription of Nkx2.1 by Sp1 and Sp3 in pulmonary epithelial cells. *Biochim Biophys Acta.* 2000;1490(3):213–224.
105 Yuan B, Li C, Kimura S, et al. Inhibition of distal lung morphogenesis in Nkx2.1$^{(-/-)}$ embryos. *Dev Dyn.* 2000;217(2):180–190.
106 Iwatani N, Mabe H, Devriendt K, et al. Deletion of NKX2.1 gene encoding thyroid transcription factor-1 in two siblings with hypothyroidism and respiratory failure. *J Pediatr.* 2000;137(2): 272–276.
107 Naltner A, Ghaffari M, Whitsett JA, Yan C. Retinoic acid stimulation of the human surfactant protein B promoter is thyroid transcription factor 1 site-dependent. *J Biol Chem.* 2000;275(1):56–62.
108 Cassel TN, Berg T, Suske G, Nord M. Synergistic transactivation of the differentiation-dependent lung gene Clara cell secretory protein (secretoglobin 1a1) by the basic region leucine zipper factor CCAAT/enhancer-binding protein alpha and the homeodomain factor Nkx2.1/thyroid transcription factor-1. *J Biol Chem.* 2002;277(40):36970–36977.
109 Li C, Zhu NL, Tan RC, et al. Transforming growth factor-beta inhibits pulmonary surfactant protein B gene transcription through SMAD3 interactions with NKX2.1 and HNF-3 transcription factors. *J Biol Chem.* 2002;277(41):38399–38408.
110 Lin CR, Kioussi C, O'Connell S, et al. Pitx2 regulates lung asymmetry, cardiac positioning and pituitary and tooth morphogenesis. *Nature.* 1999;401(6750):279–282.
111 Martin JF, Bradley A, Olson EN. The paired-like homeo box gene MHox is required for early events of skeletogenesis in multiple lineages. *Genes Dev.* 1995;9(10):1237–1249.
112 Ohto H, Takizawa T, Saito T, et al. Tissue and developmental distribution of Six family gene products. *Int J Dev Biol.* 1998; 42(2):141–148.

Metaphors

24

Shall I Compare the Endothelium to a Summer's Day?

The Role of Metaphor in Communicating Science

Steven Moskowitz* and William C. Aird†

*Advanced Medical Graphics, Boston, Massachusetts; †Beth Israel Deaconess Medical Center, Harvard Medical School, Boston, Massachusetts

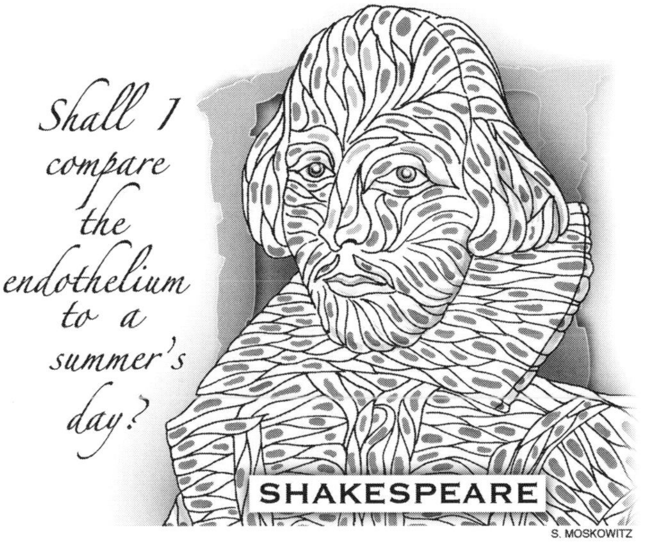

Figure 24.1. The metaphorical twist. Our Shakespeare attempts to portray endothelium with an elegant phrase, as the endothelium attempts to portray Shakespeare with individual cells.

When a complex biological system such as the endothelium is under scientific scrutiny, rigorously objective tools and techniques are used to make measurements, collect data, and analyze results. While adhering to the hallmark objectivity of science, a scientist may find it difficult to imagine a welcome role for something so fraught with subjectivity and prone to misinterpretation as metaphor. To refer to an arterial "wall" in the context of vascular medicine seems straightforward enough, because the term refers to a barrier or demarcation of the vascular structure. But we might ask: Does it hinder or further our understanding of blood vessels by likening endothelial and perivascular cells to the bricks and extracellular matrix to the mortar that make up an actual wall? At best, the term incompletely conveys the endothelial cell's functions. Might it be better to use the metaphor of a "gate" or "gatekeeper" instead of "wall" to capture the role of the endothelial cell (EC) in mediating permeability? The distinction between these two metaphorical descriptions is of paramount importance to our understanding: A permeable "wall" might be considered breached; whereas a permeable "gate" is functioning properly.

The purpose of this chapter is to recognize the role of metaphor, both past and present, in the field of endothelial biomedicine. We discuss how metaphors serve scientists on a regular basis; we consider some reasons why seemingly subjective descriptions can be a basis for scientific understanding; and we analyze the merits of both linguistic and visual expression in communicating science.

WHAT IS METAPHOR?

A metaphor is a figure of speech – typically a few words (although sometimes conveyed as a single word or as many hundreds of words) – that compares two different things (abstract or concrete) for the purpose of indicating a resemblance between them.[1] Metaphors also can be communicated through other modes of expression, such as illustration or photography. Key to the definition is that its literal meaning is bypassed as it takes on a new figurative meaning. For example, if we think of blood vessel ECs, perivascular cells, and

1 For the purposes of this chapter, we use *metaphor* as an umbrella term, even though metaphor is one of many types of expression (including simile and analogy) that draws comparisons by using words (or pictures) other than in their literal sense. Simile is generally considered a subcategory of metaphor, and metaphor itself is considered by some a subcategory of analogy. A simile overtly indicates its figurative sense to a reader by using the words *like* or *as* to compare two things. The sentence "Endothelial cells stack up *like* bricks in a wall" is a simile, although its underlying concept is metaphorical. An analogy often compares things known to be similar and builds on these known similarities as a form of reasoning.

extracellular matrix as forming a wall, the word *wall* loses its common literal meaning (i.e., a solid, upright structure) and takes on a figurative meaning derived from our common experience with walls. So, in this instance, we draw less from our understanding of the physical qualities of a wall than from the conceptual qualities; bricks and mortar become irrelevant, while the notion of separation ("barrier") or confinement ("container") is ultimately conveyed.

Such fundamental conceptual qualities commonly are conveyed in metaphors and, as Theodore Brown points out, they "arise from our most ubiquitous physical experiences, such as verticality, space, and vision" (1). Metaphors based on human sensory experience may be conveyed by language, but they are not merely a by-product of language or word play. Rather, they are tied deeply to the conceptualization of "one mental domain in terms of another" (2).

In seeking answers to how we compare and utilize concepts in a metaphor, ongoing studies in the fields of psychology, philosophy, anthropology, and linguistics have recognized some of its key features. Each metaphor consists of two domains: a source domain, and a target domain (Figure 24.2A). The *source* refers to the concept (or thing) *from* which we derive a

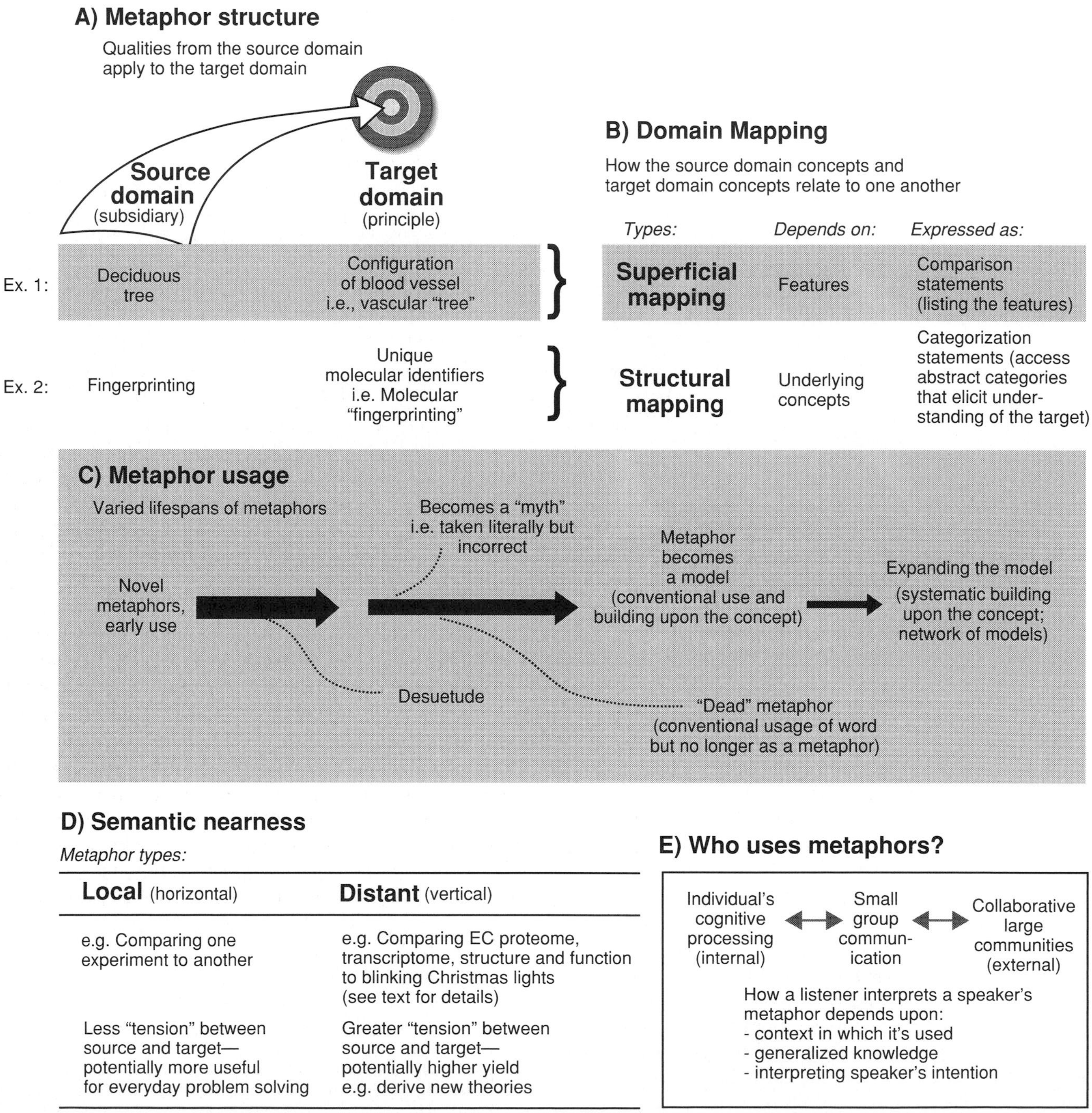

Figure 24.2. A summary of key concepts and terms to describe and analyze metaphors.

known quality; the *target* is the concept (or thing) *to* which the quality applies. For example, in the phrase "The endothelium is a gatekeeper," *endothelium* is the target domain; *gatekeeper* is the source domain. The source domain and target domain each embody properties that are being compared, some of which are emphasized by this comparison, and others that are suppressed.

Comparing properties of a given source domain and its target domain is called *mapping*. The source and target may correspond in a predictable way, whereby their superficial features are compared and matched (Figure 24.2B). As an example of superficial mapping, when vasculature is called a "tree," the salient features being compared are branching and tapering. Alternatively, an understanding may be more appropriately based on deeper, underlying concepts (termed *structural mapping*). An example is the use of the term blood vessel "wall," to describe a barrier or container. A more extreme example of a metaphor that invites structural mapping would be to describe the endothelium as an ant colony – the underlying concept being that the component parts – ECs (ants) – go about their business oblivious to the needs of the organ or system at large (colony), following local rules that generate an apparently complex global behavior.

Source and target can be further analyzed in terms of their semantic nearness (Figures 24.2D and 24.3). If the source and target are from similar fields, the metaphor is termed *local*, and the comparison deemed *horizontal*. For example, "ultramicroscopic circulation" is used as a metaphor borrowed from within its own field. Originally the term *circulation* was used to describe forward, circular flow of blood through the vasculature. In the case of ultramicroscopic circulation, the word *circulation* also is being used to convey movement of blood substances, but only in the metaphorical sense because transendothelial transport is neither forward nor circular. Here it appears in its original context:

> . . . visible flow of blood through the capillaries is, in fact, very small in comparison with the *invisible* flow of water and dissolved materials back and forth through the capillary walls. . . . this *invisible* component of the circulation takes place at a rate which is many times greater than that of the entire cardiac output. Indeed, it is by means of this **"ultramicroscopic circulation"** through the capillary wall that the circulatory system as a whole fulfills its ultimate function in the transport of materials to and from the cells of the body. (3, Emphasis added.)

When metaphors consist of a source domain *distant* to the target domain, they are said to have a greater *tension* between them.[2] The underlying concept may require explanation in order to be understood by the reader, but the payoff can be great. The use of the distant metaphors "fingerprint" and "zip code" are shown here in context:

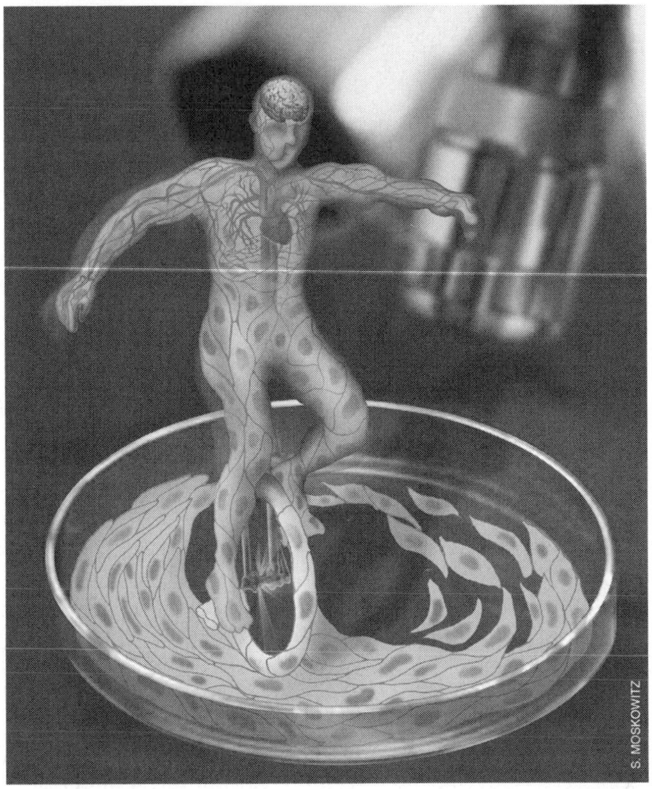

Figure 24.3. Semantic nearness. A unicyclist is used to represent the modulating effect of endothelium based on the heterogeneity of individual cells and their orchestrated effects at a systemic level. The figure conveys a reductionist–holist spectrum from single cells (in cell culture) to blood vessels to whole organism. Moreover, the endothelium, like the unicyclist, appears from a distance to be inactive or quiescent, but on closer inspection is constantly acting and reacting, making fine adjustments to its environment. Because the metaphor is presented visually, the context of the metaphor– its laboratory setting – is made apparent in an attempt to add "local" attributes to a more "distant" metaphor. As reported by Dunbar and Blanchette, scientists use those analogies more frequently that are derived from their own or similar fields.

> The importance of specific morphological changes elicited by FGF-2 [fibroblast growth factor -2] is underscored by our recent observation that major stimulators and inhibitors of angiogenesis induce distinct **"fingerprint"** patterns of vascular morphology when applied to the quail CAM [chorioallantoic membrane]. (4, Emphasis added.)

And,

> Vascular targeting exploits molecular differences that exist in blood vessels of different organs and tissues, as well as differences between normal blood vessels and angiogenic or remodeled blood vessels. Differences in

2 Tension should not be confused with the emotional response elicited by a metaphor. For example, although the metaphors of endothelium as a gatekeeper and as an ant colony are both distant in the strict sense of the word (i.e., they draw on sources outside the field of biology), they are certain to provoke different emotional responses.

> plasma-membrane proteins ("vascular zip codes") can be used to target therapeutic or imaging agents directly to a particular organ or tumor. (5, Emphasis added.)

Additional examples from published papers that utilize terms distant from the field of biomedicine abound. Because the endothelium is invisible to the human eye, it is common to apply meanings derived from the macroscopic world to the microscopic one. For example, in 1983, Thorgeirsson (6) referred to a series of metaphors (most of which had been previously introduced by other investigators):

Architecture: "The general architecture of the endothelial lining . . . . "
Constellation: ". . . in addition to the usual constellation of intracellular organelles, most endothelial cells contain a unique structure, the Weibel-Palade body . . ."
Flask, teardrop, mouth: "The flask or teardrop shaped caveola connecting to the cell surface through a mouth or soma . . ."
Hairy: "The so-called 'coated' or 'hairy vesicles' . . ."
Channels, thoroughfare: ". . . such chains might form channels through the cytoplasm of the endothelial cells and thus provide for a continuous, water-filled thoroughfare for molecular transport."
Foot: ". . . the astrocytic foot processes . . ."
Sieve, plate: ". . . were arranged in characteristic patches, termed sieve plates."

And Thorgeirsson's following passage even has a hint of a merchant marine report:

> . . . molecular **transport routes** (pores) in both large and **small vessels, migration routes** for blood cells . . . (6, Emphasis added.)

Other notable examples in the literature include:

Taps, sink, carrier: "Does this water merely follow protein as it leaks form open 'taps' in the cortex, with both protein and water flowing from the taps into the white matter 'sink' below . . . substances transported by facilitated diffusion are thought to be moved by some 'carrier'" (7).

Cobblestone: "However, tumors followed for 16, 21, 34, and 44 days retained their ovoid shape and gray cobblestoned surface . . . " (8).

Starve: "'Tumors unable to induce angiogenesis remain dormant at a microscopic size: 1–2 mm^3 in situ,' he wrote in 1971 in the *New England Journal of Medicine* (Folkman, 1971), so by attacking cancer's lifeline, it might be possible to starve tumors into regression" (9).

Previous studies have provided insight into the use of metaphor by scientists. Dunbar (10,11) reported a total of 99 analogies used by four groups of scientists during their weekly lab meetings (four meetings per group). They classified analogies into four groups, depending on how scientists used them to achieve their goals. Three of the uses (formulating a hypothesis, designing an experiment, and working out problems in an existing experiment) used analogies that were *local/horizontal* (i.e., comparing work to similar experiments or similar organisms). They write: "Rather than new theories and hypotheses being generated by making analogies to very different domains, new theories are generated by making analogies to related domains" (11). Scientists frequently used science as sources for their analogies.

The fourth use by scientists was "to explain a result." For that, the scientists sometimes used more *distant/vertical* analogies, derived from sources outside their field. If the scientists obtained a series of unexpected findings in a given organism, they would draw on other organisms for the source of their analogies. The analogies became more *distant/vertical*, and the scientists tended to make broader generalizations, linking their results to similar results obtained by others' experiments. They used *distributed reasoning*, which is noted to work better if the scientists in the group have varied backgrounds (to provide varied perspectives) (Figure 24.2E).

The lifespan of metaphors varies considerably: Some metaphors are used in an ad hoc fashion, never to be used again, whereas others provide such elucidation that they are adopted and become widely used (Figure 24.2C). The word *wall* has been used in the sense of separation (or confinement) for so long that, for most people, it is no longer considered a metaphor, but a *dead metaphor* (also called *conventional metaphor*) – one that has allowed the literal interpretation of the word to expand to include the former metaphorical meaning. This is the case for innumerable words and expressions in language. In her book, Mary B. Hesse (12) offers this thought: "Rationality consists [of] the continuous adaptation of our language to our continually expanding world, and metaphor is one of the chief means by which this is accomplished."

Bowdle and Gentner (13) describe this evolution of meaning as the *career of a metaphor*, highlighting the fact that originally a metaphor may introduce a novel concept, but as it is adopted by a culture, it becomes a conventional part of the language with a formal definition less likely to be misinterpreted. An example of this is seen with the word *virus*, which has a literal interpretation expanded from its former metaphorical meaning. The biological term *virus* was introduced in the 1700s. In the 1970s, it was first used by fiction writers in conjunction with computers and, by the early 1980s, "computer virus" was academically recognized and defined as a program that modifies ("infects") other programs "to include a possibly evolved version of itself" (14).

In endothelial biomedicine, the term *vascular labeling* was introduced, defined, and quickly adopted from its metaphorical usage. The properties of vascular labeling related to, and expanded on, existing meanings of the word *labeling*.

> Our experiments were based on a method to which we refer as "**vascular labeling**," and which permits one to identify leading vessels by means of visible accumulation of foreign particles within their walls. (15, Emphasis added.)

In some cases, the metaphor is ultimately judged to be a myth (e.g., the endothelium as an inert layer of nucleated cellophane). Other metaphors may stand the test of time by providing a completely different perspective on their targets of interest and subsequently evolve into accepted scientific models (e.g., William Harvey's metaphorical use of the word *circle* to describe the vascular system).

According to Lakoff and Johnson, our understanding of the world is derived from metaphors. They assert a pervasiveness of metaphor in our lives and refer to *conceptual metaphors*, which structure the way we think, speak, and act (16). *Container metaphors*, for example, are the underlying concept for all things that have an inside and an outside. Some containers, such as jars and boxes, literally are containers; other things have insides and outsides that have emerged metaphorically, such as blood vessels and their walls. *Orientational metaphors* are based on the way we experience the physical world. If I were to say, "I'm feeling 'up' now that I've received the NIH grant," I have expressed an abstract emotion based on concepts derived ultimately from the physical environment.

Linguistic and philosophical considerations of how words have meaning, and how meaning relates to the world, impact any discussion of metaphor. The issues surrounding our natural ability to learn and use language are intertwined with the methods used for analyzing the issues. Is it the nature of the world that structures our language, or do we apply language (concepts and words) to structure it? How is it that we can derive meaning from language? How are our thoughts related to the words we use? The work of a number of philosophers exemplifies these issues: Gottlob Frege (1848–1925) put forth the *context principle*, which states that the meaning of words can be derived only within the context of an entire expression. Ferdinand de Saussure (1857–1913) stated that language depends on an entire community – not on individuals. Lev Vygotsky (1896–1934) asserted that thought and language are a "unity" and therefore dependent on one another; in his view, speech is not a secondary result of thought. Alfred Tarski (1902–1983) believed that "truth" is something that can be said of sentences because of properties of language, and that it can indeed correspond to the things in the world. Noam Chomsky (1928–) professed that people are born with an innate ability to understand the *universal grammar* of language, which itself possesses a "universal structure" to be understood. Ludwig Wittgenstein (1889–1951) argued that meaning is not simply derived from words in the context of a sentence but also depends on what the expressions are being used. Meaning, therefore, includes the activities of people and the circumstances in which language is being used.

Wittgenstein also proposed that different uses of one word share a *family resemblance*, similar to the way siblings and parents share a resemblance, or how baseball and checkers can both be considered games without any well-delineated similarity. He asserted that no boundaries exist for words, so a radish can be bitter, just as a memory can. A lemon can be sour; my car can be a lemon, and my experience can be sour due to my lemon of a car. Identical words can be used to convey different meanings depending on context, a feature of language that denies meaning is tied to the world in a set way.

We create analogy and metaphor with ease, but their *affordances* – what their features provide for us – also vary depending on use. Dawkins wrote: "The trick is to strike a balance between too much indiscriminate analogizing on the one hand, and a sterile blindness to fruitful analogies on the other"(17). Ancient Greeks used analogical descriptions in a quest to define all that couldn't be seen – gods, soul, elemental workings – and in describing nature and human behavior. Likewise, according to Gombrich, the visual arts developed as a means to produce "plausible narration of sacred events" (18). Anaximenes' and Heraclitus' early use of analogy to describe cause-and-effect changes opened the way for scientific explanation. In an effort to distinguish knowledge from opinion, Plato warned that the wrong message could be derived from the use of analogy if its use was not a part of carefully constructed discourse. Although the use of unbiased reason is necessary to gain knowledge of the world, Plato's dialogues demonstrate that analogy and metaphor are part of the art of discovery.

METAPHOR IN ENDOTHELIAL BIOMEDICINE

Modern vascular science began with a splash of metaphor by William Harvey. Harvey's scientific revolution hinged upon a single word: *circle*. This pivotal word is part of a metaphor in his passage describing the course of blood flow:

> ... the blood should somehow find its way from the arteries into the veins, and so return to the right side of the heart, I began to think whether there might not be a Motion, As It Were, In A **Circle** (19). (Emphasis added.)

Harvey used analogy to compare blood circulation with the hydrological cycle recognized by Aristotle:

> Now, this I afterwards found to be true; and I finally saw that the blood, forced by the action of the left ventricle into the arteries, was distributed ... This motion we may be allowed to call circular, in the same way as Aristotle says that the air and the rain emulate the circular motion of the superior bodies; for the moist earth, warmed by the sun, evaporates; the vapours drawn upwards are condensed, and descending in the form of rain, moisten the earth again (19).

And here, multiple use of metaphor and analogy in an effusive letter from Harvey to Charles I, King of Great Britain at the time (1628):

> The heart of animals is the foundation of their life, the sovereign of everything within them, the sun of their microcosm, that upon which all growth depends, from which all power proceeds. The King, in like manner, is the foundation of his kingdom, the sun of the world around him, the heart of the republic, the fountain whence all power, all grace doth flow. What I have here written of the motions of the heart I am the more emboldened to present to your Majesty, according to the custom of the present age, because almost all things human are done after human examples, and many things in a King are after the pattern of the heart (19).

Here "heart" metaphorically is sovereign (like the King) and the sun. Likewise, the "King" metaphorically is sun, heart, and fountain. Harvey's lavish use of metaphor demonstrates several of its key functions: (a) to interpret observations, (b) to produce knowledge and further understanding, and (c) to convey this understanding to others.

Drawing on the brain's natural associative tendencies, we use metaphor to describe something novel in terms of something familiar, and for that reason their use is common among those at the leading edge of a scientific field. Arthur I. Miller writes: "Metaphors are an essential part of scientific creativity because they provide a means for seeking literal descriptions of the world . . . metaphors underscore the continuity of theory change . . . " (20). Some of the most common terms for vascular structures were derived from superficial similarities to the easily visible natural world. Robert Hooke first used the word "cell" (1665), because the structures of cork slices he saw resembled the familiar cells in which monks lived. Arteries were named for their capacity to carry air ([Gr.] *aer*) and spirits, not blood. The term *capillary* is derived from the word *hair* (capillary: 1656, from L. *capillaris* "of hair," from *capillus* "hair" [of the head]).

According to Thorgeirsson and Roberston, the word *endothelium* was:

> . . . introduced in 1865 by the German anatomist His, combing the Greek words "end," meaning within, and "thele," meaning nipple. Its literal meaning is thus: "within the nipple." It was proposed as a complementary anatomic term to "epithelium" ("over the nipple") first used by Ruysch around 1700 to describe the cell layer covering the tongue and other areas with nipple-like papillae. Henle expanded the domain of the latter term to include all specialized cell surface layers, including those linking the respiratory, gastrointestinal, and urinary tracts (21).

In addition to the plethora of metaphors already cited (architecture, constellation, flask, teardrop, mouth, hairy, channels, thoroughfare, foot, sieve, plate, taps, sink, carrier, cobblestone, fingerprint, zip code, and so on), a review of early papers reveals common uses of analogical comparison, which are easy to spot due to the direct reference to it:

> **Similar, striking parallel**: "Some of our data is strikingly similar to that of Kontos who employs acute hypertension to injure the endothelium. The hypertension is either initiated indirectly by . . . We see a striking parallel between these data and those of our light/dye model of injury" (7).

In the first few paragraphs of an abstract by Simionescu and colleagues, the authors employed no less than five analogical comparisons, using the following phrases:

> " . . . **comparable to**, though less elaborate than"
> " . . . behave **like** 'single unit' particles"
> " . . . junctional elements are **similar to**"
> " . . . appears to be a stretched-out **version of**"
> "The other type is **reminiscent of** . . . "(22).

In a classic paper by Folkman and colleagues:

> New capillaries are elicited and the tumor then enters a phase in which nutrients arrive by perfusion. It is possible that TAF is responsible for this final stage. It is tempting to suggest that tumor growth might be arrested at a very small size if the angiogenesis activity of this factor could be blocked. **This would be analogous** to the cessation of growth of bacterial colonies when their size exceeds the diffusion of nutrients. (23, Emphasis added.)

ORGANIZING INFORMATION

The concept of analogy may be predicated on the philosophical problem of unity and multiplicity in the world. "Are things so different that they can never be classified? Or so unified that they can never be distinguished?" (24). Between these extreme notions exists the similarity and dissimilarity from which we make comparisons. Attempts to relate large, universal concepts in the "big picture" to the particulars all around us can be a lifelong philosophical quest. Although we may be short-sighted in trying the get the big picture, we have a natural ability to see relationships in the physical world. The Gestalt psychologists point out our propensity toward perceiving groupings of shapes and things (rather than isolated things – at least in certain circumstances); they note a predilection for wholeness (rather than parts) and continuity (rather than interrupted shapes). When looking at a dashed line, for example, we first perceive the cohesive line, rather than the isolated line segments and intervening spaces. Only by scrutinizing it do we note its component parts. Grasping wholes aids pattern recognition, which is an important skill in biomedicine;

as best demonstrated by the ability of an experienced cardiologist to diagnose acute myocardial infarction based on a patient's electrocardiogram, without having to attend to the minute details of the tracing.

The fact that we recognize parts within a whole, or multiplicity of wholes, relates to an ability to understand the concept of quantity. Research on preverbal infants and on people in primitive societies without arithmetic systems demonstrates that such mathematical abilities are innate or occur at a very early age.

Graphs, which are commonly used to visually convey quantitative information, rely on this innate concept of quantity. A rectangular bar (on a graph) that appears tall means "greater" compared to the shorter bar, which means "less." The analogical component – the correlation of bar height to the data it represents – takes advantage of our effortless understanding of quantity. Cortical mapping of the shapes in our visual field may provide some explanation for our ability to correlate size, but even without images, comparative forms of words also reveal our understanding of quantity. "Big, bigger, biggest" describes the relative degree of bigness. The series can be extended to include new words (e.g., slightly bigger, somewhat smaller).

If one were to use the subjective words *happy, happier*, or *happiest* – their qualities can be compared and ordered in the same way. Although no standard units for happiness exist, we can still make a qualitative distinction between degrees of the emotion. Qualitative distinctions – the ability to tell things apart and describe them without measurement – are nonetheless assessed in terms of the *degree*. The degree of a quality is tantamount to an amount of it so, in a sense, we're still measuring it in terms of similarity and dissimilarity to something else. For example, when we take a medical history, we might ask the patient: "On a scale of one to ten, how would you rate the pain you are experiencing?"

In science, and in other endeavors, technical language is useful only to the extent that it is familiar to us. For example, precise geographical coordinates of latitude and longitude are irrelevant in our daily routine. The complexity of technically accurate descriptions can conceal the obvious: "I observe one vertical member, and one horizontal, fastened perpendicularly to it at the base. Above it are two cantilevered members of the same length fastened perpendicularly. One is fastened at the top of the vertical member, and the other is halfway between the upper and lower ones."

Or I could simply say: "It looks like a capital E."

In fact, a similar description was used by Gudmundur Thorgeirsson:

> In the early stages of realignment C and L shaped nuclear forms were seen giving the impression of nuclei being caught in the act of reorienting themselves under the new conditions of flow (6).

How do analogy and metaphor fare against more conventionally objective descriptions? In lieu of epistemological formality, we turn to another metaphor to consider the informational content of a description. Paul Laffoley's "epistemic ladder" considers gradations of information arranged by degree (25). Rather than delineate between "that which we know to be true" (fact) from "that which we know to be false" (fiction), the epistemic ladder describes successive steps in a progression from partial to complete knowledge. The metaphor of a ladder imparts a sense of order without putting limits on the content. In Laffoley's model, near the base is knowledge that is mostly incomplete, such as that conveyed by arbitrary representation. As an example, we might use a highly simplified (cartoon) silhouette, such as one found on a deer-crossing sign to represent a deer. The next step advances to a closer representation: A deer's track in the snow, for example, would indicate the presence of a deer at that place and time. "Contact" has been made between knower and known in this step. Extension proceeds from contact, in the form of an *icon*, such as a photograph of the animal or a specimen preserved by taxidermy. Evidence gleaned from indirect observation, common in modern science, would be tantamount to the found deer tracks, whereas analogy and metaphor are lower on the epistemic ladder, akin to the image on a sign.

In endothelial biomedicine, the ladder is "bottom heavy." That is, we rely on metaphor and analogy (and to some extent, on indirect measures of structure and function) to represent the endothelium in vivo. Although tens of thousands of studies exist in the literature pertaining to cultured ECs, many of the data obtained from these investigations can be assigned to the lowest rungs of Laffoley's ladder, owing to their poor correlation with in vivo biology.

Wurman has listed additional ways in which to organize information logically (26). The most obvious ones appeal fundamentally to our innate sense of time and space, and to our innate abilities with quantity and language. Depending on what information needs to be conveyed, we may choose to present it chronologically, alphabetically, in numerical sequence, as a hierarchy (or continuum), by category, or by location (spatially), among others. Organizing information spatially utilizes our visual system, which encompasses innate perceptual and cognitive abilities. The visual system is able to identify objects and their positions in space and, despite myriad viewing conditions, objects can be identified in a fraction of a second, from different viewing angles, and even if partially occluded or attached in an unfamiliar way to another object.

Alternative ways to organize content can reveal or highlight relations between particular information. Some anatomy texts are organized by region of the body, whereas others are organized by system or by tissue similarity. In the case of this book, sections are divided according to time; the first section deals with the past (history of medicine and evolution), the middle section with the present, and the last section with the future. Chapters also are organized around the metaphor of the EC as an input–output device; some chapters refer to input, others to coupling (i.e., intracellular signaling), and yet others to output (i.e., cell phenotype).

VISUAL METAPHOR

The visual system is at our disposal for tasks other than direct perception. Visual mental imagery has much in common with visual perception – that is, what we picture in our minds "without looking" shares neural framework and processes with what we see with our eyes. Kosslyn found that information from mental images corresponds to the information perceived in similar real-world scenes: We can "inspect" mental images in a way similar to that in which we inspect real-world scenes, an ability that can aid in interpreting metaphors (27). For example, the endothelium-as-nucleated-cellophane metaphor conjures up a mental image of extraordinary thinness and robustness (but the metaphor is misleading when inferred to represent a nonpermeable barrier and may even distract as we actually visualize a shiny, cellophane surface). Compared with schematic representations, mental images can be far more powerful and instructive (just as reading a book often is considered to be more fulfilling than watching a movie based on that book); however, there is still debate on how many new properties can be discovered simply from the mental manipulation of imagery.

Analogy and metaphor serve to intensify and personalize the empirical and make good use of our capacity to visualize and conceptualize. Scientists, in turn, occasionally devise thought experiments that rely on such ability. With relative ease, we picture a scene, perhaps calling on exemplars or archetypes, and, with basic language skills, come to an understanding of the problem at hand. Both logical and analogical components are part of the thought experiments, which indirectly enlightens the reader to science, as art, literature, and philosophy do, without relying on data or experience. Theodore L. Brown (1) relates metaphor to "implicit learning," which can occur outside awareness and in contrast to explicit, overt learning. We make decisions and act based on unconscious mental processes. Brown quotes David Gooding: "Thought experiments work because they are distillations of practice, including material world experience" (1).

Consider the following to describe EC heterogeneity in its vast array of (about 60 trillion) cells. It has a touch of metaphor, thought experiment, and mental imagery:

> For purposes of discussion, let's formulate a hypothesis that endothelial cells are heterogeneous, and that phenotypic differences between endothelial cells are governed by spatial and temporal differences in the local microenvironment – just as individual differences in human physical and behavioral traits are shaped by different personal experiences. We could design an experiment in which we map the phenotypes of endothelial cells from different sites of mouse vasculature by using immunohistochemistry to monitor different proteins – just as we might determine protein expression in two different cell types in culture – and then determine whether these patterns change when the mice are administered endotoxin. The study would show that at any given moment, different vascular beds express different proteins. Moreover, at any one site, endotoxemia would result in a change in the expression pattern. We would conclude that indeed the endothelium displays marked heterogeneity and that endotoxin results in phenotypic alteration. In explaining the results, we would argue that endothelial cells are highly sensitive and responsive to extracellular signals. Like a barcode reader, the endothelium is constantly taking inventory of its surrounding environment. Similar to a chameleon, the endothelium is molding itself to the requirements of the underlying tissue. If we could somehow color-code the phenotype of the endothelium – for example, by assigning a shade of color based on that cell's proteome, transcriptome and function – then at any given moment, the endothelium would display a remarkably rich color palette. If we were then to "roll the film" and observe the color palette in real time the colors would fade in and out or even blink on and off like lights on a Christmas tree (insofar as the microenvironment changes from one moment to the next).

As the above example illustrates, we have transitioned from local to distant analogies and metaphors to explain endothelial cell heterogeneity in terms that most biomedical practitioners will understand.

The ubiquity of images and graphical representations reveals the confidence we have in constructing and interpreting them. Organizing information by location, referred to as mapping, is the correlation of size or position of something to its representation. We not only analogize three-dimensional (3D) space into two-dimensional (2D) representations, but the representation can be a metaphor for something else, if need be. Willats distinguishes between "drawing" and "denoting"; denoting assigns meaning to the points, lines, and shapes that comprise a picture, whereas drawing refers to the position (and size) of elements. Our understanding of the schematic diagrams used to represent biochemical pathways depends on such denotation. Arbitrary shapes, connected by arrows, analogize the chain of cause and effect in a cascade of biochemical events, as exemplified by many of the figures in this volume. The words, shapes, and color in diagrams provide us with a conceptual understanding of cellular processes not directly observable. We infer an objective resemblance between process and representation but, in fact, the assigned elements and their spatial relation are analogical at best and therefore somewhat subjective. Ephemeral processes can be slowed down and stabilized in diagrams, just as lengthy processes can be sped up and summarized. A potentially wide range of interpretations for a diagram is narrowed by the viewer's ability to discern, the author's ability to explain, and the designer's ability to utilize principles of design.

On the printed page, the spatial arrangement of information must be utilized to convey time. One of the most prevalent metaphors in this regard is seen in the face of a clock, which

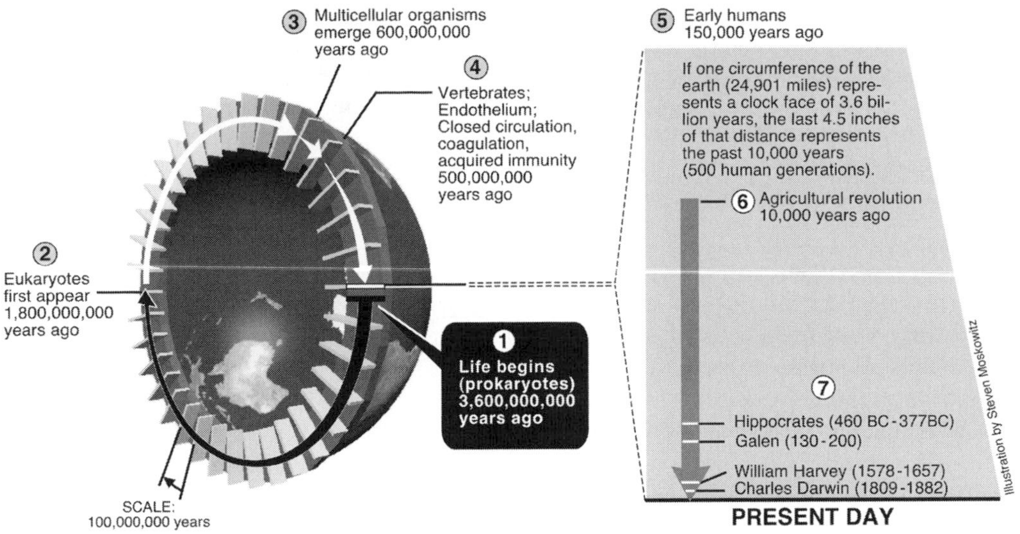

Figure 24.4. Earth as a timeline. The metaphor of the clock is changed in scale to dramatize the relative incipience of the human species. The human species is 150,000 years old, whereas the transition from the Stone Age (Paleolithic) to the agricultural revolution (Neolithic) occurred a mere 10,000 years ago. If one circumference of Earth (24,901 miles) represents a clock face of 3.6 billion years, the last 4.5 inches of that distance represents the past 10,000 years. The history of medicine (particularly as it relates to the cardiovascular system) is a mere blink of the eye in evolutionary terms.

has a unit of time (an hour) measured by a unit of space (the distance around the clock) (Figure 24.4).

We are just as easily able to apply relative quantities to time as we are to space, and we readily understand graphical representation of temporal events such as time lines, sequences of individual movie frames, and other graphical cues for time, such as directionally blurred edges to indicate motion. Indeed, an important goal in endothelial biomedicine is to develop improved tools for representing the complex temporal dynamics of this cell layer.

An artist's rendering attempts to analogize percepts and concepts to visually express them to an observer. Linear perspective, overlap, shadow and light, and tints of color are among the artistic devices used to provide visual cues to an observer when seeing spatial relationships on a flat page. Radiographs, radio-labeling, and innumerable techniques used to formulate scientific understanding all depend on the ultimate placement of abstracted data into the human realm for observation. The invisible must be made visible through photographic or illustrative means.

An understated and unadorned artistic style is concordant with the objectivity sought by science. If a color can edify, or an outline correctly delineate, or a pattern distinguish one thing from another without being superfluous – then it's deemed acceptable. But, given the numerous options for representation, an analogical approach may be most logical. For example, consider the rendering of blood vessels. The arterial wall is not red, nor is the venous wall blue. But they are often illustrated that way for a resemblance to the oxygenated or deoxygenated blood they carry, respectively. In other words, the color metaphor is relevant for the state of oxygenation of the blood, not the physical appearance of arterial or venous tissue. To be useful to an observer, a salient graphic distinction between arteries and veins has been made based on analogy.

The precise color, texture, or line thickness used in a diagram can be selected by any number of practical or aesthetic criteria. When Copernicus postulated an orbiting earth, a concept that refuted Ptolemaic geometry, religious doctrine, and common sense, any model that placed the sun (instead of the Earth) at the center of the solar system would have sufficed. Galileo preferred circular orbits for aesthetic reasons, but Kepler, more accurately, used ellipses. At the time, one's own eyes could not be trusted – empirical accounts of the sun moving across the sky were deemed illusory – but the analogical motions of little mechanical orreries (i.e., mechanical models of the solar system) were deemed truth.

By way of technology our vision is extended through the use of microscopes and telescopes, and our computational ability is augmented with computers. Instrumentation allows for degrees of precision beyond the normal threshold of experience. Innate abilities are built on in a structured way to gain understanding of the world. To this extent, scientists become pilots of a windowless aircraft, only able to directly observe the instrument panel. With in silico research, biologists delve further into analogical and metaphorical means of observation, another step removed from living tissue.

After completing all the machine processing, we are allowed periodic glimpses at the resulting visual forms translated from the data. Our innate mathematical abilities have limited prowess in statistical computation, but when the results are graphically displayed, we can more easily digest or "explore" the data – using the same visual and cognitive skills that have evolved in response to stimuli posed by our environment, which allowed us to avoid predators, hunt for food, and find mates. We compare and contrast the appearance of points and lines on a line graph, for example, and easily find trends and patterns among the shapes that may reveal trends and patterns in data. We find coherence between the big picture and the minutiae, and distinguish similarities and differences among these analogical forms.

As scientific descriptions lengthen, and the relationships become more complex, it becomes advantageous for a reader to view a diagram, table, or other graphic to address the ambiguity inherent in linguistic description. Spatial arrangement of the type found in tables and other graphic elements found in diagrams might present additional forms of ambiguity that must be avoided as well. By taking into consideration the noted principles of Gestalt psychology on human visual perception, graphic representation can communicate information more reliably. These principles include: *Proximity*, which refers to the fact that we perceive a relation between elements that are close to one another. If two curves occur on a graph, we'd expect the descriptive labels to be in proximity to their corresponding curves. If that isn't possible, (e.g., due to space constraints) we may utilize the principle of *Connectedness* and draw a line between the curve and its corresponding label, to visually connect them. If that doesn't work, we might employ the principle of *Similarity*, which finds greater relatedness between similar elements, and make the curve and its corresponding label the same color. *Closure, Continuity*, and *Common Fate* are described in Figure 24.5.

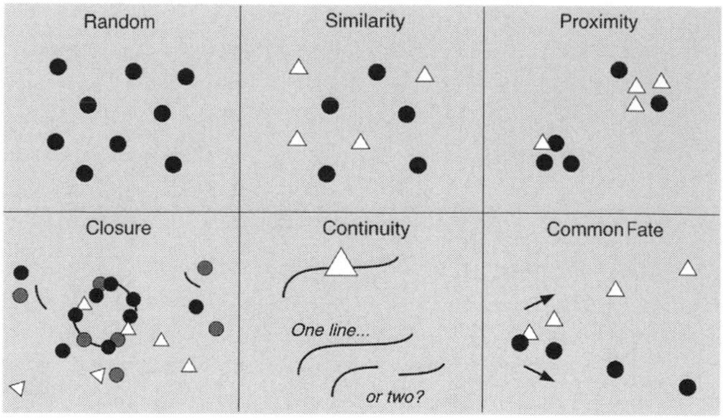

Figure 24.5. Gestalt principles of perceptual organization. Visual perception allows us to interpret the shapes on a page in a fashion similar to how we interpret our physical environment. We group objects according to some of the "rules" that Gestalt psychologists have identified, which have evolved to allow us to make sense of visual stimuli. This figure appears in a gray box, which identifies it as separate from the background of the white page, thus demonstrating figure–ground relationship. We are able to distinguish figures from backgrounds in usual circumstances, although camouflage, for example, may be deliberately designed to undermine this ability. *In the top row*, random dots appear in the first panel; they become visually separated into two, easily recognizable groups in the second panel by using two distinct shapes (Similarity). The principle of grouping by proximity is demonstrated in the third panel (as well as in the overall layout): Labels are deliberately placed closest to the designated part of the diagram. (The proximity of individual letters within a printed word allows us to consider it to be a word). *In the bottom row*, the principle of Closure readily demonstrates that the elements linked together are interpreted as one group, despite the proximity or similarity to other nearby elements. Continuity allows us to interpret a partially occluded line as a single line rather than two segments. The last panel, Common Fate, refers to the fact that objects moving in the same direction are interpreted as being grouped together. This static view is reminiscent of line graphs in which the trends are made obvious by their "trajectory." It's this trajectory that allows the data points to be easily identified as part of the group, and not just the fact they may be similar symbols.

Once it's understood how we see shapes on a page, an optimized presentation, which might incorporate a mixture of words and pictures, can be arranged in such a way that an observer's perception of the forms can be anticipated. The resulting composition can maximize the conveyance of information by removing ambiguity and overall reinforcing the intended message. The Gestalt principles expose how designers use fonts, sizes, alignments, color, contrast, and graphic elements in concert to better present information.

Whether by looking at a table, diagram, or visualization of data generated by computer, the observer's own analysis proceeds from a natural ability to see in ways described by Gestalt psychology.

An undisputed role exists for visualization in communication, but a varying degree of usefulness is found in its use for advancing science. A 3D protein structure exquisitely rendered by a computer brings the unseen realm into view. We acquire instant familiarity as we observe the protein structure as readily as we would if it were right here sitting on our desk – but press a button and the balls and stick model changes to a ribbon diagram, then a smooth-surfaced structure with planetary-like topography – then from blue to red, large to small, rotating, or in cross-section. We get a glimpse of the data, and the accumulation of these glimpses forms our understanding. The representations are not molecules, not even pictures of molecules, but pictures of icons that we believe represent some aspects of the molecule's properties. We don't immediately grasp the scale or other important characteristics, but nevertheless the model satisfies us by its naturalistic appearance. Our machines make the analogical representations, utilizing the same aforementioned artistic devices – linear perspective, overlap, shadows and light, tints of color – to provide visual cues to an observer in the same way they have done throughout the history of science. Our interest is piqued by imagery and, as Martin Kemp wrote: "how something works acquires a different status when it can be visualized as a working model – regardless of however schematic the representation may be with respect to reality of the phenomenon" (28). Key concepts in biology, such as Darwinian evolution, can be understood with words alone (although as Figure 24.6 shows, they may be embellished by schematic representation), but the complex processes of biology that support the theory, at any level – the cell, tissue, organ, individual, population, species – often require visualization from all but the most reductionist point of view. The role of imagery is as integral to the modern scientific process as it has been for any role in the past.

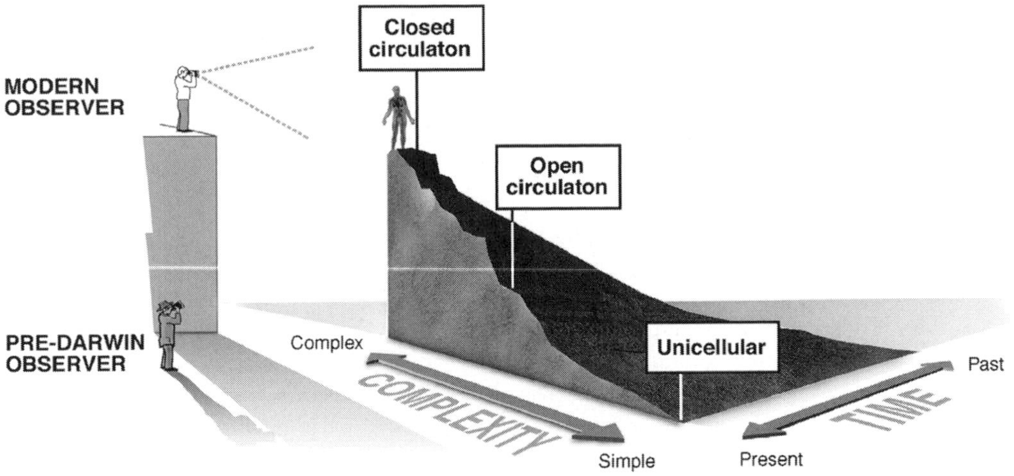

Figure 24.6. Multiple uses of metaphor. The modern observer, looking on an evolutionary "landscape" has a vantage point high above a pre-Darwin observer. Conceptually the modern observer has a better "perspective," which is represented as elevation above the ground where the pre-Darwin observer stands. The pre-Darwin observer can see only the present façade of the landscape, whereas the modern observer is informed by a view of the past. The evolutionary landscape (itself a metaphor) has a spatial axis that serves as a metaphor for "time." The summit of the landscape marked by "Closed circulation" represents a greater evolutionary complexity. Finally, the angular presentation of the arrows, along with the tapering of the labels "Complexity" and "Time" are among the graphical cues that serve as a metaphor for receding space. The modern observer might utilize Darwin's "tree of descent" metaphor, which contrasted with the "ladder of life" metaphor held by the pre-Darwin observer which held that a linear, unbroken relationship existed between organisms culminating in humans rather than a diversification from common ancestry.

The use of image, and the artistic devices to create them, may have originated historically in the same way that linguistic analogy and metaphor have – to satisfy "demand for the plausible narration of sacred events" (18). Through trial and error, and an interaction with science and technology, artistic devices have "evolved" toward better means of representation, culminating in various modes of film and digital photography.

But even the photographic image has its problems in terms of verisimilitude. First, each photographic image is mediated by lens, camera setting, film or digital capture resolution, and light intensity. It is subsequently modulated by digital or chemical means. Recording the scene from a different angle, with a longer exposure, a different lens, a brighter light, alters the result. The photographic end product may be considered an unbiased representation at a particular place and time but, due to a host of intervening factors, it is only objective to the point that the parameters of the recording process are known. In addition, when looking at the image, "truth" becomes what the viewer infers, based on a host of psychological factors. To understand photographic images, the viewer must make assumptions about what is being seen, such as the size of the subject, its 3D form, or the brightness of the scene, all of which can be misinterpreted without sufficient visual cues or explanations.

We rightfully place heavy reliance on visual representations and common-sense (innate) ability to interpret them. It's nevertheless notoriously difficult for people, including scientists, to abandon erroneous viewpoints. Eric Mazur, who teaches physics at Harvard University, was queried by one student as to how questions on a physics survey should be answered: "According to what you taught us, or by the way I think about these things?" (29). This student, who had seen firsthand demonstrations of physics principles, may well have asked: "Shall I believe my own eyes, or shall I believe yours?" When visual representation fails to convey, linguistic metaphor may be just what is needed to bridge the communication gap.

KEY POINTS

- Analogy and metaphor have played a pivotal role in not only building scientific terminology, but also in making sense of data and creating models for understanding.
- Metaphors utilize a type of reasoning that allows scientists to view novel phenomena in terms of that which has already been encountered, and to build on that knowledge. Barbara Stafford writes: "Analogizing has the virtue of making distant people, other periods, and even diverse contemporary contexts part of our world" (30).

- The benefits to an objective science are no different from the benefits to culture in general. "The very nature of science," notes Earl R. MacCormac, "is such that scientists need the metaphor as a bridge between old and new theories" (31).
- Metaphors utilize the similarities between their target and source. The meaning that emerges from the relationship doesn't rely simply on the meaning of words, but on our thoughts, perceptions, and underlying core concepts, which are understood in terms of one another.
- In the final analysis, metaphors structure our thinking, reasoning, and understanding of science in very fundamental ways.

REFERENCES

1 Brown TL. *Making Truth.* Urbana and Chicago: University of Illinois Press, 2003.
2 Lakoff G. *Metaphor and Thought*. Ortony A, ed. Cambridge: Cambridge University Press, 1992.
3 Pappenheimer JR. Passage of molecules through capillary walls. *Physiol Rev*. 1953;33:387–423.
4 Parsons-Wingerter P. Fibroblast growth factor-2 selectively stimulates angiogenesis of small vessels in arterial tree. *Arterioscler Thromb Vasc Biol.* 2000;20:1250–1256.
5 Pasqualini R, Arap W, McDonald DM. Probing the structural and molecular diversity of tumor vasculature. *Trends Mol Med.* 2002;8:563–571.
6 Thorgeirsson G. *Structure and morphological features of vascular endothelium.* Cryer A, ed. Amsterdam: Elsevier Science Publishers, 1983.
7 Rosenblum WI. Aspects of endothelial malfunction and function in cerebral microvessels. *Lab Invest*. 1986;55:252–268.
8 Gimbrone MA, Leapman SB, Cotran RS, et al. Tumor dormancy in vivo by prevention of neovascularization. *J Exp Med.* 1972;136:261–276.
9 Brower V. Less is more. *EMBO Rep.* 2003;4:831–834.
10 Dunbar K. *Model-based reasoning in scientific discovery*. Magnani L, Nersessian P, Thagard P, eds. New York: Plenum Press, 2001.
11 Dunbar K, Blanchette I. The in vivo/in vitro approach to cognition: the case of analogy. *Trends Cogn Sci.* 2001;5:334–339.
12 Hesse MB. *Models and Analogies in Science.* Notre Dame: University of Notre Dame Press, 1966.
13 Bowdle BF, Gentner D. The career of a metaphor. *Psychol Rev*. 2005;112:193–216.
14 Cohen F. *How Does a Virus Spread through a System? A Short Course on Computer Viruses.* Pittsburgh: ASP Press, 1990.
15 Majno G, Palade GE, Schoefl GI. Studies on inflammation II. The site of action of histamine and serotonin along the vascular tree: a topographic study. *J Biophys Biochem Cytol.* 1961;11:607–626.
16 Lakoff G, Johnson M. *Metaphors We Live By*. Chicago and London: The University of Chicago Press, 1980.
17 Dawkins R. *The Blind Watchmaker*. New York: W. W. Norton & Co., 1986.
18 Gombrich EH. *The Image and the Eye. Further Studies in the Psychology of Pictorial Representation.* London: Phaidon Press Ltd., 1982.
19 Harvey W. Movement of the heart and blood in animals. (1628). In: Franklin KJ, trans. *The Circulation of the Blood and Other Writings*. London: Everyman's Library, 1963.
20 Miller AI. *Metaphor and Analogy in the Sciences.* Hallyn F, ed. London: Kluwer Academic Publishers, 2000.
21 Thorgeirsson G, Robertson AL. The vascular endothelium – pathobiologic significance. *Am J Pathol*. 1978;93:803–848.
22 Simionescu M, Simionescu N, Palade GE. Segmental differentiation of cell junctions in the vascular endothelium. *J Cell Biol.* 1976;68:705–723.
23 Folkman J, Merler E, Abernathy C, et al. Isolation of tumor factor responsible for angiogenesis. *J Exp Med*. 1971;133:275–288.
24 Chesterton GK. St. Thomas Aquinas. eBook 0100331.txt. http://gutenberg.net.au/ebooks01/0100331.txt First posted 2001.
25 Laffoley P. *The Phenomenology of Revelation.* New York: Kent Fine Art, 1989.
26 Lidwell W, Holden K, Butler J. *Universal Principles of Design: A Cross-Disciplinary Reference*. Gloucester: Rockport Publishers, 2003.
27 Kosslyn SM. *Image and Brain: The Resolution of the Imagery Debate*. Cambridge: The MIT Press, 1994.
28 Kemp M. *Visualizations: The Nature Book of Art and Science.* Berkeley: University of California Press, 2000.
29 Mazur E. Qualitative vs. quantitative thinking: are we teaching the right thing? *Optics Photonics News*. 1992;3:38.
30 Stafford BM. *Visual Analogy: Consciousness as the Art of Connecting.* Cambridge: The MIT Press, 2001.
31 MacCormac ER. *Metaphor and Myth in Science and Religion.* Durham, NC: Duke University Press, 1976.

25

The Membrane Metaphor

Urban Design and the Endothelium

Kenneth L. Kaplan* and Daniel L. Schodek†

**Patient Stroke Treatment Module Project; †Harvard Design School, Cambridge, Massachusetts*

Scholars, from scientists to artists, have been drawing analogies between the human body and the built environment for centuries.[1] The functioning of major transportation networks within a city has frequently been explained in terms of how the circulatory system in a body works, and vice-versa, with major roadways compared with arteries and veins. The comparisons are surprisingly interesting. In both systems, major and minor pathways exist, with varying capacities and roles that connect and/or service different functional components of the overall city or of the body. In both systems, clogging, with its unfortunate consequences is caused by either temporary blockage or longer-term capacity reduction due to narrowing of pathways. A blockage at one point can affect far-removed locations. Mathematical models that bear some similarities have been developed to describe each of these systems. In the urban context, for example, the effects of closing a major pathway on traffic redistribution throughout the whole system of roadways can be predicted. These models find everyday use in our cities.

Although analogies between urban functions and human bodily functions are common, we should carefully think about their appropriate use. The two systems are obviously radically different for innumerable reasons, and this in turn undermines the usefulness of the analogy. It is axiomatic that nothing can ever be proven by use of analogies or metaphors to begin with. Analogies or metaphors, however, do often prompt the scholars or researchers in each field to view their own problems in a different light. New approaches to thorny issues are occasionally suggested (4). It is in this spirit that we push forward with the risky business of speculating how some aspects of a city's functioning might be relevant to studies of the endothelium (and perhaps vice versa).

We begin by noting that the models described in the opening paragraph are intrinsically limited. In the urban context, these transportation network models deal only with the ebb and flow of primary traffic on major roadways (analogous to the conduit vessels of the human body). They deal with surrounding areas of the built environment in only the most rudimentary ways, if at all. An urban area, however, consists of a highly complex environment made up of an intense fabric of streets (analogous to the tens of thousands of miles of invisible capillaries in the human body) and other primary transportation networks, utility and service networks (including wired and wireless communications), pedestrian pathways, open spaces, a variety of building forms and types serving many different functions, and many other features. Some elements in a city provide control or regulatory functions, some are service- or transactionally oriented (including common retail areas), others are essentially productive in nature (industrial zones, power plants) or deal with pragmatic issues of waste disposal, and yet others deal with habitation or with the mind and even the human spirit. The list goes on. These many functions are housed in physical constructions that assume an astonishing array of specific forms. Although the forms individually assumed may vary widely, distinct patterns of land use can be perceived in most cities (e.g., residential, retail, industrial). The character of the supporting networks (including transportation) is both influenced by these patterns and, in turn, helps shape them.

At a far more specific level, we see that marvelous interaction zones occur in an urban area at the interface zones between streets, pedestrian walkways, and adjacent buildings.[2] It is here that writers such as Jane Jacobs have noted that the

1 From the architects Vitruvius and Alberti, to Michelangelo (1), to William Harvey, to contemporary author Peter Ackroyd (2) (to name but a few) many have anthropomorphized buildings and cities; researchers in the emerging area of urban metabolism view cities as biological systems (3).

2 In the circulation, blood is delivered by arteries to the tiny capillaries. The arteries are large and have a really thick wall – much like a superhighway with guard rails. These arteries serve only as conduits. It is at the level of the capillary where (as in the case of the inner city streets) all the exchange of nutrients, wastes, and gases takes place between blood and underlying tissue.

truly dynamic "life of the city" actually takes place. Pedestrians move back and forth on their own pathways, entering and leaving shops and housing, but also interacting with and affecting traffic flows in adjoining streets. Specific city nodes act as large-scale attractors or sources (e.g., cultural centers, major retail or office centers, train stations) that in turn influence pedestrian and vehicular flows between them in a way that is dependent on their spatial relationships. The accompanying pedestrian flows between major nodes also infiltrate through an interface zone comprised of entry portals and other ancillary path entries directly into a multitude of small shops and other elements that line city streets. The complex set of human actions that occur at these interface zones between buildings and networks contributes much to the life of a city.

If we step back a moment, we can conceptualize the interface zone just described as a kind of porous membrane that surrounds a major flow network component and separates it from functional volumes. The membrane provides access, as well as control, along its length to passage from the major flow networks (e.g., streets and major sidewalks) into adjacent spatial zones (e.g., building interiors) having different functions and utilities. It is a zone of intense and highly complex activity, and it is here where an analogy with the endothelium – admittedly slightly strained – is suggested.

Curious parallels do indeed exist. In the urban example, the varying interior functions of adjacent buildings have both needs for goods and demands for services that are at least partly provided for via passages through the portals that comprise the putative membrane from the adjacent network elements. Different functions provide different kinds of draws. The very acts of passage, as well as activities inside, in turn influence the behavior of not only the adjacent network, but of the whole system itself.

Thinking in these terms brings up a number of questions. What are the factors that control the level of porosity of the membrane, and hence movement rates and volumes of passage? In the urban example, a whole host of complex factors enter into the discussion. Many relatively steady-state influences stem from the need for a normal level of services. Unusually high levels of porosity also can be externally generated and stem from many broad cognitively, culturally, or psychologically based rationales (e.g., holiday shopping days cause more stores to be open longer hours). On the other hand, high porosity levels can be generated internally (e.g., a store might deliberately have a sale and extended hours to make itself more attractive to individuals on adjacent flow networks, and nearby competing stores quickly follow suit). In view of the present attempt to draw broad parallels with the endothelium, the latter observation raises the curious question of whether any human body element adjacent to an endothelium layer ever attempts to make itself "more attractive" in a comparable way. Although this proposition may be completely absurd, it does suggest the power of analogy – truly odd questions can be raised.

As we look closer at the buildings within an urban area, we see that the outside envelope of a building (often referred to as its "skin") serves many functions that are surprisingly analogous to its counterparts in living organisms. Building skins traditionally have served to provide a protective covering (in this case from the elements), but also have served as a mediating system between the inside building functions and the outside environment, for example, operable windows that control air flows or selectively admit light. Exciting new research into "smart skins" has pushed the roles that building skins can play even further. One new role is that of a wall serving to gather, send, receive, and display information via various embedded sensor-based technologies. Here, large-area light, thermal, and sound sensors can be used to gather information about surrounding environmental conditions, and tactile sensors can be used to gather information about forces acting on the wall. When coupled with other communication technologies, this information can be used for a wide variety of monitoring and control functions. These same walls also can be used to display a wide variety of information, typically visual but also audio. Many are interactive. Individuals simply either passing by or directly interacting with the wall can cause different responses. Intriguingly, the potential for a kind of information-based system intelligence becomes possible. Smart skins can go beyond the role of dealing with sensor-based information alone, and can be used to mediate surrounding environments via intrinsic smart material characteristics related to properties such as porosity (so-called "breathable" walls) and to absorbing, storing, distributing, and releasing heat. Here, properties intrinsically provide variable responses to external stimuli via material properties alone. All these different approaches make the "smart skin" a truly active and interactive responsive boundary surface.[3]

We now return from the realm of largely idle speculation to harder questions. We began by noting that there do indeed exist mathematical modeling techniques for simulating the behaviors of major transportation networks. Are there any such modeling techniques developed in relation to urban design and planning that even begin to reflect the complexity of actions within the interface zones (at the "membrane") described above? At this point, the answer is largely negative. Current network transportation models only deal with surrounding environments in only instrumental ways and are structured in largely deterministic forms that are only doubtfully conceptually valid for the kinds of interactions described. Various cellular automata approaches that remain largely undeveloped might ultimately prove to be of some value here (5). Some hope exists in this modeling approach, in that it appears that the behavior of any specific member on our streetscape is indeed affected by the immediate actions of surrounding members, and that the behavior of the whole is in turn markedly influenced by the nuances of the actions of individual members. It is not impossible to imagine influences associated with the idea of attractors, sources, and sinks that also affect the behavior of the whole. Could corresponding

3 Perhaps the endothelial lining of tubular blood vessels may be viewed as a smart skin that senses and responds to the extracellular environment.

computational rule structures be established that have any validity? Perhaps, but we have a long way to go. Does the medical and biology community have valid computational models for their systems that might help us in understanding urban activities? Understandings can flow in both directions.

If computational models are not yet mature enough for our purposes, let's look at conceptual models. Of particular interest to architects and urban designers is the definition of the endothelium as a "distributed spatial system" that interfaces with adjoining biological systems. Although the definitional parameters of "system" as applied to the endothelium and to other complex biological systems is a topic of great debate among the life-science disciplines, its relevance to architectural design thinking is the notion that the endothelium is a spatial continuum within the context of the interior spatial volume of the human body. Space is a key concept for architects and urban designers, along with proportion, symmetry, harmony, ornament, color, and, of course, function. As well, they must incorporate culture, history, and the interaction of the built and natural environments. Designers are taught to calibrate the relationship of the detail to the whole. They study spatial phenomena and have the conceptual tools to understand formal principles of system growth or demise.

With regard to cities, architects – or more specifically urban and landscape designers – must be able to collect data and then analyze the data at various spatial scales. For example, the recent re-evaluation of lower Manhattan after 9/11 required a precise understanding of the specific site of the Twin Towers in relationship to the surrounding city blocks at one scale, and then a larger spatial measure of the neighborhoods of Manhattan and the other boroughs of the city, to an even larger scale of the regional impact to the surrounding states, because the site also served as a critical transportation hub for commuters from the tri-state area of New Jersey, Connecticut, and metropolitan New York. Designers have the ability to manipulate and calibrate spatial perspective, both conceptually and then with an assortment of tools ranging from clay urban-site models to data-rich GIS mapping software at multiple scales. Although these tools give the designer some flexibility in understanding the dynamic interplay of human and urban space, which in real terms is a constantly shifting landscape that embodies a vast assortment of levels of data from the economic to the sociopolitical, they do not provide the level of detail to robustly address the "membrane" issue. Complex human interaction, communication, and movement patterns in the city are not well understood by designers or their counterparts in the various specialized disciplines that comprise the larger field of urbanism.

It appears that, traditionally, biologists and clinicians have not approached the biological systems from a spatial perspective. Most biologists don't seem to use spatial concepts, and math used in biology doesn't measure "geometric" space. Most models of biological systems illustrate their two-dimensional (2-D) aspects or physical and chemical properties. An exception is the interest in spatial and temporal concepts relative to cell differentiation (6).

Yet medicine and science have been advanced recently by an assortment of imaging and digital technologies that now translate the body into a rich three-dimensional (3-D) spatial geography that is stimulating these disciplines to think more broadly about the body from a perspective that is in fact suitable to the insights of architects, who are much more comfortable with the notion of space as an observable and measurable phenomenon. This suggests an opportunity that harkens back to the Renaissance, when artists and architects were viewed as valuable contributors to scientific and medical knowledge.

Underpinning and sometimes confusing the issue is the struggle to comprehend the vast amounts of data made available by recent discoveries related to the molecular structure and functioning of discrete aspects of biological systems. This drive for data collection is defined by a necessary obsession for "microscopic" detail that is both the strength and weakness of the current biomedical community modus operandi. What seems to have been lost in the process of specialization is how to think about the relationship of the parts to the larger conceptual bedding of the whole. The design fields have long been occupied with what constitutes a system –including complex anatomy, typology, and functionality – for the primary reason that the designer's principal task is to conceive of and build new systems (7). Accomplishing this task requires a specialized knowledge and toolkit, but more importantly, a uniquely holistic perspective of systems at multiple scales. This conceptual perspective could be of value as life scientists puzzle the relationship of the part to the whole.

Let's further complicate this exercise and again summon the power of analogy, even stretch the notion of metaphor and look at a city as an actual function, or expression, of human biological systems. Cities are built by people, often without much apparent planning. Perhaps we humans do pattern our built environment, consciously or unconsciously, on the systems from which we are built. Human behavior creates the city: Just as DNA evolves as a result of external environment and internal chemistry, so does the city evolve.

Although the city as organism is not a new idea, as noted earlier, the relationships of humans (and the biological systems that comprise them) to their built environments is still largely misunderstood. Could designers better understand the city by integrating some reductionist thinking into the holistic, urbanist approach? Biologists seem to be further along in understanding the parts. As life scientists learn more about the systems behavior of cells at the micro-environmental level, they discover striking parallels between human social behavior and cell social behavior. If designers were to perceive the city as the ultimate expression of these two populations – populations that are intertwined, yet operating blindly of each other – might that transform how they respond to the pressing problems of our urban centers?

Recent research programs (e.g., funded by NSF, NIH, CDC) demonstrate a growing (albeit slowly) recognition that we need truly multidisciplinary teams to study cities and their impacts on humans and the environment – teams that integrate urban designers and planners, public health experts, and social scientists with scientists such as ecologists and biologists. But the future of our cities may depend on our understanding of the biological mechanisms underpinning the self-organizing tendencies of cells and cities. Will the designers of the future be trained in cell biology, as well as social sciences, or will the nature of their work be profoundly impacted by collaborations with life scientists? Could we work together to better understand the "life of the city" as well as the workings of the human body?

KEY POINTS

- Endothelial biologists and urban designers and architects are all interested in distributed spatial systems. Sharing concepts, perspectives, and tools may help us learn how some aspects of a city's functioning might be relevant to studies of the endothelium (and vice versa).
- Although we should carefully think about their appropriate use, metaphors allow the freedom to raise speculative questions. Researchers in each field may view their own problems in a different light, and new approaches to thorny issues are occasionally suggested.
- In an urban area, the interface zones between streets, pedestrian walkways, and adjacent buildings serve as a type of porous membrane that surrounds a major flow network component and separates it from functional volumes. It is a zone of intense and highly complex activity, with curious parallels to the endothelium.
- Although design tools provide some flexibility in understanding the dynamic interplay of human and urban space, they do not provide the level of detail to robustly address the "membrane" issue.
- Discoveries regarding the discrete aspects of biological systems could help designers solve complex problems of urban life.
- The designer's holistic perspective of systems at multiple scales could be of value as life scientists puzzle the relationship of the part to the whole.

Future Goals

- To undertake collaborations that enable micro- and macro-scale researchers to exploit converse perspectives and to develop and share new approaches and tools

REFERENCES

1 Hale JR. *Renaissance Fortification: Art or Engineering?* London: Thames and Hudson; 1977:43.
2 Ackroyd P. *London: The Biography*. New York: Doubleday; 2000: 1–2.
3 Royal Commission on Environmental Pollution Study on Urban Environments, Well-being and Health. *Urban Metabolism Desk Study*. p 6. www.rcep.org.uk/urbanenvironment.htm.
4 Keller EF. *Making Sense of Life: Explaining Biological Development with Models, Metaphors, and Machines*. Cambridge MA: Harvard University Press; 2002:117–119.
5 Johnson S. *Emergence*. New York: Simon & Schuster; 2001:87–89.
6 Gerhart J, Kirschner M. *Cells, Embryos, and Evolution*. Boston: Blackwell Science; 1997:238–243.
7 Perez-Gomez A. *Architecture and the Crisis of Modern Science*. Cambridge, MA: The MIT Press; 1983:298–326.

26

Computer Metaphors for the Endothelium

Dexter Pratt

Genstruct, Cambridge, Massachusetts

"It was a dark and stormy night. The Weibel-Palade bodies surged to the surface, releasing their witch's brew into the turbulent surf downstream of the rupture. Neutrophils shuddered to a stop as selectins grappled, tore away, and then finally caught."

INTRODUCTION

"Endothelial biology is weather." That is a metaphor, and a bold one, because it compares two systems of enormous complexity and implies that they share fundamental features despite an utterly different scale and mechanism. But is that a "computer" metaphor? Shouldn't this chapter find its metaphors in circuits, signal processing, and the Internet? My answer is "No," because in the mechanics of computation, the details of technology, we see an entirely artificial landscape, a world dominated by well-defined protocols, regimented structures, and perfect repeatability. It is a world in which "crosstalk," the secondary interaction between components or processes, is scrupulously avoided.[1] By contrast, crosstalk is the norm in biological systems. Signaling "protocols" exist in biology, but their effect is nuanced and context-dependent, shifting the behavior of systems that are the dynamically balanced product of many competing and overlapping processes.

To find meaningful computer metaphors, we need to look at computer applications that attempt to deal with natural, emergent complexity. The metaphors derived from these applications are not just "computer" metaphors, but are "computable" – they provide insight into the pragmatic construction of systems that can encode biological theories and hypotheses. The scientific process is essentially the discovery and articulation of rules that describe and predict the events we can observe in the world around us. When those rules are expressed in a computable form, algorithms can perform useful inferences about systems and datasets too large for a scientist to reason about. In that spirit, the current chapter explores the following three metaphors for the endothelium:

- Endothelial biology is weather.
- The endothelium is the vegetation covering a vast terrain.
- Endothelial biology is a rule-based system.

The comparison of endothelial biology to weather is an important computer metaphor because a rich field exists in which computational models of weather and climate have been employed in short-term forecasting, long-term climate prediction, and paleo-climatology. The limitations of weather modeling highlight how difficult it is to make specific predictions about the behavior of a complex system and the need for copious and detailed measurements to initialize and test the model. But the successes of weather forecasting show that a partial solution to an important problem can still have great value.

The computer models of vegetation patterns considered in the second section are interesting because they demonstrate techniques in which complicated behavior emerges from relatively small and simple sets of rules. But the hope that these techniques can help us find and articulate simple rules for complicated endothelial behavior is tempered by the need for detailed measurement and testing. Many rule sets – *hypotheses* by another name – may generate patterns of simulated behavior that resemble the behavior of the natural system, but the proof of any hypothesis still relies on specific measurement and experiment. Complex predictions by a hypothesis can be evaluated only when enough observations are made to distinguish them from the results of competing hypotheses.

In the third section, the comparison of endothelial biology to a rule-based system seems more like a computer metaphor, drawing on the jargon of artificial intelligence. In each of the three metaphors, biology is made computable by reduction

1 The basic kind of crosstalk – in which signals in adjacent wires interfere with each other – is avoided in any sort of computer. In principle, one *could* make a parallel system that had fuzzy secondary interactions between processors in order to run some sort of probabilistic agent interaction faster, but you would be far better off doing that in a software simulation. Basically, computer hardware is made to perfectly execute exact instructions trillions and trillions of times without error, *and* to be debuggable: to be cleanly broken down into well-articulated separate pieces so that it's possible for humans to design, program, and repair.

to rules, but the term "rule-based system" will be used in the sense of systems characterized by large databases of IF-THEN rules, compared to the small rule sets in the previous section. Viewed in a broad sense, rule-based systems provide a way to define hypothesis and deduction in the face of systems too large to be encompassed by the mind of any single scientist. Although the behavior of the endothelium emerges from the interactions of trillions of independent cells, it is nevertheless the case that important aspects of our knowledge can be described by simple, qualitative rules: Rules that can be manipulated mechanically, rules that can be processed by algorithms to derive conclusions. Any given deduction processed by such a biological rule-based system may seem trivial, but great value exists in the straightforward consideration of tens of thousands of deductions at once.

The theme that emerges from these three metaphors is that systems biology, and specifically endothelial systems biology, can attempt to deal with the enormous complexity of the living systems by using diverse computational techniques. No technique will be a free lunch: Very large models are hard to build, require the input of many high-quality observations, and face fundamental limitations in their ability to predict the specific behavior of complex systems. But practical value can be realized from partial solutions.

ENDOTHELIAL BIOLOGY IS WEATHER

In weather forecasting, computational models must grapple with an enormous system exactly as it occurs in nature. They must make specific, local predictions in a timely manner. Research in this field is highly rewarded for successful "snow in Chicago starting 6 PM Tuesday" predictions. In endothelial biology, we too, must contend with a naturally occurring system of great – probably much greater – complexity. We desire to make specific predictions for individual humans, such as "Lydia Jones will suffer a rupture of one or more large plaques in her coronary arteries within the next 60 days, whereas Janet Wu, despite greater stenosis and higher levels of low-density lipoprotein (LDL), will have stable arterial plaques for at least the next decade." Unlike weather forecasting, we also want to make predictions about possible interventions: "Statin therapy will prevent a predicted stroke for Peter Simpson, but will be of no value for Robert Stein."

The obvious fact about computer models for weather forecasting is that they are extremely successful, but only up to a point. The accuracy of a specific forecast declines rapidly with time: Beyond a week, forecasts are not much better than a statement of "typical weather for this time of year in this location." The models that produce these forecasts require huge data inputs, including measurements of temperature, pressure, wind velocity, partial pressure of water vapor, cloud cover, and length of day. During the 1960s, when computer models were grossly limited by the speed and memory size of the existing computers, it was assumed that forecasts would improve as better computers were available. And although that assumption turned out to be absolutely true, the hard fact is that exponential increases in computer speed and memory produced small, linear gains in the forecast horizon. According to the American Meteorology Society, the accuracy of forecasts made in 1998 more than doubled compared to the forecasts of the 1970s; this stands in sharp contrast to the computers on which the models ran, in which the number of transistors in a processor doubled roughly every *18 months* during that same time (1,2). Specific predictions about the weather are intrinsically difficult because small differences in the initial conditions make ever larger differences in the resulting behavior. Computer forecasting of weather was one of the first fields in which this kind of dependency was studied, resulting in the popular term "butterfly effect," based on the idea that even an event as small as the flapping of a butterfly's wings could have large effects on subsequent weather.

If we apply the example of weather modeling to the endothelium, we should first notice the tremendous disparity in the data available for the predictions that we would like to make. In the case of the weather, only one individual – the planet Earth –is measured, moment to moment, by an enormous array of sensors, including satellite photography, radar systems, and ground-based weather stations. In endothelial biology, we have many individuals to observe, including humans, experimental animals, and derived systems such as cell cultures. In any given experiment, we may measure the expression of thousands of genes in several tissues, but only at that single moment in time when a pathology sample is taken or when an animal is sacrificed. Typically, only system parameters, such as heart rate, blood pressure, and the levels of plasma proteins and metabolites, may be measured multiple times in the course of a study. Imagine if this was the type of data available for weather forecasting, treating each individual animal like an entire planet: Meteorologists would have a snapshot – single time point data for the weather on many different earthlike planets – but never a second set of measurements for any planet. On a given planet – Earth, for example – they might have time sequence data, but these would be only for general parameters, such as "mean ocean temperature over time." The weather models that those forecasters could develop might draw interesting conclusions about climate, but it would be unlikely that they would ever successfully predict "snow in Chicago at 6 PM."

So the first point is that, in modeling a complex system such as the weather or the endothelium, we need detailed observations over time, preferably observations of the specific system – the individual – for which we wish to make predictions.

The second point is that sharp limits are imposed on the specific predictions that we can make, because the systems we want to model are sufficiently complex that their behavior over time is dependent on small differences in initial conditions. At short time frames, good predictions can be made by determining the effects of dominant factors – straightforward chains of logic can predict the immediate reaction to a perturbation. At medium time frames, sophisticated models can make useful predictions by considering secondary effects and feedback processes. But in the long term, our attempts to predict or affect biology are at the level of climate.

THE ENDOTHELIUM IS THE VEGETATION COVERING A VAST TERRAIN

The endothelium is a vast surface – 4,000 to 7,000 m^3 in humans. On that surface, the local conditions control the behavior of the endothelium – and the endothelium alters its local environment in turn. Endothelial cells (ECs) are very plastic, changing their behavior in response to the local conditions.

We can compare this to the land area of our planet: We live on a vast surface on which most areas tend to become vegetated, but the nature of the vegetation depends on the local conditions. Moreover, the growth, resource use, shading, reflectivity, and decomposition of plants in an area continually alter its conditions, making way for further changes in vegetation. For example, if a forest is burned or harvested, it regrows as a succession of species over time, some species successful as early colonizers of open areas, others as later inhabitants of a shaded forest floor. The species and their progression reflect the local climate, topography, and surrounding ecosystems: the sum of many, many individuals growing well or poorly in the prevailing conditions.

In an analogous manner, when the endothelium is damaged or removed in an area of the vasculature, it regenerates by the response of neighboring ECs and/or by the recruitment of endothelial progenitor cells to the site. In both cases, the landscape is populated by dynamic processes in which individual plants or cells grow, colonize, and interact with each other. Given these similarities, I will develop the story of some computer models of vegetation and then return to the endothelium.

A simplistic model of vegetation might predict the flora of an area based on local parameters such as climate, latitude, and geography. This model would essentially be a large table based on ecosystem surveys. Although valuable, such a model merely predicts a typical state for each set of inputs, not the process by which the system achieves that state: It does not provide insight into the response of the system to change or stress. There is no representation of the mechanisms that produce the outcome. And without general mechanisms, it is unlikely to predict the result of novel changes, such as the application of a selective herbicide.

Moreover, these models fail at a level of granularity. If vegetation were composed of perfectly adaptable plants, one might expect smooth transitions as one moves from one climate to another. Each plant would adopt the appropriate intermediate form for the conditions at each point in the transition. But although real plants are quite malleable in their growth and form, discontinuities arise as the dominance of species changes. For any small area in the transition between grassland and forest, we expect to find either a tree, a shrub, or grass. One might handle this in the simple model by treating the state of that patch of ground as a probability (i.e., 40% of the unit areas are grass, 50% have a shrub, 10% have a tree). But the patterns of growth predicted by random assignment do not match the patterns observed in actual ecosystems. Plants of similar function and species array themselves in a patchy manner, sometimes forming spots, sometimes stripes or loops. In desert areas, shrubs are spaced with characteristic frequencies. These patterns mean that the plants interact, altering the conditions in the adjacent areas of terrain and in their own area. A better vegetation model should therefore represent time, so changes can occur over time. And it should represent each location, describing it with parameters that will allow predictions for the location and its neighborhood.

Cellular automata (CA) are a class of computer models that have been applied to the modeling of vegetation. A CA is perhaps the simplest type of agent-based model (ABM) in which a surface is represented as a grid of squares, and time is divided into a series of *turns*, as in a game. Each square may be considered an *agent*, an entity defined by the values of specific parameters and by the rules for its interaction with other agents in the system. At each turn, each square is assigned new values for its parameters, based on its previous state and the previous states of nearby squares. For example, in a very simple CA, squares might have only one parameter, such as color, and limited values, such as white or black. Their subsequent color is the result of their previous color and the colors of the eight adjacent squares. One set of rules of this sort comprised "the Game of Life," so named because it had some inspiration in the behavior of cells growing on a surface. Too few or too many neighbors and the cell in the square died; just the right number and a new cell would be generated in an empty square.

These rules were not a serious attempt to model cellular behavior, but "Life" was important because these simple rules produced surprisingly complex behavior: Patterns of black and white could grow, shrink, oscillate, and even move across the grid. Patterns were found that formed stable oscillators that spawned moving progeny at regular intervals. So "Life" seemed very "alive," even though it was the result of very simple rules. Over the next decades, many researchers developed ideas of emergent behavior – complexity arising from simple rules that specified the interaction of simple components, especially as an approach to explain complex biological behaviors.

Complex biological behavior is daunting: It simultaneously encourages reductionist research to study the mechanisms of component systems and descriptive, statistical studies of larger systems. But emergent behavior viewed in computer models suggests – holds out hope – that some systems demonstrating complexity can be modeled by simple rules, indicating that a manageable number of parameters dominate the behavior.

In that spirit, CA have been applied to a number of biological processes that unfold spatially, including vegetation on a landscape. Rietkerk and colleagues (3) compared the output of a simple CA model to the observed patterns of patchy vegetation in various ecosystems. The model is a two-dimensional grid, in which each square represents an arbitrary area of

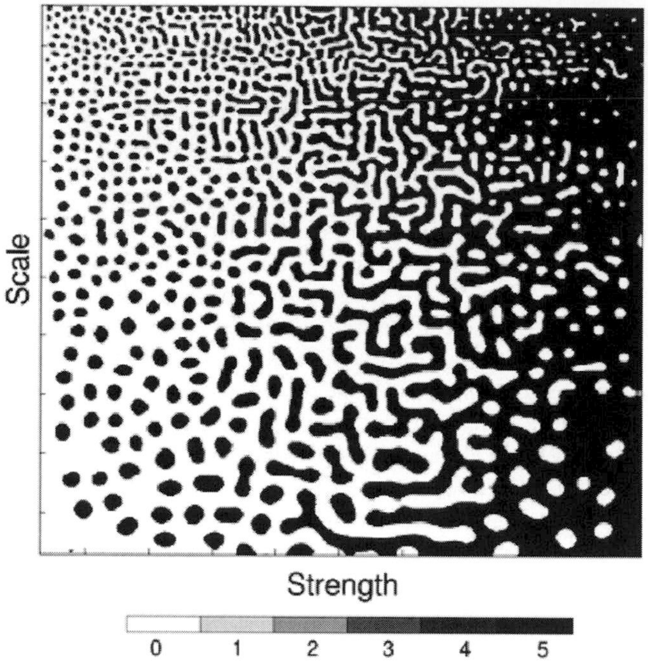

Figure 26.1. Changes in patterns produced by the model as the strength of the positive feedback is varied (x-axis) versus the scale of its influence (y-axis).

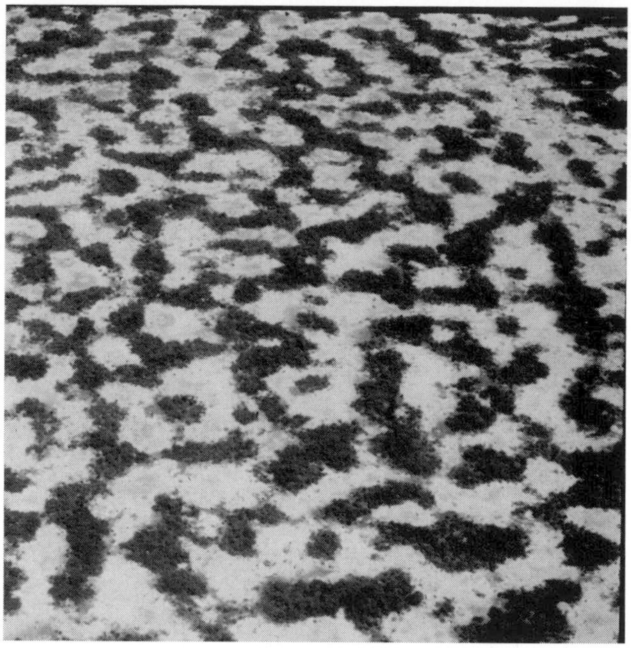

Figure 26.2. Example of a vegetation pattern in an arid ecosystem.

ground and is colored in proportion to the vegetation in that area. Vegetation in each square changes over time based on two rules: It is both promoted and inhibited by nearby vegetation. The promotion (positive feedback) is stronger at short range, whereas the inhibition (negative feedback) can act over longer distances. Figure 26.1 shows the changes in patterns produced by the model as the strength of the positive feedback is varied (x-axis) versus the scale of its influence (y-axis).

The patterns produced by these rules are in fact similar to observed patterns of vegetation in some ecosystems, as shown in the aerial photograph in Figure 26.2. The hypothesis is that the simple rules produce these recognizable patterns because the rules correspond to dominant processes in these ecosystems: Plants compete for resources and limit each other's growth (negative feedback), but they also alter the local environment in ways that conserve resources (short-range positive feedback). In the case of an arid ecosystem, the critical parameters for the system might be rainfall, soil permeability, and evaporation rate. The future of each square depends on the state of its hydration, which in turn depends on the primary parameters:

- Increased vegetation decreases evaporation and increases permeability.
- Decreased vegetation increases evaporation, but benefits adjacent squares by increasing runoff of rainwater due to decreased permeability.

With sufficient rainfall and moderate temperature, the system tends toward a state of uniform vegetation. Vice versa, with very low rainfall and high temperatures, all squares become barren. But between these extremes patterns develop that mimic the observed patterns of patchy vegetation at transitions between grassland and desert.

This sounds very appealing, but to consider a model as a theory, we need a better evaluation than "produces patterns that seem similar to observed behavior of vegetation." For any particular set of data, different rule sets (i.e., different hypotheses) might produce results indistinguishable to the human eye.

To develop the model as a theory requires the development of testable hypotheses, the design and execution of experiments to test these hypotheses, and a comparison of the results with predictions over wide ranges of parameters. Indeed, Rietkerk and colleagues conclude their paper by citing a need to move from qualitative to quantitative prediction. In the world of vegetation, such experiments can be performed. In addition to photographic surveys of vegetation patterns, it is possible to measure rainfall, runoff, permeability, and hydration over time. Moreover, it is possible to perturb these systems – changing water input, removing vegetation, adding artificial shade – and then compare the observed changes to the predictions of the hypothesis.

So how does this type of model apply to the endothelium? ECs generally form an unbroken layer of "vegetation" on the vasculature. Unlike vegetation, only one genotype of EC exists, but many phenotypes exist that the cells adopt in response to their changing environment. One can then imagine the creation of a CA model for the behavior of one or more aspects of areas of the endothelium. In that model, the surface of some segment of the endothelium would be represented as a large grid, each square corresponding to an area roughly comparable to a cell. The granularity of time in the model could range from fractions of seconds to hours, depending on what aspects of behavior are to be captured. For each "turn," the status of each square in the grid would be determined by the status of neighboring squares (presence of cells, alignment, shape, expression of certain receptors) and by the climate-like parameters of blood flow, shear stress, cytokines, and lipoprotein levels. It is easy to imagine the use of these techniques to create a model of the endothelium that would generate detailed predictions of multicellular behavior.

But perhaps first we should ask what measurements are available to us. To what extent can we measure the state of a patch of endothelium over time? Can we even come close to the measurements of water movement, temperature, and plant growth that ecologists would make? If our measurements are too crude, too sparse, how can we assess a model that generates many complex predictions? For example, for a given parameter of the endothelium – P-selectin expression – we might ask if our model predicts smooth transitions in time and space. One set of choices in defining the parameters of the model might predict P-selectin expression varying continuously across the surface of the endothelium, whereas another set of choices might predict discontinuous levels of expression, in which individual cells switch abruptly between distinct activation states. To evaluate which choices make a better model, to evaluate *any* model that predicts patterns of P-selectin expression, we must find a way to measure the microscopic variation in P-selectin expression over the surface of the endothelium under varying stimuli. We need enough data so that the model's predictions can be tested over a wide range of conditions, establishing specific evidence beyond "familiar patterns." For example, a study might involve imaging layers of cultured ECs for the expression of multiple surface proteins (P-selectin included) and for cell shape and alignment. The imaging would be repeated under variations in flow, after mechanical damage, or following stimulation with cytokines or growth factors.

Recent applications of CA to systems of interacting cells have included models of microvascular growth. Peirce and colleagues (4) developed a model that used over 50 rules integrating epigenetic stimuli, molecular signals, and cellular behaviors to predict microvascular network patterning events. Their model simulates the behaviors of individual cells during new vessel formation, vessel length extension, and recruitment of contractile perivascular cells. Moreover, it produced successful quantitative predictions of changes in vascular density in response to an exogenous growth factor and changes in contractile vessel lengths in response to increases in circumferential wall strain.

Our first two metaphors indicate that systems complex in space and time can be modeled if sufficient measurement is available of the observed system (weather), or by the use of simple rules (CA). However, those models still require extensive datasets to be tested. And our expectations for these models should be limited: Detailed predictions may be inherently difficult when small variations in initial conditions can cause large variations in later behavior.

ENDOTHELIAL BIOLOGY IS A RULE-BASED SYSTEM

The phrase "rule-based system" conveys an impression of order, but little else. After all, any model is in some sense a set of rules, and the same can be said for any computer program. The more specific meaning of this term is a class of programs characterized by large databases of IF-THEN rules, in which thousands of rules can be manipulated mechanically by algorithms to derive conclusions concerning objects and processes in the real world. In the previous section, we examined ways in which small, simple sets of rules could generate complexity, but now we look for ways to cope with inherently complex models in which there exist many, many genes, proteins, locations, and processes and a large but incomplete body of knowledge specifying their relationships.

Rule-based systems are sometimes optimistically referred to as *expert systems*, when they are intended to emulate the reasoning of human experts. Saying "endothelial biology is a rule-based system" is a way to approach a fundamental idea: Important aspects of biology can be addressed by very large models composed of nonmathematical, qualitative rules. Embracing this metaphor – biology as many abstract concepts linked by many simple rules – permits the development of pragmatic systems in which algorithms manipulate biological knowledge. It is valuable to enable algorithms to reason about endothelial biology because we need a way to pursue biological research, to explore hypotheses and deductions when too many facts and observations exist for any human researcher to grasp or process. Any given deduction made by a biological rule-based system may be simple compared to the insight of a human biologist, but there is power in the straightforward consideration of tens of thousands of deductions at once.

I start my discussion with the most important point: Success is possible. Systems of this kind can be constructed and used with diverse experimental data; they can produce significant scientific insight about the biological changes observed in an experiment and inform the design of subsequent experiments. At Genstruct, Inc., we have developed a working biological rule-based system that has both a very large set of qualitative rules encoding biological information largely derived from the scientific literature (roughly 150,000 cause-and-effect relationships) and a suite of reasoning algorithms that are used to develop hypotheses based on those rules as constrained by the data from specific experiments. This system has been used successfully in both academic and commercial projects, using experimental data from areas as diverse as cancer biology, diabetes research (5), lipid biology, and inflammation studies.

Figure 26.3 shows a causal network that can explain 104 of 694 RNA expression changes observed in an experiment in which the LNCaP prostate cancer cell line was stimulated by a synthetic androgen (6). This network was composed by algorithms that used the RNA expression changes induced by androgen in the experiment to interrogate a causal rule base (Genstruct's Knowledge Assembly model of human biology). One algorithm selected chains of reasoning – hypotheses – that

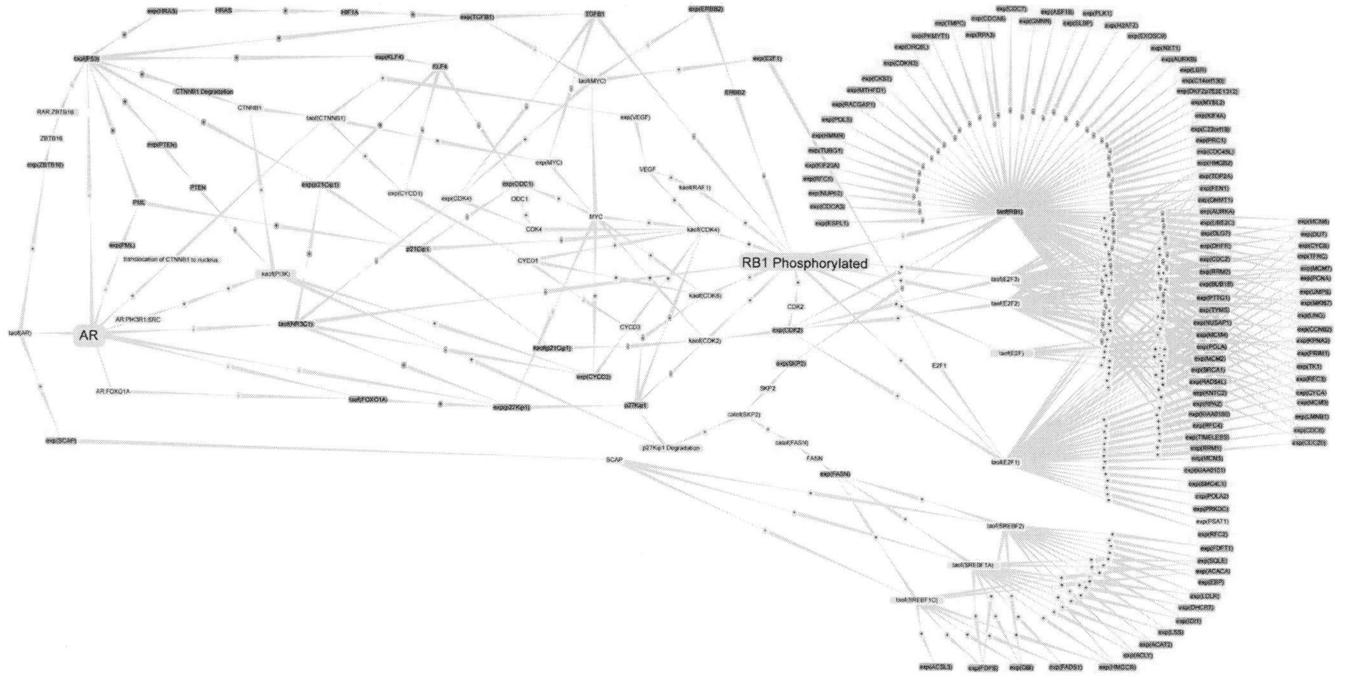

Figure 26.3. Causal network explaining 104 gene expression changes observed in an experiment. The details of the network are not readable at this scale, but its structure makes the important point: A network element toward the left of the diagram representing the experimental perturbation is causally connected to elements representing transcriptional evidence in the fan-like structure at the right of the diagram. Each causal path through this network is a testable hypothesis for a mechanism of action in the experiment.

were explanations of multiple observed RNA expression changes. Another algorithm identified possible causal chains that link those hypotheses to the downstream effects of androgen treatment and of androgen receptor activation. The diagram does not show the entire Knowledge Assembly model; it shows only the small fraction of that model that comprises an explanation of 104 of the gene changes observed in the experiment.

This diagram invites the question "But where are the rules?" The rules are represented by the edges of the graph and the nodes they connect. They express causal relationships, such as "An increase in the amount of phosphorylated retinoblastoma 1 (RB1) causes an increase in the transcriptional activity of E2F1." Each rule can be represented as an arrow on a pathway diagram, describing the ways in which processes and molecules control each other's behavior.

Is a biological rule-based system nothing but a giant pathway diagram, drawn in a computable format? In spirit, the answer is "Yes," but in practice the answer is strongly "No." The first issue is that in a rule-based system, knowledge must be encoded in a way that precisely preserves all the semantic distinctions necessary to perform the desired inferences. Pathway diagrams, on the other hand, are intended to be read by knowledgeable researchers, to succinctly communicate a manageable number of concepts and relationships in a single page. To that end, they exploit the ability of humans to interpret symbols based on context, to effortlessly shift levels of abstraction, and to bring to bear background information and unspoken assumptions about the biological system. So one arrow between colored boxes in a diagram may mean "activation of the kinase activity of A by B," whereas another may mean "increase in RNA expression of C due to the activation of transcriptional activity of D." A third arrow may mean "X translocates across the plasma membrane due to the activity of channel Y." One colored box may be labeled "RB1," meaning the product of a particular gene, whereas another labeled "E2F" refers to the proteins of an entire family of related genes. For an algorithm to use that information – to make inferences similar to the reasoning of a human reader of the diagram – each meaning expressed by the diagram must be represented in a way that captures those essential distinctions.

The second issue is the need for consistency of terminology across all the facts encoded in a rule-based system. Human researchers appreciate consistency, but they are adept at moving between diagrams that use different symbols for the same proteins or diagrams in which the same symbols are applied to genes, proteins, or RNA transcripts depending on the context. But to perform meaningful inference using a rule-based system, each unique biological concept must be represented in the system by a unique and unambiguous term. In this way, biological rule-based systems are less like pathway diagrams and more like circuit diagrams, in which each relationship and component type is well defined. And indeed, an excellent example of the representation of biological rules is Kohn's molecular interaction map of the mammalian cell-cycle control and DNA repair systems, which encodes over 500 causal relationships in a precise semantics using a graphic format similar to a circuit diagram (7).

Thinking of assemblages of biological knowledge about the endothelium as rule-based systems rather than pathways, networks, or circuits is an important viewpoint because it forces one to ask critical questions:

- What kind of deductions do I want to make?
- What kind of conclusions could my system potentially draw that would be worth the effort of its construction?
- Would these conclusions be testable?
- Would they advance the knowledge of the biology of the system?

These questions lead to the definition of the discrete semantics that the rules and concepts must express, the nuances that must be encoded, and the distinctions that can be omitted. These semantics distinguish systems that can directly enable the development of scientific hypotheses from those that can support only browsing and searching.

One type of question that a rule-based system can answer is "What if?" If we hypothesize a perturbation to the system, such as a decrease in a process or an increase in the amount of some component, then any rules that use that process or component in their "if" clause can be applied to draw the conclusion in their "then" clause. Those conclusions may then be treated as new perturbations – increases and decreases that are then applied to other terms in the system. This process can be repeated multiple times, propagating changes throughout the system. The result is a list of predicted changes and, for each prediction, a causal chain of events supporting that prediction.

But it may be more useful to ask "What happened?" than "What if?" In today's endothelial research, enormous datasets are generated routinely by high-throughput technologies such as microarray expression profiling, tissue proteomics, or serum metabolomics. Each individual measurement out of thousands or millions may reveal a state change that reflects the biology of the experiment, and each state change should ideally be explained by the analysis and interpretation of the experiment. However, these changes do not occur independently, but are tightly knit into a causal fabric by the rules of the biological system itself. In practice, when a dataset yields hundreds of state changes between two samples, it is intractable for a human researcher to explore all the possible causes for those changes, let alone search for commonalities. But within a biological rule-based system, it is practical to do just that: to explore the possible causes of every observed state change, the causes of those causes, and so on for multiple iterations. Each possible cause may then be evaluated to determine its fitness as an independent hypothesis, or as part of another clearly defined hypothesis. Clearly, hypotheses whose predictions are highly concordant with the set of observed state changes are among the most interesting; concordance and other parameters can be used to rank and filter hypotheses, producing a manageable list of detailed hypotheses for review and testing.

In sum, it is valuable to think about aspects of endothelial biology as sets of rules operating on abstract concepts, because finding rules – finding causal relationships – is at the root of the scientific process. Encoding biological facts in a manner that makes them computable requires the explicit choice of clear and consistent terminology and semantics, a process that encourages clarity and completeness in any biological model. When those facts are computable, meaningful inference can be performed over models and datasets too large for any individual researcher to reason about.

CONCLUSION

Embrace computable metaphors. Think of biological knowledge in terms that can lead to computational models. The biology of the endothelium is too big for the mind of any individual scientist. But when biological knowledge is manipulated by algorithms, larger phenomena can be described, explained, and predicted. Rule-based systems show how many thousands of simple rules can encode biological relationships and use them to perform scientific inference, such as the interpretation of the state changes from a high-throughput experiment. CA demonstrate how small sets of rules, when applied numerous times, can generate complex predictions of biological behavior. Weather modeling is an inspiring success in the practical prediction of the behavior of an enormous system, but it also shows the inherent limits of those predictions. Although the construction of very large computational models is not easy, those models can yield pragmatic results in situations that are otherwise intractable.

KEY POINTS

- Systems biology, and specifically endothelial systems biology, is the art of defining and implementing computable metaphors, employing diverse computational techniques to generate testable predictions and hypotheses.
- Encoding biological facts in a computable framework facilitates meaningful inference when the models and datasets become too large for the mind of any individual scientist.
- No "free lunch" exists. The effective models of biological systems require the explicit choice of clear and consistent terminology and semantics, the input of many high-quality observations, and are fundamentally limited in their ability to predict specific behaviors of complex systems.
- Practical value can be realized from partial solutions.

Future Goals

- Complex models require comprehensive measurements: to build and test large scale models, systems must be measured in detail and under many conditions.

REFERENCES

1 *Bull. Amer. Met. Soc.* 79, 2161–2163. Available at http://www.ametsoc.org/policy/statewaf.html

2 Gelsinger P, Gargini P, Parker G, Yu A. Microprocessors circa 2000. *IEEE Spectrum*. October, 1989.

3 Rietkerk M, Dekker SC, de Ruiter PC, van de Koppel J. Self-organized patchiness and catastrophic shifts in ecosystems. *Science*. 2004;305:1926–1929.

4 Peirce SM, Van Gieson EJ, Skalak TC. Multicellular simulation predicts microvascular patterning and in silico tissue assembly. *FASEB J*. 2004;18(6):731–733.

5 Pollard J, et al. A computational model to define the molecular causes of type 2 diabetes mellitus. *Diabetes Tech Therapeut*. 2005; 7(2):323–336.

6 Pratt D, et al. Causal analysis of the androgen-induced transcriptional program of human prostate cancer. Poster presented at AACR Special Conference on Basic, Translational, and Clinical Advances in Prostate Cancer. November 2004.

7 Kohn KW. Molecular interaction map of the mammalian cell cycle control and DNA repair systems. *Mol Biol Cell*. 1999;(8): 2703–2734.

PART II

ENDOTHELIAL CELL AS INPUT-OUTPUT DEVICE

Input

27

Introductory Essay

Endothelial Cell Input

Helmut G. Augustin

Medical Faculty Mannheim, University of Heidelberg and German Cancer Research Center (DKFZ), Germany

The vascular endothelium lines the inside of all blood vessels. As such, it forms one of the largest internal surfaces that mediates the compartmentalization of the body. The endothelium thereby acts as interface between the blood and the different organs. Structurally, the endothelial layer is diverse and heterogeneous. It is organ and caliber specifically differentiated in a way that best serves the functional needs of the underlying tissue. For example, barrier-forming endothelia such as the brain and lung endothelium are continuous with numerous tight junctions that act as a permeability barrier. The endothelium in the kidneys is continuous, but has numerous fenestrae that facilitate the kidneys' filtration function. Sinusoidal endothelial cells (ECs) are discontinuous, allowing easy entry and exit of fluids and solutes.

The molecular analysis of organ- and caliber-specific EC differentiation is still in its infancy. A number of organ-specific EC molecules have been identified, such as endothelial-specific molecule (ESM)-1 as a marker of lung ECs (1) and the stabilins as markers of sinusoidal ECs (2). Yet, the functional role of such organ-specific EC molecules is not understood. Similarly, several caliber-specific EC molecules have been identified in recent years. EphrinB2 is selectively expressed by arterial (and angiogenic) ECs, whereas EphB4 is preferentially expressed by venous ECs (3). Correspondingly, ephrinB2- and EphB4-deficient mice have essentially complementary embryonically lethal phenotypes characterized by perturbed arteriovenous differentiation (3). The asymmetric arteriovenous expression of ephrinB2 and EphB4 has stimulated research into the identification of EC molecules with arteriovenous asymmetric expression pattern. Some 20 arterial molecules (and much fewer venous-specific EC molecules) have been identified in the last 10 years. Correspondingly, gridlock and COUP-TFII have been characterized as transcription factors controlling arterial and venous EC differentiation, respectively (4,5).

ECs are surrounded by a complex microenvironment that controls the forces acting on the EC layer (Figure 27.1). Luminally, the polarized EC monolayer is exposed to the flow of blood. Blood flow and blood pressure exert physical forces on the endothelium (shear stress and mechanical strain). Likewise, circulating cells interact with surface molecules of the vascular endothelium (not only leukocytes, platelets, and red blood cells but also metastasizing tumor cells), and molecules in the local or systemic circulation may directly affect EC functions including permeability, coagulation, adhesiveness, and vasotonus. Laterally, ECs are in contact with neighboring ECs through adherens junctions and tight junctions. Cell contact–dependent signaling is well established as controlling many important EC functions. Abluminally, the EC layer is in contact with a caliber- and organ-specific extracellular matrix that is produced by the EC itself as well as by neighboring cells. ECs are abluminally also in contact with mural cells (smooth muscle cells and pericytes), the cell processes of which penetrate the EC extracellular matrix to exert regulatory functions on contacting ECs, particularly those related to the control of EC quiescence.

The vascular endothelium has been conceptualized throughout this book as an input-output device. Signals act on the endothelium (input), are processed within the endothelium (coupling), and enable the endothelium to exert specific functions (output). The input forces acting on ECs can be classified into milieu factors and specific molecular regulators. The mechanistic relationship between the microenvironmental milieu and specific molecular activities is, for most milieu factors, still poorly understood. For example, the identification of the hypoxia regulation of vascular endothelial growth factor (VEGF) expression is among the most important and most cited discoveries in the field of angiogenesis research (6) because it links the milieu factor hypoxia with the induction of gene expression of a key regulatory molecule of the angiogenic cascade. Yet, it took another 10 years to discover the prolyl hydroxylases as the long-sought enigmatic oxygen sensors (7) (Chapter 29). Similarly, the role of biophysical forces in the vascular system (shear stress, mechanical strain) in controlling the molecular repertoire and phenotypic behavior of the vascular

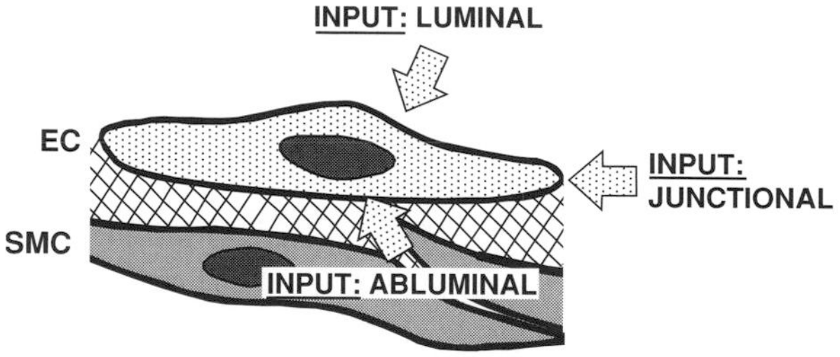

Figure 27.1. EC input. Anatomically, the vascular endothelium forms a cellular system of limited complexity. It is luminally exposed to biomechanical forces (shear stress, mechanical strain) and interacts with cells in the circulation. Laterally, ECs are in contact with each other through adherens junctions and tight junctions. Abluminally, ECs are in contact with their organ-specific extracellular matrix as well as with mural cells (smooth muscle cells [SMC], pericytes). Nevertheless, this comparatively simple anatomical unit is capable of responding to an enormous plethora of stimuli (*INPUT*) and can exert an array of different effector functions (*OUTPUT*), many of which are related to some of the most devastating human diseases.

endothelium has been solidly established. Yet, a molecular mechanosensor has not yet been identified (Chapter 28).

Milieu factors will eventually translate into the activities of specific molecular regulators. Locally and systemically acting stimulating and inhibiting molecules may act as specific or pleiotropic effectors of EC function. Some peptide growth factors, such as VEGF (Chapter 32) and the angiopoietins (Chapter 39) act almost EC-specifically through the restricted expression of their corresponding receptors by ECs. Other growth factors are pleiotropic in their action. For example, hepatocyte growth factor (HGF) (Chapter 33), the fibroblast growth factors (FGFs) (Chapter 34), and transforming growth factors (TGFs) (Chapter 35) are considered prototypic pleiotropic growth factors because they act on many cell types, but nevertheless exert specific effects on ECs.

Research performed over the last few years has identified intriguing parallels between the formation of neuronal networks and the establishment of vascular networks. Consequently, the neuronal system and the vascular system share a set of fate-determining, guidance signal–mediating, and positional information–transducing molecules. These include semaphorins with their neuropilin receptors (Chapter 37), ephrins and their corresponding Eph receptors (Chapter 38), Robo-Slit molecules (Chapter 40), and Notch molecules (Chapter 41). Mutations of many of these molecules cause specific embryonic lethal vascular phenotypes, reflecting the critical rate-limiting necessity of such guidance- and network-regulating signaling systems for proper vascular assembly.

The inflammatory response program is among the best characterized activation programs of the vascular endothelium. Inflammatory EC activation through tumor necrosis factor (TNF)-α (Chapter 31) or endotoxin (Chapter 46) causes a specific downstream signaling program that leads to the expression of inflammation-associated EC adhesion molecules that facilitate the recruitment of inflammatory cells. Similarly, a long list of other mediators and regulators of EC signaling and gene expression control EC function. These include opioids (Chapter 50), reactive oxygen species (ROS) (Chapter 42), nucleotides and nucleosides (Chapter 43), advanced glycation products (AGEs) (Chapter 47), complement (Chapter 48), the contact system (Chapter 49), thrombospondin (TSP) (Chapter 36), syndecan (Chapter 44), and sphingolipids (Chapter 45). ECs also are vulnerable to the effects of circulating toxins. Examples are environmental toxins (Chapter 59) and snake venoms (Chapter 51).

Other milieu factors are able to control phenotypic properties and functions of the vascular endothelium through receptor-independent and/or multifactorial mechanisms. These include acid–base metabolism (Chapter 30), hyperthermia (Chapter 52), hyperbaric oxygen (Chapter 53), barotrauma (Chapter 54), deep sea diving (Chapter 55), exercise (Chapter 56), high altitude (Chapter 57), and gravity (Chapter 58).

Contact with neighboring cells similarly contributes to the specific microenvironments that control EC phenotype and function. Neighboring cells can control ECs through direct contact or through paracrine-acting mediators. ECs are in most intimate contact with mural cells (smooth muscle cells and pericytes [Chapters 60 and 61]), which control the quiescent phenotype of the endothelial layer. Likewise, cells in the circulation, such as red blood cells (Chapter 62), leukocytes (Chapter 63), and platelets (Chapter 64), interact with ECs, resulting in bidirectional signaling events. In addition to contacts with abluminally and luminally contacting cells, ECs also may establish a dialogue with other cell types in select vascular beds. Such interactions include cardiomyocytes in the heart (Chapter 65), hepatocytes (Chapter 66) and stellate cells (Chapter 67) in the liver, and podocytes in the kidneys (Chapter 68).

Examples exist of molecules that contribute both to input and output. Some of these autocrine factors, such as nitric oxide and carbon monoxide, are covered in the section on Output.

Collectively, the vascular endothelium forms an enormously complex input-output system. The relative anatomical simplicity of the vessel wall stands in striking contrast to its remarkable signal transduction capacity and high level of activity. Clearly, the vascular endothelium is functionally involved in some of the most devastating human diseases. These include primary vascular diseases such as hypertension, coagulopathies, ischemic diseases, and atherosclerotic vascular changes. Furthermore, the vessel wall also is involved in many nonvascular diseases including among others tumor growth (tumor angiogenesis) and endocrine disorders (diabetic retinopathy). It may not come as a surprise that more than two-thirds of all human deaths involve the vascular system either directly or indirectly. Thus, a solid understanding of the factors and forces that act on the endothelium – as outlined in the chapters of this section – forms the foundation for a better mechanistic understanding of vascular pathophysiology.

REFERENCES

1 Wellner M, Herse F, Janke J, et al. Endothelial cell specific molecule-1 – a newly identified protein in adipocytes. *Horm Metab Res* 2003;35:217–221.
2 Politz O, Gratchev A, McCourt PA, et al. Stabilin-1 and -2 constitute a novel family of fasciclin-like hyaluronan receptor homologues. *Biochem J* 2002;362:155–164.
3 Heroult M, Schaffner F, Augustin HG. Eph receptor and ephrin ligand-mediated interactions during angiogenesis and tumor progressions. *Exp Cell Res* 2006;312:642–650.
4 You LR, Lin FJ, Lee CT, et al. Suppression of Notch signalling by the COUP-TFII transcription factor regulates vein identity. *Nature* 2005;435:98–104.
5 Zhong TP, Rosenberg M, Mohideen MA, et al. Gridlock, an HLH gene required for assembly of the aorta in zebrafish. *Science* 2000;287:1820–1824.
6 Shweiki D, Itin A, Soffer D, Keshet E. Vascular endothelial growth factor induced by hypoxia may mediate hypoxia-initiated angiogenesis. *Nature* 1992;359:843–845.
7 Epstein AC, Gleadle JM, McNeill LA, et al. C. elegans EGL-9 and mammalian homologs define a family of dioxygenases that regulate HIF by prolyl hydroxylation. *Cell* 2001;107:43–54.

28

Hemodynamics in the Determination of Endothelial Phenotype and Flow Mechanotransduction

Peter F. Davies

Institute for Medicine & Engineering, University of Pennsylvania, Philadelphia

The movement of blood in the systemic and pulmonary circulations influences every cell of those vascular systems, but its most immediate mechanical effect is on the endothelium. The endothelium is well recognized as a hemodynamic mechanotransduction interface that in turn influences the biology of the vessel and its physiological responses. Furthermore, in arteries, hemodynamic forces are associated with endothelial phenotypes that predispose specific sites to focal atherogenesis. This chapter considers the relationships between hemodynamics and endothelial phenotypes and reviews our current understanding of the broader mechanisms of endothelial mechanosensing and signal transduction as they relate to arterial physiology and pathology. The microcirculation and postcapillary venous return – regions in which flow effects on leukocyte interactions with the endothelium occur in the context of acute inflammation – are considered elsewhere in this volume (see Chapter 114).

THE ENDOTHELIUM IS HIGHLY RESPONSIVE TO LOCAL HEMODYNAMICS

A feature common to all endothelial cells (ECs) studied in a variety of vascular locations and cell cultures is the ability to respond to local changes of blood flow, particularly to acute and sustained changes of hemodynamic shear stresses that deform the cell (1–4). The endothelia of arteries and heart valves are generally subjected to the highest magnitudes and variations in hemodynamic forces. Arterial hemodynamics has received the most experimental scrutiny, in part because of the correlative relationships that exist between certain hemodynamic characteristics and atherosclerosis. However, models and measurements of flow complexities around the heart valves recently have focused on molecular investigations of the role of hemodynamics in regulating valve endothelial pathophysiology; both are considered here.

As presented at length elsewhere in this book, ECs play an important role in flow-mediated vasoregulatory control. Acute dilatation and constriction of arteries in response to changes of flow are controlled through endothelial-derived nitrovasodilators, prostaglandins, lipoxygenases, hyperpolarizing factors, and related molecules (1,3). Sustained changes in hemodynamics stimulate the structural remodeling of the arterial wall through a process that is also endothelium-dependent (4) and involves a carefully orchestrated sequence of gene and protein expressions that facilitate major structural adaptation to the new hemodynamic environment. Furthermore, hemodynamics plays a critical role in the arterial circulation at those locations where most vascular pathologies originate and develop their morbidity (5). The localized initiation and development of atherosclerosis and cardiac valve dysfunction are highly influenced by hemodynamics (6).

HISTORICAL CONTEXT

Until the Renaissance, the prevalent theory of blood movement remained that put forward by the second-century Greek physician Galen, who proposed that blood vessels were discrete noncirculating ebb-and-flow systems supplying the separate fluid compartments associated with the primary organs. During the fifteenth and sixteenth centuries, Versalius and Da Vinci highlighted some of the inconsistencies of the Galen theory (7), but it was not until the early seventeenth century that a complete circulation was demonstrated by William Harvey. However, although it had been recognized for a thousand years that large blood volumes move through complex branching networks, blood vessels were studied from a fluid dynamic perspective only as passive rigid tubes until well into the twentieth century. During the past three decades, hemodynamic forces and cellular mechanics have become firmly established in most aspects of vascular biology, and in endothelial research in particular.

The relationship between hemodynamics and the geometry of the cardiovascular system was first remarked upon during the sixteenth century by Leonardo Da Vinci, who extrapolated his interest in hydraulics to the anatomy of the human heart (7). While comparing dissected heart valves and arteries to the geometry of natural river systems, he predicted that flow would separate in regions around the valves and branches of the arteries as a matter of (physiological) course. During the nineteenth century, the pathologist Virchow (8) noted that "athero-sclerosis" favored larger arteries, and he made the important distinction that the *cellular* components of the tissue play a prominent role in mediating inflammation. However he did not comment on the role of flow itself in the process. Little further consideration was given to hemodynamics and a metabolically active endothelium, even during the early twentieth century, when a clearly nonrandom distribution of lesions was obvious at autopsy, and the first risk factor for atherosclerosis (hypercholesterolemia) had been identified (9). Renewed interest in hemodynamic patterns in atherosclerosis, but initially not particularly the endothelium, was driven by rheological engineers during the 1960s. The fluid dynamics studies of Caro and colleagues (10) in London, and Fry, Roach, and colleagues in the United States and Canada (11,12) provoked biologists to reconsider the endothelium as an active interface. This coincided with ultrastructural transmission electron microscopy studies that revealed intracellular complexity predictive of the dynamic biology of these cells (13). The isolation and culture of ECs in 1973 (14) led to an appreciation of their structural, functional, and metabolic complexity and facilitated those in vitro flow studies performed under controlled conditions that began at the end of that decade (15–17). During the same period, the central role of the endothelium in vascular pathology was established by demonstration of the retention of an intact endothelium during the initiation and development of atherosclerotic lesions, thereby connecting endothelial structure and function to lesion development (18). In pathology, therefore, the link between complex hemodynamics and heterogeneous patho-susceptible endothelial phenotype became evident.

In 1980, Furchgott made the seminal discovery that vascular relaxation is endothelial-dependent (19). It was quickly realized that Schretzenmayer's observations, made during the 1930s, of acute flow-dependent vasoregulation were mediated by the endothelium (20). The role of the endothelium in flow-regulated vessel remodeling was subsequently demonstrated (4). From the late 1970s onward, devices to impose simulated blood flow over cultured ECs were developed to investigate hemodynamic responsiveness (15–17), and this was paralleled by a renewed appreciation of in vivo hemodynamics in relation to lesion location, surgical manipulation, and vascular grafts. These inquiries have continued to the present day, with the progressive inclusion of the most modern technologies in device engineering, cell and molecular biology, computation, and imaging.

Experimental studies of hemodynamics and the endothelium primarily address two thematic questions related to arteries:

- What role does mechanotransduction play in the localization of atherosclerotic lesions and other cardiovascular pathologies?
- What are the fundamental mechanisms of mechanosignaling and transduction in normal vascular physiological function?

Endothelial mechanotransduction mechanisms are likely to underlie the susceptibility to atherosclerosis and inflammation at those arterial locations where unstable hemodynamics is prevalent and where synergisms of hemodynamics and chemical signaling (cytokines, chemokines) mediate the expression of endothelial adhesion proteins, leukocyte attachment, and cell migration.

HEMODYNAMIC FORCES

Blood, a complex inhomogeneous mixture of fluid and cells, is imperfectly modeled as a homogeneous Newtonian fluid in most experimental endothelial investigations, an approximation that, despite its limitations, has provided significant insights into endothelial biology. Forces imparted to a vessel by blood flow can be resolved into two principal vectors: (a) blood pressure perpendicular to the vessel wall and (b) shear stress that acts parallel to the vessel surface to create a frictional force at the endothelium, usually expressed as dyn/cm^2 or Pascals (10 dyn/cm^2 = 1 Pascal). Strain is the deformation that results when a stress (force/unit area) acts on a cell or tissue. Thus, both pressure and shear stress result in strain of the vessel wall.

The shear stress varies inversely with the third power of radius. Therefore, shear stresses at high flow in a wide tube (e.g., the aorta) and at lower flow in a narrow tube (e.g., the capillaries) are within a similar range. However, in the larger vessels much greater pulsatility occurs, which results in more variable flow characteristics than in the steadier flow environment of smaller arteries and the microcirculation. Intermolecular attraction between the layers within the fluid is strong, resulting in blood flow as a cohesive body of fluid slowed only by frictional contact with the vessel surface (endothelium); this is pulsatile laminar (or layered) flow. The pulsatility is simply a slowing down or speeding up of laminar flow moving in the same direction throughout each cardiac cycle. In oscillatory flow, which is pulsatile, a reversal of flow occurs during part of the pulse cycle. In arteries, oscillatory flow usually is the result of the bulk flow separating into distinct compartments of undisturbed and disturbed flow as the interfluid molecular interactions are overcome along the interface of separation,

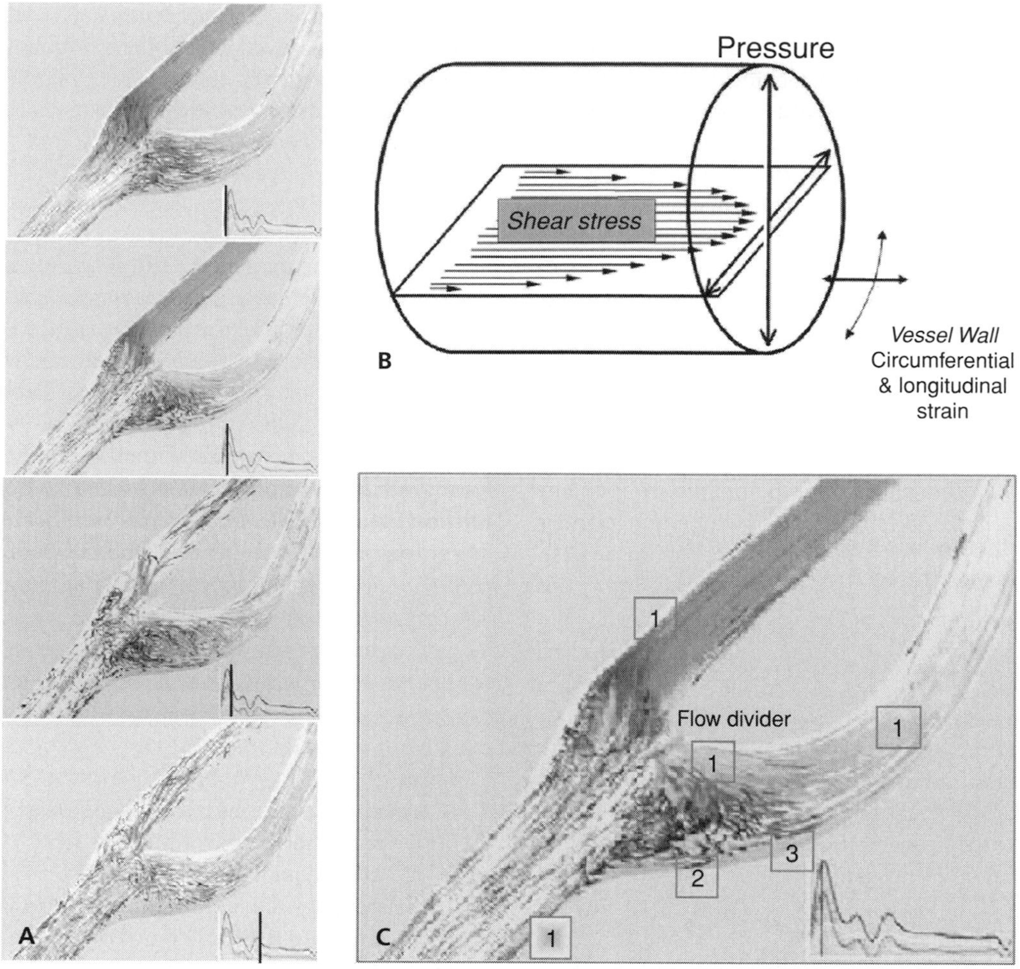

Figure 28.1. Flow characteristics in the normal human carotid artery at the bifurcation into internal and external branches illustrating some spatial and temporal characteristics of the flow. (**A**) Four frames at progressive points in the cardiac cycle illustrate the transition from unidirectional laminar flow (and shear stress) in the bulbous carotid sinus (panel 1) to the onset (panel 2) and development (panel 3) of complex disturbed flow, including flow reversal beyond peak systole and the restoration of unidirectional laminar flow in the sinus during diastole (panel 4). (**B**) Resolution of hemodynamic forces into simplified principal vectors of shear stress parallel to the vessel wall and pressure perpendicular to the wall. These stress forces result in cell and tissue deformation (strain) in the endothelium and underlying vessel wall. (**C**) Enlarged panel from 1A showing spatial separation of flow disturbance. Flow varies in velocity but is laminar and unidirectional at multiple locations (labeled 1). These include the common carotid artery proximal to the bifurcation, the entire lesser branch (internal carotid) artery, the wall of the external carotid artery adjacent to the flow divider, and the distal region of the same branch. In contrast, a transient unstable vortex containing elements of flow reversal (labeled 2) develops in the external carotid artery opposite to the flow divider. Such regions are highly susceptible to the development of atherosclerosis. At 3, the laminar unidirectional flow "reattaches" to the wall beyond the boundary of disturbance. The images in A and C were extracted and modified from a remarkable computational fluid dynamics recording of blood flow in this tissue by Professor David A. Steinman of the Department of Mechanical and Industrial Engineering, University of Toronto (image gallery URL: http://www.mie.utoronto.ca/labs/bsl/). Adapted with permission.

an event that requires a considerable input of energy. Such oscillatory disturbed flow in arteries is invariably associated with an encounter with an obstacle (e.g., flow divider of a branch artery; Figure 28.1), or a curvature in the vessel around which the blood following the outer trajectories must travel farther than that of the inner curve (e.g., the aortic arch), or a sudden expansion of the cross-sectional area of the vessel (e.g., aortic valve stenosis). The instabilities often form a vortex that may remain laminar within its boundaries or, when enough energy is imparted into the separated flow, it may break down to chaos (turbulent flow). Flow separations are difficult to define precisely because, at a single location, not only may they exhibit pulsations, oscillations, secondary helical flows, laminarity, and turbulence, but the characteristics are changing from millisecond to millisecond throughout the cardiac cycle. Collectively, the term *disturbed flow* commonly is used for such regions. Although the great majority of flow in the arterial circulation is pulsatile, unidirectional, and laminar, the disturbed flow regions are important because they predispose to atherosclerosis. Figure 28.1 outlines these general flow characteristics in a normal human carotid artery sinus.

Whatever the local flow characteristics, the endothelium is the principal recipient of shear stress forces by virtue of its direct contact with the blood flow. Although some transmural interstitial fluid movement acts on the vascular smooth muscle cells (VSMCs) to generate shear stresses, the mixture of VSMC and connective tissues (collagens, elastins, glycosaminoglycans) and four-dimensional fluid movement render it difficult to experimentally estimate the level of shear stresses with high resolution. However, they generally are considered to be low.

In most of the arterial circulation, blood velocity profiles approximate pulsatile laminar flow in a rigid tube. In the capillaries and venules, the flow is steady but subject to upstream arteriolar vasoregulation that can vary the flow rate considerably, including stasis when blood flow is diverted to other capillary beds. Thus, variations in capillary flow occur with a much longer and more irregular periodicity than in nearby precapillary arterioles.

Flow velocity is maximal at the center of the lumen of larger vessels, remains high across much of the cross-sectional area, and decreases sharply near the vessel surface, where wall shear stress acts at the endothelial surface. Shear stress is in the direction of the flow and is proportional to the fluid velocity and viscosity; in undisturbed laminar flow in the arterial circulation, the endothelium is subjected to pulsatile unidirectional shear stress typically in the range 10 to 20 dyn/cm^2. Because the cell morphology reflects the hemodynamic environment, most arterial ECs are elongated and aligned in the direction of flow. In contrast, in disturbed flow regions, where pulsatile instabilities generate transient vortices that collapse and reform with each cardiac cycle, the cells exhibit no preferred orientation or alignment. Figure 28.2 illustrates endothelial

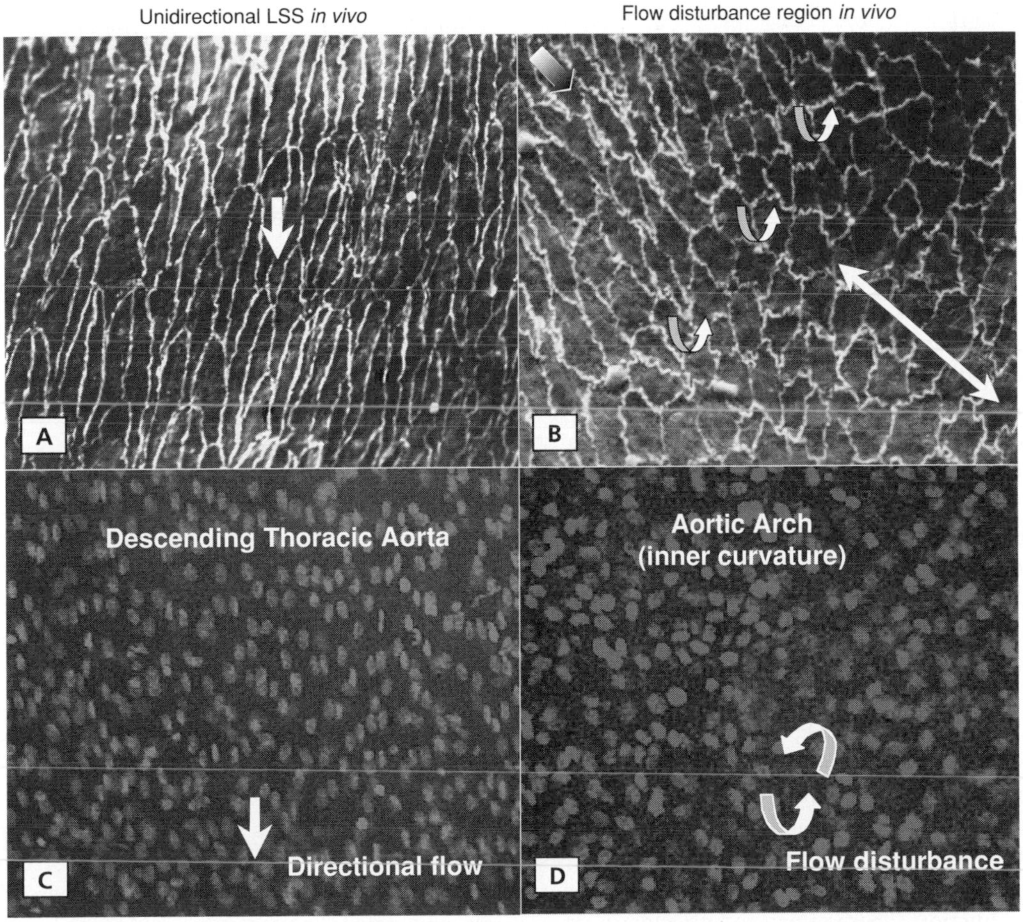

Figure 28.2. Flow-mediated EC orientation. In situ EC and nuclear orientation in undisturbed unidirectional laminar flow in the descending thoracic aorta (**A,C**) and separated disturbed flow (**B**) adjacent to an intercostal branch of the aorta. Such flow characteristics are associated with arterial branch sites, curvatures, and bifurcations. In **B**, a region of cell alignment changes abruptly to polygonal cells beyond a line of flow separation (*curved arrows*) that marks the boundary of a disturbed flow region (*double-headed arrow*). In **D**, the nuclear pattern is representative of a similar disturbed flow region in the inner curvature of the aortic arch.

morphology in regions of different flow characteristics in vivo. At high flow velocities and/or heart rate, the complex but predictable disturbed laminar flow in these regions may collapse into chaos (turbulence). Thus, both disturbed laminar flow and nonlaminar turbulent flow depart from the pattern of those unidirectional forces at the endothelium that are characteristic of most other regions, and the ECs at these locations fail to align. From a large number of fluid dynamics studies and in vivo measurements, the range of shear stress in the arterial circulation in such regions has been estimated to vary from negative values, through zero values at the edges of flow-separation regions, and up to +20 dyn/cm^2. During episodes of increased cardiac output or hypertension, these values may increase considerably; transients in excess of 100 dyn/cm^2 have been recorded (15). The spatio-temporal gradients of shear stress in flow disturbance are extraordinarily large compared with sites of undisturbed flow, and the same regions are susceptible to the development of atherosclerosis when experimental risk factors are added. The phenotypic basis of atherosusceptibility at these locations in vivo is of great interest.

HEMODYNAMICS AND ENDOTHELIAL PHENOTYPES

Recently, methods that facilitate the global interrogation of cell structure, function, and tissue organization using high-throughput platforms have been applied to studies of endothelial mechanobiology. Endothelial phenotypes can be defined by "-omics" – techniques such as genomics, proteomics, metabolomics, phenomics, and systems biology that probe the genome, proteome, physiome, and function. These technologies provide valuable insights into endothelial biology through access to large databases managed by specialized statistical and bioinformatics units. The approach embraces a great breadth of information to provide a more comprehensive overview of cell biology as well as the identification of molecular associations leading to focused investigation. In particular, when applied at multiple spatial scales, phenotyping reveals greater endothelial heterogeneity than was previously realized; substantial differences in endothelial phenotype exist not only from vessel to vessel, but also from region to region, and cell to cell within subregions. The convergence of "-omics" with endothelial biology is not only part of the forefront of vascular biology but also is driving multiscale studies of endothelium in an effort to understand the basis of heterogeneity and evaluate its importance in relation both to vascular homeostasis in general and to the hemodynamic influences on vascular function and dysfunction. These efforts are directed in vitro and, more recently, in vivo and in situ.

Phenotypes and Flow

The endothelial phenotype is malleable and can be considered to be determined by two principal components: (a) an intrinsic phenotype imprinted genetically during development that defines the essential identity of the cell as endothelial, and (b) elements of the microenvironment that locally modify some characteristics of the intrinsic phenotype. Different vascular beds express specialized morphologies and functions appropriate to their tissue or organ location (e.g., fenestrated endothelium, blood–brain barrier endothelium, discontinuous sinusoidal endothelium, etc.). All retain the essential endothelial features of blood compatibility and express molecular markers of endothelial identity determined during development. Local environmental factors are responsible for much of the plasticity of endothelial structure and function and for the cell heterogeneity that exists within vascular beds (21–23). Separation of these criteria is somewhat artificial, and the programmed and local environmental factors work together to define phenotype. Recently, Chi and colleagues have demonstrated the retention of certain in vivo vascular bed–specific endothelial characteristics for several passages in tissue culture (24).

Abundant data show hemodynamic forces to be an important environmental determinant of endothelial phenotype in vivo. For example, endothelial gene and protein expressions vary at sites of different hemodynamic characteristics (22–28), and cause–effect relationships between hemodynamics and endothelial nitric oxide synthase (eNOS) expression were recently demonstrated in a coarcted carotid artery mouse model (29). In embryonic development, a venous obstruction increased cardiac shear stress and resulted in altered expression patterns of Kruppel-like factor 2 (KLF2), endothelin (ET)-1, and NOS (30). Furthermore, more than two decades of in vitro flow experiments have demonstrated endothelial structural, functional, and adaptive changes associated with shear stress stimuli (1).

Phenotypes in Undisturbed and Disturbed Flow

Fluid dynamics complexity increases in those regions of transient unstable flow separation associated with branches, bifurcations, and curvatures in the arterial circulation and downstream reattachment (rejoining the main flow). In flow disturbance, the average shear stress is lower but is associated with steep temporal and spatial gradients of shear stress, all within a restricted region of flow disturbance. Occasional turbulence (chaos) as well as disturbed laminar flow impart multidirectional forces to the cells. Consequently, these locations, unlike elsewhere in the arterial circulation, are characterized by an absence of regular EC alignment (see Figure 28.2). Disturbed flow regions correlate closely with susceptibility to pathological change: atherosclerosis in arteries and calcific sclerosis in heart valves. Endothelium at such locations in normal swine (22) and transgenic mice (27) expresses significant differences in gene expression compared with adjacent undisturbed laminar flow regions, and these differences are proposed to be an important influence on susceptibility to, or protection from, pathological change (31).

Endothelial Multigene Expression in Aorta in Vivo

Regions of disturbed flow (DF) susceptible to lesion development were compared to regions of undisturbed laminar flow (UF) in the aortic arch and thoracic aorta, respectively, in adult male swine. We profiled gene expression by paired replicate analyses in approximately 10^4 freshly isolated ECs from DF and UF regions of the inner curvature of the aortic arch and middle descending thoracic aorta respectively (Figure 28.3) (22). Total RNA was isolated for nucleic acid amplification, labeling, and microarray analysis. To confirm that the morphologies of the endothelium in each region corresponded to the effects of directional or disturbed nondirectional flow, the endothelial nuclei were stained with fluorescent nuclear Hoechst dye (DAPI). The axial ratio of the cell nuclei, indicative of cell orientation, showed good cell alignment in the region of undisturbed flow. In contrast, no preferential alignment was present in the aortic arch, a region of disturbed flow. Linearly amplified endothelial RNA was used to synthesize labeled DNA probes to screen filter arrays of 13,824 human cDNAs. Differential expression between DF and UF regions was validated by real-time reverse transcription PCR (QRT-PCR) sampling of genes. Of the array set, 13.6% (1,880 genes) were identified as differentially expressed in DF ECs compared with the UF site at $p < .05$. The significantly different genes were analyzed for meaningful biological associations. As summarized in Table 28-1, higher proinflammatory cytokine expression occurred in the arch (DF) compared with UF, and with expression of nuclear factor (NF)-κB transcript (but not NF-κB protein), as well as the adhesion factors CD44 and von Willebrand factor (vWF). The expression of adhesion molecules vascular adhesion molecule (VCAM)-1, intercellular adhesion molecule (ICAM)-1 and -2, P-selectin, and E-selectin (all associated with the early development of atherosclerosis) was no different between regions. Furthermore, profiles of multiple genes with redox-related functions revealed a net *anti*-oxidative phenotype in DF regions indicative of *protection* against reactive oxygen species. The DF region appears to be in a phenotypic state that is primed for inflammation but, in the absence of further risk factors, is physiologically stable – possibly because

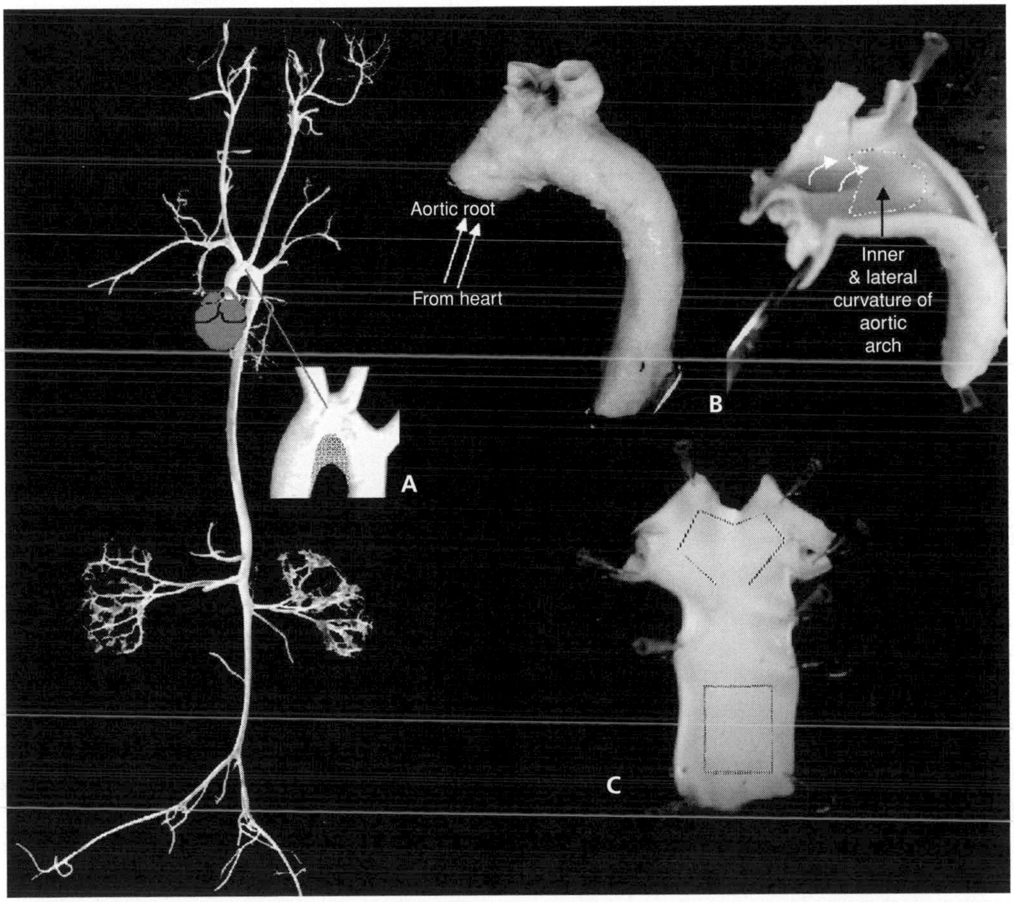

Figure 28.3. Anatomical regions of disturbed and undisturbed flow. Disturbed (inner and lateral curvature of the aortic arch) and undisturbed (descending thoracic aorta between intercostal branches) hemodynamic sites in adult swine aorta from which ECs were isolated for the study described in reference 22. An additional undisturbed flow site sampled was the common carotid artery. (Adapted with permission from Passerini AG, Polacek DC, Shi C, et al. Coexisting pro-inflammatory and anti-oxidative endothelial transcription profiles in a disturbed flow region of the adult porcine aorta. *Proc Natl Acad Sci USA*. 2004;101:2482–2487.)

Table 28-1: Coexisting Proinflammatory and Antioxidant Endothelial Gene Expression Profiles in Disturbed Flow, Atherosusceptible Regions of the Adult Pig Aorta in Vivo

Proinflammatory phenotype in disturbed flow (DF) in vivo*	
Upregulated	IL-1a, IL-1a receptor, IL-6, IL-8RB, MCP-1, NFkB, Protein C receptor, RAGE
Downregulated	IL-8, IL-10, IL-14, CXCR4, PAF receptor, c-fos,
Unchanged	ICAM-1, PECAM-1, ELAM-1, VCAM-1, ICAM-2, VE-cadherin and other molecules associated with cell-mediated onset of inflammation
Antioxidant phenotype in DF in vivo*	
Upregulated	Heme oxygenase 1, glutathione S-transferase, quinone oxidoreductase, glutathione peroxidase, SOD3. All are associated with antioxidant, protective mechanisms.
Downregulated	Vitamin D_3-upregulated protein (VDUP-1) (also known as thoredoxin-interacting protein), cytochrome c oxidase, NADH dehydrogenase. Their downregulation is consistent with antioxidant protection. eNOS was also downregulated. NAD(P)H oxidase was unchanged.
Cholesterol balance in DF in vivo*	
Upregulated	ATP-binding cassette A1, ApoA1 regulatory protein, Apo A1, Clusterin (Apo J); most are involved in reverse cholesterol transport.
Downregulated	Apo E, PPARγ, scavenger receptor A, Retinoid X receptor B
Coagulation balance in DF in vivo*	
Upregulated	vWF, Factor XIII (procoagulant), t-PA, and plasminogen (anticoagulant)
Downregulated	PAI-1, urokinase receptor

*When compared to undisturbed flow.
IL, interleukin; MCP, macrophage chemoattractant protein; RAGE, receptor for advanced glycation end product; CXCR, chemokine receptor; PAF, platelet activating factor; ICAM, intercellular adhesion molecule; PECAM, platelet-endothelial cell adhesion molecule; ELAM, endothelial leukocyte adhesion molecule; VCAM, vascular cell adhesion molecule; VE-cadherin, vascular endothelial cadherin; SOD, superoxide dysmutase; Apo, apolipoprotein; PPAR, peroxisome proliferator-activated receptor; vWF, von Willebrand factor; t-PA, tissue-type plasminogen activator; PAI, plasminogen activator inhibitor.
(Adapted with permission from Passerini AG, Polacek DC, Shi C, et al. Coexisting pro-inflammatory and anti-oxidative endothelial transcription profiles in a disturbed flow region of the adult porcine aorta. *Proc Natl Acad Sci USA*. 2004;101:2482–2487.)

of coexisting protective pathways. Thus it meets a definition of athero-susceptible, but not atherogenic. Genes assigned to other mechanisms implicated in lesion development, reverse cholesterol transport, and coagulation, also showed net *protective* profiles in the DF region (see Table 28-1). A delicate balance of pro- and antiatherosclerotic mechanisms may exist simultaneously in the endothelium of lesion-prone sites to create a setting of net vulnerability to atherogenesis but with protective measures also present. A shift from atherosusceptibility to atherogenesis may occur by *inhibition of the protection.* Thus, disturbed flow characteristics act as a local risk factor for development of atherosclerosis, but they also provide a degree of protection; additional risk factors may overcome the protection. Such a scenario is consistent with the absence of atherosclerosis in the presence of disturbed flow but in the absence of other risk factors. As noted in the section "Endothelial gene expression in heart valves in vivo," similar protective gene positioning is noted in the undiseased aortic valve (28).

Gene Clustering by Hemodynamic Location in Vivo

In a follow-up study, we compared endothelial transcript profiles from hemodynamically distinct arterial regions in 15 mature pigs: males and females fed a normal diet, and males briefly fed a high-fat diet (2 weeks) (32). Hierarchical clustering analysis showed preferential grouping of arrays by region over the risk factors of hypercholesterolemia and gender. A set of differentially expressed genes was identified that clearly distinguished regions of disturbed flow from undisturbed flow; however, few differences were observed within the same region based on gender or diet. The results implicated regional hemodynamics as a dominant determinant of endothelial phenotypic heterogeneity underlying atherosusceptibility in vivo.

Endothelial Gene Expression under Flow in Vitro

Predating the recent in vivo endothelial profiling, candidate flow-sensitive genes were identified in tissue culture (reviewed in 33), and this was quickly followed by microarray analyses of gene expression (33–36). The results, however, are not easily compared to those of in vivo studies that capture the profiles of regions adapted in situ to differential hemodynamic environments. More relevant are two in vitro microarray studies that compared flow conditions analogous to DF and UF. In the first, Garcia-Cardena and colleagues (37) compared differential expression in cultured human ECs subjected to turbulent (TSS) and laminar shear stress (LSS) – analogous to DF versus UF regions in vivo. One hundred genes were identified as differentially expressed (TSS versus LSS) after 24 h (68 upregulated and 32 downregulated). The authors identified a set of genes that was down in LSS but up in TSS as potentially the most pathologically relevant. They considered highly regulated genes of known or putative function in signaling, response to injury, or atherogenesis, and focused on linking the protective effects of LSS to matrix biology and cell cycle.

Using microarrays (588 genes) and subtraction cloning, Brooks and colleagues (38) compared gene expression in cultured human arterial ECs under pulsatile DF conditions with cells subjected to LSS. They identified more than 100 genes as differentially expressed at 24 h, many of which were upregulated genes associated with mechanisms known to be proatherosclerotic, particularly inflammatory molecules, adhesion factors, and oxidation-related molecules. Few genes

Table 28-2: Principal Endothelial Genes Regulated by Disturbed Flow in Vitro (24 h)

Upregulated: Turbulence versus LSS (From Ref. 37)
TNF receptor stimulating factor 1, CD 1D, exportin, FGF receptor 3, BRCA2, NK tumor recognition sequence, GlcNAc transferase, spastic paraplegia 7, uracil DNA glycosylase, fibrogenic lymphokine, P2X purinoceptor

Upregulated: Low shear disturbed flow versus modest LSS (From Ref. 38)
ICAM-1; PECAM-1; ELAM-1; VCAM-1; ICAM-2; IL-9 receptor; MCP-1; IL-2 receptor α; IL-3 receptor α; TNF receptor 7; MIP-2α; p53; TSP1; integrins α3, αm, α6, α5, αv, β1; VE-cadherin; N-cadherin; thrombin receptor; ET-1; endothelin converting enzyme; glutathione S-transferase; MMP-14; MMP-7; HOX-7; HOX-11; cytochrome P450; alcohol dehydrogenase

Downregulated: Turbulence versus LSS (From Ref. 37)
C-termini binding protein-2, translation initiation factor 2, follicular lymphoma variant translocation 1, FOS-like antigen, granulin, KIAA0631 protein, SDF-1, thyroid hormone receptor interactor 7, TPA-inducible protein, cytochrome P450 subfamily V, fibronectin 1, G-proteins β1 and γ5, RAGE, thioredoxin reductase, sparc/osteonectin

Downregulated: Low shear disturbed flow versus modest LSS (From Ref. 38)
IL-7; RANTES; IFg; p57kip2; cyclin b1; GADD 153; HSP-40, 47, and 70; LFA-1; PAI-2; SOD-1 and -2; MMP-1 and -8; smad 1; BMP-4; b-myb; estrogen sulfotransferase; 1,4 α-glucagon branching enzyme

LSS, laminar shear stress (steady, undisturbed unidirectional flow); TNF, tumor necrosis factor; FGF, fibroblast growth factor; BRCA, breast cancer genes; MIP, macrophage inflammatory protein; TSP, thrombospondin; ET, endothelin; MMP, matrix metalloproteinase; HOX, homeobox gene; SDF, stromal derived factor; RANTES, regulated on activation, normal T cells expressed and secreted; GADD, growth arrest and DNA damage; HSP, heat shock protein; BMP, bone morphogenic protein

expressed by these in vitro investigations were common to those identified as differentially expressed in in vivo studies of normal arteries. Furthermore, little agreement existed between the two in vitro studies. The reasons may be the nature of the disturbed flow in each study, different cell origins, sensitivities of microarray analysis, or myriad technical considerations. Despite significantly different outcomes however, both these in vitro studies demonstrated that conditions of disturbed flow resulted in significant differential endothelial gene expression compared with conditions of unidirectional steady flow. Principal gene expression differences of interest from these two studies are outlined in Table 28-2.

Differential Endothelial PKC Activity, Isoform Expression, and Posttranslational Modification at Hemodynamically Distinct Arterial Sites

As an initial investigation into proteins playing a role in complex signaling networks involved in mechanotransduction, Magid and Davies (26) focused on protein kinase C (PKC), a family of signaling kinases that have been implicated in a wide variety of cellular processes. The PKC protein family has an endothelial distribution that is affected by flow in vitro (39). Moreover, the expression pattern of PKC isoforms suggest differences between athero-susceptible and athero-protected arterial sites (22). PKC isoforms ε and ζ have been implicated in eNOS (40) and NF-κB signaling (41), respectively. Using arterial endothelial protein lysates, endothelial PKC activities were mapped to DF and UF regions of swine aorta and carotid artery, sites susceptible or resistant to atherosclerosis. Total endothelial PKC activity (all isoforms) in the DF regions was 145% to 240% of that in UF locations, whereas the UF regions were not significantly different from each other. PKC isoforms α, β, ε, ι, λ, and ζ were expressed in similar proportions in both UF and DF regions, suggesting that differences of kinase activity were not directly attributable to expression levels. The inhibition of members of the "conventional" and "novel" PKC families in the lysates had no effect on kinase activity, which remained different in UF and DF lysates. However, inhibition of PKC-ζ, a member of the "atypical" PKC family, reduced the DF lysate kinase activity to that of UF levels. Threonine 410 phosphorylation in PKC-ζ, a proxy for activated PKC-ζ, was elevated in the DF region, whereas substantially more phosphorylated threonine 560 was present in PKC-ζ in the UF region, targeting the protein for degradation. These experiments suggest that posttranslational destabilization of PKC-ζ in the UF region leads to the observed decreased kinase activity compared with the DF site.

Endothelial Gene Expression in Heart Valves in Vivo

The aortic valve is another important tissue of multiscale, site-specific endothelial phenotype influenced by hemodynamics. The complex geometry, biomechanics, dynamics, and pathology of the valve endothelium is somewhat understudied in comparison to arterial endothelium. Earlier studies identified morphological and functional differences between aortic side and ventricular side endothelium (i.e., local site specificity), and recent work from this lab has identified side-specific endothelial phenotype heterogeneity (28). During the cardiac cycle, valve endothelium is subjected to complex fluid dynamics that are distinctly different on each side of the leaflet (42). Calcific sclerosis occurs preferentially on the aortic side of the valve leaflet, where greater flow disturbances prevail. Thus, a spatial correlation exists between the localization of calcific lesions and the local biomechanical environment, similar to that observed for athero-susceptible regions in the large arteries. Local environmental factors, including biomechanics, may contribute to differential endothelial phenotypes that define the sidedness of the valve and the focal susceptibility to calcification. Furthermore, recent evidence suggests that valve endothelium may express different biological properties compared with adjacent contiguous endothelium of the ascending aorta (43). Aortic valve disease shows some characteristics of atherogenesis, suggesting that the role of the endothelium also may be critical for the maintenance of optimal valve function and may influence its dysfunction.

Differential Endothelial Transcript Profiling from Aortic- Versus Ventricular-Side Aortic Valve

Our group used microarray analyses of endothelial gene expression from either side of normal adult pig valves to identify distinct differential expression patterns representative of the steady state in vivo (28). On the aortic versus the ventricular side, 584 genes were differentially expressed in situ by the endothelium. Several over-represented biological classifications with putative relevance to endothelial regulation of valvular homeostasis and aortic side vulnerability to calcification were identified. A pattern of gene expression permissive to calcification on the disease-prone aortic-side endothelium was identified. These included osteoprotegerin (OPG; tumor necrosis factor receptor superfamily, member 11b), C-type natriuretic peptide, and parathyroid hormone, each of which has been shown to inhibit cardiovascular calcification and each of which was suppressed on the aortic side of the valve leaflet. Also underexpressed on the aortic side was the transcript for chordin, a secreted protein that inhibits the osteoinductive effects of bone morphogenetic proteins by sequestering them in latent complexes. Aortic valve sections immunostained with an anti-OPG antibody noted a striking differential expression of the OPG protein consistent with the microarray predictions, with strong staining in and around the ventricular-side endothelium, but little expression on the aortic side. However, as was noted in the DF regions of arteries, coexisting putative protective mechanisms also were expressed at a higher level on the disease-prone side. Specifically, the transcripts for several intracellular antioxidative enzymes, including microsomal glutathione S-transferase 2, glutathione S-transferase ω-1, and peroxiredoxin 2, were more highly expressed. The lack of differential expression of proinflammatory molecules on the aortic side may protect against inflammation and lesion initiation in the normal valve. These studies implicate the endothelium in regulating valvular calcific sclerosis and suggest that the spatial heterogeneity of valvular endothelial phenotypes may contribute to the focal susceptibility for lesion development.

The above-mentioned recent studies of endothelial phenotype at different locations throughout the circulation represent snapshots of gene and protein expression in complex hemodynamic environments, where the endothelium is known to be subjected to different types of shear deformation. However, it should be remembered that, in vivo, the flow not only physically deforms the cells but also transports plasma proteins and chemicals to the tissue and removes metabolic products from it. The influence of such convective transport on endothelial phenotype is difficult to assess in vivo but may be considerable when short-lived molecules are released or metabolized at or near the cell surface (e.g., adenine nucleotides from platelets or cleavable peptides such as plasma bradykinin); a change of flow can quickly alter the local concentration in the boundary layer of the endothelium (44). It is unclear how important this direct contribution to endothelial flow responses is in vivo. In ex-vivo perfused arterioles, Koller and colleagues (45) demonstrated that vessel diameter changes were similar whether shear stress was increased by flow or by viscosity (in which convective transport remained constant, but shear stress increased at constant flow rate). Their results suggested that the shear stress deformation force dominated any effects of transport, and support the widely accepted view that endothelial mechanotransduction manifested through deformation is the most important hemodynamic mechanism influencing endothelial responses and phenotype.

MECHANISMS OF FLOW-MEDIATED ENDOTHELIAL MECHANOTRANSDUCTION

Although the importance of in vivo hemodynamics experiments, such as mapping cell phenotypes, measuring physiological responses, creating transgenics, and probing in situ using immunohistochemistry, cannot be overemphasized, access to the detailed cellular mechanisms of mechanotransduction in intact vessels remains limited. A more manageable approach to the mechanisms is through reductionist experiments in vitro.

Success in endothelial culture was quickly followed in the early 1980s by the development of in vitro devices that simulate key elements of blood flow characteristics in vivo. The adaptation of cone and plate rheological instruments (46) that expose large numbers of ECs to uniform shear stresses (calculated as laminar flow velocities over a nominally flat surface) provided the first direct observations of endothelial morphological, cytoskeletal, and metabolic responses (18,19,47). The subsequent introduction of an extreme form of disturbed flow, turbulent flow, into this system revealed that the flow characteristics (disturbed/undisturbed) were as important as the magnitude of shear stress. In contrast to laminar flow, turbulent flow stimulated endothelial proliferation whether the shear stress was high or low (20).

Most in vitro studies of endothelial flow responses compare cells subjected to shear stress with cells maintained in no-flow. Criticism frequently has been directed at this experimental approach because flow stasis does not exist throughout the arterial side of the circulation in vivo. Despite this, however, such studies have revealed mechanisms subsequently shown to be highly relevant to the in vivo environment, possibly because no-flow comparisons to flow may elicit a higher signal-to-background ratio than do incremental flow rate comparisons in vitro, or because no-flow models simulate aspects of the average low–shear stress environment measured in DF regions in vivo. More recently, serious attempts have been made to establish more complicated flow characteristics in vitro (e.g., steady versus pulsatile versus oscillatory) (48), including the imposition of complex flow profiles from recordings in vivo (49). Although the cultured monolayer of endothelium in such experiments is considered to be a homogeneous population of cells, atomic-force microscopy measurements of the surface topography demonstrate that significant force variations exist between cells when flow is introduced (50). Data derived from such experiments likely reflect an averaging of the cellular response measurements.

Extensive literature is available on in vitro endothelial responses to flow ranging from ion channel activation (very fast) to morphological changes (slow) bridged through the activation of many biochemical pathways involving transcriptional, translational, and posttranslational responses. The reader is referred to several reviews (1,51–53). Below are highlighted some organizational themes of endothelial mechanotransduction that illustrate the complexity of flow-signaling cascades.

Decentralized Endothelial Mechanotransduction

Throughout the 1980s, flow mechanotransduction was viewed somewhat simplistically as a shear stress–induced deformation event at the apical endothelial surface from which arises biochemical signaling analogous to agonist-receptor signal transduction. The development of tensegrity and other theories of connected tension elements in cells, particularly the contributions of Ingber (see Chapter 192) led to a reconsideration of the spatial distribution of flow forces in the endothelium (51) and the subsequent presentation of a formalized decentralized model based on live cell imaging experiments (1). *The model proposes that, although shear stress acts initially at the apical (luminal) cell surface and immediate biochemical responses located there, surface forces are simultaneously transmitted throughout the body of the cell such that multiple elements, located away from the apical surface may, independently or in concert, transduce the mechanical signals into biological responses* (Figure 28.4). The contributions of each element may vary, and indeed a single element may dominate at a given moment. Note that these statements apply to the transduction event – that is, the site of conversion from a mechanical deformation to a localized biological response. The subsequent downstream mechano-activated biochemical pathways may not remain restricted to the same subcellular site and also may occur later, but they originate there. *Mechanotransduction* is a term that is currently used to include both the force transduction event and the downstream signaling pathways that lead to altered cellular function. Flow induces many different endothelial responses that occur at different times and that may or may not be interdependent. The subcellular spatial relationships of mechanotransduction have been understudied, and those that have been studied suggest complicated distributions and translocations (1,52–54). Examples of spatio-temporal signaling cascades in flow responses are outlined below, but first it is instructive to consider an organizational basis for these mechanisms.

The finer distinctions within flow-mediated endothelial mechanotransduction might be considered as follows:

1. *Physical deformation.* The displacement of one or more cellular elements required for a flow response. This is the most immediate of events and is required to initiate any subsequent responses. Because the shear stress acts at the apical cell surface, earlier models considered endothelial mechano-sensing systems to be confined to this region; some investigators maintain that view. However, while no doubt remains that signaling elements are located at the apical cell surface (e.g., activation of membrane ion channels [55] or membrane phospholipids [56]), strong evidence suggests simultaneous force transmission by the cytoskeleton (57) causing deformation of cellular elements widely dispersed throughout the cell (with consequent dispersal of mechanotransduction), as outlined in step 2.
2. *Force transmission (mechanotransmission).* The deformation of a connected system under tension – the adherent EC in a confluent monolayer – results in virtually instantaneous transmission of the forces throughout the connected elements. The transmission conduit is principally the cytoskeletal fibers distributed throughout the cell body and the submembranous, spectrin-like cortical cytoskeleton around the cell. Both systems provide rigidity, shape, and structure to the cell and are under tension. Demonstrations of flow-initiated intermediate filament displacement (57), actin filament deformation (58), and directed motion of mitochondria attached to microtubules (59) support the view that apical forces are transmitted to "remote" sites via cytoskeletal displacements. Interference with cytoskeletal structure inhibits flow responses, as defined for various experimental systems. Thus, as proposed in the decentralization model, the actual mechanotransduction event may occur not only at the site of fluid shear interaction with the cell but also at considerable distances (microns) from the apical surface. Furthermore, the deformation effects may extend to adjacent cells, communicated through cell junctional structures (57).
3. *Force conversion to an initial biological response ("true" mechanotransduction).* It follows from step 2 that this critical event may occur at a number of locations simultaneously involving parallel, convergent, or divergent mechanisms and be associated with a variety of different proteins, but all are considered to involve an initial (or simultaneous) physical displacement or deformation in which the plasma fluid forces meet the cell surface. Such an interface is the Holy Grail of mechanobiology because it represents the first step in biological signaling resulting from flow force change. It is therefore the prime target for investigative and/or potential therapeutic intervention. An initial single canonical mechanism has not yet been defined and indeed seems unlikely. Suggested mechanisms include force-induced conformational changes in membrane proteins, physical effects on the lateral mobility of membrane molecules, direct force effects on ion channels, separation of assembled cell–cell oligomeric proteins at junctions, deformation of caveolar structures, and physical inhibition of integrin dynamics and clustering, especially at the basal sites of cell adhesion to extracellular matrices. Physical deformation also may displace local cofactors engaged

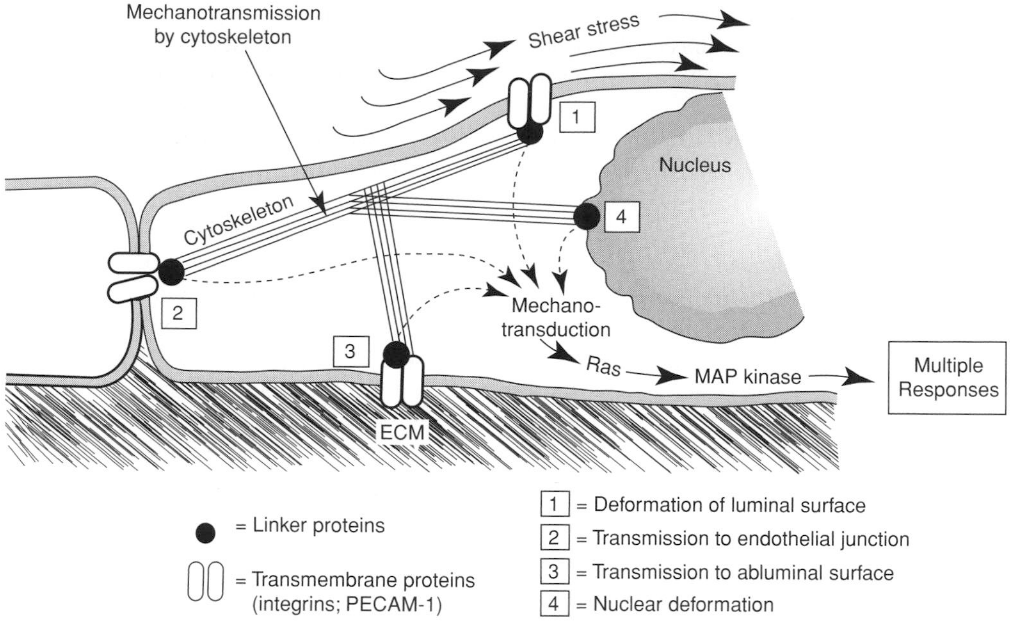

Figure 28.4. Decentralized model of endothelial mechanotransduction by shear stress. The cytoskeleton plays a central role in mechanotransmission of tension changes throughout the cell. Examples of direct signaling at the luminal surface (1), junctional signaling (2), and adhesion site signaling (3) are shown. Nuclear deformation also likely results in mechanosignaling (4). The locations are based on direct or indirect experimental evidence, are not mutually exclusive, and are likely interconnected. At 1, deformation of the luminal cell surface, possibly via the glycocalyx. Examples include localized activation of K^+, Na^+, and Cl^- ion channels, phospholipase activity leading to calcium signaling, and caveolar signaling. At 2, transmission of forces to the intercellular junction protein complexes via the cortical and/or filamentous cytoskeleton. VEGFR2 located at the luminal surface (site 1) or near the junction (site 2) may associate with VE-cadherin, β-catenin, and phosphatidylinositol 3-kinase to phosphorylate Akt and the primary transmembrane protein at this location, PECAM-1. At 3, cytoskeletal force transmission to adhesion sites. Transmembrane integrins bound to extracellular matrix serve as a focus for deformation that results in autophosphorylation of FAK, which binds the SH2 domain of c-Src, a kinase family that phosphorylates paxillin and p130 Cas and leads (via Ras G-proteins) to integrin-dependent activation of MAPK. In a variation of a similar assembly cascade, PYK2 found in multiple locations within the cell contributes to phosphorylation upon translocation to adhesion sites, the cytoskeleton itself, or the nucleus. A second, parallel integrin-mediated pathway involves the activation of Shc, which binds Src family kinases through SH2 domains. Shc phosphorylation leads to Ras-MAPK activity. A third integrin-mediated pathway is via RhoA activation that profoundly influences actin assembly. At (4), recent evidence suggests nuclear deformation or displacement that may result in mechanotransduction, possibly via lamins in the nuclear membrane.

in homeostatic regulation, including soluble and bound forms.

4. *Immediate and downstream signaling responses.* The fastest responses are likely to be closer to the force conversion event than are delayed responses. They include apical membrane ion channel activation, membrane lipid cleavage, membrane fluidity changes, and protein phosphorylation events, all of which lead to signaling pathways. These "immediate" responses occur at multiple locations throughout the EC, including the basal surface and the cell junctions, consistent with decentralized responses. Many different slower downstream events have been reported in response to shear stress in vitro. The database does not lend itself to a neat, interpretable series of responses, and indeed it is more likely that multiple parallel responses to flow allow the cell to quickly adapt to the local vascular environment. However, compelling evidence suggests the existence of multistep mechanotransduction pathways, and structural elements that are common to most responses.

The Central Role of the Endothelial Cytoskeleton in Flow Signaling

A unifying component in endothelial mechanotransduction is the cytoskeleton, which in all anchorage-dependent cells is in a state of intracellular tension arising from the association of cytoskeletal elements with each other and with cellular membrane proteins and organelles. When ECs are subjected to flow, the hemodynamic external forces are opposed by intrinsic cytoskeletal tension. Thus, although the apical surface itself is a primary site of deformation, forces communicated throughout the cytoskeleton influence abluminal adhesion sites, cell–cell junctions, the nuclear membrane, and other organelles. The almost instantaneous displacement of some of these organelles (cytoskeletal filaments, nucleus, mitochondria) can be mapped by two- and three-dimensional imaging in living ECs (57,58). Small areas of cell adhesion to the substratum (close and focal adhesions and their associated integrins) are dynamic, and some of these undergo directional remodeling within a few minutes after exposure to flow, even though the shear stress is acting on the opposite (luminal) side of the cell (1,60). At the luminal surface, Wang and colleagues (59) have demonstrated the transfer of twisting forces across the luminal endothelial membrane via transmembrane integrins and its resistance by cytoskeletal components. At adhesion sites on the abluminal surface, accessory proteins link cytoskeletal elements to integrins that bind extracellular adhesion proteins. Thus, continuity of cytoskeleton structural elements is present throughout the cell. Biochemical pathways associated with adhesion site integrins (reviewed in 52) show that they are important locations for the mechanotransduction elicited by cytoskeletal force transmission.

The cytoskeleton is involved in shear stress mechanotransduction at multiple levels. It determines cell geometry, including the luminal topography and thus the fine distribution of flow-generated surface forces. It maintains cell tension at: (a) adhesion sites, critical locations regulating many aspects of cell function; (b) cell junctions, which regulate contact inhibition of endothelial growth and cell–cell communication; and (c) the nuclear membrane, which may be physically important for gene regulation. It interacts with biochemical pathways throughout the cell, including those regulating its own turnover.

SPATIO-TEMPORAL COMPLEXITY IN FLOW-MEDIATED ENDOTHELIAL MECHANOTRANSDUCTION

The spatial organization of intracellular signaling may result in the stimulation of multiple parallel, convergent, and/or divergent mechanotransduction pathways. Although the mechanism(s) of initial transduction of a purely mechanical signal is unknown, efforts to understand endothelial mechanotransduction in terms of the generation of second messengers, the activation of transcription factors, and altered gene and protein expression that lead to structural and functional consequences are of great value. Recently, the spatial elements of mechanotransduction are receiving attention, in part because of a better appreciation of the four-dimensional aspects of mechanosignaling pathways (1,57–59). Rather than attempt to list many, often apparently unconnected, flow-induced endothelial responses, it is more instructive to summarize examples in which the experiments permit an integrated spatio-temporal interpretation of flow mechanosignaling. Several investigations have led to proposed mechanotransduction mechanisms that pay attention to the role played by the spatial structure of the cell.

Protein-Rich Tyrosine Kinase 2

A role for the spatially versatile protein-rich tyrosine kinase (PYK2) in flow mechanotransduction was recently reported (53). This kinase is an example of a molecule that is an important player in multiple signaling pathways, of which mechanotransduction is only one component. PYK2 is spatially interesting because it usually is found in the cytosol but has the versatility to relocate to focal adhesions, the perinuclear cytoplasm, or the nucleus. It also shares homology with focal adhesion kinase (FAK) that is localized to adhesion plaques at the basal side of the cell. In non-ECs it has been shown to exist downstream of G–protein-coupled receptors and able to link G-protein activation to NFκB activation through phosphoinositide 3-kinase (PI3K), the serine/threonine kinase Akt, and IκB kinase. PYK2 is phosphorylated in response to a variety of agonists and hormones and by ion channel activation. In endothelium, it binds Crk-associated substrate p130 (Cas), which itself is tyrosine phosphorylated through a calcium-dependent c-Src activation initiated by flow. FAK and PYK2 share their association with Cas through SH domains. Tyrosine phosphorylation of PYK2 in ECs by shear stress is dependent on the generation of reactive oxygen species (ROS) (53). The obligatory role of intracellular calcium mobilization for PYK2 phosphorylation, mediated through phospholipase C (PLC) activation and inositol triphosphate (IP_3), showed PYK2 to be a calcium-dependent kinase. Although stimulation of PLC activity by shear stress has been shown to lead to PKC activation, inhibition of PKC directly had no effect on flow-mediated PYK2 phosphorylation, suggesting a PLC-calcium-PYK2 pathway. The correlation of calcium dependence, flow-activation through ROS, and shared SH domain suggests that tyrosine phosphorylation of Cas depends on PYK2 activation. To test this hypothesis more directly, ECs were transfected with a kinase-inactive PYK2, the overexpression of which blocked flow- and ROS-mediated phosphorylation of both PYK2 and Cas. Transfection with a kinase-inactive Src did not inhibit PYK2 phosphorylation, suggesting that Src-dependent Cas phosphorylation occurred downstream of PYK2. Although transfection with a kinase-inactive c-src had no effect on PYK2 phosphorylation by flow, other unknown Src family kinases appear to regulate PYK2 activation, as demonstrated by the effectiveness of the global Src family kinase inhibitor PP2 to inhibit flow-mediated PYK2 activation (53). Both Cas and PYK2 phosphorylation were inhibited by depletion of ROS by the antioxidant N-acetylcysteine and, conversely, were stimulated by superoxide generating agents. Overall, the study links together several elements in mechanical signaling: (a) flow activation of ROS (possibly through G-protein activation), (b) PLC-mediated intracellular calcium mobilization, and (c) Src family kinase (but not c-src)-mediated PYK2 phosphorylation, which appears to be obligatory for p130 Cas phosphorylation.

The protein p130Cas, with which PYK2 associates, is reported to be localized to focal adhesion sites, including association with zyxin and LIM proteins, and is known to be critical for cell spreading and motility. PYK2 can also physically associate with cytoskeleton linker molecules, such as paxillin. These associations are consistent with its reported localization at adhesion sites, where its autophosphorylation is regulated by FAK through the FAK focal adhesion-targeting domain. Although phosphorylation of PYK2 by flow was unaffected by treatment with the actin microfilament destabilizing agent cytochalasin D, suggesting that an intact actin filament network may not be critical for flow-mediated PYK2 phosphorylation, binding of p130 Cas, FAK, and paxillin to actin microfilaments occurs in response to stretch of L929 cells. Paradoxically, however, PYK2 localizes to the perinuclear region in fibroblasts and becomes exclusively nuclear upon modification of one of its Src homology binding sites, suggesting a possible direct role in the regulation of transcription. The cytosolic distribution and rapid association with structural and adapter proteins suggest a kinase with great versatility of location and function, and highlight the complex interplay between mechanosignaling and other signaling pathways.

Integrin-Mediated Mechanotransduction

The location and dynamics of abluminal focal adhesion sites were first observed to be responsive to shear stress through the use of tandem-scanning confocal microscopy (60). These observations were followed by demonstrations of the shear-responsive phosphorylation of proteins localized to adhesion sites, including FAK and paxillin. The experiments suggested that shear stress at the luminal surface can stimulate abluminal mechanotransduction at the adhesion sites (1). The physical association of transmembrane integrins with the extracellular matrix (ECM) on the outside of the cell and signaling molecule complexes, linker proteins, and the cytoskeleton inside the cell (61), confer an important role for integrins in the regulation of mechanotransduction.

Shyy and Chien have investigated the role of adhesion site integrins in mechanotransduction (52). They integrated the results of endothelial flow studies with data obtained from platelet activation and growth-factor signaling (both being integrin-mediated). In their proposed mechanism, flow activates integrins, increasing their avidity for the ECM. Integrin activation by many different stimuli is influenced by extracellular or intracellular factors, particularly small GTPases. If activation is blocked with RGD peptides or by monoclonal antibodies against integrin-β subunits, typical downstream responses to flow and shear stress are inhibited (e.g., mitogen-activated protein kinase (MAPK) phosphorylation, NF-κB activation). One mechanism of integrin involvement in flow mechanotransduction is proposed to be through the assembly of signaling complexes on the cytosolic side of the membrane, at which integrin-dependent autophosphorylation of FAK leads to its association with the SH2 domain of c-Src, which in turn phosphorylates paxillin and p130 Cas. This mechanism is similar to that of PYK2 phosphorylation (discussed earlier) and results in the assembly of similar molecules. These act as a scaffold to assemble C3G (a guanine nucleotide exchange factor for Ras G-protein family member Rap1) and adapter protein Crk, which leads to the activation of MAPK. Like PYK2, spatial evidence suggests (using green fluoresence protein [GFP] as a reporter) the recruitment of activated FAK to new focal adhesion sites when ECs are subjected to shear stress, suggesting that the scaffold assembly is likely to be highly compartmentalized to portions of the plasma membrane.

A similar spatial relationship may occur in a second mechanism of integrin activation by shear stress. In this case, the protein Shc is activated and binds Src family tyrosine kinases through SH2 binding. Fyn is one such kinase that, together with its associated caveolar protein caveolin (Cav)-1, binds to Shc, phosphorylates it, and recruits an adaptor protein growth factor receptor binding protein 2 (Grb2). This leads to the activation of the Ras–MAPK pathway prominent in endothelial flow responses.

RhoA also is activated by shear stress through mechanisms that require integrin-ECM binding. Given the association of Rho with the regulation of the cytoskeleton through Rho-associated kinase (ROCK), and LIM kinase via cofilin (62), it is likely that the flow-responsive adhesion site complexes involving integrins are intimately related to the binding of cytoskeletal elements at adhesion sites and cell junctions. Integrins are not restricted to the abluminal surface; flow activation via integrins may occur without the engagement of the ECM. Furthermore, it also has been proposed that nonintegrin mechanisms of Ras pathway activation occur by direct G-protein activation in the plasma membrane, and as a direct result of deformation of membrane lipids without cytoskeletal or integrin involvement (63). The cell must either filter or integrate the spatially and temporally variable signaling events of mechanotransduction.

Primary Transduction of Shear Through Junctional Mechanosensors

Fujiwara and colleagues (64) first demonstrated the phosphorylation of the endothelial junction adhesion protein platelet-endothelial adhesion molecule (PECAM)-1, its binding to SH2 tyrosine phosphatase, and the subsequent accumulation of signaling molecules at the junction in response to shear stress; they proposed PECAM-1 to be a mechanosensor. Shay-Salit and colleagues (65) demonstrated flow activation of vascular endothelial growth factor receptor 2 (VEGFR2) and its assembly with phosphorylated vascular endothelial (VE)-cadherin, β-catenin, and PI3K, leading to the phosphorylation of Akt-1. Building on these findings, Tzima and colleagues (54) have recently reported that this mechanosensory complex, involving junctional proteins, regulates a subset of mechanotransduction pathways through integrins. In transfection studies of non-ECs, they showed that PECAM-1, VE-cadherin, and VEGFR2 are sufficient to transduce shear stress in the presence of an intact cytoskeleton and its associated proteins. Conversely, their selective inhibition or deletion in ECs inhibited important flow-related transduction; flow-induced activation of VEGFR2 did not occur in *PECAM-1*$^{-/-}$ and

VE-cadherin$^{-/-}$ ECs, nor did pathways dependent on integrin activation. In vivo evidence in support of the mechanism was provided by local regions of endothelial activation (NF-κB nuclear translocation) that correlated with disturbed hemodynamic sites in wild-type mice but were not activated in *PECAM-1*$^{-/-}$ mice.

The above examples support the transfer of shear stress forces to the adhesion sites and junctional regions of the cell consistent with a decentralized model of mechanotransduction.

INVESTIGATIVE AND THERAPEUTIC VALUE

Other than the acquisition of fundamental knowledge about cell and tissue responses, what is the value of endothelial phenotyping and the pursuit of detailed investigations of mechanotransduction?

Diversity of phenotype is important because it may permit specific therapeutic targeting of endothelium in different vascular beds and, for more refined phenotype biomarkers, targeting of those discrete regions within the arterial system that are susceptible to atherosclerosis. As multiscale studies progress, it may be possible to identify a highly susceptible phenotype for endothelial dysfunction, restricted to a focal location, thus permitting intervention early in the development of pathological change. The interplay of hemodynamics and endothelial phenotype also likely plays a central role in the protective effects of exercise against cardiovascular diseases; a better understanding of the mechanisms involved may be exploitable.

The importance of a comprehensive understanding of mechanotransduction at the single-cell and multicellular levels cannot be overstated. All cells in the body exist in local biomechanical environments that are subject to frequent or occasional change that requires the tissue to adapt accordingly. Although certain mechanoresponse mechanisms are common to many cell types, other such mechanisms are cell- and tissue-type specific. For example, endothelial cells are particularly responsive to flow forces, bone cells to compressive and fluid forces, and other cells to strain. Elucidation of the similarities and differences in these mechanisms will provide a general model for mechanotransduction as well as pathways specific to a tissue, respectively.

KEY POINTS

- Flow characteristics and associated hemodynamic forces are regulators of endothelial phenotype with important implications for the physiology and pathology of the cardiovascular system.
- Heterogeneity of endothelial phenotype must be understood over multiple-length scales with a hierarchical organization of endothelial physiology. Global analyses in parallel with candidate gene and protein analyses are being effectively applied to define heterogeneity in vivo and in vitro.
- Hemodynamics is a powerful predictor of vascular pathological susceptibility, initiation, and development. In vitro (reductionist) experiments to determine detailed mechanisms are most valuable when they simulate physiological flow in vivo. Site-specific investigations in vivo are essential to determine the spatial relevance of endothelial phenotype both in the susceptibility to cardiovascular diseases and in the pathogenesis itself.
- Both atherogenesis and calcific sclerosis of heart valves originate focally, not diffusely. If the endothelium is a determinant of focal susceptibility and pathogenesis, short-length scale studies of endothelial phenotype are essential. Therapeutic targets (based on phenotypes) may be highly restricted spatially.
- When hemodynamic forces deform the EC surface, the mechanics are communicated throughout the cell, most directly via the cytoskeleton to cell–cell and cell–matrix adhesion sites and the nuclear membrane. Thus, mechanotransduction is *decentralized*, and the potential exists for multiple mechanosensors remote from the site of interface with blood. In effect, the entire cell is a mechanotransducer.

Future Goals

- To refine hemodynamic imaging techniques to resolution levels that match the multiscale biology of endothelial heterogeneity
- To further develop of multiscale high-fidelity cell phenotyping methods at the levels of RNAs (including microRNAs, etc), proteins, and function applied to ever smaller numbers of cells (and eventually single cells) with an emphasis on in vivo expression

ACKNOWLEDGMENTS

The author acknowledges support by grants from the NIH (NHLBI: HL36049 MERIT, HL62250, HL64388, HL70128) and from the National Space Biomedical Research Institute (NASA) SMST-102.

REFERENCES

1 Davies PF. Flow-mediated endothelial mechanotransduction. *Physiol Rev*. 1995;75:519–560.

2 Resnick N, Gimbrone MA, Jr. Hemodynamic forces are complex regulators of endothelial gene expression. *FASEB J*. 1995;9:874–882.

3 Griffith TM. Endothelial control of vascular tone by nitric oxide and gap junctions: A haemodynamic perspective. *Biorheology*. 2002;39:307–318.

4 Langille BL, O'Donnel F. Reductions in arterial diameter produced by chronic decreases in blood flow are endothelium-dependent. *Science*. 1986;231:405–407.

5 Glagov S, Zarins CK, Giddens DP, Ku DN. Hemodynamics and atherosclerosis. Insights and perspectives gained from studies of human arteries. *Arch Pathol Lab Med*. 1988;112:1018–1031.

6 Mohler ER. Are atherosclerotic processes involved in aortic valve calcification? *Lancet*. 2000;356:524–525.

7 O'Malley CD. Leonardo Da Vinci. In: *Leonardo on the Human Body*. Saunders translation. New York: Dover Publications, 1983.

8 Virchow R. *Cellular Pathology as based upon Physiological and Pathological Histology*, 2nd edition. Chance F, trans. New York: Dover Publications, 1971.

9 Anitschow N. Experimentelle atherosklerose der aorta beim neerschweinchen. *Ziegler's Beitrage*. 1922;70:1–23.

10 Caro CG, Fitzgerald JM, Schroter RC. Arterial wall shear and distribution of early atheroma in man. *Nature*. 1969;223:1159–1161.

11 Fry DL. *Atherogenesis: Initiating factors*. CIBA Foundation Symposium. 1973;12:96–118.

12 Cornhil JF, Roach MR. A quantitative study of the localization of atherosclerotic lesions in the rabbit aorta. *Atherosclerosis*. 1976;23:489–499.

13 Florey H. A series of electron micrographs of the intima of blood vessels. *Proc R Soc Med*. 1960;53:13–14.

14 Jaffe EA, Nachman RL, Becker CG, Minick CR. Culture of human ECs derived from umbilical veins. Identification by morphologic and immunologic criteria. *J Clin Invest*. 1973;52:2745–2756.

15 Dewey CF, Gimbrone MA, Jr., Bussolari SR, Davies PF. The dynamic response of vascular ECs to fluid shear stress. *J Biomech Eng*. 1981;103:177–185.

16 Davies PF, Dewey CF, Bussolari SR, Gordon EJ, Gimbrone MA, Jr. Influence of hemodynamic forces on vascular endothelial function: In vitro studies of shear stress and pinocytosis in cultured bovine aortic ECs. *J Clin Invest*. 1984;73:1121–1129.

17 Davies PF, Remuzzi A, Gordon ES, Dewey CF, Gimbrone MA, Jr. Turbulent fluid shear stress induces vascular EC turnover in vitro. *Proc Nat Acad Sci USA*. 1986;83:2114–2118.

18 Davies PF, Reidy MA, Goode TB, Bowyer DE. Scanning electron microscopy in the evaluation of endothelial integrity of the fatty streak lesion of atherosclerosis. *Atherosclerosis*. 1976;25:125–130.

19 Furchgott RF, Zawadzki JV. The obligatory role of ECs in the relaxation of arterial smooth muscle by acetylcholine. *Nature*. 1980;288:373–376.

20 Pohl U, Holtz J, Busse R, Bassenge E. Crucial role of endothelium in the vasodilator response to increased flow in vivo. *Hypertension*. 1986;8:37–44.

21 Aird WC, Edelberg JM, Weiler-Guettler H, Simmons WW, Smith TW, Rosenberg RD. Vascular bed-specific expression of an EC gene is programmed by the tissue microenvironment. *J Cell Biol*. 1997;138:1117–1124.

22 Passerini AG, Polacek DC, Shi C, et al. Coexisting pro-inflammatory and anti-oxidative endothelial transcription profiles in a disturbed flow region of the adult porcine aorta. *Proc Natl Acad Sci USA*. 2004;101:2482–2487.

23 Durr E, Yu J, Krasinska KM, Carver LA, Yates JR, Testa JE, Oh P, Schnitzer JE. Direct proteomic mapping of the lung microvascular EC surface in vivo and in cell culture. *Nat Biotechnol*. 2004;8:985–992.

24 Chi JT, Chang HY, Haraldsen G, et al. Endothelial cell diversity revealed by global expression profiling. *Proc Natl Acad Sci USA*. 2003;100:10623–10628.

25 Porat RM, Grunewald M, Globerman A, Itin A, Barshtein G, Alhonen L, Alitalo K, Keshet E. Specific induction of tie1 promoter by disturbed flow in atherosclerosis-prone vascular niches and flow-obstructing pathologies. *Circ Res*. 2004;94:394–401.

26 Magid R, Davies PF. Endothelial protein kinase C isoform identity and differential activity of PKCζ in an athero-susceptible region of porcine aorta. *Circ Research*. 2005;97:443–449.

27 Hajra L, Evans AI, Chen M, Hyduk SJ, Collins T, Cybulsky MI. The NF-kappa B signal transduction pathway in aortic ECs is primed for activation in regions predisposed to atherosclerotic lesion formation. *Proc Natl Acad Sci USA*. 2000;97:9052–9057.

28 Simmons CA, Manduchi E, Grant G, Davies PF. Spatial heterogeneity of endothelial phenotypes correlates with side-specific vulnerability to calcification in normal porcine aortic valves. *Circ Research*. 2005;96:792–799.

29 Cheng C, van Haperen R, de Waard M, et al. Shear stress affects the intracellular distribution of eNOS: direct demonstration by a novel in vivo technique. *Blood*. 2005.

30 Groenendijk BC, Hierck BP, Vrolijk J, et al. Changes in shear stress-related gene expression after experimentally altered venous return in the chicken embryo. *Circ Res*. 2005;96:1291–1298.

31 Gimbrone MA Jr., Anderson KR, Topper JN, et al. Special communication: the critical role of mechanical forces in blood vessel development, physiology and pathology. *J Vasc Surg*. 1999; 29:1104–1151.

32 Passerini AG, Shi C, Francesco NM, et al. Regional determinants of arterial endothelial phenotype dominate the impact of gender or short-term exposure to a high-fat diet. *Biochem Biophys Res Comm*. 2005;332:142–148.

33 Topper JN, Gimbrone MA Jr. Blood flow and vascular gene expression: fluid shear stress as a modulator of endothelial phenotype. *Mol Med Today*. 1999;5:40–46.

34 Lin K, Hsu PP, Chen BP, et al. Molecular mechanism of endothelial growth arrest by laminar shear stress. *Proc Natl Acad Sci USA*. 2000;97:9385–9389.

35 McCormick SM, Eskin SG, McIntire LV, et al. DNA microarray reveals changes in gene expression of shear stressed human umbilical vein ECs. *Proc Natl Acad Sci USA*. 2001;98:8955–8960.

36 Wasserman SM, Mehraban F, Komuves LG, Yang RB, Tomlinson JE, Zhang Y, Spriggs F, Topper JN. Gene expression profile of human ECs exposed to sustained fluid shear stress. *Physiol Genomics*. 2002;12:13–23.

37 Garcia-Cardena G, Comander J, Anderson KR, Blackman BR, Gimbrone, MA Jr. Biomechanical activation of vascular endothelium as a determinant of its functional phenotype. *Proc Natl Acad Sci USA*. 2001;98:4478–4485.

38 Brooks AR, Lelkes PI, Rubanyi GM. Gene expression profiling of human aortic ECs exposed to disturbed flow and steady laminar flow. *Physiol Genomics*. 2002;9:27–41.

39 Hu YL, Chien S. Effects of shear stress on protein kinase C distribution in ECs. *J Histochem Cytochem*. 1997;45:237–249.

40 Traub O, Monia BP, Dean NM, Berk BC. PKCε is required for mechano-sensitive activation of ERK1/2 in ECs. *J Biol Chem*. 1997;272:31251–31257.

41 Leitges M, Sanz L, Martin P, et al. Targeted disruption of the PKCζ gene results in the impairment of the NFkB pathway. *Mol Cell.* 2001;8:771–780.

42 Sacks MS, He Z, Baijens L, Wanant S, Shah P, Sugimoto H, Yoganathan AP. Surface strains in the anterior leaflet of the functioning mitral valve. *Ann Biomed Eng*. 2002;30:1281–1290.

43 Butcher JT, Penrod AM, Garcia AJ, Nerem RM. Unique morphology and focal adhesion development of valvular ECs in static and fluid flow environments. *Arterioscler Thromb Vasc Biol.* 2004;24:1429–1434.

44 Dull RO, Davies PF. Flow modulation of agonist (ATP)-response (Ca^{++}) coupling in vascular ECs. *Am J Physiol.* 1991;261:H149–H156.

45 Koller A, Sun D, Kaley G. Role of shear stress and endothelial prostaglandins in flow- and viscosity-induced dilation of arterioles in vitro. *Circ Research.* 1993;72:1276–1284.

46 Bussolari SR, Dewey CF Jr, Gimbrone MA Jr. Apparatus for subjecting living cells to fluid shear stress. *Rev Sci Instrum.* 1982; 53:1851–1854.

47 Levesque MJ, Nerem RM. The study of rheological effects on vascular ECs in culture. *Biorheology*. 1989;26:345–357.

48 Sorescu GP, Sykes M, Weiss D, et al. Bone morphogenic protein 4 produced in ECs by oscillatory shear stress stimulates an inflammatory response. *J Biol Chem.* 2003;278:31128–31135.

49 Dai G, Kaazempur-Mofrad MR, Natarajan S, et al. Distinct endothelial phenotypes evoked by arterial waveforms derived from atherosclerosis-susceptible and -resistant regions of human vasculature. *Proc Natl Acad Sci USA.* 2004;101:14871–14876.

50 Barbee KA, Davies PF, Lal R. Shear stress-induced reorganization of the surface topography of living endothelial cells imaged by atomic force microscopy. *Circ Research.* 1994;74:163–171.

51 Davies PF, Tripathi S. Mechanical stress mechanisms in cells: An endothelial paradigm. *Circ Research.* 1993;72:239–245.

52 Shyy JY, Chien S. Role of integrins in endothelial mechanosensing of shear stress. *Circ Res.* 2002;91:769–775.

53 Tai L-K, Okuda M, Abe J, et al. Fluid shear stress activates Pyk2 via a reactive oxygen species dependent pathway. *Arterioscl Thromb Vasc Biol.* 2002;22:1790–1796.

54 Tzima E, Irani-Tehrani M, Kiosses WB, et al. A mechanosensory complex that mediates the EC response to fluid shear stress. *Nature.* 2005;437:426–431.

55 Olesen SP, Clapham DE, Davies PF. Hemodynamic shear stress activates a K^+ current in vascular ECs. *Nature.* 1988;331:168–170.

56 Gudi S, Nolan JP, Frangos JA. Modulation of GTPase activity of G proteins by fluid shear stress and phospholipid composition. *Proc Natl Acad Sci USA.* 1998;95:2515–2519.

57 Davies PF, Helmke BP. The cytoskeleton under external fluid mechanical forces. *Annals Biomed Eng*. 2002;30:1–13.

58 Helmke BP. Molecular control of cytoskeletal mechanics by hemodynamic forces. *Physiology*. 2005;20:43–53.

59 Wang N, Ingber DE. Probing transmembrane mechanical coupling and cytomechanics using magnetic twisting cytometry. *Biochem Cell Biol.* 1995;73:327–335.

60 Davies PF, Robotewsky A, Griem ML. Quantitative studies of EC adhesion: Directional remodeling of focal adhesion sites in response to flow forces. *J. Clin Invest*. 1994;93:2031–2038.

61 Giancotti FG, Ruoslahti E. Integrin signaling. *Science.* 1999;285: 1028–1032.

62 Maekawa M, Ishizaki T, Boku S, et al. Signaling from Rho to the actin cytoskeleton through protein kinases ROCK and LIM-kinase. *Science.* 1999;285:895–898.

63 Gudi S, Huvar I, White CR, et al. Rapid activation of Ras by fluid flow is mediated by Galpha(q) and Gbetagamma subunits of heterotrimeric G proteins in human ECs. *Arterioscler Thromb Vasc Biol.* 2003;23:994–1000.

64 Fujiwara K, Masuda M, Osawa M, et al. Is PECAM-1 a mechanoresponsive molecule? *Cell Struct Funct*. 2001;26:11–17.

65 Shay-Salit A, Shushy M, Wolfovitz E, et al. VEGF receptor 2 and the adherens junction as a mechanical transducer in vascular ECs. *Proc Natl Acad Sci USA.* 2002;99:9462–9466.

29

Hypoxia-Inducible Factor 1

Gregg L. Semenza

The Johns Hopkins University School of Medicine, Baltimore, Maryland

The existence of multicellular organisms is based on the efficient capture of solar energy by plants through photosynthesis, a process by which carbon dioxide (CO_2) and water are converted into glucose and oxygen (O_2). These are subsequently utilized by all eukaryotic organisms to generate ATP and, as by-products, CO_2 and water, thus completing the circle of life on our planet (Figure 29.1). The highly efficient recovery of the energy contained within the chemical bonds of glucose through the process of oxidative phosphorylation provides the power necessary to assemble and maintain complex multicellular machines such as *Homo sapiens*, in which more than 100,000,000,000,000 parts (i.e., cells) are assembled and organized into a functional unit (i.e., organism). A requirement for the efficient generation of ATP is the continuous delivery of O_2 to every cell in the body. In postnatal life, this requirement is met through the concerted action of the lungs, blood, heart, and vessels. The lungs of an adult human take in 5 to 6 liters of air per minute or approximately 8,000 liters per day. Within the lungs, O_2 is bound by hemoglobin present within erythrocytes, which are pumped by the heart through blood vessels that represent the transportation infrastructure for O_2 and glucose delivery to the tissues and for the removal of CO_2, hydrogen ions (H^+), potassium ions (K^+), and other toxic metabolites.

Whereas the major blood vessels are generated in a stereotypical pattern during development, the capillaries through which O_2 and CO_2 are exchanged in the tissues develop not through a hardwired program but rather based on physiological signals generated by individual cells. The basis for this signaling involves the ability of every nucleated cell in the body to sense the local O_2 concentration and to respond to imbalances between O_2 supply and demand. Thus, reduced O_2 availability (hypoxia) triggers the production of angiogenic growth factors, which are secreted proteins that bind to cell surface receptors on endothelial cells (ECs) and their progenitors, thereby stimulating the growth of new blood vessels through the processes of angiogenesis and vasculogenesis. It is not clear what factors determine whether new blood vessels in a tissue are generated by angiogenesis or vasculogenesis, although the more severe the hypoxic stimulus (as in a rapidly growing tumor), the greater contribution of bone marrow–derived progenitor cells (i.e., vasculogenesis) to new blood vessel formation. However, in either case, hypoxia is likely to be a major physiological stimulus for vascularization. Delivery of increased O_2 via these new vessels corrects intracellular hypoxia, thus eliminating the stimulus for angiogenic growth factor production. Blood flow through major vessels can be increased via the process of arteriogenesis, in which the luminal diameter of conduit vessels is increased. These homeostatic mechanisms ensure that each cell in a healthy organism is provided with adequate O_2 to meet its metabolic demands. O_2 levels must be maintained within narrow limits because inadequate O_2 may impair cell function or survival due to ATP depletion, and excess O_2 may impair cell function or survival due to the damaging effect of reactive oxygen species (ROS) on cellular macromolecules. Failure of these homeostatic mechanisms contributes to the pathogenesis of all major causes of mortality in the United States, including heart disease, cancer, stroke, and chronic lung disease, which account for over two-thirds of all deaths annually.

Although hypoxia due to rapidly increasing tissue mass and increasing O_2 consumption is likely to represent a major physiological signal during embryonic and fetal development, in many adults, vessel stenosis due to atherosclerotic plaque formation and rupture leads to ischemia, in which cells are subjected to low levels of O_2, glucose, and other nutrients and to high levels of CO_2, K^+, H^+, and other toxic metabolites. Although ischemia represents a more complex pathophysiological state, it is the deprivation of O_2 (i.e., hypoxia) that appears to be the major stimulus capable of inducing an adaptive angiogenic response.

HYPOXIA-INDUCIBLE FACTOR 1 IS A MASTER REGULATOR OF OXYGEN HOMEOSTASIS

Physiological responses to hypoxia occur over a broad time scale. The activity of pre-existing proteins can be modified within seconds as a result of post-translational modification, whereas the synthesis of new proteins requires minutes to

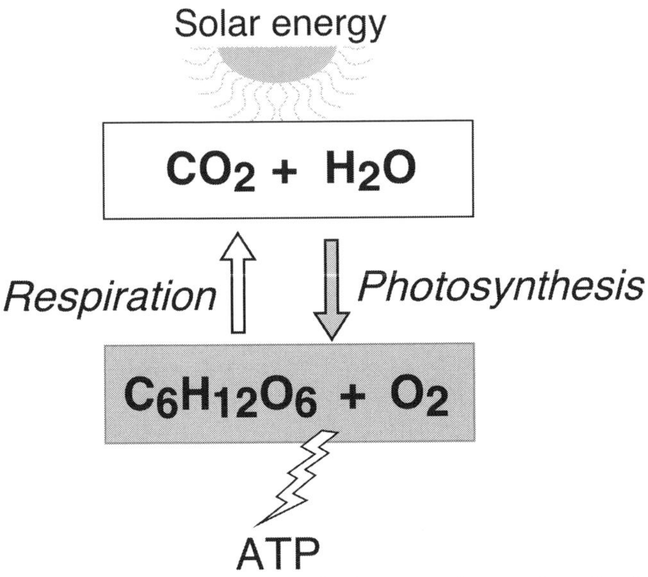

Figure 29.1. Energy trapping and substrate cycling on planet Earth. Solar energy is trapped in the chemical bonds of glucose by photosynthesis in plants and transferred to the high-energy phosphate of ATP during cellular respiration in all multicellular organisms. O_2, which is generated as a side product during photosynthesis, is required to generate ATP by oxidative phosphorylation. CO_2 and H_2O are produced as side products.

hours. Often, the expression of new proteins is associated with changes in messenger RNA (mRNA) levels that reflect altered transcriptional regulation. The regulation of gene transcription by hypoxia-inducible factor (HIF)-1 represents the most well-defined molecular mechanism for maintaining O_2 homeostasis in metazoans. HIF-1 was identified through an analysis of the *cis*-acting sequences and *trans*-acting factors required for O_2-regulated expression of the human *EPO* gene, which encodes erythropoietin. EPO is a glycoprotein hormone that stimulates the proliferation and survival of red blood cell progenitors and thus determines the blood O_2-carrying capacity. Analysis of the *EPO* gene hypoxia-response element, the *cis*-acting nucleotide sequence that is required for increased transcription of the gene in response to hypoxia, led to the identification (1), biochemical purification (2), and molecular cloning (3) of HIF-1. HIF-1 has been identified in all metazoan species that have been analyzed, from *Caenorhabditis elegans* to *Homo sapiens* (organisms whose cell number differs by more than 10 orders of magnitude). In contrast, HIF-1 is not present in yeast, suggesting that the appearance of HIF-1 represented an adaptation that was essential to metazoan evolution. The battery of genes regulated by HIF-1 is different in each cell type and, for some genes, expression can be induced or repressed by HIF-1 depending upon the cell type (4). Among the critical physiological processes regulated by HIF-1 target genes are erythropoiesis, angiogenesis, and glycolysis, which represent examples of systemic, local tissue, and intracellular adaptive responses to hypoxia, respectively.

HIF-1 is a heterodimeric protein that is composed of HIF-1α and HIF-1β subunits. The amino-terminal half of each subunit consists of basic helix-loop-helix and PER-ARNT-SIM homology (PAS) domains that mediate heterodimerization and DNA binding. The carboxyl terminal half of HIF-1α contains two transactivation domains that mediate interactions with coactivators such as cyclic AMP response element-binding protein (CBP) and p300 (5). Coactivators interact with both sequence-specific DNA binding proteins such as HIF-1 and with the general transcription factors associated with RNA polymerase II. Coactivators also have histone acetyltransferase activity that is required to make the DNA embedded in chromatin accessible to the polymerase complex for transcription into RNA.

The HIF-1β subunit is constitutively expressed, whereas the expression and activity of the HIF-1α subunit are precisely regulated by the cellular O_2 concentration. HIF-1α accumulates instantaneously under hypoxic conditions and, upon reoxygenation, is rapidly degraded, with a half-life of less than 5 minutes in posthypoxic tissue culture cells (3,6). This represents an overestimation of the half-life, because it includes the time required for O_2 to diffuse out of the culture medium. In isolated perfused and ventilated lung preparations subjected to hypoxia and reoxygenation, the half-life HIF-1α is less than 1 minute (7). No protein has been shown to have a shorter half-life.

In addition to HIF-1α, a structurally and functionally related protein designated HIF-2α, which is the product of the *EPAS1* gene, can also heterodimerize with HIF-1β (8). HIF-1α:HIF-1β and HIF-2α:HIF-1β heterodimers appear to have overlapping but distinct target-gene specificities (9,10). Unlike HIF-1α, HIF-2α is not expressed in all cell types and, when expressed, can be inactive as a result of cytoplasmic sequestration (11). Although HIF-2α was originally identified in ECs, it is expressed in several other cell types, including parenchymal cells in kidney, lung, and a variety of carcinomas. A third protein, designated HIF-3α, also has been identified and a splice variant, designated inhibitory PAS domain protein (IPAS), has been shown to bind to HIF-1α and inhibit its activity (12).

In mice, complete deficiency of HIF-1α results in embryonic lethality at midgestation that is associated with dramatic vascular regression due to extensive EC death (13–15). O_2 delivery to cells of the developing embryo becomes limited by diffusion, such that the establishment of a functioning circulatory system is required for embryonic survival beyond embryonic day E(10) in the mouse. In wild-type mouse embryos, HIF-1α expression increases dramatically between E8.5 and E9.5, whereas embryos that lack HIF-1α expression die between E9.5 and E10.5 with cardiac malformations, vascular regression, and massive cell death (13–17). Conditional knockout mice lacking HIF-1α expression in neural cells have marked cerebral atrophy associated with vascular regression (18). Vascularization of tumor xenografts derived from *HIF-1α*–null mouse embryonic stem cells also is impaired severely (15,16). The phenotypes of various germline and conditional *HIF-1*–knockout mice are summarized in Table 29-1.

Table 29-1: Phenotype of Germline and Conditional HIF-1–Knockout Mice

Germline knockouts

Gene	*Protein*	*Phenotype*	*References*
Arnt	HIF-1β	Embryonic lethality by E10.5	65
		Vascular and placental defects	
Epas1	HIF-2α	Embryonic lethality at E9.5–13.5	66
		Vascular defects	
		Embryonic lethality at E12.5–16.5	67
		Impaired catecholamine production	
		Survivors develop multiorgan failure	68
Hif1a	HIF-1α	Developmental arrest by E8.5	13–17
		Embryonic lethality by E10.5	
		Vascular regression; cardiac defects; Impaired erythropoiesis	
		Neural tube defects	

Conditional knockouts of HIF-1α

Cell Type	*Phenotype*	*References*
Breast epithelium	Impaired mammary differentiation and lipid secretion	69
Cardiomyocyte	Altered Ca^{2+} flux and reduced myocyte contractility	70
	Reduced myocardial vascularization	
Cartilage	Impaired chondrocyte survival	71
EC	Impaired tumor vascularization, wound healing	55
Intestinal epithelium	Impaired epithelial barrier protection	72
Mononuclear leukocytes	Impaired inflammatory and bactericidal responses	73–74

A variety of other transcription factors have been implicated in the O_2-dependent regulation of specific genes in specific cell types. For example, in vascular and mononuclear phagocytic cells of the lung, early growth response factor (EGR)-1 has been shown to mediate the hypoxia-inducible expression of tissue factor, intercellular adhesion molecule (ICAM)-1, plasminogen activator inhibitor (PAI)-1, and the cytokines interleukin (IL)-1β, interferon-inducible protein 10 (IP-10), macrophage inflammatory protein (MIP)-2, monocyte chemoattractant protein (MCP)-1, RANTES (Regulated on Activation, Normal T-cell Expressed and Secreted), and vascular endothelial growth factor (VEGF) (19). However, only in the case of HIF-1 has the molecular mechanism by which changes in O_2 concentration are transduced to changes in transcription factor activity been elucidated.

MOLECULAR MECHANISMS OF OXYGEN SENSING

The O_2-dependent degradation of HIF-1α involves ubiquitination and degradation by the 26*S* proteasome (20). The von Hippel-Lindau tumor suppressor protein (VHL) is required for this process (Figure 29.2), because renal carcinoma cells lacking functional VHL constitutively express HIF-1α and HIF-1 target genes under nonhypoxic conditions (21). VHL forms a complex with elongin B, elongin C, cullin 2, and RBX1 to form an E3 ubiquitin–protein ligase capable of functioning with E1 ubiquitin-activating and E2 ubiquitin-conjugating enzymes to mediate the ubiquitination of HIF-1α (22). Proline (Pro) residue 564 is hydroxylated in an O_2-dependent manner, and this modification is required for VHL binding (23–25). Pro-402 represents a second site of hydroxylation and VHL binding (26). Pro-402 and 564 are each contained within a similar amino acid sequence (LXXLAP; A, alanine; L, leucine; P, Pro; X, any amino acid). HIF-2α and HIF-3α expression also are regulated by prolyl hydroxylation and VHL binding (21,26).

Three prolyl hydroxylases were identified in mammalian cells and shown to utilize O_2 as a substrate to generate 4-hydroxyproline at residue 402 and/or 564 of HIF-1α (26,27). These proteins are homologues of EGL-9, which was identified through genetic studies as the HIF-1α prolyl hydroxylase in *C. elegans* (26). Alternative designations for the three mammalian homologues include EGLN (*EGL*-nine homologue), PHD (prolyl hydroxylase domain protein), and HPH (HIF-1α prolyl hydroxylase) 1–3. The hydroxylation reaction also requires 2-oxoglutarate (α-ketoglutarate) as a substrate and generates succinate as a side product. Ascorbate is required as a cofactor. The prolyl hydroxylase catalytic site contains an iron (Fe [II]) ion that is coordinated by one aspartate and two histidine residues. Unlike heme-containing proteins, the Fe (II) in 2-oxoglutarate-dependent oxygenases can be chelated or substituted by cobalt (Co [II]), rendering the enzyme inactive. Most importantly, these prolyl hydroxylases have a relatively high K_m for O_2 that is slightly above its atmospheric concentration, such that O_2 is rate limiting for enzymatic activity under physiological conditions (26). As a result, changes in the cellular O_2 concentration are directly transduced into changes in the rate at which HIF-1α is hydroxylated, ubiquitinated, and degraded.

Remarkably, HIF-1α transactivation domain function is regulated by the O_2-dependent hydroxylation of asparagine residue 803 (see Figure 29.2), which blocks the binding of the coactivators CBP and p300 (28). Factor-inhibiting HIF-1 (FIH-1), which was identified in a yeast two-hybrid screen as a protein that interacts with and inhibits the activity of the HIF-1α transactivation domain (29), functions as the asparaginyl hydroxylase (30). As in the case of the prolyl hydroxylases, FIH-1 appears to utilize O_2 and α-ketoglutarate, and contains Fe (II) in its active site, although it has a K_m for O_2 that is three times lower than the prolyl hydroxylases (31).

Spectroscopic analyses of a peptide from the HIF-1α transactivation domain complexed with the interacting domain

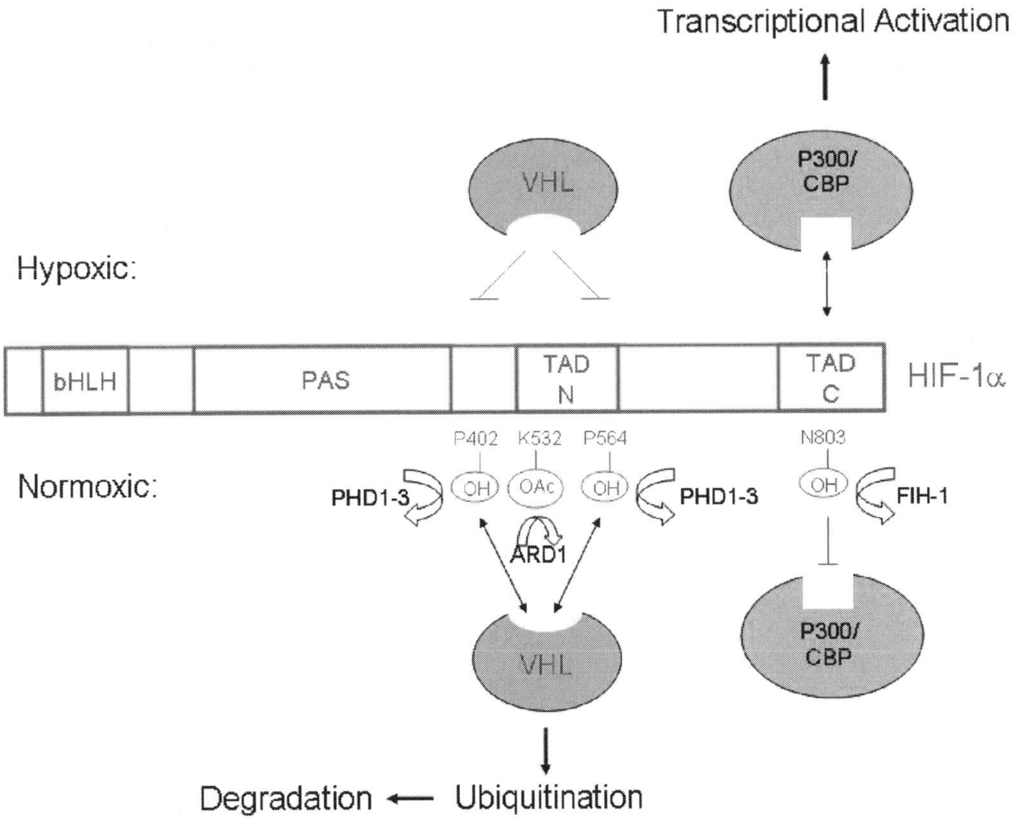

Figure 29.2. O_2-dependent regulation of HIF-1 activity. O_2 is a substrate for the hydroxylation of HIF-1α. Residues Pro-402 and/or Pro-564 are modified by the prolyl hydroxylases PHD1–3, leading to VHL-mediated ubiquitination and proteasomal degradation. Asn-803 in the carboxy-terminal transactivation domain (TAD-C) is modified by the asparaginyl hydroxylase FIH-1, which blocks HIF-1α interaction with the coactivators CBP and p300, thus inhibiting transcriptional activation. HIF-1α is also post-translationally modified by the acetyltransferase ARD1 at Lys-532, which promotes ubiquitination and degradation of HIF-1α.

of CBP or p300 revealed that asparagine 803 (Asn-803) is present within an α-helix buried deep within the protein interface, where it participates in multiple hydrogen bonding interactions that are predicted to stabilize the complex (32,33). Hydroxylation of Asn-803 is predicted to disrupt these protein–protein interactions. Similarly, hydroxylation of proline 564 has been shown to function also as a molecular switch to positively regulate the interaction of HIF-1α and VHL (34,35). Thus, hydroxylation provides a mechanism for regulating protein–protein interactions, similar to the effect of phosphorylation and other post-translational modifications. However, hydroxylation is unique because the modification occurs in an O_2-dependent manner, thus establishing a direct link between cellular oxygenation and HIF-1 activity.

OXYGEN-INDEPENDENT EXPRESSION OF HYPOXIA-INDUCIBLE FACTOR 1

Humans, like other metazoans, are constant and obligate consumers of O_2. The more cells present in a tissue, the more O_2 consumed (although the absolute amount of O_2 consumed differs among cell types due to differences in rates of ATP turnover, oxidative phosphorylation, and glycolysis). Thus, when one cell begets two daughter cells, O_2 consumption increases. It is not surprising that major pathways that transduce proliferative and survival signals from growth factor receptors also induce HIF-1α expression (Figure 29.3) in what can be viewed as a preemptive strategy for maintaining O_2 homeostasis. Proliferating cells express VEGF, which stimulates angiogenesis to provide additional perfusion required to maintain the oxygenation of an increased number of cells. In addition, proliferating cells preferentially utilize glycolytic rather than oxidative metabolism to generate ATP. The concomitant induction of angiogenesis and glycolysis with cell proliferation is mediated in part via the activation of HIF-1.

The increase in HIF-1α levels in response to growth factor stimulation differs in two important respects from the increase in HIF-1α levels in response to hypoxia. First, whereas hypoxia increases HIF-1α levels in all cell types, growth factor stimulation induces HIF-1α expression in a cell-type–specific manner. The rules governing this specificity have not been elucidated. Second, whereas hypoxia is associated with decreased degradation of HIF-1α, growth factors, cytokines, and other signaling

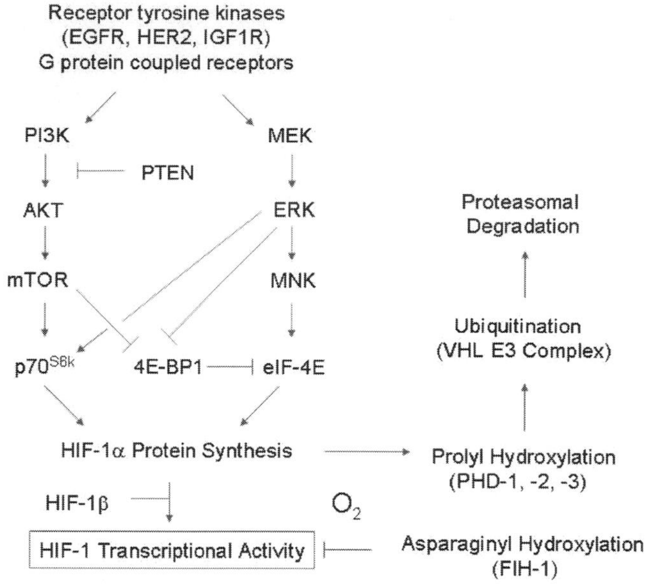

Figure 29.3. O_2-independent regulation of HIF-1 activity. Activation of receptor tyrosine kinases and G-protein–coupled receptors leads to signal transduction via the PI3K and MAPK (ERK) pathways, leading to phosphorylation of eIF-4E and its inhibitor 4E-BP1, events that increase the rate of HIF-1α protein synthesis (*left side of figure*), whereas O_2-dependent regulation of HIF-1α affects its stability and transcriptional activity (*right side of figure*).

molecules stimulate HIF-1α synthesis via activation of the phosphoinositide 3-kinase (PI3K) or mitogen-activated protein kinase (MAPK) pathways (36,37). Pulse-chase studies of MCF-7 breast cancer cells stimulated with heregulin demonstrated increased HIF-1α protein synthesis that was inhibited by treatment with rapamycin, a macrolide antibiotic which inhibits mammalian target of rapamycin (mTOR), a kinase that functions downstream of PI3K and Akt (37). The effect of heregulin was mediated via the 5′-untranslated region of *HIF-1α* mRNA (37). The known targets for phosphorylation by mTOR are regulators of protein synthesis (see Figure 29.3). The translation of several dozen different mRNAs are known to be regulated by this pathway, and it is likely that specific sequences in the 5′-untranslated region may determine the degree to which the translation of any particular mRNA can be modulated by mTOR signaling. HIF-1α protein expression is likely to be particularly sensitive to changes in the rate of synthesis because of its extremely short half-life under non-hypoxic conditions. In addition to effects on HIF-1α synthesis, activation of the Raf/MEK extracellular signal-regulated kinase (ERK) signaling pathway also has been shown to stimulate HIF-1α transactivation domain function (38). This effect is due at least in part to ERK phosphorylation of the coactivator p300, with which the transactivation domains interact (39). Unlike hypoxia, which induces HIF-1α protein stability and transcriptional activity in all cell types, the regulation of HIF-1 activity by growth factor signaling is cell-type specific.

Finally, although great progress has occurred in elucidating the molecular mechanisms by which HIF-1α protein levels are regulated, *HIF-1α* mRNA expression also is increased in response to hypoxia or ischemia in both experimental animals and humans (7,40,41). In tissue culture cells, O_2-dependent regulation of *HIF-1α* mRNA expression is not observed and thus, the mechanisms underlying this regulation have not been determined. However, *HIF-1α* mRNA contains multiple copies of the sequence 5′-AUUUA-3′, which has been implicated the regulation of mRNA stability (3).

HYPOXIA-INDUCIBLE FACTOR 1 REGULATES ANGIOGENIC GROWTH FACTOR EXPRESSION

Ischemic heart disease, a major cause of mortality, is treated by pharmacologic agents, balloon angioplasty, and coronary artery bypass graft surgery. Therapeutic angiogenesis aims to stimulate neovascularization by delivery of an angiogenic factor or its protein-coding DNA sequence. Phase I clinical trials have demonstrated safety, but phase II trials have not demonstrated efficacy. This outcome reflects the complexity of angiogenesis, which involves the temporally and spatially orchestrated expression by multiple cell types of multiple angiogenic factors, including VEGF, angiopoietin 1 (Ang1), Ang2, fibroblast growth factor 2 (FGF2), placental growth factor (PlGF), platelet-derived growth factor (PDGF)-B, transforming growth factor-β_1, and others. In ischemic tissue, angiogenic factors are produced by a variety of cell types, including myocytes, fibroblasts, and vascular cells.

VEGF is produced early in the angiogenic cascade and is responsible for the initial activation of ECs. Transgenic expression of VEGF in mouse skin results in increased numbers of blood vessels that manifest excessive permeability, whereas expression of both VEGF and Ang1 results in increased vessels without excessive permeability (42). However, transgenic coexpression of Ang1 in the heart blocks the angiogenic effect of VEGF (43). PlGF plays a critical role in ischemia-induced angiogenesis and has synergistic effects with VEGF in some tissues (44). Synergistic effects of combined treatment with PDGF-BB and FGF2 also have been reported (45). These data suggest that increased expression of a single angiogenic factor is not sufficient for functional vascularization, and that the effects of angiogenic factors are tissue-specific.

HIF-1 plays a critical role in angiogenesis by activating the transcription of genes encoding angiogenic growth factors including VEGF, Ang1, Ang2, PlGF, and PDGF-B (4). HIF-1 directly activates the transcription of the *VEGF* gene by binding to a hypoxia-response element located approximately 1 kb 5′ to the gene (46), whereas it is not known whether regulation of *Ang1, Ang2, PlGF,* and *PDGF-B* is direct or indirect.

The regulation of these genes is remarkably cell-type–specific. *Ang2* expression is induced by hypoxia in arterial ECs, repressed in arterial smooth muscle cells, and unchanged in cardiac fibroblasts and myocytes, whereas VEGF is induced by hypoxia in all four cell types (4). These results underscore the importance of determining the subset of HIF-1 target genes regulated by hypoxia in a particular cell type. In embryonic stem cells, *VEGF, Ang1, PlGF,* and *PDGF-B* mRNA expression was induced by hypoxia, whereas in *HIF-1α*–null cells, the expression of all four mRNA species was markedly reduced under both hypoxic and nonhypoxic conditions (4).

Consistent with the effect of HIF-1α loss of function, infection of cardiac fibroblasts, cardiac myocytes, arterial ECs, and arterial smooth muscle cells under nonhypoxic conditions with AdCA5, an adenovirus encoding a mutant form of HIF-1α that is resistant to O_2-dependent degradation, induces the same pattern of angiogenic-factor gene expression as that observed in hypoxic cells (4). HIF-1 can function as either an activator or repressor of *Ang1* and *Ang2* gene expression in a cell-type–specific manner. The molecular mechanism underlying this surprising observation remains to be established. These results are consistent with published data indicating that these factors can either induce or inhibit angiogenesis depending on the cellular context. This remarkable cell-type–specific regulation of gene expression underscores the general principle that HIF-1 functions as a messenger to the nucleus signaling hypoxia. The response to this signal depends on prior programming of the cell (i.e., the presence or absence of other transcriptional regulators of potential HIF-1 target genes).

The injection of AdCA5 into mouse eyes in the absence of hypoxia or ischemia induced neovascularization in multiple capillary beds (4), including those not responsive to VEGF alone (47). Lectin and anti–α-smooth muscle actin antibody staining demonstrated that these vessels contain both ECs and smooth muscle cells or pericytes. A link between HIF-1 expression and VEGF expression in ischemic retina has been well-established in ischemia-induced retinal neovascularization (48). However, the ability of AdCA5 to induce the sprouting of new vessels that extend into the vitreous cavity from the superficial capillary bed, as occurs in patients with diabetic retinopathy, contrasts with the inability of high levels of VEGF alone to do so (47). These studies indicate that increased levels of a single VEGF isoform are not sufficient to cause new vessels to sprout from the superficial vessels. Neovascularization induced by AdCA5 was associated with a dramatic increase in *PlGF* mRNA and protein expression. PlGF has been shown to act synergistically with VEGF to stimulate neovascularization, and both factors are required for neovascularization in the ischemic retina (44). Thus, the combined effect of increased PlGF and VEGF expression may underlie neovascularization in the superficial capillary bed, induced by the intravitreous injection of AdCA5.

An analysis of gene expression revealed increased expression of *Ang1, Ang2, PDGF-B, PlGF,* and *VEGF* mRNA in AdCA5-injected eyes. These results indicate that HIF-1 functions as a master regulator of angiogenesis by controlling the expression of multiple angiogenic growth factors, and that adenovirus-mediated expression of a constitutively active form of HIF-1α is sufficient to induce angiogenesis in the nonischemic tissue of an adult animal. These results may provide a mechanistic basis for the observation that transgenic mice expressing a constitutively active form of HIF-1α in the skin manifest increased vascularization without the increased

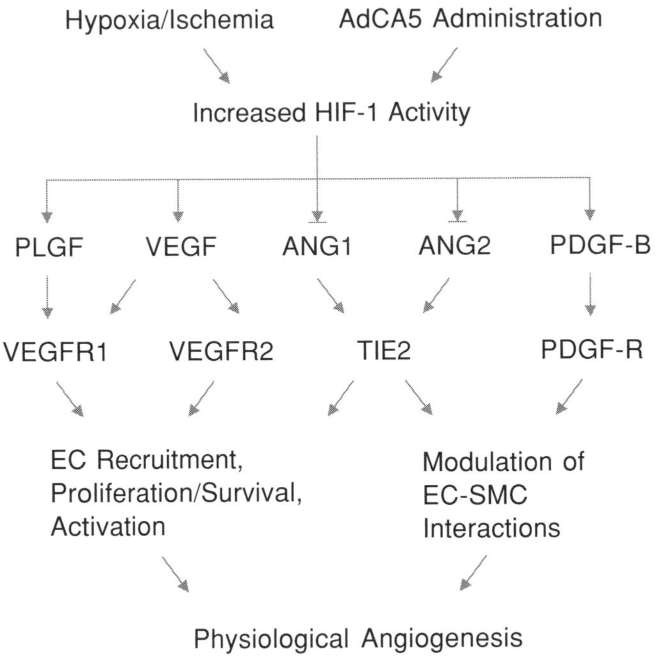

Figure 29.4. Regulation of angiogenic growth factor expression by HIF-1. Increased HIF-1 activity due to either the activation of the endogenous protein in response to hypoxia/ischemia or due to administration of AdCA5, an adenovirus encoding a constitutively active form of HIF-1α, induces cell-type–specific changes in the expression of genes encoding angiogenic growth factors. The protein products of these genes bind to receptors on the surface of ECs and smooth muscle cells and activate angiogenic programs in these cells.

vascular permeability observed in transgenic mice expressing VEGF from the same promoter (49). Because *cis*-acting hypoxia-response elements containing HIF-1 binding sites have not been demonstrated for the genes encoding angiogenic factors (other than VEGF), it remains to be determined whether HIF-1–dependent regulation of these genes is direct or indirect. However, whether by direct or indirect mechanisms, the data indicate that HIF-1 is a pleiotropic mediator of the angiogenic program (Figure 29.4).

Whereas the studies described above provided evidence for the role of HIF-1 in promoting the growth of microvasculature, as occurs in diabetic retinopathy, they did not address the issue of whether HIF-1 could promote the remodeling of arteries following the occlusion of a large, conduit vessel. To address this issue, the femoral artery of rabbits was occluded by intravascular coil placement, and the animals received an intramuscular injection of AdCA5 or AdLacZ in the region of the occluded vessel. Two weeks after occlusion, the animals that received AdCA5 showed a significantly greater recovery of blood pressure in the affected limb than did animals that received AdLacZ (50). In addition, perfusion was significantly improved, as determined by angiography, and histological analysis revealed a significant increase in the arterial luminal area in sections of muscle from the affected limb. Thus, in addition to stimulating angiogenesis, HIF-1 also stimulates arteriogenesis. As in the mouse eye, AdCA5 injection in the rabbit leg (in the absence of femoral occlusion) induced the expression of mRNAs encoding multiple angiogenic cytokines, including VEGF, PlGF, PDGF-B, MCP-1, and stromal-derived factor 1α (50).

Two other constitutively active forms of HIF-1α expressed via either transgenic (49) or gene therapy (51,52) approaches have been shown to increase vascularization during development or in response to ischemia, respectively. Cardiomyocyte-specific expression in transgenic mice of PR39, a peptide that selectively inhibits the degradation of HIF-1α, also was associated with increased vascularization and protection against ischemia (53). The coordinate activation of VEGF and PlGF and downregulation of *Ang1* mRNA expression by HIF-1 in cardiomyocytes (4) is an interesting finding, as recent studies suggest that these changes in angiogenic factor expression may stimulate both angiogenesis and arteriogenesis/collateralization in the ischemic heart. The demonstration that neovascularization is rapidly induced in AdCA5-injected eyes in the absence of ischemia (4) is noteworthy, because most patients with coronary artery disease do not have ischemia at rest. Further studies are required to determine whether the administration of AdCA5, other modified forms of HIF-1α, or peptides that inhibit HIF-1α ubiquitination or degradation may be of therapeutic utility in patients with ischemic cardiovascular disease.

CELL-AUTONOMOUS EFFECTS OF HYPOXIA-INDUCIBLE FACTOR 1 IN ENDOTHELIAL CELLS

The studies presented in the preceding sections established the paradigm that O_2 deprivation within the parenchymal cells of each organ induces HIF-1 activity. This leads to the expression of angiogenic growth factors such as VEGF that are secreted and bind to receptors present on the surface of ECs, leading to their activation. This paradigm has undergone an important shift with the recent demonstration that HIF-1 also has critical cell-autonomous functions within ECs.

Gene expression profiles were compared in arterial ECs cultured under nonhypoxic versus hypoxic conditions and in nonhypoxic cells infected with AdLacZ, an adenovirus encoding β-galactosidase, versus AdCA5, which encodes a constitutively active form of HIF-1α (as described earlier) (54). By analyzing the union of the independently derived data sets, 245 gene probes showed a statistically significant increase in expression of at least 1.5-fold in response to hypoxia and in response to AdCA5. Using the same analytic strategy, 325 gene probes showed a statistically significant decrease in expression of at least 1.5-fold in response to hypoxia and in response to AdCA5. The largest category of genes downregulated by both hypoxia and AdCA5 encoded those proteins involved in cell growth or proliferation. Many genes upregulated by both hypoxia and AdCA5 encoded cytokines, growth factors, receptors, and other signaling proteins. Transcription factors accounted for the largest group of HIF-1–regulated genes, indicating that HIF-1 controls a network of transcriptional responses to hypoxia in ECs. The infection of ECs with AdCA5 under nonhypoxic conditions was sufficient to induce increased basement membrane invasion and tube formation similar to the responses induced by hypoxia, indicating that HIF-1 mediates the cell-autonomous activation of ECs.

These gain-of-function studies were complemented by loss-of-function studies, in which HIF-1α expression in ECs was selectively eliminated in mice that were homozygous for a floxed *Hif1a* allele and expressed Cre recombinase from the *Tie2* gene promoter (55). Loss of HIF-1α expression impaired EC proliferation, chemotaxis, and extracellular matrix invasion. Most strikingly, loss of HIF-1α expression in ECs resulted in a profound impairment of tumor angiogenesis. All the observed defects were associated with decreased VEGF expression and loss of autocrine signaling via VEGF receptor 2 (VEGFR2) in *HIF-1α–null* ECs. Thus, both gain-of-function and loss-of-function studies indicate that HIF-1 plays a cell-autonomous role in ECs that is critical for tumor angiogenesis but less critical for angiogenesis during development, which was apparently normal in the *Hif1a$^{flox/flox}$;Tie2-Cre* mice. Although the studies involving these mice suggest that an important function of HIF-1 in ECs is to mediate autocrine VEGF–VEGFR2 signaling, the microarray analysis suggests that this is probably an oversimplification of the large network of gene expression controlled by HIF-1, and it is likely to contribute to many aspects of EC biology.

INHIBITORS OF ANGIOGENESIS AND HYPOXIA-INDUCIBLE FACTOR 1 IN CANCER THERAPY

Just as HIF-1 is a candidate target for therapeutic angiogenesis in ischemic cardiovascular disease, HIF-1 appears to be a target (direct or indirect) of many novel antiangiogenic cancer therapies. Many of the novel signal transduction inhibitors that are being tested in clinical trials have antiangiogenic properties that relate in part to their effect on HIF-1α expression (Table 29-2). These include receptor tyrosine kinase inhibitors such as Herceptin and Gleevec, which block HER2^{neu} and BCR-ABL signaling, which leads to HIF-1α expression in breast cancer and chronic myelogenous leukemia cells, respectively, via activation of the PI3K signal transduction pathway. Overexpression of cyclooxygenase 2 (COX2) has been shown to promote colon cancer angiogenesis. Prostaglandin E_2, the product of the COX2 reaction, induces HIF-1α expression via activation of G-protein–coupled receptors and PI3K and MAPK pathways, leading to increased synthesis of HIF-1α protein (see Figure 29.3). Thus, COX2 inhibitors, such as celecoxib, reduce HIF-1α levels in cancer cells. The heat shock protein HSP90 is a chaperone for HIF-1α, and inhibitors of HSP90 induce the degradation of HIF-1α through a VHL-independent pathway.

To the extent that HIF-1α synthesis depends on signaling via the PI3K/Akt/mTOR and RAF/MEK/ERK pathways, inhibitors of these pathways will reduce the steady-state

Table 29-2: Novel Anticancer Agents that Inhibit HIF-1 Activity

Mechanism of Action	*Agent(s)**
Receptor tyrosine kinase inhibitors	**Herceptin, Gleevec**
Cyclooxygenase-2 inhibitors	**Celecoxib**
HSP90 inhibitors	Geldanamycin, *17-AAG*
mTOR inhibitors	Rapamycin, *CCI-779*
RAF/MEK inhibitors	*BAY43–9006*, PD98059
HDAC inhibitors	*FK228, NVP-LAQ824*
Angiogenesis inhibitors	*Endostatin*
Proteasome inhibitors	**Bortezomib**, MG132
Topoisomerase I inhibitors	**Topotecan**
Microtubule-targeting agents	*2-methoxyestradiol*
Thioredoxin inhibitors	Pleurotin, PX-12
p300 CH1 domain inhibitor	Chetomin
Unknown	PX-478, YC-1

*Approved drugs are indicated by bold font; drugs in clinical trials are indicated by italics; all others are in preclinical development.

levels of HIF-1α. Because the inhibition of these pathways is predicted to have protean effects in cancer cells, the extent to which the therapeutic efficacy of agents such as rapamycin can be attributed to its inhibition of HIF-1 is unclear. A recent DNA microarray analysis, which attempted to determine the consequences of mTOR inhibition in a transgenic mouse prostate cancer model, identified two functional groups of genes that were upregulated in neoplasms driven by Akt gain-of-function and downregulated in response to mTOR inhibition. One group consisted of antiapoptotic B-cell leukemia/lymphoma 2 (BCL2) family members and the other group consisted of HIF-1 target genes, principally glucose transporters and glycolytic enzymes (56). Microarray studies have provided evidence that the HIF-1 transcriptome represents a major pathway in ECs that is inhibited by endostatin treatment (57). This is consistent with the data presented in the previous section. Histone deacetylase inhibitors also have been shown to have antiangiogenic effects that appear to be attributable to their inhibition of HIF-1 activity, although their precise mechanism of action in this context has not been delineated. The proteasome inhibitor MG132 has been shown to block both the degradation and nuclear import of HIF-1α and thus inhibits HIF-1 transcriptional activity. Whether bortezomib (Velcade), a proteasome inhibitor currently in clinical trials, has this same effect remains to be determined.

A cell-based high-throughput screen for the identification of small molecule inhibitors of the HIF-1 pathway using the National Cancer Institute "Diversity Set," a collection of approximately 2,000 compounds selected to represent the greater chemical diversity of the National Cancer Institute chemical repository, led to the identification of topoisomerase I inhibitor camptothecin and several analogues as inhibitors of HIF-1α expression (58). Preclinical studies suggest that daily administration of the drug results in significant inhibition of tumor growth and angiogenesis (59). The endogenous angiogenesis inhibitor 2-methoxyestradiol inhibits HIF-1α expression, an effect that depends on the compound's activity as a microtubule-disrupting agent (60). Thioredoxin is known to interact with and positively regulate HIF-1 activity, and thioredoxin inhibitors also inhibit HIF-1α expression (61). A screen for low-molecular-weight compounds that block the interaction of the HIF-1α transactivation domain with the CH1 domain of p300 led to the identification of chetomin as an inhibitor of HIF-1 transcriptional activity (62). Because chetomin binds to p300 rather than to HIF-1α, it is likely to interfere with the binding of other transcription factors to the CH1 domain. Two low-molecular-weight compounds that potently block HIF-1α expression, tumor growth, and angiogenesis via unknown mechanisms are PX-478 and YC-1 (63,64). Thus, the understanding of the critical role of HIF-1 in tumor angiogenesis has led to considerable interest in targeting this factor for inhibition as a novel antiangiogenic anticancer therapy.

KEY POINTS

- HIF-1 mediates angiogenesis and other adaptive responses to hypoxia.
- HIF-1 controls the expression of multiple angiogenic growth factors.
- HIF-1 controls cell-autonomous responses of ECs to hypoxia/ischemia.

Future Goals

- To determine whether increased HIF-1α expression, as a result of gene therapy or pharmacologic induction, will promote therapeutic angiogenesis
- To determine whether small-molecule inhibitors of HIF-1 will be useful components of novel anticancer therapeutic regimens

REFERENCES

1 Semenza GL, Wang GL. A nuclear factor induced by hypoxia via de novo protein synthesis binds to the human erythropoietin gene enhancer at a site required for transcriptional activation. *Mol Cell Biol.* 1992;12:5447–5454.

2 Wang GL, Semenza GL. Purification and characterization of hypoxia-inducible factor 1. *J Biol Chem.* 1995;270:1230–1237.

3 Wang GL, Jiang BH, Rue EA, Semenza GL. Hypoxia-inducible factor 1 is a basic-helix-loop-helix-PAS heterodimer regulated by cellular O_2 tension. *Proc Natl Acad Sci USA.* 1995;92:5510–5514.

4 Kelly BD, Hackett SF, Hirota K, et al. Cell type-specific regulation of angiogenic growth factor gene expression and induction of angiogenesis in nonischemic tissue by a constitutively active form of hypoxia-inducible factor 1. *Circ Res.* 2003;93:1074–1081.

5 Jiang BH, Zheng JZ, Leung SW, et al. Transactivation and inhibitory domains of hypoxia-inducible factor 1 alpha: modulation of transcriptional activity by oxygen tension. *J Biol Chem.* 1997272:19253–19260.

6 Jewell UR, Kvietikova I, Scheid A, et al. Induction of HIF-1 alpha in response to hypoxia is instantaneous. *FASEB J.* 2001;15:1312–1314.

7 Yu AY, Frid MG, Shimoda LA, et al. Temporal, spatial, and oxygen-regulated expression of hypoxia-inducible factor-1 in the lung. *Am J Physiol.* 1998;275:L818–L826.

8 Tian H, McKnight SL, Russell DW. Endothelial PAS domain protein 1 (EPAS1), a transcription factor selectively expressed in endothelial cells. *Genes Dev.* 1997;11:72–82.

9 Hu CJ, Wang LY, Chodosh LA, et al. Differential roles of hypoxia-inducible factor 1 alpha (HIF-1 alpha) and HIF-2 alpha in hypoxic gene regulation. *Mol Cell Biol.* 2003;23:9361–9374.

10 Sowter HM, Raval R, Moore J, et al. Predominant role of hypoxia-inducible transcription factor (HIF)-1 alpha versus HIF-2 alpha in regulation of the transcriptional response to hypoxia. *Cancer Res.* 2003;63:6130–6134.

11 Park SK, Dadak AM, Haase VH, et al. Hypoxia-induced gene expression occurs solely through the action of hypoxia-inducible factor 1 alpha (HIF-1 alpha): role of cytoplasmic trapping of HIF-2 alpha. *Mol Cell Biol.* 2003;23:4959–4971.

12 Makino Y, Cao R, Svensson K, et al. Inhibitory PAS domain protein is a negative regulator of hypoxia-inducible gene expression. *Nature.* 2001;414:550–554.

13 Iyer NV, Kotch LE, Agani F, et al. Cellular and developmental control of O_2 homeostasis by hypoxia-inducible factor 1 alpha. *Genes Dev.* 1998;12:149–162.

14 Kotch LE, Iyer NV, Laughner E, Semenza GL. Defective vascularization of HIF-1 alpha-null embryos is not associated with VEGF deficiency but with mesenchymal cell death. *Dev Biol.* 1999;209:254–267.

15 Ryan HE, Lo J, Johnson RS. HIF-1 is required for solid tumor formation and embryonic vascularization. *EMBO J.* 1998;17:3005–3015.

16 Carmeliet P, Dor Y, Herbert JM, et al. Role of HIF-1 alpha in hypoxia-mediated apoptosis, cell proliferation, and tumor angiogenesis. *Nature.* 1998;394:485–490.

17 Compernolle V, Brusselmans K, Franco D, et al. Cardia bifida, defective heart development and abnormal neural crest migration in embryos lacking hypoxia-inducible factor-1 alpha. *Cardiovasc Res.* 2003;60:569–579.

18 Tomita S, Ueno M, Sakamoto M, et al. Defective brain development in mice lacking the Hif-1 alpha gene in neural cells. *Mol Cell Biol.* 2003;23:6739–6749.

19 Yan SF, Fujita T, Lu J, et al. Egr-1, a master switch coordinating upregulation of divergent gene families underlying ischemic stress. *Nat Med.* 2000;6:1355–1361.

20 Salceda S, Caro J. Hypoxia-inducible factor 1 alpha (HIF-1 alpha) protein is rapidly degraded by the ubiquitin-proteasome system under normoxic conditions. Its stabilization by hypoxia depends on redox-induced changes. *J Biol Chem.* 1997;272:22642–22647.

21 Maxwell PH, Wiesener MS, Chang GW, et al. The tumour suppressor protein VHL targets hypoxia-inducible factors for oxygen-dependent proteolysis. *Nature.* 1999;399:271–275.

22 Kamura T, Sato S, Iwai K, et al. Activation of HIF1 alpha ubiquitination by a reconstituted von Hippel-Lindau (VHL) tumor suppressor complex. *Proc Natl Acad Sci USA.* 2000;97:10430–10435.

23 Ivan M, Kondo K, Yang H, et al. HIF alpha targeted for VHL-mediated destruction by proline hydroxylation: implications for O_2 sensing. *Science.* 2001;292:464–468.

24 Jaakkola P, Mole DR, Tian YM, et al. Targeting of HIF-alpha to the von Hippel-Lindau ubiquitylation complex by O_2-regulated prolyl hydroxylation. *Science.* 2001;292:468–472.

25 Yu F, White SB, Zhao Q, Lee FS. HIF-1 alpha binding to VHL is regulated by stimulus-sensitive proline hydroxylation. *Proc Natl Acad Sci USA.* 2001;98:9630–9635.

26 Epstein AC, Gleadle JM, McNeill LA, et al. C. elegans EGL-9 and mammalian homologs define a family of dioxygenases that regulate HIF by prolyl hydroxylation. *Cell.* 2001;107:43–54.

27 Bruick RK, McKnight SL. A conserved family of prolyl-4-hydroxylases that modify HIF. *Science.* 2001;294:1337–1340.

28 Lando D, Peet DJ, Whelan DA, et al. Asparagine hydroxylation of the HIF transactivation domain a hypoxic switch. *Science.* 2002;295:858–861.

29 Mahon PC, Hirota K, Semenza GL. FIH-1: a novel protein that interacts with HIF-1 alpha and VHL to mediate repression of HIF-1 transcriptional activity. *Genes Dev.* 2001;15:2675–2686.

30 Lando D, Peet DJ, Gorman JJ, et al. FIH-1 is an asparaginyl hydroxylase enzyme that regulates the transcriptional activity of hypoxia-inducible factor. *Genes Dev.* 2002;16:1466–1471.

31 Koivunen P, Hirsila M, Gunzler V, et al. Catalytic properties of the asparaginyl hydroxylase (FIH) in the oxygen sensing pathway are distinct from those of its prolyl-4-hydroxylases. *J Biol Chem.* 2004;279:9899–9904.

32 Dames SA, Martinez-Yamout M, De Guzman RN, et al. Structural basis for HIF-1 alpha/CBP recognition in the cellular hypoxic response. *Proc Natl Acad Sci USA.* 2002;99:5271–5276.

33 Freedman SJ, Sun ZY, Poy F, et al. Structural basis for recruitment of CBP/p300 by hypoxia-inducible factor-1 alpha. *Proc Natl Acad Sci USA.* 2002;99:5367–5372.

34 Hon WC, Wilson MI, Harlos K, et al. Structural basis for the recognition of hydroxyproline in HIF-1 alpha by pVHL. *Nature.* 2002;417:975–978.

35 Min JH, Yang H, Ivan M, et al. Structure of an HIF-1 alpha-pVHL complex: hydroxyproline recognition in signaling. *Science.* 2002;296:1886–1889.

36 Fukuda R, Hirota K, Fan F, et al. Insulin-like growth factor 1 induces hypoxia-inducible factor 1-mediated vascular ECs growth factor expression, which is dependent on MAP kinase and phosphatidylinositol 3-kinase signaling in colon cancer cells. *J Biol Chem.* 2002;277:38205–38211.

37 Laughner E, Taghavi P, Chiles K, et al. HER2 (neu) signaling increases the rate of hypoxia-inducible factor 1 alpha (HIF-1 alpha) synthesis: novel mechanism for HIF-1-mediated vascular endothelial growth factor expression. *Mol Cell Biol.* 2001;21:3995–4004.

38 Richard DE, Berra E, Gothie E, et al. p42/p44 mitogen-activated protein kinases phosphorylate hypoxia-inducible factor 1 alpha (HIF-1 alpha) and enhance the transcriptional activity of HIF-1. *J Biol Chem.* 1999;274:32631–32637.

39 Sang N, Stiehl DP, Bohensky J, et al. MAPK signaling up-regulates the activity of hypoxia-inducible factors by its effects on p300. *J Biol Chem.* 2003;278:14013–14019.

40 Bergeron M, Yu AY, Solway KE, et al. Induction of hypoxia-inducible factor-1 (HIF-1) and its target genes following focal ischaemia in rat brain. *Eur J Neurosci.* 1999;11:4159–4170.

41 Lee SH, Wolf PL, Escudero R, et al. Early expression of angiogenesis factors in acute myocardial ischemia and infarction. *N Engl J Med.* 2000;342:626–633.

42 Thurston G, Suri C, Smith K, et al. Leakage-resistant blood vessels in mice transgenically overexpressing angiopoietin-1. *Science.* 1999;286:2511–2514.

43 Visconti RP, Richardson CD, Sato TN. Orchestration of angiogenesis and arteriovenous contribution by angiopoietins and vascular endothelial growth factor (VEGF). *Proc Natl Acad Sci USA.* 2002;99:8219–8224.

44 Carmeliet P, Moons L, Luttun A, et al. Synergism between vascular endothelial growth factor and placental growth factor contributes to angiogenesis and plasma extravasation in pathological conditions. *Nat Med.* 2001;7:575–583.

45 Cao R, Brakenhielm E, Pawliuk R, et al. Angiogenic synergism, vascular stability and improvement of hind-limb ischemia by a combination of PDGF-BB and FGF-2. *Nat Med.* 2003;9:604–613.

46 Forsythe JA, Jiang BH, Iyer NV, et al. Activation of vascular endothelial growth factor gene transcription by hypoxia-inducible factor 1. *Mol Cell Biol.* 1996;16:4604–4613.

47 Ohno-Matsui K, Hirose A, Yamamoto S, et al. Inducible expression of vascular endothelial growth factor in adult mice causes severe proliferative retinopathy and retinal detachment. *Am J Pathol.* 2002;160:711–719.

48 Ozaki H, Yu AY, Della N, et al. Hypoxia inducible factor-1 alpha is increased in ischemic retina: temporal and spatial correlation with VEGF expression. *Invest Ophthalmol Vis Sci.* 1999;40:182–189.

49 Elson DA, Thurston G, Huang LE, et al. Induction of hypervascularity without leakage or inflammation in transgenic mice overexpressing hypoxia-inducible factor-1 alpha. *Genes Dev.* 2001;15:2520–2532.

50 Patel TH, Kimura H, Weiss CR, et al. Constitutively active HIF-1alpha improves perfusion and arterial remodeling in an endovascular model of limb ischemia. *Cardiovasc Res.* 2005;68:144–154.

51 Shyu KG, Wang MT, Wang BW, et al. Intramyocardial injection of naked DNA encoding HIF-1 alpha/VP16 hybrid to enhance angiogenesis in an acute myocardial infarction model in the rat. *Cardiovasc Res.* 2002;54:576–583.

52 Vincent KA, Shyu KG, Luo Y, et al. Angiogenesis is induced in a rabbit model of hindlimb ischemia by naked DNA encoding an HIF-1 alpha/VP16 hybrid transcription factor. *Circulation.* 2000;102:2255–2261.

53 Li J, Post M, Volk R, et al. PR39, a peptide regulator of angiogenesis. *Nat Med.* 2000;6:49–55.

54 Manalo DJ, Rowan A, Lavoie T, et al. Transcriptional regulation of vascular endothelial cell responses to hypoxia by HIF-1. *Blood.* 2005;105:659–669.

55 Tang N, Wang L, Esko J, et al. Loss of HIF-1 alpha in endothelial cells disrupts a hypoxia-driven VEGF autocrine loop necessary for tumorigenesis. *Cancer Cell.* 2004;6:485–495.

56 Majumder PK, Febbo PG, Bikoff R, et al. mTOR inhibition reverses prostate intraepithelial neoplasia through regulation of apoptotic and HIF-1-dependent pathways. *Nat Med.* 2004;10:594–601.

57 Abdollahi A, Hahnfeldt P, Maercker C, et al. Endostatin's antiangiogenic signaling network. *Mol Cell.* 2005;13:649–663.

58 Rapisarda A, Uranchimeg B, Scudiero DA, et al. Identification of small molecule inhibitors of hypoxia-inducible factor 1 transcriptional activation pathway. *Cancer Res.* 2002;62:4316–4324.

59 Rapisarda A, Zalek J, Hollingshead M, et al. Schedule-dependent inhibition of hypoxia-inducible factor-1 alpha protein accumulation, angiogenesis, and tumor growth by topotecan in U251-HRE glioblastoma xenografts. *Cancer Res.* 2004;64:6845–6848.

60 Mabjeesh NJ, Escuin D, LaVallee TM, et al. 2ME2 inhibits tumor growth and angiogenesis by disrupting microtubules and dysregulating HIF. *Cancer Cell.* 2003;3:363–375.

61 Welsh SJ, Williams RR, Birmingham A, et al. The thioredoxin redox inhibitors 1-methylpropyl 2-imidazolyl disulfide and pleurotin inhibit hypoxia-induced factor 1 alpha and vascular endothelial growth factor formation. *Mol Cancer Ther.* 2003;2:235–243.

62 Kung AL, Zabludoff SD, France DS, et al. Small molecule blockade of transcriptional coactivation of the hypoxia-inducible factor pathway. *Cancer Cell.* 2004;6:33–43.

63 Welsh S, Williams R, Kirkpatrick L, et al. Antitumor activity and pharmacodynamic properties of PX-478, an inhibitor of hypoxia-inducible factor-1 alpha. *Mol Cancer Ther.* 2004;3:233–244.

64 Yeo EJ, Chun YS, Cho YS, et al. YC-1: a potential anticancer drug targeting hypoxia-inducible factor 1. *J Natl Cancer Inst.* 2003;95:516–525.

65 Maltepe E, Schmidt JV, Baunoch D, et al. Abnormal angiogenesis and responses to glucose and oxygen deprivation in mice lacking the protein ARNT. *Proc Natl Acad Sci USA.* 1996;93:6731–6736.

66 Peng J, Zhang L, Drysdale L, Fong GH. The transcription factor EPAS-1/hypoxia-inducible factor 2 alpha plays an important role in vascular remodeling. *Proc Natl Acad Sci USA.* 2000;97:8386–8391.

67 Tian H, Hammer RE, Matsumoto AM, et al. The hypoxia-responsive transcription factor EPAS1 is essential for catecholamine homeostasis and protection against heart failure during embryonic development. *Genes Dev.* 1998;12:3320–3324.

68 Scortegagna M, Ding K, Oktay Y, et al. Multiple organ pathology, metabolic abnormalities and impaired homeostasis of reactive oxygen species in *Epas1O*$^{-/-}$ mice. *Nat Genet.* 2003;35:331–340.

69 Seagroves TN, Hadsell D, McManaman J, et al. HIF-1 alpha is a critical regulator of secretory differentiation and activation, but not vascular expansion, in the mouse mammary gland. *Development.* 2003;130:1713–1724.

70 Huang Y, Hickey RP, Yeh JL, et al. Cardiac myocyte-specific HIF-1 alpha deletion alters vascularization, energy availability, calcium flux, and contractility in the normoxic heart. *FASEB J.* 2004;18:1138–1140.

71 Schipani E, Ryan HE, Didrickson S, et al. Hypoxia in cartilage: HIF-1 alpha is essential for chondrocyte growth arrest and survival. *Genes Dev.* 2001;15:2865–2876.

72 Karhausen J, Furuta GT, Tomaszewski JE, et al. Epithelial hypoxia-inducible factor-1 is protective in murine experimental colitis. *J Clin Invest.* 2004;114:1098–1106.

73 Cramer T, Yamanishi Y, Clausen BE, et al. HIF-1 alpha is essential for myeloid cell-mediated inflammation. *Cell.* 2003;112:645–657.

74 Peyssonnaux C, Datta V, Cramer T, et al. HIF-1 alpha expression regulates the bactericidal capacity of phagocytes. *J Clin Invest.* 2005;115:1806–1815.

30

Integrative Physiology of Endothelial Cells

Impact of Regional Metabolism on the Composition of Blood-Bathing Endothelial Cells

Mitchell L. Halperin and Kamel S. Kamel

St. Michael's Hospital, University of Toronto, Ontario, Canada

This chapter has a different emphasis from others in this book because it is written from the perspective of integrative physiology. Our objective is to discuss the conditions that endothelial cells (ECs) are exposed to in the body from the luminal side and to raise a number of questions that may influence how vascular biologists think about their subject. In this context, new data may be required to determine whether these are just theoretical considerations or whether they are, in fact, actual perturbations for these cells. If the latter were true, one must consider how these perturbations are dealt with and whether they have an impact on EC function.

Our first task will be to define the composition of blood in the arterial tree. The description will be expanded to veins draining specific regions of individual organs where large differences in blood composition occur. Perhaps the best example of this latter point is the marked difference in the composition of venous blood draining the renal cortex as compared with that draining the hypoxic, hypertonic, and hypercarbic renal medulla.

Rather than focusing on many compounds, we emphasize the major volatile gases (oxygen [O_2], carbon dioxide [CO_2], and ammonia [NH_3]); ions such as hydrogen (H^+), potassium (K^+), calcium (Ca^{2+}) and citrate; as well as water because of their biological importance and implications for normal cell function. Where applicable, the discussion in each section is divided into three parts. First, we describe the composition of blood in each vascular location. Second, we ask: "What demands might be placed on ECs because of their exposure to the composition of blood in their lumen?" Third, bearing the above in mind, we ask, "What questions should scientists working with ECs consider when they examine endothelial function in vivo and when experiments are performed in vitro?"

ENDOTHELIAL CELLS IN THE ARTERIAL TREE[1]

ECs in all the arteries are exposed to a similar, if not identical, luminal composition, because these blood vessels serve primarily as conduits of blood. Certain exceptions can be made to this statement, including the pulmonary artery and the efferent renal arteriole. Little exchange of gases and nutrients occurs in arterial segments. Thus, any differences in blood composition between the proximal and distal ends of the artery depend on the degree to which arterial ECs synthesize and release substances into the blood. On the abluminal side of the arterial endothelium, the input is likely similar throughout the arterial tree, consisting of signals derived from extracellular matrix and vascular smooth muscle cells. Arterial blood has all the metabolic constituents needed to support ECs, including a high P_{O_2} and a physiologically appropriate P_{CO_2} (40 mm Hg).[2]

The relatively high intracellular P_{O_2} in arterial ECs has potential disadvantages. O_2 is toxic when it reacts with an electron (e^-) to form superoxide (O_2^-) (Equation 1). This O_2^-, if it is not removed as quickly as it is formed, will rise in concentration and react with a number of reactive groups

1 In addressing differences in the composition of blood between and within arteries and veins, it is important to consider variation across space and over time. In arteries, few differences in composition of blood occur in space, but there can certainly be changes over time (e.g., with exercise, or with pathophysiology); in veins, large differences occur across space (in terms of what each organ the vein is draining) and differences also occur in some beds over time.

2 Because O_2 is sparingly soluble in water (1), a limitation exists to the diffusion of O_2. Nevertheless, O_2 diffuses through capillaries to most cells, and the P_{O_2} in their mitochondria is only a few mm Hg. Although this is high enough to regenerate ATP at extremely rapid rates because of the high affinity of mitochondrial enzymes that bind O_2 in the electron transport system (2), the low P_{O_2} may diminish the formation of superoxide (vide infra).

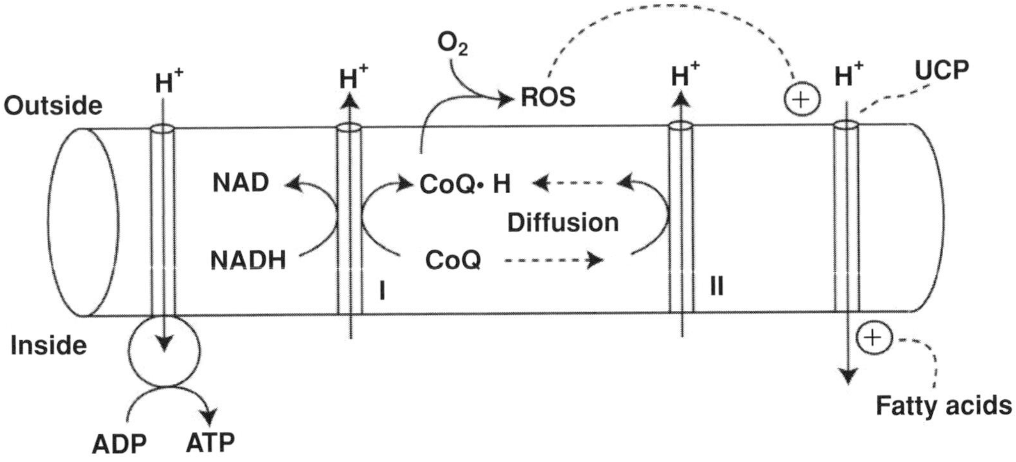

Figure 30.1. Uncoupling of oxidative phosphorylation. The horizontal structure represents the inner mitochondrial membrane. On the far left, the diffusion of H^+ into mitochondria drives ATP regeneration (coupled oxidative phosphorylation). In the central part of the illustration, only complexes I and II are shown – they pump out H^+ and are linked by the diffusion of coenzyme Q. The semialdehyde form of reduced coenzyme Q reacts with O_2 to produce ROS. The UCP is shown on the right. When ROSs rise and long-chain fatty acids are present, flux of H^+ through UCPs rises and coenzyme Q becomes less reduced, but ATP is not regenerated.

in proteins, lipids, and nucleic acids, which could compromise cell functions (3,4). The chemistry of this type of O_2^- formation is a nonenzymatic reaction; its rate is directly proportional to the concentration of its substrates. With respect to the concentration of e^-, when the rate of flux in the electron transport system is not rapid, the state of electron carriers becomes more reduced. The level of one of these electron carriers, the semialdehyde form of coenzyme Q, is most important in this context because, when its concentration and the Po_2 are both high, an accelerated rate of formation of reactive oxygen species (ROS) occurs (Equation 1):

$$O_2 + e^- \rightarrow O_2^- \text{ (ROS)} \qquad \text{(Equation 1)}$$

It is important to emphasize that, although low concentrations of ROS have useful physiological functions, high levels of ROS may cause cellular damage (5). Hence, the metabolic pathways that "detoxify" ROS are very important, including catalase and superoxide dismutase. One example will be mentioned to illustrate this point: Uncoupling of mitochondrial oxidative phosphorylation is one mechanism to diminish the rate of production of ROS (Figure 30.1) (6). Mitochondria have uncoupler proteins (UCPs) that can dissipate the large electrical and small H^+ gradient across the inner mitochondrial membrane (3,7). For these UCPs to have a higher conductance for H^+ in the short term, several factors must be present. One stimulus for ROS removal is a higher ROS concentration. In addition, high physiological concentrations of long-chain fatty acids lead to a more rapid H^+ flux (3,4,6,8). One possible reason for this requirement for long-chain fatty acids is deducible from the larger picture of the integrative physiology of prolonged fasting in Paleolithic times (see Chapter 16). In more detail, the Po_2 and the concentration of the semialdehyde form of coenzyme Q are both high during prolonged fasting (9). Hence, uncoupling is advantageous to defend against high concentrations of ROS. Having a requirement, however, for long-chain fatty acids ensures that this uncoupling does not lead to the oxidation of glucose, the fuel needed for the brain during early fasting (10). In contrast, after eating a high carbohydrate meal, the level of long-chain fatty acids should be low, and this could increase the risk of damage secondary to ROS accumulation in arterial ECs, especially if other pathways for ROS removal were not proceeding at an adequate rate.

Thus, questions to be addressed by vascular biologists include: "Do arterial ECs have special characteristics to allow them to handle an ongoing oxidative stress?" and "Under what circumstances and/or diseases might the defenses against ROS be diminished?" These considerations apply when experiments are carried out in vitro, because the use of a high Po_2 in the incubation system and variations in the fuel mixture used may have implications for the validity of the conclusions drawn in these experimental settings.

ENDOTHELIAL CELLS IN THE VENOUS TREE

In contrast to arterial blood, the composition of venous blood varies from organ to organ, and at times within different regions of the same organ. Indeed, if one were to compare venous blood in two different organs, large differences would be seen in composition, depending on what that organ extracted from and secreted into the blood of the upstream capillary bed. Some organs have a relatively constant composition of venous blood. Other organs demonstrate significant temporal changes in the composition (e.g., Po_2 and Pco_2) of venous blood.

Organs with a Relatively Constant Extraction of Oxygen and Formation of Carbon Dioxide

To have a constant composition of Po_2 and Pco_2 in the venous blood, as a general principle, a tight correlation must exist between the biological work of each of these organs (O_2 consumption and CO_2 production) and their rate of blood flow. In three major organs the composition of venous blood remains constant over time, but with very different individual (organ-specific) Po_2 and Pco_2 levels.

Brain

The brain has the ability to autoregulate its blood flow, and it also has a metabolic rate that is relatively constant throughout the day (11). Hence, the jugular venous Po_2 and Pco_2 should be relatively constant. Because the brain extracts close to one-third of the O_2 from each liter of arterial blood, the venous Pco_2 is close to 46 mm Hg. This constancy over time is valuable, because certain areas of the brain use changes in their Pco_2 as a signal. The best example is the control of alveolar ventilation via changes in the Pco_2 that are sensed by chemoreceptors.

Kidney

The kidneys are the most important site at which a decrease in Po_2 provides the signal for the synthesis and release of the hormone erythropoietin (EPO) (12). Because of the shape of the O_2–hemoglobin dissociation curve (Figure 30.2), the largest change in the Po_2 per mmol of O_2 extracted from arterial blood (or decrease in the content of arterial O_2) occurs when blood is closest to being 100% saturated with O_2. Moreover, because the signal for the Po_2 should be large, it is important to have the blood flow high enough to ensure that O_2 extraction is quite small per liter of blood. This, in effect, means that kidney cells and renal cortical efferent arterial ECs must exist in an area having a very high Po_2. As outlined for arterial blood, the pertinent questions are: "Will these cells be more at risk to O_2 toxicity than other venous ECs?" and "Do they possess efficient defense mechanisms to deal with the formation of ROS?"

Heart

Myocardial cells extract between 65% and 75% of the O_2 from each liter of blood (13). This strategy provides a number of physiological advantages. First, it is easier to have signals related to a low Po_2 and/or high Pco_2 in the short term. For example, the low Po_2 and high Pco_2 make vasodilatory stimuli more effective. Hence, as soon as a demand occurs for an increased cardiac output, blood flow rises abruptly, and cells may continue to have rapid rates of oxidative metabolism. Second, living in an environment with a low Po_2 can stimulate angiogenesis. This ensures a very rich cardiac blood flow rate and a shorter distance between capillaries and mitochondria, to permit O_2 to diffuse faster.

Because of this high O_2 extraction, cardiac venous ECs exist in a low-Po_2 and high-Pco_2 environment (10). In exercise, the Po_2 might be even lower and the Pco_2 even higher in coronary sinus blood. Although this low Po_2 diminishes the risk of ROS formation, a disadvantage exists for venous ECs, because they will exist in an environment with a high Pco_2. If these cells were sufficiently permeable to CO_2, their Pco_2 would be high. This would drive the bicarbonate buffer system equation to the right, which raises the concentration of H^+ in these cells (Equation 2). This higher concentration of H^+ forces H^+ to bind to intracellular proteins, changing their charge, shape, and possibly function (Equation 3, Figure 30.3) (14).

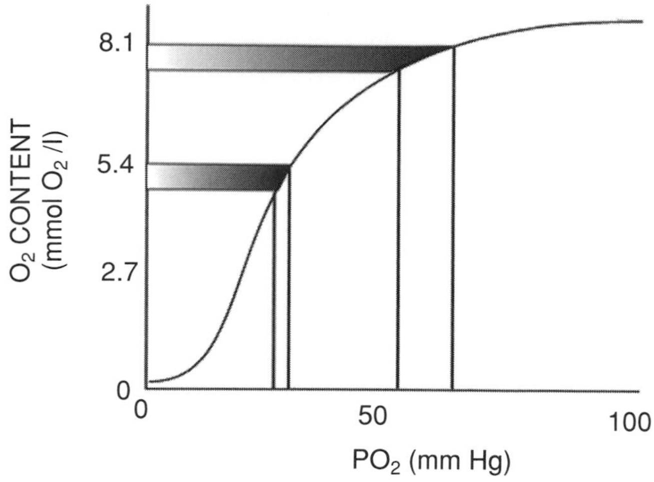

Figure 30.2. Hemoglobin/oxygen dissociation curve. The shaded horizontal rectangle represents the extra O_2 extracted per L when there is a small reduction in the concentration of hemoglobin in blood and the same quantity of O_2 must be extracted. The solid vertical lines indicate the change in arterial Po_2 due to this extra extraction of O_2 per L of blood. There is a larger change in the arterial Po_2 for this increment in O_2 extraction and a much more sensitive signal for the release of EPO when on the flatter portion of the O_2–hemoglobin dissociation curve (12).

$$CO_2 + H_2O \leftrightarrow H_2CO_3 \leftrightarrow H^+ + HCO_3^- \qquad \text{(Equation 2)}$$

$$H^+ + \text{Protein}^o \leftrightarrow H \cdot \text{Protein}^+ \qquad \text{(Equation 3)}$$

The questions for vascular biologists include: "Are ECs sufficiently permeable to CO_2?" and "How do ECs in cardiac veins adapt to this difference in charge on their proteins?" and "Are these ECs more vulnerable to a H^+ load (e.g., during a sprint) because they might not have physiological buffering by their bicarbonate buffer system if their Pco_2 is high?"

Organs with a Variable Extraction of Oxygen and Thereby Formation of Carbon Dioxide per Liter of Blood Flow

Skeletal Muscle

At rest, skeletal muscle extracts a small amount of O_2 per liter of blood flow. The venous Pco_2 in these cells is close to 46 mm Hg, but a large variation occurs in normal

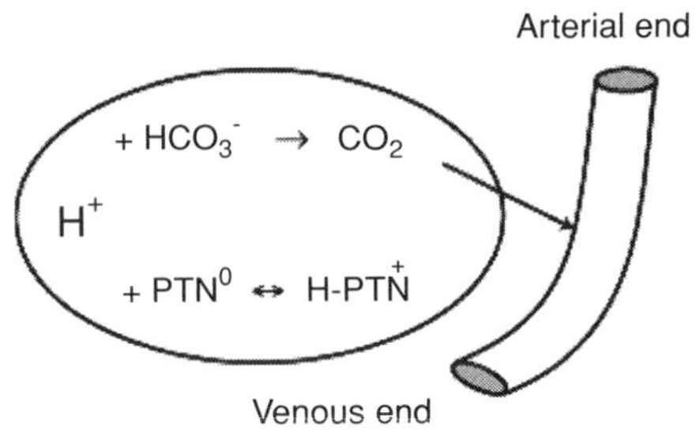

Figure 30.3. Role of EC P_{CO_2} in the selection of intracellular buffer systems. The aim of buffering is to prevent H^+ from binding to proteins in cells (PTN^o) – this can occur if the tissue P_{CO_2} falls. For this to happen, a low venous P_{CO_2} must be present. In contrast, if the venous P_{CO_2} is high, the P_{CO_2} will be high in ECs, the $[H^+]$ in these cells will rise, and H^+ will be bound to intracellular proteins ($H \cdot PTN^+$), changing their charge, shape, and possibly their function.

individuals (15). During exercise and in patients with a severe degree of extracellular fluid volume (ECF) volume contraction and metabolic acidosis, muscle cells extract more O_2 from each liter of blood flow; hence, the P_{O_2} falls and the P_{CO_2} rises in venous blood (16). This high venous P_{CO_2} exposes these cells to the risk mentioned earlier for cardiac venous ECs, with a possible titration of intracellular proteins with H^+ (see Figure 30.3).

This raises the following questions for the endothelial biologist: "Are ECs permeable to CO_2?" and "Is there a change in their function caused by H^+ binding to ICF proteins?"

Intestinal Tract

The major feature here is that venous ECs are exposed to all the nutrients absorbed in the intestinal tract, as well as to the products of enterocyte metabolism. A few examples are listed to illustrate the extremes of conditions that these cells face.

WATER. When water is ingested rapidly, the concentration of sodium (Na^+) in plasma (P_{Na}) of arterial blood is 4 ± 1 mmol/L lower than the P_{Na} in simultaneously drawn peripheral brachial venous blood (17). With respect to potential threats, extrapolating back to portal venules draining the early small intestine (where most of this water is absorbed) implies that the P_{Na} here might be dramatically lower in this setting. Hence, if these ECs have aquaporin-1 water channels, they could swell acutely and dilute the concentrations of important regulatory molecules in cells. Of great interest, aquaporin-1 also functions as one of the systems that facilitates the diffusion of P_{CO_2}. Therefore questions arise for the endothelial biologist: "Do these ECs have water channels?" and "Does a large and acute fall in the P_{Na} in luminal fluid change important properties of these ECs?"

CITRUS FRUITS. Two abundant constituents of oranges are K^+ cations and citrate anions. When absorbed, the concentration of K^+ in the plasma (P_K) of portal venous blood should rise. With respect to potential threats, if these ECs have K^+ channels that are open in their luminal membrane, the voltage across this membrane will decrease. As a result, one might expect alterations in voltage-gated calcium ion (Ca^{2+}) channels with secondary effects on the concentration of ionized Ca^{2+} in these cells. Acute episodic changes in this important regulator of metabolism could have important implications for the function of these cells. Therefore, endothelial biologists might ask: "Are there K^+ channels in these cells?" and "Are there changes in the $[Ca^{2+}]$ in ECs exposed to high K^+?"

In a similar "vein," the absorbed citrate anions chelate Ca^{2+} in plasma. Therefore, these ECs will be faced with acute falls in the concentration of Ca^{2+} in venous blood. Again, the endothelial biologist might want to examine what implications this may have for the function of these cells. In fact, episodic absorption and changes in the composition of venous blood-draining enterocytes were the rule in early human evolution because prehistoric diets were rich in fruits and berries (18).

AMMONIUM. The liver must be supplied with close to one mmol of ammonium (NH_4^+) per mmol of amino acid to permit it to convert amino acids to glucose plus urea (19). The intestinal tract supplies the bulk of this NH_4^+. The sources of this NH_4^+ are largely glutamine metabolized in enterocytes (20) and the hydrolysis of urea by colonic bacterial urease (21). In fact, the concentration of NH_4^+ in the portal vein is several hundred μmol/L (19), a concentration of NH_4^+ that is almost an order of magnitude higher than in most other venous systems.

To determine potential threats, it is important to recognize that NH_4^+ and NH_3 do not cross lipid membranes readily. The transport system is recognized as an NH_3 channel (22). Because its outer region has a very hydrophobic area in the luminal membrane of ECs in the portal vein, the apparent P_K for NH_4^+ falls by close to three units (Figure 30.4). This results in H^+ returning to plasma while NH_3 enters the pore that traverses the plasma membrane. Once NH_3 enters the cytosol

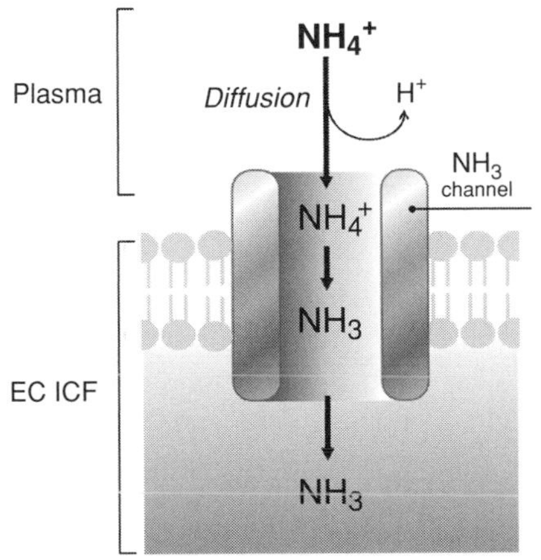

Figure 30.4. Transfer of NH_4^+ from the plasma to the ICF of the EC. The plasma is shown on the top and the ICF compartment of the EC in the portal vein is shown on the bottom of this illustration. The NH_3 channel in the center is greatly exaggerated in size. NH_4^+ diffuses into a hydrophobic area of the mouth of the NH_3 channel where NH_3 is formed. The entry of NH_3 into the cell raises its pH and NH_4^+ concentration.

of ECs, it is converted back to NH_4^+, and the amount formed is a function of the local [H^+]. In organelles with a high [H^+], if they too had these NH_3 channels, an enormous concentration of NH_4^+ would be produced. Hence, it is important to ask: "Do these ECs have Rh-glycoproteins of the classes that form NH_3 channels (23)?" and "What impact might they have on EC function?"

KEY POINTS

- Some demands placed on ECs in the arterial and venous systems related to the composition of arterial and venous blood have been identified.
- With changes in each constituent, potential disadvantages have been identified. Data are needed concerning the pathways for these chemicals to cross EC membranes.
- A consideration of these compositional differences in time and location leads to a number of questions that should be considered when the functions of ECs are investigated in vivo and in vitro.

REFERENCES

1 Voet D, Voet JG. *Biochemistry.* New York: John Wiley and Sons, 1990.

2 Mitchell P. Coupling of phosphorylation to electron and hydrogen transfer by a chemi-osmotic type of mechanism. *Nature.* 1961;191:144–148.

3 Brand MD. Uncoupling to survive? The role of mitochondrial inefficiency in ageing. *Exp Gerontol.* 2000;35:811–820.

4 Lowell BB, Spiegelman BM. Towards a molecular understanding of adaptive thermogenesis. *Nature.* 2000;404:652–660.

5 Hildeman DA, Mitchell T, Kappler J, Marrack P. T cell apoptosis and reactive oxygen species. *J Clin Invest.* 2003;111:575–581.

6 Echtay KS, Brand MD. Coenzyme Q induces GDP-sensitive conductance in kidney mitochondria. *Biochem Soc Transactions.* 2001;29:763–768.

7 Porter RK. Mitochondrial proton leak: a role for uncoupling proteins 2 and 3? *Biochimica Biophysica Acta.* 2001;1504:120–127.

8 Harper M-E, Dent RM, Bezaire V, et al. UCP3 and its putative function: consistencies and controversies. *Biochem Soc Transactions.* 2001;29:768–773.

9 Boss O, Samec S, Dulloo A, et al. Tissue-dependent upregulation of rat uncoupling protein 2-expression in response to fasting or cold. *FEBS Lett.* 1997;412:111–114.

10 Davids MR, Eduote Y, Jungas RL, et al. Facilitating an understanding of integrative physiology: Emphasis on the composition of body fluid compartments. *Can J Physiol Pharmacol.* 2002;80:835–850.

11 Kety SS, Polis BD, Nadler CS, Schmidt C. The blood flow and oxygen consumption of the human brain in diabetic acidosis and coma. *J Clin Invest.* 1948;27:500–510.

12 Halperin ML, Cheema-Dhodli S, Lin S-H, Kamel KS. Properties that permit the renal cortex to be the oxygen sensor for the release of erythropoetin: clinical implications. *Clin J Am Soc Nephrol.* 2006;1:1049–1053.

13 Goodale WT, Olson RE, Hackel DB. The effects of fasting and diabetes mellitus on myocardial metabolism in man. *Am J Med.* 1959;27:212–220.

14 Vasuvattakul S, Warner LC, Halperin ML. Quantitative role of the intracellular bicarbonate buffer system in response to an acute acid load. *Am J Physiol.* 1992;262:R305–R309.

15 Halperin ML, Rosenbaum D, Kamel K. Approach to the patient with metabolic acidosis: Emphasis on the ECF volume and the venous P_{CO_2}. In: Andreoli TA, Ritz E, Rosivall L, eds. *Budapest Nephrology School: Nephrology, Hypertension, Dialysis, Transplantation.* Vol. 1. Budapest, Hungary, 2005.

16 Napolova O, Urbach S, Davids MR, Halperin ML. How to assess the degree of extracellular fluid volume contraction in a patient with a severe degree of hyperglycemia. *Nephrol Dial Trans.* 2003;18:2674–2677.

17 Shafiee MA, Charest AF, Cheema-Dhadli S, et al. Defining conditions that lead to the retention of water: The importance of the arterial sodium concentration. *Kidney Internat.* 2005;67:613–621.

18 Eaton SB, Konner M. Paleolithic nutrition. *N Engl J Med.* 1985; 312:283–289.

19 Jungas RL, Halperin ML, Brosnan JT. Lessons learnt from a quantitative analysis of amino acid oxidation and related gluconeogenesis in man. *Physiol Reviews.* 1992;72:419–448.

20 Yang D, Hazey JW, David F, et al. Integrative physiology of splanchnic glutamine and ammonium metabolism as revealed by balance techniques in conjunction with a sensitive assay for $^{13}NH^{4+}$ enrichment. *Am J Physiol.* 2000;278:E469–E476.

21 Jackson AA. Salvage of urea-nitrogen in the large bowel: functional significance in metabolic control and adaptation. *Biochem Soc Trans.* 1998;26:281–286.

22 Khademii S, O'Connell III J, Remis J, et al. Mechanism of ammonia transport by AMT/MEP/Rh: Structure of AMTB at 135A. *Science.* 2004;305:1587–1594.

23 Weiner ID, Verlander JW. Renal and hepatic expression of the ammonium transporter proteins, Rh B Glycoprotein and Rh C Glycoprotein. *Acta Physiol Scand.* 2003;179:331–338.

31

Tumor Necrosis Factor

Jordan S. Pober

Yale University School of Medicine, New Haven, Connecticut

Tumor necrosis factor (TNF) was originally identified as an endogenous mediator of hemorrhagic necrosis of experimental tumors in rabbits. This hemorrhagic necrosis is caused by bacterial products, especially lipopolysaccharide derived from the walls of gram-negative bacteria (1,2). Human TNF protein was purified to homogeneity, and its complete amino acid sequence was determined (3,4); a complementary DNA (cDNA) was subsequently cloned and expressed (4). At the time of cloning, the molecule was renamed as TNF-α to emphasize its close structural relationship to lymphotoxin (LT)-α, then designated as TNF-β. (This latter name has gone out of use, so the names TNF and TNF-α are now synonymous, and the designation "α" is superfluous.) Cachectin was independently identified as a mediator of inflammation and as a mediator of cachexia in parasitic infections (2). Following its purification, amino acid sequencing established murine cachectin as identical to TNF (5).

TUMOR NECROSIS FACTOR FAMILY

TNF and LT-α define the TNF superfamily of secreted and cell-surface proteins that bind to structurally related receptors (the TNFR superfamily) (6). The TNF superfamily is largely confined to vertebrates and probably coevolved with the adaptive immune system, although a single TNF-like molecule recently has been identified in *Drosophila* (6). The largest cluster of TNF superfamily molecules, including both TNF and LT-α, is encoded within the major histocompatibility complex. Knockout of *TNF* gene in mice leads to abnormal formation of Peyer's patches in the small intestine but otherwise normal development (7).

The primary cellular sources of TNF are leukocytes, especially mononuclear phagocytes, T lymphocytes, and mast cells (8). There is a report that cultured human umbilical arterial endothelial cells (ECs) may synthesize TNF (9), but we have not observed TNF expression under similar conditions by multiple other cultured human ECs types (J. Pober, unpublished observations). Mononuclear phagocytes synthesize TNF largely in response to toll-like receptor microbial ligands, such as lipopolysaccharide. T cells secrete TNF in response to antigen recognition. Mast cells constitutively synthesize TNF and store it for rapid release following antigen or anaphylotoxin binding. Circulating TNF levels are generally low, less than 1 pg/mL in normal circumstances, but can rise to over 100 pg/mL in settings such as gram-negative sepsis (10). TNF is synthesized as a 24-kDa type II membrane protein and may function as a cell-surface ligand on activated macrophages and perhaps T cells. It is released by proteolysis, predominantly mediated by TNF-α converting enzyme (TACE), a metalloproteinase. The biologically active form of TNF is a homotrimer formed from three 17-kDa processed polypeptides.

TUMOR NECROSIS FACTOR RECEPTORS

TNF interactions with cellular targets involve binding to either of two cell surface receptors: TNFR1 (CD120a) or TNFR2 (CD120b). Although their external domains are homologous (and define a TNFR superfamily), these two receptors differ significantly in their intracellular domains. Trimeric ligand binding either induces receptor trimer formation or induces conformation changes in preformed receptor trimers, initiating signaling. Both receptors bind TNF (or LT) homotrimers with similar affinities (Kd = 1–10 nM), although TNFR2 may be preferentially activated by the transmembrane form of TNF expressed on activated macrophages. It has been proposed that TNFR2 may capture TNF and pass it to TNFR1, but little direct evidence suggests heterotrimeric receptors containing a mixture of TNFR1 and TNFR2 polypeptides.

Cultured human umbilical vein ECs express both TNF receptors (11). TNFR1 is present on resting ECs, although the majority of receptors are confined to the Golgi apparatus (12). In situ, surface receptors may be concentrated within caveolae, the loss of which results in a diminution of surface expression in cultured ECs (13). Certain signals, such as histamine, result both in mobilization of Golgi receptors to the cell surface and TACE-mediated shedding of receptors from the cell surface (14). (Recall that TACE is the same enzyme that liberates

TNF from the surface of activated mononuclear phagocytes.) TNFR2 molecules, although present in some cultured ECs, are absent from most ECs in situ but can be rapidly (within hours) induced in response to TNF (15). TNFR2, when expressed, is primarily a cell-surface protein and, like TNFR1, may be shed following proteolysis by TACE. Soluble TNFRs act as natural inhibitors of TNF signaling (16).

TUMOR NECROSIS FACTOR SIGNALING

Signaling by TNFR1 involves assembly and disassembly of interacting protein complexes (17,18). The intracellular carboxyl-terminus of TNFR1 contains a protein–protein interaction motif known as a death domain (DD). DDs are found on a subset of TNFR superfamily members in addition to TNFR1 (sometimes called death receptor 1 or DR-1), including Fas (DR-2), DR-3, TNF-related apoptosis-inducing ligand (TRAIL)-R1 (DR-4), TRAIL-R2 (DR-5), and DR-6. Ligand-occupied TNFR1 rapidly (within minutes) recruits an adaptor protein called TNF receptor–associated death domain (TRADD) protein to its cytosolic DD, displacing another DD-containing protein known as silencer of death domains (SODD). The TNFR1–TRADD complex rapidly recruits two other adaptor proteins, receptor interacting protein (RIP)-1 and TNF receptor–associated factor (TRAF) 2. (In some cells, TRAF2 may be replaced by TRAF5.) This signaling complex (sometimes called a signalosome) undergoes three changes within a matter of minutes, namely internalization, dissociation of TRADD-containing complexes from TNFR1, and activation of cytosolic kinases. One such kinase is MEKK-3, a mitogen-activated protein (MAP) kinase kinase kinase (MAP-KKK) that phosphorylates and activates the I-κ B-kinase complex, leading to activation of nuclear factor (NF)-κB (19). The internalized signalosome also may recruit a different MAPKKK, known as apoptotic signaling kinase (ASK)-1 (20). Binding of ASK-1 leads to activation of kinase cascades that activate c-jun N-terminal kinase (JNK) and p38 MAP kinase (MAPK), which also are known as stress-activated protein (SAP) kinases for their roles in responding to injurious stimuli. ASK-1 recruitment and activation may be facilitated by an adaptor protein known as ASK-1 interacting protein (AIP)-1 (21). AIP-1 and ASK-1 binding to TRAF2 may prevent interactions with MEKK-3, switching signaling from NF-κB to SAP kinases. It is important to note that these signalosome complexes likely are associated with internal membrane structures, but the precise cellular compartments involved are not known.

After a lag period of a few hours, the signalosome further evolves to recruit Fas-associated DD-containing protein (FADD) (22). The recruitment of FADD converts the signalosome to a death-inducing signaling complex (DISC) that binds and promotes the autocatalytic activation of procaspases 8 and 10. Caspase recruitment and/or activation may be blocked by cellular-FLICE inhibitory protein (cFLIP; FLICE is an older name for caspase 8). Activated caspase 8 (or 10) initiates death pathways through proteolytic activation of caspase 3 and/or through mitochondrial signaling initiated by cleavage of a BH-3 domain-only member of the B-cell leukemia/lymphoma (Bcl)-2 family known as Bid (23). Cleaved Bid binds to Bax on the mitochondrial membrane, triggering release of several death-inducing proteins including cytochrome c, SMAC/Diablo, and apoptosis-inducing factor. The death response initiated by the DISC can be antagonized at several levels by NF-κB–induced proteins such as c-FLIP, Bcl-2, and members of the inhibitor of apoptosis protein (IAP) family. Under normal circumstances, JNK and p38 MAPK activation in response to TNF is transient, but if these signals persist beyond 60 minutes, they can also initiate mitochondrial-dependent death via caspase-independent activation of Bid and its molecular target Bax (24). In summary, the three primary signaling pathways activated by ligand binding to TNFR1 in ECs are NF-κB–mediated gene transcription (involving proteins that are largely antiapoptotic and proinflammatory/procoagulant), SAP kinase–mediated gene transcription (involving proteins that are primarily proinflammatory/procoagulant), and caspase-mediated cell death (which normally is effectively antagonized by NF-κB–induced proteins). However, if NF-κB–induced gene expression is blocked (by drugs or toxins), or if procaspase 8 levels are enhanced (e.g., by interferon [IFN]-γ treatments), then TNF may cause EC apoptosis.

Several additional TNFR1-activated signaling responses are less well understood. These include Ras-dependent activation of raf and p42/44 MAPK (25) and phosphinositide 3-kinase (PI3K)–dependent activation of Akt and its downstream effectors (26). These two potentially progrowth/antiapoptotic responses may involve receptor cross-talk between TNFR1 and other cellular receptors. Cross-talk between different receptors is facilitated by colocalization of different receptors into segregated regions of the plasma membrane, such as those created by cholesterol- or sphingolipid-rich patches, commonly called lipid rafts. In ECs, most lipid rafts are further sequestered into membrane invaginations known as caveolae, which are organized by the cholesterol binding proteins known as caveolins 1 and 2. Caveolae are abundant in ECs in situ and, although most ECs rapidly lose their caveolae in culture, these structures are maintained in the immortalized cell line EA.hy.926. Recent studies have shown that TNFR1 is indeed localized to caveolae in this cell line, and that activation of Akt, but not of NF-κB is lost when caveolae are disrupted, consistent with the cross-talk hypothesis (13). Yet another type of response is initiated when ligand-occupied TNFR1 activates neutral sphingomyelinase via a adaptor protein known as factor for activation of neutral sphingomyelinase (FAN) (27); human ECs appear to express little if any FAN and do not seem to activate this pathway (W. Min and J. Pober, unpublished observations). Finally, TNFR1 can activate cytosolic cathepsin B, leading to mitochondria-dependent death responses independent of Bid (28). This response, which may involve RIP-1 (Li and Pober, unpublished observations) is only seen when cathepsin B escapes from lysosomes, a condition observed in human ECs that have been treated with IFN-γ (29).

TNFR2 signaling in ECs is less well understood. In some cell types, TNFR2 may directly recruit TRAF2, and receptor overexpression in ECs leads to activation of NF-κB (30). However, it is not clear that this response can be initiated in human ECs by ligand binding. In cultured bovine aortic ECs, TNFR2 signaling may activate PI3K and Akt. This response depends on epithelial/endothelial tyrosine kinase (Etk) and involves cross-talk with the vascular endothelial growth factor receptor 2 (VEGFR2, also known as KDR or FLK-1) (31). The TNFR2-mediated activation of Etk also may be observed in human glomerular ECs in situ (15). TNFR2 signaling via this pathway has been proposed to result in angiogenic responses (32).

Although most TNF responses of ECs may be described as a matter of induced or increased protein expression, some responses involve the reduced expression of proteins. For example, TNF treatment of ECs can reduce expression of the anticoagulant protein thrombomodulin (33) or the EC enzyme nitric oxide synthase-3 (also known as endothelial or eNOS) (34). The mechanisms of these effects are unclear, but may involve reduced transcription due to diversion of coactivator proteins like CBP/p300 and/or due to mRNA destabilization.

ENDOTHELIAL CELL PHENOTYPES

At the cellular level, the TNF responses of ECs vary depending on their location within the vascular system. Some proinflammatory proteins, such as the leukocyte adhesion molecules E-selectin and vascular cell adhesion molecule (VCAM)-1, are restricted in their expression to those ECs lining postcapillary venules (35). This may arise from selective inhibition of their transcription by the laminar flow–induced protein Krüppel-like factor 2 (KLF2) in arterial ECs, although it does not explain lack of expression in true capillaries (36). Not all TNF-induced proteins are inhibited by KLF 2, and TNF can induce increased expression of intercellular adhesion molecule (ICAM)-1 as well as major histocompatibility complex (MHC) class I molecules on ECs throughout the vascular system. TNF also causes a rearrangement in the cytoskeleton of ECs, altering cell shape and disrupting cell–cell junctions. These actions increase paracellular leakiness (37). Such changes in EC organization depend on new protein synthesis rather than direct receptor-initiation of signaling responses (38). TNF-induced leakage across the vasculature is being exploited as a means of enhancing the delivery of chemotherapeutic agents to isolated sites of tumor growth, for example, in the limbs (39). At the organ level, TNF responses in ECs lead to leukocyte recruitment and intravascular coagulation (thrombosis) (40), which are the two fundamental mechanisms of host defense. When these responses become systemic (as opposed to local), they can contribute to multiorgan failure and to disseminated intravascular coagulation. IFN-γ is a key cofactor in such systemic responses, and it may potentiate both TNF-mediated gene expression and EC death. The potentiation of gene expression (often called synergy) largely involves the activation of parallel transcription factors (e.g., by IFN-γ–mediated signal transducers and activators of transcription [STAT]-1 signaling and TNF-mediated NF-κB signaling) (41), whereas the enhanced EC death may involve potentiation of TNF-activated pathways (e.g., by induction of pro-caspase 8 to exceed the capacity of inhibition by cFLIP) (42) or lysosomal release of cathepsin B into the cytosol, where it may be activated by TNF (29).

KEY POINTS

- TNF is the prototypic proinflammatory cytokine, and vascular ECs are a principal target of its actions. Anti-TNF therapies are utilized increasingly to control tissue damage, but the efficacy of this approach varies among different inflammatory settings. This variability may relate the activation of different TNF responses in the local ECs. Defining the signaling pathways activated within ECs by TNF may permit more precise therapeutic interventions to reduce inflammation and tissue damage.
- TNF is a pleiotropic inflammatory mediator, synthesized predominantly by mononuclear phagocytes, T cells, and mast cells, that acts on cellular targets by binding to one of two different receptors, commonly designated as TNFR1 and TNFR2.
- Vascular ECs are a principal target of TNF actions and may differentially express TNFR1 and TNFR2.
- TNFR1 signals result in new transcription of proinflammatory and antiapoptotic genes via NF-κB and SAP kinase signaling pathways, and also may activate programmed death through caspase- and cathepsin-dependent signaling pathways.
- The proinflammatory actions of TNF on ECs include expression of leukocyte adhesion molecules and chemokines, as well as expression of procoagulant proteins and induction of vascular leakiness.
- Other cytokines, especially IFN-γ, may potentiate TNF actions.
- Locally, such responses contribute to host defense, but systemically may underlie syndromes such as sepsis and multiorgan failure.
- TNFR2 signals may be proangiogenic and may involve cross-talk with VEGFR2.

Future Goals

- To selectively blockade or stimulate TNFR1 versus TNFR2 responses in specific disease settings to achieve desired therapeutic outcomes
- To protect ECs from TNF-initiated programmed cell death

REFERENCES

1 Carswell EA, Old LJ, Kassel RL, et al. An endotoxin-induced serum factor that causes necrosis of tumors. *Proc Natl Acad Sci USA.* 1975;72:3666–3670.

2 Beutler B, Cerami A. The biology of cachectin/TNF – a primary mediator of the host response. *Ann Rev Immunol.* 1989;7:625–655.

3 Aggarwal BB, Kohr WJ, Hass PE, et al. Human tumor necrosis factor. Production, purification, and characterization. *J Biol Chem.* 1985;260:2345–2354.

4 Aggarwal BB, Aiyer RA, Pennica D, et al. Human tumour necrosis factors: structure and receptor interactions. *Ciba Found Symp.* 1987;131:39–51.

5 Beutler B, Greenwald D, Hulmes JD, et al. Identity of tumour necrosis factor and the macrophage-secreted factor cachectin. *Nature.* 1985;316:552–554.

6 Bodmer JL, Schneider P, Tschopp J. The molecular architecture of the TNF superfamily. *Trends Biochem Sci.* 2002;27:19–26.

7 Korner H, Cook M, Riminton DS, et al. Distinct roles for lymphotoxin-alpha and tumor necrosis factor in organogenesis and spatial organization of lymphoid tissue. *Eur J Immunol.* 1997;27:2600–2609.

8 Tracey KJ, Cerami A. Tumor necrosis factor: a pleiotropic cytokine and therapeutic target. *Ann Rev Med.* 1994;45:491–503.

9 Neuhaus T, Totzke G, Gruenewald E, et al. Tumour necrosis factor-alpha gene expression and production in human umbilical arterial endothelial cells. *Clin Sci (Lond).* 2000;98:461–470.

10 Debets JM, Kampmeijer R, van der Linden MP, et al. Plasma tumor necrosis factor and mortality in critically ill septic patients. *Crit Care Med.* 1989;17:489–494.

11 Slowik MR, De Luca LG, Fiers W, Pober JS. Tumor necrosis factor activates human endothelial cells through the p55 tumor necrosis factor receptor but the p75 receptor contributes to activation at low tumor necrosis factor concentration. *Am J Pathol.* 1993;143:1724–1730.

12 Bradley JR, Thiru S, Pober JS. Disparate localization of 55-kd and 75-kd tumor necrosis factor receptors in human endothelial cells. *Am J Pathol.* 1994;146:27–32.

13 D'Alessio D, Al-Lamki RS, Bradley JR, Pober JS. Caveolae participate in tumor necrosis factor receptor 1 signaling and internalization in a human endothelial cell line. *Am J Pathol.* 2005;166:1273–1282.

14 Wang J, Al-Lamki RS, Zhang H, et al. Histamine antagonizes tumor necrosis factor (TNF) signaling by stimulating TNF receptor shedding from the cell surface and Golgi storage pool. *J Biol Chem.* 2003;278:21751–21760.

15 Al-Lamki RS, Wang J, Vandenabeele P, et al. TNFR1- and TNFR2-mediated signaling pathways in human kidney are cell type-specific and differentially contribute to renal injury. *FASEB J.* 2005;19:1637–1645.

16 Wallach D, Engelmann H, Nophar Y, et al. Soluble and cell surface receptors for tumor necrosis factor. *Agents Actions Suppl.* 1991;35:51–57.

17 Wu H. Assembly of post-receptor signaling complexes for the tumor necrosis factor receptor superfamily. *Adv Protein Chem.* 2004;68:225–279.

18 Darnay BG, Aggarwal BB. Early events in TNF signaling: A story of associations and dissociations. *J Leukoc Biol.* 1997;61:559–566.

19 Schmidt C, Peng B, Li Z, et al. Mechanisms of proinflammatory cytokine-induced biphasic NF-kappaB activation. *Mol Cell.* 2003;12:1287–1300.

20 Hoeflich KP, Yeh WC, Tao Z, et al. Mediation of TNF receptor-associated factor effector functions by apoptosis signal-regulating kinase-1 (ASK1). *Oncogene.* 1999;18:5814–5820.

21 Zhang H, Zhang R, Luo Y, et al. AIP1/DAB2IP, a novel member of the Ras-GAP family, transduces TRAF2-induced ASK1-JNK activation. *J Biol Chem.* 2004;279:44955–44965.

22 Micheau O, Tschopp J. Induction of TNF receptor 1-mediated apoptosis via two sequential signaling complexes. *Cell.* 2003;114:181–190.

23 Yin XM. Signal transduction mediated by Bid, a pro-death Bcl-2 family proteins, connects the death receptor and mitochondria apoptosis pathways. *Cell Res.* 2000;10:161–167.

24 Deng Y, Ren X, Yang L, et al. A JNK-dependent pathway is required for TNF alpha-induced apoptosis. *Cell.* 2003;115:61–70.

25 Hildt E, Oess S. Identification of Grb2 as a novel binding partner of tumor necrosis factor (TNF) receptor I. *J Exp Med.* 1999;189:1707–1714.

26 Madge LA, Pober JS. A phosphatidylinositol 3-kinase/Akt pathway, activated by tumor necrosis factor or interleukin-1, inhibits apoptosis but does not activate NFkappaB in human endothelial cells. *J Biol Chem.* 2000;275:15458–15465.

27 Adam-Klages S, Adam D, Wiegmann K, et al. FAN, a novel WD-repeat protein, couples the p55 TNF-receptor to neutral sphingomyelinase. *Cell.* 1996;86:937–947.

28 Madge LA, Li JH, Choi J, Pober JS. Inhibition of phosphatidylinositol 3-kinase sensitizes vascular endothelial cells to cytokine-initiated cathepsin-dependent apoptosis. *J Biol Chem.* 2003;278:21295–21306.

29 Li JH, Pober JS. The cathepsin B death pathway contributes to TNF plus IFN-gamma-mediated human endothelial injury. *J Immunol.* 2005;175:1858–1866.

30 Feng X, Gaeta ML, Madge LA, et al. Caveolin-1 associates with TRAF2 to form a complex that is recruited to tumor necrosis factor receptors. *J Biol Chem.* 2001;276:8341–8349.

31 Zhang R, Xu Y, Ekman N, et al. Etk/Bmx transactivates vascular endothelial growth factor 2 and recruits phosphatidylinositol 3-kinase to mediate the tumor necrosis factor-induced angiogenic pathway. *J Biol Chem.* 2003;278:51267–51276.

32 Pan S, An P, Zhang R, et al. Etk/Bmx as a tumor necrosis factor receptor type 2-specific kinase: role in endothelial cell migration and angiogenesis. *Mol Cell Biol.* 2002;22:7512–7523.

33 Lentz SR, Tsiang M, Sadler JE. Regulation of thrombomodulin by tumor necrosis factor-alpha: comparison of transcriptional and posttranscriptional mechanisms. *Blood.* 1991;77:542–550.

34 Yoshizumi M, Perrella MA, Burnett JC Jr., Lee ME. Tumor necrosis factor downregulates an endothelial nitric oxide synthase mRNA by shortening its half-life. *Circ Res.* 1993;73:205–209.

35 Messadi DV, Pober JS, Murphy GF. Effects of recombinant gamma-interferon on HLA-DR and DQ expression by skin cells in short-term organ culture. *Lab Invest.* 1988;58:61–67.

36 SenBanerjee S, Lin Z, Atkins GB, et al. KLF2 Is a novel transcriptional regulator of endothelial proinflammatory activation. *J Exp Med.* 2004;199:1305–1315.

37 Stolpen AH, Guinan EC, Fiers W, Pober JS. Recombinant tumor necrosis factor and immune interferon act singly and in combination to reorganize human vascular endothelial cell monolayers. *Am J Pathol.* 1986;123:16–24.

38 Clark P, Pober JS, Kluger MS. *J Invest Dermatol.* 2007; in press.

39 Lejeune FJ. Clinical use of TNF revisited: improving penetration of anti-cancer agents by increasing vascular permeability. *J Clin Invest*. 2002;110:433–435.
40 Pober JS. Cytokine-mediated activation of vascular endothelium: physiology and pathology. *Am J Pathol*. 1988;133:426–433.
41 Johnson DR, Pober JS. HLA class I heavy chain gene promoter elements mediating synergy between tumor necrosis factor and interferons. *Mol Cell Biol*. 1994;14:1322–1332.
42 Li JH, Kluger MS, Madge LA, et al. Interferon-gamma augments CD95(APO-1/Fas) and pro-caspase-8 expression and sensitizes human vascular endothelial cells to CD95-mediated apoptosis. *Am J Pathol*. 2002;161:1485–1495.

32

Vascular Permeability Factor/Vascular Endothelial Growth Factor and Its Receptors

Evolving Paradigms in Vascular Biology and Cell Signaling

Debabrata Mukhopadhyay, Resham Bhattacharya, and Deborah A. Hughes

Mayo Clinic Cancer Center, Rochester, Minnesota

VASCULAR ENDOTHELIAL GROWTH FACTORS

Background and Discovery

The pioneering observations of Dvorak and his group first showed that tumor ascites fluids from guinea pigs, hamsters, and mice contained an activity that rapidly increased microvascular permeability. The permeability-increasing activity purified from either the culture medium or ascites fluid of one tumor – the guinea pig line 10 hepatocarcinoma – was a 34,000- to 42,000-dalton protein that was distinct from other known permeability factors (1). The vascular permeability factor (VPF) was also secreted by five human tumor cell lines and could be eluted from immobilized heparin. Two tumorigenic (in nude mice) human cell lines were found to secrete at least 14-fold more VPF than their directly matched, nontumorigenic counterparts, suggesting that this factor was important for tumor growth (2).

Later Ferrara and colleagues, and others, isolated a heparin-binding growth factor from the conditioned media of bovine pituitary follicular cells (FCs) and described it as vascular endothelial growth factor (VEGF). This factor was specific for vascular endothelial cells (ECs) and was able to induce angiogenesis in vivo. Complementary DNA (cDNA) clones for bovine and human *VEGF* were isolated from cDNA libraries prepared from FC and HL60 leukemia cells, and it was subsequently recognized that VPF and VEGF were encoded by a single gene (3–7).

Gene and Splice Variants

The human *VEGF* gene, located on the short arm of chromosome 6, is comprised of eight exons (8) and gives rise to five isoforms containing 121, 145, 165, 189, and 206 amino acids, respectively, that are produced as a result of alternate splicing (9–11). The full transcript encodes a 189-amino acid isoform. A longer molecular species, the 206 isoform, which contains an additional 17 codons following the 24 codons encoded by exon 6, appears to be expressed only in embryonal tissue (12). The transcript encoding the 165-amino acid form, in which exon 6 is omitted, is expected to generate the 45-kDa peptide following signal peptide cleavage. A fourth transcript, lacking exon 7, encodes a 145-amino acid isoform (11). A shorter transcript, lacking both exons 6 and 7, encodes a 121-amino acid isoform that does not bind heparin, whereas the four larger forms, VEGF-A_{145}, VEGF-A_{165}, VEGF-A_{189}, and VEGF-A_{206}, all bind heparin with increasing affinity. VEGF-A_{189} and VEGF-A_{206} remain cell-associated despite having a signal peptide, whereas the smaller isoforms are secreted (13,14).

Vascular Endothelial Growth Factor-Related Genes

The VEGF family currently comprises seven members: VEGF-A (five isoforms as discussed above), VEGF-B (15,16), VEGF-C (17), VEGF-D (18) or VEGF-related protein (VRP), VEGF-E (also called *orf* virus) (19), svVEGF (identified in snake venom) (20), and placental growth factor (PlGF) (21,22). These VEGF-related proteins form glycosylated dimers, which are secreted after cleavage of their signal peptide. They might exist as homo- or heterodimers. They contain the characteristic eight cysteine residues that are highly conserved within the VEGF family. An eighth family member, endocrine-gland–derived vascular endothelial growth factor (EG-VEGF) was recently identified as an angiogenic mitogen selective for ECs of endocrine glands (23).

Expression

VEGF is expressed by a variety of normal adult organs including lung, kidney, adrenal gland, heart, liver, and stomach mucosa (24), and by many cell types including megakaryocytes (25,26) peritoneal macrophages, glomerular podocytes (27), glial cells (28), cardiac myocytes, and human placental as well

Table 32-1: VEGF and VEGF Receptor Knockout and Transgenic Phenotypes

Knockout	*Organism*	*Phenotype*	*References*
VEGF-A ($^{+/-}$), ($^{-/-}$)	Mouse	Embryonic difficulty due to severe cardiovascular defect.	62 63
Flt-1 ($^{-/-}$)	Mouse	Lethal at E8.5 Abnormal vasculature Increased hemangioblast committment Defect in vascular sprouting Rescued by soluble FLT-1	100 101 104
Flt-1 tyrosine kinase ($^{-/-}$)	Mouse	Normal embryonic development	103
Flk-1 ($^{-/-}$)	Mouse Angioblasts/quail	Lethal at E8.5–9.5 Defect vasculogenesis and hematopoiesis Lack mature ECs and HCs No blood islands Failure to migrate	130 133

HCs, hematopoietic cells

as glomerular ECs (29,30). VEGF-A is overexpressed by the vast majority of solid human cancers (reviewed in 3, 31–33) and, increasingly, has been found to be of importance in lymphomas and a variety of hematological malignancies (34–40).

VEGF-A is expressed not only by invasive cancer cells but also by at least some premalignant lesions (e.g., precursor lesions of breast, cervix, and colon cancers); furthermore, expression levels increase in parallel with malignant progression (41–43).

Regulation

VEGF-A expression is regulated by a number of mechanisms. Hypoxia upregulates VEGF-A expression by stabilizing both hypoxia-inducible factor (HIF)-1α (thus inducing transcription of the *VEGF-A* gene) and *VEGF* mRNA (44–48). Oncogenes like *src* and *ras*, and tumor suppressor genes such as *p53*, *p73*, and von Hippel Lindau (*VHL*), all play important roles in regulating VEGF-A expression in different tumors (49–51). VHL comprises a protein complex that targets HIF-1α for ubiquitination and proteolysis. Therefore, when VHL is absent or mutated, HIF-1α is stabilized, causing upregulation of VEGF-A (52,53). p53 and p73 suppress *VEGF-A* transcription, and thus their deficiency in tumor cells results in increased VEGF-A levels (54–56). Numerous growth factors, cytokines, and lipid mediators upregulate VEGF-A expression in different cells, including epidermal growth factor (EGF), transforming growth factor (TGF)-α, fibroblast growth factor 2 (FGF2), TGF-β, keratinocyte growth factor, tumor necrosis factor, interleukin (IL)-1 and IL-6, insulin-like growth factor 1, hepatocyte growth factor (HGF), and prostaglandins E1 and E2 (57–59). VEGF-A also is expressed by steroidogenic cells in response to hormonal stimulation in the adrenal cortex, corpus luteum, and Leydig cells (e.g., the cycling uterus and ovary) (24,60,61).

Physiological Role

VEGF is such a potent regulator of vascular development that its dosage must be tightly regulated. Disruption of even a single allele of the *VEGF* gene in mice results in embryonic lethality due to severe cardiovascular defects, thus demonstrating the importance of VEGF-A for appropriate vascular development (62,63) (Table 32-1). VEGF is a key regulator of physiological angiogenesis, which is required during wound healing (64,65), ovulation, menstruation (66), and pregnancy (67).

Pathological Role

High levels of VEGF may lead to pathological angiogenesis in a wide variety of diseases, including chronic inflammation, psoriasis (68), arthritis (69), macular degeneration, diabetic retinopathy (70,71), and tumors. Recent genetic studies have revealed that reduced VEGF levels cause neurodegeneration by impairing neural tissue differentiation and are associated with an increased risk of amyotrophic lateral sclerosis in European populations (72,73). Low VEGF levels have been implicated in Alzheimer disease and Huntington disease, in which neural perfusion deficits lead to chronic ischemia (74,75).

Vascular Endothelial Growth Factor and Therapy

VEGF and its functional inhibitors have been used for the treatment of a variety of diseases. Efforts to inhibit VEGF-induced tumor angiogenesis include the development of humanized neutralizing anti-VEGF monoclonal antibodies (76), inhibitory soluble VEGF receptors (66,77,78), antisense *VEGF* mRNA-expressing constructs (79,80), VEGF toxin conjugates (81), antagonistic VEGF mutants (82), and inhibitors of VEGF receptors (83,84). Avastin, a recombinant humanized antibody to VEGF, was approved by the U.S. Food and Drug Administration (FDA) in February 2004 for use in combination with intravenous 5-fluorouracil-based chemotherapy as a treatment for patients with first-line or previously untreated metastatic cancer of the colon or rectum (85).

The angiogenic properties of VEGF also have been exploited to induce angiogenesis for the treatment of diseases characterized by impaired blood supply. Intramuscular *VEGF* gene transfer prevented axonal loss and myelin degeneration in a rabbit model of ischemic neuropathy (86), and showed significant improvement in the vascular ankle–brachial index in treated legs in patients with critical lower limb ischemia (87). Successful attempts aimed at the induction of collateral blood vessels in ischemic heart disease and re-endothelization by VEGF recently have been reported (88–90). Collectively, these studies provide sufficient rationale for evaluating the potential therapeutic utility of VEGF.

The remaining part of this chapter focuses on the regulation and function of the VEGF receptors.

VASCULAR ENDOTHELIAL GROWTH FACTOR RECEPTORS

Background and Discovery

Several years after the discovery of VEGF as a permeability factor, EC-specific mitogen, and angiogenic cytokine, the search for a putative VEGF receptor was underway. Vaisman and colleagues, using both binding and cross-linking techniques followed by Scatchard analysis, defined two binding sites for ^{125}I-VEGF on bovine ECs: (a) a high-affinity binding site with a dissociation constant of 10^{-12} M and a density of 3×10^3 receptors/cell, and (b) a low-affinity binding site with a dissociation constant of 10^{-11} M and density of 4×10^4 receptors/cell (91). The receptors were shown to have high specificity, because the addition of other growth factors – such as FGF, platelet-derived growth factor (PDGF), EGF, and insulin – were not shown to inhibit VEGF binding.

Simultaneously, other investigators in pursuit of discovery of novel receptor tyrosine kinase (RTK) genes utilized sequences derived from proto-oncogenes or conserved sequences of tyrosine kinases to screen cDNA libraries. Shibuya and colleagues isolated two novel mammalian kinase genes by weak cross-hybridization with the v-*ras* oncogene (92). One, obtained from a human placenta cDNA library, contained a predicted amino acid sequence of 1,338 residues with seven extracellular immunoglobulin-like domains and a tyrosine kinase domain with a long peptide insertion, indicating that this new gene was closely related to the *fms* family. Consequently, the gene was designated as *FLT* (fms-like tyrosine kinase). It was expressed in several human cell lines derived from placenta, liver, muscle, kidney, and in embryo kidney–derived 293 cells. However, it was undetectable in 20 human tumor cell lines examined. De Vries and colleagues, using oligonucleotide sequences common to tyrosine kinases, isolated a clone identical to *FLT* (93). They demonstrated that, when expressed in COS cells, FLT specifically bound with high affinity to VEGF. Furthermore, *FLT* cDNA, when expressed in *Xenopus laevis*, demonstrated a functional response, with release of calcium upon binding to VEGF. FLT is also designated as FLT-1 and VEGF receptor 1 (VEGFR1).

Similar screening approaches resulted in the discovery of VEGFR2. Mathews and colleagues cloned a RTK cDNA from mouse cell populations enriched for totipotent hematopoietic stem cell and progenitor cells (94). The deduced sequence contained 1,367 amino acids and, similar to FLT, the extracellular region contained seven immunoglobulin-like domains, with a cytoplasmic region containing a long peptide insertion in its tyrosine kinase domain, suggesting that it belonged to the RTK subclass type III. This gene was designated fetal liver kinase 1 (*Flk-1*). Terman and colleagues used polymerase chain reaction (PCR) to amplify DNA segments corresponding to the kinase insert domain of type III RTKs in a human EC cDNA library (95). A clone was isolated with a predicted protein sequence structurally similar to FLK-1, and was designated kinase insert domain-containing receptor (KDR).

Since these initial findings, the known biological functions of VEGF, VEGF-related proteins, and their respective receptors have expanded dramatically to involve a wide spectrum of physiological and pathological entities.

VEGFR1/FLT-1: Role in Development

Although VEGFR1 was discovered more than a decade ago, the precise function of this receptor is still controversial. In early murine development, VEGFR1 is expressed by embryonic day (E)7.5 in the extraembryonic yolk sac mesoderm (96). A bimodal level of expression was observed during development, with a moderate level of expression during organogenesis (E9.5–E12.5), a very low level of expression during fetal growth (E14.5–E16.5), and a high level of expression in newborn and adult mice. In all tissues examined, which included brain, heart, kidney, and lung, expression was restricted to ECs (97). Studies examining human fetal tissue showed high levels of expression of both VEGFR1 and VEGFR2 throughout the developing endothelia, with the highest levels of expression in the heart, lungs and kidney (98,99); however, careful studies examining the exact time course have not been carried out in the same detail as in animal models.

Knockout studies reveal a crucial role of VEGFR1 in early angiogenesis. A homozygous null mutation of the *Flt-1* gene in mice results in embryonic lethality at mid-somite stage (E8.5) (100). ECs are present in embryonic and extraembryonic structures, but form abnormal vascular structures. Fong and colleagues subsequently showed that this defect resulted from increased hemangioblastic commitment among mesenchymal cells prior to the formation of blood islands in *Flt-1*$^{-/-}$ mice, resulting in increased population density and vascular disorganization (101). These data suggest that FLT-1 plays a critical role in determining cell fate. Others have shown that ECs lacking FLT-1 have a higher rate of cell division than do wild-type controls, suggesting that FLT-1 negatively modulates EC proliferation during embryogenesis (102). In contrast to the full-length receptor, targeted mutation of the tyrosine kinase domain of FLT-1 receptor (TK$^{-/-}$) does not result in embryonic lethality or abnormalities in embryonic development and angiogenesis (103). Furthermore, ECs isolated from FLT-1 (TK$^{-/-}$) animals display a normal mitogenic response to VEGF. These data argue that FLT-1 has an important role in development that is not dependent on the tyrosine kinase domain. Dynamic image analysis of mouse embryonic stem cells suggests that FLT-1 is a positive modulator of vascular sprout formation and branching (104). In the mouse embryo, the absence of FLT-1 resulted in reduced branching of the vascular plexus. This defect was rescued by transgenic expression of soluble FLT-1, suggesting the soluble, rather than membrane-bound form, influences vascular patterning at this stage (104).

VEGFR1/FLT-1: Expression in the Adult and Ligands

In the adult, VEGFR1/FLT-1 is expressed primarily on ECs, but also on monocytes/macrophages (105,106), smooth muscle cells (SMCs) (107–110), mesangial cells (111), osteoblasts (112), and hematopoietic stem cells (113). Known ligands for VEGFR1 are VEGF (93), PlGF (114), and VEGF-B (15). VEGFR1 has the highest affinity for recombinant human VEGF, with a dissociation constant (Kd) of 10 to 20 pM (93) (Figure 32.1). The binding site for VEGF-A and PlGF has been mapped to the second and third Ig-like domains, with loop two being sufficient for VEGF binding (66,115–117). VEGF binding results in weak tyrosine autophosphorylation of VEGFR1 (93,118). VEGFR1 has been considered traditionally as a decoy receptor of VEGF in that it competes for binding and prevents VEGF from interacting with other receptors (77). PlGF has been shown to potentiate the effects of VEGF, which is thought to occur through competitive inhibition with VEGFR1, thus increasing the amount of VEGF available to bind with VEGFR2. Interestingly, the affinity of PlGF for VEGFR1 is somewhat lower (Kd 170 pM) than it is for VEGF (119,120). Different amino acid residues within the same domains are involved in the ligand–receptor interface between VEGFR1 and, respectively, VEGF and PlGF (117,121). More recently, a synergism between VEGF and PlGF in pathological angiogenesis has been demonstrated in vivo (122,123). Binding of PlGF by VEGFR1 was shown to result in the transactivation of VEGFR2, and enhanced VEGF-driven angiogenesis. The latter effect was dependent on the tyrosine kinase activity of VEGFR1 (124). The upregulation of specific genes, in response to VEGF and PlGF versus VEGF alone, suggests that unique signaling pathways are involved. Further studies are needed to delineate these pathways and will provide further insight into the role of VEGFR1 in VEGF signaling and biology.

Soluble FLT-1

Soluble FLT-1 (sFLT-1), a splice variant of the VEGFR1 that contains the first six Ig-like domains but lacks the transmembrane and cytoplasmic domain, was first identified in conditioned media of human umbilical vein ECs (HUVECs) (77,125). sFLT-1 acts as an antiangiogenic factor by binding to and acting as a sink for VEGF and PlGF, and also in a dominant negative fashion, by forming inactive heterodimers with membrane-spanning VEGF receptors (126). The role of sFLT-1 in physiology and pathophysiology is discussed below.

VEGFR2/KDR/FLK-1: Role in Development

During development, VEGFR2 (FLK-1, KDR) is one of the earliest markers for endothelial progenitors (127,128). In mice, FLK-1 is expressed by E7 in the yolk sac mesoderm and proximal lateral embryonic mesoderm, a region fated to become heart. At later stages, FLK-1 expression is restricted to endothelial precursors in the yolk sac blood islands, endocardial tubes, and allantois within the embryo. By E8.5, FLK-1 expression is abundant in the proliferating ECs of vascular sprouts and branching vessels and is later reduced in regions where proliferation has ceased. Within germ layers and organs throughout development, the expression of FLK-1 mirrors that of VEGF.

FLK-1 plays a critical role in embryonic vasculogenesis, angiogenesis, and vascular patterning (129–131). FLK-1–deficient mice show in utero demise between E8.5 and E9.5, and exhibit severe defects in vasculogenesis and hematopoiesis, with lack of mature endothelial and hematopoietic cells and failure to form blood islands (130). In zebrafish, delivery of a VEGFR2 kinase inhibitor at the one-cell stage abolishes axial and intersegmental vessels (ISV), whereas administration at 24 hours inhibits only ISV formation (131,132). These results are consistent with the known temporal sequence of development in the zebrafish, in which axial vessels form first via vasculogenesis, and later, ISV through angiogenesis. FLK-1 also plays a crucial role in vascular patterning. Ambler and colleagues showed that *Flk-1*$^{-/-}$ angioblasts, when transplanted into the presomatic mesoderm cavity of quail, fail to migrate to specific embryonic locations, whereas wild-type angioblasts migrate and respond normally to these

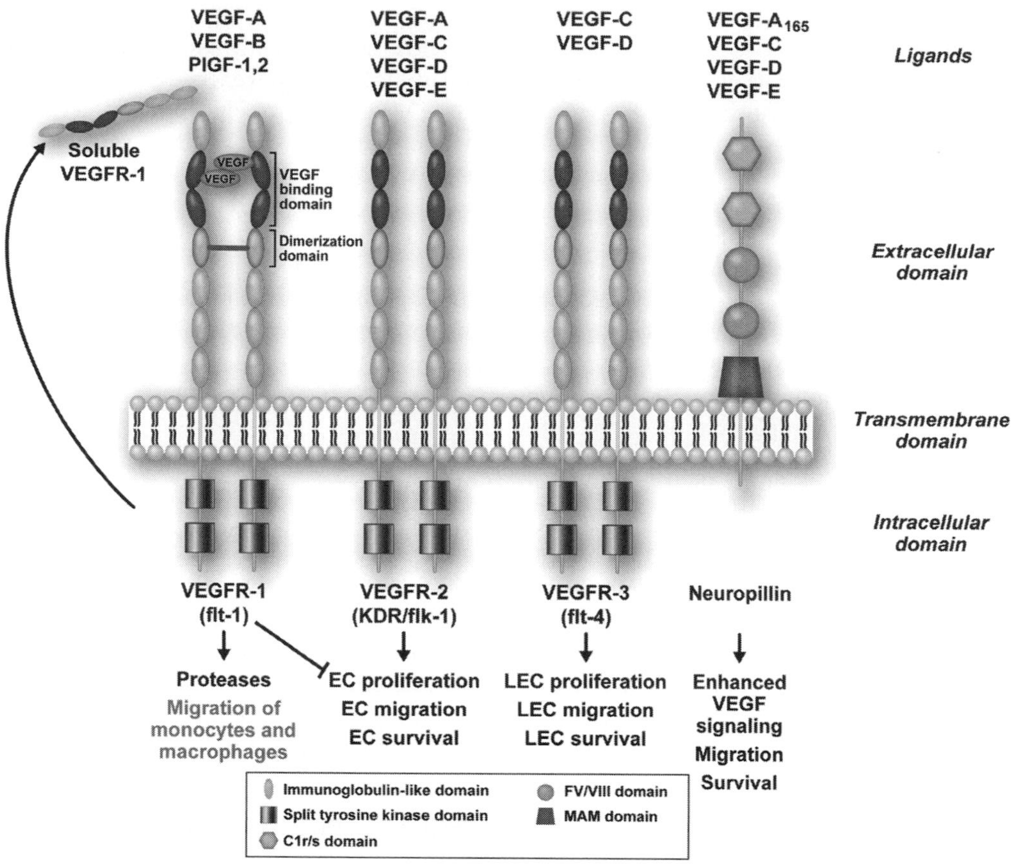

Figure 32.1. The VEGF receptor family. The three signaling tyrosine kinase receptors VEGFR1/FLT-1, VEGFR2/KDR/FLK-1, and VEGFR3/FLT-4 consist of seven immunoglobulin-like structures in the extracellular domain, a single transmembrane region, and a consensus tyrosine kinase domain interrupted by a kinase insert domain. Binding of ligand to VEGFR1 promotes the production of proteases in ECs, and the migration of monocytes and macrophages. Binding of ligand and activation of VEGFR2 promotes EC proliferation, migration, and survival. Similarly, ligand binding to VEGFR3 promotes proliferation, migration, and survival in lymphatic ECs (LECs). The receptors neuropilin-1 and neuropilin-2 possess no kinase activity but enhance VEGFR2 signaling. For explanation of the individual neuropilin domains, see Ref 226.

patterning cues (133). These findings demonstrate that FLK-1 has several distinct and overlapping functions throughout development.

VEGFR2/KDR/FLK-1: Expression in Adult and Ligands

VEGFR2 is the main mediator of the mitogenic, angiogenic, and permeability-enhancing effects of VEGF-A in the adult. It is expressed primarily on vascular ECs (134,135) and endothelial progenitor cells (136). Low levels have been detected in SMCs (110). The ligands for VEGFR2 are VEGF-A (and its isoforms 121, 145, 165) (13), VEGF-C (17), VEGF-D (137), and VEGF-E (138) (see Figure 32.1). VEGFR2 is a 200- to 230-kDa, high-affinity receptor for VEGF-A with a Kd of 75 to 760 pM (95,120,139) and maps to human chromosome 13q12–13. Upon binding by VEGF-A, VEGFR2 undergoes dimerization and ligand-dependent tyrosine phosphorylation. Interestingly, a mutant form of VEGFR2, lacking the Ig-like domains four through seven, was shown to bind to VEGF-A_{165} with equal or better affinity as compared to the native receptor, suggesting that these domains may have an important biological function in blocking unwanted signaling events (140). The precise signaling pathways by which VEGFR2 mediates its actions with regards to EC function are discussed later.

Soluble VEGFR2

A truncated form of VEGFR2 that lacks the C-terminal half of the kinase domain as a result of alternative splicing was reported in normal rat retina (141). More recently, Ebos and colleagues described a soluble VEGFR2 (s-VEGFR2) with an apparent molecular size of 160 kDa, detected in mouse and human plasma (142). This truncated form of sVEGFR2 could be detected in the conditioned media from various EC lines from both human and mouse. More studies are needed to define the precise biological functions of this newly discovered receptor variant.

Role of VEGFR1/FLT-1 and VEGFR2/KDR/FLK-1 in Physiology

Wound Healing

Cutaneous injury initiates a series of events to control hemorrhage. Platelet activation occurs at the site of injury; these platelets secrete several cytokines and growth factors, including VEGF, and subsequently attract circulating neutrophils and monocytes (143). Many cell types contribute to the source of VEGF in wounds, including keratinocytes, macrophages, fibroblasts, ECs, and T lymphocytes (144). Microvascular hyperpermeability induced by VEGF is followed by EC proliferation and migration, which in turn leads to increased angiogenesis at the wound edge (145). At 3 days after wounding, prior to neovascularization, FLT-1 was not detected within the wound, but in ectatic vessels bordering the wound (97). At 7 days, high expression of FLT-1 was observed in large vessels bordering the wound, as well as in small vessels within the wound, representing neovascularization. These data suggest that FLT-1 plays a role in directing new vessel growth and recovery of the endothelium to a quiescent phenotype after neovascularization has occurred.

Female Reproductive Cycle

Follicular growth, the development and endocrine function of the ovarian corpus luteum, and cyclic regeneration of the endometrium during the female reproductive years requires a highly regulated angiogenic response (146–148). Menstrual bleeding itself is brought about by tissue breakdown and damage of superficial endometrial vessels, followed by repair of the vessels within 5 days after its onset. *VEGF* mRNA has been shown to be expressed by endometrial stromal cells, significantly increased at the onset of menstruation and mediated by hypoxia, TGF-α, and IL-1β. *VEGF, VEGFR1,* and *VEGFR2* mRNA levels were highest, and sFLT levels lowest, during peak EC proliferation in the endometrium (149). VEGFR2 expression correlated with activation of the receptor, as demonstrated by phosphorylation assays.

Other Roles of FLT-1

As mentioned in previous sections, FLT-1 is expressed on several different cell types and organ systems. Other functions mediated by this receptor in response to VEGF demonstrate the complexity of the actions of VEGF at the level of organ system and organism. One well established function of FLT-1 is that it mediates the migration of mononuclear phagocytes across the EC monolayer in response to binding of VEGF-A and PlGF (106). VEGF-E, which binds exclusively to VEGFR2 and induces autophosphorylation, acts as a potent mitogen- and permeability-enhancing factor, but does not stimulate the migration of monocytes, demonstrating that FLT-1 is necessary for the expression of the entire spectrum of VEGF-related events in the adult vasculature (150,151).

Interestingly, LeCouter and colleagues recently showed that selective activation of FLT-1 induced liver sinusoidal ECs to produce HGF and IL-6 (152). This was shown to increase hepatocyte growth but not endothelial proliferation in vitro and in vivo. Additionally, it reduced hepatotoxin-induced liver damage in mice. This study describes a novel role that extends beyond traditional angiogenic mechanisms for VEGFR1 in adult endothelium. It also provides potential new therapies utilizing FLT-1 agonists for the possible preservation of organ function in liver disease.

VEGFR1 and VEGFR2 in Pathophysiology

Hypoxia-Induced Expression of VEGF Receptors

Hypoxia is an important stimulus for the expression of VEGF and its receptors in physiologic and pathological processes. Given the heterogeneity of EC and vascular bed phenotypes, it is not surprising that different studies have yielded varying results regarding hypoxic responses. Despite this, some general findings are consistent across studies. With hypoxia, *VEGFR1* is transcriptionally upregulated through the interaction of HIF-1α with a hypoxia-inducible enhancer element located in upstream promoter of *VEGFR1* (118,153). Both transcriptional upregulation (118,153–155) and increased protein expression (153) have been demonstrated in vitro and in vivo, in a variety of EC types. Interestingly, during development hypoxia is a key stimulatory factor for the formation of vascular tubes. During this process, coordinate upregulation of VEGF-A, VEGF-B, and VEGFR1 (156) is essential for the establishment of normal coronary circulation.

Nilsson and colleagues (157) illustrated the heterogeneity among EC types by characterizing responses to hypoxia in ECs derived from vessels within bovine adrenal cortex, human foreskin, human umbilical vein, human saphenous vein, and human pulmonary artery. Hypoxia induced HIF-1α in all EC types; however, the related HIF-2α protein was induced in dermal microvascular ECs, whereas weak or no induction occurred in other EC types. *VEGF-A* was transcriptionally upregulated in all cell types, *VEGFR1* in all cell types with the exception of HUVECs, and *VEGFR2* in dermal microvascular cells alone. Consistent with others studies that have implicated HIF-2α in the transcriptional regulation of *VEGF-A* and *VEGFR2* genes (158–160), increased gene expression of *VEGF* and *VEGFR2* correlated with clear induction of HIF-2α in microvascular ECs. With regards to protein expression, hypoxia downregulated VEGFR1 expression at the cell surface in dermal microvascular ECs. Incubation with neutralizing antibodies against VEGF-A during hypoxia prevented downregulation of VEGFR1 receptor expression, suggesting a negative autocrine loop. Additionally, venous ECs were found to transcriptionally upregulate the lymphangiogenic genes *VEGFR3* and both *VEGF-C* and *VEGF-D* under hypoxic conditions (157). Others have reported that the protein expression of VEGFR2 is upregulated in ECs in vitro (107) and in vivo (161) in response to hypoxia. This is thought to occur through posttranscriptional modification resulting in enhanced mRNA stabilization (162).

Preeclampsia

During normal pregnancy, placental cytotrophoblasts, which help establish the fetomaternal circulation through invasion of the uterine wall, produce and release sFLT-1 into the maternal circulation (163,164). sFLT-1 has been shown to be elevated in pregnant women with preeclampsia (165), a common disorder of pregnancy characterized by hypertension and proteinuria in the third trimester and which is associated with significant morbidity and mortality (see Chapter 161). Elevated plasma concentration of sFLT-1 are detectable about 5 weeks before the manifestation of this disorder (166). Evidence suggests that inadequate invasion of the cytotrophoblasts during placental development results in underperfusion of the placenta (167). Tissue hypoxia stimulates the expression and release of sFLT-1, which binds to VEGF and PlGF, resulting in a functional deficiency of these growth factors. The net effect is attenuation of vasomotor relaxation and angiogenesis, thus promoting a cycle of endothelial dysfunction and increased placental hypoxia. It appears that increased sFLT-1 is due to increased production by cytotrophoblasts, rather than ECs or villous fibroblasts. sFLT-1 may be an important diagnostic tool in this disease and others.

Tumor Angiogenesis

Tumor growth and metastasis is angiogenesis-dependent. The progression from a neoplastic process to carcinogenesis involves an "angiogenic switch" in which tumors, once they have reached a certain size (approximately 2 mm^3), begin to recruit their own blood supply (168). This occurs through the production of growth factors and cytokines, including VEGF. As mentioned previously, VEGF is overexpressed by the vast majority of solid tumors (31,33) and several hematological malignancies (34–36). Although tumor cells are the main source of VEGF-A, stromal cells and vascular endothelium have been reported to express VEGF-A, albeit at lower concentration, especially at sites of hypoxia (41,169–171). Tumors studied thus far that express a high level of VEGF are supplied by microvessels lined by ECs that overexpress both VEGFR1 and VEGFR2 (31,33,41,42,172–174). The mechanisms involved the upregulation of VEGFR1 and VEGFR2 are not well understood, but experimental evidence suggests that chronic exposure to VEGF is a major determinant of this change in expression. Transgenic mice that overexpress VEGF in the skin under the control of a keratin promoter, or mice injected with adenovirus engineered to overexpress VEGF, were both shown to have upregulation of VEGFR1 and VEGFR2 in the EC of microvessels adjacent to these regions (175). This appears to contradict in vitro findings of Nilsson and colleagues (157) but underscores the differences that may exist in culture systems versus in vivo.

Angiogenesis-related genes and gene products may serve as prognostic factors in cancer. For example, circulating levels VEGF-A and sFLT-1 (176,177) have been shown to be potential predictors of clinical outcomes in breast cancer. Cell lines from patients with multiple myeloma have been shown to express VEGFR1 and, in some cases, VEGFR2; in these patients, angiogenic markers such as microvessel density correlate with outcome (178). High tissue expression of VEGF-C and VEGFR2 in ovarian carcinoma portends poor prognosis (179), whereas VEGF-A and VEGFR1 levels are associated with lymphoma progression (180). Numerous receptor tyrosine kinase inhibitors have been developed as antiangiogenic agents and are being studied in preclinical and clinical trials (181). Other strategies have included VEGF-A neutralizing antibodies, toxins, endogenous inhibitors of angiogenesis, and DNA vaccines. Despite these tremendous advances, many questions regarding basic mechanisms governing tumor growth and metastasis remain. Further research focused on the delineation of the molecular determinants and pathways involved will help to identify unique and specific therapeutic targets.

Pulmonary Hypertension

Plexiform lesions, composed of ECs, develop in the lumen of medium and small precapillary pulmonary arteries and are considered to be the characteristic lesion of primary pulmonary hypertension and severe pulmonary hypertension secondary to left-to-right heart shunt, HIV infection, CREST (calcinosis, Raynaud phenomenon, esophageal dysmotility, sclerodactyly, and telangiectasia) syndrome, and liver cirrhosis (182). The mechanisms involved bear a striking resemblance to that of tumor angiogenesis (183) in which localized hypoxia stimulates upregulation of VEGF through HIF-1α. VEGF, VEGFR2, HIF-1α, and HIF-1β have been shown to be expressed at higher levels within plexiform lesions (184), as compared to unaffected regions within the same lung, or in controls. Upregulation of these factors stimulates an angiogenic response and vascular remodeling that proceeds in a highly uncoordinated fashion. The lesions consist of mass of disorganized capillaries that eventually obliterate the vessel lumen, restricting flow and promoting further progression of the disease. The precise molecular mechanisms have yet to be elucidated, but decreased expression of phosphoinositide 3-kinase (PI3K), src, and Akt within the plexiform lesions, as compared with controls and despite increased expression of VEGFR2, suggests downregulation of important activating signals within these cells (184). This likely results in profound endothelial dysfunction, because these pathways are known stimulate endothelial nitric oxide synthase (eNOS) and prostacyclin production, which are important mediators of vascular tone and function (185).

Ocular Microvascular Proliferative Disorders

VEGF has emerged as a major mediator of intraocular neovascularization in diabetic retinopathy (186,187). Diffuse microvascular disease with resultant capillary closure and tissue hypoxia stimulates the production of VEGF and neovascularization. This leads to increased vascular permeability, tissue edema, hemorrhage, and fibrosis. VEGF production was found not only in retinal ECs, but also in retinal capillary pericytes, retinal pigment cells, glial cells, and ganglion cells, all of which potentially participate in the pathogenesis of retinopathy. In animal models of ischemic retinopathy, a variety of therapeutic approaches have been explored, including chimeric proteins that bind VEGF (188), monoclonal antibodies (189), antisense oligonucleotides (190), and VEGF receptor tyrosine kinase inhibitors (191). All of these approaches demonstrated a significant reduction in retinal neovascularization, suggesting that this is a promising strategy for this disease.

Atherosclerosis and Restenosis

The specific biological effects of VEGF and its receptors with respect to atherosclerosis and restenosis have yet to be clarified. VEGF-A has been proposed to have several protective effects in the arterial wall including increasing the production of nitric oxide (NO) and prostacyclin (192), reducing low-density lipoprotein oxidation (193), and mediating anti-apoptotic effects in the endothelium (194). Conversely, several lines of evidence suggest that angiogenesis and VEGF-A play an important role in atherogenesis. An association between increased vasa vasorum and atherosclerotic plaques in diseased arteries, as compared with normal arteries, has been demonstrated (195). In human coronary arteries, VEGF-A expression was detected in macrophages, SMCs, and ECs within atheromatous plaques, whereas expression within normal coronaries was not detectable (196). Immunostaining for FLT-1 and FLK-1 demonstrated increased expression in aggregating macrophages within these lesions and ECs within microvessels in totally occlusive lesions, but not in normal coronary arteries (196). Furthermore, animals treated with VEGF exhibit accelerated plaque development (197).

Neointimal formation is a major cause of restenosis after coronary intervention (198,199). Neointima formation following coronary interventions arises from a response to arterial injury, which induces a number of changes in the vessel wall that promote vascular SMC proliferation and migration into the site of injury. Angioplasty induces mechanical stretching of the vessel. This disrupts internal elastic lamina and media and denudes the endothelium, subsequently exposing the blood vessel wall to circulating mitogens that promote the release of cytokines from ECs, vascular SMCs, platelets, and inflammatory cells. The vasculoprotective versus the atherogenic and neointima-promoting effects of VEGF-A in arteries is still a matter of debate (200). Increased protein expression VEGF-A and FLT-1 in a porcine model of coronary restenosis has been demonstrated within SMCs of the neointima, and in ECs and macrophages in microvessels around stent struts following stent implantation (201).

Prior studies have suggested that acceleration of re-endothelization can prevent restenosis and inhibit in stent thrombosis (202,203). Local delivery of VEGF after arterial injury has been shown to promote endothelial regeneration, accelerate endothelium-dependent relaxation, and reduce neointimal formation (204). In injury models, VEGF, through its receptor VEGFR1, induces the migration and activation of monocytes (106), adhesion molecules (205), and monocyte chemoattractant protein (MCP)-1. In vitro experiments have shown that VEGF-A increased the production of matrix metalloproteinases 1 and 9 in human SMCs (109), suggesting a possible role for VEGF-A in enhancing SMC migration. Other investigators have shown that VEGF-A promotes migration but not proliferation of human SMCs, which were shown to express FLT-1 and VEGFR2 (206). Furthermore, VEGF-E, which is known to bind specifically to VEGFR2, induced migration as well (108) Gene transfer of soluble FLT-1 has been shown to prevent neointimal formation after periadventitial injury in mice (207) and to attenuate neointimal formation in animal models after intraluminal injury, with concurrent downregulation of VEGF and inflammatory markers (208). Conversely, phVEGF plasmid-coated polymer stents were shown to improve re-endothelization versus uncoated stents, with reduced neointimal formation in hypercholesterolemic rabbits. Local drug delivery of an antiproliferative agent with drug-eluting stents is a promising strategy for the prevention of restenosis (209). However, this strategy has the potential for impairing endothelial recovery with an associated risk of in-stent thrombosis. Further studies are needed to define the appropriate "VEGF balance" to promote re-endothelization and prevent restenosis.

Other VEGF Receptors

VEGFR3/FLT-4

VEGFR3, first cloned from human erythroleukemia cells (210), is a highly glycosylated receptor tyrosine kinase, approximately 180-kDa in size. The cDNA was cloned from placental libraries with localization to q35 on human chromosome 5 (211). Two isoforms of VEGFR3 are known: The short form lacks a 65–amino acid sequence found in the cytoplasmic tail of the long form (212). Two ligands are known to bind to VEGFR3: VEGF-C (17) and VEGF-D (137,213). The activation of VEGFR3 leads to protein kinase C (PKC)-dependent activation of extracellular signal-related kinase (ERK)-1 and -2, and activation of protein kinase B/Akt, which supports cell proliferation and survival (214).

VEGFR3 Expression in Development and in the Adult

Several studies have shown that VEGFR3 is widely expressed on ECs in several species, including human, early in development prior to and during vessel formation (99,135). Unlike VEGFR1 and VEGFR2, VEGFR3 is not expressed in blood islands. As development continues, VEGFR3 is more prominent in developing veins, in vascular networks adjacent to developing skin, and vasculature within the liver. After organogenesis is complete, the expression of VEGFR3 is confined primarily to the endothelium of lymphatic vessels (98). More recent work has shown that, during development, VEGFR3 also is expressed on fenestrated endothelium within cartilage channels, endocrine organs, cells lining sinusoids of bone marrow, endothelial lining the splenic sinusoids, and glomerular capillaries within the kidney (215). In the normal adult, VEGFR3 is expressed in placenta, lung, heart, and kidney (215) in splenic sinusoid, capillaries of the glomerulus, within the vasa vasorum of the aorta, and in endocrine organs (adenohypophysis, thyroid, parathyroid and adrenal glands) (215).

Role of VEGFR3 in Physiology and Pathology

VEGFR3 knockout in mice results in embryonic lethality at E9.5. Although vasculogenesis and angiogenesis occur, these embryos possess large, disorganized vessels with defective lumen formation, leading to fluid accumulation in the pericardial cavity and cardiovascular failure, thus demonstrating a critical role of VEGFR3 in cardiovascular development prior to the formation of lymphatics (216,217).

Missense point mutations in one *VEGFR3* allele leads to the inactivation of tyrosine kinase activity and results in chronic lymphedema (218). VEGFR3 has been shown to be expressed in a wide variety of tumors and hematological malignancies including Kaposi sarcoma (219), angiosarcoma (220), hemangiomas (221), lymphangiomas (98), and undifferentiated teratocarcinoma cells, with downregulation of expression upon differentiation (212), and in many human leukemic cells lines (222,223). High expression also is found in the neovasculature of benign and malignant vascular tumors (99), breast cancer (224), and on SMCs in atherosclerotic arteries (110).

Neuropilin

Neuropilin (NRP), a 130- to 140-kDa membrane glycoprotein, was first identified in the optic tectum of *Xenopus laevis* and subsequently in the developing brain (225). Two NRP genes are known: *NRP1* and *NRP2* (226–228). NRP1 localizes to neuronal axons (227), but is also abundant on ECs and developing embryonic blood vessels, and is important in normal developmental angiogenesis (62,63) and tumor angiogenesis (229). *NRP1*-knockout mice display abnormal axonal networks, deficiencies in neuronal vascularization, aortic arch malformations, and diminished and disorganized yolk sac vascularization (230). NRP1 is not a tyrosine kinase receptor, but acts as a coreceptor for VEGFR2, enhancing VEGF-A_{165} activity (231) cell survival and migration (232,233).

Signaling Pathways of VEGFR1 and VEGFR2 in the Vascular Endothelium Leading to Biological Responses

Most of the biological activities of VEGF are mediated primarily by interaction with the two major receptors, VEGFR1 and VEGFR2 (Figure 32.2). At present, the signaling cascades following VEGF interaction with cultured ECs are only partially understood but are known to involve a series of protein phosphorylations. Tyrosine residues known to be phosphorylated on VEGFR1 are Tyr^{1213}, Tyr^{1333}, Tyr^{1242}, and Tyr^{1327}. Three SH2 domain–containing proteins bind to the phosphorylation site Tyr^{1213} namely, SHP-2, phospholipase C (PLC)γ-1, and Grb-2. The site at Tyr^{1333} allows binding of PLCγ-1 and the adaptor molecules Crk and Nck. Tyr^{1242} and Tyr^{1327} are poorly phosphorylated and have not been shown to associate with signaling molecules.

Binding of VEGF or PlGF to VEGFR1 induces cell migration in monocytes (106,234). In ECs, VEGFR1 demonstrates an inhibitory effect on VEGFR2-induced proliferation but not migration (235,236). It is not yet completely clear why VEGFR1 does not induce cell proliferation in response to VEGF, but this could be due to low kinase activity of the receptor resulting from a small number of tightly regulated phosphorylation sites. Accordingly, a repressor sequence has been identified in the VEGFR1 juxtamembrane domain, and replacement of this by VEGFR2 allows VEGFR1 to respond to stimulation by VEGF-A. Furthermore, the repressor sequence may regulate interactions with the phosphatases (237,238). Studies with the receptor chimera of EGF receptor (EGFR) and VEGFR1 (EGLT) or VEGFR2 (EGDR) have suggested a role for PI3K in mediating this inhibitory function through Tyr^{794}. Inhibitors of PI3K as well as the dominant negative mutant of p85 (PI3K subunit) reversed the inhibition of proliferation, whereas a constitutive active mutant of p110 introduced the inhibition to HUVEC-EGDR (235).

The binding of VEGF to VEGFR2 results in autophosphorylation of Tyr^{951} and Tyr^{996} present in the kinase insert domain, Tyr^{1054} and Tyr^{1059} in the kinase domain, and Tyr^{1175} and Tyr^{1214} in the C-terminal tail. Phosphorylated Tyr^{951} binds VRAP (239), and Tyr^{1175} binds Sck (240) and PLCγ-1 (241). Binding of PLC γ-1 activates PKC, which in turn activates *ras*. This is followed by the induction of the ERK pathway (p42/44 mitogen-activated protein kinase [MAPK]). ERK then phosphorylates and activates transcription factors in the nucleus. This pathway is important for VEGFR2–mediated DNA synthesis and proliferation (120,242–246).

U73122, a specific inhibitor of the PLC family, inhibited VEGF-stimulated HUVEC proliferation and migration (235). The catalytic activities of PLC-β1, -β2, -β3, and also PLCγ were measured in cellular extracts prepared from a HUVEC suspension stimulated with VEGF for 20 or 40 seconds, and 1, 2, and 5 minute. The activity of PLC-β3 peaked at 40 seconds, and dropped to baseline within 1 minute after stimulation, whereas no significant change occurred in the catalytic activity of other PLCs, except PLCγ, which started increasing after 1 minute (247). These results indicated that G-protein–regulated PLC-β3, rather than receptor tyrosine kinase–regulated PLCγ-1, was mediating the VEGF-induced increase in $[Ca^{2+}]_i$. The role of G-proteins in VEGF signaling is novel, considering the receptors that lead to signaling are receptor tyrosine kinases. The ability to block VEGF-stimulated HUVEC proliferation and migration with a Gq-specific antagonist fusion peptide and β-ARK1 minigene suggests the existence of a novel signaling pathway. Inhibition of Gq activation blocked VEGFR2 phosphorylation (248,249). This suggests a unique pathway by which Gq modulates RTK activation. The free G-$\beta\gamma$ subunit activates PLC-β3, followed by an increase in $[Ca^{2+}]_i$ that results in proliferation and migration of ECs and thus angiogenesis. These results thus suggest that Gq, G-$\beta\gamma$, and LC-β3 are required for VEGFR2

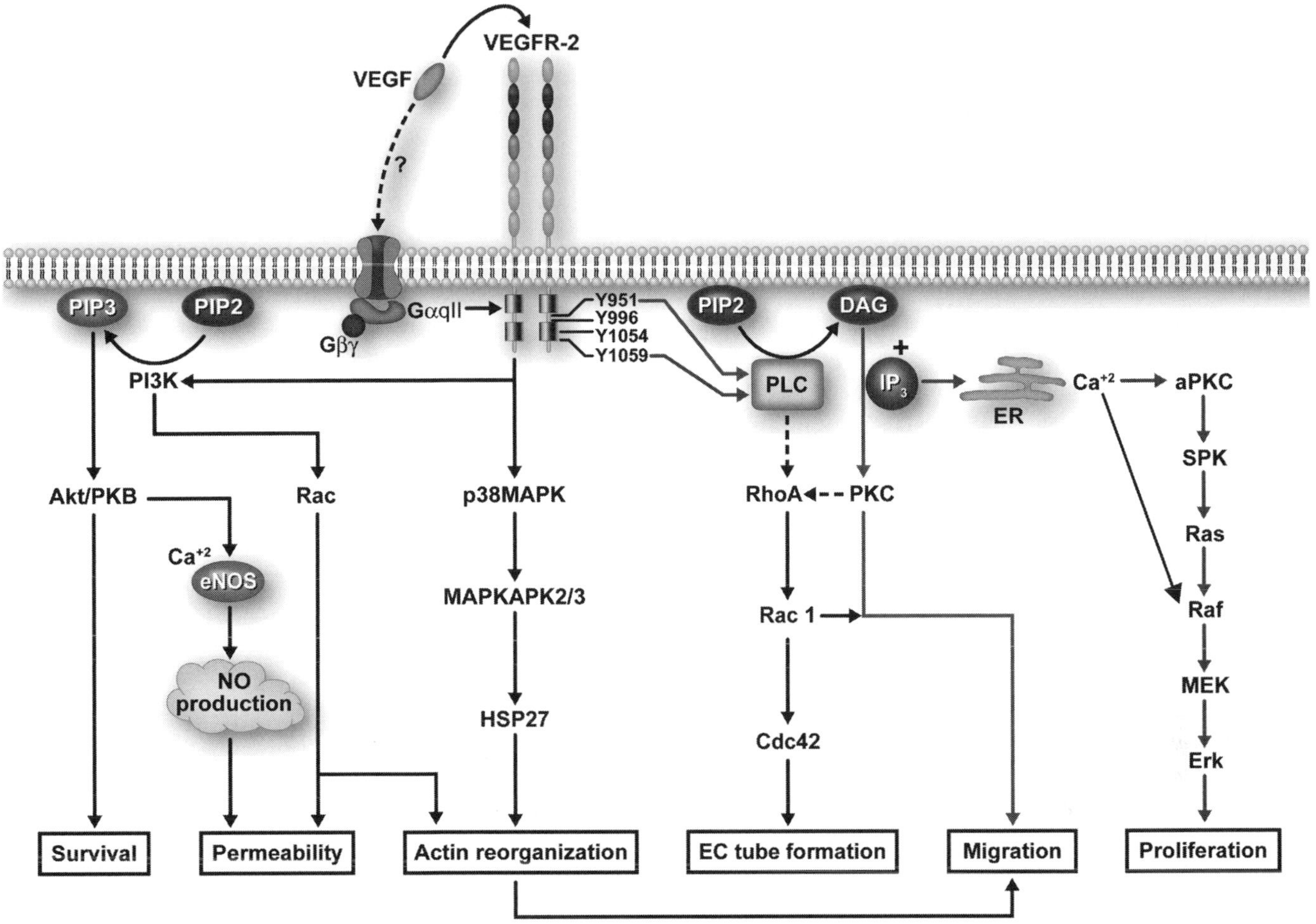

Figure 32.2. Scheme for VEGFR2–mediated signaling. Ligand binding induces dimerization and activation of Ga-proteins resulting in autophosphorylation of specific tyrosine residues on the receptor. Several SH2 domain–containing proteins, including PLCs, bind to the tyrosine phosphorylated residues on the receptor, leading to their own phosphorylation and activation. PLC activation results in hydrolysis of membrane phospholipid PIP2, which leads to generation of DAG and IP3. DAG activates PKCs, whereas IP3 binds to IP3 receptors on the ER that results in release from Ca^{2+} stores, thus raising the intracellular concentration of $[Ca^{2+}]_i$. The exact mechanism by which PI3K and p38MAPK are activated by VEGFR2 is not known. Downstream signal propagation from these aforementioned molecules leads to different endothelial functions such as survival, permeability, tube formation, migration, and proliferation. SH2, Src homology-2 domains; PLC, phospholipase C; PIP2, phosphatidylinositol (4,5)-bisphosphate; DAG, 1,2-diacylglycerol; IP3, inositol (1,4,5)-triphosphate; aPKC, atypical protein kinase C; PI3K, phosphoinositide 3-kinase; p38MAPK, p38 mitogen-activated protein kinase; PIP3, phosphatidylinositol (3,4,5)-triphosphate; Akt/PKB, protein kinase B; eNOS, endothelial nitric oxide synthase; NO, nitric oxide; MAPKAPK2/3, MAPK-activating protein kinase-2 and 3; HSP27, heat shock protein 27, ER, endoplasmic reticulum; SPK, sphingosine kinase; Erk, extracellular-signal-regulated kinase; VEGFR2, vascular endothelial growth factor receptor 2; VEGF, vascular endothelial growth factor.

function and indicate for the first time that G-protein can be activated by VEGF (249,250).

SHP-1 and SHP-2, two SH2 protein-tyrosine phosphatases, have been shown to physically associate with VEGFR2 secondary to VEGF stimulation. These molecules are potentially capable of functionally counteracting the endothelial response to VEGF (244). VEGFR2 also activates PI3K, which causes an increase in IP3 and leads to activation of the PKB (Akt/PKB)-induced survival pathway (251), activation of endothelial nitric oxide synthase resulting in the release of nitric oxide (252–254), and a mediated increase in vascular permeability and migration that is further accentuated by the small GTP binding protein Rac (255). The VEGFR2 pathway also leads to cell migration and actin reorganization that result from activation of p38 MAPK and focal adhesion kinase (FAK) along with paxillin (245,256,257).

VEGFR2 signaling also regulates tube formation by aortic ECs (258). The Rho GTPases represent an important class of molecules involved in EC morphogenesis. Cdc42 and Rac1 are required for the process of EC intracellular vacuole formation and coalescence that regulates EC lumen formation in three-dimensional (3D) extracellular matrices, whereas RhoA appears to stabilize capillary tube networks. Once EC tube networks are established, supporting cells, such as pericytes, are recruited to further stabilize these networks, perhaps by regulating EC basement membrane matrix assembly. Furthermore,

EC morphogenesis is balanced by a tendency for newly formed tubes to regress. This morphogenesis–regression balance is controlled by differential gene expression of such molecules as VEGF, angiopoietin-2, and plasminogen activator inhibitor (PAI)-1, as well as a plasmin- and matrix metalloproteinase–dependent mechanism that induces tube regression through degradation of extracellular matrix (ECM) scaffolds that support EC-lined tubes (259–262).

CONCLUSION

The discovery of VEGF and its receptors on ECs initiated a new era of intense research that focused primarily on endothelial biology. However, the subsequent discovery of these receptors on a number of other cell types indicated that highly coordinated complex interactions exist. Furthermore, the variation among VEGF family members with respect to their properties adds to the intricacy of the biology of VEGF and its receptors. These provide new tools that can be utilized to dissect signaling pathways and to develop prognostic markers and potential therapies. Proangiogenic therapy, including gene therapy using VEGF-A, has shown some promise in patients with critical limb ischemia and ischemic heart disease. Antiangiogenic therapy has had some degree of success not only in cancer patients but also in nononcologic disorders such as rheumatoid arthritis, psoriasis, diabetic retinopathy, and age-dependent macular degeneration. Nonetheless, these advances should be viewed with cautious optimism, because such therapy may be associated with significant side effects, including hypertension, thrombosis, proteinuria, and even fatal hemorrhage. Understanding the role of VEGF and its receptors in physiological and pathological processes should lead to the development of targeted therapies with improved efficacy. To achieve maximal therapeutic benefit, the timing of administration must be optimized and the drugs used in conjunction with other treatment modalities, such as chemotherapy and radiotherapy.

Even though the molecular determinants of VEGF actions in the endothelium have been characterized in great detail over the past few years, different areas of VEGF receptor signal transduction still remain to be elucidated. Because both VEGFR1 and VEGFR2 are expressed on the vascular endothelium, it has been difficult to define their respective roles in signaling. This is particularly true for VEGFR1. Nonetheless, the use of chimeric receptor approaches, antisense strategies, and siRNA techniques has shed some light on the distinct signaling properties of these receptors and demonstrates "cross-talk" between the two receptors. Clearly, an emerging paradigm in VEGF receptor signaling has been the shift from VEGF-based linear signaling to "cross-talk" among the different VEGF receptors and the involvement of other receptors such as G-protein–coupled receptors in the initiation of signaling. Although we have shown the requirement of G-αq11 and G-$\beta\gamma$ in the initiation of VEGF-mediated endothelial signaling, many questions remain unanswered, such as: Are these G-proteins coupled to a receptor at the plasma membrane? If so, then how are they activated by VEGF? Pathways initiated by the VEGF receptors have been well studied, but the molecular determinants causing receptor compartmentalization and endocytosis leading to signaling remain poorly understood. Another important goal is to identify signals relayed from VEGF receptors to the nucleus to generate information required for EC differentiation during angiogenesis and development. Finally, a more in-depth understanding of the regulatory functions of VEGF and its receptors remains essential for realizing potential promising breakthroughs and challenging questions in the pathophysiology of the endothelium.

KEY POINTS

- VPF/VEGF and its receptors control physiological and pathological angiogenesis.
- The regulation of VPF/VEGF and its receptors is complex and context dependent.

Future Goals

- To better understand of VPF/VEGF biology to help us design selective targets that can be useful for therapeutics for several pathological states including cardiovascular diseases and cancer

ACKNOWLEDGMENT

This work is partly supported by NIH grants CA78383, HL072178, and HL70567, and also by a grant from American Cancer Society to DM. We greatly appreciate the secretarial help of Denise Lecy.

REFERENCES

1 Senger DR, Galli SJ, Dvorak AM, et al. Tumor cells secrete a vascular permeability factor that promotes accumulation of ascites fluid. *Science.* 1983;219(4587):983–985.
2 Senger DR, Perruzzi CA, Feder J, Dvorak HF. A highly conserved vascular permeability factor secreted by a variety of human and rodent tumor cell lines. *Cancer Res.* 1986;46(11):5629–5632.
3 Ferrara N, Henzel WJ. Pituitary follicular cells secrete a novel heparin-binding growth factor specific for vascular endothelial cells. *Biochem Biophys Res Commun.* 1989;161(2):851–858.
4 Leung DW, Cachianes G, Kuang WJ, et al. Vascular endothelial growth factor is a secreted angiogenic mitogen. *Science.* 1989; 246(4935):1306–1309.
5 Gospodarowicz D, Abraham JA, Schilling J. Isolation and characterization of a vascular endothelial cell mitogen produced by pituitary-derived folliculo stellate cells. *Proc Natl Acad Sci USA.* 1989;86(19):7311–7315.
6 Gospodarowicz D, Lau K. Pituitary follicular cells secrete both vascular endothelial growth factor and follistatin. *Biochem Biophys Res Commun.* 1989;165(1):292–298.

7 Plouet J, Schilling J, Gospodarowicz D. Isolation and characterization of a newly identified endothelial cell mitogen produced by AtT-20 cells. *EMBO J*. 1989;8(12):3801–3806.

8 Tischer E, Mitchell R, Hartman T, et al. The human gene for vascular endothelial growth factor. Multiple protein forms are encoded through alternative exon splicing. *J Biol Chem*. 1991; 266(18):11947–11954.

9 Ferrara N, Houck K, Jakeman L, Leung DW. Molecular and biological properties of the vascular endothelial growth factor family of proteins. *Endocr Rev*. 1992;13(1):18–32.

10 Houck KA, Leung DW, Rowland AM, et al. Dual regulation of vascular endothelial growth factor bioavailability by genetic and proteolytic mechanisms. *J Biol Chem*. 1992;267(36):26031–26037.

11 Poltorak Z, Cohen T, Sivan R, et al. VEGF145, a secreted vascular endothelial growth factor isoform that binds to extracellular matrix. *J Biol Chem*. 1997;272(11):7151–7158.

12 Houck KA, Ferrara N, Winer J, et al. The vascular endothelial growth factor family: identification of a fourth molecular species and characterization of alternative splicing of RNA. *Mol Endocrinol*. 1991;5(12):1806–1814.

13 Robinson CJ, Stringer SE. The splice variants of vascular endothelial growth factor (VEGF) and their receptors. *J Cell Sci*. 2001;114(Pt 5):853–865.

14 Neufeld G, Cohen T, Gengrinovitch S, Poltorak Z. Vascular endothelial growth factor (VEGF) and its receptors. *FASEB J*. 1999;13(1):9–22.

15 Olofsson B, Pajusola K, Kaipainen A, et al. Vascular endothelial growth factor B, a novel growth factor for endothelial cells. *Proc Natl Acad Sci USA*. 1996;93(6):2576–2581.

16 Grimmond S, Lagercrantz J, Drinkwater C, et al. Cloning and characterization of a novel human gene related to vascular endothelial growth factor. *Genome Res*. 1996;6(2):124–131.

17 Joukov V, Pajusola K, Kaipainen A, et al. A novel vascular endothelial growth factor, VEGF-C, is a ligand for the FLT4 (VEGFR-3) and KDR (VEGFR-2) receptor tyrosine kinases. *EMBO J*. 1996;15(2):290–298.

18 Yamada Y, Nezu J, Shimane M, Hirata Y. Molecular cloning of a novel vascular endothelial growth factor, VEGF-D. *Genomics*. 1997;42(3):483–488.

19 Maglione D, Guerriero V, Viglietto G, et al. Isolation of a human placenta cDNA coding for a protein related to the vascular permeability factor. *Proc Natl Acad Sci USA*. 1991;88(20):9267–9271.

20 Takahashi H, Hattori S, Iwamatsu A, et al. A novel snake venom vascular endothelial growth factor (VEGF) predominantly induces vascular permeability through preferential signaling via VEGF receptor-1. *J Biol Chem*. 2004;279(44):46304–46314.

21 Jiang BH, Rue E, Wang GL, et al. Dimerization, DNA binding, and transactivation properties of hypoxia-inducible factor 1. *J Biol Chem*. 1996;271(30):17771–17778.

22 Hauser S, Weich HA. A heparin-binding form of placenta growth factor (PlGF-2) is expressed in human umbilical vein endothelial cells and in placenta. *Growth Factors*. 1993;9(4):259–268.

23 LeCouter J, Kowalski J, Foster J, et al. Identification of an angiogenic mitogen selective for endocrine gland endothelium. *Nature*. 2001;412(6850):877–884.

24 Berse B, Brown LF, Van de Water L, et al. Vascular permeability factor (vascular endothelial growth factor) gene is expressed differentially in normal tissues, macrophages, and tumors. *Mol Biol Cell*. 1992;3(2):211–220.

25 Banks RE, Forbes MA, Kinsey SE, et al. Release of the angiogenic cytokine vascular endothelial growth factor (VEGF) from platelets: significance for VEGF measurements and cancer biology. *Br J Cancer*. 1998;77(6):956–964.

26 Mohle R, Green D, Moore MA, et al. Constitutive production and thrombin-induced release of vascular endothelial growth factor by human megakaryocytes and platelets. *Proc Natl Acad Sci USA*. 1997;94(2):663–668.

27 Kretzler M, Schroppel B, Merkle M, et al. Detection of multiple vascular endothelial growth factor splice isoforms in single glomerular podocytes. *Kidney Int Suppl*. 1998;67:S159–S161.

28 Ogunshola OO, Stewart WB, Mihalcik V, et al. Neuronal VEGF expression correlates with angiogenesis in postnatal developing rat brain. *Brain Res Dev Brain Res*. 2000;119(1):139–153.

29 Bocci G, Fasciani A, Danesi R, et al. In-vitro evidence of autocrine secretion of vascular endothelial growth factor by endothelial cells from human placental blood vessels. *Mol Hum Reprod*. 2001;7(8):771–777.

30 Uchida K, Uchida S, Nitta K, et al. Glomerular endothelial cells in culture express and secrete vascular endothelial growth factor. *Am J Physiol*. 1994;266(1 Pt 2):F81–F88.

31 Brown LF, Detmar M, Claffey K, et al. Vascular permeability factor/vascular endothelial growth factor: A multifunctional angiogenic cytokine. *Exs*. 1997;79:233–269.

32 Dvorak HF, Brown LF, Detmar M, Dvorak AM. Vascular permeability factor/vascular endothelial growth factor, microvascular hyperpermeability, and angiogenesis. *Am J Pathol*. 1995;146(5): 1029–1039.

33 Stiver SI, Dvorak HF. Vascular Permeability Factor/Vascular Endothelial Growth Factor (VPF/VEGF). *J Clin Ligand Assay*. 2000;23(3):193–205.

34 Anderson KC. Multiple Myeloma. Advances in disease biology: therapeutic implications. *Semin Hematol*. 2001;38(2 Suppl 3): 6–10.

35 Aoki Y, Tosato G. Vascular endothelial growth factor/vascular permeability factor in the pathogenesis of primary effusion lymphomas. *Leuk Lymphoma*. 2001;41(3–4):229–237.

36 Bellamy WT, Richter L, Frutiger Y, Grogan TM. Expression of vascular endothelial growth factor and its receptors in hematopoietic malignancies. *Cancer Res*. 1999;59(3):728–733.

37 Chen H, Treweeke AT, West DC, et al. In vitro and in vivo production of vascular endothelial growth factor by chronic lymphocytic leukemia cells. *Blood*. 2000;96(9):3181–3187.

38 Foss HD, Araujo I, Demel G, et al. Expression of vascular endothelial growth factor in lymphomas and Castleman's disease. *J Pathol*. 1997;183(1):44–50.

39 Hussong JW, Rodgers GM, Shami PJ. Evidence of increased angiogenesis in patients with acute myeloid leukemia. *Blood*. 2000;95(1):309–313.

40 Kini AR, Peterson LA, Tallman MS, Lingen MW. Angiogenesis in acute promyelocytic leukemia: induction by vascular endothelial growth factor and inhibition by all-trans retinoic acid. *Blood*. 2001;97(12):3919–3924.

41 Brown LF, Berse B, Jackman RW, et al. Expression of vascular permeability factor (vascular endothelial growth factor) and its receptors in breast cancer. *Hum Pathol*. 1995;26(1):86–91.

42 Guidi AJ, Abu-Jawdeh G, Berse B, et al. Vascular permeability factor (vascular endothelial growth factor) expression and

angiogenesis in cervical neoplasia. *J Natl Cancer Inst*. 1995;87 (16):1237–1245.
43 Wong MP, Cheung N, Yuen ST, et al. Vascular endothelial growth factor is up-regulated in the early pre-malignant stage of colorectal tumour progression. *Int J Cancer*. 1999;81(6):845–850.
44 Claffey KP, Robinson GS. Regulation of VEGF/VPF expression in tumor cells: consequences for tumor growth and metastasis. *Cancer Metastasis Rev*. 1996;15(2):165–176.
45 Shih SC, Claffey KP. Hypoxia-mediated regulation of gene expression in mammalian cells. *Int J Exp Pathol*. 1998;79(6): 347–357.
46 Levy AP, Levy NS, Wegner S, Goldberg MA. Transcriptional regulation of the rat vascular endothelial growth factor gene by hypoxia. *J Biol Chem*. 1995;270(22):13333–13340.
47 Forsythe JA, Jiang BH, Iyer NV, et al. Activation of vascular endothelial growth factor gene transcription by hypoxia-inducible factor 1. *Mol Cell Biol*. 1996;16(9):4604–4613.
48 Kallio PJ, Wilson WJ, O'Brien S, et al. Regulation of the hypoxia-inducible transcription factor 1alpha by the ubiquitin-proteasome pathway. *J Biol Chem*. 1999;274(10):6519–6525.
49 Mukhopadhyay D, Tsiokas L, Zhou XM, et al. Hypoxic induction of human vascular endothelial growth factor expression through c-Src activation. *Nature*. 1995;375(6532):577–581.
50 Rak J, Filmus J, Finkenzeller G, et al. Oncogenes as inducers of tumor angiogenesis. *Cancer Metastasis Rev*. 1995;14(4):263–277.
51 Rak J, Kerbel RS. Ras regulation of vascular endothelial growth factor and angiogenesis. *Methods Enzymol*. 2001;333:267–283.
52 Ohh M, Kaelin WG, Jr. The von Hippel-Lindau tumour suppressor protein: new perspectives. *Mol Med Today*. 1999;5(6):257–263.
53 Clifford SC, Maher ER. Von Hippel-Lindau disease: clinical and molecular perspectives. *Adv Cancer Res*. 2001;82:85–105.
54 Mukhopadhyay D, Tsiokas L, Sukhatme VP. Wild-type p53 and v-Src exert opposing influences on human vascular endothelial growth factor gene expression. *Cancer Res*. 1995;55(24):6161–6165.
55 Pal S, Datta K, Mukhopadhyay D. Central role of p53 on regulation of vascular permeability factor/vascular endothelial growth factor (VPF/VEGF) expression in mammary carcinoma. *Cancer Res*. 2001;61(18):6952–6957.
56 Zhang L, Yu D, Hu M, et al. Wild-type p53 suppresses angiogenesis in human leiomyosarcoma and synovial sarcoma by transcriptional suppression of vascular endothelial growth factor expression. *Cancer Res*. 2000;60(13):3655–3661.
57 Claffey KP, Abrams K, Shih SC, et al. Fibroblast growth factor 2 activation of stromal cell vascular endothelial growth factor expression and angiogenesis. *Lab Invest*. 2001;81(1):61–75.
58 Wojta J, Kaun C, Breuss JM, et al. Hepatocyte growth factor increases expression of vascular endothelial growth factor and plasminogen activator inhibitor-1 in human keratinocytes and the vascular endothelial growth factor receptor flk1 in human endothelial cells. *Lab Invest*. 1999;79(4):427–438.
59 Stavri GT, Zachary IC, Baskerville PA, et al. Basic fibroblast growth factor upregulates the expression of vascular endothelial growth factor in vascular smooth muscle cells. Synergistic interaction with hypoxia. *Circulation*. 1995;92(1):11–14.
60 Cullinan-Bove K, Koos RD. Vascular endothelial growth factor/vascular permeability factor expression in the rat uterus: rapid stimulation by estrogen correlates with estrogen-induced increases in uterine capillary permeability and growth. *Endocrinology*. 1993;133(2):829–837.
61 Garrido C, Saule S, Gospodarowicz D. Transcriptional regulation of vascular endothelial growth factor gene expression in ovarian bovine granulosa cells. *Growth Factors*. 1993;8(2):109–117.
62 Carmeliet P, Ferreira V, Breier G, et al. Abnormal blood vessel development and lethality in embryos lacking a single VEGF allele. *Nature*. 1996;380(6573):435–439.
63 Ferrara N, Carver-Moore K, Chen H, et al. Heterozygous embryonic lethality induced by targeted inactivation of the VEGF gene. *Nature*. 1996;380(6573):439–442.
64 Parikh AA, Ellis LM. The vascular endothelial growth factor family and its receptors. *Hematol Oncol Clin North Am*. 2004;18 (5):951–971, vii.
65 Brown LF, Yeo KT, Berse B, et al. Expression of vascular permeability factor (vascular endothelial growth factor) by epidermal keratinocytes during wound healing. *J Exp Med*. 1992;176(5): 1375–1379.
66 Davis-Smyth T, Presta LG, Ferrara N. Mapping the charged residues in the second immunoglobulin-like domain of the vascular endothelial growth factor/placenta growth factor receptor FLT-1 required for binding and structural stability. *J Biol Chem*. 1998;273(6):3216–3222.
67 Pauli SA, Tang H, Wang J, et al. The vascular endothelial growth factor (VEGF)/VEGF receptor 2 pathway is critical for blood vessel survival in corpora lutea of pregnancy in the rodent. *Endocrinology*. 2005;146(3):1301–1311.
68 Young HS, Summers AM, Bhushan M, et al. Single-nucleotide polymorphisms of vascular endothelial growth factor in psoriasis of early onset. *J Invest Dermatol*. 2004;122(1):209–215.
69 Clavel G, Bessis N, Boissier MC. Recent data on the role for angiogenesis in rheumatoid arthritis. *Joint Bone Spine*. 2003;70 (5):321–326.
70 Alon T, Hemo I, Itin A, et al. Vascular endothelial growth factor acts as a survival factor for newly formed retinal vessels and has implications for retinopathy of prematurity. *Nat Med*. 1995;1 (10):1024–1028.
71 Caldwell RB, Bartoli M, Behzadian MA, et al. Vascular endothelial growth factor and diabetic retinopathy: pathophysiological mechanisms and treatment perspectives. *Diabetes Metab Res Rev*. 2003;19(6):442–455.
72 Storkebaum E, Carmeliet P. VEGF: a critical player in neurodegeneration. *J Clin Invest*. 2004;113(1):14–18.
73 Lambrechts D, Storkebaum E, Morimoto M, et al. VEGF is a modifier of amyotrophic lateral sclerosis in mice and humans and protects motoneurons against ischemic death. *Nat Genet*. 2003;34(4):383–394.
74 Deckel AW, Duffy JD. Vasomotor hyporeactivity in the anterior cerebral artery during motor activation in Huntington's disease patients. *Brain Res*. 2000;872(1–2):258–261.
75 Kalaria RN. Small vessel disease and Alzheimer's dementia: pathological considerations. *Cerebrovasc Dis*. 2002;13(Suppl 2):48–52.
76 Presta LG, Chen H, O'Connor SJ, et al. Humanization of an anti-vascular endothelial growth factor monoclonal antibody for the therapy of solid tumors and other disorders. *Cancer Res*. 1997;57(20):4593–4599.
77 Kendall RL, Thomas KA. Inhibition of vascular endothelial cell growth factor activity by an endogenously encoded soluble receptor. *Proc Natl Acad Sci USA*. 1993;90(22):10705–10709.

78 Lin P, Sankar S, Shan S, et al. Inhibition of tumor growth by targeting tumor endothelium using a soluble vascular endothelial growth factor receptor. *Cell Growth Differ*. 1998;9(1):49–58.
79 Cheng SY, Huang HJ, Nagane M, et al. Suppression of glioblastoma angiogenicity and tumorigenicity by inhibition of endogenous expression of vascular endothelial growth factor. *Proc Natl Acad Sci USA*. 1996;93(16):8502–8507.
80 Saleh M, Stacker SA, Wilks AF. Inhibition of growth of C6 glioma cells in vivo by expression of antisense vascular endothelial growth factor sequence. *Cancer Res*. 1996;56(2):393–401.
81 Ramakrishnan S, Olson TA, Bautch VL, Mohanraj D. Vascular endothelial growth factor-toxin conjugate specifically inhibits KDR/flk-1-positive endothelial cell proliferation in vitro and angiogenesis in vivo. *Cancer Res*. 1996;56(6):1324–1330.
82 Siemeister G, Schirner M, Reusch P, et al. An antagonistic vascular endothelial growth factor (VEGF) variant inhibits VEGF-stimulated receptor autophosphorylation and proliferation of human endothelial cells. *Proc Natl Acad Sci USA*. 1998;95(8):4625–4629.
83 Skobe M, Rockwell P, Goldstein N, et al. Halting angiogenesis suppresses carcinoma cell invasion. *Nat Med*. 1997;3(11):1222–1227.
84 Strawn LM, McMahon G, App H, et al. Flk-1 as a target for tumor growth inhibition. *Cancer Res*. 1996;56(15):3540–3545.
85 Ellis LM. Bevacizumab. *Nat Rev Drug Discov*. 2005;(Suppl):S8–S9.
86 Schratzberger P, Schratzberger G, Silver M, et al. Favorable effect of VEGF gene transfer on ischemic peripheral neuropathy. *Nat Med*. 2000;6(4):405–413.
87 Simovic D, Isner JM, Ropper AH, et al. Improvement in chronic ischemic neuropathy after intramuscular phVEGF165 gene transfer in patients with critical limb ischemia. *Arch Neurol*. 2001;58(5):761–768.
88 Asahara T, Bauters C, Pastore C, et al. Local delivery of vascular endothelial growth factor accelerates reendothelialization and attenuates intimal hyperplasia in balloon-injured rat carotid artery. *Circulation*. 1995;91(11):2793–2801.
89 Tsurumi Y, Kearney M, Chen D, et al. Treatment of acute limb ischemia by intramuscular injection of vascular endothelial growth factor gene. *Circulation*. 1997;96(9 Suppl):II-382–388.
90 Baumgartner I, Pieczek A, Manor O, et al. Constitutive expression of phVEGF165 after intramuscular gene transfer promotes collateral vessel development in patients with critical limb ischemia. *Circulation*. 1998;97(12):1114–1123.
91 Vaisman N, Gospodarowicz D, Neufeld G. Characterization of the receptors for vascular endothelial growth factor. *J Biol Chem*. 1990;265(32):19461–19466.
92 Shibuya M, Matsushime H, Yamane A, et al. Isolation and characterization of new mammalian kinase genes by cross hybridization with a tyrosine kinase probe. *Princess Takamatsu Symp*. 1989;20:103–110.
93 de Vries C, Escobedo JA, Ueno H, et al. The fms-like tyrosine kinase, a receptor for vascular endothelial growth factor. *Science*. 1992;255(5047):989–991.
94 Hamlyn JM, Blaustein MP, Bova S, et al. Identification and characterization of a ouabain-like compound from human plasma. *Proc Natl Acad Sci USA*. 1991;88(14):6259–6263.
95 Terman BI, Dougher-Vermazen M, Carrion ME, et al. Identification of the KDR tyrosine kinase as a receptor for vascular endothelial cell growth factor. *Biochem Biophys Res Commun*. 1992;187(3):1579–1586.
96 Breier G, Clauss M, Risau W. Coordinate expression of vascular endothelial growth factor receptor-1 (flt-1) and its ligand suggests a paracrine regulation of murine vascular development. *Dev Dyn*. 1995;204(3):228–239.
97 Peters KG, De Vries C, Williams LT. Vascular endothelial growth factor receptor expression during embryogenesis and tissue repair suggests a role in endothelial differentiation and blood vessel growth. *Proc Natl Acad Sci USA*. 1993;90(19):8915–8919.
98 Kaipainen A, Korhonen J, Mustonen T, et al. Expression of the fms-like tyrosine kinase 4 gene becomes restricted to lymphatic endothelium during development. *Proc Natl Acad Sci USA*. 1995;92(8):3566–3570.
99 Partanen TA, Makinen T, Arola J, et al. Endothelial growth factor receptors in human fetal heart. *Circulation*. 1999;100(6):583–586.
100 Fong GH, Rossant J, Gertsenstein M, Breitman ML. Role of the FLT-1 receptor tyrosine kinase in regulating the assembly of vascular endothelium. *Nature*. 1995;376(6535):66–70.
101 Fong GH, Zhang L, Bryce DM, Peng J. Increased hemangioblast commitment, not vascular disorganization, is the primary defect in flt-1 knock-out mice. *Development*. 1999;126(13):3015–3025.
102 Kearney JB, Ambler CA, Monaco KA, et al. Vascular endothelial growth factor receptor FLT-1 negatively regulates developmental blood vessel formation by modulating endothelial cell division. *Blood*. 2002;99(7):2397–2407.
103 Hiratsuka S, Minowa O, Kuno J, et al. FLT-1 lacking the tyrosine kinase domain is sufficient for normal development and angiogenesis in mice. *Proc Natl Acad Sci USA*. 1998;95(16):9349–9354.
104 Kearney JB, Kappas NC, Ellerstrom C, et al. The VEGF receptor flt-1 (VEGFR-1) is a positive modulator of vascular sprout formation and branching morphogenesis. *Blood*. 2004;103(12):4527–4535.
105 Yamane A, Seetharam L, Yamaguchi S, et al. A new communication system between hepatocytes and sinusoidal endothelial cells in liver through vascular endothelial growth factor and FLT tyrosine kinase receptor family (FLT-1 and KDR/Flk1). *Oncogene*. 1994;9(9):2683–2690.
106 Barleon B, Sozzani S, Zhou D, et al. Migration of human monocytes in response to vascular endothelial growth factor (VEGF) is mediated via the VEGF receptor flt-1. *Blood*. 1996;87(8):3336–3343.
107 Brogi E, Schatteman G, Wu T, et al. Hypoxia-induced paracrine regulation of vascular endothelial growth factor receptor expression. *J Clin Invest*. 1996;97(2):469–476.
108 Ishida A, Murray J, Saito Y, et al. Expression of vascular endothelial growth factor receptors in smooth muscle cells. *J Cell Physiol*. 2001;188(3):359–368.
109 Wang H, Keiser JA. Vascular endothelial growth factor upregulates the expression of matrix metalloproteinases in vascular smooth muscle cells: role of flt-1. *Circ Res*. 1998;83(8):832–840.
110 Belgore F, Blann A, Neil D, et al. Localisation of members of the vascular endothelial growth factor (VEGF) family and their receptors in human atherosclerotic arteries. *J Clin Pathol*. 2004;57(3):266–272.
111 Amemiya T, Sasamura H, Mifune M, et al. Vascular endothelial growth factor activates MAP kinase and enhances collagen synthesis in human mesangial cells. *Kidney Int*. 1999;56(6):2055–2063.

112 Gerber HP, Vu TH, Ryan AM, et al. VEGF couples hypertrophic cartilage remodeling, ossification and angiogenesis during endochondral bone formation. *Nat Med.* 1999;5(6):623–628.

113 Choi K, Kennedy M, Kazarov A, et al. A common precursor for hematopoietic and endothelial cells. *Development.* 1998;125(4):725–732.

114 Park JE, Chen HH, Winer J, et al. Placenta growth factor. Potentiation of vascular endothelial growth factor bioactivity, in vitro and in vivo, and high affinity binding to FLT-1 but not to Flk1/KDR. *J Biol Chem.* 1994;269(41):25646–25654.

115 Davis-Smyth T, Chen H, Park J, et al. The second immunoglobulin-like domain of the VEGF tyrosine kinase receptor FLT-1 determines ligand binding and may initiate a signal transduction cascade. *EMBO J.* 1996;15(18):4919–4927.

116 Muller YA, Christinger HW, Keyt BA, de Vos AM. The crystal structure of vascular endothelial growth factor (VEGF) refined to 1.93 A resolution: multiple copy flexibility and receptor binding. *Structure.* 1997;5(10):1325–1338.

117 Wiesmann C, Fuh G, Christinger HW, et al. Crystal structure at 1.7 A resolution of VEGF in complex with domain 2 of the FLT-1 receptor. *Cell.* 1997;91(5):695–704.

118 Gerber HP, Condorelli F, Park J, Ferrara N. Differential transcriptional regulation of the two vascular endothelial growth factor receptor genes. FLT-1, but not Flk1/KDR, is up-regulated by hypoxia. *J Biol Chem.* 1997;272(38):23659–23667.

119 Sawano A, Takahashi T, Yamaguchi S, et al. FLT-1 but not KDR/Flk1 tyrosine kinase is a receptor for placenta growth factor, which is related to vascular endothelial growth factor. *Cell Growth Differ.* 1996;7(2):213–221.

120 Waltenberger J, Claesson-Welsh L, Siegbahn A, et al. Different signal transduction properties of KDR and FLT1, two receptors for vascular endothelial growth factor. *J Biol Chem.* 1994;269(43):26988–26995.

121 Iyer S, Leonidas DD, Swaminathan GJ, et al. The crystal structure of human placenta growth factor-1 (PlGF-1), an angiogenic protein, at 2.0 A resolution. *J Biol Chem.* 2001;276(15):12153–12161.

122 Carmeliet P, Moons L, Luttun A, et al. Synergism between vascular endothelial growth factor and placental growth factor contributes to angiogenesis and plasma extravasation in pathological conditions. *Nat Med.* 2001;7(5):575–583.

123 Luttun A, Tjwa M, Moons L, et al. Revascularization of ischemic tissues by PlGF treatment, and inhibition of tumor angiogenesis, arthritis and atherosclerosis by anti-FLT1. *Nat Med.* 2002;8(8):831–840.

124 Autiero M, Waltenberger J, Communi D, et al. Role of PlGF in the intra- and intermolecular cross talk between the VEGF receptors FLT1 and Flk1. *Nat Med.* 2003;9(7):936–943.

125 Kendall RL, Wang G, Thomas KA. Identification of a natural soluble form of the vascular endothelial growth factor receptor, FLT-1, and its heterodimerization with KDR. *Biochem Biophys Res Commun.* 1996;226(2):324–328.

126 Goldman CK, Kendall RL, Cabrera G, et al. Paracrine expression of a native soluble vascular endothelial growth factor receptor inhibits tumor growth, metastasis, and mortality rate. *Proc Natl Acad Sci USA.* 1998;95(15):8795–8800.

127 Millauer B, Wizigmann-Voos S, Schnurch H, et al. High affinity VEGF binding and developmental expression suggest Flk1 as a major regulator of vasculogenesis and angiogenesis. *Cell.* 1993;72(6):835–846.

128 Yamaguchi TP, Dumont DJ, Conlon RA, et al. flk1, an flt-related receptor tyrosine kinase is an early marker for endothelial cell precursors. *Development.* 1993;118(2):489–498.

129 Flamme I, Breier G, Risau W. Vascular endothelial growth factor (VEGF) and VEGF receptor 2 (flk1) are expressed during vasculogenesis and vascular differentiation in the quail embryo. *Dev Biol.* 1995;169(2):699–712.

130 Shalaby F, Rossant J, Yamaguchi TP, et al. Failure of blood-island formation and vasculogenesis in Flk1-deficient mice. *Nature.* 1995;376(6535):62–66.

131 Chan J, Bayliss PE, Wood JM, Roberts TM. Dissection of angiogenic signaling in zebrafish using a chemical genetic approach. *Cancer Cell.* 2002;1(3):257–267.

132 Lee P, Goishi K, Davidson AJ, et al. Neuropilin-1 is required for vascular development and is a mediator of VEGF-dependent angiogenesis in zebrafish. *Proc Natl Acad Sci USA.* 2002;99(16):10470–10475.

133 Ambler CA, Schmunk GM, Bautch VL. Stem cell-derived endothelial cells/progenitors migrate and pattern in the embryo using the VEGF signaling pathway. *Dev Biol.* 2003;257(1):205–219.

134 Eichmann A, Marcelle C, Breant C, Le Douarin NM. Two molecules related to the VEGF receptor are expressed in early endothelial cells during avian embryonic development. *Mech Dev.* 1993;42(1–2):33–48.

135 Kaipainen A, Korhonen J, Pajusola K, et al. The related FLT4, FLT1, and KDR receptor tyrosine kinases show distinct expression patterns in human fetal endothelial cells. *J Exp Med.* 1993;178(6):2077–2088.

136 Asahara T, Masuda H, Takahashi T, et al. Bone marrow origin of endothelial progenitor cells responsible for postnatal vasculogenesis in physiological and pathological neovascularization. *Circ Res.* 1999;85(3):221–228.

137 Achen MG, Jeltsch M, Kukk E, et al. Vascular endothelial growth factor D (VEGF-D) is a ligand for the tyrosine kinases VEGF receptor 2 (Flk1) and VEGF receptor 3 (FLT4). *Proc Natl Acad Sci USA.* 1998;95(2):548–553.

138 Shibuya M. Vascular endothelial growth factor receptor-2: its unique signaling and specific ligand, VEGF-E. *Cancer Sci.* 2003;94(9):751–756.

139 Quinn TP, Peters KG, De Vries C, et al. Fetal liver kinase 1 is a receptor for vascular endothelial growth factor and is selectively expressed in vascular endothelium. *Proc Natl Acad Sci USA.* 1993;90(16):7533–7537.

140 Tao Q, Backer MV, Backer JM, Terman BI. Kinase insert domain receptor (KDR) extracellular immunoglobulin-like domains 4–7 contain structural features that block receptor dimerization and vascular endothelial growth factor-induced signaling. *J Biol Chem.* 2001;276(24):21916–21923.

141 Wen Y, Edelman JL, Kang T, et al. Two functional forms of vascular endothelial growth factor receptor-2/Flk1 mRNA are expressed in normal rat retina. *J Biol Chem.* 1998;273(4):2090–2097.

142 Ebos JM, Bocci G, Man S, et al. A naturally occurring soluble form of vascular endothelial growth factor receptor 2 detected in mouse and human plasma. *Mol Cancer Res.* 2004;2(6):315–326.

143 Martin P. Wound healing – aiming for perfect skin regeneration. *Science.* 1997;276(5309):75–81.

144 Schaffer CJ, Nanney LB. Cell biology of wound healing. *Int Rev Cytol.* 1996;169:151–181.

145 Dvorak HF, Detmar M, Claffey KP, et al. Vascular permeability factor/vascular endothelial growth factor: An important mediator of angiogenesis in malignancy and inflammation. *Int Arch Allergy Immunol.* 1995;107(1–3):233–235.

146 Reynolds LP, Killilea SD, Redmer DA. Angiogenesis in the female reproductive system. *FASEB J.* 1992;6(3):886–892.

147 Kamat BR, Brown LF, Manseau EJ, et al. Expression of vascular permeability factor/vascular endothelial growth factor by human granulosa and theca lutein cells. Role in corpus luteum development. *Am J Pathol.* 1995;146(1):157–165.

148 Goodger AM, Rogers PA. Blood vessel growth in the endometrium. *Microcirculation.* 1995;2(4):329–343.

149 Graubert MD, Ortega MA, Kessel B, et al. Vascular repair after menstruation involves regulation of vascular endothelial growth factor-receptor phosphorylation by sFLT-1. *Am J Pathol.* 2001;158(4):1399–1410.

150 Meyer M, Clauss M, Lepple-Wienhues A, et al. A novel vascular endothelial growth factor encoded by Orf virus, VEGF-E, mediates angiogenesis via signalling through VEGFR2 (KDR) but not VEGFR1 (FLT-1) receptor tyrosine kinases. *EMBO J.* 1999;18(2):363–374.

151 Ogawa S, Oku A, Sawano A, et al. A novel type of vascular endothelial growth factor, VEGF-E (NZ-7 VEGF), preferentially utilizes KDR/Flk1 receptor and carries a potent mitotic activity without heparin-binding domain. *J Biol Chem.* 1998;273(47): 31273–31282.

152 LeCouter J, Moritz DR, Li B, et al. Angiogenesis-independent endothelial protection of liver: role of VEGFR-1. *Science.* 2003; 299(5608):890–893.

153 Tang N, Wang L, Esko J, et al. Loss of HIF-1alpha in endothelial cells disrupts a hypoxia-driven VEGF autocrine loop necessary for tumorigenesis. *Cancer Cell.* 2004;6(5):485–495.

154 Barleon B, Siemeister G, Martiny-Baron G, et al. Vascular endothelial growth factor up-regulates its receptor fms-like tyrosine kinase 1 (FLT-1) and a soluble variant of FLT-1 in human vascular endothelial cells. *Cancer Res.* 1997;57(23):5421–5425.

155 Marti HH, Risau W. Systemic hypoxia changes the organ-specific distribution of vascular endothelial growth factor and its receptors. *Proc Natl Acad Sci USA.* 1998;95(26):15809–15814.

156 Tomanek RJ, Lund DD, Yue X. Hypoxic induction of myocardial vascularization during development. *Adv Exp Med Biol.* 2003; 543:139–149.

157 Nilsson I, Shibuya M, Wennstrom S. Differential activation of vascular genes by hypoxia in primary endothelial cells. *Exp Cell Res.* 2004;299(2):476–485.

158 Akeno N, Czyzyk-Krzeska MF, Gross TS, Clemens TL. Hypoxia induces vascular endothelial growth factor gene transcription in human osteoblast-like cells through the hypoxia-inducible factor-2alpha. *Endocrinology.* 2001;142(2):959–962.

159 Akeno N, Robins J, Zhang M, et al. Induction of vascular endothelial growth factor by IGF-I in osteoblast-like cells is mediated by the PI3K signaling pathway through the hypoxia-inducible factor-2alpha. *Endocrinology.* 2002;143(2):420–425.

160 Favier J, Kempf H, Corvol P, Gasc JM. Coexpression of endothelial PAS protein 1 with essential angiogenic factors suggests its involvement in human vascular development. *Dev Dyn.* 2001; 222(3):377–388.

161 Kremer C, Breier G, Risau W, Plate KH. Up-regulation of flk1/ vascular endothelial growth factor receptor 2 by its ligand in a cerebral slice culture system. *Cancer Res.* 1997;57(17):3852–3859.

162 Waltenberger J, Mayr U, Pentz S, Hombach V. Functional upregulation of the vascular endothelial growth factor receptor KDR by hypoxia. *Circulation.* 1996;94(7):1647–1654.

163 Banks RE, Forbes MA, Searles J, et al. Evidence for the existence of a novel pregnancy-associated soluble variant of the vascular endothelial growth factor receptor, FLT-1. *Mol Hum Reprod.* 1998;4(4):377–386.

164 Clark DE, Smith SK, He Y, et al. A vascular endothelial growth factor antagonist is produced by the human placenta and released into the maternal circulation. *Biol Reprod.* 1998;59(6): 1540–1548.

165 Maynard SE, Min JY, Merchan J, et al. Excess placental soluble fms-like tyrosine kinase 1 (sFLT1) may contribute to endothelial dysfunction, hypertension, and proteinuria in preeclampsia. *J Clin Invest.* 2003;111(5):649–658.

166 Levine RJ, Maynard SE, Qian C, et al. Circulating angiogenic factors and the risk of preeclampsia. *N Engl J Med.* 2004;350 (7):672–683.

167 Roberts JM. Endothelial dysfunction in preeclampsia. *Semin Reprod Endocrinol.* 1998;16(1):5–15.

168 Hanahan D, Folkman J. Patterns and emerging mechanisms of the angiogenic switch during tumorigenesis. *Cell.* 1996;86(3): 353–364.

169 Hlatky L, Tsionou C, Hahnfeldt P, Coleman CN. Mammary fibroblasts may influence breast tumor angiogenesis via hypoxia-induced vascular endothelial growth factor up-regulation and protein expression. *Cancer Res.* 1994;54(23): 6083–6086.

170 Detmar M, Brown LF, Berse B, et al. Hypoxia regulates the expression of vascular permeability factor/vascular endothelial growth factor (VPF/VEGF) and its receptors in human skin. *J Invest Dermatol.* 1997;108(3):263–268.

171 Fukumura D, Xavier R, Sugiura T, et al. Tumor induction of VEGF promoter activity in stromal cells. *Cell.* 1998;94(6):715–725.

172 Hatva E, Bohling T, Jaaskelainen J, et al. Vascular growth factors and receptors in capillary hemangioblastomas and hemangiopericytomas. *Am J Pathol.* 1996;148(3):763–775.

173 Plate KH, Breier G, Millauer B, et al. Up-regulation of vascular endothelial growth factor and its cognate receptors in a rat glioma model of tumor angiogenesis. *Cancer Res.* 1993;53(23): 5822–5827.

174 Wizigmann-Voos S, Breier G, Risau W, Plate KH. Up-regulation of vascular endothelial growth factor and its receptors in von Hippel-Lindau disease-associated and sporadic hemangioblastomas. *Cancer Res.* 1995;55(6):1358–1364.

175 Detmar M, Brown LF, Schon MP, et al. Increased microvascular density and enhanced leukocyte rolling and adhesion in the skin of VEGF transgenic mice. *J Invest Dermatol.* 1998;111(1):1–6.

176 Hoar FJ, Lip GY, Belgore F, Stonelake PS. Circulating levels of VEGF-A, VEGF-D and soluble VEGF-A receptor (sFIt-1) in human breast cancer. *Int J Biol Markers.* 2004;19(3):229–235.

177 Toi M, Bando H, Ogawa T, et al. Significance of vascular endothelial growth factor (VEGF)/soluble VEGF receptor-1 relationship in breast cancer. *Int J Cancer.* 2002;98(1):14–18.

178 Kumar S, Witzig TE, Timm M, et al. Expression of VEGF and its receptors by myeloma cells. *Leukemia.* 2003;17(10):2025–2031.

179 Nishida N, Yano H, Komai K, et al. Vascular endothelial growth factor C and vascular endothelial growth factor receptor 2 are related closely to the prognosis of patients with ovarian carcinoma. *Cancer.* 2004;101(6):1364–1374.

180 Zhao WL, Mourah S, Mounier N, et al. Vascular endothelial growth factor-A is expressed both on lymphoma cells and endothelial cells in angioimmunoblastic T-cell lymphoma and related to lymphoma progression. *Lab Invest.* 2004;84(11): 1512–1519.

181 Caponigro F, Basile M, Rosa VD, Normanno N. New Drugs in Cancer Therapy, National Tumor Institute, Naples, 17–18 June 2004. *Anticancer Drugs.* 2005;16(2):211–221.

182 Tuder RM, Groves B, Badesch DB, Voelkel NF. Exuberant endothelial cell growth and elements of inflammation are present in plexiform lesions of pulmonary hypertension. *Am J Pathol.* 1994;144(2):275–285.

183 Voelkel NF, Cool C, Taraceviene-Stewart L, et al. Janus face of vascular endothelial growth factor: the obligatory survival factor for lung vascular endothelium controls precapillary artery remodeling in severe pulmonary hypertension. *Crit Care Med.* 2002;30(5 Suppl):S251–S256.

184 Tuder RM, Chacon M, Alger L, et al. Expression of angiogenesis-related molecules in plexiform lesions in severe pulmonary hypertension: evidence for a process of disordered angiogenesis. *J Pathol.* 2001;195(3):367–374.

185 Jeffery TK, Morrell NW. Molecular and cellular basis of pulmonary vascular remodeling in pulmonary hypertension. *Prog Cardiovasc Dis.* 2002;45(3):173–202.

186 Aiello LP, Avery RL, Arrigg PG, et al. Vascular endothelial growth factor in ocular fluid of patients with diabetic retinopathy and other retinal disorders. *N Engl J Med.* 1994;331(22): 1480–1487.

187 Malecaze F, Clamens S, Simorre-Pinatel V, et al. Detection of vascular endothelial growth factor messenger RNA and vascular endothelial growth factor-like activity in proliferative diabetic retinopathy. *Arch Ophthalmol.* 1994;112(11):1476–1482.

188 Aiello LP, Pierce EA, Foley ED, et al. Suppression of retinal neovascularization in vivo by inhibition of vascular endothelial growth factor (VEGF) using soluble VEGF-receptor chimeric proteins. *Proc Natl Acad Sci USA.* 1995;92(23):10457–10461.

189 Adamis AP, Shima DT, Tolentino MJ, et al. Inhibition of vascular endothelial growth factor prevents retinal ischemia-associated iris neovascularization in a nonhuman primate. *Arch Ophthalmol.* 1996;114(1):66–71.

190 Robinson GS, Pierce EA, Rook SL, et al. Oligodeoxynucleotides inhibit retinal neovascularization in a murine model of proliferative retinopathy. *Proc Natl Acad Sci USA.* 1996;93(10):4851–4856.

191 Gehlbach P, Demetriades AM, Yamamoto S, et al. Periocular gene transfer of sFLT-1 suppresses ocular neovascularization and vascular endothelial growth factor-induced breakdown of the blood-retinal barrier. *Hum Gene Ther.* 2003;14(2):129–141.

192 Laitinen M, Zachary I, Breier G, et al. VEGF gene transfer reduces intimal thickening via increased production of nitric oxide in carotid arteries. *Hum Gene Ther.* 1997;8(15):1737–1744.

193 Kuzuya M, Ramos MA, Kanda S, et al. VEGF protects against oxidized LDL toxicity to endothelial cells by an intracellular glutathione-dependent mechanism through the KDR receptor. *Arterioscler Thromb Vasc Biol.* 2001;21(5):765–770.

194 Zachary I, Mathur A, Yla-Herttuala S, Martin J. Vascular protection: a novel nonangiogenic cardiovascular role for vascular endothelial growth factor. *Arterioscler Thromb Vasc Biol.* 2000;20(6):1512–1520.

195 Barger AC, Beeuwkes R, 3rd, Lainey LL, Silverman KJ. Hypothesis: vasa vasorum and neovascularization of human coronary arteries. A possible role in the pathophysiology of atherosclerosis. *N Engl J Med.* 1984;310(3):175–177.

196 Inoue M, Itoh H, Ueda M, et al. Vascular endothelial growth factor (VEGF) expression in human coronary atherosclerotic lesions: possible pathophysiological significance of VEGF in progression of atherosclerosis. *Circulation.* 1998;98(20):2108–2116.

197 Celletti FL, Waugh JM, Amabile PG, et al. Vascular endothelial growth factor enhances atherosclerotic plaque progression. *Nat Med.* 2001;7(4):425–429.

198 Libby P, Ganz P. Restenosis revisited – new targets, new therapies. *N Engl J Med.* 1997;337(6):418–419.

199 Topol EJ, Serruys PW. Frontiers in interventional cardiology. *Circulation.* 1998;98(17):1802–1820.

200 Isner JM. Still more debate over VEGF. *Nat Med.* 2001;7(6):639–641.

201 Shibata M, Suzuki H, Nakatani M, et al. The involvement of vascular endothelial growth factor and flt-1 in the process of neointimal proliferation in pig coronary arteries following stent implantation. *Histochem Cell Biol.* 2001;116(6):471–481.

202 Asahara T, Chen D, Tsurumi Y, et al. Accelerated restitution of endothelial integrity and endothelium-dependent function after phVEGF165 gene transfer. *Circulation.* 1996;94(12):3291–3302.

203 Van Belle E, Tio FO, Chen D, et al. Passivation of metallic stents after arterial gene transfer of phVEGF165 inhibits thrombus formation and intimal thickening. *J Am Coll Cardiol.* 1997;29(6): 1371–1379.

204 Baumgartner I, Isner JM. Somatic gene therapy in the cardiovascular system. *Annu Rev Physiol.* 2001;63:427–450.

205 Kim I, Moon SO, Kim SH, et al. Vascular endothelial growth factor expression of intercellular adhesion molecule 1 (ICAM-1), vascular cell adhesion molecule 1 (VCAM-1), and E-selectin through nuclear factor-kappa B activation in endothelial cells. *J Biol Chem.* 2001;276(10):7614–7620.

206 Grosskreutz CL, Anand-Apte B, Duplaa C, et al. Vascular endothelial growth factor-induced migration of vascular smooth muscle cells in vitro. *Microvasc Res.* 1999;58(2):128–136.

207 Zhao Q, Egashira K, Hiasa K, et al. Essential role of vascular endothelial growth factor and FLT-1 signals in neointimal formation after periadventitial injury. *Arterioscler Thromb Vasc Biol.* 2004;24(12):2284–2289.

208 Ohtani K, Egashira K, Hiasa K, et al. Blockade of vascular endothelial growth factor suppresses experimental restenosis after intraluminal injury by inhibiting recruitment of monocyte lineage cells. *Circulation.* 2004;110(16):2444–2452.

209 van der Hoeven BL, Pires NM, Warda HM, et al. Drug-eluting stents: results, promises and problems. *Int J Cardiol.* 2005;99(1):9–17.

210 Aprelikova O, Pajusola K, Partanen J, et al. FLT4, a novel class III receptor tyrosine kinase in chromosome 5q33-qter. *Cancer Res.* 1992;52(3):746–748.

211 Galland F, Karamysheva A, Mattei MG, et al. Chromosomal localization of FLT4, a novel receptor-type tyrosine kinase gene. *Genomics.* 1992;13(2):475–478.

212 Pajusola K, Aprelikova O, Armstrong E, et al. Two human FLT4 receptor tyrosine kinase isoforms with distinct carboxy terminal tails are produced by alternative processing of primary transcripts. *Oncogene.* 1993;8(11):2931–2937.

213 Orlandini M, Marconcini L, Ferruzzi R, Oliviero S. Identification of a c-fos-induced gene that is related to the platelet-derived growth factor/vascular endothelial growth factor family. *Proc Natl Acad Sci USA*. 1996;93(21):11675–11680.

214 Makinen T, Veikkola T, Mustjoki S, et al. Isolated lymphatic endothelial cells transduce growth, survival and migratory signals via the VEGF-C/D receptor VEGFR-3. *EMBO J*. 2001;20(17):4762–4773.

215 Partanen TA, Arola J, Saaristo A, et al. VEGF-C and VEGF-D expression in neuroendocrine cells and their receptor, VEGFR-3, in fenestrated blood vessels in human tissues. *FASEB J*. 2000; 14(13):2087–2096.

216 Dumont DJ, Jussila L, Taipale J, et al. Cardiovascular failure in mouse embryos deficient in VEGF receptor-3. *Science*. 1998; 282(5390):946–949.

217 Hamada K, Oike Y, Takakura N, et al. VEGF-C signaling pathways through VEGFR2 and VEGFR-3 in vasculoangiogenesis and hematopoiesis. *Blood*. 2000;96(12):3793–3800.

218 Karkkainen MJ, Ferrell RE, Lawrence EC, et al. Missense mutations interfere with VEGFR-3 signalling in primary lymphoedema. *Nat Genet*. 2000;25(2):153–159.

219 Jussila L, Valtola R, Partanen TA, et al. Lymphatic endothelium and Kaposi's sarcoma spindle cells detected by antibodies against the vascular endothelial growth factor receptor-3. *Cancer Res*. 1998;58(8):1599–1604.

220 Breiteneder-Geleff S, Soleiman A, Kowalski H, et al. Angiosarcomas express mixed endothelial phenotypes of blood and lymphatic capillaries: podoplanin as a specific marker for lymphatic endothelium. *Am J Pathol*. 1999;154(2):385–394.

221 Fanburg-Smith JC, Michal M, Partanen TA, et al. Papillary intralymphatic angioendothelioma (PILA): a report of twelve cases of a distinctive vascular tumor with phenotypic features of lymphatic vessels. *Am J Surg Pathol*. 1999;23(9):1004–1010.

222 Galland F, Karamysheva A, Pebusque MJ, et al. The FLT4 gene encodes a transmembrane tyrosine kinase related to the vascular endothelial growth factor receptor. *Oncogene*. 1993;8(5):1233–1240.

223 Armstrong E, Kastury K, Aprelikova O, et al. FLT4 receptor tyrosine kinase gene mapping to chromosome band 5q35 in relation to the t(2;5), t(5;6), and t(3;5) translocations. *Genes Chromosomes Cancer*. 1993;7(3):144–151.

224 Valtola R, Salven P, Heikkila P, et al. VEGFR-3 and its ligand VEGF-C are associated with angiogenesis in breast cancer. *Am J Pathol*. 1999;154(5):1381–1390.

225 Kawakami A, Kitsukawa T, Takagi S, Fujisawa H. Developmentally regulated expression of a cell surface protein, neuropilin, in the mouse nervous system. *J Neurobiol*. 1996;29(1):1–17.

226 Chen H, Chedotal A, He Z, et al. Neuropilin-2, a novel member of the neuropilin family, is a high affinity receptor for the semaphorins Sema E and Sema IV but not Sema III. *Neuron*. 1997;19(3):547–559.

227 He Z, Tessier-Lavigne M. Neuropilin is a receptor for the axonal chemorepellent Semaphorin III. *Cell*. 1997;90(4):739–751.

228 Rossignol M, Gagnon ML, Klagsbrun M. Genomic organization of human neuropilin-1 and neuropilin-2 genes: identification and distribution of splice variants and soluble isoforms. *Genomics*. 2000;70(2):211–222.

229 Kim KJ, Li B, Winer J, et al. Inhibition of vascular endothelial growth factor-induced angiogenesis suppresses tumour growth in vivo. *Nature*. 1993;362(6423):841–844.

230 Kitsukawa T, Shimizu M, Sanbo M, et al. Neuropilin-semaphorin III/D-mediated chemorepulsive signals play a crucial role in peripheral nerve projection in mice. *Neuron*. 1997; 19(5):995–1005.

231 Soker S, Takashima S, Miao HQ, et al. Neuropilin-1 is expressed by endothelial and tumor cells as an isoform-specific receptor for vascular endothelial growth factor. *Cell*. 1998;92(6):735–745.

232 Wang L, Zeng H, Wang P, et al. Neuropilin-1-mediated vascular permeability factor/vascular endothelial growth factor-dependent endothelial cell migration. *J Biol Chem*. 2003; 278(49):48848–48860.

233 Bachelder RE, Crago A, Chung J, et al. Vascular endothelial growth factor is an autocrine survival factor for neuropilin-expressing breast carcinoma cells. *Cancer Res*. 2001;61(15): 5736–5740.

234 Clauss M, Weich H, Breier G, et al. The vascular endothelial growth factor receptor FLT-1 mediates biological activities. Implications for a functional role of placenta growth factor in monocyte activation and chemotaxis. *J Biol Chem*. 1996;271(30):17629–17634.

235 Zeng H, Dvorak HF, Mukhopadhyay D. Vascular permeability factor (VPF)/vascular endothelial growth factor (VEGF) peceptor-1 down-modulates VPF/VEGF receptor-2-mediated endothelial cell proliferation, but not migration, through phosphatidylinositol 3-kinase-dependent pathways. *J Biol Chem*. 2001;276(29):26969–26979.

236 Yoshida A, Anand-Apte B, Zetter BR. Differential endothelial migration and proliferation to basic fibroblast growth factor and vascular endothelial growth factor. *Growth Factors*. 1996;13 (1–2):57–64.

237 Gille H, Kowalski J, Yu L, et al. A repressor sequence in the juxtamembrane domain of FLT-1 (VEGFR-1) constitutively inhibits vascular endothelial growth factor-dependent phosphatidylinositol 3'-kinase activation and endothelial cell migration. *EMBO J*. 2000;19(15):4064–4073.

238 Claesson-Welsh L. Signal transduction by vascular endothelial growth factor receptors. *Biochem Soc Trans*. 2003;31(Pt 1):20–24.

239 Wu LW, Mayo LD, Dunbar JD, et al. VRAP is an adaptor protein that binds KDR, a receptor for vascular endothelial cell growth factor. *J Biol Chem*. 2000;275(9):6059–6062.

240 Warner AJ, Lopez-Dee J, Knight EL, et al. The Shc-related adaptor protein, Sck, forms a complex with the vascular-endothelial-growth-factor receptor KDR in transfected cells. *Biochem J*. 2000;347(Pt 2):501–509.

241 Takahashi T, Yamaguchi S, Chida K, Shibuya M. A single autophosphorylation site on KDR/Flk1 is essential for VEGF-A-dependent activation of PLC-gamma and DNA synthesis in vascular endothelial cells. *EMBO J*. 2001;20(11):2768–2778.

242 English J, Pearson G, Wilsbacher J, et al. New insights into the control of MAP kinase pathways. *Exp Cell Res*. 1999;253(1):255–270.

243 Pedram A, Razandi M, Levin ER. Extracellular signal-regulated protein kinase/Jun kinase cross-talk underlies vascular endothelial cell growth factor-induced endothelial cell proliferation. *J Biol Chem*. 1998;273(41):26722–26728.

244 Kroll J, Waltenberger J. The vascular endothelial growth factor receptor KDR activates multiple signal transduction pathways in porcine aortic endothelial cells. *J Biol Chem*. 1997;272(51):32521–32527.

245 Abedi H, Zachary I. Vascular endothelial growth factor stimulates tyrosine phosphorylation and recruitment to new focal adhesions of focal adhesion kinase and paxillin in endothelial cells. *J Biol Chem.* 1997;272(24):15442–15451.

246 Zeng H, Sanyal S, Mukhopadhyay D. Tyrosine residues 951 and 1059 of vascular endothelial growth factor receptor-2 (KDR) are essential for vascular permeability factor/vascular endothelial growth factor-induced endothelium migration and proliferation, respectively. *J Biol Chem.* 2001;276(35):32714–32719.

247 Zeng H, Zhao D, Yang S, et al. Heterotrimeric G alpha q/G alpha 11 proteins function upstream of vascular endothelial growth factor (VEGF) receptor-2 (KDR) phosphorylation in vascular permeability factor/VEGF signaling. *J Biol Chem.* 2003;278(23):20738–20745.

248 Zeng H, Zhao D, Mukhopadhyay D. KDR stimulates endothelial cell migration through heterotrimeric G protein Gq/11-mediated activation of a small GTPase RhoA. *J Biol Chem.* 2002;277(48):46791–46798.

249 Mukhopadhyay D, Zeng H, Bhattacharya R. Complexity in the vascular permeability factor/vascular endothelial growth factor (VPF/VEGF)-receptors signaling. *Mol Cell Biochem.* 2004;264 (1–2):51–61.

250 Mukhopadhyay D, Zeng H. Involvement of G proteins in vascular permeability factor/vascular endothelial growth factor signaling. *Cold Spring Harbor Symp Quant Biol.* 2002;67:275–283.

251 Gerber HP, McMurtrey A, Kowalski J, et al. Vascular endothelial growth factor regulates endothelial cell survival through the phosphatidylinositol 3'-kinase/Akt signal transduction pathway. Requirement for Flk1/KDR activation. *J Biol Chem.* 1998;273(46):30336–30343.

252 Fulton D, Gratton JP, McCabe TJ, et al. Regulation of endothelium-derived nitric oxide production by the protein kinase Akt. *Nature.* 1999;399(6736):597–601.

253 Kroll J, Waltenberger J. VEGF-A induces expression of eNOS and iNOS in endothelial cells via VEGF receptor-2 (KDR). *Biochem Biophys Res Commun.* 1998;252(3):743–746.

254 Kroll J, Waltenberger J. A novel function of VEGF receptor-2 (KDR): rapid release of nitric oxide in response to VEGF-A stimulation in endothelial cells. *Biochem Biophys Res Commun.* 1999;265(3):636–639.

255 Cross MJ, Dixelius J, Matsumoto T, Claesson-Welsh L. VEGF-receptor signal transduction. *Trends Biochem Sci.* 2003;28(9): 488–494.

256 Rousseau S, Houle F, Landry J, Huot J. p38 MAP kinase activation by vascular endothelial growth factor mediates actin reorganization and cell migration in human endothelial cells. *Oncogene.* 1997;15(18):2169–2177.

257 Qi JH, Claesson-Welsh L. VEGF-induced activation of phosphoinositide 3-kinase is dependent on focal adhesion kinase. *Exp Cell Res.* 2001;263(1):173–182.

258 Endo A, Fukuhara S, Masuda M, et al. Selective inhibition of vascular endothelial growth factor receptor-2 (VEGFR-2) identifies a central role for VEGFR2 in human aortic endothelial cell responses to VEGF. *J Recept Signal Transduct Res.* 2003; 23(2–3):239–254.

259 Davis GE, Bayless KJ, Mavila A. Molecular basis of endothelial cell morphogenesis in three-dimensional extracellular matrices. *Anat Rec.* 2002;268(3):252–275.

260 Davis GE, Camarillo CW. An alpha 2 beta 1 integrin-dependent pinocytic mechanism involving intracellular vacuole formation and coalescence regulates capillary lumen and tube formation in three-dimensional collagen matrix. *Exp Cell Res.* 1996; 224(1):39–51.

261 Bell SE, Mavila A, Salazar R, et al. Differential gene expression during capillary morphogenesis in 3D collagen matrices: regulated expression of genes involved in basement membrane matrix assembly, cell cycle progression, cellular differentiation and G-protein signaling. *J Cell Sci.* 2001;114(Pt 15): 2755–2773.

262 Bayless KJ, Davis GE. The Cdc42 and Rac1 GTPases are required for capillary lumen formation in three-dimensional extracellular matrices. *J Cell Sci.* 2002;115(Pt 6):1123–1136.

33

Function of Hepatocyte Growth Factor and Its Receptor c-Met in Endothelial Cells

Xue Wang, Augustine M. K. Choi, and Stefan W. Ryter

The University of Pittsburgh School of Medicine, Pennsylvania

Hepatocyte growth factor (HGF), also known as scatter factor (SF), exerts a spectrum of biological activities in epithelial and endothelial cells (ECs), including the stimulation of cell growth or motility, and the promotion of matrix invasion. This chapter introduces the molecular characteristics of HGF/SF and its receptor c-Met, with special emphasis given to the functional activities of this system in the endothelium.

HISTORY

HGF was originally identified in 1984, in the serum of partially hepatectomized rats as a mitogenic factor for cultured rat hepatocytes (1), and also has been isolated from rat platelets (2), human serum (3), and rat liver (4). This factor was independently discovered as an embryo fibroblast-derived molecule and termed "scatter factor" for its effects on epithelial cell dissociation and motility (5). HGF and SF were later determined to be identical molecules on the basis of functional, structural, genetic, and immunological characteristics (6–8). In the remainder of the chapter, we will refer to the protein as HGF.

STRUCTURAL AND FUNCTIONAL CHARACTERISTICS OF HGF

The *HGF* gene localizes as a single copy to chromosome 7q11.1–21 in humans, and consists of 18 exons and 17 introns spanning approximately 70 kb (9). HGF originates from a primary transcript of 6-kb as a biologically inactive single-chain precursor molecule of 728 amino acids (pro-HGF) (10–11). The mature form of HGF generated by proteolytic cleavage consists of a 69-kD α-subunit and a 34 kDa β-subunit (12). The α-chain contains an N-terminal hairpin loop and four kringle domains, which are 80-amino acid structures stabilized by intramolecular disulfide bridges that mediate protein–protein interactions (13). The β-chain resembles the catalytic domain of serine proteases but with amino acid substitutions in the active site that render it nonfunctional with respect to protease activity (12–13). Studies on structure–function relationships indicate that the N-terminal hairpin loop and second kringle domains of HGF form a low-affinity binding site for heparin sulfate proteoglycans, whereas the first kringle domain acts as a high-affinity binding site to the HGF receptor, c-Met; thus, these domains are essential for HGF membrane localization and activity. This structural organization highly resembles that of factors implicated in blood coagulation and fibrinolysis, such as plasminogen and prothrombin. The extracellular urokinase-type plasminogen activator (u-PA) complexes with the biologically inactive pro-HGF both in vitro and in vivo, leading to conversion of pro-HGF to mature HGF by the hydrolysis of the Arg494–Val495 peptide bond (14). u-PA cannot act on pro-HGF bearing a mutated cleavage site in vitro. Macrophages provide the primary source of u-PA, which is produced locally in response to environmental stimuli such as tissue injury (14).

In addition to the 6-kb full-length *HGF* transcript, which encodes for the 728-amino acid form of the growth factor, a variety of truncated cDNAs (4.4, 3, 2.2, 1.3 kb) have been cloned (10,15–19). In leukocyte and fibroblast cDNA libraries, a naturally occurring HGF variant arising from alternate splicing of the primary 6-kb transcript was discovered. This transcript encodes a 15-base pair deletion, resulting in a 723-amino acid form, termed dHGF. dHGF bears a five amino acid truncation in the first N-terminal Kringle domain, which alters the solubility and immunological properties of the molecule. dHGF, however, displays similar mitogenic, cytotoxic, and scattering activities as the full-length form of the molecule (15–17). Two additional splice variants of HGF (NK1 and NK2), arising from 1.3- and 2.2-kb transcripts, which contain the N-terminus and one or two intact Kringle domains, respectively, have been described (18–19). Although both variants were initially characterized as antagonists of HGF in vitro, in transgenic models NK1 acts as a partial agonist, whereas NK2 acts as an antagonist for mitogenic activity (19–20).

HGF is produced by a number of cells of mesenchymal origin including fibroblasts, Kupffer cells, and ECs (21), as well as neutrophils, and cancer cells (22). HGF expression increases in various tissues, including liver, lung, and kidney, in response to tissue injury, and is implicated in tissue repair (21). HGF transcripts are expressed in vascular tissue of rats and humans in vitro and in vivo, including vascular smooth muscle cells (VSMCs) and ECs (23). Analysis of liver-cell populations has revealed that stellate cells are the major source of HGF production. In the lung, HGF also is produced by alveolar macrophages and vascular ECs (24).

THE HGF RECEPTOR, c-MET

HGF binds with high affinity to its transmembrane receptor, c-Met, which belongs to a family of heterodimeric tyrosine kinases (25). The c-Met receptor tyrosine kinase (RTK) was originally identified as an oncogene activated in vitro after treatment of a human osteogenic sarcoma cell line with a chemical carcinogen, *N*-methyl-*N*′-nitro-*N*-nitrosoguanidine (26). The human *c-Met* gene localizes to the chromosome 7 band 7q21–q31 and spans more than 120-kb in length, consisting of 21 exons separated by 20 introns. In wild-type cells, the primary c-Met transcript produces a 150-kDa polypeptide that is partially glycosylated to produce a 170-kDa precursor protein. This 170-kDa precursor is further glycosylated and then cleaved into a highly glycosylated extracellular 50-kDa α–chain, and a 140-kDa β-chain consisting of a large extracellular region, a membrane-spanning segment, and an intracellular tyrosine kinase domain. The α- and β-chains are linked together by disulfide bonds (26).

EVOLUTION AND DEVELOPMENT

From limited studies to date, the presence of homologues of c-Met is believed to be restricted to vertebrates, including various species of birds, rodents, reptiles, and fish. No equivalent c-Met homologues were found in worms or flies (27). The HGF/c-Met system appears to serve important functions during embryogenesis, by regulating epithelial and mesenchyme interactions (28). HGF/c-Met signaling has been implicated in the development of various tissues, including the nervous system, muscle, liver, and kidney (27). *HGF*-knockout mice display midterm embryonic lethality due to placental defects (29).

c-MET SIGNAL TRANSDUCTION

c-Met is expressed in a variety of cells, including epithelial cells and ECs. The downstream signaling pathways activated by receptor–ligand interactions have been elucidated largely from studies in epithelial cells. The activation of signal transduction pathways in response to HGF stimulation is mediated in part by autophosphorylation of specific tyrosine residues within the intracellular region of c-Met (Figure 33.1). Autophosphorylation of tyrosines (Tyr^{1234} and Tyr^{1235}) located within the activation loop of the tyrosine kinase domain activates an intrinsic receptor kinase activity (27,30–31). Subsequent autophosphorylation of two C-terminal tyrosine residues (Tyr^{1349} and Tyr^{1356}) of c-Met activates

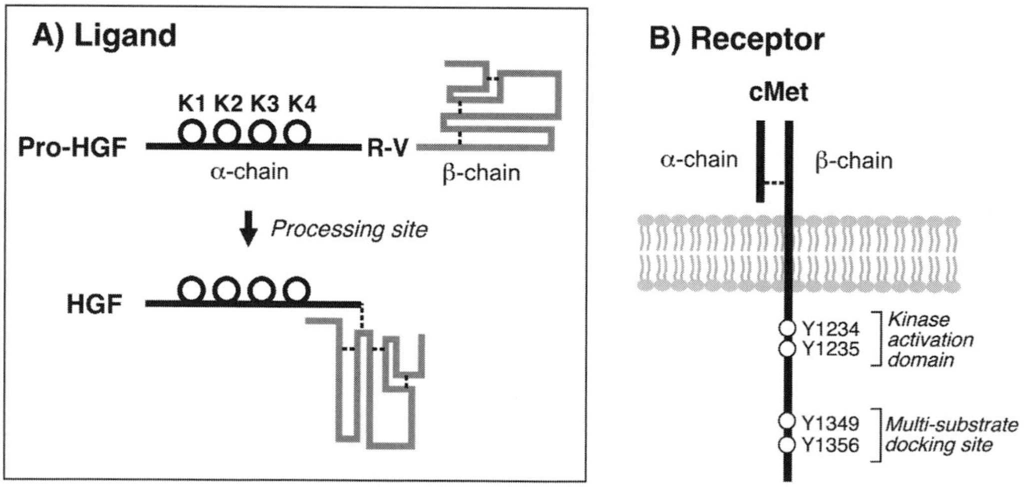

Figure 33.1. Scheme of HGF and receptor c-MET processing. HGF originates from a single-chain polypeptide precursor by proteolytic cleavage of the Arginine 494-Valine 495 peptide bond (R–V). Mature HGF consists of a dilsulfide-linked heterodimer of a 69-kD α-subunit and a 34-kDa β-subunit. The α-chain contains an N-terminal hairpin loop and four kringle domains, designated K1–K4. The HGF receptor c-Met originates from a glycosylated 170-kDa precursor that is cleaved into a highly glycosylated extracellular 50-kDa α-chain, and a 140-kDa β-chain. The α- and β-chains are linked together by disulfide bonds. Following stimulation with the ligand, c-Met is autophosphorylated at specific tyrosines (Tyr^{1234} and Tyr^{1235}) in the tyrosine kinase activation domain and (Tyr^{1349} and Tyr^{1356}) in the multisubstrate docking site.

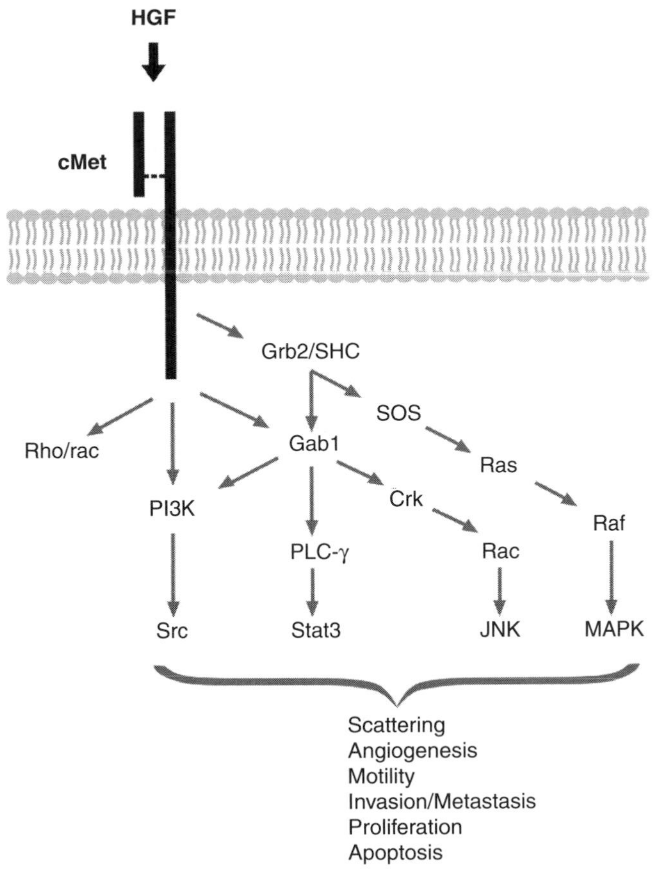

Figure 33.2. Activation of the c-Met receptor. Activation of the c-Met receptor tyrosine kinase by binding of the ligand HGF leads to the recruitment of multiple adaptor molecules and activation of signaling pathways leading to functional effects on EC growth and motility as well as other pleiotropic functions.

a conserved multisubstrate docking site (32). Sequences surrounding Tyr^{1349} and Tyr^{1356} ($Y^{1349}VHVNATY^{1356}VNV$) in mouse c-Met form the docking site that binds Src homology-2 (SH2) domain-, phosphotyrosine binding (PTB) domain-, and the c-Met binding domain (MBD)-containing signal transducers (26,32).

The adapter protein Grb2 consists of an SH2 domain sandwiched between two SH3 domains and is well known for recruiting SOS to activated RTKs to induce Ras- mitogen-activated protein kinase (MAPK) signaling (Figure 33.2). In mouse c-Met, the amino acids surrounding phosphorylated Y^{1356} ($pY^{1356}VNV$) form a consensus SH2 binding site for Grb2 (pYXN) (33). Mutations in c-Met adjacent to the upstream tyrosine Y^{1349} do not significantly affect Grb2 binding; however, mutations of Y^{1356} or N^{1358} disrupt the Met–Grb2 interaction. Although the Y^{1356F} mutation in c-Met disrupts the association of several signal transducers (growth factor receptor-bound protein [Grb]2, phosphoinositide 3-kinase [PI3K], phospholipase C [PLC]-γ, and SHP2), the N^{1358H} mutation specifically disrupts the Met–Grb2 association. Grb2 also may be recruited to c-Met indirectly via the Shc adapter protein. The PTB domain of Shc can associate with the c-Met receptor via phosphotyrosines Y^{1349} and Y^{1356}. Subsequent to c-Met activation, Shc is tyrosine phosphorylated and forms a Grb2 binding site ($pY^{317}VNV$) that is similar to the Grb2 binding site of c-Met ($pY^{1356}VNV$) (33). Therefore, activation of the Ras pathway in response to c-Met stimulation may be the result of the Shc–Grb2–SOS interaction.

Grb2 also has been implicated in recruiting the large adapter protein Grb2-associated binding protein-1 (Gab1) to the c-Met signaling complex. c-Met mutants that contain both Y^{1349} and Y^{1356} mutations are unable to associate with Gab1, whereas c-Met Y^{1356F} mutants, in which Grb2 binding is eliminated, show significantly decreased Met–Gab1 association (27). Therefore, whereas Gab1 may preferentially bind c-Met via a Grb2 linkage, the ability of Gab1 to bind c-Met both directly using Y^{1349}, and indirectly via a Grb2 linkage at Y^{1356} may allow for a more specific and a higher-affinity interaction. After HGF stimulation, Gab1 becomes strongly tyrosine phosphorylated. Overexpression of Gab1 is sufficient to induce branching morphogenesis in some cell types. Gab1 binds to various signaling molecules including PI3K, PLC-γ, and the SHP2 phosphatase (27). Although the p85 regulatory subunit of PI3K can associate directly with c-Met via the multisubstrate binding site, more PI3K activity coprecipitates with Gab1 than coprecipitates with c-Met, suggesting that PI3K, like PLC-γ, associates primarily with Gab1 (27).

The adapter proteins of the Crk family consist of SH2 and SH3 domain(s). Gab1/c-Met contains multiple Tyr-X-X-Pro (YXXP) sequences that represent potential Crk family member SH2-binding sites. Both Crk and the related Crk-like (CRKL) proteins bind to phosphorylated Gab1 and thus add an array of signal transducers that become activated on HGF stimulation. Other downstream effectors of Crk include the c-jun N-terminal kinase (JNK). Crk family members can signal through Rac to activate JNK. Dominant-negative forms of Rac block c-Met-mediated JNK activation and raise the possibility of a Met-Gab1-Crk-Rac-JNK signaling pathway (34).

Two other important signal transducers, the Src kinase and the Stat3 transcription factor, also associate with c-Met following receptor activation (35). Src can associate directly with c-Met at Y^{1349} and Y^{1356}. Similar to other receptor systems, Src activation is important for HGF-mediated cell migration and cell transformation (35). In response to HGF/c-Met stimulation, Stat3 activation and subsequent nuclear translocation/gene transcription is required for branching morphogenesis (35).

THE EFFECTS OF HGF/c-MET SIGNALING ON ENDOTHELIAL CELLS

Many studies on HGF signaling have been performed in epithelial cells, which express the HGF receptor c-Met. However, recent studies indicate that HGF/c-Met also plays an important role in vascular cells. mRNAs corresponding to both HGF and c-Met have been detected in ECs and VSMCs of rats, as assessed by reverse transcriptase polymerase chain reaction (RT-PCR) (23). *HGF* and *c-Met* mRNA also were detected in human aortic ECs (HAECs) and human VSMCs. Cultured ECs and VSMCs of both rat and human produce and secrete immunoreactive HGF (23).

Scattering and Motility

HGF can promote the "scattering" of a variety of adherent cells, including epithelial cells and ECs, as defined by spatial separation or dissociation and movement in three-dimensional support matrices (36–38). By promoting scattering, HGF activates a cellular differentiation program called "branching morphogenesis," whereby cells differentiate into branched, tube-like structures (39).

The process of cell scattering can be divided into three phases, namely cell spreading, cell–cell dissociation, and cell migration. Therefore, in order for epithelial cells to "scatter," the attenuation or dissolution of cell–cell adhesions is required. Under normal conditions, the assembly and maintenance of intercellular junctions is tightly regulated. Disassembly of these junctions occurs during certain physiological and pathological conditions, such as normal development and tumor cell invasion and metastasis. Cell motility plays a key role in both normal physiology and various disease processes. Many studies have shown that HGF/c-Met signaling increases the motility of ECs. In bovine retinal microvascular ECs, HGF-induced cell scattering and motility and growth can be blocked by inhibiting protein kinase C (PKC), MAPK, and PI3K (40).

Angiogenesis

The induction of angiogenesis (formation of new blood vessels) may be an important mechanism that permits tumor cell proliferation and eventually metastasis. Tumor-derived angiogenic factors lead to the growth of new capillaries with fragmented basement membranes, allowing tumor cells to more rapidly enter the circulation. Transfection of breast cancer cell lines with *HGF* cDNA increased tumor outgrowth when implanted into host animals. The increased growth rate of the HGF-transfected tumor cells was attributable to increased tumor angiogenesis (41).

HGF can act as a potent angiogenic molecule by acting directly on vascular ECs in vitro (37,42). HGF stimulation of vascular ECs promotes migration, proliferation, protease production, and invasion. HGF also promotes the expression of angiogenic factors such as vascular endothelial growth factor (VEGF) (43) and decreases the transcription and protein expression of vascular endothelial (VE)-cadherin in human ECs (44). Initial experiments on the angiogenic activity of HGF in vivo demonstrated that the implantation of rabbit corneas with purified HGF promoted neovascularization in vivo (38,45). In the rat diabetic ischemic hind limb model and in rat infarcted myocardium, HGF gene therapy improved blood flow by promoting angiogenesis (46,47). In such models, the proangiogenic effect of HGF was associated with activation of the transcription factor ETS (47).

Proliferation and Apoptosis

Cell proliferation plays an important role in a variety of physiological processes, including embryogenesis, growth, and wound healing. HGF/c-Met signaling can affect both proliferation and apoptosis, depending on the cell context. HGF stimulates vascular EC growth in vitro and in vivo (48). The application of recombinant HGF reduced neointimal hyperplasia in an in vivo balloon injury model of the rabbit iliac artery, by promoting re-endothelialization (49). In cultured murine lung ECs, HGF inhibited hypoxia/reoxygenation-induced apoptosis by blocking both death receptor- (extrinsic) and mitochondria- (intrinsic) dependent pathways (50). Additionally, HGF exerts antiapoptotic action in ECs by modulating several antiapoptotic factors, including Flip and Bcl-X_L (50).

Invasion and Metastasis

Tumor cell metastasis requires neoplastic cells to interact with the extracellular matrix (ECM) and to break the barrier to the circulation. The critical event of invasion is the interaction of the neoplastic cell with the basement membrane, which is mainly composed of type IV collagen, laminin, and heparin sulfate proteoglycan. ECM degradation is a highly dynamic event requiring the production of proteinases, namely plasminogen activator and matrix metalloproteinase (MMP). HGF-mediated activation of c-Met has been strongly implicated in the promotion of invasion and metastasis through MMP activity in human ECs (51). An HGF-stimulated pathway involving MAPK signaling can upregulate the expression of the serine protease u-PA and its receptor (uPAR) (52), resulting in an increase of u-PA at the cell surface. Certain components of the ECM can be directly degraded by u-PA. More importantly, u-PA cleaves plasminogen into the broader-specificity protease plasmin, which is able to efficiently degrade several ECM and basement membrane components.

Anti-inflammatory Effects

HGF inhibited the EC expression of the adhesion molecule E-selectin in human EC culture and in rats in response to tumor necrosis factor (TNF)-α treatment (53). The anti-inflammatory effect of HGF in a kidney model was associated with reduced E-selectin–mediated macrophage adhesion to the endothelium and reduced renal macrophage sequestration after TNF-α challenge (53).

THERAPEUTIC IMPLICATIONS

In the context of tumorigenesis, downregulation c-Met has been implicated as a possible therapeutic strategy to reduce tumor cell migration and invasion (30,54–56). This may be achieved through pharmacological application of small-molecule inhibitor compounds, or antagonists such as the artificial 4-Kringle HGF mutant NK4 (55) or, alternatively, by antisense/hammerhead ribozyme strategies, as illustrated by Davies and colleagues (56).

On the other hand, HGF can be beneficial in limiting tissue and organ damage in a number of models in which inflammatory and/or apoptotic processes are implicated in the injury. For example, HGF inhibits inflammation in various animal models of kidney disease (53). The proliferative effects of HGF on endothelium may be exploited for the treatment of vascular diseases (46,47,49). For example, application of recombinant HGF, which inhibits neointimal hyperplasia after balloon injury, may represent a possible therapeutic strategy for limiting restenosis after vascular injury (49). The proangiogenic effects of HGF in gene therapy approaches have potential therapeutic application in diabetes and myocardial injury (46,47).

KEY POINTS

- HGF, through its receptor c-Met, acts as a potent mitogen, motogen, and morphogen in a number of cell types, and plays important roles in EC growth, development, and angiogenesis.
- The mechanism(s) by which HGF/c-Met affect cellular proliferation, apoptosis, and cancer invasion/metastasis remain unclear.
- HGF, by activating its receptor tyrosine kinase c-Met, activates signaling pathways composed of multiple adaptor and effector molecules, including RAS-MAPK pathway, PI3K, and PLC-γ.
- The elucidation of critical molecule(s) in HGF/c-Met-signaling pathways is important for understanding the biological function of HGF, and may lead to the development of antiapoptotic, anti-inflammatory, and antiproliferative strategies.
- Modulation of the HGF/c-Met system may have potential therapeutic implications in limiting tumor growth, in organ regeneration, and in the treatment of vascular ischemic diseases.

REFERENCES

1 Michalopoulus GK, Houck KA, Dolan ML, Luetteke NC. Control of hepatocyte replication by two serum factors. *Cancer Res.* 1984;44:4414–4419.

2 Nakamura T, Teramoto H, Ichihara A. Purification and characterization of a growth factor from rat platelets for mature parenchymal hepatocytes in primary cultures. *Proc Natl Acad Sci USA.* 1986;83:6489–6493.

3 Zarnegar R, Michalopoulos GK. Purification and biological characterization of human hepatopoietin A, a polypeptide growth factor for hepatocytes. *Cancer Res.* 1989;49:3314–3320.

4 Asami O, Ihara I, Shimidzu N, et al. Purification and characterization of hepatocyte growth factor from injured liver of carbon tetrachloride-treated rats. *J Biochem.* 1991;109:8–13.

5 Stoker M, Gherardi E, Perryman M, Grey J. Scatter factor is a fibroblast-derived modulator of epithelial cell motility. *Nature.* 1987;327:239–242.

6 Weidner KM, Behrens J, Vandekerckhove J, Birchmeier W. Scatter factor: molecular characteristics and effect on the invasiveness of epithelial cells. *J Cell Biol.* 1990;111:2097–2108.

7 Furlong RA, Takehara T, Taylor WG, et al. Comparison of biological and immunochemical properties indicate that scatter factor and hepatocyte growth factor are indistinguishable. *J Cell Sci.* 1991;100:173–177.

8 Naldini L, Weidner KM, Vigna E, et al. Scatter factor and hepatocyte growth factor are indistinguishable ligands for the MET receptor. *EMBO J.* 1991;10:2867–2878.

9 Seki T, Hagiya M, Shimonishi M, et al. Organization of the human hepatocyte growth factor-encoding gene. *Gene.* 1991; 102:213–219.

10 Nakamura T, Nishizawa T, Hagiya M, et al. Molecular cloning and expression of human hepatocyte growth factor. *Nature.* 1989; 342:440–443.

11 Bardelli A, Ponzetto C, Comoglio PM. Identification of functional domains in the hepatocyte growth factor and its receptor by molecular engineering. *J Biotechnol.* 1994;37:109–122.

12 Matsumoto K, Nakamura T. Hepatocyte growth factor: molecular structure and implications for a central role in liver regeneration. *J Gastroenterol Hepatol.* 1991;6:509–519.

13 Cioce V, Csaky KG, Chan AM, et al. Hepatocyte growth factor (HGF)/NK1 is a naturally occurring HGF/scatter factor variant with partial agonist/antagonist activity. *J Biol Chem.* 1996;271: 13110–13115.

14 Naldini L, Vigna E, Bardelli A, et al. Biological activation of pro-HGF (hepatocyte growth factor) by urokinase is controlled by a stoichiometric reaction. *J Biol Chem.* 1995;270:603–611.

15 Shima N, Tsuda E, Goto M, et al. Hepatocyte growth factor and its variant with a deletion of five amino acids are distinguishable in their biological activity and tertiary structure. *Biochem Biophys Res Commun.* 1994;200:808–815.

16 Rubin JS, Chan AML, Bottaro DP, et al. A broad-spectrum human lung fibroblast-derived mitogen is a variant of hepatocyte growth factor. *Proc Natl Acad Sci USA.* 1991;88:415–419.

17 Seki T, Ihara I, Sugimura A, et al. Isolation and expression of cDNA for different forms of hepatocyte growth factor from human leukocyte. *Biochem Biophys Res Commun.* 1990;172:321–327.

18 Chan AM, Rubin JS, Bottaro DP, et al. Identification of a competitive HGF antagonist encoded by an alternative transcript. *Science.* 1991;254:1382–1385.

19 Jakubczak JL, LaRochelle WJ, Merlino G. NK1, a natural splice variant of hepatocyte growth factor/scatter factor, is a partial agonist in vivo. *Mol Cell Biochem.* 1998;18:1275–1283.

20 Otsuka T, Jakubczak J, Vieira W, et al. Disassociation of met-mediated biological responses in vivo: the natural hepatocyte growth factor/scatter factor splice variant NK2 antagonizes growth but facilitates metastasis. *Mol Cell Biol.* 2000;20:2055–2065.

21 Noji S, Tashiro K, Koyama E, et al. Expression of hepatocyte growth factor gene in endothelial and Kupffer cells of damaged rat livers, as revealed by in situ hybridization. *Biochem Biophys Res Commun.* 1990;173:42–47.

22 Jiang GJ, Martin TA, Parr C, et al. Hepatocyte growth factor, its receptor, and their potential value in cancer therapies. *Crit Rev Oncol Hematol.* 2005;53:35–69.

23 Nakamura Y, Morishita R, Higaki J, et al. Expression of local hepatocyte growth factor system in vascular tissues. *Biochem Biophys Res Commun.* 1995;215:483–488.

24 Yanagita K, Nagaike M, Ishibashi H, et al. Lung may have an endocrine function producing hepatocyte growth factor in response to injury of distal organs. *Biochem Biophys Res Commun.* 1992;182:802–809.

25 Boccaccio C, Gaudino G, Gambarotta G, et al. Hepatocyte growth factor (HGF) receptor expression is inducible and is part of the delayed-early response to HGF. *J Biol Chem.* 1994;269:12846–12851.

26 Gonzatti-Haces M, Park M, Dean M, et al. The human met oncogene is a member of the tyrosine kinase family. *Princess Takamatsu Symposium.* 1986;17:221–232.

27 Furge KA, Zhang YW, Vande Woude GF. Met receptor tyrosine kinase: enhanced signaling through adapter proteins. *Oncogene.* 2000;19:5582–5589.

28 Tsarfaty I, Rong S, Resau JH, et al. The Met proto-oncogene mesenchymal to epithelial cell conversion. *Science.* 1994;263:98–101.

29 Uehara Y, Mori C, Noda T, et al. Rescue of embryonic lethality in hepatocyte growth factor/scatter factor knockout mice. *Genesis.* 2000;27:99–103.

30 Ma PC, Maulik G, Christensen J, Salgia R. c-Met: structure, functions and potential for therapeutic inhibition. *Cancer Metastasis Rev.* 2003;22:309–325.

31 Naldini L, Vigna E, Ferracini R, et al. The tyrosine kinase encoded by the MET proto-oncogene is activated by autophosphorylation. *Mol Cell Biol.* 1991;11:1793–1803.

32 Ponzetto C, Bardelli A, Zhen Z, et al. A multifunctional docking site mediates signaling and transformation by the hepatocyte growth factor/scatter factor receptor family. *Cell.* 1994;77:261–271.

33 Maffe A, Comoglio PM. HGF controls branched morphogenesis in tubular glands. *Eur J Morphol.* 1998;36 Suppl:74–81.

34 Garcia-Guzman M, Dolfi F, Zeh K, Vuori K. Met-induced JNK activation is mediated by the adapter protein Crk and correlates with the Gab1-Crk signaling complex formation. *Oncogene.* 1999;18:7775–7786.

35 Boccaccio C, Ando M, Tamagnone L, et al. Induction of epithelial tubules by growth factor HGF depends on the STAT pathway. *Nature.* 1998;391:285–288.

36 Montesano R, Matsumoto K, Nakamura T, Orci L. Identification of a fibroblast-derived epithelial morphogen as hepatocyte growth factor. *Cell.* 1986;67:901–908.

37 Morimoto A, Okamura K, Hamanaka R, et al. Hepatocyte growth factor modulates migration and proliferation of human microvascular ECs in culture. *Biochem Biophys Res Commun.* 1991;179:1042–1049.

38 Bussolino F, Di Renzo MF, Ziche M, et al. Hepatocyte growth factor is a potent angiogenic factor which stimulates EC motility and growth. *J Cell Biol.* 1992;119:629–641.

39 Brinkmann V, Foroutan H, Sachs M, et al. Hepatocyte growth factor/scatter factor induces a variety of tissue-specific morphogenic programs in epithelial cells. *J Cell Biol.* 1995;131:1573–1586.

40 Cai W, Rook SL, Jiang ZY, et al. Mechanisms of hepatocyte growth factor-induced retinal EC migration and growth. *Invest Ophthalmol Vis Sci.* 2000;41:1885–1893.

41 Rosen EM, Lamszus K, Laterra J, et al. HGF/SF in angiogenesis. *Ciba Foundation Symposia.* 1997;212:215–226.

42 Ding S, Merkulova-Rainon T, Han ZC, Tobelem G. HGF receptor up-regulation contributes to the angiogenic phenotype of human ECs and promotes angiogenesis in vitro. *Blood.* 2003;101:4816–4822.

43 Van Belle E, Witzenbichler B, Chen D, et al. Potentiated angiogenic effect of scatter factor/hepatocyte growth factor via induction of vascular endothelial growth factor: the case for paracrine amplification of angiogenesis. *Circulation.* 1998;97:381–390.

44 Martin TA, Mansel R, Jiang WG. Hepatocyte growth factor modulates vascular endothelial-cadherin expression in human ECs. *Clin Cancer Res.* 2001;7:734–737.

45 Rosen EM, Grant DS, Kleinman HK, et al. Scatter factor (hepatocyte growth factor) is a potent angiogenesis factor in vivo. *Symp Soc Exp Biol.* 1993;47:227–234.

46 Taniyama Y, Morishita R, Hiraoka K, et al. Therapeutic angiogenesis induced by human hepatocyte growth factor gene in rat diabetic hind limb ischemia model: molecular mechanisms of delayed angiogenesis in diabetes. *Circulation.* 2001;104:2344–2350.

47 Aoki M, Morishita R, Taniyama Y, et al. Angiogenesis induced by hepatocyte growth factor in non-infarcted myocardium and infarcted myocardium: up-regulation of essential transcription factor for angiogenesis, ets. *Gene Ther.* 2000;7:417–427.

48 Mori T, Shimizu A, Masuda Y, et al. Hepatocyte growth factor-stimulating EC growth and accelerating glomerular capillary repair in experimental progressive glomerulonephritis. *Nephron Exp Nephrol.* 2003;94:E44–E54.

49 Yasuda S, Noguchi T, Gohda M, et al. Single low-dose administration of human recombinant hepatocyte growth factor attenuates intimal hyperplasia in a balloon-injured rabbit iliac artery model. *Circulation.* 2000;101:2546–2549.

50 Wang X, Zhou Y, Kim H, et al. Hepatocyte growth factor protects against hypoxia/reoxygenation-induced apoptosis in ECs. *J Biol Chem.* 2004;279:5237–5243.

51 Wang H, Keiser JA. Hepatocyte growth factor enhances MMP activity in human ECs. *Biochem Biophys Res Commun.* 2000;272: 900–905.

52 Lee KH, Hyun MS, Kim JR. Growth factor-dependent activation of the MAPK pathway in human pancreatic cancer: MEK/ERK and p38 MAP kinase interaction in uPA synthesis. *Clin Exp Metastasis.* 2003;20:499–505.

53 Gong R, Rifai A, Dworkin LD. Hepatocyte growth factor suppresses acute renal inflammation by inhibition of endothelial E-selectin. *Kidney Int.* 2006;69:1166–1174.

54 Maulik G, Shrikhande A, Kijima T, et al. Role of hepatocyte growth factor receptor, c-Met, in oncogenesis and potential for therapeutic inhibition. *Cytokine Growth Factor Rev.* 2002;13:41–59.

55 Parr C, Hiscox S, Nakamura T, et al. Nk4, a new HGF/SF variant, is an antagonist to the influence of HGF/SF on the motility and invasion of colon cancer cells. *Int J Cancer.* 2000;85:563–570.

56 Davies G, Watkins G, Mason MD, Jiang WG. Targeting the HGF/SF receptor c-met using a hammerhead ribozyme transgene reduces in vitro invasion and migration in prostate cancer cells. *Prostate.* 2004;60:317–324.

34

Fibroblast Growth Factors

Masahiro Murakami, Arye Elfenbein, and Michael Simons

Dartmouth Medical School, Hanover, New Hampshire

HISTORICAL BACKGROUND

The discovery of fibroblast growth factors (FGFs) in the 1970s arose from observations that a protein with a basic isoelectric point of 9.6, derived from bovine brain and pituitary gland, was able to induce morphological changes in BALB/c 3T3 fibroblast cells (1). Early limitations in purifying the growth factor led to the belief that the FGF derived from bovine brains was a component of myelin basic protein (2). This claim was countered by several reports of nonassociation between the two proteins (3,4), yet the definitive purification did not occur until 1984, when heparin Sepharose chromatography was used to effectively isolate FGFs (5). An acidic fibroblast growth factor (isoelectric point 5.6) was identified several years after the discovery of basic fibroblast growth factor (6–8).

In 1991, when seven FGFs had been identified, a nomenclature committee renamed aFGF and bFGF as FGF1 and FGF2, respectively (9).

EVOLUTION

In *Caenorhabditis elegans*, two FGF genes (*egg laying defective 17* [*egl-17*] and *lethal-75* [*let-75*]), and one FGF receptor (FGFR) gene (*egl-15*) are identified. The fruit fly, *Drosophila melanogaster*, has three FGF genes, *branchless*, *pyramus*, and *thisbe* and two FGFR genes, *breathless* and *heartless*. Early in evolution, *Fgfs* have undergone a gene duplication expanding from two or three to six genes; thereafter, a second gene duplication occurred during evolution of the early vertebrates, resulting in 22 FGF genes in mice and humans (FGF1–23) (human *FGF15* and mouse *Fgf19* have not been identified in the genome, and mouse *Fgf15* and human *FGF19* are now thought to be orthologous genes [10]). In mice and humans, FGFs range in molecular weight from 17 to 34 kDa and share a conserved amino acid sequence derived from evolutionarily related FGF genes.

In contrast to FGF genes, the *FGFR* gene family has expanded only later in evolution. However, the acquisition of alternative splicing in *FGFRs* has greatly increased their functional diversity, enabling the FGF system's involvement in a variety of biological processes (11). Phylogenetic analyses suggest that the FGF family can be arranged into seven subfamilies (Table 34-1). FGFs within a subfamily have similar receptor-binding properties, tissue expression patterns, and target preferences, possibly providing FGFs with wide functional redundancy.

OVERVIEW OF FIBROBLAST GROWTH FACTORS

FGFs and their receptors comprise a multifunctional system, the biological effects of which encompass diverse physiological and pathological processes including embryonic development, angiogenesis, wound healing, tissue regeneration, and tumor development (Figure 34.1) (12–14). Most members of the FGF family have an amino-terminal signal peptide and are secreted from cells. Although FGFs 1, 2, 9, 16, and 20 lack signal sequences for processing through the endoplasmic reticulum–Golgi secretory pathway, they are found in the bloodstream, extracellular space, and at the cell surface, implying the existence of a noncanonical secretion pathway. FGF2-induced cell motility is efficiently blocked by addition of neutralizing FGF2 antibody in the culture medium, implying that FGF2 acts through an autocrine, not intracrine, mechanism to promote cell motility (15).

Some FGFs genes (*FGF2* and *FGF3*) transcribe high-molecular-weight isoforms using upstream, in-frame CUG start codons to create FGFs containing an amino-terminal nuclear localization signal (NLS). In the case of FGF1, a molecule that has only one mRNA form, the presence of an amino-terminal NLS ensures its nuclear association, and this sequence is thought to be required for its mitogenic effects (16); a putative second, carboxy-terminal NLS sequence also has been characterized recently for this molecule (17). The FGFs, originally presumed to be exclusive cell-surface ligands, now are understood to have a wider impact on cell signaling events from the cell surface to the nucleus.

Studies detailing the presence and activity of FGFs in the nucleus have revealed that FGFs endocytosed from the

Table 34-1: FGF Family

	Receptor	*Expression/Target*	*Function*
FGF1 subfamily			
FGF1 Acidic FGF ECGF HBGF-1	All FGF receptors	Mesenchymal, mesodermal, neuroectodermal, EC	Angiogenesis, wound healing, organogenesis
FGF2 Basic FGF HBGF-2	1b, 1c, 2c, 3c, 4	Mesenchymal, mesodermal, neuroectodermal, endothelial, epithelial cell	Angiogenesis, vasculogenesis, wound healing hematopoiesis, neurosurvival, limb and nervous system development, tumor development
FGF4 subfamily			
FGF4	1c, 2c, 3c, 4	Limb bud	Bone morphogenesis and limb development, angiogenesis
FGF5	1c, 2c	Hair follicule, primitive ectoderm	Hair growth inhibition
FGF6	1c, 2c, 4	Myogenic lineage	Muscle regeneration or differentiation
FGF7 subfamily			
FGF3	2b	Epithelial cell	Neoplastic transformation and tumor progression, inner ear formation
FGF7 KGF	2b	Keratinocyte Epithelial cell	Epithelial cell growth, morphogenesis of epithelium, reepithelialization of wounds, hair development, and early lung organogenesis.
FGF10	2b	Keratinizing epidermal cell	Embryonic epidermal morphogenesis, brain development, lung morphogenesis, and initiation of limb bud formation; wound healing
FGF22		Hair follicle-keratinocyte, cerebellum	Hair development (?), presynaptic differentiation
FGF9 subfamily			
FGF9	1c, 2c, 3b, 3c, 4	Neuron/glial cell, astrocyte	Motor neuron survival, testicular embryogenesis, lung development
FGF16	2c, 3c	Embryonic brown adipose tissue	Proliferation of embryonic brown adipose tissue
FGF20	1c, 2c, 3c	Normal brain, cerebellum	Survival of midbrain dopaminergic neurons
FGF8 subfamily			
FGF8	(1c), 2c, 3c, 4	Testes and ovaries, adult hindbrain, primitive streak	Midbrain and limb development, organogenesis, embryo gastrulation, and left–right axis determination; tumor growth and angiogenesis
FGF17	1c, 2c, 3c, 4	Cerebellum forebrain, mid-hindbrain junction, developing skeleton and arteries	Central nervous system, bone, and vascular development
FGF18	1c, 2c, 3c, 4	Developing bone	Neurite outgrowth proliferation of the liver and small intestine; skeletal development
FGF11 subfamily*			
FGF11		Neurons	Nervous system development
FGF12		Neurons Developing heart and connective tissue	
FGF13		Neurons, developing heart	
FGF14		Neurons	
FGF19 subfamily			
FGF19	4	Fetal but not adult brain tissue, developing pharyngeal arches	Initiation of inner ear development, cardiac outflow tract alignment
FGF21		Liver	Glucose and triglyceride metabolism
FGF23	3c	Brain, thymus/renal proximal epithelial cell	Inhibits renal tubular phosphate transport

*FGFs 11–14 (FHF, Fibroblast growth factor homologous factors) lack signal peptides, remain intracellular, and function within cells in a receptor-independent manner. FHFs are known to bind to intracellular domains of voltage-gated sodium channels (VGSCs) and to a neuronal MAP kinase scaffold protein, islet-brain-2 (IB2) (106).

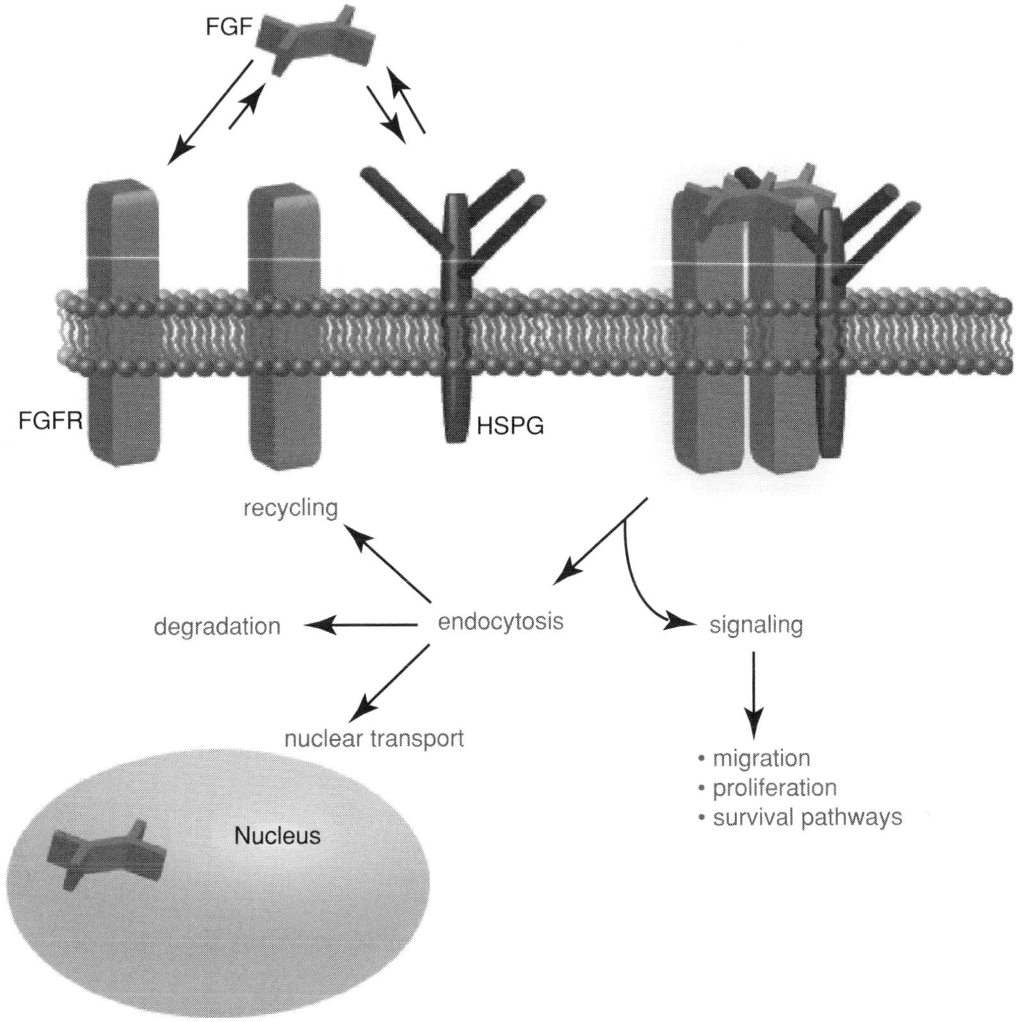

Figure 34.1. Schemata of FGF signaling. Active FGFs in the extracellular matrix exist in the monomeric form. They bind with high affinity to their tyrosine kinase receptors (FGFR1–4) and with lower affinity to heparan sulfate proteoglycans (HSPGs), such as syndecans and glypicans. FGF binding to FGFR and HSPG forms a ternary complex that involves the dimerization of FGFR and induces several parallel signaling events primarily through the cytoplasmic tails of FGFR, although certain HSPGs (including syndecan-4) also are able to signal. The complex is endocytosed, and evidence exists for subsequent recycling, degradation, or translocation to the nucleus. The differential effects of nuclear and extracellular FGFs remain incompletely elucidated, as do the roles of nuclear versus cytoplasmic FGF signals in cell proliferation, survival, and migration.

extracellular space, including those lacking a NLS, partially accumulate in the nucleus (18). Whether that results in signaling has not been determined to date. Early insights into the intracellular effects of FGFs were provided by observations that high-molecular-weight FGF2 enables cell survival under low serum conditions irrespective of a functional FGF receptor (19,20). This finding lends credence to the notion that FGFs can signal via an intracrine pathway (perhaps promoting cell survival) that is independent of classically defined FGF pathways.

More recently, low-molecular-weight FGF2 has been shown by nuclear fractionation to activate casein kinase-2, a nuclear protein that modulates cell proliferation, in the cytoplasm and in the nucleus (21). FGF accumulation in the nucleus also appears to be regulated in a cell cycle–specific manner (22,23). Finally, the nuclear effects of FGFs extend beyond the vascular system, exemplified by the recent observation that tumor cells expressing high-molecular-weight FGF2 have a metastatic advantage over those expressing low-molecular-weight FGF2 (24).

FGFs are broad-spectrum mitogens and stimulate various cellular functions including migration, proliferation, and differentiation (25,26). Some FGFs, especially FGF2, are recognized as cell survival factors that block apoptosis in many cell types, including endothelial cells (ECs) (27). Although it is reported that human umbilical vein ECs express FGF1, 2, 5, 7, 8, 11, 12, 16, and 18 in vitro (28), the angiogenic potential of most FGFs is not well characterized. ECs are one of the main target cell types for FGFs, especially for "angiogenic FGFs" (FGF1, 2, and 5) that have been shown to promote new blood formation in vivo and to bind FGFR1IIIc.

The expression patterns of FGFs vary considerably, ranging from nearly ubiquitous (FGF2) to highly restricted within particular cell subsets at specific developmental stages (FGF3, 4, 8, 17, and 19). FGF3, for example, plays a role in mesoderm induction, whereas FGF8 is associated with limb, central nervous system, and cardiac outflow development.

In pathological conditions, the expression of some FGFs is upregulated in inflammatory and angiogenic tissues, including healing wounds and tumors. The cellular sources of FGFs in these settings are thought to be mainly tissue macrophages, ECs, stromal cells, and tumor cells (13).

Several FGFs have been identified as oncogenes by their transforming capacity, and involvement of the FGF system is implicated in the tumorigenesis of several common human cancers including breast, prostate, skin, urothelial, and hematological malignancies (29). In murine mammary cancer, some FGFs are associated with proviral insertions of the mouse mammary tumor virus (MMTV) (30). A detailed analysis of proviral integration sites identified virally activated proto-oncogenes that include members of the *Wnt* and *Notch* gene family and three members of the *Fgf* family: *Int-2/Fgf-3, hst-1/Fgf-4*, and *Fgf-8*. Expression of FGF4 (HST) has been reported in human testicular germ cell tumors, including seminomas and embryonal carcinomas (31).

FIBROBLAST GROWTH FACTOR SECRETION AND ENDOCYTOSIS

Beyond their role in FGF signaling, heparan sulfate proteoglycans (HSPGs) in particular have been implicated in the process of FGF endocytosis and subsequent nuclear localization (18). Of the HSPGs, syndecans (32) and perlecans (33) have been individually implicated as critical mediators of FGF endocytosis via a ternary complex involving the FGF receptor. Endocytosis of FGF in association with its tyrosine kinase receptors also has been shown to be a requisite step in nuclear translocation. This has been reported to occur in conjunction with the concomitant endocytosis of E-cadherin, a protein most concentrated in the adherent junctions of epithelial cells (34).

Our understanding of FGF signaling complexes and their connection to FGF endocytosis remains relatively tenuous, particularly in the distinction between the effects of FGFs in the nucleus and their classically defined role in tyrosine kinase activation. In addition, FGFs have been shown to localize to the lipid rafts (35) following internalization. It is, therefore, possible that lipid rafts serve to concentrate FGFs and enhance signaling.

FGF release into the extracellular milieu remains an incompletely characterized process, although it has become apparent that FGF secretion likely involves nonclassical routes of export. Of the FGF family, FGF1 and FGF2 secretion has been studied most thoroughly, yet their respective paths from the cellular cytosol to the outside are entirely unrelated. For example, the release of FGF1 can be induced by particular conditions of stress, including heat shock, which does not affect the secretion of FGF2 (36,37). Likewise, FGF2 secretion is diminished by inhibitors of Na^+/K^+ ATPase (38), whereas FGF1 release is not similarly affected (39). FGF1 is secreted as a covalently linked homodimer, a state that confers a low heparan-binding affinity on the molecule (40). Consequently, it has been hypothesized that FGF dimer secretion requires an additional step to activate the growth factor in the extracellular space.

Although both FGF1 and FGF2 are presumed to be secreted via translocation across the cell membrane (41,42), plasma membrane–derived exovesicles containing FGF2 also have been isolated (43). It is therefore unclear whether FGFs are released via specific transporters or through vesicular shedding, although in the case of FGF1, the formation of a protein release complex under the plasma membrane argues against intravesicular involvement (41). As in the case of FGF endocytosis and signaling, cell-surface HSPGs have been associated with FGFs undergoing secretion, and they appear to be critical in establishing the complexes required for the FGF release (44).

FIBROBLAST GROWTH FACTOR INTERACTION WITH FIBROBLAST GROWTH FACTOR RECEPTORS

FGFs produce their biological effects in target cells in part by signaling via cell-surface tyrosine kinase receptors and in part – at least in the case of FGF2, but probably other FGFs as well – via syndecan-4, a membrane-spanning HSPG (45,46). Four known high-affinity receptors exist for FGFs, FGFR1 through FGFR4. These share 55% to 72% amino acid sequence identity. FGF receptors are transmembrane proteins containing two or three extracellular immunoglobulin (Ig)-like domains, a heparin-binding motif, a transmembrane domain, and intracellular tyrosine kinase domains.

Through alternative splicing, FGFR can produce numerous isoforms. Although the functional significance of every splice variant is not fully understood, of particular importance is the splicing of the carboxy-terminal half of the Ig domain III, with the exception of FGFR4, which determines the binding specificity for different FGFs. As a result of this splicing event, IIIa through IIIc isoforms are generated. Whereas IIIb and IIIc isoforms are type I transmembrane receptor tyrosine kinases, the IIIa splice variant encodes a truncated protein that, because it is secreted and not a transmembrane protein, cannot independently transduce extracellular signals, although it may act to sequester released FGFs and inhibit FGF signaling. This alternative splicing event is regulated in a tissue-specific manner and dramatically affects ligand-receptor binding specificity (47). The expression of the IIIb form is restricted to epithelial lineages, whereas the IIIc form is mainly expressed in mesenchymal lineages. In ECs, the major form of the FGF receptor is considered to be FGFR1IIIc. However, the expression of the R2IIIc and R3IIIc forms also has been reported (28), with an in vitro observation demonstrating that FGFR2 signaling in ECs only contributes to motility through the mitogen-activated protein kinase (MAPK) pathway (48).

FIBROBLAST GROWTH FACTOR INTERACTION WITH HEPARIN SULFATE PROTEOGLYCAN

In addition to tyrosine kinase FGFR, the highly complex glycosaminoglycan heparan sulfate has been shown to be critical for FGF function. FGFs' binding to heparan sulfate stabilizes FGFs in the extracellular space and facilitates their binding to FGF receptors, resulting in the formation of dimers and higher-order oligomers. FGFs fail to bind and activate FGFR in cells lacking endogenous heparan sulfate (49,50). Structural studies suggest that heparan sulfate forms a ternary complex with FGF and FGFR by binding to a groove formed by the heparan-binding sites of both the ligand and the receptor (51). The oligosaccharide sequence needed to generate a mitogenic signal with FGF2 appears to require both 2-*O*-sulfated iduronic acid and 6-*O*-sulfated *N*-acetylglucosamine groups (52).

Independent of FGF tyrosine kinase receptors, syndecan-4 can transmit a signal via its cytoplasmic tail in response to FGF2 by activating protein kinase Cα (PKCα) and potentially other intracellular signaling cascades (45,53). FGF tyrosine kinase signaling is probably not required for these events, but there clearly is a poorly understood interplay between the FGFR and syndecan-4 signaling pathways. The importance of heparan sulfates in FGF2 signaling is such that cellular responsiveness to FGF2 is diminished by the removal of syndecan-4 heparan sulfate chains (46).

FIBROBLAST GROWTH FACTOR RECEPTOR SIGNALING

The activation of FGFR leads to phosphorylation on tyrosine residues of FRS2 that in turn results in its binding to FGFR (54). FRS2 serves as a docking protein for the assembly of a signaling complex that includes Grb2, Gab1, and Shp2. This complex recruits the guanine nucleotide exchange factor SOS, through binding to Grb2, and activates Ras, resulting in the activation of MAPK pathways (extracellular signal-regulated kinase [ERK]1/2, p38, c-jun N-terminal kinase [JNK]) (55). FGFR also can activate phospholipase C (PLC)-γ, thereby stimulating production of diacylglycerol (DAG) and inositol 1,4,5-triphosphate (IP_3). In turn, this releases intracellular calcium and activates Ca^{2+}-dependent PKC (56). However, the functional significance of this pathway is not yet established. Although the MAPK (ERK1/2) pathway is considered to be involved in stimulation of proliferation, p38 and JNK MAPKs are generally associated with inflammatory or stress responses (57). In ECs, p38 MAPK seems to act as an essential mediator of FGF activity. It is involved in the FGF2-induced tubular morphogenesis of ECs by negatively regulating EC survival, proliferation, and differentiation. Interestingly, inhibition of p38 MAPK increases tube formation, decreases apoptosis, and stimulates DNA synthesis with the concomitant induction of Jagged1 expression (58).

The activation of the phosphoinositide 3-kinase (PI3K)-Akt cell survival pathway seems to be one of the important biological responses induced by FGF2 in ECs. In addition to promoting MAPK pathways, Grb2 recruits the docking protein Gab1, which is tyrosine phosphorylated by FGFR, leading to the recruitment and activation of PI3K (59). Other signaling molecules involved in FGF-driven angiogenesis, especially cell migration and tubulogenesis, are members of the Src kinase family (60,61).

Membrane-associated gangliosides recently have been implicated as modulators of various receptor tyrosine kinases by regulating their ligand binding, receptor dimerization, and receptor subcellular localization (62). FGF interacts directly with a ganglioside (GM1) (Kd = 3 nM), and FGFR activity is inhibited by both GM3 and GM1. Although it was initially thought that gangliosides can be competitive inhibitors of FGF, a recent study has demonstrated that exogenously added GM1 can restore the ability of heparin sulfate–deficient cells to respond to FGF2 (63). This suggests that gangliosides can act as a functional FGF coreceptor, by presenting FGF to its receptor in a way analogous to that of HSPGs.

Regulation of the FGF signaling system includes feedback systems by which FGF signaling is subjected to fine tuning through a negative or positive feedback loop (64). Members of the Sprouty (Spry), Sef, and MAPK phosphatase families are negative modulators of FGF signaling, whereas positive regulators that promote FGF signaling include the Ets transcription factors Erm and Pea3, low-molecular-weight protein tyrosine phosphatase (LMW-PTP) (65), and the transmembrane protein XFLRT3 (66). Although Spry is thought to be a general inhibitor of Ras-MAPK signaling, it behaves differently depending on the receptor tyrosine kinase that is activated. Overexpression of Spry-1 or Spry-2 in ECs inhibits FGF- or vascular endothelial growth factor (VEGF)-induced proliferation and differentiation by repressing the MAPK pathway.

FIBROBLAST GROWTH FACTOR AND VASCULAR FORMATION

Although the ability of FGFs to promote angiogenesis in vitro is well established, the degree of FGF contribution to angiogenesis in vivo is a more complex issue. In the former case, many studies have shown that various FGFs, including FGF2, stimulate EC chemotaxis and proliferation. In addition, FGFs also can induce ECs to invade a three-dimensional collagen matrix and organize themselves to form characteristic tubules that resemble blood capillaries (67).

However, the demonstration of angiogenic properties of FGFs in vivo has been more challenging. Unlike FGFR, it is possible to knockout many individual FGFs without lethality and, in fact, with little impact. Although FGF1 and FGF2 are known to induce new capillary formation in various in vivo angiogenesis models, *Fgf2*$^{-/-}$ mice demonstrate a mild delay in wound repair (68), but no alteration in the vessel repair following mechanical injury (69). Moreover, the double knockout of FGF1 and FGF2 failed to show abnormalities in angiogenesis, leaving the distinct role of FGF1 or FGF2 in the vascular development unclear (70). Because FGFs, especially within each subfamily, have similar receptor-binding properties and overlapping patterns of expression, these results may be explained by a functional redundancy of the FGF system.

Although the importance of the FGF system in embryonic development, especially in organogenesis and limb genesis, is well established, its role in vascular system development is much less well understood. The embryonic vasculature forms via a combination of vasculogenesis and angiogenesis. The term *vasculogenesis* refers to the segregation, migration, and assembly of angioblasts (EC precursors) from the mesoderm into the primary vascular plexus, whereas *angiogenesis* refers to the sprouting growth of capillaries from this primary plexus.

Fgfr1 knockout studies demonstrate that FGF signaling is essential for mesodermal patterning; however, a study using embryoid bodies (EBs) showed that VEGF-induced vascular development is independent of FGFR1, as demonstrated by the induction of capillaries by VEGF-A in the *Fgfr1*$^{-/-}$ EBs, contrasting with the results from the *Vegfr2*$^{-/-}$ EBs, in which the vascular development was completely arrested (71). Although FGF2 appears to play an important role in the development of the hemangioblast, the common progenitor of hematopoietic and ECs (72), and FGFR1 is expressed by a subpopulation of hematopoietic progenitor cells that give rise to endothelial progenitor cells (73), the expression of FGFR1 in EC precursors does not seem to be required for vascular development. In contrast, hematopoietic development is severely disturbed by the inactivation of FGFR1, with reduced expression of markers for primitive and definitive hematopoiesis (74).

FGF2 bead implantation in somites can induce angioblast differentiation in the epithelial somite, whereas beads lateral to the somitic mesoderm induce the formation of ectopic vessels, which suggests that FGF2 plays a role in the induction of angioblasts and their assembly into the initial vasculature (75). Furthermore, expression of dominant-negative FGFR, capable of blocking the signaling of all FGFRs, in the vasculature of embryonic day (E)9 mouse embryos leads to disrupted embryonic and extraembryonic vasculature (76).

FGF2 appears to promote collateral vessel development (arteriogenesis), which is thought to occur by the growth and remodeling of the preexisting arteries and by de novo arterialization of the capillary network (77). The initiation of arteriogenesis does not seem to be associated with an increased expression of VEGF or hypoxia-inducible genes such as hypoxia-inducible factor (HIF)-1α, and the speed of collateral development is not accelerated by infusion of VEGF (78). Both angiogenesis and collateral vessel development can occur in the heart as compensatory processes in ischemic heart disease. An upregulation of the FGF system in collateral development has been reported recently (79), raising the possibility that it plays a role in the process. In support of this hypothesis is the observation that transgenic myocardial expression of FGF1 in mice results in increased vascular branching and coronary arterial density, whereas the capillary density is unchanged (80). In a canine model of myocardial infarction, the administration of FGF2 enhances collateral development, an effect that is not observed in VEGF-treated animals (81). Moreover, administration of FGF2 in a gradual coronary occlusion model in pigs results in increase of coronary flow and reduction in infarct size in the compromised territory, leading to an improvement of cardiac function (82,83).

FGF4 and FGF5 also have been reported to induce neovascularization in experimental animal models of ischemia. The FGF4-mediated response consists mainly of the enlargement of preexisting vessels rather than an increase in capillary density. Adenovirus-mediated FGF4 expression upregulates endogenous VEGF, which may be associated with an increase in vascular permeability (84). In a model of myocardial ischemia, the intracoronary injection of recombinant adenovirus expressing FGF5 was reported to improve regional abnormalities in stress-induced function and blood flow (85).

FGF2 in combination with another angiogenic factor, platelet-derived growth factor (PDGF)-BB, administered in a protein form, synergistically induces vascular networks, which remain stable for more than a year even after depletion of angiogenic factors. The coordination of FGF2 and PDGF-BB appears to enhance both endothelial and vascular smooth muscle cell survival, conferring vascular stability after the withdrawal of growth factors (86). Interestingly, this effect is not observed with combinations of FGF2 and VEGF or PDGF-BB and VEGF.

The evidence implicating FGFs' ability to induce angiogenesis and collateral vessel growth, bolstered by some favorable results in animal studies of myocardial ischemia and peripheral arterial diseases, provides ample motivation to further investigate the functional roles of the FGF system in vivo and pursue clinical applications of FGFs as regulators of vascular formation.

FIBROBLAST GROWTH FACTOR AND VASOMOTOR REGULATION

Nitric oxide (NO) production plays a critical role in mediating vascular remodeling and vasomotor control of collateral vessels following obstruction of the main arterial supply (87).

FGF2 is a potent inducer of NO production and can induce vasodilation of both normal and collateral vessels in vitro and in vivo (88–90). The vasodilation induced by FGF is not observed in endothelium-denuded vessels or following pretreatment with L-NMMA, an inhibitor of NO synthesis, implying that FGFs modulate blood flow by acting on the endothelium of arteries in an NO-dependent manner

Table 34-2: FGF Knockout Mice

Gene	*Survival of Null Mutant*	*Phenotype*	*References*
Fgf1	Viable	None identified	70
Fgf2	Viable	Delayed wound healing, neuronal abnormalities, reduced hypertrophic response in heart, hypotension, decreased vascular smooth muscle contractility	68,69,107–109
Fgf3	Lethal, P0*	Mild inner ear and tail abnormalities	110
Fgf4	Lethal at E4–5	Arrest of growth of inner cell mass	111
Fgf5	Viable	Abnormally long hair	112
Fgf6	Viable	Normal, accelerated differentiation in regenerating muscle	113–115
Fgf7	Viable	Hair follicle growth, ureteric bud growth, fewer nephrons	116,117
Fgf8	Lethal at E7	Gastrulation defect, defects in CNS and limb development	118,119
Fgf9	Lethal, P0	Lung hypoplasia, XY sex reversal	120,121
Fgf10	Lethal, P0	Developmental defects of multiple organs, including limb, lung, thymus, pituitary, and prostate	122–125
Fgf11	NA		
Fgf12	Viable	Disturbed neuromuscular functions	12,106
Fgf13	NA		
Fgf14	Viable	Neurological phenotypes including ataxia and paroxysmal dyskinesia	126
Fgf15	Lethal at E12.5	Malalignment of aorta and pulmonary trunk	10,127
Fgf16	NA		
Fgf17	Viable	Disturbed cerebellar development	128
Fgf18	Lethal, P0	Delay of osteogenic differentiation, impairment of lung alveolar development	129,130
Fgf20	NA		
Fgf21	NA		
Fgf22	NA		
Fgf23	Viable**	Hyperphosphatemia and impaired skeletogenesis	131

CNS, central nervous system; NA, not available.
*Most of the homozygous mutants die at or soon after birth, but several have survived to adulthood.
**Death at week 13 is observed.

(91,92). The mechanisms underlying FGF-induced NO production seem to involve PKCα. PKCα stimulates NO production by activating endothelial NO synthase (eNOS) (by increasing eNOS phosphorylation at Ser^{1179}) in ECs. The overexpression of PKCα in rat femoral arteries results in a significant increase in the resting blood flow (93).

FGF2 is thought to function in the regulation of normal vascular tone. *Fgf2*-knockout mice, although morphologically normal, display decreased vascular smooth muscle contractility, low blood pressure, and thrombocytosis (69). Although the mechanism of hypotension is not clear, it is thought to be of central origin.

KNOCKOUT STUDIES

In striking contrast to the early lethality of both *Fgfr1*- and *Fgfr2*-knockout mice (94–96), the disruption of individual FGF genes in general presented much milder phenotypes, implying possible functional redundancy occurring between

Table 34-3: FGFR Knockout Mice

Gene	*Mutation*	*Phenotype*	*References*
Fgfr1	Disruption/null	Growth retardation, defect of mesodermal patterning lethal at E7.5–9.5	94,95
	Fgfr1$^{-/-}$ ES chimeras	Defective cell migration through primitive streak, malformation of chimeric limb buds	133,134
	Disruption of α(three-Ig domain) isoform	Distal truncation of limb bud, lethal at E9.5–12.5 due to posterior embryonic defects	135
	Disruption of IIIb	Homozygous mutant mice appeared normal at weaning	136
	Disruption of IIIc	Gastrulation defects	136
	Cartilage/Col2a1-Cre	No obvious defects	137
	Telencephalon/Foxg1-Cre	Malformation of olfactory bulbs	138
	Mid- and hindbrain/ En-Cre or Wnt-Cre	Cerebellar and midbrain defects, ataxia	139
	Radial glial lineage/ hGFAP-Cre	Hippocampal atrophy	140
Fgfr2	Deletion/null	Lethal at E4.5–5.5; tetraploid rescued embryos show abrogated limb outgrowth and lung development	96,141
	Deletion of IIIb and IIIc	Failure of limb bud initiation and placenta formation lethal at E10.5; tetraploid rescued embryos die at birth without limbs	142,143
	Deletion of IIIb	Agenesis or severe dysgenesis of multiple organs; lethality at birth	144
	Deletion of IIIc	Delayed ossification in the sphenoid region of the skull base, dwarfism in the long bones and axial skeleton	145
	Osteoblast and chondrocyte/Dermo1-Cre	Skeletal dwarfism and decreased bone density	146
	Ureteric bud/Hoxb 7-Cre	Aberrant ureteric bud branching and stromal mesenchyme patterning	147
	Lens/Le-Cre	Impaired lens terminal differentiation, increased apoptosis in lens epithelial cells	148
Fgfr3	Disruption/null	Bone overgrowth and deafness (inner ear defect)	149,150
Fgfr4	Disruption/null	Morphologically normal; decrease of liver function (depleted gallbladders, elevated bile acid pool, and excretion of bile acids)	151,152
Fgfr3/Fgfr4	Cross of FGFR3 and FGFR4 mutants	Dwarfism, lung abnormalities	151

FGFs. FGF knockout studies available to date are summarized in Table 34-2. FGFR knockout studies are summarized in Table 34-3.

THERAPEUTIC APPLICATIONS

The possible involvement in all three forms of vascular growth – namely vasculogenesis, angiogenesis, and arteriogenesis – positions FGFs as potential therapeutic agents to promote vascular growth in conditions such as ischemic heart disease and peripheral artery disease. Despite the angiogenic potential seen in animal models, intracoronary delivery of FGF2 protein has failed to provide clear-cut evidence of therapeutic effectiveness, as measured by long-term symptomatic improvement in patients with severe ischemic heart disease. This suggests that the time of exposure and the mode of delivery may be key elements for the success of the therapeutic use of the growth factor (77,97).

In contrast to the single bolus injection, polymer-released FGF2 implanted at the time of coronary artery bypass surgery in the unrevascularized myocardium demonstrated a reduction in the size of the ischemic defect and symptoms after 90 days in the high-dose FGF2 group when compared with the low-dose FGF2 and placebo group (98).

The attempts at FGF gene therapy have not been successful to date. Intracoronary infusion of adenovirus encoding FGF4 in patients with chronic stable angina showed a trend toward increased exercise duration and improved myocardial ischemia in small early studies (99), but a large double-blind, randomized trial failed to show benefits (T. Henry, ACC 2005 presentation).

Moreover, because FGFs are multifunctional, with the ability to modify many biological processes, additional therapeutic applications are being investigated. FGFs are involved in tissue protection by activating antiapoptotic pathways, and in tissue repair after ischemic, metabolic, or traumatic injury (100). After ischemic injury, FGF2 is able to influence neurogenesis in the hippocampus and enhance skeletal muscle regeneration (101,102). Independent of its angiogenic effect, FGF2 recently has received considerable attention with respect to its cardioprotective potential (103).

FGF7 and FGF20 are known to enhance the regenerative capacity of epithelial tissues and protect them from a variety of harmful exposures, such as chemotherapy and radiation for cancer treatment. Clinical trials that treat patients with oral mucositis resulting from chemoradiotherapy for hematologic malignancies demonstrated that FGF7 significantly reduced both the incidence and duration of severe oral mucositis (104).

Targeting the FGF system for inhibition of angiogenesis has a therapeutic potential in cancer, on the premise that tumor progression in general is angiogenesis-dependent. However, increasing evidence suggests that FGFs play crucial roles in tumors of the prostate, bladder, kidney, and testis in a manner independent of their angiogenic effect (105). Other diseases, such as diabetic retinopathy and rheumatic arthritis, also characterized by excessive neovascular formation, are good targets for antiangiogenic therapy. Inhibitors that can be used in the antiangiogenic therapy include endogenous inhibitors, chemical inhibitors, neutralizing antibodies, and soluble receptors.

KEY POINTS

- The FGF system is a multifunctional signaling system that has diverse biological activities in early embryonic development, organogenesis, and postdevelopmental pathophysiology.
- FGFs stimulate endothelial proliferation, migration, and differentiation through cell surface tyrosine kinase receptors and syndecan-4. HSPG is an essential component for FGF binding to FGFR and subsequent signaling events. Although FGFs act in an autocrine manner, high-molecular-weight FGFs that contain NLS and endocytosed FGFs accumulate in the nucleus, with increasing evidence of distinct functional properties.
- FGFs are widely viewed as angiogenic factors; however, their involvement in the processes of vascular formation also may include vasculogenesis and arteriogenesis. Beyond their role in vascular formation, FGFs' participation in vascular tone control, tissue regeneration, tissue protection, and tumor progression also is recognized.

Future Goals

- To identify and characterize (a) the secretory pathway of FGFs that lack signal peptides, (b) the function of high-molecular-weight FGFs and nuclear-localized FGFs, (c) the role of individual FGFs, (d) the detailed signaling mechanism of the FGF system, (e) the role of the FGF system in the postdevelopmental physiology and pathology
- To undertake, in the therapeutic angiogenesis, further investigation to understand critical issues for successful therapy: identification of the most effective delivery approach, proper selection of patients, timing and dosage of the angiogenic factor
- To answer the fundamental question: Is augmentation of neovascularization by a single protein or gene in the ischemic tissue a scientifically rational approach?

REFERENCES

1 Gospodarowicz D, Moran J. Effect of a fibroblast growth factor, insulin, dexamethasone, and serum on the morphology of BALB/c 3T3 cells. *Proc Natl Acad Sci USA*. 1974;71:4648–4652.

2 Westall FC, Lennon VA, Gospodarowicz D. Brain-derived fibroblast growth factor: identity with a fragment of the basic protein of myelin. *Proc Natl Acad Sci USA*. 1978;75:4675–4678.

3 Thomas KA, Riley MC, Lemmon SK, et al. Brain fibroblast growth factor: nonidentity with myelin basic protein fragments. *J Biol Chem*. 1980;255:5517–5520.

4 Westall FC, Seil FJ, Woodward WR, Gospodarowicz D. Brain fibroblast growth factors do not stimulate myelination or remyelination in tissue culture. *J Neurol Sci*. 1981;52:239–243.

5 Shing Y, Folkman J, Sullivan R, et al. Heparin affinity: purification of a tumor-derived capillary endothelial cell growth factor. *Science*. 1984;223:1296–1299.

6 Maciag T, Cerundolo J, Ilsley S, et al. An endothelial cell growth factor from bovine hypothalamus: identification and partial characterization. *Proc Natl Acad Sci USA*. 1979;76:5674–5678.

7 Lemmon SK, Riley MC, Thomas KA, et al. Bovine fibroblast growth factor: comparison of brain and pituitary preparations. *J Cell Biol*. 1982;95:162–169.

8 Thomas KA, Rios-Candelore M, Fitzpatrick S. Purification and characterization of acidic fibroblast growth factor from bovine brain. *Proc Natl Acad Sci USA*. 1984;81:357–361.
9 The fibroblast growth factor family. Nomenclature meeting report and recommendations. January 17, 1991. *Ann NY Acad Sci*. 1991;638:xiii–xvi.
10 Wright TJ, Ladher R, McWhirter J, et al. Mouse FGF15 is the ortholog of human and chick FGF19, but is not uniquely required for otic induction. *Dev Biol*. 2004;269:264–275.
11 Itoh N, Ornitz DM. Evolution of the Fgf and Fgfr gene families. *Trends Genet*. 2004;20:563–569.
12 Ornitz DM, Itoh N. Fibroblast growth factors. *Genome Biol*. 2001;2: 2001;2(3):REVIEWS3005. Epub 2001 Mar 9.
13 Powers CJ, McLeskey SW, Wellstein A. Fibroblast growth factors, their receptors and signaling. *Endocr Relat Cancer*. 2000;7: 165–197.
14 Auguste P, Javerzat S, Bikfalvi A. Regulation of vascular development by fibroblast growth factors. *Cell Tissue Res*. 2003;314:157–166.
15 Mignatti P, Morimoto T, Rifkin DB. Basic fibroblast growth factor released by single, isolated cells stimulates their migration in an autocrine manner. *Proc Natl Acad Sci USA*. 1991;88:11007–11011.
16 Imamura T, Engleka K, Zhan X, et al. Recovery of mitogenic activity of a growth factor mutant with a nuclear translocation sequence. *Science*. 1990;249:1567–1570.
17 Wesche J, Malecki J, Wiedlocha A, et al. Two nuclear localization signals required for transport from the cytosol to the nucleus of externally added FGF-1 translocated into cells. *Biochemistry*. 2005;44:6071–6080.
18 Hsia E, Richardson TP, Nugent MA. Nuclear localization of basic fibroblast growth factor is mediated by heparan sulfate proteoglycans through protein kinase C signaling. *J Cell Biochem*. 2003;88:1214–1225.
19 Bikfalvi A, Klein S, Pintucci G, et al. Differential modulation of cell phenotype by different molecular weight forms of basic fibroblast growth factor: possible intracellular signaling by the high molecular weight forms. *J Cell Biol*. 1995;129:233–243.
20 Arese M, Chen Y, Florkiewicz RZ, et al. Nuclear activities of basic fibroblast growth factor: potentiation of low-serum growth mediated by natural or chimeric nuclear localization signals. *Mol Biol Cell*. 1999;10:1429–1444.
21 Bailly K, Soulet F, Leroy D, et al. Uncoupling of cell proliferation and differentiation activities of basic fibroblast growth factor. *FASEB J*. 2000;14:333–344.
22 Bouche G, Gas N, Prats H, et al. Basic fibroblast growth factor enters the nucleolus and stimulates the transcription of ribosomal genes in ABAE cells undergoing G0—G1 transition. *Proc Natl Acad Sci USA*. 1987;84:6770–6774.
23 Malecki J, Wesche J, Skjerpen CS, et al. Translocation of FGF-1 and FGF-2 across vesicular membranes occurs during G1-phase by a common mechanism. *Mol Biol Cell*. 2004;15:801–814.
24 Thomas-Mudge RJ, Okada-Ban M, Vandenbroucke F, et al. Nuclear FGF-2 facilitates cell survival in vitro and during establishment of metastases. *Oncogene*. 2004;23:4771–4779.
25 Terranova VP, DiFlorio R, Lyall RM, et al. Human endothelial cells are chemotactic to EC growth factor and heparin. *J Cell Biol*. 1985;101:2330–2334.
26 Gospodarowicz D, Ferrara N, Schweigerer L, Neufeld G. Structural characterization and biological functions of fibroblast growth factor. *Endocr Rev*. 1987;8:95–114.
27 Chen CH, Poucher SM, Lu J, Henry PD. Fibroblast growth factor 2: from laboratory evidence to clinical application. *Curr Vasc Pharmacol*. 2004;2:33–43.
28 Antoine M, Wirz W, Tag CG, et al. Expression pattern of fibroblast growth factors (FGFs), their receptors and antagonists in primary ECs and vascular smooth muscle cells. *Growth Factors*. 2005;23:87–95.
29 Grose R, Dickson C. Fibroblast growth factor signaling in tumorigenesis. *Cytokine Growth Factor Rev*. 2005;16:179–186.
30 Dickson C, Spencer-Dene B, Dillon C, Fantl V. Tyrosine kinase signalling in breast cancer: fibroblast growth factors and their receptors. *Breast Cancer Res*. 2000;2:191–196.
31 Yoshida T, Tsutsumi M, Sakamoto H, et al. Expression of the HST1 oncogene in human germ cell tumors. *Biochem Biophys Res Commun*. 1988;155:1324–1329.
32 Tkachenko E, Lutgens E, Stan RV, Simons M. Fibroblast growth factor 2 endocytosis in ECs proceed via syndecan-4-dependent activation of Rac1 and a Cdc42-dependent macropinocytic pathway. *J Cell Sci*. 2004;117:3189–3199.
33 Deguchi Y, Okutsu H, Okura T, et al. Internalization of basic fibroblast growth factor at the mouse blood-brain barrier involves perlecan, a heparan sulfate proteoglycan. *J Neurochem*. 2002;83:381–389.
34 Bryant DM, Wylie FG, Stow JL. Regulation of endocytosis, nuclear translocation, and signaling of fibroblast growth factor receptor 1 by E-cadherin. *Mol Biol Cell*. 2005;16:14–23.
35 Chu CL, Buczek-Thomas JA, Nugent MA. Heparan sulphate proteoglycans modulate fibroblast growth factor-2 binding through a lipid raft-mediated mechanism. *Biochem J*. 2004;379: 331–341.
36 Jackson A, Friedman S, Zhan X, et al. Heat shock induces the release of fibroblast growth factor 1 from NIH 3T3 cells. *Proc Natl Acad Sci USA*. 1992;89:10691–10695.
37 Shi J, Friedman S, Maciag T. A carboxyl-terminal domain in fibroblast growth factor (FGF)-2 inhibits FGF-1 release in response to heat shock in vitro. *J Biol Chem*. 1997;272:1142–1147.
38 Dahl JP, Binda A, Canfield VA, Levenson R. Participation of Na,K-ATPase in FGF-2 secretion: rescue of ouabain-inhibitable FGF-2 secretion by ouabain-resistant Na,K-ATPase alpha subunits. *Biochemistry*. 2000;39:14877–14883.
39 Mandinova A, Soldi R, Graziani I, et al. S100A13 mediates the copper-dependent stress-induced release of IL-1alpha from both human U937 and murine NIH 3T3 cells. *J Cell Sci*. 2003; 116:2687–2696.
40 Engleka KA, Maciag T. Inactivation of human fibroblast growth factor-1 (FGF-1) activity by interaction with copper ions involves FGF-1 dimer formation induced by copper-catalyzed oxidation. *J Biol Chem*. 1992;267:11307–11315.
41 Prudovsky I, Bagala C, Tarantini F, et al. The intracellular translocation of the components of the fibroblast growth factor 1 release complex precedes their assembly prior to export. *J Cell Biol*. 2002;158:201–208.
42 Schafer T, Zentgraf H, Zehe C, et al. Unconventional secretion of fibroblast growth factor 2 is mediated by direct translocation across the plasma membrane of mammalian cells. *J Biol Chem*. 2004;279:6244–6251.
43 Taverna S, Ghersi G, Ginestra A, et al. Shedding of membrane vesicles mediates fibroblast growth factor-2 release from cells. *J Biol Chem*. 2003;278:51911–51919.

44 Engling A, Backhaus R, Stegmayer C, et al. Biosynthetic FGF-2 is targeted to non-lipid raft microdomains following translocation to the extracellular surface of CHO cells. *J Cell Sci.* 2002; 115:3619–3631.
45 Horowitz A, Tkachenko E, Simons M. Fibroblast growth factor-specific modulation of cellular response by syndecan-4. *J Cell Biol.* 2002;157:715–725.
46 Tkachenko E, Rhodes JM, Simons M. Syndecans: new kids on the signaling block. *Circ Res.* 2005;96:488–500.
47 Ornitz DM, Xu J, Colvin JS, et al. Receptor specificity of the fibroblast growth factor family. *J Biol Chem.* 1996;271:15292–15297.
48 Nakamura T, Mochizuki Y, Kanetake H, Kanda S. Signals via FGF receptor 2 regulate migration of ECs. *Biochem Biophys Res Commun.* 2001;289:801–806.
49 Mason IJ. The ins and outs of fibroblast growth factors. *Cell.* 1994;78:547–552.
50 Yayon A, Klagsbrun M, Esko JD, et al. Cell surface, heparin-like molecules are required for binding of basic fibroblast growth factor to its high affinity receptor. *Cell.* 1991;64:841–848.
51 Schlessinger J, Plotnikov AN, Ibrahimi OA, et al. Crystal structure of a ternary FGF-FGFR-heparin complex reveals a dual role for heparin in FGFR binding and dimerization. *Mol Cell.* 2000;6:743–750.
52 Guimond S, Maccarana M, Olwin BB, et al. Activating and inhibitory heparin sequences for FGF-2 (basic FGF). Distinct requirements for FGF-1, FGF-2, and FGF-4. *J Biol Chem.* 1993;268:23906–23914.
53 Volk R, Schwartz JJ, Li J, Rosenberg RD, Simons M. The role of syndecan cytoplasmic domain in basic fibroblast growth factor-dependent signal transduction. *J Biol Chem.* 1999;274:24417–24424.
54 Kouhara H, Hadari YR, Spivak-Kroizman T, et al. A lipid-anchored Grb2-binding protein that links FGF-receptor activation to the Ras/MAPK signaling pathway. *Cell.* 1997;89:693–702.
55 Schlessinger J. Common and distinct elements in cellular signaling via EGF and FGF receptors. *Science.* 2004;306:1506–1507.
56 Mohammadi M, Dionne CA, Li W, et al. Point mutation in FGF receptor eliminates phosphatidylinositol hydrolysis without affecting mitogenesis. *Nature.* 1992;358:681–684.
57 Johnson GL, Lapadat R. Mitogen-activated protein kinase pathways mediated by ERK, JNK, and p38 protein kinases. *Science.* 2002;298:1911–1912.
58 Matsumoto T, Turesson I, Book M, et al. p38 MAP kinase negatively regulates endothelial cell survival, proliferation, and differentiation in FGF-2-stimulated angiogenesis. *J Cell Biol.* 2002;156:149–160.
59 Lamothe B, Yamada M, Schaeper U, et al. The docking protein Gab1 is an essential component of an indirect mechanism for fibroblast growth factor stimulation of the phosphatidylinositol 3-kinase/Akt antiapoptotic pathway. *Mol Cell Biol.* 2004;24:5657–5666.
60 LaVallee TM, Prudovsky IA, McMahon GA, et al. Activation of the MAP kinase pathway by FGF-1 correlates with cell proliferation induction while activation of the Src pathway correlates with migration. *J Cell Biol.* 1998;141:1647–1658.
61 Tsuda S, Ohtsuru A, Yamashita S, et al. Role of c-Fyn in FGF-2-mediated tube-like structure formation by murine brain capillary ECs. *Biochem Biophys Res Commun.* 2002;290: 1354–1360.
62 Miljan EA, Bremer EG. Regulation of growth factor receptors by gangliosides. *Sci STKE.* 2002;(160):RE15.
63 Rusnati M, Urbinati C, Tanghetti E, et al. Cell membrane GM1 ganglioside is a functional coreceptor for fibroblast growth factor 2. *Proc Natl Acad Sci USA.* 2002;99:4367–4372.
64 Tsang M, Dawid IB. Promotion and attenuation of FGF signaling through the Ras-MAPK pathway. *Sci STKE.* 2004;(228): pe17.
65 Park EK, Warner N, Mood K, et al. Low-molecular-weight protein tyrosine phosphatase is a positive component of the fibroblast growth factor receptor signaling pathway. *Mol Cell Biol.* 2002;22:3404–3414.
66 Bottcher RT, Pollet N, Delius H, Niehrs C. The transmembrane protein XFLRT3 forms a complex with FGF receptors and promotes FGF signalling. *Nat Cell Biol.* 2004;6:38–44.
67 Montesano R, Vassalli JD, Baird A, et al. Basic fibroblast growth factor induces angiogenesis in vitro. *Proc Natl Acad Sci USA.* 1986;83:7297–7301.
68 Ortega S, Ittmann M, Tsang SH, et al. Neuronal defects and delayed wound healing in mice lacking fibroblast growth factor 2. *Proc Natl Acad Sci USA.* 1998;95:5672–5677.
69 Zhou M, Sutliff RL, Paul RJ, et al. Fibroblast growth factor 2 control of vascular tone. *Nat Med.* 1998;4:201–207.
70 Miller DL, Ortega S, Bashayan O, et al. Compensation by fibroblast growth factor 1 (FGF1) does not account for the mild phenotypic defects observed in FGF2 null mice. *Mol Cell Biol.* 2000;20:2260–2268.
71 Magnusson P, Rolny C, Jakobsson L, et al. Deregulation of Flk-1/vascular endothelial growth factor receptor-2 in fibroblast growth factor receptor-1-deficient vascular stem cell development. *J Cell Sci.* 2004;117:1513–1523.
72 Faloon P, Arentson E, Kazarov A, et al. Basic fibroblast growth factor positively regulates hematopoietic development. *Development.* 2000;127:1931–1941.
73 Burger PE, Coetzee S, McKeehan WL, et al. Fibroblast growth factor receptor-1 is expressed by endothelial progenitor cells. *Blood.* 2002;100:3527–3535.
74 Magnusson PU, Ronca R, Dell'Era P, et al. Fibroblast growth factor receptor-1 expression is required for hematopoietic but not endothelial cell development. *Arterioscler Thromb Vasc Biol.* 2005;25:944–949.
75 Cox CM, Poole TJ. Angioblast differentiation is influenced by the local environment: FGF-2 induces angioblasts and patterns vessel formation in the quail embryo. *Dev Dyn.* 2000;218:371–382.
76 Lee SH, Schloss DJ, Swain JL. Maintenance of vascular integrity in the embryo requires signaling through the fibroblast growth factor receptor. *J Biol Chem.* 2000;275:33679–33687.
77 Annex BH, Simons M. Growth factor-induced therapeutic angiogenesis in the heart: protein therapy. *Cardiovasc Res.* 2005; 65:649–655.
78 Deindl E, Buschmann I, Hoefer IE, et al. Role of ischemia and of hypoxia-inducible genes in arteriogenesis after femoral artery occlusion in the rabbit. *Circ Res.* 2001;89:779–786.
79 Deindl E, Hoefer IE, Fernandez B, et al. Involvement of the fibroblast growth factor system in adaptive and chemokine-induced arteriogenesis. *Circ Res.* 2003;92:561–568.
80 Fernandez B, Buehler A, Wolfram S, et al. Transgenic myocardial overexpression of fibroblast growth factor-1 increases coronary artery density and branching. *Circ Res.* 2000;87:207–213.
81 Lazarous DF, Shou M, Scheinowitz M, et al. Comparative effects of basic fibroblast growth factor and vascular endothelial

growth factor on coronary collateral development and the arterial response to injury. *Circulation*. 1996;94:1074–1082.
82 Harada K, Grossman W, Friedman M, et al. Basic fibroblast growth factor improves myocardial function in chronically ischemic porcine hearts. *J Clin Invest*. 1994;94:623–630.
83 Post MJ, Laham R, Sellke FW, Simons M. Therapeutic angiogenesis in cardiology using protein formulations. *Cardiovasc Res*. 2001;49:522–531.
84 Rissanen TT, Markkanen JE, Arve K, et al. Fibroblast growth factor 4 induces vascular permeability, angiogenesis and arteriogenesis in a rabbit hind limb ischemia model. *FASEB J*. 2003;17:100–102.
85 Giordano FJ, Ping P, McKirnan MD, et al. Intracoronary gene transfer of fibroblast growth factor-5 increases blood flow and contractile function in an ischemic region of the heart. *Nat Med*. 1996;2:534–539.
86 Cao R, Brakenhielm E, Pawliuk R, et al. Angiogenic synergism, vascular stability and improvement of hind-limb ischemia by a combination of PDGF-BB and FGF-2. *Nat Med*. 2003;9:604–613.
87 Prior BM, Lloyd PG, Ren J, et al. Arteriogenesis: role of nitric oxide. *Endothelium*. 2003;10:207–216.
88 Sellke FW, Wang SY, Friedman M, et al. Basic FGF enhances endothelium-dependent relaxation of the collateral- perfused coronary microcirculation. *Am J Physiol*. 1994;267:H1303–H1311.
89 Sellke FW, Wang SY, Stamler A, et al. Enhanced microvascular relaxations to VEGF and bFGF in chronically ischemic porcine myocardium. *Am J Physiol*. 1996;271:H713–H720.
90 Unger EF, Banai S, Shou M, et al. Basic fibroblast growth factor enhances myocardial collateral flow in a canine model. *Am J Physiol*. 1994;266:H1588–H1595.
91 Tiefenbacher CP, Chilian WM. Heterogeneity of coronary vasomotion. *Basic Res Cardiol*. 1998;93:446–454.
92 Wu HM, Yuan Y, McCarthy M, Granger HJ. Acidic and basic FGFs dilate arterioles of skeletal muscle through a NO-dependent mechanism. *Am J Physiol*. 1996;271:H1087–H1093.
93 Partovian C, Zhuang Z, Moodie K, et al. PKCalpha activates eNOS and increases arterial blood flow in vivo. *Circ Res*. 2005; 97:482–487.
94 Deng CX, Wynshaw-Boris A, Shen MM, et al. Murine FGFR-1 is required for early postimplantation growth and axial organization. *Genes Dev*. 1994;8:3045–3057.
95 Yamaguchi TP, Harpal K, Henkemeyer M, Rossant J. fgfr-1 is required for embryonic growth and mesodermal patterning during mouse gastrulation. *Genes Dev*. 1994;8:3032–3044.
96 Arman E, Haffner-Krausz R, Chen Y, et al. Targeted disruption of fibroblast growth factor (FGF) receptor 2 suggests a role for FGF signaling in pregastrulation mammalian development. *Proc Natl Acad Sci USA*. 1998;95:5082–5087.
97 Simons M, Ware JA. Therapeutic angiogenesis in cardiovascular disease. *Nat Rev Drug Discov*. 2003;2:863–871.
98 Laham RJ, Sellke FW, Edelman ER, et al. Local perivascular delivery of basic fibroblast growth factor in patients undergoing coronary bypass surgery: results of a phase I randomized, double-blind, placebo-controlled trial. *Circulation*. 1999;100: 1865–1871.
99 Grines C, Rubanyi GM, Kleiman NS, et al. Angiogenic gene therapy with adenovirus 5 fibroblast growth factor-4 (Ad5FGF-4): a new option for the treatment of coronary artery disease. *Am J Cardiol*. 2003;92:24N–31N.
100 Hampton TG, Amende I, Fong J, et al. Basic FGF reduces stunning via a NOS2-dependent pathway in coronary-perfused mouse hearts. *Am J Physiol Heart Circ Physiol*. 2000;279:H260–H268.
101 Reuss B, von Bohlen und Halbach O. Fibroblast growth factors and their receptors in the central nervous system. *Cell Tissue Res*. 2003;313:139–157.
102 Doukas J, Blease K, Craig D, et al. Delivery of FGF genes to wound repair cells enhances arteriogenesis and myogenesis in skeletal muscle. *Mol Ther*. 2002;5:517–527.
103 Detillieux KA, Sheikh F, Kardami E, Cattini PA. Biological activities of fibroblast growth factor-2 in the adult myocardium. *Cardiovasc Res*. 2003;57:8–19.
104 Spielberger R, Stiff P, Bensinger W, et al. Palifermin for oral mucositis after intensive therapy for hematologic cancers. *N Engl J Med*. 2004;351:2590–2598.
105 Cronauer MV, Schulz WA, Seifert HH, et al. Fibroblast growth factors and their receptors in urological cancers: basic research and clinical implications. *Eur Urol*. 2003;43:309–319.
106 Goldfarb M. Fibroblast growth factor homologous factors: evolution, structure, and function. *Cytokine Growth Factor Rev*. 2005;16:215–220.
107 Tobe T, Ortega S, Luna JD, et al. Targeted disruption of the FGF2 gene does not prevent choroidal neovascularization in a murine model. *Am J Pathol*. 1998;153:1641–1646.
108 Schultz JE, Witt SA, Nieman ML, et al. Fibroblast growth factor-2 mediates pressure-induced hypertrophic response. *J Clin Invest*. 1999;104:709–719.
109 Dono R, Texido G, Dussel R, et al. Impaired cerebral cortex development and blood pressure regulation in FGF-2-deficient mice. *EMBO J*. 1998;17:4213–4225.
110 Mansour SL. Targeted disruption of int-2 (fgf-3) causes developmental defects in the tail and inner ear. *Mol Reprod Dev*. 1994;39:62–67; discussion 67–68.
111 Feldman B, Poueymirou W, Papaioannou VE, et al. Requirement of FGF-4 for postimplantation mouse development. *Science*. 1995;267:246–249.
112 Hebert JM, Rosenquist T, Gotz J, Martin GR. FGF5 as a regulator of the hair growth cycle: evidence from targeted and spontaneous mutations. *Cell*. 1994;78:1017–1025.
113 Armand AS, Pariset C, Laziz I, et al. FGF6 regulates muscle differentiation through a calcineurin-dependent pathway in regenerating soleus of adult mice. *J Cell Physiol*. 2005;204(1):297–308.
114 Fiore F, Planche J, Gibier P, et al. Apparent normal phenotype of Fgf6–/– mice. *Int J Dev Biol*. 1997;41:639–642.
115 Fiore F, Sebille A, Birnbaum D. Skeletal muscle regeneration is not impaired in Fgf6 –/– mutant mice. *Biochem Biophys Res Commun*. 2000;272:138–143.
116 Guo L, Degenstein L, Fuchs E. Keratinocyte growth factor is required for hair development but not for wound healing. *Genes Dev*. 1996;10:165–175.
117 Qiao J, Uzzo R, Obara-Ishihara T, et al. FGF-7 modulates ureteric bud growth and nephron number in the developing kidney. *Development*. 1999;126:547–554.
118 Abu-Issa R, Smyth G, Smoak I, et al. Fgf8 is required for pharyngeal arch and cardiovascular development in the mouse. *Development*. 2002;129:4613–4625.
119 Sun X, Meyers EN, Lewandoski M, Martin GR. Targeted disruption of Fgf8 causes failure of cell migration in the gastrulating mouse embryo. *Genes Dev*. 1999;13:1834–1846.

120 Colvin JS, White AC, Pratt SJ, Ornitz DM. Lung hypoplasia and neonatal death in Fgf9-null mice identify this gene as an essential regulator of lung mesenchyme. *Development*. 2001;128:2095–2106.

121 Colvin JS, Green RP, Schmahl J, et al. Male-to-female sex reversal in mice lacking fibroblast growth factor 9. *Cell*. 2001;104: 875–889.

122 Min H, Danilenko DM, Scully SA, et al. Fgf-10 is required for both limb and lung development and exhibits striking functional similarity to Drosophila branchless. *Genes Dev*. 1998;12: 3156–3161.

123 Sekine K, Ohuchi H, Fujiwara M, et al. Fgf10 is essential for limb and lung formation. *Nat Genet*. 1999;21:138–141.

124 Ohuchi H, Tao H, Ohata K, et al. Fibroblast growth factor 10 is required for proper development of the mouse whiskers. *Biochem Biophys Res Commun*. 2003;302:562–567.

125 Donjacour AA, Thomson AA, Cunha GR. FGF-10 plays an essential role in the growth of the fetal prostate. *Dev Biol*. 2003; 261:39–54.

126 Wang Q, Bardgett ME, Wong M, et al. Ataxia and paroxysmal dyskinesia in mice lacking axonally transported FGF14. *Neuron*. 2002;35:25–38.

127 Vincentz JW, McWhirter JR, Murre C, et al. Fgf15 is required for proper morphogenesis of the mouse cardiac outflow tract. *Genesis*. 2005;41:192–201.

128 Xu J, Liu Z, Ornitz DM. Temporal and spatial gradients of Fgf8 and Fgf17 regulate proliferation and differentiation of midline cerebellar structures. *Development*. 2000;127:1833–1843.

129 Ohbayashi N, Shibayama M, Kurotaki Y, et al. FGF18 is required for normal cell proliferation and differentiation during osteogenesis and chondrogenesis. *Genes Dev*. 2002;16:870–879.

130 Usui H, Shibayama M, Ohbayashi N, et al. Fgf18 is required for embryonic lung alveolar development. *Biochem Biophys Res Commun*. 2004;322:887–892.

131 Shimada T, Kakitani M, Yamazaki Y, et al. Targeted ablation of Fgf23 demonstrates an essential physiological role of FGF23 in phosphate and vitamin D metabolism. *J Clin Invest*. 2004;113: 561–568.

132 Coumoul X, Deng CX. Roles of FGF receptors in mammalian development and congenital diseases. *Birth Defects Res C Embryo Today*. 2003;69:286–304.

133 Ciruna BG, Schwartz L, Harpal K, et al. Chimeric analysis of fibroblast growth factor receptor-1 (Fgfr1) function: a role for FGFR1 in morphogenetic movement through the primitive streak. *Development*. 1997;124:2829–2841.

134 Deng C, Bedford M, Li C, et al. Fibroblast growth factor receptor-1 (FGFR-1) is essential for normal neural tube and limb development. *Dev Biol*. 1997;185:42–54.

135 Xu X, Li C, Takahashi K, et al. Murine fibroblast growth factor receptor 1alpha isoforms mediate node regression and are essential for posterior mesoderm development. *Dev Biol*. 1999;208: 293–306.

136 Partanen J, Schwartz L, Rossant J. Opposite phenotypes of hypomorphic and Y766 phosphorylation site mutations reveal a function for Fgfr1 in anteroposterior patterning of mouse embryos. *Genes Dev*. 1998;12:2332–2344.

137 Xu X, Qiao W, Li C, Deng CX. Generation of Fgfr1 conditional knockout mice. *Genesis*. 2002;32:85–86.

138 Hebert JM, Lin M, Partanen J, et al. FGF signaling through FGFR1 is required for olfactory bulb morphogenesis. *Development*. 2003;130:1101–1111.

139 Trokovic R, Trokovic N, Hernesniemi S, et al. FGFR1 is independently required in both developing mid- and hindbrain for sustained response to isthmic signals. *EMBO J*. 2003;22:1811–1823.

140 Ohkubo Y, Uchida AO, Shin D, et al. Fibroblast growth factor receptor 1 is required for the proliferation of hippocampal progenitor cells and for hippocampal growth in mouse. *J Neurosci*. 2004;24:6057–6069.

141 Arman E, Haffner-Krausz R, Gorivodsky M, Lonai P. Fgfr2 is required for limb outgrowth and lung-branching morphogenesis. *Proc Natl Acad Sci USA*. 1999;96:11895–11899.

142 Xu X, Weinstein M, Li C, et al. Fibroblast growth factor receptor 2 (FGFR2)-mediated reciprocal regulation loop between FGF8 and FGF10 is essential for limb induction. *Development*. 1998;125:753–765.

143 Li C, Guo H, Xu X, et al. Fibroblast growth factor receptor 2 (Fgfr2) plays an important role in eyelid and skin formation and patterning. *Dev Dyn*. 2001;222:471–483.

144 Revest JM, Spencer-Dene B, Kerr K, et al. Fibroblast growth factor receptor 2-IIIb acts upstream of Shh and Fgf4 and is required for limb bud maintenance but not for the induction of Fgf8, Fgf10, Msx1, or Bmp4. *Dev Biol*. 2001;231:47–62.

145 Eswarakumar VP, Monsonego-Ornan E, Pines M, et al. The IIIc alternative of Fgfr2 is a positive regulator of bone formation. *Development*. 2002;129:3783–3793.

146 Yu K, Xu J, Liu Z, et al. Conditional inactivation of FGF receptor 2 reveals an essential role for FGF signaling in the regulation of osteoblast function and bone growth. *Development*. 2003;130: 3063–3074.

147 Zhao H, Kegg H, Grady S, et al. Role of fibroblast growth factor receptors 1 and 2 in the ureteric bud. *Dev Biol*. 2004;276:403–415.

148 Garcia CM, Yu K, Zhao H, et al. Signaling through FGF receptor-2 is required for lens cell survival and for withdrawal from the cell cycle during lens fiber cell differentiation. *Dev Dyn*. 2005;233:516–527.

149 Colvin JS, Bohne BA, Harding GW, et al. Skeletal overgrowth and deafness in mice lacking fibroblast growth factor receptor 3. *Nat Genet*. 1996;12:390–397.

150 Deng C, Wynshaw-Boris A, Zhou F, et al. Fibroblast growth factor receptor 3 is a negative regulator of bone growth. *Cell*. 1996;84:911–921.

151 Weinstein M, Xu X, Ohyama K, Deng CX. FGFR-3 and FGFR-4 function cooperatively to direct alveogenesis in the murine lung. *Development*. 1998;125:3615–3623.

152 Yu C, Wang F, Kan M, et al. Elevated cholesterol metabolism and bile acid synthesis in mice lacking membrane tyrosine kinase receptor FGFR4. *J Biol Chem*. 2000;275:15482–15489.

35

Transforming Growth Factor-β and the Endothelium

Barbara J. Ballermann

University of Alberta, Edmonton, Canada

The discovery that mutations in the genes encoding two endothelium-restricted transforming growth factor (TGF)-β receptors – endoglin (1) and activin-like kinase 1 (ALK1) (2) – account for most, if not all cases of the human vascular disorder hereditary hemorrhagic telangiectasia (HHT) (3) firmly established the significance of direct TGF-β signaling in endothelial cells (ECs) and in vascular development. Indeed, embryonically lethal vascular phenotypes result from homozygous deletions in no less than eight components of the TGF-β signaling system, namely TGF-β1 (4), ALK1 (5), endoglin (6), ALK5 (7), TGF-β receptor II (TβRII) (8), Smad5 (9), TGF-β–activated kinase (TAK)-1 (10), and furin (11) (Table 35-2). TGF-βs are key regulators of three-dimensional blood vessel structure, site-specific EC differentiation, and interactions between ECs and their immediate microenvironments. TGF-β and other superfamily members, influence ECs through direct activation and through stimuli that result in the release of endothelium-directed mediators from neighboring cells. Moreover, TGF-β—elicited EC-derived signals stimulate recruitment and differentiation of pericytes and vascular smooth muscle cells (VSMCs). TGF-β pathway activation is necessary for endothelial–mesenchymal transdifferentiation during formation of the cardiac cushion and, consequently, the cardiac valves. TGF-β derived from ECs regulates the phenotype of surrounding VSMCs.

More than 40 members of the TGF-β superfamily (12), which includes TGF-β1, TGF-β2, TGF-β–activated kinase (TAK)-1 and TGF-β3, activins, bone morphogenetic proteins (BMPs), growth differentiation factors (GDFs), inhibin, and muellerian inhibitory substance (MIS), have been identified so far. Pathways for TGF-βs, BMPs, and activins, all known to stimulate ECs, are shown in Table 35-1. For all TGF-β superfamily ligands, the signaling cascade is triggered by activation of transmembrane receptor serine threonine kinases, causing downstream phosphorylation and activation of Smad signaling molecules, which then translocate to the nucleus to induce or inhibit the transcription of a large number of genes (13). A remarkable array of regulatory molecules and pathways (Table 35-1), including ligand traps, ligand enhancers, inhibitory pseudoreceptors, inhibitory Smads, and the regulated activation of the TGF-βs, control the location, timing, and degree of activation of these cascades. This chapter emphasizes those components of the TGF-β signaling systems that directly or indirectly influence EC functions. A complementary chapter (Chapter 122) focuses on the clinical syndrome associated with loss of TGF-β signaling via ALK1 in ECs, HHT.

TRANSFORMING GROWTH FACTOR-β ACTIVATION AND LIGAND TRAPS

Activation

All ligands in the TGF-β superfamily are produced as prepropeptides processed by furin-like convertases that cleave mature disulfide-linked homodimeric ligand from the propeptide (Figure 35.1). In the vasculature, the unprocessed prepro-TGF-β can be secreted and then processed by extracellular furin. Extracellular processing in turn is regulated by the inhibitory action of emilin, a cysteine-rich secreted glycoprotein expressed in the vascular tree (14). In the case of TGF-β1, -2, and -3, once processed, the cleaved homodimer remains noncovalently bound to its own N-terminal propeptide domain, referred to as latency-associated peptide (LAP), rendering the TGF-β ligand incompetent to bind to receptor (15) (see Figure 35.1).

Cells release TGF-β bound to LAP either as the small latent complex (SLC) or alternatively bound via LAP to latent TGF-β binding protein (LTBP) as the large latent complex (LLC) (16). The LAP in the LLC inhibits TGF-β activity, whereas the LTBPs orchestrate localization of the LLC to specific sites in the extracellular matrix (ECM) and also facilitate activation (see Figure 35.1). Four distinct LTBP genes exist (*LTBP-1* through *LTBP-4*), each capable of giving rise to alternatively spliced short and long transcripts, producing proteins that differ at the C-terminus. Based on in silico analysis of serial analysis of gene expression (SAGE) libraries, all LTBPs are expressed

Table 35-1: TGF-β Superfamily Signaling Pathways

TGF-β Cyokine Group	*TGF-βs*			*Activins*	*Bone Morphogenetic Proteins*		
Pathway	*I*	*II*	*III*	*IV*	*V*	*VI*	*VII*
Type I recpeptor	ALK1*	ALK5	ALK5	ALK4	ALK3, -6	ALK3, -6	ALK2
Type II receptor	TβRII	TβRII	TβRII	ActRII, ActRIIB	BMPRII	BMPRII	ActRIIB
Type III coreceptor	Endoglin*	Endoglin*	Betaglycan	Betaglycan	DRAGON		
Agonist ligand	TGFβ-1, -3	TGFβ-1, -3	TGFß-1, -2, -3	Activin A	BMP-2, -4	BMP-7	BMP-4, GDF5
Propeptide ligand trap	LAP	LAP	LAP				
Matrix targeting molecule	LTBP-1, -3, -4	LTBP-1, -3, -4	LTBP-1, -3, -4	LTBP-1			
Physiological ligand activator	Tsp-1 Plasmin?	Tsp-1 Plasmin?	Tsp-1 Integrin $\alpha V\beta 6$				
Secreted ligand trap	Decorin Biglycan α2-M	Decorin Biglycan α2-M	Decorin Biglycan α2-M	Follistatin α2-M	BMPER* Noggin/Chordin Gremlin	Sclerostin/USAG-1 Gremlin	BMPER* Noggin/Chordin Gremlin
Secreted ligand enhancer					TSG	KCP	
Antagonist ligand				Inhibin		Myostatin	
Inhibitory type I-like receptor		BAMBI	BAMBI	BAMBI	BAMBI	BAMBI	
Smad anchor		SARA	SARA	SARA			
Receptor-Smad (R-Smad)	Smad1, -5, ?-8	SMAD2, -3	SMAD2, -3	SMAD2, -3	Smad1, -5, -8	Smad1, -5, -8	Smad1, -5, -8
Co-smad	Smad4	Smad4	Smad4	Smad4	Smad4	Smad4	Smad4
Inhibitory-smad (I-Smad)	Smad6	Smad7	Smad7	Smad7	Smad6	Smad6	Smad6
Endothelium-related function	Vasculogenesis Angiogenesis	Growth inhibition Apoptosis Matrix synthesis Cardiac cushion Pericyte recruitment	VEGF synthesis PAI-1 synthesis		Cardiac cushion		Endothelial precursor Cardiac cushion
Hereditary vascular disease	HHT-1, HHT-2				PAH		
Pathway active in endothelium	Yes	Yes	No	Yes	Yes	No	Yes
Pathway restricted to endothelium	Yes	Yes	No	No	No	No	No

*Indicates those components of the signaling cascades that are highly restricted to ECs.

ALK, activin like kinase; TGF-β, transforming growth factor beta; TβRII, transforming growth factor type II receptor; ActRII, activin type II receptor; BMPRII, bone morphogenetic protein type II receptor; BMP, bone morphogenetic protein; GDF, growth differentiation factor; LAP, latency associated protein; LTBP, latent transforming growth factor β binding protein; Tsp-1, thrombospondin-1; BMPER, BMP binding endothelial cell precursor derived regulator; α2-M, α2 macroglobulin; USAG-1, uterine sensitization-associated gene-1; TSG, twisted gastrulation; KCP, kielin/chordin-like protein; BAMBI, BMP activin membrane bound inhibitor; SARA, Smad anchor for receptor activation; HHT, hereditary hemorrhagic telangiectasia; PAH, pulmonary arterial hypertension.

Table 35-2: Gene Deletions in the TGF-β Superfamily Causing Vascular Defects

Deletion	*Function*	*Vascular Phenotype*	*References*
TGF-β1	Ligand	Defective vasculogenesis and hematopoeisis	4
ALK1	EC restricted type I TGF-β receptor	AV shunts; fusion of arteries and veins	5, 101
ALK5	Ubiquitous type I TGF-β receptor	Defective yolk sac vasculogenesis	7
TßRII	Type II TGF-β receptor	Defective VSMC/pericyte recruitment	8, 101
Endoglin (heterozygous)	EC Type III TGF-β coreceptor	Mouse model of HHT	185
Endoglin (homozygous)	EC Type III TGF-β coreceptor	Defective angiogenesis	6, 101
Smad1	Intracellular signaling molecule	Disorganized allantoic blood vessels	194
Smad3	Intracellular signaling molecule	Vascular inflammation	182
Smad5	Intracellular signaling molecule	Defective yolk sac vasculature	9, 195
TAK1	Smad-independent signaling	Defective yolk sac vasculogenesis, defective VSMC recruitment	10
Furin convertase	pre-pro-TGF-β processing	Defective yolk sac vasculogenesis, cardiac abnormalities	11
Emilin	Inhibits activation of pre-pro-TGF-β1 by furin	Excess arterial wall matrix, hypertension	14

in ECs, with the frequencies for LTBP-3 and -4 transcripts being higher than those for LTBP-1 and -2 (B. Ballermann, unpublished observations). The LTBPs contain 17 EGF-like motifs and four eight-cysteine repeats, a domain structure they share with the matrix protein fibrillin (see the later section on Marfan syndrome). LTBP-1, -3, and -4, but not LTBP-2, form disulfide-linked heteromers with LAP and hence bind the SLC (see Figure 35.1). Transglutaminase-2, a protein highly expressed by ECs, catalyses covalent crosslinking of LTBPs to specific binding sites on collagen, fibronectin, and fibrillin-1 (17), thereby sequestering a reservoir of latent TGF-β in the ECM, ready to signal upon activation (15). LTBP-2, which does not appear to bind LAP, regulates adhesion of some cells to fibronectin (18).

Release of TGF-β from LAP can be achieved by proteases like plasmin and calpain, by transient acidification, chaotropic agents, thrombospondin (TSP)-1, and $\alpha V\beta 6$ integrin (19). Activation of TGF-β in cocultures of ECs and VSMCs or pericytes (20) was long held to result from plasmin activity (21) similar to the process in blood clots, in which platelets release the SLC from which active TGF-β is formed through cleavage by plasmin (22). The LAP is N-glycosylated at three sites, two of which contain mannose-6 phosphate residues that bind to the mannose-6-phosphate/IGF-II receptor, which in turn facilitates plasmin-mediated activation of TGF-β in cocultures of bovine aortic ECs and VSMCs (23). Hence, spatial restriction of TGF-β activation in endothelial-pericyte or endothelial-VSMC cocultures may be provided, in part, by association of the LAP with the mannose-6-phosphate receptors on ECs. Plasmin generation is inhibited by plasminogen activator inhibitor (PAI)-1. Because PAI-1 expression is strongly activated by TGF-β1, a shift in the balance of tissue-type plasminogen activator (t-PA)/urokinase-type plasminogen activator (u-PA) to PAI-1 may serve a negative feedback function to dampen TGF-β1 action on ECs by reducing TGF-β activation (24). In this regard, activation of ECs by inflammatory cytokines also can result in enhanced PAI-1 release, in turn associated with reduced TGF-β1 activation (25). Nonetheless, because mice deficient in plasminogen do not display a phenotype consistent with TGF-β1 deficiency, it has been argued that plasmin cannot be the physiological activator of TGF-β1 (15). Even so, because the evidence favoring plasmin-mediated

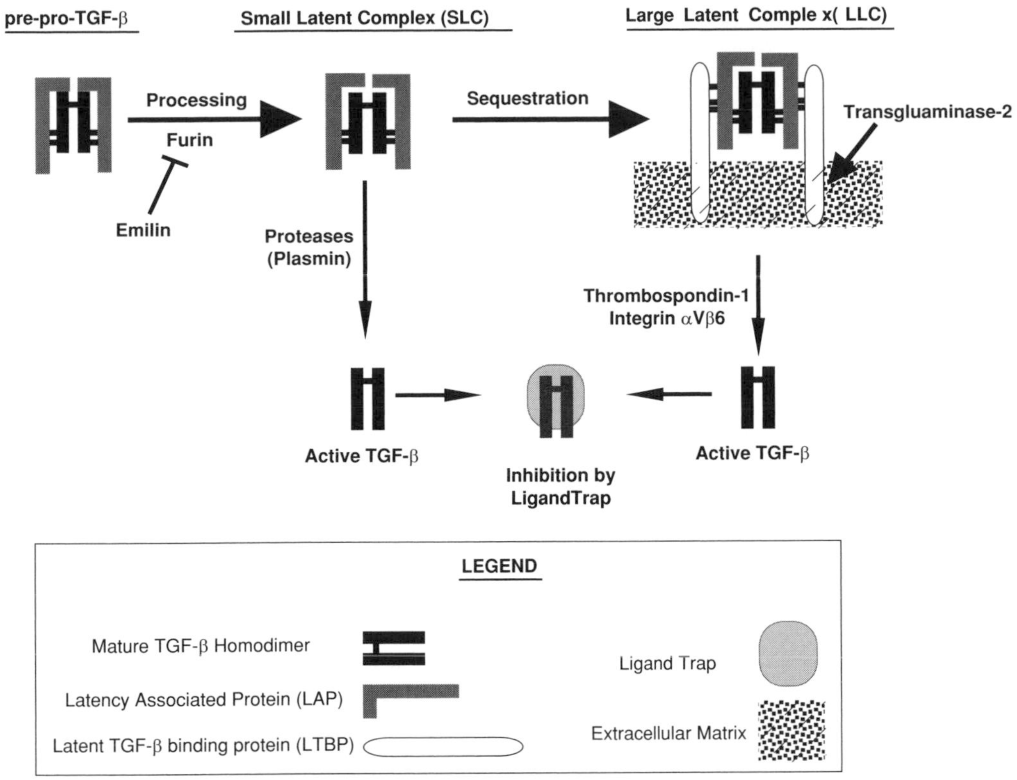

Figure 35.1. Schematic representation of TGF-β activation and subsequent silencing by ligand traps. The dual, interacting signaling pathways via ALK1 and ALK5 are unique to endothelium and are described fully in the text.

TGF-β1 activation is plentiful, and deletion of PAI-1 enhances TGF-β1 activation, at least in a model of glomerulonephritis (26), it seems likely that activation of TGF-β by plasmin can occur where plasmin is abundant, for example in blood clots, wounds, and inflammatory lesions.

Spatially restricted TGF-β activation by LTBP-dependent mechanisms plays a physiologically important role (16,15). Antibodies to LTBP-1 inhibit TGF-β1 activation in endothelial-VSMC cocultures (27) and suppress differentiation of embryonic stem (ES) cells into ECs, a process that is rescued with mature TGF-β1 (28). LTBP-1 is also necessary for TGF-β1 activation by $\alpha V\beta 6$ integrin in epithelial cells in response to shear stress, suggesting that spatial restriction of TGF-β by $\alpha V\beta 6$ integrin is mediated by LTBP-1, and that activation could result from traction on the integrin-LTBP1-LAP link (29,30). LAP also can associate with $\alpha V\beta 1$ integrin (31), highly expressed by angiogenic sprouts, but so far $\alpha V\beta 1$ integrin–dependent TGF-β activation in ECs has not been reported.

The extracellular protein fibrillin-1, mutated in Marfan syndrome, binds LTBP-1 (32) and, through this interaction, influences TGF-β1 activation. Fibrillin-1–deficient mice have much higher levels of active TGF-β1 than do controls, resulting in excessive pulmonary apoptosis and abnormal pulmonary septation as well as mitral valve prolapse that is rescued by TGF-β1 antagonism (33,34). It has therefore been suggested that LTBP-1 linkage of SLC to fibrillin-1 renders TGF-β1 inactive, and that enhanced activation results from absence of fibrillin-1.

TSP-1 probably is the main physiological activator of latent TGF-β. TSP-1 is a prominent platelet α-granule protein also produced by ECs, and is found in the ECM (35). TSP-1 is a powerful inhibitor of EC proliferation and angiogenesis (35). Latent TGF-β1 secreted by bovine aortic ECs is activated by TSP-1 without requiring localization to the cell surface (36). The activation of TGF-β1 by TSP-1 from both SLC and LLC in EC supernatants depends on its association with the LAP (37), and is due a discrete four–amino acid sequence, KRFK, in the type I repeats of TSP-1 (38). Major evidence that TSP-1 is a key physiological activator of TGF-β1 comes from the observation that TGF-β1– and TSP-1–deficient mice have similar phenotypes, that organs of the TSP-1–deficient mice contain very little active TGF-β1, and that their phenotype is rescued by TSP-1 peptides that can activate TGF-β1 (39). Binding of TSP-1 to VLAL domains of both LAP and TGF-β1 in the SLC is mediated by the WSXW region of TSP-1 and is required for TGF-β1 activation (40).

Hence, for TGF-β ligands, sequestration in the ECM is orchestrated by LAP and LTBPs, and activation is mediated by TSP-1 and potentially also by matrix-cleaving proteases. Having latent TGF-β1 ready for release by proteases at key sites of tissue remodeling is akin to placing a remodeling sensor into the extracellular compartment that can then participate in the appropriate angiogenesis and matrix repair processes (15).

Ligand Traps

For all ligands in the TGF-β superfamily, small, secreted proteoglycans act as ligand traps (see Table 35-1). Many are members of the small leucine-rich repeat proteoglycan (SLRP) family (41), and they often contain a conserved arrangement of cysteines that form intramolecular bonds to produce a "cystine knot" (42). These proteins directly bind specific ligands of the TGF-β superfamily, inhibiting their interaction with receptors and therefore providing control over ligand activity. For example, the cystine knot proteoglycan decorin binds and inhibits TGF-β (43); follistatin is a critical local regulator of activin (44); Noggin, Chordin, Gremlin, Sclerostin, and others bind and inhibit members of the BMP (45) family; and ligand traps in the DAN/Cerberus family bind and inhibit Nodal (46). The ligand trap BMPER (bone morphogenetic protein-binding endothelial cell precursor-derived regulator), which can bind BMP2, -4, and -6, is produced and secreted by FLK-1–expressing EC precursors and can inhibit their differentiation into ECs (47). Expression and secretion of TGF-β ligand traps is highly regulated spatially and temporally, serving to regulate the degree of activation of these signaling cascades.

OVERVIEW OF TRANSFORMING GROWTH FACTOR-β SIGNALING

Signal Initiation

The mature form of nearly all TGF-β cytokines contains six to nine conserved cysteine residues that form inter- and intramolecular disulfide bonds. Six of the cysteines form disulfide bonds within each monomer subunit, creating a characteristic "cystine knot" necessary for receptor binding and shared by several other growth factors, for example, platelet-derived growth factor (PDGF) and nerve growth factor (NGF) (48), as well as the ligand traps described earlier. The remainder of the cysteines form intermolecular disulfide bonds that help stabilize the hydrophobic interactions that produce the ligand homodimers (13).

TGF-β superfamily ligands bind to type I and type II receptor serine threonine kinases in the context of coreceptors, also referred to as type III TGF-β receptors (49,50). The coreceptor betaglycan, expressed by most cells, is required for activation of ALK5 by TGF-β1, -2, and -3, whereas endoglin, expressed predominantly by ECs, promotes activation of both ALK1 and ALK5 pathways by TGF-β1 and -3 in ECs (see Table 35-1). The glycophospholipid-anchored protein Dragon similarly acts as coreceptor for BMP2 and -4 binding to type II activin receptors (AcRII or AcRIIB) and ALK2, -3, and -6 (51). Coreceptor function has also been ascribed to the soluble proangiogenic and profibrotic EC product connective tissue growth factor (CTGF), which potentiates receptor activation by TGF-β1 (52). Similarly, Cripto potentiates Nodal binding to AcRII and ALK4 (46), and Kielin/chordin-like protein (KCP) increases the affinity of BMP7 for BMPRII as well as ALK2 and ALK6 (53). To what extent each of these coreceptors participates in TGF-β signaling in EC remains to be elucidated.

Signaling by the TGF-β cytokines is triggered by clustering and activation of membrane-associated receptor serine threonine kinases, grouped into the type I and type II receptor subtypes (13). These receptor serine threonine kinases contain an extracellular ligand binding domain, a single transmembrane spanning domain, and a cytoplasmic serine threonine kinase domain. The type II receptors are constitutively active serine threonine kinases that exhibit ligand family specificity (see Table 35-1). The seven type I receptor serine threonine kinases, also known as ALK, display specificity for ligand families due to their preference for specific type II receptors. The ALKs are inactive in the absence of ligand. Upon ligand-induced clustering with their respective type II receptors, the conserved TTSGSGSG domain (the GS box) of the type I receptors is phosphorylated by the type II receptor serine threonine kinase, resulting in activation of the type I receptor kinase activity (13).

Signal Propagation

Activated type I TGF-β receptor serine threonine kinases transiently associate with a subgroup of Smad proteins (Smad1, -2, -3, -5, and -8), also referred to as receptor Smads (rSmads), through conserved regions in the type I receptors and the rSmad proteins (54). This association facilitates phosphorylation of the rSmads at a conserved carboxyterminal SXS motif. ALK1, -2, -3, and -6 all interact with and phosphorylate Smad1 and -5, and potentially also Smad8, whereas ALK4, -5, and -7 activate Smad2 and -3. Smad2 phosphorylation by ALK5 requires the cosignaling protein SARA (Smad anchor for receptor activation), a FYVE domain protein (55,56). FYVE domains mediate the recruitment of signaling and membrane-trafficking proteins to early endosomes. Once Smad2 is phosphorylated, it dissociates from SARA and associates with Smad4, the common co-Smad required for nuclear translocation of the Smad signaling complex. Binding of Smad2 to SARA (55) and association of SARA with endosomes by the FYVE motif are indispensable for Smad2 phosphorylation by ALK5 (57). The cytoplasmic isoform of the promyelocytic leukaemia (cPML) tumor suppressor also participates in this cascade (58). *PML*$^{-/-}$ mouse embryo fibroblasts are resistant to TGF-β1, failing to stimulate the phosphorylation and nuclear translocation of Smad2 and -3 and gene transcription in these cells. The cPML is required for association of Smad2 and -3 with SARA and for the accumulation of SARA and TGF-β receptors in the early endosomes. Thus, Smad2 and -3 phosphorylation triggered by TGF-β receptor activation probably occurs in early endosomes. Because SARA association is lost upon Smad2 phosphorylation, SARA also may serve to retain inactive Smad2 in the cytoplasm, thus preventing its translocation to the nucleus (59). In this regard, phosphorylation of Smad2 and translocation to the nucleus of ECs can be induced in a ligand-independent fashion with disruption of endosomes by dominant negative Rab5 (57), strongly suggesting that endosome-controlled retention and release of Smad2 is critical for its function. Whether endosomal mechanisms involving SARA and cPML (or their homologs) are similarly required for all ALK-mediated rSmad phosphorylation, and in particular for the endothelium-restricted phosphorylation of Smad1 and -5 by ALK1, is as yet unknown.

Phosphorylated rSmads interact with the common co-Smad4 to form a heteromeric complex that translocates to the nucleus to positively and negatively regulate transcription of a large number of genes (13). Nucleocytoplasmic shuttling of Smads depends not only on their phosphorylation status but also on nuclear import and export domains in the Smads themselves and their interactions with the nuclear import and export machinery (60–62).

Signal Termination

Several mechanisms negatively regulate type I TGF-β cytokine receptor signaling. As already discussed, several ligand traps prevent interaction of the cytokines with receptor and therefore prevent its activation. Within the cell, type I TGF-β receptor phosphorylation is inhibited by FKBP12 (FK506 binding protein 12), which binds to the GS domain of type I receptors, and an inhibitory type-I receptor-like protein named BAMBI (BMP activin membrane-bound inhibitor), which can bind ligand but lacks kinase domain (63).

Activated type I receptors are also regulated through dephosphorylation, proteosomal degradation, and inhibition of Smad phosphorylation by inhibitory Smads. For ALK5, protein phosphatase-1 (PP1) is targeted to the receptor by the PP1 targeting subunit GADD34 in a docking cluster with Smad7 and SARA, returning ALK5 to the unphosphorylated, inactive state (64). Upon activation of type I receptors, inhibitory Smads, Smad6, and -7 interact with type 1 receptors and interrupt TGF-β cytokine signaling via distinct pathways (13). Both Smad7 and Smad6 are strongly induced by laminar shear stress in ECs (65), transcription of Smad7 is strongly induced by TGF-β1 (66), and in ECs both inhibitory Smads are induced by overexpression of constitutively active ALK1 (67). Smad6 is inhibitory primarily to the Smad1 through -5 pathways, whereas Smad7 interferes with activation of Smad2 and -3. Smad6 associates with type I receptors to competitively inhibit binding of rSmads (68), although a role at the transcriptional level also has been described (69). Smad7 similarly associates with type I receptors, and can inhibit binding of rSmads, but it also serves to target its associated receptor for destruction through association with Smurf2. Smurfs are *S*mad *u*biquitin *r*egulatory *f*actors that target Smads as well as Smad-associated proteins, including type I TGF-β receptors and SARA for ubiquitination and proteosomal degradation, thus controlling levels of these proteins in the cells (70). Interaction of Smad7 with Smurf2 facilitates ubiquitination by E3 ubiquitin ligase not only of Smad7, but also of ALK5 and SARA and results in their endocytosis via caveolae, followed by proteosomal degradation.

To sum up, signaling by TGF-β cytokines is initiated through type III coreceptor-facilitated binding of the ligand to signaling receptors. Clustering of the signaling receptors brings the constitutively active type II receptor serine threonine kinase into the correct conformational proximity to phosphorylate its principal substrate, the type I receptor. Phosphorylation of rSmads by the type I receptor serine threonine kinase then facilitates association of r- and co-Smads and their translocation to the nucleus, where target genes are activated or silenced. Negative modulation of the signal occurs through inhibitory Smads, dephosphorylation of the type I receptors, and Smurf-regulated ubiquitination and proteosomal degradation of receptors, Smads, and associated proteins.

TGF-β SIGNALING IN THE ENDOTHELIUM

Two TGF-β Signaling Pathways in Endothelial Cells

In vitro, direct stimulation of ECs with TGF-β1 generally inhibits proliferation (71), can induce apoptosis (72), and stimulates migration and assembly of ECs into capillary-like tubes in the presence of ECM (73). Concentration-dependent angiogenic responses to TGF-β are observed both in vivo (74,75) and in three-dimensional matrix in vitro (76), with a threshold effect at the lower, and an inhibitory response at the upper end of the concentration range (76). The seemingly contradictory findings that TGF-β can both stimulate and inhibit angiogenesis were difficult to explain until the presence of two distinct and interacting signaling cascades were discovered in ECs (77), one stimulating proliferation and migration, the other responsible for blocking these responses. Like most other cells, ECs express the ALK5 type I TGF-β receptor; but unlike non-ECs, they also express TGF-β–responsive ALK1. Furthermore, ECs express little if any betaglycan and instead use endoglin as their type III TGF-β coreceptor (Table 35-1). In ECs, ALK1 and ALK5 elicit distinct, but interacting signals in the context of endoglin.

Endoglin and ALK1 are Endothelial Cell–Predominant Proteins

Endoglin is a highly glycosylated homodimeric, single transmembrane spanning protein, conserved between species and expressed in a highly restricted fashion by ECs (78,79). Recognition that its structure is highly homologous to betaglycan led to the demonstration that, like betaglycan (80), endoglin serves as a TGF-β coreceptor. Unlike betaglycan however, endoglin fails to bind TGF-β2 (50), making ECs unresponsive to TGF-β2. This resistance of ECs to TGF-β2 can be rescued by betaglycan (81). Hence, one function of endoglin in ECs is to constrain their responses to two of the three TGF-β isoforms. Endoglin also can interact with activin and BMPs and, in those cases, it couples to ActRII or BMP receptors, respectively (82). The physiological role of endoglin in activin and BMP signaling is as yet poorly understood.

Expression of ALK1 also is highly restricted to endothelium (83). Both the endogenous *ALK1* promoter (84) and a fragment thereof (85) drive expression of the LacZ reporter only in ECs, with much higher levels in developing and mature arterial, compared to venous and capillary endothelium. This ALK1 expression pattern is essentially identical to that of endoglin (86). TGF-β–stimulated Smad1–5 activation also is largely restricted to ECs (77), whereas Smad1–5 in most other cells is phosphorylated in response to BMPs (see Table 35-1).

ALK1 and ALK5 are Activated by TGF-β1 and -3 in Endothelial Cells

In ECs, Smad1 interacts with ALK1 via a domain unique for type I TGF-β superfamily receptors, and is phosphorylated by activated ALK1 (54). Goumans and colleagues (77) showed that ALK1 serves as a receptor for TGF-β1 and -3 in ECs and signals via Smads1 and -5. In non-ECs that lack ALK1, Smad1 and -5 is only responsive to BMP and is not activated by the TGF-βs. Still, TGF-β1 and -3 induce not only Smad1 and -5, but also Smad2 phosphorylation in ECs. Taken together with the observations that constitutively active ALK1 phosphorylates Smad1 and -5 but not Smad2 and -3, and constitutively active ALK5 phosphorylates Smad2, but not Smad1 and -5, the data indicate that TGF-β1 and -3 can simultaneously activate two distinct signaling pathways in ECs. These two pathways have distinct biological effects in ECs. For example, by microarray analysis of human umbilical vein ECs overexpression of constitutively active ALK1, but not ALK5, strongly induces Smad6 and Smad7, as well as Id1 and Id2 (67). The kinetics of TGF-β1 and -3 in ALK1–Smad1/5 pathway activation differ from that of ALK5–Smad2/3 pathway activation, the former being activated transiently and inhibited at high concentrations of TGF-β, whereas activation of the latter occurs even at high TGF-β concentrations and is more sustained (77). The proliferative and migratory response of ECs in response to TGF-β1 and -3 is dependent on ALK1–Smad1/5-mediated Id1 induction, and this set of responses is inhibited by the ALK5 pathway. Interestingly, knockdown of ALK5 strongly inhibits both Smad2 and -3 and Smad1 and -5 responsive promoters in ECs, suggesting that signaling by ALK1 is dependent on ALK5 (77). Indeed, in *ALK5*$^{-/-}$ECs, or in wild-type ECs in which ALK5 expression is knocked down, both TGF-β–stimulated Smad2 and Smad1 and -5 pathways are blocked, and Smad1 and -5 signaling cannot be rescued simply by overexpressing ALK1, thus proving that ALK5 is required for ALK1 signaling in ECs (87). By contrast, reduced expression of ALK1 abrogates Smad1 and -5 but not Smad2 activation, showing that ALK1 is not required for ALK5 signaling.

ALK5 and ALK1 are both found in heteromeric complexes with TβRII and endoglin, and their association is strongly induced by TGF-β ligand (88). The ALK1 association with endoglin and TβRII does not occur in *ALK5*$^{-/-}$cells, but can be rescued by mutant ALK5 deleted in the Smad2 binding/activation domain, re-establishing Smad1 and -5, but not Smad2, pathway activation (87). Although ALK5-dependent ALK1 activation does not depend on Smad2 activation, it does require ALK5 kinase activity. These findings strongly support a signaling model whereby ALK1 is activated not only by TβRII but also by ALK5 (Figure 35.2). To add further complexity, ALK1/Smad5 directly inhibit ALK5 pathway activation through an effect that is downstream of Smad2 (87,88). In ECs, therefore, ALK5 is necessary for the activation of the ALK1 pathway, which in turn serves to dampen ALK5 signaling. Conversely, in the presence of high concentrations of TGF-β1, ALK5 signaling predominates, and the ALK1 pathway is suppressed.

Endoglin Functions as the TGF-β Coreceptor in Endothelial Cells

The *endoglin* gene is transcribed as two distinct splicing variants, differing only at the carboxyterminus; the cytoplasmic domain of human short (S)-endoglin being 14 amino acids long, that of long (L)-endoglin 47 amino acids long (78). The cytoplasmic domain of L-endoglin is rich in serine and threonine amino-acid residues that can become phosphorylated by ALK5 and TβR-II (89). Endoglin interacts directly with ALK5 and TβRII, and in non-ECs lacking ALK1, endoglin potentiates TGF-β–stimulated signaling via ALK5 (89). In ECs, endoglin interacts directly with TβRII and ALK1 as well as ALK5, although endoglin seems to prefer ALK1 as its partner over ALK5 (87,88). Endoglin overexpression promotes, and endoglin knockdown abrogates EC proliferation and migration (90). The proangiogenic effect of endoglin has been ascribed, at least in part, to ALK1 pathway–dependent blockade of ALK5-mediated cyclin-dependent kinase inhibitor p15^{ink4A} and p21$^{waf/cip}$ activation, whereas induction of the proangiogenic transcription factor Id has not been linked to endoglin so far. Hence, endoglin is a necessary component of TGF-β–mediated ALK1 signaling in ECs, and endoglin expression determines the growth potential of ECs (90). These findings are consistent with numerous reports of increased endoglin expression in the endothelium of tumor blood vessels undergoing angiogenesis and with inhibition of tumor angiogenesis by antiendoglin antibodies (91).

Smad-Independent Signals in Endothelial Cells

Activation of TGF-β receptors also stimulates signaling via Smad-independent mechanisms (92) that include mitogen-activated protein kinase (MAPK) pathways. TAK-1 (93) activates the stress-activated kinases p38 MAPK through MAP kinase kinase (MKK)-6 or MKK3, and c-jun N-terminal kinases (JNKs) via MKK4. This effect is rapid and does not require Smad phosphorylation (94). In ECs, TAK1 and p38 MAPK activation may play a role in altering monolayer permeability (94), cell-matrix interaction, and apoptosis (95). In non-ECs, TGF-β–induced p38 MAPK activation, likely via TAK-1, is involved in activating vascular endothelial growth

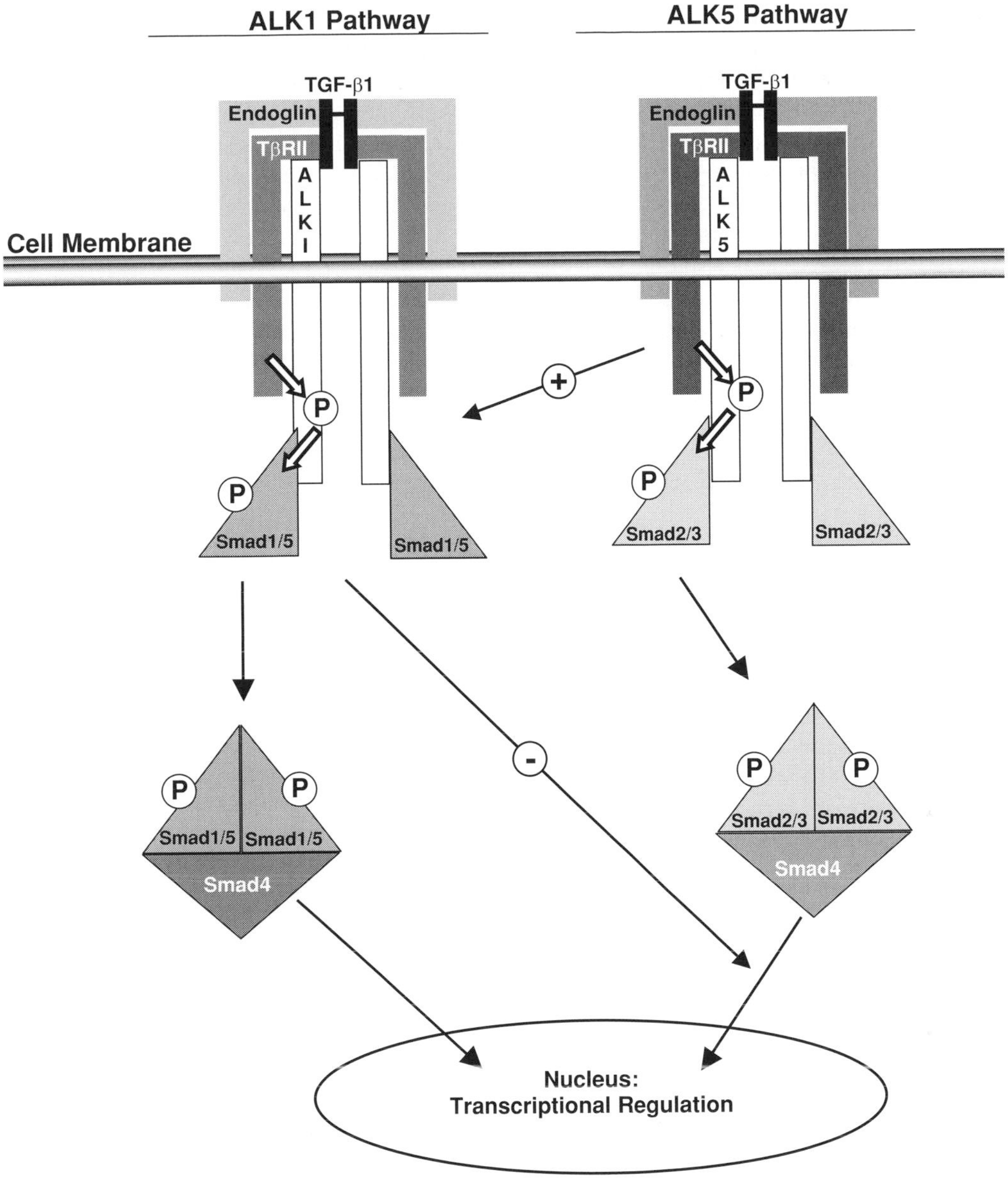

Figure 35.2. Schematic representation of TGF-β signaling in ECs. The dual, interacting signaling pathways via ALK1 and ALK5 are unique to endothelium and are described fully in the text.

factor (VEGF) expression (96), which indirectly influences EC proliferation. *TAK-1* deletion in mice results in abnormalities of yolk sac vasculogenesis that resemble those seen in ALK1- and endoglin-deficient mice (10), suggesting that TAK1 is a necessary downstream mediator of ALK1 action in ECs. Activation of small GTPases in response to TGF-β may play a role in cytoskeletal rearrangements stimulated by TGF-β. For example, in cultured pulmonary ECs, activation of the RhoA/Rho-kinase pathway is a component of the TGF-β response, and includes myosin light-chain phosphorylation, cytoskeletal reorganization, and reduction in barrier integrity (97). Other TGF-β–responsive signaling pathways involving protein phosphatases and the phosphoinositide 3-kinase (PI3K) cascade have been described, but so far insufficient work has been done in ECs to reach any firm conclusions about their importance in EC biology. Similarly, in T lymphocytes (98), endoglin may participate in signaling through TGF-β type I and II receptor-independent mechanisms, but such activity has not been reported in ECs thus far.

NECESSITY OF TGF-β SIGNALING IN VASCULAR DEVELOPMENT

Embryonic Vasculogenesis and Vascular Patterning

During mammalian embryogenesis, endothelial and hematopoietic cells develop from a common “hemangioblast” precursor cell type that is first recognized as FLK-1$^+$ at embryonic day 7 (E7) in the mouse. The first vascular structures, EC-lined blood islands, appear in the mouse embryo yolk sac at day

E8. In addition, recognizable angioblasts within the embryo mesoderm coalesce into EC tubes that form the primitive cardinal vein and dorsal aorta between E7.5 and E8. The blood islands, primitive dorsal aorta, and cardinal vein are already surrounded by VSMCs by E9.5. The common E7 endothelial/hematopoietic progenitor expresses the VEGF receptor FLK-1 (also known as VEGFR2) but not the transcription factor SCL (FLK-1^+/SCL^-), or proteins characteristic for differentiated ECs or hematopoietic cells. These cells then give rise to both EC and hematopoietic lineages through sequential steps first involving upregulation of SCL (FLK-1^+/SCL^+) followed by downregulation of FLK-1 (FLK-1^-/SCL^+). Loss of both FLK-1 and SCL (FLK-1^-/SCL^-) then marks the onset of hematopoietic and EC differentiation. Differentiation of EC and hematopoietic cell lineages can be recapitulated by embryonic stem (ES) cells in vitro stimulated to form embryoid bodies that contain blood islands lined with endothelium and filled with hematopoietic cells. Smad6 overexpression blocks the formation of the earliest FLK-1^+ precursor cell, and application of BMP4, associated with Smad1 and -5 activation, can induce formation of FLK-1^+/SCL^+ cells (99). Conversely, application of the ligand trap BMPER, which inhibits BMP4 signaling, inhibits differentiation of FLK-1^+ cells into ECs (47). Unfortunately, *BMP4*-null mouse embryos die well before the initial emergence of EC/hematopoietic precursors and are therefore uninformative in regard to EC differentiation. Evidence that TGF-β signaling is required for early vasculogenesis comes from the observation that inhibition of TGF-β activation by neutralizing LTBP1 antibodies blocks the development of capillary tubes in the ES cell assay (28). In keeping with a predominant effect of TGF-β on vasculogenesis, overexpression of TGF-β1 in mouse ES cells stimulates formation of capillary structures in embryoid bodies, but does not alter hematopoietic cell development (100). There also have been many reports of EC assembly into capillary-like structures in response to TGF-β1 in vitro (73,72). Thus, BMP4 signaling is required for the early differentiation of ECs from the FLK-1^+ hemangioblast precursor, and TGF-β activation and signaling are necessary for assembly of the ECs into tubes.

In mice deficient in TGF-β1 (4) or TβRII (8), nonvascular development appears normal through E8.5, but development of the early yolk sac vasculature is highly abnormal and leads to embryonic demise in midgestation. Although ECs are present, yolk sac capillaries are markedly distended, they are detached from underlying cell layers, and they contain fewer blood cells than those found in the wild-type embryos (8). In *ALK5*-null mice, a similar defect in yolk sac vasculogenesis is observed, but differentiation of hematopoietic precursors is not affected (7). Finally, mice with EC-restricted deletion of TβRII or ALK5 show the same disruption in yolk sac vascular architecture and embryonic lethality in midgestation (101) as that in the mice with unrestricted TβRII and ALK5 deletions. Later in development, vasculogenesis from angioblasts in renal glomerular capillaries also is dependent on TGF-β1 (102).

Endoglin is expressed in the earliest recognizable FLK-1^+/Scl^- precursors (103) and in ECs surrounding the blood islands at E8.5 (86). The ES cells from endoglin-deficient mice (*Eng*$^{-/-}$) display impaired generation of FLK-1^+/Scl^- precursors with a marked reduction in myelopoiesis and erythropoiesis but essentially intact lymphopoiesis (103). In *Eng*$^{-/-}$ mice, initial differentiation of ECs from mesenchymal progenitors appears normal, although subsequent remodeling of the endothelium into yolk sac blood vessels and their investment with VSMCs is defective (6). Like endoglin, ALK1 is highly expressed in ECs at the earliest stages of blood island formation and continues to be expressed through embryogenesis into adulthood (104). In *ALK1*$^{-/-}$ mice, no discernable defect is apparent in the differentiation of mesenchymal precursors into ECs, and coalescence of ECs to form the early dorsal aorta and cardinal vein occurs normally, albeit with several connecting shunts that eventually fuse to form a single vessel (5). As with *Eng*$^{-/-}$ mice, investment of primitive blood vessels with VSMCs is arrested at E9.5. Deletion of ALK1 in zebrafish similarly results in a highly abnormal blood vessel pattern, with very dilated arteries and veins (105). Hence, endoglin and ALK1 are not essential for EC formation from precursors or their initial coalescence into blood vessels, but they are necessary for the remodeling of the primitive vessels into distinct arteries and veins and for the normal development of the VSMC layer (5).

Defective TGF-β Signaling in Endothelial Cells Leads to Defective Vascular Smooth Muscle Cell/Pericyte Recruitment

All blood vessels begin as endothelial tubes that then recruit and stimulate assembly and differentiation of their VSMC/pericyte coat. Data from endoglin- and *ALK1*-null mice indicates that direct TGF-β signaling in ECs is required to stimulate the VSMC/pericyte coat recruitment response. Because endothelium-restricted deletion of ALK5 and TβRII in mice also results in reduced investment of vessels with differentiated VSMCs, TGF-β pathway activation in ECs is a key step that directs assembly of the VSMC/pericyte coat.

A key mediator of VSMC/pericyte recruitment is endothelium-derived PDGF-B (106). In the embryo, endothelial expression of PDGF-B is restricted to immature capillary ECs and to the endothelium of arteries undergoing angiogenesis. Proliferating VSMC/pericyte progenitors expressing receptors for PDGF-B are found at sites of endothelial PDGF-B expression (107). Furthermore, recruitment of pericytes to brain, kidney glomerular, and other capillaries is critically dependent on PDGF-B synthesis by ECs and on expression of PDGFR-β receptors by the pericyte precursors (106,108). It is known that TGF-β strongly induces *PDGF-B* gene transcription in ECs (109) through a process that is Smad3 and -4 dependent (110). Taken together with the findings that PDGF-B stimulates VSMC/pericyte migration and proliferation (106) it seems likely that investment of EC tubes with VSMCs and/or pericytes involves TGF-β–dependent PDGF-B synthesis by the developing endothelium. Nevertheless, reduced PDGF-B expression by ECs in mice with defective TGF-β signaling pathways, or rescue of VSMC/pericyte recruitment in these mice by exogenous PDGF-B have not been shown, so it remains to be seen whether reduced or absent paracrine signaling via PDGF-B explains the defect in blood vessel assembly observed with interruption of the EC TGF-β signal. Defective nitric oxide (NO) generation by ECs lacking TGF-β signaling systems also may contribute to the defect in VSMC/pericyte recruitment. TGF-β stimulates expression of endothelial NO synthase (eNOS) by ECs and thereby increases the release of NO locally (111). Both, Smad2 (112) and endoglin (113) are necessary for induction of eNOS expression by TGF-β. Furthermore, the TGF-β–stimulated assembly of ECs into tubes in vitro can be blocked by L-arginine methyl ester (L-NAME), an inhibitor of NO synthesis (114), NO is a stimulus for angiogenesis in vivo (115), and EC–derived NO induces directional migration of mural cell precursors toward ECs in vitro (116). Again, it is not known whether eNOS expression and NO synthesis are altered in mice with EC-specific TGF-β signaling defects, but *eNOS*-null mice do not develop defects in vasculogenesis or VSMC/pericyte recruitment. Hence, reduced eNOS activity in mice with defective TGF-β signaling pathways may contribute to, but is not a critical component of this process.

Differentiated VSMCs are characterized by expression of smooth muscle (SM) contractile genes *SM22*, SM-myosin heavy chain (*SM-MHC*), *SM actin*, and *calponin*. These genes are strongly upregulated by TGF-β1 in cultured VSMCs, in pluripotent C3H10T1/2, and neural crest stem cells through a TGF-β control element (TCE) that shares sequence similarity with the SP1 recognition site (117) and binds the transcription factor Kruppel-like factor4 (KLF4) (118). So far, an interaction between Smads and KLF4 has not been reported. EC-restricted deletion of TβRII or ALK5, or haploinsufficiency of endoglin all are associated with a profound reduction in EC production of TGF-β1, with consequent failure of Smad2 activation in adjacent mesothelial cells (101). Because TGF-β1 stimulates VSMC differentiation, these studies suggest that investment of the early vasculature with VSMCs results, at least in part, from a paracrine TGF-β signal that in turn stimulates differentiation of VSMC precursors (101). Findings that deletion of emilin, a furin inhibitor that reduces extracellular activation of TGF-β, results in excessive TGF-β signaling in the vascular wall, thickening of arterial walls due to excessive matrix deposition, and an increase in the systemic vascular resistance and hypertension in mice (14), strongly support the notion that TGF-β plays a central role in blood vessel maturation by defining not only the EC and VSMC phenotypes, but also the volume of matrix in the vascular wall.

Taken together, the current evidence suggests that initial differentiation of ECs from the mesothelial hematopoietic/endothelial precursor requires BMP4-mediated signals, whereas assembly of ECs into nascent blood vessels and normal vascular pattern formation with distinct arteries and veins, as well as the recruitment of VSMCs and pericytes, all requires an intact TGF-β signaling cascade. The data furthermore suggest that the ALK5 pathway may be more important than ALK1 signaling for the assembly of ECs into blood vessels, whereas the ALK1 pathway seems critical for the establishment of the normal vascular architecture and proper investment of the blood vessels with VSMCs or pericytes. Vessel assembly and pericyte recruitment also involve indirect actions of TGF-β mediated through enhanced PDGF-B and possibly NO synthesis, and TGF-β appears to be critical in determining the volume of ECM of the vascular wall.

Developmental Remodeling Through Endothelial Apoptosis

Remodeling of EC monolayers in vitro in response to TGF-β is associated with apoptosis of a subset of the cells (72,119). Apoptosis of ECs in response to TGF-β is dependent on p38 MAPK activation (95), has been ascribed to reduced bcl-2 expression (120), and is associated with reduced expression of the cyclin-dependent kinase inhibitors p21$^{waf/cip}$ (121). Similar mechanisms underlie TGF-β–induced apoptosis in a number of other cell types, and resistance to apoptosis is thought to contribute, in some instances, to tumor growth (122). The sensitivity of ECs to the apoptotic actions of TGF-β is greater than that of VSMCs or pericytes (123). Because endoglin overexpression protects ECs from TGF-β–stimulated apoptosis (124,125), and the overexpression of constitutively active ALK5 (but not ALK1) induces apoptosis in human umbilical vein ECs (67), it seems likely the apoptotic response of ECs to TGF-β results from activation of the ALK/Smad2/3 pathway. In vivo, TGF-β–dependent apoptosis is thought to play a role in renal glomerular capillary morphogenesis. The glomerular capillary bed develops through assembly of capillary cords initially lacking lumens. Because both apoptosis and lumen formation are inhibited by neutralizing TGF-β antibody, capillary lumen formation in these capillaries is thought to involve TGF-β–dependent removal of superfluous cells by apoptosis (126).

THE ANGIOGENIC RESPONSE

In Vitro and Experimental Angiogenesis

Many studies show that TGF-β can stimulate assembly of ECs into capillary-like cords in vitro (127), and sprouting angiogenesis with proliferation of cultured ECs in response to TGF-β has been described (128). Application of TGF-β in vivo, for example, in chick chorioallantoic membrane (129), in rabbit ischemic hindlimb (130), in mice implanted subcutaneously with TGF-β impregnated disks (75), or in TGF-β1 producing tumors (74) all result in angiogenesis.

Angiogenesis requires dissolution of the EC basement membrane and the perivascular matrix at the tip of a new vessel sprout (131,132), followed by stabilization of the new vessel and its investment with VSMCs or pericytes. As already discussed above, very strong evidence now suggests that TGF-β can stimulate proliferation and migration of ECs at low concentrations and that this signal is carried by endoglin, ALK1, and Smad1 and -5. The migration response is not well understood but may result, in part, from alterations in the cellular repertoire of integrins, because TGF-β increases synthesis of $\alpha 2$, $\alpha 5$, and $\beta 1$ integrins in ECs during angiogenesis (133). TGF-β also regulates the process of matrix dissolution via the plasminogen–plasmin system. Both the u-PA and t-PA are essential for in vivo angiogenic responses (134,135). Because PAI-1 is produced in large amounts by ECs stimulated with high dose TGF-β1, suggesting that the angiogenic process is dampened by TGF-β1 via ALK5–Smad2/3 pathway stimulated PAI-1 production (136). The role of the uPA/plasmin system in the process of angiogenesis has, however, been questioned because invasion of fibrin gels by new angiogenic sprouts is not inhibited in *uPA*- or *plasminogen*-null mice. Instead, the potent fibrinolytic metalloproteinase of endothelium, membrane type-1 matrix metalloproteinase (MMP) (MT1-MMP) accounts for the invasiveness of sprouting vessels (131,137). This context is of interest because decorin, a matrix protein that acts as a ligand trap for TGF-β, is a substrate for MT1-MMP (137). Degradation of decorin by MT1-MMP would tend to augment availability of TGF-β during matrix dissolution. TGF-β itself can induce MT1-MMP expression in non-ECs (138), but data for ECs are not yet available. In retinal ECs, TGF-β increases the production of MMP-9 (139), and in bovine ECs, the expression of gelatinase A/MMP-2 and B/MMP-9 is upregulated by TGF-β1. It therefore is likely that TGF-β participates in the angiogenic invasiveness by augmenting MMP synthesis by ECs at the tip of the sprout, but conclusive work still must be done.

Proliferation of ECs during the angiogenic response is stimulated not only by TGF-β, via the ALK1 pathway, but also through VEGF and fibroblast growth factor-2 (FGF2) derived from neighboring cells (140). VEGF synthesis is induced in response to TGF-β in several cell types, thus indirectly stimulating angiogenesis (141). Because TGF-β1 acts synergistically with VEGF and FGF2 to promote invasion by angiogenic sprouts (76,142), these mediators act in concert to stimulate growth and invasiveness by angiogenic vessels.

The angiogenic response is followed by a regulated process of vessel stabilization, with investment of the EC tube with VSMCs or pericytes, the formation of basement membrane, and the generation of ECM. The antiproliferative effect of TGF-β mediated by ALK5 seems to play a dominant role during this phase of the angiogenic response (143), and the induction by TGF-β of PAI-1 may antagonize the invasiveness of already formed vascular sprouts (135). Furthermore, TGF-β, acting via ALK5, stimulates synthesis of matrix proteins in most cells, including ECs (144). In cultured pulmonary artery ECs, for example, TGF-β stimulates production of type IV procollagen (145), and induction of fibronectin by TGF-β seems almost universal (127,146). In addition, TGF-β likely plays a central role in the recruitment and differentiation of VSMCs/pericytes during angiogenesis. It seems plausible that the two opposing TGF-β signaling systems in ECs, ALK1-Smad1/5 and ALK5-Smad2/3, respectively, orchestrate the initial proliferative and then the resolution phase of the angiogenic response. Autoinduction of TGF-β in ECs (147) may further raise TGF-β concentrations to levels at which a predominant ALK5/Smad2/3 response is elicited during the resolution phase.

Tumor Angiogenesis and the TGF-β System

The TGF-β signaling cascade, acting via ALK5 in non-ECs, is a major tumor suppressor system. Indeed, Smad4 was first recognized as a tumor suppressor in colon epithelium (148), and many tumors cells, including those derived from breast and colon, harbor somatic mutations that interrupt TGF-β signaling (149,150). By contrast, the type III TGF-β coreceptor endoglin is strongly expressed in the blood vessels of many tumor tissues (151). Indeed, endoglin seems to be a much better marker for tumor endothelium than are other markers of differentiated endothelium (91). A high microvessel density (MVD) in tumors, as determined by endoglin antibody labeling, is associated with shortened patient survival (91,152). Endoglin is shed into the circulation and, although it is unable to bind TGF-β in this form, high circulating endoglin levels in patients correlate with the presence of tumor metastasis (153). Endoglin antibodies have been used to image tumor microvessels in mice (154), dogs (155), and human patients (91). Therapeutic targeting of ECs in tumors using endoglin antibodies in experimental models are positive thus far. For example, synergy between antiendoglin antibodies and cyclophosphamide was observed in SCID mice bearing human tumors (156), and immunoradioisotopes suppress tumor growth in murine models bearing breast and colon carcinoma, without apparent systemic side effects (157). Nonetheless, a provocative paper by Pece-Barbara and colleagues (158) reported that absence of endoglin in ECs derived from *Eng*$^{-/-}$ mice resulted in powerful enhancement of TGF-β–stimulated Smad1 and -5 phosphorylation and cell proliferation, whereas the presence of endoglin was associated with greater TGF-β–mediated suppression of EC proliferation. Hence, it is not clear whether TGF-β–mediated signaling in ECs contributes to tumor angiogenesis or not. Still, endoglin is a promising tool for tumor vessel imaging and could potentially serve as a therapeutic target for antiangiogenic therapy in tumors.

ENDOTHELIAL–MESENCHYMAL TRANSDIFFERENTIATION DURING CARDIAC CUSHION DEVELOPMENT

The vascular tube that eventually forms the heart, also referred to as the atrioventricular (AV) canal, contains an inner lining of ECs surrounded by cell-free cardiac jelly composed of ECM, and an outer layer of myocardial cells. The formation of the cardiac cushions from which the heart valves arise and which is necessary for normal cardiac septation, involves endothelial–mesenchymal transdifferentiation (EMT) of a subset of ECs that then invade and proliferate in the cardiac jelly (159).

During the process of transdifferentiation, the cells begin to express mesenchymal markers, for example tenacin and α-SM actin, while endothelial markers, for example platelet-endothelial cell adhesion molecule (PECAM)-1, are lost (160). Abnormalities in this process can lead to defective heart valve formation or abnormal septation of the heart. Such defects occur in approximately 1% of human live births and carry with them a high morbidity.

Disruption of TGF-β and BMP signaling pathways results in cardiac cushion defects. TGF-β1 and TGF-β3 are highly expressed in the AV canal where the cardiac cushion forms (161–163). In human embryos, TβRII and ALK5 are ubiquitously expressed in all cells of the AV canal, with endoglin expression being restricted to the ECs in that region. The high level of endoglin expression is regulated temporally and associated with heart septation and valve formation (164). In the presence of myocardium, TGF-β1 can induce the transdifferentiation of AV canal ECs (161). Also, in AV cushion explants, antibodies directed against TβRII, or against TGF-β3 (165), inhibit transdifferentiation. In *endoglin*-null mice, heart development is arrested at E9, and the atrioventricular canal endocardium fails to undergo mesenchymal transformation and cushion-tissue formation (166). Taken together, these observations indicate that TGF-β1 and -3, likely acting directly on the ECs of the AV canal in an endoglin-dependent fashion, induce transdifferentiation of ECs to form the mesenchymal cells of the cardiac cushion.

In the developing mouse, BMP2 and -4 also are expressed in the cells of the developing cardiac cushion, but they are not restricted to the endothelium (167), and BMP2 has a synergistic effect with TGF-β1 on endothelial–mesenchymal transdifferentiation (168). Finally, constitutively active ALK2, but not ALK5, can induce transdifferentiation of AV canal ECs, a process that is inhibited by Smad6 (169). Because Smad6 inhibits Smad1 and -5 but not Smad2 and -3 activation (see Table 35-1), it seems likely that in the ECs of the primitive AV canal, either TGF-β1 or -3 acting on ALK1, and/or BMP2 acting via ALK2, signal the transdifferentiation of the ECs into mesenchymal cells.

Of great interest for this review, patients with Marfan syndrome, due to mutations in the *fibrillin-1* gene, develop cardiac valve abnormalities quite often, most commonly mitral valve prolapse. As already discussed, fibrillin-1 binds LTBPs, and their interaction with LAP play an important role in the local regulation of TGF-β activation. In mice with *fibrillin-1* mutations, excess activation of TGF-β not only leads to a defect in the lung that is similar to the pulmonary defect observed in patients with Marfan syndrome (33), but they also develop progressive mitral valve prolapse resembling that found in human patients (34). Moreover, a family with Marfan syndrome has been identified in whom mutations in *TβRII*, not *fibrillin-1*, produce the disorder (170). It is therefore attractive to speculate that the cardiac and vascular abnormalities in human patients with Marfan syndrome relate to the role of TGF-β signaling during the formation of the cardiac cushion or later valve remodeling.

ROLE OF TGF-β IN ATHEROSCLEROSIS

In VSMCs, TGF-β signals via the ALK5–Smad2/3 pathway induce differentiation and inhibit proliferation. If the same mechanisms hold in the media of mature blood vessels, TGF-β would tend to oppose expansion of the VSMC population and neointima formation. Several lines of evidence support the theory that TGF-β indeed acts as a protective cytokine in the mature vascular wall, and that loss of TGF-β activity or interruption of TGF-β signaling can participate in the formation of atherosclerotic plaques (171). Furthermore, because TGF-β tends to inhibit inflammatory cell adhesion and transmigration across the endothelium, and is critically important in suppressing T cell proliferation (172), it may also protect blood vessels from injury by inflammatory cells.

Clonal expansion of VSMC populations in atherosclerotic plaque has been linked to decreased expression and somatic mutation of the *TβRII* gene and consequently lower levels of receptor expression (173). Loss of TGF-β signaling, as described in tumors, could therefore result in the VSMC proliferation so characteristic in atherosclerotic plaques (173). Moreover, in mice, activation of TGF-β from the SCL (see Figure 35.1) is inhibited by high levels of apolipoprotein A and has been linked to VSMC proliferation and development of fatty streaks (171). Also, the human lipoprotein variant Lp(a), which is strongly associated with increased risk of atherosclerosis, can inhibit TGF-β activation both in vitro and when overexpressed in transgenic mice (171). Reduced availability of TGF-β due to inhibition of activation and reduced TβRII receptor expression may therefore contribute to VSMC proliferation in atherosclerosis.

Suppression of inflammation by TGF-β also has been postulated to protect the vascular wall from formation of atherosclerotic plaque (171). Deletion of TGF-β1 in mice results in a multifocal, mixed inflammatory cell response, lethal within the first 2 months of life (174), an effect that relates to its role in T-cell differentiation (172). In addition, TGF-β acting on ECs suppresses cytokine-stimulated adhesion and transmigration of monocytes (175), T lymphocytes (176), and neutrophils (177). Inflammatory cell adhesion to ECs is reduced at least in part through inhibition of cytokine-stimulated E-selectin expression (178) as well as inhibition of chemoattractant interleukin (IL)-8 (177) and monocyte chemoattractant protein (MCP)-1 (179) expression by TGF-β. Little or no effect of TGF-β on EC vascular cell adhesion molecule (VCAM)-1 or intercellular adhesion molecule (ICAM)-1 expression has been observed (178,180). Similarly, *P-selectin* gene expression is not changed, but the rapid thrombin-stimulated translocation of Weibel-Palade contents, P-selectin, and von Willebrand factor (vWF) in ECs is markedly suppressed by TGF-β (181). Therefore, both the cytokine-stimulated recruitment of leukocytes by chemokines and their initial adhesion to endothelium via E- and P-selectin are generally inhibited by TGF-β. It has been postulated that reduced TGF-β action, perhaps because of ineffective activation, can lead to more aggressive inflammatory lesions in the vasculature.

In cardiac allografts, which represent an immune-dependent model of atherosclerosis, Smad3 deficiency is associated with accelerated intimal hyperplasia with increased infiltration of adventitial macrophages expressing MCP-1, suggesting that TGF-β, acting via Smad3, is important in regulating MCP-1 expression in vivo and that its absence can augment vascular injury in this model (182). Similarly, TGF-β1 exerts a significant cardioprotective effect in a feline model of myocardial ischemia and reperfusion that appears related TGF-β1–mediated inhibition of neutrophil adhesion to the endothelium (183).

Overall therefore, accumulating evidence suggests that TGF-β has an anti-inflammatory effect in blood vessels largely due to inhibition of the leukocyte–EC interaction, and that it can inhibit neointima formation through inhibition of VSMC proliferation. Loss of TGF-β function in the vessel wall could therefore contribute to the development of atherosclerotic lesions.

HUMAN HEREDITARY VASCULAR DISORDERS DUE TO MUTATIONS IN HUMAN HEREDITARY TGF-β SIGNALING PATHWAYS

Hemorrhagic Telangiectasia

Hemorrhagic telangiectasia (HHT) is an autosomal disorder in humans also known as Osler-Weber-Rendu syndrome (discussed in detail in Chapter 122). Mutations in the *endoglin* gene located on chromosome 9q34 (1) and in the *ALK1* gene located on chromosome 12.q11 (2) cause HHT-1 and HHT-2, respectively. A number of different mutations for both genes have been reported, all associated with loss of function (3). A third locus for HHT had been proposed but the family in question was later found to have mutations in the *ALK1* gene (3). Patients with *endoglin* mutations develop more pulmonary AV malformations, and they experience epistaxis and telangiectasias earlier in life than do those with *ALK1* mutations (184).

Analysis of mice with heterozygous mutations in *endoglin* serves as a good model for HHT. They develop generalized dilation of veins with an absence of investing VSMCs in more than 70% of vessels, leading to dilation and the characteristic telengiectatic lesions (185). On the 129/Ola or mixed C57BL/6–129/Ola background, heterozygous *endoglin* mutations sometimes are associated with epistaxis, as in human patients (166). The penetrance of the HHT1 phenotype in mice is strongly dependent on the genetic background, thus indicating that several modifier genes influence the function of endoglin in the vasculature (186). Heterozygous mutations in the *ALK1* gene also lead to vascular dysplasia reminiscent of the human disorder (187). These mice display variable penetrance of the phenotype, which consists of vascular lesions in skin and oral mucosa, as well as lesions in the lung, gut, spleen, brain, and liver. Like humans, mice with mutations in *ALK1* gene suffer from gastrointestinal hemorrhage at a much greater rate than expected (187).

The mechanism by which vascular abnormalities develop in humans and in mice with heterozygous *endoglin* or *ALK1* mutations still needs much clarification. Certainly, a simple defect in EC proliferation and migration cannot account for the phenotype, nor can abnormal VSMC/pericyte recruitment. The progressive development of AV malformations implies defective vascular remodeling over time. It has been suggested that a somatic mutation in the normal allele could explain the progressive, focal vascular malformations (3), although this has yet to be confirmed.

Hereditary Pulmonary Arterial Hypertension

Hereditary pulmonary arterial hypertension (PAH) is a vascular disorder characterized by a progressive reduction in the lumen diameter of pulmonary arteries. This leads to a sustained increase in pulmonary vascular resistance that impedes ejection of blood by the right ventricle and causes subsequent right ventricular failure (188). PAH generally can be ascribed to nongenetic causes. In patients without an identifiable underlying cause for PAH, the term primary pulmonary hypertension (PHH) is also used. The prevalence of hereditary PAH is estimated to be 1 to 2 per 1,000,000 live births, and the disease can be sporadic or familial.

Two independent groups identified heterozygous mutations in the gene encoding the BMPRII (BMP receptor II) as the underlying cause of hereditary PAH (189,190) and, in a substantial portion of patients with sporadic PAH similar mutations of the *BMPRII* gene were observed (191). As is the case with ALK1 and *endoglin* mutations, incomplete penetrance may result from as-yet-unknown modifier genes. Indeed, in some families with HHT-2 without *BMPRII* mutations, severe PAH has been observed and ascribed to mutations in *ALK1* (192). The observation that mutations in two different but mechanistically related genes, *ALK1* and *BMPRII*, can produce the same clinical phenotype points to TGF-β receptors as an important molecular pathway at the origin of pulmonary vascular remodeling. Whereas *ALK1* mutations can lead to hereditary PAH, HHT-2, or both, *BMPRII* mutations appear to lead only to the PAH phenotype. It has been suggested that *BMPRII* mutations are involved mainly in VSMC proliferation and ALK1 in endothelial or VSM proliferation. Nonetheless, in patients with hereditary PAH, monoclonal expansion ECs from the pulmonary lesions has been shown and suggest that EC proliferation may underlie the compromise in blood vessel lumen in these patients (193).

KEY POINTS

- The TGF-β signaling system is unique in ECs, in that two distinct type I TGF-β receptors, ALK1 and ALK5, are activated by two of the three TGF-β

isoforms, TGF-β1 and TGF-β3. This isoform selectivity is conferred by the endothelium-restricted TGF-β coreceptor, endoglin. The two signaling pathways in ECs provide for opposing biological effects: activation of EC development and angiogenesis via the ALK1 pathway, and inhibition of angiogenesis with stabilization of the new vessels through activation of the ALK5 pathway.

- TGF-β signaling is essential for normal vascular development because deletion of any component of the cascade in ECs – the convertase furin that cleaves TGF-β1 from its propeptide, TGF-β1 itself, the coreceptor endoglin, TβRII, both TGF-β type I receptors ALK1 and ALK5, as well as Smad1 and Smad5 – all produce major defects in embryonic vascular development resulting in death at midgestation.
- Vascular disorders in human patients are produced by mutations in *TGF-β* and *BMP* receptor genes. Haploinsufficiency of endoglin or ALK1 is responsible for the human vascular disorder HHT, and haploinsufficiency of BMPRII results in the hereditary form of PAH.
- A unique and highly regulated process of activation controls the spatial and temporal actions of TGF-β. The mature form of the three TGF-β isoforms remain latent after synthesis through noncovalent association with a portion of their own propeptide, and this latent TGF-β is sequestered in the ECM. Defects in TGF-β sequestration likely underlie the cardiac abnormalities in Marfan syndrome. The sequestered TGF-β serves as a sensor of tissue remodeling during angiogenesis, because it is released in the active form upon matrix dissolution by proteases at the tip of newly forming vessels.

Future Goals

- The relatively recent discovery of two TGF-β signaling pathways in ECs begs the re-examination of a whole host of EC responses to TGF-β, as much of the in vitro work to date has been done with concentrations of TGF-β that would tend to activate ALK5 but inhibit the ALK1 pathway. Because ALK1 is so essential in vascular development, this is an important problem to be solved. In this regard, the exact mechanism by which endoglin acts in ECs, and importantly, in tumor vessels, is not well understood at this time.
- The role of sequestered latent TGF-β is fascinating and still poorly understood. Biologists often propose that cytokine concentration gradients in tissues produce distinct cellular responses. It seems likely that sequestered TGF-β could form such gradients, depending on the degree of activation during angiogenesis and development. Methods to study the precise spatial and temporal activation of the TGF-β system would open new avenues to study vascular development, angiogenesis, and vessel maturation.
- Although much is currently known about prominent EC responses to TGF-β, such as proliferation, migration, apoptosis, vessel assembly, and matrix synthesis, and many EC genes positively or negatively regulated by TGF-β have been identified, it is evident from the human vascular disorders HHT and PAH that TGF-β plays a major role in vessel remodeling throughout life. These long-term effects are not understood at all. Indeed, the three-dimensional patterning of blood vessels seems powerfully influenced by TGF-β, but we currently have only a very rudimentary understanding of this process. From work in mice and zebrafish, it is evident that TGF-β is required to keep ECs of the venous and arterial beds separate. How this is accomplished, and how TGF-β effects differ in distinct EC subpopulations, remain open questions.
- Also, as can be gleaned from the section on atherosclerosis, plausible mechanisms by which TGF-β deficiency could play a role in atherosclerotic plaque development exist, but whether they actually participate in human disease remains to be proved. If ineffective TGF-β activation indeed leaves the vasculature unprotected, this would open a real opportunity for the development of therapeutic agents that mimic TGF-β. On the other hand, if effects of low-dose TGF-β have been overlooked so far, who knows what would happen if TGF-β or its mimics were used as a therapeutic tool. Conversely, it is known that excess TGF-β plays a role in the fibrotic injury response in many organs, and inhibitors of it are being studied as therapeutic tools for those disorders. Will we unmask cardiovascular disease when we begin using such agents clinically? Similarly, in the process of tumor angiogenesis, it is unclear whether inhibition or stimulation of TGF-β action would have therapeutic value.
- Selective ALK5 kinase inhibitors already have been produced, but because ALK5 is necessary for ALK1 action, both pathways would be affected in ECs. It seems that development of ALK1 agonists and antagonists will be more relevant for EC biologists, and potentially also as human therapeutic tools.
- Finally, TGF-β does not act in isolation. Positive and negative influences by TGF-β on other mediators that regulate blood vessels are known, and other mediators influence TGF-β synthesis,

activation, and action. Over time, we will develop an ever more complete understanding of the vast array of interacting mechanisms on the vasculature. Bioinformatics tools may well be needed to help predict how a single stimulus might alter the whole network of interacting responses in the endothelium.

ACKNOWLEDGMENTS

This work was supported by grant DK50764 from the National Institutes of Health, and by grants MOP641814 and CEG63108 from the Canadian Institutes of Health Research. The author holds a Tier 1 Canada Research Chair in Endothelial Cell Biology.

REFERENCES

1 McAllister KA, Grogg KM, Johnson DW, et al. Endoglin, a TGF-beta binding protein of endothelial cells, is the gene for hereditary haemorrhagic telangiectasia type 1. *Nat Genet*. 1994;8:345–351.
2 Johnson DW, Berg JN, Baldwin MA, et al. Mutations in the activin receptor-like kinase 1 gene in hereditary haemorrhagic telangiectasia type 2. *Nat Genet*. 1996;13:189–195.
3 Marchuk DA, Srinivasan S, Squire TL, Zawistowski JS. Vascular morphogenesis: tales of two syndromes. *Hum Mol Genet*. 2003; 12 Spec No 1:R97–R112.
4 Dickson MC, Martin JS, Cousins FM, et al. Defective haematopoiesis and vasculogenesis in transforming growth factor-beta 1 knock out mice. *Development*. 1995;121:1845–1854.
5 Urness LD, Sorensen LK, Li DY. Arteriovenous malformations in mice lacking activin receptor-like kinase-1. *Nat Genet*. 2000; 26:328–331.
6 Li DY, Sorensen LK, Brooke BS, et al. Defective angiogenesis in mice lacking endoglin. *Science*. 1999;284:1534–1537.
7 Larsson J, Goumans MJ, Sjöstrand LJ, et al. Abnormal angiogenesis but intact hematopoietic potential in TGF-beta type I receptor-deficient mice. *EMBO J*. 2001;20:1663–1673.
8 Oshima M, Oshima H, Taketo MM. TGF-beta receptor type II deficiency results in defects of yolk sac hematopoiesis and vasculogenesis. *Dev Biol*. 1996;179:297–302.
9 Chang H, Huylebroeck D, Verschueren K, et al. Smad5 knockout mice die at mid-gestation due to multiple embryonic and extraembryonic defects. *Development*. 1999;126:1631–1642.
10 Jadrich JL, O'Connor MB, Coucouvanis E. The TGF{beta} activated kinase TAK1 regulates vascular development in vivo. *Development*. 2006;133:1529–1541.
11 Roebroek AJ, Umans L, Pauli IG, et al. Failure of ventral closure and axial rotation in embryos lacking the proprotein convertase Furin. *Development*. 1998;125:4863–4876.
12 Chang H, Brown CW, Matzuk MM. Genetic analysis of the mammalian transforming growth factor-beta superfamily. *Endocr Rev*. 2002;23:787–823.
13 Shi Y, Massagué J. Mechanisms of TGF-beta signaling from cell membrane to the nucleus. *Cell*. 2003;113:685–700.
14 Zacchigna L, Vecchione C, Notte A, et al. Emilin1 links TGF-beta maturation to blood pressure homeostasis. *Cell*. 2006; 124:929–942.
15 Annes JP, Munger JS, Rifkin DB. Making sense of latent TGFbeta activation. *J Cell Sci*. 2003;116:217–224.
16 Rifkin DB. Latent transforming growth factor-beta (TGF-beta) binding proteins: orchestrators of TGF-beta availability. *J Biol Chem*. 2005;280:7409–7412.
17 Nunes I, Gleizes PE, Metz CN, Rifkin DB. Latent transforming growth factor-beta binding protein domains involved in activation and transglutaminase-dependent cross-linking of latent transforming growth factor-beta. *J Cell Biol*. 1997;136: 1151–1163.
18 Hyytiäinen M, Keski-Oja J. Latent TGF-beta binding protein LTBP-2 decreases fibroblast adhesion to fibronectin. *J Cell Biol*. 2003;163:1363–1374.
19 Munger JS, Huang X, Kawakatsu H, et al. The integrin alpha v beta 6 binds and activates latent TGF beta 1: a mechanism for regulating pulmonary inflammation and fibrosis. *Cell*. 1999; 96:319–328.
20 Antonelli-Orlidge A, Saunders KB, Smith SR, D'Amore PA. An activated form of transforming growth factor beta is produced by cocultures of endothelial cells and pericytes. *Proc Natl Acad Sci USA*. 1989;86:4544–4548.
21 Sato Y, Tsuboi R, Lyons R, et al. Characterization of the activation of latent TGF-beta by co-cultures of endothelial cells and pericytes or smooth muscle cells: a self-regulating system. *J Cell Biol*. 1990;111:757–763.
22 Grainger DJ, Wakefield L, Bethell HW, et al. Release and activation of platelet latent TGF-beta in blood clots during dissolution with plasmin. *Nat Med*. 1995;1:932–937.
23 Dennis PA, Rifkin DB. Cellular activation of latent transforming growth factor beta requires binding to the cation-independent mannose 6-phosphate/insulin-like growth factor type II receptor. *Proc Natl Acad Sci USA*. 1991;88:580–584.
24 Petzelbauer E, Springhorn JP, Tucker AM, Madri JA. Role of plasminogen activator inhibitor in the reciprocal regulation of bovine aortic endothelial and smooth muscle cell migration by TGF-beta 1. *Am J Pathol*. 1996;149:923–931.
25 Kojima S, Vernooy R, Moscatelli D, et al. Lipopolysaccharide inhibits activation of latent transforming growth factor-beta in bovine endothelial cells. *J Cell Physiol*. 1995;163:210–219.
26 Hertig A, Berrou J, Allory Y, et al. Type 1 plasminogen activator inhibitor deficiency aggravates the course of experimental glomerulonephritis through overactivation of transforming growth factor beta. *FASEB J*. 2003;17:1904–1906.
27 Flaumenhaft R, Abe M, Sato Y, et al. Role of the latent TGF-beta binding protein in the activation of latent TGF-beta by co-cultures of endothelial and smooth muscle cells. *J Cell Biol*. 1993;120:995–1002.
28 Gualandris A, Annes JP, Arese M, et al. The latent transforming growth factor-beta-binding protein-1 promotes in vitro differentiation of embryonic stem cells into endothelium. *Mol Biol Cell*. 2000;11:4295–4308.
29 Annes JP, Chen Y, Munger JS, Rifkin DB. Integrin alphaVbeta6-mediated activation of latent TGF-beta requires the latent TGF-beta binding protein-1. *J Cell Biol*. 2004;165: 723–734.
30 Keski-Oja J, Koli K, von Melchner H. TGF-beta activation by traction? *Trends Cell Biol*. 2004;14:657–659.

31 Munger JS, Harpel JG, Giancotti FG, Rifkin DB. Interactions between growth factors and integrins: latent forms of transforming growth factor-beta are ligands for the integrin alphavbeta1. *Mol Biol Cell.* 1998;9:2627–2638.
32 Isogai Z, Ono RN, Ushiro S, et al. Latent transforming growth factor beta-binding protein 1 interacts with fibrillin and is a microfibril-associated protein. *J Biol Chem.* 2003;278:2750–2757.
33 Neptune ER, Frischmeyer PA, Arking DE, et al. Dysregulation of TGF-beta activation contributes to pathogenesis in Marfan syndrome. *Nat Genet.* 2003;33:407–411.
34 Ng CM, Cheng A, Myers LA, et al. TGF-beta-dependent pathogenesis of mitral valve prolapse in a mouse model of Marfan syndrome. *J Clin Invest.* 2004;114:1586–1592.
35 Lawler J. Thrombospondin-1 as an endogenous inhibitor of angiogenesis and tumor growth. *J Cell Mol Med.* 2002;6:1–12.
36 Schultz-Cherry S, Murphy-Ullrich JE. Thrombospondin causes activation of latent transforming growth factor-beta secreted by endothelial cells by a novel mechanism. *J Cell Biol.* 1993;122: 923–932.
37 Schultz-Cherry S, Ribeiro S, Gentry L, Murphy-Ullrich JE. Thrombospondin binds and activates the small and large forms of latent transforming growth factor-beta in a chemically defined system. *J Biol Chem.* 1994;269:26775–26782.
38 Ribeiro SM, Poczatek M, Schultz-Cherry S, et al. The activation sequence of thrombospondin-1 interacts with the latency-associated peptide to regulate activation of latent transforming growth factor-beta. *J Biol Chem.* 1999;274:13586–13593.
39 Crawford SE, Stellmach V, Murphy-Ullrich JE, et al. Thrombospondin-1 is a major activator of TGF-beta1 in vivo. *Cell.* 1998;93:1159–1170.
40 Young GD, Murphy-Ullrich JE. The tryptophan-rich motifs of the thrombospondin type 1 repeats bind VLAL motifs in the latent transforming growth factor-beta complex. *J Biol Chem.* 2004;279:47633–47642.
41 Ameye L, Young MF. Mice deficient in small leucine-rich proteoglycans: novel in vivo models for osteoporosis, osteoarthritis, Ehlers-Danlos syndrome, muscular dystrophy, and corneal diseases. *Glycobiology.* 2002;12:107R–116R.
42 Avsian-Kretchmer O, Hsueh AJ. Comparative genomic analysis of the eight-membered ring cystine knot-containing bone morphogenetic protein antagonists. *Mol Endocrinol.* 2004;18:1–12.
43 Hildebrand A, Romarís M, Rasmussen LM, et al. Interaction of the small interstitial proteoglycans biglycan, decorin and fibromodulin with transforming growth factor beta. *Biochem J.* 1994;302 (Pt 2):527–534.
44 Lin SY, Morrison JR, Phillips DJ, de Kretser DM. Regulation of ovarian function by the TGF-beta superfamily and follistatin. *Reproduction.* 2003;126:133–148.
45 Yanagita M.BMP antagonists: Their roles in development and involvement in pathophysiology. *Cytokine Growth Factor Rev.* 2005;16:309–317.
46 Branford WW, Yost HJ. Nodal signaling: CrypticLefty mechanism of antagonism decoded. *Curr Biol.* 2004;14:R341–R343.
47 Moser M, Binder O, Wu Y, et al. BMPER, a novel endothelial cell precursor-derived protein, antagonizes bone morphogenetic protein signaling and endothelial cell differentiation. *Mol Cell Biol.* 2003;23:5664–5679.
48 Sun PD, Davies DR. The cystine-knot growth-factor superfamily. *Annu Rev Biophys Biomol Struct.* 1995;24:269–291.
49 López-Casillas F, Wrana JL, Massagué J. Betaglycan presents ligand to the TGF beta signaling receptor. *Cell.* 1993;73:1435–1444.
50 Cheifetz S, Bellón T, Calés C, et al. Endoglin is a component of the transforming growth factor-beta receptor system in human endothelial cells. *J Biol Chem.* 1992;267:19027–19030.
51 Samad TA, Rebbapragada A, Bell E, et al. DRAGON, a bone morphogenetic protein co-receptor. *J Biol Chem.* 2005;280:14122–14129.
52 Abreu JG, Ketpura NI, Reversade B, De Robertis EM. Connective-tissue growth factor (CTGF) modulates cell signalling by BMP and TGF-beta. *Nat Cell Biol.* 2002;4:599–604.
53 Lin J, Patel SR, Cheng X, et al. Kielin/chordin-like protein, a novel enhancer of BMP signaling, attenuates renal fibrotic disease. *Nat Med.* 2005;11:387–393.
54 Chen YG, Massagué J. Smad1 recognition and activation by the ALK1 group of transforming growth factor-beta family receptors. *J Biol Chem.* 1999;274:3672–3677.
55 Tsukazaki T, Chiang TA, Davison AF, et al. SARA, a FYVE domain protein that recruits Smad2 to the TGFbeta receptor. *Cell.* 1998;95:779–791.
56 Wu G, Chen YG, Ozdamar B, et al. Structural basis of Smad2 recognition by the Smad anchor for receptor activation. *Science.* 2000;287:92–97.
57 Panopoulou E, Gillooly DJ, Wrana JL, et al. Early endosomal regulation of Smad-dependent signaling in endothelial cells. *J Biol Chem.* 2002;277:18046–18052.
58 Lin HK, Bergmann S, Pandolfi PP. Cytoplasmic PML function in TGF-beta signalling. *Nature.* 2004;431:205–211.
59 Xu L, Chen YG, Massagué J. The nuclear import function of Smad2 is masked by SARA and unmasked by TGFbeta-dependent phosphorylation. *Nat Cell Biol.* 2000;2:559–562.
60 Xu L, Alarcón C, Cöl S, Massagué J. Distinct domain utilization by Smad3 and Smad4 for nucleoporin interaction and nuclear import. *J Biol Chem.* 2003;278:42569–42577.
61 Chen HB, Rud JG, Lin K, Xu L. Nuclear targeting of transforming growth factor-beta-activated Smad complexes. *J Biol Chem.* 2005;280:21329–21336.
62 Xiao Z, Brownawell AM, Macara IG, Lodish HF. A novel nuclear export signal in Smad1 is essential for its signaling activity. *J Biol Chem.* 2003;278:34245–34252.
63 Onichtchouk D, Chen YG, Dosch R, et al. Silencing of TGF-beta signalling by the pseudoreceptor BAMBI. *Nature.* 1999;401: 480–485.
64 Shi W, Sun C, He B, et al. GADD34-PP1c recruited by Smad7 dephosphorylates TGFbeta type I receptor. *J Cell Biol.* 2004;164: 291–300.
65 Topper JN, Cai J, Qiu Y, et al. Vascular MADs: two novel MAD-related genes selectively inducible by flow in human vascular endothelium. *Proc Natl Acad Sci USA.* 1997;94:9314–9319.
66 Nakao A, Afrakhte M, Morén A, et al. Identification of Smad7, a TGFbeta-inducible antagonist of TGF-beta signalling. *Nature.* 1997;389:631–635.
67 Ota T, Fujii M, Sugizaki T, et al. Targets of transcriptional regulation by two distinct type I receptors for transforming growth factor-beta in human umbilical vein endothelial cells. *J Cell Physiol.* 2002;193:299–318.
68 Imamura T, Takase M, Nishihara A, et al. Smad6 inhibits signalling by the TGF-beta superfamily. *Nature.* 1997;389:622–626.

69 Lin X, Liang YY, Sun B, et al. Smad6 recruits transcription corepressor CtBP to repress bone morphogenetic protein-induced transcription. *Mol Cell Biol.* 2003;23:9081–9093.
70 Izzi L, Attisano L. Regulation of the TGFbeta signalling pathway by ubiquitin-mediated degradation. *Oncogene.* 2004;23:2071–2078.
71 Heimark RL, Twardzik DR, Schwartz SM. Inhibition of endothelial regeneration by type-beta transforming growth factor from platelets. *Science.* 1986;233:1078–1080.
72 Choi ME, Ballermann BJ. Inhibition of capillary morphogenesis and associated apoptosis by dominant negative mutant transforming growth factor-beta receptors. *J Biol Chem.* 1995;270:21144–21150.
73 Madri JA, Pratt BM, Tucker AM. Phenotypic modulation of endothelial cells by transforming growth factor-beta depends upon the composition and organization of the extracellular matrix. *J Cell Biol.* 1988;106:1375–1384.
74 Ueki N, Nakazato M, Ohkawa T, et al. Excessive production of transforming growth-factor beta 1 can play an important role in the development of tumorigenesis by its action for angiogenesis: validity of neutralizing antibodies to block tumor growth. *Biochim Biophys Acta.* 1992;1137:189–196.
75 Fajardo LF, Prionas SD, Kwan HH, et al. Transforming growth factor beta1 induces angiogenesis in vivo with a threshold pattern. *Lab Invest.* 1996;74:600–608.
76 Pepper MS, Vassalli JD, Orci L, Montesano R. Biphasic effect of transforming growth factor-beta 1 on in vitro angiogenesis. *Exp Cell Res.* 1993;204:356–363.
77 Goumans MJ, Valdimarsdottir G, Itoh S, et al. Balancing the activation state of the endothelium via two distinct TGF-beta type I receptors. *EMBO J.* 2002;21:1743–1753.
78 Gougos A, Letarte M. Primary structure of endoglin, an RGD-containing glycoprotein of human endothelial cells. *J Biol Chem.* 1990;265:8361–8364.
79 St-Jacques S, Cymerman U, Pece N, Letarte M. Molecular characterization and in situ localization of murine endoglin reveal that it is a transforming growth factor-beta binding protein of endothelial and stromal cells. *Endocrinology.* 1994;134:2645–2657.
80 López-Casillas F, Cheifetz S, Doody J, et al. Structure and expression of the membrane proteoglycan betaglycan, a component of the TGF-beta receptor system. *Cell.* 1991;67:785–795.
81 Sankar S, Mahooti-Brooks N, Centrella M, et al. Expression of transforming growth factor type III receptor in vascular endothelial cells increases their responsiveness to transforming growth factor beta 2. *J Biol Chem.* 1995;270:13567–13572.
82 Barbara NP, Wrana JL, Letarte M. Endoglin is an accessory protein that interacts with the signaling receptor complex of multiple members of the transforming growth factor-beta superfamily. *J Biol Chem.* 1999;274:584–594.
83 Panchenko MP, Williams MC, Brody JS, Yu Q. Type I receptor serine-threonine kinase preferentially expressed in pulmonary blood vessels. *Am J Physiol.* 1996;270:L547–L558.
84 Seki T, Yun J, Oh SP. Arterial endothelium-specific activin receptor-like kinase 1 expression suggests its role in arterialization and vascular remodeling. *Circ Res.* 2003;93:682–689.
85 Seki T, Hong KH, Yun J, et al. Isolation of a regulatory region of activin receptor-like kinase 1 gene sufficient for arterial endothelium-specific expression. *Circ Res.* 2004;94:E72–E77.
86 Jonker L, Arthur HM. Endoglin expression in early development is associated with vasculogenesis and angiogenesis. *Mech Dev.* 2002;110:193–196.
87 Goumans MJ, Valdimarsdottir G, Itoh S, et al. Activin receptor-like kinase (ALK)1 is an antagonistic mediator of lateral TGF-beta/ALK5 signaling. *Mol Cell.* 2003;12:817–828.
88 Blanco FJ, Santibanez JF, Guerrero-Esteo M, et al. Interaction and functional interplay between endoglin and ALK-1, two components of the endothelial transforming growth factor-beta receptor complex. *J Cell Physiol.* 2005;204:574–584.
89 Guerrero-Esteo M, Sanchez-Elsner T, Letamendia A, Bernabeu C. Extracellular and cytoplasmic domains of endoglin interact with the transforming growth factor-beta receptors I and II. *J Biol Chem.* 2002;277:29197–291209.
90 Lebrin F, Goumans MJ, Jonker L, et al. Endoglin promotes endothelial cell proliferation and TGF-beta/ALK1 signal transduction. *EMBO J.* 2004;23:4018–4028.
91 Duff SE, Li C, Garland JM, Kumar S. CD105 is important for angiogenesis: evidence and potential applications. *FASEB J.* 2003;17:984–992.
92 Derynck R, Zhang YE. Smad-dependent and Smad-independent pathways in TGF-beta family signalling. *Nature.* 2003;425:577–584.
93 Yamaguchi K, Shirakabe K, Shibuya H, et al. Identification of a member of the MAPKKK family as a potential mediator of TGF-beta signal transduction. *Science.* 1995;270:2008–2011.
94 Goldberg PL, MacNaughton DE, Clements RT, et al. p38 MAPK activation by TGF-beta1 increases MLC phosphorylation and endothelial monolayer permeability. *Am J Physiol Lung Cell Mol Physiol.* 2002;282:L146–L154.
95 Hyman KM, Seghezzi G, Pintucci G, et al. Transforming growth factor-beta1 induces apoptosis in vascular endothelial cells by activation of mitogen-activated protein kinase. *Surgery.* 2002;132:173–179.
96 Wang L, Kwak JH, Kim SI, et al. Transforming growth factor-beta1 stimulates vascular endothelial growth factor 164 via mitogen-activated protein kinase kinase 3-p38alpha and p38delta mitogen-activated protein kinase-dependent pathway in murine mesangial cells. *J Biol Chem.* 2004;279:33213–33219.
97 Clements RT, Minnear FL, Singer HA, et al. RhoA and Rho-kinase dependent and independent signals mediate TGF-beta-induced pulmonary endothelial cytoskeletal reorganization and permeability. *Am J Physiol Lung Cell Mol Physiol.* 2005;288:L294–L306.
98 Schmidt-Weber CB, Letarte M, Kunzmann S, et al. TGF-{beta} signaling of human T cells is modulated by the ancillary TGF-{beta} receptor endoglin. *Int Immunol.* 2005;17:921–930.
99 Park C, Afrikanova I, Chung YS, et al. A hierarchical order of factors in the generation of FLK1- and SCL-expressing hematopoietic and endothelial progenitors from embryonic stem cells. *Development.* 2004;131:2749–2762.
100 Zhang XJ, Tsung HC, Caen JP, et al. Vasculogenesis from embryonic bodies of murine embryonic stem cells transfected by Tgf-beta1 gene. *Endothelium.* 1998;6:95–106.
101 Carvalho RL, Jonker L, Goumans MJ, et al. Defective paracrine signalling by TGFbeta in yolk sac vasculature of endoglin mutant mice: a paradigm for hereditary haemorrhagic telangiectasia. *Development.* 2004;131:6237–6247.
102 Liu A, Dardik A, Ballermann BJ. Neutralizing TGF-beta1 antibody infusion in neonatal rat delays in vivo glomerular capillary formation 1. *Kidney Int.* 1999;56:1334–1348.

103 Cho SK, Bourdeau A, Letarte M, Zúñiga-Pflücker JC. Expression and function of CD105 during the onset of hematopoiesis from Flk1(+) precursors. *Blood*. 2001;98:3635–3642.
104 Roelen BA, van Rooijen MA, Mummery CL. Expression of ALK-1, a type 1 serine/threonine kinase receptor, coincides with sites of vasculogenesis and angiogenesis in early mouse development. *Dev Dyn*. 1997;209:418–430.
105 Roman BL, Pham VN, Lawson ND, et al. Disruption of acvrl1 increases endothelial cell number in zebrafish cranial vessels. *Development*. 2002;129:3009–3019.
106 Betsholtz C. Insight into the physiological functions of PDGF through genetic studies in mice. *Cytokine Growth Factor Rev*. 2004;15:215–228.
107 Hellström M, Kalén M, Lindahl P, et al. Role of PDGF-B and PDGFR-beta in recruitment of vascular smooth muscle cells and pericytes during embryonic blood vessel formation in the mouse. *Development*. 1999;126:3047–3055.
108 Levéen P, Pekny M, Gebre-Medhin S, et al. Mice deficient for PDGF B show renal, cardiovascular, and hematological abnormalities. *Genes Dev*. 1994;8:1875–1887.
109 Daniel TO, Gibbs VC, Milfay DF, Williams LT. Agents that increase cAMP accumulation block endothelial c-sis induction by thrombin and transforming growth factor-beta. *J Biol Chem*. 1987;262:11893–11896.
110 Taylor LM, Khachigian LM. Induction of platelet-derived growth factor B-chain expression by transforming growth factor-beta involves transactivation by Smads. *J Biol Chem*. 2000;275:16709–16716.
111 Inoue N, Venema RC, Sayegh HS, et al. Molecular regulation of the bovine endothelial cell nitric oxide synthase by transforming growth factor-beta 1. *Arterioscler Thromb Vasc Biol*. 1995; 15:1255–1261.
112 Saura M, Zaragoza C, Cao W, et al. Smad2 mediates transforming growth factor-beta induction of endothelial nitric oxide synthase expression. *Circ Res*. 2002;91:806–813.
113 Jerkic M, Rivas-Elena JV, Prieto M, et al. Endoglin regulates nitric oxide-dependent vasodilatation. *FASEB J*. 2004;18:609–611.
114 Papapetropoulos A, Desai KM, Rudic RD, et al. Nitric oxide synthase inhibitors attenuate transforming-growth-factor-beta 1-stimulated capillary organization in vitro. *Am J Pathol*. 1997; 150:1835–1844.
115 Ziche M, Morbidelli L, Masini E, et al. Nitric oxide mediates angiogenesis in vivo and endothelial cell growth and migration in vitro promoted by substance P. *J Clin Invest*. 1994;94:2036–2044.
116 Kashiwagi S, Izumi Y, Gohongi T, et al. NO mediates mural cell recruitment and vessel morphogenesis in murine melanomas and tissue-engineered blood vessels. *J Clin Invest*. 2005; 115(7):1816–1827
117 Hautmann MB, Madsen CS, Owens GK. A transforming growth factor beta (TGFbeta) control element drives TGFbeta-induced stimulation of smooth muscle alpha-actin gene expression in concert with two CArG elements. *J Biol Chem*. 1997;272:10948–10956.
118 King KE, Iyemere VP, Weissberg PL, Shanahan CM. Krüppel-like factor 4 (KLF4/GKLF) is a target of bone morphogenetic proteins and transforming growth factor beta 1 in the regulation of vascular smooth muscle cell phenotype. *J Biol Chem*. 2003; 278:11661–11669.
119 Pollman MJ, Naumovski L, Gibbons GH. Endothelial cell apoptosis in capillary network remodeling. *J Cell Physiol*. 1999;178: 359–370.
120 Tsukada T, Eguchi K, Migita K, et al. Transforming growth factor beta 1 induces apoptotic cell death in cultured human umbilical vein endothelial cells with down-regulated expression of bcl-2. *Biochem Biophys Res Commun*. 1995;210:1076–1082.
121 Yan Q, Sage EH. Transforming growth factor-beta1 induces apoptotic cell death in cultured retinal endothelial cells but not pericytes: association with decreased expression of p21waf1/cip1. *J Cell Biochem*. 1998;70:70–83.
122 Siegel PM, Massagué J. Cytostatic and apoptotic actions of TGF-beta in homeostasis and cancer. *Nat Rev Cancer*. 2003;3:807–821.
123 Pollman MJ, Naumovski L, Gibbons GH. Vascular cell apoptosis: cell type-specific modulation by transforming growth factor-beta1 in endothelial cells versus smooth muscle cells. *Circulation*. 1999;99:2019–2026.
124 Li C, Issa R, Kumar P, et al. CD105 prevents apoptosis in hypoxic endothelial cells. *J Cell Sci*. 2003;116:2677–2685.
125 Li C, Hampson IN, Hampson L, et al. CD105 antagonizes the inhibitory signaling of transforming growth factor beta1 on human vascular endothelial cells. *FASEB J*. 2000;14:55–64.
126 Fierlbeck W, Liu A, Coyle R, Ballermann BJ. Endothelial cell apoptosis during glomerular capillary lumen formation in vivo. *J Am Soc Nephrol*. 2003;14:1349–1354.
127 Merwin JR, Anderson JM, Kocher O, et al. Transforming growth factor beta 1 modulates extracellular matrix organization and cell-cell junctional complex formation during in vitro angiogenesis. *J Cell Physiol*. 1990;142:117–128.
128 Iruela-Arispe ML, Sage EH. Endothelial cells exhibiting angiogenesis in vitro proliferate in response to TGF-beta 1. *J Cell Biochem*. 1993;52:414–430.
129 Yang EY, Moses HL. Transforming growth factor beta 1-induced changes in cell migration, proliferation, and angiogenesis in the chicken chorioallantoic membrane. *J Cell Biol*. 1990;111:731–741.
130 van Royen N, Hoefer I, Buschmann I, et al. Exogenous application of transforming growth factor beta 1 stimulates arteriogenesis in the peripheral circulation. *FASEB J*. 2002;16:432–434.
131 Hiraoka N, Allen E, Apel IJ, et al. Matrix metalloproteinases regulate neovascularization by acting as pericellular fibrinolysins. *Cell*. 1998;95:365–377.
132 Carmeliet P. Angiogenesis in health and disease. *Nat Med*. 2003;9:653–660.
133 Enenstein J, Waleh NS, Kramer RH. Basic FGF and TGF-beta differentially modulate integrin expression of human microvascular endothelial cells. *Exp Cell Res*. 1992;203:499–503.
134 Saksela O, Moscatelli D, Rifkin DB. The opposing effects of basic fibroblast growth factor and transforming growth factor beta on the regulation of plasminogen activator activity in capillary endothelial cells. *J Cell Biol*. 1987;105:957–963.
135 Oh CW, Hoover-Plow J, Plow EF. The role of plasminogen in angiogenesis in vivo. *J Thromb Haemost*. 2003;1:1683–1687.
136 Pepper MS, Belin D, Montesano R, et al. Transforming growth factor-beta 1 modulates basic fibroblast growth factor-induced proteolytic and angiogenic properties of endothelial cells in vitro. *J Cell Biol*. 1990;111:743–755.
137 Sato H, Takino T, Miyamori H. Roles of membrane-type matrix metalloproteinase-1 in tumor invasion and metastasis. *Cancer Sci*. 2005;96:212–217.
138 Munshi HG, Wu YI, Mukhopadhyay S, et al. Differential regulation of membrane type 1-matrix metalloproteinase activity

by ERK 1/2- and p38 MAPK-modulated tissue inhibitor of metalloproteinases 2 expression controls transforming growth factor-beta1-induced pericellular collagenolysis. *J Biol Chem.* 2004;279:39042–39050.

139 Behzadian MA, Wang XL, Windsor LJ, et al. TGF-beta increases retinal endothelial cell permeability by increasing MMP-9: possible role of glial cells in endothelial barrier function. *Invest Ophthalmol Vis Sci.* 2001;42:853–859.

140 Ferrara N, Gerber HP, LeCouter J. The biology of VEGF and its receptors. *Nat Med.* 2003;9:669–676.

141 Pertovaara L, Kaipainen A, Mustonen T, et al. Vascular endothelial growth factor is induced in response to transforming growth factor-beta in fibroblastic and epithelial cells. *J Biol Chem.* 1994;269:6271–6274.

142 Sánchez-Elsner T, Botella LM, Velasco B, et al. Synergistic cooperation between hypoxia and transforming growth factor-beta pathways on human vascular endothelial growth factor gene expression. *J Biol Chem.* 2001;276:38527–38535.

143 Lebrin F, Deckers M, Bertolino P, Ten Dijke P. TGF-beta receptor function in the endothelium. *Cardiovasc Res.* 2005;65:599–608.

144 Leask A, Abraham DJ. TGF-beta signaling and the fibrotic response. *FASEB J.* 2004;18:816–827.

145 Shanker G, Olson D, Bone R, Sawhney R. Regulation of extracellular matrix proteins by transforming growth factor beta1 in cultured pulmonary endothelial cells. *Cell Biol Int.* 1999;23:61–72.

146 Müller G, Behrens J, Nussbaumer U, et al. Inhibitory action of transforming growth factor beta on endothelial cells. *Proc Natl Acad Sci USA.* 1987;84:5600–5604.

147 Van Obberghen-Schilling E, Roche NS, Flanders KC, et al. Transforming growth factor beta 1 positively regulates its own expression in normal and transformed cells. *J Biol Chem.* 1988;263:7741–7746.

148 Thiagalingam S, Lengauer C, Leach FS, et al. Evaluation of candidate tumour suppressor genes on chromosome 18 in colorectal cancers. *Nat Genet.* 1996;13:343–346.

149 Benson JR. Role of transforming growth factor beta in breast carcinogenesis. *Lancet Oncol.* 2004;5:229–239.

150 Arends JW. Molecular interactions in the Vogelstein model of colorectal carcinoma. *J Pathol.* 2000;190:412–416.

151 Burrows FJ, Derbyshire EJ, Tazzari PL, et al. Up-regulation of endoglin on vascular endothelial cells in human solid tumors: implications for diagnosis and therapy. *Clin Cancer Res.* 1995;1:1623–1634.

152 Kumar S, Ghellal A, Li C, et al. Breast carcinoma: vascular density determined using CD105 antibody correlates with tumor prognosis. *Cancer Res.* 1999;59:856–861.

153 Li C, Guo B, Wilson PB, et al. Plasma levels of soluble CD105 correlate with metastasis in patients with breast cancer. *Int J Cancer.* 2000;89:122–126.

154 Bredow S, Lewin M, Hofmann B, et al. Imaging of tumour neovasculature by targeting the TGF-beta binding receptor endoglin. *Eur J Cancer.* 2000;36:675–681.

155 Fonsatti E, Altomonte M, Nicotra MR, et al. Endoglin (CD105): a powerful therapeutic target on tumor-associated angiogenetic blood vessels. *Oncogene.* 2003;22:6557–6563.

156 Takahashi N, Haba A, Matsuno F, Seon BK. Antiangiogenic therapy of established tumors in human skin/severe combined immunodeficiency mouse chimeras by anti-endoglin (CD105) monoclonal antibodies, and synergy between anti-endoglin antibody and cyclophosphamide. *Cancer Res.* 2001;61:7846–7854.

157 Tabata M, Kondo M, Haruta Y, Seon BK. Antiangiogenic radioimmunotherapy of human solid tumors in SCID mice using (125)I-labeled anti-endoglin monoclonal antibodies. *Int J Cancer.* 1999;82:737–742.

158 Pece-Barbara N, Vera S, Kathirkamathamby K, et al. Endoglin null endothelial cells proliferate faster, and more responsive to TGFbeta 1 with higher affinity receptors and an activated ALK1 pathway. *J Biol Chem.* 2005;280:27800–27809

159 Armstrong EJ, Bischoff J. Heart valve development: endothelial cell signaling and differentiation. *Circ Res.* 2004;95:459–470.

160 Sugi Y, Yamamura H, Okagawa H, Markwald RR. Bone morphogenetic protein-2 can mediate myocardial regulation of atrioventricular cushion mesenchymal cell formation in mice. *Dev Biol.* 2004;269:505–518.

161 Potts JD, Runyan RB. Epithelial-mesenchymal cell transformation in the embryonic heart can be mediated, in part, by transforming growth factor beta. *Dev Biol.* 1989;134:392–401.

162 Akhurst RJ, Lehnert SA, Faissner A, Duffie E. TGF beta in murine morphogenetic processes: the early embryo and cardiogenesis. *Development.* 1990;108:645–656.

163 Potts JD, Vincent EB, Runyan RB, Weeks DL. Sense and antisense TGF beta 3 mRNA levels correlate with cardiac valve induction. *Dev Dyn.* 1992;193:340–345.

164 Qu R, Silver MM, Letarte M. Distribution of endoglin in early human development reveals high levels on endocardial cushion tissue mesenchyme during valve formation. *Cell Tissue Res.* 1998;292:333–343.

165 Ramsdell AF, Markwald RR. Induction of endocardial cushion tissue in the avian heart is regulated, in part, by TGFbeta-3-mediated autocrine signaling. *Dev Biol.* 1997;188:64–74.

166 Bourdeau A, Dumont DJ, Letarte M. A murine model of hereditary hemorrhagic telangiectasia. *J Clin Invest.* 1999;104:1343–1351.

167 Keyes WM, Logan C, Parker E, Sanders EJ. Expression and function of bone morphogenetic proteins in the development of the embryonic endocardial cushions. *Anat Embryol (Berl).* 2003;207:135–147.

168 Yamagishi T, Nakajima Y, Miyazono K, Nakamura H. Bone morphogenetic protein-2 acts synergistically with transforming growth factor-beta3 during endothelial-mesenchymal transformation in the developing chick heart. *J Cell Physiol.* 1999;180:35–45.

169 Desgrosellier JS, Mundell NA, McDonnell MA, et al. Activin receptor-like kinase 2 and Smad6 regulate epithelial-mesenchymal transformation during cardiac valve formation. *Dev Biol.* 2005;280:201–210.

170 Mizuguchi T, Collod-Beroud G, Akiyama T, et al. Heterozygous TGFBR2 mutations in Marfan syndrome. *Nat Genet.* 2004;36:855–860.

171 Grainger DJ. Transforming growth factor beta and atherosclerosis: so far, so good for the protective cytokine hypothesis. *Arterioscler Thromb Vasc Biol.* 2004;24:399–404.

172 Schmidt-Weber CB, Blaser K. Regulation and role of transforming growth factor-beta in immune tolerance induction and inflammation. *Curr Opin Immunol.* 2004;16:709–716.

173 McCaffrey TA, Du B, Consigli S, et al. Genomic instability in the type II TGF-beta1 receptor gene in atherosclerotic and restenotic vascular cells. *J Clin Invest.* 1997;100:2182–2188.

174 Shull MM, Ormsby I, Kier AB, et al. Targeted disruption of the mouse transforming growth factor-beta 1 gene results in multifocal inflammatory disease. *Nature*. 1992;359:693–699.

175 Cai JP, Falanga V, Chin YH. Transforming growth factor-beta regulates the adhesive interactions between mononuclear cells and microvascular endothelium. *J Invest Dermatol*. 1991;97: 169–174.

176 Gamble JR, Vadas MA. Endothelial cell adhesiveness for human T lymphocytes is inhibited by transforming growth factor-beta 1. *J Immunol*. 1991;146:1149–1154.

177 Smith WB, Noack L, Khew-Goodall Y, et al. Transforming growth factor-beta 1 inhibits the production of IL-8 and the transmigration of neutrophils through activated endothelium. *J Immunol*. 1996;157:360–368.

178 Gamble JR, Khew-Goodall Y, Vadas MA. Transforming growth factor-beta inhibits E-selectin expression on human endothelial cells. *J Immunol*. 1993;150:4494–4503.

179 Weiss JM, Cuff CA, Berman JW. TGF-beta downmodulates cytokine-induced monocyte chemoattractant protein (MCP)-1 expression in human endothelial cells. A putative role for TGF-beta in the modulation of TNF receptor expression. *Endothelium*. 1999;6:291–302.

180 Bereta J, Bereta M, Cohen S, Cohen MC. Regulation of VCAM-1 expression and involvement in cell adhesion to murine microvascular endothelium. *Cell Immunol*. 1993;147:313–330.

181 Harris H, Kirschenlohr H, Szabados N, Metcalfe J. Transforming growth factor-beta1 inhibits thrombin activation of endothelial cells. *Cytokine*. 2004;25:85–93.

182 Feinberg MW, Shimizu K, Lebedeva M, et al. Essential role for Smad3 in regulating MCP-1 expression and vascular inflammation. *Circ Res*. 2004;94:601–608.

183 Lefer AM, Ma XL, Weyrich AS, Scalia R. Mechanism of the cardioprotective effect of transforming growth factor beta 1 in feline myocardial ischemia and reperfusion. *Proc Natl Acad Sci USA*. 1993;90:1018–1022.

184 Berg J, Porteous M, Reinhardt D, et al. Hereditary haemorrhagic telangiectasia: a questionnaire based study to delineate the different phenotypes caused by endoglin and ALK1 mutations. *J Med Genet*. 2003;40:585–590.

185 Torsney E, Charlton R, Diamond AG, et al. Mouse model for hereditary hemorrhagic telangiectasia has a generalized vascular abnormality. *Circulation*. 2003;107:1653–1657.

186 Bourdeau A, Faughnan ME, McDonald ML, et al. Potential role of modifier genes influencing transforming growth factor-beta1 levels in the development of vascular defects in endoglin heterozygous mice with hereditary hemorrhagic telangiectasia. *Am J Pathol*. 2001;158:2011–2020.

187 Srinivasan S, Hanes MA, Dickens T, et al. A mouse model for hereditary hemorrhagic telangiectasia (HHT) type 2. *Hum Mol Genet*. 2003;12:473–482.

188 Eddahibi S, Morrell N, d'Ortho MP, et al. Pathobiology of pulmonary arterial hypertension. *Eur Respir J*. 2002;20:1559–1572.

189 Lane KB, Machado RD, Pauciulo MW, et al. Heterozygous germline mutations in BMPR2, encoding a TGF-beta receptor, cause familial primary pulmonary hypertension. The International PPH Consortium. *Nat Genet*. 2000;26:81–84.

190 Deng Z, Morse JH, Slager SL, et al. Familial primary pulmonary hypertension (gene PPH1) is caused by mutations in the bone morphogenetic protein receptor-II gene. *Am J Hum Genet*. 2000;67:737–744.

191 Thomson JR, Machado RD, Pauciulo MW, et al. Sporadic primary pulmonary hypertension is associated with germline mutations of the gene encoding BMPR-II, a receptor member of the TGF-beta family. *J Med Genet*. 2000;37:741–745.

192 Trembath RC, Thomson JR, Machado RD, et al. Clinical and molecular genetic features of pulmonary hypertension in patients with hereditary hemorrhagic telangiectasia. *N Engl J Med*. 2001;345:325–334.

193 Lee SD, Shroyer KR, Markham NE, et al. Monoclonal endothelial cell proliferation is present in primary but not secondary pulmonary hypertension. *J Clin Invest*. 1998;101:927–934.

194 Lechleider RJ, Ryan JL, Garrett L, et al. Targeted mutagenesis of Smad1 reveals an essential role in chorioallantoic fusion. *Dev Biol*. 2001;240:157–167.

195 Yang X, Castilla LH, Xu X, et al. Angiogenesis defects and mesenchymal apoptosis in mice lacking SMAD5. *Development*. 1999;126:1571–1580.

36

Thrombospondins

Sareh Parangi and Jack Lawler

Beth Israel Deaconess Medical Center, Harvard Medical School, Boston, Massachusetts

The thrombospondins (TSPs) are a family of proteins that regulate cellular phenotype and matrix structure during tissue remodeling and genesis (1,2). Of the five proteins that comprise the family, TSP-1 and -2 have been shown to be involved in vascular development and angiogenesis (3,4). The endothelial cell (EC) integrates positive and negative signals for proliferation, migration, survival, and apoptosis during physiological and pathological angiogenesis. TSP-1 and -2 suppress the angiogenic response by inhibiting migration and antagonizing vascular endothelial growth factor (VEGF)-induced survival. In addition, TSP-1 and -2 stimulate apoptotic pathways in ECs (5). The importance of TSP-1 as a negative regulator of angiogenesis is underscored by the fact that genetic mutations that occur in tumor cells often lead to reduced TSP-1 expression (2). Furthermore, a TSP-1–based therapeutic is currently in clinical trials for the treatment of cancer. Although generally an inhibitor of angiogenesis, TSP-1 has been shown to support angiogenesis in some contexts. This seemingly paradoxical effect stems from the fact that various cell types respond differently to the protein. Whereas TSP-1 inhibits EC migration, it supports vascular smooth muscle cell (VSMC) and inflammatory cell migration. Thus, if the induction of angiogenesis occurs in the presence of inflammation, TSP-1 can appear to promote angiogenesis. In addition, the N-terminal heparin-binding domain of TSP-1 assayed in isolation reportedly stimulates angiogenesis (6). These complex responses arise because TSP-1 and -2 are multifunctional proteins with a wide range of cell surface receptors. The cellular response to these proteins in part depends on which receptors are expressed by a given cell type and which receptors are engaged by the various domains of TSP-1 or -2.

HISTORY OF THE THROMBOSPONDINS

TSP-1 was first identified in 1971 as a 190,000-dalton protein that appeared in the supernatant of thrombin-treated platelets (7). Although it was originally thought to be cleaved from the surface of platelets, it was soon appreciated that the protein was secreted instead from α-granules as part of the platelet release reaction (8). Subsequent studies revealed that TSP-1 was synthesized by megakaryocytes as a trimer of identical subunits with a molecular weight of 142,000 (9,10). TSP-1 electrophoreses at an anomalously high molecular weight because it has an unusually high content of aspartic acid. Early studies focused on platelet TSP-1 but, in 1981 and 1982, two groups showed that TSP-1 is synthesized and secreted by fibroblasts and ECs in culture (11,12). These observations led to the cloning and sequencing of full-length human *TSP-1* from an EC cDNA library in 1986 (Table 36-1) (13). The comparison of cDNA and genomic clones that were isolated from murine libraries revealed the existence of TSP-2, a protein with the same domain structure as TSP-1 (14). Genomic cloning of the mucin *MUC-1* led to the identification of mouse *TSP-3*; the isolation of thrombospondins from a frog cDNA library led to the identification of *TSP-4*; and the cloning and sequencing of rat cartilage oligomeric matrix protein (*COMP*) or *TSP-5* resulted in its identification as the fifth member of the thrombospondin gene family (15–17). To explore the functions of the thrombospondins, several groups have determined the expression of mRNA and protein for each family member, and knockout mice have been generated (see Table 36-1). These studies reveal that most tissues express at least one family member, with many tissues synthesizing multiple thrombospondins (see later).

In 1971, Judah Folkman proposed that tumors must induce angiogenesis in order to grow beyond a few millimeters in size (18). To stimulate angiogenesis, tumor cells express growth factors like VEGF to stimulate the growth of capillaries. These observations led to the concept that antiangiogenic approaches could inhibit tumor growth. In 1987, when working with one of the first antiangiogenic compounds, TNP-470, Ingber and Folkman made the observation that whereas TNP-470 inhibited primary tumor growth, it led to the growth of metastases (19). In addition, it was noted that the upregulation of VEGF in pancreatic tumors that form in *Rip-Tag* mice occurs well before the induction of angiogenesis (20). These observations suggested that angiogenesis was being suppressed by naturally occurring inhibitors. In 1990, Noel Bouck's laboratory reported the identification of the first natural

Table 36-1: Cloning and Sequencing of Full-Length Human TSP-1

	Cloned	*Receptors*	*Embryo*	*Adult*	*EC Signaling*	*−/− Phenotype*
TSP-1	1986	PGs, integrins, CD36, calreticulin, LPR, CD47	Cardiac myocytes, neuroepithelium, cartilage and bone, liver, muscle, various epithelia, capillaries (late)	Platelets, monocytes, lung epithelium, cardiac and skeletal muscle	✓	Pneumonia, delayed wound closure, increased vessel density in some tissues
TSP-2	1991	PGs, integrins, CD36, LRP	Connective tissue, cartilage and bone, heart, kidney, large blood vessels and capillaries (early)	Skin, brain, cardiac muscle, brain	✓	Abnormal collagen fibrils, accelerated wound closure, increased vessel density in some tissues
TSP-3	1992	PGs	Neurons, lung, cartilage, brain	Lung, bone, cardiac muscle	–	Variations in postnatal bone maturation
TSP-4	1993	PGs	Somatic mesoderm, muscle, eye	Cardiac and skeletal muscle	✓	Not published
TSP-5/ COMP	1992	Integrins	Chondrocytes	Cartilage, tendon, ligament, synovium VSMCs (restenosis)	–	None

This table includes data from studies on human, mouse, rat, chicken, cow, and frog.

inhibitor of angiogenesis, TSP-1 (21). This group assayed the ability of culture supernatants from fibroblasts with a mutation in a tumor suppressor gene to inhibit EC migration. N-terminal amino acid sequencing of a protein in the active fraction revealed that a proteolytic fragment of TSP-1 was the key component. The identification of TSP-1 as an endogenous inhibitor of angiogenesis supported the hypothesis that angiogenesis is controlled by a balance between stimulators and inhibitors. Angiogenesis occurs when this balance is shifted in favor of the stimulators. Subsequent studies showed that the nondegraded protein also is active and that the effect is mediated by the interaction of the thrombospondin type 1 repeats (TSRs) with CD36, one of several TSP-1 receptors on the surface of ECs (discussed later) (2,22). Because TSP-2 also contains the TSRs, it is also a potent endogenous inhibitor of angiogenesis. In this chapter, we focus on molecular mechanisms underlying the inhibition of angiogenesis by TSP-1 and -2, and discuss recent data on TSP-1–based cancer therapies. The functions of TSP-3, TSP-4, and COMP are poorly understood, but it is unlikely that they inhibit angiogenesis because they do not contain TSRs. A role for TSP-4 in vascular function has been suggested by the observation that a polymorphism leads to a proline at position 387 instead of the more common alanine. This polymorphism reportedly is associated with an increased risk of early-onset cardiovascular disease (23). TSP-4 has been shown to be expressed by brain capillary ECs and aortic tissue. Although the two forms of TSP-4 do have differential effects on EC proliferation and migration in vitro, no clear link is known between TSP-4 and cardiovascular disease at this time (23).

EVOLUTIONARY CONSIDERATIONS

The five members of the thrombospondin family can be divided into two groups on the basis of their domain and tertiary structure. All the thrombospondins contain type-2 (epidermal growth factor [EGF]-like) repeats, type-3 repeats, and a highly conserved C-terminal domain. The type-3 repeats are composed of low- and high-affinity calcium binding sites that collaborate with the C-terminal to bind as many as 26 calcium atoms per subunit (24). The C-terminal is a 15-stranded concave β-sandwich structure that interacts with the type-3 repeats at multiple points (24). The five proteins differ in that TSP-1 and -2 (subgroup A) contain two domains that are not present in TSP-3, -4, and COMP (subgroup B): (a) a region of homology with procollagen and (b) the TSRs. The TSRs are found in numerous proteins (approximately 100 in the human genome), some of which are involved in the regulation of angiogenesis and axon guidance. The tertiary structure of the two subgroups differs in that the subgroup A proteins are trimers and the subgroup B proteins are pentamers.

Most vertebrates probably express all five TSP genes. Progressive sequence alignment algorithms suggest that the gene duplications that produced these proteins occurred approximately 500,000 years ago, during the vertebrate radiation (16). The primitive chordate, *Ciona intestinalis*, has two TSPs, one similar to the subgroup A and one with the domain structure of the subgroup B TSPs (25). A TSP gene also is found in invertebrate genomes. *Drosophila melanogaster*, *Anopheles gambiae*, and the silk moth *Bombyx mori* have a TSP that has the subgroup B molecular architecture. Like the vertebrate

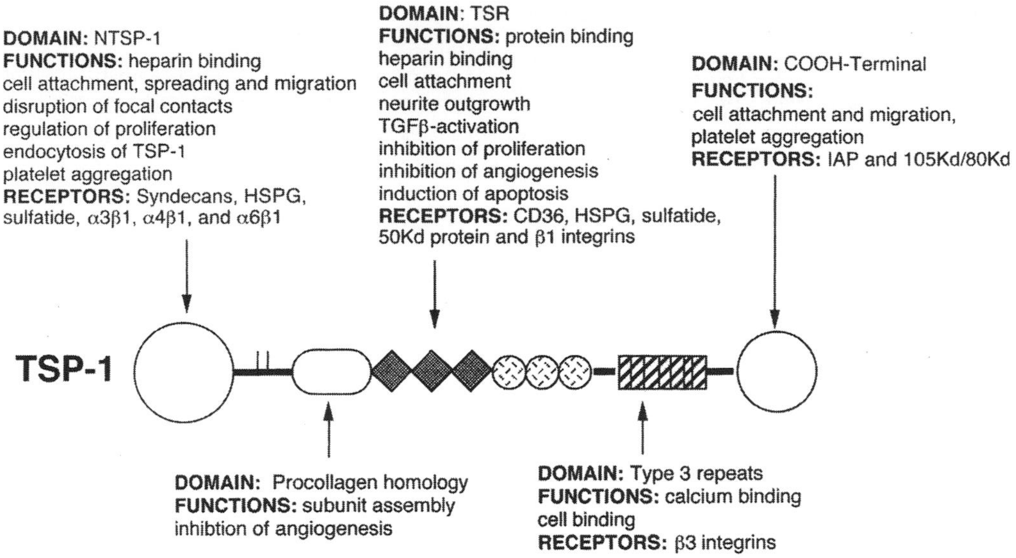

Figure 36.1. A schematic representation of the structural and functional organization of TSP-1.

subgroup B thrombospondins, the *Drosophila* protein is pentameric (26). Interestingly, the *Caenorhabditis elegans* genome does not have a TSP gene.

In TSP-1, the N-terminal and C-terminal domains are similar to each other in terms of both their structures and functions (Figure 36.1). Both domains contain cell binding sites, and antibodies to either domain inhibit platelet aggregation. Thus, the earliest forms of the protein may have been involved in cell-to-cell adhesion or communication. This assertion is consistent with the fact that RGD sequences are present in both the N-terminal and C-terminal domain of insect TSP. The insertion of the four type-2 repeats may have ensured the proper distance between the N- and C- terminal, and the multimerization may have increased valency and thus affinity of the protein for all surface receptors. Early during vertebrate development, a gene duplication may have occurred, with one copy of the gene having the opportunity to acquire new properties. One of the type-2 repeats was replaced by three TSRs by exon shuffling from another gene. Eventually, this new protein gave rise to the modern versions of TSP-1 and -2. The model suggests that the subgroup A TSPs have specialized functions that support vertebrate tissue genesis and remodeling.

DEVELOPMENTAL CONSIDERATIONS

The TSPs are widely expressed in developing vertebrates, with most tissues expressing at least one member of the family (see Table 36-1). Northern blot analysis indicates that human adult cardiac muscle expresses all the TSPs except COMP. TSP-4 and COMP are expressed primarily in skeletal muscle and cartilage, respectively. In addition, COMP has been detected in neointimal VSMCs in restenotic vessel walls (27). Only TSP-1 appears to be present in human blood platelets. In tissues that express multiple TSP genes, distinct temporal patterns of expression are often observed during embryonic development. Whereas TSP-2 is expressed in large vessels and capillaries throughout most of murine development, TSP-1 is only present in capillaries late in vascular development. As a result of the differences in the spatial and temporal patterns of expression of TSP-1 and -2, very limited opportunities may exist for these proteins to compensate for one another. The observation that the phenotype of the *TSP-1* and *-2* double null mice is, for the most part, equivalent to the sum of the phenotypes of the *TSP-1* and *-2* single knockout mice is consistent with this view (28). However, the two proteins do appear to compensate for each other in the formation of neural synapses; in vitro data indicate that both TSP-1 and -2 can support synaptogenesis, and the number of synapses in the double-null mice is significantly lower than that seen in the wild-type mice (29). The vasculature appears to develop normally in *TSP-1*– and *TSP-2*–null mice; however, increased vessel density in some tissues has been observed (30,31). The single- and double-null mice demonstrate increased vascular density in the same tissues with the extent of the increase in the double-null being equivalent to either one of the single knockouts. For example, vessel density in the iris in the double-null mice is significantly higher than in the wild-type and *TSP-1*–null mice, but not significantly different from that of the *TSP-2*–null mice (31). By contrast, aberrant wound healing in the double-null mice resembles that of the *TSP-1*–null mice and is significantly different from the wild-type and *TSP-2*–null mice (28). This seems to be due to the fact that the TSP-1 is deposited by platelets early in wound healing is essential for normal wound closure. Thus, in tissues in which both proteins are expressed, differences in the spatial and temporal patterns of expression may lead to one of them being more important for the determination of the phenotype.

Table 36-2: TSP-1 Receptors on ECs

	TSP-1 Binding Site	*Downstream Signals*	*Function*
Calreticulin	NTSP-1	PI3K, ERK	Disrupts focal contact assembly
PGs	NTSP-1, TSRs, TC31		Promotes clearance of TSP-1 and -2
LRP	NTSP-1		Promotes clearance of TSP-1 and -2
Integrins $\alpha 3\beta 1$, $\alpha 4\beta 1$	NTSP-1		Promotes EC proliferation and angiogenesis
Integrin $\alpha 6\beta 1$	NTSP-1		Promotes EC migration
$\beta 1$ integrins	TSRs	PI3K	Inhibits VEGF-induced migration and adhesion
Integrin $\alpha v\beta 3$	RGD		
CD36	TSRs	Fyn-caspase-3-p38 MAPK	Inhibits EC migration and tube formation in vitro and neovascularization in vivo; promotes apoptosis in vitro and in vivo
CD47	C-terminal domain	G proteins	Supports adhesion and extravasation of immune cells

LIGAND-RECEPTOR INTERACTIONS AND DOWNSTREAM SIGNALING

The term *matricellular protein* has been used to describe the TSPs, tenascins, osteopontin, and other proteins (30). Matricellular proteins are predominantly expressed at the interface between the cell membrane and the extracellular matrix. TSP-1 also regulates extracellular proteolysis through direct interactions with plasmin and matrix metalloproteinases (MMPs). To localize TSP-1 at its surface, most cells express a wide array of receptors for the protein (Table 36-2). These include proteoglycans (PGs), sulfatides, integrins, CD36, calreticulin, low-density lipoprotein receptor-related protein (LRP), and CD47. Most of these receptors are widely expressed on ECs; however, CD36 is expressed at low to undetectable levels on large-vessel ECs. Every domain of TSP-1 has a binding motif for at least one cell surface receptor. In many cases, the receptor binding sites have been identified using synthetic peptides, and the authenticity of these sites remains to be established using mutant forms of the intact protein or correctly folded domains.

ECs attach, spread, and form focal contacts on fibronectin-coated tissue culture plates. Addition of intact TSP-1, the N-terminal domain of TSP (NTSP-1), or a peptide designated hep I of TSP-1 (amino acids 19–36 of TSP-1) results in focal contact disassembly and increased EC migration (32). This process may be important during the tissue remodeling that is associated with angiogenesis. Calreticulin reportedly binds to the hep I sequence of TSP-1 (33). Although the majority of calreticulin is found in the endoplasmic reticulum, some is expressed on the cell surface, where it functions as a receptor for TSP-1. The complex of TSP-1 and calreticulin then associates with LRP and heterotrimeric G-protein to disrupt focal contacts through a pathway that involves phosphoinositide 3-kinase (PI3K), focal adhesion kinase, and extracellular-signal-regulated kinase (ERK)1/2 (34). This appears to be a CD36-independent pathway.

PGs are composed of a core protein and at least one glycosaminoglycan (GAG) chain. Proteoglycans or GAGs have been reported to bind NTSP-1, the TSRs, and the type-3 repeat/C-terminal (T3Cl) domain (1). Because the GAG-binding properties of NTSP-1 and the intact protein are very similar, this domain probably is the predominant GAG-binding site in the intact molecule. NTSP-1 binds to heparin, heparan sulfate, and dermatan sulfate but not chondroitin-4-sulfate. In some cases, specific PGs have been shown to bind TSPs. For example, syndecan-1 and -4, perlecan, and decorin are capable of binding TSP-1 (32).

The binding of NTSP-1 and -2 to LRP mediates their uptake and clearance (35). PGs and LRP may function as coreceptors because the internalization of TSP-1 is inhibited by heparin. Because MMPs can bind to TSP-1 and –2, the uptake and clearance of TSP bound to MMPs occurs through an LRP-dependent mechanism (30). Varying effects of the isolated NTSP-1 on EC behavior have been reported in the presence and absence of growth factors. Vogel and colleagues (36) reported that a recombinant version of NTSP-1 inhibits EC migration toward basic fibroblast growth factor 2

(FGF2). This protein also inhibits the mitogenesis and proliferation of ECs and the binding of FGF2 to heparin (36). Based on these data, the authors concluded that NTSP-1 competes with FGF2 for binding sites on cell surface PGs. By contrast, Taraboletti and colleagues (37) have found that NTSP-1 stimulates angiogenesis in the rabbit cornea in the absence of FGF2. NTSP-1 also stimulates EC invasion through Matrigel by upregulating MMP-2 and -9 and reducing tissue inhibitors of metalloproteinase 2 (TIMP-2) expression (6). Three receptors for NTSP-1 have been reported to mediate the effect of this domain on EC function: syndecan-4, and $\alpha 3\beta 1$ and $\alpha 4\beta 1$ integrins. An antibody to syndecan-4 specifically blocks the binding of NTSP-1 to human umbilical vein ECs (HUVECs) (38). In addition, synthetic peptide studies indicate that short sequences in NTSP-1 mediate binding to the integrin $\alpha 3\beta 1$ (amino acids 190–201) and $\alpha 4\beta 1$ (amino acids 151–164) (39,40). The interaction with $\alpha 3\beta 1$ and $\alpha 4\beta 1$ reportedly stimulates EC proliferation and angiogenesis. The physiological significance of these observations hinges on whether NTSP-1 is produced in vivo in sufficient quantities to elicit proangiogenic effects. The production of NTSP-1 by proteolytic cleavage in human and mouse platelets has been reported (41,42). In addition, the antiangiogenic activity of TSP-1 was first identified in a 140,000-dalton proteolytic fragment produced by cells in culture. Thus, it is likely that some NTSP-1 is produced in vivo by proteolysis.

NTSP-1 also binds to the $\alpha 6\beta 1$ integrin on human microvascular ECs and HT-1080 fibrosarcoma cells (43). The binding of TSP-1 to $\alpha 6\beta 1$ supports chemotaxis of human dermal microvascular ECs (HDMECs). The $\alpha 6\beta 1$ binding site has been mapped to amino acids 87–96 of NTSP-1. A role for $\alpha 6\beta 1$ in angiogenesis is further established by the observation that antibodies to $\alpha 6\beta 1$ inhibit tube formation in the Matrigel assay (44). Like $\alpha 4\beta 1$, $\alpha 6\beta 1$ supports the attachment and spreading of ECs. However, these integrins appear to vary in their level of activation, with $\alpha 6\beta 1$ displaying greater activity than $\alpha 4\beta 1$ in HUVECs. It is important to note that the $\alpha 3\beta 1$, $\alpha 4\beta 1$, and $\alpha 6\beta 1$ binding sites of NTSP-1 have been identified using synthetic peptides. The activity of these sites has not been confirmed using site-directed mutagenesis within the context of the correctly folded domain or intact molecule.

$\beta 1$ Integrins also bind to the TSRs and the type-2 repeats of TSP-1 (40). Adhesion of cells that express a single $\beta 1$ integrin to a recombinant protein containing all three type-1 repeats of TSP-1 (3TSR) is specifically inhibited by an antagonist to that integrin. Adhesion of cells that express several $\beta 1$ integrins is partially inhibited by each α-subunit–specific antagonist, and completely inhibited by a combination of these antagonists. $\beta 1$ Integrins recognize both the second and third TSRs, and each TSR shows pan-$\beta 1$ specificity and divalent cation dependence for promoting cell adhesion. Furthermore, $\beta 1$ integrins bind in a divalent cation-dependent manner to a 3TSR affinity column. The binding of cells to 3TSR is not inhibited by synthetic peptides that encompass the RFK, WSHWSPW, VTCG, and GVITRIR sequences, suggesting that $\beta 1$ integrins bind to a site that is distinct from the CD36 binding site. However, it is possible that these linear peptides have a much lower affinity than the correctly folded domain and therefore do not compete effectively for binding.

The TSRs reportedly inhibit VEGF-induced HUVEC migration in a $\beta 1$-dependent fashion (45). Large-vessel ECs reportedly do not express CD36. Although an antibody to $\beta 1$- or $\beta 1$-siRNA blocks the ability of 3TSR to inhibit VEGF-induced migration, $\beta 3$-siRNA is without effect. Blocking either the $\alpha 3$ or $\alpha 5$ integrin subunit prevents 3TSR from inhibiting VEGF-induced HUVEC migration, whereas antibodies to the $\alpha 2$ and αv subunits do not have a significant effect. These cells express low levels of $\alpha 1$ and $\alpha 4$, and moderate levels of $\alpha 6$. 3TSR-coated beads selectively bind $\alpha 5\beta 1$ from HUVEC lysates. Although PI3K is essential for TSR-mediated inhibition of HUVEC migration, phospholipase C (PLC)-γ and Akt are not required. These data indicate that $\alpha 5\beta 1$ mediates the inhibition of migration of ECs that lack CD36.

The interaction of TSP-1 with CD36 plays a significant role in the antiangiogenic activity of TSP-1 both in vitro and in vivo (46–48). In vitro, CD36 mediates TSP-1 inhibition of EC migration and tube formation (46,47). Both GST fusion proteins containing the TSP-1-binding region of CD36 and antibodies against CD36 block the inhibition of EC migration caused by TSP-1. Transfection of CD36 into HUVECs renders them more sensitive to inhibition of migration by TSP-1. Furthermore, molecules that bind to CD36, including collagen, oxidized low-density lipoprotein (LDL), and an anti-CD36 IgM antibody, can also inhibit EC migration (46). Although TSP-1 is a potent inhibitor of FGF2–induced corneal neovascularization in wild-type mice, it is not active in *CD36*-null mice (48). In addition, upregulation of CD36 in vivo with ligands for peroxisome proliferators-activated receptor-γ (PPARγ) increases the antiangiogenic activity of the TSP-1–based therapeutic, ABT-510 (discussed in the next section) (49). Jimenez and colleagues (48) have shown that TSP-1 activation of a CD36–Fyn–caspase-3–p38 mitogen-activated protein kinase (MAPK) cascade is essential for the antiangiogenic effect of TSP-1, as well as its induction of EC apoptosis (Figure 36.2). Activation of this pathway leads to increased EC expression of Fas ligand, which sensitizes the cells to the increased levels of Fas that are induced by factors that initiate angiogenesis (50). Nor and colleagues (51) have reported that TSP-1 inhibits angiogenesis by decreasing expression of Bcl-2 and increasing expression of Bax. Taken together, the data indicate that the antiangiogenic effect of TSP-1 depends, at least in part, on its ability to induce apoptosis of ECs (48,51).

CD36-independent mechanisms for the induction of EC apoptosis also have been identified (52–54). Guo and coworkers (52) reported that TSP-1 or peptides that comprise TSR sequences induce apoptosis of bovine aortic ECs. As described earlier, recent data indicate that the TSRs can bind to $\beta 1$ integrins, and this interaction can inhibit EC migration. We have found that CD36, CD9, and $\beta 1$ integrins form a complex on platelets and ECs (55). CD9 is a member of the tetraspanin family of integrin-associated membrane proteins (56). Although the function of these proteins remains to be

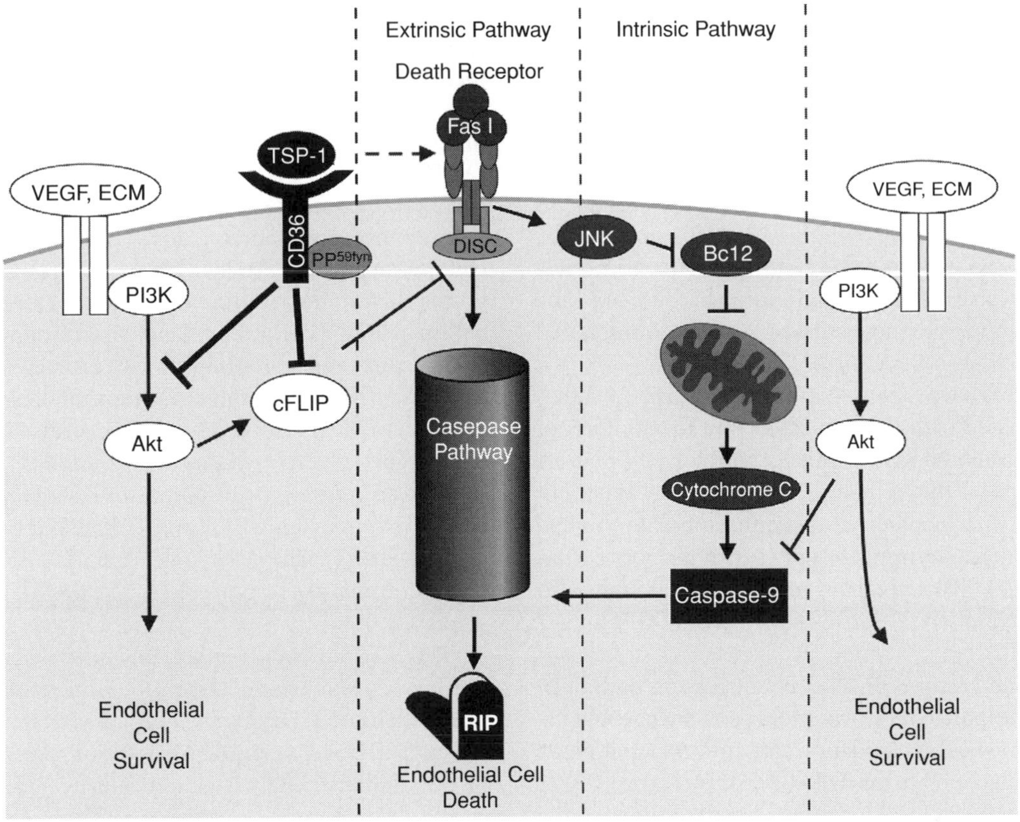

Figure 36.2. TSP-1 affects EC signaling. Complex EC signaling allows for cell survival or cell death. Both VEGF and TSP-1 affect Akt-dependent EC survival pathways as well as caspase-dependent and -independent apoptotic pathways. The balance of VEGF and TSP appears to be important in EC signaling.

fully defined, they appear to form membrane microdomains and modulate integrin function. The TSRs may engage a multiprotein complex to inhibit EC migration and induce EC apoptosis. Increased expression of CD47 also has been reported to correlate with increased EC apoptosis (53). In addition, a peptide that contains the CD47 binding sequence of the C-terminal domain of TSP-1 has been reported to inhibit angiogenesis (54). This peptide inhibits in vitro tube formation of brain capillary ECs and FGF2-induced angiogenesis in the cornea. Peptide studies have identified the sequences RFYVVMWK and IRVVM in the C-terminal domain of TSP-1 as binding sites for CD47 (57,58). However, recent structural data indicate that these sequences are not on the surface of the C-terminal domain and therefore are not available to interact with other proteins (24). Although the role of CD47 in TSP-1–induced apoptosis has not been determined, antibodies to CD47 induce apoptosis of B cells derived from patients with B-cell chronic lymphocytic leukemia (59). This apoptotic pathway does not involve caspases.

The molecular cloning of TSP-1 revealed the presence of an RGD sequence within the type-3 repeats. Subsequent studies have shown that the integrins $\alpha IIb\beta 3$ on blood platelets and $\alpha v\beta 3$ on a variety of cell types bind to the RGD sequence of TSP-1 (60–62). The availability of the RGD sequence is regulated by calcium-dependent and disulfide bond–dependent structures. An EC surface protein disulfide isomerase reportedly catalyzes disulfide interchange in TSP-1, resulting in the exposure of the RGD sequence (63).

EFFECTS ON ENDOTHELIAL CELL PHENOTYPE

Of all five members of the TSP family, only TSP-1 and -2 have been shown to exert influence on EC behavior, including adhesion, migration, proliferation, and apoptosis. The effects of TSP-1 have been well documented in both in vivo and in vitro models. Although in vitro experiments from many laboratories show that TSP-1 modulates adhesion, suppresses migration, inhibits proliferation, and induces apoptosis of ECs (64), some of the data are confusing and conflicting. Some of these disparate findings may be the result of differences in cultured ECs, such as source of origin of the ECs, species, and culture conditions, as well as passage numbers. However, some of the data undoubtedly point to the difficulty in seeing each part of the equation separately in the dynamic balance that controls angiogenesis. This is especially true for the TSPs, when compared to positive regulators of angiogenesis such as FGF1, because they are important regulators of EC function (65).

Under in vitro conditions, TSP-1 was shown to inhibit FGF2–induced murine lung capillary EC proliferation as well as FGF2–induced proliferation of bovine adrenal capillary ECs (see section on Ligand Receptor Interactions) (66,67). However, other studies have demonstrated that TSP-1, in the presence of other factors such as FGF2, lipopolysaccharide, fibrin, and collagen, is capable of stimulating EC proliferation (68,69). TSP-1 inhibited the mitogenic activity of VEGF-A_{165} both in the presence and absence of heparin, perhaps due to an interaction of VEGF-A_{165} and TSP-1 directly on ECs. The TSRs, the type-3 repeats, and the N-terminal heparin binding domain may have mediated this binding because blocking monoclonal antibodies to these regions significantly inhibited the binding of TSP-1 to VEGF-A_{165} (70).

Several authors have shown that TSP-1 increases EC adhesion to plates coated with this molecule, and the addition of a peptide containing the RGD sequence in the type-3 repeats of TSP-1 will inhibit this cellular adhesion (71). The amino terminal also may participate, because adhesion was inhibited by blocking the amino terminal of TSP-1 with the polysaccharide fucoidan (66). TSP-1 regulation of focal adhesions also appears to modulate cell responsiveness to both FGF1 and -2 (72).

Differential effects of TSP-1 on EC migration have been reported. TSP-1 induced EC chemotaxis, which could be inhibited using heparin, fucoidan, or an antibody that binds to the N-terminal domain of TSP-1 (66). However, TSP-1 inhibits VEGF-induced EC migration when it is added to the cells prior to placing them in the Boyden chamber apparatus. This effect could be reversed with anti-TSP-1 or anti-CD36 antibodies (73). In addition, recent work by Short and colleagues (45) has shown that the TSRs, the previously identified antiangiogenic domain of TSP-1, inhibits VEGF-induced HUVEC migration via a CD36-independent pathway. In this study, $\beta 1$ integrin activation was shown to be necessary for this TSR-induced inhibition of HUVEC migration. Binding of TSR to the $\alpha 5 \beta 1$ integrins outside of their RGD-binding site (since TSR does not contain RGD) resulted in a conformational change that then blocked the activation of PI3K and resulted in inhibition of EC migration (45). Endothelial cord formation also appears to be modulated by TSP-1; cord formation could be increased when anti TSP-1 antibody was added to cultures of both aortic and microvascular ECs (14).

Interestingly, TSP-1 has been reported to induce apoptosis of ECs (see the discussion on Ligand Receptor Interactions and Figure 36.2) (48,52,74). In addition, investigators have shown that the TSRs of TSP-1 are capable of antiangiogenic activity via apoptosis (52). Treatment of mice harboring experimental pancreatic tumors with intraperitoneal doses of 3TSR results in a 2.5-fold increase in EC apoptosis in the tumor (75). 3TSR also induced microvessel EC apoptosis in vitro in a CD36-dependent manner, and both the intrinsic and extrinsic apoptotic pathways activate executioner caspase-3 in 3TSR-induced apoptosis (our unpublished observations). Moreover, 3TSR decreased endogenous and VEGF-induced c-Akt activity and induced EC apoptosis in the presence of VEGF. Recombinant type-1 repeats of TSP-1 (3 TSR) have been shown to activate endothelial caspases between 8 and 12 hours of treatment with an effect similar to Tie2 blockade (76). In these same experiments, PI3K blockade using LY294002 inhibitor resulted in a loss of Akt signaling over 3 hours as well as a gradual increase in TSP-1 expression in the cultured ECs.

TSP also may play an important role in intimal hyperplasia and atherosclerosis. TSP-1 is not present in undamaged subendothelial membranes (77), but injury in the form of balloon angioplasty results in strong TSP-1 staining in the luminal surface as well as the media of the injured vessels within 1 hour (78). TSP-1 may delay healing of the injured endothelium by promoting VSMC proliferation and migration. The presence of anti–TSP-1 antibodies directed against the C-terminal of TSP-1 resulted in decreased neointimal formation in a rat carotid artery injury model (79).VSMCs are induced to migrate in response to TSP-1 through signaling pathways that include PI3K, p21 Ras, ERK-1 and -2, and p38 MAPK (80–82). The C-terminal domain of TSP-1 appears to be predominantly responsible for VSMC migration in these studies (83).

Fewer studies of the effects of TSP-2 on endothelial function exist, given the difficulty in obtaining pure material in sufficient quantities, because it cannot be isolated from platelets, as can TSP-1. A recombinant N-terminal 80-kDa fragment of TSP-2 that contains the procollagen homology domain and three type-1 repeats of TSP-2 was able to inhibit VEGF-induced migration of HDMECs in vitro (84). This fragment also inhibited Matrigel tube formation in vitro, as well as in an in vivo Matrigel assay. No effect of TSP-2 was seen on EC apoptosis.

DIAGNOSTIC AND THERAPEUTIC IMPLICATIONS

Despite remarkable progress over the past 25 years in understanding TSP structure and function, the diagnostic or therapeutic potential of these complex multifunctional molecules remains largely untapped. A number of studies have examined plasma levels of TSP-1 expression in patients with tumors. Plasma TSP-1 levels in patients with metastatic gastrointestinal, breast, or lung carcinomas were two- to threefold higher than those in healthy volunteers or in nonmetastatic malignancies (85). A direct correlation was observed between plasma TSP-1 levels and the stage of disease in patients with colorectal or gynecologic malignant tumors (86,87). Larger studies are needed to determine the prognostic value of plasma TSP-1 levels in cancer. The source of plasma TSP-1 in cancer patients has yet to be identified, and could include any combination of platelet, ECs, or tumor cells. Furthermore, high plasma levels of TSP-1 could result in beneficial effects, such as reduction in endothelial proliferation and thus tumor growth suppression at local or metastatic sites, or potentially less beneficial effects, such as increased invasiveness.

Several studies have shown that TSP-1 protein expression is inversely correlated with tumor grade and survival rate in thyroid, colon, and bladder cancers (88–90). In a study of human thyroid cancers, TSP-1 inversely correlated with the degree of invasion of the primary tumor into adjacent organs and with microvessel count per high power field (MVC) (89). A higher MVC correlated with poorer survival. The ratios of VEGF/TSP-1, VEGF-C/TSP-1, and Ang-2/TSP-1 significantly correlated with a higher microvascular density. Furthermore, the ratios VEGF/TSP-1 and angiopoietin 2 (Ang2)/TSP-1 significantly correlated with the degree of infiltration (91). Interestingly, studies of mouse skin after chronic or acute UVB irradiation also show that the balance of VEGF and TSP-1 expression plays a critical role in the shift from vascular quiescence to a proangiogenic environment (92). This switch results in dermal angiogenesis, a finding that may have potential application in the development of novel medications or topical creams to upregulate skin TSP-1, and thus combat the onset of cutaneous photoaging that results in wrinkles (92). In contrast, breast tumors and cholangiocarcinomas show increased stromal expression of TSP-1 in tumors compared to normal tissue, and in esophageal and gallbladder carcinomas TSP-1 expression directly correlated to occurrence of metastases (93,94). In a study of pancreatic cancer patients, diffuse expression of TSP-1 was significantly correlated with lymph node metastasis ($p < 0.01$), neural invasion ($p < 0.05$), and TNM stage ($p < 0.01$) (95).

Because upregulation of TSP-2 expression in the peritumoral stroma throughout the consecutive stages of chemically induced multistep carcinogenesis in mice has been observed (96), it will be interesting to look at TSP-2 levels in human malignancies. *TSP-2*–null mice display enhanced susceptibility to chemically induced skin carcinogenesis (96). A greater number of papillomas formed in the *TSP-2*–null mice; however, the rate of malignant conversion to squamous cell carcinoma was comparable in the *TSP-2*–null and wild-type mice.

The effects of TSP-1 on tumor growth and angiogenesis are summarized in Figure 36.3. In the tumor microenvironment, TSP-1 inhibits EC function directly through CD36 and probably other cell surface receptors, and indirectly through inhibition of MMP-mediated VEGF mobilization. In addition, TSP-1 can inhibit tumor cell growth and survival through the activation of tumor growth factor-β (TGF-β) Inclusion of sequences that are necessary for the activation of TGF-β in TSR-based recombinant proteins produced greater inhibition of B16F10 melanoma and A431 squamous cell carcinoma tumor growth (55,97). This effect was abrogated by systemic injections of an anti–TGF-β antibody (55). To circumvent the inhibitory effects of TSP-1, the mutations that occur in tumor suppressor genes and oncogenes frequently lead to downregulation of this protein. For example, hyperactivation of Ras and Myc leads to decreased TSP-1 expression

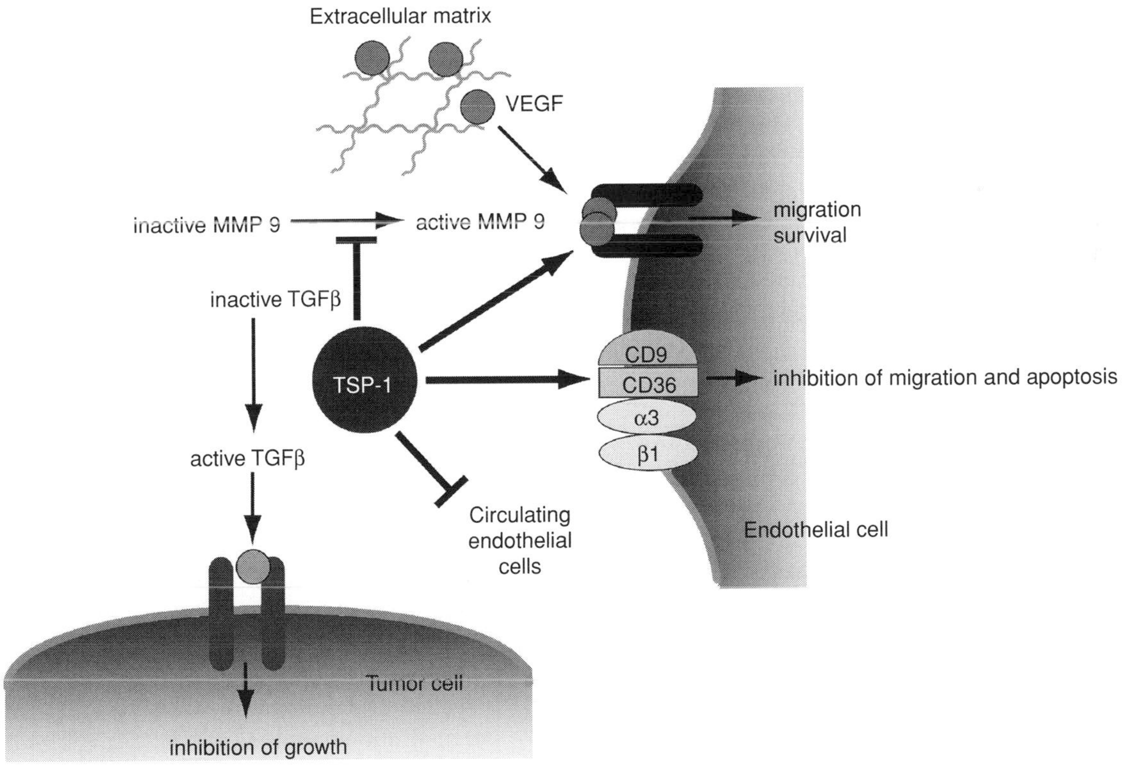

Figure 36.3. Inhibition of tumor growth and angiogenesis by TSP-1. TSP-1 is secreted in tumor tissue by stromal cells and, in some cases, by tumor cells. Through interactions with CD36 and β1 integrins, it has direct effects on the migration and survival of ECs in the vessel wall and in circulating ECs (CECs). TSP-1 also suppresses VEGF activity by inhibiting its mobilization from the extracellular matrix by matrix metalloproteinase (MMP)-9 and by binding directly to it. TSP-1 also suppresses the growth of TGF-β–responsive tumors by activating TGF-β.

and the acquisition of an angiogenic phenotype (98). These studies have led to the concept that reinstating TSP-1 expression could be used to inhibit tumor progression, and several strategies have been developed to accomplish this goal. Cell-based gene therapy and systemic delivery of synthetic peptides and recombinant proteins have been used to inhibit tumor angiogenesis (50,99,100). In addition, the antiangiogenic activity of low doses of the chemotherapeutic cyclophosphamide is mediated by systematic increases in TSP-1 levels (101,102). Low doses of cyclophosphamide inhibit angiogenesis in Matrigel plug assays performed in wild-type mice, but not in *TSP-1*–null mice (101). Although the efficacy of low-dose chemotherapeutics is diminished in *TSP-1*–null mice, it is equally effective in wild-type, endostatin-null, and tumstatin-null mice (102). The source of increased systemic levels of TSP-1 is unclear. One study presented data indicating that ECs increase TSP-1 expression in response to low-dose chemotherapeutics, whereas another study suggested that the increased TSP-1 is derived from tumor cells and perivascular stromal cells (101,102).

Circulating levels of TSP-2 have been increased by implanting a biodegradable polymer containing fibroblasts that overexpress TSP-2 (99). These increased levels persisted for at least 5 weeks and were sufficient to inhibit experimental squamous cell carcinomas, Lewis lung carcinomas, and malignant melanomas. As expected, these tumors exhibited decreased angiogenesis as measured by decreased tumor vessel size. In addition, systemic treatment of mice with a recombinant protein that contains the N-terminal through the TSRs of TSP-2 inhibits experimental squamous cell carcinoma growth (84).

Systemic delivery of recombinant versions of the TSRs of TSP-1 inhibits the growth of experimental ASPC-1 orthotopically placed pancreatic tumors, B16F10 melanomas, and Lewis lung carcinomas (75,103). Daily doses of 1 to 2.5 mg/kg of recombinant TSR protein inhibits tumor growth by 60% to 80%. For B16F10 melanoma, but not Lewis Lung carcinoma, a portion of the inhibition is associated with activation of TGF-β by the TSRs (100). Thus, for tumor cells that retain the ability to respond to TGF-β, additional therapeutic efficiency can be achieved with recombinant TSRs that include the TGF-β–activating sequence.

A synthetic TSR peptide, designated ABT-510, is currently in Phase II clinical trials for the treatment of cancer (50,104,105). This peptide is a substituted derivative of the sequence GVITRIR, and the two arginine residues are essential for activity. The first phase of clinical trials established that ABT-510 is well tolerated (104). A Phase II trial is currently under way to evaluate the safety and efficacy of ABT-510 for the treatment of soft-tissue sarcoma. The structure of the TSRs suggests that the ABT-510 sequence falls within a positively charged groove that functions as the CD36 binding site (28). In addition, anti-CD36 antibodies reportedly blocked the activity of ABT-510 (106). These data indicate that the antiangiogenic activity of ABT-510 is mediated by CD36. Because CD36 is upregulated by PPARγ, a reagent that activates PPARγ has been used systemically to sensitize ECs to ABT-510 (107).

In general, it appears that single-agent antiangiogenic strategies will not be sufficient to block tumor growth. Several studies have sought to combine TSP-based therapeutics with other treatment modalities. The combination of downregulation of VEGF by RNA interference and upregulation of TSP-1 significantly delayed fibrosarcoma formation and growth rate (108). Pretreatment with TSP-1 potentiates the effects of radiation by rendering the ECs in D-12 melanoma experimental tumors more likely to become apoptotic (109). Continued delivery of TSP-1 to mice after radiation treatment inhibited the outgrowth of metastases. A combination of TSP-1 with carboplatin also inhibited metastases in human nonseminomatous germ cell tumor xenografts (110).

Further intense investigation of the molecular effects of the TSPs on ECs and tumor cells will undoubtedly yield results that will be directly translatable to the treatment of human diseases such as cancer. Translational research protocols addressing prospectively the use of TSP-1 and TSP-2 as prognostic markers in tumor or plasma samples, as well as treatment of malignancies with TSP-1 or -2 type-1 repeat peptides, will be necessary for further advancement in this area. Thus far, the data point to the fact that TSP-based reagents are well tolerated and effective with little toxicity, but should be combined with other agents to best combat illnesses such as cancer. Ultimately, a complete understanding of the effects of TSPs on the varied cells in the tumor microenvironment will lead to a more targeted use of this protein therapeutically.

KEY POINTS

- The five members of the TSP family can be divided into two groups on the basis of their domain and tertiary structure. All the TSPs contain type-2 (EGF-like) repeats, type-3 repeats, and a highly conserved C-terminal domain.
- Of the five proteins that comprise the family, TSP-1 and -2 have been shown to be involved in vascular development and angiogenesis. Only TSP-1 and TSP-2 have been shown to exert influence on EC behavior, including adhesion, migration, proliferation, and apoptosis.
- The interaction of TSP-1 with CD36 plays a significant role in the antiangiogenic activity of TSP-1 both in vitro and in vivo. TSP-1 also acts indirectly through inhibition of MMP-mediated VEGF mobilization. In addition, TSP-1 can inhibit tumor cell growth and survival through the activation of TGF-β.
- TSP also may play an important role in intimal hyperplasia and atherosclerosis.

- Several studies have shown that TSP-1 protein expression is inversely correlated with tumor grade and survival rate in certain malignancies.
- Systemic delivery of recombinant versions of the TSRs of TSP-1 inhibits the growth of several experimental tumors including ASPC-1 orthotopically placed pancreatic tumors, B16F10 melanomas, and Lewis lung carcinomas.
- A synthetic TSR peptide, designated ABT-510, is currently in Phase II clinical trials for the treatment of cancer.
- Further intense investigation of the molecular effects of the TSPs on endothelial and tumor cells will open new avenues for therapeutic intervention.

REFERENCES

1 Chen H, Herndon ME, Lawler J. The cell biology of thrombospondin-1. *Matrix Biol.* 2000;19:597–614.

2 Lawler J, Detmar M. Tumor progression: the effects of thrombospondin-1 and -2. *Int J Biochem Cell Biol.* 2004;36:1038–1045.

3 Lawler J. Thrombospondin-1 as an endogenous inhibitor of angiogenesis and tumor growth. *J Cell Mol Med.* 2002;6:1–12.

4 de Fraipont F, Nicholson A C, Feige JJ, Van Meir EG. Thrombospondins and tumor angiogenesis. *Trends Mol Med.* 2001;7:401–407.

5 Jimenez B, Volpert OV, Crawford SE, et al. Signals leading to apoptosis-dependent inhibition of neovascularization by thrombospondin-1. *Nature Med.* 2000;6:41–48.

6 Donnini S, Morbidelli L, Taraboletti G, Ziche M. ERK1-2 and p38 MAPK regulate MMP/TIMP balance and function in response to thrombospondin-1 fragments in the microvascular endothelium. *Life Sci.* 2004;74:2975–2985.

7 Baezinger NL, Brodie GN, Majerus PW. A thrombin-sensitive protein of human platelet membranes. *Proc Natl Acad Sci USA.* 1971;68:240–249.

8 Lawler JW, Slayter HS, Coligan JE. Isolation and characterization of a high molecular weight glycoprotein from human blood platelets. *J Biol Chem.* 1978;253:8609–8616.

9 Lawler J, Derick LH, Connolly JE, et al. The structure of human platelet thrombospondin. *J Biol Chem.* 1985;260:3762–3772.

10 Galvin NJ, Dixit VM, O'Rourke KM, et al. Mapping of epitopes for monoclonal antibodies against human platelet thrombospondin with electron microscopy and high sensitivity amino acid sequencing. *J Cell Biol.* 1985;101:1434–1441.

11 McPherson J, Sage H, Bornstein P. Isolation and characterization of a glycoprotein secreted by aortic endothelial cells in culture. Apparent identity with platelet thrombospondin. *J Biol Chem.* 1981;256:11330–11336.

12 Mosher DF, Doyle MJ, Jaffe EA. Synthesis and secretion of thrombospondin by cultured human endothelial cells. *J Cell Biol.* 1982;93:343–348.

13 Lawler J, Hynes RO. The structure of human thrombospondin, an adhesive glycoprotein with multiple calcium binding sites and homologies with several different proteins. *J Cell Biol.* 1986;103:1635–1648.

14 Iruela-Arispe ML, Bornstein P, Sage H. Thrombospondin exerts an antiangiogenic effect on cord formation by endothelial cells in vitro. *Proc Natl Acad Sci USA.* 1991;88:5026–5030.

15 Vos HL, Devarayalu S, De Vries Y, Bornstein P. Thrombospondin-3 (Thbs3), a new member of the thrombospondin gene family. *J Biol Chem.* 1992;267:12192–12196.

16 Lawler J, McHenry K, Duquette M, Derick L. Characterization of human thrombospondin-4. *J Biol Chem.* 1995;270:2809–2814.

17 Oldberg A, Antonsson P, Lindblom K, Heinegard D. COMP is structurally related to the thrombospondins. *J Biol Chem.* 1992;267:22346–22350.

18 Folkman J. Tumor angiogenesis: therapeutic implications. *N Engl J Med.* 1971;285:1182–1186.

19 Ingber D, Fujita T, Kishimoto S, et al. Synthetic analogues of fumagillin that inhibit angiogenesis and suppress tumour growth. *Nature.* 1990;348:555–557.

20 Hanahan D, Christofori G, Naik P, Arbeit J. Transgenic mouse models of tumour angiogenesis: the angiogenic switch, its molecular controls, and prospects for preclinical therapeutic models. *Eur J Cancer.* 1996;32A:2386–2393.

21 Good DJ, Polverini PJ, Rastinejad F, et al. A tumor suppressor-dependent inhibitor of angiogenesis is immunologically and functionally indistinguishable from a fragment of thrombospondin. *Proc Natl Acad Sci USA.* 1990;87:6624–6628.

22 Tolsma SS, Volpert OV, Good DJ, et al. Peptides derived from two separate domains of the matrix protein thrombospondin-1 have anti-angiogenic activity. *J Cell Biol.* 1993;122:497–511.

23 Stenina OI, Desai SY, Krukovets I, et al. Thrombospondin-4 and its variants: expression and differential effects on endothelial cells. *Circulation.* 2003;108:1514–1519.

24 Kvansakul M, Adams JC, Hohenester E. Structure of a thrombospondin C-terminal fragment reveals a novel calcium core in the type 3 repeats. *EMBO J.* 2004;23:1223–1233.

25 Adams JC, Lawler J. The thrombospondins. *Int J Biochem Cell Biol.* 2004;36:961–968.

26 Adams JC, Monk R, Taylor AL, et al. Characterisation of *Drosophila* thrombospondin defines an early origin of pentameric thrombospondins. *J Mol Biol.* 2003;328:479–494.

27 Riessen R, Fenchel M, Chen H, et al. Cartilage oligomeric matrix protein (thrombospondin-5) is expressed by human vascular smooth muscle cells. *Arterioscler Thromb Vasc Biol.* 2001;21:47–54.

28 Agah A, Kyriakides TR, Lawler J, Bornstein P. The lack of thrombospondin-1 (TSP1) dictates the course of wound healing in double-TSP1/TSP2-null mice. *Am J Pathol.* 2002;161:831–839.

29 Christopherson K, Ullian E, Stokes C, et al. Thrombospondins are astrocytes-secreted proteins that promote CNS synaptogenesis. *Cell.* 2005;120:421–433.

30 Bornstein P, Agah A, Kyriakides TR. The role of thrombospondins 1 and 2 in the regulation of cell-matrix interactions, collagen fibril formation, and the response to injury. *Int J Biochem Cell Biol.* 2004;36:1115–1125.

31 Cursiefen C, Masli S, Ng TF, et al. Roles of thrombospondin-1 and -2 in regulating corneal and iris angiogenesis. *Invest Ophthalmol Vis Sci.* 2004;45:1117–1124.

32 Elzie CA, Murphy-Ullrich JE. The N-terminus of thrombospondin: the domain stands apart. *Int J Biochem Cell Biol.* 2004;36:1090–1101.

33 Goicoechea S, Pallero MA, Eggleton P, et al. The anti-adhesive activity of thrombospondin is mediated by the N-terminal domain of cell surface calreticulin. *J Biol Chem.* 2002; 277:37219–37228.

34 Baker CH, Solorzano CC, Fidler IJ. Blockade of vascular endothelial growth factor receptor and epidermal growth factor receptor signaling for therapy of metastatic human pancreatic cancer. *Cancer Res.* 2002;62:1996–2003.

35 Wang S, Herndon ME, Ranganathan S, et al. Internalization but not binding of thrombospondin-1 to low density lipoprotein receptor-related protein-1 requires heparan sulfate proteoglycans. *J Cell Biochem.* 2004;91:766–776.

36 Vogel T, Guo NH, Krutzsch HC, et al. Modulation of endothelial cell proliferation, adhesion, and motility by recombinant heparin-binding domain and synthetic peptides from the type I repeats of thrombospondin. *J Cell Biochem.* 1993;53: 74–84.

37 Taraboletti G, Morbidelli L, Donnini S, et al. The heparin binding 25 kDa fragment of thrombospondin-1 promotes angiogenesis and modulates gelatinase and TIMP-2 production in endothelial cells. *FASEB J.* 2000;14:1674–1676.

38 Ferrari do Outeiro-Bernstein MA, Nunes SS, et al. A recombinant NH(2)-terminal heparin-binding domain of the adhesive glycoprotein, thrombospondin-1, promotes endothelial tube formation and cell survival: a possible role for syndecan-4 proteoglycan. *Matrix Biol.* 2002;21:311–324.

39 Chandrasekaran L, He C-Z, Al-Barazi H, et al. Cell contact-dependent activation of a3b1 integrin modulates endothelial cell responses to thrombospondin-1. *Mol Biol Cell.* 2000; 11:2885–2900.

40 Calzada MJ, Zhou L, Sipes JM, et al. Alpha4beta1 integrin mediates selective endothelial cell responses to thrombospondins 1 and 2 in vitro and modulates angiogenesis in vivo. *Circ Res.* 2004;94:462–470.

41 Adams J, Seed B, Lawler J. Muskelin, a novel mediator of cell adhesive and cytoskeletal responses to thrombospondin-1. *EMBO J.* 1998;17:4954–4974.

42 Damas C, Vink T, Nieuwenhuis HK, Sixma JJ. The 33-kDa platelet alpha-granule membrane protein (GMP-33) is an N-terminal proteolytic fragment of thrombospondin. *Thromb Haemost.* 2001;86:887–893.

43 Calzada MJ, Sipes JM, Krutzsch HC, et al. Recognition of the N-terminal modules of thrombospondin-1 and thrombospondin-2 by alpha6beta1 integrin. *J Biol Chem.* 2003;278:40679–40687.

44 Davis GE, Camarillo CW. Regulation of endothelial cell morphogenesis by integrins, mechanical forces, and matrix guidance pathways. *Exp Cell Res.* 1995;216:113–123.

45 Short SM, Derrien A, Narsimhan RP, et al. Inhibition of endothelial cell migration by thrombospondin-1 type-1 repeats is mediated by beta1 integrins. *J Cell Biol.* 2005;168:643–653.

46 Dawson DW, Pearce SF, Zhong R, et al. CD36 mediates the in vitro inhibitory effects of thrombospondin-1 on endothelial cells. *J Cell Biol.* 1997;138:707–717.

47 Dawson DW, Bouck NP. Thrombospondin as an Inhibitor of Angiogenesis. In: Teicher BA, ed. *Antiangiogenic Agents in Cancer Therapy*. Totowa, NJ: Humana Press;1999:185–203.

48 Jimenez B, Volpert OV, Crawford SE, et al. Signals leading to apoptosis-dependent inhibition of neovascularization by thrombospondin-1 [In Process Citation]. *Nat Med.* 2000;6: 41–48.

49 Abe K, Kurakin A, Mohseni-Maybodi M, et al. The complexity of TNF-related apoptosis-inducing ligand. *Ann NY Acad Sci.* 2000;926:52–63.

50 Reiher FK, Volpert OV, Jimenez B, et al. Inhibition of tumor growth by systemic treatment with thrombospondin-1 peptide mimetics. *Int J Cancer.* 2002;98:682–689.

51 Nor JE, Mitra RS, Sutorik MM, et al. Thrombospondin-1 induces endothelial cell apoptosis and inhibits angiogenesis by activating the caspase death pathway. *J Vasc Res.* 2000;37: 209–218.

52 Guo N, Krutzsch HC, Inman JK, Roberts DD. Thrombospondin 1 and type I repeat peptides of thrombospondin 1 specifically induce apoptosis of endothelial cells. *Cancer Res.* 1997;57: 1735–1742.

53 Freyberg MA, Kaiser D, Graf R, et al. Integrin-associated protein and thrombospondin-1 as endothelial mechanosensitive death mediators. *Biochem Biophys Res Commun.* 2000.271:584–588.

54 Kanda S, Shono T, Tomasini-Johansson B, et al. Role of thrombospondin-1-derived peptide, 4N1K, in FGF-2-induced angiogenesis. *Exp Cell Res.* 1999;252:262–272.

55 Miao W, Seng WL, Duquette M, et al. Thrombospondin-1 type 1 repeat recombinant proteins inhibit tumor growth. *Cancer Res.* 2001;97:1689–1696.

56 Stipp CS, Kolesnikova TV, Hemler ME. Functional domains in tetraspanin proteins. *Trends Biochem Sci.* 2003;28:106–112.

57 Kosfeld MD, Frazier WA. Identification of active peptide sequences in the carboxyl-terminal cell binding domain of human thrombospondin-1. *J Biol Chem.* 1992;267:16230–16236.

58 Kosfeld MD, Frazier WA. Identification of a new cell adhesion motif in two homologous peptides from the COOH-terminal cell binding domain of human thrombospondin. *J Biol Chem.* 1993;268:8808–8814.

59 Mateo V, Lagneaux L, Bron D, et al. CD47 ligation induces caspase-independent cell death in chronic lymphocytic leukemia. *Nat Med.* 1999;5:1277–1284.

60 Lawler J, Hynes RO. An integrin receptor on normal and thrombasthenic platelets that binds thrombospondin. *Blood.* 1989;74:2022–2027.

61 Sun X, Skorstengaard K, Mosher DF. Disulfides modulate RGD-inhibitable cell adhesive activity of thrombospondin. *J Cell Biol.* 1992;118:693–701.

62 Adams JC, Lawler J. Diverse mechanisms for cell attachment to platelet thrombospondin. *J Cell Sci.* 1993;104:1061–1071.

63 Hotchkiss KA, Matthias LJ, Hogg PJ. Exposure of the cryptic Arg-Gly-Asp sequence in thrombospondin-1 by protein disulfide isomerase [published erratum appears in *Biochim Biophys Acta* 1999 Sep 14;1434(1):210]. *Biochim Biophys Acta.* 1998; 1388:478–488.

64 Armstrong LC, Bornstein P. Thrombospondins 1 and 2 function as inhibitors of angiogenesis. *Matrix Biol.* 2003;22:63–71.

65 Esemuede N, Lee T, Pierre-Paul D, et al. The role of thrombospondin-1 in human disease. *J Surg Res.* 2004;122:135–142.

66 Taraboletti G, Roberts D, Liotta LA., Giavazzi R. Platelet thrombospondin modulates endothelial cell adhesion, motility, and growth: a potential angiogenesis regulatory factor. *J Cell Biol.* 1990;111:765–772.

67 Bagavandoss P, Kaytes P, Vogeli G, et al. Recombinant truncated thrombospondin-1 monomer modulates endothelial cell plasminogen activator inhibitor 1 accumulation and proliferation in vitro. *Biochem Biophys Res Commun.* 1993;192:325–332.

68 Nicosia RF, Tuszynski GP. Matrix-bound thrombospondin promotes angiogenesis in vitro. *J Cell Biol.* 1994;124:183–193.

69 BenEzra D, Griffin BW, Maftzir G, Aharonov O. Thrombospondin and in vivo angiogenesis induced by basic fibroblast growth factor or lipopolysaccharide. *Invest Ophthalmol Vis Sci.* 1993;34:3601–3608.

70 Gupta K, Gupta P, Wild R, et al. Binding and displacement of vascular endothelial growth factor (VEGF) by thrombospondin: effect on human microvascular endothelial cell proliferation and angiogenesis. *Angiogenesis.* 1999;3:147–158.

71 Lawler J, Weinstein R, Hynes RO. Cell attachment to thrombospondin: the role of ARG-GLY-ASP, calcium, and integrin receptors. *J Cell Biol.* 1988;107:2351–2361.

72 Orr AW, Elzie CA, Kucik DF, Murphy-Ullrich JE. Thrombospondin signaling through the calreticulin/LDL receptor-related protein co-complex stimulates random and directed cell migration. *J Cell Sci.* 2003;116:2917–2927.

73 Dameron KM, Volpert OV, Tainsky MA, Bouck N. Control of angiogenesis in fibroblasts by p53 regulation of thrombospondin-1. *Science.* 1994;265:1582–1584.

74 Nor JE, Mitra RS, Sutorik MM, et al. Thrombospondin-1 induces endothelial cell apoptosis and inhibits angiogenesis by activating the caspase death pathway. *J Vasc Res.* 2000;37: 209–218.

75 Zhang X, Galardi E, Duquette M, et al. Antiangiogenic treatment with three thrombospondin-1 type 1 repeats recombinant proteins in an orthotopic human pancreatic cancer model. *Clin Cancer Res.* 2005;11:1–8.

76 Niu Q, Perruzzi C, Voskas D, et al. Inhibition of Tie-2 signaling induces endothelial cell apoptosis, decreases Akt signaling, and induces endothelial cell expression of the endogenous anti-angiogenic molecule, thrombospondin-1. *Cancer Biol Ther.* 2004;3:402–405.

77 Lawler J. The structural and functional properties of thrombospondin. *Blood.* 1986;67:1197–1209.

78 Raugi GJ, Mullen JS, Bark DH, et al. Thrombospondin deposition in rat carotid artery injury. *Am J Pathol.* 1990;137:179–185.

79 Chen D, Asahara T, Krasinski K, et al. Antibody blockade of thrombospondin accelerates reendothelialization and reduces neointima formation in balloon-injured rat carotid artery. *Circulation.* 1999;100:849–854.

80 Gahtan V, Wang XJ, Ikeda M, et al. Thrombospondin-1 induces activation of focal adhesion kinase in vascular smooth muscle cells. *J Vasc Surg.* 1999;29:1031–1036.

81 Gahtan V, Wang XJ, Willis AI, et al. Thrombospondin-1 regulation of smooth muscle cell chemotaxis is extracellular signal-regulated protein kinases 1/2 dependent. *Surgery.* 1999;126: 203–207.

82 Willis AI, Fuse S, Wang XJ, et al. Inhibition of phosphatidylinositol 3-kinase and protein kinase C attenuates extracellular matrix protein-induced vascular smooth muscle cell chemotaxis. *J Vasc Surg.* 2000;31:1160–1167.

83 Nesselroth SM, Willis AI, Fuse S, et al. The C-terminal domain of thrombospondin-1 induces vascular smooth muscle cell chemotaxis. *J Vasc Surg.* 2001;33:595–600.

84 Noh YH, Matsuda K, Hong YK, et al. An N-terminal 80 kDa recombinant fragment of human thrombospondin-2 inhibits vascular endothelial growth factor induced endothelial cell migration in vitro and tumor growth and angiogenesis in vivo. *J Invest Dermatol.* 2003;121:1536–1543.

85 Tuszynski GP, Smith M, Rothman VL, et al. Thrombospondin levels in patients with malignancy [published erratum appears in *Thromb Haemost* 1992 Oct 5;68(4):485]. *Thromb Haemost.* 1992;67:607–611.

86 Nathan FE, Hernandez E, Dunton CJ, et al. Plasma thrombospondin levels in patients with gynecologic malignancies. *Cancer.* 1994;73:2853–2858.

87 Yamashita Y, Kurohiji T, Tuszynski GP, et al. Plasma thrombospondin levels in patients with colorectal carcinoma. *Cancer.* 1998;82:632–638.

88 Grossfeld GD, Ginsberg DA, Stein JP, et al. Thrombospondin-1 expression in bladder cancer: association with p53 alterations, tumor angiogenesis, and tumor progression. *J Natl Cancer Inst.* 1997;89:219–227.

89 Bunone G, Vigneri P, Mariani L, et al. Expression of angiogenesis stimulators and inhibitors in human thyroid tumors and correlation with clinical pathological features. *Am J Pathol.* 1999;155:1967–1976.

90 Tokunaga T, Nakamura M, Oshika Y, et al. Thrombospondin 2 expression is correlated with inhibition of angiogenesis and metastasis of colon cancer. *Br J Cancer.* 1999;79:354–359.

91 Tanaka K, Sonoo H, Kurebayashi J, et al. Inhibition of infiltration and angiogenesis by thrombospondin-1 in papillary thyroid carcinoma. *Clin Cancer Res.* 2002;8:1125–1131.

92 Yano K, Kadoya K, Kajiya K, et al. Ultraviolet B irradiation of human skin induces an angiogenic switch that is mediated by upregulation of vascular endothelial growth factor and by downregulation of thrombospondin-1. *Br J Dermatol.* 2005; 152:115–121.

93 Oshiba G, Kijima H, Himeno S, et al. Stromal thrombospondin-1 expression is correlated with progression of esophageal squamous cell carcinomas. *Anticancer Res.* 1999;19:4375–4378.

94 Ohtani Y, Kijima H, Dowaki S, et al. Stromal expression of thrombospondin-1 is correlated with growth and metastasis of human gallbladder carcinoma. *Int J Oncol.* 1999;15:453–457.

95 Tobita K, Kijima H, Dowaki S, et al. Thrombospondin-1 expression as a prognostic predictor of pancreatic ductal carcinoma. *Int J Oncol.* 2002;21:1189–1195.

96 Hawighorst T, Velasco P, Streit M, et al. Thrombospondin-2 plays a protective role in multistep carcinogenesis: a novel host anti-tumor defense mechanism. *EMBO J.* 2001;20:2631–2640.

97 Yee KO, Streit M, Hawighorst T, et al. Expression of the type-1 repeats of thrombospondin-1 inhibits tumor growth through activation of transforming growth factor-beta. *Am J Pathol.* 2004;165:541–552.

98 Watnick RS, Cheng YN, Rangarajan A, et al. Ras modulates Myc activity to repress thrombospondin-1 expression and increase tumor angiogenesis. *Cancer Cell.* 2003;3:219–231.

99 Streit M, Stephen AE, Hawighorst T, et al. Systemic inhibition of tumor growth and angiogenesis by thrombospondin-2 using cell-based antiangiogenic gene therapy. *Cancer Res.* 2002; 62:2004–2012.

100 Miao WM, Seng WL, Duquette M, et al. Thrombospondin-1 type 1 repeat recombinant proteins inhibit tumor growth through transforming growth factor-beta-dependent and -independent mechanisms. *Cancer Res.* 2001;61:7830–7839.

101 Bocci G, Francia G, Man S, et al. Thrombospondin-1, a mediator of the antiangiogenic effects of low-dose metronomic chemotherapy. *Proc Natl Acad Sci USA.* 2003; In press.

102 Hamano Y, Sugimoto H, Soubasakos MA, et al. Thrombospondin-1 associated with tumor microenvironment contributes to low-dose cyclophosphamide-mediated endothelial cell apoptosis and tumor growth suppression. *Cancer Res.* 2004;64:1570–1574.
103 Miao W, Seng WL, Duquette M, et al. Thrombospondin-1 Type 1 repeat recombinant proteins inhibit tumor growth through transforming growth factor beta dependent and independent mechanisms. *Cancer Research.* 2001;61.7830–7839.
104 De Vos F, Hoekstra R, Gietema J, et al. A phase I dose escalating study of the angiogenesis inhibitor thrombospondin-mimetic (ABT-510) in patients with advanced cancer. *Proc Am Soc Clin Oncol.* 2002; Abstract no. 324.
105 Gordon M, Mendelson D, Guirguis M, et al. ABT-510, an anti-angiogenic, thrombospondin-1 (TSP-1) mimetic peptide, exhibits favorable safety profile and early signals of activity in a randomized phase IB trial. *Proc Am Soc Clin Oncol.* 2003; Abstract no. 780, 22:195.
106 Dawson DW, Volpert OV, Pearce SF, et al. Three distinct D-amino acid substitutions confer potent antiangiogenic activity on an inactive peptide derived from a thrombospondin-1 type 1 repeat. *Mol Pharmacol.* 1999;55:332–338.
107 Huang Z, Bao SD. Roles of main pro- and anti-angiogenic factors in tumor angiogenesis. *World J Gastroenterol.* 2004;10:463–470.
108 Filleur S, Courtin A, Ait-Si-Ali S, et al. SiRNA-mediated inhibition of vascular endothelial growth factor severely limits tumor resistance to antiangiogenic thrombospondin-1 and slows tumor vascularization and growth. *Cancer Res.* 2003;63:3919–3922.
109 Rofstad EK, Henriksen K, Galappathi K, Mathiesen B. Antiangiogenic treatment with thrombospondin-1 enhances primary tumor radiation response and prevents growth of dormant pulmonary micrometastases after curative radiation therapy in human melanoma xenografts. *Cancer Res.* 2003;63: 4055–4061.
110 Abraham D, Abri S, Hofmann M, et al. Low dose carboplatin combined with angiostatic agents prevents metastasis in human testicular germ cell tumor xenografts. *J Urol.* 2003;170:1388–1393.

37

Neuropilins

Receptors Central to Angiogenesis and Neuronal Guidance

Diane Bielenberg, Peter Kurschat, and Michael Klagsbrun

Children's Hospital, Harvard Medical School, Boston, Massachusetts

HISTORICAL CONTEXT

Neuropilins (NRPs) are type I membrane glycoproteins first described by Fujisawa and colleagues in 1987 as antigens for antibodies that bind to neuropiles and plexiform layers in the *Xenopus* tadpole nervous system (1,2). NRP function was demonstrated 10 years later when these proteins were found to be receptors for the class 3 semaphorins (SEMA). The first class 3 semaphorin identified was chick collapsin-1 (now known as Sema3A), described by Raper and colleagues as an inhibitory axonal guidance signal that collapsed growth cones and repelled axons in chick dorsal root ganglia (DRG) (1–4). Two NRPs (NRP1 and NRP2), and six class 3 semaphorins (A-F) exist (5,6).

Unexpectedly, NRPs were later shown to be regulators of angiogenesis. The first indication came in 1995, when overexpression of *Nrp1* in chimeric mice was associated with vascular defects (7). In 1998, NRP1 was detected in endothelial cells (ECs) for the first time (8). ECs express both NRP1 and NRP2, and these NRPs are functional receptors for vascular endothelial growth factor (VEGF), a potent angiogenesis factor (5). NRPs do not appear to be high-affinity receptor tyrosine kinases but may serve as coreceptors with high-affinity VEGF receptors. A complex of VEGF-A_{165}, NRP1, and VEGF receptor 2 (VEGFR2) is formed on ECs that increases VEGF binding and enhances EC chemotaxis (8,9). ECs that express NRPs also bind semaphorins. Plexins (PLXNs), a large family of transmembrane proteins, form complexes with SEMAs and NRPs and function as signaling coreceptors (10). As an example of the effect of SEMA on ECs, SEMA3A inhibits EC migration and sprouting (11), and SEMA3F inhibits tumor angiogenesis and metastasis (12). These findings are discussed in more detail in later sections.

BIOLOGICAL AND MOLECULAR SIMILARITIES BETWEEN THE NERVOUS AND VASCULAR SYSTEMS

In retrospect, it is not so surprising that similar cues regulate neuronal guidance and angiogenesis and that these two processes have much more in common than was previously thought. Structurally, neurons and blood vessels exist as hierarchically ordered networks. Both neurogenesis and angiogenesis are processes involving a high degree of branching and migration over long distances. Axons sense their environment through their growth cones, sensory structures containing filopodia at the leading tip of the axons. The growth cones react to repulsive and attractive signals in their environment, which ultimately determine axonal direction. Similarly, developing blood vessels have specialized tip cells that resemble axonal growth cones and sense their environment (13). For example, ECs are attracted by VEGF gradients. Thus, path finding (migration to an appropriate target) in both axons and blood vessels is mediated by chemo-repulsive and -attractive signals.

In addition, nerves and blood vessels are in close proximity and follow similar routes of migration. Nerves and blood vessels also are interdependent. For example, nerves are templates for blood vessel branching and arterial differentiation. Arteries, but not veins, are specifically aligned with peripheral nerves in embryonic mouse limb skin (14). These sensory nerves and Schwann cells provide the VEGF needed for arteriogenesis. In another example, DRG are highly vascular. In DRG, capillaries and neurons are in close apposition to each other, and the neurons produce VEGF that is required for capillary survival (15). The proximity of nerves and blood vessels suggests that there may be molecular cross-talk and common cues between nerves and blood vessels. Several proteins, which were first described as regulators of neuronal migration, have now been shown to guide blood vessels as well. Four pairs of neuronal guidance factors and their cognate receptors have been implicated in angiogenesis. These include Sema/NRP, ephrin/Eph, Slit/Robo, and Netrin/UNC or DCC (13,16). The four receptors are expressed on both neurons and ECs. Thus, it is apparent that a number of ligands and their receptors are involved in the dual regulation of neuronal guidance and angiogenesis.

The congruence of nerve and blood vessel interactions is a rapidly expanding field. The role of ephrin/Eph, Slit/Robo, and Netrin/UNC or DCC in endothelial biology (13,16,17) is covered elsewhere in this volume (see Chapters 38 and 40). This chapter emphasizes the role of NRPs and their semaphorin ligands in developmental and pathological angiogenesis.

NEUROPILIN STRUCTURE, LIGAND-BINDING SPECIFICITIES, AND FUNCTION

NRP is a 130- to 140-kDa membrane glycoprotein first identified in the optic tectum of *Xenopus laevis* and subsequently in the developing vertebrate brain (1). Currently, two NRP genes are known, designated as *NRP1* and *NRP2* (1,5). Human *NRP1* and *NRP2* map to 10p12 and 2q34, respectively (5). NRPs have five extracellular domains (a1, a2, b1, b2, and c), a transmembrane domain, and a short cytoplasmic domain of 40 amino acids (Figure 37.1). VEGF-A_{165} binds to the NRP1 b1b2 domain, whereas SEMA3A binds to both the NRP1 a1a2 and b1b2 domains. SEMA3A and VEGF compete for binding to the NRP1 b1b2 domain (see Figure 37.1) (11). The binding sites on NRP2 for VEGF and SEMA3F have not been identified, but are likely similar to that of NRP1. The NRP c domain is involved in dimerization (1). It is possible that NRP1 and NRP2 can heterodimerize. The cytoplasmic region of either NRP contains no apparent kinase domain, therefore its role in signaling is unclear. Yet, the three most carboxy-terminal amino acids of both NRP1 and NRP2 are serine-glutamic acid-alanine (SEA) and bind a ~40-kDa protein containing a PDZ domain. This NRP interacting protein (NIP) binds members of the GTPase activating protein (GAP) family but its function in ECs and its importance in NRP signaling remain unclear (18). In addition, there are naturally occurring soluble NRP (sNRP) isoforms containing only the a1a2 and b1b2 domains, and these sNRPs are capable of binding ligands (19).

NRP1 and NRP2 share 47% sequence homology and contain similar domain structures, yet their ligand binding specificities are somewhat different (Table 37-1). NRP1 binds SEMA3A, NRP2 binds SEMA3F, and both NRPs bind SEMA3B. Functionally, most NRP/SEMA interactions result in the chemorepulsion of NRP-expressing cells. In the neuronal system, for example, NRP1 is expressed on DRG. SEMA3A binding to NRP1 results in signaling via PLXNs and the repulsion of DRG axons and the collapse of growth cones on the axonal tip of DRG. Neurons in the superior cervical

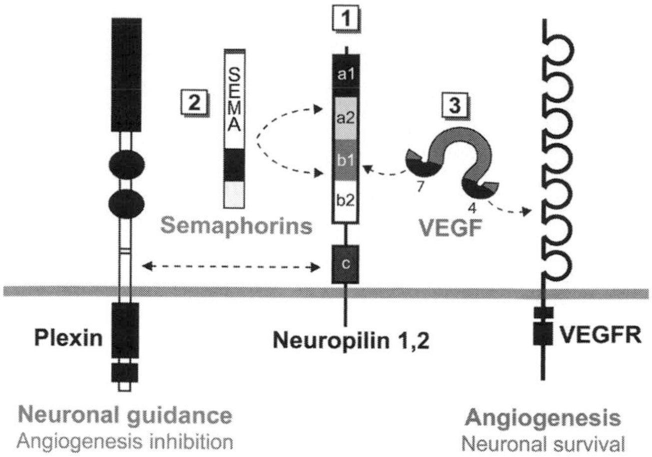

Figure 37.1. Neuropilins (NRPs) bind semaphorins and VEGF. (**1**) There are two neuropilins NRP1 and NRP2. NRPs have five extracellular subdomains (a1, a2, b1, b2, c), a transmembrane domain, and a short cytoplasmic domain. NRPs function as homo- or heterodimers (not shown) that bind via the c domain. (**2**) Class 3 semaphorins are composed of a Sema domain (*white*), an immunoglobulin loop (*black*), and a basic domain (*gray*). Semaphorins bind to NRP via the a1a2 and b1b2 domains. Plexins are transmembrane proteins containing putative cytoplasmic phosphorylation sites. Upon ligand binding, semaphorins, NRPs and plexins form a complex that mediates axon repulsion or growth cone collapse. Semaphorins also inhibit angiogenesis. (**3**) VEGF binds to NRP through the protein subunit encoded by exon 7 and to VEGFR2 (high affinity receptor) through the protein subunit encoded by exon 4. VEGF enhances angiogenesis by bridging NRP and VEGFR. VEGF can also stimulate neuronal cell survival.

ganglia (SCG) express NRP2 and can be repelled by SEMA3F. In vitro, primary ECs express both NRP1 and NRP2 and can be repelled by both SEMA3A and SEMA3F (discussed in more detail later) (12).

The NRPs also differ in their ability to bind various VEGF family members (Table 37-1). NRP1 binds VEGF-A, VEGF-B, VEGF-E, and placenta growth factor-2 (PlGF2, a splice isoform of PlGF), whereas NRP2 binds VEGF-A and VEGF-C (5,6). VEGF-A binding to NRPs is isoform-specific. VEGF-A_{165} binds to NRP1 and NRP2, whereas VEGF-A_{121} does not bind either NRP. VEGF-A_{165} (containing exon 7) is a heparin-binding isoform, whereas VEGF-A_{121} (lacking exon 6 or 7) does not bind heparin. The exon 7 domain of VEGF-A_{165} binds to the b1b2 domain of NRP1 (5,8). The exact binding site for VEGF-A and VEGF-C on NRP2 has not been elucidated.

Functionally, NRPs appear to augment ligand-induced signals by acting as coreceptors for other VEGF receptor tyrosine kinases such as VEGFR1 and VEGFR2. Because VEGF-A_{165} binds to VEGFR2 via its exon 4 domain, and since coimmunoprecipitation of NRP1 and VEGFR2 is only possible in the presence of VEGF-A_{165} ligand, it appears likely that VEGF-A_{165} forms a bridge between NRP1 and VEGFR2 (9). ECs that express NRP1 in addition to VEGFR2 show increased chemotactic activity in response to VEGF-A_{165} (8). When

Table 37-1: Ligand-Binding Specificities and EC Expression Patterns for NRPs and PLXN D1

Receptor	*Ligand*	*EC Expression*
NRP1	VEGF-A_{165} VEGF-B VEGF-E PLGF-2 SEMA3A SEMA3B	Arterial ECs
NRP2	VEGF-A_{165} VEGF-A_{145} VEGF-C SEMA3B SEMA3F	Venous ECs Lymphatic ECs
PLXN D1	SEMA3E*	ECs

*SEMA3E may signal via an NRP-independent pathway.

NRP1 is inhibited using siRNA in human umbilical vein ECs (HUVECs), VEGF-A–induced mitogenic activity (including tritiated thymidine uptake and phospho- extracellular-signal-regulated kinase [ERK] activation) is decreased (20). NRP1 also regulates VEGF-mediated EC permeability (21). Porcine aortic endothelial (PAE) cells expressing NRP1 and VEGFR2 show decreased transendothelial electrical resistance (a measurement of permeability) in response to VEGF-A_{165}, whereas PAE cells expressing NRP1 or VEGFR2 alone do not (21). In addition, blocking of NRP1, but not VEGFR2, using receptor-specific antibodies, attenuated VEGF-A_{165}-mediated permeability of human pulmonary artery endothelial (HPAE) cells (21). NRP2's coreceptor functions have not been studied as rigorously as those of NRP1. NRP2 has a similar domain structure to NRP1 and binds VEGF-A_{165}, therefore it is suspected that NRP2 may serve as a coreceptor for VEGFR2. It is possible that NRP2 may also act as a coreceptor with VEGFR3, because it binds VEGF-C, whereas NRP1 does not. NRPs also may function independently of other VEGFRs to regulate the attachment and adhesion of ECs to extracellular matrix proteins (20).

NEUROPILIN EXPRESSION IN ENDOTHELIAL CELLS

NRPs were first found on HUVECs and later on aortic ECs and capillary ECs (5,8,11). NRP expression in blood vessels has been confirmed in vivo in several organs including the skin, heart, kidney, and bone, but its expression is not ubiquitous (5,7,19). In the avian vascular system, blood islands, which are the earliest vascular structures, express both NRP1 and NRP2. However, following differentiation, NRP1 is expressed preferentially in arterial ECs, whereas NRP2 is expressed preferentially in venous ECs (Table 37-1). When quail-chick chimeras were examined, grafts composed of arterial ECs (expressing NRP1) were found to contribute to both arteries and veins in the host (22). These studies suggest that NRPs contribute toward the determination of vascular identity but are insufficient to determine vascular fate. NRPs are expressed differentially in embryonic blood vessels and lymphatic vessels. For example, NRP2 is expressed primarily in veins and lymphatic vessels (Table 37-1) (13,23). Whether this pattern remains in all adult vessels is unclear, but lymphatic ECs purified from newborn human foreskins express NRP2 and not NRP1 (12).

NEUROPILINS AND THE VASCULATURE

The importance of NRPs in vascular development has been shown in a series of transgenic mouse studies (Table 37-2) (5,24). Overexpression of *Nrp1* in mice was embryonic-lethal and displayed vascular defects (7). These mice die with various morphological deformities, including excess numbers of blood vessels, dilated blood vessels, frequent hemorrhage, and malformed hearts and limbs. *Nrp1*-knockout mice die in utero at embryonic day (E)12.5 to E13.5. In addition to abnormal axonal networks, these mice exhibited vascular defects throughout the embryo, including abnormal yolk sac and neuronal vascularization (25). Both large and small vessels were disorganized and lacked normal branching, while capillary networks were often missing. The NRP1-deficient mice had major cardiovascular defects including agenesis of the branchial arch-related great vessels, dorsal aorta, and transposition of aortic arches (25). When the *Nrp1* a1a2 domain was mutated (knockin strategy changing seven amino acids in this domain) so that SEMA–NRP1 interactions were abolished but VEGF–NRP1 interactions were retained, blood vessel development was normal in the transgenic mice, whereas axonal path finding and cardiovascular development were not (26). Thus, VEGF–NRP1, but not SEMA3A–NRP1 interactions are necessary for establishing the normal vascular architecture. EC-specific *Nrp1*-null mice (*Tie-2 Nrp1*$^{-/-}$ mice) have severe vascular and cardiac defects and survive only until mid-to-late gestation (26). NRP2-deficient mice are viable but have increased mortality following birth (27,28). These *Nrp2* mutant mice (with a targeted deletion in the *Nrp2* locus) have disorganized cranial nerves, and axonal guidance defects in the peripheral nervous system and the central nervous system (27,28). The loss of *Nrp2* results in normal development of arteries and veins, but mice display a severe reduction of small lymphatic vessels and a paucity of vascular capillaries (23). These studies suggest that NRP2 is required for the formation of small lymphatic vessels.

Double *Nrp1/Nrp2*-knockout mice die very early (E8.5) and have a more severe vascular phenotype than do either *Nrp1*- or *Nrp2*-null mice (29). These double mutants exhibit avascular yolk sacs and marked growth retardation. The vascular defects included lack of vessel development, capillary formation, and branching. The vascular phenotype of double *Nrp1/Nrp2*-knockout mice resembles that of *Vegf-a*– and

Table 37-2: Neuropilin-Related Transgenic Mouse Studies

Genotype	*Phenotype*	*References*
Nrp1$^{-/-}$	Embryonic lethal at E12.5–13.5; vascular and axonal defects	25, 44, 45
Nrp1^{sema-}	Survive to birth; normal blood vessel development, aberrant axonal path finding	26
Tie-2 Nrp1$^{-/-}$	Embryonic lethal mid to late gestation; dramatic systemic vascular deficiencies	26
Nrp2$^{-/-}$	Viable; abnormal capillaries and lymphatics; neuronal defects; abnormal neural crest migration	23, 27, 28, 46
Nrp1$^{-/-}$ *Nrp2* $^{-/-}$	Embryonic lethal at E8.5; more severe vascular phenotype than either$^{-/-}$ alone	29
Sema3A$^{-/-}$	Embryonic lethal; abnormal vascular remodeling; cardiac defects; abnormal neuronal patterning	31, 32
Sema3F$^{-/-}$	Viable; no vascular defects reported; abnormal neural crest migration; abnormal axonal guidance	46, 47
Sema3E$^{-/-}$	Viable and fertile; vascular defects	36
Plxn D1$^{-/-}$	Die after birth; vascular defects	36
Nrp1 overexpression	Embryonic lethal at E12.5–17.5; vascular, cardiovascular, and neuronal defects	7

Vegfr2–null mice, and suggests that *Nrp*s are requisite early genes in vascular development.

NRPs contribute to zebrafish vascular development as well (24). Zebrafish genes often are duplicated. There are two nrp1 (*nrp1a* and *b*) and two nrp2 (*nrp2a* and *b*) genes (24). During embryogenesis, zebrafish nrp1 and 2 genes are expressed in neuronal and vascular tissues. During vasculogenesis, *nrp1a* is first expressed by tail angioblasts and, by 48 hours post fertilization (hpf), its expression is found in the major trunk vessels and intersegmental vessels (ISV). When *nrp1* was knocked down using antisense morpholinos (MO), the zebrafish embryos had abnormal connections between the artery and vein, which led to premature return of blood circulation back to the heart (30). Other vascular defects included a lack of circulation in the ISV, in the dorsal longitudinal anastomotic vessels (DLAV), and in the caudal vein plexus. On the other hand, the axial vessels appeared normal, indicating that NRP1 is a regulator of angiogenesis but not vasculogenesis. Results from double knockdown experiments showed that VEGF and NRP1 cooperate in maintaining trunk axial and ISV blood vessel circulation (30).

SEMAPHORINS AND THE VASCULATURE

Eight classes of semaphorins are known. Of these, the class 3 semaphorins are secreted proteins capable of binding NRPs (1,2,5,6). Six class 3 semaphorins are known: SEMA3A, 3B, 3C, 3D, 3E, and 3F. Class 3 semaphorins are 100-kDa proteins, originally described as regulators of neuronal guidance capable of repelling axons and collapsing axonal growth cones (Table 37-3). Soon after the discovery of NRP in ECs, semaphorins were found to regulate vascular functions (Table 37-3). SEMA3A inhibited EC migration and sprouting of capillaries from aortic rings (even in the absence of VEGF) (11). The inhibition of EC migration was associated with the depolymerization of F-actin and retraction of lamellipodia, reminiscent of growth cone collapse in neuronal cells. The effects of semaphorins on EC functions may be due in part to competition with VEGF and in part due to direct signaling events via NRPs and PLXNs. SEMA3A and VEGF are competitive inhibitors of one another, possibly due to overlapping binding regions in the NRP1 b1b2 domain (see Figure 37.1). SEMA3A inhibits VEGF-induced EC motility, and VEGF inhibits SEMA3A-induced DRG collapse (11).

Subsequently, SEMA3A was found to inhibit angiogenesis and vascular remodeling in vivo, presumably by inhibiting integrin activation and integrin-mediated adhesion of EC to fibronectin and vitronectin (31). *Sema3A*-knockout mice have variable vascular phenotypes depending on their genetic background. In a 129/SV background, most of the mutant mice died in utero, whereas the few that survived to birth died within the first 3 days due to pronounced hypertrophy of the right ventricle of the heart (32). In a CD-1 background, *Sema3A*$^{-/-}$ mice displayed abnormal head and trunk blood vessels. At E9.5, cranial blood vessels of mutant embryos maintained a primitive capillary plexus appearance and were not remodeled (31).

Table 37-3: Diverse Activities Associated with Neuropilins and Semaphorins

Cell Type/Activity	*SEMA3/NRP**
Neuronal Cells	
Neuronal patterning	SEMA3A/NRP1
Neuritogenesis	SEMA3C/NRP1,2
Cardiac neural crest migration	SEMA3C/NRP1,2
Axonal fasciculation	SEMA3F/NRP2
Collapse of axonal growth cones	SEMA3A/NRP1
Repulsion of axons	SEMA3A/NRP1; SEMA3C/NRP1,2; SEMA3F/NRP2
Neuronal cell survival	SEMA3C/NRP1,2
ECs	
Vascular patterning	SEMA3A/NRP1; SEMA3F/NRP2; SEMA3E/PlexinD1
Inhibition of sprouting	SEMA3A/NRP1
Inhibition of tumor angiogenesis	SEMA3F/NRP2
Inhibition of adhesion/ migration	SEMA3A/NRP1; SEMA3F/NRP2
Repulsion of vascular ECs	SEMA3A/NRP1; SEMA3F/NRP2
Inhibition of lymphatic ECs	SEMA3F/NRP2
Tumor Cells	
Inhibition of proliferation	SEMA3B/NRP1,2; SEMA3F/NRP2
Inhibition of survival	SEMA3A/NRP1
Inhibition of adhesion/ migration	SEMA3F/NRP2
Inhibition of tumorigenicity	SEMA3B/NRP1,2; SEMA3F/NRP2
Inhibition of metastasis	SEMA3F/NRP2
Induction of apoptosis	SEMA3B/NRP1,2
Tumor encapsulation	SEMA3F/NRP2

*Omission of a Sema3 family member and its activity may not mean that it does not mediate this function, but rather that it has not yet been tested.

SEMA3F, which signals through NRP2, also inhibits EC function and vascular development (12,31,33). SEMA3F inhibits EC proliferation and ERK1/2 kinase activity in vitro and angiogenesis in vivo (33). In addition to its effects on proliferation, SEMA3F acts as a chemorepellant for vascular ECs and lymphatic ECs (12). When *Sema3F* was overexpressed in ECs of the perineural vascular network in the head of developing chick embryos, vascular remodeling was dramatically inhibited (with a similar phenotype to *Sema3A* overexpression) (31). *Sema3F*-null mice are viable and fertile, and vascular defects have not yet been reported (see Table 37-2). Overexpression of *SEMA3F* in melanoma cells resulted in dramatically reduced tumor angiogenesis in mouse xenograft models (12). To date, other semaphorins (including SEMA3A) have not been investigated in tumor angiogenesis models.

Semaphorins also have been implicated in the regulation of vascular patterning in zebrafish. Zebrafish possess two copies of the sema3a gene, *sema3a1* (also known as *sema1a*, *sema3aa*, or *semaZ1a*) and *sema3a2* (also known as *sema1b*, *sema3ab*, or *semaZ1b*) (24). During and after gastrulation stages, angioblasts (vascular endothelial precursors) are located in the trunk ventrally and laterally to the somites. They migrate medially and dorsally along the somites to form the dorsal aorta just ventral to the notochord. Normally, *sema3a1* is expressed by the dorsal and ventral regions of the somites and guides the migration of angioblasts. When *sema3a1* was overexpressed or knocked down in zebrafish embryos, angioblast migration was inhibited, and dorsal aorta formation was retarded (34).

SEMA3E is an unusual member of the class 3 semaphorin family because it is the only member thus far reported to bind directly to a PLXN (see Table 37-1). PLXNs are a large family of transmembrane proteins, which are kinase substrates. They bind NRPs in complex with SEMAs (10). PLXNs are expressed in neurons, ECs, and tumor cells, in a way similar to NRPs. In neurons, signaling via NRP1 is mediated primarily by PLXN A4, and signaling via NRP2 is mediated principally by PLXN A3 (35). Whether this specificity exists in ECs is unknown, but PLXNs are required for semaphorin activity in developmental vascularization (10). PLXN D1 is an unusual member of the PLXN family. Its expression is restricted to ECs. SEMA3E binds PLXN D1 directly and acts as a repulsive cue. In mice, *Sema3E*- and *PlxnD1*-knockouts have similar vascular defects, and SEMA3E appears to regulate vascular patterning by repelling endothelial tip cells (see Table 37-2) (36). SEMA3E-PLXN D1 signaling in this process does not require NRPs, contrary to previous studies showing that class 3 semaphorins need to bind to NRPs for activity (13). In another study, $PlxnD1^{-/-}$ mice were born in Mendelian fashion, but died within the first day due to congenital heart disease with a perturbed outflow tract (37). Zebrafish *plxnd1* is expressed throughout the vasculature. Knockdown of *plxnd1* in zebrafish results in highly abnormal intersegmental vessels (38). Taken together, these studies demonstrate the importance of PLXN D1 and SEMA3E in proper blood vessel path finding and angiogenesis.

NEUROPILINS AND SEMAPHORINS IN CANCER

It is now recognized that NRPs are expressed on many cell types, including tumor cells. Initially, NRP1 was discovered in the prostate cancer cell line, PC3, and the breast cancer cell line, MDA-MB-231 (8). Subsequently, NRP1 and NRP2 were found on many tumor types, especially but not limited to those of epithelial origin. These include prostate, colon, pancreas, mammary, skin, bladder, ovarian, and lung carcinomas, as well as melanomas, neuroblastomas, and osteosarcomas (5,6,39). In most tumor cells, NRP1 or NRP2 is the only VEGF receptor expressed by these cells, because VEGFR1 and VEGFR2 are seldom expressed in tumor cells. Therefore, VEGF functional activity in tumor cells is likely mediated through NRPs.

In PC3 prostate tumor cells, 1-2 $\times$ 10^5 NRP1 receptors are present per cell that bind VEGF-A_{165} with a Kd of approximately 2.8 $\times$ 10^{-10} M (SEMA3A binds NRP1 with similar affinity). Functionally, $VEGF_{165}$ stimulated tumor cell migration but not proliferation (40). In breast cancer cells expressing NRP1, VEGF-A_{165} promoted chemotaxis and survival (41).

NRP expression has been found to correlate with disease progression in several human cancers (5,6,39). Increased NRP1 expression correlated with advanced disease and high Gleason grade in prostate cancer. In addition, NRP1 was expressed preferentially in metastatic breast tumors and cell lines compared with nonmetastatic tumors and cell lines. Aggressive tumors of the gastrointestinal tract showed elevated expression of NRP1, with highest expression at the invasive leading-edge. NRP1 gene expression was upregulated by epidermal growth factor (EGF) family ligands through the EGF receptor in colon, gastric, and pancreatic cancer cell lines (42). NRP2 was differentially expressed in advanced-stage bladder cancers as compared with early-stage bladder tumors using cDNA microarrays (43).

The consequence of increased NRP expression in tumor cells was directly assessed in rat Dunning prostate cancer cells transfected with NRP1 under the control of a tetracycline-inducible promoter (40). Rats were injected with AT2.1/tetNRP1 prostate carcinoma cells and subsequently given doxycycline in their drinking water to induce NRP1 protein synthesis in the tumor cells. In vitro, the NRP1-overexpressing tumor cells had three- to fourfold increased basal motility as compared with control cells; in vivo, the NRP1-overexpressing tumors were 2.5 to 7 times larger than control tumors and had increased vascularity (microvessel density) and decreased apoptosis (40). In similar experiments, human colon carcinoma cells engineered to express increased levels of NRP1 exhibited enhanced tumor growth and angiogenesis in nude mice (42). In contrast, when prostate cancer cells were transfected with a soluble form of NRP1 (sNRP1, containing the a1a2 and b1b2 domains), the resulting tumors were severely apoptotic and contained mainly damaged and hemorrhagic tumor vessels, ostensibly by inhibiting VEGF function (19).

Because tumor cells broadly express NRPs, it came as no surprise that their ligands, the class 3 semaphorins, play a role in tumor progression (see Table 37-3). SEMA3B and SEMA3F were first isolated from regions of chromosome 3p21.3, commonly deleted in small cell lung cancers (5,6,39). Subsequently, several studies have investigated the role of SEMA3B and SEMA3F as tumor suppressors. Overexpression of SEMA3B in lung and breast cancer cell lines induced apoptosis in vitro. In nude mice, ovarian carcinoma cells expressing SEMA3B resulted in smaller tumors, whereas the same cells overexpressing SEMA3F failed to form tumors. On the other hand, overexpression of SEMA3F in a lung carcinoma cell line (GLC45) did not affect tumorigenicity. Taken together, these results suggest that SEMA3B and SEMA3F may affect the growth of some tumors.

Recently, many tumor cell types, including prostate carcinoma cells, bladder carcinoma cells, and melanoma cells, were found to express SEMA3F mRNA. This expression was downregulated as the cells became increasingly metastatic (12). This observation was consistent with previous reports of weak SEMA3F-immunostaining in advanced-stage lung cancers (6). When SEMA3F was overexpressed in metastatic melanoma cells, it completely inhibited spontaneous metastasis to lymph nodes and lungs (12). The resulting tumors expressing SEMA3F were highly encapsulated and poorly vascularized. SEMA3F expression reduced $\beta 1$ integrin levels in melanoma cells overexpressing NRP2, thereby inhibiting their adhesion and migration on fibronectin. Because SEMA3F affects multiple steps in the metastatic cascade, including adhesion, invasion, migration, and angiogenesis (12), it might be considered a new member of a metastasis suppressor family, which includes other genes such as KiSS-1, BRMS1, or KAI-1.

In summary, NRP is a receptor that plays a central role in two distinct developmental pathways – angiogenesis and neuronal guidance. Further studies may reveal other receptor/ligand families common to the nervous and vascular systems. The semaphorins may represent a new class of antiangiogenic agents capable of inhibiting tumor angiogenesis by repulsive mechanisms mediated by NRPs. Future studies will be needed to determine whether semaphorins can be used clinically to inhibit tumor angiogenesis and metastasis.

KEY POINTS

- Semaphorins and NRPs have a dual role as regulators of angiogenesis and neuronal guidance.
- ECs, neurons, and tumor cells express NRPs.
- ECs bind VEGF via NRPs to promote angiogenesis.
- ECs bind semaphorins via NRPs to inhibit angiogenesis or to guide vascular patterning.

Future Goals

- To determine whether semaphorins can be used clinically as antiangiogenic or antimetastatic agents
- To understand the complex signaling networks that govern the regulation of neuronal guidance and angiogenesis through a common receptor, NRP

REFERENCES

1 Fujisawa H. Discovery of semaphorin receptors, neuropilin and plexin, and their functions in neural development. *J Neurobiol.* 2004;59:24–33.

2 Raper JA. Semaphorins and their receptors in vertebrates and invertebrates. *Curr Opin Neurobiol.* 2000;10:88–94.

3 Fiore R, Puschel AW. The function of semaphorins during nervous system development. *Front Biosci.* 2003;8:S484–S499.

4 Bagri A, Tessier-Lavigne M. Neuropilins as Semaphorin receptors: in vivo functions in neuronal cell migration and axon guidance. *Adv Exp Med Biol.* 2002;515:13–31.
5 Klagsbrun M, Takashima S, Mamluk R. The role of neuropilin in vascular and tumor biology. *Adv Exp Med Biol.* 2002;515:33–48.
6 Neufeld G, Shraga-Heled N, Lange T, et al. Semaphorins in cancer. *Front Biosci.* 2005;10:751–760.
7 Kitsukawa T, Shimono A, Kawakami A, et al. Overexpression of a membrane protein, neuropilin, in chimeric mice causes anomalies in the cardiovascular system, nervous system and limbs. *Development.* 1995;121:4309–4318.
8 Soker S, Takashima S, Miao HQ, et al. Neuropilin-1 is expressed by endothelial and tumor cells as an isoform-specific receptor for vascular endothelial growth factor. *Cell.* 1998;92:735–745.
9 Soker S, Miao HQ, Nomi M, et al. VEGF165 mediates formation of complexes containing VEGFR-2 and neuropilin-1 that enhance VEGF165-receptor binding. *J Cell Biochem.* 2002;85: 357–368.
10 Puschel AW. The function of neuropilin/plexin complexes. *Adv Exp Med Biol.* 2002;515:71–80.
11 Miao HQ, Soker S, Feiner L, et al. Neuropilin-1 mediates collapsin-1/semaphorin IIIion of endothelial cell motility: functional competition of collapsin-1 and vascular endothelial growth factor-165. *J Cell Biol.* 1999;146:233–242.
12 Bielenberg DR, Hida Y, Shimizu A, et al. Semaphorin 3F, a chemorepulsant for endothelial cells, induces a poorly vascularized, encapsulated, nonmetastatic tumor phenotype. *J Clin Invest.* 2004;114:1260–1271.
13 Eichmann A, Noble FL, Autiero M, et al. Guidance of vascular and neural network formation. *Curr Opin Neurobiol.* 2005;15:108–115.
14 Mukouyama YS, Gerber HP, Ferrara N, et al. Peripheral nerve-derived VEGF promotes arterial differentiation via neuropilin 1-mediated positive feedback. *Development.* 2005;132:941–952.
15 Kutcher ME, Klagsbrun M, Mamluk R. VEGF is required for the maintenance of dorsal root ganglia blood vessels but not neurons during development. *FASEB J.* 2004;18:1952–1954.
16 Klagsbrun M, Eichmann A. A role for axon guidance receptors and ligands in blood vessel development and tumor angiogenesis. *Cytokine Growth Factor Rev.* 2005;16:535–548.
17 Vogel G. Developmental biology. The unexpected brains behind blood vessel growth. *Science.* 2005;307:665–667.
18 Cai H, Reed RR. Cloning and characterization of neuropilin-1-interacting protein: a PSD-95/Dlg/ZO-1 domain-containing protein that interacts with the cytoplasmic domain of neuropilin-1. *J Neurosci.* 1999;19:6519–6527.
19 Gagnon ML, Bielenberg DR, Gechtman Z, et al. Identification of a natural soluble neuropilin-1 that binds vascular endothelial growth factor: In vivo expression and antitumor activity. *Proc Natl Acad Sci USA.* 2000;97:2573–2578.
20 Murga M, Fernandez-Capetillo O, Tosato G. Neuropilin-1 regulates attachment in human endothelial cells independently of vascular endothelial growth factor receptor-2. *Blood.* 2005;105: 1992–1999.
21 Becker PM, Waltenberger J, Yachechko R, et al. Neuropilin-1 regulates vascular endothelial growth factor-mediated endothelial permeability. *Circ Res.* 2005;96:1257–1265.
22 Moyon D, Pardanaud L, Yuan L, et al. Plasticity of endothelial cells during arterial-venous differentiation in the avian embryo. *Development.* 2001;128:3359–3370.
23 Yuan L, Moyon D, Pardanaud L, et al. Abnormal lymphatic vessel development in neuropilin 2 mutant mice. *Development.* 2002; 129:4797–4806.
24 Goishi K, Klagsbrun M. Vascular endothelial growth factor and its receptors in embryonic zebrafish blood vessel development. *Curr Top Dev Biol.* 2004;62:127–152.
25 Kawasaki T, Kitsukawa T, Bekku Y, et al. A requirement for neuropilin-1 in embryonic vessel formation. *Development.* 1999; 126:4895–4902.
26 Gu C, Rodriguez ER, Reimert DV, et al. Neuropilin-1 conveys semaphorin and VEGF signaling during neural and cardiovascular development. *Dev Cell.* 2003;5:45–57.
27 Chen H, Bagri A, Zupicich JA, et al. Neuropilin-2 regulates the development of selective cranial and sensory nerves and hippocampal mossy fiber projections. *Neuron.* 2000;25:43–56.
28 Giger RJ, Cloutier JF, Sahay A, et al. Neuropilin-2 is required in vivo for selective axon guidance responses to secreted semaphorins. *Neuron.* 2000;25:29–41.
29 Takashima S, Kitakaze M, Asakura M, et al. Targeting of both mouse neuropilin-1 and neuropilin-2 genes severely impairs developmental yolk sac and embryonic angiogenesis. *Proc Natl Acad Sci USA.* 2002;99:3657–3662.
30 Lee P, Goishi K, Davidson AJ, et al. Neuropilin-1 is required for vascular development and is a mediator of VEGF-dependent angiogenesis in zebrafish. *Proc Natl Acad Sci USA.* 2002;99: 10470–10475.
31 Serini G, Valdembri D, Zanivan S, et al. Class 3 semaphorins control vascular morphogenesis by inhibiting integrin function. *Nature.* 2003;424:391–397.
32 Behar O, Golden JA, Mashimo H, et al. Semaphorin III is needed for normal patterning and growth of nerves, bones and heart. *Nature.* 1996;383:525–528.
33 Kessler O, Shraga-Heled N, Lange T, et al. Semaphorin-3F Is an Inhibitor of Tumor Angiogenesis. *Cancer Res.* 2004;64:1008–1015.
34 Shoji W, Isogai S, Sato-Maeda M, et al. Semaphorin3a1 regulates angioblast migration and vascular development in zebrafish embryos. *Development.* 2003;130:3227–3236.
35 Yaron A, Huang PH, Cheng HJ, et al. Differential requirement for Plexin-A3 and -A4 in mediating responses of sensory and sympathetic neurons to distinct class 3 Semaphorins. *Neuron.* 2005;45:513–523.
36 Gu C, Yoshida Y, Livet J, et al. Semaphorin 3E and plexin-D1 control vascular pattern independently of neuropilins. *Science.* 2005;307:265–268.
37 Gitler AD, Lu MM, Epstein JA. PlexinD1 and semaphorin signaling are required in endothelial cells for cardiovascular development. *Dev Cell.* 2004;7:107–116.
38 Torres-Vazquez J, Gitler AD, Fraser SD, et al. Semaphorin-plexin signaling guides patterning of the developing vasculature. *Dev Cell.* 2004;7:117–123.
39 Bielenberg DR, Pettaway CA, Takashima S, et al. Neuropilins in neoplasms: expression, regulation, and function. *Exp Cell Res.* 2006;312:584–593.
40 Miao HQ, Lee P, Lin H, et al. Neuropilin-1 expression by tumor cells promotes tumor angiogenesis and progression. *FASEB J.* 2000;14:2532–2539.
41 Bachelder RE, Lipscomb EA, Lin X, et al. Competing autocrine pathways involving alternative neuropilin-1 ligands regulate chemotaxis of carcinoma cells. *Cancer Res.* 2003;63:5230–5233.

42 Parikh AA, Fan F, Liu WB, et al. Neuropilin-1 in human colon cancer: expression, regulation, and role in induction of angiogenesis. *Am J Pathol.* 2004;164:2139–2151.

43 Sanchez-Carbayo M, Socci ND, Lozano JJ, et al. Gene discovery in bladder cancer progression using cDNA microarrays. *Am J Pathol.* 2003;163:505–516.

44 Kitsukawa T, Shimizu M, Sanbo M, et al. Neuropilin-semaphorin III/D-mediated chemorepulsive signals play a crucial role in peripheral nerve projection in mice. *Neuron.* 1997;19:995–1005.

45 Kawasaki T, Bekku Y, Suto F, et al. Requirement of neuropilin 1-mediated Sema3A signals in patterning of the sympathetic nervous system. *Development.* 2002;129:671–680.

46 Gammill LS, Gonzalez C, Gu C, et al. Guidance of trunk neural crest migration requires neuropilin 2/semaphorin 3F signaling. *Development.* 2006;133:99–106.

47 Sahay A, Molliver ME, Ginty DD, et al. Semaphorin 3F is critical for development of limbic system circuitry and is required in neurons for selective CNS axon guidance events. *J Neurosci.* 2003;23:6671–6680.

38

Vascular Functions of Eph Receptors and Ephrin Ligands

Helmut G. Augustin

German Cancer Center, Research Center (DKFZ) and Center for Biomedicine and Medical Technology (CBTM), Mannheim, Germany

Eph receptors comprise the largest family of receptor tyrosine kinases, consisting of eight EphA receptors (with five corresponding glycosylphosphatidylinisotol [GPI]-anchored ephrinA ligands) and six EphB receptors (with three corresponding transmembrane ephrinB ligands) (Figure 38.1). They were originally identified as neuronal path finding molecules (1,2). Yet, genetic loss-of-function experiments have surprisingly revealed that EphB receptors and ephrinB ligands act also as critical regulators of vascular assembly and arteriovenous differentiation. This chapter summarizes the current knowledge about the vascular functions of Eph receptors and ephrin ligands and discusses some emerging roles of these molecules in the regulation of vascular homeostasis, leukocyte trafficking, and tumor progression.

CHARACTERISTICS OF EPH RECEPTORS AND EPHRIN LIGANDS

Eph receptors and their corresponding ephrin ligands are grouped into A and B subfamilies based on distinct structural properties of the ephrin ligands (see Figure 38.1). EphrinA ligands are GPI-anchored peripheral membrane molecules. EphrinB ligands are transmembrane molecules whose cytoplasmic domain is capable of engaging in various signaling activities. The corresponding Eph receptors act as classical transmembrane tyrosine kinases. Correspondingly, the activation of Eph receptors by ephrin ligands is referred to as "forward signaling," whereas the EphB receptor-mediated activation of ephrinB ligands is designated as "reverse signaling."

Eph receptors and ephrin ligands were originally described as neuronal path finding molecules. Given that both receptor and ligand are transmembrane molecules, *trans* Eph/ephrin signaling is dependent on the juxtapositional contact of neighboring cells. As such, they elicit propulsive and repulsive activities on outgrowing axons and are, thus, critical mediators of neuronal network formation (1,3,4).

VASCULAR EPH RECEPTORS AND EPHRIN LIGANDS

Although the Eph/ephrin system was originally identified in the nervous system, gene targeting experiments in mice have revealed important functions during vascular development (5–7) (discussed later). EphrinB2 is an early marker of arterial ECs, and EphB4 reciprocally marks venous ECs (5–8). These genetic experiments have opened a whole new field of research in vascular biology, because EphB4 and ephrinB2 were the first molecules with a clearly defined asymmetric expression pattern in arteries and veins. These findings have subsequently stimulated research focusing on the identification of other molecules with an asymmetric expression pattern in the vascular system. This work has led to a still-growing list of molecules with distinct asymmetric arteriovenous expression patterns that include the arterial markers (DeltaC [zebrafish], Delta-like 4 (Dll4) [mouse], gridlock [zebrafish], vascular endothelial growth factor (VEGF) receptor (VEGFR)-2 [zebrafish], notch1 [mouse], notch3 [mouse], notch5 [zebrafish], tbx20 [zebrafish], BMX [mouse], neuropilin-1 [mouse, chick], and the venous EC markers (neuropilin-2 [mouse, chick], Tie2 [mouse, chick], VEGFR3 [zebrafish, mouse]) (9). Secondly, in addition to having opened the field of arteriovenous differentiation research, the identification of vascular functions of Eph receptors and ephrin ligands also has stimulated work into the study of the functional commonalities between neuronal patterning and vascular patterning. An increasing list of receptor–ligand systems including the Eph/ephrin system, the Semaphorin/Neuropilin/Plexin system, the Robo/Slit system, and the UNC5/Netrin system is being recognized for its capacity to exert distinct guidance functions on neuronal cells as well as on ECs and vascular smooth muscle cells (VSMCs) (10,11).

Prior to the identification of vascular EphB4 and ephrinB2 functions, ephrinA1 (previously called B61) was characterized as a tumor necrosis factor (TNF)-α–inducible gene in

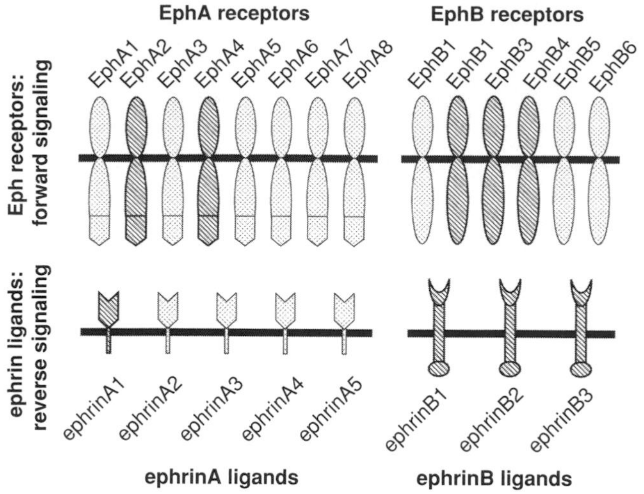

Figure 38.1. Schematic presentation of Eph receptors and ephrin ligands. Eph receptors are classical receptor tyrosine kinases. EphrinA ligands are GPI-anchored peripheral membrane molecules. EphrinB receptors are transmembrane molecules with a cytoplasmic signaling domain. Considerable receptor ligand promiscuity occurs, but little cross-talk between the A and the B subfamilies. Eph receptors and ephrin ligands whose expression has been demonstrated in the vascular system are striped and outlined in bold.

ECs (12). Yet, these early experiments did not lead to a systematic analysis of Eph and ephrin molecules in the vascular system. This is largely due to the fact that research on vascular Eph/ephrin molecules has to this date been hampered by the limited availability of adequate protein reagents to study these molecules.

The hitherto most relevant vascular Eph receptor and ephrin ligand expression data originate from LacZ staining in heterozygous mice in which the *β-galactosidase* gene is knocked into the endogenous locus (13,14). These studies have substantiated the embryonic reciprocal expression of arterial ephrinB2 and venous EphB4. They also have indicated that vascular EphB/ephrinB expression is not limited to ECs, because LacZ staining also was observed in VSMCs and pericytes. Likewise, these studies have demonstrated intense expression of ephrinB2 by angiogenic ECs in tumor microvessels and during reproductive angiogenesis.

REGULATION OF ENDOTHELIAL EPH RECEPTOR AND EPHRIN LIGAND EXPRESSION

The asymmetric arteriovenous expression of EphB4 and ephrinB2 has raised the conceptually important question of whether EphB4 and ephrinB2 are intrinsic markers of arterial and venous EC differentiation, or if their expression is controlled by microenvironmental cues. Lineage tracing experiments have provided some evidence that asymmetric EC EphB4/ephrinB2 expression may be established prior to the formation of arteries and veins, suggesting that the Eph/ephrin system may indeed act as a lineage-determining factor in arteriovenous differentiation rather than simply representing a set of marker molecules (15). This view recently has been challenged by several lines of evidence. First, cell transplantation experiments in the chick embryo indicate that arterial expression of ephrinB2 is under the control of local microenvironmental cues rather than being a lineage-determined intrinsic property of ECs (16). Indeed, manipulatory experiments in the chick chorioallantoic membrane (CAM) have demonstrated that flow is a major determinant of arterial EC ephrinB2 expression (17). Second, in cell culture studies, VEGF has been identified as a major regulator of ephrinB2 expression in angiogenic ECs (18). These latter studies are consistent with the genetically established role of VEGF as an arterializing molecule (19). Finally, a number of other microenvironmental milieu factors have been shown to induce expression of the arterial marker ephrinB2, including hypoxia (20) and shear stress (21).

Little is known about the expression of Eph receptors and ephrin ligands in the adult. ECs in the adult maintain their asymmetric arteriovenous expression pattern of EphB4 and ephrinB2, suggesting that the EphB/ephrinB system may play a role in controlling vascular homeostasis (13,14). Luminal expression of ephrinB2 by quiescent, polarized ECs in the adult vasculature (18,22) suggests a role of EphB/ephrinB interactions of ECs with circulating EphB receptor-positive hematopoietic cells.

ROLE OF THE EPH/EPHRIN SYSTEM DURING DEVELOPMENTAL ANGIOGENESIS

Genetic experiments have laid the groundwork for the identification of EphB receptors and ephrinB ligands as mediators of vascular assembly and differentiation (Table 38-1). EphrinB2-deficient mice die around E10.5 as a consequence of a grossly perturbed vascular differentiation and arteriovenous remodeling. These mice are not capable of orchestrating the positioning of arterial and venous ECs and fail to form a properly structured capillary network (5).

EphrinB2 interacts with EphB2, EphB3, and EphB4. *EphB4*$^{-/-}$ mice essentially phenocopy the phenotype of ephrinB 2-deficient mice, suggesting that the EphB4/ephrinB2 axis is the primary EphB/ephrinB interaction controlling vascular morphogenesis (6,7). Correspondingly, null mice for the other ephrinB2 receptors, *EphB2* and *EphB3*, have no apparent phenotype (6). Yet, a proportion (30%) of *EphB2/EphB3* double null mice have a variable embryonic vascular phenotype, indicating that the EphB4/ephrinB2 interaction is not fully capable of compensating for the loss of EphB2 and EphB3 (6).

Genetic experiments also have provided evidence that reverse ephrinB2 signaling is required for proper arteriovenous differentiation. This was shown by the observation that mice lacking the cytoplasmic domain of ephrinB2 (ΔephrinB2 mice) essentially have the same vascular phenotype as the full-length *ephrinB2*-null mice (8). However, another mouse strain expressing cytoplasmically truncated ΔephrinB2

Table 38-1: Vascular Phenotypes of Mice with Mutations of Eph Receptors and Ephrin Ligands

Molecule	*Genetic Manipulation*	*Phenotype*	*References*
EphA2	Deletion	No developmental phenotype; perturbed angiogenic response in ephrinA1-induced postnatal angiogenesis	51
EphB2/EphB3	Simultaneous deletion of EphB2 and EphB3	Embryonic lethality (E11.5) in some double null mice	6
EphB4	Deletion	Embryonic lethality as a consequence of perturbed arteriovenous differentiation; phenocopies *ephrinB2*-null phenotype	7
ephrinB2	Deletion	Embryonic lethality as a consequence of perturbed arteriovenous differentiation; phenocopies *EphB4*-null phenotype	5, 6, 8
ephrinB2	Deletion of cytoplasmic domain (ΔephrinB2)	Embryonic lethality as a consequence of perturbed arteriovenous differentiation; phenocopies *ephrinB2* full length–null phenotype	8
ephrinB2	Deletion of cytoplasmic domain (ΔephrinB2)	No overt developmental vascular phenotype; later defects including cardiac valve maturation	23
ephrinB2	Mutation of cytoplasmic PDZ domain	Lymphatic remodeling defect	24
ephrinB2	Deletion in mural cells	Perinatal lethality; perturbed mural association with microvessels and abnormal association with lymphatic capillaries	26
ephrinB2	Transgenic overexpression (pleiotropically or endothelial)	Neonatal death; perturbed EC – mural cell interactions	27

exhibited a different phenotype, with defects in cardiac valve formation but not in early vascular development (23). Additional support for a critical rate-limiting role of reverse ephrinB2 signaling for the normal development of the vascular system was provided by minimal-mutagenesis mouse genetic experiments of the cytoplasmic domain of ephrinB2. Mice expressing a mutant of ephrinB2 that lacks only the C-terminal PDZ interaction site have no overt arteriovenous assembly defects, but have major defects in lymphatic vessel remodeling (24).

The phenotypic manifestation of ephrinB2 mutations in different cellular compartments (e.g., vascular versus lymphatic) has stressed the need for spatially and temporally restricted genetic manipulations of *ephrinB2*. Recently, *floxed ephrinB2* mice have been generated as a versatile tool to study ephrinB2 functions in specific cell types (25). Breeding these mice to mice expressing *cre* recombinase under the control of the platelet-derived growth factor receptor-β promoter was used to genetically ablate ephrinB2 in mural cells (pericytes, VSMCs) (26). Mural-cell deletion of ephrinB2 demonstrated a role of ephrinB2 for normal association of mural cells with small-diameter blood vessels, leading to perinatal lethality as a consequence of vascular defects in skin, lung, gastrointestinal tract, and kidney glomeruli as well as abnormal association of VSMCs with lymphatic capillaries (26). Defects of cultured ephrin-B2–deficient VSMCs in spreading, focal-adhesion formation, and polarized migration also indicate cell–autonomous roles of EphB/ephrinB interactions that may result from signalling in *cis* (26).

Transgenic overexpression of ephrinB2 under the control of either the ubiquitously and constitutively active *CAG* promoter or the EC-specific *Tie2* promoter similarly provided evidence that the EphB/ephrinB system not only is involved in arteriovenous EC positioning, but also in EC interactions with mural cells (27). *CAGp-ephrinB2* transgenic mice show sudden death at neonatal stages from aortic dissecting aneurysms due to defective recruitment of VSMCs to the ascending aorta, indicating that EphB/ephrinB2 signaling between ECs and surrounding mesenchymal cells is critical for normal embryonic angiogenesis and vessel maturation. The authors concluded from these findings the existence of three distinctly different cellular compartments, namely EphB4-positive cells, ephrinB2-positive cells, and EphB/ephrinB2-negative cells (27).

ROLE OF THE EPH/EPHRIN SYSTEM DURING TUMOR ANGIOGENESIS

Eph receptors and ephrin ligands do not just regulate the migratory and networking capacity of neuronal and vascular cells. EphB receptors and ephrinB ligands also have been shown to exert positional guidance cues during intestinal epithelial cell migration and differentiation (28). Similarly, circulating leukocytes and bone marrow cells express Eph/ephrin molecules (29,30). A growing list of Eph receptors and ephrin ligands also has been identified in tumors, including colon and lung tumors (31,32). Eph/ephrin expression by tumor cells points toward a potential role of the Eph/ephrin system in controlling malignant cell interactions with the host microenvironment that controls tumor progression and metastatic dissemination. EphA2, as well as its ligand ephrinA1, was found to be expressed by both ECs and various human tumor cells, thereby establishing a microenvironment that stimulates tumor neoangiogenesis by activating EphA2 receptors expressed on angiogenic ECs (32). Blocking of EphA-class receptor activation by competitively acting soluble EphA receptors has been shown to inhibit tumor angiogenesis in two different tumor models (33). Interestingly, stimulation of EphA2 phosphorylation by an EphA2-specific antibody inhibits the malignant behavior of breast tumor cells, indicating that EphA2 signaling may induce different phenotypes in ECs and tumor cells (34).

In contrast to the EphA/ephrinA receptor ligand system (35), much less is known about the functions of the B-class Eph/ephrin molecules during tumor angiogenesis. Several studies, including tumor experiments in heterozygous *ephrinB2* mutant mice expressing LacZ in the *ephrinB2* locus, have demonstrated expression of B-class Eph/ephrin molecules in different tumors and suggested a functional relationship between Eph/ephrin expression and tumor progression (13,36–39). Corresponding functional experiments in an experimental A375 melanoma model indicated that perturbation of bidirectional EphB4/ephrinB2 signaling by overexpression of dominant-negatively acting soluble monomeric EphB4 (sEphB4) dramatically inhibits tumor growth (40). Similar findings have been reported for MCF-7 mammary tumors and SCC-15 squamous cell carcinomas grown in nude mice (41). The observed tumor growth inhibition in these experiments resulted from a combined antiangiogenic and antitumorigenic effect and suggested that EphB/ephrinB interactions may control multiple cell–cell interactions within the tumor and the vascular compartment that collectively control tumor progression (40). Similarly, complex interactions between tumor cell EphB4 and EC ephrinB2 also have been observed in a breast tumor model in which angiogenic induction through reverse signaling activation of ephrinB2 was shown to be induced by direct cell contact with EphB4-expressing tumor cells (42).

CELLULAR FUNCTIONS OF EPH RECEPTORS AND EPHRIN LIGANDS IN ENDOTHELIAL CELLS

Little progress has been made in translating the vascular phenotypes of mice with targeted mutations in Eph receptors and ephrin ligands in mechanistic experiments on the cellular level. This is largely due to the limited availability of suitable cellular models to study arteriovenous differentiation, EC positioning, and vascular guidance as it occurs during later stages of (43) the angiogenic cascade.

A number of cell–cell interaction studies involving cells of the vascular system have been performed in recent years. Corresponding to the transgenic overexpression of ephrinB2, coculture experiments of ephrinB2-expressing OP9 stromal cells and para-aortic splanchnopleura revealed a permissive role of stromal-cell ephrinB2 on vascular network formation (43). EphrinB2-positive ECs proliferate in the presence of ephrinB2-positive OP9 cells and actively induce the recruitment and proliferation of α-smooth muscle actin (α-SMA)-positive cells. In contrast, stromal cells expressing EphB4 inhibit vascular network formation, proliferation of ephrinB2-positive ECs, and recruitment of α-SMA–positive cells (43). These findings would suggest that mural-cell EphB4 acts in a negative, repulsive manner on ephrinB2-positive ECs.

Cell–cell interaction experiments with ECs support a model of reverse ephrinB2-mediated propulsive and EphB4-mediated repulsive activities during angiogenesis (44). Activation of EC ephrinB2 promotes adhesion, migration, chemotaxis, capillary network formation, and sprouting angiogenesis. In turn, forward endothelial EphB4 signaling acts in an antiadhesive, antimigratory manner, and inhibits sprouting angiogenesis (44–46). Forward EphB4 activation is associated with rapid receptor turnover, as evidenced by the internalization of the EphB4/ephrinB2 receptor complex on ephrinB2-Fc receptor body binding. Given the angiogenic and arteriogenic expression of ephrinB2, the antagonistic functions of EphB4 and ephrinB2 support an artery-to-vein push-and-pull model of invasive angiogenesis. According to this model, angiogenic activation (e.g., by VEGF induction) is associated with ephrinB2 expression, which supports a proangiogenic and proinvasive phenotype with an artery-to-vein orientation. Surrounding EphB4-positive cells may then guide invasive outgrowth of ephrinB2-positive ECs. In turn, activation of EC EphB4 transduces negative signals and leads to cellular repulsion. Consequently, propulsive and repulsive ephrinB2- and EphB4-mediated effects during capillary network formation limit cellular intermingling and control boundary formation (44).

EMERGING ROLES OF THE EPH/EPHRIN SYSTEM AS ADHESION REGULATING MOLECULAR SYSTEM IN THE VESSEL WALL

Vascular Eph/ephrin research is still in its infancy. The genetic experiments have focused on the role of EphB receptors and ephrinB ligands as critical determinants of vascular assembly and arteriovenous identity. However, bidirectional signaling through the Eph/ephrin system is increasingly emerging as a rather universal cell–cell interaction and traffic controlling molecular system. Eph/ephrin molecules are expressed by neuronal cells, ECs, VSMCs, epithelial cells, leukocytes, tumor cells, and multiple other cell types. The vascular endothelium forms one of the largest surfaces within the body, acting as a critical interface that controls the trafficking of cells, including the recruitment of circulating leukocytes as well as the metastatic dissemination of malignant tumor cells. As such, Eph receptors in interacting with their ephrin ligands are strategically positioned and suited to either exert direct adhesive functions or to control the adhesive properties of other surface molecules on ECs and interacting cells in the circulation. EphA1- and EphA4-expressing T lymphocytes are capable of binding to EC ephrinA1 (47). Similarly, Eph receptors and ephrin ligands have been shown to support thrombus growth and stability by regulating integrin outside-in signaling in platelets (48). Likewise, EC ephrinB2 has been shown to act as an entry receptor for the Nipah virus (49,50). Thus, further insights into complementary expression patterns of EphB receptors and ephrinB ligands in the vascular endothelium and the multiple cell types interacting with ECs will contribute to the better understanding of the Eph/ephrin system as a cell trafficking and vascular homeostasis–controlling cell–cell signaling system.

CONCLUSION

Eph receptors and ephrin receptors represent a unique bidirectional signaling system that controls multiple cell–cell interactions in the vascular system that are required for developmental vascular morphogenesis, vascular remodeling, and tumor angiogenesis, as well as for vascular homeostatic maintenance function in the adult.

KEY POINTS

- Bidirectional signaling through EphB receptors and ephrinB ligands is required for developmental vascular morphogenesis, transducing guidance signals that are associated with arteriovenous differentiation and positioning.
- Vascular positional guidance cues elicited by Eph/ephrin signaling parallel many aspects of neuronal guidance and positioning.
- A-class and B-class Eph receptors and ephrin ligands control tumor angiogenesis and tumor progression by controlling multiple tumor and vascular cell interactions.

Future Goals

- To undertake the functional analysis of the Eph/ephrin system as a versatile cellular guidance and positioning system controlling multiple cell–cell interactions
- To elucidate the role of endothelial Eph receptors and ephrin ligands as adhesion-regulating molecules of circulating hematopoietic cells and metastasizing tumor cells
- To unravel the differential reverse ephrinB signaling mechanisms in cells of the vessel wall and the analysis of their functional consequences

REFERENCES

1 Flanagan JG, Vanderhaeghen P. The ephrins and Eph receptors in neural development. *Annu Rev Neurosci.* 1998;21:309–345.

2 Kullander K, Klein R. Mechanisms and functions of Eph and ephrin signalling. *Nat Rev Mol Cell Biol.* 2002;3:475–486.

3 Wilkinson DG. Multiple roles of EPH receptors and ephrins in neural development. *Nat Rev Neurosci.* 2001;2:155–164.

4 Palmer A, Klein R. Multiple roles of ephrins in morphogenesis, neuronal networking, and brain function. *Genes Dev.* 2003;17:1429–1450.

5 Wang HU, Chen ZF, Anderson DJ. Molecular distinction and angiogenic interaction between embryonic arteries and veins revealed by ephrin-B2 and its receptor Eph-B4. *Cell.* 1998;93:741–753.

6 Adams RH, Wilkinson GA, Weiss C, et al. Roles of ephrinB ligands and EphB receptors in cardiovascular development: demarcation of arterial/venous domains, vascular morphogenesis, and sprouting angiogenesis. *Genes Dev.* 1999;13:295–306.

7 Gerety SS, Wang HU, Chen ZF, et al. Symmetrical mutant phenotypes of the receptor EphB4 and its specific transmembrane ligand ephrin-B2 in cardiovascular development. *Mol Cell.* 1999;4:403–414.

8 Adams RH, Diella F, Hennig S, et al. The cytoplasmic domain of the ligand ephrinB2 is required for vascular morphogenesis but not cranial neural crest migration. *Cell.* 2001;104:57–69.

9 Lawson ND, Weinstein BM. Arteries and veins: making a difference with zebrafish. *Nat Rev Genet.* 2002;3:674–682.

10 Klagsbrun M, Eichmann A. A role for axon guidance receptors and ligands in blood vessel development and tumor angiogenesis. *Cytokine Growth Factor Rev.* 2005;6:535–548.

11 Eichmann A, Makinen T, Alitalo K. Neural guidance molecules regulate vascular remodeling and vessel navigation. *Genes Dev.* 2005;19:1013–1021.

12 Pandey A, Shao H, Marks RM, et al. Role of B61, the ligand for the Eck receptor tyrosine kinase, in TNF-alpha-induced angiogenesis. *Science.* 1995;268:567–569.

13 Gale NW, Baluk P, Pan L, et al. Ephrin-B2 selectively marks arterial vessels and neovascularization sites in the adult, with expression in both endothelial and smooth-muscle cells. *Dev Biol.* 2001;230:151–160.

14 Shin D, Garcia-Cardena G, Hayashi SI, et al. Expression of ephrinB2 identifies a stable genetic difference between arterial and venous vascular smooth muscle as well as endothelial cells, and marks subsets of microvessels at sites of adult neovascularization. *Dev Biol.* 2001;230:139–150.

15 Zhong TP, Childs S, Leu JP, et al. Gridlock signalling pathway fashions the first embryonic artery. *Nature.* 2001;414:216–220.

16 Othman-Hassan K, Patel K, Papoutsi M, et al. Arterial identity of endothelial cells is controlled by local cues. *Dev Biol.* 2001; 237:398–409.

17 le Noble F, Moyon D, Pardanaud L, et al. Flow regulates arterial-venous differentiation in the chick embryo yolk sac. *Development.* 2004;131:361–375.

18 Korff T, Dandekar G, Pfaff D, et al. Endothelial ephrinB2 is controlled by microenvironmental determinants and associates context-dependently with CD31. *Arterioscler Thromb Vasc Biol.* 2006;26:468–474.

19 Mukouyama Y, Shin D, Britsch S, et al. Sensory nerves determine the pattern of arterial differentiation and blood vessel branching in the skin. *Cell.* 2002;109:693–705.

20 Claxton S, Fruttiger M. Oxygen modifies artery differentiation and network morphogenesis in the retinal vasculature. *Dev Dyn.* 2005;233:822–828.

21 Goettsch W, Augustin HG, Morawietz H. Down-regulation of endothelial ephrinB2 expression by laminar shear stress. *Endothelium.* 2004;11:259–265.

22 Augustin HG, Reiss Y. EphB receptors and ephrinB ligands: regulators of vascular assembly and homeostasis. *Cell Tissue Res.* 2003;314:25–31.

23 Cowan CA, Yokoyama N, Saxena A, et al. Ephrin-B2 reverse signaling is required for axon pathfinding and cardiac valve formation but not early vascular development. *Dev Biol.* 2004;271:263–271.

24 Makinen T, Adams RH, Bailey J, et al. PDZ interaction site in ephrinB2 is required for the remodeling of lymphatic vasculature. *Genes Dev.* 2005;19:397–410.

25 Grunwald IC, Korte M, Adelmann G, et al. Hippocampal plasticity requires postsynaptic ephrinBs. *Nat Neurosci.* 2004;7: 33–40.

26 Foo SS, Turner CJ, Adams S, et al. Ephrin-B2 controls cell motility and adhesion during blood-vessel-wall assembly. *Cell.* 2006; 124:161–173.

27 Oike Y, Ito Y, Hamada K, et al. Regulation of vasculogenesis and angiogenesis by EphB/ephrin-B2 signaling between endothelial cells and surrounding mesenchymal cells. *Blood.* 2002;100:1326–1333.

28 Batlle E, Henderson JT, Beghtel H, et al. Beta-catenin and TCF mediate cell positioning in the intestinal epithelium by controlling the expression of EphB/ephrinB. *Cell.* 2002;111:251–263.

29 Luo H, Yu G, Wu Y, et al. EphB6 crosslinking results in costimulation of T cells. *J Clin Invest.* 2002;110:1141–1150.

30 Munoz JJ, Alonso CL, Sacedon R, et al. Expression and function of the Eph A receptors and their ligands ephrins A in the rat thymus. *J Immunol.* 2002;169:177–184.

31 Dodelet VC, Pasquale EB. Eph receptors and ephrin ligands: embryogenesis to tumorigenesis. *Oncogene.* 2000;19:5614–5619.

32 Ogawa K, Pasqualini R, Lindberg RA, et al. The ephrin-A1 ligand and its receptor, EphA2, are expressed during tumor neovascularization. *Oncogene.* 2000;19:6043–6052.

33 Brantley DM, Cheng N, Thompson EJ, et al. Soluble Eph A receptors inhibit tumor angiogenesis and progression in vivo. *Oncogene.* 2002;21:7011–7026.

34 Carles-Kinch K, Kilpatrick KE, Stewart JC, et al. Antibody targeting of the EphA2 tyrosine kinase inhibits malignant cell behavior. *Cancer Res.* 2002;62:2840–2847.

35 Nakamoto M, Bergemann AD. Diverse roles for the Eph family of receptor tyrosine kinases in carcinogenesis. *Microsc Res Tech.* 2002;59:58–67.

36 Vogt T, Stolz W, Welsh J, et al. Overexpression of Lerk-5/Eplg5 messenger RNA: a novel marker for increased tumorigenicity and metastatic potential in human malignant melanomas. *Clin Cancer Res.* 1998;4:791–797.

37 Tang XX, Evans AE, Zhao H, et al. High-level expression of EPHB6, EFNB2, and EFNB3 is associated with low tumor stage and high TrkA expression in human neuroblastomas. *Clin Cancer Res.* 1999;5:1491–1496.

38 Tang XX, Zhao H, Robinson ME, et al. Implications of EPHB6, EFNB2, and EFNB3 expressions in human neuroblastoma. *Proc Natl Acad Sci USA.* 2000;97:10936–10941.

39 Stephenson SA, Slomka S, Douglas EL, et al. Receptor protein tyrosine kinase EphB4 is up-regulated in colon cancer. *BMC Mol Biol.* 2001;2:15.

40 Martiny-Baron G, Korff T, Schaffner F, et al. Inhibition of tumor growth and angiogenesis by soluble EphB4. *Neoplasia.* 2004;6: 248–257.

41 Kertesz N, Krasnoperov V, Reddy R, et al. The soluble extracellular domain of EphB4 (sEphB4) antagonizes EphB4-EphrinB2 interaction, modulates angiogenesis and inhibits tumor growth. *Blood.* 2006;107:2330–2338.

42 Noren NK, Lu M, Freeman AL, et al. Interplay between EphB4 on tumor cells and vascular ephrin-B2 regulates tumor growth. *Proc Natl Acad Sci USA.* 2004;101:5583–5588.

43 Zhang XQ, Takakura N, Oike Y, et al. Stromal cells expressing ephrin-B2 promote the growth and sprouting of ephrin-B2($^+$) endothelial cells. *Blood.* 2001;98:1028–1037.

44 Füller T, Korff T, Kilian A, et al. Forward EphB4 signaling in endothelial cells controls cellular repulsion and segregation from ephrinB2 positive cells. *J Cell Sci.* 2003;116:2461–2470.

45 Hamada K, Oike Y, Ito Y, et al. Distinct roles of ephrin-B2 forward and EphB4 reverse signaling in endothelial cells. *Arterioscler Thromb Vasc Biol.* 2003;23:190–197.

46 Hayashi S, Asahara T, Masuda H, et al. Functional ephrin-B2 expression for promotive interaction between arterial and venous vessels in postnatal neovascularization. *Circulation.* 2005; 111:2210–2218.

47 Aasheim HC, Delabie J, Finne EF. Ephrin-A1 binding to $CD4^+$ T lymphocytes stimulates migration and induces tyrosine phosphorylation of PYK2. *Blood.* 2005;105:2869–2876.

48 Prevost N, Woulfe DS, Jiang H, et al. Eph kinases and ephrins support thrombus growth and stability by regulating integrin

outside-in signaling in platelets. *Proc Natl Acad Sci USA*. 2005; 102:9820–9825.

49 Negrete OA, Levroney EL, Aguilar HC, et al. EphrinB2 is the entry receptor for Nipah virus, an emergent deadly paramyxovirus. *Nature*. 2005;436:401–405.

50 Bonaparte MI, Dimitrov AS, Bossart KN, et al. Ephrin-B2 ligand is a functional receptor for Hendra virus and Nipah virus. *Proc Natl Acad Sci USA*. 2005;102:10652–10657.

51 Brantley-Sieders DM, Caughron J, Hicks D, et al. EphA2 receptor tyrosine kinase regulates endothelial cell migration and vascular assembly through phosphoinositide 3-kinase-mediated Rac1 GTPase activation. *J Cell Sci*. 2004;117:2037–2049.

39

Endothelial Input from the Tie1 and Tie2 Signaling Pathways

Daniel Dumont

The Centre for Proteomic Studies, University of Toronto, Ontario, Canada

DISCOVERY OF TIE1, TIE2, AND THE ANGIOPOIETINS

In the early 1990s, numerous groups used an reverse transcription polymerase chain reaction (RT-PCR) approach to identify new tyrosine kinases. This approach, based on the use of degenerate primers complementary to highly conserved regions of the kinase domain, was first described by the lab of Andrew Wilks (1,2). The small RT-PCR fragments produced using this approach and RNA isolated from several sources, such as K-562 cells, embryonic mouse heart tissue, differentiated embryonic stem cells and hematopoietic stem cells (3–10), provided the starting material for the cloning and characterization of many tyrosine kinases including new families of receptor tyrosine kinases (RTK). One of these families has just two members, Tie1 and Tie2. These receptors were first identified by RT-PCR of RNA isolated from the megakaryoblastic cell line K-562 (11). cDNAs coding for these receptors were described in 1992 and 1993, respectively, and were originally named *Tie* (*Tie1*) and *Tek/Hyk* (*Tie 2*) (4,6,11).

The Tie receptors remained orphan receptors until 1996 when angiopoietin-1 (Ang1) was identified as a ligand for Tie2 (12). *Ang1* was cloned using a unique secretion trap cloning method, in which a receptor body was used to probe cells transfected with a cDNA expression library. Subsequent to the identification of Ang1, several other angiopoietins (*Ang2, 3,* and *4*) were cloned by low-stringency hybridization (13,14). *Ang3*, which is isolated from mouse, and *Ang4*, which is of human origin, are thought to be interspecies orthologues (14). Both Ang3 and Ang4 bind to and activate Tie2 within their respected species (15).

EVOLUTIONARY CONSERVATION

Genes for both the Tie1 and Tie2 signaling systems are conserved from fish to humans (16,17), suggesting that these receptors and their ligands play a pivotal role in the development of the endothelium and hematopoietic systems. Similarly, *Ang1* and *-2* have been found in the same lower vertebrates, further emphasizing their importance in vascular development (17,18). Currently, it is not known whether *Ang3* or *-4* are found in these lower vertebrates; however, an angiopoietin-related gene, *angiopoietin-like-3*, which is found in humans, was also identified in zebrafish (18), further suggesting that these ligands are important in the biology of blood vessels in diverse organisms.

ROLE IN DEVELOPMENT

Endothelial and Hematopoietic Expression and Function

Expression studies during embryogenesis revealed that the Tie-family of RTKs is expressed in endothelial cells (ECs) (5,9,11,19–21) and in hematopoietic stem cells (8), suggesting that these receptors may play a role in development of the endothelium and in hematopoiesis.

The functions of these receptors in these cell types have been examined with gene-targeted mice. Both Tie1 and Tie2 are needed for embryonic development. *Tie1*-null embryos die around embryonic day (E)13.5 of gestation with marked abnormalities in the vasculature and edema (22,23). In contrast, *Tie2*-null embryos die considerably earlier, around E9.5 of development with defects in the cardiovascular system (23,24). A role in EC survival was demonstrated in *Tie2*-null embryos that were conditionally rescued using a *Tie2*–promoter-driven tetracycline-regulated transgenic system (25). Tie2-rescued embryos in which the expression of Tie2 was subsequently silenced had a dramatic increase in apoptotic ECs, suggesting that Tie2 signaling maintains EC survival in vivo (25).

The early embryonic lethality of *Tie2*$^{-/-}$ embryos precluded the analysis of the role of this receptor in late stages of vascular development. To get around this, Puri and colleagues analyzed the ability of *Tie1*- and *Tie2*-null embryonic stem cells to contribute to the vasculature in chimeric embryos and whether any contribution was sustained throughout development (26). Tie1 and Tie2 were dispensable in the early phases of vessel development, but Tie2 was absolutely required for the microvasculature in the adult. Whether this requirement was for all components of microvasculature, capillaries, arterioles, or venules was not determined. Furthermore, similar studies focusing on the development of the hematopoietic lineage revealed that both Tie1 and -2 were required for postnatal bone marrow hematopoiesis (27).

Expression and Function in Other Cell Types

As better reagents were developed and as other cell types were examined in more detail, Tie2 expression was found in other cells including keratinocyte precursors (28), eosinophils (29), neutrophils (30), perivascular mesenchymal cells (31–33), hematopoietic stem cells (34–36) fibroblast-like cells (37), and endometrial epithelial and stromal cells of the endometrium (38). The physiological significance of this expression is not understood; however, as keratinocytes precursor cells differentiate, they downregulate the expression of Tie2, suggesting that the expression of Tie2 in these cells may play a role in the maintenance of an undifferentiated state (28).

GENE REGULATION

The *Tie1* and *Tie2* genes have been examined to determine the DNA elements in these genes that confer tissue specific expression. Early studies of the *Tie2* gene suggested much of this specificity was directed by a rather large region that was 5′ of the coding sequence. These sequences, however, were not sufficient to regulate expression in all ECs in the adult (24). Subsequently, an enhancer element was discovered in the first intron that significantly increased the expression in ECs of the adult (39).

Similar studies were initiated with the *Tie1* gene, and it was found that a smaller region of DNA 5′ from the coding sequence was able to direct endothelial expression in the developing embryo and adult (40). Further detailed analysis of these DNA sequences revealed several octomer binding sequences that were thought to be responsible for the endothelial-specific expression (41,42). Ets binding sequences were found within this region. Furthermore, Ets binding sites have been found in both *Tie* promoters, suggesting that the Ets family of proteins play a role in the expression from these promoters in the vasculature (43–47).

Hypoxia is known to drive angiogenesis, suggesting that the expression of many angiogenic factors is controlled by this stimulus. Presently, conflicting results have surfaced regarding the expression of Tie2 in response to hypoxia (48–52). It is possible that these discrepancies reflect the differences in experimental paradigms or cell types that have been used. It was recently shown that the 5′ end of the *Tie2* mRNA, which is quite long and complex, has an internal ribosome entry site (IRES) that allows for cap-independent translation. This sequence is required for efficient translation of *Tie2* mRNA under the hypoxic state (53), suggesting that Tie2 expression is enhanced during hypoxia.

ECs are at the interface between blood and the vessel wall and are therefore subjected to hemodynamic stresses. The expression of Tie1 and Tie2 was reported to be upregulated by shear (54,55), and the kinase activity of Tie2 also was elevated (56). This response of Tie1 expression to shear remains controversial, because one study examining the expression of Tie1 in cultured ECs found a downregulation (57), whereas another group examining Tie1 expression in vivo found an upregulation (55). Curiously, the extracellular domain of Tie1 is cleaved when cells in culture are exposed to shear, and the free Tie1-extracellular domain binds to Tie2 on the EC surface (57). The physiological significance of this is unclear, but it may serve to downregulate Tie2 signaling and lead to vessel remodeling. ECs also can be subjected to other stresses, including physical or mechanical stress, which has been shown to result in the increased expression of both Tie1 and Tie2 (58,59).

In addition, biochemical mediators are thought to play a pivotal role in controlling the expression of numerous angiogenic factors including Tie2 (60). In 2005, Fathers and colleagues (60) demonstrated considerable Tie2 expression heterogeneity in xenografts, illustrating that the tumor microenvironment can greatly influence the expression of Tie2.

LIGAND–RECEPTOR INTERACTIONS

Initial studies suggested that Ang1 acts as an agonist of Tie2 activity, whereas Ang2 behaves as an antagonist (13). However, Ang2, either at higher concentrations in vitro or on exposure of ECs for prolonged periods, also can act as an agonist (61). Whether these experimental conditions recapitulate true endogenous levels or environments encountered by ECs remains to be determined. However, expression of Ang2 systemically using a conditional binary transgenic mouse system suggests that Ang2 may have VEGF-independent activities in certain vascular beds (62). Furthermore, Ang1 and Ang2 are thought to be functionally redundant for lymphatic ECs, but may have this antagonist role in different vascular beds (63,64). The response of ECs to the angiopoietins may depend on the microenvironment or "context" in which the EC finds itself. Table 39-1 summarizes numerous studies examining stimulation of cells in vitro or the growth of tumors in xenograft models. The considerable inconsistencies in the data ultimately suggest that the EC type, growth conditions, and location where a particular tumor is placed in animal xenograft studies have a dramatic impact on effects of the angiopoietins (65).

Table 39-1: In Vitro and Xenograft Results in Angiopoietin Signaling

Condition	*Outcome*
Ang1 in HeLa Cells	Promotes tumor growth in the cervix
Ang1 in A431 Cells	Inhibits tumor growth in skin
Ang1 antisense in HeLa cells	Inhibits tumor growth
Ang2 expression in gliomas	Increased angiogenesis
Ang2 can act as agonist	Promotes cell survival, differentiation
Ang1,2 and VEGF cooption	Ang2 destabilizes, VEGF rescues, Ang1 stabilizes
ExTek	Soluble Tie2 can inhibit tumor growth Inhibition of laser induced retinal neovascularization
Ang2 antibody and peptide-body	Inhibit xenografts growth

In an attempt to circumvent the problem of isolating cells from different organs and to alleviate the known fact that ECs grown in culture lose many of their organ-specific characteristics, research teams have engineered transgenic animals to produce angiopoietins and have found that the vascular beds within different organs do indeed respond differently to these factors, suggesting that context is very important (66,67) (Table 39-2).

Ang3 and *Ang4* are interspecies orthologues – *Ang3* is murine and *Ang4* human (14,68). In the initial studies of the effect of these ligands using human ECs, it was thought that Ang4 was an agonist and Ang3 was an antagonist for Tie2 activation. Recent studies, however, have shown that these ligands exhibit species-specific restrictions in their ability to activate Tie2, and it now seems that both Ang3 and -4 are agonists (15).

The angiopoietins are multidomain ligands that have a coiled-coiled domain at the amino terminus that is responsible for clustering ligands, and fibrinogen-like domains at the carboxyl-terminus that bind the receptor. The angiopoietins are found as homo-oligomerized proteins held together by disulfide bonds (69–71). This oligomerization is required for the angiopoietins to cluster Tie2 and cause its activation (69,71,72). In fact, monomeric angiopoietin is inhibitory for Tie2 activation (73).

The ligand for Tie1 remains unknown but, under some circumstances, Tie1 and Tie2 heterodimerize, leading to phosphorylation of Tie1. Ang1, Ang4, or an engineered form of Ang1, COMP-Ang1 (74,75), can enhance this phosphorylation (76). Although there is a basal level of Tie1 stimulation by angiopoietins in the absence of Tie2, these ligands do not appear to bind directly to Tie1 (76), suggesting that the activation of Tie1 is through an alternative route.

DOWNSTREAM SIGNALING PATHWAYS

In an effort to uncover the mechanisms of Tie2 receptor signaling, several groups performed yeast two-hybrid assays with a truncated intracellular active kinase region (77–79). This approach proved very successful in identifying binding partners that are now thought to play an important role in Tie2 signaling. These include p85 (the regulatory subunit of phosphoinositide 3-kinase [PI3K]), Grb2, Grb7, Grb14, Shp2, ABIN-2, and Dok-R. Using a candidate-molecule approach, ShcA was also shown to be recruited to Tie2 (80) (Figure 39.1).

PI3K plays an important role in Tie2 receptor signaling. Activation of PI3K by Ang1 leads to activation of Akt, which promotes EC survival. Akt activation also leads to the phosphorylation of endothelial nitric oxide synthase (eNOS) resulting in the production of nitric oxide (NO) (61,81), an important extracellular signaling molecule. p85 interacts with tyrosine 1100 (Y1101 in humans), a residue that also recruits Grb7 (82). The importance of this site in Tie2-mediated signaling in vivo was demonstrated by genetically engineered mice that carried a mutant form of Tie2 in which tyrosine 1100 was converted to phenylalanine (83). These animals had defects in cardiac and hematopoietic development, whereas pericyte recruitment was unaltered, suggesting that multiple signals emanate from Tie2 and that only certain aspects of Tie2-mediated biology requires signaling through tyrosine 1100. Of further interest is the fact that a mouse mutant that carries a deletion in the catalytic portion of PI3K, p110α, also exhibits a vascular defect similar to the defect seen in *Tie2*-null embryos (84). Moreover, study of transgenic animals with an activated form of Akt also demonstrated that the development of the vasculature is controlled by a fine-tuning of the signals that emanate from Akt (85). Collectively, these results support an important role for PI3K signaling in vascular remodeling and maintenance.

Dok-R is another molecule that plays a role in Tie2 signaling. Recruitment of Dok-R to phosphotyrosine 1106 on Tie2 is required for Ang1-driven EC migration (86,87). Dok-R, upon binding to Tie2, is tyrosine phosphorylated and recruits the signaling molecule Nck, which is bound to the serine-threonine kinase, Pak1 (87). Pak1 initiates a phosphorylation cascade that ultimately results in reorganization of the actin cytoskeleton and in cell migration.

Another molecule thought to play an important role in Tie2 signaling is the ABIN-2 (79). ABIN-2 rescues ECs from apoptosis induced by growth factor deprivation, and a truncated form suppresses Tie2-mediated rescue of this apoptosis, suggesting that ABIN-2 also may play a role in angiopoietin-mediated cell survival (88).

Table 39-2: In Vivo Models of Angiopoietin Signaling

Gene	*Phenotype*	*Lethality*
Tie2 KO	Decreased sprouting	E9.5–12.5
Conditional rescue	Heart development Vessel branching EC survival Recruitment of pericytes Definitive hematopoiesis	
Ang1 KO	Endocardial development Trabeculation Recruitment of pericytes	E12.5
Ang2 transgenic (*Tie2*-promoter)	Endocardial/myocardial Vessel loss	E9-5–10.5
Ang2 KO	Lymphatic development Chylous ascites Hyaloid vessel	Partial-P14
Ang1 transgenic (*K14* promoter)	Large vessels Resistance to VEGF leak	NA
Ang1 transgenic (*LAP* promoter)	Large vessels in liver ???	NA
Ang1 transgenic (*MHCA* promoter)	Lack of coronary vessel development	90% dead E12.5–15.5
Ang1 transgenic (*αMHC* promoter)	Blocks VEGF-induced vessel formation	NA
Ang2 transgenic (*αMHC* promoter)	Collaborates with VEGF to induce angiogenesis edema and fibrosis	NA
Ang1-CC-transgenic (*MHCA* promoter*)	Ang1 and Ang2 inhibitor. Altered coronary vessel patterning	NA
Tie2-transgenic (*Tie2* promoter)	Psoriasis/atopic dermatitis phenotype	NA
Ang1 transgenic (*CamK* promoter)	Alteration in neuronal processes	NA
Ang2 transgenic (*LAP* promoter)	Changes in lower vessel growth Increased lymphatic density in gastrointestinal tract Increased congestion in lungs	NA

*Different MHC promoter. KO, knockout; NA, not available.

As mentioned earlier, other molecules may have a role in Tie2 receptor signaling because they bind to the intracellular domain of Tie2, but the roles of these molecules have not been elucidated.

EFFECT ON ENDOTHELIAL CELL PHENOTYPE

Level of Cell

Ang1 protects ECs from apoptosis induced by serum withdrawal, radiation, mannitol, and from detachment (anoikis) (77,82,89), suggesting that this pathway mediates EC survival. Examination of phosphorylation status of Tie2 in quiescent endothelium in the rat revealed that Tie2 was tyrosine phosphorylated, suggesting that it is active in resting ECs, and that it promotes the survival of resting ECs (89). In support of this role, Jones and colleagues (25) used a conditional rescue of *Tie2* mutant mice to demonstrate that Tie2 provides an important survival signal during the latter stages of embryonic growth. In addition to the antiapoptotic activity of Tie2 signaling, the angiopoietins also induce endothelial tubule formation in vitro. Interestingly, Ang1 and Ang2 had similar activities in this assay suggesting that, within this context, these two ligands have similar activities (90).

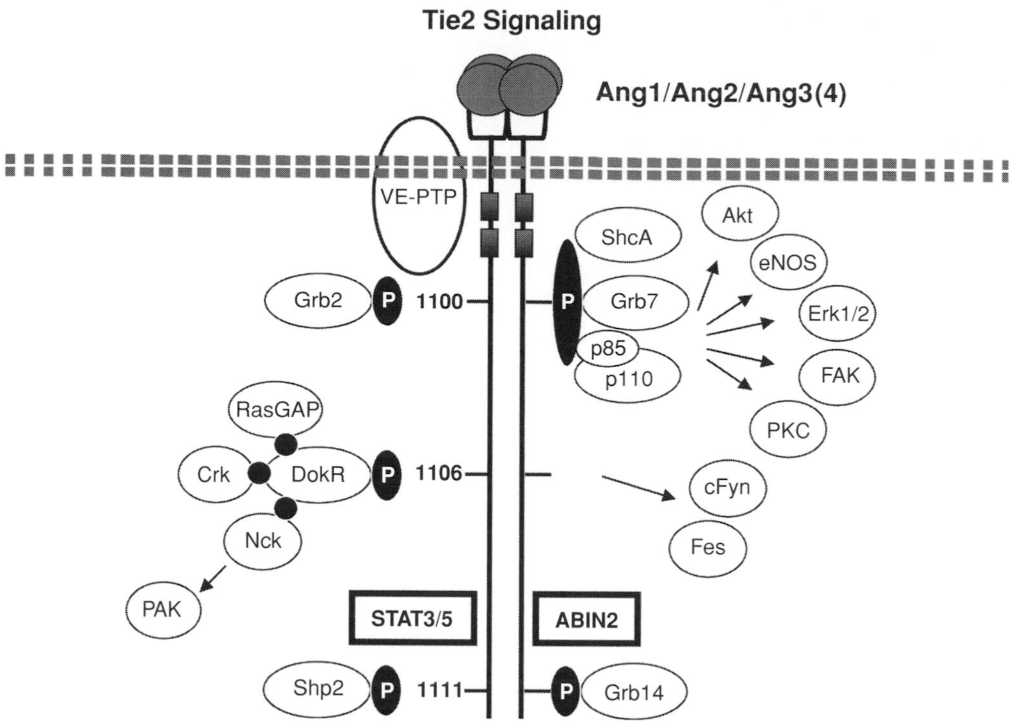

Figure 39.1. Signaling pathways modulated by Tie2. Proteins known to interact with Tie2 are shown next to the receptor (VE-PTP, ShcA, Grb2, Grb7, Grb14, p85, Shp2, Fes, STAT3, STAT5, ABIN2, and Dok-R).

Level of Blood and Lymphatic Vessels

Several groups have hypothesized that angiopoietins have an effect on vascular tone through the modulation of NO via the Tie-2 receptor signaling, but this role is not well understood (61).

Within the context of tumor angiogenesis, Ang2 seems to provide a destabilizing function that, by antagonizing Tie2 activity, in combination with VEGF, causes ECs within a vessel to detach and to start to migrate toward a chemotactic signal. *Ang2* knockout studies have suggested that this may be the case for the hyaloid vasculature, which during development supplies the lens with nutrients, but regresses after birth. In the *Ang2*-null animal, this regression does not occur (64) (see Table 39-2). Curiously, the rest of the vasculature seemed normal, suggesting that this effect may be very dependent upon the microenvironment of the ECs. Furthermore, the *Ang2*-null animal presents with defects in the developing lymphatics that can be rescued by *Ang1*, thus demonstrating a functional redundancy for these ligands in the lymphatic vasculature.

Several studies that examined the role of these ligands in xenograft tumor growth have yielded contradictory results, suggesting that the context in which the angiopoietins are expressed may dramatically influence the EC response (see Table 39-1).

Level of Whole Organism

A role for angiopoietin signaling in vascular remodeling has been suggested by the identification of missense mutations in the *Tie2* gene in several families with inherited vascular malformations (IVMs) (91). The mutated gene encodes a Tie2 receptor with elevated tyrosine kinase activity (91). It is thought this elevated kinase activity in combination with other epigenetic factors results in a defect in the recruitment and/or retention of support cells that ultimately lead to IVMs. Interestingly, one of the mutant Tie2 proteins that was isolated had increased kinase activity and also recruited an additional repertoire of signaling molecules, suggesting that manifestation of the phenotype may be caused by an increase in signaling and/or aberrant signaling (92).

Studies using conditional transgenic mice, in which Ang1 is produced in the liver and delivered systemically, also suggest that angiopoietins play a role in the development of this vascular bed. In these mice, in the absence of any increase in liver size, an increase occurred in the size of the vessels, and remodeling of the liver vasculature was disrupted (93). The response of the ECs in different vascular beds seemed to be different, further supporting the notion that the response of ECs to angiopoietins depends on the context.

DIAGNOSTIC AND THERAPEUTIC IMPLICATIONS

One of the first indications that the angiopoietin signaling system might be a good target for the design of drugs to treat cancer was a xenograft study in which tumors regressed in size when the mice were treated with a soluble form of

Tie2, exTek (94–97). Further studies have suggested that exTek also may decrease the degree of metastasis (98). Furthermore, decreasing available Ang2 with specific antibodies or peptide-bodies resulted in decreased angiogenesis and decreased tumor growth in mice (99). It is not clear whether targeting the angiopoietins will work on all types of tumors because of the context-dependent effect of the angiopoietins on the vasculature.

KEY POINTS

- Tie2 signaling plays a pivotal role in the maintenance and integrity of the endothelium.
- Tie2 is expressed on virtually all endothelium.
- The type of endothelium dramatically alters the response of ECs to the angiopoietins.
- To realize the full potential of the angiopoietins signaling system for cancer therapeutics and regenerative medicine, we will need to understand the "context-specific" response of the endothelium in different organ systems to the angiopoietins.

REFERENCES

1 Wilks AF. Cloning members of protein-tyrosine kinase family using polymerase chain reaction. *Methods Enzymol.* 1991;200: 533–546.
2 Wilks AF, Kurban RR, Hovens CM, et al. The application of the polymerase chain reaction to cloning members of the protein tyrosine kinase family. *Gene.* 1989;85:67–74.
3 Partanen J, Makela TP, Alitalo R, et al. Putative tyrosine kinases expressed in K-562 human leukemia cells. *Proc Natl Acad Sci USA.* 1990;87:8913–8917.
4 Dumont DJ, Yamaguchi TP, Conlon RA, et al. tek, a novel tyrosine kinase gene located on mouse chromosome 4, is expressed in endothelial cells and their presumptive precursors. *Oncogene.* 1992;7:1471–1480.
5 Dumont DJ, Gradwohl GJ, Fong GH, et al. The endothelial-specific receptor tyrosine kinase, tek, is a member of a new subfamily of receptors. *Oncogene.* 1993;8:1293–1301.
6 Horita K, Yagi T, Kohmura N, et al. A novel tyrosine kinase, hyk, expressed in murine embryonic stem cells. *Biochem Biophy Res Comm.* 1992;189:1747–1753.
7 Yamaguchi TP, Dumont DJ, Conlon RA, et al. flk-1, an flt-related receptor tyrosine kinase is an early marker for endothelial cell precursors. *Development.* 1993;118:489–498.
8 Iwama A, Hamaguchi I, Hashiyama M, et al. Molecular cloning and characterization of mouse TIE and TEK receptor tyrosine kinase genes and their expression in hematopoietic stem cells. *Biochem Biophy Res Comm.* 1993;195:301–309.
9 Schnurch H, Risau W. Expression of tie-2, a member of a novel family of receptor tyrosine kinases, in the endothelial cell lineage. *Development.* 1993;119:957–968.
10 Ziegler SF, Bird TA, Schneringer JA, et al. Molecular cloning and characterization of a novel receptor protein tyrosine kinase from human placenta. *Oncogene.* 1993;8:663–670.
11 Partanen J, Armstrong E, Makela TP, et al. A novel endothelial cell surface receptor tyrosine kinase with extracellular epidermal growth factor homology domains. *Mol Cell Biol.* 1992;12:1698–1707.
12 Davis S, Aldrich TH, Jones PF, et al. Isolation of angiopoietin-1, a ligand for the TIE2 receptor, by secretion-trap expression cloning. *Cell.* 1996;87:1161–1169.
13 Maisonpierre PC, Suri C, Jones PF, et al. Angiopoietin-2, a natural antagonist for Tie2 that disrupts in vivo angiogenesis. *Science.* 1997;277:55–60.
14 Valenzuela DM, Griffiths JA, Rojas J, et al. Angiopoietins 3 and 4: diverging gene counterparts in mice and humans. *Proc Natl Acad Sci USA.* 1999;96:1904–1909.
15 Lee HJ, Cho CH, Hwang SJ, et al. Biological characterization of angiopoietin-3 and angiopoietin-4. *FASEB J.* 2004;18:1200–1208.
16 Lyons MS, Bell B, Stainier D, et al. Isolation of the zebrafish homologues for the tie-1 and tie-2 endothelium-specific receptor tyrosine kinases. *Dev Dyn.* 1998;212:133–140.
17 Jones PF, McClain J, Robinson DM, et al. Identification and characterisation of chicken cDNAs encoding the endothelial cell-specific receptor tyrosine kinase Tie2 and its ligands, the angiopoietins. *Angiogenesis.* 1998;2:357–364.
18 Pham VN, Roman BL, Weinstein BM. Isolation and expression analysis of three zebrafish angiopoietin genes. *Dev Dyn.* 2001;221: 470–474.
19 Sato TN, Qin Y, Kozak CA, et al. Tie-1 and tie-2 define another class of putative receptor tyrosine kinase genes expressed in early embryonic vascular system [Published erratum appears in *Proc Natl Acad Sci USA* 1993;90(24):12056]. *Proc Natl Acad Sci USA.* 1993;90:9355–9358.
20 Maisonpierre PC, Goldfarb M, Yancopoulos GD, et al. Distinct rat genes with related profiles of expression define a TIE receptor tyrosine kinase family. *Oncogene.* 1993;8:1631–1637.
21 Korhonen J, Partanen J, Armstrong E, et al. Enhanced expression of the tie receptor tyrosine kinase in endothelial cells during neovascularization. *Blood.* 1992;80:2548–2555.
22 Puri MC, Rossant J, Alitalo K, et al. The receptor tyrosine kinase TIE is required for integrity and survival of vascular endothelial cells. *EMBO J.* 1995;14:5884–5891.
23 Sato TN, Tozawa Y, Deutsch U, et al. Distinct roles of the receptor tyrosine kinases Tie-1 and Tie-2 in blood vessel formation. *Nature.* 1995;376:70–74.
24 Dumont DJ, Gradwohl G, Fong GH, et al. Dominant-negative and targeted null mutations in the endothelial receptor tyrosine kinase, tek, reveal a critical role in vasculogenesis of the embryo. *Genes and Dev.* 1994;8:1897–1909.
25 Jones N, Voskas D, Master Z, et al. Rescue of the early vascular defects in Tek/Tie2 null mice reveals an essential survival function. *EMBO Rep.* 2001;2:438–445.
26 Puri MC, Partanen J, Rossant J, et al. Interaction of the TEK and TIE receptor tyrosine kinases during cardiovascular development. *Development.* 1999;126:4569–4580.
27 Puri MC, Bernstein A. Requirement for the TIE family of receptor tyrosine kinases in adult but not fetal hematopoiesis. *Proc Natl Acad Sci USA.* 2003;100:12753–12758.
28 Voskas D, Jones N, Van Slyke P, et al. A cyclosporine-sensitive psoriasis-like disease produced in Tie2 transgenic mice. *Am J Pathol.* 2005;166:843–855.

29 Feistritzer C, Mosheimer BA, Sturn DH, Bijuklic K, Patsch JR, Wiedermann CJ. Feirstritzer, et al. Expression and function of the angiopoietin receptor Tie-2 receptor in human eosinophils. *J Allergy Clin Immunol.* 2004;114:1077–1084.
30 Lemieux C, Maliba R, Favier J, et al. Angiopoietins can directly activate endothelial cells and neutrophils to promote proinflammatory responses. *Blood.* 2005;105:1523–1530.
31 Iurlaro M, Scatena M, Zhu WH, et al. Rat aorta-derived mural precursor cells express the Tie2 receptor and respond directly to stimulation by angiopoietins. *J Cell Sci.* 2003;116:3635–3643.
32 Tian S, Hayes AJ, Metheny-Barlow LJ, et al. Stabilization of breast cancer xenograft tumour neovasculature by angiopoietin-1. *Br J Cancer.* 2002;86:645–651.
33 Shahrara S, Volin MV, Connors MA, et al. Differential expression of the angiogenic Tie receptor family in arthritic and normal synovial tissue. *Arthritis Res.* 2002;4:201–208.
34 Hattori K, Dias S, Heissig B, et al. Vascular endothelial growth factor and angiopoietin-1 stimulate postnatal hematopoiesis by recruitment of vasculogenic and hematopoietic stem cells. *J Exp Med.* 2001;193:1005–1014.
35 Takakura N, Huang XL, Naruse T, et al. Critical role of the TIE2 endothelial cell receptor in the development of definitive hematopoiesis. *Immunity.* 1998;9:677–686.
36 Batard P, Sansilvestri P, Scheinecker C, et al. The Tie receptor tyrosine kinase is expressed by human hematopoietic progenitor cells and by a subset of megakaryocytic cells. *Blood.* 1996;87:2212–2220.
37 Otani A, Takagi H, Oh H, et al. Expressions of angiopoietins and Tie2 in human choroidal neovascular membranes. *Invest Ophthalmol Vis Sci.* 1999;40:1912–1920.
38 Hirchenhain J, Huse I, Hess A, et al. Differential expression of angiopoietins 1 and 2 and their receptor Tie-2 in human endometrium. *Mol Hum Reprod.* 2003;9:663–669.
39 Schlaeger TM, Qin Y, Fujiwara Y, et al. Vascular endothelial cell lineage-specific promoter in transgenic mice. *Development.* 1995;121:1089–1098.
40 Korhonen J, Lahtinen I, Halmekyto M, et al. Endothelial-specific gene expression directed by the tie gene promoter in vivo. *Blood.* 1995;86:1828–1835.
41 Fadel BM, Boutet SC, Quertermous T. Octamer-dependent in vivo expression of the endothelial cell-specific TIE2 gene. *J Biol Chem.* 1999;274:20376–20383.
42 Boutet SC, Quertermous T, Fadel BM. Identification of an octamer element required for in vivo expression of the TIE1 gene in endothelial cells. *Biochem J.* 2001;360:23–29.
43 Dube A, Akbarali Y, Sato TN, et al. Role of the Ets transcription factors in the regulation of the vascular-specific Tie2 gene. *Circ Res.* 1999;84:1177–1185.
44 Dube A, Thai S, Gaspar J, et al. Elf-1 is a transcriptional regulator of the Tie2 gene during vascular development. *Circ Res.* 2001;88:237–244.
45 Christensen RA, Fujikawa K, Madore R, et al. NERF 2, a member of the Ets family of transcription factors, is increased in response to hypoxia and angiopoietin-1: a potential mechanism for Tie2 regulation during hypoxia. *J Cell Biochem.* 2002;85:505–515.
46 Gaspar J, Thai S, Voland C, et al. Opposing functions of the Ets factors NERF and ELF-1 during chicken blood vessel development. *Arterioscler Thromb Vasc Biol.* 2002;22:1106–1112.
47 Minami T, Kuivenhoven JA, Evans V, et al. Ets motifs are necessary for endothelial cell-specific expression of a 723-bp Tie-2 promoter/enhancer in Hprt targeted transgenic mice. *Arterioscler Thromb Vasc Biol.* 2003;23:2041–2047.
48 Park YS, Kim NH, Jo I. Hypoxia and vascular endothelial growth factor acutely up-regulate angiopoietin-1 and Tie2 mRNA in bovine retinal pericytes. *Microvasc Res.* 2003;65:125–131.
49 Takagi H, Koyama S, Seike H, et al. Potential role of the angiopoietin/tie2 system in ischemia-induced retinal neovascularization. *Invest Ophthalmol Vis Sci.* 2003;44:393–402.
50 Ding H, Roncari L, Wu X, et al. Expression and hypoxic regulation of angiopoietins in human astrocytomas. *Neuro Oncol.* 2001;3:1–10.
51 Willam C, Koehne P, Jurgensen JS, et al. Tie2 receptor expression is stimulated by hypoxia and proinflammatory cytokines in human endothelial cells. *Circ Res.* 2000;87:370–377.
52 Mandriota SJ, Pepper MS. Regulation of angiopoietin-2 mRNA levels in bovine microvascular endothelial cells by cytokines and hypoxia. *Circ Res.* 1998;83:852–859.
53 Park E, Lee JM, Dlais JD, et al. Internal translation initiation mediated by the aniogenic factor Tie2. *J Cell Biochem.* 2005;280:20945–20953.
54 Chen BP, Li YS, Zhao Y, et al. DNA microarray analysis of gene expression in endothelial cells in response to 24-h shear stress. *Physiol Genomics.* 2001;7:55–63.
55 Porat RM, Grunewald M, Globerman A, et al. Specific induction of tie1 promoter by disturbed flow in atherosclerosis-prone vascular niches and flow-obstructing pathologies. *Circ Res.* 2004;94:394–401.
56 Lee HJ, Koh GY. Shear stress activates Tie2 receptor tyrosine kinase in human endothelial cells. *Biochem Biophys Res Commun.* 2003;304:399–404.
57 Chen-Konak L, Guetta-Shubin Y, Yahav H, et al. Transcriptional and post-translation regulation of the Tie1 receptor by fluid shear stress changes in vascular endothelial cells. *FASEB J.* 2003;17:2121–2123.
58 Zheng W, Christensen LP, Tomanek RJ. Stretch induces upregulation of key tyrosine kinase receptors in microvascular endothelial cells. *Am J Physiol Heart Circ Physiol.* 2004;287:H2739–H2745.
59 Chang H, Wang BW, Kuan P, et al. Cyclical mechanical stretch enhances angiopoietin-2 and Tie2 receptor expression in cultured human umbilical vein endothelial cells. *Clin Sci (Lond).* 2003;104:421–428.
60 Fathers KE, Stone CM, Minhas K, et al. Heterogeneity of Tie2 expression in tumor microcirculation: influence of cancer type, implantation site, and response to therapy. *Am J Pathol.* 2005;167:1753–1762.
61 Babaei S, Teichert-Kuliszewska K, Zhang Q, et al. Angiogenic actions of angiopoietin-1 require endothelium-derived nitric oxide. *Am J Pathol.* 2003;162:1927–1936.
62 Bureau W, VanSlyke P, Jones J, et al. Chronic systemic delivery of angiopoietin 2 reveals a possible independent angiogenic effect. *Am J Pathol.* 2006; In Press.
63 Holash J, Maisonpierre PC, Compton D, et al. Vessel cooption, regression, and growth in tumors mediated by angiopoietins and VEGF. *Science.* 1999;284:1994–1998.
64 Gale NW, Thurston G, Hackett SF, et al. Angiopoietin-2 is required for postnatal angiogenesis and lymphatic patterning, and only the latter role is rescued by angiopoieting-1. *Dev Cell.* 2002;3:411–423.

65 Ward NL, Dumont DJ. The angiopoietins and Tie2/Tek: adding to the complexity of cardiovascular development. *Semin Cell Dev Biol*. 2002;13:19–27.

66 Ward NL, Van Slyke P, Sturk C, et al. Angiopoietin 1 expression levels in the myocardium direct coronary vessel development. *Dev Dyn*. 2004;229:500–509.

67 Oshima Y, Oshima S, Nambu H, et al. Different effects of angiopoietin-2 in different vascular beds: new vessels are most sensitive. *FASEB J*. 2005;19:963–965.

68 Kim I, Kwak HJ, Ahn JE, et al. Molecular cloning and characterization of a novel angiopoietin family protein, angiopoietin-3. *FEBS LETTERS*. 1999;443:353–356.

69 Ward NL, Van Slyke P, Dumont DJ. Functional inhibition of secreted angiopoietin: a novel role for angiopoietin 1 in coronary vessel patterning. *Biochem Biophys Res Comm*. 2004;323:937–946.

70 Davis S, Papadopoulos N, Aldrich TH, et al. Angiopoietins have distinct modular domains essential for receptor binding, dimerization and superclustering. *Nat Struct Biol*. 2003;10:38–44.

71 Procopio WN, Pelavin PI, Lee WM, et al. Angiopoietin-1 and -2 coiled coil domains mediate distinct homo- oligomerization patterns, but fibrinogen-like domains mediate ligand activity. *J Bio Chem*. 1999;274:30196–30201.

72 Kim KT, Choi HH, Steinmetz MO, et al. Oligomerization and multimerization are critical for angiopoietin-1 to bind and phosphorylate Tie2. *J Biol Chem*. 2005;280:20126–20131.

73 Weber CC, Cai H, Ehrbar M, et al. Effects of Protein and Gene Transfer of the Angiopoietin-1 Fibrinogen-like Receptor-binding Domain on Endothelial and Vessel Organization. *J Biol Chem*. 2005;280:22445–22453.

74 Cho CH, Kammerer RA, Lee HJ, et al. Designed angiopoietin-1 variant, COMP-Ang1, protects against radiation-induced endothelial cell apoptosis. *Proc Natl Acad Sci USA*. 2004;101:5553–5558.

75 Cho CH, Kammerer RA, Lee HJ, et al. COMP-Ang1: a designed angiopoietin-1 variant with nonleaky angiogenic activity. *Proc Natl Acad Sci USA*. 2004;101:5547–5552.

76 Saharinen P, Kerkela K, Ekman N, et al. Multiple angiopoietin recombinant proteins activate the Tie1 receptor tyrosine kinase and promote its interaction with Tie2. *J Cell Biol*. 2005;169:239–243.

77 Jones N, Master Z, Jones J, et al. Identification of Tek/Tie2 binding partners. Binding to a multifunctional docking site mediates cell survival and migration. *J Biol Chem*. 1999;274:30896–30905.

78 Kontos CD, Stauffer TP, Yang WP, et al. Tyrosine 1101 of Tie2 is the major site of association of p85 and is required for activation of phosphatidylinositol 3-kinase and Akt. *Mol Cell Biol*. 1998;18:4131–4140.

79 Hughes DP, Marron MB, Brindle NP. The antiinflammatory endothelial tyrosine kinase Tie2 interacts with a novel nuclear factor-kappaB inhibitor ABIN-2. *Circ Res*. 2003;92:630–636.

80 Audero E, Cascone I, Maniero F, et al. Adaptor ShcA protein binds tyrosine kinase Tie2 receptor and regulates migration and sprouting but not survival of endothelial cells. *J Biol Chem*. 2004;279:13224–13233.

81 Chen JX, Lawrence ML, Cunningham G, et al. HSP90 and Akt modulate Ang-1-induced angiogenesis via NO in coronary artery endothelium. *J Appl Physiol*. 2004;96:612–620.

82 Jones N, Iljin K, Dumont DJ, et al. Tie receptors: New modulators of angiogenic and lymphangiogenic responses. *Nat Rev Mol Cell Biol*. 2001;2:257–267.

83 Tachibana K, Jones N, Dumont DJ, et al. Selective role of a distinct tyrosine residue on Tie2 in heart development and early hematopoiesis. *Mol Cell Biol*. 2005;25:4693–4702.

84 Lelievre E, Bourbon PM, Duan LJ, et al. Deficiency in the p110alpha subunit of PI3K results in diminished Tie2 expression and Tie2($^{-/-}$)-like vascular defects in mice. *Blood*. 2005;105:3935–3938.

85 Sun JF, Phung T, Shiojima I, et al. Microvascular patterning is controlled by fine-tuning the Akt signal. *Proc Natl Acad Sci USA*. 2005;102:128–133.

86 Jones N, Dumont DJ. The Tek/Tie2 receptor signals through a novel Dok-related docking protein, Dok-R. *Oncogene*. 1998;17:1097–1108.

87 Master Z, Jones N, Tran J, et al. Dok-R plays a pivotal role in angiopoietin-1-dependent cell migration through recruitment and activation of Pak. *EMBO J*. 2001;20:5919–5928.

88 Tadros A, Hughes DP, Dunmore BJ, et al. ABIN-2 protects endothelial cells from death and has a role in the antiapoptotic effect of angiopoietin-1. *Blood*. 2003;102:4407–4409.

89 Papapetropoulos A, Garcia-Cardena G, Dengler TJ, et al. Direct actions of angiopoietin-1 on human endothelium: evidence for network stabilization, cell survival, and interaction with other angiogenic growth factors. *Lab Invest*. 1999;79:213–223.

90 Teichert-Kuliszewska K, Maisonpierre PC, Jones N, et al. Biological action of angiopoietin-2 in a fibrin matrix model of angiogenesis is associated with activation of Tie2. *Cardiovasc Res*. 2001;49:659–670.

91 Calvert JT, Riney TJ, Kontos CD, et al. Allelic and locus heterogeneity in inherited venous malformations. *Hum Mol Gen*. 1999;8:1279–1289.

92 Korpelainen EI, Karkkainen M, Gunji Y, et al. Endothelial receptor tyrosine kinases activate the STAT signaling pathway: mutant Tie-2 causing venous malformations signals a distinct STAT activation response. *Oncogene*. 1999;18:1–8.

93 Ward NL, Haninec AL, Van Slyke P, et al. Angiopoietin-1 causes reversible degradation of the portal microcirculation in mice: implications for treatment of liver disease. *Am J Pathol*. 2004;165:889–899.

94 Lin P, Buxton JA, Acheson A, et al. Antiangiogenic gene therapy targeting the endothelium-specific receptor tyrosine kinase Tie2. *Proc Natl Acad Sci USA*. 1998;95:8829–8834.

95 Lin P, Polverini P, Dewhirst M, et al. Inhibition of tumor angiogenesis using a soluble receptor establishes a role for Tie2 in pathologic vascular growth. *J Clin Invest*. 1997;100:2072–2078.

96 Shan S, Robson ND, Cao Y, et al. Responses of vascular endothelial cells to angiogenic signaling are important for tumor cell survival. *FASEB J*. 2004;18:326–328.

97 Peters KG, Kontos CD, Lin PC, et al. Functional significance of Tie2 signaling in the adult vasculature. *Recent Prog Horm Res*. 2004;59:51–71.

98 Melani C, Stoppacciaro A, Foroni C, et al. Angiopoietin decoy secreted at tumor site impairs tumor growth and metastases by inducing local inflammation and altering neoangiogenesis. *Cancer Immunol Immunother*. 2004;53:600–608.

99 Oliner J, Min H, Leal J, et al. Suppression of angiogenesis and tumor growth by selective inhibition of angiopoietin-2. *Cancer Cell*. 2004;6:507–516.

40

Slits and Netrins in Vascular Patterning

Taking Cues from the Nervous System

Kye Won Park, Lisa D. Urness, and Dean Y. Li

School of Medicine, University of Utah, Salt Lake City

The formation of a functional blood vessel depends on endothelial cell (EC) proliferation and migration, tube formation, and the differentiation and recruitment of vascular smooth muscle cells (VSMCs). Some of the growth factors responsible for vascular development were first identified using cell biological assays that evaluated angiogenic properties. These assays include in vitro EC proliferation, adhesion, and migration assays, and in vivo aortic ring, chorioallantoic membrane (CAM), and corneal micropocket assays. These studies identified vascular endothelial growth factor (VEGF), fibroblast growth factor (FGF), transforming growth factor (TGF)-β, angiopoietins, platelet-derived growth factor (PDGF), and their homologues and isoforms, in addition to their multiple cognate receptors (1,2).

Gene targeting experiments demonstrated that many of these growth factors and their receptors are essential for vascular development and provided insight into how sprouting is initiated and a mature network is sculpted. VEGF and its receptors play an essential role in EC proliferation and differentiation. The critical requirement for VEGF at the earliest stage of vasculogenesis is demonstrated by the lack of EC development even in *Vegf* hemizygous mice (3). Once ECs aggregate and undergo tubulogenesis, VSMCs are recruited by PDGF to surround the mature vessel. In the absence of PDGF or its receptors, a subset of vessels fails to recruit VSMCs and pericytes (4). Finally, gene targeting studies have revealed that stabilization and maturation of blood vessels, through the interaction of ECs and VSMCs, is mediated by angiopoietins, TGF-β, and extracellular matrix (ECM) proteins (2). Thus, gene ablation experiments have identified which signaling pathways play essential roles in murine vascular development and have delineated the fundamental blueprint for the sequence of molecular events that govern nascent blood vessel formation. Attention has begun to shift from the identification and functional characterization of endothelial growth factors to an appreciation of the molecular mechanisms that *pattern* the mature vascular tree. Recent evidence suggests that patterning of the vascular system is genetically programmed and that the molecular mechanisms that guide vessel outgrowth are similar to those utilized by the nervous system.

DIFFERENTIATION OF ARTERIAL AND VENOUS CIRCUITS

A second level of vascular organization, overlying the primary plexus, is the differentiation and partitioning of separate arterial and venous networks. A functional vascular network requires distinct arterial and venous circuits that only connect via the capillary beds in distal target organs. Whereas it was once believed that hemodynamic forces were responsible for mediating phenotypic differences between arterial and venous ECs (5), evidence has accumulated in the last several years to indicate that genetic programs govern arterial-venous identity prior to the onset of circulation. The ephrins and their receptors, the Ephs, were identified on the basis of their roles in contact-mediated axonal guidance (6). However, the Anderson lab showed that the ligand, ephrinB2, is specifically expressed in arteries, while its receptor, EphB4, is primarily expressed in veins (7). *EphrinB2* and *EphB4* gene-targeted mice show defects in remodeling of the vascular plexus in arterial and venous beds. These data suggest that the ephrin–Eph signaling pathway may function to maintain segregation of arterial and venous domains during angiogenesis via cell–cell interactions, similar to their functions in the nervous system. Indeed, Hellmut Augustin (Chapter 38) describes cellular studies supporting a role for ephrin–Eph signaling in the establishment of capillary arterial–venous network boundaries. Taken together, these studies presaged the existence of genetic pathways that play a role in the patterning of specific arterial and venous EC populations. These observations have since paved the way for studies addressing the molecular nature of vascular guidance.

VASCULAR GUIDANCE AS A PATTERNING MECHANISM

Once the central vessels and capillary plexi are formed during vasculogenesis, they extend sprouts that must navigate throughout the embryo to nourish all the organs. Large arteries and veins branch into smaller diameter arterioles and venules, respectively, and eventually into the smallest vessels, the capillaries. In addition to a hierarchy of vessels of different diameter, arterial and venous networks must develop in parallel and interconnect in the capillary beds to form a seamless circulation. During embryonic vascular development, vessel trajectories are by no means random; rather, growth is constrained along specific routes. The intricate wiring and predictable trajectories of these vessels suggest that active guidance mechanisms exist to direct this process. It is unlikely that signaling by the vascular growth factors identified to date can account for the complexity of the embryonic vascular pattern. Although vascular mitogens, hemodynamic inputs, hypoxia, and random sprouting and pruning play pivotal roles in sculpting the vascular tree, recent evidence points to the existence of additional genetic programs that guide vessel outgrowth along specific trajectories. A forward genetic screen in zebrafish identified a gene, *out of bounds* (*Obd*), that functions to constrain angiogenic growth. *Obd* mutants exhibit anomalous migration of intersegmental vessels into the adjacent somites, suggesting that a chemorepulsive signal has been lost that normally constrains vessel trajectories between somites. This strongly suggests that vascular patterning is controlled by genetic guidance programs (8). *Obd* recently has been identified by the Weinstein group as harboring a mutation in the locus of the semaphorin receptor, *PlexinD1* (9). These data suggest that, although vascular mitogens such as VEGF have the capacity to stimulate angiogenesis, the growth response can be further refined, both positively and negatively, by other guidance factors in the local environment.

SHARED NEURAL AND VASCULAR GUIDANCE MECHANISMS

Many similarities exist between vessel and nerve networks at the anatomical and molecular levels. Blood vessels often follow parallel paths to those of nerves, such as the neurovascular bundles that course toward the distal limbs (10,11). Both organ systems must infiltrate all tissues in a highly ordered and coordinated manner, and both employ specialized filopodial extensions on their tip cells to sample their respective microenvironments for directional growth cues. Finally, like the arterial and venous subdivisions of the vascular system, the nervous system is divided into sensory and motor relays that must be precisely coordinated. The directional movement of neuronal growth cones upon exiting the central nervous system is coordinated by multiple genetic programs, which guide the growth cones by integrating a combinatorial code of attractive and repulsive, and long- and short-range cues. The most extensively characterized classes of proteins that regulate neural guidance are the ephrins, semaphorins, Slits, and netrins, and their cognate families of receptors (12). The strikingly similar repertoire of proteins expressed on vascular ECs and axonal growth cones suggests that the molecular mechanisms that guide axonal path finding also regulate endothelial sprouting and angiogenesis.

The Ephrin-Eph and Semaphorin-Neuropilin/PlexinD1 Signaling Pathways

The ephrins are membrane-bound ligands for the Eph family of receptor tyrosine kinases. Interaction between receptor-bearing axons and ephrins expressed by surrounding cells mediates contact–repulsion by inducing a collapse of the growth cone cytoskeleton. Wang and colleagues observed that axons expressing Eph receptors are restricted from extending into the caudal half of somites that express ephrin ligands (13). This juxtacrine interaction effectively channels axonal growth through permissive intersomitic corridors. Eph receptors and their ephrin ligands are also essential for organizing arterial and venous domains in the embryonic vasculature (7). This activity is likely a result of reciprocal repulsive cues between arterial and venous ECs that inhibit the fusion of arterial and venous sprouts. Similarly, juxtacrine interactions between intersomitic vessels expressing Eph receptors, and the adjacent somitic mesenchyme expressing ephrin ligands, appear to constrain vessel growth to intersomitic corridors via a repulsive guidance mechanism. Disruption of endogenous Eph/ephrin expression in the somites of mice and *Xenopus* results in aberrant invasion of intersomitic vessels into the developing somitic compartment (14,15), further suggesting that ephrin–Eph signaling may mediate repulsive guidance cues that restrict vessel growth to specific trajectories. Thus, analogous to the role of these factors in nervous system development, complementary ephrin–Eph domains of expression serve to delineate spatial boundaries for permissive growth in the vascular system. These data were among the first to suggest that the nervous and vascular systems may utilize the same array of guidance factors to govern patterning.

Semaphorins are a large family of both secreted and membrane-associated proteins. More than 30 semaphorins have been identified to date that are capable of mediating both repulsive and attractive axon guidance (16). Neuropilins are transmembrane proteins that can mediate semaphorin class 3 repulsive cues in association with plexin receptors expressed on axons. Activation of the neuropilin–plexin complex by semaphorin 3 restricts growth-cone extension to specific paths (17). The semaphorin family and their receptors, the neuropilins and PlexinD1, have been shown to regulate vascular patterning and angiogenesis in addition to their fundamental roles in neural guidance (18). Semaphorins are capable of disrupting the cytoskeleton of vascular ECs, thereby compromising migration and sprouting, analogous to semaphorin-induced axon growth cone collapse (19,20). Mice lacking semaphorin-3A (sema-3A) exhibit a gamut of neural and vascular defects that includes ectopic sprouting of intersomitic vessels into the somites (21). However, these vascular defects exhibit highly variable penetrance and are not observed in some strains. Most recently, Torres-Vasquez and colleagues and Gitler and colleagues reported that the first EC-specific semaphorin receptor, PlexinD1, regulates vascular patterning in zebrafish (*out of bounds* mutant) and mice (9,22). Targeted inactivation of *PlexinD1* in mice results in severe cardiac defects and intersomitic vessel patterning defects. The latter is most likely a result of abrogated signaling from sema-3E, or perhaps sema-3A, ligands that are both highly expressed in the adjacent somites. Gu and colleagues provide compelling evidence that the sema-3E ligand, secreted by the somites, specifically interacts with the endothelial PlexinD1 receptor (23). The activation of this signaling complex transduces inhibitory directional cues to intersomitic vessels. Genetic ablation of *Sema-3E* phenocopies the intersomitic vascular patterning defects in *PlexinD1*$^{-/-}$ mice, specifically the ingression of intersomitic vessels into the somitic mesoderm. These data define a new semaphorin signaling pathway that affects vascular patterning.

Neuropilin-1 was first identified as a neuronal cell surface marker and subsequently shown by Klagsbrun and coworkers to be expressed on the surface of vascular ECs (24). In addition to functioning as a nonsignaling accessory receptor to facilitate VEGF binding to VEGF receptor 2 (VEGFR2), neuropilin-1 also binds semaphorins. Evidence for the involvement of neuropilins in vascular development has accumulated substantially in the last several years. In addition to defects in peripheral nerve growth, neuropilin-1–deficient embryos exhibit cardiovascular defects and reduced capillary branching and neural tube vascularization abnormalities (25). Interestingly, Gu and colleagues showed that the sema-3E–PlexinD1 signaling complex functions independently of neuropilin-1 and that sema-3E does not bind neuropilin-expressing COS cells (23). The remaining semaphorin class-3 ligands, with the exception of sema-3B, interact with neuropilin and may function to block the association of neuropilin with the VEGF ligand–receptor complex. The relatively minor vascular defects associated with ablation of sema-3A may reflect the sequestration of neuropilin-1 from the VEGF–VEGFR2 complex, thereby resulting in reduced VEGF signaling.

Thus, many studies have implicated ephrin and semaphorin cues in vascular development over the past decade. These findings immediately begged the question of the potential involvement of the other two primary neural guidance pathways in vascular patterning, namely the Slits and netrins. Most recently, our laboratory and others have addressed this question by analyzing the effects of these factors in cell biological studies, as well as in vivo animal model studies.

The Slit-Robo Signal Pathway

Three Slit ligands have been identified in mammals, as well as three Roundabout (Robo) receptors that mediate axonal guidance. The Slits are large ECM proteins that control axonal and dendritic branching, regulate the migration of muscle precursor cells, and inhibit neuronal and leukocyte cell migration (26,27). In the nervous system, Slit protein is expressed in the floor plate of the neural tube and has been shown to act predominantly, although not exclusively, as a repellant guidance cue. Slit also is detected in the somites of mouse and zebrafish embryos and likely restricts the sprouting of vessels to the intersomitic corridors (28,29). A fourth member of the Robo receptor signaling pathway recently was identified in zebrafish and mice, dubbed *Magic Roundabout* or *Robo4* (30,31). This Robo receptor is unique among the Robo family due both to its divergent sequence and, more significantly, its expression in embryonic ECs. The identification of a new vascular-specific Robo family member strongly suggests a conserved role for Robo-Slit signaling in vascular development. Moreover, we identified Robo4 on the basis of its altered expression in *Alk1*$^{-/-}$ mice, which exhibit severe defects in vascular sprouting (31).

Our laboratory has conducted cell biological studies in adult ECs and heterologous cells overexpressing Robo4. Immunoprecipitation studies revealed binding of Myc-tagged Slit2 to HA-tagged Robo4. Mena, a known effector of Slit–Robo signaling in neuronal cells, also was bound to the Robo4 receptor. It has not been determined whether Slit1 or Slit3 bind to Robo4. Finally, Robo4-expressing HEK cells were decorated with *anti*-Myc antibodies in the presence of Myc-tagged Slit2-conditioned media. In functional assays, we found that Slit2–Robo4 signaling negatively regulates cell migration, but has no effect on proliferation (31). This negative read-out mirrors that of repulsive guidance observed with Slit-Robo signaling in the nervous system. These data led us to hypothesize that Slit-Robo signaling may have an inhibitory effect on vessel sprouting in vivo. Indeed, Robo4 knockdown in zebrafish demonstrated that *Robo4* is essential for vascular development (32). *Robo4* morphants display asynchronous sprouting of the intersomitic vessels from the dorsal aorta, and this defect can be rescued by coinjection of wild-type human *Robo4* mRNA. Characterization of *Robo4*-targeted mice generated in our laboratory is underway to address this hypothesis. Taken together, Slit-Robo signaling inhibits EC migration and may be critical for vascular development. Interestingly, no vascular defects have been reported in *Slit1/Slit2* double knockout (33,34) or in *Slit3*-targeted mice (35), although rigorous examination of vascular development may not have been conducted in these mice.

In contrast to our findings, others report that Robo4 stimulates EC migration in a Slit-independent manner using a soluble truncated Robo4 ectodomain reagent (Robo4-Fc) (36). These investigators observed stimulation of EC migration and proliferation, as well as in vivo angiogenesis, in rodent subcutaneous sponge and aortic ring assays (36). They were unable to detect binding of this ectodomain fusion to any of the three Slit proteins, suggesting that the proangiogenic effect of Robo4 is mediated by an unknown ligand. Taken together, the contradictory results of our two studies may be explained in several ways. The use of different probe reagents, Robo4-Fc versus Robo4-HA, may affect binding efficiency. Moreover, it is difficult to compare experiments in which Slit activity is assessed directly versus treatment with a Robo4-Fc. Finally, the functionality of these proteins may be contextual and therefore have different activities in different biological assays, depending on the expression of other cell surface receptors.

A second group reports that Slit has a proangiogenic effect mediated by Robo1 (37). They show that Slit2 is expressed by tumor cells and attracts tumor endothelium in in vitro assays and in xenografts. It would be interesting to determine if Robo4 is expressed in these tumor cells and if the activity these authors observe may be due to signaling through Robo4 in addition to Robo1. One could envision that treatment with a soluble Robo1 ectodomain protein could compromise signaling via Robo4 assuming they share reactivity to the same Slit ligand. Taken as a whole, the experimental studies on Slit-Robo signaling in vascular development are subject to the reservation that differences in functionality may be ascribed to widely dissimilar cellular contexts and experimental design. Loss-of-function and overexpression studies in animal models will ultimately be informative in regard to teasing apart the various roles of Slit-Robo signaling in vascular development.

The Netrin-Unc/DCC Signaling Pathways

Netrin 1 and -2 were discovered as chemotropic ligands that attract axons and neuronal cells. Studies by Tessier-Lavigne and his colleagues revealed that netrins attract commissural axons toward the ventral midline of the developing spinal cord (38). Unc5, neogenin, and Deleted in Colorectal Cancer (DCC) receptor families mediate netrin signaling and, depending on the combinations of these receptors expressed on the axon, may induce growth cone extension or collapse. An attractive response toward netrin is mediated by DCC family members. This signal can be silenced by DCC–Robo1 interaction in the presence of Slit (39). The conversion of netrin-attractive activity to a repulsive response also can be achieved by direct interaction between DCC and a second class of receptor, the Unc5h2 receptor (40).

Completing the picture of complementary activities for all four primary classes of axonal guidance factors in the vascular system, recent evidence suggests that this final guidance factor class, the netrins, also plays a role in vascular development. At least three netrin ligands are found in mammals, as well as several receptor partners, including DCC, neogenin, and members of the Unc family of receptors. Several groups have observed EC expression of Unc5h2 receptor in mouse embryos, suggesting that netrins may play a role in early vascular development. To assess the effect of netrin stimulation on ECs, our laboratory conducted assays of migration and proliferation in cultured adult ECs, both human aortic ECs (HAECs) and human microvascular ECs (HMVECs) (41). In all assays (including in vitro angiogenic corneal micropocket and tube formation assays), it was found that netrin functioned as a positive angiogenic factor in a dose-dependent fashion in both EC lines, with activity comparable to VEGF and PDGF. Moreover, in vivo chorioallantoic membrane and corneal micropocket assays corroborated these findings. Netrin also induced migration and proliferation of VSMCs. Interestingly, VSMCs, but not ECs, were able to adhere equally well to netrin and fibronectin in our adhesion assays. Reverse transcription polymerase chain reaction (RT-PCR) and Western blot analyses of netrin receptors in VSMCs and ECs suggested that the unique ability of VSMCs to adhere to netrin may be mediated by neogenin, given that it was the only netrin-binding receptor type identified in VSMCs but not in ECs. Furthermore, experiments with a neogenin-blocking antibody abrogated VSMC migration and adhesion to netrin. Surprisingly, none of the canonical netrin receptors were readily detected in our cultured adult ECs, although Unc5h2 is detected in the embryonic vasculature via RNA in situ hybridization analyses. These receptors may be expressed at a low level, or an unidentified netrin receptor may exist in adult ECs that mediates the proangiogenic activities we observed. A screen for additional netrin-binding factors is currently underway. Irrespective of the netrin receptor identity, these are the first data to establish netrin as a new angiogenic factor with an activity comparable to VEGF and PDGF. Given these data, one would predict that embryos harboring a hypomorphic allele of netrin 1 would display angiogenic defects. However, examination of *netrin 1* gene-trap mice revealed no obvious vascular abnormalities, owing perhaps to functional redundancy, or to variable potency of netrin on embryonic versus adult ECs.

Reports by others also support a role for netrin in mediating angiogenesis. Examination of vascular development in *Unc5b* (*Unc5h2*)$^{-/-}$ mice led Lu and colleagues to conclude that netrin–Unc5b signaling retards vessel growth during embryonic development (42). They confirm that Unc5b is expressed predominantly in the murine arterial embryonic vasculature, particularly in the growing tip cells of vessel sprouts. Ablation of Unc5b in the mouse led to excessive vessel branching, increased tip cell filopodial extensions, and aberrant vessel navigation. Application of netrin to the growing angiogenic front in cultured embryo hindbrains caused filopodial retraction, whereas this response was lost in *Unc5b*$^{-/-}$ hindbrains. These authors observed similar effects in cultured human umbilical artery ECs (HUAECs) and in rat aortic ring sprouting assays. Finally, they report increased intersegmental vessel sprouting in Unc5b and netrin 1a morpholino knockdown studies in zebrafish. Thus, similar to the Slit–Robo story, there is agreement that netrin signaling modulates vascular guidance, but controversy exists over whether the predominant effect is pro- or antiangiogenic.

How might we reconcile these data? It is possible that the effect of netrin signaling on EC biology may differ dramatically in terms of cultured cells versus vessel guidance in the animal (i.e., although netrin may be a proangiogenic factor for adult ECs, this activity may not model its function during embryologic development). It has been shown that netrin acts bifunctionally in the nervous system depending on the specific combination of receptors that predominate in the growth cone. If indeed netrin has bifunctional activities in ECs, our data would suggest that an unknown receptor exists that mediates its proangiogenic activity. In regard to the discrepancies in netrin effects on HUAECs, Lu and colleagues employed significantly higher amounts of netrin in their studies, which may account for the different cellular responses (42). Finally, evidence stemming from axonal guidance studies suggests that a cellular response to netrin may be altered depending on the extent of laminin incorporation into the ECM, as well as on the intracellular state (cAMP metabolism) of the target cells (43). In this light, our respective assays may be assessing functionality based on two very different assay conditions – those addressing netrin activity in isolation (cells treated with purified netrin) versus assays using conditioned media from an active netrin-producing monolayer. In summary, it will be important to take cellular context into account in the interpretation of angiogenic and guidance assays utilizing the netrin ligands.

FUTURE DIRECTIONS IN VASCULAR PATTERNING

Given the apparent dual roles of guidance genes in both the neuronal and vascular systems, targeted mice that lack specific neuronal guidance genes must to be revisited to assess the possible impact of these proteins on vascular development. Mice lacking DCC, neogenin, netrin, Robos, and Slits have been generated and exhibit defects in nervous system development, as reported by many groups. However, examination of these mice for possible vascular defects remains to be investigated. It is entirely possible that vascular defects may be masked due to redundant functions within the gene families, given that all the neuronal guidance factor classes have at least three family members – and up to 30 in some groups. The impact of redundancy on gene function is apparent in ablation studies of the Slit and Robo proteins. *Slit1*$^{-/-}$–*Slit2*$^{-/-}$ double knockouts exhibited limited neural guidance defects, whereas the commissural axon patterning phenotype most representative of Slit ablation first described in *Drosophila* was observed only after ablating all three murine Slit genes. Similarly, multiple Robo gene ablations will likely be necessary to recapitulate the prototypic *Drosophila* phenotype. Thus, the generation of multiple mutant mice (preferably via endothelial-specific conditional deletion) will be required to assess the roles of all family members in embryonic vascular development. In addition, it will be necessary to address the activity of each member in cell biological assays. Figure 40.1 provides a circumspect comparison of the activities of the four major guidance-cue classes in nervous system and vascular development. Although the data thus far suggest a rough conservation in the inhibitory or attractive activities for these factors between the two systems, this observation must be qualified by the difficulty in deriving a common functional theme given the diversity of assays undertaken by different investigators. This is further challenged by the possibility that a given cue may have a binary function as both an inhibitory and attractive stimulus in different cellular contexts. We have just begun to define the range of activities of these guidance factors, and more extensive studies of function, particularly in animal models, will be required to further refine our understanding of the specific roles of these factors in vascular patterning.

Ligands	Receptors	Neuronal guidance	Angiogenesis
Netrin Netrin	DCC Neogenin Unc5h1, 2, 3, 4	Attraction Attraction	Netrin1 Netrin3 Unc5h2 Neogenin
Netrin 1, 2, 3	Robo1, 2, 3, 4	Repulsion	Slit2 Robo1 Robo4
Semaphorins	Plexins Neuropilins	Repulsion	Sema3A, 3E, 3C 6D Neuropilin Plexin D1
Class A and B ephrins	Class A and B eph receptors	Repulsion	Ephrin B2 EphB4, B2, B3

Figure 40.1. Four major signaling pathways contribute to neuronal guidance and vascular patterning mechanisms. The netrin, Slit, Semaphorin, and ephrin–Eph signaling pathways either attract or repulse growth cones or, in certain contexts, can be bifunctional (e.g., the netrins). Each ligand and receptor class is comprised of multiple family members. A subset of the factors that were first described in nervous system development has been shown to function in a similar manner during angiogenesis and the patterning of the vascular system. The precise roles of many of these ligands and their cognate receptors in regard to angiogenesis have not been established.

Gaining a better understanding of the activity of guidance factors in vessel outgrowth presents a unique set of challenges for the design of informative "vascular guidance assays." Nondirected vascular sprouting and directional vascular guidance are difficult to tease apart due to the limited array of vascular assays currently employed. Guidance in the nervous system addresses the extension of a single neuron, whereas guidance in the vascular system depends on both proliferation and migration of the component vascular ECs to generate tubes. Thus, one might envision that some vascular growth factors modulate vessel patterning by regulating survival, while other growth factors and guidance factors affect proliferation and sprouting by providing attractive cues. These two mechanisms may be employed coordinately to guide vessels to an appropriate target. Alternatively, guidance factors may dictate direction only and not regulate proliferation; that is to say, once sprouting is initiated by growth factors, guidance factors may direct vessel trajectory with combinations of attractive and repulsive cues. As an example of this, Lu and colleagues (42) showed that, although vessel navigation was perturbed in *Unc5b*$^{-/-}$ mice, the proliferation state and apoptotic indices were similar to wild-type littermates, suggestive of a "pure guidance" role for Unc5b signaling. To address the function of putative vascular guidance factors, vascular sprouting assays such as the aortic ring assay can be performed with combinations of attractive and repulsive cues. It is anticipated that future studies will utilize modified angiogenic assays to assess whether vessels can be directed to sprout along engineered corridors of cues. Finally, the identification of endocrine gland–selective vascular endothelial growth factor (EG-VEGF) implies that new factors specifically regulating a subpopulation of ECs await discovery (44).

THERAPEUTIC APPLICATIONS

Without the ability to attract a blood supply, tumor growth is arrested. As such, much attention is being directed to developing new therapeutic strategies aimed at the disruption of vessel growth in cancerous tissue. These strategies have focused on modulating known vascular growth factors, such as VEGF, FGF, and PDGF, and have proven useful in animal and human trials. The realization that neural guidance factors also modulate vascular sprouting significantly expands the number of ligand–receptor signaling pathways available for clinical manipulation. Whether targeting these new factors provides a therapeutic benefit awaits further investigation. Applying repulsive vascular guidance cues to tumor masses may slow the infiltration of new vessels. Alternatively, it may be possible to induce vessel growth toward tissues in ischemic regions by treating with attractive vascular guidance cues or dominant negative receptors that could antagonize repulsive guidance cues. Importantly, attempts to induce vessel formation by providing sustained levels of VEGF alone has, in some cases, led to disorganized vessel formations similar to angiomas, rather than functional vascular networks (45). This stands as further evidence that, in addition to the known vascular cytokines and growth factors, guidance factors will be required for the generation of mature and highly organized vessels in a therapeutic setting. Considerable interest exists in identifying additional positive and negative regulators of vessel growth, because expanding our tool kit of available angiogenic factors will be critical in designing strategies to treat different angiogenic pathological states.

The first steps have been taken toward evaluating the efficacy of these new classes of vascular guidance cues in the therapeutic arena. Recent studies have shown that Slit proteins are expressed in tumor cells, whereas Robo receptors (Robo1 and Robo4) are detected in tumor endothelium (30,37,46). Using a truncated Robo1 soluble receptor, Wang and colleagues observed reduced microvessel density and tumor size in Slit-expressing melanoma xenograft models of tumorigenesis, thereby providing evidence that disruption of Slit-Robo1 signaling can block tumor growth (37). Such studies not only reveal potential strategies for the treatment of different disease states, but also expand our knowledge of the biology of the different angiogenic factors, because their expression levels and activities may vary between tumor types. Thus, clinical observations may, in turn, contribute to biological insight.

CONCLUSION

Our conceptual framework for understanding the development of the vascular system has been advanced dramatically during the last decade. Elucidating the basic signaling pathways that regulate the activity of the primary components of blood vessels (the ECs) was the first focus. The focus then shifted to understanding the interaction of these cells to form patent vessels and a seamless circulatory network. Most recently, it has become clear that guidance mechanisms are critical for directing this process, and vascular biologists have turned to the nervous system for practical insight. It is clear that the endothelial network is an intricate organ in its own right and responds in a dynamic fashion to specific cues in the environment. Because development makes reiterative use of successful regulatory mechanisms, it is, in hindsight, not surprising that the primary pathways that regulate axonal guidance have been similarly employed by the vascular system to pattern the vascular tree. The same array of ligands and receptors play roles in both systems to pattern these functionally unique, but structurally analogous systems. Finally, although it is clear that the four principal classes of neural guidance factors are shared between the nervous and vascular systems, it is not unrealistic to anticipate that unique molecules may exist to pattern specific subpopulations of ECs. The identification of these activities may similarly provide interesting therapeutic opportunities.

KEY POINTS

- The vascular and nervous systems share molecular mechanisms that govern complex patterning and functional coordination of their respective networks.
- The intricate organization of mature vascular networks likely requires coordination between vascular mitogens and guidance factors that provide directional information.
- The four well-characterized classes of axonal guidance factors, the Slits, semaphorins, netrins, and ephrins/Ephs, appear to have similar roles as guidance factors in the developing vasculature.

Future Goals

- To elucidate the activities of putative vessel guidance factors, which will require the development of informative vascular guidance assays that can distinguish directionality of growth from general mitogenic stimulation of proliferation
- To identify these new endothelial guidance factors significantly augments the tool kit of available vascular growth regulators; presumably these will provide important targets for therapeutic strategies to

control or engineer vessel growth in pathological conditions

- To overcome redundancy issues and to discern the fundamental roles of these factors in vascular development, it will be necessary to generate animal models in which multiple genes of a given family are ablated
- To determine the extent to which both networks cross-regulate their development and function, given that vessels and nerves exhibit many structural, functional, and regulatory parallels

REFERENCES

1 Urness LD, Li DY. Wiring the vascular circuitry: from growth factors to guidance cues. *Curr Top Dev Biol.* 2004;62:87–126.

2 Carmeliet P. Mechanisms of angiogenesis and arteriogenesis. *Nat Med.* 2000;6(4):389–395.

3 Ferrara N, Carver-Moore K, Chen H, et al. Heterozygous embryonic lethality induced by targeted inactivation of the VEGF gene. *Nature.* 1996;380(6573):439–442.

4 Betsholtz C, Karlsson L, Lindahl P. Developmental roles of platelet-derived growth factors. *Bioessays.* 2001;23(6):494–507.

5 Glagov S, Zarins C, Giddens DP, et al. Hemodynamics and atherosclerosis. Insights and perspectives gained from studies of human arteries. *Arch Pathol Lab Med.* 1988;112(10):1018–1031.

6 Flanagan JG, Vanderhaeghen P. The ephrins and Eph receptors in neural development. *Annu Rev Neurosci.* 1998;21:309–345.

7 Wang HU, Chen ZF, Anderson DJ. Molecular distinction and angiogenic interaction between embryonic arteries and veins revealed by ephrin-B2 and its receptor Eph-B4. *Cell.* 1998;93(5):741–753.

8 Childs S, Chen JN, Garrity DM, et al. Patterning of angiogenesis in the zebrafish embryo. *Development.* 2002;129(4):973–982.

9 Torres-Vazquez J, Gitler AD, Fraser SD, et al. Semaphorin-plexin signaling guides patterning of the developing vasculature. *Dev Cell.* 2004;7(1):117–123.

10 Martin P, Lewis J. Origins of the neurovascular bundle: interactions between developing nerves and blood vessels in embryonic chick skin. *Int J Dev Biol.* 1989;33(3):379–387.

11 Mukouyama YS, Shin D, Britsch S, et al. Sensory nerves determine the pattern of arterial differentiation and blood vessel branching in the skin. *Cell.* 2002;109(6):693–705.

12 Yu TW, Bargmann CI. Dynamic regulation of axon guidance. *Nat Neurosci.* 2001;4:S1169–S1176.

13 Wang HU, Anderson DJ. Eph family transmembrane ligands can mediate repulsive guidance of trunk neural crest migration and motor axon outgrowth. *Neuron.* 1997;18(3):383–396.

14 Helbling PM, Saulnier DM, Brandli AW. The receptor tyrosine kinase EphB4 and ephrin-B ligands restrict angiogenic growth of embryonic veins in *Xenopus laevis. Development* 2000;127(2):269–278.

15 Adams RH, Wilkinson GA, Weiss C, et al. Roles of ephrinB ligands and EphB receptors in cardiovascular development: demarcation of arterial/venous domains, vascular morphogenesis, and sprouting angiogenesis. *Genes Dev.* 1999;13(3):295–306.

16 Raper JA. Semaphorins and their receptors in vertebrates and invertebrates. *Curr Opin Neurobiol.* 2000;10(1):88–94.

17 Nakamura F, Tanaka M, Takahashi T, et al. Neuropilin-1 extracellular domains mediate semaphorin D/III-induced growth cone collapse. *Neuron.* 1998;21(5):1093–1100.

18 Carmeliet P. Blood vessels and nerves: common signals, pathways and diseases. *Nat Rev Genet.* 2003;4(9):710–720.

19 Miao HQ, Soker S, Feiner L, et al. Neuropilin-1 mediates collapsin-1/semaphorin III inhibition of endothelial cell motility: functional competition of collapsin-1 and vascular endothelial growth factor-165. *J Cell Biol.* 1999;146(1):233–242.

20 Bielenberg DR, Hida Y, Shimizu A, et al. Semaphorin 3F, a chemorepulsant for endothelial cells, induces a poorly vascularized, encapsulated, nonmetastatic tumor phenotype. *J Clin Invest.* 2004;114(9):1260–1271.

21 Serini G, Valdembri D, Zanivan S, et al. Class 3 semaphorins control vascular morphogenesis by inhibiting integrin function. *Nature.* 2003;424(6947):391–397.

22 Gitler AD, Lu MM, Epstein JA. PlexinD1 and semaphorin signaling are required in endothelial cells for cardiovascular development. *Dev Cell.* 2004;7(1):107–116.

23 Gu C, Yoshida Y, Livet J, et al. Semaphorin 3E and Plexin-D1 Control Vascular Pattern Independently of Neuropilins. *Science.* 2005;307(5707):265–268. Epub 2004 Nov 18.

24 Soker S, Takashima S, Miao HQ, et al. Neuropilin-1 is expressed by endothelial and tumor cells as an isoform-specific receptor for vascular endothelial growth factor. *Cell.* 1998;92(6):735–745.

25 Kawasaki T, Kitsukawa T, Bekku Y, et al. A requirement for neuropilin-1 in embryonic vessel formation. *Development.* 1999;126(21):4895–4902.

26 Rao Y, Wong K, Ward M, et al. Neuronal migration and molecular conservation with leukocyte chemotaxis. *Genes Dev.* 2002;16(23):2973–2984.

27 Kramer SG, Kidd T, Simpson JH, et al. Switching repulsion to attraction: changing responses to slit during transition in mesoderm migration. *Science.* 2001;292(5517):737–740.

28 Yuan W, Zhou L, Chen JH, et al. The mouse SLIT family: secreted ligands for ROBO expressed in patterns that suggest a role in morphogenesis and axon guidance. *Dev Biol.* 1999;212(2):290–306.

29 Hutson LD, Jurynec MJ, Yeo SY, et al. Two divergent slit1 genes in zebrafish. *Dev Dyn.* 2003;228(3):358–369.

30 Huminiecki L, Gorn M, Suchting S, et al. Magic roundabout is a new member of the roundabout receptor family that is endothelial specific and expressed at sites of active angiogenesis. *Genomics.* 2002;79(4):547–552.

31 Park KW, Morrison CM, Sorensen LK, et al. Robo4 is a vascular-specific receptor that inhibits endothelial migration. *Dev Biol.* 2003;261(1):251–267.

32 Bedell VM, Yeo SY, Park KW, et al. Roundabout4 is essential for angiogenesis in vivo. *Proc Natl Acad Sci USA.* 2005;102(18):6373–6378.

33 Bagri A, Marin O, Plump AS, et al. Slit proteins prevent midline crossing and determine the dorsoventral position of major axonal pathways in the mammalian forebrain. *Neuron.* 2002;33(2):233–248.

34 Plump AS, Erskine L, Sabatier C, et al. Slit1 and Slit2 cooperate to prevent premature midline crossing of retinal axons in the mouse visual system. *Neuron.* 2002;33(2):219–232.

35 Yuan W, Rao Y, Babiuk RP, et al. A genetic model for a central (septum transversum) congenital diaphragmatic hernia in mice lacking Slit3. *Proc Natl Acad Sci USAM*. 2003;100(9):5217–5222.

36 Suchting S, Heal P, Tahtis K, et al. Soluble Robo4 receptor inhibits in vivo angiogenesis and endothelial cell migration. *FASEB J*. 2005;19(1):121–123.

37 Wang B, Xiao Y, Ding BB, et al. Induction of tumor angiogenesis by Slit-Robo signaling and inhibition of cancer growth by blocking Robo activity. *Cancer Cell*. 2003;4(1):19–29.

38 Tessier-Lavigne M, Placzek M, Lumsden AG, et al. Chemotropic guidance of developing axons in the mammalian central nervous system. *Nature*. 1988;336(6201):775–778.

39 Stein E, Tessier-Lavigne M. Hierarchical organization of guidance receptors: silencing of netrin attraction by slit through a Robo/DCC receptor complex. *Science*. 2001;291(5510):1928–1938.

40 Hong K, Hinck L, Nishiyama M, et al. A ligand-gated association between cytoplasmic domains of UNC5 and DCC family receptors converts netrin-induced growth cone attraction to repulsion. *Cell*. 1999;97(7):927–941.

41 Park KW, Crouse D, Lee M, et al. The axonal attractant Netrin-1 is an angiogenic factor. *Proc Natl Acad Sci USA*. 2004;101(46):16210–16215.

42 Lu X, Le Noble F, Yuan L, et al. The netrin receptor UNC5B mediates guidance events controlling morphogenesis of the vascular system. *Nature*. 2004;432(7014):179–186.

43 Weinstein BM. Vessels and nerves: marching to the same tune. *Cell*. 2005;120(3):299–302.

44 LeCouter J, Kowalski J, Foster J, et al. Identification of an angiogenic mitogen selective for endocrine gland endothelium. *Nature*. 2001;412(6850):877–884.

45 Carmeliet P. VEGF gene therapy: stimulating angiogenesis or angioma-genesis? *Nat Med*. 2000;6(10):1102–1103.

46 Prasad A, Fernandis AZ, Rao Y, et al. Slit protein-mediated inhibition of CXCR4-induced chemotactic and chemoinvasive signaling pathways in breast cancer cells. *J Biol Chem*. 2004;279(10):9115–9124.

41

Notch Genes
Orchestrating Endothelial Differentiation

Yasuhiro Funahashi, Carrie J. Shawber, and Jan Kitajewski

Columbia University Medical Center, New York, New York

Vascular development entails multiple cell fate decisions that specify a diverse array of vascular structures, which include veins, arteries, capillaries, lymphatics, and specialized vascular beds of organs. Genetic studies have revealed that Notch plays a central role in the specification of arterial versus venous vasculature. Notch function is critical for remodelling and patterning of the vasculature, suggesting an additional role for Notch in regulating angiogenesis, a multistep process that requires endothelial cells (ECs) to respond to a variety of angiogenic stimuli. Notch proteins are receptors that modulate the ability of cells to respond to external cues, thus making them ideal regulators of angiogenesis. This chapter reviews in vivo and in vitro studies that have provided insights into Notch functions in the endothelium during development and in postnatal life. These studies show that Notch not only mediates specialization of the endothelium, but also has both positive and negative influences on the process of angiogenesis.

NOTCH AND *NOTCH LIGAND* GENES

The study of Notch started in 1917, when the famed geneticist, Thomas Hunt Morgan, first described a strain of *Drosophila* with "notched" wings. Notch genes were first molecularly identified during the 1980s, through genetic analysis of invertebrate organisms *Caenorhabditis elegans* (1) and *Drosophila* (2). Mutations in the Notch genes were described as affecting "cell fate determination," a process by which cells use molecular signals to specify distinct cellular fates. During the 1990s, the identification of Notch genes expressed in the vasculature (3,4) and both the functional analysis of Notch ligand Jagged1 in mice (5) and Notch gene target gridlock in zebrafish (6,7) began an era of vascular study of Notch. In principle, this period gave birth to studies of the molecular mechanisms of cell fate determination in vascular development, such as during arterial–venous specification.

In multicellular organisms, from worms to humans, Notch signaling regulates a variety of cell-fate decisions. Depending on the cellular context, Notch has been found to promote differentiation, proliferation, and cell survival as well as to inhibit these same processes (8–10). In mammals, the Notch family consists of four receptors (Notch1–4). Two families of ligands exist for Notch, Jagged (Jagged1 and -2), and Delta-like (Dll1, -3, and -4). Consistent with Notch functioning to regulate cell-fate differentiation via direct cell–cell interactions, both receptors and ligands are cell surface proteins with single-pass transmembrane domains (Figure 41.1).

THE NOTCH MOLECULAR PATHWAY

The activation of Notch is regulated at the level of receptor proteolysis followed by participation of the intracellular domain of Notch in a transcriptional complex (8,11). As Notch proteins transit to the cell surface, a furin-like convertase cleaves the receptor, at or near the plasma membrane, generating a heterodimeric receptor at the surface (see Figure 41.1). Upon ligand-binding, the cytoplasmic domain of Notch is released from the cell-surface by two sequential proteolytic cleavages. The first cleavage occurs just N-terminal to the transmembrane domain by a disintegrin and metalloprotease domain-like metalloprotease, whereas the second cleavage occurs within the transmembrane domain by the γ-secretase/presenilin complex. The released intracellular Notch peptide (NotchIC) translocates to the nucleus and forms a complex with the CBF, Su (H), Lag-2 (CSL) transcriptional repressor, converting it to a transcriptional activator. The *Hairy/Enhancer of Split* (*HES*) and HES-related (*Hey*, also named *CHF, HRT, HESR, gridlock*) genes are the direct targets of Notch/CSL-dependent signaling. The HES and Hey proteins are bHLH-Orange-domain transcriptional repressors. During neurogenesis and myogenesis, these transcription factors transduce Notch signaling by inhibiting tissue-specific bHLH transcription factors of the MASH and MyoD families, respectively. The direct targets of HES- and Hey-mediated transcriptional repression in ECs are unknown.

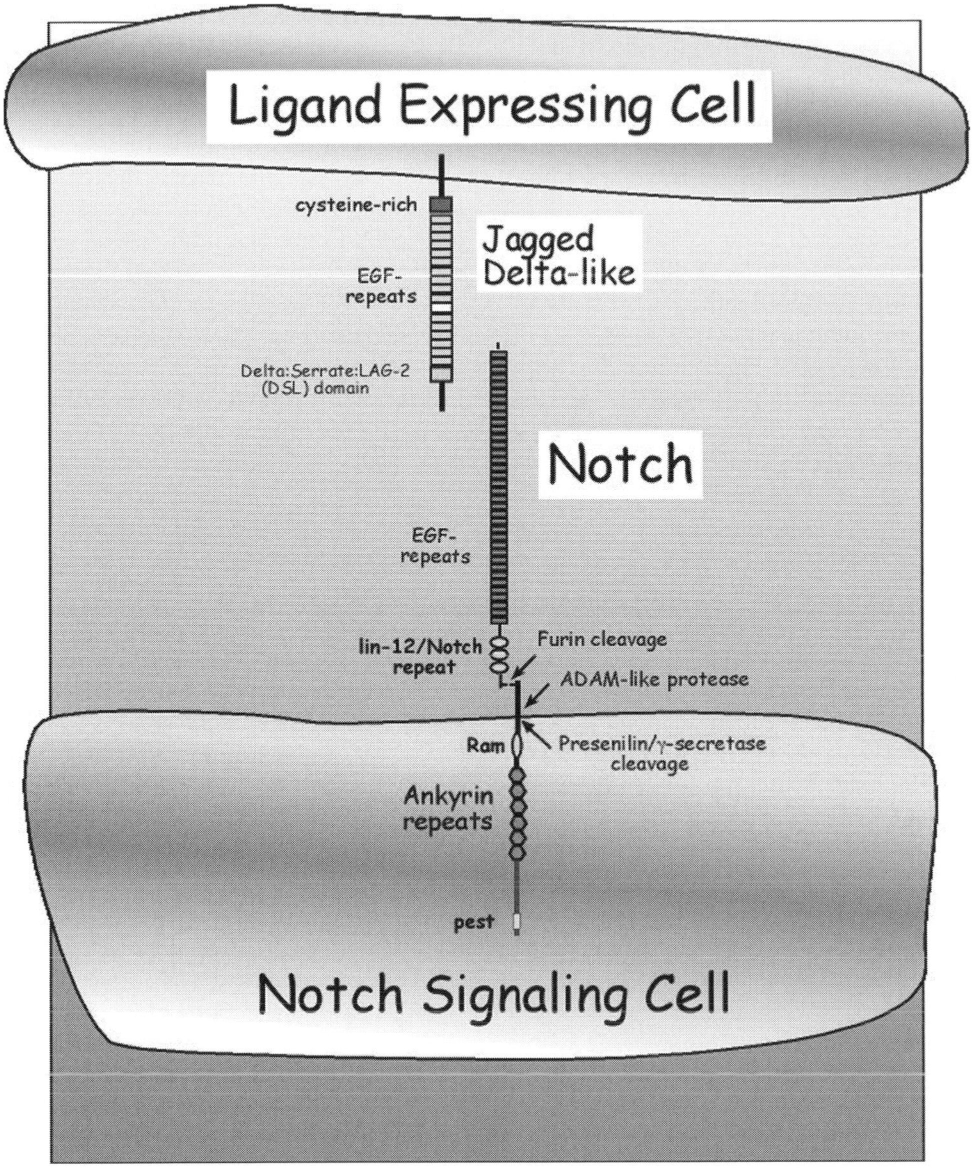

Figure 41.1. Schematic of Notch and Notch ligands. Two classes of Notch ligands, Jagged and Delta-like, are presented by cells as membrane-bound proteins that are characterized by a cysteine-rich domain, a series of EGF-like repeats, and a DSL domain unique to Notch ligands. Notch is a single-pass transmembrane protein, with the Notch extracellular domain consisting of a series of EGF-like repeats and LNR-lin-12/Notch repeats followed by a furin-cleavage site. The Notch intracellular domain contains protein interaction domains, RAM, Ankryin repeat, and PEST. The RAM is the CSL binding domain, the ankyrin repeat domain is a protein–protein interaction domain required for Notch function, and PEST a sequence important to target proteins for degradation. The Adam-like convertase cleavage occurs between the LNR and the transmembrane domain, and the γ-secretase/presenilin cleavage occurs within the transmembrane domain.

Notch also signals through a poorly defined CSL-independent pathway.

NOTCH EXPRESSION IN ARTERIAL ENDOTHELIUM AND CAPILLARIES

During mammalian development, Notch proteins and Notch ligands are dynamically expressed within the developing vasculature (12). At embryonic days (E)8 to E9 in mice, Dll4 is expressed in ECs of the dorsal aorta and developing heart tube, while Jagged1 expression begins in the aortic arch arteries and expands to the developing heart. At this same time, Notch4 is expressed in the endothelium of the anterior cardinal vein (13). The vascular expression of other Notch genes during this embryonic stage has not been reported. At E12.5, Notch1 and Notch2 are both expressed in the ECs of the arterial outflow tract, atrioventricular canal, and the aorta. Notch2 expression also is observed in the pulmonary artery. By the following day, dramatic changes occur in the expression pattern of the Notch family members. In the aorta, Notch2 expression is lost, while Notch1 expression is maintained (14). In addition, Notch4 expression is now present in ECs of the aorta but absent in the vena cava, which develops from the anterior cardinal vein (13,14).

In mouse embryos at stages E12.5 to 13.5, a diverse pattern of expression is also observed for Notch ligands. Jagged1 is coexpressed with Notch2 in the endothelium of the atrioventricular canal and the pulmonary artery. In the ECs of the aorta, both Jagged1 and Jagged2 expression overlap with that of Notch1 and Notch4 (14). Unlike Jagged ligands, Dll4 is no longer expressed in the endothelium of the aorta, but is expressed in the microvascular endothelium of the embryo (15). Expression of Jagged1 is not restricted to ECs, but extends into the surrounding vascular smooth muscle cells (VSMCs) of the aorta (14). In the aortic VSMCs, Jagged1 expression overlaps with that of Notch3 (16,17). From these limited expression studies, it appears that the expression of Notch1, -2, and -4 and Jagged2 are restricted to ECs; and Notch3 to VSMCs; whereas Jagged1 is expressed in both ECs and VSMCs. Several aspects of the expression patterns point to functional redundancies for receptors and ligands.

As embryogenesis proceeds, Notch proteins and ligands become restricted to the developing arterial endothelium. This expression is consistent with zebrafish studies in which Notch signaling was shown to regulate arterial specification of the dorsal aorta (12,18). However, expression of Jagged1 also has been observed in the endothelium of the vitelline veins, portal veins, and the anterior cardinal vein. These observations in mice suggest that Notch may have a function in venous ECs as well as regulating arterial EC differentiation.

In addition to embryogenesis, Notch proteins and ligands have been proposed to function during vascular remodeling and growth in the adult. Both Dll4 and Notch4 are expressed in the vessels of the lung, heart, and kidney of the adult mouse (13,15). It has been proposed that these highly vascularized tissues undergo high EC turnover. In the mature ovary, new blood vessel formation is necessary for follicle growth and maturation. Dll4 expression has been observed in these rapidly developing small vessels and capillaries of the mouse follicles (15). Notch1, Notch4, and Jagged1 also are expressed in the mature and angiogenic vessels of the ovary, although they have unique and overlapping patterns of expression (19). In a murine wound model, an increase in Jagged1 expression has been observed in proliferating ECs, suggesting that Notch may regulate neovascularization during wound repair. Similarly, Dll4 expression is upregulated in the invading vasculature of human tumor xenografts in mice and in the blood vessels in primary cancers of the kidney and breast (15). Thus, the patterns of Notch and Notch ligand expression indicate that Notch signaling may participate in both embryonic and adult physiological angiogenesis as well as pathological angiogenesis.

MUTATIONS IN *NOTCH* REVEAL A ROLE IN VENOUS/ARTERIAL SPECIFICATION

In zebrafish and mouse embryos, the observation that Notch and Notch ligand expression becomes predominant in the arterial endothelium suggests that Notch signaling functions to regulate arterial–venous specification. Zebrafish embryos deficient for the Hey2 homologue *gridlock*, or those that ectopically express a dominant negative form of CSL, have coarctation of the dorsal aortae, a failure of the anterior lateral dorsal aortae to merge into a single midline aorta, and shunting between the dorsal aortae and the posterior cardinal vein (20). In these loss-of-function embryos, ECs are recruited to the aortae but do not express the arterial EC marker, ephrinB2. Similarly, *Notch1*$^{-/-}$ and *Notch1*$^{-/-}$/*Notch4*$^{-/-}$,*Dll4*$^{+/-}$, *Dll4*$^{-/-}$, *Hey1*$^{-/-}$/*Hey2*$^{-/-}$, *rbpsuh*$^{-/-}$ (CSL), and Tek-Cre *Rbpsuh*$^{fl/null}$ (endothelial-specific loss of CSL) mouse embryos die around E9.5, and all display arterial vessel defects (20). In these Notch mutant embryos, coarctation of the vitelline arteries, umbilical artery, and dorsal aorta; a loss of arterial branching in the yolk sac vasculature; and arterial venous malformations are all observed. *Notch1*$^{-/-}$ and *Notch1*$^{-/-}$/*Notch4*$^{-/-}$ embryos also display defects in the anterior cardinal veins, although this defect has been proposed to be secondary to the aortic defects. In contrast to the loss-of-function embryos, the dorsal aortae are dilated in Notch gain-of-function murine embryos, *Flk1-N4/int-3* (21). Expression of arterial markers, such as Connexin37, Connexin40, CD44, and ephrinB2, is reduced or absent in the dorsal aortae of *Dll4*$^{+/-}$, *Dll4*$^{-/-}$, *Hey1*$^{-/-}$/*Hey2*$^{-/-}$, *rbpsuh*$^{-/-}$ (CSL), and Tek-Cre *Rbpsuh*$^{fl/null}$ murine embryos, whereas the dorsal aortae of *Dll4*$^{-/-}$ embryos misexpress the venous endothelial marker, EphB4 (20). Thus, in the embryonic vasculature, Notch signaling functions to suppress the venous fate in the arterial endothelium.

The genes that regulate Notch expression within the vasculature are beginning to be elucidated. In zebrafish, vascular endothelial growth factor (VEGF)-A, through the induction of Notch5 and DeltaC, regulates ephrinB2 expression within arteries (18). Consistent with VEGF-A–induced arterial specification being mediated by Notch, exogenous VEGF-A induced both Notch1 and Dll4 in cultured human iliac arterial ECs (HIAECs) and human femoral arterial ECs (HFAECs), whereas hypoxic conditions induced both VEGF-A and Dll4 in human umbilical vein ECs (HUVECs) (15,18). In mice treated with the VEGF blocking agent VEGF-Trap, Dll4 was found to be downregulated within tumor endothelium (22). Thus, VEGF-A likely regulates arterial specification through the upregulation of both Notch proteins and Notch ligands, which leads to the activation of Notch signaling. A recent study of COUP-TFII suggests that this orphan receptor functions to maintain venous EC identity by repressing Jagged1 and Notch1 expression (23). Loss of *COUP-TFII* in mice correlated with the mis-expression of Notch1, Jagged1, Hey1, and the arterial markers, ephrinB2 and neuropilin 1 (NRP1) within the veins, whereas ectopic endothelial expression of COUP-TFII resulted in a loss of Notch1, Jagged1, and NP-1 in the arterial endothelium.

MUTATIONS IN *NOTCH* UNCOVER VASCULAR REMODELING FUNCTIONS

Zebrafish and murine studies suggest that Notch signaling also may regulate other aspects of vascular development, in addition to arterial–venous specification (12,20). In zebrafish, ectopic sprouting of intersomitic vessels was observed in the *Notch Mindbomb* mutant (Mib$^{ta52b/ta52b}$) embryos, in which ligand–Notch signal activation is perturbed, suggesting that disruption of Notch results in defective patterning of microvessels. In addition, gridlock activity was found to be essential during the period of angioblast cell fate determination and migration, suggesting a role for Notch signaling in these processes (24). *Notch1*$^{-/-}$, *Notch1*$^{-/-}$/*Notch4*$^{-/-}$, *Jagged1*$^{-/-}$, *Dll4*$^{-/-}$, *Rpbsuh*$^{-/-}$ (CSL), *Hey1*$^{-/-}$/*Hey2*$^{-/-}$, or *presenilin1*$^{-/-}$/*presenilin2*$^{-/-}$ mutant mice die in utero between E9.5 and 10.5 and display capillary patterning defects. Expression of an activated form of Notch4 within the endothelium (*Flk1-N4/int-3*) or disruption of Fbw7/Sel-10, a negative regulator of Notch signaling, also results in embryonic lethality at E10.5, with vascular patterning defects. Similar to the zebrafish mutants, the intersomitic vessels of these Notch mutant embryos recruit ECs but fail to differentiate into a branching capillary network. These results further demonstrate that branching morphogenesis of the endothelium is exquisitely sensitive to the dosage of Notch signaling. In Notch mutant embryos, this failure in branching morphogenesis may be due to altered EC survival and proliferation. In *Dll4*$^{-/-}$ embryos, ECs accumulate at the apical end of the intersomitic vessels, rather than migrating and differentiating to form capillaries. In embryos that have lost one copy of *VEGFR2* and both copies of *presenilin1* (*Flk1*$^{+/lacZ}$;*PS1*$^{-/-}$), a reduction in capillary sprouting was associated with increased EC proliferation as well as increased apoptosis, suggesting that Notch signaling inhibits proliferation and promotes the survival of angiogenic ECs.

NOTCH FUNCTIONS DEFINED BY IN VITRO ANGIOGENESIS ASSAYS

Angiogenesis is a multistep process that requires ECs to respond to a variety of angiogenic stimuli. Angiogenesis initiates with the degradation of extracellular matrix (ECM), followed by budding of ECs at sites of active angiogenesis. The ECs then transition to a proliferative mode coincident with the development of new sprouts. The combined steps of budding, proliferation, and migration result in the formation of lumen-containing sprouts. The multicellular sprouts then remodel to form a network of capillary vessels. In vitro angiogenesis assays, using primary ECs have proved useful to model distinct steps of angiogenesis. Using these methodologies, Notch signaling has been described as both a positive and negative effector of angiogenesis (Figure 41.2).

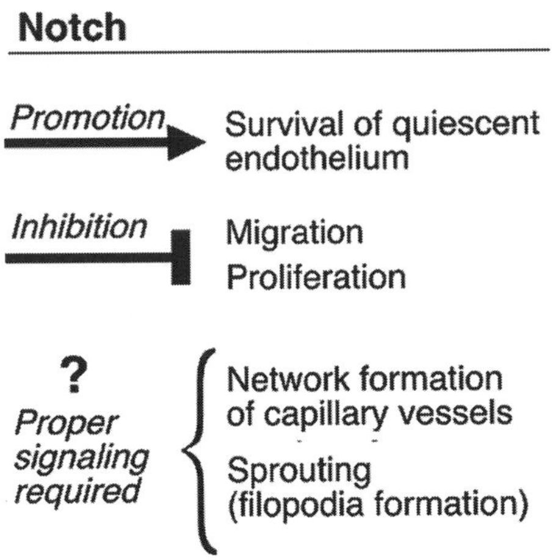

Figure 41.2. Summary of the effects of Notch signaling on distinct steps in angiogenesis, as determined by in vitro assays.

Activation of Notch signaling inhibits proliferation of human primary ECs (25–27). Notch signaling inhibits mitogen-induced expression of p21^{Cip1} and delays cyclin D-cdk4–mediated retinoblastoma (Rb) phosphorylation. In ECs, upregulation of p21^{Cip1} is necessary for nuclear localization of cyclin D and cdk4, resulting in the phosphorylation of retinoblastoma protein (Rb). Thus, Notch might use such a mechanism to inhibit EC proliferation, because an activated form of Notch is not able to induce cell-cycle arrest of immortalized ECs expressing SV40 T antigen, which inhibits the phosphorylation of Rb (28). Notch signaling also may function in the regulation of contact inhibition, because confluent cultures of ECs exhibit Notch signal activation (28). Notch signaling also controls EC apoptosis (26,29). In immortalized ECs, inflammatory-mediated cell death by lipopolysaccharide (LPS) is inhibited by Notch signal activation, accompanied by inhibition of the proapoptotic c-jun N-terminal kinase (JNK) pathway and upregulation of antiapoptotic bcl-2 protein (29). Notch signaling inhibits the migration of both human primary and immortalized ECs (25,30). The action of Notch against migration of immortalized EC line is detected in collagen-coated, but not fibrinogen-coated plates, possibly via regulation of β1 integrin signaling (30).

The role of Notch in endothelial morphogenesis, such as angiogenic sprouting and network formation of capillary vessels, is still controversial. Activation of Notch signaling can promote budding of HUVECs into the underlying collagen matrix (31). However, activation of Notch signaling in human immortalized dermal microvascular ECs decreased endothelial sprouting (30). In three-dimensional (3D) collagen gels, Notch signaling is activated during network formation of HUVECs, whereas a dominant negative form of Notch inhibits network formation (27). Either ectopic expression of activated Notch1 or overexpression of HES1, one of the primary

Table 41-1: Vascular Expression/Phenotypes of Notch Proteins

Receptor/ Ligand	*Cultured ECs*	*E8.5–9.5*	*E11.5–12.5*	*E13.5*	*Knockout Phenotype*
Notch1	Induced by VEGF in arterial ECs Induced by VEGF and bFGF in venous ECs	ND	ND	Aorta	Lethal at E9.5; defects in yolks sac plexus, placental labyrinthine, intersomitic vessels, and arterial EC specification; collapsed dorsal aorta; pericardial swelling
Notch2		ND	ND	Heart outflow tract	50% lethal at E16.5; pericardial effusion and widespread hemorrhaging. 50% viable lack of glomerular capillary tufts or glomerular capillary aneurysms; defects in hyaloid vessels
Notch3		ND	ND	Arterial VSMCs	Viable; adult-onset arteriopathy associated with failed arterial VSMC differentiation, VSMC and EC death
Notch4	Induced by VEGF and bFGF in venous ECs	E9.5 – Anterior cardinal vein ECs		Aortic, aortic tract and pulmonary artery, capillary ECs	Viable; augments Notch1 loss-of-function phenotype
Jagged1		E9.5 – Aortic arch arteries	E12.5-Aortic and aortic arch, pulmonary, coronary, and intervertebral artery ECs and VSMCs	Aortic ECs and VSMCs	Lethal at E10.5; defects in yolk sac plexus, cranial vasculature
Jagged2				Aortic ECs	Viable; no vascular defects reported
Dll4	Induced by VEGF or bFGF in arterial ECs	E8.5 – ECs of dorsal aorta umbilical artery and heart tube. E9.5 – ECs of aorta, umbilical artery, mesenteric arteries, intersomitic vessels, and yolk sac	E11.5-ECs of arteries	Capillary ECs	Lethal at E9.5; haploinsufficient: defects in yolk sac plexus, intersomitic vessels, cranial vasculature, and arterial EC specification; coarctation of dorsal aorta; pericardial swelling; AV malformations

Dll1 and Dll3 are not expressed in the vasculature, and knockout mice do not have vascular defects.
ND, No data.

targets of Notch signaling, also enhances network formation of human iliac artery ECs (26). Moreover, overexpression of HESR1 (Hey1), a primary target gene of Notch signaling, or inhibition of HESR1, blocks network formation of capillary-like vessels of primary adipose ECs (25). It is clear that these diverse results, carried out in different types of ECs, require further clarification. It is important to consider that these findings may be highlighting the necessity of the proper levels of Notch signaling to regulate endothelial morphogenesis during angiogenic progression. In support of the concept that proper Notch activity is required for vascular development is that fact that either endothelial-specific deletion or activation of Notch signaling causes vascular remodeling defects and embryonic lethality (20,21,32).

It may be that the most direct links between Notch and angiogenesis come from the finding that Notch functions downstream of VEGF-A signaling (18,22,26). An intriguing example of such a connection may be apparent in retinal vasculature. VEGF-A functions to induce angiogenic sprouting via the production of filopodia at the tip of ECs (endothelial tip cells) (33). Dll4 is strongly expressed in the tip ECs of the developing retinal vascular plexus (34). As VEGF-A induces both Notch1 and Dll4 expression in cultured ECs (26), VEGF-A might induce Dll4 expression to initiate angiogenic sprouting. Therefore, Notch signaling might assist in determining the site for angiogenesis initiation in the growing vasculature, in concert with VEGF-A. However, it remains to be determined how Notch signaling participates in defining the tip cell phenotype and in the regulation of sprouting.

NOTCH AND ENDOTHELIAL/VASCULAR DISEASE

In humans, misregulation of Notch signaling is clearly linked to disruption of vascular integrity. Cerebral autosomal dominant arteriopathy with subcortical infarcts and leukoencephalopathy (CADASIL) is an adult-onset arteriopathy caused by mutations within the EGF-like repeats of Notch3 (12,20). In CADASIL patients, dysregulation of Notch3 function results in reduced adhesion and death of arterial VSMCs, leading to recurrent strokes and (ultimately) death. This syndrome is consistent with Notch3 expression being restricted to VSMCs during embryogenesis and in the adult (16,17,34–38) (Table 41-1). The human CADASIL phenotype has been recapitulated in transgenic mice that express the CADASIL Notch3 R90C mutation in VSMCs. Ten-month-old transgenics that express Notch3 R90C display an impaired flow- and pressure-induced arterial dilation as well as compromised cerebrovasculature (39,40). These phenotypes were associated with VSMC regression and death, suggesting Notch3 function is essential for VSMC adhesion-dependent survival. In fact, constitutive activation of Notch3 in rat aortic VSMCs correlated with the induction of cFlip, an antagonist of Fas-dependent apoptosis (41). Recent analyses of *Notch3*$^{-/-}$ mice also indicate that Notch3 function is essential for the differentiation of arterial VSMCs (42,43).

Dysregulation of Notch4 signaling has been implicated in a human vascular homeostasis syndrome. In patients with mutations in *cerebral cavernous malformation 1* (*CCM1*) and a *CCM1*-knockout mouse model, development of thin-walled, hemorrhaging vessels in the brain correlates with a decrease in Notch4 expression in both arterial endothelial and VSMCs (44). Finally, retrospective analysis of Alagille patients, a complex developmental disorder with characteristic organ-specific vascular defects caused by mutations in Jagged1, revealed a significant mortality from anomalous vasculature, and that up to a third of patients' deaths correlated with the presence of defective, aneurysmal arteries.

KEY POINTS

- Vascular development entails multiple cell fate decisions in which Notch might be involved. Arterial–venous differentiation is one such cell fate decision in which Notch is a key regulator.
- Notch has been observed to have both pro- and antiangiogenic functions; thus, a proper balance or timing of Notch activity may be required during angiogenic sprouting and vascular remodeling.
- Notch also functions in vascular homeostasis and possibly in interactions between VSMCs and ECs. Human mutations in human *NOTCH3* and *JAGGED1* or reduction of NOTCH4 expression are linked to hereditary vasculopathies.

Future Goals

- To define the interaction between Notch signaling and that of other known angiogenic cascades, such as the VEGF/VEGF receptor pathway
- To identify other cell fate determination steps that occur during vascular development and endothelial differentiation that utilize Notch signaling as a regulator
- To explore further roles for Notch in pathologies of the human vasculature, such as tumor angiogenesis
- To determine the mechanisms by which Notch functions as both a pro- and antiangiogenic regulator, and determining its role in sprouting angiogenesis

ACKNOWLEDGMENTS

We are grateful to Christin Lennon for critical reading of the manuscript. Our work is supported by a grant from the National Institutes of Health (RO1HL62454). Carrie Shawber is supported by a DOD breast cancer fellowship DAMD17–031–0218.

REFERENCES

1 Greenwald IS, Sternberg PW, Horvitz HR. The lin-12 locus specifies cell fates in Caenorhabditis elegans. *Cell*. 1983;34:435–444.
2 Artavanis-Tsakonas S, Muskavitch MA, Yedvobnick B. Molecular cloning of Notch, a locus affecting neurogenesis in *Drosophila melanogaster. Proc Natl Acad Sci USA*. 1983;80:1977–1981.
3 Uyttendaele H, Marazzi G, Wu G, et al. Notch4/int-3, a mammary proto-oncogene, is an EC-specific mammalian Notch gene. *Development*. 1996;122:2251–2259.
4 Del Amo FF, Smith DE, Swiatek PJ, et al. Expression pattern of Motch, a mouse homolog of *Drosophila* Notch, suggests an important role in early postimplantation mouse development. *Development*. 1992;115:737–744.
5 Xue Y, Gao X, Lindsell CE, et al. Embryonic lethality and vascular defects in mice lacking the Notch ligand Jagged1. *Hum Mol Genet*. 1999;8:723–730.
6 Weinstein BM, Stemple DL, Driever W, et al. Gridlock, a localized heritable vascular patterning defect in the zebrafish. *Nat Med*. 1995;1:1143–1147.
7 Zhong TP, Rosenberg M, Mohideen MA, et al. Gridlock, an HLH gene required for assembly of the aorta in zebrafish. *Science*. 2000; 287:1820–1824.

8 Artavanis-Tsakonas S, Rand MD, Lake RJ. Notch signaling: cell fate control and signal integration in development. *Science*. 1999; 284:770–776.
9 Lewis J. Notch signaling and the control of cell fate choices in vertebrates. *Semin Cell Dev Biol*. 1998;9:583–589.
10 Weinmaster G. The ins and outs of Notch signaling. *Mol Cell Neurosci*. 1997;9:91–102.
11 Weinmaster G. Notch signal transduction: a real rip and more. *Curr Opin Genet Dev*. 2000;10:363–369.
12 Shawber C, Kitajewski J. Notch function in the vasculature: insights from zebrafish, mouse, and man. *BioEssays*. 2004;26:225–234.
13 Uyttendaele H, Marazzi G, Wu G, et al. Notch4/int-3, a mammary proto-oncogene, is an EC-specific mammalian Notch gene. *Development*. 1996;122:2251–2259.
14 Villa N, Walker L, Lindsell CE, et al. Vascular expression of Notch pathway receptors and ligands is restricted to arterial vessels. *Mech Dev*. 2001;108:161–164.
15 Mailhos C, Modlich U, Lewis J, et al. Delta4, an endothelial specific Notch ligand expressed at sites of physiological and tumor angiogenesis. *Differentiation*. 2001;69:135–144.
16 Leimeister C, Schumacher N, Steidl C, et al. Analysis of HeyL expression in wild-type and Notch pathway mutant mouse embryos. *Mech Dev*. 2000;98:175–178.
17 Villa N, Walker L, Lindsell CE, et al. Vascular expression of Notch pathway receptors and ligands is restricted to arterial vessels. *Mech Dev*. 2001;108:161–164.
18 Lawson ND, Vogel AM, Weinstein BM. Sonic Hedgehog and vascular endothelial growth factor act upstream of the Notch pathway during arterial endothelial differentiation. *Dev Cell*. 2002;3: 127–136.
19 Vorontchikhina MA, Zimmermann RC, Shawber CJ, et al. Unique patterns of Notch1, Notch4 and Jagged1 expression in ovarian vessels during folliculogenesis and corpus luteum formation. *Gene Expr Patterns*. 2005;5:701–709.
20 Shawber C, Kandel JJ, Kitajewski J. Notch: cell fate determination from vascular development to human vasculopathy. *Drug Discovery Today: Disease Models*. 2004;1:351–358.
21 Uyttendaele H, Ho J, Rossant J, et al. Vascular patterning defects associated with expression of activated Notch4 in embryonic endothelium. *Proc Natl Acad Sci USA*. 2001;98:5643–5648.
22 Gale NW, Dominguez MG, Noguera I, et al. Haploinsufficiency of delta-like 4 ligand results in embryonic lethality due to major defects in arterial and vascular development. *Proc Natl Acad Sci USA*. 2004;101:15949–15954.
23 You LR, Lin FJ, Lee CT, et al. Suppression of Notch signalling by the COUP-TFII transcription factor regulates vein identity. *Nature*. 2005;435:98–104.
24 Peterson RT, Shaw SY, Peterson TA, et al. Chemical suppression of a genetic mutation in a zebrafish model of aortic coarctation. *Nat Biotechnol*. 2004;22:595–599.
25 Henderson AM, Wang SJ, Taylor AC, et al. The basic helix-loop-helix transcription factor HESR1 regulates EC tube formation. *J Biol Chem*. 2001;276:6169–6176.
26 Liu Z-J, Shirakawa T, Li Y, et al. Regulation of Notch1 and Dll4 by vascular endothelial growth factor in arterial ECs: implication for modulating arteriogenesis and angiogenesis. *Mol Cell Biol*. 2003;23:14–25.
27 Taylor KL, Henderson AM, Hughes CC. Notch activation during EC network formation in vitro targets the basic HLH transcription factor HESR-1 and downregulates VEGFR-2/KDR expression. *Microvasc Res*. 2002;64:372–383.
28 Noseda M, Chang L, McLean G, et al. Notch activation induces EC cycle arrest and participates in contact inhibition: role of p21Cip1 repression. *Mol Cell Biol*. 2004;24:8813–8822.
29 MacKenzie F, Duriez P, Wong F, et al. Notch4 inhibits endothelial apoptosis via RBP-Jkappa-dependent and -independent pathways. *J Biol Chem*. 2004;279:11657–11663.
30 Leong KG, Hu X, Li L, et al. Activated Notch4 inhibits angiogenesis: role of beta 1-integrin activation. *Mol Cell Biol*. 2002;22: 2830–2841.
31 Das I, Craig C, Funahashi Y, et al. Notch oncoproteins depend on gamma-secretase/presenilin activity for processing and function. *J Biol Chem*. 2004;279:30771–30780.
32 Limbourg FP, Takeshita K, Radtke F, et al. Essential role of endothelial Notch1 in angiogenesis. *Circulation*. 2005;111:1826–1832.
33 Gerhardt H, Golding M, Fruttiger M, et al. VEGF guides angiogenic sprouting utilizing endothelial tip cell filopodia. *J Cell Biol*. 2003;161:1163–1177.
34 Claxton S, Fruttiger M. Periodic Delta-like 4 expression in developing retinal arteries. *Gene Expr Patterns*. 2004;5:123–127.
35 Joutel A, Corpechot C, Ducros A, et al. Notch3 mutations in CADASIL, a hereditary adult-onset condition causing stroke and dementia. *Nature*. 1996;383:707–710.
36 Joutel A, Andreux F, Gaulis S, et al. The ectodomain of the Notch3 receptor accumulates within the cerebrovasculature of CADASIL patients. *J Clin Invest*. 2000;105:597–605.
37 Joutel A, Favrole P, Labauge P, et al. Skin biopsy immunostaining with a Notch3 monoclonal antibody for CADASIL diagnosis. *Lancet*. 2001;358:2049–2051.
38 Loomes KM, Taichman DB, Glover CL, et al. Characterization of Notch receptor expression in the developing mammalian heart and liver. *Am J Med Genet*. 2002;112:181–189.
39 Dubroca C, Lacombe P, Domenga V, et al. Impaired vascular mechanotransduction in a transgenic mouse model of CADASIL arteriopathy. *Stroke*. 2005;36:113–117.
40 Lacombe P, Oligo C, Domenga V, et al. Impaired cerebral vasoreactivity in a transgenic mouse model of cerebral autosomal dominant arteriopathy with subcortical infarcts and leukoencephalopathy arteriopathy. *Stroke*. 2005;36:1053–1058.
41 Wang W, Prince CZ, Mou Y, et al. Notch3 signaling in vascular smooth muscle cells induces c-FLIP expression via ERK/MAPK activation. Resistance to Fas ligand-induced apoptosis. *J Biol Chem*. 2002;277:21723–21729.
42 Domenga V, Fardoux P, Lacombe P, et al. Notch3 is required for arterial identity and maturation of vascular smooth muscle cells. *Genes Dev*. 2004;18:2730–2735.
43 Krebs LT, Xue Y, Norton CR, et al. Characterization of Notch3-deficient mice: normal embryonic development and absence of genetic interactions with a Notch1 mutation. *Genesis*. 2003;37:139–143.
44 Whitehead KJ, Plummer NW, Adams JA, et al. Ccm1 is required for arterial morphogenesis: implications for the etiology of human cavernous malformations. *Development*. 2004;131:1437–1448.

42

Reactive Oxygen Species

Kaikobad Irani

Cardiovascular Institute, University of Pittsburgh Medical Center, Pennsylvania

The endothelium plays a pivotal role in vascular homeostasis. Its strategic location between the vascular wall and the circulation allows it to serve as a bridge between these two compartments. A dynamic interaction exists between the endothelium and hormonal and cellular mediators, both from the circulation as well as from the vascular wall. This same strategic location of the endothelium also makes it a prime target for injury in the setting of vascular disease.

Accumulating evidence suggests that oxidative stress, by resulting in a maladaptive endothelial cell (EC) phenotype, is one of the common means by which conditions such as hyperlipidemia, hypertension, advanced age, hyperhomocysteinemia, and diabetes, all of which are associated with an increased risk of vascular diseases, lead to a dysfunctional endothelium. However, the term *oxidative stress* is loosely used and applied, and it would be simplistic to conclude that oxidants or reactive oxygen species (ROS) always play a pathophysiological role in the endothelium.

HISTORICAL PERSPECTIVE

ROS have been regarded historically as deleterious due to free radical–induced oxidation and damage of macromolecules such as DNA, proteins, and lipids. More recently, however, it has been appreciated that ROS can exert more subtle modulatory effects. This evolution in our understanding of the roles of ROS is exemplified by the nicotinamide adenine dinucleotide phosphate (NADPH) oxidase, a source of ROS in many cell types. The NADPH oxidase was first described in professional phagocytic cells (macrophages and neutrophils) of the innate immune system, where it is responsible for generating a large amount of the ROS superoxide as part of the oxidative burst during the process of phagocytosis (1). In this capacity, the NADPH oxidase and the ROS generated by it has long been associated with pathophysiological outcomes such as injury of cardiovascular tissue perpetrated by infiltrating neutrophils in the setting of ischemia–reperfusion. In contrast to this injurious role of ROS generated by the NADPH oxidase in phagocytes, the role of ROS derived from the NADPH oxidase (and homologues of it) in nonphagocytic cells can be more appropriately described as one of physiological signaling. This latter function of ROS as biologically relevant signaling intermediaries is primarily mediated by chemically reversible, post-translational modifications of specific amino acid residues on proteins that result in changes in protein function. A classical example of this is the progressive oxidation of specific cysteine residues by H_2O_2 to give rise to sulfenic acid, sulfinic acid, and sulfonic acid derivatives (2), which can mediate discrete and diverse regulatory outcomes.

DEFINING REACTIVE OXYGEN SPECIES AND OXIDATIVE STRESS

ROS are oxygen-based molecules that are characterized by their high chemical reactivity and are generated by all aerobic organisms. In health, ROS generation is counteracted by enzymatic and nonenzymatic antioxidant systems that scavenge or reduce ROS levels, thereby maintaining reduction-oxidation (redox) homeostasis in cells and tissues. Perturbation of this balance secondary to an increase in ROS production and/or a decrease in antioxidant reserve leads to a state of oxidative stress.

By definition, ROS are chemical species containing oxygen atoms with unpaired electrons that may be charged or uncharged. The unpaired electrons render the molecule unstable, allowing the ROS to react with electrons, other chemical species, or macromolecules such as proteins and nucleic acids. Molecular oxygen serves as the primary electron acceptor yielding the superoxide anion. Subsequent univalent reductive steps can lead to the generation of a whole range of ROS (Figure 42.1). The distinct physical and chemical properties of each of these ROS determine its toxicity as well as its suitability as a signaling intermediary.

ECs and, for that matter, almost all cell types have multiple mechanisms for the production and removal of ROS. Multiple enzymatic systems can convert molecular oxygen to superoxide (O_2^-). Those notable in the endothelium are xanthine oxidoreductase, NADPH oxidases, uncoupled endothelial

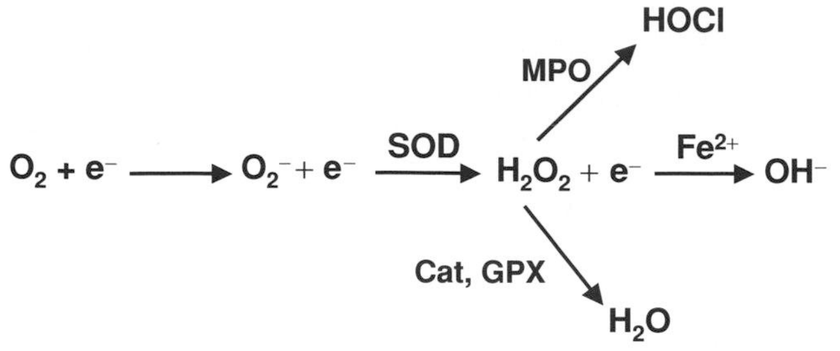

Figure 42.1. Chemical derivation of main reactive oxygen species. O_2, molecular oxygen; e^-, free electron; O_2^-, superoxide; H_2O_2, hydrogen peroxide; SOD, superoxide dismutase; MPO, myeloperoxidase; Cat, catalase; GPX, glutathione peroxidase; H_2O, water; HOCl, hypochlorous acid; Fe^{2+}, ferrous ions; OH^-, hydroxyl radical.

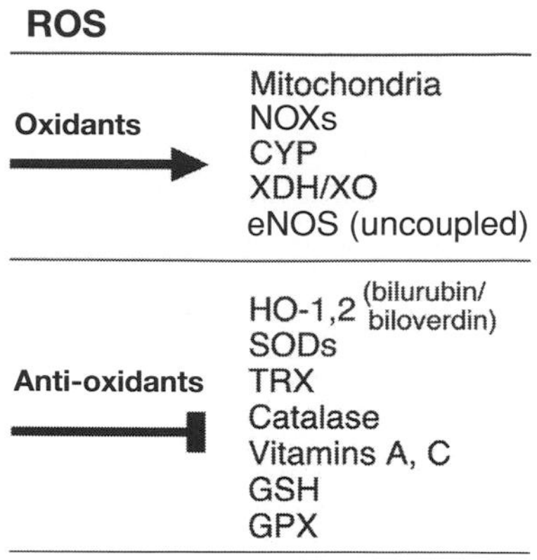

Figure 42.2. Major oxidant and antioxidant systems in endothelial cells. HO, heme oxygenase; SOD, superoxide dismutase; TRX, thioredoxin; GSH, glutathione; GPX, glutathione peroxidase; ROS, reactive oxygen species; eNOS, endothelial nitric oxide synthase; NOXs, NADPH oxidases; CYP, cytochrome P450; XDH/XO, xanthine dehydrogenase/xanthine oxidase.

nitric oxide synthase (eNOS), the cytochrome P450-based enzymes, and mitochondrial respiration. Other mechanisms for the generation of ROS include: (a) catalytic conversion of other ROS, such as H_2O_2 by dismutation of O_2^-; (b) conversion of H_2O_2 to hypochlorous acid (HOCl); and (c) reduction of ROS by transition metals such as the ferrous ion to the highly reactive hydroxyl (OH^-) radical. As an example of ROS removal, hydrogen peroxide (H_2O_2) is enzymatically converted to water (H_2O) by glutathione peroxidase (GPX) and catalase (see Figure 42.1).

SOURCES OF REACTIVE OXYGEN SPECIES IN THE ENDOTHELIUM

ROS may be produced within or without the endothelium, and may exert both autocrine and paracrine effects.

Nonendothelial Sources

As the interface between flowing blood and the vascular wall, the endothelium is exposed to ROS produced in both luminal and abluminal compartments. The classical example of a blood-borne source of ROS external to the endothelium is circulating cells of the innate immune system. In conditions in which macrophages and neutrophils have infiltrated the vascular wall (such as in an atherosclerotic plaque), their proximity to the endothelium allows the membrane-permeable ROS produced within them to affect endothelial function. When activated, infiltrating neutrophils and macrophages (which now form part of the vessel wall) are rich sources of H_2O_2 and, because they express the enzyme myeloperoxidase, also can produce large quantities of another cell-permeable ROS, hypochlorous acid (HOCl) (3). On the abluminal side of the endothelium, vascular smooth muscle cells (VSMCs) and adventitial fibroblasts can generate ROS (4).

Endothelial Sources

Although the endothelium is subject to influence by diffusible ROS produced in neighboring cells, it is itself endowed with several enzymatic systems with the capacity to generate ROS (Figure 42.2).

The NADPH Oxidase(s)

Phagocytic cells such as neutrophils and macrophages express a multimolecular oxidase that transfers an electron from NADPH to molecular O_2 to generate O_2^-, which plays an important role in the innate immune response (1). The *ph*agocyte *ox*idase (phox) comprises a heterodimer flavocytochrome consisting of a 22-kDa unit ($p22^{phox}$) and a glycosylated 91-kDa unit ($gp91^{phox}$, also termed NOX2 for *N*ADPH *ox*idase). In addition to this membrane-associated catalytic core, four cytosolic subunits ($p40^{phox}$, $p47^{phox}$, $p67^{phox}$, and the small GTPases Rac1 or Rac2) are essential for the regulation of oxidase activity. Although initially thought to be limited to phagocytes, it is now appreciated that many components of this oxidase (including the catalytic NOX2 flavoprotein) are expressed in nonphagocytes. For example, NOX2, $p47^{phox}$, $p22^{phox}$, and $p67^{phox}$, and the ubiquitous pac1, are all expressed in ECs (5–10). Although ECs may express a functional phagocytic NADPH oxidase, the regulation of oxidase activity is likely to be different than that in phagocytes. In ECs, the oxidase generates constitutive low levels of ROS even in the absence of extrinsic stimulation (in contrast to phagocytes, in which ROS production is an all-or-none phenomenon), but can increase ROS production in response to specific stimuli. This might be because in ECs, unlike in phagocytes, NOX2 is present in a preassembled complex with its regulatory components (9,11).

The observation that NOX2 is not ubiquitously expressed led to a search for homologues of NOX2, adding new members to the NOX family. The first of these, *NOX1*, was originally cloned from a colon cDNA library, and is expressed mainly in colon, VSMCs, and prostate (12–14). Since then, several other homologues, *NOX1, -2, -3, -4, and -5,* and *Duox1* and *-2* (for *du*al *ox*idase, because they possess a second N-terminal peroxidase homology domain), have been cloned, and been shown to have tissue-specific expression. In addition to NOX2, ECs express NOX1 and NOX4, albeit at much lower levels (7,15).

Endothelial NOX activity is subject to regulation by many physiologic and pathophysiological stimuli, and this activity mediates diverse EC phenotypes. Vascular endothelial growth factor (VEGF), a potent angiogenic agent that specifically targets ECs, leads to a NOX2-dependent increase in ROS levels in ECs (16,17). This effect is important for VEGF signaling, vis-à-vis EC proliferation and migration (16,17). In addition, VEGF induces antioxidant defenses in ECs, such as the expression of manganese superoxide dismutase (MnSOD) in mitochondria, which also depends on the activation of NOX (18). NOXs also are activated by cytokines such as tumor necrosis factor (TNF)-α and interleukin (IL)-1β, and may play a part in survival mechanisms when ECs are challenged with these stimuli (19,20). Both angiotensin II and endothelin-1 also activate endothelial NOX (10,21), which may be an important mechanism for the endothelial dysfunction seen in heart failure. Endothelial NOX activity also is stimulated by high insulin (22) or glucose (23) and by advanced glycation end products (24), as well as by oxidized low-density lipoprotein (LDL) (25,26), which may contribute to diabetes- and hyperlipidemia-associated endothelial dysfunction. Environmental stressors such as ischemia (27), hypoxia–reoxygenation (28), and nutrient deprivation (29) also have been reported to upregulate endothelial NOX activity, implicating endothelial NOXs in ischemic and postischemic vascular dysfunction. Finally, laminar and oscillatory shear enhance endothelial NOX activity (30).

Endothelial Nitric Oxide Synthase

eNOS is responsible for the production of nitric oxide (NO), a reactive nitrogen species that exerts many effects on the vascular system that counteract those of ROS. However, under certain circumstances, eNOS can paradoxically become a source of ROS by transferring an electron to molecular oxygen rather than its usual substrate, L-arginine, resulting in the generation of superoxide (31,32). This process, termed *uncoupling*, occurs when there is paucity or oxidation of tetrahydrobiopterin (BH4), a cofactor for eNOS (31,33,34). Interestingly, ROS derived from NADPH oxidase(s) can themselves lead to oxidation of BH4, thereby uncoupling eNOS and further promoting superoxide generation. This is an example of a positive feedback loop between ROS and different enzymatic sources. The uncoupling of NOS has been implicated as a source of ROS in atherosclerosis (35), hypertension (36), hypercholesterolemia (37), and diabetes (38).

Xanthine Dehydrogenase/Xanthine Oxidase

Catalysis of xanthine to urate by xanthine oxidase (XO) results in the generation of superoxide (39). XO is derived from oxidative conversion of xanthine dehydrogenase (XDH). XO is expressed on the luminal side of the endothelium, and its activity is upregulated – both at a transcriptional and posttranscriptional level – by inflammatory stimuli such as interferon (INF)-γ (40) or in response to phagocyte adhesion to the endothelium (41). Furthermore, observations that the inhibition of XO improves the endothelial dysfunction seen with hypercholesterolemia (42,43), smoking (44), and in atherosclerosis (45), imply that XDH/XO is an important source of endothelial ROS in these pathologies. In addition, exposure of the endothelium to oscillatory shear augments conversion of xanthine hydrogenase to XO (46).

The Mitochondria

Mitochondria make a significant contribution to overall ROS production in the endothelium. Mitochondria generate superoxide as a by-product of electron transport during ATP synthesis. Under normal physiological conditions, up to 2% of consumed molecular oxygen is converted to superoxide by the mitochondria. This fraction is increased by specific stimuli and under certain pathological conditions. Similar to the functional interaction between NADPH oxidase–derived ROS and eNOS uncoupling, there exists an amplification loop in which ROS generation in the mitochondria leads to NADPH oxidase activation (47).

Mitochondrial ROS generation is increased in response to many stimuli and has various effects. EC death, whether due to cell detachment or cytokines such as TNF-α, is associated with an increase in mitochondrial ROS generation (19,48). Both hypoxia and hyperoxia lead to an increase in mitochondrial-derived ROS in ECs (49,50). In addition, hypoxia followed by reoxygenation also increases the mitochondrial generation of ROS (28,51). Interestingly, mitochondrial ROS also play a part in the proinflammatory response of the endothelium to mechanical forces (52–54). With regards to known clinical cardiovascular risk factors, both diabetes and oxidized LDL are associated with an increase in mitochondrial ROS flux (55,56), whereas LDL, which is inversely related to cardiovascular risk, suppresses mitochondrial generation of ROS (57). Indeed, an increase in mitochondrial superoxide production due to hyperglycemia can lead to activation of proinflammatory and biochemical pathways associated with vascular complications of diabetes, and normalization of mitochondrial ROS production prevents hyperglycemic damage in vascular ECs (58). Finally, leptin, a satiety hormone reported to be elevated in patients with insulin resistance, leads to an increase in mitochondrial ROS generation, which mediates leptin-induced inflammatory response of ECs (59). Thus, it is becoming increasingly apparent that the mitochondria are a rich source of ROS in ECs, and that these ROS mediate primarily proinflammatory signals, particularly in response to metabolic abnormalities such as those seen in diabetes and insulin-resistance syndromes.

CYTOCHROME P450

The cytochrome P450 (CYP) family are flavoprotein enzymes that can oxidize a variety of substrates using NADPH and O_2 as cofactors and, in the process, also can produce ROS such as O_2^- and H_2O_2 (60). CYP2, a member of this family, is expressed in ECs and converts arachidonic acid to epoxyeicosatrienoic acids, which have vasodilatory properties (61). CYP2 also has been identified as a source of ROS in ECs (62). Because of the potential to generate both vasodilators and ROS, it is not entirely clear whether CYP2 plays a salutary or detrimental role in endothelial function. Nonetheless, CYP2 impairs endothelium-dependent vasorelaxation and promotes the expression of EC leukocyte adhesion molecules (62). The cholesterol-lowering statins reduce endothelial CYP2 activity (61), whereas oxidized LDL downregulates its expression (63).

Antioxidant Systems in Endothelial Cells

The endothelium is endowed with several antioxidant systems that play an important part in regulating not only the quantities of specific ROS, but also the compartmentalization of these ROS (see Figure 42.2). These antioxidant systems can be broadly divided into two categories.

Enzymatic Systems

Enzymatic systems within ECs that act on ROS include superoxide dismutases (SODs), glutathione peroxidases, catalase, heme oxygenase (HO), and the thioredoxin/thioredoxin reductase system. The predominant SOD in ECs is copper-zinc SOD (CuZnSOD or SOD1) (64). The importance of CuZnSOD in maintaining redox homeostasis in ECs is illustrated by the endothelial dysfunction seen with genetic knockout or pharmacologic inhibition of this enzyme (65,66). The physiologic role of SOD1 is evident by the fact that it is upregulated in response to laminar shear forces (67). SOD2 or MnSOD is primarily expressed in the mitochondria and is upregulated in response to physiologic agents such as VEGF (18). Finally, the heparan sulfate–bound extracellular SOD (ecSOD or SOD3), expressed on the EC surface, is also important in maintaining the vascular redox balance, because mice lacking this enzyme are prone to angiotensin II-induced hypertension (68).

The HO system consists of HO-2 and an inducible form (HO-1). HO-2 is constitutively expressed in ECs (69), whereas HO-1 is transcriptionally induced by hypoxia, NO, H_2O_2, heme, oxidized LDL (69), and other stimuli. Both enzymes, by generating bilirubin and biliverdin from heme, increase the antioxidant capacity of cells.

ECs also express catalase and GPX, both of which catalytically convert H_2O_2 to water. GPX utilizes the thiol glutathione (GSH) for this reaction, which in turn is oxidized to a disulfide form (GSSG). GSH then is regenerated from GSSG by glutathione reductase.

Thioredoxin (TRX), a ubiquitously expressed low-molecular-weight selenoprotein, possesses antioxidant properties. The thioredoxin/thioredoxin reductase system is an important antioxidant defense in ECs (70). TRX, by virtue of a critical dithiol motif (C-X-X-C) modulates the redox state of important cysteine residues in a number of proteins and transcription factors. For example, in ECs, TRX binds to and inactivates apoptosis signal-regulating kinase 1(ASK1) (71). TNF-α–induced ROS target the di-thiol motif of TRX, leading to dissociation of TRX from ASK1 and promoting EC apoptosis (71). The C-X-X-C motif also can act as a direct antioxidant. Levels of reduced TRX are replenished in the cell by TRX reductase.

Interestingly, the reductive activity of TRX also is regulated by NO, through the S-nitrosylation of a distinct cysteine residue (72). This modification may be an important mechanism for the antioxidative properties of the HMG-CoA reductase inhibitor (statin) class of clinical pharmaceuticals (73). TRX, because of its dithiol CXXC moiety, can act both as a direct antioxidant and also maintain thiol residues in many protein and transcription factors in its reduced form (70).

Nonenzymatic Systems

ECs possess a full armamentarium of chemicals that, through direct scavenging of ROS, participate in redox homeostasis. These include the aforementioned GSH, TRX, and heme byproducts, as well as vitamins A and C (74,75).

TISSUE- AND VASCULAR BED-SPECIFIC DIFFERENCES IN ENDOTHELIAL REACTIVE OXYGEN SPECIES

It is well recognized that ECs from different vascular beds have different properties (76). This difference also extends to ECs from macrovessels versus microvessels (77). However, a dearth of information exists about whether there exist differences in primary mechanisms of ROS generation, ROS elimination, or both in ECs from different tissues, ECs from venous versus arterial beds, and in ECs from large versus small vessels. Nevertheless, studies do show differences in the activity of endogenous enzymatic sources of ROS between ECs from macro- and microvessels, and tissue-specific differences in endothelial expression of receptors for ligands that induce ROS. For example, experimental evidence shows that cultured microvascular ECs have significantly higher NADPH oxidase activity than do human umbilical vein ECs (8), suggesting that mechanisms of ROS generation do differ in ECs of different origin and implying that such differences may contribute to the different properties of these cells. Similarly, expression of chemokine receptors on ECs that mediate chemokine-stimulated ROS production varies greatly between tissues (78).

THE ENDOTHELIAL RESPONSE TO REACTIVE OXYGEN SPECIES

Cultured ECs and the intact endothelium exhibit varied responses when challenged with ROS, whether they are derived from internal enzymatic sources or externally.

Impaired Endothelium-Dependent Relaxation

Perhaps the best-studied response of the endothelium to an increase in ROS flux is its ability to govern vasorelaxation. Endothelium-dependent agonists such as bradykinin and acetylcholine, changes in oxygen tension, and hemodynamic forces all affect vessel diameter and blood flow (79,80). This effect is mediated predominantly by eNOS-derived NO. Impairment of endothelium-dependent vasorelaxation is a striking manifestation of many vascular disorders including hypertension and atherosclerosis (81,82). An increase in ROS (in particular superoxide), derived from within the endothelium or without, by chemically reacting with endothelium-derived NO to form the secondary species peroxynitrite (which itself is highly reactive) leads to a decrease in its availability (83,84) and impaired endothelium-dependent vasorelaxation.

Inflammation

The endothelium can be transformed from an antiinflammatory surface to one that is proinflammatory. This change, a feature of atherosclerosis, is induced by a variety of inflammatory stimuli including cytokines such as TNF-α and IL-1β and oscillatory shear forces, and involves the expression of endothelial–leukocyte adhesion molecules such as vascular cell adhesion molecule (VCAM)-1 and intercellular adhesion molecule (ICAM)-1 (85,86). That antioxidants suppress the expression of these adhesion molecules on ECs suggests that ROS mediate this response.

Angiogenesis

Because of the important role of ROS in EC migration, proliferation, organization into primitive tube-like structures in vitro (87), as well as the expression of the angiogenic factor VEGF in tumor cells (88), it follows that ROS are likely to be involved in the in vivo growth of new blood vessels. Indeed, ROS, in particular those generated by NOX2, do mediate signaling and angiogenesis induced by VEGF (17,89). Because angiogenesis plays a prominent role in the development of atherothrombosis, ROS within ECs also may contribute to the progression of atheromatous lesions by this mechanism.

MOLECULAR TARGETS OF REACTIVE OXYGEN SPECIES WITHIN ENDOTHELIAL CELLS

ROS have long been known to target macromolecules such as lipids and nucleic acids. This "oxidative damage" has been well recognized as a marker – if not the cause – of cellular injury induced by ROS. However, as a role for ROS as "second messengers" has emerged in recent years, so has the understanding that they can target specific classes of intracellular proteins, many reversibly.

Tyrosine Kinases

The role of ROS as intracellular second messengers involved in receptor tyrosine kinase–mediated signaling was first shown in platelet-derived growth factor (PDGF)–treated VSMCs (90). A large body of evidence has shown that ROS serve a very similar role in other cell types by linking the membrane event of ligand-induced receptor tyrosine kinase activation to effectors of such receptors. The addition of H_2O_2 to ECs mimics ligand-induced receptor tyrosine kinase by leading to phosphorylation of the PDGF receptor (91). ROS, particularly those derived from NOX2, also participate in phosphorylation of the VEGF tyrosine kinase receptor KDR (17). Moreover, through activation of the epidermal growth factor (EGF) receptor tyrosine kinase, H_2O_2 also stimulates c-jun N-terminal kinases (JNKs) (92) and eNOS (93) in ECs. "Transactivation," or ligand-independent activation of receptor tyrosine kinases, now has been described to occur in response to a number of other stimuli (94).

Because transactivation of receptor tyrosine kinases in ECs by ROS involves tyrosine phosphorylation of these kinases, this prompted a search for tyrosine-kinase targets of ROS capable of phosphorylating receptor tyrosine kinases. Several studies suggest that such kinases may be, in fact, of the nonreceptor variety. For example, the Src tyrosine kinase is responsible for H_2O_2–induced tyrosine phosphorylation of the EGF receptor in ECs (92). In addition, endogenous ROS triggered in ECs subjected to fluid shear forces or cyclic strain stimulate activity of Src kinase and other Src-related tyrosine kinases, such as the proline-rich tyrosine kinase Pyk2 (95,96). Src kinase activation stimulated by VEGF, and in response to ligation of ICAM-1 in ECs, also is mediated by ROS (97,98). Finally, the addition of exogenous sources of ROS can activate p125 focal adhesion kinase (FAK) (99).

Tyrosine Phosphatases

One means by which ROS may lead to tyrosine phosphorylation is their likely effect on tyrosine phosphatases. Tyrosine phosphatases tonically inhibit tyrosine kinases, both of the receptor and nonreceptor variety. All tyrosine phosphatases share an active site that includes an oxidant-sensitive cysteine residue (100). Oxidation of this cysteine residue can occur in cells (101) and in vitro (102). This provides an attractive molecular mechanism by which H_2O_2 and other ROS can result in the tyrosine phosphorylation of a host of cellular proteins. Indeed, in ECs, chemical inhibition of tyrosine phosphatases mimics VEGF-induced signaling mediated by ROS (103). In concert with this observation, the overexpression of low-molecular-weight protein tyrosine phosphatases (LMW-PTP) in ECs enhances stimulated cell migration (104). Furthermore, the inhibition of the specific tyrosine phosphatases src homology-2–containing protein-tyrosine phosphatases-2 (SHP-2) prevents Src kinase activation induced by cross-linking of ICAM-1 (98). Consistent with this, SHP-2 is inactivated in ECs treated with H_2O_2 (105).

Serine/Threonine Kinases

The mitogen-activated protein kinase (MAPK) family is the largest family of serine/threonine kinases. This family consists of three kinase subfamilies: the extracellular-signal-regulated kinases (ERKs), the JNKs, and p38 mitogen-activated protein kinase (MAPK). All the MAPKs share a common T-X-Y motif, and their activation is generally dependent on phosphorylation of the threonine and tyrosine residues in this motif.

MAPKs are activated in ECs in response to a variety of stimuli and, in many cases, this activation depends on ROS. The inflammatory stimulus TNF-α requires ROS to stimulate p38MAPK-mediated IL-8 production in ECs (106). Furthermore, the direct addition of H_2O_2 stimulates p38 MAPK activity in ECs (107). JNKs also are activated by H_2O_2 (92), and vasculoprotective stimuli such as laminar flow suppress ROS-stimulated JNK activation (108). In contrast, certain anti-inflammatory exogenous stimuli, such as atrial natriuretic factor (ANP), also lead to an increase in intracellular ROS and JNK activity in ECs (109). Interestingly, this increase in JNK activity requires the induction of mitogen-activated protein kinase phosphatase (MKP)-1 (109), suggesting that a dephosphorylation event is required to trigger JNK activity, and that JNK has a role in the anti-inflammatory effect of ANP in ECs.

Finally, ERK activation in ECs subjected to various stimuli, including laminar flow (110) and cyclic strain (111), is mediated by ROS. ERK activation in ECs in response to H_2O_2 has been shown to lead to an increase in endothelial NO production by stimulating endothelial NOS activity (112), suggesting a counter-intuitive, yet important role of ROS in promoting endothelial function.

THERAPEUTIC IMPLICATIONS

With the central role that ROS play in endothelial pathophysiology, the potential is great for ameliorating endothelium-dependent manifestations of diseases such as hypertension and atherosclerosis by manipulating ROS. Unfortunately, this potential has been largely unrealized. Clinical trials using antioxidant vitamins in patients at risk of cardiovascular disease have shown no benefit in reducing cardiovascular events (113). This reality suggests that ROS play a much more complex role in endothelial biology than previously imagined. The myriad sources, functions, effects, and targets – only a fraction of which are known and reviewed here – and many of which may be disease-specific, attest to this. This complexity demands that therapies to manipulate endothelial ROS must be finely targeted at the molecular level in order to be successful.

CONCLUSION

The endothelium is uniquely poised both as a sensor and regulator of ROS production in the vascular wall. Endothelium from different vascular beds may have different capacities to achieve these functions, which may be suited to their own specific location.

KEY POINTS

- Endothelial ROS are produced by a number of intracellular enzymatic sources that are subject to regulation by many different environmental, chemical, and mechanical stimuli.
- Endothelial ROS levels also are determined by the endothelium's endogenous antioxidant defenses.
- Endothelial ROS can have various effects on endothelial function, both physiological and pathophysiological.
- The endothelium attempts to maintain redox homeostasis.
- Redox homeostasis is perturbed in many cardiovascular disease states.

REFERENCES

1 Babior BM, Lambeth JD, Nauseef W. The neutrophil NADPH oxidase. *Arch Biochem Biophys*. 2002;397(2):342–344.
2 Paget MS, Buttner MJ. Thiol-based regulatory switches. *Annu Rev Genet*. 2003;37:91–121.
3 Winterbourn CC, Vissers MC, Kettle AJ. Myeloperoxidase. *Curr Opin Hematol*. 2000;7(1):53–58.
4 Rey FE, Pagano PJ. The reactive adventitia: fibroblast oxidase in vascular function. *Arterioscler Thromb Vasc Biol*. 2002;22(12): 1962–1971.
5 Bayraktutan U, Blayney L, Shah AM. Molecular characterization and localization of the NAD(P)H oxidase components gp91-phox and p22-phox in endothelial cells. *Arterioscler Thromb Vasc Biol*. 2000;20(8):1903–1911.
6 Gorlach A, Brandes RP, Nguyen K, et al. A gp91phox containing NADPH oxidase selectively expressed in endothelial cells is a major source of oxygen radical generation in the arterial wall. *Circ Res*. 2000;87(1):26–32.
7 Jones SA, O'Donnell VB, Wood JD, et al. Expression of phagocyte NADPH oxidase components in human endothelial cells. *Am J Physiol*. 1996;271(4 Pt 2):H1626–H1634.
8 Li JM, Shah AM. Differential NADPH- versus NADH-dependent superoxide production by phagocyte-type endothelial cell NADPH oxidase. *Cardiovasc Res*. 2001;52(3):477–486.
9 Li JM, Shah AM. Intracellular localization and preassembly of the NADPH oxidase complex in cultured endothelial cells. *J Biol Chem*. 2002;277(22):19952–19960.
10 Li JM, Shah AM. Mechanism of endothelial cell NADPH oxidase activation by angiotensin II. Role of the p47phox subunit. *J Biol Chem*. 2003;278(14):12094–12100.
11 Li JM, Mullen AM, Yun S, et al. Essential role of the NADPH oxidase subunit p47(phox) in endothelial cell superoxide

production in response to phorbol ester and tumor necrosis factor-alpha. *Circ Res.* 2002;90(2):143–150.

12 Cheng G, Cao Z, Xu X, et al. Homologs of gp91phox: cloning and tissue expression of Nox3, Nox4, and Nox5. *Gene.* 2001;269 (1–2):131–140.

13 Lassegue B, Sorescu D, Szocs K, et al. Novel gp91(phox) homologues in vascular smooth muscle cells: nox1 mediates angiotensin II-induced superoxide formation and redox-sensitive signaling pathways. *Circ Res.* 2001;88(9):888–894.

14 Suh YA, Arnold RS, Lassegue B, et al. Cell transformation by the superoxide-generating oxidase Mox1. *Nature.* 1999;401 (6748):79–82.

15 Sorescu D, Weiss D, Lassegue B, et al. Superoxide production and expression of nox family proteins in human atherosclerosis. *Circulation.* 2002;105(12):1429–1435.

16 Colavitti R, Pani G, Bedogni B, et al. Reactive oxygen species as downstream mediators of angiogenic signaling by vascular endothelial growth factor receptor-2/KDR. *J Biol Chem.* 2002; 277(5):3101–3108.

17 Ushio-Fukai M, Tang Y, Fukai T, et al. Novel role of gp91(phox)-containing NAD(P)H oxidase in vascular endothelial growth factor-induced signaling and angiogenesis. *Circ Res.* 2002;91(12):1160–1167.

18 Abid MR, Tsai JC, Spokes KC, et al. Vascular endothelial growth factor induces manganese-superoxide dismutase expression in endothelial cells by a Rac1-regulated NADPH oxidase-dependent mechanism. *FASEB J.* 2001;15(13):2548–2550.

19 Deshpande SS, Angkeow P, Huang J, et al. Rac1 inhibits TNF-alpha-induced endothelial cell apoptosis: dual regulation by reactive oxygen species. *FASEB J.* 2000;14(12):1705–1714.

20 Gu Y, Xu YC, Wu RF, et al. p47phox participates in activation of RelA in endothelial cells. *J Biol Chem.* 2003;278(19):17210–17217.

21 Duerrschmidt N, Wippich N, Goettsch W, et al. Endothelin-1 induces NAD(P)H oxidase in human endothelial cells. *Biochem Biophys Res Commun.* 2000;269(3):713–717.

22 Kashiwagi A, Shinozaki K, Nishio Y, et al. Endothelium-specific activation of NAD(P)H oxidase in aortas of exogenously hyperinsulinemic rats. *Am J Physiol.* 1999;277(6 Pt 1):E976–E983.

23 Inoguchi T, Li P, Umeda F, et al. High glucose level and free fatty acid stimulate reactive oxygen species production through protein kinase C – dependent activation of NAD(P)H oxidase in cultured vascular cells. *Diabetes.* 2000;49(11):1939–1945.

24 Wautier MP, Chappey O, Corda S, et al. Activation of NADPH oxidase by AGE links oxidant stress to altered gene expression via RAGE. *Am J Physiol Endocrinol Metab.* 2001;280(5):E685–E694.

25 Rueckschloss U, Galle J, Holtz J, et al. Induction of NAD(P)H oxidase by oxidized low-density lipoprotein in human endothelial cells: antioxidative potential of hydroxymethylglutaryl coenzyme A reductase inhibitor therapy. *Circulation.* 2001;104 (15):1767–1772.

26 Stokes KY, Clanton EC, Russell JM, et al. NAD(P)H oxidase-derived superoxide mediates hypercholesterolemia-induced leukocyte-endothelial cell adhesion. *Circ Res.* 2001;88(5):499–505.

27 Al-Mehdi AB, Zhao G, Dodia C, et al. Endothelial NADPH oxidase as the source of oxidants in lungs exposed to ischemia or high K^+. *Circ Res.* 1998;83(7):730–737.

28 Kim KS, Takeda K, Sethi R, et al. Protection from reoxygenation injury by inhibition of rac1. *J Clin Invest.* 1998;101(9):1821–1826.

29 Lopes NH, Vasudevan SS, Gregg D, et al. Rac-dependent monocyte chemoattractant protein-1 production is induced by nutrient deprivation. *Circ Res.* 2002;91(9):798–805.

30 Hwang J, Ing MH, Salazar A, et al. Pulsatile versus oscillatory shear stress regulates NADPH oxidase subunit expression: implication for native LDL oxidation. *Circ Res.* 2003;93(12): 1225–1232.

31 Vasquez-Vivar J, Kalyanaraman B, Martasek P, et al. Superoxide generation by endothelial nitric oxide synthase: the influence of cofactors. *Proc Natl Acad Sci USA.* 1998;95(16):9220–9225.

32 Xia Y, Tsai AL, Berka V, et al. Superoxide generation from endothelial nitric-oxide synthase. A Ca^{2+}/calmodulin-dependent and tetrahydrobiopterin regulatory process. *J Biol Chem.* 1998;273(40):25804–25808.

33 Bec N, Gorren AC, Voelker C, et al. Reaction of neuronal nitric-oxide synthase with oxygen at low temperature. Evidence for reductive activation of the oxy-ferrous complex by tetrahydrobiopterin. *J Biol Chem.* 1998;273(22):13502–13508.

34 Gorren AC, Bec N, Schrammel A, et al. Low-temperature optical absorption spectra suggest a redox role for tetrahydrobiopterin in both steps of nitric oxide synthase catalysis. *Biochemistry.* 2000;39(38):11763–11770.

35 Laursen JB, Rajagopalan S, Galis Z, et al. Role of superoxide in angiotensin II-induced but not catecholamine-induced hypertension. *Circulation.* 1997;95(3):588–593.

36 Landmesser U, Dikalov S, Price SR, et al. Oxidation of tetrahydrobiopterin leads to uncoupling of endothelial cell nitric oxide synthase in hypertension. *J Clin Invest.* 2003;111(8):1201–1209.

37 Pritchard KA, Jr., Groszek L, Smalley DM, et al. Native low-density lipoprotein increases endothelial cell nitric oxide synthase generation of superoxide anion. *Circ Res.* 1995;77(3):510–518.

38 Hink U, Li H, Mollnau H, et al. Mechanisms underlying endothelial dysfunction in diabetes mellitus. *Circ Res.* 2001;88(2): E14–E22.

39 Harrison R. Structure and function of xanthine oxidoreductase: where are we now? *Free Radic Biol Med.* 2002;33(6):774–797.

40 Dupont GP, Huecksteadt TP, Marshall BC, et al. Regulation of xanthine dehydrogenase and xanthine oxidase activity and gene expression in cultured rat pulmonary endothelial cells. *J Clin Invest.* 1992;89(1):197–202.

41 Wakabayashi Y, Fujita H, Morita I, et al. Conversion of xanthine dehydrogenase to xanthine oxidase in bovine carotid artery endothelial cells induced by activated neutrophils: involvement of adhesion molecules. *Biochim Biophys Acta.* 1995;1265(2–3): 103–109.

42 Cardillo C, Kilcoyne CM, Cannon RO III, et al. Xanthine oxidase inhibition with oxypurinol improves endothelial vasodilator function in hypercholesterolemic but not in hypertensive patients. *Hypertension.* 1997;30(1 Pt 1):57–63.

43 Ohara Y, Peterson TE, Harrison DG. Hypercholesterolemia increases endothelial superoxide anion production. *J Clin Invest.* 1993;91(6):2546–2551.

44 Guthikonda S, Sinkey C, Barenz T, et al. Xanthine oxidase inhibition reverses endothelial dysfunction in heavy smokers. *Circulation.* 2003;107(3):416–421.

45 Spiekermann S, Landmesser U, Dikalov S, et al. Electron spin resonance characterization of vascular xanthine and NAD(P)H oxidase activity in patients with coronary artery disease: relation to endothelium-dependent vasodilation. *Circulation.* 2003;107 (10):1383–1389.

46 McNally JS, Davis ME, Giddens DP, et al. Role of xanthine oxidoreductase and NAD(P)H oxidase in endothelial superoxide production in response to oscillatory shear stress. *Am J Physiol Heart Circ Physiol.* 2003;285(6):H2290–H2297.

47 Schafer M, Schafer C, Ewald N, et al. Role of redox signaling in the autonomous proliferative response of endothelial cells to hypoxia. *Circ Res.* 2003;92(9):1010–1015.

48 Li AE, Ito H, Rovira II, et al. A role for reactive oxygen species in endothelial cell anoikis. *Circ Res.* 1999;85(4):304–310.

49 Pagano A, Barazzone-Argiroffo C. Alveolar cell death in hyperoxia-induced lung injury. *Ann NY Acad Sci.* 2003;1010: 405–416.

50 Pearlstein DP, Ali MH, Mungai PT, et al. Role of mitochondrial oxidant generation in endothelial cell responses to hypoxia. *Arterioscler Thromb Vasc Biol.* 2002;22(4):566–573.

51 Therade-Matharan S, Laemmel E, Duranteau J, et al. Reoxygenation after hypoxia and glucose depletion causes reactive oxygen species production by mitochondria in HUVEC. *Am J Physiol Regul Integr Comp Physiol.* 2004;287(5):R1037–R1043.

52 Ali MH, Pearlstein DP, Mathieu CE, et al. Mitochondrial requirement for endothelial responses to cyclic strain: implications for mechanotransduction. *Am J Physiol Lung Cell Mol Physiol.* 2004;287(3):L486–L496.

53 Ichimura H, Parthasarathi K, Quadri S, et al. Mechano-oxidative coupling by mitochondria induces proinflammatory responses in lung venular capillaries. *J Clin Invest.* 2003;111(5):691–699.

54 Parthasarathi K, Ichimura H, Quadri S, et al. Mitochondrial reactive oxygen species regulate spatial profile of proinflammatory responses in lung venular capillaries. *J Immunol.* 2002; 169(12):7078–7086.

55 Srinivasan S, Yeh M, Danziger EC, et al. Glucose regulates monocyte adhesion through endothelial production of interleukin-8. *Circ Res.* 2003;92(4):371–377.

56 Zmijewski JW, Moellering DR, Goffe CL, et al. Oxidized LDL induces mitochondrially associated reactive oxygen/nitrogen species formation in endothelial cells. *Am J Physiol Heart Circ Physiol.* 2005;289(2):H852–H861.

57 Nofer JR, Levkau B, Wolinska I, et al. Suppression of endothelial cell apoptosis by high density lipoproteins (HDL) and HDL-associated lysosphingolipids. *J Biol Chem.* 2001;276(37):34480–34485.

58 Nishikawa T, Edelstein D, Du XL, et al. Normalizing mitochondrial superoxide production blocks three pathways of hyperglycaemic damage. *Nature.* 2000;404(6779):787–790.

59 Yamagishi SI, Edelstein D, Du XL, et al. Leptin induces mitochondrial superoxide production and monocyte chemoattractant protein-1 expression in aortic endothelial cells by increasing fatty acid oxidation via protein kinase A. *J Biol Chem.* 2001;276(27):25096–25100.

60 Fleming I. Cytochrome p450 and vascular homeostasis. *Circ Res.* 2001;89(9):753–762.

61 Fisslthaler B, Michaelis UR, Randriamboavonjy V, et al. Cytochrome P450 epoxygenases and vascular tone: novel role for HMG-CoA reductase inhibitors in the regulation of CYP 2C expression. *Biochim Biophys Acta.* 2003;1619(3):332–339.

62 Fleming I, Michaelis UR, Bredenkotter D, et al. Endothelium-derived hyperpolarizing factor synthase (Cytochrome P450 2C9) is a functionally significant source of reactive oxygen species in coronary arteries. *Circ Res.* 2001;88(1):44–51.

63 Thum T, Borlak J. Mechanistic role of cytochrome P450 monooxygenases in oxidized low-density lipoprotein-induced vascular injury: therapy through LOX-1 receptor antagonism? *Circ Res.* 2004;94(1):E1–E13.

64 Cai H, Harrison DG. Endothelial dysfunction in cardiovascular diseases: the role of oxidant stress. *Circ Res.* 2000;87(10):840–844.

65 Didion SP, Ryan MJ, Didion LA, et al. Increased superoxide and vascular dysfunction in CuZnSOD-deficient mice. *Circ Res.* 2002;91(10):938–944.

66 Wambi-Kiesse CO, Katusic ZS. Inhibition of copper/zinc superoxide dismutase impairs NO.-mediated endothelium-dependent relaxations. *Am J Physiol.* 1999;276(3 Pt 2):H1043–H1048.

67 Inoue N, Ramasamy S, Fukai T, et al. Shear stress modulates expression of Cu/Zn superoxide dismutase in human aortic endothelial cells. *Circ Res.* 1996;79(1):32–37.

68 Jung O, Marklund SL, Geiger H, et al. Extracellular superoxide dismutase is a major determinant of nitric oxide bioavailability: in vivo and ex vivo evidence from ecSOD-deficient mice. *Circ Res.* 2003;93(7):622–629.

69 Perrella MA, Yet SF. Role of heme oxygenase-1 in cardiovascular function. *Curr Pharm Des.* 2003;9(30):2479–2487.

70 Yamawaki H, Haendeler J, Berk BC. Thioredoxin: a key regulator of cardiovascular homeostasis. *Circ Res.* 2003;93(11):1029–1033.

71 Liu Y, Min W. Thioredoxin promotes ASK1 ubiquitination and degradation to inhibit ASK1-mediated apoptosis in a redox activity-independent manner. *Circ Res.* 2002;90(12):1259–1266.

72 Haendeler J, Hoffmann J, Tischler V, et al. Redox regulatory and anti-apoptotic functions of thioredoxin depend on S-nitrosylation at cysteine 69. *Nat Cell Biol.* 2002;4(10):743–749.

73 Haendeler J, Hoffmann J, Zeiher AM, et al. Antioxidant effects of statins via S-nitrosylation and activation of thioredoxin in endothelial cells: a novel vasculoprotective function of statins. *Circulation.* 2004;110(7):856–861.

74 Dickinson DA, Forman HJ. Cellular glutathione and thiols metabolism. *Biochem Pharmacol.* 2002;64(5–6):1019–1026.

75 Droge W. Free radicals in the physiological control of cell function. *Physiol Rev.* 2002;82(1):47–95.

76 Chi JT, Chang HY, Haraldsen G, et al. Endothelial cell diversity revealed by global expression profiling. *Proc Natl Acad Sci USA.* 2003;100(19):10623–10628.

77 King J, Hamil T, Creighton J, et al. Structural and functional characteristics of lung macro- and microvascular endothelial cell phenotypes. *Microvasc Res.* 2004;67(2):139–151.

78 Hillyer P, Mordelet E, Flynn G, et al. Chemokines, chemokine receptors and adhesion molecules on different human endothelia: discriminating the tissue-specific functions that affect leucocyte migration. *Clin Exp Immunol.* 2003;134(3):431–441.

79 Ignarro LJ, Buga GM, Wood KS, et al. Endothelium-derived relaxing factor produced and released from artery and vein is nitric oxide. *Proc Natl Acad Sci USA.* 1987;84(24):9265–9269.

80 Perrella MA, Edell ES, Krowka MJ, et al. Endothelium-derived relaxing factor in pulmonary and renal circulations during hypoxia. *Am J Physiol.* 1992;263(1 Pt 2):R45–R50.

81 Reddy KG, Nair RN, Sheehan HM, et al. Evidence that selective endothelial dysfunction may occur in the absence of angiographic or ultrasound atherosclerosis in patients with risk factors for atherosclerosis. *J Am Coll Cardiol.* 1994;23(4):833–843.

82 Werns SW, Walton JA, Hsia HH, et al. Evidence of endothelial dysfunction in angiographically normal coronary arteries of patients with coronary artery disease. *Circulation.* 1989;79(2): 287–291.
83 Munzel T, Sayegh H, Freeman BA, et al. Evidence for enhanced vascular superoxide anion production in nitrate tolerance. A novel mechanism underlying tolerance and cross-tolerance. *J Clin Invest.* 1995;95(1):187–194.
84 Rajagopalan S, Kurz S, Munzel T, et al. Angiotensin II-mediated hypertension in the rat increases vascular superoxide production via membrane NADH/NADPH oxidase activation. Contribution to alterations of vasomotor tone. *J Clin Invest.* 1996;97 (8):1916–1923.
85 Chappell DC, Varner SE, Nerem RM, et al. Oscillatory shear stress stimulates adhesion molecule expression in cultured human endothelium. *Circ Res.* 1998;82(5):532–539.
86 Marui N, Offermann MK, Swerlick R, et al. Vascular cell adhesion molecule-1 (VCAM-1) gene transcription and expression are regulated through an antioxidant-sensitive mechanism in human vascular endothelial cells. *J Clin Invest.* 1993;92(4): 1866–1874.
87 Maulik N, Das DK. Redox signaling in vascular angiogenesis. *Free Radic Biol Med.* 2002;33(8):1047–1060.
88 Arbiser JL, Petros J, Klafter R, et al. Reactive oxygen generated by Nox1 triggers the angiogenic switch. *Proc Natl Acad Sci USA.* 2002;99(2):715–720.
89 Kuroki M, Voest EE, Amano S, et al. Reactive oxygen intermediates increase vascular endothelial growth factor expression in vitro and in vivo. *J Clin Invest.* 1996;98(7):1667–1675.
90 Sundaresan M, Yu ZX, Ferrans VJ, et al. Requirement for generation of H_2O_2 for platelet-derived growth factor signal transduction. *Science.* 1995;270(5234):296–299.
91 Chen K, Albano A, Ho A, et al. Activation of p53 by oxidative stress involves platelet-derived growth factor-beta receptor-mediated ataxia telangiectasia mutated (ATM) kinase activation. *J Biol Chem.* 2003;278(41):39527–39533.
92 Chen K, Vita JA, Berk BC, et al. c-Jun N-terminal kinase activation by hydrogen peroxide in endothelial cells involves SRC-dependent epidermal growth factor receptor transactivation. *J Biol Chem.* 2001;276(19):16045–16050.
93 Thomas SR, Chen K, Keaney JF Jr. Hydrogen peroxide activates endothelial nitric-oxide synthase through coordinated phosphorylation and dephosphorylation via a phosphoinositide 3-kinase-dependent signaling pathway. *J Biol Chem.* 2002;277(8): 6017–6024.
94 Zwick E, Hackel PO, Prenzel N, et al. The EGF receptor as central transducer of heterologous signalling systems. *Trends Pharmacol Sci.* 1999;20(10):408–412.
95 Cheng JJ, Chao YJ, Wang DL. Cyclic strain activates redox-sensitive proline-rich tyrosine kinase 2 (PYK2) in endothelial cells. *J Biol Chem.* 2002;277(50):48152–48157.
96 Tai LK, Okuda M, Abe J, et al. Fluid shear stress activates proline-rich tyrosine kinase via reactive oxygen species-dependent pathway. *Arterioscler Thromb Vasc Biol.* 2002;22(11):1790–1796.
97 Lin MT, Yen ML, Lin CY, et al. Inhibition of vascular endothelial growth factor-induced angiogenesis by resveratrol through interruption of Src-dependent vascular endothelial cadherin tyrosine phosphorylation. *Mol Pharmacol.* 2003;64(5):1029–1036.
98 Wang Q, Pfeiffer GR II, Gaarde WA. Activation of SRC tyrosine kinases in response to ICAM-1 ligation in pulmonary microvascular endothelial cells. *J Biol Chem.* 2003;278(48):47731–47743.
99 Gozin A, Franzini E, Andrieu V, et al. Reactive oxygen species activate focal adhesion kinase, paxillin and p130cas tyrosine phosphorylation in endothelial cells. *Free Radic Biol Med.* 1998; 25(9):1021–1032.
100 Barford D, Jia Z, Tonks NK. Protein tyrosine phosphatases take off. *Nat Struct Biol.* 1995;2(12):1043–1053.
101 Lee SR, Kwon KS, Kim SR, et al. Reversible inactivation of protein-tyrosine phosphatase 1B in A431 cells stimulated with epidermal growth factor. *J Biol Chem.* 1998;273(25):15366–15372.
102 Caselli A, Marzocchini R, Camici G, et al. The inactivation mechanism of low molecular weight phosphotyrosine-protein phosphatase by H_2O_2. *J Biol Chem.* 1998;273(49):32554–32560.
103 Tao Q, Spring SC, Terman BI. Comparison of the signaling mechanisms by which VEGF, H_2O_2, and phosphatase inhibitors activate endothelial cell ERK1/2 MAP-kinase. *Microvasc Res.* 2005;69(1–2):36–44.
104 Shimizu H, Toyama O, Shiota M, et al. Protein tyrosine phosphatase LMW-PTP exhibits distinct roles between vascular endothelial and smooth muscle cells. *J Recept Signal Transduct Res.* 2005;25(1):19–33.
105 Maas M, Wang R, Paddock C, et al. Reactive oxygen species induce reversible PECAM-1 tyrosine phosphorylation and SHP-2 binding. *Am J Physiol Heart Circ Physiol.* 2003;285(6): H2336–H2344.
106 Hashimoto S, Gon Y, Matsumoto K, et al. N-acetylcysteine attenuates TNF-alpha-induced p38 MAP kinase activation and p38 MAP kinase-mediated IL-8 production by human pulmonary vascular endothelial cells. *Br J Pharmacol.* 2001;132(1): 270–276.
107 Huot J, Houle F, Marceau F, et al. Oxidative stress-induced actin reorganization mediated by the p38 mitogen-activated protein kinase/heat shock protein 27 pathway in vascular endothelial cells. *Circ Res.* 1997;80(3):383–392.
108 Hojo Y, Saito Y, Tanimoto T, et al. Fluid shear stress attenuates hydrogen peroxide-induced c-Jun NH2-terminal kinase activation via a glutathione reductase-mediated mechanism. *Circ Res.* 2002;91(8):712–718.
109 Furst R, Brueckl C, Kuebler WM, et al. Atrial natriuretic peptide induces mitogen-activated protein kinase phosphatase-1 in human endothelial cells via Rac1 and NAD(P)H oxidase/Nox2-activation. *Circ Res.* 2005;96(1):43–53.
110 Yeh LH, Park YJ, Hansalia RJ, et al. Shear-induced tyrosine phosphorylation in endothelial cells requires Rac1-dependent production of ROS. *Am J Physiol.* 1999;276(4 Pt 1):C838–C847.
111 Cheng JJ, Chao YJ, Wung BS, et al. Cyclic strain-induced plasminogen activator inhibitor-1 (PAI-1) release from endothelial cells involves reactive oxygen species. *Biochem Biophys Res Commun.* 1996;225(1):100–105.
112 Cai H, Li Z, Davis ME, et al. Akt-dependent phosphorylation of serine 1179 and mitogen-activated protein kinase kinase/extracellular signal-regulated kinase 1/2 cooperatively mediate activation of the endothelial nitric-oxide synthase by hydrogen peroxide. *Mol Pharmacol.* 2003;63(2):325–331.
113 Jialal I, Devaraj S. Antioxidants and atherosclerosis: don't throw out the baby with the bath water. *Circulation.* 2003;107(7):926–928.

43

Extracellular Nucleotides and Nucleosides as Autocrine and Paracrine Regulators within the Vasculature

Silvia Deaglio and Simon C. Robson

Beth Israel Deaconess Medical Center, Harvard Medical School, Boston, Massachusetts

The normal vascular endothelium provides a barrier that separates blood cells and plasma factors from highly reactive elements of the deeper layer of the vessel wall. It maintains blood fluidity and flow by inhibiting coagulation and platelet activation and promoting fibrinolysis (1). These properties are governed by important, specific thromboregulatory mechanisms that include the release of prostacyclin (2), the generation of nitric oxide (NO) (3), and the expression of heparan sulfate (4), together with the expression of natural anticoagulants such as tissue factor pathway inhibitor or thrombomodulin and fibrinolytic mechanisms involving tissue plasminogen activator (1). These properties of the endothelium and antithrombotic pathways are addressed elsewhere in this volume. This chapter focuses on those key biological activities of the vasculature that have been identified recently and shown to be ectonucleotide catalysts that generate the respective nucleosides through phosphohydrolysis (5,6).

Extracellular nucleotides (e.g., ATP, ADP, UTP, and UDP) are released by leukocytes, platelets, and endothelial cells (ECs) in the blood, where they serve as extracellular signals (7). These mediators bind the multiple type-2 purinergic/pyrimidinergic (P2Y1–14 and P2X1–7) receptors on platelets, endothelium, vascular smooth muscle cells (VSMCs), and leukocytes (8). The 15 defined and characterized P2-receptors of the P2Y and P2X families have different specificities and have been shown to transmit signals from extracellular nucleotides, as discussed later. These receptors trigger and mediate short-term (acute) processes affecting cellular metabolism, NO release, adhesion, activation, and migration together with other more protracted developmental responses, such as cell proliferation, differentiation, and apoptosis (9–11).

These P2-receptor–mediated effects are closely modulated by ectoenzymes, termed *ectonucleotidases* (e.g., ecto-ADPases, ecto-ATPases), which bind and then hydrolyze extracellular nucleotides, ultimately to the respective nucleosides (that in turn activate a series of adenosine or P1 receptors) (6,11). The dominant ectoenzymes or ectonucleotidases of the vasculature are characterized more fully as the ectonucleoside triphosphate diphosphohydrolases (E-NTPDases) of the CD39 family. This important biological property expressed by the endothelium and associated cells is responsible for the regulation of extracellular levels of nucleotides (5,12–15).

Extracellular nucleotide stimulation of purinergic/pyrimidinergic (P2) receptors is associated with activation of platelets, leukocytes, and ECs, and may culminate in vascular thrombosis and inflammation in vivo. This chapter considers extracellular nucleotide or purinergic/pyrimidinergic signaling and its associated nucleoside-mediated effects, and chiefly addresses the role of the ectonucleotidases of the vascular endothelium and accessory cells in modulating these signals (6).

HISTORICAL PERSPECTIVES

Over 70 years ago, Drury and Szent-Györgyi (16) first described the circulatory effects of injected purine nucleotides and nucleosides – ATP and adenosine – on the cardiovascular system. The concept of a purinergic signaling system, using extracellular nucleotides and nucleosides as mediators, was introduced by Burnstock approximately 30 years ago, as an initial hypothesis dealing with neurotransmission (17). Subsequent work detailed how extracellular nucleotides could serve as extracellular messengers that modulate exocrine and endocrine systems, the vasculature and hemostatic mechanisms, and musculoskeletal, immune, and inflammatory cells, as reviewed in Burnstock and Knight (11).

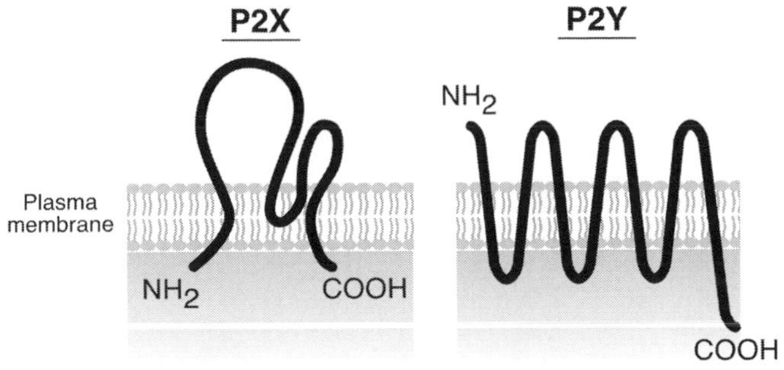

Figure 43.1. The two families of membrane-bound P2-type receptors that recognize extracellular nucleotides. Representative elements of the two families of nucleotide type-P2 receptors are depicted here. P2X are "rapid" ligand-gated ion channels permeable to Na^+, K^+, and also Ca^{2+} (subtypes P2X1–7). P2Y are the "slow" metabotropic receptors (P2Y1–14) and are seven-transmembrane G_q- or G_i-protein linked receptors.

Initially, this concept was controversial, in that it was unclear how such commonplace metabolites and ubiquitous intracellular components could be specifically involved in selective extracellular signaling. However, given the protean roles of nucleotides in the intracellular compartment, it seemed plausible that these chemicals also could function as extracellular mediators. Theoretically, these pathways also could be employed to provide nucleosides and other substrates for intracellular biochemical salvage pathways, which would provide for the rapid resynthesis of nucleotides and avoid redundancy of nucleoside biosynthetic pathways in endothelium. Such a dual role would require the origination of interrelated proteins to shuttle these purine and pyrimidine compounds between the extra- and intracellular compartments and to provide for specific recognition and ultimate clearance. Furthermore, nucleotide-mediated cellular signals might well be coordinated in a manner analogous to that described for the acetylcholinesterase system, as would occur during neurotransmission (18). Hence, in purinergic responses, there might well have evolved regulatory steps at the level of nucleotide release, receptor expression and/or desensitization, and phosphohydrolysis of the nucleotide mediators to the specific derivatives.

Indeed, subsequent studies demonstrated that each of the following factors is involved in mediating the specificity of purinergic/pyrimidinergic signaling: (a) the derivation or source of the extracellular nucleotides (7,8,19); (b) the expression of specific receptors for these molecular transmitters (or the nucleotide and nucleoside derivatives) (20–23){8521}[1] (Figure 43.1); and (c) the existence of ectonucleotidases that dictate the cellular responses by hydrolyzing the nucleotides (to nucleosides) (6,15,24).

1 Molecular Recognition Section of National Institutes of Health http://mgddk1.niddk. nih.gov: 8000/nomenclature.html.

Table 43-1: Biochemical Names of Bases, Nucleosides, and Nucleotides

	Base	*Nucleosides (= Base + Pentose)*	*Nucleotides (= Nucleoside + Phosphates)*
Purines	Adenine	Adenosine	AMP, ADP, ATP
	Guanine	Guanosine	GMP, GDP, GTP
Pyrimidines	Cytosine	Cytidine	CMP, CDP, CTP
	Thymine	Thymidine	TMP, TDP, TTP
	Uracil	Uridine	UMP, UDP, UTP

NUCLEOSIDES AND NUCLEOTIDES AS PRIMORDIAL SIGNALING MOLECULES AND MEDIATORS: BIOCHEMISTRY AND EVOLUTIONARY RELEVANCE

Nucleosides are glycosylamines made by attaching a nucleobase to a pentose sugar ring. Examples of these include cytidine, uridine, adenosine, guanosine, thymidine, and inosine (Table 43-1). Nucleosides can be phosphorylated by specific kinases in the cell and produce nucleotides. Typically, nucleotides are considered the monomeric, structural unit of those nucleotide chains that form nucleic acids (RNA and DNA). Nucleotides also play important roles in cellular energy transport and transformation (nucleoside triphosphates such as ATP are the energy-rich end-products of the majority of biochemical energy-releasing pathways), in enzyme regulation, and as intracellular second messengers.

Cellular injury causes a release of nucleosides and nucleotides. Hence, it is possible that purinergic/pyrimidinergic signaling evolved as a sensor for environmental stresses and that later, receptors for neuroactive agonists in more advanced vertebrates may have evolved from external chemoreceptors. In this respect, organization of the purinergic/pyrimidinergic system presumably involved a stepwise evolution of new gene families, gene duplication, and conservation of adaptive genes to facilitate the specificity of nucleotide metabolism and responses. This would involve the development of the many P2 and adenosine P1 receptors, and the various ectonucleotidases such as diadenosine polyphosphate hydrolases, the CD39/NTPDase family, nucleotide pyrophosphatases/phosphodiesterases (E-NPP), 5′-ectonucleotidase, alkaline phosphatases, NAD-glycohydrolase, CD38/NAD and nucleosidase activity, and adenosine deaminase.

NTPDases, in particular, might be able to serve as molecular switches at the membrane, each having differing functions according to the context of the signal. Such complexities must have been dictated by the ubiquitous nature of the signaling molecules (e.g., ATP). This situation concerning extracellular nucleotide signaling can be suitably contrasted with the

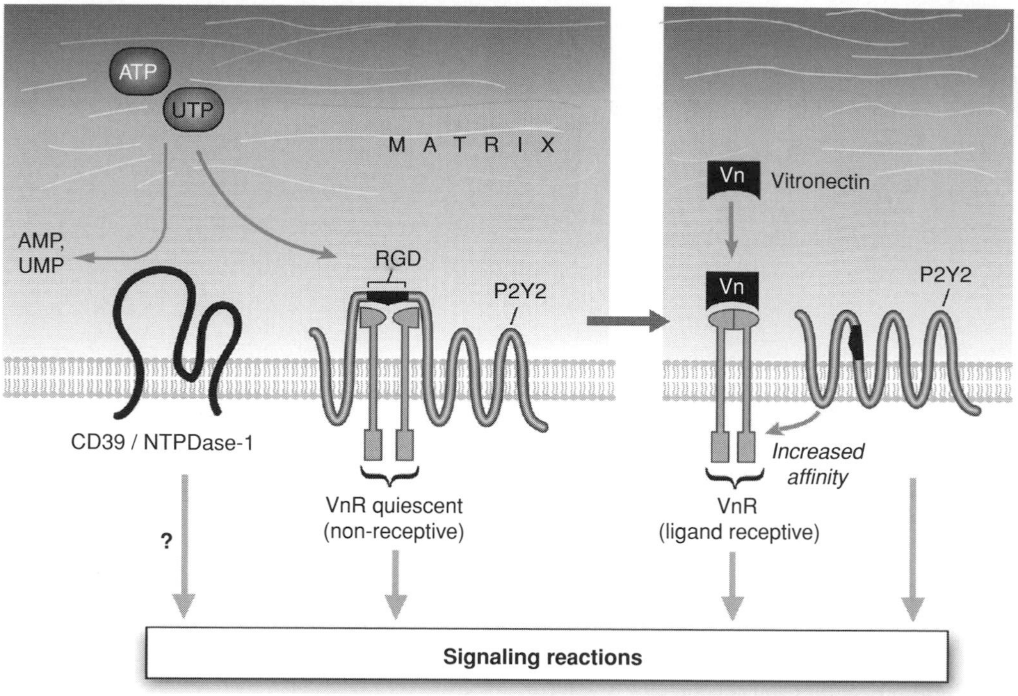

Figure 43.2. Proposed and speculative mechanisms for nucleotide-mediated signaling and linked integrin activation. The P2Y2 nucleotide receptor contains the integrin-binding domain arginine-glycine-aspartic acid (RGD) in the first extracellular loop, suggesting that this G-protein–coupled receptor might interact directly with the vitronectin receptor, the $\alpha v\beta 3$ integrin (*1*). Interactions between P2Y2 and endothelial integrins (e.g., $\alpha v\beta 3$ and the ectonucleotidase CD39/NTPDase1) with other membrane proteins in a macromolecular complex may be important for coordinating P2Y2-mediated responses triggered by nucleotides (*2*), as discussed in the text. Specifically, nucleotides can induce expression of the $\alpha v\beta 3$ integrin, and positively influence avidity and affinity of the receptor with matrix proteins such as vitronectin in vascular endothelial matrix tissues via interactions with the P2Y2 receptor, as modulated by CD39/NTPDase1. Ligation of the integrins by vitronectin and other matrix proteins results in "outside-in" signaling and post-ligand binding alteration (*3*). Whether any potential direct signaling components of CD39 modulate these cellular activation responses remains speculative at this time.

unique specificity of peptide hormones or vasoactive factors, which often are designated for single, defined receptors (25,26).

Thus, the potential of these and other surface-located enzymes involved with nucleotide metabolism would also extend to ADP-ribosylation and phosphorylation of extracellular proteins (27).

EXTRACELLULAR NUCLEOTIDES IN THE VASCULATURE

The metabolic roles of nucleotides are well established. Over the past decade, extracellular nucleotides have been increasingly recognized as important mediators of a variety of processes, including vascular inflammation and thrombosis, with varying impacts in different experimental systems and models of human disease (6). At least in part, vascular biology can be understood by considering endothelial and accessory cells as having a certain degree of plasticity and behaving as adaptive input-output devices. Input arises from the extracellular milieu and may include biochemical signals triggered by extracellular nucleotides (or nucleosides) and the resulting biomechanical responses that are transduced by adhesion receptors in response to P2-mediated signals, such as the affinity changes in integrins (Figure 43.2). The output is manifested as alterations in cellular phenotype, such as in activation responses, and includes a number of structural and functional changes implicated in diverse pathophysiological processes and thrombotic disease.

Adenosine- and ATP-mediated effects have been implicated in the local control of vessel tone as well as in vascular cell migration, proliferation, and differentiation. As an example, ATP may be released from sympathetic nerves and result in a constriction of VSMCs through effects mediated by P2X receptors. In contrast, ATP released from ECs during changes in flow (shear stress) or following exposure to hypoxic conditions activates the P2Y receptors expressed by these cells as well as VSMCs. This activation occurs in an autocrine and paracrine manner and releases NO, resulting in vessel relaxation as a purinergic event. Any nucleotide released will be ultimately hydrolyzed to adenosine and will result in vasodilatation via the effects of VSMC adenosine P1 receptors. P2

Table 43-2: Purinergic Receptors Expressed by the Vasculature

Receptor	*Agonist*	*Cell Type*	*Phenotype/Effects*
P1	**Adenosine/AMP**		
A1	Adenosine	Nerves, cardiac cells	Cardiac rate, preconditioning
A2a	Adenosine	Immune cells, platelets and ECs	Vasodilatation and renin effects Anti-inflammatory and antithrombotic
A2b	Adenosine	Mast cells, platelets	Vasodilatation
A3	Adenosine	Mast cells	Janus-like and varied Cell proliferation and apoptosis
P2	**ADP, UDP, ATP, ATP, and UDP-glucose**		
P2Y1	ADP	ECs, platelets, stem cells (mouse)	Platelet activation Apoptosis
P2Y2	ATP/UTP	ECs/VSMCs	Vasodilation
P2Y4	UTP/ATP	?ECs/VSMCs	Calcium mobilization
P2Y6	UDP	VSMCs	Calcium mobilization
P2Y11	ATP	ECs	
P2Y12	ADP	Platelets	Platelet activation
P2Y13	ADP	Leukocytes	?
P2Y14	UDP-glucose	?	?
P2X1	ATP	VSMCs, platelets	Vasoconstriction; platelet activation, cell adhesion. and vascular permeability; calcium mobilization
P2X4	ATP	ECs	EC activation and calcium mobilization
P2X7	ATP	ECs, monocytes	Apoptosis, EC activation responses

receptors also appear on vascular cells and are thought to be associated with changes in cell adhesion and permeability (11).

These cellular processes and nucleotide-triggered events are modulated during angiogenesis and influence the development of atherosclerosis and restenosis following angioplasty (11,28–31).

RELEASE OF NUCLEOTIDES WITHIN THE VASCULATURE

Extracellular nucleotides, and potentially nucleosides, are continuously released from cells (e.g., following the exocytosis of ATP/UTP-containing vesicles, facilitated diffusion by putative ABC transporters, or by poorly understood electrodiffusional movements through ATP/nucleotide channels). It has been shown that rates of increase are higher in injured or stressed cells (7,32–35). Several mechanisms account for the presence of nucleotides or nucleosides in plasma (36). As alluded to earlier, these include aggregating platelets, degranulating macrophages, excitatory neurons, injured cells, and cells undergoing mechanical or oxidative stress resulting in lysis, selective permeabilization of cellular membranes, and exocytosis of secretory vesicles, such as from platelet-dense bodies (7,37). It is important to note here that many processes of arterial vascular injury are associated with the release of adenine nucleotides that exert a variety of inflammatory effects on the endothelium, platelets and, leukocytes (7,19).

PURINERGIC RECEPTORS IN THE VASCULATURE

P1 (adenosine) receptors can be differentiated from P2 (ATP/ADP) receptors by direct pharmacological and molecular means (Table 43-2). The endogenous ligand adenosine (and potentially inosine at A3 receptors) activates adenosine receptors.[2] Adenosine signaling and uptake inhibits many intracellular ATP-utilizing enzymes, including the adenylyl cyclase systems. To date, four subtypes of P1 receptors have been cloned – namely A1, A2A, A2B, and A3 – with substantial interspecies differences (20–23). A pseudogene also exists for the A2B adenosine receptor (ENSG00000182537) with 79% identity to the A2B adenosine receptor cDNA coding sequence but which is unable to encode a functional receptor. The adenosine receptors are classified according to their affinities for adenosine and variant coupling to adenylate cyclase (23).

P2 receptors bind extracellular nucleotides and initiate cellular signaling. Burnstock and Kennedy originally distinguished between two major types of P2 nucleotide receptors, the P2X and P2Y subtypes, again based largely on pharmacological criteria (38). These are expressed by the endothelium, VSMCs, platelets, and leukocytes. Further studies of transduction mechanisms followed by cloning of both P2X and P2Y

2 The nomenclature is determined by the NC-IUPHAR Subcommittee on Adenosine Receptors.

receptors have been carried out, resulting (in 1994) in a new nomenclature system that is generally accepted (8) (see Figure 43.1 and Table 43.2).

In summary, two main families of nucleotide receptors exist (Table 43-2): P2X and P2Y. P2X are "rapid" ligand-gated ion channels permeable to Na^+, K^+, and also Ca^{2+} (subtypes P2X1–7) (21). The P2X receptors (nomenclature determined by the NC-IUPHAR Subcommittee on P2X Receptors) are putative trimeric transmitter-gated channels, with two putative transmembrane domains, in which the endogenous ligand is ATP. In P2X receptors, structural features should be the initial criteria for nomenclature. The relationships of many of the cloned receptors to endogenously expressed receptors have not been established to date. Functional P2X receptors exist as polymeric transmitter-gated channels. The native receptors may occur as homopolymers (e.g., P2X1 in VSMCs) or heteropolymers (e.g., P2X2:P2X3 in nervous tissues). P2X7 receptors have been shown to form functional homopolymers that initiate pores permeable to low-molecular-weight solutes.

P2Y are the so-called "slow" metabotropic receptors (P2Y1–14) (20).[3] P2Y receptors (provisional nomenclature as agreed by NC-IUPHAR Subcommittee on P2Y Receptors) are activated by the endogenous ligands ATP, ADP, UTP, UDP, and UDP-glucose. These receptors are seven-transmembrane G_q- or G_i-protein linked receptors that initiate signal transduction coupled to activation of phospholipase C (PLC) or to inhibition of adenylate cyclase, respectively (20) (see Figure 43.1). The relationship of many of the cloned receptors to endogenously expressed receptors is not yet fully established. To prevent additional problems with the accepted terminology, it is recommended that terms such as "UTP-preferring (or ATP-) P2Y receptor" or "P2Y1-like" be used until further classifications can be applied. For example, P2Y7 is not a true nucleotide-receptor but rather encodes a leukotriene receptor; P2Y8 is an LPA receptor, and no functional data exist for P2Y10. Finally, the recently proposed and described P2Y15 receptor is not a genuine P2 receptor in that it recognizes dicarboxylic acids (39).

The distribution of the families of validated receptors in the vascular system and elsewhere, using their pharmacological properties, recently has been collated exhaustively and defined in depth by Burnstock and Knight (11). P2X1, P2Y2, and P2Y6 are the most highly expressed P2 receptors in VSMCs, and they mediate the contractile and mitogenic actions of extracellular nucleotides. P2X4, P2Y11, P2Y1, and P2Y2 are the most highly expressed P2 receptors in ECs (11).

In the blood, extracellular nucleotides/nucleosides can influence cardiac function, vasomotor responses, platelet activation, thrombosis, and inflammatory processes (19,35,40, 41). ATP and ADP appear to regulate hemostasis through the activation of platelet P2 receptors, most notably P2Y12 and P2Y1. P2X1 is rapidly desensitized and likely important under very high shear stress. ADP is a major platelet-recruiting factor originating from platelet-dense granules during activation, whereas ATP derived from the same sources has been considered a competitive antagonist of ADP for platelet P2Y receptors and a putative agonist for P2X1 receptors (7,37,42). ATP (and UTP) also stimulate endothelial P2Y1 and P2Y2 receptors to release prostacyclin (PGI2) and NO, two vasodilators and inhibitors of platelet aggregation (43–47). This latter protective action of ATP may limit the extent of intravascular platelet aggregation and help localize thrombus formation to areas of vascular damage (45,48,49).

Interestingly, the P2Y2 nucleotide receptor contains the integrin-binding domain arginine-glycine-aspartic acid (RGD) in the first extracellular loop, suggesting that this G-protein–coupled receptor might interact directly with an integrin. Subsequent work by Erb and colleagues has detailed that this RGD sequence is required for P2Y2-mediated activation of G_o, but not G_q signaling pathways. Further work has indicated that interactions between P2Y2, integrins such as the vitronectin receptor $\alpha v\beta 3$, and other membrane proteins may be important for coordinating P2Y2-mediated responses (50) (see Figure 43.2). Other, more recent work has detailed that P2Y2 and possibly P2Y4 receptors mediate the vascular responses important during inflammation and upregulate integrins crucial in angiogenic responses (51).

In addition, ATP also stimulates P2X receptors to cause plasma membrane permeabilization, induction of apoptosis, organic anion transport, and stimulation of Ca^{2+} mobilization (19,52). The active form of ATP that causes membrane permeabilization is a fully ionized tetrabasic ion (or ATP^{4-}) that interacts with the unique multimeric P2Z/P2X7 receptors known to be expressed on the endothelium and monocyte-macrophages (53). In fact, this nucleotide is a relative inhibitor of NTPDases, because these are divalent cation-dependent enzymes (54) (S.C. Robson, unpublished observations). Such P2X7 receptors also are stimulated by ATP-γ-S but are unresponsive to UTP (35,55). Activation of P2X7 receptors can induce cellular activation and apoptosis (55,56).

Pathways of nucleotide-mediated signaling are further complicated by P2-receptor desensitization phenomena. Thus, adenine nucleotides may also directly limit the ADP-mediated reactivity of platelets (19,37,57,58). As in other members of the G-protein–coupled receptor family, some P2Y receptors readily undergo agonist-induced desensitization (59). This may involve phosphorylation of the receptor by multiple protein kinases (10). For unclear reasons, P2X- and P2Y-type receptors differ widely with respect to desensitization rates. Rapidly desensitizing receptors and/or channels (P2Y1, P2Y2, P2X1, and P2X3) may be contrasted with slow desensitizing receptors such as the P2Y6, P2X2, and P2X7 receptors (the last very important in inflammatory reactions). P2X4 is intermediate in this regard (60). Resistance to ATP and other nucleotides in the vasculature also may be explained by downregulation or absolute deficiencies of defined P2 receptors on cell membranes, because cell intracellular transduction

3 Molecular Recognition Section of National Institutes of Health, http://mgddk1.niddk.nih.gov:8000/nomenclature.html.

Table 43-3: Vascular Ectonucleotidases: NTPDases of CD39 Family

Protein Names	*Previous Nomenclature and Vascular Localization*	*Gene Human Mouse*	*Chromosome Location Human Mouse*	*Accession Number*
NTPDase1	CD39, ATPDase, ecto-apyrase. ECs and VSMCs	*ENTPD1* *Entpd1*	10q24 19C3	U87967 NM009848
NTPDase2	CD39L1, ecto-ATPase Adventitial cells and pericytes, nerves	*ENTPD2* *Entpd2*	9q34 2A3	AF144748 AY376711
NTPDase3	CD39L3, HB6 ECs	*ENTPD3* *Entpd3*	3p21.3 9F4	AF034840 AY376710
NTPDase8	Liver canalicular ecto-ATPase, hATPDase ?Central sinusoidal ECs	*ENTPD8* *Entpd8*	9q34 2A3	AY430414 AY364442

mechanisms differ, or by the presence of ectonucleotidases that hydrolyze the nucleotide mediators (61).

Select examples exist of how the differential expression of P2-receptors affects pathological outcomes. Saphenous vein, internal mammary, and other arteries have been used for coronary grafting and bypass surgery. Levels of endothelial P2Y2 receptors are comparable in all three vessels, but endothelial P2X4 receptors are expressed at much higher levels in the saphenous vein samples (11). P2X4 receptors are involved in calcium influx into ECs, they modulate blood vessel contractility, and they are upregulated in situations involving intima proliferation. These P2X4 receptors possibly play a more significant role in the higher frequency of restenosis in vein grafts, as reflected by the susceptibility of saphenous grafts to the development of atherosclerosis (62).

A novel role for P2Y2 receptors in atherosclerosis also has been suggested, because this receptor and the potential agonists (either ATP or UTP) induce EC vascular adhesion molecule (VCAM)-1 expression and consequent intimal hyperplasia in animal studies (63). Recent data indicate that this P2Y2 nucleotide receptor also can transactivate receptor tyrosine kinases in human coronary artery ECs (HCAECs) to generate inflammatory responses, which is considered associated with atherosclerosis (64). These long-term (trophic) effects of extracellular nucleotides in VSMC or EC proliferation and the linkage to ectonucleotidases in associated vascular lesions suggest the exploration of therapeutic strategies in relation to atherosclerosis and vascular inflammatory conditions associated with transplantation (15,65).

ECTONUCLEOTIDASES AND THE NUCLEOSIDE TRIPHOSPHATE DIPHOSPHOHYDROLASE (NTPDase/CD39 FAMILY)

Ectonucleotidases are a subset of ectoenzymes that hydrolyze extracellular purine and pyrimidine nucleotides (15,24). As detailed previously, these enzymes differ from intracellular nucleotidases in molecular structure and are also distinct from alkaline phosphatases and ecto-5′-nucleotidase; the latter are enzymes capable of releasing phosphate from a large variety of organic compounds, including the hydrolysis of nucleotides (66,67). These ectoenzymes are ubiquitous, broadly distributed in tissues, and well defined at the biochemical level. However, because specific inhibitors are not available, their pathophysiological relevance has remained largely elusive. The recent development of genetic techniques to modulate the levels of defined members of these families promises to advance our understanding of the biology of the purinergic/pyrimidinergic system.

One important function of ectonucleotidases might be the modulation of P2-receptor–mediated signaling by the removal of extracellular ATP, ADP, and related nucleotides. The ultimate generation of extracellular adenosine not only abrogates or terminates nucleotide-mediated effects but also activates adenosine receptors, with often opposing pathophysiological effects. Ectonucleotidases also produce the key molecules for purine salvage and the consequent replenishment of ATP stores within multiple cell types. Indeed, although nucleotides appear not to be taken up by cells, their dephosphorylated nucleoside derivatives interact with several specific transporters to enable membrane passage. The regulated dephosphorylation of extracellular nucleotides by ectonucleotidases may be critical for appropriate purinergic/pyrimidinergic signaling and metabolic homeostasis (6,15,24,59).

CD39 was characterized originally as an activation marker, identified on B cells, monocytes, subsets of activated NK cells, and T lymphocytes (68–70). Once the enzymatic function of CD39 was elucidated, work proceeded on genes that share considerable sequence identity, including the five apyrase conserved regions (71,72). Some of these eight enzymes are membrane-attached, with external-oriented active sites, whereas others are soluble (73–75) (Table 43-3). The most prevalent ectonucleotidases expressed by vascular cells (e.g.,

ECs, pericytes, VSMCs, and adventitial cells) are members of the CD39 family. These members include NTPDase1/CD39, ecto-ATPase and ecto-ADPase, and CD39L1/NTPDase2 (Table 43-3) (74,75).

NTPDases are subject to multiple post-translational changes that influence both tertiary and quaternary structure and that have major implications for enzymatic function (72,76). Interestingly, the structures of CD39 and P2X receptors appear remarkably similar (20,74), and both can form multimers (i.e., tetramers) (77,78).

The endothelial membrane–expressed CD39/NTPDase-1 is the major ectonucleotidase in the vasculature (59). Other NTPDases associated with the vasculature are the cell-associated ecto-ATPases (CD39L1 or NTPDase-2) and soluble ecto-ADPases (CD39L2 or NTPDase-6; akin to the monocyte-expressed CD39L4 or NTPDase-5) (66,79,80). The ectoenzyme CD39/NTPDase1 can be shown to efficiently bind and hydrolyze extracellular ADP (and ATP) to AMP; the product AMP does activate select P1 receptors, but is preferentially hydrolyzed to adenosine by the ubiquitous CD73 and ecto-5′-nucleotidases. This phosphohydrolytic reaction limits the platelet activation response, which depends on the autocrine and paracrine release of ADP with activation of specific purinergic receptors (5,12,81). In contrast, CD39L1/NTPDase2, a preferential nucleoside triphosphatase, activates platelets by converting the competitive antagonist (ATP) of platelet ADP-receptors to the specific agonist (ADP) of the P2Y1 and P2Y12 receptors. In keeping with these biochemical properties, CD39 is mainly expressed by ECs and VSMCs, where it serves as a thromboregulatory factor. In contrast, CD39L1 is associated with the adventitial surfaces of muscularized vessels, microvascular pericytes, and stromal cells, where it could potentially serve as a hemostatic factor (82).

OTHER ECTOENZYMES IN THE VASCULATURE

NTPDases of the CD39 family are Ca^{2+}/Mg^{2+}-dependent ectonucleotidases that hydrolyze nucleoside 5′-triphosphates and nucleoside 5′-diphosphate (6,15,66). This ectonucleotidase chain initiated by NTPDases is terminated by ecto-5′-nucleotidase (CD73; EC 3.1.3.5) (83).

Adenosine deaminase (ADA; EC 3.5.4.4) is another ectoenzyme that binds to CD26 and is involved in the generation of inosine from adenosine. Together, CD73 and ADA closely regulate local and pericellular extracellular and plasma concentrations of adenosine (26,84).

The CD38 family of ectoenzymes also is involved in recycling extracellular nucleotides by metabolizing NAD through the generation of cyclic ADP ribose and ADP ribose (85). The surface expression levels of CD38 are tightly regulated in vivo in response to inflammatory signals (86). Furthermore, a clear role for CD38 in the interactions taking place between circulating lymphocytes and the vessel wall was recognized several years ago (87). The latter finding was the starting point of the search for an EC-bound ligand, which culminated in the identification of platelet-endothelial cell adhesion molecule (PECAM)/CD31 as a counter-receptor for CD38 (88).

The CD38 paradigm is also representative of a more general trend involving several nucleotide-metabolizing enzymes (89). CD38 and CD157 (a member of the CD38 family) also may serve as receptors, controlling intracellular signaling pathways apparently independently of the ectoenzymatic activity. The two functional activities might represent two separate evolutionary developments resulting in the final multifunctional molecule. Phylogenetic studies support the view that the enzymatic activity is the "older" conserved function, whereas at a later point CD38 may have acquired membrane anchorage and the ability to function as a receptor (90).

These common traits also involve localization in membrane lipid microdomains and the physical and functional association with partners specialized in signal transduction (91–93). It has been shown recently that the NH_2 terminus of the molecule CD39 is physically associated with components of the Ras signaling pathway. These data suggest that receptor-like function also might have been acquired by this ectonucleotidase.[4]

HOMEOSTASIS

Depending on the repertoire of receptors and signaling components, P2 and P1 receptors influence those cellular pathways crucial for homeostasis. Although in regard to nucleotide receptors, platelet P2Y12 has received the most attention for its role in ADP-mediated thrombosis, recently obtained evidence suggests that P2Y1, P2Y2, and P2X4 in vascular ECs, VSMCs, and monocytes are important receptors in the mediation of vascular responses. Nucleotide-mediated effects are in turn modulated by ectonucleotidases, as discussed earlier.

The various biological functions of these P2 and adenosine receptors, together with NTPDases, may be influenced by differential expression in various vascular beds, certain unique biochemical characteristics and the effects of their relevant substrate nucleotides, or by products on various receptor subtypes expressed in the local environment. It has been proposed that, in concert with the nearly 20 described P2 or P1 or adenosine receptors, combinations of NTPDases have the capacity to terminate signaling, alter specificities of the response, or even generate signaling molecules (e.g., ADP) from precursors (e.g., ATP) (6). However, little is currently known of the structural or functional associations of NTPDases with one another or with P2-receptors. It is proposed that, under certain conditions, NTPDases may protect the integrity of any future response by preventing receptor desensitization reactions (59,94). In contrast, coexpression of ectonucleotidases

4 AASLD meeting 2004; Wu Y, Sun X, Keiichi Enjyoji K, Robson SC. CD39 Interacts with Ran Binding Protein M (RanBPM) to directly modulate Ras activation and cellular proliferation in liver regeneration following partial hepatectomy. *Hepatology* 2004;135;40, 222A.

may be essential for the survival of cells that express P2X7 receptors, because these receptors do not readily autoregulate by desensitization responses and can induce apoptosis (56).

PATHOBIOLOGY

To test these how extracellular nucleotide–mediated signaling influences pathophysiological events, we and others have developed and validated several techniques to manipulate CD39 expression in the vasculature and study effects under conditions of inflammatory stress. For example, we have globally deleted *Cd39* in mice. The mutant mice null for this dominant ectonucleotidase exhibit major perturbations of P2-receptor–mediated signaling in the vasculature and immune systems (6,95,96).

These phenomena manifest as hemostatic defects, thromboregulatory disturbances, and heightened acute inflammatory responses with a failure to generate those cellular immune responses that are associated with vascular endothelium, monocyte, dendritic cell, and platelet integrin dysfunction (28,67). The therapeutic potential of NTPDase1 to regulate P2-receptor function in the vasculature and to mitigate against thrombotic or inflammatory stress has been further established. This has been studied by the generation of CD39 transgenic mice (15,97), the use of adenoviral vectors to upregulate CD39 in cardiac grafts (98), and the use of soluble derivatives of CD39 and apyrases (14,99). All these studies show beneficial effects for the overexpression of NTPDases within the vasculature, or by their pharmacological administration (14,15).

Other mutant mice models are summarized in Table 43-4.

Table 43-4: Mutant Mice Models of Pertinent Purinergic Receptors and Ectoenzymes

Genetic Model	*Phenotype (Vascular)*	*References*
Knockout		
P2Y1	Bleeding, resistance to thrombosis	110
P2Y2	Susceptibility to sepsis (with P2Y1 deletion)	111
P2Y12		112
	Bleeding, platelet defect	113
P2X1	Platelet thrombosis defect under high shear	115 114
P2X7	Inflammation; EC unclear sequelae	116
A2A	Neuroprotective vascular phenotype; but discordant effects with respect to liver vasculature and bone marrow derived cells	117 118
CD39	Hemostatic defects, heightened acute inflammatory responses, endothelial dysfunction.	59, 95, 96 82
CD39L1	Pending (S.C. Robson)	
ADA	Perinatal hepatic injury and lethality Rescue possible – pulmonary injury and immunodeficiency	119, 120
CD38	Immunodeficiency	121
Transgenic (select)		
P2X4	Increased cardiac contractility	122
CD39	Vascular protection against thrombosis and humoral allograft rejection	97

THERAPEUTICS

Platelets are known to express P2Y1, P2Y12, and P2X1 receptors. The Clopidogrel versus Aspirin in Patients at Risk of Ischaemic Events (CAPRIE) and Clopidogrel in Unstable Angina to Prevent Recurrent Events (CURE) studies have provided definitive evidence that clopidogrel and ticlopidine, antagonists of the platelet P2Y12 receptor, decrease the clinical sequelae of recurrent strokes and myocardial infarcts, especially when combined with aspirin (100,101). More recently developed drugs hold even greater promise. MRS 2500, a potent, highly selective antagonist for the human P2Y1 receptor, also has been shown to have substantive antiaggregatory potential for platelets (102). Synergism occurs in the inhibition of ADP-induced platelet activation via the P2Y1 and P2Y12 receptors, and also for antithrombin approaches using P2Y receptor inhibition or ectonucleotidases (15,103,104). A synergistic interaction also occurs between ATP and noradrenaline in stimulating platelet aggregation, which suggests a prothrombotic role for ATP in stress (14).

Genetic variations occur in P2Y1 and P2Y12 receptor DNA in apparently normal subjects; these variations may explain variations in the platelet responses and possibly individual variation in atherothrombotic risk and bleeding manifestations. These discoveries also may provide a basis for the understanding of interpatient variation and complication rates in response to standard doses of antiplatelet drugs (105).

The beneficial effects of administered NTPDases have been determined in several animal models of vascular inflammation (99,106). Exogenous infusions of soluble NTPDases were able to rescue *cd39*-null mice from toxicity induced by ischemia–reperfusion injury and after stroke induction (96,107). Clinical studies of these soluble thromboregulatory factors are planned (15,108,109).

KEY POINTS

- Purinergic signaling by extracellular nucleotides impacts on vascular injury and inflammation and is modulated by the CD39/NTPDase family of ectonucleotidases.
- Extracellular nucleotide-mediated events, with inflammatory stress and/or immune reactions, have

important consequences for platelet activation, thrombogenesis, angiogenesis, vascular remodeling, and the metabolic milieu of the vasculature.

- Modulated, distinct NTPDase expression appears to regulate nucleotide-mediated signaling in the vasculature.
- Current pharmacological interventions to block platelet microthrombi are associated with limited therapeutic success, adverse clinical events, and bleeding.
- Available experimental data suggest that therapeutic modalities targeting nucleotide-mediated platelet activation and the evolution of rational drug delivery systems may facilitate control of the thrombotic component of inflammatory vascular disease while preserving adequate hemostasis.

Future Goals

- To determine whether increased CD39/NTPDase1 expression as a result of gene therapy or pharmacological induction will provide effective treatment modalities for vascular thrombotic disorders, atherosclerosis, and the vascular inflammation seen in transplant-related diseases, such as rejection
- To study selective P2-receptor inhibitors and effects in thrombotic human disease and the disorders tabulated above
- To determine whether targeted inhibition of CD39/NTPDase1 expression or pharmacological blockade using small molecule inhibitors will provide a useful component of novel anticancer and antiangiogenesis regimens
- To determine the biology of the other vascular NTPDases of the CD39 family

ACKNOWLEDGMENTS

Grant support from NIH: HL63972, HL57307, and HL076540. Scientific consultations included Dr. Jean Sevigny (Table 43-3).

REFERENCES

1 Ross R. Cell biology of atherosclerosis. [Review]. *Ann Rev of Physiol.* 1995;57(791):791–804.
2 Sinzinger H, Virgolini I, Gazso A, et al. Review – Eicosanoids in atherosclerosis. *Exp Pathol.* 1991;43(2):2–19.
3 Cooke JP, Dzau VJ. Nitric oxide synthase – role in the genesis of vascular disease. [Review]. *Ann Rev of Med.* 1997;48(489):489–509.
4 Ihrcke NS, Wrenshall LE, Lindman BJ, et al. Role of heparan sulfate in immune system-blood vessel interactions. *Immunol Today.* 1993;14(500):500–505.
5 Robson SC, Kaczmarek E, Siegel JB, et al. Loss of ATP diphosphohydrolase activity with EC activation. *J Exp Med.* 1997; 185(1):153–163.
6 Robson SC, Enjyoji K, Goepfert C, et al. Modulation of extracellular nucleotide-mediated signaling by CD39/nucleoside triphosphate diphosphohydrolase-1. *Drug Dev Res.* 2001;53(2–3): 193–207.
7 Luthje J. Origin, metabolism and function of extracellular adenine nucleotides in the blood [published erratum appears in *Klin Wochenschr* 1989 May 15;67(10):558]. [Review]. *Klin Wochenschrift.* 1989;67(6):317–327.
8 Abbracchio MP, Burnstock G. Purinoceptors – are there families of P2X and P2Y purinoceptors. [Review]. *Pharm and Therapeutics.* 1994;64(3):445–475.
9 Harden TK, Lazarowski ER, Boucher RC. Release, metabolism and interconversion of adenine and uridine nucleotides: implications for G protein-coupled P2 receptor agonist selectivity. *Trends Pharmacol Sci.* 1997;18(2):43–46.
10 Weisman GA, Erb L, Garrad RC, et al. P2Y nucleotide receptors in the immune system: Signaling by a P2Y(2) receptor in U937 monocytes. *Drug Dev Res.* 1998;45(3–4):222–228.
11 Burnstock G, Knight G. Cellular distribution and functions of P2 receptor subtypes in different systems. *Int Rev Cytol.* 2004; 240:31–304.
12 Marcus AJ, Safier LB, Hajjar KA, et al. Inhibition of platelet function by an aspirin-insensitive EC ADPase. Thromboregulation by ECs. *J Clin Invest.* 1991;88(5):1690–1696.
13 Kaczmarek E, Koziak K, Sevigny J, et al. Identification and characterization of CD39 Vascular ATP diphosphohydrolase. *J Bio Chem.* 1996;271(51):33116–33122.
14 Marcus AJ, Broekman MJ, Drosopoulos JH, et al. Metabolic control of excessive extracellular nucleotide accumulation by CD39/ectonucleotidase-1: implications for ischemic vascular diseases. *J Pharmacol Exp Ther.* 2003;305(1):9–16.
15 Robson SC, Wu Y, Sun X, et al. Ectonucleotidases of CD39 family modulate vascular inflammation and thrombosis in transplantation. *Semin Thromb Hemost.* 2005;31(2):217–233.
16 Drury AN, Szent-Györgyi A. The physiological activity of adenine compounds with special reference to their action upon the mammalian heart. *J Physiol (Lond).* 1929;68:213–237.
17 Burnstock G. Purinergic nerves. *Pharmacol Rev.* 1972;24:509–581.
18 Pepeu G. Cholinergic neurotransmission in the central nervous system. *Arch Int Pharmacodyn Ther.* 1972;196(Suppl):229–243.
19 Dubyak GR, Elmoatassim C. Signal transduction via p2-purinergic receptors for extracellular ATP and other nucleotides. [Review]. *Am J Physiol.* 1993;265(3 Part 1):C577-C606.
20 Fredholm BB, Abbracchio MP, Burnstock G, et al. Nomenclature and classification of purinoreceptors. *Pharmacol Rev.* 1994; 46:143–512.
21 Buell G, Collo G, Rassendren F. P2X receptors – an emerging channel family. [Review]. *Eur J Neurosci.* 1996;8(10):2221–2228.
22 Roman RM, Fitz JG. Emerging roles of purinergic signaling in gastrointestinal epithelial secretion and hepatobiliary function. *Gastroenterology.* 1999;116:964–979.
23 Palmer TM, Stiles GL. Adenosine receptors. *Neuropharmacology.* 1995;34:683–694.
24 Plesner L. Ecto-ATPases: identities and functions. *Int Rev of Cytol.* 1995.;158:141–214.

25 Sasamura H, Dzau VJ, Pratt RE. Desensitization of angiotensin receptor function. *Kidney Int*. 1994;46(6):1499–1501.
26 Goding JW, Howard MC. Ectoenzymes of lymphoid cells. *Immunol Rev*. 1998;161:5–10.
27 Zimmermann H. Extracellular purine metabolism. [Review]. *Drug Dev Res*. 1996;39(3–4):337–352.
28 Goepfert C, Sundberg C, Sevigny J, et al. Disordered cellular migration and angiogenesis in cd39-null mice. *Circulation*. 2001;104(25):3109–3115.
29 Goepfert C, Imai M, Brouard S, et al. CD39 modulates EC activation and apoptosis. *Mol Med*. 2000;6(7):591–603.
30 Wihlborg AK, Wang L, Braun OO, et al. ADP receptor P2Y12 is expressed in vascular smooth muscle cells and stimulates contraction in human blood vessels. *Arterioscler Thromb Vasc Biol*. 2004;24(10):1810–1815. Epub 2004 Aug 12.
31 Wang L, Andersson M, Karlsson L, et al. Increased mitogenic and decreased contractile P2 receptors in smooth muscle cells by shear stress in human vessels with intact endothelium. *Arterioscler Thromb Vasc Biol*. 2003;23(8):1370–1376. Epub 2003 Jun 5.
32 Abraham EH, Prat AG, Gerweck L, et al. The multidrug resistance (mdr1) gene product functions as an ATP channel. *Proc Natl Acad Sci USA*. 1993;90(1):312–316.
33 Grierson JP, Meldolesi J. Shear stress-induced [$Ca2^{+}$]i transients and oscillations in mouse fibroblasts are mediated by endogenously released ATP. *J Biol Chem*. 1995;270(9):4451–4456.
34 Tsujimoto Y. Apoptosis and necrosis – intracellular ATP level as a determinant for cell death modes. [Review]. *Cell Death Differ*. 1997;4(6):429–434.
35 Franceschi C, Abbracchio MP, Barbieri D, et al. Purines and cell death. *Drug Dev Res*. 1996;39(3–4):442–449.
36 Traut TW. Physiological concentrations of purines and pyrimidines. [Review]. *Mol Cell Biochem*. 1994;140(1):1–22.
37 Fijnheer R, Boomgaard MN, van den Eertwegh AJ, et al. Stored platelets release nucleotides as inhibitors of platelet function. *Thromb Haemost*. 1992;68(5):595–599.
38 Burnstock G, Kennedy C. Is there a basis for distinguishing two types of P2-purinoceptor? *Gen Pharmacol*. 1985;16:433–440.
39 Abbracchio MP, Burnstock G, Boeynaems JM, et al. The recently deorphanized GPR80 (GPR99) proposed to be the P2Y15 receptor is not a genuine P2Y receptor. *Trends Pharmacol Sci*. 2005;26(1):8–9.
40 Boeynaems J-M, Pearson JD. P2 purinoreceptors on vascular ECs: physiological significance and transduction mechanisms. *Trends Pharmacol Sci*. 1990;11:34–37.
41 Brake A, Schumacher M, Julius D. ATP receptors in sickness, pain and death. *Chem Biol*. 1996;3(4):229–232.
42 Hechler B, Vigne P, Leon C, et al. ATP derivatives are antagonists of the P2Y(1) receptor – similarities to the platelet ADP receptor. *Mol Pharmacol*. 1998;53(4):727–733.
43 Yang S, Cheek DJ, Westfall DP, et al. Purinergic axis in cardiac blood vessels. Agonist-mediated release of ATP from cardiac ECs. *Circ Res*. 1994;74(3):401–407.
44 Fisette PL, Denlinger LC, Proctor RA, et al. Modulation of macrophage function by p2y-purinergic receptors. [Review]. *Drug Dev Res*. 1996;39(3–4):377–387.
45 Marcus AJ, Safier LB. Thromboregulation: multicellular modulation of platelet reactivity in hemostasis and thrombosis. [Review]. *FASEB J*. 1993;7(6):516–522.
46 Juul B, Plesner L, Aalkjaer C. Effects of ATP and related nucleotides on the tone of isolated rat mesenteric resistance arteries. *J Pharmacol Exp Ther*. 1993;264(3):1234–1240.
47 Motte S, Communi D, Pirotton S, et al. Involvement of multiple receptors in the actions of extracellular ATP: the example of vascular ECs. [Review]. *Int J Biochem Cell Biol*. 1995;27(1):1–7.
48 Cote YP, Picher M, St-Jean JP, et al. Identification and localization of ATP-diphosphohydrolase (apyrase) in bovine aorta: relevance to vascular tone and platelet aggregation. *Biochim Biophys Acta*. 1991;1078(2):187–191.
49 Cote YP, Filep JG, Battistini B, et al. Characterization of ATP-diphosphohydrolase activities in the intima and media of the bovine aorta: evidence for a regulatory role in platelet activation in vitro. *Biochim Biophys Acta*. 1992;1139(1–2):133–142.
50 Erb L, Liu J, Ockerhausen J, et al. An RGD sequence in the P2Y(2) receptor interacts with alpha(V)beta(3) integrins and is required for G(0)-mediated signal transduction. *J Cell Biol*. 2001;153(3):491–501.
51 Kaczmarek E, Erb L, Koziak K, et al. Modulation of EC migration by extracellular nucleotides: involvement of focal adhesion kinase and phosphatidylinositol 3-kinase-mediated pathways. *Thromb Haemost*. 2005;93(4):735–742.
52 Chow SC, Kass G, Orrenius S. Purines and their roles in apoptosis. *Neuropharmacology*. 1997;36(9):1149–1156.
53 Koziak E, Sevigny J, Robson SC, et al. Analysis of CD39/ATP diphosphohydrolase expression in ECs, platelets and leukocytes. *Thromb Haemost*.1999;82:1538–1544.
54 Frassetto SS, Dias RD, Sarkis JJ. Characterization of an ATP diphosphohydrolase activity (APYRASE, EC 3.6.1.5) in rat blood platelets. *Mol Cell Biochem*. 1993;129(1):47–55.
55 Ferrari D, Chiozzi P, Falzoni S, et al. ATP-mediated cytotoxicity in microglial cells. *Neuropharmacology*. 1997;36(9):1295–1301.
56 Von Albertini M, Palmetshofer A, Kaczmarek E, et al. Extracellular ATP and ADP activate transcription factor NF-kappa-B and induce EC apoptosis. *Biochem Biophys Res Commun*. 1998;248(3):822–829.
57 Hourani SMO, Hall DA. Receptors for ADP on human blood platelets. *Trends Pharmacol Sci*. 1994;15:103–107.
58 Lages B, Weiss HJ. Enhanced increases in cytosolic Ca^{2+} in ADP-stimulated platelets from patients with delta-storage pool deficiency – a possible indicator of interactions between granule-bound ADP and the membrane ADP receptor. *Thromb Haemost*. 1997;77(2):376–382.
59 Enjyoji K, Sevigny J, Lin Y, et al. Targeted disruption of cd39/ATP diphosphohydrolase results in disordered hemostasis and thromboregulation. *Nat Med*. 1999;5(9):1010–1017.
60 Koshimizu T, Tomic M, Koshimizu M, et al. Identification of amino acid residues contributing to desensitization of the P2x(2) receptor channel. *J Biol Chem*. 1998;273(21):12853–12857.
61 Clifford EE, Martin KA, Dalal P, et al. Stage-specific expression of p2y receptors, ecto-apyrase, and ecto-5′-nucleotidase in myeloid leukocytes. *Am J Physiol Cell Physiol*. 1997;42(3):C973-C987.
62 Ray FR, Huang W, Slater M, et al. Purinergic receptor distribution in ECs in blood vessels: a basis for selection of coronary artery grafts. *Atherosclerosis*. 2002;162(1):55–61.
63 Seye CI, Kong Q, Erb L, et al. Functional P2Y2 nucleotide receptors mediate uridine 5′-triphosphate-induced intimal hyperplasia in collared rabbit carotid arteries. *Circulation*. 2002;106(21):2720–2726.

64 Seye CI, Yu N, Gonzalez FA, et al. The P2Y2 nucleotide receptor mediates vascular cell adhesion molecule-1 expression through interaction with VEGF receptor-2 (KDR/Flk-1). *J Biol Chem.* 2004;279(34):35679–35686. Epub 2004 Jun 2.

65 Gendron FP, Benrezzak O, Krugh BW, et al. Purine signaling and potential new therapeutic approach: possible outcomes of NTPDase inhibition. *Curr Drug Targets.* 2002;3(3):229–245.

66 Zimmermann H. Two novel families of ectonucleotidases: molecular structure, catalytic properties and a search for function. *Trends Pharmacol Sci.* 1999;20:231–236.

67 Zimmermann H. Nucleotides and CD39: principal modulatory players in hemostasis and thrombosis. *Nat Med.* 1999;5:987–988.

68 Kansas GS, Wood GS, Tedder TF. Expression, distribution, and biochemistry of human CD39. Role in activation-associated homotypic adhesion of lymphocytes. *J Immunol.* 1991;146(7):2235–2244.

69 Maliszewski CR, Delespesse GJ, Schoenborn MA, et al. The CD39 lymphoid cell activation antigen. Molecular cloning and structural characterization. *J Immunol.* 1994;153(8):3574–3583.

70 Favaloro EJ. Differential expression of surface antigens on activated endothelium. *Immunol Cell Biol.* 1993;71(571):571–581.

71 Handa M, Guidotti G. Purification and cloning of a soluble ATP-diphosphohydrolase (apyrase) from potato tubers (*Solanum tuberosum*). *Biochem Biophys Res Commun.* 1996;218(3):916–923.

72 Schulte am Esch J, Sevigny J, Kaczmarek E, et al. Structural elements and limited proteolysis of CD39 influence ATP diphosphohydrolase activity. *Biochemistry.* 1999;38(8):2248–2258.

73 Wang TF, Guidotti G. Golgi localization and functional expression of human uridine diphosphatase. *J Biol Chem.* 1998;273(18):11392–11399.

74 Zimmermann H, Kirley T, Robson SC, et al. Nomenclature for two families of novel ectonucleotidases. In: Van Duffel L, ed. *Proceedings of the Second International Workshop on Ecto-ATPases and Related Ectonucleotidases.* Maastricht, The Netherlands; Shaker Publishing, 2000:1–8.

75 Zimmermann H, Braun N, Heine P, et al. Molecular and functional properties of E-NTPDase-1, E-NTPDase-2 and 5′-ectonucleotidase in nervous tissue. In: Van Duffel L, ed. *Proceedings of the Second International Workshop on Ecto-ATPases and Related Ectonucleotidases.* Maastricht, The Netherlands; Shaker Publishing, 2000:18–25.

76 Wang TF, Handa M, Guidotti G. Structure and function of ectoapyrase (CD39). *Drug Dev Res.* 1998;45(3–4):245–252.

77 Kim M, Yoo OJ, Choe S. Molecular assembly of the extracellular domain of P2X2, an ATP-gated ion channel. *Biochem Biophys Res Commun.* 1997;240(3):618–622.

78 Nicke A, Rettinger J, Buttner C, et al. Evolving view of quaternary structures of ligand-gated ion channels. *Prog Brain Res.* 1999;120:61–80.

79 Chadwick BP, Frischauf AM. The CD39-like gene family – identification of three new human members (CD39L2, CD39L3, and Cd39L4), their murine homologues, and a member of the gene family from Drosophila melanogaster. *Genomics.* 1998;50(3):357–367.

80 Mulero JJ, Yeung G, Nelken ST, et al. CD39-L4 is a secreted human apyrase, specific for the hydrolysis of nucleoside diphosphates. *J Biol Chem.* 1999;274(29):20064–20067.

81 Robson S, Sevigny J, Imai M, et al. Thromboregulatory potential of endothelial cd39/nucleoside triphosphate diphosphohydrolase: modulation of purinergic signalling in platelets. *Emerging Therapeutic Targets.* 2000;4:155–171.

82 Sevigny J, Sundberg C, Braun N, et al. Differential catalytic properties and vascular topography of murine nucleoside triphosphate diphosphohydrolase 1 (NTPDase1) and NTPDase2 have implications for thromboregulation. *Blood.* 2002;99(8):2801–2809.

83 Zimmermann H. 5′-nucleotidase: molecular structure and functional aspects. *Biochem J.* 1992;285:345–365.

84 Resta R, Yamashita Y, Thompson LF. Ectoenzyme and signaling functions of lymphocyte cd73. *Immunol Rev.* 1998;161:95–109.

85 Howard M, Grimaldi JC, Bazan JF, et al. Formation and hydrolysis of cyclic ADP-ribose catalyzed by lymphocyte antigen CD38. *Science.* 1993;262(5136):1056–1059.

86 Malavasi F, Funaro A, Roggero S, et al. Human CD38: a glycoprotein in search of a function. *Immunol Today.* 1994;15(3):95–97.

87 Deaglio S, Dianzani U, Horenstein AL, et al. Human CD38 ligand. A 120-KDA protein predominantly expressed on ECs. *J Immunol.* 1996;156(2):727–734.

88 Deaglio S, Morra M, Mallone R, et al. Human CD38 (ADP-ribosyl cyclase) is a counter-receptor of CD31, an Ig superfamily member. *J Immunol.* 1998;160(1):395–402.

89 Deaglio S, Mehta K, Malavasi F. Human CD38: a (r)evolutionary story of enzymes and receptors. *Leuk Res.* 2001;25(1):1–12.

90 Ferrero E, Malavasi F. The metamorphosis of a molecule: from soluble enzyme to the leukocyte receptor CD38. *J Leukoc Biol.* 1999;65(2):151–161.

91 Koziak K, Kaczmarek E, Kittel A, et al. Palmitoylation targets CD39/endothelial ATP diphosphohydrolase to caveolae. *J Biol Chem.* 2000;275(3):2057–2062.

92 Deaglio S, Zubiaur M, Gregorini A, et al. Human CD38 and CD16 are functionally dependent and physically associated in natural killer cells. *Blood.* 2002;99(7):2490–2498.

93 Pacheco R, Martinez-Navio JM, Lejeune M, et al. CD26, adenosine deaminase, and adenosine receptors mediate costimulatory signals in the immunological synapse. *Proc Natl Acad Sci USA.* 2005;102(27):9583–9588. Epub 2005 Jun 27.

94 Vigne P, Breittmayer JP, Frelin C. Analysis of the influence of nucleotidases on the apparent activity of exogenous ATP and ADP at P2Y(1) receptors. *Br J Pharmacol.* 1998;125(4):675–680.

95 Mizumoto N, Kumamoto T, Robson SC, et al. CD39 is the dominant Langerhans cell associated ecto-NTPDase: Modulatory roles in inflammation and immune responsiveness. *Nat Med.* 2002;8(4):358–365.

96 Pinsky DJ, Broekman MJ, Peschon JJ, et al. Elucidation of the thromboregulatory role of CD39/ectoapyrase in the ischemic brain. *J Clin Invest.* 2002;109(8):1031–1040.

97 Dwyer KM, Robson SC, Nandurkar HH, et al. Thromboregulatory manifestations in human CD39 transgenic mice and the implications for thrombotic disease and transplantation. *J Clin Invest.* 2004;113(10):1440–1446.

98 Imai M, Takigami K, Guckelberger O, et al. Recombinant adenoviral mediated CD39 gene transfer prolongs cardiac xenograft survival. *Transplantation.* 2000;70(6):864–870.

99 Koyamada N, Miyatake T, Candinas D, et al. Apyrase administration prolongs discordant xenograft survival. *Transplantation.* 1996;62(12):1739–1743.

100 Hirsh J, Bhatt DL. Comparative benefits of clopidogrel and aspirin in high-risk patient populations: lessons from the CAPRIE and CURE studies. *Arch Intern Med.* 2004;164(19): 2106–2110.

101 Savi P, Herbert JM. Clopidogrel and ticlopidine: P2Y12 adenosine diphosphate-receptor antagonists for the prevention of atherothrombosis. *Semin Thromb Hemost.* 2005;31(2):174–183.

102 Cattaneo M, Lecchi A, Ohno M, et al. Antiaggregatory activity in human platelets of potent antagonists of the P2Y 1 receptor. *Biochem Pharmacol.* 2004;68(10):1995–2002.

103 Nylander S, Mattsson C, Ramstrom S, et al. Synergistic action between inhibition of P2Y12/P2Y1 and P2Y12/thrombin in ADP- and thrombin-induced human platelet activation. *Br J Pharmacol.* 2004;142(8):1325–1331. Epub 2004 Jul 20.

104 Robson SC. Acute vascular rejection/delayed xenograft rejection and consumptive coagulopathy in xenotransplantation. *Curr Opin Organ Transplant.* 2003;8(1):76–82.

105 Hetherington SL, Singh RK, Lodwick D, et al. Dimorphism in the P2Y1 ADP receptor gene is associated with increased platelet activation response to ADP. *Arterioscler Thromb Vasc Biol.* 2005;25(1):252–257. Epub 2004 Oct 28.

106 Candinas D, Koyamada N, Miyatake T, et al. Loss of rat glomerular ATP diphosphohydrolase activity during reperfusion injury is associated with oxidative stress reactions. *Thromb Haemost.* 1996;76(5):807–812.

107 Guckelberger O, Sun XF, Sevigny J, et al. Beneficial effects of CD39/ecto–nucleoside triphosphate diphosphohydrolase-1 in murine intestinal ischemia-reperfusion injury. *Thromb Haemost.* 2004;91(3):576–586.

108 Sampram ES, Ouriel K. In vitro verification of antithrombotic effect of recombinant soluble nucleotide triphosphate diphosphohydrolase 1. *J Vasc Interv Radiol.* 2004;15(4):379–384.

109 Hatakeyama K, Hao H, Imamura T, et al. Relation of CD39 to plaque instability and thrombus formation in directional atherectomy specimens from patients with stable and unstable angina pectoris. *Am J Cardiol.* 2005;95(5):632–635.

44

Syndecans

Eugene Tkachenko, John M. Rhodes, and Michael Simons

Dartmouth Medical School, Hanover, New Hampshire

Proteoglycans consist of negatively charged sulfated unbranched glycosaminoglycan (GAG) chains that are covalently attached to transmembrane- or membrane-anchored proteins. Such proteins carrying various GAG chains are referred to as *core proteins.* GAGs consist of a repeating disaccharide unit. The major GAGs are hyaluronic acid, dermatan sulfate, chondroitin sulfate (CS), heparan sulfate (HS), and keratin sulfate. Different GAG moieties are found on different core proteins, including syndecans, glypicans, betaglycan, lumican, and perlecan. This chapter discusses the syndecan family of core proteins.

Syndecans are a family of transmembrane proteins that perform several distinct functions (Table 44-1). By virtue of their wide expression and their ability to interact with a large number of ligands, ranging from circulating growth factors to extracellular matrix (ECM) proteins, syndecans can integrate and coordinate various cellular signaling inputs. Furthermore, they can activate specific signaling cascades and induce profound changes in cell structure and behavior. This ability to integrate and modulate various signaling inputs, as well as to signal in their own right, has earned this family of cell surface receptors the name "tuners of transmembrane signaling" (1).

In addition, due to shedding of their extracellular domains that include large HS and CS chains, syndecans have the ability to alter the protein and water context of the surrounding ECM, greatly affecting storage, distribution, and transport of many extracellular proteins.

SYNDECAN STRUCTURE

The syndecan family in all vertebrates is composed of four genes (syndecan-1, -2, -3, -4). Each syndecan possesses a protein core which, in its extracellular domain, has attachment sites for three to five HS and/or CS chains (Figure 44.1). Among different syndecans, the chains are structurally diverse and their exact composition determines the extent of interactions with various HS- or CS-binding proteins. This structural diversity results from a series of post-translational modifications in which newly synthesized chains are modified by epimerization of glucuronic to iduronic acid residues and sulfation of the C2 position of glucuronic and/or iduronic acid (for further details Ref. 2).

Various combinations of enzymes involved in post-translational chain modifications produce unique binding motifs that selectively recognize different extracellular proteins. Thus, a specific combination of sulfotransferase-mediated 2-*O* and 6-*O* sulfation is necessary for synthesis of the fibroblast growth factor 2 (FGF2)-binding site, whereas 3-*O*-sulfotransferase 1 activity is needed for generation of the antithrombin-III–binding site. These modifications can play an important biological role as demonstrated, for example, in the case of 2-*O*-sulfotransferase (2-OST) deficiency, which results in marked abnormalities in FGF signaling, whereas augmentation of 2-OST expression leads to enhanced FGF responsiveness (3). It has been generally assumed that HS and CS chains lack preference for different syndecans. However, a recent study has suggested that HS and CS chains on syndecan-1 and syndecan-4 are structurally different (4). Another layer of complexity has been added by the discovery of HS-specific 6-*O*-sulfatases (Sulf 1 and Sulf2), which can act extracellularly and have the ability to modify heparin binding growth factor signaling (5).

The syndecan core protein is composed of an extracellular, single-span transmembrane domain and a short intracellular domain. The extracellular domains are widely divergent in their amino acid composition among the four syndecans but all have the nucleation sites for attachment of HS and/or CS chains. In contrast, the transmembrane domains of syndecans are highly conserved. A unique feature of these domains is their high affinity for self-association. In fact, all syndecans probably exist as homo- and heterodimers, with higher-order aggregates forming upon binding to partner proteins.

The cytoplasmic domain in all syndecans has three distinct regions – conserved regions 1 and 2 (C1 and C2) and a variable region (V). The C1 domain is virtually identical in all four mammalian syndecans. It is involved in syndecan dimerization and has the ability to bind several intracellular proteins including ezrin, tubulin, Src kinase, and cortactin (6,7). The highly conserved C2 domain contains a

Table 44-1: Syndecan Family

Syndecan	*Expression*	*Cell Signal Intermediates*	*Function*
S-1	Epithelial and mesenchymal tissues; VSMCs	Syntenin, synectin, synbindin, and CASK	Inhibition of cell growth; cell–cell and cell–ECM interactions; establishing of epithelial cells polarity; role in ephrin signaling
S-2	Epithelial, mesenchymal, and neuronal tissues	Syntenin, synectin, synbindin, and CASK	Stimulation of cell growth; important in vascular development (zebrafish); capillary sprouting; regulation of TGF-β signaling; role in ephrin signaling
S-3	Neuronal and MSK tissue	Syntenin, synectin, synbindin, and CASK	Inhibition of cell growth; feeding behavior control; role in muscle differentiation
S-4	All cells	Syntenin, synectin, synbindin and CASK; PKCα, Src, syndesmos, ezrin	Stimulation of cell growth; control of postnatal angiogenesis, mediation of FGF2 signaling; cell spreading, cell migration, and focal adhesion formation

PDZ2-binding site that can bind four syndecan-interacting PDZ domain-containing proteins: syntenin, synectin, synbindin, and CASK (8–11). The specificity of binding of the various syndecans to these PDZ-binding partners has not been established.

The V domain is, as its name implies, highly heterogeneous among the four syndecans. In syndecan-4, the V-region sequence includes a phosphatidylinositol-(4,5)-bisphosphate (PIP_2) binding site that is involved in syndecan-4 dimerization, in which two syndecan-4 cytoplasmic domains are linked by two PIP_2 molecules in antiparallel fashion (see Figure 44.1) (12).

SYNDECAN EXPRESSION

Syndecan-1 is expressed predominantly in epithelial and mesenchymal tissues; syndecan-2 in cells of mesenchymal origin, neuronal and epithelial cells; syndecan-3 almost exclusively in neuronal and musculoskeletal tissue; and syndecan-4 in virtually every cell type (6). Although endothelial cells (ECs) express both syndecan-2 and -4, syndecan-4 is the predominant species in adult ECs. Growth factors and mechanical stress play an important role in the regulation of syndecan expression. Thus, in aortic vascular smooth muscle cells (VSMCs), FGF2 induces syndecan-4, but not syndecan-1 or -2 expression. Likewise, mechanical stress is a prominent inducer of syndecan-4 expression in VSMCs during arteriogenesis (13). A number of inflammatory conditions such as acute myocardial infarction (14) or wounds (15) also are associated with increased syndecan-4 expression in VSMCs and endothelium. One interesting mediator of syndecan expression is an inflammatory cell-derived peptide, PR39, that has a unique ability to markedly upregulate syndecan-1 and -4 expression in vitro and in vivo in a number of cell types (16).

THE ROLE OF SYNDECANS IN DEVELOPMENT

Several syndecans play an important role in the development of the vascular system. Exposure of zebrafish embryos to syndecan-2 morpholino oligonucleotides resulted in suppression of intersegmental vessels, although formation of dorsal vessels (aorta, cardinal vein) was not affected (17). Furthermore, application of moderate doses of syndecan-2 and vascular endothelial growth factor (VEGF) morpholinos demonstrated synergistic inhibition of angiogenesis, whereas coexpression of syndecan-2 and VEGF-A_{165} resulted in an increase of vessel formation compared with VEGF alone (17). These results suggest that syndecan-2 potentiates VEGF-induced capillary sprouting, but the precise mode of interactions between syndecan-2 and VEGF has not been established.

Homozygous disruption of syndecan-4 expression in mice leads to frequent thrombus formation in the vessels of the placental labyrinth, resulting in high embryo loss and suggesting a role for syndecan-4 in blood clotting (18). The adult *syndecan-4*$^{-/-}$ mice have impaired skin wound healing that is thought to be secondary to defective angiogenesis (19) and demonstrate increased sensitivity to lipopolysaccharide-induced shock.

SYNDECAN-DEPENDENT SIGNALING

Syndecans engage in signal transduction in several distinct ways. Due to the ability of their HS and CS chains to bind growth factors, thereby restricting their presence to the membrane surface and facilitating their subsequent interactions with corresponding high-affinity receptors, syndecans can serve as coreceptors for these growth factors. Indeed, inhibition of HS chain synthesis grossly affects signaling (the role of CS chains in syndecan-dependent signaling has not been established). For example, heparinase treatment of cultured VSMCs decreases their responsiveness to FGF2 (20), whereas heparinase treatment in vivo inhibits angiogenesis (21). Similarly, a homozygous deletion of 2-*OST* results in mice that die perinatally because of the complete failure of kidney formation (22), whereas a knockdown of the 6-*O*-sulfotransferase in *Drosophila* severely perturbs tracheal development (23),

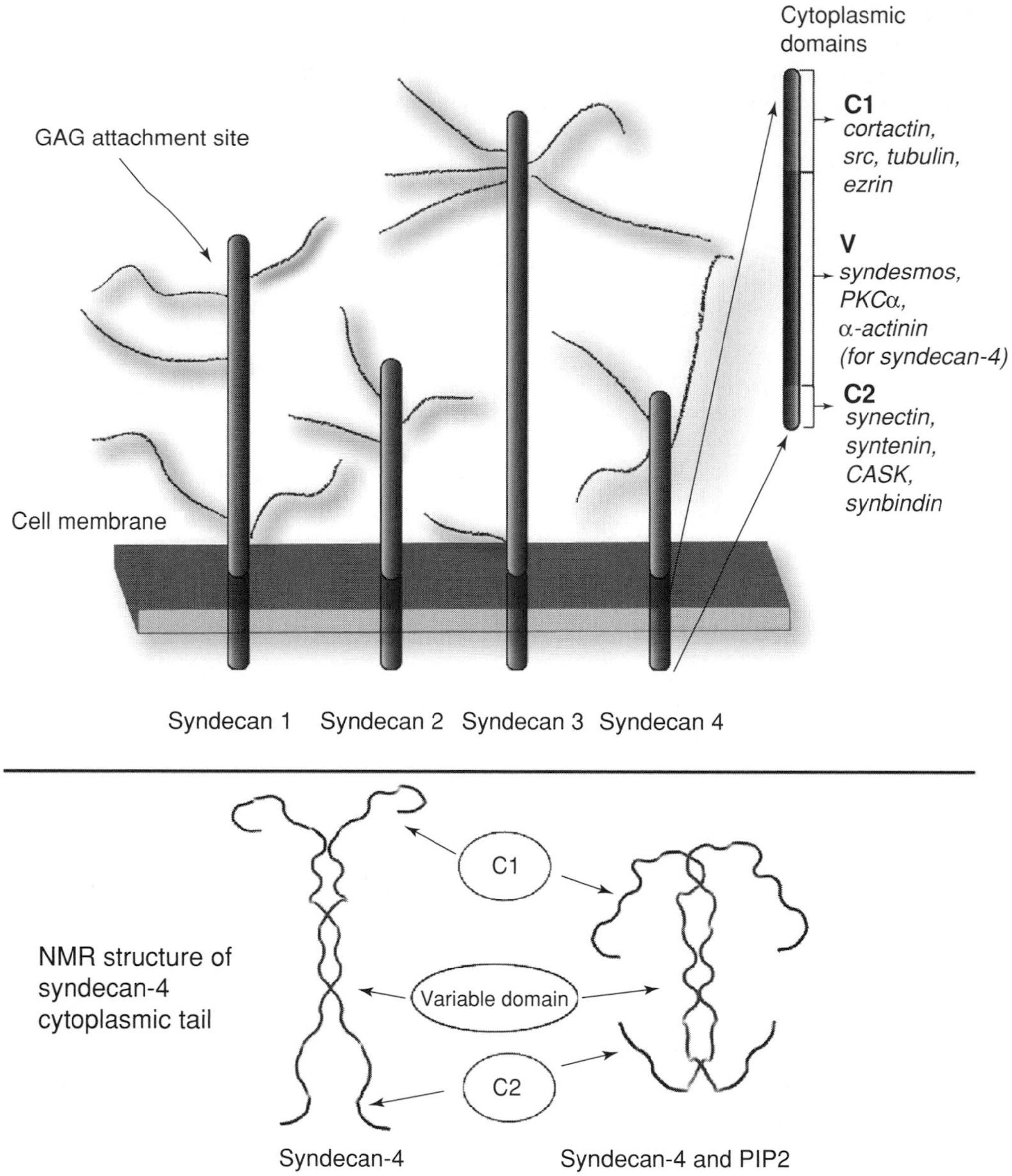

Figure 44.1. Syndecan structure. *Upper panel.* Cartoon representation of the vertebrate syndecan structure. *Lower panel.* NMR structure of the syndecan-4 dimer in the presence and absence of PIP2. Note that interaction with PIP2 results in reorientation of PDZ-binding domains.

an FGF-dependent process. Deletions of other genes involved in HS biosynthesis have equally profound phenotypic effects.

In addition to HS chain-dependent signaling, syndecans can initiate signaling via their protein core domains by activating a variety of intracellular signaling pathways, including mitogen-activated protein kinase (MAPK), Src, Rac1, and protein kinase C (PKC)α. Each of the syndecans has its own characteristic biology that will be briefly reviewed in the following sections.

In approaching syndecan-mediated signaling it is useful to think of syndecan-1 and -3 as one subfamily and syndecan-2 and -4 as another. This reflects both the homology of variable domains between corresponding syndecans as well as functional similarities. Thus, although expression of both syndecan-1 and -3 is associated with the inhibition of cell growth, that of syndecan-2 and -4 is associated with its stimulation.

SYNDECAN-1

Syndecan-1 plays a role in the regulation of cell–cell and cell–ECM interactions. Because its expression is predominantly limited to the cells of ectoderm-derived tissues, it is not clear how important a role it plays in the vascular system, although its expression in certain types of VSMCs and endothelium during inflammation has been noted. In epithelial cells, downregulation of syndecan-1 expression results in loss of cell polarity, associated with a reduced level of E-cadherin on the cell surface, suggesting involvement in the epithelial–mesenchymal switch during development and in wound healing (24). Overexpression of syndecan-1 leads to the increased shedding of its ectodomain, which results in inhibition of FGF2–induced proliferation due to sequestering of the growth factor (25). Consequently, mice overexpressing syndecan-1 have delayed dermal wound repair (26).

Syndecan-1 plays an important role in the regulation of inflammation, as suggested by increased leukocyte–endothelial interactions in *syndecan-1*–null mice. It also plays an important role in chemokine gradient formation for the transendothelial and transepithelial migration of neutrophils (27).

Syndecan-1 also plays a role in regulating ephrin signaling in venous–arterial differentiation. During this process, EphB4 is expressed in venule endothelium and a homozygous disruption of EphB4 or one of its ligands, ephrinB2, results in fatal vascular defects. Activation of EphB4-positive ECs causes upregulation of syndecan-1 (28), resulting in suppression of FGF2 signaling.

SYNDECAN-2

Syndecan-2 is the predominant syndecan expressed during embryonic development. In adult cells, it is involved in regulation of transforming growth factor (TGF)-β signaling. Unlike other syndecans that bind to growth factors via HS and CS chains, syndecan-2 directly binds TGF-β via a protein–protein interaction. Furthermore, syndecan-2 coimmunoprecipitates with the TGF-β as well as with the type III TGF-β receptor (betaglycan) and expression of a syndecan-2 mutant with a truncated cytoplasmic domain results in impaired response to TGF-β (29).

Similar to syndecan-1, syndecan-2 also may play a role in ephrin signaling. Cell surface ephrin–Eph signaling in ECs induces clustering of syndecan-2 and recruitment of cytoplasmic molecules, which leads to localized actin polymerization via Rho family GTPases, N-WASP, and the Arp2/3 complex (30).

SYNDECAN-3

Syndecan-3 does not play an important role in the vascular system. However, it plays an important role in skeletal muscle differentiation. A knockout of syndecan-3 results in expression of myogenin, a master transcription factor for muscle differentiation, and in accelerated skeletal muscle differentiation and myoblast fusion. Syndecan-3 also appears to be involved in the regulation of hedgehog signaling by means of proper spatial and temporal regulation of Indian hedgehog (Ihh) (31).

SYNDECAN-4

Syndecan-4 has become the most extensively studied member of the syndecan family, and it plays an important role in vascular biology due to its ability to modulate FGF2 signaling, regulate cell migration via cross-talk with β1 integrin, and control adhesion via cytoskeletal modifications (7). Some of these functions are probably shared with other syndecans, but most derive from the presence of syndecan-4–specific cytoplasmic V domain.

Syndecan-4 directly initiates a number of intracellular signaling events including activation of PKCα, Src, and Rac1. The initiation of syndecan-4 signaling is prompted through the clustering of its extracellular domains by a ligand that can be a soluble growth factor, a matrix protein such as fibronectin, or an antibody. The importance of this signaling pathway is further emphasized by the fact that introduction of a syndecan-4 construct with the mutated PDZ or PIP_2 binding regions that, respectively, eliminated syndecan-4 ability to bind PDZ proteins or PIP_2, inhibited FGF2– but not serum-induced EC growth, migration, and the ability to form vascular structures on Matrigel (32).

Although it is clear that syndecan-4 can modulate FGF2 signaling via its cytoplasmic tail interactions, the precise mechanism of this effect is uncertain. A sustained activation of FGF2 signaling requires internalization and perhaps nuclear transport of the ligand. Syndecan-4 plays an important role in FGF2 endocytosis in ECs that proceeds in a clathrin- and dynamin-independent fashion, requires syndecan-4-dependent activation of Rac1, and strongly resembles macropinocytosis (33).

A unique feature of syndecan-4 signaling is activation of PKCα in the absence of Ca^{2+}. This ability of syndecan-4 to activate a calcium-dependent PKC, first shown by Oh and colleagues, requires core-protein multimerization and the presence of PIP_2 (34). This event depends on dephosphorylation of the Ser^{183} site in its cytoplasmic domain (32), whereas its phosphorylation markedly reduces its affinity for PIP_2 (35) and prevents PKCα activation in vitro and in vivo (35). Remarkably, PKCδ phosphorylates the Ser^{183} site, thus setting up a system that allows one PKC to regulate the activity of another (Figure 44.2) (36).

The molecular details of syndecan-4–dependent PKCα activation have not yet been established. The results of yeast two-hybrid screening demonstrated that although a full-length PKCα weakly bound the variable domain of syndecan-4 (4V), the PKCα constructs that lack the pseudosubstrate region or that encode only the whole catalytic domain interacted more strongly (37). A mutation of Tyr^{192} in the syndecan-4 cytoplasmic domain abolished this interaction (37). Interestingly, the corresponding binding site in the catalytic domain of PKCα (amino acid sequence 513–672) encompasses the regulatory autophosphorylation sites implicated in control of PKC activation and stability (37). On the other hand, in vitro surface plasmon resonance studies suggested that binding affinity of a full length PKCα to the

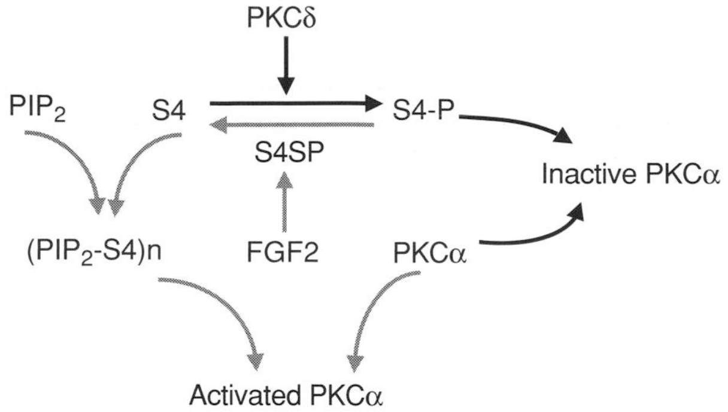

Figure 44.2. Syndecan-4-dependent modulation of PKC activity. Under normal circumstances, the Ser[183] site is phosphorylated by PKCδ. Exposure to FGF2 results in dephosphorylation of the site, which in turn allows syndecan-4 oligomerization, binding of PIP2, and activation of PKCα. Any signal that would activate PKCδ would impair syndecan-4 dependent PKCα activation.

syndecan-4 cytoplasmic domain is very low in the absence of PIP_2 (35).

Syndecan-4 colocalizes with PKCα in focal adhesions, suggesting a role in cell adhesion, spreading, and stress fiber formation. Indeed, studies in cultured fibroblasts have shown that overexpression of syndecan-4, but not of a mutant lacking its cytoplasmic domain, specifically increases the level of endogenous PKCα and enhances the translocation of PKCα in the plasma cell membrane in general and in the membrane raft fractions in particular, and increases the activity of membrane PKCα (38).

CELL ADHESION AND THE CYTOSKELETON

One of the important roles played by syndecans is the modulation of cytoskeletal rearrangements. For example, syndecan HS binding to fibronectin leads to cell spreading, and overexpression of syndecan-4 results in cell flattening as well as increase in focal adhesion formation and a decrease in cell migration in fibroblasts (39). In contrast, *syndecan-4*–null fibroblasts plated on fibronectin exhibit enhanced lamellipodia formation and increased level of Rac1 (40) and low Rho (41) activities, compared with wild-type cells. Interestingly, in ECs also plated on fibronectin, clustering of syndecan-4 constructs by antibodies results in upregulation of Rac1 activity (33). These differences in Rac activation might be due to the differences in syndecan-4 localization following binding to matrix versus soluble ligand. In this way, growth factors might modulate cell–matrix interactions by removal of the syndecan component via endocytosis (33).

In VSMCs, shear stress causes syndecan-4 dissociation from the focal adhesions (13). Overexpression of syndecan-4 blocks this dissociation and also results in reduced mechanical stress–induced cell migration. This observation suggests that syndecan-4 might be involved in regulation of VSMC migration during arteriogenesis.

Syndecan-4, but not syndecan-1 or -2, coimmunoprecipitates with CXCR4. This syndecan-4/CXCR4 complex is likely a functional unit involved in chemokine stromal cell-derived factor (SDF)-1 binding that may play a role in SDF-1-dependent stem cell recruitment, a process that can be involved in angiogenesis, atherogenesis, and vascular repair, among other processes.

Finally, syndecans may regulate the cytoskeleton via their interactions with members of the ezrin-radixin-moesin (ERM) family. The ERM proteins have NH_2- and COOH-terminal domains that associate with the plasma membrane and the actin cytoskeleton, respectively. Association of syndecan-2 with actin cytoskeleton through ezrin binding was observed in COS-1 cells following expression of a constitutively active form of RhoA (42). Thus, ezrin provides RhoGTPases with a regulated link between syndecan-2 and actin. The ezrin binding motif is identical between syndecan-2 and -4, implying the possibility of syndecan-4/ezrin interactions (42).

CONCLUSION

Syndecans are a family of four single-span transmembrane proteoglycans with a divergent amino terminal extracellular domain decorated with heparan and chondroitin glycosaminoglycans and a conserved intracellular domain. These proteins modulate signaling shown to be critical in angiogenesis and development.

KEY POINTS

- Signaling of heparin-binding growth factors via syndecans is mediated primarily via the structural diversity of their glycosaminoglycan chains. This diversity is governed by changes in the pattern and sulfation of the GAG's "building-block" monosaccharides.
- Syndecan-1 is primarily limited to cells of ectoderm-derived tissues, where it is involved in regulation of cell–cell and cell–matrix interactions. It also plays an important role in the regulation of inflammation and in chemokine gradient formation in the transendothelial and transepithelial migration of neutrophils, as revealed by *syndecan-1*–null mice experiments. Ephrin signaling in venous–arterial differentiation is regulated partially by syndecan-1, because activation of EphB4 causes upregulation of syndecan-1 and thus suppressing FGF2 signaling.
- Syndecan-2 is expressed in mesenchymal and epithelial as well as neuronal tissues but mainly during embryonic development. Direct protein–protein

interaction between TGF-β and syndecan-2 as well as immunoprecipitation of the type III TGF-β receptor (betaglycan) suggests participation in the TGF-β pathway. A role in ephrin signaling also is suggested by ephrin-Eph signaling-induced clustering of syndecan-2.

- Syndecan-3, although not appearing to play an important role in the vascular system, does have a role in skeletal muscle differentiation, which correlates with its primary expression in neuronal and musculoskeletal tissue. Knockouts of syndecan-3 result in myoblast fusion and accelerated skeletal muscle differentiation as a result of improper expression of myogenin. Indian hedgehog's proper temporal and spatial regulation also appears to depend on syndecan-3 expression.
- Syndecan-4 is expressed almost ubiquitously in all tissues. Through its ability to "tune" FGF2 signaling, syndecan-4 regulates cell migration via cross-talk with β1 integrin and controls adhesion via cytoskeletal modifications. Initiation of signaling events is prompted by clustering of its extracellular domain via soluble or matrix proteins. Using its intracellular domain, syndecan-4 has been shown to activate PKCα, Src, and Rac1. Although the exact mechanism of syndecan-4 modulation of FGF2 signaling is unknown, it has been demonstrated that syndecan-4 is important in FGF2 endocytosis in ECs by a clathrin- and dynamin-independent process requiring syndecan-4–dependent activation of Rac1. It also has been revealed to interact with CXCR4 and the ERM family of proteins, suggesting roles in angiogenesis and cytoskeletal regulation, respectively.

Future Goals

- To further elucidate syndecan-4's role in the FGF2 signaling pathway: Specifically, (how) does endocytosis mediated by syndecan-4 regulate the temporal signaling of the FGF2 pathway? Similarly, is syndecan-4 involved in signaling by other heparin-binding growth factors?
- To explore of the role of syndecan-4 in the activation/inactivation of G-proteins and the actin cytoskeleton: What other proteins are involved in the signaling pathway, and how are they regulated in the context of syndecan-4 mediated migration?
- To examine the role of syndecans as scaffold proteins: What growth factors or morphogens are temporally and spatially localized by syndecans, and how is this process regulated in development?

ACKNOWLEDGMENTS

We would like to thank members of the Simons' laboratory and the Angiogenesis Research Center colleagues for helpful discussions and incisive comments. Supported in part by NIH grants HL62289 and HL63609 (MS).

REFERENCES

1 Zimmermann P, David G. The syndecans, tuners of transmembrane signaling. *FASEB J.* 1999;13:S91–S100.
2 Esko JD, Selleck SB. Order out of chaos: assembly of ligand binding sites in heparan sulfate. *Annu Rev Biochem.* 2002;71:435–471.
3 Li J, Shworak NW, Simons M. Increased responsiveness of hypoxic ECs to FGF2 is mediated by HIF-1alpha-dependent regulation of enzymes involved in synthesis of heparan sulfate FGF2-binding sites. *J Cell Sci.* 2002;115(Pt 9):1951–1959.
4 Deepa SS, Yamada S, Zako M, et al. Chondroitin sulfate chains on syndecan-1 and syndecan-4 from normal murine mammary gland epithelial cells are structurally and functionally distinct and cooperate with heparan sulfate chains to bind growth factors. A novel function to control binding of midkine, pleiotrophin, and basic fibroblast growth factor. *J Biol Chem.* 2004;279(36):37368–37376.
5 Ai X, Do AT, Lozynska O, et al. QSulf1 remodels the 6-O sulfation states of cell surface heparan sulfate proteoglycans to promote Wnt signaling. *J Cell Biol.* 2003;162(2):341–351.
6 Couchman JR. Syndecans: proteoglycan regulators of cell-surface microdomains? *Nat Rev Mol Cell Biol.* 2003;4(12):926–937.
7 Tkachenko E, Rhodes JM, Simons M. Syndecans: new kids on the signaling block. *Circ Res.* 2005 2005;96(5):488–500.
8 Gao Y, Li M, Chen W, et al. Synectin, syndecan-4 cytoplasmic domain binding PDZ protein, inhibits cell migration. *J Cell Physiol.* 2000;184(3):373–379.
9 Grootjans JJ, Zimmermann P, Reekmans G, et al. Syntenin, a PDZ protein that binds syndecan cytoplasmic domains. *Proc Natl Acad Sci USA.* 1997;94(25):13683–13688.
10 Hsueh YP, Yang FC, Kharazia V, et al. Direct Interaction of CASK/LIN-2 and Syndecan Heparan Sulfate Proteoglycan and Their Overlapping Distribution in Neuronal Synapses. *J Cell Biol.* 1998;142(1):139–151.
11 Ethell IM, Hagihara K, Miura Y, et al. Synbindin, A novel syndecan-2-binding protein in neuronal dendritic spines. *J Cell Biol.* 2000;151(1):53–68.
12 Shin J, Lee W, Lee D, et al. Solution structure of the dimeric cytoplasmic domain of syndecan-4. *Biochemistry.* 2001;40(29):8471–8478.
13 Li L, Chaikof EL. Mechanical stress regulates syndecan-4 expression and redistribution in vascular smooth muscle cells. *Arterioscler Thromb Vasc Biol.* 2002;22(1):61–68.
14 Li J, Brown LF, Laham RJ, et al. Macrophage-dependent regulation of syndecan gene expression. *Circ Res.* 1997;81(5):785–796.
15 Geary RL, Koyama N, Wang TW, et al. Failure of heparin to inhibit intimal hyperplasia in injured baboon arteries. The role of heparin-sensitive and -insensitive pathways in the stimulation of smooth muscle cell migration and proliferation. *Circulation.* 1995;91(12):2972–2981.
16 Li J, Post M, Volk R, et al. PR39, a peptide regulator of angiogenesis. *Nat Med.* 2000;6(1):49–55.

17 Chen E, Hermanson S, Ekker SC. Syndecan-2 is essential for angiogenic sprouting during zebrafish development. *Blood*. 2004; 103(5):1710–1719.
18 Ishiguro K, Kadomatsu K, Kojima T, et al. Syndecan-4 deficiency impairs the fetal vessels in the placental labyrinth. *Dev Dyn*. 2000;219(4):539–544.
19 Echtermeyer F, Streit M, Wilcox-Adelman S, et al. Delayed wound repair and impaired angiogenesis in mice lacking syndecan-4. *J Clin Invest*. 2001;107(2):R9–R14.
20 Chu CL, Buczek-Thomas JA, Nugent MA. Heparan sulphate proteoglycans modulate fibroblast growth factor-2 binding through a lipid raft-mediated mechanism. *Biochem J*. 2004;379(Pt2):331–341.
21 Sasisekharan R, Moses MA, Nugent MA, et al. Heparinase inhibits neovascularization. *Proc Natl Acad Sci USA*. 1994;91(4): 1524–1528.
22 Wilson VA, Gallagher JT, Merry CL. Heparan sulfate 2-O-sulfotransferase (Hs2st) and mouse development. *Glycoconj J*. 2002;19(4–5):347–354.
23 Kamimura K, Fujise M, Villa F, et al. Drosophila heparan sulfate 6-O-sulfotransferase (dHS6ST) gene. Structure, expression, and function in the formation of the tracheal system. *J Biol Chem*. 2001;276(20):17014–17021.
24 Woods A, Couchman JR. Syndecan-4 and focal adhesion function. *Curr Opin Cell Biol*. 2001;13(5):578–583.
25 Mali M, Elenius K, Miettinen HM, et al. Inhibition of basic fibroblast growth factor-induced growth promotion by overexpression of syndecan-1. *J Biol Chem*. 1993;268(32):24215–24222.
26 Elenius V, Gotte M, Reizes O, et al. Inhibition by the soluble syndecan-1 ectodomains delays wound repair in mice overexpressing syndecan-1. *J Biol Chem*. 2004;279(40):41928–41935.
27 Gotte M. Syndecans in inflammation. *FASEB J*. 2003;17(6):575–591.
28 Yuan K, Hong TM, Chen JJ, et al. Syndecan-1 up-regulated by ephrinB2/EphB4 plays dual roles in inflammatory angiogenesis. *Blood*. 2004;104(4):1025–1033.
29 Chen L, Klass C, Woods A. Syndecan-2 regulates transforming growth factor-beta signaling. *J Biol Chem*. 2004;279(16):15715–15718.
30 Irie F, Yamaguchi Y. EPHB receptor signaling in dendritic spine development. *Front Biosci*. 2004;9:1365–1373.
31 Shimo T, Gentili C, Iwamoto M, et al. Indian hedgehog and syndecans-3 coregulate chondrocyte proliferation and function during chick limb skeletogenesis. *Dev Dyn*. 2004;229(3):607–617.
32 Horowitz A, Tkachenko E, Simons M. Fibroblast growth factor-specific modulation of cellular response by syndecan-4. *J Cell Biol*. 2002;157(4):715–725.
33 Tkachenko E, Lutgens E, Stan RV, et al. Fibroblast growth factor 2 endocytosis in ECs proceed via syndecan-4-dependent activation of Rac1 and a Cdc42-dependent macropinocytic pathway. *J Cell Sci*. 2004;117(Pt 15):3189–3199.
34 Oh ES, Woods A, Couchman JR. Syndecan-4 proteoglycan regulates the distribution and activity of protein kinase C. *J Biol Chem*. 1997;272(13):8133–8136.
35 Horowitz A, Murakami M, Gao Y, et al. Phosphatidylinositol-4,5-bisphosphate Mediates the Interaction of Syndecan-4 with Protein Kinase C. *Biochemistry*. 1999;38(48):15871–15877.
36 Murakami M, Horowitz A, Tang S, et al. PKC-delta regulates PKC-alpha activity In a syndecan-4 dependent manner. *J Biol Chem*. 2002;277(23):20367–20371.
37 Lim ST, Longley RL, Couchman JR, et al. Direct binding of syndecan-4 cytoplasmic domain to the catalytic domain of protein kinase C alpha (PKC alpha) increases focal adhesion localization of PKC alpha. *J Biol Chem*. 2003;278(16):13795–13802.
38 Keum E, Kim Y, Kim J, et al. Syndecan-4 regulates localization, activity and stability of protein kinase C-alpha. *Biochem J*. 2004; 378(Pt 3):1007–1014.
39 Longley RL, Woods A, Fleetwood A, et al. Control of morphology, cytoskeleton and migration by syndecan-4. *J Cell Sci*. 1999;112 (Pt 20):3421–3431.
40 Saoncella S, Calautti E, Neveu W, et al. Syndecan-4 regulates ATF-2 transcriptional activity in a Rac1-dependent manner. *J Biol Chem*. 2004;279(45):47172–47176.
41 Wilcox-Adelman SA, Denhez F, Goetinck PF. Syndecan-4 modulates focal adhesion kinase phosphorylation. *J Biol Chem*. 2002; 277(36):32970–32977.
42 Granes F, Berndt C, Roy C, et al. Identification of a novel Ezrin-binding site in syndecan-2 cytoplasmic domain. *FEBS Lett*. 2003;547(1–3):212–216.

45

Sphingolipids and the Endothelium

Timothy Hla

Center for Vascular Biology, University of Connecticut Health Center, Farmington

Structural diversity in the membrane lipids is essential for proper function of eukaryotic membranes. Sphingolipids, named after the mythical Sphinx due to their enigmatic physicochemical properties, are critical for specialized membrane microdomains such as rafts and caveolae. In higher organisms, sphingolipid metabolites, such as sphingosine, sphingosine 1-phosphate (S1P), and ceramide are utilized in intracellular metabolism, signaling, and extracellular cell–cell communication events. Sphingolipids are particularly important for the mammalian vascular system, in which plasma levels of S1P are several orders of magnitude higher than tissue levels, thereby constituting a vascular S1P gradient. S1P is a major regulator of vital endothelial cell (EC) functions including vascular permeability, vascular stabilization/maturation, angiogenesis, vascular tone control, nitric oxide (NO) synthesis, and survival. The endothelium produces and responds to S1P. Indeed S1P receptors in ECs are dynamically induced and regulated. Thus, sphingolipids (sphingomyelin [SM], sphingosine, ceramide, S1P, as well as numerous glycosphingolipid [GSL] species) constitute a major class of molecules involved in the structure and function of the endothelium (Table 45-1). Further knowledge in this area promises to provide novel therapeutic approaches in the restoration of endothelial health, a vital phenomenon compromised in many human maladies.

SPHINGOLIPID METABOLISM

Sphingolipids, which are built from the sphingosine base, are major structural components of the biological membranes of all eukaryotes. Both GSLs and SM belong to this class of lipids. Sphingolipids are localized on specialized domains of the membranes (1). For example, SM and GSL are essential constituents of membrane rafts. SM contains the sphingosine backbone, a fatty-acid side chain that is linked via the amide bond, and the amphipathic phosphocholine headgroup. Most SM species contain saturated or *trans*-monounsaturated fatty acids with 16 to 24 carbons. This contrasts with the *cis*-polyunsaturated fatty acids present in the *sn2* position of glycerol-based phospholipids. In addition, sphingolipids are generally found associated with cholesterol, which is thought to be essential in the formation of plasma membrane microdomains and of the detergent-insoluble domains commonly referred to as *rafts*.

In addition to their structural role, sphingolipids are metabolized to intermediates that are utilized as intracellular regulatory lipids, metabolic intermediates and extracellular mediators and receptors. For example, sphingosine and ceramide are studied extensively as intracellular regulatory lipids (2). Hexadecanal and phosphoethanolamine, which are products of the S1P lyase enzyme, are utilized as intermediates in the phospholipid biosynthetic pathway (3). S1P is recognized as an extracellular mediator that activates cell surface G-protein–coupled receptors (4), and GSLs are recognized as ligand–receptor pairs that are involved in extracellular cell–cell recognition (5).

The metabolic pathway of sphingolipid metabolism in mammals is shown in Figure 45.1. The de novo pathway is present in all cells and produces SM. Many of the pathway's enzymes have been characterized molecularly from the yeast *Saccharomyces cerevisiae* (6). The biosynthesis of SM takes place in the intracellular membranes, such as the Golgi complex. Once formed, SM is shuttled between specific membrane compartments. The transport systems of sphingolipids is not well understood, even though specific transport proteins, such as the ceramide transporter, have been described. SM also is degraded in the sphingomyelinase (SMase) pathway. The sequential action of SMase, ceramidase, and sphingosine kinase (Sphk) results in the formation of S1P. The formation of S1P occurs at several locations in the cell, such as the cytosol, the intracellular face of the plasma membrane, and the extracellular compartment. S1P is degraded by the S1P phosphatase or the S1P lyase, which are found on the endoplasmic reticulum (ER) membrane (7). The subcellular topology, localization, and transport of various sphingolipid mediators, as well as the metabolic enzymes, are poorly understood at present. It is likely that site-specific formation of sphingolipid mediators is important to their bioactivity and is coupled tightly to the respective effector systems in cells.

Table 45-1: S1P Receptors and Endothelial Cell Expression

Receptor	*Expression in ECs*	*EC Phenotype*	*Induction*	*Expression in VSMCs*	*VSMC Phenotype*	*–/–*
$S1P_1$ (also known as EDG-1)	YES	Induces migration, proliferation, cell–cell adhesion; tube formation (S1P3 needed as well); inhibits permeability; enhances survival and angiogenesis; induces NO	Shear stress, phorbol esters, and angiogenic growth factors; increased in tumor endothelium	Induced (activated, proliferative VSMCs)	Promotes cell migration	Embryonic lethal at E12.5–14.5; hemorrhage and vascular leak
$S1P_2$ (also known as EDG-5)	+/–	Inhibits migration and proliferation	Not known	YES (quiescent, synthetic VSMCs)	Inhibits cell migration; promotes contraction	Viable, but cooperates with $S1P_1$ in vascular development
$S1P_3$ (also known as EDG-3)	YES	Functions with $S1P_1$	TNF-α	YES (quiescent, synthetic VSMCs)	Promotes contraction	Viable; impaired endothelium-dependent vasorelaxation

The vascular endothelium is enriched in sphingolipids and participates actively in sphingolipid metabolism and signaling. The strategic anatomic position of the endothelium enables access to the abundant sphingolipids present in plasma, blood-borne cells, and lipoproteins. Indeed, S1P and sphingomyelin are abundant in plasma. In addition, ECs express S1P receptors and therefore respond to S1P (8). Moreover, expression of Sphk isoenzymes as well as ceramidase and SMase by ECs

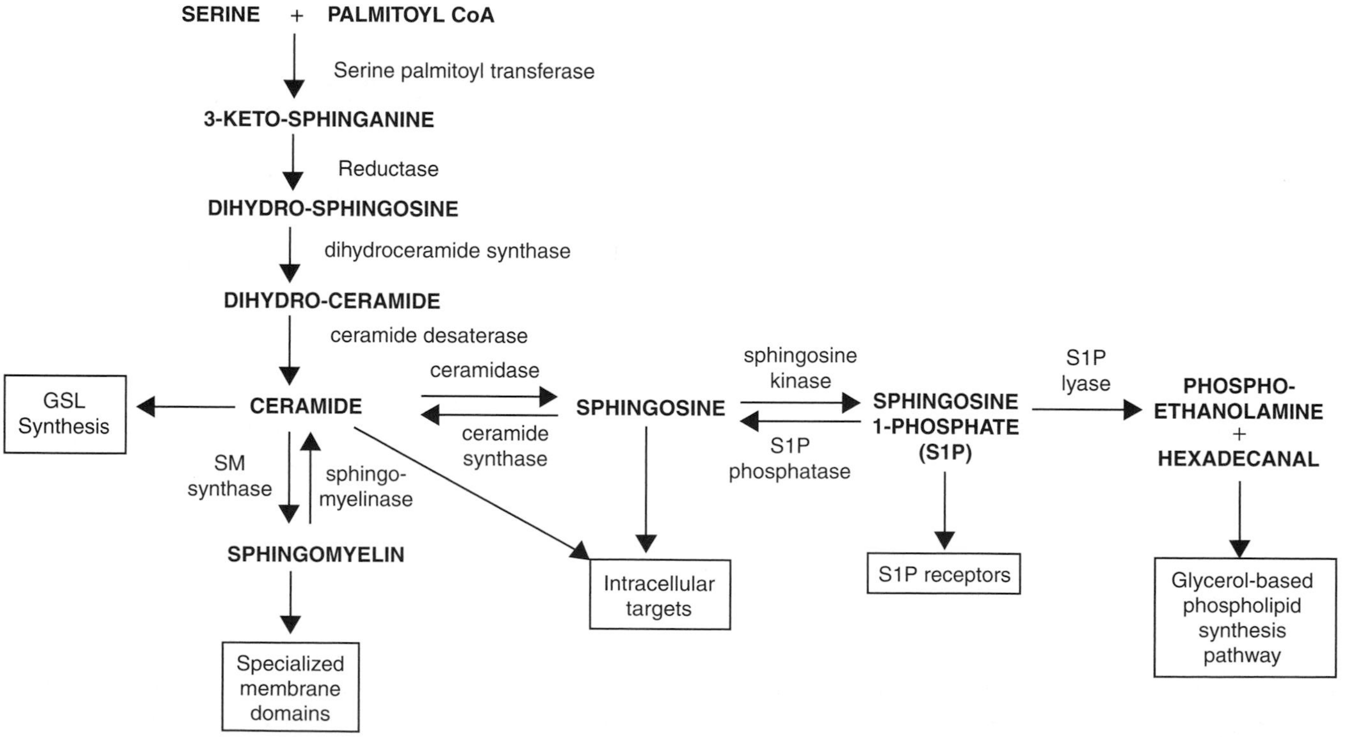

Figure 45.1. Sphingolipid metabolism in signaling.

makes this cell type a contributor of S1P in plasma (9). In this chapter, we discuss the role of sphingolipid metabolism in vascular biology, especially as it pertains to EC functions.

STRUCTURAL ROLES OF SPHINGOLIPIDS

Although much controversy surrounds the concept of membrane lipid rafts, it is clear that sphingolipids are critical for the formation and function of these membrane microdomains. The structural properties of SM and GSL enable this family of lipids to associate with cholesterol and thus possess differential lateral mobility within the plasma membrane (10). Rafts are enriched in sphingolipids and cholesterol as well as phosphatidyl inositol-linked proteins. In ECs, caveolae or flask-shaped endocytotic vesicles, are common and are observed by transmission electron microscopy methods. In some cases, such vesicles, referred to as vesiculo-vacuolar organelles (VVOs), exist as transcytotic channels (11). They have been implicated in the transcellular transport of plasma proteins and plasma. In addition, caveolae are well known to cluster signaling proteins such as kinases, NO synthase (NOS), and G-proteins. Sphingolipids may be essential for the structure and function of caveolae and VVOs. Important transmembrane proteins such as platelet-endothelial cell adhesion molecule (PECAM)-1/CD31 is found on vesicular membrane invaginations that resemble bunches of grapes (12). These represent the plasma membrane reserve pool that allows leukocyte trafficking through the endothelial monolayer. In addition, discontinuous endothelium in secretory organs contains specialized membrane patches called *fenestrae*, which are essential for the secretion of fluid from plasma into these tissues and the secretion of hormones from tissue into plasma. The role of sphingolipids in the formation and maintenance of complex membrane domains is not well understood and must be defined.

SPHINGOLIPID MEDIATORS AND THE ENDOTHELIUM

Recent work has clearly established S1P as a bona fide lipid mediator that acts via G-protein–coupled receptors (4). In contrast, sphingosine and ceramide, even though they were considered previously as intracellular signaling lipids, act in a poorly defined manner (13). Much of the work in the literature involves treatment of cells with nonphysiological lipids such as C2-ceramide, which may have nonspecific effects on the membranes. However, modulation of enzymes involved in ceramide and sphingosine metabolism in unicellular and multicellular organisms using genetic approaches have clearly shown that sphingosine, S1P, and ceramide all have intracellular effects, suggesting that they are utilized in intracellular signaling pathways (1). The challenge at hand is to sort out generalized effects on membranes from specific receptor-mediated effects.

S1P, like lysophosphatidic acid (LPA), acts on plasma membrane G-protein–coupled receptors of the endothelial differentiation gene (EDG) family, and it regulates intracellular signaling events in a wide variety of cells (4). As discussed in the next section, S1P action on ECs is of critical importance to the normal development and physiological function of the vascular system. In addition, dysregulated S1P signaling may be important in several disease states.

THE VASCULAR S1P GRADIENT

S1P is found generally associated with carrier proteins or membrane lipids. Although comprehensive knowledge of S1P binding proteins is not known, it is found in association with albumin and high-density lipoprotein fractions in plasma (14). Indeed, total serum concentration of S1P is very high, estimated to be in the range of 0.4 to 2 μM in mammals. However, plasma concentration is somewhat lower (0.1 to 1 μM), suggesting that S1P is produced and/or released by hematopoietic cells upon blood clotting. In contrast, tissue S1P levels are extremely low (nM concentrations). Thus, a large vascular gradient of S1P exists in mammals. The mechanisms involved in the establishment and/or maintenance of the S1P gradient are not known. Whether this occurs also in other vertebrates, chordates, and invertebrates is not known. S1P-like lipids are produced by all eukaryotes. The receptors for S1P appear to be restricted to vertebrates, although close homologues have been shown to exist in the chordate genome (15). The evolutionary significance of the vascular S1P gradient must be further explored.

Because platelets appear to express high levels of Sphk, lack the S1P lyase, and platelet activation leads to the appearance of S1P in the serum, it is generally assumed that platelets are a major cellular source of S1P (14). However, other hematopoietic cells (e.g., monocytes, macrophages, mast cells, neutrophils) also secrete S1P (16). Thus, release of S1P from activated and aggregated hematopoietic cells is likely to be a major source of S1P in serum. This mechanism may come into play during thrombosis and inflammation. However, under physiological conditions, the cellular source of S1P in the vascular system is not known. Although many cells are capable of secreting S1P from intracellular sources, S1P also may be formed in the extracellular environment. For example, vascular ECs secrete Sphk-1, which uses extracellular sphingosine and ATP to produce S1P (9). In addition, ceramidase and SMase enzymes also are secreted, suggesting that S1P can be formed in both intracellular and extracellular locations (7).

The S1P gradient in the closed circulatory system of mammals suggests that vascular and hematopoietic cells are constantly exposed to high concentrations of S1P. Given that the dissociation constant for S1P receptors is in the nanomolar range, cells bearing the receptors are constantly stimulated with maximal S1P concentrations. $S1P_1$ receptor is readily downregulated by ligand-induced internalization via the endosomal pathway (17). Even though receptor recycling takes place, a very small fraction of receptors is present on the cell surface when excess ligand is present. These data suggest that S1P receptors are likely to be sequestered in the cytosol of cells in the vascular system, and this may limit receptor function for certain processes such as cell migration. Indeed, this was shown to be functionally important for lymphocyte egress from secondary lymphoid organs and the thymus. For example, interference with the $S1P_1$ receptor through the immunomodulatory compound FTY720 resulted in the inhibition of T-cell egress and immunosuppression (18). Similarly, vascular EC and vascular smooth muscle cell (VSMC) migration, which require S1P receptor on the plasma membrane, is likely to be transient in the presence of high extracellular S1P concentrations. However, the S1P gradient is likely to be reversed during thrombosis and angiogenesis, when platelet aggregation and other cell activation would release S1P in the extracellular milieu, such as the clot and the provisional fibrin matrix. The changes in S1P gradient during physiological and pathophysiological conditions may be extremely critical to the biological responses.

S1P FUNCTION IN VASCULAR DEVELOPMENT AND ANGIOGENESIS

The first S1P receptor to be characterized was originally isolated from vascular ECs as an immediate-early gene induced during in vitro angiogenesis (19). Cell culture studies show that S1P is a potent inducer of EC migration, survival, proliferation, cell–cell adhesion, and cell–matrix adhesion. It potently induces morphogenesis and tube formation in in vitro assays of angiogenesis (4). Vascular ECs express multiple S1P receptors. Most cell lines and strains express $S1P_1$ and $S1P_3$, whereas some also express $S1P_2$. For example, human umbilical vein ECs express $S1P_1$ and $S1P_3$, whereas bovine lung microvascular ECs express $S1P_1$ and $S1P_2$. Collaboration between the two receptors is necessary for EC morphogenesis (20). The $S1P_2$ receptor, however, may inhibit cell migration and other angiogenic phenomena, suggesting a complex regulatory mechanism in this system of angiogenesis (21). Indeed, in models of in vivo angiogenesis, either by itself or in concert with polypeptide angiogenic factors, S1P potently influences angiogenesis (20).

S1P also has potent effects on VSMCs. Normal rat adult VSMCs express $S1P_2$ and $S1P_3$ receptors, which are presumed to regulate G_q- and $G_{12/13}$-dependent vascular contraction (22). In these cells, activation of Rho GTPase by $S1P_2$ receptor results in the profound inhibition of cell migration. In the fetal or intimal phenotype of VSMCs, $S1P_1$ receptor is induced, which activates the Rac and $p70^{S6K}$ kinase pathways (22). These events are important for cell migration and proliferation. $S1P_1$ receptor is overexpressed in atheromatous plaques in humans, suggesting that the balanced function of different S1P receptors is important for vascular physiology and disease (22).

Knockout studies in mice have indicated the essential role of $S1P_1$ receptor in the embryonic vascular maturation. *S1p*$^{-/-}$ mice die in utero between embryonic days 12.5 to 14.5 (E12.5–14.5) of development due to hemorrhage and vascular permeability. Incomplete coverage of various blood vessels with mural cells, particularly in the arterial system was observed. This study provided evidence for a critical role of $S1P_1$ during the later phases of vascular system development (23). This process can be recapitulated in vitro; ECs derived from *S1p1*$^{-/-}$ mice show defective interaction with mural cells (24). It appears that activation of $S1P_1$ in vascular ECs triggers G_i/Rac-dependent activation of microtubule assembly, which is essential for the proper trafficking of the cell–cell adhesion molecule N-cadherin to heterotypic cell–cell junctions. N-cadherin itself is activated by complex formation with catenins, lateral clustering, and phosphorylation. This mechanism of the N-cadherin–dependent activation of vascular maturation also may be important in the adult. Indeed, siRNA-mediated knockdown of the *$S1P_1$* mRNA, which is induced in tumor microvessels, resulted in diminished vascular maturation, reduced tumor microvessel density, and attenuated tumor growth rate, suggesting the importance of the S1P/$S1P_1$ signaling system in tumor angiogenesis (25).

In addition to $S1P_1$, other S1P receptors may also play a role in vascular development. Although single knockouts of $S1P_2$ and $S1P_3$ are viable, double knockout or triple knockouts of the three receptors led to earlier lethality with defects in the vascular system (26). This suggests that some vascular functions are regulated in a redundant manner by these S1P receptors.

S1P AND VASCULAR PERMEABILITY

The potent ability of S1P to induce homotypic cell–cell junction assembly (VE-cadherin–based) and heterotypic cell–cell junctions (N-cadherin–based), suggested a role for this mediator in regulating vascular permeability. Indeed, S1P treatment of cultured ECs results in cortical actin changes, which are necessary for the formation of cell–cell junctions (20). Moreover, S1P inhibits paracellular permeability in monolayer cultures of ECs (27). Importantly, vascular endothelial growth factor (VEGF)-induced vascular permeability in vivo is potently inhibited by acute treatment with FTY720, which is phosphorylated and acts on the $S1P_1$ and $S1P_3$ receptors as a potent agonist (28). In addition, in murine and canine models of lipopolysaccharide-induced pulmonary edema, S1P and FTY720 potently reduced EC injury and edema (29). The role of other S1P receptors, particularly the $S1P_2$ receptor, on vascular permeability is not understood. It is possible that high S1P levels in plasma are important for the physiological maintenance of vascular permeability. Importantly, pharmacologic modulation of this signaling system may provide novel therapeutic targets in the control of vascular permeability disorders, such as sepsis, Dengue hemorrhagic fever, and others.

ROLE OF S1P IN VASCULAR TONE CONTROL

$S1P_1$- and $S1P_3$-mediated signaling in ECs has been linked to activation of the G_i pathway, induction of intracellular Ca^{2+} levels, and phosphorylation of mitogen-activated protein kinase (MAPK) and phosphoinositide 3-kinase (PI3K)/Akt (30), all of which are important for the activation of the endothelial NOS (eNOS) system. Indeed, S1P was shown to potently induce the phosphorylation and activation of eNOS, resulting in increased NO production, which is essential for vasorelaxation (31). In *S1p3*$^{-/-}$ mice, endothelium-dependent vasorelaxation responses are impaired (32).

In contrast, basal expression of $S1P_2$ and $S1P_3$ in the VSMCs of resistance arteries would be expected to counteract this system. Both receptors can couple to the G_q and $G_{12/13}$ system, thus activating phospholipase C-β/Ca^{2+}/protein kinase C (PKC) pathway and Rho/Rho-associated kinase (ROCK) pathway, respectively (30). Both of these pathways ultimately impact the actin/myosin-based system and result in VSMC contraction. Thus, in some vascular beds, such as in the basilar artery in the brain and in renal arteries, S1P acts as a potent vasoconstrictor (33,34).

Differential S1P receptor expression in ECs, injury of ECs, and S1P receptor expression in VSMCs may all contribute to the role of S1P in vascular tone control. As stated earlier, the ligand is abundant in the plasma; therefore, EC receptors are likely to be continuously activated. However, upon injury to the ECs, VSMC receptors are likely to be engaged. The role of receptor internalization is likely to limit the vasoactive responses, in particular those that depend on the NO pathway. S1P receptor expression is inducible. For example, $S1P_1$ is upregulated by shear stress, phorbol esters, and angiogenic growth factors, whereas $S1P_3$ is induced by tumor necrosis factor (TNF)-α in ECs (7). Therefore, under the influence of inflammation, thrombosis, and angiogenesis, the receptor expression profile is likely to be altered, and thus may change the vascular contractile response to S1P.

ROLE OF S1P IN ENDOTHELIAL INJURY

Stimulation of ECs with S1P in vitro results in activation of cell survival pathways, such as those regulated by PI3K and extracellular signal-regulated kinase (ERK). Thus, S1P protects ECs from cell death induced by chemicals such as ceramide and 15-deoxy-$\Delta^{12,14}$-prostaglandin J_2. The survival effects of S1P requires the $S1P_1$ receptor expression and involves the G_i pathway (20). In many other cells, S1P protects against cell death – via both receptor-mediated and other pathways (16). For example, in addition to G-protein–dependent antiapoptotic pathways, metabolic conversion of S1P may be important for antiapoptosis. Therefore, plasma S1P may be involved in normal maintenance of EC function and health.

As discussed earlier, S1P is a potent inducer of endothelial cell–cell junctions. The proper maintenance of cell–cell junctions is critical for proper barrier formation and thus prevents excessive vascular permeability in continuous endothelia. Thus, S1P receptor function may be critical for physiologic maintenance of EC transport and barrier properties. Because abnormal vascular permeability is a precursor to chronic tissue changes, such as abnormal fibrin deposition, inflammation, angiogenesis, and fibrosis, alterations in the S1P signaling system may contribute to a variety of chronic pathological changes. The interaction of S1P with other noxious stimuli, such as oxidant stress has not been described. However, given that S1P is a pro-survival factor in nonvascular cells treated with radiation and cytotoxic agents, it likely protects vascular cells against various injurious insults.

S1P AND VASCULAR PATHOLOGY

In addition to the role of S1P in tumor angiogenesis, S1P signaling pathway may contribute to a number of vascular diseases, for example, restenosis after vascular injury, transplant-associated vasculopathy, angiogenesis post myocardial infarction, wound healing, and atherosclerosis.

VSMC migration into the neointima of injured vessels, followed by proliferation of VSMCs and matrix deposition, are important characteristics of restenosis after vascular injury, which is a chronic health problem in patients who receive revascularization therapy. $S1P_1$ receptor induces VSMC migration and proliferation and is expressed in pathologic lesions in atheromatous plaques (22). It also activates the critical signaling pathway regulated by ribosomal protein S6kinase ($p70^{S6K}$), which is a target of the drug rapamycin. Thus, $S1P_1$ function may be important in this disease condition, and drugs that target this receptor may be useful in controlling restenosis.

Similarly, VSMC proliferation and restenosis are important complications that limit organ function in transplants. In this scenario, immune pathways are predominant and contribute to VSMC expansion and stenosis of the vessel. Importantly, many transplant therapies, such as rapamycin and FTY720, also target the effect of S1P receptors or the receptors themselves, respectively. Thus, consideration of this issue may lead to better therapeutic approaches in the control of transplant-associated vasculopathy (35).

Atherosclerosis is a complex disease that affects the majority of the population in the modern society. Although the role of S1P and sphingolipids in the pathogenesis of atherosclerosis is poorly understood, available data suggest their significant involvement. Sphingomyelin, ceramide, sphingosine, and S1P are constituents of lipoproteins. The inhibition of de novo sphingomyelin synthesis with myriocin, a potent inhibitor of the rate-limiting enzyme serine palmitoyl transferase, inhibited atherosclerosis in mouse models without affecting lipoprotein levels (36–38). This suggests that sphingolipid metabolites may play critical roles in atherosclerosis. High-density lipoprotein (HDL) fraction contains significant levels of S1P, and this may confer the vasculoprotective and NO-releasing functions of HDL (39). In addition, S1P may regulate monocyte/macrophage functions, such as survival and inflammatory gene expression (40,41). The ability to regulate EC functions also may be important for initial phases, as well as for the progression of atherosclerosis. Further studies are needed to fully define the role and impact of sphingolipids in atherosclerosis.

PHARMACOLOGIC APPROACHES IN THE CONTROL OF SPHINGOLIPID MEDIATORS

Given the ubiquity and potency of sphingolipid mediators in various organ systems, novel pharmacologic opportunities exist to control various disease processes in which these lipids are implicated. This is particularly true for extracellular receptors for S1P, which are G-protein–coupled and therefore are commonly referred to in the pharmaceutical industry as "druggable." Although the current state of the pharmacologic tools for S1P receptors can be classified as primordial, a number of initiatives suggest that more refined tools will be forthcoming. $S1P_1$- and $S1P_2$-specific tools have been described that have immunologic effects (42,43). The vascular effects of such reagents remain to be investigated.

The developing FTY720 story suggests that S1P receptor modulation may have a potentially significant impact on human health in the future. It is a derivative of myriocin, and was developed due to its potent property to induce reversible lymphopenia without causing the myeloid-suppressive side effects of classical immunosuppressants (35). Interestingly, FTY720 is phosphorylated by sphingosine kinase-2 and the resultant FTY720-P acts as an agonist in S1P receptors (all except $S1P_2$). FTY720-P binding does not result in receptor recycling; therefore, this compound causes the sustained internalization of $S1P_1$ receptors. This presumably interferes with the chemotaxis-sensing property of T-cell receptors, which is important for cellular egress from the lymph nodes and other peripheral lymphoid tissues (18). Currently, FTY720 is being tested for its immunomodulatory actions in transplant rejection and in multiple sclerosis. Some side effects, such as bradycardia, were noted – which is perhaps to be expected, because S1P receptor systems are involved in many organ systems. The utility of FTY720 or S1P receptor-modulatory drugs in vascular and angiogenic indications should be of interest.

CONCLUSION

Sphingolipid biology is anticipated to yield important novel insights into the workings of the endothelium. Recent work has illustrated the important role played by sphingolipid mediators in the structure and function of the ECs. The powerful and broad-spectrum nature of the S1P signaling system in the vascular ECs is illustrative of this concept. That structurally critical lipids are mobilized in processes of signal transduction points to the economy by which evolutionary mechanisms have selected the players. The insights from the study of these simple lipids will not only lead to better understanding of physiology and pathology in the vasculature, but also will likely bring forth novel opportunities for therapeutic intervention, which will add to our armamentarium in vascular medicine.

KEY POINTS

- Sphingolipids are important constituents of ECs.
- S1P is a major signaling lipid in ECs.
- ECs secrete S1P.
- ECs respond to S1P via its G-protein–coupled receptors.
- S1P regulates EC migration, survival, proliferation, adherens junction assembly, and NO synthesis in vitro.
- S1P regulates angiogenesis, vascular maturation, and vascular permeability in vivo.
- S1P receptors in ECs provide a novel therapeutic opportunities.

ACKNOWLEDGMENTS

This work is supported by NIH grants HL67330 and HL70694.

REFERENCES

1 Futerman AH, Hannun YA. The complex life of simple sphingolipids. *EMBO Rep.* 2004;5(8):777–782.
2 Hannun YA, Linardic CM. Sphingolipid breakdown products: anti-proliferative and tumor-suppressor lipids. *Biochim Biophys Acta.* 1993;1154(3–4):223–236.
3 Dobrosotskaya IY, Goldstein JL, Brown MS, et al. Reconstitution of sterol-regulated endoplasmic reticulum-to-Golgi transport of SREBP-2 in insect cells by co-expression of mammalian SCAP and Insigs. *J Biol Chem.* 2003;278(37):35837–35843.
4 Hla T, Lee MJ, Ancellin N, et al. Lysophospholipids – receptor revelations. *Science.* 2001;294(5548):1875–1878.
5 Proia RL. Glycosphingolipid functions: insights from engineered mouse models. *Philos Trans R Soc Lond B Biol Sci.* 2003;358 (1433):879–883.
6 Hannun YA, Luberto C, Argraves KM. Enzymes of sphingolipid metabolism: from modular to integrative signaling. *Biochemistry.* 2001;40(16):4893–4903.
7 Saba JD, Hla T. Point-counterpoint of sphingosine 1-phosphate metabolism. *Circ Res.* 2004;94(6):724–734.
8 Lee MJ, Van Brocklyn JR, Thangada S, et al. Sphingosine-1-phosphate as a ligand for the G protein-coupled receptor EDG-1. *Science.* 1998;279(5356):1552–1555.
9 Ancellin N, Colmont C, Su J, et al. Extracellular export of sphingosine kinase-1 enzyme. Sphingosine 1-phosphate generation

and the induction of angiogenic vascular maturation. *J Biol Chem.* 2002;277(8):6667–6675. Epub 2001 Dec 10.

10 Rietveld A, Simons K. The differential miscibility of lipids as the basis for the formation of functional membrane rafts. *Biochim Biophys Acta.* 1998;1376(3):467–479.

11 Dvorak AM, Kohn S, Morgan ES, et al. The vesiculo-vacuolar organelle (VVO): a distinct EC structure that provides a transcellular pathway for macromolecular extravasation. *J Leukoc Biol.* 1996;59(1):100–115.

12 Mamdouh Z, Chen X, Pierini LM, et al. Targeted recycling of PECAM from endothelial surface-connected compartments during diapedesis. *Nature.* 2003;421(6924):748–753.

13 Kolesnick R, Fuks Z. Radiation and ceramide-induced apoptosis. *Oncogene.* 2003;22(37):5897–5906.

14 Yatomi Y, Ohmori T, Rile G, et al. Sphingosine 1-phosphate as a major bioactive lysophospholipid that is released from platelets and interacts with ECs. *Blood.* 2000;96(10):3431–3438.

15 Hla T. Genomic insights into mediator lipidomics. *Prostaglandins Other Lipid Mediat.* 2005;77(1–4):197–209.

16 Spiegel S, Milstien S. Sphingosine-1-phosphate: an enigmatic signalling lipid. *Nat Rev Mol Cell Biol.* 2003;4(5):397–407.

17 Liu CH, Thangada S, Lee MJ, et al. Ligand-induced trafficking of the sphingosine-1-phosphate receptor EDG-1. *Mol Biol Cell.* 1999;10(4):1179–1190.

18 Matloubian M, Lo CG, Cinamon G, et al. Lymphocyte egress from thymus and peripheral lymphoid organs is dependent on S1P receptor 1. *Nature.* 2004;427(6972):355–360.

19 Hla T, Maciag T. An abundant transcript induced in differentiating human ECs encodes a polypeptide with structural similarities to G-protein-coupled receptors. *J Biol Chem.* 1990; 265(16):9308–9313.

20 Lee MJ, Thangada S, Claffey KP, et al. Vascular EC adherens junction assembly and morphogenesis induced by sphingosine-1-phosphate. *Cell.* 1999;99(3):301–312.

21 Sanchez T, Thangada S, Wu MT, et al. PTEN as an effector in the signaling of antimigratory G protein-coupled receptor. *Proc Natl Acad Sci USA.* 2005;102(12):4312–4317.

22 Kluk MJ, Hla T. Role of the sphingosine 1-phosphate receptor EDG-1 in vascular smooth muscle cell proliferation and migration. *Circ Res.* 2001;89(6):496–502.

23 Liu Y, Wada R, Yamashita T, et al. Edg-1, the G protein-coupled receptor for sphingosine-1-phosphate, is essential for vascular maturation. *J Clin Invest.* 2000;106(8):951–961.

24 Paik JH, Skoura A, Chae SS, et al. Sphingosine 1-phosphate receptor regulation of N-cadherin mediates vascular stabilization. *Genes Dev.* 2004;18(19):2392–2403.

25 Chae SS, Paik JH, Furneaux H, et al. Requirement for sphingosine 1-phosphate receptor-1 in tumor angiogenesis demonstrated by in vivo RNA interference. *J Clin Invest.* 2004;114(8):1082–1089.

26 Kono M, Mi Y, Liu Y, et al. The sphingosine-1-phosphate receptors S1P1, S1P2, and S1P3 function coordinately during embryonic angiogenesis. *J Biol Chem.* 2004;279(28):29367–29373.

27 Garcia JG, Liu F, Verin AD, et al. Sphingosine 1-phosphate promotes EC barrier integrity by Edg-dependent cytoskeletal rearrangement. *J Clin Invest.* 2001;108(5):689–701.

28 Sanchez T, Estrada-Hernandez T, Paik JH, et al. Phosphorylation and action of the immunomodulator FTY720 inhibits vascular EC growth factor-induced vascular permeability. *J Biol Chem.* 2003;278(47):47281–47290.

29 McVerry BJ, Peng X, Hassoun PM, et al. Sphingosine 1-phosphate reduces vascular leak in murine and canine models of acute lung injury. *Am J Respir Crit Care Med.* 2004;170(9):987–993.

30 Ancellin N, Hla T. Differential pharmacological properties and signal transduction of the sphingosine 1-phosphate receptors EDG-1, EDG-3, and EDG-5. *J Biol Chem.* 1999;274(27):18997–19002.

31 Morales-Ruiz M, Lee MJ, Zollner S, et al. Sphingosine 1-phosphate activates Akt, nitric oxide production, and chemotaxis through a Gi protein/phosphoinositide 3-kinase pathway in ECs. *J Biol Chem.* 2001;276(22):19672–19677.

32 Nofer JR, van der Giet M, Tolle M, et al. HDL induces NO-dependent vasorelaxation via the lysophospholipid receptor S1P3. *J Clin Invest.* 2004;113(4):569–581.

33 Alewijnse AE, Peters SL, Michel MC. Cardiovascular effects of sphingosine-1-phosphate and other sphingomyelin metabolites. *Br J Pharmacol.* 2004;143(6):666–684.

34 Tosaka M, Okajima F, Hashiba Y, et al. Sphingosine 1-phosphate contracts canine basilar arteries in vitro and in vivo: possible role in pathogenesis of cerebral vasospasm. *Stroke.* 2001;32(12):2913–2919.

35 Brinkmann V, Cyster JG, Hla T. FTY720: sphingosine 1-phosphate receptor-1 in the control of lymphocyte egress and endothelial barrier function. *Am J Transplant.* 2004;4(7):1019–1025.

36 Hojjati MR, Li Z, Zhou H, et al. Effect of myriocin on plasma sphingolipid metabolism and atherosclerosis in apoE-deficient mice. *J Biol Chem.* 2005;280(11):10284–10289.

37 Park TS, Panek RL, Mueller SB, et al. Inhibition of sphingomyelin synthesis reduces atherogenesis in apolipoprotein E-knockout mice. *Circulation.* 2004;110(22):3465–3471.

38 Tabas I. Sphingolipids and atherosclerosis: a mechanistic connection? A therapeutic opportunity? *Circulation.* 2004;110(22): 3400–3401.

39 Kimura T, Sato K, Malchinkhuu E, et al. High-density lipoprotein stimulates EC migration and survival through sphingosine 1-phosphate and its receptors. *Arterioscler Thromb Vasc Biol.* 2003;23(7):1283–1288.

40 Garg SK, Volpe E, Palmieri G, et al. Sphingosine 1-phosphate induces antimicrobial activity both in vitro and in vivo. *J Infect Dis.* 2004;189(11):2129–2138.

41 Lee H, Liao JJ, Graeler M, et al. Lysophospholipid regulation of mononuclear phagocytes. *Biochim Biophys Acta.* 2002;1582(1–3):175–177.

42 Jo E, Sanna MG, Gonzalez-Cabrera PJ, et al. S1P1-selective in vivo-active agonists from high-throughput screening: off-the-shelf chemical probes of receptor interactions, signaling, and fate. *Chem Biol.* 2005;12(6):703–715.

43 Rosen H, Goetzl EJ. Sphingosine 1-phosphate and its receptors: an autocrine and paracrine network. *Nat Rev Immunol.* 2005; 5(7):560–570.

46

Endothelium

A Critical Detector of Lipopolysaccharide

Jaswinder Kaur and Paul Kubes

University of Calgary, Alberta, Canada

Endotoxin is part of the outer membrane of the cell wall of gram-negative bacteria. It also is referred to as lipopolysaccharide (LPS), owing to its historical discovery. During the 1800s, it was understood that bacteria could secrete toxins into their environment, which were broadly known as "exotoxins." The term *endotoxin* came about following the discovery that portions of the gram-negative bacteria itself can cause toxicity. Over the next 50 years, studies focused on endotoxin revealed that the actual molecule responsible for the effects of endotoxin was in fact LPS. Endotoxins largely are responsible for clinical manifestations of the infections induced by pathogenic gram-negative bacteria such as *E. coli*, *Salmonella*, *Shigella*, *Pseudomonas*, *Neisseria*, *Haemophilus*, and others. The goal of this chapter is to discuss the receptor, signaling pathway, and effect of LPS on endothelium.

LIPOPOLYSACCHARIDE STRUCTURE

LPS is comprised of a polysaccharide ("sugar") chain and a lipid moiety, known as lipid A, the portion responsible for toxic effects (Figure 46.1). Lipid A is found in the exterior leaflet of the outer membrane. Lipid A is the most conserved portion of the molecule; however, it can vary according to the nature of fatty acid chains and its degree of phosphorylation. This portion is the biologically active component of LPS, because isolated lipid A or synthetic lipid A (free of contaminants) displays the same bioactivity as intact LPS in vitro and in vivo (1). Indeed, in bovine pulmonary artery endothelial cell (BPAEC) monolayers, lipid A has been shown to be the bioactive portion of LPS responsible for inducing changes in endothelial barrier function, with a dose-dependent increase in EC monolayer permeability comparable to that of native LPS (2). Incubation of BPAECs with LPS in the presence of polymyxin B, a derivative of the bacteria *Bacillus polymyxa*, which binds to the lipid A portion of LPS (3), blocked LPS-induced endothelial cell (EC) barrier dysfunction (2).

TOLL-LIKE RECEPTORS

For a long time, the mechanism by which endotoxin initiated a signal remained elusive. Then, in 1998, positional cloning revealed Toll-like receptor 4 (TLR4) as the central component of the LPS receptor (4). TLR4 is expressed predominantly by the myeloid lineages (monocytes, macrophages, microglia, myeloid dendritic cells, and granulocytes) (5–7) in mice and humans. In human tissues, the strongest expression of TLR4 is observed in the spleen and on peripheral blood leukocytes, with weak expression apparent in other tissues (8,9). On the other hand, murine *TLR4* mRNA is expressed most strongly in the lung, heart, and spleen, but is also present in the muscle, liver, and kidney (10).

The term *Toll-like receptor* was adopted following the observed similarity between the *Drosophila* Toll receptor and mammalian TLRs. The *Drosophila* Toll receptor is a regulator of dorso-ventral polarity in fly embryos. The homology between the intracellular domains of *Drosophila* Toll and the interleukin (IL)-1 receptor was recognized by Gay and Keith in 1991 (11). The similarities of these type-I transmembrane proteins extends to both the signaling domains, which contain the Toll/IL-1 Receptor (TIR) motifs, and the ectodomains, which have multiple leucine-rich repeats. The first sequence of the mammalian TLR homolog of *Drosophila* was obtained in 1994 (12). Lemaitre and colleagues, in 1996 (13), demonstrated its importance in regulating antifungal immunity in adult flies. The first functional role for a mammalian TLR was established by Medzhitov and colleagues (14), who demonstrated that coupling intracellular domains of a TLR with the extracellular domain of CD4 resulted in the activation of the transcription factor, nuclear factor (NF)-κB. This induced the expression of NF-κB–regulated inflammatory genes, including IL-1, IL-6, and IL-8, and the costimulatory molecule B7.1, which is required for the activation of naïve T cells. Although no ligand for a TLR(s) was known, these novel observations were suggestive of a role for TLR in innate as well as

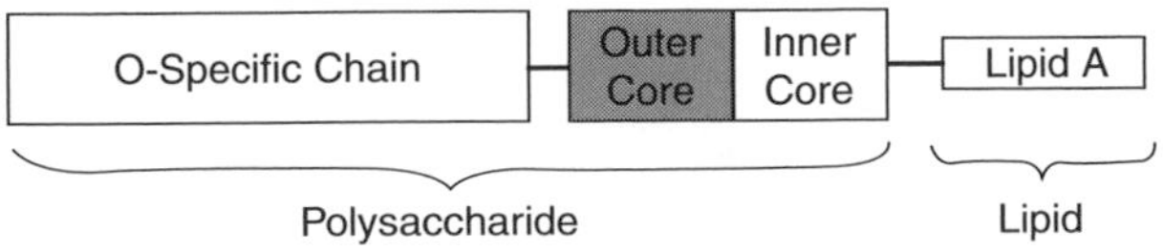

Figure 46.1. General structure of LPS. The biological activity of endotoxin generally is associated with the LPS. In particular, toxicity is a result of lipid component (lipid A) and immunogenicity with the polysaccharide components.

adaptive immunity. In 1998, it was revealed that a whole family of mammalian TLRs exist that bind to a number of different microbial ligands – and that the long-sought bacterial receptor for LPS was indeed TLR4 (4,9,15).

Mammalian TLRs have evolved in a process that began prior to the divergence of plants and animals. The TLR multigene family encodes conserved pattern recognition receptors (PRRs) that sense a variety of conserved pathogen-associated molecular patterns (PAMPs) that are unique to pathogens and are absent from the host cells in both the invertebrate and vertebrate lineages (16). TLRs are type-I transmembrane proteins with ectodomains consisting of leucine-rich repeat motifs that recognize the various microbial ligands. The cytoplasmic domain contains TIR motifs, which are highly conserved across the TLR families as well as through evolution, because they are present not only in animals but plants as well. TIR motifs mediate the bimolecular interactions necessary for downstream signal transduction from the TLR receptors. These germline-encoded receptors initiate signaling cascades to transcribe a diverse repertoire of genes, including antimicrobial peptides, cytokines, inflammatory mediators, and regulators of phagocytosis (17–19).

As noted earlier, the Toll signaling pathway was initially identified in *Drosophila* (13,20) and characterized for its role in initiating the innate immune response against fungal and gram-positive bacterial infections (13,21). TLR family members recognize several classes of pathogens and initiate appropriate innate and adaptive immune responses. TLR2 in combination with TLR1 recognizes lipoproteins with triacylated N-terminal cysteines, while in combination with TLR6, it recognizes diacylated lipoproteins (22,23). TLR3 recognizes double-stranded RNA, whereas LPS is recognized by TLR4 and flagellin by TLR5. Compounds such as nucleic acids and heme are recognized by the TLR7–9 family (24–26). Although the specific ligand for TLR10 remains to be identified, it has been demonstrated that TLR10 directly associates with MyD88 (myeloid differentiation primary response gene 88), the common Toll IL-1 receptor domain adapter (27). TLR11 receptor is present in mice, but only represented as a pseudogene in humans; it is known to recognize uropathogenic *Escherichia coli* (28). The ligand for TLR11 has been described as a profilin-like protein from *Toxoplasma gondii* (29). Not only are the coding sequences and function of the vertebrate TLRs highly conserved, but so too are the signaling pathways initiated by the TLRs.

TOLL-LIKE RECEPTOR SIGNALING PATHWAYS

As discussed earlier, Toll receptors are class I transmembrane proteins that have a leucine-rich extracellular domain, a short transmembrane domain, and an intracellular signaling domain known as the TIR domain (26). TLRs have TIR motifs in their cytoplasmic tails, and signal through cytoplasmic adapter proteins that also contain TIR motifs. To date, three TIR motif-containing proteins, MyD88, Tirap (Toll-IL-1 receptor [TIR] domain-containing adapter protein), and TIR domain containing adaptor inducing interferon (IFN)-β (Trif) have been identified. These are involved in mediating LPS signaling. Genetic evidence does suggest the existence of another adapter protein, "X," that perhaps partially substitutes for Trif (30). More recently, an essential accessory protein for signaling, MD2, a 20- to 30-kDa glycoprotein, which is secreted and binds to an extracellular domain of TLRs, has been described (31).

TLR signaling is initiated from the cytoplasmic TIR domains. The crucial role of these domains was first demonstrated in the C3H/HeJ mice, which have a point mutation of a proline residue in the TIR domain of TLR4 (4,32). The next signaling molecule characterized downstream from the TIR domains was the TIR domain-containing adaptor protein, MyD88. MyD88 contains a TIR domain in its C-terminal portion and a death domain in its N-terminal portion. Following stimulation, MyD88 recruits a serine/threonine kinase, IL-1 receptor-associated kinase (IRAK) to TLRs via the death domains of both molecules. *MyD88*$^{-/-}$ mice are unable to produce inflammatory cytokines, such as tumor necrosis factor (TNF)-α, IL-6, and IL-12, in response to any of the TLR ligands (33).

The recruitment and activation of IRAK subsequently leads to the recruitment and activation of TNF receptor–associated factor 6 (TRAF6), which in turn activates NF-κB either through the IκB kinase complex and the kinases TAB-1 and TAK-1 (34), or through evolutionarily conserved signaling intermediates in Toll pathways (ECSIT) and mitogen-activated protein kinase (MAPK) (35). These pathways converge on NF-κB, which can now disassociate from its inhibitory proteins that retain NF-κB in the cytoplasm in a quiescent state. The liberated NF-κB then freely moves into the nucleus to induce the transcription of NF-κB–responsive genes (Figure 46.2).

To date, eleven TLR subtypes have been identified, but TLR4 is the only TLR that mediates responses to LPS (36). Interestingly, one study had suggested an additional LPS-induced response via TLR2 (37). However, these data have been discounted, because repurification of commercially available LPS resulted in abrogation of TLR2 responsiveness (indicating previous LPS samples were contaminated) (38). Furthermore, Weis and colleagues clearly demonstrated that non-LPS ligands (e.g., peptidoglycan [PGN], zymosan and lipoteichoic acid [LTA]) mediate TLR2 signaling (38). This controversy was further resolved by studies of TLR2-deficient

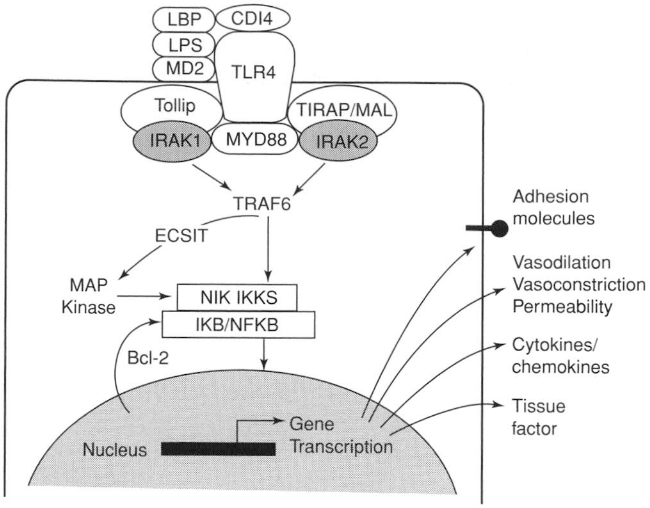

Figure 46.2. TLR4 signaling-mediated effects on ECs. LPS in the bloodstream, released by infectious gram-negative bacteria, overactivates ECs to release a number of cytokines, chemokines, and vasoregulatory mediators and to express adhesion molecules and procoagulant factors. Increased permeability also results from activation. LPS binds endothelial-expressed TLR4 in the presence of CD14 and MD2. Activation of this receptor complex mediates the recruitment of the cytoplasmic proteins MyD88, TIRAP, Tollip, IRAK 1 and 2, and TRAF6. TRAF6 ultimately activates the MAPK pathway and NF-κB. In quiescent cells, NF-κB is retained in the cytoplasm by IκB. Incubation of ECs with LPS leads to degradation of IκB and thus translocation of NF-κB into the nucleus. Translocated NF-κB binds to DNA sequences of NF-κB regulated target genes, such as IL-1, IL-6, and IL-8, tissue factor, and adhesion molecules.

mice, which showed robust responses to LPS administration. Although alternative signaling pathways have been suggested (39,40), most LPS preparations are principally recognized by TLR4. Possible exceptions are LPS from *Porphyromonas gingivalis* and *Leptospira interrogans*, which appear to be recognized by TLR2 (41,42).

Further evidence in support of LPS-TRL–mediated effects comes from the recent identification of spontaneous mutations in the murine *Tlr4* gene, which for many years were called "LPS-resistant mice." These mutations render two distinct strains of mice, C3H/HeJ and C57BL/10ScCr, resistant to LPS (4). In the C3H/HeJ mouse, a missense mutation was identified in the third exon of *Tlr4*, whereas C57BL/10ScCr mice were identified as homozygous for a null mutation of *Tlr4*. A genetically engineered strain has been described that has the *Tlr4* gene deleted by homologous recombination and is also unresponsive to LPS challenges (32).

ENDOTOXIN, TLR4, AND ENDOTHELIUM

The expression of TLR receptors on endothelium is regulated by inflammatory mediators, such as LPS, TNF-α, and IFN-γ (43). LPS-stimulated ECs also release high levels of MD2 in vitro (44). Interestingly, MD2 may aid the direct binding of LPS to TLR4 (45). Using an LPS inhibitor, E5531, it was demonstrated that TLR4 and MD2 were critical for signaling in human umbilical ECs (HUVECs), similar to observations in phagocytic cells (5). Immortalized human dermal microvessel ECs also express MD2 and respond to LPS (44). However, it has not been determined if primary ECs express MD2, use soluble MD2, or are activated independently of MD2. MD2 recently was reported to also colocalize with TLR2 and to amplify the response to TLR2 ligands (46).

Although the majority of reports suggest TLR4 is expressed by cultured ECs, one report suggests otherwise. Foxwell and colleagues (47) were unsuccessful in detecting surface TLR4 expression on HUVECs. In contrast, others reported the presence of TLR4 in HUVECs using reverse transcription polymerase chain reaction (RT-PCR) and Northern blot analysis (48). More recently, TLR4 expression on endothelium also has been shown at the protein level (49). LPS-induced NF-κB activation in ECs can be blocked using dominant negative mutant proteins involved in downstream Toll signaling (50). It was subsequently shown that the LPS-resistant mutant mice cannot transmit signals across the EC membrane in response to LPS (4,10,32,48). In addition, it also has been demonstrated that LPS induces the mRNA expression of *TNF-α* and *IL-1β*, and the release of IL-6, IL-8, and monocyte chemoattractant protein (MCP)-1 from human coronary artery ECs via TLR4. In the same study, LPS-induced and TLR4-mediated expression of endothelial adhesion molecules was also observed (51). However, it is unclear whether LPS directly binds to TLR4, or whether LPS induces an unidentified ligand of TLR4 that transmits the signals. Much less is known about the other TLRs in the endothelium. TLR2 is thought to be expressed in very low levels in ECs or not at all (47). However, it is upregulated by various stimuli in ECs (52).

ROLE OF ENDOTHELIUM IN INFECTION

The strategic positioning of ECs at the interface between circulating blood and tissue allows these cells to engage in diverse physiological functions essential for organ homeostasis. ECs have been shown to regulate a number of physiological events, such as organ perfusion, permeability, adhesion of blood cells, pro-coagulation, and anticoagulation (53). Because of their prime location, ECs can encounter bacterial toxins both in the circulation and the extravascular space. However, it remains unknown whether LPS can be detected abluminally by endothelium or whether sentinel cells (macrophage, mast cells, and the like) detect the LPS in tissue and activate the endothelium. Upon their activation, ECs release vasoactive molecules to affect blood flow, and they recruit effector cells of the immune system to the site of infection. The expression of adhesion molecules allows leukocytes to transmigrate into the infected tissue. Furthermore, the activated endothelium presents a procoagulant surface and promotes the formation of clots to enclose the infected site. Therefore, these endothelial events are absolutely essential for an efficient inflammatory response that eliminates the infectious agent. However, if the endothelium is overactivated systemically, as in sepsis and septic shock, the result may be devastating (see Chapter 141).

Sepsis and severe sepsis are caused by inappropriate inflammatory and immunological host responses to infections. Gram-negative bacterial sepsis, a common life-threatening event, often is complicated by systemic vascular collapse, disseminated intravascular coagulation, and vascular leak syndromes. One component common to all these complications is EC injury and dysfunction. It is widely believed that the dysregulation of the endothelium plays a significant role in mediating the sepsis phenotype. Aside from direct activation of the endothelium by LPS, myriad cytokines, chemokines, and other mediators also affect the endothelium and lead to vascular dysfunction.

LIPOPOLYSACCHARIDE, CD14, AND LPS-BINDING PROTEIN

To respond to LPS directly, ECs must express CD14. Indeed, it has been shown that activation of ECs and other immune cells occurs by LPS binding to CD14, with subsequent interaction with TLR4 (54). CD14 is a glycosylphosphatidylinositol (GPI)-linked protein that is constitutively expressed on monocytes and macrophages and, to a lesser extent, on neutrophils. In addition to the membrane-bound isoform of sCD14, an alternatively spliced soluble form of CD14 (sCD14) is released from various cells and is found in normal serum or plasma (55,56). Interestingly, early studies suggested that ECs do not express CD14 (56,57) but require the soluble form to respond to LPS. However, Jersmann and colleagues (58) have demonstrated constitutive expression of membrane CD14 (mCD14) on HUVECs, where its expression was further enhanced by LPS activation. These in vitro studies revealed that passaging of ECs led to a loss of endothelial CD14 (58). Membrane CD14 has a short transmembrane portion consisting of 21 amino acids. Both sCD14 and mCD14 lack an intracytoplasmic region and signaling domain, suggesting that these molecules must interact with a receptor (e.g., TLR4, which is capable of transducing the LPS signal). Indeed, the primary function (59) of these proteins is to bind LPS and to facilitate the interaction of endotoxins with TLRs.

The ability of endothelium to interact with LPS via sCD14 is well established. In vitro studies show that ECs respond to very low levels of LPS only in the presence of serum containing sCD14. The addition of sCD14 to serum-free cultures sensitizes ECs to LPS up to 100-fold. For example, LPS-mediated inflammatory EC responses such as the release of cytokines (55) and vasoactive molecules (60), the expression of adhesion molecules, and disruption of endothelial barrier function occurs at picogram quantities of sCD14 in the presence of plasma and in nanogram quantities in the absence of plasma (61). Interestingly, it has been shown that CD14-deficient cells do not respond to low levels of LPS even in the presence of sCD14, suggesting an essential role for mCD14 for sCD14 function. ECs, smooth muscle cells, or phagocytes from CD14-deficient mice can respond to LPS if the concentration of LPS is high enough. Recently, using *CD14*$^{-/-}$ mice, we observed that some vascular beds were completely resistant to LPS, whereas endothelium from other vascular beds still reacted with adhesion molecule expression. Clearly this reflected the heterogeneity of different ECs rather than simply a concentration effect, and it raises the possibility that other molecules can replace CD14. Further evidence that other LPS-recognizing molecules may exist is that anti-CD14 antibodies are incapable of completely inhibiting cell activation in response to high levels of LPS (62,63). Alternatively, high doses of LPS could bypass CD14 and directly activate downstream receptors (e.g., TLR4).

Other candidates that could potentially recognize LPS include the integrin family of adhesion molecules, CD11/CD18, and LPS-binding protein (LBP), which also bind LPS (59). Binding of LPS to CD14 is facilitated by LBP, a 60-kDa acute-phase serum protein primarily synthesized by hepatocytes (64–66). LBP transfers LPS to mCD14 as well as to sCD14. The importance of LBP in LPS-induced effects has been demonstrated in mice lacking the LBP gene; these mice are partially resistant to LPS (67). Also, administering anti-LBP in tandem with LPS protected mice from LPS-induced shock (68). The role that LBP plays in the transfer of LPS to sCD14 is unclear. However, addition of recombinant sCD14 to serum-free media restores LPS-related cytotoxicity, whereas addition of recombinant LBP has no effect, suggesting that sCD14 is the prerequisite molecule for LPS-induced endothelial activation independent of LBP.

LIPOPOLYSACCHARIDES AND ENDOTHELIAL PERMEABILITY

A breakdown of vascular barrier function, also referred to as plasma leakage, is a classic feature of severe sepsis and septic shock. The endothelial barrier regulates the transport of cellular components, macromolecules, and fluid from the bloodstream into the surrounding tissue. Cytoskeletal proteins located along the EC borders maintain intercellular tightness in association with adherence and tight junctions (69). In response to mediators that increase permeability, ECs contract, driven by reorganization of the cytoskeleton. This contraction results in endothelial intercellular contacts opening and contributing to increases in paracellular permeability of the endothelium. The increase in endothelial permeability can be caused by direct effects of LPS on the endothelium or by a variety of inflammatory mediators, including cytokines, platelet-activating factor (PAF), and leukotrienes (LTs). These latter mediators are released from surrounding resident macrophages and mast cells or by adhering and emigrating neutrophils.

LPS has been shown to directly enhance endothelial permeability. LPS-mediated disruption of endothelial barrier function depends on CD14 and LBP (70). LPS augments

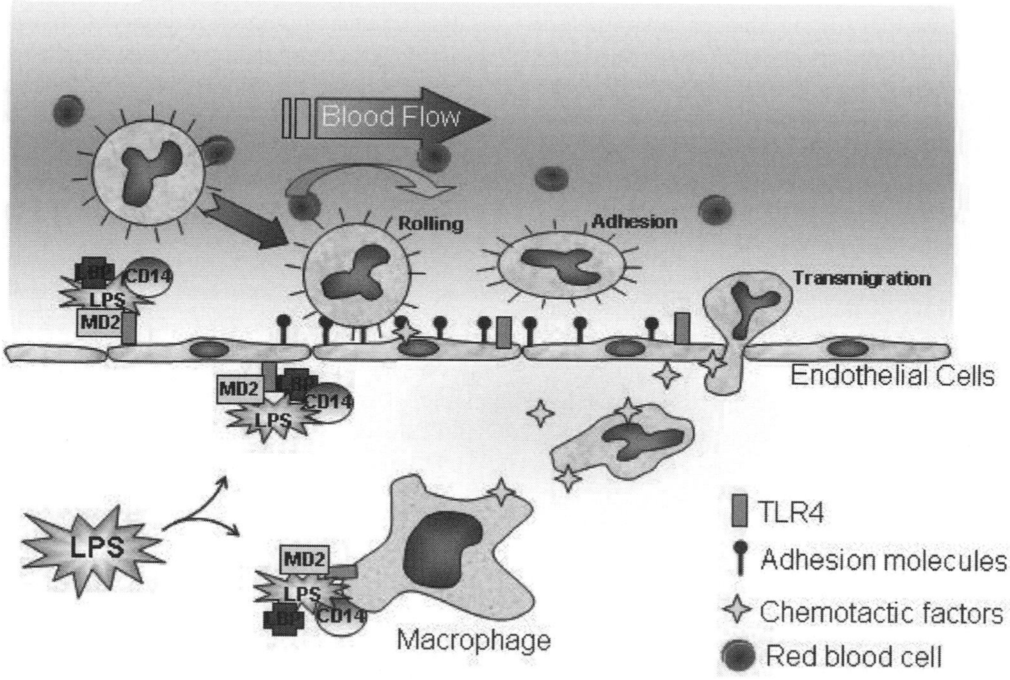

Figure 46.3. Studies have shown that the stimulation of TLR4 on the endothelium leads to the rapid surface expression of adhesion molecules. As a consequence, increased leukocyte rolling, adhesion, and transmigration occurs. Additionally, TLR4-expressing cells, such as tissue-resident macrophages, also release a number of inflammatory factors, such as cytokines and chemokines, which also act on the endothelium.

the movement of radiolabeled tracer molecules across EC monolayers in the absence of hydrostatic or osmotic pressure gradients, granulocyte effector cells, or a number of other host-derived nonendothelial mediators (2). Studies in vitro have demonstrated that LPS induces F-actin depolymerization and intercellular gap formation, processes that coincide with endothelial barrier dysfunction (2). This sequence of events can be inhibited by using an endotoxin-neutralizing protein to prevent the rapid LPS-induced protein tyrosine phosphorylation events. LPS-induced changes in endothelial barrier function are not dependent on new protein synthesis, because the endothelial responses occur very quickly and protein synthesis inhibitors do not decrease LPS-induced transendothelial flux (61). LPS-induced endothelial permeability is augmented by the cleavage or disengagement of the adherens junction proteins β- and γ-catenin, which are involved in maintaining cell–cell contacts. Studies strongly suggest that LPS-induced endothelial permeability is associated with enzymatic cleavage of adherens junction proteins (71).

LIPOPOLYSACCHARIDES AND ENDOTHELIAL–LEUKOCYTE INTERACTIONS

The recruitment of leukocytes from the circulation to the interstitium is central to the inflammatory process. Leukocyte recruitment involves a sequence of events constituting rolling, adhesion, and transmigration. These processes utilize a number of inflammatory mediators and adhesion molecules expressed on ECs and leukocytes. Figure 46.3 illustrates the cascade of LPS-induced leukocyte recruitment on activated ECs. TLR4 activation induces not only the expression of adhesion molecules, but also the production of chemotactic agents, including chemokines and other inflammatory molecules such as PAF and LTB4. The local production of these molecules leads to the accumulation of inflammatory cells within a tissue following LPS administration (72,73). This entire cascade of leukocyte recruitment can be recapitulated on endothelial monolayers, suggesting direct effects of LPS on endothelium.

It has been demonstrated that TLR4 activation through administration of LPS to a local site induces a localized inflammation within that tissue (72,74). Using intravital microscopy to observe leukocyte recruitment, it was shown that, under baseline conditions, very few leukocytes rolled and no adhesion or emigration into the tissue occurred. However, upon local administration of LPS, significant rolling, adhesion, and subsequent leukocyte transmigration into the tissue was observed. Notably, these events were not observed in *Tlr4*$^{-/-}$ mice, thus confirming the involvement of the LPS–TLR4 complex in inducing inflammation in this model (74). Chimeric mice lacking TLR4 on endothelium but having normal TLR4 on circulating leukocytes had impaired leukocyte recruitment. This rapid leukocyte recruitment in response to local administration of LPS involved P-selectin, which is prestored in the Weibel-Palade bodies of ECs and is upregulated immediately by agonists such as LPS, thrombin, and oxidants. P-selectin–blocking antibodies completely inhibited the

LPS-induced leukocyte–endothelial interactions observed within the first 90 minutes (73). The LPS response was further bolstered by a second wave of recruitment after 90 minutes that was not blocked with anti-P-selectin antibodies. These delayed interactions were, however, completely blocked using E-selectin antibodies in combination with P-selectin, as well as in mice deficient in both selectins (72).

ENDOTOXEMIA AND LEUKOCYTE RECRUITMENT

The systemic administration of LPS mechanistically is very different from a local tissue exposure to LPS. Systemic LPS administered intraperitoneally or intravenously results in widespread activation of TLR4 (75). This leads to a profound expression of adhesion molecules including E- and P-selectin, intercellular adhesion molecule (ICAM)-1 and vascular cell adhesion molecule (VCAM)-1 on endothelium in every organ. However, this does not lead to leukocyte recruitment into all tissues. In fact, a disproportionately large number of leukocytes enter into the lungs. We do know that neither selectins nor integrins support the recruitment of leukocytes into lungs (76–79). Although the underlying mechanism for the preferred recruitment of leukocytes into lungs is unknown, chimeric mice having TLR4 on leukocytes but lacking TLR4 on endothelium did not express adhesion molecules on endothelium and did not recruit leukocytes into lungs. This suggests that the endothelial adhesive profile in lungs stimulated by TLR4 is different and dominant for subsequent leukocyte recruitment.

The liver also recruits some leukocytes during endotoxemia. Endothelial heterogeneity becomes very evident when the liver microcirculation is visualized in response to LPS using intravital microscopy. This organ constitutes a unique vascular architecture. Systemic administration of LPS has been shown to induce a significant amount of leukocyte adhesion in the collecting or central venules of the liver (80), and this adhesion was prevented in mice deficient in both E- and P-selectin. A large amount of leukocyte adhesion also was observed within the sinusoids; however, this was not inhibited by blocking selectins (80). Although we have proposed that leukocyte recruitment into the liver is a trapping phenomenon, this is a default position, because no adhesion molecule to date has been identified to be important in this recruitment process. These trapping phenomena are observed following the administration of LPS or bacteria to mice, but do not seem to be prominent in mice treated with TNF-α or IL-1.

As a whole, these data suggest that the mechanisms encompassing leukocyte recruitment in response to LPS can be quite different depending on the specific tissue and its vascular architecture. Several questions remain: a) is TLR4 temporally regulated by vascular bed-specific mechanisms, b) is the spatial expression (basal vs. apical) of TLR4 variable among tissues, and c) if so, does this also contribute to local versus systemic LPS responses and the adhesion molecules involved? Finally, there clearly exist some unidentified adhesion mechanisms independent of selectins and integrins evoked by LPS; therefore, standard antiadhesion therapy may not successfully limit leukocyte sequestration in endotoxemia.

LIPOPOLYSACCHARIDE-INDUCED PROCOAGULANT ACTIVITY

One important function of ECs is the maintenance of an antithrombotic surface. However, a consistent observation is that LPS induces tissue factor expression in cultured ECs, altering the endothelial surface from an anti- to a procoagulant state. This occurs via activation of NF-κB and the suppression of anticoagulant thrombomodulin (81,82). The release of tissue plasminogen activator and the expression of plasminogen activator inhibitor 1 also is increased by LPS (83). Platelet recruitment to the endothelium can cause vascular dysfunction by releasing various inflammatory molecules (84). In addition, the platelets could express adhesion molecules such as P-selectin, thus allowing neutrophil binding and activation and leading to further uncontrolled inflammation. This latter event is further underscored by the recent discovery that platelets also express TLR4 (85).

LIPOPOLYSACCHARIDE PRIMING PHENOMENON

Much like in preconditioning in ischemia–reperfusion, studies have demonstrated that repeated LPS exposure causes a diminished inflammatory response to LPS in vitro, including reduced TNF-α expression by monocytes, macrophages, and ECs (76,77,86). In addition, it has been documented that preconditioning with LPS leads to a reduction of tissue factor, ICAM-1, urokinase-type plasminogen activator (u-PA), E-selectin, and VCAM-1 expression in response to thrombin stimulation (87). Using DNA microarray analyses it was observed that pretreatment with LPS alone resulted in induction and suppression of many other genes (5). This pretreatment with LPS consequently inhibited thrombin-induced activation of NF-κB as well (5).

LIPOPOLYSACCHARIDE-INDUCED ENDOTHELIAL CELL INJURY AND APOPTOSIS

Apoptosis is another event that may be involved in the pathogenesis of severe sepsis and septic shock. A number of in vitro studies show apoptotic cell death of ECs in response to LPS (88), although convincing in vivo data is lacking. Research suggests that endothelium derived from different sites displays varying degrees of resistance to apoptosis in response to LPS. Moreover, species differences may exist. Bovine ECs derived from distinct anatomical origins are highly sensitive to LPS-induced apoptosis, whereas human ECs are resistant (89).

It is well established that the activation of NF-κB is important for EC survival, and that disruption of this pathway enhances the cytotoxic effects of apoptotic proteins. Hull and colleagues (90) have shown that the C-terminal portion of TRAF6 (TRAF6-C) inhibits LPS-induced NF-κB nuclear translocation and c-jun N-terminal kinase (JNK) activation in ECs and hence inhibits LPS-initiated endothelial apoptosis. TRAF6-C also was shown to block LPS-altered mitochondrial transmembrane potential, cytochrome C release, and caspase activation.

OTHER EFFECTS OF LIPOPOLYSACCHARIDES ON ENDOTHELIUM

Endothelium exposed to LPS transcriptionally upregulates nitric oxide synthase (NOS)-2 (also known as inducible NOS). In addition, experimental evidence suggests that LPS downregulates constitutive NOS-3 (also known as endothelial NOS) in endothelium (91). The intracellular signaling events involved in endothelial NOS-2 and cyclooxygenase (COX)-2 expression in response to LPS are yet undetermined. It is known that induction of NOS-2 in mononuclear cells requires binding of the LPS/CD14 complex to TLR4, with subsequent activation of NF-κB (1). Likely similar mechanisms are active in the endothelium as well. LPS also may contribute to endothelial dysfunction by increasing the vasoconstrictor tone of vascular smooth muscle cells, because ECs can release endothelin in response to LPS (1). A connection between endothelin and iNOS has been suggested. Hence, ECs respond to microbial toxins by hypersecreting vasoactive mediators, including vasoconstrictors as well as vasodilators. Additionally, endothelium expresses the inducible COX-2 protein (1). Similar to the isoform shift of NOS, experimental evidence suggests that a switch from COX-1 to COX-2 also may occur in ECs in response to LPS (1).

KEY POINTS

- The endothelium requires CD14 for activation; however, it remains unclear whether the membrane or soluble isoform is essential.
- TLR4 is the critical endothelial LPS detector.
- Endothelial TLR4 may be a critical molecule that mediates the sepsis phenotype.

Future Goals

- To identify additional LPS signaling pathways (independent of CD14)
- To identify the adhesion molecules involved in leukocyte recruitment into lungs and liver

REFERENCES

1 Grandel U, Grimminger F. Endothelial responses to bacterial toxins in sepsis. *Crit Rev Immunol.* 2003;23(4):267–299.
2 Bannerman DD, Fitzpatrick MJ, Anderson DY, et al. Endotoxin-neutralizing protein protects against endotoxin-induced endothelial barrier dysfunction. *Infect Immun.* 1998;66(4):1400–1407.
3 Cooperstock MS. Inactivation of endotoxin by polymyxin B. *Antimicrob Agents Chemother.* 1974;6(4):422–425.
4 Poltorak A, He X, Smirnova I, et al. Defective LPS signaling in C3H/HeJ and C57BL/10ScCr mice: mutations in Tlr4 gene. *Science.* 1998;282(5396):2085–2088.
5 Kawata T, Bristol JR, Rossignol DP, et al. E5531, a synthetic non-toxic lipid A derivative blocks the immunobiological activities of lipopolysaccharide. *Br J Pharmacol.* 1999;127(4):853–862.
6 Hornung V, Rothenfusser S, Britsch S, et al. Quantitative expression of toll-like receptor 1–10 mRNA in cellular subsets of human peripheral blood mononuclear cells and sensitivity to CpG oligodeoxynucleotides. *J Immunol.* 2002;168(9):4531–4537.
7 Muzio M, Bosisio D, Polentarutti N, et al. Differential expression and regulation of Toll-like receptors (TLR) in human leukocytes: selective expression of TLR3 in dendritic cells. *J Immunol.* 2000; 164(11):5998–6004.
8 Zarember KA, Godowski PJ. Tissue expression of human Toll-like receptors and differential regulation of Toll-like receptor mRNAs in leukocytes in response to microbes, their products, and cytokines. *J Immunol.* 2002;168(2):554–561.
9 Rock FL, Hardiman G, Timans JC, et al. A family of human receptors structurally related to *Drosophila* Toll. *Proc Natl Acad Sci USA.* 1998;95(2):588–593.
10 Qureshi ST, Lariviere L, Leveque G, et al. Endotoxin-tolerant mice have mutations in Toll-like receptor 4 (Tlr4). *J Exp Med.* 1999;189(4):615–625.
11 Gay NJ, Keith FJ. *Drosophila* Toll and IL-1 receptor. *Nature.* 1991; 351(6325):355–356.
12 Nomura N, Miyajima N, Sazuka T, et al. Prediction of the coding sequences of unidentified human genes. I. The coding sequences of 40 new genes (KIAA0001-KIAA0040) deduced by analysis of randomly sampled cDNA clones from human immature myeloid cell line KG-1. *DNA Res.* 1994;1(1):27–35.
13 Lemaitre B, Nicolas E, Michaut L, et al. The dorsoventral regulatory gene cassette spatzle/Toll/cactus controls the potent antifungal response in *Drosophila* adults. *Cell.* 1996;86(6):973–983.
14 Medzhitov R, Preston-Hurlburt P, Janeway CA Jr. A human homologue of the *Drosophila* Toll protein signals activation of adaptive immunity. *Nature.* 1997;388(6640):394–397.
15 Yang RB, Mark MR, Gurney AL, Godowski PJ. Signaling events induced by lipopolysaccharide-activated toll-like receptor 2. *J Immunol.* 1999;163(2):639–643.
16 Roach JC, Glusman G, Rowen L, et al. The evolution of vertebrate Toll-like receptors. *Proc Natl Acad Sci USA.* 2005;102(27): 9577–9582.
17 Medzhitov R, Janeway C Jr. Innate immunity. *N Engl J Med.* 2000;343(5):338–344.
18 Janeway CA Jr, Medzhitov R. Innate immune recognition. *Annu Rev Immunol.* 2002;20:197–216.
19 Brennan CA, Anderson KV. Drosophila: the genetics of innate immune recognition and response. *Annu Rev Immunol.* 2004;22: 457–483.

20 Wasserman SA. A conserved signal transduction pathway regulating the activity of the rel-like proteins dorsal and NF-kappa B. *Mol Biol Cell.* 1993;4(8):767–771.
21 Wasserman SA. Toll signaling: the enigma variations. *Curr Opin Genet Dev.* 2000;10(5):497–502.
22 Takeda K, Kaisho T, Akira S. Toll-like receptors. *Annu Rev Immunol.* 2003;21:335–376.
23 Ozinsky A, Underhill DM, Fontenot JD, et al. The repertoire for pattern recognition of pathogens by the innate immune system is defined by cooperation between toll-like receptors. *Proc Natl Acad Sci USA.* 2000;97(25):13766–13771.
24 Aderem A, Hume DA. How do you see CG? *Cell.* 2000;103(7): 993–996.
25 Diebold SS, Kaisho T, Hemmi H, et al. Innate antiviral responses by means of TLR7-mediated recognition of single-stranded RNA. *Science.* 2004;303(5663):1529–1531.
26 Heil F, Hemmi H, Hochrein H, et al. Species-specific recognition of single-stranded RNA via toll-like receptor 7 and 8. *Science.* 2004;303(5663):1526–1529.
27 Hasan U, Chaffois C, Gaillard C, et al. Human TLR10 is a functional receptor, expressed by B cells and plasmacytoid dendritic cells, which activates gene transcription through MyD88. *J Immunol.* 2005;174(5):2942–2950.
28 Zhang D, Zhang G, Hayden MS, et al. A toll-like receptor that prevents infection by uropathogenic bacteria. *Science.* 2004; 303(5663):1522–1526.
29 Yarovinsky F, Zhang D, Andersen JF, et al. TLR11 activation of dendritic cells by a protozoan profilin-like protein. *Science.* 2005;308(5728):1626–1629.
30 Hoebe K, Janssen EM, Kim SO, et al. Upregulation of costimulatory molecules induced by lipopolysaccharide and double-stranded RNA occurs by Trif-dependent and Trif-independent pathways. *Nat Immunol.* 2003;4(12):1223–1229.
31 Shimazu R, Akashi S, Ogata H, et al. MD-2, a molecule that confers lipopolysaccharide responsiveness on Toll-like receptor 4. *J Exp Med.* 1999;189(11):1777–1782.
32 Hoshino K, Takeuchi O, Kawai T, et al. Cutting edge: Toll-like receptor 4 (TLR4)-deficient mice are hyporesponsive to lipopolysaccharide: evidence for TLR4 as the Lps gene product. *J Immunol.* 1999;162(7):3749–3752.
33 Feng CG, Scanga CA, Collazo-Custodio CM, et al. Mice lacking myeloid differentiation factor 88 display profound defects in host resistance and immune responses to Mycobacterium avium infection not exhibited by Toll-like receptor 2 (TLR2)- and TLR4-deficient animals. *J Immunol.* 2003;171(9):4758–4764.
34 Ninomiya-Tsuji J, Kishimoto K, Hiyama A, et al. The kinase TAK1 can activate the NIK-I kappaB as well as the MAP kinase cascade in the IL-1 signalling pathway. *Nature.* 1999;398(6724): 252–256.
35 Kopp E, Medzhitov R, Carothers J, et al. ECSIT is an evolutionarily conserved intermediate in the Toll/IL-1 signal transduction pathway. *Genes Dev.* 1999;13(16):2059–2071.
36 Beutler B. Tlr4: central component of the sole mammalian LPS sensor. *Curr Opin Immunol.* 2000;12(1):20–26.
37 Yoshioka T, Morimoto Y, Iwagaki H, et al. Bacterial lipopolysaccharide induces transforming growth factor beta and hepatocyte growth factor through toll-like receptor 2 in cultured human colon cancer cells. *J Int Med Res.* 2001;29(5):409–420.
38 Hirschfeld M, Ma Y, Weis JH, et al. Cutting edge: repurification of lipopolysaccharide eliminates signaling through both human and murine toll-like receptor 2. *J Immunol.* 2000;165(2):618–622.
39 Tohme ZN, Amar S, Van Dyke TE. Moesin functions as a lipopolysaccharide receptor on human monocytes. *Infect Immun.* 1999;67(7):3215–3220.
40 Triantafilou K, Triantafilou M, Dedrick RL. A CD14-independent LPS receptor cluster. *Nat Immunol.* 2001;2(4):338–345.
41 Hirschfeld M, Weis JJ, Toshchakov V, et al. Signaling by toll-like receptor 2 and 4 agonists results in differential gene expression in murine macrophages. *Infect Immun.* 2001;69(3):1477–1482.
42 Werts C, Tapping RI, Mathison JC, et al. Leptospiral lipopolysaccharide activates cells through a TLR2-dependent mechanism. *Nat Immunol.* 2001;2(4):346–352.
43 Faure E, Thomas L, Xu H, et al. Bacterial lipopolysaccharide and IFN-gamma induce Toll-like receptor 2 and Toll-like receptor 4 expression in human ECs: role of NF-kappa B activation. *J Immunol.* 2001;166(3):2018–2024.
44 Abreu MT, Vora P, Faure E, et al. Decreased expression of Toll-like receptor-4 and MD-2 correlates with intestinal epithelial cell protection against dysregulated proinflammatory gene expression in response to bacterial lipopolysaccharide. *J Immunol.* 2001;167(3):1609–1616.
45 da Silva CJ, Ulevitch RJ. MD-2 and TLR4 N-linked glycosylations are important for a functional lipopolysaccharide receptor. *J Biol Chem.* 2002;277(3):1845–1854.
46 Dziarski R, Wang Q, Miyake K, et al. MD-2 enables Toll-like receptor 2 (TLR2)-mediated responses to lipopolysaccharide and enhances TLR2-mediated responses to Gram-positive and Gram-negative bacteria and their cell wall components. *J Immunol.* 2001;166(3):1938–1944.
47 Andreakos E, Sacre SM, Smith C, et al. Distinct pathways of LPS-induced NF-kappa B activation and cytokine production in human myeloid and nonmyeloid cells defined by selective utilization of MyD88 and Mal/TIRAP. *Blood.* 2004;103(6):2229–2237.
48 Faure E, Equils O, Sieling PA, et al. Bacterial lipopolysaccharide activates NF-kappaB through toll-like receptor 4 (TLR-4) in cultured human dermal ECs. Differential expression of TLR-4 and TLR-2 in ECs. *J Biol Chem.* 2000;275(15):11058–11063.
49 Tavener SA, Kubes P. Cellular and molecular mechanisms underlying LPS-associated myocyte impairment. *Am J Physiol Heart Circ Physiol.* 2005;290(2):H800–H8006. Epub 2005 Sep 19.
50 Zhang FX, Kirschning CJ, Mancinelli R, et al. Bacterial lipopolysaccharide activates nuclear factor-kappaB through interleukin-1 signaling mediators in cultured human dermal ECs and mononuclear phagocytes. *J Biol Chem.* 1999;274(12):7611–7614.
51 Zeuke S, Ulmer AJ, Kusumoto S, et al. TLR4-mediated inflammatory activation of human coronary artery ECs by LPS. *Cardiovasc Res.* 2002;56(1):126–134.
52 Mullaly SC, Kubes P. Toll gates and traffic arteries: from endothelial TLR2 to atherosclerosis. *Circ Res.* 2004;95(7):657–659.
53 Bassenge E. Endothelial function in different organs. *Prog Cardiovasc Dis.* 1996;39(3):209–228.
54 Wright SD, Ramos RA, Tobias PS, et al. CD14, a receptor for complexes of lipopolysaccharide (LPS) and LPS binding protein. *Science.* 1990;249(4975):1431–1433.
55 Pugin J, Ulevitch RJ, Tobias PS. A critical role for monocytes and CD14 in endotoxin-induced EC activation. *J Exp Med.* 1993; 178(6):2193–2200.

56 Frey EA, Miller DS, Jahr TG, et al. Soluble CD14 participates in the response of cells to lipopolysaccharide. *J Exp Med.* 1992; 176(6):1665–1671.

57 Pugin J, Schurer-Maly CC, Leturcq D, et al. Lipopolysaccharide activation of human endothelial and epithelial cells is mediated by lipopolysaccharide-binding protein and soluble CD14. *Proc Natl Acad Sci USA.* 1993;90(7):2744–2748.

58 Jersmann HP, Hii CS, Hodge GL, et al. Synthesis and surface expression of CD14 by human ECs. *Infect Immun.* 2001;69(1): 479–485.

59 Fenton MJ, Golenbock DT. LPS-binding proteins and receptors. *J Leukoc Biol.* 1998;64(1):25–32.

60 Meyrick B, Hoover R, Jones MR, et al. In vitro effects of endotoxin on bovine and sheep lung microvascular and pulmonary artery ECs. *J Cell Physiol.* 1989;138(1):165–174.

61 Goldblum SE, Ding X, Brann TW, et al. Bacterial lipopolysaccharide induces actin reorganization, intercellular gap formation, and endothelial barrier dysfunction in pulmonary vascular ECs: concurrent F-actin depolymerization and new actin synthesis. *J Cell Physiol.* 1993;157(1):13–23.

62 El Samalouti VT, Schletter J, Brade H, et al. Detection of lipopolysaccharide (LPS)-binding membrane proteins by immuno-coprecipitation with LPS and anti-LPS antibodies. *Eur J Biochem.* 1997;250(2):418–424.

63 Flaherty SF, Golenbock DT, Milham FH, et al. CD11/CD18 leukocyte integrins: new signaling receptors for bacterial endotoxin. *J Surg Res.* 1997;73(1):85–89.

64 Tobias PS, Soldau K, Ulevitch RJ. Isolation of a lipopolysaccharide-binding acute phase reactant from rabbit serum. *J Exp Med.* 1986;164(3):777–793.

65 Tobias PS, Mathison JC, Ulevitch RJ. A family of lipopolysaccharide binding proteins involved in responses to gram-negative sepsis. *J Biol Chem.* 1988;263(27):13479–13481.

66 Schumann RR, Leong SR, Flaggs GW, et al. Structure and function of lipopolysaccharide binding protein. *Science.* 1990; 249(4975):1429–1431.

67 Jack RS, Fan X, Bernheiden M, et al. Lipopolysaccharide-binding protein is required to combat a murine gram-negative bacterial infection. *Nature.* 1997;389(6652):742–745.

68 Gallay P, Heumann D, Le Roy D, et al. Lipopolysaccharide-binding protein as a major plasma protein responsible for endotoxemic shock. *Proc Natl Acad Sci USA.* 1993;90(21):9935–9938.

69 Lum H, Malik AB. Regulation of vascular endothelial barrier function. *Am J Physiol.* 1994;267(3 Pt 1):L223–L241.

70 Goldblum SE, Brann TW, Ding X, et al. Lipopolysaccharide (LPS)-binding protein and soluble CD14 function as accessory molecules for LPS-induced changes in endothelial barrier function, in vitro. *J Clin Invest.* 1994;93(2):692–702.

71 Bannerman DD, Sathyamoorthy M, Goldblum SE. Bacterial lipopolysaccharide disrupts endothelial monolayer integrity and survival signaling events through caspase cleavage of adherens junction proteins. *J Biol Chem.* 1998;273(52):35371–35380.

72 Yipp BG, Andonegui G, Howlett CJ, et al. Profound differences in leukocyte-EC responses to lipopolysaccharide versus lipoteichoic acid. *J Immunol.* 2002;168(9):4650–4658.

73 Johnston B, Walter UM, Issekutz AC, et al. Differential roles of selectins and the alpha4-integrin in acute, subacute, and chronic leukocyte recruitment in vivo. *J Immunol.* 1997;159(9):4514–4523.

74 Andonegui G, Goyert SM, Kubes P. Lipopolysaccharide-induced leukocyte-EC interactions: a role for CD14 versus toll-like receptor 4 within microvessels. *J Immunol.* 2002;169(4): 2111–2119.

75 Danner RL, Elin RJ, Hosseini JM, et al. Endotoxemia in human septic shock. *Chest.* 1991;99(1):169–175.

76 Lush CW, Cepinskas G, Kvietys PR. LPS tolerance in human ECs: reduced PMN adhesion, E-selectin expression, and NF-kappaB mobilization. *Am J Physiol Heart Circ Physiol.* 2000;278(3): H853-H861.

77 Zuckerman SH, Evans GF, Snyder YM, et al. Endotoxin-macrophage interaction: post-translational regulation of tumor necrosis factor expression. *J Immunol.* 1989;143(4):1223–1227.

78 Lonnemann G, Bechstein M, Linnenweber S, et al. Tumor necrosis factor-alpha during continuous high-flux hemodialysis in sepsis with acute renal failure. *Kidney Int Suppl.* 1999;72:S84–S87.

79 Astiz M, Saha D, Lustbader D, et al. Monocyte response to bacterial toxins, expression of cell surface receptors, and release of anti-inflammatory cytokines during sepsis. *J Lab Clin Med.* 1996;128(6):594–600.

80 Wong J, Johnston B, Lee SS, et al. A minimal role for selectins in the recruitment of leukocytes into the inflamed liver microvasculature. *J Clin Invest.* 1997;99(11):2782–2790.

81 Mackman N. Regulation of the tissue factor gene. *Thromb Haemost.* 1997;78(1):747–754.

82 Moore KL, Andreoli SP, Esmon NL, et al. Endotoxin enhances tissue factor and suppresses thrombomodulin expression of human vascular endothelium in vitro. *J Clin Invest.* 1987;79 (1):124–130.

83 Matsumoto H, Ueshima S, Fukao H, et al. Effects of lipopolysaccharide on the expression of fibrinolytic factors in an established cell line from human ECs. *Life Sci.* 1996;59(2):85–96.

84 Vincent JL, Yagushi A, Pradier O. Platelet function in sepsis. *Crit Care Med.* 2002;30(5 Suppl):S313-S317.

85 Andonegui G, Kerfoot SM, McNagny K, et al. Platelets express functional Toll-like receptor-4. *Blood.* 2005;106(7):2417–2423.

86 Haas JG, Meyer N, Riethmuller G, et al. Inhibition of lipopolysaccharide-induced in vitro desensitization by interferon-gamma. *Eur J Immunol.* 1990;20(5):1181–1184.

87 Wada Y, Otu H, Wu S, et al. Preconditioning of primary human ECs with inflammatory mediators alters the "set point" of the cell. *FASEB J.* 2005;19(13):1914–1916.

88 Frey EA, Finlay BB. Lipopolysaccharide induces apoptosis in a bovine EC line via a soluble CD14 dependent pathway. *Microb Pathog.* 1998;24(2):101–109.

89 Harlan JM, Harker LA, Reidy MA, et al. Lipopolysaccharide-mediated bovine EC injury in vitro. *Lab Invest.* 1983;48(3):269–274.

90 Hull C, McLean G, Wong F, et al. Lipopolysaccharide signals an endothelial apoptosis pathway through TNF receptor-associated factor 6-mediated activation of c-Jun NH2-terminal kinase. *J Immunol.* 2002;169(5):2611–2618.

91 Walter R, Schaffner A, Schoedon G. Differential regulation of constitutive and inducible nitric oxide production by inflammatory stimuli in murine ECs. *Biochem Biophys Res Commun.* 1994;202(1):450–455.

47

Receptor for Advanced Glycation End-Products and the Endothelium

A Path to the Complications of Diabetes and Inflammation

Jean-Luc Wautier* and Ann Marie Schmidt†

** University Lariboisiere-Saint Louis and Institut National de la Transfusion Sanguine, Paris, France; † College of Physicians and Surgeons, Columbia University, New York, New York*

Nonenzymatically modified adducts of proteins or lipids engage the vascular endothelium by mechanisms distinct from those of the native, unmodified species. These modified adducts, termed *advanced glycation end-products* (AGEs), are particularly prevalent in diabetes, but accumulate as well in diseases such as renal failure, in states of intense inflammation, and in aging. Extracellular AGEs mediate their cellular activity by binding to the receptor for age (RAGE). Increasing evidence suggests that disease-associated ligands other than AGE may bind and activate RAGE and thus contribute to diverse tissue-damaging complications. Studies in this area have provided a framework for targeting the ligand–RAGE axis as a novel therapeutic opportunity in diabetes.

THE PROBLEM OF GLYCATION

A range of physiological and pathophysiological states provides a ripe environment for the post-translational modification of proteins and lipids that eventuate in the formation of AGEs. Such modified species bind to and activate specialized receptors (RAGE) present on the surface of multiple cell types, including endothelial cells (ECs). AGE–RAGE interactions result in EC dysfunction, and appear to play an important pathophysiological role in several diseases, including type 1 and type 2 diabetes.

Maillard Reaction and the Formation of Advanced Glycation Endproducts

The nonenzymatic glycation of proteins was first described in 1912, by Louis-Camille Maillard (1). This reaction begins when the carbonyl group (either aldehyde or ketone) of the reducing sugar forms a reversible Schiff base with the amino group of the molecule. Schiff bases may undergo subsequent intramolecular rearrangements to form Amadori products (2). A series of further rearrangements may occur, including dehydration and condensation reactions to form irreversible endproducts, or the AGEs (Figure 47.1). For example, the initial Schiff base adducts formed from glucose, lysine, and N-terminal amino acid residues may rearrange to form the key intermediate, fructosamine. Fructosamine degradation and the direct reaction of α-oxoaldehydes with proteins may form many types of AGEs in vivo, including bis(lysyl)imidazolium cross-links, hydroimidazolones, and monolysyl adducts (2,3).

AGEs may form in both the intracellular and extracellular environment, and are a heterogeneous class of compounds. Some may be fluorescent and of yellow-brown color; others may form stable intermolecular and intramolecular cross-links (2,4). Furthermore, the degree of AGE formation on any given species will vary depending on a range of other factors. For example, in the case of proteins, AGE depends on the endogenous reactivity of specific amino groups, and it is directly influenced by distinct factors within the environment, such as increased oxidative stress, the surrounding glucose concentration (both in the circulation and in the microenvironment), and the half-life of the protein (4,5).

It was the latter property that led to the identification in vivo of the importance of glycation in human subjects with diabetes. Specifically, Rahbar and colleagues first reported a distinct form of hemoglobin in subjects with diabetes. This species came to be known as glycosylated hemoglobin, or hemoglobin A1c, a naturally occurring minor human hemoglobin that is elevated in human diabetic subjects (6). Because of the relatively long half-life of red blood cells, the degree of modification of this hemoglobin species in diabetes is a barometer of the effectiveness of long-term glycemic control. At present, glycosylated hemoglobin is a critical tool in the armamentarium of diabetes control in the clinic and in

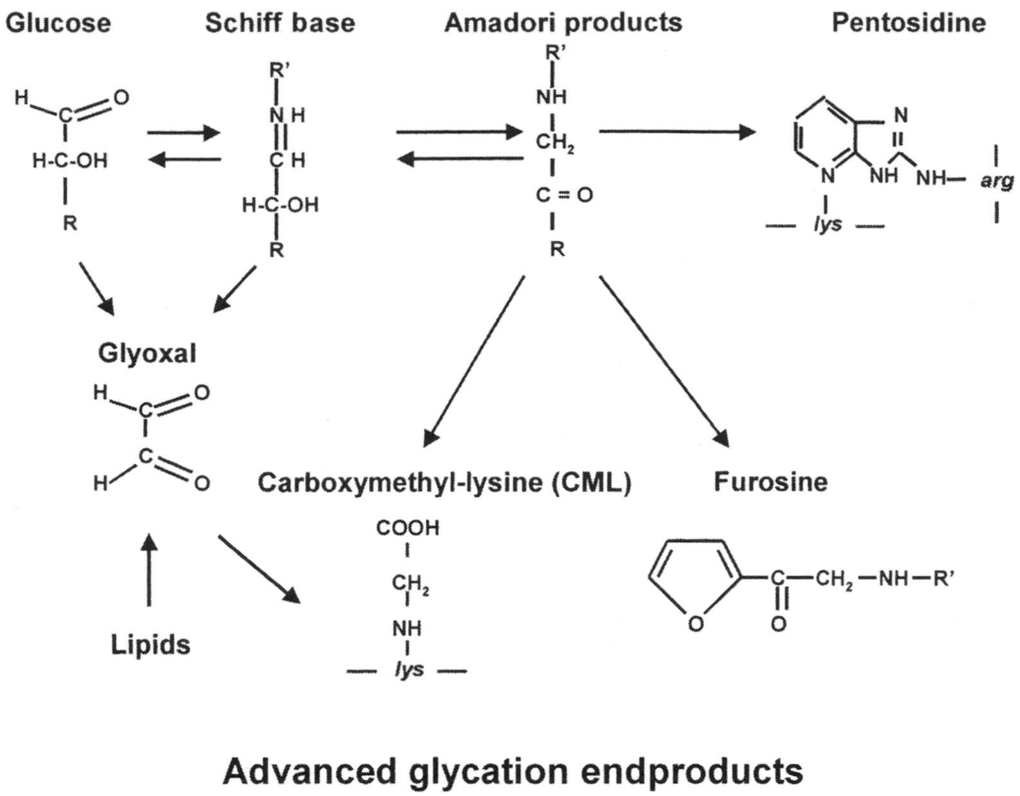

Figure 47.1. Generation of AGEs. Pathways by which glucose may yield generation of Schiff bases and Amadori products and, after further molecular rearrangements, AGEs.

long-term epidemiological studies, such as the Diabetes Control and Complications Trial Research Group (DCCT) (7).

Importantly, beyond the impact of hyperglycemia, AGEs may form via other mechanisms as well. For example, it has been shown that myeloperoxidase pathways may lead to the direct generation of carboxymethyl lysine (CML)-AGEs (8). In addition, it is established that enhanced oxidative stress, either by activation of nicotinamide adenine dinucleotide phosphate (NADPH) oxidase and/or mitochondrial dysfunction, which typifies the diabetic milieu (9–11), may favor further AGE generation (12). Based on the existence of hyperglycemia-independent mechanisms of AGE production, it is not surprising that enhanced AGE accumulation occurs in settings beyond diabetes. For example, AGEs may form in renal failure, chronic neurodegenerative disorders, and with natural aging (13–18).

In Vivo Modifiers of Advanced Glycation Endproducts Formation and Detoxification

In addition to direct forces that stimulate and favor AGE generation in the intracellular and extracellular compartments, a number of enzymatic reactions may further propel AGE formation or, indeed, modulate AGE removal. In the former case, the aldose reductase (AR) or polyol pathway may generate species that "feed forward" into AGE formation. In this pathway, glucose is reduced to sorbitol by AR; fructose generated by this pathway is converted into fructose-3-phosphate by the action of 3-phosphokinase (3-PK). This leads to the generation of 3-deoxyglucosone (3-DG), a central precursor in the generation of an array of AGEs, in particular, CML adducts and others (19–20).

In other settings, the glyoxalase pathway, particularly glyoxalase I, mediates AGE detoxification and removal. In ECs stably transfected to express human glyoxalase I, incubation in high-glucose conditions failed to increase AGE formation compared with nontransfected control cells (21). In natural aging, it has been found that glyoxalase activity is reduced, thereby providing a mechanism by which AGE accumulation may be even further enhanced (22). Thus, in addition to gradually enhanced generation of these species in euglycemic aging, the impaired removal or detoxification of these adducts may further exacerbate their accumulation.

CONSEQUENCES OF ADVANCED GLYCATION ENDPRODUCTS

Advanced Glycation Endproducts and Effects on Structural Integrity of the Vessel Wall

AGEs may directly affect the structural integrity of the vessel wall and underlying basement membrane. For example, extensive cross-linking of AGE modifications within matrix molecules such as collagen may disrupt matrix–matrix and matrix–cell interactions (23,24). In the intracellular milieu, nonenzymatic glycation of basic fibroblast growth factor (bFGF) may impair its function (25). Other studies indicate that AGEs quench nitric oxide (NO), thereby potentially altering vascular relaxation and function (26). When AGEs are formed directly on enzymes, such modification may modulate their function (27).

Advanced Glycation Endproducts and Receptors

In addition to the effects of AGEs on the integrity of the vessel wall and basement membrane, AGEs have been shown to interact with specific cellular receptors. A number of cell surface interaction sites for AGEs have been identified, including macrophage scavenger receptor (MSR) type II, OST-48, 80K-H, galectin-3, CD36, and RAGE (28–32). Most of these receptors appear to remove and detoxify AGEs. One exception is RAGE, which, when bound to ligand, triggers intracellular signaling (32,33). RAGE is a member of the immunoglobulin superfamily of cell surface molecules (33). The extracellular domain of RAGE is composed of one V-type immunoglobulin domain, followed by two C-type immunoglobulin domains (33). A single hydrophobic transmembrane domain is followed by a short, but highly charged cytosolic domain that is essential for RAGE-mediated signal transduction (33). Although RAGE has subsequently been shown to be present on a variety of cell types, the very first studies identified this receptor on ECs. A plethora of evidence suggests that AGE–RAGE interactions critically modulate the clinical phenotype of diseases associated with increased ligand, particularly diabetes.

ADVANCED GLYCATION ENDPRODUCTS, RECEPTOR FOR AGE, AND ENDOTHELIUM: BINDING AND OUTCOMES

In Vitro–Prepared and in Vivo–Derived Advanced Glycation Endproducts Bind Receptor for AGE

RAGE was first identified in ECs (32). In those experiments, bovine serum albumin (BSA) was incubated with reducing sugars such as D-glucose or D-ribose and, after incubation, was dialyzed free of remaining sugars. The resulting product, AGE-BSA, or its unmodified counterpart was then radiolabeled with I^{125}. When immobilized lung extract or bovine aortic ECs were incubated with I^{125}-AGE-BSA at 4°C, dose-dependent and saturable binding was noted, with a K_d of approximately 70 nM (32). Pretreatment of the monolayers with trypsin or detergent effectively eliminated AGE binding, strongly suggesting that the interaction site was, at least in part, composed of protein. Definitive evidence that RAGE was a receptor for AGE was noted by pretreatment of the monolayers with blocking antibodies to RAGE, but not nonimmune IgG. Further, transient transfection of cells largely devoid of RAGE with constructs expressing full-length functional RAGE resulted in AGE binding, compared with the absence of AGE binding in the mock vector-transfected cells (33). Finally, immunohistochemistry performed with antibodies raised to RAGE demonstrated that the receptor indeed was present on the surface of immobilized, nonpermeabilized cultured ECs (32).

A chief concern in these first experiments was that the AGEs prepared and tested in vitro were artificially modified and, thus, not likely to reflect AGE species present in human fluids or tissues in diabetes. To address this issue, cultured ECs were incubated with an in vivo source of AGE, namely diabetic erythrocytes. In these studies, binding of diabetic erythrocytes was reduced by incubation with excess AGE-BSA, anti-AGE IgG, or in the presence of anti-RAGE IgG, but not nonimmune IgG (34,35).

Efforts to elucidate specific AGEs that might interact with RAGE led to the identification of CML-AGE products as signal transduction ligands of RAGE (36). CML-AGEs, nonfluorescent and non–cross-linked adducts, are prevalent in vivo and have been suggested to be the most common class of AGEs found in human diabetic tissues (37,38). CML-AGEs were radiolabeled with I^{125} and binding studies performed both on plastic tubes and in cultured cells. When CML-ovalbumin was incubated with cultured human umbilical vein ECs (HUVECs), dose-dependent and saturable binding occurred. In addition, incubation of ECs with CML-AGE resulted in increased expression of adhesion molecules, such as vascular cell adhesion molecule (VCAM)-1, and activation of nuclear factor (NF)-κB (36). These effects were blocked by the addition of neutralizing anti-RAGE antibodies or the overexpression of dominant negative RAGE (36).

Thus, support for the role of AGE-EC RAGE stimulation was procured by using a variety of in vitro– and in vivo–derived AGEs, as well as specific AGE structures, such as CML adducts. It is important to note that these findings do not exclude that distinct AGEs beyond CML adducts may bind to RAGE.

AGE–RAGE Interactions Result in Endothelial Cell Perturbation

The diabetic erythrocyte AGE–RAGE interactions discussed earlier were shown to have functional consequences. Under in vitro conditions, these included increased oxidant stress, as assessed by the generation of thiobarbituric acid reactive substances (TBARS) and activation of NF-κB (35). To assay for effects in vivo, erythrocytes from diabetic or control rats were infused into syngeneic nondiabetic rats. Consistent with the premise that diabetic erythrocytes bound cell surface RAGE, ^{51}Cr-labeled diabetic erythrocytes displayed shorter survival compared with nondiabetic erythrocytes, an effect that was blocked by infusion of anti-RAGE IgG (35). In vivo, liver tissue of rats infused with diabetic erythrocytes displayed increased oxidant stress, a process blocked by anti-RAGE IgG (35). Thus, these studies demonstrated for the first time that in vivo–derived AGEs bound RAGE and brought about specific consequences in the cells, such as enhanced oxidant stress.

In addition to enhancing oxidant stress in ECs, AGE–RAGE signaling has been shown to alter the transcription and translation of key molecules involved in the recruitment and activation of inflammatory cells. As indicated above, both in vitro– and in vivo–derived AGEs were reported to upregulate VCAM-1 on the surface of ECs in a RAGE-dependent manner (36,39). In other studies, infusion of AGEs into rabbits resulted in upregulation of both VCAM-1 and intercellular adhesion molecule (ICAM)-1 (40). Such findings implicate key roles for RAGE in the earliest steps initiating acceleration of vascular disease, particularly those in AGE-enriched environments as occur in diabetes and renal failure.

Ongoing studies suggested that RAGE was not solely a receptor for AGEs. As new ligands of the receptor were identified, efforts were expanded to test the premise that these molecules bound RAGE on ECs and modulated their properties.

RECEPTOR FOR AGE IS A MULTILIGAND RECEPTOR: IMPLICATIONS FOR THE INFLAMMATORY RESPONSE

Proinflammatory Ligands of Receptor for AGE

S100/Calgranulins

In addition to AGEs, recent studies have shown that S100/calgranulin molecules are RAGE ligands (41). The S100/calgranulins are a family of at least 20 distinct polypeptide members, characterized by their ability to bind calcium. S100/calgranulins are expressed by a wide array of cell types, including polymorphonuclear leukocytes, dendritic cells, mononuclear phagocytes (MPs), and lymphocytes (42–44), and at one time they were believed to function exclusively within the intracellular compartment. However, S100A12 (calgranulin C or extracellular newly identified RAGE binding protein, EN-RAGE) was shown to bind RAGE in a saturable and dose-dependent manner on cultured ECs, as well as on other cells, such as vascular smooth muscle cells (VSMCs), MPs, and lymphocytes (44). Initially, it was difficult to reconcile a role for these species in receptor-mediated signaling with their established intracellular location and function. However, it has been shown clearly that discrete mechanisms exist by which these molecules may be released by activated cells (45,46), thereby allowing them to engage the cell surface receptor RAGE in an autocrine and paracrine manner.

Similar to the effects of AGEs, incubation of a prototypic S100/calgranulin, S100A12, with cultured HUVECs resulted in RAGE-dependent activation of NF-κB, upregulation of VCAM-1 antigen and activity, and induction of ICAM-1 protein (41). In addition to S100A12, further studies indicated that another member of the S100 family, S100b, also bound and activated RAGE in ECs. Thus, it is plausible that multiple members of the S100/calgranulin family may interact with RAGE.

Amphoterin or High-Mobility Group Box-1

Amphoterin (also known as high-mobility group box 1 (HMGB1) was originally described and characterized as a DNA-binding protein (47). In addition to its role in transcriptional regulation, HMGB1 recently was shown to function as a soluble inflammatory mediator. Wang and colleagues demonstrated that HMGB1 is released from activated monocytes and macrophages, and contributes to the propagation of inflammation (48,49). Compared with other inflammatory mediators such as interleukin (IL)-6 and tumor necrosis factor (TNF)-α, HMGB1 is a relatively late marker of endotoxemia. Such findings suggest that lipopolysaccharides (LPS) may exert its often devastating effects due to multiple waves of cellular perturbation, with HMGB1 initiating a key later wave. Additional studies showed that HMGB1 release from monocytes is triggered by a vesicle-mediated pathway (50). Moreover, release of HMGB1 from necrotic, but not apoptotic cells, stimulated inflammatory mechanisms in affected tissues (51).

In subsequent studies, HMGB1 was shown to directly activate HUVECs, as evidenced by upregulation of VCAM-1, ICAM-1, and E-selectin, in parallel with the release of IL-8 and granulocyte colony stimulating factor (G-CSF) (52). Key roles for Elk and NF-κB pathways, at least in part via RAGE, were shown to underlie these effects of amphoterin on the modulation of EC gene expression (52). Together, these data suggest that even in the absence of diabetes and hyperglycemia, RAGE ligands may stimulate ECs to trigger vascular and inflammatory dysfunction.

Mac-1

In addition to proinflammatory S100/calgranulins and amphoterin, recent studies have indicated that RAGE is a counter-receptor for leukocyte integrin Mac-1 (also known as CD11b/CD18) (53). Further, these studies indicated that, in the presence of S100b, the RAGE–mac-1 interaction is augmented (53). Such findings suggest that a scaffold of RAGE ligands amplifies inflammatory mechanisms in the tissues. If left unchecked, RAGE-mediated ongoing and sustained inflammation, coupled with failure of regeneration mechanisms, may lead to irreversible tissue injury.

Receptor for AGE and Euglycemic Inflammatory Responses: In Vivo Studies

Two key observations, discussed earlier, raised the possibility that RAGE plays a pathophysiological role in *euglycemic* inflammatory conditions: (a) factors other than high glucose are able to promote AGE formation, and (b) non-AGE ligands associated with inflammation may bind to and activate RAGE. Consistent with this notion, studies in nondiabetic mice demonstrated that blockade of RAGE suppressed delayed type hypersensitivity, decreased colonic inflammation in *IL-10*$^{-/-}$ mice (41), suppressed experimental autoimmune encephalomyelitis, suppressed joint inflammation and destruction in a model of type II collagen sensitization, and enhanced survival and decreased inflammation in a murine model of sepsis induced by cecal ligation and puncture (54–57).

Based on the above findings, we hypothesized that RAGE might be linked to autoimmunity and inflammation in pancreatic islets. To test this hypothesis, we employed a model of adoptive transfer of diabetogenic spleen cells into NOD/scid mice (56). Both RAGE and S100 were expressed on islet cells with an inflammatory infiltrate in sections of pancreata from NOD/scid mice that had received a transfer of diabetogenic splenocytes, but not from control NOD/scid mice that were not subjected to transfer of these splenocytes. RAGE was expressed in key populations of T cells ($CD4^+$ and $CD8^+$) and B cells within the islet, as well as monocytes. Blockade of ligand–RAGE interactions, using soluble RAGE (sRAGE; comprising the extracellular V-C-C′ Ig-like domains), significantly prolonged the time to development of hyperglycemia in NOD/scid mice compared with vehicle-treated mice subjected to adoptive transfer of diabetogenic splenocytes (56). Taken together, these findings highlight key roles for RAGE in the inflammatory response and suggest that release of these proinflammatory ligands from activated monocytes and lymphocytes may play a key role in amplification of inflammation.

THE ROLE OF RECEPTOR FOR AGE IN MODULATING DIABETES-ASSOCIATED COMPLICATIONS

Hyperpermeability

Endothelial dysfunction accompanies diabetes and has been linked to the incidence and severity of cardio- and cerebrovascular complications in this disorder (58–64). Diabetes-associated EC dysfunction is manifested, at least in part, by abnormal response to acetylcholine and impaired flow-induced vasodilation. In addition, diabetes is associated with increased vascular leakage in both animal models of disease and in human subjects (65–69). To determine the role of RAGE in mediating endothelial barrier dysfunction, we employed a model of streptozotocin-induced, insulin-deficient diabetes in rats. Compared with control animals, diabetic rats demonstrated increased vascular permeability (particularly in the intestine, skin, and kidney). Importantly, this effect was inhibited in a dose-dependent manner by a single intravenous injection of sRAGE (70). Soluble RAGE was administered in a single dose after 9 to 11 weeks of established hyperglycemia in these animals. These experiments were the first to demonstrate that in vivo blockade of RAGE modulates an established vascular complication of diabetes (70). In the same study, the infusion of diabetic (compared with nondiabetic) red blood cells into syngeneic nondiabetic rats resulted in increased permeability in the intestine, skin, kidney, vena cava, and heart (70). The latter effect was attenuated by pretreatment with the antioxidant, probucol. These data support a key role for RAGE and oxidative stress as essential factors mediating vascular dysfunction in vivo.

Atherosclerosis

Receptor for AGE and Human Diabetic Atherosclerosis

Cipollone and colleagues examined endarterectomy specimens from diabetic and nondiabetic patients. They demonstrated that: (a) RAGE expression was increased in diabetic atherosclerosis compared with nondiabetic controls; (b) RAGE expression colocalized with cyclooxygenase (COX)-2, type 1/type 2 microsomal prostaglandin E_2, and matrix metalloproteinases (MMPs), particularly in macrophages at the vulnerable regions of the atherosclerotic plaques; and (c) increased expression of RAGE in the atherosclerotic plaques linearly correlated with the level of glycosylated hemoglobin (71). Others have reported increased expression of RAGE, in parallel with enhanced tissue factor antigen/activity, in peripheral blood mononuclear cells from type 2 diabetic patients with vascular complications compared with those without complications (72). Taken together, these data raise the possibility that the ligand–RAGE axis plays a pathophysiological role in diabetic atherosclerosis.

Receptor for AGE Blockade and Atherosclerosis in Apolipoprotein E-Null Mice

To address the question of pathophysiology, an animal model of diabetic atherosclerosis was developed, in which 6-week-old male apolipoprotein E (*ApoE*)-null mice (73,74) were injected with streptozotocin. After approximately 6 weeks of established diabetes, aortas displayed increased lesions at aortic branch points and the lesser curvature compared with nondiabetic controls (75). Quantitative analysis of multiple sections demonstrated significantly increased area and number of lesions in diabetic versus euglycemic *ApoE*-null controls (75). Histological analysis of oil red O-stained sections revealed more complex lesions (fibrous caps, calcification, necrosis, cholesterol clefts) in the diabetic specimens. Importantly, diabetic lesions displayed increased expression of RAGE protein and enhanced accumulation of AGEs, the latter as measured by enzyme-linked immunosorbent assay (ELISA) and immunohistochemistry (75).

Next, diabetic mice were treated with once-daily intraperitoneal administration of sRAGE beginning immediately at the time of documentation of diabetes and continued for an additional 6 weeks. Analysis of aortas revealed a marked dose-dependent suppression of atherosclerotic lesion area, number, and complexity in samples from diabetic *ApoE*-null mice treated with sRAGE compared with controls (75). Taken together, these findings provide evidence for a pathogenic role of RAGE in accelerated diabetic atherosclerosis.

RAGE blockade did not alter the glucose or lipid profiles, nor did it affect levels of glycosylated hemoglobin (75). Unexpectedly, however, sRAGE resulted in reduced levels of AGEs in the diabetic animals (75,76). One explanation for these observations is that suppression of RAGE-mediated oxidant stress in the vessel wall reduces subsequent formation and accumulation of AGEs. Consistent with this concept, we observed diminished susceptibility to copper-induced oxidation of low-density lipoprotein (LDL) retrieved from sRAGE-treated diabetic *ApoE*-null mice compared with LDL obtained from vehicle-treated diabetic *ApoE*-null mice (75).

To promote an increased disease burden, and thus more accurately mimic the clinical setting, diabetic *ApoE*-null mice were left untreated for approximately 6 weeks. Mice then received daily sRAGE or vehicle for an additional 6 weeks. sRAGE administration blocked interval progression of lesion area and complexity compared with control animals (77). Treatment also resulted in decreased infiltration of MPs and VSMCs into atherosclerotic plaques. Most sRAGE-treated mice demonstrated decreased oxidant stress, as evidenced by decreased COX-2 and nitrotyrosine epitopes, and decreased inflammation, as evidenced by JE/monocyte chemoattractant protien (MCP)-1 and tissue factor antigens, MMP9 antigen/activity, and phosphorylated p38 mitogen-activated protein kinase (MAPK) (77). Again, RAGE blockade did not affect glucose or lipid profiles (77).

Recent studies have extended these findings to a murine model of type 2 diabetes. *ApoE*-null mice were bred into the *db/db* background, thereby generating hypercholesterolemic mice with leptin receptor deficient signaling. Compared with *ApoE*-null mice without diabetes, the *ApoE*-null *db/db* mice displayed significantly increased atherosclerosis. When sRAGE was administered to these diabetic mice once daily from age 9 to 11 weeks, a significant decrease in atherosclerosis at the aortic root was noted (78). In parallel, decreased vascular inflammation was observed in the animals treated with sRAGE versus vehicle (78). These findings expanded the impact of RAGE in diabetic atherosclerosis to include both models of insulin-deficient and insulin-resistant diabetes.

Restenosis

In addition to its key role in modulating atherosclerosis, RAGE also has been implicated in exaggerated neointimal expansion after acute vascular injury. In a previous study, carotid balloon injury resulted in accelerated neointimal expansion in diabetic, insulin-resistant fatty Zucker rats compared with age-matched euglycemic controls (79). The administration of sRAGE beginning just prior to arterial injury and continuing for 1 week postoperatively resulted in a significant decrease in the intima:media ratio (I:M ratio) compared with vehicle-treated rats (serum albumin) (79). In parallel, the incorporation of bromodeoxyuridine (BrdU) was significantly decreased in VSMCs in the presence of sRAGE, thereby strongly suggesting key roles for RAGE in the response of VSMCs to arterial injury.

To further explore the pathogenic role of RAGE in arterial remodeling, we employed a well-described model of acute femoral artery endothelial denudation in euglycemic C57BL/6 mice (80). RAGE and its ligands, CML-AGEs and S100/calgranulins, were upregulated in the vessel wall consequent to injury, particularly in VSMCs (80). These studies highlight the fact that, even in the absence of hyperglycemia, AGE generation may be triggered by intense inflammatory stimulation. In this model, consequent to endothelial denudation, rapid recruitment of polymorphonuclear leukocytes to the injured vessel wall likely triggers the generation of CML, and perhaps other AGEs, by inflammatory pathways such as myeloperoxidase (8,81). In this model, RAGE blockade yielded significantly reduced I:M ratios (80). In other experiments, homozygous *RAGE*-null mice or transgenic mice expressing dominant negative (DN) RAGE selectively in VSMCs (driven by the *SM22α* promoter) were employed. Compared to their respective littermates, homozygous *RAGE*-null mice or transgenic mice expressing DN RAGE in VSMCs displayed decreased I:M ratio on day 28 (80). Together, these data suggest that acute endothelial injury triggers the generation of RAGE ligands, which in turn are linked to RAGE-dependent VSMC proliferation, migration, and restenosis.

Nephropathy

Yamamoto and colleagues generated transgenic mice overexpressing RAGE in EC using the *Flk-1* promoter. In that report, the authors indicated that some transgenic expression of RAGE occurred in macrophages as well. These mice were crossed with mice that develop insulin-dependent diabetes shortly after birth. Compared to single transgenic diabetic mice lacking the RAGE transgene, double transgenic mice displayed enlargement of the kidney, glomerular hypertrophy, increased albuminuria, mesangial expansion and glomerulosclerosis, and increased serum creatinine (82). Moreover, serum AGE levels were increased in the double transgenic mice. Treatment with an AGE inhibitor, (+/–)-2-isopropylidene-hydrazono-4-oxo-thiazolidine-5-ylacetanilide (OPB-195), inhibted advanced nephropathy in the double transgenic mice, strongly supporting a role for the ligand–RAGE signaling axis in the pathogenesis of experimental diabetic nephropathy.

We have carried out studies in the *db/db* mouse. In the long term, these animals develop renal changes that parallel, at least in part, those evident in human diabetic kidney (83). Consistent with observations in the human kidney, we showed that the podocyte is the principal site of RAGE expression in the glomerulus of *db/db* mice. Podocyte RAGE expression was higher in diabetic compared with nondiabetic controls. Moreover, RAGE also was detected in glomerular ECs (83). Also, CML-AGE adducts and S100/calgranulins were upregulated in *db/db* kidneys (83). Compared with nondiabetic controls, administration of sRAGE to *db/db* mice resulted in reduced expression of vascular endothelial growth factor (VEGF) antigen and transforming growth factor (TGF)-β mRNA, decreased glomerular and mesangial area, reduced thickness of the GBM, and lower urinary albumin excretion (83). Importantly, the protection against renal cortical VEGF and TGF-β induction was confirmed in *RAGE*-null mice rendered diabetic with streptozotocin (83).

RECEPTOR FOR AGE AND SIGNAL TRANSDUCTION

The signal transduction pathways downstream of RAGE activation in ECs and other cell types are diverse. Depending on the acuteness or chronicity of ligand stimulation, a range of signaling pathways may be activated. In the context of EC dysfunction in diabetes or other inflammatory states,

AGEs, S100/calgranulins, and amphoterin may activate ECs, as well as other cell types intimately involved in vascular disease initiation and progression, such as MPs and VSMCs. Signaling cascades activated upon ligand–RAGE interaction include pathways such as p21ras, extracellular signal-regulated kinase (ERK)-1/2 (p44/p42) MAPKs, p38 and SAPK/C-jun N-terminal kinase (JNK) MAPKs, Rho GTPases, phosphoinositide-3 kinase (PI3K), JNK/STAT pathway, and $p21^{waf}$; downstream consequences such as activation of the key transcription factors NF-κB and cAMP-response element-binding protein (CREB) also have been reported (36,90,84–94).

Importantly, ligand-stimulated RAGE activation results in the generation of reactive oxygen species (ROS), at least in part via activation of NADPH oxidase (88). Definitive roles for RAGE-mediated activation of NADPH oxidase were demonstrated in studies in *gp91*^{phox}-null mice. Specifically, monocytes from wild-type, but not NADPH oxidase–deficient animals, displayed increased generation of tissue factor upon incubation with AGEs (95). Studies are underway to delineate if additional mechanisms, distinct from NADPH oxidase, underlie the potent ability of RAGE to generate ROS.

The cytosolic domain of RAGE appears to be critical for RAGE-dependent signal transduction and modulation of gene expression and cellular phenotype. The deletion of the cytosolic domain of RAGE (DN RAGE) abrogates ligand-mediated RAGE signaling and function both in vitro and in vivo. For example, when DN RAGE is expressed in vivo in VSMCs, cells of MP lineage, neurons of the central or peripheral nervous system, or CD4 lymphocytes, RAGE-mediated signaling is effectively suppressed, thereby strikingly modulating injury-triggered outcomes in diverse settings (36,41,55,80,91, 96–98).

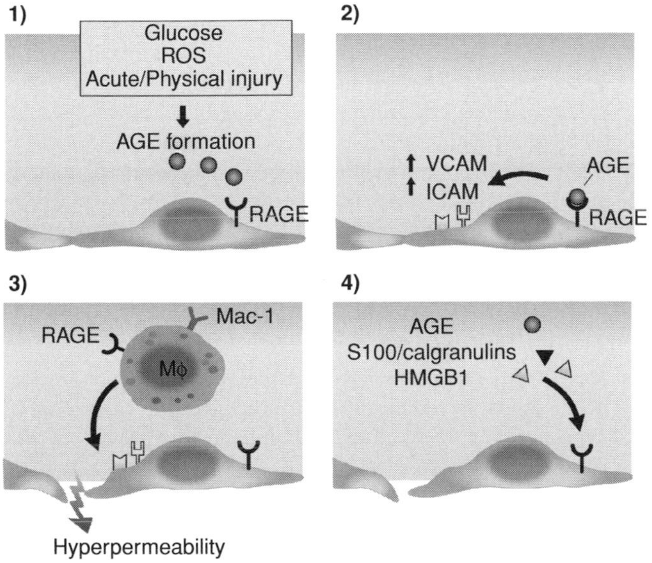

Figure 47.2. Endothelial RAGE: sparking a cascade of events linked to cellular perturbation and tissue injury. It is our hypothesis that endothelial RAGE is a key priming step for the activation of multiple proinflammatory mechanisms that amplify cellular perturbation in a range of disorders. We propose that in hyperglycemia, inflammation, or settings characterized by oxidative stress, acute, rapid, or sustained and chronic generation and accumulation of AGEs occurs. The interaction of AGEs with EC RAGE stimulates the generation of ROS and upregulation of proinflammatory mechanisms, such as the expression of adhesion molecules and chemokines, thereby leading to the recruitment and activation of inflammatory cells such as MP and lymphocytes. These processes may be further amplified by Mac-1–RAGE-dependent interactions. RAGE-dependent hyperpermeability enhances inflammatory cell entry into the injured vessel. Once such inflammatory cells are recruited to perturbed ECs, we posit that the generation or release of S100/calgranulins and amphoterin by these activated cells augment and sustain RAGE-activation. Then, by both autocrine and paracrine interactions, S100/calgranulins, amphoterin, and AGEs may amplify inflammation in the vessel wall and surrounding milieu, and multiple other cell types within the tissue. We suggest that initial injury triggered by EC RAGE may set the stage for the development of the diverse complications of diabetes in the macrovessels, retina, glomerulus, and central and peripheral nerves. Indeed, even in euglycemic inflammation, such as that triggered by acute EC injury or autoimmunity, we posit that EC RAGE plays a critical role in augmenting and amplifying cellular distress.

ENDOTHELIAL CELL RAGE: IGNITING A CASCADE OF INFLAMMATORY AND TISSUE DESTRUCTIVE PATHWAYS

The last decade has witnessed many advances in our understanding of AGEs and RAGE. However, several key questions remain unanswered. How do RAGE-dependent mechanisms contribute to diverse complications in diabetes and inflammation? What is the role of EC RAGE in these processes? As illustrated in Figure 47.2, we hypothesize that, in settings of hyperglycemia, inflammation, or oxidative stress, a transient and in some cases sustained accumulation of AGEs occurs. In the vessel wall, circulating AGEs bind and activate EC RAGE, which in turn stimulates generation of ROS, increased permeability, leukocyte adhesion, and transendothelial migration. These proinflammatory reactions may lead to the local release of S100/calgranulins and amphoterin/HMGB1, which serve to perpetuate RAGE-mediated signaling and vessel injury. In the final analysis, the elucidation of the role for endothelial-derived RAGE in pathophysiology will depend on the generation and analysis of mice in which RAGE has been conditionally deleted in the endothelium.

KEY POINTS

- The activation of endothelial RAGE by ligands such as AGEs, S100/calgranulins, and/or amphoterin/HMGB1 is a key priming step for the activation of multiple proinflammatory mechanisms that amplify cellular perturbation in a range of disorders, such as diabetes and inflammation.

- RAGE is a multiligand receptor of the immunoglobulin superfamily expressed by ECs, other vascular cells, inflammatory cells, and target cells. Examples of target cells include ECs and vascular cells themselves, podocytes, central and peripheral neurons, and islet cells.
- RAGE-dependent signal transduction, triggered by ligands, is a key first step in activating ECs and other cells, thereby stimulating the transcription and translation of a host of genes linked to perturbation and inflammatory mechanisms.

Future Goals

- To dissect the precise proximal signaling mechanisms initiated in ECs and other RAGE-bearing cells by which ligand–RAGE interaction modulates a host of genes that regulate inflammatory and other critical stress responses in cells
- To test the impact of blocking RAGE in clinical settings; specifically, antagonists of the receptor. Such studies are the only means by which to definitively address the role of RAGE in the pathogenesis of inflammation and diabetes and its complications.

ACKNOWLEDGMENTS

This work was supported by grants from the Universite Paris 7, Institut National de la Transfusion Sanguine (JLW), and the USPHS and the Juvenile Diabetes Research Foundation (AMS). AMS is a recipient of a Burroughs Wellcome Fund Clinical Scientist Award in Translational Research.

REFERENCES

1 Maillard LC. Formation d'humus et de combustibles mineraux sans intervention de l'oxygene atmospherique, des microorganismes, des hautes terperatures ou fortes pressions. *C R Acad Sci (Paris)*. 1912;154:1554–1556.
2 Ahmed N, Thornalley PJ. Quantitative screening of protein biomarkers of early glycation, advanced glycation, oxidation and nitrosation in cellular and extracellular proteins by tandem mass spectroscopy multiple reaction monitoring. *Biochem Soc Trans*. 2003;31:1417–1422.
3 Wautier J-L, Schmidt AM. Protein Glycation: a firm link to endothelial cell dysfunction. *Circ Res*. 2004;95:233–238.
4 Ulrich P, Cerami A. Protein glycation, diabetes and aging. *Recent Prog Horm Res*. 2001;56:1–21.
5 Brownlee M. Advanced glycation endproducts in diabetic complications. *Curr Opin Endocrinol Diabetes*. 1996;3:291–297.
6 Rahbar S, Blumenfeld O, Ranney HM. Studies of an unusual hemoglobin in patients with diabetes mellitus. *Biochem Biophys Res Commun*. 1969;36:838–843.
7 The Diabetes Control and Complications Trial Research Group. The effect of intensive treatment of diabetes on the development and progression of long-term complications in insulin-dependent diabetes mellitus. *New Engl J Med*. 1993;329:977–986.
8 Anderson MM, Requena JR, Crowley JR, et al. The myeloperoxidase system of human phagocytes generates Nepsilon-(carboxymethyl)lysine on proteins: a mechanism for producing advanced glycation endproducts at sites of inflammation. *J Clin Invest*. 1999;104:103–113.
9 Niskikawa T, Edelstein D, Du XL, et al. Normalizing mitochondrial superoxide production blocks three pathways of hyperglycemic damage. *Nature*. 2000;404:787–790.
10 Brownlee M. Biochemistry and molecular cell biology of diabetic complications. *Nature*. 2001;414:813–820.
11 Schmidt AM, Hori O, Brett J, et al. Cellular receptors for advanced glycation endproducts. Implications for induction of oxidant stress and cellular dysfunction in the pathogenesis of vascular lesions. *Arterioscler Thromb*. 1994;14:1521–1528.
12 Traverso N, Menini S, Maineri EP, et al. Malondialdehyde, a lipoperoxidation-derived aldehyde, can bring about secondary oxidative damage to proteins. *J Gerontol Biol Sci Med Sci*. 2004;59:B890-B895.
13 Makita Z, Yanagisawa K, Kuwajima S, et al. Advanced glycation endproducts and diabetic nephropathy. *J Diabetes Complications*. 1995;9:265–268.
14 Horie K, Miyata T, Maeda K, et al. Immunohistochemical colocalization of glycoxidation products and lipid peroxidation products in diabetic renal glomerular lesions. *J Clin Invest*. 1997;100:2995–3004.
15 Sousa MM, Du Yan S, Fernandes R, et al. Familial amyloid polyneuropathy: receptor for advanced glycation end products-dependent triggering of neuronal inflammatory and apoptotic pathways. *J Neurosci*. 2001;21:7576–7586.
16 Smith MA, Taneda S, Richey P, et al. Advanced Maillard reaction end products are associated with Alzheimer disease pathology. *Proc Natl Acad Sci USA*. 1994;91:5710–5714.
17 Shibata N, Hirano A, Kato S, et al. Advanced glycation endproducts are deposited in neuronal hyaline inclusions: a study of familial amyotrophic lateral sclerosis with superoxide dismutase-1 mutation. *Acta Neuropathol*. 1999;97:240–246.
18 Schleicher E, Wagner E, Nerlich A. Increased accumulation of glycoxidation product carboxymethyllysine in human tissues in diabetes and aging. *J Clin Invest*. 1997;99:457–468.
19 Niwa T. 3-deoxyglucosone metabolism, analysis, biological activity, and clinical implication. *J Chromatogr B Biomed Sci Appl*. 1999;731:23–36.
20 Hasuike Y, Nakanishi T, Otaki Y, et al. Plasma 3-deoxyglucosone elevation in chronic renal failure is associated with increased aldose reductase in erythrocytes. *Am J Kid Dis*. 2002;40:464–471.
21 Shinohara M, Thornalley PJ, Giardino I, et al. Overexpression of glyoxalase-I in bovine endothelial cells inhibits intracellular advanced glycation endproduct formation and prevents hyperglycemia-induced increases in macromolecular endocytosis. *J Clin Invest*. 1998;101:1142–1147.
22 Thornalley PJ. Glyoxalase I structure, function and a critical role in the enzymatic defence against glycation. *Biochem Soc Trans*. 2003;31:1343–1348.
23 Tanaka S, Avigad G, Brodsky B, et al. Glycation induces expansion of the molecular packing of collagen. *J Molec Biol*. 1988;203:495–505.
24 Haitoglou CS, Tsilbary ECs, Brownlee M, et al. Altered cellular interactions between endothelial cells and nonenzymatically

glycosylated laminin/type IV collagen. *J Biol Chem.* 1992;267: 12404–12407.

25 Giardino I, Edelstein D, Brownlee M. Nonenzymatic glycosylation in vitro and in bovine endothelial cells alters basic fibroblast growth factor activity. A model for intracellular glycosylation in diabetes. *J Clin Invest.* 1994;94:110–117.

26 Bucala R, Tracey K, Cerami, A. AGEs quench nitric oxide and mediate defective endothelium-dependent vasodilation in experimental diabetes. *J Clin Invest.* 1991;87:432–438.

27 Wautier JL, Guillausseau PJ. Advanced glycation endproducts, their receptors and diabetic angiopathy. *Diabetes Metab.* 2001;27: 535–542.

28 El Khoury J, Thomas CA, Loike JD, et al. Macrophages adhere to glucose-modified basement membrane via their scavenger receptors. *J Biol Chem.* 1994;269:10197–10200.

29 Vlassara H, Li YM, Imani F, et al. Galectin-3 as a high affinity binding protein for AGE: a new member of the AGE-Receptor complex. *Molec Med.* 1995;1:634–646.

30 Li YM, Mitsuhashi T, Wojciechowicz D, et al. Molecular identity and cellular distribution of advanced glycation endproduct receptors: relationship of p60 to OST-48 and p90 and 80K-H membrane proteins. *Proc Natl Acad Sci U S A.* 1996;93:11047–11052.

31 Ohgami N, Nagai R, Ikemoto M, et al. CD36 serves as a receptor for AGEs. *J Diabetes Complicat.* 2002;16:56–59.

32 Schmidt AM, Vianna M, Gerlach M, et al. Isolation and characterization of binding proteins for advanced glycosylation endproducts from lung tissue which are present on the endothelial cell surface. *J Biol Chem.* 1992;267:14987–14997.

33 Neeper M, Schmidt AM, Brett J, et al. Cloning and expression of RAGE: a cell surface receptor for advanced glycosylation end products of proteins. *J Biol Chem.* 1992;267:14998–15004.

34 Wautier JL, Paton RC, Wautier MP, et al. Increased adhesion of erythrocytes to endothelial cells in diabetes mellitus and its relation to vascular complications. *N Engl J Med.* 1981;305:237–242.

35 Wautier JL, Wautier MP, Schmidt AM, et al. Advanced glycation end products (AGEs) on the surface of diabetic erythrocytes bind to the vessel wall via a specific receptor inducing oxidant stress in the vasculature: a link between surface-associated AGEs and diabetic complications. *Proc Natl Acad Sci USA.* 1994;91:7742–7746.

36 Kislinger T, Fu C, Huber B, et al. Nε(carboxymethyl)lysine modifications of proteins are ligands for RAGE that activate cell signaling pathways and modulate gene expression. *J Biol Chem.* 1999;274:31740–31749.

37 Ikeda K, Higashi T, Sano H, et al. Carboxymethyllysine protein adduct is a major immunological epitope in proteins modified with AGEs of the Maillard reaction. *Biochemistry.* 1996;35:8075–8083.

38 Reddy S, Bichler J, Wells-Knecht K, et al. Carboxymethyllysine is a dominant AGE antigen in tissue proteins. *Biochemistry.* 1995; 34:10872–10878.

39 Schmidt AM, Hori O, Chen J, et al. Advanced glycation endproducts interacting with their endothelial receptor induce expression of vascular cell adhesion molecule-1 (VCAM-1): a potential mechanism for the accelerated vasculopathy of diabetes. *J Clin Invest.* 1995;96:1395–1403.

40 Vlassara H, Fuh H, Donnelly T, et al. Advanced glycation endproducts promote adhesion molecule (VCAM-1, ICAM-1) expression and atheroma formation in normal rabbits. *Mol Med.* 1995;1:447–456.

41 Hofmann MA, Drury S, Fu C, et al. RAGE mediates a novel proinflammatory axis: a central cell surface receptor for S100/calgranulin polypeptides. *Cell.* 1999;97:889–901.

42 Donato R. S100: a multigenic family of calcium-modulated proteins of the EF-hand type with intracellular and extracellular functional roles. *Intl J Biochem Cell Biol.* 2001;33:637–668.

43 Schafer BW, Heinzmann CW. The S100 family of EF-hand calcium-binding proteins: functions and pathology. *Trends Biochem Sci.* 1996;21:134–140.

44 Zimmer DB, Cornwall EH, Landar A, et al. The S100 protein family: history, function, and expression. *Brain Res Bull.* 1995;37: 417–429.

45 Frosch M, Strey A, Vogl T, et al. Myeloid-related proteins 8 and 14 are specifically secreted during interaction of phagocytes and activated endothelium and are useful markers for monitoring disease activity in pauciarticular-onset juvenile rheumatoid arthritis. *Arthritis Rheum.* 2000;43:628–637.

46 Rammes A, Roth J, Goebeler M, et al. Myeloid-related protein (MRP) 8 and MRP14, calcium-binding proteins of the S100 family, are secreted by activated monocytes via a novel, tubulin-dependent pathway. *J Biol Chem.* 1997;272:9496–9502.

47 Rauvala H, Pihlaskari R. Isolation and some characteristics of an adhesive factor of brain that enhances neurite outgrowth in central neurons. *J Biol Chem.* 1987;262:16625–16635.

48 Wang H, Bloom O, Zhang M, et al. HMG-1 as a late mediator of endotoxin lethality in mice. *Science.* 1999;285:248–251.

49 Andersson U, Wang H, Palmblad K, et al. High mobility group 1 protein (HMG-1) stimulates proinflammatory cytokine synthesis in human monocytes. *J Exp Med.* 2000;192:565.

50 Gardella S, Andrei C, Ferrera D, et al. The nuclear protein HMGB1 is secreted by monocytes via a non-classical, vesicle-mediated secretory pathway. *EMBO Rep.* 2002;3:995–1001.

51 Scaffidi P, Misteli T, Bianchi ME. Release of chromatin protein HMGB1 by necrotic cells triggers inflammation. *Nature.* 2002; 418:191–195.

52 Treutiger CJ, Mullins GE, Johansson AS, et al. High mobility group 1 B-box mediates activation of human endothelium. *J Intern Med.* 2003;254:375–385.

53 Chavakis T, Bierhaus A, Al-Fakhri N, et al. The pattern recognition receptor RAGE is a counterreceptor for leukocyte integrins: a novel pathway for inflammatory cell recruitment. *J Exp Med.* 2003;198:1507–1515.

54 Hofmann MA, Drury S, Hudson BI, et al. RAGE and arthritis: The G82S polymorphism amplifies the inflammatory response. *Genes Immun.* 2002;3:123–135.

55 Yan SSD, Wu ZY, Zhang HP, et al. Suppression of experimental autoimmune encephalomyelitis by selective blockade of encephalitogenic T-cell infiltration of the central nervous system. *Nat Med.* 2003;9:287–293.

56 Chen Y, Yan SS, Colgan J, et al. Blockade of late stages of autoimmune diabetes by inhibition of the receptor for advanced glycation end products. *J Immunol.* 2004;173:1399–1405.

57 Liliensiek B, Weigand MA, Bierhaus A, et al. Receptor for advanced glycation endproducts (RAGE) regulates sepsis but not the adaptive immune response. *J Clin Invest.* 2004;113:1641–1650.

58 Williams SB, Cucso JA, Roddy MA, et al. Impaired nitric oxide-mediated vasodilation in patients with non-insulin-dependent diabetes mellitus. *J Am Coll Cardiol.* 1996;27:567–574.

59 Johnstone MT, Creager SJ, Scales KM, et al. Impaired endothelium-dependent vasodilation in patients with insulin-dependent diabetes mellitus. *Circulation.* 1993;88:2510–2516.

60 De Vriese AS, Verbeuren TJ, Van de Voorde J, et al. Endothelial dysfunction in diabetes. *Br J Pharmacol.* 2000;130:963–974.
61 Caballero AE, Arora S, Saouaf R, et al. Microvascular and macrovascular reactivity is reduced in subjects at risk for type 2 diabetes. *Diabetes.* 1999;48:1856–1862.
62 Veves A, Saouaf R, Donaghue VM, et al. Aerobic exercise capacity remains normal despite impaired endothelial function in the Micro- and macrocirculation of physically active IDDM patients. *Diabetes.* 1997;46:1846–1852.
63 Kostouros GD, Olgart L, Gazeliu B, et al. Facilitated diffusion by iontophoresis of certain chemical compounds to the rat incisor pulp. *Eur J Oral Sci.* 1996;104:570–576.
64 Oldenburg KR, Vo KT, Smith GA, et al. Iontophoretic delivery of oligonucleotides across full-thickness hairless mouse skin. *J Pharmacol Sci.* 1995;84:915–921.
65 Viberti G. Increased capillary permeability in diabetes mellitus and its relationship to microvascular angiopathy. *Am J Med.* 1983;146:688–694.
66 Mattock M, Morrish N, Viberti G, et al. Prospective study of microalbuminuria as predictor of mortality in NIDDM. *Diabetes.* 1992;41:736–741.
67 Feldt-Rasmussen B. Increased transcapillary escape rate of albumin type I diabetic patients with microalbuminuria. *Diabetologia.* 1996;29:282–286.
68 Nannipieri M, Rizzo L, Rapuano A, et al. Increased transcapillary escape rate of albumin in a microalbuminuric type II diabetic subject. *Diabetes Care.* 1995;18:1–9.
69 Williamson J, Chang K, Tilton R, et al. Increased vascular permeability in spontaneously diabetic BB/W rats and in rats with mild versus severe streptozotocin-induced diabetes. *Diabetes.* 1987;36:813–821.
70 Wauter JL, Zoukourian C, Chappey O, et al. Receptor-mediated endothelial cell dysfunction in diabetic vasculopathy. Soluble receptor for Advanced Glycation Endproducts blocks hyperpermeability in diabetic rats. *J Clin Invest.* 1996;97:238–243.
71 Cipollone F, Iezzi A, Fazia M, et al. The Receptor RAGE as a progression factor amplifying arachidonate-dependent inflammatory and proteolytic response in human atherosclerotic plaques: role of glycemic control. *Circulation.* 2003;108:1070–1077.
72 Buchs AE, Kornberg A, Zahavi M, et al. Increased expression of tissue factor and receptor for advanced glycation end-products in peripheral blood mononuclear cells of patients with type 2 diabetes with vascular complications. *Exp Diabetes Res.* 2004;5:163–169.
73 Plump AS, Smith JD, Hayek T, et al. Severe hypercholesterolemia and atherosclerosis in apolipoprotein E-deficient mice created by homologous recombination in ES cells. *Cell.* 1992;71:343–353.
74 Zhang SH, Reddick RL, Piedrahita JA, et al. Spontaneous hypercholesterolemia and arterial lesions in mice lacking apolipoprotein E. *Science.* 1992;258:468–471.
75 Park L, Raman KG, Lee KJ, et al. Suppression of accelerated diabetic atherosclerosis by soluble Receptor for AGE (sRAGE). *Nat Med.* 1998;4:1025–1031.
76 Kislinger T, Tanji N, Wendt T, et al. RAGE mediates inflammation and enhanced expression of tissue factor in the vasculature of diabetic apolipoprotein E null mice. *Arterioscler Thromb Vasc Biol.* 2001;21:905–910.
77 Bucciarelli LG, Wendt T, Qu W, et al. RAGE blockade stabilizes established atherosclerosis in diabetic apolipoprotein E null mice. *Circulation.* 2002;106:2827–2835.
78 Wendt T, Harja E, Bucciarelli LG, et al: RAGE modulates vascular inflammation and atherosclerosis in a murine model of type 2 diabetes. *Atherosclerosis.* 2005. In press.
79 Zhou Z, Wang K, Penn MS, et al. Receptor for AGE (RAGE) mediates neointimal formation in response to arterial injury. *Circulation.* 2003;107:2238–2243.
80 Sakaguchi T, Yan SF, Yan SD, et al. Arterial restenosis: central role of RAGE-dependent neointimal expansion. *J Clin Invest.* 2003; 111:959–972.
81 Roque M, Fallon J, Badimon J, et al. Mouse model of femoral artery denudation injury associated with the rapid accumulation of adhesion molecules on the luminal surface and recruitment of neutrophils. *Arterioscler Thromb Vasc Biol.* 2000;20:335–342.
82 Yamamoto Y, Kato I, Doi T, et al. Development and prevention of advanced diabetic nephropathy in RAGE-overexpressing mice. *J Clin Invest.* 2001;108:261–268.
83 Wendt TM, Tanji N, Guo J, et al. RAGE drives the development of glomerulosclerosis and implicates podocyte activation in the pathogenesis of diabetic nephropathy. *Am J Pathol.* 2003;162:1123–1137.
84 Huang JS, Guh JY, Chen HC, et al. Role of receptor for advanced glycation end-product (RAGE) and the JAK/STAT-signaling pathway in AGE-induced collagen production in NRK-49F cells. *J Cell Biochem.* 2001;81:102–113.
85 Yeh CH, Sturgis L, Haidacher J, et al. Requirement for p38 and p44/42 mitogen-activated protein kinases in RAGE-mediated nuclear factor-kappa B transcriptional activation and cytokine secretion. *Diabetes.* 2001;50:1495–1504.
86 Deora AA, Win T, Vanhaesebroeck B, et al. A redox-triggered ras-effector interaction. Recruitment of phosphatidylinositol 3'-kinase to ras by redox stress. *J Biol Chem.* 1998;273:29923–29928.
87 Huttunen HJ, Fages C, Rauvala H. Receptor for Advanced Glycation Endproducts (RAGE)-mediated neurite outgrowth and activation of NF-kB require the cytoplasmic domain of the receptor but different downstream signaling pathways. *J Biol Chem.* 1999;274:19919–19924.
88 Yan SD, Schmidt AM, Anderson G, et al. Enhanced cellular oxidant stress by the interaction of advanced glycation endproducts with their receptors/binding proteins. *J Biol Chem.* 1994;269: 9889–9897.
89 Lander HL, Tauras JM, Ogiste JS, et al. Activation of the receptor for advanced glycation endproducts triggers a MAP kinase pathway regulated by oxidant stress. *J Biol Chem.* 1997;272:17810–17814.
90 Huttunen HJ, Kuja-Panula J, Rauvala H. RAGE signaling induces CREB-dependent chromogranin expression during neuronal differentiation. *J Biol Chem.* 2002;277:38635–38646.
91 Taguchi A, Blood DC, del Toro G, et al. Blockade of amphoterin/RAGE signaling suppresses tumor growth and metastases. *Nature.* 2000;405:354–360.
92 Brizzi MF, Dentelli P, Rosso A, et al. RAGE- and TGF-beta receptor-mediated signals converge on STAT5 and p21waf to control cell cycle progression of mesangial cells: a possible role in the development and progression of diabetic nephropathy. *FASEB J.* 2004;18:1249–1251.
93 Li JH, Wang W, Huang XR, et al. Advanced glycation endproducts induce tubular epithelial-myofibroblast transition through the RAGE-ERK 1/2 MAP kinase signaling pathway. *Am J Pathol.* 2004;164:1389–1397.

94 Shaw SS, Schmidt AM, Banes AK, et al. S100B-RAGE mediated augmentation of angiotensin II-induced activation of JAK2 in vascular smooth muscle cells is dependent on PLD2. *Diabetes.* 2003;52:2381–2388.
95 Wautier MP, Chappey O, Corda S, et al. Activation of NADPH Oxidase by Advanced Glycation Endproducts (AGEs) links oxidant stress to altered gene expression via RAGE. *Am J Physiol Endocrinol Metab.* 2001;280:E685-E694.
96 Rong LL, Yan SF, Wendt T, et al. RAGE modulates peripheral nerve regeneration via recruitment of both inflammatory and axonal outgrowth pathways. *FASEB J.* 2004;18:1818–1825.
97 Arancio O, Zhang HP, Chen X, et al. RAGE potentiates Abeta-induced perturbation of neuronal function in transgenic mice. *EMBO J.* 2004;23:4096–4105.
98 Cataldegirmen G, Zeng S, Feirt N, et al. RAGE limits regeneration after massive liver injury by coordinated suppression of TNF-alpha and NF-kappaB. *J Exp Med.* 2005;201:473–484.

48

Complement

Anne Nicholson-Weller

Beth Israel Deaconess Medical Center, Harvard Medical School, Boston, Massachusetts

COMPLEMENT FUNCTIONS IN INNATE AND ADAPTIVE IMMUNITY

The complement system evolved as part of innate immunity and as such it has an important role in host defense. Complement exhibits the four fundamental properties of innate immunity: (a) its components are encoded by the germ line, (b) its genes are invariant within a species, (c) it is constitutively poised for activation, and (d) it recognizes and is activated by large molecular motifs associated with viral, bacterial, and fungal pathogens – the so-called pathogen-associated molecular patterns, or PAMPs (1). Pathogens do not express PAMPs by choice. PAMPs are associated with critical prokaryote function, such as the barrier functions of the outer membrane lipopolysaccharide (LPS) of gram-negative bacteria and the cell wall lipoproteins of gram-positive bacteria. Many of the most virulent bacterial pathogens have evolved polysaccharide capsules as a counterdefense mechanism to cover and hide their PAMPs and thereby evade recognition by complement.

Adaptive immunity, of a type similar to that of modern mammals, evolved in the ancestors of present day jawed fish. In contrast to innate immunity, adaptive immunity (a) encodes its genes in peripheral cells (lymphocytes); (b) allows mutation and selection of genes to adapt to the antigenic experience of the individual, as opposed to the species; (c) takes 1 to 3 weeks to generate a new response; and (d) recognizes molecular details of antigens. Animals with adaptive immunity retain innate immunity, including complement, to provide a first line of defense during the interval (1–3 wks) necessary to mount a new specific adaptive immune response. In addition, when innate immunity recognizes what is dangerous, it activates antigen-presenting cells and thereby provides critical priming for an adaptive immune response (2). For example, an antigen is potentiated 1,000- to 10,000-fold for a B-cell response when it is tagged with a C3 fragment (3). Much of the data for the participation of innate immunity in adaptive immunity were derived from studies of knockout mice, and will be discussed later. Many of the components of innate immunity also are involved in responses to tissue damage, and complement is no exception. Thus, altered host proteins that are recognized by innate immunity can become antigens for an inappropriate adaptive immune response and result in autoimmunity.

Ancient Origins of C3 and Its Pivotal Role in Modern Complement Systems

From homology studies, it is proposed that C3 arose some 700 million years ago from gene duplication of α_2-*macroglobulin* (α_2-*M*), which is widely distributed in metazoans (4). α_2-M has an internal thioester bond. When α_2-M is cleaved by proteases, its thioester bond transacylates to the protease and thereby entraps the protease. Because pathogens routinely use proteases as virulence factors, α_2-M is considered an ancient host defense molecule of innate immunity. After entrapping a protease, the complex of activated α_2-M–entrapped-protease is cleared by receptors that recognize activated α_2-M. In humans, the α_2-M receptor is CD91/low-density-lipoprotein-receptor–related protein (5,6).

A C3-like molecule has been identified in coral, which diverged from the main trunk of evolution during the Precambrian period (7). Thus, the α_2-*M* gene duplication must have occurred even earlier than the divergence of coral. Consistent with the coral data, homologs of C3 and C3-activating proteases are found in the ancient protostome *Carcinoscorpius rotundicauda* (horseshoe crab) (8), as well as in some deuterostomes, including Echinodermata (sea urchins) and early Chordata (sea squirts, lancelets) (9).

C3 is a 180-kDa heterodimer molecule with its internal thioester bond within the α-chain. Normally, the thioester bond is protected within a hydrophobic pocket (10). When C3 is activated by specific cleavage, releasing a N-terminal fragment (C3a) from the α-chain. The residual C3b molecule undergoes a conformational change that exposes the labile thioester bond, allowing C3b to dock on adjacent OH-groups by ester linkage or, less commonly, on NH groups by amide linkage (11). The acceptor molecules are not random – for example, glycophorin A is a preferred target on red blood cells (RBCs) (12), whereas thr^{144} on human IgG1 is a specific target (13). As the universal tag of innate immunity, C3b

Table 48-1: Inhibitors of Complement Activation

Inhibitor	*Prevents Activation of*	*Plasma (P) or Membrane Protein*	*Mode of Inhibition*
C1 INH	C4	P	SERPIN*, binding C1rC1s
C4 binding protein†	C3	P	Accelerates decay of C4b2a; cofactor for factor I cleavage
Factor H†	C3	P and peripheral membrane protein	Accelerates decay of C3bBb; cofactor for factor I cleavage of C3b
CD46†	C3	Transmembrane	Cofactor for factor I cleavage of C3b
Factor I	C4, C3	P	Protease cleavage of C4b and C3b
DAF (CD55)†	C3, C5	GPI-membrane	Accelerates decay of C4b2a and C3bBb
CR1 (CD35)†	C3	Transmembrane	Cofactor for factor I; cleavage of C3b
CD59	C5b-8, C5b-9	GPI-membrane	Blocks addition of C8 to C5b67; blocks addition of C9 to C5b, -6, -7, -8
Clusterin (apolipoprotein J)	C5b7	P	Competes with membrane for binding C5b67
Vitronectin	C5b7	P	Competes with membrane for binding C5b67

*Serine protease inhibitor or SERPIN.
†Extracellular domain consists of short consensus repeats (SCR), or complement control protein (CCP) repeats, or Sushi domains.

marks harmful molecules associated with PAMPs. Specific internalization (phagocytic or endocytic) receptors exist for C3b and each of C3b's sequentially proteolyzed membrane-bound derivatives (Table 48-1). Specificity for tagging PAMPs, as opposed to host cells, is governed by multiple attributes of the complement system, including substrate specificity for each of the complement enzymes, need for sequential activation of components, lability of active fragments, preferred acceptors of C3b, and by the presence of inhibitors of C3 activation and C5b-9 insertion at the membrane of host cells (Table 48-2).

COMPLEMENT PATHWAYS

C3 Activation

Three pathways evolved for efficient activation of C3: the lectin, classical, and alternative pathways (Figure 48.1). Each pathway is specialized to recognize different types of PAMPs. The lectin pathway, which is evolutionarily the oldest of the three pathways, uses mannan binding lectin (MBL) or ficolins to directly recognize PAMP sugars, such as D-mannose and *N*-acetyl-D-glucosamine. After binding, these lectins activate MBL-associated serine protease (MASP)-2, which in turn, sequentially activates C4 and C2. C4 was derived from C3 by gene duplication during evolution. Cleavage-activation of C4 exposes its internal thioester bond, which is within the major cleavage fragment (C4b), and thereby allows C4b to dock on adjacent $^{-}$OH or $^{-}$NH groups. The activation and docking of C4b is completely analogous to the cleavage-activation of C3 and docking of C3b (see Figure 48.1). C2 is only efficiently activated if it is bound to C4b, and the resulting C4b2a enzyme cleaves C3 and C3bC4b2a cleaves C5.

The classical pathway for C3 activation evolved in the ancestors of jawed fish coincident with the origins of modern mammalian adaptive immunity. This pathway is activated when IgM and IgG bind their specific antigens and become conformationally modified. C1q recognizes the bound immunoglobulin and initiates the sequential activation of C1r and C1s, serine proteases homologous to MASP-2 of the lectin pathway. C1s, in turn, activates C4 and C2, as described for the lectin pathway (see Figure 48.1). Both MBL and C1q, once they shed their associated proteases, act as opsonins (14,15). This suggests that these recognition molecules functioned independently of their proteases and the rest of complement early in evolution and then later became incorporated into pathways to activate C3. Consistent with this hypothesis, C1q recognizes some PAMPs, such as the membranes of retroviruses and lipid A of endotoxin (16,17).

The alternative pathway is comprised of three activation proteins: C3b, factor D, and factor B; and three regulator proteins: properdin, factor I, and factor H. This pathway also recognizes PAMP carbohydrates, but by a type of "negative selection." Without any specific stimulus for activation, C3 is

Table 48-2: Complement-Binding Proteins and Receptors

Binding Proteins	*Complement Ligands*	*Functions*
Calreticulin (CRT)*	C1q	Spreading of ECs, activation of platelets†
p33*	C1q	Spreading of ECs, activation of platelets†
Receptors		
CR1 (CD35)*,‡	C3b, C4b, C1q, MBL	Phagocytic receptor on leukocytes, immune adhesion receptor on erythrocytes, upregulated in/on ECs after ischemia–reperfusion injury
CR3 (CD11b/CD18)	iC3b	Phagocytic receptor on leukocytes
CR4 (CD11/CD18)	iC3b	Endocytic receptor on dendritic cells
CR2 (CD21)‡	C3dg	Endocytic receptor on B cells
C3a receptor*	C3a	Modulates inflammation†
C5a receptor*	C5a	Modulates inflammation†

*Expressed on ECs and other cells.
†See text.
‡Receptor contains SCR domains.

cleaved by the slow turnover of both the classical and lectin pathways (18). In addition, the thioester bond in C3 can be attacked by water, converting C3 into C3b-like C3. Both C3b and C3b-like C3 can bind factor B, a zymogen (19). Factor D, which is a constitutively active circulating serine protease with unique specificity for factor B in the C3bB complex, generates the C3-cleaving C3bBb convertase and C5-cleaving convertase $(C3b)_2Bb$.

Properdin, a positive regulator, binds the alternative pathway C3- and C5-cleaving multimeric enzymes and enhances their activity by stabilizing them. In contrast, factors H and I act as negative regulators: factor H binds to the C3b and dislodges the enzymatic Bb subunit; and then factor I, using H as a cofactor, cleaves the C3b to iC3b, which is unable to participate in further complement activation because it cannot bind factor B. In addition to its C3b binding site, factor H contains sites that directly bind the polyanions and sialic acids of host cells, or to small integrin-binding ligand, N-linked glycoproteins (SIBLINGs), which in turn bind to CD44 or RGD-binding integrins on the cell surface (20,21). These binding properties allow factor H to concentrate as a peripheral membrane protein on the surface of host cells, and thereby protect the cells from autologous complement damage. Microbes lack heparin-like molecules (polyanions) and SIBLING receptors and therefore cannot concentrate factor H at their surface. As a consequence, these microbes are vulnerable to unregulated C3 and C5 activation at their surface. A few pathogens, such as Group B Streptococci and type B *Neisseria meningitidis*, have sialic acid–containing capsules and therefore can bind factor H and circumvent factor H control.

The importance of factor H for protection of endothelial cells (ECs) is underscored by two pathologic conditions resulting from factor H mutations, namely atypical hemolytic uremic syndrome (aHUS) and age-related macular degeneration (AMD). These conditions are discussed later. Genetic deficiency of factor I leads to complete consumption of C3 due to unregulated alternative pathway activity. Factor I–deficient patients present with pyogenic infections in infancy, in a syndrome identical to that of patients with genetically determined C3 deficiency (Table 48-3).

Activation of the Terminal Complement Sequence

C5 also was derived from C3 by gene duplication but, in contrast to C3 and C4, it lacks an internal thioester bond. C6, C7, C8, and C9 share domains distinct from those of C3 and C5, and appear to have evolved separately, but in parallel with perforin, the nongated pore-forming protein of cytotoxic lymphocytes and natural killer cells (44).

C5 is the last component to be activated by proteolytic cleavage. C5 binds to C3b adjacent to C4b2a of the lectin and classical pathway or the C3bBb enzyme of the alternative pathway, and it is cleaved to generate C5a (10 kDa), an anaphylactic fragment, and C5b (167 kDa), which initiates the terminal complement pathway that is shared by the lectin, classical, and alternative pathways. C5b sequentially condenses with C6 and then C7. C5b7 formation exposes a nascent lipid binding site and, if a membrane is adjacent, C5b7 will insert itself (45). No specific receptors exist for C5b7 binding. In the absence of an adjacent membrane, C5b7 may bind hydrophobic sites in clusterin or vitronectin and thereby become inactivated. The addition of C8 to membrane-inserted C5b7 allows a floppy transmembrane pore to form, whereas the subsequent addition of C9 forms a rigid C5b-9 pore (46). Adding more C9 to the pore widens its diameter.

REGULATORS OF COMPLEMENT ACTIVATION

Despite the presence of numerous complement serine proteases, C1r, C1s, factor D, factor Bb, C2a, and the inhibitor factor I, only one serine protease inhibitor functions in the complement system, namely C1 inhibitor (C1 INH). Within the complement system, C1 INH is specific for C1r and C1s, although it also inhibits clotting factors XIIa, Xia, and kallikrein. Deficiency of C1 INH leads to hereditary angioedema; the local, but sometimes fatal edema that occurs with this syndrome is now known to result from kallikrein cleavage of high-molecular kininogen to produce the mediator bradykinin (47) (see Table 48-3).

Except for CD59, which inhibits complement channel formation (discussed later), the rest of the complement inhibitors are focused on limiting C3 and C5 activation. Of the C3 and C5

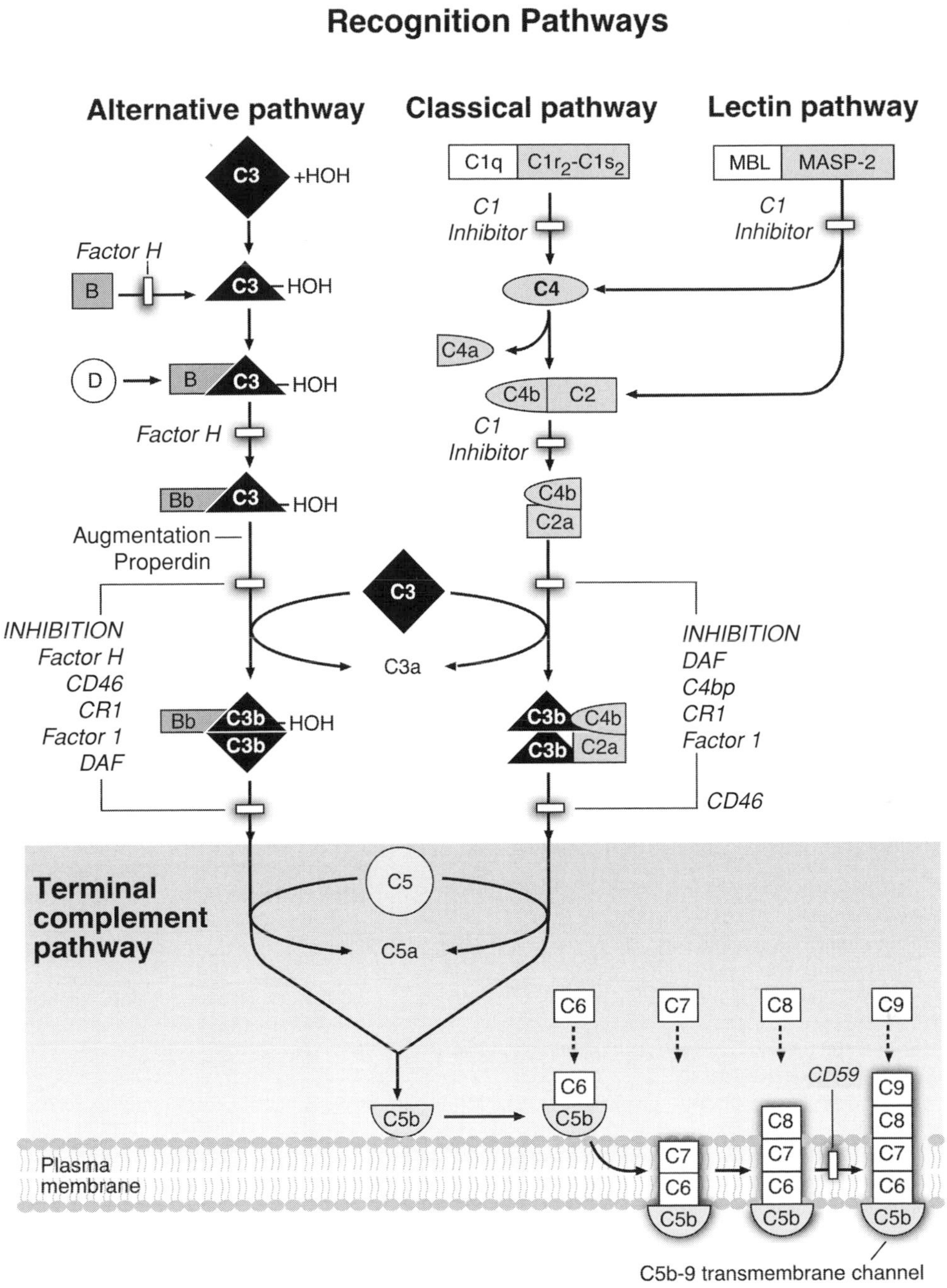

Figure 48.1. The three pathways of complement activation and common final pathway to generate the complement pore. Proteins sharing structural homology are shown in the same shade; enzymes are denoted by triangles; and major inhibitors are shown in boxes that interrupt the activation pathways. The lectin pathway is activated when MBL binds sugars associated with pathogen-associated molecular patterns (PAMPs), especially N-acetylglucosamine, mannose, and N-acetylmannosamine. The classical pathway is most commonly activated when the C1q subcomponent of macromolecular C1 binds the heavy chains of IgG and IgM after these immunoglobulins have bound their specific antigens. The alternative pathway slowly ticks over in the fluid phase as C3 reacts with water {C3b(HOH)} and becomes "C3b-like." However, this pathway is only efficiently activated when C3b is deposited on a surface that fails to bind factor H, thereby allowing the affinity of factor B to exceed that factor H. C3 (HOH) or "C3b-like C3" is uncleaved C3 in which the internal thioester bond has reacted with water, and the molecule assumes the conformation of C3b in terms of allowing factor B to bind and initiate further complement activation; C4bp, C4 binding protein; CR1, complement receptor 1 or CD35; DAF, decay-accelerating factor or CD55; MBL, mannan-binding lectin; MASP-2, MBL-associated serine protease-2.

Table 48-3: Phenotypes of Complement-Deficient Humans and Knockout Mice

Affected Protein	*Humans (Refs.)*	*Mice (Refs.)*
Mannan binding lectin-1 (MBL1)	Normally not expressed (22)	MBL1 is at least partially redundant with MBL2, see below.
MBL2	Deficiency is common, up to 10% in some populations. Associated with predisposition to infection, especially in altered host (23)	*MBL1*$^{-/-}$/*MBL2*$^{-/-}$ protects mice from ischemia reperfusion injury (24).
MBL-associated serine protease-2 (Masp-2)	One case report: autoimmune phenomena (25); may also have mild phenotype*	ND†
Properdin	Meningococcal infection (26)	ND
Factor B	No phenotype (27)	Normal; if backcrossed with MRL/lpr lupus-prone mice, ameliorates autoimmune disease (28)
Factor D	Meningococcal infection (29,30)	Normal; if backcrossed with MRL/lpr lupus-prone mice, ameliorates autoimmune disease (31)
C1q	93% prevalence of severe lupus-like illness (32)	Renal disease, strain-dependent (33)
C1r and/or C1s	*C1r* and *C1s* are linked genes and usually subjects are deficient in both. Deficiency is rare, and about 57% of subjects have severe autoimmune disease (32)	ND
C4	75% prevalence of SLE (32)	Decreased clearance of apoptotic cells in vitro (34)
C2	10% prevalence of SLE (32)	ND; knockout resulted in loss of adjacent genes including factor B (35)
C3	Pyogenic infections in infancy and childhood ± glomerulonephritis (26)	Decreased opsonic clearance, abnormal adaptive response, decreased inflammation; see Mouse Genome Informatics (MGI) of Jackson Labs‡
C5, C6, C7, or C8	Increased prevalence of *Neisseria* infections (26)	Very common, found in several inbred strains of mice, including AKR, DBA, NOD, NZB, and SWR; phenotypes are summarized in MGI
C9	Mild increase in prevalence of *Neisseria* infections (26)	ND
C3aR	ND	Decreased Th2-mediated inflammation, see MGI
C5aR	ND	Increased susceptibility to *Pseudomonas aeruginosa* pneumonia (36); modulates inflammation, see MGI
C1 INH	Hereditary angioedema, see OMIM**	Increased vascular permeability, see MGI
C4bp	In one family: two asymptomatic, and one with aggressive autoimmune syndrome (37)	ND
Factor H	Hemolytic uremic syndrome (HUS), see OMIM	C3 deficiency, spontaneous glomerulonephritis, increased mortality; see MGI
Factor I	Factor B and C3 catabolism, pyogenic infections in infancy and childhood, see OMIM	ND
DAF (CD55) (GPI anchored)	Genetic deficiency on RBCs has no phenotype, see OMIM; acquired deficiency associated with GPI-anchor deficiency contributes to paroxysmal hemoglobinuria (60,61).	No phenotype for knockout (38); widely expressed in wt including RBCs (39)
TM-DAF	Not expressed	ND; expression limited to mature sperm in wt (39)
Crry	Not expressed	Lethal during embryogenesis due to unrestricted complement activation in the placenta (40)
CD46	HUS and age-related macular degeneration, see OMIM	Increased fertility due to more efficient sperm; expression limited to sperm; see MGI
CR1 (CD35)	No known genetic deficiencies; acquired deficiencies occur in immune complex disease	See below
CR2 (CD21)	ND	In mice, CR1 and CR2 are alternatively spliced from the same gene; knockouts to date have resulted in deficiency of both proteins; see OMIM
CR3 (Mac-1 or CD11b/CD18	Usually results from CD18 mutation and therefore affects all β_2 integrins and causes leukocyte adhesion deficiency. See OMIM.	Abnormal leukocyte trafficking, see MGI.
CD59a	Episodic hemolysis, stroke. See OMIM; acquired deficiency with GPI-anchor deficiency contributes to paroxysmal hemoglobinuria. See OMIM	Widely expressed, including RBCs; knockout results in mild hemolytic anemia (41)
CD59b	Not expressed	Expression predominately on testes, with low levels on RBCs; effects of knockout are controversial with respect to hemolysis (42,43)

*Jens Jensenius, personal communication
†No data
‡Mouse Genone Informatics (MGI) of Jackson Labs: <http://www.informatics.jax.org>
**Online Mendelian Inheritance in Man (OMIM): <http://www.ncbi.nlm.nih.gov/entrez/query.fcgi?db=OMIM>

inhibitors, only factor I has enzymatic activity, which cleaves C4b and C3b so that they cannot function for the further complement activation. The remaining C3 and C5 regulators are fluid-phase or membrane molecules that bind C4b or C3b and dissociate the catalytic subunit C2a or Bb of the classical pathway or alternative-pathway C3- and C5-activating enzymes, respectively. They also may act as cofactors for factor I cleavage (see Figure 48.1). Genes for the complement inhibitors are clustered in a region known as the regulators of complement activation (RCA) at 1q32. These regulators include C4 binding protein (C4bp) and factor H in the fluid phase, and complement receptor 1 (CR1 or CD35), decay-accelerating factor (DAF) (CD55), and CD46 in the membrane. Complement receptor 2 (CR2 or CD21) is in the same gene cluster and, although it does not inhibit complement, it binds the C3dg fragment of C3 and positively potentiates signaling of the B cell's antigen receptor (3). The common structural motif of the RCA proteins and their functions are discussed later and summarized in Tables 48-1 and 48-2.

The extracellular domains of the RCA C4b/C3b binding molecules are comprised of short consensus repeats (SCRs, also known as complement control protein [CCP] repeats or Sushi domains). SCRs are about 60 AA, with commonly four cysteines forming disulfide bonds between the first and third Cys residues and second and fourth Cys residues. SCRs, which are ancient scaffolding modules found in sponges (48), have varying flexibility depending on the amino acid sequences between the disulfide-bonded loops. Although the extracellular domains of RCA proteins are comprised almost exclusively of SCRs, only a few of the SCRs are involved with binding C4b or C3b (49). For example, the common allotype of CR1 (CD35) has an extracellular domain comprised of 30 SCRs. The NH-terminal 28 SCRs are organized as four long, homologous repeats (LHRs) – A, B, C, and D – of seven SCRs each (50). Mapping by substitution mutagenesis has shown that it is primarily the first three SCRs of LHR-A, -B, and -C that are responsible for binding. These are C4b in LHR-A, C4b or C3b in LHR-B, and C3b in LHR-C (51,52). The residues in LHR-D responsible for binding C1q and MBL have not been mapped (14,15). Amino acid analysis of SCRs from complement regulators have identified 11 SCR types and found that identical sequences of SCR types are repeated in proteins, providing further evidence that complement regulators and inhibitors comprised of SCRs arose from a process of gene duplication (53). SCR also are found in some noncomplement proteins, such as factor XIII, β-2 glycoprotein 1, and the interleukin (IL)-2 receptor, in which the SCRs function as scaffolding modules (54).

The SCR-containing membrane regulators DAF (CD55) and CD46 inhibit C3 and C5 activation: DAF causes the dissociation and decay of the C4b2a and C3bBb C3/C5 cleaving enzymes (55); whereas CD46 acts as a cofactor for factor I cleavage of C4b and C3b to iC4b and iC3b (56), respectively. Factor H, acting both as a fluid-phase inhibitor and a peripheral membrane protein inhibitor, dissociates and decays the C3bBb enzyme and acts a cofactor for the factor I–mediated proteolysis of C3b. The extracellular domain of factor H is comprised of 20 SCRs and includes distinct binding sites for the following ligands: three for C3b, two for heparin, and one for C-reactive protein (57–59).

The non–SCR-containing membrane protein CD59 inhibits the assembly of the terminal C5b-9 complex in autologous cells. CD59 binds to C8 and C9 as the terminal complex is forming in the plasma membrane and thereby prevents the addition of the first C9 and additional C9s that widen the transmembrane channel. Both DAF and CD59 are linked to the plasma membrane by glycan phosphatidylinositol (GPI) anchors. Deficiency of GPI-anchored proteins, specifically the lack of DAF and CD59 in the affected erythrocytes of patients with paroxysmal nocturnal hemoglobinuria, leads to the abnormal deposition of C3b (60,61) and unregulated membrane insertion of C5b-9 that results in intravascular hemolysis (62).

Although the complement components of mammals are largely interchangeable, the receptors (Table 48-1), as well as regulators (Table 48-2), have species specificity. This is illustrated by the fact that human complement does not normally lyse human RBCs, but it readily lyses sheep and rabbit RBCs, which are used to assay human classical and alternative pathways, respectively. Finally, although most of the human inhibitors and receptors are homologous to mouse proteins, in the case of CR1 and DAF their functions have diverged (Table 48-3).

COMPLEMENT EFFECTORS: C1q EFFECTS, OPSONINS, B-CELL SIGNALING, ANAPHYLOTOXINS, AND C5b-9

Two intracellular proteins find their way to the external face of the plasma membrane and are able to bind C1q. They are calreticulin (CRT), a chaperone of the endoplasmic reticulum, and p33, a mitochondrial protein. CRT binds a collagen domain of C1q, whereas p33 binds the globular domain. Because CRT and p33 lack a membrane domain, it is assumed that they are tethered to the plasma membrane by associating with another membrane protein and that any cell signaling occurring with C1q binding is due to the associated membrane protein (63). CRT is associated with CD59 on the surface of human neutrophils (64), whereas the pattern of signaling in ECs (to be discussed later) suggests that CRT and some of p33 may be associated with β1 integrin (65). Evidence also suggests that p33 is associated with the receptor for urokinase-type plasminogen activator (uPAR) on ECs (63) (Figure 48.2). Platelets, which express both CRT and p33, are activated by binding C1q (66).

Complement is the major source of opsonins for innate immunity. In adaptive immunity, complement opsonins function synergistically with specific immunoglobulin to trigger uptake by phagocytic cells. Although C1q, MBL, and C4b are all opsonins, the most effective ligands are cleavage fragments

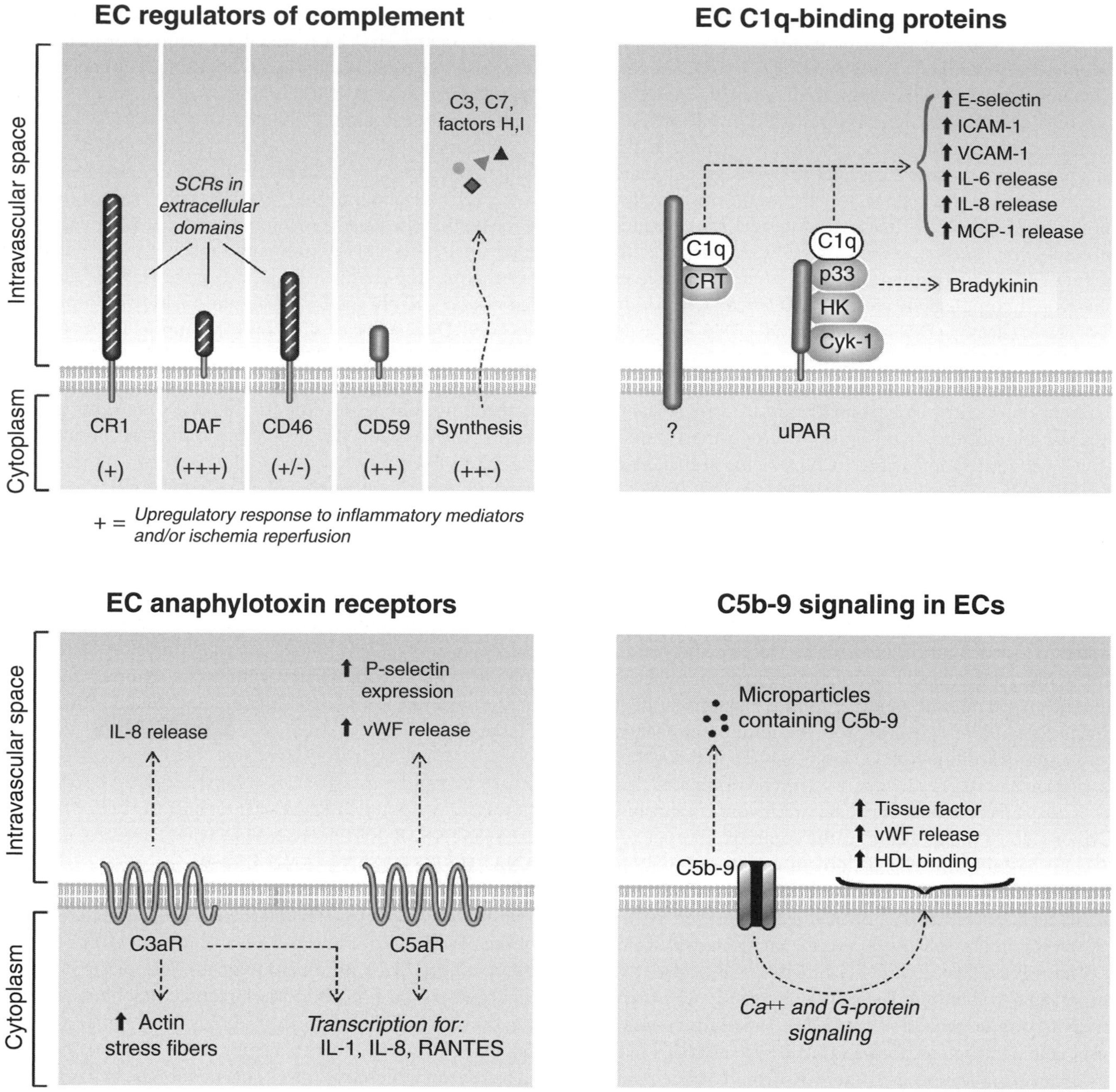

Figure 48.2. EC responses to complement agonists and noncomplement inflammatory mediators. The C1q binding proteins calreticulin (CRT) and p33 are anchored to EC plasma membrane by membrane protein thought to be a $\beta 1$ integrin. p33 also is found in association with cytokeratin (Cyk)-1 and urinary plasminogen-activator receptor (uPAR). C3a and C5a react with their respective receptors and stimulate the immediate release of some mediators and the transcription of inflammatory cytokines. C5b-9 activates G-proteins and allows a calcium flux that results in some immediate phenotypic changes and vesiculation to remove the C5b-9 complexes (microparticles). Ischemia-reperfusion injury in vitro leads to the expression of complement receptor 1 (CR1 or CD35). Inflammatory mediators can upregulate the expression of complement regulatory molecules and lead to the synthesis of C3, C7, factor H, and factor I.

of C3. The importance of C3 results from its relatively high serum concentration (1.5 mg/mL) and the amplification of its activation by the C3-cleaving enzymes, C4b2a, and especially C3bBb, which participates in a positive feedback loop for more C3 activation (see Figure 48.1). CR1 is the receptor for C1q, MBL, C4b, and C3b of neutrophils, monocytes, and dendritic cells, where it has an important role in phagocytosis. It also is the receptor for C1q on glomerular podocytes, where it may be involved in endocytosis. RBC CR1 has an important role in clearing complement-opsonized particles from the intravascular space (discussed later). Finally, CR1 functions as a cofactor for factor I cleavage of C3b and iC3b. When C3b is cleaved to iC3b, it no longer can participate in complement activation but it becomes an opsonin for the integrins $\alpha_2\beta_2$ and $\alpha_3\beta_2$ (complement receptors 3 and 4, respectively), which are also phagocytic receptors expressed on neutrophils, macrophages, and dendritic cells.

Using CR1 as a cofactor, iC3b is cleaved to C3dg, the ligand for CR2 (complement receptor 2 or CD21), which is primarily expressed on B cells. When antigen bearing C3dg binds to the B-cell antigen receptor, the attached C3dg ligates CR2, which then signals through associated CD19 to amplify B-cell antigen receptor signaling (3). Thus, particles recognized as "foreign" by the complement system may have their C3b processed to C3dg, which has been shown in one case to potentiate 10,000-fold the threshold for B-cell activation (3). This experiment provides an explanation as to why adjuvants enhance the immunogenicity of purified antigens: Adjuvants activate innate immunity and thereby prime the adaptive immune response. Extrapolating from in vitro studies done with the Raji B-cell line, B cells without a committed antigen receptor can use CR2 to endocytose C3dg-antigen for antigen presentation (67). Studies in mice have corroborated other roles for C4 and C3, as well as their receptors, in enhancing adaptive immunity (2). Of note, although human CR1 and CR2 are encoded by distinct genes, mouse CR2 and CR1 are alternatively spliced gene products. Mouse CR2 is 67% homologous to human CR2, whereas mouse CR1 consists of the entire mouse CR2 molecule plus six additional NH-terminal SCRs that contain a C3b binding site and factor I cofactor site (68). To date, all murine knockouts target both molecules, which together are referred to as CR2/CR1.

The NH-terminal cleavage-activation of the α-chains of C4, C3, and C5 generate the anaphylotoxins, C4a, C3a, and C5a, respectively. Knockouts of C4, C3, and C5 are confounded because, not only do they inhibit complement activation, they also eliminate the role of these anaphylotoxins. A specific receptor for C4a has not been identified, but receptors for C3a and C5a are well characterized: They are G-coupled and are widely and differentially distributed on leukocytes and other tissues, including endothelium. Specific knockouts of the C3aR and the C5aR have underscored complex roles for these receptors in modulating inflammation (Table 48-3). C4a and C3a, which are strongly positively charged, also can act in a receptor-independent manner to potentiate the chemokine CXCL12 (or SDF-1) (69).

Although C5b-9 often is referred to as the membrane attack complex (MAC), except in pathologic conditions, this is a misnomer. In vivo only *Neisseria* species are killed by C5b-9 (26). C5b-9, which lacks a specific receptor on autologous cells, acts like a cytokine for host cells because it permits a flux of calcium, and it stimulates heterotrimeric G-proteins of the G_i/G_o family (70). Signaling by C5b-9 must be redundant, however, because deficiency in a terminal complement component in normal humans and mice, which prevents the formation of C5b-9, has no phenotype. C5b-9 does participate in human and experimental inflammation (discussed later). Because C5b-9 is an ungated channel, it is necessary for the cell to eliminate the channel or the cell will lyse. Both vesiculation and endocytosis have been described. When ECs are subjected to C5b-9, the cells vesiculate to eliminate the C5b-9 pore from their plasma membrane (71).

Role of Complement in Immune Adherence-Mediated Clearance

All animals with a closed circulation face the problem of clearing their vascular space of noxious particles, such as microbes, apoptotic debris, and immune complexes. The mechanism, known as immune adherence-mediated clearance, has been conserved during evolution and involves three steps. First, particles are recognized by the complement system and tagged with opsonins. Second, using complement receptors, opsonized particles are ligated to the surface of cells. In the case of primates, erythrocytes are the cell and the receptor is CR1 (72). Finally, the particles are removed from the surface of the RBCs by phagocytic cells in the liver and spleen, and the RBCs return to the circulation (73). Nonhuman primates have a GPI-anchored CR1-like molecule on RBCs (74), whereas in humans, CR1 is a transmembrane molecule (72). Other vertebrate animals ligate complement-opsonized particles to platelets, and in mice the complement receptor is platelet-associated factor H (75).

Normal functioning of immune adherence clearance prevents inflammation by both avoiding leukocyte activation in the intravascular space and preventing blood-borne noxious particles (microbes or immune complexes) from lodging in vulnerable organs, such as the brain and kidney, and becoming a nidus of intravascular inflammation. In the absence of immune adherence, immune complexes are cleared more rapidly in vivo (76,77), but the EC becomes the target. In vitro modeling has shown that the presence of immune complexes leads to activation of neutrophils and increased permeability of an endothelial monolayer. When human RBCs are added to this mixture the inflammatory stimulus of the immune complexes is abrogated because they bind RBCs in preference to neutrophils (78).

THE ENDOTHELIUM AND COMPLEMENT

Effect of Inflammatory Mediators on Complement Proteins

Of the three membrane complement regulatory proteins, DAF expression is most responsive to upregulation by transcription or translation in response to inflammatory cytokines, whereas CD59 is minimally affected, and CD46 is not affected (Table 48-4) (79,80).

In contrast, incubation of coronary artery endothelial or saphenous vein ECs with C-reactive protein leads to the upregulation of DAF, CD59, and CD46 (84). Ischemia–reperfusion injury of cultured human umbilical vein ECs (HUVECs) leads to the upregulation of CR1 (86). LPS, tumor necrosis factor (TNF)-α, and interferon (IFN)-γ all cause the upregulation of

Table 48-4: Effect of Cytokines/Drugs on Complement Regulatory Molecules Expressed on Endothelial Cells

Stimulus	*DAF*	*CD46*	*CD59*	*References*
IL-1β	No Δ	No Δ	No Δ	79
TNF-α	↑	No Δ	No Δ	80
Thrombin	↑	—	—	81
IFN-γ	↑	No Δ	No Δ	80
bFGF	↑↑	No Δ	No Δ	82
VEGF	↑↑	—	—	83
CRP*	↑	↑↑	↑	84
Statins	↑↑	No Δ	↑	85

↑ – significant increase
↑↑ – ≥100% increase
*C-reactive protein

the C1q binding protein, p33 (87). Inflammatory agonists also can induce ECs to synthesize complement proteins including C3 and alternative pathway components (88,89), C7 (90), and the fluid phase regulators, factors H and I (91,92).

Complement Signaling to Endothelial Cells

Signaling Through C1q Binding Proteins

C1q immobilized in plastic wells causes seeded human dermal microvascular ECs to spread. The spreading can be partially inhibited by pretreating the cells with either antibody against CRT or p33, and can be completely inhibited by an RGD-containing peptide. The ability of an RGD-containing peptide to completely inhibit spreading has led to the hypothesis that $\beta 1$ integrins are associated with one or both of the C1q binding proteins and may be mediating the signaling in ECs (65).

C3a and C5a Signaling in Endothelial Cells

Receptors for C3a and C5a are constitutively expressed on HUVECs and ECs in vivo. Although C3 and C5 are homologous proteins, their N-terminal fragments, liberated after the cleavage-activation of the parent molecules, show complete specificity for their respective receptors. They are both known as anaphylotoxins because of their ability to induce smooth muscle contraction. Because C3 is cleaved more readily than is C5, the molar ratio of C3a will exceed that of C5a in vivo. Their receptors are G-protein–coupled and, in the case of leukocytes, both are pertussis toxin inhibitable, whereas the C3aR of ECs is not inhibited by pertussis toxin (93). In cultured ECs, both C3a and C5a induce the transcription of mRNA for *IL-1β*, *IL-8*, and regulated on activation normal T-cell expressed and secreted (RANTES), and the rapid release of IL-8 (94). C3a induces actin stress fibers in HUVECs, whereas C5a causes cell retraction over days, with resultant increased paracellular permeability (93). C5a also induces the rapid release of von Willebrand factor (vWF) and upregulation of P-selectin, which results in increased endothelial adhesiveness for neutrophils (95). Thus, both C3a and C5a stimulate ECs to modulate cytokine production and affect leukocyte trafficking.

C5b-9 Signaling to Endothelial Cells

Signaling pathways downstream of C5b-9 have been reviewed (96). C5b-9 stimulates HUVEC degranulation, secretion of multimers of vWF, vesiculation, and binding of high-density lipoprotein (HDL) particles and their protein adaptors apolipoproteins A-I and A-II (apo A-I and A-II). C5b-9-induced degranulation is mimicked by other agonists that induce an increase in intracellular calcium, such as ionophore (97,98), but vesiculation is a specific response to C5b-9. Vesicles, which contain the C5b-9, are avid docking sites for clotting factor Va. Subsequently, Va binds factor Xa to assemble the Va,Xa prothombinase complex at the vesicle surface (71). C5b-9 also induces tissue factor expression on ECs, an effect that occurs over days and depends on secreted IL-1 as an intermediary stimulus. Thus, at least three mechanisms are in place whereby sublytic complement stimulation can lead to procoagulant signals. It is likely that C5b-9-stimulated procoagulant activity participates in syndromes involving complement fixing antiendothelial antibodies. Such syndromes include thrombotic thrombocytopenia purpura, Kawasaki disease, systemic vasculitis, and acute graft rejection (99–101).

C5b-9-induced binding of HDL and apo A-I and A-II to HUVECs depends on a cytosolic calcium flux, but cannot be reproduced by calcium ionophore (102). The receptor for HDL and apoA-I and A-II on endothelium has been identified as scavenger receptor B type I (103). One explanation for increased ligand binding is that the scavenger receptor is upregulated by C5b-9, as is DAF (80).

Complement-Induced Endothelial-Cell Pathology

Dysregulation of complement and damage to ECs may be due to either potent stimuli for activation and/or mutations in the complement regulatory proteins. Examples of potent complement-activating stimuli are reperfusion injury, allotransplantation, and to a greater degree, xenotransplantation. Transplant surgeons have demonstrated that long-term organ survival is adversely affected by reperfusion injury (104), illustrating that transplants are not only targets of the host adaptive immune response, but also of innate immunity's response to ischemia, which is particularly severe in organs retrieved from cadavers. Rodent models of reperfusion injury also demonstrate a central role for complement, although the mechanism of complement activation is controversial. It is likely that the target organ and the degree of ischemia affects how complement is activated because different models activate either the lectin pathway through mannan-binding lectin, or the classical pathway with the participation of natural IgM (105,106).

Biopsies of transplanted ischemically damaged human kidneys show mannan-binding lectin deposition (107).

AMD is an inflammatory condition that is characterized by focal extracellular debris, known as *drusen* (German for stones), that develop between the retinal pigment epithelial layer and its specialized basement membrane known as the Bruch membrane. Drusen contain multiple elements including vitronectin, C3 cleavage products, and C5b-9 complexes (108,109). Familial forms of AMD had been genetically linked to a region near the factor H gene, and recent data implicates a single nucleotide polymorphism (SNP) (Y404H) as highly associated with the familial form of AMD in five different patient groups (110–113). Although drusen are above the Bruch membrane, and the choroid layer containing the vasculature lies just beneath the Bruch membrane, this particular SNP is most closely associated with the variant of AMD that features neovascularization of the macula. The SNP is in the factor H domain involved in binding to host cells and C-reactive protein. It is not clear whether AMD is initiated by complement activation, or associated with complement activation. In either case, the SNP potentially compromises the ability of factor H to bind to host cells, including ECs, and protect them from errant C3 and C5 activation. Retinal pigmented epithelial cells are sources of both proangiogenic vascular endothelial growth factor (VEGF) and antiangiogenic pigmented epithelium derived factor (PEDF). In vitro, disruption of cell–cell contact, which might result from drusen formation in vivo, induces *VEGF* mRNA and inhibits *PEDF* mRNA transcription by pigmented epithelial cells (114).

In diabetic retinal microangiopathy, which is characterized by microinfarcts and neovascularization, C5b-9 is deposited in the capillaries just beneath the Bruch membrane. The absence of C4 antigen in the vessels is consistent with activation of complement by the alternative pathway (115). GPI-anchored complement regulators DAF and CD59 are relatively deficient in the lysates of retinas from diabetic subjects as compared with those from normal subjects (116). DAF and CD59 deficiency would allow unregulated C3 activation at the cell membrane and unregulated insertion of C5b-9. Further compromise of CD59 function may occur from glycation secondary to hyperglycemia (117). The membrane complement regulator CD46, which is transmembrane, is not decreased in diabetic retinas (116).

aHUS, which accounts for about 10% of all HUS cases, is not associated with infections due to Shiga toxin–producing bacteria. The microangiopathic hemolytic anemia, thrombocytopenia, and renal failure of aHUS is pathologically similar to that of HUS, but aHUS is often more severe and may be recurrent and familial. About one third of aHUS patients have functional mutations in either factor H, factor I, or CD46, which are all regulators of C3 activation (118). aHUS due to factor H mutations has been associated with recurrence after renal transplantation (119). Variable penetrance of the aHUS susceptibility has led to searches for gene modifiers, and aHUS patients are more likely to share SNPs in other complement regulator genes when compared with a control population (120).

Atheromata are associated with complement deposition (121). Whether complement has a role in initiating the lesion is debatable, but it almost certainly has a role in sustaining the inflammation. Complement is activated, both in vitro and in vivo, when exposed to the modified lipid deposits characteristic of the atheromatous lesion (122). C-reactive protein can bind to the lipids and further activate C1. Several complement-deficient animal models have been used to investigate the roles of complement in the pathogenesis of atherosclerosis. If C3 is knocked out in atherosclerosis prone $Apoe^{-/-}$ mice, the serum lipids do not change, but two changes occur in the aortic lesions, compared with the wt $Apoe^{-/-}$ animals. First, the lesion size is increased in $Apoe^{-/-}$/C3-deficient mice as compared with the $Apoe^{-/-}$/C3-normal mice. Second, in the absence of C3, the aortic lesion does not mature beyond the foamy macrophage stage (123). Specifically, no smooth muscle cell proliferation occurs. C5b-9 is thought to have a role in smooth muscle proliferation (121), and C5b-9 cannot form in the absence of C3 activation. In another animal model, C6 deficiency in rabbits protects them from diet-induced aortic atherosclerosis (124). However, not all the animal data are consistent: C5 deficiency does not protect the $Apoe^{-/-}$ mouse from aortic atherosclerotic lesions (125). Only additional studies will clarify the role of C5b-9 in atherosclerosis.

Compromise in immune adherence clearance can result from the inability to opsonize particles adequately. This is one explanation for the association of a deficiency of an early classical pathway component and the risk of developing lupus erythematosus. Interestingly, the risk is greatest for developing lupus the earlier the deficient component is in the activation sequence, with a >90% risk associated with C1q deficiency (126) and only a 10% risk associated with C2 deficiency (32). It is not known whether the noxious particle that is not cleared adequately in lupus is apoptotic debris or immune complexes. Immune adherence clearance can be overwhelmed when a large burden of antigen results in the continued production of immune complexes that cannot be appropriately cleared. This occurs in subacute bacterial endocarditis and serum sickness, and the uncleared immune complexes injure the endothelium and cause clinical vasculitis.

KEY POINTS

- ECs are bathed in complement. High levels of membrane regulators of complement activation normally protect ECs from direct complement damage. The effective immobilization of complement opsonized noxious particles and clearance by immune adherence further protect ECs from bystander complement damage.

- Acquired and genetic deficiencies of complement regulators are linked to diabetic retinopathy, aHUS, and age-related macular degeneration.
- Compromise of complement-dependent immune adherence clearance predisposes to vasculitis.

Future Goals

- To gain acute control over complement activation at the time of organ transplant to dampen acute and chronic rejection
- To undertake additional studies of complement component deficiencies in atherosclerotic prone mice to delineate if complement has a role in the initiation of atheromata, or just in their progression

REFERENCES

1 Medzhitov R, Janeway CA Jr. Innate immunity: the virtues of a nonclonal system of recognition. *Cell.* 1997;91(3):295–298.

2 Carroll MC. The complement system in regulation of adaptive immunity. *Nat Immunol.* 2004;5(10):981–986.

3 Dempsey PW, Allison ME, Akkaraju S, et al. C3d of complement as a molecular adjuvant: bridging innate and acquired immunity. *Science.* 1996;271:348–350.

4 Sottrup-Jensen L, Stepanik TM, Kristensen T, et al. Common evolutionary origin of alpha 2-macroglobulin and complement components C3 and C4. *Proc Natl Acad Sci USA.* 1985;82(1):9–13.

5 Kristensen T, Moestrup SK, Gliemann J, et al. Evidence that the newly cloned low-density-lipoprotein receptor related protein (LRP) is the alpha 2-macroglobulin receptor. *FEBS Lett.* 1990;276(1–2):151–155.

6 Strickland DK, Ashcom JD, Williams S, et al. Sequence identity between the alpha 2-macroglobulin receptor and low density lipoprotein receptor-related protein suggests that this molecule is a multifunctional receptor. *J Biol Chem.* 1990;265(29):17401–17404.

7 Dishaw LJ, Smith SL, Bigger CH. Characterization of a C3-like cDNA in a coral: phylogenetic implications. *Immunogenetics.* 2005;57(7):535–548.

8 Zhu Y, Thangamani S, Ho B, et al. The ancient origin of the complement system. *EMBO J.* 2005;24(2):382–394.

9 Fujita T, Matsushita M, Endo Y. The lectin-complement pathway – its role in innate immunity and evolution. *Immunol Rev.* 2004;198:185–202.

10 Janssen BJ, Huizinga EG, Raaijmakers HC, et al. Structures of complement component C3 provide insights into the function and evolution of immunity. *Nature.* 2005;437(7058):505–511.

11 Sahu A, Lambris JD. Structure and biology of complement protein C3, a connecting link between innate and acquired immunity. *Immunol Rev.* 2001;180:35–48.

12 Parker CJ, Soldato CM, Rosse WF. Abnormality of glycophorin-alpha on paroxysmal nocturnal hemoglobinuria erythrocytes. *J Clin Invest.* 1984;73(4):1130–1143.

13 Sahu A, Kozel TR, Pangburn MK. Specificity of the thioester-containing reactive site of human C3 and its significance to complement activation. *Biochem J.* 1994;302(Pt 2):429–436.

14 Klickstein LB, Barbashov S, Liu T, et al. Complement Receptor Type 1 (CR1, CD35) is a Receptor for C1q. *Immunity.* 1997;7: 345–355.

15 Ghiran I, Barbashov SF, Klickstein LB, et al. Complement Receptor 1/CD35 is a receptor for mannan-binding lectin. *J Exp Med.* 2000;192:1797–1808.

16 Bartholomew RM, Esser AF. Mechanism of antibody-independent activation of the first component of complement (C1) on retrovirus membranes. *Biochemistry.* 1980;19:2847–2853.

17 Cooper NR, Morrison DC. Binding and activation of the first component of human complement by the lipid A region of lipopolysaccharides. *J Immunol.* 1978;120:1862–1868.

18 Manderson AP, Pickering MC, Botto et al. Continual low-level activation of the classical complement pathway. *J Exp Med.* 2001; 194(6):747–756.

19 Pangburn MK, Schreiber RD, Muller-Eberhard HJ. Formation of the initial C3 convertase of the alternative complement pathway. Acquisition of C3b-like activities by spontaneous hydrolysis of the putative thioester in native C3. *J Exp Med.* 1981; 154(3):856–867.

20 Pangburn MK, Pangburn KL, Koistinen V, et al. Molecular mechanisms of target recognition in an innate immune system: interactions among factor H, C3b, and target in the alternative pathway of human complement. *J Immunol.* 2000;164(9):4742–4751.

21 Jain A, Karadag A, Fohr B, et al. Three SIBLINGs (small integrin-binding ligand, N-linked glycoproteins) enhance factor H's cofactor activity enabling MCP-like cellular evasion of complement-mediated attack. *J Biol Chem.* 2002;277(16): 13700–13708.

22 Seyfarth J, Garred P, Madsen HO. The "involution" of mannose-binding lectin. *Hum Mol Genet.* 2005;14(19):2859–2869.

23 Foster CB, Lehrnbecher T, Mol F, et al. Host defense molecule polymorphisms influence the risk for immune- mediated complications in chronic granulomatous disease. *J Clin Invest.* 1998;102(12):2146–2155.

24 Walsh MC, Bourcier T, Takahashi K, et al. Mannose-binding lectin is a regulator of inflammation that accompanies myocardial ischemia and reperfusion injury. *J Immunol.* 2005;175(1): 541–546.

25 Stengaard-Pedersen K, Thiel S, Gadjeva M, et al. Inherited deficiency of mannan-binding lectin-associated serine protease 2. *N Engl J Med.* 2003;349(6):554–560.

26 Figueroa J, Andreoni J, Densen P. Complement deficiency states and meningococcal disease. *Immunol Res.* 1993;12(3):295–311.

27 Densen P, Weiler J, Ackerman L, et al. Functional and antigenic analysis of human factor B deficiency. *Mol Immnol.* 1996;33(Suppl1):68(Abstract 270).

28 Watanabe H, Garnier G, Circolo A, et al. Modulation of renal disease in MRL/lpr mice genetically deficient in the alternative complement pathway factor B. *J Immunol.* 2000;164(2):786–794.

29 Hiemstra PS, Langeler E, Compier B, et al. Complete and partial deficiencies of complement factor D in a Dutch family. *J Clin Invest.* 1989;84(6):1957–1961.

30 Biesma DH, Hannema AJ, van Velzen-Blad H, et al. A family with complement factor D deficiency. *J Clin Invest.* 2001; 108(2):233–240.

31 Elliott MK, Jarmi T, Ruiz P, et al. Effects of complement factor D deficiency on the renal disease of MRL/lpr mice. *Kidney Int.* 2004;65(1):129–138.
32 Walport MJ. Complement. Second of two parts. *N Engl J Med.* 2001;344(15):1140–1144.
33 Botto M, Dell'Agnola C, Bygrave AE, et al. Homozygous C1q deficiency causes glomerulonephritis associated with multiple apoptotic bodies [see comments]. *Nat Genet.* 1998;19(1):56–59.
34 Taylor PR, Carugati A, Fadok VA, et al. A hierarchical role for classical pathway complement proteins in the clearance of apoptotic cells in vivo. *J Exp Med.* 2000;192(3):359–366.
35 Taylor PR, Nash JT, Theodoridis E, et al. A targeted disruption of the murine complement factor B gene resulting in loss of expression of three genes in close proximity, factor B, C2, and D17H6S45. *J Biol Chem.* 1998;273(3):1699–1704.
36 Hopken UE, Lu B, Gerard NP, et al. The C5a chemoattractant receptor mediates mucosal defence to infection. *Nature.* 1996;383(6595):86–89.
37 Trapp RG, Fletcher M, Forristal J, et al. C4 binding protein deficiency in a patient with atypical Behçet's disease. *J Rheumatol.* 1987;14(1):135–138.
38 Sun X, Funk CD, Deng C, et al. Role of decay-accelerating factor in regulating complement activation on the erythrocyte surface as revealed by gene targeting. *Proc Natl Acad Sci USA.* 1999;96(2):628–633.
39 Miwa T, Sun X, Ohta R, et al. Characterization of glycosylphosphatidylinositol-anchored decay accelerating factor (GPI-DAF) and transmembrane DAF gene expression in wild-type and GPI-DAF gene knockout mice using polyclonal and monoclonal antibodies with dual or single specificity. *Immunology.* 2001;104(2):207–214.
40 Xu C, Mao D, Holers VM, et al. A critical role for murine complement regulator crry in fetomaternal tolerance. *Science.* 2000;287(5452):498–501.
41 Holt DS, Botto M, Bygrave AE, et al. Targeted deletion of the CD59 gene causes spontaneous intravascular hemolysis and hemoglobinuria. *Blood.* 2001;98(2):442–449.
42 Qin X, Krumrei N, Grubissich L, et al. Deficiency of the mouse complement regulatory protein mCd59b results in spontaneous hemolytic anemia with platelet activation and progressive male infertility. *Immunity.* 2003;18(2):217–227.
43 Baalasubramanian S, Harris CL, Donev RM, et al. CD59a is the primary regulator of membrane attack complex assembly in the mouse. *J Immunol.* 2004;173(6):3684–3692.
44 Mondragon-Palomino M, Pinero D, Nicholson-Weller A, et al. Phylogenetic analysis of the homologous proteins of the terminal complement complex supports the emergence of C6 and C7 followed by C8 and C9. *J Mol Evol.* 1999;49(2):282–289.
45 Hammer CH, Nicholson A, Mayer MM. On the mechanism of cytolysis by complement: evidence on the insertion of the C5b and C7 subunits of the C5b,6,7 complex into the phospholipid bilayer of the erythrocyte membrane. *Proc Natl Acad Sci USA.* 1975;72:5076–5080.
46 Michaels DW, Abramovitz AS, Hammer CH, et al. Increased ion permeability of planar lipid bilayer membranes after treatment with the C5b-9 cytolytic attack mechanism of complement. *Proc Natl Acad Sci USA.* 1976;73:2852–2856.
47 Davis AE III. The pathophysiology of hereditary angioedema. *Clin Immunol.* 2005;114(1):3–9.
48 Blumbach B, Pancer Z, Diehl-Seifert B, et al. The putative sponge aggregation receptor. Isolation and characterization of a molecule composed of scavenger receptor cysteine-rich domains and short consensus repeats. *J Cell Sci.* 1998;111(Pt 17):2635–2644.
49 Soares DC, Gerloff DL, Syme NR, et al. Large-scale modelling as a route to multiple surface comparisons of the CCP module family. *Protein Eng Des Sel.* 2005;18(8):379–388.
50 Klickstein LB. The complete primary structure of the human C3b/C4b receptor: Tandem long homologous repeats contain distinct C3b and C4b recognition sites. [PhD diss.]. Cambridge, MA: Harvard University, 1989.
51 Klickstein LB, Bartow TJ, Miletic B, et al. Identification of distinct C3b and C4b recognition sites in the human C3b/C4b receptor (CR1, CD35) by deletion mutagenesis. *J Exp Med.* 1988; 168:1699–1717.
52 Krych M, Hourcade D, Atkinson JP. Sites within the complement C3b/C4b receptor important for the specificity of ligand binding. *Proc Natl Acad Sci USA.* 1991;88(10):4353–4357.
53 McLure CA, Dawkins RL, Williamson JF, et al. Amino acid patterns within short consensus repeats define conserved duplicons shared by genes of the RCA complex. *J Mol Evol.* 2004;59(2): 143–157.
54 Gaboriaud C, Rossi V, Fontecilla-Camps JC, et al. Evolutionary conserved rigid module-domain interactions can be detected at the sequence level: the examples of complement and blood coagulation proteases. *J Mol Biol.* 1998;282(2):459–470.
55 Nicholson-Weller A, Burge J, Fearon DT, et al. Isolation of a human erythrocyte membrane glycoprotein with decay-accelerating activity for C3 convertases of the complement system. *J Immunol.* 1982;129(1):184–189.
56 Liszewski MK, Post TW, Atkinson JP. Membrane cofactor protein (MCP or CD46): Newest member of the regulators of complement activation gene cluster. *Ann Rev Immunol.* 1991;9:431–455.
57 Sharma AK, Pangburn MK. Identification of three physically and functionally distinct binding sites for C3b in human complement factor H by deletion mutagenesis. *Proc Natl Acad Sci USA.* 1996;93(20):10996–11001.
58 Jarva H, Jokiranta TS, Hellwage J, et al. Regulation of complement activation by C-reactive protein: targeting the complement inhibitory activity of factor H by an interaction with short consensus repeat domains 7 and 8–11. *J Immunol.* 1999;163(7): 3957–3962.
59 Giannakis E, Jokiranta TS, Male DA, et al. A common site within factor H SCR 7 responsible for binding heparin, C-reactive protein and streptococcal M protein. *Eur J Immunol.* 2003;33(4):962–969.
60 Nicholson-Weller A, March JP, Rosenfeld SI, et al. Affected erythrocytes of patients with paroxysmal nocturnal hemoglobinuria are deficient in the complement regulatory protein, decay accelerating factor. *Proc Natl Acad Sci USA.* 1983;80(16):5066–5070.
61 Pangburn MK, Schreiber RD, Müller-Eberhard HJ. Deficiency of an erythrocyte membrane protein with complement regulatory activity in paroxysmal nocturnal hemoglobinuria. *Proc Natl Acad Sci USA.* 1983;80:5430–5434.
62 Parker CJ, Wiedmer T, Sims PJ, et al. Characterization of the complement sensitivity of paroxysmal nocturnal hemoglobinuria erythrocytes. *J Clin Invest.* 1985;75(6):2074–2084.
63 Ghebrehiwet B, Peerschke EI. cC1q-R (calreticulin) and gC1q-R/p33: ubiquitously expressed multi-ligand binding cellular

proteins involved in inflammation and infection. *Mol Immunol.* 2004;41(2–3):173–183.

64 Ghiran I, Klickstein LB, Nicholson-Weller A. Calreticulin is at the surface of circulating neutrophils and uses CD59 as an adaptor molecule. *J Biol Chem.* 2003;278(23):21024–21031.

65 Feng X, Tonnesen MG, Peerschke EI, et al. Cooperation of C1q receptors and integrins in C1q-mediated EC adhesion and spreading. *J Immunol.* 2002;168(5):2441–2448.

66 Peerschke EI, Ghebrehiwet B. Human blood platelet gC1qR/p33. *Immunol Rev.* 2001;180:56–64.

67 Hess MW, Schwendinger MG, Eskelinen EL, et al. Tracing uptake of C3dg-conjugated antigen into B cells via complement receptor type 2 (CR2, CD21). *Blood.* 2000;95(8):2617–2623.

68 Molina H, Kinoshita T, Webster CB, et al. Analysis of C3b/C3d binding sites and factor I cofactor regions within mouse complement receptors 1 and 2. *J Immunol.* 1994;153(2):789–795.

69 Honczarenko M, Ratajczak MZ, Nicholson-Weller A, et al. Complement C3a enhances CXCL12 (SDF-1)-mediated chemotaxis of bone marrow hematopoietic cells independently of C3a receptor. *J Immunol.* 2005;175(6):3698–3706.

70 Niculescu F, Rus H, Shin ML. Receptor-independent activation of guanine nucleotide-binding regulatory proteins by terminal complement complexes. *J Biol Chem.* 1994;269:4417–4423.

71 Hamilton KK, Hattori R, Esmon CT, et al. Complement proteins C5b-9 induce vesiculation of the endothelial plasma membrane and expose catalytic surface for assembly of the prothrombinase enzyme complex. *J Biol Chem.* 1990;265(7):3809–3814.

72 Fearon DT. Identification of the membrane glycoprotein that is the C3b receptor of the human erythrocyte, polymorphonuclear leukocyte, B lymphocyte and monocyte. *J Exp Med.* 1980; 152:20–30.

73 Nelson RAJ. The immune adherence phenomenon: an immunologically specific reaction between microorganisms and erythrocytes leading to enhanced phagocytosis. *Science.* 1953;118: 733–737.

74 Chen W, Logar CM, Shen XP, et al. The chimpanzee and cynomolgus monkey erythrocyte immune adherence receptors are encoded by CR1-like genes. *Immunogenetics.* 2000;52(1–2): 46–52.

75 Alexander JJ, Hack BK, Cunningham PN, et al. A protein with characteristics of factor H is present on rodent platelets and functions as the immune adherence receptor. *J Biol Chem.* 2001;276(34):32129–32135.

76 Halma C, Daha MR, Camps JA, et al. Deficiency of complement component C3 is associated with accelerated removal of soluble ^{123}I-labelled aggregates of IgG from the circulation. *Clin Exp Immunol.* 1992;90(3):394–400.

77 Davies KA, Peters AM, Beynon HLC, et al. Immune complex processing in patients with systemic lupus erythematosus. *J Clin Invest.* 1992;90:2075–2083.

78 Beynon HL, Davies KA, Haskard DO, et al. Erythrocyte complement receptor type 1 and interactions between immune complexes, neutrophils, and endothelium. *J Immunol.* 1994;153(7): 3160–3167.

79 Moutabarrik A, Nakanishi I, Namiki M, et al. Cytokine-mediated regulation of the surface expression of complement regulatory proteins, CD46(MCP), CD55(DAF), and CD59 on human vascular ECs. *Lymphokine Cytokine Res.* 1993;12(3): 167–172.

80 Mason JC, Yarwood H, Sugars K, et al. Induction of decay-accelerating factor by cytokines or the membrane-attack complex protects vascular ECs against complement deposition. *Blood.* 1999;94(5):1673–1682.

81 Lidington EA, Haskard DO, Mason JC. Induction of decay-accelerating factor by thrombin through a protease-activated receptor 1 and protein kinase C-dependent pathway protects vascular ECs from complement-mediated injury. *Blood.* 2000; 96(8):2784–2792.

82 Mason JC, Lidington EA, Ahmad SR, et al. bFGF and VEGF synergistically enhance endothelial cytoprotection via decay-accelerating factor induction. *Am J Physiol Cell Physiol.* 2002;282(3):C578–C587.

83 Mason JC, Lidington EA, Yarwood H, et al. Induction of EC decay-accelerating factor by vascular endothelial growth factor: a mechanism for cytoprotection against complement-mediated injury during inflammatory angiogenesis. *Arthritis Rheum.* 2001;44(1):138–150.

84 Li SH, Szmitko PE, Weisel RD, et al. C-reactive protein upregulates complement-inhibitory factors in ECs. *Circulation.* 2004;109(7):833–836.

85 Mason JC, Ahmed Z, Mankoff R, et al. Statin-induced expression of decay-accelerating factor protects vascular endothelium against complement-mediated injury. *Circ Res.* 2002;91(8): 696–703.

86 Collard CD, Bukusoglu C, Agah A, et al. Hypoxia-induced expression of complement receptor type 1 (CR1, CD35) in human vascular ECs. *Amer J Physiol.* 1999;276:C450–C458.

87 Guo WX, Ghebrehiwet B, Weksler B, et al. Up-regulation of EC binding proteins/receptors for complement component C1q by inflammatory cytokines. *J Lab Clin Med.* 1999;133(6):541–550.

88 Ripoche J, Mitchell JA, Erdei A, et al. Interferon gamma induces synthesis of complement alternative pathway proteins by human ECs in culture. *J Exp Med.* 1988;168(5):1917–1922.

89 Coulpier M, Andreev S, Lemercier C, et al. Activation of the endothelium by IL-1 alpha and glucocorticoids results in major increase of complement C3 and factor B production and generation of C3a. *Clin Exp Immunol.* 1995;101(1):142–149.

90 Langeggen H, Pausa M, Johnson E, et al. The endothelium is an extrahepatic site of synthesis of the seventh component of the complement system. *Clin Exp Immunol.* 2000;121(1):69–76.

91 Brooimans RA, Hiemstra PS, van der Ark AA, et al. Biosynthesis of complement factor H by human umbilical vein ECs. Regulation by T cell growth factor and IFN-gamma. *J Immunol.* 1989;142(6):2024–2030.

92 Schlaf G, Demberg T, Beisel N, et al. Expression and regulation of complement factors H and I in rat and human cells: some critical notes. *Mol Immunol.* 2001;38(2–3):231–239.

93 Schraufstatter IU, Trieu K, Sikora L, et al. Complement C3a and C5a induce different signal transduction cascades in ECs. *J Immunol.* 2002;169(4):2102–2110.

94 Monsinjon T, Gasque P, Chan P, et al. Regulation by complement C3a and C5a anaphylatoxins of cytokine production in human umbilical vein ECs. *FASEB J.* 2003;17(9):1003–1014.

95 Foreman KE, Vaporciyan AA, Bonish BK, et al. C5a-induced expression of P-selectin in ECs. *J Clin Invest.* 1994;94(3):1147–1155.

96 Niculescu F, Rus H. Mechanisms of signal transduction activated by sublytic assembly of terminal complement complexes on nucleated cells. *Immunol Res.* 2001;24(2):191–199.

97 Hattori R, Hamilton KK, McEver RP, et al. Complement proteins C5b-9 induce secretion of high molecular weight multimers of endothelial von Willebrand factor and translocation of granule membrane protein GMP-140 to the cell surface. *J Biol Chem.* 1989;264(15):9053–9060.
98 Hattori R, Hamilton KK, Fugate RD, et al. Stimulated secretion of endothelial von Willebrand factor is accompanied by rapid redistribution to the cell surface of the intracellular granule membrane protein GMP-140. *J Biol Chem.* 1989;264(14):7768–7771.
99 Cines DB. Disorders associated with antibodies to ECs. *Rev Infect Dis.* 1989;11Suppl 4:S705–S711.
100 Brasile L, Kremer JM, Clarke JL, et al. Identification of an autoantibody to vascular EC-specific antigens in patients with systemic vasculitis. *Am J Med.* 1989;87(1):74–80.
101 Leventhal JR, Matas AJ, Sun LH, et al. The immunopathology of cardiac xenograft rejection in the guinea pig-to-rat model. *Transplantation.* 1993;56(1):1–8.
102 Hamilton KK, Sims PJ. The terminal complement proteins C5b-9 augment binding of high density lipoprotein and its apolipoproteins A-I and A-II to human ECs. *J Clin Invest.* 1991; 88(6):1833–1840.
103 Acton S, Rigotti A, Landschulz KT, et al. Identification of scavenger receptor SR-BI as a high density lipoprotein receptor. *Science.* 1996;271(5248):518–520.
104 Laskowski I, Pratschke J, Wilhelm MJ, et al. Molecular and cellular events associated with ischemia/reperfusion injury. *Ann Transplant.* 2000;5(4):29–35.
105 Hart ML, Ceonzo KA, Shaffer LA, et al. Gastrointestinal ischemia-reperfusion injury is lectin complement pathway dependent without involving C1q. *J Immunol.* 2005;174(10): 6373–6380.
106 Chan RK, Ding G, Verna N, et al. IgM binding to injured tissue precedes complement activation during skeletal muscle ischemia-reperfusion. *J Surg Res.* 2004;122(1):29–35.
107 de Vries B, Walter SJ, Peutz-Kootstra CJ, et al. The mannose-binding lectin-pathway is involved in complement activation in the course of renal ischemia-reperfusion injury. *Am J Pathol.* 2004;165(5):1677–1688.
108 Mullins RF, Russell SR, Anderson DH, et al. Drusen associated with aging and age-related macular degeneration contain proteins common to extracellular deposits associated with atherosclerosis, elastosis, amyloidosis, and dense deposit disease. *FASEB J.* 2000;14(7):835–846.
109 Johnson LV, Leitner WP, Staples MK, et al. Complement activation and inflammatory processes in Drusen formation and age related macular degeneration. *Exp Eye Res.* 2001;73(6):887–896.
110 Hageman GS, Anderson DH, Johnson LV, et al. A common haplotype in the complement regulatory gene factor H (HF1/CFH) predisposes individuals to age-related macular degeneration. *Proc Natl Acad Sci USA.* 2005;102(20):7227–7232.
111 Edwards AO, Ritter R III, Abel KJ, et al. Complement factor H polymorphism and age-related macular degeneration. *Science.* 2005;308(5720):421–424.
112 Haines JL, Hauser MA, Schmidt S, et al. Complement factor H variant increases the risk of age-related macular degeneration. *Science.* 2005;308(5720):419–421.
113 Klein RJ, Zeiss C, Chew EY, et al. Complement factor H polymorphism in age-related macular degeneration. *Science.* 2005;308(5720):385–389.
114 Wang XF, Cui JZ, Prasad SS, et al. Altered gene expression of angiogenic factors induced by calcium-mediated dissociation of retinal pigment epithelial cells. *Invest Ophthalmol Vis Sci.* 2005; 46(4):1508–1515.
115 Gerl VB, Bohl J, Pitz S, et al. Extensive deposits of complement C3d and C5b-9 in the choriocapillaris of eyes of patients with diabetic retinopathy. *Invest Ophthalmol Vis Sci.* 2002;43(4): 1104–1108.
116 Zhang J, Gerhardinger C, Lorenzi M. Early complement activation and decreased levels of glycosylphosphatidylinositol-anchored complement inhibitors in human and experimental diabetic retinopathy. *Diabetes.* 2002;51(12):3499–3504.
117 Qin X, Goldfine A, Krumrei N, et al. Glycation inactivation of the complement regulatory protein CD59: a possible role in the pathogenesis of the vascular complications of human diabetes. *Diabetes.* 2004;53(10):2653–2661.
118 Zipfel PF, Heinen S, Jozsi M, et al. Complement and diseases: Defective alternative pathway control results in kidney and eye diseases. *Mol Immunol.* 2005;43(1–2):97–106.
119 Olie KH, Goodship TH, Verlaak R, et al. Posttransplantation cytomegalovirus-induced recurrence of atypical hemolytic uremic syndrome associated with a factor H mutation: successful treatment with intensive plasma exchanges and ganciclovir. *Am J Kidney Dis.* 2005;45(1):E12–E15.
120 Esparza-Gordillo J, Goicoechea de Jorge E, Buil A, et al. Predisposition to atypical hemolytic uremic syndrome involves the concurrence of different susceptibility alleles in the regulators of complement activation gene cluster in 1q32. *Hum Mol Genet.* 2005;14(8):1107.
121 Niculescu F, Rus H. The role of complement activation in atherosclerosis. *Immunol Res.* 2004;30(1):73–80.
122 Seifert PS, Kazatchkine MD. Generation of complement anaphylatoxins and C5b-9 by crystalline cholesterol oxidation derivatives depends on hydroxyl group number and position. *Mol Immunol.* 1987;24(12):1303–1308.
123 Buono C, Come CE, Witztum JL, et al. Influence of C3 deficiency on atherosclerosis. *Circulation.* 2002;105(25):3025–3031.
124 Schmiedt W, Kinscherf R, Deigner HP, et al. Complement C6 deficiency protects against diet-induced atherosclerosis in rabbits. *Arterioscler Thromb Vasc Biol.* 1998;18(11):1790–1795.
125 Patel S, Thelander EM, Hernandez M, et al. ApoE(-/-) mice develop atherosclerosis in the absence of complement component C5. *Biochem Biophys Res Commun.* 2001;286(1):164–170.
126 Bowness P, Davies KA, Norsworthy PJ, et al. Hereditary C1q deficiency and systemic lupus erythematosus. *QJM.* 1994;87(8): 455–464.

49

Kallikrein-Kinin System

Robert Colman

Sol Sherry Thrombosis Research Center, Temple University School of Medicine, Philadelphia, Pennsylvania

Three proteins – factor XII, prekallikrein (PK), and high-molecular-weight kininogen (HK) – comprise the contact system, currently referred to as the plasma kallikrein-kinin system (KKS) (Table 49-1). HK was originally identified as a substrate (1) that yields a bioactive peptide, bradykinin (BK), when cleaved. BK in vivo stimulates microvascular permeability and arteriolar vasodilation, resulting in hypotension. Two enzymes are required for the hydrolysis of HK. In 1955, factor XII was discovered by Oscar Ratnoff in a study of a patient, John Hageman, who lacked the protein and had a very prolonged coagulation time (2). Prekallikrein (PK) was first purified from human plasma as the activated enzyme, kallikrein, after being identified by Hathaway in 1965, in a subject with modestly prolonged activated partial thromboplastin time (aPTT) (3). Factor XII autoactivates to factor XIIa, which cleaves plasma PK to kallikrein. Kallikrein then feeds back to cleave factor XII to factor XIIa. Plasma kallikrein and factor XIIa (to a lesser extent) both hydrolyze HK to form cleaved HK (HKa) and BK. The importance of HK was first apparent when three different research groups described patients deficient in HK with prolonged aPTT. One such patient, Williams (4), was identified by our laboratory to have absent BK, even in the urine. All three proteins – factor XII, PK, and HK – were cloned in the mid-1980s.

FACTOR XII

Factor XII is encoded by a single gene of 12 kb that maps to chromosome 5, comprising 13 introns and 14 exons (5). A putative signal peptide sequence on the NH_2 terminus of the zymogen is located on the first exon, followed by a region of unknown homology encoded by the second exon. A type II region homologous with collagen-binding properties in fibronectin is represented by exons II and IV. Exon V codes for the epidermal growth factor (EGF)-like domain, followed by the type I homology or the fibrin finger in fibronectin, coded by exon XI. Exon VII encodes the second EGF-like domain, preceding a kringle structure and a proline-rich region on exons VIII through IX. The light chain of factor XIIa is encoded by five exons (X–XIV). Exon XIV is the largest, consisting of 165 nucleotides encoding 55 amino acid residues of the serine protease region and 150 base pairs (bp) of the 3′untranslated end of the mRNA (6).

The *factor XII* gene is similar in structure to the serine protease family of *tissue-type plasminogen activator (t-PA)* and *urokinase-type plasminogen activator (u-PA)* genes. The mRNA codes for a 596-amino acid, single polypeptide chain with a molecular mass of 80 kDa and an isoelectric point of 6.3 ± 0.1. Factor XII concentration in plasma is 31 ± 8 μg/mL (0.375 μM). Human liver has been shown to express *factor XII* mRNA, and rat hepatocytes in culture synthesize factor XII. In isolated livers of estrogen- and prolactin-treated rats, factor XII is elevated. Rat *factor XII* DNA has been shown to have a functional estrogen promoter element that binds 17β-estradiol (7). Estrogens in premenopausal women in the context of oral contraception, replacement therapy in postmenopausal women, or pregnancy all elevate factor XII (8).

Factor XII is cleaved to the two-chain factor XIIa at R353-V354, remaining connected by a disulfide bond. Further cleavage at three peptide bonds in the catalytic light chain yields the 28-kDa factor XII_F, which lacks the heavy chain. Because all the cell binding sites are in the heavy chain, factor XIIa binds to platelets, neutrophils, and endothelial cells (ECs), but factor XII_F does not. Targeted deletion of murine *factor XII* gene results in complete plasma factor XII deficiency. The mice do not have spontaneous or excessive injury related bleeding but they are protected against collagen- and epinephrine-induced thromboembolism (9).

PREKALLIKREIN

The gene structure for human *PK* has only recently been determined (10); it is composed of 15 exons and 14 introns. The results of chromosomal localization for human *factor XI* and *PK* (both chromosome 4) genes strongly corroborate a gene duplication event from a common ancestor for both *PK* and

Table 49-1: Biologic Functions of Contact Proteins

	Receptors	*Function*	*Human Deficiency*	*Mouse*$^{-/-}$	*References*
Prekallikrein	Binds to cells only in presence of HK	Cleaves pro-urokinase to u-PA (increased plasmin formation); kallikrein activates XII and hydrolyzes HK to HKa and BK	High aPTT; no excessive bleeding; not protected from thrombosis	Not yet generated	10,14
XII	gC1qR GP1b	Auto-activates to XIIa; XIIa cleaves prekallikrein to kallikrein; kallikrein hydrolyzes HK to HKa and BK	High aPTT; no excessive bleeding; not protected from thrombosis	Protected from platelet-related thrombosis but does not cause excessive hemorrhage	6,9
HK	GP1b, Mac-1, uPAR, gC1qR, CK-1	Antiadhesive displaces fibrinogen from platelets and leukocytes; VN blocks HKa binding to uPAR by forming complex with VN, thus blocking VN binding to $\alpha_v\beta_3$ on ECs. HK is proangiogenic (via BK); HKa inhibits FGF2-mediated EC proliferation (via induction of apoptosis) and VN-mediated EC migration	High aPTT; reduced/absent BK in plasma and urine; no excessive bleeding; not protected from thrombosis	Not yet generated, probably because two functional genes exist in mouse; rat deficiency results in decreased angiogenesis and increased thrombosis	21,24, 27–31, 33–35, 46,47

factor XI (11), about 270 million years ago. *PK* mRNA codes for a mature human protein of 609 amino acids with 58% identity to human factor XI. Plasma PK has an isoelectric point of 8.7 and an estimated plasma concentration of 42 ± 3 μg/mL (0.49 μM). Approximately 75% circulates bound to HK (12,13), and only 25% circulates as free PK. When analyzed on a nonreduced polyacrylamide gel electrophoresis in SDS, plasma PK is shown to consist of two components of Mr 85 and 88 kDa, which differ in the degree of glycosylation. During the conversion of PK to kallikrein, its active form, a single bond (Arg371–Ile372) is split, generating a heavy chain of 371 amino acids (53 kDa) and a light chain of 248 amino acids (36 and 33 kDa), held together by a single disulfide bond (14). PK has four tandem repeats in the amino-terminal portion of the molecule due to the linking of the first and sixth, second and fifth, and third and fourth half-cysteine residues present in each repeat. This arrangement results in four groups of 90 or 91 amino acids that are arranged in "apple" domains (15) which, in addition to PK, are only seen in factor XI. The fourth repeat contains two additional 1/2-Cys residues, forming an additional disulfide bridge. A short connecting region of nine amino acids follows repeat 4, culminating at the cleavage site between the light and heavy chains. The light chain contains the catalytic triad His415, Asp464, and Ser559.

KININOGENS

The two plasma kininogens, HK and low-molecular-weight kininogen (LK), are the products of a single-copy gene (16) in humans that maps to 3q26-qter. The human *kininogen* gene (27 kb) consists of 11 exons and produces unique mRNA, respectively, for HK and LK by alternative splicing. *HK* and *LK* both contain the coding region for exons 1 to 9, a part of exon 10 containing the *BK* sequence, and the first 12 amino acids after the carboxy-terminal amino acid of BK. Exon 11 codes for the 4-kDa light chain of LK, D5. Exon 10 contains the coding sequence for the 56-kDa light chain of HK, including domain 5_H and domain 6. The mRNA for human *HK* and *LK* are 3.5 and 1.7 kb, respectively. In mice, two functional genes exist (17). This situation explains why a knockout has yet to be reported. Recent studies of expression of the human *kininogen* gene have shown that it is strongly upregulated by agonists of the farnesoid X receptor (18).

The heavy chain of HK and LK consists of three tandem repeat units related to cystatin-designated domains 1, 2, and 3 (19,20), which are coded for by exons 1 to 3, 4 to 6, and 7 to 9, respectively. Domain 1 contains a low-affinity calcium-binding site, and domain 2 inhibits calpain and cathepsin L. Domain 3 can inhibit cathepsin L, B, and H but not calpain. More importantly, domain 3, which has been expressed in

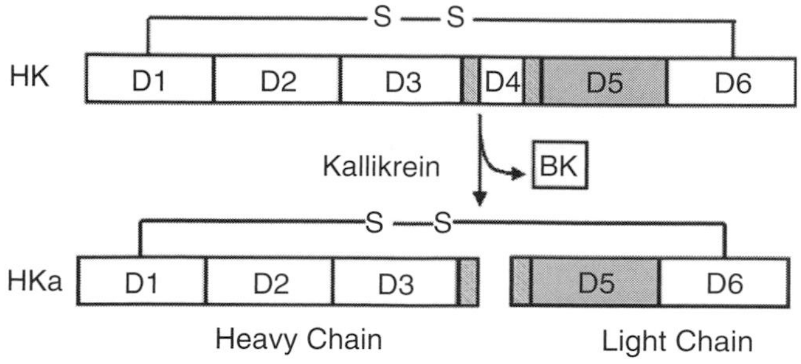

Figure 49.1. Domain organization of HK and HKa. HK is divided into six domains, D1 through D6. Plasma kallikrein cleaves HK, releasing the nonapeptide BK. The resulting protein (HKa) is a two-chain molecule held together by a single disulfide bond between D1 and D6.

Escherichia coli, binds to neutrophils, monocytes, platelets, and ECs with high (nM) affinity. Domain 4 contains BK, and its first five amino acids comprise a low-affinity binding site to platelets. Domain 5_H (unique to HK) also has been expressed in bacteria, and is a major cell binding structure that interacts with the same cells as domain 3 but with different receptors or at different sites in the same receptor. Domain 6 has a high-affinity site for binding PK, which mediates the formation of the binary complex of HK and PK. The latter complex circulates in plasma. When HK is cleaved by kallikrein, BK is released from D4, resulting in a heavy chain (D1–D3) and light chain (D5–D6) that remain covalently linked by a single disulfide bond (Figure 49.1).

KININOGEN RECEPTORS

HK interacts with platelets, ECs, monocytes, and neutrophils to exert discrete physiological effects. No other cells have yet been shown to bind kininogen. On platelets, two receptors have been identified. HK binding to the GPIb/IX/V complex (21) is primarily due to D3 and prevents high-affinity thrombin binding (22), thus exerting an antithrombotic effect both in vitro in human platelets and in vivo in rats deficient in HK (23). HK also binds to globular C_1q receptor (gC_1qR) on platelets. On neutrophils and monocytes, a major receptor is Mac-1 ($\alpha_m\beta_2$) (24) to which both D3 and D5 bind (25). HK can effectively inhibit fibrinogen binding, because the K_D is 5 to 15 nM, compared to that of fibrinogen, at 600 nM (26). The reactions of the contact system and fibrinolytic proteins on the endothelial surface are summarized in Figure 49.2.

HK and/or HKa binds to three major receptors on ECs: urokinase plasminogen activator receptor (uPAR) (27), gC1qR (28,29), and cytokeratin-1 (CK-1) (30), which colocalize on the EC surface. CK-1 interacts with either uPAR or gC1qR, but uPAR and gC1qR do not directly bind to each other. This pairing is consistent with the observation that CK-1 interacts with HK through D3 and D5, whereas uPAR (31) and gC1qR bind primarily HKa through D5. The binding to ECs is tight (K_d – 7 nM), requires Zn^{2+}, and results in two important reactions. The first is the cleavage of HK to HKa and BK by plasma kallikrein. Although the cleavage can occur in the fluid phase, it is limited by the presence of the protease inhibitors C1-inhibitor and α_2-macroglobulin. When kallikrein is bound to HK

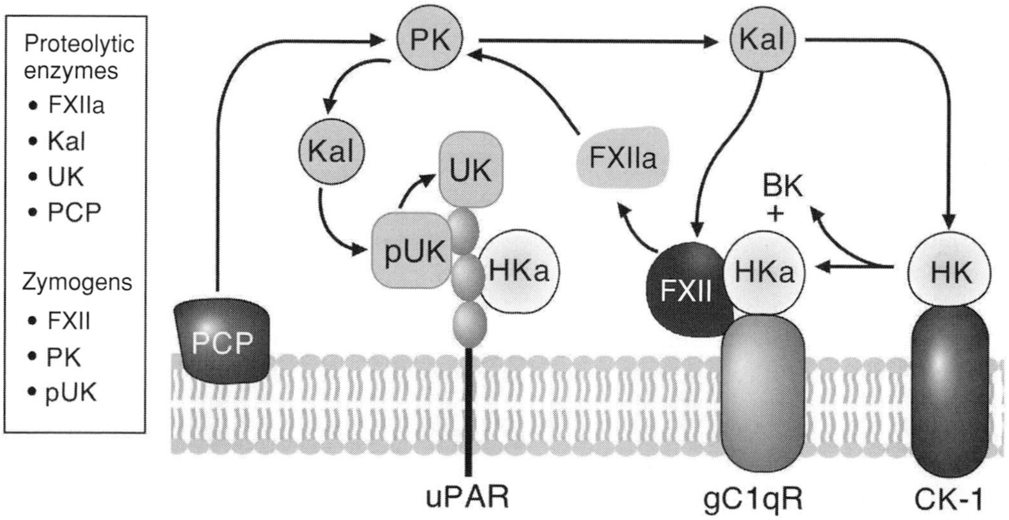

Figure 49.2. Interactions of contact system proteins on ECs – role in angiogenesis. Factor XII (FXII) binds to globular C1q receptor (gC1qR), which allows its autoactivation to FXIIa. FXIIa activates prekallikrein (PK) to kallikrein (Kal). Kal feeds back to cleave FXII to $FXII_a$. HK binds to cytokeratin (CK)-1. Kal cleaves HK to release bradykinin (BK), a stimulator of angiogenesis; HKa, an angiogenic inhibitor, remains. HKa binds to gC1qR and uPAR but not to CK-1. HK binds only to CK-1. Prolylcarboxypeptidase (PCP) activates PK to Kal. Kal cleave prourokinase (pUK) to urokinase (UK). pUK and UK bind to domain 1 of uPAR.

or receptors on ECs or leukocytes, the protease inhibitors are much less potent. Therefore HK is less susceptible to cleavage.

ROLE OF CONTACT SYSTEM IN FIBRINOLYSIS

Plasma kallikrein is brought to the endothelial surface by HK or HKa, because it binds to D6. Two enzymes can activate the zymogen, PK, to the active enzyme, plasma kallikrein. The first already discussed is factor XIIa, which binds separately to gC1qR on the endothelial surface (32). PK also is cleaved by a membrane-associated serine protease, prolylcarboxypeptidase (33), to the active enzyme. Plasma PK cleaves prourokinase into the two-chain activated form, a reaction independent of factor XIIa. Prourokinase then hydrolyzes plasminogen to form plasmin on the endothelial surface. Because HK binds to uPAR, as does prourokinase, this is an extremely efficient kinetic process; the cell binding also protects the enzyme against plasminogen activator inhibitor (PAI)-1. This cell surface plasmin formation also is enhanced by BK, which is a potent agonist for t-PA release. Plasmin can directly hydrolyze basement membrane proteins, such as laminin and fibronectin, and cleaves collagen indirectly by activating matrix metalloproteases.

HKa: INHIBITOR OF ANGIOGENESIS

HKa is a potent antiadhesive protein. In addition to displacing fibrinogen from neutrophils and platelets (24), HKa inhibits the binding of vitronectin (VN) to uPAR (27) and thus cell adhesion. In addition, HKa forms a complex with VN that inhibits VN binding to the integrin $\alpha_v\beta_3$ on ECs (34). Because detachment of ECs and penetration of the basement membrane are two of the early steps of angiogenesis, the role of HK and HKa in angiogenesis has been further investigated. We observed that HK-D5 inhibited the EC proliferation stimulated by fibroblast growth factor 2 (FGF2) as well as EC migration toward VN (35). These effects are separable, because two different peptides derived from HK-D5 have separate actions – HK440 to 455 inhibits proliferation, whereas HK486 to 502 inhibits migration. Kawasaki and colleagues (36) compared these results in vitro and demonstrated that peptides that contain the sequence HGK were active in inhibiting EC migration. The synergism of these two peptides predicted that both D5 and HKa would exhibit potent inhibition of angiogenesis. This prediction was confirmed in ovo in the chick chorioallantoic membrane (CAM), where D5 had an $IC_{50} = 100$ nM and HKa an $IC_{50} = 30$ nM. The mechanism by which proliferation was inhibited proved to be the induction of apoptosis (37). D5 decreased cyclin D1, a critical protein in the cell cycle required for the G1–S transition. HKa, which also led to apoptosis, produced upregulation of Cdc2 and cyclin A (38). Furthermore, the induction of apoptosis was regulated by extracellular matrix proteins. HKa-induced apoptosis occurs when cells are grown on VN but not on fibronectin (39). These inhibitory effects are mediated by HKa binding to uPAR and dissociation of the integrins $\alpha_v\beta_3$ or $\alpha_5\beta_1$. This reaction then disrupts the downstream intracellular signaling, including phosphorylation of focal adhesion kinase by the Syk kinase, Yes, and the subsequent phosphorylation of paxillin (Figure 49.3). Both ECs and cancer cells must invade tissues to form vessels and to metastasize, respectively. Recently, we showed that GST-D5 can inhibit the growth of human colon cancer in vitro by preventing the G1–S transition in the cell cycle (40). Browder and colleagues (41) have shown that proteins such as collagen and plasminogen, when proteolytically cleaved, give rise to the angiogenic inhibitors endostatin and angiostatin, respectively.

BRADYKININ: A STIMULATOR OF ANGIOGENESIS

Collagen is needed structurally for new vessel formation. Plasminogen, when activated, gives rise to plasmin, which enhances migration and angiogenesis. Because HKa and D5 were both antiangiogenic, we postulated and subsequently demonstrated that HK would be required for angiogenesis itself (42). BK was shown to stimulate neovascularization directly, consistent with our finding that inhibiting plasma kallikrein prevented the formation of BK and thus inhibits angiogenesis. Several other investigators, including Parenti and colleagues (43), also implicated BK as an angiogenic agonist which, by stimulating the B1 receptor, upregulated endogenous endothelial FGF2 by stimulating nitric oxide synthase (NOS). BK, through the B1 receptor, also stimulated human vascular smooth muscle cells (VSMCs) to secrete vascular endothelial growth factor (VEGF) (44). In contrast, Seegers and colleagues (45) indicated that the B2 receptor was involved in EC proliferation during acute synovitis. One mechanism for this effect could be the transactivation of the VEGF receptor, KDR/FLK-1 (46).

USE OF MONOCLONAL ANTIBODY TO KININOGEN TO INHIBIT ANGIOGENESIS

HK binding to ECs is required for cleavage, which produces the angiogenic stimulator, BK. Although HK could be cleaved on the surface of leukocytes, the rapid destruction of BK by kininases ensure that it acts as a local mediator. The finding has been exploited to "treat" experimental angiogenesis using a monoclonal antibody to human HK (C11C1). C11C1 recognizes orthinokininogen in chicken plasma and thus in the CAM model (42). C11C1 can inhibit FGF2 or VEGF-induced neovascularization or even BK stimulation of angiogenesis in the CAM. Mab C11C1 inhibits the growth of human fibrosarcoma on the CAM (42). These results suggested the possibility that mAb C11C1 could be used in the treatment of human tumors by targeting angiogenesis. We have recently shown that this monoclonal antibody blocks the binding to human

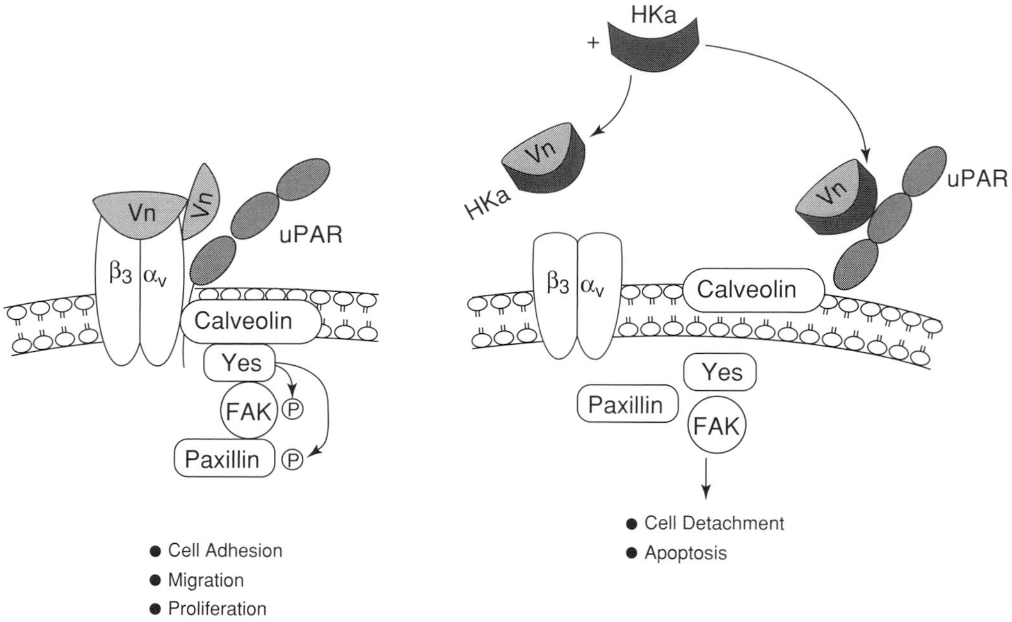

Figure 49.3. Action model illustrates interference with endothelial extracellular interactions by HKa and intracellular signaling required for angiogenesis. *Left:* Vitronectin (Vn) binds to both the integrin $\alpha_v\beta_3$ and uPAR, which also bind to each other. uPAR recruits caveolin, which in turn provides an environment for activation of Yes, which phosphorylates focal adhesion kinase (FAK), which further phosphorylates paxillin. These signals lead to EC adhesion, migration, and proliferation. *Right:* HKa competes with Vn for a binding site on uPAR (domain 2/3). HKa also forms a stoichiometric complex with Vn to prevent its binding to $\alpha_v\beta_3$. Both these steps disrupt the downstream signaling complex seen in the left panel.

and murine ECs and inhibits the growth of human colon carcinoma as a xenograft in an immunoincompetent (nude) mouse (47). Angiogenesis is not confined to supporting cancer cell growth. Chronic reactive arthritis in rats injected intraperitoneally with proteoglycan-polysaccharide for group A streptococcus is associated with angiogenesis as well, similar to that of human inflammatory arthritis. MAb C11C1 inhibited this reactive arthritis in Lewis rats by blocking the binding of HK and the subsequent proteolytic release of BK (48).

BRADYKININ AND HKa: PATHOPHYSIOLOGICAL ANTAGONISTS

Because the cleavage of HK on the surface of ECs gives rise to BK, an angiogenic stimulator, and HKa, a potent angiogenic inhibitor, it was important to determine which one predominates. Hayashi and colleagues (49) solved this conundrum by testing angiogenesis ex vivo and in vivo in Brown Norway Katholique rats, which are deficient in both HK and LK. They found that angiogenesis was suppressed in the kininogen-deficient rats compared with Brown Norway Kitasato rats, which have normal plasma kininogen concentrations. Because BK has a short (30 sec) half-life but is very potent, whereas HKa has a long (9 hrs) half-life but requires higher molar concentrations, it is likely that BK turns on angiogenesis while HKa turns it off (50).

KEY POINTS

- The interactions of the contact system with ECs can best be illustrated by their effect on angiogenesis.
- Cleavage of HK by plasma kallikrein on the EC surface gives rise to two mediators, BK, which turns on angiogenesis, and HKa, which turns it off.
- BK, by binding to B2 receptors, initiates proliferation by intracellular signaling pathways that stimulate VEGF secretion and bind to B1 receptors, which upregulate FGF2.
- HKa binds to uPAR and ligands of selected interacting integrins. HKa physically displaces these proteins, thereby disrupting phosphorylation of Syk kinases such as Yes and focal adhesion kinase and its substrate, paxillin, thereby inhibiting migration.

Future Goals

- To map downstream signaling pathways that would reveal nuclear effects
- To delineate further pathways involving extracellular matrix proteins

REFERENCES

1 Jacobsen S. Substrates for plasma kinin-forming enzymes in human, dog and rabbit plasmas. *Br J Pharmacol.* 1966;26:403–411.

2 White CA, Rees TD, Hurt WC. Factor XII (Hageman factor) deficiency in a periodontal surgery patient. *J Oral Med.* 1986;41:105.

3 Hathaway WE, Belhasen LP, Hathaway HS. Evidence for a new plasma thromboplastin factor. I. Case report, coagulation studies and physicochemical properties. *Blood.* 1965;26:521.

4 Colman RW, Bagdasarian A, Talamo RC, et al. Williams trait. Human kininogen deficiency with diminished levels of plasminogen proactivator and prekallikrein associated with abnormalities of the Hageman factor-dependent pathways. *J Clin Invest.* 1975;56:1650–1662.

5 Cool DE, MacGillivray RT. Characterization of the human blood coagulation factor XII gene. Intron/exon gene organization and analysis of the 5′-flanking region. *J Biol Chem.* 1987;262:13662–13673.

6 Cool DE, Edgell CJ, Louie GV, et al. Characterization of human blood coagulation factor XII cDNA: prediction of the primary structure of factor XII and the tertiary structure of beta-factor XIIa. *J Biol Chem.* 1985;260:13666–13676.

7 Farsetti A, Misiti S, Citarella F, et al. Molecular basis of estrogen regulation of Hageman factor XII gene expression. *Endocrinology.* 1995;136:5076–5083.

8 Citarella F, Mistri S, Felice A, et al. Estrogen induction and contact phase activation of human factor XII. *Steroids.* 1996;61:270–276.

9 Renne T, Pozgajova M, Gruner S, et al. Defective thrombus formation in mice lacking coagulation factor XII. *J Exp Med.* 2005;18:202(2):271–281.

10 Yu H, Anderson PJ, Freedman BI, et al. Genomic structure of the human plasma prekallikrein gene, identification of allelic variants, and analysis in end-stage renal disease. *Genomics.* 2000;69(2):225–234.

11 Veloso D, Shilling J, Shine J, et al. Recent evolutionary divergence of plasma prekallikrein and factor XI. *Thromb Res.* 1986;43:153–160.

12 Mandle R, Jr., Colman RW, Kaplan AP. Identification of prekallikrein and high molecular weight kininogen as a complex in human plasma. *Proc Natl Acad Sci USA.* 1976;73:4179–4183.

13 Scott CF, Colman RW. Function and immunochemistry of prekallikrein-high molecular weight kininogen complex in plasma. *J Clin Invest.* 1980;65:413–421.

14 Mandle R, Jr., Kaplan AP. Hageman factor substrates. Human plasma prekallikrein: mechanism of activation by Hageman factor and participation in Hageman factor-dependent fibrinolysis. *J Biol Chem.* 1977;252:6097–6104.

15 McMullen BA, Fujikawa K, Davie EW. Location of the disulfide bonds in human plasma prekallikrein: the presence of four novel apple domains in the amino-terminal portion of the molecule. *Biochemistry.* 1991;30:2050–2056.

16 Kitamura N, Kitagawa H, Fukushima D, et al. Structural organization of the human kininogen gene and a model for its evolution. *J Biol Chem.* 1985;260:8610–8617.

17 Cardoso CC, Garrett T, Cayla C, et al. Structure and expression of two kininogen genes in mice. *Biol Chem.* 2004;385(3–4):295–301.

18 Zhao A, Lew JL, Huang L, et al. Human kininogen gene is transactivated by the farnesoid X receptor. *J Biol Chem.* 2003;278(31):28765–28770.

19 Kellermann J, Lottspeich F, Henschen A, et al. Completion of the primary structure of human high-molecular-mass kininogen. The amino acid sequence of the entire heavy chain and evidence for its evolution by gene triplication. *Eur J Biochem.* 1986;154:471–478.

20 Salvesen G, Parkes C, Abrahamson M, et al. Human low-Mr kininogen contains three copies of a cystatin sequence that are divergent in structure and in inhibitory activity for cysteine proteinases. *Biochem J.* 1986;234:429–434.

21 Bradford HN, DeLa Cadena RA, Kunapuli SP, et al. Human kininogens regulate thrombin binding to platelets through the GPIb-IX–V complex. *Blood.* 1997;90:1508–1515.

22 Puri RN, Zhou F, Hu CJ, et al. High molecular weight kininogen inhibits thrombin-induced platelet aggregation and cleavage of aggregin by inhibiting binding of thrombin to platelets. *Blood.* 1991;77:500–507.

23 Colman RW, White JV, Scovell S, et al. Kininogens are antithrombotic proteins in vivo. *Arterioscler Thromb Vasc Biol.* 1999;19:2245–2250.

24 Gustafson EJ, Lukasiewicz H, Wachtfogel YT, et al. High molecular weight kininogen inhibits fibrinogen binding to cytoadhesins of neutrophils and platelets. *J Cell Biol.* 1989;109:377–387.

25 Wachtfogel YT, DeLa Cadena RA, Kunapuli SP, et al. High molecular weight kininogen binds to Mac-1 on neutrophils by its heavy chain (domain 3) and its light chain (domain 5). *J Biol Chem.* 1994;269:19307–19312.

26 Gustafson EJ, Schmaier AH, Wachtfogel YT, et al. Human neutrophils contain and bind high molecular weight kininogen. *J Clin Invest.* 1989;84:28–35.

27 Colman RW, Pixley RA, Najamunnisa S, et al. Binding of high molecular weight kininogen to human ECs is mediated via a site within domains 2+3 of the urokinase receptor. *J Clin Invest.* 1997;100:1481–1487.

28 Joseph K, Ghebrehiwet B, Peerschke EI, et al. Identification of the zinc-dependent EC binding protein for high molecular weight kininogen and factor XII: identity with the receptor that binds to the globular "heads" of C1q (gC1q-R). *Proc Natl Acad Sci USA.* 1996;93:8552–8557.

29 Joseph K, Ghebrehiwet B, Kaplan AP. Cytokeratin 1 and gC1qR mediate high molecular weight kininogen binding to ECs. *Clin Immunol.* 1999;92(3):246–255.

30 Hasan AAK, Zisman T, Schmaier AH. Identification of cytokeratin 1 as a binding protein and presentation receptor for kininogens on ECs. *Proc Natl Acad Sci USA.* 1998;95(7):3615–3620.

31 Mahdi F, Shariat-Madar Z, Kuo A, et al. Mapping the interaction between high molecular weight kininogen and the urokinase plasminogen activator receptor. *J Biol Chem.* 2004;279:16621–16628.

32 Joseph K, Shibayama Y, Ghebrehiwet B, et al. Factor XII-dependent contact activation on ECs and binding proteins gC1qR and cytokeratin 1. *Thromb Haemost.* 2001;85(1):119–124.

33 Shariat-Madar Z, Mahdi F, Schmaier AH. Identification and characterization of prolylcarboxypeptidase as an EC prekallikrein activator. *J Biol Chem.* 2002;277(20):17962–17969.

34 Chavakis T, Kanse SM, Lupu F, et al. Different mechanisms define the antiadhesive function of high molecular weight kininogen in integrin- and urokinase receptor-dependent interactions. *Blood.* 2000;96(2):514–522.

35 Colman RW, Jameson BA, Lin Y, et al. Domain 5 of high molecular weight kininogen (kininostatin) down-regulates EC

proliferation and migration and inhibits angiogenesis. *Blood.* 2000;95(2):543–550.

36 Kawasaki M, Maeda T, Hanasawa K, et al. Effect of His-Gly-Lys motif derived from domain 5 of high molecular weight kininogen on suppression of cancer metastasis both in vitro and in vivo. *J Biol Chem.* 2003;278(49):49301–49307.

37 Guo YL, Wang S, Colman RW. Kininostatin, an angiogenic inhibitor, inhibits proliferation and induces apoptosis of human ECs. *Arterioscler Thromb Vasc Biol.* 2001;21(9):1427–1433.

38 Wang S, Hasham MG, Isordia-Salas I, et al. Upregulation of Cdc2 and cyclin A during apoptosis of ECs induced by cleaved high-molecular-weight kininogen. *Am J Physiol Heart Circ Physiol.* 2003;284(6):H1917–H1923.

39 Guo YL, Wang S, Cao DJ, et al. Apoptotic effect of cleaved high molecular weight kininogen is regulated by extracellular matrix proteins. *J Cell Biochem.* 2003;89(3):622–632.

40 Bior A, Pixley RA, Colman RW. Domain 5 of kininogen inhibits proliferation of human colon cancer cell line (HCT-116) by interfering with G1/S in the cell cycle. *J Thromb Haemost.* 2006; [Epub ahead of print].

41 Browder T, Folkman J, Pirie-Shepherd S. The hemostatic system as a regulator of angiogenesis. *J Biol Chem.* 2000;275(3):1521–1524.

42 Colman RW, Pixley RA, Sainz I, et al. Inhibition of angiogenesis by antibody blocking the action of proangiogenic high-molecular-weight kininogen. *J Thromb Haemost.* 2003;1:164–170.

43 Parenti A, Morbidelli L, Ledda F, et al. The bradykinin/B1 receptor promotes angiogenesis by up-regulation of endogenous FGF-2 in endothelium via the nitric oxide synthase pathway. *FASEB J.* 2001;15(8):1487–1489.

44 Knox AJ, Corbett L, Stocks J, et al. Human airway smooth muscle cells secrete vascular endothelial growth factor: up-regulation by bradykinin via a protein kinase C and prostanoid-dependent mechanism. *FASEB J.* 2001;15(13):2480–2488.

45 Seegers HC, Avery PS, McWilliams DF, et al. Combined effect of bradykinin B2 and neurokinin-1 receptor activation on EC proliferation in acute synovitis. *FASEB J.* 2004;18(6):762–764.

46 Thuringer D, Maulon L, Frelin C. Rapid transactivation of the vascular endothelial growth factor receptor KDR/Flk-1 by the bradykinin B2 receptor contributes to endothelial nitric-oxide synthase activation in cardiac capillary ECs. *J Biol Chem.* 2002; 277(3):2028–2032.

47 Song JS, Sainz IM, Cosenza SC, et al. Inhibition of tumor angiogenesis in vivo by a monoclonal antibody targeted to domain 5 of high molecular weight kininogen. *Blood.* 2004;104(7):2065–2072.

48 Espinola RG, Uknis A, Sainz IM, et al. A monoclonal antibody to high-molecular weight kininogen is therapeutic in a rodent model of reactive arthritis. *Am J Pathol.* 2004;165(3):969–976.

49 Hayashi I, Amano H, Yoshida S, et al. Suppressed angiogenesis in kininogen-deficiencies. *Lab Invest.* 2002;82(7):871–880.

50 Guo YL, Colman RW. Two faces of high molecular weight kininogen (HK) in angiogenesis: Bradykinin turns it on and cleaved HK (HKa) turns it off. *J Thromb Haemost.* 2005;3:670–676.

50

Opioid Receptors in Endothelium

Kalpna Gupta and Elliot J. Stephenson

Vascular Biology Center, University of Minnesota Medical School, Minneapolis

Opioids are well known for their highly effective analgesic properties and for their addiction-causing tendencies. The term *opioid* is derived from the Greek word *opion*, meaning "poppy juice." Opium is obtained from the green pods of the opium poppy plant. The alkaloids morphine, codeine, and thebaine are derived from opium. Animals ranging from worms to humans produce endogenous opioid peptides. The term *opioid* encompasses both the naturally occurring (endogenous) opioids from opium poppy and animals and semisynthetic and synthetic alkaloid drugs that interact with opioid receptors. To date, four different opioid receptors (ORs) have been cloned: μ, δ, and κ (also known as MOR, DOR, and KOR, respectively), and nociceptin/orphanin FQ (ORL1) (1). All four receptors are expressed in the vascular endothelium, where they are uniquely positioned to transduce signals from circulating opioids (synthesized centrally or by peripheral non-neuronal cells) before they reach their target tissue, the brain. Compared to our advanced understanding of the opioid–neural axis, remarkably little is known about opioid signaling in the endothelium. Recent studies indicate that opioid–opioid receptor interactions in endothelial cells (ECs) result in increased production of nitric oxide (NO), pointing to a potential role for opioid signaling in mediating vasodilation, cell adhesion, cell survival, and angiogenesis (Figure 50.1). In this chapter, we discuss the nature and mechanisms of these interactions and explore the utility of this signaling axis as a therapeutic target in vascular diseases.

HISTORY: FROM MESOPOTAMIA TO MODERN MEDICINE

Opioids are inextricably linked to the history of mankind. In 3400 BC, the Sumerians of Mesopotamia called the opium poppy *Hul Gil*, the "joy plant" for its euphoric effect. In 460 BC, Hippocrates dismissed the magical attributes of opium but acknowledged its usefulness in treating internal and gynecological diseases as well as epidemics. Opium first began to traverse the globe when Alexander the Great conquered India and introduced opium there in 330 BC. The first description of opium's properties and medical use for treating diarrhea and sexual debility appeared in the ancient East Indian medical treatises *The Shodal Gadanigrah*, *Sharangdhar Samahita*, and *The Dhanvantri Nighantu* in 1200 AD. Starting in 1750, the British East India Company dominated the lucrative export of opium from India to China and, in 1800, the British Levant Company purchased nearly half of Turkey's opium harvest strictly for importation to Europe and the United States. According to the United Nations, Afghanistan currently supplies 87% of the world's opium.

In 1803, Friedrich Sertürner of Paderborn, Germany discovered the active ingredient in opium by dissolving it in acid, then neutralizing it with ammonia. The result: the alkaloid morphine. Morphine was lauded as "God's own medicine" for its reliability, long-lasting effects, and safety; physicians believed that opium had finally been perfected and tamed. Ironically, this global drug that ancient and medieval humans used wisely is now abused recklessly. Modern research has resurrected the ancient status of these drugs to only a small extent because of the predominantly narrow focus on analgesia, addiction, and tolerance. In contrast, in blood vessels where opioids have a close encounter with the endothelium, they remain inadequately studied.

OPIOID RECEPTORS

Overview

Although opioid binding sites were first proposed in the early 1950s and identified in mammalian brain in 1973, the first opioid receptor DOR was cloned in 1992, followed by the cloning of the other three receptors (1). All four opioid receptors are present on the endothelium (2,3). Each opioid receptor can be activated by several endogenous or exogenous opioids, although a degree of selectivity exists, as summarized in Table 50-1. Although the activity of all four receptors can be antagonized by naloxone, specific antagonists exist for each individual opioid receptor.

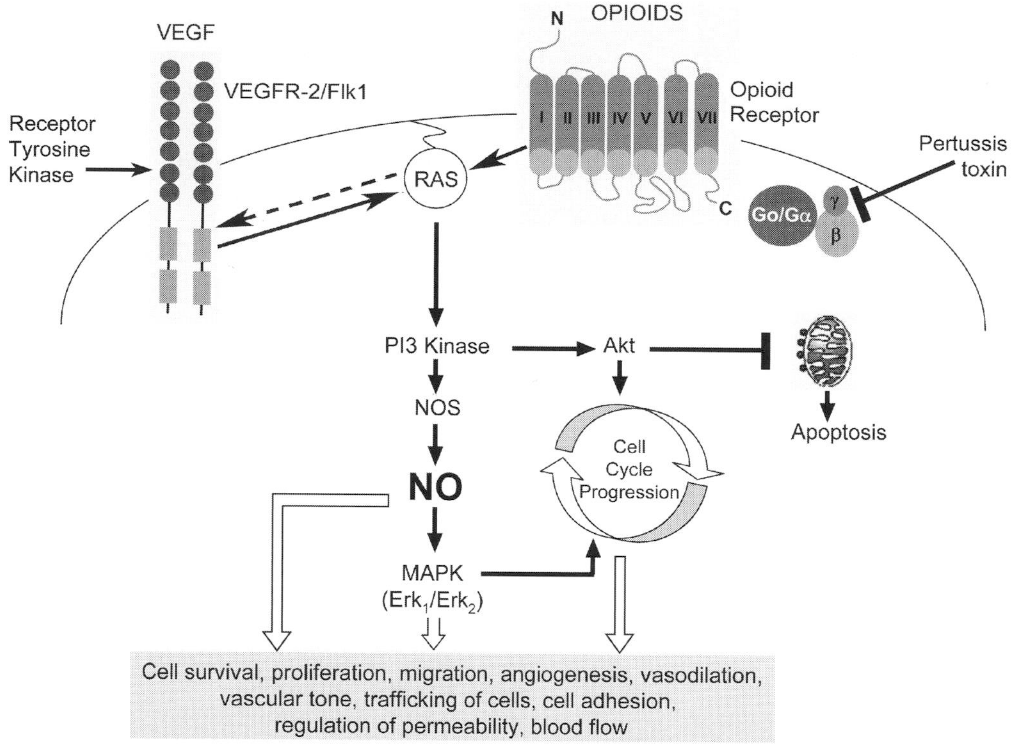

Figure 50.1. Schematic representation of opioid signaling in endothelium. Opioids interact with 7-transmembrane domain receptors coupled to G-proteins and stimulate the MAPK/ERK pathway via a NO-dependent mechanism. On the other hand, opioid receptors activate Akt phosphorylation and promote cell cycling. Opioid receptor signaling resembles FLK-1 signaling, which is in part caused by their transactivation of FLK-1. Activation of these intracellular signaling events leads to endothelial proliferation, survival, angiogenesis, and other vascular effects that can have opposite effects on vascular pathobiology.

Opioid receptors are seven transmembrane domain G-protein–coupled receptors (GPCRs) which are coupled to the pertussis toxin-dependent G_i/G_o type of G-proteins. Upon receptor activation, G-proteins interact with multiple effector systems, leading to the inhibition of adenyl cyclase and voltage-gated calcium (Ca^{2+}) channels and stimulation of G-protein–activated inwardly rectifying potassium (K^+) channels. However, chronic activation of opioid receptors leads to adenyl cyclase superactivation and increased levels of cyclic adenosine monophosphate (cAMP). Because cAMP is a survival factor for ECs, acute and chronic activation of opioid receptors can be a matter of cell death and survival, respectively (4,5). Thus, it is important to consider the duration of exposure when studying the role of opioids in endothelial biomedicine.

Regulation of Opioid Receptor Expression

Receptor densities and activity are critical determinants of opioid effects. Opioid receptor expression is regulated by agonist exposure and growth factors in a heterogeneous manner (6,7). Nerve growth factor (NGF) stimulates DOR expression in PC12 pheochromocytoma cells (8); basic fibroblast growth factor (bFGF) (also known as FGF2) and platelet-derived growth factor (PDGF)-BB induce KOR in oligodendrocytes (9). Emerging data on ECs show that interleukin (IL)-1β and vascular endothelial growth factor (VEGF) upregulate *MOR* mRNA in human umbilical vein ECs (HUVECs) and mouse retinal ECs (mRECs) (10–12). Cocaine, ethanol, and naloxone alter opioid receptor gene expression in NG108–15 neuroblastoma hybrid cells (11). *MOR* mRNA and protein are barely detectable in normal rodent skin, but are highly upregulated in wounds (Gupta, unpublished observations). In these wounds, MOR is localized in epidermal keratinocytes and in vascular and lymphatic ECs. Because keratinocytes produce high levels of VEGF in newly formed wound epidermis (13), these data suggest that injury-mediated secretion of VEGF results in autocrine and paracrine induction of opioid receptors in keratinocytes and ECs, respectively. In pathophysiological conditions, a vascular microenvironment replete with proinflammatory cytokines and growth factors may similarly modulate opioid receptors on endothelium (Figure 50.2).

Regulation of Opioid Receptor Activity

Like classical GPCRs, opioid receptors are regulated by a cascade of events involving receptor phosphorylation by GPCR kinases (GRKs) and subsequent β-arrestin recruitment, thus uncoupling the receptor from its G protein, followed by receptor endocytosis. Endocytosis may lead to degradation and secondary downregulation of the receptor. However, not all endocytosed opioid receptors are degraded; some are recycled

Table 50-1: Opioid Receptors and Their Functions

Opioid Receptor	*Exogenous Agonists*	*Endogenous Opioids*	*Inducers*	*Signaling Pathways and Phenotype*	*Cellular Expression*	*Putative Functions*
μ (MOR)	Morphine Hydromorphone Fentanyl Meperidine Methadone Oxycodone Sufentanil DAMGO Etorphine	Endomorphins Endorphins Demorphin	IL-1 (ECs) VEGF (ECs) IL-6 (neuroblastoma cells SH SY5Y)	PI3K/AKT, NF-κB (neurons) Ras/Rac PI3K/AKT, MAPK, NO in endothelial cells (assuming this is the receptor morphine was working through) Transactivation of FLK-1, PDGF-β and EGFR receptors Vasodilatation STAT3 GATA4	ECs Cardiac myocytes Keratinocytes Neuronal cells Mesangial cells MCF7 tumor cells Monocytes	Wound healing Heart: cardioprotective Kidney: ? Lung: respiratory depression Brain Tumors Skin Blood vessels: angiogenesis Lymphatic: lymphatic vessels Peripheral nerves: pain Retina/CNS: neovascularization/ analgesia HIV-1 Tat-induced death of glial precursors and astrocytes
δ (DOR)	DADLE DPDPE BW373U86 SNC 80	Enkephalins Deltorphins	NGF (PC12) IL-4 (immune neuronal cells)	STAT3, Akt K^+ATP channels Ras/Rac/MAPK survival PKC, CaMKII Heterodimerization with β-adrenergic receptor	ECs Adult rat ventricular myocardium Mesangial cells	Cardioprotection Cardiogenesis IPC induced antiarrhythmic Ischemia–reperfusion
κ (KOR)	Dynorphin 1 U-50,488 U-69,593 ICI 199,441 ICI 197,067 6′-GNTI	Dynorphin	PDGF, bFGF (oligodendrocytes)	PI3K K^+ATP channels PKC Survival (cardioprotective) MAPK; GATA4 Heterodimerization with β-adrenergic receptor	ECs Adult rat ventricular myocardium Embryonal pluripotent stem cells Neonatal hearts	Cardioprotection Kidney
Nociceptin/ orphanin FQ (ORL1)	[Arg^{14}, Lys^{15}] nociceptin ZP 120 Ro 64–6198	Nociceptin/ orphanin FQ	Ciliary neurotrophic factor (CNTF)	Akt Ras/Rac/MAPK survival PKC, CaMKII		

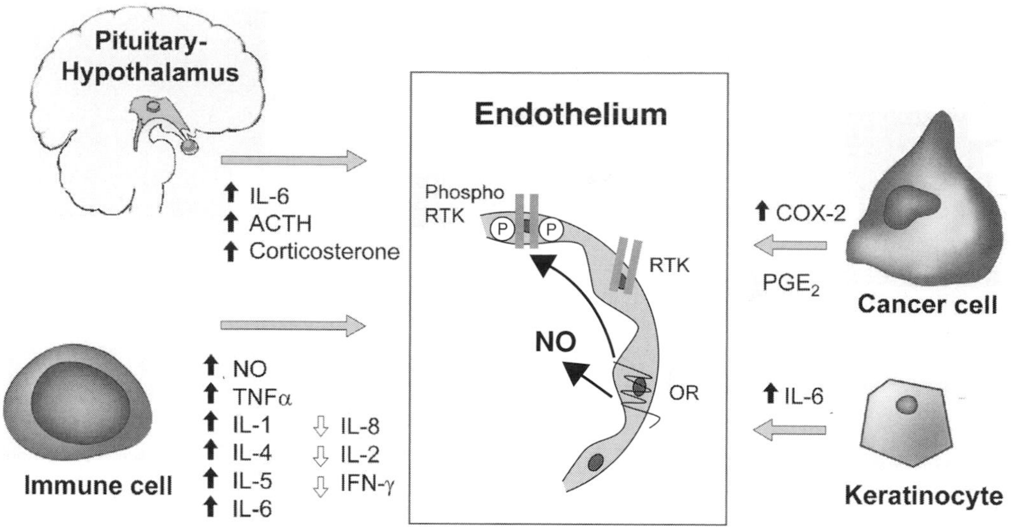

Figure 50.2. Opioid receptors on different cells in the neighborhood of endothelium, as well as distant cells in the pituitary-hypothalamus, modulate the endothelial microenvironment by secreting cytokines, hormones, NO, and other substances in response to external stimuli or exogenously administered or endogenously produced opioids. ACTH, adrenocorticotropic hormone.

in a cell-, agonist-, and receptor-specific manner. For example, whereas DOR is transported deeper into the endocytotic pathway and rapidly degraded by lysosomes, MOR is recycled following endocytosis (14,15). The "net signal" for relative activity of the receptor versus endocytosis is termed *relative activity versus endocytosis* (RAVE) (16). RAVE defines the relative ability of an opioid agonist to induce cell signaling. In the case of morphine, high RAVE values seem to reflect the relative inability of this agent to induce receptor endocytosis.

Endothelial–Opioid Interactions

MOR was cloned in 1993; its presence on the endothelium was shown by binding studies in 1995 (17), and its expression in ECs was recently confirmed by sequencing (18). The presence of DOR on ECs has been inferred from studies examining the binding of OR agonists to human aortic and saphenous vein ECs (19). In contrast, other studies found that the activity of DOR agonist peptides on porcine aortic ECs was naloxone-insensitive (20), that specific opioid receptor binding of ^{3}H-morphine on bovine aortic ECs was not detectable, and that morphine was located almost exclusively within the nucleus (21). We demonstrated that MOR, DOR, and KOR are expressed on HUVECs, human dermal microvascular ECs (HDMECs), and mRECs using reverse transcription polymerase chain reaction (RT-PCR) with MOR-, DOR-, and KOR-specific primers (2,12). Importantly, we and others have shown that activation of MOR on ECs stimulates NO-mediated signaling, vasodilatation, and angiogenesis (2,17).

Pasi and colleagues (22) demonstrated that both β-endorphin (BEP) and morphine inhibit angiogenesis in a chorioallantoic membrane (CAM) assay. However, the concentrations of morphine tested (5–50 μg/4 μL; equivalent to 1.65–16.5 mM) were several log-fold higher than those found in the serum/plasma of patients being treated for pain (2 nM–3.5 μM). We observed that morphine at concentrations up to 100 μM stimulates EC proliferation, but is cytotoxic at concentrations $\geq$1 mM (2). The biological effects of opioids on the endothelium may therefore be highly dose-dependent. In any case, it has become imperative to gain a better understanding of endothelial–opioid interactions because opioids are clinically used in many diseases in which endothelial activity, angiogenesis, and blood vessel wall pathophysiology play a central role.

OPIOID SIGNALING

Opioids and Opioid Receptors Promote Growth and Survival

Opioids directly induce cell proliferation, as shown in CHO, COS, and C6 glioma cells transfected with MOR, DOR, or KOR, as well as in primary adult hippocampal progenitors (1,2,23). In lymphocytes, morphine suppresses apoptosis by blocking p53-mediated death signaling via MOR Bcl-2/Bax signaling (24). In general, ORs stimulate phosphoinositide-3-kinase (PI3K)-dependent signaling, leading to proliferation and survival. For example, in CHO cells, MOR agonists stimulate a PI3K-dependent signaling cascade in a manner similar to VEGF signaling in ECs (25). DAMGO, a MOR agonist, stimulates Akt via PI3K, which inhibits apoptosis in neurons (23). Long-term treatment with DAMGO increases nuclear factor (NF)-κB activity in cortical neurons (26). NF-κB is known to be activated by G-protein heterodimers ($G_{\beta\gamma}$) through the PI3-Kinase/Akt pathway (27). Opioid receptors also can modulate signals generated by classical growth factors. DOR activation potentiates the proliferation of serum-induced neuroblastoma cells (28) and serum- and growth factor-induced proliferation in CHO cells stably transfected with the receptor (29).

Our laboratory was the first to identify an opioid-induced intracellular signaling pathway in ECs (Figure 50.1) (2). Morphine and other agonists of MOR, DOR, and KOR, which are pertussis toxin-dependent GPCRs, stimulate mitogen-activated protein kinase/extracellular signal-regulated kinase (MAPK/ERK) phosphorylation and proliferation of HDMECs. The MOR agonist DAMGO also potentiates serum-induced proliferation of HDMECs. Morphine stimulates MAPK/ERK phosphorylation and HDMEC proliferation via PI3K and NO-dependent mechanisms. Furthermore, morphine also stimulates Akt and promotes EC survival and cell cycling by upregulating cyclin D1. Among the growth factors, VEGF alone has been shown to stimulate angiogenesis via NO-dependent MAPK phosphorylation (30). Thus, opioids act like growth factors in promoting proliferation, migration, and survival via the same signaling pathway utilized by VEGF in endothelium (see Figure 50.1). These in vitro proangiogenic observations are further strengthened by the in vivo promotion of angiogenesis by morphine in tumors and wounds, described in the next section. More recently, we confirmed that morphine signaling is similar to that of VEGF in mRECs (12). In these cells, morphine also stimulates signal transducer and activator of transcription (STAT)3 directly as well as via transactivation of VEGF receptor 2 (VEGFR2)/FLK-1.

Transactivation

Although opioids act via typical GPCRs, growth factors act via phosphorylation of their receptor protein tyrosine kinases (PTKs). GPCRs transactivate PTK receptors through multilevel crosstalk between intermediate signaling pathways (31). Therefore, it is likely that our observation of opioid-induced proangiogenic, growth- and survival-promoting signaling is in part caused by the transactivation of growth factor receptors (see Figure 50.1). Indeed, our recent studies show that VEGFR2/FLK-1 and PDGF-β receptors are transactivated by MOR in mRECs (12). These observations are consistent with the finding that MOR transactivates the epidermal growth factor receptor (EGFR) in astrocytes (32,33). This also raises the possibility of constitutive activation of PTKs by opioid receptors in the regulation of the normal endothelial phenotype. Opioid receptors may therefore represent therapeutic targets for indirectly modulating the effect of PTK receptors such as VEGF receptors in angiogenesis.

Adhesion and Immunomodulation

In response to stress, leukocytes express opioids (predominantly BEP) that bind to peripheral opioid receptors, leading to local pain relief in animals and humans (34). This study showed that in complete Freund's adjuvant-induced inflammation in hind paws, 43% to 58% of CD45$^+$ hematopoietic cells express opioid peptides. At the inflammatory site, activated ECs expressing P-selectin, E-selectin, and platelet-endothelial cell adhesion molecule (PECAM)/CD31 are juxtaposed with BEP-expressing leukocytes. This model suggests that endothelial selectins and integrins mediate the recruitment of opioid-expressing inflammatory cells.

Experiments from our laboratory show that morphine by itself does not stimulate significant adhesion molecule expression by HDMECs in vitro, but modulates tumor necrosis factor (TNF)-α–induced vascular cell adhesion molecule (VCAM)-1 and E-selectin expression (Figure 50.3). TNF-α promotes leukocyte rolling, firm adhesion and transmigration, in part via the increased expression and/or function of endothelial P-selectin, E-selectin, and VCAM-1. Incubation of HDMECs with TNF-α for 4 hours resulted in 15- and 10-fold increases in E-selectin and VCAM-1 expression, respectively, as compared to unstimulated cells ($p < 0.0001$, Tukey-Kramer method; unstimulated (US) vs. TNF-α or VCAM-1, at 4 hours). After 18 hours of incubation, VCAM-1 expression had returned to baseline (US) levels, whereas E-selectin had decreased by 50% compared with 4 hours. This time-dependent reduction in VCAM-1 and E-selectin expression was blocked by coincubation with morphine ($p < 0.0001$). These findings suggest that chronic exposure of ECs to morphine results in sustained TNF-α–induced expression of adhesion molecules.

Long-term exposure to morphine alone increases the adhesion of monocytes to saphenous vein and internal thoracic artery endothelium by 30%, the same extent as that induced by human immunodeficiency virus (HIV) gp120 (35). Prolonged exposure to both morphine and gp120 resulted in a 100% increase in adhesion, indicating that morphine acts synergistically with gp120. Thus, the effects of morphine on the endothelium depend on duration of exposure and microenvironmental conditions. Chronic exposure to morphine also potentiates the leukocyte–endothelial adhesion response to IL-1β in rat mesentery (36). In contrast, short-term exposure to morphine induces an NO-dependent conformational change (rounding of cells) in spontaneously activated human monocytes, granulocytes, and arterial ECs, rendering them less adhesive (37).

Adhesion molecule expression is regulated by NF-κB. Opioids modulate NF-κB expression in several cell types. Morphine stimulates NF-κB activation in rat heart and human NT2-N neurons (38,39). The long-term treatment of cortical neurons with the MOR agonist DAMGO increases NF-κB activity (26). However, Roy and colleagues found that at μM-concentrations, morphine inhibited NF-κB activation in macrophages (40).

Opioids also are believed to have immunomodulatory effects (41). Morphine induces apoptosis in macrophages by activating transforming growth factor (TGF)-β. Data cited above and our own observations on the direct upregulation of VCAM-1 and E-selectin by morphine favor the thesis that chronic morphine treatment stimulates EC adhesion in a proinflammatory microenvironment.

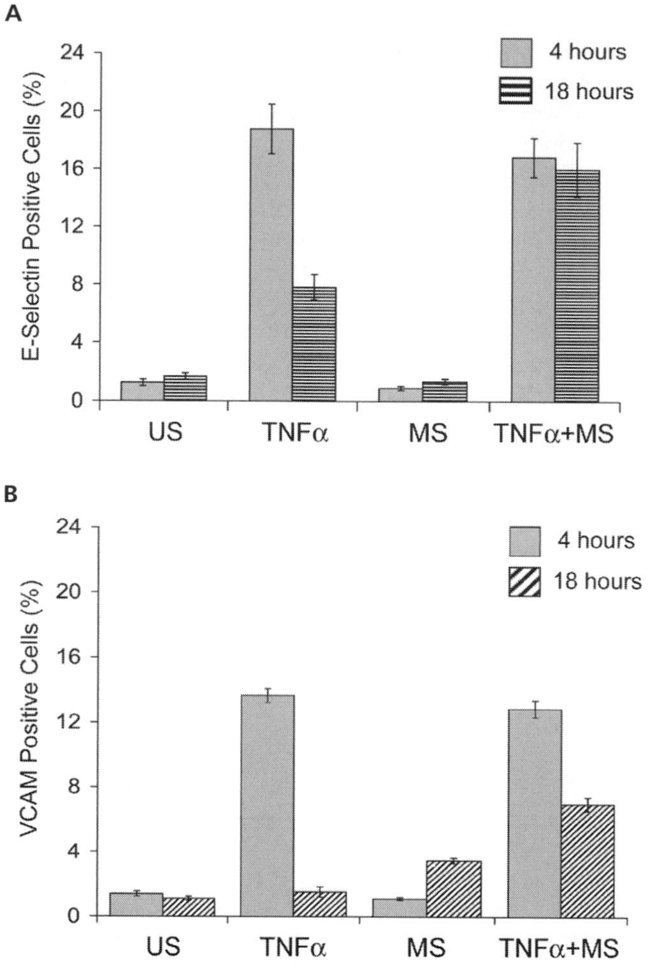

Figure 50.3. Morphine upregulates adhesion molecule expression on human dermal microvascular ECs (HDMECs) in a proinflammatory microenvironment. HDMECs were incubated with 1 μM morphine sulfate (MS) in the presence or absence of 100 ng/mL TNF-α for 4 or 18 hours in serum-deplete conditions, then FACS analyzed for E-selectin and VCAM-1 expression. TNF-α stimulated a significant increase in both E-selectin and VCAM-1 expression after 4 hours of incubation versus unstimulated (US) cells ($p < 0.0001$). After 18 hours of incubation, TNF-α-induced stimulation of E-selectin was inhibited by about 60% and VCAM-1 by 100%. MS did not have any effect on E-selectin or VCAM-1 except for a modest increase in VCAM-1 expression after 18 hours versus US ($p < 0.0001$). However, in the presence of MS, TNF-α-induced downregulation of both adhesion molecules at 18 hours was significantly inhibited (TNF vs. TNF + MS at 18 hours for E-selectin or VCAM-1). $n = 6$; Mean ± SEM; Statistical analysis: Multiple comparisons by Tukey-Kramer method.

MEDICAL IMPLICATIONS

Ischemia–Reperfusion Injury and Cardiovascular Disease

In cardiovascular disease, ischemia and reperfusion injury are directly associated with inflammatory changes, vasoregulation, and vessel wall physiology. We found that 3 weeks of morphine treatment in clinically relevant doses upregulates endothelial NO synthase (eNOS), inducible NOS (iNOS), cyclooxygenase (COX)-2, and heme oxygenase (HO)-1, and increases perfusion and vasodilatation in the mouse kidney (42). These opioid-induced mechanisms also mediate cardioprotection in animal models (43,44). Opioids and opioid receptors have been shown to provide cytoprotection following ischemia reperfusion injury and hypoxic insult in models ranging from murine cardiac ischemia to hibernating polar bears (45). A dramatic increase in the circulating levels of opioid peptides in hibernating mammals, and a reversal of hibernation by opioid receptor antagonists, provides an exciting example of the protective role of opioids in a hypoxia-tolerant state (45).

The induction of vasodilatation via vascular smooth muscle cell (VSMC) relaxation is one of the principal roles of endothelial NO. Opioids can directly stimulate eNOS as well as iNOS (17,46). Although eNOS-derived NO stimulates the relaxation of VSMCs, iNOS-derived NO from immune cells can cause desensitization of VSMCs to NO. Interferon (IFN)-γ has been proposed to stimulate iNOS in T cells and to inhibit eNOS (47). Morphine negatively regulates *IFN-γ* promoter activity and inhibits IFN-γ secretion by the activated T cells via MOR (48). Therefore, opioids can stimulate endothelial NO production on one hand and impede iNOS-derived IFN-γ–dependent NO production on the other. Thus, opioids and opioid receptors may be important in preventing endothelial dysfunction and even arteriosclerosis.

Opioids, through their different receptors, provide preconditioning and cytoprotection against ischemia in the heart (49). Opioids activate DOR and KOR, which leads to protein kinase C (PKC) activation. Activated PKC acts as an amplifier of the preconditioning stimulus and stabilizes, through phosphorylation, the open state of the mitochondrial K_{ATP} channel and sarcolemmal K_{ATP} channel. The opening of K_{ATP} channels ultimately elicits cytoprotection by decreasing cytosolic and mitochondrial Ca^{2+} overload. Chronic opioid stimulation opens K_{ATP} channels and also prevents apoptosis by activating the survival signal Akt. Studies also demonstrate that KOR stimulates cardioprotective signaling by activating mitochondrial-K_{ATP} and PKC (50). In vivo, morphine reduces ischemia–reperfusion injury in isolated rat heart. The cardioprotective effects of morphine are mediated by a local opioid receptor-K_{ATP} channel-linked mechanism in rat hearts. After ischemia and reperfusion, coronary flow (CF), heart rate (HR), left ventricular pressure (LVP), and the first derivative left ventricular pressure (LVP/dt_{max}) of isolated hearts decreased significantly in this study (51). After morphine preconditioning, CF, HR, LVP, and LVP/dt_{max} increase and infarct size decreases significantly ($p < 0.01$). These studies suggest that opioids, via their different receptors, provide protection against ischemic tissue injury.

All three opioid receptors appear to be involved in cardioprotection, based on the effect of specific agonist-induced activity in the heart. Stimulation of DOR after hemorrhage, during the recompensatory period, leads to improved mean arterial pressure (MAP) (52). In freely moving hemorrhaged rats, the average HR gain (bpm/mm Hg) after 2 mg/kg

Delt-Dvar (Deltorphin-Dvariant, a selective DOR agonist) treatment was greater and the BP_{50} (BP at one-half the HR range) was significantly lower than after saline treatment. The KOR agonist U50,488, along with other KOR agonists, shows antiarrhythmic actions in rat hearts, which appears to be mediated by the blockade of Na^+, K^+, or Ca^{2+} channels, independent of KOR (53). Endomorphin-2, a MOR agonist, elicits depressor and bradycardic responses and inhibits baroreflex function when injected into the rostral ventrolateral medullary pressor area (RVLM) (54). The presence of MOR and endomorphins has been demonstrated in the general area encompassing the RVLM. Unilateral microinjections of endomorphin-2 (0.0125–0.5 mmol/L) into the RVLM elicited decreases in MAP (16–30 mm Hg) and HR (12–36 bpm) that lasted for 2 to 4 minutes. The effects of endomorphin-2 were mediated via MOR because preadministration of naloxonazine (1 mmol/L), an antagonist of MOR, abolished these responses. These effects are consistent with a known hyperpolarizing effect of opioid peptides on RVLM neurons. Furthermore, two chimeric peptides YGGFMKKKFMRFamide (YFa) and [D-Ala2] YAGFMKKKFMRFamide ([D-Ala2] YFa), consisting of a sequence of met-enkephalin (55), showed a dose-dependent fall in MAP in the dose-range of 13 to 78 μmol/kg in rats. After naloxone treatment (5 mg/kg), the vasodepressor effect of [D-Ala2] YFa and YFa was only partially blocked, when compared to met-enkephalin. Partial blockade of the vasodepressive effect of YFa and [D-Ala2] YFa by naloxone may be attributed to interaction of these chimeric peptides with opioid receptors and others, such as adrenergic receptors and D-analogue receptors. A short-term, high-dose infusion of BEP improved left ventricular ejection fraction (LVEF), reduced systemic vascular resistance, blunted neurohormonal activation, and stimulated the growth hormone (GH)/insulin-like growth factor (IGF)-1 axis in patients with mild to moderate congestive heart failure (56). These changes were paralleled by a significant increase in plasma levels of glucagons, GH, and IGF-1, and a significant decrease in plasma levels of endothelin and catecholamines. In another human trial, long-term opiate exposure was suggested to mitigate coronary artery disease severity and its fatal consequences (57). These data indicate that opioid receptors provide cardioprotection directly as well as indirectly by modulating the balance between vasodilators and vasoconstrictors. Moreover, their short-term versus long-term effects are different, which warrants a systematic examination of their activity in pathological states to exploit their potential for therapy in cardiovascular disease.

In the Brain

Hypoxia releases NO/cGMP from glial cells, which in turn increases opioid release. Thus, activation of neuronal NOS (nNOS) contributes to increased cerebrospinal fluid concentration of the endogenous opioids met-enkephalin and leu-enkephalin during hypoxia in the newborn pig. Together NO, opioids, and ATP-sensitive K^+ channels contribute to hypoxia-induced pial artery dilation (58). In vitro, morphine preconditions Purkinje cells against cell death under in vitro simulated ischemia–reperfusion conditions (59). Therefore, endogenous opioids and opioid receptors may be critical in vasoregulation and ischemia–reperfusion in the brain, an area that has not been explored yet for its therapeutic potential.

In Sexual Debility

Our current understanding of opioid-induced NO and vasodilatation may be the basis for the use of opium for sexual debility as far back as 1200, described in an ancient East Indian medical treatise. (Coincidentally, the drug name Viagra is derived from the East Indian Sanskrit word *viaghra*, meaning tiger.) Penile erection results from an arousal-induced synthesis of NO in nonadrenergic-noncholinergic nerves (NANC), ECs, and cavernosal VSMCs. We speculate that opioids may stimulate nNOS and eNOS, leading to vasodilation and relaxation of cavernosal VSMCs, which in turn engorge the corpora cavernosa with blood. The subcellular mechanism by which tumescence occurs also involves NO-induced activation of soluble guanylate cyclase, increased cyclic guanosine monophosphate (cGMP) levels, and activation of cGMP-dependent protein kinase (PKG). PKG phosphorylates numerous ion channels and pumps, each promoting a reduction in cytosolic calcium. In particular, PKG activates high-conductance calcium-sensitive K^+ channels, which hyperpolarize the arterial and cavernosal VSMC membranes, causing relaxation. Morphine directly stimulates the arginine/NO/cGMP/K^+_{ATP}/PKG/ pathway (60). These potential effects of opioids on the genitourinary system provide the rationale for investigation of their possible therapeutic role in sexual debility.

From Bench to Bedside – Wound Healing

It appears that injury turns on a natural switch to provide pain relief and healing by secreting BEP and upregulating MOR. The MOR agonist BEP is upregulated at the wound site in the epidermal keratinocytes and the proinflammatory cells (61). We found that MOR is upregulated several-fold in the wound margins and endothelium, as compared with undamaged skin in mice and rats (unpublished observations). Furthermore, we showed that topical application of the MOR agonist drugs morphine, hydromorphone, and fentanyl promotes healing of ischemic wounds in rats (46). These effects were not seen on systemic administration of the same drugs. Although opioids promote wound healing, the opioid receptor antagonist naloxone inhibits it and also antagonizes the healing effect of fentanyl, lending credence to the argument that MOR mediates the healing process. Opioid-induced wound healing was associated with increased cellular proliferation of both ECs and non-ECs, angiogenesis, and increased collagen formation in opioid-treated as compared with vehicle-treated wounds. These cellular changes are accompanied by upregulation of iNOS, eNOS, and VEGFR2/FLK-1 in opioid-treated wounds.

We hypothesize that topical opioids can be safely used to treat wounds and nonhealing ulcers, such as those observed in diabetes and sickle cell disease. The small amounts of opioids applied peripherally on ischemic wounds remain undetectable in the circulation and are therefore unlikely to cause dependence or addiction. We have initiated a Phase 1 clinical trial at the University of Minnesota to evaluate the capability of topical opioids to heal wounds (62).

From Bedside to Bench – Tumor Progression

Tumor progression and metastasis are angiogenesis-dependent. Our observations that the chronic administration of clinically relevant doses of morphine stimulate tumor angiogenesis and MCF7 breast tumor growth in mice (2) suggest that opioid receptor activity is critical in cancer progression. We observed that naloxone impedes tumor growth in this model by antagonizing estrogen receptor (ER) activity directly and also via crosstalk between ER and MOR in MCF7 cells (63). ER also is expressed on ECs, and its activation stimulates angiogenesis. Thus, naloxone-like compounds may inhibit tumor growth by targeting both endothelial and tumor cells. Opioids are currently the drugs of choice for the treatment of severe pain in cancer. Therefore, it is imperative to determine the clinical effect of opioids on tumor progression and metastasis. If confirmed in human studies, strategies to antagonizing the inadvertent promotion of tumor growth by opioids will need to be explored.

Lacunae

Opioid pharmacology and biochemistry in the endothelium, including tolerance, internalization, and receptor dimerization and recycling, remain underexplored. Activation of MOR may be critical in vascular function in both normal physiology and pathophysiological conditions. Understanding these mechanisms will help create a safe vascular environment for use of exogenous opioids.

Our observations that opioid-induced activation of opioid receptors can transactivate various other receptors, including those for hormones such as estrogen and cytokines such as PDGF and VEGF, may provide an explanation for the limited success of many antiangiogenic therapies that target growth factors, their receptors, and PTKs. More effective blockage of endothelial proliferation and angiogenesis may require concurrent inhibition of both growth factor receptors and opioid receptors.

CONCLUSION

This century is the beginning of a new era for appreciating the endothelial effects of opioids and their receptors, until recently believed to be confined to the behavioral and neuronal world. The existence of all four defined opioid receptors on endothelium is well documented. Of these, MOR appears to play a more critical role in vascular biology for two reasons: (a) MOR stimulates NO, which plays a key role in vascular function and (b) opioid analgesics are MOR agonists, and are used in diseases where vascular pathophysiology and angiogenesis are critical in disease prevention or progression. The proangiogenic, survival-promoting, and NO-mediated signaling and modulation of adhesion molecule expression by opioids in the endothelium suggest that opioid receptors provide novel therapeutic targets for endothelial dysfunction and angiogenesis-dependent diseases. Results from our laboratory and others are beginning to show the therapeutic use of opioids in wound healing and cardiovascular disease, which provide a driving force to unravel the mysteries of opioid receptor activity in the endothelium.

KEY POINTS

- Opioids may influence diverse angiogenesis-dependent diseases by activating opioid receptors on endothelium.
- Opioid receptors stimulate MAPK/ERK and PKB/Akt signaling and promote endothelial proliferation and survival.
- MOR stimulates vasoregulatory and cytoprotective signaling and mitigates ischemia–reperfusion injury.

Future Goals

- To elucidate the role of each individual opioid receptor in the endothelium
- To develop opioid receptor antagonists to inhibit angiogenesis in cancer and proliferative retinopathy. Stimulate angiogenesis using opioid receptor agonists to promote healing of chronic ischemic ulcers
- To mitigate the consequences of tissue ischemia by opioid receptor mediated vasoregulation

ACKNOWLEDGMENTS

The authors thank Philip M. Portoghese, Ph.D. and Robert P. Hebbel, MD for their helpful discussions from time to time, and Pankaj Gupta, MD for critical review of the manuscript. I remain indebted to Tasneem Poonawala, Chunsheng Chen, Marc L. Weber, Mariya Farooqui, and others in my lab for their dedication to opioid-endothelial research. We thank Carol Taubert for preparing the illustrations and word processing the manuscript, and Tyson Rogers for the statistical analysis. This work is supported by National Institutes of Health grants HL68802 and CA109582, the Susan G. Komen Breast Cancer Foundation, and funding from the Division of Hematology, Oncology and Transplantation, University of Minnesota Medical School.

REFERENCES

1 Waldhoer M, Bartlett SE, Whistler JL. Opioid receptors. *Annu Rev Biochem.* 2004;73:953–990.
2 Gupta K, Kshirsagar S, Chang L, et al. Morphine stimulates angiogenesis by activating proangiogenic and survival-promoting

signaling and promotes breast tumor growth. *Cancer Res.* 2002; 62:4491–4498.

3 Granata F, Potenza RL, Fiori A, et al. Expression of OP4 (ORL1, NOP1) receptors in vascular endothelium. *Eur J Pharmacol.* 2003;482:17–23.

4 Polte T, Schroder H. Cyclic AMP mediates endothelial protection by nitric oxide. *Biochem Biophys Res Commun.* 1998;251:460–465.

5 Arai M, Thurman RG, Lemasters JJ. Contribution of adenosine A(2) receptors and cyclic adenosine monophosphate to protective ischemic preconditioning of sinusoidal ECs against storage/reperfusion injury in rat livers. *Hepatology.* 2000;32:297–302.

6 Law PY, Loh HH. Regulation of opioid receptor activities. *J Pharmacol Exp Ther.* 1999;289:607–624.

7 Stiene-Martin A, Zhou R, Hauser KF. Regional, developmental, and cell cycle-dependent differences in mu, delta, and kappa-opioid receptor expression among cultured mouse astrocytes. *Glia.* 1998;22:249–259.

8 Hellewell SB, Bowen WD. A sigma-like binding site in rat pheochromocytoma (PC12) cells: decreased affinity for (+)-benzomorphans and lower molecular weight suggest a different sigma receptor form from that of guinea pig brain. *Brain Res.* 1990;527:244–253.

9 Tryoen-Toth P, Gaveriaux-Ruff C, Maderspach K, et al. Regulation of kappa-opioid receptor mRNA level by cyclic AMP and growth factors in cultured rat glial cells. *Brain Res Mol Brain Res.* 1998;55:141–150.

10 Chang SL, Moldow RL, Wu G-d, et al. The association between opiates and cytokines. *Adv Exp Med Biol.* 1998;437:4–6.

11 Jenab S, Inturrisi CE. Ethanol and naloxone differentially upregulate delta opioid receptor gene expression in neuroblastoma hybrid (NG108–15) cells. *Brain Res Mol Brain Res.* 1994;27:95–102.

12 Chen C, Farooqui M, Gupta K. Morphine stimulates VEGF-like signaling in mouse retinal ECs. *Curr Neurovas Res.* 2006;3:171–180.

13 Brown LF, Yeo KT, Berse B, et al. Expression of vascular permeability factor (vascular endothelial growth factor) by epidermal keratinocytes during wound healing. *J Exp Med.* 1992;176:1375–1379.

14 Finn AK, Whistler JL. Endocytosis of the mu opioid receptor reduces tolerance and a cellular hallmark of opiate withdrawal. *Neuron.* 2001;32:829–839.

15 Whistler JL, Enquist J, Marley A, et al. Modulation of postendocytic sorting of G protein-coupled receptors. *Science.* 2002;297: 615–620.

16 Whistler JL, Chuang HH, Chu P, et al. Functional dissociation of mu opioid receptor signaling and endocytosis: implications for the biology of opiate tolerance and addiction. *Neuron.* 1999; 23:737–746.

17 Stefano GB, Hartman A, Bilfinger TV, et al. Presence of the mu3 opiate receptor in ECs. Coupling to nitric oxide production and vasodilation. *J Biol Chem.* 1995;270:30290–30293.

18 Cadet M, Mantione KJ, Stefano GB. Molecular identification and functional expression of mu3, a novel alternatively spliced variant of the human mu opiate receptor gene. *J Immunol.* 2003; 170:5118–5123.

19 Stefano GB, Salzet M, Hughes TK, et al. Delta-2 opioid receptor subtype on human vascular endothelium uncouples morphine stimulated nitric oxide release. *Int J Cardiol.* 1998;64(Suppl 1): S43–S51.

20 Arendt RM, Schmoeckel M, Wilbert-Lampen U, et al. Bidirectional effects of endogenous opioid peptides on endothelin release rates in porcine aortic EC culture: mediation by delta opioid receptor and opioid receptor antagonist-insensitive mechanisms. *J Pharmacol Exp Ther.* 1995;272:1–7.

21 Melzig MF, Heder G, Siems WE, et al. Stimulation of endothelial angiotensin-converting enzyme by morphine via non-opioid receptor mediated processes. *Pharmazie.* 1998;53:634–637.

22 Pasi A, Qu BX, Steiner R, et al. Angiogenesis: modulation with opioids. *Gen Pharmacol.* 1991;22:1077–1079.

23 Tegeder I, Geisslinger G. Opioids as modulators of cell death and survival – unraveling mechanisms and revealing new indications. *Pharmacol Rev.* 2004;56:351–369.

24 Suzuki S, Chuang LF, Doi RH, et al. Morphine suppresses lymphocyte apoptosis by blocking p53-mediated death signaling. *Biochem Biophys Res Commun.* 2003;308:802–808.

25 Polakiewicz RD, Schieferl SM, Gingras AC, et al. Mu opioid receptor activates signaling pathways implicated in cell survival and translational control. *J Biol Chem.* 1998;273:23534–23541.

26 Hou YN, Vlaskovska M, Cebers G, et al. A mu-receptor opioid agonist induces AP-1 and NF-kappa B transcription factor activity in primary cultures of rat cortical neurons. *Neurosci Lett.* 1996;212:159–162.

27 Xie P, Browning DD, Hay N, et al. Activation of NF-kappa B by bradykinin through a Galpha(q)- and Gbeta gamma-dependent pathway that involves phosphoinositide 3-kinase and Akt. *J Biol Chem.* 2000;275:24907–24914.

28 Law PY, Bergsbaken C. Properties of delta opioid receptor in neuroblastoma NS20Y: receptor activation and neuroblastoma proliferation. *J Pharmacol Exp Ther.* 1995;272:322–332.

29 Law PY, McGinn TM, Campbell KM, et al. Agonist activation of delta-opioid receptor but not mu-opioid receptor potentiates fetal calf serum or tyrosine kinase receptor-mediated cell proliferation in a cell-line-specific manner. *Mol Pharmacol.* 1997;51:152–160.

30 Murohara T, Asahara T, Silver M, et al. Nitric oxide synthase modulates angiogenesis in response to tissue ischemia. *J Clin Invest.* 1998;101:2567–2578.

31 Wetzker R, Bohmer F-D. Transactivation joins multiple tracks to the ERK/MAPK cascade. *Nat Rev.* 2003;4:651–657.

32 Belcheva MM, Szucs M, Wang D, et al. Mu-opioid receptor-mediated ERK activation involves calmodulin-dependent epidermal growth factor receptor transactivation. *J Biol Chem.* 2001; 276:33847–33853.

33 Belcheva MM, Tan Y, Heaton VM, et al. Mu opioid transactivation and down-regulation of the epidermal growth factor receptor in astrocytes: Implications for mitogen-activated protein kinase signaling. *Mol Pharmacol.* 2003;64:1391–1401.

34 Machelska H, Brack A, Mousa SA, et al. Selectins and integrins but not platelet-EC adhesion molecule-1 regulate opioid inhibition of inflammatory pain. *Br J Pharmacol.* 2004;142:772–780.

35 Stefano GB, Salzet M, Bilfinger TV. Long-term exposure of human blood vessels to HIV gp120, morphine, and anandamide increases endothelial adhesion of monocytes: uncoupling of nitric oxide release. *J Cardiovasc Pharmacol.* 1998;31:862–868.

36 Chang SL, Moldow RL, House SD, et al. Morphine affects the brain-immune axis by modulating an interleukin-1 dependent pathway. In: Friedman H, Eisenstein T, Madden J, Sharp B, eds. *AIDS, Drugs of Abuse and the Neuroimmune Axis.* New York: Plenum Publishing Corp., 1996:35–42.

37 Magazine HI, Liu Y, Bilfinger TV, et al. Morphine-induced conformational changes in human monocytes, granulocytes, and

ECs and in invertebrate immunocytes and microglia are mediated by nitric oxide. *J Immunol.* 1996;156:4845–4850.
38 Wang X, Douglas SD, Commons KG, et al. A non-peptide substance P antagonist (CP-96,345) inhibits morphine induced NF-kappa B promoter activation in human NT2-N neurons. *J Neurosci Res.* 2004;75:544–553.
39 Frassdorf J, Weber NC, Obal D, et al. Morphine induces late cardioprotection in rat hearts in vivo: the involvement of opioid receptors and NF-kappa B. *Anesth Analg.* 2005;101:934–941.
40 Roy S, Cain KJ, Chapin RB, et al. Morphine modulates NF kappa B activation in macrophages. *Biochem Biophys Res Commun.* 1998;245:392–396.
41 Vallejo R, de Leon-Casasola O, Benyamin R. Opioid therapy and immunosuppression: a review. *Am J Ther.* 2004;11:354–365.
42 Arerangaiah R, Chalasani N, Udager AM, et al. Opioids induce renal abnormalities in tumor bearing mice. *Nephron Exp Nephrol.* 2007:In press.
43 Kodani E, Xuan YT, Shinmura K, et al. Delta opioid receptor induced lat preconditioning is mediated by cyclo-oxygense-2 in conscious rabbits. *Am J Physiol Heart Circ Physiol.* 2002;283: H1943–H1957.
44 Patel HH, Hsu AK, Gross GJ. COX-2 and iNOS in opioid induced delayed cardioprotection in the intact rat. *Life Sci.* 2004;75:129–140.
45 Peart JN, Gross ER, Gross GJ. Opioid induced preconditioning: recent advances and future perspectives. *Vasc Pharmacol.* 2005; 42:211–218.
46 Poonawala T, Levay-Young BK, Hebbel RP, et al. Opioids heal ischemic wounds in the rat. *Wound Repair and Regen.* 2005;13: 165–174.
47 Mitchell RN, Lichtman AH. The link between IFN-gamma and allograft arteriopathy: is the answer NO? *J Clin Invest.* 2004;114: 762–764.
48 McBride SM, Smith-Sonneborn J, Oeltgen P, et al. Delta2 opioid receptor agonist facilitates mean arterial pressure recovery after hemorrhage in conscious rats. *Shock.* 2005;23:264–268.
49 Zaugg M, Lucchinetti E, Uecker M, et al. Anaesthetics and cardiac preconditioning. Part I. Signalling and cytoprotective mechanisms. *Br J Anaesth.* 2003;91:551–565.
50 Cao CM, Xia Q, Tu J, et al. Cardioprotection of interleukin-2 is mediated via kappa-opioid receptors. *J Pharmacol Exp Ther.* 2004;309:560–567.
51 Shi E, Jiang X, Bai H, et al. Cardioprotective effects of morphine on rat heart suffering from ischemia and reperfusion. *Chin Med J (Engl).* 2003;116:1059–1062.
52 Wang J, Barke RA, Charboneau R, et al. Morphine negatively regulates interferon-gamma promoter activity in activated murine T cells through two distinct cyclic AMP-dependent pathways. *J Biol Chem.* 2003;278:37622–37631.
53 Pugsley MK, Walker MJ, Wong TM. Cardiovascular actions of the kappa-agonist U-50,488H in the absence and presence of opioid receptor blockade. *Br J Pharmacol.* 1992;105:521–526.
54 Kasamatsu K, Sapru HN. Attenuation of aortic baroreflex responses by microinjections of endomorphin-2 into the rostral ventrolateral medullary pressor area of the rat. *Am J Physiol Regul Integr Comp Physiol.* 2005;[Epub ahead of print].
55 Hanif K, Fahim M, Pavar MC, et al. Hypotensive effect of novel chimeric peptides of met-enkephalin and FMRFa. *Regul Pept.* 2005;125:155–161.
56 Cozzolino D, Sasso FC, Salvatore T, et al. Acute effects of beta-endorphin on cardiovascular function in patients with mild to moderate chronic heart failure. *Am Heart J.* 2004;148:E13.
57 Marmor M, Penn A, Widmer K, et al. Coronary artery disease and opioid use. *Am J Cardiol.* 2004;15:1295–1297.
58 Armstead WM. Relationship among NO, the KATP channel, and opioids in hypoxic pial artery dilation. *Am J Physiol.* 1998;275: H988–H994.
59 Lim YJ, Zheng S, Zuo Z. Morphine preconditions Purkinje cells against cell death under in vitro simulated ischemia-reperfusion conditions. *Anesthesiol.* 2004;100:562–568.
60 Sachs D, Cunha FQ, Ferreira SH. Peripheral analgesic blockade of hypernociception: activation of arginine/NO/cGMP/protein kinase G/ATP-sensitive K+ channel pathway. *Proc Natl Acad Sci USA.* 2004;101:3680–3685.
61 Bigliardi PL, Sumanovski LT, Buchner S, et al. Different expression of mu-opiate receptor in chronic and acute wounds and the effect of beta-endorphin on transforming growth factor beta type II receptor and cytokeratin 16 expression. *J Invest Dermatol.* 2003;120:145–152.
62 Gupta K, Zhang J. Angiogenesis: a curse or cure? *Postgrad Med J.* 2005; 81:236–42.
63 Farooqui M, Geng ZH, Stephenson EJ, et al. Naloxone acts as an antagonist of estrogen receptor in MCF7 cancer cells. *Mol Cancer Ther.* 2006;5:611–620.

51

Snake Toxins and Endothelium

Jay W. Fox* and Solange M.T. Serrano[†]

* *University of Virginia School of Medicine, Charlottesville;*
[†] *Instituto Butantan, São Paulo, Brazil*

HISTORY OF SNAKE TOXINS AND THEIR EFFECTS ON ENDOTHELIUM

Venomous snakes are by far the most feared of venomous animals. Four families of venomous snakes exist: the Viperidae (which includes vipers, adders, and pit vipers), the Elapidae (which includes the cobras, mambas, kraits, coral snakes, and sea snakes), the Colubridae (which includes the venomous rear-fanged snakes such as the boomslang and the twig snake), and the Atractaspididae – the family termed the *burrowing snakes*. In most developed countries, fatalities associated with snake envenoming are rare; however, in many tropical regions in places such as Africa, South America, and Southeast Asia, snake envenomation often is associated with a relatively high mortality rate. By definition, venomous snakes are snakes that produce venom. Approximately 2,700 different snake species exist, of which nearly one-fifth are venomous (1). Interestingly, many of the venomous snakes rarely, if ever, envenomate humans and hence are falsely considered nonvenomous. From the earliest observations of viperid snake envenomations, it was obvious that severe coagulopathic consequences occurred at both the local and systemic level. These coagulopathies, although not necessarily absent in envenomations by other snakes from other venomous families, were certainly more pronounced with snakes from Viperidae. In 1664, Francisco Redi (1626–1697) discovered that it was the venom from the snake that caused the "direful effects" associated with the bite of a snake (2). The Italian scientist Felice Fontana (1720–1805) was the first to study the effect of snake venom on blood and recognized that the effect of venom on blood coagulation properties was directly related to the toxicity of the venom (3). More recently, S. Weir Mitchell (1829–1914) focused his attention on the viperid group of rattlesnake venoms. Mitchell and colleagues demonstrated that venoms are complex mixtures of proteins whose various compositions in different snakes give rise to the different potencies observed (4). Further Mitchell and Reichert were the first to show that viperid venoms were capable of causing local and in severe cases systemic ecchymosis, necrosis, edema, and hemorrhage owing to alterations in the capillary wall with subsequent loss of blood and hypotension (5). Thus, beginning with the first correlation of snakebite-associated, venom-induced pathologies to the association of complex venom proteins being responsible for the dramatic symptoms associated with viperid envenoming, it has been clear that viperid venoms are potent factors for the perturbation of the prey's circulatory system with the endothelium being one of many tissues affected by the venom.

VIPERID VENOMS, TOXINS, AND ENVENOMATION

Snake venoms represent a complex mixture of proteins, peptides, nucleic acids, and salts that vary based on snake species, populations of the same species, environment, and even sex. The toxins that are contained in the venom function for prey acquisition, digestion, and defense. With regard to prey acquisition, clearly, the ability of the venom to rapidly immobilize the prey by paralysis or circulatory system collapse is highly advantageous. Recent proteomic analyses of several individual viperid venoms indicate the presence of more than 1,000 proteins (6,7).

Most toxins in venoms can be categorized as belonging to the following groups: (a) peptides that function as neurotoxins, membrane destabilizers, or proteinase inhibitors; and (b) hydrolases and amino acid oxidases including phospholipases, hyaluronidases, acetylcholinesterases, and proteinases. This last subgroup, proteinases, likely plays the greatest role in the disruption of hemostasis. Both serine and metalloproteinases are abundant in viperid venoms (8,9), and function to degrade vessel walls and affect coagulation, respectively. The serine proteinases that function as kininogenases also may be important in this process in light of their hypotensive properties.

The pathological consequences of viperid snakebite are varied and reflect the complexity of the venom. Historically, the study of venom has been approached by using isolated toxins from the venom to characterize their biochemical and biological properties. Although this strategy has proven helpful in delineating specific properties of many toxins, it falls short in revealing the role of a specific toxin in the context of the whole venom. The pathological effects observed with whole venom are attributable to both additive and synergistic functions of individual toxins, and hence it is experimentally difficult to know how an individual toxin functions in the context of the hundreds of other biologically active proteins and peptides in the venom. Furthermore, using that approach one only understands what one is assaying for; in other words, you will only find what you look for. Thus, when discussing individual toxins in this chapter, the reader must keep in mind two important caveats: (a) most of the studies described are with isolated toxins, and hence the biological activities ascribed to the isolated toxin may have different outcomes in the presence of the whole venom; and (b) unless specifically noted, the lack of mention of a particular activity for a protein does not imply that it has been tested for and deemed absent.

Often, snakebites, including those of the viperid snake, are "dry" bites, meaning no venom is injected. However, in productive bites by viperid snakes large quantities of venom can be injected with severe consequences, including death. Although not as predominant in viperid venoms as in elapid venoms, nevertheless neurotoxic symptoms can occur. Generally, damage of skeletal muscle occurs, including muscle pain and myoglobinuria, which in severe cases can lead to renal failure. Hemostatic alterations are most common with viperid envenoming, which often leads to hypocoagulable blood owing to direct hydrolysis of fibrinogen and/or the activation of the endogenous fibrinolytic system by venom proteinases, which may lead to systemic bleeding. Other hallmarks of viperid envenomation are local swelling, hemorrhage, edema, and necrosis. Generally, these features are caused by venom metalloproteinases, which act on vessel walls and the basement membranes and stromal extracellular matrix that surround them, to give rise to blood extravasion into the tissues. Large blisters can occur along with the subcutaneous bleeding, together contributing to local tissue necrosis. Hypotensive shock is a usual consequence of viperid envenoming because of the loss of fluids from edema and hemorrhage and hypotensive/vasodilatory peptide production. As one can imagine, many of these symptoms are interrelated and thus underscore the complex interplay of venom toxins as they give rise to the pathologies associated with viperid envenoming.

SNAKE TOXINS THAT DIRECTLY AFFECT THE ENDOTHELIUM[1]

Toxins found in viperid venom that have been shown to directly affect the endothelium include (a) snake venom metalloproteinases (SVMPs), (b) snake venom serine proteinases (SVSPs), (c) natriuretic peptides, (d) sarafotoxins (SRTXs), and (e) vascular endothelial growth factor (VEGF)–like proteins (Table 51-1). Toxins that play an indirect role in affecting the endothelium include (a) SVMPs, (b) disintegrins, (c) SVSPs, (d) C-type lectins, and (e) bradykinin potentiating peptides.

Snake Venom Metalloproteinases

One of the major groups of toxins present in viperid venoms is the SVMPs, which have been rather thoroughly characterized as to their structure and function (8). The SVMPs are classified into four subgroups (PI–PIV) that vary from one another based on the presence of additional nonmetalloproteinase domains on the C-terminal side of the metalloproteinase domain (Figure 51.1; for color reproduction, see Color Plate 51.1). The nonmetalloproteinase domains (disintegrin, disintegrin-like, cysteine-rich, and C-type lectin) likely play roles in targeting the SVMPs to specific substrates. For example, the disintegrin and disintegrin-like domains may target the PII and PIII SVMPs to integrins (10), and the cysteine-rich domain may target the PIII and PIV SVMPs to von Willebrand factor (vWF) A domains (9). In the case of the PI, PII, and PIII subgroups, some members of each have been demonstrated to cause hemorrhage (11). A variety of both in vivo and in vitro studies have indicated that the primary cause of hemorrhage is the degradation or disruption of basement membrane structure around capillaries (12). Microscopically, there appears to be a clearing of regions of capillary basement membrane with a subsequent dissolution of the endothelial lining, ultimately allowing escape of capillary contents into the stroma (Figure 51.2). In vitro studies with the hemorrhagic SVMPs using basement membranes, isolated basement membrane proteins, and extracellular matrix proteins demonstrated that these metalloproteinases are effective agents for matrix degradation (13–15). The question as to why some SVMPs are hemorrhagic and others are not is still open; however, it is generally thought that the nonproteinase domains in the PIII hemorrhagic toxins subgroup are important for selectively targeting these toxins to key substrates, which, when hydrolyzed, allow for the escape of capillary contents into the stroma. Recently, the cysteine-rich domain of atrolysin A, a PIII hemorrhagic SVMP from *Crotalus atrox* venom, was shown to support binding of the toxin to a vWF A domain found in integrins (16) as well as vWF itself (9).

In addition to causing gaps in the integrity of the endothelium, SVMP-mediated disruption/degradation of adjacent basement membranes and stromal extracellular matrix may have additional effects on endothelial cells (ECs). For example, in a recent study, two structurally similar PIII SVMPs, jararhagin (from *Bothrops jararaca* venom) and berythractivase (from *Bothrops erythromelas* venom), with differing hemostatic properties were shown to exert different effects on ECs. Both toxins caused the release of nitric oxide, prostacyclin, and interleukin (IL)-8 from human umbilical vein ECs

1 For purposes of discussion, we have divided the effects of various snake venom toxins on the endothelium into "direct" and "indirect." However, separating direct from indirect effects is not always straightforward. Moreover, many of these toxins have both direct and indirect effects on the endothelium.

Table 51-1: Snake Venom Toxin Activities on Endothelium and Vasculature

Toxin	*Activity*	*References*
Direct Effects		
Metalloproteinases	Hemorrhagic	8, 10, 41, 42
Serine proteinases	Increase of capillary permeability	19, 43–45
Natriuretic peptides	Relaxation of coronary arteries	20, 21, 46–48
Sarafotoxins	Vasoconstriction; cardiotoxic	49–53
VEGF-like proteins	Increase of capillary permeability	23, 54–56
Indirect Effects		
Metalloproteinases	Fibrin(ogen)olytic; inhibition of platelet-aggregation; apoptosis of ECs; cleavage of vWF; binding to cell surfaces and subendothelium	25, 34, 57–62
Disintegrins	Inhibition of platelet aggregation; inhibition of angiogenesis; blocking of function of EC integrin receptors; antiangiogenic	61, 63–66
Serine proteinases	Kinin release; plasma coagulant; platelet activation; plasminogen activation; protein C activation; factor V activation	35, 67–72
C-type lectin-like proteins	Inhibition or activation of platelet functions; anticoagulant, procoagulant	73–76
Bradykinin potentiating peptides	Inhibition of ACE; potentiation of bradykinin	20, 77, 78

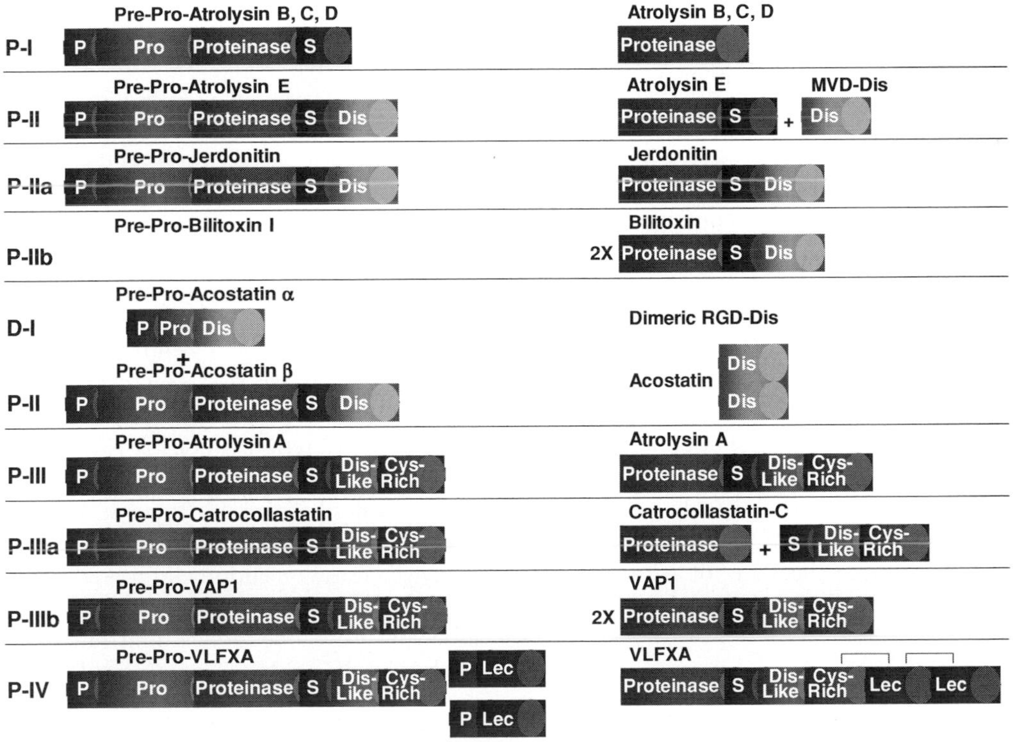

Figure 51.1. Schematic of SVMP structure. **P:** signal sequence; **Pro:** latency domain; **S:** spacer region; **Dis:** disintegrin domain; **Cys:** cysteine-rich domain; **Lec:** lectin-like domain. (Reproduced with permission from Fox JW, Serrano SMT. Structural considerations of the snake venom metalloproteinases, key members of the M12 reprolysin family of metalloproteinases. *Toxicon.* 2005;45:969–985.) For color reproduction, see Color Plate 51.1.

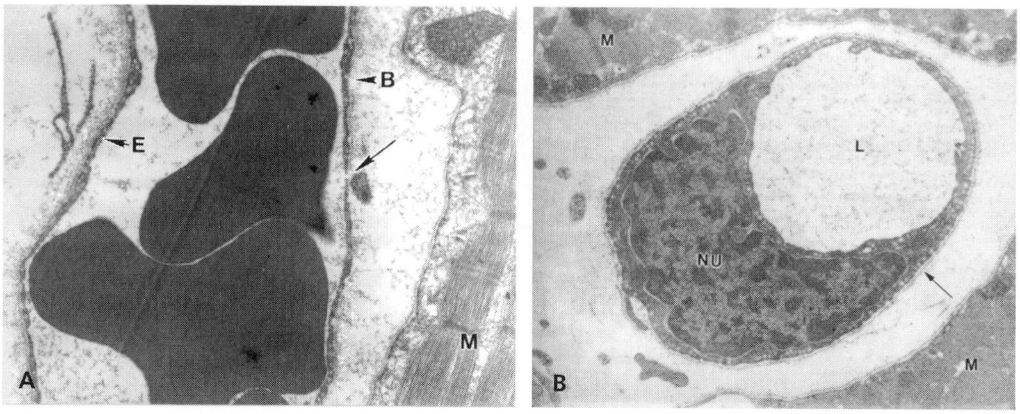

Figure 51.2. Ultrastructural changes. (**A**) Electron micrograph of a capillary in mouse gastrocnemius muscle 5 min following injection with 50 μg of BaP1, a PI SVMP isolated from *Bothrops asper* venom. The EC (E) shows a significant reduction in its thickness and in the number of pinocytotic vesicles. The continuity of the EC also is interrupted (*arrow*). The integrity of the capillary basement membrane is also affected (**B**). Contiguous skeletal muscle cell is also seen (M). (Image was generously provided by Dr. José Gutiérrez, University of Costa Rica). (**B**) Electron micrograph of a normal capillary from skeletal muscle tissue of a mouse that was injected intramuscularly with saline solution. The capillary lumen (L) is surrounded by an EC that has a prominent nucleus (NU). The basement membrane (*arrow*) surrounds the EC. Portions of two muscle cells (M) are located in the vicinity of the capillary vessel. Magnification: ×17,000. (Image provided by J. Gutierrez.)

(HUVECs), but in the case of jararhagin increased EC apoptosis also occurred (17). The mechanism by which these effects occurred is not clear; one could speculate that they could be the result of disrupted matrix-cell signaling, shedding of cell surface proteins, or both. Using DNA microarray studies, we demonstrated that treatment of HUVECs with snake venoms resulted in significant changes in gene expression (18). The pattern of response was consistent with a proapoptotic phenotype. In recent GeneChip studies, we showed that, in primary human fibroblasts, the PIII SVMP jararhagin activated p38 mitogen-activated protein kinase (MAPK) and triggered signaling pathways involved in integrin signaling and cellular metabolism. Similarly, two functional classes of upregulated proteins, namely those related to cell death and inflammation, were observed as being significantly populated (18). Interestingly, toxin-mediated changes in gene expression were dependent on the toxin's proteolytic activity and not on its engagement of cell surface receptors. Thus, it seems that the SVMPs can function to perturb endothelium by proteolytically targeting basement membrane and stromal extracellular matrix as well as directly functioning as a "sheddase" of EC surfaces to disrupt matrix–cell signaling and lead to altered EC function and, in some cases, death.

Snake Venom Serine Proteinases

The SVSPs are structurally homologous with serine proteinases of the chymotrypsin family, with molecular weights ranging from approximately 26 kDa to 67 kDa. Like other chymotrypsin family members, SVSPs are inhibited by typical active-site serine alkylation agents. They have been shown to increase capillary permeability, an effect that is generally attributable to their kallikrein-like enzymatic activity (see references listed in Table 51-1).[2] Interestingly, one serine proteinase, so-called capillary permeability-increasing enzyme-2, from *Agkistrodon caliginousus*, did not appear to have bradykinin-releasing activity, and its capillary permeability–inducing ability was thought to be the result of proteolytic release of peptides from fibrinogen (19).

Natriuretic Peptides

Natriuretic peptides are a group of naturally occurring substances that counter the activity of the renin-angiotensin system. In humans three classes of natriuretic peptides exist: (a) atrial natriuretic peptide (ANP), (b) B-type natriuretic peptide (BNP), and (c) C-type natriuretic peptide (CNP). All these peptides have vasodilator and hypotensive effects and generally function via stimulation of the membrane-bound guanylyl cyclases on endothelium. ANP and BNP cause natriuresis and a reduction in intravascular volume; CNP, although also hypotensive, does not demonstrate significant diuretic or natriuretic effects. Natriuretic peptides have been isolated

2 Kallikrein digests high-molecular-weight kininogen (HK) to release bradykinin. Bradykinin binds to receptors on nearby ECs, which liberates vasoactive prostaglandins or nitric oxide, resulting in, for example, vasodilatation and increased capillary permeability (see Chapter 49).

from a variety of snake venoms, with C-type natriuretic peptides being the most common. Typically, venom CNPs have a 17-amino acid ring structure connected by disulfide bonds with various numbers of amino acids extending from the N- and C-termini. Analysis of a cDNA clone from the *Bothrops jararaca* venom gland showed that a CNP was present in a transcript containing seven tandem repeats for bradykinin-potentiating peptides followed by the sequence of the CNP (20). Venom from *Dendroaspis angusticeps*, the green mamba, contains unusual natriuretic peptides that recognize ANP-A receptors but not ANP-B receptors, for which venom CNPs are usually specific (21).

Sarafotoxins

SRTXs are vasoconstrictive peptides isolated from the venoms of the Atractaspididae family of elapid snakes. These peptide toxins are homologous to the endothelins and affect the vasoconstriction of cardiac and smooth muscle. STRXs have 21 amino acids and two disulfide bonds. A wide variety of isotoxins have been isolated from this family of snakes, which reflect the subtle bioactivity requirements for various snakes with regard to typical prey species. The observation that these "toxins" are orthologs to the vertebrate endothelins underscores one evolutionary process by which poisonous snakes develop toxins. In this case, the protein scaffolds of normally occurring biologically active proteins and peptides are utilized and modified such that, when presented at relevant locations in appropriate concentrations, they produce the toxic effect required for the snake's survival.

Vascular Endothelial Growth Factor-Like Proteins

VEGF-like proteins have been isolated from a variety of snake venoms and appear to function by increasing capillary permeability and/or EC proliferation. Some snake venom VEGF-like proteins (svVEGF) appear to be specific for the kinase insert domain-containing VEGF receptor (KDR/VEGFR2). Two VEGF-like proteins, Vamin and VR-1 isolated from *Viper ammodytes ammodytes* venom, show preferential binding to VEGFR2 (22). The VEGF-like protein Tfsv-VEGF from *Trimeresurus flavorviridis* venom shows preferential binding to VEGFR1 and therefore elicits potent capillary permeability–inducing activity rather than EC mitogenesis (23). Structural studies of the various svVEGF-like proteins have yielded insights into the requirements for binding interactions with the different VEGF receptors and the subsequent biological effects of such interactions (22). Regardless of the individual specificity of the various svVEGF-like proteins, their capillary permeability–inducing activity has a profound effect on the vascular endothelium and likely contributes in a synergistic manner to the production of hemorrhage, edema, and necrosis observed following envenoming. Whether the mitogenic properties of certain of the svVEGF-like proteins actually play a role in venom toxicity or pathogenesis is unclear.

TOXINS THAT INDIRECTLY AFFECT THE ENDOTHELIUM

Five classes of toxins exist that can indirectly affect the endothelium in a manner that contributes to the overall pathological effects of the venom. These include (a) SVMPs, (b) disintegrin and disintegrin-like peptides and proteins, (c) SVSPs, (d) C-type lectin-like proteins, and (e) bradykinin-potentiating peptides. It is interesting to note that the SVMPs and the SVSPs are capable of both directly and indirectly affecting endothelium and thus have a very significant combined effect on that tissue.

Snake Venom Metalloproteinases

The indirect effect of the SVMPs stems primarily from their capacity to induce fibrin(ogen)olysis, inhibit platelet aggregation, cause apoptosis of ECs, and/or cleave vWF. Those toxins that promote fibrin(ogen)olysis do so by cleaving the α-chain of fibrinogen, giving rise to non- or poorly clottable blood (24). Many of the SVMPs also can cleave fibrin clots, thus attenuating any resolution of venom-induced hemorrhage by fibrin clots. Most PIII-type SVMPs have been demonstrated to inhibit platelet aggregation (8,16, and Figure 51.1) by virtue of the ability of their disintegrin-like/cysteine-rich domain to bind the platelet $\alpha_2\beta_1$ integrin (25). Similarly, these domains in the PIII toxins can target the SVMP to subendothelial collagen, where they can cause further proteolytic damage to tissue structure (9).

Several SVMPs have been demonstrated to cause EC apoptosis with the mechanism associated with the proteolytic activity of proteinase. Hypothetically, one can envision endothelial apoptosis to occur because of (a) proteolysis of extracellular matrix/basement membrane adjacent to adherent endothelium; (b) proteolysis of EC surface proteins required for maintenance of cellular viability, such as receptors for extracellular matrix proteins; (c) or proteolysis of cell surface proteins involved in cell-cell interaction. SVMPs have been demonstrated to be capable of cleaving all such proteins. You and colleagues (26) isolated a PIII SVMP from the venom of *Gloydius halys* and demonstrated that its ability to cause EC apoptosis in vitro was caused by both the protein's metalloproteinase activity as well activities associated with the nonproteinase disintegrin-like/cysteine-rich domains. It is likely that these results represent a general functional feature of the PII, PIII, and PIV SVMPs in that the nonproteinase domain(s) serve as exosites to target these classes of SVMPs to specific substrates, which are then degraded by the metalloproteinase domain (7). Wu and Peng (27) demonstrated that the PIII SVMP graminelysin isolated from *Trimeresurus gramineus* venom caused EC apoptosis in vitro by a combination of proteolysis of extracellular matrix and adherens junctions comprising vascular endothelial (VE)-cadherin and β-, γ- and α-catenin, resulting in cell detachment followed by activation of caspase-3 and a decrease of the bcl-2/Bax ratio.

Clearly most of the PIII SVMPs, under appropriate conditions, can cause EC apoptosis via variable sequences of proteolytic events in vitro. In vivo experiments examining EC apoptosis by SVMPs are limited. However, as mentioned previously, Gallagher and colleagues (28) have recently shown, using laser-capture microscopy of mouse muscle tissue treated in vivo with the PIII jararhagin from *Bothrops jararaca* venom, that the gene expression profiles of the affected myocytes, ECs, and fibroblasts indicate alterations of integrin-signaling pathways and upregulation of cell-death pathways reflecting disruption of cell adhesion signal transduction and the entrance of the cells into apoptosis. Under the experimental conditions used, apo-jararhagin, lacking metalloproteinase activity, did not show these properties, underscoring the role of proteolysis in affecting cellular homeostasis.

Disintegrins and Disintegrin-Like Proteins

Disintegrins are polypeptides found in viperid venoms that bind selectively to integrin receptors to functionally block the receptor. The disintegrin polypeptides range from approximately 40 to 100 amino acid residues in length with numerous disulfide bonds. The original report of a disintegrin from snake venom showed that the peptide containing the RGD motif was capable of blocking the platelet fibrinogen receptor, αIIbβ3 to prevent platelet aggregation (29). Since then, numerous disintegrins and dimeric disintegrins, both with and without RGD motifs, have been isolated and their integrin specificities explored (30). The disintegrin peptides from snake venoms are primarily derived by proteolytic processing from precursor PII SVMPs (8,31). Evidence exists that disintegrins also can be produced from specific gene products in addition to proteolytic processing from PII SVMPs.

In addition to and/or in lieu of platelet aggregation inhibitory activity, some disintegrins have been shown to inhibit cell migration, cell adhesion, and angiogenesis, depending on the specific integrins to which the disintegrins bind (30). With particular relevance to this chapter, the ability to inhibit angiogenesis was demonstrated for the disintegrin obtustatin isolated from the venom of *Vipera lebetina obtusa* (32). Obtustatin is the shortest disintegrin discovered to date; it is a somewhat unusual disintegrin in that its inhibitory loop is two residues shorter than that of other disintegrins. The Lys-Thr-Ser sequence in the loop is critical for binding to the $\alpha 1\beta 1$ integrin, which is important in angiogenesis (33).

PIII SVMPs contain disintegrin-like sequences, which differ from the disintegrin domain proper in that the RGD motif is substituted with other sequences. Moreover, the disulfide bond organization differs between disintegrin-like and disintegrin proteins. Although it is generally thought that the disintegrin-like domain also can support integrin binding, direct evidence for this is lacking.

Some venom contains disintegrin-like/cysteine-rich domains proteolytically processed from PIII SVMP structures. These domains have been demonstrated to inhibit collagen-induced platelet aggregation via inhibition of platelet αIIbβ3 integrin (34). Interestingly, recombinant cysteine-rich domain from atrolysin A, a PIII SVMP from *Crotalus atrox* venom, has also been shown to inhibit collagen-stimulated platelet aggregation (16).

Thus, potential exists for many disintegrins and disintegrin-like proteins in venom to interact with EC surfaces, giving rise to a variety of biological outcomes depending on the receptor repertoire on the ECs.

Snake Venom Serine Proteinases

Proteomic studies on viperid venoms show that these venoms are a rich source of serine proteinases (7). Many of the SVSPs in venoms have been demonstrated to affect coagulation (35). Virtually all SVSPs isolated to date are single-chain glycoproteinases with molecular masses between 26 kDa and 67 kDa – the single exception being brevinase, a heterodimeric proteinase with fibrinolytic activity isolated from the venom of *Agkistrondon blomhoffi brevicaudus*. In general, the indirect effects of SVSPs on the endothelium are owing to their abilities to act on blood components in the coagulation cascade, fibrinolysis, and/or platelet aggregation. Many SVSPs also can defibrinogenate blood but, unlike the SVMPs, these proteinases generally are specific for the β-chain of fibrinogen (24). Like the SVMPs, many of the SVSPs also can cleave fibrin clots. An example of venom fibrin(ogen)olytic snake venom proteinases currently in use in the clinic includes applications for fibrin clot dissolution, attesting to their potency and specificity (36).

More specifically, SVSPs have been shown to activate protein C, cleave fibrinogen Aα and Bβ chains, release fibrinogen peptides A and/or B, release kinin, activate receptors protease-activated receptor (PAR)-1 and PAR-4, cleave the α chain of complement C3, activate Factor V and Factor XIII, and activate plasminogen. These effects can produce mass disseminated intravascular coagulation to incapacitate the prey. The so-called kallikrein-like SVSPs act by releasing hypotensive lys-bradykinin or bradykinin from kininogen and thereby indirectly affect the endothelium. Given the numerous serine proteinases in venoms and the wide variety of proteolytic specificities, and hence biological activities that they display, it is clear that, upon envenomation, the SVSPs have a profound effect on hemostasis. This is supported by the observed coagulopathies in animal models of envenomation and many snake envenomation cases (37). Such disturbances no doubt impinge upon endothelium and endothelial function, which then further contributes to the overall pathogenesis of the venom.

Snake Venom C-Type Lectin-Like Proteins

This family of venom proteins is distinguished by structures that have a C-type lectin fold, characteristic of the selectins and mannose-binding proteins (38). The snake venom C-type lectin-like proteins are heterodimeric proteins comprising an α and β chain that can oligomerize to form larger structures. All the venom C-type lectin-like proteins indirectly affect the endothelium by altering platelet aggregation. The largest class of venom C-type lectins is the platelet GPIb receptor-binding proteins. Within this class, most of the venom C-type lectins bind to GPIb and inhibit vWF binding to the receptor, whereas others bind at a different site on GPIb and inhibit thrombin-mediated activation of platelets. Another class of C-type lectins, represented by the proteins botrocetin and bitiscetin, binds to vWF and causes induction of binding of vWF to platelet GPIb to promote platelet aggregation. A third class of snake venom C-type lectins binds to platelet collagen receptor GPVI to activate the receptor and give rise to hyper-aggregation. Finally, a fourth class of venom C-type lectins has been demonstrated to bind not only to GPIb but also to the integrin $\alpha 2\beta 1$. Because $\alpha 2\beta 1$ integrin is expressed by ECs, venom C-type lectins of this class, as represented by rhodocetin and aggretin, may directly influence EC phenotype (e.g., by inhibiting integrin interaction with collagen).

In summary, the snake venom C-type lectin-like proteins is a complex class of venom toxins with both pro- and anti-coagulation properties, all of which indirectly and in some instances directly affect the endothelium.

Bradykinin-Potentiating Peptides

One of the most therapeutically significant discoveries stemming from the study of snake venoms was of the bradykinin potentiating peptides found in viperid venoms. The bradykinin potentiating peptides (BPPs) are short (5–14 residues) proline-rich oligopeptides beginning with an amino terminal pyroglutamyl residue (39). These peptides are inhibitors of angiotensin converting enzyme (ACE) and thus potentiate the activity of bradykinin to give rise to an antihypertensive effect. Elucidation of BPP structures and activities played a critical role in the development of the antihypertensive drug captopril, an active site ACE inhibitor used to treat patients with hypertension. During viperid envenomation, the BPPs function in a synergistic manner with the bradykinin-producing SVSPs in the venom to give rise to vascular shock in the prey. Also, the presence of the BPPs in the venom when in the venom gland are at sufficient concentration to function as reversible inhibitors of the SVMPs and thereby minimize toxin proteolysis/degradation in the venom prior to envenomation (40). The BPPs are synthesized as a precursor protein containing tandem repeats of the BPPs in conjunction with a C-type natriuretic peptide (20).

Thus, the BPPs play an indirect role in affecting endothelium by virtue of their antihypertensive activities; they thus function in the overall coagulopathies associated with viper envenomation.

IN VITRO VERSUS IN VIVO CONSIDERATIONS ON THE FUNCTIONS OF SNAKE TOXINS AND THE ENDOTHELIUM

The issue of in vitro versus in vivo effects of toxins in biological systems is a longstanding issue. Traditionally, snake venom toxins have been isolated based on a particular activity they display in vitro or in vivo. Studies are then performed in an attempt to correlate and understand the in vitro activities compared to the in vivo activities. However, this is not always successful. As an example, in the early days of SVMP isolation and investigation it was shown that these protein toxins were very effective in producing hemorrhage in animal models. However, owing to limitations in substrates and assays used for proteolysis, the proteolytic activity of many of these venoms was overlooked. Later, using different substrates and assay methods, it became clear that these toxins were proteolytic and, in fact, their hemorrhagic activities were dependent on metalloproteinase activity. A second issue in the studies of venoms, whether in vivo or in vitro, is that venoms are complex, with many different biologically active proteins present that undoubtedly act in coordinated and synergistic manners to produce the symptomatic pathologies associated with envenomation by a particular snake species. Given the complexity of venom, it is almost impossible to completely understand the role of an individual toxin in the context of the whole venom, although this is clearly a goal for investigators in the field.

KEY POINTS

- Snake venoms are complex, with many protein toxins present in the venom that can directly and/or indirectly affect endothelium function.
- Many of the toxins that affect endothelium function do so in a coordinated or synergistic manner with other toxins present in the venom to give rise to the overall effect on the endothelium and subsequently the pathologies associated with snake envenoming.
- Due to the issues noted in the previous bullets, the in vitro activities/outcomes of noted toxins on ECs may be quite different in the in vivo context of whole venom in the prey.
- Snake toxins, owing to their abilities to both indirectly and directly affect ECs and endothelium, are fascinating, useful tools for the investigation of endothelium function.

REFERENCES

1 Mebs D. *Venomous and Poisonous Animals: a handbook for Biologists, Toxicologists and Toxinologists, Physicians and Pharmacists.* Boca Raton FL: CRC Press, 2002.
2 Redi F. *Osservazioni Intorno alle Vipere.* Firenze, Italy, 1664.
3 Fontana F. *Richerche Fisiche Sopra il Veleno Della Vipera.* Lucca, Italy, 1767.

4 Mitchell SW. Researches upon the venom of the rattlesnake. *Smithsonian Contributions to Knowledge*. 1860;12:145.

5 Mitchell SW, Reichert ET. Researches upon the venoms of poisonous serpents. *Smithsonian Contributions to Knowledge*. 1886; 26:186.

6 Fox JW, Shannon JD, Stefansson B, et al. Role of discovery science in toxinology: Examples, in venom proteomics. In: Menez A, ed. *Perspectives in Toxinology*. New York: John Wiley & Sons, 2002.

7 Serrano SMT, Shannon JD, Wang D, et al. A multifaceted analysis of Viperid snake venoms by two-dimensional electrophoresis: an approach to understanding venom proteomics. *Proteomics*. 2005; 5(2):501–510.

8 Fox JW, Serrano SMT. Structural considerations of the snake venom metalloproteinases, key members of the M12 reprolysin family of metalloproteinases. *Toxicon*. 2005;45(8):969–985.

9 Serrano SMT, Jia LG, Wang D, et al. Function of the cysteine-rich domain of the hemorrhagic metalloproteinase atrolysin A: targeting adhesion proteins collagen I and von Willebrand factor. *Biochem J*. 2005; In press.

10 Bjarnason JB, Fox JW. Snake venom metalloproteinases: reprolysins. *Methods Enzymol*. 1995;24:345–368.

11 Gutierrez JM, Rucavado A, Escalante T, Diaz C. Hemorrhage induced by snake venom metalloproteinases: biochemical and biophysical mechanisms involved in microvessel damage. *Toxicon*. 2005;45(8):997–1011.

12 Ownby CL, Bjarnason J, Tu AT. Hemorrhagic toxins from rattlesnake (*Crotalus atrox*) venom. Pathogenesis of hemorrhage induced by three purified toxins. *Am J Pathol*. 1978;93(1):201–218.

13 Baramova EN, Shannon JD, Bjarnason JB, Fox JW. Identification of the cleavage sites by a hemorrhagic metalloproteinase in type IV collagen. *Matrix* 1990;10:91–97.

14 Baramova EN, Shannon JD, Bjarnason JB, Fox JW. Degradation of extracellular matrix proteins by hemorrhagic metalloproteinases. *Arch Biochem Biophys*. 1989b;275:63–71.

15 Bjarnason JB, Hamilton D, Fox JW. Studies on the mechanism of hemorrhagic production by five proteolytic hemorrhagic toxins from *Crotalus atrox* venom. *Biol Chem Hoppe-Seyler*. 1988;369: 121–129.

16 Jia LG, Wang XM, Shannon JD, et al. Inhibition of platelet aggregation by the recombinant cysteine-rich domain of the hemorrhagic snake venom metalloproteinase, atrolysin A. *Arch Biochem Biophys*. 2000;373(1):281–286.

17 Schattner M, Fritzen M, Ventura J de S, et al. The snake venom metalloproteases berythractivase and jararhagin activate endothelial cells. *Biol Chem*. 2005;386(4):369–374.

18 Gallagher PG, Bao Y, Serrano SMT, et al. Use of microarrays for investigating the subtoxic effects of snake venoms: insights into venom-induced apoptosis in human umbilical vein endothelial cells. *Toxicon*. 2003;41(4):429–440.

19 Shimokawa K, Takahashi H. Capillary permeability-increasing enzyme-2 from the venom of *Agkistrodon caliginosus* (kankoku-mamushi): activity resulting from the release of peptides from fibrinogen. *Toxicon*. 1997;35:597–605.

20 Murayama N, Hayashi MA, Ohi H, et al. Cloning and sequence analysis of a *Bothrops jararaca* cDNA encoding a precursor of seven bradykinin-potentiating peptides and a C-type natriuretic peptide. *Proc Natl Acad Sci USA*. 1997;94(4):1189–1193.

21 Schweitz H, Vigne P, Moinier D, et al. A new member of the natriuretic peptide family is present in the venom of the green mamba (*Dendroaspis angusticeps*). *J Biol Chem*. 1992;267(20):13928–13932.

22 Suto K, Yamazaki Y, Morita T, Mizuno H. Crystal structures of novel vascular endothelial growth factors (VEGF) from snake venoms: insight into selective VEGF binding to kinase insert domain-containing receptor but not to fms-like tyrosine kinase-1. *J Biol Chem*. 2005;280(3):2126–2131.

23 Takahashi H, Hattori S, Iwamatsu A, et al. A novel snake venom vascular endothelial growth factor (VEGF) predominantly induces vascular permeability through preferential signaling via VEGF receptor-1. *J Biol Chem*. 2004;279:46304–46314.

24 Swenson S, Markland FS Jr. Snake venom fibrin(ogen)olytic enzymes. *Toxicon*. 2005;45(8):1021–1039.

25 Kamiguti AS, Hay CRM, Theakston RDG, Zuzel M. Insights into the mechanism of haemorrhage caused by snake venom metalloproteinases. *Toxicon*. 1996;34:627–642.

26 You WK, Seo HJ, Chung KH, Kim DS. A novel metalloprotease from *Gloydius halys* venom induces endothelial cell apoptosis through its protease and disintegrin-like domains. *J Biochem (Tokyo)*. 2003;134(5):739–749.

27 Wu WB, Peng HC, Huang TF. Disintegrin causes proteolysis of beta-catenin and apoptosis of endothelial cells. Involvement of cell-cell and cell-ECM interactions in regulating cell viability. *Exp Cell Res*. 2003;286:115–127.

28 Gallagher P, Bao Y, Serrano SMT, et al. Role of the snake venom toxin jararhagin in proinflammatory pathogenesis: in vitro and in vivo gene expression analysis of the effects of the toxin. *Arch Biochem Biophys*. 2005;441:1–15.

29 Huang TF, Holt JC, Lukasiewicz H, Niewiarowski S. Trigramin. A low molecular weight peptide inhibiting fibrinogen interaction with platelet receptors expressed on glycoprotein IIb-IIIa complex. *J Biol Chem*. 1987;262(33):16157–16163.

30 Calvete JJ, Marcinkiewicz C, Monleon D, et al. Snake venom disintegrins: evolution of structure and function. *Toxicon*. 2005; 45(8):1063–1074.

31 Shimokawa K, Jia LG, Shannon JD, Fox JW. Isolation, sequence analysis, and biological activity of atrolysin E/D, the non-RGD disintegrin domain from *Crotalus atrox* venom. *Arch Biochem Biophys*. 1998;354(2):239–246.

32 Marcinkiewicz C, Weinreb PH, Calvete JJ, et al. Obtustatin: a potent selective inhibitor of alpha1beta1 integrin in vitro and angiogenesis in vivo. *Cancer Res*. 2003;63(9):2020–2023.

33 Moreno-Murciano MP, Monleon D, Calvete JJ, et al. Amino acid sequence and homology modeling of obtustatin, a novel non-RGD-containing short disintegrin isolated from the venom of *Vipera lebetina obtusa*. *Protein Sci*. 2003;12(2):366–371.

34 Jia LG, Wang XM, Shannon JD, et al. Function of disintegrin-like/cysteine-rich domains of atrolysin A. Inhibition of platelet aggregation by recombinant protein and peptide antagonists. *J Biol Chem*. 1997;272(20):13094–13102.

35 Serrano SMT, Maroun RC. Snake venom serine proteinases: sequence homology vs. substrate specificity, a paradox to be solved. *Toxicon*. 2005;45(8):1115–1132.

36 Toombs CF. New directions in thrombolytic therapy. *Curr Opin Pharmacol*. 2001;1(2):164–168.

37 Zhang Y, Wisner A, Xiong Y, Bon C. A novel plasminogen activator from snake venom. Purification, characterization, and molecular cloning. *J Biol Chem*. 1995;270(17):10246–10255.

38 Lu Q, Navdaev A, Clemetson JM, Clemetson KJ. Snake venom C-type lectins interacting with platelet receptors.

Structure-function relationships and effects on haemostasis. *Toxicon.* 2005;45(8):1089–1098

39 Hayashi MA, Camargo AC. The Bradykinin-potentiating peptides from venom gland and brain of *Bothrops jararaca* contain highly site specific inhibitors of the somatic angiotensin-converting enzyme. *Toxicon.* 2005;45(8):1163–1170.

40 Robeva A, Politi V, Shannon JD, et al. Synthetic and endogenous inhibitors of snake venom metalloproteinases. *Biomed Biochim Acta.* 1991;50(4–6):769–773.

41 Hite LA, Shannon JD, Bjarnason JB, Fox JW. Sequence of a cDNA clone encoding the zinc metalloproteinase hemorrhagic toxin e from *Crotalus atrox*: evidence for signal, zymogen, and disintegrin-like structures. *Biochemistry.* 1992;31:6203–6211.

42 Paine MJ, Desmond HP, Theakston RD, Crampton JM. Purification, cloning, and molecular characterization of a high molecular weight hemorrhagic metalloprotease, jararhagin, from *Bothrops jararaca* venom. Insights into the disintegrin gene family. *J Biol Chem.* 1992;267(32):22869–22876.

43 Shimokawa K, Takahashi H. Purification of a capillary permeability-increasing enzyme-2 from the venom of *Agkistrodon caliginosus* (Kankoku-Mamushi). *Toxicon.* 1993;31:1213–1219.

44 Shimokawa K, Takahashi H. Some properties of a capillary permeability-increasing enzyme-2 from the venom of *Agkistrodon caliginosus* (Kankoku-Mamushi). *Toxicon.* 1993;31:1221–1227.

45 Sugihara H, Nikai T, Kito R, Sato H. Purification and characterization of arginine ester hydrolases from the venom of the Chinese habu snake (*Trimeresurus mucrosquamatus*). *Toxicon.* 1984;22:63–73.

46 Amininasab M, Elmi MM, Endlich N, et al. Functional and structural characterization of a novel member of the natriuretic family of peptides from the venom of *Pseudocerastes persicus. FEBS Lett.* 2004;557(1–3):104–108.

47 Collins E, Bracamonte MP, Burnett JC Jr., Miller VM. Mechanism of relaxations to dendroaspis natriuretic peptide in canine coronary arteries. *J Cardiovasc Pharmacol.* 2000;35(4):614–618.

48 Ho PL, Soares MB, Maack T, et al. Cloning of an unusual natriuretic peptide from the South American coral snake *Micrurus corallinus. Eur J Biochem.* 1997;250(1):144–149.

49 Ambar I, Kloog Y, Kochva E, et al. Characterization and localization of a novel neuroreceptor for the peptide sarafotoxin. *Biochem Biophys Res Commun.* 1988;157(3):1104–1110.

50 Hayashi MA, Ligny-Lemaire C, Wollberg Z, et al. Long-sarafotoxins: characterization of a new family of endothelin-like peptides. *Peptides.* 2004;25(8):1243–1251.

51 Sokolovsky M. Endothelins and sarafotoxins: physiological regulation. *Trends Biochem Sci.* 1991;16:261–264.

52 Takasaki C, Tamiya N, Bdolah A, et al. Sarafotoxins S6: several isotoxins from *Atractaspis engaddensis* (burrowing asp) venom that affect the heart. *Toxicon.* 1988;26(6):543–548.

53 Takasaki C, Itoh Y, Onda H, Fujino M. Cloning and sequence analysis of a snake, *Atractaspis engaddensis* gene encoding sarafotoxin S6c. *Biochem Biophys Res Commun.* 1992;189(3):1527–1533.

54 Gasmi A, Abidi F, Srairi N, Oijatayer A, et al. Purification and characterization of a growth factor-like which increases capillary permeability from *Vipera lebetina* venom. *Biochem Biophys Res Commun.* 2000;268:69–72.

55 Junqueira de Azevedo IL, Farsky SH, Oliveira ML, Ho PL. Molecular cloning and expression of a functional snake venom vascular endothelium growth factor (VEGF) from the *Bothrops insularis* pit viper. A new member of the VEGF family of proteins. *J Biol Chem.* 2001;276:39836–39842.

56 Yamazaki Y, Takani K, Atoda H, Morita T. Snake venom vascular endothelial growth factors (VEGFs) exhibit potent activity through their specific recognition of KDR (VEGF receptor 2). *J Biol Chem.* 2003;278:51985–51988.

57 Kamiguti AS, Markland FS, Zhou Q, et al. Proteolytic cleavage of the β_1 subunit of platelet $\alpha_2\beta_1$ integrin by the metalloproteinase jararhagin compromises collagen- stimulated phosphorylation of pp72 Syk. *J Biol Chem.* 1997;272:32599–32605.

58 Markland FS. Fibrolase, an active thrombolytic enzyme in arterial and venous thrombosis model systems. *Adv Exp Med Biol.* 1996;391:427–438.

59 Masuda S, Hayashi H, Araki S. Two vascular apoptosis-inducing proteins from snake venom are members of the metalloprotease/disintegrin family. *Eur J Biochem.* 1998;253:36–41.

60 Rucavado A, Lomonte B, Ovadia M, Gutierrez JM. Local tissue damage induced by BaP1, a metalloproteinase isolated from *Bothrops asper* (Terciopelo) snake venom. *Exp Mol Pathol.* 1995; 63(3):186–199.

61 Wu WB, Chang SC, Liau MY, Huang TF. Purification, molecular cloning and mechanism of action of graminelysin I, a snake-venom-derived metalloproteinase that induces apoptosis of human ECs. *Biochem J.* 2001;357:719–728.

62 Wu WB, Huang TF. Activation of MMP-2, cleavage of matrix proteins, and adherens junctions during a snake venom metalloproteinase-induced EC apoptosis. *Exp Cell Res.* 2003;288(1): 143–157.

63 Calvete JJ, Moreno-Murciano MP, Theakston RDG, et al. Snake venom disintegrins: novel dimeric disintegrins and structural diversification by disulfide bond engineering. *Biochem J.* 2003; 372:725–734.

64 Golubkov V, Hawes D, Markland FS. Anti-angiogenic activity of contortrostatin, a disintegrin from *Agkistrodon contortrix contortrix* snake venom. *Angiogenesis.* 2003;6(3):213–224.

65 Gould RJ, Polokoff MA, Friedman PA, et al. Disintegrins: a family of integrin inhibitory proteins from viper venoms. *Proc Soc Exp Biol Med.* 1990;195:16871.

66 McLane MA, Marcinkiewicz C, Vijay-Kumar S, et al. Viper venom disintegrins and related molecules. *Proc Soc Exp Biol Med.* 1998;219:109–119.

67 Markland FS Jr. Snake venoms and the hemostatic system. *Toxicon.* 1998;36:1749–1800.

68 Serrano SMT, Mentele R, Sampaio CA, Fink E. Purification, characterization, and amino acid sequence of a serine proteinase, PA-BJ, with platelet-aggregating activity from the venom of *Bothrops jararaca. Biochemistry.* 1995;34:7186–7193.

69 Serrano SMT, Hagiwara Y, Murayama N, et al. Purification and characterization of a kinin-releasing and fibrinogen-clotting serine proteinase (KN-BJ) from the venom of *Bothrops jararaca*, and molecular cloning and sequence analysis of its cDNA. *Eur J Biochem.* 1998;251:845–853.

70 Stocker K, Barlow GH. The coagulant enzyme from *Bothrops atrox* venom (batroxobin). *Methods Enzymol.* 1976;45:214–223.

71 Stocker K, Fischer H, Meier J, et al. Characterization of the protein C activator Protac from the venom of the southern copperhead (*Agkistrodon contortrix*) snake. *Toxicon.* 1987;25:239–252.

72 Tokunaga F, Nagasawa K, Tamura S, et al. The factor V-activating enzyme (RVV-V) from Russell's viper venom. Identification of

isoproteins RVV-V alpha, -V beta, and -V gamma and their complete amino acid sequences. *J Biol Chem.* 1988;263:17471–17481.

73 Clemetson KJ, Navdaev A, Dormann D, et al. Multifunctional snake C-type lectins affecting platelets. *Haemostasis.* 2001;31 (3–6):148–154.

74 Matsui T, Hamako J, Matsushita T, et al. Binding site on human von Willebrand factor of bitiscetin, a snake venom-derived platelet aggregation inducer. *Biochemistry.* 2002;41(25):7939–7946.

75 Morita T. Use of snake venom inhibitors in studies of the function and tertiary structure of coagulation factors. *Int J Hematol.* 2004;79(2):123–129.

76 Sekiya F, Atoda H, Morita T. Isolation and characterization of an anticoagulant protein homologous to botrocetin from the venom of *Bothrops jararaca. Biochemistry.* 1993;32(27):6892–6897.

77 Ferreira SH, Bartelt DC, Greene LJ. Isolation of bradykinin-potentiating peptides from *Bothrops jararaca* venom. *Biochemistry.* 1970;9(13):2583–2593.

78 Ferreira SH, Rocha E, Silva M. Potentiation of bradykinin and eledoisin by BPF (bradykinin potentiating factor) from *Bothrops jararaca* venom. *Experientia.* 1965;21(6):347–349.

79 Ondetti MA, Cushman DW. Angiotensin converting enzyme inhibitors: biochemical properties and biological activities. In: Soffer RL, ed. *Biochemical Regulation of Blood Pressure.* New York: Wiley; 1981:165–204.

52

Inflammatory Cues Controlling Lymphocyte–Endothelial Interactions in Fever-Range Thermal Stress

Qing Chen, Kristen Clancy, Wan-Chao Wang, and Sharon S. Evans

Roswell Park Cancer Institute, Buffalo, New York

A longstanding question in immunology revolves around the physiological benefit of the ancient fever response. Despite the fact that fever occurs at great metabolic cost, it is associated with improved survival during infection in endothermic and ectothermic vertebrate species (1). The prevailing paradigm with regard to leukocyte trafficking has been that febrile temperatures influence leukocyte delivery to tissues principally through bystander effects on hemodynamic parameters (i.e., vasodilation and increased blood flow). This chapter focuses on emerging evidence for a proactive role of temperatures in the range of natural fever (i.e., 38–40°C) in regulating the molecular events that support lymphocyte adhesion in high-endothelial venules (HEVs). HEVs are major sites of lymphocyte extravasation, serving as a locus for recirculation of blood-borne lymphocytes through peripheral lymphoid organs. Continuous entry of naïve and central memory lymphocytes across HEVs is crucial for immune homeostasis and immune surveillance. Notably, the mechanisms by which fever-range thermal stress promote lymphocyte–endothelial interactions are tightly regulated with respect to the type of vessels involved. Sustained exposure to fever-range thermal stress selectively targets adhesion in HEVs, whereas squamous endothelial cells (ECs) of noninflamed extralymphoid organs are not responsive. The physiological fever-range thermal stimulus described here is distinct from heat shock conditions (e.g., ≥43°C) that globally promote adhesion in nonactivated EC. These observations support the concept that HEVs act as sentinels during febrile inflammatory responses by heightening the delivery of naïve and central memory lymphocytes to secondary lymphoid organs.

OVERVIEW OF THE MOLECULAR MECHANISMS ORCHESTRATING LYMPHOCYTE TRAFFICKING ACROSS HIGH-ENDOTHELIAL VENULES

HEVs function as gatekeepers controlling the egress of lymphocytes out of the peripheral blood compartment and into secondary lymphoid organs where pathogens and cognate antigens are encountered (2–4). These specialized postcapillary venules are localized in T cell–enriched zones of all secondary lymphoid organs except the spleen. The ECs lining HEVs are morphologically and biochemically distinct from the flat, squamous ECs of the majority of the vessels throughout the body. These so-called high ECs (HECs) exhibit a cuboidal morphology, allowing them to extrude into the lumen of vessels and provide an irregular surface topography. These biophysical parameters theoretically contribute to turbulent blood flow, facilitating margination of leukocytes along the vessel wall. The vasculature in vertebrate species provides an extensive surface area for potential sites of extravasation. However, under noninflammatory steady-state conditions, the majority of constitutive lymphocyte extravasation occurs preferentially across HEVs that compose a relatively small percentage of the total vasculature. The mechanisms controlling lymphocyte extravasation in HEVs of lymphoid organs is briefly discussed in the next section. Additional details regarding the molecular basis of trafficking across lymphoid organ HEVs or HEV-like vessels of inflamed tissues are described in Chapters 155 and 170.

Lymphocyte migration across HEVs is coordinated by adhesion molecules and chemokine/chemokine receptor partners. These molecules participate in a stepwise sequence of reversible adhesion events that include (a) initial tethering and rolling, (b) chemokine-mediated activation and firm adhesion, and (c) transendothelial migration (2,5,6). Extravasation of naïve and central memory lymphocytes in peripheral lymph nodes (PLN) is initiated by the L-selectin leukocyte homing receptor. Positioning of L-selectin on microvillous projections enables this molecule to initiate contact between free-flowing lymphocytes and the walls of HEVs. L-selectin binds transiently to counter-receptors on HEVs collectively termed PLN addressins (PNAd) that include CD34, glycosylation-dependent cell adhesion molecule (GlyCAM)-1, podocalyxin, endomucin, and sgp200 (3). The N-terminal calcium-binding lectin domain of L-selectin interacts directly with negatively charged mucin domains of PNAd molecules generated by post-translational enzymatic modifications (i.e., glycosylation, fucosylation, sulfation, sialylation) (3). The sulfation determinant on PNAd recognized by MECA-79 monoclonal antibody (mAb) is absolutely required to support lymphocyte–HEV interactions in PLN (2,3,5). MECA-79-reactive PNAd molecules are not expressed on the lumenal surface of Peyer patches (PP) HEVs. Instead, L-selectin binds to the mucin stalk of mucosal addressin cell adhesion molecule (MAdCAM)-1 in PP HEVs (5). The efficiency of lymphocyte tethering and rolling in PP HEVs is enhanced by binding of a second leukocyte homing receptor, $\alpha 4\beta 7$ integrin, to a negatively charged aspartic acid residue in the N-terminal immunoglobulin domain of MAdCAM-1 (5). Lymphocyte homing in mesenteric lymph nodes (MLN) is mediated by combined interactions of L-selectin with PNAd and MAdCAM-1 as well as $\alpha 4\beta 7$ integrin binding to MAdCAM-1 (5).

Intravital microscopy studies have elegantly demonstrated that the process of tethering and rolling dramatically reduces the velocity of lymphocytes in HEVs (5,6) (see also Chapter 170). This is most evident in higher-order venules that are located downstream from the capillary bed. Here, free-flowing lymphocytes, moving at a rate of >400 to 500 μm/sec, transition to slow rolling cells with a speed of $\leq$50 μm/sec (7,8). The increased transit time allows lymphocytes to sample the chemokine microenvironment on the surface of HEVs. A single chemokine synthesized by HECs and stromal cells, the CC chemokine ligand (CCL)21 (TCA-4/SLC/6C-kine/exodus 2) plays a predominant role in supporting the progression from slow-rolling to firm-sticking cells in LN and PP HEVs (4,6).

Engagement of CCL21 by CCR7, a 7-transmembrane-spanning chemokine receptor on naïve and central memory lymphocytes, triggers G-protein-dependent conformational changes in the $\beta 2$ integrin, leukocyte function adhesion molecule-1 (LFA)-1. This enables LFA-1 to bind with high affinity to its constitutively expressed counter-receptors, intercellular adhesion molecule (ICAM)-1 and ICAM-2, on the walls of HEVs (2,4,6). ICAM-1 and ICAM-2 appear to have redundant functions in HEVs based on evidence that either molecule can substitute during homeostatic trafficking under conditions in which the other molecule is not operative (i.e., during mAb blockade or in genetically deficient mice) (9). In PLN HEVs, firm adhesion of lymphocytes is primarily mediated by LFA-1/ICAM-1–2 interactions, whereas in MLN and PP, both LFA-1/ICAM-1–2 and $\alpha 4\beta 7$ integrin/MAdCAM-1 contribute to firm adherence (5). LFA-1/ICAM-1–2 interactions have also been implicated in supporting transendothelial migration although the molecular mechanisms controlling this process in HEVs are not well understood (4–6). At extralymphoid sites of injury or inflammation, ICAM-1 is highly induced by inflammatory cytokines on cuboidal, HEVs-like vessels, where it predominates over ICAM-2 in recruitment of neutrophils, macrophages, and lymphocyte subsets (9).

FEVER-RANGE THERMAL STRESS AMPLIFIES LYMPHOCYTE–HEV ADHESION AND LYMPHOCYTE HOMING

Local increases in temperature at sites of inflammation and, in some cases, systemic fever are cardinal features of a host response to infection or inflammation. A recent series of studies has shown that fever-range temperatures alter the tissue distribution of lymphocytes in vivo. In this regard, significant decreases in the number of lymphocytes are observed in the circulating pool following elevation of the core temperature of mice (10–12). This was accomplished experimentally using whole body hyperthermia (WBH) protocols developed by Repasky and colleagues to simulate the temperature and duration of physiologic fever (13). Transient decreases in the number of peripheral blood lymphocytes have also been reported in patients with advanced cancer undergoing clinical fever-range WBH therapy (12,13). In mice, it was shown that these cells redistribute to HEV-bearing organs (LN and PP) but not to organs that lack HEVs, such as spleen, liver, or pancreas (8,10,14). Investigation of the underlying mechanisms has revealed that fever-range thermal stress influences trafficking by independently stimulating adhesion in two distinct cellular targets – lymphocytes (i.e., B and T cells) and HEC (Figure 52.1). Collectively, these findings support the notion that fever-range thermal stress provides a danger signal during inflammation to proactively regulate lymphocyte egress across HEVs in secondary lymphoid organs (1,8).

Thermal Stress Stimulates Lymphocyte Homing Receptor Function

To investigate the effect of fever-range thermal stress on lymphocyte adhesion, primary lymphocyte populations or lymphocyte cell lines expressing a defined profile of adhesion molecules (Table 52-1) were cultured in vitro under fever-like temperature conditions (i.e., 40°C) for 6 hours (1,8,10,15–19). This reductionist approach of solely modifying temperature conditions allowed for the identification of a role for the *thermal component* of the febrile inflammatory response in regulating lymphocyte trafficking. Changes in adhesion were evaluated under shear in frozen-section in vitro adherence assays where lymphoid tissue HEVs serve as substrates. Alternatively, lymphocytes treated in vitro with heat were evaluated for their ability to traffic to various tissues in short-term (1 hour) in vivo homing assays. These studies established that marked increases in lymphocyte adhesion to HEVs and homing to HEV-bearing organs (e.g., PLN, MLN, and PP) are observed following lymphocyte culture at physiologic fever-range temperatures (i.e., 2–4°C above normal temperature for 6 hours) (Table 52-1) (1,8,10,15–19). Notably, fever-range thermal stress does not improve short-term homing of lymphocytes to organs that do not express HEVs such as the spleen (Table 52-1) (1,8, 10,15).

Evidence for organ-specific trafficking of heat-treated lymphocytes raised the possibility that thermal stress targets

Table 52-1: Fever-Range Thermal Stress Improves Lymphocyte–HEV Adhesion and Lymphocyte Trafficking to LN and PP Organs

	Adhesion Molecule Expression			*HEV Adhesion after HT*		*Homing after HT*		
Cells	*L-Selectin*	*α4β7*	*LFA-1*	*LN*	*PP*	*LN*	*PP*	*Spleen*
Primary Lymphocytes								
Human PBL	+	+	+	↑	↑	ND	ND	ND
Mouse splenocytes	+	+	+	↑	↑	↑	↑	↔
Mouse LN–derived cells	+	+	+	↑	ND	↑*	↑*	↔*
Cell Lines								
300.19/L-selectin transfectant†	+	–	–	↑	↑	↑	↑	↔
300.19/Δcyto transfectant††	+	–	–	↔	↔	↔	↔	↔
TK1 cells‡	–	+	+	↔	↑	↔	↑	↔

HT, hyperthermia treatment (6 hours at 40°C); ND, not determined; + indicates positive expression of adhesion molecules; – indicates lack of expression of adhesion molecules; ↑ indicates the increased adhesion or homing to the indicated tissues after heat treatment; ↔ indicates no change in adhesion or homing after heat treatment.

*Q. Chen and S. Evans, personal observations.

†Murine B lymphoma cell line transfected with full-length human L-selectin (300.19/L-selectin).

††Murine B lymphoma cell line transfected with a nonfunctional form of human L-selectin (300.19/Δcyto; lacking C-terminal 11 amino acids).

‡Murine CD8$^+$ T lymphoma cell line; L-selectin is expressed at levels below the threshold to support adhesion.

References: 15,17–21.

the function of known lymphocyte homing receptors. This hypothesis was confirmed using L-selectin and $\alpha 4\beta 7$ integrin-specific function-blocking mAb or cell lines expressing a nonfunctional form of L-selectin (i.e., murine B lymphoma 300.19/Δcyto transfectants that lack the C-terminal 11 amino acid cytoplasmic tail) (Table 52-1) (20). Fever-range thermal stress increases L-selectin–mediated adhesion and homing of lymphocytes via engagement of PNAd on PLN and MLN HEVs as well as MAdCAM-1 on PP HEVs (Table 52-1, Figure 52.1) (15–19). Parallel increases in $\alpha 4\beta 7$ integrin-mediated adhesion to MAdCAM-1 on PP and MLN HEVs were demonstrated (Table 52-1, Figure 52.1) (10,21). Exquisite selectivity in thermal regulation of integrin-binding activity is suggested by results that heat treatment of lymphocytes fails to increase $\alpha 4\beta 7$ integrin-mediated adhesion to the extracellular matrix protein, fibronectin, or LFA-1–dependent binding to ICAM-1 (15,19). An important finding is that equivalent increases in L-selectin–dependent adhesion are observed in

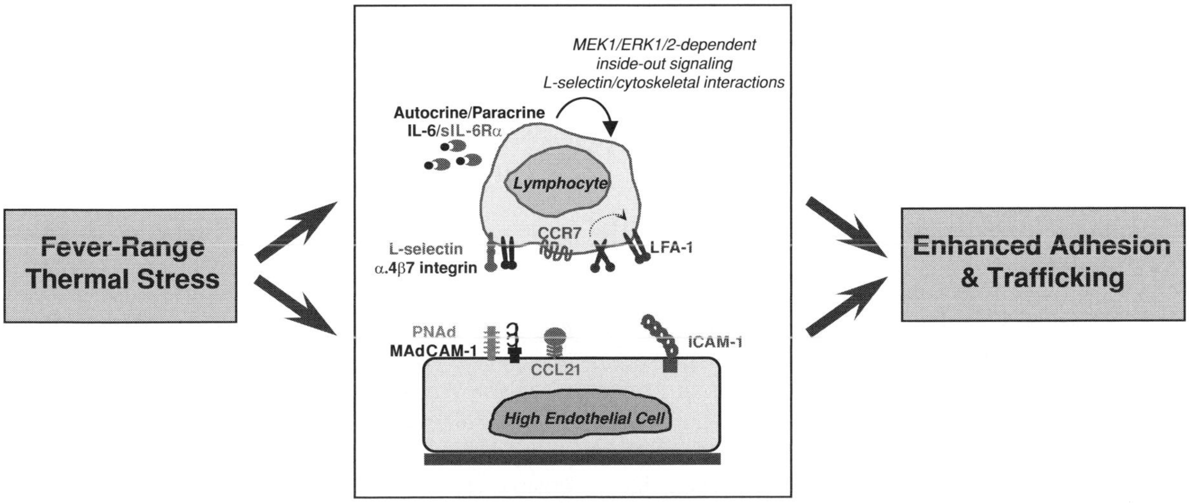

Figure 52.1. Fever-range thermal stress proactively stimulates lymphocyte–HEV adhesion. Fever-range temperatures enhance lymphocyte trafficking to lymphoid organs by exerting independent effects on the molecular events culminating in enhanced adhesion in lymphocytes and ECs of HEVs.

lymphocytes whether they are exposed to fever-range thermal stress in vitro or in vivo (i.e., during heat treatment of cultured cells or WBH treatment, respectively) (10,13,18). Multiple lymphocyte subsets were found to increase L-selectin–based adhesion to HEVs in response to thermal stress (i.e., $CD4^+$ and $CD8^+$ T cells, $CD19^+$ B cells, $CD56^+$ NK cells, $CD45RA^+$ naïve lymphocytes, and $CD45RO^+$ memory cells) (17–19). Moreover, thermal stress improves adhesion to PNAd or MAdCAM-1 substrates in lymphocyte populations of endothermic species that diverged during evolution 300 million years ago (e.g., mouse, human, chicken) (1). Collectively, these findings support the notion that thermal regulation of lymphocyte homing receptor activity confers a survival benefit that was maintained during the diversification of vertebrate species.

A growing body of evidence indicates that conventional mechanisms are not responsible for thermal control of L-selectin and $\alpha 4\beta 7$ integrin adhesion. Thermal stress does not alter the cell surface density of these homing receptors on lymphocyte subsets (15,17–19). Moreover, fever-range temperatures do not affect the lectin-binding activity of the N-terminal domain of L-selectin (15). Electron microscopy revealed that heat also does not influence the overall distribution of L-selectin on microvillous projections (15). However, these latter results do not exclude the possibility that heat enhances L-selectin clustering in membrane microdomains that cannot be resolved by immunogold-labeled antibody reagents.

Studies were undertaken to determine if thermal stress affects the association of L-selectin with the detergent-insoluble cytoskeletal matrix as a possible mechanism for regulating the binding activity of this homing receptor. In lymphocytes maintained at normothermal temperature (37°C), L-selectin is fully extracted from the detergent-insoluble subcellular fraction (16–18,22). However, upon L-selectin cross-linking by antibodies or L-selectin ligands (GlyCAM-1), this homing receptor rapidly ($\leq$5 seconds) becomes associated with the detergent-insoluble cytoskeletal matrix (17,22). L-selectin interactions with cytoskeletal elements require the 11-amino acid region within the cytoplasmic tail that contains a binding site for the cytoskeletal linker protein, α-actinin (17,23). These results are consistent with the notion that L-selectin becomes stably associated with the actin-based cytoskeleton during transient tethering and rolling along the lumenal surface of HEVs. Remarkably, treatment of lymphocytes with fever-range thermal stress causes L-selectin to preassociate with the detergent-insoluble matrix without the requirement for L-selectin cross-linking (1,8,16–18). Thus, heat-induced association of L-selectin with the cytoskeletal scaffolding underlying the lymphocyte plasma membrane is speculated to increase L-selectin tensile strength and thereby its ability to withstand shear within postcapillary HEVs (17,18).

One of the most intriguing findings is that thermal regulation of lymphocyte adhesion can be segregated into a two-step process, thereby excluding a role of heat per se in directly altering the organization or conformation of adhesion molecules in the lipid bilayer of the plasma membrane. In this regard, conditioned medium from cells treated in vitro with fever-range thermal stress (i.e., the "initiation phase") can be used in the "effector phase" to stimulate L-selectin/cytoskeletal interactions as well as the binding activity of L-selectin or $\alpha 4\beta 7$ integrin in lymphocytes maintained at normothermal temperatures (1,8,15,16,18,21). These experiments provide unequivocal evidence that soluble factors are responsible for mediating the pro-adhesive effects of thermal stress. Moreover, the source of the soluble factor appears to be remarkably cell-type specific. Hematopoietic cells (T cells, B cells, monocytes) and stromal cells (ECs, fibroblasts) are all sources of heat-induced transactivating pro-adhesive factors, whereas no activity is detected in culture supernatants of cell lines representing parenchymal cells of various organs (lung, liver, breast, brain, skin) (18,21).

These observations led to the discovery of a novel role for a well-known immunomodulatory cytokine, interleukin (IL)-6, in regulating lymphocyte homing (Figure 52.1) (1,8,18). Functional blockade of IL-6 and its receptor components – IL-6 receptor α (IL-6Rα) and the gp130 signal transducing subunit – prevent fever-range thermal stimulation of L-selectin adhesion in vitro and in vivo. In contrast, other cytokines – including tumor necrosis factor (TNF)-α, IL-1β, interferon (IFN)-α, IFN-γ, IL-8, IL-11, leukemia inhibitory factor, or oncostatin M – do not contribute to the pro-adhesive activity of thermal stress in lymphocytes under normal conditions. Thermal stimulation of lymphocyte adhesion was further found to depend on a trans-signaling mechanism that involves not only IL-6 but also a soluble form of the IL-6Rα (1,8,18). The requirement for these two components to function together as a heterodimeric cytokine provides a sophisticated level of control in this system. Combined biochemical and pharmacological inhibitor approaches positioned extracellular-signal-regulated kinase (ERK) 1/2, but not p38 mitogen-activated protein kinase (MAPK) or c-jun N-terminal kinase (JNK), in the IL-6/sIL-6Rα signaling pathway upstream of activation of L-selectin–cytoskeletal interactions and L-selectin avidity/affinity (Figure 52.1) (18). Taken together, these data enlarge on the concept that IL-6/sIL-6Rα trans-signaling actively contributes to immune responses by regulating leukocyte trafficking during both acute and chronic inflammation (24).

Fever-Range Thermal Stress Amplifies High-Endothial Venule Adhesion

An independent line of investigation revealed that fever-range thermal stress enhances adhesion in HEVs that are the major portals governing lymphocyte extravasation (Figure 52.1). These studies employed frozen-section in vitro adherence assays to compare the binding activity of HEVs in lymphoid organs of normothermal (NT) controls or mice treated 6 hours with fever-range WBH (core temperature of 39.5° ±0.5°C). Fever-range thermal stress markedly increases the ability of PLN HEVs to support lymphocyte adhesion under shear in

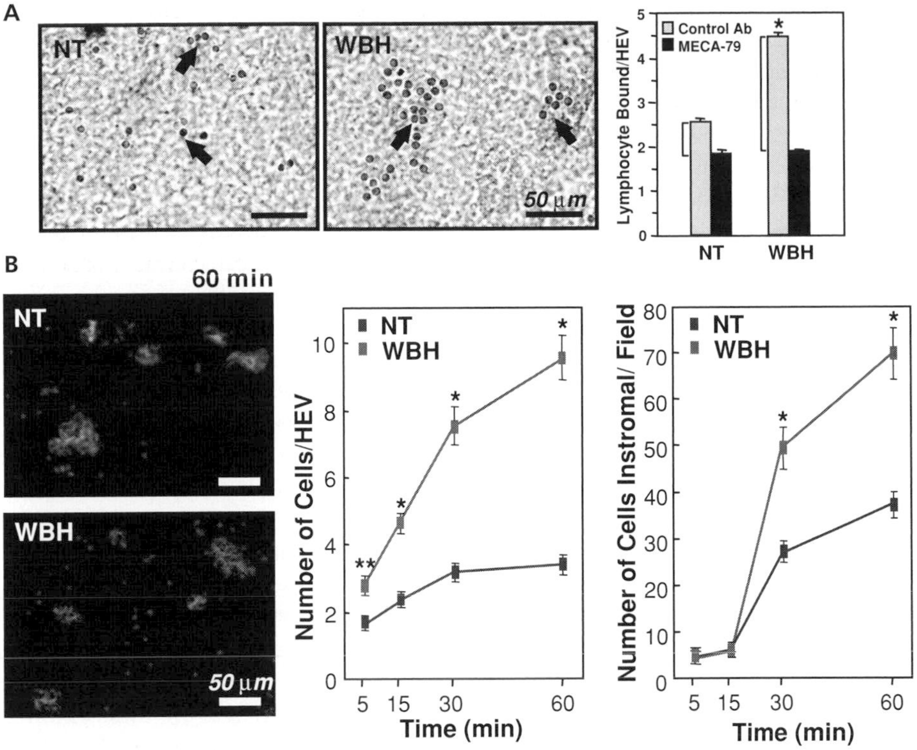

Figure 52.2. Fever-range thermal stress enhances lymphocyte adhesion to PLN HEVs (**A**) and homing to PLN organs (**B**). BALB/c mice were treated with fever-range whole body hyperthermia (WBH) for 6 hours (core temperature, 39.5 ± 0.5°C). The core temperature of normothermal control mice (NT) was 36.8 ± 0.2°C. (**A**) Adherence of mouse splenocytes to HEVs in PLN cryosections was evaluated in vitro under shear. The level of L-selectin/PNAd-specific adhesion (*brackets*) was determined by treating PLN cryosections with MECA-79 mAb. Photomicrographs show typical images of lymphocytes (*black arrows*) bound to HEVs in toluidine-stained PLN tissues. The number of adherent lymphocytes was quantified by light microscopy in a total of 300–500 HEVs. For consistency in double-blind evaluation, HEVs were quantified only if they contained ≥1 adherent cell. (**B**) Short-term homing studies were performed essentially as described. Mouse splenocytes labeled with TRITC were injected intravenously (3×10^7 cells/mouse) into NT control or WBH-treated mice and PLN were isolated at the indicated time points. Cryosections of PLN were stained with PNAd-specific MECA-79 mAb and FITC-labeled goat anti-rat IgMμ. Photomicrographs show typical images of TRITC-labeled cells associated with PNAd-positive HEVs or infiltrated into the stroma of PLN tissue at 60 minutes. The number of TRITC-labeled cells associated with HEVs (>40 HEVs) or stroma (>10 fields; 0.35 mm^2/field) in each sample was quantified (double-blind) by fluorescence microscopy. Data are the mean ± SE and are representative of ≥3 independent experiments. The differences between NT and WBH-treated mice were significant by unpaired two-tailed Student *t*-test ($^*p < 0.0001$, $^{**}p < 0.001$). Fever-range WBH did not increase the localization of fluorescent-labeled cells in splenic tissues at 60 minutes (not shown). (Further experimental details are provided in Refs. 10 and 14.).

this in vitro assay (Figure 52.2A) (1,8,14). The molecular events regulated by thermal stress appear to be remarkably stable when considering that enhanced adhesion is detected in frozen tissues stored at –20°C. Lymphocyte adhesion to HEVs in PLN cryosections is inhibited by the MECA-79 mAb, which recognizes a sulfation determinant on PNAd (see Figure 52.2A), as well as by mAb that bind to the N-terminal lectin domain of the L-selectin homing receptor (DREG-56, MEL-14) (1,8,10). These results confirm the requirement for L-selectin/PNAd adhesive partners in supporting lymphocyte adhesion to HEVs of heat-treated animals. Parallel increases in $\alpha4\beta7$ integrin/MAdCAM-1–dependent adhesion were identified following heat treatment in PP HEVs using $\alpha4\beta7^{hi}$ L-selectin^{lo} TK1 cells and neutralizing mAb specific for $\alpha4\beta7$ integrin and MAdCAM-1 (i.e., DATK32 and MECA-367, respectively)(1,8,10). Thermal stimulation of HEV adhesion is also observed in LN and PP organ cultures following incubation at fever-range temperatures in vitro (10). These data suggest that regulation of HEV adhesion occurs under local microenvironmental control and is not dependent on feedback mechanisms provided by neuronal or lymphatic systems or by the hypothalamus/pituitary/adrenal axis that orchestrates physiological responses during febrile episodes.

Thermal stimulation of adhesion at the level of the HEV correlates with improved trafficking of lymphocytes to LN and PP in short-term homing studies (1,8,14). In this series of studies, the core temperature of mice is initially raised to the range of natural fever (39.5°C ± 0.5°C) by WBH treatment for 6 hours. Mice are then allowed to revert to their normal basal temperature (36.8°C ± 0.2°C) prior to intravenous injection of fluorescent-labeled splenocytes and enumeration of labeled cells in various organs. Because fluorescent-tagged splenocytes are not subjected directly to thermal stress, this experimental design allows for the analysis of vascular responses to elevated temperatures. Flow cytometric analysis has shown that T and B lymphocytes constitute the major population of tetramethylrhodamine B isothiocynate (TRITC)-labeled cells that traffic across HEVs in normothermic mice or in mice pretreated with WBH, whereas entry of monocytes or neutrophils is limited under either temperature condition (14). An example of this type of homing study is shown in Figure 52.2B. Enhanced interactions of TRITC-labeled lymphocytes with HEVs (stained with PNAd-specific MECA-79 mAb and FITC-labeled secondary Ab) are detected in the PLN of WBH-treated mice as early as 5 minutes after lymphocyte injection. Increased lymphocyte–HEV interactions are observed in response to heat treatment at subsequent time points of 15, 30, and 60 minutes (Figure 52.2B) (8,14). Notably, lymphocyte interactions with HEVs temporally precede infiltration of fluorescent-labeled lymphocytes into the stroma of PLN organs of heat-treated animals (Figure 52.2B). These observations are consistent with the notion that HEVs are the major focal point directing lymphocyte trafficking into lymphoid organs in response to thermal stress.

The kinetics for optimal stimulation of HEV adhesion is tightly regulated. Moderate increases in HEV function are detected by in vitro adherence assays or in vivo homing studies following WBH treatment for 2 hours, whereas markedly elevated responses are observed following sustained exposure to thermal stress for 6 to 8 hours (8,14). The approximately twofold increase in lymphocyte–HEV adhesion and trafficking documented in response to fever-range thermal stress appears to represent a biologically significant amplification of lymphocyte access to lymphoid organs. Stimulation of the frequency of lymphocyte extravasation across HEVs, which is estimated to occur at a rate of 5×10^6 cells per second in humans under normothermal temperatures (2), would be expected to profoundly enhance the potential for immune surveillance. HEV adhesion returns to normal basal levels within 12 hours of cessation of thermal stress (10). Transient regulation of vascular adhesion is in line with the sequence of events in natural fever where physiological feedback loops are designed to restore biological systems to steady-state equilibrium after resolution of an infection.

A notable finding relates to the selectivity of thermal regulation of endothelial adhesion (1,8,10). Fever-range WBH preferentially amplifies adhesion in HEVs of LN and PP, whereas no increase in adhesion is detected by in vitro adherence assays in squamous ECs of noninflamed extralymphoid organs (e.g., pancreas). Moreover, WBH does not increase the localization of fluorescent-labeled lymphocytes in non–HEV-bearing organs (e.g., spleen, pancreas) in short-term homing studies (1,8,10). These findings suggest that febrile temperatures associated with infection or inflammation proactively focus the delivery of lymphocytes across HEVs in lymphoid organs, while sparing noninflamed extralymphoid sites.

The specific adhesion events targeted by thermal stress in HEVs have recently been characterized. Enhanced HEV adhesion is not accompanied by detectable increases in expression of the molecules that support primary tethering and rolling, that is, PNAd (Figure 52.2B) or MAdCAM-1 (1,10,14). Fever-range thermal stress was instead found to regulate molecular events that control secondary firm arrest of lymphocytes in HEVs. Thus, fever-range WBH treatment of mice markedly upregulates the luminal expression of ICAM-1 on HEVs of LN and PP, leading to enhanced lymphocyte trafficking at these sites, whereas no induction of ICAM-1 is observed in noninflamed vessels of extralymphoid organs (14). Direct visualization of lymphocyte–HEV interactions by intravital microscopy further revealed that WBH treatment increases firm adhesion without influencing primary tethering/rolling interactions (14). Based on these findings, it is proposed that high-density ICAM-1 on HEVs during febrile episodes increases the avidity and/or affinity of the molecular interactions with LFA-1 on apposing lymphocyte surface membranes. It remains to be determined whether heat induction of ICAM-1, in the context of infection or inflammation, might enable this trafficking molecule to cooperate with PNAd or MAdCAM-1 to stabilize primary adhesion events in HEVs, as reported for non-HEV systems in which overlapping roles have been defined for selectins and ICAM-1 (24).

Temperatures that exceed the range of natural fever appear to override the selectivity of vascular targeting. Heat shock (43°C) upregulates ICAM-1 expression as well as the ability to support lymphocyte adhesion in primary human endothelial cultures that model resting macrovascular endothelium (human umbilical vein ECs [HUVECs]) and microvascular endothelium (human dermal microvascular ECs [HMVECs]) (20,25,26). Moreover, heat shock proteins (HSPs) including HSP70, HSP60, and HSP65, which are major intracellular proteins synthesized in response to high-temperature heat shock or cellular stress, have been shown to act extracellularly to induce the expression of ICAM-1 in nonactivated primary ECs (HUVECs) in vitro (27–30). In sharp contrast, physiologic fever-range temperatures (39.5°C ± 0.5°C) have no effect on adhesion molecule expression (ICAM-1, E-selectin, vascular cell adhesion molecule [VCAM]-1, P-selectin, platelet-endothelial cell adhesion molecule [PECAM]-1, PNAd, MAdCAM-1), cytokine release (IL-1β, TNF-α, IFN-γ, IL-6, IL-11, IL-12, IL-13), or chemokine secretion (IL-8, RANTES, monocyte chemoattractant protein [MCP]-1, monokine induced by IFN-γ [MIG]) in cultured ECs (21,31). Nonactivated endothelium is not entirely refractory to fever-range thermal stress, however, because proadhesive factors can be recovered from the

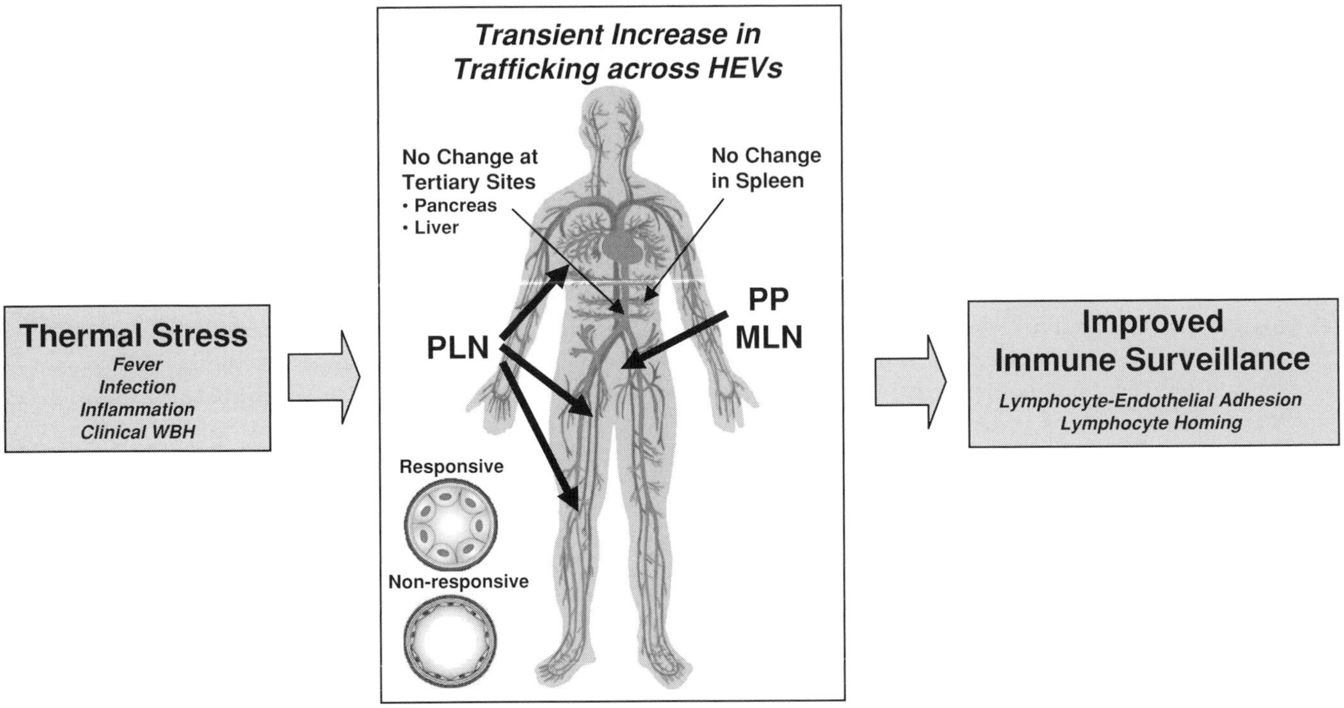

Figure 52.3. Model for fever-range thermal regulation of lymphocytes trafficking across selected vascular beds. Fever-range thermal stress in the context of inflammation or clinical thermal therapy promotes lymphocyte extravasation across cuboidal HEVs in peripheral lymph nodes (PLN), mesenteric LN (MLN), and Peyer's patches (PP). In contrast, lymphocyte–endothelial adhesion and trafficking across nonactivated squamous endothelium is not enhanced in spleen or noninflamed extralymphoid tissues.

conditioned medium of HUVECs and HMVECs following culture at 40°C (1,8,18,21). These soluble factors act in trans to stimulate the binding function of L-selectin and $\alpha4\beta7$ integrin on lymphocytes. Based on these findings, it is tempting to speculate that the vast majority of vascular beds indirectly contribute to lymphocyte delivery to LN and PP during febrile inflammatory responses by providing factors that stimulate lymphocyte homing receptor function.

KEY POINTS

- Recent studies support the concept that febrile temperatures function as a rheostat to amplify lymphocyte trafficking to secondary lymphoid organs and thereby enhance the efficacy of immune surveillance (Figure 52.3).
- Fever-range thermal stress increases L-selectin and $\alpha4\beta7$ integrin-dependent adhesion in B and T lymphocytes, thereby promoting trafficking via HEVs in LN and PP. Thermal regulation of adhesion in lymphocytes involves a trans-signaling mechanism mediated by engagement of membrane-anchored gp130 signal transduction molecules by IL-6 and a soluble form of the IL-6Rα.
- Fever-range hyperthermia transiently promotes the binding activity of gatekeeper HEVs in lymphoid organs. As a result, thermal stress induces a substantial increase in lymphocyte binding to HEVs under shear as well as transmigration across HEVs into the parenchyma of lymphoid organs. Thermal stress does not promote adhesion in squamous ECs of noninflamed extralymphoid organs either in vitro or in vivo.
- Selective targeting of primary and secondary adhesion events in specialized HEVs focuses the delivery of immune effector cells to peripheral lymphoid organs where the opportunity exists for optimal sensitization or restimulation of naïve or central memory lymphocytes, respectively (see Figure 52.3). In the absence of such tight control of lymphocyte–endothelial adhesion, inappropriate trafficking of lymphocytes could lead to extensive damage in noninflamed extralymphoid tissues.

Future Goals

- An unresolved issue relates to how site-specific vascular targeting is maintained by fever-range thermal stress. It is probable that the unique microenvironment of lymphoid organs as well as the differentiation/ activation status of HECs contribute to the specificity of vascular responses. It will be of interest in follow-up studies to determine if adhesion of HEV-like vessels at extralymphoid sites of acute or chronic inflammation is similarly regulated by febrile temperatures.

- Future studies are required to examine whether fever-range thermal stress initiates additional signaling events in ECs of different vascular beds, such as induction endothelial nitric oxide synthase, hemostatic factors, or apoptotic cascades.
- Major questions remain regarding the molecular mechanisms underlying thermal control of adhesion in HEVs. Inflammatory cytokines such as TNF-α, IL-1β, IFN-γ, or IL-6 are potential candidates because they are known to stimulate vascular expression of ICAM-1 in extralymphoid organs in the context of acute or chronic inflammation (2). Moreover, increased adhesion is detected in HEVs during fever responses induced by LPS or turpentine that are associated with high systemic levels of inflammatory cytokines (10). Notably, recent studies have implicated IL-6 as the central mediator of fever-range thermal stimulation of adhesion in HEVs as well as in lymphocytes (14,18). These findings support the emergence of IL-6 as a molecular switch controlling lymphocyte trafficking across vascular barriers during acute and chronic inflammation (24,32). An important issue that awaits further investigation is how pleiotropic cytokines, such as IL-6, selectively target lymphocyte-HEV adhesion while maintaining the tightly regulated balance between physiological and pathological responses during febrile inflammatory responses.

ACKNOWLEDGMENTS

We thank Elizabeth Repasky and Heinz Baumann for insightful discussions relating to this work and Jennifer Black and Michelle Appenheimer for the critical review of the manuscript. This work was supported by grants from the NIH, the Department of Defense, and the Roswell Park Alliance Foundation.

REFERENCES

1 Appenheimer MM, Chen Q, Girard RA, et al. Impact of fever-range thermal stress on lymphocyte-endothelial adhesion and lymphocyte trafficking. *Immunol Invest*. 2005;34(3):295–323.

2 Girard JP, Springer TA. High endothelial venules (HEVs): specialized endothelium for lymphocyte migration. *Immunol Today*. 1995;16(9):449–457.

3 Rosen SD. Ligands for L-selectin: homing, inflammation, and beyond. *Annu Rev Immunol*. 2004;22:129–156.

4 Miyasaka M, Tanaka T. Lymphocyte trafficking across high endothelial venules: dogmas and enigmas. *Nat Rev Immunol*. 2004; 4(5):360–370.

5 Butcher EC, Williams M, Youngman K, et al. Lymphocyte trafficking and regional immunity. *Adv Immunol*. 1999;72:209–253.

6 von Andrian UH, Mempel TR. Homing and cellular traffic in lymph nodes. *Nat Rev Immunol*. 2003;3(11):867–878.

7 von Andrian UH. Intravital microscopy of the peripheral lymph node microcirculation in mice. *Microcirculation*. 1996;3(3):287–300.

8 Chen Q, Fisher DT, Kucinska SA, et al. Dynamic control of lymphocyte trafficking by fever-range thermal stress. *Cancer Immunol Immunother*. 2006;55:299–311.

9 Lehmann JC, Jablonski-Westrich D, Haubold U, et al. Overlapping and selective roles of endothelial intercellular adhesion molecule-1 (ICAM-1) and ICAM-2 in lymphocyte trafficking. *J Immunol*. 2003;171(5):2588–2593.

10 Evans SS, Wang WC, Bain MD, et al. Fever-range hyperthermia dynamically regulates lymphocyte delivery to high endothelial venules. *Blood*. 2001;97(9):2727–2733.

11 Ostberg JR, Repasky EA. Use of mild, whole body hyperthermia in cancer therapy. *Immunol Invest*. 2000;29(2):139–142.

12 Kraybill WG, Olenki T, Evans SS, et al. A phase I study of fever-range whole body hyperthermia (FR-WBH) in patients with advanced solid tumours: correlation with mouse models. *Int J Hyperthermia*. 2002;18(3):253–266.

13 Pritchard MT, Ostberg JR, Evans SS, et al. Protocols for simulating the thermal component of fever: preclinical and clinical experience. *Methods*. 2004;32(1):54–62.

14 Chen G, Fisher DT, Clancy KA, et al. Fever-range thermal stress promotes lymphocyte trafficking across high endothelial venules via an interleukin-6 trans-signaling mechanism. *Nat Immunol*. 2006;7:1299–1308.

15 Wang WC, Goldman LM, Schleider DM, et al. Fever-range hyperthermia enhances L-selectin-dependent adhesion of lymphocytes to vascular endothelium. *J Immunol*. 1998;160(2):961–969.

16 Evans SS, Frey M, Schleider DM, et al. Regulation of leukocyte-EC interactions in tumor immunity. In: Croce MA, ed. *The Biology of Tumors*. New York: Plenum Press; 1998:273–86.

17 Evans SS, Schleider DM, Bowman LA, et al. Dynamic association of L-selectin with the lymphocyte cytoskeletal matrix. *J Immunol*. 1999;162(6):3615–3624.

18 Chen Q, Wang WC, Bruce R, et al. Central role of IL-6 receptor signal-transducing chain gp130 in activation of L-selectin adhesion by fever-range thermal stress. *Immunity*. 2004;20(1):59–70.

19 Evans SS, Bain MD, Wang WC. Fever-range hyperthermia stimulates alpha4beta7 integrin-dependent lymphocyte-endothelial adhesion. *Int J Hyperthermia*. 2000;16(1):45–59.

20 Kansas GS, Ley K, Munro JM, Tedder TF. Regulation of leukocyte rolling and adhesion to high endothelial venules through the cytoplasmic domain of L-selectin. *J Exp Med*. 1993;177(3):833–838.

21 Shah A, Unger E, Bain MD, et al. Cytokine and adhesion molecule expression in primary human ECs stimulated with fever-range hyperthermia. *Int J Hyperthermia*. 2002;18(6):534–551.

22 Leid JG, Steeber DA, Tedder TF, Jutila MA. Antibody binding to a conformation-dependent epitope induces L-selectin association with the detergent-resistant cytoskeleton. *J Immunol*. 2001; 166(8):4899–4907.

23 Pavalko FM, Walker DM, Graham L, et al. The cytoplasmic domain of L-selectin interacts with cytoskeletal proteins via alpha-actinin: receptor positioning in microvilli does not require interaction with alpha-actinin. *J Cell Biol.* 1995;129(4):1155–1164.

24 Jones SA, Rose-John S. The role of soluble receptors in cytokine biology: the agonistic properties of the sIL-6R/IL-6 complex. *Biochim Biophys Acta.* 2002;1592(3):251–263.

25 Steeber DA, Campbell MA, Basit A, Ley K, et al. Optimal selectin-mediated rolling of leukocytes during inflammation in vivo requires intercellular adhesion molecule-1 expression. *Proc Natl Acad Sci USA.* 1998;95(13):7562–7567.

26 Lefor AT, Foster CE 3rd, Sartor W, Engbrecht B, et al. Hyperthermia increases intercellular adhesion molecule-1 expression and lymphocyte adhesion to ECs. *Surgery.* 1994;116(2):214–220.

27 Chen Q, Evans SS. Thermal regulation of lymphocyte trafficking: hot spots of the immune response. *Int J Hyperthermia.* In press.

28 Kol A, Bourcier T, Lichtman AH, Libby P. Chlamydial and human heat shock protein 60s activate human vascular endothelium, smooth muscle cells, and macrophages. *J Clin Invest.* 1999; 103(4):571–577.

29 Galdiero M, de l'Ero GC, Marcatili A. Cytokine and adhesion molecule expression in human monocytes and ECs stimulated with bacterial heat shock proteins. *Infect Immun.* 1997; 65(2):699–707.

30 Verdegaal ME, Zegveld ST, van Furth R. Heat shock protein 65 induces CD62e, CD106, and CD54 on cultured human ECs and increases their adhesiveness for monocytes and granulocytes. *J Immunol.* 1996;157(1):369–376.

31 Hasday JD, Bannerman D, Sakarya S, et al. Exposure to febrile temperature modifies EC response to tumor necrosis factor-alpha. *J Appl Physiol.* 2001;90(1):90–98.

32 Jones SA. Directing transition from innate to acquired immunity: defining a role for IL-6. *J Immunol.* 2005;175(6):3463–3468.

53

Hyperbaric Oxygen and Endothelial Responses in Wound Healing and Ischemia–Reperfusion Injury

Bryan Belikoff, Wende R. Reenstra, and Jon A. Buras

Northeastern University, Boston, Massachusetts

Hyperbaric oxygen (HBO_2) therapy is defined as breathing 100% oxygen (O_2) at a pressure greater than 1.4 atmospheres absolute (ATA) (1). HBO_2 is currently accepted as a clinical adjunctive therapy for the following indications: wound healing, carbon monoxide (CO) poisoning, crush injury, clostridial myositis and myonecrosis, decompression sickness, osteomyelitis (refractory), skin grafts and flaps, and air or gas embolism (1–3). HBO_2 therapy is administered in either a monoplace hyperbaric chamber (where the chamber is compressed with 100% oxygen) or in a multiplace hyperbaric chamber (where the chamber is compressed with air and the patient breathes 100% oxygen via a tight-fitting mask or hood) (Figure 53.1). Many benefits of HBO_2 therapy are gained through its effects on the endothelium, including improvements in microvascular flow, induction of proangiogenic factors, inhibition of inducible intercellular adhesion molecule (ICAM)-1 expression, and regulation of endothelial derived nitric oxide (NO) production. The ubiquitous nature of the endothelium provides a gateway for many of the HBO_2 protective mechanisms to be effective at a number of injury sites. This chapter reviews the effects of HBO_2 therapy on disease states that are most dependent on endothelial responses: wound healing and ischemia–reperfusion (I/R) injury.

HISTORY OF HYPERBARIC OXYGEN THERAPY

HBO_2 therapy was born from the concept that individuals exposed to increased ambient air pressure would have improvement in many different disease states (4). Although there was no scientific basis for this concept, it was put into practice as early as 1662 by a British clergyman, Henshaw (5). Compressed-air therapy enjoyed a nonscientifically founded popularity during the 19th century in a manner similar to mineral water spas (6). The French surgeon Fontaine created a mobile air pressurized operating theater in 1879 (7). The use of nitrous oxide anesthesia was observed to be safer due to the approximately doubled percentage of inspired oxygen created by compression. Under these conditions of pressure, patients did not appear as cyanotic as usual during emergence from anesthesia (7). The benefit of compressed-air therapy for tunnel workers was first recognized by Ernest W. Moir in 1889, during construction of the Hudson River tunnel (8). During this construction project, use of a recompression chamber reduced the death rate of workers from 25% to 1.66%.

The birth of HBO_2 therapy occurred in 1955 when oxygen, rather than air, was used as the environment for compression. This concept was credited to Churchill-Davidson in an effort to enhance radiation therapy for cancer patients (9). Also in 1956, Boerema and colleagues suggested that compression of the operating room with O_2 would improve tolerance to circulatory arrest during cardiac surgery (10). As a result of promising animal experiments, hyperbaric operating theaters were constructed throughout the United States to support cardiac surgery. The development of cardiopulmonary bypass technology marked the decline of HBO_2 therapy usage in the 1970s. Unfortunately, HBO_2 was used indiscriminately and without scientific evidence for multiple medical conditions, limiting the credibility of this therapy in the eyes of the medical community. Recently, more rigorous clinical studies have demonstrated a beneficial role for HBO_2 therapy in treating CO poisoning and diabetic foot ulcers (2,3). Also, basic science research has begun to identify the mechanisms responsible for HBO_2 in hopes of providing a basis for the rational application of HBO_2 in clinical trials.

PRINCIPLES AND PRACTICE OF HYPERBARIC OXYGEN

In mammals, the delivery of oxygen into the bloodstream in exchange for metabolic gases occurs at the alveolar–blood barrier in pulmonary capillaries. At this interface oxygen exists in three different forms: (a) as a gas within the alveoli, (b) dissolved in plasma, and (c) bound by hemoglobin (Hb).

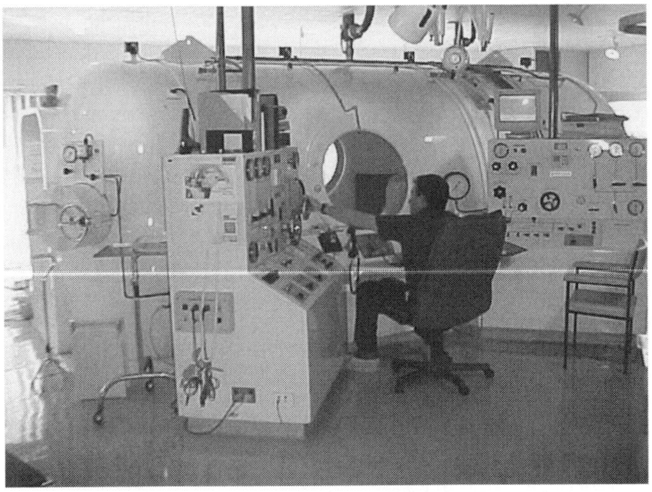

Figure 53.1. Multiplace hyperbaric chamber. The multiplace chamber is compressed with air and the patient breathes 100% oxygen via mask or hood inside the chamber. (Photo courtesy of Dick Clarke.)

O_2 and CO_2 travel between the blood–gas interface by simple diffusion, resulting in the transfer of gas from areas of high to low partial pressure. The partial pressure of a gas in the gas phase is dependent on its concentration in the gas mixture and is determined by multiplying its concentration by the total pressure (mmHg) of the vessel. The relationship between the partial pressure of a gas and the concentration of dissolved gas is given by Henry's law: $P = kC$, where P represents the pressure of the gas in the gas phase, C represents the concentration of the dissolved gas, and k is a constant characteristic of the particular liquid (solution).

The uptake of O_2 can be divided into two stages: (a) diffusion of O_2 across the alveolar–blood barrier into the interior of the red blood cell, and (b) the reversible reaction of O_2 binding with Hb to form oxyhemoglobin, $O_2 + Hb \leftrightarrow HbO_2$.

Fick's law of diffusion states that the rate of diffusion across a barrier is proportional to the area (A) of the membrane and inversely proportional to its thickness (T):

$$V_{gas} = (A/T) \times D \times (P_1 - P_2) \qquad \text{(Equation 1)}$$

where D represents gas solubility/$\sqrt{}$(molecular weight)$_{gas}$, V_{gas} represents velocity of gas, and P represents pressure.

Normal arterial blood with a P_{O2} of 100 mmHg contains only 0.3 ml O_2/100 mL or 3 mL O_2/liter by simple diffusion alone. The amount of O_2 dissolved in the blood by simple diffusion alone is not sufficient to meet the metabolic demands of the body's tissues. Nature has circumvented this problem through the production of Hb to enhance the O_2 carrying capacity of blood in vertebrates (one interesting exception is the Antarctic icefish, discussed in Chapter 7). Hb is an iron-porphyrin–containing globular protein that binds dissolved oxygen in blood. The maximum amount of O_2 that can bind with Hb is called the oxygen capacity. The %O_2 saturation is calculated by dividing the number of O_2 molecules bound to Hb by the oxygen capacity, multiplied by 100. The following equation shows the relationship between oxyhemoglobin and dissolved oxygen in blood and is used to calculate the total O_2 concentration of blood (mL O_2/100 mL):

$$\{1.39 \times [\text{Hb}] \times (\%O_2\ \text{saturation}/100) + 0.003\, P_{O2}\ \text{mmHg}\} \qquad \text{(Equation 2)}$$

As Hb becomes saturated with O_2, the blood nears O_2 capacity. Subsequently, the P_{O2} of the blood dramatically increases, and the rate of O_2 diffusion across the alveolar–blood barrier slows due to the P_{O2} in the blood reaching that of alveolus. This becomes apparent in a normal individual (whose blood contains 15 gmHb/100 mL) with an arterial blood P_{O2} above 50 mmHg, where the O_2 combined with Hb begins to near O_2 capacity. If the P_{O2} of the alveoli is increased, the P_{O2} dissolved in the blood will increase, and the converse also is true.

HBO_2 exposure at 2.5 ATA will significantly increase the O_2 concentration of the arterial blood, even though the amount of O_2 in the blood is close to O_2 capacity. The ability of HBO_2 to enhance O_2 concentration is mediated by the non–Hb-dependent direct dissolution of O_2 in blood. By increasing the atmospheric pressure to 2.5 ATA, the P_{O2} in the alveoli will be approximately three times greater than breathing 100% pure O_2 alone and approximately 18 times greater than inspiring normal air at 1 ATA (760 mm Hg). This situation results in more free O_2 (O_2 not bound by Hb) dissolved directly in the blood and thus increased O_2 delivery in the systemic circulation. Experimentally, the entire blood volume may be replaced with normal saline under HBO_2 conditions and the animal may function without impairment (11).

CHRONIC WOUNDS

Aggressive elimination of infected and nonviable tissue in combination with meticulous wound care is the foundation of successful wound healing. However, some wounds do not respond to simple wound care and progress to a chronic state. By definition, chronic wounds are recalcitrant, reoccur, and/or are progressive and refractory to a wide spectrum of treatments applied. Chronic wounds arise in a variety of situations and pathologies. The most common chronic wounds in Western medical practice are a consequence of diabetes, arterial or venous disease, sustained pressure, or therapeutic irradiation for the treatment of tumors (12).

Chronic wounds are common and constitute a significant health problem. The actual incidence and impact are difficult to assess accurately because of the wide range of disease states and pathologies associated with chronic wounds. Current therapeutic interventions for chronic wounds (beyond antibiotics and supportive care) include the use of HBO_2 and exogenous NO donors, as well as topical growth factors such as platelet-derived growth factor (PDGF)-BB (13–15). The beneficial effects of HBO_2 on wound healing are well documented clinically, but the underlying mechanisms are relatively unknown. Basic science research utilizing both in vivo and in vitro wound healing models has demonstrated that HBO_2 treatment accelerates wound healing through NO-dependent mechanisms (16).

The Effect of Hyperbaric Oxygen on Wound Healing

The rationale for using HBO_2 in wound healing is that the common denominator in many chronic wounds is tissue hypoxia. Hypoxemia, caused initially in wounds by vascular disruption, is a key factor that eventually limits wound healing (17). A majority of wounds have been shown to be hypoxic at the center of the wound bed. This initial hypoxia is thought to be a trigger for blood vessel growth. However, although hypoxia is thought to initiate neovascularization, it cannot sustain it. Wound healing is not sustainable in the hypoxic environment as many of the processes involved in wound healing, such as collagen deposition, fibroblast migration, and growth factor production, require oxygen for metabolism. This concept is further supported by the observation that supplemental oxygen administration accelerates blood vessel growth in a rabbit ear model (18).

Hyperbaric Oxygen and Angiogenesis

HBO_2 appears to exert its beneficial effects on wound healing in part by promoting the formation of new capillaries. The effect of HBO_2 exposure on the endothelium became apparent after its usage as an adjunctive therapy for the treatment of diabetic foot ulcers. Clinical studies demonstrated that HBO_2 exposure for 90 minutes at 2.5 ATA significantly increases transcutaneous pO_2 levels in the dorsal foot of patients with diabetic foot ulcers (2). At the conclusion of this study, the sustained increase in tissue pO_2 levels was attributed in part to new vessel growth.

The formation of new capillaries from preexisting vessels, such as occurs in wound healing, is termed *angiogenesis*. Clinical and experimental evidence suggests that increasing oxygen concentration in hypoxic wound beds results in increased blood vessel growth (19,20). The mechanism by which hyperoxia might induce blood vessel growth is not well understood but is thought to involve vascular endothelial growth factor (VEGF) and angiopoietin (Ang), both potent inducers of angiogenesis.

VEGF is a 21- to 49-kDa protein and is one of the most important angiogenic growth factors involved in adult wound healing. VEGF interacts with two tyrosine kinase receptors located on the endothelial cell (EC) surface, VEGFR2 and VEGFR1, and VEGFR2–VEGF binding stimulates EC proliferation. VEGFR1–VEGF binding is antagonistic to VEGFR2–VEGF-induced EC proliferation and induces tube formation and capillary development (21,22).

Ang2 is a 70- to 88-kDa protein that is selectively expressed at sites of pathological angiogenesis. Binding of the Ang2 ligand to its EC surface receptor, Tie2, loosens the endothelium for vascular remodeling and enhances the responsiveness of ECs to VEGF stimulation (23). Initial cellular events in angiogenesis during wound healing appear to be dependent on Ang2–Tie2 binding, whereas the later events of vessel formation depend on the availability of functional VEGF (23).

VEGF became an attractive mediator to explain the beneficial effects of HBO_2 therapy due to its neovasculogenic properties. Initially this association appears paradoxical, as hypoxia, not hyperoxia, was thought to be the principal inducer of VEGF. However, studies by Howdieshell and colleagues, utilizing a porcine wound model, determined that VEGF expression occurred in the absence of hypoxia (24). These authors suggested the presence of another inducer of VEGF expression. Further data suggest a role for oxygen and HBO_2 in VEGF regulation. In a study by Sheikh and colleagues utilizing an in vivo wound healing model, stainless steel wire mesh cylinders were implanted beneath the dorsal skin of rats (25). HBO_2 therapy was administered twice daily at 2.1 ATA for 90 minutes over 7 days. Oxygen electrodes were used to measure oxygen tensions from within the site of injury before, during, and after HBO_2 exposure. Baseline wound oxygen tension remained hypoxic (7.4–13.7 mm Hg) throughout the control experiment and increased significantly during HBO_2 to 373.2 to 537.2 mm Hg. Increased wound pO_2 levels were sustained at 45.6 to 150.5 mm Hg for 1 hour after the end of hyperbaric treatment. Fluid samples were collected and the results demonstrated significant increases of VEGF protein in wound fluid samples as early as 2 days after initial HBO_2 exposure, peaking at day 5 (25).

Conversely, in vitro experiments by Lin and colleagues suggest that HBO_2 instead upregulates expression of the proangiogenic factor Ang2, an Ang1 antagonist involved in vascular remodeling and vessel sprouting (23,26). In this study, human umbilical vein ECs (HUVECs) exposed to HBO_2 for 90 minutes at 2.5 ATA selectively expressed Ang2, but not VEGF (26). However, in addition to the difference between model systems, HBO_2 therapy was administered only once as opposed to twice daily for multiple days as in the Sheikh and colleagues in vivo model previously described (26).

Several possible explanations exist for the absence of HBO_2-induced VEGF in HUVECs. One possibility is that prolonged or repetitive HBO_2 exposures may be required to induce VEGF expression. A more plausible explanation is that increased levels of VEGF protein found in the wound fluid samples in vivo may be derived from an alternate cell type. For example, HBO_2 could induce VEGF production from macrophages instead of the endothelium during the inflammatory response (27). Additionally, one must consider differences in cell type and cell origin when evaluating experimental data. HUVECs, commonly used for in vitro studies, may respond differently to HBO_2 than ECs derived from cutaneous endothelium.

Despite the above discrepancies, it is possible to envision a hypothesis explaining how HBO_2 enhances wound healing through the induction of both Ang2 and VEGF. If early HBO_2 exposures are able to induce endothelial Ang2 expression, whereas prolonged HBO_2 exposures induce VEGF, then in the context of poor blood perfusion observed in wound models, HBO_2's function would be to increase oxygen tension levels, reduce metabolic cellular dysfunction, and promote an environment supporting neovascularization. The formation of new blood vessels would replenish oxygen to the ischemic tissue, and the subsequent adequate supply of oxygen would enhance wound closure.

The mechanism explaining how HBO_2 could increase VEGF levels is currently unknown, although regulation may occur via NO- and hypoxia-inducible factor (HIF)-1–dependent mechanisms. *VEGF* is transcriptionally induced in hypoxia by the binding of HIF-1 to hypoxia-response elements in the 5′ promoter region of the *VEGF* gene (28,29). Furthermore, a recent report suggests that NO is also a regulator of *VEGF* expression. Under normoxic conditions, NO donors induce *VEGF* expression by enhancing HIF-1 binding activity in human glioblastoma A-172 and hepatoma Hep3B cells (30). This may explain in part the paradox of VEGF synthesis under both hypoxic and hyperoxic conditions, as HIF-1 binding and activity is enhanced by both hypoxia and NO. Recently, it has been shown that HBO_2 may both induce NO production and stabilize HIF-1 protein, providing a mechanism whereby HBO_2 could directly influence VEGF expression (31,32).

Hyperbaric Oxygen, Nitric Oxide, and Endothelial Nitric Oxide Synthase

NO is a free radical gas produced by a number of cellular sources, including the endothelium, and exerts multiple beneficial effects on wound healing (33). Three isoforms of NO synthase are responsible for NO production. Each isoform shares approximately 50% sequence identity with N-terminal variability. All three isoforms – endothelial NO synthase (eNOS), inducible NOS (iNOS), and neuronal NOS (nNOS) – generate NO in an oxygen-dependent manner through the conversion of L-arginine to L-citrulline.

eNOS-generated NO diffuses from the EC to vascular smooth muscle cells, where it binds to the heme group within the active site of guanyl cyclase. This binding increases cyclic guanosine-3′,5′-monophosphate production, which activates an enzymatic cascade that leads to smooth muscle relaxation and vasodilation. eNOS protein is preferentially expressed in ECs. Regulation of eNOS production and activity is multifaceted. Hyperoxia, as a result of inhalation of 100% pure O_2, and shear stress within the endothelial lumen upregulate eNOS mRNA and protein expression, whereas intracellular calcium influx and calcium–calmodulin binding modulates eNOS activity (34,35).

The second NOS isoform, iNOS, is expressed in many cell types, including ECs, vascular smooth muscle cells, macrophages, polymorphonuclear leukocytes (PMNLs), and fibroblasts, following exposure to lipopolysaccharide (LPS), interleukin (IL)-1β, or tumor necrosis factor (TNF)-α. Although iNOS accounts for the majority of NO production, it is not constitutively expressed in cells. The third NOS isoform, nNOS, is expressed predominantly in the CNS.

Multiple studies support the role of NO in wound healing. Applying exogenous NO to wounds enhances wound closure in diabetic mouse models (36). Further studies by Boykin and colleagues have demonstrated a decrease in wound closure rates in diabetic patients with decreased NO production (37). In vivo studies also demonstrate that both *iNOS*- and *eNOS*-knockout mice exhibit impaired wound healing, suggesting a specific role for eNOS in wound closure (38,39). Further wound healing experiments on aortic segments harvested from eNOS-deficient mice show retarded EC migration and EC proliferation (38). Finally, recent data utilizing an in vitro HUVEC model suggest that neovascularization occurs via an *eNO* S-dependent pathway (26).

The beneficial effects of HBO_2 on wound healing may be mediated by eNOS production of NO. The contribution of eNOS-generated NO from HBO_2 exposure was assessed in an in vitro model of I/R injury. Exposure of human and bovine ECs (BAECs) to HBO_2 resulted in upregulation of eNOS but not iNOS protein, suggesting that the total contribution of endothelial NO is derived from eNOS (35). Studies using *iNOS*- and *eNOS*-knockout mice have demonstrated impaired wound closure rates despite HBO_2 treatment (40). Evaluation of wound tissue demonstrated reduced levels of PDGF receptor-β and epidermal growth factor receptor (EGFR) expression in response to HBO_2 as compared to wild-type mice (40). Taken together, the data imply that the mechanism of HBO_2-enhanced wound closure may be via the NOS-dependent upregulation of wound growth factors. Additionally, a recent in vitro study shows that HBO_2-induced Ang2 expression appears to be mediated by eNOS production of NO (26).

Unifying Mechanisms of Hyperbaric Oxygen in Wound Healing

Studies to date implicate eNOS generation of NO as a unifying mechanism of HBO_2-induced wound healing. Initial exposure of the endothelium to HBO_2 upregulates eNOS expression, leading to increases in NO levels within the microvasculature. The presence of NO decreases vasomotor tone and increases blood flow to the injured tissue. Results from wound fluid samples in the in vivo rat model suggest that the incoming blood flow will have increased pO_2 levels, preventing further ischemia-related tissue damage and promoting angiogenesis. Simultaneously, the increase in blood flow will increase the shear stress within the lumen of the endothelium and together with hyperoxia may further increase the production of eNOS protein and NO. A second effect of HBO_2-induced endothelial-derived NO may be increased production of Ang2, with subsequent sprouting of new vessels. Further investigations with Ang2 production in vivo are required to confirm the link between eNOS and NO-induced Ang2 expression. Finally, it has been demonstrated that NO may increase VEGF levels through HIF-1, leading to enhanced vessel formation.

Taken together, it appears that HBO_2 is influencing angiogenesis through NO-dependent mechanisms.

HYPERBARIC OXYGEN AND THE ENDOTHELIUM: ISCHEMIA–REPERFUSION INJURY

I/R injury is a complex biphasic process that results in metabolically induced cellular dysfunction and cell death. The initial phase of I/R injury is an acute interruption of blood flow within the vasculature that creates ischemia – defined as a cessation of blood flow – and an acute tissue environment of hypoxic and hypoglycemic conditions. The endothelium represents the first population of cells affected by acute ischemia and initiates multiple cellular, immunological, and physiological responses in an attempt to regain metabolic homeostasis. Although the initial period of ischemia may be corrected by removal of the obstructing lesion, a secondary period of tissue damage occurs during reperfusion involving impaired microvascular flow and is referred to as the "no-reflow phenomenon" (41). The secondary phase of microvascular stasis may be attributed to alterations in vessel tone and adhesion of PMNLs to the endothelium, with ensuing cellular and humoral mediator activation.

Prior to PMNL migration into ischemic tissue, microvascular PMNL plugs may exacerbate ischemic conditions through physical obstruction, EC activation and activation of other humoral inflammatory mediators such as the coagulation and complement cascade. Further damage to injured tissue is caused by the release of free radicals and destructive enzymes by PMNLs following endothelial transmigration (42). Indeed, tissue destruction caused by PMNL invasion into ischemic tissue may be responsible for the infarct extension observed in myocardial infarction and stroke (43–45). Furthermore, PMNL/endothelial-mediated reperfusion injury following myocardial infarction may contribute to greater scar formation during ventricular remodeling, which is a predictor of a poorer clinical outcome (46). Studies demonstrate that inhibition of PMNL adherence to the endothelium or the addition of free radical scavengers improves clinical outcome (47,48). Thus, interruption of PMNL adhesion to the endothelium and prevention/limitation of free radical formation represents a central focus in I/R injury prevention research.

HBO_2 has been suggested as a therapeutic intervention for I/R injury and appears to exert a beneficial effect through both disruption of PMNL adhesion and inhibition of free radical generation (49). Experimental studies have been performed evaluating the mechanisms of HBO_2 benefit in I/R injury models of muscle and brain in vivo and EC-PMNL interactions in vitro (35,50–54). These outcomes are surprising and paradoxical, as one would normally predict that reperfusion of previously ischemic tissue with supranormal pO_2 concentrations would only serve to enhance the production of reactive oxygen species (ROS) and cellular damage. However, multiple studies have shown that although HBO_2 may increase ROS formation, cellular damage, as assessed by lipid peroxidation, is actually reduced (49). The mechanism for this observation, referred to as the "oxygen paradox," remains largely undefined.

Early observations demonstrating that HBO_2 reduces edema and necrosis in ischemic skeletal muscle stimulated subsequent investigations into the beneficial mechanisms of HBO_2 in I/R injury (55). The primary benefit of HBO_2 in I/R injury appears to be inhibition of PMNL-EC interactions (35,56). HBO_2 was first noted to reduce PMNL adherence to the endothelial lining as determined by intravital microscopy in rodent gracilis muscle preparations following experimental I/R injury (57). Subsequent investigations on mechanisms of HBO_2 in I/R injury have focused on the CD11/18-ICAM-1 adhesion molecule axis.

HYPERBARIC OXYGEN AND THE PMNL–ENDOTHELIAL ADHESION AXIS

The interaction of membrane-bound PMNL CD11/18 and endothelial ICAM-1 mediates firm adhesion of PMNLs to the endothelium. Briefly, CD11/18 is a heterodimeric protein member of the integrin family of adhesion molecules that is constitutively expressed on the PMNL cell surface (48). Binding of CD11/18 to its primary ligand, ICAM-1 causes a bidirectional cellular activation and is responsible for firm adhesion of PMNLs to the endothelial surface. ICAM-1 is an inducible 90- to 114-kDa cell surface glycoprotein and a member of the immunoglobulin superfamily (48). ICAM-1 expression may be induced by a variety of proinflammatory mediators, such as cytokines and bacterial endotoxins. I/R injury is known to induce endothelial expression of ICAM-1 in vivo (48). However, ICAM-1 induction appears to require both hypoxic and hypoglycemic conditions, because hypoxia alone, or hypoxia with reoxygenation, does not induce ICAM-1 expression in human aortic ECs (HAECs), HUVECs, or BAECs (35,58). The combination of hypoxia and hypoglycemia followed by normoxic/normoglycemic conditions is required to induce ICAM-1 expression in HUVECs and BAECs (35).

Initial studies on HBO_2 interference with the CD11/18–ICAM-1 axis were performed in a model of CO poisoning. CO represents a subtype of I/R injury by creating global ischemia as opposed to the classic compartmentalized ischemic models (59). CO poisoning creates localized hypoxia through binding of CO to heme structures found in Hb and cytochromes, diminishing oxygen transfer to Hb in the capillary lung beds and preventing the dissociation of oxygen-bound Hb in the systemic capillary circulation, thus interfering with oxidative metabolism. The effects of CO poisoning after CO elimination are analogous to transient vascular interruption of cerebral blood flow and result in increased levels of lipid membrane peroxidation, leukocyte infiltration, hypotension, and cerebral hypoperfusion (59). In vivo studies of CO poisoning in rodent models demonstrated that the beneficial effect of HBO_2 is mediated by interruption of CD11/18-mediated adhesion to the endothelium and not by suppression of PMNL activation (60). Further studies on human blood samples exposed to HBO_2 confirm the inhibitory effect of HBO_2 on PMNL adherence (60). The mechanism of inhibition appears to involve inhibition of membrane guanylate cyclase (60). The inhibitory effects of HBO_2 exposure at 2.8 ATA or 3.0 ATA on human PMNL adherence and respiratory burst function were reversible after 21 hours, suggesting that HBO_2 may be nondetrimental to innate host defense mechanisms (53).

The importance of endothelial ICAM-1 in mediating I/R injury has been described in multiple models (48). Inhibition of ICAM-1 via antibody interference inhibits PMNL EC binding in vitro and blocks PMNL binding, tissue infiltration, and tissue damage in vivo (61). Further in vivo studies using mice genetically deficient in ICAM-1 support the beneficial effects of ICAM-1 inhibition during renal I/R injury; *ICAM-1*$^{-/-}$ animals demonstrated decreased PMNL infiltration and were protected from ischemic injury to the kidney (62). Additionally, ICAM-1–deficient mice prove resistant to I/R injury in cerebral I/R injury models (63). Thus, both in vitro and in vivo data suggest that endothelial ICAM-1 may represent a valuable target for therapeutic intervention in I/R injury.

I/R injury models demonstrating HBO_2 prevention of PMNL/endothelium adhesion led to recent studies investigating the effects of HBO_2 exposure on ICAM-1 expression (35,36). In an in vitro I/R injury model, HUVECs and BAECs were exposed to hypoxia and hypoglycemia (mock ischemia) for 4 hours and then exposed to normoxia and normoglycemia for 20 hours (mock reperfusion) (35). ECs exposed to 2.5 ATA HBO_2 for 90 minutes at the initiation of mock reperfusion demonstrated a significant reduction in ICAM-1 protein expression. HBO_2 treatment also prevented adhesion of PMNLs to mock I/R injured ECs, suggesting that HBO_2 treatment may also prevent PMNL adhesion through ICAM-1 regulation (35). The ability of HBO_2 to reduce ICAM-1 expression following I/R injury was recently confirmed in vivo using a rodent musculocutaneous flap model (56).

HBO_2 is similar to any other pharmacological mediator as it exhibits a clear dose-dependent effect. For example, in contrast to HBO_2's suppressive effect on endothelial ICAM-1 expression following I/R injury, other studies have shown that isolated, prolonged HBO_2 exposure of healthy mice to 3.0 ATA for 3 to 6 hours or isolated ECs 100% oxygen at 1 ATA for 24 hours or more actually upregulates ICAM-1 expression and enhances PMNL adhesion (64,65). As a result, it appears that the endothelium is sensitive to the duration of HBO_2 exposure and that the EC metabolic state may influence the ultimate cellular response.

Hyperbaric Oxygen, Nitric Oxide, and Ischemia–Reperfusion Injury

NO may play a central role in the mechanism of HBO_2 protection during I/R injury. As a result of I/R injury damage to the endothelium, NO production by eNOS is compromised, and several studies have demonstrated that NO is protective through inhibition of PMNL infiltration and PMNL-induced tissue damage during I/R injury (41,66,67). As described earlier, HBO_2 selectively induces eNOS protein in cultured HUVECs and BAECs (35). Furthermore, following I/R injury, HBO_2-mediated downregulation of ICAM-1 production is prevented by the presence of L-N-arginine methyl ester (L-NAME), a NOS inhibitor (35). Consequently, HBO_2 appears to regulate ICAM-1 production on the EC surface via NO production from eNOS. Recently, HBO_2 was shown to increase NO production by eNOS and also to increase *eNOS* mRNA expression in vivo following I/R injury in a rodent gracilis muscle model (31). These studies confirm the central role of the endothelium in mediating HBO_2 protection following I/R injury.

Unifying Mechanisms of Hyperbaric Oxygen in Ischemia–Reperfusion Injury

The mechanism of HBO_2 action in preventing I/R injury appears to be associated with an increase in eNOS and NO production, a decrease in ICAM-1 production, and a concomitant reduction in PMNL–endothelial adhesion. Interestingly, the expression of other endothelial adhesion molecules, such as E-selectin, may also be suppressed by HBO_2 following I/R injury (68). Review of the adhesion molecule transcriptional control reveals that the heterodimeric transcription factor nuclear factor (NF)-κB serves a positive, central role in gene regulation (69). Furthermore, NF-κB activity is negatively regulated by increased NO levels in ECs, as exogenously applied NO may reduce NF-κB activation and adhesion molecule gene transcription (66). These data support the possibility that the mechanism of HBO_2 involves increased production of NO from eNOS with subsequent downregulation of NF-κB activity and adhesion molecule expression. Furthermore, NO may exert a suppressive effect on PMNL membrane guanylate cyclase, which could then interfere with CD18 function (70). In this regard, HBO_2-induced NO could have an anti-adhesive effect on both components of the PMNL–endothelial axis during I/R injury (Figure 53.2).

Preliminary work suggests that HBO_2 may inhibit NF-κB DNA-binding activity and nuclear translocation of the p65 subunit of NF-κB in ECs using an in vitro I/R injury model (J. Buras, unpublished observations). Other indirect evidence supporting this hypothesis comes from in vivo studies on the HBO_2 regulation of Toll-like receptor 4 (TLR4) activation. LPS-challenged rats exposed to HBO_2 at 2.5 ATA for 60 minutes demonstrated reduced PMNL intestine mucosal infiltration and also inhibited nuclear NF-κB translocation to the nucleus (71). Based on these observations, it is plausible that HBO_2 may downregulate NF-κB activation and serve as an anti-inflammatory stimulus in the setting of I/R.

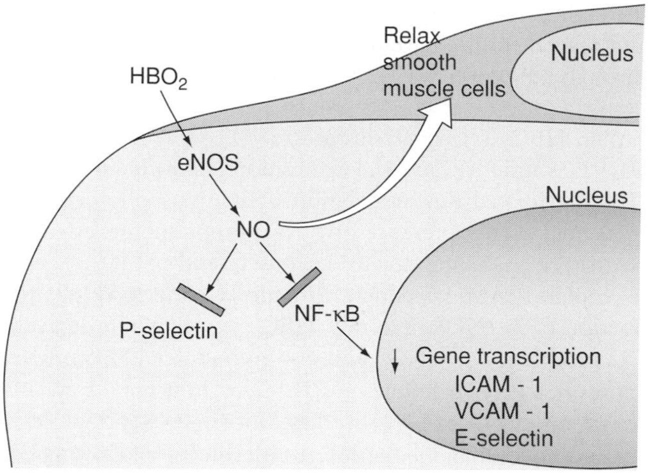

Figure 53.2. Unifying antiadhesive effects of HBO_2 on the endothelium during I/R injury. HBO_2 may affect multiple components of EC-inflammatory cell adhesion. Effects of HBO_2 may be mediated via activation of NO production from eNOS, resulting in local vessel dilation, inhibition of NF-κB activation with subsequent reduction of cell adhesion molecule expression.

CONCLUSION

The effect of HBO_2 on the cellular response is clearly linked to the exact disease state of the host (e.g., during wound healing or I/R injury). Furthermore, the effect of HBO_2 on the endothelium is dose-dependent and suggests that further emphasis be placed on the development of oxygen as a drug.

KEY POINTS

- HBO_2 therapy has a wide range of effects on the endothelium during several different disease states, including improvement of microvascular flow, induction of proangiogenic factors, augmentation of inducible cell adhesion molecule expression, and regulation of NO production.
- Exposure of the endothelium to HBO_2 upregulates eNOS expression, leading to increases in NO levels within the microvasculature.
- NO may increase VEGF levels through HIF-1.
- HBO_2-induced endothelial-derived NO may increase production of Ang2.
- HBO_2 reduces PMNL–EC interaction by reducing CD11/18 function and downregulating EC ICAM-1.
- Reduction of EC ICAM-1 by HBO_2 is dependent on NO production.
- HBO_2 may inhibit adhesive EC properties by reducing NF-κB activation.

Future Goals

- To clarify the signaling mechanisms responsible for the observed effects of HBO_2 and explain how the same elevation in Po_2 may elicit disparate cellular responses in different disease states

REFERENCES

1 Tibbles PM, Edelsberg JS. Hyperbaric-oxygen therapy. *N Engl J Med*. 1996;334(25):1642–1648.
2 Faglia E, Favales F, Aldeghi A, et al. Adjunctive systemic hyperbaric oxygen therapy in treatment of severe prevalently ischemic diabetic foot ulcer. A randomized study. *Diabetes Care*. 1996;19(12):1338–1343.
3 Weaver L. Double-blind randomized control trial in acute carbon monoxide poisoning. *Int J Technol Assess Health Care*. 2005; 21(1):151.
4 Kindwall E, Whelan H. *Hyperbaric Medicine Practice*. Flagstaff, AZ: Best Publishing Company, 1999.
5 Simpson A. *Compressed Air as a Therapeutic Agent in the Treatment of Consumption, Asthma, Chronic Bronchitis and Other Diseases*. Edinburgh: Sutherland and Knox, 1857.
6 Arntzenius A. *De Pneumatische Therapie*. Amsterdam: Scheltema and Holkemas Boekhandel, 1887.
7 Fontaine J. Emploi chirurgical de l'air comprime. *Union Med*. 1879;28:445.
8 Moir E. Tunnelling by compressed air. *J Soc Arts*. 1896;XLIVLIV:567–83.
9 Churchill-Davidson I, Sanger C, Thomlinson R. High-pressure oxygen and radiotherapy. *Lancet*. 1955;1:1091–1095.
10 Boerema I, Kroll J, Meijne N, et al. High atmospheric pressure as an aid to cardiac surgery. *Arch Chir Neerl*. 1956;8:193–211.
11 Boerema I, Meyne NG, Brummelkamp WH, et al. [Life without blood.]. *Ned Tijdschr Geneeskd*. 1960;104:949–954.
12 Kranke P, Bennett M, Roeckl-Wiedmann I, Debus S. Hyperbaric oxygen therapy for chronic wounds. *Cochrane Database Syst Rev*. 2004(2):CD004123.
13 Masterson JS, Fratkin LB, Osler TR, Trapp WG. Treatment of pneumatosis cystoides intestinalis with hyperbaric oxygen. *Ann Surg*. 1978;187(3):245–247.
14 Bennett SP, Griffiths GD, Schor AM, et al. Growth factors in the treatment of diabetic foot ulcers. *Br J Surg*. 2003;90(2):133–146.
15 Newton DJ, Bennett SP, Fraser J, et al. Pilot study of the effects of local pressure on microvascular function in the diabetic foot. *Diabet Med*. 2005;22(11):1487–1491.
16 Reenstra WR, Olrow D, Buras JA. Expression of growth factor recpetors in dermal fibroblasts in response to hyperbaric oxygen is nitric oxide dependent. *Undersea Hyperb Med*. 2005.
17 Sen CK, Khanna S, Babior BM, et al. Oxidant-induced vascular endothelial growth factor expression in human keratinocytes and cutaneous wound healing. *J Biol Chem*. 2002;277(36):33284–33290.
18 Knighton DR, Silver IA, Hunt TK. Regulation of wound-healing angiogenesis-effect of oxygen gradients and inspired oxygen concentration. *Surgery*. 1981;90(2):262–270.

19 Gibson JJ, Angeles AP, Hunt TK. Increased oxygen tension potentiates angiogenesis. *Surg Forum*. 1997;48:696–699.
20 Cianci P. Adjunctive hyperbaric oxygen therapy in the treatment of diabetic foot wounds. In: Levin ME, O'Neal LW, Bowker JH, eds. *The Diabetic Foot*. St. Louis: Mosby-Year Book Inc.; 1993: 305–20.
21 Hanahan D. Signaling vascular morphogenesis and maintenance. *Science*. 1997;277(5322):48–50.
22 Bussolati B, Dunk C, Grohman M, et al. Vascular endothelial growth factor receptor-1 modulates vascular endothelial growth factor-mediated angiogenesis via nitric oxide. *Am J Pathol*. 2001; 159(3):993–1008.
23 Maisonpierre PC, Suri C, Jones PF, et al. Angiopoietin-2, a natural antagonist for Tie2 that disrupts in vivo angiogenesis. *Science*. 1997;277(5322):55–60.
24 Howdieshell TR, Riegner C, Gupta V, et al. Normoxic wound fluid contains high levels of vascular endothelial growth factor. *Ann Surg*. 1998;228(5):707–715.
25 Sheikh AY, Gibson JJ, Rollins MD, et al. Effect of hyperoxia on vascular endothelial growth factor levels in a wound model. *Arch Surg*. 2000;135(11):1293–1297.
26 Lin S, Shyu KG, Lee CC, et al. Hyperbaric oxygen selectively induces angiopoietin-2 in human umbilical vein endothelial cells. *Biochem Biophys Res Commun*. 2002;296(3):710–715.
27 Walczak R, Joseph SB, Laffitte BA, et al. Transcription of the vascular endothelial growth factor gene in macrophages is regulated by liver X receptors. *J Biol Chem*. 2004;279(11):9905–9911.
28 Wang GL, Jiang BH, Rue EA, Semenza GL. Hypoxia-inducible factor 1 is a basic-helix-loop-helix-PAS heterodimer regulated by cellular O2 tension. *Proc Natl Acad Sci USA*. 1995;92:5510–5514.
29 Semenza GL, Agani F, Booth G, et al. Structural and functional analysis of hypoxia-inducible factor. *Kidney Int*. 1997;51(2):553–555.
30 Kimura H, Weisz A, Kurashima Y, et al. Hypoxia response element of the human vascular endothelial growth factor gene mediates transcriptional regulation by nitric oxide: control of hypoxia-inducible factor-1 activity by nitric oxide. *Blood*. 2000;95(1): 189–197.
31 Baynosa RC, Naig AL, Murphy PS, et al. The effect of hyperbaric oxygen on NOS activity and transcription in ischemia reperfusion injury. *J Am Coll Surg*. 2005;201:S57–S58.
32 Salhanick SD, Belikoff B, Holt DE, et al. Hyperbaric oxygen protects from acetaminophen toxicity and induces hypoxia-inducible factor-1 alpha. *Undersea Hyperb Med*. 2005.
33 Lowenstein CJ, Snyder SH. Nitric oxide, a novel biologic messenger. *Cell*. 1992;70(5):705–707.
34 Black SM, Johengen MJ, Ma ZD, et al. Ventilation and oxygenation induce endothelial nitric oxide synthase gene expression in the lungs of fetal lambs. *J Clin Invest*. 1997;100(6):1448–1458.
35 Buras JA, Stahl GL, Svoboda KK, Reenstra WR. Hyperbaric oxygen downregulates ICAM-1 expression induced by hypoxia and hypoglycemia: the role of NOS. *Am J Physiol Cell Physiol*. 2000; 278(2):C292–C302.
36 Masters KS, Leibovich SJ, Belem P, et al. Effects of nitric oxide releasing poly(vinyl alcohol) hydrogel dressings on dermal wound healing in diabetic mice. *Wound Repair Regen*. 2002;10(5): 286–294.
37 Boykin JV. How hyperbaric oxygen therapy helps heal chronic wounds. *Nursing*. 2002;32(6):24.
38 Lee PC, Salyapongse AN, Bragdon GA, et al. Impaired wound healing and angiogenesis in eNOS-deficient mice. *Am J Physiol*. 1999;277(4 Pt 2):H1600–H1608.
39 Yamasaki K, Edington HD, McClosky C, et al. Reversal of impaired wound repair in iNOS-deficient mice by topical adenoviral-mediated iNOS gene transfer. *J Clin Invest*. 1998; 101(5):967–971.
40 Reenstra WR, Holt DE, Orlow D, Buras JA. Hyperbaric oxygen increases growth factor receptor expression and wound healing in vivo. *Acad Emerg Med*. 2005;12(5Suppl 1):8.
41 Carden DL, Granger DN. Pathophysiology of ischaemia-reperfusion injury. *J Pathol*. 2000;190(3):255–266.
42 Weiss SJ. Tissue destruction by neutrophils. *N Engl J Med*. 1989; 320(6):365–376.
43 Engler RL, Dahlgren MD, Morris DD, Peterson MA, et al. Role of leukocytes in response to acute myocardial ischemia and reflow in dogs. *Am J Physiol*. 1986;251(2 Pt 2):H314–H323.
44 Hallenbeck JM, Dutka AJ, Tanishima T, et al. Polymorphonuclear leukocyte accumulation in brain regions with low blood flow during the early postischemic period. *Stroke*. 1986;17(2):246–253.
45 Schofer J, Montz R, Mathey DG. Scintigraphic evidence of the "no reflow" phenomenon in human beings after coronary thrombolysis. *J Am Coll Cardiol*. 1985;5(3):593–598.
46 Ito H, Maruyama A, Iwakura K, et al. Clinical implications of the "no reflow" phenomenon. A predictor of complications and left ventricular remodeling in reperfused anterior wall myocardial infarction. *Circulation*. 1996;93(2):223–228.
47 Tamura Y, Chi LG, Driscoll EM, Jr., et al. Superoxide dismutase conjugated to polyethylene glycol provides sustained protection against myocardial ischemia/reperfusion injury in canine heart. *Circ Res*. 1988;63(5):944–959.
48 Kakkar AK, Lefer DJ. Leukocyte and endothelial adhesion molecule studies in knockout mice. *Curr Opin Pharmacol*. 2004; 4(2):154–158.
49 Buras J. Basic mechanisms of hyperbaric oxygen in the treatment of ischemia-reperfusion injury. *Int Anesthesiol Clin*. 2000; 38(1):91–109.
50 Veltkamp R, Siebing DA, Heiland S, et al. Hyperbaric oxygen induces rapid protection against focal cerebral ischemia. *Brain Res*. 2005;1037(1–2):134–138.
51 Mink RB, Dutka AJ. Hyperbaric oxygen after global cerebral ischemia in rabbits reduces brain vascular permeability and blood flow. *Stroke*. 1995;26(12):2307–2312.
52 Sirsjo A, Lehr HA, Nolte D, et al. Hyperbaric oxygen treatment enhances the recovery of blood flow and functional capillary density in postischemic striated muscle. *Circ Shock*. 1993;40(1):9–13.
53 Thom SR. Functional inhibition of leukocyte B2 integrins by hyperbaric oxygen in carbon monoxide-mediated brain injury in rats. *Toxicol Appl Pharmacol*. 1993;123(2):248–256.
54 Zamboni WA, Roth AC, Russell RC, Graham B, et al. Morphologic analysis of the microcirculation during reperfusion of ischemic skeletal muscle and the effect of hyperbaric oxygen. *Plast Reconstr Surg*. 1993;91(6):1110–1123.
55 Nylander G, Lewis D, Nordstrom H, Larsson J. Reduction of postischemic edema with hyperbaric oxygen. *Plast Reconstr Surg*. 1985;76(4):596–603.
56 Hong JP, Kwon H, Chung YK, Jung SH. The effect of hyperbaric oxygen on ischemia-reperfusion injury: an experimental study in a rat musculocutaneous flap. *Ann Plast Surg*. 2003;51(5):478–487.

57 Zamboni WA, Roth AC, Russell RC, Nemiroff PM, et al. The effect of acute hyperbaric oxygen therapy on axial pattern skin flap survival when administered during and after total ischemia. *J Reconstr Microsurg*. 1989;5(4):343–347; discussion 9–50.

58 Maurus CF, Schmidt D, Schneider MK, et al. Hypoxia and reoxygenation do not upregulate adhesion molecules and natural killer cell adhesion on human endothelial cells in vitro. *Eur J Cardiothorac Surg*. 2003;23(6):976–983; discussion 83.

59 Thom SR. Leukocytes in carbon monoxide-mediated brain oxidative injury. *Toxicol Appl Pharmacol*. 1993;123(2):234–247.

60 Thom SR, Mendiguren I, Hardy K, et al. Inhibition of human neutrophil beta2-integrin-dependent adherence by hyperbaric O2. *Am J Physiol*. 1997;272(3 Pt 1):C770–C777.

61 Ioculano M, Squadrito F, Altavilla D, et al. Antibodies against intercellular adhesion molecule 1 protect against myocardial ischaemia-reperfusion injury in rat. *Eur J Pharmacol*. 1994; 264(2):143–149.

62 Kelly KJ, Williams WW Jr., Colvin RB, et al. Intercellular adhesion molecule-1-deficient mice are protected against ischemic renal injury. *J Clin Invest*. 1996;97(4):1056–1063.

63 Connolly ES, Jr., Winfree CJ, Springer TA, et al. Cerebral protection in homozygous null ICAM-1 mice after middle cerebral artery occlusion. Role of neutrophil adhesion in the pathogenesis of stroke. *J Clin Invest*. 1996;97(1):209–216.

64 Suzuki Y, Aoki T, Takeuchi O, et al. Effect of hyperoxia on adhesion molecule expression in human endothelial cells and neutrophils. *Am J Physiol*. 1997;272(3 Pt 1):L418–L425.

65 Bowman CM, Butler EN, Vatter AE, Repine JE. Hyperoxia injuries endothelial cells in culture and causes increased neutrophil adherence. *Chest*. 1983;83(5 Suppl):S33-S35.

66 Lefer AM, Lefer DJ. The role of nitric oxide and cell adhesion molecules on the microcirculation in ischaemia-reperfusion. *Cardiovasc Res*. 1996;32(4):743–751.

67 Sato H, Zhao ZQ, Jordan JE, et al. Basal nitric oxide expresses endogenous cardioprotection during reperfusion by inhibition of neutrophil-mediated damage after surgical revascularization. *J Thorac Cardiovasc Surg*. 1997;113(2):399–409.

68 Buras J, Reenstra WR. Hyperbaric oxygen decreases endothelial cell E-selectin protein expression in an in vitro model of ischemia/reperfusion. *Ann Emerg Med*. 1998;32:S17.

69 Collins T, Read MA, Neish AS, et al. Transcriptional regulation of endothelial cell adhesion molecules: NF-kappa B and cytokine-inducible enhancers. *FASEB J*. 1995;9(10):899–909.

70 Banick PD, Chen Q, Xu YA, Thom SR. Nitric oxide inhibits neutrophil beta 2 integrin function by inhibiting membrane-associated cyclic GMP synthesis. *J Cell Physiol*. 1997;172(1): 12–24.

71 Sakoda M, Ueno S, Kihara K, et al. A potential role of hyperbaric oxygen exposure through intestinal nuclear factor-kappaB. *Crit Care Med*. 2004;32(8):1722–1729.

54

Barotrauma

Deborah A. Quinn and Charles A. Hales

Massachusetts General Hospital, Harvard Medical School, Boston, Massachusetts

Clinical barotrauma to the lung induced by mechanical ventilation, including pneumothorax, has been recognized as a problem for many years, but it is only in the past 10 years that the more subtle effects of mechanical ventilation on the lung have been recognized (1). The first comprehensive study that showed mechanical ventilation with high pressures and high tidal volumes in normal lungs caused pulmonary edema was carried out by Webb and Tierney in 1974 (2). In 1988, Dreyfus and colleagues, by using thoracoabdominal strapping to limit lung stretch during mechanical ventilation with high airway pressures, showed no ensuing pulmonary edema occurred with high pressures when the lung was not allowed to stretch (3). A negative pressure extrathoracic ventilator was then used, showing that pulmonary edema occurred at large tidal volumes, even though airway pressures were low. This demonstrated that lung stretch occurring with artificial ventilation, and not airway pressure, was the mechanism causing lung injury. Positive end-expiratory pressure may modify the response to lung stretch as occurs at disproportionately high tidal volumes. Although some data in rats support this concept (1), applying positive end-expiratory pressure to humans with lung injury and lowering the tidal volume simultaneously did not protect the lung (4). Along with lung edema, mechanical ventilation with high tidal volume ventilation has been shown to cause a release of inflammatory cytokines and chemokines with a subsequent influx of neutrophils into the lung (5). This combination of pulmonary edema and lung inflammation caused by barotrauma, or, perhaps more correctly, volutrauma, secondary to mechanical ventilation, has been termed ventilator-induced lung injury (VILI).

The use of low tidal volumes to avoid VILI has been shown to decrease mortality in patients with acute lung injury (ALI) who require mechanical ventilation (6). ALI and the more severe form of ALI, the adult respiratory distress syndrome (ARDS), are associated with nonhomogeneous damage to the lung. As a result, even moderate-sized tidal volumes can cause overdistention of the less severely involved, hence more compliant, areas of the lung. These latter effects may be modeled by applying high tidal volumes to animals with normal lungs. In vitro cell stretch models have been used to study the mechanisms of stretch-induced changes in the endothelium (1).

The aim of this chapter is to show that the endothelium plays a key role in barotrauma from VILI. We present data that support a role for the endothelium in VILI-mediated pulmonary edema (Table 54-1) and lung inflammation (Table 54-2).

LUNG CAPILLARY LEAK AND ENDOTHELIAL INJURY WITH HIGH TIDAL VOLUME VENTILATION

The edema in VILI arises from alterations in permeability rather than changes in hydrostatic pressure. In contrast to hydrostatic pulmonary edema (as occurs with congestive heart failure), VILI-associated permeability-type edema results in increased dry lung weight, increased extravasation of ^{125}I-labeled fibrinogen and Evans blue dye bound to albumin, and increased protein in bronchoalveolar lavage (BAL) fluid (1). Furthermore, we have shown that high-volume mechanical ventilation for 2 hours in rats results in an increased wet/dry lung weight ratio, despite a minimal rise in pulmonary artery pressure and no change in the left atrial pressure (7). The mild increase in pulmonary artery (and presumably intracapillary) pressure was insufficient to induce pulmonary edema, thus confirming that the lung leak in VILI is associated with a disruption of lung microvascular barrier function. In animal models, the development of capillary leak is an early event, occurring as early as after 1 hour of mechanical ventilation (D.A. Quinn and C.A. Hales, unpublished observations) and continues to rise at 2 hours of ventilation (7,8). Edema formation precedes the production of cytokines and the influx of neutrophils, suggesting that cytokines and neutrophils are not necessary for edema formation in the early stages of VILI (8). However neutrophils may contribute to the edema formation in the later stages by preventing its resolution.

The endothelium has been shown to play a key role in mediating high tidal volume ventilation-induced capillary leak

Table 54-1: Mechanical Stretch-Induced EC Alterations That May Contribute to Pulmonary Capillary Leak with High Tidal Volume Ventilation

Endothelial Alteration	*Experimental Model*	*References*
Stress failure	Intact rat model	1
Calcium influx with MLC phosphorylation	EC monolayers Isolated perfused rat lungs	11, 12, 15, 16, 20
Stretch-induced susceptibility to edematogenic agents	EC monolayers	17
Aquaporin channel regulation of fluid flux	Intact rat model	7
Activation of EC tyrosine kinases	Isolated perfused rat lungs	19, 20
Activation of 5-lipoxygenase	Intact mouse model	21
Activation of phospholipase A2	Intact mouse model EC monolayers	23, 24
Activation of iNOS	Intact mouse model	22

through two mechanisms (Table 54-3): (a) direct barotrauma-induced stress failure and (b) stretch-induced changes in endothelial intracellular signaling and cell membrane leading to barrier dysfunction.

VILI is accompanied by severe morphological damage to the alveolar endothelium. Ultrastructural studies using electron microscopy have revealed detachment of endothelial cells (ECs), gaps in the endothelium and focal endothelial blebs, and accumulation of interstitial edema. The damage to the epithelium is patchy. In some areas complete destruction of Type 1 cells occurs, whereas Type II cells are preserved (1). In addition, plasma von Willebrand factor (vWF), a biochemical marker of endothelial injury, was significantly elevated in rats with acid-injured lung ventilated with 12 mL/kg as compared with rats ventilated with 6 or 3 mL/kg (9). Together these data suggest that VILI may cause direct injury to the endothelium and stress failure of the capillary membrane; inactivation of surfactant leads to increased filtration (1). The combination of increased capillary leak and increased filtration leads to flooding of the alveoli with high-protein edema. Cyclic stretch also increases transepithelial permeability in vitro and may also contribute to lung stretch–induced capillary leak (10). However, the contributions of epithelial cell permeability are less well studied. Stretch-induced transepithelial permeability was not altered by inhibition of protein kinase C (PKC) or tyrosine kinase in vitro (10).

In addition to causing mechanical stress failure, high tidal volume ventilation may promote changes in EC signaling, which in turn contribute to capillary leak in VILI, including influx of intracellular calcium and alterations of myosin light chain (MLC) kinase, phosphotyrosine phosphatase, tyrosine kinase, 5-lipoxygenase (5-LO), inducible nitric oxide synthase (iNOS), and phospholipase A2 activity. Mechanical stretch has been shown to initiate calcium entry through nonspecific cation channels of endothelial monolayers and lung epithelial cells (11–13). In isolated perfused rat lungs, Parker and colleagues found that gadolinium, a nonspecific blocker of stretch-activated cation channels, attenuated the increase in capillary filtration caused by ventilation with high pressures (14). Under in vitro conditions, stretch-induced calcium entry has been associated with activation of MLC phosphorylation and increased permeability of pulmonary artery EC monolayers (15). Inhibition of MLC phosphorylation or inhibition of calcium entry through calcium channels blocked high tidal volume ventilation-induced capillary filtration in isolated perfused rat lungs (16). Pulmonary artery ECs exposed to cyclic mechanical strain of 18% but not 5% caused an increase in MLC, p38 mitogen-activated protein kinase (MAPK), and extracellular signal-regulated kinase (ERK)1/2 phosphorylation and actomyosin remodeling (17). Cyclic mechanical strain (18%) also significantly enhanced thrombin-stimulated MLC phosphorylation and thrombin-induced pulmonary artery endothelial monolayer permeability, as measured by transendothelial electrical resistance. Mice deficient in MLC kinase had increased length of survival with lipopolysaccharide (LPS) plus VILI as compared to wild-type mice. The amount of lung edema and inflammation was not compared between knockout mice and wild-type mice in this study (18).

Lung stretch–induced capillary permeability has been reported to be dependent on activation of tyrosine kinases. Bhattacharya and colleagues found that lung expansion activated rat pulmonary endothelial tyrosine kinases in isolated blood-perfused rat lungs (19). Parker and colleagues used a phosphotyrosine phosphatase inhibitor (phenylarsine) and a tyrosine kinase inhibitor (genistein) to assess the role of these enzymes in lung stretch–induced capillary permeability in isolated perfused rat lungs (20). Inhibition of phosphotyrosine phosphatase was shown to lead to an increase in susceptibility of rat lungs to high tidal volume ventilation, whereas inhibition of tyrosine kinase attenuated capillary permeability.

The use of knockout mice has contributed to the investigation of role 5-lipoxygenase, iNOS, and phospholipase A2 activity in lung stretch–induced capillary leak. (See Table 54-4 for a summary of knockout mice used in the study of VILI.) Both mice deficient in 5-lipoxygenase and iNOS had decreased capillary leak and lung inflammation when exposed to high tidal volume ventilation as compared to wild-type litter mates (21,22). To explore the role of phospholipase A2 activity in VILI, both mice deficient in Clara cell secretory protein (CCSP), which have decreased phospholipase A2 activity, and inhibitors of phospholipase A2 activity were used (23). Increased phospholipase A2 activity had been found in bovine aortic ECs exposed to cyclic mechanical strain, and it was

Table 54-2: Mechanical Stretch-Induced EC Alterations That May Contribute to Cytokine Production and Lung Neutrophil Influx with High Tidal Volume Ventilation

Endothelial Alteration	*Experimental Model*	*References*
Increased expression of ICAM, P-selectin, and L-selectin	EC monolayers Intact mouse model	40–42, 44
Increased production of IL-8	EC monolayers	31, 32
Activation of JNK and NF-κB	EC monolayers Intact mouse model	31, 32
Stretch-induced oxidant injury	EC monolayers	36
Integrin activation dependent production of IL-6	EC monolayers	40–42
Low-molecular-weight hyaluronan-induced production of IL-8 through Toll-like receptors 2 and 4, CD44, and RHAMM binding	EC monolayers	48, 49
Activation of 5-lipoxygenase	Intact mouse model	21
Activation of phospholipase A2	EC monolayers Intact mouse model	23, 24
Activation of iNOS	Intact mouse model	22

hypothesized that lung stretch would also increase phospholipase A2 activity (24). In mice exposed to large tidal volume mechanical ventilation, phospholipase A2 activity was increased in wild-type mice as compared to nonventilated controls and even higher levels in *CCSP*$^{-/-}$ mice exposed to large tidal volume mechanical ventilation. The use of a phospholipase A2 inhibitor decreased phospholipase A2 activity in both wild-type and *CCSP*$^{-/-}$ mice exposed to large tidal volume mechanical ventilation. Increased levels of phospholipase A2 activity was correlated with increased pulmonary capillary leak and lung inflammation. Inhibition of phospholipase A2 activity decreased both lung inflammation and lung capillary leak. TNF (tumor necrosis factor)-α was found to be not critical for lung stretch–induced capillary leak in knockout mice (25).

In addition to expressing barrier properties, the normal lung microvascular endothelium regulates lung fluid osmotic water permeability through water channels. Ten types of aquaporin channels that regulate water movement exist. Aquaporin 1 channels are strongly expressed throughout the microvascular endothelial beds outside the brain (26). In the lung endothelium, they are preferentially localized in the caveolae (7). Aquaporin 5 is also found in the lung on the apical membrane epithelial cells (26). We found that $HgCl_2$, which inhibits aquaporin channels by binding to cysteine on the channel, worsened high-tidal volume-induced pulmonary edema (Figure 54.1) (7). Infusion of cysteine, by providing alternative binding sites for $HgCl_2$, blocked this effect. There was no effect of $HgCl_2$ on lung edema with low tidal volumes. Though aquaporin 1 and aquaporin 5 have been shown to regulate lung fluid osmotic water permeability, neither *aquaporin 1*- nor *aquaporin 5*-knockout mice showed evidence of impaired alveolar clearance (26). The effects of high tidal volume ventilation have not been studied in *aquaporin 1*- or *aquaporin 5*-knockout mice. *Aquaporin 1*-knockout mice have not shown alterations in lung edema in other models of ALI, including sepsis and acid aspiration (27,28). Other factors may account for the role for aquaporins in lung edema formation in VILI. It is possible that stretch may have increased aquaporin channel expression or activity, but this has not been studied. Alternatively, the stress failure of the endothelium as seen on electron microscopy may also lead to loss of aquaporin channels and decreased ability to regulate fluid osmotic permeability secondary to the high levels of protein leak found in VILI.

In summary, the endothelium plays an important role in the lung stretch–induced pulmonary capillary leak both through the direct effects of barotrauma leading to stress failure and by stretch-induced intracellular and membrane changes that cause increased endothelial layer permeability.

THE ROLE OF THE ENDOTHELIUM IN HIGH TIDAL VOLUME VENTILATION-INDUCED INFLAMMATION

In contrast to its early effects on endothelial permeability, large tidal volume ventilation results in a delayed induction of cytokine production and neutrophil influx. Studies of high tidal volume ventilation in animal models (including mice, rats, sheep, and rabbits) have demonstrated increased levels of macrophage inflammatory protein (MIP)-2 (which is analogous to interleukin [IL]-8 in humans), IL-1β, IL-6,

Table 54-3: Endothelial vs. Epithelial Contributions to VILI

	Endothelial	*Epithelial*	*Whole Lung*
Direct Injury			
Detachment	√ (VILI)		
Gaps	√ (VILI)		
Destruction		√ (VILI)	
Cell Signaling			
Ca entry	√ (CS) √ (perfusion)	√ (CS)	
MLC Pi	√ (CS)		
p38 MAPK Pi	√ (CS)	√ (CS)	√ lung parenchyma
ERK Pi	√ (CS)	√ (CS)	√ lung parenchyma
Tyrosine kinases	√ (perfusion)		√
Protein kinase C	√ (CS)		
Phospholipase C	√ (CS)		
Phosphotyrosine phosphatase	√ (perfusion)		
JNK, NF-κB	√ (CS)	√ (CS, VILI)	√ lung parenchyma
ROS	√ (CS)	√ (CS)	
Increased cytokine expression	√ (CS)	√ (CS)	√ (BAL and lung parenchyma)
Integrin activation	√ (CS)		
Increased ICAM-1/selectin expression	√ (CS)		
Low-molecular-weight hyaluronan		√ (CS)	√ lung parenchyma
5-lipoxygenase			√ lung parenchyma
iNOS			√ lung parenchyma
Phospholipase A2	√ (CS)		√ lung parenchyma

VILI refers to in vivo studies of ventilator-induced lung injury; CS refers to in vitro studies of cells under cyclical strain; perfusion refers to ex vivo studies using perfused isolated lung. Pi, phosphorylation.

and TNF-α in the BAL fluid. These changes correlated with increased expression of the corresponding mRNA in the lung parenchyma (29). The source of these cytokines has not been identified in vivo, but in vitro cyclic stretch induces production of IL-8, IL-6, and IL-1β in alveolar epithelial cells; IL-8 and IL-6 in pulmonary artery ECs; and IL-8 in macrophages (30–34). Production of chemotactic cytokines is temporally related to an influx of neutrophils into the alveolar space, as evidenced by myeloperoxidase assay (which measures influx of neutrophils in the interstitium, vessels, and airways) and by BAL (which measures movement of the neutrophils into the airways).

The mechanism of stretch-induced cytokine production has been found to involve activation of the c-jun N-terminal kinase (JNK) and nuclear factor (NF)-κB pathways. In the intact mouse treated with subcutaneous injection of SP600125, a specific inhibitor of JNK, and in mice deficient in either JNK1 or JNK2, there was an attenuation of MIP-2 production and neutrophil influx (35). Studies of in vitro cyclic mechanical stretch showed that cyclic stretch-induced IL-8 production was dependent on both activation of JNK and NF-κB in alveolar epithelial cells and JNK in pulmonary artery ECs (32,33). Similar studies examining a role for NF-κB have yet to be performed in pulmonary artery ECs exposed to cyclic stretch. However, in human umbilical vein ECs (HUVECs), laminar shear stress resulted in a dose- and time-dependent increase in NF-κB nuclear translocation and induction of *IL-8* mRNA expression (31).

Another possible mechanism of stretch-induced IL-8 expression is the production of reactive oxygen species (ROS). In aortic ECs, cyclic stretch has been shown to induce ROS levels through an nicotinamide adenine dinucleotide (NADH)/

Table 54-4: Knockout Mice and VILI

Knockout	*VILI Phenotype*		*References*
	Inflammation	Capillary Leak	
JNK	⇓	⇔	35
HAS3	⇓	⇔	47
CCSP	⇑	⇑	23
5LO	⇓	⇓	21
iNOS	⇓	⇓	22
TNF-α	⇔	⇔	25

nicotinamide adenine dinucleotide phosphate (NADPH) oxidase-dependent pathway (36). We have shown that the antioxidant N-acetylcysteine blocks stretch-induced production of IL-8 in alveolar epithelial cells in vitro (37) and attenuates high tidal volume ventilation-induced lung neutrophil influx in vivo (38). Whether a similar mechanism holds true for pulmonary artery ECs remains to be seen.

In addition to IL-8, IL-6 (another inflammatory cytokine) has been found to be increased in the serum of ARDS patients ventilated with large tidal volumes (39). In HUVECs, constant uni-axial strain[1] resulted in rapid (15 minutes) phosphorylation and degradation of IκB and translocation of NF-κB to the nucleus and subsequent induction of *IL-6* mRNA expression (2 hours). This effect was blocked by NF-κB inhibitors pyrrolidine dithiocarbamanate (PDTC) and SN50, or antisense oligodeoxynucleotides for NF-κB subunits p65 and p50 (30). The stretch-induced activation of NF-κB in HUVECs was found to be transduced by cell membrane integrins. Continuous uni-axial strain of HUVECs caused phosphorylation of FAK and paxillin, markers of integrin activation. GRGDNP, a blocker of integrin activation, abolished stretch-induced IL-6 production. Inhibitors of PKC and phospholipase C also blocked IL-6 production and NF-κB activation, suggesting that mechanical stretch results in integrin activation, causing downstream activation of PKC and phospholipase C with subsequent translocation of NF-κB to the nucleus and transcription of *IL-6* mRNA (40). We have also found in airway epithelial cells that cyclic stretch-induced IL-6 production was blocked by the inhibition of NF-κB and JNK activity with antioxidants (37).

Cyclic stretch has been shown to induce intercellular adhesion molecule (ICAM)-1 and P-selectin expression in HUVECs and kidney mesangial cells (41,42). ICAM-1 was found to be necessary for cyclic stretch-induced adhesion of monocytes to HUVECs and neutrophils to mesangial cells (41,42). In the lung, adhesion molecules are not always necessary for migration of leukocytes, especially in the capillary bed, although they may be important in the postcapillary venules (43). In mice, however, infusion of a neutralizing anti–L-selectin antibody prevented high tidal volume–induced

1 In vitro cell stretch can be performed with bi-axial strain (i.e., strain applied in two dimensions) or uni-axial strain (i.e., strain applied in one dimension).

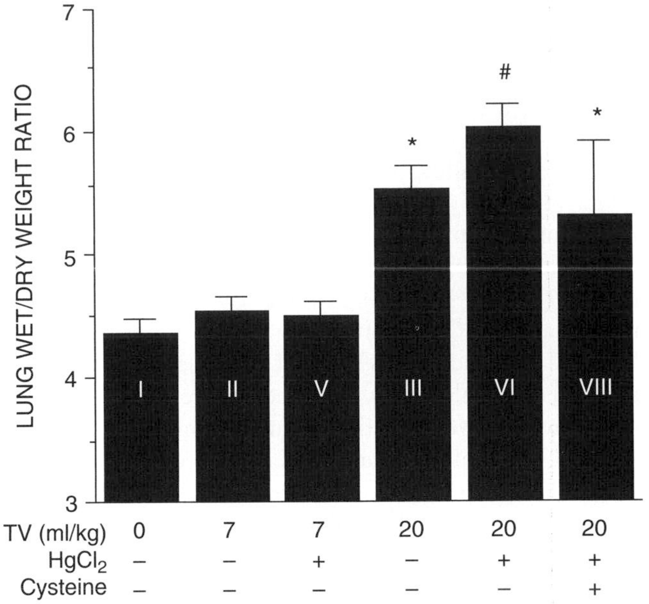

Figure 54.1. Effect of tidal volume on lung water content. Lung water content in rats as assessed by wet/dry weight ratio was not affected by volume ventilation at 85 breaths per minute for 2 hours with tidal volume (TV) of 7 mL/kg (Group II) compared to nonintubated control rats (Group I) but was increased by TV of 20 mL/kg (Group III). $HgCl_2$, a competitive inhibitor of aquaporins 1 and 5, significantly ($p < 0.05$) worsened the wet/dry weight ratio in rats ventilated at a TV of 20 mL/kg (Group VI) but not at 7 mL/kg (Group V). Equimolar infusions of cysteine 5 minutes after the $HgCl_2$ prevented the excess increase in lung wet/dry weight ratio caused by the $HgCl_2$ in the rats ventilated at TV 20 mL/kg (Group VIII). $^{*}p < 0.05$ from TV 7 mL/kg (Group II) or normals (Group I); $^{\#}p < 0.05$ from TV 20 mL/kg. (Reproduced with permission from Hales CA, Du, HK, Volokhov A, et al. Aquaporin channels may modulate ventilator-induced lung injury. *Respir Physiol.* 2001;124:159–166.)

sequestration of neutrophils (44). In another study, administration of a new anti-inflammatory agent that blocks CD11b/CD18 (the leukocyte β integrin that binds ICAM-1) inhibited neutrophil infiltration in surfactant-depleted rabbits on mechanical ventilation (45). Taken together, these data implicate an important role for the endothelium in mediating lung neutrophil influx with high tidal volume ventilation.

Changes in the extracellular matrix (ECM) may also affect EC function during mechanical ventilation. The ECM of the lung is made up of many different components, including collagens, elastin, proteoglycans (PGs), and stromal cells. PGs are a heterogeneous group of molecules that make up the ground substance of the ECM. PGs consist of a core protein to which one or more glycosaminoglycans (GAGs) are covalently attached. The GAGs in the lung include chondroitin sulfate, dermatan sulfate, heparan sulfate, and hyaluronan (HA). HA is a negatively charged, linear GAG composed of alternating N-acetyl glucosamine and glucuronic acid in repeating disaccharide units, with no covalently linked protein core. High-molecular-weight HA (>500 kDa; HMW HA) is an important component of the lung interstitium and functions to maintain the structural integrity and compliance of the normal lung. Low-molecular-weight HA (<500 kDa; LMW HA) can function as an intracellular signaling molecule in inflammation. We have found increased production of LMW HA in lung parenchyma of rats exposed to high tidal volume ventilation (46). LMW HA can be produced by de novo synthesis by HA synthase 3 (HAS3). We have found that mice lacking HAS3 do not produce LMW HA in response to high tidal volume ventilation and were resistant to high tidal volume ventilation–induced cytokine production and neutrophil influx (47). LMW HA (200–300) induced IL-8 in human pulmonary artery ECs by binding to CD44, Toll-like receptor (TLR)2, TLR4, and RHAMM (48). Low-molecular-weight HA fragments have also been found to induce IL-8 expression by human microvascular ECs isolated from neonatal dermis (49) and TLR4-blocking antibody inhibited HA fragment–induced IL-8 production. As compared to wild-type mice, TLR4-deficient mice failed to increase serum MIP-2 production in response to infusion of HA fragments.

Circulating HA has been shown to be cleared by sinusoidal ECs in the liver. In patients with hepatotoxicity, and therefore damaged sinusoidal ECs, increased levels of both HMW and LMW HA have been found along with increased levels of serum IL-8, suggesting the increased levels of LMW HA may have led to inflammation (50). Thus, another source of IL-8 production with high tidal volume ventilation may be via the production of LMW HA, which in turn induces IL-8 production by ECs.

In addition, we have found increased levels of LMW HA in the serum of rats ventilated with high tidal volumes and in the serum of patients with ARDS (D.A. Quinn and C.A. Hales, unpublished data). We have previously shown that high tidal volume ventilation of the lung caused increased capillary leak in the kidney (51). We hypothesize that lung stretch increased expression of HAS3, leading to increased production of LMW HA. The LMW HA produced in the lung can be released into the circulation, potentially leading to activation of the systemic endothelium.

KEY POINTS

- The endothelium plays a key role in lung barotrauma caused by VILI.
- The endothelium modulates lung stretch–induced pulmonary capillary leak both through the direct effects of barotrauma, leading to stress failure, and by stretch-induced intracellular and membrane changes that cause increased endothelial layer permeability.
- The endothelium modulates lung stretch–induced inflammation through upregulation of neutrophil binding sites, the production of cytokines through stretch-induced oxidant production with JNK and NF-κB activation, and through stimulation by inflammatory mediators, such as LMW HA produced during lung stretch.

Future Goals

- Many of the data here have been produced in ECs from a variety of sources and not only pulmonary artery ECs. It has been shown that ECs from different vascular beds behave differently. These data need to be confirmed in pulmonary artery ECs both in vitro and in vivo experiments.
- The role of nonprotein, noncytokine stimulants of endothelium, such as LMW HA, is a new and exciting field. These inflammatory mediators need further exploration and may provide a new target for therapeutic interventions.

REFERENCES

1 Dreyfuss D, Saumon G. Ventilator-induced lung injury: lessons from experimental studies. *Am J Respir Crit Care Med.* 1998; 157:294–323.
2 Webb HH, Tierney DF. Experimental pulmonary edema due to intermittent positive pressure ventilation with high inflation pressures. Protection by positive end-expiratory pressure. *Am Rev Respir Dis.* 1974;110:556–565.
3 Dreyfuss D, Soler P, Basset G, Saumon G. High inflation pressure pulmonary edema. Respective effects of high airway pressure, high tidal volume, and positive end-expiratory pressure. *Am Rev Respir Dis.* 1988;137:1159–1164.
4 Brower RG, Lanken PN, MacIntyre N, et al. Higher versus lower positive end-expiratory pressures in patients with the acute respiratory distress syndrome. *N Engl J Med.* 2004;351:327–336.
5 Tremblay L, Valenza F, Ribeiro SP, et al. Injurious ventilatory strategies increase cytokines and c-fos m-RNA expression in an isolated rat lung model. *J Clin Invest.* 1997;99:944–952.
6 Acute Respiratory Distress Syndrome Network. Ventilation with lower tidal volumes as compared with traditional tidal volumes for acute lung injury and the acute respiratory distress syndrome. The Acute Respiratory Distress Syndrome Network. *N Engl J Med.* 2000;342:1301–1308.
7 Hales CA, Du HK, Volokhov A, et al. Aquaporin channels may modulate ventilator-induced lung injury. *Respir Physiol.* 2001; 124:159–166.
8 Quinn DA, Moufarrej RK, Volokhov A, Hales CA. Interactions of lung stretch, hyperoxia, and MIP-2 production in. *J Appl Physiol.* 2002;93:517–525.
9 Frank JA, Gutierrez JA, Jones KD, et al. Low tidal volume reduces epithelial and endothelial injury in acid-injured rat lungs. *Am J Respir Crit Care Med.* 2002;165:242–249.
10 Cavanaugh KJ, Jr., Margulies SS. Measurement of stretch-induced loss of alveolar epithelial barrier integrity with a novel *in vitro* method. *Am J Physiol Cell Physiol.* 2002;283:C1801–C1808.

11 Naruse K, Sokabe M. Involvement of stretch-activated ion channels in $Ca2^{+}$ mobilization to mechanical stretch in endothelial cells. *Am J Physiol.* 1993;264:C1037–C1044.
12 Winston FK, Thibault LE, Macarak EJ. An analysis of the time-dependent changes in intracellular calcium concentration in endothelial cells in culture induced by mechanical stimulation. *J Biomech Eng*. 1993;115:160–168.
13 Wirtz HR, Dobbs LG. Calcium mobilization and exocytosis after one mechanical stretch of lung epithelial cells. *Science.* 1990; 250:1266–1269.
14 Parker JC, Ivey CL, Tucker JA. Gadolinium prevents high airway pressure-induced permeability increases in isolated rat lungs. *J Appl Physiol.* 1998;84:1113–1118.
15 Garcia JG, Lazar V, Gilbert-McClain LI, et al. Myosin light chain kinase in endothelium: molecular cloning and regulation. *Am J Respir Cell Mol Biol.* 1997;16:489–494.
16 Parker JC. Inhibitors of myosin light chain kinase and phosphodiesterase reduce ventilator-induced lung injury. *J Appl Physiol.* 2000;89:2241–2248.
17 Birukov KG, Jacobson JR, Flores AA, et al. Magnitude-dependent regulation of pulmonary endothelial cell barrier function by cyclic stretch. *Am J Physiol Lung Cell Mol Physiol.* 2003;285:L785–L797.
18 Wainwright MS, Rossi J, Schavocky J, et al. Protein kinase involved in lung injury susceptibility: evidence from enzyme isoform genetic knockout and in vivo inhibitor treatment. *Proc Natl Acad Sci USA.* 2003;100:6233–6238.
19 Bhattacharya S, Sen N, Yiming MT, et al. High tidal volume ventilation induces proinflammatory signaling in rat lung endothelium. *Am J Respir Cell Mol Biol.* 2003;28:218–224.
20 Parker JC, Ivey CL, Tucker A. Phosphotyrosine phosphatase and tyrosine kinase inhibition modulate airway pressure-induced lung injury. *J Appl Physiol.* 1998;85:1753–1761.
21 Caironi P, Ichinose F, Liu R, et al. 5-Lipoxygenase deficiency prevents respiratory failure during ventilator-induced lung injury. *Am J Respir Crit Care Med.* 2005;172:334–343.
22 Peng X, Abdulnour RE, Sammani S, et al. Inducible nitric oxide synthase contributes to ventilator-induced lung injury. *Am J Respir Crit Care Med.* 2005;172:470–479.
23 Yoshikawa S, Miyahara T, Reynolds SD, et al. Clara cell secretory protein and phospholipase A2 activity modulate acute ventilator-induced lung injury in mice. *J Appl Physiol.* 2005;98:1264–1271. Epub 2004 Dec 17.
24 Ikeda M, Takei T, Mills I, et al. Extracellular signal-regulated kinases 1 and 2 activation in endothelial cells exposed to cyclic strain. *Am J Physiol.* 1999;276:H614–H622.
25 Yoshikawa S, King JA, Lausch RN, et al. Acute ventilator-induced vascular permeability and cytokine responses in isolated and in situ mouse lungs. *J Appl Physiol.* 2004;97:2190–2199.
26 Verkman AS. Aquaporin water channels and endothelial cell function. *J Anat*. 2002;200:617–627.
27 Su X, Song Y, Jiang J, Bai C. The role of aquaporin-1 (AQP1) expression in a murine model of lipopolysaccharide-induced acute lung injury. *Respir Physiol Neurobiol.* 2004;142:1–11.
28 Song Y, Fukuda N, Bai C, et al. Role of aquaporins in alveolar fluid clearance in neonatal and adult lung, and in oedema formation following acute lung injury: studies in transgenic aquaporin null mice. *J Physiol.* 2000;525:771–779.
29 Dos Santos CC, Slutsky AS. Invited review: mechanisms of ventilator-induced lung injury: a perspective. *J Appl Physiol.* 2000;89:1645–1655.
30 Kobayashi S, Nagino M, Komatsu S, et al. Stretch-induced IL-6 secretion from endothelial cells requires NF-kappaB activation. *Biochem Biophys Res Commun.* 2003;308:306–312.
31 Liang F, Huang N, Wang B, et al. Shear stress induces interleukin-8 mRNA expression and transcriptional activation in human vascular endothelial cells. *Chin Med J (Engl).* 2002;115:1838–1842.
32 Matyal R, Hales CA, Quinn DA. Stretch-induced IL-8 production in human pulmonary artery endothelial cells. *Amer J Respir Crit Care Med.* 2000;161:A417.
33 Li LF, Ouyang B, Choukroun G, et al. Stretch-induced IL-8 depends on c-Jun NH2-terminal and nuclear. *Am J Physiol Lung Cell Mol Physiol.* 2003;285:L464–L475.
34 Pugin J, Dunn I, Jolliet P, et al. Activation of human macrophages by mechanical ventilation *in vitro*. *Am J Physiol.* 1998;275:L1040–L1050.
35 Li LF, Liao SK, Lee CH, et al. Ventilation-induced neutrophil infiltration and apoptosis depend on apoptosis signal-regulated kinase 1 pathway. *Crit Care Med.* 2005;33:1913–1921.
36 Howard AB, Alexander RW, Nerem RM, et al. Cyclic strain induces an oxidative stress in endothelial cells. *Am J Physiol.* 1997;272:C421–C427.
37 Jafari B, Ouyang B, Li LF, et al. Intracellular glutathione in stretch-induced cytokine release from alveolar type-2 like cells. *Respirology*. 2004;9:43–53.
38 Matthay MA, Bhattacharya S, Gaver D, et al. Ventilator-induced lung injury: in vivo and in vitro mechanisms. *Am J Physiol Lung Cell Mol Physiol.* 2002;283:L678–L682.
39 Parsons PE, Eisner MD, Thompson BT, et al. Lower tidal volume ventilation and plasma cytokine markers of inflammation in patients with acute lung injury. *Crit Care Med.* 2005;33:1–6-discussion 230–232.
40 Sasamoto A, Nagino M, Kobayashi S, et al. Mechanotransduction by integrin is essential for IL-6 secretion from endothelial cells in response to uniaxial continuous stretch. *Am J Physiol Cell Physiol.* 2005;288:C1012–C1022.
41 Yun JK, Anderson JM, Ziats NP. Cyclic-strain-induced endothelial cell expression of adhesion molecules and their roles in monocyte-endothelial interaction. *J Biomed Mater Res.* 1999;44:87–97.
42 Riser BL, Varani J, Cortes P, et al. Cyclic stretching of mesangial cells up-regulates intercellular adhesion molecule-1 and leukocyte adherence: a possible new mechanism for glomerulosclerosis. *Am J Pathol.* 2001;158:11–17.
43 Burns AR, Smith CW, Walker DC. Unique structural features that influence neutrophil emigration into the lung. *Physiol Rev.* 2003;83:309–336.
44 Choudhury S, Wilson MR, Goddard ME, et al. Mechanisms of early pulmonary neutrophil sequestration in ventilator-induced lung injury in mice. *Am J Physiol Lung Cell Mol Physiol.* 2004; 287:L902–L910. Epub 2004 Jul 16.
45 Rimensberger PC, Fedorko L, Cutz E, Bohn DJ. Attenuation of ventilator-induced acute lung injury in an animal model by inhibition of neutrophil adhesion by leumedins (NPC 15669). *Crit Care Med.* 1998;26:548–555.
46 Mascarenhas MM, Day RM, Ochoa CD, et al. Low molecular weight hyaluronan from stretched lung enhances interleukin-8 expression. *Am J Respir Cell Mol Biol.* 2004;30:51–60.
47 Bai KJ, Spicer AP, Mascarenhas MM, et al. The role of hyaluronan synthase 3 in ventilator-induced lung injury. *Am J Respir Crit Care Med.* 2005;172:92–98.

48 Mascarenhas MM, Garg HG, Hales CA, Quinn DA. Hyaluronan (HA), CD 44 and Toll-like Receptors (TLR) in ventilator-induced lung injury (VILI). *FASEB J.* 2005;19:A912.5.

49 Taylor KR, Trowbridge JM, Rudisill JA, et al. Hyaluronan fragments stimulate endothelial recognition of injury through TLR4. *J Biol Chem.* 2004;279:17079–17084.

50 Williams AM, Langley PG, Osei-Hwediah J, et al. Hyaluronic acid and endothelial damage due to paracetamol-induced hepatotoxicity. *Liver Int.* 2003;23:110–115.

51 Choi WI, Quinn DA, Park KM, et al. Systemic microvascular leak in an *in vivo* rat model of ventilator-induced. *Am J Respir Crit Care Med.* 2003;167:1627–1632.

55

Endothelium and Diving

Alf O. Brubakk, Olav S. Eftedal, and Ulrik Wisløff

Norwegian University of Science and Technology, Trondheim, Norway

Diving, as a method for exploring and exploiting the underwater world, requires equipment and methods that have been available for only about 200 years. The first practical procedures for decompression from dives were developed by Boycott and colleagues (1) at the beginning of the 1900s. With the development of the self-contained underwater breathing apparatus (SCUBA) in the 1940s, divers were able to swim freely, and saturation diving methods[1] developed in the 1950s allowed divers to stay under pressure for weeks.

Divers may reach a depth of 50 m using air as their breathing gas. Beyond 50 m, it is necessary to employ helium mixtures, owing to depth-related toxic effects of nitrogen and excessive oxygen tension. For a review of the general effects of diving on the organism, the reader is referred to Bennett and Elliott (2).

The major risk of injury associated with diving relates to decompression upon return to the surface. Gas will be taken up in solution by the tissues of the body during the dive proportional to the depth, and the uptake is exponentially related to the time spent under pressure. The gas content at the bottom is given by Henry's law:

$$C = P \times L \qquad \text{(Equation 1)}$$

where P is partial pressure of the gas and L is the solubility coefficient.

Upon returning to the surface, this excess gas must be eliminated. Gas elimination follows an exponential curve, with time constants determined by blood flow to the different tissues. If pressure is reduced faster than gas can be eliminated, the partial pressure of gas in the tissue will be higher than the environmental pressure. This supersaturation can lead to gas coming out of solution, forming bubbles. These bubbles are considered to be the cause of clinical symptoms, so-called decompression sickness (DCS). Symptoms of DCS include pain in muscles or joints or more severe symptoms from the cardiovascular and the central nervous systems.

1 Saturation diving is a method in which the diver stays at pressure for a period long enough for complete equilibration of gas tensions so that no further uptake of gas takes place (usually at least 24 hours). This method typically is used for deep commercial diving.

Vascular bubbles can be observed following nearly all decompressions (3). In the majority of dives, no acute clinical symptoms occur, and these bubbles have therefore been termed "silent bubbles" (4). Intravascular gas bubbles are more likely to form on the venous side of the circulation, both because the inert gas tensions are higher here than on the arterial side and because the blood pressure is lower (5). These bubbles may become lodged in the pulmonary vasculature and eliminated, or they may gain access to the arterial circulation via a patent foramen ovale (PFO) or vascular shunts in the lung. Such shunts seem to open even after light exercise (6). Several studies have documented that neurological DCS is related to the presence of a PFO (7,8), thus indicating that arterialization of vascular bubbles may be one important mechanism for injury. However, other mechanisms have also been proposed (9).

A working hypothesis for this chapter is that vascular bubbles play an important role in the pathophysiology of neurological DCS by altering endothelial cell (EC) phenotypes. Even if acute clinical symptoms are lacking, such alterations could form the basis for possible long-term injuries. An important goal is to identify the properties of the bubble and/or host that determine progression to DCS and to use this information to prevent injury and long-term sequelae.

A search on PubMed using the keywords *endothelium* and *diving* yielded nine hits, only one of them published after 2000. Similarly, a search for *endothelium* and *decompression* generated nine hits, including four from our own laboratory, whereas *endothelium* and *gas bubbles* gave 21 hits, only seven of these related to diving. Thus, the endothelium has not been a major focus in diving research.

MECHANISMS OF VASCULAR BUBBLE FORMATION

De novo formation of gas bubbles requires supersaturations exceeding those found in the vascular system after a dive by a factor of 50 to 100 (10). Thus, it is widely held that bubbles grow from preformed nuclei. These nuclei are likely gas filled and have been estimated to measure ≈1 μm in diameter (11). Experimental data point to their presence in the vascular system under normal conditions, although not in free-flowing blood (12–15). Harvey points out that "if bubbles come from nuclei they must arise from gas nuclei sticking to or formed on or within the endothelial linings of the vascular system or extravascular spaces, and only when they have enlarged to the point of instability do they pass into the blood stream" (15). As bubbles do not readily form intracellularly (10), there are probably no nuclei inside the ECs.

Bubbles are extremely stable on hydrophobic surfaces (16). Thus, lipid-rich microdomains on the surface of the endothelium (e.g., caveolae) may have a particular propensity for formation of and/or stable attachment to bubble nuclei (17). These are also locations where nitric oxide (NO) is produced (18). Caveolae are invaginations in the vascular wall (17), and may represent a stabilizing mechanism for bubbles (19). This indicates that caveolae are attractive sites for the formation of bubble nuclei. A reduction in surface tension of hydrophobic membranes has been shown to increase the number of stable nuclei (20). The latter study also showed a lack of correlation between bubble formation and surface tension of the fluid above the membrane; if anything, the reverse seemed to be true.

However, as we have previously demonstrated that a small reduction in surface tension of serum will significantly increase bubble production in vivo (21), we have to assume that this reduction reflects a reduction of surface tension of the vessel wall. Thus, we can hypothesize that spatial and temporal differences in surface tension of the vessel wall contribute to the large intra- and interpersonal variation in bubble formation by variation in the number and size of bubble nuclei.

Increased body fat is associated with increased vascular bubble formation and an increased risk for DCS (22,23). A correlation exists between blood lipids and obesity (24,25), and an increase in blood lipids in individuals suffering from obesity would decrease surface tension of blood and probably also of the vessel wall. Thus, it is likely that obesity will increase both the number of nucleation sites and the size of nuclei and bubbles. Because the speed of bubble growth is significantly influenced by the size of the nuclei, these individuals may be at risk for faster bubble growth during dives, increasing their susceptibility to DCS as the risk of DCS is related to the volume of gas in bubbles (26).

The growth of bubbles is determined by the LaPlace equation:

$$P_{bubble} - p_{amb} = 2\gamma/R + \delta \qquad \text{(Equation 2)}$$

where P_{bubble} is the gas tension inside the bubble, p_{amb} is the gas tension in the fluid surrounding the bubble, γ is the surface tension of the liquid-gas interface, R is bubble radius, and δ is additional deformation pressure opposing bubble expansion.

If the gas tensions in the fluid (e.g., blood) exceed the gas tension inside the bubble, the bubble will grow. As a reduction in surface tension will reduce the pressure gradients required for growth, a reduction in external deformation pressure (e.g., blood pressure) will have the same effect. Note also that an increase in bubble radius will reduce pressure inside the bubble.

Even if no direct observation of this exists, bubbles are probably formed from nuclei adhering to the endothelium. Their presence in free-flowing blood must indicate that the adhering forces have been reduced or that they have grown to a size that allows them to be washed away by the bloodstream. It is well established that an increase in venous blood flow caused by muscle contraction increases the number of bubbles observed the venous outflow tract after a dive.

FACTORS INFLUENCING BUBBLE FORMATION

In studies of rats we have shown that the number of vascular bubbles observed following a dive can be significantly reduced by performing heavy physical exercise 20 hours before the dive (Figure 55.1). This protective effect was absent when exercise was performed 48 hours before the dive, even in highly trained rats (27). Moreover, exercise sooner than 10 hours before the dive had no effect on bubble formation. Similar findings have been reported in humans. In divers performing a dive to 18 m, bubble formation was significantly reduced in those that performed heavy exercise 24 hours before the dive (28). A study by Blattau and colleagues indicated that exercise at lower intensities 2 hours before a dive might have similar effects in divers (29).

Exercise induces production of endothelial-derived NO in the vascular system through increase in blood flow and shear stress, not only in the organs directly involved in the exercise (30,31). NO is not only a potent vasodilator but also reduces the "stickiness" or adhesiveness of cellular particles to the endothelial surface (32,33). This may also reduce the number of nuclei available for bubble production. The very

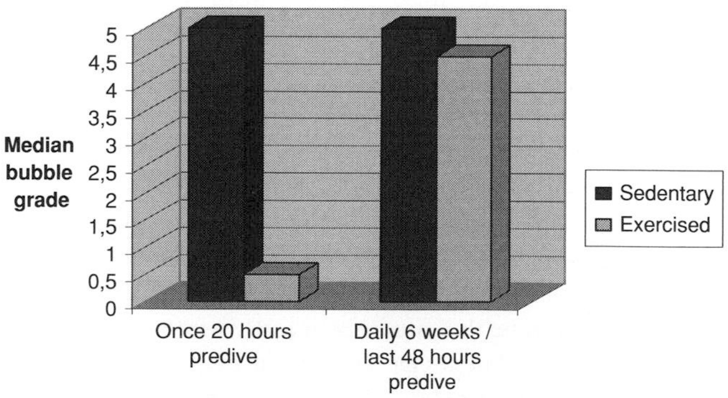

Figure 55.1. Median bubble grades in sedentary and exercised rats (90-minute intervals at 90% max). The reduction in bubble grade in the animals exercised 20 hours predive is significant. (Data used with permission from Wisloff U, Brubakk AO. Aerobic endurance training reduces bubble formation and increases survival in rats exposed to hyperbaric pressure. *J Physiol.* 2001;537(Pt 2):607–611.)

microdomain that is believed to promote adherence of bubble nuclei, namely the caveolus, is also a key regulator of NO production in ECs (18). We have shown that exogenous NO before a dive reduces bubble formation after the dive, a similar effect as was achieved by exercise 20 hours predive. NO blockade increases bubble formation (27,34,35).

Traditionally, efforts to reduce formation of bubbles, hence incidence and/or severity of DCS, have focused mainly on reducing supersaturation by slowing the rate of ascent. Although there is no question that these protocols have reduced DCS-related morbidity, there still remains considerable room for improvement (36,37). Based on the above observations, we believe that the endothelium represents a potentially valuable target for preventing serious (e.g., neurological) DCS.

Traditionally, the correlation between weight and the propensity to form gas bubbles has been attributed to the fact that nitrogen is far more soluble in fat than it is in water (38), thus resulting in a positive correlation between body weight and supersaturation of nitrogen. However, an alternative explanation relates to differences in surface tension. Surface tension is negatively correlated with the number of bubble nuclei (20). A reduction in surface tension may be brought about by an increase in lipids in the blood. We have shown that exercise 24 hours before, or NO donors given immediately before, a dive reduces the amount of bubbles in heavy rats. Conversely NO blockade promotes bubble production in sedentary light rats (27,34,35). These observations indicate that NO is an important mechanism of bubble formation. However, exercise 24 hours before the dive prevents the bubble-inducing effect of NO blockade in light rats, thus there must also be NO-independent mechanisms involved. Interestingly, this effect is not observed in heavy rats (Alf O. Brubakk, unpublished observation). We have also seen that exercise immediately before a dive reduces the protective effect of 24 hours predive exercise and NO donors in heavy, but not in light, animals (39).

It has been shown that blood lipids are increased by approximately 30% immediately after heavy exercise (37). This is followed by a decrease in total blood lipids over the next several hours (38), whereas high-density lipoprotein (HDL) is increased 24 hours after exercise (39). Elevation of blood lipids reduces surface tension, which in turn facilitates production of bubble nuclei (19). Also, low-density lipoprotein (LDL) and cholesterol have been shown to decrease NO production, whereas HDL increases NO activity (40,41). This can explain the time-dependent response of exercise on bubble formation. One might also speculate that the differences observed between light and heavy rats in effects of exercise may be caused by different levels of blood lipids, but to our knowledge this has not been studied.

The observations discussed earlier are summarized in Table 55-1. Based on these data, one might speculate that lean animals have lower basal levels of lipids and hence fewer bubble nuclei predive and fewer bubbles during/postdive. A study by Wisløff and colleagues has shown that heavy rats have higher levels of blood lipids than leaner rats (40). This will, however, not explain why NO blockade promotes bubble formation in lean animals, an effect that is prevented if exercise is performed 24 hours before the dive. One might speculate that the protective effect of exercise observed 24 hours prior to diving may be related to a time-dependent reduction in circulating lipids. The resulting effects on surface tension will promote washing out of existing bubble nuclei and will serve to inhibit formation of new bubble nuclei. This could be a mechanism to explain why exercise 24 hours before a dive protects lean but not heavy animals that are NO blocked. In the lean animals, with possible lower lipid levels, exercise will reduce blood lipids, allowing endothelial adhesiveness to be reduced sufficiently to allow washing out of bubble nuclei, although the effect is not strong enough in the fat animals with blocked NO. This mechanism could also explain why heavy exercise shortly before decompression could affect bubble formation in heavy but not lean animals. The findings above support the hypothesis that bubble nuclei and their adherence to the endothelium are critical for bubble formation in the blood. However, once bubbles get larger, a reduction in surface tension in the blood will reduce adhesiveness to the vessel wall (41), thus allowing more bubbles to enter the bloodstream.

Table 55-1: Vascular Bubble Formation after Decompression in the Rat

Body Weight	*Sedentary*	*Exercise*		*References*
		24 h pre	30 min pre	
<280 g	Few	Few	Few	27,39
NO	Few	Few	Few	35
L-NAME	Many	Few	Not studied	34
>300 g	Many	Few*	Many	27
NO	Few	Few	Many	27
L-NAME	Many	Many	Many	U. Wisløff et al., unpublished observations

*Exercise 30 minutes before a dive blocks this effect.

Other substances may also play a role in bubble formation. Bradykinin is a potent vasodilator, acting by increasing endothelial production of endothelial hyperpolarizing factor (EDHF), which acts on the smooth muscle of the vessels by an NO-independent mechanism (42). In normal endothelium, NO is apparently the main vasodilator; when the endothelium is injured, EDHF production is increased. Bradykinin is, however, also a potent proinflammatory hormone, increasing adhesiveness of cellular elements to the vessel wall (43). Some studies has shown that EDHF may have anti-inflammatory properties, reducing the adhesiveness of the endothelium, similar to the effect of NO (44), indicating that the response to endothelial injury is quite complex.

Chryssanthou and colleagues developed an experimental model in genetically obese mice and showed that the lighter

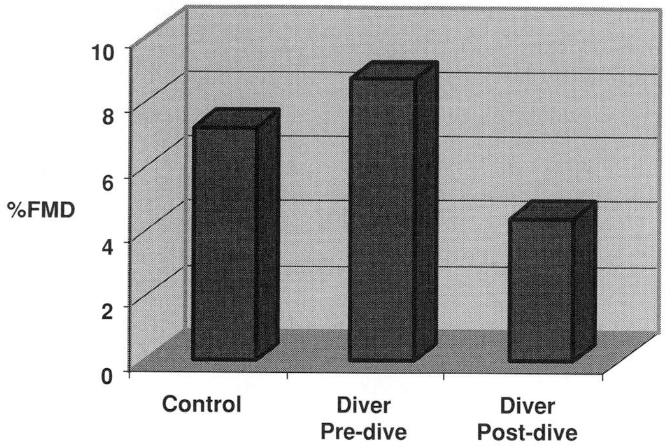

Figure 55.2. FMD in controls and after a single air dive to 280 kPa for 80 minutes. (Data used with permission from Brubakk AO, Duplancic D, Valic Z, et al. A single air dive reduces arterial endothelial function in man. *J Physiol.* 2005;566(Pt 3):901–906.)

animals did not develop lethal DCS, whereas the heavier animals died following massive bubble formation (45). However, if bradykinin was given to the animals during decompression, mortality dramatically increased. Bradykinin given prior to compression did not have this effect. His group also showed that if anti-inflammatory drugs were administered prior to decompression, mortality was dramatically reduced (46). Bubbles were not measured, but it appears that fewer bubbles seem to be formed in the control animals who did not receive bradykinin (46). This evidence indicates that properties of the endothelial surface are of considerable importance for bubble formation after diving.

EFFECT OF BUBBLES ON THE ENDOTHELIUM

As discussed earlier, the endothelium plays an important role in the formation of decompression-induced vascular bubbles. Several studies show that once formed, bubbles will damage or reduce the function of the endothelium in a dose-dependent manner (47–49). A recent study in experienced divers showed that arterial endothelial function as evaluated by flow-mediated dilatation (FMD) was reduced after a single air dive where very few venous gas bubbles could be detected in the pulmonary artery (Figure 55.2) (50). Philp and colleagues showed in rats that circulating ECs could be detected in the blood in proportion to the severity of DCS (51).

Activation of the endothelium may lead to the formation and release of so-called endothelial microparticles (EMP) (52). These may reduce endothelial function (53), possibly by increasing expression of the endothelial adhesion molecules such as vascular cell adhesion molecule (VCAM)-1, intercellular adhesion molecule (ICAM)-1, and E-selectin and by influencing NO production. Increased EMP levels have been reported in a number of cardiovascular diseases and after using a heart–lung machine (54). It is therefore not unlikely that gas bubbles from dives can lead to such activation. Oxygen radicals may also activate endothelium (55). This is probably also part of the dive effect observed above, as the reduction in endothelial function can be prevented by administering the antioxidants vitamins E and C before the dive.

Ward and colleagues (56) demonstrated that gas bubbles may activate complement in vitro. C5a is a potent molecule generated from complement activation through cleavage of C5 and is usually accompanied by formation of the membrane attack complex (MAC). C5a increases endothelial P-selectin expression, secretion of von Willebrand factor, and smooth muscle contraction. Complement activation will lead to activation of neutrophils and the formation of MAC that will lead to the destruction of nucleated cells (57). This process leads to adhesion of leukocytes to the endothelial layer and eventually to their transport through the endothelium. Activation of neutrophils has been demonstrated during decompression from saturation dives (58). In the skin, C5a will lead to erythema and edema and promote infiltration of inflammatory cells (59). This picture is very similar to what can be actually observed in the skin of divers following decompression. It is of interest to note that skin changes related to decompression are associated with the presence of a PFO (60).

Using monoclonal antibodies, Bergh and colleagues (61) demonstrated dose-dependent activation of C5a by gas bubbles in vitro. They also showed that the response was similar regardless of the content of the bubble, indicating that it was the surface of the bubble that was of importance for activation. The study also reported considerable variation in response over time (a period of 6 months) both in rabbits and man. This variation was not considered by Ward and colleagues, who claimed that individuals could be divided into sensitive and nonsensitive groups according to the degree of activation of complement and that clinical symptoms of decompression illness were related to the degree of activation (62).

In a subsequent study on divers, Hjelde and associates have been unable to verify Ward's finding (63). However, interestingly enough, the study showed that those individuals who had a low level of C5a before the dive produced many gas bubbles. Furthermore, a single air dive seemed to reduce the level of C5a. We have no explanation for this observation, but the reduction in C5a levels in the blood may indicate that diving and gas bubbles activate C5a receptors as well as C5a, thus leading to binding of C5a and hence removal from the blood in vivo. Shastri and colleagues also failed to observe any activation of C3a after a dive (64). However, activation of C5a has been demonstrated by Stevens and associates (65) in divers up to 14 hours after they had been treated for DCS. Anti-C5a failed to protect the rabbit endothelium against injury after exposure to gas bubbles (66).

The data previous indicate that the effect of bubbles on complement is complex in vivo. It has been shown that complement activation and increased production of C5a will lead to increased expression of circulating endothelial adhesion molecules about 4 hours after injury (67). This is in good agreement with our finding that a reduction in endothelial function is observed between 1 and 6 hours after exposure to gas bubbles (48). If decompression with endothelial damage leads to an inflammatory response, this may reduce NO bioavailability (68). An increase in VCAM-1 and ICAM-1 was observed in the blood of divers 5 minutes after surfacing; this effect lasted 24 hours (69).

Complement activation seems to be the main mechanism for acute lung injury not caused by gas bubbles (70). Activation and damage to ECs during acute inflammation can lead to leakage of plasma proteins and microvascular hemorrhage, which appear to be major contributing factors in the development of acute lung injury. The changes in pulmonary function seen in divers, including a reduction in carbon monoxide diffusion capacity and lung compliance (71), would support the hypothesis that inflammatory processes in the lungs are a result of the decompression process. The reduction in diffusion capacity is quite rapid and follows the development of bubbles (72). Hyperoxia can lead to an increased expression of vascular adhesion molecules (73), thus it is reasonable to suggest that exposure to increased oxygen tensions during the dive and bubbles during decompression may have additive effects.

Based on the above, the endothelium in the lung vasculature must be considered a primary target organ for gas bubbles and is probably exposed to gas bubbles to a greater or lesser degree in all decompressions. Generally, in diving the main focus on the lungs has been on their role as a filter, through which the bubbles are eliminated before they can be transmitted to the arterial side, where their potential for damage is greater. However, as pointed out above, bubbles will probably also enter the arterial circulation, either as a result of high gas load on the lungs that exceeds the filtering capability (74) or through shunts or a PFO.

Most divers have repeatedly been exposed to vascular bubbles during their career. If clinical signs of serious DCS can be avoided, exposure to vascular bubbles does not seem to injure the spinal cord (75). In a group of 10 amateur and 10 professional divers who had performed extensive diving, no histological changes could be seen. Five of the divers had a history of mild musculoskeletal DCS and several had aseptic bone necrosis (likely related to recurrent decompression). In the brain, significant loss of cells in the ependymal layer of the ventricles could be detected in a group of experienced divers when compared to controls (76). This is probably caused by gas bubbles in the spinal fluid; such bubbles will probably primarily adhere to the lining of the ventricles. Gas bubbles in the vascular system in the brain can lead to breakage of the blood–brain barrier (77). Broman and colleagues have demonstrated that even very short contact between gas bubbles and endothelium (1–2 minutes) will lead to such breakage (78). Furthermore, studies in rabbits indicate that such contact leads to endothelial damage and progressive reduction of cerebral blood flow and function (79).

PREVENTION OF ENDOTHELIAL INJURY

It has been shown that exposure to 280 kPa oxygen for 45 minutes, as in hyperbaric oxygen (HBO_2) therapy, inhibits B2-integrin–dependent adherence of neutrophils to the endothelium (80,81). Predive treatment with HBO_2 has been shown to protect against lung and CNS injury following decompression, probably through the same mechanisms (82). This anti-inflammatory effect is probably also a key factor in the response to HBO_2 for treatment of manifest decompression sickness. It is also worth noting that exercise prevented CNS injury in the pig model of decompression sickness used by Broome and associates (83). However, in the latter study the time between the last exercise bout and the dive was not reported. Thus, it is formally possible that the beneficial effects were related to a reduction in bubble production, as observed in rats and humans following exercise 24 hours before a dive (27,84).

Heat shock proteins (HSPs) are activated in response to a number of stressors, including hyperoxia or hypoxia, heat, cold, exercise, and some heavy metals or drugs. They have an important function in controlling the folding and structure of proteins and protect the cells from injury (85). Divers may be exposed to a number of stressors (hyperoxia, hard physical work, exposure to infections, etc.) that could possibly induce HSP formation. We have shown in rats that a short increase in body temperature to 42°C 24 hours before the dive reduced mortality by 50% and that this exposure increased HSP70 levels but not HSP90 or endothelial NO synthase (eNOS), and that bubble formation was not affected (A. O. Brubakk, unpublished observation). These findings are supported by a study showing increased HSP70 expression in animals with decompression sickness (86). Exercise also has an effect on HSP expression, and HSPs may represent a link between flow-mediated induction of shear stress and NO (87). A study by Siu and colleagues (88) showed that moderate exercise increased HSP70 48 hours after the last exercise bout. Other studies have shown that HSP70 increased immediately after exercise, with a new increase 24 and 48 hours later (89). Recently it has become clear that exercise (90), as well as hyperbaric oxygen (91,92), can act as a preconditioning regimen, protecting against injury, probably through the regulation of HSP and NO. Lochner and associates (93) showed that preconditioning in the presence of N(G)-nitro-L-arginine methylester (L-NAME) attenuated this effect, and an NO donor had the same effect as preconditioning by ischemia. Thom and colleagues showed that hyperoxia increases NO production by increasing HSP90 expression (94). HSP seems to be of crucial importance for preventing an increased production of radical oxygen species. HSP90 is an important modulator of eNOS production, determining the balance between NO and reactive oxygen species and may thus be of importance in preventing injury caused by increased oxygen tensions (95).

However, some data indicate that HSP activation can also increase damage to the endothelium. Certain bacteria produce HSPs, which are strongly antigenic and can trigger a significant immunoresponse (96). One study demonstrated increased endothelial damage in animals with such a response (97). Infections still are a significant problem in saturation diving operations (98). Consequently, if the bacterial flora in the saturation chamber can produce such an immunological response, saturation diving may have a higher risk of endothelial damage due to bubbles than other types of diving.

Further studies are needed to elucidate the role of HSPs in the prevention of DCS and bubble injury.

CONCLUSION

It is well documented that gas bubbles occur both in the arterial and the venous side of the circulation following decompression from a dive. Even if many questions still exist, we think it is well documented that these bubbles will reduce endothelial function. We suggest that this may be a central mechanism in the development of serious decompression injury and possibly also long-term effects of diving. Traditionally, protection against DCS and serious injury has been focused on influencing bubble formation by changing ascent procedures from a dive. The data presented in this chapter indicate that vascular bubble formation, which may be important for central nervous system injury, is significantly influenced by the properties of the endothelium. As these properties can be changed in various ways, it opens up a novel way of preventing injury from diving.

KEY POINTS

- Vascular gas bubbles leading to endothelial dysfunction are the main cause of neurological injury following decompression.
- Vascular gas bubbles are formed from nuclei adhering to the vascular wall.
- Exercise and/or NO reduce the adherence to the vessel wall and prevent vascular bubble formation.
- An increase in vascular lipids will increase the number of nuclei and vascular bubbles.
- HSP may protect against bubble-induced injury.

Future Goals

- A reduction in the incidence of DCS has occurred over the last 40 years. However, the relative number of incidents of DCS involving the central nervous system has increased. The data presented indicate that injury of the CNS may be related to vascular bubble formation and endothelial function. Although nearly all research in this field has focused on supersaturation and methods to prevent that, future improvements may be related to methods to improve endothelial function and hence influence nucleation. If the relationship between bubble formation and lipids is confirmed, then increasing obesity in the diving population will pose additional challenges. One testable hypothesis can be that individuals with increased levels of cholesterol have an increased risk of DCS and increased bubble formation.
- Age, and possibly a low aerobic capacity, is considered a risk factor for DCS. Our studies have shown that exercise performed 24 hours before a dive prevents bubble formation and that this effect is related to NO production. These studies did not show any protective effect of increased aerobic capacity. However, some human data indicate that bubble formation and the risk of DCS is related to aerobic capacity (22,99). The ability of the endothelium to produce NO is reduced in subjects, though, and this is an effect that can be improved by exercise (see Chapter 56). In a group of subjects older than 60 years, Taddei and colleagues (100) found that those who performed regular exercise (VO_2 max 63.7 mL/kg/min) had a significantly better endothelial function than sedentary individuals (VO_2 max 38 mL/kg/min). Thus, the importance of aerobic capacity and the timing of exercise in relation to the dive should be evaluated.
- One significant problem in repeated dives that has never been satisfactorily solved is the question about the safe time between dives. Generally it is assumed that the reason a second dive is considered more risky is that there remains inert gas in the body, leading to an increased supersaturation in the second dive. However, the work described in this chapter indicates that the problem also may be related to the endothelial lining being damaged by the first dive. Thus, studies of the repair process and the factors involved may change dive procedures. This can be studied both experimentally and by determining endothelial function following dives of different severity.
- Studies by Wisløff and associates have shown that differences in endothelial function probably are genetically determined and that these differences are linked to aerobic capacity (40). This could be a key to understanding the significant interindividual differences in bubble formation observed. In studies in which two groups of divers of similar age and physical characteristics performed the same dive, bubble formation varied by a factor of 100 between the groups (50). Changes in gene expression due to environmental factors may be of importance.

- Numerous studies have shown that endothelial dysfunction is independently related to future cardiovascular events (e.g., myocardial infarction, stroke, transient ischemic attack) (101). It is thus conceivable that repeated endothelial injury caused by vascular gas bubbles may contribute to the development of atherosclerotic lesions and that this may be a basis for long-term effects of diving.
- The studies described indicate that drugs may be used to modify decompression and to reduce the risk of bubble formation and hence the risk of serious neurological injury. This is of particular importance if rapid decompressions have to be performed as in evacuation from a sunken submarine or emergency evacuation from a saturation chamber. Based on the above, possible candidates for such an approach are NO donors, statins, anti-inflammatory drugs, or HSP.

ACKNOWLEDGMENTS

The studies that form the basis for this review have been supported mainly by the Norwegian University of Science and Technology (NTNU); by Phillips Petroleum Norway through the HADES program; and by the Norwegian Petroleum Directorate, Norsk Hydro, Esso Norge, and Statoil under the "dive contingency contract" (No 4600002328) with Norwegian Underwater Intervention (NUI).

REFERENCES

1 Boycott-AE, Damant CC, Haldane JS. The prevention of compressed-air illness. *J Hyg, London.* 1908;8:342–443.
2 Brubakk AO, Newman TS. *Bennett and Elliott's Physiology and Medicine of Diving.* 5th ed. London: Saunders, 2003.
3 Nishi RY, Brubakk AO, Eftedal O. Bubble detection. In: Brubakk AO, Neuman TS, eds. *Bennet & Elliott's Physiology and Medicine of Diving.* London: Saunders; 2003:501–529.
4 Behnke AR. Decompression sickness following exposure to high pressures. In: Fulton JF, ed. *Decompression Sickness.* Philadelphia, PA: Saunders; 1951:53–89.
5 Eckenhoff RG, Olstad CS, Carrod G. Human dose-response relationship for decompression and endogenous bubble formation. *J Appl Physiol.* 1990;69(3):914–918.
6 Eldridge MW, Dempsey JA, Haverkamp HC, et al. Exercise-induced intrapulmonary arteriovenous shunting in healthy humans. *J Appl Physiol.* 2004;97(3):797–805.
7 Moon RE., Camporesi EM., Kisslo JA. T1 – Patent foramen ovale and decompression sickness in divers. Patent foramen ovale and decompression sickness in divers. *Lancet North Am Ed.* 1989; 513–514.
8 Wilmshurst P, Bryson P. Relationship between the clinical features of neurological decompression illness and its causes. *Clin Sci (Lond).* 2000;99(1):65–75.
9 Francis TJR, Mitchell SJ. Pathophysiology of decompression sickness. In: Brubakk AO, Neuman TS, eds. *Bennett and Elliott's Physiology and Medicine of Diving.* London: Saunders; 2003: 530–556.
10 Hemmingsen EA, Hemmingsen BB. Lack of intracellular bubble formation in microorganisms at very high gas supersaturations. *J Appl Physiol.* 1979;47:1270–1277.
11 Yount DE. Growth of bubbles from nuclei. In: Brubakk AO, Kanwisher J, Sundnes G, eds. *Diving in Animals and Man.* Trondheim: Tapir Publishers; 1986:131–164.
12 Yount DE, Gillary EW, Hoffman DC. A microscopic investigation of bubble formation nuclei. *J Acoustical Soc Amer.* 1984; 76(5):1511–1521.
13 Yount DE, Strauss RH. On the evolution, generation and regeneration of gas cavitation nuclei. *J Acoust Soc Am.* 1982;65:1431–1439.
14 Conkin J, Foster PP, Powell MR, Waligora JM. Relationship of the time course of venous gas bubbles to altitude decompression illness. *Undersea Hyperbaric Med.* 1996;23:141–149.
15 Harvey NE. Physical factors in bubble formation. In: Fulton JF, ed. *Decompression Sickness.* Philadelphia: WB Saunders Co; 1951:90–114.
16 Liebermann L. Air bubbles in water. *J Appl Physics.* 1957;28:205–211.
17 Anderson RGW, Jacobsen K. A role for lipid shells in targeting protein in caveolae, rafts and other lipid domains. *Science.* 2002; 296:1821–1825.
18 Shaul PW. Regulation of endothelial nitric oxide synthase: location, location, location. *Annu Rev Physiol.* 2002;64:749–774.
19 Chappell MA, Payne SJ. A physiological model of gas pockets in crevices and their behavior under compression. *Respir Physiol Neurobiol.* 2005;153(2):166–180. Epub 2005 Nov 22.
20 Gaskins N, Vann RD, Hobbs E, et al. Surface tension and bubble formation in agar gelatin. *Undersea Hyperb Med.* 2001;28(Supplement):56.
21 Hjelde A, Koteng S, Eftedal O, Brubakk AO. Surface tension and bubble formation after decompression in the pig. *Appl Cardiopulm Pathophysiol.* 2000;9:47–52.
22 Carturan D, Boussuges A, Vanuxem P, et al. Ascent rate, age, maximal oxygen uptake, adiposity, and circulating venous bubbles after diving. *J Appl Physiol.* 2002;93(4):1349–1356.
23 Webb JT, Pilmanis AA, Balldin UI, Fischer JR. Altitude decompression sickness susceptibility: influence of anthropometric and physiologic variables. *Aviat Space Environ Med.* 2005;76(6): 547–551.
24 Tanner JM. The relation between serum cholesterol and physique in healthy young men. *J Physiol.* 1951;115(4):371–390.
25 Lefevre M, Champagne CM, Tulley RT, et al. Individual variability in cardiovascular disease risk factor responses to low-fat and low-saturated-fat diets in men: body mass index, adiposity, and insulin resistance predict changes in LDL cholesterol. *Am J Clin Nutr.* 2005;82(5):957–963.
26 Yount DE, Hoffman DC. On the use of a bubble formation model to calculate diving tables. *Aviat Space Environ Med.* 1986; 76:1511–1521.
27 Wisløff U, Brubakk AO. Aerobic endurance training reduces bubble formation and increases survival in rats exposed to hyperbaric pressure. *J Physiol.* 2001;537(Pt 2):607–611.
28 Dujic Z, Duplancic D, Marinovic-Terzic I, et al. Aerobic exercise before diving reduces venous gas bubble formation in humans. *J Physiol.* 2004;555(Pt 3):637–642.

29 Blatteau JE, Gempp E, Galland FM, et al. Aerobic exercise 2 hours before a dive to 30 msw decreases bubble formation after decompression. *Aviat Space Environ Med*. 2005;76(7):666–669.
30 Green D, Cheetham C, Mavaddat L, et al. Effect of lower limb exercise on forearm vascular function: contribution of nitric oxide. *Am J Physiol Heart Circ Physiol*. 2002;283(3):H899–H907.
31 Kingwell BA, Sherrard B, Jennings GL, Dart AM. Four weeks of cycle training increases basal production of nitric oxide from the forearm. *Am J Physiol*. 1997;272(3 Pt 2):H1070–H1077.
32 Kerem M, Mehmet C. Endothelial adhesivity, pulmonary hemodynamics and nitric oxide synthesis in ischemia-reperfusion. *Eur J Cardithoracic Surg*. 2000;18:248–252.
33 Buga GM, Gold ME, Fukuto JM, Ignarro LJ. Shear stress-induced release of nitric oxide from endothelial cells grown on beads. *Hypertension*. 1991;17(2):187–193.
34 Wisløff U, Richardson RS, Brubakk AO. NOS inhibition increases bubble formation and reduces survival in sedentary but not exercised rats. *J Physiol*. 2003;546(Pt 2):577–582.
35 Wisløff U, Richardson RS, Brubakk AO. Exercise and nitric oxide prevent bubble formation: a novel approach to the prevention of decompression sickness? *J Physiol*. 2004;555(Pt 3):825–829.
36 Brubakk AO, Arntzen AJ, Wienke BR, Koteng S. Decompression profile and bubble formation after dives with surface decompression: experimental support for a dual phase model of decompression. *Undersea Hyperb Med*. 2003;30(3):181–193.
37 Gutvik C, Brubakk AO. A model predictive framework for dynamic calculation of optimal decompression profiles. *Undersea Hyperb Med*. 2004;31(3):342.
38 Langø T, Mørland T, Brubakk AO. Diffusion coefficients ans solubility coefficients for gases in biological fluids and tissues; a review. *Undersea Hyperb Med*. 1996;23:247–272.
39 Berge VJ, Jorgensen A, Loset A, et al. Exercise ending 30 min pre-dive has no effect on bubble formation in the rat. *Aviat Space Environ Med*. 2005;76(4):326–328.
40 Wisløff U, Najjar SM, Ellingsen O, et al. Cardiovascular risk factors emerge after artificial selection for low aerobic capacity. *Science*. 2005;307(5708):418–420.
41 Suzuki A, Armstead SC, Eckmann DM. Surfactant reduction in embolism bubble adhesion and endothelial damage. *Anesthesiology*. 2004;101(1):97–103.
42 Halcox JP, Narayanan S, Cramer-Joyce L, et al. Characterization of endothelium-derived hyperpolarizing factor in the human forearm microcirculation. *Am J Physiol Heart Circ Physiol*. 2001;280(6):H2470–H2477.
43 Shigematsu S, Ishida S, Gute DC, Korthuis RJ. Bradykinin-induced proinflammatory signaling mechanisms. *Am J Physiol Heart Circ Physiol*. 2002;283(6):H2676–H2686.
44 Campbell WB. New role for epoxyeicosatrienoic acids as anti-inflammatory mediators. *Trends Pharmacol Sci*. 2000;21(4):125–127.
45 Chryssanthou C, Kalberer J Jr., Kooperstein S, Antopol W. Studies on dysbarism: II. Influence of bradykinin and "bradykinin antagonists" on decompression sickness in mice. *Aerospace Med*. 1964;35:741–746.
46 Chryssanthou C, Rubin L, Graber B. Amelioration of decompression sickness in mice by pretreatment with cyproheptadine. *Undersea Biomed Res*. 1980;7(4):321–329.
47 Nossum V, Brubakk AO. Endothelial damage by bubbles in the pulmonary artery of the pig. *Undersea Hyperb Med*. 1999;26:1–8.
48 Nossum V, Hjelde A, Brubakk AO. Small amounts of venous gas embolism cause delayed impairment of endothelial function and increase polymorphonuclear neutrophil infiltration. *Eur J Appl Physiol*. 2002;86(3):209–214.
49 Rosenblum WI. Biology of disease. Aspects of endothelial malfunction and function in cerebral microvessels. *Lab Invest*. 1986;55:252–268
50 Brubakk AO, Duplancic D, Valic Z, et al. A single air dive reduces arterial endothelial function in man. *J Physiol*. 2005;566(Pt 3):901–906.
51 Philp R, Inwood M, Warren B. Interactions between gas bubbles and components of the blood: implications in decompression sickness. *Aerospace Med*. 1972;43(9):946–953.
52 Jimenez JJ, Jy W, Mauro LM, et al. Endothelial cells release phenotypically and quantitatively distinct microparticles in activation and apoptosis. *Thromb Res*. 2003;109(4):175–180.
53 Brodsky SV, Zhang F, Nasjletti A, Goligorsky MS. Endothelium-derived microparticles impair endothelial function in vitro. *Am J Physiol Heart Circ Physiol*. 2004;286(5):H1910–H1915.
54 VanWijk MJ, VanBavel E, Sturk A, Nieuwland R. Microparticles in cardiovascular diseases. *Cardiovasc Res*. 2003;59(2):277–287.
55 Adamopoulos S, Parissis JT, Kremastinos DT. New aspects for the role of physical training in the management of patients with chronic heart failure. *Int J Cardiol*. 2003;90(1):1–14.
56 Ward CA, Koheil A, McCullough D, et al. Activation of complement at plasma-air or serum-air interface of rabbits. *J Appl Physiol*. 1986;60(5):1651–1658.
57 Kilgore KS, Friedrichs GS, Homeister JW, Lucchesi BR. The complement system in myocardial ischaemia/reperfusion injury. *Cardiovasc Res*. 1994;28:437–444.
58 Benestad HB, Hersleth IB, Hardersen H, Molvaer OI. Functional capacity of neutrophil granulocytes in deep-sea divers. *Scand J Clin Lab Invest*. 1990;50:9–18.
59 Swerlick RA, Yancey KB, Lawry TJ. A direct in vivo comparison of the inflammatory properties of human C5a and C5a des arg in human skin. *Immunology*. 1988;140:2376–2381.
60 Wilmshurst PT, Pearson MJ, Walsh KP, et al. Relationship between right-to-left shunts and cutaneous decompression illness. *Clin Sci (Lond)*. 2001;100(5):539–542.
61 Bergh K, Hjelde A, Iversen OJ, Brubakk AO. Variability over time of complement activation induced by air bubbles in human and rabbit sera. *J Appl Physiol*. 1993;74:1811–1815.
62 Ward CA, McCullough D, Fraser WD. Relation between complement activation and susceptibility to decompression sickness. *J Appl Physiol*. 1987;62(3):1160–1166.
63 Hjelde A, Bergh K, Brubakk AO, Iversen OJ. Complement activation in divers after repeated air/heliox dives and its possible relevance to DCS. *J Appl Physiol*. 1995;78(3):1140–1144.
64 Shastri KA, Logue GL, Lundgren CE, et al. Diving decompression fails to activate complement. *Undersea Hyperb Med*. 1997;24(2):51–57.
65 Stevens DM, Gartner SL, Pearson RR. Complement activation during saturation diving. *Undersea Hyperb Med*. 1993;20:279–288.
66 Nossum V, Hjelde A, Bergh K, Brubakk AO. Lack of effect of anti-C5a monoclonal antibody on endothelial injury by gas bubbles in the rabbit after decompression. *Undersea Hyperb Med*. 2000;27(1):27–35.
67 Albrecht EA, Chinnaiyan AM, Varambally S, et al. C5a-induced gene expression in human umbilical vein endothelial cells. *Am J Pathol*. 2004; 164(3):849–859.

68 Clapp BR, Hingorani AD, Kharbanda RK, et al. Inflammation-induced endothelial dysfunction involves reduced nitric oxide bioavailability and increased oxidant stress. *Cardiovasc Res.* 2004;64(1):172–178.

69 Laden G, Madden L, Purdy G, Greenman J. Endothelial damage as a marker of decompression stress. *Undersea Hyperb Med.* 2004;31(3):344.

70 Ward RA, Till GO, Kunkel R. Evidence for the role of hydroxyl radical in complement and neutrophil-dependent tissue injury. *J Clin Invest.* 1983;72:789–801.

71 Thorsen E, Segadal K, Kambestad BK, Gulsvik A. Divers' lung function: small airways disease? *Br J Ind Med.* 1990;47:519–523.

72 Dujic Z, Eterovic D, Denoble P, et al. Effect of a single air dive on pulmonary diffusing capacity in professional divers. *J Appl Physiol.* 1993;74(1):55–61.

73 Willam C, Schindler R, Frei U, Eckardt KU. Increases in oxygen tension stimulate expression of ICAM-1 and VCAM-1 on human endothelial cells. *Am J Physiol.* 1999;276(6 Pt 2):H2044–H2052.

74 Vik A, Brubakk AO, Hennessy TR, et al. Venous air embolism in swine: transport of gas bubbles through the pulmonary circulation. *J Appl Physiol.* 1990;69:237–244.

75 Mrk SJ, Morild I, Brubakk AO, et al. A histopathologic and immunocytochemical study of the spinal cord in amateur and professional divers. *Undersea Hyperb Med.* 1994;21(4):391–402.

76 Mrk SJ, Morild E. A neuropathological study of the ependymoventricular surface in divers brains. *Undersea Hyperb Med.* 1994;21:43–51.

77 Chryssanthou C, Springer M, Lipschitz S. Blood-brain and blood-lung barrier alterations by dysbaric exposure. *Undersea Biomed Res.* 1977;4:111–116.

78 Broman T, Branemark PI, Johansson B, Steinwell O. Intravital and post-mortem studies on air embolism damage of the blood-brain-barrier. *Acta Neur Scand.* 1966;42:146–152.

79 Helps SC, Parsons DW, Reilly PL, Gorman DF. The effect of gas emboli on rabbit cerebral blood flow. *Stroke.* 1990;21(1):94–99.

80 Thom S, Mendiguren I, Hardy K, et al. Inhibition of human neutrophil beta2-integrin-dependent adherence by hyperbaric O2. *Am J Physiol.* 1997;272(3):C770–C777.

81 Thom S, Mendiguren I, Nebolon M, et al. Temporary inhibition of human neutrophil B2 integrin function by hyperbaric oxygen (HBO). *Clin Res.* 1994;42:130.

82 Martin JD, Thom SR. Vascular leukocyte sequestration in decompression sickness and prophylactic hyperbaric oxygen therapy in rats. *Aviat Space Environ Med.* 2002;73(6):565–569.

83 Broome JR, Dutka AJ, McNamee GA. Exercise conditioning reduces the risk of neurologic decompression illness in swine. *Undersea Hyperb Med.* 1995;22(1):73–85.

84 Dujic Z, Duplancic D, Marinovic-Terzic I, et al. Aerobic exercise before diving reduces venous gas bubble formation in humans. *J Physiol.* 2004;555(Pt 3):637–642.

85 Snoeckx LH, Cornelussen RN, Van Nieuwenhoven FA, et al. Heat shock proteins and cardiovascular pathophysiology. *Physiol Rev.* 2001;81(4):1461–1497.

86 Su CL, Wu CP, Chen SY, et al. Acclimatization to neurological decompression sickness in rabbits. *Am J Physiol Regul Integr Comp Physiol.* 2004;287(5):R1214–R1218.

87 Garcia-Cardena G, Fan R, Shah V, et al. Dynamic activation of endothelial nitric oxide synthase by Hsp90. *Nature.* 1998;392 (6678):821–824.

88 Siu PM, Bryner RW, Martyn JK, Alway SE. Apoptotic adaptations from exercise training in skeletal and cardiac muscles. *FASEB J.* 2004;18(10):1150–1152. Epub 2004 May 7.

89 Gonzalez B, Manso R. Induction, modification and accumulation of HSP70s in the rat liver after acute exercise: early and late responses. *J Physiol.* 2004;556(Pt 2):369–385.

90 Paroo Z, Haist JV, Karmazyn M, Noble EG. Exercise improves postischemic cardiac function in males but not females: consequences of a novel sex-specific heat shock protein 70 response. *Circ Res.* 2002;90(8):911–917.

91 Dong H, Xiong L, Zhu Z, et al. Preconditioning with hyperbaric oxygen and hyperoxia induces tolerance against spinal cord ischemia in rabbits. *Anesthesiology.* 2002;96(4):907–912.

92 Xiong L, Zhu Z, Dong H, et al. Hyperbaric oxygen preconditioning induces neuroprotection against ischemia in transient not permanent middle cerebral artery occlusion rat model. *Chin Med J (Engl).* 2000;113(9):836–839.

93 Lochner A, Marais E, Du TE, Moolman J. Nitric oxide triggers classic ischemic preconditioning. *Ann NY Acad Sci.* 2002;962: 402–414.

94 Thom SR, Bhopale V, Fisher D, et al. Stimulation of nitric oxide synthase in cerebral cortex due to elevated partial pressures of oxygen: an oxidative stress response. *J Neurobiol.* 2002;51(2):85–100.

95 Pritchard KA Jr., Ackerman AW, Gross ER, et al. Heat shock protein 90 mediates the balance of nitric oxide and superoxide anion from endothelial nitric-oxide synthase. *J Biol Chem.* 2001;276(21):17621–17624.

96 Bonorino C, Nardi NB, Zhang X, Wysocki LJ. Characteristics of the strong antibody response to mycobacterial Hsp70: a primary, T cell-dependent IgG response with no evidence of natural priming or gamma delta T cell involvement. *J Immunol.* 1998;161(10):5210–5216.

97 George J, Greenberg S, Barshack I, et al. Accelerated intimal thickening in carotid arteries of balloon-injured rats after immunization against heat shock protein 70. *J Am Coll Cardiol.* 2001;38(5):1564–1569.

98 Ahlen C, Mandal LH, Iversen OJ. Identification of infectious Pseudomonas aeroginosa strains in an occupational saturation diving environment. *Occup Environ Med.* 1998;55(7):480–484.

99 Gernhardt ML, Pollock NW, Vann RD, et al. Development of an in-suit exercise prebreathe protocol supporting extravehicular activity in microgravity. *Undersea Hyperb Med.* 2004;31(3):338.

100 Taddei S, Galetta F, Virdis A, et al. Physical activity prevents age-related impairment in nitric oxide availability in elderly athletes. *Circulation.* 2000;101(25):2896–2901.

101 Gonzalez MA, Selwyn AP. Endothelial function, inflammation, and prognosis in cardiovascular disease. *Am J Med.* 2003; 115(Suppl 8A):S99–S106.

56

Exercise and the Endothelium

Ulrik Wisløff, Per M. Haram, and Alf O. Brubakk

Norwegian University of Science and Technology, Trondheim, Norway

Today's human body was "designed," through natural selection, to maximize fitness in the early ancestral environment, a time in which physical activity was obligatory for survival (see Chapter 16). As an important corollary, the human body is not ideally suited for the modern Western lifestyle, where inactivity is the norm. Indeed, physical inactivity is now established as an independent risk factor for cardiovascular morbidity and mortality, an effect that is similar to that of high blood pressure, high levels of blood lipids, and smoking combined (1). An important prerequisite for reversing this trend and prolonging life is to encourage a change in societal attitude such that physical activity, and not a sedentary lifestyle, is viewed as the physiological and behavioral norm. Indeed, exercise training results in reduced primary and secondary vascular events, independent of amelioration of other cardiac risk factors. Exercise is associated with significant physiological adaptations in many systems, including skeletal, muscle, heart, metabolism, and the vasculature. Central to the protective effect of exercise on endothelium is increased bioavailability of nitric oxide (NO), achieved through increased production of NO and decreased reactive oxygen species (ROS)–mediated inhibition of NO activity. The goals of this chapter are to describe some of the long- and short-term adaptations to exercise in the blood vessel endothelium, to review potential intracellular signaling pathways and other molecular mechanisms associated with these adaptations and to relate these changes to integrated performance and health effects.

ENDOTHELIAL FUNCTION

The endothelium was previously regarded as an inert cellular layer lining the blood vessels, but it has since been acknowledged as an important organ with autocrine and paracrine functions. The endothelial isoform of NO synthase (eNOS) produces NO, which is the most important endothelial-derived relaxing factor. In addition to relaxing vascular smooth muscle, NO counteracts the formation of atherosclerosis through inhibition of leukocyte adhesion and invasion, smooth muscle cell proliferation, platelet aggregation, and inflammation. In short, NO is critical for endothelial *function*. Abnormalities in one or more pathways that ultimately regulate the availability of NO may lead to endothelial *dysfunction*, which is characteristic of cardiovascular disease and coronary risk factors, including hypertension, hypercholesterolemia, cigarette smoking, diabetes mellitus, and obesity. Importantly, endothelial dysfunction, as defined by impaired endothelial-dependent vasorelaxation, has been identified as an independent risk factor and a strong prognostic marker of long-term cardiovascular morbidity and mortality in latent and manifest cardiovascular disease (2). Thus, the preservation of endothelial function should be a major therapeutic goal.

IS THERE A LINK BETWEEN AEROBIC CAPACITY AND ENDOTHELIAL FUNCTION?

It is generally accepted that there exists a close link between aerobic capacity (maximal oxygen uptake) and endothelial function in humans (3). In a recent study, we added support to these observations by demonstrating that rats with inborn low aerobic capacity had reduced endothelial function compared with high-aerobic-capacity controls, as determined by acetylcholine-mediated vascular relaxation in isolated ring segments of carotid arteries (4).

In keeping with these observations, many studies have demonstrated an ameliorating effect of exercise on endothelial function and aerobic capacity, even in rats with inborn low-aerobic capacity (Table 56-1) (5). It is widely held that these benefits are most pronounced in (and perhaps even limited to) subjects with preexisting endothelial dysfunction. For example, exercise has been shown to improve endothelial function in rats with metabolic syndrome (5) (a syndrome that includes a cluster of risk factors of cardiovascular disease such as weight gain, high blood pressure, reduced endothelial

Table 56-1: Summary of Animal and Human Exercise Studies

Health Status (Refs.)	*Acute Exercise*	*Chronic Exercise*	*Detraining*
Normal rats (23)		10 weeks of high-intensity endurance training (8-min intervals at 90% of maximal oxygen uptake for 1 h, 5 d/week) improved aerobic capacity and endothelial function.	Endothelium-dependent dilation regressed completely within 2–4 weeks after cessation of training.
Normal rats (19)	A single bout of high-intensity exercise (4 × 4 min at 90% of maximal oxygen uptake) improved endothelium-dependent dilation with peak response at 12–24 hours post-exercise.		Improvement absent 48 hours post-exercise.
Normal rats (19)		6 weeks of high-intensity exercise (4 min intervals at 90% of maximal oxygen uptake, 1 h, 5 d/week) induced larger adaptation than observed after a single bout of exercise.	Effect lasted for about 1 week.
Rats with metabolic syndrome (5)	A single bout of high-intensity exercise (4 × 4 min at 90% of maximal oxygen uptake for 1 h) improved endothelial function measured 24 hours post-exercise. No effect after a single bout of exercise at 70% of maximal oxygen uptake.		Improvement absent 48 hours post-exercise.
Rats with metabolic syndrome (5)		8 weeks of intensity-controlled endurance training improved aerobic capacity and endothelial function.	
Rats with postinfarction heart failure (6)		10 weeks of intensity-controlled endurance training improved aerobic capacity and endothelial function.	
Rats with high aerobic capacity (5)	Moderate (70% of maximal oxygen uptake) and high (90% of maximal oxygen uptake) aerobic intensity improved endothelium-dependent dilatation after a single bout.	Moderate (70% of maximal oxygen uptake) and high (90% of maximal oxygen uptake) aerobic intensity improved endothelium-dependent dilatation after 8 weeks of regular exercise.	
Pigs (18)	7 days of endurance training in pigs resulted in improved endothelium-dependent vasodilatation.		
Humans with chronic heart failure (22)		4 weeks of daily handgrip training improved flow-dependent dilatation.	Improvement absent 6 weeks after ceasing handgrip training.
Patients with stable coronary artery disease (7)		12 months of exercise training (20 min of bicycle ergometry per day at 70% of maximal heart rate) gave similar or better outcome compared to patients that underwent percutaneous coronary angioplasty.	
Patients with coronary artery disease (8)		Six times per day for 10 minutes on a bicycle ergometer at 80% of peak heart rate, 4 weeks, improved endothelial function.	

(*continued*)

Table 56-1 ***(continued)***

Health Status (Refs.)	*Acute Exercise*	*Chronic Exercise*	*Detraining*
Patients with coronary artery disease (9)		Three times daily for 10 minutes on row ergometer and three times daily for 10 minutes on bicycle ergometer, 4 weeks, improved endothelial function.	
Patients with coronary artery disease (10)		Three times daily for 10 minutes on a row ergometer and three times daily for 10 minutes on a bicycle ergometer, 4 weeks, improved endothelial function.	
Patients with coronary artery disease (12)		After 4 weeks of in-hospital training (60 min of bicycle ergometry per day), all training patients were enrolled in a 5-month home-based program of 20 minutes' ergometry training per day and one group training session per week. Home-based endurance training sustained part of the effects of hospital-based endurance on endothelium-dependent vasodilation in coronary artery disease. However, acetylcholine-induced increases in CBF were lower after home-based exercise.	
Patients with coronary artery disease (25)		8 weeks of combined strength and endurance training with large muscle groups or home-based training sessions performing continuous aerobic exercise at 70%–85% HR_{peak} for up to 45–60 minutes. Both exercise training regimens improved endothelium-dependent conduit vessel dilation in subjects with CAD.	Training-induced changes lost 8 weeks after cessation of training.
Patients with chronic heart failure (11)		4 weeks, six times daily for 10 minutes on a bicycle ergometer at 70% peak oxygen consumption gave correction of endothelial dysfunction.	
Patients with chronic heart failure (31)		12 weeks of either interval training (90% of peak heart rate) or moderate exercise (70% of peak heart rate) intensity improved flow-mediated dilatation in the brachial artery in patients with postinfarction heart failure. More robust changes after high-intensity training.	Still improved endothelial function in patients that trained interval at 1–3 years follow-up after cessation of the training project.
Patients with chronic heart failure (24)		8 weeks of three l-hour circuit sessions of whole-body exercise (both strength and endurance) each week. Exercise training improved endothelium-dependent and -independent vascular function and peak vasodilator capacity in patients with CHF.	Training-induced changes lost 8 weeks after cessation of training.
Humans with type 2 diabetes (26)		8 weeks, three 1-hour sessions of combined strength and endurance exercise as well as cycle ergometry and treadmill walking at 70% to 85% of peak HR. Endothelium-dependent vasodilation was enhanced in both conduit and resistance vessels.	Endothelial function returned to baseline 8 weeks after cessation of training.

Health Status (Refs.)	*Acute Exercise*	*Chronic Exercise*	*Detraining*
Obese adolescents (27)		8 weeks, three 1-hour sessions of combined strength- and endurance exercise involving both cycle ergometer and resistance training. Cycle ergometry was maintained at 65%–85% of maximum HR and resistance training intensity at 55%–70% of pretraining maximum strength. Exercise training normalized endothelial dysfunction.	Endothelial function returned to baseline 8 weeks after cessation of training.
Humans with metabolic syndrome (30)	Flow-mediated dilatation in the brachial improved following single-bout interval training (90% of maximal heart rate), but not single-bout moderate-intensity exercise (70% of maximal heart rate).	Flow-mediated dilatation in the brachial improved following a 16-week endurance training program at both levels of exercise intensity, but more robust changes after high intensity training.	Improved endothelial function after a single bout of exercise lasts for 72 hours post-exercise.
Healthy humans (13)		10 weeks of supervised, standardized, aerobic and anaerobic training, consisting of daily 3-mile runs and upper-body strength and endurance exercises improved endothelial function.	
Healthy humans (14)		8-week training regime consisted of three 1-hour sessions of whole body exercise each week, concentrating on the large muscle groups of the lower limbs. Selected torso and upper-body exercises, which did not involve handgripping or forearm exercise, were also included. No change in endothelial function was observed.	
Healthy humans (17)		The training program consisted of four 1-hour running sessions per week. The intensity of training was adjusted to correspond to 70%–80% of each subject's VO_2max. Endothelium-dependent vasodilatation in forearm vessels decreased by 32%–35%	

function, hyperinsulinemia, and increased triglyceride concentration in blood) and postinfarction heart failure (6), as well as in humans with chronic heart disease (7–12). In contrast, studies in healthy subjects reveal conflicting data with some showing improved (13), unaltered (14–16), or even depressed (17) endothelial function. Based on recent data from our laboratory in inactive healthy rats, we believe that the discrepancies in the literature regarding the effect of exercise in normal healthy subjects are explained in part by differences in the exercise protocol and timing of assays relative to the last exercise bout. On the other hand, it seems that endothelial function is well preserved in young, healthy women and men, and that a high aerobic training status due to long-term aerobic training does not improve the dilating capacity any further, but that athletes have larger diameter of their arteries compared to untrained counterparts and thus have a larger "functional capacity" (i.e., blood-transporting capacity) of their vessels (15,16).

TIME COURSE OF ADAPTATION AND REGRESSION

Although regular exercise is known to improve endothelium-dependent arterial relaxation, less is known about how fast the salutary effects appear after initiating a training program and how long the effects last after cessation of exercise.

ADAPTATION

The endothelium is a highly adaptable organ and only a few days of exercise training have shown beneficial effects. For example, as little as 7 days of endurance training in pigs resulted in improved endothelium-dependent arterial vasodilatation (18). Recently we carried out a detailed analysis of the effect of high-intensity endurance training and detraining upon endothelial function in rats (19). As shown in

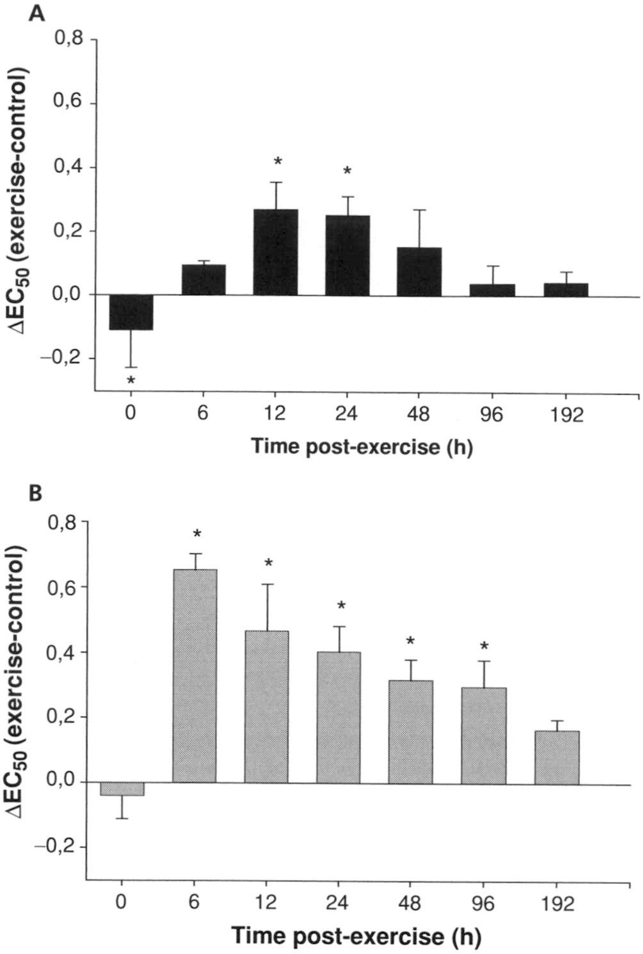

Figure 56.1. Exercise-induced endothelium-dependent vasodilatation. Endothelium-dependent vasodilatation after a single bout of exercise relative to sedentary controls (**A**) and 6 weeks of exercise training (**B**). EC_{50} values = the dose that evokes 50% of maximal dilatation. $^{*}p < 0.05$. (Data used with permission from Haram PM, Adams V, Kemi OJ, et al. Endothelial adaptation following regular and acute exercise. *Eur J Cardiovasc Prev Rehabil.* 2006;13:585–591.)

Figure 56.1, endothelial function in healthy animals improved across a full spectrum of exercise duration. Most notably, we demonstrated that even a single bout of exercise yielded significant changes in endothelium-dependent dilation (Figure 56.1), in line with a study of healthy elderly humans (20). These findings were subsequently confirmed both in rats with intrinsic high- and low aerobic capacity (21).

REGRESSION

Whereas regular exercise is known to increase endothelium-dependent arterial relaxation, little is known about the response to detraining. Current literature suggests that localized and systemic exercise-induced improvements in endothelial function are quickly lost following the cessation of training. For example, 6 weeks after ceasing 4 weeks of daily handgrip training in patients with chronic heart failure, flow-dependent dilatation had returned to pretraining levels (22), and endothelium-dependent vasorelaxation regressed completely within 2 to 4 weeks of detraining after finishing a 10-week endurance program (interval training at 85% of maximal oxygen uptake 5 days per week) in previously untrained rats (23). In line with these data, endothelial function returned to baseline 8 weeks after cessation of 8 weeks of endurance training in patients with congestive heart failure (24), coronary artery disease (25), and type 2 diabetes (26) and in obese adolescents (27,28). In healthy rats the beneficial effects of a 6-week endurance training program was lost after 1 week of inactivity (19). In our own detailed analyses of training and detraining, we found that increased endothelium-dependent dilation after a single bout of exercise peaked at 12 to 24 hours and was reversed after 2 days (Figure 56.1). In contrast, 6 weeks of interval training was associated with greater adaptation and a more gradual return to baseline (approximately 1 week) following cessation of activity (Figure 56.1) (19).

The above considerations regarding adaptation and regression provide important new insights into the effect of exercise training on endothelial function. First, exercise regimens of short duration (including single bout) and long duration (several days per week for more than 4 weeks in published studies) significantly influence endothelial function. The finding that a single bout of exercise is able to initiate a substantial improvement in endothelial function may change the way we look at exercise as a tool in the prevention and rehabilitation of cardiovascular disease. Knowing that the benefits of exercise begin after one bout of exercise can be very motivating for individuals starting in a rehabilitation program, and also for the physicians prescribing the program. Also, beneficial effects of exercise on endothelial function regress after cessation of training, the pattern of which depends on the nature of the initial exercise regimen. Furthermore, these data suggest that even highly trained individuals cannot "store" exercise-induced improvements in endothelial function for a long period of time and that regular exercise is necessary for long-term preservation of endothelial function.

EFFECTS OF HEALTH STATUS AND EXERCISE INTENSITY

As mentioned earlier, the discrepancy in the literature pertaining to the effect of exercise on endothelial function is likely related to interstudy variation in training regimens and the time lag between the last exercise bout and the measurement of endothelial function. One important variable in the training regimen, which is often overlooked, is the intensity of exercise. Exercise intensity should ideally be defined as a workload corresponding to either a given percentage of maximal oxygen uptake or maximal heart rate (29). Unfortunately, many studies employ fixed-intensity (expressed in absolute terms; e.g., km/h) exercise throughout the experiment, despite the fact that the intensity of the load required to induce physical conditioning increases as the performance improves during the course of training (29). Hambrecht and colleagues (7–12) report that moderate-to-high exercise intensity partly reverses endothelial dysfunction in patients with coronary artery disease but they do not determine the effects of different exercise intensities.

In rats with intrinsic (inborn) high aerobic capacity, both moderate (70% of maximal oxygen uptake) and high (90% of maximal oxygen uptake) aerobic intensity improve endothelium-dependent dilatation after either a single bout of exercise or 8 weeks of regular exercise (21). In contrast, in rats with metabolic syndrome, only high-intensity training improved endothelial function after a single bout of exercise, whereas both moderate exercise intensity and high-intensity interval training improved the endothelial function after 8 weeks of endurance training (21). Consistent with these findings, flow-mediated dilatation in the brachial artery in patients with metabolic syndrome was improved following a single bout of high-intensity interval training (4 × 4 minutes at 90%–95% of maximal heart rate, separated by 3-minute active pauses at 70% of maximal heart rate) but not single-bout moderate-intensity exercise, whereas the same patients benefited from a 16-week endurance training program at both levels of exercise intensity (although there was significantly larger improvement in interval-trained subjects) (30). Finally, 12 weeks of either interval training or moderate exercise intensity improved flow-mediated dilatation in the brachial artery in patients with postinfarction heart failure, but again the most robust changes were seen in interval trained subjects (4 × 4 minutes at 90% of peak heart rate, separated by 3 minutes at 60%–70% of peak heart rate) (31).

In summary, most studies suggest that patients with preexisting endothelial dysfunction respond more favorably to high-intensity training (9,23,32). However, there exists one study that shows detrimental effect on endothelial function after 12 weeks (30 minutes for five to seven times per week) of continuous high intensity bicycling (75% of maximal oxygen uptake) (33). Recent data indicate that this may be a matter of when the measurements are being done in relation to the last exercise bout (19).

MOLECULAR MECHANISMS OF EXERCISE-MEDIATED PROTECTION OF ENDOTHELIUM

Nitric Oxide

General agreement exists that exercise-induced endothelium-dependent vasodilatation is due primarily to shear stress-mediated induction of endothelial-derived NO bioactivity, brought about by increased NO production and attenuation of oxidative stress. For example, short- and long-term exercise has been shown to induce NO synthesis and vasomotor responses to acetylcholine and L-arginine but not to sodium nitroprusside, pointing to the importance of the endothelium in mediating this effect. We have shown that improved endothelial response after both a single bout of exercise and chronic exercise is inhibited by the L-arginine antagonist, N(G)-nitro-L-arginine methyl ester (L-NAME) (19).

Hambrecht and colleagues (9) showed that eNOS protein (phosphorylated and unphosphorylated) was upregulated in the mammary artery from patients subjected to exercise prior to coronary artery bypass surgery. Furthermore, compared with active controls, sedentary mice displayed reduced e*NOS* protein expression and endothelium-dependent vasorelaxation, effects that were reversed with subsequent exercise (34). Shear stress and/or exercise not only increases e*NOS* mRNA and protein but also results in increased vascular endothelial growth factors (VEGF) levels and secondary phosphorylation of Akt, which in turn phosphorylates and activates eNOS activity (Figure 56.2) (35). Importantly, the importance of eNOS in mediating adaptation to exercise was demonstrated in studies of *eNOS* $^{-/-}$ mice (36).

The bioavailability of endothelial-derived NO is dependent not only on the expression and posttranslational modification of eNOS, but also on the local redox state. Superoxide anions react with NO (37) to form the peroxynitrite anion (38), resulting in quenching and reduced bioavailability of NO. Peroxynitrite is a strong oxidant that uncouples eNOS by oxidizing the zinc-thiolate center, leading to decreased production of NO. Together, these effects promote endothelial dysfunction.

Interestingly, strenuous exercise is associated with oxidative vascular stress and impaired vascular function in ring segments of aorta from rats immediately following activity (19). These effects are reversed by L-NAME, implicating NO, hence peroxynitrite, as a central mediator of this acute response (19). Consistent with this hypothesis, decreasing the level of superoxide anions by adding superoxide dismutase (SOD) in the perfusion buffer enhanced the vasodilation, whereas disabling the extracellular superoxide dismutase by adding diethyldithiocarbamate (DETCA) further decreased vasodilation.

With chronic exercise training, the potential deleterious effect of the oxygen free radical on NO bioavailability is offset by an upregulation of antioxidant enzymes (17,37) such as cytosolic copper-and-zinc-containing superoxide dismutase (39) and the downregulation of pro-oxidative enzymes. Taken together, these changes lead to increased half-life of NO and improve eNOS function.

Endothelium-Derived Hyperpolarizing Factor and Prostaglandins

In addition to NO, several other vasodilating agents exist, the most debated being the endothelium-derived hyperpolarizing factor (EDHF). This is an L-arginine antagonist and indomethacin-resistant factor released from the endothelium, which probably diffuses into the vascular smooth muscle and causes hyperpolarization and vasorelaxation (40). Currently it seems like EDHF has a major influence in small arteries, whereas NO is the dominant vasodilator in larger arteries such as resistance and conduit arteries (40). EDHF has yet to be identified, but several possible candidates exist, such as arachidonic acid products and potassium. Conflicting results

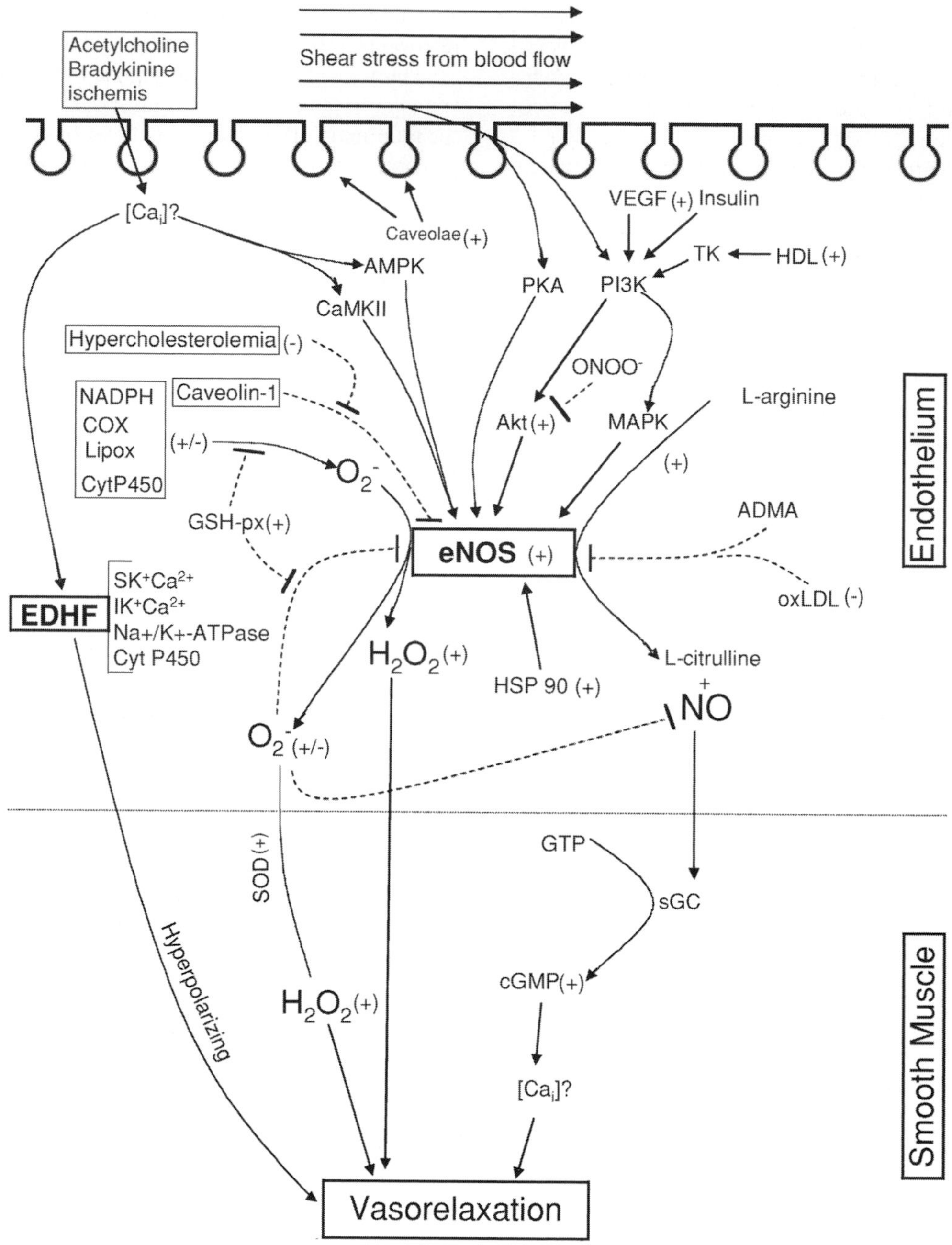

Figure 56.2. A simplified schema of signal transduction in ECs that contribute to regulation of endothelial function. VEGF, vascular endothelial growth factor; AMPK, AMP-protein kinase; CaMKII, calmodulin kinase II; NADPH, nicotinamide adenine dinucleotide phosphate; COX, cyclooxygenase; O_2^-, superoxide anion; GSH-px, glutathione peroxidase; SOD, superoxide dismutase; ADMA, N(G),N(G′)-dimethylarginine; HDL, high-density lipoprotein; Ox LDL, oxidized low-density lipoprotein; TK, tyrosine kinase; PKA, protein kinase A; Akt, protein kinase b; MAPK, mitogen-activated protein kinase; sGC, soluble guanylate cyclase; cGMP, guanosine 3,5-cyclic monophosphate; H_2O_2, hydrogen peroxide; $ONOO^-$, peroxynitrite; (+), upregulated in exercise; (−), downregulated in exercise.

exist to whether EDHF is reduced in disease, and even less evidence exists for an exercise-induced increase in EDHF.

Prostaglandins (PGI) can be potent vasodilators, but their relevance in exercise-induced vasodilation has been debated. Woodman and colleagues (41) suggest that exercise reverses the endothelial dysfunction induced by a sedentary lifestyle and high-fat diet through upregulation/restoration of both NO- and PGI-mediated vasorelaxation.

Exercise and Endothelial Caveolae

eNOS resides in 50 to 100 nm invaginations of the plasma membrane termed *caveolae*. Caveolae consist of the protein caveolin-1, phospholipids, sphingolipids, and cholesterol (42). The physiological role of caveolae is an area of intense research interest. Several functions have been proposed, including vesicular transport, cholesterol homeostasis, and, more recently, cell signaling (see Chapter 73). According to the "caveolae-signaling hypothesis" (43), caveolin-1 provides oligomeric docking sites for organizing and concentrating signaling molecules within caveolae membranes (44). In line with this hypothesis, eNOS is bound to and inhibited by caveolin-1 in caveolae (45). Studies in *caveolin-1$^{-/-}$* mice (45,46) demonstrate unstable basal vessel tone and vigorous acetylcholine-mediated NO production, indicating a lack of inhibition from caveolin-1. The interbreeding of *ApoE$^{-/-}$* and *caveolin-1$^{-/-}$* animals resulted in a significant reduction in atherosclerotic lesion area (47), suggesting an important role for caveolae in the formation of atheroma. However, in vitro experiments in which bovine aortic endothelial cells (ECs) were exposed to shear stress led to an increase in the number of caveolae, in addition to increased phosphorylation of eNOS and subsequent production of NO (48).

As discussed earlier, we have characterized rats selected for aerobic capacity over 11 generations, yielding a strain of rats with either high or low aerobic capacity (4). Those with high aerobic capacity resembled a healthy athletic phenotype, whereas those with low aerobic capacity had metabolic syndrome, including endothelial dysfunction. We determined the relationship between endothelial function and caveolae density, both in sedentary and exercising animals of both strains. The main findings in this study were (a) a low density of luminal caveolae was associated with a classical high-risk profile for cardiovascular disease including endothelial dysfunction, low maximal oxygen uptake, high adipose mass, relative glucose intolerance, and high blood pressure; and (b) exercise upregulated luminal caveolae density and reduced the cardiovascular risk profile. To our knowledge these are the first data reporting a relationship among caveolae, exercise, and health status. The data add important information to the proposed functions of caveolae in that a low luminal endothelial caveolae density reflects a high cardiovascular risk profile, whereas a high density reflects a low cardiovascular risk profile. Intervention in the form of exercise increased the number of caveolae in addition to modifying the risk profile by reducing blood pressure, increasing endothelial function and oxygen uptake, and reducing the amount of abdominal fat.

CONCLUSION

Substantial knowledge linking the EC function to clinical effects of exercise training has emerged over the last decade. In individuals with endothelial dysfunction, accumulating evidence suggests of the beneficial effects of training on quality of life and survival. As presented in this chapter, several molecular mechanisms and signaling pathways for exercise-induced changes in ECs function are being uncovered and provide the basis for modern exercise physiology and directions of future research.

KEY POINTS

- High-intensity endurance training is more beneficial for improving endothelial function than moderate exercise.
- A single bout of high-intensity exercise may improve endothelial function for up to 2 days.
- It is not possible to "store" exercise-induced improvements in endothelial function for a long period of time. Thus, regular exercise is necessary for long-term preservation of endothelial function.

Future Goals

- To determine the optimal intensity and frequency necessary for improved endothelial function, both in animal models for in-depth study of the cellular and molecular mechanisms and in humans for clinical relevance
- To seek new knowledge in cellular and molecular biology in relation to exercise and pathological physiology to improve treatment of endothelial dysfunction (including the role of EDHF and caveolae)
- To implement the increasing knowledge of basal physiological mechanisms into everyday practice of sports and medicine

REFERENCES

1 Alvær K, Dybving E, Engeland A, et al. Folkehelserapporten. Vedlegg 1 i stortingsmelding nr. 16. (2002–2003). Resept for et sunnere Norge. *Nasjonalt folkehelseinstitutt for Helsedepartementet.* 2002.

2 Schächinger V, Britten MB, Zeiher AM. Prognostic impact of coronary vasodilator dysfunction on adverse long-term outcome of coronary disease. *Circulation.* 2000;101:1899–1906.

3 Laughlin MH, Wolfe JB. Memorial lecture. Physical activity in prevention and treatment of coronary disease: the battle line is in exercise vascular cell biology. *Med Sci Sports Exerc.* 2004;36:352–362.

4 Wisløff U, Najjar SM, Ellingsen O, et al. Cardiovascular risk factors emerge after artificial selection for low aerobic capacity. *Science.* 2005;307:418–420.

5 Haram PM, Lee SJ, Al-Share QY, et al. Endurance training reduces disease risk factors in rats artificially selected for low aerobic capacity. Submitted.

6 Haram PM, Kemi OJ, Høydal MA, et al. Combination of losartan and high intensity exercise on endothelial dysfunction in heart failure. Submitted.

7 Hambrecht R, Walther C, Möbius-Winkler S, et al. Percutaneous coronary angioplasty compared with exercise training in patients with stable coronary disease. *Circulation.* 2004;109:1371–1378.

8 Hambrecht R, Wolf A, Gielen S, et al. Effect of exercise on coronary endothelial function in patients with coronary artery disease. *N Engl J Med.* 2000;342:454–460.

9 Hambrecht R, Adams V, Erbs S, et al. Regular physical activity improves endothelial function in patients with coronary artery disease by increasing phosphorylation of endothelial nitric oxide synthase. *Circulation.* 2003;107:3152–3158.

10 Adams V, Linke A, Krankel N, et al. Impact of regular physical activity on the NAD(P)H oxidase and angiotensin receptor system in patients with coronary artery disease. *Circulation.* 2005; 111:555–562.

11 Linke A, Schoene N, Gielen S, et al. Endothelial dysfunction in patients with chronic heart failure: systemic effects of lower-limb exercise training. *J Am Coll Cardiol.* 2001;37:392–397.

12 Gielen S, Erbs S, Linke A, et al. Home-based versus hospital-based exercise programs in patients with coronary artery disease: effects on coronary vasomotion. *Am Heart J.* 2003;145:E3–E6.

13 Clarkson P, Montgomery HE, Mullen MJ, et al. Exercise training enhances endothelial function in young men. *J Am Coll Cardiol.* 1999;33:1379–1385.

14 Maiorana A, O'Driscoll G, Dembo L, et al. Exercise training, vascular function, and functional capacity in middle-aged subjects. *Med Sci Sports Exerc.* 2001;33:2022–2028.

15 Moe IT, Hoven H, Hetland EV, et al. Endothelial function in highly endurance-trained and sedentary, healthy young women. *Vasc Med.* 2005;10(2):97–102.

16 Rognmo Ø, Kahrs C, Bjørnstad TH, et al. Endothelial function in highly endurance-trained and sedentary young men: effects of acute aerobic exercise. *Br J Sports Med.* In press.

17 Bergholm R, Makimattila S, Valkonen M, et al. Intense physical training decreases circulating antioxidants and endothelium-dependent vasodilation in vivo. *Atherosclerosis.* 1999;145:141–149.

18 McAllister RM, Laughlin MH. Short-term exercise training alters responses of porcine femoral and brachial arteries. *J Appl Physiol.* 1997;82:1438–1444.

19 Haram PM, Adams V, Kemi OJ, et al. Endothelial adaptation following regular and acute exercise. *Eur J Cardiovasc Prev Rehabil.* 2006;13:585–591.

20 Benjamin EJ, Larson MG, Keyes MJ, et al. Clinical correlates and heritability of flow-mediated dilation in the community: the Framingham Heart Study. *Circulation.* 2004;109(5):613–619.

21 Haram PM, Østensen Bendheim M, Kemi OJ, et al. Effects of exercise intensity and health status upon adaptation in endothelial function. In preparation.

22 Hornig B, Maier V, Drexler H. Physical training improves endothelial function in patients with chronic heart failure. *Circulation.* 1996;93:210–214.

23 Kemi OJ, Haram PM, Wisloff U, Ellingsen O. Aerobic fitness is associated with cardiomyocyte capacity and endothelial function in exercise training and detraining. *Circulation.* 2004;109:2897–2904.

24 Maiorana A, O'Driscoll G, Dembo L, et al. Effect of aerobic and resistance exercise training on vascular function in heart failure. *Am J Physiol.* 2000;279:H1999–H2005.

25 Walsh JH, Best M, Maiorana AJ, et al. Exercise improves conduit vessel endothelial function in CAD patients. *J Appl Physiol.* 2003;285:20–25.

26 Maiorana A, O'Driscoll G, Cheetham C, et al. The effect of combined aerobic and resistance exercise training on vascular function in type 2 diabetes. *J Am Coll Cardiol.* 2001;38:860–866.

27 Watts K, Beye P, Siafarikas A, et al. Exercise training normalises vascular dysfunction and improves central adiposity in obese adolescents. *J Am Coll Cardiol.* 2004;43:1823–1827.

28 Watts K, Beye P, Siafarikas A, et al. Exercise training in obese children. *J Pediatrics.* 2004;144:620–625.

29 Astrand PO, Rodahl K. Textbook of work physiology. In: *Textbook of Work Physiology.* Singapore: McGraw-Hill, 2003.

30 Tjønna AE, Rognmo Ø, Haram PM, et al. High intensity-training is superior to moderate exercise intensity in treatment of the metabolic syndrome. Manuscript in preparation.

31 Wisløff U, Støylen A, Loennechen JP, et al. Anti-remodelling effects of short-term high-intensity exercise training in patients with stable post-infarction heart failure. Manuscript in preparation

32 Kemi OJ, Haram PM, Loennechen JP, et al. Moderate vs. high intensity: differential effects on aerobic fitness, cardiomyocyte contractility and endothelial function. *Cardiovasc Res.* 2005;67: 161–172.

33 Goto C, Higashi Y, Kimura M, et al. Effect of different intensities of exercise on endothelium-dependent vasodilation in humans. Role of endothelium-dependent nitric oxide and oxidative stress. *Circulation.* 2003;108:530–535.

34 Suvorava T, Lauer N, Kojda G. Physical inactivity causes endothelial dysfunction in healthy young mice. *J Am Coll Cardiol.* 2004; 44(6):1320–1327.

35 Abid MR, Guo S, Minami T, et al. Vascular endothelial growth factor activates PI3K/Akt/forkhead signaling in endothelial cells. *Arterioscler Thromb Vasc Biol.* 2004;24(2):294–300.

36 Momken I, Lechene P, Ventura-Clapier R, Veksler V. Voluntary physical activity alterations in endothelial nitric oxide synthase knockout mice. *Am J Physiol Heart Circ Physiol.* 2004;287(2): H914–H920.

37 Rubanyi GM, Vanhoutte PM. Superoxide anions and hyperoxia inactivate endothelium-derived relaxing factor. *Am J Physiol.* 1986;250:822–827.

38 Patwell DM, Mcardle A, Morgan JE, et al. Release of reactive oxygen and nitrogen species from contracting skeletal muscle cells. *Free Radic Biol Med.* 2004;377:1064–1072.

39 Inoue N, Ramasamy S, Fukai T, et al. Shear stress modulates expression of Cu/Zn superoxide dismutase in human aortic endothelial cells. *Circ Res.* 1996;79:32–37.

40 Bryan RM Jr., You J, Golding EM, Marrelli SP. Endothelium-derived hyperpolarizing factor: a cousin to nitric oxide and prostacyclin. *Anesthesiology.* 2005;102(6):1261–1277.

41 Woodman CR, Price EM, Laughlin MH. Selected contribution: aging impairs nitric oxide and prostacyclin mediation of endothelium-dependent dilation in soleus feed arteries. *J Appl Physiol.* 2003;95(5):2164–2170.

42 Simons K, Toomre D. Lipid rafts and signal transduction. *Nat Rev Mol Cell Biol.* 2000;1:31–41.

43 Razani B, Woodman SE, Lisanti MP. Caveolae: from cell biology to animal physiology. *Pharmacol Rev.* 2002;54:431–467.

44 Sargiacomo M, Sudol M, Tang ZL, Lisanti MP. Signal transducing molecules and GPI-linked proteins form a caveolin-rich insoluble complex in MDCK cells. *J Cell Biol.* 1993;22:789–807.

45 Drab M, Verkade P, Elger M, et al. Loss of caveolae, vascular dysfunction, and pulmonary defects in caveolin-1 gene-disrupted mice. *Science*. 2001;293:2449–2453.
46 Razani B, Engelman JA, Wang XB, et al. Caveolin-1 null mice are viable but show evidence of hyperproliferative and vascular abnormalities. *J Biol Chem*. 2001;276(41):38121–38138.
47 Frank PG, Lee H, Parks DS, et al. Genetic ablation of caveolin-1 confers protection against atherosclerosis. *Arterioscler Thromb Vasc Biol*. 2004;24(1):98–105.
48 Boyd NL, Park H, Yi H, et al. Chronic shear induces caveolae formation and alters ERK and Akt responses in endothelial cells. *Am J Physiol Heart Circ Physiol*. 2003;285:H1113–H1122.

57

The Endothelium at High Altitude

Nicholas L.M. Cruden and David J. Webb

Centre for Cardiovascular Science, University of Edinburgh, United Kingdom

The endothelium is a complex organ that, besides providing a selectively permeable barrier, contributes to the regulation of vascular tone and vascular remodeling, oxidative homeostasis, and the immune response. Acute exposure to hypoxia results in a decrease in the endothelial release of vasorelaxant factors such as nitric oxide (NO) and an increase in the production of vasoconstrictors such as endothelin-1. This imbalance in mediators of vascular tone, combined with alterations in endothelial cell (EC) permeability, is thought to play a key role in the pathogenesis of acute mountain sickness (AMS) (1), high-altitude cerebral edema (HACE) (1), and high-altitude pulmonary edema (HAPE) (2). Besides changes in the regulation of vascular tone, alterations in endothelial function at high altitude may be associated with increased levels of oxidative stress and inflammation. Finally, the endothelium appears to be important in mediating the changes in blood vessel structure that occur as a result of chronic exposure to high altitude.

ACUTE RESPONSE TO HYPOXIA

The major physiological consequence of exposure to acute hypoxia and high altitude is organ-specific vasoconstriction (3). Within the pulmonary vascular bed, for example, hypoxia-induced vasoconstriction is an important mechanism for matching ventilation and perfusion. A dysfunctional endothelial response, however, may lead to exaggerated vasoconstriction, alterations in regional blood flow, and ventilation-perfusion mismatches. Moreover, an exaggerated hypoxic vasoconstrictor response may lead to capillary stress failure, alterations in EC permeability, and ultimately tissue edema. Not surprisingly, therefore, endothelial dysfunction at high altitude has been implicated in the pathogenesis of two potentially life-threatening conditions that occur in otherwise healthy subjects exposed to altitudes of greater than 3,000 m – HACE (1), intracerebral fluid accumulation characterized by ataxia and altered consciousness that may progress rapidly to coma and death, and HAPE (2), a noncardiogenic pulmonary edema. Interestingly, hypoxic vasoconstriction alone does not appear to be sufficient for the development of altitude-related disease (4) and additional aspects of EC dysfunction, or injury, have been implicated. These are discussed below.

ENDOTHELIUM-DEPENDENT VASOMOTOR FUNCTION

ECs respond to hypoxia by releasing a number of vasoconstrictor agents, including endothelin-1 and platelet-derived growth factor (PDGF)-B (5). Plasma concentrations of endothelin-1 are elevated in mountaineers exposed to high altitude and are higher still in mountaineers prone to HAPE (6–8), the increase in plasma endothelin-1 correlating with the rise in systolic pulmonary artery pressure and the degree of hypoxia (8).

In contrast, levels of endothelial NO synthase (eNOS) are reduced in ECs exposed to hypoxia and associated NO production is impaired (9). Consistent with this, exhaled NO concentrations are reduced in HAPE-prone individuals (10). Moreover, inhalation of NO reduces pulmonary artery systolic pressure in subjects with HAPE to levels found in subjects without HAPE (11). NO also suppresses the increase in endothelin-1 production by ECs exposed to hypoxia (5). Therefore, reduced NO concentrations associated with hypoxic exposure will result both in impaired vasodilatation and an increase in the production of the vasoconstrictor endothelin-1.

REACTIVE OXYGEN SPECIES

Reactive oxygen species (ROS), or oxygen-free radicals, are highly toxic molecules that contain one or two unpaired electrons. Under hypoxic conditions, ROS are generated as a result of increased xanthine conversion to uric acid by xanthine oxidase, releasing the potent ROS superoxide. Additional mechanisms implicated in the production of ROS include leukocyte activation, catecholamine oxidation, and disruption of the mitochondrial oxidative respiration (12). Increased oxidative stress impairs endothelial function (13). Evidence of increased oxidative stress has been reported in humans exposed to high altitude (14) and may result in direct damage to the EC membrane leading to increased capillary permeability, a feature of both HAPE and HACE. In addition, ROS can interact with and inactivate NO, leading to decreased NO bioavailability (13,15). Although a number of studies have reported a treatment benefit with antioxidant therapy in the prevention of AMS, a syndrome of nonspecific symptoms that is largely thought to precede the development of HACE (16–18), this has not been a universal finding (19). Further work is required to fully characterize the role of ROS and antioxidant therapy in the pathophysiology of high-altitude acclimatization and illness.

INFLAMMATION

EC expression of interleukin (IL)-1α, -6, and -8, as well as macrophage chemotactic protein, is increased under hypoxic conditions in vitro (20–22), and plasma concentrations of IL-6 are raised in subjects ascending rapidly to high altitude (23). Elevated plasma concentrations of soluble E-selectin at high altitude provide further evidence of altered EC function, or EC injury, under these conditions (6,24). It has been suggested that soluble E-selectin has a protective effect on the endothelium (25). Membrane-bound E-selectin is a glycoprotein produced exclusively by ECs that mediates endothelium–leukocyte interactions. Proteolytic cleavage of E-selectin from the EC surface results in the release of soluble E-selectin into the circulation, where it may impede the action of its membrane-bound form, blocking endothelium–leukocyte adhesion, enhancing tissue perfusion, and reducing ischemia-induced damage (26).

Additional data supporting a role for inflammation in the pathogenesis of high-altitude illness include increased cytokine concentrations in bronchoalveolar lavage (BAL) fluid from subjects with established HAPE (27–29), an association between HLA genotype and HAPE (30), and the response of high-altitude headache to anti-inflammatory drugs (31,32). However, the absence of an elevated neutrophil count or cytokine concentrations in bronchoalveolar lavage fluid early in the development of HAPE would suggest that inflammation occurs as a secondary event and is not a primary pathophysiological mechanism (33).

SIGNAL TRANSDUCTION

The Po_2 of arterial blood is approximately 150 mm Hg and in tissues is approximately 40 mm Hg (5). Under hypoxic conditions, tissue oxygen tension may fall as low as 15 to 30 mm Hg. ECs sense and respond to oxygen tensions below 70 mm Hg, but details of how this is achieved remain unclear (5). Proposed mechanisms include heme-containing proteins that undergo conformational change under hypoxic conditions (5), alterations in cellular concentrations of ROS (34), and shifts in the relative balance between ROS and NO production (35). Downstream signaling pathways implicated in the endothelial response to hypoxia include activation of protein kinases and transcription factors such as AP-1, hypoxia-inducible factor (HIF)-1, and nuclear factor (NF)-κB (5,36) (Figure 57.1).

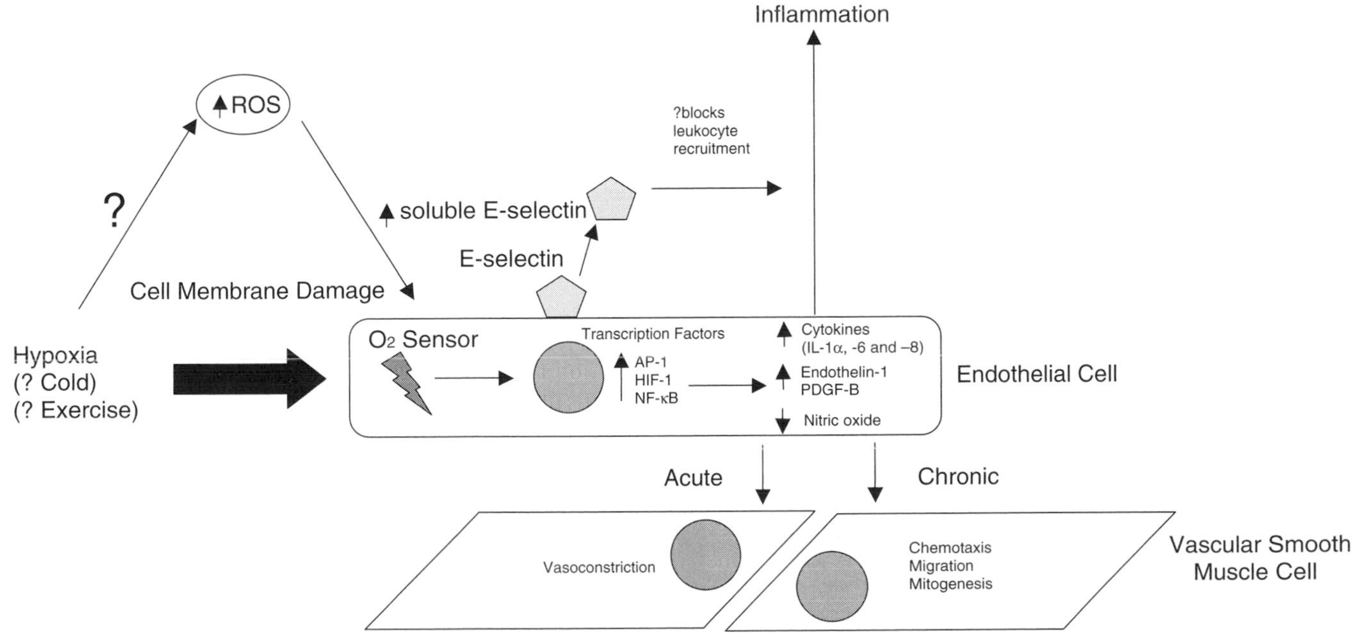

Figure 57.1. Schematic summary of effects of exposure to high altitude on EC function.

CHRONIC RESPONSE TO HYPOXIA

Chronic exposure to hypoxia and high altitude is associated with extensive vascular remodeling characterized predominantly by an increase in vascular smooth muscle (37). Chronic hypoxic exposure induces EC production of a number of molecules that have potent smooth muscle cell chemotaxic and mitogenic properties (5). These include endothelin-1, PDGF-B, and vascular endothelial growth factor (VEGF). In contrast, NO, levels of which are reduced under hypoxic conditions, inhibits fibroblast and smooth muscle cell mitogenesis. Thus hypoxia would appear not only to induce factors that promote vascular remodeling but also to inhibit the synthesis of mediators that normally keep this process in check. The consequences of an increase in vascular smooth muscle, particularly within the lungs, is not clear, but it may reflect a natural adaptive mechanism designed to protect individuals chronically exposed to high altitude from the capillary failure, characteristic of HACE and HAPE, that may occur on reentry to high altitude (37).

CONFOUNDING FACTORS

In addition to hypoxia, subjects at high altitude may be exposed to extreme cold and varying degrees of physical exercise. Elevated plasma concentrations of endothelin-1 and E-selectin have been reported following extreme, but not mild, cold exposure (6,38). Moreover, evidence suggests that moderate-intensity exercise may upregulate endothelial NO bioavailability and improve endothelium-dependent vasodilatation and antioxidant function (13,39). In contrast, high-intensity exercise may increase circulating markers of oxidative stress (40). At present, however, the interplay of cold exposure and exercise with the alterations in endothelial function observed at high altitude is poorly understood.

KEY POINTS

- Alterations in endothelial function occur on exposure to high altitude and appear to play a central role in the process of acclimatization.
- An abnormal EC response to high-altitude exposure has been implicated in the pathogenesis of altitude-related illnesses such as AMS, HACE, and HAPE.
- Some data support the use of antioxidant therapy in the treatment of AMS and even greater support for inhaled NO as a therapy for HAPE.
- Further work is required to examine the potential of the EC and its products as therapeutic targets in the treatment of high-altitude-related diseases.

REFERENCES

1 Hackett PH, Roach RC. High altitude cerebral edema. *High Alt Med Biol.* 2004;5:136–146.
2 Schoene RB. Unraveling the mechanism of high altitude pulmonary edema. *High Alt Med Biol.* 2004;5:125–135.
3 Fishman AP. Hypoxia on the pulmonary circulation. How and where it acts. *Circ Res.* 1976;38:221–231.
4 Sartori C, Allemann Y, Trueb L, et al. Exaggerated pulmonary hypertension is not sufficient to trigger high-altitude pulmonary oedema in humans. *Schweiz Med Wochenschr.* 2000;130:385–389.
5 Faller DV. Endothelial cell responses to hypoxic stress. *Clin Exp Pharmacol Physiol.* 1999;26:74–84.
6 Cruden NL, Newby DE, Ross JA, et al. Effect of cold exposure, exercise and high altitude on plasma endothelin-1 and endothelial cell markers in man. *Scott Med J.* 1999;44:143–146.
7 Morganti A, Giussani M, Sala C, et al. Effects of exposure to high altitude on plasma endothelin-1 levels in normal subjects. *J Hypertens.* 1995;13:859–865.
8 Sartori C, Vollenweider L, Loffler BM, et al. Exaggerated endothelin release in high-altitude pulmonary edema. *Circulation.* 1999; 99:2665–2668.
9 Phelan MW, Faller DV. Hypoxia decreases constitutive nitric oxide synthase transcript and protein in cultured ECs. *J Cell Physiol.* 1996;167:469–476.
10 Duplain H, Sartori C, Lepori M, et al. Exhaled nitric oxide in high-altitude pulmonary edema: role in the regulation of pulmonary vascular tone and evidence for a role against inflammation. *Am J Respir Crit Care Med.* 2000;162:221–224.
11 Scherrer U, Vollenweider L, Delabays A, et al. Inhaled nitric oxide for high-altitude pulmonary edema. *N Engl J Med.* 1996;334:624–629.
12 Roche E, Romero-Alvira D. Role of oxygen free radicals in altitude-related disorders. *Med Hypotheses.* 1994;42:105–109.
13 Ogita H, Liao J. Endothelial function and oxidative stress. *Endothelium.* 2004;11:123–132.
14 Jefferson JA, Simoni J, Escudero E, et al. Increased oxidative stress following acute and chronic high altitude exposure. *High Alt Med Biol.* 2004;5:61–69.
15 Steiner DR, Gonzalez NC, Wood JG. Interaction between reactive oxygen species and nitric oxide in the microvascular response to systemic hypoxia. *J Appl Physiol.* 2002;93:1411–1418.
16 Roncin JP, Schwartz F, D'Arbigny P. EGb 761 in control of acute mountain sickness and vascular reactivity to cold exposure. *Aviat Space Environ Med.* 1996;67:445–452.
17 Gertsch JH, Seto TB, Mor J, et al. Ginkgo biloba for the prevention of severe acute mountain sickness (AMS) starting one day before rapid ascent. *High Alt Med Biol.* 2002;3:29–37.
18 Bailey DM, Davies B. Acute mountain sickness; prophylactic benefits of antioxidant vitamin supplementation at high altitude. *High Alt Med Biol.* 2001;2:21–29.
19 Gertsch JH, Basnyat B, Johnson EW, et al. Randomised, double blind, placebo controlled comparison of ginkgo biloba and acetazolamide for prevention of acute mountain sickness among Himalayan trekkers: the prevention of high altitude illness trial (PHAIT). *BMJ.* 2004;328:797.
20 Yan SF, Tritto I, Pinsky D, et al. Induction of interleukin 6 (IL-6) by hypoxia in vascular cells. Central role of the binding site for nuclear factor-IL-6. *J Biol Chem.* 1995;270:11463–11471.
21 Shreeniwas R, Koga S, Karakurum M, et al. Hypoxia-mediated induction of endothelial cell interleukin-1 alpha. An

autocrine mechanism promoting expression of leukocyte adhesion molecules on the vessel surface. *J Clin Invest*. 1992;90: 2333–2339.
22 Karakurum M, Shreeniwas R, Chen J, et al. Hypoxic induction of interleukin-8 gene expression in human ECs. *J Clin Invest*. 1994; 93:1564–1570.
23 Bailey DM, Kleger GR, Holzgraefe M, et al. Pathophysiological significance of peroxidative stress, neuronal damage, and membrane permeability in acute mountain sickness. *J Appl Physiol*. 2004;96:1459–1463.
24 Grissom CK, Zimmerman GA, Whatley RE. Endothelial selectins in acute mountain sickness and high-altitude pulmonary edema. *Chest*. 1997;112:1572–1578.
25 Blann AD, Tse W, Maxwell SJ, et al. Increased levels of the soluble adhesion molecule E-selectin in essential hypertension. *J Hypertens*. 1994;12:925–928.
26 Menasche P, Peynet J, Haeffner-Cavaillon N, et al. Influence of temperature on neutrophil trafficking during clinical cardiopulmonary bypass. *Circulation*. 1995;92:II334–II340.
27 Schoene RB, Hackett PH, Henderson WR, et al. High-altitude pulmonary edema. Characteristics of lung lavage fluid. *JAMA*. 1986;256:63–69.
28 Schoene RB, Swenson ER, Pizzo CJ, et al. The lung at high altitude: bronchoalveolar lavage in acute mountain sickness and pulmonary edema. *J Appl Physiol*. 1988;64:2605–2613.
29 Kubo K, Hanaoka M, Yamaguchi S, et al. Cytokines in bronchoalveolar lavage fluid in patients with high altitude pulmonary oedema at moderate altitude in Japan. *Thorax*. 1996;51:739–742.
30 Hanaoka M, Kubo K, Yamazaki Y, et al. Association of high-altitude pulmonary edema with the major histocompatibility complex. *Circulation*. 1998;97:1124–1128.
31 Ferrazzini G, Maggiorini M, Kriemler S, et al. Successful treatment of acute mountain sickness with dexamethasone. *BMJ*. 1987;294:1380–1382.
32 Burtscher M, Likar R, Nachbauer W, et al. Aspirin for prophylaxis against headache at high altitudes: randomised, double blind, placebo controlled trial. *BMJ*. 1998;316:1057–1058.
33 Swenson ER, Maggiorini M, Mongovin S, et al. Pathogenesis of high-altitude pulmonary edema: inflammation is not an etiologic factor. *JAMA*. 2002;287:2228–2235.
34 Acker H. Cellular oxygen sensors. *Ann NY Acad Sci*. 1994;718:3–10; discussion 11–12.
35 Gonzalez NC, Wood JG. Leukocyte-endothelial interactions in environmental hypoxia. *Adv Exp Med Biol*. 2001;502:39–60.
36 Mortimer H, Patel S, Peacock AJ. The genetic basis of high-altitude pulmonary oedema. *Pharmacol Ther*. 2004;101: 183–192.
37 Heath D, Williams DR. The lung at high altitude. *Invest Cell Pathol*. 1979;2:147–156.
38 Hynynen M, Ilmarinen R, Saijonmaa O, et al. Plasma endothelin-1 concentration during cold exposure. *Lancet*. 1991;337:1104–1105.
39 Maiorana A, O'Driscoll G, Taylor R, et al. Exercise and the nitric oxide vasodilator system. *Sports Med*. 2003;33:1013–1035.
40 Goto C, Higashi Y, Kimura M, et al. Effect of different intensities of exercise on endothelium-dependent vasodilation in humans: role of endothelium-dependent nitric oxide and oxidative stress. *Circulation*. 2003;108:530–535.

58

Endothelium in Space

Janice V. Meck* and Ralph E. Purdy†

*University of Texas at Houston, NASA/Johnson Space Center;
† University of California, Irvine

Of all the challenges to the endothelium discussed in the chapters within this volume, perhaps the most extreme is the challenge presented when members of a bipedal species that has evolved exquisite cardiovascular controls to support life on Earth, decides to leave its natural environment to explore a world without gravity. We speak of our pioneering astronauts. This chapter describes the cardiovascular changes that the human body undergoes when it leaves Earth to reside in a zero-gravity (actually microgravity) environment for even short periods of time and discusses how the vascular endothelium (particularly the endothelial nitric oxide synthase [eNOS] enzyme) may participate in the adaptations. We also discuss how other forms of NOS might respond to spaceflight. Results from the ground-based rodent model of microgravity, hindlimb unloading, also will be discussed as they have helped determine mechanisms of cardiovascular adaptations. Finally, we speculate about possible adaptations to spaceflight that have not yet been measured.

When exposed to microgravity, the human body undergoes changes in physiological function that appear to be normal adaptations to the new environment (Table 58-1). The cardiovascular system in particular undergoes profound changes. Unfortunately, these adaptations leave the cardiovascular system ill-equipped to handle a sudden return to the gravity of Earth. The most universal symptom after landing, experienced by virtually every astronaut, is postflight orthostatic intolerance. This manifests at a minimum as an increased heart rate response to upright posture and at a maximum as frank orthostatic hypotension and presyncope (lightheadedness, nausea) or even syncope (fainting). Another universal postflight symptom is decreased exercise capacity after landing. Both of these adverse effects present substantial risks if rapid physical activity or emergency egress is required during or immediately after landing. Given the ambitious goals for future space exploration enunciated by the National Aeronautics and Space Administration (NASA), the development of effective countermeasures to microgravity-induced cardiovascular changes is a pressing need. It is also important to note that 500,000 Americans suffer from a chronic syndrome of orthostatic intolerance, and that all long-term bed-rest patients, as well as quadriplegic patients (1), are at risk for developing orthostatic intolerance. Thus, countermeasures developed for astronauts also have the potential to be effective in the treatment of hundreds of thousands of orthostatic intolerance patients.

The present state of our knowledge of microgravity-induced cardiovascular changes is based on numerous experiments, conducted both in actual microgravity of spaceflight and in simulated microgravity on the ground, over four decades. Head-down bed rest often is used as a ground-based model of microgravity for the study of humans. The most commonly used microgravity model for rodent studies has been hindlimb unloading, during which animals are suspended by their tails at an angle that raises their hind legs so that they locomote only with their forelegs (Figure 58.1).

The hindlimb-unloading rodent model simulates in many ways the cardiovascular effects of spaceflight in humans. These effects include muscle atrophy, hypovolemia, resting tachycardia, impaired baroreflex responses, and reduced ability to vasoconstrict. Studies of astronauts before, during, and after spaceflight have identified the major hemodynamic and neurohumoral adaptations that are associated with orthostatic intolerance. In turn, the rodent hindlimb-unloading studies have provided insights into the mechanisms underlying these hemodynamic and neurohumoral changes. Important to this chapter, evidence suggests that nitric oxide (NO) mechanisms may underlie or, at least contribute to, all of the major parameter changes that characterize microgravity-induced cardiovascular adaptations.

CARDIOVASCULAR ADAPTATIONS TO SPACEFLIGHT

During the normal daily activities of humans on Earth, the blood pressure gradient between the head and feet rapidly

Table 58-1: Common Physiological Effects of Spaceflight

↓	Blood volume
↓	Baroreflex function
↓	LV mass
↓	Stroke volume
↓	Orthostatic tolerance
↓	Aerobic capacity
↓	Skeletal muscle mass
↓	Bone mass
↑	Renal stone risk
↓	Postural equilibrium control
↓	Locomotor control
↓	Visual–vestibular integration

and repeatedly can change from no gradient in the supine position to a gradient of about 130 mm Hg in the upright position. A normal human is exquisitely adapted to accommodate these gradient changes and maintain constant, adequate blood flow to the brain. Most people can arise from a nap on the sofa, walk into the kitchen for a drink, and then sit down in a chair without fainting. They are unaware of the multifaceted, redundant system of monitors and controls, evolved over the millennia, which allow them to make these movements. Some of the tools that help maintain blood flow to the head are mechanical, such as the one-way valves within, and the large skeletal muscles that surround the deep veins in the legs. Others involve continual adjustments by the autonomic nervous system or adjustments to maintain intravascular volume. Still other controls exist within the vasculature itself to regulate local flow and pressure. All these mechanisms evolved within the confines of Earth's gravity, but when a human enters the unique environment of space, the influences of gravity are removed, and the body must adapt to an environment it has not previously experienced. During spaceflight changes in posture become hemodynamically irrelevant; locomotion utilizes primarily the arms, rather than the legs; blood pressure becomes uniform throughout the body; the heart has less work to do; there is no up or down; and the weight-bearing bones bear no weight. The large hemodynamic changes that occur during spaceflight were anticipated and feared by early flight surgeons. There were many who feared that the first humans who were launched into space would not survive, but over the last four decades we have learned that the human body has a remarkable ability to adapt and work efficiently within the microgravity of space. However, although the adaptations to microgravity are appropriate, they leave the body ill-prepared to return to Earth. About 20% of returning astronauts are clearly hypotensive and presyncopal on landing day after spaceflights of about 2 weeks, and about 83% are hypotensive and presyncopal after 4- to 6-month flights (2). The hallmarks of postflight orthostatic hypotension are decreased plasma volume coupled with an inappropriately vascular resistance response upon standing.

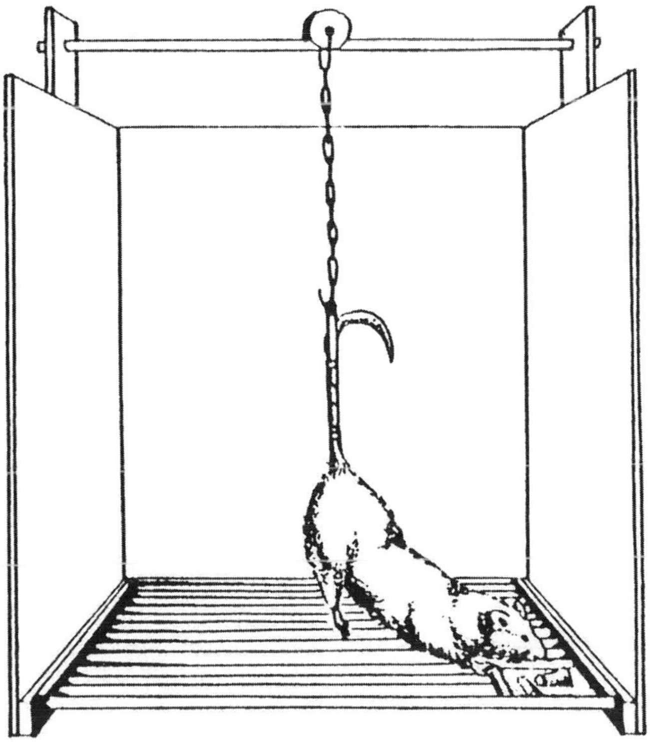

Figure 58.1. Hindlimb-unloaded rat. (Courtesy of Dr. Jin Ma.)

PLASMA VOLUME LOSS

Hypovolemia in returning astronauts is a universal finding that has been well described. When a human first becomes weightless, a rapid cephalad fluid shift occurs, which is thought to trigger an early diuresis. Atrial natriuretic peptide increases dramatically early in flight and increases renal sodium excretion (3). After several days the plasma volume stabilizes at a new level that is lower than before flight and stays at that level for at least 6 months (2). From inflight photographs, which show astronauts' "chicken legs," it is evident that the main loss of volume is in the legs. Central volume, which is monitored by the cardiopulmonary baroreceptors in the chest, is maintained. There have been many efforts to restore this loss of volume. In 1985, NASA mandated that astronauts must consume salt tablets and water equivalent to a liter of isotonic saline prior to landing, but still plasma volume is decreased an average of 9% from preflight after landing (4–6). During head-down tilt, attempts to restore plasma volume with a liter of intravenous saline failed because the extra fluid was quickly eliminated (7). Thus, the body actively maintains the plasma volume at the level it perceives to be appropriate in microgravity and resists attempts at intervention intended to restore volume to levels that are appropriate in normal gravity. The resultant hypovolemia is a primary contributor to postflight cardiovascular deconditioning.

HYPOADRENERGIC RESPONSES

Another cardiovascular adjustment to microgravity is a change in the autonomic response to baroreceptor input. During the weightlessness of space the central nervous system receives entirely different input from the arterial baroreceptors. In the absence of gravity, there is no tendency for blood to "pool" in what on Earth would be dependent limbs. Consequently, the central nervous system receives no baroreceptor input to counter upright posture. This appears to effect a central remodeling, such that the capability for proper integration and initiation of appropriate sympathetic efferent responses to hypotensive input is lost. This results in an inadequate vascular resistance increase with upright posture. Interestingly, this remodeling occurs to a greater extent in some individuals than others (4–6).

On landing day, after having spent up to 6 months in an environment where it was not necessary to respond to changes in posture, the sympathetic nervous system has approximately 1 hour from the time the engines ignite for reentry, to readapt and respond within the full force of Earth's gravity. It must not only "remember" how to respond but also compensate for the reduced blood volume. The longer the flight, the more difficult it is to "remember" (2). During and after landing, if a hypersympathetic response is not initiated during upright posture, astronauts become presyncopal. Three separate studies have shown that astronauts who do become presyncopal do not release appropriate amounts of norepinephrine during postflight upright tilt tests (4–6). The mechanism for this appears to be central. When intravenous injections of tyramine, an indirect sympathomimetic, were given before and after spaceflight, it was found that, after flight, the amount of norepinephrine released by the tyramine injections was not reduced, compared to preflight, in either presyncopal or nonpresyncopal astronauts (5). This provided strong evidence that baseline norepinephrine synthesis and availability are not reduced as a result of spaceflight, but rather that spaceflight had reduced the ability of the central nervous system to organize an appropriate sympathetic response to baroreceptor input. During this time, several studies using the hindlimb unloading in rats provided important new information in support of this idea.

The possibility of changes in central integration of baroreceptor input was pursued using the hindlimb-unloading rat model mentioned above. In a landmark study, afferent aortic depressor nerve traffic versus efferent renal sympathetic nerve traffic was measured during pharmacologically induced pressure decreases. After hindlimb unloading, baroreceptor afferent traffic decreased appropriately when arterial pressure decreased, but the resultant increase in efferent sympathetic traffic was attenuated. Thus the "gain" of the baroreflex was attenuated. This indicates a dysfunction of the ability of the central nervous system to integrate hypotensive afferent input and organize an appropriate efferent response. A schematic representation of the baroreflex is presented in Figure 58.2. In a follow-up study, the same group reported enhanced GABA-mediated inhibition of efferent sympathetic traffic in the rostral ventrolateral medulla (RVLM) after hindlimb unloading (8).

What role might NO play in hypothesized changes in central regulation of blood pressure after spaceflight? It is well known that endogenous NO plays a generalized role in autonomic regulation of blood pressure by decreasing sympathetic output (9) and also by influencing norepinephrine synthesis (10) and release (11). It can act within the paraventricular nucleus (PVN) of the hypothalamus, where parvocellular neurons project to the nucleus of the solitary tract (NTS), the caudal ventrolateral medulla (CVLM), the RVLM (12), and to sympathetic preganglionic neurons in the spinal cord (12). All these areas are important in cardiovascular control. Functional data suggest that NO inhibits these PVN neurons, as pharmacological blockade of NOS in the PVN increases efferent sympathetic traffic (13). Thus, increases in neuronal NOS (nNOS) activity in the PVN of the hypothalamus may inhibit efferent sympathetic activity in response to baroreceptor input. NO-producing neurons in many areas of the brain are activated by stress. It is reasonable to suggest that nNOS would be increased during the stress of spaceflight and landing and may inhibit sympathetic responsiveness. This line of research and reasoning provides important insight into a possible mechanism by which the central nervous system may remodel during spaceflight and lead to inadequate sympathetic responses to standing, resulting in orthostatic hypotension in astronauts after spaceflight.

VASCULAR REMODELING

We have so far discussed changes in neural regulation of blood pressure as a mechanism for the low standing vascular resistance on landing day. Another possible mechanism is that the vasculature itself may remodel during spaceflight. Data from astronauts have not been conclusive. When pressor responses to intravenous injections of phenylephrine (an α_1 adrenergic agonist) were measured, there were no differences between preflight and landing day. In addition, β_2 adrenergic receptors were not changed from preflight (5). However, before flight, pressor responses to phenylephrine were smaller in those astronauts who are destined to become presyncopal on landing day than in those who are not. These data suggest that adrenergic receptor responses are not affected by spaceflight, but that individual differences in vascular responsiveness may indicate preflight predisposition to postflight problems. Other data suggest different conclusions. Whitson and colleagues showed a marked increase in plasma norepinephrine on landing day, and a nearly fivefold increase in plasma renin activity (suggesting high levels of angiotensin , a powerful vasoconstrictor), with no parallel increase in total peripheral resistance (14). In a third study, nonpresyncopal astronauts had significantly greater norepinephrine release with standing on landing day but no increase in resistance, also suggesting impairment of vasoconstrictor responses (6).

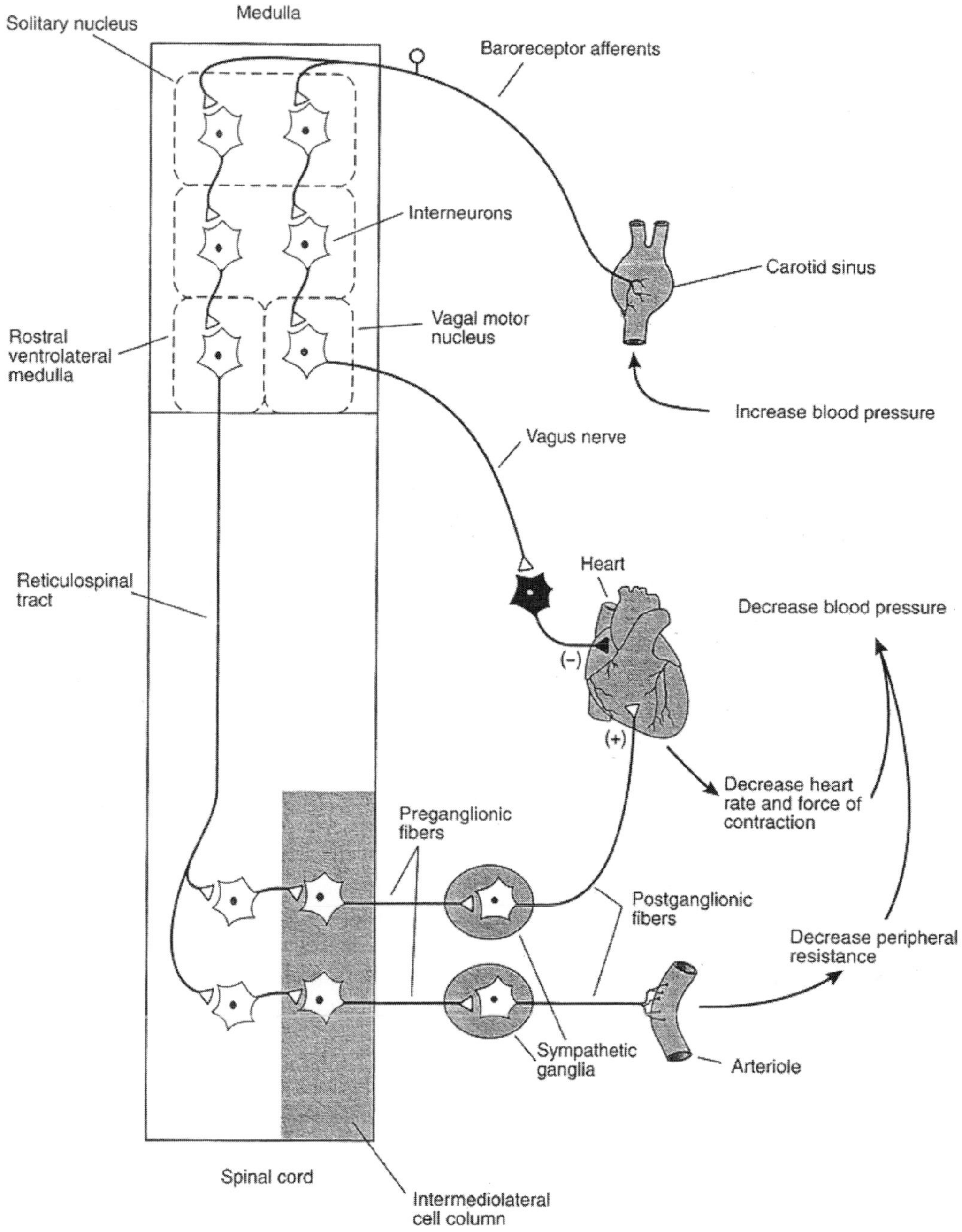

Figure 58.2. Schematic representation of the baroreflex. (Reproduced with permission from Kandel ER, Schwartz JH, Jessell TM. *Principles of Neural Science.* New York: Appleton & Lange; 1991.)

There is every reason to believe that the vasculature would remodel in response to the large changes in local hemodynamic forces experienced when the body becomes weightless. On Earth, vessels in the legs experience much higher pressures than those in the arms, as well as large changes in pressures throughout a 24-hour period. During spaceflight, those factors are missing. The legs are not used for locomotion, they do not bear weight, they become differentially dehydrated, their skeletal muscles atrophy, and interstitial pressures probably change. The arms, on the other hand, are the primary means of locomotion, particularly in crew members who perform extravehicular activities (EVAs) or spacewalks. It follows that the vasculature in the arms and the legs would experience opposite, although not equal, changes in flows, pressures, and shear forces when they go from Earth to space. One could therefore reasonably theorize that the vasculature in the arms might adapt differently to spaceflight than the vasculature in the legs. Virtually no studies have been performed on astronauts regarding this matter. Therefore, we must turn once again to our animal model.

During hindlimb unloading, flow is changed in many vascular beds. In particular, the cephalad fluid shift caused by hindlimb unloading increases flow in the upper body (15) and decreases flow in the lower body (16). The changes in flow appear to correlate with parallel changes in both structure and function. Vasculature in the forelimbs shows hypertrophic remodeling, whereas that in the hindlimb shows atrophic remodeling (16). There also are decreases in the media cross-sectional area in arteries in the hindlimb but not the forelimb (16), suggesting atrophic remodeling in the hindlimb but not the forelimb.

Changes in flow as discussed earlier change shear stress, which is a powerful stimulus for the modulation of eNOS expression. It varies directly with flow and inversely with the cube of the vessel radius. Endothelial function varies with shear stress over a time course dictated by changes in both flow and vessel radius (16). Studies of endothelial responses in hindlimb-unloaded rats support the idea of differential changes in function in the forelimbs versus the hindlimbs. In ex vivo preparations of the carotid artery and other arteries in the upper body, hindlimb unloading causes increases in flow (17), increased responses to acetylcholine (18), upregulation of eNOS (19), and a decreased contractile response, which is restored toward control levels when the endothelium is removed (18). In the pulmonary artery, also positioned to experience increased blood flow during hindlimb unloading (20), there occur enhanced responses to acetylcholine and an increase in eNOS protein after hindlimb unloading (20). In isolated pulmonary artery rings, contractile responses are smaller in hindlimb unloaded animals than in control animals, but pharmacological blockade of eNOS abolishes the difference (20). In contrast to changes in arteries in the upper body, soleus muscle-feed arteries in the hindlimb show reduced vasodilatory responses to acetylcholine injections (21) and reduced eNOS expression (22). Additional studies of the soleus muscle-feed arteries illustrate the time-dependent nature of changes in endothelial function and flow-induced artery remodeling. At the beginning of hindlimb unloading, shear stress (calculated from measured flow and internal diameter) is reduced because of the decrease in flow (16), but it is restored to pretreatment levels after 14 days of hindlimb unloading. This is because the low blood flow stimulates arterial remodeling so that lumen diameter is decreased, thereby increasing shear (16). These changes in shear stress affect endothelial function. The initial decline in shear stress is responsible for the decrease in eNOS and in acetylcholine-induced vasodilator response observed at 14 days. On the other hand, because shear stress is restored to normal at 14 days, the acetylcholine-induced vasodilator response returns to normal at 28 days. Thus, hindlimb-unloading-induced changes in shear stress are followed, after a lag time, by changes in both eNOS expression and endothelium-mediated vasodilation (16).

It is not clear if or how microgravity-induced changes in endothelial function might affect performance on landing day. If hindlimb unloading of a quadruped can initiate such significant structural and functional changes in vascular beds in the upper and lower body, it is reasonable to expect that when the head-to-foot pressure gradients are eliminated in humans the changes in vascular structure and function would be even greater. However, because responses are vascular bed–specific, it is difficult to predict overall hemodynamic effects. Even if the vascular beds in the upper body increase eNOS expression, enhance endothelial function, and diminish responses to vasoconstrictors during spaceflight, opposite changes in the leg vasculature might negate any overall hemodynamic effect. In addition, the time-dependent nature of microgravity-induced changes in endothelial function discussed earlier would limit the potential impact as flight duration increased. Flight-induced changes might be complete prior to landing. Although these changes may be entirely appropriate during flight, they could be one of many factors underlying postflight orthostatic intolerance.

This discussion of the possible contribution of vascular endothelium to orthostatic intolerance is important in the consideration of a therapeutic intervention. The localized and time-dependent nature of microgravity-induced changes in the endothelium argues against the utility of targeting this cell layer therapeutically. Moreover, inhibition of eNOS by such nonselective NOS inhibitors such as N(G)-nitro-L-arginine methyl ester (L-NAME) is known to elevate blood pressure into the hypertensive range (23). Thus, any therapeutic strategy aimed at eNOS inhibition would require the difficult task of achieving a stable, partial eNOS inhibition. On these grounds, it currently appears impractical to consider the endothelium and eNOS as therapeutic targets.

If the overproduction of NO is to be considered a possible cause of decreased vascular responsiveness and resistance, then by far the most likely candidate is the NO produced by inducible NOS (iNOS). Inducible NOS is known to be responsible for the severe hypotension associated with circulatory shock, and NOS inhibitors have been shown to be effective in those patients (24). Following induction, iNOS is active for up to 20 hours, and produces NO in nanomolar concentrations, one thousandfold greater than the amount produced by eNOS.

In vivo evidence for a role for iNOS in simulated microgravity has also been obtained. First, plasma NO is elevated after hindlimb unloading in rats (25). Second, selective inhibition of iNOS both lowers plasma NO and elevates the post-hindlimb-unloading blood pressure to normal (26). Western blot analysis has shown that, along with nNOS, iNOS is upregulated following hindlimb unloading in rats in many tissues, including the aorta, heart, and kidney (27). Blood pressure responses to injections of norepinephrine are smaller in hindlimb-unloaded rats than in control rats. In contrast, injections of the specific iNOS inhibitor aminoguanidine causes a much greater pressor response in hindlimb unloading than in control rats (27). This supports the possibility that the elevated iNOS expression is sufficiently widespread in the vasculature to exert a systemic hemodynamic effect – that is, the consequent elevated concentrations of NO would render the vasculature less responsive to endogenous vasoconstrictors. In an individual vessel, iNOS also appears to be more active after hindlimb unloading. Precontracted femoral artery rings exposed to L-arginine undergo nearly complete relaxation. This response is abolished by aminoguanidine. The contractile response of the femoral artery to norepinephrine also has been studied (18). In the presence of L-arginine, hindlimb-unloaded femoral arteries contract less in response to norepinephrine than controls. However, similar to the systemic response, aminoguanidine abolishes the differences. A presently unresolved issue is the lack of evidence for increased iNOS expression in hindlimb-unloaded femoral arteries using Western blot analysis (19).

The evidence cited above indicates that upregulation of iNOS and the consequent production of NO is sufficiently widespread among vascular beds to have important hemodynamic effects. It is possible that the increased production of NO by iNOS may contribute to low peripheral resistance, low venous return, and greater susceptibility to orthostatic hypotension on landing day. Thus, it is reasonable to suggest that specific iNOS inhibition may represent a potential countermeasure for the prevention and/or reversal of orthostatic intolerance.

CONCLUSION

In the entire history of human adventure and exploration, the decision to leave the planet and venture into outer space was by far the boldest. It required the development of a plethora of engineering and hardware systems. At the outset of space exploration, the human system was feared to be the most vulnerable, yet has proven to be wonderfully adaptive and plastic. It can adapt and function nominally in space and return to normal function on Earth in a short period of time. The vascular endothelium has great adaptability, and the studies reviewed in this chapter demonstrate that the endothelium plays a significant role in the physiological accommodation to spaceflight. Moreover, the endothelial, neuronal, and/or inducible NOS sources of NO are likely to participate in nearly every aspect of cardiovascular deconditioning and the consequent orthostatic intolerance. eNOS, nNOS, and iNOS are changed differentially by microgravity. The regulation of eNOS appears to be mediated, in part, by hemodynamic changes, but the causes of the changes in neuronal and inducible NOS are unknown. These are among the many unanswered questions that remain to be addressed. Perhaps *NOS*-knockout mice can be flown in the future.

KEY POINTS

- Spaceflight results in orthostatic hypotension.
- The etiology is multifactorial, but NO may play an important role. All current evidence resides in hindlimb-unloaded rat data.
- Upregulation of nNOS contributes to impairment of baroreceptor-mediated increases in sympathetic flow during orthostatic (hypotensive) challenge.
- eNOS expression is regulated locally by differential hemodynamics, thus overall contribution to orthostatic hypotension is unclear.
- iNOS expression is upregulated in noncardiovascular and cardiovascular tissue, has global hemodynamic effects, and constitutes a potential site for therapeutic intervention.

Future Goals

- To determine the therapeutic value of iNOS blockade to prevent and/or reverse the orthostatic hypotension associated with spaceflight, long-term bed rest, and spinal cord injury

REFERENCES

1 Vaziri ND. Nitric oxide in microgravity-induced orthostatic intolerance: relevance to spinal cord injury. *J Spinal Cord Med.* 2003;26(1):5–11.

2 Meck JV, Reyes CJ, Perez SA, et al. Marked exacerbation of orthostatic intolerance after long- vs. short-duration spaceflight in veteran astronauts. *Psychosom Med.* 2001;63(6):865–873.

3 Huntoon CSL, Cintròn NM. Endocrine system and fluid and electrolyte balance. In: Nicogossian A, Mohler SR, Gazenko OG, Grigoriev A, eds. *Space Biology and Medicine: Humans in Spaceflight.* Reston, VA: American Institute of Aeronautics and Astronautics Inc.; 1996:89–104.

4 Fritsch-Yelle JM, Whitson PA, Bondar RL, Brown TE. Subnormal norepinephrine release relates to presyncope in astronauts after spaceflight. *J Appl Physiol.* 1996;81(5):2134–2141.

5 Meck JV, Waters WW, Ziegler MG, et al. Mechanisms of postspaceflight orthostatic hypotension: low alpha1-adrenergic receptor responses before flight and central autonomic dysregulation postflight. *Am J Physiol Heart Circ Physiol.* 2004;286(4):H1486–H1495.

6 Waters WW, Ziegler MG, Meck JV. Post-spaceflight orthostatic hypotension occurs mostly in women and is predicted by low vascular resistance. *J Appl Physiol.* 2002;92:586–594.

7 Gaffney FA, Buckey JC, Lane LD, et al. The effect of a 10-day period of head-down tilt on the cardiovascular responses to intravenous saline loading. *Acta Physiol Scand.* 1992;144,S604:121–130.

8 Moffitt JA, Schadt JC, Hasser EM. Altered central nervous system processing of baroreceptor input following hindlimb unloading in rats. *Am J Physiol.* 1999;277(6 Pt 2):H2272–H2279.

9 Krukoff TL. Central actions of nitric oxide in regulation of autonomic functions. *Brain Res Brain Res Rev.* 1999;30(1):52–65.

10 Schwarz PM, Rodriguez-Pascual F, Koesling D, et al. Functional coupling of nitric oxide synthase and soluble guanylyl cyclase in controlling catecholamine secretion from bovine chromaffin cells. *Neuroscience.* 1998;82(1):255–265.

11 Sears CE, Choate JK, Paterson DJ. Effect of nitric oxide synthase inhibition on the sympatho-vagal control of heart rate. *J Auton Nerv Syst*. 1998;73(1):63–73.

12 Pyner S, Coote JH. Identification of branching paraventricular neurons of the hypothalamus that project to the rostroventrolateral medulla and spinal cord. *Neuroscience*. 2000;100(3):549–556.

13 Zhang K, Mayhan WG, Patel KP. Nitric oxide within the paraventricular nucleus mediates changes in renal sympathetic nerve activity. *Am J Physiol*. 1997;273(3 Pt 2):R864–R872.

14 Whitson PA, Charles JB, Williams WJ, Cintròn NM. Changes in sympathoadrenal response to standing in humans after spaceflight. *J Appl Physiol*. 1995;79(2):428–433.

15 Colleran PN, Wilkerson MK, Bloomfield SA, et al. Alterations in skeletal perfusion with simulated microgravity: a possible mechanism for bone remodeling. *J Appl Physiol*. 2000;89(3):1046–1054.

16 Delp MD, Colleran PN, Wilkerson MK, et al. Structural and functional remodeling of skeletal muscle microvasculature is induced by simulated microgravity. *Am J Physiol Heart Circ Physiol*. 2000;278(6):H1866–H1873.

17 Somody L, Fagette S, Blanc S, et al. Regional blood flow in conscious rats after head-down suspension. *Eur J Appl Physiol Occup Physiol*. 1998;78(4):296–302.

18 Sangha DS, Vaziri ND, Ding Y, Purdy RE. Vascular hyporesponsiveness in simulated microgravity: role of nitric oxide-dependent mechanisms. *J Appl Physiol*. 2000;88(2):507–517.

19 Ma J, Kahwaji CI, Ni Z, et al. Effects of simulated microgravity on arterial nitric oxide synthase and nitrate and nitrite content. *J Appl Physiol*. 2003;94(1):83–92.

20 Nyhan D, Kim S, Dunbar S, et al. Impaired pulmonary artery contractile responses in a rat model of microgravity: role of nitric oxide. *J Appl Physiol*. 2002;92(1):33–40.

21 Schrage WG, Woodman CR, Laughlin MH. Hindlimb unweighting alters endothelium-dependent vasodilation and ecNOS expression in soleus arterioles. *J Appl Physiol*. 2000;89(4):1483–1490.

22 Jasperse JL, Woodman CR, Price EM, et al. Hindlimb unweighting decreases ecNOS gene expression and endothelium-dependent dilation in rat soleus feed arteries. *J Appl Physiol*. 1999;87(4):1476–1482.

23 Ribeiro MO, Antunes E, de Nucci G, et al. Chronic inhibition of nitric oxide synthesis. A new model of arterial hypertension. *Hypertension*. 1992;20(3):298–303.

24 Anzueto A, Lodato RF, Lorente J, et al. Multicenter, placebo-controlled, double-blind trial of the nitric oxide synthase inhibitor 546C88 in patients with septic shock: acute hemodynamic effects. *Am J Resp Crit Care Med*. 1997;155:A263.

25 Bayorh MA, Socci RR, Watts S, et al. L-NAME, a nitric oxide synthase inhibitor, as a potential countermeasure to post-suspension hypotension in rats. *Clin Exp Hypertens*. 2001;23(8):611–622.

26 Eatman D, Walton M, Socci RR, et al. NOS inhibition attenuates post-suspension hypotension in Sprague-Dawley rats. *Clin Exp Hypertens*. 2003;25(1):11–24.

27 Vaziri ND, Ding Y, Sangha DS, Purdy RE. Upregulation of NOS by simulated microgravity, potential cause of orthostatic intolerance. *J Appl Physiol*. 2000;89(1):338–344.

59

Toxicology and the Endothelium

Howard D. Beall and J. Douglas Coffin

Center for Environmental Health Sciences, University of Montana, Missoula

Toxic substances represent important inputs for endothelial cells (ECs). EC toxicity represents an emerging research area in the discipline of toxicology. Toxic agents can be classified as toxins of biological origin such as hormones, cytokines, and plant and animal toxins. Alternatively, toxicants are of anthropogenic origin – e.g. industrial chemicals, solvents, and polycyclic aromatic hydrocarbons. Toxicants also include naturally occurring substances released by human activity that are moderately toxic or nontoxic at low levels but highly toxic when released to generate high levels. Examples of these include metals (Hg, Cd, and Pb) and metalloids (As). Table 59-1 shows examples of toxins and toxicants that have been shown to affect the endothelium.

The endothelium is the first barrier to come into contact with toxic agents in the systemic circulation. Exposure of ECs to low levels of such substances is frequent and nonharmful. High-level exposure may lead to significant toxicity. Many agents – such as cytokines (tumor necrosis factor [TNF]-α), drugs (antibiotics), hormones (corticosteroids, estrogens), ethanol, and trace metals – cause general damage to ECs or "endothelial dysfunction." At that point, the endothelium loses its ability to homeostatically adapt to the toxic insult and becomes part of the resulting pathology.

Given that the structure and function of the endothelium are heterogeneous, the effects of toxins on EC phenotype may vary between different organs. Most cardiovascular pathologies can be related to endothelial toxicology, and these include coronary artery disease or atherosclerosis, arteriosclerosis, arrhythmia, hemorrhagic or thrombogenic stroke, hypertension, thrombosis, infection, and peripheral vascular disease. Some of the best-known examples are snake venoms that cause edema, hemorrhage, or thrombosis; insect toxins that cause local edema; or drug overdoses that block vasoregulation. Endothelial toxicity can also contribute to diseases of the vascular system. Arsenic exposure has been shown to exacerbate atherosclerosis through oxidative stress and inflammation (1). On the other hand, toxic agents such as cadmium can produce a hypertensive response that, when prolonged, causes secondary damage to the endothelium (Table 59-1). Toxicity-mediated thrombosis often requires activation of the endothelium as a critical step (2).

In this chapter, general responses of ECs to toxic exposures are presented along with the pathological consequences of those exposures. Selected model toxic agents are discussed that best represent the field of endothelial toxicology. Finally, this chapter includes an important discussion on how endothelial toxicity may impair vascular development.

MECHANISMS OF TOXICITY

Toxic exposures to the endothelium generate a variety of responses that fall into homeostatic or pathological categories including cell differentiation, proliferation, adhesion, apoptosis, and metabolic changes. These manifest as changes in the vasculature such as vasodilation or vasoconstriction, inflammation, atherogenesis, angiogenesis, thrombosis, or edema. Depending on the manifestation, the outcome may be acute or chronic and either moderate or severe. The molecular mechanisms and regulatory pathways controlling these responses are often related to homeostasis or a particular disease, and they are described in detail throughout this volume. However, they have not been well characterized or defined in relation to toxicology. The molecular interactions that generate these vascular problems are better described and they include oxidative stress, lipid peroxidation, blockage of electron transport, and also many interactions in which toxic agents act as agonists or antagonists to important regulatory molecules such as cell surface (e.g., cytokine, growth factor, and hormone) receptors, ion channels, and transporters.

A common mechanism of toxicity is generation of reactive oxygen species (ROS) or reactive nitrogen species (RNS) leading to oxidative stress. Oxidative stress can directly injure the endothelium (3), which can then have an impact on other vascular cell types, such as smooth muscle cells and fibroblasts. Toxicant-induced oxidative stress may also trigger

Table 59-1: List of Toxins and Toxicants Reported to Affect Endothelium

Toxin/Toxicant	*Example*	*Effect*	*Pathology*	*References*
Biotoxins				
Endotoxin	LPS	Act, Dys, Apop	Sepsis, Edema	11,22,77–79
Snake venom	Fer-de-lance	Dys, Ox, Nec	Edema, Hem, Gangrene, Throm	17,33
Virus	Ebola, influenza	Dys, Ox, Nec	Edema, Hem	34,36,80,81
Homocysteine		Act, Dys	Athero, Inflam, Throm	2,28,82
Hormones	Corticosteroids, estrogens	Act, Dys	AMD, Ret, TA	83,84
Cytokines	TNF-α, INF	Act, Dys, Apop, Adh	Hyper, Athero, PVD	6,12,85–89
Growth factors	VEGF	Act, Prolif, Diff, Apop	Hyper, Athero, TA, Edema	14,90,91
Chemokines	IL-8	Act, Dys, Prolif, Diff, Apop	KS, Hem, TA	92,93
OxLDL		Apop	Athero	5
Cardioglycosides	Ouabain	Apop	Hyper	7
Eicosanoids		Dys	Hyper	94
Metalloids				
Arsenic		Act, Dys, Ox	Athero, Hyper, PVD, Inflam, Throm	1,24,38,39, 44,46
Metals				
Iron		Ox, Apop, Nec	Vasculopathy	29
Lead		Ox, Dys	Hyper	19
Mercury		Dys, Apop	Vasculopathy	18
Cadmium		Dys	Hyper	20
Nickel		Act	Inflam	30
Drugs				
Catecholamines	Norepinephrine, serotonin	Act, Dys, Ox	Hyper, Throm	23,95,96
Antineoplastics	Fluorouracil, bleomycin	Dys	Hyper, Edema	97,98
Antibiotics	Rapamycin, gentamicin	Dys, Apop	Athero, Edema	15,99
Antihistamines	Tranilast	Dys	AMD, Ret	31
Anesthetics	Nitric oxide	Dys	Edema, Hyper	21
Analgesics	Coxibs	Dys	Throm	25
Industrial Toxicants				
Solvents, benzene, and derivatives	Ethanol, benzopyrene, benzene	Dys, Act, Ox	Inflam, Edema, Stroke	16, 32
Halogenated hydrocarbons	PCBs	Ox, Dys	Athero, Inflam	100,101
Dioxins	TCDD	Dys	Vasculopathy	102

Toxins reported to affect endothelium by causing activation (Act), dysfunction (Dys), oxidative stress (Ox), apoptosis (Apop), proliferation (Prolif), leukocyte adhesion (Adh), or necrosis (Nec). These alterations in ECs are associated with pathologies such as atherosclerosis (Athero), hypertension (Hyper), inflammation (Inflam), thrombosis (Throm), age-related macular degeneration (AMD), tumor angiogenesis (TA), peripheral vascular disease (PVD), retinopathy (Ret), Kaposi sarcoma (KS), and hemorrhage (Hem). Other abbreviations include PCB (polychlorinated biphenyls), TCDD (2,3,7,8 tetrachlorodibenzo-*p*-dioxin), Coxibs (cyclooxygenase inhibitors), INF (interferon γ).

endothelial apoptosis or produce an inflammatory response in which initial damage to the endothelium leads to recruitment and adhesion of inflammatory cells and release of cytokines, chemokines, and growth factors (2–4). Lipid peroxidation, associated with ischemia/reperfusion injury, causes damage to cell membranes with subsequent alteration of transmembrane regulatory proteins and instability of organelles (mitochondria, Golgi, and endoplasmic reticulum). ECs damaged through antibiotic-mediated lipid peroxidation often undergo premature apoptosis (5,6).

Besides oxidative stress, endothelial damage leading to endothelial dysfunction can also occur by the direct cytotoxic action of drugs and chemicals (Table 59-1). For example, cardioglycoside such as ouabain or digoxin bind Na/K transporters, causing endothelial dysfunction (7). Alternatively, nontoxic chemicals may be bioactivated to toxic metabolites by enzymes in vascular cells. ECs express a range of enzymes capable of activating chemicals to toxic species, including amine oxidases, cytochrome P450 isozymes, and the prostaglandin synthetase complex of enzymes (8).

ENDOTHELIAL ACTIVATION/DYSFUNCTION FROM TOXIC EXPOSURES

Exposure to toxic agents can cause endothelial activation that is characterized by an increase in secretion of proteins or metabolites. Factors expressed or released by activated endothelium include adhesion molecules such as intercellular adhesion molecule (ICAM)-1 and vascular cell adhesion molecule (VCAM)-1 (9,10). Expression of cytokines such as TNF-α and chemokines including interleukin (IL)-8 are reported as responses to toxic insults related to sepsis or lipopolysaccharide (LPS) exposure (11,12). Growth factor response to endothelial toxicity is well documented with vascular endothelial growth factor (VEGF) family members most often reported (9,13–16). Toxicity-mediated thrombosis or hemorrhage is related to endothelial activation or dysfunction and is best exemplified by snakebites (17). Finally, altered hemostasis is a well-documented vascular response to toxic exposures based on changes in nitric oxide (NO) (9,18–20) or endothelin (ET)-1 (19,21–23). Patterns of activation depend not only on the toxic compound but also on the site within the vascular tree. Sustained endothelial activation can lead to endothelial dysfunction, defined by abnormalities in one or more functions, including changes in vasomotor control, barrier properties, leukocyte trafficking, inflammation, hemostasis, and apoptosis.

VASOMOTOR CONTROL

ECs are normally in a vasodilatory state due to secretion of NO and vasodilatory prostaglandins such as prostacyclin (PGI_2), and either of these can be altered in a toxic insult causing hypertension or hypotension (21,24,25). Toxicity may result in impaired response of the vasculature to endothelium-dependent relaxation (EDR). An important mediator of EDR is NO, synthesized from L-arginine, nicotinamide adenine dinucleotide phosphate (NADPH), and oxygen (O_2) by endothelial NO synthase (eNOS). Oxidative stress from exposure to toxic agents could potentially decrease the bioavailability of NO by several mechanisms. Generation of superoxide anion ($O_2^{\cdot-}$), an ROS that reacts rapidly with NO to form a powerful oxidant, peroxynitrite ($ONOO^-$), may be increased in activated ECs or surrounding tissues. Excess superoxide anion could originate from NAD(P)H oxidases on the EC membrane, the mitochondrial respiratory chain, or uncoupled eNOS (26). In addition, $ONOO^-$ can oxidize and inactivate an essential eNOS cofactor, tetrahydrobiopterin (BH_4), thereby uncoupling eNOS and decreasing NO production (27).

Endothelial activation/dysfunction may also be associated with the release of angiotensin-converting enzyme (ACE) (12,16,28–31) and ET-1 (4,19,21,23), which results in a further shift of the vasomotor balance toward the vasoconstrictive side. This can aggravate existing hypertension and ultimately increase cardiovascular morbidity and mortality (4). On the other hand, hypertension itself can cause endothelial dysfunction by directly damaging ECs.

Barrier Function

Another form of endothelial dysfunction resulting from toxic insult is loss of endothelial barrier integrity. The failure of this barrier could lead to serious pathological consequences, especially for specialized endothelium such as the blood–brain barrier (12,32). For example, certain types of venom (17,23,33) or viral proteins (34–36) may cause severe edema or hemorrhage by disrupting the integrity of, or even obliterating, the endothelial barrier. Overproduction of oxidants leading to oxidative stress can disrupt the endothelium by disorganization of critical junctional proteins, including vascular endothelial (VE)-cadherin (37).

Leukocyte Trafficking and Inflammation

ROS and RNS generation attributed to toxic exposures may activate ECs to produce cytokines, chemokines, and adhesion molecules. Leukocytes (e.g., T cells and monocytes) are thus recruited to the EC surface, attach to the surface, and are translocated to the subendothelial space (4,9,10,30,33). This is an important mechanism underlying atherosclerosis that is, for example, associated with arsenic toxicity (1,24,38,39).

Hemostasis

ECs are key mediators of coagulation, and they secrete a range of both procoagulant and anticoagulant factors. Chronic inflammation with release of proinflammatory cytokines creates a procoagulant state in the endothelium, and in the case of atherosclerotic disease it may cause weakening and rupture of plaques, potentially leading to myocardial infarction or stroke (40). Direct damage to the endothelium caused by mechanical means or administration of an irritating substance can cause formation of a thrombus at the site of the injury.

PATHOLOGICAL CONSEQUENCES OF ENDOTHELIAL INJURY

The most important pathological consequence of endothelial activation/dysfunction is the development of atherosclerosis. Atherosclerosis is the single most important contributor to other cardiovascular diseases. Endothelial dysfunction has been documented both in the early stages of atherosclerosis as well as in patients at risk for development of, but with no clinical evidence of, coronary artery disease (3). An activated or damaged endothelium can also promote thrombotic or hypertensive disease (41).

Cardiovascular disease is a complex set of disorders involving multiple cell types and organ systems. Endothelial activation/dysfunction can both contribute to and result from cardiovascular disease. The remainder of this chapter focuses on toxic agents and their specific effects on ECs.

MODEL SYSTEMS FOR STUDYING ENDOTHELIAL TOXICOLOGY

Until recently, there were very few EC model systems available for studying toxic responses of the endothelium. Aortic ECs of bovine and porcine origin were used most often in toxicology studies along with cells from the human umbilical vein. In the last few years, human arterial ECs have become available commercially, including cells of aortic, coronary artery, and pulmonary artery origin. A lingering problem is the paucity of rodent EC models to support toxicology studies. Steady progress has been made culturing murine ECs, but it is still difficult to obtain more than three or four doublings. Many tumor angiogenesis models have been adapted for studies in endothelial toxicology, including the use of Matrigel (42,43). The obvious advantage for murine models in toxicology is access to the many excellent transgenic and knockout mouse models available for functional studies.

SELECTED ENDOTHELIAL TOXIC AGENTS

Agents that have been identified as endothelial toxic agents include butadiene, acrolein, metals and metalloids, and various drugs including aspirin, cocaine, and the antineoplastic agent cyclophosphamide (8). Of course, many more drugs and chemicals exist that are cytotoxic to ECs (see Table 59-1). In addition, natural substances such as homocysteine, oxidized low-density lipoprotein (oxLDL), and bacterial endotoxins can elicit toxic responses by ECs. This section is devoted to several well-characterized toxic substances, some of which are single agents and others complex mixtures.

Arsenic

In recent years the metalloid arsenic has received considerable attention in endothelial toxicity studies. Natural arsenic levels in drinking water are very high in some locales, and it is released into the environment by human activities in other regions, increasing exposure risk in human populations (1,44). Chronic exposure to arsenic in drinking water has been identified as a risk factor for a number of cardiovascular diseases, including peripheral vascular disease (1), ischemic heart disease (1,24,38,39,45), and hypertension (46). In mouse models of atherosclerosis, significantly exacerbated lesions have been documented in mice that are chronically exposed to arsenic in their drinking water (24,38). The important pathophysiological role of the EC in atherosclerosis and other cardiovascular diseases should stimulate further research into the effects of arsenic on this cell type.

Arsenic exposure studies using a variety of cultured ECs have shown increased ROS and RNS generation, potentially leading to oxidative stress. The oxidative stress is believed to result in altered EC proliferation, adhesion, and apoptosis (47–49). Vascular NADPH oxidases have been implicated in the increased formation of superoxide anion in ECs treated with arsenic (50). Increased NO synthesis in aortic ECs in response to arsenic has also been reported (51), but this remains controversial as the response to arsenic appears to be dependent on cell type, arsenic species, and concentration (52). A plausible explanation may be that peroxynitrite is produced by the scavenging of NO by superoxide anion. Evidence for increased peroxynitrite formation in arsenic-treated aortic ECs has been reported (39). In any case, the evidence suggests that ECs exposed to arsenic produce ROS and RNS, which cause oxidative stress, activation of the endothelium, and, ultimately, endothelial dysfunction.

Direct evidence for endothelial dysfunction from arsenic exposure has recently been documented in a comprehensive study using both in vitro and in vivo assays. Treatment of rat aortic rings with arsenic inhibited relaxation induced by acetylcholine, a standard methodology for determining endothelial dysfunction (46). In the same study, live rats were administered arsenic intravenously. Arsenic pretreatment prevented blood pressure reduction from acetylcholine infusion, the first demonstration of the inhibitory effect of arsenic on vasomotor tone in an animal model (46). These findings support the epidemiological evidence showing that chronic arsenic exposure is associated with the development of hypertension (46).

Epidemiological evidence has also suggested that arsenic may increase the potential for thrombus formation (44). In response to this finding, an in vivo study was conducted in rats that showed enhanced platelet aggregation and arterial thrombus formation following chronic arsenic treatment (44). A possible mechanism behind this effect is increased expression of P-selectin, an important protein involved in the adhesion of platelets to ECs.

Metals

Mercury, lead, and cadmium are environmental contaminants that have been implicated in EC toxicity. Acute mercury (53) intoxication has been associated with hypertension in both human case reports and controlled animal studies. Inorganic mercury is known to cause injury to ECs. In a recent study using environmentally relevant exposures, mercury chloride caused structural and functional damage to rat endothelium and loss of endothelial-dependent vasoreactivity (18). Chronic low-level lead exposure has also been associated with hypertension in humans and animals through a mechanism that may include generation of ROS and RNS by ECs (2,19). Lead neurotoxicity may be due in part to alterations in the integrity of the blood–brain barrier (12,32). Cadmium also is recognized as an endothelial toxicant and may contribute to neurological diseases or stroke (54) by causing an inflammatory response that involves activation of ECs. Like lead, cadmium exposure can cause disruption of the blood–brain barrier (54). Other metals, including chromium and nickel, may activate ECs and initiate toxic mechanisms. Iron also has been implicated in endothelial dysfunction in relation to oxidative stress leading to EC apoptosis or necrosis in vasculopathy (29).

Cigarette Smoke

Cigarette smoking is a major risk factor for atherosclerosis and other cardiovascular diseases. Cigarette smoke contains thousands of chemical components, many of which are highly toxic. A major effect of inhaled cigarette smoke is induction of oxidative stress. As discussed previously, ROS (resulting from cigarette smoke) react with and inactivate NO produced by ECs. This produces other oxidants such as peroxynitrite and decreases NO availability, leading to endothelial dysfunction. Components of cigarette smoke stimulate adhesion molecule expression and leukocyte adhesion to ECs, an early step in atherogenesis (4). Cigarette smoking affects the endothelium indirectly through activation of platelets (55). Platelet activation increases their adherence to ECs. This leads to a prothrombotic state in the vasculature and increased potential for acute coronary or cerebrovascular events.

Particulate Matter

The effect of air pollutants, especially particulate matter, on cardiovascular disease is an area of recent concern and increased research activity (56). Particulate matter is composed of many potentially toxic components, including polycyclic aromatic hydrocarbons, metals, and the particles themselves. Inhaled particulate matter deposits in the lungs and causes the expected respiratory effects, but epidemiological studies also show that cardiovascular morbidity and mortality increase on high air pollution days (57). The mechanisms behind the systemic cardiovascular effects of locally deposited particles are largely unknown.

Most of the evidence points to inflammation as a common mechanism in the cardiovascular toxicity of inhaled particulate matter. The EC is at the center of this inflammatory response. Rats that were exposed to residual oil fly ash (ROFA), a model of particulate matter, by intratracheal instillation exhibited diminished endothelium-dependent dilation in the systemic microvasculature (58). Furthermore, increased rolling and adhesion of leukocytes on ECs were observed in the microvasculature, suggesting that ROFA had produced a systemic inflammatory response (58). A relationship between particulate matter exposure from air pollution and carotid intima-media thickness, an indicator of early atherosclerosis, was recently demonstrated in an epidemiological study carried out in Los Angeles, California (59). Particulate matter–induced endothelial activation/dysfunction leading to atherogenesis is one plausible explanation for this reported association. Future work should focus on the specific components of particulate matter that contribute to endothelial activation/dysfunction and downstream pathologies.

ENDOTHELIAL TOXICOLOGY IN VASCULAR DEVELOPMENT

The best known toxic agent affecting vascular development is thalidomide. Administration of thalidomide to pregnant women from 1959 to 1961 led to severe cases of amelia and phocomelia in their children (60). The mechanisms for the limb deformations were largely unknown in that era, but recent use of thalidomide as an angiogenesis inhibitor in chemotherapy (61,62) has regenerated interest in determining the molecular basis for thalidomide's effects on angiogenesis with possible applications to development. Previously published data show that, regardless of the cause, failure to vascularize limbs during development can lead to limb defects (63) similar to amelia. Based on the effects of thalidomide on tumor angiogenesis, thalidomide could be inhibiting limb angiogenesis as a means to cause amelia or phocomelia. Recent studies have revealed that thalidomide increases endothelial oxidative stress (64), providing an example of endothelial dysfunction through increased apoptosis. Based on these results, a general model for toxicity-based oxidative stress and endothelial dysfunction can be constructed (Figure 59.1) whereby a toxic exposure inhibits angiogenesis and blocks perfusion. This leads to hypoxia or anoxia that causes extensive tissue damage in the developing limb and subsequent teratogenesis. Moreover, consideration of these results in the context of developmental field models (65,66) suggests that the limb bud vessels may be necessary to ensure proper positioning of the precursors needed for formation of other tissues, such as bone, nerves, and muscle.

Arsenic is another toxin implicated in impairment of vascular development and subsequent teratogenesis, miscarriage, low birth weight, and/or premature delivery in humans (53). As mentioned previously, arsenic exposure to the endothelium in vitro results in production of ROS and RNS, such as peroxynitrite, which can alter the NO balance (Figure 59.1). Either collectively or alone, any of these arsenic stress-related effects is sufficient to cause endothelial dysfunction manifest as increased apoptosis (45), decreased proliferation (67), and possibly altered cell adhesion or migration as the key components to arsenic-mediated vascular dysmorphogenesis. These results suggest that arsenic-based teratogenesis is caused by a lack of neovascularization/perfusion of developing structures in the embryo. The subsequent hypoxia causes tissue damage in the embryo, resulting in teratogenesis and possibly miscarriage should the vascular dysmorphogenesis occur in the placenta.

Preeclampsia is one of the most common problems associated with preterm delivery, low birth weight, and miscarriage. The exact cause of preeclampsia is unknown, but it is linked to defects in placental vasculogenesis similar to our observations with arsenic exposure and placental insufficiency. Current

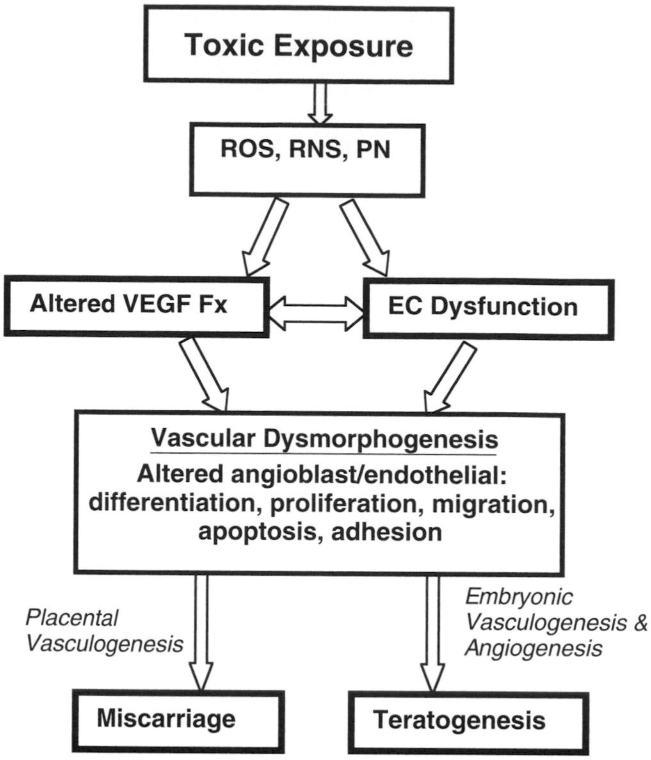

Figure 59.1. Schematic relating endothelial toxicity to oxidative stress, including production of ROS, RNS, and peroxynitrite (PN). Oxidative stress leads to alteration of the VEGF regulatory axis and EC dysfunction, affecting angioblast differentiation, proliferation, migration, apoptosis, and adhesion. These are all key processes in vascular morphogenesis, and, therefore, toxicity triggers dysmorphogenesis manifest as miscarriage if involving placental vasculogenesis, or teratogenesis if involving either vasculogenesis or angiogenesis in the embryo.

models suggest that the molecular mechanism for preeclampsia centers on the VEGF regulatory axis whereby increases in the soluble FLT-1 receptor result in an inhibitory sink for VEGF (and placental growth factor [PlGF]) (68,69). Genetic mouse models have established the importance of the VEGF signaling axis in development; deletion of a single allele of VEGF or homozygous knockout of *FLT-1* or *FLK-1* results in embryonic lethality. The results of the preeclampsia studies provide important proof of principle that abnormalities in the VEGF pathway may be linked to impaired development in humans (69,70). Hypothetically, exposure to environmental toxicants could cause or at least exacerbate vascular pathologies of unknown origin, such as preeclampsia, because hypoxia-inducible factor (HIF)-1, which regulates VEGF expression, is related to the aryl hydrocarbon receptor (71,72).

Dioxins are another candidate for triggering developmental vasculotoxicity and dysmorphogenesis. Unlike arsenic and thalidomide, no data exist associating dioxins with a human condition to show a definitive effect on the embryo or fetus – i.e., teratogenesis, miscarriage, or preterm delivery. However, TCDD exposures in zebrafish, lake trout, and chick embryos all reveal vascular dysmorphogenesis related to metabolism by the CYP1A1 cytochrome P450 and alteration of VEGF or the VEGF receptor FLT-1 (13,73,74).

Neovascularization of developing tissues is mediated by vasculogenesis and angiogenesis (75,76). The fundamental cellular and molecular bases for the two modes of vascular development have unique distinctions but intersect with growth factor and signaling pathways that regulate EC differentiation, proliferation, migration, and adhesion during morphogenesis. Although developmental vasculotoxicity has not been specifically or definitively linked to alterations in these processes, the current data show that arsenic toxicity primarily affects placental vasculognesis, whereas the effects of thalidomide are relatively specific to limb bud angiogenesis (Figure 59.1). Further studies are needed to clarify this distinction. Should these interpretations hold up to continued elucidation of the mechanisms and scrutiny of the models, vascular toxicology may prove an interesting means for not only characterizing the molecular and cellular mechanisms for vascular development but also for providing clues for therapeutic manipulations.

CONCLUSION

The endothelium is a key target for toxic events that fundamentally alter cardiovascular physiology facilitating cardiovascular disease, miscarriage, and teratogenesis.

KEY POINTS

- Toxic exposures generate stress reactions in the endothelium that alter normal physiology and developmental processes.
- Generation of ROS and RNS by toxic agents causes oxidative stress and inflammation leading to endothelial activation and dysfunction.
- Endothelial dysfunction in the adult results in exacerbation of cardiovascular diseases such as hypertension and atherosclerosis. Angioblast/endothelial dysfunction during development can result in miscarriage or teratogenesis.

Future Goals

- Toxicology should be a major consideration in human epidemiologic models applied to diseases and fecundity for both human and agricultural populations.
- Endothelial toxicology requires further investigation, consideration, and inclusion in the etiology of cardiovascular disease.

- The relationship between developmental toxic exposures and vascular dysmorphogenesis needs elucidation for practical prevention of miscarriage and teratogenesis.

REFERENCES

1 Navas-Acien A, Sharrett AR, Silbergeld EK, et al. Arsenic exposure and cardiovascular disease: a systematic review of the epidemiologic evidence. *Am J Epidemiol*. 2005;162:1037–1049.

2 Lentz SR. Mechanisms of homocysteine-induced atherothrombosis. *J Thromb Haemost*. 2005;3:1646–1654.

3 Davignon J, Ganz P. Role of endothelial dysfunction in atherosclerosis. *Circulation*. 2004;109:III27–III32.

4 Schiffrin EL. Beyond blood pressure: the endothelium and atherosclerosis progression. *Am J Hypertens*. 2002;15:S115–S122.

5 Kuzuya M, Naito M, Funaki C, et al. Probucol prevents oxidative injury to endothelial cells. *J Lipid Res*. 1991;32:197–204.

6 Deepa PR, Varalakshmi P. Influence of a low-molecular-weight heparin derivative on the NO levels and apoptotic DNA damage in Adriamycin-induced cardiac and renal toxicity. *Toxicology*. 2006;217:176–183.

7 Blaustein MP. Endogenous ouabain: role in the pathogenesis of hypertension. *Kidney Int*. 1996;49:1748–1753.

8 Ramos KS, Melchert RB, Chacon E, Acosta D. Toxic responses of the heart and vascular systems. In: Casarett & Doull's, eds. *Toxicology: The Basic Science of Poisons*. New York: McGraw-Hill, 2001.

9 Kuldo JM, Ogawara KI, Werner N, et al. Molecular pathways of endothelial cell activation for (targeted) pharmacological intervention of chronic inflammatory diseases. *Curr Vasc Pharmacol*. 2005;3:11–39.

10 Kuldo JM, Westra J, Asgeirsdottir SA, et al. Differential effects of NF-kappa B and p38 MAPK inhibitors and combinations thereof on TNF-alpha- and IL-1 beta-induced proinflammatory status of endothelial cells in vitro. *Am J Physiol Cell Physiol*. 2005;289:C1229–C1239.

11 Andreoli SP. The pathophysiology of the hemolytic uremic syndrome. *Curr Opin Nephrol Hypertens*. 1999;8:459–464.

12 Eisenhauer PB, Jacewicz MS, Conn KJ, et al. Escherichia coli Shiga toxin 1 and TNF-alpha induce cytokine release by human cerebral microvascular endothelial cells. *Microb Pathog*. 2004;36:189–196.

13 Ivnitski-Steele ID, Sanchez A, Walker MK. 2,3,7,8-tetrachlorodibenzo-p-dioxin reduces myocardial hypoxia and vascular endothelial growth factor expression during chick embryo development. *Birth Defects Res A Clin Mol Teratol*. 2004;70:51–58.

14 Zhu BQ, Heeschen C, Sievers RE, et al. Second hand smoke stimulates tumor angiogenesis and growth. *Cancer Cell*. 2003;4:191–196.

15 Goto T, Fujigaki Y, Sun DF, et al. Plasma protein extravasation and vascular endothelial growth factor expression with endothelial NO synthase induction in gentamicin-induced acute renal failure in rats. *Virchows Arch*. 2004;444:362–374.

16 Bandi N, Ayalasomayajula SP, Dhanda DS, et al. Intratracheal budesonide-poly(lactide-co-glycolide) microparticles reduce oxidative stress, VEGF expression, and vascular leakage in a benzo(a)pyrene-fed mouse model. *J Pharm Pharmacol*. 2005; 57:851–860.

17 Lu Q, Clemetson JM, Clemetson KJ. Snake venoms and hemostasis. *J Thromb Haemost*. 2005;3:1791–1799.

18 Golpon HA, Puchner A, Barth P, et al. Nitric oxide-dependent vasorelaxation and endothelial cell damage caused by mercury chloride. *Toxicology*. 2003;192:179–188.

19 Vaziri ND. Pathogenesis of lead-induced hypertension: role of oxidative stress. *J Hypertens Suppl*. 2002;20:S15–S20.

20 Kolluru GK, Tamilarasan KP, Geetha Priya S, et al. Cadmium induced endothelial dysfunction: consequence of defective migratory pattern of endothelial cells in association with poor NO availability under cadmium challenge. *Cell Biol Int*. 2006;30:427–438.

21 Lowson SM. Alternatives to NO. *Br Med Bull*. 2004;70:119–131.

22 Deaciuc IV, Spitzer JJ. Hepatic sinusoidal endothelial cell in alcoholemia and endotoxemia. *Alcohol Clin Exp Res*. 1996;20:607–614.

23 Nouira S, Elatrous S, Besbes L, et al. Neurohormonal activation in severe scorpion envenomation: correlation with hemodynamics and circulating toxin. *Toxicol Appl Pharmacol*. 2005; 208:111–116.

24 Bunderson M, Brooks DM, Walker DL, et al. Arsenic exposure exacerbates atherosclerotic plaque formation and increases nitrotyrosine and leukotriene biosynthesis. *Toxicol Appl Pharmacol*. 2004;201:32–39.

25 Fosslien E. Cardiovascular complications of non-steroidal anti-inflammatory drugs. *Ann Clin Lab Sci*. 2005;35:347–385.

26 Li JM, Shah AM. Endothelial cell superoxide generation: regulation and relevance for cardiovascular pathophysiology. *Am J Physiol Regul Integr Comp Physiol*. 2004;287:R1014–R1030.

27 Laursen JB, Somers M, Kurz S, et al. Endothelial regulation of vasomotion in apoE-deficient mice: implications for interactions between peroxynitrite and tetrahydrobiopterin. *Circulation*. 2001;103:1282–1288.

28 Lang D, Kredan MB, Moat SJ, et al. Homocysteine-induced inhibition of endothelium-dependent relaxation in rabbit aorta: role for superoxide anions. *Arterioscler Thromb Vasc Biol*. 2000; 20:422–427.

29 Balla J, Vercellotti GM, Nath K, et al. Haem, haem oxygenase and ferritin in vascular endothelial cell injury. *Nephrol Dial Transplant*. 2003;5(18 Suppl):V8–V12.

30 Wataha JC, Lockwood PE, Marek M, Ghazi M. Ability of Ni-containing biomedical alloys to activate monocytes and endothelial cells in vitro. *J Biomed Mater Res*. 1999;45:251–257.

31 Isaji M, Miyata H, Ajisawa Y, et al. Tranilast inhibits the proliferation, chemotaxis and tube formation of human microvascular endothelial cells in vitro and angiogenesis in vivo. *Br J Pharmacol*. 1997;122:1061–1066.

32 Haorah J, Knipe B, Leibhart J, et al. Alcohol-induced oxidative stress in brain endothelial cells causes blood-brain barrier dysfunction. *J Leukoc Biol*. 2005;78:1223–1232.

33 Zamuner SR, Zuliani JP, Fernandes CM, et al. Inflammation induced by Bothrops asper venom: release of proinflammatory cytokines and eicosanoids, and role of adhesion molecules in leukocyte infiltration. *Toxicon*. 2005;46:806–813.

34 Ergonul O. Crimean-Congo haemorrhagic fever. *Lancet Infect Dis*. 2006;6:203–214.

35 Hensley LE, Jones SM, Feldmann H, et al. Ebola and Marburg viruses: pathogenesis and development of countermeasures. *Curr Mol Med.* 2005;5:761–772.
36 Yang ZY, Duckers HJ, Sullivan NJ, et al. Identification of the Ebola virus glycoprotein as the main viral determinant of vascular cell cytotoxicity and injury. *Nat Med.* 2000;6:886–889.
37 Zhao X, Alexander JS, Zhang S, et al. Redox regulation of endothelial barrier integrity. *Am J Physiol Lung Cell Mol Physiol.* 2001;281:L879–L886.
38 Simeonova PP, Hulderman T, Harki D, Luster MI. Arsenic exposure accelerates atherogenesis in apolipoprotein E($^{-}/^{-}$) mice. *Environ Health Perspect.* 2003;111:1744–1748.
39 Bunderson M, Coffin JD, Beall HD. Arsenic induces peroxynitrite generation and cyclooxygenase-2 protein expression in aortic endothelial cells: possible role in atherosclerosis. *Toxicol Appl Pharmacol.* 2002;184:11–18.
40 Tousoulis D, Davies G, Stefanadis C, et al. Inflammatory and thrombotic mechanisms in coronary atherosclerosis. *Heart.* 2003;89:993–997.
41 Lip GY, Blann AD. Thrombogenesis, atherogenesis and angiogenesis in vascular disease: a new "vascular triad." *Ann Med.* 2004;36:119–125.
42 Soucy NV, Ihnat MA, Kamat CD, et al. Arsenic stimulates angiogenesis and tumorigenesis in vivo. *Toxicol Sci.* 2003;76:271–279.
43 Soucy NV, Mayka D, Klei LR, et al. Neovascularization and angiogenic gene expression following chronic arsenic exposure in mice. *Cardiovasc Toxicol.* 2005;5:29–41.
44 Lee MY, Bae ON, Chung SM, et al. Enhancement of platelet aggregation and thrombus formation by arsenic in drinking water: a contributing factor to cardiovascular disease. *Toxicol Appl Pharmacol.* 2002;179:83–88.
45 Nuntharatanapong N, Chen K, Sinhaseni P, Keaney JF Jr. EGF Receptor-dependent JNK activation is involved in arsenite-induced p21Cip1/Waf1 upregulation and endothelial apoptosis. *Am J Physiol Heart Circ Physiol.* 2005;289(1):H99–H107. Epub 2005 Feb 25.
46 Lee MY, Jung BI, Chung SM, et al. Arsenic-induced dysfunction in relaxation of blood vessels. *Environ Health Perspect.* 2003; 111:513–517.
47 Robbesyn F, Salvayre R, Negre-Salvayre A. Dual role of oxidized LDL on the NF-kappaB signaling pathway. *Free Radic Res.* 2004; 38:541–551.
48 Santilli F, Cipollone F, Mezzetti A, Chiarelli F. The role of NO in the development of diabetic angiopathy. *Horm Metab Res.* 2004;36:319–335.
49 Suematsu M, Suzuki H, Delano FA, Schmid-Schonbein GW. The inflammatory aspect of the microcirculation in hypertension: oxidative stress, leukocytes/endothelial interaction, apoptosis. *Microcirculation.* 2002;9:259–276.
50 Smith KR, Klei LR, Barchowsky A. Arsenite stimulates plasma membrane NADPH oxidase in vascular endothelial cells. *Am J Physiol Lung Cell Mol Physiol.* 2001;280:L442–L449.
51 Liu F, Jan KY. DNA damage in arsenite- and cadmium-treated bovine aortic endothelial cells. *Free Radic Biol Med.* 2000;28: 55–63.
52 Gurr JR, Yih LH, Samikkannu T, et al. Nitric oxide production by arsenite. *Mutat Res.* 2003;533:173–182.
53 Ahmad SA, Sayed MH, Barua S, et al. Arsenic in drinking water and pregnancy outcomes. *Environ Health Perspect.* 2001; 109:629–631.
54 Elliott P, Arnold R, Cockings S, et al. Risk of mortality, cancer incidence, and stroke in a population potentially exposed to cadmium. *Occup Environ Med.* 2000;57:94–97.
55 Cirillo P, De Rosa S, Pacileo M, Gargiulo A, et al. Nicotine induces tissue factor expression in cultured endothelial and smooth muscle cells. *J Thromb Haemost.* 2006;4:453–458.
56 Weinhold B. Environmental cardiology: getting to the heart of the matter. *Environ Health Perspect.* 2004;112:A880–A887.
57 Zanobetti A, Schwartz J. The effect of particulate air pollution on emergency admissions for myocardial infarction: a multicity case-crossover analysis. *Environ Health Perspect.* 2005;113:978–982.
58 Nurkiewicz TR, Porter DW, Barger M, et al. Particulate matter exposure impairs systemic microvascular endothelium-dependent dilation. *Environ Health Perspect.* 2004;112:1299–1306.
59 Kunzli N, Jerrett M, Mack WJ, et al. Ambient air pollution and atherosclerosis in Los Angeles. *Environ Health Perspect.* 2005; 113:201–206.
60 Newman CG. The thalidomide syndrome: risks of exposure and spectrum of malformations. *Clin Perinatol.* 1986;13:555–573.
61 Yabu T, Tomimoto H, Taguchi Y, et al. Thalidomide-induced anti-angiogenic action is mediated by ceramide through depletion of VEGF receptors, and antagonized by sphingosine-1-phosphate. *Blood.* 2005;106(1):125–134. Epub 2005 Mar 1.
62 D'Amato RJ, Loughnan MS, Flynn E, Folkman J. Thalidomide is an inhibitor of angiogenesis. *Proc Natl Acad Sci USA.* 1994;91:4082–4085.
63 Nolan RS, Schwartz G, Farquharson S, et al. Hematomas and limb skeletal malformations in chicken embryos following exposure to 5-fluoro-2-deoxyuridine. *Biomed Environ Sci.* 1998;11:15–30.
64 Kaicker S, McCrudden KW, Beck L, et al. Thalidomide is anti-angiogenic in a xenograft model of neuroblastoma. *Int J Oncol.* 2003;23:1651–1655.
65 Tabin CJ. Why we have (only) five fingers per hand: hox genes and the evolution of paired limbs. *Development.* 1992;116:289–296.
66 Capellini TD, Di Giacomo G, Salsi V, et al. Pbx1/Pbx2 requirement for distal limb patterning is mediated by the hierarchical control of Hox gene spatial distribution and Shh expression. *Development.* 2006;133:2263–2273.
67 Woo SH, Park MJ, An S, et al. Diarsenic and tetraarsenic oxide inhibit cell cycle progression and bFGF- and VEGF-induced proliferation of human endothelial cells. *J Cell Biochem.* 2005;95(1):120–130.
68 Luttun A, Carmeliet P. Soluble VEGF receptor Flt1: the elusive preeclampsia factor discovered? *J Clin Invest.* 2003;111:600–602.
69 Levine RJ, Maynard SE, Qian C, et al. Circulating angiogenic factors and the risk of preeclampsia. *N Engl J Med.* 2004;350:672–683.
70 Karumanchi SA, Stillman IE. In vivo rat model of preeclampsia. *Methods Mol Med.* 2006;122:393–399.
71 Nevo O, Soleymanlou N, Wu Y, et al. Increased expression of sFlt-1 in in vivo and in vitro models of human placental hypoxia is mediated by HIF-1. *Am J Physiol Regul Integr Comp Physiol.* 2006;291(4):R1085–1093. Epub 2006 Apr 20.
72 Ema M, Taya S, Yokotani N, et al. A novel bHLH-PAS factor with close sequence similarity to hypoxia-inducible factor

1alpha regulates the VEGF expression and is potentially involved in lung and vascular development. *Proc Natl Acad Sci USA.* 1997;94:4273–4278.

73 Dong W, Teraoka H, Tsujimoto Y, et al. Role of aryl hydrocarbon receptor in mesencephalic circulation failure and apoptosis in zebrafish embryos exposed to 2,3,7,8-tetrachlorodibenzo-p-dioxin. *Toxicol Sci.* 2004;77:109–116.

74 Guiney PD, Walker MK, Spitsbergen JM, Peterson RE. Hemodynamic dysfunction and cytochrome P4501A mRNA expression induced by 2,3,7,8-tetrachlorodibenzo-p-dioxin during embryonic stages of lake trout development. *Toxicol Appl Pharmacol.* 2000;168:1–14.

75 Risau W, Sariola H, Zerwes HG, et al. Vasculogenesis and angiogenesis in embryonic-stem-cell-derived embryoid bodies. *Development.* 1988;102:471–478.

76 Poole TJ, Coffin JD. Vasculogenesis and angiogenesis: two distinct morphogenetic mechanisms establish embryonic vascular pattern. *J Exp Zool.* 1989;251:224–231.

77 Evgenov OV, Liaudet L. Role of nitrosative stress and activation of poly(ADP-ribose) polymerase-1 in cardiovascular failure associated with septic and hemorrhagic shock. *Curr Vasc Pharmacol.* 2005;3:293–299.

78 Pan M, Choudry HA, Epler MJ, et al. Arginine transport in catabolic disease states. *J Nutr.* 2004;134:S2826–S2829; discussion S2853.

79 Dauphinee SM, Karsan A. Lipopolysaccharide signaling in endothelial cells. *Lab Invest.* 2006;86:9–22.

80 Klenk HD. Infection of the endothelium by influenza viruses. *Thromb Haemost.* 2005;94:262–265.

81 Hensley LE, Geisbert TW. The contribution of the endothelium to the development of coagulation disorders that characterize Ebola hemorrhagic fever in primates. *Thromb Haemost.* 2005;94:254–261.

82 Moat SJ, McDowell IF. Homocysteine and endothelial function in human studies. *Semin Vasc Med.* 2005;5:172–182.

83 Ciulla TA, Walker JD, Fong DS, Criswell MH. Corticosteroids in posterior segment disease: an update on new delivery systems and new indications. *Curr Opin Ophthalmol.* 2004;15:211–220.

84 Schumacher G, Neuhaus P. The physiological estrogen metabolite 2-methoxyestradiol reduces tumor growth and induces apoptosis in human solid tumors. *J Cancer Res Clin Oncol.* 2001;127:405–410.

85 Slungaard A, Vercellotti GM, Walker G, et al. Tumor necrosis factor alpha/cachectin stimulates eosinophil oxidant production and toxicity towards human endothelium. *J Exp Med.* 1990; 171:2025–2041.

86 Lucas R, Kresse M, Latta M, Wendel A. Tumor necrosis factor: how to make a killer molecule tumor-specific? *Curr Cancer Drug Targets.* 2005;5:381–392.

87 Mallat Z, Tedgui A. Apoptosis in the vasculature: mechanisms and functional importance. *Br J Pharmacol.* 2000;130:947–962.

88 Chen W, Thoburn CJ, Miura Y, et al. Autoimmune-mediated vasculopathy. *Clin Immunol.* 2001;100:57–70.

89 Kipshidze N, Moussa I, Nikolaychik V, et al. Influence of Class I interferons on performance of vascular cells on stent material in vitro. *Cardiovasc Radiat Med.* 2002;3:82–90.

90 de Gramont A, Van Cutsem E. Investigating the potential of bevacizumab in other indications: metastatic renal cell, non-small cell lung, pancreatic and breast cancer. *Oncology.* 2005;3 (69 Suppl):46–56.

91 Roboz GJ, Dias S, Lam G, et al. Arsenic trioxide induces dose- and time-dependent apoptosis of endothelium and may exert an antileukemic effect via inhibition of angiogenesis. *Blood.* 2000;96:1525–1530.

92 Koch AE, Polverini PJ, Kunkel SL, et al. Interleukin-8 as a macrophage-derived mediator of angiogenesis. *Science.* 1992; 258:1798–1801.

93 Bironaite D, Siegel D, Moran JL, et al. Stimulation of endothelial IL-8 (eIL-8) production and apoptosis by phenolic metabolites of benzene in HL-60 cells and human bone marrow endothelial cells. *Chem Biol Interact.* 2004;149:37–49.

94 Munzel T, Daiber A, Mulsch A. Explaining the phenomenon of nitrate tolerance. *Circ Res.* 2005;97:618–628.

95 Michelakis E. Anorectic drugs and vascular disease: the role of voltage-gated K+ channels. *Vascul Pharmacol.* 2002;38:51–59.

96 Zhang LF, Peng SQ, Wang S. Influence of lead ($Pb2^{+}$) on the reactions of in vitro cultured rat aorta to 5-hydroxytryptamine. *Toxicol Lett.* 2005;159:71–82.

97 Alter P, Herzum M, Soufi M, et al. Cardiotoxicity of 5-fluorouracil. *Cardiovasc Hematol Agents Med Chem.* 2006;4:1–5.

98 Uzel I, Ozguroglu M, Uzel B, et al. Delayed onset bleomycin-induced pneumonitis. *Urology.* 2005;66:195.

99 Ruygrok PN, Muller DW, Serruys PW. Rapamycin in cardiovascular medicine. *Intern Med J.* 2003;33:103–109.

100 Ramadass P, Meerarani P, Toborek M, et al. Dietary flavonoids modulate PCB-induced oxidative stress, CYP1A1 induction, and AhR-DNA binding activity in vascular endothelial cells. *Toxicol Sci.* 2003;76:212–219.

101 Hennig B, Slim R, Toborek M, et al. Effects of lipids and antioxidants on PCB-mediated dysfunction of vascular endothelial cells (EC). *Cent Eur J Public Health.* 2000;(8 Suppl):18–19.

102 Ivnitski-Steele I, Walker MK. Inhibition of neovascularization by environmental agents. *Cardiovasc Toxicol.* 2005;5:215–226.

60

Pericyte–Endothelial Interactions

Mark W. Majesky

Carolina Cardiovascular Biology Center, University of North Carolina at Chapel Hill

Blood vessels are built and maintained by two interdependent cell types, endothelial cells (ECs) and mural cells. Signals exchanged between these two cell populations control the formation, remodeling, maturation, and function of the vascular network. Pericytes are smooth muscle-like mural cells with unusual properties that play highly dynamic roles in the microvasculature. In response to signals produced by ECs, pericytes invest and partially cover the microvessel wall, where they act to stabilize nascent endothelial tubes, provide essential survival factors, inhibit EC proliferation, and guide vessel wall remodeling. Proper investment of vessel walls with pericytes is a critical and necessary step in vascular development and angiogenesis (Figure 60.1). The purpose of this chapter is to review the multiple roles that endothelial–pericyte interactions play in the development and function of microvessels.

PERICYTE INVESTMENT OF MICROVASCULAR ENDOTHELIUM

The acquisition of a pericyte coating around microvessels (terminal arterioles, capillaries, post-capillary venules) is referred to as *investment* (1). The process of investment includes the fundamental steps of cell migration, alignment, contact, and phenotype changes associated with mural cells interacting with ECs. These individual steps overlap in time and are controlled by multiple signaling pathways. Investment of blood vessels with pericytes is a reversible process that is highly responsive to changes in rate and direction of blood flow (2,3). Pericyte investment is mediated by a variety of different factors secreted into the local environment, including platelet-derived growth factor (PDGF)-B, angiopoietin 1 (Ang1) and angiopoietin 2 (Ang2), sphingosine-1-phosphate (S1P), transforming growth factor (TGF)-β1, and nitric oxide (NO) (4–9). The microvascular defects in mice with mutations in these secreted pericyte investment factors are summarized later. For additional details not covered here, the reader is directed to several excellent recent reviews (1,10,11).

Platelet-Derived Growth Factor-B

Possibly the most well-studied mediator of endothelial–pericyte communication is PDGF-B. ECs make PDGF-B (12–14), and its production is upregulated by elevated shear stress (15) and sprouting angiogenesis (4,16,17). The highest levels of PDGF-B expression are localized to the lead migratory tip-cell of the endothelial sprout (17). Once secreted by the endothelial sprout, PDGF-B interacts reversibly with constituents of the microvascular extracellular matrix (ECM), particularly heparan sulfate proteoglycans (18,19), via matrix-retention motifs in the C-terminus of the mature growth factor. Binding of PDGF-B to the ECM results in the establishment of a local concentration gradient around the newly formed capillary vessels. Pericytes express PDGF-receptor β (PDGF-Rβ) and respond to a local gradient of PDGF-B by proliferation to expand the pericyte pool and by directed migration toward the endothelium (4,20,21). These responses to PDGF-B also are true for smooth muscle cell (SMC) investment of developing large arteries and veins (16). Chemotaxis in response to a PDGF-B concentration gradient is normally followed by pericytes contacting ECs, extending processes around the microvessel wall, and exhibiting phenotype changes associated with enhanced vessel stability and maturation. Null mutations in PDGF-B or PDGF-Rβ result in vascular malformations associated with failure to recruit or expand the pericyte component of some but not all developing vessels leading to microaneurysms and microvascular ruptures (Table 60-1) (4,20,22,23). Of particular interest is the finding that selective deletion of only the matrix retention motif in PDGF-B also produced defects in investment of pericytes around the microvessel wall (24). These defects are seen as reduced numbers of total pericytes, poor attachment of the surviving pericytes to the endothelium, and the extension of pericyte cytoplasmic processes away from, rather than around, the microvessel wall (24). The long-term consequence of deletion of the PDGF-B matrix retention motif included severe retinal degeneration, glomerulosclerosis, and chronic renal insufficiency (24). The importance of PDGF-B

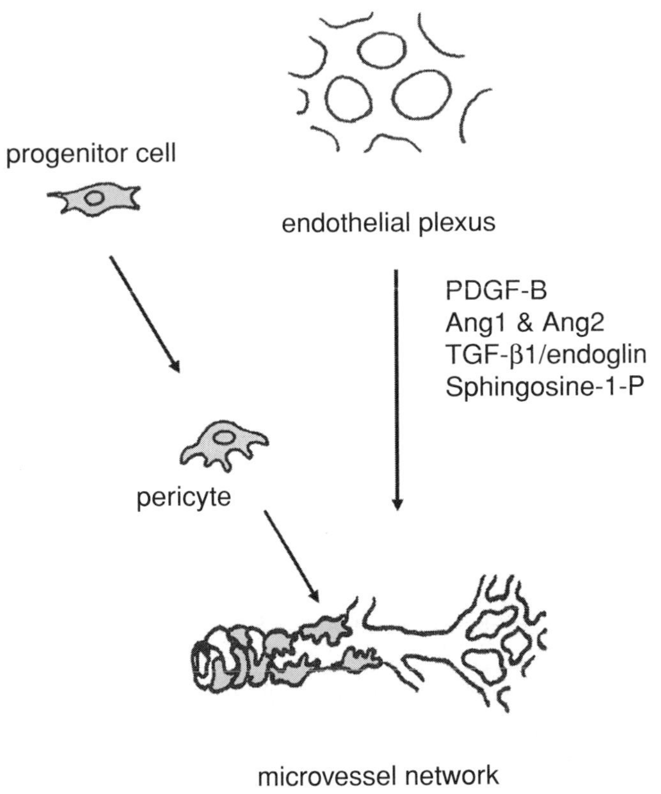

Figure 60.1. A nascent endothelial plexus recruits pericytes to form a stable microvessel network. Critical endothelial-derived factors that direct pericyte recruitment include PDGF-B, Ang1 and Ang2, TGF-β signaling via the type III TGF-β receptor endoglin, and sphingosine 1-phosphate. Pericyte–endothelial interactions are required to produce a stable, functional microvessel wall. Pericytes provide essential survival factors, promote junction formation between ECs, and inhibit EC proliferation.

for pericyte investment is underscored by the finding that even in *PDGF-B*$^{+/-}$ mice, that express one-half the wild-type levels of PDGF-B, significant reductions in pericyte numbers in brain and retinal microvasculature could be found (25).

Sphingosine-1-Phosphate

Another important pathway for mural cell recruitment is signaling via the endothelial differentiation gene 1 (Edg1)/S1P receptor (S1P1). Edg1/S1P1 is a G protein–coupled serpentine receptor that binds S1P, a secreted sphingolipid that is produced by ECs and platelets. Mice lacking Edg1-mediated signaling have pericyte investment defects that closely phenocopy PDGF-Rβ-deficient mice (6). Loss of Edg1/S1P1 signaling in ECs results in failure of microtubule polymerization and loss of proper trafficking of N-cadherin to polarized plasma membrane domains. This defect results in loss of N-cadherin–dependent cell–cell adhesion between ECs and pericytes, leading to microaneurysms, hemorrhage, and embryonic death between embryonic day (E)12.5 and E14.5 (26). A phylogenetically conserved role for S1P-signaling in the control of cell migrations important for cardiovascular

Table 60-1: Knockouts with Pericyte/Vascular Phenotypes

Gene	*Survival*	*Phenotype*	*References*
PDGF-B	Dies at birth	Mural cell recruitment defects	4,22
PDGF-Rβ	Dies at birth	Mural cell recruitment defects	16,23
Edg1/S1P1	Embryonic lethal	Failure to recruit SMCs	6
Tie2/Tek	Embryonic lethal	Failure of EC remodeling	29,31
Ang1	Embryonic lethal	Failure of EC remodeling	5
Ang2	Dies after birth	Lymphatic vessel defects	32
TGF-β1	Embryonic lethal	Defective vasculogenesis	35
Endoglin	Embryonic lethal	Failure of EC remodeling	7
FoxC2	Dies at birth	Multiple cardiovascular defects	67

development is illustrated by the zebrafish mutant *miles apart* (mil). The mil mutation was identified by positional cloning and shown to encode a G protein–coupled receptor of the lysosphingolipid receptor family that is activated on binding S1P (27). Zebrafish *mil* embryos exhibit cardia bifida due to failure of the bilateral cardiac primordia to migrate to the midline and form the tubular heart (27).

Angiopoietins

Identification of a pair of orphan tyrosine kinase receptors with immunoglobulin and epidermal growth factor–like ectodomains that are specifically expressed in ECs, called Tie1 and Tek/Tie2, led to the discovery of the angiopoietins Ang1 and Ang2 (5,28–30). Ang1 is produced by mural cells, binds to and activates Tek/Tie2 receptors on ECs, and signals maturation processes involving cell–cell and cell–matrix interactions leading to stable, leak-resistant vessels. Tie2-deficient embryos die by ED10.5 with multiple foci of hemorrhage and edema (29,31). Initial formation and patterning of early blood vessels are normal in Tie2-deficient embryos, but subsequent vascular remodeling including pericyte and SMC investment is defective. Histological examination revealed that vascular defects result from a breakdown in cell–cell communication between ECs and mural cells, leading to failure of ECs to properly adhere to underlying basement membrane and to each other. The role of Ang2 seems to depend on what other factors are present in the microenvironment. In the absence of vascular endothelial growth factor (VEGF), Ang2 acts as an antagonist of Ang1, leading to vessel destabilization, regression, endothelial apoptosis, and anti-angiogenesis (32,33). In the presence of VEGF, Ang2 is proangiogenic and facilitates vascular sprouting, branching, and elongation (33). Ang2-deficient mice have no defects in embryonic vascular development but exhibit major functional abnormalities in lymphatic vessels after birth. Unlike its previously demonstrated properties as an antagonist of Tie2 signaling, the effects of Ang2 on lymphangiogenesis are more compatible with its activity as an agonist for Tie2 in lymphatic ECs (34).

Transforming Growth Factor-β1

TGF-β1 is a multifunctional cytokine that has diverse effects within the vascular system. Studies of vascular development in knockout mice reveal the importance of TGF-β1 signaling in multiple steps of vessel formation and maturation. Approximately 50% of *TGF-β1^{-}/$^{-}$* embryos die by ED10.5 because of defects in yolk sac vasculogenesis and hematopoiesis (35). TGF-β signals through a receptor complex that consists of three different subunits, type I, II, and III (36). ECs express a specialized type III TGF-β receptor called endoglin. Endoglin-deficient mice die around E11.5 due to defects in embryonic vascular remodeling that are distinctly different from the defects seen with *TGF-β1^{-}/$^{-}$* embryos. The primary deficiencies in *endoglin^{-}/$^{-}$* embryos are in pericyte and SMC recruitment, resulting in fragile vessels that fail to sustain the increases in blood flow and pressure that accompany normal growth of the embryo (7). Early stages of vasculogenesis and hematopoiesis are normal, but arterial wall remodeling fails in *endoglin^{-}/$^{-}$* mice. Interestingly, venous ECs do not express endoglin. A study of TGF-β1 signaling in yolk sac vessels of *endoglin^{-}/$^{-}$* mice showed reduced levels of activated, phospho-Smad2 in ECs that were correlated with defects in mesenchymal cell recruitment and SMC differentiation leading to fragile, dilated, and rupture-prone vessels (8). Because endoglin is only expressed in ECs, and defects are observed in mural cells, endoglin signaling is thought to control the production of soluble factors by ECs that mediate pericyte and SMC recruitment during angiogenesis. At present, the identity of these factors is not clear.

Nitric Oxide

The diverse nature of factors that promote pericyte recruitment by ECs could reflect a requirement for distinct, nonredundant steps in investment. However, it is also true that many of these diverse signals, including PDGF-B and TGF-β, activate a common response in ECs, namely production and release of NO by endothelial NO synthase (eNOS). To test the role of NO in pericyte recruitment, Kashiwagi and colleagues studied tumor angiogenesis in mice following injections of transplantable B16 murine melanoma cells. Tumor-bearing mice that were treated with the eNOS inhibitor N^{G}-monomethyl-L-arginine (L-NMMA) exhibited reduced pericyte recruitment and decreased mural cell coverage around angiogenic vessels when compared with vehicle-treated mice (9). In cell culture assays, inhibition of eNOS resulted in greatly reduced directional migration of mural cell progenitors toward ECs (9). These studies suggest that NOS activity in the endothelium plays an important role in the communication between ECs and mural cells required to build normal vessel walls.

PERICYTE ALIGNMENT AND CONTACT WITH ENDOTHELIUM

Pericytes are not equally distributed over the surface of microvessels. Rather, they are most frequently found localized over endothelial cell–cell junctions and at microvascular branch points. Moreover, pericyte density varies considerably among different capillary beds (1). Pericyte coverage of the abluminal surface area ranges from 10% to 50% depending on the microvascular bed (11). The highest pericyte density is found in microvessels in the central nervous system (CNS), possibly because of their involvement in maintaining a blood–brain barrier. The morphology of pericyte–endothelial contacts reflects the function of the particular microvascular bed (11). For example, in tissues involved in gas exchange (lung, placenta), pericytes are only sparsely distributed in the microvasculature thus allowing for more or less unrestricted passage into and out of the circulation. In a study of the microvascular responses to histamine-induced inflammation, pericytes were observed to form umbrella-like covers over gaps between ECs, serving to restrain extravasated cells and large plasma proteins from escaping the vessel wall (37). Likewise, interleukin-2 produced a generalized leakage of capillary vessels that was accompanied by pericyte realignments to cluster along EC junctions (38). Unlike SMCs in larger arteries and veins that are separated from the endothelium by one or more layers of ECM, pericytes are located within the EC basement membrane, where they make direct contacts with ECs via tight and gap junctions (39,40). In addition, pericyte-derived products, such as Ang1, have been shown to induce the expression of the tight junction protein occludin in cultured brain microvascular endothelium (41). Moreover, immunoreactive N-cadherin and β-catenin are found in a clustered distribution in capillary ECs that matches pericyte coverage and is lost upon detachment of pericytes from microvessels (39). The importance of homotypic N-cadherin-mediated adhesion between pericytes and ECs was demonstrated by the disrupting effects of blocking antibodies to N-cadherin on brain angiogenesis in developing chick embryos (42).

PERICYTE ORIGINS AND PHENOTYPES: INFLUENCE OF ENDOTHELIUM

The embryonic origins of pericytes are unclear. Based on their expression of some of the common marker genes for SMCs, such as smooth muscle (SM) α-actin and desmin, it is often presumed that pericytes are a SMC that is localized to the microvasculature (43). However, pericytes also express markers that are not uniformly expressed by vascular SMCs, such as high-molecular-weight melanoma antigen (called NG2 in the mouse) (44). Very few lineage mapping studies have addressed whether pericytes share the same origin in development as SMCs. The little evidence that has been reported argues that pericyte origins may be at least as diverse as that of vascular SMCs. For example, a neural crest origin has been demonstrated for pericytes in the brain (11). Moreover, several reports suggest that arteriolar SMCs can serve as a progenitor pool for pericytes. Thus, upstream SMCs have been reported to detach from the arteriolar wall and migrate downstream along capillary basement membranes to adopt a pericyte-like phenotype in skeletal muscle arterioles in newborn rats (45,46). Unlike cardiac and skeletal muscle, vascular SMCs and pericytes are capable of considerable cell migration and proliferation while maintaining expression of differentiated contractile protein marker genes (47–49). A more recent study employed chimeric mice that were made by reconstituting C57Bl/6 mice with syngeneic bone marrow. In two different models of angiogenesis (B16 melanoma tumors or VEGF-A injections), the bone marrow–derived cells found in newly formed vessel walls were not ECs but rather NG2-positive pericytes that did not detectably express the smooth muscle markers SM α-actin or desmin (50). It also is possible that many pericytes are a type of multipotential mesenchymal cell that gives rise to SMCs when acted upon by endothelium but can adopt alternative cell fates when different signals are encountered. For example, pericytes have been shown to produce bone-forming cells, adipocytes, and myofibroblasts depending on environmental cues, leading to the suggestion that they are a kind of multipotential "reserve" cell that participates in wound repair, tissue remodeling, osteogenesis, and angiogenesis (51). An important question yet to be resolved is whether pericytes also can produce ECs and, if so, what roles that process may play in angiogenesis and tissue repair. Regardless of the source of pericytes, it is likely that common endothelial-derived signals are involved in their recruitment, given the phenotypes of *PDGF-B/PDGF-R*$\beta^{-/-}$ mice (20,39).

An important but poorly understood process is how ECs and pericytes become specialized to function in tissue-specific ways in different vascular beds. One critical step in this process is the determination of arterial–venous identity. Previous models, based in part on arterialization of vein grafts, held that arterial identity was conferred on ECs by high rates of blood flow, oxygen content, or intraluminal pressures that are characteristic of the arterial circulation. However, analysis of transcription of the *ephrinB2* locus in vascular development suggested that arterial–venous identity is genetically determined and is an inherent property of ECs even before the onset of blood flow in the embryo (52). Evidence in support of a genetic pathway for specification of arterial or venous identity for ECs quickly followed, particularly from studies of vascular development in zebrafish embryos where the early steps in vasculogenesis are readily accessible and observable (53). Those studies showed that midline structures, including the notochord, release the signaling protein sonic hedgehog, which stimulates the nearby somites to produce VEGF. Among the target cells for VEGF are ECs that are induced to express arterial-specific markers in a notch-dependent manner (54).

In addition to arterial–venous identity of the endothelium, studies of ephrinB2-LacZ reporter mice indicate that SMCs in arteries and arterioles also selectively express this arterial-specific marker (55). Moreover, mural cell expression of ephrinB2-LacZ can even be traced down to the level of pericytes that ensheathe capillary walls where it abruptly terminates about midway between terminal arterioles and postcapillary venules (55). During neovascularization in adult tissues, arterial-specific markers appear as early as new pericytes can be identified (56). These data show that arterial–venous identity is an early and intrinsic property of both ECs and mural cells not only in developing embryos but also in adult tissues as well (55,56).

Analysis of transcriptional regulation of the human von Willebrand factor (*hvWF*) gene provided strong evidence for a molecular basis of EC diversity within the vascular system. For example, when 487 bp of 5′-flanking sequence from the *hvWF* gene were used to drive a β-galactosidase reporter gene in transgenic mice, reporter gene activity was detected in adult brain vasculature, but not in heart, kidney, or aorta (57). By contrast, when 2182 bp of genomic sequence from the same *hvWF* locus were tested, reporter gene expression was now found in the heart, skeletal muscle, and brain of adult mice (58). Transplantation of adult cardiac tissue into subcutaneous positions in the ear resulted in induction of reporter gene expression from the 2182 bp *hvWF* transgene in auricular microvessels that were in close proximity to the cardiac graft, suggesting an important role for specific target-tissue-derived factors in directing the patterns of *hvWF* transgene expression previously described (58). Subsequent analysis showed that cardiac myocytes stimulated microvascular ECs to produce PDGF-AB, which activated the endothelial PDGF-Rα to trigger upregulation of *hvWF* reporter gene expression (59). These studies suggest that endothelial phenotypic diversity has a transcriptional basis that is, at least in part, determined by the unique properties of individual target tissues.

PERICYTE FUNCTIONS

Proper incorporation of mural cells is indispensable for blood vessel formation and function. During embryonic development, ECs form a dense network (or plexus) of vessels that is in excess of what will be needed to establish a functional circulation. Vascular pruning reduces the number of vessels to a final density required for efficient tissue perfusion. In the retina, newly formed microvessels are immature and vulnerable to regression until pericytes are recruited (60). The investment of retinal microvessels by pericytes marks the close of a "plasticity window" and the onset of vascular maturation. An important role for EC-derived PDGF-B in retinal pericyte recruitment was verified by the phenotype of mice with endothelium-specific deletion of PDGF-B (61). These mice exhibited a range of retinal microvascular defects that resemble those seen in diabetic retinopathy (61). In fact, pericyte density throughout the CNS was greatly reduced, with a corresponding increase in capillary regression and microaneurysms (61). However, the precise role of pericyte–endothelial communication in maturation of the retinal vasculature remains unclear as estimates of the number and location of pericytes depend on the markers used, and more recent studies with newer and more selective pericyte markers suggest that pericytes are abundant in both immature and mature retinal microvessels (39).

Endothelial–pericyte communication includes growth-control signals that inhibit cell proliferation. For example, although sprouting angiogenesis proceeds normally in the absence of pericytes, the diameter of pericyte-deficient microvessels in the global *PDGF-B*–knockout mice is abnormal (20). Capillary vessel diameters are either increased or decreased in the absence of pericytes, most likely due to vascular bed–specific responses to the loss of endothelial growth control signals normally provided by pericytes. Indeed, previous studies in cell culture models provided evidence that pericyte investment inhibits EC proliferation, in part via activation of latent TGF-β1 (62). This conclusion is supported by subsequent findings of EC hyperplasia in *PDGF-B*–knockout mice (4,20). Loss of pericytes also leads to defects in formation of EC junctions (20). Along with dilated vessel segments, loss of mechanical support provided by pericytes and their associated ECM makes these vessels prone to dilation and rupture (4,20).

It is commonly assumed that endothelial tubes initially form without pericyte involvement, and that subsequent acquisition of pericyte coverage then leads to vessel remodeling, maturation, and stabilization. However, new evidence suggests that pericytes can invest actively sprouting endothelium and, in some cases, may actually be pioneer cells for angiogenesis that precede the movement of EC sprouts into tissue environments. Moreover, it has been reported that pericyte tubes can be found in tumor vasculature and developing retina with no evidence of ECs in these vessels. Indeed, perfusion of tumor-bearing mice with fluorescein-tagged dextran suggested that pericyte tubes can even support sustained blood flow (63). Nevertheless, given the very small distances that may exist between pericytes and the vascular lumen with even a thin profile of endothelium still intervening, the existence of pericyte-only vessels remains to be conclusively established. Regardless of this caveat, our concepts of the function of pericytes in angiogenesis will need to include the very earliest steps in vascular sprouting and tissue invasion (11).

In addition to serving as microvascular support cells, pericytes also exhibit properties of multipotential mesenchymal cells that are capable of differentiating into myofibroblasts, adipocytes, osteoblasts, and macrophage-like phagocytic cells (64). As such, the functions of pericytes in vivo most likely extend beyond their familiar role as endothelial support cells in angiogenesis to encompass important activities in wound healing, bone and connective tissue remodeling, and as effector cells of inflammation. For example, pericytes in brain microvessels have been shown to respond to inflammatory cytokines, such as tumor necrosis factor (TNF)-α and interferon (IFN)-γ, by upregulating the expression of the leukocyte adhesion molecule intercellular cell adhesion molecule (ICAM)-1 and acquiring the activity of an antigen-presenting cell for primed T-lymphocytes via expression of MHC class II II molecules on their surface (65,66).

PERICYTE–ENDOTHELIAL CELL INTERACTIONS IN VASCULAR DISEASE

Defects in pericyte–endothelial interactions have been implicated in the development of a variety of vascular diseases including diabetic microangiopathies, particularly of the retina, Raynaud syndrome, early stages of scleroderma, various central nervous system disorders including multiple sclerosis and Alzheimer disease, vascular calcification, and fibrotic disorders (1,11). Some of the disorders in which defects in pericyte-EC communication have been suggested to play a pathogenic role are discussed later.

Lymphedema

Pericytes are normally absent from lymphatic vessels. But in a condition known as lymphedema distichiasis (LD), lymphatic vessel function is defective as a result of a mutation in the forkhead-containing transcription factor FoxC2. In mice made deficient in FoxC2 by gene targeting, lymphatic vessels were found to have an increased investment with pericytes and a lack of formation of lymphatic valves (67). The absence of FoxC2 appears to respecify lymphatic endothelium to express a more vascular endothelial phenotype, including increased expression of PDGF-B, reduced VEGFR3 expression, and increased levels of endoglin expression (67). A consequence of this switch in endothelial identity appears to be expression of a gene set for pericyte–mural cell investment.

Scleroderma

Scleroderma refers to a heterogeneous group of fibrotic disorders characterized by extensive deposition of connective tissue, mostly collagen, in either a localized (usually skin) or generalized (including internal organs) distribution. It is postulated to be an autoimmune disease with a strong microvascular component (68). In progressive systemic scleroderma, capillaries are reduced in number, exhibit perivascular infiltrates of lymphocytes and macrophages, and develop extensive fibrosis (69). Microvessels at the periphery of the sclerotic zone were found to exhibit a more than twofold increase in pericyte density, whereas in the sclerotic zone itself very few blood vessels are found at all (70). Moreover, during the early stages of systemic scleroderma, increased expression of PDGF-Rβ was found on pericytes, suggesting an unusually activated state of these perivascular cells, possibly contributing to the characteristic perivascular fibrosis found in these patients (71).

Calcification

Ectopic calcification of blood vessels is a common complication of aging and atherosclerosis. Recent studies show that angiogenesis plays an important role in atherosclerotic plaque progression (72). One of characteristic features of advanced, complicated lesions of atherosclerosis is plaque calcification. Angiogenesis is thought to promote ectopic tissue calcification via cytokines and growth factors that are released from capillary ECs and via the formation of osteoprogenitor cells from resident microvascular pericytes (51).

Hypertension

In hypertension, the number of SM α-actin-positive pericytes in brain microvessels of spontaneously hypertensive rats (SHR) was reported to be four times greater than in normotensive Wistar-Kyoto rats (73). Although this may reflect endothelial dysfunction in the SHR strain, it is possible that a systemic pattern of excess pericyte investment by hypertensive endothelium may be an important contributor to increased peripheral resistance in hypertension.

Diabetic Retinopathy

A loss of pericytes in the retinal microvasculature is one of the earliest histological changes in diabetic retinopathy (74). Pericyte "drop out" also is reported in a number of animal models of diabetes (75). Retinal microvessels have a higher density of pericytes than most other capillary beds and are very sensitive to any change in pericyte numbers. For example, in conditional *PDGF-B*–knockouts where loss of PDGF-B was restricted to the endothelium, pericyte reductions of 50% or more were associated with the development of retinal lesions in one or both eyes (61). Pericyte loss initially triggers EC death and capillary regression, followed by exuberant vascular proliferative responses, retinal fibrosis, and contraction, resulting in retinal detachment and blindness. Therefore, disruptions in the careful balance of endothelial–pericyte interactions in the retinal vasculature can set into motion a reactive sequence of events with severe consequences. The finding of retinopathy in *PDGF-B*–knockout mice suggests that changes in pericyte number or function might actually direct the course of disease rather than serve only as a reactive marker of the condition (76).

PERICYTE–ENDOTHELIAL INTERACTIONS AS TARGETS FOR ANTIANGIOGENIC THERAPY

Antiangiogenesis therapy is generally regarded as a promising adjunct to current treatments for cancer. The tumor microvasculature exhibits a number of important differences in organization and function when compared to normal microvascular beds (77). The vascular network of most solid tumors lacks a clear hierarchy of arterioles, capillaries, and venules. Although pericytes are present in tumor vessels, striking differences exist in their shape and organization that suggest abnormalities in endothelial–pericyte communication (78). Pericytes in pancreatic tumor vessels were irregular in shape, exhibited only loose contacts with capillary endothelium, and displayed frequent cytoplasmic extensions that projected abnormally into the tumor tissue (78). When tumor-bearing mice were injected with a small-molecule tyrosine kinase inhibitor that has target selectivity for the PDGF-Rβ (SU6668), pericytes were observed to detach from capillary endothelium, and the tumor vessels in pancreatic islet neoplasms were found to regress (79). When combined with an inhibitor of VEGF receptors (SU5416), the two kinase inhibitors together were more effective at arresting tumor growth at all stages of tumor development that were studied than either inhibitor alone (79). These and related findings have led to the idea that interfering with pericyte–endothelial communication would be an effective antitumor angiogenesis strategy. On the other hand, the idea that improving tumor vasculature would be desirable for cancer chemotherapy also has been proposed. According to this strategy, careful scheduling of a sequence of agents that could improve the function of tumor vasculature would allow more effective delivery of oxygen and chemotherapeutic drugs into the tumor mass to be achieved thereby making the tumor more vulnerable to cytotoxic therapy (80). Whether inhibiting or promoting tumor angiogenesis turns out to be the most effective treatment strategy may depend on the type of tumor, the stage of tumor progression, and the extent of tumor invasion or metastasis. In any case, a detailed understanding of the mechanisms by which ECs and pericytes interact to form and maintain blood vessels will provide a rich resource for new drug targets to enhance or inhibit therapeutic angiogenesis.

CONCLUSION

This brief review has emphasized the importance of cell–cell communication between ECs and pericytes in the formation, remodeling, and function of microvessels. Although the complexity of the signals exchanged between these two cell types is daunting, it is encouraging that genetic approaches are providing important and detailed insights into this intricate cross-talk. The analysis of microvascular defects in the *PDGF-B*–knockout mouse, for example, has revealed a remarkable specialization of the endothelial tip cell in angiogenic sprouts for migration and signaling. Further study of this specialized EC phenotype will no doubt identify additional secreted factors that control and coordinate pericyte responses during angiogenesis. An important area for further study concerns the origins of pericytes in adult tissues and whether pericyte deficiencies in vivo can be considered approachable by stem/progenitor cell therapy. Finally, further study of endothelial–mural cell interactions during vascular development will continue to provide important clues as to how these two cell types communicate in adult tissues to build and remodel the microvasculature.

KEY POINTS

- Pericyte–EC communication is essential for functional microvascular network formation in development and angiogenesis.
- Pericyte investment of capillary vessels provides critical survival and maturation signals for the endothelium.
- ECs produce multiple signals to recruit pericytes and mural cells, including PDGF-B, S1P, Ang1, and TGF-β1.
- Pericytes can exhibit the properties of multipotential cells, capable of differentiating into chondrogenic, osteogenic, adipogenic, and neurogenic cell types.
- Pericyte–EC communication offers multiple targets for new approaches to anti-angiogenesis therapy.

Future Goals

- To undertake a comprehensive characterization of the endothelial tip-cell phenotype in angiogenesis; this will significantly advance our understanding of how pericytes are recruited to areas of neovessel formation
- To determine what spectrum of mesenchymal cell types can be produced by pericytes in the microvasculature and to identify the signals that control pericyte fates in normal and disease tissues

ACKNOWLEDGMENTS

The author would like to acknowledge helpful discussions with Da-Zhi Wang, Vicki Bautch, and James Faber. Work in the author's laboratory was supported by the NIH (HL-19242), the American Heart Association, and by the Carolina Cardiovascular Biology Center at the University of North Carolina.

HISTORICAL OVERVIEW OF PERICYTE RESEARCH

1812 William Wood identifies painful subcutaneous glomus tumors as a vascular neoplasm. Glomus tumors would be shown later to be composed of pericytes.

1873 Rouget describes highly branching connective tissue cells associated with capillary walls. These cells, sometimes called Rouget cells, will later become commonly known as pericytes.

1923 Zimmermann identifies and names "pericytes" using a modification of Golgi's silver impregnation stain for nerve cells. This staining technique allowed the long, thin, and highly branched cell processes characteristic of microvessel-associated pericytes to be seen in tissue sections.

1924 Masson describes the normal structure of the glomus body, a small neurovascular complex sensitive to variations in temperature that regulates arteriolar blood flow. The glomus body shunts blood from arterial to venous systems without an intervening capillary network. Found in fingers, toes, and nail beds, glomus bodies contain abundant sensory nerves, accounting for the painful sensations of pericyte-rich glomus tumors.

1940 The Clarks report their observations on microvessels in rabbit ear chambers. Their studies revealed the highly dynamic and reversible nature of endothelial–pericyte interactions in response to changes in blood flow in living blood vessels.

1942 Stout and Murray identify the epithelioid cells in glomus tumors as a type of pericyte. Stout went on to identify and characterize a related but more invasive pericyte-derived neoplasm, which he named hemangiopericytoma.

1961 Majno and Palade identify a role for pericytes in control of microvascular permeability in response to histamine or serotonin.

1983 Gitlin and D'Amore describe an improved method for isolation and culture of retinal pericytes that becomes widely used to study pericyte properties in vitro.

1994 Targeted gene deletion studies in mice by Betsholtz and by Soriano identify a crucial role played by PDGF-B signaling, via the PDGF-β receptor, for pericyte recruitment in vascular development and angiogenesis.

REFERENCES

1 Sims D. Diversity within pericytes. *Clin Exp Pharmacol Physiol.* 2000;27:842–846.
2 Clark E, Clark E. Microscopic observations on the extraendothelial cells of living mammalian blood vessels. *Am J Anat*. 1940;66: 1–49.
3 Sundberg C, Ivarsson M, Gerdin B, Rubin K. Pericytes as collagen-producing cells in excessive dermal scarring. *Lab Invest*. 1996;74:452–466.
4 Lindahl P, Johansson B, Leveen P, Betsholtz C. Pericyte loss and microaneurysm formation in PDGF-B-deficient mice. *Science.* 1997;277:242–245.
5 Suri C, Jones P, Patan S, et al. Requisite role of angiopoietin-1, a ligand for the tie2 receptor, during embryonic angiogenesis. *Cell.* 1996;87:1171–1180.
6 Liu Y, Wada R, Yamashita T, et al. Edg-1, the G protein-coupled receptor for sphingosine-1-phosphate, is essential for vascular maturation. *J Clin Invest*. 2000;106:951–961.
7 Li D, Sorensen L, Brooke B, et al. Defective angiogenesis in mice lacking endoglin. *Science.* 1999;284:1534–1537.
8 Carvalho R, Jonker L, Goumans M, et al. Defective paracrine signalling by TGF-beta in yolk sac vasculature of endoglin mutant mice: a paradigm for hereditary haemorrhagic telangiectasia. *Development*. 2004;131:6237–6247.
9 Kashiwagi S, Izumi Y, Gohongi T, et al. NO mediates mural cell recruitment and vessel morphogenesis in murine melanomas and tissue-engineered blood vessels. *J Clin Invest*. 2005;115: 1816–1827.

10 Jain R. Molecular regulation of vessel maturation. *Nat Med.* 2003; 9:685–693.
11 Armulik A, Abramsson A, Betsholtz C. Endothelial/pericyte interactions. *Circ Res.* 2005;97:512–523.
12 DiCorleto P, Bowen-Pope D. Cultured endothelial cells produce a platelet-derived growth factor-like protein. *Proc Natl Acad Sci USA.* 1983;80:1919–1923.
13 Barrett T, Benditt E. Sis (platelet-derived growth factor B chain) gene transcript levels are elevated in human atherosclerotic lesions compared to normal artery. *Proc Natl Acad Sci USA.* 1987; 84:1099–1103.
14 Zerwes H, Risau W. Polarized secretion of a platelet-derived growth factor-like chemotactic factor by endothelial cells in vitro. *J Cell Biol.* 1987;105:2037–2041.
15 Resnick N, Collins T, Atkinson W, et al. Platelet-derived growth factor B chain promoter contains a cis-acting fluid shear-stress-responsive element. *Proc Natl Acad Sci USA.* 1993;90:4591–4595.
16 Hellstrom M, Kalen M, Lindahl P, et al. Role of PDGF-B and PDGFR-β in recruitment of vascular smooth muscle cells and pericytes during embryonic blood vessel formation in the mouse. *Development.* 1999;126:3047–3055.
17 Gerhardt H, Golding M, Fruttiger M, et al. VEGF guides angiogenic sprouting utilizing endothelial tip cell filopodia. *J Cell Biol.* 2003;161:1163–1177.
18 Raines E, Ross R. Compartmentalization of PDGF on extracellular binding sites dependent on exon-6-encoded sequences. *J Cell Biol.* 1992;116:533–543.
19 Ostman A, Andersson M, Betsholtz C, et al. Identification of a cell retention signal in the B-chain of platelet-derived growth factor and in the long splice version of the A-chain. *Cell Regul.* 1991;2:503–512.
20 Hellstrom M, Gerhardt H, Kalen M, et al. Lack of pericytes leads to endothelial hyperplasia and abnormal vascular morphogenesis. *J Cell Biol.* 2001;153:543–553.
21 Tallquist M, Soriano P. Cell autonomous requirement for PDGFR-alpha in population of cranial and cardiac neural crest cells. *Development.* 2003;130:507–518.
22 Leveen P, Pekny M, Gebre-Medhin S, et al. Mice deficient for PDGF B show renal, cardiovascular, and hematological abnormalities. *Genes Dev.* 1994;8:1875–1887.
23 Soriano P. Abnormal kidney development and hematological disorders in PDGF beta receptor mutant mice. *Genes Dev.* 1994; 8:1888–1896.
24 Lindblom P, Gerhardt H, Liebner S, et al. Endothelial PDGF-B retention is required for proper investment of pericytes in the microvessel wall. *Genes Dev.* 2003;17:1835–1840.
25 Hammes H, Lin J, Renner O, et al. Pericytes and the pathogenesis of diabetic retinopathy. *Diabetes.* 2002;51:3107–3112.
26 Paik J, Skoura A, Chae S, et al. Sphingosine-1-phosphate receptor regulation of N-cadherin mediates vascular stabilization. *Genes Dev.* 2004;18:2392–2403.
27 Kupperman E, An S, Osborne N, et al. A sphingosine-1-phosphate receptor regulates cell migration during vertebrate heart development. *Nature.* 2000;406:192–195.
28 Dumont D, Gradwohl G, Fong G, et al. The endothelial-specific receptor tyrosine kinase, tek, is a member of a new subfamily of receptors. *Oncogene.* 1993;8:1293–1301.
29 Sato T, Tozawa Y, Deutsch U, et al. Distinct roles of receptor tyrosine kinases tie-1 and tie-2 in blood vessel formation. *Nature.* 1995;376:70–74.
30 Davis S, Aldrich T, Jones P, et al. Isolation of angiopoietin-1, a ligand for the tie2 receptor, by secretion-trap expression cloning. *Cell.* 1996;87:1161–1169.
31 Dumont D, Gradwohl G, Fong G, et al. Dominant-negative and targeted null mutations in the endothelial receptor tyrosine kinase, tek, reveal a critical role in vasculogenesis of the embryo. *Genes Dev.* 1994;8:1897–1909.
32 Maisonpierre P, Suri C, Jones P, et al. Angiopoietin-2, a natural antagonist for tie2 that disrupts in vivo angiogenesis. *Science.* 1997;277:55–60.
33 Lobov I, Brooks P, Lang R. Angiopoietin-2 displays VEGF-dependent modulation of capillary structure and endothelial cell survival *in vivo*. *Proc Natl Acad Sci USA.* 2002;99:11205–11210.
34 Gale N, Thurston G, Hackett S, et al. Angiopoietin-2 is required for postnatal angiogenesis and lymphatic patterning, and only the latter role is rescued by angiopoietin-1. *Dev Cell.* 2002;3:411–423.
35 Dickson M, Martin J, Cousins F, et al. Defective haematopoiesis and vasculogenesis in transforming growth factor-β1 knock out mice. *Development.* 1995;121:1845–1854.
36 Shi Y, Massague J. Mechanisms of TGF-β signaling from cell membrane to the nucleus. *Cell.* 2003;113:685–700.
37 Sims D, Miller F, Donald A, Perricone M. Ultrastructure of pericytes in early stages of histamine-induced inflammation. *J Morphol.* 1990;206:333–342.
38 Sims D, Miller F, Horne M, Edwards M. Interleukin-2 alters the positions of capillary and venule pericytes in rat cremaster muscle. *J Submicrosc Cytol Pathol.* 1994;26:507–513.
39 Gerhardt H, Betsholtz C. Endothelial-pericyte interactions in angiogenesis. *Cell Tissue Res.* 2003;314:15–23.
40 Cuevas P, Gutierrez-Diaz J, Reimers D, et al. Pericyte endothelial gap junctions in human cerebral capillaries. *Anat Embryol (Berl).* 1984;170:155–159.
41 Hori S, Ohtsuki S, Hosoya K, et al. A pericyte-derived angiopoietin-1 multimeric complex induces occludin gene expression in brain capillary endothelial cells through Tie-2 activation in vitro. *J Neurochem.* 2004;89:503–513.
42 Gerhardt H, Wolburg H, Redies C. N-cadherin mediates pericytic-endothelial interaction during brain angiogenesis in the chicken. *Dev Dyn.* 2000;218:472–479.
43 Owens G, Kumar M, Wamhoff B. Molecular regulation of vascular smooth muscle cell differentiation in development and disease. *Physiol Rev.* 2004;84:767–801.
44 Ozerdem U, Grako K, Dahlin-Huppe K, et al. NG2 proteoglycan is expressed exclusively by mural cells during vascular morphogenesis. *Dev Dyn.* 2001;222:218–227.
45 Price R, Owens G, Skalak T. Immunohistochemical identification of arteriolar development using markers of smooth muscle differentiation: evidence that capillary arterialization proceeds from terminal arterioles. *Circ Res.* 1994;75:520–527.
46 Skalak T, Price R, Zeller P. Where do new arterioles come from? Mechanical forces and microvessel adaptation. *Microcirculation.* 1998;5:91–94.
47 Cook C, Weiser M, Schwartz P, et al. Developmentally timed expression of an embryonic growth phenotype in vascular smooth muscle cells. *Circ Res.* 1994;74:189–196.
48 Hungerford J, Owens G, Aargraves W, Little C. Development of the aortic vessel wall as defined by vascular smooth muscle and extracellular markers. *Dev Biol.* 1996;178:375–392.
49 Lee S, Hungerford J, Little C, Iruela-Arispe M. Proliferation and differentiation of smooth muscle cell precursors occurs

simultaneously during the development of the vessel wall. *Dev Dyn.* 1997;209:342–352.
50 Rajantie I, Ilmonen M, Alminaite A, et al. Adult bone marrow-derived cells recruited during angiogenesis comprise precursors for periendothelial vascular mural cells. *Blood.* 2004;104:2084–2086.
51 Collett G, Canfield A. Angiogenesis and pericytes in the initiation of ectopic calcification. *Circ Res.* 2005;96:930–938.
52 Wang H, Anderson D. Eph family transmembrane ligands can mediate repulsive guidance of trunk neural crest migration and motor axon outgrowth. *Neuron.* 1997;18:383–396.
53 Torres-Vazquez J, Kamei M, Weinstein B. Molecular distinction between arteries and veins. *Cell Tissue Res.* 2003;314:43–59.
54 Lawson N, Vogel A, Weinstein B. *Sonic hedgehog* and *vascular endothelial growth factor* act upstream of the notch pathway during arterial endothelial differentiation. *Dev Cell.* 2002;3:127–136.
55 Gale N, Baluk P, Pan L, et al. Ephrin-B2 selectively marks arterial vessels and neovascularization sites in the adult, with expression in both endothelial and smooth-muscle cells. *Dev Biol.* 2001;230:151–160.
56 Shin D, Garcia-Cardena G, Hayashi S, et al. Expression of ephrinB2 identifies a stable genetic difference between arterial and venous vascular smooth muscle as well as endothelial cells, and marks subsets of microvessels at sites of adult neovascularization. *Dev Biol.* 2001;230:139–150.
57 Aird W, Jahroudi N, Weiler-Guettler H, et al. Human von Willebrand factor gene sequences target expression to a subpopulation of endothelial cells in transgenic mice. *Proc Natl Acad Sci USA.* 1995;92:4567–4571.
58 Aird W, Edelberg J, Weiler-Guettler H, et al. Vascular bed-specific expression of an endothelial cell gene is programmed by the tissue microenvironment. *J Cell Biol.* 1997;138:1117–1124.
59 Edelberg J, Aird W, Wu W, et al. PDGF mediates cardiac microvascular communication. *J Clin Invest.* 1998;102:837–843.
60 Benjamin L, Hemo I, Keshet E. A plasticity window for blood vessel remodeling is defined by pericyte coverage of the preformed endothelial network and is regulated by PDGF-B and VEGF. *Development.* 1998;125:1591–1598.
61 Enge M, Bjarnegard M, Gerhardt H, et al. Endothelium-specific platelet-derived growth factor-B ablation mimics diabetic retinopathy. *EMBO J.* 2002;21:4307–4316.
62 Hirschi K, Rohovsky S, Beck L, et al. Endothelial cells modulate the proliferation of mural cell precursors via platelet-derived growth factor-BB and heterotypic cell contact. *Circ Res.* 1999;84:298–305.
63 Ozerdem U, Stallcup W. Early contribution of pericytes to angiogenic sprouting and tube formation. *Angiogenesis.* 2003;6:241–249.
64 Farrington-Rock C, Crofts N, Doherty M, et al. Chondrogenic and adipogenic potential of microvascular pericytes. *Circulation.* 2004;110:2226–2232.
65 Balabanov R, Beaumont T, Dore-Duffy P. Role of central nervous system microvascular pericytes in activation of antigen-primed splenic T-lymphocytes. *J Neurosci Res.* 1999;55:578–587.
66 Thomas W. Brain macrophages: on the role of pericytes and perivascular cells. *Brain Res Brain Res Rev.* 1999;31:42–57.
67 Petrova T, Karpanen T, Norrmen C, et al. Defective valves and abnormal mural cell recruitment underlie lymphatic vascular failure in lymphedema distichiasis. *Nat Med.* 2004;10:974–981.
68 Harris M, Rosen A. Autoimmunity in scleroderma: the origin, pathogenetic role, and clinical significance of autoantibodies. *Curr Opin Rheumatol.* 2003;15:778–784.
69 Fleischmajer R, Perlish J. Capillary alterations in scleroderma. *J Am Acad Dermatol.* 1980;2:161–170.
70 Helmbold P, Fiedler E, Fischer M, Marsch W. Hyperplasia of dermal microvascular pericytes in scleroderma. *J Cutan Pathol.* 2004;31:431–440.
71 Rajkumar V, Sundberg C, Abraham D, et al. Activation of microvascular pericytes in autoimmune Raynaud's phenomenon and systemic sclerosis. *Arthritis Rheum.* 1999;42:930–941.
72 Moulton K, Vakili K, Zurakowski D, et al. Inhibition of plaque neovascularization reduces macrophage accumulation and progression of advanced atherosclerosis. *Proc Natl Acad Sci USA.* 2003;100:4736–4741.
73 Herman I, Jacobson S. In situ analysis of microvascular pericytes in hypertensive rat brains. *Tissue Cell.* 1988;20:1–12.
74 Hirschi K, D'Amore P. Pericytes in the microvasculature. *Cardiovasc Res.* 1996;32:687–698.
75 Robison WJ, McCaleb M, Feld L, et al. Degenerated intramural pericytes ("ghost cells") in the retinal capillaries of diabetic rats. *Curr Eye Res.* 1991;10:339–350.
76 Betsholtz C, Lindblom P, Bjarnegard M, et al. Role of platelet-derived growth factor in mesangium development and vasculopathies: lessons from platelet-derived growth factor and platelet-derived growth factor receptor mutations in mice. *Curr Opin Nephrol Hypertens.* 2004;13:45–52.
77 Jain R, Booth M. What brings pericytes to tumor vessels? *J Clin Invest.* 2003;112:1134–1136.
78 Morikawa S, Baluk P, Kaido T, et al. Abnormalities in pericytes on blood vessels and endothelial sprouts in tumors. *Am J Pathol.* 2002;160:985–1000.
79 Bergers G, Song S, Meyer-Morse N, et al. Benefits of targeting both pericytes and endothelial cells in the tumor vasculature with kinase inhibitors. *J Clin Invest.* 2003;111:1287–1295.
80 Jain R. Normalization of tumor vasculature: An emerging concept in antiangiogenic therapy. *Science.* 2005;307:58–62.

61

Vascular Smooth Muscle Cells

The Muscle behind Vascular Biology

Shivalika Handa, Karolina Kolodziejska, and Mansoor Husain

Toronto General Hospital, University of Toronto, Ontario, Canada

Vascular smooth muscle cells (VSMCs) may be considered end-effector organs of the vasculature. They mediate the vasomotor responses orchestrated by the endothelium and are the principal pathogenic agents in diseases such as atherosclerosis, restenosis, and hypertension. Indeed, VSMCs represent an ideal target for the therapy of these conditions.

In atherosclerosis, for example, endothelial cells (ECs) and macrophages are primarily involved in the initiation of this disease, whereas VSMCs are typically the last to manifest pathological change (Table 61-1). However, once they do, the proliferative, synthetic, and matrix-modulating capacities of VSMCs underlie obstructive lesion formation and play a critical role in determining plaque stability (1).

VSMCs are highly responsive to their environment and are able to switch their phenotype from quiescent and contractile to migratory, synthetic, and proliferative (2). Various cues from the surrounding milieu, such as hormonal signals from overlying ECs, cytokines from invading macrophages, and mechanical stresses on the vessel wall, can both initiate and sustain the modulation of VSMC form and function.

Under normal conditions, ECs lining the arterial lumen act as sentinels and gatekeepers. By virtue of their location, ECs are the first to respond to circulating factors and hemodynamic stresses, rapidly relaying these signals to VSMCs for transduction (3). ECs also prevent circulating cells and macromolecules, such as lipids or plasma proteins, from penetrating the underlying intima and media indiscriminately. This insulates the underlying VSMCs from stimuli that might otherwise activate them or initiate their phenotypic conversion. However, when ECs become diseased or damaged this barrier function may fail (4). Under these circumstances, ECs elaborate molecular "alarm bells" that initiate inflammatory cell adhesion (5), thrombosis, vasoconstriction, and eventually VSMC proliferation. As such, a symbiotic relationship exists between ECs and VSMCs: ECs need VSMCs to carry out specific tasks, and VSMCs require the protection and guidance of ECs to do so.

In this chapter we will overview the role of VSMCs in vascular function and disease, highlighting how this cell type is the "muscle" behind vascular biology, and how learning the language spoken between the endothelium and VSMCs reveals pathophysiological insights and may lead to novel therapeutic approaches.

HISTORY

Early Descriptions of Vascular Smooth Muscle Cells: A Cell with Two Faces

As early as 1911, comparative-staining reactions described the vascular wall of elastic and muscular arteries as mostly "mucoid," showing metachromatic material that was thought to be a precursor to elastin (6). By the late 1950s, the tunica media of elastic arteries was described as containing an atypical VSMC, lacking well-defined sarcolemmal envelopes and dense masses of cytoplasm, and not as spindle shaped as the classical VSMCs described in other arteries. Rather, these cells were termed fibroblast-like but were recognized as VSMCs because of the presence of myofilaments (6). Such descriptions are some of the earliest on the presence of a synthetic phenotype of VSMCs in vivo.

Based on immunostaining and high-resolution electron microscopy, we again learned that the synthetic phenotype of VSMCs displays many characteristics similar to those of fibroblasts. Specifically, synthetic-type VSMCs harbor finger-like projections, an increased prevalence of rough endoplasmic reticulum and Golgi bodies, and underdeveloped or discontinuous basement membranes (7). However, in contrast to fibroblasts, these cells stain positively for smooth muscle (SM) myosin-1, and SM myosin heavy chain (SM-MHC) isoform B (8). Nevertheless, controversy remains as to how and whether the phenotypically modulated synthetic VSMC can be distinguished from a fibroblast or from a putative bipotential precursor known as the myofibroblast. For further details on

Table 61-1: Comparison between Vascular Smooth Muscle Cells and Endothelial Cells

	VSMCs	*ECs*
Activation	Proliferative, migratory Synthetic phenotypes Dedifferentiation, transformation	Procoagulant, proadhesive, proinflammatory phenotype
Spatial heterogeneity	Differences between VSMC and pericytes Differences between regions of large and small vessels and between intimal, medial, adventitial VSMC	Differences between large and small vessels, between same-sized vessels in different organs, and between neighboring ECs
Origins	Splanchnic and lateral plate mesoderm during early development Marrow-derived circulating and vessel resident progenitors in adults	Splanchnic and lateral plate mesoderm during early development Marrow-derived circulating progenitors in adults
Cell–cell communication	Respond to EC-derived PDGF, NO, and endothelin-1	Respond to VSMC-derived Ang1
Ca^{2+} entry or Ca^{2+}-release from internal stores	Contraction in contractile cells Migration and proliferation in synthetic cells	NO release and vasodilation
Hemodynamic forces	Strain	Shear stress and strain
Response to		
Serine proteases	√	√
Cytokines (TNF)	√	√
Chemokines	√	√
Growth factors	√	√
LPS	√	√
Cell Cycle	Vessel resident cells normally exist in G0 growth arrest Recurrent reentry into cell cycle in response to injury, mitogens, and cytokines	Vessel resident cells normally in G_0 growth arrest Proliferation and capture of circulating progenitors participating in endothelial repair
Implicated in Disease	Hypertension Atherosclerosis Restenosis Shock	Thrombosis Inflammation Hypertension Atherosclerosis Restenosis Organ rejection

TNF, tumor necrosis factor; LPS, lipopolysaccharides.

the most current molecular methods for distinguishing these two cell types, the reader is referred to a recent review (9).

DEVELOPMENT

The phylogeny of VSMCs and ECs remains elusive. Vertebrates, such as the well-studied zebrafish (*Danio rerio*) and African clawed frog (*Xenopus laevis*), possess highly organized vascular networks consisting of both cell types. These model organisms have proven to be powerful tools for understanding vascular development, including patterning, vasculogenesis, and angiogenesis. Invertebrates such as the nematode (*Caenorhabditis elegans*) and fruit fly (*Drosophila melanogaster*) display visceral smooth muscle cell (SMC) and skeletal-like muscles, as well as cardiac muscles in the case of *Drosophila* (10–14), but neither possesses VSMCs or ECs.

The vascular system of the developing vertebrate embryo is assembled via two distinct mechanisms. Initially, EC precursors (i.e., angioblasts) undergo specification from the mesoderm and proliferation. Subsequently, these migrate and come together to form a primitive network of vessels. Such de novo assembly of a vascular plexus has been termed *vasculogenesis.* In amniotes, there exist two main sites for primary vascular network formation: the yolk sac blood islands and the embryo proper (15). Successive remodeling involves sprouting and branching of new vessels from preexisting ones, transforming the primary plexus into a highly elaborate EC network now acting to recruit SMCs and pericytes. Generally, the latter step is excluded from capillaries. In the adult, well-characterized examples of angiogenesis accompany tissue growth, pregnancy, the female reproductive cycle, and wound healing. Pathologically, angiogenesis plays a critical role in diabetic retinopathy, tumor growth, and metastasis (16). The role of VSMCs or pericytes in the formation of vessels larger than capillaries – that is, where they may be required for structural integrity – is just beginning to be understood (17).

Although the source of angioblasts or hemangioblasts (bipotential precursors of endothelial and hematopoietic lineages) is the splanchnopleural and local mesoderm (15), the origins of VSMCs are more confounding. Most SMCs originate from splanchnic mesoderm and local mesoderm, however VSMCs of the coronary arteries originate from epicardium (18–21) and VSMCs of the aortic outflow tract are derived from neural crest (22,23).

At the molecular level both vasculogenesis and angiogenesis are initiated via vascular endothelial growth factor (VEGF), a powerful mitogen and chemoattractant of angioblasts. Its signaling is mediated via VEGF receptors VEGFR1 (FLT-1) and VEGFR2 (FLK-1/KDR) expressed on ECs (24–26). VEGFR2 is the earliest known marker of developing ECs and begins to be expressed at embryonic day (E)7.0 in the mouse (24). Angiogenesis, being the more complex process, requires vasodilatation and increased EC permeability, followed by destabilization of surrounding VSMCs and matrix, enabling new sprout formation (16). Remodeling depends on angiopoietin 2 (Ang2) signaling through its receptor Tie2, expressed on all ECs, as well as activation of matrix metalloproteinases (MMPs), chymases, and other proteases. Proliferating ECs migrate to assemble a new adjoining vessel and begin secreting platelet-derived growth factor (PDGF)-B, thus attracting local VSMC precursors (27). Contact between these two cell types initiates a cascade of events leading to vessel stabilization and maturation. Upon transforming growth factor (TGF)-β1 activation, ECs cease to proliferate and migrate and reduce their VEGFR2 expression, whereas SMCs acquire their mature phenotype. The expression of Ang1 by VSMCs signals back to ECs through the Tie2 receptor to ensure vessel stabilization (17).

In addition to participating in EC signaling, VSMC progenitors also require expression of a number of transcription factors to become differentiated to their typical contractile form. These include the serum response factor (SRF) and its cofactor myocardin, the MADS-box family members Smad5 and myocyte enhancer factor 2C (MEF2C), lung Kruppel-like factor (LKLF), and heart and neural crest derivatives expressed transcript (dHAND) (28,29). Coordinate expression of SM-specific contractile and cytoskeletal proteins has become a standard for assessing differentiation status. The earliest of all such markers is SMα-actin, expressed at E9.0 in the mouse, followed shortly by SM22α, SMα-actinin, α– and β-metavinculin, then h-caldesmon, h1-calponin, and SM1 isoform of SM-MHC, with the SM2 isoform of SM-MHC being a late molecular marker of maturing VSMCs (2). Similarly, the use of temporal expression patterns of proteins associated with the EC lineage, beginning with TAL1 and VEGFR2, through platelet-endothelial cell adhesion molecule (PECAM)-1 (also known as CD31), CD34, vascular endothelial (VE)-cadherin, and Tie2, has become a standard tool for tracking endothelial development (30).

Assembly of vessels must naturally reflect their future function, be it in an aorta, capillary, or a venous or arterial vessel. EphrinB2 has been found to mark future arterial, but not venous ECs, while in a complementary manner EphB4 (a receptor for ephrinB2) marks the venous endothelium. This divergence occurs at the earliest stages of the capillary plexus formation when these cell types are intermingled and virtually indistinguishable, indicating possible cross-talk between these two populations (31). Importantly, vessel-specific signals that VSMC recruitment and differentiation must also exist. This can be inferred from studies defining arterial- (and not venous- or visceral-) specific promoter elements in the *SM22α* gene (32,33).

It is worth noting that the exploitation of this knowledge base for the regeneration or bioengineering of large-sized vessels containing both ECs and VSMCs has yet to occur. Our ability to "grow" blood vessels large enough to restore impaired organ perfusion may depend on the application of such insights. On the other hand, the potential for targeting genetic therapies to specific vascular beds has been realized for both ECs (34,35) and VSMCs (36–39), at least in animal models.

PHYSIOLOGY

Cardiac, Skeletal, and Smooth Muscle: Specialized Functions More Different than Alike, with Surprising Regenerative Capacity

The adult human body contains three distinct muscle tissues: skeletal, cardiac, and smooth. Skeletal muscle cells are large, multinucleated cells that contain actin and myosin filaments arranged in parallel within functional groups or sarcomeres. This organization gives rise to striations that are a distinctive feature of these cells and visible by light microscopy (40). Skeletal muscle cells are believed to be terminally differentiated and incapable of mitosis. However, regenerative repair of this tissue is possible and appears to occur from resident satellite cells or circulating precursor cells (41,42). Skeletal muscle cells will not contract unless stimulated to do so by nearby nerve cells; therefore, this tissue also is known as striated voluntary muscle (40).

Cardiac muscle cells are comparatively smaller in size than skeletal muscle but also are striated and generally thought to contain only one centrally placed nucleus. However, these cells are capable of close alignment or fusion with each other to form myotubes and fibers, which may then appear multinucleated. Traditionally, we have also viewed cardiac myocytes as being terminally differentiated cells incapable of cell division and have thought of cardiac muscle as a tissue that cannot regenerate after injury due to the absence of satellite cells (40). These concepts have recently been challenged, both by the recognition of cardiac-resident or circulating stem cells capable of making new cardiac myocytes (43,44) and by genetic animal models in which cardiac myocytes overexpressing cell cycle proteins have been shown to retain proliferative potential (45). In the context of endothelial biomedicine, it is particularly important to highlight here recent studies demonstrating that circulating human $CD34^+$ (i.e., endothelial) progenitor cells (EPCs) can be programmed to transdifferentiate into cardiomyogenic cells (46) and that human EPCs can form direct tubular communications with cocultured cardiomyocytes (47). Indeed, both transdifferentiation and cell fusion are believed to explain the observed regenerative potential of human EPCs in mouse models of cardiac injury (48). Cardiac muscle tissue is known as striated involuntary muscle (40), as the nervous system provides only indirect control over the heart rhythm, which can be independently sustained by specialized pacemaker cells in the sinoatrial (SA) and atrioventricular (AV) nodes.

By contrast, SMCs are small, spindle-shaped mononucleated cells that do not undergo cell fusion or myotube formation. The actin and myosin filaments in these cells are organized differently from skeletal or cardiac cells such that no striations are present. Without requiring a progenitor population (although these also exist [49–54]), SMCs can self-replicate and regenerate after injury. Visceral SM contractions can be triggered by the autonomic nervous system, such as in the enteric plexus of the gut, but these cells generally contract either spontaneously or in response to coordinated hormonal signals and are consequently termed *involuntary muscle* (40).

The Physiology of Muscle Contraction

The main function of a muscle cell is to contract. Both striated and nonstriated muscle types share a common mechanism of contraction known as the sliding filament model. At the beginning of a cycle, myosin cross bridges are bound to actin with high affinity, creating the rigor state. Cross bridges are detached upon binding to ATP, which is hydrolyzed followed by a weakly bound reattachment of the myosin head. Release of inorganic phosphate (P_i) permits transition into strongly bound force-generating states that are terminated by ADP release. This, however, is where the similarities end.

In striated muscle cells, contraction is initiated by release of acetylcholine from nerve cells, which triggers an action potential by stimulating the opening of membrane calcium (Ca^{2+}) channels, allowing an influx of extracellular calcium. Contraction follows a rise in the intracellular calcium concentration ($[Ca^{2+}]_i$), and relaxation occurs when the $[Ca^{2+}]_i$ drop. In striated muscle cells, the interaction of calcium with troponin activates contraction.

In SMCs the situation is distinctly different (55). Calcium binds to calmodulin (CaM), and this complex interacts with the catalytic subunit of myosin light chain kinase (MLCK). Ca^{2+}/CaM-activated MLCK then phosphorylates serine 19 on the regulatory light chain of myosin, MLC-20. Phosphorylation of MLC-20 is the "on" switch for actin-activated ATPase activity, which follows the sliding filament pathway to contraction. Decreases in $[Ca^{2+}]_i$ inactivate MLCK, which permits dephosphorylation of MLC-20 by myosin light chain phosphatase (MLCP). Dephosphorylation of MLC-20 inhibits ATPase activity and the muscle relaxes.

Most importantly, contraction-inducing increases in $[Ca^{2+}]_i$ in VSMCs can be brought about by two very different mechanisms. The first is somewhat analogous to what occurs in striated muscle and is known as electromechanical coupling. This pathway involves changes in the resting membrane potential of smooth myocytes, leading to depolarization and influx of extracellular calcium through voltage-gated calcium channels (VDCCs). Coupled with calcium-induced calcium release (CICR) from the sarcoplasmic reticulum (SR), increases in $[Ca^{2+}]_i$ occur and contraction ensues through Ca^{2+}/CaM-mediated activation of MLCK. In electromechanical coupling, resting negative membrane potentials (typically –40 to –70 mV) are restored by potassium channels and sodium–potassium exchanger activity. This closes VDCC and stops further influx of extracellular calcium. The $[Ca^{2+}]_i$ can then be returned to basal levels by the sarcoplasmic-endoplasmic reticulum Ca^{2+}-ATPase (SERCA) pumps (56) and efflux via the plasma membrane Ca^{2+}-ATPase (PMCA) pumps and, to a much lesser extent, the sodium calcium exchanger (NCX) (57). Even after Ca^{2+} levels are lowered and Ca^{2+}-CaM-dependent MLCK activity is lost, smooth muscle relaxation will not occur until MLCP does its work.

Although only employed by a few select visceral myocytes, electromechanical coupling enables autonomic nervous system control over smooth muscle contraction. Tonic or steady-state SMCs respond to excitatory stimuli by graded depolarizations only, whereas phasic smooth muscle responds with action potentials (58). The former is an important function for SMCs that need to be maintained in a contractile or rigor state for long periods of time and at very low metabolic cost. An example of this in the human body is in the smooth muscle of certain sphincters.

The second pathway through which contraction-inducing increases in the $[Ca^{2+}]_i$ are brought about is pharmacomechanical coupling. This route is independent of any changes in resting membrane potential and does not require extracellular calcium. This pathway involves receptor–ligand interactions on the cellular membrane, which activate second messenger signaling molecules that lead to the release of sequestered compartments of intracellular calcium (55). Importantly, spatially and functionally distinct intracellular calcium stores exist in VSMCs that enable Ca^{2+}-dependent functions other than contraction to be carried out in much the same way (59–61). Although the role of calcium regulation in the control of cell cycle will be discussed below, we will focus here on the mechanisms through which hormonal signals from the endothelium achieve contraction in underlying VSMCs.

Principally, pharmacomechanical coupling in VSMCs is initiated by a hormone–receptor interaction between a specific ligand and its G-protein–coupled receptor (GPCR). This leads to activation of phospholipase C (PLC), which generates inositol 1,4,5-triphosphate (IP3) (62). IP3 binding to IP3 receptors (IP3R) located on the SR mediates release of this particular calcium store and an increase in $[Ca^{2+}]_i$ (63). It is important to recognize that in addition to producing an IP3-mediated increase in $[Ca^{2+}]_i$, GPCR-signaling also activates other molecular participants that sensitize the VSMCs to Ca^{2+}-initiated contractions through a variety of molecular mechanisms acting either directly or indirectly on MLCK and MLCP. The Rho/ RhoA/Rho-associated kinase (ROCK) pathway (64), amongst others, is a critical mediator of such added functionality, which is a dominant feature of pharmacomechanical coupling and absent from electromechanical coupling. The importance of these pathways to VSMC contractility in specific, and cardiovascular function in general (65), is highlighted by the efficacy of pharmacological agents specifically targeting this pathway (66,67). Along similar lines, specific members of the RGS family of proteins (regulators of G-protein signaling) expressed in VSMCs have been shown to dampen GPCR-dependent calcium transients, and mice deficient in RGS2 exhibit profound hypertension (68). However, much still needs to be learned about regulation of VSMC contraction. For example, in transgenic mice designed to have reduced $[Ca^{2+}]_i$ by virtue of forced overexpression of human PMCA4b in VSMCs, the resultant phenotype was one of increased vascular reactivity and elevated blood pressure (69). Although not completely elucidated, the mechanism believed to be active in this model involves the subcellular localization of this particular PMCA, and its coassociation and inhibition of neuronal nitric oxide synthase (nNOS) (69).

Because of the time needed to produce IP3 and mediate the successive steps involved in sensitization and contraction, VSMC contractions brought about by pharmacomechanical coupling may take as long 1 to 3 seconds to take effect (55). Although such a delay would be unacceptable for striated muscle functions, this putative disadvantage is more than compensated by the intricate "remote control" of VSMC tone and function offered by this pathway. Indeed, without VSMCs having evolved pharmacomechanical coupling, every contractile VSMC would need to be electrically linked to a depolarizing EC or neuron. Having said that, experimental evidence supports a physiologically important role for gap-junction–mediated connectivity between VSMCs themselves (70).

Maintaining the Vascular Wall: It's About More than Contractility

The human body contains blood vessels of many different shapes and sizes. The largest arterial blood vessels are elastic arteries. These contain many layers of VSMCs alternating with sheets of elastic laminae. The smallest artery that contains VSMCs is the precapillary arteriole, which may have only a single layer. Because it is obvious that the structures of these two types of vessels are very different, one cannot assume that their physiologies will be the same. For example, although presumably exposed to similar "risk factors" in vivo, only large or conduit arteries are associated with atherosclerotic plaque formation, whereas the smaller arterioles, or resistance arteries, are not. It is tempting to speculate that this could be due more to fundamental differences in the number and nature of VSMCs residing in these locations than to differences in EC biology.

Remodeling responses also are markedly different between these two types of vessels. In conduit arteries, the outer diameter of the vessel is often enlarged, without a reduction in lumen size. This form of remodeling is referred to as *outward hypertrophic remodeling,* a state in which an increased lumen diameter and medial cross-sectional area are present (71). By contrast, resistance vessels typically undergo *inward eutrophic remodeling,* where the lumen is decreased but the cross-sectional area of the media has not changed. The precise mechanisms mediating this type of remodeling are not yet known, but one of the theories put forward is that intimal proliferation is balanced by medial apoptosis. In some cases, resistance arteries also exhibit *inward hypertrophic remodeling,* where the lumen diameter is decreased and the medial cross-sectional area has enlarged (71).

What physical forces determine vessel wall proportions, and what laws govern remodeling? In normal arteries, the product of blood pressure and lumen size determine the thickness of the vascular wall (72). Since the mean pressure in conduit arteries is relatively constant, the determining factor for wall thickness is the radius of their lumens. By Laplace's law (Equation 1), for any given pressure, the larger a vessel's radius the greater the outward stretching force known as *wall tension* will be (73):

$$T = PR/2t \qquad \text{(Equation 1)}$$

The only way to reduce wall tension is either to lower intraluminal pressure (not physiologically feasible) or increase wall thickness. Since the endothelial and adventitial layers of an artery are structurally "weak," the media must meet most of the stress on the vessel wall. As vessels with larger radii will require more layers of medial VSMCs and more structural matrix elements, one would intuitively expect a higher prevalence of proliferative- and synthetic-type VSMCs in the media of larger elastic arteries. Indeed, culturing and cloning studies have shown that two distinct populations of VSMCs can be found in the medium-sized internal thoracic artery of humans (74). Studies of SMC-specific transgene expression in vivo provide further evidence for VSMC-subtype-selective gene regulation (75–77).

The radius of the lumen also is determined by the size and activity of the vascular bed it serves (72). Vessel branching increases as one moves away from the core of the body, which permits vascularization of distal tissues. Understandably, the lumen of the blood vessel progressively decreases as branching increases. Although the role of blood pressure may be limited in determining normal vessel diameter, the role of blood flow is critically important, especially in small resistance arteries (72). ECs sense differential shear or frictional forces imposed by varying blood velocities. Increased velocity leads to increased lumen diameter mediated by EC signaling and VSMC relaxation. Likewise, reduced flow velocity decreases lumen diameter through VSMC constriction. Understanding the communications between the EC sensor and the VSMC effector is thus central to understanding both the dynamic and structural changes to vessel size and systemic hemodynamics (3).

PATHOLOGY

Atherosclerotic and restenotic coronary and cerebrovascular diseases remain the leading causes of death and disability in our society, and VSMC proliferation is the histological hallmark of these lesions (1,78,79). Although intraluminal stents have reduced the negative remodeling that accompanies angioplasty (80,81), in-stent stenosis due to the migration and proliferation of VSMCs remained, until recently, a serious and not infrequent complication (82). Brachytherapy and drug-eluting stents have limited the incidence and extent of in-stent stenosis (83–85), however, efforts aimed at elucidating the fundamental mechanisms that govern VSMC proliferation remain highly relevant to our understanding of atherosclerosis and its consequences. Yet, despite enormous efforts aimed at understanding the molecular and cellular basis of proliferative vasculopathies, it is both remarkable and unfortunate that so little of this research has translated to the clinical arena. These difficulties are likely due to the inherent redundancies present in biological signaling systems and the consequent failure of therapies aimed at upstream pathways. Accordingly, many groups have focused their efforts on improving our understanding of the final common pathway of cell cycle progression in VSMCs. Indeed, the success of drug-eluting stents owes much to their targeting of the VSMC cell cycle (86).

It is noteworthy that the ability of differentiated VSMCs to repeatedly reenter the cell cycle is unique among mature myocytes. Indeed, this property underlies the phenotypic plasticity of VSMCs and forms the very basis of their pathogenic potential. Thus, it may be incumbent upon us to also address the molecular mechanisms involved in the development and differentiation of VSMCs, as they may improve our ability to identify and prevent the dedifferentiation of VSMCs in disease (2,9).

Phenotypic Plasticity of Vascular Smooth Muscle Cells

A remarkable feature of VSMCs is their ability to modulate their phenotype between quiescent contractile and synthetic proliferative states. This process was initially recognized in cell culture where structural changes in VSMCs during early primary cell culture were observed (87). Cells undergo phenotypic modification such that they lose their contractile apparatus and exhibit disorganization of myofilaments, show increased presence of endoplasmic reticulum and Golgi, increase the synthesis of DNA and RNA, and become highly proliferative (7). Morphologically, synthetic VSMCs show a rounded or polygonal shape verses the elongated spindle shape of the contractile phenotype. Subsequently, functional observations revealed that these cells also synthesize and secrete collagen and elastin (7), thereby earning the designation "synthetic" phenotype.

Following these important observations, several investigators have shown that culturing VSMCs from large elastic arteries from bovine (88) and human (74) sources results in phenotypically distinct cell types. Additionally, it has been shown that in vascular proliferative diseases such as atherosclerosis, VSMCs take cues from the surrounding environment, neighboring ECs, and modified macromolecules such as low-density lipoprotein (LDL) to switch to a synthetic phenotype, which are a dominant pathogenic force in neointima and plaque formation (2).

An important and interesting observation from these and other studies is that synthetic-type VSMCs downregulate expression of marker genes such as *SMα-actin*, *SM22α*, and *SM-MHC* (89) and concomitantly upregulate a set of genes required for synthetic, migratory, and proliferative functions (9). This process is reminiscent of the developmental or undifferentiated state of smooth muscle and is often referred to as *dedifferentiation.* Recent discovery of VSMC and EC progenitor cells circulating in the blood and capable of incorporating into damaged vasculature or contributing to neovascularization of tissues in need of repair have challenged and expanded our understanding of this complex system (54).

The mechanisms underlying these processes will need to be harnessed for effective regenerative therapies, and they highlight an important concept in pathogenesis: our body's two primary biological programs, development and defense, share common pathways. Indeed, this concept has provided strong rationale for studies in developmental biology and opens the door to new tools, such as stem cell technologies, to aid in the therapy of a large variety of human diseases.

Cardiovascular Disease

Cardiovascular disease is a leading cause of death in North America (90), affecting more than 70 million Americans (91). Moreover, this burden will continue to grow as people with two or more risk factors for cardiovascular disease – smoking, obesity, hypertension, diabetes, etc. – make up 28% of all adults. Of primary relevance to this chapter on VSMCs are the vascular diseases of atherosclerosis and hypertension.

The Role of Vascular Smooth Muscle Cells in Atherosclerosis

Atherosclerosis is an inflammatory disease involving lipid accumulation within the wall of muscular and elastic arteries, causing plaque formation. These plaques consist mostly of foam cells, which are macrophages and/or VSMCs that have taken up lipid particles, necrotic cell debris, cholesterol esters, and calcium deposits. Lesion formation in humans occurs as early as childhood, at which stage the lesion is often described as a fatty streak and consists mainly of macrophages and T lymphocytes (1).

The formative cell biology of this disease has long been a topic of much interest. The popular response-to-injury hypothesis implicates the ECs as the initiating culprit. Although early theories were based on endothelial denudation as the key first event, more recent attention has been given to dysfunction of *intact* endothelium as the starting point. ECs dysfunction can be caused by many factors, such as elevated modified LDL levels, free radicals formed from smoking, shear stresses in high blood pressure, genetic defects, and infections by viruses and/or bacteria (1).

Injury to the endothelium leading to dysfunction results in increased expression of adhesion molecules and chemokines, such as vascular cell adhesion molecule (VCAM)-1 and monocyte chemotactic protein (MCP)-1, and subsequent recruitment and accumulation of blood leukocytes to the area (5,92). Injury to the endothelium also increases its permeability, thereby decreasing its ability to form a protective barrier to the underlying vascular cells (4). Blood platelets, monocytes, and other immune-related cells are free to enter the vascular wall, where they proliferate and, in combination with ECs, produce cytokines and growth factors such as PDGF, TGF-β, and fibroblast growth factor 2 (FGF2) (1).

These molecules all stimulate VSMC proliferation and migration, which leads to neointima formation and thickening of the vascular wall. At this stage, the blood vessel can undergo outward hypertrophic remodeling to compensate for the increased medial cross-sectional area. Accumulation of macrophages and other blood cells within the vascular wall provides a source of hydrolytic enzymes and other immune-related responses that lead to the formation of a necrotic core. Continuous VSMC migration and proliferation combined with the inflammatory response leads to an enlarged lesion that becomes covered by a fibrous cap. Eventually, dilatory compensation by the vessel wall reaches its outward limit and the growing plaque begins to occlude the lumen area and obstruct blood flow (1).

In the coronary arteries, this process may manifest first as angina, described as an oppressive pain or tightness in the chest, and a potential warning sign of a more dangerous cardiovascular event. Continued disease progression can lead to complete occlusion of the blood vessel at the site of plaque formation or plaque rupture, leading to occlusion and embolism by thrombus. As described, these events may occur in a coronary artery, leading to myocardial infarction; in the peripheral vasculature, resulting in critical limb ischemia; or in cerebral vessels, causing stroke. Approximately three-quarters of all deaths resulting from cardiovascular disease are a consequence of such complications of atherosclerosis (91).

Known for their migratory and proliferative capacity, VSMCs are key mediators in the progression of atherosclerosis. In this disease, VSMCs respond to signals sent from the diseased endothelium and other inflammatory cell types, thereby playing the role of a late but critical accomplice. A focus on this pathogenic potential has resulted in a large number of studies on the mechanisms by which VSMCs migrate and proliferate, only some of which will be briefly reviewed in the following section.

Mitogens, Calcium, and Cell Cycle Control

VSMCs in the vasculature are primarily quiescent and rely on specific signals to initiate migration and proliferation. Such signals arise from different sources, such as activated macrophages, platelets, ECs, or other "transformed" VSMCs. One major stimulant of VSMC migration and proliferation is PDGF, which is produced by platelets and macrophages within atherosclerotic lesions; however, ECs and VSMCs have been shown to be additional important sources of this critical chemoattractant and mitogen (2). Indeed, the unique receptor and signaling pathways through which the different PDGF isoforms stimulate both migration and proliferation of VSMCs have been the focus of several studies to which the reader is referred (93–95). A recent review further details the lessons learned from mouse models of global and tissue-specific perturbations in PDGF signaling (27).

TGF-β also has been shown to be involved in VSMC proliferation, matrix deposition, and neointima formation in transgenic and injury models of atherosclerosis (96). However, studies involving this molecule have yielded contradictory results. Some have shown that TGF-β promotes VSMC differentiation, whereas others have shown that neutralizing antibodies to TGF-β promote the development of atherosclerosis (2).

The cellular transcription factors that mediate the signals of these and other mitogens have been extensively studied. For example, the proto-oncogenes *c-fos*, *c-myc*, and *c-myb* are early-response genes that follow growth factor–receptor binding in VSMCs (97). Temporally late in this sequence, the *c-myb* gene encodes the nuclear phosphoprotein c-Myb, a transcription factor that plays a key role in promoting late G1-to-S phase cell cycle progression in this cell type (98). First, antisense oligonucleotides to *c-myb* (99), and then unique dominant negative c-Myb (39) constructs were used to demonstrate that reductions in c-Myb activity abolish the normal rise in both resting and stored $[Ca^{2+}]_i$ as cells move to the G_1/S transition. This effect was partly due to a significant increase in the Ca^{2+} efflux rate of dominant negative (DN)-Myb–transfected cells (98), mediated in part by increased mRNA and protein levels of plasma membrane Ca^{2+}-ATPase 1 (PMCA1) and PMCA4 (98,100). The importance of PMCA1 as a critical regulator of Ca^{2+} efflux and $[Ca^{2+}]_i$ was specifically demonstrated in both growth-arrested and proliferating VSMCs, without manipulations of c-Myb (101).

Although other c-Myb–dependent targets, unrelated to calcium regulation, also may play a role in regulating VSMC proliferation (102–105), the importance of the Ca^{2+}-mediated effects cannot be overstated. Indeed, a number of investigators have shown that the regulation of cell cycle–associated $[Ca^{2+}]_i$ and cell cycle progression also is dependent on SERCA-maintained Ca^{2+} stores (106–110) and the IP3-responsive Ca^{2+}-release pathway (111–115). For example, treatments with Thapsigargin, a known SERCA inhibitor, have been shown to inhibit the proliferation and migration of venous (116) and arterial VSMCs (117). Of interest, preliminary data suggest that in addition to the known cell cycle–associated decreases in PMCA, expression levels of the most abundant isoforms of SERCA (SERCA2) and the IP3 receptor (IP3R1) also are regulated during G_1-to-S phase cell cycle progression of VSMCs (118) and that these may also be mediated by c-Myb (119). These observations are consistent with the ability of c-Myb to act as both a *repressor* (100,120, 121) and activator of transcription (102–105,119) and implicate c-Myb as a fundamentally important master switch for cell cycle Ca^{2+} homeostasis (122,123).

Taken together, the above data strongly support the existence of a Ca^{2+}-sensitive checkpoint in G_1/S. Although many investigators have similarly demonstrated the importance of $[Ca^{2+}]_i$ in cell cycle progression (124,125), remarkably little is known about the precise manner in which Ca^{2+} mediates this effect. Given the temporal association of cyclin E and the cyclin-dependent kinase-2 (cdk2) with the G_1/S interface (126,127), and the fact that cdk2 expression levels are Ca^{2+}/CaM-responsive (128), it seems reasonable to hypothesize that the cdk2/cyclin E complex may mediate the Ca^{2+}-sensitivity of G_1/S transitions. Indeed, preliminary data show that cdk2 and cyclin E are coexpressed at the G_1/S interface of VSMCs and that their activity at this cell cycle stage is Ca^{2+}-dependent (129,130).

In addition to studies employing VSMCs in vitro, the issue of whether c-Myb–directed therapies would be of physiological importance in vivo has been addressed (97,122). In the first tissue-specific approach described, a carotid artery denudation model of restenosis in the mouse was used to show that VSMC proliferation and neointima formation were significantly inhibited by DN-Myb expression (39). However, these experiments were conducted in young animals with presumably intact endothelial function and normal carotid arteries. Given that the biology of VSMCs from atherosclerotic vessels is distinct from those of normal vessels, the question of whether c-Myb– or other cell cycle–directed therapies targeted specifically to VSMCs will be of benefit in preventing or reducing restenosis in atherosclerotic arteries or atherosclerosis itself remains unanswered.

Notwithstanding the evidence supporting a role for c-Myb in the proliferation of VSMCs, little is known of its role in the embryonic development and differentiation of VSMCs. For that matter, little is known, in general, of the mechanisms underlying the phenotypic plasticity and ability of VSMCs, but not other myocytes, to dedifferentiate. Although one study has suggested a potential role for *c-myb* in the development of neural crest–derived SMCs (131), this may have been confounded by the poor specificity of the antisense *c-myb* oligonucleotides employed (132–134). Mouse embryos homozygous for the *c-myb* allele ($c\text{-}myb^{-/-}$) appear to perish at approximately E15.5 because of failed hematopoiesis (135). Given that the discovery of *c-myb* was based on its sequence similarity to leukemia-causing oncogenes (136), and that its critical role in myeloid differentiation had already been established (136–139), attributing the phenotype of $c\text{-}myb^{-/-}$ mice to a disorder of hematopoiesis was expected (135). Whether subtle defects in cardiovascular development were present in the *c-myb*–knockout embryos was never specifically examined. However, the normal expected frequency of pups generated in crosses to enable tissue-specific expression of a DN-Myb suggested that interference with c-Myb in a spatiotemporal pattern defined by the *SM22α* promoter employed in this study did not result in abnormalities of cardiovascular development (39). These data suggested that c-Myb–dependent genes either do not play a role in the development of cardiac and/or vascular structures at time points defined by SM22α-expression or that there exist redundant (or DN-Myb–insensitive) pathways active during differentiation and development but not during pathogenic proliferation of adult VSMCs. This would represent an extremely important distinction, with broad implications for VSMC-directed gene therapies. Indeed, the question of whether c-Myb plays a role in the development and differentiation of VSMCs through assays employing embryonic stem (ES) cells (140–142) harboring reporter genes activated by SMCs, and more specifically VSMCs, differentiation (32,75,143,144) are now of interest.

Further insight into the mechanisms underlying VSMC proliferation have been gained by studying the effects of inhibiting the action of a negative regulator of cell proliferation such as TGF-β (145) or the glucocorticoid receptor (146). In an atherosclerotic mouse model, mice that were injected with antibodies against the PDGF-β receptor exhibited decreased lesion size and reduced numbers of intimal VSMCs (147). Recently a group of investigators published findings of inhibited serum-stimulated SMC proliferation with DNAzymes, moieties that target translational start sites of specific genes such as *Egr-1* (148).

Stents: Marrying Metals, Medicines, and Marrow

As outlined earlier, atherosclerosis is a highly prevalent disease, and its complications affect hundreds of millions of people worldwide. This enormous market, and the need for effective solutions to restore and maintain the patency of narrowed and occluded atherosclerotic vessels, have driven biotechnology research for several decades. A particularly promising development has been the marriage between mechanical device manufacturers and basic biomedical scientists – producing the recent offspring known as the drug-eluting stent (DES).

A stent is composed of a thin and initially collapsed wire network designed to open as a cylindrical tube inside the lumen in which it is deployed. This inflation at very high pressures presses the stent circumferentially out against the occluding plaque and/or thrombus in the vessel. This restores blood flow to the target tissue, with the stent's tensile strength preventing elastic recoil or vasospasm. Initially, the use of bare metal stents was common; however, early thrombotic closure within the stent and late in-stent restenosis were not infrequent problems. Although the development of potent antiplatelet agents such as GpIIb/IIIa antagonists (149) and ADP receptor antagonists (150) greatly reduced the former, in-stent restenosis remained the Achilles' heel of this technology. Consequently, variations on the simple wire stent were explored to prevent the malignant VSMC proliferation that characterized this particularly troublesome lesion.

Utilizing radiotherapy, stents themselves were made radioactive by direct ion implantation of ^{32}P. At low doses (0.14 μCi), radioactivity may prevent VSMC proliferation, as shown in a pig model of coronary disease (151). However, the beneficial effect of low-dose radiotherapy appeared to be limited to short-term studies of up to 28 days postimplantation (151). When the experimental model was followed for 6 months after stent implantation, studies showed that higher doses of radiotherapy (24–48 μCi) were more effective, with the caveat that the stented area displayed delayed healing in the form of incomplete endothelialization (152). Radiotherapy also has been shown to induce medial thinning and adventitial scarring in both the human and the pig (153). Nevertheless, the subsequent use of transient intra-arterial radiation or "brachytherapy" by means of a beta emitter brought transiently to the site of a treated in-stent restenosis has shown sustained efficacy over 2 years of follow-up and remains popular in some centers (154,155). However, one study comparing radiotherapy with DES in patients with in-stent restenosis suggests the latter to be a better choice (156).

DES for the treatment of in-stent restenosis, and first-time coronary artery lesions prone to restenosis, has become the subject of several clinical studies. Commonly used drugs are paclitaxel, which prevents VSMC proliferation and migration in vivo (157); rapamycin, which has antiproliferative and anti-inflammatory effects in humans after angioplasty and has been shown to repress proinflammatory and proadhesion molecule expression (158); and sirolimus, which inhibits VSMC proliferation by blocking cell cycle progression (86). Clinical trials have shown beneficial effects of paclitaxel-eluting stents 9 months after implantation (159), and when multiple, overlapping stents are used to cover a large area, favorable effects have been observed up to 2 years postimplantation (160).

A new and exciting technology has emerged involving the use of stents that capture EPCs. Stents coated with antihuman CD34 antibodies trap circulating EPCs, which attach to the stent and subsequently form a functional endothelial layer over the stented area. These stents offer an advantage over the use of pharmacological agents in that by producing a new, functional EC layer, the vasculature is essentially rebuilt, whereas pharmacotherapy is limited by the bioavailability of the drug. Results of the first clinical trial using this stent have shown that this is a feasible and promising technique for stent-based treatment of coronary artery disease (161).

The Role of Vascular Smooth Muscle Cells in Hypertension

Approximately one in every three Americans has high blood pressure (91), which is defined as a systolic pressure of 140 mm Hg or higher and diastolic blood pressure of 90 mm Hg or higher. Hypertension affects more men than women until the age of 55, after which there is a higher prevalence of women with hypertension (91). Of those with this condition, as many as 30% are unaware and 25% are on medication but do not have optimal blood pressure control (91).

Many of the mechanisms underlying hypertension remain unknown. However, there have been a vast number of basic and clinical studies, each focused on a particular "mediator" of the disease. The initiating factors can vary from smoking and EC dysfunction, to pregnancy, renal dysfunction, genetic diseases, and so on. One common theme in hypertension is that although the initiating mechanism may be variable, progression of the disease is dependent on VSMC function and structure. Specifically, remodeling responses, hypertrophy, and contractile responses of the vessel wall all rely on VSMC biology. Therefore, studies on the intricate regulation of contraction, relaxation, and regulated growth of VSMCs will be central to therapeutic breakthroughs in this disease.

In the vasculature, shear stress, acetylcholine, and other factors induce Ca^{2+} release in EC. This in turn activates endothelial NOS (eNOS), which then produces nitric oxide (NO), otherwise known as endothelial-derived relaxing factor (EDRF). This molecule can then diffuse to underlying VSMCs, activate soluble guanylate cyclase (sGC) to produce cyclic GMP (cGMP), and mediate VSMC and overall vessel relaxation (162). Nitroglycerin, an age-old therapy for coronary artery disease, is a NO-donor that also stimulates sGC to produce cGMP. Both endogenous and exogenous NO-stimulated cGMP relax VSMCs through two main pathways: first, by decreasing $[Ca^{2+}]_I$, and second, by altering the calcium sensitivity of the contractile apparatus (163). Once formed, cGMP activates cGMP-dependent protein kinase (PKG). Via phosphorylation, this enzyme inactivates membrane calcium channels, increases potassium channel activity, decreases IP3-mediated activity of IP3R, and deactivates PLC, which reduces IP3 production. PKG acts to reduce the sensitivity of the cell's contractile system by increasing the activity of MLCP (163).

As intimated earlier, the Rho-GTPase system is important in regulating the contractile, migratory, and phenotypic status of SMC. The Rho family of proteins carries out its signaling mechanisms by binding and activating Rho-kinases (ROCK). These effector enzymes regulate, again by phosphorylation, many important molecules in the cell. ROCK phosphorylate MLC to induce nominal direct contractile effects, however, more significant is their inhibitory phosphorylation of MLCP, which dramatically increases contractility of the VSMC. In essence, the Rho/ROCK pathway increases the calcium sensitivity of VSMCs such that lower $[Ca^{2+}]_i$ are required for contraction. ROCK activity also can increase actin stress fiber and focal adhesion formation and mediate cell migration by reorganization of the actin cytoskeleton, and it also plays a role in the phenotypic modulation of VSMCs (64).

Sphingosine-1-phosphate (S1P) is released from activated platelets in the blood. This molecule acts on specialized S1P receptors (endothelial differentiation gene [Edg] receptors) located on both ECs and VSMCs. S1P has been shown to be involved in the maintenance of ECs barrier integrity, migration, and proliferation (164). Depending on which receptor subtype S1P is acting on, it can either induce or inhibit VSMC proliferation and migration (164). S1P has been shown to be an upstream mediator of Rho-GTPase pathways (165), activation of PLC/IP3-mediated calcium release from the SR, and phosphorylation of MLC by ROCK, subsequently sustaining VSMC contraction (166).

As mentioned earlier, the causes of hypertension are variable and scientists engaged in this field argue the respective importance of their particular pathways of interest. Vascular biologists believe that the root problem stems from an imbalance between contracting and relaxing forces in the ECs and VSMCs; endocrine and renal experts will implicate the renin–angiotensin system or disproportional salt and water retention in the kidneys; whereas neurophysiologists will implicate neurohormones and the sympathetic nervous system. On these specific topics, the reader is referred to recent reviews (3,167–170). Finally, there is an unequivocal genetic component to human hypertension, and animal models such as the spontaneously hypertensive rat, syngeneic rat strains, or various transgenic mouse models have implicated many individual genes or sets of genes as crucial contributors (69,171–175). Although it is clear that no single mechanism acts in isolation to perpetuate hypertension, the participation of the VSMC at either the systemic or organ-specific level is undeniable.

THERAPY

Common Cardiovascular Drug Treatments that Directly Target Vascular Smooth Muscle Cells

HMG-CoA reductase inhibitors, otherwise known as statins, are often prescribed to patients with high cholesterol or those at a significant risk for cardiovascular disease. Initially used for their lipid-lowering effects, statins are now recognized as multifunctional and of benefit in a variety of conditions. In addition to lowering plasma LDL and increasing high-density lipoprotein (HDL) concentrations, statins have been shown to inhibit VSMC proliferation, act as anti-inflammatory agents, inhibit platelet function, and improve endothelial function (176).

Angiotensin-converting enzyme inhibitors (ACEI) block the ability of ACE to convert angiotensin (Ang)-I to Ang-II, through its activation of the angiotensin receptors (AT1 and AT2), mediates potent vasoconstriction but also has proliferative effects on VSMCs (177). Ang-II stimulates the expression of other mitogens, adhesion molecules, chemotactic proteins, cytokines, oxidized LDL, and MMPs (178), all of which contribute to atherosclerotic plaque instability. An additional therapeutic effect of ACEI is their ability to prevent the ACE-mediated degradation of bradykinin (BK), an endogenous activator of eNOS, thereby capturing all the salutary benefits of NO. Indeed, ACEI have been commonly prescribed to patients with cardiovascular disease now for many years. ACEI have been shown to be beneficial by increasing plaque stability and decreasing inflammation in the vascular wall (178), and these biological findings have translated into significant clinical benefits (179).

Angiotensin-receptor blockers (ARBs) work in a slightly different manner but have similar effects to those of ACEI. The ARB competitively inhibit AT-1 and/or AT-2 receptor activation, achieving all the benefits (and side effects) of blocking Ang-II, but not those of preserving BK (180). ARB treatment can correct small artery remodeling responses in patients with hypertension (181).

Calcium channel blockers (CCBs) are often prescribed for their negative inotropic and chronotropic effects on the heart. As described earlier, cardiac muscle relies on calcium entry to induce contraction, at a rate that is carefully regulated by pacemaker cells. Cardiac contractility, heart rate (and thus myocardial oxygen demand), and arrhythmias can all be treated with CCB. Although pharmacomechanical coupling appears to be the dominant contractile pathway in VSMCs, these myocytes also harbor VDCC, and CCB may achieve their vasodilatory and antihypertensive properties in part by limiting calcium entry. CCB inhibit voltage-dependent calcium entry via T-type and L-type channels, which results in vasorelaxation and are certainly effective agents for the treatment of hypertension (182).

DIAGNOSTICS

As reviewed earlier in this chapter, the initiation of vascular disease is often due to endothelial dysfunction, disordered ECs signaling, and abnormal VSMC responses. Eventually, VSMCs receiving prolonged and specific pathological signals may transform to a migratory, synthetic, and proliferative phenotype that underlies vascular disease.

Currently, noninvasive medical diagnostic tools enabling molecular analysis of ECs or VSMCs at early or late stages of disease are few and far between. This is a fundamental limitation that requires attention, as the clinical implementation of much basic science knowledge will be difficult to achieve without such tools. Although still considered to be “in development,” physicians may soon be armed with technologies capable of detecting some characteristic molecular perturbations associated with abnormal hemodynamic forces, oxidative stress, adhesion molecule expression, and vascular cell apoptosis (see later). These developments may permit earlier and more accurate diagnoses, allowing more specific regenerative or reparative approaches. Having said that, there exists a critical need for a greater number and variation of probes designed for the molecular imaging of ECs and VSMCs. To become clinically useful, these probes will have to be nontoxic, specific, and detectable with high-resolution imaging. Meanwhile, some of the currently available noninvasive and invasive techniques utilized for studies of VSMCs are briefly reviewed.

Computed Tomography

This radiographic technique has undergone rapid redevelopment with the arrival of 64-slice cameras that can image vessel walls and provide angiographic images of remarkable clarity (183). This technique can also measure calcification in the aorta and coronary artery, a finding that correlates with obstructive coronary artery disease, and to a lesser extent with cardiac event rates (184).

Magnetic Resonance Imaging

The nonradiographic imaging offered by MRI has revolutionized many areas of clinical medicine and surgery. Its potential for application as a molecular diagnostic tool for the vasculature is reviewed elsewhere in this volume (see Chapter 176). With respect to VSMC imaging, MRI can characterize a carotid atherosclerotic plaque, defining different components of the plaque, namely the fibrous cap, necrotic core, and calcium deposition (185). This imaging technique also can quantify size or volume of plaque, detect breaks in the fibrous cap, and identify thrombotic events within the plaque (185). Of particular interest is the possibility of coupling MRI with ferromagnetic tracers composed of nanoparticles linked to antibodies targeting specific cell surface markers. Such a strategy has been used, for example, to define the expression of VCAM-1 in ECs of atherosclerotic animals (186).

Single Photon Emission Computed Tomography and Positron Emission Tomography

Single photon emission computed tomography (SPECT) and the much-higher-energy coincidence-imaging made possible with positron emission tomography (PET) have been mainstays for the noninvasive diagnosis of flow-limiting coronary stenosis and viable myocardium. However, their potential as platforms for molecular imaging has only just begun to be realized (187). Radionuclide tracers coupled to anti–Annexin-V antibodies have been shown to detect apoptosis in the myocardium (188), and similar "hot spots" in the vessel wall have correlated with inflammatory transformation of atherosclerotic plaque (189,190).

Gluteal Artery Biopsy

This powerful technique has been successfully exploited in human subjects with hypertension to explore the vasomotor physiology and molecular biology of small arteries in this condition (71). After a 1 cm horizontal incision is made in the upper external gluteal quadrant, a 1 × 0.5 × 0.5 cm cube of superficial subcutaneous tissue and immediately underlying fat are excised. In this specimen, several small arterioles can be microdissected and employed for perfusion myography (191) or biological and biochemical analyses. No skin is removed, facilitating repeat procedures and rendering the intervention small, minimally invasive, and well-tolerated one. Patients describe little residual discomfort, and practically no complications occur (71).

This technique is particularly useful, because larger arteries may not show changes in lumen diameter due to eutrophic remodeling. By contrast, small arteries typically exhibit signs of inward remodeling. As such, the effectiveness of antihypertensive medicines on arterial remodeling can be directly monitored (191) and correlations made between vessel structure and function.

KEY POINTS

- VSMCs are the muscle behind the EC, the two cell types existing in a symbiotic relationship that begins in development and continues through to disease.
- VSMCs are unique muscle cells capable of migrating, contracting, proliferating, synthesizing, and destabilizing their surrounding matrix components on remote cues from circulating plasma proteins, ECs, platelets, and inflammatory cells.
- The contractile phenotype of VSMCs has elaborate spatially and functionally distinct calcium stores that are tightly regulated and used to effect a variety of calcium-dependent functions, the foremost of which is pharmacomechanical coupling.
- The proliferative phenotype of VSMCs has equally elaborate cascades of transcription factors that regulate calcium homeostatic mechanisms, which ultimately mediate calcium-sensitive cell cycle proteins and cell cycle progression.
- Both phenotypes of VSMCs participate in disease, the contractile phenotype mediating abnormal vasomotor control in hypertension, and the proliferative phenotype mediating the remodeling of hypertension, atherosclerosis, and restenosis.
- By targeting VSMC pathobiology, the marriage of implantable stent technology with radiation, anti-inflammatory and antiproliferative medicines, and EC capture has significantly altered the long-term success of angioplasty procedures for restenosis and atherosclerosis.
- Most of the most commonly prescribed drugs in cardiovascular medicine have important mechanisms of beneficial action via their effects on the VSMCs.
- Diagnostic advances including molecular imaging and small-tissue biopsies with attendant studies on vessel structure and function will be the windows through which the next generation of insights in vascular health and disease will be observed.

REFERENCES

1 Ross R. Atherosclerosis – an inflammatory disease. *N Engl J Med.* 1999;340:115–126.

2 Owens G, Kumar M, Wamhoff B. Molecular regulation of vascular smooth muscle cell differentiation in development and disease. *Physiol Rev.* 2004;84:767–801.

3 Vanhoutte PM, Feletou M, Taddei S. Endothelium-dependent contractions in hypertension. *Br J Pharmacol.* 2005;144:449–458.

4 Johnson-Leger C, Imhof BA. Forging the endothelium during inflammation: pushing at a half-open door? *Cell Tissue Res.* 2003;314:93–105.

5 Choi J, Enis DR, Koh KP, et al. T lymphocyte-endothelial cell interactions. *Annu Rev Immunol.* 2004;22:683–709.

6 Licata R. Arterial and arteriolar systems: embryology and gross, microscopic and submicroscopic anatomy. In: Abramson D, ed. *Blood Vessels and Lymphatics.* New York: Academic Press; 1962: 3–25.

7 Sjolund M, Madsen K, von der Mark K, Thyberg J. Phenotype modulation in primary cultures of smooth-muscle cells from rat aorta. Synthesis of collagen and elastin. *Differentiation.* 1986;32: 173–180.

8 Sartore S, Chiavegato A, Faggin E, et al. Contribution of adventitial fibroblasts to neointima formation and vascular remodelling. *Circ Res.* 2001;89:1111.

9 Yoshida T, Owens GK. Molecular determinants of vascular smooth muscle cell diversity. *Circ Res.* 2005;96:280–291.

10 Fukushige T, Krause M. The myogenic potency of HLH-1 reveals wide-spread developmental plasticity in early C. elegans embryos. *Development.* 2005;132:1795–1805.

11 Bartnik E, Osborn M, Weber K. Intermediate filaments in muscle and epithelial cells of nematodes. *J Cell Biol.* 1986;102:2033–2041.

12 Paternostro G, Vignola C, Bartsch DU, et al. Age-associated cardiac dysfunction in *Drosophila melanogaster. Circ Res.* 2001;88: 1053–1058.

13 Gao Z, Joseph E, Ruden DM, Lu X. Drosophila Pkd2 is haploid-insufficient for mediating optimal smooth muscle contractility. *J Biol Chem.* 2004;279:14225–14231.

14 Baker PW, Kelly Tanaka K, Klitgord N, Cripps RM. Adult myogenesis in *Drosophila melanogaster* can proceed independently of myocyte enhancer factor-2. *Genetics.* 2005;170(4):1747–1759. Epub 2005 Jun 14.

15 Risau W, Flamme I. Vasculogenesis. *Annu Rev Cell Dev Biol.* 1995;11:73–91.

16 Distler JH, Hirth A, Kurowska-Stolarska M, et al. Angiogenic and angiostatic factors in the molecular control of angiogenesis. *Q J Nucl Med.* 2003;47:149–161.

17 Ramsauer M, D'Amore PA. Getting Tie(2)d up in angiogenesis. *J Clin Invest.* 2002;110:1615–1617.

18 Landerholm TE, Dong XR, Lu J, et al. A role for serum response factor in coronary smooth muscle differentiation from proepicardial cells. *Development.* 1999;126:2053–2062.

19 Li WE, Waldo K, Linask KL, et al. An essential role for connexin43 gap junctions in mouse coronary artery development. *Development.* 2002;129:2031–2042.

20 Majesky MW. Vascular smooth muscle diversity: insights from developmental biology. *Curr Atheroscler Rep.* 2003;5:208–213.

21 Majesky MW. Development of coronary vessels. *Curr Top Dev Biol.* 2004;62:225–259.

22 Bergwerff M, Verberne ME, DeRuiter MC, et al. Neural crest cell contribution to the developing circulatory system: implications for vascular morphology? *Circ Res.* 1998;82:221–231.

23 Li J, Zhu X, Chen M, et al. Myocardin-related transcription factor B is required in cardiac neural crest for smooth muscle differentiation and cardiovascular development. *Proc Natl Acad Sci USA.* 2005;102:8916–8921.

24 Shalaby F, Rossant J, Yamaguchi TP, et al. Failure of blood-island formation and vasculogenesis in Flk-1-deficient mice. *Nature.* 1995;376:62–66.

25 Carmeliet P, Ferreira V, Breier G, et al. Abnormal blood vessel development and lethality in embryos lacking a single VEGF allele. *Nature.* 1996;380:435–439.

26 Ferrara N, Carver-Moore K, Chen H, et al. Heterozygous embryonic lethality induced by targeted inactivation of the VEGF gene. *Nature.* 1996;380:439–442.

27 Armulik A, Abramsson A, Betsholtz C. Endothelial/pericyte interactions. *Circ Res.* 2005;97:512–523.

28 Li S, Wang DZ, Wang Z, et al. The serum response factor coactivator myocardin is required for vascular smooth muscle development. *Proc Natl Acad Sci USA.* 2003;100:9366–9370.

29 Oettgen P. Transcriptional regulation of vascular development. *Circ Res.* 2001;89:380–388.

30 Drake CJ, Fleming PA. Vasculogenesis in the day 6.5 to 9.5 mouse embryo. *Blood.* 2000;95:1671–1679.

31 Wang HU, Chen ZF, Anderson DJ. Molecular distinction and angiogenic interaction between embryonic arteries and veins revealed by ephrin-B2 and its receptor Eph-B4. *Cell.* 1998;93: 741–753.

32 Li L, Miano JM, Mercer B, Olson EN. Expression of the SM22alpha promoter in transgenic mice provides evidence for distinct transcriptional regulatory programs in vascular and visceral smooth muscle cells. *J Cell Biol.* 1996;132:849–859.

33 Moessler H, Mericskay M, Li Z, et al. The SM 22 promoter directs tissue-specific expression in arterial but not in venous or visceral smooth muscle cells in transgenic mice. *Development.* 1996;122:2415–2425.

34 Dor Y, Djonov V, Abramovitch R, et al. Conditional switching of VEGF provides new insights into adult neovascularization and pro-angiogenic therapy. *EMBO J.* 2002;21:1939–1947.

35 Dor Y, Djonov V, Keshet E. Induction of vascular networks in adult organs: implications to proangiogenic therapy. *Ann NY Acad Sci.* 2003;995:208–216.

36 Kingston PA, Sinha S, David A, et al. Adenovirus-mediated gene transfer of a secreted transforming growth factor-beta type II receptor inhibits luminal loss and constrictive remodeling after coronary angioplasty and enhances adventitial collagen deposition. *Circulation.* 2001;104:2595–2601.

37 Holtwick R, Gotthardt M, Skryabin B, et al. Smooth muscle-selective deletion of guanylyl cyclase-A prevents the acute but not chronic effects of ANP on blood pressure. *Proc Natl Acad Sci USA.* 2002;99:7142–7147.

38 Kingston PA, Sinha S, Appleby CE, et al. Adenovirus-mediated gene transfer of transforming growth factor-beta3, but not transforming growth factor-beta1, inhibits constrictive remodeling and reduces luminal loss after coronary angioplasty. *Circulation.* 2003;108:2819–2825.

39 You X, Mungrue I, Kalair W, et al. Conditional expression of a dominant-negative c-Myb in vascular smooth muscle cells inhibits arterial remodeling after injury. *Circ Res.* 2003;92:314–321.

40 Martini F, Timmons M, McKinley M. *Human Anatomy*. Upper Saddle River, NJ: Prentice Hall, 2000.

41 Goetsch SC, Martin CM, Embree LJ, Garry DJ. Myogenic progenitor cells express filamin C in developing and regenerating skeletal muscle. *Stem Cells Dev*. 2005;14:181–187.

42 Dezawa M, Ishikawa H, Itokazu Y, et al. Bone marrow stromal cells generate muscle cells and repair muscle degeneration. *Science*. 2005;309:314–317.

43 Urbanek K, Torella D, Sheikh F, et al. Myocardial regeneration by activation of multipotent cardiac stem cells in ischemic heart failure. *Proc Natl Acad Sci USA*. 2005;102:8692–8697.

44 Dawn B, Stein AB, Urbanek K, et al. Cardiac stem cells delivered intravascularly traverse the vessel barrier, regenerate infarcted myocardium, and improve cardiac function. *Proc Natl Acad Sci USA*. 2005;102:3766–3771.

45 Pasumarthi KB, Nakajima H, Nakajima HO, et al. Targeted expression of cyclin D2 results in cardiomyocyte DNA synthesis and infarct regression in transgenic mice. *Circ Res*. 2005;96:110–118.

46 Koyanagi M, Haendeler J, Badorff C, et al. Non-canonical Wnt signaling enhances differentiation of human circulating progenitor cells to cardiomyogenic cells. *J Biol Chem*. 2005;280:16838–16842.

47 Koyanagi M, Brandes RP, Haendeler J, et al. Cell-to-cell connection of endothelial progenitor cells with cardiac myocytes by nanotubes: a novel mechanism for cell fate changes? *Circ Res*. 2005;96:1039–1041.

48 Zhang S, Wang D, Estrov Z, et al. Both cell fusion and transdifferentiation account for the transformation of human peripheral blood CD34-positive cells into cardiomyocytes in vivo. *Circulation*. 2004;110:3803–3807.

49 Sata M, Saiura A, Kunisato A, et al. Hematopoietic stem cells differentiate into vascular cells that participate in the pathogenesis of atherosclerosis. *Nat Med*. 2002;8:403–409.

50 Simper D, Stalboerger PG, Panetta CJ, et al. Smooth muscle progenitor cells in human blood. *Circulation*. 2002;106:1199–1204.

51 Caplice NM, Bunch TJ, Stalboerger PG, et al. Smooth muscle cells in human coronary atherosclerosis can originate from cells administered at marrow transplantation. *Proc Natl Acad Sci USA*. 2003;100:4754–4759.

52 Hu Y, Zhang Z, Torsney E, et al. Abundant progenitor cells in the adventitia contribute to atherosclerosis of vein grafts in ApoE-deficient mice. *J Clin Invest*. 2004;113:1258–1265.

53 Zernecke A, Schober A, Bot I, et al. SDF-1alpha/CXCR4 axis is instrumental in neointimal hyperplasia and recruitment of smooth muscle progenitor cells. *Circ Res*. 2005;96:784–791.

54 Caplice NM, Doyle B. Vascular progenitor cells: origin and mechanisms of mobilization, differentiation, integration, and vasculogenesis. *Stem Cells Dev*. 2005;14:122–139.

55 Somlyo A, Somlyo A. Signal transduction and regulation in smooth muscle. *Nature*. 1994;372:231–236.

56 Adachi T, Weisbrod RM, Pimentel DR, et al. S-Glutathiolation by peroxynitrite activates SERCA during arterial relaxation by nitric oxide. *Nat Med*. 2004;10:1200–1207.

57 Furukawa K, Tawada Y, Shigekawa M. Regulation of the plasma membrane $Ca2^{+}$ pump by cyclic nucleotides in cultured vascular smooth muscle cells. *J Biol Chem*. 1988;263:8058–8065.

58 Somlyo A, Somlyo A. Electromechanical and pharmacomechanical coupling in vascular smooth muscle. *J Pharm Exp Ther*. 1968;159:129–145.

59 Tribe RM, Borin ML, Blaustein MP. Functionally and spatially distinct $Ca2^{+}$ stores are revealed in cultured vascular smooth muscle cells. *Proc Natl Acad Sci USA*. 1994;91:5908–5912.

60 Boittin FX, Galione A, Evans AM. Nicotinic acid adenine dinucleotide phosphate mediates $Ca2^{+}$ signals and contraction in arterial smooth muscle via a two-pool mechanism. *Circ Res*. 2002;91:1168–1175.

61 Fedoryak OD, Searls Y, Smirnova IV, et al. Spontaneous $Ca2^{+}$ oscillations in subcellular compartments of vascular smooth muscle cells rely on different $Ca2^{+}$ pools. *Cell Res*. 2004;14:379–388.

62 Marks A. Intracellular calcium-release channels: regulators of cell life and death. *Am J Physiol*. 1997;272:H597–H605.

63 Berridge MJ. Unlocking the secrets of cell signaling. *Annu Rev Physiol*. 2005;67:1–21.

64 Riento K, Ridley A. Rocks: multifunctional kinases in cell behaviour. *Nat Rev Mol Cell Biol*. 2003;4:446–456.

65 Shimokawa H, Takeshita A. Rho-Kinase is an important therapeutic target in cardiovascular medicine. *Arterioscler Thromb Vasc Biol*. 2005;25(9):1767–75. Epub 2005 Jul 7.

66 Nagaoka T, Fagan KA, Gebb SA, et al. Inhaled Rho kinase inhibitors are potent and selective vasodilators in rat pulmonary hypertension. *Am J Respir Crit Care Med*. 2005;171:494–499.

67 Kishi T, Hirooka Y, Masumoto A, et al. Rho-kinase inhibitor improves increased vascular resistance and impaired vasodilation of the forearm in patients with heart failure. *Circulation*. 2005;111:2741–2747.

68 Heximer S, Knutsen R, Sun X, et al. Hypertension and prolonged vasoconstrictor signaling in RGS2-deficient mice. *J Clin Invest*. 2003;111:1259.

69 Gros R, Afroze T, You X, et al. Plasma membrane calcium ATPase overexpression in arterial smooth muscle increases vasomotor responsiveness and blood pressure. *Circ Res*. 2003; 93:614–621.

70 Fanchaouy M, Serir K, Meister JJ, et al. Intercellular communication: role of gap junctions in establishing the pattern of ATP-elicited $Ca2^{+}$ oscillations and $Ca2^{+}$-dependent currents in freshly isolated aortic smooth muscle cells. *Cell Calcium*. 2005; 37:25–34.

71 Schiffrin E, Hayoz D. How to assess vascular remodelling in small and medium-sized muscular arteries in humans. *J Hypertens*. 1997;5:571–584.

72 Rodbard S. Effect of mechanical forces on structure of vascular system. In: Abramson D, ed. *Blood Vessels and Lymphatics*. New York: Academic Press; 1962:41–61.

73 Roman M, Saba P, Pini R, et al. Parallel cardiac and vascular adaptation in hypertension. *Circulation*. 1992;86.

74 Li S, Fan Y, Chow L, et al. Innate diversity of adult human arterial smooth muscle cells. *Circ Res*. 2001;89:517.

75 Madsen CS, Regan CP, Hungerford JE, et al. Smooth muscle-specific expression of the smooth muscle myosin heavy chain gene in transgenic mice requires 5'-flanking and first intronic DNA sequence. *Circ Res*. 1998;82:908–917.

76 Regan CP, Manabe I, Owens GK. Development of a smooth muscle-targeted cre recombinase mouse reveals novel insights regarding smooth muscle myosin heavy chain promoter regulation. *Circ Res*. 2000;87:363–369.

77 Manabe I, Owens GK. The smooth muscle myosin heavy chain gene exhibits smooth muscle subtype-selective modular regulation in vivo. *J Biol Chem*. 2001;276:39076–39087.

78 Nobuyoshi M, Kimura T, Ohishi H, et al. Restenosis after percutaneous transluminal coronary angioplasty: pathologic observations in 20 patients. *J Am Coll Cardiol.* 1991;17:433–439.
79 Ross R. The pathogenesis of atherosclerosis: a perspective for the 1990s. *Nature.* 1993;362:801–809.
80 Strauss BH, Umans VA, van Suylen RJ, et al. Directional atherectomy for treatment of restenosis within coronary stents: clinical, angiographic and histologic results. *J Am Coll Cardiol.* 1992;20:1465–1473.
81 Strauss BH, Serruys PW, Bertrand ME, et al. Quantitative angiographic follow-up of the coronary Wallstent in native vessels and bypass grafts (European experience – March 1986 to March 1990). *Am J Cardiol.* 1992;69:475–481.
82 Hoffmann R, Mintz GS, Dussaillant GR, et al. Patterns and mechanisms of in-stent restenosis. A serial intravascular ultrasound study. *Circulation.* 1996;94:1247–1254.
83 Ajani AE, Kim H, Waksman R. Clinical trials of vascular brachytherapy for in-stent restenosis: update. *Cardiovasc Radiat Med.* 2001;2:107–113.
84 Teirstein PS, Kuntz RE. New frontiers in interventional cardiology: intravascular radiation to prevent restenosis. *Circulation.* 2001;104:2620–2626.
85 Hiatt BL, Ikeno F, Yeung AC, Carter AJ. Drug-eluting stents for the prevention of restenosis: in quest for the Holy Grail. *Catheter Cardiovasc Interv.* 2002;55:409–417.
86 Suzuki T, Kopia G, Hayashi S, et al. Stent-based delivery of sirolimus reduces neointimal formation in a porcine coronary model. *Circulation.* 2001;104:1188–1193.
87 Nilsson J, Thyberg J. Fine structure of arterial smooth muscle cells cultured in the presence of whole blood serum or plasma-derived serum. *Cell Tissue Res.* 1982;223:87–99.
88 Frid M, Aldashev A, Dempsey E, Stenmark K. Smooth muscle cells isolated from discrete compartments of the mature vascular media exhibit unique phenotypes and distinct growth capabilities. *Circ Res.* 1997;81:940–952.
89 Regan C, Adam P, Madsen C, Owens G. Molecular mechanisms of decreased smooth muscle differentiation marker expression after vascular injury. *J Clin Invest.* 2000;106:9.
90 Statistics NCfH. Leading Causes of Death in the US. http://www.cdc.gov/nchs/fastats/lcod.htm 2002.
91 AHA. Heart Disease and Stroke Statistics – 2005 Update. http://www.americanheart.org/downloadable/heart/1105390918119HDSStats2005Update.pdf 2005.
92 Iiyama K, Hajra L, Iiyama M, et al. Patterns of vascular cell adhesion molecule-1 and intercellular adhesion molecule-1 expression in rabbit and mouse atherosclerotic lesions and at sites predisposed to lesion formation. *Circ Res.* 1999;85:199–207.
93 Blank R, Thompson M, Owens G. Cell cycle versus density dependence of smooth muscle alpha actin expression in cultured rat aortic smooth muscle cells. *J Cell Biol.* 1988;107:299–306.
94 Corjay M, Thompson M, Lynch K, Owens G. Differential effect of platelet-derived growth factor- versus serum-induced growth on smooth muscle alpha-actin and nonmuscle beta-actin mRNA expression in cultured rat aortic smooth muscle cells. *J Biol Chem.* 1989;264:10501–10506.
95 Holycross B, Blank R, Thompson M, et al. Platelet-derived growth factor-BB-induced suppression of smooth muscle cell differentiation. *Circ Res.* 1992;71:1525–1532.
96 Hautmann M, Madsen C, Owens G. A transforming growth factor beta (TGFbeta) control element drives TGFbeta-induced stimulation of smooth muscle alpha-actin gene expression in concert with two CArG elements. *J Biol Chem.* 1997;272:10948–10956.
97 Husain M, Simons M. Vascular antisense targets: c-myc, c-myb and PCNA. In: LeRoy E, Rabbani MD, eds. *Applications of Antisense Therapies to Restenosis.* Boston: Kluwer Academic, 1999:71–98.
98 Husain M, Bein K, Jiang L, et al. c-Myb-dependent cell cycle progression and $Ca2^+$ storage in cultured vascular smooth muscle cells. *Circ Res.* 1997;80:617–626.
99 Simons M, Rosenberg R. Antisense nonmuscle myosin heavy chain and c-myb oligonucleotides suppress smooth muscle cell proliferation in vitro. *Circ Res.* 1992;71:835–843.
100 Afroze T, Yang L, Wang C, et al. Calcineurin-independent regulation of plasma membrane $Ca2^+$ ATPase-4 in the vascular smooth muscle cell cycle. *Am J Physiol Cell Physiol.* 2003;285: C88–C95.
101 Husain M, Jiang L, See V, et al. Regulation of vascular smooth muscle cell proliferation by plasma membrane $Ca(2^+)$-ATPase. *Am J Physiol.* 1997;272:C1947–C1959.
102 Cogswell JP, Cogswell PC, Kuehl WM, et al. Mechanism of c-myc regulation by c-Myb in different cell lineages. *Mol Cell Biol.* 1993;13:2858–2869.
103 Travali S, Ferber A, Reiss K, et al. Effect of the myb gene product on expression of the PCNA gene in fibroblasts. *Oncogene.* 1991;6:887–894.
104 Venturelli D, Travali S, Calabretta B. Inhibition of T-cell proliferation by a MYB antisense oligomer is accompanied by selective down-regulation of DNA polymerase alpha expression. *Proc Natl Acad Sci USA.* 1990;87:5963–5967.
105 Ku DH, Wen SC, Engelhard A, et al. c-myb transactivates cdc2 expression via Myb binding sites in the 5′-flanking region of the human cdc2 gene [published erratum appears in *J Biol Chem* 1993 Jun 15;268(17):13010]. *J Biol Chem.* 1993;268:2255–2259.
106 Short AD, Bian J, Ghosh TK, et al. Intracellular $Ca2^+$ pool content is linked to control of cell growth. *Proc Natl Acad Sci USA.* 1993;90:4986–4990.
107 Waldron RT, Short AD, Meadows JJ, et al. Endoplasmic reticulum calcium pump expression and control of cell growth. *J Biol Chem.* 1994;269:11927–11933.
108 Waldron RT, Short AD, Gill DL. Thapsigargin-resistant intracellular calcium pumps. Role in calcium pool function and growth of thapsigargin-resistant cells. *J Biol Chem.* 1995;270: 11955–11961.
109 Graber MN, Alfonso A, Gill DL. $Ca2^+$ pools and cell growth: arachidonic acid induces recovery of cells growth-arrested by $Ca2^+$ pool depletion. *J Biol Chem.* 1996;271:883–888.
110 Gill DL, Waldron RT, Rys-Sikora KE, et al. Calcium pools, calcium entry, and cell growth. *Biosci Rep.* 1996;16:139–157.
111 Miyazaki S, Yuzaki M, Nakada K, et al. Block of $Ca2^+$ wave and $Ca2^+$ oscillation by antibody to the inositol 1,4,5-trisphosphate receptor in fertilized hamster eggs [published erratum appears in *Science* 1992 Oct 9;258(5080):following 203]. *Science.* 1992;257:251–255.
112 Berridge MJ. Inositol trisphosphate and calcium signalling. *Nature.* 1993;361:315–325.
113 Ciapa B, Pesando D, Wilding M, Whitaker M. Cell-cycle calcium transients driven by cyclic changes in inositol trisphosphate levels. *Nature.* 1994;368:875–878.
114 Takuwa N, Zhou W, Kumada M, Takuwa Y. Involvement of intact inositol-1,4,5-trisphosphate-sensitive $Ca2^+$ stores in cell

cycle progression at the G1/S boundary in serum-stimulated human fibroblasts. *FEBS Lett.* 1995;360:173–176.
115 Whitaker M. Regulation of the cell division cycle by inositol trisphosphate and the calcium signaling pathway. *Adv Second Messenger Phosphoprotein Res.* 1995;30:299–310.
116 George S, Johnson J, Angelini G, Jeremy J. Short-term exposure to thapsigargin inhibits neointima formation in human saphenous vein. *ATVB.* 1997;17:2500–2506.
117 Moses S, Dreja K, Lindqvist A, el al. Smooth muscle cell response to mechanical injury involves intracellular calcium release and ERK1/ERK2 phosphorylation. *Exp Cell Res.* 2001;269:88–96.
118 Afroze T, Husain M. Increased expression of the IP3R may contribute to the rising cytosolic $Ca2^{+}$ concentrations of proliferating SMC. *Can J Cardiol.* 1999;15:107D.
119 Afroze T, Husain M. c-Myb-dependent activation of the IP3R1 promoter. *FASEB J.* 2001;15:A80.
120 Afroze T, Husain M. c-Myb-binding sites mediate G(1)/S-associated repression of the plasma membrane $Ca(2^{+})$-ATPase-1 promoter. *J Biol Chem.* 2000;275:9062–9069.
121 Mizuguchi G, Kanei-Ishii C, Takahashi T, et al. c-Myb repression of c-erbB-2 transcription by direct binding to the c-erbB-2 promoter. *J Biol Chem.* 1995;270:9384–9389.
122 Afroze T, Husain M. Cell cycle dependent regulation of intracellular calcium concentration in vascular smooth muscle cells: a potential target for drug therapy. *Curr Drug Targets Cardiovasc Haematol Disord.* 2001;1:23–40.
123 Afroze T, Husain M. c-Myb-mediated transcription of multiple $Ca2^{+}$ transporters at the G1/S transition point of vascular smooth muscle cells. In: *Recent Research Developments in Biological Chemistry*. London: Research Signpost, 2002:273–285.
124 Lu KP, Means AR. Regulation of the cell cycle by calcium and calmodulin. *Endocr Rev.* 1993;14:40–58.
125 Berridge MJ. Calcium signalling and cell proliferation. *Bioessays.* 1995;17:491–500.
126 Ekholm SV, Reed SI. Regulation of G(1) cyclin-dependent kinases in the mammalian cell cycle. *Curr Opin Cell Biol.* 2000; 12:676–684.
127 Endicott JA, Noble ME, Tucker JA. Cyclin-dependent kinases: inhibition and substrate recognition. *Curr Opin Struct Biol.* 1999;9:738–744.
128 Colomer J, Lopez-Girona A, Agell N, Bachs O. Calmodulin regulates the expression of cdks, cyclins and replicative enzymes during proliferative activation of human T lymphocytes. *Biochem Biophys Res Commun.* 1994;200:306–312.
129 Choi J, Husain M. Calmodulin-mediated cell cycle regulation: new mechanisms for old observations. *Cell Cycle.* 2006;5: 2183–2186.
130 Choi J, Chiang A, Taulier N, et al. A calmodulin binding site on cyclin E mediates Ca^{2+}-sensitive G1/S transitions in vascular smooth muscle cells. *Circ Res.* 2006;98:1273–1281.
131 Gadson PF, Jr., Dalton ML, Patterson E, et al. Differential response of mesoderm- and neural crest-derived smooth muscle to TGF-beta1: regulation of c-myb and alpha1 (I) procollagen genes. *Exp Cell Res.* 1997;230:169–180.
132 Burgess TL, Fisher EF, Ross SL, et al. The antiproliferative activity of c-myb and c-myc antisense oligonucleotides in smooth muscle cells is caused by a nonantisense mechanism. *Proc Natl Acad Sci USA.* 1995;92:4051–4055.
133 Guvakova MA, Yakubov LA, Vlodavsky I, et al. Phosphorothioate oligodeoxynucleotides bind to basic fibroblast growth factor, inhibit its binding to cell surface receptors, and remove it from low affinity binding sites on extracellular matrix. *J Biol Chem.* 1995;270:2620–2627.
134 Villa AE, Guzman LA, Poptic EJ, et al. Effects of antisense c-myb oligonucleotides on vascular smooth muscle cell proliferation and response to vessel wall injury. *Circ Res.* 1995;76:505–513.
135 Mucenski ML, McLain K, Kier AB, et al. A functional c-myb gene is required for normal murine fetal hepatic hematopoiesis. *Cell.* 1991;65:677–689.
136 Weston KM. The myb genes. *Semin Cancer Biol.* 1990;1:371–382.
137 Gewirtz AM, Anfossi G, Venturelli D, et al. G1/S transition in normal human T-lymphocytes requires the nuclear protein encoded by c-myb. *Science.* 1989;245:180–183.
138 Gewirtz AM, Calabretta B. A c-myb antisense oligodeoxynucleotide inhibits normal human hematopoiesis in vitro. *Science.* 1988;242:1303–1306.
139 Ness SA, Marknell A, Graf T. The v-myb oncogene product binds to and activates the promyelocyte-specific mim-1 gene. *Cell.* 1989;59:1115–1125.
140 Drab M, Haller H, Bychkov R, et al. From totipotent embryonic stem cells to spontaneously contracting smooth muscle cells: a retinoic acid and db-cAMP in vitro differentiation model. *FASEB J.* 1997;11:905–915.
141 Kolodziejska-Baginska K, Yang LL, Nagy A, Husain M. High levels of smooth muscle-specific mRNA expression precede the contractile phenotype during vascular smooth muscle cell differentiation from mouse embryonic stem cells. *Can J Cardiol.* 2003;(Abstract).
142 Kolodziejska-Baginska K, Yang LL, Nagy A, Husain M. Transcriptional analysis of vascular smooth muscle cell differentiation in embryoid bodies. *Cardiovasc Pathol.* 2004;13: S1–S160.
143 Kuro-o M, Nagai R, Tsuchimochi H, et al. Developmentally regulated expression of vascular smooth muscle myosin heavy chain isoforms. *J Biol Chem.* 1989;264:18272–18275.
144 Mack CP, Owens GK. Regulation of smooth muscle alpha-actin expression in vivo is dependent on CArG elements within the 5′ and first intron promoter regions. *Circ Res.* 1999;84:852–861.
145 Khanna A. Concerted effect of transforming growth factor-beta, cyclin inhibitor p21, and c-myc on smooth muscle cell proliferation. *Am J Physiol Heart Circ Physiol.* 2004;286: H113–H140.
146 Roth M, Johnson P, Borger P, et al. Dysfunctional interaction of C/EBPalpha and the glucocorticoid receptor in asthmatic bronchial smooth-muscle cells. *N Engl J Med.* 2004;531:560–574.
147 Sano H, Sudo T, Yokode M, et al. Functional blockade of platelet-derived growth factor receptor-beta but not of receptor-alpha prevents vascular smooth muscle cell accumulation in fibrous cap lesions in apolipoprotein E-deficient mice. *Circulation.* 2001;103:2955–2960.
148 Fahmy R, Khachigian L. Locked nucleic acid modified DNA enzymes targeting early growth response-1 inhibit human vascular smooth muscle cell growth. *N Engl J Med.* 2004;32:2281–2285.
149 Atwater BD, Roe MT, Mahaffey KW. Platelet glycoprotein IIb/IIIa receptor antagonists in non-ST segment elevation acute coronary syndromes: a review and guide to patient selection. *Drugs.* 2005;65:313–324.
150 Mehta SR, Yusuf S. Short- and long-term oral antiplatelet therapy in acute coronary syndromes and percutaneous coronary intervention. *J Am Coll Cardiol.* 2003;41:S79–S88.

151 Carter A, Scott D, Bailey L, et al. Dose-response effects of 32-P Radioactive stents in an atherosclerotic porcine coronary model. *Circulation.* 1999;100:1548–1554.

152 Farb A, Shroff S, John M, et al. Late Arterial Responses (6 and 12 months) after 32-P B-emitting stent placement. *Circulation.* 2001;103:1912–1919.

153 Virmani R, Farb A, Carter A, Jones R. Pathology of radiation-induced coronary artery disease in human and pig. *Cardiovasc Radiat Med*. 1999;1:98–101.

154 Silber S, Popma JJ, Suntharalingam M, et al. Two-year clinical follow-up of 90Sr/90 Y beta-radiation versus placebo control for the treatment of in-stent restenosis. *Am Heart J*. 2005;149:689–694.

155 Krotz F, Sohn HY, Klauss V, Schiele TM. Intracoronary brachytherapy – clinical state and pathophysiological considerations. *Curr Pharm Des.* 2005;11:421–433.

156 Pohl T, Kupatt C, Steinbeck G, Boekstegers P. Angiographic and clinical outcome for the treatment of in-stent restenosis with sirolimus-eluting stent compared to vascular brachytherapy. *Z Kardiol.* 2005;94:405–410.

157 Drachman D, Edelman E, Seifert P, et al. Neointimal thickening after stent delivery of paclitaxel: change in composition and arrest of growth over six months. *J Am Coll Cardiol.* 2000;36:2325–2332.

158 Nuhrenberg T, Voisard R, Fahlisch F, et al. Rapamycin attenuates vascular wall inflammation and progenitor cell promoters after angioplasty. *FASEB J*. 2005;19:246–248.

159 Stone G, Ellis S, Cox D, et al. A polymer-based, paclitaxel-eluting stent in patients with coronary artery disease. *N Engl J Med.* 2004;350:221–231.

160 Tsagalou E, Chieffo A, Iakovou I, et al. Multiple overlapping drug-eluting stents to treat diffuse disease of the left anterior descending coronary artery. *J Am Coll Cardiol.* 2005;45:1570–1573.

161 Aoki J, Serruys P, van Beusekom H, et al. Endothelial progenitor cell capture by stents coated with antibody against CD34: the HEALING-FIM (Healthy Endothelial Accelerated Lining Inhibits Neointimal Growth-First In Man) registry. *J Am Coll Cardiol.* 2005;45:1574–1579.

162 Moncada S, Higgs A. The L-arginine-nitric oxide pathway. *N Engl J Med.* 1993;329:2002–2012.

163 Carvajal J, Germain A, Huidobro-Toro J, Weiner C. Molecular mechanism of cGMP-mediated smooth muscle relaxation. *J Cell Physiol.* 2000;84:409–420.

164 Tamama K, Okajima F. Sphingosine 1-phosphate signaling in atherosclerosis and vascular biology. *Curr Opin Lipidol.* 2002;13: 489–495.

165 Shirao S, Kashiwagi S, Sato M, et al. Sphingosylphosphorylcholine is a novel messenger for Rho-kinase-mediated $Ca2^{+}$ sensitization in the bovine cerebral artery: unimportant role for protein kinase C. *Circ Res.* 2002;91:112–119.

166 Zhou H, Murthy K. Distinctive G protein-dependent signaling in smooth muscle by sphingosine 1-phosphate receptors S1P1 and S1P2. *Am J Physiol Cell Physiol.* 2004;286:C1130–C1138.

167 Sakai K, Sigmund CD. Molecular evidence of tissue renin-angiotensin systems: a focus on the brain. *Curr Hypertens Rep.* 2005;7:135–140.

168 Johnson RJ, Rodriguez-Iturbe B, Kang DH, et al. A unifying pathway for essential hypertension. *Am J Hypertens.* 2005;18: 431–440.

169 Osborn JW, Jacob F, Guzman P. A neural set point for the long-term control of arterial pressure: beyond the arterial baroreceptor reflex. *Am J Physiol Regul Integr Comp Physiol.* 2005;288: R846–R855.

170 Meneton P, Jeunemaitre X, de Wardener HE, MacGregor GA. Links between dietary salt intake, renal salt handling, blood pressure, and cardiovascular diseases. *Physiol Rev*. 2005;85:679–715.

171 Safar M, Chamiot-Clerc P, Dagher G, Renaud J. Pulse pressure, endothelium function, and arterial stiffness in spontaneously hypertensive rats. *Hypertension.* 2001;38:1416–1421.

172 Zhao Y, Campbell A, Robb M, et al. Protective role of angiopoietin-1 in experimental pulmonary hypertension. *Circ Res.* 2003;92:984–991.

173 Gros R, Van Wert R, You X, et al. Effects of age, gender, and blood pressure on myogenic responses of mesenteric arteries from C57BL/6 mice. *Am J Physiol Heart Circ Physiol.* 2002;282:H380–H388.

174 Lavoie JL, Bianco RA, Sakai K, et al. Transgenic mice for studies of the renin-angiotensin system in hypertension. *Acta Physiol Scand*. 2004;181:571–577.

175 Toporsian M, Gros R, Kabir M, et al. A role for endoglin in coupling eNOS activity and regulating vascular tone revealed in hereditary hemorrhagic telangiectasia. *Circ Res.* 2005;96:684–692.

176 Wierzbicki A, Poston R, Ferro A. The lipid and non-lipid effects of statins. *Pharmacol Ther*. 2003;99:95–112.

177 Meloche S, Pelletier S, Servant M. Functional cross-talk between the cyclic AMP and Jak/STAT signaling pathways in vascular smooth muscle cells. *Mol Cell Biochem.* 2000;212:99–109.

178 Scholkens B, Landgraf W. ACE inhibition and atherogenesis. *Can J Physiol Pharmacol.* 2002;80:354–359.

179 Yusuf S, Sleight P, Pogue J, et al. Effects of an angiotensin-converting-enzyme inhibitor, ramipril, on cardiovascular events in high-risk patients. The Heart Outcomes Prevention Evaluation Study Investigators. *N Engl J Med.* 2000;342:145–153.

180 Molinaro G, Cugno M, Perez M, et al. Angiotensin-converting enzyme inhibitor-associated angioedema is characterized by a slower degradation of des-arginine(9)-bradykinin. *J Pharmacol Exp Ther*. 2002;303:232–237.

181 Park J, Schiffrin E. Effects of antihypertensive therapy on hypertensive vascular disease. *Curr Hypertens Rep.* 2000;2:280–288.

182 Inoue I, Katayama S. The possible therapeutic actions of peroxisome proliferator-activated receptor alpha (PPAR alpha) agonists, PPAR gamma agonists, 3-hydroxy-3-methylglutaryl coenzyme A (HMG-CoA) reductase inhibitors, angiotensin converting enzyme (ACE) inhibitors and calcium (Ca)-antagonists on vascular endothelial cells. *Curr Drug Targets Cardiovasc Haematol Disord.* 2004;4:35–52.

183 Leber AW, Knez A, von Ziegler F, et al. Quantification of obstructive and nonobstructive coronary lesions by 64-slice computed tomography: a comparative study with quantitative coronary angiography and intravascular ultrasound. *J Am Coll Cardiol.* 2005;46:147–154.

184 Kondos GT, Hoff JA, Sevrukov A, et al. Electron-beam tomography coronary artery calcium and cardiac events: a 37-month follow-up of 5635 initially asymptomatic low- to intermediate-risk adults. *Circulation.* 2003;107:2571–2576.

185 Rajaram V, Pandhya S, Patel S, et al. Role of surrogate markers in assessing patients with diabetes mellitus and the metabolic syndrome and in evaluating lipid-lowering therapy. *Am J Cardiol.* 2004;93:C32-C48.

186 Kelly KA, Allport JR, Tsourkas A, et al. Detection of vascular adhesion molecule-1 expression using a novel multimodal nanoparticle. *Circ Res.* 2005;96:327–336.

187 Blankenberg FG, Mari C, Strauss HW. Development of radiocontrast agents for vascular imaging: progress to date. *Am J Cardiovasc Drugs.* 2002;2:357–365.

188 Hofstra L, Liem IH, Dumont EA, et al. Visualisation of cell death in vivo in patients with acute myocardial infarction. *Lancet.* 2000;356:209–212.

189 Kolodgie FD, Petrov A, Virmani R, et al. Targeting of apoptotic macrophages and experimental atheroma with radiolabeled annexin V: a technique with potential for noninvasive imaging of vulnerable plaque. *Circulation.* 2003;108:3134–3139.

190 Kietselaer BL, Reutelingsperger CP, Heidendal GA, et al. Noninvasive detection of plaque instability with use of radiolabeled annexin A5 in patients with carotid-artery atherosclerosis. *N Engl J Med.* 2004;350:1472–1473.

191 Endemann D, Pu Q, De Ciuceis C, et al. Persistent remodeling of resistance arteries in type 2 diabetic patients on antihypertensive treatment. *Hypertension.* 2004;43:399–404.

62

Cross-Talk between the Red Blood Cell and the Endothelium

Nitric Oxide as a Paracrine and Endocrine Regulator of Vascular Tone

Sruti Shiva and Mark T. Gladwin

Vascular Medicine Branch, National Heart, Lung, and Blood Institute; National Institutes of Health; Bethesda, Maryland

The endothelium and red blood cells, although often studied in isolation, clearly interact to modulate nitric oxide (NO)-dependent vascular homeostasis, including basal blood flow, shear- or flow-mediated vasodilation, and hypoxic vasodilation. NO, produced by endothelial NO synthase (eNOS), plays a fundamental role in regulating blood flow. Recent data implicate a critical function for hemoglobin and the erythrocyte in regulating the bioavailability of NO in the vascular compartment. In fact, a delicate balance exists between endothelial NO formation, which is produced tonically and in response to shear stress (modulated by red blood cell mass), and NO scavenging by the erythrocytic hemoglobin. The scavenging of NO by hemoglobin inside an intact red blood cell is greatly diminished by diffusional barriers to NO – intrinsic to and surrounding the red cell. Compartmentalization of hemoglobin by the red cell also limits extravasation of hemoglobin dimers into the interstitial compartment. Intravascular hemolysis disrupts this balance by the release of hemoglobin from the red blood cell into plasma. This extracellular hemoglobin is then able to scavenge endothelial-derived NO approximately 600-fold faster than erythrocytic hemoglobin, thus disrupting NO homeostasis. This may lead to vasoconstriction, decreased blood flow, platelet activation, increased endothelin (ET)-1-dependent vasoconstrictor activity (because NO modulates ET receptor expression to promote a vasodilator phenotype), adhesion molecule expression, and end-organ injury (1,2). This suggests a novel mechanism of disease for hereditary and acquired hemolytic conditions such as sickle cell disease and malaria. In addition to providing an NO scavenging role in the physiological regulation of NO-dependent vasodilation, hemoglobin and the erythrocyte may deliver NO as the hemoglobin deoxygenates. Although this process has been ascribed previously to S-nitrosated hemoglobin (SNO-hemoglobin), recent data from our laboratories suggest that deoxygenated hemoglobin reduces the ubiquitous blood and tissue anion nitrite (NO_2^-) to NO, leading to vasodilation along the physiological oxygen (O_2) gradient.

In this context, the endothelium can be considered a signaling organ that generates both paracrine NO and endocrine NO_2^- (in addition to other putative endocrine NO-modified species). In the remainder of this chapter, we outline the basic principles by which NO mediates local signaling between the red cell and endothelium; describe a role for the endocrine action of nitrite, a newly uncovered storage form of NO; and describe how nitrite may potentially be used as a therapeutic agent to treat diseases characterized by endothelial dysfunction and organ ischemia.

ENDOTHELIUM-DEPENDENT VASODILATION AND PARACRINE PROPERTIES OF NITRIC OXIDE

In 1933, the German physiologist Schretzenmayer demonstrated that increased blood flow to the canine femoral artery could result in an increased blood vessel diameter (3), thus demonstrating the physiological property of endothelium-dependent flow-mediated vasodilation. Almost 50 years later, in 1981, Furchgott and colleagues proposed that the endothelium produced a vasodilator substance called *endothelium-derived relaxing factor* in response to acetylcholine stimulation. They experimentally demonstrated that endothelium-denuded aortic rings paradoxically constricted rather than dilated on exposure to acetylcholine (4). The surprising revelation that this paracrine (endothelium-to-smooth muscle) relaxing

Table 62-1: Regulators of NO Production from eNOS

Type of Regulation	*Mediator of Regulation*	*Role in Regulation*	*References*
Substrates	L-arginine Oxygen NADPH	Required for enzyme activity	10, 143, 144
Cofactors and prosthetic groups	Tetrahydrobiopterin FAD FMN Heme Zinc	Required for enzyme activity	10, 11, 145
Post-translational modifications	Phosphorylation	Several sites of phosphorylation including serine, threonine, and tyrosine residues	146, 147
	Myristoylation Palmitoylation	Targets enzyme to caveolae	148, 149
Protein–protein interaction	Calcium bound Calmodulin	Necessary for enzyme activity	150
	Caveolin	Inhibitor of activity	151, 152
	HSP90	Activator of activity	146, 153
Regulation of expression	Translational regulation	*eNOS* gene contains region for the binding of regulatory transcription factors	147, 154

NADPH, nicotinamide adenine dinucleotide phosphate; HSP90, heat shock protein 90

factor was the free radical gaseous molecule NO led to a Nobel prize for Furchgott, Ignarro, and Murad (4–9).

The endothelium expresses NO synthase enzymes that catalyze the O_2-dependent conversion of the guanidino group of L-arginine to NO, producing L-citrulline as a byproduct (10, 11) (Table 62-1). To mediate vasodilation, the NO generated diffuses to underlying vascular smooth muscle cells (VSMCs), where it binds to the heme group of soluble guanylate cyclase to activate the enzyme. Activated soluble guanylate cyclase in turn produces cyclic GMP (cGMP), a second messenger that drives a kinase cascade that ultimately results in relaxation of the smooth muscle. The discovery of this pathway cemented the vital role of the endothelium as a master organ system regulating vascular homeostasis by producing paracrine vasodilator, antioxidant, antithrombotic, and antiproliferative mediators.

We now understand that vasomotor tone is determined by an exquisite balance of vasoconstrictors, vasodilators, and receptors expressed by the endothelium. Although the list of both paracrine and endocrine vasodilators produced by the endothelium is expansive (Table 62-2), NO is responsible for maintaining approximately 25% of basal blood flow (12–14), 100% of flow- or shear-mediated blood flow (15), 40% to 50% of acetylcholine-dependent blood flow (16), and 10% of exercise-dependent blood flow (17). It also contributes importantly to heat-mediated flow (18). (The emerging role of endothelium-derived NO in the regulation of hypoxic vasodilation is discussed later in this chapter). NO further regulates the levels and biochemical fate of superoxide through a direct reaction that also regulates levels of bioactive NO and forms the potent oxidant peroxynitrite ($ONOO^-$). NO also alters the production of other vasodilators such as prostacyclin; modulates the distribution and concentration of the potent vasoconstrictor ET-1 and its receptors; inhibits thrombosis at multiple levels; is an intermediate in the vascular endothelial growth factor (VEGF) signaling pathway; and tonically downregulates endothelial adhesion molecules, such as intercellular adhesion molecule (ICAM)-1, vascular cell adhesion molecule (VCAM)-1, and P- and E-selectin (2). It is increasingly apparent that this system becomes unbalanced in diseases of the cardiovascular system, such as pulmonary hypertension and coronary artery disease. Patients with systemic hypertension, pulmonary hypertension, diabetes, obesity, and atherosclerotic vascular disease all demonstrate impaired production of endothelium-derived NO (13,16). In fact, the coronary vasculature of these patients paradoxically constricts with exposure to acetylcholine, thus recapitulating Furchgott's isolated rabbit aortic ring experiments (13). The expansive role of NO in vascular health and disease supports the assertion that NO is a central regulator of homeostatic vascular function.

RED BLOOD CELLS AND HEMOGLOBIN MODULATE THE BIOLOGICAL FATE OF ENDOTHELIUM-DERIVED NITRIC OXIDE

Although the paracrine exchange of vasoactive mediators between the endothelium and smooth muscle has been studied extensively, the role of the intravascular compartment in regulating NO homeostasis only recently has been explored. Although intravascular shear stress on the endothelium from

Table 62-2: Endothelial-Derived Vasodilators and Vasoconstrictors

Vasoactive Substance	*Effect*	*Method of Production*	*Target*	*Downstream Signaling Pathway*	*References*
ET-1	Constrictor	Expressed as propeptide and activated by endothelin-converting enzyme in the endothelium	ET_A and ET_B	Phospholipase C, IP_3	155, 156
Prostaglandin H_2	Constrictor	Cyclooxygenase and PH_2 synthase	Thromboxane receptor	Phospholipase C, IP_3	157, 158
Thromboxane A_2	Constrictor	Cyclooxygenase and PH_2 synthase	Thromboxane receptor	Phospholipase C, IP_3	159
Angiotensin II	Constrictor	Released as Ang I by the kidney and converted to Ang II by angiotensin-converting enzyme	AT_1	Adenylate cyclase, cAMP, Protein kinase A	160
Adrenomedullin	Dilator	Expressed as a propeptide and converted by the endothelium to the active form	Adrenomedullin receptors	Adenylate cyclase, cAMP, Protein kinase A	161, 162
NO	Dilator	NO synthase	Soluble guanylate cyclase	cGMP, Protein kinase G	163, 164
Bradykinin	Dilator	Kallikrein	Bradykinin (BK)1 and BK2 receptors	Phospholipase C, IP_3, increase calcium binding to calmodulin for eNOS	165, 166
Prostacyclin	Dilator	Cyclooxygenase and prostacyclin synthase	IP receptor	Adenylyl cyclase and cAMP	166, 167

blood flow stimulates eNOS and the production of NO, hemoglobin within the red blood cell exerts an opposing effect on NO homeostasis by scavenging NO and decreasing NO bioactivity (19,20). Because NO is highly diffusible, its inactivation and spatial localization depends on its reactivity with other molecules, including heme proteins and reactive oxygen species (ROS) on both sides of the endothelium (21). Because the small uncharged NO molecule is freely and randomly diffusible, in a given area where NO is produced, such as the endothelium, a spatial gradient of NO concentration occurs. Owing to this random diffusion, this gradient is constrained by scavenging reactions even if these reactions occur on only one side of the endothelium. Thus, even NO generated on the basolateral side of the endothelium is susceptible to scavenging by an intravascular NO sink (22).

In particular, NO reacts at nearly diffusion-limited rates ($10^7\ M^{-1}s^{-1}$) with both oxy and deoxyhemoglobin within the vasculature (19,23). When hemoglobin is oxygenated, reaction with NO yields the inert molecule nitrate and methemoglobin (Equation 1), while NO binds to the iron of deoxyhemoglobin to form iron-nitrosyl hemoglobin (Equation 2) (19,23–25).

$$NO + oxyHb\,(Fe^{II} - O_2) \rightarrow NO_3^- + metHb\,(Fe^{III}) \quad \text{(Equation 1)}$$

$$NO + deoxyHb\,(Fe^{II}) \rightarrow Hb - NO\,(Fe^{II} - NO) \quad \text{(Equation 2)}$$

In chemical terms, these reactions approach the diffusion limit and are essentially irreversible. These chemical equilibrium kinetics create a paradox in vascular biology. From a kinetic and equilibrium standpoint, the high concentration of hemoglobin (10 mM concentration in heme) present in blood should (theoretically) rapidly consume most available NO synthesized by the endothelium before it diffuses to the smooth muscle cell layer (26). Stated differently, the proximity of an irreversible high-affinity NO sink (hemoglobin) should severely limit the diffusional radius of NO from its endothelial source. The solution to this paradox requires compartmentalization of hemoglobin within the erythrocyte, which generates diffusional barriers for NO that decrease the reaction rates between NO and hemoglobin. In addition to an intrinsic diffusional barrier created by the erythrocytic membrane and submembrane protein scaffolding (27,28), there exists an unstirred layer around the red blood cell membrane (29,30) and an erythrocyte-free zone that forms along the endothelium as red cells centrally concentrate in laminar flowing blood (31–33). The combination of these barriers decreases the reaction rate of NO with hemoglobin by over 600-fold under physiological conditions. This allows NO to escape inactivation by hemoglobin and diffuse to the smooth muscle to mediate local paracrine activity (29,30,32,33). In addition to the formation of diffusional barriers, the compartmentalization of hemoglobin in the red cell prevents extravasation of hemoglobin dimers into the interstitial compartment (19,34).

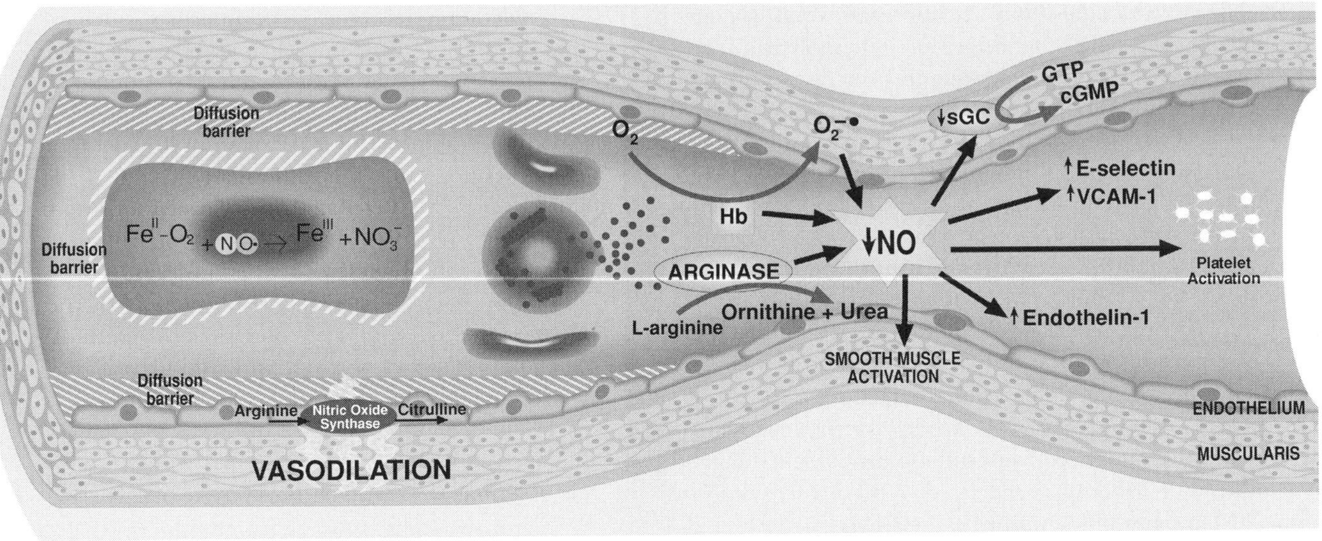

Figure 62.1. Disruption of NO-dependent vascular signaling by hemolysis. NO is produced by the enzyme NOS in the endothelium. Under physiological conditions, NO diffuses into the smooth muscle cells as well as into the lumen of the vessel to mediate vasodilation and a host of other responses important to maintaining vascular homeostasis. NO bioactivity is limited by its chemical reactivity with molecules such as intraerythrocytic hemoglobin and superoxide generated from xanthine oxidase. The reaction of NO with hemoglobin is limited by diffusional barriers around the red blood cell and along the endothelium. During intravascular hemolysis, cell free oxyhemoglobin in the plasma (Fe^{II} – O_2) rapidly scavenges NO, converting it to nitrate (NO_3^-). This decreases NO bioavailability to both smooth muscle cells and other vascular targets of NO, leading to a disruption of vascular homeostasis. For color reproduction, see Color Plate 62.1.

Although the NO–hemoglobin dioxygenation reaction (Equation 1) and iron-nitrosylation (Equation 2) are reduced by compartmentalization of hemoglobin in red cells, the scavenging reactions remain significant and contribute to vascular tone and the maintenance of NO as a paracrine signaling molecule. The half-life of NO in whole blood is less than 2 milliseconds, and increasing hematocrit reduces NO availability and raises blood pressure. This pressor effect is notable during polycythemia and with the clinical treatment of anemia using erythropoietin (35,36). The NO-scavenging effects of red cell hemoglobin are likely teleologically balanced by opposing effects of erythropoietin on inducing NOS expression and by shear-mediated NO synthase activation (37,38).

Disruption of Nitric Oxide–Dependent Vascular Homeostasis in Hemolytic Disease

During intravascular hemolysis, erythrocytes lyse and release hemoglobin into the plasma, disrupting the diffusional barriers between NO and hemoglobin (Figure 62.1 for color reproduction, see Color Plate 62.1). (1,2,20,39). This decompartmentalized hemoglobin rapidly oxidizes NO to nitrate, inhibiting NO-dependent responses in the vessel. This idea is supported by in vitro, animal, and human studies that demonstrate that cell-free hemoglobin at concentrations as low as 6 μM completely inhibits acetylcholine-mediated vasodilation and produces vasoconstriction (40–42). A reduction in NO bioavailability is characteristic of a number of pathologies that share a component of intravascular hemolysis, such as sickle cell disease and paroxysmal nocturnal hemoglobinuria (2,39). Our group has shown that the plasma from patients with sickle cell disease contains higher concentrations of cell-free hemoglobin (4.2 $\pm$ 1.1 μM; ranges from undetectable to 20 μM) than healthy controls (0.2 $\pm$ 0.1 μM), and stoichiometrically consumes NO in vitro and in vivo (39). In these patients, higher cell-free hemoglobin levels also correlated with a resistance to the vasodilatory effects of infused sodium nitroprusside, further supporting the hypothesis that NO scavenging by cell-free hemoglobin can disrupt NO-dependent vascular homeostasis (39). Vasodilation during infusions of NO donors (nitroprusside, nitroglycerin, NONOates) is blunted in patients with sickle cell disease (39,43) and in transgenic mouse models of sickle cell disease (44–46). This resistance to NO correlates with plasma hemoglobin levels (39,45). Apart from contributing to the pathogenesis of hemolytic diseases, NO scavenging by cell-free hemoglobin is thought to contribute to the toxicity of the stroma-free hemoglobin-based blood substitutes. In clinical and preclinical trials, the administration of these substitutes led to systemic hypertension (47), smooth muscle spasm (48), decreased organ perfusion (49), and risk of death, consistent with a decrease in NO bioactivity (2). In addition to adversely affecting blood flow, NO scavenging has other implications within the vasculature. Physiologically, NO tonically inhibits the transcription of soluble vascular cell adhesion molecules (VCAM-1 and ICAM-1) in endothelial cells (ECs). Thus, a decrease in NO bioactivity also leads to an increase in the expression of VCAM-1 and ICAM-1 (39,50,51). NO inhibits platelet activation, thrombogenesis, and clot stability, and it regulates ET-1 expression and the distribution of its receptors (52–55). NO also plays an intermediary role in the VEGF signaling pathway. It is therefore likely that chronic NO inactivation in hemolytic diseases has extensive downstream consequences on vascular homeostasis, thrombosis, and angiogenesis.

Indeed, hemolytic conditions such as sickle cell disease and paroxysmal nocturnal hemoglobinuria are associated with relative hypertension (compared to the hypotension of nonhemolytic anemias) (56), pulmonary hypertension (57,58), and esophageal and smooth muscle dystonias during paroxysms of hemolysis (59). In addition to the release of hemoglobin during hemolysis, the red blood cell also contains large quantities of arginase. Increases in plasma levels of this enzyme, which metabolizes arginine to ornithine, reduce the substrate for eNOS. In patients with sickle cell disease, plasma arginase activity increases in direct proportion to intravascular hemolytic rate (60). Furthermore, increased arginase activity and an associated low arginine:ornithine ratio is associated with pulmonary hypertension and risk of death in these patients (60).

Epidemiological data increasingly suggests that chronic intravascular hemolysis produces a state of hemolysis-associated endothelial dysfunction and a progressive proliferative vasculopathy. The clinical manifestations of this vasculopathy include pulmonary hypertension, cutaneous leg ulceration, priapism, and stroke (20,61,62). Pulmonary hypertension is a recognized complication of sickle cell disease, thalassemia, paroxysmal nocturnal hemoglobinuria, spherocytosis, stomatocytosis, and microangiopathic hemolytic anemias, such as hemolytic uremic syndrome (HUS) and thrombotic thrombocytopenic purpura (TTP) (57,63,64).

Multiple systems have evolved to limit the toxicity of cell-free plasma hemoglobin, including the high-molecular-weight haptoglobin system (prevents extravasation of hemoglobin and limits NO scavenging) (40,65,66); hemopexin; CD163 hemoglobin scavenger protein (which not only mediates haptoglobin-hemoglobin clearance but also upregulates interleukin [IL]-10 and heme oxygenase 1) (67); and the heme oxygenase 1/biliverdin reductase/p21 pathways, which exert anti-inflammatory, antioxidant, and antiproliferative effects (68–72).

NITRIC OXIDE AS AN ENDOCRINE MEDIATOR

Although the NO radical has a half-life of only 2 milliseconds in whole blood, a number of studies suggest that in addition to a local paracrine activity, NO may possess an endocrine bioactivity within the blood. These studies suggest that NO is somehow stabilized, transported, and bioactivated at remote sites within the vasculature:

- In humans, NO gas inhalation (80 ppm) has limited effects on basal blood flow in the forearm. However, when eNOS is inhibited, inhaled NO significantly increases regional forearm blood flow (73).
- Infusions of NO solutions into the human forearm are associated with measurable increases in blood flow in the opposing arm (74).
- In a feline model of mesenteric ischemia–reperfusion, inhaled NO increases blood flow following reperfusion and decreases leukocyte adhesion. In this model, inhaled NO also increases blood flow to the intestine during NO synthase inhibition (75,76).
- Mean systemic arterial pressure and systemic vascular resistance are decreased in sheep breathing 80 ppm NO gas (77).
- Inhaled NO (80 ppm) increases renal blood flow, urinary flow, and glomerular filtration rate in anesthetized pigs (78).
- In a recently completed, randomized, placebo-controlled trial of NO therapy for premature newborns with respiratory failure, NO treatment significantly reduced the combined end point of intracranial hemorrhage, periventricular leukomalacia, or ventriculomegaly, an effect consistent with endocrine transport of an NO intermediate in the blood to the central nervous system (79).

In many of these studies, the observed vasodilation was associated with increases in plasma levels of S-nitrosated, N-nitrosated, and iron-nitrosylated proteins as well as NO_2^-. These data suggest that NO might be stabilized in the circulation in the form of a higher oxidation product (nitrite) or as an NO-modified protein, peptide, and/or lipid (S-nitrosothiol, N-nitrosamine, or nitrated lipid) and released later at distal sites to act as an NO-dependent signaling molecule.

Although support for the concept of a circulating NO store has advanced in the last few years, much controversy surrounds the identity of the circulating species, how it forms, and its mechanism of action. The principle that NO may be stabilized in blood, and that the inactivation reactions with hemoglobin may thus be limited, was first proposed by Loscalzo and Stamler. They hypothesized that NO (after a required abstraction of an electron) could form a covalent bond with cysteine residues on albumin to form S-nitrosated albumin (SNO-albumin) (80,81). This paradigm was later extended by the Stamler group to SNO-hemoglobin (82). This field is extremely controversial, largely because of major questions about the concentrations and function of SNO-albumin and SNO-hemoglobin in the human circulation (reported values range from 7 μM to undetectable, with more modern methodologies documenting levels of less than 10 nM) (66,82–88). However, it is likely that a number of intravascular species are capable of endocrine-mediated vasodilation, including S-nitrosothiols (76,81), NO_2^- (89), N-nitrosamines (29,86,90), iron-nitrosyl (83), and recently identified nitrated lipids (91). Figure 62.2 illustrates the putative pathways involved in the formation and stabilization of NO in blood as an endocrine vasodilating species. In the following sections, we briefly review the SNO-Hb hypothesis, the major challenges to this theory, and then focus on the emerging role of the nitrite anion as a storage molecule of NO that is activated during physiological blood and tissue hypoxia and acidosis.

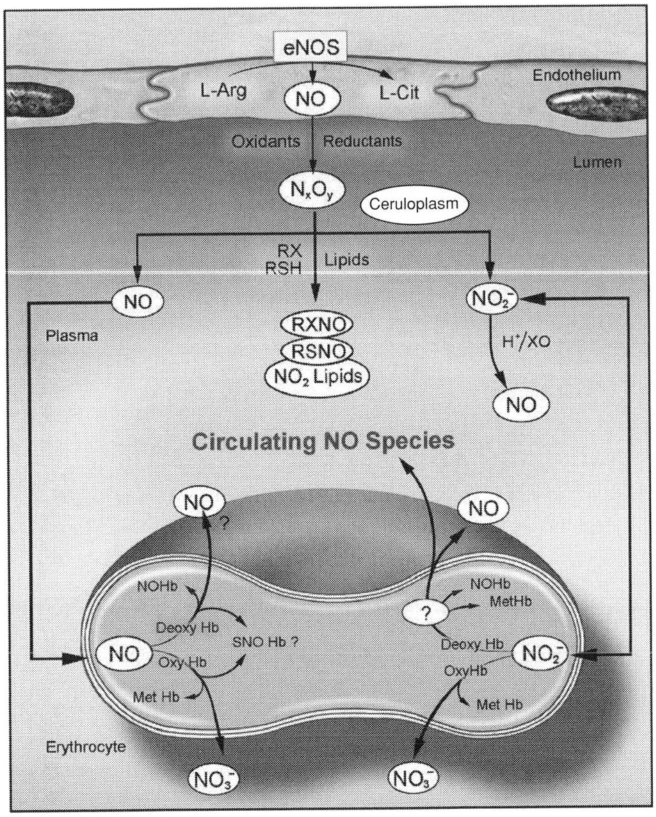

Figure 62.2. Pathways for the formation of potential circulating NO species. The majority of NO produced by the endothelium is metabolized by the erythrocyte. NO reacts with oxyhemoglobin to form nitrate or with deoxyhemoglobin to produce iron-nitrosyl hemoglobin (HbNO). The mechanisms of S-nitrosated hemoglobin (SNOHb) formation are not entirely clear, but both SNOHb and iron-nitrosyl-hemoglobin (HbNO) have been hypothesized to deliver NO to distal sites. A minor portion of endothelium-derived NO escapes erythrocytic scavenging and is converted to potential endocrine mediators (nitrite, nitrated lipids, plasma nitrosothiols, N-nitrosamines) by the plasma. The plasma component ceruloplasmin oxidizes NO to nitrite. Nitrite enters the erythrocyte, where it is oxidized to nitrate by oxyhemoglobin or is reduced to bioactive NO by deoxyhemoglobin. The mechanism for NO escape from this reaction is still unclear, but may involve an intermediate NO species (unknown intermediates denoted by question mark). (Reprinted with permission from Dejam A, Hunter CJ, Pelletier MM, et al. Erythrocytes are the major intravascular storage sites of nitrite in human blood. *Blood.* 2005;106:734–739. Epub 2005 Mar 17.) For color reproduction, see Color Plate 62.2.

The S-Nitrosated Hemoglobin Hypothesis

The SNO-hemoglobin hypothesis suggests the formation of a covalent modification of the cysteine-93 residue on the β-chain of the hemoglobin protein by NO (82,92). In this model, NO first binds cooperatively to a vacant heme on the oxygenated hemoglobin molecule to form iron-nitrosyl hemoglobin (93). This requires a preferential binding of NO to a deoxyheme on 99% oxygenated hemoglobin. A subsequent transfer of the NO group from the heme to the β-93 cysteine, to form SNO-hemoglobin occurs (94). This NO group is then delivered to tissues as an "X-NO" intermediate by a mechanism linked to the allosteric structural transition of hemoglobin between the oxy (R) state and deoxy (T) state (95). The SNO-hemoglobin concept was initially supported by data showing relatively high (2.5 μM) levels of SNO-hemoglobin in the blood and an arterial-to-venous gradient of this species (38,96). In addition, SNO-hemoglobin and NO-treated erythrocytes could vasodilate aortic rings as well as blood vessels in vivo (92,94). However, data generated by several different labs seriously challenge essential elements of the SNO-hemoglobin hypothesis (38). The cooperative or preferential binding of NO to the deoxyhemes of oxygenated hemoglobin has not been reproduced by other laboratories and has been ascribed to an artifact of bolus NO addition (82,97–99). A recent study monitoring the position of NO on the hemoglobin molecule using electron paramagnetic resonance (EPR) during oxy-to-deoxy cycling did not show a transfer of NO between the heme and the thiol on the protein (100,101). The levels of SNO-hemoglobin measured in the human circulation by other groups are one to two logs lower than the 2.5 μM concentrations obtained using photolysis-based assays, and significant artery-to-vein gradients of SNO-hemoglobin have not been detected by other laboratories (83,102,103). A resolution to this controversy will likely require carefully performed physiological studies of hypoxic vasodilation in a cysteine 93-substituted mouse model. Although the specific mechanisms of the SNO-hemoglobin model remain controversial, the principle of endocrine transport of an NO species in blood and the possibility that red cells contribute to hypoxia-dependent NO homeostasis has driven a field of research attempting to identify novel mechanisms by which NO may act as an endocrine mediator.

Nitrite as a Circulating Store of Nitric Oxide

It has recently been proposed that nitrite (NO_2^-) serves as a circulating storage pool of bioactive NO in the human blood stream (20,104–107). The small circulating anion, found at high nanomolar (300–600 nM) concentrations in blood, is generated in the body by the oxidation of NO (104). The circulating plasma metalloprotein ceruloplasmin has been shown to be responsible for oxidizing NO to nitrite in the plasma (108). In addition, plasma nitrate (NO_3^-) is concentrated into saliva and reduced to nitrite (NO_2^-) by commensal mouth bacterial flora. This nitrite is further reduced to NO in the stomach, where it regulates mucosal blood flow and mucous production (109–112). Although dietary nitrite and nitrate appears to contribute to plasma nitrite levels, studies suggest that up to 70% of plasma nitrite derives from endothelially produced NO (113). In human physiological studies, our laboratory detected arterial-to-venous gradients of nitrite in human plasma and observed increased nitrite consumption during exercise stress (106). This consumption, coupled with measured increases in plasma nitrite associated with endocrine vasodilation during NO gas inhalation in normal human volunteer studies, led to the hypothesis that the intravascular reduction of nitrite to NO might contribute to in vivo blood flow regulation (73).

Consistent with a vasodilatory role for nitrite, early in vitro studies demonstrated that nitrite was able to mediate vasodilation and the activation of soluble guanylate cyclase when applied to aortic rings at high (micro- to millimolar) concentrations (8,114,115), However, because nitrite levels are less than 1 μM in blood, the high concentrations of nitrite required to vasodilate aortic ring preparations appeared to preclude a physiological role for nitrite as an endogenous vasodilator (116). We were therefore quite surprised to find that infusions of near physiological concentrations of nitrite into the human forearm circulation produced vasodilation, a phenomenon associated with the rapid formation of NO-modified hemoglobin during artery-to-venous transit (89). (The chemical mechanism is described in the section "Hemoglobin as an Allosterically Regulated Nitrite Reductase: Physiological Roles for Deoxyhemoglobin and Myoglobin-Dependent Nitrite Reduction.") It is increasingly clear that nitrite is indeed a bioactive signaling molecule, generating NO under physiological levels of hypoxia and acidosis (117–121). A number of mechanisms for nitrite reduction to NO are being studied, including an enzymatic reduction by xanthine oxidoreductase (122–124), acidic reduction (disproportionation) (125), and heme-based nitrite reduction by hemoglobin, myoglobin, or other tissue heme proteins (89,105,119, 126,127).

Nitrite and Hemoglobin – A Long History

The reaction of nitrite with myoglobin has been harnessed by man since the time of the Greeks and Romans. In the process of meat curing, the application of nitrate derived from saltpeter and ascorbic acid in the form of citrus juice results in enzymatic and nonenzymatic nitrate reduction to nitrite and subsequently to NO by bacteria and heme proteins in meat. Haldane and Hoagland demonstrated that the reduction of nitrite to NO in a reaction with hemoglobin or myoglobin produced iron-nitrosyl heme, which was responsible for the resulting red color of cured meat (105,128). Nitrosylation of the heme inhibits iron release from the porphyrin ring, preventing iron-catalyzed lipid oxidation and deleterious Fenton chemistry that is thought to contribute to the spoiling of the meat (128). In 1923, the U.S. Food and Drug Administration (FDA) authorized the use of nitrite as the curing agent in place of higher concentrations of nitrate that were previously used.

Therapeutically, the antidotal activity of nitrite for cyanide poisoning was first demonstrated in 1888, and nitrite has since become an essential component of the cyanide antidote kit used today (129). The principle of nitrite's action for cyanide poisoning derives from the ability of nitrite to oxidize hemoglobin to ferric methemoglobin. This form of hemoglobin binds cyanide with a very high affinity, causing cyanide to dissociate from its other targets and bind to hemoglobin. With all the cyanide circulating as cyanomethemoglobin, an agent such as thiosulfate can be given to react with the cyanide to form thiocyanate, which will be excreted in the urine. Interestingly, studies performed in 1937 clearly demonstrated that doses of nitrite comparable to those used for the treatment of cyanide poisoning produced circulatory collapse (130). However, despite the fact that the reactions of nitrite with hemoglobin have long been studied and exploited therapeutically, the physiological role of this reaction and the fate of any by-products, such as NO, generated from this reaction has been reappraised only recently.

HEMOGLOBIN AS AN ALLOSTERICALLY REGULATED NITRITE REDUCTASE: PHYSIOLOGICAL ROLES FOR DEOXYHEMOGLOBIN AND MYOGLOBIN-DEPENDENT NITRITE REDUCTION

In 1981, Doyle and colleagues characterized the reduction of nitrite by purified deoxyhemoglobin in a reaction yielding NO and methemoglobin (Equation 3) (131).

$$\text{Deoxyhemoglobin}\,(\text{Fe}^{\text{II}}) + \text{nitrite} + \text{H}^+ \rightarrow \text{methemoglobin}\,(\text{Fe}^{\text{III}}) + \text{NO} + \text{OH}^- \quad (\text{Equation 3})$$

NO formed from this reaction can then bind vacant deoxyheme sites to form NO-modified hemoglobin, as shown in Equation 2.

The kinetics of this reaction have been re-examined with particular emphasis on the physiological role of hemoglobin as a nitrite reductase, able to produce bioactive NO and potentially other NOx intermediates (132,133). Three important ideas emerge from these studies: (a) the reduction of nitrite by deoxyhemoglobin is allosterically regulated by the quaternary structure of hemoglobin; (b) the reduction of nitrite generates measurable quantities of NO gas and likely other NOx intermediates; and (c) the reduction of nitrite by deoxyhemoglobin by the above reaction possesses both pH and O_2 sensor chemistry with increased rates of reduction at low Pao_2 and low pH (132).

From a physiological standpoint, consider that this reaction requires deoxyhemoglobin and a proton – providing O_2 and pH sensor chemistry – and generates NO, a potent vasodilator. Methemoglobin formed during the reaction will not autocapture and inactivate the NO formed. These conditions provide a chemistry that has the potential to generate NO and mediate vasodilation under hypoxic and acidic physiological stress.

We recently re-examined the reaction of excess nitrite with cell-free deoxyhemoglobin and found that the products of the reaction are met- and iron-nitrosyl hemoglobin in equimolar concentrations. In addition, the kinetics of the reaction deviated from second order and demonstrated a surprising sigmoidal reaction profile, characterized by a maximal reaction rate when 50% of the heme was ligated (with NO or oxidized to methemoglobin) (132,133). This sigmoidal profile

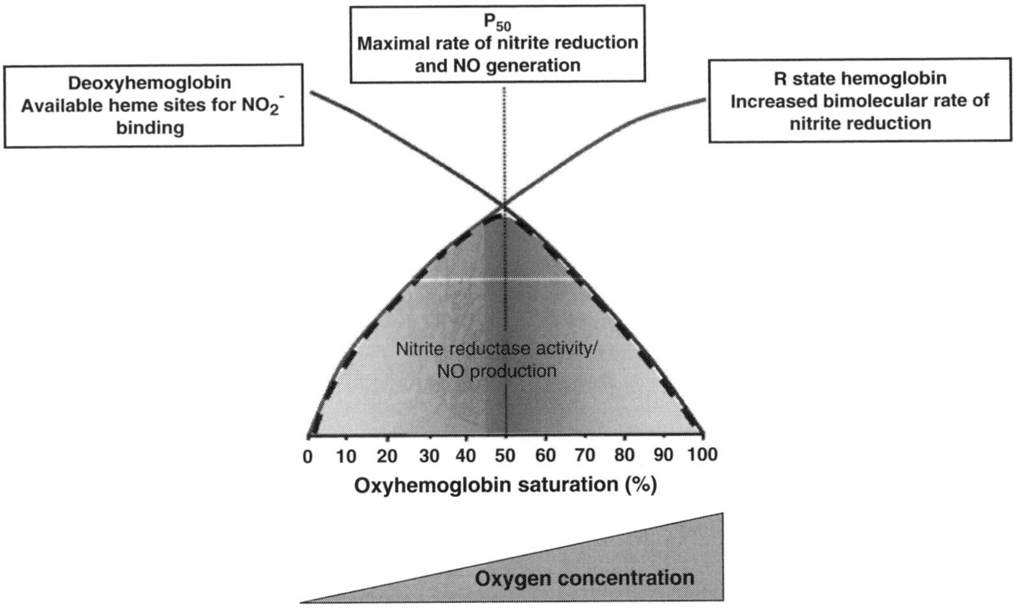

Figure 62.3. Allosteric regulation of nitrite reductase rate. The rate of hemoglobin-dependent nitrite reduction (*black dashed line*) depends on the number of available heme sites for reaction and the amount of R state tetramers. As O_2 saturation of hemoglobin increases, the number of available heme sites for nitrite binding decreases, decelerating the reaction. However, the R-state tetramer is increasingly stabilized, which accelerates reaction rate. The maximal reaction rate occurs at an optimal balance of these two subprocesses, which occurs when 50% of the hemoglobin is saturated with oxygen. The partial pressure at which this occurs is the P_{50} of hemoglobin. For color reproduction, see Color Plate 62.3.

is the result of the allosteric quaternary structural transition of the hemoglobin tetramer from the T (deoxy) to R (oxy) state during the course of the reaction. Over the course of this anaerobic reaction of nitrite with deoxyhemoglobin, the concentration of T-state deoxyhemoglobin decreases as the deoxyheme is converted to metheme and NO-heme; the depletion of deoxyheme has an effect of decelerating the reaction, because fewer heme sites are available for nitrite to react with. This is counterbalanced by an increasing concentration of R-state hemoglobin, produced by the R-state stabilizing effects of NO-heme and metheme, which are formed in this reaction (Equation 2 and Equation 3). R-state hemoglobin tetramer has a lower heme reduction potential (i.e., is a better electron donor to nitrite), which accelerates the reduction reaction of nitrite to NO. The balance of these two opposing subprocesses results in a sigmoidal reaction profile in which the maximal rate occurs when the majority of tetramers switch from the T to R state, a point that coincides with approximately 50% heme ligation (Figure 62.3; for color reproduction, see Color Plate 62.3).

Physiologically, O_2 is the major regulator of allosteric conformation and is the predominant ligand for hemoglobin. Consistent with the chemistry described earlier, the measured rate of nitrite reduction by hemoglobin is maximal at a hemoglobin-O_2 saturation between 40% and 60%, around the hemoglobin P_{50} (132). A maximal nitrite reductase activity at the hemoglobin P_{50} is biochemically consistent with a potential role in hypoxic vasodilation. Hypoxic vasodilation is a highly conserved physiological response in which blood flow and O_2 delivery increases in proportion to tissue O_2 consumption. This response requires the ability to detect changes in O_2 concentration or pH in the tissue, and to couple this sensing to the rapid elaboration of a vasodilator (134). Although the exact mechanism of this response is not yet known, it occurs when hemoglobin O_2 saturations fall between 40% and 60% (20–40 mmHg O_2), a value approximating the P_{50} of hemoglobin and the point of maximal nitrite reduction (135). Studies investigating the gradients of hemoglobin O_2 saturation across the vasculature show that arteriolar O_2 concentration drops from A1 to A4 caliber arterioles. In skeletal muscle, this drop is substantial, resulting in hemoglobin saturation values of less than 40% (Po_2 of 20 mm Hg) in A4 level arterioles (136). These data demonstrate that conditions ideal for maximal nitrite reductase activity occur within resistance vessels, consistent with a potential physiological role for nitrite-dependent hypoxic vasodilation. In addition to O_2 unloading and NO production occurring within the arterioles and arteriolized precapillaries, it is possible that NO generation at the capillary level could modulate NO-dependent hypoxic vasodilation by retrograde propagation, as characterized by Duling and Segal (137,138). A role for nitrite, hypoxia, and the red cell in hypoxic vasodilation is supported by aortic ring bioassay experiments in which preconstricted vessels were exposed to nitrite in the presence and absence of red blood cells over a range of O_2 concentrations. In these studies, vessels exposed to both nitrite and red blood cells vasodilated maximally at a Po_2 approaching the intrinsic hemoglobin P_{50} (89,139).

Within tissues, the reduction of nitrite by deoxyhemoglobin or myoglobin may contribute to the regulation of mitochondrial respiration and tissue O_2 gradients. Owing to the low redox potential of the heme group of myoglobin, the rate of reaction of nitrite with myoglobin is approximately 50-fold faster than with deoxyhemoglobin; hence, the rate of NO production is also greater (132). Under hypoxic conditions, nitrite (present in the tissue between 1 and 10 μM) has the potential to react with myoglobin to produce NO and inhibit local mitochondria by binding to cytochrome *c* oxidase (139,140,141). NO-dependent inhibition of mitochondrial respiration has been proposed to facilitate O_2 diffusion. Alternatively, hypoxia-dependent inhibition of respiration may generate a protective response to subsequent ischemic challenge (ischemic preconditioning), a concept that is discussed more thoroughly in the next section.

THERAPEUTIC IMPLICATIONS: NITRITE AS A NOVEL NITRIC OXIDE DONOR TARGETING ISCHEMIC AND ACIDIC TISSUES

Pulmonary Hypertension

Neonatal pulmonary hypertension is a condition associated with NO deficiency, right-to-left shunting of blood, profound hypoxemia, and low hemoglobin O_2 saturations (arterial hemoglobin O_2 saturations of approximately 55%). In a newborn sheep model of pulmonary hypertension, nitrite delivered as a nebulized aerosol produced a 50% reduction in hypoxia-induced pulmonary hypertension (119). Consistent with the conversion of nitrite to bioactive NO by hemoglobin, the vasodilation was accompanied by the formation of iron-nitrosyl hemoglobin and an increase in exhaled NO gas. Furthermore, the magnitude of decrease in pulmonary artery pressure in response to nitrite was dependent on blood pH, with decreased pulmonary pressures as pH was lowered, a chemistry consistent with the reduction of nitrite to NO by the deoxyhemoglobin reaction (119).

Subarachnoid Hemorrhage–Associated Cerebral Artery Vasospasm

Nitrite may also have therapeutic potential for conditions in which hemolysis plays a role in pathogenesis. In subarachnoid hemorrhage-induced cerebral vasospasm, the vasospasm of large intracranial arteries is thought to occur secondary to hemolysis of clotted erythrocytes within the subarachnoid space. In addition to free deoxy- and oxyhemoglobin scavenging NO, eNOS is inhibited, resulting in decreased regional NO bioavailability (120). In a well-characterized nonhuman primate model of subarachnoid hemorrhage-induced cerebral vasospasm, 14 animals were observed for 14 days, with half receiving a continuous infusion of nitrite and the others receiving saline. All animals in the control group developed cerebral vasospasm 7 to 14 days after aneurysm rupture, whereas vasospasm was prevented in all animals who received continuous intravenous infusions of nitrite (120). Furthermore, the concentration of nitrite in the cerebrospinal fluid of these animals correlated inversely with the degree of vasospasm (120).

Ischemia–Reperfusion Injury

In addition to pathologies involving the vasculature, nitrite may be beneficial in preventing ischemia–reperfusion injury in tissues (118,121). Tissue injury induced by ischemia and subsequent reperfusion is characterized by an increased production of ROS, leading to protein modification, mitochondrial damage, an inappropriate inflammatory response, and tissue death by both apoptosis and necrosis. In a model of ischemia–reperfusion in a Langendorf heart, nitrite at concentrations as low as 10 μM reduced infarct size (121). In a mouse model of hepatic ischemia–reperfusion, infusion of nitrite in doses as low as 1.2 nmol (blood levels of less than 200 nM) during the 45-minute ischemic period protected the liver from tissue death. This protection was replicated in a mouse model of myocardial infarction, in which 2.4 to 48 nmol of infused nitrite in the left ventricle reduced infarct size by 50% in comparison with control animals that did not receive nitrite. Interestingly, tissue protection was accompanied by the formation and maintenance of NO-modified proteins after reperfusion. In control animals, the concentration of plasma nitros(yl)ated species (RXNOs; including nitrosamines and iron-nitrosyl) was maximal 1 minute after reperfusion and decayed thereafter, whereas in nitrite-treated animals, these species remained elevated 30 minutes after reperfusion (118). Although the mechanism of this nitrite-dependent tissue cytoprotection has not yet been elucidated, and the identity of these modified proteins is unknown, several potential targets exist, including soluble guanylate cyclase and proteins within the mitochondrion.

KEY POINTS

- The endothelium and red blood cells interact to regulate NO-dependent blood flow and vascular homeostasis. The reactions of NO and hemoglobin largely determine the paracrine and endocrine properties of NO. Scavenging reactions control the paracrine properties of NO, and nitrite reductase reactions contribute to an endocrine hypoxic bioactivity. Disruption of this balance during intravascular hemolysis severely limits NO bioavailability, whereas the therapeutic amplification of these reactions with therapeutically delivered nitrite increases regional NO bioavailability.
- Intravascular hemolysis disrupts vasomotor tone through the release of hemoglobin from the red blood cell into plasma, which is then able to scavenge

endothelial-derived NO 600-fold faster than compartmentalized erythrocytic hemoglobin, thereby disrupting NO homeostasis.

- In addition to providing an NO scavenging role in the physiological regulation of NO-dependent vasodilation, hemoglobin and the erythrocyte may deliver NO as the hemoglobin deoxygenates. Recent data suggest that deoxygenated hemoglobin reduces blood nitrite to NO, leading to vasodilation along the physiological O_2 gradient.
- The endothelium can be considered a signaling organ that generates both paracrine NO and endocrine nitrite NO_2^- and other NO-modified protein species.

Future Goals

- To establish and further characterize the role of hemolysis as a mechanism for endothelial dysfunction that leads to proliferative vasculopathy and pulmonary hypertension in human disease such as the hereditary hemolytic anemias, iatrogenic hemolytic states such as cardiopulmonary bypass and red cell alloimmunization, and during infections such as malaria and hemolytic uremia syndrome
- To develop hemoglobin or red cell–based blood substitutes that bypass NO scavenging reactions (19,137)
- To determine the mechanisms by which nitrite formation and function is regulated physiologically, particularly the role of enzyme systems in this balance
- To determine the therapeutic effects of nitrite on the vasculature in the setting of hemolytic disease as well as other vascular dysfunction

REFERENCES

1 Reiter CD, Gladwin MT. An emerging role for nitric oxide in sickle cell disease vascular homeostasis and therapy. *Curr Opin Hematol.* 2003;10(2):99–107.

2 Rother RP, Bell L, Hillmen P, Gladwin MT. The clinical sequelae of intravascular hemolysis and extracellular plasma hemoglobin: a novel mechanism of human disease. *JAMA.* 2005; 293(13):1653–1662.

3 Schretzenmayer A. über Kreislaufregulatorische Vorgänge an den grossen Arterien bei der Muskelarbeit. *Pflügers Arch.* 1933; 232:S743–S748.

4 Furchgott RF, Zawadzki JV. The obligatory role of endothelial cells in the relaxation of arterial smooth muscle by acetylcholine. *Nature.* 1980;288(5789):373–376.

5 Ignarro LJ, Buga GM, Wood KS, et al. Endothelium-derived relaxing factor produced and released from artery and vein is nitric oxide. *Proc Natl Acad Sci USA.* 1987;84(24):9265–9269.

6 Ignarro LJ, Byrns RE, Buga GM, Wood KS. Endothelium-derived relaxing factor from pulmonary artery and vein possesses pharmacologic and chemical properties identical to those of nitric oxide radical. *Circ Res.* 1987;61(6):866–879.

7 Ignarro LJ, Wood KS, Wolin MS. Activation of purified soluble guanylate cyclase by protoporphyrin IX. *Proc Natl Acad Sci USA.* 1982;79(9):2870–2873.

8 Mittal CK, Arnold WP, Murad F. Characterization of protein inhibitors of guanylate cyclase activation from rat heart and bovine lung. *J Biol Chem.* 1978;253(4):1266–1271.

9 Palmer RM, Ferrige AG, Moncada S. Nitric oxide release accounts for the biological activity of endothelium-derived relaxing factor. *Nature.* 1987;327(6122):524–526.

10 Andrew PJ, Mayer B. Enzymatic function of nitric oxide synthases. *Cardiovasc Res.* 1999;43(3):521–531.

11 Stuehr DJ. Structure-function aspects in the nitric oxide synthases. *Annu Rev Pharmacol Toxicol.* 1997;37:339–359.

12 Chu A, Chambers DE, Lin CC, et al. Effects of inhibition of nitric oxide formation on basal vasomotion and endothelium-dependent responses of the coronary arteries in awake dogs. *J Clin Invest.* 1991;87(6):1964–1968.

13 Quyyumi AA, Dakak N, Andrews NP, et al. Nitric oxide activity in the human coronary circulation. Impact of risk factors for coronary atherosclerosis. *J Clin Invest.* 1995;95(4):1747–1755.

14 Rees DD, Palmer RM, Hodson HF, Moncada S. A specific inhibitor of nitric oxide formation from L-arginine attenuates endothelium-dependent relaxation. *Br J Pharmacol.* 1989;96(2):418–424.

15 Joannides R, Haefeli WE, Linder L, et al. Nitric oxide is responsible for flow-dependent dilatation of human peripheral conduit arteries in vivo. *Circulation.* 1995;91(5):1314–1319.

16 Panza JA, Casino PR, Kilcoyne CM, Quyyumi AA. Role of endothelium-derived nitric oxide in the abnormal endothelium-dependent vascular relaxation of patients with essential hypertension. *Circulation.* 1993;87(5):1468–1474.

17 Gilligan DM, Panza JA, Kilcoyne CM, et al. Contribution of endothelium-derived nitric oxide to exercise-induced vasodilation. *Circulation.* 1994;90(6):2853–2858.

18 Kellogg DL Jr., Liu Y, Kosiba IF, O'Donnell D. Role of nitric oxide in the vascular effects of local warming of the skin in humans. *J Appl Physiol.* 1999;86(4):1185–1190.

19 Olson JS, Foley EW, Rogge C, et al. No scavenging and the hypertensive effect of hemoglobin-based blood substitutes. *Free Radic Biol Med.* 2004;36(6):685–697.

20 Schechter AN, Gladwin MT. Hemoglobin and the paracrine and endocrine functions of nitric oxide. *N Engl J Med.* 2003; 348(15):1483–1485.

21 Wink DA, Mitchell JB. Chemical biology of nitric oxide: insights into regulatory, cytotoxic, and cytoprotective mechanisms of nitric oxide. *Free Radic Biol Med.* 1998;25(4–5):434–456.

22 Jeffers A, Glad Win MT, Kim-Shapiro D. Computation of plasma hemoglobin nitric oxide scavenging in hemolytic anemias. *Free Radic Biol Med.* 2006;41:1557–1565.

23 Dou Y, Maillett DH, Eich RF, Olson JS. Myoglobin as a model system for designing heme protein based blood substitutes. *Biophys Chem.* 2002;98(1–2):127–148.

24 Eich RF, Li T, Lemon DD, et al. Mechanism of NO-induced oxidation of myoglobin and hemoglobin. *Biochemistry.* 1996; 35(22):6976–6983.

25 Herold S, Exner M, Nauser T. Kinetic and mechanistic studies of the NO*-mediated oxidation of oxymyoglobin and oxyhemoglobin. *Biochemistry.* 2001;40(11):3385–3395.

26 Lancaster JR Jr. A tutorial on the diffusibility and reactivity of free nitric oxide. *Nitric Oxide.* 1997;1(1):18–30.

27 Huang KT, Han TH, Hyduke DR, et al. Modulation of nitric oxide bioavailability by erythrocytes. *Proc Natl Acad Sci USA.* 2001;98(20):11771–11776.

28 Vaughn MW, Huang KT, Kuo L, Liao JC. Erythrocytes possess an intrinsic barrier to nitric oxide consumption. *J Biol Chem.* 2000;275(4):2342–2348.

29 Coin JT, Olson JS. The rate of oxygen uptake by human red blood cells. *J Biol Chem.* 1979;254(4):1178–1190.

30 Liu X, Miller MJ, Joshi MS, et al. Diffusion-limited reaction of free nitric oxide with erythrocytes. *J Biol Chem.* 1998;273(30): 18709–18713.

31 Butler AR, Megson IL, Wright PG. Diffusion of nitric oxide and scavenging by blood in the vasculature. *Biochim Biophys Acta.* 1998;1425(1):168–176.

32 Liao JC, Hein TW, Vaughn MW, et al. Intravascular flow decreases erythrocyte consumption of nitric oxide. *Proc Natl Acad Sci USA.* 1999;96(15):8757–8761.

33 Vaughn MW, Kuo L, Liao JC. Effective diffusion distance of nitric oxide in the microcirculation. *Am J Physiol.* 1998;274(5 Pt 2):H1705–H1714.

34 Nakai K, Sakuma I, Ohta T, et al. Permeability characteristics of hemoglobin derivatives across cultured endothelial cell monolayers. *J Lab Clin Med.* 1998;132(4):313–319.

35 Ruschitzka FT, Wenger RH, Stallmach T, et al. Nitric oxide prevents cardiovascular disease and determines survival in polyglobulic mice overexpressing erythropoietin. *Proc Natl Acad Sci USA.* 2000;97(21):11609–11613.

36 Azarov I, Huang KT, Basu S, et al. Nitric oxide scavenging by red blood cells as a function of hematocrit and oxygenation. *J Biol Chem.* 2005;280(47):39024–39032.

37 Beleslin-Cokic BB, Cokic VP, Yu X, et al. Erythropoietin and hypoxia stimulate erythropoietin receptor and nitric oxide production by endothelial cells. *Blood.* 2004;104(7):2073–2080.

38 Gladwin MT, Lancaster JR Jr., Freeman BA, Schechter AN. Nitric oxide's reactions with hemoglobin: a view through the SNO-storm. *Nat Med.* 2003;9(5):496–500.

39 Reiter CD, Wang X, Tanus-Santos JE, et al. Cell-free hemoglobin limits nitric oxide bioavailability in sickle-cell disease. *Nat Med.* 2002;8(12):1383–1389.

40 Nakai K, Ohta T, Sakuma I, et al. Inhibition of endothelium-dependent relaxation by hemoglobin in rabbit aortic strips: comparison between acellular hemoglobin derivatives and cellular hemoglobins. *J Cardiovasc Pharmacol.* 1996;28(1):115–123.

41 Pohl U, Lamontagne D. Impaired tissue perfusion after inhibition of endothelium-derived nitric oxide. *Basic Res Cardiol.* 1991;86(Suppl 2):97–105.

42 Minneci PC, Deans KJ, Zhi H, et al. Hemolysis-associated endothelial dysfunction mediated by accelerated NO inactivation by decompartmentalized oxyhemoglobin. *J Clin Invest.* 2005;115(12):3409–3417. Epub 2005 Nov 17.

43 Eberhardt RT, McMahon L, Duffy SJ, et al. Sickle cell anemia is associated with reduced nitric oxide bioactivity in peripheral conduit and resistance vessels. *Am J Hematol.* 2003;74(2):104–111.

44 Kaul DK, Liu XD, Chang HY, et al. Effect of fetal hemoglobin on microvascular regulation in sickle transgenic-knockout mice. *J Clin Invest.* 2004;114(8):1136–1145.

45 Kaul DK, Liu XD, Fabry ME, Nagel RL. Impaired nitric oxide-mediated vasodilation in transgenic sickle mouse. *Am J Physiol Heart Circ Physiol.* 2000;278(6):H1799–H1806.

46 Nath KA, Shah V, Haggard JJ, et al. Mechanisms of vascular instability in a transgenic mouse model of sickle cell disease. *Am J Physiol Regul Integr Comp Physiol.* 2000;279(6):R1949–R1955.

47 Hess JR, MacDonald VW, Brinkley WW. Systemic and pulmonary hypertension after resuscitation with cell-free hemoglobin. *J Appl Physiol.* 1993;74(4):1769–1778.

48 Murray JA, Ledlow A, Launspach J, et al. The effects of recombinant human hemoglobin on esophageal motor functions in humans. *Gastroenterology.* 1995;109(4):1241–1248.

49 Vogel WM, Dennis RC, Cassidy G, et al. Coronary constrictor effect of stroma-free hemoglobin solutions. *Am J Physiol.* 1986;251(2 Pt 2):H413–H420.

50 De Caterina R, Libby P, Peng HB, et al. Nitric oxide decreases cytokine-induced endothelial activation. Nitric oxide selectively reduces endothelial expression of adhesion molecules and proinflammatory cytokines. *J Clin Invest.* 1995;96(1):60–68.

51 Gladwin MT, Schechter AN, Ognibene FP, et al. Divergent nitric oxide bioavailability in men and women with sickle cell disease. *Circulation.* 2003;107(2):271–278.

52 Radomski MW, Palmer RM, Moncada S. Endogenous nitric oxide inhibits human platelet adhesion to vascular endothelium. *Lancet.* 1987;2(8567):1057–1058.

53 Schafer A, Wiesmann F, Neubauer S, et al. Rapid regulation of platelet activation in vivo by nitric oxide. *Circulation.* 2004; 109(15):1819–1822.

54 Shao J, Miyata T, Yamada K, et al. Protective role of nitric oxide in a model of thrombotic microangiopathy in rats. *J Am Soc Nephrol.* 2001;12(10):2088–2097.

55 Shibata Y, Takaoka M, Maekawa D, et al. Involvement of nitric oxide in the suppressive effect of 17beta-estradiol on endothelin-1 overproduction in ischemic acute renal failure. *J Cardiovasc Pharmacol.* 2004;44(Suppl 1):S459–S461.

56 Rodgers GP, Walker EC, Podgor MJ. Is "relative" hypertension a risk factor for vaso-occlusive complications in sickle cell disease? *Am J Med Sci.* 1993;305(3):150–156.

57 Gladwin MT, Sachdev V, Jison ML, et al. Pulmonary hypertension as a risk factor for death in patients with sickle cell disease. *N Engl J Med.* 2004;350(9):886–895.

58 Jison ML, Gladwin MT. Hemolytic anemia-associated pulmonary hypertension of sickle cell disease and the nitric oxide/arginine pathway. *Am J Respir Crit Care Med.* 2003;168(1):3–4.

59 Hillmen P, Hall C, Marsh JC, et al. Effect of eculizumab on hemolysis and transfusion requirements in patients with paroxysmal nocturnal hemoglobinuria. *N Engl J Med.* 2004;350(6): 552–559.

60 Morris CR, Kato GJ, Poljakovic M, et al. Dysregulated arginine metabolism, hemolysis-associated pulmonary hypertension, and mortality in sickle cell disease. *JAMA.* 2005;294(1):81–90.

61 Kato GJ, McGowan VR, Machado RF, et al. Lactate dehydrogenase as a biomarker of hemolysis-associated nitric oxide resistance, priapism, leg ulceration, pulmonary hypertension and death in patients with sickle cell disease. *Blood.* 2005;107(6): 2279–85. Epub 2005 Nov 15.

62 Nolan VG, Wyszynski DF, Farrer LA, Steinberg MH. Hemolysis-associated priapism in sickle cell disease. *Blood.* 2005;106(9): 3264–3267.

63 Castro O, Gladwin MT. Pulmonary hypertension in sickle cell disease: mechanisms, diagnosis, and management. *Hematol Oncol Clin North Am.* 2005;19(5):881–896, vii.

64 Machado RF, Gladwin MT. Chronic sickle cell lung disease: new insights into the diagnosis, pathogenesis and treatment of pulmonary hypertension. *Br J Haematol.* 2005;129(4):449–464.

65 Edwards DH, Griffith TM, Ryley HC, Henderson AH. Haptoglobin-haemoglobin complex in human plasma inhibits endothelium dependent relaxation: evidence that endothelium derived relaxing factor acts as a local autocoid. *Cardiovasc Res.* 1986;20(8):549–556.

66 Wang X, Tanus-Santos JE, Reiter CD, et al. Biological activity of nitric oxide in the plasmatic compartment. *Proc Natl Acad Sci USA.* 2004;101(31):11477–11482.

67 Philippidis P, Mason JC, Evans BJ, et al. Hemoglobin scavenger receptor CD163 mediates interleukin-10 release and heme oxygenase-1 synthesis: antiinflammatory monocyte-macrophage responses in vitro, in resolving skin blisters in vivo, and after cardiopulmonary bypass surgery. *Circ Res.* 2004;94(1): 119–126.

68 Baranano DE, Rao M, Ferris CD, Snyder SH. Biliverdin reductase: a major physiologic cytoprotectant. *Proc Natl Acad Sci USA.* 2002;99(25):16093–16098.

69 Ferris CD, Jaffrey SR, Sawa A, et al. Haem oxygenase-1 prevents cell death by regulating cellular iron. *Nat Cell Biol.* 1999;1(3): 152–157.

70 Nath KA, Katusic ZS, Gladwin MT. The perfusion paradox and vascular instability in sickle cell disease. *Microcirculation.* 2004;11(2):179–193.

71 Otterbein LE, Bach FH, Alam J, et al. Carbon monoxide has anti-inflammatory effects involving the mitogen-activated protein kinase pathway. *Nat Med.* 2000;6(4):422–428.

72 Ryter SW, Otterbein LE, Morse D, Choi AM. Heme oxygenase/carbon monoxide signaling pathways: regulation and functional significance. *Mol Cell Biochem.* 2002;234–235(1–2): 249–263.

73 Cannon RO 3rd, Schechter AN, Panza JA, et al. Effects of inhaled nitric oxide on regional blood flow are consistent with intravascular nitric oxide delivery. *J Clin Invest.* 2001;108(2):279–287.

74 Rassaf T, Preik M, Kleinbongard P, et al. Evidence for in vivo transport of bioactive nitric oxide in human plasma. *J Clin Invest.* 2002;109(9):1241–1248.

75 Fox-Robichaud A, Payne D, Hasan SU, et al. Inhaled NO as a viable antiadhesive therapy for ischemia/reperfusion injury of distal microvascular beds. *J Clin Invest.* 1998;101(11):2497–2505.

76 Ng ES, Jourd'heuil D, McCord JM, et al. Enhanced S-nitrosoalbumin formation from inhaled NO during ischemia/reperfusion. *Circ Res.* 2004;94(4):559–565.

77 Takahashi Y, Kobayashi H, Tanaka N, et al. Nitrosyl hemoglobin in blood of normoxic and hypoxic sheep during nitric oxide inhalation. *Am J Physiol.* 1998;274(1 Pt 2):H349–H357.

78 Troncy E, Francoeur M, Salazkin I, et al. Extra-pulmonary effects of inhaled nitric oxide in swine with and without phenylephrine. *Br J Anaesth.* 1997;79(5):631–640.

79 Kinsella JP, CuHer GR, Walsh WF, et al. Early inhaled nitric oxide therapy in premature newborns with respiratory failure. *N Engl J Med.* 2006;355:354–364.

80 Scharfstein JS, Keaney JF, Jr, Slivka A, et al. In vivo transfer of nitric oxide between a plasma protein-bound reservoir and low molecular weight thiols. *J Clin Invest.* 1994;94(4):1432–1439.

81 Stamler JS, Jaraki O, Osborne J, et al. Nitric oxide circulates in mammalian plasma primarily as an S-nitroso adduct of serum albumin. *Proc Natl Acad Sci USA.* 1992;89(16):7674–7677.

82 Jia L, Bonaventura C, Bonaventura J, Stamler JS. S-nitrosohaemoglobin: a dynamic activity of blood involved in vascular control [see comments]. *Nature.* 1996;380(6571):221–226.

83 Gladwin MT, Ognibene FP, Pannell LK, et al. Relative role of heme nitrosylation and beta-cysteine 93 nitrosation in the transport and metabolism of nitric oxide by hemoglobin in the human circulation. *Proc Natl Acad Sci USA.* 2000;97(18):9943–9948.

84 Gladwin MT, Schechter AN. NO contest: nitrite versus S-nitroso-hemoglobin. *Circ Res.* 2004;94(7):851–855.

85 Marley R, Feelisch M, Holt S, Moore K. A chemiluminescence-based assay for S-nitrosoalbumin and other plasma S-nitrosothiols. *Free Radic Res.* 2000;32(1):1–9.

86 Rassaf T, Bryan NS, Kelm M, Feelisch M. Concomitant presence of N-nitroso and S-nitroso proteins in human plasma. *Free Radic Biol Med.* 2002;33(11):1590–1596.

87 Rassaf T, Bryan NS, Maloney RE, et al. NO adducts in mammalian red blood cells: too much or too little? *Nat Med.* 2003;9(5):481–482; author reply 2–3.

88 Stamler JS. S-nitrosothiols in the blood: roles, amounts, and methods of analysis. *Circ Res.* 2004;94(4):414–417.

89 Cosby K, Partovi KS, Crawford JH, et al. Nitrite reduction to nitric oxide by deoxyhemoglobin vasodilates the human circulation. *Nat Med.* 2003;9(12):1498–1505.

90 Gruetter CA, Barry BK, McNamara DB, Kadowitz PJ, et al. Coronary arterial relaxation and guanylate cyclase activation by cigarette smoke, N′-nitrosonornicotine and nitric oxide. *J Pharmacol Exp Ther.* 1980;214(1):9–15.

91 Lim DG, Sweeney S, Bloodsworth A, et al. Nitrolinoleate, a nitric oxide-derived mediator of cell function: synthesis, characterization, and vasomotor activity. *Proc Natl Acad Sci USA.* 2002;99(25):15941–15946.

92 Stamler JS, Jia L, Eu JP, et al. Blood flow regulation by S-nitrosohemoglobin in the physiological oxygen gradient. *Science.* 1997;276(5321):2034–2037.

93 Gow AJ, Luchsinger BP, Pawloski JR, et al. The oxyhemoglobin reaction of nitric oxide. *Proc Natl Acad Sci USA.* 1999;96(16): 9027–9032.

94 McMahon TJ, Moon RE, Luschinger BP, et al. Nitric oxide in the human respiratory cycle. *Nat Med.* 2002;8(7):711–717.

95 Pawloski JR, Hess DT, Stamler JS. Export by red blood cells of nitric oxide bioactivity. *Nature.* 2001;409(6820):622–626.

96 Singel DJ, Stamler JS. Chemical physiology of blood flow regulation by red blood cells: the role of nitric oxide and S-nitrosohemoglobin. *Annu Rev Physiol.* 2005;67:99–145.

97 Huang Z, Louderback JG, Goyal M, et al. Nitric oxide binding to oxygenated hemoglobin under physiological conditions. *Biochim Biophys Acta.* 2001;1568(3):252–260.

98 Huang Z, Ucer KB, Murphy T, et al. Kinetics of nitric oxide binding to R-state hemoglobin. *Biochem Biophys Res Commun.* 2002;292(4):812–818.

99 Zhang Y, Hogg N. Mixing artifacts from the bolus addition of nitric oxide to oxymyoglobin: implications for S-nitrosothiol formation. *Free Radic Biol Med.* 2002;32(11):1212–1219.

100 Xu X, Cho M, Spencer NY, et al. Measurements of nitric oxide on the heme iron and beta-93 thiol of human hemoglobin during cycles of oxygenation and deoxygenation. *Proc Natl Acad Sci USA.* 2003;100(20):11303–11308.

101 Doctor A, Gaston B, Kim-Shapiro D. Detecting physiologic fluctuations in the S-nitrosohemoglobin micropopulation: triiodide versus 3C. *Blood*. 2006;108:3225–3226.
102 Doctor A, Platt R, Sheram ML, et al. Hemoglobin conformation couples erythrocyte S-nitrosothiol content to O2 gradients. *Proc Natl Acad Sci USA*. 2005;102(16):5709–5714.
103 Rogers SC, Khalatbari A, Gapper PW, et al. Detection of human haemoglobin-bound nitric oxide. *J Biol Chem*. 2005;280(29): 26720–26728. Epub 2005 May 6.
104 Dejam A, Hunter CJ, Pelletier MM, et al. Erythrocytes are the major intravascular storage sites of nitrite in human blood. *Blood*. 2005;106(2):734–739. Epub 2005 Mar 17.
105 Gladwin MT, Crawford JH, Patel RP. The biochemistry of nitric oxide, nitrite, and hemoglobin: role in blood flow regulation. *Free Radic Biol Med*. 2004;36(6):707–717.
106 Gladwin MT, Shelhamer JH, Schechter AN, et al. Role of circulating nitrite and S-nitrosohemoglobin in the regulation of regional blood flow in humans. *Proc Natl Acad Sci USA*. 2000; 97(21):11482–11487.
107 Modin A, Bjorne H, Herulf M, et al. Nitrite-derived nitric oxide: a possible mediator of 'acidic-metabolic' vasodilation. *Acta Physiol Scand*. 2001;171(1):9–16.
108 Shiva S, Wang X, Ringwood LA, et al. Ceruloplasmin is a NO oxidase and nitrite synthase that determines endocrine NO homeostasis. *Nat Chem Biol*. 2006;2:486–493.
109 Bjorne HH, Petersson J, Phillipson M, et al. Nitrite in saliva increases gastric mucosal blood flow and mucus thickness. *J Clin Invest*. 2004;113(1):106–114.
110 Lundberg JO, Govoni M. Inorganic nitrate is a possible source for systemic generation of nitric oxide. *Free Radic Biol Med*. 2004;37(3):395–400.
111 Lundberg JO, Weitzberg E. NO generation from nitrite and its role in vascular control. *Arterioscler Thromb Vasc Biol*. 2005; 25(5):915–922.
112 Lundberg JO, Weitzberg E, Cole JA, Benjamin N. Nitrate, bacteria and human health. *Nat Rev Microbiol*. 2004;2(7):593–602.
113 Rassaf T, Feelisch M, Kelm M. Circulating no pool: assessment of nitrite and nitroso species in blood and tissues. *Free Radic Biol Med*. 2004;36(4):413–422.
114 Furchgott RF, Bhadrakom S. Reactions of strips of rabbit aorta to epinephrine, isopropylarterenol, sodium nitrite and other drugs. *J Pharmacol Exp Ther*. 1953(108):129–143.
115 Ignarro LJ, Gruetter CA. Requirement of thiols for activation of coronary arterial guanylate cyclase by glyceryl trinitrate and sodium nitrite: possible involvement of S-nitrosothiols. *Biochim Biophys Acta*. 1980;631(2):221–231.
116 Lauer T, Preik M, Rassaf T, et al. Plasma nitrite rather than nitrate reflects regional endothelial nitric oxide synthase activity but lacks intrinsic vasodilator action. *Proc Natl Acad Sci USA*. 2001;98(22):12814–12819.
117 Bryan NS, Rassaf T, Maloney RE, et al. Cellular targets and mechanisms of nitros(yl)ation: an insight into their nature and kinetics in vivo. *Proc Natl Acad Sci USA*. 2004;101(12):4308–4313. Epub 2004 Mar 10.
118 Duranski MR, Greer JJ, Dejam A, et al. Cytoprotective effects of nitrite during in vivo ischemia-reperfusion of the heart and liver. *J Clin Invest*. 2005;115(5):1232–1240. Epub 2005 Apr 14.
119 Hunter CJ, Dejam A, Blood AB, et al. Inhaled nebulized nitrite is a hypoxia-sensitive NO-dependent selective pulmonary vasodilator. *Nat Med*. 2004;10(10):1122–1127.
120 Pluta RM, Dejam A, Grimes G, et al. Nitrite infusions to prevent delayed cerebral vasospasm in a primate model of subarachnoid hemorrhage. *JAMA*. 2005;293(12):1477–1484.
121 Webb A, Bond R, McLean P, et al. Reduction of nitrite to nitric oxide during ischemia protects against myocardial ischemia-reperfusion damage. *Proc Natl Acad Sci USA*. 2004;101(37): 13683–13688.
122 Millar TM, Stevens CR, Benjamin N, et al. Xanthine oxidoreductase catalyses the reduction of nitrates and nitrite to nitric oxide under hypoxic conditions. *FEBS Lett*. 1998;427(2):225–228.
123 Zhang Z, Naughton D, Winyard PG, et al. Generation of nitric oxide by a nitrite reductase activity of xanthine oxidase: a potential pathway for nitric oxide formation in the absence of nitric oxide synthase activity [published erratum appears in *Biochem Biophys Res Commun* 1998 Oct 20;251(2):667]. *Biochem Biophys Res Commun*. 1998;249(3):767–772.
124 Zhang Z, Naughton DP, Blake DR, et al. Human xanthine oxidase converts nitrite ions into nitric oxide (NO). *Biochem Soc Trans*. 1997;25(3):S524.
125 Zweier JL, Wang P, Samouilov A, Kuppusamy P. Enzyme-independent formation of nitric oxide in biological tissues [see comments]. *Nat Med*. 1995;1(8):804–809.
126 Dejam A, Hunter CJ, Schechter AN, Gladwin MT. Emerging role of nitrite in human biology. *Blood Cells Mol Dis*. 2004; 32(3):423–429.
127 Nagababu E, Ramasamy S, Abernethy DR, Rifkind JM. Active nitric oxide produced in the red cell under hypoxic conditions by deoxyhemoglobin mediated nitrite reduction. *J Biol Chem*. 2003;278(47):46349–46356. Epub 2003 Sep 2.
128 Pegg RB, Shahidi F. *Nitrite Curing of Meat*. Trumbull, CT: Food & Nutrition Press, 2000.
129 Pedigo L. Antagonism between amyl nitrite and prussic acid. *Trans Med Soc Virginia*. 1888;19:124–131.
130 Weiss S, Wilkins RW, Haynes FW. The nature of circulatory collapse induced by sodium nitrite. *J Clin Invest*. 1937;16(1):73–84.
131 Doyle MP, Pickering RA, DeWeert TM, et al. Kinetics and mechanism of the oxidation of human deoxyhemoglobin by nitrites. *J Biol Chem*. 1981;256(23):12393–12398.
132 Huang ZS, Kim-Shapiro S, Patel R, et al. Enzymatic function of hemoglobin as a nitrite reductase that produces NO under allosteric control. *J Clin Invest*. 2005;115(8):2099–107.
133 Huang KT, Keszler A, Patel NK, et al. The reaction between nitrite and deoxyhemoglobin: reassessment of reaction kinetics and stoichiometry. *J Biol Chem*. 2005;280(35):31126–31131. Epub 2005 Apr 18.
134 Tune JD, Gorman MW, Feigl EO. Matching coronary blood flow to myocardial oxygen consumption. *J Appl Physiol*. 2004; 97(1):404–415.
135 Ross JM, Weldy J, Guyton AC. Autoregulation of blood flow by oxygen lack. *Am J Physiol*. 1963;202:21–24.
136 Tsai AG, Johnson PC, Intaglietta M. Oxygen gradients in the microcirculation. *Physiol Rev*. 2003;83(3):933–963.
137 Segal SS, Damon DN, Duling BR. Propagation of vasomotor responses coordinates arteriolar resistances. *Am J Physiol*. 1989; 256(3 Pt 2):H832–H837.
138 Segal SS, Duling BR. Conduction of vasomotor responses in arterioles: a role for cell-to-cell coupling? *Am J Physiol*. 1989; 256(3 Pt 2):H838–H845.

139 Crawford JH, Isbell TS, Huang Z, et al. Hypoxia, red blood cells and nitrite regulate NO-dependent hypoxic vasodilatation. *Blood*. 2005;107(2):566–574. Epub 2005 Sep 29.

140 Gladwin M, Schechter AN, Kim-Shapiro DB, et al. The emerging biology of the nitrite anion in signaling, blood flow and hypoxic nitric oxide homeostasis. *Nat Chem Biol*. 2005.

141 Shiva S, Huang Z, Grubina R, et al. Myoglobin is a nitrite reductase that generates NO and inhibits mitochondrial respiration. *Circ Res*. 2007; In press.

142 Manjula BN, Tsai A, Upadhya R, et al. Site-specific PEGylation of hemoglobin at Cys-93(beta): correlation between the colligative properties of the PEGylated protein and the length of the conjugated PEG chain. *Bioconjug Chem*. 2003;14(2):464–472.

143 Solomonson LP, Flam BR, Pendleton LC, et al. The caveolar nitric oxide synthase/arginine regeneration system for NO production in ECs. *J Exp Biol*. 2003;206(Pt 12):2083–2087.

144 Wyatt AW, Steinert JR, Mann GE. Modulation of the L-arginine/nitric oxide signalling pathway in vascular endothelial cells. *Biochem Soc Symp*. 2004(71):143–156.

145 Alp NJ, Channon KM. Regulation of endothelial nitric oxide synthase by tetrahydrobiopterin in vascular disease. *Arterioscler Thromb Vasc Biol*. 2004;24(3):413–420.

146 Fulton D, Gratton JP, Sessa WC. Post-translational control of endothelial nitric oxide synthase: why isn't calcium/calmodulin enough? *J Pharmacol Exp Ther*. 2001;299(3):818–824.

147 Wu KK. Regulation of endothelial nitric oxide synthase activity and gene expression. *Ann NY Acad Sci*. 2002;962:122–130.

148 Busconi L, Michel T. Endothelial nitric oxide synthase. N-terminal myristoylation determines subcellular localization. *J Biol Chem*. 1993;268(12):8410–8413.

149 Prabhakar P, Cheng V, Michel T. A chimeric transmembrane domain directs endothelial nitric-oxide synthase palmitoylation and targeting to plasmalemmal caveolae. *J Biol Chem*. 2000;275(25):19416–19421.

150 Michel JB, Feron O, Sase K, et al. Caveolin versus calmodulin. Counterbalancing allosteric modulators of endothelial nitric oxide synthase. *J Biol Chem*. 1997;272(41):25907–25912.

151 Goligorsky MS, Li H, Brodsky S, Chen J. Relationships between caveolae and eNOS: everything in proximity and the proximity of everything. *Am J Physiol Renal Physiol*. 2002;283(1):F1–F10.

152 Minshall RD, Sessa WC, Stan RV, et al. Caveolin regulation of endothelial function. *Am J Physiol Lung Cell Mol Physiol*. 2003; 285(6):L1179–L1183.

153 Kone BC. Protein-protein interactions controlling nitric oxide synthases. *Acta Physiol Scand*. 2000;168(1):27–31.

154 Li H, Wallerath T, Munzel T, Forstermann U. Regulation of endothelial-type NO synthase expression in pathophysiology and in response to drugs. *Nitric Oxide*. 2002;7(3):149–164.

155 D'Orleans-Juste P, Plante M, Honore JC, et al. Synthesis and degradation of endothelin-1. *Can J Physiol Pharmacol*. 2003; 81(6):503–510.

156 Mohacsi A, Magyar J, Tamas B, Nanasi PP. Effects of endothelins on cardiac and vascular cells: new therapeutic target for the future? *Curr Vasc Pharmacol*. 2004;2(1):53–63.

157 Davidge ST. Prostaglandin H synthase and vascular function. *Circ Res*. 2001;89(8):650–660.

158 Katusic ZS, Shepherd JT. Endothelium-derived vasoactive factors: II Endothelium-dependent contraction. *Hypertension*. 1991;18(5Suppl):III86–III92.

159 Wang LH, Kulmacz RJ. Thromboxane synthase: structure and function of protein and gene. *Prostaglandins Other Lipid Mediat*. 2002;68–69:409–422.

160 Vanhoutte PM, Feletou M, Taddei S. Endothelium-dependent contractions in hypertension. *Br J Pharmacol*. 2005;144(4):449–458.

161 Hay DL, Smith DM. Adrenomedullin receptors: molecular identity and function. *Peptides*. 2001;22(11):1753–1763.

162 Nagaya N, Kangawa K. Adrenomedullin in the treatment of pulmonary hypertension. *Peptides*. 2004;25(11):2013–2018.

163 Ignarro LJ. Nitric oxide as a unique signaling molecule in the vascular system: a historical overview. *J Physiol Pharmacol*. 2002; 53(4 Pt 1):503–514.

164 Krumenacker JS, Hanafy KA, Murad F. Regulation of nitric oxide and soluble guanylyl cyclase. *Brain Res Bull*. 2004;62(6):505–515.

165 Ignjatovic T, Stanisavljevic S, Brovkovych V, et al. Kinin B1 receptors stimulate nitric oxide production in endothelial cells: signaling pathways activated by angiotensin I-converting enzyme inhibitors and peptide ligands. *Mol Pharmacol*. 2004;66(5):1310–1316.

166 Shepherd JT, Katusic ZS. Endothelium-derived vasoactive factors: I. Endothelium-dependent relaxation. *Hypertension*. 1991; 18(5Suppl):III76–III85.

167 Parkington HC, Coleman HA, Tare M. Prostacyclin and endothelium-dependent hyperpolarization. *Pharmacol Res*. 2004; 49(6):509–514.

63

Leukocyte–Endothelial Cell Interactions

Volker Vielhauer, Xavier Cullere, and Tanya Mayadas

Brigham and Women's Hospital, Harvard Medical School, Boston, Massachusetts

Pathologists of the 19th century were aware that leukocytes migrate from the blood across the wall of microvessels and accumulate in inflamed tissue. The purpose of this migration, a process called *diapedesis*, remained enigmatic until the discoveries of Elias Metchnikoff in 1901. He demonstrated that leukocytes attack and kill bacteria, and he recognized diapedesis as a fundamental mechanism of host defense (1). Indeed, any immune response relies on the well-coordinated trafficking of leukocytes from the circulation into lymphoid tissue (homeostatic immune surveillance) and sites of tissue injury (inflammation). The emerging model is that leukocyte adhesion and extravasation are guided by a series of well-choreographed interactions between leukocytes and endothelial cells (ECs). These interactions are supported by temporally controlled and spatially restricted engagement of receptor–ligand pairs between cell types. This triggers bidirectional signaling events that lead to changes in cellular phenotype and function. The extravasation of a specific subset of leukocytes at selective sites during inflammation is achieved through expression of specific adhesion molecules and cell-type–selective chemoattractants (2,3).

The function of recruited neutrophils is to control and eliminate infection and to promote tissue repair. To achieve these ends during an inflammatory response, the lifespan of transmigrated neutrophils is extended, compared to circulating counterparts, partially through endothelial-derived factors. The endothelium also provides cues to those phagocytes that result in their locally contained release of cytokines, chemokines, growth factors, proteinases, and reactive oxygen species (ROS). Clark and Clark (4) recognized the importance of changes in the "consistency" of the endothelium in initiating an inflammatory response. Furthermore, they noted that a spatially localized and graded physiologic response to inflammatory stimuli was important for maintaining the integrity of the endothelium (4). It is now well accepted that excessive leukocyte accumulation contributes to the initiation and progression of a number of diseases including atherosclerosis, vasculitis, meningitis, ischemia–reperfusion injury, and autoimmune disease. Because the inflamed endothelium is responsible for coordinating the recruitment and function of phagocytes, and is also potentially the most proximal target of leukocyte-mediated signals, it is not surprising that endothelial dysfunction has been linked to the pathogenesis of multiple vascular diseases. The thesis of this chapter is that the cross-talk between leukocytes and ECs during an inflammatory response has short- and long-term effects on the functional phenotype of the vasculature. In one direction, endothelial-derived effectors modulate leukocyte activity and their survival. In the other direction, activated phagocytes release cellular factors that can moderate EC function and viability in physiological and pathophysiological settings. The bidirectional communication between leukocytes and ECs is essential for host defense and physiologic inflammation (Figure 63.1). Understanding the molecular basis of this cross-talk will provide insights into how its dysregulation can provide a nidus for endothelial dysfunction.

LEUKOCYTE ROLLING

The first step in leukocyte recruitment comprises initial tethering followed by reversible rolling. Rolling is mediated by molecules that are specialized for adhesion under shear stress, namely the selectins. The selectin family consists of leukocyte- (L-selectin, CD62L), platelet- (P-selectin, CD62P), and endothelial selectin (E-selectin, CD62E) (5). L-selectin is constitutively expressed on nonactivated neutrophils, monocytes, and eosinophils, and on the majority of lymphocytes. L-selectin is important in T-cell rolling and homing to secondary lymphoid organs as well as leukocyte rolling on inflamed endothelium. P-selectin and E-selectin are induced on the surface of the vascular endothelium activated by inflammatory stimuli. P-selectin also is expressed on the surface of activated platelets. Both P- and E-selectin support rolling of all leukocyte subsets in various inflamed tissues and are essential for rolling and homing of hematopoietic progenitor cells to the bone marrow after bone marrow transplantation (6,7). Selectins recognize and bind to specific carbohydrate moieties – the posttranscriptionally modified sialylated, fucosylated lactosamine-type glycans that decorate core molecules

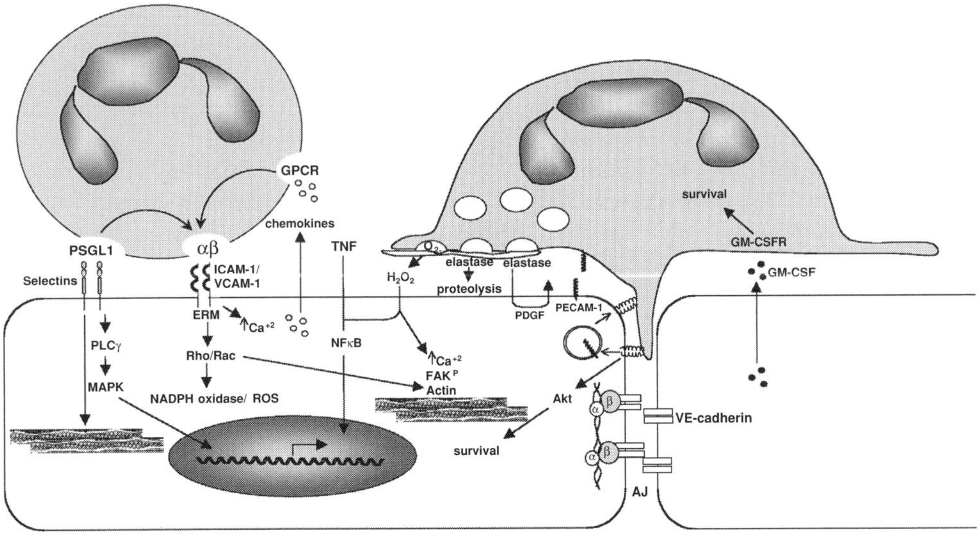

Figure 63.1. Leukocyte-derived inputs that modulate the phenotype of the endothelium. Phagocyte interaction with the endothelium occurs through adhesion molecules: PSGL-1/P- and E-selectin, and $\beta 2$ or $\beta 1$ integrin/ICAM-1 and VCAM-1, respectively. These interactions increase intracellular Ca^{+2} and ROS, and lead to changes in the actin cytoskeleton and gene transcription in the endothelium. After phagocytes undergo firm adhesion, they transmigrate. PECAM-1, constitutively recycling at the endothelial borders, is targeted to areas of leukocyte transmigration, where homophilic interactions with PECAM-1 on the leukocyte promote leukocyte egress. PECAM-1 engagement activates the antiapoptotic kinase Akt in ECs. Adherent leukocytes release superoxide and proteinases. Superoxide (O_2^-) is converted to hydrogen peroxide (H_2O_2), which penetrates the endothelium and has effects on NF-κB activation, intracellular Ca^{+2} levels, and actin dynamics. Elastase has proteolytic functions that result in cleavage of surface adhesion receptors and chemokines, as well as nondegradative signaling functions that promote PDGF release. High and/or prolonged exposure of the endothelium to leukocyte-derived products leads to EC injury. GM-CSF produced by activated ECs extends the lifespan of migrating neutrophils and thus the pool of functional neutrophils at sites of inflammation.

(mainly mucins) as a result of post-translational modification by specific glycosylation enzymes. An important receptor on leukocytes that contain these selectin ligands is the P-selectin glycoprotein ligand (PSGL)-1, originally identified as a ligand for P-selectin but subsequently shown also to bind E-selectin. Consistent with this, PSGL-1 supports both P- and E-selectin–mediated leukocyte rolling (8–12). Notably, the rolling phase is not mediated exclusively by selectins and their glycosylated ligands. The integrin $\alpha 4\beta 1$ (very late activation antigen-4, VLA-4) can support T-cell tethering and rolling on vascular cell adhesion molecule (VCAM)-1, an integrin ligand expressed on activated endothelium (13,14). CD44, a transmembrane glycoprotein present on activated T cells, facilitates leukocyte rolling on endothelial-bound hyaluronan (15,16). CD44 also associates physically with VLA-4 on T cells to facilitate VLA-4–mediated slow rolling and subsequent firm adhesion of T cells (17). Interestingly, fractalkine, a transmembrane mucin-chemokine chimeric molecule induced on the surface of activated ECs, mediates rapid capture (without rolling) and integrin-independent firm adhesion of free flowing monocytes, T cells, and NK cells through interaction with its receptor CX3CR1 (18).

One of the earliest responses of leukocyte contact with the endothelium is an elevation of intracellular free calcium that is accompanied by cytoskeletal changes in the endothelium. Mounting evidence suggests that "rolling" receptors contribute to these changes by serving a signaling function in ECs (19). For example, E-selectin engagement promotes its linkage to the cytoskeleton and results in the dephosphorylation of its cytoplasmic tail (20). Also, clustered E-selectin redistributes to caveolin-1–containing lipid rafts, where it activates signaling molecules such as phospholipase Cγ (21,22). E-selectin activates the immediate early response gene *c-fos* through the mitogen-activated protein kinase (MAPK) pathway (23), suggesting that leukocyte–EC interactions may modulate the transcriptional profile of the endothelium. Further studies are needed to demonstrate that signaling through E-selectin has functional consequences for the endothelium during an inflammatory response. Studies evaluating P-selectin–mediated signaling in the endothelium are limited, but suggest that it too may serve a signaling function. P-selectin serine and threonine phosphorylation of its cytoplasmic tail has been documented in activated ECs and platelets (24), and histidine phosphorylation is noted in stimulated platelets (25). Because the cytoplasmic tails of E- or P-selectin are not necessary for leukocyte adhesion per se (26), the role of their intracellular domains may indeed be primarily to modulate downstream biologic events that affect the phenotype of the endothelium.

Endothelial engagement by rolling leukocytes has functional consequences for the leukocyte. Transient interaction with P- and E-selectin leads to the induction of an intermediate state of leukocyte integrin activation that reduces the velocity of the rolling leukocytes. This provides an opportunity for leukocytes to sample the microenvironment for additional cues required for leukocyte arrest (27–29). On the leukocyte, occupancy of PSGL-1 induces tyrosine phosphorylation and activation of MAPKs (30), cytokine release (31–34), activation of $\beta 2$ integrins (29,35,36), and serum response factor (SRF)-mediated transcriptional activation of immediate early genes (37). The pathways by which PSGL-1 triggers such intracellular signals include its association with and phosphorylation by the syk tyrosine kinase. This association is mediated by ezrin and moesin, proteins that link to the actin cytoskeleton (37). PSGL-1 also is expressed on platelets and could possibly serve similar signaling functions in these cells (38). Finally, although a bona fide ligand for L-selectin on the inflamed endothelium has not been identified, L-selectin engagement triggers various cellular events in leukocytes, including ionized calcium mobilization, increases in protein tyrosine phosphorylation, and rearrangements of the cytoskeleton (39–41).

LEUKOCYTE ARREST

The arrest of leukocytes on the activated endothelium is a prerequisite for subsequent leukocyte spreading, transendothelial migration, and diapedesis into the subendothelial tissue. Neutrophil arrest is signaled by soluble factors released by the endothelium and/or membrane-bound signaling molecules on the endothelium. The cues for arrest are provided by a class of chemotactic cytokines, the chemokines that are secreted by activated ECs, activated platelets, and subendothelial parenchymal cells (42). During inflammation, activated ECs produce inflammatory chemokines including macrophage inflammatory proteins (MIPs) CCL3 and CCL4; monocyte chemotactic proteins (MCPs) CCL7, CCL8, and CCL13; RANTES (Regulated on Activation, Normal T-cell Expressed and Secreted)/CCL5; interleukin (IL)-8/CXCL8; and fractalkine/CXCL1. Moreover, chemokines generated in the subendothelial tissue have been shown to cross the endothelium via caveolar transcytosis following binding to glycosaminoglycans and/or Duffy receptor and be selectively presented on the apical microvilli. The activated endothelium also upregulates structures such as heparin-sulfate proteoglycans, which retain and present these chemokines to circulating leukocytes (43). These immobilized chemokines on the apical surface of the endothelium are then poised to locally stimulate the rolling leukocytes that express the relevant receptors. This ensures spatially localized activation and transmigration of leukocytes during an inflammatory response. Thus, expression of a chemokine in a tissue, and the presence of its receptor on a circulating leukocyte, contribute to the selectivity of the local leukocyte recruitment.

Chemokine signaling through specific leukocyte-expressed 7-transmembrane G-protein–coupled receptors converts the low-affinity, selectin-mediated leukocyte–EC interaction into a high-affinity, integrin-mediated interaction that leads to the firm arrest of the leukocyte on the endothelial surface. Several chemokines have been described that mediate integrin-dependent firm adhesion to activated endothelium, accordingly described as "arrest" chemokines (44). The CXC chemokine IL-8/CXCL8, produced by tumor necrosis factor (TNF)-activated endothelium activates $\beta 2$ integrins and leads to the arrest of rolling neutrophils (45). More recently, RANTES and KC (the mouse chemokine resembling human IL-8) were identified as arrest chemokines for rolling monocytes (44). Interestingly, L-selectin stimulation on the leukocyte induces surface redistribution of CXCR4, the receptor for the chemokine SDF-1 from intracellular vesicles (46). Thus, in addition to selectins mediating activation of $\beta 2$ integrins and leading to slow rolling, selectin-mediated leukocyte rolling may prime the cell for subsequent chemokine-mediated integrin activation.

Chemokine-induced integrin activation, known as *inside-out signaling*, is particularly relevant in leukocytes. In these cells, the integrin heterodimers (one α- and one β-subunit) are present in a low-affinity, low-avidity state on the cell surface but are induced to rapidly bind ligand upon chemokine stimulation (47). Activation of integrins is mediated by GTPases of the Ras and Rho family and associated signaling cascades, which are well-described and reviewed elsewhere (48,49). Increased integrin affinity and avidity are achieved through conformational changes, association of integrins with the cytoskeleton, and integrin clustering. Integrin binding to their ligands brings about the third step of the adhesion cascade, the tight adhesion of the leukocyte to integrin ligands presented on the activated endothelium. The major integrins required for this adhesion are the $\beta 2$ integrins lymphocyte function-associated antigen (LFA)-1 ($\alpha L\beta 2$) and Mac-1 ($\alpha M\beta 2$), the $\alpha 4\beta 1$ integrin VLA-4, and the $\alpha 4\beta 7$ integrin. The endothelial ligands to which the integrins bind are members of the IgG superfamily that are upregulated on the activated endothelium. Intercellular adhesion molecule (ICAM)-1 and ICAM-2 are counter-receptors for the $\beta 2$ integrins LFA-1 and Mac-1, whereas VCAM-1 interacts with VLA-4 (50) and $\alpha 4\beta 7$ additionally binds MAdCAM, which is important in lymphocyte homing to intestinal tissue (51).

Similar to the rolling receptors, engagement of the IgG superfamily of ligands on the endothelium also generates intracellular signals in ECs (52). Engagement of ICAM-1 on the endothelium induces calcium signaling, cytoskeletal rearrangements (53–54), activation of Rho GTPase, and phosphorylation of several proteins including focal adhesion kinase, paxillin, and p130Cas (55). Crosslinking of VCAM-1 also induces increases in intracellular calcium and stress fiber formation (19) as well as Rho-dependent Rac activation and ROS generation (56–58). Interestingly, inhibition of downstream signaling by Rac1 was shown to reduce the migration of leukocytes across endothelial monolayers (57). Thus,

ICAM-1–induced Rac activity may promote the opening of endothelial cell–cell contacts, thus allowing leukocytes to transmigrate. ICAM-1 and VCAM-1 clustering induces their association with the Ezrin/Radixin/Moesin (ERM) proteins moesin and ezrin (59) in a Rho-dependent manner, providing docking structures for adherent leukocytes to transmigrate. These data indicate that both RhoA and Rac1 in ECs are activated by adhesion receptors and may signal changes in the endothelium that facilitate leukocyte egress. ICAM-1–mediated signaling events during neutrophil adherence also were shown to activate xanthine oxidase in ECs, which in turn mediated the production of ROS. This led to cytoskeletal remodeling and cell stiffening, steps that may influence both neutrophil migration as well as EC permeability (60). ROS production stimulated by lymphocyte binding to VCAM-1 has been shown to result in actin coalescence in ECs and activation of EC-associated matrix metalloproteinases, with the latter being required for lymphocyte transmigration (61). Thus, leukocyte adhesion to the endothelium may result in locally generated signals that prime the endothelium for subsequent transmigration.

Shortly after arrest, most leukocytes spread and begin to migrate laterally over the apical surface of the endothelium prior to transmigration. The outside-in signaling that occurs through ligand-bound integrins leads to cell polarization and forward movement of the leukocyte, driven by the combination of leading-edge protrusion and contractility at both the front and the back of the cell. A large number of focal adhesions, as well as structural and regulatory proteins, are located at the leading edge of the migrating cell. These include chemokine receptors, integrins like VLA-4 (but not LFA-1), and regulators of the actin cytoskeleton assembly, such as the small GTPase Rac1, its effector PAK1, and the Arp2/3 complex (49,62). During movement over the endothelium, leukocytes have to form new adhesion contacts at protruding sites while breaking down old ones at the retracting end of the cell. The turnover of focal adhesions involves intracellular interactions between integrins, adapter molecules, and microtubules. For example, α4 integrins use paxillin to regulate the attachment and detachment of the leading edge (63). LFA-1 signals through myosin light-chain kinase and through p160ROCK, which regulates cell contraction and detachment (64). Binding of LFA-1 to endothelial ligands also induces actin polymerization at the leading edge of the leukocyte, possibly through the Rho family exchange factor Vav1 (65,66). Thus, the dynamic exchange of information between the migrating leukocyte and endothelium is required for directed leukocyte movement.

LEUKOCYTE TRANSMIGRATION

Usually within minutes, firmly arrested leukocytes, which spread and migrate over the endothelium, extravasate. Proactive roles for the endothelium in facilitating diapedesis of leukocytes, the final step of the adhesion cascade, have been suggested by a variety of studies. A testament to this is the early demonstration by Huang and colleagues that a rise inintracellular calcium levels in ECs, induced by leukocyte contact, was required for efficient transendothelial migration (67). The majority of transmigration occurs at junctions present at endothelial cell–cell contacts, and this is referred to as the *paracellular route.* The endothelial junctions consist primarily of two types of complexes: the adherens junctions and the tight junctions, which remodel to facilitate various endothelial processes (68). These structures transiently disassemble during leukocyte transmigration and after stimulation with growth factors and inflammatory stimuli. For example, transmigrating leukocytes induce a focal and transient redistribution of junctional vascular endothelial (VE)-cadherin. Although no role for VE–cadherin in leukocyte transmigration has been established, the dynamic changes in the location of this protein during diapedesis does suggest that active endothelial junctional remodeling occurs during the transmigration process. Molecules at endothelial cell–cell contacts that are indeed required for paracellular leukocyte transmigration include platelet-endothelial cell adhesion molecule (PECAM)-1, a member of the immunoglobulin superfamily (69,70), and junctional adhesion molecule (JAM)-A, a component of the tight junctions (71,72). The mechanisms by which these two proteins facilitate transcellular migration, and the signals that they generate upon their engagement, will be discussed.

PECAM-1 is concentrated at the intercellular junctions of ECs. It is also expressed on the surface of platelets and most leukocytes. PECAM-1 is essential for leukocyte diapedesis, and supports cell-to-cell adhesion by homophilic and heterophilic interactions. Recently, Mamdouh and colleagues found that PECAM-1 constitutively recycles evenly along EC borders (73). However, during transendothelial migration across human umbilical vein EC (HUVEC) monolayers, PECAM-1 is targeted to areas of the interendothelial junction where monocytes transmigrate (73). This may provide a source of endothelial PECAM-1 for the leukocyte PECAM-1 to engage via homophilic interactions. The continuous addition of PECAM-1 into the junction may create a gradient of cellular adhesion sites through the junction, forming a molecular zipper that provides traction to drive diapedesis at these locations. Following stimulation of ECs with various stimuli, PECAM-1 is tyrosine- and serine/threonine-phosphorylated, resulting in its interaction with catenin family members (70). Engagement of PECAM-1 modulates EC migration through its interaction with the tyrosine phosphatase SHP-2 (74) and through the small GTPase Rho (75). During leukocyte transendothelial migration, ECs are protected from apoptosis, and this may be attributed to PECAM-1–mediated activation of survival signals in the endothelium (76). The engagement of PECAM-1 with monoclonal antibodies in ECs also results in an increase in intracellular calcium (77). Whether the engagement specifically of PECAM-1 during leukocyte transmigration evokes similar effects on the endothelium is not known. Notably, PECAM-1 has been shown to transduce EC responses to hydrogen peroxide, potentially derived from neutrophils, by generating cation currents and a transient rise in intracellular calcium flux (78). Finally, engagement of PECAM-1 by migrating neutrophils results in a negative feedback on nuclear factor (NF)-κB activity in ECs (79). This could potentially provide a mechanism by which leukocyte transmigration could limit the amplification, or aid in the resolution of, the inflammatory response at the level of endothelial gene expression (79).

JAM-A is a member of the immunoglobulin superfamily proteins expressed at endothelial and epithelial cell junctions, but it also is present on leukocytes and platelets (80). In vivo studies have demonstrated an important role for this protein on neutrophils in neutrophil diapedesis (72,81). Insight into the role of this molecule on ECs has come from in vitro work. Endothelial JAM-A has been shown to bind LFA-1 and support leukocyte adhesion through this receptor (82), although the relative contribution of this interaction to transmigration in the presence of other adhesion molecules present on the inflamed endothelium has not been demonstrated. Interestingly, neutrophil LFA-1, which is evenly distributed upon neutrophil arrest on TNF-activated HUVECs, rapidly redistributes to form a ring-like structure at the neutrophil–endothelial junctional interface that coclusters with endothelial ICAM-1 and JAM-A as the neutrophil transmigrates (83). In this system, endothelial ICAM-1, but not JAM-A, appeared to contribute to LFA-1–dependent transmigration. These studies illustrate the complex roles played by junctional molecules during leukocyte recruitment, being possibly involved in both adhesion and diapedesis. JAM-A signaling has been reported to affect EC motility and fibroblast growth factor (FGF)-induced tube formation in ECs (84). Another member of the JAM family, JAM-C, has been shown to interact with the $\beta 2$ integrin Mac-1 (85) and promote leukocyte transendothelial migration in vivo and in vitro (86). Thus, JAM molecules on the endothelium have the potential to significantly alter the inflammatory response and serve as a possible conduit for leukocyte signaling into ECs.

Transmigration through adjacent ECs is not the exclusive pathway for leukocyte diapedesis. Mounting evidence suggests that leukocytes can pass through the body of an EC (i.e., the transcellular route) and does not require mechanisms for the separation of adherence junctions. Indeed some of the first electron microscopy studies on transendothelial migration provided in vivo evidence for the predominance of the transcellular route of diapedesis (87,88). Recently, a cuplike transmigratory structure was described using high-resolution fluorescence imaging techniques. This endothelial structure is composed of vertical microvilli-like projections that surround transmigrating leukocytes and induce redistribution of leukocyte integrins (LFA-1 and Mac-1) into linear tracks oriented parallel to the direction of diapedesis (89). These structures are highly enriched in ICAM-1 and VCAM-1, whereas PECAM-1 or VE-cadherin seem to be excluded. The disruption of these projections inhibited leukocyte transmigration, which suggest that this transmigratory cup represents an important mechanism of directional guidance during transendothelial migration (89). Interestingly, both trans- and paracellular diapedesis by neutrophils, monocytes, and lymphocytes were associated with transmigratory cup formation, suggesting that this endothelial structure represents a more general mechanism fundamental to transendothelial migration (89). Consistent with this concept, endothelial ICAM-1 was identified as essential for both paracellular and transcellular neutrophil transmigration in in vitro flow assays. Interestingly, interfering with ICAM-1 cytoplasmic tail function preferentially reduced transcellular transendothelial migration. Moreover, ICAM-1 endothelial surface density and distribution as well as TNF-induced changes in EC shape contributed to transcellular transendothelial migration (90). Apparently, the proper adhesive function of ICAM-1 (and VCAM-1) during leukocyte adhesion and transendothelial migration depends on the presence of endothelial tetraspanin microdomains. Tetraspanins relocalize to the contact site with transmigrating leukocytes and associate laterally with both ICAM-1 and VCAM-1. Interference with tetraspanin microdomains prevented lymphocyte transendothelial migration and increased lymphocyte detachment under shear flow (91).

In summary, leukocyte transmigration requires complex, overlapping cellular and molecular interactions between leukocytes and ECs at all stages of leukocyte transmigration. Recent studies emphasize that the engagement of adhesion molecule–ligand pairs on leukocytes and ECs results in signal transduction in both cell types, which can impact functional outcomes. The importance of the carefully choreographed interaction between leukocytes and ECs to human physiology is highlighted in human leukocyte adhesion deficiency (LAD) syndromes. These patients have missing or dysfunctional surface adhesion molecules such as the $\beta 2$ integrins and selectin ligands that are required for leukocyte egress from the vessel into tissues. The LAD syndromes result from failures of leukocytes, primarily cells of the innate immune system including neutrophils and monocytes, to transmigrate which leads to compromised host defense (92).

ROLE OF ENDOTHELIUM ON THE LIFESPAN OF LEUKOCYTES

During inflammation, the lifespan of normally short-lived peripheral blood neutrophils is prolonged in the circulation and that of transmigrated neutrophils is further delayed compared to circulating cells. This delay may serve to increase the pool of functional neutrophils at sites of inflammation. Importantly, activated ECs play a pivotal role in modulating leukocyte apoptosis. Under in vitro conditions, transmigration of neutrophils through a cytokine-activated endothelial monolayer was shown to delay neutrophil apoptosis through adhesive interactions between the two cell types, as well as endothelial-secreted cytokines such as granulocyte monocyte colony stimulating factor (GM-CSF) (93). In contrast, resting ECs may induce leukocyte apoptosis by expressing functional Fas ligand, which triggers cells death in cells carrying its receptor Fas. Fas ligand is constitutively expressed in nonlymphoid cells of the eye and testis, which may render these tissues immune-privileged by eliminating infiltrating lymphocytes and granulocytes (94). Interestingly, TNF stimulation leads to downregulation of Fas ligand, which may promote the survival and accumulation of extravasating leukocytes during the inflammatory response in these tissues (94). In conclusion, ECs present at the interface of circulating blood cells and tissue modulate the accumulation of leukocytes at sites of inflammation by potentially two mechanisms: the control of leukocyte egress and regulation of the lifespan of these cells.

LEUKOCYTE-DEPENDENT ENDOTHELIAL SIGNALING AND INJURY

As discussed, leukocyte adhesion-induced calcium signaling in ECs occurs at various stages of leukocyte transmigration and is functionally important under homeostatic conditions. However, it also may have pathophysiologic relevance in certain disease states, such as transplant rejection. Monocyte adhesion to xenogeneic but not syngeneic ECs results in calcium signals and reorganization of the F-actin cytoskeleton in the endothelium, with an associated increase in endothelial permeability (95). Earlier studies also showed that leukocyte interactions with the normal vasculature during an inflammatory response leads to tissue edema, although the mechanism involved remained undetermined (96). A key contractile event in ECs leading to barrier disruption is the phosphorylation of regulatory myosin light chains by calcium/calmodulin-mediated activation of myosin light-chain kinase and/or through the activity of the Rho/Rho kinase pathway (97–101). Thus, the location and extent of calcium release in ECs may facilitate transmigration, or provide a cue for endothelial permeability and dysfunction.

In addition to leukocyte adhesion-induced calcium signaling, leukocyte-derived soluble factors can potentially have deleterious consequences for the endothelium. Reactive oxygen and nitrogen species are formed in neutrophils following particle uptake and are normally targeted intracellularly to phagolysosomes. However, ROS are released extracellularly in response to stimuli such as cytokines and chemoattractants or following a process referred to as *frustrated phagocytosis,* wherein the target is too large to ingest. ROS are both labile and reactive, and the majority of superoxide released extracellularly is converted to H_2O_2 via xanthine oxidase (found in neutrophils and ECs) or to hypochlorous acid through the action of myeloperoxidase. H_2O_2 is highly membrane-permeable and may contribute to EC dysfunction associated with ischemia–reperfusion (102), hypertension (103,104), and sepsis (105). Although leukocyte-derived ROS and nitric oxide (NO) radicals are traditionally viewed as cytotoxic and causing frank injury to ECs, they have now gained recognition as molecules that may contribute to EC oxidant stress signaling. Neutrophil-derived ROS increase microvessel permeability and intracellular calcium levels in the endothelium (105,106). HUVECs treated with ROS exhibited increased actin stress fibers, which was associated with phosphorylation of focal adhesion kinase, paxillin, and p130cas tyrosine (107,108). ROS treatment also resulted in enhanced neutrophil adhesion (109), and the neutrophil NADPH oxidase was required for the TNF-induced NF-κB activation and ICAM-1 expression in the endothelium in vivo (110). The described ROS-mediated increases in endothelial adhesion receptor expression and change in redox state could enhance cell–cell interactions. These actions may contribute to the progression of atherosclerosis or plaque rupture in acute myocardial infarction. The CD18-dependent adhesion of neutrophils to the endothelium also results in the transmission of neutrophil NADPH oxidase-dependent signals that leads to enhanced expression of the Toll-like receptor 2 (TLR2) in the endothelium (111). This latter report suggests the intriguing possibility that leukocyte-derived inputs into the endothelium result in local amplification of innate immune responses against pathogens. Finally, oxidative stress might alter signal transduction and transcription factor activity through glutathiolation of thiol residues in proteins involved in these pathways or through H_2O_2-mediated depletion of cellular GSH reserves (103).

Neutrophils have an abundance of proteinases, which are stored in three different classes of intracellular granules (primary, secondary, and tertiary). Proteinases released from primary and secondary granules may cleave EC surface adhesion receptors, chemokines, and extracellular matrix molecules and result in a change in the endothelial phenotype at sites of inflammation. Elastase also serves a nondegradative signaling function by significantly promoting platelet-derived growth factor (PDGF) release from activated ECs (112). On the other hand, proteinases may promote EC injury. Studies to date suggest that serine proteinases, such as elastase released from neutrophils, promote EC injury in various diseases. For example, neutrophil elastase has been reported to decrease endothelial production of NO and prostaglandin I_2, thereby contributing to the development of ischemia–reperfusion-induced liver injury (113). The toxic effects of neutrophil elastase are greatly enhanced by increased oxidative stress because ROS inhibit the naturally occurring neutrophil elastase inhibitor, alpha-1-proteinase inhibitor. In addition to cell proteolysis resulting in gross changes in the endothelial integrity, elastase could also indirectly impact leukocyte transmigration, since it has been shown to cleave endothelial junctional molecules, adhesion molecules, and surface-retained chemokines (104). Furthermore, the neutrophil proteinase-mediated release of spatially restricted signaling molecules and chemokines at the endothelial surface may result in activation of leukocytes in blood and/or at remote tissue targets. This may occur in sepsis, acute respiratory distress syndrome (ARDS), and ischemia–reperfusion syndromes.

BLOCKING LEUKOCYTE–ENDOTHELIAL CELL INTERACTIONS: THERAPEUTIC IMPLICATIONS

Inflammation is a fundamental response to tissue injury and infection, but is detrimental in clinically important inflammatory disorders. The phenotypic changes in the endothelium that often occur in chronic inflammatory sites can fundamentally change the rheostat of the inflammatory response. A striking example of this is the adoption of a unique plump or tall appearance reminiscent of high endothelial venules (HEVs, specialized for recruitment of lymphocytes in secondary lymphoid organs) in ECs in many chronic inflammatory settings. These ECs express unique epitopes for L-selectin ligands, MECA-79, which are normally largely restricted to expression in HEVs (114). The implication is that lymphocyte traffic may be increased at these sites; this has ramifications for the cellular immune response and the ensuing inflammation. Blocking leukocyte recruitment through inhibition of leukocyte–endothelial interactions is an attractive therapeutic approach that could yield benefits in disease states like autoimmune disease, atherosclerosis, and ischemia–reperfusion injury. Indeed, the effect of many anti-inflammatory drugs, such as nonsteroidal anti-inflammatory agents (NSAIDs) and corticosteroids, can be ascribed, in part, to inhibition of adhesion molecules. Inhibiting either IL-1β or TNF through the use of antibodies or their respective soluble receptors reduces the expression of adhesion molecules on activated ECs. Moreover, NSAIDs, corticosteroids, and antioxidants also decrease the expression of inflammatory adhesion molecules and chemokines, in part by down-modulating the function of the inflammatory transcription factor NF-κB. In the search for more selective and potent drugs, the leukocyte and EC adhesion molecules emerged as direct targets for therapeutic interventions (115).

In many experimental models, inhibiting leukocyte rolling by blocking selectin activity affects the accumulation of leukocytes. However, although initial preclinical studies were promising, in clinical trials, antibodies against all selectins, either individually or in combination, and recombinant ligand for P- and L-selectin (rPSGL-1-Ig) did not protect against inflammation induced by ischemic conditions (115). Similarly, clinical trials blocking β2 integrin or its ligand ICAM-1 were unsuccessful in reducing either myocardial or cerebral ischemic injury (116–118). By contrast, inhibition of LFA-1 is successful in the treatment of graft-versus-host disease, transplant rejection (119), and psoriasis (115).

In contrast to the limited success of blocking the selectins and β2 integrins, targeting α4 integrins (such as VLA-4) with monoclonal antibodies has been beneficial in several human diseases, such as multiple sclerosis and inflammatory bowel disease (120,121). This success might relate to the principal leukocyte subtype targeted in these autoimmune diseases. Whereas trials of selectin and β2 integrin antiadhesion therapies have primarily involved inflammatory conditions in which neutrophil infiltration predominates, the beneficial effect of antagonizing α4 integrins was achieved in disease in which macrophages and lymphocytes play crucial roles. An advantage of selectively targeting mononuclear leukocytes by α4 integrin inhibition also can be that, at long-term therapy, serious adverse effects with regard to increased susceptibility to infections may not occur. By contrast, blocking selectins and β2 integrin adhesion pathways, which strongly affect adhesion of neutrophils, mimics the clinical LAD-I and LAD-II syndromes and could compromise innate immune responses.

CONCLUSION

In the last three decades, it became increasingly clear that leukocyte extravasation from the blood into the tissues, both under physiologic and pathologic conditions, is not a passive phenomenon, but a highly regulated and selective process. As discussed, this process involves adhesive interactions between leukocytes and ECs, resulting in the generation of intracellular signaling events in both cell types. These events are required for the temporal and spatial regulation of leukocyte diapedesis into the tissues critical for inflammation and host defense. Leukocyte-derived inputs into the endothelium may serve to modify subsequent leukocyte recruitment and responses to inflammatory stimuli. Taken a step further, repetitive leukocyte–EC cross-talk could lead to aberrant endothelial responses that contribute to chronic inflammation. Thus, one may argue that a multipronged approach is required to prevent the leukocyte–endothelial cross-talk that leads to leukocyte diapedesis and associated endothelial phenotypic alterations. A better understanding of the inputs and outputs of both cell types will provide a framework to design targeted therapeutics to achieve this objective.

KEY POINTS

- Cross-talk between leukocytes and the activated endothelium, which occurs locally via adhesive interactions and soluble mediators, results in short- and long-term modifications of the endothelial phenotype.
- Regulated phagocyte adhesion signals changes in the endothelial actin cytoskeleton, secondary messengers, and gene transcription.
- The timed, limited, and low exposure of ECs to inflammatory mediators generated by phagocytes leads to homeostatic signaling, whereas excessive concentrations of these mediators leads to EC injury.
- Blocking of leukocyte-generated inputs in the endothelium represents an attractive therapeutic target for treatment of inflammatory diseases.

Future Goals

- To develop techniques to detect spatially localized intracellular signals in the endothelium generated by leukocyte–EC interactions
- To elucidate specific intracellular signaling molecules critical for relaying particular leukocyte mediated inputs into the endothelium. This will aid in identifying therapeutic targets for inflammatory diseases

REFERENCES

1 Metchnikoff E. *L'immunité dans les maladies infectieuses.* Paris: Masson & Cie., 1901.
2 Butcher EC. Leukocyte-endothelial cell recognition: three (or more) steps to specificity and diversity. *Cell.* 1991;67:1033–1036.
3 Springer TA. Traffic signals for lymphocyte recirculation and leukocyte emigration: the multistep paradigm. *Cell.* 1994;76:301–314.
4 Clark E, Clark E. Observations on changes in blood vascular endothelium in the living animal. *Am J Anat.* 1935;57:385–438.
5 Ley K. The role of selectins in inflammation and disease. *Trends Mol Med.* 2003;9:263–268.
6 Frenette PS, Subbarao S, Mazo IB, et al. Endothelial selectins and vascular cell adhesion molecule-1 promote hematopoietic progenitor homing to bone marrow. *Proc Natl Acad Sci USA.* 1998;95:14423–14428.
7 Mazo IB, Gutierrez-Ramos JC, Frenette PS, et al. Hematopoietic progenitor cell rolling in bone marrow microvessels: parallel contributions by endothelial selectins and vascular cell adhesion molecule 1. *J Exp Med.* 1998;188:465–474.
8 Moore KL, Patel KD, Bruehl RE, et al. P-selectin glycoprotein ligand-1 mediates rolling of human neutrophils on P-selectin. *J Cell Biol.* 1995;128:661–671.
9 Yang J, Hirata T, Croce K, et al. Targeted gene disruption demonstrates that P-selectin glycoprotein ligand 1 (PSGL-1) is required for P-selectin-mediated but not E-selectin-mediated neutrophil rolling and migration. *J Exp Med.* 1999;190:1769–1782.
10 Hirata T, Merrill-Skoloff G, Aab M, et al. P-Selectin glycoprotein ligand 1 (PSGL-1) is a physiological ligand for E-selectin in mediating T helper 1 lymphocyte migration. *J Exp Med.* 2000;192:1669–1676.
11 Norman KE, Katopodis AG, Thoma G, et al. P-selectin glycoprotein ligand-1 supports rolling on E- and P-selectin in vivo. *Blood.* 2000;96:3585–3591.
12 Xia L, Sperandio M, Yago T, et al. P-selectin glycoprotein ligand-1-deficient mice have impaired leukocyte tethering to E-selectin under flow. *J Clin Invest.* 2002;109:939–950.
13 Berlin C, Bargatze RF, Campbell JJ, et al. $\alpha 4$ integrins mediate lymphocyte attachment and rolling under physiologic flow. *Cell.* 1995;80:413–422.
14 Alon R, Kassner PD, Carr MW, et al. The integrin VLA-4 supports tethering and rolling in flow on VCAM-1. *J Cell Biol.* 1995;128:1243–1253.
15 Clark RA, Alon R, Springer TA. CD44 and hyaluronan-dependent rolling interactions of lymphocytes on tonsillar stroma. *J Cell Biol.* 1996;134:1075–1087.
16 DeGrendele HC, Estess P, Siegelman MH. Requirement for CD44 in activated T cell extravasation into an inflammatory site. *Science.* 1997;278:672–675.
17 Steeber DA, Venturi GM, Tedder TF. A new twist to the leukocyte adhesion cascade: intimate cooperation is key. *Trends Immunol.* 2005;26:9–12.
18 Fong AM, Robinson LA, Steeber DA, et al. Fractalkine and CX3CR1 mediate a novel mechanism of leukocyte capture, firm adhesion, and activation under physiologic flow. *J Exp Med.* 1998;188:1413–1419.
19 Lorenzon P, Vecile E, Nardon E, et al. Endothelial cell E- and P-selectin and vascular cell adhesion molecule-1 function as signaling receptors. *J Cell Biol.* 1998;142:1381–1391.
20 Yoshida M, Szente BE, Kiely JM, et al. Phosphorylation of the cytoplasmic domain of E-selectin is regulated during leukocyte-endothelial adhesion. *J Immunol.* 1998;161:933–941.
21 Tilghman RW, Hoover RL. E-selectin and ICAM-1 are incorporated into detergent-insoluble membrane domains following clustering in endothelial cells. *FEBS Lett.* 2002;525:83–87.
22 Kiely JM, Hu Y, Garcia-Cardena G, Gimbrone MA Jr. Lipid raft localization of cell surface E-selectin is required for ligation-induced activation of phospholipase Cγ. *J Immunol.* 2003;171:3216–3224.
23 Hu Y, Szente B, Kiely JM, Gimbrone MA Jr. Molecular events in transmembrane signaling via E-selectin. SHP2 association, adaptor protein complex formation and ERK1/2 activation. *J Biol Chem.* 2001;276:48549–48553.
24 Fujimoto T, McEver RP. The cytoplasmic domain of P-selectin is phosphorylated on serine and threonine residues. *Blood.* 1993;82:1758–1766.
25 Crovello CS, Furie BC, Furie B. Histidine phosphorylation of P-selectin upon stimulation of human platelets: a novel pathway for activation-dependent signal transduction. *Cell.* 1995;82:279–286.
26 Kansas GS, Pavalko FM. The cytoplasmic domains of E- and P-selectin do not constitutively interact with alpha-actinin and are not essential for leukocyte adhesion. *J Immunol.* 1996;157:321–325.
27 Forlow SB, White EJ, Barlow SC, et al. Severe inflammatory defect and reduced viability in CD18 and E-selectin double-mutant mice. *J Clin Invest.* 2000;106:1457–1466.
28 Kunkel EJ, Dunne JL, Ley K. Leukocyte arrest during cytokine-dependent inflammation in vivo. *J Immunol.* 2000;164:3301–3308.
29 Green CE, Pearson DN, Camphausen RT, et al. Shear-dependent capping of L-selectin and P-selectin glycoprotein ligand 1 by E-selectin signals activation of high-avidity $\beta 2$-integrin on neutrophils. *J Immunol.* 2004;172:7780–7790.
30 Hidari KI, Weyrich AS, Zimmerman GA, McEver RP. Engagement of P-selectin glycoprotein ligand-1 enhances tyrosine phosphorylation and activates mitogen-activated protein kinases in human neutrophils. *J Biol Chem.* 1997;272:28750–28756.
31 Celi A, Pellegrini G, Lorenzet R, et al. P-selectin induces the expression of tissue factor on monocytes. *Proc Natl Acad Sci USA.* 1994;91:8767–8771.
32 Damle NK, Klussman K, Dietsch MT, et al. GMP-140 (P-selectin/CD62) binds to chronically stimulated but not resting $CD4^{+}$ T lymphocytes and regulates their production of proinflammatory cytokines. *Eur J Immunol.* 1992;22:1789–1793.
33 Weyrich AS, McIntyre TM, McEver RP, et al. Monocyte tethering by P-selectin regulates monocyte chemotactic protein-1 and tumor necrosis factor-alpha secretion. Signal integration and NF-κ B translocation. *J Clin Invest.* 1995;95:2297–2303.
34 Weyrich AS, Elstad MR, McEver RP, et al. Activated platelets signal chemokine synthesis by human monocytes. *J Clin Invest.* 1996;97:1525–1534.
35 Evangelista V, Manarini S, Sideri R, et al. Platelet/polymorphonuclear leukocyte interaction: P-selectin triggers protein-tyrosine phosphorylation-dependent CD11b/CD18 adhesion: role of PSGL-1 as a signaling molecule. *Blood.* 1999;93:876–885.
36 Ma YQ, Plow EF, Geng JG. P-selectin binding to P-selectin glycoprotein ligand-1 induces an intermediate state of $\alpha M\beta 2$ activation and acts cooperatively with extracellular stimuli

to support maximal adhesion of human neutrophils. *Blood.* 2004;104:2549–2556.

37 Urzainqui A, Serrador JM, Viedma F, et al. ITAM-based interaction of ERM proteins with Syk mediates signaling by the leukocyte adhesion receptor PSGL-1. *Immunity.* 2002;17:401–412.

38 Frenette PS, Denis CV, Weiss L, et al. P-Selectin glycoprotein ligand 1 (PSGL-1) is expressed on platelets and can mediate platelet-endothelial interactions in vivo. *J Exp Med.* 2000;191: 1413–1422.

39 Laudanna C, Constantin G, Baron P, et al. Sulfatides trigger increase of cytosolic free calcium and enhanced expression of tumor necrosis factor-α and interleukin-8 mRNA in human neutrophils. Evidence for a role of L-selectin as a signaling molecule. *J Biol Chem.* 1994;269:4021–4026.

40 Waddell TK, Fialkow L, Chan CK, et al. Signaling functions of L-selectin. Enhancement of tyrosine phosphorylation and activation of MAP kinase. *J Biol Chem.* 1995;270:15403–15411.

41 Brenner B, Gulbins E, Busch GL, et al. L-selectin regulates actin polymerisation via activation of the small G-protein Rac2. *Biochem Biophys Res Commun.* 1997;231:802–807.

42 Zlotnik A, Yoshie O. Chemokines: a new classification system and their role in immunity. *Immunity.* 2000;12:121–127.

43 Middleton J, Patterson AM, Gardner L, et al. Leukocyte extravasation: chemokine transport and presentation by the endothelium. *Blood.* 2002;100:3853–3860.

44 Ley K. Integration of inflammatory signals by rolling neutrophils. *Immunol Rev.* 2002;186:8–18.

45 DiVietro JA, Smith MJ, Smith BR, et al. Immobilized IL-8 triggers progressive activation of neutrophils rolling in vitro on P-selectin and intercellular adhesion molecule-1. *J Immunol.* 2001;167:4017–4025.

46 Ding Z, Issekutz TB, Downey GP, Waddell TK. L-selectin stimulation enhances functional expression of surface CXCR4 in lymphocytes: implications for cellular activation during adhesion and migration. *Blood.* 2003;101:4245–4252.

47 Hynes RO. Integrins: bidirectional, allosteric signaling machines. *Cell.* 2002;110:673–687.

48 Worthylake RA, Burridge K. Leukocyte transendothelial migration: orchestrating the underlying molecular machinery. *Curr Opin Cell Biol.* 2001;13:569–577.

49 van Buul JD, Hordijk PL. Signaling in leukocyte transendothelial migration. *Arterioscler Thromb Vasc Biol.* 2004;24:824–833.

50 Yonekawa K, Harlan JM. Targeting leukocyte integrins in human diseases. *J Leukoc Biol.* 2005;77:129–140.

51 Kummer C, Ginsberg MH. New approaches to blochade of alpha4-integrins, proven therapeutic targets in chronic inflammation. *Biochem Pharmacol.* 2006;72:1460–1468.

52 Wang Q, Doerschuk CM. The signaling pathways induced by neutrophil-endothelial cell adhesion. *Antioxid Redox Signal.* 2002;4:39–47.

53 Pfau S, Leitenberg D, Rinder H, et al. Lymphocyte adhesion-dependent calcium signaling in human endothelial cells. *J Cell Biol.* 1995;128:969–978.

54 Etienne-Manneville S, Manneville JB, Adamson P, et al. ICAM-1-coupled cytoskeletal rearrangements and transendothelial lymphocyte migration involve intracellular calcium signaling in brain endothelial cell lines. *J Immunol.* 2000;165:3375–3383.

55 Etienne S, Adamson P, Greenwood J, et al. ICAM-1 signaling pathways associated with Rho activation in microvascular brain endothelial cells. *J Immunol.* 1998;161:5755–5761.

56 Matheny HE, Deem TL, Cook-Mills JM. Lymphocyte migration through monolayers of endothelial cell lines involves VCAM-1 signaling via endothelial cell NADPH oxidase. *J Immunol.* 2000;164:6550–6559.

57 van Wetering S, van den BN, van Buul JD, et al. VCAM-1-mediated Rac signaling controls endothelial cell-cell contacts and leukocyte transmigration. *Am J Physiol Cell Physiol.* 2003;285:C343–C352.

58 Cook-Mills JM, Johnson JD, Deem TL, et al. Calcium mobilization and Rac1 activation are required for VCAM-1 (vascular cell adhesion molecule-1) stimulation of NADPH oxidase activity. *Biochem J.* 2004;378:539–547.

59 Barreiro O, Yanez-Mo M, Serrador JM, et al. Dynamic interaction of VCAM-1 and ICAM-1 with moesin and ezrin in a novel endothelial docking structure for adherent leukocytes. *J Cell Biol.* 2002;157:1233–1245.

60 Wang Q, Doerschuk CM. Neutrophil-induced changes in the biomechanical properties of endothelial cells: roles of ICAM-1 and reactive oxygen species. *J Immunol.* 2000;164:6487–6494.

61 Cook-Mills JM, Deem TL. Active participation of endothelial cells in inflammation. *J Leukoc Biol.* 2005;77:487–495.

62 Webb DJ, Parsons JT, Horwitz AF. Adhesion assembly, disassembly and turnover in migrating cells – over and over and over again. *Nat Cell Biol.* 2002;4:E97–E100.

63 Rose DM, Han J, Ginsberg MH. Alpha4 integrins and the immune response. *Immunol Rev.* 2002;186:118–124.

64 Smith A, Bracke M, Leitinger B, et al. LFA-1-induced T cell migration on ICAM-1 involves regulation of MLCK-mediated attachment and ROCK-dependent detachment. *J Cell Sci.* 2003; 116:3123–3133.

65 Porter JC, Bracke M, Smith A, et al. Signaling through integrin LFA-1 leads to filamentous actin polymerization and remodeling, resulting in enhanced T cell adhesion. *J Immunol.* 2002; 168:6330–6335.

66 Riteau B, Barber DF, Long EO. Vav1 phosphorylation is induced by $\beta 2$ integrin engagement on natural killer cells upstream of actin cytoskeleton and lipid raft reorganization. *J Exp Med.* 2003;198:469–474.

67 Huang AJ, Manning JE, Bandak TM, et al. Endothelial cell cytosolic free calcium regulates neutrophil migration across monolayers of endothelial cells. *J Cell Biol.* 1993;120:1371–1380.

68 Bazzoni G, Dejana E. Endothelial cell-to-cell junctions: molecular organization and role in vascular homeostasis. *Physiol Rev.* 2004;84:869–901.

69 Muller WA, Randolph GJ. Migration of leukocytes across endothelium and beyond: molecules involved in the transmigration and fate of monocytes. *J Leukoc Biol.* 1999;66:698–704.

70 Ilan N, Madri JA. PECAM-1: old friend, new partners. *Curr Opin Cell Biol.* 2003;15:515–524.

71 Martin-Padura I, Lostaglio S, Schneemann M, et al. Junctional adhesion molecule, a novel member of the immunoglobulin superfamily that distributes at intercellular junctions and modulates monocyte transmigration. *J Cell Biol.* 1998;142: 117–127.

72 Del Maschio A, De Luigi A, Martin-Padura I, et al. Leukocyte recruitment in the cerebrospinal fluid of mice with experimental meningitis is inhibited by an antibody to junctional adhesion molecule (JAM). *J Exp Med.* 1999;190:1351–1356.

73 Mamdouh Z, Chen X, Pierini LM, et al. Targeted recycling of PECAM from endothelial surface-connected compartments during diapedesis. *Nature.* 2003;421:748–753.

74 Gratzinger D, Barreuther M, Madri JA. Platelet-endothelial cell adhesion molecule-1 modulates endothelial migration through its immunoreceptor tyrosine-based inhibitory motif. *Biochem Biophys Res Commun.* 2003;301:243–249.
75 Gratzinger D, Canosa S, Engelhardt B, Madri JA. Platelet endothelial cell adhesion molecule-1 modulates endothelial cell motility through the small G-protein Rho. *FASEB J.* 2003;17:1458–1469.
76 Newman PJ, Newman DK. Signal transduction pathways mediated by PECAM-1: new roles for an old molecule in platelet and vascular cell biology. *Arterioscler Thromb Vasc Biol.* 2003; 23:953–964.
77 Gurubhagavatula I, Amrani Y, Pratico D, et al. Engagement of human PECAM-1 (CD31) on human endothelial cells increases intracellular calcium ion concentration and stimulates prostacyclin release. *J Clin Invest.* 1998;101:212–222.
78 Ji G, O'Brien CD, Feldman M, et al. PECAM-1 (CD31) regulates a hydrogen peroxide-activated nonselective cation channel in endothelial cells. *J Cell Biol.* 2002;157:173–184.
79 Cepinskas G, Savickiene J, Ionescu CV, Kvietys PR. PMN transendothelial migration decreases nuclear NFkappaB in IL-1β-activated endothelial cells: role of PECAM-1. *J Cell Biol.* 2003;161:641–651.
80 Dejana E. Endothelial cell-cell junctions: happy together. *Nat Rev Mol Cell Biol.* 2004;5:261–270.
81 Corada M, Chimenti S, Cera MR, et al. Junctional adhesion molecule-A-deficient polymorphonuclear cells show reduced diapedesis in peritonitis and heart ischemia-reperfusion injury. *Proc Natl Acad Sci USA.* 2005;102:10634–10639.
82 Ostermann G, Weber KS, Zernecke A, et al. JAM-1 is a ligand of the β2 integrin LFA-1 involved in transendothelial migration of leukocytes. *Nat Immunol.* 2002;3:151–158.
83 Shaw SK, Ma S, Kim MB, et al. Coordinated redistribution of leukocyte LFA-1 and endothelial cell ICAM-1 accompany neutrophil transmigration. *J Exp Med.* 2004;200:1571–1580.
84 Bazzoni G, Tonetti P, Manzi L, et al. Expression of junctional adhesion molecule-A prevents spontaneous and random motility. *J Cell Sci.* 2005;118:623–632.
85 Santoso S, Sachs UJ, Kroll H, et al. The junctional adhesion molecule 3 (JAM-3) on human platelets is a counterreceptor for the leukocyte integrin Mac-1. *J Exp Med.* 2002;196:679–691.
86 Chavakis T, Keiper T, Matz-Westphal R, et al. The junctional adhesion molecule-C promotes neutrophil transendothelial migration in vitro and in vivo. *J Biol Chem.* 2004;279:55602–55608.
87 Williamson JR, Grisham JW. Electron microscopy of leukocytic margination and emigration in acute inflammation in dog pancreas. *Am J Pathol.* 1961;39:239–256.
88 Marchesi VT, Gowans JL. The migration of lymphocytes through the endothelium of venules in lymph nodes: an electron microscopy study. *Proc R Soc Lond B Biol Sci.* 1964;159: 283–290.
89 Carman CV, Springer TA. A transmigratory cup in leukocyte diapedesis both through individual vascular endothelial cells and between them. *J Cell Biol.* 2004;167:377–388.
90 Yang L, Froio RM, Sciuto TE, et al. ICAM-1 regulates neutrophil adhesion and transcellular migration of TNF-alpha-activated vascular endothelium under flow. *Blood.* 2005;106:584–592.
91 Barreiro O, Yanez-Mo M, Sala-Valdes M, et al. Endothelial tetraspanin microdomains regulate leukocyte firm adhesion during extravasation. *Blood.* 2005;105:2852–2861.
92 Bunting M, Harris ES, McIntyre TM, et al. Leukocyte adhesion deficiency syndromes: adhesion and tethering defects involving beta2 integrins and selectin ligands. *Curr Opin Hematol.* 2002;9:30–35.
93 Mayadas TN, Cullere X. Neutrophil beta2 integrins: moderators of life or death decisions. *Trends Immunol.* 2005;26(7):388–395.
94 Walsh K, Sata M. Negative regulation of inflammation by Fas ligand expression on the vascular endothelium. *Trends Cardiovasc Med.* 1999;9:34–41.
95 Peterson MD, Vlasova E, Ciano-Oliveira C, et al. Monocyte-induced endothelial calcium signaling mediates early xenogeneic endothelial activation. *Am J Transplant.* 2005;5:237–247.
96 Wedmore CV, Williams TJ. Control of vascular permeability by polymorphonuclear leukocytes in inflammation. *Nature.* 1981;289:646–650.
97 Hixenbaugh EA, Goeckeler ZM, Papaiya NN, et al. Stimulated neutrophils induce myosin light chain phosphorylation and isometric tension in endothelial cells. *Am J Physiol.* 1997;273:H981–H988.
98 Saito H, Minamiya Y, Kitamura M, et al. Endothelial myosin light chain kinase regulates neutrophil migration across human umbilical vein endothelial cell monolayer. *J Immunol.* 1998;161:1533–1540.
99 Garcia JG, Verin AD, Herenyiova M, English D. Adherent neutrophils activate endothelial myosin light chain kinase: role in transendothelial migration. *J Appl Physiol.* 1998;84:1817–1821.
100 Yuan SY, Wu MH, Ustinova EE, et al. Myosin light chain phosphorylation in neutrophil-stimulated coronary microvascular leakage. *Circ Res.* 2002;90:1214–1221.
101 Saito H, Minamiya Y, Saito S, Ogawa J. Endothelial Rho and Rho kinase regulate neutrophil migration via endothelial myosin light chain phosphorylation. *J Leukoc Biol.* 2002;72:829–836.
102 Ichikawa H, Flores S, Kvietys PR, et al. Molecular mechanisms of anoxia/reoxygenation-induced neutrophil adherence to cultured endothelial cells. *Circ Res.* 1997;81:922–931.
103 Griendling KK, Sorescu D, Lassegue B, Ushio-Fukai M. Modulation of protein kinase activity and gene expression by reactive oxygen species and their role in vascular physiology and pathophysiology. *Arterioscler Thromb Vasc Biol.* 2000;20:2175–2183.
104 Touyz RM. Oxidative stress and vascular damage in hypertension. *Curr Hypertens Rep.* 2000;2:98–105.
105 Gao XP, Standiford TJ, Rahman A, et al. Role of NADPH oxidase in the mechanism of lung neutrophil sequestration and microvessel injury induced by gram-negative sepsis: studies in p47phox$^{-/-}$ and gp91phox$^{-/-}$ mice. *J Immunol.* 2002;168: 3974–3982.
106 Zhu L, Castranova V, He P. fMLP-stimulated neutrophils increase endothelial $[Ca^{2+}]$i and microvessel permeability in the absence of adhesion: role of reactive oxygen species. *Am J Physiol Heart Circ Physiol.* 2005;288:H1331–H1338.
107 Gozin A, Franzini E, Andrieu V, et al. Reactive oxygen species activate focal adhesion kinase, paxillin and p130cas tyrosine phosphorylation in endothelial cells. *Free Radic Biol Med.* 1998;25:1021–1032.
108 Vepa S, Scribner WM, Parinandi NL, et al. Hydrogen peroxide stimulates tyrosine phosphorylation of focal adhesion kinase in vascular endothelial cells. *Am J Physiol.* 1999;277:L150–L158.
109 Sellak H, Franzini E, Hakim J, Pasquier C. Reactive oxygen species rapidly increase endothelial ICAM-1 ability to bind neutrophils without detectable upregulation. *Blood.* 1994;83:2669–2677.

110 Fan J, Frey RS, Rahman A, Malik AB. Role of neutrophil NADPH oxidase in the mechanism of tumor necrosis factor-alpha-induced NF-kappa B activation and intercellular adhesion molecule-1 expression in endothelial cells. *J Biol Chem.* 2002;277:3404–3411.

111 Fan J, Frey RS, Malik AB. TLR4 signaling induces TLR2 expression in endothelial cells via neutrophil NADPH oxidase. *J Clin Invest*. 2003;112:1234–1243.

112 Totani L, Cumashi A, Piccoli A, Lorenzet R. Polymorphonuclear leukocytes induce PDGF release from IL-1 β-treated endothelial cells: role of adhesion molecules and serine proteases. *Arterioscler Thromb Vas Biol.* 1998;18:1534–1540.

113 Okajima K, Harada N, Uchiba M, Mori M. Neutrophil elastase contributes to the development of ischemia-reperfusion-induced liver injury by decreasing endothelial production of prostacyclin in rats. *Am J Physiol Gastrointest Liver Physiol.* 2004; 287:G1116–G1123.

114 Renkonen J, Tynninen O, Hayry P, et al. Glycosylation might provide endothelial zip codes for organ-specific leukocyte traffic into inflammatory sites. *Am J Pathol.* 2002;161: 543–550.

115 Ulbrich H, Eriksson EE, Lindbom L. Leukocyte and endothelial cell adhesion molecules as targets for therapeutic interventions in inflammatory disease. *Trends Pharmacol Sci.* 2003;24:640–647.

116 Baran KW, Nguyen M, McKendall GR, et al. Double-blind, randomized trial of an anti-CD18 antibody in conjunction with recombinant tissue plasminogen activator for acute myocardial infarction: limitation of myocardial infarction following thrombolysis in acute myocardial infarction (LIMIT AMI) study. *Circulation.* 2001;104:2778–2783.

117 Faxon DP, Gibbons RJ, Chronos NA, et al. The effect of blockade of the CD11/CD18 integrin receptor on infarct size in patients with acute myocardial infarction treated with direct angioplasty: the results of the HALT-MI study. *J Am Coll Cardiol.* 2002;40:1199–1204.

118 Enlimomab Acute Stroke Trial Investigators. Use of anti-ICAM-1 therapy in ischemic stroke: results of the Enlimomab Acute Stroke Trial. *Neurology.* 2001;57:1428–1434.

119 Dedrick RL, Bodary S, Garovoy MR. Adhesion molecules as therapeutic targets for autoimmune diseases and transplant rejection. *Expert Opin Biol Ther.* 2003;3:85–95.

120 Miller DH, Khan OA, Sheremata WA, et al. A controlled trial of natalizumab for relapsing multiple sclerosis. *N Engl J Med.* 2003; 348:15–23.

121 Ghosh S, Goldin E, Gordon FH, et al. Natalizumab for active Crohn's disease. *N Engl J Med.* 2003;348:24–32.

64

Platelet–Endothelial Interactions

Patricia B. Maguire, Orina Belton, Niaobh O'Donoghue, Sandra Austin, and Judith Coppinger

Conway Institute, University College Dublin, Ireland

Platelets are anucleate, discoid cell fragments measuring 1.5 to 3.0 μM in diameter. They are derived from bone-marrow megakaryocytes that are normally maintained in a nonadhesive state, whereby they circulate freely in blood. Anucleate platelets are unique to mammals, with nonmammalian vertebrates (such as zebrafish) possessing nucleated thrombocytes. In 1865, a German anatomist Max Schultze (1825–1874) first described platelets as "spherules" much smaller than red blood cells. A few years later, in 1882, Giulio Bizzozero (1846–1901) found that platelets played a role in coagulation because they could clump and form a blood clot at the site of vessel wall injury. It was not until 1961, however, that the platelet aggregating effect of adenosine diphosphate was discovered (1) and, in the following year, that a machine for measuring aggregation (an aggregometer) was developed (2). The most important breakthroughs surrounded the discovery of proaggregatory platelet thromboxane A2 (TXA_2) and anti-aggregatory endothelial cell (EC) prostacyclin (PGI_2), both discovered in the 1970s, along with the finding that aspirin inhibited prostaglandin synthesis and platelet activation. Indeed, finding inhibitors to platelets was exciting because it established the therapeutic possibility of preventing arterial thrombosis and ultimately vessel occlusion (the precipitating event in most myocardial infarctions and many strokes) by means of antiplatelet therapy.

The activation of platelets to arrest bleeding at sites of vascular damage results in platelet adhesion to the vessel wall, where they convert from passive, small discs into larger, flattened structures with extended pseudopods that act as a surface for the propagation of the clotting cascade. Activated platelets secrete and synthesize further platelet agonists, inflammatory mediators, and vasoactive substances. Through conformational changes in the major integrin αIIbβ3, these platelets "stick" to other platelets via fibrinogen (a process termed *platelet aggregation*) to form a hemostatic plug. Platelets, however, can inappropriately undergo activation and aggregation as a consequence of vascular disease, leading to thrombosis and ultimately vessel occlusion. In addition to this classical role in thrombosis, studies have shown that activated platelets also are involved (along with other products) in the development of atherosclerotic lesions, heparin-induced thrombocytopenia (HIT), antiphospholipid antibody syndrome (APS), and thrombotic thrombocytopenic purpura (TTP).

A critical facet of the normal and pathological platelet response is the regulatory effect that activated platelets and ECs exert on each other and on other cells in the microenvironment. As well as their primary hemostatic function, platelets can serve to nourish and support the vascular endothelium. By providing a source of growth factors such as fibroblast growth factor 2 (FGF2) and vascular endothelial growth factor (VEGF), platelets can improve EC survival and potentially contribute to vessel angiogenesis (3). Alternatively, under normal circumstances, the vascular endothelium, through secretion of mediators such as PGI_2, prostaglandin E_2, nitric oxide, nucleoside triphosphate diphosphohydrolases, tissue factor pathway inhibitor, tissue-type plasminogen activator (t-PA), and heparan sulfate, maintains a nonadhesive, nonthrombogenic vessel lining that forms a barrier to prevent the extravasation of blood cells. The activation of ECs, for example by ischemia–reperfusion and diseases such as hypertension and diabetes, leads to a change in their surface structure, with presentation of surface receptors that allow rolling and adhesion of platelets and leukocytes. Such endothelial dysfunction is attributed as one of the initiating factors in the pathogenesis of atherosclerosis and can alter these transcellular regulatory mechanisms, thereby increasing endothelial adhesiveness to platelets and leukocytes and resulting in their subsequent adherence and activation (4,5).

The mechanisms contributing to the interaction of platelets with denuded vessel wall have been well described over the past decade. Less well defined is the process by which platelets roll along and adhere to intact, but activated, endothelium. However, several recent studies have demonstrated platelet interaction with intact endothelium in vivo, even under conditions of high shear stress (6,7). It is becoming clear that many similarities exist between platelet–endothelial interactions and the multistep process by which leukocytes

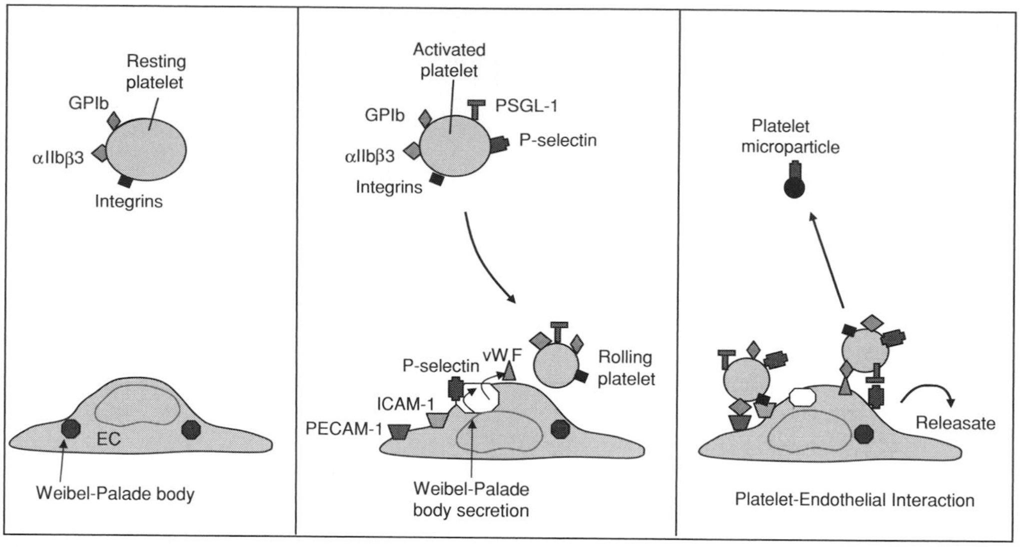

Figure 64.1. Schematic representation of the major platelet–EC interactions. Activation of ECs leads to presentation of surface receptors (P-selectin, vWF, ICAM-1, PECAM-1), which facilitate rolling and adhesion of activated platelets (increased surface expression of adhesion molecules, integrins, P-selectin and PSGL-1).

roll and adhere to ECs prior to extravasation. Little or no evidence as yet suggests that platelets transmigrate across the intact endothelium. Thus, platelet adhesion to ECs leads to a buildup of activated platelets on the vessel surface, providing a platform for leukocyte recruitment to the vessel wall and a mechanism for the delivery of proinflammatory cytokines to atherosclerotic sites.

Furthermore, several in vivo models demonstrate that platelet activation in early lesion formation contributes to the development of the lesion (8). Thus, accumulation of activated platelets on the inflamed endothelium (increased intercellular adhesion molecule [ICAM]-1, vascular cell adhesion molecule [VCAM]-1 and E-selectin expression) provides a reactive surface for the recruitment of leukocytes and additional platelets and allows the localized delivery of the platelet "releasate" on to the endothelial surface. This releasate is composed of a multitude of inflammatory and vasoactive substances and exosomes that can attract atherogenic leukocytes from the circulation, activate ECs and, by infiltrating the vessel wall, trigger vascular cell proliferation, migration, and inflammation (6,7,9,10). Thus, the participation of activated platelets in the genesis of atherosclerotic lesions centers on the products translocated to their surface and those released following adhesion and activation. It is worth noting that, in addition to an important role in atherosclerosis, platelet–EC interactions play an important role in other diseases including sickle cell disease, myeloproliferative disorders, sepsis, and nephritis.

PLATELET–ENDOTHELIAL ADHESION

The adhesion of platelets to the endothelium is a multistep process in which platelets are first tethered to the damaged blood vessel, then followed by rolling and both transient and firm adhesion. In recent years, much attention has focused on understanding the molecular intermediates on platelets and ECs that promote platelet–endothelial interactions including integrins, cell adhesion molecules, and selectins. In platelets, several of these proteins are present in α-granules and, on activation, are expressed on the platelet surface (Figure 64.1). Interestingly, the inhibition of platelet adhesion to the endothelium dramatically reduces leukocyte accumulation and attenuates atherosclerotic lesion progression in the cholesterol fed *ApoE*$^{-/-}$ mouse (10). Thus, the exact role of each receptor in mediating platelet–EC interactions must be elucidated as they may influence the progression of atherosclerosis in different ways.

Selectins, present on both platelets and ECs, mediate the initial contact between circulating platelets and the endothelium (11,12). P-selectin is stored in Weibel-Palade bodies in ECs, and is expressed on the endothelial surface in response to inflammatory stimuli. In addition, P-selectin is stored in the platelet α-granules and is translocated to the surface upon activation. Surface-bound P-selectin mediates interaction of ECs and platelets with leukocytes that express the P-selectin glycoprotein ligand (PSGL)-1 (13). Falati and colleagues examined thrombus formation in wild-type, *P-selectin-* and *PSGL-1–* knockout animals using in vivo microscopy. Mice lacking P-selectin or PSGL-1 developed platelet-rich thrombi compared to wild types, suggesting a role for P-selectin in thrombus formation (14). Specifically, it has been shown that endothelial P-selectin is crucial for the promotion of atherosclerotic lesion growth because, in its absence, only relatively small lesions develop. However, platelet P-selectin also contributes to the lesion development because lesions in wild-type recipients receiving transplants with wild-type platelets were larger than those receiving P-selectin–deficient platelets and were more frequently calcified (15). In addition, deficiency of

P-selectin dramatically protected against neointimal lesion formation after arterial injury in *ApoE*$^{-/-}$ mice and significantly attenuated macrophage recruitment into the arterial wall (16). Although it is as yet unclear as to which of the phenotypes associated with the P-selectin mouse are attributed to platelet or ECs, it has been shown recently that platelet-derived, not endothelial-derived P-selectin, is required for neointimal formation after vascular injury in the *P-selectin*–knockout mouse (17). Furthermore, blood levels of soluble P-selectin, mainly derived from platelets, were found to correlate with the progression of atherosclerosis (18). These data are consistent with earlier immunohistochemical studies showing increased expression of P-selectin protein on the endothelium overlying human atherosclerotic lesions and on the endothelium of patients with unstable angina (19).

Although platelet- and EC-derived P-selectin mediates platelet rolling in inflammation, it has been suggested that the process of platelet rolling does not necessarily require previous platelet activation, because platelets from *P-selectin*–knockout mice roll as efficiently as those from wild-type animals (11). However, it is important to note that P-selectin on both platelets and endothelium is expressed on the cell surface only on activation of the cells (20). Furthermore, although PSGL-1 mediates leukocyte–endothelium interactions and leukocyte–platelet interactions in vivo, it appears that PSGL-1 is not involved in platelet–endothelial interactions, as its expression has not been detected on the platelet membrane, and platelet rolling is not inhibited by an anti–PSGL-1 antibody in vivo (21). However, this has been disputed by other groups, which have shown that PSGL-1 on platelets mediates platelet–endothelial interactions in vivo (12). Therefore it seems feasible to suggest that the interaction of platelet PSGL-1 with endothelial P-selectin may, at least in part, mediate platelet rolling. However, it should be noted that the presence of PSGL-1 on a particular cell type does not indicate its functional relevance. For example, although lymphocytes express PSGL-1, only 15% are actually able to bind P-selectin (22).

Another platelet receptor that may mediate interaction of the platelets with the endothelium is glycoprotein (GP)Ib, which is also a ligand for P-selectin (23). Although it has been shown, using antibodies against P-selectin and GPIb, that the von Willebrand Factor (vWF) mediates platelet adhesion to the subendothelial matrix and intact ECs (23), it is still not clear if GPIb interaction with P-selectin on the endothelial surface induces platelet activation similar to that observed with GPIb-vWF interaction in thrombus formation (24). Although selectins enable platelets to rapidly bind to the endothelial layer, these interactions have high dissociation rates, resulting in platelet rolling. Consequently these P-selectin–PSGL-1 interactions are not sufficient to promote firm adhesion of platelets, and thus new bonds characterized by low dissociation rates must be formed. These interactions between platelets and the vascular wall involve endothelial integrins and cell adhesion molecules.

ICAM-1 is upregulated in leukocytes and ECs and mediates adhesion by establishing strong bonds with integrins and the firm arrest of inflammatory cells on the vascular surface. In addition, an ICAM-1 mechanism is involved in platelet adhesion and aggregation in vitro and in vivo (25). It has been shown that ICAM-2 is expressed on the surface of resting and activated platelets, and this may play an important role in leukocyte–platelet interactions in inflammation and thrombosis (26). In ICAM-1–deficient animals, reduced platelet adhesion to the endothelium is present, most likely because of decreased EC interactions with leukocytes. In addition, expression of ICAM-1 on ECs is enhanced by activated platelets via the cytokine interleukin (IL)-1β, which promotes neutrophil and monocyte recruitment to the endothelium.

Platelet-endothelial cell adhesion molecule (PECAM)-1, another member of the Ig family, is expressed on platelets and leukocytes, and is concentrated at the lateral junctions of ECs. In platelets, PECAM-1 is derived from α-granules and is thought to play a role in leukocyte migration (27). In PECAM-1–deficient mice, there is a prolonged bleeding time and, in addition, platelets from these animals show enhanced adhesion to collagen as well as enhanced platelet aggregation (28) to negatively modulate the prothrombotic platelet collagen interactions. Therefore, PECAM-1 may play a role in preventing platelet thrombus formation. In addition, PECAM-1 expression by donor ECs attenuates the development of transplant arteriosclerosis, possibly by affecting macrophage infiltration (29).

Indeed, both platelet GPIbα and αIIbβ3 have been demonstrated to contribute to platelet endothelial adhesion in vitro (30). Furthermore, in vivo studies have shown that antibodies against GPIbα reduce both transient and firm adhesion to the endothelium the *ApoE*$^{-/-}$ model. In contrast, in the same model, inhibition of αIIbβ3 had only partial effects on transient platelet adhesion but almost completely inhibited firm attachment to ECs. It is important to note that a similar reduction in transient and firm adhesion was observed in the *ApoE*$^{-/-}$ mouse with platelets deficient in αIIbβ3 (6). This study also showed that inhibition of platelet adhesion to the endothelium in the early stages of atherosclerosis not only decreased lesion formation but also attenuated inflammatory cell accumulation. This provides strong evidence for the role of platelet–endothelial interactions in early atherosclerotic lesion formation. However, it is important to note that αIIbβ3 may not be as important in the early development of atherosclerosis. Weng and colleagues have shown that in both an *ApoE*$^{-/-}$ and low-density lipoprotein receptor (*LDLR*)$^{-/-}$ mouse models of atherosclerosis, β3 integrin deficiency was associated with increased atherosclerosis (31). Increased vascular disease in the absence of members of the β3 integrin family also has been previously reported. Indeed, humans with Glanzmann thrombasthenia due to deficiency of αIIbβ3 develop carotid atherosclerosis (32). In addition, despite the unquestioned efficacy of short-term intravenous inhibition of αIIbβ3, long-term oral inhibition of αIIbβ3 increases mortality in humans (33).

Although the role for vWF in the development of atherosclerosis has not fully been elucidated and remains a

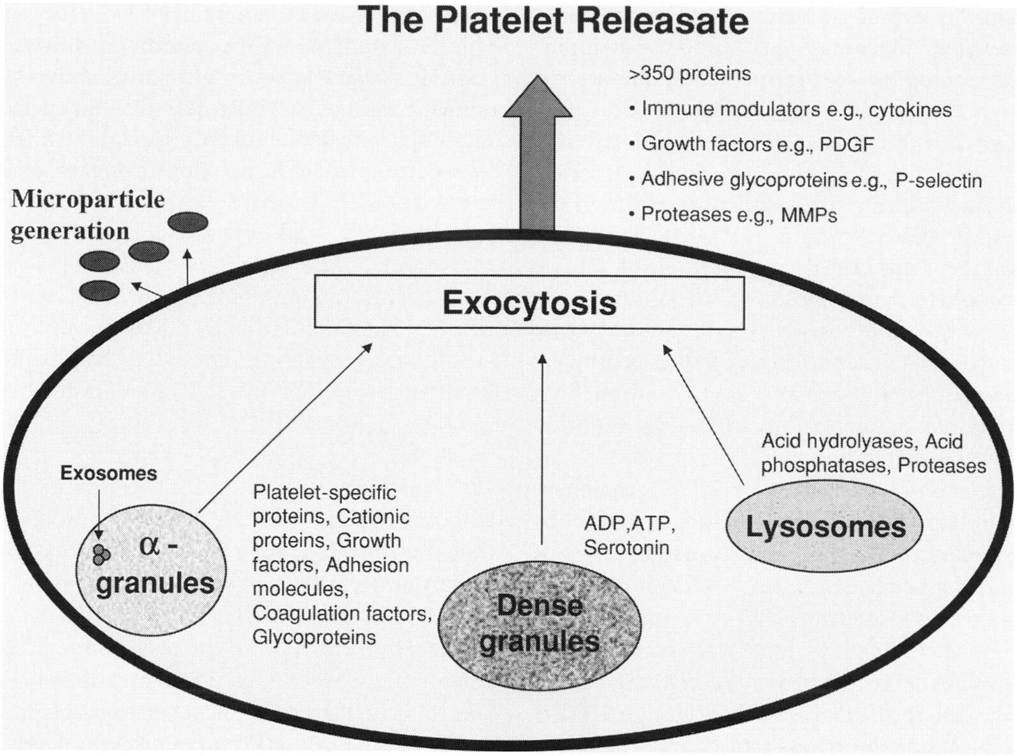

Figure 64.2. Platelet granules and vesicles. Upon platelet activation, three types of preformed granules, α-granules, dense granules, and lysosomes, fuse with the plasma membrane (in a process known as *exocytosis*); change their surface structure (express P-selectin); and release their contents into the external environment. This platelet releasate contains more than 350 soluble proteins and enzymes from α-granules and lysosomes as well as exosomal vesicles and can exert potent biological effects. In addition, microparticles are generated from the platelet plasma membrane and are involved in hemostasis and angiogenesis.

matter of debate, it has been shown that mice deficient in vWF have decreased atherosclerosis (34). This has been suggested to be due to the lack of platelet–EC interactions; however, other possible mechanisms cannot be excluded. It also has been shown that enhanced secretion of vWF in response to inflammatory stimuli can lead to local recruitment of platelets (35). Thus, inflammatory states are associated with an increased potential for EC–platelet interactions (36). This may explain in part why the deficiency of vWF protects against atherosclerosis. Stimulated ECs recently have been shown to express long string-like multimers of von Willebrand factor, known as ultralarge vWF (ULVWF), to which platelets can adhere and support leukocyte tethering under high shear conditions (37).

PLATELET–ENDOTHELIAL COMMUNICATION

Upon activation, platelets release a wealth of adhesive and proinflammatory substances that allow cross-talk between platelets and the endothelium. Physiologically, this releasate could be involved in many normal EC processes, including angiogenesis, due to the release of VEGF and matrix metalloproteinases (MMPs). Additionally, the effects of this releasate are believed to be central in the initiation, development, and exacerbation of atherosclerotic lesions, and offers a direct link between platelets, inflammation, atherosclerosis, and other platelet-associated diseases.

Platelet Granules and Vesicles

Platelets contain a number of preformed, morphologically distinguishable storage granules: $\alpha-$granules, dense granules, and lysosomes (9). In addition, platelets also release two distinct membrane vesicle populations during activation: exosomes and microparticles (Figure 64.2) (38).

α-Granules

α-Granules are suggested to arise from the trans-Golgi complex in the megakaryocyte and are the most abundant granule with about 80 present per platelet. Several subclasses of α-granule proteins exist: platelet-specific proteins, cationic proteins, growth factors, adhesion molecules, coagulation factors, and glycoproteins, which have a diverse range of functions including many chemokines and growth factors for ECs, smooth muscle cells, and fibroblasts. These proteins originate either from endogenous synthesis in the megakaryocyte or platelet or by endocytosis from the plasma; they are packaged into granules through vesicle trafficking processes. Gray platelet syndrome is associated with thrombocytopenia, enlarged platelets, and a specific absence of α-granules and their contents.

Dense Granules

Dense granules, of which there are approximately seven per platelet, contain no secretable proteins but instead store many low-molecular-weight compounds, for example 5-hydroxytryptamine (serotonin) and the nucleotides ADP and ATP, which are potent activators of platelet aggregation and vasoconstriction. Indeed, ADP plays a role in the "second wave of platelet activation" and is essential for TxA_2-induced platelet aggregation via the phosphoinositide-3-kinase (PI3K) pathway. Functional ADP receptors (P2Y and P2X1) are present on the endothelium; upon activation, these produce vasodilatation. Platelet dense granules are related to melanosomes, and such lysosome-related organelles are deficient in the autosomal recessive Hermansky-Pudlak syndrome.

Lysosomes

Lysosomes are membrane-bound vesicles equipped with an ATP-dependent proton pump; they are morphologically similar to the α-granule. Several glycosidal enzymes are released from lysosomes, as well as a number of acid phosphatase enzymes and proteases, such as cathepsin D and E. Lysosomes represent the post-multivesicular body (MVB) stage in the endocytic route and are the main site of protein and lipid degradation.

Exosomes

Studies of platelets have shown the release of α-granule membrane-derived exosomes following exocytosis (38). These membrane-bound vesicles (<100 nm in diameter) are biochemically and morphologically distinct from larger apoptotic vesicles such as microparticles (39). Exosomes are secreted by a multitude of other cells, including cytotoxic T cells and antigen-presenting cells, such as B cells and dendritic cells. No evidence as yet suggests that exosomes are released from ECs. They can be released from platelets in response to a variety of agonists and contain the lysosomal/exosomal tetraspan protein CD63; however, their function in platelet biology remains unknown (38). Indeed, platelet-derived exosomes do not readily bind annexin V, factor X, or prothrombin in vitro, suggesting that they do not play a procoagulant role (38). However, mast cell–derived exosomes activate ECs to secrete plasminogen activator inhibitor (PAI)-1, increased levels of which are associated with an increased risk of ischemic cardiovascular events (40). Early studies suggested that exosomes served to eliminate undegraded endosomal or lysosomal proteins and membranes. Recent studies indicate, however, that in different cell types, exosomes may play an immunoregulatory intercellular role. For example, B-lymphocyte–derived exosomes have been shown to stimulate $CD4^+$ T lymphocytes in vitro (41). Furthermore, B cell–derived exosomes have been shown to express functional integrins on their surfaces that are capable of mediating exosome adhesion to the extracellular matrix (42). Interestingly, dendritic cell exosomes were found to stimulate antitumor immune responses and the rejection of established tumors in mice (39). Proteomic analysis of exosomes secreted from various cell types has revealed the presence of ubiquitous proteins such as tubulin, actin and actin-binding proteins, plasma membrane proteins, proteins involved in apoptosis (e.g., 14–3–3) as well as cell-type–specific proteins (39). However, the full molecular content of proteins found on the surface and within platelet exosomes has never been investigated, although they do contain α-granular membrane proteins such as cellular prion protein (43). Future studies are necessary to define the role of platelet exosomes in mediating EC phenotype.

Microparticles

Microparticles (MPs) are membrane-bound vesicles that are formed from the platelet surface within minutes following activation by many agonists or shear stress. Their generation seems to be a regulated process distinct from granule exocytosis, requiring an increase in cytosolic calcium and subsequent disruption of the membrane cytoskeleton. The formation of platelet MPs has been demonstrated in vitro in the presence of several agonists including thrombin and collagen, and in vivo under high shear conditions (44,45). Such MP formation is closely associated with the exposure of phosphatidyl serine on the outer membrane leaflet and provides an anionic surface for the binding of coagulation factors such as coagulation factor Va (46). The majority of circulating MPs are derived from platelets, and increased numbers of platelet MPs are found in many diseases such as diabetes, sickle cell disease, cancer, and following cardiopulmonary bypass surgery, where they cause an increased risk of thrombosis (47,48). Although their exact physiological and pathophysiological functions are unknown, MPs, independent of their origin, can transfer information, mostly at the level of ECs. Platelet MPs are involved in hemostasis and thrombosis because they aid in the initiation and amplification of the coagulant cascade. They can promote angiogenesis via extracellular-signaling-receptor kinase (ERK) and PI3K activation (49). Platelet MPs also stimulate ECs through modifications of arachidonic acid metabolism and the generation of TxA_2, thereby inducing expression of cyclooxygenase (COX)-2 and the generation and release of PGI_2, suggesting the involvement of platelet MPs in the development of inflammation.

Platelet MPs have a protein content similar to the activated plasma membrane containing GPIb/IX/V, PECAM-1, αIIbβ3, annexin V, and the tetraspanin proteins Peta-3, CD9, and P-selectin. It is thought that the presence of these glycoproteins on MPs may be important in their interactions with the endothelium (50). Furthermore, IL-1β also is present in MPs, in addition to its release in soluble form, and it can induce inflammatory gene expression in ECs (51). MPs generated from ADP-activated platelets have recently been studied using a proteomics approach (52). Using 1D sodium dodecyl sulfate polyacrimide gel electrophoresis (SDS-PAGE) in combination with mass spectrometry (MS), 578 proteins were identified, including the membrane surface proteins GPIIIa, GPIIb, and P-selectin, as well other platelet proteins such as the chemokines platelet factor 4 (PF4), neutrophil-activating peptide (NAP)-2, and RANTES (Regulated upon Activation, Normal T-cell Expressed and Secreted).

The Platelet Releasate

Upon platelet activation, there ensues a series of graded responses (dependent on the degree of agonist stimulation) that include a change in cell shape, cell aggregation, and movement of preformed storage granules to the cell surface. Exocytosis of these granules involves the redistribution of soluble membrane proteins to the plasma membrane surface – such as the adhesion molecule P-selectin – as well as the release of their contents (soluble proteins and enzymes from α-granules and lysosomes, as well as exosomal vesicles). These substances are known collectively as the platelet releasate. These releasates move into the external milieu, where they exert potent biological effects. This releasate contains a range of growth factors, coagulation proteins, adhesion molecules, cell-activating molecules, cytokines, integrins, proinflammatory molecules (transforming growth factor [TGF]-β, PF4, RANTES), and angiogenic factors. Such proteins act in an autocrine or paracrine fashion to modulate cell signaling. Thus, they play a key role in hemostasis. Importantly, increasing evidence suggests that the platelet releasate plays a pathological role in vascular conditions such as thrombosis and atherosclerosis.

Despite the biological significance of such released chemokines and signaling factors from platelets, the full complement of proteins comprising the platelet releasate and the mechanisms by which these proteins exert their effects on other cell types, such as ECs, remain poorly defined. Recently, a method was developed to purify the secreted fraction of thrombin-activated human platelets. This has allowed researchers to establish the first extensive catalog of the molecular context of the platelet releasate, a fraction highly enriched for platelet granular and exosomal contents (53). It is expected that the secretion profile of platelets changes depending on the agonist, although this change may be more quantitative than qualitative. In this study, the platelet releasate was separated using a multilayered proteomic approach using a combination of two-dimensional electrophoresis and multidimensional chromatography, both coupled to mass spectrometry. In total, over 300 proteins were identified in the thrombin-activated platelet releasate, 81 of which were present in two or more experiments. A list of these released proteins, classified according to their main function (if known), is found in Table 64-1.

Of these 81, 37% were previously reported to be released from platelets, including PF4, thrombospondin (TSP), osteonectin, metalloproteinase inhibitor 1, and TGF-β. Another 35% are released from other secretory cells: for example, cofilin, profilin, 14–3–3ζ, and actin from dendritic cells; peptidyl-prolyl-cis isomerase (cyclophilin A) from smooth muscle cells; phosphoglycerate kinase from fibrosarcoma cells; and β2-microglobulin and vitamin D–binding protein from the liver. The remaining 23 (28%) recurring proteins were not reported to be released from any cell type, with several mapping to expressed sequence tags (ESTs) of unknown function. Moreover, out of the 81 proteins, 46 are detected at the mRNA level in platelets. The functions of many of these proteins, however, are consistent with the known proinflammatory role of platelets, and include the previously undescribed platelet proteins secretogranin-II and -III, monocyte chemoattractant precursors, and cyclophilin A, a vascular smooth muscle and EC growth factor and proinflammatory cytokine. Indeed, several molecules released by platelets are now known to play significant roles in platelet-to-EC communication.

Immune Modulators

Released platelet chemokines coordinate leukocyte recruitment to the inflamed endothelium. These chemokines include PF4, platelet basic protein, and RANTES from α-granules. Other proinflammatory cytokines released by platelets include those stored in the cytoplasm, such as CD40L, and those synthesized upon activation, such as IL-1β. Indeed, platelet-derived IL-1β induces EC adhesiveness by enhancing the expression of ICAM-1, with consequent leukocyte adhesion (51). Additionally, two antibacterial proteins, thrombocidin-1 and -2, which are structurally similar to chemokines such as NAP-2 and secretogranin (Sg)-II and –III, have been reported to be released from platelets (53,54).

PF4 is a lysine-rich, heparin-binding, low-molecular-weight (7 kDa) platelet-specific protein (55) that belongs to the CXC chemokine subfamily. It comprises 2% to 3% of the releasate (concentrations approaching approximately 25 μg/mL [4 μM] in the vicinity of the vessel wall). The biological role of PF4 is unclear; however, secreted PF4 binds to endothelial heparin sulfate proteoglycans, inhibiting the activity of antithrombin III (56). It also binds to the Duffy antigen/receptor for chemokines (DARC) on ECs and red blood cells (the consequences of which are unknown) (57,58), and to an alternatively spliced variant of CXCR3 found on ECs (59). Additionally, endothelial expression of E-selectin is induced by PF4 through low-density lipoprotein receptor-related protein (LRP) in an NF-κB–dependent manner (60). PF4 also may act as a anticoagulant, because it binds to thrombomodulin and, like protein S (also released from platelets), can stimulate the formation of activated protein C on the platelet surface (61). PF4 has been shown to bind to exposed subendothelium, where an internal fragment generated by proteolytic processing may function as an inhibitor of EC proliferation and angiogenesis (62). Platelets have been shown to deposit PF4 directly on the luminal surface of the endothelium in a mouse atherosclerotic model, significantly promoting leukocyte infiltration (10). Furthermore, PF4 and low-density lipoprotein (LDL) have been found to colocalize in atherosclerotic lesions, especially in macrophage-derived foam cells, because PF4 binds to oxidized LDL and may contribute to its uptake (63).

Table 64-1: Summary of 81 Proteins from the Thrombin-Activated Platelet Releasate

Protein	*Gene Name*	*Function*
Adhesion		
Thrombospondin	TSP1_HUMAN	Adhesive glycoprotein that mediates cell-to-cell and cell-to-matrix interactions; can bind to fibrinogen, fibronectin, laminin, type V collagen, and integrins $\alpha V/\beta 1$, $\alpha V/\beta 3$ and $\alpha IIb/\beta 3$; upon secretion, can bind $\alpha IIb\beta 3$, $\alpha v\beta 3$, and GPIV; can potentiate aggregation by complexing with fibrinogen and becoming incorporated into fibrin clots
SPARC protein (osteonectin)	SPRC_MOUSE	Upon secretion, forms a specific complex with thrombospondin; appears to regulate cell growth through interactions with the extracellular matrix and cytokines; binds PDGF and cell membranes; expressed in tissues undergoing morphogenesis, remodeling, and wound repair
Platelet glycoprotein V	GPV_HUMAN	Part of the GPIb-IX-V complex on the platelet surface; cleaved by the protease thrombin during thrombin-induced platelet activation
Von Willebrand Factor	VWF_HUMAN	Binds GPIb-IX-V
Glycoprotein Ibα chain	GPBA_HUMAN	Surface membrane protein of platelets that participates in formation of platelet plug by binding A1 domain of von Willebrand factor
Nidogen	NIDO_HUMAN	Glycoprotein found in basement membranes; interacts with laminin, collagen, and integrin on neutrophils
Endothelial cell multimerin precursor	ECM_HUMAN	Carrier protein for platelet factor V/Va
Inflammation		
Platelet basic protein	SZO7_HUMAN	Proteolytic cleavage yields the chemokines β-thromboglobulin and neutrophil activating peptide (NAP) 2
Secretory granule proteoglycan core protein precursor	PGSG_HUMAN	Function unknown; associates and is coreleased with inflammatory mediators such as platelet factor 4
Clusterin	CLUS_HUMAN	Not clear; possibly platelet-derived apolipoprotein J participates in short-term wound repair and chronic pathogenic processes at vascular interface
Platelet factor 4	PLF4_HUMAN	Platelet-specific chemokine with neutrophil activating properties
Latent transforming growth factor (TGF)-β binding protein isoform 1S	LTBS_HUMAN	Subunit of the TGF-β1 complex secreted from platelets
Latent TGF binding protein	O88349	Subunit of the TGF-β1 complex secreted from platelets
Platelet factor 4 variant	PF4V_HUMAN	Similar to PF4
Vitamin D binding protein	VTDB-HUMAN	Carries vitamin D sterols; prevents actin polymerization; has T-lymphocyte surface association
β2 microglobulin	B2MG_HUMAN	The β–chain of the major histocompatibility complex (MHC) class I molecule
Plasminogen	PLMN_HUMAN	Dissolves fibrin in blood clots, proteolytic factor in tissue remodeling, tumor invasion, and inflammation
αIb glycoprotein	A1BG_HUMAN	Found in plasma; role not clear, possibly involved in cell recognition as a new member of the immunoglobulin family
Apolipoprotein A1	APO1_HUMAN	Role in high-density lipoprotein binding to platelets
α1-acid glycoprotein	A1AH_HUMAN	Modulates activity of the immune system during the acute phase reaction; binds platelet surface
Gelsolin	GELS_HUMAN	Two isoforms: a cytoplasmic actin-modulating protein and a secreted isoform involved in the inflammatory response
Serotransferrin precursor	TRFE_HUMAN	Precursor to macromolecular activators of phagocytosis (MAPP), which enhance leukocyte phagocytosis via the FcγRII receptor; may stimulate cell proliferation
Secretogranin III	SG3_MOUSE	Unknown; possibly involved in secretory granule biogenesis; may be cleaved into active inflammatory peptide-like secretogranin II
Cyclophilin A	CYPH_MOUSE	Cellular protein with isomerase activity; secreted vascular smooth muscle cell growth factor and endothelial activator

(continued)

Table 64-1 (*continued*)

Protein	*Gene Name*	*Function*
Actin Binding		
α-actinin 1	AAC1_HUMAN	Actin-binding and actinin cross-linking protein found in platelet α-granules; interacts with thrombospondin on the platelet surface; plays a role in monocyte/macrophage maturation
α-actinin 2	AAC2_HUMAN	Actin-binding and actinin cross-linking protein found in platelet α-granules; interacts with thrombospondin on the platelet surface
α-actinin 4	AAC4_HUMAN	Actin-binding and actinin cross-linking protein found in platelet α-granules; interacts with thrombospondin on the platelet surface
Actin	ACTA_HUMAN	Major cytoskeletal protein; external function unknown
Actin	ACTB_HUMAN	Major cytoskeletal protein; external function unknown
14–3–3 protein ζ/δ	14–3–3Z MOUSE	External function unknown; involved intracellularly in signal transduction; however, may have a role in regulating exocytosis
Adenylyl cyclase associated protein 1	CAP1_HUMAN	Contains a WH2 actin-binding domain (as β-thymosin 4); known to regulate actin dynamics; may mediate endocytosis
Tubulin	TBA_HUMAN	Cytoskeletal protein involved in microtubule formation
Cofilin	COF_HUMAN	Actin depolymerization/regulation in cytoplasm
Profilin	PRO1_HUMAN	Actin demolymerization/regulation in cytoplasm
Filamin	FLNA_HUMAN	Actin-binding protein; essential for GP1b-α anchorage at high shear; substrate for caspase-3
Talin	TAL1_HUMAN	Actin-binding protein that binds to integrin-β3 domain
Zyxin	ZYX_HUMAN	Associates with the actin cytoskeleton near adhesion plaques; binds α-actinin and VASP
Transgelin	TAG_HUMAN	Actin-binding protein; loss of transgelin expression important in early tumor progression; may serve as a diagnostic marker for breast and colon cancer
Thymosin β4	TYB4_HUMAN	Intracellularly: G-actin binding protein; functions as an antimicrobial peptide when secreted; increases the rate of attachment and spreading of ECs on matrix components and migration of HUVECs
Vinculin	VINC_HUMAN	Actin-binding protein
WD-repeat protein	WDR1_HUMAN	Actin-binding protein
FKSG30	Q9BYX7	Actin binding protein
Titin	Q8WZ42	Anchoring protein of actomyosin filaments; role in secretion of myostatin
Protease		
Metalloproteinase inhibitor 1	TIM1_HUMAN	Interacts with metalloproteinases and inactivates them; stimulates growth and differentiation of erythroid progenitors, dependant on disulfide bonds
Amyloid β-A4 protein (Protease nexin II)	A4_HUMAN	Exhibits potent protease inhibitor and growth factor activity; may play a role in coagulation by inhibiting factors XIa and IXa
α1-antitrypsin precursor	A1AT_HUMAN	Acute phase protein, similar to complement, inhibits proteinases
α2-Macroglobulin	A2MG_HUMAN	Acute phase protein, similar to complement, inhibits proteinases
Coagulation		
Vitamin K-dependant protein S	PRTS_HUMAN	Anticoagulant; cofactor to protein C in the degradation of coagulation factors Va and VIIIA secretion; downregulated by TNF-α; deficiency associated with thrombosis; also released by endothelial and vascular smooth muscle cells
Coagulation factor V	FA5_HUMAN	Cofactor that participates with Factor Xa to activate prothrombin to thrombin
Coagulation factor XIIIA chain	F13A_HUMAN	Coagulation protein involved in the formation of the fibrin clots
Complement C3	CO3_HUMAN	Activator of the complement system; cleaved to α, β, and γ chains normally prior to secretion, and is a mediator of the local inflammatory response
Complement C4	CO4-HUMAN	Activator of the complement system; cleaved normally prior to secretion; its products mediate the local inflammatory response
78kDa glucose related protein	GR78_MOUSE	Chaperone in the ER involved in the inhibition of secreted coagulation factors
Prothrombin	THRB_HUMAN	Converts fibrinogen to fibrin and activates coagulation factors including factor V
Fibrinogen α-chain	FIBA_HUMAN	Cofactor in platelet aggregation; endocytosed into platelets from plasma
Fibrinogen β-chain	FIBB_HUMAN	Cofactor in platelet aggregation; endocytosed into platelets from plasma

Protein	*Gene Name*	*Function*
Fibrinogen γ-chain	FIBG_HUMAN	Cofactor in platelet aggregation; endocytosed into platelets from plasma;
Calumenin	CALU_MOUSE	an inhibitor of the vitamin K epoxide reductase-warfarin interaction
Tumor		
Similar to hepatocellular carcinoma-associated antigen 59	Q99JW3	Tumor marker
Mitochondrial		
Glyceraldehyde 3-phosphate dehydrogenase	G3P_HUMAN	Mitochondrial enzyme involved in glycolysis; may catalyze membrane fusion
Phosphoglycerate kinase	PKG_MOUSE	Glycotic enzyme; secreted from tumor cells and involved in angiogenesis
Superoxide dismutase	SODC_HUMAN	Involved in cellular oxygen metabolism, role for SOD-1 in inflammation
Pyruvate kinase M2 isozyme	KYP2_HUMAN	Involved in final stage of glycolysis; presented as an autoantigen by dendritic cells
Fructose bisphosphate aldolase	ALFA_MOUSE	Glycolytic enzymes that converts fructose 1,6-bis phosphate to glyceraldehyde 3-phosphate and dihydroxy acetone phosphate
Exocytosis		
Calmodulin	CALM_HUMAN	Known to regulate calcium-dependent acrosomal exocytosis in neuroendocrine cells
Rho GDP-dissociation inhibitor 2	GDIS_MOUSE	Regulates the GDP/GTP exchange reaction of Rho proteins; regulates platelet aggregation; involved in exocytosis in mast cells
Hypothetical		
Filamin fragment (Hypothetical 54-kDa protein)	Q99KQ2	Unknown
Hypothetical protein	Y586_HUMAN	Unknown
Hypothetical protein	Q9BTV9	Unknown
Fibrinogen type protein	Q8VCM7	Similar to fibrinogen
Lysosomal		
Proactivator polypeptide	SAP_HUMAN	Activator proteins for sphingolipid hydrolases (saposins) that stimulate the hydrolysis of sphingolipids by lysosomal enzymes
Miscellaneous		
RNA-binding protein	Q9UQ35	RNA binding protein
Hemoglobin α-chain	HBA_HUMAN	Oxygen transport; potentiates platelet aggregation through thromboxane receptor
Hemoglobin β-chain	HBB_HUMAN	Oxygen transport
Hemopexin	HEMO_HUMAN	Heme-binding protein with metalloproteinase domains
Transthyetin	TTHY_HUMAN	Thyroid hormone-binding protein secreted from the choroid plexus and the liver in to CSF and plasma, respectively
Pleckstrin	PLEK_HUMAN	A substrate for protein kinase C; its phosphorylation is important for platelet secretion
Intracellular hyaluronan-binding protein p 57	Q9JKS5	Unknown
Bromodomain and PHD finger-containing protein 3	BRF3_HUMAN	Unknown

(*continued*)

Table 64-1 (*continued*)

Protein	*Gene Name*	*Function*
Rho GTPase-activating protein	Q92512	Promotes the intrinsic GTP hydrolysis activity of Rho family proteins; involved in regulating myosin phosphorylation in platelets
Albumin	ALBU_HUMAN	Major plasma protein secreted from the liver into the blood; endocytosed into platelets from plasma

Proteins identified by Coppinger and colleagues, 2004, categorized according to their main function. Eighty-one proteins were identified using a proteomics approach from the thrombin-stimulated platelet supernatant fraction. Mass spectrometric spectra were identified using the SEQUEST program and a composite mouse and human database (NCBI July 2002 Release) in three replicate experiments. Gene name and known functions are provided.

Finally, heterophilic interactions of PF4 and RANTES promote monocyte arrest on the endothelium (64). HIT, a thrombotic disorder related to the administration of heparin, involves the formation of antibodies to a complex of PF4 with heparin. The pathogenesis of this disease is believed to be associated with excessive secretion of PF4. Resulting antibodies can directly activate ECs and, through FcγRII receptors, lead to platelet activation and subsequent thrombosis and inflammation (65).

Platelet basic protein has neutrophil-activating properties. It is a platelet-specific protein that is part of the chemokine protein family. Platelet basic protein is converted by proteolysis to connective tissue activating protease (CTAP-III), and by N-terminal cleavage to β-thromboglobulin (βTG), whose proteolytic products include NAP-2. NAP-2 is a potent activator of neutrophils, stimulating neutrophil chemotaxis, calcium mobilization, and exocytosis (55). Moreover, NAP-2 reacts with IL-8 receptors on neutrophils, causing degranulation and secretion of elastase. Neutrophils also release cathepsin G, which exerts a positive feedback effect by causing platelets to release and degrade CTAP-III to NAP-2, thereby activating additional neutrophils (66). Furthermore, βTG is a potent activator of smooth muscle cell proliferation and platelet–leukocyte adhesion, and its levels previously have been found to be significantly elevated in deep vein thrombosis.

The CC chemokine RANTES, first purified as a product of activated T cells, is a powerful chemoattractant for memory T lymphocytes and monocytes. This proinflammatory factor is deposited by activated platelets on the surface of monocytes and the endothelium of atherosclerotic arteries, mediated by EC P-selectin (10,67). The deposition of RANTES in vivo was shown to trigger shear-resistant monocyte arrest on inflamed or atherosclerotic endothelium, as well as the potent chemotaxis of eosinophils (10,64,68). Interestingly, antagonism of RANTES receptors with the CC chemokine antagonist Met-RANTES reduces the progression of atherosclerosis in a hypercholesterolemic mouse model (69).

Secretogranins (Sg)-II and -III are members of the chromogranin family of acidic secretory proteins and have recently been found in the platelet releasate (53). These proteins were previously shown only to be localized to storage vesicles of neuroendocrine cells (70) and are not known to be present in ECs. Upon release from these cells, SgII is proteolytically processed to yield the 33-amino acid polypeptide secretoneurin, which has a tissue distribution and function similar to the proinflammatory neuropeptides substance P and neuropeptide Y (70). In fact, secretoneurin may itself modulate inflammatory reactions, because it has been shown to increase the chemotactic migration of monocytes and eosinophils, deactivate neutrophil chemotaxis, and stimulate EC migration (71). This effect of secretoneurin to stimulate monocyte adhesion to arterial and venous ECs in vitro, followed by their transendothelial migration, was comparable to that of tumor necrosis factor (TNF)-α, lipopolysaccharide (LPS), or IL-1 under the same experimental conditions (72). Furthermore, secretoneurin stimulates proliferation and exerts antiapoptotic effects on ECs, where it activates intracellular PI3K/Akt and mitogen-activated protein kinase (MAPK) pathways (73). Whether cleavage products of SgIII play a similar role is unknown. Nevertheless, although absent in normal vasculature, SgIII was widely expressed in human atherosclerotic lesions (53). Therefore, secretoneurin and cleavage products of SgIII, such as substance P, may be implicated in expanding the inflammatory response of atherosclerosis by stimulating the replication of monocyte-derived macrophages and the entry of new monocytes into lesions.

CD40 ligand (CD40L, CD154) is a 33-kDa type-II transmembrane protein of the TNF family. It was originally identified as a surface molecule expressed on activated T cells. Although present on several cells of the vasculature (including ECs), platelets are the primary source, and following their activation, CD40L rapidly translocates from the cytoplasm to their surface. Surface-expressed CD40L can be cleaved to an 18-kDa soluble form, sCD40L; however, unlike α-granule secretion, in which the substances are fully released within 5 to 10 minutes, this is a long-lasting process that takes at least 90 minutes to reach the maximal level of extracellular sCD40L concentration. CD40L is prothrombotic. It binds to αIIbβ3 and is involved in the recruitment of inflammatory cells at the vessel wall by ligating CD40 on ECs (no evidence for subsequent rolling mediated by CD40L-CD40). It induces the endothelial expression of adhesion molecules (ICAM-1, VCAM-1, and E-selectin) and chemokines (such as tissue factor, monocyte chemoattractant protein [MCP]-1 and IL-8) (74,75). Interaction of membrane-bound or sCD40L with CD40 also may inhibit the re-endothelization of an injured vessel, thereby enhancing restenosis (76). Furthermore, CD40L appears to be particularly relevant in the pathogenesis of acute coronary syndromes, percutaneous coronary interventions, and cardiopulmonary bypass, because enhanced circulating levels of soluble and membrane-bound CD40L are found, and elevated levels of sCD40L are a reliable predictor of cardiovascular events (77). Moreover, inhibition of CD40L reduces the development of atherosclerosis in mice (78).

Growth Factors

Growth factors that derive from platelets stimulate cell proliferation and have very important roles in growth, development, wound repair, and inflammatory reactions. Growth factors released from α-granules include the smooth muscle cell mitogens platelet-derived growth factor (PDGF), TGF-β, epidermal growth factor (EGF), as well as bone-derived growth factor (BDGF) and hepatocyte growth factor (53,79). The overexpression of factors such as PDGF has been linked to diseases such as atherosclerosis and cancer (4). Normal vessels express low levels of the protein and its receptors; however, a marked increase of PDGF has been observed in the eccentric lesions of atherosclerosis (80). Additionally, the role of PDGF as a transforming and angiogenic growth factor has been demonstrated (81). VEGF also is released from platelets and plays a principal role in EC migration and proliferation, promoting angiogenesis in normal (e.g., embryonic vascular development and clotting/wound healing) and pathological conditions (e.g., thrombosis/cardiac ischemia and tumor progression/metastasis). Evidence exists that angiopoetin 1 (Ang1) is also released from platelets. Ang1 binds to Tie2 receptor-like tyrosine kinase receptors located on ECs and is essential for normal vascular development and maintaining the quiescence and stability of mature vasculature (82). This interplay between VEGF and angiopoietin is essential for vessel wall stabilization. Plasminogen also is released from platelets, and it can be proteolytically cleaved to angiostatin, a negative regulator of angiogenesis (53).

Cyclophilins are peptidyl-propyl *cis-trans*-isomerases. They are members of the immunophilin family and act intracellularly both as catalysts and chaperones in protein folding. They have extracellular signaling functions, such as the induction of chemotaxis and adhesion of memory CD4 cells. Cyclophilin A is secreted by vascular smooth muscle cells in response to oxidative stress, where it acts in an autocrine manner to stimulate ERK1/2 and vascular smooth muscle proliferation (83). Recently, cyclophilin A has been found to be secreted by ECs, and also was found to act in an autocrine manner to stimulate EC proliferation, migration, and invasive capacity and induce the expression of several genes, including CD147 (84,85). Furthermore, endothelial-derived cyclophilin A was found to have proinflammatory cytokine properties, upregulating the expression of E-selectin and VCAM-1, stimulating MAPKs, and inducing apoptosis similar to the proapoptotic effect of TNF-α, indicating the multifunctional role of cyclophilin A in the vascular system (86). Interestingly, cyclophilin A also is released from activated platelets and thus may stimulate the activation, migration, and proliferation of ECs and smooth muscle cells (53). Indeed, although absent in normal vasculature, cyclophilin A also was identified in human atherosclerotic lesions, which may be partially derived from incorporated platelets (53,86).

Adhesive Glycoproteins

The platelet releasate is rich in adhesive glycoproteins such as P-selectin (as mentioned earlier), TSP, vWF, and multimerin. As well as being present in the subendothelium, vWF is stored and secreted from platelet α-granules and EC Weibel-Palade bodies, because it is involved in binding to the GPIb-IX-V receptor under high shear. Multimerin, a carrier protein for platelet (but not plasma) factor V/Va, resembles vWF in its repeating, homomultimeric structure; it also is expressed in α-granules and Weibel-Palade bodies.

TSP is a 450-kDa Ca^{2+}-binding, disulphide-linked trimeric glycoprotein that constitutes 25% of the α-granule content; however, it is not exclusively a product of platelets and megakaryocytes, because it is present also in ECs, fibroblasts, monocytes, and macrophages. TSP can dictate the formation of multiprotein complexes at the vessel wall through specific interactions with growth factors, cytokines, other matrix components, and membrane proteins, including fibrinogen, P-selectin, osteonectin, and CD47 (87). Indeed, complexing with fibrinogen can potentiate platelet aggregation and the formation of the fibrin clot. TSP1 can modulate angiogenesis through its role in regulating EC apoptosis, protease expression, and VEGF expression by interacting with EC surface receptors such as CD36, proteoglycan receptors, or the integrins $\alpha V\beta 3$ and $\alpha 3\beta 1$ (88). Furthermore, TSP also plays a role in the modulation of EC PECAM-1 expression and function during vascular development and angiogenesis (89). TSP1 has been linked to a role as an inhibitor of tumor progression; a putative tumor cell binding protein, angiocidin, has been identified (90). Additionally, secreted TSP1 regulates the multimeric size and hemostatic activity of vWF (91).

Actin-Binding Proteins

Actin-binding proteins α-actinin and thymosin-$\beta 4$ (a 4.9-kDa polypeptide) have been documented as secreted from platelet α-granules (92), where they bind TSP and exhibit antimicrobial properties, respectively (93). Indeed, several released platelet proteins have antimicrobial properties, including PF4, RANTES, CTAP-III, platelet basic protein, and fibrinopeptide A and B (93). Following the release of thymosin-$\beta 4$, it is cross-linked to fibrin mediated by factor XIIIa, thereby increasing its local concentration near sites of clots and tissue damage, where it may contribute to wound healing, angiogenesis, and inflammatory responses. Furthermore, thymosin-$\beta 4$ increases the rate of attachment and spreading of ECs on matrix components and stimulates the migration of ECs, inducing the synthesis of PAI-1 and increasing its extracellular expression (94). Many of the other actin-binding proteins identified in the platelet releasate are released from other cells; these releasates include profilin, cofilin, actin, and tubulin from the exosomes of dendritic cells (39,53).

Proteases

In addition to lysosomal enzymes, several enzymes have been reported to be released from platelets, such as Adam-TS-5 and MMP-2 and -24 (53), which could support physiological processes such as angiogenesis as well as pathologically promote plaque instability or rupture through extracellular matrix remodeling and cell infiltration or migration. This ability to produce proteolytic enzymes as well as growth factors may be critical in the role that these cells play in the damage and repair that ensues as the lesion progresses.

THERAPEUTIC AND DIAGNOSTIC IMPLICATIONS

The prevention of platelet activation or the disruption of platelet–EC or platelet–leukocyte interactions by neutralizing adhesion receptors or soluble extracellular released products may be a means of therapeutic intervention in atherosclerosis. Indeed, the inhibition of platelet-derived proteins such as CD40 ligand and the blockade of chemokine receptor–ligand interactions reduces the development of atherosclerosis in mice (69,78). Thus, the targeting of selected secreted platelet proteins or their surface receptors may provide a novel means of modifying atherosclerosis without the risk associated with the direct inhibition of platelet adhesion. In addition, these secreted proteins could potentially be used as diagnostic markers, because no currently identified single marker is ideal for the ex vivo detection of in vivo platelet activation.

KEY POINTS

- Adhesive and signaling interactions between platelets and the endothelium play a pivotal role in vascular homeostasis and pathological conditions.
- Selectins and cell adhesion molecules mediate adhesive interactions between platelets and the endothelium.
- The platelet releasate is comprised of a multitude of inflammatory and vasoactive substances and vesicles that can attract atherogenic leukocytes from the circulation, activate ECs and, by infiltrating the vessel wall, stimulate vessel growth and repair.
- Future in vivo studies, in which certain secreted platelet proteins are enhanced or eliminated in transgenic animal models, are necessary to fully determine the biological function of these factors and their relevance to the initiation and progression of atherosclerotic lesions. Furthermore, comprehending the full complement of the platelet releasate will significantly improve our understanding of cell–cell interactions, coagulation, vessel growth, and remodeling.
- Neutralization of these platelet-derived proinflammatory factors may become an interesting means for therapeutic or preventative intervention in diseases such as atherosclerosis.

REFERENCES

1 Gaarder A, Jonsen J, Laland S, et al. Adenosine diphosphate in red cells as a factor in the adhesiveness of human blood platelets. *Nature*. 1961;192:531–532.
2 Born GV. Aggregation of blood platelets by adenosine diphosphate and its reversal. *Nature*. 1962;194:927–929.
3 Pintucci G, Froum S, Pinnell J, et al. Trophic effects of platelets on cultured endothelial cells are mediated by platelet-associated fibroblast growth factor-2 (FGF-2) and vascular endothelial growth factor, VEGF. *Thromb Haemost*. 2002;88:834–842.
4 Ross R. Atherosclerosis – an inflammatory disease. *N Engl J Med*. 1999;340:115–126.
5 Gimbrone MA Jr., Buchanan MR. Interactions of platelets and leukocytes with vascular endothelium: in vitro studies. *Ann NY Acad Sci*. 1982;401:171–183.
6 Frenette PS, Johnson RC, Hynes RO, Wagner DD. Platelets roll on stimulated endothelium in vivo: an interaction mediated by endothelial P-selectin. *Proc Natl Acad Sci USA*. 1995;92:7450–7454.
7 Massberg S, Brand K, Gruner S, et al. A critical role of platelet adhesion in the initiation of atherosclerotic lesion formation. *J Exp Med*. 2002;196:887–896.
8 Belton OA, Duffy A, Toomey S, Fitzgerald DJ. Cyclooxygenase isoforms and platelet vessel wall interactions in the apolipoprotein E knockout mouse model of atherosclerosis. *Circulation*. 2003;108:3017–3023.
9 Fukami H, Holmsen H, Kowalska M, Niewiarowski S. Platelet Secretion. In: Colman RW, Hirsh J, Marder VJ, Clowes AW, George JN, eds. *Haemostasis and Thrombosis: Basic Principles and Clinical Practice*. 4th Ed. Philadelphia, PA: Lippincott Williams and Wilkins; 2001:561–574.
10 Huo Y, Schober A, Forlow SB, et al. Circulating activated platelets exacerbate atherosclerosis in mice deficient in apolipoprotein E. *Nat Med*. 2003;9:61–67.
11 Johnson RC, Mayadas TN, Frenette PS, et al. Blood cell dynamics in P-selectin-deficient mice. *Blood*. 1995;86:1106–1114.
12 Frenette PS, Denis CV, Weiss L, et al. P-Selectin glycoprotein ligand 1 (PSGL-1) is expressed on platelets and can mediate platelet-endothelial interactions in vivo. *J Exp Med*. 2000;191:1413–1422.
13 Ley K. The role of selectins in inflammation and disease. *Trends Mol Med*. 2003;9:263–268.

14 Falati S, Liu Q, Gross P, et al. Accumulation of tissue factor into developing thrombi in vivo is dependent upon microparticle P-selectin glycoprotein ligand 1 and platelet P-selectin. *J Exp Med.* 2003;197:1585–1598.
15 Burger PC, Wagner DD. Platelet P-selectin facilitates atherosclerotic lesion development. *Blood.* 2003;101:2661–2666.
16 Manka D, Forlow SB, Sanders JM, et al. Critical role of platelet P-selectin in the response to arterial injury in apolipoprotein-E-deficient mice. *Arterioscler Thromb Vasc Biol.* 2004;24:1124–1129.
17 Wang K, Zhou X, Zhou Z, et al. Platelet, not endothelial, P-selectin is required for neointimal formation after vascular injury. *Arterioscler Thromb Vasc Biol.* 2005;25:1584–1589.
18 Ridker PM, Buring JE, Rifai N. Soluble P-selectin and the risk of future cardiovascular events. *Circulation.* 2001;103:491–495.
19 Johnson Tidey R, McGregor JL, Taylor PR, Poston RN. Increase in the adhesion molecule P-selectin in endothelium overlying atherosclerotic plaques. *Am J Pathol.* 1994;144:952–961.
20 McEver RP. Regulation of function and expression of P-selectin. *Agents Actions Suppl.* 1995;47:117–119.
21 Gawaz M. Role of platelets in coronary thrombosis and reperfusion of ischemic myocardium. *Cardiovasc Res.* 2004;61:498–511.
22 Vachino G, Chang XJ, Veldman GM, et al. P-selectin glycoprotein ligand-1 is the major counter-receptor for P-selectin on stimulated T cells and is widely distributed in non-functional form on many lymphocytic cells. *J Biol Chem.* 1995;270:21966–21974.
23 Romo GM, Dong JF, Schade AJ, et al. The glycoprotein Ib-IX-V complex is a platelet counterreceptor for P-selectin. *J Exp Med.* 1999;190:803–814.
24 Ruggeri ZM. Von Willebrand factor, platelets and endothelial cell interactions. *J Thromb Haemost.* 2003;1:1335–1342.
25 Massberg S, Enders G, Matos FC, et al. Fibrinogen deposition at the postischemic vessel wall promotes platelet adhesion during ischemia-reperfusion in vivo. *Blood.* 1999;94:3829–3838.
26 Diacovo TG, deFougerolles AR, Bainton DF, Springer TA. A functional integrin ligand on the surface of platelets: intercellular adhesion molecule-2. *J Clin Invest.* 1994;94:1243–1251.
27 Gong N, Chatterjee S. Platelet endothelial cell adhesion molecule in cell signaling and thrombosis. *Mol Cell Biochem.* 2003;253:151–158.
28 Jones KL, Hughan SC, Dopheide SM, et al. Platelet endothelial cell adhesion molecule-1 is a negative regulator of platelet-collagen interactions. *Blood.* 2001;98:1456–1463.
29 Ensminger SM, Spriewald BM, Steger U, et al. Platelet-endothelial cell adhesion molecule-1 (CD31) expression on donor endothelial cells attenuates the development of transplant arteriosclerosis. *Transplantation.* 2002;74:1267–1273.
30 Bombeli T, Schwartz BR, Harlan JM. Adhesion of activated platelets to endothelial cells: evidence for a GPIIbIIIa-dependent bridging mechanism and novel roles for endothelial intercellular adhesion molecule 1 (ICAM-1), alphavbeta3 integrin, and GPIbalpha. *J Exp Med.* 1998;187:329–339.
31 Weng S, Zemany L, Standley KN, et al. Beta3 integrin deficiency promotes atherosclerosis and pulmonary inflammation in high-fat-fed, hyperlipidemic mice. *Proc Natl Acad Sci USA.* 2003;100:6730–6735.
32 Shpilberg O, Rabi I, Schiller K, et al. Patients with Glanzmann thrombasthenia lacking platelet glycoprotein alpha(IIb)beta(3) (GPIIb/IIIa) and alpha(v)beta(3) receptors are not protected from atherosclerosis. *Circulation.* 2002;105:1044–1048.
33 Chew DP, Moliterno DJ. GP IIb/IIIa inhibitors in coronary artery disease management: what the latest trials tell us. *Cleve Clin J Med.* 2001;68:1017–1023.
34 Methia N, Andre P, Denis CV, et al. Localized reduction of atherosclerosis in von Willebrand factor-deficient mice. *Blood.* 2001;98:1424–1428.
35 Andre P, Denis CV, Ware J, et al. Platelets adhere to and translocate on von Willebrand factor presented by endothelium in stimulated veins. *Blood.* 2000;96:3322–3328.
36 Theilmeier G, Michiels C, Spaepen E, et al. Endothelial von Willebrand factor recruits platelets to atherosclerosis-prone sites in response to hypercholesterolemia. *Blood.* 2002;99:4486–4493.
37 Bernardo A, Ball C, Nolasco L, et al. Platelets adhered to endothelial cell-bound ultra-large von Willebrand factor strings support leukocyte tethering and rolling under high shear stress. *J Thromb Haemost.* 2005;3:562–570.
38 Heijnen HF, Schiel AE, Fijnheer R, et al. Activated platelets release two types of membrane vesicles: microparticles by surface shedding and exosomes derived from exocytosis of multivesicular bodies and alpha granules. *Blood.* 1999;94:3791–3799.
39 Thery C, Boussac M, Veron P, et al. Proteomic analysis of dendritic cell-derived exosomes: a secreted subcellular compartment distinct from apoptotic vesicles. *J Immunol.* 2001;166:7309–7318.
40 Al-Nedawi KN, Czyz M, Bednarek R, et al. Thymosin beta 4 induces the synthesis of plasminogen activator inhibitor 1 in cultured endothelial cells and increases its extracellular expression. *Blood.* 2004;103:1319–1324.
41 Raposo G, Nijman HW, Stoorvogel W, et al. B lymphocytes secrete antigen-presenting vesicles. *J Exp Med.* 1996;183:1161–1172.
42 Clayton AA, Turkes S, Dewitt R, et al. Adhesion and signaling by B cell-derived exosomes: the role of integrins. *FASEB J.* 2004;18:977–999.
43 Robertson C, Booth SA, Beniac DR, et al. Cellular prion protein is released on exosomes from activated platelets. *Blood.* 2006; In press.
44 Holme PA, Solum NO, Brosstad F, et al. Stimulated Glanzmann's thrombasthenia platelets produced microparticles. Microvesiculation correlates better to exposure of procoagulant surface than to activation of GPIIb-IIIa. *Thromb Haemost.* 1995;74:1533–1540.
45 Holme PA, Orvim U, Hamers MJ, et al. Shear-induced platelet activation and platelet microparticle formation at blood flow conditions as in arteries with a severe stenosis. *Arterioscler Thromb Vasc Biol.* 1997;17:646–653.
46 Sims PJ, Faioni EM, Wiedmer T, Shattil SJ. Complement proteins C5b-9 cause release of membrane vesicles from the platelet surface that are enriched in the membrane receptor for coagulation factor Va and express prothrombinase activity. *J Biol Chem.* 1988;263:18205–18212.
47 Horstman LL, Ahn YS. Platelet microparticles: a wide angle perspective. *Crit Rev Onc Hematol.* 1999;30:111–142.
48 Van Wijk MJ, VanBavel E, Sturk A, Nieuwland R. Microparticles in cardiovascular diseases. *Cardiovasc Res.* 2003;59:277–287.
49 Martinez MC, Tesse A, Zobairi F, Andriantsitohaina R. Shed membrane microparticles from circulating and vascular cells in regulating vascular function. *Am J Physiol Heart Circ Physiol.* 2005;288:H1004–H1009.
50 Owens MR, Holme S, Cardinali S. Platelet microparticles adhere to subendothelium and promote adhesion of platelets. *Thromb Res.* 1992;66:247–258.

51 Lindemann S, Tolley ND, Dixon DA, et al. Activated platelets mediate inflammatory signaling by regulated interleukin 1beta synthesis. *J Cell Biol.* 2001;154:485–490.
52 Garcia BA, Smalley DM, Cho H, et al. The platelet microparticle proteome. *J Proteome Res.* 2005;4:1516–1521.
53 Coppinger JA, Cagney G, Toomey S, et al. Proteomic characterization of the proteins released from activated platelets leads to localization of novel platelet proteins in human atherosclerotic lesions. *Blood.* 2004;103:2096–2104.
54 Krijgsveld J, Zaat SA, Meeldijk J, et al. Thrombocidins, microbicidal proteins from human blood platelets, are C-terminal deletion products of CXC chemokines. *J Biol Chem.* 2000;275:20374–20381.
55 Holt JC, Rabellino EM, Gewirtz AM, et al. Occurrence of platelet basic protein, a precursor of low affinity platelet factor 4 and beta thromboglobulin, in human platelets and megakaryocytes. *Exp Hematol.* 1998;16:302–306.
56 Stern D, Nawroth P, Marcum J, et al. Interaction of antithrombin III with bovine aortic segments. Role of heparin in binding and enhanced anticoagulant activity. *J Clin Invest.* 1985;75:272–279.
57 Hadley TJ, Lu ZH, Wasniowska K, et al. Postcapillary venule endothelial cells in kidney express a multispecific chemokine receptor that is structurally and functionally identical to the erythroid isoform, which is the Duffy blood group antigen. *J Clin Invest.* 1994;94:985–991.
58 Peiper SC, Wang ZX, Neote K, et al. The Duffy antigen/receptor for chemokines (DARC) is expressed in endothelial cells of Duffy negative individuals who lack the erythrocyte receptor. *J Exp Med.* 1995;181:1311–1317.
59 Lasagni L, Francalanci M, Annunziato F, et al. An alternatively spliced variant of CXCR3 mediates the inhibition of endothelial cell growth induced by IP-10, Mig, and I-TAC, and acts as functional receptor for platelet factor 4. *J Exp Med.* 2003;197:1537–1539.
60 Yu G, Rux AH, Ma P, et al. Endothelial expression of E-selectin is induced by the platelet specific chemokine Platelet Factor 4 through LRP in an NF-kappaB dependent manner. *Blood.* 2005;105:3545–3551.
61 Slungaard A, Fernandez JA, Griffin JH, et al. Platelet factor 4 enhances generation of activated protein C in vitro and in vivo. *Blood.* 2003;102:146–151.
62 Brill A, Elinav H, Varon D. Differential role of platelet granular mediators in angiogenesis. *Cardiovasc Res.* 2004;63:226–235.
63 Nassar T, Sachais BS, Akkawi S, et al. Platelet factor 4 enhances the binding of oxidized low-density lipoprotein to vascular wall cells. *J Biol Chem.* 2003;278:6187–6193.
64 von Hundelshausen P, Koenen RR, Sack M, et al. Heterophilic interactions of platelet factor 4 and RANTES promote monocyte arrest on endothelium. *Blood.* 2005;105:924–930.
65 Blank M, Shoenfeld Y, Tavor S, et al. Anti-platelet factor 4/heparin antibodies from patients with heparin-induced thrombocytopenia provoke direct activation of microvascular endothelial cells. *Int Immunol.* 2002;14:121–129.
66 Cohen AB, Stevens MD, Miller EJ, et al. Generation of the neutrophil-activating peptide-2 by cathepsin G and cathepsin G-treated human platelets. *Am J Physiol.* 1992;263:L249–L256.
67 Schober A, Manka D, von Hundelshausen P, et al. Deposition of platelet RANTES triggering monocyte recruitment requires P-selectin and is involved in neointima formation after arterial injury. *Circulation.* 2002;106:1523–1529.
68 von Hundelshausen P, Weber KS, Huo Y, et al. RANTES deposition by platelets triggers monocyte arrest on inflamed and atherosclerotic endothelium. *Circulation.* 2001;103:1772–1777.
69 Veillard NR, Kwak B, Pelli G, et al. Antagonism of RANTES receptors reduces atherosclerotic plaque formation in mice. *Circ Res.* 2004;94:253–261.
70 Taupenot L, Harper K, O'Connor D. The chromogranin-secretogranin family. *N Engl J Med.* 2003;348:1134–1149.
71 Wiedermann CJ, Dunzendorfer S, Kahler CM, et al. Secretoneurin and neurogenic inflammation. *Zhongguo Yao Li Xue Bao.* 1999;20:789–794.
72 Kahler C, Kaufmann G, Kahler S, Wiedermann C. The neuropeptide secretoneurin stimulates adhesion of human monocytes to arterial and venous endothelial cells in vitro. *Regul Pept.* 2002;110:65–73.
73 Kirchmair R, Gander R, Egger M, et al. The neuropeptide secretoneurin acts as a direct angiogenic cytokine in vitro and in vivo. *Circulation.* 2004;109:777–783.
74 Henn V, Slupsky JR, Gräfe M, et al. CD40 ligand on activated platelets triggers an inflammatory reaction of endothelial cells. *Nature.* 1998;391:591–594.
75 Henn V, Steinbach S, Buchner K, et al. The inflammatory action of CD40 ligand (CD154) expressed on activated human platelets is temporally limited by coexpressed CD40. *Blood.* 2001;98:1047–1054.
76 Urbich C, Dernbach E, Aicher A, et al. CD40 ligand inhibits endothelial cell migration by increasing production of endothelial reactive oxygen species. *Circulation.* 2002;106:981–986.
77 Aukrust P, Müller F, Ueland T, et al. Enhanced levels of soluble and membrane bound CD40 ligand in patients with unstable angina. Possible reflection of T lymphocyte and platelet involvement in the pathogenesis of acute coronary syndromes. *Circulation.* 1999;100:614–620.
78 Schonbeck U, Sukhova GK, Shimizu K, et al. Inhibition of CD40 signaling limits evolution of established atherosclerosis in mice. *Proc Natl Acad Sci USA.* 2000;97:7458–7463.
79 Harrison P, Cramer EM. Platelet alpha-granules. *Blood Rev.* 1993;7:52–62.
80 Raines EW. PDGF and cardiovascular disease. *Cytokine Growth Factor Rev.* 2004;15:237–254.
81 Hong T, Shimada Y, Uchida S, et al. Expression of angiogenic factors and apoptotic factors in leiomyosarcoma and leiomyoma. *Int J Mol Med.* 2001;8:141–148.
82 Li JJ, Huang YQ, Basch R, Karpatkin S. Thrombin induces the release of angiopoietin-1 from platelets. *Thromb Haemost.* 2001;85:204–206.
83 Jin Z, Melaragno M, Liao D, et al. Cyclophilin A is a secreted growth factor induced by oxidative stress. *Circ Res.* 2000;87:789–796.
84 Kim SH, Lessner SM, Sakurai Y, Galis ZS. Cyclophilin A as a novel biphasic mediator of endothelial activation and dysfunction. *Am J Pathol.* 2004;164:1567–1574.
85 Yang H, Li M, Chai H, et al. Effects of cyclophilin A on cell proliferation and gene expressions in human vascular smooth muscle cells and endothelial cells. *J Surg Res.* 2005;123:312–319.
86 Jin ZG, Lungu AO, Xie L, et al. Cyclophilin A is a proinflammatory cytokine that activates endothelial cells. *Arterioscler Thromb Vasc Biol.* 2004;24:1186–1191.
87 Chen H, Herndon ME, Lawler J. The cell biology of thrombospondin-1. *Matrix Biol.* 2000;19:597–614.

88 Chandrasekaran L, He CZ, Al-Barazi H, et al. Cell contact-dependent activation of a3b1 integrin modulates endothelial cell responses to thrombospondin-1. *Mol Biol Cell.* 2000;9:2885–2900.
89 Wang Y, Su X, Wu Z, Sheibani N. Thrombospondin-1 deficient mice exhibit an altered expression pattern of alternatively spliced PECAM-1 isoforms in retinal vasculature and endothelial cells. *J Cell Physiol.* 2005;204:352–361.
90 Zhou J, Rothman VL, Sargiannidou I, et al. Cloning and characterization of angiocidin, a tumor cell binding protein for thrombospondin-1. *J Cell Biochem.* 2004;92: 125–146.
91 Xie L, Chesterman CN, Hogg PJ. Control of von Willebrand factor multimer size by thrombospondin-1. *J Exp Med.* 2001;193: 1341–1349.
92 Dubernard V, Arbeille BB, Lemesle MB, Legrand C. Evidence for an alpha granular pool of the cytoskeletal protein alpha actinin in human platelets that redistributes with the adhesive glycoprotein thrombospondin-1 during the exocytotic process. *Arterioscler Thromb Vasc Biol.* 1997;17:2293–2305.
93 Tang YQ, Yeaman MR, Selsted ME. Antimicrobial peptides from human platelets. *Infect Immun.* 2002;70:6524–6533.
94 Al-Nedawi K, Szemraj J, Cierniewski CS. Mast cell-derived exosomes activate endothelial cells to secrete plasminogen activator inhibitor type 1. *Arterioscler Thromb Vasc Biol.* 2005;25: 1744–1749.

65

Cardiomyocyte–Endothelial Cell Interactions

Jian Li and Frank W. Sellke

Beth Israel Deaconess Medical Center, Harvard Medical School, Boston, Massachusetts

The vascular system delivers blood containing oxygen (O_2) and nutrients to tissues according to their local needs. In the finest branches of the vascular tree, capillaries consist of little more than a monolayer of endothelial cells (ECs) surrounded by an extracellular matrix and occasional pericytes. At these sites, microvascular endothelium is in intimate contact with underlying parenchymal cells.

In the heart, capillary ECs lie in close proximity to cardiomyocytes. As a result, significant cross-talk occurs between these two cell types, which involves reciprocal gene regulation, signal transduction, and energy supply (1). Changes in endothelial–myocyte cross-talk may contribute to and/or arise from numerous cardiac pathologies. In this chapter, we review in vitro and in vivo studies that provide mechanistic insight into the nature of this cell–cell communication.

MORPHOLOGICAL CONSIDERATIONS

The endothelium serves as the interface between the circulation and the vascular wall, integrating local and systemic events that regulate vascular function. During embryogenesis, the vascular endothelium is the first cardiovascular cell type to differentiate. These cells provide critical cues to all stages of cardiovascular development, and they are critically involved in myocardial trabeculation and the formation of primitive nutrient vessels.

In the normal heart, the ratio of the number of ECs to cardiac myocytes is approximately three to one (1). Three types of ECs have been identified in the heart based on their location: (a) endocardial ECs, which line the ventricles and atria; (b) myocardial microvascular ECs that line the arterioles, venules, and capillaries of the myocardium; and (c) ECs of epicardial arteries and veins (2–5). Of these, endocardial and microvascular ECs are in closest contact with cardiomyocytes.

Although the various EC types in the heart share certain common phenotypes, growing evidence suggests that they are not identical. For example, they differ in their embryological origin, cytoskeletal organization, and electrophysiological properties (6). The endocardium is formed by the process of vasculogenesis and is uniquely poised to sense and respond to cardiac performance and to circulating blood entering and leaving the heart. Endocardial (and coronary vascular) ECs release substances that modify the contraction of cardiac myocytes, most notably endothelin (ET)-1 (discussed in the next section) (7). ECs lining the coronary blood vessels and capillaries are formed by angiogenesis and receive less than 5% of cardiac output, yet they comprise the vast majority of surface area of the cardiac circulation and play a primary role in supplying the myocardial muscle mass with O_2 and nutrients (8). Endocardial and microvascular ECs are likely to signal via cell-type–specific mechanisms with underlying cardiomyocytes.

SIGNAL INPUT TO ENDOTHELIAL CELLS FROM CARDIAC MYOCYTES

Regulation of Hemostatic Balance

Previous studies have demonstrated that cardiac myocyte-derived signals are capable of modulating the expression of hemostatic genes in cardiac microvascular ECs. For example, cardiac myocytes, but not fibroblasts or hepatocytes, induce endothelial expression of endogenous von Willebrand factor (vWF) as well as *vWF* promoter activity under both in vitro and in vivo conditions (9). Expression of the endogenous gene and promoter was restricted to a minority of the cardiac microvascular ECs, suggesting the existence of a subpopulation of ECs capable of transducing the cardiomyocyte-derived signal(s).

Further studies demonstrated a role for the platelet-derived growth factor (PDGF) pathways in mediating the effect of cardiomyocytes on the expression of vWF (in addition to other genes) in microvascular ECs. In the absence of cardiomyocytes, cultured cardiac microvascular ECs were shown to constitutively express the PDGF-A isoform, leading to formation of the PDGF-AA homodimer. However, in the presence of cardiomyocytes, the microvascular ECs expressed PDGF-AB. PDGF-AB–mediated activation of the PDGF-α receptor was reported to induce the expression of vWF in ECs (10,11).

Regulation of Inflammation

In a study of graft failure in heart transplantation, Wang and colleagues reported that local expression of leukocyte adhesion receptor intercellular adhesion molecule (ICAM)-1 results in an explosive cascade of inflammatory events within the reperfused graft. The authors provided evidence for a positive-feedback loop in which local endothelial expression of ICAM-1 and myocardial expression of interleukin (IL)-1 leads to an amplification of the inflammatory response (12). IL-1 receptor blockade inhibited the expression of ICAM-1 in both the graft and remote organs, and improved graft survival, suggesting that the ICAM-1–IL-1 axis plays a pathophysiological role in primary cardiac graft failure (12,13).

Regulation of Angiogenesis

Vascular endothelial growth factor (VEGF) is involved in the growth and maintenance of blood vessels (14,15). In myocardium, VEGF expressed in cardiac myocytes is a predominant factor in controlling EC survival, proliferation, and angiogenesis. Studies have indicated that mouse embryos heterozygous for *VEGF* die around embryonic day (E)11 with malformations of the heart and major vessels. Cardiomyocyte-specific deletion of VEGF results in increased mortality between E14 and E17 (16). ECs express three endothelial-specific receptor tyrosine kinases: VEGF receptor 1 and -2 (VEGFR1 and VEGFR2) and Tie2, which bind to VEGF and angiopoietins 1 and -2 (Ang1 and Ang2), respectively. Several studies have suggested that the VEGF receptor pathway and the Tie2 pathway are independent and essential mediators of angiogenesis (17,18), leading to the hypothesis that simultaneous inhibition of both pathways should result in additive effects on angiogenesis. Indeed, blockage of both VEGFR2 and Tie2 pathways simultaneously resulted in a significant inhibition of angiogenesis (19).

The induction of VEGF expression by cardiac myocytes occurs via a hypoxia-inducible mechanism. Once released by the myocyte, VEGF activates its cognate receptors on the surface of ECs, inducing angiogenesis in ischemic myocardium. In a mouse model of myocardial ischemia, VEGF levels were shown to accumulate in the peri-infarct area (20). Under these conditions, *VEGF* mRNA expression was localized in cardiac myocytes and was associated with the formation of new capillaries that invade the infarcted myocardium. Thus, paracrine-mediated VEGF signaling may have evolved as an adaptive mechanism to improve deficient blood supply to ischemic myocardium.

Ischemic cardiac myocytes secrete other mediators that may engage in paracrine signaling pathways. In a recent study, conditioned medium from hypoxic myocytes was added to monolayers of human umbilical vein ECs (HUVECs) (21). DNA array analyses revealed more than 200 endothelial genes that were significantly induced or downregulated in the presence of conditioned medium (Table 65-1), including two widely studied markers, cyclooxygenase (COX)-2 (21), and syndecan-4 (22). COX-2 has been implicated in inflammation and angiogenesis (23,24). Syndencan-4 plays an important role in the regulation of EC growth and migration and in fibroblast growth factor 2 (FGF2) dependent signaling (25–27). Increased syndecan-4 expression augments the ability of ECs to migrate, proliferate, and form vascular structures in Matrigel in response to FGF2 (28). Further studies demonstrated that hypoxic cardiac myocytes induce endothelial expression of COX-2 (and its products, PGI_2 and PGE) by a VEGF-dependent pathway, whereas hypoxic myocyte-mediated induction of syndecan-4 involves tumor necrosis factor (TNF)-α (21).

The angiopoietin/Tek/Tie growth signaling pathway also plays an important role in the communication between cardiac myocytes and ECs. Tek/Tie receptors are expressed in ECs receiving the angiopoietin signal from cardiac myocytes and support the maturation, maintenance, and survival of ECs (29–31). Other growth factors that are involved in the angiogenic process are transforming growth factor (TGF)-β and FGF2 (32). These are also expressed abundantly in cardiac myocytes. The pathways through which these growth factors operate need further confirmation.

Mechanical Forces

Myocardial ECs are constantly exposed to mechanical forces resulting from blood flow and transmural pressure. Mechanical forces have been shown to stimulate capillary growth under conditions that produce mild-to-moderate vessel wall stretch (33). Mechanical stimulation leads to altered K^+ conductance in ECs, resulting in increased Ca^{2+} influx and the release from the ECs of vasoactive substances such as nitric oxide (NO), ET-1, and prostaglandins (34). In addition, increases in K^+ channel density are important in mediating endothelium-dependent responses to stimulating agents such as histamine and thrombin (34).

SIGNAL OUTPUT FROM ENDOTHELIAL CELLS TO CARDIAC MYOCYTES

Regulation of Myocardial Contractibility

ECs within the heart may release a number of substances that modulate myocardial function. The paracrine modulation of cardiac myocyte function by EC-derived factors is likely to be an important mechanism contributing to the overall regulation of cardiac contractile function, both in physiological and pathological states. Endothelial-derived mediators include NO (35,36), ET-1 (37,38), angiotensin converting enzyme (ACE), prostanoids (6,39), adenylpurines (38), and many other substances. A notable feature of many of these agents is that they influence contractile behavior by modifying cardiac myofilament properties rather than alternating cytosolic Ca^{2+} transients (7).

There are three isoforms of ET (40,41). Endothelium-derived ET-1 is a major determinant of cardiac contractility.

Table 65-1: Gene Regulation on HUVECs after Treatment with Cardiac Myocytes Hypoxia Conditioned Media (HMCM)

Name	*Access #*	*Fold*
Upregulation of Endothelial Genes (Selected)		
Leukocyte surface protein (CD31)	gb:NM_000442.1	28.97
von Willebrand factor (VWF)	gb:NM_000552.2	24.03
Cadherin 5, type 2, VE-cadherin (vascular epithelium) (CDH5)	gb:AB035304.1	19.87
Intercellular adhesion molecule-2 (ICAM-2)	gb:BC003097.1	19.87
Endothelin 1 (EDN1)	gb:NM_001955.1	18.10
Similar to placental growth factor, vascular endothelial growth factor-related protein	gb:S72960.1	15.66
Transcription factor EC (TFEC)	gb:NM_012252.1	9.39
EGF-containing fibulin-like extracellular matrix protein 1 (EFEMP1)	gb:NM_004105.2	9.19
Interferon, α-inducible protein 27 (IFI27)	gb:NM_005532.1	8.45
Matrix metalloproteinase 1 (interstitial collagenase) (MMP1)	gb:M13509.1	6.53
Coagulation factor II (thrombin) receptor (F2R)	gb:NM_001992.2	6.16
Kinase insert domain receptor (a type III receptor tyrosine kinase) (KDR)	gb:AF063658.1	5.95
Heparan sulfate proteoglycan (HSPG2)	gb:NM_005529.2	5.63
Rho GDP dissociation inhibitor (GDI) β (ARHGDIβ)	gb:NM_001175.1	5.55
Ras association (RalGDSAF-6) domain family 2 (RASSF2)	gb:NM_014737.1	4.75
Protein kinase C, eta (PRKCH)	gb:NM_006255.1	4.47
T-cell acute lymphocytic leukemia 1 (TAL1)	gb:NM_003189.1	4.13
Tissue factor pathway inhibitor-β (TFPI-β)	gb:AF021834.1	4.40
Endothelial differentiation, sphingolipid G-protein–coupled receptor, 1 (EDG1)	gb:AF233365.1	3.68
Integrin, α-6 (ITGA6)	gb:NM_000210.1	3.30
PTPL1-associated RhoGAP 1 (PARG1)	gb:NM_004815.1	2.96
Transforming growth factor-β-2	gb:M19154.1	2.83
Transcription factor RTEF-1 (RTEF1)	gb:U63824	2.77
Putative endothelin receptor type B-like protein	gb:U87460.1	2.65
Death-associated protein kinase 1 (DAPK1)	gb:NM_004938.1	2.56
Type XVIII collagen (COL18A1)	gb:NM_016214.1	2.49
Latent transforming growth factor-β binding protein 3 (LTBP3)	gb:AF135960.2	2.32
Angiopoietin 2 (ANGPT2)	gb:AB009865.1	2.19
Protein kinase, cAMP-dependent, catalytic, β (PRKACB)	gb:NM_002731.1	2.17
Downregulation of endothelial genes (Selected)		
Hypoxia-inducible factor 1, α-subunit (basic helix-loop-helix transcription factor) (HIF1A)	gb:U22431.1	0.20
Lysosomal membrane glycoprotein-2 (LAMP2)	gb:NM_002294.1	0.20
TGF-β activated kinase 1c	gb:AB009358.2	0.20
FLC3A mRNA for MAP1 light chain 3–related protein	gb:NM_007285.1	0.20
Ubiquitin specific protease	gb:AF350251.1	0.20
Forkhead box H1 (FOXH1)	gb:NM_003923.1	0.20
XIAP associated factor-1 (HSXIAPAF1)	gb:NM_017523.1	0.30
Integrin, β- (ITGβ5)	gb:NM_002213.1	0.30
Insulin-like growth factor 2 receptor (IGF2R)	gb:NM_000876.1	0.30
Proteolipid protein 2 (colonic epithelium-enriched) (PLP2)	gb:NM_002668.1	0.30
Breast cancer anti-estrogen resistance 3 (BCAR3)	gb:AF124250.1	0.30
Forkhead box F1 (FOXF1)	gb:NM_001451.1	0.30
Tumor necrosis factor receptor superfamily, member 1A (TNFRSF1A)	gb:NM_001065.1	0.30
Mutant p53 binding protein 1 (MBP1)	gb:AB030655.1	0.30
Vascular endothelial growth factor	gb:AF091352.1	0.30
Putative BTK-binding protein	gb:AF235049.1	0.30
Jun B proto-oncogene (JUNB)	gb:NM_002229.1	0.30
ADP-ribosylation factor-like 1 (ARL1)	gb:L28997.1	0.30
Nuclear factor, interleukin 3–regulated (NFIL3)	gb:NM_005384.1	0.40
Matrix metalloproteinase 9 (MMP9)	gb:NM_004994.1	0.40
Endothelin receptor type A (EDNRA)	gb:NM_001957.1	0.40
Mitogen-activated protein kinase kinase kinase 12 (MAP3K12)	gb:U07358.1	0.40
Angiopoietin-1 mRNA	gb:NM_001146.1	0.40
p38 mitogen activated protein kinase (MAPK)	gb:L35253.1	0.40

Name	*Access #*	*Fold*
Stem cell growth factor; lymphocyte secreted C-type lectin	gb:BC005810.1	0.40
Bcl-2-associated transcription factor	gb:AF249273.1	0.40
Mitogen-activated protein kinase kinase kinase kinase 3 (MAP4K3)	gb:NM_003618.1	0.40
Platelet derived growth factor C (PDGFC)	gb:NM_016205.1	0.40
Apoptosis-related protein PNAS-4 (PNAS-4)	gb:AF229834.1	0.40
Interferon-γ receptor 2 (interferon-γ transducer 1) (IFNGR2)	gb:NM_005534.1	0.40
Wilms tumour 1-associating protein (KIAA0105)	gb:NM_004906.1	0.50
Tissue inhibitor of metalloproteinase 2 (TIMP2)	gb:NM_003255.2	0.50
Apoptosis-related protein (APG5L)	gb:AF293841.1	0.50
Interleukin 1-β–converting enzyme isoform gamma (IL1βCE)	gb:U13698.1	0.50
G protein-coupled receptor 48 (GPR48)	gb:NM_018490.1	0.50
Caspase-9-β	gb:AB020979.1	0.50

ET-1 is known to have potent effects on contractile function and gene expression in cardiac myocytes (42,43). In addition to promoting contraction, ET-1 also induces myocyte proliferation (44,45). Expression of ET-1 is stimulated by hypoxia, shear stress, and various pharmacological receptor-dependent agonists (46–49). The role for EC–cardiomyocyte communication in mediating ET-l–dependent cardiac function is supported by many studies (50–52). For example, cardiomyocyte-derived paracrine mediators have been shown to regulate the abundance of precursor transcripts for *ET-1* in cardiac microvascular ECs (47,51,52). In another study, TGF-β was shown to induce expression of *ET-1* precursor mRNA in cardiac microvascular ECs, suggesting a potential autocrine and paracrine role for this growth factor (53).

ACE is located on the luminal surface of ECs (54,55), whereas other components of the angiotensin system exist locally in both cardiac microvascular ECs and cardiac myocytes (56,57). Moreover, a substantial ACE-independent conversion of angiotensin (Ang)-I to Ang-II occurs in the ventricles of the heart and may occur through cardiac chymase, a large proportion of which is localized in cardiac ECs (1).

Chronic ACE inhibition, initiated after acute myocardial infarction, can reduce ventricular dilatation and improve patient survival. ACE inhibition enhances coronary blood flow at the time of reperfusion and can prevent impairment of endothelium-dependent arteriolar responses and improve microvascular endothelial function and augmented postischemic coronary blood flow for ischemic myocardium (58). Therapeutically, ACE inhibitors have beneficial effects on ventricular remodeling and diastolic dysfunction in heart failure. Therefore, both cardiac myocytes and ECs are involved in mediating the effects of ACE inhibitors.

The endothelium of the heart releases several chemical mediators that influence cardiac regulatory pathways. Endothelial NO synthase (eNOS) is expressed predominantly in coronary vascular and endocardial ECs (59). NO synthesized by the activity of eNOS in the cardiac ECs directly modulates myocardial contractility (60,61). eNOS expression is followed by specific stimuli, for example, hypoxia. Shear stress and the mechanical deformation that occurs during the cardiac cycle also are relevant physiological stimuli for the release of eNOS from ECs. The fact that an increase in shear stress leads to an increase in Ca^{2+}, which in turn leads to NO release, suggests that the signal transduction is similar to that of receptor-dependent activation. Ischemia–reperfusion injury of the heart is not limited to cardiomyocytes but also extends to coronary vascular cells.

Mechanical forces modulate a number of signal transduction pathways in ECs in myocardium. These pathways include the heterotrimer G-proteins (62), cyclic adenosine monophosphate (cAMP) (63), protein kinase C (PKC) (64), and mitogen-activated protein kinases (MAPKs) (65). In addition to the direct modulatory effects on cardiac function, stimulation of these pathways leads to the activation of multiple transcription factors involved in mediating cross-talk between ECs and cardiac myocytes. The balance between NO and PGI_2 is important in regulating myocardial cAMP–cGMP ratios (NO and PGI_2 have opposing effects on myocardial relaxation) (66). eNOS activation in cardiac myocytes may suppress ouabain-induced arrhythmias through increased cGMP (67). Cultured ECs produce abundant PGI_2 and PGE_2. Interestingly, expression levels vary between endocardial ECs and coronary ECs (68). NO and PGI_2 release from ECs is coupled via cross-reaction between the NOS and COX-2 pathways (69,70).

Others have demonstrated that endothelium promotes cardiomyocyte survival, increases synchronized contraction of cardiac myocytes, and modulates expression of connexin-43 (71). These findings emphasize the functional importance of communication between ECs and cardiac myocytes and underscore the potential advantages of targeting both cell types in cardiac disease.

Regulation of Energy Supply

Two principal EC-dependent systems regulate contractile efficiency of cardiac myocytes: (a) endothelial-derived ET-1, which optimizes contraction efficiency by increasing isometric force and decreasing actomyosin ATPase activity (43,72,73);

and (b) β-adrenergic agonists, which increase ATPase activity through an EC-dependent mechanism, optimized for power of energy (6,74). The relative amounts of up- and downregulating EC-released substances are related to tissue O_2 tension (7). The advantage of a ceiling may be the prevention of increased formation of free radicals as O_2 tension rises.

The results observed with isolated cardiac myocytes confirm that these cells operate as O_2 sensors and secrete substances that induce ECs to release contractility-enhancing mediators (e.g., ET-1) and possibly an unidentified relaxant (75,76). This cross-talk mediates cardiac contraction and vascular tone. For example, endothelial-derived ET-1 enhances myofilament intracellular Ca^{2+} transients (77), decreases actomyosin ATPase activity, and decreases the velocity of unloaded shortening (43). In addition, increased ET-1 also may promote vascular tone and thereby decrease energy used by the cardiac myocytes (78).

Substances released from ECs in response to various stimuli may cause changes in several ion channels in cardiac myocytes. These substances regulate Ca^{2+} concentration, which is an important step in cellular signal transduction. Following a stimulus, the increase in Ca^{2+} occurs in two phases: (a) release of Ca^{2+} from the intracellular sources, resulting in an initial transient increase; and (b) sustained influx of Ca^{2+} from the extracellular space. This influx causes the Ca^{2+} to remain elevated and depends on the electrochemical gradient of Ca^{2+}; therefore, it is dependent on the resting membrane potential of the cardiac myocytes (6,79).

CONCLUSION

In this chapter, we have summarized some of the newer concepts regarding physiological and pathophysiological communication between ECs and cardiac myocytes. We have focused on studies that have directly examined the input of ECs, such as VEGF from cardiac myocytes targeting ECs, and output of ECs, including ET-1 from ECs affecting the myocardial contractibility of surrounding cardiac myocytes. Accordingly, the cardiovascular system is one system governing blood supply and circulation, and the cross-talk between vascular ECs and cardiac myocytes takes place consistently in almost every aspect. In reviewing the literature, it is clear that, during the last three decades, many investigators have found that there are mutual effects of these two cell types. Among these studies, the differences between endocardial ECs and myocardial capillary ECs with regard to their interactions with cardiac myocytes have been recognized and emphasized. However, because there are limitations of the techniques used, the direct experimental evidence from in vivo animal studies is less convincing than the data from cocultured cells in vitro. Nevertheless, as vascular biological research examines more fundamental questions, it will be important to consider the modulating role of ECs to their surrounding cellular environment. It is a complex system in which signaling and gene regulation pathways cross-react with one another. Future studies may discover the fundamental mechanisms that can be examined in the intact heart, including myocardium, coronary circulation, and capillaries.

KEY POINTS

- The cross-talk between vascular ECs and cardiac myocytes takes place consistently in almost every aspect of their function. Thus, we may need to consider the effect of this communication when studying gene regulation and protein alteration in either ECs or cardiac myocytes, especially in therapeutic manners.
- Signal input to ECs from cardiac myocytes: Regulation of ECs in homeostasis and angiogenesis is mediated by induction of PDGF and VEGF from cardiac myocytes. Endothelial COX-2, vWF, and TNF-α are involved in the communication as targets of the factors from cardiac myocytes.
- Signal output from ECs to cardiac myocytes: Endothelial-derived mediators including NO, ET-1, ACE, and many other substances, influence contractile behavior by modifying cardiac myofilament properties and alternating cytosolic Ca^{2+} transients.

Future Goals

- To identify and analyze novel factors that are involved in the input and output of ECs and their effects on cardiac myocytes
- To examine the cross-talk between ECs and cardiac myocytes in vivo to understand fundamental mechanisms of the cell–cell dialogue in the intact heart, including myocardium, coronary circulation, and capillaries

REFERENCES

1 Brutsaert DL. Cardiac endothelial-myocardial signaling: its role in cardiac growth, contractile performance, and rhythmicity. *Physiol Rev*. 2003;83:59–115.

2 Brutsaert DL, Fransen P, Andries LJ, et al. Cardiac endothelium and myocardial function. *Cardiovasc Res*. 1998;38:281–290.

3 Brutsaert DL, Meulemans AL, Sipido KR, Sys SU. Effects of damaging the endocardial surface on the mechanical performance of isolated cardiac muscle. *Circ Res*. 1988;62:358–366.

4 Li K, Rouleau JL, Andries LJ, Brutsaert DL. Effect of dysfunctional vascular endothelium on myocardial performance in isolated papillary muscles. *Circ Res*. 1993;72:768–777.

5 Li K, Qi X, Andries L, et al. Vascular-derived myocardial contractile factor: positive myocardial inotropic substance released from medial layer of the canine aorta. *J Mol Cell Cardiol*. 1996;28:881–892.

6 Kuruvilla L, Kartha CC. Molecular mechanisms in endothelial regulation of cardiac function. *Mol Cell Biochem.* 2003;253:113–123.

7 Winegrad S. EC regulation of contractility of the heart. *Annu Rev Physiol.* 1997;59:505–525.

8 Winegrad S, Henrion D, Rappaport L, Samuel JL. Vascular EC-cardiac myocyte crosstalk in achieving a balance between energy supply and energy use. *Adv Exp Med Biol.* 1998;453:507–514.

9 Aird WC, Edelberg JM, Weiler-Guettler H, et al. Vascular bed-specific expression of an endothelial cell gene is programmed by the tissue microenvironment. *J Cell Biol.* 1997;138:1117–1124.

10 Lindner V. Expression of platelet-derived growth factor ligands and receptors by rat aortic endothelium in vivo. *Pathobiology.* 1995;63:257–264.

11 D'Amore PA, Smith SR. Growth factor effects on cells of the vascular wall: a survey. *Growth Factors.* 1993;8:61–75.

12 Wang CY, Naka Y, Liao H, et al. Cardiac graft intercellular adhesion molecule-1 (ICAM-1) and interleukin-1 expression mediate primary isograft failure and induction of ICAM-1 in organs remote from the site of transplantation. *Circ Res.* 1998;82:762–772.

13 Salom RN, Maguire JA, Hancock WW. Endothelial activation and cytokine expression in human acute cardiac allograft rejection. *Pathology.* 1998;30:24–29.

14 Ferrara N. Vascular endothelial growth factor and the regulation of angiogenesis. *Recent Prog Horm Res.* 2000;55:15–35; discussion 35–36.

15 Ferrara N, Keyt B. Vascular endothelial growth factor: basic biology and clinical implications. *EXS.* 1997;79:209–232.

16 Gerber HP, Hillan KJ, Ryan AM, et al. VEGF is required for growth and survival in neonatal mice. *Development.* 1999;126:1149–1159.

17 Siemeister G, Schirner M, Weindel K, et al. Two independent mechanisms essential for tumor angiogenesis: inhibition of human melanoma xenograft growth by interfering with either the vascular endothelial growth factor receptor pathway or the Tie-2 pathway. *Cancer Res.* 1999;59:3185–3191.

18 Jendreyko N, Popkov M, Beerli RR, et al. Intradiabodies, bispecific, tetravalent antibodies for the simultaneous functional knockout of two cell surface receptors. *J Biol Chem.* 2003;278:47812–47819.

19 Jendreyko N, Popkov M, Rader C, Barbas CF 3rd. Phenotypic knockout of VEGF-R2 and Tie-2 with an intradiabody reduces tumor growth and angiogenesis in vivo. *Proc Natl Acad Sci USA.* 2005;102:8293–8298.

20 Li J, Brown LF, Hibberd MG, et al. VEGF, flk-1, and flt-1 expression in a rat myocardial infarction model of angiogenesis. *Am J Physiol.* 1996;270:H1803–H1811.

21 Wu G, Mannam AP, Wu J, et al. Hypoxia induces myocyte-dependent COX-2 regulation in endothelial cells: role of VEGF. *Am J Physiol Heart Circ Physiol.* 2003;285:H2420–H2429.

22 Zhang Y, Pasparakis M, Kollias G, Simons M. Myocyte-dependent regulation of EC syndecan-4 expression. Role of TNF-alpha. *J Biol Chem.* 1999;274:14786–14790.

23 Wilkinson-Berka JL. Vasoactive factors and diabetic retinopathy: vascular endothelial growth factor, cycoloxygenase-2 and nitric oxide. *Curr Pharm Des.* 2004;10:3331–3348.

24 Wu G, Luo J, Rana JS, et al. Involvement of COX-2 in VEGF-induced angiogenesis via P38 and JNK pathways in vascular endothelial cells. *Cardiovasc Res.* 2006;69:512–519.

25 Horowitz A, Simons M. Phosphorylation of the cytoplasmic tail of syndecan-4 regulates activation of protein kinase Calpha. *J Biol Chem.* 1998;273:25548–25551.

26 Simons M, Horowitz A. Syndecan-4-mediated signalling. *Cell Signal.* 2001;13:855–862.

27 Keum E, Kim Y, Kim J, et al. Syndecan-4 regulates localization, activity and stability of protein kinase C-alpha. *Biochem J.* 2004;378:1007–1014.

28 Zhang Y, Li J, Partovian C, et al. Syndecan-4 modulates basic fibroblast growth factor 2 signaling in vivo. *Am J Physiol Heart Circ Physiol.* 2003;284:H2078–H2082.

29 Partanen J, Puri MC, Schwartz L, et al. Cell autonomous functions of the receptor tyrosine kinase TIE in a late phase of angiogenic capillary growth and endothelial cell survival during murine development. *Development.* 1996;122:3013–3021.

30 Puri MC, Partanen J, Rossant J, Bernstein A. Interaction of the TEK and TIE receptor tyrosine kinases during cardiovascular development. *Development.* 1999;126:4569–4580.

31 Puri MC, Rossant J, Alitalo K, et al. The receptor tyrosine kinase TIE is required for integrity and survival of vascular endothelial cells. *EMBO J.* 1995;14:5884–5891.

32 Kardami E, Fandrich RR. Basic fibroblast growth factor in atria and ventricles of the vertebrate heart. *J Cell Biol.* 1989;109:1865–1875.

33 Price RJ, Skalak TC. Circumferential wall stress as a mechanism for arteriolar rarefaction and proliferation in a network model. *Microvasc Res.* 1994;47:188–202.

34 Fan J, Walsh KB. Mechanical stimulation regulates voltage-gated potassium currents in cardiac microvascular endothelial cells. *Circ Res.* 1999;84:451–457.

35 Heymes C, Vanderheyden M, Bronzwaer JG, et al. Endomyocardial nitric oxide synthase and left ventricular preload reserve in dilated cardiomyopathy. *Circulation.* 1999;99:3009–3016.

36 Paulus WJ, Shah AM. NO and cardiac diastolic function. *Cardiovasc Res.* 1999;43:595–606.

37 Mohacsi A, Magyar J, Tamas B, Nanasi PP. Effects of endothelins on cardiac and vascular cells: new therapeutic target for the future? *Curr Vasc Pharmacol.* 2004;2:53–63.

38 Shah AM, Grocott-Mason RM, Pepper CB, et al. The cardiac endothelium: cardioactive mediators. *Prog Cardiovasc Dis.* 1996;39:263–284.

39 Mebazaa A, Wetzel RC, Dodd-o JM, et al. Potential paracrine role of the pericardium in the regulation of cardiac function. *Cardiovasc Res.* 1998;40:332–342.

40 Levin ER. Endothelins as cardiovascular peptides. *Am J Nephrol.* 1996;16:246–251.

41 Levin ER. Endothelins. *N Engl J Med.* 1995;333:356–363.

42 Russell FD, Skepper JN, Davenport AP. Human endothelial cell storage granules: a novel intracellular site for isoforms of the endothelin-converting enzyme. *Circ Res.* 1998;83:314–321.

43 Mebazaa A, Mayoux E, Maeda K, et al. Paracrine effects of endocardial endothelial cells on myocyte contraction mediated via endothelin. *Am J Physiol.* 1993;265:H1841–H1846.

44 Suzuki E, Nagata D, Kakoki M, et al. Molecular mechanisms of endothelin-1-induced cell-cycle progression: involvement of extracellular signal-regulated kinase, protein kinase C, and phosphatidylinositol 3-kinase at distinct points. *Circ Res.* 1999;84:611–619.

45 Kramer BK, Nishida M, Kelly RA, Smith TW. Endothelins. Myocardial actions of a new class of cytokines. *Circulation.* 1992;85:350–356.

46 Pohl U, Busse R. Hypoxia stimulates release of endothelium-derived relaxant factor. *Am J Physiol.* 1989;256:H1595–H1600.
47 Katusic ZS, Shepherd JT. Endothelium-derived vasoactive factors: II. Endothelium-dependent contraction. *Hypertension.* 1991;18:III86–III92.
48 Vanhoutte PM, Luscher TF, Graser T. Endothelium-dependent contractions. *Blood Vessels.* 1991;28:74–83.
49 Pearson PJ, Lin PJ, Schaff HV. Production of endothelium-derived contracting factor is enhanced after coronary reperfusion. *Ann Thorac Surg.* 1991;51:788–793.
50 Agapitov AV, Haynes WG. Role of endothelin in cardiovascular disease. *J Renin Angiotensin Aldosterone Syst.* 2002;3:1–15.
51 Fedak PW, Rao V, Verma S, et al. Combined endothelial and myocardial protection by endothelin antagonism enhances transplant allograft preservation. *J Thorac Cardiovasc Surg.* 2005; 129:407–415.
52 Spieker LE, Noll G, Ruschitzka FT, Luscher TF. Endothelin receptor antagonists in congestive heart failure: a new therapeutic principle for the future? *J Am Coll Cardiol.* 2001;37:1493–1505.
53 Nishida M, Springhorn JP, Kelly RA, Smith TW. Cell-cell signaling between adult rat ventricular myocytes and cardiac microvascular endothelial cells in heterotypic primary culture. *J Clin Invest.* 1993;91:1934–1941.
54 Dostal DE, Baker KM. The cardiac renin-angiotensin system: conceptual, or a regulator of cardiac function? *Circ Res.* 1999;85: 643–650.
55 Dostal DE, Baker KM. Angiotensin and endothelin: messengers that couple ventricular stretch to the Na^+/H^+ exchanger and cardiac hypertrophy. *Circ Res.* 1998;83:870–873.
56 Dzau VJ. Circulating versus local renin-angiotensin system in cardiovascular homeostasis. *Circulation.* 1988;77:I4–I13.
57 Fischer TA, Ungureanu-Longrois D, Singh K, et al. Regulation of bFGF expression and ANG II secretion in cardiac myocytes and microvascular endothelial cells. *Am J Physiol.* 1997;272:H958–H968.
58 Piana RN, Wang SY, Friedman M, Sellke FW. Angiotensin-converting enzyme inhibition preserves endothelium-dependent coronary microvascular responses during short-term ischemia-reperfusion. *Circulation.* 1996;93:544–551.
59 Andries LJ, Brutsaert DL, Sys SU. Nonuniformity of endothelial constitutive nitric oxide synthase distribution in cardiac endothelium. *Circ Res.* 1998;82:195–203.
60 Smith JA, Shah AM, Lewis MJ. Factors released from endocardium of the ferret and pig modulate myocardial contraction. *J Physiol.* 1991;439:1–14.
61 Schulz R, Nava E, Moncada S. Induction and potential biological relevance of a $Ca^{(2+)}$-independent nitric oxide synthase in the myocardium. *Br J Pharmacol.* 1992;105:575–580.
62 Gudi SR, Clark CB, Frangos JA. Fluid flow rapidly activates G proteins in human endothelial cells. Involvement of G proteins in mechanochemical signal transduction. *Circ Res.* 1996;79: 834–839.
63 Cohen CR, Mills I, Du W, et al. Activation of the adenylyl cyclase/cyclic AMP/protein kinase A pathway in endothelial cells exposed to cyclic strain. *Exp Cell Res.* 1997;231:184–189.
64 Pulinilkunnil T, An D, Yip P, et al. Palmitoyl lysophosphatidylcholine mediated mobilization of LPL to the coronary luminal surface requires PKC activation. *J Mol Cell Cardiol.* 2004;37:931–938.
65 Tseng H, Peterson TE, Berk BC. Fluid shear stress stimulates mitogen-activated protein kinase in endothelial cells. *Circ Res.* 1995;77:869–878.
66 Balligand JL, Kelly RA, Marsden PA, et al. Control of cardiac muscle cell function by an endogenous nitric oxide signaling system. *Proc Natl Acad Sci USA.* 1993;90:347–351.
67 Kubota I, Han X, Opel DJ, et al. Increased susceptibility to development of triggered activity in myocytes from mice with targeted disruption of endothelial nitric oxide synthase. *J Mol Cell Cardiol.* 2000;32:1239–1248.
68 Mebazaa A, Wetzel R, Cherian M, Abraham M. Comparison between endocardial and great vessel endothelial cells: morphology, growth, and prostaglandin release. *Am J Physiol.* 1995;268:H250–H259.
69 Hendrickson RJ, Cappadona C, Yankah EN, et al. Sustained pulsatile flow regulates endothelial nitric oxide synthase and cyclooxygenase expression in co-cultured vascular endothelial and smooth muscle cells. *J Mol Cell Cardiol.* 1999;31:619–629.
70 Gerritsen ME. Physiological and pathophysiological roles of eicosanoids in the microcirculation. *Cardiovasc Res.* 1996;32: 720–732.
71 Narmoneva DA, Vukmirovic R, Davis ME, et al. Endothelial cells promote cardiac myocyte survival and spatial reorganization: implications for cardiac regeneration. *Circulation.* 2004;110: 962–968.
72 McClellan G, Weisberg A, Rose D, Winegrad S. Endothelial cell storage and release of endothelin as a cardioregulatory mechanism. *Circ Res.* 1994;75:85–96.
73 McClellan G, Weisberg A, Winegrad S. Effect of endothelin-1 on actomyosin ATPase activity. Implications for the efficiency of contraction. *Circ Res.* 1996;78:1044–1050.
74 Shah AM, Mebazaa A, Wetzel RC, Lakatta EG. Novel cardiac myofilament desensitizing factor released by endocardial and vascular ECs. *Circulation.* 1994;89:2492–2497.
75 McClellan G, Weisberg A, Kato NS, et al. Contractile proteins in myocardial cells are regulated by factor(s) released by blood vessels. *Circ Res.* 1992;70:787–803.
76 Ramaciotti C, McClellan G, Sharkey A, et al. Cardiac endothelial cells modulate contractility of rat heart in response to oxygen tension and coronary flow. *Circ Res.* 1993;72: 1044–1064.
77 Wang JX, Paik G, Morgan JP. Endothelin 1 enhances myofilament Ca^{2+} responsiveness in aequorin-loaded ferret myocardium. *Circ Res.* 1991;69:582–589.
78 Margulies KB, Hildebrand FL Jr., Lerman A, et al. Increased endothelin in experimental heart failure. *Circulation.* 1990;82: 2226–2230.
79 Brutsaert DL, De Keulenaer GW, Fransen P, et al. The cardiac endothelium: functional morphology, development, and physiology. *Prog Cardiovasc Dis.* 1996;39:239–262.

66

Interaction between Hepatocytes and Liver Sinusoidal Endothelial Cells

David Semela and Vijay Shah

Mayo Clinic, Rochester, Minnesota

The liver is the largest organ in the body and, in humans, weighs between 1,400 and 1,600 g (2.5% of total body weight). It receives 25% of the total cardiac output, which arrives via the hepatic artery (one-third of hepatic blood flow) and the portal vein (two-thirds of hepatic blood flow). Blood flows through liver sinusoids, a unique microvasculature that consists of plates of liver sinusoidal endothelial cells (LSECs) between plates of hepatocytes, before coming in contact with the liver parenchyma. LSECs account for 20% of total liver cells (an estimated 1×10^8 cells) whereas hepatocytes represent the majority of liver cells (estimated 60% or 3×10^8 cells). The remaining cell populations are (a) hepatic stellate cells (HSCs) (5%–8%), which are liver-specific pericytes; (b) biliary epithelial cells, also known as cholangiocytes, which line the bile ducts; and (c) macrophages called *Kupffer cells.* LSECs are unique in comparison to other organs in that they possess multiple fenestrae (pores) arranged in sieve plates (Figure 66.1) and lack an organized basement membrane. LSECs lie in close proximity to hepatocytes, and the interaction of these two cell types has been the subject of intense study. Cellular cross-talk between LSECs and hepatocytes plays an important role in homeostasis and physiologic processes such as liver organogenesis and liver regeneration. Moreover, abnormalities in intercellular communication underlie virtually every disease of the liver, including acute and chronic inflammatory liver disease, liver fibrosis, and hepatocarcinogenesis. Of these physiological and pathological states, liver regeneration is the most studied and best understood. Thus, this chapter focuses primarily on the latter model as a means to explore the biology of hepatocyte–endothelial cell (EC) interactions. LSECs interact with many other cell types in the liver, including HSCs, Kupffer cells, and circulating blood cells. These interactions are discussed elsewhere in this volume (see Chapters 67 and 133).

LIVER REGENERATION AFTER PARTIAL HEPATECTOMY

The extraordinary capacity of the liver to regenerate was first described in the Greek myth of Prometheus (Table 66-1). According to this tale, Prometheus was chained to a rock as punishment for stealing fire from Zeus and giving it to man. Zeus' eagle would come and eat part of Prometheus' liver every day, but it would grow back overnight. Today, surgeons routinely carry out partial resections of the liver for purposes of living donor organ transplantation or during the process of tumor excision. Following partial hepatectomy (up to 70% removal of the organ), the liver will completely regenerate and reach its original functional mass with normal microscopic architecture within weeks. This contrasts with other organs, which have no or only limited regenerative capacity. The phenomenon of liver regeneration has been described in mammals, birds, fish, reptiles, and amphibians. It has been suggested that this remarkable capacity for regeneration evolved as a means to protect the mammalian liver against food toxin–induced fulminant hepatic failure (1).

Experimental partial hepatectomy involves the removal of two-thirds of the liver, typically from rodents. Within 8 to 10 days, the liver remnant enlarges until the previous functional liver mass is restored (2). This process represents a hyperplastic response of the remaining liver, leading to a new lobar anatomy rather than true regeneration with restoration of the initial anatomy. Hepatocytes in the normal liver are quiescent, with only 0.01% to 0.001% undergoing mitosis at any given time. Partial hepatectomy activates numerous transcription factors, induces the expression of more than 70 genes, and promotes cell cycle entry with synchronous proliferation of almost all hepatocytes (2). The initial wave of hepatocyte proliferation is followed by a second wave of replication involving

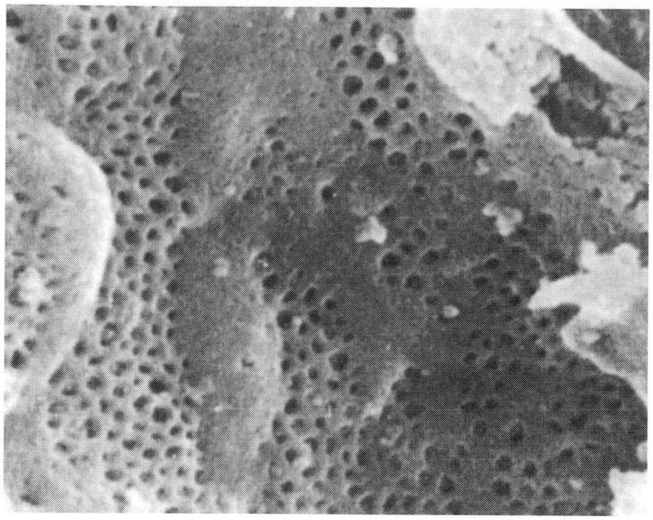

Figure 66.1. Scanning electron microscopy of normal fenestrated hepatic sinusoids. (Reproduced with permission from Motta P, Muto M, Fujita T. *The Liver: An Atlas of Scanning Electron Microscopy*. New York: Igaku-Shoin Medical Publishers, 1978:129.)

nonparenchymal liver cells, including HSCs, Kupffer cells, and LSECs, which start to proliferate on day 3 after partial hepatectomy and reach a peak on day 5 (3).

Ross and coworkers have detailed the timing and stages associated with post–partial-hepatectomy liver regeneration in mice (3) (Figure 66.2). The initial proliferation of hepatocytes leads to the formation of avascular islands of cells. Hepatocyte signaling induces the activation and proliferation of neighboring LSECs, resulting in their migration into the avascular islands (4,5). LSECs continue to proliferate and form three-dimensional structures, ultimately leading to recanalization and formation of patent sinusoids. Eight to 10 days following partial hepatectomy, liver mass is fully reconstituted, and the architecture is restored to normal. Recent evidence suggests that endothelial progenitor cells (EPCs) (mobilized from bone marrow by systemically circulating growth factors and chemokines produced in the regenerating liver) participate in this process by homing to sites of neovascularization in the liver and committing to sinusoidal ECs (6). Bone marrow cells also have been shown to commit to HSCs, which are liver-specific pericytes supporting LSECs during liver regeneration (7).

Several key signaling molecules have been implicated in the process of liver regeneration. These include the cytokines interleukin (IL)-6 and tumor necrosis factor (TNF), which play a role in the G_0/G_1 transition, and the growth factors hepatocyte growth factor (HGF) and transforming growth factor (TGF)-α, which are involved in cell cycle progression after the cells reach G_1 (2). With the exception of TGF-α (which is produced by hepatocytes), all the above mediators are secreted by nonparenchymal cells (including LSECs) but act on hepatocytes as well as on nonparenchymal cells in an autocrine and paracrine way to orchestrate the regenerative response (2,8).

Table 66-1: Milestones in the Field of Liver Regeneration/Angiogenesis

700 BC	Hesiod's *Theogony* depicts the phenomenon of liver regeneration in the myth of Prometheus.
1880–1900	The principles of liver regeneration and liver hemostasis were first studied by Ponfick, Mayer, Kousnetzoff, Pensky, and Pringle (36).
1931	Higgins and Anderson describe the now widely used partial hepatectomy model involving removal of two-thirds of the liver in rats for the study of liver regeneration (37).
1967	Moolten and Bucher show that a circulating mitogen (later identified as HGF) is present after hepatectomy inducing liver regeneration (9).
1975	Peak mitotic activity of LSECs after partial hepatectomy is found to occur on day 4, only after hepatocyte proliferation (38).
1984	Michalopoulos and Nakamura identify a serum protein from hepatectomized rats by gel permeation chromatography which is a potent stimulator of hepatocyte proliferation in vitro (39,40).
1990	Gherardi and coworkers characterize this protein as HGF (41).
1991	c-MET is found to be the transmembrane receptor for HGF (42).
1994	Yamane proposes HGF and VEGF with its receptors as a reciprocal communication system between hepatocytes and LSECs (43).
1996	Expression of VEGF in hepatocytes and of VEGF receptors on LSECs are shown to occur after partial hepatectomy (16).
2001	Ross and colleagues describe the spatiotemporal expression of angiogenic growth factor receptors during liver regeneration in different liver cell populations (3).
2003	LeCouter and Ferrara demonstrate that HGF is one of the liver sinusoidal EC-derived paracrine mediators promoting hepatocyte growth. VEGF receptor 1 is shown to stimulate hepatocyte but not EC proliferation in vivo and reduces liver damage in mice exposed to a hepatotoxin (15).

In 1967, Moolten and Bucher demonstrated that a circulating mitogen following hepatectomy in one rat causes regeneration of an intact liver in a second rat when the circulations of both animals were connected (9). During the 1980s and 1990s, this factor was identified and characterized, and ultimately named HGF (also known as scatter factor or hematopoietin A). HGF is an 80-kDa disulfide-linked heterodimeric protein with

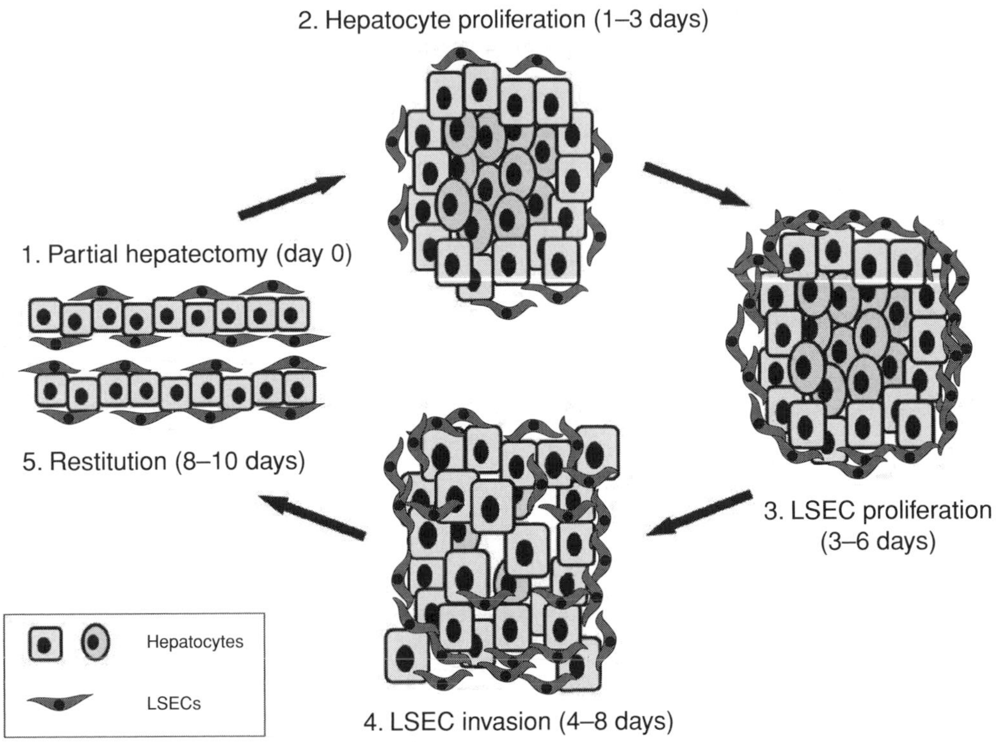

Figure 66.2. Spatiotemporal representation of proliferation and migration of LSECs and hepatocytes during liver regeneration. 1. Normal liver architecture with LSECs lining sinusoids and adjacent hepatocytes before partial hepatectomy. 2. Hepatocytes proliferate and form avascular islands composed of cell clusters of 10 to 14 hepatocytes (see Ref. 5). 3. LSECs proliferate in a second wave after hepatocytes. 4. LSECs migrate into the avascular islands of hepatocytes. 5. Reconstitution of the normal liver architecture 8 to 10 days after partial hepatectomy in rodents. (Adapted with permission from Ross MA, Sander CM, Kleeb TB, et al. Spatiotemporal expression of angiogenesis growth factor receptors during the revascularization of regenerating rat liver. *Hepatology* 2001;34:1135–1148; and Martinez-Hernandez A, Amenta PS. The extracellular matrix in hepatic regeneration. *FASEB J* 1995;9:1401–1410.)

mitogenic and motogenic properties for a number of cell types including hepatocytes and ECs. HGF, which is the most potent stimulator of hepatocyte proliferation, is produced during liver regeneration by LSECs as well as HSCs, and it signals through the tyrosine kinase receptor c-Met by activating mitogen-activated protein kinase (MAPK), phosphoinositide 3-kinase (PI3K)/Akt, Rac/Pak, and Crk/Rap1 pathways in hepatocytes (Figure 66.3). HGF/c-Met signaling is essential for hepatic recovery after partial hepatectomy or after exposure to a hepatotoxin (CCl_4), as conditional *Met* mutant mice show severe impairment of liver regeneration under these conditions (10,11).

Vascular endothelial growth factor (VEGF) is the most important growth and survival factor for endothelium. VEGF is a 45-kDa homodimeric glycoprotein that is constitutively expressed in hepatocytes at low levels in hepatocytes of the normal liver (12). VEGF production by hepatocytes and HSCs rapidly increases during liver regeneration. Upregulation of VEGF might be triggered by hypoxia in the center of the regenerative and avascular hepatocyte islands (13). Hypoxia activates the transcription factor hypoxia-inducible factor (HIF)-1, which in turn induces the expression of downstream target genes including *VEGF* and VEGF receptor (*VEGFR) 1.* Hepatocyte production of VEGF peaks 48 to 72 hours after hepatectomy and is detected mainly in periportal hepatocytes (5,12). The administration of VEGF in hepatectomized rodents increases hepatocyte and sinusoidal EC proliferation (12,14), accelerates gain in liver mass (15), and improves functional hepatic recovery (13). Neutralizing antibodies against VEGF inhibit hepatocyte and EC proliferation after partial hepatectomy (12).

VEGF production is accompanied by an increase in the expression of the VEGFR1 on hepatocytes and of VEGFR1 and VEGFR2 on LSECs (3,4,13,16). Activation of VEGFR2 stimulates LSEC proliferation. On the other hand, binding of VEGF to VEGFR1 on LSECs induces the secretion of growth and survival factors such as HGF and IL-6 from LSECs, which in turn stimulate hepatocyte proliferation (15). The role of VEGFR1 in hepatocytes is not clear.

In addition to VEGF and HGF, other angiogenic growth factors (and their receptors) are upregulated during liver regeneration, including angiopoietins (Angs), platelet-derived growth factor (PDGF), and fibroblast growth factor (FGF) (Table 66-2) (17).

Table 66-2: Expression of Growth Factors and Their Receptors in Hepatocytes, Liver ECs, Cholangiocytes, and Hepatic Stellate Cells after Partial Hepatectomy

Growth Factor/Receptor	*Expressing Cells*	*Increase after Partial Hepatectomy*
VEGF	Hepatocytes, HSCs	48–72 h
VEGFR1 (FLT-1)	LSECs, HSCs	72 h to 12 d
VEGFR2 (FLK-1/KDR)	LSECs	72 h to 10 d
HGF	LSECs	Immediate, peak 12–24 h
c-Met	LSECs, hepatocytes, cholangiocytes	No change in LSECs
Ang1	unknown	72–96 h (mRNA)
Ang2	unknown	72–168 h (mRNA)
Tie1	LSECs	Start at 48 h, peak at 96 h
Tie2	LSECs	Start at 48 h, peak at 96 h
PDGF	LSECs, HSCs	48 h
PDGF-Rβ	LSECs, HSCs	3–12 d
aFGF	HSCs	24 h (mRNA)
bFGF	HSCs	Unchanged
FGF-R1 (Flg)	Hepatocytes	Unchanged
FGF-R2 (Bek)	Hepatocytes	Not determined
EGF	Hepatocytes	Immediate, peak 12–24 h
TGF-α	Hepatocytes	12–24 h (mRNA)
EGF-R	LSECs	Constitutive, 3–14 d

Adapted with permission from Medina J, Arroyo AG, Sanchez-Madrid F, Moreno-Otero R. Angiogenesis in chronic inflammatory liver disease. *Hepatology* 2004;39(5):1185–1195.
EGF, epidermal growth factor.

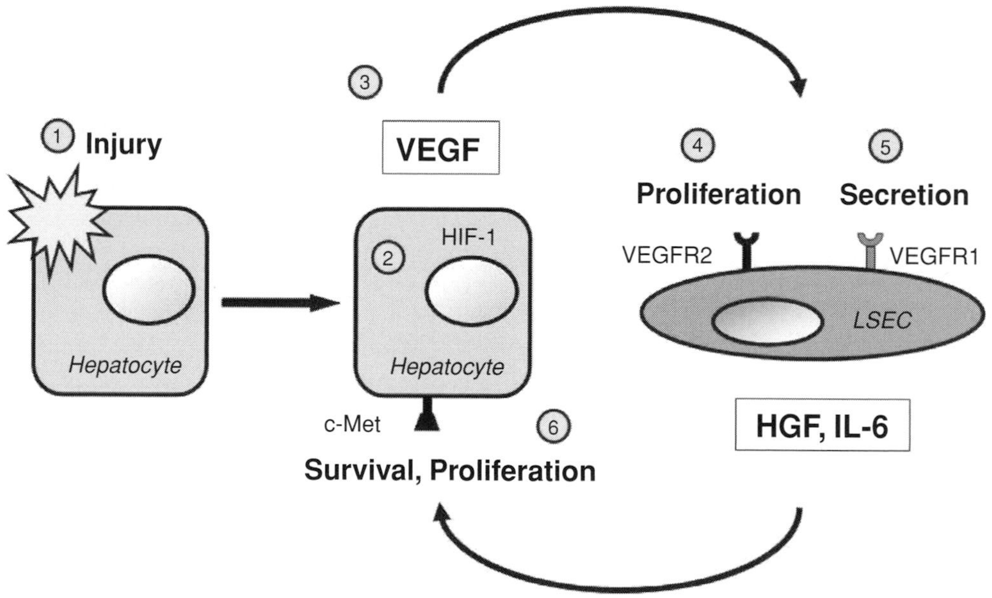

Figure 66.3. Cellular cross-talk between LSECs and hepatocytes during liver regeneration. 1. Injury of the liver leads to liver regeneration, as described in Figure 66.2, with formation of avascular hepatocyte clusters. 2. In response to local hypoxia, hepatocytes in these clusters activate the transcription factor HIF-1, which induces the expression of downstream target genes including VEGF. 3. Hepatocytes secrete VEGF. 4. Activation of VEGFR2 stimulates LSEC proliferation. 5. Activation of VEGFR1 induces the secretion of survival and growth factors such as HGF and IL-6 from LSEC. 6. These secreted factors stimulate hepatocyte proliferation (c-Met is the receptor for HGF).

Table 66-3: Liver Organogenesis in VEGF and HGF Knockout and Knockdown Mice

Gene (Refs.)	*Survival of Mutant*	*Phenotype*
VEGF (23, 24)	Lethal at E12	Vessel malformation and abnormal organ development
VEGF (liver only) (25)	Viable	Abnormal sinusoidal network, lack of LSEC fenestrae, impaired lipoprotein intake in hepatocytes
VEGF (neonatal) (26)	Increased mortality	Small liver with immature hepatic sinusoids, reduced number of LSECs with high apoptosis index
VEGFR1 (27, 28)	Lethal at E8.5	Abnormal vasculature, excessive angioblast proliferation
VEGFR2 (19)	Lethal at E9.5–10.5	Lack of mature ECs, fails to form a liver bud
HGF, c-Met (20–22)	Lethal at E13.5–16.5	Small liver, loss of hepatocytes

LIVER ORGANOGENESIS

Similar to their role in liver regeneration, HGF and VEGF signaling are required for liver organogenesis by mediating signal exchange between hepatocytes and LSECs during mouse development. Hepatic cells are induced within the endoderm by embryonic day (E)8.5 of mouse gestation (18). Interaction between these cells and surrounding ECs or angioblasts induces outgrowth of the liver bud into the septum transversum mesenchyme at E9.5 (Figure 66.4) (19). This stage is then followed by the de novo formation of a local vascular network from hemangioblasts (vasculogenesis) and the recruitment of hematopoietic cells (19). Knockout mice for HGF (20,21) or c-Met (22) fail to complete development and die in utero between days E13.5 and E16.5 (Table 66-3). These mutations affect the embryonic liver, which is reduced in size and shows extensive loss of parenchymal cells. The importance of VEGF signaling during embryonic development is highlighted by the fact that lack of a single *VEGF* gene allele results in abnormal blood vessels, abnormal organ development, and embryonic lethality in mice by E12 (23,24). Carpenter and coworkers studied the development of the hepatic vasculature using an elegant liver-specific conditional VEGF-knockdown system (25). Downregulation of VEGF during liver organogenesis resulted in an abnormal hepatic microvascular network with incomplete lining of the sinusoidal lumen, partially collapsed sinusoidal channels, lack of LSEC fenestrae, and impaired lipoprotein uptake in hepatocytes (25). But VEGF also is required during the neonatal development of the liver. The inactivation of the *VEGF* gene in newborn mice leads to a small liver with an immature hepatic sinusoidal network, focal loss of integrity of the space of Disse, and a reduced number of LSECs with high levels of apoptosis (26). The liver parenchyma in these mice shows small and rounded hepatocytes that fail to produce the typical plate architecture surrounding the sinusoids (26). Knockout of the *VEGFR1* gene in mice results in the formation of abnormal vascular channels and excessive angioblast proliferation in embryonic and extra-embryonic regions, causing lethality by day E8.5 (27,28). Homozygous loss of VEGFR2 results in lack of ECs, impaired liver organogenesis (19), and embryonic lethality at day E9.5 to E10.5, whereas heterozygous mice are normal (29). Mice with homozygous deficiency in VEGFR2 lacking mature ECs and blood vessels show normal thickening of the hepatic endoderm but absence of liver bud emergence (19). Vasculogenic ECs with intact signaling through VEGFR2 are therefore critical in the earliest stages of liver organogenesis, even prior to blood vessel function (19).

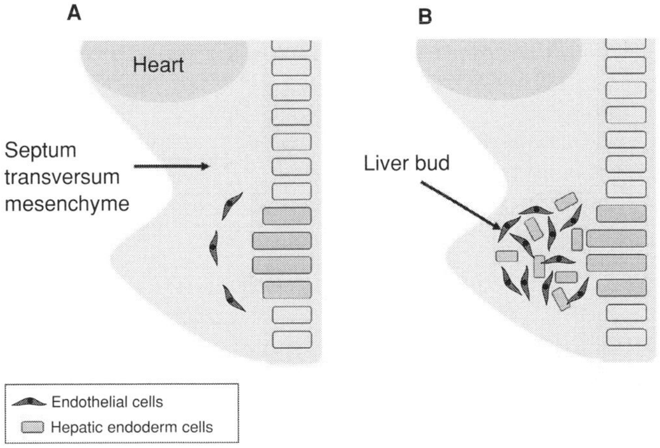

Figure 66.4. Model of liver bud formation during liver organogenesis. Hepatic endoderm cells are induced within the endoderm by day E8.5 of mouse gestation (**A**). Interaction between hepatic endoderm cells and surrounding ECs induces outgrowth of the liver bud into the septum transversum mesenchyme at E9.5 (**B**), which is followed by the generation of an extensive local vascular network with formation of the future sinusoids. (Adapted with permission Matsumoto K, Yoshitomi H, Rossant J, Zaret KS. Liver organogenesis promoted by endothelial cells prior to vascular function. *Science* 2001;294:559–563.)

DIAGNOSTIC AND THERAPEUTIC IMPLICATIONS

Studies in patients undergoing liver resection (for example, for liver metastasis), show increased serum levels of IL-6 and HGF, which is in line with Molton and Bucher's early observations in the animal regeneration model explained earlier (30). Similar studies with HGF and VEGF have been carried out in patients undergoing living donor liver transplantation (31,32). To recover, both donor and recipient rely on liver regeneration after the transplantation.

Could liver regeneration and recovery in such patients be accelerated if growth factors would be administered? Several animal studies confirm faster recovery and accelerated liver regeneration after partial hepatectomy if growth factors such as HGF and/or VEGF are administered intravenously or via recombinant adenoviral vectors 48 hours before or at the day of surgery (12–14). No clinical studies have been carried out so far. However, a recent study has investigated whether extracorporeal liver support systems, which have been developed to remove metabolites and facilitate recovery or as a bridge to liver transplantation in patients with liver failure, can influence liver regeneration. It has been shown that when not only metabolites and toxins, but also various hepatic growth factors, were simultaneously removed by the artificial liver support it potentially retarded liver regeneration (33,34).

Several proteins involved in the regulation of angiogenesis, such as VEGF and HGF, have been investigated as biomarkers and potential therapeutic targets in patients with hepatocellular carcinoma and chronic inflammatory liver disease (17,35).

KEY POINTS

- Molecular cross-talk between LSECs and hepatocytes regulates organogenesis, regeneration, and homeostasis of the liver.
- Partial hepatectomy as a model to study cellular and humoral interaction within the liver has identified intense and timely regulated paracrine cross-talk between liver sinusoidal ECs and hepatocytes during hepatic regeneration.
- VEGF secreted by hepatocytes stimulates LSEC proliferation and angiogenesis by signaling through VEGFR2, whereas activation of VEGFR1 on LSECs induces secretion of liver-specific growth factors such as HGF.
- HGF secreted by LSECs in conjunction with other growth and survival factors provides the main stimulus for hepatocyte proliferation during liver regeneration.
- HGF and VEGF signaling are also fundamental for normal liver organogenesis by mediating signal exchange between the mesenchyme and epithelium during development.

Future Goals

To develop diagnostic and therapeutic strategies for patients with liver disease based on insight from the recent molecular discoveries in the field of liver regeneration

REFERENCES

1 Michalopoulos GK, DeFrances MC. Liver regeneration. *Science.* 1997;276(5309):60–66.

2 Kumar V, Fausto N, Abbas A. *Robbins and Cotran's Pathologic Basis of Disease.* 7th ed. Boston: Elsevier Science, 2004.

3 Ross MA, Sander CM, Kleeb TB, et al. Spatiotemporal expression of angiogenesis growth factor receptors during the revascularization of regenerating rat liver. *Hepatology.* 2001;34(6):1135–1148.

4 Sato T, El-Assal ON, Ono T, et al. Sinusoidal EC proliferation and expression of angiopoietin/Tie family in regenerating rat liver. *J Hepatol.* 2001;34(5):690–698.

5 Martinez-Hernandez A, Amenta PS. The extracellular matrix in hepatic regeneration. *FASEB J.* 1995;9(14):1401–1410.

6 Fujii H, Hirose T, Oe S, et al. Contribution of bone marrow cells to liver regeneration after partial hepatectomy in mice. *J Hepatol.* 2002;36(5):653–659.

7 Baba S, Fujii H, Hirose T, et al. Commitment of bone marrow cells to hepatic stellate cells in mouse. *J Hepatol.* 2004;40(2):255–260.

8 Taub R. Liver regeneration: from myth to mechanism. *Nat Rev Mol Cell Biol.* 2004;5(10):836–847.

9 Moolten FL, Bucher NL. Regeneration of rat liver: transfer of humoral agent by cross circulation. *Science.* 1967;158(798):272–274.

10 Borowiak M, Garratt AN, Wustefeld T, et al. Met provides essential signals for liver regeneration. *Proc Natl Acad Sci USA.* 2004;101(29):10608–10613.

11 Huh C-G, Factor VM, Sanchez A, et al. Thorgeirsson SS. Hepatocyte growth factor/c-met signaling pathway is required for efficient liver regeneration and repair. *Proc Natl Acad Sci USA.* 2004;101(13):4477–4482.

12 Taniguchi E, Sakisaka S, Matsuo K, et al. Expression and role of vascular endothelial growth factor in liver regeneration after partial hepatectomy in rats. *J Histochem Cytochem.* 2001;49(1):121–130.

13 Redaelli CA, Semela D, Carrick FE, et al. Effect of vascular endothelial growth factor on functional recovery after hepatectomy in lean and obese mice. *J Hepatol.* 2004;40(2):305–312.

14 Assy N, Spira G, Paizi M, et al. Effect of vascular endothelial growth factor on hepatic regenerative activity following partial hepatectomy in rats. *J Hepatol.* 1999;30(5):911–915.

15 LeCouter J, Moritz DR, Li B, et al. Angiogenesis-independent endothelial protection of liver: role of VEGFR-1. *Science.* 2003; 299(5608):890–893.

16 Mochida S, Ishikawa K, Inao M, et al. Increased expressions of vascular endothelial growth factor and its receptors, flt-1 and KDR/flk-1, in regenerating rat liver. *Biochem Biophys Res Commun.* 1996;226(1):176–179.

17 Medina J, Arroyo AG, Sanchez-Madrid F, Moreno-Otero R. Angiogenesis in chronic inflammatory liver disease. *Hepatology.* 2004;39(5):1185–1195.

18 Gualdi R, Bossard P, Zheng M, et al. Hepatic specification of the gut endoderm in vitro: cell signaling and transcriptional control. *Genes Dev.* 1996;10(13):1670–1682.

19 Matsumoto K, Yoshitomi H, Rossant J, Zaret KS. Liver organogenesis promoted by endothelial cells prior to vascular function. *Science.* 2001;294(5542):559–563.

20 Schmidt C, Bladt F, Goedecke S, et al. Scatter factor/hepatocyte growth factor is essential for liver development. *Nature.* 1995; 373(6516):699–702.

21 Uehara Y, Minowa O, Mori C, et al. Placental defect and embryonic lethality in mice lacking hepatocyte growth factor/scatter factor. *Nature*. 1995;373(6516):702–705.
22 Bladt F, Riethmacher D, Isenmann S, et al. Essential role for the c-met receptor in the migration of myogenic precursor cells into the limb bud. *Nature*. 1995;376(6543):768–771.
23 Carmeliet P, Ferreira V, Breier G, et al. Abnormal blood vessel development and lethality in embryos lacking a single VEGF allele. *Nature*. 1996;380(6573):435–439.
24 Ferrara N, Carver-Moore K, Chen H, et al. Heterozygous embryonic lethality induced by targeted inactivation of the VEGF gene. *Nature*. 1996;380(6573):439–442.
25 Carpenter B, Lin Y, Stoll S, et al. VEGF is crucial for the hepatic vascular development required for lipoprotein uptake. *Development*. 2005;132(14):3293–3303.
26 Gerber HP, Hillan KJ, Ryan AM, et al. VEGF is required for growth and survival in neonatal mice. *Development*. 1999;126(6): 1149–1159.
27 Fong GH, Rossant J, Gertsenstein M, Breitman ML. Role of the Flt-1 receptor tyrosine kinase in regulating the assembly of vascular endothelium. *Nature*. 1995;376(6535):66–70.
28 Fong GH, Zhang L, Bryce DM, Peng J. Increased hemangioblast commitment, not vascular disorganization, is the primary defect in flt-1 knock-out mice. *Development*. 1999;126(13):3015–3025.
29 Shalaby F, Rossant J, Yamaguchi TP, et al. Failure of blood-island formation and vasculogenesis in Flk-1-deficient mice. *Nature*. 1995;376(6535):62–66.
30 Chijiiwa K, Saiki S, Tanaka M. Serum interleukin-6 and hepatocyte growth factor levels in patients after hepatectomy. *Hepatogastroenterology*. 2002;49(44):467–471.
31 Ninomiya M, Harada N, Shiotani S, et al. Hepatocyte growth factor and transforming growth factor beta1 contribute to regeneration of small-for-size liver graft immediately after transplantation. *Transpl Int*. 2003;16(11):814–819.
32 Yagi S, Iida T, Taniguchi K, et al. Impact of portal venous pressure on regeneration and graft damage after living-donor liver transplantation. *Liver Transpl*. 2005;11(1):68–75.
33 Auth MK, Kim HS, Beste M, et al. Removal of metabolites, cytokines and hepatic growth factors by extracorporeal liver support in children. *J Pediatr Gastroenterol Nutr*. 2005;40(1):54–59.
34 Mullin EJ, Metcalfe MS, Maddern GJ. Artificial liver support: potential to retard regeneration? *Arch Surg*. 2004;139(6):670–677.
35 Semela D, Dufour JF. Angiogenesis and hepatocellular carcinoma. *J Hepatol*. 2004;41(5):864–880.
36 Hardy KJ. Liver surgery: the past 2000 years. *Aust NZ J Surg*. 1990;60(10):811–817.
37 Higgins M, Anderson R. Experimental pathology of the liver. I. Restoration of the liver of the white rat following partial surgical removal. *Arch Pathol*. 1931;12:186–202.
38 Widmann J, Fahimi H. Proliferation of mononuclear phagocytes (Kupffer cells) and ECs in regenerating rat liver. *Am J Pathol*. 1975;80(3):B349–B366.
39 Michalopoulos G, Houck K, Dolan M, Leutteke N. Control of hepatocyte replication by two serum factors. *Cancer Res*. 1984;44(10):4414–4419.
40 Nakamura T, Nawa K, Ichihara A. Partial purification and characterization of hepatocyte growth factor from serum of hepatectomized rats. *Biochem Biophys Res Commun*. 1984;122(3):1450–1459.
41 Gherardi E, Stoker M. Hepatocytes and scatter factor. *Nature*. 1990;346(6281):228.
42 Naldini L, Vigna E, Narsimhan R, et al. Hepatocyte growth factor (HGF) stimulates the tyrosine kinase activity of the receptor encoded by the proto-oncogene c-MET. *Oncogene*. 1991;6(4): 501–504.
43 Yamane A, Seetharam L, Yamaguchi S, et al. A new communication system between hepatocytes and sinusoidal ECs in liver through vascular endothelial growth factor and Flt tyrosine kinase receptor family (Flt-1 and KDR/Flk-1). *Oncogene*. 1994; 9(9):2683–2690.

67

Stellate Cell–Endothelial Cell Interactions

Haruki Senoo

Akita University, School of Medicine, Japan

The hepatic lobule consists of parenchymal cells (epithelial cells) and nonparenchymal cells associated with the sinusoids: endothelial cells (ECs), Kupffer cells, pit cells, dendritic cells, and stellate cells (also known as vitamin A-storing cells, lipocytes, interstitial cells, fat-storing cells, Ito cells) (1–3) (Figure 67.1). ECs (4,5) express lymphocyte costimulatory molecules (6) and form the greater part of the extremely thin lining of the sinusoids, which are larger than ordinary capillaries and more irregular in shape. Kupffer cells are tissue macrophages and components of the diffuse mononuclear phagocyte system. They usually are situated on the endothelium, with cellular processes extending between the underlying ECs. The greater part of their irregular cell surface is exposed to the blood in the lumen of the sinusoid. Pit cells are natural killer cells. Dendritic cells (located in the portal triad in human, and in periportal and central areas in rats) capture and process antigens, migrate to lymphoid organs, and secrete cytokines to initiate immune responses (7). Hepatic stellate cells (1–3,8–11) lie in the space between sinusoidal ECs and parenchymal cells (perisinusoidal space or space of Disse) and, like ECs, are mesenchymal in origin. Recently, retinoid-storing stellate cells have been reported to exist in extrahepatic organs including the kidney, intestine, lung, and pancreas (2,12,13). The goal of this chapter is to review the structure and function of the hepatic stellate cell and to discuss its interactions with neighboring ECs.

HEPATIC STELLATE CELL

Hepatic stellate cells distribute regularly within hepatic lobules. The cell consists of a spindle-shaped or angular cell body and long, branching cytoplasmic processes that encompass the endothelial tubes of sinusoids (14,15). Some processes penetrate the hepatic cell plates to reach the subendothelial space of neighboring sinusoids. Accordingly, a single stellate cell wraps two, three, or sometimes four sinusoids along its processes. The total length of sinusoids surrounded by a single stellate cell is 60 to 140 μm in the rat liver.

The subendothelial processes are flat and have three cell surfaces (relative to the EC): inner, outer, and lateral. The inner surface is smooth and adheres to the abluminal (basal) surface of the sinusoidal endothelium. Intercalated between the stellate cell and EC is the basement membrane, consisting of components such as type IV collagen and laminin. The outer surface faces the space of Disse and is decorated with short microvillous protrusions. The lateral edges of the subendothelial processes are characteristically studded with numerous spike-like microprojections, the tips of which make contacts with the microvillous facets of the hepatic parenchymal cells (see Figure 67.1). Although the stellate cells are closely connected to the sinusoidal ECs through basement membrane components, they make only spotty contacts with parenchymal cells. Recently, the hepatic stellate cells have been demonstrated at molecular and morphological levels to adhere to each other through adherens junctions (16).

Retinoid Storage and Metabolism

Retinoids (vitamin A and its metabolites) are known to regulate diverse cellular activities such as cell proliferation, differentiation, morphogenesis, and tumorigenesis (17). Under physiological conditions, hepatic stellate cells store 80% of the total body retinoids in the form of retinyl palmitate contained within lipid droplets in the cytoplasm.

Hepatic stellate cells play a critical role in retinoid homeostasis. These cells regulate both transport and storage of retinoids, maintaining the concentration of retinoids in the bloodstream within physiological range. Retinoids circulate as a complex of retinol and a specific binding protein called retinol-binding protein (RBP) (Figure 67.2). Stellate cells take up retinol from the blood via receptor-mediated endocytosis (17). Once inside the cell, free retinol has several fates, one of which is to recomplex with RBP and return to the bloodstream (17–19).

More than 50 years ago, Rodahl reported that mammals in the Arctic were able to store a large amount of retinoids in the liver (20). More recently, Higashi and colleagues investigated the cellular and molecular mechanisms underlying the transport and storage of retinoids in Arctic animals inhabiting the Svalbard archipelago (situated at 80° N, 15° E) (21). These

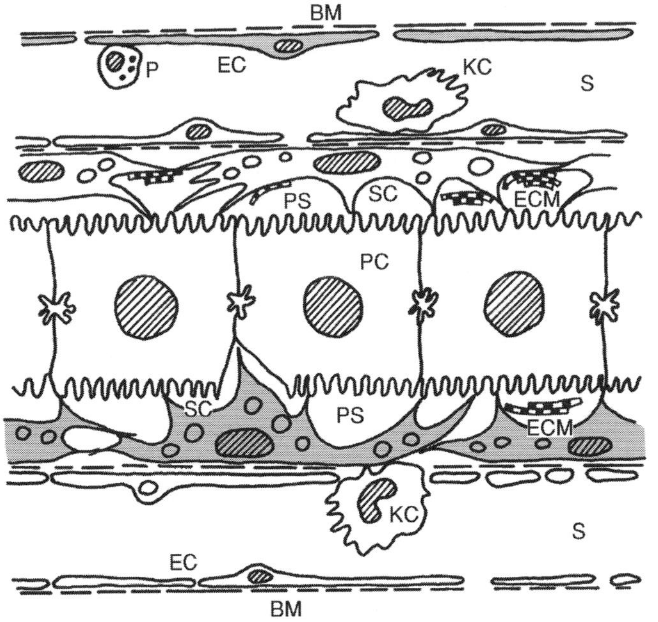

Figure 67.1. Structure of hepatic lobule. Hepatic cords of the lobule consist of parenchymal cells (*PC*). Endothelial cells (*EC*) form the thin lining of the sinusoids (*S*). Kupffer cells (*KC*) are tissue macrophages and belong to the monocyte-macrophage cell lineage. Pit cells (*P*) have natural killer activity. Stellate cells (*SC*) lie in the space between parenchymal cells and ECs, and store 80% of retinoids of the whole body as retinyl palmitate in the lipid droplets in the cytoplasm. The classical definition of the perisinusoidal space (space of Disse) was the space between PCs and ECs. However, the new concept of the perisinusoidal space (*PS*) is the space between PCs and the complex of SCs and ECs. BM. Basement membrane components; ECM, extracellular matrix.

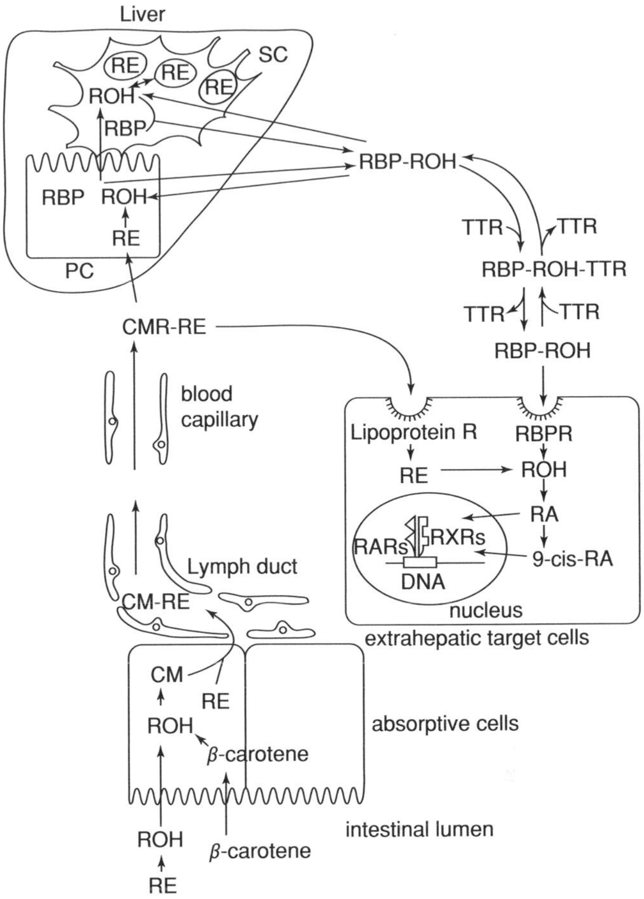

Figure 67.2. Major pathway for retinoid transport in the body. Dietary retinyl esters (*RE*) are hydrolyzed to retinol (*ROH*) in the intestinal lumen before absorption by enterocytes, and carotenoids are absorbed and then partially converted to retinol in the enterocytes. In the enterocytes, retinol reacts with fatty acid to form esters before incorporation into chylomicrons (*CM*). Chylomicrons then reach the general circulation by way of the intestinal lymph, and chylomicron remnants (*CMR*) are formed in blood capillaries. Chylomicron remnants, which contain almost all the absorbed retinol, are mainly cleared by the liver parenchymal cells (*PC*) and, to some extent, also by cells in other organs. In liver parenchymal cells, retinyl esters are rapidly hydrolyzed to retinol, which then binds to retinol-binding protein (*RBP*). A complex of retinol–RBP (*RBP–ROH*) is secreted and transported to hepatic stellate cells (*SC*). Stellate cells store retinoids mainly as retinyl palmitate and secrete retinol-RBP directly into the blood. Most retinol-RBP in the bloodstream is reversibly complexed with transthyretin (*TTR*). The uncomplexed retinol-RBP is presumably taken up in a variety of cells by cell surface receptors specific for RBP. RA, retinoic acid; 9-cis RA, 9-cis retinoic acid; RARs, retinoic acid receptors; RXR, retinoid X receptors; lipoprotein R, lipoprotein receptor; RBPR, retinol-binding protein receptor.

animals were found to store retinoids predominantly in hepatic stellate cells, with only a small amount present in other organs such as kidney, spleen, lung, and jejunum. Top predators among Arctic animals store 6 to 23 μm retinyl ester per g liver, which is 20 to 100 times higher than the levels normally found in other animals, including humans. Thus, the hepatic stellate cells in these animals have an unusually high capacity for uptake and storage of retinoids.

Xenobiotics (such as polychlorinated biphenyls and dioxins) may reduce the threshold of retinoid toxicity (17), and both retinoids and fat-soluble xenobiotics have a tendency to accumulate in the food chain (22). Total kidney retinol, which may be used as a biomarker for retinoid-related toxicity or excess, in polar bear and bearded seal was below 1% of their liver value, which is in the normal range for most animals (22). Arctic fox and glaucous gull, however, had kidney levels of about 9% and 42% of the liver values, respectively (23). This increased kidney concentration and decreased capacity for storage in hepatic stellate cells of total retinol in Arctic fox and glaucous gull are most likely signs of retinoid-toxicity that deserve attention. This observation is alarming and has not been observed previously in free-living animals.

In pathological conditions such as liver cirrhosis, hepatic stellate cells lose retinoids, proliferate vigorously, and synthesize and secrete a large amount of extracellular matrix (ECM) components such as collagen, proteoglycan, and glycoprotein (24). The appearance of the cell changes from star-shaped to fibroblast-like or myofibroblast-like, with well-developed rough-surfaced endoplasmic reticulum and Golgi apparatus (25). The dysregulation of collagen gene expression in hepatic stellate cells is a central pathogenetic step during the development of hepatic fibrosis (10).

Matrix metalloproteinases (MMPs) and tissue inhibitor of metalloproteinases (TIMPs) were reported to be synthesized by hepatic parenchymal and stellate cells (26). Recent reports indicate that the substratum used for culturing hepatic stellate cells is an important determinant of MMP activity, hence ECM remodeling (27). The three-dimensional structure of ECM can reversibly regulate hepatic stellate cell morphology, proliferation, and function (28–30). Stellate cells can take up collagen fibrils by endocytosis (31) and phagocytosis (32). Thus, hepatic stellate cells play pivotal roles in sinusoidal blood flow retinoid homeostasis, remodeling, fibrosis or cirrhosis of the liver, and so on. Additional studies have implicated a role for this cell type in liver regeneration (33).

SINUSOIDAL ENDOTHELIAL CELL

The sinusoidal wall of the liver is covered with a layer of ECs that enclose the perisinusoidal space (space of Disse) (4). The hepatic sinusoids differ from other capillaries in the body in that the ECs possess fenestrae lacking a diaphragm (5), and their basal lamina is discontinuous, even beneath the nonfenestrated regions of the cell. Fenestrae are grouped into clusters called *sieve plates.* Mean fenestration diameter in the rat is 150 to 175 nm, occupying 6% to 8% of the endothelial surface area (4,5). The fenestrae can change in size and number in response to various stimuli; agents that disrupt actin filaments can almost double the number of fenestrae within minutes (4,5).

On the basis of morphological and physiological evidence, it was reported that the grouped fenestrae act as a dynamic filter. Fenestrae filter fluids, solutes, and particles that are exchanged between the sinusoidal lumen and the perisinusoidal space, allowing only particles smaller than the fenestrae to reach the parenchymal cells or to leave the perisinusoidal space.

Another functional characteristic of the hepatic sinusoidal ECs is their high endocytic capacity (4,5). This function is reflected by the presence of numerous endocytic vesicles and by the effective uptake of a wide variety of substances from the blood by receptor-mediated endocytosis.

INTERACTIONS BETWEEN STELLATE CELLS AND SINUSOIDAL ENDOTHELIAL CELLS

Cross-talk between stellate cells and sinusoidal ECs plays a key role in homeostasis (14,15). The stellate cells are closely associated with ECs in a manner similar to that of the pericytes and ECs in the capillaries of other organs. In the normal rat liver, basement membrane components are interposed between the two cell types. In lower vertebrates, sinusoidal walls are not as well organized as they are in mammals; the stellate cells are often found in close contact with thick bundles of collagen fibrils. Interestingly, a similar phenotype is observed in fibrotic livers of mammals (25). It seems likely that the degree of stellate cell–EC communication has increased during the course of vertebrate evolution (14,15).

In mammals, stellate cells are organized into a sheath encompassing the sinusoids. Because the processes of these cells are long and branching, they are difficult to observe under light and electron microscopy. As a result, many studies underestimate the true size of stellate cells. The interactions between hepatic stellate cells and sinusoidal ECs occur via extracellular components. It is clear that the ECM plays an active and complex role in regulating the behavior of the cells they contact, influencing morphology, development, migration, proliferation, and function.

Conventionally, the perisinusoidal space has been defined as the space that lies between the sinusoids and the liver parenchyma, and which contains hepatic stellate cells. However, in the normal mammalian liver, the endothelial tubes are intimately wrapped by stellate cells. Virtually no space exists between ECs and the stellate cells. Together, these two cell types form an inseparable entity that borders the sinusoidal lumen on one side and the liver parenchyma on the other. Viewed from this perspective, the perisinusoidal space represents the limited space between the EC–stellate cell complex and parenchymol cells (14,15). Within this region are found nerve fibers, collagens (type I, III, V, and II), glycoproteins, and proteoglycans.

KEY POINTS

- Hepatic stellate cells that lie in the space between parenchymal cells and ECs play pivotal roles in the regulation of retinoid homeostasis in the whole body.
- Hepatic stellate cells are principal cells for induction of hepatic fibrosis and liver cirrhosis.
- The stellate cell consists of a spindle-shaped or angular cell body and long and branching cytoplasmic processes that encompass the endothelial tubes of sinusoids.
- The interactions between hepatic stellate cells and sinusoidal ECs occur through basement membrane components.

Future Goals

- To elucidate mutual functional regulation between ECs and stellate cells
- To develop therapies for hepatic fibrosis and liver cirrhosis

REFERENCES

1 Wake K. "Sternzellen" in the liver: perisinusoidal cells with special reference to storage of vitamin A. *Am J Anat*. 1971;132:429–462.
2 Wake K. Perisinusoidal stellate cells (fat-storing cells, interstitial cells, lipocytes), their related structure in and around the liver sinusoids, and vitamin A-storing cells in extrahepatic organs. *Int Rev Cytol*. 1980;66:303–353.
3 Senoo H, Sato M, Imai K. Hepatic stellate cells – from the viewpoint of retinoid handling and function of the extracellular matrix. *Acta Anat Nippon*. 1997;72:79–94.
4 Wanless IR. Physioanatomic considerations. In: Schiff ER, Sorrell MF, Maddrey WC, eds. *Schiff's Diseases of the Liver*. 8th ed. Philadelphia: Lippincott-Raven Publishers; 1999:3–37.
5 Braet F, Luo D, Spector I, et al. Endothelial and pit cells. In: Arias IM, Boyer JL, Chisari FV, Fausto N, Schachter D, Shafritz DA, eds. *The Liver: Biology and Pathobiology*. 4th ed. Philadelphia: Lippincott Williams & Wilkins; 2001:437–453.
6 Kojima N, Sato M, Suzuki A, et al. Enhanced expression of B7–1, B7–2, and intercellular adhesion molecule 1 in sinusoidal ECs by warm ischemia/reperfusion injury in rat liver. *Hepatology*. 2001; 34:751–757.
7 Steiniger B, Klempnauer J, Wonigeit K. Phenotype and histological distribution of interstitial dendritic cells in the rat pancreas, liver, heart, and kidney. *Transplantation*. 1984;38:169–175.
8 Blomhoff R, Wake K. Perisinusoidal stellate cells of the liver: important roles in retinol metabolism and fibrosis. *FASEB J*. 1991;5:271–277.
9 Sato M, Suzuki S, Senoo H. Hepatic stellate cells: unique characteristics in cell biology and phenotype. *Cell Struct Func*. 2003; 28:105–112.
10 Li D, Friedman SL. Hepatic stellate cells: morphology, function, and regulation. In: Arias IM, Boyer JL, Chisari FV, Fausto N, Schachter D, Shafritz DA, eds. *The Liver: Biology and Pathobiology*. 4th ed. Philadelphia: Lippincott Williams & Wilkins; 2001:456–468.
11 Senoo H. Structure and function of hepatic stellate cells. *Med Electron Microsc*. 2004;37:3–15.
12 Wold HL, Wake K, Higashi N, et al. Vitamin A distribution and content in tissues of the lamprey (*Lampetra japonica*). *Anat Rec*. 2004;276A:134–142.
13 Nagy NE, Holven KB, Roos N, et al. Storage of vitamin A in extrahepatic stellate cells in normal rats. *J Lipid Res*. 1997;38:645–658.
14 Wake K. Structure of the sinusoidal wall in the liver. In: Wisse E, Knook DL, Wake K, eds. *Cells of the Hepatic Sinusoid*. Vol. 5. Leiden, The Netherlands: Kupffer Cell Foundation, 1995;241–246.
15 Wake K. Hepatic stellate cells. *Connect Tiss*. 1998;30:245–246.
16 Hiagashi N, Kojima N, Miura M, et al. Cell-cell junctions between mammalian (human and rat) hepatic stellate cells. *Cell Tissue Res*. 2004;317:35–43.
17 Blomhoff R. *Vitamin A in Health and Disease*. New York: Marcel Dekker, 1994.
18 Senoo H, Stang E, Nilsson A, et al. Internalization of retinol-binding protein in parenchymal and stellate cells of rat liver. *J Lipid Res*. 1990;31:1229–1239.
19 Senoo H, Smeland S, Malaba L, et al. Transfer of retinol-binding protein from HepG2 human hepatoma cells to cocultured rat stellate cells. *Proc Natl Acad Sci USA*. 1993;90:3616–3620.
20 Rodahl K. Toxicity of polar bear liver. *Nature (London)*. 1949; 164:530–531.
21 Higashi N, Senoo H. Distribution of vitamin A-storing lipid droplets in hepatic stellate cells in liver lobules-A comparative study-. *Anat Rec*. 2003;(Pt A)271:240–248.
22 Barrie LA, Gregor D, Hargrave B, et al. Arctic contaminants: sources, occurrence and pathways. *Sci Total Environ*. 1992;122:1–74
23 Senoo H, Wake K, Wold HL, et al. Decreased capacity for vitamin A storage in hepatic stellate cells in arctic animals. *Comp Hepatol*. 2004;3(Suppl 1):S18. Epub at http://www.comparative-hepatology.com/content/3/S1/S18.
24 Senoo H, Hata R, Nagai Y, Wake K. Stellate cells (vitamin A-storing cells) are the primary site of collagen synthesis in non-parenchymal cells in the liver. *Biomed Res*. 1984;5:451–458.
25 Senoo H, Wake K. Suppression of experimental hepatic fibrosis by administration of vitamin A. *Lab Invest*. 1985;52:182–194.
26 Benyon RC, Arthur MJP. Extracellular matrix degradation and the role of hepatic stellate cells. *Seminar Liver Dis*. 2001;21:373–384.
27 Li Y-L, Sato M, Kojima N, et al. Regulatory role of extracellular matrix components in expression of matrix metalloproteases in cultured hepatic stellate cells. *Cell Struct Func*. 1999;24:255–261.
28 Senoo H, Hata R. Extracellular matrix regulates and L-ascorbic acid 2-phosphate further modulates morphology, proliferation, and collagen synthesis of the perisinusoidal stellate cells. *Biochem Biophys Res Commun*. 1994;200:999–1006.
29 Senoo H, Imai K, Sato M, et al. Three-dimensional structure of extracellular matrix reversibly regulates morphology, proliferation and collagen metabolism of perisinusoidal stellate cells (vitamin A-storing cells). *Cell Biol Int*. 1996;20:501–512.
30 Senoo H, Imai K, Matano Y, Sato M. Molecular mechanisms in the reversible regulation of morphology, proliferation and collagen metabolism in hepatic stellate cells by the three-dimensional structure of the extracellular matrix. *J Gastroent Hepatol*. 1998;13:S19–S32.
31 Mousavi SA, Sato M, Sporstøl M, et al. Uptake of denatured collagen into hepatic stellate cells: evidence for the involvement of urokinase plasminogen activator associated protein/Endo 180. *Biochem J*. 2005;387:39–46.
32 Higashi N, Wake K, Sato M, et al. Degradation of extracellular matrix by extrahepatic stellate cells in the intestine of the lamprey, *Lampetra japonica*. *Anat Rec*. 2005;285A:668–675.
33 Mabuchi A, Mullaney I, Sheard PA, et al. Role of hepatic stellate cells in the early phase of liver regeneration in rat: formation of tight adhesions to parenchymal cells. *J Hepatol*. 2004;40:910–916.

68

Podocyte–Endothelial Interactions

Susan E. Quaggin

Mt. Sinai Hospital, University of Toronto, Ontario, Canada

The glomerular filtration barrier (GFB) separates the blood from the urinary space. It consists of two cell types: glomerular visceral epithelial cells, also known as podocytes, and glycocalyx-coated, fenestrated endothelial cells (ECs) that form the glomerular capillary system. They are separated by a glomerular basement membrane (GBM), which is produced by both cell types (Figures 68.1 and 68.2) and consists of laminins, collagens, and heparan sulfate–containing moieties such as perlecan and agrin. In humans, each kidney contains approximately 600,000 to 1 million individual glomeruli that produce 180 L of filtrate per day. Mesangial cells sit between the capillary loops; they provide support and produce the extracellular matrix (ECM) molecules and growth factors required for proper formation of the developing capillary system. The structure of the GFB was first appreciated in electron micrographic (EM) studies that were performed during the 1970s (1–3). Using EM, it was possible to visualize fenestrations in glomerular ECs and specialized intercellular junctions between the foot processes of podocytes known as slit diaphragms (SDs), which are electron-dense structures resembling the teeth of a zipper. Until very recently, it was believed that the major function of the SD was to form the ultimate structural barrier to the blood to prevent loss of critical proteins into the urine.

The identification of human nephrin (NPHS1; nephrosis 1) by Karl Trygvasson's group in 1998 (4) led to an explosion in our understanding of the molecular basis of the SD and the biology of the podocyte (5,6). Nephrin molecules from adjacent foot processes span the SD space and, through homophilic interactions, make up a major structural component of the SD. In addition, the slit proteins reside in lipid raft microdomains of the podocyte cell membrane and function as signaling centers that relay inputs from the glomerular microenvironment to the cytoskeleton and intracellular environment of the podocyte. Nephrin interacts with the actin cytoskeleton through phosphorylated tyrosines in its cytoplasmic tail. Recruitment of Nck adaptor proteins to phosphorylated nephrin leads to actin reorganization in vitro, whereas podocyte-selective loss of Nck leads to failure of podocyte foot process formation (7,8). Three nephrin homologues have been identified in mammals: NEPH1, -2, and -3 (9); at least one of these proteins, NEPH1, is known to form both homo- and heterodimers with nephrin in vitro (10). On the heels of the discovery of nephrin, several other components of the slit and/or cytoskeleton were identified, including podocin (NPHS2), α-actinin 4 and CD2AP (11–14). Human mutations in each of the *NPHS1*, *NPHS2*, and *α-actinin 4* genes are associated with steroid-resistant forms of congenital nephrosis or focal segmental glomerulosclerosis (FSGS), emphasizing the role of the SD and cytoskeleton in podocyte and glomerular biology. Recently, mutations in the transient receptor potential cation channel, *TRPC6*, which is expressed by both podocytes and glomerular ECs, and *PLCe1*, which is expressed by podocytes, have been found to underlie other familial forms of FSGS and mesangial sclerosis, respectively (15–17). Several additional proteins have been shown to coprecipitate with nephrin in a large multimeric complex; these include ZO-1, p120 catenin, CASK (18), and IQGAP1, but their functional role is less clear (19). Still other proteins required for SD function have been identified by knockout studies, although their exact subcellular localization and interacting partners remain unclear. One example of this latter group includes the cadherin FAT (20); *FAT*-null mice die prior to birth with no foot processes or SDs.

In addition to its role in filtration, the glomerulus is a specialized high-flux capillary bed, as large volumes of water and small solutes pass from the blood to the urinary space. In this context, the podocyte may be considered as a vascular supporting cell. What is the evidence that they function as vascular supporting cells? They are in close association with adjacent ECs, and produce ECM proteins that are required for the podocyte–endothelial interface. However, most importantly, they express a number of vascular growth factors known to be essential for endothelial migration, differentiation, survival, and activity. These factors include vascular endothelial growth factor (VEGF)-A, VEGF-C, angiopoietin 1 (Angl), and ephrinB2 (21–24). Adjacent ECs express the major receptors for each of these ligands, although the Eph receptor responsible for ephrinB2 signaling within the glomerulus has not yet been identified. Reciprocal signals from the endothelial to podocyte

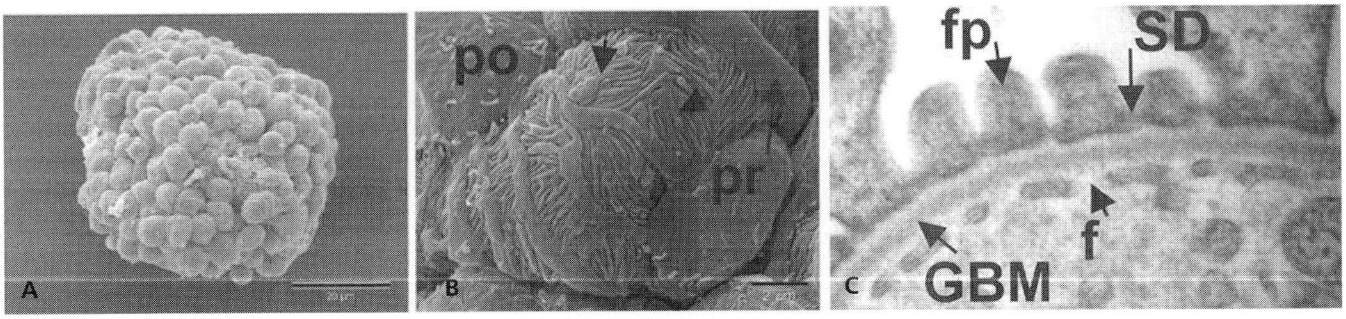

Figure 68.1. Electron micrographs of the renal glomerulus. **(A)** Scanning electron micrograph shows an isolated whole mouse glomerulus. The cells seen on the outer surface represent podocytes. **(B)** Scanning electron micrograph at higher power shows the cell body of a podocyte (*po*) with its primary processes (*pr*) and secondary foot processes (*unlabeled arrows*) that wrap around glomerular capillaries. **(C)** Transmission electron micrograph shows foot processes (*fp*) of a podocyte. The slit diaphragms (*SD*) are specialized intercellular junctions between interdigitating foot processes. Fenestrations (*f*) are present in the glomerular endothelium. Podocytes and ECs are separated by an intervening glomerular basement membrane (*GBM*). Urine is formed as it passes through the fenestrations and SDs.

compartment also likely exist, although their identities are not known.

Despite the advances that have been gained in our understanding of the molecular components of the filtration barrier, some controversy exists regarding the role of each of the components and cell types – including the podocytes, ECs, glycocalyx, glomerular basement membrane, and SDs – in providing the permeselective qualities that permit high flux and loss of small solutes and water into the urine, but prevent the loss of larger molecules, including albumin and blood clotting factors. Evidence from clinical medicine and the laboratory suggests that each of these components plays a role. Loss of SD proteins alone, as occurs in congenital nephrosis of the Finnish

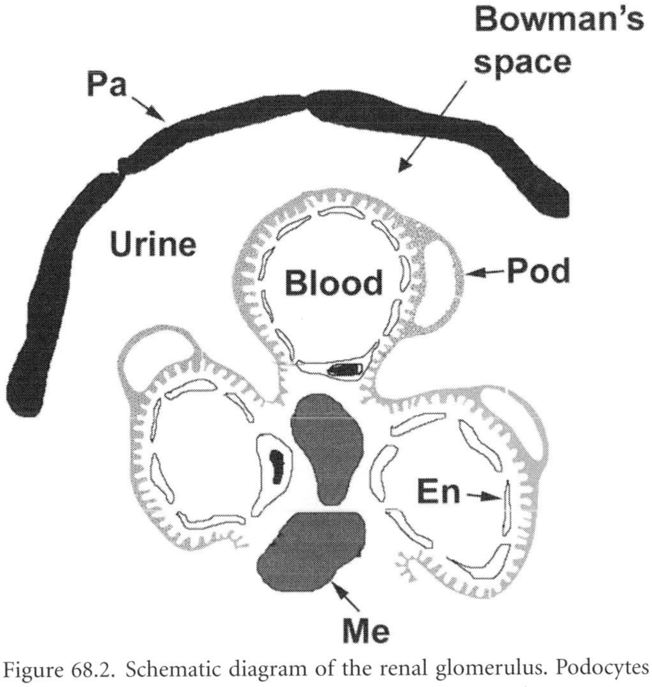

Figure 68.2. Schematic diagram of the renal glomerulus. Podocytes (*Pod*) are shown in light gray and overlie the fenestrated ECs. Mesangial cells (*Me*) sit between the capillary loops. Pa, parietal epithelial cells.

type due to mutations in *NPHS1*, results in nephrotic range proteinuria. These findings illustrate the important role that the SD plays in the barrier. Yet, how the SD prevents loss of protein in the basal state is still debated. Electron tomographic studies can resolve images of the SD to 5 to 10 nm and permit three-dimensional reconstruction (25). Together with computer modeling, these data led Wartiovaara and colleagues to argue that albumin is excluded from urine based on size exclusion within the slit. Oliver Smithies proposes a different model, in which the width of the SDs and foot processes determines the gel properties of the filtration barrier (26). In contrast to podocytes and the SD, the role of the glomerular endothelium in filtration is less well appreciated. In addition to the fenestrations, a glycocalyx is observed on the blood side of the endothelium; this is believed to provide some barrier to filtration (27). A body of evidence supports a critical role for ECs in the permselectivity of the GFB: Both the loss of fenestrations that occurs in preeclampsia and both the under- and overproduction of vascular growth factors that function at the level of the endothelium lead to rapid increases in permeability and loss of proteins into the urine (28,29). Furthermore, in several experimental models nephrotic-range proteinuria occurs with preservation of foot processes and SDs, at least ultrastructurally (30). Several interesting reviews and hypothetical argument papers have been published recently regarding these models (31).

DEVELOPMENTAL CONSIDERATIONS

Glomeruli arise from metanephric mesenchymal cells that, in turn, derive from the intermediate mesoderm (32). At approximately embryonic day (E)10.5 in mice and 4.5 to 5 weeks in humans, the ureteric bud (UB) develops from the caudal end of the Wolffian duct and grows into the metanephric mesenchyme (MM). Metanephric mesenchymal cells are induced by the adjacent UB and undergo a series of well-defined morphological changes including condensation and then epithelization to produce a vesicle, comma-shape, S-shape, capillary

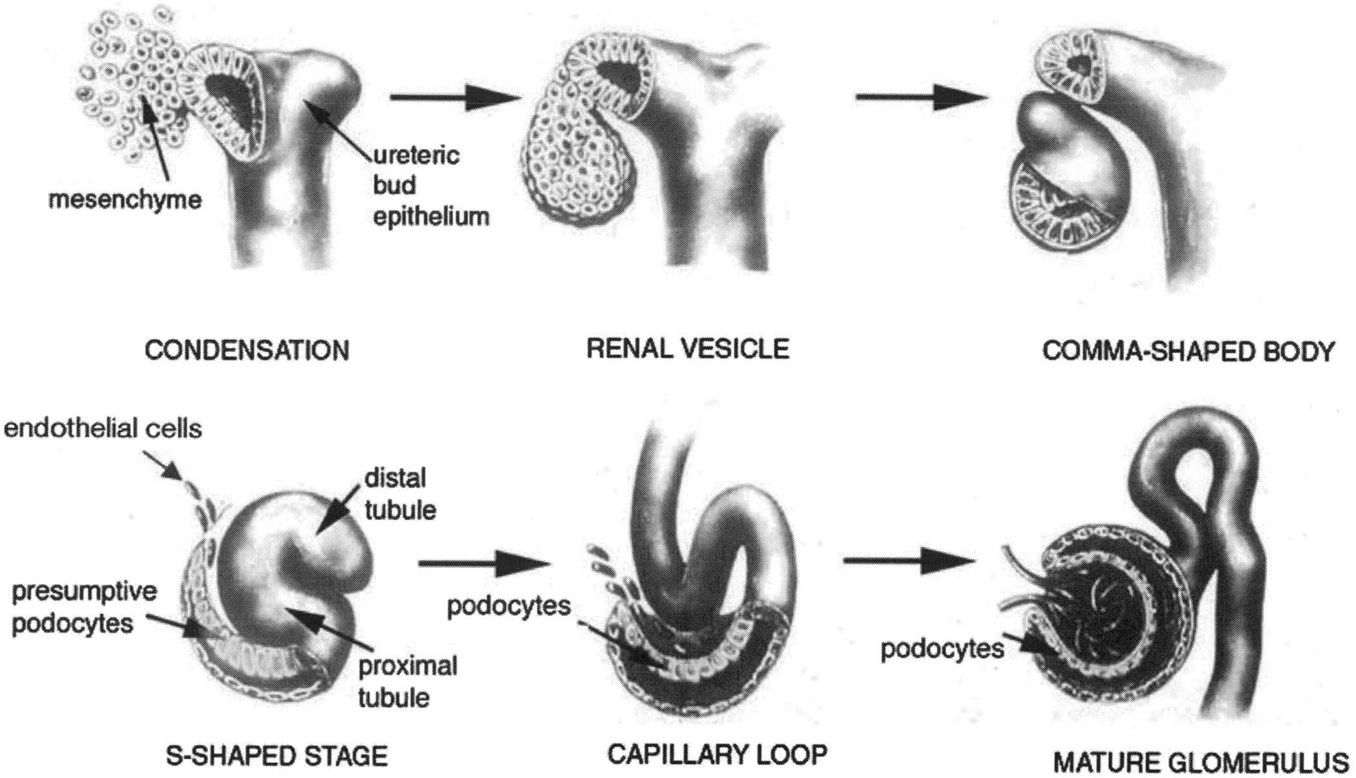

Figure 68.3. Development of the glomerular filtration barrier. Glomeruli develop from metanephric mesenchymal cells that are induced to condense and epithelialize by the adjacent ureteric bud epithelium. Presumptive podocytes are first observed at the S-shape stage of development as columnar epithelial cells. A vascular cleft forms immediately adjacent to the podocyte precursors that express VEGF-A. The VEGF receptors (VEGFR1 and VEGFR2) are expressed by incoming ECs. At the capillary loop stage, ECs develop fenestrations, and capillaries form in direct apposition to podocytes. The functional filtration barrier is seen in mature glomeruli. (Adapted with permission from Saxen L. *Organogene of the kidney*. Cambridge University Press, 1987.)

loop, and finally mature glomerulus (Figure 68.3). Podocyte precursors are first evident at the S-shape stage of development. A vascular cleft develops immediately adjacent to the columnar-shaped podocytes, which already produce high levels of the vascular mitogen VEGF-A (see Figure 68.3). FLK-1–positive ECs migrate into this cleft from the MM and set up the glomerular capillary system in direct apposition to the podocytes. By the capillary loop stage of development, fenestrations are observed within the endothelium. The mature capillary loop structure depends on the migration of mesangial cells into the center of the developing glomerulus to provide support and produce the "flowerlike" structure of the system. Interactions between glomerular ECs and mesangial cells require platelet-derived growth factor (PDGF) receptor (PDGFR)-β and PDGF-B signaling. Loss of either the PDGF-β receptor or its ligand PDGF-B in knockout mice prevents mesangial cell migration and, as a result, glomeruli form with a single ballooned capillary loop only (33,34).

LIGAND–RECEPTOR INTERACTIONS: VASCULAR ENDOTHELIAL GROWTH FACTOR-A SIGNALING

During glomerular development and in the mature glomerulus, podocytes produce high levels of VEGF-A. During this time, VEGF receptors are expressed by adjacent endothelium, suggesting a tightly regulated paracrine system for the development and maintenance of the filtration barrier. In addition, podocytes express the VEGF coreceptors – neuropilin 1 (NRP1) and NRP2 – in vivo, and immortalized podocyte cell lines express high levels of VEGFR1, NRP1 and NRP2, leaving open the possibility that an autocrine loop for VEGF-A signaling exists in the glomerular podocytes as well (21,35). In vitro, inhibition of tyrosine kinase signaling leads to apoptosis of podocytes (36). In some laboratories, these podocyte cell lines also have been shown to express VEGFR2 (37); however, to date no one has been able to demonstrate expression of this receptor by podocytes in vivo.

The angiopoietin ligand–receptor pairs are expressed by podocytes and glomerular ECs, respectively. Chimeric mouse studies demonstrated a cell-autonomous requirement for the Ang1 receptor (Tie2) in glomerular ECs, because *Tie2*-null embryonic stem cells can contribute to many vascular beds, but never to glomerular ECs (38).

Similar to other fenestrated vascular beds in the body, the adjacent vascular supporting cells – the podocytes – continue to constitutively express VEGF-A in the postnatal and adult periods. Although hypoxia-inducible regulation of VEGF-A is well established and is believed to play a major role during

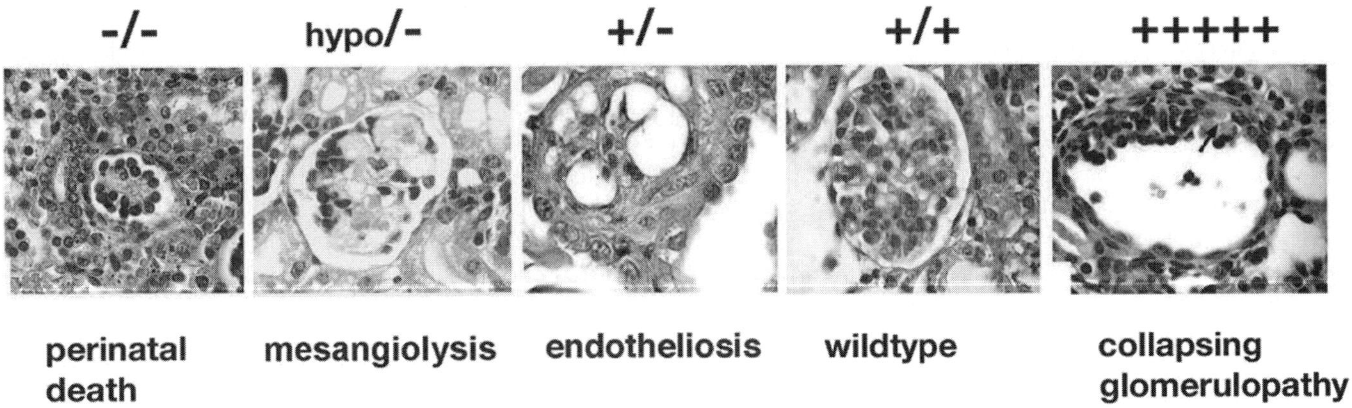

Figure 68.4. Exquisite dosage sensitivity for VEGF-A exists in the developing glomerulus. A wild-type mouse glomerulus is shown for comparison ($^{+}/^{+}$). Loss of all VEGF-A production by podocytes ($^{-}/^{-}$) results in failure of the filtration barrier to form; removal of one *VEGF-A* allele ($^{+}/^{-}$) in podocyte-specific haploinsufficient mice results in endotheliosis, whereas an intermediate dose (hypomorphic and one null allele) leads to massive mesangiolysis and death at 3 weeks. By contrast, marked upregulation of VEGF-A ($^{+++++}$) in podocytes leads to collapsing glomerulopathy, similar to the lesion observed in acquired immune deficiency syndrome (AIDS)-related nephropathy.

glomerular development (39), these mechanisms appear to be less important in the adult kidney, suggesting the presence of alternative, nonhypoxia-mediated regulatory pathways for VEGF-A production, such as integrin-mediated regulation (40).

VASCULAR ENDOTHELIAL GROWTH FACTOR-A SIGNALING AND THE GLOMERULUS

Numerous clinical studies have demonstrated an association between dysregulation of the VEGF-A protein or gene expression, and glomerular disease (41). Given the nature of association studies, it was not clear whether this dysregulation was primary or secondary. Elegant studies by Karumanchi and colleagues have identified elevations in circulating levels of soluble VEGFR1/FLT-1 (sFLT-1) and patients with preeclampsia (28,42) (see Chapter 161). Furthermore, injection of sFLT-1 into rats led to endotheliosis – the classic renal lesion observed in preeclampsia, in which ECs become swollen and lose their fenestrations. Injection of a mouse antibody to VEGFR2/FLK-1 led to redistribution of the SD protein, nephrin, and proteinuria in mice (43), suggesting a possible podocyte-dependent mechanism for proteinuria in addition to direct effects on the endothelium.

Conditional gene targeting in mice has provided additional support for the pivotal role that VEGF-A signaling plays in all aspects of glomerular endothelial function. Using the Cre-loxP system, a series of transgenic mice were generated that carried different "doses" or copy numbers of the *VEGF-A* allele (29,44–46). Deletion of all VEGF-A from podocytes led to perinatal death and failure of GFB formation because of defects in EC migration into the vascular cleft, decreased survival, and failure of fenestrations to form. Mice that were hemizygous for the *VEGF-A* gene within their podocytes (one wild-type copy only) developed endotheliosis similar to patients with preeclampsia. Because systemic VEGF-A levels were presumably normal in this genetic model of VEGF-A reduction, it suggests that circulating levels of VEGF-A are unable to rescue the local paracrine requirement for VEGF-A signaling. This endothelial defect led to global deterioration of glomerular function and glomerulosclerosis by 3 months of age. Further reduction of VEGF-A (one hypomorphic allele and one null allele in the podocyte) led to a phenotype intermediate between the null and haploinsufficient mice (Figure 68.4). Although EC migration occurred normally into the glomeruli of hypomorphic VEGF mice, they rapidly disappeared, leaving capillary ghosts devoid of all endothelium (Figure 68.5). This led to mesangiolysis (dissolution of mesangial cells), presumably due to defects in signaling from the endothelium to mesangium.

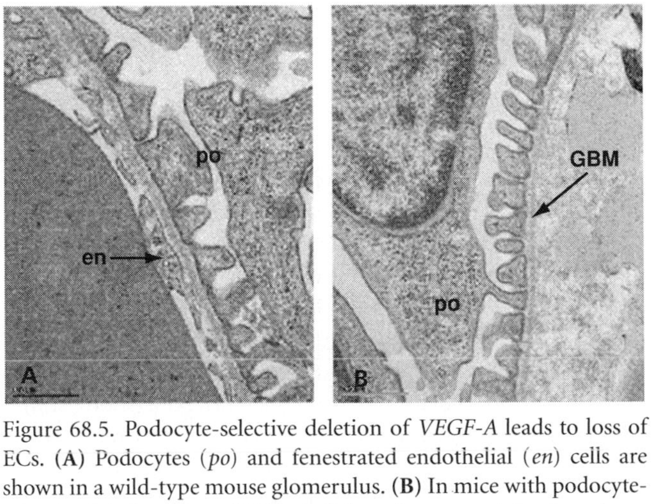

Figure 68.5. Podocyte-selective deletion of *VEGF-A* leads to loss of ECs. (**A**) Podocytes (*po*) and fenestrated endothelial (*en*) cells are shown in a wild-type mouse glomerulus. (**B**) In mice with podocyte-selective deletion of one *VEGF-A* allele and one remaining hypofunctioning allele, only capillary ghosts devoid of ECs are seen. Podocyte foot processes (*po*) and glomerular basement membrane (*GBM*) are present. (Adapted with permission from Eremina V, et al. VEGF-A signaling in the podocyte-endothelial compartment is required for mesangial cell migration and survival. *J Am Soc Nephrol.* 2006;17: 724–735.)

By contrast, upregulation of the major VEGF-A_{164} isoform in developing podocytes led to collapsing glomerulopathy and death from renal failure at 5 days of age (44). Taken together, these results emphasize the critical role for VEGF-A signaling from podocytes to ECs during glomerular development.

However, this requirement does not end after development. Although VEGF-A production by podocytes occurs against the "flow of urine" from capillaries to urinary space, it is still required by the endothelium in the adult. Podocyte-specific induction of the VEGF-A_{164} isoform using a tetracycline-inducible gene-targeting system in mice (47) leads to rapid changes in glomerular EC permeability and nephrotic-range nonselective proteinuria within 48 hours (29,47). In this situation, podocyte foot processes are intact, consistent with a central role for the endothelium in glomerular permselectivity. Deletion of VEGF-A from adult podocytes also leads to glomerular injury, disease, and renal failure with major defects observed in glomerular endothelia (29).

CLINICAL IMPLICATIONS

Only recently have podocytes come to be viewed as vascular supporting cells. A large body of work in murine models demonstrates the intimate nature of interactions between podocytes and glomerular ECs and highlights the role of vascular signaling pathways in glomerular development and disease (29,45,46).

Is this relevant to our patients in the clinic? In the simplest interpretation, it is paramount that we recognize potential safety issues for drugs and inhibitors that target these vascular pathways in diseases such as cancer and vasculopathies. Of note, the most common adverse effects observed in patients treated with VEGF-inhibitory drugs is the development of proteinuria and/or hypertension (48). Whether circulating levels of VEGF-A in these patients influence the glomerular endothelium to produce symptoms is not yet clear, but increased awareness of possible adverse events in patients treated with these drugs is required. By contrast, the emergence of prolyl hydroxylase inhibitors to stabilize hypoxia inducible factors (HIFs) will lead to increased local levels of VEGF-A; in theory, these drugs also may affect the glomerular endothelium to cause proteinuria. Future studies will be needed to determine if this is a consideration in the clinic.

Given the embryonic lethality of *VEGF-A*–knockout and haploinsufficient mice (49), it is unlikely that such major mutations underlie human disease. However, more subtle defects that alter gene expression or affect the storage and/or delivery of VEGF-A are likely to be involved in glomerular disease. Studies to determine the role of VEGF-A signaling within the glomerulus are providing new insights into glomerular filtration. For example, the upregulation of VEGF-A within podocytes rapidly leads to nonselective nephrotic-range proteinuria (45), and proteinuria in clinical medicine portends serious renal disease and is the surrogate marker used to predict progression. A lower level of proteinuria, termed microalbuminuria (albumin <300 mg/24 hour), is a prognostic indicator not only for diabetic nephropathy and glomerular disease but also for cardiovascular disease; its presence is associated with poorer clinical outcome. Given the central role of VEGF-A and glomerular endothelia in glomerular permselectivity and function, it follows that VEGF-A may contribute to microalbuminuria. The fact that microalbuminuria is associated with systemic vascular disease may reflect a more generalized endothelial defect that can be detected through urinalysis. This is simplified due to the accessibility of the glomerular capillary bed to the external environment, and suggests that urinalysis may be extended to include biomarkers of endothelial function.

KEY POINTS

- Podocyte–endothelial cross-talk is essential for the development and maintenance of the GFB – a specialized capillary bed.
- Podocytes function as vascular support cells within the glomerulus.
- Paracrine signaling from the podocyte to EC is critical for glomerular function; VEGF-A is an important example of one such signaling pathway.

Future Goals

- To identify additional signaling pathways that are important for podocyte–endothelial interactions
- To determine how VEGF-A pathway is altered in human glomerular disease and proteinuric states

REFERENCES

1 Karnovsky MJ, Ainsworth SK. The structural basis of glomerular filtration. *Adv Nephrol Necker Hosp*. 1972;2:35–60.
2 Rodewald R, Karnovsky MJ. Porous substructure of the glomerular slit diaphragm in the rat and mouse. *J Cell Biol*. 1974;60:423–433.
3 Reeves W, Caulfield JP, Farquhar MG. Differentiation of epithelial foot processes and filtration slits: sequential appearance of occluding junctions, epithelial polyanion, and slit membranes in developing glomeruli. *Lab Invest*. 1978;39:90–100.
4 Kestila M, Lenkkeri U, Mannikko M, et al. Positionally cloned gene for a novel glomerular protein – nephrin – is mutated in congenital nephrotic syndrome. *Mol Cell*. 1998;1:575–582.
5 Benzing T. Signaling at the slit diaphragm. *J Am Soc Nephrol*. 2004;15:1382–1391.
6 Huber TB, Benzing T. The slit diaphragm: a signaling platform to regulate podocyte function. *Curr Opin Nephrol Hypertens*. 2005; 14:211–216.
7 Verma R, Kovari I, Soofi A, et al. Nephrin ectodomain engagement results in Src kinase activation, nephrin phosphorylation,

Nck recruitment, and actin polymerization. *J Clin Invest*. 2006; 116(5):1346–1359.
8 Jones N, Blasutig IM, Eremina V, et al. Nck adaptor proteins link nephrin to the actin cytoskeleton of kidney podocytes. *Nature*. 2006;440:818–823.
9 Sellin L, Huber TB, Gerke P, et al. NEPH1 defines a novel family of podocin interacting proteins. *FASEB J*. 2003;17:115–117.
10 Gerke P, Huber TB, Sellin L, et al. Homodimerization and heterodimerization of the glomerular podocyte proteins nephrin and NEPH1. *J Am Soc Nephrol*. 2003;14:918–926.
11 Boute N, Gribouval O, Roselli S, et al. NPHS2, encoding the glomerular protein podocin, is mutated in autosomal recessive steroid-resistant nephrotic syndrome [In process citation]. *Nat Genet*. 2000;24:349–354.
12 Shih NY, Li J, Karpitskii V, et al. Congenital nephrotic syndrome in mice lacking CD2-associated protein [see comments]. *Science*. 1999;286:312–315.
13 Kaplan JM, Kim SH, North KN, et al. Mutations in ACTN4, encoding alpha-actinin-4, cause familial focal segmental glomerulosclerosis. *Nat Genet*. 2000;24:251–256.
14 Lenkkeri U, Mannikko M, McCready P, et al. Structure of the gene for congenital nephrotic syndrome of the Finnish type (NPHS1) and characterization of mutations. *Am J Hum Genet*. 1999;64:51–61.
15 Winn MP, Conlon PJ, Lynn KL, et al. A mutation in the TRPC6 cation channel causes familial focal segmental glomerulosclerosis. *Science*. 2005;308:1801–1804.
16 Reiser J, Polu KR, Moller CC, et al. TRPC6 is a glomerular slit diaphragm-associated channel required for normal renal function. *Nat Genet*. 2005;37:739–744.
17 Hinkes B, Wiggins RC, Gbadegesin R, et al. Positional cloning uncovers mutations in PLCE1 responsible for a nephrotic syndrome variant that may be reversible. *Nat Genet*. 2006;38:1397–1405.
18 Lehtonen S, Lehtonen E, Kudlicka K, et al. Nephrin forms a complex with adherens junction proteins and CASK in podocytes and in Madin-Darby canine kidney cells expressing nephrin. *Am J Pathol*. 2004;165:923–936.
19 Liu XL, Kilpelainen P, Hellman U, et al. Characterization of the interactions of the nephrin intracellular domain. *FEBS J*. 2005;272:228–243.
20 Inoue T, Yaoita E, Kurihara H, et al. FAT is a component of glomerular slit diaphragms. *Kidney Int*. 2001;59:1003–1012.
21 Harper SJ, Xing CY, Whittle C, et al. Expression of neuropilin-1 by human glomerular epithelial cells in vitro and in vivo. *Clin Sci (Lond)*. 2001;101:439–446.
22 Satchell SC, Harper SJ, Mathieson PW. Angiopoietin-1 is normally expressed by periECs. *Thromb Haemost*. 2001;86:1597–1598.
23 Takahashi T, Takahashi K, Gerety S, et al. Temporally compartmentalized expression of ephrin-B2 during renal glomerular development. *J Am Soc Nephrol*. 2001;12:2673–2682.
24 Satchell SC, Harper SJ, Tooke JE, et al. Human podocytes express angiopoietin 1, a potential regulator of glomerular vascular endothelial growth factor. *J Am Soc Nephrol*. 2002;13:544–550.
25 Wartiovaara J, Ofverstedt LG, Khoshnoodi J, et al. Nephrin strands contribute to a porous slit diaphragm scaffold as revealed by electron tomography. *J Clin Invest*. 2004;114:1475–1483.
26 Smithies O. Why the kidney glomerulus does not clog: a gel permeation/diffusion hypothesis of renal function. *Proc Natl Acad Sci USA*. 2003;100:4108–4113.
27 Sorensson J, Fierlbeck W, Heider T, et al. Glomerular endothelial fenestrae in vivo are not formed from caveolae. *J Am Soc Nephrol*. 2002;13:2639–2647.
28 Maynard SE, Min JY, Merchan J, et al. Excess placental soluble fms-like tyrosine kinase 1 (sFlt1) may contribute to endothelial dysfunction, hypertension, and proteinuria in preeclampsia. *J Clin Invest*. 2003;111:649–658.
29 Sison K, Eremina V, Haigh J, et al. VEGF-A is required for maintenance of glomerular function: insights from conditional knockouts in the adult kidney. *NIH, NIDDK Glomerular Workshop*, Abstract, 2005.
30 Karumanchi SA, Epstein FH, Stillman IE. Is loss of podocyte foot processes necessary for the induction of proteinuria? *Am J Kidney Dis*. 2005;45:436.
31 Deen WM. What determines glomerular capillary permeability? *J Clin Invest*. 2004;114:1412–1414.
32 Saxen L. *Organogenesis of the Kidney*. Cambridge: Cambridge University Press, 1987.
33 Lindahl P, Hellstrom M, Kalen M, et al. Paracrine PDGF-B/PDGF-Rbeta signaling controls mesangial cell development in kidney glomeruli. *Development*. 1998;125:3313–3322.
34 Leveen P, Pekny M, Gebre-Medhin S, et al. Mice deficient for PDGF B show renal, cardiovascular, and hematological abnormalities. *Genes Dev*. 1994;8:1875–1887.
35 Foster RR, Saleem MA, Mathieson PW, et al. Vascular endothelial growth factor and nephrin interact and reduce apoptosis in human podocytes. *Am J Physiol Renal Physiol*. 2005;288:F48–F57.
36 Foster RR, Hole R, Anderson K, et al. Functional evidence that vascular endothelial growth factor may act as an autocrine factor on human podocytes. *Am J Physiol Renal Physiol*. 2003;284: F1263–F1273.
37 Guan F, Villegas G, Teichman J, Tufro A. VEGF regulates podocin expression and podocyte phenotype. *J Am Soc Nephrol*. 2004; 15:A97.
38 Partanen J, Puri MC, Schwartz L, et al. Cell autonomous functions of the receptor tyrosine kinase TIE in a late phase of angiogenic capillary growth and EC survival during murine development. *Development*. 1996;122:3013–3021.
39 Freeburg PB, Robert B, St. John PL, Abrahamson DR. Podocyte expression of hypoxia-inducible factor (HIF)-1 and HIF-2 during glomerular development. *J Am Soc Nephrol*. 2003;14:927–938.
40 Datta K, Li J, Karumanchi SA, et al. Regulation of vascular permeability factor/vascular endothelial growth factor (VPF/VEGF-A) expression in podocytes. *Kidney Int*. 2004;66:1471–1478.
41 Wakelin SJ, Marson L, Howie SE, et al. The role of vascular endothelial growth factor in the kidney in health and disease. *Nephron Physiol*. 2004;98:73–79.
42 Levine RJ, Maynard SE, Qian C, et al. Circulating angiogenic factors and the risk of preeclampsia. *N Engl J Med*. 2004;350:672–683.
43 Sugimoto H, Hamano Y, Charytan D, et al. Neutralization of circulating vascular endothelial growth factor (VEGF) by anti-VEGF antibodies and soluble VEGF receptor 1 (sFlt-1) induces proteinuria. *J Biol Chem*. 2003;278:12605–12608.
44 Eremina V, Sood M, Haigh J, et al. Glomerular-specific alterations of VEGF-A expression lead to distinct congenital and acquired renal diseases. *J Clin Invest*. 2003;111:707–716.
45 Eremina V, Quaggin SE. The role of VEGF-A in glomerular development and function. *Curr Opin Nephrol Hypertens*. 2004;13:9–15.

46 Eremina V, Cui S, Gerber H, et al. VEGF-A signaling in the podocyte-endothelial compartment is required for mesangial cell migration and survival. *J Am Soc Nephrol.* 2006;17:724–735.

47 Belteki G, Haigh J, Kabacs N, et al. Conditional and inducible transgene expression in mice through the combinatorial use of Cre-mediated recombination and tetracycline induction. *Nucleic Acids Res.* 2005;33:E51.

48 Kabbinavar F, Hurwitz HI, Fehrenbacher L, et al. Phase II, randomized trial comparing bevacizumab plus fluorouracil (FU)/leucovorin (LV) with FU/LV alone in patients with metastatic colorectal cancer. *J Clin Oncol.* 2003;21:60–65.

49 Carmeliet P, Ferreira V, Breier G, et al. Abnormal blood vessel development and lethality in embryos lacking a single VEGF allele. *Nature.* 1996;380:435–439.

Coupling

69

Introductory Essay:

Endothelial Cell Coupling

Michael Simons

Dartmouth Medical School, Hanover, New Hampshire

Endothelial cells (ECs), by virtue of their location at the interface between the bloodstream and the vessel wall, must integrate bidirectional signaling inputs and generate appropriate responses. These responses in turn can affect both EC interaction with the flowing blood and the circulating blood cells, as well as affect the function of the vessel wall. Thus, these deceptively simple looking cells are charged with a highly complex task of responding to and modifying the behavior of the liquid phase (plasma) and the solid phase (vessel wall) environments.

This integrative signaling function performed by the endothelium includes a number of components, including sensors of the physical environment such as the rate of blood flow (shear stress), of the circulating growth factors and cytokines, and of the circulating cells. In addition, the type of matrix the EC rests on, and cells it is in contact with, including other ECs, provide additional inputs into the complex integration matrix. As a practical example of this challenge, let us consider that, under certain circumstances, an EC may be called on to deal with a variety of inputs, including signals from multiple soluble ligands binding to tyrosine kinase receptors (e.g., vascular endothelial growth factor [VEGF], fibroblast growth factor [FGF], platelet-derived growth factor [PDGF]); serine/threonine kinases (e.g., epidermal growth factor [EGF]); G-protein-coupled receptors (e.g., thrombin); input from binding circulating cells via cellular adhesion molecules; signals from the extracellular matrix via integrins, syndecans, and other adhesion molecules; and signals from other adjoining ECs via tight junctions. All this input is integrated to yield a variety of responses including the initiation of migration or cell division and changes in production of nitric oxide and other EC-derived vasoactive substances, such as prostaglandins. Processing of these signals thus results in the endothelium altering its own properties – such as changes in its shape, expression of various receptors, and production of signaling molecules – as well as potentially altering the appearance of the blood vessel at the site where the signal was applied or even affecting the overall state of circulation.

Thus, the endothelium is a key component of a complex signaling network operating over several orders of magnitude on the spatiotemporal scales (1). Some of the examples of these scales of operation include the endocrine signaling provided by circulating growth factors and cytokines, the paracrine interaction with smooth muscle cells directed by the production of nitric oxide, and the autocrine signaling generated by secretion of FGF1.

Shifting next to the intracellular scale of endothelial signaling, one observes multiple individual signaling networks, such as VEGF receptor or αv integrin signaling cascades, that are typically tightly compartmentalized. We have a tendency to think of these signaling cascades in linear terms and to ascribe a change in cell phenotype resulting from an alteration in function of one of the signaling cascade proteins (e.g., inhibition of Akt-1 phosphorylation) to the particular signaling cascade interfered with. Yet this hardly can be correct. It is quite clear that various signaling cascades interact and modify one another, with resulting output from signaling cascade X being dependent on the presence or absence of signals from cascades Y and Z (2). This is a very important issue that is usually ignored in discussions of individual signaling pathways.

The interactions between different signaling networks typically lead to the appearance of new functionality, described by the term *emergent properties.* One example of such emergent properties is the effect of the EGF receptor signaling cascade cross-talk with other cellular processes. A thorough analysis of these interactions demonstrates several emergent properties, including extended signal duration, activation of feedback loops, changes in thresholds for stimulation of biological processes and, perhaps most importantly, multiple signaling outputs (2).

The chapters in this section of the book discuss various microdomains and signaling pathways in ECs. As these chapters illustrate, we clearly have and continue to gain very detailed understanding of individual organelles and pathways. The section begins with a chapter that summarizes the differences and similarities between endothelium and epithelium (Chapter 70). Dvorak provides us with an overview of the seminal advances related to ultrastructure during 1950s and 1960s (Chapter 71). As discussed in chapters by Lowenstein, Schnitzer, and Stan, the last decade has witnessed remarkable progress in understanding the signaling functions and molecular basis of EC microdomains (e.g., caveolae) and organelles (e.g., Weibel-Palade bodies) (Chapters 72–74). Other structural elements discussed in this section, which contribute to signaling, include the glycocalyx (Chapter 75) and the cytoskeleton (Chapter 76). The field of water channels and ion channels has gained significance over the past years, as evidenced by the awarding of Nobel prizes in Medicine for discoveries in these areas. In this volume, Verkman and Yao apply principles of water and ion transport to an understanding of endothelial function in health and disease (Chapters 78 and 79). Finally, McCarty and Hynes discuss the critical interaction between integrins and the extracellular matrix (Chapter 77).

At the level of signaling pathways, many important intermediates occur, only some of which are discussed here. These include chapters on phosphoinositide 3-kinase (PI3K)-Akt (Chapter 80), mitogen-activated protein kinases (MAPKs) (Chapter 81), protein kinase C (PKC) (Chapter 82), Rho GTP-binding proteins (Chapter 83), phosphatases (Chapter 84), and Wnt signaling (Chapter 85).

Activation (or repression) of signaling pathways in ECs may lead to changes in proteins (posttranscriptional) or mRNA (posttranscriptional and/or transcriptional). Changes in gene expression are mediated by DNA binding proteins. Interestingly, few if any transcription factors (or for that matter target genes) are truly specific to ECs (one exception may be Vezf1, which is discussed in Chapter 95). It seems likely that EC-specific gene expression is mediated by the coordinate action of nonspecific transcription factors and coactivators. However, as recently discussed in a comprehensive review of EC gene regulation (3), there is far more to learn about the endothelium by focusing on those transcription factors and target genes whose activity is modulated by the microenvironment, regardless of their lineage specificity. Examples of transcription factors that serve as signal transducers that couple changes in the extracellular milieu to alterations in gene expression include nuclear factor (NF)-κB (Chapter 86), peroxisome proliferator-activated receptors (PPARs) (Chapter 87), GATA (Chapter 88), Ets (Chapter 89), early growth response (Egr)-1 (Chapter 90), Krüppel-like factor 2 (KLF2) (Chapter 91), nuclear factor of activated T cells (NF-AT) (Chapter 92), forkhead (Chapter 93), MEF (Chapter 94), Sox (Chapter 96), and Id (Chapter 97). As detailed in these chapters, many of these transcription factors have been implicated not only in homeostasis, but also in the pathogenesis of vascular disease states.

Yet the question remains whether this knowledge of individual components on endothelial signaling is sufficient to gain a thorough understanding of how the endothelium functions as an organ, or whether the emergent picture is more complex than the sum of its parts and cannot be deduced from the first principles – in this case, the knowledge of individual signaling pathways.

As an illustration of this complexity, let's consider two examples. A simple matter of two-pathway interaction is illustrated by endothelial expression of the $\alpha v\beta 5$ integrin. ECs expressing it are far more sensitive to VEGF stimulation, and may demonstrate an entirely different phenotypic response than those that do not (4). Another example is that of transforming growth factor (TGF)-β which, when studied in a conventional in vitro culture, produces results quite different from those obtained in a three-dimensional (3D) collagen matrix (5). This presumably occurs because the cellular inputs from a 3D matrix differ from those of the two-dimensional (2D) matrix, and this difference is enough to dramatically alter cellular response to TGF-β. It is hardly necessary to note that in vivo response is frequently, if not always, different from the effects observed in vitro.

Thus, the endothelium itself can in some sense be considered an emergent system in which the total system demonstrates functionality not obviously evident from its parts. Although the concept of emergent systems has gained much popularity lately, with everything from consciousness to Newton laws being viewed as emergent, the implication of such a nomenclature is not clear. In some cases, the underlying message is that studying constitutive components is not particularly useful to the overall understanding of the system, because the emergent nature cannot be accounted for by its individual parts. This, of course, runs counter to a reductionist approach firmly ingrained in science in general and biological science in particular. The other implication, to paraphrase Samuel Johnson, is that "one does not criticize the way endothelium functions, it is sufficient that it functions at all."

Thus, we will need new tools and new ideas to better understand the role of endothelium and the function of its numerous signaling networks. Various simulation approaches have been advanced to study the network–network interactions (6,7), and several in silico methods have been developed to study the computational properties of protein networks (8). But more traditional biological techniques will also play an important role. A better understanding of the compartmentalization of individual network components, including the roles played by caveolae, lipid rafts, and endosomes; the function of various scaffold proteins that assemble the network components together; and the organization of signaling inputs into reaction channeling (9), will be critical to a sophisticated understanding of signaling.

Furthermore, as with any complex system, stability is the key question. How stable is the phenotype being studied, and how easily can it be perturbed by changes in individual parameters or by outside influences? Remarkably, molecular biology provides wonderful tools for dealing with these issues: In vivo manipulations of individual and multiple gene expression changes in desired cell types is a routine of daily life in vascular biology laboratories. Now we need to systematically and comprehensively apply these tools to study of endothelial function under a variety of normal and abnormal conditions.

Although we have a pretty good idea how the individual signaling components mentioned here work, we have a rather incomplete understanding of how the various signals are integrated and harnessed into a coherent response. The understanding of this integration, occurring under a variety of conditions from normal to pathologic, is the next big challenge facing vascular biology. A successful solution to this riddle will provide not only the intellectual satisfaction of understanding a complex system but will allow us to design more effective treatments for a variety of illnesses – for rare is the illness that does not involve endothelium.

A practical example of this can be seen in the field of therapeutic angiogenesis. Multiple studies in vitro and in a variety of simple animal models have convinced us that a single exposure to an angiogenic growth factor such as VEGF or FGF is sufficient to induce a meaningful growth of new vasculature (10). Yet, when applied to humans or even to atherosclerotic mice, this approach utterly failed, thus demonstrating that new inputs generated by the altered disease environment have so altered the endothelium as to make it unresponsive to the stimuli to which it was responsive before (11).

To understand signal integration at the level of the individual EC, a layer of ECs and, indeed, an organ, will require new thinking and new experimental approaches. But the payoff will be worth it.

REFERENCES

1 Papin JA, Hunter T, Palsson BO, et al. Reconstruction of cellular signalling networks and analysis of their properties. *Nat Rev Mol Cell Biol.* 2005;6:99–111.

2 Bhalla US, Iyengar R. Emergent properties of networks of biological signaling pathways. *Science.* 1999;283:381–387.

3 Minami T, Aird WC. Endothelial gene regulation. *Trends in Cardiovascular Medicine.* 2005;15:174–184.

4 De S, Razorenova O, McCabe NP, et al. VEGF-integrin interplay controls tumor growth and vascularization. *Proc Natl Acad Sci USA.* 2005;102:7589–7594.

5 Madri JA, Pratt B, Tucker A. Phenotypic modulation of ECs by transforming growth factor β depends upon composition and organization of extracellular matrix. *J Cell Biol.* 1988;106:1375–1384.

6 Johnson CG, Goldman JP, Gullick WJ. Simulating complex intracellular processes using object-oriented computational modelling. *Prog Biophys Mol Biol.* 2004;86:379–406.

7 Papin JA, Palsson BO. The JAK-STAT signaling network in the human B-cell: an extreme signaling pathway analysis. *Biophys J.* 2004;87:37–46.

8 Rousseau F, Schymkowitz J. A systems biology perspective on protein structural dynamics and signal transduction. *Curr Opin Struct Biol.* 2005;15:23–30.

9 Weng G, Bhalla US, Iyengar R. Complexity in biological signaling systems. *Science.* 1999;284:92–96.

10 Simons M, Ware JA. Therapeutic angiogenesis in cardiovascular disease. *Nat Rev Drug Discov.* 2003;2:863–871.

11 Simons M. Angiogenesis: where do we stand now? *Circulation.* 2005;111:1556–1566.

70

Endothelial and Epithelial Cells

General Principles of Selective Vectorial Transport

Rolf Kinne

Max Planck Institut für molekulare Physiologie, Dortmund, Germany

One of the major prerequisites for the existence of biological organisms is compartmentalization. This process is an important early step in biological evolution, when the first organisms dissociated themselves from the surrounding medium by the generation of a selectively permeable membrane. In more and more complex organisms, not only are cells separated from their environment but also organs, such as the brain, are protected from changes in the interstitial fluid. Ultimately, the whole organism generates a *milieu interieur* whose composition is monitored with great precision and kept within narrow limits that optimize body function. Compartmentalization in complex organisms, such as mammals, is achieved by specifically organized cells, such as endothelial cells (ECs) and epithelial cells. On the one hand, both cell types act as barriers that separate compartments. On the other hand, they both mediate the vectorial transport of solutes and macromolecules. The predominance of the "wall function" versus the "gate function" varies from the blood–brain barrier endothelium to the sinusoidal endothelium and from "tight" epithelia to "leaky" epithelia. The underlying cellular and molecular mechanisms appear, however, to be quite similar. The current chapter reviews these biophysical and biochemical similarities and attempts to delineate common principles of structure and function in endothelial and epithelial cells.

Special emphasis is thereby placed on the relation between asymmetry and net movement of small solutes. Asymmetry (i.e., polarity) of cell membrane transport properties is found in both epithelial and EC layers, and provides the basis for transcellular transport. Furthermore, the asymmetry of membrane transporters is discussed with regard to their different properties at the extra- and intracellular face. These molecular asymmetries play an important role in the efficiency, direction, and regulation of transport processes across the plasma membranes in endothelial and epithelial cells.

PERMEABILITY OF ENDOTHELIA AND EPITHELIA

The pathways used in endothelia and epithelia for the penetration of cells, macromolecules, organic nutrients, inorganic salts, and water are summarized in Table 70-1. In both cell layers, para- and transcellular pathways are used, with the size of the permeating compounds spanning from cells permeating through discontinuous endothelia in liver and spleen, for example, to water permeation through the tight endothelium of the blood–brain barrier (1). Epithelia, although also categorized into leaky, intermediate, or tight, are much more restrictive with regard to their permeability to cells and macromolecules. Selectivity appears at the level of organic solutes, inorganic ions, and water. This selectivity is reflected in their transepithelial resistance and transepithelial voltage, as well as in their hydraulic conductivity (2).

Thus, the degree of similarity in transendothelial and transepithelial pathways increases with the tightness of the cell layer, with the blood–brain barrier and the tight epithelia becoming most similar in their basic transport properties.

GENERAL PRINCIPLES OF TRANSPORT THROUGH CELL LAYERS

Energetic Considerations

The net transport of any electrolyte i across a cell layer is the sum of the passive transport by solvent drag (similar to heat convection) and electrochemical potential difference, and of the active transport. This is expressed in Equation 1:

$$J_i = (1 - \sigma_i)\,\bar{c_i}\,J_v - P_i\left(\Delta c_i + \frac{zF}{RT}\Delta\varphi\bar{c_i}\right) + J_{act},$$

(Equation 1)

Table 70-1: Gateways in Endothelia and Epithelia

	Occurrence of Permeation	
Permeating Compound	*EC*	*Epithelial Cell*
Cells	Frequent, in particular in sinusoidal endothelium	Rare, more frequent in intestinal epithelium
Macromolecules (< albumin)	Frequent, except at blood–brain barrier	Rare, selective uptake in renal and intestinal epithelia
Organic nutrients	Paracellular pathways rare; transcellular pathways predominate	Paracellular pathways rare, predominantly transcellular transport
Inorganic salts	Paracellular pathways predominate, except at blood–brain barrier	Paracellular pathways predominate in leaky epithelia, transcellular pathways predominate in tight epithelia
Water	Paracellular pathways predominate; transcellular pathway in nonfenestrated endothelia	Transcellular pathways predominate

Compiled from Refs. 106–118.

where σ represents the reflection coefficient, c the mean concentration of i in the two compartments, J_v the volume flow, P the permeability coefficient, Δc the concentration difference, z the charge of the molecule, F the Faraday constant, R the gas constant, T the temperature in Kelvin degrees, $\Delta\varphi$ the electrical potential difference, and J_{act} the active transport component. This active transport component is by definition the transport observed when all physical driving forces are zero, and it is coupled to a chemical reaction (3–5).

The flux of a nonelectrolyte is described by Equation 2.

$$J_i = (1 - \sigma_i)\,\bar{c}_i J_v - P_i \Delta c_i + J_{act} \qquad \text{(Equation 2)}$$

The volume flow Jv can be formulated according to Starling (6) as:

$$J_v = L_p A \cdot [p - P] \qquad \text{(Equation 3)}$$

where L_p represents the hydraulic conductivity, A the effective cell layer surface, p the osmotic pressure difference, and P the hydrostatic pressure difference. Therefore, solute movement by solvent drag also depends on the osmotic and hydrostatic pressure differences that cause water movement across the cell layer (for a more extensive treatment of water flow across a cell layer see Ref. 7).

Active transport across a cell layer usually involves transcellular transport and, at one point or another, the coupling of an exergonic (i.e., energy yielding) chemical reaction to a vectorial flux (5). The transport across the membrane can be either primary active or secondary or even tertiary active (8). Primary active transport is mediated by ATP-dependent pumps such as the inorganic ion translocating Na^+-K^+-ATPase, Ca^{++}-ATPases (PMCA), H^+-ATPase, and the like, or the organic anions ABC transport systems. Secondary active transport systems use ion gradients established by primary active pumps across the plasma membrane as driving force. Typical examples in mammalian cells are the sodium (Na^+)-solute cotransport systems for sugar, amino acids, organic metabolites, and similar substances (8). Tertiary active transport systems use the gradients established by the secondary active transport as driving forces for transfer across the membrane. The latter systems are mostly antiport or exchange systems, such as the PAH/α-ketoglutarate antiporter (9) localized in the basolateral plasma membrane of the proximal tubule, which exchanges with a broad substrate specificity organic anions with α-ketoglutarate accumulated by a Na^+-α-ketoglutarate cotransport system (8,9). Secondary active and tertiary active transport also can be combined in one translocation step, such as the y^+L amino acid transport system in which a neutral amino acid (together with Na^+) is taken up into the cell by secondary active transport and, at the same time, arginine is transported out of the cell against its electrochemical gradient by tertiary active transport (10,11).

Paracellular Transport

The transport of molecules between cells is governed by passive driving forces such as differences in concentration, electrical potential, pressure, or temperature in the two compartments that are separated by the cell layer (5). The rate of transfer also depends on the physical and chemical properties of the paracellular pathway, such as effective pore size, geometry, surface charges, and presence of binding sites, to name a few (12–14). Paracellular transport thus can be selective but only to a limited extent. Across leaky cell layers, in general, no significant concentration differences between compartments are established. However, in the presence of even small differences in electrochemical potential, the paracellular pathway is the predominant route for the bulk flow of salts in leaky epithelia and discontinuous endothelia (14,15).

It is important to note at this point that significant differences exist in the cell-to-cell contact between epithelia and endothelia. In the former, desmosomes are found, specialized cell junctions that are formed by desmosomal cadherins and additional associated proteins into which intermediate filaments are inserted. They are also known as macula adherens junctions, and they link the intermediate filament network of one cell to its neighbor (16). Although mainly considered as imparting tensile strength and resilience to the epithelium, they also may limit the size of molecules able to pass through the paracellular pathway. Similarly, paracellin, a paracellular protein that confers specificity for paracellular calcium and magnesium permeability has also thus far only been observed in epithelial cells (in the renal thick ascending limb of Henle's loop and duct cells in salivary glands).

Transcellular Transport

Hydrophobic Molecules

The plasma membrane surrounding the endothelial and epithelial cells is composed mainly of membrane proteins and phospholipids. The latter provide an ideal solvent for small hydrophobic molecules; these molecules can be translocated from one cell side to the other (called luminal-to-abluminal translocation in endothelia and apical-to-basolateral translocation in epithelia). Thereby, diffusion of the translocated molecules within the hydrophobic solvent (17) flip-flop between the two monolayers (18), and the mobility of the inner layer of the phospholipid bilayers aid their migration (19). These pathways are known both for endothelial and epithelial cells, and they are characterized by a specificity based on the oil/water partition coefficient (20,21). The net transport across the cell layer depends strongly, however, on the mode of delivery to the one cell surface and mode of removal from the other cell surface. In this context, the role of plasma proteins and interstitial tissue in providing a reservoir and/or "sink" for hydrophobic molecules is of particular importance.

Hydrophilic Molecules

Passage of hydrophilic molecules such as inorganic salts, sugars, and amino acids through the plasma membrane requires the presence of specific mediators. When the electrochemical potentials of the transportates are equal at both sides of the cell (i.e., luminal and abluminal sides in endothelia; apical and basolateral sides in epithelia), net transcellular transport can occur only if the transport properties of the two membranes facing the *cis* and *trans* compartments are different (22). Thus, an asymmetry of the cell must exist to allow for vectorial transcellular transport (5). Such a cellular *asymmetry* has been clearly demonstrated in endothelial (blood–brain barrier) and epithelial cells (kidney tubules) for example, using indicator dilution techniques in vivo (15,23–27) and in uptake studies employing isolated capillaries or tubules or cultured cells in vitro (10,28–33). Thereby, the two transporting surfaces of the cell were functionally defined, and striking differences could be observed. The blood–brain barrier showed a strongly asymmetric behavior for the movement of potassium (K^+), glycine, and glutamate, whereas much less asymmetry was evident for sugars and large neutral amino acids (34–37). In the proximal tubule, the apical cell side avidly absorbed sugars and amino acids; at the basal-lateral side, p-aminohippuric acid (PAH) entered the cells most rapidly (25,26). The physiological significance of these findings is evident. In the blood–brain barrier, substances that interfere with the electrical activity of the neurons or act as neurotransmitters, such as K^+, glycine, and glutamate, are removed from the extracellular fluid, whereas in the kidney, the secretion of metabolic waste products or xenobiotics (PAH as representative) is initiated at the basolateral cell side. Both in the blood–brain barrier and in the proximal tubule, the first steps of nutrient transfer take place at the luminal cell side.

CELLULAR ASYMMETRY AS BASIS OF VECTORIAL TRANSCELLULAR TRANSPORT

As pointed out earlier, an asymmetry of the transport properties of the two cell membranes arranged in series is required to achieve transcellular vectorial transport. The basis for this idea was laid by Forster and Taggart in 1950, when they proposed that the coupling of metabolic energy to the active transport of organic solutes occurs only at one cell membrane – leading to a "push" mechanism when the intracellular concentration of the transported solute was higher than the concentration in the luminal medium, or to a "pull" mechanism when the concentration gradient was established across the contralumenal membrane (38). This sequence of events in molecular terms is realized in the active transcellular movement of sugars, amino acids, and phosphate, for example, where one membrane contains a secondary-active Na^+-cotransport system for the solute and the other membrane a Na^+-independent uniporter (8). In 1958, Koefoed-Johnsen and Ussing postulated an asymmetry for the frog skin to explain the active Na^+ transport across the epithelium, which involves a Na^+ channel in the cell membrane facing the outside and an active Na^+ pump at the inside (39). This is one of the typical arrangements observed, for example, for the transcellular movement of calcium and protons. A small modification for bile acids, xenobiotics, and other substances is the presence of a diffusive pathway in one membrane and an ABC transporter in the other cell membrane (see the next section).

The cellular and molecular basis of such asymmetry has been elucidated successfully mainly by two techniques. In vesicle transport studies, the transport properties of the two membranes have been determined under well-defined driving forces in vitro after their separation from the intracellular space and from each other (40–42). The other technique uses immunohistochemistry, which localizes particular transport systems based on their antigenicity. Highly specific antibodies directed against distinctive epitopes of transporters can be generated to allow the identification and localization of transport molecules. Both methods have their advantages and disadvantages. In the former, the interaction with other membrane-associated and/or modifying proteins might be lost (43–47), and transport rates (and sometimes also affinities and substrate specificities) might vary from the situation in intact cells. In the latter, only the *presence* of a transport system is observed – no information about its current state of activity is provided. Sometimes antigenic sites might also be inaccessible, as observed, for example, for the glucose transporter-1 (GLUT1) in bovine blood–brain capillaries (48).

Another major step forward in defining the transport properties of the different cell poles was the development of the patch-clamp technique, which allows the identification and characterization of channels for inorganic and organic ions (49,50).

More recently, it has been realized that the degree of cellular polarity can, however, be modified. Modification may occur via two mechanisms. First, new transporters may be inserted into one cell pole after a particular stimulus – best exemplified by aquaporin water channels (51) or the insulin-responsive GLUT glucose transporter (52). This phenomenon is increasingly realized to be of major importance in regulating the rate of transfer and also occasionally the net direction of translocation (53). Second, a modification may occur by the activation or inactivation of transporter in loco; that is, at the membrane in which they are residing in. In brain capillaries, recent studies on the regulation of the $Na^+/K^+/2Cl^-$ cotransporter (NKCC) revealed such a modification in loco. Ischemia activates a basal lateral NKCC; this may lead to a reversal of chloride fluxes and the removal of electrolytes released by damaged cell (54). Such regulatory phenomena also are affected by drugs, such as nicotine – demonstrating the flexibility of the cells generated by the possibilities of intracellular trafficking of regulatory members of the signal transduction pathways (such as protein kinases and phosphatases) or the transporters themselves (54). This kind of "flexible" polarity should always be taken into account when considering epithelial and endothelial polarity.

DISTRIBUTION OF TRANSPORT SYSTEMS IN ENDOTHELIAL CELLS

Figure 70.1 compiles data obtained by a combination of techniques on the distribution of transport systems in ECs at the blood–brain barrier. Some of the transport systems, such as those for D-glucose (GLUT1) and L-leucine (system L), are present both in the luminal and contraluminal membrane (10,55–58). Others, such as the Na^+,K^+-ATPase are present only in the abluminal membrane. This enzyme, in combination with a luminal K^+ channel, is probably mainly involved

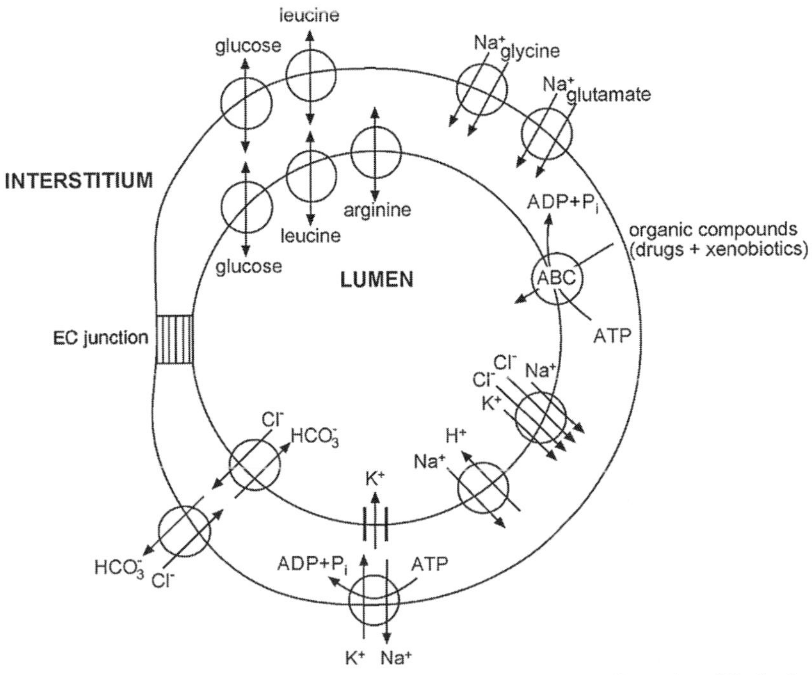

Figure 70.1. Distribution of transport systems in the EC of a brain capillary. (Modified after Risau W. Differentiation of blood–brain barrier endothelium. *News Physiol Sci.* 1989;4:151–153.)

in K^+ removal from the brain tissue into the blood. In addition a $Na^+/K^+/2Cl^-$ cotransport system has been shown to be present in the luminal membrane (54,59). Thus, the luminal K^+ channels (60,61) also may be involved in recycling of K^+ after its uptake via the $Na^+/K^+/2Cl^-$ cotransporter. The arrangement of these transporters is very similar to the one found in the distal part of the nephron (see later section), where secretion of K^+ is observed (62,63). In addition, the Na^+,K^+-ATPase maintains the electrochemical potential difference for Na^+ across the plasma membrane that is employed for the uptake of glycine, glutamate, and taurine into the cells. As mentioned earlier, the former two transport systems are apparently present only in the abluminal membrane.

The molecular identity of the transport systems in ECs for amino acids can be summarized as follows (10): System L1 (for "large" amino acids) expressed both in the luminal and abluminal membrane, is a heterodimeric transport system and transports (and exchanges) alanine, leucine, glycine, serine, and glutamine (64). The abluminal membrane contains three Na^+-dependent amino acid transport systems: the system A (65) and $B^{0,+}$, and one for anionic amino acids. System A is the major Na^+-dependent transport systems for neutral amino acids such as alanine and proline; its hallmark is that it interacts with the nonphysiological α-(methylamino)isobutyric acid (MeAB). System $B^{0,+}$ appears to be of minor importance. At the abluminal cell side, anionic amino acids such as glutamate are probably translocated both by the Na^+-dependent X^-_{AC} system and the Na^+-independent X^-_C transport system that exchanges cystine and glutamate. Depending on the ECs investigated, one or the other transport route prevails. At the luminal membrane of ECs, cationic amino acids such as arginine, which plays a crucial role in the nitric oxide production of ECs (66–69), are taken up via a Na^+-independent transport system y^+ (probably CAT1 similar). Another major luminal pathway for amino acids in some ECs is the y^+L system. This heterodimeric transport system mediates the exchange of a cationic amino acid for Na^+ and a neutral amino acid – a particularly intriguing combination.

DISTRIBUTION OF TRANSPORT SYSTEMS IN EPITHELIAL CELLS

In most epithelial cells, the asymmetry of the cell membranes is more striking than that of endothelia. For any particular transported solute, the transport systems in the apical and

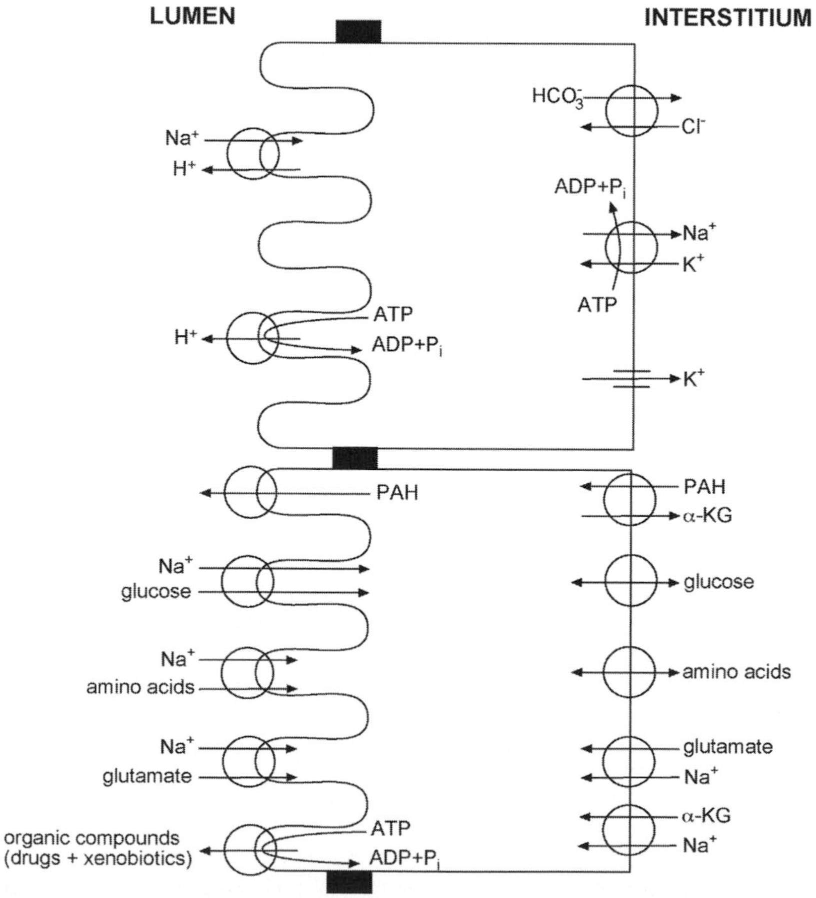

Figure 70.2. Distribution of transport systems in the epithelial cell of the renal proximal tubule.

basolateral membrane are very often completely different. Sugar entry into the cells of the proximal tubule and intestine, for example, is mediated by a Na^+-dependent cotransport system, SGLT (Figure 70.2) (8), whereas cell exit occurs via the Na^+-independent transport system, GLUT2 (70). The same holds for a large number of other organic solutes that are retrieved from the lumen of the tubule and small intestine, respectively (8). PAH is taken up at the contraluminal membrane by a specific exchange system with α–ketoglutarate (9) and released at the luminal membrane by another transporter that bears similarities to a channel. Therefore, the entry steps for these solutes are driven directly or indirectly by the electrochemical potential difference of Na^+ across the cell membrane, which is maintained by the primary active Na^+ pump, the Na^+,K^+-ATPase. As a consequence, the transport systems in the two membranes are set up in such a way that the *active* transepithelial transport of sugars and PAH is possible.

In the distal tubule, K^+ is secreted by a similar mechanism as in the brain capillary (62,63). Organic solutes, however, are taken up almost exclusively from the contraluminal side and subserve cellular metabolism. This latter phenomenon is observed for all epithelia whose luminal membranes are in contact with fluids that do not contain the appropriate metabolic substrates; other examples are the urinary bladder, the colon, and other organs.

MOLECULAR ASYMMETRY AT THE MEMBRANE LEVEL AS BASIS FOR VECTORIAL TRANSPORT

Similar to the requirement that vectorial transport across the cell requires an asymmetry of the two cell surfaces, such an asymmetry also must be postulated at the membrane level. Indeed, all transport systems cloned thus far exhibit such an asymmetry when the amino acid sequence and the (in most cases computer-predicted) topology of the transport system is considered. The asymmetry also is reflected in different properties of the two sides of the molecule. Classic examples are Na^+,K^+-ATPase, with its different accessibility of ion-binding sites and the active center for ATP hydrolysis from the two membrane surfaces (71) and ion channels that show rectification of current (i.e., a different conductivity depending on the direction and strength of the electric field applied to them) (49,72). Also, Na^+ cotransport systems have been shown to exhibit a strong functional asymmetry. In general, the affinity for Na^+ at the domains facing the inside of the cell is much lower than at the outside, allowing for a rapid dissociation of both the Na^+ and the cotransported solute. In addition, regulation of transport systems occurs mainly by modification of internally located phosphorylation, methylation, or prenylation sites. Also, uniporters for amino acids and carbohydrates show marked differences in substrate affinity and specificity when investigated in the cis or in the trans conformation.

One example that has been studied in our laboratory is the L-glutamate transport system present in the kidney as well as in the endothelium (73,74). For these experiments, brush border membrane vesicles were isolated from hog kidney. The isolated membranes are oriented right side out (i.e., in influx experiments, the properties of the transporter exhibited at the extracellular face are studied, whereas in efflux experiments the properties of the transporter at the cytoplasmic face are examined). In influx experiments, Na^+ stimulates L-glutamate influx into the vesicles. In efflux experiments, it becomes obvious that the cytoplasmic face of the transporter lacks the Na^+ dependence. This asymmetry increases strongly the efficiency of the transmembrane translocation by minimizing back leak that would reduce energetic coupling (75). Similar observations have been made with the Na^+-D-glucose cotransporter. When expressed in inside-out vesicles of yeast, the uptake of D-glucose is Na^+ independent and has a very low affinity for sugars. In addition, it lacks the sensitivity to phlorhizin found when testing the extracellular face of the carrier (76).

The Na^+-dependent glutamate transporter in the kidney also exhibits a strong asymmetry with regard to its sensitivity to changes in the intra- or extravesicular pH. L-glutamate transport is strongly stimulated when the pH of the extravesicular medium is reduced from 7.4 to 5.4 (77). A similar reduction of the intravesicular pH does not change the L-glutamate influx significantly. Furthermore, the stimulatory effect of protons at the extravesicular face is not due to an increased driving force provided for H^+-glutamate cotransport, because the identical stimulation is found when both the intra- and extracellular pH are lowered at the same time.

It can, therefore, be expected that a functional asymmetry exists for probably all transporters – including the Na^+-independent uniporters. Its characterization will have important consequences on how transmembrane and transcellular transport across endothelial and epithelial cell layers can be described.

DIRECTION OF TRANSCELLULAR ORGANIC SOLUTE TRANSPORT IN ENDOTHELIAL CELLS

D-Glucose Transport across Endothelial Cells

The consequences of the apparent cellular asymmetry of the GLUT1 distribution for transcellular transport of D-glucose depend on several assumptions concerning the kinetic properties of the transport systems in the two membranes and the glucose concentrations in the various compartments. With regard to the latter, we can assume that, under normal physiological conditions, the glucose concentration is highest in the blood, intermediate in the EC, and lowest in the brain tissue. If, under these circumstances, the two glucose transport systems behave identically, the rate-limiting step for the transport across the cell would be the transport across the luminal membrane. The basolateral transporters then could be a functional reserve that acts, in concert with an increased insertion of transporters from an intracellular pool into the luminal membrane, to increase glucose supply to the brain in case of increased demand. Such regulation of transendothelial D-glucose transport has been discussed in the literature (43,78). Another explanation for the asymmetry in the distribution of GLUT1 could, however, be a functional asymmetry of the transporter itself. For the human erythrocyte glucose transporter and GLUT1, it has been described that the affinity of the transporter for the mediation of efflux of D-glucose is much lower than for the mediation of influx (79,80). If this difference holds also for the situation in the ECs, the greater number of transporters at the basolateral membrane would compensate (at least partially) the kinetic disadvantage for the permeation of D-glucose across this cell pole.

Amino Acid Transport across Endothelial Cells

With regard to the transport of amino acids the situation is rather complex. This complexity stems from the fact that amino acid transporters often exchange one amino acid against another; therefore, the direction of the translocation across one membrane can be reversed at the same membrane when exchange systems are arranged in parallel. In this instance, the functional and kinetic data obtained in isolated capillaries, membranes, and cell cultures can be helpful. Thus, for glutamate, the basal-lateral localization can explain the observed avid uptake at the brain-oriented cell pole. Together with the high stoichiometry of the system (2 to 3 Na^+ ions per glutamate), a reversal of the direction appears to be improbable.

With regard to arginine, Na^+-independent uptake is driven by the membrane potential and the low intracellular arginine concentration. This uptake could be, however, counteracted by the exchange of arginine for Na^+ and a neutral amino acid entering the cell via the y^+L system at the same luminal cell pole.

For the large amino acids, the presence of the same transport system at the luminal and the abluminal side allows only for vectorial transcellular transport when the transporters are in the right location (81), the driving forces are in the proper direction, and the exchangeable amino acids are in the proper concentration.

TRANSPORT OF DRUGS AND XENOBIOTICS ACROSS ENDOTHELIAL CELLS

The expression and activity of transporters in the endothelium of the blood–brain barrier also critically determine the bioavailability of drugs essential for treating neurological diseases, including brain tumors and epilepsy (82,83). Similarly, the removal of neurotoxins and glutathione-S-conjugates from the ECs and/or xenobiotics from the brain into the blood depends on the presence of appropriate transport systems in the cells. In both instances, the directionality of transcellular transport is very clearly defined, as indicated by the strong unidirectional, energy-dependent efflux observed in brain capillary cell lines and isolated capillaries (84–86). The molecular bases for this transport are the ABC transporters multidrug resistance-associated protein (MRP or ABCB1), P-glycoprotein encoded by MDR1 (or ABCC1), and a homologue of the human breast cancer resistance protein breast cancer resistance protein (BCRP) (or ABCG2), which are exclusively located at the luminal pole of the ECs in the brain (87,88). The lipophilic organic compounds presented to the ECs either from the brain or blood side enter the cells via simple diffusion according to the oil partition coefficient or by the flip-flop mechanism (18) mentioned earlier. They are then, often after conjugation within the cell for detoxification, extruded across the luminal membrane into the blood. Thus, one function of the endothelium is to act as an effective barrier against the entry of these substances from the blood into the brain due to the recycling of the compounds across the luminal membrane. For substances conjugated in the cells or entering the ECs from the brain side, an effective removal from the brain into the blood takes place (89,90).

It is interesting to note that similar systems operate in a variety of epithelial cells such as the liver hepatocytes, renal proximal tubule cells, and secretory cells such as the mammary glands (91).

CONCLUSION

Since I reviewed this topic several years ago (92), the knowledge of the transport properties of the endothelium has increased significantly. Interestingly, the methods that helped in the past to decipher the chain of events comprising transcellular transport in epithelia-isolated tubules and in particular cell cultures (93–95) and isolated membranes (8,40) have now found their way to the endothelium (10). An additional gain came from the cloning of transporters and the possibility of generating specific antibodies for use in immunohistochemistry. In general, it has become clear that epithelia and endothelia are quite alike – histology defines the endothelium as "flat epithelium in contact with lymph or blood." The main differences prevail when the tight junctions become fenestrae.

Another major difference is the reactions of the ECs engendered by blood cells interacting with them (96–101). In classical epithelia such as kidney and liver, such phenomena are rare. Mechanical stress also is a potent modifier of endothelial function (102–104). Another point of difference is that the surface of the endothelium is exposed to signaling molecules present in the bloodstream more frequently than is the epithelium, although even there luminal messenger systems are becoming more and more elucidated (105). If, for example, wound healing is considered, such factors – particularly the interplay between endothelium and epithelium – are important. This mutual interplay is one area in which the specific, differing properties of both cell types become dominant over the similarities pointed out in this chapter. This is particularly important when disease states, such as inflammation and cancer, are considered (105).

KEY POINTS

- Both endothelia and epithelia fulfill wall and gate functions in simultaneously separating and connecting various body compartments. The basic principles by which the cells achieve this function are, despite their different evolutionary origins, very similar. This holds in particular for the transcellular transport pathways, in which the two membranes facing the *cis* and *trans* compartments (i.e., the luminal and abluminal side) of the endothelium are crossed in series. The paracellular pathways differ more in their composition and properties, but the same physical principles are at work to drive solutes across the cell layer.
- Cellular polarity and membrane asymmetry are the basis for vectorial transcellular transport. This polarity is generated by the different composition of the luminal and contraluminal cell membrane with regard to transport systems.
- These transport systems are either primary active transport systems, in which the translocation of the solute is directly coupled to ATP hydrolysis, or secondary or even tertiary active transport systems, in which (electrochemical) gradients of inorganic ions (mostly Na^+) or organic molecules (mostly metabolites) are coupled to the symport or antiport of other solutes.
- The membranes also differ in the presence and activity of channels for inorganic anions, inorganic cations, and organic molecules.
- The polarity between the two membranes is more evident in epithelia than in endothelia, although similar transport systems are involved.
- Transport through the paracellular pathway is governed by passive driving forces.
- The paracellular pathway exhibits major differences between endothelia and epithelia; those most evident are the fenestrae in endothelia, as contrasted to the tight junctions in epithelia, and the presence of desmosomes in the latter.
- The prediction of the direction of flow in endothelia is complicated by the low asymmetry of the cells, the presence of multiple transport systems (in particular for amino acids), the lack of knowledge concerning the chemical and electrical driving forces, and the occurrence of significant paracellular movements.

Future Goals

- To study all these parameters in parallel at the same point in time and space to make possible predictions on the transport rates and transport routes
- To understand the mechanisms of transfer of organic compounds across EC layers – in particular therapeutic drugs in the brain – which could lead to important progress in the treatment of disease

REFERENCES

1 Bennet HS. Morphological classification of vertebrate blood capillaries. *Anat Rec.* 1958;130(2):271.

2 Greger R. Epithelial transport. In: Greger R, Windhorst U, eds. *Comprehensive Human Physiology*. Berlin: Springer; 1996:1217–1232.

3 Froemter E, Rumrich G, Ullrich KJ. Phenomenologic description of Na^+, Cl- and Hco-/3 absorption from proximal tubules of rat-kidney. *Pflugers Arch.* 1973;343(3):189–220.

4 Froemter E. The electrophysiological analysis of tubular transport. *Kidney Int.* 1986;30(2):216–228.

5 Sauer F. Nonequilibrium thermodynamics of kidney tubule transport. In: Orloff J, Berliner RW, eds. *Handbook in Physiology*. Bethesda, MD.: American Physiological Society; 1973:399–414.

6 Starling EH. On the absorption of fluid from the connective tissue spaces. *J Physiol (London)*. 1896;19:312–326.

7 Schafer JA, Reeves WB, Andreoli TE. Mechanism of fluid transport. In: Windhager EE, ed. *Renal Physiology*. Oxford University Press: American Physiological Society; 1992:659–713.

8 Kinne RKH. Selectivity and direction – plasma-membranes in renal transport. *Am J Physiol.* 1991;260(2):F153–F162.

9 Burckhardt BC, Burckhardt G. Transport of organic anions across the basolateral membrane of proximal tubule cells. *Rev Physiol Biochem Pharmacol.* 2003;146:95–158.

10 Mann GE, Yudilevich DL, Sobrevia L. Regulation of amino acid and glucose transporters in endothelial and smooth muscle cells. (Review) *Physiol Rev.* 2003;83(1):183–252.

11 Cariappa R, Heath-Monnig E, Smith CH. Isoforms of amino acid transporters in placental syncytiotrophoblast: plasma membrane localization and potential role in maternal/fetal transport. *Placenta.* 2003;24(7):713–726.

12 Froemter E. Viewing the kidney through microelectrodes. *Am J Physiol.* 1984;247(5):F695–F705.

13 Crone C, Levitt DG. Capillary permeability to small solutes. In: Renkin EM, Michel CC, eds. *Handbook of Physiology*. Bethesda, MD: American Physiological Society; 1984:411–466.

14 Bradbury MWB. *The Concept of a Blood-Brain Barrier*. New York: Wiley, 1979.

15 Bradbury MWB, Patlak CS, Oldendorf WH. Analysis of brain uptake and loss of radiotracers after intracarotid injection. *Am J Physiol.* 1975;229(4):1110–1115.

16 Huber O. Structure and function of desmosomal proteins and their role in development and disease. *Cell Mol Life Sci.* 2003; 60(9):1872–1890.

17 Scow RO, Blanchettemackie EJ, Smith LC. Role of capillary endothelium in clearance of chylomicrons – model for lipid transport from blood by lateral diffusion in cell-membranes. *Circ Res.* 1976;39(2):149–162.

18 Eytan GD. Mechanism of multidrug resistance in relation to passive membrane permeation. *Biomed Pharmacother.* 2005;59(3): 90–97.
19 Dragsten PR, Blumenthal R, Handler JS. Membrane asymmetry in epithelia – is the tight junction a barrier to diffusion in the plasma-membrane. *Nature.* 1981;294(5843):718–722.
20 Panzenboeck U, Balazs Z, Sovic A, et al. ABCA1 and scavenger receptor class B, type I, are modulators of reverse sterol transport at an in vitro blood-brain barrier constituted of porcine brain capillary endothelial cells. *J Biol Chem.* 2002;277(45):42781–42789.
21 Oldendorf WH. Lipid solubility and drug penetration of the blood-brain barrier. *Proc Soc Exp Biol Med.* 1974;147:813–816.
22 Boulpaep EE. Solute-coupled water transport in the kidney. In: Solomon AK, Karnovsky M, eds. *Molecular Specialization and Symmetry in Membrane Function.* Cambridge: Harvard University Press; 1978:294–315.
23 Crone C. The permeability of capillaries in various organs as determined by used of the “indicator diffusion” method. *Acta Physiol Scand.* 1983;58:292–305.
24 Chinard FP. Transcapillary exchange of water and other substances in certain organs of the dog. *Am J Physiol.* 1955;183:221–234.
25 Silverman M, Trainor C. In vivo determination of cellular uptake in the kidney. *Fed Proc.* 1982;41(14):3054–3060.
26 Silverman M. Comparison of glucose transport mechanisms at opposing surfaces of the renal proximal tubular cell. *Biochem Cell Biol.* 1986;64(11):1092–1098.
27 Yudilevich DL, De Rose N. Blood-brain transfer of glucose and other molecules measured by rapid indicator dilution. *Am J Physiol.* 1971;220(3):841–846.
28 Joó F. The blood-brain barrier in vitro: ten years of research on microvessels isolated from the brain. *Neurochem Int.* 1985;7:1–25.
29 Goldstein GW, Betz AL, Bowman PD. Use of isolated brain capillaries and cultured endothelial-cells to study the blood-brain-barrier. *Fed Proc.* 1984;43(2):191–195.
30 Gstraunthaler G, Steinmassl D, Pfaller W. Renal cell cultures: a tool for studying tubular function and nephrotoxicity. *Toxicol Lett.* 1990;53(1–2):1–7.
31 Joó F. The cerebral microvessels in culture, an update. *J Neurochem.* 1992 58:1–17.
32 Rubin LL, Hall DE, Porter S, et al. A cell culture model of the blood-brain barrier. *J Cell Biol.* 1991;115(6):1725–1735.
33 Handler JS. Studies of kidney-cells in culture. *Kidney Int.* 1986; 30(2):208–215.
34 Hussar P, Tserentsoodol N, Koyama H, et al. The glucose transporter GLUT1 and the tight junction protein occludin in nasal olfactory mucosa. *Chem Senses.* 2002;27(1):7–11.
35 Pardridge WM. Blood-brain transport of nutrients. Introduction. *Fed Proc.* 1986;45(7):2047–2049.
36 Komura J, Tamai I, Senmaru M, et al. Sodium and chloride ion-dependent transport of beta-alanine across the blood-brain barrier. *J Neurochem.* 1996;67(1):330–335.
37 Betz AL. Transport of ions across the blood-brain barrier. *Fed Proc.* 1986;45(7):2050–2054.
38 Forster RP, Taggart JV. Use of isolated renal tubules for the examination of metabolic processes associated with active cellular transport. *J Cell Physiol.* 1950;36(2):251–270.
39 Koefoed-Johnsen V, Ussing HH. The nature of the frog skin potential. *Acta Physiol Scand.* 1958;42(3–4):298–308.
40 Kinne-Saffran E, Kinne RKH. Membrane isolation – strategy, techniques, markers. *Methods Enzymol.* 1989;172:3–17.
41 Kinne R, Kinne-Saffran E. Membrane-vesicles as tools to elucidate epithelial-cell function. *Eur J Cell Biol.* 1981;25(2):346–352.
42 Murer H, Kinne R. The Use of isolated membrane-vesicles to study epithelial transport processes. *J Membr Biol.* 1980;55(2): 81–95.
43 Busik JV, Olson LK, Grant MB, Henry DN. Glucose-induced activation of glucose uptake in cells from the inner and outer blood-retinal barrier. *Invest Ophthalmol Vis Sci.* 2002;43(7): 2356–2363.
44 Minshall RD, Sessa WC, Stan RV, et al. Caveolin regulation of endothelial function. *Am J Physiol Lung Cell Mol Physiol.* 2003; 285(6):L1179–L1183.
45 Takabe W, Kanai Y, Chairoungdua A, et al. Lysophosphatidylcholine enhances cytokine production of endothelial cells via induction of L-type amino acid transporter 1 and cell surface antigen 4F2. *Arterioscler Thromb Vasc Biol.* 2004;24(9):1640–1645.
46 Zharikov SI, Block ER. Association of L-arginine transporters with fodrin: implications for hypoxic inhibition of arginine uptake. *Am J Physiol Lung Cell Mol Physiol.* 2000;278(1):L111–L117.
47 Zharikov SI, Krotova KY, Belayev L, Block ER. Pertussis toxin activates L-arginine uptake in pulmonary endothelial cells through downregulation of PKC-alpha activity. *Am J Physiol Lung Cell Mol Physiol.* 2004;286(5):L974–L983.
48 Simpson IA, Vannucci SJ, DeJoseph MR, Hawkins RA. Glucose transporter asymmetries in the bovine blood-brain barrier. *J Biol Chem.* 2001;276(16):12725–12729.
49 Hille B. Ionic channels – evolutionary origins and modern roles. *Q J Exp Physiol Cogn Med Sci.* 1989;74(6):785–804.
50 Neher E, Sakmann B. Single-channel currents recorded from membrane of denervated frog muscle fibres. *Nature.* 1976;260 (5554):799–802.
51 Brown D. The ins and outs of aquaporin-2 trafficking. *Am J Physiol Renal Physiol.* 2003;284(5):F893–F901.
52 Thong FS, Dugani CB, Klip A. Turning signals on and off: GLUT4 traffic in the insulin-signaling highway. *Physiology (Bethesda).* 2005;20:271–284.
53 Kipp H, Khoursandi S, Scharlau D, Kinne RKH. More than apical: distribution of SGLT1 in Caco-2 cells. *Am J Physiol Cell Physiol.* 2003;285(4):C737–C749.
54 Abbruscato TJ, Lopez SP, Roder K, Paulson JR. Regulation of blood-brain barrier Na,K,2Cl-cotransporter through phosphorylation during in vitro stroke conditions and nicotine exposure. *J Pharmacol Exp Ther.* 2004;310(2):459–468.
55 Farrell CL, Pardridge WM. Blood-brain-barrier glucose transporter is asymmetrically distributed on brain capillary endothelial luminal and abluminal membranes – an electron-microscopic immunogold study. *Proc Natl Acad Sci USA.* 1991;88(13): 5779–5783.
56 Farrell CL, Yang J, Pardridge WM. Glut-1 Glucose transporter is present within apical and basolateral membranes of brain epithelial interfaces and in microvascular endothelia with and without tight junctions. *J Histochem Cytochem.* 1992;40(2):193–199.
57 Harik SI, Kalaria RN, Whitney PM, et al. Glucose transporters are abundant in cells with occluding junctions at the blood eye barriers. *Proc Natl Acad Sci USA.* 1990;87(11):4261–4264.

58 Maher F, Vannucci SJ, Simpson IA. Glucose-transporter proteins in brain. *FASEB J*. 1994;8(13):1003–1011.
59 Sun DD, Lytle C, O Donnell ME. Astroglial cell-induced expression of Na-K-Cl cotransporter in brain microvascular endothelial cells. *Am J Physiol Cell Physiol*. 1995;38(6):C1506–C1512.
60 Nilius B, Droogmans G. Ion channels and their functional role in vascular endothelium. (Review). *Physiol Rev*. 2001;81(4):1415–1459.
61 Ningaraj NS, Rao MK, Black KL. Adenosine 5′-triphosphate-sensitive potassium channel-mediated blood-brain tumor barrier permeability increase in a rat brain tumor model. *Cancer Res*. 2003;63(24):8899–8911.
62 Wang WH, Sackin H, Giebisch G. Renal potassium channels and their regulation. *Annu Rev Physiol*. 1992;54:81–96.
63 Giebisch G, Hebert SC, Wang WH. New aspects of renal potassium transport. *Pflugers Arch*. 2003;446(3):289–297.
64 O'Kane RL, Hawkins RA. Na+-dependent transport of large neutral amino acids occurs at the abluminal membrane of the blood-brain barrier. *Am J Physiol Endocrinol Metab*. 2003;285(6):E1167–E1173.
65 Takanaga H, Tokuda N, Ohtsuki S, et al. ATA2 is predominantly expressed as system A at the blood-brain barrier and acts as brain-to-blood efflux transport for L-proline. *Mol Pharmacol*. 2002;61(6):1289–1296.
66 Closs EI, Scheld JS, Sharafi M, Forstermann U. Substrate supply for nitric-oxide synthase in macrophages and endothelial cells: role of cationic amino acid transporters. *Mol Pharmacol*. 2000;57(1):68–74.
67 Dye JF, Vause S, Johnston T, et al. Characterization of cationic amino acid transporters and expression of endothelial nitric oxide synthase in human placental microvascular endothelial cells. *FASEB J*. 2004;18(1):125–127. Epub 2003 Nov 3.
68 Mcdonald KK, Zharikov S, Block ER, Kilberg MS. A caveolar complex between the cationic amino acid transporter 1 and endothelial nitric-oxide synthase may explain the arginine paradox. *J Biol Chem*. 1997;272(50):31213–31216.
69 Sala R, Rotoli BM, Colla E, et al. Two-way arginine transport in human endothelial cells: TNF-alpha stimulation is restricted to system y(+). *Am J Physiol Cell Physiol*. 2002;282(1):C134–143.
70 Thorens B. Facilitated glucose transporters in epithelial-cells. *Annu Rev Physiol*. 1993;55:591–608.
71 Kotyk A, Amler E. Na,K-adenosine-triphosphatase: the paradigm of a membrane-transport protein. *Physiol Res*. 1995;44(5):261–274.
72 Doyle DA. Structural themes in ion channels. *Eur Biophys J*. 2004;33(3):175–179.
73 Heinz E, Sommerfeld DL, Kinne RKH. Electrogenicity of sodium L-glutamate cotransport in rabbit renal brush-border membranes – a reevaluation. *Biochim Biophys Acta*. 1988;937(2):300–308.
74 Storck T, Schulte S, Hofmann K, Stoffel W. Structure, expression, and functional-analysis of a Na^+-dependent glutamate aspartate transporter from rat-brain. *Proc Natl Acad Sci USA*. 1992;89(22):10955–10959.
75 Centelles JJ, Kinne RKH, Heinz E. Energetic coupling of Na-glucose cotransport. *Biochim Biophys Acta*. 1991;1065(2):239–249.
76 Firnges MA, Lin JT, Kinne RKH. Functional asymmetry of the sodium-D-glucose cotransporter expressed in yeast secretory vesicles. *J Membr Biol*. 2001;179(2):143–153.
77 Kinne R, Sommerfeld D, Heinz E. Modulation of sodium-cotransport systems by other ions. *Biophys Chem*. 1988;29(1–2):105–109.
78 Nishizaki T, Matsuoka T. Low glucose enhances Na+/glucose transport in bovine brain artery endothelial cells. *Stroke*. 1998;29(4):844–849.
79 Carruthers A. Facilitated diffusion of glucose. *Physiol Rev*. 1990;70(4):1135–1176.
80 Gould GW, Thomas HM, Jess TJ, Bell GI. Expression of human glucose transporters in xenopus oocytes – kinetic characterization and substrate specificities of the erythrocyte, liver, and brain isoforms. *Biochemistry*. 1991;30(21):5139–5145.
81 Bauch C, Forster N, Loffing-Cueni D, et al. Functional cooperation of epithelial heteromeric amino acid transporters expressed in madin-darby canine kidney cells. *J Biol Chem*. 2003;278(2):1316–1322.
82 Fromm MF. P-glycoprotein: a defense mechanism limiting oral bioavailability and CNS accumulation of drugs. *Int J Clin Pharmacol Ther*. 2000;38(2):69–74.
83 Fromm MF. Importance of P-glycoprotein for drug disposition in humans. *Eur J Clin Invest*. 2003;33(2 Suppl):6–9.
84 Hegmann EJ, Bauer HC, Kerbel RS. Expression and functional activity of P-glycoprotein in cultured cerebral capillary endothelial cells. *Cancer Res*. 1992;52(24):6969–6975.
85 Homma M, Suzuki H, Kusuhara H, et al. High-affinity efflux transport system for glutathione conjugates on the luminal membrane of a mouse brain capillary endothelial cell line (MBEC4). *J Pharmacol Exp Ther*. 1999;288(1):198–203.
86 Kusuhara H, Suzuki H, Naito M, et al. Characterization of efflux transport of organic anions in a mouse brain capillary endothelial cell line. *J Pharmacol Exp Ther*. 1998;285(3):1260–1265.
87 Seetharaman S, Barrand MA, Maskell L, Scheper RJ. Multidrug resistance-related transport proteins in isolated human brain microvessels and in cells cultured from these isolates. *J Neurochem*. 1998;70(3):1151–1159.
88 Zhang W, Mojsilovic-Petrovic J, Andrade MF, et al. Expression and functional characterization of ABCG2 in brain endothelial cells and vessels. *FASEB J*. 2003;17(14):2085–2087. Epub 2003 Sep 4.
89 Schinkel AH. The roles of P-glycoprotein and MRP1 in the blood-brain and blood-cerebrospinal fluid barriers. *Adv Exp Med Biol*. 2001;500:365–372.
90 Sun H, Dai H, Shaik N, Elmquist WF. Drug efflux transporters in the CNS. *Adv Drug Deliv Rev*. 2003;55(1):83–105.
91 Tanigawara Y. Role of P-glycoprotein in drug disposition. *Ther Drug Monit*. 2000;22(1):137–140.
92 Kinne RKH. Endothelial and epithelial cells: general principles of selective vectorial transport. *Int J Microcirc Clin Exp*. 1997;17(5):223–230.
93 Prieto P, Blaauboer BJ, de Boer AG, et al. Blood-brain barrier in vitro models and their application in toxicology. The report and recommendations of ECVAM Workshop 49. *Altern Lab Anim*. 2004;32(1):37–50.
94 Nagy Z, Vastag M, Kolev K, et al. Human cerebral microvessel endothelial cell culture as a model system to study the blood-brain interface in ischemic/hypoxic conditions. *Cell Mol Neurobiol*. 2005;25(1):201–210.
95 Sambuy Y, De AI, Ranaldi G, et al. The Caco-2 cell line as a model of the intestinal barrier: influence of cell and culture-related factors on Caco-2 cell functional characteristics. *Cell Biol Toxicol*. 2005;21(1):1–26.

96 Rao RM, Shaw SK, Kim M, Luscinskas FW. Emerging topics in the regulation of leukocyte transendothelial migration. *Microcirculation.* 2005;12(1):83–89.
97 Hordijk P. Endothelial signaling in leukocyte transmigration. *Cell Biochem Biophys.* 2003;38(3):305–322.
98 McIntyre TM, Prescott SM, Weyrich AS, Zimmerman GA. Cell-cell interactions: leukocyte-endothelial interactions. *Curr Opin Hematol.* 2003;10(2):150–158.
99 Edens HA, Parkos CA. Modulation of epithelial and endothelial paracellular permeability by leukocytes. *Adv Drug Deliv Rev.* 2000;41(3):315–328.
100 Wolburg H, Lippoldt A. Tight junctions of the blood-brain barrier: development, composition and regulation. (Review) *Vascul Pharmacol.* 2002;38(6):323–337.
101 Parkos CA, Colgan SP, Madara JL. Interactions of neutrophils with epithelial cells:lessons from the intestine. (Review) *J Am Soc Nephrol.* 1994;5(2):138–152.
102 Weinbaum S, Zhang XB, Han YF, et al. Mechanotransduction and flow across the endothelial glycocalyx. *Proc Natl Acad Sci USA.* 2003;100(13):7988–7995.
103 Drenckhahn D, Ness W. The endothelial contractile cytoskeleton. In: Born GVR, Schwartz CJ, eds. *Vascular Endothelium: Physiology, Pathology, and Therapeutic Opportunities.* Stuttgart: Schattauer; 1997:1–25.
104 Davies PF. Flow-mediated endothelial mechanotransduction. (Review). *Physiol Rev.* 1995;75(3):519–560.
105 Mullin JM. Epithelial barriers, compartmentation, and cancer. *Sci STKE.* 2004;2004(216):E2.
106 Garlick DG, Renkin EM. Transport of large molecules from plasma to interstitial fluid and lymph in dogs. *Am J Physiol.* 1970; 219(6):1595–1605.
107 Goldstein GW, Betz AL. The blood-brain-barrier. *Sci Am.* 1986; 255(3):74–83.
108 Levick JR, Smaje LH. An analysis of the permeability of a fenestra. *Microvasc Res.* 1987;33(2):233–256.
109 Milici AJ, Watrous NE, Stukenbrok H, Palade GE. Transcytosis of albumin in capillary endothelium. *J Cell Biol.* 1987;105(6): 2603–2612.
110 Pappenheimer JR, Renkin EM, Borrero LM. Filtration, diffusion and molecular sieving through peripheral capillary membranes a contribution to the pore theory of capillary permeability. *Am J Physiol.* 1951;167(1):13–46.
111 Pardridge WM. Recent advances in blood-brain barrier transport. *Annu Rev Pharmacol Toxicol.* 1988;28:25–39.
112 Predescu D, Palade GE. Plasmalemmal vesicles represent the large-pore system of continuous microvascular endothelium. *Am J Physiol.* 1993;265(2):H725–H733.
113 Renkin EM, Crone C. Microcirculation and capillary exchange. In: Greger R, Windhorst U, eds. *Comprehensive Human Physiology*. Berlin: Springer; 1996:1965–1979.
114 Simionescu M, Simionescu N. Endothelial transport of macromolecules – transcytosis and endocytosis – a look from cell biology. *Cell Biol Rev.* 1991;25(1):5–80.
115 Gnoth MJ, Rudloff S, Kunz C, Kinne RKH. Investigations of the in vitro transport of human milk oligosaccharides by a Caco-2 monolayer using a novel high performance liquid chromatography-mass spectrometry technique. *J Biol Chem.* 2001;276 (37):34363–34370.
116 Predescu D, Vogel SM, Malik AB. Functional and morphological studies of protein transcytosis in continuous endothelia [Review]. *Am J Physiol Lung Cell Mol Physiol.* 2004;287(5):L895–L901.
117 Tamai I, Tsuji A. Transporter-mediated permeation of drugs across the blood-brain barrier. *J Pharm Sci.* 2000;89(11):1371–1388.
118 Risau W. Differentiation of blood-brain barrier endothelium. *News Physiol Sci.* 1989;4:151–153.

71

Electron Microscopic–Facilitated Understanding of Endothelial Cell Biology

Contributions Established during the 1950s and 1960s

Ann M. Dvorak

Beth Israel Deaconess Medical Center, Harvard Medical School, Boston, Massachusetts

The application of electron microscopy to biological research initially required intensive investigation directed toward improvements in instrumentation and specimen preparation. As these investigations progressed, ultrastructural descriptions of the cell, different cells, and subcellular organelles common to cells and those which are distinctive of specific cell lineages emerged. Application of ultrastructural studies to the endothelium facilitated understanding of endothelial cell (EC) biology. In this chapter, we review the contributions of electron microscopy to this understanding, which was established during the 1950s and 1960s.[1]

CONTRIBUTIONS ESTABLISHED DURING 1950–1959

The endothelium provides a physical barrier between the vascular lumen and the extravascular tissue space. All blood vessels are lined by interconnected ECs that comprise the endothelial layer. Before the electron microscope was developed, knowledge of EC structure and function was based primarily on physiological and light microscopic studies. One of the primary uses of the electron microscope was to confirm or deny the prevalent theories and findings obtained with these methods. Thus, by the late 1940s, physiologists had modeled the necessity for pores of small and large sizes in the endothelium to explain permeability data (1), and light microscopic studies proposed an intercellular cement to account for the connection between adjacent ECs (2). As well as re-examining such issues with the electron microscope, studies proceeded on the anatomic organization, heterogeneity of different vascular beds, classification of types of microvessels, and description of individual cellular components of small vessels and of the organelle content of microvascular endothelium (3). Early studies using tracers to probe permeability possibilities began during the 1950s, as well as studies of the mediator-induced leaking vessels characteristic of inflammatory changes involving small vessels (4).

Ultrastructural Description of the Endothelium

Defining *capillary* to include any small blood vessel, *endothelium* to refer to the lining of any blood or lymphatic channel, and *EC* to be the cell lining such channels (3) set the operational parameters for ultrastructural studies during the 1950s. Further refinement of these definitions included the size of vessels and their investiture with smooth muscle cells, pericytes (5), and elastic tissue, thus separating the thin-walled, small, smooth muscle cell–free capillaries from the larger, smooth muscle- and elastica-encased arterioles. Large venules with variable areas of thin and thick endothelium and pericyte investiture were also distinguished by electron microscopy.

Electron microscopists examined vessels from multiple organ sites and established that the microvasculature was of several characteristic types. These were classified as continuous (with or without large fenestrae) or discontinuous (3). These characterizations have held up over time, and such vessels typically can be found in different vascular beds. For example, discontinuous endothelium has large openings between individual ECs and is typified by sinusoid endothelium in the liver; continuous fenestrated endothelium is found in endocrine

1 This review covers two decades (1950–1969) that were instrumental to defining the ultrastructural anatomy of ECs. These studies provided the background for an explosion of information made possible by the ultrastructural studies regarding EC biology spanning 1970–2005, which are not covered here. Some later electron microscopic studies done in the author's laboratory during the past 10 years directly answer questions raised by the early studies and are referenced here. The apparent gap in citations of ultrastructural work (1970–2005) reflects only the assignment and space constraints. Electron microscopists continue to enrich our knowledge of EC biology and will certainly continue to do so in the future.

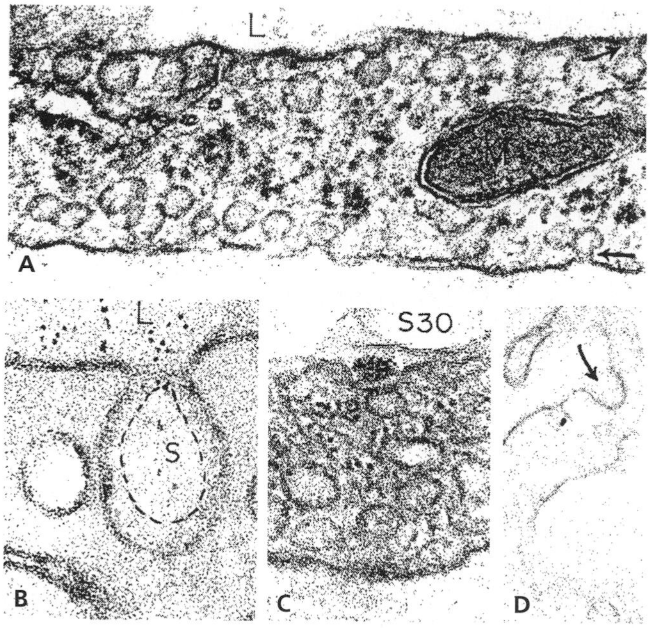

Figure 71.1. EC plasmalemmal vesicles. (**A**) Mouse skin capillary EC shows PVs attached to the luminal (*L*) and abluminal plasma membranes. Free and membrane-bound ribosomes and a mitochondria (*M*) fill in the central cytoplasm. Some PVs display elongated, narrow necks that connect them to the surfaces of this EC (*arrows*). (**B**) Guinea pig skin capillary PV (intravenous anionic ferritin) shows a thin diaphragm closing the luminal (*L*) aspect of this flask-shaped structure attached by a narrow neck to the plasma membrane. Ferritin is plentiful within the vessel lumen; several ferritin particles have entered the PV through a stoma (s) (*outlined with a dotted line in this unstained image*). (**C**) Mouse skin capillary PV (intravenous cationized ferritin, sample obtained 30 minutes after local exposure to serotonin) shows numerous electron-dense cationized ferritin particles bound to the luminal surface of a PV diaphragm, which restricts entry of the tracer into the PV. (**D**) Mouse skin microvessel EC in an induced angiogenic site and prepared with an immunonanogold method shows silver-enhanced gold particles indicating vascular endothelial growth factor receptor 2 (VEGFR2) in a PV attached to the luminal plasma membrane. A large coated pit is also similarly attached but is not labeled (*arrow*). L, lumen. (**A**) ×109,000; (**B**) ×210,000; (**C**) ×137,000; (**D**) Bar = 140 nm. (A, B, and C are reproduced with permission from Feng D, Nagy JA, Dvorak HF, Dvorak AM. Ultrastructural studies define soluble macromolecular, particulate, and cellular trans-EC pathways in venules, lymphatic vessels, and tumor-associated microvessels in man and animals. *Microsc Res Tech.* 2002;57:289–326. Figure D is reproduced with permission from Feng D, Nagy JA, Brekken RA, et al. Ultrastructural localization of the vascular permeability factor/vascular endothelial growth factor (VPF/VEGF) receptor-2 (FLK-1, KDR) in normal mouse kidney and in the hyperpermeable vessels induced by VPF/VEGF-expressing tumors and adenoviral vectors. *J Histochem Cytochem* 2000;48:545–556.)

organs and continuous nonfenestrated endothelium lines skin vessels.

The increased resolution afforded by the electron microscope allowed a description of EC organelles. Thus, the first description was possible of EC plasmalemmal vesicles (caveolae) (Figure 71.1) attached to luminal and abluminal surfaces and apparently free in the cytoplasm of ECs (6,7). Other organelles described included mitochondria, Golgi apparatus, dense bodies, endoplasmic reticulum, and intracellular fibrils (6). A basement membrane composed of a feltwork of 10-nm fine fibrils was shown to encase the abluminal endothelial surface (6).

Areas of contact between adjacent ECs were described as specialized attachments that served as barriers between the vessel lumen and the perivascular space (6), heralding future studies that definitively describe EC junctions (discussed in the next section).

Where Is the Cement?

Starling (8) thought that "cracks between ECs filled with cement" were probable leakage sites for fluid escaping from the vascular lumen. During the 1940s, intercellular cement between ECs was in vogue (2), and changes in this cement were implicated as a mechanism for vascular leakage. Electron microscopists working during the 1950s, however, did not find any intercellular cement (9), thus refuting the idea of filtration slits between ECs.

Where Are the Holes?

During the late 1940s, physiologists modeled capillary permeability and proposed a "pore theory" for capillary leakage (1). This model stipulated that endothelia, of necessity, must be perforated by pores of essentially two sizes, termed *large pores and small pores* (1). Enter the new field of electron microscopy during the 1950s, and a considerable effort to find these pores ensued. Pores of the dimensions postulated were not found by electron microscopists to perforate endothelium (3). Subsequent efforts by electron microscopists began to investigate further the possible role(s) of plasmalemmal vesicles and/or interendothelial attachments within the controlled, known permeability properties and barrier functions of the endothelium.

What about Inflammation?

As electron microscopists examined the microvasculature in search of static holes in the endothelium, to prove physiological principles and implicate the endothelium as an inert barrier to fluid exchange dictated by pore size, and failed to do so, others used the electron microscope to investigate the role of endothelial changes in inflammation typified by increased permeability (4). These early electron microscopic studies used colloidal particles (mercuric sulfide variably sized from 5 to 25 nm) as a tracer for permeability pathway(s) across dermal capillaries topically exposed to histamine (4). Although the image quality was hampered by primary osmium fixation, embedment in methacrylate, and sectioning with glass knives, some conclusions were possible. Thus, a transendothelial cell pathway for leakage of colloidal particles was found, rather than a paraendothelial cell pathway, and those EC organelles

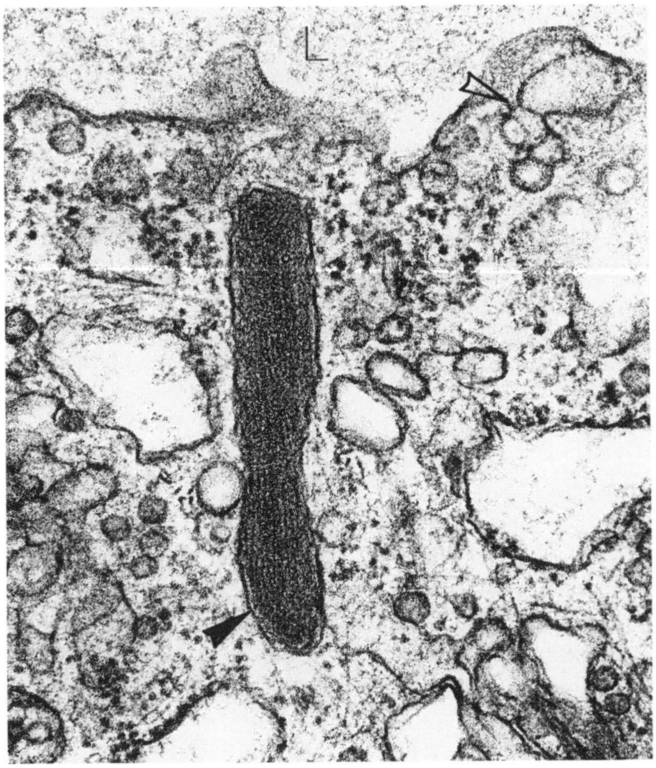

Figure 71.2. Weibel-Palade body. An elongated WPB (*closed arrowhead*) oriented at a right angle to the EC apical plasma membrane and vascular lumen (*L*) of a tumor-associated venule in a human diaphragm sample. Note that the WPB is enveloped by a membrane and is filled with small tubules. Most of the tubules are located longitudinally; a few are seen in cross-section. Also evident in this photo is a VVO (*open arrowhead*) attached to the plasma membrane. This cluster shows four vesicles connected by knobbed diaphragms, necks, and stomata, ×95,000. (Reproduced with permission from Dvorak AM. Weibel-Palade body. *Blood* 1996;87:409.)

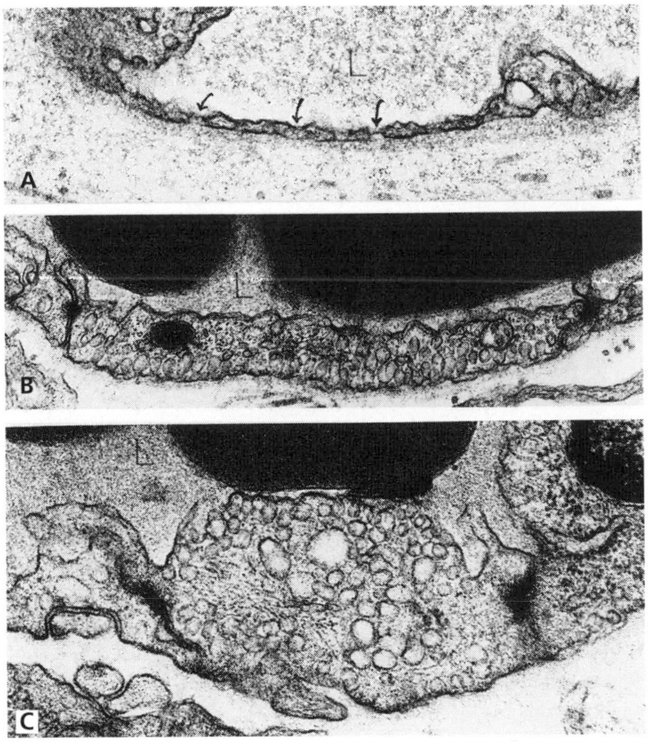

Figure 71.3. Fenestrated endothelium. In **A**, fenestrated endothelium lines a tumor-associated microvessel from the same sample imaged in Figure 71.2. Note that the thin endothelium is periodically perforated by large diaphragm-guarded fenestrae (*arrows*). By contrast, the capillary in **B**, from a mouse peritoneal sample, is not fenestrated. The EC contains numerous PVs, most of which are attached to the abluminal plasma membrane. Note the short, closely apposed lateral endothelial borders, which also show focal adherens junctions. **C**, a sample similar to **B**, shows a venule, also from a nonfenestrated microvessel, that has a thicker endothelium than that in panels **A** and **B**, filled with a VVO between two adherens junctions involving three ECs. L, lumen. (**A**) ×66,000; (**B**) ×53,000; (**C**) ×70,500.

in the cytoplasm containing tracer were all membrane-bound, including vesicles, folds, clefts, and caveolae (plasmalemmal vesicles). Of particular interest, despite exposure to histamine and the rapid onset of edema at the injection site, electron microscopic examination of inter-EC attachments showed no widening or gap formation in kinetically acquired samples (4). Moreover, no holes were found in the continuous endothelium lining these vessels.

However, complex intracytoplasmic membrane-bound structures honeycombed the endothelial cytoplasm in histamine-treated vessels. Some of the vesicles comprising these structures were as large as 200 nm in diameter. Rarely, perforating channels containing tracer were noted and interpreted to be cytoplasmic membranous structures rather than altered intercellular junctions, because no thickenings of the membrane, contacts, or attachment belts were associated with them (4). During the 1950s, these early electron microscopic descriptions of permeability pathways in inflammation involving skin microvessels served as a prelude to the electron microscopic studies of the 1960s, which discovered tracer-filled plasmalemmal vesicle transport across ECs (10). During the 1990s, discoveries were made, using sophisticated instrumentation, ultrathin serial sections, tracer technology, and computer-assisted three dimensional reconstructions to clarify EC cytoplasmic vesiculo-vacuolar organelles (VVOs) as a permeability pathway between the blood vascular space and the extravascular compartment (11,12).

CONTRIBUTIONS ESTABLISHED DURING 1960–1969

By the end of the 1950s, electron microscopic studies had largely established the ultrastructural components and organization of capillaries, arterioles, and venules. The array of endothelial organelles identified in the microcirculation included the first description of plasmalemmal vesicles (see Figure 71.1) (6), and early tracer experiments of allergic inflammation defined a transcellular pathway for leakage through ECs (4) and, as proposed earlier (6) possible transcellular transport via plasmalemmal vesicles (4). Importantly,

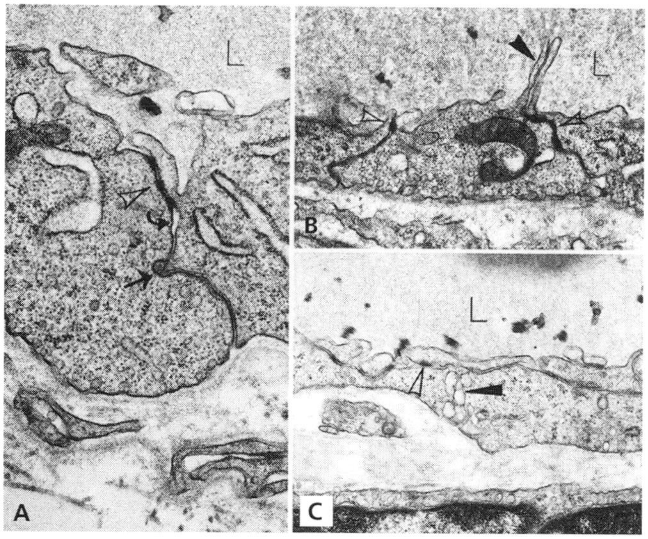

Figure 71.4. EC lateral borders. Mouse skin samples imaged using routine ultrastructural methods after intravenous injection of colloidal carbon. (Note, in all three panels, carbon particles are present in the vessel lumen but not in ECs or in extravascular tissues). The lateral plasma membranes of ECs are closely apposed and, in these medium-magnification images, areas of occludens and adherens junctions show enhanced electron density and are located in the apical one-third of the lateral borders (*open arrowheads*). In **A**, a coated vesicle is attached to the lateral plasma membrane (*straight arrow*) of one EC beneath the specialized junctional complexes and the widened intercellular cleft (*curved arrow*). In **B**, elongated EC flaps (also called marginal folds) guarding the lateral border of two ECs protrude into the vessel lumen (*closed arrowhead*). In **C**, an interconnected cluster of vesicles (*closed arrowhead*) spans the cytoplasm of one EC from immediately below the occludens junction (also called the tight junction; *open arrowhead*) from the lateral plasma membrane to the abluminal (also called basilar) plasma membrane. This structure has been termed the EC VVO. (See text and Ref. 11). L, lumen. (**A**) ×37,000; (**B**) ×22,000; (**C**) ×28,000.

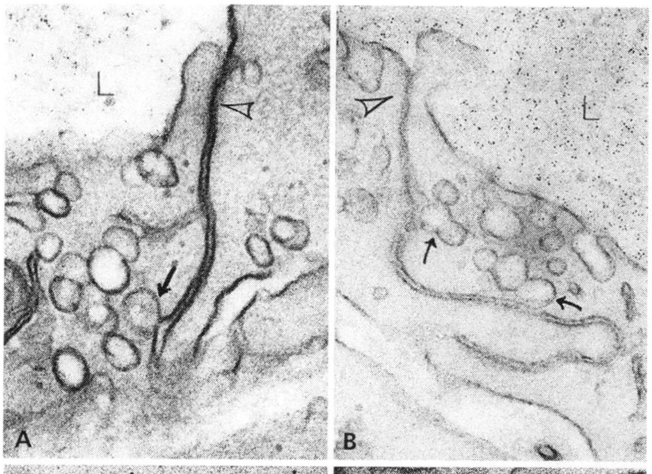

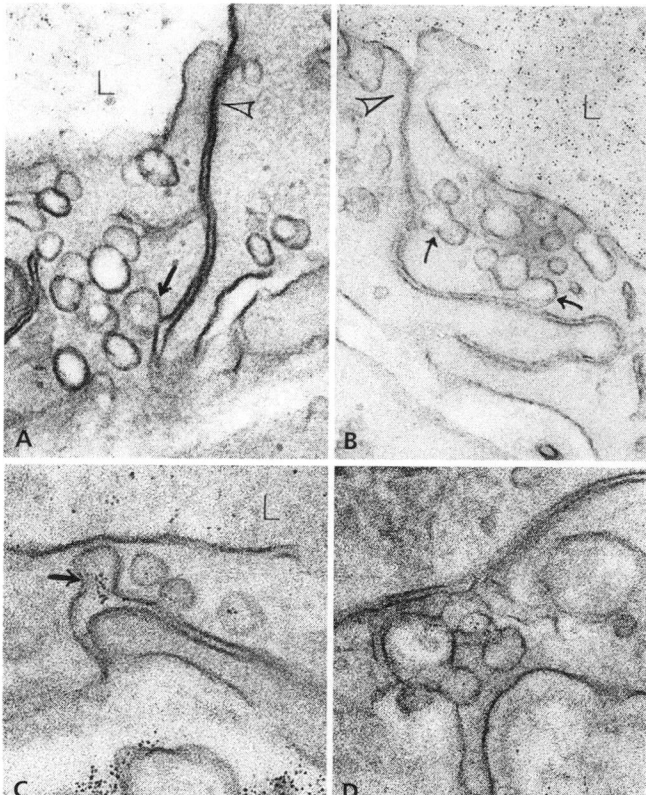

Figure 71.5. Lateral borders of unstained ECs processed using a potassium ferrocyanide-reduced osmium postfixation step. Mouse (**A**), guinea pig (**B, D**), and rat (**C**). Skin samples were taken after anionic ferritin was given intravenously in animals receiving skin injections of saline (**A**), histamine (**B, D**), or VEGF (**C**). With this method and magnification, the lateral plasma membranes are electron dense, allowing visualization of the closeness of membranes in the occludens junctions (*A, B arrowheads*). The subadjacent cytoplasmic densities associated with adherens junctions are poorly visible with this method, thus allowing better imaging of the tight junctional areas. Ferritin is visible in the lumens of vessels (**A–C**). In **A**, a single PV (*arrow*) is attached to the lateral membrane of one EC. In **B**, two clusters of vesicles (*arrows*) are similarly attached. In **C**, PVs contain ferritin – one has fused with the lateral border of one EC, and extravasated ferritin lies in the extravascular space below. In **D**, a vesicle cluster (VVO) connected to the lateral border of one EC shows diaphragms, stomata, and necks. The stomata and necks contain ferritin, and the entire structure connects the lateral and basilar (also called abluminal) plasma membranes of this EC. (**A**) ×107,000; (**B**) ×99,000; (**C**) ×155,000; (**D**) ×120,000.

the new imaging capabilities definitively denied the existence of intercellular cement (2) and static EC pores (1). During the 1960s, these findings in aggregate fueled the search (and development of tools to facilitate this search) for permeability pathways in the microcirculation (13–41). The extensive morphological studies of the 1960s also provided the first description of a new EC-specific organelle, the Weibel-Palade body (Figure 71.2) (42), a more complete description of inter-EC junctions (3,15–18,43–47) and EC suborganellar characteristics that define plasmalemmal vesicles (see Figure 71.1) (19–23,44,48), and EC fenestrae (Figure 71.3) (23,48,49). In addition, both the fluid phase and the cellular phase of acute inflammation were probed by electron microscopists to better understand possible pathways of egress from vessels during acute inflammatory reactions (24–26,50–55).

More about the General Anatomy of Endothelial Cells

As the electron microscopes of the world were turned on, more and more cellular organelles and specializations were defined generally. Many of these also were recorded in ECs. Thus, in addition to those described during the 1950s, EC structures defined during the 1960s included an endocapillary layer stained with ruthenium red (56,57); EC flaps or marginal folds, guarding primarily the luminal introit to the aligned lateral EC borders (Figure 71.4) (37,44,46,47,58,59) (structures found to contribute to ingestion by ECs [17,51]); coated vesicles (Figures 71.1 and 71.4) (22,44,47); membrane-bound lysosome-like dense bodies (56,58,60–64); multivesicular bodies (37,38,47,56,61–66); mitochondria (61,62); Golgi apparatus (61,62); endoplasmic reticulum (61,62); ribosomes (61,62); centrioles (62); cytoplasmic filaments (60,65,67,68); microtubules (47,69,70); and lipid droplets (56,71,72).

Identification of a New Endothelial Cell–Specific Organelle by Electron Microscopy

The high-resolution analyses made possible by electron microscopy allowed electron microscopists to peer inside EC organelles and, in so doing, to describe an entirely new organelle that is unique to ECs (42). This organelle, an elongated, tightly membrane-bound, small, tubule-packed structure was named the Weibel-Palade body (Figure 71.2) for its discoverers (42). Several other early microscopists also noted differences between the Weibel-Palade body and other EC organelles (47,56). This EC-specific organelle has become the ultrastructural gold standard for the identification of ECs and is now known to play a key role in EC biology (73).

Inside the Fenestra

The term *fenestra* was initially used to define the unique opening in the type of thin endothelium that obtained its name from the presence of these fenestrae – that is, fenestrated endothelium (see Figure 71.3) (14,17,20,22,23,28,33,35,39,47–49,62, 74). These relatively large endothelial openings were found generally to be guarded by a single, thin electron-dense membrane called a diaphragm and to display an electron-dense knob in the center of this diaphragm (49). Some microvascular beds were said to have open fenestrae (no diaphragms) (28) but, as technology improved, most microvascular beds (including those in the kidney, which displayed fenestrated endothelia) were shown to have fenestrae closed by diaphragms (17,22, 48,49). Good preparations imaged en face at high magnification showed radial striations proceeding from the central knobs to fenestral edges (23).

Inside the Plasmalemmal Vesicle

EC plasmalemmal vesicles (PVs), also known as pinocytotic vesicles and caveolae, could well be called Palade vesicles in recognition of Dr. Palade's original description of them in 1953 (6). These small, flask-shaped structures are located attached to all three borders of ECs, as well as apparently free in the cytoplasm (Figures 71.1 and 71.3) (6,14–23,29,33,35,44,46–49,59–62,65,75). These locations and their anatomy inspired the hypothesis that they could participate in the transfer of material from the vascular lumina to the extracellular space (6) (discussed in the next section). Ultrastructural studies have defined those unique characteristics of PVs (see Figure 71.1) (14–23,29,35,44,46–49,56,59,61) that have proven useful in the recognition of PVs as participants in chain and cluster formations with permeability properties (76,77). PVs attached to luminal, lateral, or abluminal plasma membranes show constrictions at these attachments where a thin single membrane, also called a diaphragm, closes the vesicle opening. These

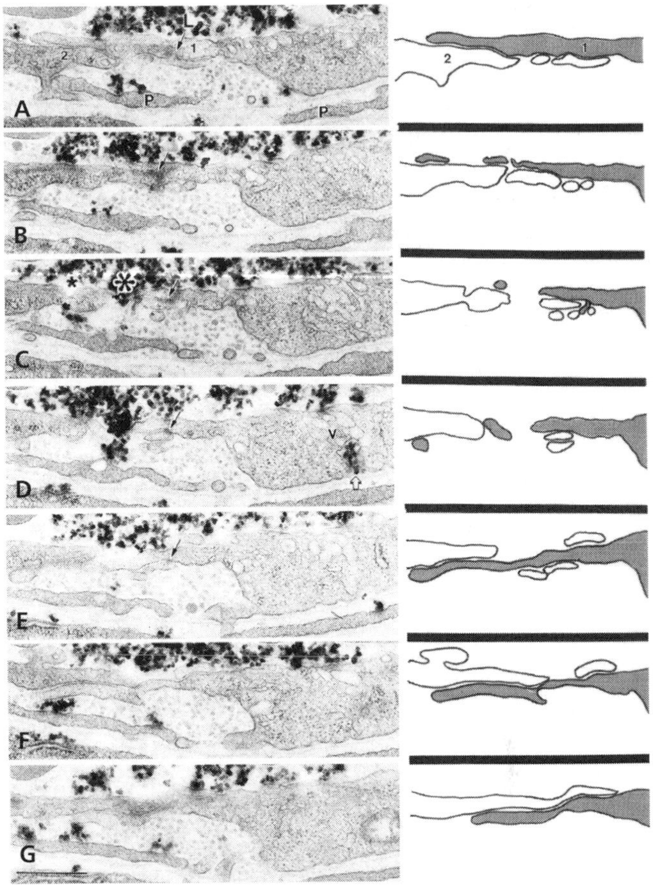

Figure 71.6. Colloidal carbon-filled trans-EC pores in a venule of rat cremaster muscle. Colloidal carbon was injected intravenously and tissue was harvested 1 min after local injection of 50 ng VPF/VEGF. *Left*, **A–G** illustrate sections 29 through 35 from a series of 41 consecutive 100-nm thick serial sections. *Right*, corresponding tracings of EC plasma membranes. Two transcellular pores are encompassed in these seven serial sections. The left and smaller pore illustrated only in **C** (*small star*) passes through a single EC (labeled *2*). The diameter of this pore was <100 nm, and it was encompassed entirely within a single section (**C**); therefore, the tracing of section **C** (*right panel*) does not indicate a complete transendothelial opening because it traces plasma membrane superficial and deep to the pore. The larger pore (*large star*, **C**) crosses through portions of two overlapping ECs (labeled *1* and *2*) in **C** and **D**. A VVO (*V*) that extends across EC 1 is followed in sections **A–F**; in **D**, carbon particles are present in a portion of this structure that opens to the endothelial ablumen (*open arrow*). The intercellular cleft and occludens-type junctions (*arrows*) joining the two ECs are closed normally at all levels. Note that, in the course of these serial sections, EC 1, which is luminal in **A–C**, dips beneath EC 2 in **D–G**; i.e., peripheral portions of the two ECs interdigitate, forming a woven pattern. L, lumen; P, pericyte. Scale bar (applies to **A–G**), 0.6 μm. (Reproduced with permission from Feng D, Nagy JA, Hipp J, et al. Reinterpretation of EC gaps induced by vasoactive mediators in guinea-pig, mouse and rat: many are transcellular pores. *J Physiol.* 1997;504:747–761.)

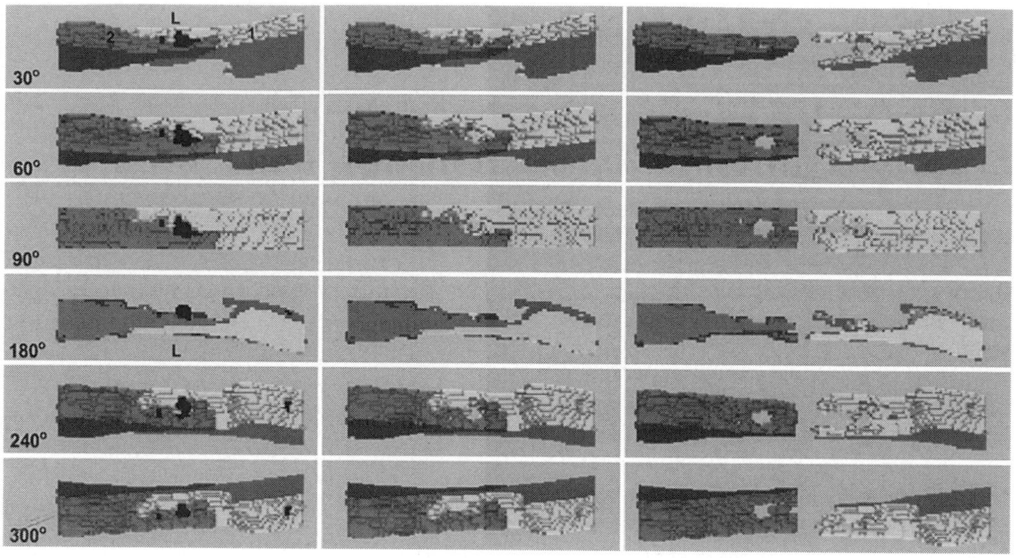

Figure 71.7. Computer-generated 3-D reconstruction of the pores illustrated in Figure 71.6. Panels represent successive rotations toward the viewer around a horizontal axis at the angles indicated. 0 degrees (not shown) corresponds to a vascular cross-section taken at right angles to the direction of blood flow and a 90-degree view looking down on the luminal surface; views ≥240 degrees illustrate the abluminal surface. ECs 1 and 2 represent cells correspondingly numbered in Figure 71.6. Left panel includes extravasating carbon, which is represented in black. Middle panel illustrates views corresponding to those in left panel, except that carbon was deleted from the reconstruction to facilitate visualization of the pore interior. In the right panel, the overlapping ECs are viewed separately. Two trans-EC pores are illustrated. The smaller (*left*) pore extends to the albumen by passing through only EC 2 and corresponds to the pore marked by the smaller star in Figure 71.6C; background is visible from both luminal (e.g., 90-degree) and abluminal (e.g., 240-degree) views. The larger pore passes through overlapping portions of both ECs 1 and 2 and corresponds to the pore designated by the large star in Figure 71.6C. Abluminal views (*left panel*, 240 degrees and 300 degrees) illustrate carbon present within abluminal portions of cell 1, corresponding to the carbon-containing VVO illustrated in Figure 71.6D. Note (*middle panels*, 240- and 300-degree views, here and for Figure 71.6D) that carbon does not extend through the entire thickness of cell 1. L, lumen. (Reproduced with permission from Feng D, Nagy JA, Hipp J, et al. Reinterpretation of EC gaps induced by vasoactive mediators in guinea-pig, mouse and rat: many are transcellular pores. *J Physiol.* 1997;504:747–761.) For color reproduction, see Color Plate 71.7.

diaphragms (also discussed in Chapter 72) also show central electron-dense knobs and radiating striations from the knobs to the vesicle edges. Variously elongated narrow necks attach vesicles to the plasma membranes.

The openings to PVs have variously been called *mouths, stomata, fenestrae,* and *apertures* (Figure 71.1). The unfortunate use of the term *fenestrae* to indicate these narrow, circular openings into PVs provides for possible confusion with their larger counterpart, also termed *fenestrae,* which typify fenestrated endothelium (see Figure 71.3) (as discussed earlier). We prefer to designate the openings into PVs as *stomata* and call attention to their similarity to the fenestrae of fenestrated endothelia, which also are guarded by knobbed diaphragms. An interesting property of PV stomata is their multiplicity. That is, individual PVs may display single, double, or many stomata, thus providing necessary connection points for the formation of small chains or large clusters of interconnected PVs in the endothelium (76,77).

This electron microscopic definition of the substructure of PVs during the 1950s and 1960s facilitated the recognition of interconnected chains of PVs during the 1970s (76) and clusters of PVs and larger vacuoles (VVOs) during the 1990s (11,77).

Junctionology

Originally, cells were thought to be surrounded by a cement that bound them together (2). The use of electron microscopy showed this not to be true and, early on, microscopists used a number of terms to allude to inter-EC connections. ECs are polarized cells with apical, basilar, and lateral borders. Intercellular connections populate the lateral borders, which in simple capillary endothelia are short, straight, and easier to study than the complex interdigitating and overlapped lateral borders of ECs in venules (78). Some of the terms used to describe lateral-border EC anatomy (see Figures 71.3 through 71.5) include attachment belts (2,3,28,59,62,79), adhesion plates (28,29), intercellular clefts (15,17,35,41,47), intercellular junctions (not otherwise specified) (39), interlocking inter-EC processes (78), quintuple-layered membrane junctions at

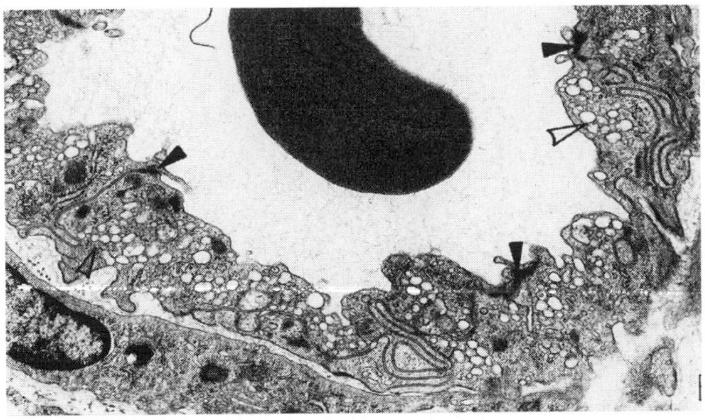

Figure 71.8. Control mouse skin venule. Numerous collections of cytoplasmic vesicles and vacuoles constitute the vesiculo-vacuolar organelle (VVO) (*open arrowheads*). Note that the inter-EC lateral borders show complex interdigitations; electron-dense tight and adherens junctions are located near the luminal aspect of the lateral borders of ECs (*closed arrowheads*), ×21,500. (Reproduced with permission from Feng D, Nagy JA, Dvorak HF, Dvorak AM. Ultrastructural studies define soluble macromolecular, particulate, and cellular trans-EC pathways in venules, lymphatic vessels, and tumor-associated microvessels in man and animals. *Microsc Res Tech.* 2002;57:289–326.)

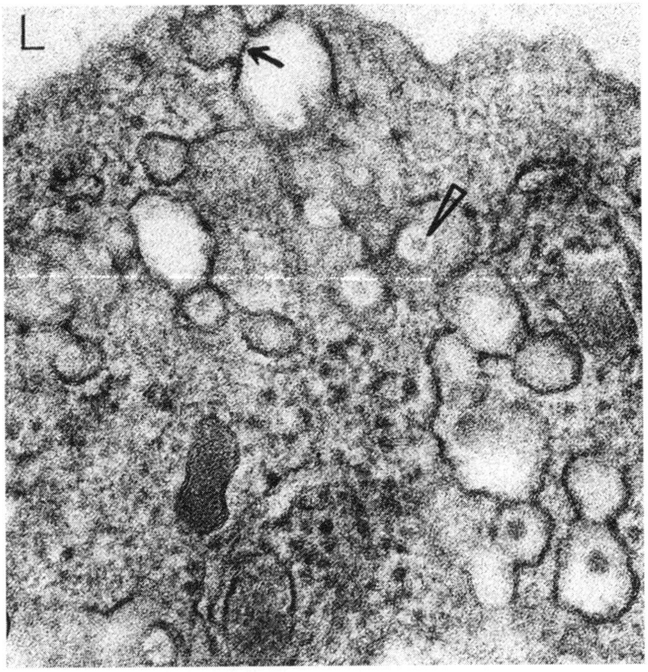

Figure 71.9. Control mouse skin venule EC shows portions of a VVO near the lumen (*L*). Individual vesicles and vacuoles are connected by thin diaphragms (*closed arrow*) and display electron-lucent stomata, many of which contain a central electron-dense knob (*arrowhead*). Numerous ribosomes fill the cytoplasm. ×109,000. (Reproduced with permission from Feng D, Nagy JA, Dvorak HF, Dvorak AM. Ultrastructural studies define soluble macromolecular, particulate, and cellular trans-EC pathways in venules, lymphatic vessels, and tumor-associated microvessels in man and animals. *Microsc Res Tech.* 2002;57:289–326.)

terminal bars (43), occludens junctions (tight junctions), and adherens junctions (15,16,23,44,46,47,61–63,65,67). Of these terms, the originally described quintuple-layered membrane junctions of Muir and Peters (43) also are known as occludens junctions and as tight junctions in EC lateral borders. They generally are located nearer to the apical surface than to the basilar surface, and they can be recognized by closely apposed membranes of variable lengths. The adherens junction, also seen between the lateral borders of ECs, has an intercellular space; straight, apposed membranes; and a dense condensation of the subjacent cytoplasm in each cell. The ultrastructural nomenclature for these specializations of EC lateral borders follows that originally defined for epithelial cells (45). The occludens (tight) and adherens junctions of ECs are structurally resemble those in epithelia. ECs (like some simple epithelial cells) do not display desmosomes.

Because inter-EC junctions (generally in capillaries) were studied using high magnifications and with a few partial series of sections, some favored that the junctions were closed by fused membranes, and others that they were narrowly open between the apposed lateral membranes (15,17,43,44). In fact, images reflecting both these situations are available that demonstrate the actual "tightness" of tight junctions.

Although electron microscopists have made great contributions to our understanding of inter-EC junctions, one should keep in mind that most studies were done in capillaries, not in the more complex venular lateral borders (78). Also, because the correct terminology has been provided for these specialized attachments, these terms should be used appropriately. That is, either an occludens (tight) (Figures 71.4 and 71.5) or adherens (Figure 71.4) junction can be called an inter-EC junction and both (as well as multiples of each) can be found separated by wider spaces between any given pair of ECs. These spaces have been referred to correctly as clefts, intercellular spaces, and sometimes as "gaps" (see Figures 71.4 and 71.5) (15,17,35,41,47). However, since the classic papers appeared (15,17,43–45), a vast literature has emerged that misuse the term *EC junction.* That is, *cell junction* is all too often used to refer to any portion or all of the EC lateral border (Figures 71.4 and 71.5). Based on the contributions of electron microscopy carried out during the 1960s, it is appropriate to use the terms *lateral borders* and *intercellular clefts,* reserving the usage of *junction* to indicate only the focal, specialized areas of adherens and occludens junctions regularly present between ECs at their lateral borders (see Figures 71.4 and 71.5).

Vesicular Transport

Plasmalemmal vesicles (also termed *pinocytotic vesicles, Palade vesicles,* or *caveolae*) are uniquely constructed and located to fill, fuse, and move between the three cell surfaces that comprise ECs (Figure 71.5) (6,21,22,29,34,44). Most reports of capillary endothelium find them attached to the plasma membranes of the basilar, apical, lateral borders. Some appeared to be free in the cytoplasm in the standard 70- to

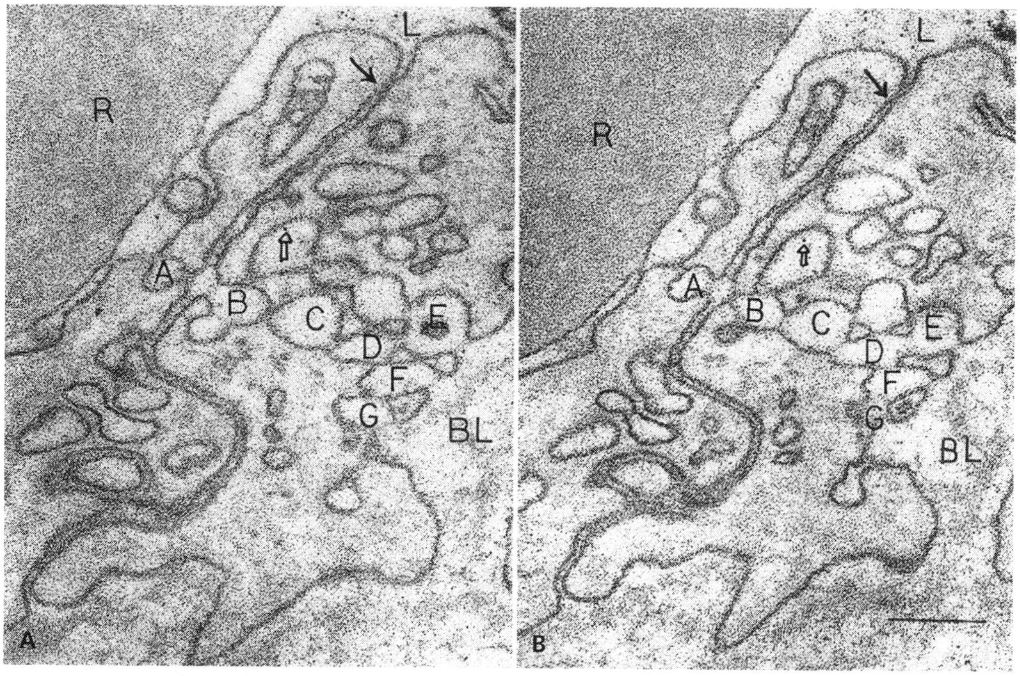

Figure 71.10. Two consecutive ultrathin (12 nm) sections, from a series of 61, of a control mouse skin venule. The mouse had been injected intravenously with ferritin 5 min prior to sacrifice and intradermally with HBSS 1 min prior to sacrifice. Together these sections demonstrate direct, open communication between VVOs present in adjacent ECs. A vesicle (**A**) in the left-hand cell connects with the intercellular cleft which, in turn, connects successively with vesicles and vacuoles (**B–G**) in the right-hand cell. Diaphragms close the stomata, linking some adjacent vesicles/vacuoles (for example, B–C, C–D, and D–F in panel B). Closed arrows indicate area of tight junction near the luminal surfaces. Scattered FE particles are present in the vascular lumen (*L*) and in the vesicle (*open arrow*) immediately above vesicle labeled B. R, red blood cell; BL, basal lamina. (**A, B**) ×88,000, scale bar = 0.2 μm. (Reproduced with permission from Feng D, Nagy JA, Hipp J, et al. Reinterpretation of EC gaps induced by vasoactive mediators in guinea-pig, mouse and rat: many are transcellular pores. *J Physiol.* 1997;504:747–761.)

80-nm single sections used for viewing by electron microscopy during the 1960s.

The utility of PVs for transendothelial transport first envisioned by Palade (6) was also first tested by Palade using 3- to 25-nm gold particles (27). He showed that colloidal gold injected intravenously entered vesicles attached to the luminal front, increased with time after injection, and later was found in the perivascular space. Gold particles were not present free in the cytoplasm of ECs and did not enter intercellular spaces (27).

Palade's group used another electron-dense tracer, ferritin (~11 nm), to image vesicle transport in capillary ECs (see Figures 71.1 and 71.5) (14). As with the gold tracer, PVs on both luminal and abluminal EC borders contained ferritin. Ferritin did not traverse interendothelial junctions (14). In an attempt to better understand vesicle connection possibilities in endothelia, Bruns and Palade (44) made a three-dimensional (3D) model in plastic of a short segment of capillary EC cytoplasm. The model was based on photographs of seven serial 50-nm sections. In this model, the authors found five vesicle pairs that were apparently fused. One set opened to the luminal front, another set to the abluminal front, and two sets did not open to either surface in this model. From this, the authors concluded that vesicles were not interconnected in chains or clusters from surface to surface (44). Later studies have, however, demonstrated such interconnectedness (11,76).

The 1960s showed numerous efforts by electron microscopists to model vesicles in their transport roles across endothelia. Thus, vesicles were said to be capable of motion by random Brownian movement (21,31,36). Others noted that another EC vesicle system – coated vesicles (Figure 71.4) – avidly took up ferritin in tracer experiments, and that multivesicular bodies in the cytoplasm of ECs stored ferritin (56,63,64).

All together, the ultrastructural tracer experiments done during the 1960s provided evidence for the vesicular transport of macromolecular colloids and electron-dense enzymes into and through capillary ECs (14,15,17–19,23,35,41). As such, the suggestion that the vesicular transport properties imaged using tracers fulfilled the criteria for the large pore system of physiologists (1) gained favor (14). Some (14,23) envisioned future studies in which interconnected chains (76) and clusters (11,77) of vesicles would be demonstrated to have constrictions corresponding to diaphragm insertions at PV-like

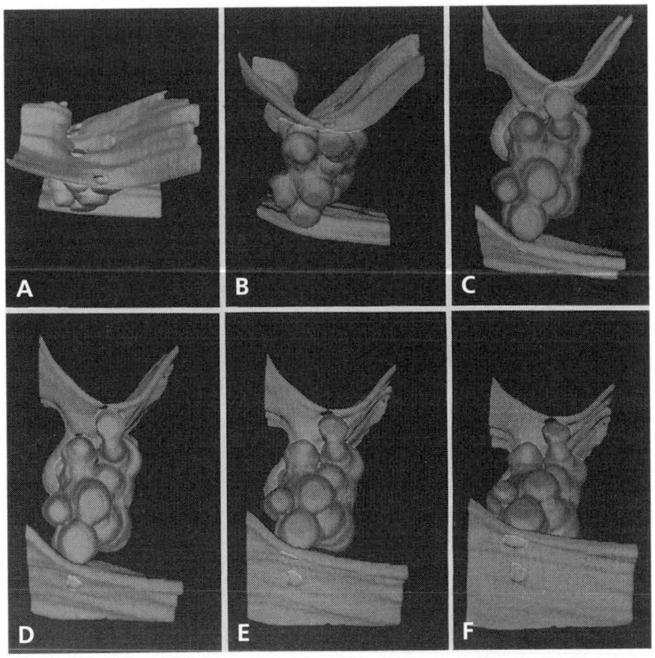

Figure 71.11. Computer-generated 3-D reconstruction of portion of a venular VVO from mouse skin injected 5 min earlier with 50 ng VPF. The interior volumes of VVO vesicles and vacuoles in successive electron micrographs were traced onto transparent overlays with reference marks to retain register. Tracings were digitized at a resolution of 5.9 nm/pixel. Because section thickness (15 nm) was greater than in-plane resolution, bicubic interpolation was performed between sections, and convolutions were used to smooth surfaces. (**A–F**) Portion of a VVO (15 consecutive serial, ultrathin EM sections illustrating 25 individual vesicles-vacuoles) reconstructed using Advanced Visual Systems (Waltham, MA) Software, here viewed in successive rotations around a horizontal axis at intervals of 30 degrees (except 15 degrees, [**C–D**]). There are two openings (**E, F**) to the vascular lumen and four to the abluminal surface (**A**). (Reproduced with permission from Feng D, Nagy JA, Hipp J, et al. Vesiculo-vacuolar organelles and the regulation of venule permeability to macromolecules by vascular permeability factor, histamine and serotonin. *J Exp Med.* 1996;183:1981–1986.) For color reproduction, see Color Plate 71.11.

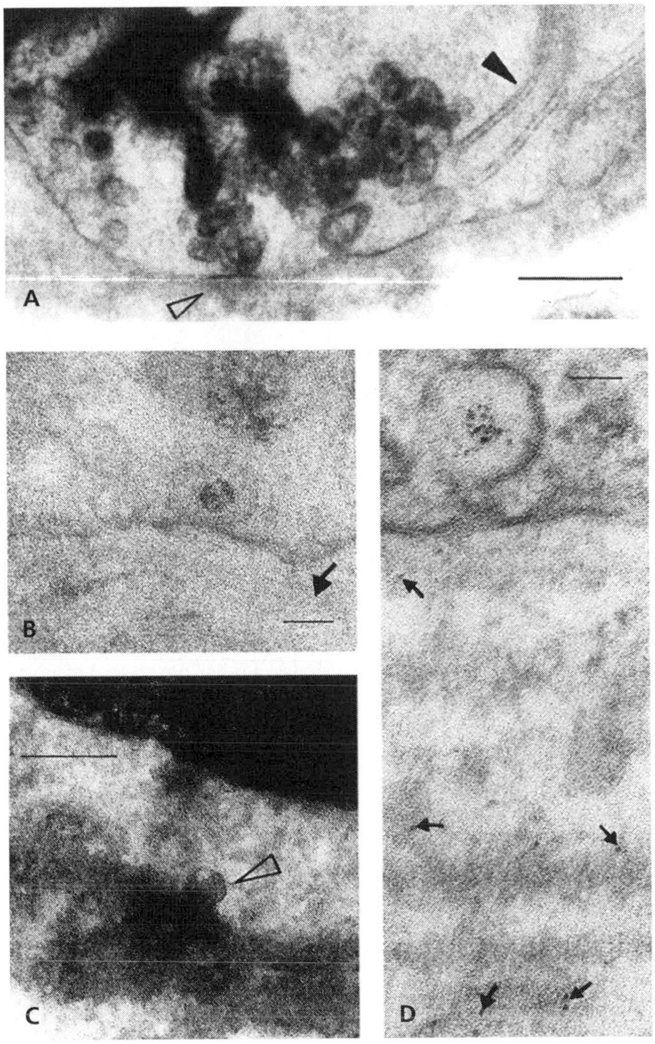

Figure 71.12. Electron microscopic tracer studies of trans-EC leakage through VVOs of mouse tumor microvascular ECs. (**A**) At 10 sec after IV injection of HRP, the electron-dense tracer fills the vessel lumen and a cytoplasmic VVO, which has released a small amount to the basal lamina focally (*open arrowhead*). The inter-EC junction (*closed arrowhead*) does not contain HRP. Bar = 260 nm. (**B**) A tumor EC VVO connected to the abluminal surface shows an open, HRP-containing stoma. The abluminal surface of the EC plasma membrane is focally stained with HRP beneath this cell; the basal lamina (*arrow*) remains unstained. Bar = 80 nm. (**C**) At 10 sec after IV HRP, several vesicles in this VVO contain electron-dense HRP, similar in intensity to the overlying HRP-filled vessel lumen. One HRP-loaded vesicle (*arrowhead*) is fused with the abluminal plasma membrane and has released a cloud of HRP focally into the underlying basal lamina and surrounding connective tissue. Bar = 154 nm. (**D**) At 30 min after IV injection of anionic ferritin, a VVO vesicle stoma at the basilar EC front is filled with ferritin particles. Individual ferritin particles (*arrows*) have dispersed widely in the extracellular matrix beneath this EC. Bar = 63 nm. (Reproduced with permission from Dvorak AM, Kohn S, Morgan ES, et al. The vesiculo-vacuolar organelle (VVO): a distinct endothelial cell structure that provides a transcellular pathway for macromolecular extravasation. *J Leukoc Biol.* 1996;59:100–115.)

necks (11,76,77). These constrictions within the fused vesicles and vacuoles could account for the small pore system of physiologists (1).

Junctional Transport

During the 1960s, a new era of ultrastructural studies was ushered in by the development of smaller probes for the investigation of EC permeability pathways (32,33). These studies made use of injected enzymes (horseradish peroxidase [HRP] was the prototype) that are rendered electron dense by an enzyme reaction and an affinity for osmium (32,3). Use of this approximately 4- to 5-nm probe with timed collection of samples after intravenous injections provided a potential probe for the physiologist's small pores. All together, this probe clearly entered the intercellular clefts, permeated the ~4-nm gaps in the EC tight junctions, and filled PVs in capillaries of muscle origin

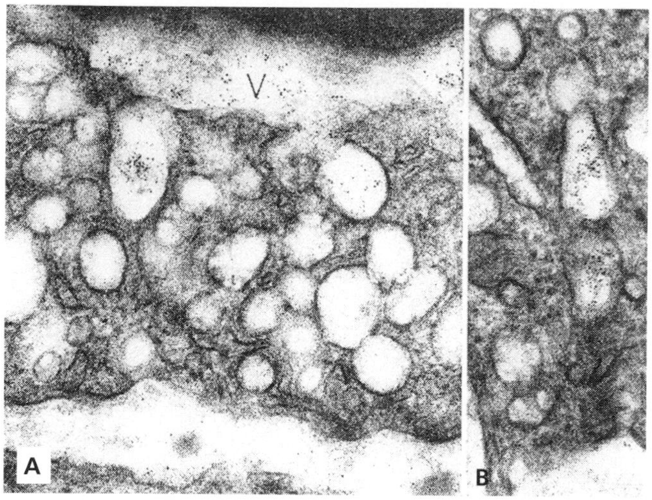

Figure 71.13. Mediator-induced VVO leakage demonstrated with FE in control skin venules. (**A**) Extravasation of FE by way of VVOs in guinea pig skin venules 20 min after intradermal injection of VPF/VEGF or (**B**) 10 min later in response to an intradermal injection of histamine. (**A**) Note FE particles entering VVO vesicles-vacuoles that open to the vascular lumen (*V*); (**B**) FE has progressed to mid-cytoplasmic VVO components. A few FE particles have spilled into the subendothelial spaces in (**A, B**). (**A**) ×109,000; (**B**) ×111,000. (Reproduced with permission from Feng D, Nagy JA, Pyne K, et al. Pathways of macromolecular extravasation across microvascular endothelium in response to VPF/VEGF and other vasoactive mediators. *Microcirculation.* 1999;6:23–44.)

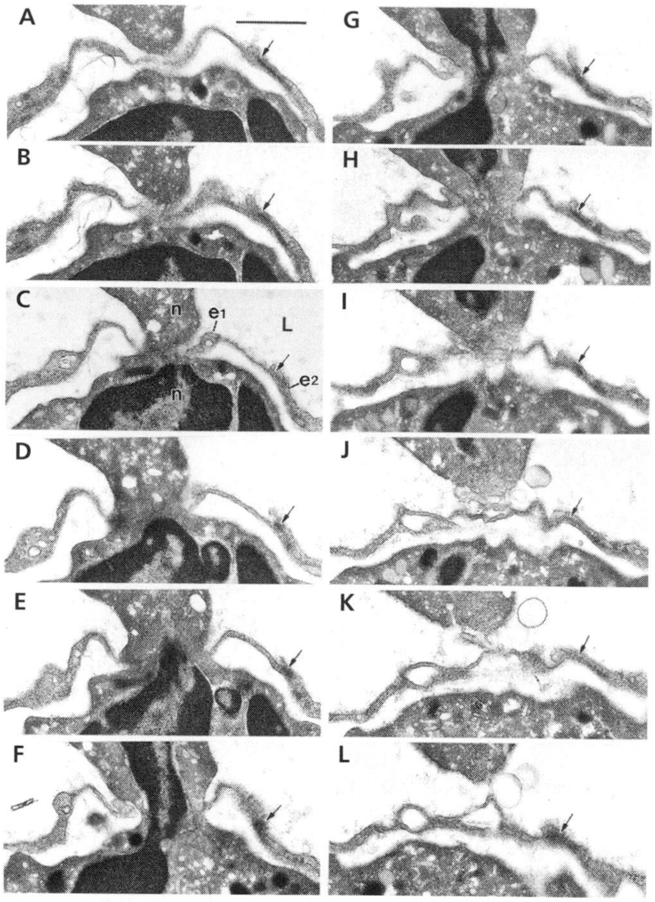

Figure 71.14. Transmigration of a neutrophil (*n*) across a thinned portion of venular endothelium at 60 min after local intradermal injection of 10^5 M FMLP. Twelve sections (numbers 13, 14, 16, 19, 21, 24, 25, 27, 29, and 30–32) of a series of 74 consecutive serial sections are illustrated. Portions of the neutrophil's nucleus (**E–G**) are included within the pore. The pore passes through a single EC (*e1*), but a junction of e1 and with a second EC, e2, is indicated (*arrow*); this junction is intact and maintained a distance of $>1\mu$m from the pore margin at all levels of sectioning. L = lumen. Bar = 1 μm. (Reproduced with permission from Feng D, Nagy JA, Pyne K, et al. Neutrophils emigrate from venules by a trans-EC pathway in response to FMLP. *J Exp Med.* 1998;187:903–915.)

(15,17,19,41). Similar studies of brain microvessels, which are normally poorly permeable, showed tracer in a vastly reduced PV population and none in tight junctions between ECs (16). In neither location in brain endothelium did HRP pass to the interstitium through the endothelial barrier (16).

During the 1960s, new tracer studies done using the prototypes ferritin and HRP lent credibility to the concept of endothelial holes that are important for permeability purposes. They also, however, drew further attention to the heterogeneity of vascular beds within an organ and between different organs, and to the necessity for future studies to probe these differences carefully.

More about Inflammation

Humoral Phase

Electron microscopists working during the 1960s tended to discount the earliest ultrastructural examination of the humoral phase of inflammation (4). In the early studies, changes involving the lateral endothelial borders or hole development through the cytoplasm of ECs were not found after direct exposure to histamine (4). Complex membrane-bound cytoplasmic intracytoplasmic structures were thought to supply a trans-EC pathway for stimulated macromolecular leakage in acute inflammation (4). Although a tracer (colloidal mercury) was used in the early study (4), during the 1950s, general technical considerations in electron microscopy sample preparations did not allow for the production of excellent images of skin. In 1961, another group initiated ultrastructural studies of acute inflammation and stated for the first time that histamine induced "dissecting aneurysms of the vessel wall," otherwise referred to as *gap formation* (20). (These gaps are not to be confused with the use of this term to describe spaces that occur normally between the lateral borders of apposed ECs.) In inflammation, the newly developed gaps are said to develop as stimulated ECs pull apart, thus providing a gap for the egress of plasma (20).

At this point, the use of colloidal carbon as a tracer for vascular injury was applied to electron microscopy (Figures 71.4, 71.6, and 71.7; for color reproduction, see Color Plate 71.7) (24–26,40). In inflammation, these particles are sufficiently

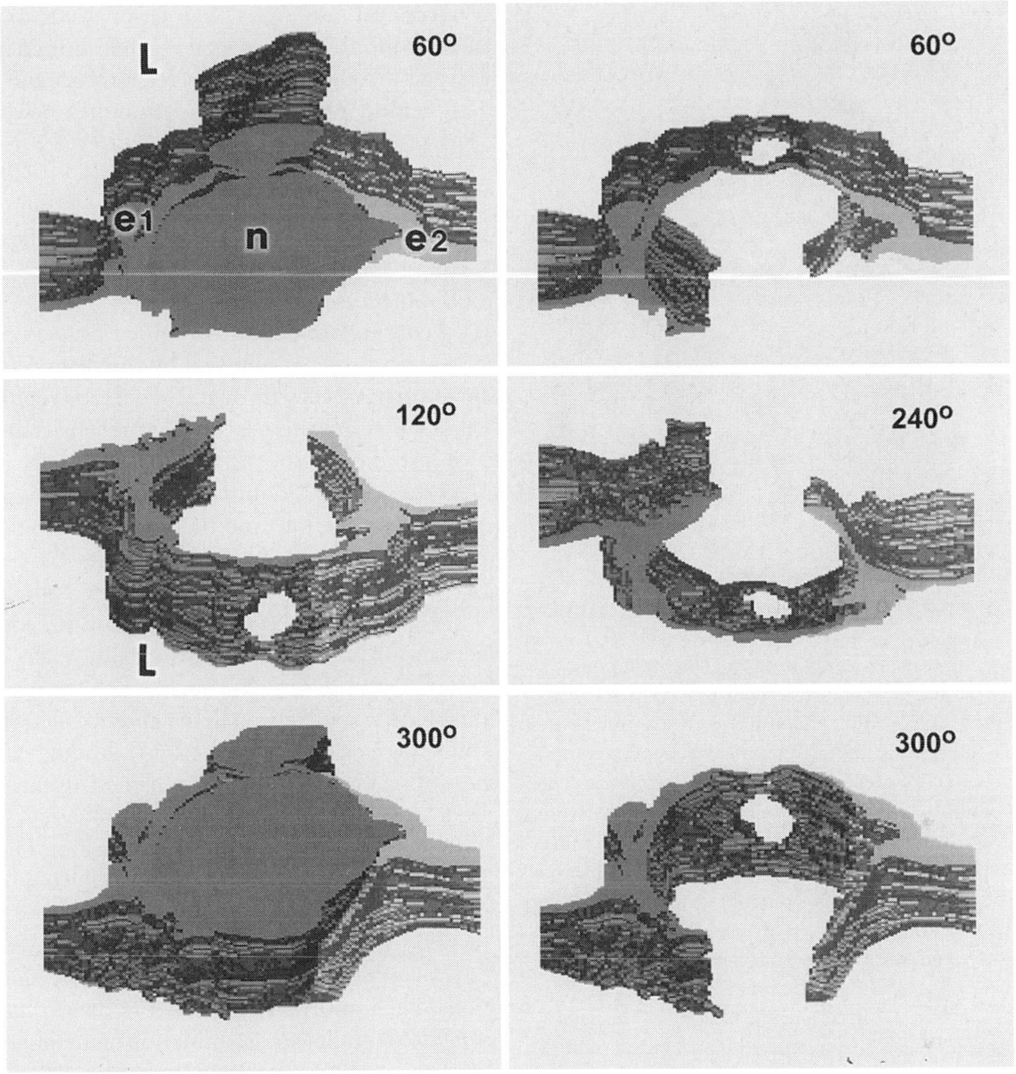

Figure 71.15. Computer-generated 3-D reconstruction of the transmigrating neutrophil illustrated in Figure 71.14. The panels portray successive rotations toward the viewer around a horizontal axis at angles of 60, 120, 240, and 300 degrees as indicated. 0 degrees (*not shown*) would represent a vascular cross-section at right angles to the direction of blood flow, and 90 degrees (*also not shown*) would represent a view looking directly down the luminal surface. Emigrating neutrophil (*n*) is in *upper and lower left*; in the other panels, the neutrophil was subtracted electronically to visualize the pore that passes cleanly through the cytoplasm of EC e1 distinctly apart from the junction of e1 with e2. Cytoplasmic arms of both e1 and e2 embrace the neutrophil luminally and, to a lesser extent, abluminally. L, lumen. (Reproduced with permission from Feng D, Nagy JA, Pyne K, et al. Neutrophils emigrate from venules by a trans-EC pathway in response to FMLP. *J Exp Med.* 1998;187:903–915.) For color reproduction, see Color Plate 71.15.

large that they accumulate at basement membranes after leaking from the vascular compartment. The carbon accumulations produce blackened vessels that are easily seen by light microscopy and can be selected for higher magnification studies by electron microscopy.

Electron microscopists, looking either at standard preparations or following intravenous injections of carbon, became convinced that ECs developed large gaps at their junctions through which carbon-containing macromolecular leakage occurred in inflammation (20,24–26,50). Use of the term *junction* in these cases generally referred to a separation at lateral EC borders, not permeation through tight junctions. However, the mechanism of such gap formation remained obscure and, in some of the published studies, visible adherens and occludens junctions nearby are not leaking carbon. Clearly another approach was necessary. One group noted that nuclear pinches formed in the reactive ECs stimulated by histamine, and they postulated that the active contraction of ECs pulled them apart at "junctions," thus allowing for the escape of vascular fluid (26). In one study, colocalized leaks of carbon and silver nitrate staining at cell "junctions" by light microscopy were said to provide proof of gap formation at intercellular junctions (24).

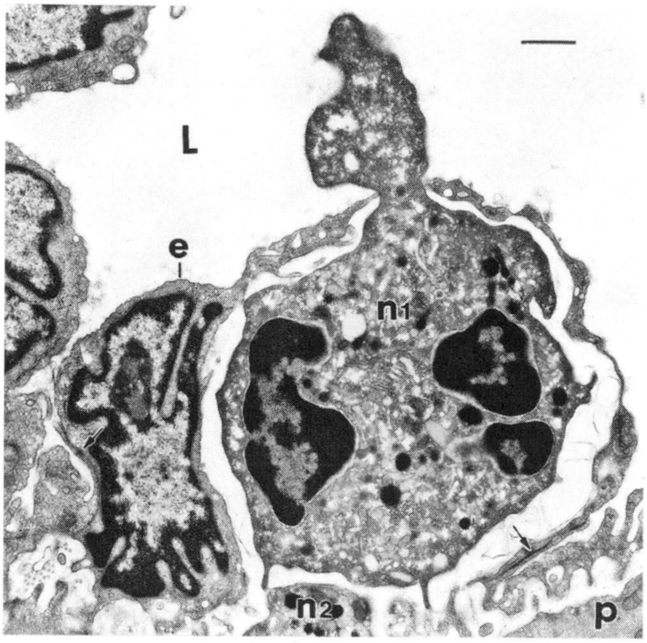

Figure 71.16. Neutrophil trans-EC migration in acute inflammation in guinea pig skin. The image represents section 19 of a consecutive series of 45 100-nm sections that encompassed the luminal pore through which the neutrophil is migrating in its entirety. Maximum pore size was 0.75 μm, 60 minutes after 10^5 M FMLP exposure. The nearest inter-EC junctions (*arrows*) are closed and are remote from the pore through which the neutrophil (*n1*) is transmigrating. A short, thin process projecting from n1 is in contact with a second neutrophil (*n2, below*), which, beyond the area of this micrograph, has progressed through the basal lamina and underlying pericyte (*p*). Note pinches of EC (*e*) nucleus that appear to result from the compressing bulk of the adjacent neutrophil, not from gap formation at junctions. The thin rim of EC cytoplasm overlying n1 and the bulging EC nucleus is relatively smooth, in contrast to surface endothelium to the right of the figure, which exhibits extensive wrinkling with extension of cytoplasmic processes into the vascular lumen and abluminally into the basal lamina. L = lumen. Bar = 1 μm.

The silver nitrate method, however, identifies the lateral borders of ECs, not just junctions. And, in the electron microscopic study, whereas nuclear pinches were quantifiably more numerous in histamine-stimulated samples than in controls, no convincing photographs of carbon-filled gap formation and nuclear pinches in the same cell were presented (26).

By the end of the 1960s, the electron microscopic studies seemed to be solidly accepted regarding the mechanism of mediator-induced vascular leaking (20,24–26,50). The ultimate resolution of where and how the leakage sites develop in ECs in acute inflammation would require large sets of serial sections and 3D computer reconstructions. These were done during the 1990s (Figures 71.6 and 71.7) and showed that carbon leaked through parajunctional and cytoplasmic trans-EC membrane-bound structures adjacent to unaltered lateral borders and adherens and occludens junctions (12,80). Also, exact replication of the cremaster model for examining histamine-induced vascular leakage (24,26) using ferritin as a tracer and numerous large series of 12- to 14-nm sections allowed for 3-D reconstructions that showed macromolecular leakage through interconnected vesicles and vacuoles spanning endothelial cytoplasm, structures newly termed *VVOs* (Figures 71.3 through 71.5, and Figures 71.8 through 71.13; for color reproduction, see Color Plate 71.11) (11,12,81). Closed junctions were present and were not passaging ferritin in this well-known model of acute inflammation (see Figure 71.5).

Cellular Phase

Acute inflammation is typified by the enhanced leakage of plasma proteins followed by the diapedesis of circulating inflammatory cells through the wall of venules (24,68). The cellular phase of this process was the subject of electron microscopic studies during the 1960s. Numerically, polymorphonuclear cells (PMNs) dominate the cellular migration process in acute inflammation, and therefore their traffic route was the primary study objective in these studies (51–55). The question that was framed, but not answered completely, was: Did PMNs pass through the cytoplasm of ECs, or did they travel between ECs? In the early studies, virtually everybody suggested that PMNs might pass through the cytoplasm, but that this route was unproven in the absence of serial sections. One study was based on serial sections, although neither the thickness of sections nor the completeness of the series was given (53).

In this study, it was decided, based on one serial set, that PMNs could "pry open the 'intercellular junction' with an extended pseudopod and travel down the junction" (53). Upon this evidence, and since the 1960s, future investigations claim, but do not prove, that PMNs migrate down intercellular junctions. The important body of work during the 1960s (51–55) served to stimulate re-examination of this issue using extensive ultrastructural sets of serial sections and computer-assisted 3-D reconstructions (Figures 71.14 through 71.16) (12,82). In one model, the new studies established a trans-EC traffic route for PMNs through EC cytoplasm in acute inflammation. Nearby intercellular junctions and lateral EC borders were intact (Figures 71.14 through 71.16; for color reproduction, see Color Plate 71.15).

CONCLUSION

Ultrastructural studies during the 1950s and 1960s greatly advanced the understanding of EC biology, including the correction of ideas developed earlier, without the use of the magnifications and resolution made possible by electron microscopes. The general anatomy of the microvasculature, described during the 1950s and 1960s, still stands today, and the identification of new organelles in ECs provided the rationale for future work. Vascular permeability routes through the endothelium in health and disease were probed with newly developed or applied tracers for electron microscopy. In all, a large body of work was accomplished with ever-improving instruments and methods, all of which provided powerful stimuli for new electron microscopic studies of the endothelium that are still in progress.

REFERENCES

1 Pappenheimer JR, Renkin EM, Borrero LM. Filtration, diffusion and molecular sieving through peripheral capillary membranes; a contribution to the pore theory of capillary permeability. *Am J Physiol.* 1951;167:13–46.
2 Chambers R, Zweifach BW. Intercellular cement and capillary permeability. *Physiol Rev.* 1947;27:436–463.
3 Bennett HS, Luft JH, Hampton JC. Morphological classifications of vertebrate blood capillaries. *Am J Physiol.* 1959;196:381–390.
4 Alksne J. The passage of colloidal particles across the dermal capillary wall under the influence of histamine. *Q J Exp Physiol Cogn Med Sci.* 1959;44:51–66.
5 Farquhar MG, Hartmann JF. Electron microscopy of cerebral capillaries. *Anat Rec.* 1956;124:288–289.
6 Palade GE. The fine structure of blood capillaries. *J Appl Physics.* 1953;24:1424.
7 Palade GE. The endoplasmic reticulum. *J Biophys Biochem Cytol.* 1956;2(Suppl):85–98.
8 Starling EH. *The Fluids of the Body.* Chicago: Keener, 1909.
9 Florey HW, Poole JCF, Meek GA. Endothelial cells and cement lines. *J Pathol Bacteriol.* 1959;77:625–636.
10 Luft JH. Capillary permeability. In: Zweifach, Grant, McCluskey, eds. *The Inflammatory Process.* Vol. 2. New York: Academic Press; 1973:47–92.
11 Dvorak AM, Feng D. The vesiculo-vacuolar organelle (VVO): a new EC permeability organelle. *J Histochem Cytochem.* 2001;49: 419–431.
12 Feng D, Nagy JA, Dvorak HF, Dvorak AM. Ultrastructural studies define soluble macromolecular, particulate, and cellular trans-EC pathways in venules, lymphatic vessels, and tumor-associated microvessels in man and animals. *Microsc Res Tech.* 2002;57:289–326.
13 Florey H. Exchange of substances between the blood and tissues. *Nature.* 1961;192:908–912.
14 Bruns RR, Palade GE. Studies on blood capillaries. II. Transport of ferritin molecules across the wall of muscle capillaries. *J Cell Biol.* 1968;37:277–299.
15 Karnovsky MJ. The ultrastructural basis of capillary permeability studied with peroxidase as a tracer. *J Cell Biol.* 1967;35:213–236.
16 Reese TS, Karnovsky MJ. Fine structural localization of a blood-brain barrier to exogenous peroxidase. *J Cell Biol.* 1967;34:207–217.
17 Karnovsky MJ. The ultrastructural basis of transcapillary exchanges. *J Gen Physiol.* 1968;52:S64–S95.
18 Jennings MA, Florey L. An investigation of some properties of endothelium related to capillary permeability. *Proc R Soc Lond B Biol Sci.* 1967;167:39–63.
19 Shea SM, Karnovsky MJ, Bossert WH. Vesicular transport across endothelium: simulation of a diffusion model. *J Theor Biol.* 1969; 24:30–42.
20 Palade GE. Blood capillaries of the heart and other organs. *Circulation.* 1961;24:368–388.
21 Casley-Smith JR. The dimensions and numbers of small vesicles in cells, endothelial and mesothelial and the significance of these for endothelial permeability. *J Microsc.* 1969;90:251–268.
22 Palade GE, Bruns RR. Structural modulations of plasmalemmal vesicles. *J Cell Biol.* 1968;37:633–649.
23 Clementi F, Palade GE. Intestinal capillaries. I. Permeability to peroxidase and ferritin. *J Cell Biol.* 1969;41:33–58.
24 Majno G, Palade GE. Studies on inflammation. I. The effect of histamine and serotonin on vascular permeability: an electron microscopic study. *J Biophys Biochem Cytol.* 1961;11:571–605.
25 Marchesi VT. The passage of colloidal carbon through inflamed endothelium. *Proc R Soc Lond B Biol Sci.* 1962;156:550–552.
26 Majno G, Shea SM, Leventhal M. Endothelial contraction induced by histamine-type mediators: an electron microscopic study. *J Cell Biol.* 1969;42:647–672.
27 Palade GE. Transport of quanta across the endothelium of blood capillaries. *Anat Rec.* 1960;136:254.
28 Farquhar MG, Wissig SL, Palade GE. Glomerular permeability. I. Ferritin transfer across the normal glomerular capillary wall. *J Exp Med.* 1961;113:47–66.
29 Jennings MA, Marchesi VT, Florey HW. The transport of particles across the walls of small blood vessels. *Proc R Soc Lond B Biol Sci.* 1962;156:14–19.
30 Casley-Smith JR. Endothelial permeability. The passage of particles into and out of diaphragmatic lymphatics. *Q J Exp Physiol Cogn Med Sci.* 1964;49:365–383.
31 Tomlin SG. Vesicular transport across ECs. *Biochim Biophys Acta.* 1969;183:559–564.
32 Graham RC Jr., Karnovsky MJ. The early stages of absorption of injected horseradish peroxidase in the proximal tubules of the mouse kidney: ultrastructural cytochemistry by a new technique. *J Histochem Cytochem.* 1966;14:291–302.
33 Graham RC Jr., Karnovsky MJ. Glomerular permeability. Ultrastructural cytochemical studies using peroxidases as protein tracers. *J Exp Med.* 1966;124:1123–1134.
34 Karnovsky MJ, Shea SM. Transcapillary transport by pinocytosis. *Microvasc Res.* 1970;2:353–360.
35 Karnovsky MJ. Morphology of capillaries with special reference to muscle capillaries. In: Crone C, Lassen NA, eds. *Capillary Permeability: The Transfer of Molecules and Ions between Capillary Blood and Tissue.* (Alfred Benzon Symp. II.) Munksgaard, Copenhagen: Academic Press; 1970:341–350.
36 Shea SM, Karnovsky MJ. Brownian motion: a theoretical explanation for the movement of vesicles across the endothelium. *Nature.* 1966;212:353–355.
37 Suter ER, Majno G. Passage of lipid across vascular endothelium in newborn rats. An electron microscopic study. *J Cell Biol.* 1965;27:163–177.
38 Schoefl GI, French JE. Vascular permeability to particulate fat: morphological observations on vessels of lactating mammary gland and of lung. *Proc R Soc Lond B Biol Sci.* 1968;169:152–165.
39 Florey HW. The structure of normal and inflamed small blood vessels of the mouse and rat colon. *Q J Exp Physiol Cogn Med Sci.* 1961;46:119–122.
40 Cotran RS, Suter ER, Majno G. The use of colloidal carbon as a tracer for vascular injury. A review. *Vasc Dis.* 1967;4:107–127.
41 Karnovsky MJ, Cotran RS. The intercellular passage of exogenous peroxidase across endothelium and mesothelium. *Anat Rec.* 1966;154:365.
42 Weibel ER, Palade GE. New cytoplasmic components in arterial endothelia. *J Cell Biol.* 1964;23:101–112.
43 Muir AR, Peters A. Quintuple-layered membrane junctions at terminal bars between ECs. *J Cell Biol.* 1962;12:443–448.

44 Bruns RR, Palade GE. Studies on blood capillaries. I. General organization of blood capillaries in muscle. *J Cell Biol.* 1968;37: 244–276.
45 Farquhar MG, Palade GE. Junctional complexes in various epithelia. *J Cell Biol.* 1963;17:375–412.
46 Fawcett DW. Surface specializations of absorbing cells. *J Histochem Cytochem.* 1965;13:75–91.
47 Stehbens WE. Ultrastructure of vascular endothelium in the frog. *Q J Exp Physiol Cogn Med Sci.* 1965;50:375–384.
48 Elfvin L-G. The ultrastructure of the capillary fenestrae in the adrenal medulla of the rat. *J Ultrastruct Res.* 1965;12:687–704.
49 Rhodin JA. The diaphragm of capillary endothelial fenestrations. *J Ultrastruct Res.* 1962;6:171–185.
50 Movat HZ, Fernando NV. Allergic inflammation. I. The earliest fine structural changes at the blood-tissue barrier during antigen-antibody interaction. *Am J Pathol.* 1963;42:41–59.
51 Hurley JV, Xeros N. Electron microscopic observations on the emigration of leucocytes. *Aust J Exp Biol Med Sci.* 1961;39:609–623.
52 Marchesi VT, Florey HW. Electron micrographic observations on the emigration of leucocytes. *Q J Exp Physiol Cogn Med Sci.* 1960;45:343–348.
53 Marchesi VT. The site of leucocyte emigration during inflammation. *Q J Exp Physiol Cogn Med Sci.* 1961;46:115–118.
54 Florey HW, Grant LH. Leucocyte migration from small blood vessels stimulated with ultraviolet light: an electron-microscope study. *J Pathol Bacteriol.* 1961;82:13–17.
55 Marchesi VT. Some electron microscopic observations on interactions between leukocytes, platelets, and ECs in acute inflammation. *Ann NY Acad Sci.* 1964;116:774–788.
56 French JE. Atherosclerosis in relation to the structure and function of the arterial intima, with special reference to the endothelium. *Int Rev Exp Pathol.* 1966;5:253–353.
57 Luft JH. The ultrastructural basis of capillary permeability. In: Zweifach, Grant, McCluskey, eds. *The Inflammatory Process.* New York: Academic Press; 1965:121–1959.
58 Pease DC, Paule WJ. Electron microscopy of elastic arteries; the thoracic aorta of the rat. *J Ultrastruct Res.* 1960;3:469–483.
59 Fernando NV, Movat HZ. The fine structure of the terminal vascular bed. III. The capillaries. *Exp Mol Pathol.* 1964;34: 87–97.
60 Rhodin JA. Fine structure of vascular walls in mammals with special reference to smooth muscle component. *Physiol Rev.* 1962; 42(Suppl 5):48–87.
61 Florey HW. The EC. *Br Med J.* 1966;2:487–490.
62 Farquhar MG. Fine structure and function in capillaries of the anterior pituitary gland. *Angiology.* 1961;12:270–292.
63 Farquhar MG, Palade GE. Functional evidence for the existence of a third type of cell in the renal glomerulus. *J Cell Biol.* 1962;13:55–87.
64 Florey HW. The uptake of particulate matter by endothelial cells. *Proc R Soc Lond B Biol Sci.* 1967;166:375–383.
65 Rhodin JA. Ultrastructure of mammalian venous capillaries, venules, and small collecting veins. *J Ultrastruct Res.* 1968;25: 452–500.
66 Rosenbluth J, Wissig SL. The distribution of exogenous ferritin in toad spinal ganglia and the mechanism of its uptake by neurons. *J Cell Biol.*1964;23:307–325.
67 Rhodin JA. The ultrastructure of mammalian arterioles and precapillary sphincters. *J Ultrastruct Res.* 1967;18:181–223.
68 Majno G. In: Hamilton WF, Dow P, eds. *Handbook of Physiology.* Vol. 3, sect. 2. Baltimore: Williams and Wilkins; 1965:2293–2375.
69 Behnke O. A preliminary report on "microtubules" in undifferentiated and differentiated vertebrate cells. *J Ultrastruct Res.* 1964;11:139–146.
70 Sanborn E, Koen PF, MacNabb JD, Moore G. Cytoplasmic microtubules in mammalian cells. *J Ultrastruct Res.* 1964;11:123–138.
71 Geer JC, McGill HC Jr., Strong JP. The fine structure of human atherosclerotic lesions. *Am J Pathol.* 1961;38:263–287.
72 Still WJS, Marriott PR. Comparative morphology of the early atherosclerotic lesion in man and cholesterol-atherosclerosis in the rabbit. An electron microscopic study. *J Atheroscler Res.* 1964; 44:373–386.
73 Wagner DD. The Weibel-Palade body: the storage granule for von Willebrand factor and P-selectin. *Thromb Haemost.* 1993;70: 105–110.
74 Luft JH. Fine structure of the diaphragms across capillary "pores" in mouse intestine. *Anat Rec.* 1964;148:307–308.
75 Moore DH, Ruska H. The fine structure of capillaries and small arteries. *J Biophys Biochem Cytol.* 1957;3:457–462.
76 Simionescu N, Simionescu M, Palade GE. Permeability of muscle capillaries to small heme-peptides. Evidence for the existence of patent transendothelial channels. *J Cell Biol.* 1975;64:586–607.
77 Kohn S, Nagy JA, Dvorak HF, Dvorak AM. Pathways of macromolecular tracer transport across venules and small veins. Structural basis for the hyperpermeability of tumor blood vessels. *Lab Invest.* 1992;67:596–607.
78 Movat HZ, Fernando NV. The fine structure of the terminal vascular bed. IV. The venules and their perivascular cells (pericytes, adventitial cells). *Exp Mol Pathol.* 1964;34:98–114.
79 Yamada E. The fine structure of the renal glomerulus of the mouse. *J Biophys Biochem Cytol.* 1955;1:551–566.
80 Feng D, Nagy JA, Hipp J, et al. Reinterpretation of EC gaps induced by vasoactive mediators in guinea-pig, mouse and rat: many are transcellular pores. *J Physiol.* 1997;504 (Pt 3):747–761.
81 Feng D, Nagy JA, Hipp J, et al. Vesiculo-vacuolar organelles and the regulation of venule permeability to macromolecules by vascular permeability factor, histamine and serotonin. *J Exp Med.* 1996;183:1981–1986.
82 Feng D, Nagy JA, Pyne K, et al. Neutrophils emigrate from venules by a trans-EC pathway in response to FMLP. *J Exp Med.* 1998;187:903–915.

72

Weibel-Palade Bodies

Vesicular Trafficking on the Vascular Highway

Charles J. Lowenstein, Craig N. Morrell, and Munekazu Yamakuchi

The Johns Hopkins University School of Medicine, Baltimore, Maryland

Edward Weibel and George Palade made a remarkable discovery in 1964. Surveying human lungs with an electron microscope, they saw a "hitherto unknown rod-shaped cytoplasmic component . . . in endothelial cells of small arteries" (1). These rods, measuring 0.1 μm thick and 3 μm in length, contained dense material organized as fibers running parallel to the long axis (Figure 72.1; for color reproduction, see Color Plate 72.1A). Weibel and Palade concluded that the "nature and significance of these cytoplasmic components are yet unknown."

Over the last four decades, we have learned an enormous amount about these endothelial-specific organelles, now called Weibel-Palade bodies (WPBs). WPBs are regulated secretory granules. Within seconds of vascular injury, endothelial cells (ECs) activate the internal exocytic machinery, driving fusion of the WPBs with the cellular membrane. WPBs contain a variety of compounds that, once released into the blood, activate vascular inflammation and thrombosis (Figure 72.2). WPB exocytosis is the initial response of the vasculature to injury.

However, a number of intriguing questions about WPBs remain unanswered. Why are specific proteins targeted to enter the WPBs, whereas other cargo is excluded? How are WPBs directed to fuse with the plasma membrane and not with other organelles? Which vascular beds contain the most endothelial granules, and why? Finally, are endothelial granules capable of plasticity? Do the contents of these WPBs change as their environment changes, giving the ECs the ability to alter their response as their own environment changes, thus reflecting a vascular memory?

CONTENTS OF WEIBEL-PALADE BODIES

WPBs contain molecules that regulate thrombosis and inflammation, as well as other vascular regulators (Table 72.1). The predominant component of WPB is von Willebrand factor (vWF) (2). Within the EC, vWF forms dimers in the endoplasmic reticulum, multimers in the Golgi apparatus, and then is either secreted from the cell via the constitutive pathway or stored in WPBs for subsequent secretion (3,4). The highest molecular weight vWF are preferentially incorporated into WPBs. Parallel bundles of vWF form the striations within endothelial granules first seen by Weibel and Palade (4). WPB exocytosis releases multimeric vWF into the blood, where it is cleaved by the metalloproteinase ADAMTS13 (a disintegrin and metalloprotease with thrombospondin type 1 repeats 13) (5). vWF mediates platelet adherence to collagen in the subendothelial space through its interactions with the platelet glycoprotein (GP) complex GPIb-IX-V; vWF also mediates platelet aggregation by interacting with activated GPIIb/IIIa on platelets (6). Humans and mice lacking vWF have defects in hemostasis (7). Although WPBs also may contain fibrinolytic mediators such as tissue-type plasminogen activator (t-PA) (8), activation of endothelial exocytosis induces immediate platelet adhesion to the endothelium mediated by vWF and its interaction with platelet P-selectin and platelet GPIb-IX-V (9,10). Thus, secretion of the WPB contents is prothrombotic.

WPBs also contain inflammatory mediators. P-selectin, the major endothelial receptor mediating leukocyte rolling, is stored in WPBs (11,12). P-selectin is a transmembrane protein, stored inside WPBs with its amino-terminal domain inside the granule, and its carboxy-terminal cytoplasmic tail in the cytoplasm. During exocytosis, P-selectin translocates from the granule to the external cellular surface. Its carboxy-terminal tail remains cytoplasmic, but its amino-terminal domain becomes extracellular, where it interacts with its ligand P-selectin glycoprotein ligand (PSGL)-1 on the surface of leukocytes, triggering leukocyte rolling along blood vessels. However, within 20 minutes of its externalization, P-selectin is removed from the cell surface by endocytosis, recycles to the lysosome or Golgi apparatus, and returns to the WPBs (13). Mice lacking P-selectin have decreased leukocyte recruitment to sites of injury (14). (In contrast to vWF release through both constitutive and secretory pathways, P-selectin is only released through granule exocytosis.) Exocytosis of WPBs is thus proinflammatory.

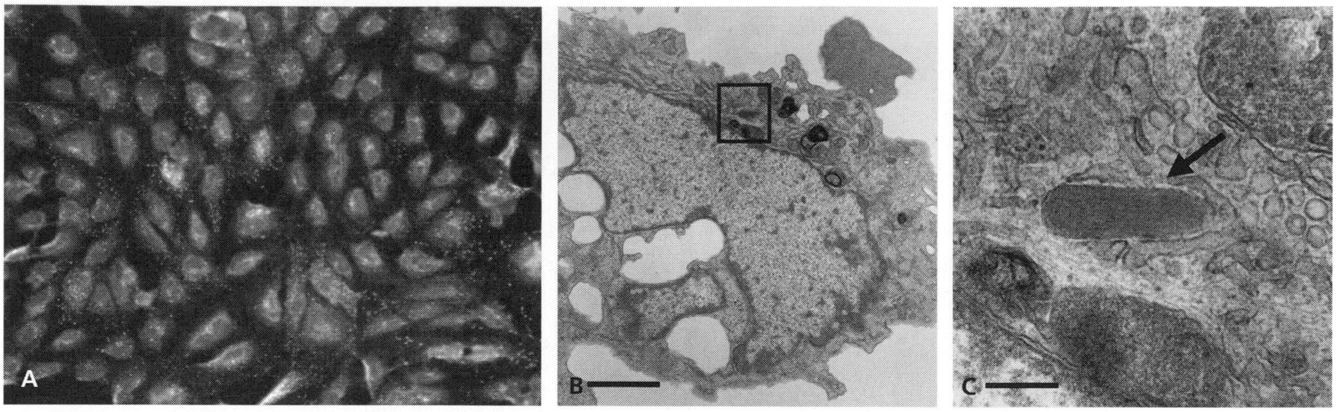

Figure 72.1. Weibel-Palade bodies in ECs. (**A**) Immunofluorescent images of ECs with nuclei stained blue and vWF stained green. (**B**) Electron microscopy of a human EC with boxed region expanded below. Bar = 200 μm. (**C**) Electron microscopy of a WPB within a human EC (*arrow* shows WPB; bar = 200 nm). For color reproduction, see Color Plate 72.1A.

Additional molecules in WPBs regulate inflammation. Interleukin (IL)-8 is a C-X-C chemokine stored within WPBs (15,16). Stimulation of ECs with IL-1β or tumor necrosis factor (TNF)-α causes the accumulation of IL-8 within WPBs. Once released by exocytosis, IL-8 can activate neutrophil adhesion. WPBs thus contain separate molecules that activate distinct stages in leukocyte trafficking: P-selectin, which mediates leukocyte rolling; and IL-8, which induces leukocyte adhesion.

Another component of WPB is a member of the tetraspanin superfamily, CD63 (17). Tetraspanins are transmembrane proteins that promote a diverse array of neuronal and immune functions. The precise role of CD63 in WPBs is unknown. CD63 may play a role in the biogenesis of WPBs, fusion of WPBs with the plasma membrane, or extracellular signaling.

PLATELET α-GRANULES VERSUS ENDOTHELIAL GRANULES

Because both platelet α-granules and endothelial WPBs contain vWF and P-selectin, it is tempting to speculate that both granules contain similar components. However, platelet α-granules may contain more types of proteins than do. WPBs. Approximately 15 proteins have been identified within WPBs (see Table 72-1). In contrast, a proteomic analysis described at least 81 proteins released from thrombin stimulated platelets – although these proteins may be components of α-granules, dense granules, lysosomal granules, or exosomes (18). The diversity of α-granule proteins may be accounted for in part by the ability of platelets to take up various compounds from the blood and store them in granules. Nonetheless, a comparison of the components of endothelial and platelet granules will not be possible until a precise proteomic description of isolated granules is performed. Until such experiments are performed, additional clues to the composition of these granules can come from understanding how cargo is selected and loaded into granules.

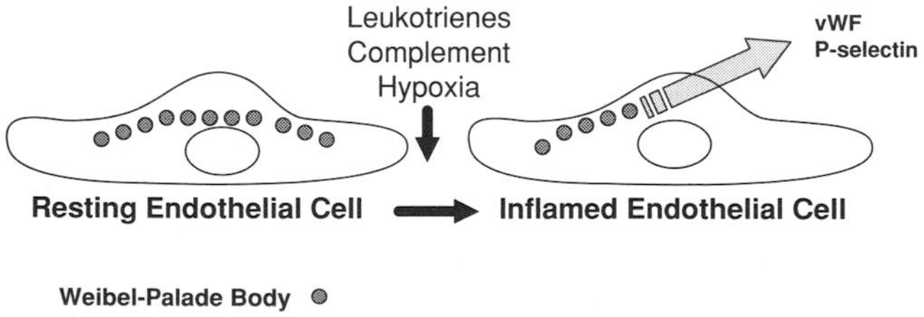

Figure 72.2. A variety of agonists trigger resting ECs to release the contents of WPB into the bloodstream.

CARGO LOADING AND BIOGENESIS OF WEIBEL-PALADE BODIES

The earliest model for molecular sorting of specific proteins into WPBs suggested that WPBs are pre-existing organelles to which proteins are directed by signal sequences. This model was supported by studies showing that the 742-amino acid residue propeptide of vWF is necessary for the targeting of vWF to WPBs (4,19–24). In contrast, expression of the 2,050-amino acid residue mature vWF lacking the propeptide sequence leads to a cytoplasmic localization of vWF. Coexpression of the vWF propeptide and the mature vWF *in trans* leads to localization of the latter in granules. Furthermore, the vWF propeptide can direct a chimeric protein, consisting of the vWF propeptide fused to complement C3, into granules (23). In fact, expression of the vWF propeptide alone, without the mature vWF, also leads to granule localization of the propeptide (4). Thus, in the revised model, vWF propeptide directs vWF into WPBs.

The model for cargo sorting into WPBs has been extended by observations suggesting that cargo loading of WPBs and formation of WPBs are linked: The process of sorting proteins into the granule drives formation of the granule itself. Coat proteins play a dual role in granule biogenesis, not only

Table 72-1: Contents of WPB.

WPB Contents	*Vascular Function*
vWF	Thrombosis
Factor XIIIa	Thrombosis
t-PA	Fibrinolysis
P-selectin (PADGEM, GMP-140)	Inflammation
IL-8	Inflammation
Eotaxin	Inflammation
α-1,3-fucosyltransferase VI	Inflammation
CD63 (Lamp3)	Unknown
Endothelin-1	Vasoconstriction
Endothelin converting enzyme	Vasoconstriction
Calcitonin gene-related peptide	Vasodilation
Angiopoietin 2	Angiogenesis

catalyzing the budding of granules from donor trans-Golgi–derived membranes, but also sorting cargo proteins into granules. Coat proteins select diverse proteins for inclusion into granules through interactions between specific cargo proteins and multiple cargo-binding sites on the coat proteins (25–27). This model is supported by studies that show the vWF propeptide is necessary for formation of WPBs; however, the coat proteins associated with WPBs have not been identified. Ectopic expression of vWF propeptide leads to the formation of novel WPB-like granules in AtT-20 neuroendocrine cells that contain adrenocorticotropin secretory granules but normally lack WPBs (21). Conversely, WPBs are absent from mice deficient in vWF (28). In contrast, the *P-selectin*–knockout mice contain WPBs in their ECs (14). Collectively, these data imply that an unidentified coat protein in the trans-Golgi of ECs interacts with the vWF propeptide, simultaneously forming the WPBs while loading vWF into the nascent granule.

What determines the distribution of WPBs throughout the vascular tree? WPBs have been found in the vascular ECs of all organs, and throughout the vascular tree in arteries, capillaries, and veins (29). Because leukocyte trafficking often is detected in postcapillary venules, it might be expected that more WPBs are found here. However, a precise description of the distribution of WPBs is not available. The factors that initiate the biogenesis of WPBs or that regulate the number of WPBs within an individual EC are unknown.

What is the molecular basis for localizing WPBs within ECs? Presumably, WPBs are transported to specific subcellular locations, as are other cellular organelles, along microtubule tracks by a kinesin isoform. Although another molecular motor, dynein, may play a role in the subcellular localization of WPBs, the specific mechanisms of WPB transport are not well defined (30). In fact, the distribution of WPBs within ECs is not well characterized. However, the secretion of WPBs appears to be polarized. Approximately 90% of vWF is released through the basolateral surface of human umbilical vein ECs (HUVECs) ex vivo (31). We have also visualized the basolateral release of vWF during acute rejection of allografts in vivo (32). The release of vWF and other components of WPBs into the subendothelial space is surprising, because many of these factors are thought to regulate leukocyte and platelet interactions with the endothelium. This polarized release of WPBs suggests that unknown mechanisms deliver WPBs to both the apical and the basolateral surfaces of ECs.

THE EXOCYTIC MACHINERY OF ENDOTHELIAL CELLS

During endothelial exocytosis, the membrane of the WPB fuses with the cellular plasma membrane, releasing the contents of the granule out to the extracellular space. The endothelial proteins that drive fusion of the WPB with the cellular membrane are a subset of the vesicular trafficking proteins conserved in yeast, *Drosophila*, and mammalian cells. Granule exocytosis is a type of vesicle trafficking that involves a series of discrete stages: loading of cargo into the nascent vesicle, vesicle budding, targeting of the vesicle to its fusion partner, priming of the vesicle, membrane fusion, and, finally, vesicle recycling (Figure 72.3) (33–36). The families of proteins that mediate these steps include the Rab and Rab effectors, soluble N-ethylmaleimide sensitive factor receptor (SNARE) proteins, members of the Sec1/Munc18 protein family, and N-ethylmaleimide sensitive factor (NSF).

Rab proteins regulate vesicle tethering to target membranes. The Rab family member RalA associates with WPBs and is phosphorylated following activation of exocytosis (37–39). Rab27a associates with mature WPBs that contain vWF, but not with nascent WPBs lacking vWF (40). However, the

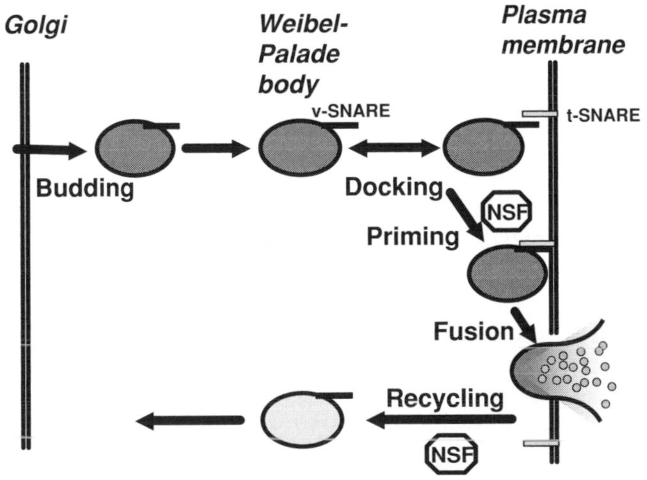

Figure 72.3. WPB and the cycle of vesicle trafficking. Granules bud off from the Golgi. Vesicle-SNARE (v-SNARE) proteins on the granule membrane interact with target-SNARE (t-SNARE) proteins on the plasma membrane, specifying the destination of the granule. An influx of calcium then triggers fusion of the granule and cytoplasmic membranes, causing exocytosis. NSF either primes the granule for fusion, facilitates empty granule recycling, or mediates both processes.

precise role that RalA, Rab27a, and other Rab family members play in WPB exocytosis is unknown.

SNAREs are transmembrane proteins localized to granule and target membranes. They play a role in directing granules to the appropriate target membrane prior to membrane fusion. SNAREs in granule membranes can interact with SNAREs in cellular membranes, forming a stable SNARE complex and placing the two membrane fusion partners in apposition prior to fusion. Experiments using RNA silencing or antibody interference have demonstrated that syntaxin-4 is one of the three SNARE molecules that targets WPBs to the plasma membrane (41,42). The identity of the other two SNARE partners of syntaxin-4 on WPBs are unknown, but are probably SNAP-23 and synaptobrevin-3 (VAMP-3), which interact with syntaxin-4 to direct granule exocytosis in other secretory cells. Syntaxin-4 is found predominantly on the inner surface of the plasma membrane; synaptobrevin-3 and SNAP-23 are usually associated with granule membranes (34). The Sec1/Munc18 family members are negative regulators of SNARE interactions that separate from the SNARE complex prior to membrane fusion. Although Munc18 is phosphorylated prior to WPB exocytosis, its precise role in EC exocytosis is unclear (42).

NSF is a critical component of the exocytic machinery in ECs. NSF associates with the SNARE complex via an accessory protein, α-soluble NSF attachment protein (α-SNAP), hydrolyzes ATP, and disassembles the SNARE complex (43–47). Precisely how NSF converts the chemical energy of ATP into physical energy to separate the three SNARE molecules is unknown. Although chemical or genetic inactivation of NSF blocks vesicle trafficking, the precise step at which NSF mediates vesicle trafficking is controversial. NSF was originally thought to prime vesicles for fusion, but subsequent studies suggested that NSF is necessary for vesicle recycling after fusion. Inhibiting NSF in ECs with intracellular antibodies blocks WPB exocytosis; adding recombinant NSF back to these cells restores exocytosis (41). Thus NSF is the molecular motor that drives endothelial exocytosis in WPBs.

TRIGGERS OF WEIBEL-PALADE BODY EXOCYTOSIS

Two aspects of WPB exocytosis are particularly notable: WPB release is both rapid and stereotypical. Agonists trigger endothelial exocytosis within seconds to minutes. WPB release is thus an immediate early response of ECs – a rapid response to injury independent of gene transcription (48). (In contrast, vascular injury also can activate a delayed endothelial response that involves transcription of genes, but this response may take hours.) The endothelial response to thrombin is immediate: Thrombin activates WPB exocytosis within minutes, leading to the rapid externalization of P-selectin, which mediates leukocyte rolling along the endothelium (49). In contrast, the endothelial response to hypoxia is both immediate and delayed. Hypoxia not only triggers WPB exocytosis within minutes, leading to rapid externalization of P-selectin, but also activates the transcription of intercellular adhesion molecule (ICAM)-1 within hours, which eventually leads to leukocyte adherence to the vessel wall (50,51). Thus, WPB exocytosis is an immediate endothelial response to injury.

Table 72-2: Triggers of WPB Exocytosis

Class	*Agonist*
Physical	Hypoxia Radiation Trauma Shear stress
Endogenous chemicals	ATP ADP Reactive oxygen species Leukotrienes Histamine Epinephrine Serotonin
Polypeptides	Thrombin Fibrin Oxidized LDL Complement VEGF

WPB exocytosis is also a stereotypical response to a wide variety of agonists, physical, chemical, and biological (Table 72-2). Physical injury, such as hypoxia, ischemia, radiation, and trauma activates WPB secretion (50,52–57). Endogenous chemicals, including ADP, ATP, epinephrine, serotonin, and histamine, also trigger endothelial exocytosis (12,58–62). Proteins such as thrombin, fibrin, complement, vascular endothelial growth factor (VEGF), and oxidized low-density lipoprotein (LDL) induce WPB release (49,63–65). All these diverse agents not only activate unique receptors and specific signal transduction cascades, but also trigger WPB exocytosis, leading to vascular inflammation and thrombosis. WPB exocytosis is thus a final common pathway by which ECs respond to injury.

NEGATIVE REGULATORS OF ENDOTHELIAL EXOCYTOSIS

Although many activators of WPB exocytosis have been identified, the only inhibitor of exocytosis identified thus far is nitric oxide (NO). Several lines of evidence suggest that NO inhibits exocytosis by regulating NSF (41). First, NO chemically modifies recombinant NSF, S-nitrosylating three of its nine cysteine residues. Second, S-nitrosylated NSF cannot disassemble the SNARE complex. Third, exocytosis can be restored by adding NSF back to ECs previously inhibited with NO. Finally, inhibition of NO production by pharmacologic or genetic tools increases endothelial exocytosis and platelet adherence in vivo

(41,66). NO thus inhibits vascular inflammation in part by blocking endothelial exocytosis.

If NO inhibits endothelial inflammation, then lack of NO permits excessive vascular inflammation. This may explain why endothelial dysfunction is a risk factor for atherosclerosis. The hallmark of endothelial dysfunction is the inability of the endothelium to produce NO. In healthy patients, NO may limit endothelial exocytosis in response to vascular injury. However, in patients with endothelial dysfunction and decreased NO production, perhaps NSF is not inhibited, thus permitting an exaggerated endothelial response to injury, excessive release of inflammatory and thrombotic mediators, and vascular inflammation.

ENDOTHELIUM AS VASCULAR MEMORY

If memory is the ability to store and retrieve information, then ECs function as a vascular memory. Just as neurons integrate presynaptic signals and secrete neurotransmitters from neurovesicles, so do ECs respond to external agonists by exocytosis of vascular messengers from WPBs. Neurons and ECs share components of the exocytic machinery: Both cell types rely on SNAREs and NSF to regulate exocytosis. Furthermore, the strength of the exocytic response of both neurons and ECs can be modulated. Long-term potentiation (LTP) and long-term depression (LTD) describe the ability of neurons to increase and decrease, respectively, exocytosis of neurotransmitters in response to repetitive stimulation. The EC response to agonists is also plastic: ECs modulate exocytosis in response to the external environment. For example, WPBs normally do not contain IL-8. However, prolonged treatment with TNF-α or IL-1β induces ECs to synthesize IL-8 and store it inside WPBs. Thus, ECs alter the contents of WPBs in response to inflammatory stimuli. After exposure to cytokines, ECs can release IL-8 from WPBs, modulating neutrophil activation. ECs also modulate exocytosis in response to anti-inflammatory signals. NO inhibits endothelial exocytosis for 6 hours. However, after exposure to NO ceases, ECs gradually recover the ability to activate exocytosis (41). ECs thus exhibit both LTP and LDP. Proinflammatory and anti-inflammatory signals modulate endothelial exocytosis – in a sense, ECs "remember" their inflammatory environment.

KEY POINTS

Upon discovering the rod-shaped organelles that now bear their name, Weibel and Palade noted that these granules "must be a structure of some functional significance which for the moment remains obscure . . . It is hoped that further . . . studies may shed some light on the nature of the rods and their components."(1) In the decades since the discovery of WPB, investigators have learned that:

- Endothelial exocytosis of WPBs is a rapid response to vascular injury.
- WPB exocytosis activates vascular inflammation and thrombosis.
- NO inhibits vascular inflammation by regulating WPB exocytosis.

Future Goals

- To identify WPB coat proteins that simultaneously direct granule budding and select proteins for inclusion in WPB
- To define the molecular components of the exocytic machinery in ECs
- To invent novel drugs that inhibit endothelial exocytosis for the treatment of inflammatory diseases

ACKNOWLEDGMENTS

Supported by grants from the NIH (R01 HL63706–04, R01 HL074061, P01 HL65608, P01 HL56091), the Ciccarone Center, and the John and Cora H. Davis Foundation to CJL; and by grants RR07002 and HL074945 from the NIH to CM.

REFERENCES

1 Weibel ER, Palade GE. New cytoplasmic components in arterial endothelia. *J Cell Biol.* 1964;23:101–112.
2 Wagner DD, Olmsted JB, Marder VJ. Immunolocalization of von Willebrand protein in Weibel-Palade bodies of human endothelial cells. *J Cell Biol.* 1982;95(1):355–360.
3 Wagner DD, Mayadas T, Marder VJ. Initial glycosylation and acidic pH in the Golgi apparatus are required for multimerization of von Willebrand factor. *J Cell Biol.* 1986;102(4):1320–1324.
4 Wagner DD, Saffaripour S, Bonfanti R, et al. Induction of specific storage organelles by von Willebrand factor propolypeptide. *Cell.* 1991;64(2):403–413.
5 Levy GG, Motto DG, Ginsburg D. ADAMTS13 Turns 3. *Blood.* 2005;106(1):11–17. Epub 2005 Mar 17.
6 Ruggeri ZM. Platelets in atherothrombosis. *Nat Med.* 2002;8(11): 1227–1234.
7 Denis C, Methia N, Frenette PS, et al. A mouse model of severe von Willebrand disease: defects in hemostasis and thrombosis. *Proc Natl Acad Sci USA.* 1998;95(16):9524–9529.
8 Huber D, Cramer EM, Kaufmann JE, et al. Tissue-type plasminogen activator (t-PA) is stored in Weibel-Palade bodies in human endothelial cells both in vitro and in vivo. *Blood.* 2002; 99(10):3637–3645.
9 Padilla A, Moake JL, Bernardo A, et al. P-selectin anchors newly released ultralarge von Willebrand factor multimers to the endothelial cell surface. *Blood.* 2004;103(6):2150–2156.

10 Dong JF, Moake JL, Nolasco L, et al. ADAMTS-13 rapidly cleaves newly secreted ultralarge von Willebrand factor multimers on the endothelial surface under flowing conditions. *Blood.* 2002;100(12):4033–4039.
11 Bonfanti R, Furie BC, Furie B, Wagner DD. PADGEM (GMP140) is a component of Weibel-Palade bodies of human endothelial cells. *Blood.* 1989;73(5):1109–1112.
12 McEver RP, Beckstead JH, Moore KL, et al. GMP-140, a platelet alpha-granule membrane protein, is also synthesized by vascular endothelial cells and is localized in Weibel-Palade bodies. *J Clin Invest.* 1989;84(1):92–99.
13 Subramaniam M, Koedam JA, Wagner DD. Divergent fates of P- and E-selectins after their expression on the plasma membrane. *Mol Biol Cell.* 1993;4(8):791–801.
14 Mayadas TN, Johnson RC, Rayburn H, et al. Leukocyte rolling and extravasation are severely compromised in P selectin-deficient mice. *Cell.* 1993;74(3):541–554.
15 Wolff B, Burns AR, Middleton J, Rot A. Endothelial cell "memory" of inflammatory stimulation: human venular endothelial cells store interleukin 8 in Weibel-Palade bodies. *J Exp Med.* 1998; 188(9):1757–1762.
16 Utgaard JO, Jahnsen FL, Bakka A, et al. Rapid secretion of prestored interleukin 8 from Weibel-Palade bodies of microvascular endothelial cells. *J Exp Med.* 1998;188(9):1751–1756.
17 Vischer UM, Wagner DD. CD63 is a component of Weibel-Palade bodies of human endothelial cells. *Blood.* 1993;82(4):1184–1191.
18 Coppinger JA, Cagney G, Toomey S, et al. Characterization of the proteins released from activated platelets leads to localization of novel platelet proteins in human atherosclerotic lesions. *Blood.* 2004;103(6):2096–2104.
19 Journet AM, Saffaripour S, Cramer EM, et al. von Willebrand factor storage requires intact prosequence cleavage site. *Eur J Cell Biol.* 1993;60(1):31–41.
20 Voorberg J, Fontijn R, Calafat J, et al. Biogenesis of von Willebrand factor-containing organelles in heterologous transfected CV-1 cells. *EMBO J.* 1993;12(2):749–758.
21 Haberichter SL, Fahs SA, Montgomery RR. von Willebrand factor storage and multimerization: 2 independent intracellular processes. *Blood.* 2000;96(5):1808–1815.
22 Haberichter SL, Jacobi P, Montgomery RR. Critical independent regions in the VWF propeptide and mature VWF that enable normal VWF storage. *Blood.* 2003;101(4):1384–1391.
23 Haberichter SL, Jozwiak MA, Rosenberg JB, et al. The von Willebrand factor propeptide (VWFpp) traffics an unrelated protein to storage. *Arterioscler Thromb Vasc Biol.* 2002;22(6):921–926.
24 Haberichter SL, Merricks EP, Fahs SA, et al. Re-establishment of VWF-dependent Weibel-Palade bodies in VWD endothelial cells. *Blood.* 2005;105(1):145–152.
25 Barlowe C. Molecular recognition of cargo by the COPII complex: a most accommodating coat. *Cell.* 2003;114(4):395–397.
26 Miller EA, Beilharz TH, Malkus PN, et al. Multiple cargo binding sites on the COPII subunit Sec24p ensure capture of diverse membrane proteins into transport vesicles. *Cell.* 2003;114(4): 497–509.
27 Mossessova E, Bickford LC, Goldberg J. SNARE selectivity of the COPII coat. *Cell.* 2003;114(4):483–495.
28 Denis CV, Andre P, Saffaripour S, Wagner DD. Defect in regulated secretion of P-selectin affects leukocyte recruitment in von Willebrand factor-deficient mice. *Proc Natl Acad Sci USA.* 2001;98(7):4072–4077.
29 Hoyer LW, De los Santos RP, Hoyer JR. Antihemophilic factor antigen. Localization in endothelial cells by immunofluorescent microscopy. *J Clin Invest.* 1973;52(11):2737–2744.
30 Rondaij MG, Bierings R, Kragt A, et al. Dynein-dynactin complex mediates protein kinase A-dependent clustering of Weibel-Palade bodies in endothelial cells. *Arterioscler Thromb Vasc Biol.* 2006;26(1):49–55.
31 Sporn LA, Marder VJ, Wagner DD. Differing polarity of the constitutive and regulated secretory pathways for von Willebrand factor in endothelial cells. *J Cell Biol.* 1989;108(4):1283–1289.
32 Qian Z, Gelzer-Bell R, Yang SX, et al. Inducible nitric oxide synthase inhibition of Weibel-Palade body release in cardiac transplant rejection. *Circulation.* 2001;104(19):2369–2375.
33 Mellman I, Warren G. The road taken: past and future foundations of membrane traffic. *Cell.* 2000;100(1):99–112.
34 Jahn R, Sudhof TC. Membrane fusion and exocytosis. *Annu Rev Biochem.* 1999;68:863–911.
35 Jahn R, Lang T, Sudhof TC. Membrane fusion. *Cell.* 2003;112(4): 519–533.
36 Jahn R. Principles of exocytosis and membrane fusion. *Ann NY Acad Sci.* 2004;1014:170–178.
37 Rondaij MG, Sellink E, Gijzen KA, et al. Small GTP-binding protein Ral is involved in cAMP-mediated release of von Willebrand factor from endothelial cells. *Arterioscler Thromb Vasc Biol.* 2004;24(7):1315–1320.
38 de Leeuw HP, Fernandez-Borja M, Reits EA, et al. Small GTP-binding protein Ral modulates regulated exocytosis of von Willebrand factor by endothelial cells. *Arterioscler Thromb Vasc Biol.* 2001;21(6):899–904.
39 de Leeuw HP, Wijers-Koster PM, van Mourik JA, Voorberg J. Small GTP-binding protein RalA associates with Weibel-Palade bodies in endothelial cells. *Thromb Haemost.* 1999;82(3):1177–1181.
40 Hannah MJ, Hume AN, Arribas M, et al. Weibel-Palade bodies recruit Rab27 by a content-driven, maturation-dependent mechanism that is independent of cell type. *J Cell Sci.* 2003;116 (Pt 19):3939–3948.
41 Matsushita K, Morrell CN, Cambien B, et al. Nitric oxide regulates exocytosis by S-nitrosylation of N-ethylmaleimide-sensitive factor. *Cell.* 2003;115(2):139–150.
42 Fu J, Naren AP, Gao X, et al. Protease-activated receptor-1 activation of endothelial cells induces protein kinase Calpha-dependent phosphorylation of syntaxin 4 and Munc18c: role in signaling p-selectin expression. *J Biol Chem.* 2005;280(5):3178–3184.
43 Fayos BE, Wattenberg BW. Regulated exocytosis in vascular endothelial cells can be triggered by intracellular guanine nucleotides and requires a hydrophobic, thiol-sensitive component. Studies of regulated von Willebrand factor secretion from digitonin permeabilized endothelial cells. *Endothelium.* 1997;5(4): 339–350.
44 Malhotra V, Orci L, Glick BS, et al. Role of an N-ethylmaleimide-sensitive transport component in promoting fusion of transport vesicles with cisternae of the Golgi stack. *Cell.* 1988;54(2):221–227.
45 Block MR, Glick BS, Wilcox CA, et al. Purification of an N-ethylmaleimide-sensitive protein catalyzing vesicular transport. *Proc Natl Acad Sci USA.* 1988;85(21):7852–7856.
46 Whiteheart SW, Rossnagel K, Buhrow SA, et al. N-ethylmaleimide-sensitive fusion protein: a trimeric ATPase

whose hydrolysis of ATP is required for membrane fusion. *J Cell Biol*. 1994;126(4):945–954.

47 Whiteheart SW, Schraw T, Matveeva EA. N-ethylmaleimide sensitive factor (NSF) structure and function. *Int Rev Cytol.* 2001; 207:71–112.

48 Pober JS, Cotran RS. The role of endothelial cells in inflammation. *Transplantation.* 1990;50(4):537–544.

49 Sporn LA, Marder VJ, Wagner DD. Inducible secretion of large, biologically potent von Willebrand factor multimers. *Cell.* 1986; 46(2):185–190.

50 Pinsky DJ, Naka Y, Liao H, et al. Hypoxia-induced exocytosis of endothelial cell Weibel-Palade bodies. A mechanism for rapid neutrophil recruitment after cardiac preservation. *J Clin Invest*. 1996;97(2):493–500.

51 Shreeniwas R, Koga S, Karakurum M, et al. Hypoxia-mediated induction of endothelial cell interleukin-1 alpha. An autocrine mechanism promoting expression of leukocyte adhesion molecules on the vessel surface. *J Clin Invest*. 1992;90(6):2333–2339.

52 Zizzi HC, Zibari GB, Granger DN, et al. Quantification of P-selectin expression after renal ischemia and reperfusion. *J Pediatr Surg*. 1997;32(7):1010–1013.

53 Yamada S, Mayadas TN, Yuan F, et al. Rolling in P-selectin-deficient mice is reduced but not eliminated in the dorsal skin. *Blood*. 1995; 86(9): 3487–3492.

54 Hallahan DE, Virudachalam S. Ionizing radiation mediates expression of cell adhesion molecules in distinct histological patterns within the lung. *Cancer Res.* 1997;57(11):2096–2099.

55 Hallahan DE, Staba-Hogan MJ, Virudachalam S, Kolchinsky A. X-ray-induced P-selectin localization to the lumen of tumor blood vessels. *Cancer Res.* 1998;58(22):5216–5220.

56 Hallahan DE, Virudachalam S. Accumulation of P-selectin in the lumen of irradiated blood vessels. *Radiat Res.* 1999;152(1):6–13.

57 Reidy MA, Chopek M, Chao S, et al. Injury induces increase of von Willebrand factor in rat endothelial cells. *Am J Pathol.* 1989;134(4):857–864.

58 Vischer UM, Wollheim CB. Purine nucleotides induce regulated secretion of von Willebrand factor: involvement of cytosolic $Ca2^{+}$ and cyclic adenosine monophosphate-dependent signaling in endothelial exocytosis. *Blood*. 1998;91(1):118–127.

59 Palmer DS, Aye MT, Ganz PR, et al. Adenosine nucleotides and serotonin stimulate von Willebrand factor release from cultured human endothelial cells. *Thromb Haemost*. 1994;72(1):132–139.

60 Vischer UM, Barth H, Wollheim CB. Regulated von Willebrand factor secretion is associated with agonist-specific patterns of cytoskeletal remodeling in cultured endothelial cells. *Arterioscler Thromb Vasc Biol*. 2000;20(3):883–891.

61 Schluter T, Bohnensack R. Serotonin-induced secretion of von Willebrand factor from human umbilical vein endothelial cells via the cyclic AMP-signaling systems independent of increased cytoplasmic calcium concentration. *Biochem Pharmacol*. 1999;57(10):1191–1197.

62 Kagawa H, Fujimoto S. Electron-microscopic and immunocytochemical analyses of Weibel-Palade bodies in the human umbilical vein during pregnancy. *Cell Tissue Res.* 1987;249(3):557–563.

63 Ribes JA, Francis CW, Wagner DD. Fibrin induces release of von Willebrand factor from endothelial cells. *J Clin Invest.* 1987; 79(1):117–123.

64 Ota H, Fox-Talbot K, Hu W, et al. Terminal complement components mediate release of von Willebrand factor and adhesion of platelets in arteries of allografts. *Transplantation.* 2005;79(3): 276–281.

65 Matsushita K, Yamakuchi M, Morrell CN, et al. Vascular endothelial growth factor regulation of Weibel-Palade-body exocytosis. *Blood*. 2005;105(1):207–214.

66 Iafrati MD, Vitseva O, Tanriverdi K, et al. Compensatory mechanisms influence hemostasis in the setting of eNOS-deficiency. *Am J Physiol Heart Circ Physiol*. 2005;288(4):H1627–1632. Epub 2004 Nov 24.

73

Multiple Functions and Clinical Uses of Caveolae in Endothelium

Lucy A. Carver and Jan E. Schnitzer

Sidney Kimmel Cancer Center, San Diego, California

Caveolae were first identified half a century ago by electron microscopy (EM) as small, flask-shaped, smooth plasmalemmal invaginations. They are most abundant in certain endothelial cells (ECs) and adipocytes but exist in many other cell types. For the next 40 years, exploration into these mysterious organelles was limited primarily to morphological study by EM, including transendothelial transport studies using electron-dense tracers. Until recently, when caveolae were considered at all, they were thought at best to function only in fluid-phase endocytosis (the uptake of small molecules and/or fluids surrounding the cell, also known as *pinocytosis*) and thus were frequently called *pinocytic* or "drinking" vesicles. During the last decade, the development of new molecular tools, markers, and purification techniques that permit selective functional and architectural analysis of caveolae has led to a renaissance in the research of caveolae and to new insights into their structure, composition, physiology, and function. This renewal has been fueled further by key fundamental discoveries of their role in important cellular processes, including signaling and mechanotransduction, cholesterol trafficking, cell growth, and membrane trafficking. Selective targeting of caveolae in vivo can mediate tissue-specific delivery of substances to the endothelium and even across the endothelial layer (via transcytosis) into underlying tissue. In this chapter, we discuss the structure, purification, and composition of caveolae, their relation to lipid rafts, and their role in trafficking and signaling, with emphasis on functions pertinent mostly to ECs, such as acute mechanotransduction and transcytosis. The possible clinical utility of caveolae in the vascular targeting of drugs, genes, and nanomedicines also is presented.

DEFINING CAVEOLAE

Initially, the terms *noncoated plasmalemmal invagination* and *caveolae* referred to the smooth invaginations at the plasma membrane with bipolar-oriented, thin striations around the bulb observed by EM (1,2) (Figure 73.1). These structures were noted to be distinct from the thick, electron-dense coat characteristic of clathrin-coated vesicles (Table 73-1). Caveolae are most abundant (up to 60% of the plasma membrane) in the continuous microvascular endothelia of the lung, skeletal muscle, and heart; moderate in endothelia of large blood vessels; relatively rare in the highly restrictive continuous microvascular endothelia of brain and testes; and apparently absent in sinusoidal endothelia (3), such as in liver (where ECs express abundant clathrin-coated vesicles). They are most abundant in the arteriolar and capillary segments of the microvasculature and less so in the postcapillary venules.

Caveolae appear to require for their existence the expression of caveolin-1 (also called VIP21) (4,5). The caveolin gene family has three members in mammals. Caveolin-1 is a 21-kDa integral membrane protein that exists as two isoforms produced by alternative initiation, α and β, and which oligomerize to form the structural coat of the caveolar bulb (5,6). Caveolin-1 is an integral membrane protein whereby both the NH_2 and COOH termini face the cytoplasm with the intervening hydrophobic domain forming a hairpin loop embedded within the membrane bilayer (5,7). Caveolin-2 is a 20-kDa protein that can form stable hetero-oligomeric complexes with caveolin-1 in the caveolar bulb (8). Both caveolin-1 and -2 are expressed in some ECs, adipocytes, epithelial cells, smooth muscle cells, and fibroblasts (8). Caveolin-3 appears to be expressed mainly in muscle cells, including smooth, skeletal, and cardiac myocytes, and is present in the caveolae of these cell types (9).

Caveolin-1 appears to be a major structural component of caveolae. The expression of caveolin-1 is both sufficient and necessary to drive the de novo formation of caveolae under in vitro and in vivo conditions (10–14). In human umbilical vein ECs (HUVECs), the reduction of *caveolin-1* mRNA expression levels using antisense oligonucleotides ablates caveolae formation at the cell surface (14). Caveolin-2 requires caveolin-1 for proper membrane targeting and for protein stability (11,15), but caveolin-1 may not necessarily rely on caveolin-2 for the formation of caveolae, its overall protein stability, and/or its

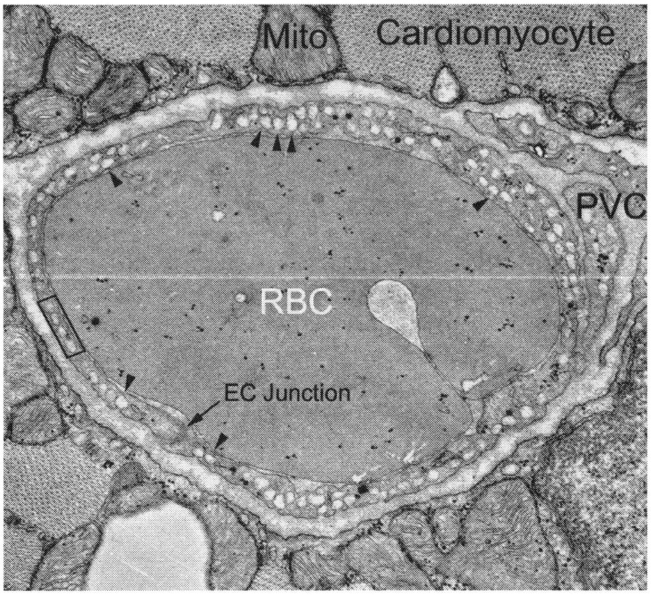

Figure 73.1. Electron micrograph of cross-section of a capillary in mouse myocardium. A red blood cell (RBC) is cramped in the middle of the capillary and next to the very attenuated ECs, with the luminal and abluminal surfaces less 100-nm apart in some regions (*box*). Arrowheads point to some of the caveolae in the EC. Arrow points to EC junction (route for paracellular transport). Beyond the EC is basement membrane (extracellular matrix), some perivascular cells (PVC), and the main tissue cell of the heart, a cardiomyocyte (rich in mitochondria and myofibrils). Mito, mitochondria.

localization in the plasma membrane (13,16). Yet, the coexpression of both caveolin-1 and -2 appears necessary for the formation of stable, deep, plasma membrane–attached caveolae (16). Recent studies have shown that caveolin-2–regulated caveolae biogenesis is dependent on the phosphorylation state of caveolin-2 (16). In addition, the formation and maintenance of caveolae seems to require the presence of cholesterol, which binds tightly to caveolin-1 (17,18). The depletion or removal of membrane cholesterol using cholesterol binding agents, such as filipin, results in the disassembly of caveolae, dispersal of caveolar proteins diffusely across the plasma membrane, and loss of specific trafficking and signaling functions (4,17,19–23).

Caveolin-1 was initially discovered by analysis of the material comprising the caveolar structural coat. Antibodies recognizing VIP-21, a 22-kd substrate for v-src tyrosine kinase in virus-transformed chick embryo fibroblasts, bound the filaments that comprise the striated coat, and the substance was named the caveolin (4). With the discovery of caveolin-1 as a marker for caveolae (4), many methods were introduced to isolate caveolin-containing membranes by subcellular fractionation. The earliest methods were unidimensional, relying on detergent-resistance and buoyant density or solely on buoyant density for isolation of caveolar membranes after cell fragmentation (24). It soon became clear that both detergent and detergent-free methods were problematic. First, caveolin-1 can exist in other locations in the cell, including Golgi apparatus (25) and endoplasmic reticulum (ER) (6,18), leading to questions about its specificity as a sole marker for following caveolae isolation. Second, these preparations are heavily contaminated with low-density, detergent-resistant membranes derived from other intracellular organellar structures, such as nuclei (26,27). The isolation of plasma membranes from fragmented cells using Percoll gradients first, before detergent or sonication treatment, can help but, at least so far, the isolates still contain ample markers for Golgi and ER membranes (24). Such contamination is expected for unidimensional isolation based almost solely on buoyant density for the separation of membrane compartments.

In addition to nuclei, Golgi, and ER, unidimensional isolation of low-density membranes also yields lipid rafts. Lipid rafts are thought to exist as very small plasma membrane microdomains that share certain biochemical properties with the canonical, invaginated caveolae. Because lipid rafts cannot be visualized by light or EM techniques, they are defined biochemically as dynamic, detergent-resistant regions of highly saturated sphingolipids that self-assemble with cholesterol to pack into a highly ordered lipid phase and which are enriched in a wide variety of glycosylphosphatidyl inositol-anchored proteins (GPI-AP) (28). Caveolae and lipid rafts may be found on the cell surface as discrete entities, but they also can be closely associated with each other so that, in some cases, lipid rafts are essentially attached to a caveola at the annulus (26).

Using Nanotechnology to Improve Microdomain Isolation

The development of more rigorous, higher-dimension separation techniques for isolating caveolae has resulted in a greater understanding of the relationship between caveolae and lipid rafts by permitting purification of caveolae from tissue or cultured cells (26,29). One technique perfused positively charged colloidal silica nanoparticles (NPs) to selectively coat the cell surface membrane before crosslinking to form a stable, strong membrane pellicle, then employed tissue/cell homogenization to break the cells or tissue into subcellular components. Because the NP has a density greater than any component of the cell or tissue, a series of centrifugations through high-density gradients yields silica-coated plasma membranes as a membrane pellet. Intracellular membranes or membranes of other cell types (when processing tissue) are not NP-coated so they float on the gradient and are nearly eliminated from the pellet. For example, biochemical and EM analyses show that, in rat organs, this technique provides reasonably purified EC plasma membranes ($\geq$20-fold enrichment of key EC surface marker proteins) with minimal contamination ($\geq$20-fold depletion of markers of other tissue cells as well as nuclei, Golgi, ER, and other intracellular membranes).

As depicted in Figure 73.2, caveolae have been purified from silica-coated luminal EC plasma membranes isolated from rat lung either by physical homogenization (26) or by physiological GTP induction of caveolae budding (30). The silica coating serves to stabilize the flat parts of the plasma

Table 73-1: Comparison of Properties between Caveolae and Lipid Rafts

	Caveolae	*Lipid Rafts*
Other names	Pinocytic vesicles; noncoated plasmalemmal invagination	
Distribution	Continuous microvascular endothelia of the lung, skeletal muscle, and heart; moderate in endothelia of large blood vessels; relatively rare in the highly restrictive continuous microvascular endothelia of brain and testes. Absent in liver sinusoids; presence in lymphatic ECs not yet verified	All ECs
Constituent proteins/enriched constituents	Caveolin-1, -2, and -3; G-protein G_q; PDGF-R; eNOS; Ras, Raf, dynamin, SNAP, vSNARE, VAMP-2, small and large GTP binding proteins; select calcium-dependent, lipid binding annexins; NSF; intersectin; albumin binding proteins	Immunoglobulin E receptors, T-cell receptors, and GPI-AP; G-proteins G_i and G_s
Abrogation	Caveolin-1 deficiency; removal of membrane cholesterol	
Description		Dynamic, detergent-resistant regions of highly saturated sphingolipids that self-assemble with cholesterol to pack into a highly ordered lipid phase and which are enriched in a wide variety of glycosylphosphatidyl inositol-anchored proteins (GPI-APs)
In vivo:in vitro abundance	10–100:1	
Enriched	Caveolin, Ca^{2+}-ATPase, the glycolipid GM_1, VAMP, eNOS, dynamin, and the IP_3 receptor	
Trafficking	Cholera toxin; albumin-gold complexes; SV40, ebolavirus, and polyomavirus.	

membrane, so that the invaginations can be detached selectively from the plasma membrane, leaving behind the plasma membrane proper (including lipid rafts, pits coated by clathrin, or other microdomains) still attached to the silica NPs. The buoyant caveolae are collected at a specific low density after being floated away from the high-density silica membrane pellicle by sucrose gradient centrifugation. Other possible invaginations include clathrin-coated structures, which must be fully developed as invaginations rather than large pits to avoid silica coating. These are much fewer at the lung EC surface in vivo than are caveolae, and, regardless, are bigger and much more dense than caveolae. More than 95% of the caveolae on the silica-coated plasma membranes can be removed to yield a homogeneous population of morphologically distinct, smooth, flask-shaped vesicles of 60 to 80 nm in diameter as visualized by EM (26) (Figure 73.2). This reasonable homogeneity is confirmed biochemically by caveolin antibodies binding >95% of the isolate to magnetic beads. Biochemical analysis also reveals that the isolated caveolae represent specific microdomains of the cell surface with their own unique molecular topography (26,31). They are enriched in seven caveolar markers established by immunogold EM: caveolin-1, calcium (Ca^{2+})-ATPase, the glycolipid GM_1, vesicle-associated membrane protein (VAMP), endothelial nitric oxide synthase (eNOS), dynamin, and the IP_3 receptor (26,29,31–35). By contrast, clathrin, angiotensin-converting enzyme (ACE), Band 4.1, and β-actin are not detected or at least markedly depleted ($\geq$15-fold) from the isolated caveolae, despite ACE, Band 4.1, and β-actin being amply present in the silica-coated plasma membranes. Alternatively, caveolae also can be isolated from low-density membrane subfractions by immunoaffinity techniques using magnetic nanoparticles and select caveolin antibodies (29).

Once free of caveolae, the stable attachment of silica particles to the remaining silica-coated plasma membranes provides a means to isolate the remaining flat detergent-resistant membranes, the so-called lipid rafts, away from both caveolae and the plasmalemma proper (26). After a high-salt wash to remove the silica beads, most of the remaining flat plasma membrane subjected to Triton solubilization will be "melted

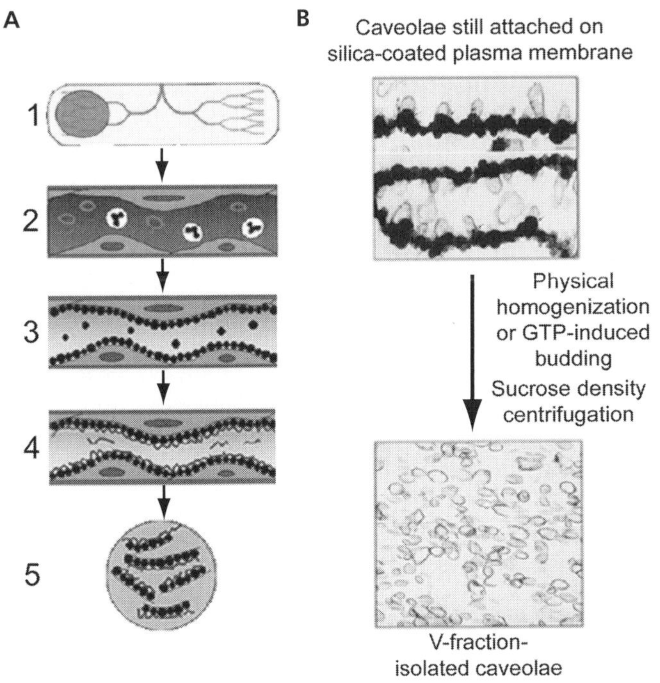

Figure 73.2. Subcellular fractionation to isolate luminal EC plasma membranes normal or tumor tissue in vivo. (**A**) The vasculature of an organ or solid tumor (*1*) is flushed free of blood, then perfused with a positively charged colloidal silica solution to (*2* → *3*) coat specifically the luminal EC surface membrane directly exposed to the circulating blood. The monolayer of silica particles stably adherent to this plasma membrane of interest are cross-linked with polyacrylic acid to establish intraparticle cohesion and to neutralize excess positive charges (*4*). This coating increases the density of the plasma membrane significantly above any other tissue component, so that after tissue homogenization, large sheets of silica-coated luminal EC plasma membranes (*5*) with their attached caveolae are readily isolated away from other membranes and debris by ultracentrifugation through a high-density medium. Further subcellular fractionation to isolate caveolae (V-fraction) can be performed by homogenization or GTP-induced budding (both of which release the caveolae from the silica-coated plasma membranes), followed by flotation via sucrose density centrifugation. (**B**) Electron micrographs of silica-coated rat lung EC plasma membranes (*top*) and caveolae (*bottom*) after isolation.

away" to leave behind the detergent-resistant microdomains, which vesiculate and can be isolated by sucrose density centrifugation as a homogeneous population of rather large vesicles (very similar to those described next). These new NP-based preparations reveal molecular distinctions between caveolae and putative lipid rafts, both isolated from the same EC plasma membranes (26). Caveolae appear concentrated in caveolin-1 but not so for several GPI-APs, which instead exist quite concentrated in lipid rafts devoid of caveolin-1 (26,29).

Lipid Rafts

In contrast to the multidimensional silica NP-based methods for isolating caveolae, unidimensional methods isolate caveolin-rich fractions containing lipid rafts, the obvious smaller, omega-shaped caveolae, and even intact large vesicles with a caveola attached (26). The attached caveola usually is oriented inside the larger vesicle but also can be seen sometimes with the bulb oriented outside. Immunogold EM reveals GPI-AP and carbonic anhydrase on the larger vesicles but not the free or attached caveolae, suggesting that these larger vesicles are, or contain, lipid rafts (26). This finding of coisolation using unidimensional methods is underscored by studies showing that similar, large, detergent-resistant vesicles rich in GPI-AP are isolated from cells lacking caveolin or caveolae (36,37). Moreover, only approximately 40% of these isolates can be bound to, and isolated by, magnetic beads coated with caveolin antibodies (much less than the 98% binding observed for the silica-based preparations) (29). Thus, it is quite apparent that unidimensional methods based primarily on physical separation by membrane buoyant density are unable to discriminate caveolae from lipid rafts. This heterogeneity is compounded further by the presence of additional microdomains (of similarly low buoyant density and usually detergent-resistant) derived from contaminating membranes of intracellular organelles (as discussed earlier).

In retrospect, it is not surprising that, as a consequence of early reports showing that caveolin-containing, detergent-resistant membrane fractions were rich in both GPI-AP and caveolin, many investigators concluded that caveolae constituted the GPI-AP-enriched lipid rafts in the plasma membrane. This notion was supported further by the findings that antibodies to GPI-AP can be found in caveolae using immunocytochemistry (38) and EM (23,39), and that GPI-AP can be internalized via caveolae (39). Later, however, it also was reported that several GPI-APs are not normally clustered in caveolae but rather artifactually induced to localize in caveolae by antibody sequestration resulting from crosslinking by secondary reporter antibodies (40). These early conflicting data led to confusion about the exact definition and composition of caveolae that remains to this day.

It was even proposed that caveolin interacts with and traps GPI-AP in caveolae (41), but this seems unlikely given the structural nature of caveolin and the opposite orientation of these proteins in the bilayer. This does not mean that some GPI-AP may not exist concentrated in caveolae in some cell types. Many GPI-APs, when sequestered, may move into caveolae. Thus, natural crosslinking ligands to GPI-APs, yet to be identified, could potentially sequester select GPI-APs and concentrate them in caveolae. Alternatively, an intermediate linker between caveolin and select GPI-APs eventually may be found. But for now it is apparent that GPI-APs appear to reside preferentially as a group in rafts. Also note that clathrin-coated vesicles can endocytose GPI-APs (42), so the relationship between the molecular inventory of caveolae and lipid raft is not a simple continuum.

In the end, caveolae and lipid rafts (as defined by detergent-resistant low-density membranes) appear to be distinct microdomains containing a unique set of proteins, including signaling molecules. For example, caveolae contain growth factor receptors (e.g., platelet-derived growth factor receptor) whereas lipid rafts contain immunoglobulin E receptors, T-cell receptors, and GPI-APs (43). In cells expressing caveolin and caveolae, the heterotrimeric G-protein, G_q, tends to concentrate in caveolae but not in lipid rafts, whereas G_i and G_s exist in lipid rafts preferentially over caveolae (44). The apparent ability of caveolin-1 to form a complex in the plasma membrane with G_q, but not G_i or G_s, appears to trap G_q effectively in caveolae and minimize its presence in lipid rafts (44). In cells that don't normally express caveolae, G_q partitions into lipid rafts with other G-proteins (44). These findings discovered by tissue/cell subfractionation were confirmed by immunofluorescence microscopy on intact cells (44).

If they exist, lipid rafts are probably very small, flat, plasmalemmal microdomains (<50 nm) enriched in GPI-AP. As such, they lack a distinct morphology easily detected by EM and thus require sophisticated biophysical detection that, to date, has not definitively established their existence. They currently are defined most often operationally through biochemical isolation and detergent resistance (45). Such analysis is complicated by the fact that many cellular components can be detergent-resistant and have low buoyant densities, including cytoskeletally tethered membrane proteins, protein–protein complexes, and adhesion sites. Also, one must not forget that, during the 1980s, nonionic detergent insolubility of a protein was considered good evidence for cytoskeletal association. In contrast to rafts, the existence of caveolae is certain. They have a distinct omega-shaped invaginated form easily detected by EM, are readily labeled at the cell surface with caveolin antibodies, and can be isolated to homogeneity as appropriately sized and shaped vesicles coated with caveolin. Much more research is required to buttress so far suggestive evidence and to establish clearly the existence and even proposed function of lipid rafts. Although much suggestive evidence has accumulated to support a function for rafts in signaling, until raft targeting is shown to be functionally obligatory and protein–protein interactions auxiliary, this hypothesis will remain rather speculative and unproven.

Many still wish to consider rafts and caveolae as near equivalents. Some even describe caveolae as a subset of lipid rafts. If so, then nearly all unidimensional isolation procedures become, by definition, more homogeneous. Also the lipid raft field becomes more grounded and not reliant on establishing the existence of lipid rafts, which currently appears close to impossible. In the end, it should not be forgotten that finding distinction is fundamental to the philosophy and progress of science.

FUNCTIONS OF CAVEOLAE

Role of Caveolae in Ligand-Based Cell Surface Signaling

The compartmentalization of key signaling molecules in caveolae appears to be necessary to provide the rapid, efficient, and specific propagation of extracellular stimuli to intracellular targets. This key concept has emerged from studies performed over the last decade. Caveolae are rich in many signaling molecules including platelet-derived growth factor receptor (PDGFR) (21,46), eNOS (35,47,48), select heterotrimeric G-proteins, and nonreceptor tyrosine kinases (21). Many of these proteins appear to interact with caveolin (49,50), and may function in acute mechanotransduction (20,35) and cell surface receptor signaling (50–52). Caveolae are rich in cholesterol, and the removal of cholesterol from the plasma membrane by cholesterol binding agents such as filipin affects microdomain organization or compartmentalization (4,23). These compounds disassemble caveolae, which results in a dispersal of caveolar molecules to a more random distribution over the cell surface, thereby disrupting both transport and signaling functions (19,21). For example, filipin treatment can prevent PDGF-induced signaling downstream of the initial receptor autophosphorylation (21). Knockdown of caveolin by antisense oligonucleotides eliminates caveolae and inhibits angiogenesis (which is dependent on specific signaling factors) both in vitro and in vivo (14). *Caveolin-1*–knockout mice lack detectable cell surface caveolae and exhibit alterations in signaling, including changes in the functions of eNOS, p42/44 mitogen-activated protein kinase (MAPK), and M_3 muscarinic cholinergic and insulin signaling (53–57).

G-protein–coupled receptors (GPCR) may utilize cellular localization and internalization to regulate receptor–effector signaling specificity and receptor desensitization, respectively. Upon ligand engagement, many GPCR are rapidly internalized via multiple endocytic pathways not yet fully defined. Claing and colleagues detected three distinct pathways for GPCR internalization based on their sensitivity to overexpression of β-arrestin and dynamin mutants and the small G-protein–coupled receptor kinase (GRK)-interacting protein, GIT1 (58). Pathway I (β_1 adrenergic receptor, β_2 adrenergic receptor, μ-opioid receptor, adenosine 2B receptor, and M_1 muscarinic receptor) is inhibited by overexpressed wild-type GIT1, mutant β-arrestin, and K44A dynamin. Pathway II mediates internalization of ETB and vasoactive intestinal peptide-1 receptor and is inhibited by K44A dynamin but not by GIT1 or mutant β-arrestin, whereas pathway III (angiotensin 1A receptor and M_2 muscarinic receptor) is not affected by overexpression of the three proteins. Clathrin-mediated endocytosis clearly mediates pathway I, but the mechanisms underlying the others have been much less clear. It has been speculated that caveolae likely mediate pathway II (58), consistent with observed the dynamin-mediated fission of caveolae (33), but this still requires detailed experimentation and rigorous proof. Various GPCR appear to sequester in caveolae and/or lipid rafts upon activation by ligand, (e.g., endothelial differentiation gene [EDG]-1, B2 Bk, and thyrotropin receptors) (59–61). Other GPCRs, such as ETA, may constitutively reside in cell surface "caveolae-like rafts" (51). It appears likely that caveolae promote efficient signal propagation into the cell not only through effective compartmentalization but also through regulated and directed trafficking to intracellular compartments for continued signaling or desensitization. This internalization may also facilitate subsequent desensitization of select pathways to further ligand or other stimulation.

Role of Caveolae in Mechanotransduction

A major physiological function of the endothelium is the conversion of hemodynamic forces into a series of adaptive biological responses that minimize mechanical stress as well as injury to the blood vessel and the tissue itself. Physical forces generated by vascular blood flow appear to play a central role in the acute and chronic regulation of the structural and functional phenotype of the endothelium. By being directly exposed to the circulating blood, the vascular endothelium is constantly subjected to fluid mechanical forces, such as frictional fluid shear from blood flow parallel to the luminal EC surface and vascular pressure transmitted across the entire blood vessel surface into the tissue. These mechanical forces induce compensatory protective responses by ECs that ultimately bring about alterations in cellular structure and function, including the release of vasoactive substances that function to alter vascular tone and thus mitigate hemodynamic stress. Normal compensatory endothelial responses to mechanical stress include the rapid release of a number of vasoactive substances, such as nitric oxide (NO), prostacyclins, and coagulation factors, as well as the induction or upregulation of a number of "atheroprotective" genes including cyclooxygenase-2, superoxide dismutase, and eNOS (62). Disturbances in this process may play a significant role in the pathogenesis of cardiovascular diseases such as atherosclerosis, hypertension, diabetic microangiopathy, and coagulopathies (63). Disturbed flow is considered a dominant initiator of atherosclerotic lesions and correlates with lesion localization at vessel bifurcations. ECs in these regions exhibit an altered phenotype with enhanced proliferation, increased permeability, and impaired endothelial vasoresponsiveness. Fluid stressors with time promote EC adhesion and atherogenesis (62,64).

Because the luminal EC surface is exposed directly to blood flow and all resultant force vectors, it appears logical that acute mechanotransduction occurs at this plasma membrane. In 1995, it was proposed that caveolae may be acute mechanosensing organelles at the surface of ECs (65). Evidence in support of this hypothesis has been accumulating (20,22,35,66–68). Molecular mapping shows that caveolae are enriched in various signaling molecules known to be activated in response to shear stress, including specific G-proteins, select nonreceptor tyrosine kinases, Ras, Raf, and eNOS (20,21,35). Increased vascular flow and pressure in situ very rapidly induce the tyrosine phosphorylation of EC surface proteins located primarily in caveolae (20). Caveolae have 50-fold more kinase activity than the rest of the plasma membrane (21). Disassembly of caveolae disperses the molecular constituents of this specialized compartment and prevents acute flow-activation of cell surface tyrosine kinases as well as downstream cytosolic kinases such as extracellular signal-regulated kinase (ERK) but not c-jun N-terminal kinase (JNK) (20,22). Increased fluid stress also rapidly stimulates eNOS, which is concentrated at the EC surface in caveolae (35). Finally, select G-proteins are highly concentrated in caveolae (44), and shear stress appears to induce G-protein activation, which in turn stimulates the production of NO and activates ERK and JNK (66). In cultured ECs, physiological levels of shear stress redistribute caveolin from the Golgi to the cell surface, induce caveolae formation, and regulate ERK and Akt activity (67).

Caveolin-1 may act as a mechanosensitive scaffold for attracting and concentrating a variety of signaling molecules to caveolae, thus forming a specialized mechanosensing compartment (20,21,35). Many laboratories have used various assays in different systems to detect an interaction between caveolin and eNOS, which may inhibit enzymatic activity and be regulated physiologically (35,49,69,70). EM has confirmed eNOS concentration in caveolae in endothelium (35). Increased flow in situ rapidly dissociates eNOS from caveolin and concomitantly increases eNOS association with key positive modulators such as calmodulin (35). *Caveolin-1*–null mice lack caveolae and provide further evidence of a regulatory role for caveolae and caveolin in eNOS (12). They exhibit an inability to respond properly to mechanostressors and thus display reduced ability to maintain steady contractile tone, as well as having defects in endothelium-dependent vasodilation and contractility (12,57).

It has been proposed that caveolin, as a highly oligomerized, proteinaceous coat anchored directly in the inner lipid membrane around the bulb of caveolae, may function like a tension-bearing spring, sensing acute changes in mechanical forces at the cell surface to regulate the activity of key signaling molecules (20,35). Caveolin, possibly through conformational sensitivities of the oligomer, may act as an inhibitory clamp for eNOS that is released to permit enzymatic activation (35). The injection of a cell-permeable caveolin scaffolding domain peptide into mice uncouples caveolin-1-eNOS complexes through competitive binding to keep eNOS inactive, thereby eliminating physiological regulation by caveolin-1, preventing activation of eNOS, and inhibiting NO-mediated vascular hyperpermeability and vasodilation (71). Other signaling molecules (G-proteins, NRTK, Ras) also may interact with caveolin to concentrate in caveolae and regulate their signaling (44,50,52). Altogether, these data strongly implicate plasmalemmal caveolae as an important component of the acute mechanosensing and signaling mechanism in the vascular endothelium.

Neutral sphingomyelinase (N-SMase), an enzyme exposed externally on the plasma membrane, is activated to produce ceramide in an acute and transient manner by increase in flow rate and pressure in rat lung vasculature. This acute mechanoactivation occurred primarily in EC surface caveolae enriched in sphingomyelin and neutral sphingomyelinase, but not acid sphingomyelinase (72). Specific inhibition of N-SMase but not acid SMase activity prevented mechanoactivation of N-SMase and downstream tyrosine and MAPKs. Cell-permeable ceramide analogues mimicked the rapid mechanoinduced tyrosine phosphorylation of cell surface proteins. Both ceramide and mechanical stress-induced mechanosignaling were inhibited by genistein, Herbimycin A, and PP2, indicating a role of Src-like kinases in ceramide-mediated mechanotransduction. Ceramides also induced serine/threonine phosphorylation to activate the Akt/endothelial NO synthase pathway. Thus, N-SMase in EC surface caveolae may be the upstream initiating mechanosensor, which acutely triggers mechanotransduction by generation of the lipid second messenger ceramide. Further discussion of EC mechanosensation can be found in Chapter 28.

Role of Caveolae in Transvascular Exchange and Capillary Permeability

ECs form an attenuated monolayer lining all blood vessels and creating a critical barrier to control the exchange of molecules from the tissue to the blood, as well as from the blood to the inside of each tissue (perivascular space, interstitium, tissue parenchyma, and tissue cells). The lipid membrane tends to restrict transmembrane transport– in both directions – of both small and large circulating molecules, including water, ions, and proteins because of electrostatic, steric, and/or hydrophobic exclusion forces (73,74). Highly regulated paracellular pathways through the intercellular junctions mediate the filtration of water and small solutes driven by concentration and pressure gradients (75). Caveolae can provide transcellular pathways to transport macromolecules, apparently trafficking their cargo as a vesicular shuttle from the luminal to abluminal aspect of the endothelium (76–78). They also can form transient transendothelial channels in very attenuated regions of the cell through the fusion of two or more caveolae, each located on apposing plasma membranes, to provide a direct conduit for the exchange of both small and large plasma molecules (79). These transcellular pathways may include both fluid-phase and receptor-mediated transport.

Although studied for half a century, defining the exact function of caveolae in transport has been somewhat problematic. In fact, Palade, after his discovery of caveolae in endothelium, had the insight to propose that caveolae transport molecules as quanta (specific packets of a finite number of molecules, as opposed to simply a stream of molecules.) across the EC into the tissue (80,81). Subsequent decades of research using EM show that various macromolecules such as albumin, orosomucoid, ferritin, and even gold-tagged blood proteins can enter luminal caveolae and, with time, are found in abluminal caveolae and the tissue interstitium (76,82,83). In all these cases, the labeling within caveolae, especially those apparently free in the cell, was interpreted as proof for caveolar-mediated transport. Yet, these probes may be difficult to follow under native conditions in vivo, and do not appear to bind in caveolae with a specificity, affinity, and level of exclusivity that rules out transport by other potentially less specific pathways. Moreover, because of the racemose structure of caveolae, wherein many vesicles link to each other in a chain to form a grape-like branching "cave" penetrating deep into the cell, conventional sections for EM can be problematic and quite misleading for interpreting transport studies with these less-than-specific markers. Conventional sections show that about 50% of the noncoated vesicles are apparently free in the cytoplasm, and studies using serial ultrathin sectioning reveal that $\leq$1% are truly free and unattached to other membranes (84–86). That very few vesicles are found free in the cytoplasm has been interpreted as proof that caveolae are not dynamic, but rather static elements that do not bud from the cell surface to function in endocytosis or transcytosis (84–86). An alternative interpretation is that, within the time frame of specimen processing, few free vesicles are found because the caveolar budding and fusion process, once initiated, is very rapid, and the thermodynamic tendency favors fusion with cell membranes. Yet it is also possible that any observed transendothelial transport of probes to the interstitium in past studies may be primarily mediated by nonspecific paracellular transport, including retrograde entry of the probe into the abluminal caveolae (the "back-filling" model). This alternative pathway cannot be dismissed simply because the lack of both high-affinity probes targeting caveolae specifically and physically identical control probes (not targeting caveolae) precludes definitive proof that caveolae function in transport as discrete carrier vesicles capable of delivery. The idea that caveolae function in any transport (including transcytosis) has been the subject of a very contentious 40-year-old debate (63,87).

The last decade has witnessed the development of new assays for measuring endothelial transport and studying the mechanism of caveolae trafficking, including intact and permeabilized cultured cell assays, in vitro reconstituted cell-free systems, and immunofluorescence confocal microscopy. The development of a cell-free system that reconstituted caveolar budding in vitro demonstrated that caveolae fission can be induced and regulated by GTP hydrolysis (30). Simply adding GTP (but not UTP, CTP, nonhydrolyzable GTP [GTPγS], GDP, or ATP alone) to the isolated EC plasma membranes greatly stimulates the budding of caveolae (30). These budded vesicles are collected away from the remaining plasma membrane by centrifugation in a sucrose gradient and are concentrated in caveolin and other known caveolar proteins but not in several GPI-APs. This GTP effect has been confirmed in cultured ECs. After mild permeabilization to get the GTP into the cells, EM shows the loss of nearly all caveolae at the cell surface, and confocal immunofluorescence microscopy shows much greater internalization of probes entering caveolae. In fact, nonhydrolyzable GTPγS seems to "fix" at the cell surface, and prevents both caveolar fission in vitro and the EC endocytosis of various probes entering caveolae (30). The GTPase mediating this budding has been identified as dynamin (33) (see the following discussion). These data provided the first concrete evidence that at least some caveolae can indeed bud off the plasma membrane to form discrete carrier vesicles that participate in membrane trafficking, and thus put to rest a decades-old contention that all caveolae are fixed plasmalemmal invaginations incapable of budding.

Primarily through studying cultured cells, other evidence has been accumulating in favor of caveolae functioning as dynamic vesicular carriers. Caveolae appear to traffic select ligands, such as cholera toxin and albumin–gold complexes, to specific destinations within the cell, such as endosomes and lysosomes, in many cell types including ECs (39,88,89). Caveolae also have been implicated in the infectious process of SV40 (90,91), ebolavirus (92), and polyomavirus (93,94). SV40 binds to cell surface major histocompatibility class (MHC) class I receptors and associates with caveolae to be internalized via a relatively slow, dynamin-dependent mechanism, with each vesicle containing a single virus to be found approximately 2 hours later in the ER (90,95,96). Polyomavirus follows a different, Brefeldin A-sensitive trafficking pathway than does SV40 (97). Furthermore, select pharmacological agents, such as filipin and N-ethylmaleimide (NEM), have been identified to inhibit the endocytosis and transcytosis of select ligands that preferentially bind within caveolae. Cholesterol-binding agents, such as filipin, reduce caveolae number and significantly decrease caveolae-mediated cellular trafficking of such molecules as albumin and insulin (4,19,98). NEM is a thioalkylating agent that inhibits the fusion of carrier vesicles to target membranes and thus reduces multiple cellular vesicular transport processes (99–102). As with other vesicular pathways, including synaptic neurotransmitter release (103), apparent caveolae-mediated transport appears to occur via NEM-sensitive vesicle-membrane fusion (104).

More recently, the ability to isolate EC caveolae to a high degree of purity directly from tissue has enabled the discovery of specific molecular machinery in caveolae that mediates the budding, docking, and fusion of transport vesicles (77). Several key proteins are concentrated in caveolae, including dynamin; SNAP; vSNARE; VAMP; small and large GTP binding proteins; select Ca^{2+}-dependent, lipid binding annexins; and the NEM-sensitive membrane fusion factor (NSF) (31,33,34). In cultured ECs, selective cleavage studies show that caveolae require intact VAMP-2 for efficient, targeted transport of cholera toxin B subunit to intracellular organelles such as endosomes (34). Immunogold EM on ultrathin frozen sections of lung tissue reveals VAMP-2 at the bulb of caveolae (34) and dynamin concentrated at the neck of caveolae (33). Overexpression of a dominant negative mutant form of dynamin, which does not hydrolyze GTP normally because of a single amino acid change (K44A), inhibits the GTP-induced budding of caveolae in the reconstituted cell-free assay and prevents caveolae-mediated internalization in both intact and permeabilized ECs (33,105). Dynamin forms an oligomer around the neck of caveolae, where it can hydrolyze GTP to drive neck extension and result in the fission of caveolae to form free transport vesicles. Recently, intersectin, a known component of the endocytic machinery in non-ECs, was found at the caveolar neck in ECs in a complex with dynamin and SNAP-23, where it may regulate caveolar fission and internalization (106). Thus, caveolae appear to have the key molecular machinery necessary to function as endocytic and possibly transcytotic vesicles.

To study in greater detail the transport function of caveolae, it is critical to have caveolae-specific probes that can target the caveolae from outside of the cell. Currently used probes, such as cholera toxin, albumin, cross-linked GPI-AP, and even viruses such as SV40, target caveolae neither directly nor specifically. They can also be internalized by other endocytic pathways (40,98,107). Recently, a few such high-affinity probes have been created via monoclonal antibody hybridoma technology using as immunogen the isolated silica-coated EC plasma membranes and caveolae (78,108). After intravenous administration of antibody conjugated to colloidal gold NPs (Au-NPs), EM of lung tissue reveals specific and rapid targeting of caveolae followed by transendothelial transport (78). Within 3 minutes, Au-NPs appear bound at the EC surface, specifically inside luminal caveolae, with little to no Au-NPs detected in caveolae located further inside the cell nor in clathrin-coated pits or in the subendothelial space. By 10 minutes, more Au-NPs are observed inside each caveola, with more caveolae containing Au-NPs, including those obviously connected at the luminal surface as well as those further inside the cell. Perhaps more importantly, caveolae at the abluminal surface contain Au-NPs and can be captured rather frequently with Au-NPs emanating from the caveolar neck into the perivascular space. By 15 minutes, in areas where the attenuated endothelium is in close apposition to the alveolar epithelium, some of the Au-NPs accumulating in the subendothelial space percolates through the basement membrane and is even taken up by epithelial caveolae for transport across the cell to the alveolar space. Control Au-NPs conjugated to isotyped matched IgG or antibodies to GPI-AP do not accumulate in caveolae and is not transported across the endothelium into the tissue, thereby ruling out any nonspecific paracellular mechanism. Thus, caveolae can indeed mediate the transcytosis of select molecular cargo to overcome both the endothelial and epithelial cell barriers in tissue (78).

Insights into the role of caveolae in capillary permeability also have been gained from studies of *caveolin-1*–knockout mice. Normally, albumin and even albumin-conjugated to gold particles are thought to bind specific albumin-binding proteins in caveolae (88,109–112), and appear to be transported by caveolae into and even across the EC barrier into the tissue (76,83). But in *caveolin-1*–knockout mice, the caveolae in microvascular endothelium are absent, and this transport pathway is lost (113). Although albumin-gold particles bound to the EC surface in vivo, they did not bind caveolae and demonstrated negligible transport. However, given that *caveolin-1*–knockout mice are viable, other mechanisms must be responsible for mediating the transendothelial transport of essential nutrients and other molecules to underlying tissues. Indeed, these animals demonstrate increased paracellular transport. They have an overall increased microvascular permeability in vivo (12,56,57,114) resulting in the rapid, nonselective transvascular transport of both large and small molecules from the bloodstream, apparently due to significant changes to EC tight junctions (56). Because eNOS influences permeability by producing NO (56,71,115–118) that can greatly increase transvascular exchange, eNOS activity without caveolin lacks appropriate regulation and becomes sufficiently elevated to cause the vascular hyperpermeability in the *caveolin-1*–knockout mice. Injection of the NOS inhibitor

N(G)-nitro-L-arginine methyl ester (L-NAME) into *caveolin-1*–knockout mice reverses most of the microvascular hyperpermeability (56). Injection of a cell-permeable caveolin scaffolding domain peptide into mice prevents the activation of eNOS and inhibits NO-mediated vascular hyperpermeability and vasodilation (71). The physiological consequences of the apparently widespread EC dysfunction caused by a lack of cell surface caveolae are severe. *Caveolin-1*–knockout mice suffer from edematous lungs, cardiac abnormalities, low tolerance to physical stress, aberrations in lipid metabolism, and decreased life span (12,57,114,119). Future, more conclusive, experiments are needed, using inducible *caveolin* knockdown methods in adults under more acute conditions that are likely to avoid this apparently paradoxical, yet partially compensatory, effect on capillary permeability (56,71,115–118).

Recently, an interesting structure potentially related to caveolae has been discovered. Vesiculo-vacuolar organelles (VVOs) are grape-like clusters of vesicles and vacuoles found in EC cytoplasm that interconnect with each other and with the luminal and abluminal plasma membranes via stomata that may be open or closed by a diaphragm (120). VVOs extend across ECs, represent sites of transendothelial hyperpermeability in tumor microvasculature and in normal and inflammatory venules, and appear to be regulated by vascular permeability factor/vascular endothelial growth factor (VPF/VEGF), serotonin, and histamine. The possible localization of caveolin and VAMP in vivo suggests a close relationship between these structures and caveolae (120). Because of their large size, they cannot be present in the highly attenuated ECs normally found in most microvessels, but rather reside primarily in the thicker endothelium of postcapillary venules and tumors. It can be speculated that factors that induce caveolae internalization and reduce EC membrane tension (lower pressure) may reduce cell attenuation, decrease caveolae cell surface density, and possibly even redistribute caveolae to VVOs, thereby increasing not only their number and/or size but also nonspecific transport and overall apparent vessel permeability, while reducing the permselectivity of the EC barrier.

The current data demonstrating the presence of key molecular machinery for the budding, docking, and fusion of caveolae as vesicular carriers, and the selective transendothelial transport of probes targeting caveolae versus identical control probes, clearly establish that caveolae are capable of mediating transcytosis in endothelium. The key questions remaining are: To what degree does this occur in vivo? How active is this process? And, is it regulated (if so, how and what induces it)?

To get answers, we must move beyond current methods. Static imaging by EM only records data points at specific time points; this lacks continuity in the observation and thus, may miss key details. In addition, the electron-dense gold particles used to tag antibodies or other molecules for EM detection are rather large and can alter both the trafficking pathway and kinetics of trafficking (e.g., retard the rate at which the process occurs). Thus, it will be necessary to develop better imaging probes coupled to sensitive dynamic live imaging techniques to define the kinetics and degree of active transport possible by caveolae under physiological conditions found in tissue in vivo. Intravital microscopy and multiphoton microscopy may be useful in this regard. Higher-intensity fluorophores and NPs, such as quantum dots, may eventually permit the trafficking of a single caveola (especially in cell culture) or at least small clusters in vivo. Last, a more comprehensive, detailed proteomic mapping of caveolae using mass spectrometry (see Chapter 99) to define proteins concentrated selectively in this microdomain may elucidate not only new functions but also novel targets for developing new probes to study caveolar functions.

Other Functions of Caveolae

In addition to the functions discussed earlier for caveolae in ECs, caveolae play key roles in cellular pathways in other cell types. For example, caveolae and caveolin have regulatory roles in cholesterol metabolism and in cholesterol-dependent cell growth and adhesion, as well as in glucose metabolism and multidrug resistance. These aspects of caveolae function have been previously well-reviewed (121–123).

CLINICAL UTILITY OF CAVEOLAE: A PATHWAY INTO TISSUE FOR DRUG AND GENE THERAPY

The design of traditional therapeutic agents has focused primarily on the immediate interaction between a drug and its intended pharmacological site of action, usually a specific tissue. For many drugs that are small molecules, accessibility into tissue normally does not pose a problem. After entering the circulation by injection or absorption, these drugs readily move across the EC barrier (primarily via paracellular transport) from the bloodstream across intercellular junctions to gain access to underlying tissue cells. The ability of drugs to nonspecifically diffuse throughout many tissues of the body can in fact be quite disadvantageous, because access to many tissues increases the effective volume of distribution within the body, which in turn results in the dilution of drug potency and unwanted pharmacological side effects. On the other hand, larger molecules, such as proteins, antibodies, NPs, or gene vectors, lack access to many organs because they are unable to readily cross the blood vessel endothelium to reach their targets (124–127). For example, although initially promising in selectively killing tumor cells in culture, monoclonal antibodies have not been very successful in targeting drugs to extravascular sites, in part because they poorly cross the blood vessel endothelium to reach their antigen targets deep inside the solid tumor tissue (126,127). In fact, dose escalation becomes necessary to achieve some efficacy but also unfortunately increases systemic toxicity. Thus, for both large and small pharmaceutical agents, the ability to rapidly target the endothelium and promote transendothelial passage to the tissue of interest may greatly improve pharmaco-efficacy at lower dosages and limit systemic side effects caused by drug uptake in healthy tissue and cells (125,128).

To achieve "site-specific drug delivery," most drug discovery efforts have focused on designing drugs as well as NP- and antibody-based delivery systems that specifically interact with diseased or malfunctioning target cells, with little consideration for the challenges and opportunities inherent in transendothelial transport (125,128). For example, high throughput, in vitro assays that screen for pharmacological actions on the desired cell type are being used extensively in the design of new drugs. Most in vivo "omic"-based target discovery efforts are aimed at the whole tissue level, and are thus biased toward the more numerous parenchymal cell types, whose targets are poorly accessible to most biological agents injected into the circulation. Thus, these drugs frequently perform much less effectively in vivo, where the agent must reach its target cells in a tissue in sufficient quantities to be potent while sparing normal organs (126,127).

Classic gene therapy aims to deliver genetic material to a specific diseased tissue using viral and nonviral vectors to achieve the sustained expression of a specific protein product. The new and rapidly growing field of nanomedicine seeks to apply nanotechnology for the diagnosis, monitoring, and treatment of disease. A major goal of nanomedicine is to use NPs as carriers to selectively deliver, upon injection into the bloodstream, a drug, imaging agent, or gene as a probe to its site of action within specific organs or solid tumors. Most attempts to use an antibody, gene vectors, or NPs to target solid tumors or other tissues in vivo have met with modest success. NPs (e.g., pegylated cationic liposomes, latex spheres, quantum dots) and viral vectors are much larger than standard drugs and even biologics (20 to >300 nm in diameter). When injected intravenously, they show very limited uptake into most organs and tumors (129,130). They appear to enter solid tumors mostly through "passive" targeting, and their uptake is attributable mainly to tumor vessel leakiness, which occurs with considerable heterogeneity in small tumor regions. As large foreign particles, they are rapidly removed from the circulation by the reticulo-endothelial system (RES), primarily in liver and spleen. Yet, if selective targeting of gene vectors or NPs is to be effective, it will be necessary to obtain not only accessible targets but an active means to overcome the restrictive EC barrier to reach their site of action, the underlying tissue or tumor cells. Although not yet realized, active targeting is predicted to mediate the faster, more efficient targeting of specific cell types (129). The benefits of more rapid targeting are twofold: (a) by decreasing the dosages necessary to attain the desired effect, potential drug-, gene-, and NP-induced toxicities can be minimized; and (b) quick removal of the targeting gene vectors or NPs from the bloodstream may greatly reduce or avoid their destruction by scavenging macrophages or sequestration by the RES. Until these critical issues are explored and understood, the exciting promise of gene therapy and/or nanomedicine may not be attained.

Vascular targeting strategy is a new site-directed delivery strategy that seeks to facilitate tissue-, organ-, or tumor-specific drug and gene delivery by targeting the inherently accessible EC surface of blood vessels feeding the target tissue, rather than the relatively inaccessible targets sites located on the cells residing further inside the tissue (77,131). Taking this concept one step further, a selective EC caveolae targeting strategy may confer an additional advantage by providing both targeted delivery and tissue penetration (131). Caveolae at the luminal EC surface may permit access inside the EC (endocytosis) and even across the EC barrier (transcytosis) to reach the underlying cells inside the tissue, thereby providing a theoretical means for transporting molecular and functional imaging agents as well as drugs, nanomedicines, and gene vectors across the normally restrictive EC barriers to reach underlying tissue cells, the normal desired targets of pharmacotherapies (Figure 73.3). Targeting caveolae may provide additional benefits beyond just targeting the EC surface by providing greater accumulation and improved penetration throughout the tissue. In addition, caveolae-targeted

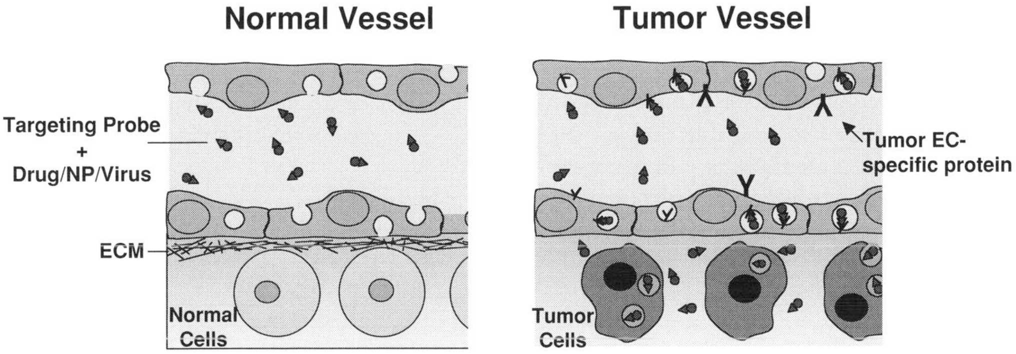

Figure 73.3. Caveolae vascular targeting strategy. The molecular heterogeneity of the EC surface caveolae may allow specific targeting of imaging agents, drugs, nanomedicines, and even viral vectors. Probes, such as an antibody, that specifically target tumor-induced EC caveolar proteins may direct linked cargo (drug/NP/virus) to enhance tumor-specific delivery into and/or across the tumor EC barrier to the underlying tumor cells (*right*). Blood vessels that supply normal tissue do not express tumor-induced antigens and are spared exposure to tumor-targeted cytotoxic drugs or genes (*left*). This novel strategy, with the discovery of reasonably specific caveolar targets for tumor endothelium, may provide the means to greatly improve tumor imaging and therapy. This obviously may extend to other organs and diseases.

molecules may exhibit relatively nonsaturable accumulation in contrast to antibodies that merely target the static EC surface to saturate available binding sites. The potential utility of caveolae in cancer therapy has been reviewed recently (132).

Proof-of-principle experiments utilizing antibodies specific for EC caveolar proteins of the lung have shown that these agents can display the adequate molecular specificity, affinity, and level of exclusivity to selectively overcome the EC barrier to deliver drugs or imaging agents to the underlying tissue cells. Gold-conjugated antibodies that specifically bind their target protein in lung caveolae have been visualized by EM after intravascular administration to specifically target the lung endothelial caveolae (78,108). As noted earlier, uptake into the lung tissue occurs within minutes, followed by rapid transport across the endothelium for release by caveolae now on the abluminal surface (78). Once at the abluminal side of the endothelial barrier, the molecules percolate through the tissue parenchyma to gain access to, and be taken up by, the tissue cells. Future efforts will no doubt focus on extending this work to live dynamic imaging in vivo and to include caveolae on other vascular beds, including tumors.

The utility of caveolar targeting need not be limited to overcoming the vascular EC barrier. The localization of select signaling and mechanotransducing molecules in caveolae creates an additional impetus to develop caveolae-targeted delivery agents. For example, caveolae-targeted small molecules, siRNAs, peptides, or NPs may be used to inhibit specific signaling components concentrated in caveolae, such as eNOS, G-proteins, N-SMase, and the various signaling kinases, to treat such diseases as atherosclerosis and hypertension. One also can imagine using such agents to induce signaling or other events, such as caveolar budding, to bring about the desensitization of specific signaling pathways. Caveolin itself may be a useful target. Sessa and colleagues have generated a chimeric peptide consisting of the caveolin-1 scaffolding domain fused to a cellular internalization sequence that, when injected into mice, selectively inhibited eNOS activity and resulted in decreased inflammation-induced vascular permeability (71). This peptide also slowed tumor growth after intravenous injection, apparently by inhibiting eNOS (and probably other EC signaling molecules), which is required for VEGF effects on vessel permeability (71). Last, it is also probable that caveolae are not homogeneous organelles that all carry the same targets and signaling constituents, so that it may be possible to selectively target specific subpopulations of caveolae.

CONCLUSION

In the 40 years since their discovery, caveolae have undergone a renaissance in terms of interest and discovery of their unique functions in signalling and transport. New techniques to isolate them directly from ECs in vivo, new reagents that allow visualization of their activity in cells, and new methods for identifying their molecular components have led to an explosion in our understanding of their biology. Clinically, caveolae may represent an important pathway for realizing true site-directed delivery of antibodies, NPs, and gene vectors to the endothelium itself or to cells inside the tissue. Without the caveolae portal system to penetrate into tissue, nanomedicine and gene therapy so far lack efficient access to their sites of action within the tissue and thus may have limited clinical utility. In the coming decade, it appears interesting and potentially clinically fruitful to define all potential caveolar molecules and targets, to understand caveolar function and trafficking mechanisms, and to elucidate the potential of caveolae vascular targeting as a strategy for the site-specific delivery of drugs, NPs, gene vectors, and imaging agents. This strategy may also be applied to past drugs that can be highly effective but were tabled by significant systemic toxicity. Re-engineering such drugs to include caveolae targeting capability may improve efficacy and reduce side effects via selective delivery to the desired site of action inside one or more select tissues.

KEY POINTS

- Caveolae are smooth invaginations at the plasma membrane with bipolar-oriented, thin striations around the bulb consisting of polymers of caveolin. They are abundant in continuous ECs, moderate in ECs of large blood vessels, relatively rare in the highly restrictive continuous ECs of brain and testes, and apparently absent in sinusoidal endothelia.
- Many methods have been used to isolate caveolin-containing membranes by subcellular fractionation. The most effective of these methods involves silica NP perfusion or antibody-based methods to isolate caveolae away from lipid rafts.
- Caveolae have key functions in ligand-activated cell surface signalling, mechanotransduction, transvascular exchange, endocytosis, and vascular permeability.
- Caveolae may be an important mediator of the site-directed delivery and penetration of antibodies, NPs, and gene vectors into the endothelium itself or to cells inside a selected tissue.

REFERENCES

1 Palade GE. Fine structure of blood capillaries. *J Appl Physics.* 1953;24:1424.
2 Yamada E. The fine structure of the gall bladder epithelium of the mouse. *J Biophys Biochem Cytol.* 1955;1:445–458.
3 Ogi M, Yokomori H, Oda M, et al. Distribution and localization of caveolin-1 in sinusoidal cells in rat liver. *Med Electron Microsc.* 2003;36:33–40.
4 Rothberg KG, Heuser JE, Donzell WC, et al. Caveolin, a protein component of caveolae membrane coats. *Cell.* 1992;68:673–682.

5 Sargiacomo M, Scherer PE, Tang Z, et al. Oligomeric structure of caveolin: implications for caveolae membrane organization. *Proc Natl Acad Sci USA*. 1995;92:9407–9411.
6 Monier S, Parton RG, Vogel F, et al. VIP21-caveolin, a membrane protein constituent of the caveolar coat, oligomerizes in vivo and in vitro. *Mol Biol Cell*. 1995;6:911–927.
7 Dupree P, Parton RG, Raposo G, et al. Caveolae and sorting in the trans-Golgi network of epithelial cells. *EMBO J*. 1993;12: 1597–1605.
8 Scherer PE, Lewis RY, Volonte D, et al. Cell-type and tissue-specific expression of caveolin-2. Caveolins 1 and 2 co-localize and form a stable hetero-oligomeric complex in vivo. *J Biol Chem*. 1997;272:29337–29346.
9 Tang Z, Scherer PE, Okamoto T, et al. Molecular cloning of caveolin-3, a novel member of the caveolin gene family expressed predominantly in muscle. *J Biol Chem*. 1996;271:2255–12561.
10 Fra AM, Williamson E, Simons K, Parton RG. Detergent-insoluble glycolipid microdomains in lymphocytes in the absence of caveolae. *J Biol Chem*. 1994;269:30745–30748.
11 Parolini I, Sargiacomo M, Galbiati F, et al. Expression of caveolin-1 is required for the transport of caveolin-2 to the plasma membrane. Retention of caveolin-2 at the level of the Golgi complex. *J Biol Chem*. 1999;274:25718–25725.
12 Drab M, Verkade P, Elger M, et al. Loss of caveolae, vascular dysfunction, and pulmonary defects in caveolin-1 gene-disrupted mice. *Science*. 2001;293:2449–2452.
13 Razani B, Wang XB, Engelman JA, et al. Caveolin-2-deficient mice show evidence of severe pulmonary dysfunction without disruption of caveolae. *Mol Cell Biol*. 2002;22:2329–2344.
14 Griffoni C, Spisni E, Santi S, et al. Knockdown of caveolin-1 by antisense oligonucleotides impairs angiogenesis in vitro and in vivo. *Biochem Biophys Res Commun*. 2000;276:756–761.
15 Mora R, Bonilha VL, Marmorstein A, et al. Caveolin-2 localizes to the golgi complex but redistributes to plasma membrane, caveolae, and rafts when co-expressed with caveolin-1. *J Biol Chem*. 1999;274:25708–25717.
16 Sowa G, Pypaert M, Fulton D, Sessa WC. The phosphorylation of caveolin-2 on serines 23 and 36 modulates caveolin-1-dependent caveolae formation. *Proc Natl Acad Sci USA*. 2003; 100:6511–6516.
17 Murata M, Peranen J, Schreiner R, et al. VIP21/caveolin is a cholesterol-binding protein. *Proc Natl Acad Sci USA*. 1995;92: 10339–10343.
18 Smart EJ, Ying Y, Donzell WC, Anderson RG. A role for caveolin in transport of cholesterol from endoplasmic reticulum to plasma membrane. *J Biol Chem*. 1996;271:29427–29435.
19 Schnitzer JE, Oh P, Pinney E, Allard J. Filipin-sensitive caveolae-mediated transport in endothelium: reduced transcytosis, scavenger endocytosis, and capillary permeability of select macromolecules. *J Cell Biol*. 1994;127:1217–1232.
20 Rizzo V, Sung A, Oh P, Schnitzer JE. Rapid mechanotransduction in situ at the luminal cell surface of vascular endothelium and its caveolae. *J Biol Chem*. 1998;273:26323–26329.
21 Liu J, Oh P, Horner T, et al. Organized endothelial cell surface signal transduction in caveolae distinct from glycosylphosphatidylinositol-anchored protein microdomains. *J Biol Chem*. 1997;272:7211–7222.
22 Park H, Go YM, St. John PL, et al. Plasma membrane cholesterol is a key molecule in shear stress-dependent activation of extracellular signal-regulated kinase. *J Biol Chem*. 1998;273:32304–32311.
23 Rothberg KG, Ying YS, Kamen BA, Anderson RG. Cholesterol controls the clustering of the glycophospholipid-anchored membrane receptor for 5-methyltetrahydrofolate. *J Cell Biol*. 1990;111:2931–2938.
24 Smart EJ, Ying YS, Mineo C, Anderson RG. A detergent-free method for purifying caveolae membrane from tissue culture cells. *Proc Natl Acad Sci USA*. 1995;92:10104–10108.
25 Kurzchalia TV, Dupree P, Monier S. VIP21-Caveolin, a protein of the trans-Golgi network and caveolae. *FEBS Lett*. 1994;346: 88–91.
26 Schnitzer JE, McIntosh DP, Dvorak AM, et al. Separation of caveolae from associated microdomains of GPI-anchored proteins [see comments]. *Science*. 1995;269:1435–1439.
27 Oh P, Carver LA, Schnitzer JE. Isolation and subfractionation of plasma membranes to purify caveolae separately from lipid rafts. In: Celis JE, ed. *Cell Biology: A Laboratory Handbook*. 3rd ed. Vol. 2. Amsterdam: Elsevier, 2004.
28 Brown DA, London E. Structure and function of sphingolipid- and cholesterol-rich membrane rafts. *J Biol Chem*. 2000;275: 17221–17224.
29 Oh P, Schnitzer JE. Immunoisolation of caveolae with high affinity antibody binding to the oligomeric caveolin cage. Toward understanding the basis of purification [published erratum appears in *J Biol Chem*. 1999;274(41):29582]. *J Biol Chem*. 1999;274:23144–23154.
30 Schnitzer JE, Oh P, McIntosh DP. Role of GTP hydrolysis in fission of caveolae directly from plasma membranes [published erratum appears in *Science* 1996;274(5290):1069]. *Science*. 1996;274:239–242.
31 Schnitzer JE, Liu J, Oh P. Endothelial caveolae have the molecular transport machinery for vesicle budding, docking, and fusion including VAMP, NSF, SNAP, annexins, and GTPases. *J Biol Chem*. 1995;270:14399–14404.
32 Schnitzer JE, Oh P, Jacobson BS, Dvorak AM. Caveolae from luminal plasmalemma of rat lung endothelium: microdomains enriched in caveolin, Ca(2+)-ATPase, and inositol trisphosphate receptor. *Proc Natl Acad Sci USA*. 1995;92:1759–1763.
33 Oh P, McIntosh DP, Schnitzer JE. Dynamin at the neck of caveolae mediates their budding to form transport vesicles by GTP-driven fission from the plasma membrane of endothelium. *J Cell Biol*. 1998;141:101–114.
34 McIntosh DP, Schnitzer JE. Caveolae require intact VAMP for targeted transport in vascular endothelium. *Am J Physiol*. 1999; 277:H2222–H2232.
35 Rizzo V, McIntosh DP, Oh P, Schnitzer JE. In situ flow activates endothelial nitric oxide synthase in luminal caveolae of endothelium with rapid caveolin dissociation and calmodulin association. *J Biol Chem*. 1998;273:34724–34729.
36 Fra AM, Williamson E, Simons K, Parton RG. De novo formation of caveolae in lymphocytes by expression of VIP21-caveolin. *Proc Natl Acad Sci USA*. 1995;92:8655–8659.
37 Gorodinsky A, Harris DA. Glycolipid-anchored proteins in neuroblastoma cells form detergent-resistant complexes without caveolin. *J Cell Biol*. 1995;129:619–627.
38 Rothberg KG, Ying YS, Kolhouse JF, et al. The glycophospholipid-linked folate receptor internalizes folate without entering the clathrin-coated pit endocytic pathway. *J Cell Biol*. 1990;110:637–649.
39 Parton RG, Joggerst B, Simons K. Regulated internalization of caveolae. *J Cell Biol*. 1994;127:1199–1215.

40 Mayor S, Rothberg KG, Maxfield FR. Sequestration of GPI-anchored proteins in caveolae triggered by cross-linking. *Science*. 1994;264:1948–1951.

41 Lisanti MP, Tang Z, Sargiacomo M. Caveolin forms a hetero-oligomeric protein complex that interacts with an apical GPI-linked protein: implications for the biogenesis of caveolae. *J Cell Biol*. 1993;123:595–604.

42 Rijnboutt S, Jansen G, Posthuma G, et al. Endocytosis of GPI-linked membrane folate receptor-alpha. *J Cell Biol*. 1996;132:35–47.

43 Brown DA, London E. Functions of lipid rafts in biological membranes. *Annu Rev Cell Dev Biol*. 1998;14:111–136.

44 Oh P, Schnitzer JE. Segregation of heterotrimeric G proteins in cell surface microdomains: Gq binds caveolin to concentrate in caveolae whereas Gi and Gs target lipid rafts by default. *Mol Biol Cell*. 2001;12:685–698.

45 Carver LA, Schnitzer JE, Anderson RG, Mohla S. Role of caveolae and lipid rafts in cancer: workshop summary and future needs. *Cancer Res*. 2003;63:6571–6574.

46 Liu P, Ying Y, Ko YG, Anderson RG. Localization of platelet-derived growth factor-stimulated phosphorylation cascade to caveolae. *J Biol Chem*. 1996;271:10299–10303.

47 Garcia-Cardena G, Oh P, Liu J, et al. Targeting of nitric oxide synthase to endothelial cell caveolae via palmitoylation: implications for nitric oxide signaling. *Proc Natl Acad Sci USA*. 1996;93:6448–6453.

48 Feron O, Belhassen L, Kobzik L, et al. Endothelial nitric oxide synthase targeting to caveolae. Specific interactions with caveolin isoforms in cardiac myocytes and endothelial cells. *J Biol Chem*. 1996;271:22810–22814.

49 Garcia-Cardena G, Martasek P, Masters BS, et al. Dissecting the interaction between nitric oxide synthase (NOS) and caveolin. Functional significance of the nos caveolin binding domain in vivo. *J Biol Chem*. 1997;272:25437–25440.

50 Li S, Okamoto T, Chun M, et al. Evidence for a regulated interaction between heterotrimeric G proteins and caveolin. *J Biol Chem*. 1995;270:15693–15701.

51 Chun M, Liyanage UK, Lisanti MP, Lodish HF. Signal transduction of a G protein-coupled receptor in caveolae: colocalization of endothelin and its receptor with caveolin. *Proc Natl Acad Sci USA*. 1994;91:11728–11732.

52 Li S, Couet J, Lisanti MP. Src tyrosine kinases, G-alpha subunits, and H-Ras share a common membrane-anchored scaffolding protein, caveolin. Caveolin binding negatively regulates the auto-activation of Src tyrosine kinases. *J Biol Chem*. 1996;271:29182–29190.

53 Cohen AW, Park DS, Woodman SE, et al. Caveolin-1 null mice develop cardiac hypertrophy with hyperactivation of p42/44 MAP kinase in cardiac fibroblasts. *Am J Physiol Cell Physiol*. 2003;284:C457–C474.

54 Lai HH, Boone TB, Yang G, et al. Loss of caveolin-1 expression is associated with disruption of muscarinic cholinergic activities in the urinary bladder. *Neurochem Int*. 2004;45:1185–1193.

55 Cohen AW, Razani B, Wang XB, et al. Caveolin-1-deficient mice show insulin resistance and defective insulin receptor protein expression in adipose tissue. *Am J Physiol Cell Physiol*. 2003;285:C222–C235.

56 Schubert W, Frank PG, Woodman SE, et al. Microvascular hyperpermeability in caveolin-1 (-/-) knock-out mice. Treatment with a specific nitric-oxide synthase inhibitor, L-name, restores normal microvascular permeability in Cav-1 null mice. *J Biol Chem*. 2002;277:40091–40098.

57 Razani B, Engelman JA, Wang XB, et al. Caveolin-1 null mice are viable but show evidence of hyperproliferative and vascular abnormalities. *J Biol Chem*. 2001;276:38121–38138.

58 Claing A, Perry SJ, Achiriloaie M, et al. Multiple endocytic pathways of G protein-coupled receptors delineated by GIT1 sensitivity. *Proc Natl Acad Sci USA*. 2000;97:1119–1124.

59 Drmota T, Novotny J, Gould GW, et al. Visualization of distinct patterns of subcellular redistribution of the thyrotropin-releasing hormone receptor-1 and Gq alpha /G11alpha induced by agonist stimulation. *Biochem J*. 1999;340 (Pt 2):529–538.

60 Igarashi J, Michel T. Agonist-modulated targeting of the EDG-1 receptor to plasmalemmal caveolae. eNOS activation by sphingosine 1-phosphate and the role of caveolin-1 in sphingolipid signal transduction. *J Biol Chem*. 2000;275:32363–32370.

61 de Weerd WF, Leeb-Lundberg LM. Bradykinin sequesters B2 bradykinin receptors and the receptor-coupled G-alpha subunits G-alpha-q and G-alpha-i in caveolae in DDT1 MF-2 smooth muscle cells. *J Biol Chem*. 1997;272:17858–17866.

62 Gimbrone MA Jr. Endothelial dysfunction, hemodynamic forces, and atherosclerosis. *Thromb Haemost*. 1999;82:722–726.

63 Schnitzer JE. The endothelial cell surface and caveolae in health and disease. In: Born GVR, Schwartz CJ, eds. *Vascular Endothelium: Physiology, Pathology, and Therapeutic Opportunities*. Stuttgart: Schattauer;1997:77–95.

64 Topper JN, Gimbrone MA Jr. Blood flow and vascular gene expression: fluid shear stress as a modulator of endothelial phenotype. *Mol Med Today*. 1999;5:40–46.

65 Schnitzer JE. Molecular architecture of endothelial caveolae: possible stress-sensing organelles. *Ann Biomed Eng*. 1995;23:S34.

66 Park H, Go YM, Darji R, et al. Caveolin-1 regulates shear stress-dependent activation of extracellular signal-regulated kinase. *Am J Physiol Heart Circ Physiol*. 2000;278:H1285–H1293.

67 Boyd NL, Park H, Yi H, et al. Chronic shear induces caveolae formation and alters ERK and Akt responses in endothelial cells. *Am J Physiol Heart Circ Physiol*. 2003; 285:H1113–H1122.

68 Czarny M, Liu J, Oh P, Schnitzer JE. Transient mechanoactivation of neutral sphingomyelinase in caveolae to generate ceramide. *J Biol Chem*. 2003;278:4424–4430.

69 Ju H, Zou R, Venema VJ, Venema RC. Direct interaction of endothelial nitric-oxide synthase and caveolin-1 inhibits synthase activity. *J Biol Chem*. 1997;272:18522–18525.

70 Michel JB, Feron O, Sacks D, Michel T. Reciprocal regulation of endothelial nitric-oxide synthase by Ca2+-calmodulin and caveolin. *J Biol Chem*.1997;272:15583–15586.

71 Bucci M, Gratton JP, Rudic RD, et al. In vivo delivery of the caveolin-1 scaffolding domain inhibits nitric oxide synthesis and reduces inflammation. *Nat Med*. 2000;6:1362–1367.

72 Czarny M, Schnitzer JE. Neutral Sphingomyelinase inhibitor scyphostatin prevents and ceramide mimics mechanotransduction in vascular endothelium. *Am J Physiol Heart Circ Physiol*. 2004;287:H1344–H1352.

73 Schnitzer JE. Analysis of steric partition behavior of molecules in membranes using statistical physics. Application to gel chromatography and electrophoresis. *Biophys J*. 1988;54:1065–1076.

74 Schnitzer JE, Carley WW. Electrostatic and steric partition function of the endothelial glycocalyx. *Fed Proc*. 1986;45:A1152.

75 Huxley VH, Curry FE. Albumin modulation of capillary permeability: test of an adsorption mechanism. *Am J Physiol*. 1985;248:H264–H273.

76 Milici AJ, Watrous NE, Stukenbrok H, Palade GE. Transcytosis of albumin in capillary endothelium. *J Cell Biol.* 1987;105:2603–2612.
77 Schnitzer JE. Caveolae: from basic trafficking mechanisms to targeting transcytosis for tissue-specific drug and gene delivery in vivo. *Adv Drug Deliv Rev.* 2001;49:265–280.
78 McIntosh DP, Tan X-Y, Oh P, Schnitzer JE. Targeting endothelium and its dynamic caveolae for tissue-specific transcytosis in vivo: a pathway to overcome cell barriers to drug and gene delivery. *Proc Natl Acad Sci USA.* 2002;99:1996–2001.
79 Wagner RC, Chen S-C. Transcapillary transport of solute by the endothelial vesicular system: evidence from thin serial section analysis. *Microvasc Res.* 1991;42:139–150.
80 Bruns RR, Palade GE. Studies on blood capillaries. II. Transport of ferritin molecules across the wall of muscle capillaries. *J Cell Biol.* 1968;37:277–299.
81 Palade GE. Transport in quanta across the endothelium of blood capillaries. *Anat Rec.* 1960;136:254.
82 Clough G, Michel CC. The role of vesicles in the transport of ferritin through frog endothelium. *J Physiol.* 1981;315:127–142.
83 Ghitescu L, Fixman A, Simionescu M, Simionescu N. Specific binding sites for albumin restricted to plasmalemmal vesicles of continuous capillary endothelium: receptor-mediated transcytosis. *J Cell Biol.* 1986;102:1304–1311.
84 Bundgaard M, Frokjaer-Jensen J, Crone C. Endothelial plasmalemmal vesicles as elements in a system of branching invaginations from the cell surface. *Proc Natl Acad Sci USA.* 1979;76:6439–6442.
85 Frokjaer-Jensen J, Wagner RC, Andrews SB, et al. Three-dimensional organization of the plasmalemmal vesicular system in directly frozen capillaries of the rete mirabile in the swim bladder of the eel. *Cell Tissue Res.* 1988;254:17–24.
86 Severs NJ. Caveolae: static inpocketings of the plasma membrane, dynamic vesicles or plain artifact? *J Cell Sci.* 1988;90:341–348.
87 Renkin EM. Capillary transport of macromolecules: pores and other endothelial pathways. *J Appl Physiol.* 1985;58:315–325.
88 Schnitzer JE, Carley WW, Palade GE. Albumin interacts specifically with a 60-kDa microvascular endothelial glycoprotein. *Proc Natl Acad Sci USA.* 1988;85:6773–6777.
89 Tran D, Carpentier J-L, Sawano F, et al. Ligands internalized through coated or noncoated invaginations follow a common intracellular pathway. *Proc Natl Acad Sci USA.* 1987;84:7957–7961.
90 Pelkmans L, Kartenback J, Helenius A. Caveolar endocytosis of Simian virus 40 reveals a novel two-step vesicular transport pathway to the ER. *Nature Cell Biol.* 2001;3:473–483.
91 Norkin LC. Simian virus 40 infection via MHC class I molecules and caveolae. *Immunol Rev.* 1999;168:13–22.
92 Empig CJ, Goldsmith MA. Association of the caveola vesicular system with cellular entry by filoviruses. *J Virol.* 2002;76:5266–5270.
93 Mackay RL, Consigli RA. Early events in polyoma virus infection: attachment, penetration, and nuclear entry. *J Virol.* 1976;19:620–636.
94 Richterova Z, Liebl D, Horak M, et al. Caveolae are involved in the trafficking of mouse polyomavirus virions and artificial VP1 pseudocapsids toward cell nuclei. *J Virol.* 2001;75:10880–10891.
95 Kartenbeck J, Stukenbrok H, Helenius A. Endocytosis of simian virus 40 into the endoplasmic reticulum. *J Cell Biol.* 1989;109:2721–2729.
96 Pelkmans L, Puntener D, Helenius A. Local actin polymerization and dynamin recruitment in SV40-induced internalization of caveolae. *Science.* 2002;296:535–539.
97 Mannova P, Forstova J. Mouse polyomavirus utilizes recycling endosomes for a traffic pathway independent of COPI vesicle transport. *J Virol.* 2003;77:1672–1681.
98 Orlandi PA, Fishman PH. Filipin-dependent inhibition of cholera toxin: evidence for toxin internalization and activation through caveolae-like domains. *J Cell Biol.* 1998;141:905–915.
99 Goda Y, Pfeffer SR. Identification of a novel, N-ethylmaleimide-sensitive cytosolic factor required for vesicular transport from endosomes to the trans-Golgi network in vitro. *J Cell Biol.* 1991;112:823–831.
100 Rothman JE, Orci L. Movement of proteins through the Golgi stack: a molecular dissection of vesicular transport. *FASEB J.* 1990;4:1460–1468.
101 Rothman JE, Orci L. Molecular dissection of the secretory pathway. *Nature.* 1992;355:409–415.
102 Wessling-Resnick M, Braell WA. Characterization of the mechanism of endocytic vesicle fusion in vitro. *J Biol Chem.* 1990;265:16751–16759.
103 Sollner T, Bennett MK, Whiteheart SW, et al. A protein assembly-disassembly pathway in vitro that may correspond to sequential steps of synaptic vesicle docking, activation, and fusion. *Cell.* 1993;75:409–418.
104 Schnitzer JE, Allard J, Oh P. NEM inhibits transcytosis, endocytosis, and capillary permeability: implication of caveolae fusion in endothelia. *Am J Physiol.* 1995;268:H48–H55.
105 Oh P, Schnitzer JE. Dynamin-mediated fission of caveolae from plasma membranes. *Mol Biol Cell.* 1996;7:A83.
106 Predescu SA, Predescu DN, Timblin BK, et al. Intersectin regulates fission and internalization of caveolae in endothelial cells. *Mol Biol Cell.* 2003;14:4997–5010.
107 Villaschi S, Johns L, Cirigliano M, Pietra GG. Binding and uptake of native and glycosylated albumin-gold complexes in perfused rat lungs. *Microvasc Res.* 1986;32:190–199.
108 Oh P, Li Y, Yu J, et al. Subtractive proteomic mapping of the endothelial surface in lung and solid tumours for tissue-specific therapy. *Nature.* 2004;429:629–635.
109 Ghinea N, Fixman A, Alexandru D, et al. Identification of albumin-binding proteins in capillary endothelial cells. *J Cell Biol.* 1988;107:231–239.
110 Schnitzer JE, Carley WW, Palade GE. Specific albumin binding to microvascular endothelium in culture. *Am J Physiol.* 1988;254:H425–H437.
111 Schnitzer JE, Sung A, Horvat R, Bravo J. Preferential interaction of albumin-binding proteins, gp30 and gp18, with conformationally modified albumins. Presence in many cells and tissues with a possible role in catabolism. *J Biol Chem.* 1992;267:24544–24553.
112 Schnitzer JE. gp60 is an albumin-binding glycoprotein expressed by continuous endothelium involved in albumin transcytosis. *Am J Physiol.* 1992;262:H246–H254.
113 Schubert W, Frank PG, Razani B, et al. Caveolae-deficient endothelial cells show defects in the uptake and transport of albumin in vivo. *J Biol Chem.* 2001;276:48619–48622.
114 Zhao YY, Liu Y, Stan RV, et al. Defects in caveolin-1 cause dilated cardiomyopathy and pulmonary hypertension in knockout mice. *Proc Natl Acad Sci USA.* 2002;99:11375–11380.

115 Feng Y, Venema V, Venema R, et al. VEGF-induced permeability increase is mediated by caveolae. *Invest Ophthalmol Vis Sci.* 1999;40:157–167.

116 Fukumura D, Gohongi T, Kadambi A, et al. Predominant role of endothelial nitric oxide synthase in vascular endothelial growth factor-induced angiogenesis and vascular permeability. *Proc Natl Acad Sci USA.* 2001;98:2604–2609.

117 Gratton J-P, Lin MI, Yu J, et al. Selective inhibition of tumor microvascular permeability by cavtratin blocks tumor progression in mice. *Cancer Cell.* 2003;4:31–39.

118 Zhu L, Schwegler-Berry D, Castranova V, Pingnian H. Internalization of caveolin-1 scaffolding domain facilitated by Antennapedia homeodomain attenuates PAF-induced increase in microvessel permeability. *Am J Physiol Heart Circ Physiol.* 2004; 286:195–201.

119 Park DS, Cohen AW, Frank PG, et al. Caveolin-1 null (−/−) mice show dramatic reductions in life span. *Biochemistry.* 2003; 42:15124–15131.

120 Dvorak AM, Feng D. The vesiculo-vacuolar organelle (VVO). A new endothelial cell permeability organelle. *J Histochem Cytochem.* 2001;49:419–432.

121 Fielding CJ, Fielding PE. Caveolae and intracellular trafficking of cholesterol. *Adv Drug Deliv Rev.* 2001;49:251–264.

122 Cohen AW, Combs TP, Scherer PE, Lisanti MP. Role of caveolin and caveolae in insulin signaling and diabetes. *Am J Physiol Endocrinol Metab.* 2003;285:E1151–E1160.

123 Lavie Y, Fiucci G, Liscovitch M. Upregulation of caveolin in multidrug resistant cancer cells: functional implications. *Adv Drug Deliv Rev.* 2001;49:317–323.

124 Schnitzer JE. Vascular targeting as a strategy for cancer therapy. *N Engl J Med.* 1998;339:472–474.

125 Jain RK. The next frontier of molecular medicine: delivery of therapeutics. *Nat Med.* 1998;4:655–657.

126 Tomlinson E. Theory and practice of site-specific drug delivery. *Adv Drug Deliv Rev.* 1987;1:87–198.

127 Dvorak HF, Nagy JA, Dvorak AM. Structure of solid tumors and their vasculature: implications for therapy with monoclonal antibodies. *Cancer Cells.* 1991;3:77–85.

128 Miller N, Vile R. Targeted vectors for gene therapy. *FASEB J.* 1995;9:190–199.

129 Moghimi SM, Hunter AC, Murray JC. Nanomedicine: current status and future prospects. *FASEB J.* 2005;19:311–330.

130 Ferrari M. Cancer nanotechnology: opportunities and challenges. *Nat Rev Cancer.* 2005;5:161–171.

131 Carver LA, Schnitzer JE. Tissue-specific pharmacodelivery and overcoming key cell barriers in vivo: vascular targeting of caveolae. In: Muzykantov V, Torchilin B, eds. *Biomedical Aspects of Drug Targeting.* Boston: Kluwer Academic Publishers; 2002: 107–128.

132 Carver LA, Schnitzer JE. Caveolae: mining little caves for new cancer targets. *Nat Rev Cancer.* 2003;3:571–381.

74

Endothelial Structures Involved in Vascular Permeability

Radu V. Stan

Dartmouth Medical School, Hanover, New Hampshire

Vascular endothelium is a highly differentiated cellular monolayer with the organization of a simple squamous epithelium. It lines the entire cardiovascular system and thus constitutes a quasi-ubiquitous presence in organs and tissues throughout the body. One of the most significant aspects of the vascular endothelium is its diversity of phenotype and function, which endows it with the ability to regulate the exchange of myriad substances between blood plasma and interstitial fluid in all tissues of the body. The ability of the endothelium to regulate the transendothelial exchange or vascular permeability is critically important in the growth, maintenance, and survival of all tissues, as well as in the delivery of therapies to correct locations.

Within the circulation, exchanges occur across the endothelium at all levels of the vascular tree. However, two functional segments are recognized with respect to the role of the vessel wall in the transendothelial exchanges between blood and tissues served: the conduit vessels and the exchange vessels. The *conduit vessels* take blood from heart to tissues (large arteries down to the smaller muscular arterioles upstream of the precapillary sphincter) or from tissues to heart (small muscular veins collecting from venules and draining into large veins). In these vessels (with large sectional diameter but low aggregated exchange surface), the transendothelial exchanges are thought to have impact only on the vessel wall and not on the organs/tissues crossed by these vessels. The *exchange vessels* consist of capillaries and venules, and are thought to be involved in active exchanges between blood and tissues. This segment, which has low sectional surface per vessel (i.e., <100 μm diameter for venules and <10–20 μm for capillaries) yet the largest aggregated surface, is characterized by a very thin wall consisting of endothelium (with an average thickness of less than 0.3 μm) overlying basement membrane and occasional pericytes (situated mostly in venules). The term *microvascular permeability* as used extensively in the literature (1) (and the current chapter), describes the transendothelial exchange of fluids and solutes.

It is generally recognized that endothelium mediates "basal" microvascular permeability, which refers to the steady, continuous transverse flow of fluid and hydrophilic solutes (both small and large molecules) across the normal healthy microvascular endothelium, and "inducible" (i.e., increased) microvascular permeability, which refers to the increase in permeability to fluid and plasma proteins that occurs in inflamed tissues or tumors. Although the mechanisms underlying basal permeability remain poorly understood, previous investigations have provided important insights into the molecular basis of inducible permeability. An interesting yet unresolved question is whether increased permeability in inflammation and tumors represents an exaggeration of the normal response or is mediated by separate pathway(s).

Compared with other epithelia, the endothelium is far more permeable – by two to three orders of magnitude – to water and solutes (both small and large molecules), with the exception of brain endothelium. The relatively high basal permeability was determined to have both a diffusive and convective component. Based on physiological studies, the best fitting model for basal permeability, known as the *pore theory*, predicted that exchange takes place through two types of patent (open), water-filled, cylindrical or slit-like channels: small pores with diameters of <11 nm and approximately 15.1 units/μm^2, and large pores with diameters of approximately 50 nm and approximately 0.2 to 0.02 units/μm^2 (1–3).

During the 1950s and 1960s, cell biologists employed electron microscopy (EM) alone or combined with in vivo tracers to elucidate the structural basis for fluid and solute exchange across the endothelium (4–7). The latter studies – which employ probes of variable chemical and physical properties – suggested that the paracellular pathway represents the equivalent of the physiologically defined small pores. Also in agreement with the pore theory, EM and tracer studies indicated that the transendothelial channels (TECs) and fenestrae are structural equivalents of large pores in fenestrated

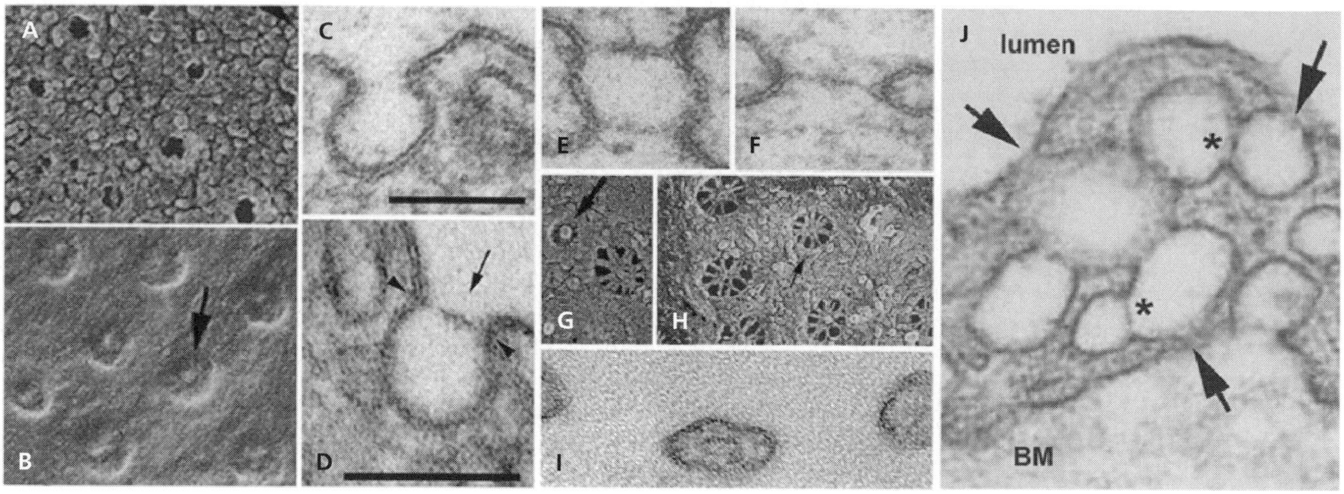

Figure 74.1. Structures involved in transendothelial permeability. Caveolae with (**B,D**) and without (**A,C**) SDs. Fenestrae subtended by a FD (**E–H**) or kidney glomerular fenestrae without FDs (**I**). Vesiculo-vacuolar organelle (VVO) from the heart endothelium of a *caveolin-1*–null mouse (**J**).

and discontinuous endothelia (Figure 74.1).[1] However, ultrastructural studies failed to identify pore-like equivalents for large pores in continuous endothelia (where most of the data supporting the "pore theory" had been obtained), with the exception of vesiculo-vacuolar organelles (VVOs) in the venular endothelium (4).

The failure to identify the structural equivalent of large pores in continuous endothelium presaged a longstanding controversy between capillary physiologists and cell biologists/morphologists regarding the molecular basis for the transendothelial passage of macromolecules. In the final analysis, tracer studies have provided compelling evidence for the existence of "functional pores" that mediate transfer across endothelia (a process termed *transcytosis*). Transcytosis – which appears to be a feature of all epithelia (8) – describes the process of macromolecular transport across cellular barriers via membrane-bounded vesicular carriers, including plasmalemmal vesicles (caveolae), clathrin-mediated pathways, and nonclathrin, noncaveolar pathways (9).

HISTORY

The earliest contributions to our understanding of the exchange of water and solutes across permselective barriers date from the 19th century, with the work of physiologists such as C. Bernard and E. H. Starling (10). As discussed earlier, the seminal work of physiologists in the last century led to the formulation and development of the "pore theory" of capillary permeability (11–16). The introduction of the electron microscope during the 1950s ushered in an era of elegant structure-function studies. George Palade, a pioneer in the ultrastructural field, was the first to describe what appeared by transmission electron microscopy (TEM) as "uncoated"[2] plasma membrane invaginations and vesicles in the cytoplasm of endothelial cells (ECs); he named these structures plasmalemmal vesicles (17). Later studies uncovered the presence of a diaphragm overlying the opening or stomata of the plasmalemmalvesicles (so-called *stomatal diaphragms* [SDs]) (18–20). Very soon after their discovery, by employing electron opaque tracers (e.g., colloidal gold, ferritin), Palade showed that macromolecules use the plasmalemmal vesicles to cross the endothelium "in quanta," that is, in discrete vesicular carriers that shuttle between the luminal and abluminal sides (21,22). Subsequent work by Palade's and other groups have demonstrated the vesicular transport for several physiologically relevant macromolecules in endothelium, which crystallized under the concept of *transcytosis* (term coined by N. Simionescu, one of Palade's collaborators) (23). A few years after the description of the plasmalemmal vesicles, E. Yamada described similar uncoated vesicles in other epithelia, which he called *caveolae* (24). Until the discovery of caveolin-1 as marker of caveolae, the terms *plasmalemmal vesicles* and *caveolae* were used interchangeably based simply on similar morphology. At the beginning of the 1990s, when caveolin-1 was shown to stain most of the plasmalemmal vesicles in endothelium (25–29), it was assumed that all uncoated vesicles are

1 Endothelia have been classically defined into three main structural types, depending on their content of specific structures: the continuous, fenestrated, and discontinuous endothelium (schematized in Figure 74-2). The *continuous endothelium* forms an uninterrupted barrier between the blood and tissues. It features a large population of caveolae or plasmalemmal, extremely few TECs, and virtually no fenestrae. In addition to relatively fewer caveolae, the *fenestrated endothelium* features pores such as fenestrae, transendothelial channels, and endothelial pockets. *Discontinuous* or *sinusoidal endothelium* lines the sinusoids in the liver and bone marrow. It has extremely few caveolae and displays sinusoidal gaps.

2 As opposed to the clathrin-coated vesicles that have a characteristic fuzzy coat by EM.

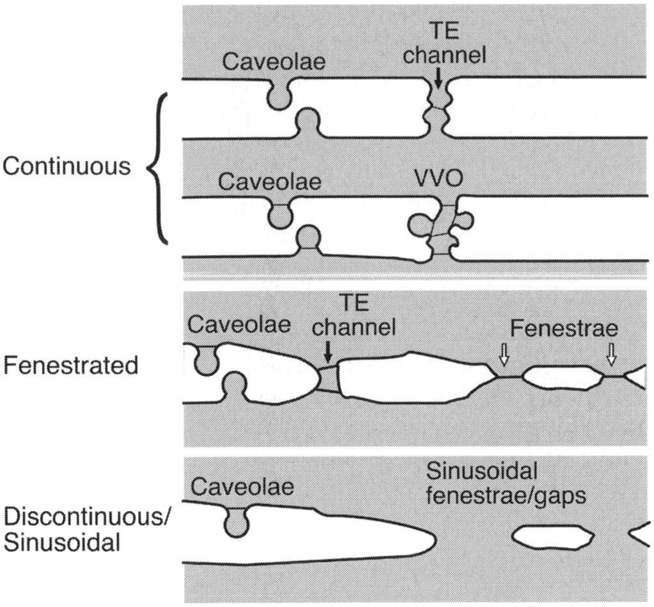

Figure 74.2. Morphological types of endothelium. Schematic cross-sections of EC segments of the continuous, fenestrated, and discontinuous type and their relevant specific organelles. Continuous endothelium has caveolae with or without SDs and few TECs. VVOs have been described in the continuous ECs of the venules. The fenestrated endothelium features caveolae, always provided with a SD, TEC, and fenestrae. A FD spans the fenestrae in all cases except kidney glomerulus. The discontinuous endothelium has extremely few caveolae provided with SDs and sinusoidal fenestrae.

the same, and the term *caveolae* became truly synonymous with *plasmalemmal vesicles.* However, during the last few years, a large number of alternative endocytotic pathways that use uncoated vesicles have been demonstrated in many cell types, endothelium included (30–33). Some of these pathways seem to participate in transcytosis, along with caveolae and clathrin-coated pits (9,34,35).

Early EM studies revealed the presence of fenestrae in certain vascular beds, giving rise to the time-honored classification of fenestrated versus nonfenestrated endothelium (Figure 74.2). Oberling first reported the existence of diaphragms in the fenestrae of kidney endothelium (36,37). Subsequent studies demonstrated fenestral diaphragms (FDs) in other types of endothelium (38–42). In 1970, E. Wisse (43) documented in the liver sinusoids an endothelial lining that lacked a basement membrane and that showed many discontinuities of varied sizes, which he named *sinusoidal fenestrae* or *gaps.* Thus, a third endothelial morphological type was recognized, namely the discontinuous (or sinusoidal) endothelium. Scanning electron microscopy (SEM) studies showed that the discontinuities of this type of endothelium are not gaps between ECs, because the intercellular junctions were fully formed, but they are circular pores through the EC (44). Thus, the term *fenestrae* is much more appropriate to describe them. However, their relationship with the fenestrae of the fenestrated endothelium (smaller, constant in size, subtended by a diaphragm) is not clear and creates some confusion in the field. TECs were first described as "fenestrae with two diaphragms" in fenestrated endothelium, and were considered the same entity as fenestrae. Their diaphragms were later shown to have different chemical properties in terms of lectin and cationic probe binding (45–47), and were thus proposed to represent a separate structure. Also, the presence of TECs formed by a chain of interconnected vesicles with or without diaphragms was revealed by electron-opaque tracers in continuous endothelium (48). As a variant to fenestrae and TECs, in 1986, Milici and Palade described "endothelial pockets" (or "fenestrated domes"), a very little known morphological entity consisting of a vacuole-like space communicating via fenestrae with the luminal and abluminal space (49). Finally, in 1992, H. Dvorak's group reported a novel structure in ECs, the so-called VVOs (50).

ONTOGENETIC CONSIDERATIONS

No systematic studies have documented the appearance of caveolae, fenestrations, TECs, pockets, VVOs, or diaphragms and their specific protein components during development. Most of the data available concern the formation of the fenestrae and their diaphragms.

In the blood island and vascular plexus stage of vascular development, examination of micrographs published in various papers shows a continuous endothelium devoid of fenestrae or TECs. Unfortunately, the magnification shown is usually too low for a qualitative assessment of the presence of caveolae or VVOs and their respective diaphragms, most of the studies being more concerned with the organization of the vasculature than the ultrastructural detail of ECs.

The rat brain starts developing as an avascular mass or primordium surrounded by vascularized tissue called the perineurium, whence the developing brain will attract sinusoid-like vessels by sprouting angiogenesis between approximately embryonic day (E)12 and E16. At E12, the vessels in perineurium have fenestrated endothelia (51), but it is not known when precisely fenestrae appear in these vessels, E12 being the earliest time point reported in this paper. Interestingly, fenestrae subtended by diaphragms are present in the vessels that enter the nervous matter from E12 to E16 by sprouting angiogenesis but are lost upon the formation of the blood–brain barrier (BBB), beginning with E16 (51).

The capillaries of rat intestinal mucosa are initially continuous and express caveolae; it is not clear whether SDs are present. These continuous capillaries become fenestrated late in gestation, at E15 to E16, when they start to express fenestrae subtended by diaphragms (52,53). The reason for the delayed induction of the fenestrae in the intestine is not known. A possible explanation may be found by examining the situation in the adult, in which the vessels in the intestinal tunica muscularis are continuous, and the ECs have caveolae without SDs. As the vessels enter the mucosa (i.e., the villus), their endothelium remains continuous in the central bundle vessels but their caveolae have SDs. The capillary loops that come in contact with the epithelium are fenestrated, and their fenestrae have SDs. The intestinal mucosa does not fully develop in the embryo but requires the postnatal establishment of the intestinal flora, which leads to the elongation of the vili and formation of fenestrated vessels by angiogenesis.

Interestingly, in rat kidney development, the fenestrae that appear first in the kidney glomerular capillaries are subtended by diaphragms. Moreover, these diaphragms bind cationic ferritin, a property common to all fenestrae in the adult embryo, conferred by heparan sulfate proteoglycans (54). It is not known, however, at which precise stage fenestrae appear in the kidney. The diaphragms of glomerular fenestrae are gradually lost within a week or two after birth, resulting in the unique phenotype of the adult glomerulus with fenestrae lacking diaphragms (37,54–56).

In the liver at E10 to E12, during the hematopoietic stage, the capillaries have a basement membrane and are fenestrated, but the fenestrae resemble those of the adult fenestrated endothelium: circular pores, regular in size (average 60–70 nm), and subtended by FDs. Starting at E17, these fenestrae are progressively replaced by larger, more size-heterogeneous fenestrae that are typical of the adult liver and other sinusoidal ECs (57,58).

In summary, it is not known when fenestrae, TECs, caveolae, VVOs, and their respective diaphragms first appear in different vascular beds. Drawing a parallel between the appearance of the glomerular fenestrae and those of the sinusoids is tempting, because both types of fenestrae lacking diaphragms seem to be preceded by fenestrae subtended by diaphragms.

Plasmalemma vesicle protein 1 (PV1) (also known as plasmalemma vesicle-associated protein [PLVAP]) is the only known marker (discussed later) of endothelial SDs and FDs. The mRNA is definitively expressed in both mouse and rat embryos by E7 (59), corresponding to the formation of blood islands. PV1 protein is detectable by E12 in most organs, and is lost by E16 to E18 in the brain at time of BBB formation (60). Moreover, the loss of PV1 expression correlates with the loss of fenestrae on BBB formation (61). In mice at E11.5, PV1 is expressed in hindlimb buds, but not forelimbs (62). PV1 also is expressed on bone marrow endothelial progenitor cells (63) as well as on the ECs that spontaneously differentiate on embryoid bodies after 5 days outgrowth (64).

ULTRASTRUCTURE OF ORGANELLES INVOLVED IN PERMEABILITY

The transendothelial exchange of plasma and solutes takes place at intercellular junctions (paracellular pathway) and through the EC (transcellular route) via fenestrae, caveolae, TECs, and VVOs. Here, we discuss the structures, organelles, and microdomains involved in the transcellular pathway. Although their exquisite architecture has been compared to works of art (65), their molecular components have remained elusive for decades.

Caveolae and Their Stomatal Diaphragms

Caveolae (also termed *plasmalemmal vesicles*) occur in most mammalian cell types, including ECs (for reviews see Refs. 6 and 66, and Chapter 73). They are morphologically defined as spherical invaginations of plasma membrane of regular shape and size (~70 nm) (Figure 74.1A–D) (18–20). They occur singly or in grape-like clusters attached to either the apical or basolateral domain of the endothelial plasma membrane. In fenestrated and discontinuous endothelia, and in select endothelia of continuous type (e.g., in lung, tongue, kidney vasa recta), caveolar necks are subtended by a SD, as evidenced by rapid-freeze deep-etch techniques (Figure 74.1B (67,68). By TEM, the SD is a thin (~5–7 nm) protein barrier that bridges the neck or stoma of the caveolae at the surface of the cell or between vesicles that form part of an intracellular cluster (Figure 74.1B and D). Because of the dynamic nature of caveolae that could undergo internalization, their necks have a variable diameter, which results in variable diameter SDs (<40 nm). At the site of SD insertion, the plasma membrane forms a sharp angle. Electron-dense material usually is present on the intracellular aspect of the membrane within this angle, as shown in orthogonal TEM sections (Figure 74.1D). Both orthogonal and en face views demonstrate a central density or knob on the SDs (20,38). Favorable en face views of the SDs (e.g., in the lung capillaries) suggest the central knob to be connected to the vesicular neck by very thin fibrils. Also, in rapidly frozen deeply etched specimens, a hint of the radial fibrils pattern also is detectable (68). The diaphragm is apparently lipid-free, because it does not have the trilaminar appearance of the plasma membrane bilayer. It is destroyed by proteases (45), which implies its proteinaceous nature (66).

Endothelial Fenestrae and Their Diaphragms

Endothelial fenestrae (Figure 74.1F–I) are circular windows resembling "boat portholes" that cut through the cell body. They are arranged in ordered linear arrays within large planar clusters called *sieve plates.* Individual fenestral pores have a remarkably constant diameter (~62–68 nm) (28,39,69). The fenestral rim delineates their circumference, where the luminal aspect of plasma membrane continues with the abluminal one under a sharp angle filled with electron-dense material, as in the case of the SD. The circumference of the fenestral pore has an octagonal symmetry (68,70,71).

With the exception of glomerular endothelium, all endothelial fenestrae posses a diaphragm (37,54–56). Compared with SDs, FDs have a slightly larger diameter. However, both types of diaphragms are structurally similar, consisting of a thin protein barrier anchored at the fenestral rim and provided with a central knob or density connected to the fenestral rim by thin fibrils (41). Bearer and Orci exquisitely demonstrated that FDs consist of radial fibrils, starting at the rim and interweaving in a central mesh, the equivalent of the central density or knob seen by TEM (Figure 74.1G–H) (68). Unfortunately, the SDs of caveolae and TECs could not be resolved in the same detail owing to their smaller size and technical limitations (i.e., the angle of shadowing) (68). SDs of caveolae resemble those of TECs, with both featuring a central particle (68) (Figure 74.2G) and hints of FD-like radial fibrils (Figure 74.2B).

Recent studies employing special perfusion techniques have revealed an interesting difference between fenestral and SDs; FDs, but not those of caveolae or TECs, contain large (up to 400 nm) tufts (termed *fascinae fenestrae*) on the their luminal side (72). The fibers forming these tufts have been interpreted as the morphologic equivalent of the heparan sulfate proteoglycans shown to reside on the luminal surface of fenestrae (73).

Transendothelial Channels

TECs are patent pores spanning the EC body from their apical (luminal) to the basolateral (abluminal) front (Figure 74.1E). TECs are rarely found in continuous endothelia, where they seem to be formed by the fusion of either one caveolae/plasmalemmal vesicle with both luminal and abluminal aspects of the plasmalemma or by chains of usually two to four vesicles (20,48,74). In fenestrated endothelia, TECs occur in the attenuated part of the EC and are provided with two SDs (one luminal and one abluminal) (39,69,75,76). As mentioned, TEC diaphragms display the same features as the SDs of caveolae.

Vesiculo-Vacuolar Organelles

VVOs are morphologically defined as chains of interconnected vesicles and vacuoles of variable size that form intricate TECs, spanning the cytoplasm of the EC from the apical to the basal side of the cell (77). They normally possess SDs at the connection points between vesicles and vacuoles, as well as at the level of their stoma or point of communication with the extracellular space (78). The SDs of the VVOs closely resemble those of caveolae by TEM. VVOs have been described in postcapillary venules as well as tumor blood vessels (50,77,79). It is formally possible that VVOs represent clusters of caveolae (20,69), and thus do not represent a bona fide, novel organelle (76). However, in support of its status as a "novel organelle," EM studies in mice null for caveolin-1 (which lack caveolae in all nonmuscle cell types) have demonstrated the presence of structures resembling VVOs in venular endothelium (Figure 74.1J) (9,80,81)]. The relationship between these structures and VVOs remains to be clarified following biochemical evaluation.

Endothelial Pockets

By EM, these infrequently seen structures resemble a pocket or a large vacuole formed by cellular processes that contain fenestrae with the usual structure (76). The information on these is scarce (49) and, so far, they seem to occur in very low numbers only in the fenestrated endothelia.

MOLECULAR STRUCTURE OF ORGANELLES INVOLVED IN PERMEABILITY

Probes with Broad Specificity

The chemical components of SDs and FDs have been investigated using nonspecific probes such as lectins (46,82–84) and charged molecules (45,47,85–87). SDs bind lectins avidly and lack anionic sites, whereas FDs bind lectins poorly if at all (46,84), and have multiple anionic sites on the luminal side conferred by heparan sulfate proteoglycans (45,72,86). The presence of lectin-binding sites on SDs and FDs, together with the sensitivity of the diaphragms to protease degradation (45), suggest that these structures contain glycoproteins.

Plasmalemmal Vesicle–Associated Protein

Recently, SDs and FDs were shown to share at least one biochemical marker, namely PV1 (PLVAP) (59,88–90). The PV1 antigen is recognized by a novel endothelial antibody (21D5 mAb) (91), which colocalizes strictly on endothelial caveolae from rat lung immunoisolated with anti–caveolin-1 antibodies (92). Endothelial specificity is further supported by data obtained with two additional anti-PV1 monoclonal antibodies, MECA-32 (60,93) and PAL-E (94,95), which were shown to recognize the PV1 antigen (68). PV1 is encoded by the plasmalemmal vesicle-associated protein (*PLVAP*) gene in humans, and its presence has been clearly documented only in mammals (Homologene record #10578) (59,87–90; and R. Stan, unpublished observations). The gene or significant orthologues are absent from yeast (*Saccharomyces cerevisiae*), insects (*Drosophila melanogaster*), and nematodes (*Caenorhabditis elegans*), in which SDs and FDs are not present. No significant orthologue was found in lower vertebrates such as ascids, amphibians, or fish. Thus PLVAP genes encode for a family of proteins (Pfam 6637) that are highly conserved in mammals, consistent with a function of mammalian closed cardiovascular systems. *PV1* mRNA is expressed in most organs and tissues, with highest levels in lung, kidney, spleen, endocrine glands, and digestive tract, followed by modest levels in large vessels, heart, skeletal muscle, skin, brain, and testis (59,73,88,89).

PV1 is a 50-kDa single-span, type II membrane N-glycoprotein (88,91) that forms homodimers in situ (89). The monomer has a very basic calculated isoelectric point (pI) ($\sim$9.1), and the determined pI of the dimer in glycosylated form is still shifted to the basic (pI $\sim$7.8). PV1 binds avidly to heparin at physiological pH (96). PV1 has a short (27-aa) N-terminal intracellular tail and a long (358-aa) C-terminal extracellular domain (73,88). Although the intracellular domain lacks known consensus motifs, it contains two short stretches of amino acids, one next to the transmembrane domain (9-aa) and the other one at the extreme N-terminus (7-aa), that are perfectly conserved across species and that may play a functional role in PV1 biology. The conserved sequence next to the transmembrane domain might contain a putative caveolin-1 interacting motif (ØxxØxxxxØ, where x is any aminoacid (aa) and Ø is a hydrophobic aa such as Trp, Phe, or Tyr) (97). The extracellular domain contains four N-glycosylation sites near the transmembrane domain, a proline-rich region near the C–terminus, and two large coiled-coil domains conserved in all mammals. The secondary structure is predicted to be mostly α-helical (88,89), sustained by the presence of two large consensus coiled-coil domains that are obligate α-helix formers (98). Every seventh amino acid of the α-helix of the coiled-coil domain is hydrophobic, which results in a spiral hydrophobic interface through which one coiled-coil domain interacts with the hydrophobic interface on the cognate coiled-coil to facilitate the formation of an intermolecular superhelix (98). Moreover, seven of the cysteines are situated within the predicted coiled-coil domains. This would suggest that the coiled-coil mediated interaction between two monomers of a dimer would be "reinforced" by several disulfide bonds. All these data suggest that PV1 might adopt a rod-like shape. The fact that the PV1 dimers can be solubilized in SDS suggests that the putative protein–protein interactions between adjacent dimers and other putative interacting proteins (e.g., other PV1 dimers or other protein[s]) are noncovalent in nature.

PV1 is expressed in a subset of ECs (73, 88–90) (e.g., capillaries of the lung, choroid plexus, retina, adrenals, pancreas, intestinal villi, and peritubular capillaries in the kidney, liver, and spleen). PV1 is absent from the continuous endothelia of large vessels of the lung, aorta, and coronary artery as well as from capillaries in the heart, skin, skeletal muscle, and intestinal smooth muscle (where caveolae do not have SDs), and from the fenestrated endothelium of the kidney glomerulus (where fenestrae do not have FDs). All in all, the pattern of expression of PV1 correlates well with the pattern of expression of SDs and FDs in situ (73, and our unpublished data), and the observation that PV1 is present in cultured ECs that contain diaphragms (90).

By immunocytochemistry, PV1 was found to be specifically associated with the SDs of caveolae and TECs, and FDs, at both the apical and basal side of ECs (73,88,89), this being the first demonstration of a protein with such localization. The label was absent from other microdomains of the ECs (e.g., clathrin-coated pits, plasmalemma proper, intercellular junctions), as well as from any other cell type in the organs investigated, thus bolstering the claims of endothelial specificity.

PV1 Is Necessary and Sufficient for Formation of Stomatal Diaphragms and Fenestral Diaphragms

PV1 could function either as a structural and/or functional component of diaphragms. Recent data from our laboratory strongly suggest that PV1 is a key structural component of SDs and FDs, and is both necessary and sufficient for diaphragm expression (90). This is based on several lines of evidence: (a) PV1 forms homodimers in situ, as demonstrated by cotransfection of PV1 tagged with two different epitopes, followed by immunoprecipitation and immunoblotting with anti-tag antibodies (89); (b) several PV1 homodimers reside in close proximity within the same diaphragm, as shown by cross-linking experiments of PV1 carried out in situ in rat lung and kidneys (89); (c) treatment of ECs in culture, lacking both PV1 and diaphragms, with reagents that induce de novo formation of SDs and FDs, upregulates PV1 expression – as expected, PV1 could be found in the newly formed SDs and FDs; (d) *PV1* mRNA silencing, using an siRNA approach, prevented the formation of both SDs and FDs (actually, this approach prevented the formation of fenestrae and TECs as a whole, which suggests that both TECs and fenestrae require PV1 for their biogenesis); and (e) overexpression of tagged PV1 in either ECs or non-ECs (e.g., fibroblasts, COS7, HeLa, HEK293) cell types lacking PV1 expression and diaphragms, led to the formation of caveolar SDs. This result shows that either PV1 forms the diaphragms by itself or the other components of the diaphragms are ubiquitously expressed (at least in the cell types studied), PV1 being the limiting factor (90). Based on these data, a model of PV1 integration in the diaphragms was proposed in which PV1 dimers would form or participate in the formation of the diaphragm fibrils (66).

FUNCTION OF ORGANELLES INVOLVED IN PERMEABILITY

In this section, we describe the functional implications of diaphragms in caveolae, fenestrations, TECs, and VVOs. By virtue of their location in these structures, diaphragms likely act as sieves in transendothelial exchange. The passage of select molecules from the blood plasma to the interstitial fluid could be either inhibited or facilitated depending on chemical (i.e., binding to the diaphragm or not) or biophysical (i.e., size, geometry, charge) criteria.

Stomatal Diaphragms of Caveolae in Continuous Endothelium

No studies directly compare the permeability of continuous endothelium whose caveolae either possess or lack SDs. One approach to this question might be the analysis of lymph composition in different organs (99). However, because the relative contribution of the cell types present in a given organ to the lymph is not known, and the basement membrane composition is expected to be of importance in mediating permeability, differences in lymph composition may provide few clues about the function of the SDs.

Such studies also are limited by microheterogeneity in SD distribution. For example, in perfusion studies, some but not all caveolar invaginations in the microvascular endothelium of the lung are provided with SDs (R.V. Stan, unpublished observations). In EM studies of the lung microcirculation, tracers appeared to gain access to some but not all caveolae, suggesting an uneven distribution of SDs (100–105). It is formally possible that the variable presence of diaphragms in a given vascular bed may be explained on technical grounds (e.g., perfusion-related disruption of SDs), or on the basis of misidentification (e.g., similarly sized lipid raft–based endocytotic vesicle systems) (106,107).

Finally, endothelial caveolae have been involved in several functions (see Ref. 6 and references therein); therefore, the SDs may have other roles in addition to that of a possible permeability modulator.

Diaphragms of Fenestrated Endothelia

Fenestrated capillaries occur in organs that are involved in the reabsorption of water and small solutes (e.g., kidney and intestine) or hormones (e.g., endocrine glands) from the interstitium into the blood. Most of the data in the literature argue that the fenestrated capillaries are highly permeable to water and small hydrophilic solutes, but their permeability to macromolecules, including albumin, is basically the same as that of the continuous endothelium (108). The fenestrae are the structural equivalents of the large pores described by the pore theory (1). The relative contribution of other organelles, such as caveolae and their SDs, TECs, and VVOs in mediating permeability in fenestrated endothelium is not clear.

Fenestrae demonstrate low albumin penetrance. This was explained by the contribution of the proteoglycan tufts, as well as the basement membranes, to overall permeability. It has been calculated that the permeability to water and small solutes could be explained with only a fraction of the fenestrae being open, therefore the function of the proteoglycans would be that of a permeability-reducing plug (109). Tracer experiments show that all fenestrae are permeable to horseradish peroxidase (molecular diameter (Ø) 4 nm) (39), whereas only a few permit the exit of ferritin (Ø 11 nm) (39), dextrans (Ø 12.5–22.5 nm), and glycogen (Ø 22 and 30 nm) (110). As in the case of the fenestrae, all these tracers gained access to few caveolae past their SDs (53,110). As shown by Bearer and Orci (68), the pores between the fibrils of fenestrae have dimensions of 5 to 6 nm; therefore, the passage of the larger tracers must be accommodated by fenestrae with larger pores. An interpretation of these data may be that the diaphragms are dynamic structures, and the degree of their permeability could be modulated by yet unknown factors. Another possibility that cannot be discarded at present is artifact creation in perfusion experiments or during processing in experiments in which the tracers were injected into circulation.

Vesiculo-Vacuolar Organelles

The functions of VVOs have been extensively reviewed recently (4). They have been involved in increased permeability related to inflammation and tumors, permeability to macromolecules, and particulate and immune-cell extravasation. Interestingly, their SDs seem to play a restrictive role in the permeability of particulates (78). The results obtained in cells in culture (111) bolster the claims for VVOs as an organelle involved in the permeability induced by cytokines such as vascular endothelial growth factor (VEGF). However, it is not clear how their restricted localization to the venular endothelium would contribute to the overall permeability in a particular vascular bed, accepting the hypothesis that the bulk of transendothelial exchanges would occur at the level of the capillaries.

KEY POINTS

- Two types of vascular permeability occur: "basal" and "increased." The molecular mechanisms of these types of vascular permeability are not clearly understood.
- Several endothelial structures, such as caveolae, fenestrae, TECs, and VVOs, have been involved in different aspects of permeability. There is little information about the components and the regulation of these structures, especially of fenestrae, TECs and VVOs. PLVAP (PV1) protein is the only protein shown to specifically reside within these structures.
- Work on the components and regulation of these structures should lead to the elucidation of the molecular mechanisms of permeability.

REFERENCES

1 Michel CC, Curry FE. Microvascular permeability. *Physiol Rev*. 1999;79:703.
2 Renkin EM. Cellular and intercellular transport pathways in exchange vessels. *Am Rev Respir Dis*. 1992;146:S28.
3 Rippe B, Rosengren BI, Carlsson O, Venturoli D. Transendothelial transport: the vesicle controversy. *J Vasc Res*. 2002;39:375.
4 Feng D, Nagy JA, Dvorak HF, Dvorak AM. Ultrastructural studies define soluble macromolecular, particulate, and cellular transendothelial cell pathways in venules, lymphatic vessels, and tumor-associated microvessels in man and animals. *Microsc Res Tech*. 2002;57:289.
5 Predescu D, Vogel SM, Malik AB. Functional and morphological studies of protein transcytosis in continuous endothelia. *Am J Physiol Lung Cell Mol Physiol*. 2004;287:L895.
6 Stan RV. Structure and function of endothelial caveolae. *Microsc Res Tech*. 2002;57:350.
7 Stan RV. Endothelial diaphragms: nanogates of vascular permeability. *Anat Rec*. 2005.
8 Tuma PL, Hubbard AL. Transcytosis: crossing cellular barriers. *Physiol Rev*. 2003;83:871.
9 Stan RV, ed. *Channels Across Endothelial Cells*. Georgetown, TX: Landes Bioscience; 2006:271–278. Epub Jan 2006.
10 Starling EH. On the absorption of fluid from the connective tissue spaces. *J Physiol (London)*. 1896;19:312–332.
11 Chinard FP, Flexner LB. Capillary permeability. *Bull Johns Hopkins Hosp*. 1951;88:489.

12 Crone C. The permeability of capillaries in various organs as determined by use of the 'indicator diffusion' method. *Acta Physiol Scand.* 1963;58:292.

13 Pappenheimer JR, Renkin EM, Borrero LM. Filtration, diffusion and molecular sieving through peripheral capillary membranes; a contribution to the pore theory of capillary permeability. *Am J Physiol.* 1951;167:13.

14 Pappenheimer JR. Capillary permeability: deductions concerning the number and dimensions of ultramicroscopic openings in the capillary walls. *Ann N Y Acad Sci.* 1952;55:465.

15 Staverman AJ. The physics of the phenomena of permeability. *Acta Physiol Pharmacol Neerl.* 1954;3:522.

16 Vargas F, Johnson JA. An estimate of reflection coefficients for rabbit heart capillaries. *J Gen Physiol.* 1964;47:667.

17 Palade GE. Fine structure of blood capillaries. *J Applied Physics.* 1953;24:1424.

18 Bruns RR, Palade GE. Studies on blood capillaries. II. Transport of ferritin molecules across the wall of muscle capillaries. *J Cell Biol.* 1968;37:277.

19 Bruns RR, Palade GE. Studies on blood capillaries. I. General organization of blood capillaries in muscle. *J Cell Biol.* 1968;37:244.

20 Palade GE, Bruns RR. Structural modulations of plasmalemmal vesicles. *J Cell Biol.* 1968;37:633.

21 Palade GE. Transport in quanta across endothelium in blood capillaries. *Anat Rec.* 1960;136:254.

22 Palade GE. Blood capillaries of the heart and other organs. *Circulation.* 1961;24:368.

23 Simionescu N, ed. *Transcytosis and Traffic of Membranes in ECs.* Berlin: Springer-Verlag; 1981:657–672.

24 Yamada E. The fine structure of the gall bladder epithelium of the mouse. *J Biophys Biochem Cytol.* 1955;1:445.

25 Dupree P, Parton RG, Raposo G, et al. Caveolae and sorting in the trans-Golgi network of epithelial cells. *EMBO J.* 1993;12:1597.

26 Glenney JR Jr. The sequence of human caveolin reveals identity with VIP21, a component of transport vesicles. *FEBS Lett.* 1992;314:45.

27 Glenney JR Jr., Soppet D. Sequence and expression of caveolin, a protein component of caveolae plasma membrane domains phosphorylated on tyrosine in Rous sarcoma virus-transformed fibroblasts. *Proc Natl Acad Sci USA.* 1992;89:10517.

28 Kurzchalia TV, Dupree P, Parton RG, et al. VIP21, a 21-kD membrane protein is an integral component of trans-Golgi-network-derived transport vesicles. *J Cell Biol.* 1992;118:1003.

29 Rothberg KG, Heuser JE, Donzell WC, et al. Caveolin, a protein component of caveolae membrane coats. *Cell.* 1992;68:673.

30 Mamdouh Z, Chen X, Pierini LM, et al. Targeted recycling of PECAM from endothelial surface-connected compartments during diapedesis. *Nature.* 2003;421:748.

31 Muro S, Wiewrodt R, Thomas A, et al. A novel endocytic pathway induced by clustering endothelial ICAM-1 or PECAM-1. *J Cell Sci.* 2003;116:1599.

32 Takizawa T, Anderson CL, Robinson JM. A novel Fc gamma R-defined, IgG-containing organelle in placental endothelium. *J Immunol.* 2005;175:2331.

33 Tkachenko E, Lutgens E, Stan RV, Simons M. Fibroblast growth factor 2 endocytosis in ECs proceed via syndecan-4-dependent activation of Rac1 and a Cdc42-dependent macropinocytic pathway. *J Cell Sci.* 2004;117:3189.

34 Kirkham M, Parton RG. Clathrin-independent endocytosis: new insights into caveolae and non-caveolar lipid raft carriers. *Biochim Biophys Acta.* 2005;1745:273.

35 Muro S, Koval M, Muzykantov V. Endothelial endocytic pathways: gates for vascular drug delivery. *Curr Vasc Pharmacol.* 2004;2:281.

36 Gautier A, Bernhard W, Oberling C. Sur l'existence d'un appareil lacunaire pericapillaire du glomerule de Malpighi, revele par la microscopie electronique. *C R Seances Soc Biol Fil.* 1950;144:1605.

37 Oberling C, Gautier A, Bernhardt W. La structure des capillaires glomerulairevue au microscope electronique. *Presse Med.* 1951;59:938.

38 Clementi F, Palade GE. Intestinal capillaries. II. Structural effects of EDTA and histamine. *J Cell Biol.* 1969;42:706.

39 Clementi F, Palade GE. Intestinal capillaries. I. Permeability to peroxidase and ferritin. *J Cell Biol.* 1969;41:33.

40 Elfvin LG. The ultrastructure of the capillary fenestrae in the adrenal medulla of the rat. *J Ultrastruct Res.* 1965;12:687.

41 Friederici HH. On the diaphragm across fenestrae of capillary endothelium. *J Ultrastruct Res.* 1969;27:373.

42 Rhodin JA. The diaphragm of capillary endothelial fenestrations. *J Ultrastruct Res.* 1962;6:171.

43 Wisse E. An electron microscopic study of the fenestrated endothelial lining of rat liver sinusoids. *J Ultrastruct Res.* 1970;31:125.

44 Braet F, Wisse E. Structural and functional aspects of liver sinusoidal endothelial cell fenestrae: a review. *Comp Hepatol.* 2002;1:1.

45 Simionescu M, Simionescu N, Silbert JE, Palade GE. Differentiated microdomains on the luminal surface of the capillary endothelium. II. Partial characterization of their anionic sites. *J Cell Biol.* 1981;90:614.

46 Simionescu M, Simionescu N, Palade GE. Differentiated microdomains on the luminal surface of capillary endothelium: distribution of lectin receptors. *J Cell Biol.* 1982;94:406.

47 Simionescu N, Simionescu M, Palade GE. Differentiated microdomains on the luminal surface of the capillary endothelium. I. Preferential distribution of anionic sites. *J Cell Biol.* 1981;90:605.

48 Simionescu N, Siminoescu M, Palade GE. Permeability of muscle capillaries to small heme-peptides. Evidence for the existence of patent transendothelial channels. *J Cell Biol.* 1975;64:586.

49 Milici AJ, Peters KR, Palade GE. The endothelial pocket. A new structure in fenestrated endothelia. *Cell Tissue Res.* 1986;244:493.

50 Kohn S, Nagy JA, Dvorak HF, Dvorak AM. Pathways of macromolecular tracer transport across venules and small veins. Structural basis for the hyperpermeability of tumor blood vessels. *Lab Invest.* 1992;67:596.

51 Yoshida Y, Yamada M, Wakabayashi K, Ikuta F. Endothelial fenestrae in the rat fetal cerebrum. *Brain Res Dev Brain Res.* 1988;44:211.

52 Milici AJ, Bankston PW. Fetal and neonatal rat intestinal capillaries: a TEM study of changes in the mural structure. *Am J Anat.* 1981;160:435.

53 Milici AJ, Bankston PW. Fetal and neonatal rat intestinal capillaries: permeability to carbon, ferritin, hemoglobin, and myoglobin. *Am J Anat.* 1982;165:165.

54 Reeves WH, Kanwar YS, Farquhar MG. Assembly of the glomerular filtration surface. Differentiation of anionic sites in glomerular capillaries of newborn rat kidney. *J Cell Biol.* 1980;85:735.

55 Pease DC. Electron microscopy of the vascular bed of the kidney cortex. *Anat Rec.* 1955;121:701.
56 Yamada E. The fine structure of the mouse kidney glomerulus. *J Biophys Biochem Cytol.* 1955;1:551.
57 Bankston PW, Pino RM. The development of the sinusoids of fetal rat liver: morphology of ECs, Kupffer cells, and the transmural migration of blood cells into the sinusoids. *Am J Anat.* 1980;159:1.
58 Naito M, Wisse E. *Kupffer Cells and Other Liver Sinusoidal Cells.* E Wisse, Ed. Amsterdam: Elsevier 1977,497–505.
59 Stan RV, Arden KC, Palade GE. cDNA and protein sequence, genomic organization, and analysis of cis regulatory elements of mouse and human PLVAP genes. *Genomics.* 2001;72:304.
60 Hallmann R, Mayer DN, Berg EL, et al. Novel mouse endothelial cell surface marker is suppressed during differentiation of the blood brain barrier. *Dev Dyn.* 1995;202:325.
61 Bradbury MW. The structure and function of the blood-brain barrier. *Fed Proc.* 1984;43:186.
62 Margulies EH, Kardia SL, Innis JW. A comparative molecular analysis of developing mouse forelimbs and hindlimbs using serial analysis of gene expression (SAGE). *Genome Res.* 2001;11:1686.
63 Penn PE, Jiang DZ, Fei RG, et al. Dissecting the hematopoietic microenvironment. IX. Further characterization of murine bone marrow stromal cells. *Blood.* 1993;81:1205.
64 Vittet D, Prandini MH, Berthier R, et al. Embryonic stem cells differentiate in vitro to ECs through successive maturation steps. *Blood.* 1996;88:3424.
65 Orci L, Pepper MS. Microscopy: an art? *Nat Rev Mol Cell Biol.* 2002;3:133.
66 Stan RV. Structure of caveolae. *Biochim Biophys Acta.* 2005; 1746(3):334–348. Epub 2005 Sep 16.
67 Noguchi Y, Shibata Y, Yamamoto T. Endothelial vesicular system in rapid-frozen muscle capillaries revealed by serial sectioning and deep etching. *Anat Rec.* 1987;217:355.
68 Bearer EL, Orci L. Endothelial fenestral diaphragms: a quick-freeze, deep-etch study. *J Cell Biol.* 1985;100:418.
69 Simionescu M, Simionescu N, Palade GE. Morphometric data on the endothelium of blood capillaries. *J Cell Biol.* 1974;60:128.
70 Apkarian RP. The fine structure of fenestrated adrenocortical capillaries revealed by in-lens field-emission scanning electron microscopy and scanning transmission electron microscopy. *Scanning.* 1997;19:361.
71 Maul GG. Structure and formation of pores in fenestrated capillaries. *J Ultrastruct Res.* 1971;36:768.
72 Rostgaard J, Qvortrup K. Electron microscopic demonstrations of filamentous molecular sieve plugs in capillary fenestrae. *Microvasc Res.* 1997;53:1.
73 Stan RV, Kubitza M, Palade GE. PV-1 is a component of the fenestral and stomatal diaphragms in fenestrated endothelia. *Proc Natl Acad Sci USA.* 1999;96:13203.
74 Milici AJ, Furie MB, Carley WW. The formation of fenestrations and channels by capillary endothelium in vitro. *Proc Natl Acad Sci USA.* 1985;82:6181.
75 Milici AJ, L'Hernault N, Palade GE. Surface densities of diaphragmed fenestrae and transendothelial channels in different murine capillary beds. *Circ Res.* 1985;56:709.
76 Roberts WG, Palade GE. In: Rissau W, Rubanyi GM, eds. *Morphogenesis of Endothelium.* Vol. 8. Amsterdam: Hardwood Academic Publishers; 2000:23–41.
77 Dvorak AM, Kohn S, Morgan ES, et al. The vesiculo-vacuolar organelle (VVO): a distinct endothelial cell structure that provides a transcellular pathway for macromolecular extravasation. *J Leukoc Biol.* 1996;59:100.
78 Feng D, Nagy JA, Pyne K, et al. Pathways of macromolecular extravasation across microvascular endothelium in response to VPF/VEGF and other vasoactive mediators. *Microcirculation.* 1999;6:23.
79 Vasile E, Dvorak AM, Stan RV, Dvorak HF. Isolation and characterization of caveolae and vesiculo-vacuolar organelles from ECs cultured with VPF/VEGF and from human lung. *Mol Biol Cell.* 2000;11:121.
80 Drab M, Verkade P, Elger M, et al. Loss of caveolae, vascular dysfunction, and pulmonary defects in caveolin-1 gene-disrupted mice. *Science.* 2001;293:2449.
81 Zhao YY, Liu Y, Stan RV, et al. Defects in caveolin-1 cause dilated cardiomyopathy and pulmonary hypertension in knockout mice. *Proc Natl Acad Sci USA.* 2002;99:11375.
82 Bankston PW, Porter GA, Milici AJ, Palade GE. Differential and specific labeling of epithelial and vascular ECs of the rat lung by *Lycopersicon esculentum* and *Griffonia simplicifolia* I lectins. *Eur J Cell Biol.* 1991;54:187.
83 Furuya S. Ultrastructure and formation of diaphragmed fenestrae in cultured ECs of bovine adrenal medulla. *Cell Tissue Res.* 1990;261:97.
84 Pino RM. The cell surface of a restrictive fenestrated endothelium. I. Distribution of lectin-receptor monosaccharides on the choriocapillaris. *Cell Tissue Res.* 1986;243:145.
85 Bankston PW, Milici AJ. A survey of the binding of polycationic ferritin in several fenestrated capillary beds: indication of heterogeneity in the luminal glycocalyx of fenestral diaphragms. *Microvasc Res.* 1983;26:36.
86 Simionescu M, Simionescu N, Palade GE. Preferential distribution of anionic sites on the basement membrane and the abluminal aspect of the endothelium in fenestrated capillaries. *J Cell Biol.* 1982;95:425.
87 Simionescu M, Simionescu N, Palade GE. Partial chemical characterization of the anionic sites in the basal lamina of fenestrated capillaries. *Microvasc Res.* 1984;28:352.
88 Stan RV, Ghitescu L, Jacobson BS, Palade GE. Isolation, cloning, and localization of rat PV-1, a novel endothelial caveolar protein. *J Cell Biol.* 1999;145:1189.
89 Stan RV. Multiple PV1 dimers reside in the same stomatal or fenestral diaphragm. *Am J Physiol Heart Circ Physiol.* 2004;286:H1347.
90 Stan RV, Tkachenko E, Niesman IR. PV1 is a key structural component for the formation of the stomatal and fenestral diaphragms. *Mol Biol Cell.* 2004;15:3615.
91 Ghitescu LD, Crine P, Jacobson BS. Antibodies specific to the plasma membrane of rat lung microvascular endothelium. *Exp Cell Res.* 1997;232:47.
92 Stan RV, Roberts WG, Predescu D, et al. Immunoisolation and partial characterization of endothelial plasmalemmal vesicles (caveolae). *Mol Biol Cell.* 1997;8:595.
93 Duijvestijn AM, Kerkhove M, Bargatze RF, Butcher EC. Lymphoid tissue- and inflammation-specific endothelial cell differentiation defined by monoclonal antibodies. *J Immunol.* 1987;138:713.
94 Niemela H, Elima K, Henttinen T, et al. Molecular identification of PAL-E, a widely used endothelial cell marker. *Blood.* 2005;106(10):3405–3409. Epub 2005 Aug.

95 Schlingemann RO, Dingjan GM, Emeis JJ, et al. Monoclonal antibody PAL-E specific for endothelium. *Lab Invest*. 1985;52: 71.
96 Hnasko R, McFarland M, Ben-Jonathan N. Distribution and characterization of plasmalemma vesicle protein-1 in rat endocrine glands. *J Endocrinol*. 2002;175:649.
97 Couet J, Li S, Okamoto T, et al. Identification of peptide and protein ligands for the caveolin-scaffolding domain. Implications for the interaction of caveolin with caveolae-associated proteins. *J Biol Chem*. 1997;272:6525.
98 Lupas A. Prediction and analysis of coiled-coil structures. *Methods Enzymol*. 1996;266:513.
99 Aukland K, Reed RK. Interstitial-lymphatic mechanisms in the control of extracellular fluid volume. *Physiol Rev*. 1993;73: 1.
100 Ghitescu L, Fixman A, Simionescu M, Simionescu N. Specific binding sites for albumin restricted to plasmalemmal vesicles of continuous capillary endothelium: receptor-mediated transcytosis. *J Cell Biol*. 1986;102:1304.
101 John TA, Vogel SM, Minshall RD, et al. Evidence for the role of alveolar epithelial gp60 in active transalveolar albumin transport in the rat lung. *J Physiol*. 2001;533:547.
102 McIntosh DP, Tan XY, Oh P, Schnitzer JE. Targeting endothelium and its dynamic caveolae for tissue-specific transcytosis in vivo: a pathway to overcome cell barriers to drug and gene delivery. *Proc Natl Acad Sci USA*. 2002;99:1996.
103 Nistor A, Simionescu M. Uptake of low density lipoproteins by the hamster lung. Interactions with capillary endothelium. *Am Rev Respir Dis*. 1986;134:1266.
104 Popov D, Simionescu M. Alterations of lung structure in experimental diabetes, and diabetes associated with hyperlipidaemia in hamsters. *Eur Respir J*. 1997;10:1850.
105 Simionescu M. Ultrastructural organization of the alveolar-capillary unit. *Ciba Found Symp*. 1980;78:11.
106 Lamaze C, Dujeancourt A, Baba T, et al. Interleukin 2 receptors and detergent-resistant membrane domains define a clathrin-independent endocytic pathway. *Mol Cell*. 2001;7:661.
107 Sabharanjak S, Sharma P, Parton RG, Mayor S. GPI-anchored proteins are delivered to recycling endosomes via a distinct cdc42-regulated, clathrin-independent pinocytic pathway. *Dev Cell*. 2002;2:411.
108 Granger DN, Granger JP, Brace RA, et al. Analysis of the permeability characteristics of cat intestinal capillaries. *Circ Res*. 1979;44:335.
109 Levick JR, Smaje LH. An analysis of the permeability of a fenestra. *Microvasc Res*. 1987;33:233.
110 Simionescu N, Simionescu M, Palade GE. Permeability of intestinal capillaries. Pathway followed by dextrans and glycogens. *J Cell Biol*. 1972;53:365.
111 Chen J, Braet F, Brodsky S, et al. VEGF-induced mobilization of caveolae and increase in permeability of ECs. *Am J Physiol Cell Physiol*. 2002;282:C1053.

75

Endothelial Luminal Glycocalyx

Protective Barrier between Endothelial Cells and Flowing Blood

Bernard M. van den Berg*, Max Nieuwdorp†, Erik Stroes†, and Hans Vink†

*Beth Israel Deaconess Medical Center and Harvard Medical School, Boston, Massachusetts;
†Academic Medical Center and University of Amsterdam, Amsterdam, The Netherlands

All cells – from single-cell microorganisms to highly organized mammalian cells – are shielded from their surrounding milieu by a membranous, carbohydrate-rich layer, or glycocalyx. The glycocalyx also is involved in nutrient uptake, and it facilitates the binding and concentrating of factors necessary for proper cell function. In the vasculature, the endothelial glycocalyx protects the vascular wall from direct exposure to flowing blood, contributes to the vascular permeability barrier and its antiadhesive properties, and stimulates the endothelial release of nitric oxide in response to fluid shear stress.

Historically, studies aimed at understanding mechanisms of endothelial permeability led to the concept that vessel walls are lined with an extracellular layer of membrane-bound substances (1,2). Danielli (1), and Chambers and Zweifach (2) hypothesized the existence of a thin, noncellular layer on the endothelial surface, termed the *endocapillary layer*. In the latter study, the perfusion of frog mesentery with an Evans blue–containing solution revealed blue-colored thin strands and sheets of translucent material on the inner surface of the capillary (2). In subsequent experiments involving intravenous injections of pontamine sky blue and intravital microscopy of the hamster cheek pouch, Copley and Staple observed an unstained plasmatic zone adjacent to the endothelial surface, giving rise to the notion that the endothelial surface is covered by a thin molecular layer (the endoendothelial fibrin lining) and an adjacent immobile plasma region (3).

To date, although both concepts, in essence, still hold true, novel data on the structural and compositional properties of the endothelial glycocalyx implicate a highly active role for this noncellular layer in vascular wall homeostasis. Like so many other properties of the endothelium, this role is likely to vary between different sites of the vascular tree. Indeed, efforts to correlate the structure and composition of the glycocalyx with underlying physiology and pathophysiology should take into account site- and time-specific differences in vascular permeability and blood flow.

STRUCTURAL PROPERTIES OF THE ENDOTHELIAL GLYCOCALYX

Electron Microscopy

A combination of ruthenium red staining and conventional electron microscopy (EM) provided the first direct visualization of the endothelial glycocalyx (4). Ruthenium red is a cationic dye that has a high affinity for acidic mucopolysaccharides and, together with osmium tetroxide fixation, generates electron density. The electron micrographs revealed a small, irregularly shaped layer extending approximately 50 to 100 nm into the vessel lumen. Subsequent EM studies, in which varying perfusates or fixatives were used to stabilize anionic carbohydrate structures, demonstrated the presence of a glycocalyx (of heterogeneous dimension and appearance) on continuous endothelium throughout the vasculature (5–10). Fenestrated endothelium, in addition, was found to have a combination of surface-bound stained structures (50–100 nm thick) and distinct filamentous plugs (20–40 filaments, with a length of about 350 nm) on the surface of the fenestrae (11). Collectively, these investigations provided compelling evidence for a thick endothelial surface layer (ESL) throughout the vascular tree (Figure 75.1). Colocalization of lectins to the observed stained structures confirmed their saccharine nature (5,7,9).

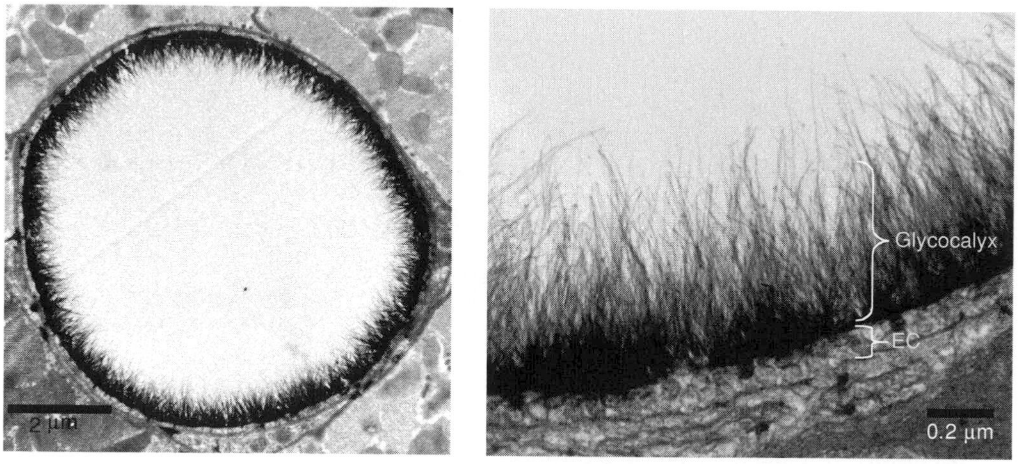

Figure 75.1. Electron micrograph of a goat coronary capillary stained with Alcian blue.

Intravital Microscopy

Improvements in intravital microscopy, such as the introduction of quantitative methods and fluorescence techniques, have yielded additional insights into the glycocalyx. Among the more important discoveries is that the glycocalyx is much thicker than originally believed based on EM studies. For example, using a hamster cremaster muscle model, Klitzman and Duling demonstrated dramatic differences in the microvascular and systemic hematocrit (12). Moreover, these differences were abrogated upon enzymatic treatment of the microvascular network with heparinase (13). Based on these findings, it was hypothesized that there exists a 300- to 1,000-nm thick, slow-moving plasma layer on the endothelial cell (EC) surface consisting principally of heparan sulfate (HS) proteoglycans.

As an alternative approach to characterize the glycocalyx, we labeled the free-flowing plasma using FITC-dextran. In hamster cremaster capillaries, the width of the labeled plasma column was compared with the anatomical diameter of the lumen. The anatomical diameter was consistently greater than that of the plasma column, revealing a continuous EC surface layer with a thickness of 400 to 500 nm (Figure 75.2) (14). Subsequent theoretical considerations predicted a glycocalyx thickness of 500 to 1,000 nm, accounting for the observed variations in red cell motion through the microvessels and the discrepancy between in vivo and in vitro estimates of resistance to blood flow (15–17).

The functional relevance of the glycocalyx was demonstrated in a canine model of coronary reactive hyperemia (18). Reactive hyperemia following a 15-second occlusion of the left circumflex artery was compared in dogs in which coronary arteries were infused with or without hyaluronidase. Treatment with hyaluronidase resulted in increased coronary conductance (the inverse of resistance). Heat-inactivated hyaluronidase had no such effect. Further in vivo evidence for the role of the hyaluronan in mediating glycocalyx properties was provided by Henry and Duling, who demonstrated that hyaluronidase treatment of hamster microvessels in the cremaster muscle resulted in increased access of 70- to 145-kDa FITC dextrans, but not higher molecular weight FITC dextrans or red blood cells (19).

Taken together, these findings are consistent with the presence of a hyaluronan-containing glycocalyx whose volume and thickness provide normal resistance to flowing blood. Indeed, the unexpectedly large dimension of the glycocalyx exceeds – by several times – the dimensions of the endothelium and its associated receptors, including leukocyte adhesion molecules. It is tempting to speculate that glycocalyx dimension per se plays an important role in blood vessel physiology. Indeed, various studies have demonstrated a link between altered glycocalyx dimension and pathophysiology, as in ischemia–reperfusion (6), hypoxia (10), high-density lipoprotein (HDL) (20) and low-density lipoproteins (LDL) (21,22), and variations in wall shear stress (7,23).

COMPOSITION OF THE ENDOTHELIAL GLYCOCALYX

The endothelial glycocalyx is composed of a negatively charged mesh consisting of proteoglycans (core proteins containing long, unbranched carbohydrate side-chains termed *glycosaminoglycans*), glycoproteins (proteins containing short carbohydrate side-chains), and glycolipids (24). The most common core proteins are syndecans, glypicans, and perlecans. Glycosaminoglycans include HS, chondroitin sulfate (CS), and hyaluronan. Examples of glycoproteins include selectins, integrins, and members of the immunoglobulin superfamily. In addition to cell surface–bound components, the composition of the glycocalyx also is influenced by absorption of soluble plasma components, such as albumin, orosomucoid, fibrinogen, and fibronectin. The endothelial glycocalyx also harbors a wide array of enzymes and proteins that play a role in leukocyte–EC and platelet–EC adhesion, including extracellular superoxide dismutase

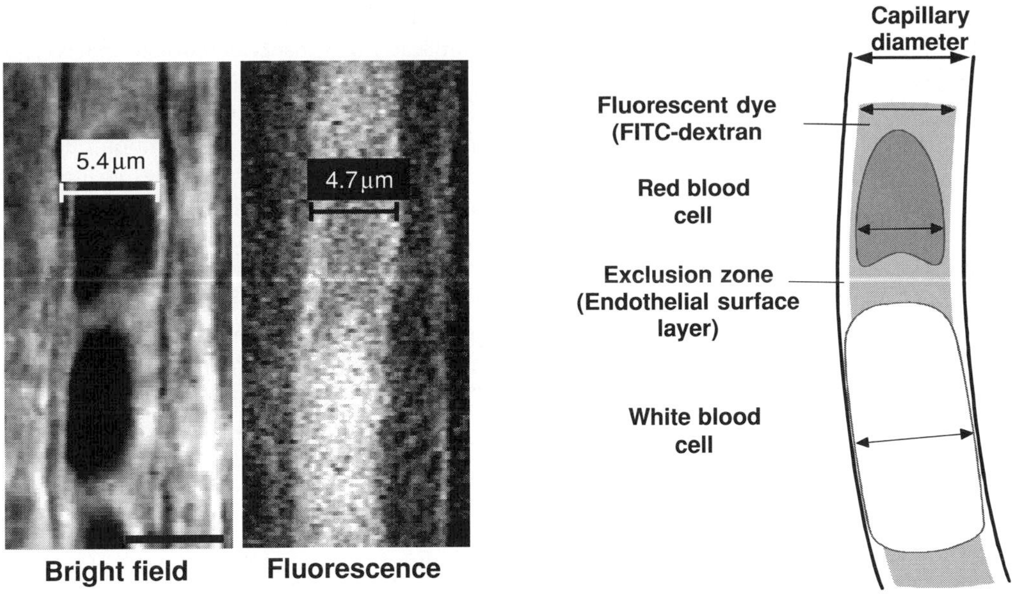

Figure 75.2. Combined bright-field and fluorescence intravital microscopy. These studies reveal a gap of about 200 to 800 nm between passing red blood cells, and a fluorescent column of an infused fluorescent dye coupled to a large molecule (Dextran 70), and the EC boundary. (Reproduced with permission from Vink H, Duling BR. Identification of distinct luminal domains for macromolecules, erythrocytes, and leukocytes within mammalian capillaries. *Circ Res.* 1996;79(3):581–589.)

(ec-SOD), angiotensin-converting enzyme (ACE), antithrombin III (ATIII), lipoprotein lipase (LpL), hepatic endothelial lipase (HEL), apolipoproteins, growth factors, and chemokines.

Constituents of the glycocalyx, such as hyaluronan, have been found to be intimately involved in mediating vascular homeostasis (7,9,19). For example, the glycocalyx serves as a mechano-shear sensor for nitric oxide (NO) release (25–27). Certain enzymes present in the glycocalyx contribute to its vasculoprotective effect. In particular, ec-SOD, which dismutates oxygen radicals to hydrogen peroxide and is bound to proteoglycans within the glycocalyx, is shed upon damage to the glycocalyx and results in an unbalance in favor of a pro-oxidant state (28).

ENDOTHELIAL GLYCOCALYX AND VASCULAR PERMEABILITY

Vascular barrier properties determine the transport of fluid between blood and interstitial space according to the Starling-Landis equation (29):

$$J_c = L_p A\left[(P_c - P_t) - \sigma_p(\pi_c - \pi_t)\right]$$

where J_c = net transcapillary fluid shift, L_p = hydraulic conductivity of capillary wall, A = capillary membrane filtration area, P_c = capillary blood pressure, P_t = tissue fluid pressure, σ_p = reflection coefficient for plasma proteins, π_c = capillary blood colloid osmotic pressure, and π_t = tissue fluid colloid osmotic pressure.

A continuous turnover of fluid occurs in the body. Fluid is filtered from blood to tissues at the arterial end of the circulation and is reabsorbed at the venous end. This pattern is largely dictated by differences in the hydrostatic pressure between the two ends of the vascular segment. Excess fluid not taken up at the venous end is removed from the tissue by the lymphatics.

The hydraulic conductivity reflects vessel wall permeability to water. Hydraulic conductivity varies between different segments of the vascular tree (e.g., arterioles, capillaries, and venules) and between microvascular beds from different organs. These differences, which presumably reflect site-specific demands of the underlying tissue, are explained – at least in part – by the variable existence of transendothelial channels, fenestrae, and intercellular gaps.

In 1980, Curry and Michel proposed the fiber matrix concept (30). According to this model, the molecular sieving properties of the capillary wall are determined by a fiber matrix, covering all endothelial channels and filling intracellular boundaries. It was suggested that the local protein concentration gradient that generates the colloid osmotic pressure (or oncotic pressure) is localized across the glycocalyx and not between the plasma and tissue underlying the endothelium. Thus, the endothelial surface glycocalyx maintains fluid balance between blood and tissue and results in a balance between the absorption and filtration of water. This hypothesis was confirmed by the remarkable similarity of protein permeability and reflection coefficient (i.e., the rejection of molecular passage) in fenestrated and continuous capillaries despite their differences in filtration coefficient and ultrastructure (31). Recently, this was further illustrated by very similar estimates (between 34 and 45 mEq/L) of the permeability parameter related to the presence of charge-selective properties (ω) between peripheral and glomerular capillaries, although the renal glomeruli are the body's most active filtration units, producing about 180 liters of primary urine per day, with a minimal loss of proteins (32).

Because glomerular capillaries consist of fenestrated ECs, whereas peripheral ECs are predominantly of the continuous type, it appears that, in most vascular beds, the capillary wall behaves as a molecular filter that allows the free exchange (by convection and diffusion) of water, ions, and small hydrophilic solutes between the plasma and tissue spaces, but limits the passage of highly charged macromolecules. These findings argue for similar selective permeability properties throughout the various vascular networks and organs, predominantly dependent on the high plasma concentration of negatively charged albumin, the main contributor to oncotic pressure.

ENDOTHELIAL GLYCOCALYX IN MICRO- VERSUS MACROVASCULATURE

Studies in both micro- and macrovessels demonstrated similarity in glycocalyx constituents such as hyaluronan (7,9,19), the release of NO (25–27), and the presence of ec-SOD (28), which are all involved in vascular homeostasis and protection against damage. These findings are of particular interest because altered vascular permeability, attenuated NO bioavailability, and redox dysregulation are among the earliest characteristics of atherogenesis (33).

Despite these observations, it has proved difficult to show direct relevance of the glycocalyx as a vasculoprotective paradigm for larger vessels. The latter is predominantly due to the fact that glycocalyx research has traditionally focused on the microvasculature, in which atherogenesis does not occur. However, several studies have emphasized the relevance of the glycocalyx in macrovessels (6,23). van Haaren and colleagues recently visualized a thick endothelial glycocalyx in the larger arteries of rats (34). Interestingly, the reduced thickness of the glycocalyx at high-shear compared to low-shear regions of the murine carotid artery lends direct support to a potential role of glycocalyx perturbation in rendering low-shear regions more susceptible to atherogenesis (23,35). The glycocalyx in larger vessels also has been shown to decrease extravasation of LDL particles into the subendothelial space (36,37). Together, these data imply that, in addition to its functional importance in small vessels, the glycocalyx serves a vasculoprotective role in large vessels.

ROLE OF ENDOTHELIAL GLYCOCALYX IN HEMOSTASIS

The endothelium is involved intimately in the regulation of coagulation. Under physiological conditions, the generation of thrombin is tightly regulated by the inhibitory actions of ATIII, thrombomodulin, endothelial protein C receptor, and tissue factor pathway inhibitor (TFPI), all of which are bound to the luminal endothelium (38,39). Denudation of endothelium leads to exposure of blood to the tissue factor–rich vessel wall. Activation or dysfunction of *intact* endothelium results in downregulation of natural anticoagulant mechanisms, including reduced expression of thrombomodulin and the endothelial protein C receptor. Whether structural or functional in nature, these changes may result in a shift in the hemostatic balance toward the procoagulant side. Interestingly, a specific disruption of the glycocalyx itself results in thrombin generation and platelet adhesion as early as 10 minutes after glycocalyx removal (22). Moreover, activation agonists such as proinflammatory cytokines have profound effects on the synthesis of glycocalyx constituents, such as HS and hyaluronan, thus contributing to impaired natural anticoagulant defenses (40,41).

ROLE OF ENDOTHELIAL GLYCOCALYX IN LEUKOCYTE ADHESION

Leukocyte trafficking is regulated by a highly orchestrated multistep pathway that involves the inducible expression of cell adhesion molecules on the surface of the endothelium (42). The dimension of the glycocalyx greatly exceeds the length of the extracellular domain of adhesion molecules such as P- and E-selectin, intercellular adhesion molecule (ICAM)-1, and vascular cell adhesion molecule (VCAM)-1.

Thus, new concepts have been developed to explain how leukocytes interact with endothelial adhesion molecules (43,44). According to one model, leukocytes develop cooperative weak interactions with hyaluronan, which exhibits electrostatic attraction properties (43). The subsequent fate of adhesion depends on modification of the glycocalyx layer such that leukocyte-bound ligands are exposed to EC surface adhesion molecules, for example, as might occur with the directed lateral diffusion of active transport of the hyaluronan and the attached membrane receptors, or the hyaluronidase-mediated degradation of hyaluronan. The second model implicates a role for long L-selectin–containing leukocyte microvilli in contacting the EC through the glycocalyx (44,45). Once the microvilli penetrate the glycocalyx, they may initiate rolling along the endothelial surface. Again, implicit in this paradigm is the subsequent modification of the glycocalyx to allow for firm adhesion, spreading, and transmigration.

It is interesting to speculate that disease-associated modification of the glycocalyx results in dysregulated leukocyte trafficking. Exposure of ECs to ox-LDL, as occurs in atherosclerosis, has been shown to decrease the amount of HS proteoglycans associated with the luminal cell surface (46). Furthermore, inflammation and ischemia-induced shedding of glycocalyx in activated endothelium may be coupled to vascular inflammation (47). Consistent with its inhibitory (barrier) role in leukocyte adhesion (48), glycocalyx preservation (via the infusion of major glycocalyx constituents) was shown to abolish leukocyte rolling to the endothelial surface in response to potent proatherogenic challenges (49).

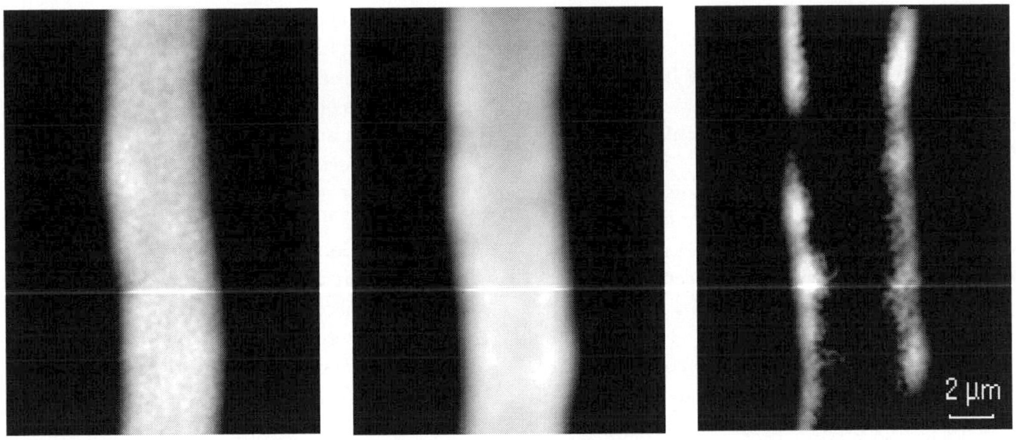

Figure 75.3. Three-dimensional microscopic reconstruction of fluorescent tracer distributions in mouse cremaster tissue capillaries. (Reproduced with permission from Vink H, Stace TM, Damiano ER. 2003.)

ASSESSMENT OF SYSTEMIC GLYCOCALYX DIMENSION IN HUMANS

To date, direct visualization of endothelial glycocalyx in humans has been unsuccessful, mainly due to the fact that the endothelial glycocalyx is a very delicate structure that is critically dependent on the presence of flowing blood (24). Because the endothelial glycocalyx provides limited access to plasma macromolecules and erythrocytes, the best way to measure the endothelial glycocalyx in humans is to compare intravascular volumes using a glycocalyx-permeable tracer, such as neutral Dextran 40 (MW 40 kDa) and a glycocalyx-impermeable tracer, such as anionic sulfated dextran or labeled erythrocytes (14) (Figure 75.3). Based on the electromechanical nature of interactions of erythrocytes and anionic plasma molecules with the sulfated glycosaminoglycans, high-resolution three-dimensional (3-D) intravital fluorescence microscopy reveals a partial exclusion of intravenously injected sulfated dextran molecules (anionic) within a 1-micron thick capillary EC glycocalyx, in comparison to Texas red–labeled dextran molecules (neutral). To make this method feasible in humans, intravascular distribution of flowing blood volume can be quantified using labeled autologous erythrocytes (50). In combination with total intravascular volume measurements using the neutral tracer Dextran 40, it is possible to distinguish between flowing plasma volume and total plasma volume and hence systemic glycocalyx volume.

CONCLUSION

Located at the interface of flowing blood and the vascular endothelial lining, the endothelial glycocalyx displays a wide array of functions, including maintenance of interstitial fluid balance, hemostasis, leukocyte adhesion, and mechanotransduction. Structural and/or functional disruption of the glycocalyx may arise from and/or contribute to a number of pathophysiological conditions, including atherosclerosis. Important considerations are the composition and the physical dimensions of the glycocalyx, and how these properties differ between different vascular beds in health and disease. Whereas glycocalyx disruption is accompanied by enhanced sensitivity of the vasculature toward atherogenic stimuli, glycocalyx "restoration" attenuates these abnormalities, suggesting a potential benefit of glycocalyx-based therapies. From a diagnostic standpoint, an important goal is to determine the feasibility and value of systemic glycocalyx measurement as a surrogate marker for cardiovascular disease.

KEY POINTS

- The glycocalyx is the carbohydrate rich matrix that lines the luminal surface of the vasculature and forms the true interface between endothelium and flowing blood.
- Novel intravital techniques indicate that the glycocalyx is considerably thicker than originally believed.
- The glycocalyx consists of endothelial-derived and plasma-derived components.
- The glycocalyx is vasculoprotective and plays an important role in homeostasis.

Future Goals

- To develop and test novel therapies aimed at restoring glycocalyx in disease
- To develop diagnostic tools for measuring glycocalyx dimensions and constituents

REFERENCES

1 Danielli JF. Capillary permeability and edema in the perfused frog. *J Physiol (Lond)*. 1940;98:109–129.
2 Chambers R, Zweifach BW. Intercellular cement and capillary permeability. *Physiol Rev*. 1947;27:436–463.
3 Copley AL. The endothelial fibrin lining, fibrinogen gel clotting, and the endothelium-blood interface. *Ann NY Acad Sci*. 1983; 416:377–396.
4 Luft JH. Fine structure of capillary and endocapillary layer as revealed by ruthenium red. *Microcirc Symp Fed Proc*. 1966;25: 1773–1783.
5 Baldwin AL, Winlove CP. Effects of perfusate composition on binding of ruthenium red and gold colloid to glycocalyx of rabbit aortic endothelium. *J Histochem Cytochem*. 1984;32:259–266.
6 Beresewicz A, Czarnowska E, Maczewski M. Ischemic preconditioning and superoxide dismutase protect against endothelial dysfunction and endothelium glycocalyx disruption in the postischemic guinea-pig hearts. *Mol Cell Biochem*. 1998;186:87–97.
7 Haldenby KA, Chappell DC, Winlove CP, et al. Focal and regional variations in the composition of the glycocalyx of large vessel endothelium. *J Vasc Res*. 1994;31:2–9.
8 Sims DE, Horne MM. Non-aqueous fixative preserves macromolecules on the endothelial cell surface: an in situ study. *Eur J Morphol*. 1993;32:59–64.
9 van den Berg BM, Vink H, Spaan JAE. The endothelial glycocalyx protects against myocardial edema. *Circ Res*. 2003;92:592–594.
10 Ward BJ, Donnelly JL. Hypoxia induced disruption of the cardiac endothelial glycocalyx: implications for capillary permeability. *Cardiovasc Res*. 1993;27:384–389.
11 Rostgaard J, Qvortrup K. Electron microscopic demonstrations of filamentous molecular sieve plugs in capillary fenestrae. *Microvasc Res*. 1997;53:1–13.
12 Klitzman B, Duling BR. Microvascular hematocrit and red cell flow in resting and contracting striated muscle. *Am J Physiol*. 1979;237:H481-H490.
13 Desjardins C, Duling BR. Heparinase treatment suggests a role for the endothelial cell glycocalyx in regulation of capillary hematocrit. *Am J Physiol*. 1990;258:H647-H654.
14 Vink H, Duling BR. Identification of distinct luminal domains for macromolecules, erythrocytes, and leukocytes within mammalian capillaries. *Circ Res*. 1996;79(3):581–589.
15 Damiano ER. The effect of the endothelial-cell glycocalyx on the motion of red blood cells through capillaries. *Microvasc Res*. 1998;55:77–91.
16 Feng J, Weinbaum S. Lubrication theory in highly compressible porous media: the mechanics of skiing, from red cells to humans. *J Fluid Mech*. 2000;422:281–317.
17 Secomb TW, Hsu R, Pries AR. Motion of red blood cells in a capillary with an endothelial surface layer: effect of flow velocity. *Am J Physiol*. 2001;281:H629–H636.
18 Van Teeffelen JWGE, Dekker S, Fokkema DS, et al. Hyaluronidase treatment of coronary glycocalyx increases reactive hyperemia but not adenosine hyperemia in dog hearts. *Am J Physiol*. 2005; 289:H2508–H2513.
19 Henry CB, Duling BR. Permeation of the luminal capillary glycocalyx is determined by hyaluronan. *Am J Physiol*. 1999;277(46): H508–H514.
20 Paka L, Kako Y, Obunike JC, Pillarisetti S. Apolipoprotein E containing high density lipoprotein stimulates endothelial production of heparan sulfate rich in biologically active heparin-like domains: a potential mechanism for the anti-atherogenic actions of vascular apolipoprotein E. *J Biol Chem*. 1999;274:4816–4823.
21 Constantinescu AA, Vink H, Spaan JA. Elevated capillary tube hematocrit reflects degradation of endothelial cell glycocalyx by oxidized LDL. *Am J Physiol*. 2001;280:H1051–H1057.
22 Vink H, Constantinescu AA, Spaan JA. Oxidized lipoproteins degrade the endothelial surface layer: implications for platelet-endothelial cell adhesion. *Circulation*. 2000; 101(13):1500–1502.
23 van den Berg BM, Spaan JAE, Rolf TM, Vink H. Atherogenic region and diet diminish glycocalyx dimension and increase intima-to-media ratios at the murine carotid artery bifurcation. *Am J Physiol*. 2006;290:H915–H920.
24 Pries AR, Secomb TW, Gaehtgens P. The endothelial surface layer. *Pflugers Arch*. 2000;440(5):653–666.
25 Weinbaum S, Zhang X, Han Y, et al. Mechanotransduction and flow across the endothelial glycocalyx. *Proc Natl Acad Sci USA*. 2003;100(13):7988–7995.
26 Mochizuki S, Vink H, Hiramatsu O, et al. Role of hyaluronic acid in shear induced endothelium derived nitric oxide release. *Am J Phys*. 2003;285(2):H722–H726.
27 Florian JA, Kosky JR, Ainslie K, et al. Heparan sulfate proteoglycan is a mechanosensor on ECs. *Circ Res*. 2003;93(10):E136–E142.
28 Maczewski M, Duda M, Pawlak W, Beresewicz A. Endothelial protection from reperfusion injury by ischemic preconditioning and diazoxide involves a SOD-like anti-O2- mechanism. *J Physiol Pharmacol*. 2004;55(3):537–550.
29 Starling EH. On the absorption of fluids from the connective tissue spaces. *J Physiol*. 1896;19:312–326.
30 Curry FE, Michel CC. A fibre matrix model of capillary permeability. *Microvasc Res*. 1980;20:96–99.
31 Renkin EM. Multiple pathways of capillary permeability. *Circ Res*. 1977;41:735–743.
32 Ohlson M, Sörensson J, Haraldsson B. A gel-membrane model of glomerular charge and size selectivity in series. *Am J Physiol*. 2001;280:F396–F405.
33 Libby P. Inflammation in atherosclerosis. *Nature*. 2002;420 (6917):868–874.
34 van Haaren PM, van Bavel E, Vink H, Spaan JA. Localization of the permeability barrier to solutes in isolated arteries by confocal microscopy. *Am J Physiol*. 2003;285(6):H2848–H2856.
35 Wang S, Okano M, Yoshida Y. Ultrastructure of ECs and lipid deposition on the flow dividers of brachiocephalic and left subclavian arterial bifurcations of the rabbit aorta. *J Jpn Atheroscler Soc*. 1991;19:1089–1100.
36 Adamson RH. Permeability of frog mesenteric capillaries after partial pronase digestion of the endothelial glycocalyx. *J Physiol*. 1990;428:1–13.
37 Huxley VH, Williams DA. Role of a glycocalyx on coronary arteriole permeability to proteins: evidence from enzyme treatments. *Am J Physiol*. 2000;278(4):H1177–H1185.
38 Esmon CT. Inflammation and thrombosis. *J Thromb Haemost*. 2003;1(7):1343–1348.
39 Levi M, van der Poll T, Buller HR. Bidirectional relation between inflammation and coagulation. *Circulation*. 2004;8;109 (22):2698–2704.
40 Rosenberg RD, Shworak NW, Liu J, et al. Heparan sulfate proteoglycans of the cardiovascular system. Specific structures

emerge but how is synthesis regulated? *J Clin Invest*. 1997;100(11 Suppl):S67–S75.

41 Henry CB, Duling BR. TNF-alpha increases entry of macromolecules into luminal endothelial cell glycocalyx. *Am J Physiol*. 2000;279(6):H2815–H2823.

42 Lin SJ, Shyue SK, Hung YY, et al. Superoxide dismutase inhibits the expression of vascular cell adhesion molecule-1 and intracellular cell adhesion molecule-1 induced by tumor necrosis factor-alpha in human ECs through the JNK/p38 pathways. *Arterioscler Thromb Vasc Biol*. 2005;25(2):334–340.

43 Cohen M, Joester D, Geiger B, Addadi L. Spatial and temporal sequence of events in cell adhesion: from molecular recognition to focal adhesion assembly. *Chembiochem*. 2004;5: 1393–1399.

44 Zhao Y, Chien S, Weinbaum S. Dynamic contact forces on leukocyte microvilli and their penetration of endothelial glycocalyx. *Biophys J*. 2001;80:1124–1140.

45 Bruehl RE, Springer TA, Bainton DF. Quantitation of L-selectin distribution on human leukocyte microvilli by immunogold labeling and electron microscopy. *J Histochem Cytochem*. 1996; 44:835–844.

46 Pillarisetti S, Paka L, Obunike JC, et al. Subendothelial retention of lipoprotein (a). Evidence that reduced heparan sulfate promotes lipoprotein binding to subendothelial matrix. *J Clin Invest*. 1997;100(4):867–874.

47 Mulivor AW, Lipowsky HH. Inflammation- and ischemia-induced shedding of venular glycocalyx. *Am J Physiol*. 2004; 286(5):H1672–H1680.

48 Mulivor AW, Lipowsky HH. Role of glycocalyx in leukocyte-endothelial cell adhesion. *Am J Physiol*. 2002;283(4):H1282–H1289.

49 Constantinescu AA, Vink H, Spaan JA. Endothelial cell glycocalyx modulates immobilization of leucocytes at the endothelial surface. *Arterioscler Thromb Vasc Biol*. 2003;23(9):1541–1547.

50 Orth VH, Rehm M, Thiel M, et al. First clinical implications of perioperative red cell volume measurement with a nonradioactive marker (sodium fluorescein). *Anesth Analg*. 1998;87(6): 1234–1238.

76

The Endothelial Cytoskeleton

Christopher V. Carman

Beth Israel Deaconess Medical Center, Harvard Medical School, Boston, Massachusetts

The vascular endothelium represents an enormous and heterogeneous organ faced with diverse challenges. The individual endothelial cells (ECs) that make up this organ must organize a polarized and discrete monolayer, with a range of tissue-specific specializations. The endothelium must be capable of carefully balancing and dynamically regulating both barrier function and selective permeability to solutes and immune cells. ECs must resist significant, and in some settings extreme, mechanical forces including fluid shear, hydrostatic pressure, and cyclical stretch. Finally, these cells must be able to efficiently migrate and, indeed, invade tissue matrices during the formation of new vessels. The cytoskeleton of the endothelium is central to meeting all these challenges. Our growing knowledge of the roles and regulation of cytoskeletal components in endothelium provides us with an improved understanding of endothelial function in both health and disease.

AN OVERVIEW OF THE CYTOSKELETON

The cytoskeleton is formed by three kinds of protein filaments, which together provide cells with shape, mechanical strength, spatial organization/polarity, and movement. They also serve to connect protein complexes and organelles in distinct parts of the cell, and can provide tracks for transport between them. These filaments include microfilaments (or "actin filaments"), microtubules, and intermediate filaments. Each of these is formed by the polymerization of separate sets of proteins with distinct dynamics and stability. The dynamics, stability, and function of these filaments are highly dependent on a large repertoire of accessory and regulatory proteins that control the localized assembly, connect filaments to each other and other cellular components, and provide motors that move organelles (or other filaments) along their length.

Microfilaments

Microfilaments are approximately 5 to 9 nm in diameter and are composed of polymers of globular ATP-binding/hydrolyzing proteins termed *actin* (G-actin; each ~42 kDa) (1). Microfilaments are polarized, with distinct slow (*minus*) and fast (*plus* or *barbed*) growing ends, and form either as bundles of linear filaments (*fibers*) or highly branched lattices. The degree of branching is regulated by the Arp2/3 complex, which associates laterally with individual microfilaments to create branch points at a distinct 70-degree angle (2).

Basic structures formed by microfilaments include actin stress fibers, cortical actin lattices, and various distinct protrusive structures including lamellipodia, filopodia, pseudopodia, podosomes, and invadopodia (3,4). Actin stress fibers represent bundles of microfilaments that are attached to the cytoplasmic domain of cell surface adhesion receptors, such as integrins, typically arranged into clusters termed *focal contacts* and *focal adhesions* (1). Microfilaments, in combination with the actin-associated motor protein complex myosin, serve to generate centripetal contractile force. Cortical actin lattices form at the periphery of the cell underlying the plasma membrane, where they contribute to cell shape and mechanical stability. The various protrusive structures, such as lamellipodia, are critical for cell shape change, spreading, and migration. It is widely appreciated that the specific architecture and dynamics of the actin cytoskeleton are predominantly regulated by the Rho family of small GTPases including Rho, Rac, and Cdc42 (5). More recently, the small GTPase, Rap has been added as a key actin regulatory protein (6).

In endothelium, individual ECs form a monolayer stabilized by specialized interendothelial contacts – termed *adherens junctions* – rich in the homophilic adhesion molecule, vascular endothelial (VE)-cadherin (7). Actin in endothelial monolayers forms specialized dense cortical actin fibers in a Rac- and Rap-dependent manner (8). These provide peripheral bands of actin that help to organize and stabilize the adherens junctions, partly by association with VE-cadherin through the cytoplasmic adaptor proteins, the catenins (7). Actin stress fibers, stimulated by Rho signaling, function antagonistically to cortical actin fibers in endothelium, thereby weakening intercellular junctions (Figure 76.1), as discussed later (8).

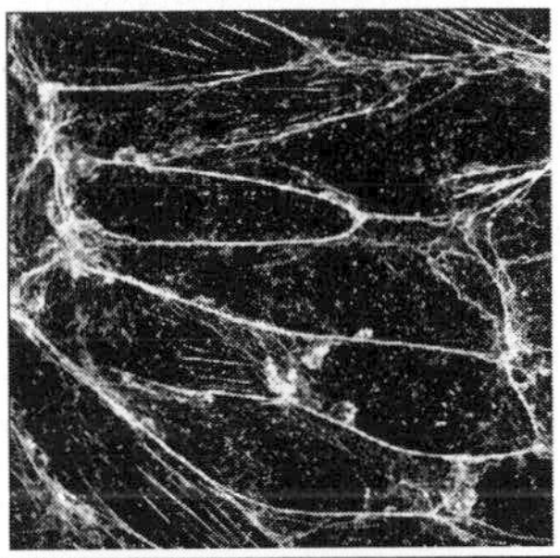

ENDOTHELIAL ACTIN PHENOTYPES

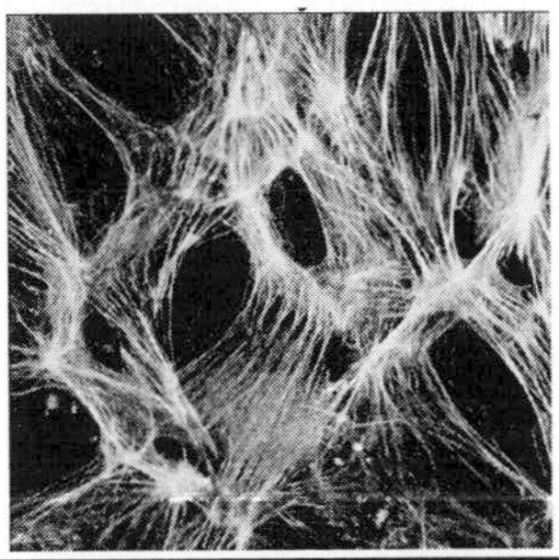

Cortical Actin Fibers	Actin Stress Fibers
Features	
Enhanced barrier	Increased permeability
Rac-, Rap- & cAMP-dependent	Rho- & Calcium-dependent
Signaling inputs	
Chronic laminar shear	Acute or disturbed shear
Mural cells (pericytes, VSMCs, astrocytes) interactions	Leukocyte interaction (ICAM-1 and VCAM-1)
Matrix: Laminin	Matrix: Collagen & Fibrin
Sphingosine-1 phosphate	TNF-α
Catecholamines	VEGF
Serotonin	Thrombin
Adenosine/ATP	Histamine
Adrenomedullin	Bradykinin
NO (variable)	LPA
Angiopoietin 1	Endothelin-1
Hepatocyte growth factor	

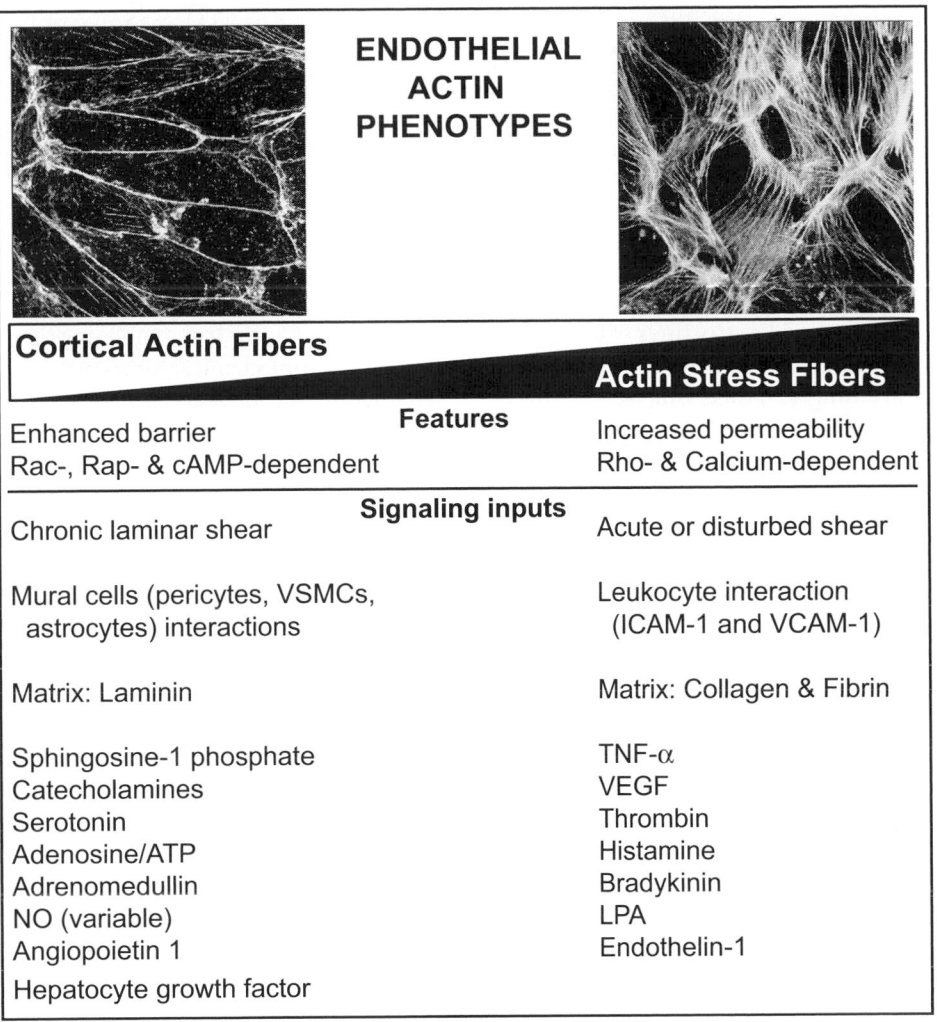

Figure 76.1. Antagonistic stress fiber–dominant and cortical actin fiber–dominant endothelial phenotypes. Shown are micrographs of human umbilical vein ECs stained with phalloidin-Alexa488 to visualize actin microfilaments. The image on the left depicts an example of a strong cortical actin fiber phenotype, in which the vast majority of the microfilaments are tightly associated with the interendothelial junctions. The image on the right depicts an example of a strong stress fiber phenotype, in which stress fibers dominate the central portion of each cell, and intercellular gaps are apparent. These images show relative extremes of what is believed to be a continuum of opposing and antagonistic phenotypes (8). The lower table compares the functional consequences, signaling pathways, and signaling inputs associated with each of the two phenotypes.

Microtubules

Microtubules are filaments of approximately 25 nm in diameter, which are derived from the globular GTP-binding/hydrolyzing proteins termed tubulins (~55 kDa each) (9). The fundamental building block of microtubules is a stable heterodimer of single α- and β-tubulin subunits, which undergoes both linear and lateral associations to form rigid, hollow tubes. This polymerization occurs in a polarized fashion such that the addition of new heterodimers occurs predominantly at the plus end, whereas the minus end is associated with, and stabilized by, the microtubule-organizing center (MTOC) rich in specialized γ-tubulin subunits. Microtubules are extremely dynamic structures, continuously alternating between phases of growth and shrinkage at the plus end (depending on the hydrolysis of GTP to GDP) in a process termed *dynamic instability* (10). A wide range of microtubule-associated proteins (MAPs) and posttranslational modifications (including phosphorylation and acetylation) modulate the stability of microtubules in response to extracellular stimuli (11).

In cells, arrays of microtubules emanate from the perinuclear MTOC with the plus ends extending outward toward the cell periphery. Microtubules play a major role in spatial organization and polarization. A critical means by which this is achieved is through the specialized MAPs, kinesin and dynein, that serve as molecular motors to conduct movement along microtubules tracks (10). These are capable of attaching to various cellular structures, allowing the controlled positioning of organelles and directed vesicular trafficking (10).

In endothelium, microtubules are thought to function primarily in establishing polarity and, in collaboration with the microfilament system, in providing cell shape and stability (12–14).

Intermediate Filaments

Intermediate filaments are rope-like structures of 8 to 10 nm in diameter formed by a large and heterogeneous family of fibrous proteins, including keratins, vimentin, neurofilament proteins, and nuclear lamins (~40–200 kDa each) (15). The primary intermediate filament proteins found in ECs are vimentin and cytokeratin (16,17). Individual intermediate filament subunits form dimeric parallel coiled-coils that undergo antiparallel association with each other to form tetramers (the basic building block unit of intermediate filaments). These tetramers aggregate into helical bundles to form intermediate filaments. The antiparallel nature of the tetramers imparts symmetry to intermediate filaments, distinguishing them from microfilaments and microtubules, which are both polarized and highly dependent on polarity for their function. In general, the vast majority of cellular intermediate filament tetramers are assembled into filaments, with very little tetramer found free in the cytoplasm, suggesting relatively limited dynamics for these structures. However, their assembly is able to be regulated. For example, serine phosphorylation within the N-terminus of intermediate filament proteins, triggered by various extracellular stimuli, can induce rapid depolymerization of intermediate filaments.

In most cell types, intermediate filaments form a network surrounding the nucleus and extending throughout the cytoplasm, with a variable degree of plasma membrane interaction. In addition, the nuclear lamins provide a dense meshwork that underlies the nuclear envelope. The intermediate filaments are the most stable of the three cytoskeletal filaments, and their primary function is to provide mechanical stability and maintain cell shape (14). Indeed, intermediate filaments are most prominent in cells that are subject to significant mechanical stress, including muscle cells, epithelial cells, and ECs. Moreover, among different endothelia, the amount of intermediate filament proteins expressed tends to be proportional to the magnitude of mechanical stress experienced (18).

In addition, evidence has emerged for roles of intermediate filaments in stabilizing adhesions in endothelia. Vimentin has been shown to colocalize with $\alpha1\beta2$ integrin– and $\beta3$ integrin–positive focal adhesions in ECs, where they function to stabilize these structures (19,20). In addition, the laminin receptor, $\alpha4\beta6$ integrin has been observed to form both fibrillar and hemidesmosome-like adhesions enriched in vimentin in endothelium (21).

THE CYTOSKELETON AS AN EFFECTOR OF ENDOTHELIAL MORPHOLOGY AND HETEROGENEITY

As discussed earlier, the cytoskeleton plays critical roles in establishing cellular shape, organization, and dynamics. Endothelia in vivo are dispersed ubiquitously throughout the body, where they take on distinct morphologies and functions that suit the diverse needs of the local environments. The local environment itself provides essential contextual cues, including the specific matrix, soluble mediators, cell–cell interactions, and biomechanical forces, which drive endothelial differentiation. A wide range of signaling and adhesion receptors exist to receive and communicate this information intracellularly, as discussed in detail throughout many of the chapters of this book. Here, we focus on the cytoskeleton as a translator of this information into specific endothelial morphologies and functions.

Matrix and Mural Cells

The minimal architecture of the endothelium consists of the ECs and the underlying matrix to which they are attached. However, in many endothelia, "mural cells" provide a third critical component that localizes in the outer (i.e., abluminal) circumference of the vessels. Mural cells include pericytes, vascular smooth muscle cells (VSMCs), and astrocytes, which are heterogeneously associated with distinct vascular structures. Pericytes predominate in the microvasculature, whereas VSMCs are associated most often with the macrovasculature, and astrocytes are restricted to the vasculature of the central nervous system (CNS) (22,23). The ECs themselves contribute importantly to their own basement membranes by secreting specific matrix components (24). However, mural cells also contribute, both directly and indirectly, to matrix composition. Pericytes secrete specific matrix components and, upon interaction with ECs, they also upregulate the production of matrix by the endothelium (23,24).

The specific composition of the resulting matrix profoundly affects the endothelial cytoskeleton and the overall EC morphology. For example, collagens and fibronectin/fibrin are recognized by endothelial $\alpha5\beta1$ and $\alpha v\beta3$ integrins, which in turn trigger concomitant activation of Rho and inhibition of Rac, leading to stress fiber formation, disruption of adherens junctions and, in some circumstances, the morphogenesis of new capillary sprouts (24). In contrast, laminin-1, acting through integrin $\alpha6\beta1$ and $\alpha3\beta1$, stabilizes the endothelium by activating protein kinase A (PKA) and Rac1 signaling, while concomitantly inhibiting Rho and stress fiber formation (24).

In addition to matrix-dependent effects, pericytes and VSMCs also mediate direct contacts with and supply soluble factors, such as transforming growth factor (TGF)-β, to endothelium that promote cortical actin fibers and increased adherens junction stability (22,23). Similarly, in the CNS, astrocytes as well as glial cells, secrete soluble factors that elevate intraendothelial cyclic adenosine monophosphate (cAMP) levels, thereby inducing strong formation of cortical actin fibers and tight junctions (i.e., the blood–brain barrier) (25).

Mechanical Stress

ECs are subjected to an extremely dynamic range of mechanical stresses, including fluid shear, hydrostatic pressure, and cyclic stretch. ECs are exquisitely sensitive to such stress, and modify their cytoskeletal composition and architecture according to the different magnitude and type of stress experienced.

Fluid shear forces range from <1 dyne/cm^2 in the microvasculature to >90 dyne/cm^2 in the large conduit vessels of the macrovasculature. Moreover, shear forces can be laminar, turbulent, or "disturbed," and either steady state or pulsatile (26,27). Exposure of in vitro endothelial cultures (grown initially under static conditions) to acute shear force results in an almost immediate activation of Rho signaling coupled to stress-fiber formation and contraction (28). The application of long-term (>12 hr) steady-state laminar shear (akin to the situation found in the microvasculature in most physiologic settings) results in an alignment of cells with the direction of shear flow and a concentration of actin into cortical fibers that stabilize the adherens junctions (26,29). In addition, steady-state laminar shear stress induces significant increases in levels of microfilament proteins, including β-actin, Arp2/3 complex subunit 3, and destrin (26). The nature and degree of cytoskeletal rearrangements elicited depend on the magnitude and type of the shear stress (26,27). For example, a switch from laminar to disturbed or turbulent flow alters the actin phenotypes toward distinct morphologies that are thought to contribute to the increased susceptibility to atherosclerosis (30).

Although the shear mechanosensors in endothelium are incompletely understood, evidence suggests that the integrin adhesion receptors, which link the extracellular matrix with the cytoskeleton, as well as VE-cadherin and platelet-endothelial cell adhesion molecule (PECAM)-1, which link interendothelial contacts to the cytoskeleton, function in this capacity (28,31).

Cyclic stretch is particularly important for the endocardium (the endothelium of the heart chambers), the large conduit vessels, and the pulmonary vasculature. The application of distinct cyclic stretch regimes to pulmonary artery ECs elicits distinct cytoskeletal rearrangements. Whereas physiologic regimes have been shown to induce Rac- and cortactin (an actin binding protein that stimulates Arp2/3-dependent actin polymerization)-dependent enhancement of barrier function, supraphysiologic and pathologic cyclic stretch regimes (similar to those associated with mechanical ventilation–induced lung injury) induced Rho-dependent increases in cell susceptibility to barrier-perturbing stimuli (i.e., thrombin) (32). Other studies also have shown that the amount and orientation of stress fibers induced by cyclic stretch was determined by the interplay between the Rho pathway and the magnitude of stretch (33).

Microtubules also respond to fluid shear flow. Indeed, shear alters the planar cell polarity of ECs and causes the MTOC to orient downstream of the nucleus in the direction of flow in an integrin-, Cdc42-, and glycogen synthase kinase (GSK)-3β–dependent manner (28,34). Moreover, shear induces rapid and sustained stabilization in microtubules, as inferred by tubulin acetylation (34).

Similarly, the onset of laminar shear flow in bovine aortic ECs expressing vimentin-GFP was shown to induce a rapid redistribution of intermediate filaments, suggesting that they participate in mechanotransduction and play roles in shear adaptation (35,36). Vimentin also regulates focal contact size and helps stabilize cell matrix adhesions in response to flow. Shear induces thickening of vimentin intermediate filaments and growth of focal contacts (20). Moreover vimentin knockdown reduces EC adhesion to the substrate (20). Finally, in vivo/in situ evidence suggests that vimentin expression levels may be modulated according to the shear regime experienced by specific ECs (18).

Thus, all three elements of the cytoskeleton exhibit mechanosensitivity and adjust their architecture according to their mechanical stress regime. In many cases, the cytoskeletal arrangements seem to function in optimizing the integrity of the endothelium. However, it is likely that many more subtle consequences of such modifications also exist.

Soluble Factors

ECs rely on a variety of soluble factors for cell formation, survival, and differentiation, including vascular endothelial growth factor (VEGF), platelet-derived growth factor (PDGF), basic fibroblast growth factor (bFGF), TGF-β, and hepatocyte growth factor (HGF). These exhibit distinct spatial and temporal patterns of expression, thereby contributing to local differences in endothelial morphology and function. Many of these do so through influences on the endothelial cytoskeleton (8). For example, VEGF, an angiogenic and survival factor, has been shown to stimulate Rho-, Rho kinase- (ROCK; a Rho effector), and myosin-dependent increases in permeability in cultured coronary venular ECs (37). Alternatively, HGF, an angiogenic and barrier-stablizing hormone, drives the formation of cortical actin and organization of VE-cadherin in adherences junctions, through a Rac- and GSK-3β–dependent signaling mechanism (38).

Sphingosine-1 phosphate (S1P) has emerged as a factor with particularly important function in the formation and maintenance of vascular endothelium. S1P has been shown to bind endothelial G-protein–coupled receptors and activate endothelial Rac signaling, leading to the formation of cortical actin fibers and organization of the adherens junctions (39,40). Under basal conditions in vivo, S1P exists at micromolar concentrations, and a depletion of platelets, the primary producers of circulating S1P, leads to significant vascular leak and edema (41). Thus, S1P is thought to represent a constitutive signaling input facilitating the organization and barrier function of the vascular endothelium through its effects on the actin cytoskeleton. That lymphatic endothelia, which are exposed to many-fold lower S1P concentrations (42), constitutively maintain, and are functionally reliant on, loosely organized junctions and high permeability, argues that one role for S1P is to act as a general differentiation cue helping to distinguish vascular and lymphatic endothelium.

Heterogeneity in the Endothelial Cytoskeleton

Details of the in vivo/in situ architecture of the endothelial cytoskeleton are generally lacking. However, a few vascular structures are amenable to direct in situ analysis of the cytoskeleton; the geometry of the heart and the associated large conduit vessels provides relatively flat surfaces that allow en face microscopic imaging. Comparative analysis of the microfilaments, microtubules, and intermediate filament systems in various regions of the endocardium and in the vascular endothelium of the various connected large vessels provides clear demonstrations of cytoskeletal heterogeneity. For example, throughout most of the outflow tract of the left ventricle, endocardial actin forms highly organized, tightly packed cortical actin fibers, with virtually no stress fibers. However, on ridges that traverse the outflow tract, actin stress fibers dominate with few cortical actin fibers observable (43). Vimentin forms distinct perinuclear rings of density and an intricate network of intermediate filaments dispersed throughout the cytoplasm in all regions of the endocardium examined. However, in aortic endothelium, intermediate filaments are highly aligned with the direction of blood flow and lack perinuclear rings. Moreover, the levels of total vimentin expression vary dramatically in different regions of the endocardium and associated vessels (18). Microtubules also exhibit highly distinct architectures in endocardium, aortic endothelium, and pulmonary valve endothelium (43). Such studies provide a clear demonstration of local heterogeneity in endothelial cytoskeleton, with dramatic differences seen even in spatially juxtaposed regions of endothelium. Differential mechanical forces (shear and cyclic stretch) probably represent the dominant environmental differences driving specialization in this example. However, it can be envisioned that the sum of all the local environmental factors discussed together determine the specific cytoskeletal architectures of distinct endothelia.

ENDOTHELIAL CELL PERMEABILITY

One of the most critical functions of the endothelium is balancing barrier function and selective permeability to fluids, solutes, and immune cells. A wide range of vasoactive and inflammatory mediators impinge on the endothelial cytoskeleton to dynamically modulate this property.

Microfilaments and Endothelial Permeability

The actin cytoskeleton plays a dominant role in the modulation of EC permeability. As discussed earlier, ECs in the context of a confluent endothelial monolayer exhibit antagonistic Rho-dependent stress fiber–dominant and Rac-dependent cortical actin fiber–dominant phenotypes (see Figure 76.1) (8,29,44).

The stress fiber–dominant phenotype is stimulated by signals that elevate intracellular calcium and activate Rho. The Rho effector ROCK and myosin light chain (MLC; a component of the myosin motor complex) kinase (MLCK; a calcium-dependent kinase that phosphorylates MLC and positively regulates contractility) together promote stress fiber formation and myosin-dependent cellular contractions that weaken the intercellular junctions and increase paracellular permeability (45). In addition, Rho signals directly interfere with Rac signaling. Key inflammatory mediators that stimulate this phenotype include tumor necrosis factor (TNF)-α, histamine, thrombin, VEGF, lysophosphatidic acid (LPA), and endothelin (see Figure 76.1) (8).

The cortical actin fiber–dominant phenotype is stimulated by two small GTPases: Rac and Rap. Rac signaling in general is coupled to cell spreading and lamellipodia formation (5), contrasting with the contractility initiated by Rho. In endothelium, Rac stimulates p21-activated kinase (PAK), which in turn promotes cortical actin fibers, translocation of cortactin to the cell periphery, and junctional assembly. PAK also functions to negatively modulate myosin-mediated contractility. The best known stimulators of this pathway in endothelium are S1P and ATP (39,46,47). It has been known for years that elevated intracellular cAMP promotes endothelial barrier function (48). cAMP is understood to act both through PKA, which inhibits Rho signaling (49), and through exchange protein activated by cAMP (EPAC)-1, a cAMP-responsive guanine nucleotide exchange factor that activates the small GTPase Rap. Rap, like Rac, enhances cortical actin fibers and reinforces adherence junctions (50,51). Barrier-enhancing agents that act through cAMP include β-adrenergic agonists (catecholamines), serotonin, and adrenomedullin (see Figure 76.1).

Microfilaments and Fenestrae

Some vascular beds, such as the kidney glomerulus and liver sinusoid, are specialized for relatively high constitutive levels of permeability. These beds contain specialized transcellular pores of regulated diameters, termed *fenestrae.* Studies suggests that the actin cytoskeleton plays an important role in establishing the architecture, and therefore permeability properties, of these structures (52). Moreover, increasing evidence suggests that the permeability properties of these structures are dynamically regulated in vivo (53). In vitro studies show that disruption of actin microfilaments (54,55) or inhibition of Rho signaling (56) reduces stress fiber content and augments the number and/or diameter of endothelial fenestrae.

Moreover, the targeted delivery of the actin-disrupting agent cytochalasin B to hepatocytes also increased the number of fenestrae in vivo (57). Alternatively, activation of Rho signaling, for example with LPA (a vasoactive mediator) promotes stress fiber formation and decreases the size and numbers of fenestrae in liver sinusoidal endothelium (56). Thus, whereas Rho activation enhances paracellular permeability in many endothelia, in fenestrated endothelia transcellular permeability is attenuated by Rho.

Microtubules and Endothelial Permeability

Microtubules have been shown to contribute to endothelial barrier function. LIM kinase (LIMK) is a Rho-activated kinase that functions, in part, to deactivate the actin-severing protein cofilin, thereby facilitating growth of stress fibers and increased paracellular permeability. Recently, LIMK also was shown to destabilize microtubules in response to thrombin, suggesting a possible role for microtubules in the modulation of permeability (58). Indeed, depolymerization of microtubules with the pharmacologic agents vinblastine and nocodazole promoted increased endothelial permeability (59,60). This effect seems to be at least partially due to the release of microtubule-sequestered RhoA and LIM kinase (58,61–68). Alternatively, cAMP-mediated barrier protective effects have been shown to act, in part, through PKA-dependent stabilization of microtubules (69). In addition, the MAP kinesin associates with p120 catenin, suggesting that trafficking of VE-cadherin could be mediated by microtubules, and implying a role for microtubules in the organization of adherens junctions (70–72). Similarly, the kinesin KIF 5 has been implicated in the trafficking of those focal adhesion components necessary for stabilizing matrix contacts (73).

In capillary endothelium, solutes such as albumin cross the endothelium transcellularly (in addition to the paracellular route) through a process of internalization and trafficking of specialized vesicles (caveolae) from the luminal to abluminal surfaces of the endothelium (*transcytosis*) (74). Although it remains to be investigated in detail, some preliminary studies suggest a role for the microtubule system in this process. For example, microtubule-disrupting agents (74) or the knockdown of specific microtubule motor proteins (kinesins HIF 3A or 3B) (75) in endothelium reduces transcytosis of albumin.

Intermediate Filaments and Endothelial Permeability

In endothelium, the most prominent functional roles associated with intermediate filaments are the maintenance of mechanical integrity and cell shape (14). In response to shape-changing vasoactive stimuli, cytokeratin intermediate filaments undergo dynamic rearrangement in pulmonary microvascular endothelium (16). Moreover, thrombin and phorbol esters induce phosphorylation of endothelial vimentin, suggesting a potential role of protein kinase C (PKC) in the modulation of vimentin intermediate filaments (76,77). In addition, the adherens junction protein VE-cadherin has been demonstrated, through biochemical studies, to associate with vimentin, and this interaction is modulated by the permeability-stimulating agent histamine (78). Taken together, these studies imply potential modulation of EC permeability by vimentin and cytokeratin. Yet, *vimentin*-knockout mice do not display any overt changes in EC architecture or function (79). However, responses to permeability-enhancing agents in these mice have yet to be directly investigated.

Permeability to Immune Cells

In addition to fluid and solute, ECs interact with blood leukocytes, as well as pathogenic bacteria, that cross the endothelial barrier. The endothelial cytoskeleton plays an important role in modulating or facilitating these processes. In addition to the wide range of inflammatory and vasoactive mediators that influence these processes, as described earlier, the migrating cells themselves also significantly affect endothelial cytoskeletal dynamics.

The process of leukocyte extravasation is central to both normal and pathologic inflammatory responses. As leukocytes accumulate on the luminal surface of the endothelium, adhesion receptors, such as intercellular adhesion molecule (ICAM)-1, vascular cell adhesion molecule (VCAM)-1, and PECAM-1, trigger intraendothelial calcium fluxes, activation of Rho, and increased myosin contractility. These events are believed to weaken the intercellular junctions facilitating leukocyte migration at these locations (paracellular diapedesis) (80–87). Conversely, the activation of Rap-dependent cortical actin-fibers through elevated cAMP both strengthened adherence junctions and reduced monocyte diapedesis in one study (50). However, in a similar investigation, cAMP increased barrier function to solutes, but not neutrophils (51).

Recently, we and others have demonstrated that the endothelium proactively generates actin-, ICAM-1-, and VCAM-1–enriched upright microvilli-like projections that surround adherent and transmigrating leukocytes (termed *docking structures* or *transmigratory-cups*) (88–90). These are formed proactively by the endothelium through the recruitment of ezrin-radixin-moesin (ERM) family proteins ezrin and moesin (actin-binding linker proteins) to the cytoplasmic domain of clustered ICAM-1 and VCAM-1, which in turn promotes local actin polymerization (88). Additional studies demonstrate an important role for vimentin intermediate filament reorganization in the formation of transmigratory cups (91). A key function for these structures is providing a physical basis (i.e., vertical traction structures) for leukocytes to exert forces against the surface of the endothelium, thereby facilitating efficient transmigration (90,91). Another recent study demonstrates that the recruitment of the actin binding/regulator protein cortactin to sites of leukocyte diapedesis somehow functions to facilitate this process (92). Thus, the endothelium seems to respond to the presence of luminally bound leukocytes with cytoskeletal rearrangements that act to support leukocyte trafficking.

Interestingly, certain bacteria have apparently subverted these endothelial responses to cross endothelial barriers and gain access to underlying tissues. Remarkably, the *Neisseria meningitidis* has adopted cell surface molecules (type IV pili) that engage ezrin-interacting endothelial receptors CD44 and ICAM-1. This induces actin-dependent microvilli-like structures, similar to those formed in transmigratory cups, which facilitate bacterial transcytosis across the blood brain barrier (93,94). The similarity between the *N. meningitidis*- and leukocyte-initiated structures is underscored by the finding that bacteria present on the endothelial surface can effectively "compete" for (and sequester) endothelial ezrin, thereby preventing leukocyte transmigratory cup formation and diapedesis (95).

As with solutes, leukocytes also have been found, both in vivo and in vitro, to use transcellular routes (i.e., via pores formed directly through individual ECs) for diapedesis (90,91,96–99). Very recently it was suggested that the actin-rich domains are important for this process (97). However, our studies suggest that the actin stress fiber–dominant phenotype reduced transcellular diapedesis (by creating a steric barrier to transcellular pores), whereas either directly driving the cortactin fiber–dominant phenotype or inhibiting Rho dramatically upregulated the efficiency of transcellular (but not paracellular) diapedesis (C. Carman, unpublished observations). This idea that Rho negatively modulates transcellular permeability is consistent with the findings, described earlier, that Rho activity reduces the number and diameter of fenestrae in liver sinusoid endothelium (53). Interestingly, further support for this idea is provided by the recent finding that *Staphylococcus aureus* can induce transcellular "macroapertures" (remarkably similar to the transcellular pore formed during leukocyte diapedesis) that facilitate extravasation of these bacteria via secreted Rho-inactivating toxins (100).

ANGIOGENESIS

In healthy adults ECs, are, for the most part, highly differentiated and relatively quiescent. However, under certain settings, for example during wound healing, ECs must rapidly mobilize to form new blood vessels. This requires cell matrix degradation, cell shape change, polarization, migration, and invasion into tissue matrix, all of which depend on the endothelial cytoskeleton. Initially, in response to angiogenic stimuli, ECs begin to degrade their underlying basement membrane, exposing increased ECs to collagen I. This induces Rho signaling, actin stress–fiber formation, and myosin-dependent contraction (dissociating VE-cadherin from intercellular junctions) and supports morphogenesis of new capillary sprouts (24). In this setting, the microtubules play a critical role in establishing cell polarization (28,34). In addition to its role in maintaining barrier function, S1P represents an important and potent chemotactic agent. Stimulation of endothelium with S1P in the correct setting promotes Rac- and cortactin-depended lamellar protrusions that form a leading edge for migration and chemotaxis (40,42). In addition, studies have revealed that novel Cdc42-dependent protrusive structures, termed *podosomes* and *invadopodia*, function in this setting. These, represent dynamic micron-scale actin-rich cylindrical protrusions found on highly migratory and invasive cells, including blood leukocytes and tumor cells (4). Such structures recently have been shown to be formed by endothelium to allow degradation and protrusion/invasion into the underlying matrix, suggesting a potentially critical role in angiogenesis (4,101,102).

THE ENDOTHELIAL CYTOSKELETON AS A BIOMEDICAL TARGET?

As a result of many basic investigations, several potential therapeutic approaches targeting the endothelial cytoskeleton have emerged.

As outlined earlier, Rho signaling and resulting changes in the actin cytoskeleton play critical roles in barrier function, inflammation, and leukocyte extravasation. Small-molecule strategies that have been developed to disrupt these pathways show potential for therapeutic benefit in inflammatory pathology. Cholesterol-lowering drugs (statins) function to block cholesterol synthesis by inhibiting 3-hydroxy-3-methylglutaryl (HMG) CoA reductase. However, the resulting depletion of isoprenoids also prevents efficient posttranslational prenylation of Rho, which is a requirement for its proper function. Accumulating evidence suggests that the cholesterol-independent or "pleiotropic" anti-inflammatory therapeutic effects of statins likely result from inhibition of Rho signaling in endothelium, as well as in vascular smooth muscle and immune cells (103,104). Rho-dependent benefits in cardiovascular disease have been associated with the inhibition of stress-fiber formation, which leads to improved barrier function, reduced leukocyte trafficking, and increased endothelial nitric oxide synthase (eNOS) signaling (103,104). Studies also have suggested that statins may improve blood–brain barrier function in a murine model of multiple sclerosis (i.e., experimental autoimmune encephalomyelitis [EAE]) in a Rho- and actin-dependent manner, which thereby limits pathologic lymphocyte trafficking to the CNS (105). Similar anti-inflammatory and barrier-protective effects have been seen with two small-molecule inhibitors of Rho kinase, fasudil and Y-27632, that act as selective competitive inhibitors for ATP binding (104,106).

As discussed earlier, S1P acts to promote endothelial barrier function through the activation of Rac1 signaling and cortical actin fibers. Recently, Merck developed the small-molecule prodrug FTY720, which is phosphorylated in vivo to form FTY720-P, a stable circulating analogue of S1P. FTY720 has profound immunosuppressive activity and is currently in clinical trials for kidney graft rejection and multiple sclerosis. Immunosuppression results from the accumulation of lymphocytes in secondary lymphoid organs due to blocked egress from these locations. FTY720 acts partially through direct effects on lymphocytes. However, because FTY720 has been shown to enhance cortical actin fiber formation and barrier function in vascular endothelium, it is hypothesized that enhanced barrier function of sinus-lining endothelium of lymphoid organs contributes to immunosuppression by precluding leukocyte diapedesis (107). The recent demonstration in vivo that FTY720 promotes the organization of lymphatic endothelial junctions, as evidenced by the appearance of dense accumulations of PECAM-1, β-catenin, and zona occludens-1 in the junctions between sinus ECs supports this idea (108). Interestingly, other studies have shown barrier protective effects of FTY720 on vascular endothelium in inflammatory settings such as acute lung injury (109).

Recently, FTY720 also was examined as an inhibitor of angiogenesis. Because S1P plays an important role in endothelial chemotaxis, it was hypothesized that FTY720 might perturb this functions in ways that alter angiogenesis. Indeed, FTY720 has proven efficacious in blocking tumor-associated angiogenesis in several models (110,111).

Like statins, the chemotherapeutic drug Taxol, commonly used to treat breast cancer, recently has been revealed to have pleiotropic mechanisms of action that involve the endothelial cytoskeleton. As a microtubule stabilizing agent, Taxol was originally developed to block tumor cell proliferation, specifically by trapping cells during the process of mitosis. However, recent studies have shown secondary effects that result from a perturbation of the EC polarization and migration necessary for angiogenesis (112–114).

KEY POINTS

- The major functional role of the microfilament system in endothelium is to establish and modulate interendothelial junctions.
- The major role of the microtubule system in endothelium is to establish cell polarity and facilitate trafficking.
- The major role of the intermediate filament system in endothelium is to provide structural and mechanical stability.
- Microenvironmental inputs drive distinct cytoskeletal architectures that help establish location-specific endothelial morphology and function.
- The reciprocal modulation of actin stress and cortical actin fibers is critical for the dynamic modulation of endothelial barrier function to solutes and immune cells.
- Emerging understanding of the roles and regulation of endothelial cytoskeleton in health and disease has led to several cytoskeleton-targeted therapeutic strategies.

Future Goals

- Endothelial morphology and function display great diversity in vivo as a consequence of distinct microenvironmental inputs. Thus, a key goal for creating new understanding of the endothelium is to develop improved approaches to study endothelial morphology and function, and especially cytoskeletal architecture and dynamics, in diverse endothelia both in situ and in vivo.

REFERENCES

1 Burridge K, Chrzanowska-Wodnicka M. Focal adhesions, contractility, and signaling. *Annu Rev Cell Dev Biol.* 1996;12:463–518.
2 Pollard TD, Beltzner CC. Structure and function of the Arp2/3 complex. *Curr Opin Struct Biol.* 2002;12:768–774.
3 Ridley AJ, Schwartz MA, Burridge K, et al. Cell migration: integrating signals from front to back. *Science* 2003;302:1704–1709.
4 Linder S, Aepfelbacher M. Podosomes: adhesion hot-spots of invasive cells. *Trends Cell Biol.* 2003;13:376–385.
5 Hall A. Rho GTPases and the actin cytoskeleton. *Science.* 1998; 279:509–514.
6 Bos JL. Linking Rap to cell adhesion. *Curr Opin Cell Biol.* 2005; 17:123–128.
7 Lampugnani MG, Resnati M, Raiteri M, et al. A novel endothelial-specific membrane protein is a marker of cell-cell contacts. *J. Cell Biol.* 1992;118:1511–1522.
8 Wojciak-Stothard B, Ridley AJ. Rho GTPases and the regulation of endothelial permeability. *Vascul Pharmacol.* 2002;39:187–199.
9 Wade RH, Hyman AA. Microtubule structure and dynamics. *Curr Opin Cell Biol.* 1997;9:12–17.
10 Mandelkow EM, Mandelkow E. Microtubule oscillations. *Cell Motil Cytoskeleton.* 1992;22:235–244.
11 Jessell TM. Adhesion molecules and the hierarchy of neural development. *Neuron.* 1998;1:3–13.
12 Ingber DE. Cellular tensegrity: defining new rules of biological design that govern the cytoskeleton. *J Cell Sci.* 1993;104:613–627.
13 Ingber DE. Tensegrity I. Cell structure and hierarchical systems biology. *J Cell Sci.* 2003;116:1157–1173.
14 Ingber DE. Tensegrity II. How structural networks influence cellular information processing networks. *J Cell Sci.* 2003;116: 1397–1408.
15 Strelkov SV, Herrmann H, Aebi U. Molecular architecture of intermediate filaments. *Bioessays.* 2003;25:243–251.
16 Alexander JS, Patton WF, Yoon MU, Shepro D. Cytokeratin filament modulation in pulmonary microvessel endothelial cells by vasoactive agents and culture confluency. *Tissue Cell.* 1991;23:141–150.
17 Patton WF, Yoon MU, Alexander JS, et al. Expression of simple epithelial cytokeratins in bovine pulmonary microvascular endothelial cells. *J Cell Physiol.* 1990;143:140–149.

18 Schnittler HJ, Schmandra T, Drenckhahn D. Correlation of endothelial vimentin content with hemodynamic parameters. *Histochem Cell Biol.* 1998;110:161–167.
19 Kreis S, Schonfeld HJ, Melchior C, et al. The intermediate filament protein vimentin binds specifically to a recombinant integrin alpha2/beta1 cytoplasmic tail complex and co-localizes with native alpha2/beta1 in endothelial cell focal adhesions. *Exp Cell Res.* 2005;305:110–121.
20 Tsuruta D, Jones JC. The vimentin cytoskeleton regulates focal contact size and adhesion of endothelial cells subjected to shear stress. *J Cell Sci.* 2003;116:4977–4984.
21 Homan SM, Mercurio AM, LaFlamme SE. Endothelial cells assemble two distinct alpha6beta4-containing vimentin-associated structures: roles for ligand binding and the beta4 cytoplasmic tail. *J Cell Sci.* 1998;111:2717–2728.
22 Hirschi KK, D'Amore PA. Pericytes in the microvasculature. *Cardiovasc Res.* 1996;32:687–698.
23 Shepro D, Morel NM. Pericyte physiology. *FASEB J.* 1993;7: 1031–1038.
24 Davis GE, Senger DR. Endothelial extracellular matrix: biosynthesis, remodeling, and functions during vascular morphogenesis and neovessel stabilization. *Circ Res.* 2005;97:1093–1107.
25 Rist RJ, Romero IA, Chan MW, et al. F-actin cytoskeleton and sucrose permeability of immortalised rat brain microvascular endothelial cell monolayers: effects of cyclic AMP and astrocytic factors. *Brain Res.* 1997;768:10–18.
26 Garcia-Cardena G, Comander J, Anderson KR, et al. Biomechanical activation of vascular endothelium as a determinant of its functional phenotype. *Proc Natl Acad Sci USA.* 2001;98:4478–4485.
27 Davies PF. Flow-mediated endothelial mechanotransduction. *Physiol Rev.* 1995;75:519–560.
28 Tzima E, Kiosses WB, del Pozo MA, Schwartz MA. Localized cdc42 activation, detected using a novel assay, mediates microtubule organizing center positioning in endothelial cells in response to fluid shear stress. *J Biol Chem.* 2003;278:31020–31023.
29 Wojciak-Stothard B, Ridley AJ. Shear stress-induced endothelial cell polarization is mediated by Rho and Rac but not Cdc42 or PI 3-kinases. *J Cell Biol.* 2003;161:429–439.
30 Gimbrone MAJ, Topper JN, Nagel T, et al. Endothelial dysfunction, hemodynamic forces, and atherogenesis. *Ann NY Acad Sci.* 2000 l920:230–239.
31 Tzima E, Irani-Tehrani M, Kiosses WB, et al. A mechanosensory complex that mediates the endothelial cell response to fluid shear stress. *Nature* 2005;437:426–431.
32 Birukova AA, Chatchavalvanich S, Rios A, et al. Differential regulation of pulmonary endothelial monolayer integrity by varying degrees of cyclic stretch. *Am J Pathol.* 2006;168:1749–1761.
33 Kaunas R, Nguyen P, Usami S, Chien S. Cooperative effects of Rho and mechanical stretch on stress fiber organization. *Proc Natl Acad Sci USA.* 2005;102:15895–15900.
34 McCue S, Dajnowiec D, Xu F, et al. Shear stress regulates forward and reverse planar cell polarity of vascular endothelium in vivo and in vitro. *Circ Res.* 2006;98:939–946.
35 Helmke BP, Thakker DB, Goldman RD, Davies PF. Spatiotemporal analysis of flow-induced intermediate filament displacement in living endothelial cells. *Biophys J.* 2001;80:184–194.
36 Helmke BP, Rosen AB, Davies PF. Mapping mechanical strain of an endogenous cytoskeletal network in living endothelial cells. *Biophys J.* 2003;84:2691–2699.
37 Sun H, Breslin JW, Zhu J, et al. Rho and ROCK signaling in VEGF-induced microvascular endothelial hyperpermeability. *Microcirculation.* 2006;13:237–247.
38 Liu F, Schaphorst KL, Verin AD, et al. Hepatocyte growth factor enhances endothelial cell barrier function and cortical cytoskeletal rearrangement: potential role of glycogen synthase kinase-3beta. *FASEB J.* 2002;16:950–962.
39 Garcia JG, Liu F, Verin AD, et al. Sphingosine 1-phosphate promotes endothelial cell barrier integrity by Edg-dependent cytoskeletal rearrangement. *J. Clin. Invest.* 2001;108:689–701.
40 McVerry BJ, Garcia JG. In vitro and in vivo modulation of vascular barrier integrity by sphingosine 1-phosphate: mechanistic insights. *Cell Signal.* 2005;17:131–139.
41 Lo SK, Burhop KE, Kaplan JE, Malik AB. Role of platelets in maintenance of pulmonary vascular permeability to protein. *Am J Physiol.* 1998;254:H763–H771.
42 Hla T. Physiological and pathological actions of sphingosine 1-phosphate. *Semin Cell Dev Biol.* 2004;15:513–520.
43 Andries LJ, Brutsaert DL. Endocardial endothelium in the rat: cell shape and organization of the cytoskeleton. *Cell Tissue Res.* 1993;273:107–117.
44 Wojciak-Stothard B, Tsang LY, Paleolog E, et al. Rac1 and RhoA as regulators of endothelial phenotype and barrier function in hypoxia-induced neonatal pulmonary hypertension. *Am J Physiol Lung Cell Mol Physiol.* 2006;290:L1173–L1182.
45 Bogatcheva NV, Garcia JG, Verin AD. Molecular mechanisms of thrombin-induced endothelial cell permeability. *Biochemistry (Mosc).* 2002;67:75–84.
46 Lee MJ, Thangada S, Claffey KP, et al. Vascular endothelial cell adherens junction assembly and morphogenesis induced by sphingosine-1-phosphate. *Cell* 1999;99:301–312.
47 Jacobson JR, Dudek SM, Singleton PA, et al. Endothelial cell barrier enhancement by ATP is mediated by the small GTPase Rac and cortactin. *Am J Physiol Lung Cell Mol Physiol.* 2006; 291:L289–L295.
48 Langeler EG, van Hinsbergh VW. Norepinephrine and iloprost improve barrier function of human endothelial cell monolayers: role of cAMP. *Am. J. Physiol.* 1991;260:C1052–C1059.
49 Qiao J, Huang F, Lum H. PKA inhibits RhoA activation: a protection mechanism against endothelial barrier dysfunction. *Am J Physiol Lung Cell Mol Physiol.* 2003;284:L972–L980.
50 Wittchen ES, Worthylake RA, Kelly P, et al. Rap1 GTPase inhibits leukocyte transmigration by promoting endothelial barrier function. *J Biol Chem.* 2005;280:11675–11682.
51 Cullere X, Shaw SK, Andersson L, et al. Regulation of vascular endothelial barrier function by Epac, a cAMP-activated exchange factor for Rap GTPase. *Blood.* 2005;105:1950–1955.
52 Nagai T, Yokomori H, Yoshimura K, et al. Actin filaments around endothelial fenestrae in rat hepatic sinusoidal endothelial cells. *Med Electron Microsc.* 2004;37:252–255.
53 Braet F. How molecular microscopy revealed new insights into the dynamics of hepatic endothelial fenestrae in the past decade. *Liver Int.* 2004;24:532–539.
54 Braet F, Spector I, Shochet N, et al. The new anti-actin agent dihydrohalichondramide reveals fenestrae-forming centers in hepatic endothelial cells. *BMC Cell Biol.* 2002;3:7.

55 Steffan AM, Gendrault JL, Kirn A. Increase in the number of fenestrae in mouse endothelial liver cells by altering the cytoskeleton with cytochalasin B. *Hepatology* 1987;7:1230–1238.

56 Yokomori H, Yoshimura K, Funakoshi S, et al. Rho modulates hepatic sinusoidal endothelial fenestrae via regulation of the actin cytoskeleton in rat endothelial cells. *Lab. Invest*. 2004;84.

57 Braet F, Vekemans K, Morselt H, et al. The effect of cytochalasin B – Loaded liposomes on the ultrastructure of the liver sieve. *Comp Hepatol*. 2004;14:S27-S29.

58 Gorovoy M, Niu J, Bernard O, et al. LIM kinase 1 coordinates microtubule stability and actin polymerization in human endothelial cells. *J Biol Chem*. 2005;280:26533–26542.

59 Petrache I, Birukova A, Ramirez SI, et al. The role of the microtubules in tumor necrosis factor-alpha-induced endothelial cell permeability. *Am J Respir Cell Mol Biol*. 2003;28:574–581.

60 Verin AD, Birukova A, Wang P, et al. Microtubule disassembly increases endothelial cell barrier dysfunction: role of MLC phosphorylation. *Am J Physiol Lung Cell Mol Physiol*. 2001;281:L565–L574.

61 Glaven JA, Whitehead I, Bagrodia S, et al. The Dbl-related protein, Lfc, localizes to microtubules and mediates the activation of Rac signaling pathways in cells. *J Biol Chem*. 1999;274:2279–2285.

62 Krendel M, Zenke FT, Bokoch GM. Nucleotide exchange factor GEF-H1 mediates cross-talk between microtubules and the actin cytoskeleton. *Nat Cell Biol*. 2002;4:294–301.

63 Ren Y, Li R, Zheng Y, Busch H. Cloning and characterization of GEF-H1, a microtubule-associated guanine nucleotide exchange factor for Rac and Rho GTPases. *J Biol Chem*. 1998;273:34954–34960.

64 van Horck FP, Ahmadian MR, Haeusler LC, et al. Characterization of p190RhoGEF, a RhoA-specific guanine nucleotide exchange factor that interacts with microtubules. *J Biol Chem*. 2001;276:4948–4356.

65 Birukov KG, Csortos C, Marzilli L, et al. Differential regulation of alternatively spliced endothelial cell myosin light chain kinase isoforms by p60(Src). *J Biol Chem*. 2001;276:8567–8573.

66 Birukova AA, Birukov KG, Smurova K, et al. Novel role of microtubules in thrombin-induced endothelial barrier dysfunction. *FASEB J*. 2004;18:1879–1890.

67 Birukova AA, Birukov KG, Gorshkov B, et al. MAP kinases in lung endothelial permeability induced by microtubule disassembly. *Am J Physiol Lung Cell Mol Physiol*. 2005;289:L75–L84.

68 Birukova AA, Birukov KG, Adyshev D, et al. Involvement of microtubules and Rho pathway in TGF-beta1-induced lung vascular barrier dysfunction. *J Cell Physiol*. 2005;204:934–947.

69 Birukova AA, Liu F, Garcia JG, Verin AD. Protein kinase A attenuates endothelial cell barrier dysfunction induced by microtubule disassembly. *Am J Physiol Lung Cell Mol Physiol*. 2004;287:L86–L93.

70 Chen X, Kojima S, Borisy GG, Green KJ. p120 catenin associates with kinesin and facilitates the transport of cadherin-catenin complexes to intercellular junctions. *J Cell Biol*. 2003;163:547–557.

71 Yanagisawa M, Kaverina IN, Wang A, et al. A novel interaction between kinesin and p120 modulates p120 localization and function. *J Biol Chem*. 2004;279:9512–9521.

72 Vincent PA, Xiao K, Buckley KM, Kowalczyk AP. VE-cadherin: adhesion at arm's length. *Am J Physiol Cell Physiol*. 2004;286:C987–C997.

73 Krylyshkina O, Kaverina I, Kranewitter W, et al. Modulation of substrate adhesion dynamics via microtubule targeting requires kinesin-1. *J Cell Biol*. 2002;156:349–359.

74 Tuma PL, Hubbard AL. Transcytosis: crossing cellular barriers. *Physiol Rev*. 2003;83:871–932.

75 Mehta D, Malik AB. Signaling mechanisms regulating endothelial permeability. *Physiol Rev*. 2006;86:279–367.

76 Bormann BJ, Huang CK, Lam GF, Jaffe EA. Thrombin-induced vimentin phosphorylation in cultured human umbilical vein endothelial cells. *J Biol Chem*. 1986;261:10471–10474.

77 Stasek JEJ, Patterson CE, Garcia JG. Protein kinase C phosphorylates caldesmon77 and vimentin and enhances albumin permeability across cultured bovine pulmonary artery endothelial cell monolayers. *J Cell Physiol*. 1992;153:62–75.

78 Shasby DM, Ries DR, Shasby SS, Winter MC. Histamine stimulates phosphorylation of adherens junction proteins and alters their link to vimentin. *Am J Physiol Lung Cell Mol Physiol*. 2002;6:L1330–L1338.

79 Colucci-Guyon E, Portier MM, Dunia I, et al. Mice lacking vimentin develop and reproduce without an obvious phenotype. *Cell*. 1994;79:679–694.

80 Etienne S, Adamson P, Greenwood J, et al. ICAM-1 signaling pathways associated with Rho activation in microvascular brain endothelial cells. *J. Immunol*. 1998;161:5755–5761.

81 Etienne-Manneville S, Manneville JB, Adamson P, et al. ICAM-1-coupled cytoskeletal rearrangements and transendothelial lymphocyte migration involve intracellular calcium signaling in brain endothelial cell lines. *J. Immunol*. 2000;165:3375–3383.

82 Wojciak-Stothard B, Williams L, Ridley AJ. Monocyte adhesion and spreading on human endothelial cells is dependent on Rho-regulated receptor clustering. *J. Cell Biol*. 1999;145:1293–1307.

83 Adamson P, Etienne S, Couraud PO, et al. Lymphocyte migration through brain endothelial cell monolayers involves signaling through endothelial ICAM-1 via a rho-dependent pathway. *J. Immunol*. 1999;162:2964–2973.

84 Huang AJ, Manning JE, Bandak TM, et al. Endothelial cell cytosolic free calcium regulates neutrophil migration across monolayers of endothelial cells. *J. Cell Biol*. 1993;120:1371–1380.

85 Muller WA. Leukocyte-endothelial-cell interactions in leukocyte transmigration and the inflammatory response. *Trends Immunol*. 2003;24:327–334.

86 Curry FE. Modulation of venular microvessel permeability by calcium influx into endothelial cells. *FASEB J*. 1992;6:2456–2466.

87 Strey A, Janning A, Barth H, Gerke V. Endothelial Rho signaling is required for monocyte transendothelial migration. *FEBS Lett*. 2002;517:261–266.

88 Barreiro O, Yanez-Mo M, Serrador JM, et al. Dynamic interaction of VCAM-1 and ICAM-1 with moesin and ezrin in a novel endothelial docking structure for adherent leukocytes. *J. Cell. Biol*. 2002;157:1233–1245.

89 Carman CV, Jun C-D, Salas A, Springer TA. Endothelial cells proactively form microvilli-like membrane projections upon ICAM-1 engagement of leukocyte LFA-11. *J. Immunol*. 2003;171:6135–6144.

90 Carman CV, Springer TA. A transmigratory cup in leukocyte diapedesis both through individual vascular endothelial cells and between them. *J. Cell Biol*. 2004;167:377–388.

91 Nieminen M, Henttinen T, Merinen M, et al. Vimentin function in lymphocyte adhesion and transcellular migration. *Nat Cell Biol.* 2006;8:156–162.

92 Yang L, Kowalski JR, Zhan X, et al. Endothelial cell cortactin phosphorylation by Src contributes to polymorphonuclear leukocyte transmigration in vitro. *Circ Res.* 2006;98:394–402.

93 Eugene E, Hoffmann I, Pujol C, et al. Microvilli-like structures are associated with the internalization of virulent capsulated Neisseria meningitidis into vascular endothelial cells. *J Cell Sci.* 2002;115:1231–1241.

94 Nassif X, Bourdoulous S, Eugene E, Couraud PO. How do extracellular pathogens cross the blood-brain barrier? *Trends Microbiol.* 2002;10:227–232.

95 Doulet N, Donnadieu E, Laran-Chich MP, et al. Neisseria meningitidis infection of human endothelial cells interferes with leukocyte transmigration by preventing the formation of endothelial docking structures. *J Cell Biol.* 2006;173:627–637.

96 Yang L, Froio RM, Sciuto TE, et al. ICAM-1 regulates neutrophil adhesion and transcellular migration of TNF-alpha-activated vascular endothelium under flow. *Blood* 2005;106:584–592.

97 Millan J, Hewlett L, Glyn M, et al. Lymphocyte transcellular migration occurs through recruitment of endothelial ICAM-1 to caveola- and F-actin-rich domains. *Nat. Cell Biol.* 2006;8: 113–123.

98 Cinamon G, Shinder V, Shamri R, Alon R. Chemoattractant signals and beta 2 integrin occupancy at apical endothelial contacts combine with shear stress signals to promote transendothelial neutrophil migration. *J Immunol.* 2004;173:7282–7291.

99 Feng D, Nagy JA, Pyne K, et al. Neutrophils emigrate from venules by a transendothelial cell pathway in response to FMLP. *J. Exp. Med.* 1998;187:903–915.

100 Boyer L, Doye A, Rolando M, et al. Induction of transient macroapertures in endothelial cells through RhoA inhibition by Staphylococcus aureus factors. *J Cell Biol.* 2006;173:809–819.

101 Moreau V, Tatin F, Varon C, Genot E. Actin can reorganize into podosomes in aortic endothelial cells, a process controlled by Cdc42 and RhoA. *Mol Cell Biol.* 2003;23:6809–6822.

102 Ghersi G, Zhao Q, Salamone M, et al. The protease complex consisting of dipeptidyl peptidase IV and seprase plays a role in the migration and invasion of human endothelial cells in collagenous matrices. *Cancer Res.* 2006;66:4652–4661.

103 Rikitake Y, Liao JK. Rho GTPases, statins, and nitric oxide. *Circ Res.* 2005;97:1232–1235.

104 Rolfe BE, Worth NF, World CJ, et al. Rho and vascular disease. *Atherosclerosis.* 2005;183:1–16.

105 Greenwood J, Walters CE, Pryce G, et al. Lovastatin inhibits brain endothelial cell Rho-mediated lymphocyte migration and attenuates experimental autoimmune encephalomyelitis. *FASEB J.* 2003;17:905–907.

106 Shimokawa H, Takeshita A. Rho-kinase is an important therapeutic target in cardiovascular medicine. *Arterioscler Thromb Vasc Biol.* 2005;25:1767–1775.

107 Brinkmann V, Cyster JG, Hla T. FTY720: sphingosine 1-phosphate receptor-1 in the control of lymphocyte egress and endothelial barrier function. *Am J Transplant.* 2004;4:1019–1025.

108 Singer II, Tian M, Wickham LA, et al. Sphingosine-1-phosphate agonists increase macrophage homing, lymphocyte contacts, and endothelial junctional complex formation in murine lymph nodes. *J Immunol.* 2005;175:7151–7161.

109 Peng X, Hassoun PM, Sammani S, et al. Protective effects of sphingosine 1-phosphate in murine endotoxin-induced inflammatory lung injury. *Am. J. Respir. Crit. Care Med.* 2004;169:1245–1251.

110 LaMontagne K, Littlewood-Evans A, Schnell C, et al. Antagonism of sphingosine-1-phosphate receptors by FTY720 inhibits angiogenesis and tumor vascularization. *Cancer Res.* 2006;66: 221–231.

111 Ho JW, Man K, Sun CK, et al. Effects of a novel immunomodulating agent, FTY720, on tumor growth and angiogenesis in hepatocellular carcinoma. *Mol Cancer Ther.* 2005;4:1430–1438.

112 Pasquier E, Honore S, Pourroy B, et al. Antiangiogenic concentrations of paclitaxel induce an increase in microtubule dynamics in endothelial cells but not in cancer cells. *Cancer Res.* 2005; 65:2433–2440.

113 Lu H, Murtagh J, Schwartz EL. The microtubule binding drug laulimalide inhibits vascular endothelial growth factor-induced human endothelial cell migration and is synergistic when combined with docetaxel (taxotere). *Mol Pharmacol.* 2006;69:1207–1215.

114 Pourroy B, Honore S, Pasquier E, et al. Antiangiogenic concentrations of vinflunine increase the interphase microtubule dynamics and decrease the motility of endothelial cells. *Cancer Res.* 2006;66:3256–3263.

77

Endothelial Cell Integrins

Joseph H. McCarty* and Richard O. Hynes†

*M.D. Anderson Cancer Center, Houston Texas; †Center for Cancer Research, Massachusetts Institute of Technology, Cambridge

The extracellular matrix (ECM) is an insoluble network of proteins that provides structural support to multicellular tissues and regulates a variety of cell behaviors. Most metazoan cells express transmembrane proteins, known as integrins, which serve as receptors for many ECM protein ligands (1). Integrins connect the extracellular environment to the intracellular cytoskeleton, and also regulate the signal transduction cascades leading to gene regulatory events. Integrin-mediated adhesion and signaling are essential for the proper formation of most tissues, and they play particularly important roles in vascular development and homeostasis (2).

HISTORY OF INTEGRINS

The discovery of integrins centered around efforts to identify a cell surface receptor for the fibrillar ECM protein fibronectin (FN) (3,4). Biochemical purification of the FN receptor revealed a complex of two proteins of differing molecular weights (5). Subsequent cloning and sequencing of cDNAs encoding components of the FN receptor complex identified two transmembrane proteins, named *integrins* for their postulated role in linking the ECM to the intracellular cytoskeleton. Soon thereafter, several investigators working in areas ranging from human immunology to fly genetics published the cDNA sequences encoding other transmembrane proteins with striking homologies to the FN receptor (3). Thus, it was realized that integrins represented a multigene family with adhesive functions on diverse cell types. We now know that integrins are heterodimeric proteins consisting of two noncovalently associated subunits, termed α and β. Most integrins recognize multiple ECM ligands, and many ECM proteins can bind to more than one integrin receptor. Integrins and their various ECM ligands play essential roles in virtually every aspect of physiological and pathological blood vessel function.

EVOLUTIONARY AND DEVELOPMENTAL CONSIDERATIONS

Integrins are present in all metazoans, ranging from sponges to nematodes, flies, and mammals. In general, the expansion of integrin genes correlates with body plan complexity (6). For example, *Caenorhabditis elegans* contains only three integrin genes: two α subunits and one β subunit that can combine to form two integrin heterodimers. The *Drosophila* genome has seven different integrin genes: two β subunits and five α subunits that can yield at least five integrin heterodimers. In vertebrates, 26 integrin genes exist, 18 encoding α subunits and eight encoding β subunits, yielding 24 distinct integrin heterodimers (Figure 77.1). Similar to flies and worms, vertebrates also contain laminin- and RGD-binding integrins, as well as an expanded set of collagen- and FN-binding integrins, and several leukocyte-specific integrins that also mediate heterotypic cell–cell adhesion (1,6).

Genetic analyses reveal important functional roles for integrins in most developmental processes (Table 77-1). Mutant screens in flies and worms reveal essential roles for integrins in cell differentiation, organ morphogenesis, and tissue homeostasis (7). Most of the 26 integrin genes have been ablated in mice, revealing a diverse range of phenotypes that reveal little, if any, functional redundancy (1,8). The phenotypes of the mouse integrin mutants range from peri-implantation defects, to lethality at various embryonic and neonatal ages, to postnatal viability with organ-specific abnormalities.

INTEGRIN GENE REGULATION

Mechanical stress and infectious agents, as well as cytokines and growth factors, all regulate integrin expression at both the transcriptional and post-translational levels (9). A detailed description of these processes is beyond the scope of this chapter, so we refer the reader to other excellent articles that discuss

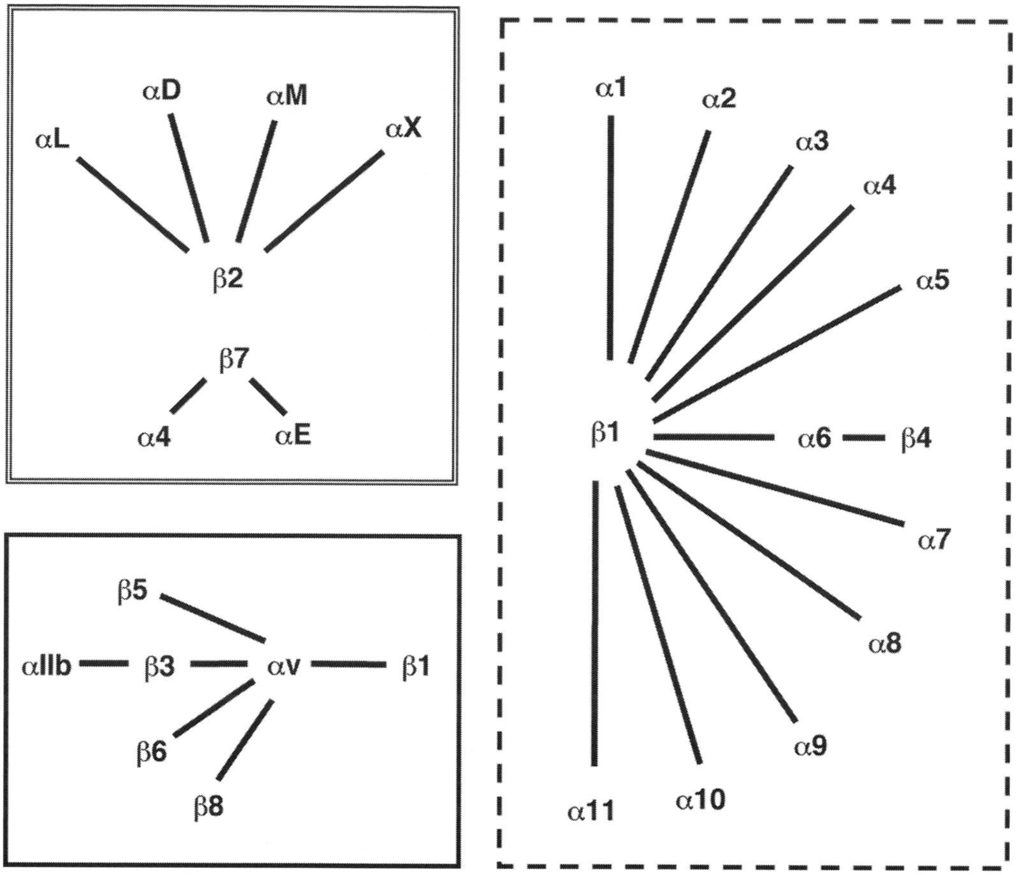

Figure 77.1. The integrin family of ECM receptors. The 26 integrin α- and β-subunits, separated into three groups, are shown. Lines connecting subunits indicate the 24 known heterodimeric pairs. The leukocyte-specific integrins are shown in the box bordered by a double line (*upper left*). The αv-subfamily is highlighted in the box with a solid line (*lower left*), and the β1-subfamily is shown in the box with a dashed line (*right*).

these various regulators in depth (9,10). Instead, we focus here on the regulation of endothelial integrin gene expression by pro- and antiangiogenic factors.

Angiogenic growth factors and their receptors play essential roles in physiological and pathological neovascularization, and regulate these events in part via modulation of integrin expression and function. For example, vascular endothelial growth factor (VEGF) treatment of cultured endothelial cells (ECs) stimulates transcriptional upregulation of multiple integrin subunits including αv, β3, β5, α1, and α2 (11). Additionally, the stimulation of cultured ECs with basic fibroblast growth factor (bFGF) leads to upregulation of αvβ5 expression, with little effect on the expression of other integrin gene products. Stimulation of ECs with transforming growth factor (TGF)-β1 also leads to upregulation of multiple integrin subunits, including αv, β3, β5, α5, and β1 (9).

Alternative splicing of exons encoding portions of the extracellular and intracellular domains also regulates integrin functions (12). For example, many β subunit cytoplasmic domains contain NPXY signaling motifs that play important roles in recruiting signaling proteins. Splice variants of β1, β3, and β5 integrins that lack this motif display altered cellular localization and signal transduction. Thus, the regulation of integrin splice variants may generate functional differences in EC adhesion, migration, and survival.

INTEGRIN PROTEIN STRUCTURE

The three-dimensional structures of integrin extracellular and intracellular portions have been solved, with the EC integrin αvβ3 serving as the prototype for many of these studies (13). Here we provide a general description of integrin structure and activation, and refer the reader to other papers for more details (1,14–16).

Unlike most transmembrane receptors, high-affinity ligand binding by integrins – termed *activation* – can be regulated via intracellular (inside-out) signals. In their inactive state, integrins recognize ECM ligands with low affinity, and the crystal structure of αvβ3 reveals the extracellular domains folded into a "bent" conformation. Inactive integrins are maintained in part by the α and β cytoplasmic domains, which interact to form a "molecular clasp." Inside-out activation disrupts the α–β cytoplasmic interaction and causes

Table 77-1: Integrin Mutant Mice with Angiogenesis Phenotypes

Integrin Subunit (Refs.)	*Mutant Phenotype*
β1 (8)	Peri-implantation lethality at E6.5. β1-null ES cells and teratocarcinomas display a poorly developed vasculature.
β3/β5 (24)	β3/β5 double-knockout mice are viable and fertile. They display normal physiological angiogenesis, but enhanced pathological angiogenesis. Some mutants develop hemorrhage due to defects in platelet clotting.
β4 (36)	Mutants expressing a β4 gene product that lacks the cytoplasmic signaling portion are viable and fertile. However, mutants display defective pathological angiogenesis.
β8(27)	~70% of embryos dies by E11. These embryos have defective placental layering. The remaining 30% die shortly after birth owing to cerebral hemorrhage.
α1 (34)	Viable and fertile; reduced tumor growth owing to defective pathological angiogenesis. Elevated levels of circulating angiostatin.
α5 (1,30,31)	Lethality at E10.5. Defective embryonic and extraembryonic blood vessel formation. Somitic and neural crest defects.
αv (25,26,28)	~70% of embryos dies by E11. These embryos have defective placental layering. The remaining 30% die shortly after birth owing to cerebral hemorrhage, a cleft palate, and abnormal intestinal development.

Mouse genetic data identify several integrin subunits with roles during neovascularization. Listed are only those integrin mutants reported to display physiological and/or pathological angiogenesis defects.

a switchblade-like conformational change in the extracellular domains, leading to enhanced ECM ligand affinity. The cytoskeletal protein talin, which binds to an NPXY sequence in the integrin β-cytoplasmic tail, plays an essential role in inside-out integrin activation. Integrins activated by outside-in and inside-out mechanisms recruit various intracellular signal transduction molecules, which leads to protein phosphorylation cascades and gene regulatory events that cause major effects on cell behavior.

INTEGRINS AND RECEPTOR TYROSINE KINASES

Integrins physically associate with a variety of transmembrane receptors, including G-protein–coupled receptors, ion channels, and other cell adhesion molecules. On ECs, angiogenic growth factors and their cognate receptor tyrosine kinases (RTKs) play well-established synergistic roles in integrin activation (17). Indeed, proper microvascular development and homeostasis depend on precise regulation of cross-talk events between integrin and RTK signaling components (18). For example, the integrin αvβ3 can physically interact with FLK-1, the VEGF-A receptor tyrosine kinase. αvβ3–FLK-1 association augments VEGF-A–induced FLK-1 phosphorylation and enhances EC migration on vitronectin, an αvβ3 ligand. Physical association between αvβ3 and FLK-1 occurs via the αvβ3 extracellular domain, and is independent of growth factor stimulation (19). Interestingly, vascular development occurs normally in β3-null mice; however, in cultured β3-null ECs elevated FLK-1 phosphorylation and enhanced downstream signaling events occur, suggesting a compensatory mechanism by FLK-1 in the absence of β3 integrin (20).

INTEGRINS AND THEIR INFLUENCES ON ENDOTHELIAL CELL FUNCTIONS

Integrin adhesion to vascular basement membrane ligands regulates many intracellular signaling events, affecting EC responses such as migration and survival. ECs in multiple vascular beds and at varying developmental and postnatal time points express several integrins, including α1β1, α2β1, α3β1, α4β1, α5β1, α6β4, αvβ3, and αvβ5 (2). Here we summarize pertinent biochemical and genetic data linking these eight integrins and their ECM ligands to various vascular events.

αv Integrins

There are five members of the αv subfamily of integrins: αv integrin can pair with β1, β3, β5, β6, and β8 (see Figure 77.1). αvβ3 and αvβ5 have been implicated in various types of physiological and pathological angiogenesis. αvβ3 or αvβ5 protein expression is upregulated on cultured ECs or on angiogenic blood vessels in response to different angiogenic growth factors or cytokines (21). αv Integrin–mediated neovascularization in these various systems is inhibited upon addition of function-blocking antibodies or peptide antagonists, which ultimately cause EC death (22). Hence, αvβ3 has been pursued intensely as a potential antiangiogenic target, and αvβ3 antagonists are currently being tested in human clinical trials (see next section).

Genetic ablation of the $\beta 3$ and/or $\beta 5$ genes does not lead to overt angiogenic abnormalities. Indeed, mice null for αv β3 and/or αv$\beta 5$ are viable and fertile, and blood vessel development proceeds normally in these mutants (23). Physiological angiogenesis occurs normally in mice null for $\beta 3$ or both $\beta 3$ and $\beta 5$. Interestingly, pathological retinal angiogenesis as well as tumor growth and neovascularization are actually enhanced in $\beta 3^{-/-}$ or $\beta 3/\beta 5$-null mice (24). Thus, the inferred significance of αvβ3 and αvβ5 in neovascularization based on the antibody and peptide inhibition results versus the genetic ablation data are quite conflicting. Interestingly, in $\beta 3$- or

$\beta 3/\beta 5$-null animals upregulation of signaling events occurs, mediated by the FLK-1 receptor tyrosine kinase, suggesting that, in the absence of $\beta 3$- and $\beta 5$-integrin expression, a compensatory response occurs involving enhanced VEGF/FLK-1 signal transduction.

Lastly, mice harboring an ablated αv gene, and thus lacking expression of all five αv integrins, do not develop widespread vascular defects (25). Approximately 70% of αv-null embryos survive to midgestation, having developed a normal vasculature, but die by embryonic day (E)11, owing to placental layering abnormalities. The remaining αv-null embryos that survive beyond midgestation actually are carried to term and develop a largely normal vasculature. However, they do develop cerebral hemorrhage, a cleft palate, and die within a day after birth. The hemorrhage is not due to primary endothelial or pericyte defects, but rather involves defective associations between angiogenic cerebral blood vessels and central nervous system (CNS) glia, where αv protein is expressed (26). Interestingly, the cerebral hemorrhage that develops in the αv-nulls is due to the loss of one specific integrin, $\alpha v\beta 8$, which is a receptor for the latency-associated peptide of TGF-$\beta 1$ (27). Mice null for the $\beta 8$ gene develop phenotypes that are very similar to those seen in the αv-nulls (28). Recently, we have used Cre/lox methodology to selectively ablate αv integrin expression in neural cells of the developing and postnatal CNS (29). Similar to the phenotype observed in the complete αv-null mice, conditional mutant embryos and neonates also develop CNS-specific hemorrhage. Unlike the complete αv-nulls, most conditional αv mutants survive for several months, but develop progressive spinocerebellar neurodegeneration, leading to symptoms such as hindlimb rigidity, ataxia, and urinary dysfunction. Similar phenotypes develop in conditional $\beta 8$–knockout animals (30). This conditional knockout approach confirms a role for glial-expressed αv integrins, particularly αvβ8, in regulating proper blood vessel development and cerebrovascular homeostasis.

$\alpha 5\beta 1$ Integrin

$\alpha 5\beta 1$ Integrin is a major cell surface receptor for the ECM protein FN, and various biochemical and genetic data support important functional roles for $\alpha 5\beta 1$-FN adhesion in blood vessel formation (2,31). Ablation of $\alpha 5$ gene expression in mouse embryos leads to lethality by embryonic day E10.5, owing to severe embryonic and extraembryonic vascular defects. These mutants also develop posterior trunk and somitic abnormalities. Additionally, teratocarcinomas formed from $\alpha 5$-null embryonic stem (ES) cells form a poorly developed vasculature and display reduced tumor size. Similarly, embryoid bodies derived from $\alpha 5$-null ES cells also display delayed vascular differentiation and organization (32,33). In concordance with the $\alpha 5$ genetic data, deletion of the *FN* gene leads to embryonic lethality associated with severe vascular defects, and similar angiogenic abnormalities are observed in *FN*-null embryoid bodies (34).

Other experimental models also support a necessary role for $\alpha 5\beta 1$ and FN in angiogenesis. For example, $\alpha 5$ integrin and FN protein levels are upregulated on angiogenic tumor blood vessels, and vascular growth factors upregulate $\alpha 5\beta 1$ expression on angiogenic blood vessels in the chick chorioallantoic membrane (CAM). In both cases, function-blocking antibodies or peptides directed against $\alpha 5\beta 1$ or FN reduce neovascularization (35).

$\alpha 1\beta 1$ and $\alpha 2\beta 1$ Integrins

The collagen-binding integrins $\alpha 1\beta 1$ and $\alpha 2\beta 1$ are implicated in angiogenesis. Both integrins are upregulated on angiogenic blood vessels following VEGF stimulation (2). Additionally, function-blocking anti-$\alpha 1$ and -$\alpha 2$ antibodies inhibit angiogenesis in vivo using Matrigel plugs or transplanted tumor models. Interestingly, $\alpha 1$-null mice are viable and fertile, yet are less susceptible to tumor growth (36). This resistance correlates with increased levels of circulating matrix metalloproteinase activity. Elevated proteolysis in $\alpha 1$-null mice leads to the conversion of plasminogen to angiostatin, a potent inhibitor of EC neovascularization. Mice harboring a deletion of the $\alpha 2$ gene are viable and fertile and do not display any gross vascular abnormalities (37). A potential contribution of $\alpha 2\beta 1$ to pathological neovascularization has yet to be reported. Obviously, analysis of mice null for both $\alpha 1\beta 1$ and $\alpha 2\beta 1$ expression should be useful in understanding the exact functions of these integrins during physiological and pathological angiogenesis.

$\beta 1$ Integrins

The $\beta 1$ integrin subunit can pair with at least twelve α-subunits (see Figure 77.1). $\beta 1$-containing integrins are expressed in virtually all vertebrate cells and are involved in most aspects of development. Thus, $\beta 1$ genetic ablation leads to embryonic lethality at E7.5, which is prior to the onset of vascular development (8). However, analyses of $\beta 1$-null ES cells, embryoid bodies, and teratocarcinomas reveal an essential role for $\beta 1$ integrins in blood vessel formation. Although $\beta 1$ integrins are dispensable for vasculogenesis, they are absolutely necessary for developmental angiogenesis. Interestingly, ES cells or embryoid bodies null for $\beta 1$ integrin, $\alpha 5$ integrin, or FN display many strikingly similar defects. Thus, many of the $\beta 1$-null defects can probably be explained by loss of integrin $\alpha 5\beta 1$.

$\alpha 3$, $\alpha 4$, and $\alpha 6$ Integrins

Endothelial laminin receptors also are involved in regulating angiogenesis. Genetic ablation of the signal transduction portion of $\alpha 6\beta 4$ integrin leads to defects in growth factor–induced angiogenesis as well as pathological neovascularization (38). The laminin-binding integrins $\alpha 3\beta 1$ and $\alpha 6\beta 1$ also play cooperative roles in EC adhesion in vitro, as well as various angiogenesis assays in vivo (39). Lastly, the $\alpha 4\beta 1$ integrin binds to a variety of ECM ligands, including the vascular basement membrane proteins thrombospondin-1 and -2 (40). Disruption of $\alpha 4\beta 1$-mediated adhesion to the thrombospondins leads to abnormal EC functions in various in vitro and in vivo angiogenesis assays (41).

ANGIOSTATIC PROTEINS ACT VIA INTEGRINS

The *angiogenic switch* model (42) proposes that a balance exists between pro- and antiangiogenic factors. Changes in the balance regulate whether a blood vessel remains in a quiescent state or becomes angiogenic. Many ECM proteins within the vascular basement membrane, acting via integrins, are regulators of the angiogenic switch. The pro- or antiangiogenic functions of a particular ECM protein, as well as the integrin receptor(s) it binds, can be greatly affected by proteolysis (43).

Type IV collagen, a major component of the vascular basement membrane, is normally a ligand for the integrins $\alpha 1\beta 1$ and $\alpha 2\beta 1$. Matrix metalloprotease (MMP)-9–mediated cleavage of the NC1 domain of collagen IV ($\alpha 3$) yields the potent angiogenesis inhibitor tumstatin. Tumstatin binds to $\alpha v\beta 3$ integrin and inhibits the protein translation machinery, leading to EC apoptosis and reduced tumor progression (43). Proteolysis of the C-terminal NC1 domain of collagens XV or XVIII generates the antiangiogenic factor endostatin. Murine endostatin inhibits VEGF- or bFGF-induced EC proliferation and migration in vitro, and also represses pathological angiogenesis in some transplantable tumor models (44). Endostatin is primarily a ligand for both $\alpha 5\beta 1$ (45). Importantly, although the exact molecular mechanisms by which endostatin exerts its antiangiogenic effects remain unknown, it is clear that they are different from those regulated by tumstatin and do not require $\alpha v\beta 3$ (46).

Endorepellin, an 80-kDa C-terminal proteolytic fragment of the ubiquitous basement membrane protein, perlecan, also has antiangiogenic activity (47). Endorepellin blocks EC migration and tube formation in vitro, and inhibits growth factor-induced angiogenesis in Matrigel plugs and the CAM assay. Endorepellin binds to $\alpha 2\beta 1$ integrin, which leads to EC actin cytoskeletal disassembly and focal contact disruption. Interestingly, endorepellin also can bind to endostatin and counteract its antiangiogenic activity, suggesting that a balance exists between angiostatic proteins and also affects angiogenesis.

The proteolysis of many other ECM proteins also generates antiangiogenic fragments, although the mechanistic details of how they exert these effects remain unclear. For example, when proteolytically cleaved, thrombospondin-1 "cryptic fragments" bind to the scavenger receptor CD36 and serve as potent repressors of angiogenesis (40). Similarly, degradation of the proangiogenic factor FN generates the angiogenesis inhibitor anastellin (48), and degradation of plasminogen yields angiostatin (49). Interestingly, many of these ECM cryptic fragments require the circulating forms of plasma FN and vitronectin for their antiangiogenic activities (50), thus identifying a new complexity to angiogenesis regulation.

DIAGNOSTIC AND THERAPEUTIC IMPLICATIONS

There is no doubt that modulating integrin function can have beneficial effects in treating human disease. For example the peptide antagonist eptifibatide (Integrilin), which targets the platelet-specific integrin $\alpha IIb\beta 3$, has proven efficacious in treating patients with acute coronary artery syndromes (51). Furthermore, natalizumab (Tysabri/Antegren), a humanized monoclonal blocking antibody directed against the $\alpha 4$ integrin, has proven effective in inhibiting the T cell–mediated autoimmunity associated with multiple sclerosis and Crohn disease (52).

The fact that tumor growth and metastasis depend on neovascularization has spearheaded efforts to develop angiogenesis inhibitors to treat cancer, and some of these strategies have shown promising clinical results (53). Accordingly, $\alpha v\beta 3$ integrin has been a target of intense focus for neovascular intervention. This integrin is highly expressed on angiogenic ECs and structural data (see the previous section) will allow for the rational design of antagonists with exquisite molecule specificities and biomimetic properties. To date, a humanized anti-$\alpha v\beta 3$ antibody has yielded varied results in human clinical trials, although approaches using small-molecule antagonists or combinatorial radiotherapy may prove more effective (54). Additionally, small-molecule inhibitors targeting $\alpha 5$ integrin have yielded encouraging results in mouse models of colorectal cancer (55); whether these results correlate with tumor regression in human clinical trials remains to be determined.

KEY POINTS

- Integrin-mediated adhesion and signal transduction play important roles during physiologic and pathologic angiogenesis.
- Integrin expression is regulated at the transcriptional and post-translational levels by various angiogenic growth factors.
- Cell surface integrins form physical complexes with different endothelial receptor tyrosine kinases, which regulate EC functions.
- Many endogenous angiostatic proteins – derived from proteolysis of ECM components – are ligands for EC integrins.

Future Goals

- To develop EC integrins as attractive targets for antiangiogenesis therapies
- To understand the molecular pathways by which different integrins regulate angiogenesis, which will prove useful in determining strategies to more effectively modulate their function and may identify other protein targets for neovascular inhibition

REFERENCES

1 Hynes RO. Integrins: bidirectional, allosteric signaling machines. *Cell.* 2002;110(6):673–687.

2 Hynes RO, Lively JC, McCarty JH, et al. The diverse roles of integrins and their ligands in angiogenesis. *Cold Spring Harb Symp Quant Biol.* 2002;67:143–153.

3 Hynes RO. The emergence of integrins: a personal and historical perspective. *Matrix Biol.* 2004;23(6):333–340.

4 Ruoslahti E. The RGD story: a personal account. *Matrix Biol.* 2003;22(6):459–465.

5 Pytela R, Pierschbacher MD, Ruoslahti E. Identification and isolation of a 140 kd cell surface glycoprotein with properties expected of a fibronectin receptor. *Cell.* 1985;40(1):191–198.

6 Hynes RO, Zhao Q. The evolution of cell adhesion. *J Cell Biol.* 2000;150(2):F89–F96.

7 Bokel C, Brown NH. Integrins in development: moving on, responding to, and sticking to the extracellular matrix. *Dev Cell.* 2002;3(3):311–321.

8 Bouvard D, Brakebusch C, Gustafsson E, et al. Functional consequences of integrin gene mutations in mice. *Circ Res.* 2001;89(3):211–223.

9 Kim LT, Yamada KM. The regulation of expression of integrin receptors. *Proc Soc Exp Biol Med.* 1997;214(2):123–131.

10 Katsumi A, Orr AW, Tzima E, Schwartz MA. Integrins in mechanotransduction. *J Biol Chem.* 2004;279(13):12001–12004.

11 Senger DR, Claffey KP, Benes JE, et al. Angiogenesis promoted by vascular endothelial growth factor: regulation through alpha1beta1 and alpha2beta1 integrins. *Proc Natl Acad Sci USA.* 1997;94(25):13612–13617.

12 de Melker AA, Sonnenberg A. Integrins: alternative splicing as a mechanism to regulate ligand binding and integrin signaling events. *Bioessays.* 1999;21(6):499–509.

13 Xiong JP, Stehle T, Zhang R, et al. Crystal structure of the extracellular segment of integrin alpha Vbeta3 in complex with an Arg-Gly-Asp ligand. *Science.* 2002;296(5565):151–155.

14 Vinogradova O, Velyvis A, Velyviene A, et al. A structural mechanism of integrin alpha(IIb)beta(3) "inside-out" activation as regulated by its cytoplasmic face. *Cell.* 2002;110(5):587–597.

15 Liddington RC, Ginsberg MH. Integrin activation takes shape. *J Cell Biol.* 2002;158(5):833–839.

16 Shimaoka M, Springer TA. Therapeutic antagonists and conformational regulation of integrin function. *Nat Rev Drug Discov.* 2003;2(9):703–716.

17 Giancotti FG, Ruoslahti E. Integrin signaling. *Science.* 1999;285 (5430):1028–1032.

18 Hood JD, Frausto R, Kiosses WB, et al. Differential {alpha}v integrin-mediated Ras-ERK signaling during two pathways of angiogenesis. *J Cell Biol.* 2003;162(5):933–943.

19 Byzova TV, Goldman CK, Pampori N, et al. A mechanism for modulation of cellular responses to VEGF: activation of the integrins. *Mol Cell.* 2000;6(4):851–860.

20 Reynolds AR, Reynolds LE, Nagel TE, et al. Elevated Flk1 (vascular endothelial growth factor receptor 2) signaling mediates enhanced angiogenesis in beta3-integrin-deficient mice. *Cancer Res.* 2004;64(23):8643–8650.

21 Friedlander M, Brooks PC, Shaffer RW, et al. Definition of two angiogenic pathways by distinct integrins. *Science.* 1995;270: 1500–1502.

22 Cheresh DA. Death to a blood vessel, death to a tumor. *Nat Med.* 1998;4(4):395–396.

23 Hodivala-Dilke KM, Reynolds AR, Reynolds LE. Integrins in angiogenesis: multitalented molecules in a balancing act. *Cell Tissue Res.* 2003;314(1):131–144.

24 Reynolds LE, Wyder L, Lively JC, et al. Enhanced pathological angiogenesis in mice lacking beta3 integrin or beta3 and beta5 integrins. *Nat Med.* 2002;8(1):27–34.

25 Bader BL, Rayburn H, Crowley D, Hynes RO. Extensive vasculogenesis, angiogenesis, and organogenesis precede lethality in mice lacking all alpha v integrins. *Cell.* 1998;95(4):507–519.

26 McCarty JH, Monahan-Earley RA, Brown LF, et al. Defective associations between blood vessels and brain parenchyma lead to cerebral hemorrhage in mice lacking alphav integrins. *Mol Cell Biol.* 2002;22(21):7667–7677.

27 Cambier S, Gline S, Mu D, et al. Integrin alpha(v)beta8-mediated activation of transforming growth factor-beta by perivascular astrocytes: an angiogenic control switch. *Am J Pathol.* 2005; 166(6):1883–1894.

28 Zhu J, Motejlek K, Wang D, et al. Beta8 integrins are required for vascular morphogenesis in mouse embryos. *Development.* 2002;129(12):2891–2903.

29 McCarty JH, Lacy-Hulbert A, Charest A, et al. Selective ablation of alphav integrins in the central nervous system leads to cerebral hemorrhage, seizures, axonal degeneration and premature death. *Development.* 2005;132(1):165–176.

30 Proctor JM, Zang K, Wang D, et al. Vascular development of the brain requires beta8 integrin expression in the neuroepithelium. *J Neurosci.* 2005;25(43):9940–9948.

31 Hynes RO, Bader BL, Hodivala-Dilke KM. Integrins in vascular development. *Braz J Med Biol Res.* 1999;32:501–510.

32 Taverna D, Hynes RO. Reduced blood vessel formation and tumor growth in alpha5-integrin- negative teratocarcinomas and embryoid bodies. *Cancer Res.* 2001;61(13):5255–5261.

33 Francis SE, Goh KL, Hodivala-Dilke K, et al. Central roles of alpha5beta1 integrin and fibronectin in vascular development in mouse embryos and embryoid bodies. *Arterioscler Thromb Vasc Biol.* 2002;22(6):927–933.

34 Hynes RO. A reevaluation of integrins as regulators of angiogenesis. *Nat Med.* 2002;8(9):918–921.

35 Kim S, Bell K, Mousa SA, Varner JA. Regulation of angiogenesis in vivo by ligation of integrin alpha5beta1 with the central cell-binding domain of fibronectin. *Am J Pathol.* 2000;156(4):1345–1362.

36 Pozzi A, Moberg PE, Miles LA, et al. Elevated matrix metalloprotease and angiostatin levels in integrin alpha 1 knockout mice cause reduced tumor vascularization. *Proc Natl Acad Sci USA.* 2000;97(5):2202–2207.

37 Mercurio AM. Lessons from the alpha2 integrin knockout mouse. *Am J Pathol.* 2002;161(1):3–6.
38 Nikolopoulos SN, Blaikie P, Yoshioka T, et al. Integrin beta4 signaling promotes tumor angiogenesis. *Cancer Cell.* 2004;6(5): 471–483.
39 Gonzalez AM, Gonzales M, Herron GS, et al. Complex interactions between the laminin alpha 4 subunit and integrins regulate EC behavior in vitro and angiogenesis in vivo. *Proc Natl Acad Sci USA.* 2002;99(25):16075–16080.
40 Lawler J. Thrombospondin-1 as an endogenous inhibitor of angiogenesis and tumor growth. *J Cell Mol Med.* 2002;6(1):1–12.
41 Calzada MJ, Zhou L, Sipes JM, et al. Alpha4beta1 integrin mediates selective EC responses to thrombospondins 1 and 2 in vitro and modulates angiogenesis in vivo. *Circ Res.* 2004;94(4):462–470.
42 Hanahan D, Folkman J. Patterns and emerging mechanisms of the angiogenic switch during tumorigenesis. *Cell.* 1996;86(3): 353–364.
43 Kalluri R. Basement membranes: structure, assembly and role in tumour angiogenesis. *Nat Rev Cancer.* 2003;3(6):422–433.
44 O'Reilly MS, Boehm T, Shing Y, et al. Endostatin: an endogenous inhibitor of angiogenesis and tumor growth. *Cell.* 1997; 88(2):277–285.
45 Bix G, Iozzo RV. Matrix revolutions: "tails" of basement-membrane components with angiostatic functions. *Trends Cell Biol.* 2005;15(1):52–60.
46 Hamano Y, Zeisberg M, Sugimoto H, et al. Physiological levels of tumstatin, a fragment of collagen IV alpha3 chain, are generated by MMP-9 proteolysis and suppress angiogenesis via alphaV beta3 integrin. *Cancer Cell.* 2003;3(6):589–601.
47 Bix G, Fu J, Gonzalez EM, et al. Endorepellin causes EC disassembly of actin cytoskeleton and focal adhesions through alpha2beta1 integrin. *J Cell Biol.* 2004;166(1):97–109.
48 Yi M, Ruoslahti E. A fibronectin fragment inhibits tumor growth, angiogenesis, and metastasis. *Proc Natl Acad Sci USA.* 2001;98(2): 620–624.
49 O'Reilly MS, Holmgren L, Shing Y, et al. Angiostatin: a novel angiogenesis inhibitor that mediates the suppression of metastases by a Lewis lung carcinoma. *Cell.* 1994;79(2):315–328.
50 Akerman ME, Pilch J, Peters D, Ruoslahti E. Angiostatic peptides use plasma fibronectin to home to angiogenic vasculature. *Proc Natl Acad Sci USA.* 2005;102(6):2040–2045.
51 Rossi ML, Zavalloni D. Inhibitors of platelets glycoprotein IIb/IIIa (GP IIb/IIIa) receptor: rationale for their use in clinical cardiology. *Mini Rev Med Chem.* 2004;4(7):703–709.
52 Elices MJ. Natalizumab. Elan/Biogen. *Curr Opin Investig Drugs.* 2003;4(11):1354–1362.
53 Folkman J. Endogenous angiogenesis inhibitors. *APMIS.* 2004; 112(7–8):496–507.
54 Tucker GC. Alpha v integrin inhibitors and cancer therapy. *Curr Opin Investig Drugs.* 2003;4(6):722–731.
55 Stoeltzing O, Liu W, Reinmuth N, et al. Inhibition of integrin alpha5beta1 function with a small peptide (ATN-161) plus continuous 5-FU infusion reduces colorectal liver metastases and improves survival in mice. *Int J Cancer.* 2003;104(4):496–503.

78

Aquaporin Water Channels and the Endothelium

Alan S. Verkman

Cardiovascular Research Institute, University of California, San Francisco

The aquaporins (AQPs) are a family of water transporting channels that are expressed in many mammalian tissues including epithelial, endothelial, and other cell types. At least ten AQPs are present in mammals, and more exist in amphibians, plants, and lower organisms. Functional measurements suggest that mammalian AQPs 1, 2, 4, 5, and 8 are primarily water-selective, whereas AQPs 3, 7, and 9 (called *aquaglyceroporins*) also transport glycerol and possibly other small solutes. The AQPs function as pores to increase water transport across cell membranes in response to an osmotic gradient. Structural studies of AQP1 indicate a tetrameric assembly in membranes in which each monomer contains six tilted helical segments surrounding a putative aqueous pore. Human mutations exist for several AQPs. Humans with mutations in AQP0, the major intrinsic protein of lens fiber (also called MIP) develop cataracts (1). Humans with mutations in the *AQP1* gene (Colton blood group antigen) manifest a urinary concentrating defect (2), and humans with mutations in the *AQP2* gene, the vasopressin-regulated water channel, have autosomal hereditary nephrogenic diabetes insipidus (3).

AQP1 is expressed strongly throughout microvascular endothelial beds outside the brain, such as in kidney (vasa recta), lung and airways, skin, secretory glands, skeletal muscle, pleura, and peritoneum (4,5). AQP1 also is expressed in the endothelium of proliferating tumor microvessels (6) and in nonvascular endothelial cells (ECs) in cornea, trabecular meshwork in the canal of Schlemm, and central lacteals in small intestine (4,5,7). Figure 78.1A shows immunolocalization of AQP1 in ECs in a variety of microvascular and nonvascular endothelia. AQPs have not been identified in ECs in the brain, although a different AQP (AQP4) is expressed in astrocyte foot processes that comprise the blood–brain barrier (BBB) in close contact with ECs. The expression of AQP1 in various vascular and nonvascular endothelia suggests a possible role in transendothelial water movement. Our laboratory generated a series of knockout mice to examine the role of AQPs in mammalian physiology and disease. This chapter summarizes the general paradigms that have emerged from phenotype analysis of *AQP*-knockout mice, followed by an analysis of AQP1 function in endothelial biology.

NONENDOTHELIAL ROLES OF AQUAPORINS REVEALED BY KNOCKOUT MICE

Phenotype analysis of transgenic mice lacking AQPs has been useful in defining their physiological roles. The role of AQPs in transepithelial fluid transport is fairly well-understood. Mice lacking AQP1 manifest defective urinary concentrating ability (8) as a consequence of impaired isosmolar fluid absorption in the proximal tubule (9) and countercurrent multiplication (10,11). Mice lacking water channels of the kidney collecting duct, AQPs 2, 3, or 4, have variable impairment of urinary concentrating ability (12–14) because of reduced transepithelial water permeability in the collecting duct and consequent impaired lumen-to-interstitial water transport. The impaired fluid absorption in the kidney proximal tubule in AQP1 deficiency indicates the need for high transepithelial water permeability for rapid, near-isosmolar fluid transport (9). Impairment of near-isosmolar fluid transport also was found for fluid secretion in salivary (15) and airway submucosal (16) glands in AQP5 deficiency, and for AQP1-dependent secretion of cerebrospinal fluid by choroid plexus (17) and aqueous fluid by the ocular ciliary epithelium (18). High, AQP-facilitated transepithelial water permeability permits rapid, passive water transport in response to active transepithelial salt transport. However, active fluid transport in many tissues does not appear to be AQP-dependent, such as alveolar fluid absorption in AQP1 or AQP5 deficiency (19,20), and sweat secretion in AQP5 deficiency (21). The rate of active fluid absorption/secretion per unit epithelial surface area in alveolus and sweat gland is much lower than that in kidney proximal tubule or salivary gland, suggesting that high AQP-dependent water permeability is not required to support relatively slow fluid absorption or secretion.

A number of interesting, less predictable phenotypes were found for AQP deletion in nonepithelial tissues. AQP4 appears to play an important role in the central nervous system,

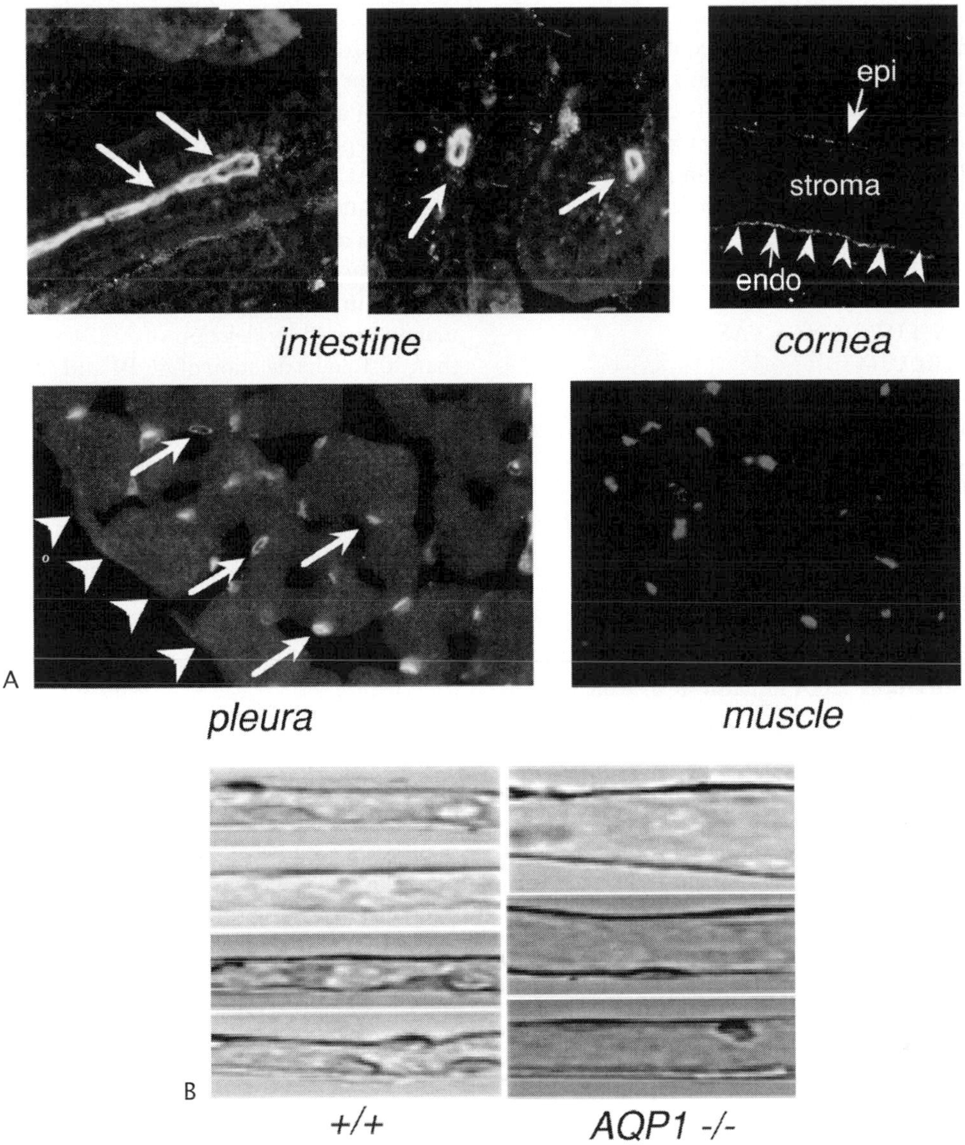

Figure 78.1. AQP1 expression in ECs in mice. (**A**) Tissues stained with purified polyclonal anti-AQP1 antibody. Arrows point to central lacteals in duodenum, corneal endothelium, pleural microvessels, and skeletal muscle microvessels. (**B**) Light micrographs of renal vasa recta microdissected from mouse kidney of wild-type and *AQP1*-null mice. (Reproduced with permission from Pallone TL, Edwards A, Ma T, et al. Requirement of aquaporin-1 for NaCl driven water transport across descending vasa recta. *J Clin Invest.* 2000;105:215–222; Binder D, Oshio K, Ma T, et al. Increased seizure threshold in mice deficient in aquaporin-4 water channels. *Neuroreport.* 2004;15:259–262; and Thiagarajah JR, Verkman AS. Aquaporin deletion in mice reduces corneal water permeability and delays restoration of transparency after swelling. *J Biol Chem.* 2001;277:19139–19144.)

both for fluid movement into and out of the brain, and for rapid neural signal transduction. Mice lacking AQP4 have reduced swelling in models of cytotoxic (cellular) brain edema, including water intoxication, ischemic stroke, and meningitis (22,23). However, brain swelling and clinical outcome were worse in AQP4-deficient mice in models of vasogenic (leaky vessel) brain edema, including cortical freeze injury and brain tumor, where it was concluded that the AQP4-dependent exit of water from brain tissue is impaired. *AQP4*-null mice also have impaired evoked potential responses to sound (24) and light (25), as well as altered seizure threshold (26), which has been proposed to be related to impaired AQP4-dependent Kir4.1 K^+ channel activity (27) and/or altered extracellular space dynamics (28).

The aquaglyceroporins AQP3 and AQP7 transport both water and glycerol, leading to interesting phenotypes in *AQP3*- and *AQP7*-null mice involving impaired glycerol transport. AQP3 is expressed in the basal layer of epidermal keratinocytes in skin, where its deletion in mice produces dry skin with impaired biosynthetic function as a consequence of reduced glycerol content in epidermis and stratum corneum (29). AQP7 is expressed in adipocytes, where its deletion in mice produces progressive adipocyte hypertrophy as a consequence of glycerol and triglyceride accumulation in adipocytes (30), as well as an impaired serum glycerol response to fasting and hormonal stimuli (31).

AQUAPORIN 1 IN THE RENAL VASA RECTA ENDOTHELIUM

AQP1 is expressed in the microvascular endothelium of outer medullary descending vasa recta (OMDVR). The formation of a concentrated urine requires the trapping of NaCl and urea in the renal medulla, as a well as high water permeability across the collecting duct epithelium. In portions of OMDVR, water efflux from the vascular space into the renal interstitium occurs despite the existence of Starling forces (hydrostatic and oncotic forces) that favor volume influx, suggesting that water efflux involves a water-only (AQP1) pathway in which NaCl and urea osmotic gradients drive water movement. To investigate the role of AQP1 in vasa recta function, OMDVR were perfused in vitro (11). Interestingly, OMDVR from *AQP1*-null mice were much larger in diameter than those from wild-type mice (see Figure 78.1B). Osmotic water permeability in response to a 200-mM NaCl gradient (bath > lumen) was reduced by more than tenfold in *AQP1*-null mice, and by approximately twofold when driven by a raffinose gradient. These data, together with mercurial inhibition measurements, suggested that most NaCl-driven water transport occurs by a transcellular route through AQP1. In a mathematical model of the medullary microcirculation, deletion of AQP1 resulted in diminished concentrating ability due to enhancement of medullary blood flow, partially accounting for the observed urinary concentrating defect. Therefore, AQP1 in OMDVR is an important component of the urinary concentrating mechanism. Mechanistically, NaCl and urea gradients drive water transport from the OMDVR lumen to the medullary interstitium across the AQP1-containing luminal and abluminal membranes, thereby shunting blood flow from descending to ascending vasa recta in the outer medulla. The reduced blood flow to the deep inner medullary portions of the vasa recta is predicted to enhance countercurrent exchange by limiting solute washout. The enlarged vasa recta in AQP1 deficiency may be a kidney-specific effect involving chronically increased vasa recta blood flow, because neither microvascular structure nor density differ in extrarenal vascular beds, such as in skin and skeletal muscle (32).

AQUAPORIN 1 IN LUNG MICROVESSELS

The lung and airways participate in fluid movement during airway hydration, reabsorption of alveolar fluid in the neonatal period, and formation and resolution of pulmonary edema resulting from heart failure or lung injury. The barriers to water movement between the airspace and capillary compartments consist of an epithelium, interstitium and endothelium. The alveolar epithelium is composed mainly of type 1 cells, that express AQP5 at their apical membrane. The microvascular endothelium expresses AQP1. AQP3 and AQP4 are expressed mainly in airway epithelia. Osmotic water transport between the airspace and capillary compartments, as measured by a pleural surface fluorescence method, was reduced approximately tenfold by deletion of AQP1 or AQP5, and by more than 30-fold by deletion of AQP1 and AQP5 together (19,20). However, there was no effect of AQP1 or AQP5 deletion on alveolar fluid absorption in the adult or neonatal lung, or on the response of the adult lung to epithelial or EC injury (33). Also, AQP1 deletion in mice did not impair airway humidification (34) or lung carbon dioxide transport (35). Together these results indicate that, although AQP1 in the lung and airway microvascular endothelium facilitates osmotic water transport, its deletion does not impair physiologically relevant lung fluid or gas transport physiology.

AQUAPORIN 1 FUNCTION AT THE PLEURAL AND PERITONEAL BARRIERS

Fluid is continuously secreted into and cleared from the pleural space. Pleural fluid can accumulate in congestive heart failure, lung infection, lung tumor, and the acute respiratory distress syndrome. Fluid entry into the pleural space involves filtration across microvascular endothelia near the pleural surface and movement across a mesothelial barrier lining the pleural space, whereas fluid clearance occurs primarily through lymphatic drainage. AQP1 is expressed in microvascular endothelia near the visceral and parietal pleura as well as in mesothelial cells in visceral pleura (36). Osmotic water permeability was measured in anesthetized, ventilated mice from the kinetics of pleural fluid osmolality after instillation of hypertonic or hypotonic fluid into the pleural space. Osmotic equilibration of pleural fluid was rapid in wild-type mice (50% equilibration in <2 min) and slowed by approximately fourfold in *AQP1*-null mice. However, the clearance of isosmolar saline instilled in the pleural space (~4 mL/kg/hour) was not affected by AQP1 deletion, nor was the accumulation of pleural fluid (~0.035 mL/hour) in a volume overload model of congestive heart failure. Thus, although AQP1 facilitates rapid osmotic equilibration across the pleural surface, AQP1 does not appear to be a physiologically important mechanism of pleural fluid accumulation or clearance.

A similar conclusion was reached for the role of AQP1 in fluid movement into and out of the peritoneal cavity (37). The peritoneal cavity is lined by a membranous barrier that provides a large surface for fluid movement between peritoneal capillaries and the peritoneal cavity. Ascites can accumulate in conditions associated with decreased serum oncotic pressure, increased portal venous pressure, or peritoneal cavity inflammation or infection. The large peritoneal surface is exploited in peritoneal dialysis, where water, electrolytes, urea, and uremia-causing toxins are extracted from blood by repeated infusion and removal of dialysate solutions into the peritoneal cavity. AQP1 is expressed in the capillary endothelium and mesangium near the peritoneal luminal surface. Osmotically induced water transport in *AQP1*-null mice was approximately 2.5-fold reduced, compared with wild-type mice, suggesting a role for AQP1 in osmotic water removal during peritoneal dialysis. However, fluid absorption after intraperitoneal saline infusion was not affected by AQP1 deletion, suggesting that AQP1 is not an important determinant of peritoneal fluid movement under physiological conditions.

AQUAPORIN 1 IN LACTEAL ENDOTHELIUM IN SMALL INTESTINE

AQP1 is expressed at those sites in the proximal gastrointestinal tract that play a role in dietary fat processing, including cholangiocytes in liver (bile production), pancreatic microvascular endothelium (pancreatic fluid production), gallbladder microvascular endothelium (bile storage), and intestinal lacteal endothelium (chylomicron absorption). The hypothesis was tested that AQP1 facilitates dietary fat processing (38). It was found that relatively young *AQP1*-null mice fed a high-fat (50%) diet failed to gain weight and developed steatorrhea and serum hypertriglyceridemia, whereas litter-matched wild-type mice gained weight rapidly. The *AQP1*-null mice thrived when returned to a normal low-fat diet. The *AQP1*-null mice on a high-fat diet had elevated concentrations of pancreatic enzymes in their small intestine and stool, normal pH in duodenal fluid, and normal bile/pancreatic fluid production, suggesting a defect in fat absorption rather than digestion. The mechanisms by which AQP1 acts in lacteal endothelium to facilitate chylomicron absorption remain to be determined.

AQUAPORIN 1 IN CORNEAL ENDOTHELIAL FUNCTION

AQP1 is expressed in corneal ECs, and AQP5 in corneal epithelial cells. Corneal thickness, water permeability, and response to experimental swelling was compared in wild-type mice and mice lacking AQP1 or AQP5 (39). Corneal thickness was significantly reduced in *AQP1*-null mice (101 μm) and increased in *AQP5*-null mice (144 μm) compared with wild-type mice (123 μm). After exposure of the corneal endothelial surface to hypotonic saline by anterior chamber perfusion, the rate of corneal swelling (7.1 μm/min in wild-type mice) was greatly reduced by AQP1 deletion (1.6 μm/min), indicating that AQP1 provides the principal route for osmotically driven water transport across the corneal endothelium. Although baseline corneal transparency was not impaired by AQP1 deletion, the recovery of corneal transparency and thickness after hypotonic swelling (10 min exposure of corneal surface to distilled water) was remarkably delayed in *AQP1*-null mice, with approximately 75% recovery at 7 minutes in wild-type mice compared to 5% recovery in *AQP1*-null mice. The impaired recovery of corneal transparency in *AQP1*-null mice provides evidence for the involvement of AQP1 in the active extrusion of fluid from the corneal stroma across the corneal endothelium. Measurements in ECs cultured from the corneas of wild-type and *AQP1*-null mice confirmed the reduced water permeability in AQP1 deficiency and suggested a potential role for AQP1 in corneal EC volume regulation (40).

NOVEL ROLE OF AQUAPORIN 1 IN ANGIOGENESIS AND CELL MIGRATION

AQP1 is expressed strongly in the ECs of proliferating microvessels in malignant brain tumors, bone marrow microvessels in multiple myeloma, and proliferating microvessels in chick embryo chorioallantoic membrane. We recently found remarkably impaired tumor growth in *AQP1*-null mice after subcutaneous or intracranial tumor cell implantation, with reduced tumor vascularity and extensive necrosis (Figures 78.2A and 78.2B; for color reproduction, see Color Plate 78.2) (32). Reduced microvessel growth also was found in implanted pellets of Matrigel containing endothelial growth factors. A novel mechanism for the impaired angiogenesis was established from cell culture studies. Although adhesion and proliferation were similar in primary cultures of aortic endothelium from wild-type versus *AQP1*-null mice, cell migration was greatly impaired in AQP1-deficient cells (Figure 78.2C), with abnormal vessel formation in vitro. Stable transfection of non-ECs with AQP1, or a structurally different water-selective transporter (AQP4), accelerated cell migration and in vitro wound healing. Interestingly, motile AQP1-expressing cells had prominent membrane ruffles at the leading edge, with polarization of AQP1 protein to lamellipodia, where rapid water fluxes occur. As a possible mechanism for the dependence of cell migration on AQPs, it has been proposed that cell membrane protrusions are formed as a consequence of actin cleavage and ion uptake at the tip of a lamellipodium, creating local osmotic gradients that drive the influx of water through the cell membrane (41,42). Water entry is thought to then increase local hydrostatic pressure to cause cell membrane protrusion. Aquaporins may thus provide an important pathway for water entry into lamellipodia in some cell types.

CONCLUSION

Phenotype studies in AQP-deficient mice indicate a variety of diverse roles for AQPs in mammalian physiology, ranging from urinary concentrating function and exocrine gland fluid secretion, to water balance in brain, and glycerol/triglyceride regulation in adipocytes. AQP1 is widely expressed in microvascular endothelia outside the brain; however, only in renal

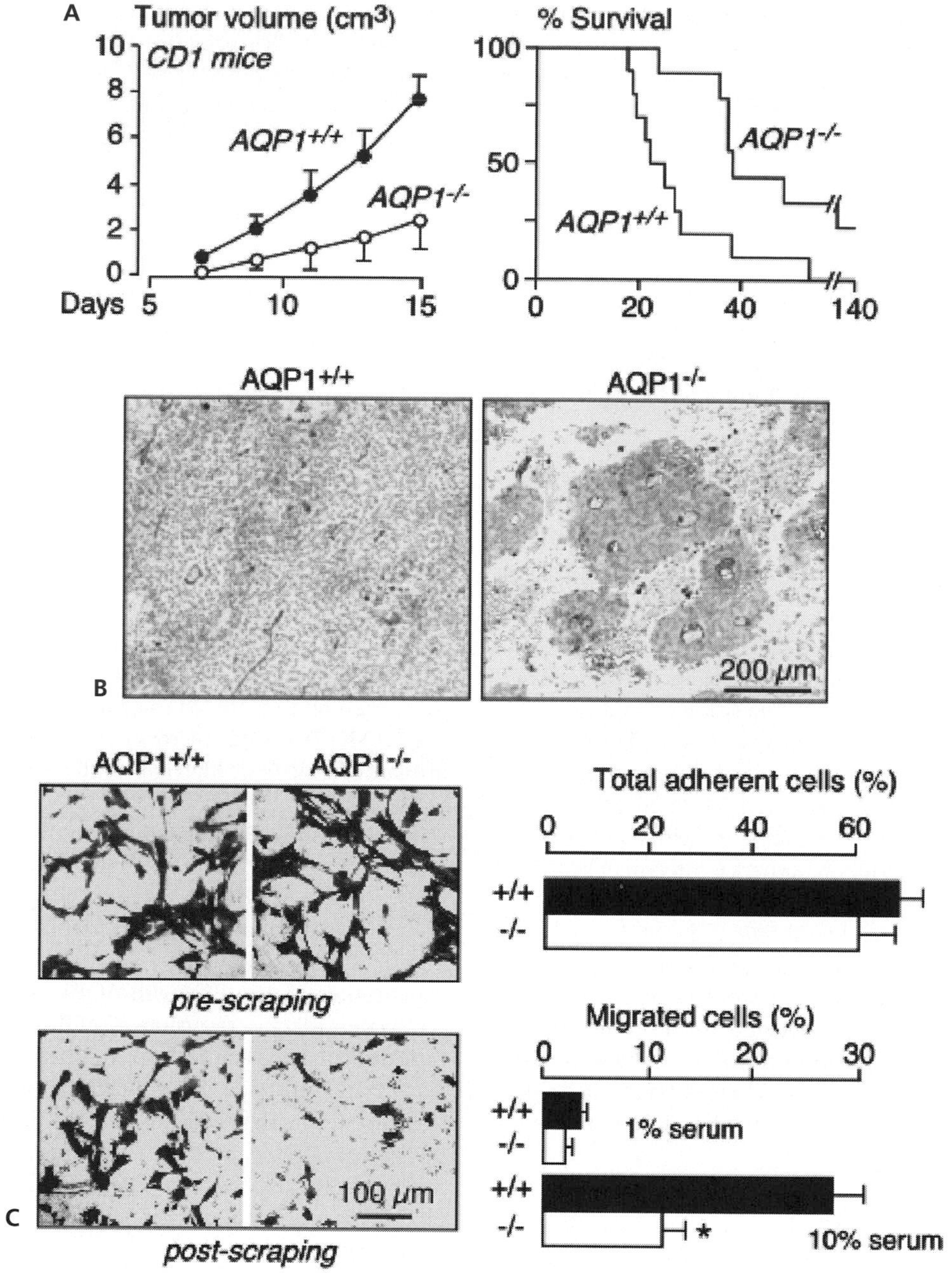

Figure 78.2. Impaired tumor angiogenesis and EC migration in AQP1 deficiency. (**A**) Reduced tumor growth in *AQP1*-null mice. Growth of subcutaneous melanoma in wild-type versus *AQP1*-null mice (*left*). Survival of mice with subcutaneous melanoma (*right*). (**B**) Tumor stained with isolectin-B4. Note islands of tumor surrounded by necrotic tissue in AQP1-deficient mice. (**C**) Impaired migration of ECs lacking AQP1. Adhesion and migration (*left*) of ECs quantified by counting cells in Boyden chamber assay before and after scraping. Summary of percentage adherent and migrated cells (*right*). (Reproduced with permission from Saadoun S, Papadopoulos MC, Hara-Chikuma M, Verkman AS. Targeted disruption of aquaporin-1 gene impairs angiogenesis and cell migration. *Nature*. 2005;434:786–792.) For color reproduction, see Color Plate 78.2.

vasa recta, where transendothelial osmotic transport driven by gradients of small solutes occurs, is AQP1 of clear physiological significance. The reasons for the expression of aquaporins in microvascular endothelia without demonstrable functional significance remain unclear. Reduced angiogenesis in *AQP1*-null mice and the impaired migration of AQP1-deficient ECs suggests a novel role for AQP1 in rapid vessel proliferation in tumors and after injury. Indeed, AQP-dependent cell migration may be a general phenomenon of relevance to tumor spread, wound healing, and organ regeneration. In nonmicrovascular endothelium, AQP1 in intestinal lacteals may facilitate intestinal absorption, and AQP1 in corneal endothelium appears to be quite important for corneal fluid balance. The phenotype studies suggest that the pharmacological inhibition of AQP1 may provide new therapies in human diseases, such as in diuretics, regulators of intraocular pressure and swelling in brain and cornea, and inhibitors of tumor angiogenesis.

KEY POINTS

- AQP1 provides the major water transport pathway in the endothelium.
- AQP1 is required for normal EC function in renal vasa recta and cornea; AQP1 facilitates tumor vessel angiogenesis by enhancing EC migration.

Future Goals

- To develop AQP1 inhibitors as diuretics and antiangiogenesis agents

ACKNOWLEDGMENTS

Supported by grants DK35124, HL59198, EY13574, EB00415, and HL73854 from the National Institutes of Health, and grant R613 from the Cystic Fibrosis Foundation.

REFERENCES

1 Berry V, Francis P, Kaushal S, et al. Missense mutations in MIP underlie autosomal dominant "polymorphic" and lamellar cataracts linked to 12q. *Nat Genet*. 2000;25:15–17.
2 King LS, Choi M, Fernandez PC, et al. Defective urinary-concentrating ability due to a complete deficiency of aquaporin-1. *New Engl J Med*. 2001;345:175–179.
3 Deen PM, Verkijk MA, Knoers NV, et al. Requirement of human renal water channel aquaporin-2 for vasopressin-dependent concentration of urine. *Science*. 1994;264:92–95.
4 Nielsen S, Smith BL, Christensen EI, Agre P. Distribution of the aquaporin CHIP in secretory and resorptive epithelia and capillary endothelia. *Proc Natl Acad Sci USA*. 1993;90:7275–7279.
5 Hasegawa H, Lian SC, Finkbeiner WE, Verkman AS. Extrarenal tissue distribution of CHIP28 water channels by in situ hybridization and antibody staining. *Am J Physiol*. 1994;266:C893–C903.
6 Endo M, Jain RK, Witwer B, Brown D. Water channel (aquaporin 1) expression and distribution in mammary carcinomas and glioblastomas. *Microvasc Res*. 1999;58:89–98.
7 Wen Q, Diecke FP, Iserovich P, et al. Immunocytochemical localization of aquaporin-1 in bovine corneal endothelial cells and keratocytes. *Exp Biol Med*. 2001;226:463–467.
8 Ma T, Yang B, Gillespie A, et al. Generation and phenotype of a transgenic knock-out mouse lacking the mercurial-insensitive water channel aquaporin-4. *J Clin Invest*. 1997;100:957–962.
9 Schnermann J, Chou CL, Ma T, et al. Defective proximal tubular fluid reabsorption in transgenic aquaporin-1 null mice. *Proc Natl Acad Sci USA*. 1998;95:9660–9664.
10 Chou CL, Knepper MA, Van Hoek AN, et al. Reduced water permeability and altered ultrastructure in thin descending limb of Henle in aquaporin-1 null mice. *J Clin Invest*. 1999;103:491–496.
11 Pallone TL, Edwards A, Ma T, et al. Requirement of aquaporin-1 for NaCl driven water transport across descending vasa recta. *J Clin Invest*. 2000;105:215–222.
12 Ma T, Yang B, Gillespie A, et al. Severely impaired urinary concentrating ability in transgenic mice lacking aquaporin-1 water channels. *J Biol Chem*. 1998;273:4296–4299.
13 Ma T, Song Y, Yang B, et al. Nephrogenic diabetes insipidus in mice deficient in aquaporin-3 water channels. *Proc Natl Acad Sci USA*. 2000;97:4386–4391.
14 Yang B, Gillespie A, Carlson EJ, et al. Early neonatal mortality in a transgenic AQP2 knock-in model of nephrogenic diabetes insipidus. *J Biol Chem*. 2001;276:2775–2779.
15 Ma T, Song Y, Gillespie A, et al. Defective secretion of saliva in transgenic mice lacking aquaporin-5 water channels. *J Biol Chem*. 1999;274:20071–20074.
16 Song Y, Verkman AS. Aquaporin-5 dependent fluid secretion in airway submucosal glands. *J Biol Chem*. 2001;276:41288–41292.
17 Oshio K, Song Y, Verkman AS, Manley GT. Reduced intraventricular pressure and cerebrospinal fluid production in mice lacking aquaporin-1 water channels. *FASEB J*. 2005;18:76–78.
18 Zhang, D, Vetrivel L, Verkman AS. Aquaporin deletion in mice reduces intraocular pressure and aqueous fluid production. *J Gen Physiol*. 2002;119:561–569.
19 Bai C, Fukuda CX, Song Y, et al. Role of aquaporin water channels in lung fluid transport: Phenotype analysis of aquaporin 1 and 4 knockout mice. *J Clin Invest*. 1999;103:555–561.
20 Ma T, Fukuda N, Song Y, et al. Lung fluid transport in aquaporin-5 knockout mice. *J Clin Invest*. 2000;105:93–100.
21 Song Y, Sonawane N, Verkman AS. Localization of aquaporin-5 in sweat glands and functional analysis using knockout mice. *J Physiol*. 2002;541:561–568.
22 Manley GT, Fujimura M, Ma T, et al. Aquaporin-4 deletion in mice reduces brain edema following acute water intoxication and ischemic stroke. *Nature Med*. 2000;6:159–163.
23 Papadopoulos MC, Krishna S, Manley GT, Verkman AS. Aquaporin-4 facilitates the reabsorption of excess fluid in vasogenic brain edema. *FASEB J*. 2004;18:1291–1293.
24 Li J, Verkman AS. Impaired hearing in mice lacking aquaporin-4 water channels. *J Biol Chem*. 2001;276:31233–31237.
25 Da T, Verkman AS. Aquaporin-4 gene disruption in mice protects against impaired retinal function and cell death after ischemia. *Invest Opthalmol Vis Sci*. 2004;45:4477–4483.
26 Binder DK, Oshio K, Ma T, et al. Increased seizure threshold in mice deficient in aquaporin-4 water channels. *Neuroreport*. 2004;15:259–262.
27 Nagelhus EA, Mathiisen TM, Ottersen OP. Aquaporin-4 in the central nervous system: cellular and subcellular distribution and coexpression with KIR4.1. *Neuroscience*. 2004;129:905–913.
28 Binder D, Oshio K, Ma T, et al. Increased seizure threshold in mice deficient in aquaporin-4 water channels. *Neuroreport*. 2004;15:259–262.
29 Hara M, Verkman AS. Glycerol replacement corrects defective skin hydration, elasticity and barrier function in aquaporin-3-deficient mice. *Proc Natl Acad Sci USA*. 2003;100:7360–7365.
30 Hara-Chikuma M, Sohara E, Rai T, et al. Progressive adipocyte hypertrophy in aquaporin-7 deficient mice: adipocyte glycerol

permeability as a novel regulator of fat accumulation. *J Biol Chem.* 2005;280:15493–15496.
31 Maeda N, Funahashi T, Hibuse T, et al. Adaptation to fasting by glycerol transport through aquaporin 7 in adipose tissue. *Proc Natl Acad Sci USA.* 2004;101:17801–17806.
32 Saadoun S, Papadopoulos MC, Hara-Chikuma M, Verkman AS. Targeted disruption of aquaporin-1 gene impairs angiogenesis and cell migration. *Nature.* 2005;434:786–792.
33 Song Y, Fukuda N, Bai C, et al. Role of aquaporins in alveolar fluid clearance in neonatal and adult lung, and in edema formation following lung injury. *J Physiol.* 2000;525:771–779.
34 Song Y, Jayaraman S, Yang B, et al. Role of aquaporin water channels in airway fluid transport, humidification, and surface liquid hydration. *J Gen Physiol.* 2001;117:573–582.
35 Yang B, Fukuda N, van Hoek AN, et al. Carbon dioxide permeability of aquaporin-1 measured in erythrocytes and lung of aquaporin-1 null mice and in reconstituted proteoliposomes. *J Biol Chem.* 2000;275:2686–2692
36 Song Y, Yang B, Matthay MA, et al. Role of aquaporin water channels in pleural fluid dynamics. *Am J Physiol.* 2000;279:C1744–C1750.
37 Yang B, Folkesson HG, Yang J, et al. Reduced water permeability of the peritoneal barrier in aquaporin-1 knockout mice. *Am J Physiol.* 1999;276:C76–C81.
38 Ma T, Jayaraman, S, Wang, KS, et al. Defective dietary fat processing in transgenic mice lacking aquaporin-1 water channels. *Am J Physiol.* 2000;280:C126–C134.
39 Thiagarajah JR, Verkman AS. Aquaporin deletion in mice reduces corneal water permeability and delays restoration of transparency after swelling. *J Biol Chem.* 2001;277:19139–19144.
40 Kuang K, Yiming M, Wen Q, et al. Fluid transport across cultured layers of corneal endothelium from aquaporin-1 null mice. *Exp Eye Res.* 2004;78:791–798.
41 Condeelis J. Life at the leading edge: the formation of cell protrusions. *Annu Rev Cell Biol.* 1993;9:411–444.
42 Lauffenburger DA, Horwitz AF. Cell migration: a physically integrated molecular process. *Cell.* 1996;84:359–369.

79

Ion Channels in Vascular Endothelium

Xiaoqiang Yao

Faculty of Medicine, Chinese University of Hong Kong

Among the innumerable proteins that support life in its many forms, ion channels are some of the most fascinating, partly for what they do, and partly for how they do it. Endothelial cells (ECs) express a great variety of ion channels. The most obvious function of these channels is to sense chemical and mechanical stimuli, by which they are activated, and to transform these stimuli into changes in ion fluxes. Simply put, in response to stimuli, the ion channels undergo an extremely rapid conformational change that converts an impermeable structure into highly permeable holes in the membrane through which ions can pass.

It appears that the main functions of these channels are to control two vital parameters: Calcium (Ca^{2+}) influx and membrane potential (Figure 79.1). ECs generally are regarded as nonexcitable. Little functional role exists for voltage-gated Ca^{2+} channels such as the L-type and T-type Ca^{2+} channels in ECs; these channels play more important roles in myocytes and neurons. In ECs, Ca^{2+} enters the cells via nonselective cation channels. Ca^{2+} influx results in a rise in cytosolic Ca^{2+} level ($[Ca^{2+}]_i$), which in turn controls the production and/or release of numerous vasoactive agents from endothelium. These agents include nitric oxide (NO), endothelium-derived hyperpolarizing factor (EDHF), vasodilatory and vasoconstrictive prostaglandins, endothelins, and tissue-type plasminogen activator (t-PA), and they serve to regulate multiple vascular functions, such as vessel tone, vascular permeability, blood coagulation, and EC growth. In addition to controlling the endothelial $[Ca^{2+}]_i$, endothelial ion channels, especially potassium (K^+) and chloride (Cl^-) channels, also act to maintain endothelial membrane potential. Membrane potential, together with the transmembrane concentration gradient for Ca^{2+}, provides the electrochemical driving force for transmembrane Ca^{2+} influx into ECs. Furthermore, endothelial membrane hyperpolarization may spread to underlying vascular smooth muscle layers through myoendothelial gap junctions and directly hyperpolarize the smooth muscle, causing vascular relaxation (1).

Endothelial ion channels originally were studied by measuring radioisotope (such as $^{45}Ca^{2+}$ and $^{86}Rb^+$) uptake and, more recently, with the patch clamp technique and by measuring changes in the fluorescence of ion-sensitive indicator dyes. Rapid developments in molecular biology have led to the cloning of a large number of ion channels, many of which are found to be expressed in ECs. However, it still remains difficult to directly correlate a cloned channel with endogenous channel activity recorded from ECs. This is attributable, in part, to a lack of specific compounds with which to block or activate the different ion channels and especially different channel isoforms, as well as the inherent complexity of EC ion channels. Many important ion channels in ECs such as transient receptor potential (TRP) channels and K^+ channels are composed of several subunits, which can exist as homo- or heteromultimers. Unfortunately, the precise subunit composition and relative stoichiometry of many of these channels are still unclear. As a result, the electrophysiological and pharmacological properties of native channels and their regulation may be very different from those of the cloned channels, which are mostly studied in the form of functional homomultimeric channel complexes.

EVOLUTIONARY CONSIDERATIONS

Some of the most important channels in ECs are TRP, cyclic nucleotide-gated channels (CNG), inwardly rectifying K^+ channels (K_{ir}), Ca^{2+}-sensitive K^+ channels (K_{Ca}), adenosine triphosphate (ATP)-sensitive K^+ channels(K_{ATP}), voltage-gated Cl^- channels (ClC), and Ca^{2+}-sensitive Cl^- channels (CLCA) (Table 79-1). Most TRP channels and K^+ channels are composed of four principal subunits, each of which contains either six transmembrane segments (6-TMS) (TRP, K_{Ca}) or two transmembrane segments (2-TMS) (K_{ir}, K_{ATP}). The evolutionary precursor of these channels appears to be the basic building block of 2-TMS, which constitutes the entire protein in many prokaryotic channels. An additional 4-TMS is added to 2-TMS in the process of evolution to create 6-TMS channels (2). A likely ancestor for endothelial K^+ channels and cation channels (TRP, CNG) appears to be those channels akin to bacterial K^+ channel KcsA (3). Interestingly, the structure of Cl^- channels is very different from that of cation channels. ClC channels are found

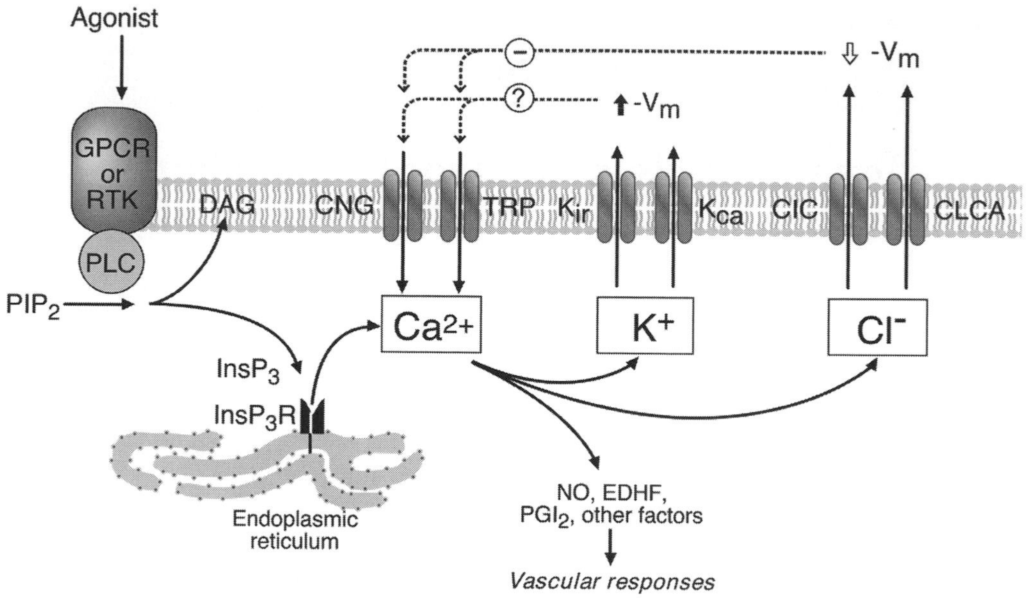

Figure 79.1. A general scheme showing the role of endothelial ion channels in the control of Ca^{2+} influx and membrane potentials. Agonists bind to G-protein–coupled receptors (GPCR) or tyrosine kinase–linked receptors (RTK), causing activation of phospholipase C (PLC), leading to the production of diacylglycerol (DAG) and $InsP_3$. DAG activates some TRP channels. $InsP_3$ binds to its receptor (InsP3R), resulting in store-Ca^{2+} release and store-operated Ca^{2+} influx via TRP channels. K^+ channels and Cl^- channels control membrane potential, which is the driving force for Ca^{2+} influx. Endothelial $[Ca^{2+}]_i$ regulates the production and release of vasoactive agents from endothelium, causing subsequent vascular responses.

in bacteria and contain two identical subunits. Each subunit consists of 18 helices, and together they form two topologically related domains, which span the membrane in opposite directions in an arrangement called *antiparallel architectures* (4).

Ca^{2+}-PERMEABLE NONSELECTIVE CATION CHANNELS

Early studies showed that binding of agonists such as bradykinin, histamine, ATP, and endothelin-1 to their membrane receptors stimulates the activity of Ca^{2+}-permeable nonselective cation channels (5,6) and thus enhances Ca^{2+} influx (5) in vascular ECs. The single channel conductance of these channels is between 20 and 30 picosiemen (pS) for monovalent cations and 4 to 12 pS for Ca^{2+}. In general, these channels do not discriminate between K^+ and sodium (Na^+), and the permeability ratio of $P_{Ca^{2+}}/P_{K^+,Na^+}$ varies from 0.03 to 2 (6).

It is widely accepted that, as in many other cell types, one of the predominant mechanisms for the above-mentioned agonist-induced Ca^{2+} influx in ECs is the capacitative Ca^{2+} entry (CCE), which was first proposed by Putney (7). According to this model, binding of agonists to their respective membrane receptors activates G-proteins, which in turn stimulate phospholipase C (PLC), leading to the generation of inositol 1,4,5-triphosphate ($InsP_3$) and diacylglycerol (DAG) (7,8). $InsP_3$ binds to its receptor ($InsP_3R$), causing Ca^{2+} release from intracellular Ca^{2+} stores within endoplasmic reticulum (8). The resultant decrease in Ca^{2+} content within endoplasmic reticulum triggers Ca^{2+} influx from the extracellular milieu. In addition to CCE, agonists also may stimulate Ca^{2+} entry in ECs via noncapacitative mechanisms.

The molecular identity of endothelial CCE is still not well understood. TRP channels, which were first cloned in *Drosophila melanogaster* (9), provide the molecular candidates for CCE. Tremendous advances in recent years have resulted in the isolation of at least 28 unique mammalian TRP homologues from human, mouse, rat, rabbit, and bovine tissues (10). These mammalian TRP homologues have been divided into six subfamilies: canonical (TRPC), vanilloid (TRPV), melastatin (TRPM), mucolipin (TRPML), polycystin (TRPP), and ankyrin transmembrane proteins (TRPA) (10). With the exception of some polycystins, all TRPs possess 6-TMS and a pore region between TMS5 and TMS6. Functional TRP channels are believed to be composed of four subunits (11). The charged residues in the putative TMS4 helix, which usually underlie voltage gating in voltage-gated channels, are not present in TRP. Vascular ECs at a minimum express TRPC, TRPV, TRPP, and TRPM.

Canonical Transient Receptor Potential

TRPC is the most extensively studied TRP. TRPC contains three highly conserved ankyrin domains at the NH_2 terminus and a TRP box, which contains a conserved stretch of 25 amino acids at its COOH terminal to TMS6. TRPC is activated

Table 79-1: Properties of EC Ion Channels

	Ions	*EC Expression*	*Activation*	*−/−*
Voltage-gated Ca^{2+} channels	Ca^{2+}	L-type? T-type?	Membrane depolarization	
Nonselective cation channels				
TRPC	Ca^{2+}, Na^+, K^+	All 7 isoforms	TRPC1-C7: CCE? TRPC3,C6,C7: DAG	*TRCP4*$^{-/-}$: impaired store-operated currents; impaired EC-dependent vascular smooth muscle cell relaxation; lack of thrombin-induced actin stress fiber formation; reduction in thrombin-induced permeability
TRPV	V1–V4: Ca^{2+}, Na^+, K^+ V5–V6: Ca^{2+}	TRPV1, V2, V4	TRPV1-V4: hot or warm temperature. TRPV1: vanilloids, anandamide, eicosanoid.TRPV4: 5′,6′-EET, osmotic swelling, shear stress.	
TRPM (except M4,M5)	Ca^{2+}, Na^+, K^+ M6,M7: Mg, Ca^{2+}, Na^+, K^+	TRPM3, M6–M8	TRPM2: ADP-ribose, NAD, H_2O_2 TRPM3: decrease in osmolarity.TRPM7: Mg-ATP, phosphorylation,TRPM8: Menthol, icillin, cool temperature	
TRPP	Ca^{2+}, Na^+, K^+	TRPP1, P2	Flow shear stress	
CNG	Ca^{2+}, Na^+, K^+	CNGA1, A2, A4	cGMP, cAMP	
Membrane Potential				
K channels				
K_{ir}	K^+	$K_{ir2.1,\,2.2,\,2.4}$	K, ATP	
K_{ca}	K^+	BK_{ca}, IK_{ca}, SK2, SK3	All K_{ca}: Ca BK_{ca}: membrane depolarization	Conditional transgenic mice: low SK3 expression enhances EC-dependent pressure- and phenylephrine-induced vascular constrictions, elevates blood pressure
K_{ATP}	K^+	$K_{ir6.1,\,6.2}$	Reduction in ATP	
Cl channels				
CLCA	Cl^-	CLCA1,2,5	Ca	
VRAC	Cl^-, HCO_3^-, NO_3^-	VRAC	Osmotic swelling, low ionic strength	
CLC	Cl^-	CLC2–7	All CLC: membrane depolarization CLC2, CLC3: osmotic swelling	

either by binding of agonists to plasma membrane G-protein–coupled receptors, leading to the activation of PLC-β, or by binding of growth factors to tyrosine kinase–linked receptors, resulting in the activation of PLC-γ (12,13). Activity of either PLC-β or PLC-γ hydrolyses the lipid precursor phosphatidylinositol-4,5-bisphosphate (PIP_2) to yield $InsP_3$ and diacylglycerol (12,13). Although it is well accepted that stimulation of PLC is essential for TRPC activation, the activation steps following PLC stimulation are diverse and sometimes controversial. The proposed mechanisms of TRPC activation include the capacitative mechanism mediated by $InsP_3$-$InsP_3R$ (12,14) and a noncapacitative mechanism that involves diacylglycerol (15). DAG is a product of PLC activity, and it can activate some TRPC isoforms (TRPC3, -C6, -C7), but it is unable to activate other isoforms (TRPC4, -C5) (15,16). Ca^{2+} may also modulate the activity of TRPC (12,17). A calmodulin (CaM) binding site is found in the COOH terminus of TRPC that overlaps with the $InsP_3R$ binding domain. The activity of TRPC also is regulated by protein phosphorylation. Protein kinase G and protein kinase C can both phosphorylate multiple TRPC isoforms, causing a decrease in channel activity (18,19), thereby providing negative feedback regulation to allow fine control for TRPC-mediated Ca^{2+} influx.

ECs have been shown to express seven isoforms of TRPC (as well as TRPV1, -2, and -4; all TRPM except TRPM5; and TRPP1, and -2) (1,6,20). Most of these data concerning the expression of TRP were collected from cultured ECs based on reverse transcription polymerase chain reaction (RT-PCR) assays (1,6,20). A recent study utilizing both in situ hybridization and immunohistochemistry confirmed the expression of all seven TRPC isoforms in ECs of different-sized human arteries (20). Unfortunately, there is still little information about the relative abundance of different TRPC isoforms in ECs in vivo. Moreover, the extent to which expression of the isoforms varies between different vascular beds remains unknown.

Functional studies show that TRPC1 mediates CCE in ECs (17,21). Transfection of an antisense oligonucleotide against TRPC1 effectively reduces *TRPC1* mRNA levels and, at the same time, decreases the CCE induced by the endoplasmic reticulum Ca^{2+}-ATPase inhibitor, thapsigargin, and it also reduces store-operated Ca^{2+} currents in human pulmonary artery ECs (21). Evidence also shows that TRPC1-mediated Ca^{2+} entry may be important for arrangement of the microfilamentous cytoskeleton (17). TRPC4 is another channel that may play a crucial role in vascular function. *TRPC4*-knockout ($TRPC4^{-/-}$) mice display impaired CCE and store-operated currents in aortic and lung ECs. Endothelium-dependent smooth muscle relaxation in response to the vasoactive agonists ATP and acetylcholine also is impaired in these mice (22). Furthermore, ECs of $TRPC4^{-/-}$ mice exhibit lack of actin stress fiber formation and cell retraction in response to thrombin (23), and the knockout mice demonstrate reduced thrombin-mediated lung microvascular permeability (23). TRPC1 also may participate in the control of vascular permeability. Inhibition of TRPC1 activity either by using an anti-TPC1 antibody or by interfering with the interaction of TRPC1 with $InsP_3R$, reduces the vascular endothelial growth factor (VEGF)-induced increase in transendothelial permeability (24). Evidence also suggests that TRPC is involved in oxidative stress-induced increases in membrane conductance. The oxidant tert-butylhydroperoxide (tBHP) activates cation conductance in porcine aortic ECs. The tBHP-induced currents resemble TRP-related currents in terms of cation selectivity, sensitivity to La^{3+} inhibition, and lack of voltage dependence. Expression of the N-terminal fragment of human TRPC3 (NTRP3), which interferes with the channel assembly, abolishes the oxidant-induced cation currents, thus suggesting that TRPC3 proteins may be the molecular basis of endothelial oxidant-activated cation channels (25). Taken together, these data provide compelling evidence for an important role of TRPC in regulating vascular tone and controlling vascular permeability.

Another important consideration for the physiological function of TRPC is the coassembly mechanism. Although TRPC is predicted to form tetrameric structures (11), the stoichiometry of the functional channels has not been determined. Several studies have established the ability of TRPC to coassemble to form heteromultimers (12,26). Coassembly could occur at least between TRPC1 and C3 (12); TRPC1 and -C4 and -C5 (12); and between the members within the same TRPC subgroups (e.g., TRPC4 and -C5 or TRPC3, -6, and -7) (26). Because multiple TRPC homologues are expressed within a single EC type (17,20), it is likely that ECs contain multiple forms of TRPC heterotetramers, which may confer on the ECs a variety of different Ca^{2+}-permeable channels with different properties. At present, however, no structural analysis data are available concerning the oligomeric states of endothelial TRPC channels. Resolving this puzzle is an important priority.

Vanilloid Transient Receptor Potential

At least three members of TRPV are expressed in vascular ECs: TRPV1, -V2, and -V4 (1). TRPV1 and -V4 are activated by endogenous vanilloids such as anandamide and 2-arachidonoyl-glycerol (2-AG), both of which have vascular effect in their own right (27,28). Functionally, TRPV1 is involved in endothelium-dependent dilation in rat mesenteric arteries (29). Another important property of TRPV4 is its mechanosensitivity. TRPV4 is activated by cell swelling or flow shear str ess (30). Cell swelling stimulates the phospholipase A_2 (PLA_2)-dependent formation of arachidonic acid, which, through its metabolic product 5′,6′-EET, activates TRPV4 (30,31).

All three EC TRPV channels are temperature-sensitive. TRPV4 is activated by warm temperature, with its low threshold at around 25°C to 27°C; therefore the channels may be constitutively open in ECs at body temperature, and this may contribute to the elevated intracellular Ca^{2+} concentration in native endothelium, causing a steady-state release of NO. TRPV1 and -V2 are activated at ≥43°C and ≥53°C, respectively (32), but these threshold values can be modulated by chemical ligands and the phosphorylation state of the channels. This plasticity potentially may confer a broad range of temperature sensitivity to cells expressing TRPV1 (32). The temperature sensitivity of TRPV channels may contribute to peripheral vasoconstriction at low temperature and vasodilation at high temperature.

Polycystin Transient Receptor Potential

The best-known TRPP channels are TRPP1 (PKD1) and TRPP2 (PKD2). Mutations in either the TRPP1 or TRPP2 gene, which encodes respectively for polycystin 1 (PC1) and polcystin-2 (PC2) proteins, result in a common genetic disease known as autosomal dominant polycystic kidney disease. Both PC1 and PC2 are integral membrane proteins. PC1 contains a large extracellular NH_2 terminus, and the proteins act as a G-protein–coupled receptor. PC2 shares sequence similarity with the TRP channel family and functions as a Ca^{2+}-permeable cation channel. In renal epithelial cells, PC1 and PC2 physically interact with each other to form a mechanosensor to detect fluid flow (33). Both PC1 and PC2 are required for flow-induced Ca^{2+} influx in renal epithelial cells (33).

TRPP1 and TRPP2 are found to be expressed in cultured vascular ECs by immunohistochemical studies (1,34). In cultured human umbilical vein ECs, TRPP1 is localized to the lateral membrane of cells in contact with one another, whereas no expression is found at the apical and basal membrane, suggesting a role of polycystin proteins in cell differentiation and cell–cell interactions. In addition, the fact that TRPP1 acts as the mechanosensor in renal epithelial cells to sense flow shear stress also raises the possibility that TRPP channels may play a similar role in vascular ECs to sense flow shear stress.

Cyclic Nucleotide-Gated Channels

At least six CNG-related genes have been cloned. CNGA1 through -A3 are referred to as principal subunits, because they can form functional channels on their own. CNGA4, -B1, and -B3 are referred to as modulatory subunits, because unlike the A1 to A3 subunits, they do not form functional channels on their own (35). ECs express at least CNGA1, -A2, and -A4 (36,37). Inhibition of CNGA2 channels reduces the capacitative Ca^{2+} entry as well as thapsigargin-induced cation currents in pulmonary artery ECs (36).

THE ION CHANNELS CONTROLLING MEMBRANE POTENTIAL

The measured resting membrane potential of vascular ECs varies considerably between –10 mV to –80 mV due to differences in seal resistances, culture conditions, and interspecies and arterial–venous variations (6,38). The endothelium of the intact rat aorta has an average membrane potential of –58 mV (6). K^+ and Cl^- channels are the two most important classes of channels controlling endothelial membrane potential. ECs in culture consist of two population of cells, one with a resting potential between –70 and –60 mV that is mainly determined by inwardly rectifying K^+ channels (K^+-type ECs) and the other with resting potential between –40 mV to –10 mV that is controlled by Cl^- conductance (Cl^--type ECs) (6).

Potassium Ion Channels

Inwardly Rectifying Potassium Ion Channels

K_{ir} is considered to be the major determinant for the resting membrane potential of ECs (6). The predominant K_{ir}s expressed in ECs are $K_{ir2.1}$, but $K_{ir2.2}$ also may have an important role in resting conditions (39). K_{ir} channels conduct large inward currents at potentials more negative than the K^+ equilibrium potentials (E_K), but permit very small outward currents at potentials slightly more positive than E_K (6,38). Strong inward rectification of $K_{ir2.1}$ reportedly is due to the blockage of outward currents by cytoplasmic Mg^{2+} and polyamine, which is known to be determined partly by the negatively charged amino acid residues Asp172, Glu224, and Glu299. K_{ir} currents are blocked by external divalent cations with $Ba^{2+} >> Sr^{2+} >> Mg^{2+} > Mn^{2+} = Ca^{2+}$ (38). K_{ir} may also act as a K^+ sensor. A small increase in extracellular K^+, such as that resulting from the activation of large-conductance K_{Ca} channels, may shift the reversal potential of K_{ir} toward a more positive value, thereby increasing the conductance of the channels (6).

Functionally, endothelial K_{ir} is inhibited by vasoactive agonists such as angiotensin II, vasopressin, vasoactive intestinal polypeptide, and histamine (6,38). K_{ir} also seems to be metabolically regulated. ATP and channel phosphorylation appear to be essential for maintaining the activity of channels (6). Reduction of intracellular ATP and induction of hypoxic conditions dramatically downregulate $K_{ir2.1}$. More recently, it has been shown that $K_{ir2.1}$ expressed in *Xenopus* oocytes and HEK293 can be activated by flow shear stress via a tyrosine phosphorylation–dependent mechanism, suggesting that $K_{ir2.1}$ channels in ECs can act as a mechanotransducer in response to flow shear stress (38).

Calcium Ion-Sensitive Potassium Ion Channels

The influx of Ca^{2+} in stimulated ECs exerts a depolarizing effect that is offset by activation of a variety of K_{Ca} channels. K_{Ca} channels can be classified into large (BK_{Ca}), intermediate (IK_{Ca}), and small conductance channels (SK_{Ca}) based on their single channel conductance. All K_{Ca} channels are comprised of four functional α-subunits, each of which exhibits six transmembrane segments and an intracellular COOH and NH_2 terminus. BKca, IKca, and at least two SK_{Ca} channels (SK2, SK3) are found to be expressed in ECs in vivo (6,40). BK_{Ca} has a conductance between 165 and 220 pS, and is activated by the concerted actions of cytosolic Ca^{2+} and membrane depolarization. The channel is blocked by tetraethylammonium, charybdotoxin, iberiotoxin, and D-turbocurarine. The benzimidazolone compounds NS004 and NS1619 increase the activity of BK_{Ca}. Unlike BK_{Ca}, the activity of IK_{Ca} and SK_{Ca} is not voltage-dependent. Instead, they are activated solely by cytosolic Ca^{2+}, with a unit conductance of 2 to 20 and 20 to 85 pS, respectively, and they are more sensitive to Ca^{2+} than is BK_{Ca}. IK_{Ca} is blocked by charybdotoxin, clotrimazole, and tetrabutylammonium, but is not affected by iberiotoxin (41). IK_{Ca} is activated by $InsP_3$-sensitive Ca^{2+} release (6). SK_{Ca} is blocked by tetrabutylammonium, apamin, and clotrimazole.

Functionally, it has been shown that SK_{Ca} and IK_{Ca} in ECs play a pivotal role in the endothelium-dependent smooth muscle hyperpolarization. K^+ efflux via IK_{Ca} and SK_{Ca} may activate K^+-sensitive K_{ir} channels in smooth muscle cells and induce hyperpolarization (41). A recent study, using a transgenic mouse in which SK3 expression levels can be manipulated using dietary components, provides compelling evidence that SK3 in ECs exerts a profound and tonic hyperpolarizing influence in resistance arteries (40).

ATP-Sensitive K^+ Channels

K_{ATP} is believed to exist as an octameric structure composed of four α-subunits and four β-subunits. The α-subunits belong to the inward rectifier family ($K_{ir6.1/6.2}$), whereas the β-subunits (SUR1, SUR2A, SUR2B) are the site of sulfonylurea binding. The activity of K_{ATP} channels increases as the concentration of intracellular ATP or glucose availability decreases (38). K_{ATP} is blocked by tetraethylammonium, the sulfonylurea drugs glibenclamide and tolbutamide, and an elevation of intracellular pH. Cromakalim, pinacidil, and diazoxide can increase the activity of K_{ATP}. The role of K_{ATP} in ECs has attracted significant interest because it has the potential to provide a mechanism for coupling blood flow to the metabolic requirements of the tissue. It is possible that a high metabolic rate decreases intracellular ATP, which in turn activates K_{ATP}, causing membrane hyperpolarization and smooth muscle relaxation.

Cl^- Channels

ECs express volume-sensitive anion channels (VRAC). These channels are mainly permeable to Cl^- under physiological conditions, but amino acids and organic osmolytes also may permeate through the channels. The single-channel conductance of VRAC is approximately 40 to 50 pS at positive potentials and 10 to 20 pS at negative potentials, indicating that the channels are outward rectifiers. The channels are inhibited by the classical Cl^- channel blockers DIDS, SITS, 9-AC, and niflumic acid. Cell shrinkage reduces the basal current in ECs below its values in control conditions, indicating that VRAC is partially activated in resting cells and therefore important for the resting membrane potential. On the other hand, cell swelling or low intracellular ionic strength may activate tyrosine kinase, leading to the activation of VRAC (6). The molecular nature of the VRAC channel is still unknown. Several putative extracellular candidates have been proposed, including P-glycoprotein, pIcln, and ClC3, but all these candidates have been questioned (6).

ClC channels are voltage-gated Cl^- channels and many of them (ClC2 through ClC7) are expressed in ECs (42). CLCA channels are Ca^{2+}-dependent Cl^- channels and at least CLCA1, -2, and -5 are expressed in ECs (43). Although all CLCA channels show strong outward rectification, the current–voltage relationship may vary among different channel subtypes. Some ClC channels such as ClC2 and ClC3 are volume-sensitive, and thus may play a role in volume regulation (6). Different Cl^- channels regulate endothelial anion efflux under different conditions.

FUNCTIONAL ROLE OF ENDOTHELIAL ION CHANNELS

Vascular Tone Control

One of the most important functions of endothelial ion channels is to regulate vascular tone. Vascular ECs are exposed to circulating blood that contains numerous Ca^{2+}-mobilizing agents including metabolites, local paracrine agents, neurotransmitters, growth factors, and cytokines. Some of these factors, such as serotonin, ATP, and adenosine diphosphate (ADP), are released from platelets and erythrocytes into circulating blood. Others are mostly produced in ECs themselves, such as substance P, ATP, vasopressin, angiotensin II, histamine, and endothelin (44). These compounds bind to their respective membrane receptors and elicit a rise of endothelial $[Ca^{2+}]_i$ via $InsP_3$-mediated store-Ca^{2+} release, and via capacitative and noncapacitative Ca^{2+} influx. Examples of noncapacitative Ca^{2+} influx include diacylglycerol-mediated activation of TRPC3, -C6, and -C7 (15). The rise in endothelial $[Ca^{2+}]_i$ also may be elicited by some other physical and chemical stimuli, such as flow shear stress, cell swelling, and oxidative stress.

A rise in $[Ca^{2+}]_i$ triggers the ECs to produce and release vasorelaxants such as NO, EDHF, and prostacyclin, causing a subsequent reduction in vascular tone. Basal release of NO from ECs appears to be a Ca^{2+}-independent process, but a rise in $[Ca^{2+}]_i$ is certainly able to stimulate endothelial NO synthase, thereby increasing the production and release of NO in ECs (45). A rise in $[Ca^{2+}]_i$ may also stimulate PLA_2, an enzyme required for prostacyclin production. EDHF is another relaxing factor, the production and release of which is Ca^{2+}-dependent. The chemical identity of EDHF is not fully resolved. Candidates for EDHF may vary with vascular beds and vessel sizes. The proposed candidates for EDHF include the cytochrome P450 metabolites of arachidonic acid such as 5,6-EET, K^+ ions released from ECs due to activity of K_{Ca}, and hydrogen peroxide (5,41,44).

TRP and CNG channels play pivotal roles in endothelial Ca^{2+} signaling, because they allow Ca^{2+} to enter ECs in response to various stimuli. K^+ and Cl^- channels are also important, because they maintain endothelial membrane potential, which is one of the driving forces for transmembrane Ca^{2+} influx. EC hyperpolarization also may spread to underlying vascular smooth muscle layers through gap junctions, directly causing smooth muscle hyperpolarization and subsequent vascular relaxation (6,41). At present, at least two channels, TRPC4 and SK3, have been convincingly demonstrated to play critical roles in the control of vascular tone and blood pressure. Freichel and colleagues showed that knocking out the *TRPC4* gene impairs endothelium-dependent vascular relaxation (22), and Taylor and colleagues demonstrated that transgenic mice in which SK3 expression is suppressed by dietary doxycycline display an enhanced vascular constriction and a pronounced elevation of blood pressure (40).

Blood Coagulation

ECs also release factors that influence blood coagulation. NO, prostacyclin, t-PA, and tissue factor pathway inhibitor (TFPI) function as natural anticoagulants, whereas von Willebrand factor (vWF), platelet activating factor (PAF), and plasminogen activator inhibitor (PAI)-1 serve as procoagulants. The release of these factors from ECs is Ca^{2+}-dependent. Many of these factors, such as t-PA, TFPI, vWF, are stored in granules and then released by Ca^{2+}-dependent exocytosis.

Control of Vascular Permeability

Inflammatory mediators such as thrombin and histamine induce vascular leakage, which is defined as increased endothelial permeability to plasma proteins and other solutes. These inflammatory mediators induce an $[Ca^{2+}]_i$ increase in ECs. The rise in $[Ca^{2+}]_i$ activates key signaling pathways, which lead to myosin light chain–dependent EC contraction and disassembly of vascular endothelial (VE)-cadherin at the adherens junctions, causing a subsequent increase in vascular permeability. Evidence shows that TRPC1 and -C4 channels play important roles in the control of vascular permeability (23,24). In *TRPC4*$^{-/-}$ mice, thrombin-induced increase in vascular permeability was significantly impaired (23).

Endothelial Cell Growth and Angiogenesis

Angiogenesis inducers such as VEGF, basic fibroblast growth factor (bFGF), platelet-derived growth factor, and others can stimulate the proliferation and migration of ECs, and regulate the formation of new blood vessels. Evidence shows that VEGF, by binding to its receptors, activates a signaling transduction cascade involving autophosphorylation, activation of PLC-γ, and induction of CCE. TRP channels, especially TRPC1, TRPC6, and VRAC, appear to be particularly important for Ca^{2+} influx induced by angiogenesis factors (6,24,46). Blockers of VRAC such as tamoxifen, 4,4′ diisothiocyanatostilbene-2,2′-disulfonic acid (DIDS), and flufenamic acid, suppress angiogenesis (6). The mechanism of action for VRAC in angiogenesis is still unknown, but it may be linked to the role of VRAC in volume regulation or in the modulation of endothelial Ca^{2+} levels, which is important for cell cycle control.

Mechanosensing

Blood flow exerts a viscous drag, or shear stress, on the surface of the vascular ECs. Furthermore, pulsatile stretch on the vascular wall stretches the cell membrane of the ECs. In response, ECs undergo diverse biochemical and physiological changes, including the release of vasodilators and alterations of gene expression. Many of these responses depend on mechanosensitive Ca^{2+} influx into the ECs. ECs express several flow shear stress–activated channels including TRPV4 (28), the complex of TRPP1 and TRPP2 (1), and $K_{ir2.1}$ (6). Several stretch-activated channels also are expressed in ECs. These include TRPC1 and TRPV2, which are directly activated by membrane stretch, and TRPM4 and TRPC6, which are indirectly activated by endogenous ligands produced after membrane stretch. However, note that the mechanosensitivity of these channels has only been demonstrated in other cell types, and that no experimental evidence directly links any of these channels to the global increase of $[Ca^{2+}]_i$ in ECs.

KEY POINTS

- Ion channel controls endothelial Ca^{2+} influx, thereby influencing diverse vascular functions such as vascular tone, vascular permeability, secretion, EC proliferation and angiogenesis.
- TRP channels (except TRPM4 and -M5) and CNG channels allow Ca^{2+} influx in ECs.
- K^+ channels and Cl^- channels control membrane potential, which is one of the driving force for Ca^{2+} influx.
- Experiments with transgenic mice provide compelling evidence for crucial roles of several ion channels such as TRPC4 and SK3 in the regulation of vascular permeability and vascular tone.

Future Goals

- To develop selective drugs targeting different isoforms of TRP channels, K^+ channels and Cl^- channels
- To elucidate the mechanism of regulation for endothelial ion channels
- To resolve the oligomeric states of endothelial TRP channels

REFERENCES

1 Yao X, Garland CJ. Recent developments in vascular endothelial cell transient receptor potential channels. *Circ Res.* 2005;97: 853–863.
2 Anderson PAV, Greenberg RM. Phylogeny of ion channels: clues to structure and function. *Comp Biochem Physiol B.* 2001;129: 17–28.
3 Lu Z, Klem AM, Ramu Y. Ion conduction pore is conserved among potassium channels. *Nature.* 2001;413:809–813.
4 Jentsch TJ, Stein V, Weinreich F, Zdebik AA. Molecular structure and physiological function of chloride channels. *Physiol Rev.* 2002;82:503–568.

5 Adams DJ, Barakeh J, Laskey R, van Breemen C. Ion channels and regulation of intracellular calcium in vascular endothelial cells. *FASEB J*. 1989;3:2389–2400.
6 Nilius B, Droogman G. Ion channels and their functional role in vascular endothelium. *Physiol Rev*. 2001;81:1415–1459.
7 Putney JW Jr. A model for receptor-regulated calcium entry. *Cell Calcium*. 1986;7:1–12.
8 Parekh AB, Penner R. Store depletion and calcium influx. *Physiol Rev*. 1997;77:901–930.
9 Montell C, Rubin GM. Molecular characterization of the Drosophila trp locus: a putative integral membrane protein required for phototransduction. *Neuron*. 1989;2:1313–1323.
10 Moran MM, Xu H, Clapham DE. TRP ion channels in the nervous system *Curr Opin Neurobiol*. 2004;14:362–369.
11 Birnbaumer L, Zhu X, Jiang M, et al. On the molecular basis and regulation of cellular capacitative calcium entry: roles for Trp proteins. *Proc Natl Acad Sci USA*. 1996;93:15195–15202.
12 Minke B, Cook B. TRP channel proteins and signal transduction. *Physiol Rev*. 2002;82:429–472.
13 Venneken R, Voets T, Bindels RJM, et al. Current understanding of mammalian TRP homologues. *Cell Calcium*. 2002;31:253–264.
14 Kiselyov K, Xu X, Mozhayeva G, et al. Functional interaction between InsP3 receptors and store-operated Htrp3 channels. *Nature*. 1998;396:478–482.
15 Hofmann T, Obukhov AG, Schaefer M, et al. Direct activation of human TRPC6 and TRPC3 channels by diacylglycerol. *Nature*. 1999; 397:259–263.
16 Plant TD, Schaefer M. TRPC4 and TRPC5: receptor-operated Ca^{2+}-permeable nonselective cation channels. *Cell Calcium*. 2003;33:441–450.
17 Nilius B, Droogman G, Wondergem R. Transient receptor potential channels in endothelium: solving the calcium entry puzzle. *Endothelium*. 2003;10:5–15.
18 Kwan HY, Huang Y, Yao X. Regulation of canonical transient receptor potential isoform 3 (TRPC3) channel by protein kinase G. *Proc Natl Acad Sci USA*. 2004;101:2625–2630.
19 Trebak M, Hempel N, Wedel BJ, et al. Negative regulation of TRPC3 channels by protein kinase C-mediated phosphorylation of serine 712. *Mol Pharmacol*. 2005;67:558–563.
20 Yip H, Chan WY, Leung PC, et al. The expression of TRPC homologs in endothelial cells and smooth muscle layers of human arteries. *Histochem Cell Biol*. 2004;122:553–561.
21 Brough GH, Wu S, Cioffi D, et al. Contribution of endogenously expressed Trp1 to a Ca^{2+}-selective, store-operated Ca^{2+} entry pathway. *FASEB J*. 2001;15:1727–1738.
22 Freichel M, Suh SH, Pfeifer A, et al. Lack of an endothelial store-operated Ca^{2+} current impairs agonist-dependent vasorelaxation in $TRP4^{-/-}$ mice. *Nat Cell Biol*. 2001;3:121–127.
23 Tiruppathi C, Minshall RD, Paria BC, et al. Role of Ca signaling in the regulation of endothelial permeability. *Vascul Pharmacol*. 2003;39:173–185.
24 Jho D, Mehta D, Ahmmed G, et al. Angiopoietin-1 opposes VEGF-induced increase in endothelial permeability by inhibiting TRPC1-dependent Ca2 influx. *Circ Res*. 2005;96:1282–1290.
25 Groschner K, Rosker C, Lukas M. Role of TRP channels in oxidative stress. *Novartis Found Symp*. 2004;258:222–230.
26 Hofmann T, Schaefer M, Schultz G, Gudermann T. Subunit composition of mammalian transient receptor potential channels in living cells. *Proc Natl Acad Sci USA*. 2002;99:7461–7466.
27 Ross RA. Anandamide and vanilloid TRPV1 receptors. *Br J Pharmacol*. 2003;140:790–801.
28 Nilius B, Vriens J, Prenen J, et al. TRPV4 calcium entry channel: a paradigm for gating diversity. *Am J Physiol*. 2004;286:C195–C205.
29 Domenicali M, Ros J, Fernandez-Varo G, et al. Increased anandamide induced relaxation in mesenteric arteries of cirrhotic rats: role of cannabinoid and vanilloid receptors. *Gut*. 2005;54:522–527.
30 Watanabe H, Vriens J, Prenen J, et al. Anandamide and arachidonic acid use epoxyeicosatrienoic acids to activate TRPV4 channels. *Nature*. 2003;424:434–438.
31 Vriens J, Watanabe H, Janssens A, et al. Cell swelling, heat, and chemical agonists use distinct pathways for the activation of the cation channel TRPV4. *Proc Natl Acad Sci USA*. 2004;101:396–401.
32 Tominaga M, Caterina MJ. Thermosensation and pain. *J Neurobiol*. 2004;61:3–12.
33 Nauli SM, Alenghat FJ, Luo Y, et al. Polycystins 1 and 2 mediate mechanosensation in the primary cilium of kidney cells. *Nature Genet*. 2003;33:129–137.
34 Ibraghimov-Beskrovnaya O, Dackowski WR, Foggensteiner L, et al. Polycystin: in vitro synthesis, in vivo tissue expression, and subcellular localization identifies a large membrane-associated protein. *Proc Natl Acad Sci USA*. 1997;94:6397–6402.
35 Kaupp UB, Seifert R. Cyclic nucleotide-gated ion channels. *Physiol Rev*. 2002;82:769–824.
36 Cioffi DL, Wu S, Stevens T. On the endothelial cell I_{soc}. *Cell Calcium*. 2003;33:323–336.
37 Yao X, Leung PS, Kwan HY, et al. Rod-type cyclic nucleotide-gated cation channel is expressed in vascular endothelium and vascular smooth muscle cells. *Cardiovasc Res*. 1999;41:282–290.
38 Adams DJ, Hill MA. Potassium channels and membrane potential in the modulation of intracellular calcium in vascular endothelial cells. *J Cardiovasc Electrophysiol*. 2004;15:598–610.
39 Fang Y, Schram G, Romanenko VG, et al. Functional expression of Kir2.x in human aortic endothelial cells: the dominant role of Kir2.2. *Am J Physiol Cell Physiol*. 2005;289:C1134–C1144.
40 Taylor MS, Bonev AD, Gross TP, et al. Altered expression of small-conductance Ca^{2+}-activated K^+ (SK3) channels modulates arterial tone and blood pressure. *Circ Res*. 2003;93:124–131.
41 Busse R, Edwards G, Feletou M, et al. EDHF: bringing the concepts together. *Trends Pharmacol Sci*. 2002;23:374–380.
42 Davies N, Akhtar S, Turner HC, et al. Chloride channel gene expression in the rabbit cornea. *Mol Vis*. 2004;10:1028–1037.
43 Abdel-Ghany M, Cheng HC, Elble RC, et al. The interacting binding domains of the beta(4) integrin and calcium-activated chloride channels (CLCAs) in metastasis. *J Biol Chem*. 2003;278:49406–49416.
44 Himmel HM, Whorton AR, Strauss HC. Intracellular calcium, currents, and stimulus-response coupling in endothelial cells. *Hypertension*. 1993;21:112–127.
45 Dimmeler S, Fleming I, Fisslthaler B, et al. Activation of nitric oxide synthase in endothelial cells by Akt-dependent phosphorylation. *Nature*. 1999;399:601–605.
46 Antoniotti S, Lovisolo D, Fiorio Pla A, Munaron L. Expression and functional role of bTRPC1 channels in native endothelial cells. *FEBS Lett*. 2002;510:189–195.

80

Regulation of Angiogenesis and Vascular Remodeling by Endothelial Akt Signaling

Ichiro Shiojima and Kenneth Walsh

Whitaker Cardiovascular Institute, Boston University School of Medicine, Boston, Massachusetts

Since the initial identification of several classes of receptor tyrosine kinases and their ligands as crucial mediators of vascular development, considerable progress has been made toward understanding the process of angiogenesis at sites of tissue growth and/or repair (1,2). A number of clinical trials are currently evaluating angiogenic ligands for their ability to induce neovascularization in ischemic tissues (3,4), and the intracellular signaling pathways that mediate the proangiogenic effects of these growth factors are being extensively investigated. Among a number of signaling pathways activated by angiogenic growth factors, the phosphoinositide 3-kinase (PI3K)/Akt pathway is of particular interest because it regulates downstream target molecules that are potentially involved in blood vessel growth and homeostasis, and thus seems to be a major mediator that couples "inputs" with "outputs" in the endothelium.

PI3K-AKT SIGNALING AXIS: UPSTREAM ACTIVATORS AND DOWNSTREAM TARGETS

Akt originally was identified as a cellular counterpart of the oncogene derived from the murine AKT8 retrovirus; the same gene product was independently isolated as a protein kinase related to protein kinase A and C and was therefore named protein kinase B (PKB) or RAC (related to protein kinase A and C) (5–8). Mammalian genomes contain three *Akt* genes, *Akt1/PKBα*, *Akt2/PKBβ*, and *Akt3/PKBγ*, whereas *Drosophila melanogaster* and *Caenorhabditis elegans* contain one and two *Akt* genes, respectively (9,10). These genes encode proteins containing a pleckstrin homology (PH) domain in the amino terminus, a central kinase domain, and a carboxy terminal regulatory domain. All three mammalian *Akt* genes are widely expressed in various tissues, but Akt1 is most abundant in brain, heart, and lung, while Akt2 is expressed primarily in skeletal muscle and embryonic brown fat, and Akt3 is predominantly expressed in brain, kidney, and embryonic heart (9,10). In unstimulated cells, Akt protein exists in cytoplasm, and the two regulatory phosphorylation sites (T308 and S473 in the case of Akt1) are in an unphosphorylated state. Upon growth factor stimulation, the PH domain binds to the lipid products of PI3K, and Akt is recruited to the plasma membrane. Akt is then sequentially phosphorylated at T308 and S473 by upstream kinases referred to as 3-phosphoinositide-dependent protein kinase 1 (PDK1) and PDK2, respectively, to yield a fully activated kinase (Figure 80.1) (11). PDK1 has been isolated and characterized, but the identity of PDK2 is still controversial. Several candidate molecules have been suggested to be a potential S473-kinase, including integrin-linked kinase (ILK) and Akt itself (autophosphorylation) (12). Fully activated Akt becomes available to phosphorylate its downstream substrates, and a portion of these molecules detach from the plasma membrane and translocate to various subcellular locations including the nucleus. Akt then is dephosphorylated and inactivated by protein phosphatases such as protein phosphatase 2A (PP2A).

Akt is a critical regulator of PI3K-mediated cell survival (9). A large number of studies have demonstrated in various cell types that constitutive activation of Akt signaling is sufficient to block cell death induced by a variety of apoptotic stimuli, and that transduction of dominant-negative Akt inhibits growth factor–induced cell survival. Several downstream targets of Akt are recognized to be apoptosis-regulatory molecules including Bad, the FOXO family of forkhead transcription factors, IKKα, MDM2, and Yes-associated protein (YAP), and these findings are consistent with the notion that Akt functions as a survival kinase (13). However, other downstream effectors of Akt are involved in different aspects of cellular regulation. For example, Akt (a) mediates the metabolic effects of insulin by inducing membrane translocation of the glucose transporter GLUT4 or through the phosphorylation and inactivation of glycogen synthase kinase (GSK)-3 (14); (b) regulates cell cycle and cellular senescence, at least in part, through modulating the activities of downstream

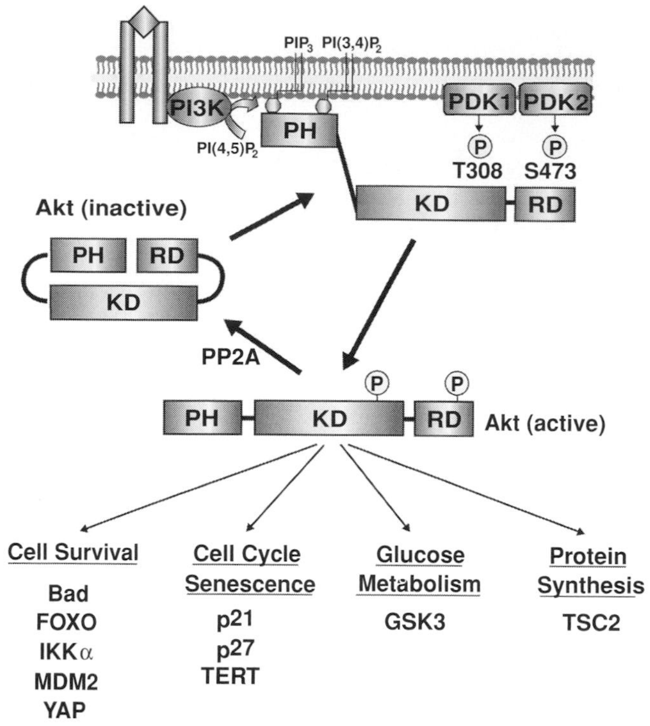

Figure 80.1. Mechanism of Akt activation and partial list of downstream molecules. Akt is activated by growth factors or cytokines in a PI3K-dependent manner, and phosphorylation of two residues by PDK1 (T308) and PDK2 (S473) is required for its full activation. Downstream target molecules are grouped according to their function.

substrates such as $p21^{cip1/waf1}$, $p27^{kip1}$, and human telomerase reverse transcriptase subunit (hTERT) (15,16); and (c) enhances protein synthesis by activating the mammalian target of rapamycin (mTOR) pathway through phosphorylation/inactivation of the tuberous sclerosis complex gene product TSC2 (see Figure 80.1) (17). Taken together, it would be more appropriate to classify Akt as a multifunctional protein kinase rather than a simple regulator of cell survival.

The functions of Akt gene products also have been studied in vivo using transgenic or knockout mice (Table 80-1). Overexpression of Akt1 in the heart results in cardiac hypertrophy (18–21), and Akt1 overexpression in pancreatic β-cells leads to both hypertrophy and hyperplasia of β-cells (22,23), consistent with the idea that Akt promotes cell growth and proliferation. Expression of Akt1 in T lymphocytes resulted in the development of lymphomas (24,25), and prostate cancer formation was observed in transgenic mice overexpressing Akt1 in prostate epithelial cells (26). However, Akt1 overexpression in mammary glands failed to induce tumor formation (27–30). Thus, sustained Akt activation is sufficient to induce malignant transformation in some, but not all cell types. Targeted disruption of the *Akt1* gene in mice results in a growth retardation phenotype (31,32). *Akt2*-null mice exhibit insulin resistance and, depending on genetic background, mild growth retardation (33,34). In contrast, *Akt3*-null mice do not show the general growth retardation phenotype, but brain size and weight are reduced in adult *Akt3*-null mice due to the reduction in both cell size and cell number (35,36). Mutual compensations among three *Akt* gene products also were demonstrated by combined deletion of these genes. *Akt1/Akt2* double mutant mice die shortly after birth due to defects in multiple tissues including skeletal muscle, bone, and skin (37), and *Akt1/Akt3*-double knockout mice die in utero (38). Thus, the phenotypes of these compound mutants are more severe than single knockout mice, pointing to the functional redundancies among three *Akt* genes in mammals.

Recently, the role of Akt signaling in the vascular endothelium was examined in transgenic and knockout mice. Endothelial cell (EC)-specific overexpression of an activated form of Akt1 resulted in embryonic lethality, associated with abnormal vascular remodeling and patterning. Transient expression of Akt in ECs also blocked hyperoxia-induced capillary clearing in retina (39). On the other hand, ischemia or vascular endothelial growth factor (VEGF)-induced angiogenesis and mobilization of endothelial progenitor cells (EPCs) were impaired in *Akt1*-null mice (40), and newly formed vessels in *Akt1*-null mice were immature and leaky (41). These data suggest that Akt signaling plays an important role in promoting angiogenesis and maintaining vascular homeostasis, and that fine-tuning of the Akt signal in ECs is required for normal vascular patterning and remodeling.

AKT-DEPENDENT SURVIVAL SIGNALS IN ENDOTHELIAL CELLS

Although originally identified as a factor that induces vascular permeability, VEGF exhibits multiple biological activities in ECs, including the enhancement of EC survival. This prosurvival effect of VEGF is mediated by the FLK-1/PI3K/Akt pathway (42–44). Subsequently it also was shown that several other EC stimuli including angiopoietin 1 (Ang1), insulin, insulin-like growth factor (IGF)-I, sphingosine-1-phosphate (S1P), hepatocyte growth factor (HGF), the small proteoglycan decorin, leptin, adiponectin, and fluid shear stress also activate PI3K–Akt signaling and inhibit apoptosis, illustrating the central role of this pathway in controlling EC viability (43,45–47).

Growth factor activation of angiogenesis depends on proper EC–extracellular matrix (ECM) attachment and, in the absence of matrix attachment, cells undergo apoptotic cell death through a process termed *anoikis* (a Greek word for "homelessness") (48). VEGF activation of Akt in ECs is dependent on matrix attachment, and constitutively active Akt blocks cell detachment–induced apoptosis (42). These findings suggest that matrix attachment is required for growth factors to activate Akt and maintain EC viability. Cell attachment is mediated mainly through the engagement of ECM with integrin molecules. When integrins bind to ECM, they

Table 80-1: Summary of Genetic Mouse Models

Gene	*TG/KO*	*Phenotype*	*References*
Akt1	TG (heart)	Cardiac hypertrophy, heart failure	18–21
	TG (β-cells)	Hypertrophy and hyperplasia of beta-cells	22, 23
	TG (T lymphocytes)	T-cell lymphoma	24, 25
	TG (prostate)	Prostate neoplasia	26
	TG (mammary glands)	Delayed involution, fatty milk synthesis	27–30
	TG (ECs)	Abnormal vascular patterning	39
Akt1	KO	Growth retardation, apoptosis in thymus	31, 32
Akt2	KO	Diabetes, mild growth retardation, lipoatrophy	33, 34
Akt3	KO	Reduced brain size	35, 36
Akt1/Akt2	Double KO	Perinatal lethality, severe growth retardation	37
Akt1/Akt3	Double KO	Embryonic lethality, placental defect	38

TG, transgenic; KO, knockout.

become clustered and associate with the actin cytoskeleton through adaptor/signaling molecules. This further promotes integrin clustering and the assembly of actin filaments and leads to the formation of focal adhesion and activation of intracellular signaling (49). The αv integrin combinations have been most extensively investigated in terms of their roles in angiogenesis (50). ECs stimulated with angiogenic growth factors or those in newly formed vessels express high levels of $\alpha v\beta 3$ integrin, and antagonists against $\alpha v\beta 3$ or $\alpha v\beta 5$ integrin block the growth factor–induced angiogenesis. It also has been shown that $\alpha v\beta 3$ integrin associates with VEGF and platelet-derived growth factor (PDGF) receptors and potentiates VEGF or PDGF signaling, respectively (50). Because several integrin signaling molecules, including focal adhesion kinase (FAK), ILK, and Shc, have been implicated in Akt activation (48), downregulation of Akt activity induced by cell detachment is likely caused by the decrease in integrin-dependent Akt activation. Caspase-mediated cleavage of Akt also is implicated in the downregulation of Akt protein levels during long-term suspension culture (51). Collectively, these findings suggest that integrin signaling induced by cell attachment (outside-in signal) is an important regulator of growth factor–dependent EC survival and angiogenesis through PI3K/Akt pathways. Furthermore, VEGF induction of inside-out signals has also been shown to activate integrins (52), suggesting that integrin and growth factor signaling are cooperative and synergistic with regard to activation of Akt signaling.

Currently, relatively little is known about the downstream mediators of the Akt-dependent survival pathway in ECs, although several candidate signaling molecules have been identified including survivin (53), death domain (FADD)–like interleukin (IL)-1-converting enzyme–inhibitory protein (FLIP) (54), mitogen-activated protein kinase (MAPK) kinase kinase 3 (MEKK3) (55), FOXO3a (56), GSK-3, and β-catenin (57,58). Thus, possible combinations of these and other unidentified Akt target molecules may control EC survival, depending on the context of pro- and antiapoptotic stimuli encountered in the cellular environment.

REGULATION OF ENDOTHELIAL NITRIC OXIDE SYNTHASE ACTIVITY BY AKT

In addition to its antiapoptotic effects, VEGF induces nitric oxide (NO) release from ECs. It was shown that this effect of VEGF is mediated by Akt-dependent endothelial NO synthase (eNOS) phosphorylation at Ser 1179 (in bovine eNOS, equivalent to Ser 1177 in human eNOS), which results in an increase in eNOS activity (59). It also was reported that production of NO in response to fluid shear stress in cultured ECs is controlled by Akt-dependent phosphorylation of eNOS (60), although another study has shown that shear stress induces eNOS phosphorylation at Ser 1179 predominantly through a protein kinase A (PKA)–dependent, Akt-independent mechanism (46). Subsequently, it was found that Ser 617 (in bovine eNOS) phosphorylation by Akt in part mediates shear stress–induced activation of eNOS (61). Studies in intact animals have shown that overexpression of constitutively active Akt in the vascular endothelium results in an increase in resting diameter and blood flow that is blocked by an eNOS inhibitor N-nitro-L-arginine methyl ester (L-NAME), and transduction of dominant negative Akt attenuates endothelium-dependent vasodilation induced by acetylcholine (62). These findings demonstrate that Akt functions as a regulator of vasomotor tone in vivo. Furthermore, a possible link between Akt and endothelial dysfunction in hypertension was suggested by a recent study showing that agonist-induced eNOS activation is reduced in spontaneously hypertensive rats (SHR), and this defect is normalized by *Akt* gene transfer into the endothelium (63).

The activity of eNOS also is regulated by subcellular localization and/or protein–protein interactions. Of note, it has been shown that eNOS is localized in a specific domain of the plasma membrane, the caveolae, and interacts with caveolin-1 through the caveolin-1 scaffolding domain, and that the interaction between eNOS and caveolin-1 results in the inhibition of eNOS activity (64,65). Although originally implicated in transmembrane trafficking of macromolecules, the findings that caveolae contain a variety of signaling molecules

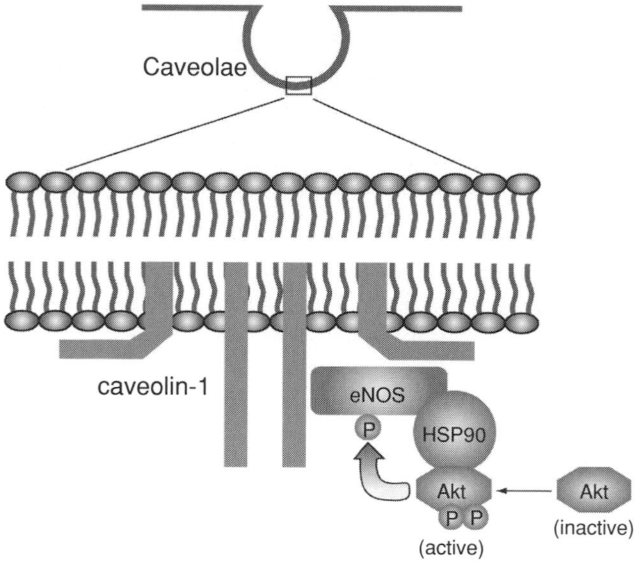

Figure 80.2. Schematic illustration of the Akt–eNOS interaction at caveolae. Caveolin-1 is localized to caveolae and associates with a number of regulatory molecules including eNOS. Association of eNOS with caveolin-1 negatively regulates eNOS activity, although targeting of eNOS to caveolae is required for proper eNOS function. Activated Akt and eNOS also associate with Hsp90. Hsp90 is believed to function as a scaffold protein for activation of eNOS by Akt-mediated phosphorylation.

and that caveolin-1 interacts directly with those caveolae-associated proteins has suggested that caveolae and caveolin-1 are involved in the compartmentalization and integration of signal transduction pathways at the cell membrane. Consistent with the inhibitory effect of caveolin-1 on eNOS activity, administration of the caveolin-1 scaffolding domain fused to cellular internalization sequences in vivo attenuates eNOS activity (65). Acetylcholine-induced vasorelaxation and NO production are enhanced in caveolin-1-deficient mice (66), and endothelial-specific overexpression of caveolin-1 impairs eNOS activation (67). The targeting of eNOS to caveolae, however, seems to be required for efficient and proper activation of eNOS upon stimulation, because conditions that inhibit the localization of eNOS in caveolae also attenuate eNOS activity (68). It also has been shown that eNOS interacts with heat shock protein 90 (Hsp90) upon stimulation with VEGF or shear stress, and this interaction enhances eNOS activity (59). Interestingly, Akt also interacts with Hsp90 upon stimulation, and this interaction enhances Akt enzymatic activity (69), suggesting that Hsp90 may serve as a scaffold protein for the efficient phosphorylation of eNOS by Akt at caveolae (Figure 80.2).

REGULATION OF ENDOTHELIAL CELL MIGRATION BY AKT

The ability of ECs to migrate and form capillary-like structures is essential for angiogenesis in vivo. VEGF enhances EC migration and capillary-like structure formation in vitro, and these activities of VEGF are PI3K/Akt–dependent (44). S1P also has been shown to enhance EC migration and capillary formation in vitro through the activation of the endothelial differentiation gene (EDG) family of G-protein–coupled receptors and PI3K/Akt–dependent pathways (70,71). Conversely, oxidized low-density lipoprotein (LDL) inhibits EC migration toward VEGF by promoting the dephosphorylation of Akt (72).

Studies in other cell types also have implicated PI3K and Akt in the control of directional cell migration and the sensing of chemoattractant gradients by the cell. For example, it has been shown that Akt transiently localizes to the leading edge membrane of migratory cells in a PI3K-dependent manner (73,74), and gene ablation studies in mice have demonstrated that PI3Kγ is required for chemotaxis and chemoattractant-dependent activation of Akt in macrophages and neutrophils (75). Akt has been shown to be required for chemotaxis in *Dictyostelium* cells as well (73,74).

Cellular movement requires the reorganization of the actin cytoskeleton, and distinct patterns of actin reorganization are required as cells establish a leading edge and then generate contractile force to migrate forward (76). Previous studies have implicated the Rho family of small G-proteins as one of the major regulators of actin reorganization. Among Rho family members, Rho, Rac, and Cdc42 are most widely studied, and each regulates specific aspects of cytoskeletal reorganization. Rho stimulates cytoplasmic stress fiber formation and actomyosin-based contractility, Rac induces membrane ruffling and extension of lamellipodia, and Cdc42 induces the extension of membrane protrusions (filopodia) and is also involved in chemoattractant gradient sensing (77,78). Although it has been shown that VEGF-induced EC migration is dependent on Rho family GTPases, the relationship between Akt and the Rho family of G-proteins is complicated and controversial (43). On one hand, Akt negatively regulates Rac1 activity by phosphorylating Rac1 and inhibiting its GTP-binding activity. On the other hand, Akt phosphorylates S1P receptor EDG-1 and induces Rac activation and cell migration in ECs. Other reports show that Rac and Cdc42 are situated upstream of Akt and that they promote Akt signaling. It is possible that the upstream–downstream relationship between Akt and Rho/Rac/Cdc42 differs depending on cell type and external stimuli.

Another possible downstream effector of Akt that regulates cell motility is p21-activated protein kinase (PAK). PAK was originally identified as a Rac1-binding protein that specifically interacts with the GTP-bound form of Rac (79). Subsequently, it was shown that PAK is activated by Rac or Cdc42 and that it regulates polarized cytoskeletal reorganization (79). It was shown in *Dictyostelium* cells that Akt regulates cell polarity and chemotaxis through the regulatory phosphorylation of PAK (80), suggesting a direct functional link between Akt and PAK in the regulation of cytoskeletal reorganization. In mammalian fibroblasts, it also was shown that Akt stimulates PAK1 activation, and that dominant-negative Akt inhibits Ras-induced activation of PAK1 (81). However, the Akt phosphorylation site in *Dictyostelium* PAK is not conserved in mammalian PAK1, suggesting an indirect activation of mammalian PAK1 by Akt. Nonetheless, the PAK family of protein kinases are attractive candidates for Akt effectors in the regulation of EC migration, and may be a convergence point of signals from Rac/Cdc42 and Akt.

STATINS AND AKT SIGNALING

The 3-hydroxyl-3-methylglutaryl coenzyme A (HMG-CoA) reductase inhibitors, or statins, are widely prescribed for the treatment of hypercholesterolemia, and several clinical trials have demonstrated that statins are effective for both primary and secondary prevention of coronary artery diseases (82). It also has been shown that the cardioprotective effects of statins are partly independent of their serum lipid-lowering effects (82,83). Subsequently, it was found that statins rapidly promote the activation of Akt in ECs, leading to eNOS phosphorylation (at Ser 635 and Ser 1179) and increased NO production (84,85). Statins also protect ECs from serum deprivation–induced apoptosis and promote capillary-like structure formation on Matrigel in an Akt-dependent manner, and statin treatment enhances angiogenesis in hindlimb ischemia of normocholesterolemic animals through an eNOS-dependent mechanism (84,86). In addition to its effect on mature ECs, it was shown that statins enhance the mobilization of EPCs from bone marrow in a PI3K/Akt–dependent manner both in mice and humans (87–89), suggesting another mechanism of Akt-dependent proangiogenic effects in statins. Although numerous lines of evidence suggest that statins promote EC function and angiogenesis, currently no evidence in clinical studies links statin treatment to an increase in cancer risk (90).

Activation of Akt by statins is blocked by treatment with inhibitors of PI3K (wortmannin or LY294002) (84), and statin treatment rapidly induces translocation of Akt to the plasma membrane in a PI3K-dependent manner (91). However, the mechanisms by which statins activate PI3K are unknown at present. In this regard, statins have been shown to decrease caveolin-1–eNOS interaction and enhance the formation of eNOS-Hsp90-Akt complex in ECs (92), although it is not clear whether these effects of statins are secondary to Akt activation. Another recent report has shown that statins have biphasic effects on angiogenesis and that low doses of statins (relevant to the serum concentrations observed in patients treated with statins) enhance, whereas high doses inhibit, angiogenesis (93). This may explain the antiangiogenic effects of statins observed in studies using higher concentrations of statin treatment (94).

In addition to their proangiogenic effects, statins also have been shown to exhibit antithrombotic actions in humans, which appear to be independent of their serum cholesterol-lowering effects (95). Recent studies have shown that the PI3K/Akt pathway inhibits the expression of tissue factor (96,97), which is the primary cellular initiator of blood coagulation and whose expression is induced in ECs and macrophages by a number of stimuli, including IL-1β and tumor necrosis factor (TNF)-α (98). Although VEGF activates both tissue factor expression and PI3K/Akt signaling, administration of inhibitors of PI3K/Akt signaling further enhances VEGF-induced tissue factor expression (96,97). Taken together, these data suggest that statins may inhibit blood coagulation, at least in part, through a selective activation of PI3K/Akt signaling in ECs, leading to an inhibition of tissue factor expression.

CONCLUSION

The PI3K/Akt signaling axis is activated by a variety of stimuli in ECs, and it regulates multiple critical steps in angiogenesis, including EC survival, migration, and capillary-like structure formation. Furthermore, this signaling pathway also regulates cardiovascular homeostasis and vessel integrity at least, in part, by controlling NO synthesis. Angiogenesis has been implicated in the pathophysiology of a number of diseases, and regulation of angiogenesis, both its increase and decrease, could be an important therapeutic strategy for those disease states. Thus, further dissection of the PI3K/Akt pathway and elucidation of downstream effector molecules will lead to a better understanding of blood vessel growth and may provide avenues for the development of novel therapeutic interventions.

KEY POINTS

- Akt signaling in ECs regulates angiogenesis and vascular remodeling.
- Akt mediates EC survival and migration.
- Akt mediates eNOS activity and vascular tone.
- Statins activate Akt in ECs and promotes angiogenesis.

Future Goals

- To determine the role of Akt signaling in physiological and pathological angiogenesis
- To determine the downstream Akt substrates in ECs that promotes survival, migration, and angiogenesis
- To develop small-molecule Akt activators or inhibitors for therapeutic angiogenesis or antiangiogenesis therapies

REFERENCES

1 Yancopoulos GD, Davis S, Gale NW, et al. Vascular-specific growth factors and blood vessel formation. *Nature*. 2000;407: 242–248.

2 Carmeliet P. Angiogenesis in health and disease. *Nat Med.* 2003; 9:653–660.
3 Isner JM, Asahara T. Angiogenesis and vasculogenesis as therapeutic strategies for postnatal neovascularization. *J Clin Invest.* 1999;103:1231–1236.
4 Losordo DW, Dimmeler S. Therapeutic angiogenesis and vasculogenesis for ischemic disease. Part I: angiogenic cytokines. *Circulation.* 2004;109:2487–2491.
5 Staal SP. Molecular cloning of the Akt oncogene and its human homologues AKT1 and AKT2: amplification of AKT1 in a primary human gastric adenocarcinoma. *Proc Natl Acad Sci USA.* 1987;84:5034–5037.
6 Bellacosa A, Testa JR, Staal SP, Tsichlis PN. A retroviral oncogene, akt, encoding a serine-threonine kinase containing an SH2-like region. *Science.* 1991;254:274–277.
7 Coffer PJ, Woodgett JR. Molecular cloning and characterisation of a novel putative protein-serine kinase related to the cAMP-dependent and protein kinase C families. *Eur J Biochem.* 1991; 201:475–481.
8 Jones PF, Jakubowicz T, Pitossi FJ, et al. Molecular cloning and identification of a serine/threonine protein kinase of the second-messenger subfamily. *Proc Natl Acad Sci USA.* 1991;88:4171–4175.
9 Datta SR, Brunet A, Greenberg ME. Cellular survival: a play in three Akts. *Genes Dev.* 1999;13:2905–2927.
10 Scheid MP, Woodgett JR. Pkb/akt: functional insights from genetic models. *Nat Rev Mol Cell Biol.* 2001;2:760–768.
11 Downward J. Mechanisms and consequences of activation of protein kinase B/Akt. *Curr Opin Cell Biol.* 1998;10:262–267.
12 Scheid MP, Woodgett JR. Unravelling the activation mechanisms of protein kinase B/Akt. *FEBS Lett.* 2003;546:108–112.
13 Downward J. PI 3-kinase, Akt and cell survival. *Semin Cell Dev Biol.* 2004;15:177–182.
14 Whiteman EL, Cho H, Birnbaum MJ. Role of Akt/protein kinase B in metabolism. *Trends Endocrinol Metab.* 2002;13:444–451.
15 Liu JP. Studies of the molecular mechanisms in the regulation of telomerase activity. *FASEB J.* 1999;13:2091–2104.
16 Liang J, Slingerland JM. Multiple roles of the PI3K/PKB (Akt) pathway in cell cycle progression. *Cell Cycle.* 2003;2:339–345.
17 Inoki K, Corradetti MN, Guan KL. Dysregulation of the TSC-mTOR pathway in human disease. *Nat Genet.* 2005;37:19–24.
18 Shioi T, McMullen JR, Kang PM, et al. Akt/protein kinase B promotes organ growth in transgenic mice. *Mol Cell Biol.* 2002;22: 2799–2809.
19 Matsui T, Li L, Wu JC, et al. Phenotypic spectrum caused by transgenic overexpression of activated Akt in the heart. *J Biol Chem.* 2002;9:9.
20 Condorelli G, Drusco A, Stassi G, et al. Akt induces enhanced myocardial contractility and cell size in vivo in transgenic mice. *Proc Natl Acad Sci USA.* 2002;99:12333–12338.
21 Shiojima I, Sato K, Izumiya Y, et al. Disruption of coordinated cardiac hypertrophy and angiogenesis contributes to the transition to heart failure. *J Clin Invest.* 2005;115:2108–2118.
22 Tuttle RL, Gill NS, Pugh W, et al. Regulation of pancreatic beta-cell growth and survival by the serine/threonine protein kinase Akt1/PKBalpha. *Nat Med.* 2001;7:1133–1137.
23 Bernal-Mizrachi E, Wen W, Stahlhut S, et al. Islet beta cell expression of constitutively active Akt1/PKB alpha induces striking hypertrophy, hyperplasia, and hyperinsulinemia. *J Clin Invest.* 2001;108:1631–1638.
24 Malstrom S, Tili E, Kappes D, et al. Tumor induction by an Lck-MyrAkt transgene is delayed by mechanisms controlling the size of the thymus. *Proc Natl Acad Sci USA.* 2001;98:14967–14972.
25 Rathmell JC, Elstrom RL, Cinalli RM, Thompson CB. Activated Akt promotes increased resting T cell size, CD28-independent T cell growth, and development of autoimmunity and lymphoma. *Eur J Immunol.* 2003;33:2223–2232.
26 Majumder PK, Yeh JJ, George DJ, et al. Prostate intraepithelial neoplasia induced by prostate restricted Akt activation: the MPAKT model. *Proc Natl Acad Sci USA.* 2003;100:7841–7846.
27 Hutchinson J, Jin J, Cardiff RD, et al. Activation of Akt (protein kinase B) in mammary epithelium provides a critical cell survival signal required for tumor progression. *Mol Cell Biol.* 2001; 21:2203–2212.
28 Schwertfeger KL, Richert MM, Anderson SM. Mammary gland involution is delayed by activated Akt in transgenic mice. *Mol Endocrinol.* 2001;15:867–881.
29 Ackler S, Ahmad S, Tobias C, et al. Delayed mammary gland involution in MMTV-AKT1 transgenic mice. *Oncogene.* 2002;21: 198–206.
30 Schwertfeger KL, McManaman JL, Palmer CA, et al. Expression of constitutively activated Akt in the mammary gland leads to excess lipid synthesis during pregnancy and lactation. *J Lipid Res.* 2003;44:1100–1112.
31 Chen WS, Xu PZ, Gottlob K, et al. Growth retardation and increased apoptosis in mice with homozygous disruption of the Akt1 gene. *Genes Dev.* 2001;15:2203–2208.
32 Cho H, Thorvaldsen JL, Chu Q, et al. Akt1/pkbalpha is required for normal growth but dispensable for maintenance of glucose homeostasis in mice. *J Biol Chem.* 2001;276:38349–38352.
33 Cho H, Mu J, Kim JK, et al. Insulin resistance and a diabetes mellitus-like syndrome in mice lacking the protein kinase Akt2 (PKB beta). *Science.* 2001;292:1728–1731.
34 Garofalo RS, Orena SJ, Rafidi K, et al. Severe diabetes, age-dependent loss of adipose tissue, and mild growth deficiency in mice lacking Akt2/PKB beta. *J Clin Invest.* 2003;112:197–208.
35 Easton RM, Cho H, Roovers K, et al. Role for Akt3/protein kinase B{gamma} in attainment of normal brain size. *Mol Cell Biol.* 2005;25:1869–1878.
36 Tschopp O, Yang ZZ, Brodbeck D, et al. Essential role of protein kinase B gamma (PKB gamma/Akt3) in postnatal brain development but not in glucose homeostasis. *Development.* 2005; 132:2943–2954.
37 Peng XD, Xu PZ, Chen ML, et al. Dwarfism, impaired skin development, skeletal muscle atrophy, delayed bone development, and impeded adipogenesis in mice lacking Akt1 and Akt2. *Genes Dev.* 2003;17:1352–1365.
38 Yang ZZ, Tschopp O, Di-Poi N, et al. Dosage-dependent effects of Akt1/protein kinase Balpha (PKBalpha) and Akt3/PKBgamma on thymus, skin, and cardiovascular and nervous system development in mice. *Mol Cell Biol.* 2005;25:10407–10418.
39 Sun JF, Phung T, Shiojima I, et al. Microvascular patterning is controlled by fine-tuning the Akt signal. *Proc Natl Acad Sci USA.* 2005;102:128–133.
40 Ackah E, Yu J, Zoellner S, et al. Akt1/protein kinase Balpha is critical for ischemic and VEGF-mediated angiogenesis. *J Clin Invest.* 2005;115:2119–2127.
41 Chen J, Somanath PR, Razorenova O, et al. Akt1 regulates pathological angiogenesis, vascular maturation and permeability in vivo. *Nat Med.* 2005;11:1188–1196.

42 Fujio Y, Walsh K. Akt mediates cytoprotection of endothelial cells by vascular endothelial growth factor in an anchorage-dependent manner. *J Biol Chem.* 1999;274:16349–16354.
43 Shiojima I, Walsh K. Role of Akt signaling in vascular homeostasis and angiogenesis. *Circ Res.* 2002;90:1243–1250.
44 Zachary I. VEGF signalling: integration and multi-tasking in endothelial cell biology. *Biochem Soc Trans.* 2003;31:1171–1177.
45 Goetze S, Bungenstock A, Czupalla C, et al. Leptin induces endothelial cell migration through Akt, which is inhibited by PPARgamma-ligands. *Hypertension.* 2002;40:748–754.
46 Boo YC, Jo H. Flow-dependent regulation of endothelial nitric oxide synthase: role of protein kinases. *Am J Physiol Cell Physiol.* 2003;285:C499–C508.
47 Ouchi N, Kobayashi H, Kihara S, et al. Adiponectin stimulates angiogenesis by promoting cross-talk between AMP-activated protein kinase and Akt signaling in endothelial cells. *J Biol Chem.* 2004;279:1304–1309.
48 Frisch SM, Screaton RA. Anoikis mechanisms. *Curr Opin Cell Biol.* 2001;13:555–562.
49 Giancotti FG, Ruoslahti E. Integrin signaling. *Science.* 1999;285:1028–1032.
50 Eliceiri BP. Integrin and growth factor receptor crosstalk. *Circ Res.* 2001;89:1104–1110.
51 Bachelder RE, Wendt MA, Fujita N, et al. The cleavage of Akt/protein kinase B by death receptor signaling is an important event in detachment-induced apoptosis. *J Biol Chem.* 2001;276:34702–34707.
52 Byzova TV, Goldman CK, Pampori N, et al. A mechanism for modulation of cellular responses to VEGF: activation of the integrins. *Mol Cell.* 2000;6:851–860.
53 Papapetropoulos A, Fulton D, Mahboubi K, et al. Angiopoietin-1 inhibits endothelial cell apoptosis via the Akt/survivin pathway. *J Biol Chem.* 2000;275:9102–9105.
54 Suhara T, Mano T, Oliveira BE, Walsh K. Phosphatidylinositol 3-kinase/Akt signaling controls endothelial cell sensitivity to Fas-mediated apoptosis via regulation of FLICE-inhibitory protein (FLIP). *Circ Res.* 2001;89:13–19.
55 Gratton JP, Morales-Ruiz M, Kureishi Y, et al. Akt down-regulation of p38 signaling provides a novel mechanism of vascular endothelial growth factor-mediated cytoprotection in endothelial cells. *J Biol Chem.* 2001;276:30359–30365.
56 Skurk C, Maatz H, Kim HS, et al. The Akt-regulated forkhead transcription factor FOXO3a controls endothelial cell viability through modulation of the caspase-8 inhibitor FLIP. *J Biol Chem.* 2004;279:1513–1525.
57 Kim HS, Skurk C, Thomas SR, et al. Regulation of angiogenesis by glycogen synthase kinase-3beta. *J Biol Chem.* 2002;277:41888–41896.
58 Skurk C, Maatz H, Rocnik E, et al. Glycogen-synthase kinase3beta/beta-catenin axis promotes angiogenesis through activation of vascular endothelial growth factor signaling in endothelial cells. *Circ Res.* 2005;96:308–318.
59 Fulton D, Gratton JP, Sessa WC. Post-translational control of endothelial nitric oxide synthase: why isn't calcium/calmodulin enough? *J Pharmacol Exp Ther.* 2001;299:818–824.
60 Dimmeler S, Fleming I, Fisslthaler B, et al. Activation of nitric oxide synthase in endothelial cells by Akt-dependent phosphorylation. *Nature.* 1999;399:601–605.
61 Michell BJ, Harris MB, Chen ZP, et al. Identification of regulatory sites of phosphorylation of the bovine endothelial nitric-oxide synthase at serine 617 and serine 635. *J Biol Chem.* 2002;277:42344–42351.
62 Luo Z, Fujio Y, Kureishi Y, et al. Acute modulation of endothelial Akt/PKB activity alters nitric oxide-dependent vasomotor activity in vivo. *J Clin Invest.* 2000;106:493–499.
63 Iaccarino G, Ciccarelli M, Sorriento D, et al. AKT participates in endothelial dysfunction in hypertension. *Circulation.* 2004;109:2587–2593.
64 Minshall RD, Sessa WC, Stan RV, et al. Caveolin regulation of endothelial function. *Am J Physiol Lung Cell Mol Physiol.* 2003;285:L1179–L1183.
65 Gratton JP, Bernatchez P, Sessa WC. Caveolae and caveolins in the cardiovascular system. *Circ Res.* 2004;94:1408–1417.
66 Hnasko R, Lisanti MP. The biology of caveolae: lessons from caveolin knockout mice and implications for human disease. *Mol Interv.* 2003;3:445–464.
67 Bauer PM, Yu J, Chen Y, et al. Endothelial-specific expression of caveolin-1 impairs microvascular permeability and angiogenesis. *Proc Natl Acad Sci USA.* 2005;102:204–209.
68 Govers R, Rabelink TJ. Cellular regulation of endothelial nitric oxide synthase. *Am J Physiol Renal Physiol.* 2001;280:F193–F206.
69 Sato S, Fujita N, Tsuruo T. Modulation of Akt kinase activity by binding to Hsp90. *Proc Natl Acad Sci USA.* 2000;97:10832–10837.
70 Kluk MJ, Hla T. Signaling of sphingosine-1-phosphate via the S1P/EDG-family of G-protein-coupled receptors. *Biochim Biophys Acta.* 2002;1582:72–80.
71 Radeff-Huang J, Seasholtz TM, Matteo RG, Brown JH. G protein mediated signaling pathways in lysophospholipid induced cell proliferation and survival. *J Cell Biochem.* 2004;92:949–966.
72 Chavakis E, Dernbach E, Hermann C, et al. Oxidized LDL inhibits vascular endothelial growth factor-induced endothelial cell migration by an inhibitory effect on the Akt/endothelial nitric oxide synthase pathway. *Circulation.* 2001;103:2102–2107.
73 Stephens L, Ellson C, Hawkins P. Roles of PI3Ks in leukocyte chemotaxis and phagocytosis. *Curr Opin Cell Biol.* 2002;14:203–213.
74 Merlot S, Firtel RA. Leading the way: directional sensing through phosphatidylinositol 3-kinase and other signaling pathways. *J Cell Sci.* 2003;116:3471–3478.
75 Wymann MP, Bjorklof K, Calvez R, et al. Phosphoinositide 3-kinase gamma: a key modulator in inflammation and allergy. *Biochem Soc Trans.* 2003;31:275–280.
76 Lauffenburger DA, Horwitz AF. Cell migration: a physically integrated molecular process. *Cell.* 1996;84:359–369.
77 Ridley AJ. Rho proteins, PI 3-kinases, and monocyte/macrophage motility. *FEBS Lett.* 2001;498:168–171.
78 van Nieuw Amerongen GP, van Hinsbergh VW. Cytoskeletal effects of rho-like small guanine nucleotide-binding proteins in the vascular system. *Arterioscler Thromb Vasc Biol.* 2001;21:300–311.
79 Daniels RH, Bokoch GM. p21-activated protein kinase: a crucial component of morphological signaling? *Trends Biochem Sci.* 1999;24:350–355.
80 Chung CY, Potikyan G, Firtel RA. Control of cell polarity and chemotaxis by Akt/PKB and PI3 kinase through the regulation of PAKa. *Mol Cell.* 2001;7:937–947.
81 Tang Y, Zhou H, Chen A, et al. The Akt proto-oncogene links Ras to Pak and cell survival signals. *J Biol Chem.* 2000;275:9106–9109.

82 Maron DJ, Fazio S, Linton MF. Current perspectives on statins. *Circulation.* 2000;101:207–213.
83 Skaletz-Rorowski A, Walsh K. Statin therapy and angiogenesis. *Curr Opin Lipidol.* 2003;14:599–603.
84 Kureishi Y, Luo Z, Shiojima I, et al. The HMG-CoA reductase inhibitor simvastatin activates the protein kinase Akt and promotes angiogenesis in normocholesterolemic animals. *Nat Med.* 2000;6:1004–1010.
85 Harris MB, Blackstone MA, Sood SG, et al. Acute activation and phosphorylation of endothelial nitric oxide synthase by HMG-CoA reductase inhibitors. *Am J Physiol Heart Circ Physiol.* 2004;287:H560–H566.
86 Sata M, Nishimatsu H, Suzuki E, et al. Endothelial nitric oxide synthase is essential for the HMG-CoA reductase inhibitor cerivastatin to promote collateral growth in response to ischemia. *FASEB J.* 2001;15:2530–2532.
87 Dimmeler S, Aicher A, Vasa M, et al. HMG-CoA reductase inhibitors (statins) increase endothelial progenitor cells via the PI 3-kinase/Akt pathway. *J Clin Invest.* 2001;108:391–397.
88 Llevadot J, Murasawa S, Kureishi Y, et al. HMG-CoA reductase inhibitor mobilizes bone marrow – derived endothelial progenitor cells. *J Clin Invest.* 2001;108:399–405.
89 Vasa M, Fichtlscherer S, Adler K, et al. Increase in circulating endothelial progenitor cells by statin therapy in patients with stable coronary artery disease. *Circulation.* 2001;103:2885–2890.
90 Davidson MH. Safety profiles for the HMG-CoA reductase inhibitors: treatment and trust. *Drugs.* 2001;61:197–206.
91 Skaletz-Rorowski A, Lutchman M, Kureishi Y, et al. HMG-CoA reductase inhibitors promote cholesterol-dependent Akt/PKB translocation to membrane domains in endothelial cells. *Cardiovasc Res.* 2003;57:253–264.
92 Brouet A, Sonveaux P, Dessy C, et al. Hsp90 and caveolin are key targets for the proangiogenic nitric oxide-mediated effects of statins. *Circ Res.* 2001;89:866–873.
93 Urbich C, Dernbach E, Zeiher AM, Dimmeler S. Double-edged role of statins in angiogenesis signaling. *Circ Res.* 2002;90:737–744.
94 Vincent L, Soria C, Mirshahi F, et al. Cerivastatin, an inhibitor of 3-hydroxy-3-methylglutaryl coenzyme a reductase, inhibits endothelial cell proliferation induced by angiogenic factors in vitro and angiogenesis in in vivo models. *Arterioscler Thromb Vasc Biol.* 2002;22:623–629.
95 Dangas G, Smith DA, Unger AH, et al. Pravastatin: an antithrombotic effect independent of the cholesterol-lowering effect. *Thromb Haemost.* 2000;83:688–692.
96 Blum S, Issbruker K, Willuweit A, et al. An inhibitory role of the phosphatidylinositol 3-kinase-signaling pathway in vascular endothelial growth factor-induced tissue factor expression. *J Biol Chem.* 2001;276:33428–33434.
97 Kim I, Oh JL, Ryu YS, et al. Angiopoietin-1 negatively regulates expression and activity of tissue factor in endothelial cells. *FASEB J.* 2002;16:126–128.
98 Moons AH, Levi M, Peters RJ. Tissue factor and coronary artery disease. *Cardiovasc Res.* 2002;53:313–325.

81

Mitogen-Activated Protein Kinases

Natalia V. Bogatcheva and Alexander D. Verin

University of Chicago, Illinois

The family of mitogen-activated protein kinases (MAPKs) comprises several protein kinases, united by their ability to phosphorylate serine or threonine residues followed by proline; therefore, these often are termed *proline-directed protein kinases.* Another characteristic feature of the MAPKs is the presence of specific phosphoacceptor site(s) in the regulatory loop of the enzyme catalytic core.

HISTORY AND EVOLUTION

MAPKs were originally named after the protein they were discovered to phosphorylate in insulin- or growth factor–treated mammalian cells, namely microtubule-associated protein 2 (MAP-2). Later, with the realization that virtually all mitogens lead to the rapid activation of these protein kinases, the enzymes were renamed *mitogen-activated protein kinases.* Now, it is clear that MAPK function is not limited to the initially described induction of cell division, but also extends to control over cell differentiation, survival, motility, and gene expression. Because MAPKs orchestrate the physiological processes fundamental for both single- and multicellular organisms, it is not surprising that homologous enzymes are found in eukaryotes from budding yeast to mammals (1). The deletion of certain MAPK genes in mammals leads to early embryonic lethality, proving the essential role of MAPK signaling in the developmental processes (Table 81-1).

MITOGEN-ACTIVATED PROTEIN KINASE STRUCTURE AND FUNCTION

To date, five different groups of MAPKs are known in mammals. Enzymes within groups are generated from a number of gene products; additional isoforms were shown to derive from the alternative splicing of pre-mRNAs (12). Mammalian MAPKs include extracellular-signal-regulated protein kinase (ERK)1/2; p38 MAPKs (α, β, γ, and δ); c-jun N-terminal kinases (JNK1, JNK2, JNK3); ERK5, also known as BMK1 (13); and the recently identified ERK3 subfamily, comprising the products of two genes, *MAPK4* and *MAPK6* (14). All kinases of the MAPK family display approximately a 40% to 50% amino acid sequence identity within their catalytic domains (13–15). ERK5 and ERK3 carry certain structural differences from the conventional MAPKs ERK1/2, p38, and JNK (13,14).

The three-dimensional (3D) structure of the earliest described MAPK representative, ERK2, reveals that this enzyme consists of two domains, connected by a linker region. Interface between the two domains forms an ATP-binding site, buried deep inside the enzyme; the protein substrate-binding site, on the contrary, is located on the surface. Because the two domains of ERK2 are rotated apart in the inactive state, activation requires the closure of the 3D structure to form an active substrate-binding site (Figure 81.1). Thus, unlike kinases activated by the displacement of the autoinhibitory domain (e.g., protein kinase C [PKC], protein kinase A [PKA], myosin light-chain kinase [MLCK]), MAPKs acquire activity after conformational changes within the region positioned within the catalytic core of the enzyme (16). These conformational changes are induced by the phosphorylation of the specific regulatory motif that is conserved between different members of the MAPK family. ERK1/2, p38 MAPK, JNK, and ERK5/BMK1 possess the double phosphoacceptor site TXY in the activation loop, whereas ERK3 is regulated by the phosphorylation of a single site in the SEG motif (14). Another unique feature of ERK3 is its regulation by protein turnover (17), in contrast to the other relatively stable members of the MAPK family.

Early study of the ERK1/2 activation mechanism revealed that these enzymes require both tyrosine and threonine phosphorylation of the phosphoacceptor site to display activity. Initially, the integration of two signaling pathways controlling dual MAPK phosphorylation was considered. However, MAPK “activating factor” isolated from cells behaved rather as a single component, capable of phosphorylating both tyrosine and threonine residues in the structure of MAPKs. To be active, this factor itself required serine/threonine phosphorylation. Thus, activation of MAPK “activating factor” (named MAPK kinase [MAPKK]) by receptor tyrosine kinases implied the presence of at least one more upstream protein kinase,

Table 81-1: MAPK-Knockout Phenotypes

Genes	*Phenotypes (Refs.)*
ERK1	Decreased T-cell responses Skin lesions and hyperkeratosis, inhibition of keratinocytes apoptosis and proliferation (3) Enhanced cocaine-evoked behavioral plasticity (4)
ERK2	Lack of mesoderm differentiation, defects in placenta, embryonic death
p38α	Defects in placental angiogenesis, embryonic death Survivors exhibit defects in erythropoiesis Increased adipogenesis (5) Cardiac-specific knockouts exhibit dedifferentiation and proliferation of neonatal cardiomyocytes (6)
p38β	No obvious phenotype (7)
p38γ	No obvious phenotype
JNK1	Defects in T-cell differentiation, proliferation, and apoptosis Low susceptibility to insulin resistance in diabetes models Marked deterioration in cardiac function following transverse aortic constriction (8) Reduced development of steatohepatitis in mice fed with methionine- and choline-deficient diet (9) Reduced inhibition of neovascularization by thrombospondin (10)
JNK2	Defects in T-cell differentiation, proliferation, and apoptosis
JNK1+JNK2	Defects in neural tube closure, embryonic death Increased proliferation and IL-2 production in T cells Resistance to UV-induced apoptosis in embryonic fibroblasts
JNK3	Resistance to kainate-induced neural damage
ERK5	Defects in angiogenesis and cardiovascular development, embryonic death
ERK3	Mice phenotype not yet described Reduced activity of MAPKAP5 (MK5) in −/− embryo fibroblasts (11)

As reviewed in Kuida K, Boucher DM. Function of MAP kinases: insights from gene-targeting studies. *J Biochem.* 2004;135:653–656, with recent data addition.

specific for serine/threonine residues. The proto-oncogene, Raf-1, has been proposed to function as MAPK kinase kinase (MAPKKK). All three enzymes were combined in the so-called *MAPK cascade*, which also was characterized in yeast and frogs, and later discovered in fruit flies, nematodes, slime mold, and plants (1). Thus, functionally, the MAPK cascade appears to be evolutionarily conserved, although the identity of cascade members in the sequence may vary significantly.

Generally, each group of MAPKs possesses specific MAPKK, which, in turn, are activated by their MAPKKK, although some MAPKKKs do not show strict selectivity. Coupling within the cascade is ensured by special docking motifs, present in the cascade enzymes as well as in their substrates. In addition, the recognition and in some cases nonenzymatic activation of cascade members may be provided by certain scaffolding proteins (18).

Upstream of the three-tiered cascade lies a plethora of signals, converging to the activation of MAPKKK. The mechanism of regulation of the most studied MAPKKK, Raf, is known to require binding of the small G-protein Ras (19). "Classical" MAPKKK activation via receptor tyrosine kinases (RTK) is achieved via a sequence of events, including membrane recruitment of the Ras guanine nucleotide exchange factor Sos, with participation of the adapter protein Grb2 (Figure 81.2). Activation of MAPKKK via G-protein–coupled receptors (GPCRs) involves Pyk2/Src-, phosphoinositide 3-kinase (PI3K)-, and PKC-dependent mechanisms (see Figure 81.2), but also may occur via the β-arrestin–dependent pathway. Among other external stimuli leading to MAPK cascade activation are reactive oxygen species (ROS), phorbol esters, proinflammatory cytokines, and shear stress (Figure 81.2). It is generally believed that growth factors and phorbol esters mostly promote ERK1/2 activation, whereas cytokines and stress stimuli induce the activation of p38 MAPK and JNK. However, it would be more correct to say that many of the factors mentioned activate all three types of conventional MAPKs, although the relevance of such activation to the cell physiological response may vary.

Less is known about stimuli regulating unconventional MAPKs. ERK5 was shown to be activated by both mitogens and stress stimuli (13), whereas increased ERK3 activity was noted in response to certain growth factors (11). Inactivation of conventional MAP kinases and ERK5 is achieved mostly with the help of dual specificity phosphatases (DSP), which, in turn, are upregulated by many mitogens and stress treatments. Certain members of the DSP family display specificity to different MAPKs, thereby providing an additional level of control over cascade regulation (20).

The physiological effects of MAPK activation are elicited via the phosphorylation of numerous protein substrates (Figure 81.3). Substrates may be phosphorylated either in the cytoplasm or the nucleus, where MAPKs translocate upon activation. The exception is ERK5, which is constantly localized within the nucleus (21). Which targets are preferentially phosphorylated depends on the intensity and duration of activation and the localization of activated MAPK, determined in turn by the nature of the extracellular signal. MAPKs directly phosphorylate cytoskeletal proteins, signaling enzymes (phospholipases and MAPK-activated protein kinases), proteins involved in control of apoptosis, and transcriptional regulators. Downstream MAPK-activated kinases compose a family of kinases with various degree of structural relativity. They include RSK (MAKAPK1A-D); MSK1 and -2; MNK1 and -2; and MAPKAP2, -3, and -5 (22). Those kinases,

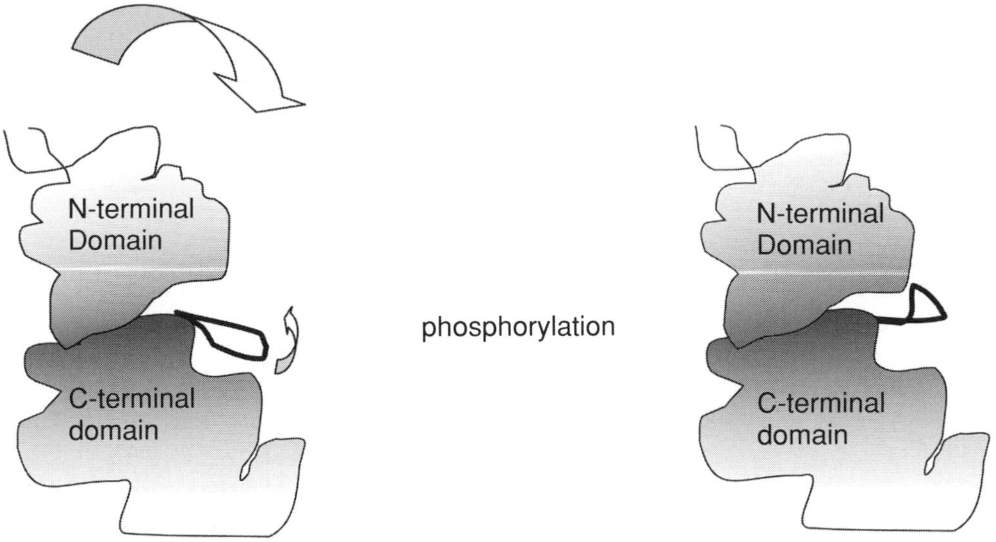

Figure 81.1. Schematic of conformational changes in ERK2 after phosphorylation. The enzyme consists of smaller N-terminal (residues 1–109 and 320–358) and larger C-terminal (residues 110–319) domains. The phosphorylated region, located within the C-terminal domain of ERK2, changes it position to promote domain closure and formation of proline-directed substrate-binding site. (Reproduced with permission from Cangarajah BJ, Khokhlatchev A, Cobb MH, Goldsmith EJ. Activation mechanism of the MAP kinase ERK2 by dual phosphorylation. *Cell.* 1997;90:859–869.)

activated by certain members of the MAPK family, phosphorylate proteins associated with cytoskeleton, ribosomal proteins, signaling or metabolic enzymes such as Akt, glycogen synthase 3β, 5-lipoxygenase, the Na/H exchanger NHE-1, and transcriptional regulators (19). The diversity of MAPKs and downstream kinases substrates allows the MAPK pathway to modulate various functions, such as cell contractility, cell metabolism, and cell fate. It is conceivable that MAPK-induced alteration of cytoskeleton and homeostatic/signaling enzymatic activity by phosphorylation may provide immediate cell response, whereas changes in gene expression would be relatively belated.

Alteration of cell contractility, often seen within minutes of agonist treatment, is achieved via phosphorylation of several actin-binding proteins (see Figure 81.2). Among direct MAPK and MAPK-activated protein kinase substrates are the regulatory actin-binding protein caldesmon, actin-stabilizing chaperone Hsp27, membrane-associated actin-binding proteins filamin and cortactin, and proteins mediating cell adhesion. Control over these protein functions allows MAPK to adjust actomyosin contractility and F-actin stability, remodel membrane structures, and regulate cell contact formation. Among noncytoskeletal MAPK targets not directly involved in the regulation of gene expression are enzymes like phospholipases and nitric oxide synthases (NOS), which allow MAPK to modulate the release of bioactive substances and thereby to coordinate the response of whole tissues or organs (19,23).

Targets of MAPK pathways affecting gene expression include numerous transcription factors and transcriptional regulators. To manipulate the cell transcription machinery, MAPK utilize several pathways. Phosphorylation by MAPK or MAPK-activated protein kinases can stimulate the translocation of transcriptional regulators to the nucleus or, conversely, from the nucleus, rendering factor activation or inactivation, respectively. Similarly, phosphorylation might alter the ability of some transcription factors to bind DNA. In addition, MAPK pathways control the level of expression of certain transcription factors, modulating their synthesis and stability. Last transcriptional regulation is achieved through the control of nucleosomal structure (12).

Transcriptional regulation often is linked to the alteration of synthesis of several bioactive substances by the cell, but it also determines cell fate, targeting the cell toward proliferation, differentiation, or apoptosis. Cell proliferation is thought to be regulated by MAPK pathways via activation of Elk-1, c-Myc, c-Fos, Ets-1, serum response factor (SRF), and other transcription factors. Cell survival signals may be conveyed by MAPK via cross-talk with the PI3K/Akt and nuclear factor (NF)-κB pathway, and the regulation of nitric oxide (NO) release. Apoptosis is likely controlled by the phosphorylation and downregulation of antiapoptotic proteins such as Bcl-2 and upregulation of caspase activity (19,23,24). In summary, the consequences of MAPK activation on a cellular level depend on the balance between different MAPK activities and the activities of other signaling pathways, which, in turn, will vary from cell to cell and from stimulus to stimulus.

MITOGEN-ACTIVATED PROTEIN KINASE CASCADE IN THE ENDOTHELIUM

The importance of MAPK for vascular functions was demonstrated in experiments using knockout animals in which *ERK2*-, *p38α*-, and *ERK5*-null animals displayed impaired

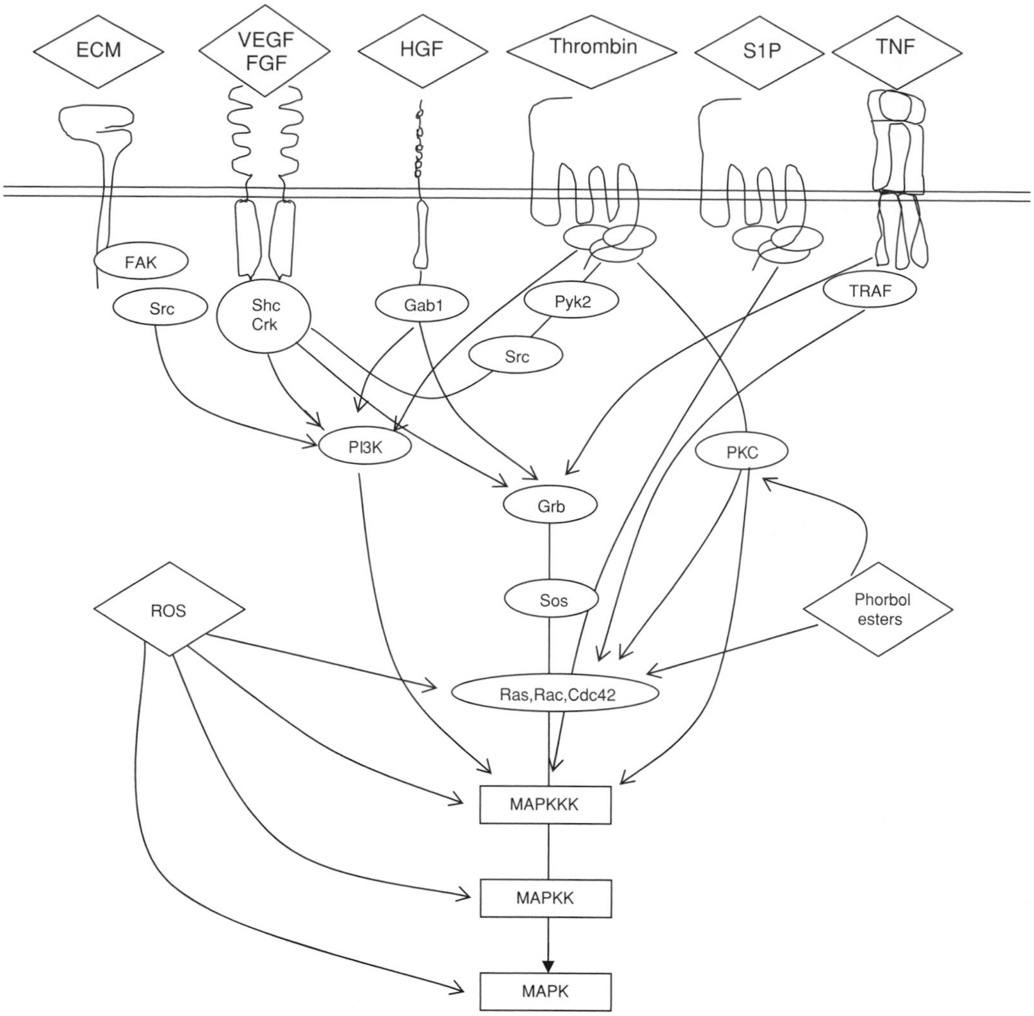

Figure 81.2. Main signaling pathways, leading to the activation of MAPK cascade in ECs cells. Numerous signals, originating from membrane receptors, utilize several mechanisms to achieve MAPK activation. One of the mechanisms involves the activation of various adapter proteins, such as Shc, Crk, Gab1, and Grb2. Other mechanisms depend on the activation of PKC or PI3K. All of them converge at the activation of small G-proteins (Ras, Rac, and Cdc42) or downstream MAPKK. Membrane-penetrating substances, such as reactive oxygen species (ROS) or phorbol esters, act through the PKC or small G-protein activation, but also suppress tyrosine dephosphorylation and deactivation of the MAPK module members. ECM, extracellular matrix proteins; FAK, focal adhesion kinase; FGF, fibroblast growth factor; HGF, hepatocyte growth factor; TNF, tumor necrosis factor; TRAF, TNF receptor associated factor; VEGF, vascular endothelial growth factor.

placental or embryonic angiogenesis (see Table 81-1). Although deficits of ERK1, JNK1, JNK2, JNK3, p38β, or p38γ were not shown to lead to abrupt vascular aberration (2), these data do not preclude the participation of the aforementioned MAPKs in prenatal angiogenesis, because a certain functional redundancy exists between isoforms, and double or triple knockouts may be necessary to exhibit the mutant phenotype.

Migration

On the cellular level, the process of vascularization is driven by endothelial cell (EC) migration and differentiation, which allows capillary tube formation and is, by the timely shift between proliferative and antiproliferative signals, necessary for the formation of a functional vasculature. EC migration, induced by growth factors or sphingosine-1 phosphate (S1P), critically depends on ERK1/2, p38 MAPK, and/or JNK activation (23,25). Because motile activity requires the cell to contract and establish contacts with the surrounding environment (24), it is hardly surprising that migrating cells display increased stress fiber formation and vinculin-mediated membrane anchoring, evidence of integrin-mediated contacts with extracellular matrices (26). The potential targets of MAPK, leading to actin network remodeling, are likely to be caldesmon and Hsp27 (Figure 81.3). Caldesmon is a known substrate for ERK1/2 and p38 MAPK, whereas Hsp27 is

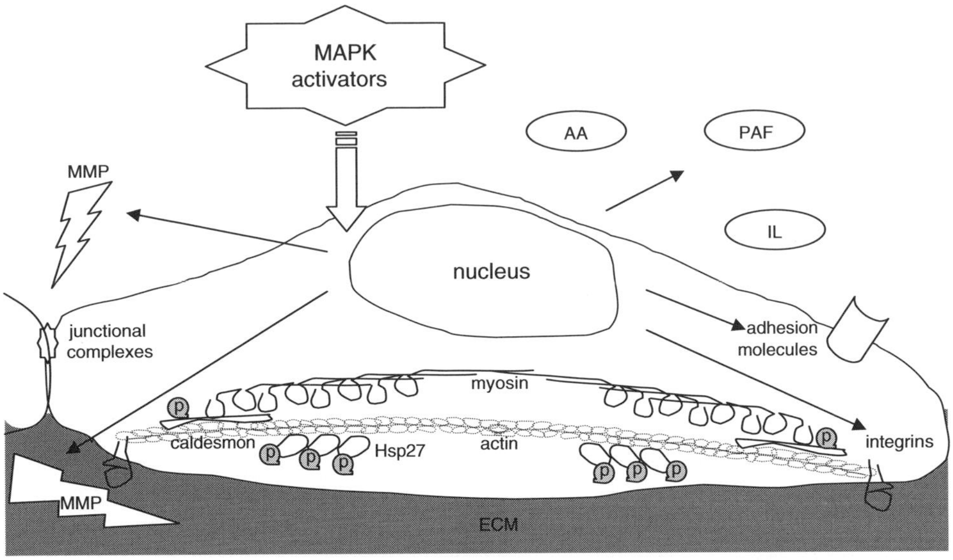

Figure 81.3. MAPK involvement in proinflammatory and proangiogenic endothelial responses. In endothelium, MAPK activation often is followed by the loss of EC monolayer integrity, increased leukocyte adhesion, and transmigration and release of bioactive mediators. These responses are a result of numerous phosphorylation events, leading to the alteration of cytoskeletal arrangement, enzymatic activity, or the expression level of several important gene products. Direct protein phosphorylation by MAPK or MAPKAP is depicted here as grey circles. Synthesis regulation by MAPK cascade is shown with arrows. PAF, platelet-activating factor; AA, arachidonic acid; IL, interleukin; MMP, matrix metalloprotease; HSP, heat shock protein; ECM, extracellular matrix.

phosphorylated via a p38 MAPK/MAPKAP-2–dependent mechanism (23). Both caldesmon and Hsp27 affect the stability of actin filaments under certain conditions. Moreover, caldesmon inhibits actomyosin interaction in quiescent cells and requires phosphorylation or interaction with Ca^{2+}-dependent protein to allow contraction. Indeed, involvement of caldesmon in EC migration was recently proved (27), although the question of whether MAPK-induced alteration of caldesmon properties is critical for the migration remains open. Similarly, the role of Hsp27 phosphorylation seems somewhat controversial, because both proangiogenic and antiangiogenic factors were shown to induce Hsp27 phosphorylation (28).

Migrating cell attachment to the substrate is likely to be regulated via a protein synthesis–dependent mechanism. Expression of one of the membrane anchors, integrin $\beta 3$, was shown to be controlled by the transcription factor Ets-1 in ECs (26,29). The level of *Ets* mRNA is known to be upregulated by activated ERK1/2 (26). Among Ets-1–controlled proteins are also matrix metalloproteinases (MMP), extracellular matrix (ECM)-degrading enzymes that are necessary for cell migration (29) (see Figure 81.3). The induction of matrix degradation activity depends on the activation of the urokinase-type plasminogen activator (u-PA) pathway. Both ERK1/2 and p38 MAPK were shown to regulate u-PA expression in ECs (30,31), implicating those kinases in control over ECM stability. The targets of the JNK pathway, which mediates migration, have been less thoroughly characterized to date. So far, JNK has been shown to regulate the level of MMP2 and MMP14 expression (32). In summary, ERK1/2, p38 MAPK, and JNK actively control EC migration, altering both the cellular organization of contractile and membrane proteins and the ECM degradation rate.

Tubular Morphogenesis, Endothelial Cell Proliferation and Survival

The formation of tubular structures is an intrinsic feature of ECs that allows them to create new capillaries. Grown in 3D matrices, ECs elongate and coalesce; the latter process is followed by the appearance of lumens within tubules. The signaling mechanisms controlling tubular morphogenesis in vitro have been studied only sporadically. Endothelial tube formation was found to depend on ERK1/2 or JNK activation (23,26). The role of p38 MAPK in tubular morphogenesis appears more contradictory. Although the inhibition of p38 MAPK in spleen ECs repressed tube formation (26), in other ECs, the ablation of p38 MAPK activity has an opposite effect. The latter event was noted to be associated with the increased expression of Jagged1, a ligand activating the Notch pathway (33). This pathway was earlier implicated in the regulation of EC differentiation and vascular remodeling.

MAPKs are known to regulate the proliferation and survival of many cell types. The general paradigm holds that activated ERK1/2 upregulate cell growth, whereas p38 MAPK and JNK inhibit proliferation and/or induce apoptosis. However, experiments using cultured ECs do not necessarily support that paradigm. The induction of cell growth in ECs treated with vascular endothelial growth factor (VEGF), basic fibroblast growth factor (bFGF), hepatocyte growth factor

(HGF), or S1P is indeed dependent on ERK1/2 (23,24,26). However, shear stress–exposed ECs, despite elevated ERK1/2 activity, display an inhibition of proliferation (23). Similarly, although p38 MAPK and JNK mediate EC antiproliferative or proapoptotic responses to high glucose and tumor necrosis factor (TNF)-α, in other studies the inhibition of p38 MAPK or JNK led to the downregulation of EC proliferation rates (23,26). Of note, p38 MAPK and JNK were shown to initiate both pro- and antiapoptotic signaling cascades, depending on cell and signal context. For instance, p38 MAPK induces cell death via downregulation of Bcl and activation of caspases (34), whereas the antiapoptotic mechanism of p38 MAPK action is linked to the NFκB pathway (35) or involves Hsp27 phosphorylation (36).

So far, little is known about the role of ERK5/BMK1 in endothelial functions. Experiments using ectopic expression of constitutively active MAPKK for ERK5 have demonstrated that ERK5 activation protects ECs from apoptosis in a growth factor–deprivation model (37). Targeted deletion of ERK5 in mice leads to the abnormal vascular leakage observed in several organs, mainly due to EC morphological alteration, and their eventual death (13). Future experiments are needed to assess the role of this member of the MAPK family in other endothelial responses.

Angiogenesis

As discussed earlier, in vitro experiments suggest that all processes critical for angiogenesis are tightly regulated by MAPK. Consistent with these results, ERK1/2 inhibition was shown to suppress retinal neovascularization in the rat retinopathy model (38), VEGF-induced angiogenesis in fetal rat cortical explants (39), and thrombin-induced angiogenesis in the chick chorioallantoic membrane (CAM) assay (40). The in vivo data regarding p38 MAPK inhibition, however, are conflicting and indicate, along with the in vitro data, that primary mechanisms, regulated by p38 MAPK in different models of angiogenesis, may vary. For example, p38 MAPK inhibition decreased vascularization in an air pouch granuloma model (41); this effect was attributed to the suppressed synthesis of proangiogenic inflammatory cytokines. By contrast, in another study, p38 MAPK inhibition abrogated the effect of antiangiogenic factor thrombospondin on mouse cornea vessel formation, suggesting a negative role of p38 MAPK in the regulation of angiogenesis. Indeed, thrombospondin was shown to induce EC apoptosis in a p38 MAPK-dependent manner (42). Along with those data, VEGF- or bFGF-induced CAM neoangiogenesis was significantly potentiated by p38 MAPK inhibitors (33,43). The most intriguing finding was that vessels formed in the presence of p38 MAPK inhibitor were characterized by EC hyperplasia and abnormal vascular structure (33). Thus, p38 MAPK activation serves to balance the processes of EC proliferation, differentiation, and apoptosis necessary for the formation of functional vessels. Data on the involvement of JNK in the process of vascularization are only now beginning to emerge. Experiments with *JNK-1*–null animals show that, along with p38 MAPK, JNK-1 plays an important role in the abrogation of angiogenesis by thrombospondin-1 (10). Future research is needed to clarify the role of other JNK enzymes in vascularization.

Endothelial Barrier

Once a functional vessel is created, it starts to act as a selective barrier to control the access of cells and fluids to the underlying tissues. The regulation of endothelial barrier properties is critical for the proper inflammatory response, because increased transendothelial permeability is required to allow mobile luminal cell migration toward the source of the tissue disturbance. Various edemagenic substances, such as H_2O_2, phorbol ester, thrombin, VEGF, and TNF-α were noted to evoke barrier dysfunction in cultured ECs in a MAPK-dependent manner (23). Because two events are of major importance for the disruption of the EC barrier – namely a decrease in cell–cell attachment and an increase in actomyosin contractility (23) – proteins associated with cellular junctions or contractile filaments are likely to be the targets of MAPK phosphorylation in activated ECs (see Figure 81.3). Indeed, endothelial junctions lose VE-cadherin, occludin, and ZO-1 as a result of ERK1/2, p38 MAPK, or JNK activation (23). These proteins do not seem to be phosphorylated by MAPKs directly, but rather by intermediate protein kinases whose identity remains unknown. One of the mechanisms, however, may involve MMP-dependent degradation of tight junctional proteins (Figure 81.3), because ERK1/2 inhibitors were shown to suppress ZO-1 proteolysis concomitant with barrier protection (44). On the other hand, MAPK and MAPKAP are thought to induce contractile response via direct phosphorylation of the actin-binding proteins Hsp27 and caldesmon (Figure 81.3). Both proteins display an increased level of phosphate incorporation coincidental with the loss of EC monolayer integrity (23). However, the functional importance of these events for endothelial barrier dysfunction remains to be confirmed. Another possible mechanism of MAPK-dependent barrier compromise involves destabilization of microtubules. The inhibition of p38 MAPK was shown to attenuate a TNF-α–induced decrease in stable tubulin content and to ameliorate endothelial barrier dysfunction (45). Last, some edemagenic agents may employ MAPK-dependent regulation of protein synthesis to increase paracellular permeability. As has been shown recently, the increase in myosin light-chain (MLC) phosphorylation in TGF-β–treated ECs occurs in a cycloxehimide-dependent manner (46). p38 MAPK seems to be the kinase responsible for this event, because both TGF-β–induced EC barrier dysfunction and MLC phosphorylation are blocked by inhibition of p38 MAPK. The protein synthesis–dependent mechanism appears to be unique for TGF-β producing delayed contractile response; it is not shared by agents that induce immediate contraction, such as thrombin (46).

Interestingly, augmentation of monolayer integrity in cultured ECs may also be MAPK-dependent (23). For example, the restoration of barrier function after ischemia–reperfusion requires ERK1/2 activity. One might expect that an increase in transendothelial resistance must be associated either with the tightening of cell–cell junctions or thickening of cortical actin rings. Indeed, the tight junction protein claudin-1 was shown to require ERK1/2–dependent phosphorylation to enhance EC barrier function (47). Barrier augmentation in HGF-treated cells was attributed to regulation of the level of the peripheral actin-binding protein, β-catenin. β-Catenin degradation depends on its phosphorylation by GSK3-β, which activity, in turn, is regulated by MAPK pathways (23). What exactly switches the MAPK cascade from barrier-disruptive to barrier-protective functions is a question for future research. The fact that MAPK activation may lead to both augmentation and deterioration of endothelial barrier function indicates that an understanding of specifics of cascade activation is required to manipulate EC barrier properties.

Inflammation

The role of MAPKs in inflammation is not limited to the regulation of EC monolayer integrity. Because leukocyte adhesion to ECs is a prerequisite for their later transmigration, the presence of specific adhesion proteins on the EC surface determines the rate of inflammatory cell infiltration to surrounding tissues. Leukocyte adhesion to ECs is regulated via a p38 MAPK- and/or JNK-dependent increase in the expression of the surface adhesion molecules such as intercellular adhesion molecule (ICAM)-1 and vascular adhesion molecule (VCAM)-1 (48), achieved via transactivation of the NF-κB pathway. Moreover, MAPKs control the subsequent process of leukocyte transmigration. ICAM-1 ligation-induced changes in the EC cytoskeleton were shown to be p38 MAPK-dependent (49). In addition, neutrophil extravasation also depends on ERK1/2 activation in ECs. Soluble neutrophil factor, inducing ERK1/2 activation in endothelium, was recently characterized as a 30-kDa protein (50). Thus, MAPKs control leukocyte transendothelial migration by altering barrier function, modulating EC expression of cell adhesion molecules, and reorganizing the EC cytoskeleton.

Release of Soluble Mediators

MAPKs modulate the release of several bioactive substances by ECs, including factors controlling inflammatory response, angiogenesis, blood coagulation, and vascular tone (see Figure 81.3). The enzymes producing proinflammatory lipid mediators, namely platelet-activating factor (PAF), arachidonic acid (AA), and its metabolites, are controlled by MAPK pathways at both the transcriptional and post-translational level. Both p38 MAPK and ERK1/2 contribute to the activation of phospholipase A2 (cPLA2) and arachidonic acid and PAF release by ECs (23). EC prostaglandin synthesis is regulated via p38 MAPK or ERK1/2–dependent control over cyclooxygenase (COX)-2 expression. p38 MAPK and JNK govern the induction of interleukin (IL)-8, coagulation factor III, and angiogenic mediator ephrinA1 (23,48). Finally, NO synthesis is controlled by both ERK1/2 and p38 MAPK via the phosphorylation and/or transcriptional regulation of inducible NOS (iNOS) and eNOS (51–53). Such control over the release of bioactive substances allows MAPKs to adjust the response of several cell types present at the site of disturbance.

IN VIVO APPLICATIONS

These data suggest that factors inducing MAPK activation affect vascular function via several mechanisms. They alter endothelial permeability or EC ability to recognize leukocytes and change the level of bioactive mediators released into the vascular lumen. Therefore, the response of blood vessels as a whole depends on the type of EC and nature of the initial signal, as well as on the pattern of secondary signals originating from mobile luminal elements and resident cells. Thus, distinct signaling pathways may dominate in specific models of inflammation, suggesting the necessity for different pharmacological inhibitors to suppress edema and neutrophil infiltration and, correspondingly, the duration and severity of inflammation on the tissue level.

For example, the MEK/ERK inhibitor U0126 was shown to reduce phorbol-ester–induced ear edema and trauma-induced brain edema (44,54). p38 MAPK inhibitors FR167653 and SB203508 suppressed carrageenan-induced paw edema, LPS-induced subcutaneous vascular leakage, or lung edema (55–57). In certain models, p38 MAPK inhibitors were shown to decrease the level of TNF-α, prostanoids (55), or IL-1β, whereas in other experiments no significant changes in the level of proinflammatory mediators were found. Surprisingly, in several studies, p38 MAPK inhibition aggravated lung edema (58) and cerebral vascular permeability (59), suggesting that the use of p38 MAPK inhibitors as antiedemagenic agents might be limited to certain types of inflammation.

CONCLUSION

MAPKs represent the essential components of those signal transduction pathways implicated in vascular development, vascular remodeling, or inflammatory responses. Given the multiple physiological or pathological processes induced by MAPK activation, it is hardly surprising that significant organ dysfunction may result from the derangement of these processes. As MAPK-mediated pathways are further uncovered, our potential to develop methods to prevent the morbidity associated with vascular leakage–dependent organ failure or vascularization-dependent tumor development may be greatly improved. However, we must keep in mind that the pharmacological inhibition of MAPK would probably affect numerous physiological functions as well. Local intervention at the site of disturbance or specific targeting of certain downstream proteins may be needed to suppress the pathological rather than physiological responses to MAPK activation.

KEY POINTS

- MAPKs are critically involved in the regulation of vascular endothelial functions.
- The physiological effects of MAPK activation are elicited via the phosphorylation of numerous specific protein substrates.
- The final consequences of MAPK activation on a cellular level depend on the balance between different MAPK activities and the activities of other signaling pathways, which, in turn, vary from cell to cell and from stimulus to stimulus.
- The manipulation of specific MAPK activities may represent a powerful tool in the treatment of vascular diseases.

REFERENCES

1 Lewis TS, Shapiro PS, Ahn NG. Signal transduction through MAP kinase cascades. *Adv Cancer Res.* 1998;74:49–139.
2 Kuida K, Boucher DM. Function of MAP kinases: insights from gene-targeting studies. *J Biochem.* 2004;135:653–656.
3 Bourcier C, Jacquel A, Hess J, et al. p44 mitogen-activated protein kinase (extracellular signal-regulated kinase 1)-dependent signaling contributes to epithelial skin carcinogenesis. *Cancer Res.* 2006;66:2700–2707.
4 Ferguson SM, Fasano S, Yang P, et al. Knockout of ERK1 enhances cocaine-evoked immediate early gene expression and behavioral plasticity. *Neuropsychopharmacology*. 2006; In press.
5 Aouadi M, Laurent K, Prot M, et al. Inhibition of p38MAPK increases adipogenesis from embryonic to adult stages. *Diabetes.* 2006;55:281–289.
6 Engel FB, Schebesta M, Duong MT, et al. p38 MAP kinase inhibition enables proliferation of adult mammalian cardiomyocytes. *Gen Devel.* 2005;19:1175–1187.
7 Beardmore VA, Hinton HJ, Eftychi C, et al. Generation and characterization of p38β (MAPK11) gene-targeting mice. *Mol Cell Biol.* 2005;25:10454–10464.
8 Tachibana H, Perrino C, Takaoka H, et al. JNK1 is required to preserve cardiac function in the early response to pressure overload. *Biochem Biophys Res Commun.* 2006;343:1060–1066.
9 Schattenberg JM, Singh R, Wang Y, et al. JNK1 but not JNK2 promotes the development of steatohepatitis in mice. *Hepatology*. 2006;43:163–172.
10 Jimenez B, Volpert OV, Reiher F, et al. c-Jun N-terminal kinase activation is required for the inhibition of neovascularization by thrombospondin-1. *Oncogene.* 2001;20:3443–3448.
11 Seternes OM, Mikalsen T, Johansen B, et al. Activation of MK5/PRAK by the atypical MAP kinase ERK3 defines a novel signal transduction pathway. *EMBO J.* 2004;23:4780–4791.
12 Yang SH, Sharrocks AD, Whitmarsh AJ. Transcriptional regulation by the MAP kinase signaling cascades. *Gene.* 2003;320:3–21.
13 Hayashi M, Lee JD. Role of BMK1/ERK5 signaling pathway: lessons from knockout mice. *J Mol Med.* 2004;82:800–808.
14 Turgeon B, Lang BF, Meloche S. The protein kinase ERK3 is encoded by a single functional gene: genomic analysis of the ERK3 gene family. *Genomics.* 2002;80:673–680.
15 Miyata Y, Nishida E. Distantly related cousins of MAP kinase: biochemical properties and possible physiological functions. *Biochem Biophys Res Commun.* 1999;266:291–295.
16 Cangarajah BJ, Khokhlatchev A, Cobb MH, Goldsmith EJ. Activation mechanism of the MAP kinase ERK2 by dual phosphorylation. *Cell.* 1997;90:859–869.
17 Coulombe P, Rodier G, Pelletier S, et al. Rapid turnover of extracellular signal-regulated kinase 3 by the ubiquitin-proteasome pathway defines a novel paradigm of mitogen-activated protein kinase regulation during cellular differentiation. *Mol Cell Biol.* 2003;23:4542–4558.
18 Johnson GL, Lapadat R. Mitogen-activated protein kinase pathways mediated by ERK, JNK, and p38 protein kinases. *Science.* 2002;298:1911–1912.
19 Roux PP, Blenis J. ERK and p38 MAPK-activated protein kinases: a family of protein kinases with diverse biological functions. *Microbiol Mol Biol Rev.* 2004;68:320–344.
20 Camps M, Nichols A, Arkinstall S. Dual specificity phosphatases: a gene family for control of MAP kinase function. *FASEB J.* 2000;14:6–16.
21 Robinson MJ, Xu BE, Stippec S, Cobb MH. Different domains of the mitogen-activated protein kinases ERK3 and ERK2 direct subcellular localization and upstream specificity in vivo. *J Biol Chem.* 2002;277:5094–5100.
22 Gaestel M. MAPKAP kinases – MKs – two's company, three's a crowd. *Nat Rev Mol Cell Biol.* 2006;7:120–130.
23 Bogatcheva NV, Dudek SM, Garcia JG, Verin AD. Mitogen-activated protein kinases in endothelial pathophysiology. *J Invest Medicine.* 2003;51:341–351.
24 Munoz-Chapuli R, Quesada AR, Angel Medina M. Angiogenesis and signal transduction in endothelial cells. *Cell Mol Life Sci.* 2004;61:2224–2243.
25 Liu F, Verin AD, Wang P, et al. Differential regulation of sphingosine-1-phosphate- and VEGF-induced endothelial cell chemotaxis. Involvement of G(ialpha2)-linked Rho kinase activity. *Am J Respir Cell Mol Biol.* 2001;24:711–719.
26 Tanaka K, Abe M, Sato Y. Roles of extracellular signal-regulated kinase 1/2 and p38 mitogen-activated protein kinase in the signal transduction of basic fibroblast growth factor in endothelial cells during angiogenesis. *Jpn J Cancer Res.* 1999;90:647–654.
27 Mirzapoiazova T, Kolosova IA, Romer L, et al. The role of caldesmon in the regulation of endothelial cytoskeleton and migration. *J Cell Physiol.* 2005;203:520–528.
28 Keezer SM, Ivie SE, Krutzsch HC, et al. Angiogenesis inhibitors target the endothelial cell cytoskeleton through altered regulation of heat shock protein 27 and cofilin. *Cancer Res.* 2003;63:6405–6412.
29 Oda N, Abe M, Sato Y. ETS-1 converts endothelial cells to the angiogenic phenotype by inducing the expression of matrix metalloproteinases and integrin beta3. *J Cell Physiol.* 1999;178:121–132.
30 Yu J, Bian D, Mahanivong C, et al. p38 Mitogen-activated protein kinase regulation of endothelial cell migration depends on urokinase plasminogen activator expression. *J Biol Chem.* 2004; 279:50446–50454.
31 Bein K, Odell-Fiddler ET, Drinane M. Role of TGF-beta1 and JNK signaling in capillary tube patterning. *Am J Physiol Cell Physiol.* 2004;287:C1012–C1022.

32 Wang BW, Chang H, Lin S, et al. Induction of matrix metalloproteinases-14 and -2 by cyclical mechanical stretch is mediated by tumor necrosis factor-alpha in cultured human umbilical vein endothelial cells. *Cardiovasc Res.* 2003;59:460–469.

33 Matsumoto T, Turesson I, Book M, et al. p38 MAP kinase negatively regulates endothelial cell survival, proliferation, and differentiation in FGF-2-stimulated angiogenesis. *J Cell Biol.* 2002; 156:149–160.

34 Grethe S, Ares MP, Andersson T, Porn-Ares MI. p38 MAPK mediates TNF-induced apoptosis in endothelial cells via phosphorylation and downregulation of Bcl-x(L). *Exp Cell Res.* 2004; 298:632–642.

35 Brouard S, Berberat PO, Tobiasch E, et al. Heme oxygenase-1-derived carbon monoxide requires the activation of transcription factor NF-kappa B to protect endothelial cells from tumor necrosis factor-alpha-mediated apoptosis. *J Biol Chem.* 2002;277:17950–17961.

36 Clermont F, Adam E, Dumont JE, Robaye B. Survival pathways regulating the apoptosis induced by tumour necrosis factor-alpha in primary cultured bovine endothelial cells. *Cell Signal.* 2003;15:539–546.

37 Pi X, Yan C, Berk BC. Big mitogen-activated protein kinase (BMK1)/ERK5 protects endothelial cells from apoptosis. *Circ Res.* 2004;94:362–369.

38 Bullard LE, Qi X, Penn JS. Role for extracellular signal-responsive kinase-1 and -2 in retinal angiogenesis. *Inv Ophth Visual Sci.* 2003;44:1722–1731.

39 Mani N, Khaibullina A, Krum JM, Rosenstein JM. Activation of receptor-mediated angiogenesis and signaling pathways after VEGF administration in fetal rat CNS explants. *J Cereb Blood Flow Metab.* 2003;23:1420–1429.

40 Caunt M, Huang YQ, Brooks PC, Karpatkin S. Thrombin induces neoangiogenesis in the chick chorioallantoic membrane. *J Thromb Haemost.* 2003;1:2097–2102.

41 Jackson JR, Bolognese B, Hillegass L, et al. Pharmacological effects of SB 220025, a selective inhibitor of P38 mitogen-activated protein kinase, in angiogenesis and chronic inflammatory disease models. *J Pharmacol Exp Ther.* 1998;284:687–692.

42 Jimenez B, Volpert OV, Crawford SE, et al. Signals leading to apoptosis-dependent inhibition of neovascularization by thrombospondin-1. *Nat Med.* 2000;6:41–48.

43 Issbrucker K, Marti HH, Hippenstiel S, et al. p38 MAP kinase – a molecular switch between VEGF-induced angiogenesis and vascular hyperpermeability. *FASEB J.* 2003;17:262–264.

44 Mori T, Wang X, Aoki T, Lo EH. Downregulation of matrix metalloproteinase-9 and attenuation of edema via inhibition of ERK mitogen activated protein kinase in traumatic brain injury. *J Neurotrauma.* 2002;19:1411–1419.

45 Petrache I, Birukova A, Ramirez SI, et al. The role of the microtubules in tumor necrosis factor-alpha-induced endothelial cell permeability. *Am J Respir Cell Mol Biol.* 2003;28:574–581.

46 Goldberg PL, MacNaughton DE, Clements RT, et al. p38 MAPK activation by TGFbeta-1 increases MLC phosphorylation and endothelial monolayer permeability. *Am J Physiol.* 2002;282: L146–L154.

47 Fujibe M, Chiba H, Kojima T, et al. Thr203 of claudin-1, a putative phosphorylation site for MAP kinase, is required to promote the barrier function of tight junctions. *Exp Cell Res.* 2004;295:36–47.

48 Viemann D, Goebeler M, Schmid S, et al. Transcriptional profiling of IKK2/NF-kappa B- and p38 MAP kinase-dependent gene expression in TNF-alpha-stimulated primary human endothelial cells. *Blood.* 2004;103:3365–3373.

49 Wang Q, Yerukhimovich M, Gaarde WA, et al. MKK3 and -6-dependent activation of p38{alpha} MAP kinase is required for cytoskeletal changes in pulmonary microvascular endothelial cells induced by ICAM-1 ligation. *Am J Physiol.* 2005;288:L359–L369.

50 Stein BN, Gamble JR, Pitson SM, et al. Activation of endothelial extracellular signal-regulated kinase is essential for neutrophil transmigration: potential involvement of a soluble neutrophil factor in endothelial activation. *J Immunol.* 2003;171:6097–6104.

51 Kuhlmann CR, Trumper JR, Abdallah Y, et al. The K+-channel opener NS1619 increases endothelial NO-synthesis involving p42/p44 MAP-kinase. *Thromb Haemost.* 2004;92:1099–1107.

52 Huang H, Rose JL, Hoyt DG. p38 Mitogen-activated protein kinase mediates synergistic induction of inducible nitric-oxide synthase by lipopolysaccharide and interferon-gamma through signal transducer and activator of transcription 1 Ser727 phosphorylation in murine aortic endothelial cells. *Mol Pharmacol.* 2004;66:302–311.

53 Anter E, Thomas SR, Schulz E, et al. Activation of endothelial nitric-oxide synthase by the p38 MAPK in response to black tea polyphenols. *J Biol Chem.* 2004;279:46637–46643.

54 Jaffee BD, Manos EJ, Collins RJ, et al. Inhibition of MAP kinase kinase (MEK) results in anti-inflammatory response in vivo. *Biochem Biophys Res Commun.* 2000;268:647–651.

55 Nishikori T, Irie K, Ozaki M, Yoshikoa T. Anti-inflammatory potency of FR167653, a p38 mitogen-activated protein kinase inhibitor, in mouse models of acute inflammation. *Eur J Pharmacol.* 2002;451:327–333.

56 Yoshinari D, Takeyoshi I, Koibuchi Y, et al. Effects of a dual inhibitor of tumor necrosis factor-alpha and interleukin-1 on lipopolysaccharide-induced lung injury in rats: involvement of the p38 mitogen-activated protein kinase pathway. *Crit Care Med.* 2001;29:628–634.

57 Birrell M, Hele D, McCluskie K, et al. Effect of the p38 kinase inhibitor, SB 203580, on Sephadex induced airway inflammation in the rat. *Eur Respir J.* 2000;16:947–950.

58 Arcaroli J, Yum HK, Kupfner J, et al. Role of p38 MAP kinase in the development of acute lung injury. *Clin Immunol.* 2001;2:211–219.

59 Lennmyr F, Ericsson A, Gerwins P, et al. Increased brain injury and vascular leakage after pretreatment with p38-inhibitor SB203580 in transient ischemia. *Acta Neurol Scand.* 2003;108: 339–345.

82

Protein Kinase C

Alex Toker

Beth Israel Deaconess Medical Center, Harvard Medical School, Boston, Massachusetts

The discovery of protein kinase C (PKC) by Y. Nishizuka and colleagues in the late 1970s was a landmark event in biology and reinforced the emerging concept at the time that extracellular agonists that stimulate receptor-mediated phospholipid hydrolysis transduce signals through the activation of a variety of serine/threonine protein kinases (1). In this context, the discovery of PKC was particularly illuminating because it also became evident that this kinase serves as the receptor for potent tumor-promoting phorbol esters, which mimic the natural allosteric activator of PKC, diacylglycerol (DAG) (2). These findings propelled the nascent PKC field into one of the most widely studied lipid second-messenger signaling pathways, which has occupied researchers in the last two decades (3). Studies to date have painted a detailed molecular picture of the mechanisms that control PKC activation, localization, and function. Thus, at the dawn of the new millennium, the spatial and temporal controls of PKC regulation by lipid activators and by phosphorylation are now well understood. Although this knowledge has allowed researchers to foray into more complex questions about how PKC transduces signals to modulate cellular responses, this area of investigation lags significantly behind.

This chapter reviews the mechanisms by which PKC functions as a signal-relay enzyme in the endothelium. Although our understanding of the role of PKC in vascular biology and pathology is still in its infancy, studies have clearly identified this kinase as a key regulator of signal coupling in endothelial cells (ECs) and have thus provided exciting prospects for translating this information into therapeutic value. The knowledge that phorbol esters such as phorbol myristate acetate (PMA) are potent activators of PKC in cells has proven to be a powerful tool for studying the role of this kinase in EC function. However, it recently has become clear that there exist numerous cellular targets of PMA and its cellular counterpart DAG other than PKCs, so it is therefore likely that EC responses originally attributed to PKCs may actually be regulated by other PMA/DAG effectors. These include chimaerins, Ras guanine nucleotide releasing proteins (GRPs), protein kinase D (PKD), Munc13 proteins, and DAG-kinases (4). Therefore, attributing the effects of PMA on cellular responses to PKC alone can be misleading. This is further complicated by the fact that up to 10 distinct isoforms of PKC, which share similar regulatory mechanisms and functional output, can exist in any one cell or tissue. Clearly, more sophisticated methods are required to accurately probe specific functions of this key signaling enzyme. The development of PKC knockouts, transgenes, and RNA interference has begun to shed some light on the role of PKC in EC biology.

THE PROTEIN KINASE C FAMILY

Although originally discovered as a single enzymatic entity, purification and cloning efforts during the 1980s resulted in the identification of 10 distinct PKC gene products in mammalian cells (5). These are subdivided into conventional PKCs (cPKC: PKC-α, PKC-βI, PKC-βII, and PKC-γ), novel PKCs (nPKC: PKC-δ, PKC-ε, PKC-η, and PKC-θ), and atypical PKCs (aPKC: PKC-ζ and PKC-ι/λ). Although originally characterized as a PKC, PKC-μ has been reclassified into its own family of kinases, the PKDs, which although distantly related, do share some similarities in regulation and function with their cousins, the PKCs. In the human Kinome, PKCs are founding members of the AGC superfamily (protein kinases A, G, and C). The classification of PKCs is shown in Table 82-1, and is largely based on their domain structure, which in turn defines their individual modes of regulation by either calcium, DAG, and/or the anionic phospholipid phosphatidylserine. Conventional and novel PKCs contain two copies of the C1 domain at the DAG/PMA binding site (Figure 82.1). In atypical PKCs, only one C1 domain is present, and so these PKCs are unresponsive to PMA and DAG. The function of the single C1 domain in these PKCs is not known. The C2 domain houses the calcium-binding site in cPKCs, and calcium binding to these enzymes promotes interaction with anionic phospholipids. An atypical C2 domain is found in nPKCs, and the lack of critical residues necessary to promote interaction with calcium renders these PKCs unresponsive

Table 82-1: Summary of PKC Classes

PKC Isoform	*DAG Binding*	*Calcium Responsive*	*Activation Mechanism*	*Vascular Phenotypes in Knockout Mice*
Conventional PKCs: α, βI, βII, γ	Yes	Yes	DAG binding	*PKCβII*$^{-/-}$: reduced retinal neovascularization
Novel PKCs: δ, ε, η, θ	Yes	No	DAG binding	No data
Atypical PKCs: ι/λ, ζ	No	No	Phosphorylation	No data

to calcium. aPKCs also are unresponsive to calcium, because they lack a C2 domain. All PKCs also are regulated by interaction with phosphatidylserine on membranes.

The catalytic domain of PKCs is highly conserved, and the secondary and tertiary structure reveals a similar mode of regulation for all enzymes. Although it was appreciated for almost a decade that PKC is itself phosphorylated in cells, it was not until the discovery of phosphoinositide-dependent kinase (PDK)-1 as the upstream kinase for PKCs' close cousin Akt/protein kinase B (PKB) that it became apparent that PDK-1 also phosphorylates PKCs at a crucial residue in a conserved region in the catalytic cleft, the activation loop (6). In fact, it is now known that before PKC is able to bind to DAG, it must first be phosphorylated at the activation loop and two other conserved residues in the carboxyl-terminus. This suggests that phosphorylation of PKC is the rate-limiting step in its activation cycle, such that phosphorylation locks the enzyme in a conformation that is able to bind to DAG and calcium. PDK-1, which is regulated by the phosphoinositide 3-kinase (PI3K) pathway, is the master regulator of PKCs and indeed many other AGC kinases. In the case of cPKCs, phosphorylation by PDK-1 is a processive event that occurs in the absence of extracellular stimuli and serves to maintain the enzyme in a conformation capable of DAG binding. In the case of nPKCs, although some evidence suggests that phosphorylation by PDK-1 serves to activate the kinase in response to ligand stimulation, the main activating mechanism is DAG binding. Once PKCs are phosphorylated at the activation loop, the two carboxyl-terminal sites, termed the *turn-motif* and the *hydrophobic site*, are phosphorylated (see Figure 82.1). In the case of cPKCs, this occurs through autophosphorylation. Although this also is true for novel and atypical PKCs, evidence suggests that a heterologous upstream kinase for the hydrophobic motif exists.

The mechanisms leading to activation of PKC that are ultimately responsible for positioning the enzyme next to its effector substrates have been elucidated in exquisite detail. Numerous excellent reviews discuss these models, and the reader is referred to these articles for more in-depth discussions (6–8). Briefly, newly synthesized PKC is phosphorylated by PDK-1, which stimulates autophosphorylation in the carboxyl-terminus as part of the maturation of the enzyme. A pseudosubstrate sequence in the PKC amino terminus plugs into the catalytic cleft, effectively preventing access to substrates. Extracellular signals promote the hydrolysis of phospholipids such as phosphatidylinositol 4,5-biphosphate thus producing DAG and inositol 1,4,5-triphosphate (IP_3), and the latter induces release of calcium from intracellular stores. Newly

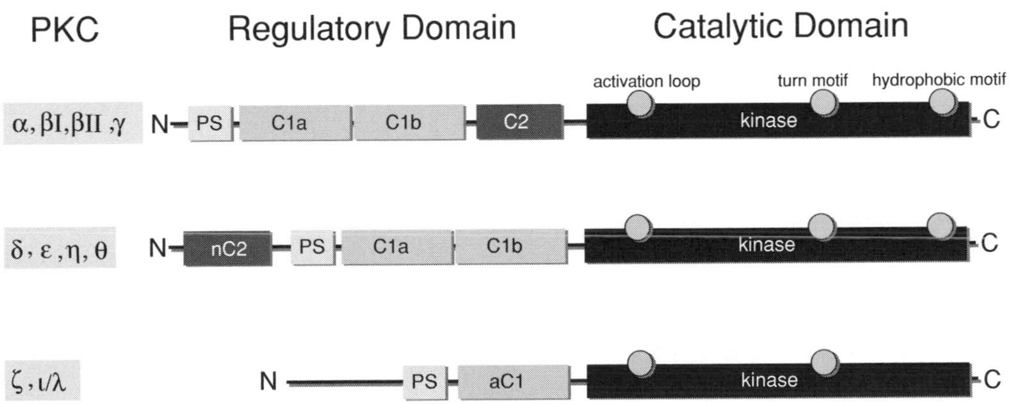

Figure 82.1. The PKC family. Schematic showing the domain structure of conventional (PKC-α, PKC-βI, PKC-βII, PKC-γ), novel (PKC-δ, PKC-ε, PKC-η, PKC-θ), and atypical (PKC-ζ, PKC-ι/λ) PKCs. Indicated are the pseudosubstrate (*PS*), C1a and C1b, C2, and kinase domains. Note that novel PKCs have a novel C2 (*nC2*) domain that does not bind calcium, and atypical PKCs have one copy of an atypical C1 (*aC1*) domain that does not bind DAG/PMA. The phosphorylation sites at the activation loop, turn motif, and hydrophobic site in the catalytic domain are also depicted.

synthesized DAG serves to recruit cPKCs and nPKCs to membranes, along with scaffolding proteins that position the enzyme at discrete cellular locations. Binding of calcium (in the case of cPKC) and phosphatidylserine locks the enzyme on membranes and promotes release of the pseudosubstrate domain. In the presence of nucleotide (adenosine triphosphate [ATP]), substrates are now phosphorylated by PKC, and the signal is transduced. In this model, PMA can substitute for DAG, although it is important to note that phorbol esters bind to C1 domains with an affinity of at least two orders higher than does DAG. Thus PMA does not accurately reproduce the kinetics of agonist-stimulated PKC activation. The life cycle of all PKC isoforms is terminated when the enzyme is dephosphorylated by protein phosphatases. Dephosphorylation of PKC is the signal for its degradation, and there is evidence that the ubiquitin/proteasome pathway is responsible.

It is important to note that variations on this model do exist. For example, although phosphorylation of conventional PKCs is part of the maturation process, for atypical and even novel PKCs, evidence suggests that some agonists actually stimulate phosphorylation of PKC at the activation loop, an event that may therefore be coupled to receptor-initiated signaling (7). In the case of aPKCs that are not PMA/DAG responsive, activation is best understood in the context of phosphorylation, which is mediated by PI3K and PDK-1 and which occurs in response to the stimulation of cells with many agonists, including vascular endothelial growth factor (VEGF). This locks the kinase in the active conformation in the absence of interaction with any lipids, although some evidence suggests that phosphatidylserine still is required for maximal aPKC activity. Similarly, other phospholipids, such as the PI3K product PtdIns-3,4,5-P_3 also may moderately affect aPKC activation through allosteric mechanisms. These differences are important in the context of understanding the function of PKC in EC responses.

PROTEIN KINASE C IN THE ENDOTHELIUM

General Considerations

Expression patterns of signaling enzymes in adult tissues can provide useful information concerning their potential functions, especially in the case of multigene families such as the PKCs. Although PKCs are essentially ubiquitous in adult mammalian tissues, some isozymes reveal restricted expression profiles. For example, PKC-θ is restricted to immune cells and skeletal muscle, whereas PKC-γ is exclusively expressed in the central nervous system. Interestingly, PKC-β shows highest expression in most vascular tissues including retina, heart, and kidney (9). Atypical PKCs are ubiquitous and are expressed in all tissues studied.

When considering a function for PKCs in EC responses, at least two requirements must be fulfilled if PKC is to play a physiologically relevant function: (a) agonist-stimulated DAG synthesis (in the case of cPKCs and nPKCs) and calcium flux (in the case of cPKCs) must occur subsequent to receptor-initiated signaling and, (b) processive (cPKC) or stimulated (nPKC and aPKC) phosphorylation of PKC by PDK-1 also must occur (Figure 82.2). In the case of VEGF, these requirements appear to be met. Stimulation of cultured ECs with VEGF leads to activation of the receptor VEGFR2 (FLK-1/KDR), leading to tyrosine phosphorylation of phospholipase C (PLC)-γ, DAG synthesis, and calcium release, as well as activation of the PI3K/PDK-1 signaling axis (10). Interestingly, this is not observed with placental growth factor (PlGF), which signals through VEGFR1 (FLT-1), suggesting that EC signaling through PKC may be restricted to VEGF-derived inputs. The activation of several PKCs in ECs stimulated with VEGF has been reported. In bovine aortic ECs, VEGF stimulates the activation of PKC-α and PKC-βII (11). In human umbilical vein ECs (HUVECs), VEGF stimulates PKC-α, PKC-ε, and PKC-ζ activation (12–14). PKC-δ activation by VEGF also has been shown in VEGF-stimulated HUVECs (15). Given that PI3K is potently activated by VEGF, it is not surprising that atypical PKC-ζ activation by VEGFR2 has been extensively documented (12,16). It should be noted, however, that the complexity of the PKC gene family has necessitated the use of chemical inhibitors to identify the activation profiles of distinct isozymes in cells and, unfortunately, these inhibitors are not sufficiently specific to accurately probe the activation or function of a given PKC. Regardless, clear evidence shows that multiple PKCs are activated by VEGF in ECs in culture.

Other agonists that stimulate PKC activation in ECs include thrombin, which promotes EC leakiness. Thrombin, which stimulates PLC-β–induced DAG synthesis, has been shown to promote PKC-α, PKC-δ, and PKC-ζ activation in a variety of EC types (17). Exposure of ECs to H_2O_2 stimulates phosphorylation of VEGFR, and this has been shown to be modulated by PKC-δ (18). Finally, ECs are continuously subjected to hemodynamic forces, and studies on ECs under mechanical strain have revealed activation of PKC-α and PKC-ε (19). Thus, the overwhelming evidence points to a role for EC stimuli in the activation of various PKCs.

Unfortunately, evidence for the activation of PKCs in the endothelium of whole tissues or organisms is lacking. This is due to technical limitations in our ability to visualize PKC activity in intact cells. For example, phosphorylation state-specific antibodies to monitor PKC activity cannot be used because phosphorylation of PKC is not an accurate measure of activation. More recent advances in imaging technology have permitted the visualization of PKC activity in real-time in transfected cells (20) and, when these techniques are refined to permit imaging in whole organisms, a wealth of important new information is likely to emerge. Until then, we are confined to in vitro studies. Similar hurdles are encountered when considering PKC knockouts in mice. All nine PKC genes have been targeted by homologous recombination, and all but eight are viable and show no obvious developmental phenotypes. The only embryonically lethal mouse knockout is PKC-ι/λ. As discussed later, some of the PKC knockouts have revealed some vascular phenotypes, such as loss of retinal vascularization.

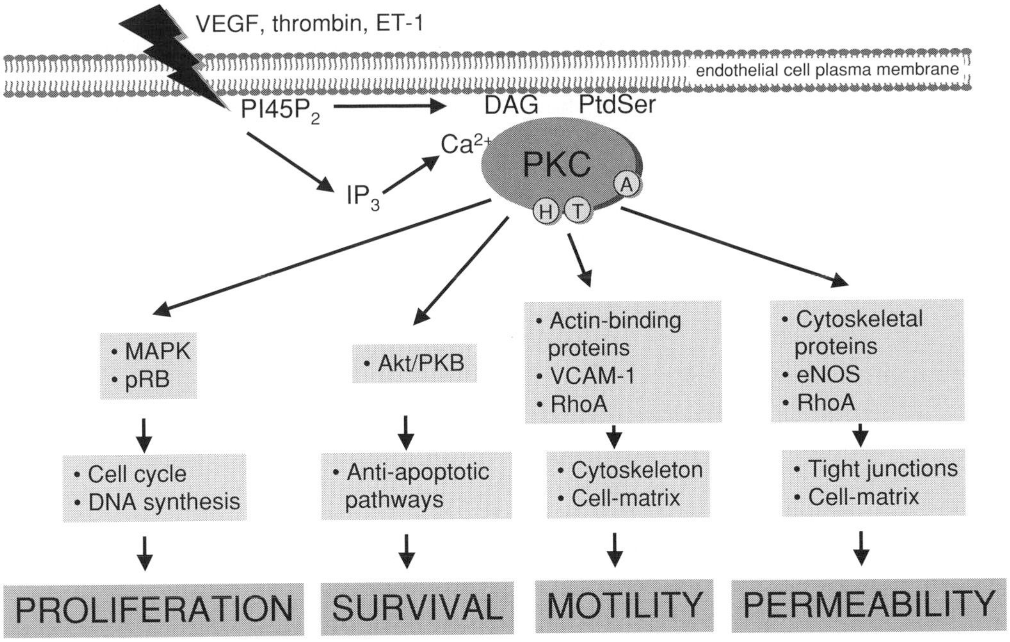

Figure 82.2. EC responses modulated by PKC. Diagrammatic representation of activation of PKC by stimuli that activate ECs, including VEGF, thrombin, and endothelin (ET)-1. All these agonists stimulate phospholipid ($PI4,5P_2$) hydrolysis, leading to synthesis of DAG and release of calcium (Ca^{2+}). Phosphatidylserine (PtdSer) also is required as a cofactor for PKC activation. Phosphorylation of PKC at the activation loop (*A*), turn motif (*T*), and hydrophobic site (*H*) is rate-limiting for PKC activation. Once activated, PKC transduces the signal in ECs by coupling to secondary signaling cascades including MAPK, phosphorylation of retinoblastoma (pRB), synergy with Akt/PKB, and phosphorylation of actin-regulatory proteins as well as eNOS. This in turn modulates endothelial phenotypes that manifest in altered proliferation, survival, motility, or permeability.

However, given the functional redundancy of the PKC family, double and triple knockouts may be required to reveal more essential roles for PKCs in other aspects of EC function. Although PKC transgenes have been developed, none to date have interrogated a specific function for PKCs in the vasculature. Based on these considerations, it is therefore not surprising that our knowledge on the role of PKCs in EC biology has been limited to cell culture experiments. Despite the simplistic approach, some important advances have been made.

Endothelial Cell Growth and Survival

Although VEGF originally was discovered as a permeability factor, it also functions as a bona fide mitogen by promoting cell-cycle progression and DNA synthesis. The use of various chemical inhibitors or dominant negative mutants of various isozymes has suggested a role for PKCs in EC proliferation and growth. For example, inhibition of PKC-θ, PKC-α, PKC-ζ, PKC-ε, and PKC-η results in inhibition of DNA synthesis and EC proliferation in response to VEGF in vitro (12,13). Can it be that simultaneous signaling through all these PKCs is required for EC growth? This seems unlikely, and until these observations are confirmed using more specific tools such as small interfering RNA (siRNA) or knockouts, the role of a given PKC in vascular proliferation will remain unclear. However, a critical role for PKC-β in EC proliferation has been demonstrated convincingly. Studies have shown that the specific PKC-β inhibitor LY333531, currently in clinical trials for the treatment of diabetes and cardiac ventricular hypertrophy, attenuates VEGF-stimulated mitogenesis (11). Even more compelling are the findings that transgenic mice overexpressing *PKC-βII* in vascular tissues show a dramatic increase in the angiogenic response to oxygen-induced retinal ischemia, whereas *PKC-βII*-null mice show a decrease in retinal neovascularization (21). Similarly, mitogenesis induced by VEGF is increased in ECs expressing PKC-βI/II, and it is attenuated in cells expressing dominant negative PKC-β mutants (21). A role for retinoblastoma phosphorylation, which is essential for cell-cycle progression and entry into S phase, was also suggested in PKC-β–expressing cells (21). Finally, myristoylated peptides that specifically blunt PKC-β activity also block VEGF-induced EC proliferation in vitro and angiogenesis in vivo (22). These data make a strong case for an important role for PKC-β in modulating EC growth and angiogenesis, although the specific substrates and transcriptional events induced by PKC-β signaling to promote these responses have yet to be defined.

By analogy with other receptor tyrosine kinases that stimulate cell growth as well as cellular survival, an important function of VEGF is its prosurvival activity in ECs. Although the PI3K/Akt/PKB signaling pathway plays a major role in promoting survival, some studies have suggested a role for PKCs in modulating this response in ECs. As discussed earlier, the exposure of ECs to oxidative stress stimulates VEGFR phosphorylation and activation, leading to enhanced survival (18). Blocking PKC-δ signaling attenuates this response. A role for PKC-δ in modulating the survival capacity of Akt/PKB also has been proposed (15). To date, no data address an in vivo role for PKCs in EC survival, either in tumor angiogenesis or developmental angiogenesis.

Endothelial Cell Adhesion and Motility

Subsequent to local ischemia, wounding, or inflammation, new blood vessels develop, and the normally quiescent ECs become activated and motile as part of the angiogenic response. Studies have implicated PKC in controlling those EC cytoskeletal rearrangements leading to increased adhesion and migration. Inhibition of PKC-α using either chemical inhibitors or antisense oligonucleotides blocks VEGF-stimulated EC migration and capillary tube formation in vitro (23). Conversely, overexpression of PKC-α increases EC motility. Similar approaches also have implicated PKC-θ in EC tube formation (24). Although the mechanisms by which PKCs modulate the motile apparatus of ECs remain obscure, studies have addressed the role of adhesion molecules in this pathway. Vascular adhesion molecule (VCAM)-1 is an adhesion molecule expressed in activated ECs, and PKC signaling controls its expression in cells stimulated with thrombin. In this pathway, the induction of VCAM-1 expression occurs through two transcriptional events modulated by p65 nuclear factor (NF)-κB and GATA. Studies have shown that PKC-δ signaling controls NF-κB induction, whereas PKC-ζ controls GATA, and both transcription factors synergize to induce VCAM-1 expression in ECs (17). Moreover, high glucose also stimulates VCAM-1 expression in ECs, and selective inhibition of PKC-βII blocks this response (25). The relevance of these studies is that enhanced expression of adhesion molecules such as VCAM-1 in the endothelium may account for vascular damage in vivo, especially during stress hyperglycemia. Therefore PKC-selective inhibitors such as LY333531 may hold promise for the treatment of endothelial dysfunction during hyperglycemia and diabetes.

Vascular Permeability

Arguably the most widely studied role for PKCs in modulating EC function is the regulation of vascular permeability. As with other EC responses, a major role has emerged for conventional PKC-β and possibly for PKC-α. Both clinical and experimental evidence has implicated PKC signaling in events leading to microvascular barrier dysfunction, which has obvious implications for the inception and progression of human vascular disorders including diabetes, atherosclerosis, and ischemia–reperfusion injury. Studies in vitro have identified a role for PKC-α, PKC-βI/II, PKC-δ, and PKC-ζ in promoting VEGF- or thrombin-induced EC permeability. For example, the inhibition of PKC-α leads to attenuated thrombin-stimulated transendothelial electrical resistance, a measure of increased endothelial permeability, and the effects of PKC-α have been shown to be modulated, at least in part, through the small GTPase Rho (26). Inhibitors of PKC-β also blunt VEGF-stimulated transendothelial albumin flux (27), and similar results have been reported for PKC-δ, albeit in response to PMA stimulation of pulmonary microvascular ECs (28). In this context, studies also have shown that phorbol esters stimulate the permeability of cultured ECs to albumin, an effect that can be reversed in the presence of PKC inhibitors (29). Gain-of-function studies, using overexpression of PKC-βI in human dermal ECs, for example, augment the PMA response (30). Interestingly, angiopoietin 1, which opposes the effects of VEGF on permeability, also blocks the activation of PKC-β induced by VEGF, further supporting that signaling through PKC-β is crucial for EC permeability (31). Although PKC-ζ is not PMA-responsive, its overexpression does result in increased thrombin-stimulated HUVEC permeability, whereas a dominant negative PKC-ζ mutant blunts the thrombin effect (32). Therefore, multiple PKC isoforms appear to positively regulate EC permeability, at least in vitro. It is worth noting that at least one study has suggested the opposite, in which the inhibition of PKC-α and PKC-β using cell-permeant inhibitory peptides actually increased vascular permeability in vivo (22). It is rather difficult to rationalize a dual role for PKCs in both promoting and blocking vascular permeability, so clearly additional data with more selective approaches to target individual PKCs are required to establish a clear-cut role for distinct isozymes in permeability.

In vivo studies also have provided compelling evidence for a role for PKC (particular PKC-β) in regulating vascular permeability. The oral administration of LY333531 or LY290181, both reportedly PKC-β–specific inhibitors, blocked endothelial abnormality and microvascular leakage in the retina and kidney of diabetic mice (33,34). This is consistent with detectable increases in PKC-β expression in the heart and aorta of diabetic animals, concomitant with increased basal permeability of coronary venules. Again, this could be reversed with PKC-β–selective, but not PKC-α–selective, inhibitors (35).

An obvious question that arises is: How do PKCs such as PKC-β alter endothelial permeability? The cellular structures that define permeability in ECs include the actin cytoskeleton, cell–adherens junctions, and cell–matrix interactions. Signaling through PKC has been shown to modulate all these structures, although the precise mechanisms have yet to be elucidated. PKCs phosphorylate several cytoskeletal proteins, including myosin light chains (MLC), the myristoylated alanine-rich C kinase substrate (MARCKS) protein, actin-binding proteins, as well as intermediate filaments. As discussed earlier, PKC-dependent activation of Rho in ECs has been shown to be required for altered permeability. Disassembly of cadherin junctions also has been shown to occur in a PKC-dependent manner (31). Finally, enhanced nitric oxide (NO) signaling and induction of endothelial NO synthase (eNOS) also play a crucial role in vascular permeability, and there is evidence that PKCs interface with NO. Exposure of ECs to phorbol esters leads to increased eNOS phosphorylation and NO production. Consistent with this, pharmacological inhibition of eNOS blocks the permeability effects stimulated by PKC activators, suggesting an intimate relationship between PKC and NO production in the regulation of vascular permeability (36).

THERAPEUTIC INTERVENTION

The only current targeted therapy aimed at ameliorating vascular complications caused by EC dysfunction specifically due to alterations in PKC signaling is the LY333531 inhibitor from Eli Lilly & Co. The in vitro and in vivo studies that have made use of this compound to specifically block PKC-β signaling have been convincing and suggest that it may be viable to target PKCs in human dysfunctions such as retinopathy and other vascular complications. Recent studies in patients with type 1 or 2 diabetes with minimal retinopathy assessed the safety and vascular effects of the drug in the retina. The results were encouraging, revealing statistically significant improvements in retinal blood flow, along with good tolerance for 30 days and no measurable adverse effects (9). Phase III trials are currently under way. Studies such as these rest on the specificity of inhibitors such as LY333531, and clearly this is the rate-limiting feature of all inhibitors aimed at blocking protein kinases. In this context, recent studies have actually demonstrated that LY333531 is not as uniquely selective as previously thought, because it can inhibit other protein kinases such as PKA and PDK-1, albeit at slightly higher doses than the IC_{50} for PKC-β. Clearly needed at this stage is a more rigorous exploration of the specific roles of distinct PKC isozymes in modulating EC function. Once this is achieved, it may be possible to design more specific and potent inhibitors of a given PKC with the goal of targeting specific vascular abnormalities.

CONCLUSION

Despite the relative infancy of the PKC field in EC biology, it is readily apparent that PKC signaling regulates multiple vascular phenotypes, ranging from cell growth and survival, to microvascular permeability. The in vitro studies discussed here should serve as an ideal springboard for more accurately dissecting the role of individual PKC isozymes in ECs in vivo. It is clear thus far that conventional PKC-β is an important signal transduction enzyme that is activated by most stimuli that control EC responses. This does not rule out equally important roles for other PKCs in controlling one or more aspects of EC responses, but clearly more work is required to confirm the function of novel and atypical PKCs in the vasculature. Now that more specific tools are available to investigate these questions, including transgenes and RNA interference, there is reason to be excited about future progress in this field.

KEY POINTS

- PKC-β is activated by most stimuli that promote EC responses, leading to alterations in growth, survival, motility, and permeability.
- Other PKCs such as PKC-δ and PKC-ζ also appear to play important roles in specific EC functions, and may thus prove to be useful molecular targets for therapeutic intervention in vascular diseases.

Future Goals

- Future exploration of the role of specific PKC isozymes in EC function should make use of RNA interference as a loss-of-function approach, and activated alleles as a gain-of-function approach. Both approaches are amenable to in vitro and in vivo manipulation.
- Double and triple knockouts of PKCs may reveal important developmental phenotypes in the vasculature and reaffirm the notion that PKCs are important for angiogenesis.

ACKNOWLEDGMENTS

I would like to thank members of my laboratory for their critical evaluation of this chapter. Work in the laboratory is supported by National Institutes of Health grants CA075134 and CA096710.

REFERENCES

1 Takai Y, Kishimoto A, Iwasa Y, et al. Calcium-dependent activation of a multifunctional protein kinase by membrane phospholipids. *J Biol Chem.* 1979; 254(10):3692–3695.
2 Castagna M, Takai Y, Kaibuchi K, et al. Direct activation of calcium-activated, phospholipid-dependent protein kinase by tumor-promoting phorbol esters. *J Biol Chem.* 1982;257(13): 7847–7851.
3 Nishizuka Y. Studies and perspectives of protein kinase C. *Science.* 1986;233(4761):305–312.
4 Yang C, Kazanietz MG. Divergence and complexities in DAG signaling: looking beyond PKC. *Trends Pharmacol Sci.* 2003;24 (11):602–608.
5 Mellor H, Parker PJ. The extended protein kinase C superfamily. *Biochem J.* 1998;332(Pt 2):281–292.
6 Newton AC. Regulation of the ABC kinases by phosphorylation: protein kinase C as a paradigm. *Biochem J.* 2003;370(Pt 2):361–371.
7 Parekh DB, Ziegler W, Parker PJ. Multiple pathways control protein kinase C phosphorylation. *EMBO J.* 2000;19(4):496–503.
8 Toker A, Newton AC. Cellular signaling: pivoting around PDK-1. *Cell.* 2000;103(2):185–188.
9 Way KJ, Katai N, King GL. Protein kinase C and the development of diabetic vascular complications. *Diabet Med.* 2001;18 (12):945–959.

10 He H, Venema VJ, Gu X, et al. Vascular endothelial growth factor signals endothelial cell production of nitric oxide and prostacyclin through flk-1/KDR activation of c-Src. *J Biol Chem.* 1999;274(35):25130–25135.

11 Xia P, Aiello LP, Ishii H, et al. Characterization of vascular endothelial growth factor's effect on the activation of protein kinase C, its isoforms, and endothelial cell growth. *J Clin Invest.* 1996;98(9):2018–2026.

12 Wellner M, Maasch C, Kupprion C, et al. The proliferative effect of vascular endothelial growth factor requires protein kinase C-alpha and protein kinase C-zeta. *Arterioscler Thromb Vasc Biol.* 1999;19(1):178–185.

13 Wu LW, Mayo LD, Dunbar JD, et al. Utilization of distinct signaling pathways by receptors for vascular endothelial cell growth factor and other mitogens in the induction of endothelial cell proliferation. *J Biol Chem.* 2000;275(7):5096–5103.

14 Mason JC, Steinberg R, Lidington EA, et al. Decay-accelerating factor induction on vascular endothelium by vascular endothelial growth factor (VEGF) is mediated via a VEGF receptor-2 (VEGF-R2)- and protein kinase C-alpha/epsilon (PKCalpha/epsilon)-dependent cytoprotective signaling pathway and is inhibited by cyclosporin A. *J Biol Chem.* 2004;279(40):41611–41618.

15 Gliki G, Abu-Ghazaleh R, Jezequel S, et al. Vascular endothelial growth factor-induced prostacyclin production is mediated by a protein kinase C (PKC)-dependent activation of extracellular signal-regulated protein kinases 1 and 2 involving PKC-delta and by mobilization of intracellular $Ca2^{+}$. *Biochem J.* 2001; 353(Pt 3): 503–512.

16 Pal S, Datta K, Khosravi-Far R, Mukhopadhyay D. Role of protein kinase Czeta in Ras-mediated transcriptional activation of vascular permeability factor/vascular endothelial growth factor expression. *J Biol Chem.* 2001;276(4):2395–2403.

17 Minami T, Sugiyama A, Wu SQ, et al. Thrombin and phenotypic modulation of the endothelium. *Arterioscler Thromb Vasc Biol.* 2004;24(1):41–53.

18 Wang JF, Zhang X, Groopman JE. Activation of vascular endothelial growth factor receptor-3 and its downstream signaling promote cell survival under oxidative stress. *J Biol Chem.* 2004;279(26):27088–27097.

19 Cheng JJ, Wung BS, Chao YJ, Wang DL. Sequential activation of protein kinase C (PKC)-alpha and PKC-epsilon contributes to sustained Raf/ERK1/2 activation in endothelial cells under mechanical strain. *J Biol Chem.* 2001;276(33):31368–31375.

20 Violin JD, Zhang J, Tsien RY, Newton AC. A genetically encoded fluorescent reporter reveals oscillatory phosphorylation by protein kinase C. *J Cell Biol.* 2003;161(5):899–909.

21 Suzuma K, Takahara N, Suzuma I, et al. Characterization of protein kinase C beta isoform's action on retinoblastoma protein phosphorylation, vascular endothelial growth factor-induced endothelial cell proliferation, and retinal neovascularization. *Proc Natl Acad Sci USA.* 2002;99(2):721–726.

22 Spyridopoulos I, Luedemann C, Chen D, et al. Divergence of angiogenic and vascular permeability signaling by VEGF: inhibition of protein kinase C suppresses VEGF-induced angiogenesis, but promotes VEGF-induced, NO-dependent vascular permeability. *Arterioscler Thromb Vasc Biol.* 2002;22(6):901–906.

23 Wang A, Nomura M, Patan S, Ware JA. Inhibition of protein kinase Calpha prevents endothelial cell migration and vascular tube formation in vitro and myocardial neovascularization in vivo. *Circ Res.* 2002;90(5):609–616.

24 Tang S, Morgan KG, Parker C, Ware JA. Requirement for protein kinase C theta for cell cycle progression and formation of actin stress fibers and filopodia in vascular endothelial cells. *J Biol Chem.* 1997;272(45):28704–28711.

25 Kouroedov A, Eto M, Joch H, et al. Selective inhibition of protein kinase Cbeta2 prevents acute effects of high glucose on vascular cell adhesion molecule-1 expression in human endothelial cells. *Circulation.* 2004;110(1):91–96.

26 Mehta D, Rahman A, Malik AB. Protein kinase C-alpha signals rho-guanine nucleotide dissociation inhibitor phosphorylation and rho activation and regulates the endothelial cell barrier function. *J Biol Chem.* 2001;276(25):22614–22620.

27 Idris I, Gray S, Donnelly R. Protein kinase C-beta inhibition and diabetic microangiopathy: effects on endothelial permeability responses in vitro. *Eur J Pharmacol.* 2004;485(1–3):141–144.

28 Tinsley JH, Teasdale NR, Yuan SY. Involvement of PKCdelta and PKD in pulmonary microvascular endothelial cell hyperpermeability. *Am J Physiol Cell Physiol.* 2004; 286(1):C105–C111.

29 Lynch JJ, Ferro TJ, Blumenstock FA, et al. Increased endothelial albumin permeability mediated by protein kinase C activation. *J Clin Invest.* 1990;85(6):1991–1998.

30 Nagpala PG, Malik AB, Vuong PT, Lum H. Protein kinase C beta 1 overexpression augments phorbol ester-induced increase in endothelial permeability. *J Cell Physiol.* 1996;166(2):249–255.

31 Wang Y, Pampou S, Fujikawa K, Varticovski L. Opposing effect of angiopoietin-1 on VEGF-mediated disruption of endothelial cell interactions requires activation of PKC beta. *J Cell Physiol.* 2004;198(1):53–61.

32 Li X, Hahn CN, Parsons M, et al. Role of protein kinase Czeta in thrombin-induced endothelial permeability changes: inhibition by angiopoietin-1. *Blood.* 2004;104(6):1716–1724.

33 Aiello LP, Bursell SE, Clermont A, et al. Vascular endothelial growth factor-induced retinal permeability is mediated by protein kinase C in vivo and suppressed by an orally effective beta-isoform-selective inhibitor. *Diabetes.* 1997;46(9):1473–1480.

34 Koya D, Haneda M, Nakagawa H, et al. Amelioration of accelerated diabetic mesangial expansion by treatment with a PKC beta inhibitor in diabetic db/db mice, a rodent model for type 2 diabetes. *FASEB J.* 2000;14(3):439–447.

35 Yuan SY, Ustinova EE, Wu MH, et al. Protein kinase C activation contributes to microvascular barrier dysfunction in the heart at early stages of diabetes. *Circ Res.* 2000;87(5):412–417.

36 Duran WN, Seyama A, Yoshimura K, et al. Stimulation of NO production and of eNOS phosphorylation in the microcirculation in vivo. *Microvasc Res.* 2000;60(2):104–111.

83

Rho GTP-Binding Proteins

Allan Murray

University of Alberta, Edmonton, Canada

Remodeling of the endothelial cell (EC) cytoskeleton is central to many functions of the endothelium. The Rho family of small GTP-binding proteins have been identified as key regulators of F-actin cytoskeletal dynamics in a variety of cell types. They integrate signals from soluble mediators interacting with cytokine, growth factor tyrosine kinase, and G-protein–coupled receptors (GPCRs); as well as signals from cell–cell, and cell–matrix protein adhesion molecules. Recently, it has become appreciated that effector molecules downstream of Rho GTP-binding proteins also modulate several other well-described cell signaling pathways. We review the role these molecules play in the cell, with a particular focus on the EC.

HISTORY

The Rho family of small GTP-binding proteins, which consists of 22 members, is part of the larger Ras GTP-binding protein superfamily. These approximately 21-kDa proteins cycle between inactive GDP- and active GTP-bound forms to act as a molecular switch in signal transduction pathways. The members of this family are grouped by virtue of a shared structural motif, the Rho insert loop, that is present in the GTPase domain and contributes to the binding specificity for downstream effector molecules (1,2). In addition to this shared structural feature, most Rho family members undergo post-translational modification to link farnesyl or geranylgeranyl groups to the cysteine in a CAAX motif at the C-terminus of the molecule. Subcellular localization of the molecule is directed by the lipid moiety and, in some family members, is also influenced by additional domains in the C-terminus.

The first members of the Rho family of small GTP-binding proteins were discovered in the late 1980s by their similarity to Ras, and today the original family members RhoA, Rac1, and Cdc42 remain the best characterized functionally (3–5). These molecules are the first representatives of three of the six subfamilies of Rho GTP-binding proteins now identified.

Pivotal work done during the early 1990s identified characteristic cellular responses to Rho GTP-binding protein activation. Activation of RhoA, either by the administration of the bacterial toxin cytotoxic necrotizing factor-1 or overexpression of dominant active mutants in Swiss 3T3 fibroblasts resulted in prominent F-actin stress fiber formation in serum-starved cells (Figure 83.1) (6,7). In contrast, the expression of active Rac1 stimulated the extension of lamellaepodia, and active Cdc42 expression induced the formation of filopodia (6,7). Common to each of these changes in cell morphology is reorganization of the cell F-actin cytoskeleton.

The importance of Rho-dependent signaling in development is highlighted by observations of Rho-deficient mice (Table 83-1). Mouse embryos deficient in Rac1 fail to progress to gastrulation, apparently because of a failure of cell migration, and hence die in utero (8). Similarly, the conditional deletion mutation of Rac1 in myeloid lineage cells demonstrates defects in cell polarization and cell movement in vitro, consistent with failure to extend lamellae at the advancing edge of the cell (9,10). Deletion of Cdc42 expression in mice also results in an embryonically lethal phenotype at a slightly earlier point in gestation, prior to gastrulation, with small embryos lacking primary embryonic ectoderm (11). RhoA deletion mutations have not been reported, but expression of a RhoA-selective bacterial toxin that inactivates Rho function in the thymus induces a dramatic decrease in cellularity, indicating profound effects on cell division and apoptosis (12). Taken together, these observations identify a critical nonredundant role for each of these molecules in the regulation of cell positioning in the embryo. Later developmental abnormalities are evident in mice lacking key proteins that terminate RhoA signals. For example, *RhoGAP*$^{-/-}$ mice manifest abnormal neural development, whereas *RhoGAP-B*$^{-/-}$ mice have less severe abnormalities in neuron patterning, but are remarkably small with reduced cell volumes (13). Neither loss-of-function mutation is compatible with survival.

MEMBERS OF THE Rho FAMILY

The Rho subfamily consists of three isoforms, RhoA, -B, and -C. Most experimental evidence is available for the effects of RhoA, however, the RhoB isoform also is expressed in vascular

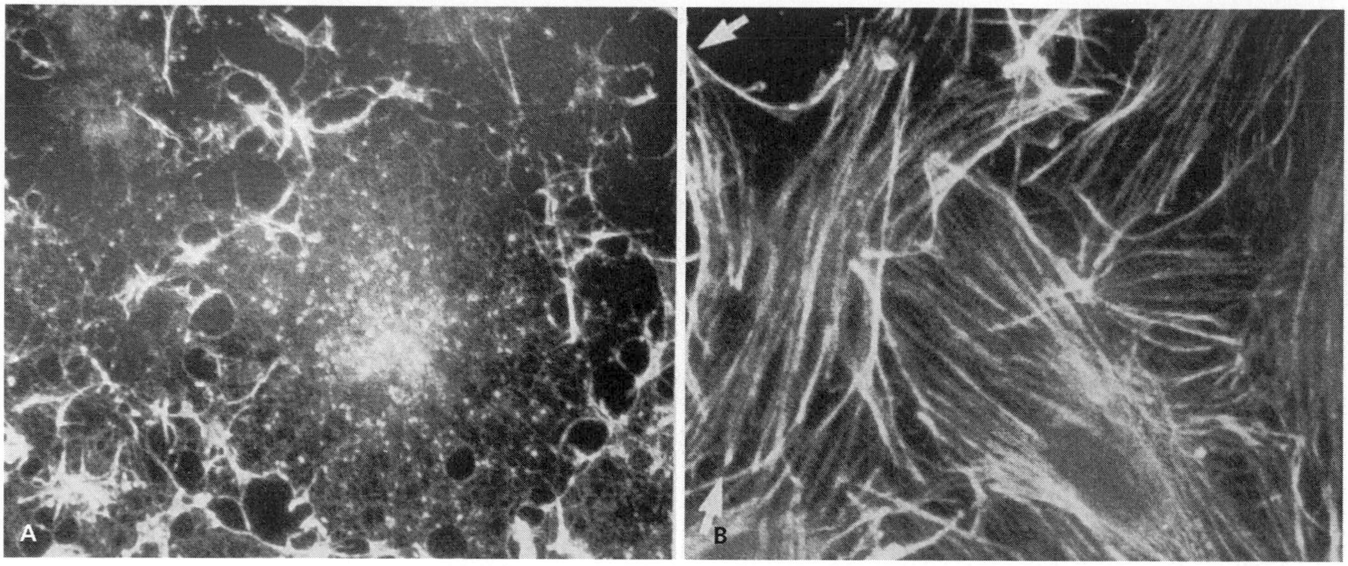

Figure 83.1. Active RhoA induces actin filament stress fibers in serum-starved Swiss 3T3 cells. The cells were fixed without (**A**) or after injection with V^{14}RhoA (**B**) then stained with FITC-conjugated phalloidin to visualize F-actin structures. The small GTP-binding protein RhoA regulates the assembly of focal adhesions (i.e., the plaques of matrix receptors and associated molecules assembled at the base of the cell) and actin stress fibers (i.e., the cables of filamentous actin [F-actin] traversing the cell body) in response to growth factors. (Reproduced with permission from Ridley AJ, Hall J. The small GTP-binding protein rho regulates the assembly of focal adhesions and actin stress fibers in response to growth factors. *Cell.* 1992;70:389–399.)

cells (14). Unlike RhoA, RhoB is a short-lived protein that is highly regulated at the transcriptional level. Most of the primary amino acid residue differences between the subfamily members occur in the C-terminus, and this likely accounts for differences in intracellular localization of the isoforms. When overexpressed in cells, RhoA is largely cytosolic, whereas RhoB is enriched at the plasma membrane. The Rac subfamily includes four members, of which functional information for Rac1 and Rac2 is most developed. Rac2 is expressed in bone-marrow–derived cells, whereas Rac1 is ubiquitous in distribution. Of the five Cdc42-related subfamily members, the function of Cdc42 has been investigated most thoroughly. Cdc42 is expressed in endothelium. Rho-dependent pathway activation can be regulated by tissue or cell type–specific expression of a specific member within a Rho subfamily, but each tissue requires Rho GTP-binding protein expression.

REGULATION OF Rho ACTIVITY

Rho-dependent pathway activation is controlled by GTP binding, which in turn is closely regulated by the association of GDP dissociation inhibitors (RhoGDI), GTPase activating proteins (RhoGAP), and GDP/GTP exchange factors (RhoGEFs) with the target Rho protein (Figure 83.2). RhoGDI binds to Rho proteins to inhibit GDP dissociation and membrane localization, and therefore pathway activation. RhoGAPs accelerate the intrinsic GTPase activity of the Rho proteins, and therefore limit interaction of the activated form with downstream effector molecules. RhoGEFs act to release GDP and facilitate Rho protein binding with GTP present in the cytosol. On the basis of sequence data, there are estimated to be a total of three RhoGDIs, 80 RhoGAPs, and 80 RhoGEFs, but most candidates have not been characterized functionally.

GDP/GTP Exchange Factors

Much interest has focused on pathway regulation through the RhoGEFs, but examples of pathway regulation achieved by influencing RhoGDI and RhoGAP activity also are known. RhoGEFs contain a characteristic tandem arrangement of a diffuse B cell lymphoma (Dbl) oncogene, and a plekstrin homology (PH) domain. Activation of Rho family members is achieved by the exchange of GTP for GDP in the nucleotide binding pocket. The Dbl domain of the RhoGEF interacts with the "switch" regions of the Rho protein and so repositions the switch to cause intramolecular restructuring and disruption of the GDP and Mg^{2+} interactions within the binding pocket (15). Because GTP is in excess in the cell, the GDP is then replaced by GTP. The PH domain may function to localize the RhoGEF to phospholipid-rich membrane locations where the Rho molecule is positioned through its prenyl moiety, or to orient the GEF at such sites to facilitate efficient GDP/GTP exchange activity (16). However, RhoGEF PH domains also appear to mediate protein–protein interactions, notably with actin-binding proteins (17,18). RhoGEF specificity for one or more target Rho proteins may be achieved by spatial localization within the cell and conformational constraints determined by the characteristics of the binding interface between the RhoGEF and Rho target protein. However, it appears that

Table 83-1: Rho GTP-Binding Proteins and Knockout Phenotypes

Gene	*Phenotype*
Rac1	Early embryonic lethal at E8.5
Rac2	Defect in leukocyte function
Cdc42	Early embryonic lethal at E6.5
RhoGAP	Early postnatal death, defect in neural development

many RhoGEFs are promiscuous in their ability to exchange GTP on several Rho subfamilies.

RhoGEFs act to couple the Rho GTP-binding protein pathways to upstream receptors. Many signals originate at the cell surface from heterotrimeric GPCRs. Well characterized Rho-dependent signal transduction pathways coupled to heterotrimeric $G_{\alpha 12/13}$ via RhoGEFs are described. For example, p115-RhoGEF, PDZ-RhoGEF, and LARG each promote GTP exchange on RhoA via a Dbl domain, but also exhibit GAP activity for $G_{\alpha 12/13}$ via a tandem RGS domain. In this way, the downstream activation of Rho signaling is coupled to the means to extinguish the heterotrimeric G-protein signal (19,20). Recently, a redundant mechanism to activate Rho GTP-binding proteins via $G_{\alpha q}$-linked GPCRs has been described (21). Rho signaling also can be initiated by protein tyrosine kinase activity. For example, tyrosine phosphorylation of Vsm-RhoGEF serves to increase RhoA activity, likely by release of N-terminal autoinhibition of the Dbl domain by an intramolecular conformation change (22). In other cases, such as Vav1–3, promiscuous exchange factors for Rho, Rac, and Cdc42, the GEF may be localized by interaction between the Vav SH2 domain and phosphotyrosine residues (23). The large number of possible combinations between 22 Rho proteins and approximately 70 Dbl domain-containing or approximately 10 DOCK-family RhoGEFs represents a significant challenge to fully characterize biologically functional interactions, and the complexity is compounded by heterogeneity in the use of targeting domains among different cell types.

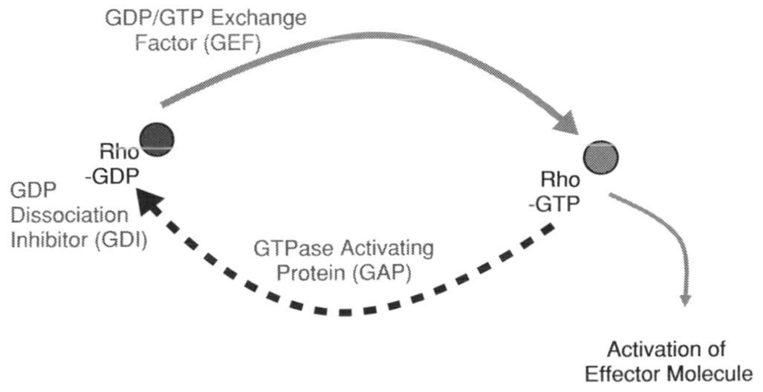

Figure 83.2. The Rho GTP-binding protein switch.

GTPase Activating Proteins

Although up to 80 RhoGAP proteins are predicted in the human genome, most candidates have yet to be characterized functionally. Structural analysis of p50RhoGAP indicates that a conserved arginine is inserted into the Rho protein GTP-binding pocket to promote GTP hydrolysis (24). Specificity of RhoGAP activities is achieved by selective expression among cell types as well as the characteristics of the complementary interacting surfaces of the GAP with the target Rho protein. Regulation of RhoGAP activity is best illustrated by the ubiquitously expressed p190RhoGAP. GAP activity is enhanced by tyrosine phosphorylation under the influence of Src family kinases, and inhibited by protein phosphatase activity (25,26).

DOWNSTREAM EFFECTOR MOLECULES

The control of downstream effector molecules is mediated by interaction of the binding surface of the Rho protein with characteristic binding motifs in the effector molecule, the Rho binding domains (RBD), or Cdc42/Rac-interactive binding (CRIB) domains. Rho pathway signaling depends on the specific activation conferred by the presence of GTP in the nucleotide binding pocket of the Rho molecule. Recent work has confirmed that interaction between the Rho effector molecule and the switch region of the Rho protein is conserved among several Rho/effector pairs studied (27). This is consistent with the observations that the switch region undergoes pronounced conformational change upon GTP loading of the Rho protein. However, additional sites of interaction also have been demonstrated to participate in the binding of various Rho effector molecules. For example, the insert loop, characteristic of the Rho GTP-binding protein family, is involved in Rho binding of Rho-dependent kinase (ROCK) and Rac binding of p67^{phox}, hence the insert loop may be a structural feature common to the family-specific binding or activation of additional effector proteins (28,29).

Rho-MEDIATED REGULATION OF THE CELL CYTOSKELETON

Modification of the cell cytoskeleton is achieved by the coordinated activity of particular Rho effector molecules. The activation of RhoA activity, for example, leads to prominent stress fiber formation in fibroblasts and ECs in culture. Stress fibers (see Figure 83.1) in turn enable cells to contract. The formation of stress fibers depends on the levels of myosin phosphorylation, which are governed by the relative activity of myosin light-chain (MLC) kinase and MLC phosphatase. RhoA activates the downstream effector molecule, ROCK, which then promotes serine phosphorylation and downregulation of MLC phosphatase activity, leading to increased myosin phosphorylation and stress fiber formation (30). ROCK also activates LIM kinase, which phosphorylates and inactivates the

F-actin–severing activity of cofilin (31–33). Similarly, a third kinase regulated by Rho GTP binding proteins through Rho kinase, phosphoinositide 5-kinase (PI5K), participates in actin assembly by generating phosphatidylinositol 4,5-bisphosphate (PIP_2) that promotes the dissociation of actin capping protein from the barbed end of F-actin to facilitate further actin polymerization (34). In this way, both net actin polymerization and F-actin bundling are enhanced, because F-actin depolymerization is inhibited by the cooperative activity of several downstream effectors of Rho GTP-binding proteins.

In addition to signaling through kinase activation to regulate cytoskeletal dynamics, Rho GTP-binding proteins also stimulate de novo actin nucleation and the polymerization of F-actin filaments. The most intensively studied pathway involves a Cdc42-binding–dependent conformation change in the actin-binding Wiscott-Aldrich syndrome protein (WASP) and subsequent downstream activation of actin polymerization through recruitment of the arp2/3 complex (35). A similar interaction between Rac and WASP family verprolin-homologous protein (WAVE), a WASP-related protein, has been identified (36). Rho proteins also interact with diaphanous-related formins (DRF). Mammalian diaphanous (mDia) has a Rho interacting motif that, on binding the cognate RhoA or Cdc42, releases autoinhibition of the DRF activity (37,38). Formins directly bind to F-actin and enhance actin polymerization; of note, they act independently of the arp2/3 complex. Furthermore, formins may be important to establish cell polarization by coordinating both F-actin and microtubule orientation (39).

Rho-MEDIATED REGULATION OF CELL MIGRATION

In motile cells such as Dictyostelium or motile mammalian leukocytes, Rho and Rac have been found localized to distinct regions of the cell membrane. At the leading edge, Rac-GTP is enriched in lamellipodia (Figure 83.3) in association with phosphoinositide 3-kinase (PI3K) and PIP_3 at the plasma membrane, whereas phosphatase and tensin homologue deleted on chromosome 10 (PTEN) is excluded from this region (40). The mechanism of this localized activity is thought to be by direct interaction of the RacGEF PH domain with the PIP_3-rich membrane. Recent work in neutrophils has established

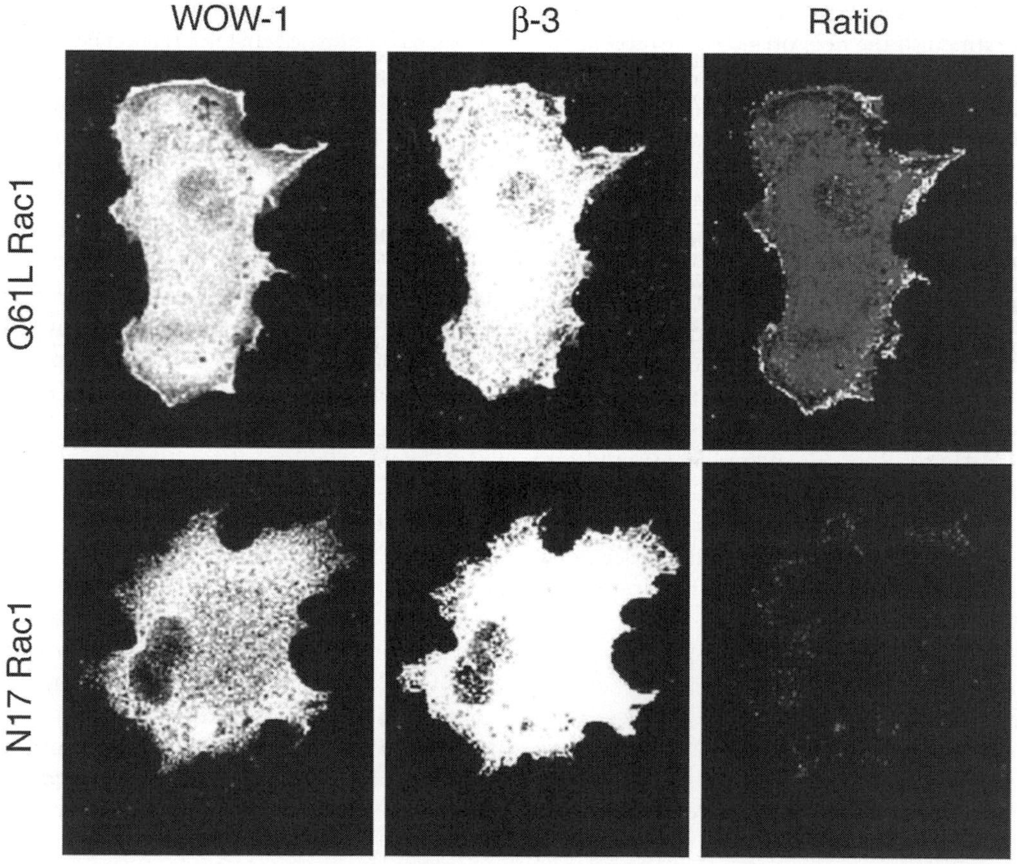

Figure 83.3. Rac1 regulates integrin activation in ECs. Bovine aortic ECs were injected with cDNA coding for dominant active L61Rac1 (*upper panels*) or dominant negative N17Rac1 (*lower panels*). The cells were fixed and stained for active (WOW-1) or total $\beta 3$ integrin distribution. The ratio of active to total $\beta 3$ integrin is shown in the right panels. (Reproduced with permission from Kiosses WB, et al. Rac recruits high-affinity integrin $\alpha v\beta 3$ to lamellipodia in EC migration. *Nature Cell Biol.* 2001;3:316–320.)

that Rac1 and Rac2 have distinct functions in polymorphonuclear leukocyte (PMN) migration to chemoattractant. Rac1 and PI3K are necessary for chemoattractant gradient sensing, whereas Rac2 acting upstream of Cdc42 functions to mediate actin polymerization (9). Another group has proposed a similar function for Cdc42 and the Cdc42GEF PIXα, acting distinctly from Rac activity (40). Cdc42 and Rac also are implicated in control of the direction of movement in monocytes and nerve cells in *Drosophila* (41,42). These observations illustrate exquisite functional specificity even among Rho subfamily members and the degree of conservation across species.

RhoA activity in the migrating cell is localized to the rear, where myosin-based contractility is required. The inhibition of RhoA or ROCK blocks retraction of the trailing uropod, increases integrin-mediated adhesiveness for matrix, and ultimately decreases leukocyte migration (43,44). Moreover, inhibition of Rho or ROCK activity results in a loss of cell polarity, such that lamellipodia form at numerous points around the cell periphery (45). The same authors identify distinct Rho GTP-binding protein coupling to heterotrimeric G-protein–coupled chemoattractant receptor: $G_{\alpha i}$ governs leading edge Rac activity and $G_{\alpha 13}$ regulates Rho. Further work is needed to identify the mechanisms of subcellular compartmentalization in Rho family members and the exchange factors used among different cell types and growth factor receptors to govern localized Rho family activation.

Rho-MEDIATED REGULATION OF GENE EXPRESSION

Rho GTP-binding proteins are involved in the regulation of gene expression. Early work identified the modulation of genes using the serum response element by Rho protein activities (46). However, this appears to be regulated indirectly by levels of G-actin in the cell, sensed by G-actin–dependent translocation of a coactivator for the serum response factor (SRF), megakaryocytic acute leukemia (MAL) (47). Several studies have demonstrated stress-activated protein kinase (SAPK), p38 MAPK, and c-jun N-terminal kinase (JNK) activation by expression of dominant-active Rac and Cdc42 (48–51). Conversely, JNK activity and mast cell protease 7 expression is impaired in *Rac2*-knockout mast cells stimulated with stem cell factor, and reconstituted by Rac2 re-expression (52). Other reports implicate the nuclear factor (NF)-κB (49,53), and Jak/STAT pathway activation (54,55) downstream of dominant active Rho family member expression.

Rho-MEDIATED REGULATION OF REDOX POTENTIAL

A third general effect of Rho GTP-binding protein activation is the generation of reactive oxygen species (ROS) under the control of Rac. Neutrophil activation, an intensively studied model for the generation of superoxide, promotes the assembly of a multimeric nicotinamide adenine dinucleotide phosphate (NADPH) oxidase complex on cell membranes. Myeloid cells deficient in Rac2 lack the ability to assemble the NADPH oxidase (NOX) complex and fail to generate a respiratory burst (56). One subunit, $p67^{phox}$/NOXA1, interacts directly with the switch region, and a second subunit, $gp91^{phox}$/NOX2, interacts with the Rho insert loop of GTP-bound Rac2. Remarkably, Rac1 is unable to complement a deficiency in Rac2 to facilitate the generation of ROS, and indeed, it appears Cdc42 may act as a competitive antagonist to Rac2 binding to $p67^{phox}$ (56). Differences in subcellular localization conferred by the polybasic residues in the C terminus of Rac1, which direct Rac1 to the plasma membrane rather than granule or lysosomal membranes, probably account in part for this difference in Rac isoform function. In other nonmyeloid cell types, NADPH oxidase-generated ROS acts in signal transduction pathways to inactivate tyrosine phosphatases. For example, angiotensin II stimulation of vascular smooth muscle cells (VSMCs) is associated with biphasic generation of free oxygen radicals, dependent in the first phase on protein kinase C (PKC) and in the second phase on Rac and tyrosine kinase activity. The NOX family members involved in nonmyeloid cell ROS generation remain to be clarified in most cases.

THE ROLE OF Rho GTP-BINDING PROTEINS IN THE ENDOTHELIUM

Vasculogenesis and Angiogenesis

Because mice null for Rho family members die in utero before their vasculature develops, it is difficult to directly examine the function of these molecules in endothelial lineage development or vasculogenesis in vivo. At present, hints to the critical role the Rho family is likely to play can be taken from the observation that knockout mice lacking the $G_{\alpha 13}$ heterotrimeric G-protein that couples these receptors to Rho family members die at about day E10 with marked abnormalities in vascular patterning by primitive ECs (57). On the other hand, angiogenesis is a feature of both wound repair and tumor development in the adult, and is more amenable to study. Expression of a dominant negative RhoA mutant in invading ECs using retroviral transduction markedly reduced tumor neovascularization in vivo, confirming that RhoA activity is involved in microvessel formation (58). Growth factors that play a critical role in angiogenesis signal through vascular endothelial growth factor (VEGF) receptor (VEGFR2/KDR), Tie2, transforming growth factor (TGF)-β, and Eph receptors. Rho signaling has been implicated downstream of each of these receptors.

Angiogenesis involves both cell migration and changes in cell shape to form tubes. These events have been studied individually using in vitro cell culture models. EC migration into a denuded area in vitro is characterized by cell polarization and lamellipodia formation. Early investigations documented that Rho GTP-binding protein activity was required to migrate and fill the defect in a wounded EC monolayer (59,60). Further work identified a role for EC Rac and Cdc42 acting downstream of VEGF-stimulated cell migration (61). Cell movement requires the coordinated remodeling of matrix adhesion structures, apparent as focal adhesion complexes in ECs cultured in vitro. The adaptor proteins, Crk and Nck, are complexed to VEGFR2/KDR after VEGF stimulation and participate in the activation of Rac and the Rac effector kinase, PAK, to remodel EC focal adhesions (62). However, PAK activation may occur independently of Rac- or Cdc42-GTP binding.

RhoA activity also is increased after VEGFR2/KDR stimulation, but interestingly appears to be dependent on the transactivation of a $G_{\alpha q}$ heterotrimeric GPCR (63). Similarly, stimulation of EC in vitro with angiopoetin 1 (Ang1) elicits cell migration, and like VEGF, stimulates both RhoA and Rac activity. Further, following Ang1 stimulation, Nck is found complexed with PAK and is associated with increased PAK activity in a signal transduction pathway analogous to VEGFR2/KDR stimulation (64,65).

The role played by Rho proteins also has been analyzed in capillary tube formation. In vitro, ECs cultured on Matrigel spontaneously reorganize their shape to form an elongated spindle morphology, assemble to form short multicellular capillary-like segments, then develop more complicated branching networks. Rac and PAK, but not Rho, activity is implicated in capillary tube formation (66,67). The assembly of capillary-like segments involves the establishment of interendothelial cell contacts and the assembly of vascular endothelial (VE)-cadherin–based adherens junctions between the cells. Rac recruits high-affinity integrin to the lamellipodia extensions formed during the change to the spindle shape (68). Dominant negative Rac1 expression in ECs further inhibits the development of VE-cadherin interendothelial cell junction structures in vitro (67), and inactivation of Rac disorders the established VE-cadherin junctions (69,70). In part, this is related to remodeling of the association of VE-cadherin to the cortical F-actin cytoskeleton (71).Thus, Rac participates in the events leading to the formation of primitive microvessels and maintenance of microvessel integrity.

Coordination between Rho subfamilies is nevertheless important for capillary network maturation. Cdc42 regulation of WASP activity is needed for lumen development of primitive capillary tubes in vitro (72). Inhibition of Rho activity appears to be involved in maintaining capillary lumen patency (73). However, inhibition of Rho or ROCK activity markedly decreases branching cord formation in vitro and in vivo, indicating that Rho activity is involved in the development of complex microvessel networks, but not the development of a patent lumen (58,66). Together, these observations illustrate that Rac, Cdc42, and Rho each participate in the formation of microvessel networks, but differential regulation of distinct aspects of microvessel assembly and maturation is evident in the in vitro experiments.

Endothelial Response to Shear Stress

ECs at different sites in the vasculature are specialized for functions at each location. In the arterial circulation, the EC is subjected to mechanical stresses during each cardiac cycle that contribute to a characteristic phenotype of arterial endothelium aligned in the direction of blood flow. Fluid shear stress sensed at the luminal surface by structural deformations in the plasma membrane or underlying cytoskeleton, at the interendothelial cell junctions, or at the matrix adhesion sites is translated into chemical signals that affect cytoskeletal organization and gene transcription.

In vitro, randomly oriented ECs subjected to shear stress realign in the direction of flow in a process requiring cytoskeletal remodeling, suggesting that Rho GTP-binding proteins may be under the influence of shear stress receptors. The initial application of shear induces a rapid increase in Rho-GTP, followed by suppression from basal values, then a late increase several hours later. Rac and Cdc42 activity, in contrast, are increased in a reciprocal pattern to Rho-GTP (74). The suppression of Rho-GTP is necessary for remodeling of cell shape, and is preceded by changes in cell matrix adhesion structure morphology and integrin affinity (75).

The mechanotransduction of shear stress to the matrix adhesion sites may directly regulate integrin affinity, which in turn appears to signal suppression of Rho through outside-inside signaling. In addition, reorientation of the EC requires the re-establishment of cell polarity. Rac activity regulates directional lamellipodia extension along the axis of flow (74). Cdc42 participates in EC polarization by governing the characteristic positioning of the microtubule organizing center upstream of the nucleus (76). Shear stress activation of Rho also is linked to the transcription of genes (such as the low-density lipoprotein [LDL] receptor) through effects on sterol regulatory element–binding proteins (77).

Reactive Oxygen Species Production by the Endothelium

Vascular ECs express $p47^{phox}$ and $p67^{phox}$ in addition to both NOX1 and NOX4 (which are homologues of leukocyte $gp91^{phox}$/ NOX2). Hence, vascular ECs are competent to generate superoxide from this enzyme system. However, NOX4 is reported to be the principal component to generate superoxide in ECs (78). Indeed quiescent ECs in culture appear to maintain assembled NADPH oxidase complexes and constitutively produce oxygen-free radicals at low levels (79). NADPH oxidase activity in ECs can be upregulated by heterotrimeric G-protein receptor activation (80,81). For example, angiotensin II stimulation of the AT-1 receptor rapidly induces EC NADPH oxidase activity (80,82) and also can induce NOX expression in vascular ECs (83). In VSMCs, it has been demonstrated that similar activity is dependent on Rac GTP. Superoxide also is generated in response to receptor tyrosine kinase stimulation in ECs. The best characterized pathway is downstream of the VEGF receptor VEGFR2/KDR (84,85). VEGF-stimulation of coronary or venous ECs elicits a transient pulse of ROS regulated by RacGEF activity that can be inhibited by dominant negative Rac1 expression. This signal appears to promote Mn-dependent superoxidase transcription in a complicated adaptive response. Interestingly VEGF-stimulated membrane translocation of the endothelial WASP homologue WAVE1 also has been demonstrated to depend on the Rac activation of PAK1-mediated phosphorylation of $p47^{phox}$, an adaptor component of the NADPH oxidase complex (86). Thus, Rac activity links VEGFR2/KDR stimulation of EC to effects such as changes in the cytoskeleton, EC motility, and gene transcription.

The activation of the NADPH oxidase system is proposed to be a critical event in repair phenomenon such as endothelial migration and angiogenesis (85). However, NADPH oxidase–dependent oxidative stress of the endothelium is also linked to endothelial dysfunction and later arteriosclerosis (87). Diverse clinically relevant noxious stimuli including hyperglycemia, stimulation by advanced glycosylation end products, or oxidized LDL have been demonstrated to activate Rac GTP and generate ROS in cultured cells (88–90). Indeed, evidence has accumulated that the inhibition of endothelial Rac prenylation may be an important mechanism of 3-hydroxy-3-methylglutaryl-coenzyme A (HMG CoA) reductase inhibitor–mediated vascular protection (90,91).

Endothelial Barrier Function and Inflammation

The endothelium plays an active role in inflammation by changing capillary permeability and facilitating the tethering and transendothelial migration of leukocytes to the inflamed vessel surface. The development of small pores in confluent ECs and increased permeability of an endothelial monolayer in vitro can be elicited by GPCR agonists such as thrombin or histamine. The thrombin receptor is known to couple to $G_{\alpha 12/13}$ heterotrimeric G-proteins and, through p115RhoGEF, to the Rho GTP-binding proteins (92). Stimulation of EC monolayers with thrombin increases Rho GTP and myosin phosphorylation, consistent with increased cell contractility. Inhibition of RhoA blocks this response and the loss of the barrier function, demonstrating a critical role for Rho activation to elicit a cardinal feature of inflammation – vascular permeability (69,93). A similar signaling pathway is activated by tumor necrosis factor (TNF)-induced monolayer permeability in vitro (94).

Proinflammatory signals that decrease EC monolayer barrier function are accompanied by a decrease in Rac activity and redistribution of adherens junction proteins such as VE-cadherin (95). However, other signals generated during an inflammatory response act to counter the vascular leak and modulate Rho GTP-binding protein signaling. For example, sphingosine-1-phosphate (S1P), a mediator derived from platelet activation that decreases capillary permeability, acts to counter the thrombin-induced increase vascular permeability (96,97). Mechanistic investigations into the signal transduction pathways recruited by EC S1P signaling have demonstrated that Rac activation downstream of the S1P1 receptor mediates the effect (96,98). Moreover other mechanisms that attenuate the effect of proinflammatory agents on Rho-dependent signal pathways and vascular permeability have been identified. Protein kinase A (PKA)-dependent increases in endothelial monolayer barrier characteristics are mediated by RhoGDI inhibition of RhoGTP relocation to the membrane (99). In contrast, Cc42 activity is linked to restoration of monolayer barrier function after edemagenic stimulus (100).

Endothelial display of adhesion molecules provides a critical cue for the recruitment of leukocytes to sites of inflammation. Mobilization of stored P-selectin – a key adhesion molecule involved in the rapid recruitment of neutrophils – from endothelial Weibel-Palade bodies after thrombin receptor stimulation is inhibited by dominant negative Cdc42 expression (101). Further, Rac and Cdc42 modulate TNF-stimulated transcription of *E-selectin*, an adhesion molecule that participates in the later recruitment of neutrophils (102). Therapeutic intervention using HMG-CoA reductase inhibitors to block prenylation of Rho molecules inhibits TNF-induced E-selectin expression and may represent a clinically relevant effect of this class of drugs (103).

Endothelial cytoskeletal reorganization to form a "docking structure" has been identified at the sites of tight adhesion of the leukocyte to inflamed EC and correlates with the efficiency of subsequent transendothelial migration (104,105). These structures involve clustering of endothelial adhesion molecules intracellular adhesion molecule (ICAM)-1 and vascular cell adhesion molecule (VCAM)-1 in association with the extension of F-actin rich fingers or ridges around the adherent leukocyte (105). In vitro clustering of ICAM-1 has been found to upregulate endothelial Rho GTP, and is critical for lymphocyte migration across a brain EC monolayer in vitro (106,107). However, it is not clear that ICAM-1–stimulated Rho-GTP initiates docking structure formation or if this is the Rho-dependent event that limits leukocyte transendothelial migration in vitro or extravasation in vivo.

DIAGNOSTIC AND THERAPEUTIC IMPLICATIONS

Therapeutic targeting of Rho GTP-binding protein activity may be a useful approach in several clinical settings. In the management of cancer, for example, the inhibition of Rho activity in the transformed cell might impair tumor growth or metastasis. Moreover, neovascularization of the tumor may be impaired by Rho inhibition (108). Rho inhibition may be useful in the management of cardiovascular diseases such as hypertension (109) or angina related to coronary artery vasospasm (110). Indeed some of the beneficial therapeutic effects attributed to HMG CoA reductase inhibitors used in the management of dyslipidemia may be related to the effects of these drugs on small GTP-binding proteins, including Rho family members. In addition to desirable effects on cholesterol biosynthesis, these agents inhibit the production of farnesyl and geranylgeranyl pyrophosphates used to isoprenylate Ras and Rho GTP-binding proteins, which facilitate membrane localization of the small GTP-binding proteins (111).

Pharmacologic agents that specifically inhibit geranylgeranyl transferase activity are currently under investigation by the pharmaceutical industry.

KEY POINTS

- Rho GTP-binding proteins are critical checkpoints that regulate key functions of the EC including angiogenesis, endothelial dysfunction, and inflammation.
- Identification of endothelial RhoGEFs and functional characterization of the participation of these molecules under particular pathological conditions is needed.
- Inhibition of these cell signaling systems may provide novel opportunities for therapeutic intervention.

REFERENCES

1 Karnoub AE, Campbell SL. Structural and biophysical insights into the role of the insert region in Rac1 function. *Biochemistry*. 2002;41(12):3875–3883.

2 Valencia A, Chardin P, Wittinghofer A, Sander C. The ras protein family: evolutionary tree and role of conserved amino acids. *Biochemistry*. 1991;30(19):4637–4648.

3 Shinjo K, Koland JG, Hart MJ, et al. Molecular cloning of the gene for the human placental GTP-binding protein Gp (G25K): identification of this GTP-binding protein as the human homolog of the yeast cell-division-cycle protein CDC42. *Proc Natl Acad Sci USA*. 1990;87(24):9853–9857.

4 Didsbury J, Weber RF, Bokoch GM, et al. Rac, a novel ras-related family of proteins that are botulinum toxin substrates. *J Biol Chem.* 1989;264(28):16378–16382.

5 Aktories K, Braun U, Rosener S, et al. The Rho gene product expressed in E. coli is a substrate of botulinum ADP-ribosyltransferase C3. *Biochem Biophys Res Commun.* 1989;158(1): 209–213.

6 Ridley AJ, Paterson HF, Johnston CL, et al. The small GTP-binding protein rac regulates growth factor-induced membrane ruffling. *Cell.* 1992;70(3):401–410.

7 Ridley AJ, Hall A. The small GTP-binding protein Rho regulates the assembly of focal adhesions and actin stress fibers in response to growth factors. *Cell.* 1992;70(3):389–399.

8 Sugihara K, Nakatsuji N, Nakamura K, et al. Rac1 is required for the formation of three germ layers during gastrulation. *Oncogene.* 1998;17(26):3427–3433.

9 Sun CX, Downey GP, Zhu F, et al. Rac1 is the small GTPase responsible for regulating the neutrophil chemotaxis compass. *Blood*. 2004;104(12):3758–3765.

10 Benvenuti F, Hugues S, Walmsley M, et al. Requirement of Rac1 and Rac2 expression by mature dendritic cells for T cell priming. *Science*. 2004;305(5687):1150–1153.

11 Chen F, Ma L, Parrini MC, et al. Cdc42 is required for PIP(2)-induced actin polymerization and early development but not for cell viability. *Curr Biol.* 2000;10(13):758–765.

12 Henning SW, Galandrini R, Hall A, Cantrell DA. The GTPase Rho has a critical regulatory role in thymus development. *EMBO J.* 1997;16(9):2397–2407.

13 Sordella R, Classon M, Hu KQ, et al. Modulation of CREB activity by the Rho GTPase regulates cell and organism size during mouse embryonic development. *Dev Cell.* 2002;2(5):553–565.

14 Adini I, Rabinovitz I, Sun JF, et al. RhoB controls Akt trafficking and stage-specific survival of endothelial cells during vascular development. *Genes Dev.* 2003;17(21):2721–2732.

15 Rossman KL, Worthylake DK, Snyder JT, et al. A crystallographic view of interactions between Dbs and Cdc42: PH domain-assisted guanine nucleotide exchange. *EMBO J.* 2002; 21(6):1315–1326.

16 Rossman KL, Cheng L, Mahon GM, et al. Multifunctional roles for the PH domain of Dbs in regulating Rho GTPase activation. *J Biol Chem.* 2003;278(20):18393–18400.

17 Bellanger JM, Astier C, Sardet C, et al. The Rac1- and RhoG-specific GEF domain of Trio targets filamin to remodel cytoskeletal actin. *Nat Cell Biol.* 2000;2(12):888–892.

18 Vanni C, Parodi A, Mancini P, et al. Phosphorylation-independent membrane relocalization of ezrin following association with Dbl in vivo. *Oncogene*. 2004;23(23):4098–4106.

19 Kozasa T, Jiang X, Hart MJ, et al. p115 RhoGEF, a GTPase activating protein for Galpha12 and Galpha13. *Science*. 1998;280 (5372):2109–2111.

20 Gohla A, Schultz G, Offermanns S. Role for G(12)/G(13) in agonist-induced vascular smooth muscle cell contraction. *Circ Res.* 2000;87(3):221–227.

21 Vogt S, Grosse R, Schultz G, Offermanns S. Receptor-dependent RhoA activation in G12/G13-deficient cells: genetic evidence for an involvement of Gq/G11. *J Biol Chem.* 2003;278(31):28743–28749.

22 Ogita H, Kunimoto S, Kamioka Y, et al. EphA4-mediated Rho activation via Vsm-RhoGEF expressed specifically in vascular smooth muscle cells. *Circ Res.* 2003;93(1):23–31.

23 Zugaza JL, Lopez-Lago MA, Caloca MJ, et al. Structural determinants for the biological activity of Vav proteins. *J Biol Chem.* 2002;277(47):45377–45392.

24 Nassar N, Hoffman GR, Manor D, et al. Structures of Cdc42 bound to the active and catalytically compromised forms of Cdc42GAP. *Nat Struct Biol.* 1998;5(12):1047–1052.

25 Haskell MD, Nickles AL, Agati JM, et al. Phosphorylation of p190 on Tyr1105 by c-Src is necessary but not sufficient for EGF-induced actin disassembly in C3H10T1/2 fibroblasts. *J Cell Sci.* 2001;114(Pt 9):1699–1708.

26 Roof RW, Haskell MD, Dukes BD, et al. Phosphotyrosine (p-Tyr)-dependent and -independent mechanisms of p190 RhoGAP-p120 RasGAP interaction: Tyr 1105 of p190, a substrate for c-Src, is the sole p-Tyr mediator of complex formation. *Mol Cell Biol.* 1998;18(12):7052–7063.

27 Dvorsky R, Ahmadian MR. Always look on the bright site of Rho: structural implications for a conserved intermolecular interface. *EMBO Rep.* 2004;5(12):1130–1136.

28 Zong H, Kaibuchi K, Quilliam LA. The insert region of RhoA is essential for Rho kinase activation and cellular transformation. *Mol Cell Biol.* 2001;21(16):5287–5298.

29 Nisimoto Y, Freeman JL, Motalebi SA, et al. Rac binding to p67(phox). Structural basis for interactions of the Rac1 effector region and insert region with components of the respiratory burst oxidase. *J Biol Chem.* 1997;272(30):18834–18841.

30 Kimura K, Ito M, Amano M, et al. Regulation of myosin phosphatase by Rho and Rho-associated kinase (Rho-kinase). *Science.* 1996;273(5272):245–248.
31 Arber S, Barbayannis FA, Hanser H, et al. Regulation of actin dynamics through phosphorylation of cofilin by LIM-kinase. [See comment]. *Nature.* 1998;393(6687):805–809.
32 Maekawa M, Ishizaki T, Boku S, et al. Signaling from Rho to the actin cytoskeleton through protein kinases ROCK and LIM-kinase. *Science.* 1999;285(5429):895–898.
33 Sumi T, Matsumoto K, Takai Y, Nakamura T. Cofilin phosphorylation and actin cytoskeletal dynamics regulated by rho- and Cdc42-activated LIM-kinase 2. *J Cell Biol.* 1999;147(7):1519–1532.
34 Tolias KF, Hartwig JH, Ishihara H, et al. Type Ialpha phosphatidylinositol-4-phosphate 5-kinase mediates Rac-dependent actin assembly. *Curr Biol.* 2000;10(3):153–156.
35 Rohatgi R, Ma L, Miki H, et al. The interaction between N-WASP and the Arp2/3 complex links Cdc42-dependent signals to actin assembly. *Cell.* 1999;97(2):221–231.
36 Eden S, Rohatgi R, Podtelejnikov AV, et al. Mechanism of regulation of WAVE1-induced actin nucleation by Rac1 and Nck.[See comment]. *Nature.* 2002;418(6899):790–793.
37 Li F, Higgs HN. The mouse Formin mDia1 is a potent actin nucleation factor regulated by autoinhibition. *Curr Biol.* 2003; 13(15):1335–1340.
38 Li F, Higgs HN. Dissecting requirements for auto-inhibition of actin nucleation by the formin, mDia1. *J Biol Chem.* 2005;280 (8):6986–6992.
39 Ishizaki T, Morishima Y, Okamoto M, et al. Coordination of microtubules and the actin cytoskeleton by the Rho effector mDia1. *Nat Cell Biol.* 2001;3(1):8–14.
40 Li Z, Hannigan M, Mo Z, et al. Directional sensing requires G beta gamma-mediated PAK1 and PIX alpha-dependent activation of Cdc42. *Cell.* 2003;114(2):215–227.
41 Allen WE, Zicha D, Ridley AJ, Jones GE. A role for Cdc42 in macrophage chemotaxis. *J Cell Biol.* 1998;141(5):1147–1157.
42 Sepp KJ, Auld VJ. RhoA and Rac1 GTPases mediate the dynamic rearrangement of actin in peripheral glia. *Development.* 2003;130(9):1825–1835.
43 Worthylake RA, Lemoine S, Watson JM, Burridge K. RhoA is required for monocyte tail retraction during transendothelial migration. *J Cell Biol.* 2001;154(1):147–160.
44 Worthylake RA, Burridge K. RhoA and ROCK promote migration by limiting membrane protrusions. *J Biol Chem.* 2003; 278(15):13578–13584.
45 Xu J, Wang F, Van Keymeulen A, et al. Divergent signals and cytoskeletal assemblies regulate self-organizing polarity in neutrophils. *Cell.* 2003;114(2):201–214.
46 Hill CS, Wynne J, Treisman R. The Rho family GTPases RhoA, Rac1, and CDC42Hs regulate transcriptional activation by SRF. *Cell.* 1995;81(7):1159–1170.
47 Miralles F, Posern G, Zaromytidou AI, Treisman R. Actin dynamics control SRF activity by regulation of its coactivator MAL. *Cell.* 2003;113(3):329–342.
48 Minden A, Lin A, Claret FX, et al. Selective activation of the JNK signaling cascade and c-Jun transcriptional activity by the small GTPases Rac and Cdc42Hs. *Cell.* 1995;81(7):1147–1157.
49 Murphy GA, Solski PA, Jillian SA, et al. Cellular functions of TC10, a Rho family GTPase: regulation of morphology, signal transduction and cell growth. *Oncogene.* 1999;18(26):3831–3845.
50 Chowdhury I, Chaqour B. Regulation of connective tissue growth factor (CTGF/CCN2) gene transcription and mRNA stability in smooth muscle cells. Involvement of RhoA GTPase and p38 MAP kinase and sensitivity to actin dynamics. *Eur J Biochem.* 2004;271(22):4436–4450.
51 Marinissen MJ, Chiariello M, Tanos T, et al. The small GTP-binding protein RhoA regulates c-jun by a ROCK-JNK signaling axis. *Mol Cell.* 2004;14(1):29–41.
52 Gu Y, Byrne MC, Paranavitana NC, et al. Rac2, a hematopoiesis-specific Rho GTPase, specifically regulates mast cell protease gene expression in bone marrow-derived mast cells. *Mol Cell Biol.* 2002;22(21):7645–7657.
53 Frost JA, Swantek JL, Stippec S, et al. Stimulation of NFkappa B activity by multiple signaling pathways requires PAK1. *J Biol Chem.* 2000;275(26):19693–19699.
54 Pelletier S, Duhamel F, Coulombe P, et al. Rho family GTPases are required for activation of Jak/STAT signaling by G protein-coupled receptors. *Mol Cell Biol.* 2003;23(4):1316–1333.
55 Turkson J, Bowman T, Adnane J, et al. Requirement for Ras/Rac1-mediated p38 and c-Jun N-terminal kinase signaling in Stat3 transcriptional activity induced by the Src oncoprotein. *Mol Cell Biol.* 1999;19(11):7519–7528.
56 Diebold BA, Fowler B, Lu J, et al. Antagonistic cross-talk between Rac and Cdc42 GTPases regulates generation of reactive oxygen species. *J Biol Chem.* 2004;279(27):28136–28142.
57 Offermanns S, Mancino V, Revel JP, Simon MI. Vascular system defects and impaired cell chemokinesis as a result of Galpha13 deficiency. *Science.* 1997;275(5299):533–536.
58 Hoang MV, Whelan MC, Senger DR. Rho activity critically and selectively regulates endothelial cell organization during angiogenesis. *Proc Natl Acad Sci USA.* 2004;101(7):1874–1879.
59 Aepfelbacher M, Essler M, Huber E, et al. Bacterial toxins block endothelial wound repair. Evidence that Rho GTPases control cytoskeletal rearrangements in migrating endothelial cells. *Arterioscler Thromb Vasc Biol.* 1997;17(9):1623–1629.
60 van Nieuw Amerongen GP, Koolwijk P, Versteilen A, van Hinsbergh VW. Involvement of RhoA/Rho kinase signaling in VEGF-induced endothelial cell migration and angiogenesis in vitro. *Arterioscler Thromb Vasc Biol.* 2003;23(2):211–217.
61 Soga N, Namba N, McAllister S, et al. Rho family GTPases regulate VEGF-stimulated endothelial cell motility. *Exp Cell Res.* 2001;269(1):73–87.
62 Stoletov KV, Gong C, Terman BI. Nck and Crk mediate distinct VEGF-induced signaling pathways that serve overlapping functions in focal adhesion turnover and integrin activation. *Exp Cell Res.* 2004;295(1):258–268.
63 Zeng H, Zhao D, Mukhopadhyay D. KDR stimulates endothelial cell migration through heterotrimeric G protein Gq/11-mediated activation of a small GTPase RhoA. *J Biol Chem.* 2002; 277(48):46791–46798.
64 Master Z, Jones N, Tran J, et al. Dok-R plays a pivotal role in angiopoietin-1-dependent cell migration through recruitment and activation of Pak. *EMBO J.* 2001;20(21):5919–5928.
65 Cascone I, Audero E, Giraudo E, et al. Tie-2-dependent activation of RhoA and Rac1 participates in endothelial cell motility triggered by angiopoietin-1. *Blood.* 2003;102(7):2482–2490.
66 Connolly JO, Simpson N, Hewlett L, Hall A. Rac regulates endothelial morphogenesis and capillary assembly. *Mol Biol Cell.* 2002;13(7):2474–2485.
67 Cascone I, Giraudo E, Caccavari F, et al. Temporal and spatial modulation of Rho GTPases during in vitro formation of

capillary vascular network. Adherens junctions and myosin light chain as targets of Rac1 and RhoA. *J Biol Chem.* 2003;278(50): 50702–50713.
68 Kiosses WB, Shattil SJ, Pampori N, Schwartz MA. Rac recruits high-affinity integrin alphavbeta3 to lamellipodia in endothelial cell migration. *Nat Cell Biol.* 2001;3(3):316–320.
69 Wojciak-Stothard B, Potempa S, Eichholtz T, Ridley AJ. Rho and Rac but not Cdc42 regulate endothelial cell permeability. *J Cell Sci.* 2001;114(Pt 7):1343–1355.
70 Waschke J, Drenckhahn D, Adamson RH, Curry FE. Role of adhesion and contraction in Rac 1-regulated endothelial barrier function in vivo and in vitro. *Am J Physiol Heart Circ Physiol.* 2004;287(2):H704–H711.
71 Baumgartner W, Adamson RH, Zeng M, et al. Requirement of Rac activity for maintenance of capillary endothelial barrier properties. *Am J Physiol Heart Circ Physiol.* 2004;286(1):H394–H401.
72 Bayless KJ, Davis GE. The Cdc42 and Rac1 GTPases are required for capillary lumen formation in three-dimensional extracellular matrices. *J Cell Sci.* 2002;115(Pt 6):1123–1136.
73 Bayless KJ, Davis GE. Microtubule depolymerization rapidly collapses capillary tube networks in vitro and angiogenic vessels in vivo through the small GTPase Rho. *J Biol Chem.* 2004; 279(12):11686–11695.
74 Wojciak-Stothard B, Ridley AJ. Shear stress-induced endothelial cell polarization is mediated by Rho and Rac but not Cdc42 or PI 3-kinases. *J Cell Biol.* 2003;161(2):429–439.
75 Tzima E, del Pozo MA, Shattil SJ, et al. Activation of integrins in endothelial cells by fluid shear stress mediates Rho-dependent cytoskeletal alignment. *EMBO J.* 2001;20(17):4639–4647.
76 Tzima E, Kiosses WB, del Pozo MA, Schwartz MA. Localized cdc42 activation, detected using a novel assay, mediates microtubule organizing center positioning in endothelial cells in response to fluid shear stress. *J Biol Chem.* 2003;278(33):31020–31023.
77 Lin T, Zeng L, Liu Y, et al. Rho-ROCK-LIMK-cofilin pathway regulates shear stress activation of sterol regulatory element binding proteins. *Circ Res.* 2003;92(12):1296–1304.
78 Ago T, Kitazono T, Ooboshi H, et al. Nox4 as the major catalytic component of an endothelial NAD(P)H oxidase. *Circulation.* 2004;109(2):227–233.
79 Li JM, Shah AM. Intracellular localization and preassembly of the NADPH oxidase complex in cultured endothelial cells. *J Biol Chem.* 2002;277(22):19952–19960.
80 Zhang H, Schmeisser A, Garlichs CD, et al. Angiotensin II-induced superoxide anion generation in human vascular endothelial cells: role of membrane-bound NADH-/NADPH-oxidases. *Cardiovasc Res.* 1999;44(1):215–222.
81 Lang D, Mosfer SI, Shakesby A, et al. Coronary microvascular endothelial cell redox state in left ventricular hypertrophy: the role of angiotensin II. *Circ Res.* 2000;86(4):463–469.
82 Pueyo ME, Gonzalez W, Nicoletti A, et al. Angiotensin II stimulates endothelial vascular cell adhesion molecule-1 via nuclear factor-kappaB activation induced by intracellular oxidative stress. *Arterioscler Thromb Vasc Biol.* 2000;20(3):645–651.
83 Rueckschloss U, Quinn MT, Holtz J, Morawietz H. Dose-dependent regulation of NAD(P)H oxidase expression by angiotensin II in human endothelial cells: protective effect of angiotensin II type 1 receptor blockade in patients with coronary artery disease. *Arterioscler Thromb Vasc Biol.* 2002;22(11): 1845–1851.
84 Abid MR, Tsai JC, Spokes KC, et al. Vascular endothelial growth factor induces manganese-superoxide dismutase expression in endothelial cells by a Rac1-regulated NADPH oxidase-dependent mechanism. *FASEB J.* 2001;15(13):2548–2550.
85 Ushio-Fukai M, Tang Y, Fukai T, et al. Novel role of gp91(phox)-containing NAD(P)H oxidase in vascular endothelial growth factor-induced signaling and angiogenesis. *Circ Res.* 2002; 91(12):1160–1167.
86 Wu RF, Gu Y, Xu YC, et al. Vascular endothelial growth factor causes translocation of p47phox to membrane ruffles through WAVE1. *J Biol Chem.* 2003;278(38):36830–36840.
87 Guzik TJ, Sadowski J, Kapelak B, et al. Systemic regulation of vascular NAD(P)H oxidase activity and nox isoform expression in human arteries and veins. [see comment]. *Arterioscler Thromb Vasc Biol.* 2004;24(9):1614–1620.
88 Uemura S, Matsushita H, Li W, et al. Diabetes mellitus enhances vascular matrix metalloproteinase activity: role of oxidative stress. *Circ Res.* 2001;88(12):1291–1298.
89 Wautier MP, Chappey O, Corda S, et al. Activation of NADPH oxidase by AGE links oxidant stress to altered gene expression via RAGE. *Am J Physiol Endocrinol Metab.* 2001;280(5):E685–E694.
90 Shin HK, Kim YK, Kim KY, et al. Remnant lipoprotein particles induce apoptosis in endothelial cells by NAD(P)H oxidase-mediated production of superoxide and cytokines via lectin-like oxidized low-density lipoprotein receptor-1 activation: prevention by cilostazol. *Circulation.* 2004;109(8):1022–1028.
91 Vecchione C, Brandes RP. Withdrawal of 3-hydroxy-3-methylglutaryl coenzyme A reductase inhibitors elicits oxidative stress and induces endothelial dysfunction in mice. *Circ Res.* 2002;91(2):173–179.
92 Gohla A, Offermanns S, Wilkie TM, Schultz G. Differential involvement of Galpha12 and Galpha13 in receptor-mediated stress fiber formation. *J Biol Chem.* 1999;274(25):17901–17907.
93 Carbajal JM, Schaeffer RC Jr. RhoA inactivation enhances endothelial barrier function. *Am J Physiol.* 1999;277(5 Pt 1): C955–C964.
94 Nwariaku FE, Rothenbach P, Liu Z, et al. Rho inhibition decreases TNF-induced endothelial MAPK activation and monolayer permeability. *J Appl Physiol.* 2003;95(5):1889–1895.
95 Vouret-Craviari V, Bourcier C, Boulter E, van Obberghen-Schilling E. Distinct signals via Rho GTPases and Src drive shape changes by thrombin and sphingosine-1-phosphate in endothelial cells. *J Cell Sci.* 2002;115(Pt 12):2475–2484.
96 Garcia JG, Liu F, Verin AD, et al. Sphingosine 1-phosphate promotes endothelial cell barrier integrity by Edg-dependent cytoskeletal rearrangement. *J Clin Invest.* 2001;108(5):689–701.
97 Schaphorst KL, Chiang E, Jacobs KN, et al. Role of sphingosine-1 phosphate in the enhancement of endothelial barrier integrity by platelet-released products. *Am J Physiol Lung Cell Mol Physiol.* 2003;285(1):L258–L267.
98 Dudek SM, Jacobson JR, Chiang ET, et al. Pulmonary endothelial cell barrier enhancement by sphingosine 1-phosphate: roles for cortactin and myosin light chain kinase. *J Biol Chem.* 2004; 279(23):24692–24700.
99 Qiao J, Huang F, Lum H. PKA inhibits RhoA activation: a protection mechanism against endothelial barrier dysfunction. *Am J Physiol Lung Cell Mol Physiol.* 2003;284(6):L972–L980.
100 Kouklis P, Konstantoulaki M, Vogel S, et al. Cdc42 regulates the restoration of endothelial barrier function. *Circ Res.* 2004; 94(2):159–166.

101 Klarenbach SW, Chipiuk A, Nelson RC, et al. Differential actions of PAR2 and PAR1 in stimulating human endothelial cell exocytosis and permeability: the role of Rho-GTPases. *Circ Res.* 2003; 92(3):272–278.

102 Min W, Pober JS. TNF initiates E-selectin transcription in human endothelial cells through parallel TRAF-NF-kappa B and TRAF-RAC/CDC42-JNK-c-Jun/ATF2 pathways. *J Immunol.* 1997;159(7):3508–3518.

103 Nubel T, Dippold W, Kleinert H, et al. Lovastatin inhibits Rho-regulated expression of E-selectin by TNFalpha and attenuates tumor cell adhesion. *FASEB J.* 2004;18(1):140–142.

104 Barreiro O, Yanez-Mo M, Serrador JM, et al. Dynamic interaction of VCAM-1 and ICAM-1 with moesin and ezrin in a novel endothelial docking structure for adherent leukocytes. *J Cell Biol.* 2002;157(7):1233–1245.

105 Carman CV, Jun CD, Salas A, Springer TA. Endothelial cells proactively form microvilli-like membrane projections upon intercellular adhesion molecule 1 engagement of leukocyte LFA-1. *J Immunol.* 2003;171(11):6135–6144.

106 Adamson P, Wilbourn B, Etienne-Manneville S, et al. Lymphocyte trafficking through the blood-brain barrier is dependent on endothelial cell heterotrimeric G-protein signaling. *FASEB J.* 2002;16(10):1185–1194.

107 Adamson P, Etienne S, Couraud PO, et al. Lymphocyte migration through brain endothelial cell monolayers involves signaling through endothelial ICAM-1 via a rho-dependent pathway. *J Immunol.* 1999;162(5):2964–2973.

108 Xue Y, Bi F, Zhang X, et al. Inhibition of endothelial cell proliferation by targeting Rac1 GTPase with small interference RNA in tumor cells. *Biochem Biophys Res Commun.* 2004;320(4):1309–1315.

109 Masumoto A, Hirooka Y, Shimokawa H, et al. Possible involvement of Rho-kinase in the pathogenesis of hypertension in humans. *Hypertension.* 2001;38(6):1307–1310.

110 Mohri M, Shimokawa H, Hirakawa Y, et al. Rho-kinase inhibition with intracoronary fasudil prevents myocardial ischemia in patients with coronary microvascular spasm. *J Am Coll Cardiol.* 2003;41(1):15–19.

111 Cicha I, Schneiderhan-Marra N, Yilmaz A, et al. Monitoring the cellular effects of HMG-CoA reductase inhibitors in vitro and ex vivo. *Arterioscler Thromb Vasc Biol.* 2004;24(11):2046–2050.

84

Protein Tyrosine Phosphatases

Arne Östman and Kai Kappert

Cancer Center Karolinska, Karolinska Institute, Stockholm, Sweden

Tyrosine phosphorylation is a fundamental regulatory mechanism, occurring exclusively in multicellular eukaryotes. This signaling mechanism is regulated by the balanced action of tyrosine kinases and protein tyrosine phosphatases (PTPs). The first PTP was purified in 1988 (1), approximately 10 years after the discovery of tyrosine kinases. The following years witnessed a rapid development that established PTPs as a family of highly specific and tightly regulated signaling proteins. It is now well established that PTPs equal tyrosine kinases in importance for the regulation of tyrosine phosphorylation (2–5). It is also predicted that, like tyrosine kinases, PTPs may emerge as drug targets for many major diseases, including conditions associated with endothelial pathology (6).

THE FAMILY OF PROTEIN TYROSINE PHOSPHATASES

Protein Tyrosine Phosphatase Subfamilies

A recent survey identified a total of 107 genes encoding PTPs in the human genome, of which 105 have mouse orthologues (3). Of these 107 genes, 38 belong to the subtype of classical PTPs, which display a specificity for dephosphorylation of tyrosine residues. Other major subgroups include the dual-specificity phosphatases (61 genes), the myotubularins (16 genes), and the non–cysteine-based Eya family of phosphatases (4 genes). This chapter is restricted to discussing the classical PTPs, hereafter only referred to as PTPs (Figure 84.1).

Structure of the Protein Tyrosine Phosphatase Domain and Catalytic Mechanism

The conserved catalytic domain of PTPs consists of approximately 280 amino acid residues, of which 22 are perfectly conserved among the 38 human PTPs (7). Six of the conserved residues are part of the active-site sequence V/I H C S X G X G R. In addition to this PTP signature motif, three conserved regions contribute to the formation of the catalytic site. The WPD loop closes in over the active site after substrate binding and provides an aspartic acid necessary for catalysis. The phospho-tyrosine (pTyr) recognition loop contains a tyrosine residue that defines the depth of the active site and thereby gives PTPs their specificity for phospho-tyrosine. Finally, the Q-loop contributes with a glutamine residue, which is also necessary for the activity. The PTP domains also share a common overall fold composed of a mixed β-sheet flanked by α-helices.

Catalysis proceeds in a two-step manner (detailed in Ref. 8). In the first step, a covalent PTP-phosphate intermediate is formed in which the phosphate group is covalently linked to the active site cysteine. In the second step of catalysis, this cysteine-phosphate is hydrolyzed. The detailed understanding of the catalytic mechanism and its structural basis has already been exploited for the structure-based design of PTP inhibitors (6,9–11). It also has provided the rationale for the design of "substrate-trapping" mutants used to identify the substrates of individual PTPs (12).

Domain Organization

PTPs are broadly categorized into receptor-like PTPs (RPTPs) and nonreceptor cytosolic variants (NRPTPs). In addition to the conserved PTP domain, PTPs display a striking variation with regard to associated domains (see Figure 84.1) (3).

The RPTPs all share the features of an extracellular domain, a single transmembrane segment, and an intracellular part composed of one or two PTP domains. In many cases, the extracellular parts are composed of multiple fibronectin type III or Ig-like domains, and thus share structural features with cell adhesion molecules. Based on these characteristics, RPTPs have been grouped into eight subtypes (3). Features distinguishing these subtypes include the presence of one (subtypes VI–VIII) or two (subtypes I–V) intracellular PTP domains, and the composition of extracellular domains, which are composed (for example) of fibronectin type III domains only in subgroup VI, whereas members of subgroup III have an

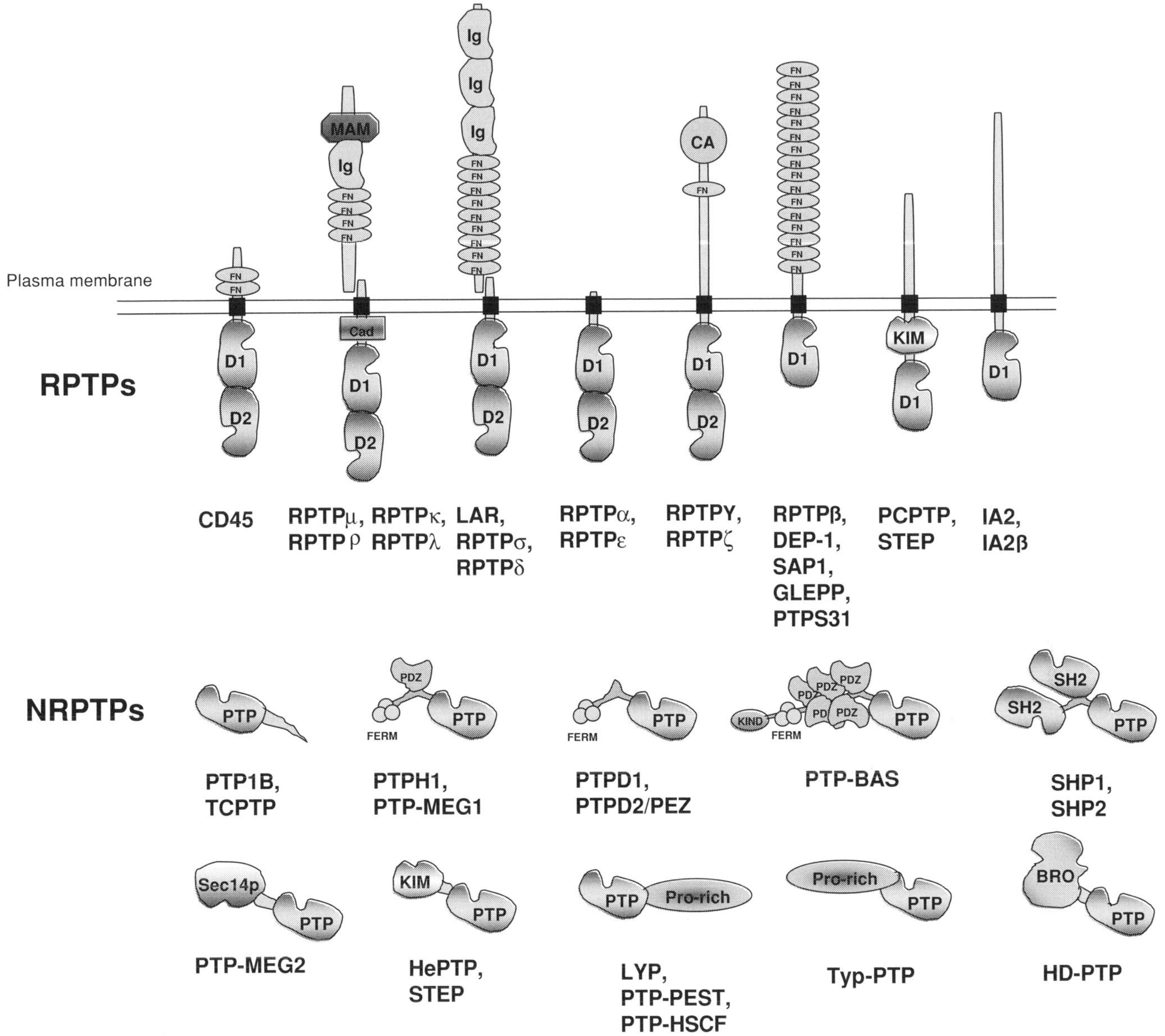

Figure 84.1. Schematic of domain structure of classical PTPs. Black box indicates transmembrane part of receptor-like PTPs. RPTP, receptor-like PTPs; NRPTP non–receptor-like PTP; D1, domain 1; D2, domain 2; FN, fibronectin-like; Ig, immunoglobulin-like; KIM, kinase interaction motif; PDZ, postsynaptic density-95/discs large/Z01 homology; SH2, src homology 2; BRO, baculovirus BRO homology; Pro-rich, proline-rich; FERM, band 4.1/ezrin/radixin/moesin homology; KIND, kinase N lobe-like domain; Sec14p, Sec14p homology (or CRAL/TRIO). (Modified with permission from Alonso A, Sasin J, Bottini N, et al. Protein tyrosine phosphatases in the human genome. *Cell* 2004;117:699–711.)

extracellular domain composed of both Ig-like and fibronectin type III domains. The significance of two tandem PTP domains in most RPTPs is still incompletely understood. However, recent evidence suggests that the second PTP domain, which in most cases has low or no catalytic activity, predominantly exerts a regulatory function.

Among the cytosolic PTPs, a common feature is additional domains involved in directing the proteins to specific subcellular locations or in mediating specific protein–protein interactions (see Figure 84.1). Such domains include the carboxy-terminal endoplasmic reticulum–targeting sequence of PTP-1B, the FERM domains of PTP-HI, which is involved in mediating interactions with the cytoskeleton, and the SH2 domains of SHP-1 and SHP-2 that mediate interactions with tyrosine-phosphorylated proteins.

Regulation of Protein Tyrosine Phosphatase Activity

PTPs are subject to multiple mechanisms that control their specific activity. Phosphorylation of PTPs on tyrosine, serine, or threonine residues has been shown to either increase or decrease their activity (13–16). More recently, an additional reversible modification, namely inhibitory oxidation of the active site cysteine residue, has emerged as a potentially general mechanism for controlling PTP activity (17,18). This phenomenon has been described both after treatment with pathological oxidizing stimuli such as hydrogen peroxide (H_2O_2) or ultraviolet (UV) irradiation, and also after treatment with physiological inducers of reactive oxygen species (ROS), like tyrosine kinase receptors (19–22). These findings provide intriguing indications of PTPs as mediators of cross-talk between redox signaling and tyrosine phosphorylation.

Although a series of reports has identified various putative ligands of RPTPs, there is still very limited information available concerning regulatory PTP ligands. The few examples include reports of pleiotrophin as an inhibitory ligand of PTP-γ and the demonstration of a yet unidentified agonistic density-enhanced phosphatase (DEP)-1 ligand present in a crude extracellular matrix preparation (23,24). In addition to ligand binding, regulated dimerization also has been implicated as a mechanism for negative regulation of RPTP activity. This was first suggested based on the structure of the catalytic domain of RPTP-α. This domain occurred as a symmetric dimer, in which the active sites were blocked by the insertion of a "wedge-like" segment of the other subunit (25). Although subsequent functional studies have provided some support for the structural findings, recent studies also indicate that this mode of regulation might not apply to all RPTPs (26).

A final mode of regulation, which has been observed both for NRPTPs and RPTPs, is stimulus-triggered proteolytic cleavage. This has been shown, for example, with the demonstration of a calpain-mediated release of PTP-1B from the endoplasmic reticulum, and the calcium-dependent shedding of the extracellular domain of LAR and PTP-σ (27,28). The latter was also associated with a translocation of the cleaved forms to intracellular compartments. Finally, proteolytic cleavage and degradation of PTP-1B, SHP-1, and LAR also recently was shown to occur after UV irradiation of cells (29).

Substrate Specificity of Protein Tyrosine Phosphatases

Early biochemical studies indicated a more than 1,000-fold difference in the K_m when different tyrosine phosphorylated peptides as PTP substrates were compared (30). These indications of stringent substrate specificity were confirmed by the use of substrate-trapping mutants of various PTPs, which showed capture of specific substrates by different PTPs (12). It also recently was demonstrated that within proteins with multiple sites of tyrosine phosphorylation, individual PTPs preferentially dephosphorylate certain subsets of these sites (31,32).

The observed selectivity has at least two determinants. First, as revealed by cocrystallization of peptide substrates and PTPs, significant interactions occur between side chains of residues near the phospho-tyrosine residues of substrates and residues flanking the active site of PTPs between substrate and enzyme (33). Second, subcellular compartmentalization restricts access to substrates and thereby also contributes to substrate specificity (34).

PROTEIN TYROSINE PHOSPHATASES AND REGULATION OF ENDOTHELIAL CELL FUNCTION

Endothelial Protein Tyrosine Phosphatases and Pathways Affected by Them

It is noteworthy that only a subset of PTPs has been described to be expressed in endothelial cells (ECs). A recent review summarizing available data tabulated 10 RPTPs and eight NRPTPs that have been demonstrated to be expressed in ECs, using different methods to detect either mRNA or proteins (35) (Table 84-1). As indicated in more detail later, some of these enzymes have been implicated as important modulators of tyrosine kinase receptor signaling, adhesion-modulated signaling, and nitric oxide (NO) signaling. In addition, PTPs are also candidate mediators of signaling triggered by shear stress and hypoxia.

Protein Tyrosine Phosphatases and Vascular Endothelial Growth Factor Receptor 2 or Tie2 Signaling

The major receptor tyrosine kinases controlling ECs are vascular endothelial growth factor (VEGF) receptor 2 (VEGFR2) and Tie2. Whereas VEGF receptors are strongly linked to EC proliferation and migration, Tie2 appears to be more important for functional maturation of ECs.

Two PTPs that have been identified as negative regulators of VEGFR2 signaling are DEP-1 and SHP-1. siRNA-mediated downregulation of DEP-1, or overexpression of a dominant-negative form of DEP-1, led to enhanced VEGFR2 phosphorylation and activation of downstream signaling (36). Interestingly, this effect was most prominent under conditions of high cell density. Association between DEP-1 and vascular endothelial (VE)-cadherin was shown to be necessary for this effect, emphasizing the importance of a localized recruitment of PTPs to their substrates. SHP-1 has been linked to VEGFR2 signaling based on evidence of direct physical associations between SHP-1 and VEGFR2, and the finding that tumor necrosis factor (TNF)-α–mediated downregulation of VEGFR2 phosphorylation requires SHP-1 expression (37,38). Also, enhanced SHP-1 association with VEGFR2 appears to be one component of tissue inhibitor of metalloproteinase (TIMP)-2–mediated inhibition of angiogenesis (39).

Concerning Tie2, special interest has been paid to the receptor-like protein PTP-β, also designated vascular endothelial (VE)-PTP (40). PTP-β was implicated as a Tie2–relevant PTP based on interaction in a yeast two-hybrid screen. Subsequently, a substrate-trapping mutant of PTP-β was found to interact with Tie2, but not with VEGFR2. Interestingly, like DEP-1, PTP-β has been shown to interact also with VE-cadherin (41).

Preliminary animal studies with PTP inhibitors also support the notion that PTPs negatively control VEGFR2 signaling. For example, enhanced vascularization has been observed in various models after treatment with PTP inhibitors (42–44).

Table 84-1: Overview of Functions of Certain PTPs Expressed in ECs and Their Knockout Phenotype

	Function in ECs	*−/− Phenotype*	*References*
RPTPs			
DEP-1/CD148	Inhibits VEGF function; associates with VE-cadherin	Die at midgestation; enlarged vessels with highly proliferative ECs	57
RPTP-α	Not described	Viable, fertile, no gross abnormalities	59
PTP-β/VE-PTP	Interacts with Tie2 (not VEGFR2); associates with VE-cadherin, promotes barrier function	Embryonic lethality, vascular malformations, enlarged cavities in vessels of yolk sac	60
PTP-δ	Not described	Insufficient food intake (consecutive semilethal), impaired learning	61
PTP-ϵ	Enhances EC migration and survival	Myelination defects	62
PTP-μ	Binds to VE-cadherin, regulates barrier integrity	Viable, fertile, no aberrant phenotype	47, 63
PTP-σ	Not described	Increased neonatal mortality, retarded growth, altered nerve regeneration, pituitary, pancreatic and gut neuroendocrine defects	64–66
LAR	Not described	Normal growth, no histological abnormalities, increased neonatal death rate for mice from −/− females, altered mammary gland development and function	67
Cytosolic PTPs			
PTP-1B	Alters VEGF-induced proliferation/migration, regulates neovascularization	Enhanced insulin sensitivity and resistance to weight gain, increased basal metabolic rate	42, 68, 69
SHP-1	Inhibits VEGF function; dephosphorylates PECAM-1	Shortened lifespan, immune deficiency*	70, 71
SHP-2	Dephosphorylates PECAM-1; associates with and mediates downstream signaling of E-selectin	Embryonic lethality	72
TC-PTP	Not described	Splenomegaly, lymphadenopathy, premature death	73

In addition to these 12 PTPs, expression in ECs also has been described for PTP-γ, PTP-κ, FAP-1, PTP-D1, PTP-D2, and PTP-PEST (see Ref. 35).

*Mice exist in the form of two natural mutations. Homozygous mutations in the SHP-1 gene cause motheaten (*me/me*) and motheaten viable (me^v/me^v) mice, which differ in phenotype severity (see Ref. 70).

Although most available information thus emphasizes the role of PTPs as negative regulators of RTK signaling in ECs, it should be cautioned that this view might be simplified. It is well established from other RTK pathways that some PTPs, particularly SHP-2, actually act as positive mediators of signaling (45). In the context of EC signaling, this also has been demonstrated by enhanced migration and survival after overexpression of the receptor-like variant of PTP-ϵ, presumably caused by dephosphorylation of the negative regulatory element pY527 of c-Src (46).

Endothelial Cell Adhesion and Permeability

ECs adhere to each other as well as to the extracellular matrix of the vessel wall. These aspects of endothelial biology are directly linked to vascular permeability. In addition, ECs express surface molecules, such as selectins, to which circulating cells adhere. These adhesive processes all involve reversible tyrosine phosphorylation and, as indicated by a few examples given here, PTPs are actively involved in the regulation of these processes.

As mentioned earlier, both PTP-β and DEP-1 have been shown to associate with VE-cadherin. In the case of PTP-β, it was shown that this interaction, analyzed in transfected cells, promoted the barrier function of VE-cadherin (41). Similar functions of PTP-μ have been demonstrated (47). An involvement of PTPs in the maintenance of barrier function also is supported by the observation that treatment of cells with a general PTP inhibitor increased serum albumin flux across cultured monolayers of ECs (48). Furthermore, evidence for PTP-mediated control of VE-cadherin phosphorylation and function in vivo was shown by increased phosphorylation of VE-cadherin after systemic treatment with peroxovanadate (49).

The receptor platelet-endothelial cell adhesion molecule (PECAM)-1 is mainly expressed at endothelial intercellular sites. PECAM-1 activity is subjected to control by integrins in a manner that involves PECAM-1 dephosphorylation (50). One aspect of this cross-talk appears to be the recruitment and dephosphorylation of PECAM-1 by SHP-1 and SHP-2 (51). More recently, ROS have been implicated in these processes, both as regulators of PECAM-1 phosphorylation and of SHP-2 catalytic activity (52).

E-selectin is an inducible adhesion molecule that is expressed at relatively high density on the surface of cultured ECs that have been activated by proinflammatory cytokines, such as interleukin (IL)-1β, TNF-α, or bacterial endotoxin. It is also detectable in vivo at sites of inflammation. In addition to supporting the rolling and stable arrest of leukocytes on activated endothelium, there is increasing evidence that E-selectin can transduce outside-in signals. Hu and colleagues (53) demonstrated that SHP-2 associates with E-selectin and subsequently acts as a positive mediator for downstream signaling in ECs. This interaction might have functional implications for the pathology of inflammation.

Protein Tyrosine Phosphatase Inhibition in Association with Shear Stress and Hypoxia in Endothelial Cells

Hypoxia–reoxygenation and shear stress are known stimuli for the production of ROS in the vessel wall. As discussed earlier, PTPs recently have been implicated as major targets for ROS signaling and are thereby candidate mediators of hypoxia and shear stress signaling in ECs. Although still sparse, some experimental support for this notion has been gathered.

ECs are constantly exposed to blood flow. A well-balanced laminar flow is considered atheroprotective, whereas a turbulent flow pattern might lead to endothelial dysfunction and preatherosclerotic conditions. That flow conditions can influence signaling is indicated by the rapid phosphorylation of PECAM-1 after shear stress exposure (54). A possible role of reversible oxidation of PTPs in this response is suggested by the observed reduction of SHP-2 activity following shear stress (55).

Hypoxia–reoxygenation affects EC signaling, and increased ROS production is one intracellular mediator of this process. Although direct demonstration of PTP inhibition under these conditions is still lacking, some indirect observations support this possibility. Waltenberger and colleagues (56) showed that VEGFR2 was upregulated in human umbilical ECs after hypoxia. Moreover, an increase of VEGF-induced receptor phosphorylation due to hypoxia was observed, and VEGF-dependent mitogenesis was increased. The possibility that PTP inhibition contributes to the increased RTK signaling in ECs under hypoxia deserves further investigations. An interesting therapeutic implication of this scenario is treatment with antioxidants for prevention of pathological angiogenesis.

Protein Tyrosine Phosphatases in the Context of Endothelial Cell Biology

As is clear from this review of PTP involvement in EC signaling, PTPs have an impact on a broad range of pathways that affect fundamental aspects of EC biology. Some of the findings discussed earlier are summarized in a cellular context in Figure 84.2 and Table 84-1.

As negative regulators of VEGFR2 and Tie2 signaling, the PTPs DEP-1, PTP-β/VE-PTP, and SHP-1 are central negative regulators of endothelial proliferation, migration, and maturation (36–38,40,41). Also, it is anticipated that hypoxia -induced endothelial proliferation involves ROS-mediated attenuation of PTPs. Concerning cell–cell interactions between ECs, the observed functional cooperation between VE-cadherin and PTP-β/VE-PTP points to the role of this PTP, and possibly others, in maintaining the barrier integrity of the endothelial layer (41). Also, SHP-2 negatively regulates the adhesive and antimigratory effects of PECAM-1 by dephosphorylation of PECAM-1 (50,51). Finally, the requirement of SHP-2 for E-selectin–mediated signaling points to a role for this PTP in the signaling events triggered by neutrophil–endothelial interactions (53).

Analyses of vascular phenotypes in PTP knockouts represent an alternative strategy for investigating the role of PTPs in EC biology. In this context, the phenotype of DEP-1 depletion provides the strongest evidence for PTP involvement in vascular biology (57). *DEP-1* $^{-/-}$ embryos die at midgestation (see Table 84-1). They display enlarged vessels with highly proliferative ECs expressing markers of low differentiation. Vessels also are characterized by reduced pericyte coverage. Although the biochemical basis for this phenotype has not been characterized extensively, it is likely that they at least partially are caused by hyperactive VEGFR2 signaling.

THERAPEUTIC OPPORTUNITIES

ECs are obvious targets for pro- and antiangiogenic strategies. As important enzymatic regulators of endothelial biology, PTPs are emerging as interesting drug targets. The development of PTP antagonists is envisioned to be a more realistic venture than the development of specific activators. However, reactivation of oxidized PTPs by antioxidants might be an

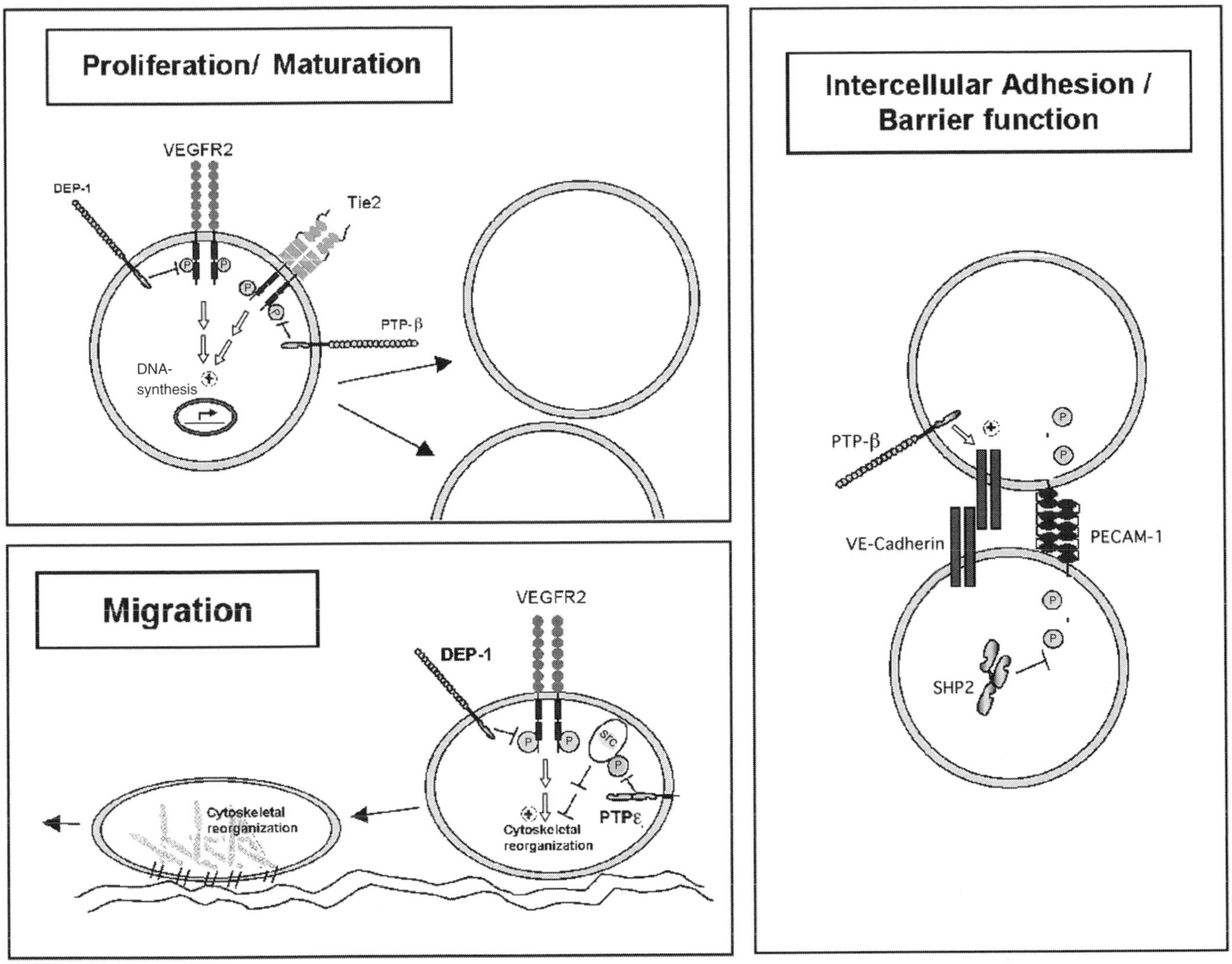

Figure 84.2. Schematic illustration of PTP involvement in EC biology. Some cellular processes that involve PTPs as signaling molecules are illustrated. The figure highlights the roles of some individual PTPs as presently understood. It is likely that the true situation involves multiple members of the PTP family.

option in situations where high specificity is not crucial. In general terms, the inhibition of PTPs that antagonize VEGFR2 and Tie2, such as DEP-1 and PTP-β, holds promise for proangiogenic approaches, whereas the inhibition of PTPs acting as positive regulators of VEGFR2 signaling, such as SHP-2, can be hypothesized to have antiangiogenic effects.

Given the structural similarity of the active site of different PTPs, the issue of inhibitor specificity has been a concern. However, a series of recent studies have indicated the possibility of generating inhibitors with submicromolar potency that display significant specificity. In the context of endothelial biology, recent PTP-β inhibitors developed at Novo Nordisk are of special importance (58). These compounds, generated by structure-based drug design, show a more than 100-fold selectivity for PTP-β when compared with (for example) PTP-1B and RPTP-α. In addition to the issue of specificity, poor bioavailability has been a concern in the development of PTP inhibitors. This mainly derives from the fact that most PTP inhibitors, developed as phospho-tyrosine mimics, display a high polarity. Prodrug strategies are therefore being tested to overcome this problem.

CONCLUSION

During the last decade, improved understanding of the general aspects of PTP biology is being integrated in more detailed studies on the roles of PTPs in EC biology. Although clear progress has been made, it is also obvious that the present knowledge is still scarce and incomplete. It is envisioned that applications of modern methods, leading to endothelial-specific depletion of individual PTPs, will reveal yet unrecognized roles for PTPs in the context of vascular biology. Furthermore, specific PTP-inhibitors, useful for in vivo applications, will ultimately be required to evaluate the therapeutic potential of PTPs.

NOTE ADDED IN PROOF

Concerning the regulation of PTPs, it has been shown that PTP-1B and possibly other PTPs are also regulated by sumolation (74). Inactivation of VE-PTP has confirmed important functions of this enzyme in ECs (60). Also, a DEP-1 monoclonal antibody has been described that also indicates a regulatory role of DEP-1 in EC biology (75). Finally, a second *DEP-1*-knockout mouse lacking an obvious vascular phenotype has been described (76).

KEY POINTS

- PTPs constitute a structurally diverse, tightly regulated enzyme family with important regulatory roles in the control of vessel biology, particularly EC proliferation, migration, and adhesion, and capillary permeability.
- DEP-1 and SHP-1 act as negative regulators of VEGFR2 receptor-induced EC proliferation.
- PTP-β and PTP-μ promote the VE-cadherin–mediated barrier function of ECs.

Future Goals

- To continue a systematic determination of the expression and activity of different PTPs in ECs under various physiological and pathological conditions
- To continue analyses of the involvement of ROS-mediated PTP inhibition in EC signaling, with emphasis on signaling triggered by shear stress or hypoxia
- To explore the potential of PTP inhibition, and reactivation of oxidized PTPs, for pro- and antiangiogenic purposes

REFERENCES

1 Tonks NK, Diltz CD, Fischer EH. Purification of the major protein-tyrosine-phosphatases of human placenta. *J Biol Chem.* 1988;263(14):6722–6730.

2 Östman A, Bohmer FD. Regulation of receptor tyrosine kinase signaling by protein tyrosine phosphatases. *Trends Cell Biol.* 2001; 11(6):258–266.

3 Alonso A, Sasin J, Bottini N, et al. Protein tyrosine phosphatases in the human genome. *Cell.* 2004;117(6):699–711.

4 Chiarugi P, Cirri P. Redox regulation of protein tyrosine phosphatases during receptor tyrosine kinase signal transduction. *Trends Biochem Sci.* 2003;28(9):509–514.

5 Tonks NK, Neel BG. Combinatorial control of the specificity of protein tyrosine phosphatases. *Curr Opin Cell Biol.* 2001;13(2): 182–195.

6 van Huijsduijnen RH, Bombrun A, Swinnen D. Selecting protein tyrosine phosphatases as drug targets. *Drug Discov Today.* 2002;7(19):1013–1019.

7 Andersen JN, Mortensen OH, Peters GH, et al. Structural and evolutionary relationships among protein tyrosine phosphatase domains. *Mol Cell Biol.* 2001;21(21):7117–7136.

8 Pannifer AD, Flint AJ, Tonks NK, Barford D. Visualization of the cysteinyl-phosphate intermediate of a protein-tyrosine phosphatase by x-ray crystallography. *J Biol Chem.* 1998;273(17): 10454–10462.

9 Johnson TO, Ermolieff J, Jirousek MR. Protein tyrosine phosphatase 1B inhibitors for diabetes. *Nat Rev Drug Discov.* 2002; 1(9):696–709.

10 Taylor SD. Inhibitors of protein tyrosine phosphatase 1B (PTP1B). *Curr Top Med Chem.* 2003;3(7):759–782.

11 Harley EA, Levens N. Protein tyrosine phosphatase 1B inhibitors for the treatment of type 2 diabetes and obesity: recent advances. *Curr Opin Investig Drugs.* 2003;4(10):1179–1189.

12 Flint AJ, Tiganis T, Barford D, Tonks NK. Development of "substrate-trapping" mutants to identify physiological substrates of protein tyrosine phosphatases. *Proc Natl Acad Sci USA.* 1997; 94(5):1680–1685.

13 Wang Y, Guo W, Liang L, Esselman WJ. Phosphorylation of CD45 by casein kinase 2. Modulation of activity and mutational analysis. *J Biol Chem.* 1999;274(11):7454–7461.

14 Garton AJ, Tonks NK. PTP-PEST: a protein tyrosine phosphatase regulated by serine phosphorylation. *EMBO J.* 1994;13(16): 3763–3771.

15 Liu F, Chernoff J. Protein tyrosine phosphatase 1B interacts with and is tyrosine phosphorylated by the epidermal growth factor receptor. *Biochem J.* 1997;327 (Pt 1):139–145.

16 Uchida T, Matozaki T, Noguchi T, et al. Insulin stimulates the phosphorylation of Tyr538 and the catalytic activity of PTP1C, a protein tyrosine phosphatase with Src homology-2 domains. *J Biol Chem.* 1994;269(16):12220–12228.

17 Salmeen A, Barford D. Functions and mechanisms of redox regulation of cysteine-based phosphatases. *Antioxid Redox Signal.* 2005;7(5–6):560–577.

18 Rhee SG, Kang SW, Jeong W, et al. Intracellular messenger function of hydrogen peroxide and its regulation by peroxiredoxins. *Curr Opin Cell Biol.* 2005;17(2):183–189.

19 Lee SR, Kwon KS, Kim SR, Rhee SG. Reversible inactivation of protein-tyrosine phosphatase 1B in A431 cells stimulated with epidermal growth factor. *J Biol Chem.* 1998;273(25):15366–15372.

20 Denu JM, Tanner KG. Specific and reversible inactivation of protein tyrosine phosphatases by hydrogen peroxide: evidence for a sulfenic acid intermediate and implications for redox regulation. *Biochemistry.* 1998;37(16):5633–5642.

21 Meng TC, Fukada T, Tonks NK. Reversible oxidation and inactivation of protein tyrosine phosphatases in vivo. *Mol Cell.* 2002; 9(2):387–399.

22 Persson C, Sjoblom T, Groen A, et al. Preferential oxidation of the second phosphatase domain of receptor-like PTP-alpha revealed by an antibody against oxidized protein tyrosine phosphatases. *Proc Natl Acad Sci USA.* 2004;101(7):1886–1891.

23 Meng K, Rodriguez-Pena A, Dimitrov T, et al. Pleiotrophin signals increased tyrosine phosphorylation of beta beta-catenin through inactivation of the intrinsic catalytic activity of the receptor-type protein tyrosine phosphatase beta/zeta. *Proc Natl Acad Sci USA.* 2000;97(6):2603–2608.

24 Sorby M, Sandstrom J, Ostman A. An extracellular ligand increases the specific activity of the receptor-like protein tyrosine phosphatase DEP-1. *Oncogene.* 2001;20(37):5219–5224.

25 Jiang G, den Hertog J, Su J, et al. Dimerization inhibits the activity of receptor-like protein-tyrosine phosphatase-alpha. *Nature.* 1999;401(6753):606–610.
26 Nam HJ, Poy F, Krueger NX, et al. Crystal structure of the tandem phosphatase domains of RPTP LAR. *Cell.* 1999;97(4):449–457.
27 Frangioni JV, Oda A, Smith M, et al. Calpain-catalyzed cleavage and subcellular relocation of protein phosphotyrosine phosphatase 1B (PTP-1B) in human platelets. *EMBO J.* 1993;12(12): 4843–4856.
28 Aicher B, Lerch MM, Muller T, et al. Cellular redistribution of protein tyrosine phosphatases LAR and PTPsigma by inducible proteolytic processing. *J Cell Biol.* 1997;138(3):681–696.
29 Gulati P, Markova B, Gottlicher M, et al. UVA inactivates protein tyrosine phosphatases by calpain-mediated degradation. *EMBO Rep.* 2004;5(8):812–817.
30 Zhang ZY, Thieme-Sefler AM, Maclean D, et al. Substrate specificity of the protein tyrosine phosphatases. *Proc Natl Acad Sci USA.* 1993;90(10):4446–4450.
31 Kovalenko M, Denner K, Sandstrom J, et al. Site-selective dephosphorylation of the platelet-derived growth factor beta-receptor by the receptor-like protein-tyrosine phosphatase DEP-1. *J Biol Chem.* 2000;275(21):16219–16226.
32 Persson C, Savenhed C, Bourdeau A, et al. Site-selective regulation of platelet-derived growth factor beta receptor tyrosine phosphorylation by T-cell protein tyrosine phosphatase. *Mol Cell Biol.* 2004;24(5):2190–2201.
33 Jia Z, Barford D, Flint AJ, Tonks NK. Structural basis for phosphotyrosine peptide recognition by protein tyrosine phosphatase 1B. *Science.* 1995;268(5218):1754–1758.
34 Haj FG, Verveer PJ, Squire A, et al. Imaging sites of receptor dephosphorylation by PTP1B on the surface of the endoplasmic reticulum. *Science.* 2002;295(5560):1708–1711.
35 Kappert K, Peters KG, Bohmer FD, Ostman A. Tyrosine phosphatases in vessel wall signaling. *Cardiovasc Res.* 2005;65(3):587–598.
36 Grazia Lampugnani M, Zanetti A, Corada M, et al. Contact inhibition of VEGF-induced proliferation requires vascular endothelial cadherin, beta-catenin, and the phosphatase DEP-1/CD148. *J Cell Biol.* 2003;161(4):793–804.
37 Kroll J, Waltenberger J. The vascular endothelial growth factor receptor KDR activates multiple signal transduction pathways in porcine aortic endothelial cells. *J Biol Chem.* 1997;272(51): 32521–32527.
38 Guo DQ, Wu LW, Dunbar JD, et al. Tumor necrosis factor employs a protein-tyrosine phosphatase to inhibit activation of KDR and vascular endothelial cell growth factor-induced endothelial cell proliferation. *J Biol Chem.* 2000;275(15):11216–11221.
39 Seo DW, Li H, Guedez L, et al. TIMP-2 mediated inhibition of angiogenesis: an MMP-independent mechanism. *Cell.* 2003; 114(2):171–180.
40 Fachinger G, Deutsch U, Risau W. Functional interaction of vascular endothelial-protein-tyrosine phosphatase with the angiopoietin receptor Tie-2. *Oncogene.* 1999;18(43):5948–5953.
41 Nawroth R, Poell G, Ranft A, et al. VE-PTP and VE-cadherin ectodomains interact to facilitate regulation of phosphorylation and cell contacts. *EMBO J.* 2002;21(18):4885–4895.
42 Soeda S, Shimada T, Koyanagi S, et al. An attempt to promote neo-vascularization by employing a newly synthesized inhibitor of protein tyrosine phosphatase. *FEBS Lett.* 2002;524(1–3):54–58.
43 Carr AN, Davis MG, Eby-Wilkens E, et al. Tyrosine phosphatase inhibition augments collateral blood flow in a rat model of peripheral vascular disease. *Am J Physiol Heart Circ Physiol.* 2004; 287(1):H268–H276.
44 Sugano M, Tsuchida K, Makino N. A protein tyrosine phosphatase inhibitor accelerates angiogenesis in a rat model of hindlimb ischemia. *J Cardiovasc Pharmacol.* 2004;44(4):460–465.
45 Neel BG, Gu H, Pao L. The "Shp"ing news: SH2 domain-containing tyrosine phosphatases in cell signaling. *Trends Biochem Sci.* 2003;28(6):284–293.
46 Nakagawa Y, Yamada N, Shimizu H, et al. Tyrosine phosphatase epsilonM stimulates migration and survival of porcine aortic endothelial cells by activating c-Src. *Biochem Biophys Res Commun.* 2004;325(1):314–319.
47 Sui XF, Kiser TD, Hyun SW, et al. Receptor protein tyrosine phosphatase micro regulates the paracellular pathway in human lung microvascular endothelia. *Am J Pathol.* 2005;166(4):1247–1258.
48 Young BA, Sui X, Kiser TD, et al. Protein tyrosine phosphatase activity regulates endothelial cell-cell interactions, the paracellular pathway, and capillary tube stability. *Am J Physiol Lung Cell Mol Physiol.* 2003;285(1):L63–L75.
49 Lambeng N, Wallez Y, Rampon C, et al. Vascular endothelial-cadherin tyrosine phosphorylation in angiogenic and quiescent adult tissues. *Circ Res.* 2005;96(3):384–391.
50 Lu TT, Yan LG, Madri JA. Integrin engagement mediates tyrosine dephosphorylation on platelet-endothelial cell adhesion molecule 1. *Proc Natl Acad Sci USA.* 1996;93(21):11808–11813.
51 Hua CT, Gamble JR, Vadas MA, Jackson DE. Recruitment and activation of SHP-1 protein-tyrosine phosphatase by human platelet endothelial cell adhesion molecule-1 (PECAM-1). Identification of immunoreceptor tyrosine-based inhibitory motif-like binding motifs and substrates. *J Biol Chem.* 1998;273(43): 28332–28340.
52 Maas M, Wang R, Paddock C, et al. Reactive oxygen species induce reversible PECAM-1 tyrosine phosphorylation and SHP-2 binding. *Am J Physiol Heart Circ Physiol.* 2003;285(6):H2336–H2344.
53 Hu Y, Szente B, Kiely JM, Gimbrone MA Jr. Molecular events in transmembrane signaling via E-selectin. SHP2 association, adaptor protein complex formation and ERK1/2 activation. *J Biol Chem.* 2001;276(51):48549–48553.
54 Osawa M, Masuda M, Kusano K, Fujiwara K. Evidence for a role of platelet endothelial cell adhesion molecule-1 in endothelial cell mechanosignal transduction: is it a mechanoresponsive molecule? *J Cell Biol.* 2002;158(4):773–785.
55 Lerner-Marmarosh N, Yoshizumi M, Che W, et al. Inhibition of tumor necrosis factor-[alpha]-induced SHP-2 phosphatase activity by shear stress: a mechanism to reduce endothelial inflammation. *Arterioscler Thromb Vasc Biol.* 2003;23(10):1775–1781.
56 Waltenberger J, Mayr U, Pentz S, Hombach V. Functional upregulation of the vascular endothelial growth factor receptor KDR by hypoxia. *Circulation.* 1996;94(7):1647–1654.
57 Takahashi T, Takahashi K, St. John PL, et al. A mutant receptor tyrosine phosphatase, CD148, causes defects in vascular development. *Mol Cell Biol.* 2003;23(5):1817–1831.
58 Lund IK, Andersen HS, Iversen LF, et al. Structure-based design of selective and potent inhibitors of protein-tyrosine phosphatase {beta}. *J Biol Chem.* 2004;279(23):24226–24235. Epub 2004 Mar 15.

59 Ponniah S, Wang DZ, Lim KL, Pallen CJ. Targeted disruption of the tyrosine phosphatase PTPalpha leads to constitutive downregulation of the kinases Src and Fyn. *Curr Biol.* 1999;9(10): 535–538.

60 Baumer S, Keller L, Holtmann A, et al. Vascular endothelial cell-specific phosphotyrosine phosphate (VE-PTP) activity is required for blood vessel development. *Blood.* 2006;107:4754–4762.

61 Uetani N, Kato K, Ogura H, et al. Impaired learning with enhanced hippocampal long-term potentiation in PTPdelta-deficient mice. *EMBO J.* 2000;19(12):2775–2785.

62 Peretz A, Gil-Henn H, Sobko A, et al. Hypomyelination and increased activity of voltage-gated $K(^+)$ channels in mice lacking protein tyrosine phosphatase epsilon. *EMBO J.* 2000;19(15): 4036–4045.

63 Koop EA, Lopes SM, Feiken E, et al. Receptor protein tyrosine phosphatase mu expression as a marker for endothelial cell heterogeneity; analysis of RPTPmu gene expression using LacZ knock-in mice. *Int J Dev Biol.* 2003;47(5):345–354.

64 Batt J, Asa S, Fladd C, Rotin D. Pituitary, pancreatic and gut neuroendocrine defects in protein tyrosine phosphatase-sigma-deficient mice. *Mol Endocrinol.* 2002;16(1):155–169.

65 Elchebly M, Wagner J, Kennedy TE, et al. Neuroendocrine dysplasia in mice lacking protein tyrosine phosphatase sigma. *Nat Genet.* 1999;21(3):330–333.

66 McLean J, Batt J, Doering LC, et al. Enhanced rate of nerve regeneration and directional errors after sciatic nerve injury in receptor protein tyrosine phosphatase sigma knock-out mice. *J Neurosci.* 2002;22(13):5481–5491.

67 Schaapveld RQ, Schepens JT, Robinson GW, et al. Impaired mammary gland development and function in mice lacking LAR receptor-like tyrosine phosphatase activity. *Dev Biol.* 1997; 188(1):134–146.

68 Elchebly M, Payette P, Michaliszyn E, et al. Increased insulin sensitivity and obesity resistance in mice lacking the protein tyrosine phosphatase-1B gene. *Science.* 1999;283(5407):1544–1548.

69 Klaman LD, Boss O, Peroni OD, et al. Increased energy expenditure, decreased adiposity, and tissue-specific insulin sensitivity in protein-tyrosine phosphatase 1B-deficient mice. *Mol Cell Biol.* 2000;20(15):5479–5489.

70 Chen HE, Chang S, Trub T, Neel BG. Regulation of colony-stimulating factor 1 receptor signaling by the SH2 domain-containing tyrosine phosphatase SHPTP1. *Mol Cell Biol.* 1996; 16(7):3685–3697.

71 Green MC, Shultz LD. Motheaten, an immunodeficient mutant of the mouse. I. Genetics and pathology. *J Hered.* 1975;66(5):250–258.

72 Saxton TM, Henkemeyer M, Gasca S, et al. Abnormal mesoderm patterning in mouse embryos mutant for the SH2 tyrosine phosphatase Shp-2. *EMBO J.* 1997;16(9):2352–2364.

73 You-Ten KE, Muise ES, Itie A, et al. Impaired bone marrow microenvironment and immune function in T cell protein tyrosine phosphatase-deficient mice. *J Exp Med.* 1997;186(5):683–693.

74 Dadke S, Cotteret S, Yip SC, et al. Regulation of protein tyrosine phosphatase 1B by sumolation. *Nat Cell Biol.* 2007;9(1): 80–85.

75 Takahashi T, Takahashi K, Mernaugh RL, et al. A monoclonal antibody against CD148, a receptor-like tyrosine phosphate, inhibits endothelial-cell growth and angiogenesis. *Blood.* 2006; 108(4):1234–1242.

76 Trapasso F, Drusco A, Costinean S, et al. Genetic ablation of Ptprj, a mouse cancer susceptibility gene, results in normal growth and development and does not predispose to spontaneous tumorigenesis. *DNA Cell Biol.* 2006;25(6):376–382.

85

Role of β-Catenin in Endothelial Cell Function

Anna Cattelino* and Stefan Liebner†

*IFOM-Fondazione Istituto FIRC di Oncologia Molecolare, Milan, Italy;
†Institute of Neurology, University of Frankfurt, Germany

Originally, a homologue of the vertebrate β-catenin was identified as a segment polarity gene involved in the wingless pathway in *Drosophila melanogaster.* Because of the structural organization of the gene, it was named *armadillo* (ARM), showing several repeats of a 42-amino acid motif in its central region (ARM-repeats) flanked by N-terminal and C-terminal regulatory domains. Its vertebrate homologues, β-catenin and γ-catenin (also named *plakoglobin*), were first characterized as structural proteins involved in cell adhesion, linking the cytoplasmic tail of type 1 cadherins (E, N-, P-, VE-cadherin) via α-catenin to the actin cytoskeleton. Although β-catenin is specific for classical cadherins of the adherens junction (AJ) complex, γ-catenin also associates with the desmosomal cadherins, desmocollin, and desmoglein, linking them via desmoplakin to the vimentin intermediate filament system. Due to its high sequence similarity to armadillo, a signaling function was also suggested for β-catenin and γ-catenin and was subsequently demonstrated by loss- and gain-of-function experiments in *Xenopus laevis*, leading to ventralized embryos and axis duplication, respectively (1). Furthermore, similar to the *wingless* growth factor in *Drosophila*, vertebrate Wnts, on binding to frizzled (Fz) receptors, stabilize cytoplasmic β-catenin or γ-catenin, leading to their activation in nuclear signaling. To accomplish this function, β-catenin interacts with HMG-box transcription factors of the lymphoid enhancer binding factor/ T-cell factor (Lef/TCF) family, which mediate DNA binding and thus enable the transactivating function of β-catenin.

The Wnt signaling pathway was initially characterized in development, where it was shown to be involved in processes as diverse as somitogenesis, and brain, limb, and vascular differentiation. In the adult, the Wnt-pathway seemed to be less important until it was shown that adenomatous polyposis coli (APC), a tumor suppressor gene, serves as a negative regulator of β-catenin stability and that a lack-of-function mutation of APC is responsible for familial adenomatous polyposis (FAP) (2). Affected patients show a predisposition for colon cancer due to increased β-catenin signaling. So far, increased β-catenin signaling has been demonstrated in various kinds of tumors, such as mammary carcinoma, hair follicle tumors, gallbladder carcinoma, and neuroblastoma.

In addition to β- and γ-catenin, other armadillo family members have been identified in vertebrates, such as p120^{ctn}, plakophilins, neural plakophilin-related armadillo protein (NPRAP/δ-catenin), APC, SRP1, p0071, and smgGDS. The common ARM-repeat is believed to mediate protein–protein interaction similar to those of SH2- or ankyrin-domains in other proteins. Although all these proteins share the ARM-repeat as a structural motif in their core region, functionally, several subfamilies can be defined, underlining the involvement of ARM family proteins in diverse cellular processes.

FROM ASCIDIANS TO HUMANS

The ability of unicellular organisms to communicate and share nutrients and information is considered the key molecular event for the evolution of multicellular organisms. Therefore, cell-to-cell adhesion structures such as AJs and β-catenin must have been evolved very early. In fact, β-catenin, and ARM family proteins in general, are highly conserved, and have been found in phylogenetically very distant species, from ascidians to humans.

The importance of Wnt signaling is underlined by the fact that mutations of this pathway have been shown to cause severe defects in numerous cell specification events during development, including formation of endomesoderm in *Caenorhabditis elegans*, sea urchins, and ascidians; segmentation in *Drosophila*; and dorsal-ventral axis formation in vertebrates. In a strict sense, Wnt signaling is limited to metazoan animals. It has been postulated that a single β-catenin gene fulfilled both adhesion and signaling functions in the last common ancestor of metazoans some 700 million years ago. The diploblastic Cnidaria (e.g., *Hydra*) possess genes encoding most of the Wnt pathway, including a Wnt growth factor, a Fz receptor, β-catenin, and Lef/TCF. The Wnt and Lef/TCF genes show interesting expression patterns during head formation in *Hydra*. In the sea cnidarian *Nematostella vectensis*,

β-catenin is needed to specify the endoderm, indicating an evolutionarily ancient role for this protein in early pattern formation (3). Both a Wnt family member and a bona fide member of the Fz family have been identified Platyhelminthes in Desmosponges, respectively. Proteins with the ARM repeats are also widespread and present in plants and yeast. However, no evidence for a Wnt growth factor has been obtained in any of these latter organisms.

The most divergent parts of the metazoan β-catenin proteins and other ARM family members are the terminal domains. Strikingly, each ARM repeat from cnidarian β-catenin aligns with its respective ARM repeat from mouse, suggesting a strong, phylogenetic conservation of each of these motifs.

Whereas β-catenin is represented by a single gene in most species, in the *C. elegans* genome three β-catenin genes – *hmp-2*, *bar-1*, and *wrm-1* – have been detected. Interestingly, one of their gene products (HMP-2) is dedicated to adhesion only, whereas BAR-1 and WRM-1 act in Wnt signaling. The biochemical and genetic functions of β-catenin have been most extensively studied in vertebrates (human, mouse, frog, zebrafish) and *Drosophila*.

Vertebrates have two Armadillo/β–catenin homologues, β-catenin and γ-catenin. Although the former remained remarkably conserved over most of metazoan evolution, the latter probably developed through a gene duplication of the bifunctional β-catenin, yielding the ancestor of γ-catenin. This duplication is probably attributable to one of the two genome duplications that occurred in the vertebrate branch 600 to 490 million years ago. γ-Catenin evolved relatively rapidly from the ancestral protein, giving rise to novel functions. Differences in γ-catenin's cellular sequestration, its accessibility for the protein degradation machinery, its affinity to Lef/TCF, and its level of transactivating ability, suggest that γ-catenin is unlikely to play a β-catenin–like role in the canonical Wnt signaling pathway. However, γ-catenin may indirectly modify or attenuate β-catenin–mediated Wnt signaling. Interestingly, it has recently been reported that β-catenin and γ-catenin associate with distinct TCF-4 subdomains, causing opposite effects on TCF-4 DNA binding: Whereas β-catenin does not affect this binding, γ-catenin prevents it (4). Although its presumed ancestral role in Wnt signaling has been reduced or modified, γ-catenin has acquired a novel role in desmosomal adhesion.

THE ROLE OF β-CATENIN IN ADHERENS JUNCTIONS ORGANIZATION

β-Catenin is a multifunctional protein known to interact with more than 20 different molecules (Figure 85.1). Similar to epithelial cells, endothelial cells (ECs) have specialized junctional regions that are functionally defined into AJs and tight junctions (TJs). However, in contrast to epithelial cells, endothelial junctions are less rigidly organized. Whereas in most epithelia TJs are located at the most apical side of the lateral plasma membrane, in the endothelium they are regularly intermingled with AJs along the lateral membrane.

As in epithelia, in endothelial AJs β-catenin acts as a bridge connecting cadherin C-termini to cytoplasmic proteins such as α-catenin, which in turn couples the cadherin–catenin complex to the actin cytoskeleton (Figure 85.2, upper panel). The main endothelial cadherins are vascular endothelial (VE)-cadherin and neuronal (N)-cadherin. Only VE-cadherin is organized in AJs, whereas the role of N-cadherin remains elusive and is probably related to the interaction between the endothelium and perivascular pericytes. Like other cadherins, VE-cadherin mediates calcium (Ca^{2+})-dependent, homophilic adhesion. Structural studies of β-catenin revealed that the ARM repeats are tightly packed and form a superhelix of helices, resulting in a compact structure that forms a binding interface for several interacting partners, including cadherins (5). Both β-catenin and its cognate γ-catenin bind to the distal region of the cadherin cytoplasmic tail in a highly conserved

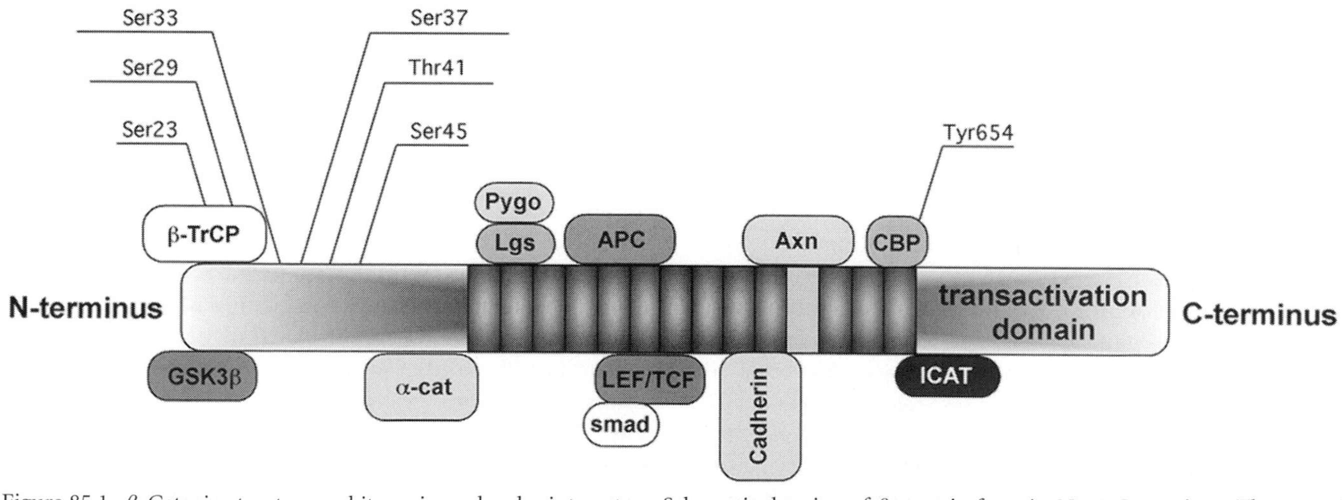

Figure 85.1. β-Catenin structure and its main molecular interactors. Schematic drawing of β-catenin from its N- to C-terminus. The central region containing the ARM repeats is indicated together with the main Ser/Thr and Tyr phosphorylation sites. The diagram displays some interactions between β-catenin and proteins of the Wnt-pathway, showing also their appropriate binding sites.

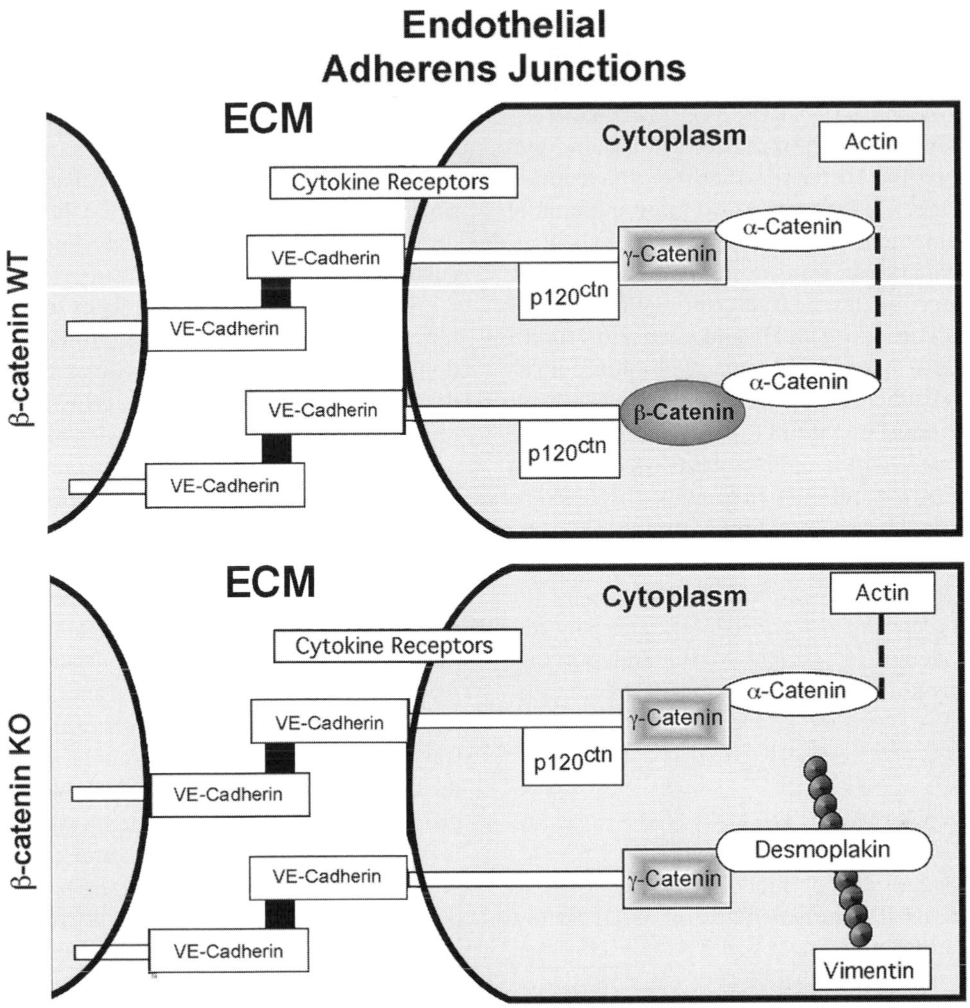

Figure 85.2. Schematic model of endothelial AJs and their modification in the absence of β-catenin. *Upper panel*: In the presence of β-catenin (WT), α-catenin is normally expressed at cell–cell junctions and connects VE-cadherin/β-catenin or VE-cadherin/γ-catenin to the actin cytoskeleton. *Lower panel*: In the absence of β-catenin (KO), the decrease in α-catenin is accompanied by an increase in desmoplakin, leading to a shift from α-catenin and actin based AJs to desmoplakin- and vimentin-based *complexus adherens*.

region termed the "catenin binding domain." Analysis of the cadherin–β-catenin complex revealed that the cadherin cytoplasmic tail is unstructured in the absence of β-catenin and is sensitive to proteolytic degradation (6). Whereas the central arm region of β-catenin is involved in cadherin binding, the amino-terminal region is involved in α-catenin interaction (see Figure 85.1).

By modulating the interaction of cadherins and the actin cytoskeleton, catenins – and in particular β-catenin – have a well-established role in the regulation of cadherin-mediated adhesion. In fact, in a variety of cellular contexts, cadherin mutants, lacking the ability to bind catenins, exhibit a compromised ability to mediate tight cell–cell adhesion. In line with this observation is that phosphorylation of the cadherin–catenin complex at tyrosine residues increases vascular permeability. Histamine and vascular endothelial growth factor (VEGF), which are potent inducers of vascular permeability, led to tyrosine phosphorylation of both cadherins and catenins in long-term confluent EC cultures, causing loosening of cell-to-cell contacts and dissociation of VE-cadherin from the actin cytoskeleton (7). It has been suggested that platelet-endothelial cell adhesion molecule (PECAM)-1 can interact with β-catenin and may function as a reservoir for, and modulator of, tyrosine phosphorylated β-catenin under conditions in which junctions are destabilized (8,9).

At the membrane, β-catenin may participate also in other signaling pathways. For example, it has been demonstrated that β-catenin is required for VE-cadherin–mediated attenuation of VEGF-induced phosphorylation of the VEGF receptor 2 (VEGFR2), thus participating in contact inhibition of ECs (7).

In epithelial cells, E-cadherin associates with actin and assembles into AJs, whereas the desmosomal cadherins desmocollin and desmoglein interact with intermediate filament-binding proteins and are organized into desmosomes. As opposed to most epithelial cells, ECs lack classical desmosomes. Nevertheless some types of ECs have *complexus adherens* formed by VE-cadherin linked to plakoglobin or p0071, which mediate the anchorage to vimentin intermediate filaments through desmoplakin (Figure 85.2, lower panel).

As described later in detail, the stabilized cytoplasmic pool of β-catenin can enter the nucleus and activate transcription of various target genes. This raises the important question of whether β-catenin interaction with cadherins at the cell membrane may limit its nuclear translocation. Because in the nucleus β-catenin upregulates the transcription of important cell cycle regulators, such as cyclin D1 and c-myc, this model would account, at least in part, for contact inhibition of cell growth mediated by cadherin–catenin clustering in confluent cells. However, this model might not explain all observations. For example, in sparse cells, β-catenin remains associated with cadherins, but cells are not inhibited in growth. This could be explained by the fact that even small, and in some cases undetectable, levels of β-catenin may be sufficient to achieve transcriptional activation. Alternatively, there is evidence for the presence of functionally distinct and conformationally different pools of β-catenin that are involved in adhesion and signaling respectively, and which might be regulated independently (10).

β-CATENIN IN WNT SIGNALING

Our current knowledge of Wnt signaling refers mostly to the so-called *canonical* pathway, depending on β-catenin as a bipartite nuclear transcription factor with Lef/TCF. However, so far two other, less well characterized (*noncanonical*) signaling events downstream of Fz receptors involving disheveled (Dvl) but not β-catenin have been demonstrated: (a) the planar cell polarity pathway (PCP), which is important for the planar organization of epithelial cells; and (b) a Ca^{2+}-dependent pathway controlling dorso-ventral axis formation in vertebrates (11).

In ECs, Wnt signaling has not yet been systematically characterized. However, some evidence points to a role for the canonical as well as the noncanonical pathway in vascular development, regression, and pathology. Here we give a general state-of-the-art description of the current understanding of Wnt signaling from the outside-in, starting with the Wnt growth factors. If available, the known situation in ECs is indicated specifically.

Wnt growth factors/morphogens are secreted, strongly glycosylated, and palmitoylated proteins. Currently 19 different vertebrate Wnt genes have been identified and their signaling roles have been characterized only in few cases. Wnt-1 is the best characterized family member, known to induce the canonical pathway via β-catenin, but induction of canonical signaling also has been demonstrated by Wnt-2, Wnt-3, Wnt-3a, Wnt-7a, Wnt-7b, and Wnt-8a. In contrast, Wnt-4, Wnt-5a, and Wnt-11 failed to induce or poorly induced β-catenin–dependent target gene expression, but inhibited canonical signaling via a Ca^{2+}-dependent pathway.

Fetal vessels of the murine placenta were shown to express Wnt-2, and vessels of the mouse embryonic yolk sac were tested positive for Wnt-5a and Wnt-10b (12). In vitro, expression of Wnt-7a and -10b could be demonstrated in mouse brain microvascular ECs (MBMECs) and Wnt-5a in human umbilical vein ECs (HUVECs) (13). Furthermore, overexpression of Wnt-4 in a mouse model for the human chromosomal duplication of Wnt-4 leads to lower and altered testicular vascularization (14).

The members of the Fz family of serpentine receptors are seven-transmembrane spanning proteins with a cysteine-rich domain (CRD) in their extracellular N-terminus. The CRD domain serves to bind the Wnt growth factors and is therefore essential for mediating Wnt signaling. Fz-receptors alone are not sufficient to activate the canonical signaling pathway. This can be achieved only by a complex of Fz with the coreceptors LRP-5/-6/*arrow*. Indeed, inherited diseases in humans and gene deletion of LRP-5 in mice demonstrate that LRP-5 is involved and necessary for canonical Wnt signaling. The receptor complex activates, through a yet unknown mechanism, the intracellular phosphoprotein disheveled (Dvl) and further downstream signaling events (Figure 85.3).

Several inhibitors/modulators of the Wnt signaling pathway exist; these act at the extracellular and membrane level of the signaling cascade. Probably the best described are the gene products of the soluble Fz-related proteins (sFRP). Although their name refers to their structural and functional relation to Fz-receptors, they are lacking the transmembrane domain of the latter and, except for the CRD domain, they share no structural homology with the FZ receptors. As a consequence, sFRP are released in the extracellular space, sequestering Wnt growth factors and thereby inhibiting or modulating their availability for signaling. In ECs, first evidence points to a major inhibiting role of FrzA in angiogenesis during vascular development and neovascularization (15).

In addition to the sequestration of Wnts, the receptor complex of Fz and LRPs can be functionally inactivated by the extracellular, cysteine-rich, protein Dickkopf (Dkk), which directly binds to LRP. In addition to this, Dkk interacts with the membrane receptor Kremen, resulting in the reduction of LRP at the membrane.

Although endothelial expression of Fz receptors has not been explored in detail, expression of Fz-1, Fz-2, Fz-3, and Fz-5 has been reported (13). As various sites of the vascular tree have been investigated and different species have been used, the present results may support a site-specific expression of Fz receptors in the vascular tree. This would also help to explain why the knockout for Fz-5 shows vascular defects only in the yolk sac and in the placenta, whereas other sites of vascular development in the embryo proper seem to be unaffected. In the absence of Wnts, free cytoplasmic β-catenin is readily degraded in the 26S proteasome system, involving axin (Axn), APC, glycogen synthase kinase 3β (GSK3β), and casein kinase 1 (CK1), which form a functional complex to target β-catenin for ubiquitination and subsequent degradation (Figure 85.3). Axn has mostly been considered to be a

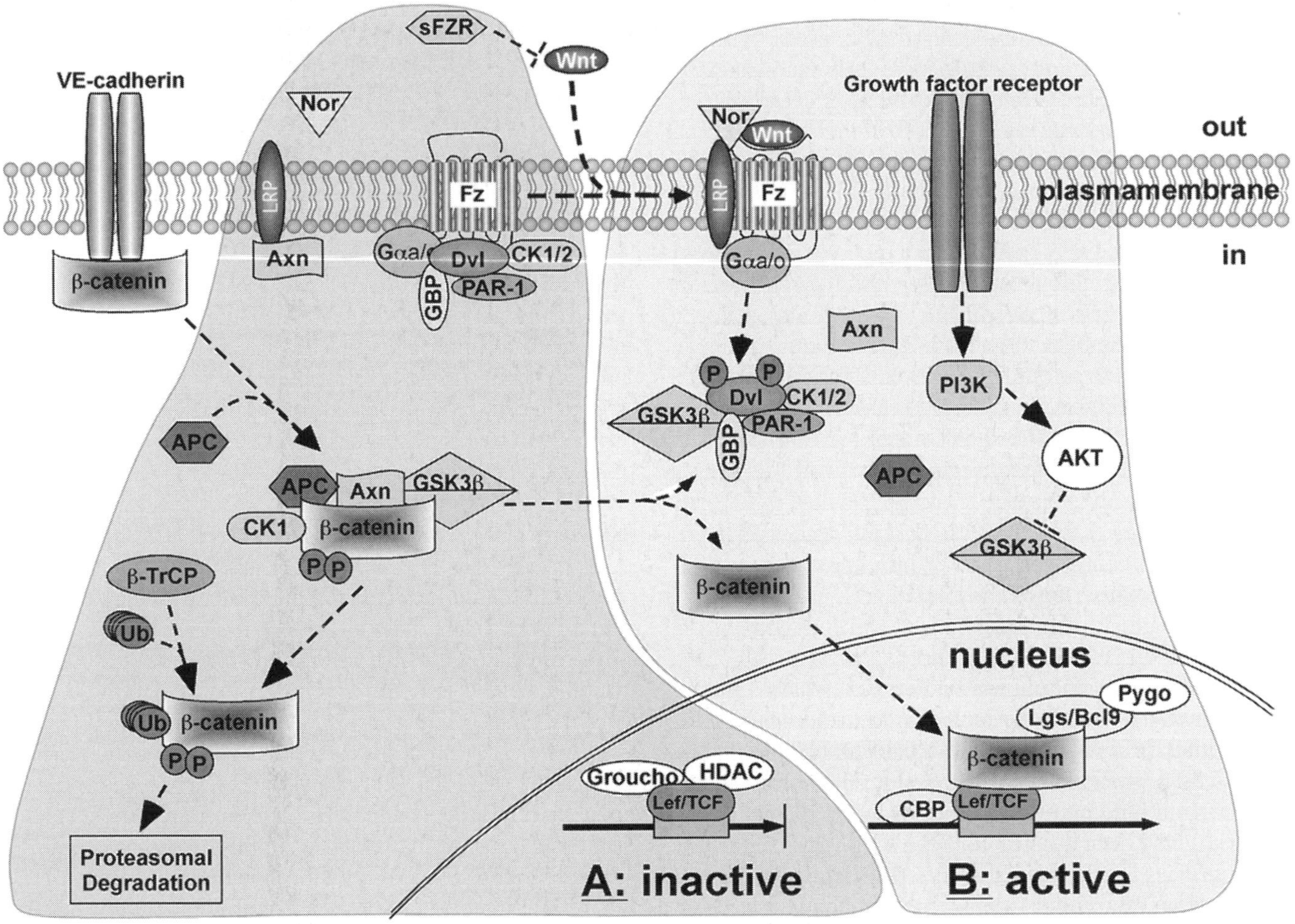

Figure 85.3. Schematic representation of the canonical Wnt-pathway. (**A**) Inactive state. In the absence of a Wnt signal, or when Wnt is sequestered by a soluble inhibitor such as sFRP from binding to the Fz receptor, LRP, and Norrin (Nor), the free pool of β-catenin (which is not bound to VE-cadherin) is phosphorylated by the APC/GSK3β/Axn/CK1 destruction complex. The Ser/Thr phosphorylated β-catenin is recognized by β-TrCP, ubiquitinated (Ub) and rapidly degraded in the proteasomal degradation complex. Consequently, Lef/TCF acts in concert with Groucho and HDAC as a transcriptional inhibitor. (**B**) Active state. When Wnt binds to Fz and LRP, Dvl becomes phosphorylated, recruiting and inactivating GSK3β in a complex with PAR-1, GBP, and CK1, and thereby inhibiting β-catenin phosphorylation and degradation. Active β-catenin then enters the nucleus, binds to Lef/TCF, CBP, Lgs/Bcl9, and Pygo, and leads to target gene transcription. GBP, GSK3β-binding protein; CBP, CREB binding protein. In an alternative, or parallel activation pathway, GSK3β is inhibited through the PI3K/AKT pathway following Tyr growth factor receptor stimulation.

scaffolding protein, providing the basis for complex formation of β-catenin with APC, CK1, and GSK3β. In a first step, β-catenin receives an N-terminal priming phosphorylation by CK1 in residue Ser45; subsequently GSK3β becomes phosphorylated by CK1 and in turn phosphorylates β-catenin in residues Thr41, Ser37, and Ser33. The N-terminal phosphorylated β-catenin can now be recognized by the E3 ubiquitin ligase β-transducin repeat-containing protein (βTrCP), and the ubiquitinated protein is degraded.

The initial steps in the Wnt signaling cascade are the least understood. After binding of the Wnt growth factor to the Fz-LRP-receptor complex, the intracellular phospho-protein Dvl is activated at the Fz-C-terminus by a yet-unknown mechanism. Dvl in turn inactivates the β-catenin destruction pathway, which is probably achieved through the destabilization of the destruction complex. Evidence points to a fundamental role of Axn in this scenario, in which, upon stimulation of the Wnt-pathway, Axn is relocated to the plasma membrane by active Dvl, leading to its dephosphorylation and inactivation. As a consequence, the β-catenin degradation complex cannot form, thus shifting the equilibrium toward the free β-catenin pool. In addition to the previously mentioned proteins involved in the Wnt pathway, many others have been identified as modulators and regulators of the Wnt-cascade; these are described in more complete detail at http://www.stanford.edu/~rnusse/wntwindow.html (see Figure 85.3).

To accomplish its transactivating function, β-catenin must enter the nucleus and interact with members of the Lef/TCF transcription factors in order to bind to DNA. The β-catenin protein itself contains no nuclear localization signal (NLS) and is dependent neither on the nuclear transport proteins

Table 85-1: Effect of Wnt Target Genes Expressed in the Vascular System

Gene	*Organism/Cell Type*	*Function*
Cyclin D1	Human/mouse/ECs	Proliferation
c-myc	Human/SMCs	Proliferation
uPAR	Human	Invasion
MMP-7	Human/HUVECs	Migration
Fibronectin	Rat/SMCs	Migration
VEGF-A/-C	Human/HUVECs	Proliferation/ angiogenesis
Connexin 43	Rat/ECs/SMCs	Gap junctions
Cyclooxygenase-2	Human/HMECs	Angiogenesis

ECs, endothelial cells; SMCs, smooth muscle cells; HUVECs, human umbilical vein ECs; HMECs, human microvascular ECs; uPAR, urokinase plasminogen activator receptor; MMP-7, metalloproteinase-7.

importin-β/β-karyopherin, nor on Lef/TCF, which contains an NLS. Instead β-catenin can bind directly to the nuclear pore and becomes, by a yet-unknown mechanism, translocated to the nucleus, a process proven to require GTP. Recently, the two β-catenin binding proteins legless (Lgs/Bcl9) and pygopus (Pygo) have been implicated in the shuttling of β-catenin to the nucleus (16). In particular, Lgs/Bcl9, which serves as a linker between Pygo and β-catenin, is essential for the translocation process. However, the point at which GTP is needed and the manner by which the protein complex accomplishes the translocation remains to be elucidated. In the nucleus, β-catenin forms a complex with the HMG-box transcription factors Lef/TCF and, in vertebrates, with other coactivators (such as p300/CBP) to initiate target gene transcription (see Figure 85.3). More than 200 genes have been described as targets for Wnt/β-catenin signaling so far (for a complete list see http://www.stanford.edu/~rnusse/wntwindow.html). The most relevant for the vascular system are summarized in Table 85-1. These include regulators of proliferation, such as cyclin D1 and c-myc, extracellular matrix (ECM) components and remodeling factors, and proangiogenic cytokines such as VEGF. Wnt signaling ceases if the transcription factor complex disaggregates and β-catenin is guided out of the nucleus by Axn and APC, which both contain nuclear export sequences (NES); this leads to cytoplasmic destruction of β-catenin.

Not all cells undergo canonical Wnt signaling even in the presence of Wnt growth factors. As known so far, ECs show a strictly time- and site-restricted pattern of Wnt signaling, as revealed by BAT-gal reporter mice for the canonical pathway. These mice show activation of the Wnt pathway exclusively in vessels growing into the brain during embryonic and early postnatal development (17). This might reflect the missing expression of specific Fz-receptors on some ECs, but also the nuclear disposition to respond to pathway activation. When Wnt signaling is inactive, Lef/TCF still can bind to the specific DNA sites, but acts under these circumstances as an inhibitory transcription factor together with the transcriptional repressor groucho, which specifically activates histone deacetylase (HDAC). Moreover, several lines of evidence point to the existence of different molecules (e.g., β-catenin interacting protein, ICAT) binding β-catenin and preventing its interaction with Lef/TCF. Also, other nuclear proteins such as Reptin52 (repressor of Pontin52) inhibit β-catenin–dependent transactivation. Most of our understanding of β-catenin–dependent transcriptional regulation has been acquired in vitro, making it difficult to predict the specific situation in vivo. Furthermore, it is impossible to extrapolate from the current state of knowledge on ECs, owing to a lack of data on the expression of core pathway genes in the endothelium.

NONCANONICAL WNT SIGNALING

Besides the canonical signaling pathway involving β-catenin, certain Wnt growth factors and Fz receptors elicit a downstream signaling through Fz and Dvl on Rho, Rac, c-Jun N-terminal kinase (JNK), and Ca^{2+}. This type of Wnt signaling was first identified in *Drosophila* as PCP, shown to be crucial for the polarization of some epithelia within a plane perpendicular to the apical–basal axis of the cell (11). The core genes involved in PCP are Fz and Dsh, which overlap with the canonical Wnt signaling pathway; in addition to these common genes, PCP requires also the atypical cadherin Flamingo/Starry night (Fmi) and the putative transmembrane protein Strabismus/Van Gogh (Stbm), as well as the negative regulator Prickl (Pk). In mammals, the best understood process in which PCP is known to be involved is the organization of the hair cells in the inner ear.

In vertebrates, in addition to the PCP pathway a second, noncanonical pathway has been identified, and is shown to signal through protein kinase C (PKC). Also, this pathway is dependent on Fz and Dvl, but acts through PKC and an upregulation of the intracellular Ca^{2+}-level, leading to stimulated cell migration.

Although striking functional similarity exists, it is not yet clear if the PCP and the PKC pathways are homologous. In vivo and in vitro, it has been demonstrated that at least Wnt-5a predominantly signals through the PKC pathway. This is of particular interest, since mice deficient for Fz-5, which is a designated Wnt-5a receptor, show major defects in vascularization and blood vessel remodeling in the yolk sac (18). This result is in line with the observation that in a reporter mouse for canonical Wnt signaling, no signaling events could be found in yolk-sac blood vessels (17; A. Cattelino, unpublished observations). Instead, the alterations of blood vessels found in mouse embryos bearing an endothelial conditional deletion of β-catenin might be due to the lack of the protein at the plasma membrane (19). As for canonical Wnt signaling, the exact function of the noncanonical pathway in ECs is still to be determined.

CROSS-TALK WITH OTHER SIGNALING PATHWAYS

As mentioned earlier, Wnt signaling is known to play a crucial role during invertebrate and vertebrate embryogenesis. Obviously, aside from this pathway, many other pathways such as transforming growth factor (TGF)-β, hedgehog, nuclear factor (NF)-κB, Notch, and others are necessary for proper embryonic development. During the last several years, an increasing amount of information has become available on the cross-talk of pathways previously thought to be distinct. In particular, the Wnt pathway was shown to be tightly connected with the TGF-β pathway, sharing also common target genes, which are cooperatively regulated (20,21). The interaction of the two pathways has been demonstrated at the membrane, cytoplasmic, and nuclear level. Although bound to E-cadherin at the plasma membrane, β-catenin can interact with the TGF-β receptor II and, upon TGF-β stimulation, elicit downstream signaling interacting with the transcription factors smad3 and smad4 (22). Although in ECs no direct interaction of Wnt- and TGF-β signaling has been demonstrated yet, evidence for cooperation between the two pathways during endothelial–mesenchymal transformation (EMT) derives from the endothelial-specific knockout for β-catenin (23).

The hedgehog growth factors, Sonic hedgehog (Shh), Indian hedgehog (Ihh), and Desert hedgehog (Dhh) activate downstream signaling through the receptors patched (Ptch) and smoothened (Smo) and activate the Gli transcription factors through protein kinase A (PKA), CK1, and GSK3β (24). The hedgehog pathway shares the proteins CK1, GSK3β, and β-TrCP with the Wnt pathway. Considerable cross-talk between these pathways has been demonstrated in the differentiation of the sebaceous lineage in the mammalian epidermis, in the differentiation of neuronal progenitor cells in the subventricular zone, in the response of diencephalic cells to dorsal Wnt-stimuli, and in somitogenesis. In the endothelium, Hedgehog signaling was shown to be involved in yolk-sac angiogenesis and arterial endothelial differentiation (25,26). Interestingly, VEGF was demonstrated to be a direct target of the Shh/Gli/pathway, suggesting a possible cooperative regulation of VEGF by Wnt and Shh signaling, although direct evidence for this is still lacking.

Furthermore, in somitogenesis, Wnt signaling induced by Wnt-3a acts upstream of the Notch pathway. Notch1 is an integral membrane receptor, which binds to the transmembrane ligands Delta and Jagged. The Notch/Delta pathway is mainly involved in cell fate decisions during development. Notch and Wnt signaling have been shown to interact at several levels: at the plasma membrane through the binding of Wnts to Notch, at the cytoplasmic level through an interaction of Dvl with the intracellular domain of Notch, and in the cytoplasm through an interaction between presenilin-1 with β-catenin (27). Although considerable evidence indicates that Notch and Wnt signaling play an important role in vascular development and differentiation of ECs, notably in EMT during heart valve development (23,28), it still remains to be elucidated whether and to what extent the two pathways interact in the endothelium.

It should be mentioned that the Wnt/β-catenin pathway also interacts with the NF-κB pathway of inflammatory cytokines. The first evidence for cross-talk between the two pathways came from the observation that the phosphorylation motif on β-catenin, which is recognized by GSK3β, is identical to a motif in the inhibitor of NF-κB (IκB). Recently, it was shown that the phosphoinositide 3-kinase (PI3K)/AKT/IkB pathway can simultaneously activate NF-κB and β-catenin/Lef/TCF signaling in colon carcinoma cell lines, leading to the cooperative upregulation of angiogenic genes like VEGF-A, -B, and -C and urokinase-type plasminogen activator (u-PA), therefore facilitating tumor angiogenesis and growth (29). Because it is known that AKT inactivates GSK3β, it might be possible that every signaling pathway involving an activation of PI3K/AKT could participate in the activation of β-catenin signaling (see Figure 85.3).

Finally, tyrosine phosphorylated β-catenin also associates with the receptor tyrosine kinase ErbB2 and with the receptor tyrosine phosphatase LAR in human melanoma cell lines. If ErbB2 is blocked pharmacologically, β-catenin–dependent signaling in the nucleus is decreased, leading to reduced cell motility (30,31).

(A complete and interactive overview of the connection map of Wnt/β-catenin signaling is available at *Science* magazine's website, Signal Transduction Knowledge Environment: http://stke.sciencemag.org/cgi/cm/stkecm;CMP_5533.)

EFFECT ON ENDOTHELIAL CELL PHENOTYPE

In some cell types, Wnt signaling has been associated with an inhibition of apoptosis, an increase in proliferation, and cell cycle progression. Nevertheless, its role in ECs is still poorly understood.

Induction of canonical Wnt signal through stimulation with Wnt-1 leads to proliferation of HUVECs (32). In the same cellular model, it has recently been reported that the stabilization of β-catenin is related to an induction of angiogenesis in Matrigel, both in vitro and in vivo, and this effect is mediated by the transcriptional activation of VEGF by β-catenin in ECs (33). Another study demonstrated an accumulation of cytosolic β-catenin in ECs during neovascularization after myocardial infarction (34). Also the overexpression of a non-degradable form of β-catenin caused enhanced cell growth, suggesting a role for the canonical Wnt signaling in EC growth and morphogenesis (35,36). A list of mice transgenic for Wnt pathway components and exhibiting a vascular phenotype is given in Table 85.2.

The role of β-catenin in the endothelium has been investigated by the conditional inactivation of β-catenin using the Cre/loxP system (19). Mice with floxed β-catenin were crossed with transgenic mice expressing Cre recombinase under the control of the endothelial-specific *Tie2* promoter.

Table 85-2: Mutation or Deletion of Wnt Components in Mice Exhibiting a Vascular Phenotype

Gene	*Phenotype*	*References*
Wnt-2	Placental defects	41
Wnt-4	Testicular vascularization defect (overexpression of Wnt-4)	14
Wnt-7b	Placental defects (probably not through β-catenin)	42
LRP5	Defective eye vascularization	43
Fz-4	Retinal and inner ear vascularization	40
Fz-5	Defects in yolk sac and placental angiogenesis	18
FrzA	Inhibition of physiological and pathological angiogenesis	15, 34
β-catenin	Altered vascular organization and heart cushion formation	19, 23

Table 85-3: Mutated Wnt Components in Human Genetic Diseases Affecting the Vascular System

Gene	*Disease*	*References*
LRP5	FEVR	44
Fz-4	FEVR	45
Norrin	Norrie disease/FEVR	40
Wnt-4	Intersex-phenotype (chromosomal duplication)	46

Surprisingly, ECs do not require β-catenin for the early phases of vasculogenesis and angiogenesis, but do require the molecule for correct vascular patterning and maintenance of vascular integrity at later stages of development. The absence of β-catenin in ECs leads to lethality at midgestation, due to an altered vasculature in the head, vitelline umbilical vessels, placental defects, and frequent hemorrhages (19). Moreover, in many regions, the vascular lumen is irregular, with lacunae and bifurcations and fluid extravasation in the pericardial cavity. In EC lines derived from mutant mouse embryos, we found a different organization of intercellular junctions with a decrease in α-catenin in favor of desmoplakin (19). These molecular changes lead to an overall alteration of the balance between the recruitment of α-catenin and desmoplakin in endothelial intercellular junctions, with a decrease of the number of strong, actin-based AJs to weak, more dynamic, vimentin-based *complexus adherens* (see Figure 85.2). Interestingly, this junctional reorganization correlates with an increase in paracellular permeability measured in vitro. Although desmoplakin and vimentin are present in other types of endothelia, *complexus adherens* are preferentially observed in lymphatics, where junctions are relatively weak, permitting a dynamic passage of cells and solutes. It has consistently been reported previously that β-catenin is reduced in lymphatics as compared with vascular endothelium (37). We hypothesize that the weakness of cell-to-cell junctions can account for the high occurrence of vascular dilatation and hemorrhages found in vivo in mutant embryos.

It is not clear to what extent the phenotype observed in the absence of β-catenin can be attributed to a block of Wnt signaling rather than to an alteration of AJ organization. By using a transgenic reporter mouse strain for canonical Wnt signaling (17), only a transient and localized activation of β-catenin transcriptional activity in the developing vasculature was observed (S. Liebner, unpublished observations).

Interestingly, the conditional deletion of β-catenin in the endothelium resulted not only in a severe defect in placental development and organization and remodeling of blood vessels in the yolk sac and perineural plexus, but also in the heart, revealing in particular a lack of heart cushion formation. The heart cushion evolves during ontogeny from an acellular structure called the cardiac jelly, which consists of ECM components; it is located around the atrioventricular canal and between endo- and myocardium. The cardiac jelly is overrun by cells derived from the endocardium, which undergo EMT during this process, leading to the formation of the heart cushion and, later in development, to the valves. EMT does not occur in endocardial cells or ECs lacking β-catenin, explaining the observed phenotype in the heart. Heart EMT has been shown to depend upon several pathways, including TGF-β, bone morphogenic protein (BMP), VEGF, and Notch, raising the intriguing possibility that the Wnt pathway was also involved. The use of a reporter mouse for canonical Wnt signaling (17) showed that cells do, in fact, undergo Wnt signaling during EMT (23). Further support for an involvement of the Wnt signaling pathway in heart EMT came from work in zebrafish, demonstrating that a loss-of-function mutation of APC leads to increased β-catenin signaling and excessive cellularity of heart cushions. Although it seems confirmed that canonical Wnt signaling is involved in endocardial EMT, it remains to be elucidated at which time point during EMT the Wnt cascade becomes active, and whether and how cooperation with other pathways is necessary for proper heart cushion formation.

The involvement of Wnt signaling in endothelial EMT might have further implications. It was reported previously that, during the formation of arteriosclerotic plaques, the proliferation of vascular smooth muscle cells (VSMCs) is dependent on canonical Wnt signaling, which cooperates with the NF-κB pathway (Table 85.3) (38). Furthermore, embryonic as well as adult ECs can transform into VSMCs through an EMT-related process. Collectively, these data suggest a role for β-catenin in the etiology of the arteriosclerotic phenotype, acting on ECs and VSMCs.

CLINICAL AND THERAPEUTIC IMPLICATIONS

As already mentioned, the Wnt pathway regulates proliferation, migration, differentiation, and apoptosis in a variety of cell types. Therefore, it is not surprising that the development of numerous organ systems depends on Wnt signaling, including the central nervous system, the female reproductive tract, mammary glands, kidney, limb, placenta, hair, and teeth. Wnt signaling also has been implicated in cardiogenesis and in directing hematopoietic cell fate. Uncontrolled Wnt signaling, resulting from mutations in β-catenin, inactivation of β-catenin phosphorylation, or the destruction complex has been shown to directly cause cell transformation and lead to different types of tumors. Increased Wnt signaling may also play a role in the pathologies of rheumatoid arthritis and Alzheimer disease. An increasing body of evidence also indicates that the Wnt pathway plays an important role in the correct development and in the pathological state of the vascular system. During human embryonic development, when vessels proliferate, nuclear and/or cytoplasmic β-catenin can be detected in placental villus capillaries, fetal capillaries (particularly in the brain), and in arteries and veins, whereas cells of the normal adult vasculature rarely if ever accumulate β-catenin in the cytoplasm or nucleus (13). On the other hand, cytoplasmic and nuclear β-catenin is frequently observed during vessel remodeling in disease states, such as myocardial infarction, where it correlates with neovascularization and increased endothelial proliferation.

β-catenin accumulation has been associated with angiogenesis in cancer, in particular in the brain, both in a mouse model of gliomas, and in neovascular ECs of human glioblastoma multiforme, medulloblastoma, and other tumors of the central nervous system (39). Several human genetically inherited diseases appear to be caused by mutations in Wnt signaling (see Table 85-2). Chromosomal duplication of Wnt-4 in humans, for example, leads to an intersex phenotype (14). Consistent with this clinical finding, Wnt-4 transgenic mice have abnormal testicular vasculature, and their testosterone synthesis is inhibited. Although the mechanism of action of Wnt-4 in this system is not clear yet, it is likely that Wnt-4 acts as an anti-male factor by interfering with β-catenin functions.

Loss of function mutations in Fz-4 and the Wnt coreceptor LRP-5 are linked to familiar exudative retinopathy (FEVR), an inherited disorder of the retinal vasculature in which peripheral capillaries fail to grow, resulting in compensatory neovascularization, leakage, bleeding, and scarring, leading ultimately to blindness. Recently Norrin, a new ligand for Fz/LRP complex, has been identified and shown to be mutated in the hereditary ocular Norrie syndrome, an X-linked retinal dysplasia that presents similar features to FEVR (40). Wnt signaling has been implicated in the maintenance and self-renewal of various pluripotent stem and progenitor cells. Whether Wnt signaling also may play a role in the regulation of EC progenitors, as well as in endothelial specification in arterial, venous, and lymphatic endothelium is an important unsolved issue.

Given the deleterious consequences of uncontrolled Wnt/β-catenin signaling in humans, it is no surprise that a considerable number of inhibitors of this pathway are in consideration or under development for clinical studies. Candidate approaches include small-molecule inhibitors that block the interaction of β-catenin with TCF or CBP, as well as small interfering RNAs (siRNAs) and the therapeutic use of antibodies against Wnts.

KEY POINTS

- The current understanding of β-catenin in the endothelium points to a fundamental role in vascular development and maintenance.
- In particular, β-catenin is crucial for the organization, regulation, and stability of AJ architecture, and for the control of vascular permeability.
- Growing evidence suggests an important involvement of β-catenin as a key player in the Wnt pathway, in normal vascular development, as well as in the pathological state of the endothelium.
- Although a challenging task, the growing scientific interest in β-catenin and Wnt signaling, mirrored by the numerous recent publications, fosters a rapid increase of our knowledge.

Future Goals

- To determine which components of the Wnt pathway become active in the endothelium during vasculo- and angiogenesis, remodeling, and pathological neovascularization, and whereas and when these events take place
- To determine if Wnt signaling in the endothelium create cross-talk with other signaling pathways in order to exert a site-specific effect on vascular differentiation
- To determine whether insights into Wnt signaling accumulated over the past 20 years might be translated into tangible therapies for some of the diseases discussed here

REFERENCES

1 Peifer M, McCrea PD, Green KJ, et al. The vertebrate adhesive junction proteins beta-catenin and plakoglobin and the *Drosophila* segment polarity gene armadillo form a multigene family with similar properties. *J Cell Biol.* 1992;118:681–691.
2 Polakis P. The oncogenic activation of beta-catenin. *Curr Opin Genet Dev.* 1999;9:15–21.
3 Holstein TW, Hobmayer E, Technau U. Cnidarians: an evolutionarily conserved model system for regeneration? *Dev Dyn.* 2003; 226:257–267.

4 Solanas G, Miravet S, Casagolda D, et al. beta-Catenin and plakoglobin N- and C-tails determine ligand specificity. *J Biol Chem.* 2004;279:49849–49856.

5 Huber AH, Nelson WJ, Weis WI. Three-dimensional structure of the armadillo repeat region of β-catenin. *Cell.* 1997;90:871–882.

6 Huber AH, Weis WI. The structure of the beta-catenin/E-cadherin complex and the molecular basis of diverse ligand recognition by beta-catenin. *Cell.* 2001;105:391–402.

7 Dejana E. Endothelial cell-cell junctions: happy together. *Nat Rev Mol Cell Biol.* 2004;5:261–270.

8 Ilan N, Madri JA. PECAM-1: old friend, new partners. *Curr Opin Cell Biol.* 2003;15:515–524.

9 Biswas P, Zhang J, Schoenfeld JD, et al. Identification of the regions of PECAM-1 involved in beta- and gamma-catenin associations. *Biochem Biophys Res Commun.* 2005;329:1225–1233.

10 Gottardi CJ, Gumbiner BM. Distinct molecular forms of beta-catenin are targeted to adhesive or transcriptional complexes. *J Cell Biol.* 2004;167:339–349.

11 Fanto M, McNeill H. Planar polarity from flies to vertebrates. *J Cell Sci.* 2004;117:527–533.

12 Austin TW, Solar GP, Ziegler FC, et al. A role for the Wnt gene family in hematopoiesis: expansion of multilineage progenitor cells. *Blood.* 1997;89:3624–3635.

13 Goodwin AM, D'Amore PA. Wnt signaling in the vasculature. *Angiogenesis.* 2002;5:1–9.

14 Jordan BK, Shen JH, Olaso R, et al. Wnt4 overexpression disrupts normal testicular vasculature and inhibits testosterone synthesis by repressing steroidogenic factor 1/beta-catenin synergy. *Proc Natl Acad Sci USA.* 2003;100:10866–10871.

15 Ezan J, Leroux L, Barandon L, et al. FrzA/sFRP-1, a secreted antagonist of the Wnt-Frizzled pathway, controls vascular cell proliferation in vitro and in vivo. *Cardiovasc Res.* 2004;63:731–738.

16 Townsley FM, Cliffe A, Bienz M. Pygopus and legless target Armadillo/beta-catenin to the nucleus to enable its transcriptional co-activator function. *Nat Cell Biol.* 2004;6:626–633.

17 Maretto S, Cordenonsi M, Dupont S, et al. Mapping Wnt/beta-catenin signaling during mouse development and in colorectal tumors. *Proc Natl Acad Sci USA.* 2003;100:3299–3304.

18 Ishikawa T, Tamai Y, Zorn AM, et al. Mouse Wnt receptor gene Fzd5 is essential for yolk sac and placental angiogenesis. *Development.* 2001;128:25–33.

19 Cattelino A, Liebner S, Gallini R, et al. The conditional inactivation of the beta-catenin gene in ECs causes a defective vascular pattern and increased vascular fragility. *J Cell Biol.* 2003;162:1111–1122.

20 Labbe E, Letamendia A, Attisano L. Association of smads with lymphoid enhancer binding factor 1/T cell- specific factor mediates cooperative signaling by the transforming growth factor-beta and wnt pathways. *Proc Natl Acad Sci USA.* 2000;97:8358–8363.

21 Nishita M, Hashimoto MK, Ogata S, et al. Interaction between Wnt and TGF-beta signalling pathways during formation of Spemann's organizer. *Nature.* 2000;403:781–785.

22 Tian YC, Fraser D, Attisano L, Phillips AO. TGF-beta1-mediated alterations of renal proximal tubular epithelial cell phenotype. *Am J Physiol Renal Physiol.* 2003;285:F130–F142.

23 Liebner S, Cattelino A, Gallini R, et al. Beta-catenin is required for endothelial-mesenchymal transformation during heart cushion development in the mouse. *J Cell Biol.* 2004;166:359–367.

24 Ogden SK, Ascano M Jr., Stegman MA, Robbins DJ. Regulation of Hedgehog signaling: a complex story. *Biochem Pharmacol.* 2004;67:805–814.

25 Kanda S, Mochizuki Y, Suematsu T, et al. Sonic hedgehog induces capillary morphogenesis by ECs through phosphoinositide 3-kinase. *J Biol Chem.* 2003;278:8244–8249.

26 Pola R, Ling LE, Silver M, et al. The morphogen Sonic hedgehog is an indirect angiogenic agent upregulating two families of angiogenic growth factors. *Nat Med.* 2001;7:706–711.

27 De Strooper B, Annaert W. Where Notch and Wnt signaling meet. The presenilin hub. *J Cell Biol.* 2001;152:F17–F20.

28 Noseda M, Chang L, McLean G, et al. Notch activation induces EC cycle arrest and participates in contact inhibition: role of p21Cip1 repression. *Mol Cell Biol.* 2004;24:8813–8822.

29 Agarwal A, Das K, Lerner N, et al. The AKT/IkappaB kinase pathway promotes angiogenic/metastatic gene expression in colorectal cancer by activating nuclear factor-kappaB and beta-catenin. *Oncogene.* 2005;24:1021–1031.

30 Shin HS, Lee HJ, Nishida M, et al. Betacellulin and amphiregulin induce upregulation of cyclin D1 and DNA synthesis activity through differential signaling pathways in vascular smooth muscle cells. *Circ Res.* 2003;93:302–310.

31 Bonvini P, An WG, Rosolen A, et al. Geldanamycin abrogates ErbB2 association with proteasome-resistant beta-catenin in melanoma cells, increases beta-catenin-E-cadherin association, and decreases beta-catenin-sensitive transcription. *Cancer Res.* 2001;61:1671–1677.

32 Wright M, Aikawa M, Szeto W, Papkoff J. Identification of a Wnt-responsive signal transduction pathway in primary ECs. *Biochem Biophys Res Commun.* 1999;263:384–388.

33 Skurk C, Maatz H, Rocnik E, et al. Glycogen-Synthase Kinase3beta/beta-catenin axis promotes angiogenesis through activation of vascular endothelial growth factor signaling in ECs. *Circ Res.* 2005;96:308–318.

34 Barandon L, Couffinhal T, Dufourcq P, et al. Frizzled A, a novel angiogenic factor: promises for cardiac repair. *Eur J Cardiothorac Surg.* 2004;25:76–83.

35 Choi JH, Hur J, Yoon CH, et al. Augmentation of therapeutic angiogenesis using genetically modified human endothelial progenitor cells with altered glycogen synthase kinase-3beta activity. *J Biol Chem.* 2004;279:49430–49438.

36 Masckauchan TN, Shawber CJ, Funahashi Y, et al. Wnt/beta-catenin signaling induces proliferation, survival and interleukin-8 in human ECs. *Angiogenesis.* 2005;8:43–51.

37 Petrova TV, Makinen T, Makela TP, et al. Lymphatic endothelial reprogramming of vascular ECs by the Prox-1 homeobox transcription factor. *EMBO J.* 2002;21:4593–4599.

38 Wang X, Adhikari N, Li Q, et al. The role of [beta]-transducin repeat-containing protein ([beta]-TrCP) in the regulation of NF-[kappa]B in vascular smooth muscle cells. *Arterioscler Thromb Vasc Biol.* 2004;24:85–90.

39 Yano H, Hara A, Takenaka K, et al. Differential expression of beta-catenin in human glioblastoma multiforme and normal brain tissue. *Neurol Res.* 2000;22:650–656.

40 Xu Q, Wang Y, Dabdoub A, et al. Vascular development in the retina and inner ear: control by Norrin and Frizzled-4, a high-affinity ligand-receptor pair. *Cell.* 2004;116:883–895.

41 Monkley SJ, Delaney SJ, Pennisi DJ, et al. Targeted disruption of the Wnt2 gene results in placentation defects. *Development.* 1996;122:3343–3353.

42 Parr BA, Cornish VA, Cybulsky MI, McMahon AP. Wnt7b regulates placental development in mice. *Dev Biol*. 2001;237:324–332.
43 Kato M, Patel MS, Levasseur R, et al. Cbfa1-independent decrease in osteoblast proliferation, osteopenia, and persistent embryonic eye vascularization in mice deficient in Lrp5, a Wnt coreceptor. *J Cell Biol*. 2002;157:303–314.
44 Toomes C, Bottomley HM, Jackson RM, et al. Mutations in LRP5 or FZD4 underlie the common familial exudative vitreoretinopathy locus on chromosome 11q. *Am J Hum Genet*. 2004;74:721–730.
45 Robitaille J, MacDonald ML, Kaykas A, et al. Mutant frizzled-4 disrupts retinal angiogenesis in familial exudative vitreoretinopathy. *Nat Genet*. 2002;32:326–330.
46 Biason-Lauber A, Konrad D, Navratil F, Schoenle EJ. A WNT4 mutation associated with Mullerian-duct regression and virilization in a 46,XX woman. *N Engl J Med*. 2004;351:792–798.

86

Nuclear Factor-κB Signaling in Endothelium

Kaiser M. Bijli and Arshad Rahman

School of Medicine and Dentistry, University of Rochester, New York

Nuclear factor (NF)-κB is perhaps the most intensely studied eukaryotic transcription factor, mainly because of its pivotal role in controlling varied biological effects ranging from inflammatory-, immune-, and stress-induced responses to cell fate decisions such as proliferation, differentiation, tumorigenesis, and apoptosis. The mammalian NF-κB family consists of five members: RelA (p65), RelB, c-Rel, NF-κB1 (p50 and its precursor p105), and NF-κB2 (p52 and its precursor p100). These proteins share a conserved N-terminal 300-amino acid Rel homology domain (RHD) that contains a nuclear localization signal (NLS) and is responsible for dimerization, sequence-specific DNA binding, and interaction with inhibitory IκB proteins (Figure 86.1A). A critical feature of RelA, RelB, and c-Rel that distinguishes them from p50 and p52 is the presence of a transactivation domain (TAD) within the carboxy-terminal region of these proteins (Figure 86.1A).

The diverse biological effects of NF-κB are mediated, in part, by the ability of NF-κB proteins to form numerous homo- and heterodimers that differentially regulate target genes (1). For example, p50 and p52 homodimers serve as repressors, whereas dimers containing RelA or c-Rel are transcriptional activators. Heterodimers of RelB with either p50 or p52 display a greater regulatory flexibility, and function both as an activator and a repressor (1,2). Accumulating evidence suggests that activation of specific dimers is mediated by distinct upstream signaling pathways, which in turn are activated in a stimulus- and cell-specific manner (2,3). This chapter begins with an overview of the current state of knowledge about NF-κB, and then discusses the role for this fascinating transcription factor in the endothelium.

HISTORY

NF-κB was first identified in 1986 by Sen and Baltimore (4) as part of an investigation that involved identification of proteins that bind to the immunoglobulin (Ig) heavy chain and κ-light chain enhancers in the nucleus of B cells. The nuclear factor found to be necessary for transcription of immunoglobulin κ-light chain in B cells was therefore named NF-κB (nuclear factor-κB). The inability to detect NF-κB by gel-shift assay in other cell types, including pre-B cells, led to the initial belief that NF-κB was restricted only to mature B cells. Later, identification of IκB proteins revealed the cytoplasmic retention of inactive NF-κB, explaining the absence of DNA-binding of NF-κB in other cell types. In 1991, the cloning of IκBα (5) showed that IκBα retains NF-κB in the cytoplasm through masking of the nuclear localization sequences. Further studies showed that stimulation of cells by a wide variety of agonists results in phosphorylation of IκBα (at Ser32 and Ser36), which then leads to proteasome-mediated degradation of IκBα, with release and nuclear translocation of NF-κB. The realization that protein synthesis–independent inducibility of NF-κB and its ability to translocate to the nucleus may allow rapid transmission of messages to their nuclear targets generated intense interest in this transcription factor. Investigations during the past 10 years have led to the discovery of IκB kinase (IKK) complex, alternate and classical pathways of NF-κB activation, and the post-translational modifications (phosphorylation and acetylation) of NF-κB subunits as well as histones surrounding the NF-κB target genes and their roles in nuclear regulation of NF-κB. These studies have established NF-κB as a central regulator of a wide spectrum of genes with critical functions in cellular physiology and pathology (Tables 86-1 and 86-2).

THE CLASSICAL AND ALTERNATE PATHWAYS OF NF-κB ACTIVATION

NF-κB dimers are mostly sequestered in the cytoplasm by IκBs, a family of inhibitory proteins that mask the NLS by virtue of their interaction with the RHD. These proteins, which include IκBα, IκBβ, BCL3, IκBε, IκBγ (Figure 86.1B), and the precursor proteins NF-κB2/p100 and NF-κB1/p105 (Figure 86.1A), contain five to seven ankyrin repeats that mediate association with the RHD. The presence of these repeats allows the precursors (NF-κB2/p100 and NF-κB1/p105) also to function

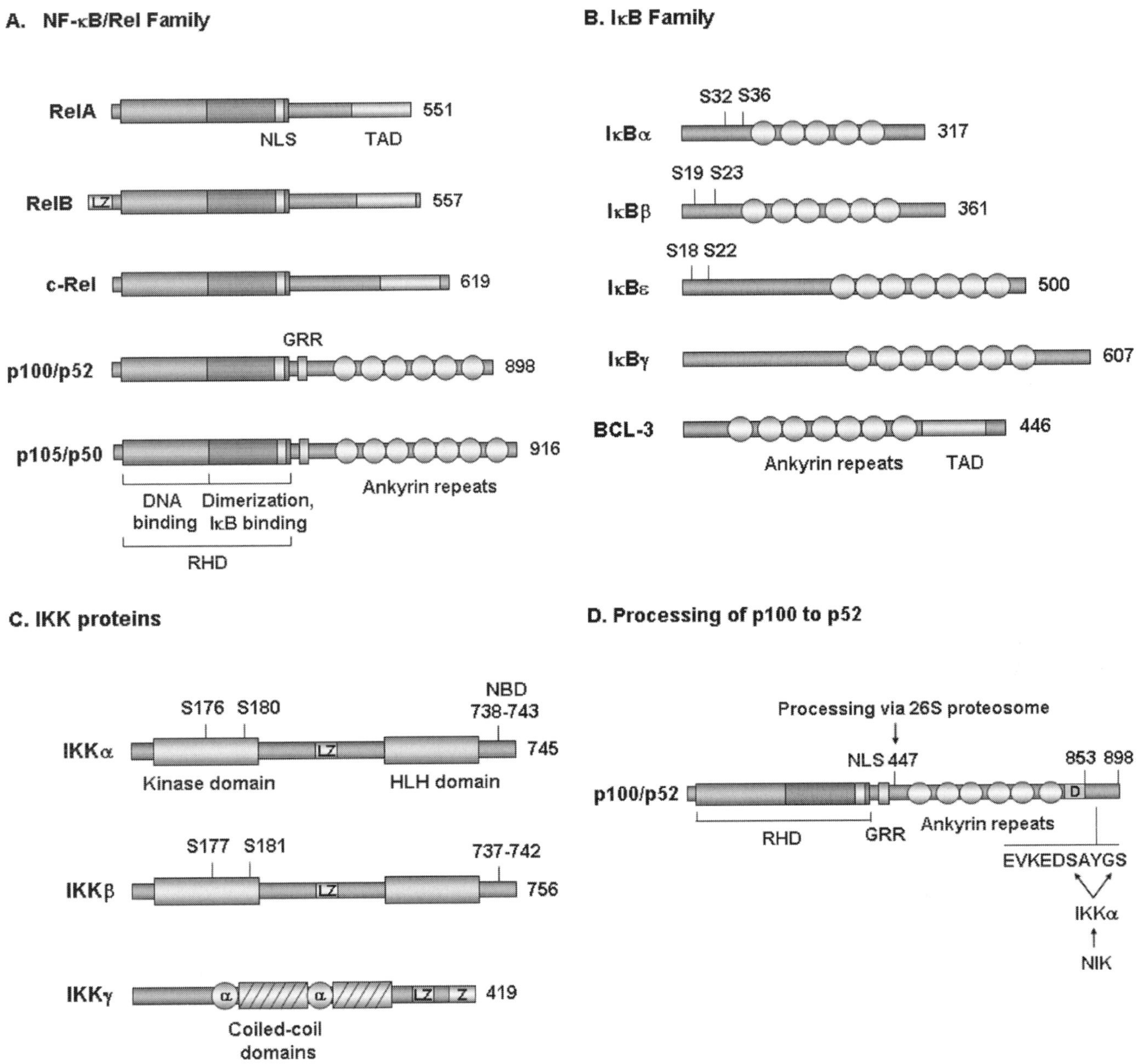

Figure 86.1. Schematic representation of NF-κB family members and their regulators. The number of amino acids in each protein is indicated on the *right.* (**A**) The N-terminal region (~300 amino acids) of NF-κB proteins has a Rel homology domain (RHD). The RHD contains a DNA binding domain and dimerization/IκBα binding domains, and a nuclear localization sequence (NLS) located on the C-terminal side of the RHD. In addition to an RHD, RelA, RelB, and c-Rel contain transactivation domains (TAD). The proteins p105 and p100 contain ankyrin repeats (*circles*) and a glycine-rich region (GRR). The ankyrin repeats render these proteins inactive, and the GRR is required for cotranslational processing of p105 to p50 and post-translational processing of p100 to p52 as indicated in (**D**). (**B**) Members of the IκB family contain ankyrin repeats (*circles*) that mediate their interaction with the RHD. Phosphorylation of indicated serines in these proteins is required for their degradation by the proteasome. Bcl3, by virtue of possessing ankyrin repeats, is frequently considered a member of the IκB family. However, Bcl3 also contains a TAD, and therefore complexes resulting from the interaction of Bcl3 with either p50 or p52 are transcriptionally active. (**C**) All three members of the IKK complex possess a leucine zipper (*LZ*) motif. Of these, only IKKα and IKKβ possess kinase and helix-loop-helix (*HLH*) domains. Phosphorylation of indicated serines is required for the activation of IKKα and IKKβ. IKKγ (*NEMO*), which is unrelated to IKKα and IKKβ, possesses a zinc finger domain (*Z*), coiled-coil domains, and two α-helical (α) domains. These domains are considered to mediate the regulatory function of NEMO and its association with NEMO binding domain (*NBD*) of IKK. (**D**) Processing of p100 to p52. The processing is initiated by phosphorylation of two serine residues at the C-terminus of p100 by IKKα, secondary to its activation by NF-κB inducing kinase (*NIK*). The presumed site of the cleavage for p100 (amino acid 447) is indicated. The sequences in GRR prevent further progression of the 26S proteasome, which digests the C-terminal half of p100 following its ubiquitination.

Table 86-1: Phenotypes of IKK/IκB/NF-κB Knockout Mice

Targeted Gene (Refs.)	*Dominant Phenotypes*
RelA/p65 (59)	Embryonic lethality, liver cell apoptosis, granulopoiesis
c-Rel (60)	Defective in lymphocyte proliferation, humoral immunity, and IL-2 expression
RelB (61)	Lethal; multiple organ inflammation
NF-κ B1/p50 (62)	Multiple deficiencies in B-cell function
NF-κ B2/p52 (63)	Altered architecture of spleen and lymph nodes, deficiency in B cell function
Iκ Bα (64)	Lethal; skin defects, atrophic spleen and thymus, granulopoiesis
Iκ Bε (65)	Reduction of CD44-CD25$^+$ T-cell subspecies, enhanced Ig isotype switching, and cytokine synthesis
Bcl-3 (66)	Altered microarchitecture of secondary lymphoid organs
IKKα (67)	Limb and skin abnormalities
IKKβ (8)	Embryonic lethality; severe liver damage due to apoptosis
IKKγ/NEMO (68)	Embryonic lethality; severe liver damage due to apoptosis
NIK (69, 70)	Impaired processing of p100 to p52, altered architecture of peripheral lymphoid organs

Table 86-2: Some Diseases Associated with NF-κB

Atherosclerosis
Rheumatoid arthritis
Inflammatory bowel disease
Asthma
Systemic inflammatory response syndrome
Ischemia–reperfusion injury
Cancer

as IκBs and retain their partners, the Rel proteins, in the cytoplasm (1,2). Crystallographic studies have revealed that IκB proteins mask only the NLS of RelA/p65, whereas the NLS of p50 remain accessible (1,6). The accessibility of the NLS on p50 and the presence of a nuclear export signal (NES) on IκBα and RelA/p65 allow constant shuttling of IκBα/NF-κB complexes between the nucleus and the cytoplasm. Despite this dynamic shuttling, the steady-state localization of IκBα/NF-κB is in the cytosol (1,7). Following signal-induced degradation of IκBα, which results in the removal of IκB NES and unmasking of RelA/p65 NLS, the steady-state localization of NF-κB is altered in favor of the nucleus.

Degradation of IκB is initiated through phosphorylation of IκBs on specific serine residues by a macromolecular cytoplasmic IKK complex composed of the catalytic subunits IKKα and IKKβ and the regulatory subunit NEMO/IKKγ (Figure 86.1C) (7). Phosphorylation marks IκBs for polyubiquitination by the E3-SCFβ-TrCP ubiquitin ligase, which in turn targets them to degradation by the 26 S-proteasome (1,2). Degradation of IκB results in translocation of the released NF-κB dimer to the nucleus, where it activates transcription following its binding to the κB-enhancer consensus sequences (5′-GGGRNYYYCC-3′, where R is a purine, Y is a pyrimidine, and N is any nucleotide) in the promoter region of target genes.

There are two major signaling pathways by which extracellular stimuli mediate IκB degradation–dependent translocation of NF-κB dimers to the nucleus. In the classical NF-κB signaling pathway, the key event is the activation of IKKβ induced by proinflammatory cytokines and pathogen-associated molecular patterns (PAMPs) following engagement of receptors belonging to the tumor necrosis factor (TNF) receptor (TNFR) and Toll-like receptor (TLR)-interleukin-1 receptor (IL-1R) superfamilies, respectively (Figure 86.2A). Activated IKKβ (in association with IKKα and IKKγ) catalyzes the phosphorylation of IκBs on two N-terminal serine residues (at sites equivalent to Ser32 and Ser36 of IκBα, and Ser19 and Ser23 of IκBβ), polyubiquination (at sites equivalent to Lys21 and Lys22 of IκBα), and subsequent degradation by the 26 S-proteasome. The liberated NF-κB dimers (generally the prototypical heterodimer p50-p65/RelA in this pathway) undergo nuclear translocation and DNA binding to activate the transcription of genes involved in innate immunity and inflammatory responses (1,2,7). It should be noted that activated IKKα also can phosphorylate IκBs; however, biochemical and genetic studies indicate that IKKβ is the dominant kinase involved in the phosphorylation of IκB proteins and, thus, is a crucial regulator of the classical NF-κB pathway (8,9).

The recently described alternative NF-κB signaling pathway has provided a second evolutionarily conserved mechanism of NF-κB activation. This pathway is unique in that it does not require participation of IKKβ and IKKγ and is strictly dependent on IKKα (Figure 86.2B). The IKKα homodimers selectively phosphorylate the two C-terminal serines of NF-κB2/p100 associated with RelB (2,3) (Figure 86.1D). The phosphorylation of these sites is essential for p100 processing to p52. Thus, processing of p100 releases a subset of transcriptionally active NF-κB dimers, consisting mainly of p52 and RelB (3,10). Processing of p52 also is dependent on ubiquitination and proteasomal degradation of p100 (11). However, unlike complete degradation of IκBs, the phosphorylation-dependent ubiquitination of p100 leads only to degradation of its inhibitory C-terminal half. This releases the p52 polypeptide containing the N-terminal RHD associated with RelB, and

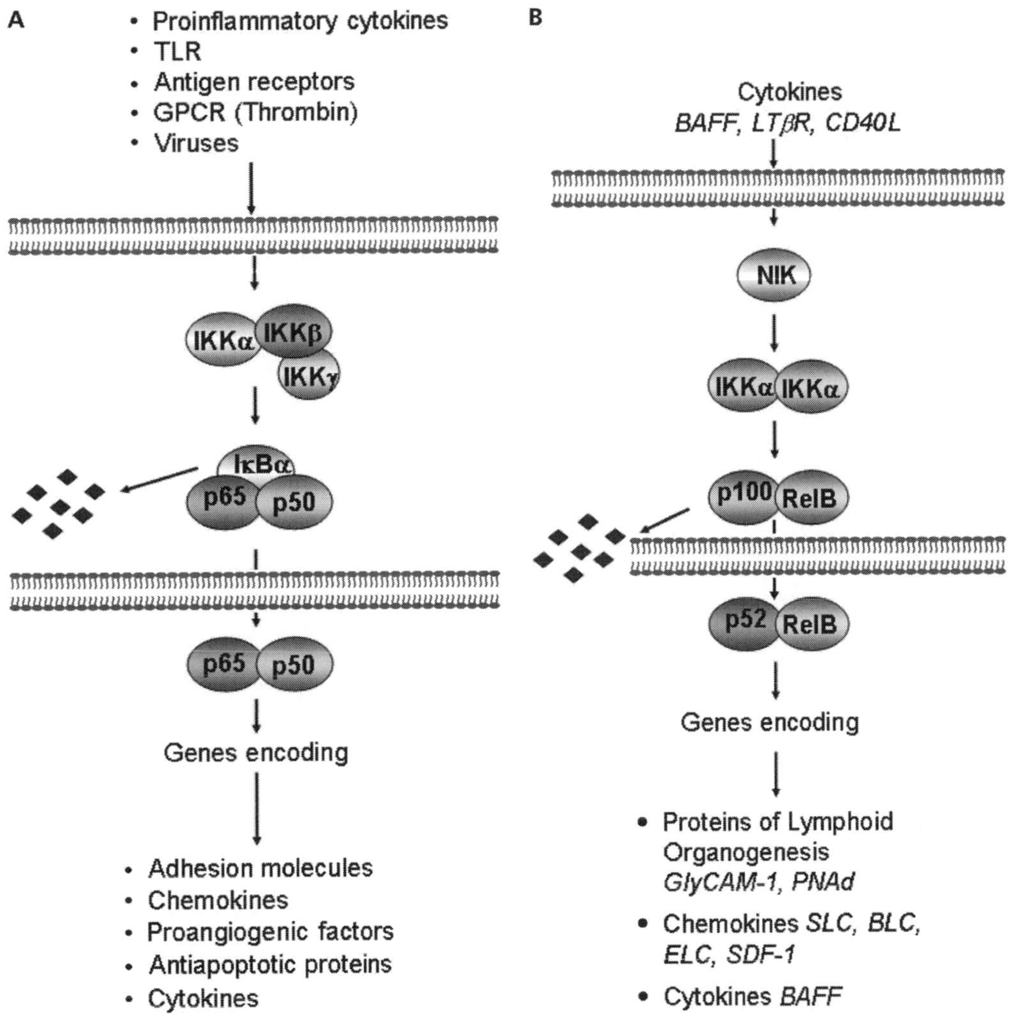

Figure 86.2. The classical and alternative pathway of NF-κB activation. (**A**) The classical pathway is activated by a host of stimuli including viruses, proinflammatory cytokines, antigen receptors Toll-like receptors (*TLR*), and G-protein–coupled receptors (*GPCR*). This pathway is mediated by IKKβ, which catalyzes the phosphorylation of IκBα and leads to its ubiquitination and degradation. The released NF-κB translocates to the nucleus and activates expression of multiple inflammatory and innate immune genes. (**B**) The alternative pathway is mediated by IKKα following its activation by NIK in response to cytokines such as LTβR, BAFF, and CD40L. Activation of this pathway leads to phosphorylation and processing of p100, generating p52:RelB heterodimers. Following nuclear translocation, p52:RelB activates the genes mainly involved in the development and maintenance of secondary lymphoid organs. BAFF, B-cell–activating factor; BLC, B-lymphocyte chemoattractant; ELC, Epstein-Barr virus–induced molecule-1 ligand CC chemokine; GlyCAM, glycosylation-dependent cell adhesion molecule; LTβR, lymphotoxin-β receptor; PNAd, peripheral lymph node addressin; SLC, secondary lymphoid tissue chemokine; SDF-1, stromal cell–derived factor-1.

the p52-RelB heterodimer translocates to the nucleus, where it activates the transcription of genes involved in the development and maintenance of lymphoid organs (2).

REGULATION OF IκB KINASE: THE POINT OF CONVERGENCE IN SIGNALING TO NF-κB

Activation of IKK is the central event in the mechanism of NF-κB activation. (One possible exception is the casein kinase 2 [CK2], which has been shown to mediate ultraviolet [UV]-induced NF-κB activation by catalyzing the phosphorylation of IκBα at a cluster of C-terminal sites [12]). The majority of extracellular signals causing activation of NF-κB converge on this high-molecular-weight kinase IKK complex, which is composed of at least three subunits: IKKα, IKKβ, and IKKγ (Figure 86.1C). The IKKα and IKKβ are the catalytic subunits and share 52% overall amino-acid sequence identity in their primary structures and 65% identity in their catalytic domains (7). The characteristic features of these kinases include an N-terminal kinase domain, a C-terminal helix-loop-helix (HLH) domain, and leucine zipper (LZ) domain (Figure 86.1C). The

HLH domain is required for full IKKβ activity, because deletion mutants lacking this domain possess diminished kinase activity following overexpression or stimulation (1,7). IKKα and IKKβ dimerization depends on a leucine zipper domain, which is therefore required for kinase activity (1,7). Unlike IKKα and IKKβ, IKKγ (also termed NEMO, or NF-κB essential modifier) has no catalytic activity but plays a critical regulatory role in protein–protein interaction by virtue of its long stretches of coiled-coil sequence and C-terminal leucine zipper domain. Gene targeting experiments in mice have shown that the IKK subunits are differentially required for NF-κB activation, depending on the stimulus and the cellular context (1,2,7) (Figures 86.2A and 86.2B).

Activation of the IKK complex involves the phosphorylation of two serine residues in the activation loop within the kinase domain of IKKα (Ser 176 and Ser 180) or IKKβ (Ser 177 and Ser 181) (Figure 86.1C). IKKβ also is phosphorylated at multiple serines in the C-terminal of the HLH motif, most likely through autophosphorylation following its activation. Deletion mutants lacking the C-terminal serine-rich segment show normal activation upon challenge with TNF-α, but remain active for longer periods. This suggests that autophosphorylation or *trans*-autophosphorylation of IKKβ may be a negative regulatory mechanism (1).

The molecular mechanisms by which signaling from cell surface receptors lead to phosphorylation and thereby activation of IKK are still poorly understood. Among the multiple upstream kinases suggested to act as IKK kinases, most have failed to yield the expected results in knockout mice. NF-κB inducing kinase (NIK), a critical component of alternative pathway of NF-κB signaling provides the strongest proof of being an upstream IKK kinase. NIK, a mitogen-activated protein kinase kinase kinase (MAP3K) family member, directly phosphorylates and activates IKKα following activation of the alternative pathway through select members of the TNFR family including lymphotoxin β receptor (LTβR) and B cell–activating factor receptor (BAFFR) (Figure 86.2B) (1,2,11). Despite the reports that NIK associates with IKKβ and NEMO, and thus participates in classical signaling pathway (11–13), the phenotype of *NIK*$^{-/-}$ mice, namely severe deficiency in lymph node development, is consistent with NIK functioning only in the alternative pathway. These findings have led to the prevailing view that each receptor system utilizes a distinct set of adaptor molecules and signaling enzymes to construct a unique pathway that mediates IKK activation in a signal- and cell-specific manner.

REGULATION OF NUCLEAR FACTOR-κB TRANSCRIPTIONAL ACTIVITY

Role of RelA Phosphorylation

Apart from IκBα degradation and nuclear translocation of NF-κB, the transcriptional activity of NF-κB also is regulated by post-translational modifications, particularly phosphorylation and acetylation (14,15). Studies have shown that phosphorylation of RelA/p65 at serine 276 and 311 in the RHD or serine 529 and 536 in the TAD increases the transcriptional capacity of NF-κB in the nucleus (14,15). Moreover, phosphorylation of RelA also decreases its affinity for its negative regulator, IκBα (14,15). Unlike IκBα phosphorylation, the RelA phosphorylation site and the kinase involved vary in a stimulus- and cell type-specific manner (14,15). For example, phosphorylation of RelA at Ser 276 following lipopolysaccharide (LPS) challenge is mediated by the catalytic subunit of protein kinase A (PKA$_c$), whereas TNF-α–induced phosphorylation at this residue is catalyzed by mitogen- and stress-activated kinase (MSK)-1 (Figure 86.3A). Interestingly, PKA$_c$ phosphorylates RelA in the cytoplasm, whereas MSK1 functions in the nucleus. TNF-α also induces phosphorylation of serine 311 within the RHD of RelA, and this phosphorylation requires the participation of yet another kinase, protein kinase (PKC)-ζ (Figure 86.3). Phosphorylation of serine 529 is mediated by CK2, whereas serine 536 is phosphorylated by IKKα and IKKβ. Other kinases that phosphorylate RelA include glycogen synthase kinase (GSK)-3β,

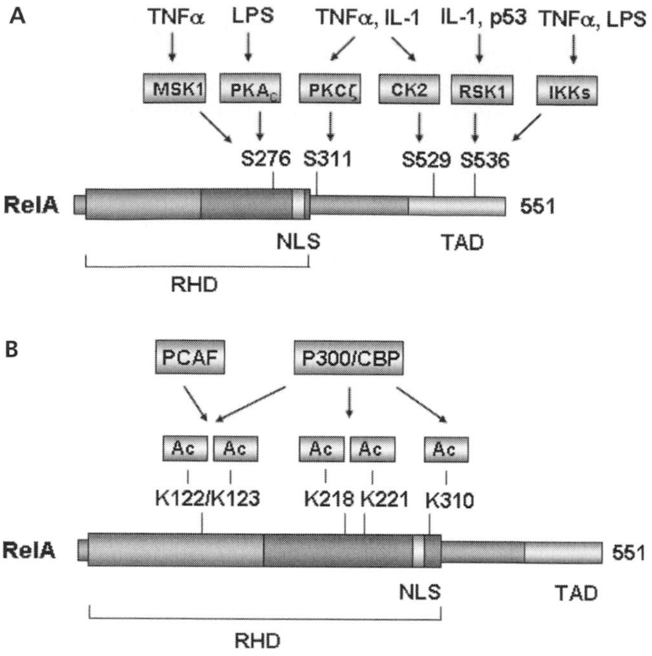

Figure 86.3. RelA phosphorylation and acetylation sites. (**A**) RelA phosphorylation sites. The phosphorylation sites in RelA include serines 276 and 311 in RHD, and serines 529 and 536 in TAD. These phosphorylation sites are targeted by multiple kinases following activation by distinct stimuli as indicated. (**B**) RelA acetylation sites. The acetylation sites in RelA include lysines 122/123, 218/221, and 310. These acetylation sites regulate distinct functions of RelA. The DNA-binding affinity of RelA to κB enhancer is increased upon acetylation (*ac*) of Lys221 by p300 or CREB-binding protein (*CBP*), and simultaneous acetylation of Lys218 prevents the association of RelA with IκBα. Acetylation of Lys310 is required for full transcriptional activity but has no effect on the DNA binding and IκBα assembly. Acetylation of Lys122 and Lys123 by p300/CBP- and p300/CBP-associated factor reduces the transcriptional activity of RelA by interfering with its binding to κB enhancer.

phosphoinositide 3-kinase (PI3K), and NF-κB–activating kinase (NAK; also known as TANK binding kinase [TBK]-1, and TRAF2-associated kinase [T2K]) (14,15). Recently, an additional phosphorylation event at serine 254 in the RHD has been implicated in the regulation of nuclear NF-κB function (Figure 86.3A) (16). This phosphorylation facilitates the binding of peptidyl-prolyl isomerase PIN-1 to the Thr254-Pro motif in RelA. The binding of PIN-1 by this mechanism inhibits RelA binding to IκBα and enhances RelA nuclear localization and protein stability. The biological significance of PIN-1 regulation of RelA was evidenced by the findings that PIN-1–deficient mice and cells are refractory to NF-κB activation by cytokine signals and that the RelA-T254A mutant that cannot act as a PIN-1 substrate is extremely unstable and fails to transactivate NF-κB target genes (16). Given the multiplicity of phosphorylation sites and the kinases involved, it is likely that concurrent phosphorylation of multiple sites has cooperative functional effects on the transcriptional activity of NF-κB.

Phosphorylation of RelA also serves an important function in promoting the recruitment of various transcriptional coactivators. Studies have established that phosphorylation of RelA at serine 276 by PKA_c, MSK1, or serine 311 by PKC-ζ, facilitates the interaction of cAMP-response-element-binding protein (CREB)-binding protein (CBP) and p300 to RelA (1,14). The RelA-CBP complex effectively displaces the transcriptionally repressive histone deacetylase complexes, especially p50-histone deacetylase (HDAC)-1 complexes that frequently occupy κB enhancers of target genes under unstimulated conditions (1,14,15). It remains unclear whether phosphorylation of serines 529 and 536 within TAD also promotes association of RelA with CBP and p300. Nevertheless, it appears likely that the phosphorylation of RelA at these residues facilitates its interaction with other components of basal transcription machinery, thereby controlling the transcriptional responses.

Role of RelA Acetylation

Acetylation of RelA, in addition to phosphorylation, is implicated in regulating the nuclear function of NF-κB. Acetylation of RelA by CBP/p300 prolongs NF-κB DNA binding, and this effect is the result of impaired interaction of acetylated RelA with IκBα (14,17). Deacetylation of RelA by HDAC-1, -2, and -3 restores the interaction with IκBα, which results in a rapid chromosomal-region maintenance 1 (CRMI)-dependent nuclear export of the NF-κB complex, thereby terminating the transcriptional response. It has been suggested that the reversible acetylation of RelA serves as an intramolecular switch critical for determining the strength and duration of the NF-κB response (17). Although the sites of acetylation in RelA have been identified (14), they seem to have differential effects on NF-κB transcriptional activity (Figure 86.3B). Although acetylation at Lys 221 stabilizes RelA binding to the κB site and interferes with IκB binding, acetylation at Lys 310 enhances transcriptional activity without influencing binding to DNA or IκB (14). In contrast to these transcriptional-promoting effects, acetylation of RelA by both CBP and p300/CBP-associated factor (PCAF) on Lys 122/123 was shown to exert a negative effect on RelA transcriptional activity by reducing its binding to the κB site (14).

Role of Histone Acetylation and Deacetylation

Acetylation and deacetylation of histone, like RelA, has important implications in gene regulation (14,18). Histone proteins (histone-H1, -H2, -H3, and -H4), which form nucleosomes associated with DNA, are the fundamental components of chromatin. In general, histones in the transcriptionally active segments of chromatin are acetylated, whereas deacetylated forms are concentrated in the transcriptionally repressed regions of chromatin. Acetylation by histone acetyltransferase (HAT) of specific lysine residues on the N-terminal tail of core histones is associated with uncoiling of the DNA that renders the promoters accessible for binding to RNA polymerase II and other cofactors. Conversely, histone deacetylation by HDAC represses gene transcription by promoting DNA winding, thereby limiting promoter accessibility to transcription factors. Chromatin remodeling associated with the acetylation or /deacetylation of histone plays an important role in regulating NF-κB responses. For example, acetylation of histone-H3 and -H4 is associated with increased NF-κB signaling and activation of genes encoding interleukin (IL)-6, IL-8, IκBα, manganese superoxide dismutase (MnSOD), granulocyte macrophage colony stimulating factor (GM-CSF), macrophage inflammatory protein 1α (MIP1α), and MIP2 (19,20).

In contrast, histone deacetylation renders the cells refractory to NF-κB activation. Inhibition of NF-κB–dependent gene expression by glucocorticoids is primarily ascribed to the ability of these hormones to induce histone deacetylation by facilitating recruitment of HDACs or by direct inhibition of HAT activity (14,21). Other mechanisms involving induction of IκBα synthesis and interaction between glucocorticoid receptor and NF-κB subunits are believed to contribute to the inhibitory effects of glucocorticoids (22). Similarly, the function of p50 homodimers in repressing NF-κB–dependent genes is consistent with its ability to recruit the HDAC1-containing complex (1,14). Recruitment of HDAC1 by this mechanism leads to histone deacetylation in regions bound by p50 homodimers. These findings highlight the importance of chromatin remodeling secondary to the acetylation or deacetylation of histones in controlling the expression of NF-κB target genes.

Role of Histone Phosphorylation

Recent studies have identified phosphorylation at serine 10 of histone H3 as a novel regulatory step in the mechanism of NF-κB activation (23,24). The kinase implicated in this phosphorylation is nuclear IKKα. However, it remains to be determined whether IKKα catalyzes histone phosphorylation directly or requires the participation of a downstream kinase.

Chromatin immunoprecipitation (ChIP) studies showed that stimulation of cells with TNF-α resulted in the recruitment of IKKα to the *IL-6* promoter (23,24). The kinetics of IKKα recruitment was similar to phosphorylation of histone H3 at serine 10. Consistent with the observation that several inflammatory stimuli that induce phosphorylation at serine 10 also promote acetylation at lysine 14 of histone H3, CBP also was recruited to the promoter. RelA appears to have a key nucleating role in the assembly of this complex and in recruiting the IKKα and CBP components to the promoter. The sequence of events include translocation of IKKα to the nucleus, where it interacts with CBP and in conjunction with RelA; these proteins are collectively recruited to NF-κB–responsive promoters and mediate the cytokine-induced phosphorylation and subsequent acetylation of specific residues in histone H3. These studies define a new nuclear function of IKKα in facilitating the recruitment of CBP in a histone H3/serine-10 phosphorylation-dependent manner, which in turn catalyzes the acetylation of histone H3, and thereby promotes transcription of NF-κB target genes. However, given that activation of *TNF-α* and *MIP-1α* genes occur independently of histone H3 phosphorylation, this does not appear to be a general mechanism in controlling NF-κB–dependent responses (25).

NUCLEAR FACTOR-κB AS AN IMPORTANT REGULATOR OF ENDOTHELIAL CELL RESPONSES

By serving as sensors and transducers of signals within the circulatory microenvironment, endothelial cells (ECs) play an integral role in several physiological processes, including the trafficking of cells and nutrients, the control of vasomotor tone, the maintenance of blood fluidity, the regulation of permeability, and the formation of new blood vessels (26). The components of the NF-κB/IκB system, as well as the IKK complex (as described earlier), are present in ECs. The role of NF-κB in ECs has been well documented, especially in the context of inflammation, apoptosis, and angiogenesis. The following section summarizes the key events regulated by NF-κB in ECs; it should, however, be noted that most studies (unless otherwise stated) are based on cell cultures.

Expression of Adhesion Molecules

Adhesion of leukocytes to the endothelium is a crucial step in transendothelial trafficking of leukocytes during inflammation and involves the sequential activation of adhesive proteins on ECs and their counter-receptors on the surface of leukocytes (27). Leukocyte adhesion begins with "capture" and "rolling" of leukocytes along the vessel wall, followed by firm adhesion to the endothelium. Rolling is mediated by binding of endothelial-bound E-selectin and P-selectin to their respective ligands on the surface of leukocytes, whereas firm adhesion involves the binding of vascular cell adhesion molecule (VCAM)-1 and intercellular adhesion molecule (ICAM)-1 to very late activation antigen (VLA)-4 and β2-integrins, respectively (27–30).

Previous studies have shown that activation of NF-κB is essential for the expression of these adhesion molecules in response to a variety of proinflammatory mediators including thrombin, TNF-α, IL-1β, LPS, phorbol esters (PMA), vascular endothelial growth factor (VEGF), interferon (IFN)-γ, and shear stress. Promoter analyses have revealed that the number and sequence of NF-κB sites, as well as the DNA sequence context in which they are present, are different in these genes. While the *E-selectin* and *P-selectin* promoters each contains three NF-κB sites, the promoters of *ICAM-1* and *VCAM-1* are characterized by the presence of one and two functional NF-κB sites, respectively (31,32). Consistent with sequence variation in the NF-κB sites, these promoters are occupied by different NF-κB dimers following TNF-α or thrombin challenge of ECs (31,33,34). For example, NF-κB sites in *E-selectin* and *VCAM-1* promoters are preferentially bound by p50/RelA heterodimer, whereas *ICAM-1* promoter is predominantly activated by RelA homodimer. Moreover, NF-κB sites in these genes either overlap with or are in close proximity to the binding sites for other transcription factors. These features are consistent with the findings that NF-κB despite being essential for the expression of these adhesion molecules, requires the cooperation of other transcription factors for maximal response, and it also explains why these genes have distinct patterns of expression in ECs (31,32). Thus, the activation of these genes is influenced by the number and arrangement of NF-κB binding sites as well as by codependency on other transcription factors, such as CAAT enhancer binding protein (C/EBP) in case of ICAM-1, interferon response factor (IRF) in case of VCAM-1, and high-mobility group-1[Y] (HMG-1[Y]) in case of E-selectin (31,32).

Unlike the requirement of NF-κB for the expression of human as well as murine *ICAM-1*, *VCAM-1*, and *E-selectin* genes, only the murine *P-selectin* gene relies on NF-κB activation. Murine *P-selectin* – but not the human analogue – contains two tandem κB elements and a reverse-oriented κB site. As expected, TNF-α or LPS stimulation results in *P-selectin* expression in murine, but not human ECs. Agonist-mediated induction of the murine promoter is mediated by the binding of NF-κB p50/p65 heterodimers and p65 homodimers to the two tandem κB elements, and binding of p65 homodimers to the reverse-oriented κB site (35).

Expression of Chemokines

Chemokines are small (8–14 kDa), structurally related molecules that interact with seven-transmembrane–spanning G-protein–coupled receptors and, in conjunction with cell adhesion molecules, play an important role in the process of leukocyte recruitment and activation at sites of inflammation (36). IL-8 and monocyte chemoattractant protein (MCP)-1 are among the chemokines predominantly induced in ECs following activation of NF-κB by inflammatory mediators such as TNF-α, IL-1β, LPS, VEGF, and thrombin (36). IL-8, a potent activator and chemoattractant of neutrophils belonging to the CXC chemokine family, was the first chemokine to be characterized. The promoter of *IL-8* is reminiscent of *ICAM-1* in that the NF-κB site is in close proximity to the C/EBP binding site and therefore requires the cooperation of C/EBP and NF-κB for maximal IL-8 induction by TNF-α. MCP-1, the prototype C-C chemokine, is a major regulator of monocyte and T-lymphocyte recruitment to sites of inflammation. In addition, MCP-1 is associated with many biological and disease processes, such as atherosclerosis and angiogenesis (36). Studies by Denk and colleagues (37) used retroviral-mediated expression of transdominant (TD) mutants of IκBα, IκBβ, or IκBε and dominant negative (DN) versions of IκB kinases (IKK)-α or -β, as well as constitutively active version of IKKβ to address the role of NF-κB in mediating the expression of *IL-8* and *MCP-1* in ECs. Using this approach, expression of TD-IκBα, -IκBβ, or -IκBε inhibited the activation of NF-κB and expression of *IL-8* and *MCP-1* in response to TNF-α challenge of ECs. Expression of DN-IKKβ prevented TNF-α–mediated upregulation, whereas DN-IKKα showed only partial inhibition of expression of these molecules. In related experiments, expression of a constitutively active IKKβ mutant induced the expression of *IL-8* and *MCP-1* in the absence of TNF-α challenge. In another study, selective inhibition of NF-κB by adenoviral-mediated transfer of IκBα impaired the expression of these chemokines as well as transendothelial migration of leukocytes by shear flow (38). Similarly, overexpression of RelA/p65 in ECs resulted in the transactivation of the *MCP-1* promoter. Collectively, these data indicate that the IKK/IκB/NF-κB pathway is not only necessary but also sufficient to induce *IL-8* and *MCP-1* expression in activated endothelium.

Expression of Cytokines

Cytokines are small secreted proteins that mediate and regulate immunity, inflammation, and hematopoiesis. The ability of ECs to produce cytokines, in addition to adhesion molecules and chemokines, further underscores the critical involvement of these cells in orchestrating the inflammatory processes. ECs, upon stimulation with such diverse stimuli as hypoxia and bacterial infection, produce a wide range of cytokines including TNF-α, IL-1, IL-5, IL-6, IL-18; and colony stimulating factor (CSF); granulocyte CSF (G-CSF); macrophage CSF (M-CSF); and GM-CSF (39,40). Activation of NF-κB is a critical determinant of the expression of these cytokines (40–42). TNF-α and IL-1 thus released or secreted by infiltrating inflammatory cells can act on ECs in an autocrine or paracrine manner to induce the expression of several of these cytokines, in addition to the adhesion molecules and chemokines, which act in concert to promote adhesion and the migration of leukocytes across endothelial barrier. Recent studies have demonstrated that the adhesive interactions between ECs and recruited inflammatory cells also can signal the activation of NF-κB in ECs, resulting in the secondary secretion of inflammatory cytokines and further induction of adhesion molecules (43,44), thus creating a positive feedback loop that may serve to amplify the inflammatory responses. Indeed, cross-talk between inflammatory cells and the endothelium may be critical to the development of chronic inflammatory states. EC–derived cytokines have been implicated in cellular chemotaxis and recruitment, hematopoiesis, coagulation, and acute-phase protein synthesis. The involvement of these processes in the maturation of an inflammatory and reparative state suggest a crucial role for endothelial-derived cytokines in several disease states such as atherosclerosis, graft rejection, asthma, vasculitis, and sepsis. These findings highlight the importance of NF-κB in controlling the production of multifunctional cytokines from activated endothelium.

Expression of Antiapoptotic Genes

Several lines of evidence have suggested that NF-κB is an important cell survival factor. Targeted deletion of the *RelA/p65* gene in mice leads to embryonic lethality at 15 to 16 days of gestation, concomitant with a massive degeneration of the liver by programmed cell death or apoptosis (45). In addition, cells with functionally inactive NF-κB undergo apoptosis when challenged with a host of stimuli (46). The antiapoptotic effect of NF-κB primarily is ascribed to its ability to activate the transcription of antiapoptotic genes. In ECs, NF-κB has been shown to induce the expression of antiapoptotic molecules such as Bcl2 homologue A1, the zinc finger protein A20, cellular inhibitor of apoptosis 1 and 2 (c-IAP1 and c-IAP2), TNF receptor-associated factor (TRAF)-1, and MnSOD (46,47). Interestingly, Levkau and colleagues (48) showed that NF-κB is activated soon after growth factor deprivation, and that the ability of an EC to maintain this NF-κB activity determines whether it survives or undergoes apoptosis. These studies demonstrate that surviving viable cells exhibit increased NF-κB activity, whereas in apoptotic cells, NF-κB is inactivated by a mechanism involving caspase-mediated cleavage of RelA/p65. This cleavage results in the loss of the carboxy-terminal transactivation domain, rendering RelA/p65 transcriptionally incompetent. The truncated RelA/p65 thus generated promotes apoptosis by acting as a dominant negative inhibitor of NF-κB, whereas the intact RelA/p65 rescues cells from undergoing apoptosis. The caspase-dependent cleavage of RelA/p65 may serve as a regulatory switch that commits the cells to apoptosis by turning off the antiapoptotic genes (48). However, it remains to be determined whether this is a general mechanism of apoptosis or whether its activation is stimulus- and/or cell-type–specific.

Intriguingly, NF-κB activation also is implicated in facilitating EC apoptosis. Recently, Chandrasekhar and colleagues (49) reported that IL-18 induces apoptosis of human cardiac microvascular ECs (HCMECs) by promoting the expression of proapoptotic genes *Fas* and *FasL* in a manner dependent on NF-κB activation. Interfering with NF-κB signaling by siRNA-mediated knockdown of RelA/p65 or expression of dominant negative IKKα or IKKβ mutant inhibited IL-18–induced HCMEC death (49). The paradoxical effects of NF-κB on apoptosis raises the following important questions: (a) What determines whether the activated NF-κB mediates expression of pro- or antiapoptotic genes and thereby promotes or inhibits EC apoptosis? and (b) Are there specific adaptor proteins that are activated in a stimulus-specific manner to link NF-κB to pro- or antiapoptotic pathways?

Expression of Proangiogenic Genes

Clues for a role of NF-κB in angiogenesis were provided by the protective effect of NF-κB on EC apoptosis. Evidence indicating the involvement NF-κB in mediating angiogenesis came from the finding that the expression of several proangiogenic genes depends on the activation of NF-κB. Studies by Yoshida and colleagues (50) showed that expression of proangiogenic factors, IL-8 and VEGF, and their receptors was associated with activation of NF-κB following TNF-α challenge of human microvascular ECs. Neutralizing IL-8 and VEGF by their respective antibodies inhibited TNF-α–induced tubular morphogenesis in vascular ECs plated onto collagen gels as well as neovascularization in the rabbit cornea in vivo. Another important angiogenic mediator induced by TNF-α in an NF-κB–dependent manner is endothelium-specific receptor tyrosine kinase, Tie2, which has been shown to mediate TNF-α–induced angiogenesis in a mouse corneal assay (51). Inhibition of NF-κB activation also prevented angiogenesis as determined by tubular morphogenesis in human microvascular ECs in response to hypoxia–reoxygenation. Angiogenic activities of such diverse stimuli as LPS, estrogen, reactive oxygen species (ROS), and 12(R)-hydroxyeicosatrienoic acid [12(R)-HETErE], an arachidonic acid metabolite, involve activation of NF-κB. These studies identify NF-κB as an important mediator of angiogenesis. Consistent with this, the administration of NF-κB inhibitor pyrrolidine dithiocarbamate (PDTC) resulted in suppression of retinal neovascularization in mice (52).

Cell Growth, Inflammation, and Tumorigenesis

Recent studies have provided evidence for an important role of IKK/NF-κB in promoting cell growth and tumorigenesis and also in linking inflammation to tumorigenesis (53). In an elegant study, Greten and colleagues (53) used a colitis-associated cancer (CAC) model to demonstrate that tissue-specific deletion of IKKβ in intestinal epithelial cells and macrophages was associated with a dramatic decrease in the formation of inflammation-associated tumors. The mechanism by which IKKβ exerts its tumorigenic effect differs between cell types. For example, in enterocytes, IKKβ contributes to tumor initiation and promotion by suppressing apoptosis, but is not required for inflammation per se, whereas in myeloid cells IKKβ is involved in the production of inflammatory mediators that promote tumor growth (53). Similarly, using a mouse model of hepatocellular carcinoma, a prototype of inflammation-associated cancer, Pikarsky and colleagues (54) have shown that NF-κB plays an essential role in promoting such cancers. These studies further demonstrate that activation of NF-κB in hepatocytes in this model depends on TNF-α secreted by endothelial and recruited inflammatory cells. The identification of IKK/NF-κB as a key link between inflammation and tumorigenesis points to the existence of a mechanism that ensures proper growth of normal cells by tightly controlling IKK/NF-κB activation.

Mammalian target of rapamycin (mTOR), alternatively termed FKBP12 and rapamycin associated protein (FRAP) or rapamycin and FKBP12 target (RAFT), is a central controller of cell growth. Minhajuddin and colleagues (55) have recently demonstrated that inhibition of mTOR by rapamycin, an immunosuppressive drug that complexes with high affinity with its cellular receptor FK506-binding protein 12 (FKBP12) with the resulting rapamycin–FKBP12 complex specifically binding to mTOR and inhibiting mTOR-dependent downstream signaling, promoted thrombin-induced NF-κB activation in ECs. Conversely, overexpression of mTOR inhibited NF-κB activation by thrombin. Analysis of the NF-κB signaling pathway revealed that inhibition of mTOR potentiated IKK activation and resulted in a rapid and persistent phosphorylation and degradation of IκBα. Consistent with these data, inhibition of mTOR resulted in a more efficient and stable nuclear localization of RelA/p65 and subsequently, the DNA binding activity of NF-κB in response to thrombin. As expected, mTOR inhibition was associated with augmented ICAM-1 expression by thrombin. These data define a novel role of mTOR in downregulating thrombin-induced ICAM-1 expression in ECs by controlling a delayed and transient activation of IKK/NF-κB. Given the role of IKK/NF-κB and ICAM-1 in tumor promotion and metastasis, the negative regulation of IKK/NF-κB activation by mTOR may be a mechanism of ensuring proper cell growth.

DIAGNOSTIC AND THERAPEUTIC IMPLICATIONS

The activation of NF-κB is implicated in the pathogenesis of a variety of human diseases (Table 86-2). The apparent association of NF-κB activation with various diseases raises the possibility that detection of "expression levels" or "activation states" of the IKK/IκB/NF-κB system can serve as an important tool to predict or diagnose the onset of these disease states. Indeed, studies by Hajra and colleagues (56) have shown the presence of activated NF-κB and induction of NF-κB–dependent genes in atherosclerotic lesions using a mouse model of atherosclerosis. These studies showed that expression of RelA/p65 and IκBα was five- to 18-fold higher in ECs in regions of proximal aorta with high probability (HP) for atherosclerotic lesion development. Moreover, in response to systemic inflammatory stimuli, NF-κB was preferentially activated in the HP region. The increased expression of NF-κB/IκB components suggests that NF-κB in aortic ECs is primed for activation in regions predisposed to atherosclerotic lesion formation. These studies also provide a rationale for the development of strategies for determining the steady-state expression of IKK/IκB/NF-κB components to localize atherosclerotic lesions or regions prone to developing such lesions.

Consistent with discrete steps involved in NF-κB activation, different agents have been shown to block NF-κB activity by targeting different events in the NF-κB activation pathway (57). For example, nonsteroidal anti-inflammatory agents such as aspirin, sodium salicylate, and sulindac exert their effects, at least in part, by inhibiting IKKβ activity. Similarly, naturally occurring flavonoids such as quercetin, resveratrol, and myricetin are thought to mediate their various biological effects including suppression of inflammation, cancer chemoprevention, and protection from vascular disease through inhibition of IKK activity. The anti-inflammatory effects of glucocorticoids such as dexamethasone and prednisone are attributed to their ability to inhibit NF-κB activation by promoting the expression of IκBα to increase cytosolic retention of NF-κB, disturbing the interaction of RelA/p65 with the basal transcription machinery, or competing for the limiting amounts of the coactivators CBPs, depending on the cellular context (57). Other compounds such as peptide aldehyde MG132, lactacystin, and the immunosuppressive drug cyclosporine A (CsA) prevent NF-κB activation by inhibiting the protease activity of the proteasome. Tacrolimus (FK506), another immunosuppressive drug, which, like CsA, is used to prevent the rejection of organ transplants, inhibits NF-κB activation by preventing translocation of c-Rel from cytoplasm to the nucleus. Thus, a number of widely used anti-inflammatory, anticancer, and immunosuppressive agents inhibit the NF-κB pathway, at least in part, as one of their targets.

One of the concerns about inhibiting several of these components of the NF-κB pathway is the specificity of such drugs and the potential for toxicity because prolonged NF-κB blockade could interfere with host defense response and result in liver apoptosis. In view of these challenges, an effective strategy for the therapeutic interventions of diseases associated with chronic or dysregulated NF-κB activation depends on achieving a delicate balance between suppressing NF-κB activity and interfering with normal cellular functions. Conditional targeting of specific NF-κB subunits, IκB proteins, or IKKs in a cell-specific manner will be a useful strategy to achieve this goal. For example, using *E-selectin* promoter, it will be possible to direct the expression of super repressor IκBα, dominant negative mutant of IKKs, or short hairpin RNA (shRNA) in targeting specific NF-κB subunits in an EC- and inflammation-specific manner, thus providing a control over the strength and duration of NF-κB signals and thereby ensuring therapeutic efficacy and minimizing systemic toxicity. Such a strategy has been used to induce conditional and EC-specific expression of neutrophil inhibitory factor (NIF), a β_2-integrin antagonist, which in turn prevents lung neutrophils infiltration and vascular injury in a mouse model of gram-negative sepsis (58).

KEY POINTS

- Endothelial NF-κB plays a key role in controlling varied biological effects ranging from inflammation, angiogenesis, and cell fate decisions.
- NF-κB contributes to inflammation by inducing the expression of cytokines, adhesion molecules, and chemokines in ECs.
- NF-κB contributes to angiogenesis by promoting the expression of proangiogenic genes such as *VEGF*, *IL-8*, and their receptors in ECs.
- Activation of NF-κB in ECs influences cell fate decisions by mediating expression of pro- or anti-apoptotic genes.

Future Goals

- To identify the proximal signaling components of the IKK/IκB/NF-κB pathway in ECs
- To determine the mechanisms responsible for dysregulated NF-κB activation in EC in disease states
- To conditionally target NF-κB in endothelium to limit the pathological effects of dysregulated NF-κB without interfering with host defense response

ACKNOWLEDGMENTS

This work was supported by NHLBI grant HL67424.

REFERENCES

1 Bonizzi G, Karin M. The two NF-κB activation pathways and their role in innate and adaptive immunity. *Trends Immunol.* 2004;25:280–288.
2 Hayden MS, Ghosh S. Signaling to NF-κB. *Genes Dev.* 2004;18: 2195–2224.
3 Senftleben U, Cao Y, Xiao G, et al. Activation of IKKα of a second, evolutionary conserved, NF-κB signaling pathway *Science.* 2001; 293:1495–1499.
4 Sen R, Baltimore D. Multiple nuclear factors interact with the immunoglobulin enhancer sequences. *Cell.* 1986;46:705–716.
5 Haskill S, Beg A, Tompkins S, et al. Characterization of an immediate early gene induced in adherent monocytes that encodes an IκB-like activity. *Cell.* 1991;65:1281–1289.
6 Huxford T, Huang DB, Malek S, et al. The crystal structure of IκBα/NF-κB complex reveals mechanisms of NF-κB inactivation. *Cell.* 1998;95:759–770.
7 Ghosh S, Karin M. Missing pieces in the NF-κB puzzle. *Cell.* 2002;109:S81–S86.
8 Li Q, Van Antwerp D, Mercurio F, et al. Severe liver degeneration in mice lacking the IKK2 gene. *Science.* 1999;284:321–325.
9 Tanaka M, Fuentes ME, Yamaguchi K, et al. Embryonic lethality, liver degeneration, and impaired NF-κB activation in IKKβ-deficient mice. *Immunity.* 1999;10:421–429.

10 Regnier CH, Song HY, Gao X, et al. Identification and characterization of an IκB kinase. *Cell.* 1997;90:357–383.
11 Xiao G, Fong A, Sun SC. Induction of p100 processing by NIK involves docking IKKα to p100 and IKKα-mediated phosphorylation. *J Biol Chem.* 2004;279:30099–30105.
12 Kato T Jr., Delhase M, Hoffmann A, et al. CK2 is a C-terminal IKK responsible for NF-κB activation during the UV response. *Mol Cell.* 2003;12:829–839.
13 Bouwmwester T, Bauch A, Ruffner H, et al. A physical and functional map of the human TNFα/NF-κB signal transduction pathway. *Nat Cell Biol.* 2004;6:97–105.
14 Chen LF, Greene WC. Shaping the nuclear activation of NF-κB. *Nat Mol Cell Biol.* 2004;5:392–401.
15 Viatour P, Merville MP, Bours V, et al. Phosphorylation of NF-κB and IκB proteins: implications in cancer and inflammation. *Trends Biochem Sci.* 2005;30:43–52.
16 Ryo A, Suizu F, Yoshida Y. Regulation of NF-κB signaling by PIN 1-dependent prolyl isomerization and ubiquitination-mediated proteolysis of p65/RelA. *Mol Cell.* 2003;12:1413–1426.
17 Chen L, Fischle W, Verdin E, et al. Duration of nuclear NF-κB action regulated by reversible acetylation. *Science.* 2001;293: 1653–1657.
18 Imhof A, Wolffe AP. Transcription: gene controlled by targeted histone acetylation. *Curr Biol.* 1998;8:R422–R424.
19 Saccani S, Pantano S, Natoli G. Two waves of NF-κB recruitment to target promoters. *J Exp Med.* 2001;193:1351–1359.
20 Chen LF, Greene WC. Regulation of distinct biological activities of the NF-κB transcription factor complex by acetylation. *J Mol Med.* 2003;81:549–557.
21 Rahman I, Marwick J, Kirkham P. Redox modulation of chromatin remodeling: impact on histone acetylation and deacetylation, NF-κB and pro-inflammatory gene expression. *Biochem Pharmacol.* 2004;68:1255–1267.
22 Almawi WY, Melemedjian OK. Negative regulation of NF-κB activation and function by glucocorticoids. *J Mol Endocrinol.* 2002;28:69–78.
23 Yamamoto Y, Verma UN, Prajapati S, et al. Histone H3 phosphorylation by IKKα is critical for cytokine-induced gene expression. *Nature.* 2003;423:655–659.
24 Anest V, Hanson JL, Cogswell PC, et al. A nucleosomal function for IKKα in NF-κB dependent gene expression. *Nature.* 2003; 423:659–663.
25 Saccani S, Pantano S, Natoli G. p38-dependent marking of inflammatory genes for increased NF-κB recruitment. *Nature Immunol.* 2002;3:69–75.
26 Aird WC. Endothelium as an organ system. *Crit Care Med.* 2004; 32:S271–S279.
27 Springer TA. Traffic signals for lymphocyte recirculation and leukocyte emigration: the multistep paradigm. *Cell.* 1994;76: 301–314.
28 Brown EJ. Adhesive interactions in the immune system. *Trends Cell Biol.* 1997;7:289–295.
29 Rahman A, Anwar KN, Uddin S, et al. PKC-δ regulates thrombin-induced ICAM-1 gene expression in ECs via activation of p38 mitogen-activated protein kinase. *Mol Cell Biol.* 2001;21:5554–5565.
30 Minami T, Abid MR, Zhang J, et al. Thrombin stimulation of vascular adhesion molecule-1 in ECs is mediated by PKC-δ-NF-κB and PKC-ζ-GATA signaling pathways. *J Biol Chem.* 2003;278: 6976–6984.
31 Collins T, Read MA, Neish AS, et al. Transcriptional regulation of EC adhesion molecules: NF-κB and cytokine-inducible enhancers. *FASEB J.* 1995;9:899–909.
32 Rahman A, Roebuck KA, Malik AB. Transcriptional regulation of endothelial adhesion molecule gene expression by oxidants and cytokines. In: Weir EK, Archer SL, Reeves JT, eds. *Nitric Oxide and Radicals in the Pulmonary Vasculature.* Armonk, New York: Futura Publishing Company, Inc.; 1996:63–85.
33 Rahman A, Anwar KN, True AL, et al. Thrombin-induced p65 homodimer binding to downstream NF-κB site of the promoter mediates endothelial ICAM-1 expression and neutrophil adhesion. *J Immunol.* 1999;162:5466–5476.
34 Minami T, Aird WC. Thrombin stimulation of the vascular cell adhesion molecule-1 promoter in ECs is mediated by tandem NF-κB and GATA motifs. *J Biol Chem.* 2001;276:47632–47641.
35 Pan J, Xia L, Yao L, et al. TNFα or lipopolysaccharide-induced expression of the murine P-selectin gene in ECs involves novel κB sites and a variant activating transcription factor/cAMP response element. *J Biol Chem.* 1998;273:10068–10077.
36 Murdoch C, Finn A. Chemokine receptors and their role in inflammation and infectious diseases. *Blood.* 2000;95:3032–3043.
37 Denk A, Goebeler M, Schmid S, et al. Activation of NF-κB via the IKK complex is both essential and sufficient for proinflammatory gene expression in primary ECs. *J Biol Chem.* 2001;276: 28451–28458.
38 Zoja C, Angioletti S, Donadelli R, et al. Shiga toxin-2 triggers endothelial leukocyte adhesion and transmigration via NF-κB dependent up-regulation of IL-8 and MCP-1. *Kidney Int.* 2002; 62:846–856.
39 Krishnaswamy G, Kelley J, Yerra L, et al. Human endothelium as a source of multifunctional cytokines: molecular regulation and possible role in human disease. *J Interferon Cytokine Res.* 1999;19:91–104.
40 Chandrasekar B, Marelli-Berg FM, Tone M, et al. β-adrenergic stimulation induces IL-18 expression via beta2-AR, PI3K, Akt, IKK and NF-κB. *Biochem Biophys Res Commun.* 2004;319:304–311.
41 Kobayashi S, Nagino M, Komatsu S, et al. Stretch-induced IL-6 secretion from endothelial cells requires NF-κB activation. *Biochem Biophys Res Commun.* 2003;308:306–312.
42 Peng HB, Rajavashisth TB, Libby P, et al. Nitric oxide inhibits macrophage-colony stimulating factor gene transcription in vascular endothelial cells. *J Biol Chem.* 1995;270: 17050–17055.
43 Fan J, Frey RS, Malik AB. TLR4 signaling induces TLR2 expression in endothelial cells via neutrophil NADPH oxidase. *J Clin Invest.* 2003;112:1234–1243.
44 Takahashi M, Kitagawa S, Masuyama JI, et al. Human monocyte-endothelial cell interaction induces synthesis of GM-CSF. *Circulation.* 1996;93:1185–1193.
45 Beg AA, Baltimore D. An essential role for NF-κB in preventing TNFα-induced cell death. *Science.* 1996;274:782–784.
46 Brouard S, Berberat PO, Tobiasch E, et al. Heme oxygenase-1-derived carbon monoxide requires the activation of transcription factor NF-κB to protect endothelial cells from TNFα-mediated apoptosis. *J Biol Chem.* 2002;277:17950–17961
47 Abid MR, Schoots IG, Spokes KC, et al. Vascular endothelial growth factor-mediated induction of manganese superoxide dismutase occurs through redox-dependent regulation of forkhead and IκB/NF-κB. *J Biol Chem.* 2004;279:44030–44038.

48 Levkau B, Scatena M, Giachelli CM, et al. Apoptosis overrides survival signals through a caspase-mediated dominant-negative NF-κB loop. *Nat Cell Biol.* 1999;1:227–233.
49 Chandrasekar B, Vemula K, Surabhi RM, et al. Activation of intrinsic and extrinsic proapoptotic signaling pathways in IL-18-mediated human cardiac endothelial cell death. *J Biol Chem.* 2004;279:20221–20233.
50 Yoshida S, Ono M, Shono T, et al. Involvement of IL-8, vascular endothelial growth factor, and basic fibroblast growth factor in TNFα-dependent angiogenesis. *Mol Cell Biol.* 1997;17:4015–4023.
51 DeBusk LM, Chen Y, Nishishita T, et al. Tie2 receptor tyrosine kinase, a major mediator of TNFα-induced angiogenesis in rheumatoid arthritis. *Arthritis Rheum.* 2003;48:2461–2471.
52 Yoshida A, Yoshida S, Ishibashi T, et al. Suppression of retinal neovascularization by the NF-κB inhibitor pyrrolidine dithiocarbamate in mice. *Invest Ophthalmol Vis Sci.* 1999;40:1624–1629.
53 Greten FR, Eckmann L, Greten TF, et al. IKKβ links inflammation and tumorigenesis in a mouse model of colitis-associated cancer. *Cell.* 2004;118:285–296.
54 Pikarsky E, Porat RM, Stein I, et al. NF-κB functions as a tumor promoter in inflammation-associated cancer. *Nature.* 2004;431:461–466.
55 Minhajuddin M, Fazal F, Bijli KM, et al. Inhibition of mTOR potentiates thrombin-induced ICAM-1 expression by accelerating and stabilizing NF-κB activation in ECs. *J Immunol.* 2005;174:5823–5829.
56 Hajra L, Evans AI, Chen M, et al. The NF-kB signal transduction pathway in aortic endothelial cells is primed for activation in regions predisposed to atherosclerotic lesion formation. *Proc Natl Acad Sci USA.* 2000;97:9052–9057.
57 Yamamoto Y, Gaynor RB. Therapeutic potential of inhibition of the NF-κB pathway in the treatment of inflammation and cancer. *J Clin Invest.* 2001;107:135–142.
58 Xu N, Gao XP, Minshall RD, et al. Time-dependent reversal of sepsis-induced PMN uptake and lung vascular injury by expression of CD18 antagonist. *Am J Physiol Lung Cell Mol Physiol.* 2002;282:L796–L802.
59 Beg AA, Sha WC, Bronson RT, et al. Embryonic lethality and liver degeneration in mice lacking the RelA component of NF-κB. *Nature.* 1995;376:167–170.
60 Kontgen F, Grumont RJ, Strasser A, et al. Mice lacking the c-rel proto-oncogene exhibit defects in lymphocyte proliferation, humoral immunity, and interleukin-2 expression. *Genes Dev.* 1995;9:1965–1977.
61 Weih F, Carrasco D, Durham SK, et al. Multiorgan inflammation and hematopoietic abnormalities in mice with targeted disruption of RelB, a member of the NF-κB/Rel family. *Cell.* 1995;80:331–340.
62 Sha WC, Liou HC, Toumanen E, et al. Targeted disruption of the p50 subunit of NF-κB leads to multifocal defects in immune responses. *Cell.* 1995;80:321–330.
63 Caamano JH, Rizzo CA, Durham SK, et al. NF-κB2 (p100/p52) is required for normal splenic microarchitecture and B cell-mediated immune responses. *J Exp Med.* 1998;187:185–196.
64 Beg AA, Sha WC, Bronson RT, et al. Constitutive NF-κB activation, enhanced granulopoiesis, and neonatal lethality in IκBα-deficient mice. *Genes Dev.* 1995;9:2736–2746.
65 Memet S, Laouini D, Epinat JC, et al. IκBε-deficient mice: reduction of one T cell precursor subspecies and enhanced Ig isotype switching and cytokine synthesis. *J Immunol.* 1999;163:5994–6005.
66 Franzoso G, Carlson L, Scharton-Kersten T, et al. Critical roles for the Bcl-3 oncoprotein in T cell-mediated immunity, splenic microarchitecture, and germinal center reactions. *Immunity.* 1997;6:479–490.
67 Takeda K, Takeuchi O, Tsujimura T, et al. Limb and skin abnormalities in mice lacking IKKα. *Science.* 1999;284:313–316.
68 Rudolph D, Yeh WC, Wakeham A, et al. Severe liver degeneration and lack of NF-κB activation in NEMO/IKKγ-deficient mice. *Genes Dev.* 2000;14:854–862.
69 Xiao G, Harhaj EW, Sun SC. NF-κB-inducing kinase regulates the processing of NF-κB2 p100. *Mol Cell.* 2001;7:401–409.
70 Miyawaki S, Nakamura Y, Suzuka H, et al. The new mutation, aly, that induces a generalized lack of lymph nodes accompanied by immunodeficiency in mice. *Eur J Immunol.* 1994;24:429–434.

87

Peroxisome Proliferator-Activated Receptors and the Endothelium

Jonathan D. Brown and Jorge Plutzky

Brigham and Women's Hospital, Harvard Medical School, Boston, Massachusetts

The endothelium is a biologically active, dynamic transducer of diverse inputs to the arterial wall and indeed the entire organism. In perhaps an analogous manner, peroxisome proliferator-activated receptors (PPARs) are biological integrators linking various proximal signals to multiple, specific cellular responses through transcriptional regulation (1,2). PPARs perform this function as nuclear receptors that are ligand-activated transcription factors (3). Through PPAR activation, inputs such as lipid metabolism, dietary intake, and drug therapy are coupled to nuclear responses, thus regulating entire cassettes of PPAR target genes (4). The complexity of the PPAR system allows this programmed response to be specific, carefully controlled, and integrated into feedback loops that coordinate systemic responses. The evidence for PPAR expression and activity in the endothelium identifies these biological transcriptional integrators as important contributors to endothelial biology (5). Given known PPAR control of pathways such as lipid metabolism, insulin sensitivity, adipogenesis, and energy balance, as well as recent data for PPAR involvement in inflammation and atherosclerosis (6), their relevance to endothelial cell (EC) transcriptional responses is apparent. After briefly reviewing PPARs in general, this chapter considers more recent data regarding endothelial PPAR responses.

HISTORY AND EVOLUTIONARY CONSIDERATIONS

The discovery of the nuclear hormone receptor family, beginning with the identification of the growth hormone receptor (GHR) in 1985, unearthed a critical molecular link between hormonal signaling cascades and the transcriptional control of specific gene cassettes (7). The following decade was marked by an explosion in the field of nuclear hormone receptor biology due to the realization that this superfamily of receptors regulates a broad array of physiologic processes ranging from reproduction to bone metabolism to nutrient control of carbohydrate and lipid metabolism. Moreover, interest in this arena was fueled by the realization that these nuclear receptors could be targeted for drug therapy. In 1990, shortly after the discovery of the GHR by Evans and his laboratory, Issemann and Green cloned the first PPAR nuclear receptor (now known as PPARα) as part of their search for the molecular target of chemicals known to induce hypolipidemia and peroxisome proliferation in rodents (8). Interestingly, the rodent peroxisome organelle contains the enzymatic machinery required for β-oxidation of lipids. Homology mapping between this first PPAR and other known steroid nuclear receptor family members, including the retinoic acid receptor, thyroid receptor, and vitamin D3 receptor, revealed significant similarity spanning a wide range of eukaryotic organisms including *Drosophila melanogaster* and *Homo sapiens* (8). Functional studies confirmed that the activation of this PPAR directly upregulates the expression of acetyl co-A oxidase (ACO), an enzyme involved in the β-oxidation of fatty acids (9). The subsequent observation that fatty acids can activate PPARs revealed a link between nutritional input and the transcriptional regulation of genes responsible for crucial metabolic processes. These findings were of obvious relevance to many areas such as diabetes and cardiovascular disease, an intersection that was heightened by the discovery of many of these nuclear receptors in other tissues – for example, the estrogen receptor in vascular smooth muscle cells (VSMCs) and the PPARs in ECs, the latter being the focus here. The conserved homology among PPARs in eukaryotes suggests a mechanism has likely evolved for the dynamic molecular adaptation to nutritional states in these organisms.

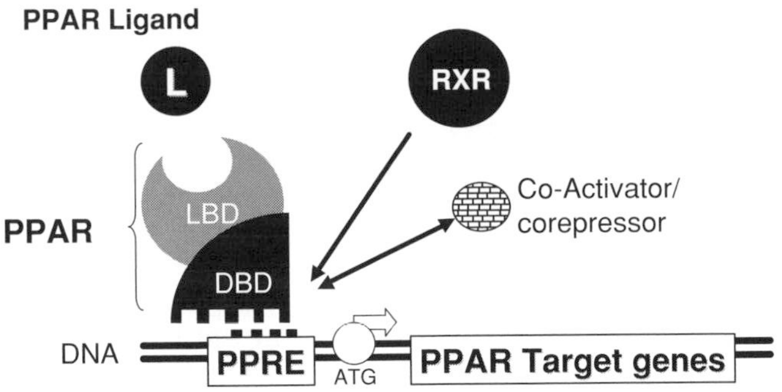

Figure 87.1. PPARs are members of the steroid hormone nuclear receptor family. Like other such nuclear receptors, PPARs are ligand-activated transcription factors with five main domains including the ligand-binding domain (*LBD*) and the DNA-binding domain (*DBD*). In response to specific ligands, the PPAR is activated and forms a heterodimeric complex with another nuclear receptor retinoid X receptor (*RXR*). This functional transcriptional complex, which also involves the release (corepressors) or recruitment (coactivators) of small accessory molecules, can then bind to specific PPAR response elements (PPRE) in the promoter region of target genes.

PEROXISOME PROLIFERATOR-ACTIVATED RECEPTORS – LIGAND-ACTIVATED NUCLEAR RECEPTORS REGULATING GENE EXPRESSION

The three known PPAR isoforms – α, γ, and δ – are all members of the steroid hormone nuclear receptor family (10). Like other steroid hormone receptors, such as the estrogen receptor and the thyroid hormone receptor, PPARs contain both ligand-binding (LBD) and DNA-binding domains (DBD) (1) (Figure 87.1). The binding of a cognate agonist to the PPAR LBD induces a conformational shift that activates the receptor, allowing formation of a heterodimeric complex between the PPAR and another nuclear receptor, the retinoic X receptor (RXR), which is activated by its own distinct ligand, namely 9-*cis*-retinoic acid. Like many other steroid hormone receptors, PPARs are obligate heterodimeric partners with RXR, requiring the presence of RXR for functional activity. The cognate interaction between a specific PPAR agonist and its PPAR partner regulates PPAR-responsive genes in vitro and in vivo. The exact mechanics through which PPAR and RXR heterodimerization actually occur in vivo remain incompletely defined. For example, insight into the order of steps, the action of unliganded receptors, and the control of RXR homodimerization versus its heterodimerization with its multiple other nuclear receptor partners all remains largely obscure. Regardless, ligand binding induces a conformational change involving the PPAR AF2 domain that allows the ligand/PPAR/RXR complex to recognize and bind to specific PPAR response elements (PPREs) in the promoter regions of specific target genes. PPREs are classically a direct DNA repeat separated by a single nucleotide. More recently, the PPAR field has focused on the importance of small accessory molecules, known as *coactivators* and *corepressors*, that must be respectively recruited or released in order for functional PPAR responses to occur (11). Importantly, PPAR activation can either induce or repress gene expression. Although the mechanism for so-called transrepression remains incompletely understood, one recent line of investigation implicates the stabilization of a corepressor-bound state through sumoylation may contribute to anti-inflammatory effects of PPARγ (11). Recent scientific evidence suggests PPARs can activate transcription in a ligand-independent manner through stochastic changes in LBD conformation (12). Although many advances in the PPAR field have been derived from the serendipitous discovery of synthetic PPAR ligands, the precise identity of natural PPAR agonists remains somewhat obscure (see the following sections for further discussion). Furthermore, the fate of these ligands, once bound to the PPAR, is unclear. Although the three PPARs share common mechanisms of action, each isoform has unique biological attributes and functional roles. We review current insight into each PPAR isoform, focusing on its role in endothelial biology.

PEROXISOME PROLIFERATOR-ACTIVATED RECEPTOR-γ – TRANSCRIPTIONAL REGULATOR OF ADIPOGENESIS AND INSULIN SENSITIVITY

PPARγ was initially identified in adipocytes with subsequent work establishing a critical role for PPARγ in the adipogenic differentiation program, for example through induction of aP2, leptin, and adiponectin (2,13–15). The discovery that synthetic PPARγ ligands identified through chemical screening could improve insulin sensitivity and lower glucose established the role of PPARγ in metabolic homeostasis (1,15). Thiazolidinediones (TZDs), a class of drugs in clinical use as antidiabetic medications in the form of pioglitazone (Actos) and rosiglitazone (formerly BRL49653, now Avandia), were found to exert their effects by binding to PPARγ (16). PPARγ also participates in lipid metabolism through the regulation of target genes such as cholesterol transporters and lipoprotein lipase (17). Homozygous PPARγ–deficient mice die in utero due to placental malformation (18–20) (Table 87-1). Although the identity of naturally occurring ligands remains elusive, several candidate endogenous PPARγ ligands have been proposed, including oxidized linoleic acid, in the form of 9- or 13-hydroxyoctadecadienoic acid (HODE) derivatives (21,22) (which also activates PPARα)(23); nitrolineoleic acid; and the prostaglandin metabolite 15-deoxy-D12,14-prostaglandin J_2 (15d-PGJ_2) (24,25). The enzyme 15-lipooxygenase may be a way of generating PPARγ–activating HODEs and is a general pathway for the production of endogenous PPAR ligands (26).

PEROXISOME PROLIFERATOR-ACTIVATED RECEPTOR-γ IN THE ENDOTHELIUM

PPARγ transcript and protein is expressed in human ECs both in vitro and in vivo, as demonstrated by Northern blot and

Table 87-1: Genetic Mouse Models

Gene	$^{-/-}$*(Refs.)*	*Synthetic Agonists (Refs.)*	*Endogenous Agonists (Refs.)*	*Expression and Function Role in ECs (Refs.)*
PPARγ	Embryonic lethal (18)	TZD (rosiglitazone, pioglitazone) (16)	HODE,15d-PGJ_2 (24)	Inhibits IFN-8-mediated induction of CXC chemokines, but not MCP-1 (29); inhibits CCR2; ET-1 (34, 35); variable results with PAI-1 in vitro and in vivo (46); $^{+/-}$ effects on VCAM-1, ICAM-1 (69); CuZn-SOD; decreased $p22^{phox}$ and $p47^{phox}$ (71,72); Increased nitric oxide bioavailability (41); inhibits angiogenesis (43)
PPARα	Viable; altered lipid profile (56)	Fibric acid derivatives, the research compound WY14643 (1)	HODE, eicosanoids, long-chain fatty acids; leukotriene B4; LPL-treated VLDL (1)	Inhibits IL-1β and TNFα-induced VCAM-1 expression (65); CuZn-SOD; decreased $p22^{phox}$ and $p47^{phox}$ (71); activation in vivo reduces atherosclerosis in mice (83)
PPARδ	Predominantly embryonic lethal; surviving animals with less adiposity (76)	Research compound GW501516 (19)	LPL-treated VLDL, prostanoid derivatives (89)	Present in ECs (19); function unclear

immunohistochemical staining of human carotid and coronary artery specimens (27,28). Subsequent work has identified various PPARγ target genes and secondary responses important to endothelial biology (5). For example, PPARγ agonists have been shown to repress chemokine expression (29). Chemokines – short for chemoattractant cytokines – are a large complex family of proteins produced by various tissues including ECs that can recruit inflammatory cells to sites of injury (30). Chemokines, strongly implicated in the pathophysiology of atherosclerosis (31,32), are categorized into subfamilies based on the primary sequence relationship of the first two of four invariant cysteine residues (30): so called "CC" or "CXC" chemokines. CXC chemokines are further defined by the presence or absence of an amino-terminal ELR (Glu-Leu-Arg) sequence. ELR-containing CXC chemokines include interleukin (IL)-8 and platelet factor 4 (PF4). Non-ELR CXC chemokines include interferon (IFN)-inducible protein of 10 kDa (IP-10), monokine induced by γ-IFN (Mig), and IFN-inducible T-cell α-chemoattractant (I-TAC). Inflammatory cytokines such as IFN-γ or tumor necrosis factor (TNF)-α can induce the expression of both CC chemokines, such as monocyte chemoattractant protein (MCP)-1, and CXC chemokines. Consequently, this network of chemokines can recruit numerous and distinct types of inflammatory cells to sites of injury, including T cells, monocytes or macrophages, and neutrophils (31).

Experimentally, synthetic PPARγ activators were shown to repress certain cytokine-induced chemokine expression in ECs cultured from human saphenous vein specimens. PPARγ activation inhibited mRNA expression in a subset of IFN-γ–induced CXC chemokines including IP-10, ITAC, and Mig. In contrast, in this experimental system, IFN-γ induction of the CC chemokine *MCP-1* mRNA (29) was unchanged by PPARγ stimulation, suggesting a selective effect of PPARγ activation. Extended further, PPARγ activation inhibited *IP-10* promoter activity in vitro in response to IFN-γ via a nuclear factor (NF)-κB–dependent mechanism. Functionally, the transcriptional effect on chemokine expression also attenuated chemotactic responses in cell culture (29,33). Despite the lack of effect on MCP-1, the receptor for MCP-1 (known as CCR2) is repressed by PPARγ agonists, which would still be consistent with a net repression of MCP-1 signaling. Although complete elucidation of all PPAR regulation of chemokines and their receptors remains to be worked out, other studies have implicated PPARs in chemokine regulation in the gut, suggesting possible benefits in inflammatory bowel disease. To date, ligand-mediated trans-repression of transcription remains poorly understood, but may involve the upregulation of cytoplasmic NF-κB inhibitors such as IκK or the direct recruitment of corepressors to the transcriptional apparatus, as noted earlier (11).

PPARγ has other effects in ECs. PPARγ activation can suppress both basal levels and induced levels of insulin-stimulated or thrombin-induced expression of the potent vasoconstrictor endothelin (ET)-1 in cultured ECs (34,35). Other investigators have found that PPARγ can limit the hypertension induced by angiotensin II infusion in rats and restore EC-dependent vasodilation in response to acetylcholine (36).

These mechanisms may contribute to the modest but consistent decrease in blood pressure observed in patients treated with thiazolidinediones (37). Although causation has not been established, angiotensin receptor blockade using some of the currently available pharmacologic antagonists can bind and activate PPARγ within physiologic dose ranges (38). To define more precisely the role of PPARγ in blood pressure regulation in ECs, mice with a targeted disruption of the *PPARγ* gene in ECs (EC-γ KO) have been engineered. At baseline, these EC-γ KO animals possessed no significant phenotypical, biochemical, or hemodynamic differences as compared with wild-type animals. However, after exposure to high fat diet, the EC-γ KO mice developed an exaggerated hypertensive response (39). Treatment with the TZD rosiglitazone reversed hyperinsulinemia in the EC-γ KO animals, but did not lower their blood pressure. These data implicate EC PPARγ specifically in the pathophysiology of high-fat diet induced hypertension. In contrast, a high-salt diet provoked hypertension in both knockout and wild-type mice to the same degree. Taken together, these findings suggest that the direct activation of EC-derived PPARγ modulates the hypertensive phenotype in the setting of insulin resistance; and the data also may explain why TZD therapy in humans can modestly lower blood pressure, independently of their well-described insulin-sensitizing effects. Further studies in these genetically altered mice will likely provide other important mechanistic insights into PPARγ–dependent endothelial biology.

Other in vitro studies have revealed enhanced nitric oxide (NO) bioavailability after treating cultured ECs with various PPARγ agonists (40,41), although PPARγ stimulation did not change gene expression or protein levels of endothelial NO synthase (eNOS), which is the enzyme responsible for basal and shear stress–mediated NO production in EC. Thus, the precise mechanism for PPARγ/NO interaction remains obscure and may occur indirectly through eNOS cofactor expression, activation, or degradation. Intriguingly, laminar shear stress (LSS) activates PPARγ in ECs in vitro (42). Moreover, the conditioned media from ECs exposed to LSS also can activate PPARγ, correlating with the upregulation of *CD36* gene expression – a known PPARγ target. Shear stress represents an important anti-inflammatory stimulus in vascular EC. In these in vitro experimental systems, pretreatment of cultured ECs with the conditioned media from LSS antagonized the IL-1-β induction of intercellular adhesion molecule (ICAM)-1. Targeted knockdown of PPARγ by small interfering RNA (siRNA) partially abrogated this effect, which implicates PPARγ as one anti-inflammatory mediator of shear stress in ECs. Other, variable PPARγ effects on adhesion molecule expression have been reported (discussed later, in conjunction with PPARα). PPARγ agonists may have antiangiogenic effects (43,44), but they also have been reported to induce vascular endothelial growth factor (VEGF) in certain cells such as VSMCs, and the results in other angiogenic models have varied (45). We have found that putative, naturally occurring PPARγ ligands induce in vitro mRNA and protein levels of plasminogen activator inhibitor (PAI)-1, an important endogenous inhibitor of fibrinolysis that has been found to be predictive of risk for myocardial infarction. PPARα ligands had no such effect (28). In contrast, other in vitro reports have found that synthetic PPARγ agonists modestly downregulate basal and cytokine-induced *PAI-1* mRNA and protein expression (46–48). Finally, synthetic PPARγ agonists clearly decrease serum PAI-1 levels in patients, although these changes also may derive from the metabolic effects of these drugs, including improved insulin sensitivity, decreased glucose levels, or lowered triglycerides levels, which may contribute to these responses. This discordance in the scientific literature likely reflects the use of distinct experimental systems, the use of different PPARγ agonists (synthetic versus natural), and the complexity of dissecting the multiplicity of potential PPARγ molecular targets from in vivo models. Translational studies in humans also have reported PPARγ–mediated improvements in endothelium-dependent vasoreactivity, assessed by forearm blood flow measurements with bradykinin stimulation, although it is not clear if this is through direct endothelial versus smooth muscle cell effects (49,50). Taken together, these endothelial PPARγ effects are congruent with reports of PPARγ action in other vascular and inflammatory cells and may account for a putative role of PPARγ in limiting inflammation and atherosclerosis (51–53).

PEROXISOME PROLIFERATOR-ACTIVATED RECEPTOR-α – TRANSCRIPTIONAL SENSOR FOR FATTY ACID METABOLISM

Issemann and Grossman first identified PPARα in 1990, while screening a murine cDNA library (8). They were searching for the molecular target of a group of chemical compounds that induced peroxisomal proliferation and carcinogenesis in rodents. Peroxisomes are intracellular organelles that participate in detoxification and fatty acid oxidation in rodents. This proliferation of peroxisomes does not occur in humans (54). PPARα is expressed in a variety of tissues, including heart, liver, skeletal muscle, and kidney. PPARα plays an important role in lipid metabolism and energy balance (55). PPARα target genes include enzymes important for β-oxidation of fatty acids. Moreover, PPARα regulates many metabolic pathways, including activation of fatty acids to acyl coenzyme A derivatives, peroxisome $\beta-$oxidation, and apolipoprotein expression (A1, AII, and CIII) (1). Endogenous PPARα activators likely include eicosanoids and certain long-chain fatty acids, although as with PPARγ, many aspects of endogenous PPARα activation remain unclear. Synthetic PPARα agonists include lipid-lowering fibric acids – fenofibrate, gemfibrozil, and the research compound WY14643 (6). The relative potency among various synthetic ligands for binding to PPARα differs, which may explain the nonuniform effects of different fibric acids, as well as general biological differences among distinct PPAR agonists.

The PPARα–deficient mouse developed by Gonzalez and colleagues has been an invaluable tool for studying PPARα (56). Of note, PPARα agonists fail to induce peroxisome proliferation in PPARα–deficient animals (see earlier discussion). PPARα–deficient mice do manifest altered lipid and lipoprotein profiles including increased total cholesterol, elevated Apo-AI, and mildly increased total high-density lipoprotein (HDL) levels, the latter apparently caused by decreased HDL metabolism (57). Triglyceride levels were not significantly different between wild-type and knockout animals. Although PPARα activators lowered triglycerides in wild-type mice, they had no effect on *PPARα*-null mice, implicating PPARα directly in triglyceride lowering by fibrates. Extended to humans, the decrease in cardiovascular events with fibrate therapy observed in clinical trials such as the Veterans Affairs High-Density Lipoprotein Cholesterol Intervention Trial Study Group (VA-HIT) suggests that PPARα activation in the vasculature, including the endothelium, may contribute to benefits in patient outcomes (58,59). More recent data from the Fenofibrate Intervention and Event Lowering in Diabetes (FIELD) study with fenofibrate were less clear, perhaps in part due to a significant degree of statin drop-in treatment that occurred disproportionately more in the placebo group (60).

PEROXISOME PROLIFERATOR-ACTIVATED RECEPTOR-α IN THE ENDOTHELIUM

PPARα is expressed in vitro and in vivo in ECs, as shown by several approaches, including northern blotting of cultured ECs and immunohistochemistry of human carotid specimens (61). Multiple PPARα–regulated target genes relevant to endothelial function have been identified (62). Endothelial adhesion molecule expression is an early step in atherogenesis. We and others have found that synthetic PPARα activators (WY and fenofibrate) inhibit TNF-α– or IL-1β–mediated induction of vascular adhesion molecule (VCAM)-1 expression in cultured ECs, as well as adhesion of fluorescently labeled monocyte-like cells (61,62). PPARα–mediated repression of adhesion molecule expression may derive from NF-κB inhibition (61). The molecular mechanism underlying PPAR antagonism of NF-κB signaling remains elusive, but may involve the altered expression of the cytoplasmic inhibitor IκBα, which is known to occur in cultured VSMCs (63,64). Importantly, we have shown that these PPARα agonist responses do not occur in microvascular ECs isolated from the hearts of PPARα–deficient mice (23,65). PPARα may regulate transcription in ECs through nutritional inputs (4,66). For example, certain forms of ω-3 fatty acids can repress adhesion molecule expression and leukocyte adhesion in a PPARα–dependent manner (67). Although our group identified no effect of PPARγ activators on VCAM-1, other investigators have reported inhibition of inducible VCAM-1 and ICAM-1 in various models (68–70). Not all PPARγ agonists limit adhesion molecule expression, an observation that underscores the potential complexity of PPARs, including the impact of specific agonists and variables that may differ among experimental systems (68). Consistent with our in vitro findings, although synthetic PPARγ agonists reduced atherosclerosis in low-density lipoprotein (LDL) receptor–deficient mice fed a high-fat diet, *VCAM-1* gene expression was unchanged in control and treated animals (53). These variable findings highlight the potential impact that the specific PPAR agonists used, the concentrations employed, cell type being studied, and the methodological approach and model system used can have on results.

PPARs also have been implicated in the endothelium's defense against oxidative stress (71,72). Inoue and colleagues found that both PPARα and -γ activators induced the expression of the superoxide scavenger enzyme Cu^{2+}, Zn2^{+}-superoxide dismutase (CuZn-SOD) in cultured vascular ECs (71). PPARα and -γ activators also decreased induction of $p22^{phox}$ and $p47^{phox}$, two subunits of the superoxide generating enzyme nicotinamide adenine dinucleotide phosphate (reduced form) (NADPH) oxidase (71). PPARα also has been suggested to play a part in feedback loops that help limit oxidative stress (73,74).

PEROXISOME PROLIFERATOR-ACTIVATED RECEPTOR-δ – TRANSCRIPTIONAL SENSOR FOR FATTY-ACID METABOLISM

Although less is known about PPARδ, this is rapidly changing. Studies into PPARδ may have been influenced by the absence of a clinical PPARδ agonist in current clinical use. PPARδ is expressed widely in essentially all tissues. It was originally characterized in the skin, where it may enhance wound healing during acute inflammation by influencing keratinocyte differentiation and cell survival pathways (75). PPARδ–deficient mice are in general not viable, although occasional mice do survive to birth. These PPARδ-deficient mice demonstrate significantly smaller body habitus, less adiposity, and abnormal myelination of the nervous system (76,77). Investigators have subsequently shown that PPARδ regulates genes involved in fatty-acid metabolism in adipose tissue, skeletal muscle, cardiac muscle, and liver (78). PPARδ also has been implicated in regulating inflammation, although these data suggested distinct roles for PPARδ in the presence versus absence of ligand. PPARδ is expressed in human ECs, but its function in this setting remains obscure (78).

POTENTIAL UNTOWARD EFFECTS OF PEROXISOME PROLIFERATOR-ACTIVATED RECEPTOR-α AND -γ ACTIVATION

In contrast to these various anti-inflammatory and antiatherosclerotic responses, limited reports do suggest some potential untoward effects on the endothelium resulting from PPAR activation. Certain specific oxidized lipids might act through PPARα to induce inflammation (79,80). In this work, minimally modified LDL (MM-LDL) and oxidation products of 1-palmitoyl-2-arachidonyl-sn-glycero-3-phosphocholine (PAPC) activated cultured human ECs to synthesize MCP-1 and IL-8. The authors suggest that this effect may be mediated partly through PPARα. MM-LDL, Ox-PAPC, or its component phospholipids were capable of activating the PPARα ligand-binding domain, as well as a consensus peroxisome proliferator-activated receptor response element (PPRE) in transfected ECs. In contrast to earlier studies, these investigators found that the PPARα activator Wy14643 stimulated the synthesis of IL-8 and MCP-1 by human arterial ECs (HAECs). By contrast, troglitazone, a PPARγ agonist, blunted the induction of IL-8 and MCP-1 gene and protein expression. Ox-PAPC and MM-LDL no longer upregulated MCP-1 in *PPARα*-null mice. The reproducibility and in vivo relevance of these findings or of the compounds employed remains unclear but certainly worthy of further study. In clinical practice, the use of TZDs is associated with weight gain, peripheral edema, and heart failure in a small subset of patients, in particular those also taking insulin therapy for diabetes. However, the endothelial effects of PPARγ agonists have not been specifically implicated in the fluid retention that can develop in some patients administered these medications (81).

Interestingly, despite the extensive PPARα–mediated anti-inflammatory and antiatherosclerotic effects, studies crossing PPARα–deficient mice with ApoE-deficient mice found less, not more, atherosclerosis (82). Others report that PPARα agonists could decrease atherosclerosis in LDL receptor–deficient mice (83,84). The reasons for these differences are not clear. One contributing factor may be inherent differences between PPARs in mice versus humans. Indeed, studies of PPARα agonists in ApoE-deficient mice found a more impressive reduction in atherosclerosis when these mice also expressed a transgene for human apoA1 (83).

ENDOGENOUS PEROXISOME PROLIFERATOR-ACTIVATED RECEPTOR ACTIVATION?

As is apparent from the preceding discussion, considerable insight into PPAR biology has been derived from the identification of the synthetic PPAR agonists for each isoform. The specificity of individual agonists for each PPAR is dictated by the unique structure of both the LBD as well as the chemical structure of the agonist. The large LBD of each PPAR has been exploited in the development of therapeutic agents that can activate two (dual agonists) or all three (pan-agonists) PPARs, simply as a function of steric forces and the available space for agonist occupancy (19).

One of the major unresolved issues in the PPAR field has been the identity of specific natural PPAR agonists that certainly must exist to activate PPARs in vivo (1). Insight into these endogenous PPAR agonists would have major implications, suggesting mechanisms that might protect against the very problems synthetic PPAR agonists are used to treat, such as diabetes mellitus and dyslipidemia (19). Likewise, defects in pathways that generate natural PPAR agonists also might lead to pathological conditions. Although the prostaglandin derivative 15d-PGJ_2 has been suggested to bind to PPARγ, the presence of this molecule in vivo and its PPARγ–independent effects has tempered enthusiasm for the relevance of this interaction. Leukotriene B4 has been proposed as an agonist for PPARα (85), although again the physiological relevance of this has been unclear. Examples of other recent work in this area suggest that lysophosphatidic acid (LPA) (86), tetradecylthioacetic acid (87), and nitrolinoleic acid (88) might also be PPARγ agonists.

An alternative approach to the question of natural PPAR agonists is to identify pathways that might generate natural PPAR ligands, as opposed to pursuing specific candidate molecules. Using this tack, both our group as well as that of Ron Evans, found that lipoprotein lipase (LPL), a key enzyme in triglyceride metabolism, could act on triglyceride-rich lipoproteins to activate PPARs (65,89). Studying murine macrophages, Chawla and colleagues found that LPL interacted with very-low-density lipoproteins (VLDLs) to activate PPARδ (89). We found that LPL hydrolyzes VLDL to generate preferentially PPARα ligands, with significant but lesser effects on PPARδ and even less on PPARγ (65) (Figure 87.2). These relative responses were seen both in cell-based as well as more rigorous cell-free radioligand displacement assays. Additional studies demonstrated that LPL-mediated PPARα activation depended on intact LPL catalysis and did not occur with other lipases tested, despite the release of similar amounts of total free fatty acids (65). Extended further, LPL-treated VLDL repressed TNF-α–induced VCAM-1 induction, but not in PPARα–deficient ECs (65). This recapitulates the effects of synthetic PPARα agonists as described earlier. The novel role for LPL in limiting inflammation is an example of the insight that can be derived from probing nuclear receptors through endogenous mechanisms, as opposed to synthetic agonists. Subsequent work has provided additional support for the notion of a transcriptional network through different lipase–lipoprotein substrate interactions, for example as seen with the ability of LPL action to release not only known PPARα ligands from electronegative LDL, but also to completely reverse electronegative LDL's potent inflammatory effects (23).

Recently, this model for lipolytic PPAR activation has been extended to also include endothelial lipase (90). We initially observed that the widely reported ability of HDL to repress leukocyte adhesion and adhesion molecule expression was lost in the presence of the general lipase inhibitor tetrahydrolipstatin. Subsequent studies led to the demonstration that HDL hydrolysis by endothelial lipase was capable of PPARα activation and repression of VCAM-1 and leukocyte adhesion. These findings, although in contrast with other lines of evidence in mice that endothelial lipase may promote atherosclerosis (91), support the notion that probing PPARs with more physiologic stimuli like lipase–lipoprotein interactions may shed light on PPARs and lipases, and their role in health and disease.

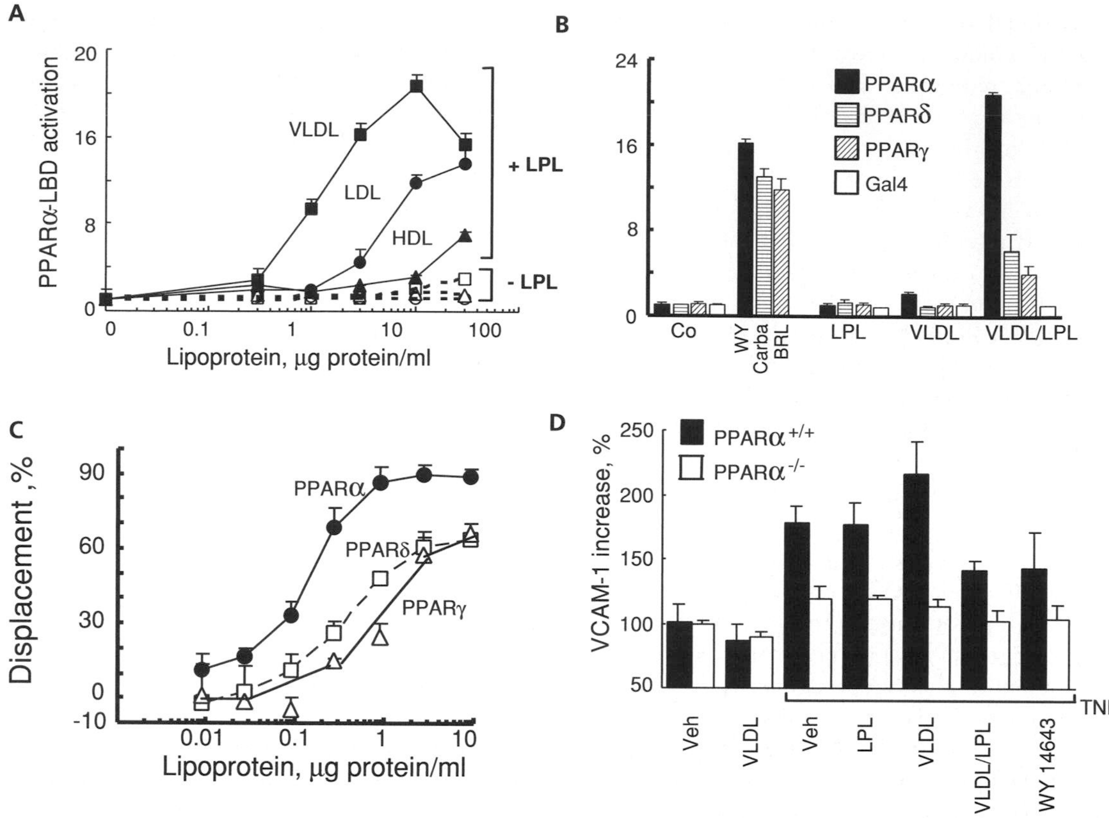

Figure 87.2. (**A**) LPL treatment of VLDL activates the PPARα ligand-binding domain (*LBD*). Standard PPARLBD assays were performed in ECs cotransfected with PPARα–LBD, the luciferase response pUASx4-TK-luc, and β-galactosidase constructs and stimulated with the lipoproteins shown in either the presence or absence of LPL (30 U/ml). (Details here and in Rossi A, Kapahi P, Natoli G, et al. Anti-inflammatory cyclopentenone prostaglandins are direct inhibitors of IkappaB kinase. *Nature.* 2000;403:103–108.) (**B**) LPL-treated VLDL (10 μg/mL) preferentially activates PPARα. LBD activation for all three known PPAR isotypes (-α, -γ,- δ) by VLDL in ECs was measured in the presence or absence of LPL using standard LBD-GAL4 assays (as in Figure 87.1), and compared to responses induced by known PPAR isotype agonists as follows: PPARγ—BRL49653 (BRL, 1 μM); PPARα—WY14643 (WY, 100 μM), PPARδ—carbaprostacyclin (carba, 10 μM). (**C**) LPL-treated VLDL generates PPAR agonists. LPL-treated VLDL was used in radioligand displacement assays. Competition curves were generated by incubating the LPL/VLDL reaction mixture with specific radiolabeled PPAR activators: 5 nM $^3H_2$ compound A with GST-hPPARα (*circles*) or GST-hPPARγ (*triangles*), or 2.5 nM $^3H_2$ compound B with GST-hPPARδ (*squares*). Radioligand displacement in the presence of the indicated concentration of VLDL (0.003–10 μg protein/mL)/LPL (200 U/mL LPL, 37°C, 1h) is plotted. (**D**) Similar to synthetic PPARα agonists, LPL-treated VLDL represses cytokine-induced VCAM-1 in wild-type, but not PPARα–deficient ECs. VCAM-1 surface expression in ECs obtained from wild-type (*black bars*) or *PPAR*$\alpha^{-/-}$ mice (*white bars*) was measured by ELISA using standard protocols. ECs cultured in 96 well plates were stimulated with murine TNF-α (10 ng/mL, 18h). Data represent absorbance at 410 nm after protein concentration normalization. Both WY14643 (100 μM) and LPL/VLDL significantly decreased VCAM-1 levels (*, p <.005) in wild-type, but not PPARα–deficient ECs. Basal VCAM-1 level in *PPAR*$\alpha^{-/-}$ is significantly increased as compared to wild-type ECs (#, p <.001), suggesting PPARα plays a basal role in repressing inflammation.

PEROXISOME PROLIFERATOR-ACTIVATED RECEPTORS – KEY MEDIATORS OF ENDOTHELIAL TRANSCRIPTIONAL RESPONSES?

PPARs can be considered transcriptional nodal points, controlling the expression of multiple gene targets in a host of different pathways. This characteristic, combined with the fact that all major PPAR isoforms are expressed in vascular ECs, suggest that PPAR regulation is potentially critical to the roles the endothelium plays in both health and disease. Thus, in addition to the possibility of PPAR activators in clinical use having indirect vascular effects – through improved insulin sensitivity or lower triglycerides and higher HDL – they may also directly influence EC function by activating PPARs present in ECs. Whether these direct effects in ECs as well as other non-metabolic settings are relevant to clinical responses remains a critical and unsettled issue. Fortunately, some insight into this area may soon be obtained, given ongoing clinical trials examining the effects of PPARγ and PPARα agonists on cardiovascular events.

KEY POINTS

- PPARs, as ligand-activated nuclear receptors or transcription factors that control cassettes of target genes involved in energy balance, lipid metabolism, insulin sensitivity, and inflammation, may play a central role in directing transcriptional responses in the endothelium.
- All three PPARs – α, γ, and δ – are expressed in the endothelium. Considerable evidence suggests the involvement of PPARα and -γ in regulating the expression of target genes important to roles the endothelium is known to play in health and disease.
- Agonists for both PPARα (lipid-lowering fibrates) and PPARγ (antidiabetic thiazolidinediones) are in current clinical use; some of the proposed benefits of these agents on the cardiovascular system, including anti-inflammatory and antiatherosclerotic effects, may derive from the direct activation of PPARs expressed in the endothelium.
- The action of both lipoprotein lipase and endothelial lipase, respectively, on VLDL and HDL, may represent endogenous mechanisms for PPAR activation, with functional consequences for endothelial responses.

Future Goals

- To establish if direct PPAR activation in the endothelium, as well as in vascular cells and inflammatory cells, can contribute to clinical cardiovascular benefits, as is currently under study
- To better understand the nature of endogenous PPAR ligands, and the clinical effects and safety of novel synthetic PPAR agonists that are in development

REFERENCES

1 Willson TM, Brown PJ, Sternbach DD, Henke BR. The PPARs: from orphan receptors to drug discovery. *J Med Chem.* 2000; 43(4):527–550.

2 Evans RM, Barish GD, Wang YX. PPARs and the complex journey to obesity. *Nat Med.* 2004;10(4):355–361.

3 Mangelsdorf D, Thummel C, Beato M, et al. The nuclear receptor superfamily: the second decade. *Cell.* 1995;83:835–839.

4 Ziouzenkova O, Plutzky J. Lipolytic PPAR activation: new insights into the intersection of triglycerides and inflammation? *Curr Opin Clin Nutr Metab Care.* 2004;7(4):369–375.

5 Plutzky J. Peroxisome proliferator-activated receptors in endothelial cell biology. *Curr Opin Lipidol.* 2001;12(5):511–518.

6 Chinetti-Gbaguidi G, Fruchart JC, Staels B. Role of the PPAR family of nuclear receptors in the regulation of metabolic and cardiovascular homeostasis: new approaches to therapy. *Curr Opin Pharmacol.* 2005;5:177–183.

7 Evans RM. The nuclear receptor superfamily: a Rosetta Stone for physiology. *Mol Endocrinol.* 2005;19(6):1429–1438.

8 Issemann I, Green S. Activation of a member of the steroid hormone receptor superfamily by peroxisome proliferators. *Nature.* 1990;347(6294):645–650.

9 Tugwood JD, Issemann I, Anderson RG, et al. The mouse peroxisome proliferator activated receptor recognizes a response element in the 5′ flanking sequence of the rat acyl CoA oxidase gene. *EMBO J.* 1992;11(2):433–439.

10 Lee CH, Olson P, Evans RM. Minireview: lipid metabolism, metabolic diseases, and peroxisome proliferator-activated receptors. *Endocrinology.* 2003;144(6):2201–2207.

11 Pascual G, Fong AL, Ogawa S, et al. A SUMOylation-dependent pathway mediates transrepression of inflammatory response genes by PPARgamma. *Nature.* 2005;437(7059):759–763.

12 Molnar F, Matilainen M, Carlberg C. Structural determinants of the agonist-independent association of human peroxisome proliferator-activated receptors with coactivators. *J Biol Chem.* 2005;280(28):26543–26556.

13 Rosen ED, Spiegelman BM. PPAR gamma: a nuclear regulator of metabolism, differentiation, and cell growth. *J Biol Chem.* 2001;17:17.

14 Tontonoz P, Graves RA, Budavari AI, et al. Adipocyte-specific transcription factor ARF6 is a heterodimeric complex of two nuclear hormone receptors, PPAR gamma and RXR alpha. *Nucleic Acids Res.* 1994;22(25):5628–5634.

15 Chawla A, Schwarz EJ, Dimaculangan DD, Lazar MA. Peroxisome proliferator-activated receptor (PPAR) gamma: adipose-predominant expression and induction early in adipocyte differentiation. *Endocrinology.* 1994;135(2):798–800.

16 Henry RR. Thiazolidinediones. *Endocrinol Metab Clin North Am.* 1997;26(3):553–573.

17 Kliewer SA, Willson TM. The nuclear receptor PPARgamma – bigger than fat. *Curr Opin Genet Dev.* 1998;8(5):576–581.

18 Barak Y, Nelson MC, Ong ES, et al. PPAR gamma is required for placental, cardiac, and adipose tissue development. *Mol Cell.* 1999;4(4):585–595.

19 Berger JP, Akiyama TE, Meinke PT. PPARs: therapeutic targets for metabolic disease. *Trends Pharmacol Sci.* 2005;26(5):244–251.

20 Kubota N, Terauchi Y, Miki H, et al. PPAR gamma mediates high-fat diet-induced adipocyte hypertrophy and insulin resistance. *Mol Cell.* 1999;4(4):597–609.

21 Tontonoz P, Nagy L. Regulation of macrophage gene expression by peroxisome-proliferator-activated receptor gamma: implications for cardiovascular disease. *Curr Opin Lipidol.* 1999;10(6): 485–490.

22 Delerive P, Furman C, Teissier E, et al. Oxidized phospholipids activate PPARalpha in a phospholipase A2- dependent manner. *FEBS Lett.* 2000;471(1):34–38.

23 Ziouzenkova O, Asatryan L, Sahady D, et al. Transcriptional PPAR responses to electronegative LDL: dual roles for lipolysis and oxidation. *J Biol Chem.* 2003;278(41):39874–39881. Epub 2003 Jul 23.

24 Kliewer SA, Lenhard JM, Willson TM, et al. A prostaglandin J2 metabolite binds peroxisome proliferator-activated receptor

gamma and promotes adipocyte differentiation. *Cell.* 1995;83(5): 813–819.
25 Forman BM, Tontonoz P, Chen J, et al. 15-Deoxy-delta 12, 14-prostaglandin J2 is a ligand for the adipocyte determination factor PPAR gamma. *Cell.* 1995;83(5):803–812.
26 Huang JT, Welch JS, Ricote M, et al. Interleukin-4-dependent production of PPARgamma ligands in macrophages by 12/15-lipoxygenase. *Nature.* 1999;400(6742):378–382.
27 Itoh H, Doi K, Tanaka T, et al. Hypertension and insulin resistance: role of peroxisome proliferator-activated receptor gamma. *Clin Exp Pharmacol Physiol.* 1999;26(7):558–560.
28 Marx N, Bourcier T, Sukhova GK, et al. PPARgamma activation in human endothelial cells increases plasminogen activator inhibitor type-1 expression: PPARgamma as a potential mediator in vascular disease. *Arterioscler Thromb Vasc Biol.* 1999;19(3): 546–551.
29 Marx N, Mach F, Sauty A, et al. Peroxisome proliferator-activated receptor-gamma activators inhibit IFN- gamma-induced expression of the T cell-active CXC chemokines IP-10, Mig, and I-TAC in human endothelial cells. *J Immunol.* 2000;164(12):6503–6508.
30 Luster AD. Chemokines – chemotactic cytokines that mediate inflammation. *N Engl J Med.* 1998;338(7):436–445.
31 Kim CH. Chemokine-chemokine receptor network in immune cell trafficking. *Curr Drug Targets Immune Endocr Metabol Disord.* 2004;4(4):343–361.
32 Terkeltaub R, Boisvert WA, Curtiss LK. Chemokines and atherosclerosis. *Curr Opin Lipidol.* 1998;9(5):397–405.
33 Kintscher U, Goetze S, Wakino S, et al. Peroxisome proliferator-activated receptor and retinoid X receptor ligands inhibit monocyte chemotactic protein-1-directed migration of monocytes. *Eur J Pharmacol.* 2000;401(3):259–270.
34 Delerive P, Martin-Nizard F, Chinetti G, et al. Peroxisome proliferator-activated receptor activators inhibit thrombin-induced endothelin-1 production in human vascular endothelial cells by inhibiting the activator protein-1 signaling pathway. *Circ Res.* 1999;85(5):394–402.
35 Satoh H, Tsukamoto K, Hashimoto Y, et al. Thiazolidinediones suppress endothelin-1 secretion from bovine vascular endothelial cells: a new possible role of PPARgamma on vascular endothelial function. *Biochem Biophys Res Commun.* 1999;254(3): 757–763.
36 Schiffrin EL, Amiri F, Benkirane K, et al. Peroxisome proliferator-activated receptors. *Hypertension.* 2003;42(4):664–668. Epub 2003 Jul 21.
37 Parulkar AA, Pendergrass ML, Granda-Ayala R, et al. Non-hypoglycemic effects of thiazolidinediones. *Ann Intern Med.* 2001;134(1):61–71.
38 Benson SC, Pershadsingh H, Ho CI, et al. Identification of telmisartan as a unique angiotensin II receptor antagonist with selective PPARgamma-modulating activity. *Hypertension.* 2004;43:993–1002.
39 Nicol CJ, Adachi M, Akiyama TE, Gonzalez FJ. PPARgamma in endothelial cells influences high fat diet-induced hypertension. *Am J Hypertens.* 2005;18(4Pt 1):549–556.
40 Calnek DS, Mazzella L, Roser S, et al. Peroxisome proliferator-activated receptor gamma ligands increase release of nitric oxide from endothelial cells. *Arterioscler Thromb Vasc Biol.* 2003;23(1): 52–57.
41 Polikandriotis JA, Mazzella LJ, Rupnow HL, Hart CM. Peroxisome proliferator-activated receptor gamma ligands stimulate endothelial nitric oxide production through distinct peroxisome proliferator-activated receptor gamma-dependent mechanisms. *Arterioscler Thromb Vasc Biol.* 2005;25(9):1810–1816.
42 Liu Y, Zhu Y, Rannou F, et al. Laminar flow activates peroxisome proliferator-activated receptor-gamma in vascular endothelial cells. *Circulation.* 2004;110(9):1128–1133.
43 Xin X, Yang S, Kowalski J, Gerritsen ME. Peroxisome proliferator-activated receptor gamma ligands are potent inhibitors of angiogenesis in vitro and in vivo. *J Biol Chem.* 1999;274(13):9116–9121.
44 Murata T, He S, Hangai M, et al. Peroxisome proliferator-activated receptor-gamma ligands inhibit choroidal neovascularization. *Invest Ophthalmol Vis Sci.* 2000;41(8):2309–2317.
45 Yamakawa K, Hosoi M, Koyama H, et al. Peroxisome proliferator-activated receptor-gamma agonists increase vascular endothelial growth factor expression in human vascular smooth muscle cells. *Biochem Biophys Res Commun.* 2000;271(3):571–574
46 Kato K, Satoh H, Endo Y, et al. Thiazolidinediones down-regulate plasminogen activator inhibitor type 1 expression in human vascular endothelial cells: a possible role for PPARgamma in endothelial function. *Biochem Biophys Res Commun.* 1999; 258(2):431–435.
47 Ma LJ, Mao SL, Taylor KL, et al. Prevention of obesity and insulin resistance in mice lacking plasminogen activator inhibitor 1. *Diabetes.* 2004;53(2):336–346.
48 Zirlik A, Leugers A, Lohrmann J, et al. Direct attenuation of plasminogen activator inhibitor type-1 expression in human adipose tissue by thiazolidinediones. *Thromb Haemost.* 2004;91(4):674–682.
49 Voytovich MH, Simonsen C, Jenssen T, et al. Short-term treatment with rosiglitazone improves glucose tolerance, insulin sensitivity and endothelial function in renal transplant recipients. *Nephrol Dial Transplant.* 2005;20(2):413–418. Epub 2004 Dec 22.
50 Campia U, Matuskey LA, et al. Peroxisome proliferator-activated receptor-gamma activation with pioglitazone improves endothelium-dependent dilation in nondiabetic patients with major cardiovascular risk factors. *Circulation.* 2006;113(6):867–875.
51 Haffner SM, Greenberg AS, Weston WM, et al. Effect of rosiglitazone treatment on nontraditional markers of cardiovascular disease in patients with type 2 diabetes mellitus. *Circulation.* 2002; 106(6):679–684.
52 Mohanty P, Aljada A, Ghanim H, et al. Evidence for a potent antiinflammatory effect of rosiglitazone. *J Clin Endocrinol Metab.* 2004;89(6):2728–2735. 2005;20(2):413–418.
53 Li AC, Brown KK, Silvestre MJ, et al. Peroxisome proliferator-activated receptor gamma ligands inhibit development of atherosclerosis in LDL receptor-deficient mice. *J Clin Invest.* 2000; 106(4):523–531.
54 Cattley RC, DeLuca J, Elcombe C, et al. Do peroxisome proliferating compounds pose a hepatocarcinogenic hazard to humans? *Regul Toxicol Pharmacol.* 1998;27(1 Pt 1):47–60.
55 Fruchart JC, Duriez P, Staels B. Peroxisome proliferator-activated receptor-alpha activators regulate genes governing lipoprotein metabolism, vascular inflammation and atherosclerosis. *Curr Opin Lipidol.* 1999;10(3):245–257.
56 Lee SS, Pineau T, Drago J, et al. Targeted disruption of the alpha isoform of the peroxisome proliferator-activated receptor gene in mice results in abolishment of the pleiotropic effects of peroxisome proliferators. *Mol Cell Biol.* 1995;15(6):3012–3022.

57 Peters JM, Hennuyer N, Staels B, et al. Alterations in lipoprotein metabolism in peroxisome proliferator- activated receptor alpha-deficient mice. *J Biol Chem.* 1997;272(43):27307–27312.
58 Rubins HB, Robins SJ, Collins D, et al. Gemfibrozil for the secondary prevention of coronary heart disease in men with low levels of high-density lipoprotein cholesterol. Veterans Affairs High-Density Lipoprotein Cholesterol Intervention Trial Study Group. *N Engl J Med.* 1999;341(6):410–418.
59 Plutzky J. Peroxisome proliferator-activated receptors as therapeutic targets in inflammation. *J Am Coll Cardiol.* 2003;42(10): 1764–1766.
60 The FIELD Investigators. Effects of long-term fenobrate therapy on cardiovascular events in 9795 people with type 2 diabetes mellitus (the FIELD study): randomized controlled trial. *Lancet.* 2005;366:1849–1861.
61 Marx N, Sukhova GK, Collins T, et al. PPARalpha activators inhibit cytokine-induced vascular cell adhesion molecule-1 expression in human endothelial cells. *Circulation.* 1999;99(24): 3125–3131.
62 Duez H, Fruchart JC, Staels B. PPARS in inflammation, atherosclerosis and thrombosis. *J Cardiovasc Risk.* 2001;8(4): 187–194.
63 Delerive P, De Bosscher K, Vanden Berghe W, et al. DNA binding-independent induction of IkappaBalpha gene transcription by PPARalpha. *Mol Endocrinol.* 2002;16(5):1029–1039.
64 Rossi A, Kapahi P, Natoli G, et al. Anti-inflammatory cyclopentenone prostaglandins are direct inhibitors of IkappaB kinase. *Nature.* 2000;403(6765):103–108.
65 Ziouzenkova O, Perrey S, Asatryan L, et al. Lipolysis of triglyceride-rich lipoproteins generates PPAR ligands: Evidence for an antiinflammatory role for lipoprotein lipase. *Proc Natl Acad Sci USA.* 2003;100(5):2730–2735. Epub 2003 Feb 26.
66 Jump DB, Clarke SD. Regulation of gene expression by dietary fat. *Annu Rev Nutr.* 1999;19:63–90.
67 Sethi S, Ziouzenkova O, Ni H, et al. Oxidized omega-3 fatty acids in fish oil inhibit leukocyte-endothelial interactions through activation of PPAR alpha. *Blood.* 2002;100(4):1340–1346.
68 Jackson SM, Parhami F, Xi XP, et al. Peroxisome proliferator-activated receptor activators target human endothelial cells to inhibit leukocyte-endothelial cell interaction. *Arterioscler Thromb Vasc Biol.* 1999;19(9):2094–2104.
69 Pasceri V, Wu HD, Willerson JT, Yeh ET. Modulation of vascular inflammation in vitro and in vivo by peroxisome proliferator-activated receptor-gamma activators. *Circulation.* 2000;101(3): 235–238.
70 Nakajima A, Wada K, Miki H, et al. Endogenous PPAR gamma mediates anti-inflammatory activity in murine ischemia-reperfusion injury. *Gastroenterology.* 2001;120(2):460–469.
71 Inoue I, Goto S, Matsunaga T, et al. The ligands/activators for peroxisome proliferator-activated receptor alpha (PPARalpha) and PPARgamma increase Cu2+,Zn2+-superoxide dismutase and decrease p22phox message expressions in primary endothelial cells. *Metabolism.* 2001;50(1):3–11.
72 Teissier E, Nohara A, Chinetti G, et al. Peroxisome proliferator-activated receptor alpha induces NADPH oxidase activity in macrophages, leading to the generation of LDL with PPAR alpha activation properties. *Circ Res.* 2004;95(12):1174–1182.
73 Kunsch C, Medford RM. Oxidative stress as a regulator of gene expression in the vasculature. *Circ Res.* 1999;85(8):753–766.
74 Devchand PR, Ziouzenkova O, Plutzky J. Oxidative stress and peroxisome proliferator-activated receptors: reversing the curse? *Circ Res.* 2004;95(12):1137–1139.
75 Michalik L, Desvergne B, Wahli W. Peroxisome proliferator-activated receptors beta/delta: emerging roles for a previously neglected third family member. *Curr Opin Lipidol.* 2003;14(2): 129–135.
76 Barak Y, Liao D, He W, et al. Effects of peroxisome proliferator-activated receptor delta on placentation, adiposity, and colorectal cancer. *Proc Natl Acad Sci USA.* 2002;99:303–308.
77 Peters JM., Lee SS, Li W, et al. Growth, adipose, brain, and skin alterations resulting from targeted disruption of the mouse peroxisome proliferator-activated receptor β. *Mol Cell Biol.* 2000; 20:5119–5128.
78 Berger JP, Akiyama TE, Meinke PT. PPARs: therapeutic targets for metabolic disease. *Trends Pharmacol Sci.* 2005;26(5):244–251.
79 Han KH, Chang MK, Boullier A, et al. Oxidized LDL reduces monocyte CCR2 expression through pathways involving peroxisome proliferator-activated receptor gamma. *J Clin Invest.* 2000;106(6):793–802.
80 Lee H, Shi W, Tontonoz P, et al. Role for peroxisome proliferator-activated receptor alpha in oxidized phospholipid-induced synthesis of monocyte chemotactic protein-1 and interleukin-8 by endothelial cells. *Circ Res.* 2000;87(6):516–521.
81 Renings AJ, Smits P, Stewart MW, Tack CJ. Fluid retention and vascular effects of rosiglitazone in obese, insulin-resistant, nondiabetic subjects. *Diabetes Care.* 2006;29:581–587.
82 Tordjman K, Bernal-Mizrachi C, Zemany L, et al. PPARalpha deficiency reduces insulin resistance and atherosclerosis in apoE-null mice. *J Clin Invest.* 2001;107(8):1025–1034.
83 Duez H, Chao YS, Hernandez M, et al. Reduction of atherosclerosis by the PPARalpha agonist fenofibrate in mice. *J Biol Chem.* 2002;107(1):1.
84 Li AC, Binder CJ, Gutierrez A, et al. Differential inhibition of macrophage foam-cell formation and atherosclerosis in mice by PPARalpha, beta/delta, and gamma. *J Clin Invest.* 2004;114(11): 1564–1576.
85 Devchand PR, Keller H, Peters JM, et al. The PPARalpha-leukotriene B4 pathway to inflammation control. *Nature.* 1996; 384(6604):39–43.
86 McIntyre TM, Pontsler AV, Silva AR, et al. Identification of an intracellular receptor for lysophosphatidic acid (LPA): LPA is a transcellular PPARgamma agonist. *Proc Natl Acad Sci USA.* 2003; 100(1):131–136.
87 Dyroy E, Yndestad A, Ueland T, et al. Antiinflammatory effects of tetradecylthioacetic acid involve both peroxisome proliferator-activated receptor alpha-dependent and -independent pathways. *Arterioscler Thromb Vasc Biol.* 2005;25(7):1364–1369.
88 Schopfer FJ, Lin Y, Baker PR, et al. Nitrolinoleic acid: an endogenous peroxisome proliferator-activated receptor gamma ligand. *Proc Natl Acad Sci USA.* 2005;102(7):2340–2345.
89 Chawla A, Lee CH, Barak Y, et al. PPARdelta is a very low-density lipoprotein sensor in macrophages. *Proc Natl Acad Sci USA.* 2003;100(3):1268–1273.
90 Ahmed W, Orasanu G, Nehra V et al. High-density lipoprotein hydrolysis by endothelial lipase activates PPARalpha: a candidate mechanism for high-density lipoprotein-mediated repression of leukocyte adhesion. *Circ Res.* 2006;98:490–498.
91 Ishida T, Choi SY, Kundu RK et al. Endothelial lipase modulates susceptibility to atherosclerosis in apolipoprotein-E-deficient mice. *J Biol Chem.* 2003;279:45085–45092.

88

GATA Transcription Factors

Takashi Minami

The Research Center for Advanced Science and Technology, The University of Tokyo, Japan

The GATA family of transcription factors are zinc finger proteins that bind directly to the consensus motif (T/A)GATA (A/G). GATA factors are highly conserved throughout evolution. The mammalian family consists of six members, GATA-1 to -6. The proteins have been classified into two subgroups, according to sequence homology and expression patterns: (a) GATA-1, -2, and -3, which are expressed predominantly in hematopoietic cells; and (b) GATA-4, -5, and -6, which are primarily expressed in heart, gastrointestinal tract, and embryonic endoderm (1,2). Endothelial cells (ECs) have been shown to express GATA-2, -3, and -6. With the exception of GATA-5, disruption of GATA factors in mice results in embryonic lethality. This chapter reviews the role for these proteins in EC biology (summarized in Table 88-1).

GATA-2

GATA-2 is widely considered to be a critical mediator of hematopoietic cells. Although it is expressed in ECs, its role in this cell type is less well defined. Several in vitro studies have demonstrated a role for GATA-2 in mediating basal gene expression in ECs. In the early 1990s, two groups independently cloned human GATA-2 and reported that GATA-2 transactivates the *pre-proendothelin-1* promoter in bovine aortic ECs (3,4). Additional studies have implicated GATA-2 in the basal regulation of von Willebrand factor (vWF) (5), utrophin B (6), platelet-endothelial cell adhesion molecule (PECAM)-1 (7,8), Tie2 (9), and endothelial nitric oxide synthase (eNOS) (10).

In addition to controlling basal expression of EC-specific genes, GATA-2 also serves as an immediate early gene, coupling changes in the extracellular environment to an alteration in EC phenotype. For example, GATA-2 is involved in mediating (a) vascular endothelial growth factor (VEGF)- and thrombin-mediated induction of the angiogenesis-related negative feedback factor, Down syndrome critical region (DSCR)-1 (11); (b) extracellular matrix-mediated induction of matrix metalloprotease (MMP)-2 (12); (c) transforming growth factor (TGF)-β1–inhibitable KDR/FLK-1 (also known as VEGFR2) expression (13); and (d) thrombin- and native LDL-mediated induction of vascular cell adhesion molecule (VCAM)-1 (14–16).

Null mutation of *GATA-2* is embryonically lethal (E10.5) owing to failure of expansion of early hematopoietic progenitor cells (17). Further evidence for the role of GATA-2 in the hematopoietic stem cell pool was derived from studies of chimeric mice and embryoid bodies (17,18). A recent study suggests an important role for GATA-2 in the induction and expansion of the first hematopoietic stem cells in the aorta-gonads-mesonephros (AGM) region. Although GATA-2 is expressed in the aortic endothelium and neighboring mesenchymal cells in the AGM at E11.5, the role for GATA-2 in the developing vasculature and intact endothelium remains to be determined.

GATA-3

GATA-3 is most often implicated in T-cell development (19). However, recent studies from our laboratory point to a potentially important role for this transcription factor in ECs. Using DNA microarrays and RNase protection assays, we have shown that *GATA-3* mRNA is expressed at similar or higher levels than GATA-2 in human umbilical vein ECs (HUVECs). Moreover, incubation of HUVECs with inflammatory mediators, such as thrombin and tumor necrosis factor (TNF)-α, resulted in a downregulation of GATA-3, but not GATA-2, expression (20). Interestingly, GATA-3 is differentially expressed in different types of cultured ECs. For example, compared with HUVECs, GATA-3 expression is lower in human pulmonary artery ECs (HPAECs), although it is undetectable in dermal microvascular ECs (HDMECs) (Figure 88.1).

To date, no published reports exist of GATA-3 target genes in ECs. To gain insights into the GATA-3–dependent transcriptional program, we carried out DNA microarrays of HUVECs infected with adenovirus overexpressing GATA-2 or GATA-3. A total of 60 genes were selectively altered (more than twofold change in duplicate arrays) by GATA-3 (Figure 88.2; for color reproduction, see Color Plate 88.2). Nine of these genes encode

Table 88-1: Putative GATA Target Genes in ECs

	Gene Name	*Regulation Type*	*Stimulus*
GATA-2	*ET-1*	Constitutive	
	vWF	Constitutive	
	utrophin B	Constitutive	
	PECAM-1 (CD31)	Constitutive	
	eNOS (NOS3)	Inducible	Glucocorticoid
	DSCR-1	Inducible	VEGF or thrombin
	MMP-2	Inducible	Collagen gel culture
	KDR/FLK-1	Inducible	TGF-β
	VCAM-1	Inducible	Thrombin
	ICAM-2	Constitutive	
	Tie2	Constitutive	
GATA-6	*vWF*	Constitutive	
	VCAM-1	Inducible	TNF-α
	u-PA	Inducible	Laminar shear

proteins related to cell morphology (e.g., cytoskeleton). Interestingly, the angiogenic inducer Cyr 61 and placental growth factor were upregulated significantly with GATA-3 but not GATA-2. A need exists for further studies (e.g., small interfering RNA [siRNA]-mediated knockdown of GATA-3 in vitro and conditional EC-specific knockout of GATA-3 in vivo) to elucidate the role for GATA-3 in EC biology.

GATA-6

GATA-6 plays an important role in endoderm differentiation (21). In adults, GATA-6 is expressed predominantly in adult heart, aorta, stomach, small intestine, and bladder (21). GATA-6 has been localized in myocardium, lung epithelium, and vascular smooth muscle cells (VSMCs), and has been shown to regulate lung- and heart-specific gene expression. A paucity of data describes a role for GATA-6 in ECs. In our microarray and real-time PCR analyses, GATA-6 is expressed constitutively in cultured ECs (HUVECs, HPAECs, and human coronary artery ECs [HCAECs]), but not in HDMECs (database from http://www.lsbm.org). Indeed, GATA-6 was shown to mediate basal expression of vWF (22) in HUVECs. In HCAECs, laminar flow (fluid shear stress) was reported to upregulate the *urokinase-type plasminogen activator (u-PA)* gene via inducible binding of GATA-6 (23). Others have shown that TNF-α treatment increases the binding of GATA-6 to the *VCAM-1* promoter in cultured ECs (24).

POST-TRANSCRIPTIONAL MODIFICATION

Extracellular signals, such as thrombin, promote the transcriptional activity of GATA-2 without inducing GATA-2 mRNA expression. Thus, GATA-2 must be subject to post-transcriptional modification. Inducible phosphorylation has been described for GATA-1 in hematopoietic progenitor cells (25), GATA-2 in hematopoietic cells (26), GATA-3 in T cells (25), and GATA-4 in cardiac myocytes (27,28). The extent to which agonist-mediated induction of GATA-2 activity in ECs is mediated by phosphorylation of the transcription factor has not been determined. We have found that treatment of GATA-2 with phosphatase inhibits thrombin-mediated induction of the transcription factor to the *VCAM-1* promoter (T. Minami and W.C. Aird, unpublished observations), arguing against a role for GATA phosphorylation in the thrombin–VCAM-1 signaling axis.

GATA factors also may undergo acetylation of lysine residues. For example, it has been reported that CREB-binding protein (CBP/p300)-mediated acetylation of GATA-1, -2, and -3 results in increased DNA binding activity and transactivation (29). It will be interesting to determine whether acetylation of GATA and associated chromatin remodeling plays a role in EC signal transduction.

More recently, it was shown that GATA-2 is covalently modified by small ubiquitin-like modifier (SUMO)-1 and -2 (30). In that study, PIASy, a member of the Protein Inhibitor of Activated Stat (PIAS) family of regulators of STAT function, was shown to enhance SUMO (particularly SUMO-2) conjugation to GATA-2, leading to suppression of GATA-2–dependent endothelin (*ET*)-*1* promoter activity. Expression of

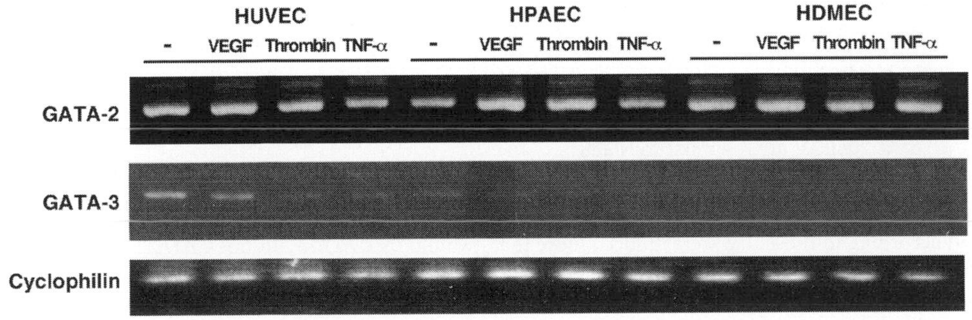

Figure 88.1. Cell subtype-specific differences in constitutive and inducible GATA expression. Reverse transcription PCR was performed with human GATA-2 or GATA-3–specific primers. Total RNA was purified from each designated ECs in the presence of VEGF (50 ng/mL), thrombin (1.5 μ/mL) or TNF-α (10 ng/mL). Arrows indicate amplified *GATA-2* or *GATA-3* gene. Cyclophilin was used as a control, showing the integrity and concentration of the isolated total RNA.

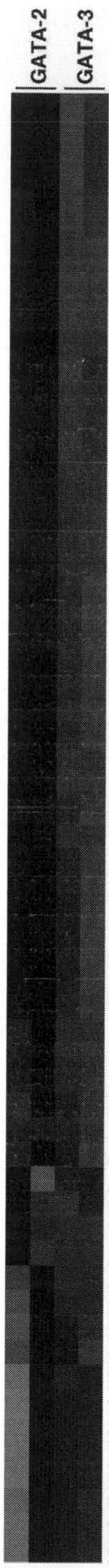

probe	gene description
221258_s_at	kinesin family member 18A
215506_s_at	ras homolog gene family, member I
204011_at	sprouty homolog 2 (Drosophila)
225081_s_at	transcription factor RAM2
212719_at	pleckstrin homology domain containing family E member 1
235399_at	kinesin light chain 2-like
201896_s_at	CDC28 protein kinase regulatory subunit 1B
222396_at	hematological and neurological expressed 1
243405_at	homo sapiens transcribed sequences
208776_at	proteasome 26S subunit, non-ATPase, 11
218412_s_at	GTF2I repeat domain containing 1
209652_s_at	placental growth factor, PIGF
219544_at	hypothetical protein FLJ22624
209403_at	TBC1 domain family, member 3
232027_at	tight junction protein 2 (zona occludiens 2)
210252_s_at	MAP-kinase activating death domain
203734_at	FOXJ2 forkhead factor
202788_at	MAP-kinase-activated protein kinase 3
218606_at	zinc finger, DHHC domain containing 7
200838_at	cathepsin B
214946_x_at	hypothetical protein FLJ10824
212370_x_at	hypothetical protein FLJ10633
211068_x_at	KIAA0592 protein
204444_at	kinesin family member 11
235545_at	hypothetical protein FLJ20354
226374_at	homo sapiens transcribed sequence with weak similarity ref:NP_055301
211926_s_at	myosin, heavy polypeptide 9, non-muscle
225655_at	ubiquitin-like, containing PHD and RING finger domain, 1
227455_at	chromosome 6 open reading frame 136
205235_s_at	M-phase phosphoprotein 1
219058_x_at	lipocalin 7
212021_s_at	antigen identified by monoclonal antibody Ki-67
203145_at	sperm associated antigen 5
208051_s_at	polyadenylate binding protein-interacting protein 1
226936_at	homo sapiens cDNA clone IMAGE:4448513
208808_s_at	high-mobility group box 2
242260_at	matrin 3
227249_at	nudE nuclear distribution gene E homolog 1
228069_at	DUF729 domain containing 1
201675_at	A kinase (PRKA) anchor protein 1
221754_s_at	coronin, actin binding protein, 1B
227918_s_at	hypothetical protein FLJ13456
204887_s_at	serine/threonine kinase 18
225608_at	similar to RIKEN cDNA 4933437K13
210764_s_at	cysteine-rich, angiogenic inducer, 61, Cyr 61
224580_at	solute carrier family 38, member 1
219004_s_at	chromosome 21 open reading frame 45
219105_x_at	origin recognitioncomplex, subunit 6 homolog-like (yeast)
211913_s_at	c-mer proto-oncogene tyrosine kinase
208613_s_at	filamin B, beta
201663_s_at	SMC4 structural maintenance of chromosome 4-like 1
225687_at	chromosome 20 open reading frame 129
244324_at	hypothetical protein MGC33382
212714_at	c-Mpl binding protein
215691_x_at	homo sapiens, clone IMAGE:5303990
229442_at	hypothetical protein MGC33382
227801_at	tumor suppressor TSBF1
212929_s_at	KIAA0592 protein
203235_at	thimet oligopeptidase 1
208903_at	similar to NG28

Figure 88.2. Color mapping with GATA-3–selective inducible genes. Microarray (Affymetrix U133 plus) analyses were performed in duplicate using HUVECs overexpressing GATA-2 or GATA-3. Genes were filtered according to those that were induced more than twofold by GATA-3 but not GATA-2. Selected genes were analyzed using the GeneTree program. Dark and light panels represent higher and lower expression, respectively, compared with the median for that particular gene. For color reproduction, see Color Plate 88.2.

PlASy is higher in ECs compared with VSMCs and is induced by fibroblast growth factor (FGF) and VEGF (30).

PROTEIN–PROTEIN INTERACTIONS

The protein–protein interactions of GATA are summarized in Table 88-2.

Nonendothelial Cells

GATA-1, -2, and -3

GATA-1 may bind CBP, resulting in transcriptional synergy (31). GATA-1–LMO2 interactions have been shown to play a key role in regulating erythroid gene expression (32). In megakaryocytes and erythroid cells, GATA-1 physically interacts with the Ets family member, Fli-1, and results in the synergistic activation of megakaryocyte-specific promoters (33). A family of proteins, called Friend of GATA (FOG) have been described that mediate both activation and repression of GATA-1, -2, and -3 target genes (1).

In thyrotropes, Pit-1 (a homeodomain protein) and GATA-2 form stable protein–protein interactions, bind to the *TSH* promoter in a ternary complex, and synergistically activate transcription (34). Pit-1 physically interacts with the second zinc finger of GATA-2. Yet another coactivator, MED220, recently has been shown to interact with the Pit-1–GATA-2 complex (35). In hematopoietic cells, GATA-2 interacts directly with the retinoic acid (RA) receptor-α, rendering the transcriptional activity of GATA RA-responsive (29).

GATA-4, -5, and -6

Nkx2.1 and Nkx2.5 interact with members of the GATA-4, -5, and -6 subfamily and synergistically act to induce expression of cardiac genes (36,37). GATA-4 and -6 interact with the

Table 88-2: Transcription Factors or Cofactors Associated with GATA Proteins

Category		*Name*	*Effect of the Target Genes*
Transcription factors	Zinc finger type family	SP1/3	Positive
		EKLF	Positive
		RBTN-2/LMO-2	Positive
		Estrogen receptor	Positive
		RARα	Positive
		RXR	Negative
		GATA itself	Positive
	Homeobox family	Hex	Negative
		HNF1a	Positive
		Nkx2.5	Positive
		Pit-1	Positive
		TTF-1	Positive
	Ets family	PU.1	Negative
		Elf-1	Positive
	MADS box	SRF	Positive
	Runt domain	AML-1	Positive
	Other	AP-1	Positive
		NF-ATc	Positive
		Smad3	Positive
		Nef (HIV)	Unknown
Cofactors	HDAC	HDAC3	Negative
		HDAC5	Negative
	FOG	FOG-1	Negative
		FOG-2	Negative
	CBP	CBP/p300	Positive
	Basal	Basal factors	Negative

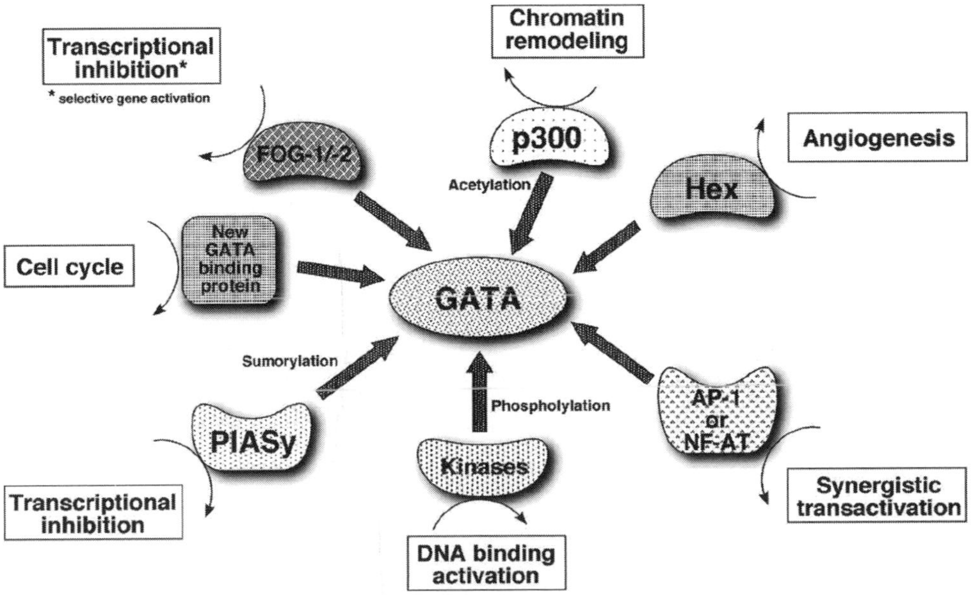

Figure 88.3. Model of GATA function.

cysteine-rich protein (CRP) family of LIM domain proteins to regulate expression in cardiac and vascular smooth muscle (38). GATA-6 interacts with LMCD1/Dyxin. In contrast to most other protein–protein interactions involving GATA factors, which synergistically activate transcription, GATA-6–LMCD1/Dyxin interactions repress GATA-6–dependent gene expression in cardiac cells (39). In VSMCs, p300 is a coactivator for GATA-6 (27). Thyroid transcription factor-1 interacts with the zinc fingers of GATA-6 to regulate the *surfactant protein-C* promoter (40).

Endothelial Cells

From a functional standpoint, GATA-mediated transactivation of genes in ECs has been shown to be enhanced or repressed by a number of proteins (Figure 88.3) . For example, GATA-2 synergizes with AP-1 to induce ET-1 expression in ECs (41). VEGF- and thrombin-mediated induction of DSCR-1 involves the cooperative binding of NF-ATc and GATA-2 and -3 to neighboring consensus motifs in the upstream promoter (11). Recently, we reported that GATA-2 physically interacts with Hex and this interaction inhibits GATA-2–mediated expression of *KDR/FLK-1* gene expression incultured ECs (42).

KEY POINTS

- GATA transcription factors play an important role in mediating basal and modulatable gene expression in ECs.
- ECs express GATA-2, -3, and -6.
- Different vascular beds indicated separate GATA expression patterns.
- GATA proteins are signal transducers, coupling changes in the extracellular environment to changes in gene expression.
- GATA transcription factors are modulated at a post-transcriptional level.

Future Goals

- To determine the roles for GATA factors in EC biology in vivo
- To elucidate the mechanisms of post-transcriptional regulation of GATA proteins in ECs

REFERENCES

1 Cantor AB, Orkin SH. Coregulation of GATA factors by the friend of GATA (FOG) family of multitype zinc finger proteins. *Semin Cell Dev Biol.* 2005;16(1):117–128.

2 Temsah R, Nemer M. GATA factors and transcriptional regulation of cardiac natriuretic peptide genes. *Regul Pept.* 2005; 128(3):177–185.

3 Dorfman DM, Wilson DB, Bruns GA, Orkin SH. Human transcription factor GATA-2. Evidence for regulation of preproendothelin-1 gene expression in ECs. *J Biol Chem.* 1992; 267(2):1279–1285.

4 Lee ME, Bloch KD, Clifford JA, Quertermous T. Functional analysis of the endothelin-1 gene promoter. Evidence for an endothelial cell-specific cis-acting sequence. *J Biol Chem.* 1990;265(18):1046–1050.

5 Jahroudi N, Lynch DC. Endothelial-cell-specific regulation of von Willebrand factor gene expression. *Mol Cell Biol.* 1994;14(2): 999–1008.

6 Perkins KJ, Davies KE. Ets, Ap-1 and GATA factor families regulate the utrophin B promoter: potential regulatory mechanisms for endothelial-specific expression. *FEBS Lett.* 2003;538(1–3): 168–172.

7 Almendro N, Bellon T, Rius C, et al. Cloning of the human platelet endothelial cell adhesion molecule-1 promoter and its tissue-specific expression. Structural and functional characterization. *J Immunol.* 1996;157(12):5411–5421.

8 Gumina RJ, Kirschbaum NE, Piotrowski K, Newman PJ. Characterization of the human platelet/endothelial cell adhesion molecule-1 promoter: identification of a GATA-2 binding element required for optimal transcriptional activity. *Blood.* 1997;89(4):1260–1269.

9 Minami T, Kuivenhoven JA, Evans V, et al. Ets motifs are necessary for endothelial cell-specific expression of a 723-bp tie-2 promoter/enhancer in hprt targeted transgenic mice. *Arterioscler Thromb Vasc Biol.* 2003;23(11):2041–2047.

10 Zhang R, Min W, Sessa WC. Functional analysis of the human endothelial nitric oxide synthase promoter. Sp1 and GATA factors are necessary for basal transcription in endothelial cells. *J Biol Chem.* 1995;270(25):15320–15326.

11 Minami T, Horiuchi K, Miura M, et al. Vascular endothelial growth factor- and thrombin-induced termination factor, down syndrome critical region-1, attenuates endothelial cell proliferation and angiogenesis. *J Biol Chem.* 2004;279(48):50537–50554.

12 Han X, Boyd PJ, Colgan S, et al. Transcriptional up-regulation of endothelial cell matrix metalloproteinase-2 in response to extracellular cues involves GATA-2. *J Biol Chem.* 2003;278(48):47785–47791.

13 Minami T, Rosenberg RD, Aird WC. Transforming growth factor-beta 1-mediated inhibition of the flk-1/KDR gene is mediated by a 5′-untranslated region palindromic GATA site. *J Biol Chem.* 2001;276(7):5395–5402.

14 Minami T, Aird WC. Thrombin stimulation of the vascular cell adhesion molecule-1 promoter in endothelial cells is mediated by tandem nuclear factor-kappa B and GATA motifs. *J Biol Chem.* 2001;276(50):47632–47641.

15 Minami T, Abid MR, Zhang J, et al. Thrombin stimulation of vascular adhesion molecule-1 in endothelial cells is mediated by protein kinase C (PKC)-delta-NF-kappa B and PKC- zeta-GATA signaling pathways. *J Biol Chem.* 2003;278(9):6976–6984.

16 Lin JH, Zhu Y, Liao HL, et al. Induction of vascular cell adhesion molecule-1 by low-density lipoprotein. *Atherosclerosis.* 1996;127(2):185–194.

17 Tsai FY, Keller G, Kuo FC, et al. An early haematopoietic defect in mice lacking the transcription factor GATA-2. *Nature.* 1994; 371(6494):221–226.

18 Tsai FY, Orkin SH. Transcription factor GATA-2 is required for proliferation/survival of early hematopoietic cells and mast cell

formation, but not for erythroid and myeloid terminal differentiation. *Blood.* 1997;89(10):3636–3643.
19 Ting CN, Olson MC, Barton KP, Leiden JM. Transcription factor GATA-3 is required for development of the T-cell lineage. *Nature.* 1996;384(6608):474–478.
20 Minami T, Sugiyama A, Wu SQ, et al. Thrombin and phenotypic modulation of the endothelium. *Arterioscler Thromb Vasc Biol.* 2004;24(1):41–53.
21 Maeda M, Ohashi K, Ohashi-Kobayashi A. Further extension of mammalian GATA-6. *Dev Growth Differ.* 2005;47(9):591–600.
22 Peng Y, Jahroudi N. The NFY transcription factor inhibits von Willebrand factor promoter activation in non-endothelial cells through recruitment of histone deacetylases. *J Biol Chem.* 2003;278(10):8385–8394.
23 Sokabe T, Yamamoto K, Ohura N, et al. Differential regulation of urokinase-type plasminogen activator expression by fluid shear stress in human coronary artery ECs. *Am J Physiol Heart Circ Physiol.* 2004;287(5):H2027–H2034.
24 Umetani M, Mataki C, Minegishi N, et al. Function of GATA transcription factors in induction of endothelial vascular cell adhesion molecule-1 by tumor necrosis factor-alpha. *Arterioscler Thromb Vasc Biol.* 2001;21(6):917–922.
25 Chen CH, Zhang DH, LaPorte JM, Ray A. Cyclic AMP activates p38 mitogen-activated protein kinase in Th2 cells: phosphorylation of GATA-3 and stimulation of Th2 cytokine gene expression. *J Immunol.* 2000;165(10):5597–5605.
26 Towatari M, May GE, Marais R, et al. Regulation of GATA-2 phosphorylation by mitogen-activated protein kinase and interleukin-3. *J Biol Chem.* 1995;270(8):4101–4107.
27 Morimoto T, Hasegawa K, Kaburagi S, et al. Phosphorylation of GATA-4 is involved in alpha 1-adrenergic agonist-responsive transcription of the endothelin-1 gene in cardiac myocytes. *J Biol Chem.* 2000;275(18):13721–13726.
28 Kitta K, Day RM, Kim Y, et al. Hepatocyte growth factor induces GATA-4 phosphorylation and cell survival in cardiac muscle cells. *J Biol Chem.* 2003;278(7):4705–4712.
29 Hayakawa F, Towatari M, Ozawa Y, et al. Functional regulation of GATA-2 by acetylation. *J Leukoc Biol.* 2004;75(3):529–540.
30 Chun TH, Itoh H, Subramanian L, et al. Modification of GATA-2 transcriptional activity in endothelial cells by the SUMO E3 ligase PIASy. *Circ Res.* 2003;92(11):1201–1208.
31 Hung HL, Lau J, Kim AY, et al. CREB-Binding protein acetylates hematopoietic transcription factor GATA-1 at functionally important sites. *Mol Cell Biol.* 1999;19(5):3496–3505.
32 Wadman IA, Osada H, Grutz GG, et al. The LIM-only protein Lmo2 is a bridging molecule assembling an erythroid, DNA-binding complex which includes the TAL1, E47, GATA-1 and Ldb1/NLI proteins. *EMBO J.* 1997;16(11):3145–3157.
33 Eisbacher M, Holmes ML, Newton A, et al. Protein-protein interaction between Fli-1 and GATA-1 mediates synergistic expression of megakaryocyte-specific genes through cooperative DNA binding. *Mol Cell Biol.* 2003;23(10):3427–3441.
34 Gordon DF, Lewis SR, Haugen BR, et al. Pit-1 and GATA-2 interact and functionally cooperate to activate the thyrotropin beta-subunit promoter. *J Biol Chem.* 1997;272(39):24339–24347.
35 Gordon DF, Tucker EA, Tundwal K, et al. MED220/thyroid receptor-associated protein 220 functions as a transcriptional coactivator with Pit-1 and GATA-2 on the thyrotropin-beta promoter in thyrotropes. *Mol Endocrinol.* 2006;20(5):1073–1089.
36 Durocher D, Charron F, Warren R, et al. The cardiac transcription factors Nkx2–5 and GATA-4 are mutual cofactors. *EMBO J.* 1997;16(18):5687–5696.
37 Sepulveda JL, Belaguli N, Nigam V, et al. GATA-4 and Nkx-2.5 coactivate Nkx-2 DNA binding targets: role for regulating early cardiac gene expression. *Mol Cell Biol.* 1998;18(6):3405–3415.
38 Chang DF, Belaguli NS, Iyer D, et al. Cysteine-rich LIM-only proteins CRP1 and CRP2 are potent smooth muscle differentiation cofactors. *Dev Cell.* 2003;4(1):107–118.
39 Rath N, Wang Z, Lu MM, Morrisey EE. LMCD1/Dyxin is a novel transcriptional cofactor that restricts GATA6 function by inhibiting DNA binding. *Mol Cell Biol.* 2005;25(20):8864–8873.
40 Liu C, Glasser SW, Wan H, Whitsett JA. GATA-6 and thyroid transcription factor-1 directly interact and regulate surfactant protein-C gene expression. *J Biol Chem.* 2002;277(6):4519–4525.
41 Kawana M, Lee ME, Quertermous EE, Quertermous T. Cooperative interaction of GATA-2 and AP1 regulates transcription of the endothelin-1 gene. *Mol Cell Biol.* 1995;15(8):4225–4231.
42 Minami T, Murakami T, Horiuchi K, et al. Interaction between hex and GATA transcription factors in vascular endothelial cells inhibits flk-1/KDR-mediated vascular endothelial growth factor signaling. *J Biol Chem.* 2004;279(20):20626–20635.

89

Coupling

The Role of Ets Factors

Peter Oettgen

Beth Israel Deaconess Medical Center, Harvard Medical School, Boston, Massachusetts

Transcription factors play a pivotal role in the coupling of endothelial gene expression to endothelial function. The tightly controlled regulation of endothelial gene expression is required for several aspects of endothelial biology. During embryogenesis, for example, the normal development of a primary vascular network, or vasculogenesis, requires a series of carefully orchestrated events that are spatially and temporally regulated. These events are largely coordinated by the precisely timed expression of selected sets of transcription factors. In the adult, endothelial cell (EC)-specific genes are differentially regulated in different body organs or tissues. These differences are in large part dependent on paracrine interactions between tissue-specific cell types and the endothelium. These paracrine interactions depend on the coupling of one cell type with another and are largely mediated by specific transcriptional programs. Under changing environmental or pathological conditions, the endothelium in the affected organ or tissue undergoes distinct changes in gene expression. The identification of specific transcription factors involved in the regulation of the endothelium under normal and pathological conditions not only has led to a better understanding of basic underlying mechanisms of endothelial gene regulation, but also may provide novel therapeutic targets for treating patients with a variety of diseases. Indeed, recent studies have demonstrated the feasibility of targeting transcription factors using small molecules (discussed in the section Therapeutic Implications).

We and other investigators have recently demonstrated the importance of selected members of the Ets transcription factor family in regulating endothelial-specific gene expression during normal development, in mediating vascular-bed specific expression in adult endothelium, and in modulating the expression of activation markers under pathological conditions. The goal of this chapter is to review how transcription factors play a central role in coupling endothelial gene expression to endothelial function under normal and pathological conditions, and to explore their potential as therapeutic targets for drug discovery, with a particular focus on the role of selected members of the Ets transcription factor family in this process.

ETS FACTORS

The Ets factors are a family of approximately 30 transcription factors that have been identified in species ranging from flies to humans, and that share a highly conserved DNA binding domain consisting of a helix-loop-helix structure (Figure 89.1; for color reproduction, see Color Plate 89.1) (1). Particular conservation occurs in the third helix, which binds to a minor groove of the DNA, in the region of the core GGAA/T region of the DNA backbone. Additional specificity is mediated by the flanking wings on either side, which bind to and recognize structural elements within the flanking DNA nucleotide sequences. The name *ETS* is derived from a sequence that was detected in an avian erythroblastosis virus, E26, where it forms a transforming gene together with Δgag and c-myb (2). This sequence was called E26 (E Twenty Six) transformation-specific sequence or Ets. The first cellular homologue to be identified was the human Ets-1 protein, which was shown to function as a proto-oncogene in chromosomal translocations from chromosome 1 to 11 in one form of acute monocytic leukemia and, later, acute lymphoblastic leukemia (3,4). The first functional role for Ets-1 was shown to be in thymocytes, where Ets-1 regulates a number of T cell–specific genes. More recently, Ets has been shown to be involved in a wide variety of developmental processes and cell growth. Furthermore, as proto-oncogenes, aberrant expression of selected members of the Ets transcription factor family is associated with the development of a variety of cancers. Over the past several years, a role for different members of the Ets transcription factor family has been demonstrated in the regulation of a wide variety of biological and pathobiological processes including cancer, angiogenesis, and inflammation.

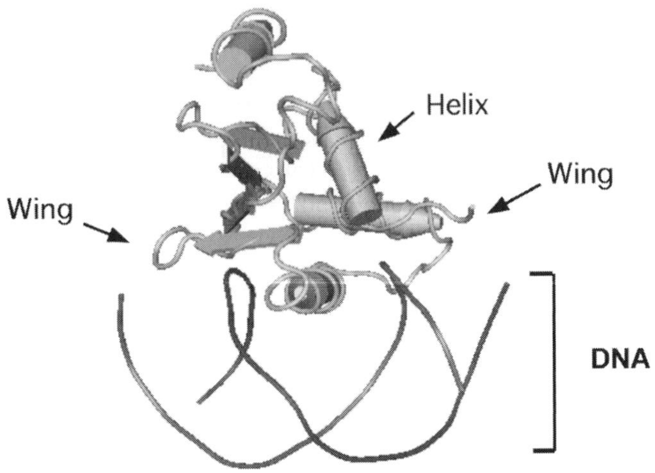

Figure 89.1. Ets DNA binding domain. Schematic of the highly conserved Ets DNA binding domain with a helix-loop-helix structure. Depicted is the structure of the Ets-1 DNA binding domain. Helix 3 is shown in close proximity to the DNA, which is depicted as an overlapping wire-like structure. The flanking wings provide DNA binding specificity among the individual family members. A fourth helix (*top*) can function to inhibit DNA binding. (Reproduced with permission from Garvie CW, Pufall MA, Graves BJ, Wolberger C. Structural analysis of the autoinhibition of Ets-1 and its role in protein partnerships. *J Biol Chem.* 2002;277:45529–45536.) For color reproduction, see Color Plate 89.1.

REGULATION OF ENDOTHELIAL-SPECIFIC GENE EXPRESSION DURING DEVELOPMENT

Vascular development requires the tightly coordinated expression of vascular-specific genes in a temporal and spatial fashion. Examples of endothelial-specific genes that have been shown to be required for vascular development include several receptor tyrosine kinases, such as the vascular endothelial growth factor (VEGF) receptors, VEGFR1, VEGFR2, and the Tie receptors, Tie1 and Tie2. Targeted disruption of the genes encoding these receptors leads to embryonal lethality at various stages of vascular development (5). Recent studies have supported a role for selected members of the Ets factor family in the regulation of these endothelial-specific tyrosine kinases. Whereas the VEGF receptors VEGFR1 and VEGFR2 are regulated by Ets-1 and Ets-2, the *Tie1* and *Tie2* genes appear to be regulated by a different subset of Ets factors including ELF-1 and NERF2 (Elf-2) (6) (Table 89-1). Further support for a role for these factors during vascular development comes from studies demonstrating increased expression of ELF-1 and ELF-2 in developing embryonic blood vessels (7,8).

Several other Ets factors are involved in regulating endothelial-specific gene expression and function during development. For example, the Ets transcription factor Fli-1 is enriched in the developing blood vessels of zebrafish (9). Targeted disruption of *Fli-1* in mice results in a loss of vascular integrity accompanied by bleeding and embryonic lethality at day E11.5 (10). The Ets factor Tel was originally identified as a proto-oncogene in the development of human leukemias (11). Interestingly, targeted disruption of this factor led not only to defects in hematopoiesis, but also to defects in extraembryonic angiogenesis that are embryonically lethal (12).

VASCULAR BED–SPECIFIC GENE REGULATION

Vascular-specific gene expression is also regulated at the level of a particular organ or tissue. Distinct differences in the expression of endothelial-specific genes exist from one organ to another (13–15). For example, the endothelial-specific gene von Willebrand factor (vWF), thrombomodulin (TM), and tissue factor pathway inhibitor (TFPI) exhibit significant differences in expression levels in normal human tissue samples (16). In the heart, vWF and TM are highly expressed, whereas only small amounts of TFPI are detected. The brain expresses undetectable amounts of TFPI and TM, and only small amounts of vWF. TFPI is highly expressed in the placenta, whereas TM and vWF are only weakly expressed. Several endothelial-specific genes, including intercellular adhesion molecule (ICAM)-2, vWF, thrombospondin, and the extracellular matrix protein *SPARC* depend on the Ets factor Erg for their expression in ECs (17). Ets binding sites also have been identified that are critical for the expression of endothelial-specific molecule (ESM)-1 that is expressed primarily in ECs of the lung, kidney, and gut (18). One particular promising strategy for identifying the molecular basis of vascular bed–specific gene expression has been to study the activity of wild-type and, in some cases mutant, DNA promoter sequences coupled to a reporter gene in transgenic mice. Using this approach, a 1.8-kb fragment of the mouse *Tie1* gene was shown to direct highly reproducible expression of LacZ within the intact endothelium (19). A mutation of a single Ets binding site in the proximal region of the *Tie1* promoter completely abolished the ability of the promoter to direct endothelial-specific gene expression in vivo. Mutation of a second, more distal Ets site leads to a marked reduction of expression in certain vascular beds such as the heart, but not in other organs such as the brain. These findings suggest that selected members of the Ets transcription factor family may be involved in mediating vascular bed–specific expression of the *Tie1* gene. The specific Ets family members that regulate the *Tie1* gene at these sites have not been determined. In addition to their role in regulating the expression of vascular-specific genes, selected members of the Ets transcription factor family also regulate the expression of transcription factors involved in regulating endothelial-specific genes. For example, the gene for transcription factor myocyte enhancer factor 2C (MEF2C) contains an intronic enhancer that directs the expression of this transcription factor in a vascular-specific manner during development. This enhancer contains highly conserved binding sites for Ets-1 (20).

Another approach that has been used more recently to overcome the problems associated with random integration

Table 89-1: Role of Selected Ets Factors in Endothelial Gene Expression

	ECs/Vascular Beds	*Targets*	*Inducers*	*−/− Phenotype*	*References*
Ets-1	Arterial and venous ECs in vitro	VEGFR1, VEGFR2 MMP-1, MMP-3, MMP-9 vWF, urokinase, MEF2	VEGF, HGF, FGF2	Reductions in the number of natural killer cells; no vascular defects	20, 34–37
Ets-2	Arterial and venous ECs in vitro	VEGFR1 VEGFR2, urokinase			6, 38
Elf-1	Arterial ECs in vitro developing embryonic vessels	Tie1, Tie2			6, 8
Elf-2 (NERF-2)	Arterial and venous ECs in vitro; developing embryonic vessels	Tie1, Tie2	Hypoxia		6, 39
Elf-3 (ESE-1)	Not detectable in quiescent ECs in vitro or in vivo, is induced in activated ECs	NOS2, COX2	IL-1β, TNF-α, LPS	Abnormalities in development of small intestine in some strains	29, 40–42
Fli-1	Developing embryonic vessels of zebrafish			Embryonic lethal at E11.5 with loss of vascular integrity and hemorrhage	10
Tel	Yolk sac ECs, broadly expressed in other cells			Defect in extra-embryonic angiogenesis	12
ERG	Arterial and venous ECs	ICAM-2, thrombospondin, SPARC			17

COX2, cyclooxygenase 2; LPS, lipopolysaccharide

observed using standard transgenic approaches is to target the transgene into the *HPRT* locus. This approach has been used to define the role of Ets and other transcription factor binding sites within the *Tie2* promoter (21). Similar studies also have been conducted with the *VEGFR1* and *vWF* genes, which have facilitated the identification of specific regulatory elements that are required for directing the expression of these genes in different vascular beds (22).

Differences in the expression of selected Ets family members also have been observed in different types of primary ECs. For example, whereas the Ets factor ELF-2 (NERF2) is expressed in both arterial (i.e., human aortic ECs [HAECs] and venous ECs [HUVECs]), ELF-1 is only expressed in arterial EC (i.e., HAEC) (6).

PROTEIN–PROTEIN AND POST-TRANSLATIONAL MODIFICATIONS

Several other mechanisms exist by which gene expression can be further regulated in addition to the binding of specific transcription factors to regulatory elements within these genes. One of the major mechanisms by which the activity of transcription factors can be modified is through protein–protein interactions. For example, Ets-1 has been shown to interact with several different proteins, including the transcription factors AML-1, nuclear factor (NF)-κB, GATA3, hypoxia-inducible factor (HIF)-2α, Pax5, and STAT5 (Figure 89.2) (23). Another mechanism, by which the activity of a transcription factor can be modified, is through post-translational modifications, such as phosphorylation. For example, phosphorylation of Ets-1 by the mitogen-activated protein kinase (MAPK) pathway leads to increased transcriptional activity, whereas phosphorylation through Cam Kinase II decreases Ets-1 activity (24). Other post-translational modifications by which the activity of Ets factors can be modified include sumoylation, ubiquitination, and acetylation (25).

ANGIOGENESIS

In addition to the regulation of the VEGF receptors during development, Ets-1 also has been shown to be upregulated in the setting of angiogenesis associated with tumor growth. Ets-1 expression in ECs is induced by VEGF, hepatocyte growth factor (HGF), and basic fibroblast growth factor (bFGF). Some of the gene targets for Ets-1 include several members of the matrix metalloproteinase family including matrix metalloproteinase (MMP)-1, MMP-3, and MMP-9 (26). The therapeutic potential of blocking Ets-1 in the setting of angiogenesis recently has been shown. In two animal models of angiogenesis, including a tumor model and a model in which

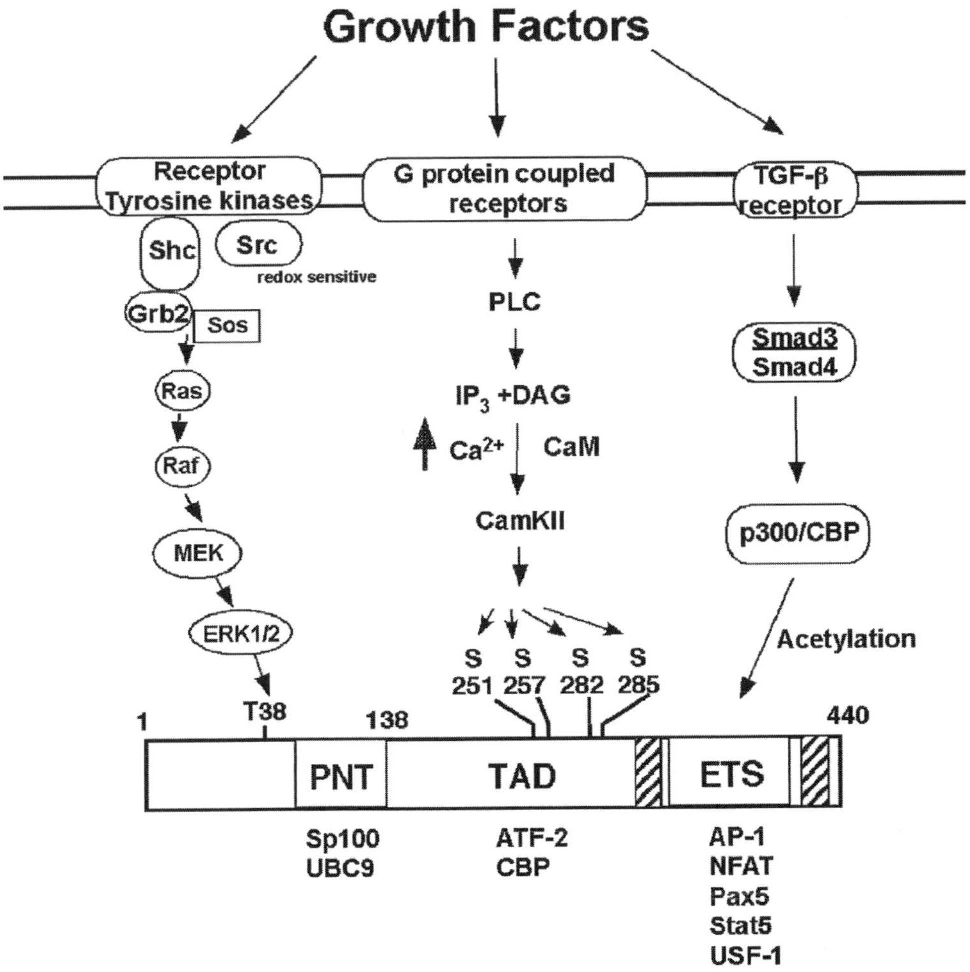

Figure 89.2. Post-translational modifications of Ets factors. An example of the multiple mechanisms by which the activity of Ets factors can be modulated at a post-translational level. The activity of Ets-1 can be positively (MAPK) or negatively (CamKII) regulated through protein–protein interactions with several domains of the protein including the pointed domain (PNT) transactivation domain (TAD) and Ets domain (Ets). (Reproduced with permission Tootle TL, Rebay I. Post-translational modifications influence transcription factor activity: a view from the Ets superfamily. *Bioessays.* 2005;27:285–298.)

bFGF was subcutaneously implanted in the mouse ear, systemic administration of a retrovirus expressing a dominant negative form of Ets-1 resulted in marked inhibition of angiogenesis and tumor growth (27).

ENDOTHELIAL ACTIVATION

A number of inflammatory mediators can lead to endothelial activation, including proinflammatory cytokines and endotoxin. We recently have identified a novel member of the Ets transcription factor, ESE-1, that has the unique property of having two DNA binding domains: an A/T hook domain and a classical Ets domain (28). ESE-1 is not expressed under normal conditions in cells of vascular origin, such as ECs or vascular smooth muscle cells (VSMCs). However in the presence of proinflammatory cytokines such as interleukin (IL)-1β or tumor necrosis factor (TNF)-α, ESE-1 expression is rapidly induced in ECs and VSMCs (29). One of the main targets of ESE-1 is the inducible form of nitric oxide synthase (NOS2). In a mouse model of endotoxemia, ESE-1 was induced within the endothelium and first layer of smooth muscle cells within the media (29). In parallel, a similar spatial and temporal induction of NOS2 was observed. Furthermore, ESE-1 functions synergistically with NF-κB to transactivate the *NOS2* gene. To further define additional proteins that might interact with ESE-1 to modulate its function, we recently identified CBP/p300 as a protein that cooperates with ESE-1 to upregulate its function (30). More interestingly, we identified another set of proteins, the Ku proteins 70 and 86 that bind to ESE-1 and inhibit its function. The Ku proteins have previously been shown to act as architectural proteins and bind to DNA, but have not been shown to modulate the function of transcription factors.

THERAPEUTIC IMPLICATIONS

The identification of selected transcription factors that are enriched in the setting of vascular inflammation or angiogenesis not only has implications regarding the understanding of the fundamental underlying molecular mechanisms of these processes, but also has important therapeutic implications. In general, transcription factors do not regulate a single gene, but more often whole sets of genes that may be involved in a particular process or cell type. The inhibition of the function of a particular factor may therefore be therapeutically useful because the activation of whole sets of genes can be blocked, as opposed to a single receptor or enzyme.

In addition to blocking peptides and the use of dominant negative proteins, the identification of small molecules that specifically inhibit selected transcription factors and exhibit therapeutic potential has recently been demonstrated. The transcription factor HIF-1α is constitutively expressed in many cell types but becomes activated in the setting of hypoxia. HIF-1α induces the expression of VEGF and is a critical mediator of angiogenesis (31). Two groups have recently identified small molecules that are capable of inhibiting the function of HIF-1α (32,33). These small molecules not only block the function of HIF-1α, but have been shown to inhibit the induction of VEGF in the setting of hypoxia and lead to marked reduction in tumor growth of human cancer cell lines in mice, with associated inhibition of tumor angiogenesis in mouse tumor models. The HIF-1α small-molecule inhibitor was administered after the injected tumor cells reached a size of 120 mm^3 (33). When the HIF-1α small-molecule inhibitor was administered to animals with the small cell lung cancer cell line SHP-77, it completely abolished the growth of the tumor, with no evidence of any tumors visible after several months. In summary, the results of these and other recent studies clearly demonstrate the therapeutic potential of targeting transcription factors that act as central mediators of the activated endothelium in the setting of pathological processes such as angiogenesis or vascular inflammation. Interactions between basic and clinical scientists, and computational and medicinal chemists, will promote the identification of novel therapeutic compounds that can be used to target the endothelium in a number of human diseases.

KEY POINTS

- Ets factors are critical transcriptional regulators of endothelial-specific gene expression.
- Several Ets factors, including Ets-1, Ets-2, Elf-1, Elf-2, Fli-1, Erg-1, and Tel regulate vasculogenesis and angiogenesis.
- Critical regulators of endothelial activation in response to inflammatory mediators include the Ets factors Elf-3 and Ets-1.

Future Goals

- To determine whether inhibition of the function of selected Ets family members in the setting of pathological angiogenesis, using peptide inhibitors or small molecules, can be used therapeutically in a variety of human diseases
- To determine whether inhibition of the function of selected Ets family members in the setting of vascular inflammation can be used therapeutically to treat diseases associated with vascular inflammation, including hypertension, atherosclerosis, and diabetes mellitus

REFERENCES

1 Wasylyk B, Hahn SL, Giovane A. The Ets family of transcription factors. *Eur J Biochem.* 1993;211:7–18.

2 Nunn MF, Seeburg PH, Moscovici C, Duesberg PH. Tripartite structure of the avian erythroblastosis virus E26 transforming gene. *Nature.* 1983;306:391–395.

3 Caubet JF, Gegonne A, Stehelin D, Berger R. Abnormal localization of proto-oncogene c-ets 1 in acute leukemia with translocation t(1:11)(q21: q23). *CR Acad Sci III.* 1986;302:589–691.

4 Diaz MO, Le Beau MM, Pitha P, Rowley JD. Interferon and c-ets-1 genes in the translocation (9;11)(p22;q23) in human acute monocytic leukemia. *Science.* 1986;231:265–267.

5 Sato TN, Qin Y, Kozak CA, Audus KL. Tie-1 and Tie-2 define another class of putative receptor tyrosine kinase genes expressed in early embryonic vascular system [published erratum appears in *Proc Natl Acad Sci USA.* 1993;90(24):12056]. *Proc Natl Acad Sci USA.* 1993;90:9355–9358.

6 Dube A, Akbarali Y, Sato TN, et al. Role of the Ets transcription factors in the regulation of the vascular- specific Tie2 gene [See comments]. *Circ Res.* 1999;84:1177–1185.

7 Dube A, Thai S, Gaspar J, et al. Elf-1 is a transcriptional regulator of the Tie2 gene during vascular development. *Circ Res.* 2001;88:237–244.

8 Gaspar J, Thai S, Voland C, et al. Opposing functions of the Ets factors NERF and ELF-1 during chicken blood vessel development. *Arterioscler Thromb Vasc Biol.* 2002;22:1106–1112.

9 Brown LA, Rodaway AR, Schilling TF, et al. Insights into early vasculogenesis revealed by expression of the Ets-domain transcription factor Fli-1 in wild-type and mutant zebrafish embryos. *Mech Dev.* 2000;90:237–252.

10 Hart A, Melet F, Grossfeld P, et al. Fli-1 is required for murine vascular and megakaryocytic development and is hemizygously deleted in patients with thrombocytopenia. *Immunity.* 2000;13: 167–177.

11 Lacronique V, Boureux A, Valle VD, et al. A TEL-JAK2 fusion protein with constitutive kinase activity in human leukemia. *Science.* 1997;278:1309–1312.

12 Wang LC, Kuo F, Fujiwara Y, et al. Yolk sac angiogenic defect and intra-embryonic apoptosis in mice lacking the Ets-related factor TEL. *EMBO J.* 1997;16:4374–4383.

13 Aird WC. Endothelial cell heterogeneity. *Crit Care Med.* 2003; 31:S221–S230.
14 Aird WC. Endothelial cell dynamics and complexity theory. *Crit Care Med.* 2002;30:S180–S185.
15 Aird WC, Edelberg JM, Weiler-Guettler H, et al. Vascular bed-specific expression of an EC gene is programmed by the tissue microenvironment. *J Cell Biol.* 1997;138:1117–1124.
16 Bajaj MS, Kuppuswamy MN, Manepalli AN, Bajaj SP. Transcriptional expression of tissue factor pathway inhibitor, thrombomodulin and von Willebrand factor in normal human tissues. *Thromb Haemost.* 1999;82:1047–1052.
17 McLaughlin F, Ludbrook VJ, Cox J, et al. Combined genomic and antisense analysis reveals that the transcription factor Erg is implicated in EC differentiation. *Blood.* 2001;98:3332–3339.
18 Tsai JC, Zhang J, Minami T, et al. Cloning and characterization of the human lung endothelial-cell-specific molecule-1 promoter. *J Vasc Res.* 2002;39:148–159.
19 Iljin K, Dube A, Kontusaari S, et al. Role of ets factors in the activity and EC specificity of the mouse Tie gene promoter. *FASEB J.* 1999;13:377–386.
20 De Val S, Anderson JP, Heidt AB, et al. Mef2c is activated directly by Ets transcription factors through an evolutionarily conserved EC-specific enhancer. *Dev Biol.* 2004;275:424–434.
21 Minami T, Kuivenhoven JA, Evans V, et al. Ets motifs are necessary for EC-specific expression of a 723-bp Tie-2 promoter/enhancer in Hprt targeted transgenic mice. *Arterioscler Thromb Vasc Biol.* 2003;23:2041–2047.
22 Minami T, Donovan DJ, Tsai JC, et al. Differential regulation of the von Willebrand factor and Flt-1 promoters in the endothelium of hypoxanthine phosphoribosyltransferase-targeted mice. *Blood.* 2002;100:4019–4025.
23 Dittmer J. The biology of the Ets1 proto-oncogene. *Mol Cancer.* 2003;2:29.
24 Pufall MA, Lee GM, Nelson ML, et al. Variable control of Ets-1 DNA binding by multiple phosphates in an unstructured region. *Science.* 2005;309:142–145.
25 Tootle TL, Rebay I. Post-translational modifications influence transcription factor activity: a view from the Ets superfamily. *Bioessays.* 2005;27:285–298.
26 Hashiya N, Jo N, Aoki M, et al. In vivo evidence of angiogenesis induced by transcription factor Ets-1: Ets-1 is located upstream of angiogenesis cascade. *Circulation.* 2004;109:3035–3041.
27 Pourtier-Manzanedo A, Vercamer C, Van Belle E, et al. Expression of an Ets-1 dominant-negative mutant perturbs normal and tumor angiogenesis in a mouse ear model. *Oncogene.* 2003;22:1795–1806.
28 Oettgen P, Alani RM, Barcinski MA, et al. Isolation and characterization of a novel epithelium-specific transcription factor, ESE-1, a member of the ets family. *Mol Cell Biol.* 1997;17:4419–4433.
29 Rudders S, Gaspar J, Madore R, et al. ESE-1 is a novel transcriptional mediator of inflammation that interacts with NF-kappa B to regulate the inducible nitric-oxide synthase gene. *J Biol Chem.* 2001;276:3302–3309.
30 Wang H, Fang R, Cho JY, et al. Positive and negative modulation of the transcriptional activity of the Ets factor ESE-1 through interaction with p300, CREB-binding protein, and Ku 70/86. *J Biol Chem.* 2004;279:25241–25250.
31 Wartenberg M, Donmez F, Ling FC, et al. Tumor-induced angiogenesis studied in confrontation cultures of multicellular tumor spheroids and embryoid bodies grown from pluripotent embryonic stem cells. *FASEB J.* 2001;15:995–1005.
32 Rapisarda A, Uranchimeg B, Scudiero DA, et al. Identification of small molecule inhibitors of hypoxia-inducible factor 1 transcriptional activation pathway. *Cancer Res.* 2002;62:4316–4324.
33 Welsh S, Williams R, Kirkpatrick L, et al. Antitumor activity and pharmacodynamic properties of PX-478, an inhibitor of hypoxia-inducible factor-1alpha. *Mol Cancer Ther.* 2004;3:233–244.
34 Schwachtgen JL, Janel N, Barek L, et al. Ets transcription factors bind and transactivate the core promoter of the von Willebrand factor gene. *Oncogene.* 1997;15:3091–3102.
35 Iwasaka C, Tanaka K, Abe M, Sato Y. Ets-1 regulates angiogenesis by inducing the expression of urokinase- type plasminogen activator and matrix metalloproteinase-1 and the migration of vascular endothelial cells. *J Cell Physiol.* 1996;169:522–531.
36 Wakiya K, Begue A, Stehelin D, Shibuya M. A cAMP response element and an Ets motif are involved in the transcriptional regulation of flt-1 tyrosine kinase (vascular endothelial growth factor receptor 1) gene. *J Biol Chem.* 1996;271:30823–30828.
37 Oda N, Abe M, Sato Y. Ets-1 converts endothelial cells to the angiogenic phenotype by inducing the expression of matrix metalloproteinases and integrin beta3. *J Cell Physiol.* 1999;178:121–132.
38 Watabe T, Yoshida K, Shindoh M, et al. The Ets-1 and Ets-2 transcription factors activate the promoters for invasion-associated urokinase and collagenase genes in response to epidermal growth factor. *Int J Cancer.* 1998;77:128–137.
39 Christensen RA, Fujikawa K, Madore R, et al. NERF2, a member of the Ets family of transcription factors, is increased in response to hypoxia and angiopoietin-1: a potential mechanism for Tie2 regulation during hypoxia. *J Cell Biochem.* 2002;85:505–515.
40 Ng AY, Waring P, Ristevski S, et al. Inactivation of the transcription factor Elf3 in mice results in dysmorphogenesis and altered differentiation of intestinal epithelium. *Gastroenterology.* 2002;122:1455–1466.
41 Grall FT, Prall WC, Wei W, et al. The Ets transcription factor ESE-1 mediates induction of the COX-2 gene by LPS in monocytes. *FEBS J.* 2005;272:1676–1687.
42 Garvie CW, Pufall MA, Graves BJ, Wolberger C. Structural analysis of the autoinhibition of Ets-1 and its role in protein partnerships. *J Biol Chem.* 2002;277:45529–45536.

90

Early Growth Response-1 Coupling in Vascular Endothelium

Levon M. Khachigian and Valerie C. Midgley

Centre for Vascular Research, University of New South Wales, Prince of Wales Hospital, Sydney, Australia

Changes in the extracellular environment can trigger alterations in gene expression and eventually phenotype through the activity of transcription factors. One such "coupler" is early growth response (Egr)-1 (also known as Krox-24, zif268, NGFI-A, and TIS8), a 60- to 80-kDa zinc finger transcription factor of the C_2H_2 subtype and the product of an immediate-early gene (1). It belongs to a family of nuclear regulators that include Egr-2, Egr-3, and Egr-4, and its expression can be induced by a variety of pathophysiologic stimuli in endothelial cells (ECs) and other vascular cell types such as smooth muscle cells (SMCs), monocytes, and macrophages, as well as in a diverse range of nonvascular cells. In 1987, Egr-1 was independently cloned by a number of groups using differential screening strategies (2–5). It is comprised of 533 amino acids and contains an amino-terminal and a carboxyl-terminal activation domain, a repression domain, and a DNA-binding domain consisting of three conserved zinc fingers. Egr-1 also contains a bipartite nuclear localization signal (residues 315–330 and a section of the DNA-binding domain) (Figure 90.1) (1,6).

Egr-1 has been implicated in a panoply of cardiovascular pathologic processes, which include atherosclerosis, restenosis, ischemia, angiogenesis, allograft rejection, and cardiac hypertrophy (7). Triggers such as injury (8), altered biomechanical environment (9), growth factor/hormone and cytokine exposure (10–13), chemical challenge (such as lysophosphatidylcholine) (14), and hypoxia (15,16) are capable of rapidly inducing Egr-1 transcription. Once induced, Egr-1 regulates the expression of many genes through specific interactions with GC-rich motifs in the promoter regions of responsive genes. These include transcription factors, signaling molecules, cell cycle regulatory proteins, cytokines, growth factors, coagulation factors, and even Egr-1 itself. Transcriptional profiling following adenoviral gene transfer in ECs indicates that several hundred genes are differentially expressed in response to Egr-1 overexpression by at least threefold (17) (Table 90-1). We and others have demonstrated that, at least in ECs, Egr-1 controls platelet-derived growth factor (PDGF)-A (9,11,18), PDGF-B (8), fibroblast growth factor (FGF)-2 (19), tissue factor (20), and angiotensin-converting enzyme (21) expression. Moreover, insulin-inducible EC proliferation and regrowth after injury is an Egr-1–dependent process (22). Egr-1 therefore transcriptionally integrates changes in the extracellular environment with altered gene expression and cell phenotype.

It is unlikely that Egr-1 acts alone in its regulation of gene expression; cofactors of Egr-1 have been identified that influence Egr-1 transcriptional activity. For example, cAMP-response-element-binding-protein-binding protein (CBP) and p300 boost Egr-1 transactivation of the *5-lipoxygenase* promoter whereby the N- and C-terminal domains of CBP interact with the transcriptional activation domain of Egr-1 (23). NAB1 and NAB2, on the other hand, serve as Egr-1 corepressors by blocking Egr-1 activity through protein–protein interaction with the inhibitory domain of Egr-1 (24–26). Lessons from our own work investigating PDGF-A (18), PDGF-B-chain (8), and more recently PDGF-C-chain (27) transcription indicate that Egr-1's interaction with DNA involves the displacement of other transcription factors, such as Sp1, from overlapping binding sites in the promoters. However, such interplay need not necessarily have a positive influence on gene expression. Suppression of the expression of hepatocyte growth factor (HGF) receptor (the product of the c-met proto-oncogene) by oxidative stress, for example, is mediated through interplay between Egr-1 and Sp1 (28).

The inducible transcription of Egr-1, like that of c-Fos (29), is under the control of mitogen-activated protein kinase (MAPK) phosphorylation cascades that phosphorylate ternary complex factor (TCF)[1]/serum response factor (SRF)-dependent gene expression (30), which binds to SRFs to activate serum-response elements (SREs) (31). Several SREs occur in the *Egr-1* promoter (32,33). These interactions are

1 For example, the Ets factors Elk-1, Sap-1a, and Net.

Table 90-1: Representative Egr-1–Dependent Genes in ECs

Gene Class	*Upregulated Genes*	*Downregulated Genes*
Transcriptional regulatory proteins	*NAB1, AP2, Egr-1, GATA-2, p54*	*Smad6*
Signaling proteins	*Rad, Notch3, small G protein, bcl-1, RAMP1*	*BMP2, BMP4, NOS*
Growth factors, cytokines, and chemokines	*IGF-2, TGF-β, PDGF-A, PDGF-B, PDGF-C*	*TRAIL, CTGF*
Cell cycle regulators	*p57, Cyclin D1*	
Extracellular matrix proteins	*VCAM-1, TIMP-1, fibronectin, ostepontin, collagen III α 1*	*VE-Cadherin, PECAM-1*
Ion channels	*Na^+-Ca^{2+} exchanger, Ca^{2+} channel α 2*	
Other	*Apolipoprotein E & D, H0–1, BAI1, hPD-1, tissue factor*	*Connexin 40, HSP40*

(Compiled from Fu M, Zhu X, Zhang J, et al. Egr-1 target genes in human ECs identified by microarray analysis. *Gene.* 2003;315:33–41, and other sources cited in this chapter.)

responsible for regulating the inducible expression of Egr-1 in response to agonists and conditions as diverse as thrombin (34), lipopolysaccharide (35), and hypoxia (16).

Our understanding of Egr-1's role in ECs and other vascular settings in vivo has, in part, been derived from the use of gene-targeting strategies. The weaponry of synthetic nucleic acid–based gene silencing agents includes antisense oligonucleotides, triplex-forming oligonucleotides, ribozymes, riboswitches, aptamers, silent RNA (siRNA), and RNA-cleaving DNA enzymes (or DNAzymes) (36). Our lab has employed the use of DNAzymes (bearing 9 + 9 nt arms, with a 3′-linked inverted T at the 3′-end) targeting Egr-1 to inhibit vascular EC (19) and SMC (37) growth in culture and to impair the capacity of these cells to regenerate after wounding in vitro. DNAzymes are single-stranded DNA molecules that bind to their target RNA through Watson-Crick base-pairing and cleave between an unpaired purine and a paired pyrimidine by virtue of a cation-dependent catalytic domain (e.g., the 15 nt long "10–23" domain) (38). Unlike alternative, more costly approaches, DNAzymes are stable in serum and may inhibit through the use of both classic noncatalytic antisense and catalytic mechanisms (39). DNAzymes have provided insights into the roles of Egr-1 in microvascular EC growth, neovascularization, tumor angiogenesis, and tumor growth. For example, we found Egr-1 DNAzymes to perturb spontaneous endothelial tubule formation within hours of plating (19), to inhibit host neovascularization of subcutaneous Matrigel plugs in mice, and to reduce limbal angiogenesis in rat cornea after vascular endothelial growth factor (VEGF) implantation. Furthermore, DNAzymes impaired MCF7 solid tumor (human breast carcinoma) growth in athymic nude mice and reduced vessel density in the tumors, without adversely affecting body weight, wound healing, blood coagulation, or reproduction (19). DNAzyme inhibition of solid tumor

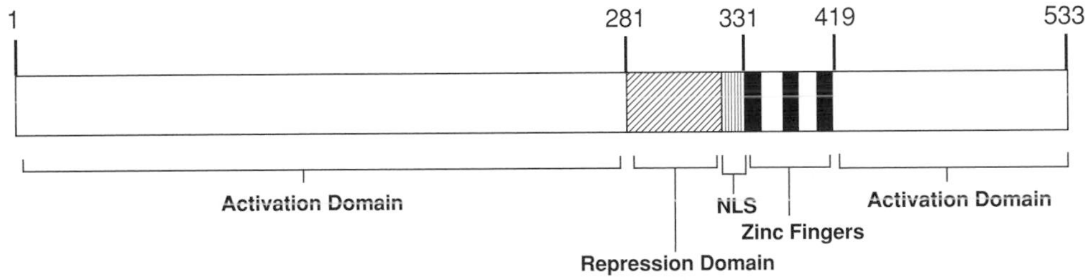

Figure 90.1. Functional domains in Egr-1. Amino acid residues spanning functional domains are indicated by numbering in the figure. NLS denotes nuclear localization domain. (Modified with permission from Gashler A, Sukhatme V. Early growth response protein 1 (Egr-1): prototype of a zinc-finger family of transcription factors. *Prog Nucleic Acid Res Mol Biol.* 1995;50:191–224; and Gashler AL, Swaminathan S, Sukhatme VP. A novel repression module, an extensive activation domain, and a bipartite nuclear localization signal defined in the immediate-early transcription factor Egr-1. *Mol Cell Biol.* 1993;13:4556–4571.)

growth in that study was indirect; that is, growth suppression was achieved via inhibition of host EC growth, consistent with Folkman's hypothesis from 30 years ago (40,41). Egr-1 inhibition blocked the expression of FGF2 (19). Interestingly, the injury-inducible endothelial expression of Egr-1 is itself dependent on the release and paracrine activity of endogenous FGF2 (12). The inhibitory properties of DNAzymes illustrated by our findings demonstrate the key role Egr-1 plays in EC growth, neovascularization, and tumor growth. Our earlier work demonstrates that Egr-1 DNAzymes also inhibit other vascular settings. When delivered adventitially (from the outside of the vessel in) to the carotid arteries of rats, Egr-1 DNAzymes inhibited intimal thickening involving aggressive SMC growth in the arterial intima, 2 weeks after balloon catheter angioplasty (19) and 3 weeks after ligation injury (42). Moreover, when delivered intraluminally in pig arteries using a special catheter, Egr-1 DNAzymes blocked in-stent restenosis 1 month after coronary stenting (43). These studies with Egr-1 DNAzymes provided the first proof-of-principle evidence that this class of molecule can be used as gene-specific and potentially versatile therapeutic tools (37). DNAzymes targeting other "couplers" such as, for example, c-Jun, have been used to inhibit microvascular endothelial growth and angiogenesis, tumor growth (39) as well as SMC growth, and other vascular proliferative settings including intimal thickening in injured arteries (44). Other groups have generated catalytic DNA targeting bcr-abl (45), c-myc (46), integrins β_1 and β_3 (47,48), protein kinase C (PKC)α (49), PKCϵ (50), Twist (51), VEGF receptor-2 (52), laminin γ_1 chain (53), human immunodeficiency virus (HIV)-1 (54,55), and platelet-type 12-lipoxygenase (56).

Since its discovery almost two decades ago, many lines of investigation have shown that Egr-1 is indeed a "master-regulator" that controls the expression of multiple growth-regulatory genes and can couple the extracellular environment with cardiovascular pathologic settings.

ACKNOWLEDGMENTS

Work in the author's laboratory was supported by grants from the National Health and Medical Research Council of Australia (NHMRC) and National Heart Foundation of Australia. LMK is a Senior Principal Research Fellow of the NHMRC.

REFERENCES

1 Gashler A, Sukhatme V. Early growth response protein 1 (Egr-1): prototype of a zinc-finger family of transcription factors. *Prog Nucleic Acid Res Mol Biol.* 1995;50:191–224.
2 Lau LF, Nathans D. Expression of a set of growth-related immediate early genes in BALB/c 3T3 cells: coordinate regulation with c-fos or c-myc. *Proc Natl Acad Sci USA.* 1987;84:1182–1186.
3 Lim RW, Varnum BC, Herschman HR. Cloning of tetradecanoyl phorbol ester-induced "primary response" sequences and their expression in density-arrested Swiss 3T3 cells and a TPA non-proliferative variant. *Oncogene.* 1987;1:263–270.
4 Milbrandt J. A nerve growth factor-induced gene encodes a possible transcriptional regulatory factor. *Science.* 1987;238:797–799.
5 Sukhatme VP, Cao X, Chang LL, et al. A zinc-finger encoding gene coregulated with c-Fos during growth and differentiation and after depolarization. *Cell.* 1988;53:37–43.
6 Gashler AL, Swaminathan S, Sukhatme VP. A novel repression module, an extensive activation domain, and a bipartite nuclear localization signal defined in the immediate-early transcription factor Egr-1. *Mol Cell Biol.* 1993;13:4556–4571.
7 Khachigian LM. Egr-1 in cardiovascular pathobiology. *Circ Res.* 2006;98:186–191.
8 Khachigian LM, Lindner V, Williams AJ, Collins T. Egr-1-induced endothelial gene expression: a common theme in vascular injury. *Science.* 1996;271:1427–1431.
9 Khachigian LM, Anderson KA, Halnon NJ, et al. Egr-1 is activated in endothelial cells exposed to fluid shear stress and interacts with a novel shear-stress response element in the PDGF A-chain promoter. *Arterioscl Thromb Vasc Biol.* 1997;17:2280–2286.
10 Silverman ES, Khachigian LM, Lindner V, et al. Inducible PDGF A-chain transcription in vascular smooth muscle cells is mediated by Egr-1 displacement of Sp1 and Sp3. *Am J Physiol.* 1997;42: H1415–H1426.
11 Delbridge GJ, Khachigian LM. FGF-1-induced PDGF A-chain gene expression in vascular endothelial cells involves transcriptional activation by Egr-1. *Circ Res.* 1997;81:282–288.
12 Santiago FS, Lowe HC, Day FL, et al. Early growth response factor-1 induction by injury is triggered by release and paracrine activation by fibroblast growth factor-2. *Am J Pathol.* 1999;154: 937–944.
13 Day FL, Rafty LA, Chesterman CN, Khachigian LM. Angiotensin II (ATII)-inducible platelet-derived growth factor A-chain gene expression is p42/44 extracellular signal-regulated kinase-1/2 and Egr-1 dependent and modulated via the ATII type 1 but not type 2 receptor – induction by ATII antagonized by nitric oxide. *J Biol Chem.* 1999;274:23726–23733.
14 Morimoto M, Kume N, Miyamoto S, et al. Lysophosphatidylcholine induces early growth response factor-1 expression and activates the core promoter of PDGF-A chain in vascular endothelial cells. *Arterioscler Thromb Vasc Biol.* 2001;21:771–776.
15 Lo LW, Cheng JJ, Chiu JJ, et al. Endothelial exposure to hypoxia induces Egr-1 expression involving PKCalpha-mediated Ras/Raf-1/ERK1/2 pathway. *J Cell Physiol.* 2001;188:304–312.
16 Yan SF, Lu J, Zou YS, et al. Hypoxia-associated induction of early growth response-1 gene expression. *J Biol Chem.* 1999;274: 15030–15040.
17 Fu M, Zhu X, Zhang J, et al. Egr-1 target genes in human endothelial cells identified by microarray analysis. *Gene.* 2003;315:33–41.
18 Khachigian LM, Williams AJ, Collins T. Interplay of Sp1 and Egr-1 in the proximal PDGF-A promoter in cultured vascular endothelial cells. *J Biol Chem.* 1995;270:27679–27686.
19 Fahmy RG, Dass CR, Sun LQ, et al. Transcription factor Egr-1 supports FGF-dependent angiogenesis during neovascularization and tumor growth. *Nat Med.* 2003;9:1026–1032.
20 Bavendiek U, Libby P, Kilbride M, et al. Induction of tissue factor expression in human endothelial cells by CD40 ligand is mediated via activator protein 1, nuclear factor kappa B, and Egr-1. *J Biol Chem.* 2002;277:25032–25039.
21 Eyries M, Agrapart M, Alonso A, Soubrier F. Phorbol ester induction of angiotensin-converting enzyme transcription is mediated by Egr-1 and AP-1 in human endothelial cells via ERK1/2 pathway. *Circ Res.* 2002;91:899–906.

22 Gousseva N, Kugathasan K, Chesterman CN, Khachigian LM. Early growth response factor-1 mediates insulin-inducible vascular endothelial cell proliferation and regrowth after injury. *J Cell Biochem.* 2001;81:523–534.

23 Silverman ES, Du J, Williams AJ, et al. cAMP-response-element-binding-protein-binding protein (CBP) and p300 are transcriptional co-activators of early growth response factor-1 (Egr-1). *Biochem J.* 1998;336(Pt 1):183–189.

24 Russo MW, Sevetson BR, Milbrandt J. Identification of NAB-1, a repressor of NGFI-A- and Krox20-mediated transcription. *Proc Natl Acad Sci USA.* 1995;92:6873–6877.

25 Svaren J, Sevetson BR, Apel ED, et al. NAB2, a corepressor of NGFI-A (Egr-1) and Krox20, is induced by proliferative and differentiative stimuli. *Mol Cell Biol.* 1996;16:3545–3553.

26 Svaren J, Sevetson BR, Golda T, et al. Novel mutants of NAB corepressors enhance activation by Egr transactivators. *EMBO J.* 1998;17:6010–6019.

27 Midgley VC, Khachigian LM. Fibroblast growth factor-2 induction of platelet-derived growth factor-C chain transcription in vascular smooth muscle cells is ERK-dependent but not JNK-dependent and mediated by Egr-1. *J Biol Chem.* 2004;279:40289–40295.

28 Zhang X, Liu Y. Suppression of HGF receptor gene expression by oxidative stress is mediated through the interplay between Sp1 and Egr-1. *Am J Physiol Renal Physiol.* 2003;284:F1216–F1225.

29 Treisman R. Journey to the surface of the cell: Fos regulation and the SRE. *EMBO J.* 1995;14:4905–4913.

30 Buchwalter G, Gross C, Wasylyk B. Ets ternary complex transcription factors. *Gene.* 2004;324:1–14.

31 Treisman R. The SRE: a growth factor responsive transcriptional regulator. *Sem Cancer Biol.* 1990;1:47–58.

32 Sakamoto KM, Bardeleben C, Yates KE, et al. 5′ upstream sequence and genomic structure of the human primary response gene, EGR-1/TIS-8. *Oncogene.* 1991;6:867–871.

33 McMahon SB, Monroe JG. A ternary complex factor-dependent mechanism mediates induction of egr-1 through selective serum response elements following antigen receptor cross-linking in B lymphocytes. *Mol Cell Biol.* 1995;15:1086–1093.

34 Wu SQ, Minami T, Donovan DJ, Aird WC. The proximal serum response element in the Egr-1 promoter mediates response to thrombin in primary human endothelial cells. *Blood.* 2002;100:4454–4461.

35 Guha M, O'Connell MA, Pawlinski R, et al. Lipopolysaccharide activation of the MEK-ERK1/2 pathway in human monocytic cells mediates tissue factor and tumor necrosis factor alpha expression by inducing Elk-1 phosphorylation and Egr-1 expression. *Blood.* 2001;98:1429–1439.

36 Breaker RR. Natural and engineered nucleic acids as tools to explore biology. *Nature.* 2004;432:838–845.

37 Santiago FS, Lowe HC, Kavurma MM, et al. New DNA enzyme targeting Egr-1 mRNA inhibits vascular smooth muscle proliferation and regrowth factor injury. *Nature Med.* 1999;5:1264–1269.

38 Santoro SW, Joyce GF. A general purpose RNA-cleaving DNA enzyme. *Proc Natl Acad Sci USA.* 1997;94:4262–4266.

39 Zhang G, Dass CR, Sumithran E, et al. Effect of deoxyribozymes targeting c-Jun on solid tumor growth and angiogenesis in rodents. *J Natl Cancer Inst.* 2004;96:683–696.

40 Folkman J. Tumor angiogenesis: therapeutic implications. *N Engl J Med.* 1971;285:1182–1186.

41 Folkman J. Anti-angiogenesis: new concept for therapy of solid tumors. *Ann Surg.* 1972;175:409–416.

42 Lowe HC, Chesterman CN, Khachigian LM. Catalytic antisense DNA molecules targeting Egr-1 inhibit neointima formation following permanent ligation of rat common carotid arteries. *Thromb Haemost.* 2002;87:134–140.

43 Lowe HC, Fahmy RG, Kavurma MM, et al. Catalytic oligodeoxynucleotides define a key regulatory role for early growth response factor-1 in the porcine model of coronary in-stent restenosis. *Circ Res.* 2001;89:670–677.

44 Khachigian LM, Fahmy RG, Zhang G, et al. c-Jun regulates vascular smooth muscle cell growth and neointima formation after arterial injury: inhibition by a novel DNAzyme targeting c-Jun. *J Biol Chem.* 2002;277:22985–22991.

45 Wu Y, McMahon R, Rossi JJ, et al. Inhibition of bcr-abl oncogene by novel deoxyribozymes (DNAzymes). *Hum Gene Ther.* 1999;10:2847–2857.

46 Sun L-Q, Cairns MJ, Gerlach WL, et al. Suppression of smooth muscle cell proliferation by a c-myc RNA-cleaving deoxyribozyme. *J Biol Chem.* 1999;274:17236–17241.

47 Cieslak M, Szymanski J, Adamiak RW, Cierniewski CS. Structural rearrangements of the 10–23 DNAzyme to beta 3 integrin subunit mRNA induced by cations and their relations to the catalytic activity. *J Biol Chem.* 2003;278:47987–47996.

48 Cieslak M, Niewiarowska J, Nawrot M, et al. DNAzymes to beta 1 and beta 3 mRNA down-regulate expression of the targeted integrins and inhibit EC capillary tube formation in fibrin and matrigel. *J Biol Chem.* 2002;277:6779–6787.

49 Sioud M, Leirdal M. Design of nuclease resistant protein kinase C-alpha DNA enzymes with potential therapeutic application. *J Mol Biol.* 2000;296:937–947.

50 Nunamaker EA, Zhang HY, Shirasawa Y, et al. Electroporation-mediated delivery of catalytic oligodeoxynucleotides for manipulation of vascular gene expression. *Am J Physiol Heart Circ Physiol.* 2003;285:H2240–H2247.

51 Hjiantoniou E, Iseki S, Uney JB, Phylactou LA. DNazyme-mediated cleavage of Twist transcripts and increase in cellular apoptosis. *Biochem Biophys Res Commun.* 2003;300:178–181.

52 Zhang L, Gasper WJ, Stass SA, et al. Angiogenic inhibition mediated by a DNAzyme that targets vascular endothelial growth factor receptor 2. *Cancer Res.* 2002;62:5463–5469.

53 Grimpe B, Dong S, Duller C, et al. The critical role of basement membrane-independent laminin gamma chain during axon regeneration in the CNS. *J Neurosci.* 2002;22:3144–3160.

54 Unwalla H, Banerjea AC. Inhibition of HIV-1 gene expression by novel macrophage-tropic DNA enzymes targeted to cleave HIV-1 TAT/Rev RNA. *Biochem J.* 2001;357:147–155.

55 Unwalla H, Banerjea AC. Novel mono- and di-DNA-enzymes targeted to cleave TAT or TAT-REV RNA inhibit HIV-1 gene expression. *Antiviral Res.* 2001;51:127–139.

56 Liu C, Cheng R, Sun LQ, Tien P. Suppression of platelet-type 12-lipoxygenase activity in human erythroleukemia cells by an RNA-cleaving DNAzyme. *Biochem Biophys Res Commun.* 2001;284:1077–1082.

91

KLF2

A "Molecular Switch" Regulating Endothelial Function

Zhiyong Lin and Mukesh K. Jain

Case Cardiovascular Research Institute, Case Western Reserve University, Cleveland, Ohio

Krüppel-like factors (KLFs) are a subclass of the zinc-finger family of transcriptional regulators implicated in the regulation of cellular growth and differentiation and tissue development. The term *Krüppel* is a German word meaning "cripple." This is based on the observation that *Drosophila* embryos homozygous for *Krüppel* die due to altered thoracic and anterior abdominal segments (1–3). Over the past 13 years, 17 mammalian KLFs have been identified and found to play important roles in diverse cell types. For example, the first mammalian KLF, termed *erythroid Krüppel-like factor (EKLF/KLF1)*, was found to play a key role in β-globin gene synthesis and erythrocyte development (4,5). Other family members have been shown to play critical roles in diverse processes ranging from epithelial cell differentiation, tumor cell growth, and bone formation (6–10). Recent studies also implicate an important role for these factors in cardiovascular biology (11,12). In this chapter, we discuss published studies to date that support a critical role of lung Krüppel-like factor 2 (LKLF/KLF2) in endothelial biology.

IDENTIFICATION AND INITIAL CHARACTERIZATION OF KRÜPPEL-LIKE FACTOR-2

KLF2 was originally cloned in the Lingrel laboratory using a homology screening strategy (13). Subsequent gene targeting studies indicate an essential role for this factor in programming the quiescent phenotype of single-positive T cells and in lung development (14,15). In addition, *KLF2-/-* mice exhibit abnormal blood vessel formation due to insufficient smooth muscle cell recruitment, which results in embryonic hemorrhage and death between embryonic (E) day 12.5 and 14.5(16). The latter observation suggested that KLF2 has an important role in vessel biology. Interestingly, it was shown that within the blood vessel wall, KLF2 expression was limited to endothelial cells (ECs) (16). However, until recently, the function of KLF2 in this cell type remained poorly understood.

FUNCTIONS OF KRÜPPEL-LIKE FACTOR-2 IN ENDOTHELIAL BIOLOGY

The vascular endothelium, which forms the inner lining of cells in all blood vessels, is a critical integrator and transducer of various physiological stimuli (17). The endothelium is a "quiescent" tissue that regulates vascular permeability, blood coagulation, and homing of immune cells to specific sites of the body. Dysfunction of the endothelium is a key pathophysiologic event in the development and progression of diverse vascular disease states. Characteristic abnormalities include impaired endothelium-dependent vasodilation, enhanced expression of leukocyte adhesion molecules, increased elaboration of procoagulant factors, and abnormal vascular remodeling.

Various stimuli can modulate the phenotype of ECs. For example, laminar shear stress (LSS) induces various factors, such as endothelial nitric oxide synthase (eNOS) and thrombomodulin (TM), that confer potent antithrombotic, antiadhesive, and anti-inflammatory properties (18,19). In contrast, noxious stimuli, such as proinflammatory cytokines or advanced glycation end products, can render the endothelium dysfunctional. As a consequence, eNOS expression is reduced while the expression of adhesion molecules (e.g., vascular cell adhesion molecule [VCAM]-1) and prothrombotic factors (e.g., tissue factor) is induced (20,21). Given the critical function of endothelium in homeostasis, the identification of the regulatory factors that mediate the effects of these stimuli on endothelial (dys)function is of considerable interest. Studies from our laboratory, coupled with others, implicate KLF2 as a molecular regulator of endothelial function (Figure 91.1).

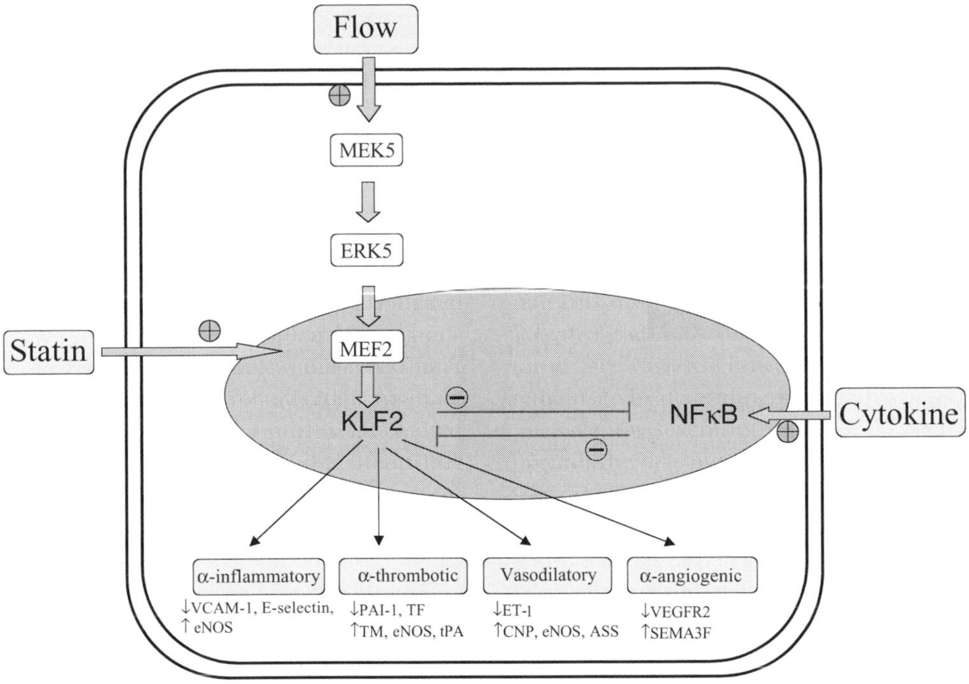

Figure 91.1. Functions and regulation of KLF2 in ECs. KLF2 differentially affects the expression of factors that confer anti-inflammatory, antithrombotic, antiproliferative effects in ECs. Laminar shear stress induces KLF2 expression through the MEK5/ERK5/MEF2 pathway. Cytokines, such as TNF-α, reduce KLF2 expression. Statins induce KLF2 expression.

Krüppel-Like Factor-2 Regulates Endothelial Proinflammatory Activation

In response to inflammatory stimuli, ECs are induced to express adhesion molecules and chemokines that recruit immune cells to the blood vessel wall. This early event then sets the stage for a series of complex interactions between these immune cells and nonimmune cells that dictate disease development and progression (22). Our laboratory first reported KLF2 as a novel transcriptional regulator of endothelial proinflammatory activation. We observed that KLF2 can potently inhibit the expression of VCAM-1 and E-selectin in ECs, resulting in significant attenuation of immune cell adhesion. The effect of KLF2 is specific because, in contrast to VCAM-1 and E-selectin, intercellular adhesion molecule (ICAM)-1 expression was not affected (23).

It is noteworthy that KLF2 inhibits the effects of multiple proinflammatory stimuli including interleukin (IL)-1β, tumor necrosis factor (TNF)-α, lipopolysaccharide (LPS), and thrombin (23,24). This raised the possibility that inhibition of a common inflammatory pathway may account for KLF2's favorable effects. Indeed, subsequent studies indicate that KLF2 can inhibit the nuclear factor (NF)-κB pathway at multiple levels. In the case of stimuli such as IL-1β and TNF-α, inhibition occurs via recruitment of transcriptional coactivators (e.g., cyclic adenosine monophosphate [cAMP] response element–binding protein CBP/p300) (23). However, in the case of thrombin, a second mechanism is also operative. We found that KLF2 potently inhibits the thrombin-mediated induction of multiple cytokines/chemokines (e.g., monocyte chemoattractant protein [MCP]-1, IL–6, and IL-8) by inhibiting the expression of its principal receptor, protease-activated receptor (PAR)-1. As a consequence, NF-κB nuclear accumulation and DNA-binding are strongly reduced (24). Thus, KLF2 can inhibit endothelial proinflammatory activation by multiple mechanisms that affect the NF-κB pathway.

Krüppel-Like Factor-2 Regulates Endothelial Thrombotic Function

Under normal circumstances, the vascular endothelium maintains blood fluidity by producing inhibitors of coagulation and platelet aggregation, modulating vascular tone, and providing a protective barrier separating hemostatic blood components from subendothelial structures (25). Hemostasis may occur if the capacity of inhibitory pathways is impaired or if the capacity of endogenous anticoagulant systems is overwhelmed by a noxious stimulus (25). Recent work indicates that KLF2 can regulate the expression of key factors that regulate vessel thrombosis. Overexpression of KLF2 strongly induced TM, a potent antithrombotic factor. In parallel, KLF2 overexpression also inhibited expression of the prothrombotic proteins plasminogen activator inhibitor (PAI)-1 and tissue factor (in the setting of cytokine activation). Conversely, small interfering RNA (siRNA)-mediated knockdown of KLF2 resulted in antithetical effects. Consistent with these gene expression effects, blood clotting time and flow rates were increased in the presence of KLF2 and reduced by KLF2-deficiency (26).

Krüppel-Like Factor-2 Regulates Expression of Factors Implicated in Regulating Vasoreactivity

Under physiologic conditions, the endothelium regulates vessel tone through the elaboration of several vasodilatory factors such as eNOS or C-natriuretic peptide (CNP). In disease states, the expression of these factors is reduced, whereas that of vasoconstrictive molecules (e.g., endothelin-1) is induced. Several recent studies implicate KLF2 in the regulation of these gene products. KLF2 strongly induces the expression of both eNOS and CNP (27). Furthermore, the expression of argininosuccinate synthase (ASS), a limiting enzyme in eNOS substrate bioavailability (28), also is induced. On the other hand, KLF2 reduced the expression of genes regulating vasoconstriction, such as endothelin-1. In addition, KLF2 decreased the expression of caoveolin-1, a cell membrane protein that serves as a negative regulator of eNOS activity (29). The importance of KLF2 was verified by siRNA studies; reduction of KLF2 significantly attenuated flow-mediated induction of eNOS and CNP while increasing endothelin-1 expression (27).

Krüppel-Like Factor-2 Regulates Endothelial Proliferation and Migration

Under normal conditions, the endothelium retains a highly quiescent phenotype. However, in response to certain stimuli (e.g., vascular endothelial growth factor [VEGF]), ECs can be induced to proliferate, migrate, and establish new blood vessels. Although angiogenesis is of physiologic importance in the adult (e.g., during the menstrual cycle), inappropriate angiogenesis can contribute to serious pathology, such as tumor formation. Studies from our group and others indicate that KLF2 has potent antiangiogenic effects. Bhattacharya and colleagues revealed that forced expression of KLF2 retards VEGF-mediated calcium flux, proliferation, and proinflammatory factor expression. Furthermore, in a nude mouse ear model, KLF2 overexpression dramatically attenuated VEGF-induced new vessel formation and tissue edema. Mechanistically, this reduction was at least in part due to KLF2's ability to inhibit the expression of the key VEGF receptor, VEGFR2/KDR (30). Dekker and colleagues also have provided evidence that KLF2 inhibits EC migration by virtue of its ability to induce the expression of semaphorin 3F (SEMA3F) (31). Thus, by reducing EC proliferation and migration, KLF2 can inhibit angiogenesis.

MECHANISMS REGULATING KRÜPPEL-LIKE FACTOR-2 EXPRESSION IN ENDOTHELIAL CELLS

As noted previously, KLF2 expression in the vasculature is limited to ECs. In this regard, it is noteworthy that expression in the endothelium is not uniform. Indeed, as elegantly demonstrated by Dekker and colleagues, KLF2 was robustly expressed in linear segments of the human aorta, but expression was notably lower at branch points of aorta (and near the bifurcation in the iliac and carotid arteries) (32,33). Linear segments of the aorta experience laminar flow – a stimulus known to induce KLF2. In contrast, branch points experience disturbed flow and are the regions most susceptible to the development of atherosclerotic lesions (17). Furthermore, branch-point areas also are characterized by heightened NF-κB activity and the expression of VCAM-1 (34). These observations are particularly noteworthy in light of studies discussed in the following sections, regarding the regulation of KLF2 expression.

Flow-Mediated Induction of Krüppel-Like Factor-2 Expression

We and others have demonstrated that LSS can strongly induce KLF2 expression in ECs (23,32,35). The mechanistic basis for this effect has been addressed recently in a series of studies. Promoter deletion studies revealed that the proximal region of *KLF2* promoter can be induced by flow (36). A breakthrough observation first reported by Kumar and colleagues was that this region contains a functional consensus binding site for a family of transcription factors termed myocyte enhancer factor 2 (MEF2) proteins (37). MEF2 factors are members of the MADS box (MCM1, Agamous, Deficiens, Serum response factor) family of transcription factors that bind to A/T-rich sequences (38). Although best known for their role in muscle development, an emerging literature implicates MEF2A and MEF2C as critical regulators of endothelial biology (39–41). For example, Wang and colleagues identified mutations in MEF2A in an inherited disorder with features of coronary artery disease (41) (see Chapter 94). Furthermore, MEF2C has been implicated as a regulator of endothelial integrity and permeability (39). The basis for the favorable effects of MEF2 factors in ECs is not understood, but the link to KLF2 provides a potential explanation. Indeed, this link has been substantiated by two additional studies. First, loss-of-function studies by Sohn and colleagues showed that extracellular signal-regulated kinase 5 (ERK5), a kinase known to activate MEF2 factors, is essential for embryonic KLF2 expression (42). This observation is particularly noteworthy because ERK5 also is known to be activated by laminar flow (43), raising the possibility that LSS-mediated activation of ERK5 stimulates MEF2, which in turn induces KLF2. Consistent with this hypothesis, Parmar and colleagues showed that the overexpression of a dominant negative MEF2 (or mutant MEK5 – an upstream activator of ERK5) prevented flow-mediated induction of KLF2 expression in ECs (27).

In addition to MEF2 factors, observations largely derived from promoter analyses under flow by the Lingrel laboratory implicate a number of additional factors that may contribute KLF2 expression. These factors include p300/CBP-associated factor (PCAF), heterogeneous nuclear ribonucleoprotein D (hnRNP D), and nucleolin (44,45). However, the importance of these factors in regulating endogenous KLF2 expression under basal conditions or in the context of flow remained to be defined.

Cytokine-Mediated Inhibition of Krüppel-Like Factor-2 Expression

Proinflammatory cytokines such as TNF-α and IL-1β have been shown to repress KLF2 expression in ECs (23,37). This is an important observation, because reduction in KLF2 expression may lead to unopposed NF-κB activity and endothelial proinflammatory activation. In this regard, a recent study has provided novel insights. Using chemical and genetic inhibitors, Kumar and colleagues determined that the TNF-α–mediated reduction of KLF2 was dependent on both the NF-κB and histone deacetylase (HDAC) pathways. Next, using a combination of promoter deletion and mutation analyses, chromatin immunoprecipitation (ChIP) assays, and siRNA-mediated knockdown studies, evidence was provided that p65 (a component of NF-κB) and HDAC4 cooperate to inhibit the ability of MEF2 factors to induce KLF2 expression (37). These studies, coupled with the observation that KLF2 can inhibit NF-κB, raise the possibility that the balance of these two transcriptional pathways may regulate the state of endothelial activation.

THERAPEUTIC IMPLICATIONS

Given the favorable properties that KLF2 confers to the endothelium, one would in principle like to identify mechanisms to exploit this in a therapeutically meaningful manner. In this regard, studies from our group and others have identified a novel link between KLF2 and a commonly prescribed class of medications, the statins (46,47). Originally designed to lower cholesterol, these 3-hydroxy-3-methylglutaryl coenzyme A (HMG-CoA) reductase inhibitors are increasingly recognized as agents that confer favorable properties to the vasculature. However, accumulating information from clinical and experimental studies also indicates that the magnitude of the favorable effects of statins exceed what may be anticipated by lipid lowering alone.

A review of studies assessing the effects of statins on endothelial gene expression revealed a marked overlap with those observed in studies of KLF2. Consistent with this observation, we and others have reported that multiple statins can induce KLF2 expression (23,47). This induction was abrogated by geranylgeranyl pyrophosphate (GGPP) but not farnesyl pyrophosphate (FPP), thereby implicating the Rho pathway. Consistent with this observation, we found that overexpression of a constitutively active form of Rho strongly reduced KLF2 expression. Furthermore, we demonstrated that induction of KLF2 by statins occurs through the MEF2 site in the proximal promoter region (46). Finally, siRNA-mediated knockdown studies demonstrated that the statin-mediated induction of *eNOS* and *TM* mRNA and protein accumulation within ECs is KLF2-dependent. Taken together, these data identify KLF2 as an important nuclear effector of statin effects in ECs.

ROLE OF OTHER KRÜPPELS IN THE ENDOTHELIUM

Several other members of the KLF family also may have important roles in endothelial biology. Previous studies indicate KLF6 (also called zf9), a ubiquitously expressed factor, is induced in arterial ECs after vascular injury (48). Other studies show that KLF6 can induce urokinase-type plasminogen activator (u-PA) – a key enzyme implicated in tissue remodeling, tumor metastasis, and apoptosis (49). This induction also increased the amount of bioactive transforming growth factor (TGF)-β1 – an important regulator of the vessel injury response. A second KLF factor expressed in ECs is KLF4 (50). Interestingly, KLF4 is induced by both laminar flow as well as thrombin stimulation. However, the role of this factor in endothelial gene expression and function remains unknown.

KEY POINTS

- Compelling evidence implicates KLF2 as a "molecular switch" that regulates endothelial gene expression and function.
- KLF2 differentially affects the expression of factors that confer anti-inflammatory, antithrombotic, antiproliferative effects in ECs.
- KLF2 is induced by laminar flow. This occurs via a MEK5/ERK5/MEF2 signaling pathway.
- Cytokines, such as TNF-α, inhibit KLF2 expression through the recruitment of p65/HDAC4 complex to prevent MEF2-mediated induction of KLF2.
- Statins induce KLF2 expression. This induction of KLF2 is critical for the ability of statins to induce key endothelial factors such as eNOS and TM.

Future Goals

- To determine the exact nature of KLF2 function in the vasculature
- To develop in vivo gain- and loss-of-function studies to understand the role of KLF2 in vessel biology in vivo
- To determine whether statin-mediated induction of KLF2 is operative and functionally important in vivo
- To determine whether current statins be modified to develop agents that can more potently induce KLF2
- To determine the existence of a role for other members of this family beyond KLF2

REFERENCES

1 Jackle H, Rosenberg UB, Preiss A, et al. Molecular analysis of Kruppel, a segmentation gene of *Drosophila melanogaster*. *Cold Spring Harb Symp Quant Biol*. 1985;50:465–473.

2 Preiss A, Rosenberg UB, Kienlin A, et al. Molecular genetics of Kruppel, a gene required for segmentation of the *Drosophila* embryo. *Nature*. 1985;313(5997):27–32.

3 Nusslein-Volhard C, Wieschaus E. Mutations affecting segment number and polarity in *Drosophila*. *Nature*. 1980;287(5785): 795–801.

4 Nuez B, Michalovich D, Bygrave A, et al. Defective haematopoiesis in fetal liver resulting from inactivation of the EKLF gene. *Nature*. 1995;375(6529):316–318.

5 Perkins AC, Sharpe AH, Orkin SH. Lethal beta-thalassaemia in mice lacking the erythroid CACCC-transcription factor EKLF. *Nature*. 1995;375(6529):318–322.

6 Wei D, Kanai M, Huang S, Xie K. Emerging role of KLF4 in human gastrointestinal cancer. *Carcinogenesis*. 2006;27(1):23–31.

7 Kimmelman AC, Qiao RF, Narla G, et al. Suppression of glioblastoma tumorigenicity by the Kruppel-like transcription factor KLF6. *Oncogene*. 2004;23(29):5077–5083.

8 Narla G, Heath KE, Reeves HL, et al. KLF6, a candidate tumor suppressor gene mutated in prostate cancer. *Science*. 2001;294 (5551):2563–2566.

9 Segre JA, Bauer C, Fuchs E. Klf4 is a transcription factor required for establishing the barrier function of the skin. *Nat Genet*. 1999; 22(4):356–360.

10 Subramaniam M, Gorny G, Johnsen SA, et al. TIEG1 null mouse-derived osteoblasts are defective in mineralization and in support of osteoclast differentiation in vitro. *Mol Cell Biol*. 2005; 25(3):1191–1199.

11 Feinberg MW, Lin Z, Fisch S, Jain MK. An emerging role for Kruppel-like factors in vascular biology. *Trends Cardiovasc Med*. 2004;14(6):241–246.

12 Suzuki T, Aizawa K, Matsumura T, Nagai R. Vascular implications of the Kruppel-like family of transcription factors. *Arterioscler Thromb Vasc Biol*. 2005;25(6):1135–1141.

13 Anderson KP, Kern CB, Crable SC, Lingrel JB. Isolation of a gene encoding a functional zinc finger protein homologous to erythroid Kruppel-like factor: identification of a new multigene family. *Mol Cell Biol*. 1995;15(11):5957–5965.

14 Wani MA, Wert SE, Lingrel JB. Lung Kruppel-like factor, a zinc finger transcription factor, is essential for normal lung development. *J Biol Chem*. 1999;274(30):21180–21185.

15 Kuo CT, Veselits ML, Leiden JM. LKLF: a transcriptional regulator of single-positive T cell quiescence and survival [See comments]. *Science*. 1997;277(5334):1986–1990.

16 Kuo CT, Veselits ML, Barton KP, et al. The LKLF transcription factor is required for normal tunica media formation and blood vessel stabilization during murine embryogenesis. *Genes Dev*. 1997;11(22):2996–3006.

17 Gimbrone MA Jr., Topper JN, Nagel T, et al. Endothelial dysfunction, hemodynamic forces, and atherogenesis. *Ann NY Acad Sci*. 2000;902:230–9; discussion 9–40.

18 Topper JN, Cai J, Falb D, Gimbrone MA Jr. Identification of vascular endothelial genes differentially responsive to fluid mechanical stimuli: cyclooxygenase-2, manganese superoxide dismutase, and endothelial cell nitric oxide synthase are selectively up-regulated by steady laminar shear stress. *Proc Natl Acad Sci USA*. 1996;93(19):10417–10422.

19 Takada Y, Shinkai F, Kondo S, et al. Fluid shear stress increases the expression of thrombomodulin by cultured human endothelial cells. *Biochem Biophys Res Commun*. 1994;205(2):1345–1352.

20 Neish AS, Williams AJ, Palmer HJ, et al. Functional analysis of the human vascular cell adhesion molecule 1 promoter. *J Exp Med*. 1992;176(6):1583–1593.

21 Bevilacqua MP, Pober JS, Majeau GR, et al. Recombinant tumor necrosis factor induces procoagulant activity in cultured human vascular endothelium: characterization and comparison with the actions of interleukin 1. *Proc Natl Acad Sci USA*. 1986;83 (12):4533–4537.

22 Libby P. Inflammation in atherosclerosis. *Nature*. 2002;420 (6917):868–874.

23 SenBanerjee S, Lin Z, Atkins GB, et al. KLF2 is a novel transcriptional regulator of endothelial proinflammatory activation. *J Exp Med*. 2004;199(10):1305–1315.

24 Lin Z, Hamik A, Jain R, et al. Kruppel-like factor 2 inhibits protease activated receptor-1 expression and thrombin-mediated endothelial activation. *Arterioscler Thromb Vasc Biol*. 2006;26 (5):1185–1189.

25 Esmon CT. Crosstalk between inflammation and thrombosis. *Maturitas*. 2004;47(4):305–314.

26 Lin Z, Kumar A, SenBanerjee S, et al. Kruppel-like factor 2 (KLF2) regulates endothelial thrombotic function. *Circ Res*. 2005;96(5):E48–E57.

27 Parmar KM, Larman HB, Dai G, et al. Integration of flow-dependent endothelial phenotypes by Kruppel-like factor 2. *J Clin Invest*. 2006;116(1):49–58.

28 Goodwin BL, Solomonson LP, Eichler DC. Argininosuccinate synthase expression is required to maintain nitric oxide production and cell viability in aortic endothelial cells. *J Biol Chem*. 2004;279(18):18353–18360.

29 Razani B, Engelman JA, Wang XB, et al. Caveolin-1 null mice are viable but show evidence of hyperproliferative and vascular abnormalities. *J Biol Chem*. 2001;276(41):38121–38138.

30 Bhattacharya R, Senbanerjee S, Lin Z, et al. Inhibition of vascular permeability factor/vascular endothelial growth factor-mediated angiogenesis by the Kruppel-like factor KLF2. *J Biol Chem*. 2005;280(32):28848–28851.

31 Dekker RJ, Boon RA, Rondaij MG, et al. KLF2 provokes a gene expression pattern that establishes functional quiescent differentiation of the endothelium. *Blood*. 2006;107(11):4354–4363. Epub 2006 Feb.

32 Dekker RJ, van Soest S, Fontijn RD, et al. Prolonged fluid shear stress induces a distinct set of endothelial cell genes, most specifically lung Kruppel-like factor (KLF2). *Blood*. 2002;100(5):1689–1698.

33 Dekker RJ, van Thienen JV, Rohlena J, et al. Endothelial KLF2 links local arterial shear stress levels to the expression of vascular tone-regulating genes. *Am J Pathol*. 2005;167(2):609–618.

34 Hajra L, Evans AI, Chen M, et al. The NF-kappa B signal transduction pathway in aortic endothelial cells is primed for activation in regions predisposed to atherosclerotic lesion formation. *Proc Natl Acad Sci USA*. 2000;97(16):9052–9057.

35 Wang N, Miao H, Li YS, et al. Shear stress regulation of Kruppel-like factor 2 expression is flow pattern-specific. *Biochem Biophys Res Commun*. 2006;341(4):1244–1251.

36 Huddleson JP, Srinivasan S, Ahmad N, Lingrel JB. Fluid shear stress induces endothelial KLF2 gene expression through a defined promoter region. *Biol Chem.* 2004;385(8):723–729.

37 Kumar A, Lin Z, SenBanerjee S, Jain MK. Tumor necrosis factor alpha-mediated reduction of KLF2 is due to inhibition of MEF2 by NF-kappaB and histone deacetylases. *Mol Cell Biol.* 2005;25(14):5893–5903.

38 McKinsey TA, Zhang CL, Olson EN. Activation of the myocyte enhancer factor-2 transcription factor by calcium/calmodulin-dependent protein kinase-stimulated binding of 14–3–3 to histone deacetylase 5. *Proc Natl Acad Sci USA.* 2000;97(26): 14400–14405.

39 Hayashi M, Lee JD. Role of the BMK1/ERK5 signaling pathway: lessons from knockout mice. *J Mol Med.* 2004;82(12):800–808.

40 Olson EN. Undermining the endothelium by ablation of MAPK-MEF2 signaling. *J Clin Invest.* 2004;113(8):1110–1112.

41 Wang L, Fan C, Topol SE, et al. Mutation of MEF2A in an inherited disorder with features of coronary artery disease. *Science.* 2003;302(5650):1578–1581.

42 Sohn SJ, Li D, Lee LK, Winoto A. Transcriptional regulation of tissue-specific genes by the ERK5 mitogen-activated protein kinase. *Mol Cell Biol.* 2005;25(19):8553–8566.

43 Yan C, Takahashi M, Okuda M, et al. Fluid shear stress stimulates big mitogen-activated protein kinase 1 (BMK1) activity in endothelial cells. Dependence on tyrosine kinases and intracellular calcium. *J Biol Chem.* 1999;274(1):143–150.

44 Ahmad N, Lingrel JB. Kruppel-like factor 2 transcriptional regulation involves heterogeneous nuclear ribonucleoproteins and acetyltransferases. *Biochemistry.* 2005;44(16):6276–6285.

45 Huddleson JP, Ahmad N, Lingrel JB. Upregulation of the KLF2 transcription factor by fluid shear stress requires nucleolin. *J Biol Chem.* 2006;281:15121–15128.

46 Sen-Banerjee S, Mir S, Lin Z, et al. Kruppel-like factor 2 as a novel mediator of statin effects in endothelial cells. *Circulation.* 2005;112(5):720–726.

47 Parmar KM, Nambudiri V, Dai G, et al. Statins exert endothelial atheroprotective effects via the KLF2 transcription factor. *J Biol Chem.* 2005;280(29):26714–26719.

48 Botella LM, Sanchez-Elsner T, Sanz-Rodriguez F, et al. Transcriptional activation of endoglin and transforming growth factor-beta signaling components by cooperative interaction between Sp1 and KLF6: their potential role in the response to vascular injury. *Blood.* 2002;100(12):4001–4010.

49 Kojima S, Hayashi S, Shimokado K, et al. Transcriptional activation of urokinase by the Kruppel-like factor Zf9/COPEB activates latent TGF-beta1 in vascular endothelial cells. *Blood.* 2000;95(4):1309–1316.

50 Yet SF, McA'Nulty MM, Folta SC, et al. Human EZF, a Kruppel-like zinc finger protein, is expressed in vascular endothelial cells and contains transcriptional activation and repression domains. *J Biol Chem.* 1998;273(2):1026–1031.

92

NFAT Transcription Factors

Takashi Minami

The Research Center for Advanced Science and Technology, The University of Tokyo, Japan

The nuclear factor of activated T cells (NFAT) family of transcription factors consists of five family members, four of which are regulated by calcium/calcineurin (NFATc1, -2, -3, and -4) and one of which is calcium/calcineurin independent (NFAT5).[1] Although calcineurin is highly conserved from yeast and humans, NFATc1 through 4 transcription factors appeared later during evolution, with the rise of chordates. The NFAT signaling pathway was first described in T cells. However, most cell types express one or more isoforms of NFAT. Indeed, these proteins have been shown to play a critical role in such diverse processes as the development of skeletal muscle, cartilage, the central nervous system (CNS), and the cardiovascular system. In this chapter, we review evidence for the role of NFAT in endothelial cell (EC) biology. Of particular interest is the recent finding that NFAT signaling in ECs is tightly regulated by a negative feedback inhibitor, the Down syndrome critical region (DSCR)-1.

STRUCTURE OF NUCLEAR FACTOR OF ACTIVATED T CELLS

NFAT proteins contain a highly conserved DNA-binding domain that is similar to the DNA-binding domain of the Rel family of transcription factors (Figure 92.1). Indeed, all five members of the NFAT family may be classified as members of the extended nuclear factor (NF)-κB/Rel family (1). In contrast to canonical Rel-containing factors, the Rel binding domain of NFAT binds weakly as a monomer or dimer. Instead, NFAT cooperates with other transcription factors to upregulate target genes (discussed in the next section). In addition to the DNA-binding domain, NFAT proteins contain a regulatory domain, which mediates transactivation, and docking sites for calcineurin (CnA) and NFAT kinases. The calcium-dependent NFATs (NFATc1–4) contain 14 conserved phosphorylation sites, all but one of which is dephosphorylated by CnA (discussed next).

1 NFAT5 (also known as TonEBP, OREBP), the primordial member of the NFAT family, functions in immune response and kidney homeostasis.

REGULATION OF NUCLEAR FACTOR OF ACTIVATED T CELLS SIGNALING

The activity of NFAT is mediated by its phosphorylation state (1,2). The phosphorylation state is controlled by the opposing actions of the calcium (Ca^{2+})-dependent serine/threonine protein phosphatase, CnA[2], and a number of so-called "maintenance" and "export" serine/threonine kinases, including GSK-3β, casein kinase-1, p38 mitogen-activated protein kinase (MAPK), and c-jun N-terminal kinase (JNK). In resting cells, NFAT is phosphorylated and localized in the cytoplasm. Upon ligand binding to cell surface receptors, a series of linked steps leads to the induction of NFAT transcriptional activity:

1. Activation of phospholipase C (PLC)
2. Release of inositol triphosphate (IP3)
3. Transient release of calcium from intracellular stores through IP3 receptors on the endoplasmic reticulum
4. Influx of calcium through a specialized calcium release–activated calcium (CRAC) channel
5. Ca^{2+} binding to calmodulin
6. Calmodulin-dependent activation of CnA

2 Calcineurin was purified during the 1970s, and was named owing to its regulation by calcium and its presence in neural tissue. Mammalian calcineurin consists of three components: calmodulin, a catalytic subunit (CnA), and a regulatory subunit (CnB). There are three isoforms of CnA (CnAα, CnAβ and CnAγ), and two isoforms of CnB (CNB1 and CNB2). A rise in intracellular calcium concentration leads to increased binding of calmodulin to CnA, and a conformation shift such that the active site of CnA is exposed. Calcineurin interacts with many proteins, including scaffold proteins (e.g., Cabin1), endogenous inhibitors (e.g., Csp1, DSCR-1), and substrates (e.g., NFAT). The classic pharmacological inhibitors of calcineurin, CsA and FK506, both derived from fungi, bind to endogenous immunophilin proteins (cyclophilin in the case of CsA, FKBP12 for FK 506), which then inhibit the catalytic activity of calcineurin, thus blocking both NFAT and non-NFAT calcineurin substrates.

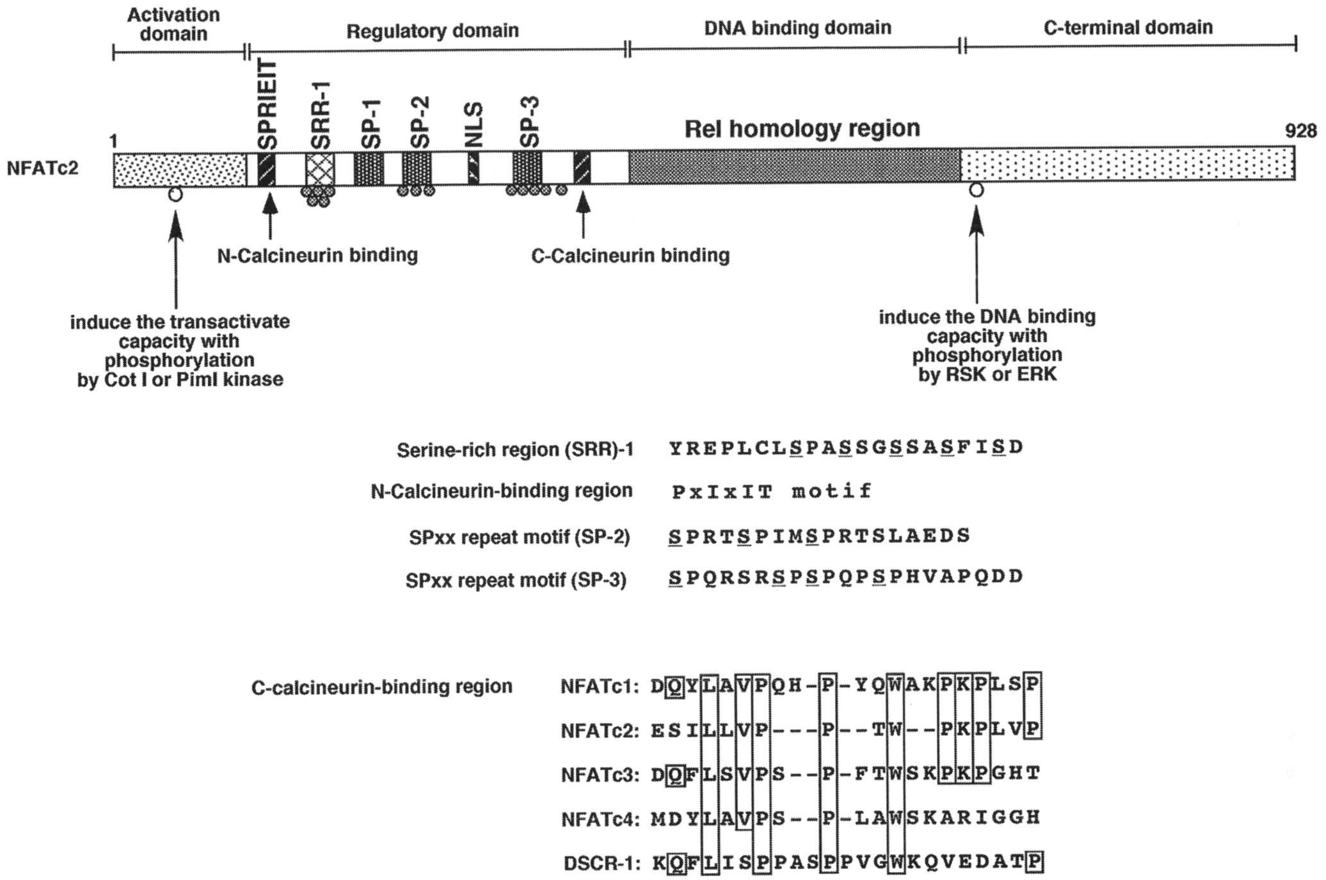

Figure 92.1. Schematic diagram of NFAT structure. *(Top)* Shown is NFATc2. Hatched circles indicate phosphorylated serine residues that are dephosphorylated by calcineurin. Open circles indicate serine residues that are phosphorylated by Cot I and Pim I (*left*) or RSK and ERK (*right*). *(Middle)* Sequences of phosphorylated motifs in NFATc2 (phosphorylated serine residues are underlined). *(Bottom)* Sequences of the C-calcineurin binding region aligned with the corresponding sequences of each NFAT family protein and the similar region in DSCR-1. NLS, nuclear localization signal.

7. CnA-mediated dephosphorylation of NFAT, which exposes the nuclear localization signal
8. Translocation of dephosphorylated NFAT to the nucleus[3]
9. Binding of nuclear localized NFAT to promoters and activation of target gene expression

As mentioned earlier, NFATs may bind as monomers, homodimers, or heterodimers, but more commonly cooperate with other transcription factors, including AP1, GATA, or myocyte enhancer factor 2 (MEF2) families.[4] The nature of the binding partners is determined – at least in part – by the NFAT isoform and the cell type.

CnA is potently inhibited by the immunosuppressive drugs cyclosporin and FK506, and by a number of endogenous inhibitors, collectively known as calcipressins. These include calcineurin-binding protein 1 (CABIN1), A-kinase anchor protein (AKAP79), and DSCR-1.

ROLE FOR NUCLEAR FACTOR OF ACTIVATED T CELLS IN THE VASCULATURE

In Vivo Studies

Targeted disruption of NFAT has demonstrated a role for these transcription factors not only in immunity, but also in a

3 Although transcriptional activity of NFAT is regulated primarily by its subcellular localization and by the relative activity of coactivators such as AP1, evidence suggests that NFAT may be post-transcriptionally modified, for example by phosphorylation of the transactivation domain by Cot1/Tpl-2 or Pim1 kinases, or phosphorylation of the DNA-binding region by p60 RSK and MAPK.

4 The classical paradigm in the immunology literature involves cooperative interactions between NFAT and AP1. Concomitant activation of the Ca^{2+}/calcineurin/NFAT and PKC/Ras/Fos/Jun pathways leads to synergistic activation of many target genes in T cells. GATA–NFAT interactions have been described in T cells, cardiomyocytes and, more recently, ECs. Dephosphorylated NFATc2 (but not other NFAT family members) associates with MEF2 and the coactivator p300, resulting in transcriptional synergy in T cells and skeletal muscle.

number of other critical developmental processes, including the formation of the cardiovascular system.

NFATc1$^{-/-}$ mice are lethal at embryonic day (E)13.5 to 17.5 and demonstrate abnormal aortic and pulmonary valve development, as well as malformation of the interventricular septum (3,4). The heart valves forms through a unique developmental process termed *endocardial-mesenchymal transformation* (EMT), in which ECs within the endocardial cushions delaminate from the endocardial layer and transdifferentiate to become mesenchymal cells (5). These mesenchymal cells proliferate and invade the myocyte-derived basement membrane-like cardiac jelly in the endocardial cushion area. The cushion undergoes a series of morphogenic steps that lead to the formation of the valve. During mouse embryogenesis, NFATc1 expression is restricted to the heart (3) and possibly the thymus (4). Expression in the heart is limited to the endocardium. NFATc1 is localized in the endocardial tubes at E7.5; the endocardium and vitelline vein at E8.5; the endocardium of the ventricles, atria, and outflow tract between E9 and E11.5; and in the lining of the aortic, pulmonary, and AV valves, and endocardial cushion between E11.5 and E13.5 (3). In contrast to the developing heart, studies of newborn and adult hearts did not reveal evidence of NFATc1 expression, suggesting that the gene is turned off at some point during development (3). Immunofluorescence of embryonic hearts revealed nuclear localization of NFATc1 in many, but not all endocardial cells. This pattern was reversed by administration of CsA or FK506 to embryos. These data suggest that NFAT1c1 activity is mediated by short-range paracrine signals. Vascular endothelial growth factor (VEGF) may represent one such signal; VEGF is produced by cardiomyocytes, and was shown to induce nuclear translocation of NFATc1 in pulmonary valve ECs (6).

NFAT1c$^{-/-}$ embryos displayed multiple abnormalities, including hypertrophied ventricular walls, narrowed/occluded outflow tract, abnormal/underdeveloped valve structure, and persistent interventricular foramen/defective septum formation (3). Vascular endothelium appeared unaffected by loss of NFATc1, as indicated by the normal expression of platelet-endothelial cell adhesion molecule (PECAM)-1. The authors hypothesized that death at E14.5 was caused by valvular incompetence or stenosis, and concluded that the Ca^{2+}/calcineurin/NFATc signaling pathway is essential for normal cardiac valves and septum morphogenesis. These findings, along with those of another group (4), suggest that ubiquitous calcium–CnA signaling is channeled into a valve-promoting pathway via the time- and site-specific expression of NFATc1 in endocardial cells.

In a more recent study, the EC-specific deletion of *Cnb1* resulted in cytoplasmic localization of NFATc1 in endocardial cells and valvular defects similar to those observed in *NFATc1*$^{-/-}$ mice (7). Tie2–mediated overexpression of NFATc1 rescued the valvular defects and viability of *NFATc1*$^{-/-}$ mice, but not the mice deficient in endocardial Cnb1. The administration of CsA to embryos at E11 phenocopied the *NFATc1*$^{-/-}$ and endocardial-specific *Cnb1*-knockout. Together, these data support a critical role for endocardial Ca^{2+}/CnA/NFATc1 signaling in valve formation. Of note, these genetic and pharmacologic manipulations did not result in defective EMT. Rather, a triple deletion of NFATc2/c3/c4 was required to reduce the numbers of mesenchymal cells in the cushion areas (7).

Combined *NFATc3/c4*-knockout resulted in embryonic lethality (E11.5) owing to impaired blood vessel assembly (8). The phenotype included underdeveloped yolk sac vasculature with enlarged and disordered capillary plexus and poorly developed vitelline vessels, and abnormalities in the major vessels of the embryo proper. Differentiation, proliferation, and formation of the initial vascular plexus were not affected. A reduction of supporting cells occurred, leading to the conclusion that NFATc3 and -c4 are required for the recruitment of vascular smooth muscle and pericyte precursors to the developing vessel wall. A survey of genes involved in vascular development revealed elevated expression VEGF and FLT-1, consistent with the notion that NFAT signaling represses these genes or activates a repressor of their production (a recent study favors the former mechanism [7]). NFATc4 was shown to be expressed preferentially in the neural tube and somites, with minimal expression in ECs. The double mutant embryos demonstrated increased expression of VEGF and blood vessel invasion in the neural tube and somites, suggesting that NFAT signaling normally inhibits vascular overgrowth at these sites (7). Similarly, suppression of myocardial VEGF production by calcineurin was shown to be essential for EMT (7).

In Vitro Studies

NFAT transcription factors have been localized in cultured ECs. Several NFAT agonists were identified, including phorbol 12-myristate 13-acetate (PMA), VEGF, and histamine (9,10). Many putative NFAT target genes have been described, including *granulocyte monocyte colony stimulating factor* (*GM-CSF*), *tissue factor*, *interleukin (IL)-8*, *E-selectin*, *COX-2*, and *ATP2A3* (9,11).

We recently have shown that thrombin and VEGF, but not tumor necrosis factor (TNF)-α, result in activation of NFATc. In that study, VEGF and thrombin treatment of human umbilical vein ECs (HUVECs) resulted in the induction of DSCR-1 via the coordinate binding NFATc1 or c2, and GATA-2/3 on the promoter (12). To further delineate NFAT target genes in ECs, we carried out DNA microarrays of HUVECs infected with adenovirus overexpressing either GFP (Ad-Control) or a constitutively active form of NFAT (Ad-CA-NFAT). In duplicate arrays, a total of 59 defined genes were upregulated more than fourfold in both experiments (Figure 92.2; for color reproduction, see Color Plate 92.2). These included genes involved in inflammation (*IL-8*, *monocyte chemoattractant protein [MCP]-]*, and *fractalkine*), coagulation (*tissue factor*), leukocyte trafficking (*E-selectin*, *vascular cell adhesion molecule [VCAM]-1*, and *intercellular adhesion molecule [ICAM]-1*), transcription factors (*Egr-1* and nuclear hormone receptor, *RAR*α), and autoinhibition (*DSCR-1*).

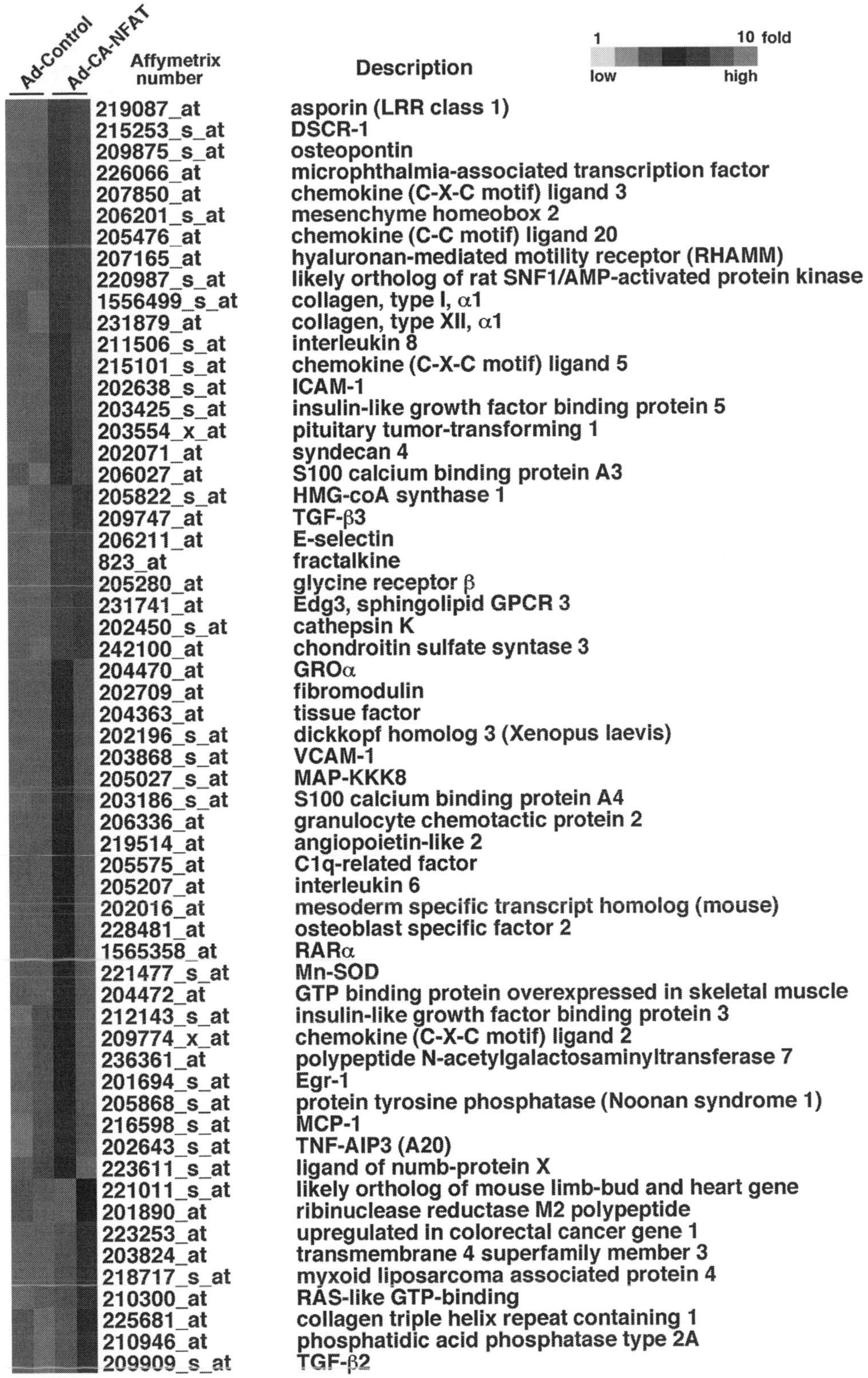

Figure 92.2. NFAT target genes in primary ECs. Microarray (Affymetrix U133 plus) analyses were performed in duplicate using HUVECs overexpressing constitutively active NFAT (CA-NFAT). Genes were filtered according to those induced fourfold by CA-NFAT versus control. Selected genes were analyzed by GeneTree program. Shaded panels represent higher and lower expression, respectively, compared with the median for that particular gene. For color reproduction, see Color Plate 92.2.

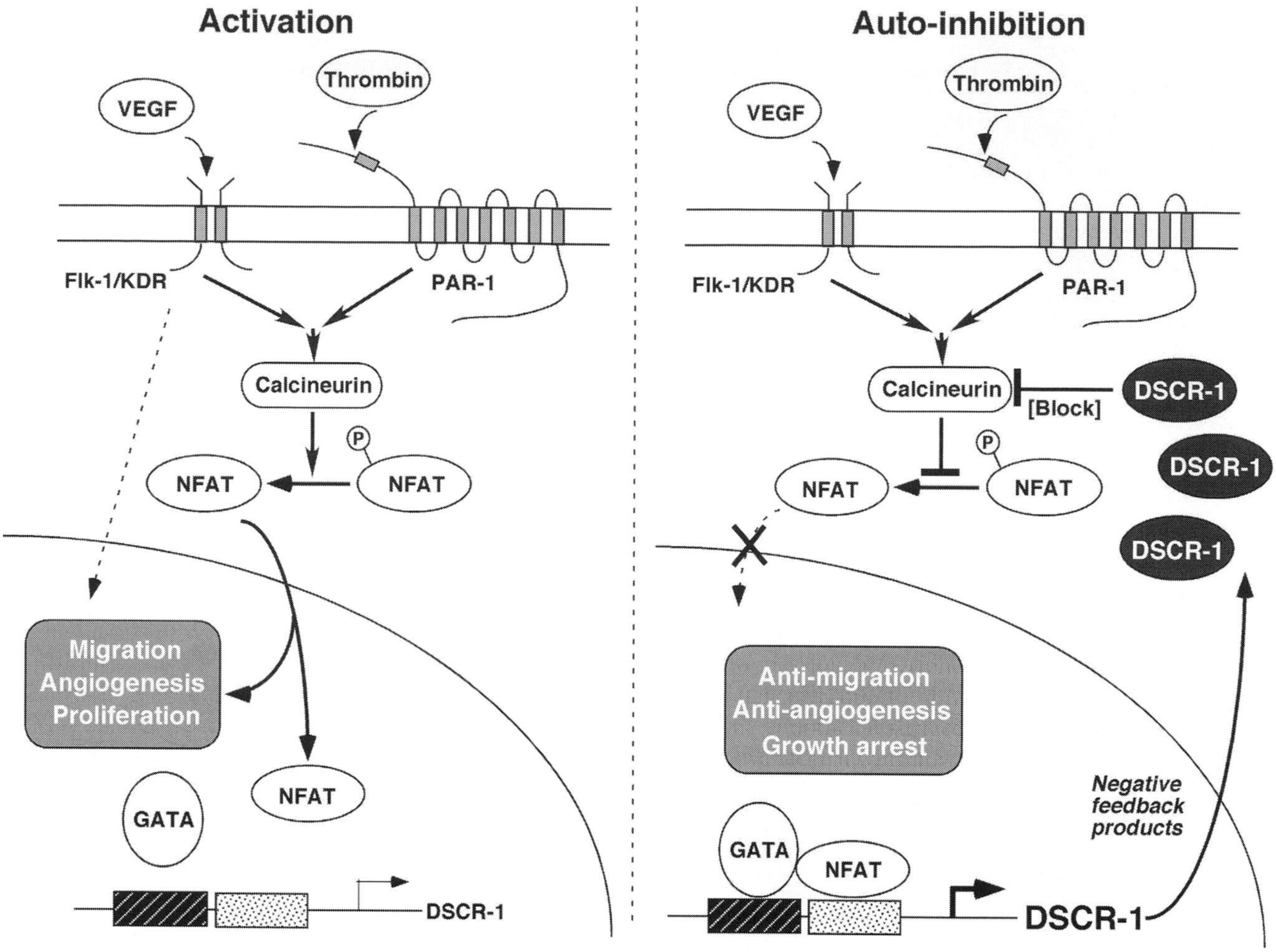

Figure 92.3. Schematic model shows the VEGF- or thrombin-calcineurin-NFAT signaling and the DSCR-1–mediated autoinhibition system in ECs.

DOWN SYNDROME CRITICAL REGION 1

The *DSCR-1* gene (also known as MCIP-1), designated as such because it resides within the Down syndrome critical region of human chromosome 21, encodes a small (∼25 kDa) protein that binds to and inhibits the catalytic subunit of calcineurin (13). DSCR-1 is highly expressed in the CNS in the developing embryo, and preferentially expressed in heart, brain, and skeletal muscle in adults. Overexpression of the DSCR-1 occurs in patients with Down syndrome (who have an extra copy of the gene), and in patients with Alzheimer disease.[5] The *DSCR-1* gene includes seven exons and six introns. The first four exons are alternative and code for four different isoforms. A 5′ promoter regulates expression of the first three isoforms, which are derived from alternative splicing. The most common of these contains exons 1, 5, 6, and 7. An intragenic region between exons 3 and 4 contains an alternative promoter that initiates transcription of the fourth isoform (exons 4, 5, 6, and 7).

Using DNA microarrays, we showed that, of the various transcripts that were responsive both to VEGF and thrombin, DSCR-1 was the most highly induced at 1 hour (22-fold with VEGF and 18-fold with thrombin) (12). The inducible DSCR-1 proved to be the isoform encoded by exons 4 through 7. In contrast to VEGF and thrombin, TNF-α had little effect on DSCR-1 levels.

Adenovirus-mediated overexpression of DSCR-1, but not control EGFP, inhibited VEGF- and thrombin-mediated nuclear localization of NFATc at 1 hour, and inhibited VEGF-mediated tube formation (12). Overexpression of DSCR-1 increased the percentage of cells in G_0/G_1 phase and decreased the fraction in S phase, an effect that was reversed by coinfection with Ad-CA-NFAT. DSCR-1 overexpression in primary ECs resulted in the downregulation of multiple proangiogenic, proinflammatory, and procoagulant genes, including *ICAM-1*, *VCAM-1*, *tissue factor*, *IL-8*, and *E-selectin*. Finally, overexpression of DSCR-1 in mice blocked matrix neovascularization and tumor progression.

5 Patients with Down syndrome have decreased incidences of diabetic retinopathies, atheromas, and solid tumors. These conditions are linked to angiogenesis and/or EC activation. It is interesting to speculate that this protection is mediated – in part – by increased DSCR-1 expression (thus autoinhibition of angiogenesis and EC activation).

Taken together, these data suggest that DSCR-1 serves in a negative feedback loop to inhibit EC proliferation and activation, as well as angiogenesis (Figure 92.3). Such a mechanism is reminiscent of the NF-κB–IκBα autoinhibitory loop. Cellular activation results in phosphorylation-dependent degradation and subsequent ubiquitination of IκBα. As a result, RelA translocates to the nucleus and partners with other members of the NF-κB family to transactivate a multitude of target genes, including proinflammatory mediators. In addition, the NF-κB family induces the early expression of its inhibitor, IκB, which serves to dampen further RelA activity. The NF-AT–DSCR and NF-κB–IκBα negative feedback loops may function to "fine tune" the desired downstream effect of the transcription factor and signal transducer.

KEY POINTS

- NFAT transcription factors play an important role in EC biology during development and in the postnatal period.
- NFATc1 is restricted in time and space to developing endocardium, and plays key role in valve formation.
- NFATc3/c4 transcription factors are important for blood vessel development.
- NFATc family members play an important role in adult endothelium, responding to extracellular signals such VEGF and thrombin, and contributing to multiple phenotypes including angiogenesis and EC activation.
- DSCR-1 is an inducible autoinhibitory protein that turns off NFAT signaling in the endothelium.

Future Goals

- To determine the relative roles for NFAT isoforms in physiology and pathophysiology of adult endothelium
- To elucidate the potential for DSCR-1 as a novel therapeutic target

REFERENCES

1 Hogan PG, Chen L, Nardone J, Rao A. Transcriptional regulation by calcium, calcineurin, and NFAT. *Genes Dev.* 2003; 17(18):2205–2232.

2 Macian F. NFAT proteins: key regulators of T-cell development and function. *Nat Rev Immunol.* 2005;5(6):472–484.

3 de la Pompa JL, Timmerman LA, Takimoto H, et al. Role of the NF-ATc transcription factor in morphogenesis of cardiac valves and septum. *Nature.* 1998;392(6672):182–186.

4 Ranger AM, Grusby MJ, Hodge MR, et al. The transcription factor NF-ATc is essential for cardiac valve formation. *Nature.* 1998; 392(6672):186–190.

5 Eisenberg LM, Markwald RR. Molecular regulation of atrioventricular valvuloseptal morphogenesis. *Circ Res.* 1995;77(1): 1–6.

6 Johnson EN, Lee YM, Sander TL, et al. NFATc1 mediates vascular endothelial growth factor-induced proliferation of human pulmonary valve ECs. *J Biol Chem.* 2003;278(3):1686–1692.

7 Chang CP, Neilson JR, Bayle JH, et al. A field of myocardial-endocardial NFAT signaling underlies heart valve morphogenesis. *Cell.* 2004;118(5):649–663.

8 Graef IA, Chen F, Chen L, et al. Signals transduced by Ca($^{2+}$)/calcineurin and NFATc3/c4 pattern the developing vasculature. *Cell.* 2001;105(7):863–875.

9 Hadri L, Pavoine C, Lipskaia L, et al. Transcription of the sarcoplasmic/endoplasmic reticulum Ca^{2+}-ATPase type 3 gene, ATP2A3, is regulated by the calcineurin/NFAT pathway in endothelial cells. *Biochem J.* 2006;394(Pt 1):27–33.

10 Boss V, Wang X, Koppelman LF, et al. Histamine induces nuclear factor of activated T cell-mediated transcription and cyclosporin A-sensitive interleukin-8 mRNA expression in human umbilical vein endothelial cells. *Mol Pharmacol.* 1998;54(2):264–272.

11 Hesser BA, Liang XH, Camenisch G, et al. Down syndrome critical region protein 1 (DSCR1), a novel VEGF target gene that regulates expression of inflammatory markers on activated endothelial cells. *Blood.* 2004;104(1):149–158.

12 Minami T, Horiuchi K, Miura M, et al. Vascular endothelial growth factor- and thrombin-induced termination factor, down syndrome critical region-1, attenuates endothelial cell proliferation and angiogenesis. *J Biol Chem.* 2004;279(48):50537–50554.

13 Klee CB, Ren H, Wang X. Regulation of the calmodulin-stimulated protein phosphatase, calcineurin. *J Biol Chem.* 1998;273 (22):13367–13370.

93

Forkhead Signaling in the Endothelium

Md. Ruhul Abid and William C. Aird

Beth Israel Deaconess Medical Center, Harvard Medical School, Boston, Massachusetts

Forkhead proteins, members of the winged helix family of transcription factors, have been implicated in a multitude of biological processes. According to conventional views, mammalian forkhead proteins are death-promoting/cell cycle arrest factors whose function must be inhibited to ensure cell survival and growth. In the absence of serum or growth factors, forkhead is dephosphorylated and localized in the nucleus, where it activates proapoptotic and/or antigrowth/cell cycle genes including p27 kinase inhibitor protein ($p27^{kip1}$), Bcl-2–interacting mediator of cell death (Bim), Fas ligand (FasL), growth-arrest and DNA damage inducible gene 45A (GADD45A), and B-cell translocation gene (BTG)-1. Serum or growth factor signaling results in phosphoinositide 3-kinase (PI3K)/Akt-dependent phosphorylation and nuclear translocation of forkhead proteins, with subsequent cell survival and/or proliferation.

Recent studies suggest that this model of mammalian forkhead function is overly simplistic. In addition to their role in promoting cell death and cell cycle arrest, forkhead proteins play a critical role in mediating embryonic angiogenesis, cell differentiation, metabolism, redox state, and immunity.

Most of our current knowledge about forkhead function is derived from studies in the fruit fly (*Drosophila melanogaster*), roundworm (*Caenorhabditis elegans*), and nonendothelial mammalian cell types. In the first part of this chapter, we draw on this extensive literature to review the complex signaling mechanisms underlying forkhead regulation of metabolism, survival, proliferation, longevity, and tumorigenesis. In the second part, we discuss more recent evidence implicating a role for forkhead proteins in endothelial cell (EC) biology.

FROM FRUIT FLY TO HUMAN

The name *forkhead* was first derived in 1989 from a homeotic gene (*fkh*) that encodes a nuclear protein involved in the development of the terminal regions of the *Drosophila melanogaster* embryo (1). In the following year, a group of liver-specific forkhead-like transcription factors – hepatic nuclear factor (HNF)-3A, -B, and -C – was discovered (2). HNF3 and *fkh* have been shown to share a highly conserved DNA-binding domain consisting of approximately 100 amino acids. Since their original discovery, forkhead genes have been identified in a wide variety of species ranging from yeast to humans.

Initial clues about forkhead signaling pathways and the molecular mechanisms of forkhead function were derived from studies in *C. elegans*. The *C. elegans* forkhead transcription factor, DAF-16 (the orthologue of mammalian FoxO [*F*orkhead b*ox*-containing protein, *O* subfamily]), was shown to be downstream of an intracellular signaling cascade involving DAF-2 (insulin/insulin growth factor [IGF]-1 receptor homologue)-AGE-1 (PI3K homologue)-Akt1 and Akt2 (Akt homologues), and to regulate lifespan, reproduction, and metabolism, including response to oxidative stress (3–5).

Since these early reports, more than 100 members of the forkhead family have been identified in different species. Recently, a new nomenclature, based on the structure of these transcription factors, has been adopted in which all FOX proteins are organized into 15 subfamilies or classes. The three mammalian homologues of DAF-16 – namely FKHR/FoxO1, FKHRL1/FoxO3a, and AFX/FoxO4 – belong to the 'O' subfamily of FOX proteins. All the members contain a highly conserved DNA-binding domain (FKH), the relative location of which may vary among different members. Forkhead proteins are often described as "winged helix" due to the presence of helical sections flanked by a double-wing structure that resembles a butterfly, as resolved by crystal structure analysis. Although commonly known as transcription factors due to their ability to bind DNA, forkhead proteins may act as activators (transactivators) or inhibitors (repressors) of gene transcription (discussed next).

A role for forkhead in human cancer was suspected during the cloning of the mammalian counterparts of forkhead transcription factors AFX/FoxO4, FKHR/FoxO1, and FKHRL1/FoxO3a. AFX (*ALL1* fused gene from chromosome X, 1; new name, FoxO4) was originally identified as a fusion partner of the mixed-lineage leukemia (*MLL*) gene in acute leukemia, whereas FKHR/FoxO1 (forkhead homolog in rhabdomyosarcoma; new name, FoxO1) was shown to be fused with PAX3 (paired box 3) or PAX7 in rhabdomyosarcoma (6–9). Another mammalian homolog of forkhead, FKHRL1 (forkhead homolog in rhabdomyosarcoma-like 1; new name, FoxO3a) was also identified (10).

More recently, all three mammalian orthologues of DAF-16 (FKHR/FoxO1, FKHRL1/FoxO3a and AFX/FoxO4) were shown to be substrates for Akt (11–17). These and subsequent studies in mammalian cells uncovered a critical role for the forkhead family of transcription factors in coupling extracellular signals (particularly insulin and IGF-1) to downstream changes in gene expression in primary nonendothelial cells and cell lines originated from fibroblast, muscle, liver, and adipose tissues. As mentioned earlier, the conventional model of forkhead function holds that, when cells are exposed to growth factors or serum, there occurs the activation of PI3K/Akt, secondary phosphorylation, and nuclear exclusion of forkhead transcription factors; downregulation of target gene expression; and enhanced cell survival and proliferation (11–15,17). Putative forkhead-responsive genes include insulin-like growth factor binding protein (IGFBP)-1, glucose-6-phosphatase, phosphoenolpyruvate carboxykinase, pancreatic duodenal homeobox (Pdx)-1, mitochondrial 3-hydroxy-3-methylglutaryl-CoA synthase, FasL, tumor necrosis factor apoptosis related ligand (TRAIL), GADD45A, transforming growth factor (TGF)-β, B-cell lymphoma (BCL)-6, Bim, and $p27^{kip1}$ (18–27). In contrast, the withdrawal of serum or growth factors leads to nuclear translocation of forkhead proteins, transcriptional activation of target genes, and subsequent cell cycle arrest and/or apoptosis.

REGULATION OF FORKHEAD PROTEINS

Phosphorylation and Subcellular Localization

Forkhead proteins are regulated by two major post-translational events: phosphorylation and acetylation. Phosphorylation is initiated upstream by PI3K, which in turn is activated by G-proteins and tyrosine kinases, including the receptors for insulin, IGF-1, platelet-derived growth factor (PDGF), and vascular endothelial growth factor (VEGF) (28–31). On activation, PI3K triggers a signaling cascade that includes downstream serine-threonine kinases, PI3K-dependent kinase (PDK1), Akt, and serum and glucocorticoid-regulated kinase (SGK) (28). Akt is the main downstream signaling molecule that phosphorylates and inactivates proapoptotic and cell cycle inhibitory proteins, including BAD (displaces Bax from binding to Bcl-2 resulting in cell death) and forkhead proteins. Akt-mediated phosphorylation occurs at Thr32, Ser253, and Ser315 in FKHRL1/FoxO3a; at Thr24, Ser256, and Ser319 in FKHR/FoxO1; and at Thr28, Ser193, and Ser258 in AFX/FoxO4 (13,32,33). Each of these sites in forkhead contains the highly conserved Akt-recognition motif, RXRXXS/T, where *R* denotes arginine, *S* and *T* denote serine and threonine, respectively, and *X* denotes any residue.

All forkhead transcription factors require phosphorylation of the N-terminal (T24 in FKHR/FoxO1, T32 in FKHRL1/FoxO3a, and T28 in AFX/FoxO4) and the DNA-binding domain/nuclear localization sequence (NLS) (S256 in FKHR/FoxO1, S253 in FKHRL1/FoxO3a, and S193 in AFX/FoxO4) for translocation from nucleus to the cytoplasm. Phosphorylation of the N-terminal threonine and the NLS serine residues also is required for recognition by and association with 14–3–3 proteins that retain forkhead in the cytosol. The second Akt phosphorylation site (within the NLS) is thought to disrupt the nuclear localization signals (34). The third or C-terminal phosphorylation site (S319 in FKHR/FoxO1, S315 in FKHRL1/FoxO3a, and AFX/FoxO4) is a substrate for both Akt and SGK. Although all the Akt motifs in forkhead can be phosphorylated by Akt and SGK (because they have similar substrate recognition motifs), only the third or C-terminal site appears to be preferred by the kinase, SGK.

Two additional serine residues exist immediately after the third (C-terminal) Akt/SGK serine phosphorylation site, which are phosphorylated by casein kinase 1 (CK1) (Figure 93.1) (35). These two serine residues are recognized and phosphorylated by CK1 only when the third Akt motif (S315 in FKHR/FoxO1; S319 in FKHRL1/FoxO3a; S258 in AFX/FoxO4) is already phosphorylated – i.e., the phosphorylation by CK1 appears to be a “primed” event. Another serine phosphorylation site exists, immediately after the CK1 sites, which is phosphorylated by dual-specificity tyrosine-phosphorylated and –regulated kinase 1A (DYRK1A) (36). Together with the C-terminal Akt/SGK and CK1-substrate serine residues, this DYRK1A phosphorylation site appears to form a patch of four acidic residues in the vicinity of the nuclear export sequence (NES) of the forkhead proteins. The precise functional importance of these phosphorylation events requires further study.

A small GTPase, Ras, also has been implicated in the regulation of forkhead function (37). Ras is recruited to the receptors of growth factors by adaptor proteins and is loaded with GTP. Ras activates GEF (guanine exchange factor) that activates Ral, and Ral modulates Ras-mediated induction of cellular proliferation. Ras–Ral signaling has been shown to phosphorylate threonine residues at 447 and 451 in AFX/FoxO4, which appears to influence transactivational activity of the forkhead protein (18,38). It is not clear at the moment whether other forkhead proteins require this Ras–Ral activity, because this phenomenon was studied only with AFX/FoxO4. In addition, phosphorylation of T447 and T451 did not influence subcellular localization of AFX/FoxO4 (Figure 93.2).

The nuclear export and import of forkhead proteins are active processes, because these proteins have an estimated molecular mass of more than 50 kDa (ranging from 75 kDa to 125 kDa). To date, several *cis-* and *trans-*acting factors that are involved in the regulation of subcellular localization of forkhead proteins have been identified. In addition to phosphorylation events, two important *cis*-acting elements exist that have been demonstrated to be involved in the relocation of forkhead. As shown in Figure 93.1, all FoxOs contain a

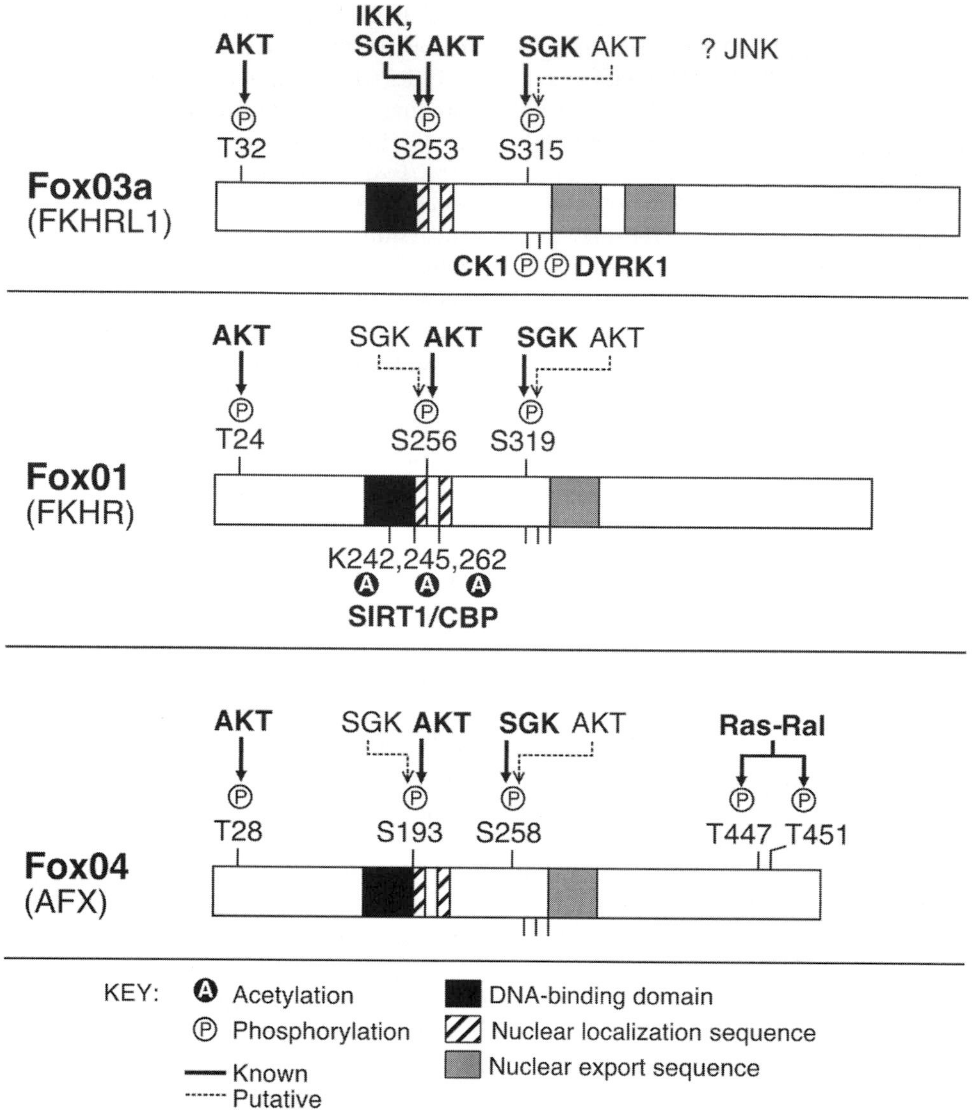

Figure 93.1. The mammalian forkhead proteins, FKHRL1/FoxO3a, FKHR/FoxO1 and AFX/FoxO4, and their functional sites. Shown are the putative phosphorylation sites and their respective kinases (AKT/PKB; SGK, serum glucocorticoid kinase; IKK, IκB kinase; JNK, c-jun N-terminal kinase; CK1, casein kinase 1; DYRK1, dual-specificity tyrosine-phosphorylated and –regulated kinase 1), the acetylation sites and the proteins responsible for acetylation (CBP, CEBP binding protein) and deacetylation (SITR1, silent information regulator). T, threonine; S, serine; K, lysine.

lysine-rich NLS situated in the C-terminal end of the DNA-binding domain. This NLS region has been shown to be responsible for nuclear localization (34). The second Akt-substrate phosphorylation site (e.g., S256 in FKHR/FoxO1) is believed to alter the conformation and function of this NLS motif. In addition, all three mammalian forkhead contain another *cis*-acting element in the C-terminal NES (34,39).

Importins and exportins are *trans*-acting factors involved in mediating the nuclear–cytoplasmic distribution of forkhead proteins. Importins recognize the NLS and mediate the translocation of forkhead to the nucleus. Ran-GTP binds to importin and releases the transported protein in the nucleus. Exportins, such as chromosomal region maintenance 1 (CRM1) protein, bind to the NES and facilitate the export of forkhead from the nucleus to the cytosol. It has been demonstrated that a phosphorylation-resistant mutant of AFX/FoxO4 is amenable to nuclear export (34), suggesting that the association between forkhead and CRM1 is independent of forkhead phosphorylation status (34,40).

Additional *trans*-acting factors involved in the regulation of subcellular localization of forkhead are 14–3–3 proteins that have a molecular mass of 30 kDa. 14–3–3 proteins are homodimers and can also form heterodimers with other members of the same family of proteins, assuming a U shape.

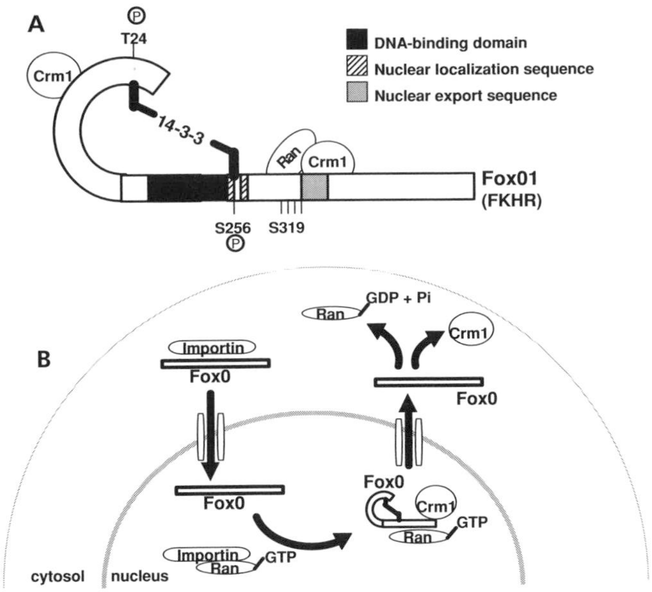

Figure 93.2. Mechanism of subcellular localization of the FoxO proteins. (**A**) Proteins mediating active transport of FKHR/FoxO1 from the nucleus to the cytoplasm. 14–3–3 dimers bind to the Akt-phosphorylation sites in FKHR, threonine 24 (T24) and serine 256 (S256). Chromosomal region maintenance protein 1 (CRM1) binds to the nuclear export sequence (NES). CRM1 and Ran-GTP mediate nuclear export of forkhead. (**B**) Nuclear and cytoplasmic shuttling of FoxO proteins. Importins bind to the nuclear localization sequence FoxO and mediate transport of FoxO from the cytosol to the nucleus. Once in the nucleus, Ran-GTP binds importin and releases forkhead which then becomes available for CRM1 binding. CRM1 binds to the NES of FKHR/FoxO1 in the nucleus. CRM1, in conjunction with Ran-GTP, mediates nuclear exclusion of forkhead.

The phosphorylated serine residues (pS) followed by an X (any residue)-proline in RSXpSXP and RXXXpSXP motifs are recognized by 14–3–3. The N-terminal (threonine) Akt-substrate sites of the mammalian forkhead proteins contain an overlapping optimal site for 14–3–3 binding. It has been demonstrated that phosphorylation by Akt of the N-terminal threonine and DNA-binding–domain serine is essential for 14–3–3-mediated cytosolic sequestration of the transcription factor (34,39). It is plausible that the two "arms" of the U of 14–3–3 recognize and bind to these two Akt phosphorylation sites in forkhead proteins. In addition, phosphorylated forkhead was reported to be degraded by ubiquitin-proteasome (41), suggesting another level of complexity in the regulation of forkhead translocation and stability.

IκB kinase has also been found to cause phosphorylation and cytoplasmic translocation of FKHRL1/FoxO3a (42). Hu and colleagues demonstrated that, in some phospho-Akt-negative tumors, forkhead exclusion from the nucleus was associated with higher levels of IκB kinase β (42). This group further showed that IκB-mediated phosphorylation was followed by ubiquitination and proteosomal degradation of FKHRL1/FoxO3a. In addition, overexpression of forkhead could inhibit IκB-induced cellular proliferation in vitro and tumor growth in mice.

Recently, Oh and associates showed that the c-jun N-terminal kinase (JNK) pathway acts in parallel with the insulin-like signaling pathway to regulate lifespan in *C. elegans,* and both JNK and insulin pathways converge onto DAF-16 (43). They also showed that JNK-1 directly interacts with and phosphorylates DAF-16 in *C. elegans,* and in response to heat stress, JNK-1 promotes the translocation of DAF-16 into the nucleus. These findings raise the possibility that a similar pathway exists in mammals where interaction between two well-conserved proteins, JNK-1 and forkhead, may provide a mechanism by which JNK regulates longevity and stress resistance.

Transcriptional Activity

Transcriptional activity of forkhead proteins is primarily dependent on their subcellular localization – i.e., whether forkhead is present in the nucleus. In addition, regulation of transcriptional activity by modification of the DNA-binding domain and/or transactivational domain has been reported. Other transcriptional proteins and coactivators have been shown to interact with forkhead, thereby altering its function. Finally, recent reports of acetylation of forkhead protein have uncovered another level of complexity in the regulation of the activity of these transcription factors.

The DNA-binding domain of forkhead proteins recognizes and binds to a conserved forkhead-responsive element (FHRE) that is present in forkhead target genes (12,13,22,23). The second Akt phosphorylation site that is located adjacent to the DNA-binding domain has been shown to be critical for forkhead's ability to bind DNA. DNA binding is hindered by phosphorylation of this serine residue (12,17).

The C-terminus was reported to be involved in the transactivational activity of the forkhead transcription factor (FKHR/FoxO1) in rat hepatoma cells (44). In this study, a series of serine residues in the C-terminus (at 319, 499, and residues within a stretch of 350 to 364) were found to be phosphorylated by insulin, which increased the transactivation potential of FKHR/FoxO1. The DAF-16 Ser-Ala mutant, in which the Akt phosphorylation sites are mutaed, was also demonstrated to have increased PIK3-dependent, 14-3-3 independent transcriptional activity (45), suggesting the presence of a PI3K-dependent transactivational domain in the C-terminus besides the classical Akt-dependent sites in forkhead proteins.

Forkhead proteins have been found to interact with other proteins that modulate their activity. FKHR/FoxO1, FKHRL1/FoxO3a, and DAF-16 interact with p300/CREB binding protein (CBP) and enhance the glucocorticoid-stimulated transcription by CBP (46–48). FKHR/FoxO1 was found to interact with CCAAT/enhancer binding protein (C/EBP)-β in a DNA-independent manner (49). However, DNA-binding was essential for the additive action of C/EBP-β

and FKHR/FoxO1 on the decidual prolactin (*dPRL*) promoter (50). DYRK1A and DYRK1B, in addition to DYRK1A's kinase activity on forkhead, were shown to increase the transcriptional potential of FKHR/FoxO1 in inducing expression of glucose-6-phosphatase via a direct physical interaction (50–52). Additionally, DNA-binding-independent interaction of FKHR/FoxO1 with proliferation-activated receptor gamma coactivator (PGC)-1α was demonstrated to enhance transcription of glucose-6-phosphatase and phosphoenolpyruvate carboxykinase 1 genes by PGC-1α (53).

The nuclear receptors interact with and act as inhibitors of forkhead transcription factors (54). Forkhead transcription factors contain an LXXLL domain in the C-terminus that plays an important role in their interaction with the nuclear receptors. It has been shown that estrogen receptor (ER) and androgen receptor (AR) repress the function of FKHR/FoxO1 (55). In turn, interaction of FKHR/FoxO1, FKHRL1/FoxO3a, and AFX/FoxO4 steroid and nonsteroid nuclear receptors may either activate or repress the transcriptional activity of these receptors. Such interactions have been described for ER, AR, HNF-4, progesterone receptor, thyroid hormone receptor, retinoic acid receptor (RAR), glucocorticoid receptor, and peroxisome-proliferator-activated receptor (PPAR) (55–59). FKHR/FoxO1 binds to and negatively regulates HNF-4 activity (59). Akt-mediated phosphorylation of FKHR/FoxO1 by insulin decreases physical interaction of HNF-4 and FKHR/FoxO1 (59). Interaction of forkhead with the nuclear receptors can be independent or dependent of nuclear receptor ligand binding. For example, ligand binding of RAR is not required for RAR-FKHR/FoxO1 interaction, whereas estrogen facilitates binding of ER to FKHR/FoxO1 (56).

Very recently, SIRT1, a human homolog of silent information regulator (Sir) 2, has been demonstrated to deacetylate mammalian FoxO proteins, resulting in either potentiation (60) or deactivation (61) of forkhead activity. Daitoku and colleagues showed that Sir2 deacetylates the CBP-induced acetylation of FKHR/FoxO1 at the lysine residues K242, K245, and K262 and potentiates transcriptional activity of forkhead (62). Sir2 also serves as a transcriptional coactivator of FKHR/FoxO1. In contrast, Motta and associates demonstrated that SIRT1-mediated deacetylation of FKHRL1/FoxO3a reduced both Bim and $p27^{kip1}$ expression and enhanced cell survival (61). On the other hand, Brunet and colleagues showed that deacetylation of forkhead by SIRT1 shifted the balance from proapoptosis to a cell-cycle arrest and stress-resistance phenotype with a corresponding reduction in Bim but increase in $p27^{kip1}$ expression (62). These findings are in accordance with the previous findings in *C. elegans* that overexpression of *Sir2* gene could increase lifespan in the roundworm. The discrepancy between the functional activities of forkhead upon deacetylation by SIRT1 demonstrated by these three groups indicates that further study will be required to understand the effect of deacetylation on forkhead at the functional level.

Transcriptional Repression

Forkhead proteins have also been reported to act as transcriptional repressors. Two members of the murine forkhead gene family that belong to a different subclass P, Foxp1 and Foxp2, were shown to be expressed in the lung and to act as transcriptional repressors (63). *Xenopus* forkhead protein, FoxD5a, acts as a transcriptional repressor in axis formation and neural plate expansion (64). The latter study showed that expression of *en2*, *Krox20*, proneural genes (*Xnrgn1*, *neuroD*), and a neural differentiation gene (*n-tubulin*) is repressed in FoxD5a-expressing cells. Recently, cell cycle inhibition by FKHRL1/FoxO3a was shown to be mediated by transcriptional repression of cyclin D1 and D2 by forkhead (65). In addition, a forkhead transcription factor of subclass L, FoxL2, known to be associated with premature ovarian follicle in a subset of women, was shown to be a repressor of the human steroidogenic acute regulatory (StAR) and other follicle differentiation genes (66–68). Mutation or disruption of FoxL2 was demonstrated to accelerate follicle development, resulting in increased recruitment of dormant follicles and subsequent premature ovarian failure. Thus, an increase in the expression or activity of forkhead proteins may not necessarily be translated into an increase in a generalized expression of forkhead target genes (69).

PATHOPHYSIOLOGICAL ROLE OF FORKHEAD PROTEINS

Although most of our earlier knowledge about forkhead functions was derived from the studies in the roundworm, *C. elegans*, forkhead has been implicated in many cell-type-specific and signal-specific functions. To examine forkhead functions in mammals, several groups have focused on different cell lines and primary cells in culture. Currently, the physiological functions of forkhead are being studied using loss-of-function (by ablation of one or both alleles) or gain-of-function mutants (overexpression of wild-type or constitutively active) in mice.

Role in Metabolism

Adenovirus-mediated overexpression of FKHR/FoxO1 in mice increases gluconeogenesis (glucose production) in liver through PPAR-γ coactivator PGC-1α (53). Because hepatic glucose production is a key issue in the development and treatment of diabetes, forkhead may be a therapeutic target in this disease. In contrast to the in vitro findings in cell-based systems, overexpression of a constitutively active FKHR/FoxO1 did not result in any observable increase in apoptosis in transgenic mice (70). On the other hand, diabetes due to ablation of IRS2 (insulin receptor substrate 2) was rescued by haploinsufficiency (deletion of one allele) of FKHR/FoxO1 (27), suggesting that a decrease in forkhead gene dosage can alleviate the effects of a lack of insulin-mediated downstream signaling. These findings support the notion that insulin resistance or failure of propagation of inhibitory signaling of insulin to downstream forkhead results in a phenotype that resembles overexpression or constitutive activation of forkhead in mice – i.e., increase in hepatic gluconeogenesis. Accordingly, Nakae and colleagues showed that haploinsufficiency of FKHR/FoxO1 protected against insulin resistance (70). On the contrary, *FKHR/FoxO1*-null mice die at E10.5 due to incomplete vascular development (71,72).

Deletion of FKHRL1/FoxO3a showed age-dependent infertility due to premature ovarian failure as a result of accelerated differentiation and consequent depletion of primary ovarian follicles (71). In contrast, *AFX/FoxO4*-null mice did not show any observable abnormalities (71). These findings suggest that the forkhead gene products, FKHR/FoxO1, FKHRL1/FoxO3a, and AFX/FoxO4 are functionally diverse (i.e., nonredundant) in mammals.

Role in Cell Differentiation

A role for forkhead in cellular differentiation was derived from the in vitro (using cell culture) and in vivo (transgenic mice) observations. FoxOs were shown to couple extracellular cues with intracellular signals that regulate differentiation of myoblasts (73,74), preadipocytes (75), pancreatic β cells (27), and thymocytes (76). Recently, two independent studies demonstrated that the ubiquitin ligases, Atrogin-1 (MAFbx) and MuRF1, that cause muscular atrophy are FoxO targets, and that inhibition of forkhead activity prevents the induction of these ligases (77,78).

Role in Neoplasia

Forkhead's involvement in tumorigenesis and cancer was first suspected from its role in cell cycle regulation. Forkhead modulates several cell cycle checkpoint genes that regulate G_0/G_1 (e.g., $p27^{kip1}$ [18,79], BTG1 [80], cyclin D [25]) and G_2/M (e.g., GADD45A [25,81]). However, a role for forkhead proteins in cancer was confirmed with the discovery of two chromosomal translocations associated with human cancer. A chromosomal translocation resulting in the fusion of the coding region of either PAX3 or PAX7 to FKHR/FoxO1 was shown to be involved in the pathogenesis of rhabdomyosarcoma (82). In addition, AFX/FoxO4 was shown to be fused with the transcription factor MLL in leukemia (83). To explain the molecular mechanisms of forkhead-mediated tumorigenesis, two opposing models have been proposed. The first model assumes that a gain-of-function of forkhead predisposes to tumor formation. The finding that the PAX3-FoxO1 fusion is a more potent transcriptional activator than PAX3 alone supports this model (82). On the other hand, failure of the PAX3-FoxO1 transgenic mice to develop tumors (84–86) called for an alternate model. In this model, a loss of function mutation of forkhead is thought to be responsible for tumorigenesis, suggesting that a critical forkhead target gene with roles in both cell cycle arrest and apoptosis may contribute to cell survival and transformation (87). In addition, PTEN, an endogenous negative regulator of PI3K and a tumor suppressor, was shown to positively regulate forkhead through the PI3K/Akt pathway, resulting in induction of apoptosis or cell cycle arrest by upregulating $p27^{kip1}$ expression (18,79,88). More than 60% of prostate cancer tumors are known to have loss of function mutations in PTEN. Modur and associates demonstrated that forkhead overexpression in these PTEN mutant tumors resulted in apoptosis (and thereby inhibition of tumor progression) through direct induction of a proapoptotic member of the tumor necrosis factor (TNF) family, TRAIL (89).

In breast cancer, overexpression of FKHRL1/FoxO3a was found to increase expression of Bim, an apoptotic protein. An antineoplastic drug, paclitaxel, appears to increase the expression of FKHRL1/FoxO3a, which subsequently decreases cell survival through induction of Bim expression and thus contributes to the tumor suppressive actions of the drug (90). In contrast, estrogen enhances survival in mammary cancer cell lines by stimulating p21-activated kinase (Pak)-1 that induces phosphorylation and nuclear exclusion of forkhead (91).

Seoane and colleagues demonstrated that in mammalian cells, forkhead coordinates signals at the intersection of Smad, FoxG1, and PI3K pathways (92). A complex of forkhead and Smad proteins activates $p21^{Cip1}$, a growth arrest gene; Smad proteins are mediators of TGFβ signaling. This Smad- and forkhead-mediated upregulation of $p21^{Cip1}$ is inhibited by PI3K and positively regulated by FoxG1, a member of a different subfamily of forkhead. Seoane and colleagues also showed that the presence of increased PI3K/Akt activity and FoxG1 expression in neuroepithelial and glioblastoma cells resulted in the reduction of $p21^{Cip1}$ levels and concomitant increase in cell survival and proliferation. Moreover, Nakae and associates reported that the FoxO target gene, $p21^{Cip1}$, is involved in preadipocyte differentiation (75).

Taken together, these findings suggest that inactivation of forkhead proteins is crucial in the development and/or progression of a variety of cancers, raising the possibility that forkhead activity may be targeted therapeutically. Recently, Kau and colleagues devised several chemical molecules that act to inhibit nuclear export of FKHR/FoxO1 in PTEN-deficient (thus, with a high level of PI3K/Akt activity) tumor cells (93). Interestingly, many of the molecules identified in this study are inhibitors of nuclear export receptor CRM1, an PI3K/Akt-mediated export pathway.

FORKHEAD PROTEINS IN ENDOTHELIAL BIOMEDICINE

The quest to elucidate the role of forkhead in endothelium has just begun. Most of our understanding about FoxOs in endothelium is currently derived from in vitro studies using primary human ECs. A small number of studies have addressed in vivo functions by either overexpressing or deleting one or more of the mammalian isoforms of FoxO proteins in mice. The first evidence suggesting a role for forkhead in the vasculature came from a study of non-FoxO forkhead proteins, Foxc1 (Mf1) and Foxc2 (Mfh1). Mice null for *Foxc1/Mf1* had severe defects in the anterior segment of the eye with a complete failure of differentiation of the inner corneal endothelial layer (94). Deletion of another forkhead homolog in mice, Foxc2/Mfh1, resulted in severe cardiovascular defects (95). Kalinichenko and associates showed that the newborn mice heterozygously null for the forkhead box f1 (*Foxf1*) transcription factor developed lethal alveolar hemorrhage due to defective mesenchyme–epithelial interaction during lung morphogenesis (96). Mutations in *FoxC2/Mfh1* were also reported to be responsible for the lymphatic vascular disorder, lymphedema-distichiasis (97). However, until recently, evidence was lacking for a role of FoxO proteins in EC biology.

In Vitro

The first studies to demonstrate FoxO expression in ECs were published in 2003 and 2004 (30,98,99). FKHR/FoxO1/, FKHRL1/FoxO3a, and AFX/FoxO4 were shown to be expressed in human coronary artery ECs (HCAECs) and umbilical vein ECs (HUVECs) (30). This study also demonstrated for the first time that VEGF induced phosphorylation of FKHR/FoxO1 and AFX/FoxO4, nuclear export of FKHR/FoxO1 and FKHRL1/FoxO3a, and downregulation of transcriptional activity of FKHRL1/FoxO3a in ECs in a PI3K/Akt-dependent manner. VEGF-mediated cell survival, promotion of DNA synthesis, and cell cycle progression from G1 to S were shown to be forkhead-dependent in ECs (30). Forkhead-mediated induction of the cell cycle inhibitory gene *p27*^{kip1} was partly implicated in all these VEGF effects in ECs. This study clearly demonstrated that VEGF inhibited expression of p27^{kip1} at the mRNA and protein levels through a forkhead-dependent pathway. Interestingly, the forkhead isoforms, FKHR/FoxO1 and AFX/FoxO4, but not FKHRL1/FoxO3a, were detected in HCAECs and HUVECs at the protein level by western blots; FKHRL1/FoxO3a was only detectable by RT-PCR in HCAECs, HUVECs, and human pulmonary artery ECs (30). Other studies have shown that FoxO factors lie downstream of epoxyeicosatrienoic acid (99), and angiopoietin 1 (Ang1) (100) in ECs and modulate the expression of several forkhead-target genes including *p27*^{kip1}, *survivin*, and *TRAIL*. An elegant study by Skurk and colleagues showed for the first time that FKHRL1/FoxO3a could also act as a transcriptional repressor in ECs (98). They found that FKHRL1/FoxO3a inhibited transcription of FLICE inhibitory protein (FLIP), a dominant-negative inhibitor of caspase-8 function (101), resulting in the activation of the "extrinsic" or caspase-dependent apoptotic pathway in HUVECs (98). Thus, as was observed in non-ECs, forkhead may function as a transcriptional activator or repressor in ECs.

A recent study from our lab has demonstrated that VEGF-PI3K/Akt-mediated regulation of phosphorylation and transcriptional activity of forkhead in ECs is dependent on NADPH oxidase-derived redox signaling (102). Hepatocyte growth factor (HGF) also induced phosphorylation and inactivation of FKHR/FoxO1 in HCAECs in a PI3K-dependent manner. However, HGF-mediated induction of forkhead phosphorylation was found to be redox-independent in these cells (102). These findings suggested that the requirement of NADPH oxidase/redox activity in ECs is specific to the growth factor/agonist and not to the forkhead protein per se.

In contrast to non-ECs, where insulin suppresses manganese superoxide dismutase (MnSOD) levels via a PI3K/Akt-forkhead-dependent signaling pathway, VEGF was shown to induce expression of MnSOD in ECs. The importance of forkhead as a positive regulator in this pathway was evidenced by the observation that VEGF-mediated induction of MnSOD was enhanced by inhibition of PI3K or Akt or by overexpression of TM-FKHRL1/FoxO3a. Because both insulin and VEGF phosphorylate and inhibit forkhead, these data suggested that VEGF-mediated induction of MnSOD must involve an overriding positive pathway. Indeed, we demonstrated that VEGF-mediated induction of PKC-nuclear factor (NF)-κB was responsible for the net positive effect of VEGF on MnSOD expression. We have recently identified a number of other VEGF-responsive genes whose net response depends on the relative activities of positive signaling pathways (e.g., NF-κB) and negative signaling pathways (namely, the PI3K/Akt-mediated exclusion of forkhead from the nucleus) (103).

Potente and associates (104) recently reported that FKHR/FoxO1 and FKHRL1/FoxO3a are abundantly expressed in HUVECs and that overexpression of constitutively active TM-FKHR/FoxO1 and TM-FKHRL1/FoxO3a significantly inhibited EC migration and tube formation in vitro. This study also demonstrated that FKHR/FoxO1 and FKHRL1/ FoxO3a regulate overlapping yet distinct (i.e., nonredundant) target genes, including those involved in angiogenesis and vascular remodeling. For example, FKHR/FoxO1, but not FKHRL1/FoxO3a, was implicated in transcriptional induction of *Ang2*, whereas both FoxO proteins were shown to downregulate endothelial nitric oxide synthase (eNOS) expression. In addition, TM-FKHRL1/FoxO3a deficiency increased eNOS expression and enhanced postnatal vessel formation in mice.

In Vivo

In 2004, two groups independently demonstrated that mice null for *FKHR/FoxO1*, but not *FKHRL1/FoxO3a* or *AFX/FoxO4*, died on embryonic day 10.5 as a consequence of defects in the branchial arches and impaired vascular development of the embryo and yolk sac (71,72). Deletion of *FKHR/FoxO1* did not appear to affect EC differentiation or vasculogenesis but rather impaired vascular sprouting (i.e., angiogenesis). Whereas *FKHRL1/FoxO3a*-null mice exhibited an age-dependent infertility due to ovarian failure, *AFX/FoxO4*-null mice did not show any obvious phenotype (71,105).

Furuyama and associates (72) examined *FKHR/FoxO1*$^{-/-}$ versus *FKHR/FoxO1*$^{+/+}$ mice for differences in gene expression. In quantitative PCR, transcripts of *VEGF, FLT-1, FLK-1, Ang1, Ang2, Tie1, Tie2, EphB2, EphB3,* and *EphB4* were comparable in mutant and wild-type E9.5 yolk sacs. In contrast, mRNA levels of *connexin-37, connexin-40,* and *ephrinB2* were significantly lower in the mutant samples. In embryoid bodies, comparable differentiation was noted of mutant and wild-type ES cells into ECs and, consistent with the in vivo data, no difference in expression of FLT-1, FLK-1, Tie1, and Tie2. VEGF-mediated induction of proliferation and migration of was not impaired in *FKHR/FoxO1*$^{-/-}$ ECs. VEGF induced an elongated or spindle-shaped phenotype in polygonal ECs isolated from wild-type mice and cultured in vitro. In contrast, a significant percentage of ECs isolated from the *FKHR/FoxO1*-null mice remained flat and polygonal even after treatment with VEGF. These observations led to the conclusion that the defective vascular development was attributable to either a failure of response or an abnormal response of ECs to growth factors or angiogenic stimuli in the *FKHR/FoxO1*-null embryos (72).

As noted earlier, our current understanding of forkhead physiology in vivo is heavily based on the observations made from the studies in the roundworm, *C. elegans.* In this metazoan, regulation of the mammalian forkhead homologue, DAF-16, by the insulin-receptor homologue, DAF-2, has been shown to modulate several physiological functions (4,5,106). It is not clear whether insulin/IGF-1, the major signaling pathway that regulates the *C. elegans* forkhead (DAF-16), plays a similar role in the regulation of the mammalian vascular system. A study using vascular EC-specific *insulin receptor (IR)*-knockout mouse demonstrated that inactivation of the IR on EC had no major consequences on vascular development or glucose homeostasis under basal conditions but altered expression of vasoactive mediators (eNOS, endothelin-1) and may play a role in maintaining vascular tone and regulation of insulin sensitivity to dietary salt intake (107). These results contrast with the embryonic lethality and abnormal vascular development of *FKHR/FoxO1*$^{-/-}$ mice (71,72). Taken together, these results suggest that, unlike the metazoan DAF-2/DAF-16 signaling, the link between insulin and FKHR/FoxO1 signaling pathways in the mammalian vascular development has been modified during evolution.

CONCLUSION

Forkhead signaling has been implicated in many cellular and organismal functions including cell survival, proliferation, differentiation, metabolism, redox regulation, and longevity.

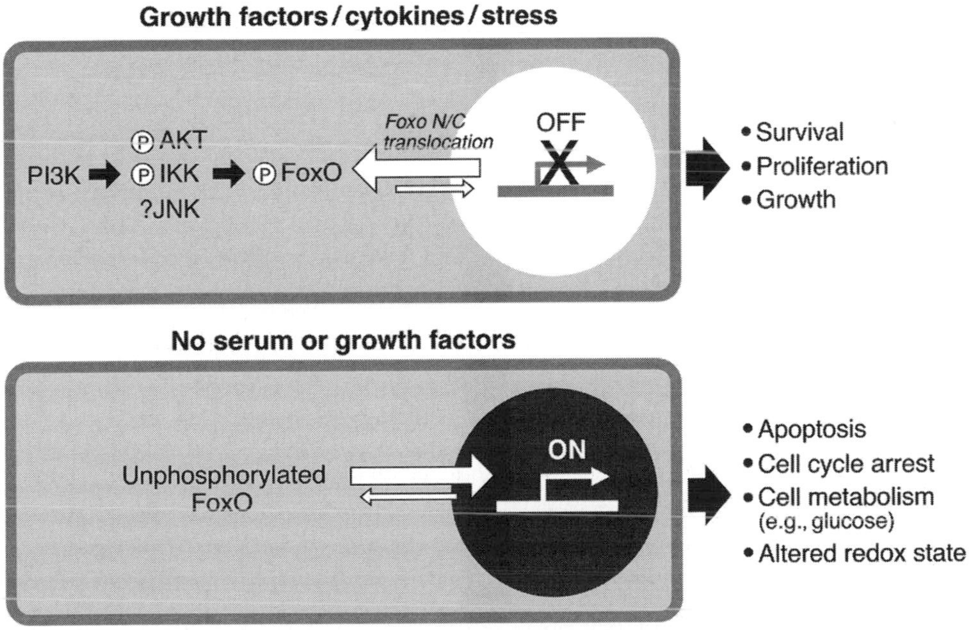

Figure 93.3. Regulation of FoxO proteins by extracellular agonists. *Upper panel,* exposure of ECs to growth factors, cytokines or stress results in PI3K/Akt-dependent phosphorylation of FoxO, nuclear exclusion of phospho-FoxO, and attenuation of FoxO-dependent gene expression. The net effect is increased EC survival, proliferation, and/or growth. *Lower panel,* in the absence of these agonists (e.g., in serum-free medium), there is no stimulation of the PI3K/Akt pathway. Thus, FoxO remains unphosphorylated and is localized to the nucleus where it exerts transcriptional activity (shown is FoxO-mediated target gene activation, although FoxO may repress certain target genes). The net outcome is apoptosis, cell cycle arrest, and/or altered cell metabolism/redox state. N/C, nuclear cytoplasmic.

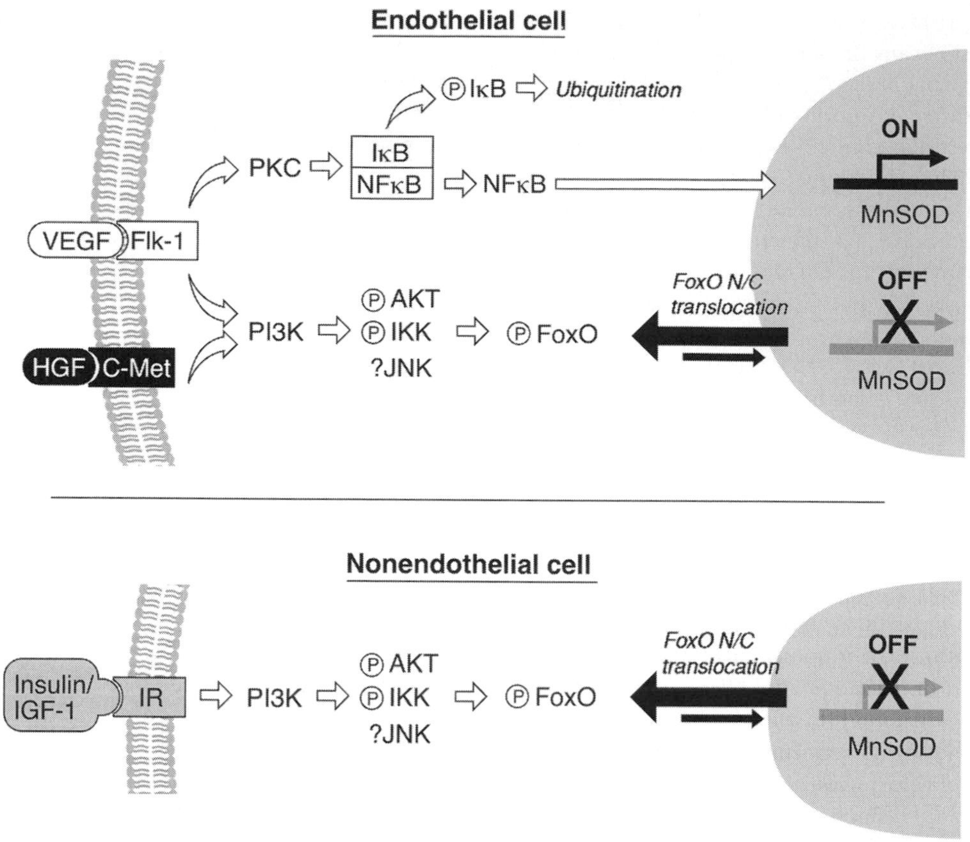

Figure 93.4. Role for FoxO proteins in cell signaling is agonist- and cell type–specific. In ECs, VEGF and HGF result in P13K/Akt-depedent phosphorylation and inactivation of FoxO. In the case of HGF, nuclear exclusion of FoxO results in reduced expression of the FoxO target gene, MnSOD. In the case of VEGF, concomitant activation of NF-κB overrides the negative effect of FoxO translocation, leading to net induction of MnSOD. In non-ECs, the effect of insulin/IGF-1 parallels that of HGF in ECs, with phosphorylation-mediated inactivation of FoxO and downregulation of MnSOD expression. N/C, nuclear cytoplasmic.

Forkhead performs all these important cellular functions by serving as an intracellular signal transducer that couples information in the extracellular environment through cell surface receptors for growth factors, cytokines, hormones, and stress, to downstream changes in phenotype.

Existing evidence supports a critical role for forkhead proteins in ECs. In addition to the classical model in which agonist-mediated phosphorylation and nuclear exclusion of FoxO proteins results in downregulation of death-promoting/cell cycle arrest genes, these transcription factors also participate in complex transcriptional regulation of health-promoting genes (e.g., MnSOD) whose net expression depends on interacting positive and negative influences (Figure 93.3). In vitro and in vivo studies suggest that FoxO proteins are involved in multiple EC functions, including survival/apoptosis, cell cycle progression, proliferation, migration, and blood vessel formation. Whether forkhead is involved in other functions of ECs (e.g., metabolism) awaits further study. Moreover, the extent to which endothelial FoxOs contribute to vascular hemostasis will require the generation and analysis of lineage-specific knockouts.

KEY POINTS

- FoxO transcription factors play an important role in EC biology (Figure 93.4).
- In vitro, several agonists have been shown to phosphorylate FoxO, including VEGF, epoxyeicosatrienoic acid, and Ang1.
- Phosphorylation of FoxO leads to nuclear exclusion of the transcription factor and hence reduced transcriptional activity (which may be activating or repressing in nature).
- FoxO target genes in the ECs include proapoptotic and cell cycle arrest genes, as well as genes involved in metabolism and health.
- Under in vivo conditions, mammalian forkhead family members FKHR/FoxO1, FKHRL1/FoxO3a, and AFX/FoxO4 have nonredundant roles in vascular development.

Future Goals

- To understand the pathophysiological roles of FoxO proteins in adult endothelium
- To determine the potential for FoxO proteins as therapeutic targets in cardiovascular diseases

ACKNOWLEDGMENT

This work was supported by NIH grant HL077348.

REFERENCES

1 Weigel D, Jurgens G, Kuttner F, et al. The homeotic gene fork head encodes a nuclear protein and is expressed in the terminal regions of the *Drosophila* embryo. *Cell.* 1989;57(4):645–658.

2 Lai E, Prezioso VR, Smith E, et al. HNF-3A, a hepatocyte-enriched transcription factor of novel structure is regulated transcriptionally. *Genes Dev.* 1990;4(8):1427–1436.

3 Dorman JB, Albinder B, Shroyer T, Kenyon C. The age-1 and daf-2 genes function in a common pathway to control the lifespan of *Caenorhabditis elegans.* Genetics. 1995;141(4):1399–1406.

4 Ogg S, Paradis S, Gottlieb S, et al. The Fork head transcription factor DAF-16 transduces insulin-like metabolic and longevity signals in *C. elegans. Nature.* 1997;389(6654):994–999.

5 Paradis S, Ruvkun G. *Caenorhabditis elegans* Akt/PKB transduces insulin receptor-like signals from AGE-1 PI3 kinase to the DAF-16 transcription factor. *Genes Dev.* 1998;12(16):2488–2498.

6 Hillion J, Le Coniat M, Jonveaux P, et al. AF6q21, a novel partner of the MLL gene in t(6;11)(q21;q23), defines a forkhead transcriptional factor subfamily. *Blood.* 1997;90(9):3714–3719.

7 Fredericks WJ, Galili N, Mukhopadhyay S, et al. The PAX3-FKHR fusion protein created by the t(2;13) translocation in alveolar rhabdomyosarcomas is a more potent transcriptional activator than PAX3. *Mol Cell Biol.* 1995;15(3):1522–1535.

8 Davis RJ, Bennicelli JL, Macina RA, et al. Structural characterization of the FKHR gene and its rearrangement in alveolar rhabdomyosarcoma. *Hum Mol Genet.* 1995;4(12):2355–2362.

9 Borkhardt A, Repp R, Haas OA, et al. Cloning and characterization of AFX, the gene that fuses to MLL in acute leukemias with a t(X;11)(q13;q23). *Oncogene.* 1997;14(2):195–202.

10 Anderson MJ, Viars CS, Czekay S, et al. Cloning and characterization of three human forkhead genes that comprise an FKHR-like gene subfamily. *Genomics.* 1998;47(2):187–199.

11 Biggs WH, 3rd, Meisenhelder J, Hunter T, et al. Protein kinase B/Akt-mediated phosphorylation promotes nuclear exclusion of the winged helix transcription factor FKHR1. *Proc Natl Acad Sci U S A.* 1999;96(13):7421–7426.

12 Brunet A, Bonni A, Zigmond MJ, et al. Akt promotes cell survival by phosphorylating and inhibiting a forkhead transcription factor. *Cell.* 1999;96(6):857–868.

13 Kops GJ, Burgering BM. Forkhead transcription factors: new insights into protein kinase B (c- akt) signaling. *J Mol Med.* 1999; 77(9):656–665.

14 Nakae J, Park BC, Accili D. Insulin stimulates phosphorylation of the forkhead transcription factor FKHR on serine 253 through a Wortmannin-sensitive pathway. *J Biol Chem.* 1999; 274(23):15982–15985.

15 Tang ED, Nunez G, Barr FG, Guan KL. Negative regulation of the forkhead transcription factor FKHR by Akt. *J Biol Chem.* 1999;274(24):16741–16746.

16 Rena G, Guo S, Cichy SC, et al. Phosphorylation of the transcription factor forkhead family member FKHR by protein kinase B. *J Biol Chem.* 1999;274(24):17179–17183.

17 Guo S, Rena G, Cichy S, et al. Phosphorylation of serine 256 by protein kinase B disrupts transactivation by FKHR and mediates effects of insulin on insulin-like growth factor-binding protein-1 promoter activity through a conserved insulin response sequence. *J Biol Chem.* 1999;274(24):17184–17192.

18 Medema RH, Kops GJ, Bos JL, Burgering BM. AFX-like forkhead transcription factors mediate cell-cycle regulation by Ras and PKB through p27kip1. *Nature.* 2000;404(6779):782–787.

19 Stahl M, Dijkers PF, Kops GJ, et al. The forkhead transcription factor FoxO regulates transcription of p27Kip1 and Bim in response to IL-2. *J Immunol.* 2002;168(10):5024–5031.

20 Dijkers PF, Medema RH, Lammers JW, et al. Expression of the pro-apoptotic Bcl-2 family member Bim is regulated by the forkhead transcription factor FKHR-L1. *Curr Biol.* 2000;10(19): 1201–1204.

21 Samatar AA, Wang L, Mirza A, et al. Transforming growth factor-beta 2 is a transcriptional target for Akt/protein kinase B via forkhead transcription factor. *J Biol Chem.* 2002;277(31): 28118–28126.

22 Schmoll D, Walker KS, Alessi DR, et al. Regulation of glucose-6-phosphatase gene expression by protein kinase Balpha and the forkhead transcription factor FKHR. Evidence for insulin response unit-dependent and -independent effects of insulin on promoter activity. *J Biol Chem.* 2000;275(46):36324–36333.

23 Hall RK, Yamasaki T, Kucera T, et al. Regulation of phosphoenolpyruvate carboxykinase and insulin-like growth factor-binding protein-1 gene expression by insulin. The role of winged helix/forkhead proteins. *J Biol Chem.* 2000;275(39):30169–30175.

24 Nadal A, Marrero PF, Haro D. Down-regulation of the mitochondrial 3-hydroxy-3-methylglutaryl-CoA synthase gene by insulin: the role of the forkhead transcription factor FKHRL1. *Biochem J.* 2002;366(Pt 1):289–297.

25 Furukawa-Hibi Y, Yoshida-Araki K, Ohta T, et al. FOXO forkhead transcription factors induce G(2)-M checkpoint in response to oxidative stress. *J Biol Chem.* 2002;277(30):26729–26732.

26 Tang TT, Dowbenko D, Jackson A, et al. The forkhead transcription factor AFX activates apoptosis by induction of the BCL-6 transcriptional repressor. *J Biol Chem.* 2002;277(16):14255–14265.

27 Kitamura T, Nakae J, Kitamura Y, et al. The forkhead transcription factor Foxo1 links insulin signaling to Pdx1 regulation of pancreatic beta cell growth. *J Clin Invest.* 2002;110(12):1839–1847.

28 Cantley LC. The phosphoinositide 3-kinase pathway. *Science.* 2002;296(5573):1655–1657.

29 Ghosh Choudhury G, Lenin M, Calhaun C, et al. PDGF inactivates forkhead family transcription factor by activation of Akt in glomerular mesangial cells. *Cell Signal.* 2003;15(2):161–170.

30 Abid MR, Guo S, Minami T, et al. Vascular endothelial growth factor activates PI3K/Akt/forkhead signaling in endothelial cells. *Arterioscler Thromb Vasc Biol.* 2004;24(2):294–300.

31 Claesson-Welsh L. Signal transduction by vascular endothelial growth factor receptors. *Biochem Soc Trans.* 2003;31(Pt 1):20–24.

32 del Peso L, Gonzalez-Garcia M, Page C, et al. Interleukin-3-induced phosphorylation of BAD through the protein kinase Akt. *Science.* 1997;278(5338):687–689.

33 Tran H, Brunet A, Griffith EC, Greenberg ME. The many forks in FOXO's road. *Sci STKE.* 2003;2003(172):RE5.

34 Zhao X, Gan L, Pan H, et al. Multiple elements regulate nuclear/cytoplasmic shuttling of FOXO1: characterization of phosphorylation- and 14-3-3-dependent and -independent mechanisms. *Biochem J.* 2004;378(Pt3):839–849.

35 Rena G, Bain J, Elliott M, Cohen P. D4476, a cell-permeant inhibitor of CK1, suppresses the site-specific phosphorylation and nuclear exclusion of FOXO1a. *EMBO Rep.* 2004;5(1):60–65.

36 Woods YL, Rena G, Morrice N, et al. The kinase DYRK1A phosphorylates the transcription factor FKHR at Ser329 in vitro, a novel in vivo phosphorylation site. *Biochem J.* 2001;355 (Pt 3):597–607.

37 Kops GJ, de Ruiter ND, De Vries-Smits AM, et al. Direct control of the Forkhead transcription factor AFX by protein kinase B. *Nature.* 1999;398(6728):630–644.

38 De Ruiter ND, Burgering BM, Bos JL. Regulation of the forkhead transcription factor AFX by Ral-dependent phosphorylation of threonines 447 and 451. *Mol Cell Biol.* 2001;21(23):8225–8235.

39 Brunet A, Kanai F, Stehn J, et al. 14-3-3 transits to the nucleus and participates in dynamic nucleocytoplasmic transport. *J Cell Biol.* 2002;156(5):817–828.

40 Brownawell AM, Kops GJ, Macara IG, Burgering BM. Inhibition of nuclear import by protein kinase B (Akt) regulates the subcellular distribution and activity of the forkhead transcription factor *AFX. Mol Cell Biol.* 2001;21(10):3534–3546.

41 Matsuzaki H, Daitoku H, Hatta M, et al. Insulin-induced phosphorylation of FKHR (Foxo1) targets to proteasomal degradation. *Proc Natl Acad Sci USA.* 2003;100(20):11285–11290.

42 Hu MC, Lee DF, Xia W, et al. IkappaB kinase promotes tumorigenesis through inhibition of forkhead FOXO3a. *Cell.* 2004;117(2):225–237.

43 Oh SW, Mukhopadhyay A, Svrzikapa N, et al. JNK regulates lifespan in Caenorhabditis elegans by modulating nuclear translocation of forkhead transcription factor/DAF-16. *Proc Natl Acad Sci U S A.* 2005;102(12):4494–4499. Epub 2005 Mar 14.

44 Perrot V, Rechler MM. Characterization of insulin inhibition of transactivation by a C-terminal fragment of the forkhead transcription factor Foxo1 in rat hepatoma cells. *J Biol Chem.* 2003;278(28):26111–26119.

45 Cahill CM, Tzivion G, Nasrin N, et al. Phosphatidylinositol 3-kinase signaling inhibits DAF-16 DNA binding and function via 14-3-3-dependent and 14-3-3-independent pathways. *J Biol Chem.* 2001;276(16):13402–13410.

46 Nasrin N, Ogg S, Cahill CM, et al. DAF-16 recruits the CREB-binding protein coactivator complex to the insulin-like growth factor binding protein 1 promoter in HepG2 cells. *Proc Natl Acad Sci U S A.* 2000;97(19):10412–10417.

47 Mahmud DL, G-Amlak M, Deb DK, Platanias LC, Uddin S, Wickrema A. Phosphorylation of forkhead transcription factors by erythropoietin and stem cell factor prevents acetylation and their interaction with coactivator p300 in erythroid progenitor cells. *Oncogene.* 2002;21(10):1556–1562.

48 Chan HM, La Thangue NB. p300/CBP proteins: HATs for transcriptional bridges and scaffolds. *J Cell Sci.* 2001;114(Pt 13): 2363–2373.

49 Christian M, Zhang X, Schneider-Merck T, et al. Cyclic AMP-induced forkhead transcription factor, FKHR, cooperates with CCAAT/enhancer-binding protein beta in differentiating human endometrial stromal cells. *J Biol Chem.* 2002; 277(23):20825–20832.

50 von Groote-Bidlingmaier F, Schmoll D, Orth HM, et al. DYRK1 is a co-activator of FKHR (FOXO1a)-dependent glucose-6-phosphatase gene expression. *Biochem Biophys Res Commun.* 2003;300(3):764–769.

51 Himpel S, Panzer P, Eirmbter K, et al. Identification of the autophosphorylation sites and characterization of their effects in the protein kinase DYRK1A. *Biochem J.* 2001;359(Pt 3): 497–505.

52 Kentrup H, Becker W, Heukelbach J, et al. Dyrk, a dual specificity protein kinase with unique structural features whose activity is dependent on tyrosine residues between subdomains VII and VIII. *J Biol Chem.* 1996;271(7):3488–3495.

53 Puigserver P, Rhee J, Donovan J, et al. Insulin-regulated hepatic gluconeogenesis through FOXO1-PGC-1alpha interaction. *Nature.* 2003;423(6939):550–555.

54 Van Der Heide LP, Hoekman MF, Smidt MP. The ins and outs of FoxO shuttling: mechanisms of FoxO translocation and transcriptional regulation. *Biochem J.* 2004;380(Pt 2): 297–309.

55 Li P, Lee H, Guo S, et al. AKT-independent protection of prostate cancer cells from apoptosis mediated through complex formation between the androgen receptor and FKHR. *Mol Cell Biol.* 2003;23(1):104–118.

56 Zhao HH, Herrera RE, Coronado-Heinsohn E, et al. Forkhead homologue in rhabdomyosarcoma functions as a bifunctional nuclear receptor-interacting protein with both coactivator and corepressor functions. *J Biol Chem.* 2001;276(30):27907–27912.

57 Schuur ER, Loktev AV, Sharma M, et al. Ligand-dependent interaction of estrogen receptor-alpha with members of the forkhead transcription factor family. *J Biol Chem.* 2001;276(36): 33554–33560.

58 Dowell P, Otto TC, Adi S, Lane MD. Convergence of peroxisome proliferator-activated receptor gamma and Foxo1 signaling pathways. *J Biol Chem.* 2003;278(46):45485–45491.

59 Hirota K, Daitoku H, Matsuzaki H, et al. HNF-4 is a novel downstream target of insulin via FKHR as a signal-regulated transcriptional inhibitor. *J Biol Chem.* 2003;7:7.

60 Daitoku H, Hatta M, Matsuzaki H, et al. Silent information regulator 2 potentiates Foxo1-mediated transcription through its deacetylase activity. *Proc Natl Acad Sci U S A.* 2004;101(27): 10042–10047.

61 Motta MC, Divecha N, Lemieux M, et al. Mammalian SIRT1 represses forkhead transcription factors. *Cell.* 2004;116(4):551–563.

62 Brunet A, Sweeney LB, Sturgill JF, et al. Stress-dependent regulation of FOXO transcription factors by the SIRT1 deacetylase. *Science.* 2004;303(5666):2011–2015.

63 Shu W, Yang H, Zhang L, et al. Characterization of a new subfamily of winged-helix/forkhead (Fox) genes that are expressed

in the lung and act as transcriptional repressors. *J Biol Chem.* 2001;276(29):27488–27497.
64 Sullivan SA, Akers L, Moody SA. foxD5a, a Xenopus winged helix gene, maintains an immature neural ectoderm via transcriptional repression that is dependent on the C-terminal domain. *Dev Biol.* 2001;232(2):439–457.
65 Schmidt M, Fernandez de Mattos S, van der Horst A, et al. Cell cycle inhibition by FoxO forkhead transcription factors involves downregulation of cyclin D. *Mol Cell Biol.* 2002;22(22):7842–7852.
66 Pisarska MD, Bae J, Klein C, Hsueh AJ. Forkhead l2 is expressed in the ovary and represses the promoter activity of the steroidogenic acute regulatory gene. *Endocrinology.* 2004;145(7):3424–3433.
67 Schmidt D, Ovitt CE, Anlag K, et al. The murine winged-helix transcription factor Foxl2 is required for granulosa cell differentiation and ovary maintenance. *Development.* 2004;131(4):933–942.
68 Uda M, Ottolenghi C, Crisponi L, et al. Foxl2 disruption causes mouse ovarian failure by pervasive blockage of follicle development. *Hum Mol Genet.* 2004;13(11):1171–1181.
69 Accili D, Arden KC. FoxOs at the crossroads of cellular metabolism, differentiation, and transformation. *Cell.* 2004; 117(4):421–426.
70 Nakae J, Biggs WH, 3rd, Kitamura T, et al. Regulation of insulin action and pancreatic beta-cell function by mutated alleles of the gene encoding forkhead transcription factor FoxO1. *Nat Genet.* 2002;32(2):245–253.
71 Hosaka T, Biggs WH, 3rd, Tieu D, et al. Disruption of forkhead transcription factor (FOXO) family members in mice reveals their functional diversification. *Proc Natl Acad Sci U S A.* 2004;101(9):2975–2980.
72 Furuyama T, Kitayama K, Shimoda Y, et al. Abnormal angiogenesis in Foxo1 (Fkhr)-deficient mice. *J Biol Chem.* 2004;279 (33):34741–34749.
73 Hribal ML, Nakae J, Kitamura T, et al. Regulation of insulin-like growth factor-dependent myoblast differentiation by FoxO forkhead transcription factors. *J Cell Biol.* 2003;162(4):535–541.
74 Bois PR, Grosveld GC. FKHR (FOXO1a) is required for myotube fusion of primary mouse myoblasts. *EMBO J.* 2003;22(5):1147–1157.
75 Nakae J, Kitamura T, Kitamura Y, et al. The forkhead transcription factor Foxo1 regulates adipocyte differentiation. *Dev Cell.* 2003;4(1):119–129.
76 Leenders H, Whiffield S, Benoist C, Mathis D. Role of the forkhead transcription family member, FKHR, in thymocyte differentiation. *Eur J Immunol.* 2000;30(10):2980–2990.
77 Sandri M, Sandri C, Gilbert A, et al. Foxo transcription factors induce the atrophy-related ubiquitin ligase atrogin-1 and cause skeletal muscle atrophy. *Cell.* 2004;117(3):399–412.
78 Stitt TN, Drujan D, Clarke BA, et al. The IGF-1/PI3K/Akt pathway prevents expression of muscle atrophy-induced ubiquitin ligases by inhibiting FOXO transcription factors. *Mol Cell.* 2004;14(3):395–403.
79 Nakamura N, Ramaswamy S, Vazquez F, et al. Forkhead transcription factors are critical effectors of cell death and cell cycle arrest downstream of PTEN. *Mol Cell Biol.* 2000;20(23):8969–8982.
80 Bakker WJ, Blazquez-Domingo M, Kolbus A, et al. FoxO3a regulates erythroid differentiation and induces BTG1, an activator of protein arginine methyl transferase 1. *J Cell Biol.* 2004; 164(2):175–184.
81 Tran H, Brunet A, Grenier JM, et al. DNA repair pathway stimulated by the forkhead transcription factor FOXO3a through the Gadd45 protein. *Science.* 2002;296(5567): 530–534.
82 Xia SJ, Pressey JG, Barr FG. Molecular pathogenesis of rhabdomyosarcoma. *Cancer Biol Ther.* 2002;1(2):97–104.
83 So CW, Cleary ML. Common mechanism for oncogenic activation of MLL by forkhead family proteins. *Blood.* 2003;101(2): 633–639.
84 Anderson MJ, Shelton GD, Cavenee WK, Arden KC. Embryonic expression of the tumor-associated PAX3-FKHR fusion protein interferes with the developmental functions of Pax3. *Proc Natl Acad Sci U S A.* 2001;98(4):1589–1594.
85 Lagutina I, Conway SJ, Sublett J, Grosveld GC. Pax3-FKHR knock-in mice show developmental aberrations but do not develop tumors. *Mol Cell Biol.* 2002;22(20):7204–7216.
86 Relaix F, Polimeni M, Rocancourt D, et al. The transcriptional activator PAX3-FKHR rescues the defects of Pax3 mutant mice but induces a myogenic gain-of-function phenotype with ligand-independent activation of Met signaling in vivo. *Genes Dev.* 2003;17(23):2950–2965.
87 Burgering BM, Kops GJ. Cell cycle and death control: long live forkheads. *Trends Biochem Sci.* 2002;27(7):352–360.
88 Graff JR, Konicek BW, McNulty AM, et al. Increased AKT activity contributes to prostate cancer progression by dramatically accelerating prostate tumor growth and diminishing p27Kip1 expression. *J Biol Chem.* 2000;275(32):24500–24505.
89 Modur V, Nagarajan R, Evers BM, Milbrandt J. FOXO proteins regulate tumor necrosis factor-related apoptosis inducing ligand expression. Implications for PTEN mutation in prostate cancer. *J Biol Chem.* 2002;277(49):47928–47937.
90 Sunters A, Fernandez de Mattos S, Stahl M, et al. FoxO3a transcriptional regulation of Bim controls apoptosis in paclitaxel-treated breast cancer cell lines. *J Biol Chem.* 2003;278(50): 49795–49805.
91 Birkenkamp KU, Coffer PJ. Regulation of cell survival and proliferation by the FOXO (Forkhead box, class O) subfamily of Forkhead transcription factors. *Biochem Soc Trans.* 2003;31(Pt 1):292–297.
92 Seoane J, Le HV, Shen L, et al. Integration of Smad and forkhead pathways in the control of neuroepithelial and glioblastoma cell proliferation. *Cell.* 2004;117(2):211–223.
93 Kau TR, Schroeder F, Ramaswamy S, et al. A chemical genetic screen identifies inhibitors of regulated nuclear export of a forkhead transcription factor in PTEN-deficient tumor cells. *Cancer Cell.* 2003;4(6):463–476.
94 Kidson SH, Kume T, Deng K, et al. The forkhead/winged-helix gene, Mf1, is necessary for the normal development of the cornea and formation of the anterior chamber in the mouse eye. *Dev Biol.* 1999;211(2):306–322.
95 Winnier GE, Kume T, Deng K, et al. Roles for the winged helix transcription factors MF1 and MFH1 in cardiovascular development revealed by nonallelic noncomplementation of null alleles. *Dev Biol.* 1999;213(2):418–431.
96 Kalinichenko VV, Lim L, Stolz DB, et al. Defects in pulmonary vasculature and perinatal lung hemorrhage in mice heterozygous null for the Forkhead Box f1 transcription factor. *Dev Biol.* 2001;235(2):489–506.

97 Irrthum A, Devriendt K, Chitayat D, et al. Mutations in the transcription factor gene SOX18 underlie recessive and dominant forms of hypotrichosis-lymphedema-telangiectasia. *Am J Hum Genet.* 2003;72(6):1470–1478.
98 Skurk C, Maatz H, Kim HS, et al. The Akt-regulated forkhead transcription factor FOXO3a controls endothelial cell viability through modulation of the caspase-8 inhibitor FLIP. *J Biol Chem.* 2004;279(2):1513–1525.
99 Potente M, Fisslthaler B, Busse R, Fleming I. 11,12-Epoxyeicosatrienoic acid-induced inhibition of FOXO factors promotes endothelial proliferation by down-regulating p27Kip1. *J Biol Chem.* 2003;278(32):29619–29625.
100 Daly C, Wong V, Burova E, et al. Angiopoietin-1 modulates endothelial cell function and gene expression via the transcription factor FKHR (FOXO1). *Genes Dev.* 2004;18(9):1060–1071.
101 Irmler M, Thome M, Hahne M, et al. Inhibition of death receptor signals by cellular FLIP. *Nature.* 1997;388(6638):190–195.
102 Abid MR, Schoots IG, Spokes KC, et al. Vascular endothelial growth factor-mediated induction of manganese superoxide dismutase occurs through redox-dependent regulation of forkhead and IkappaB/NF-kappaB. *J Biol Chem.* 2004;279(42): 44030–44038.
103 Abid MR, Shih SC, Otu HH, et al. A novel class of vascular endothelial growth factor-responsive genes that require forkhead activity for expression. *J Biol Chem.* 2006;281(46):35544–35553. Epub 2006 Sep 15.
104 Potente M, Urbich C, Sasaki KI, et al. Involvement of FoxO transcription factors in angiogenesis and postnatal neovascularization. *J Clin Invest.* 2005;115(9):2382–2392.
105 Castrillon DH, Miao L, Kollipara R, et al. Suppression of ovarian follicle activation in mice by the transcription factor Foxo3a. *Science.* 2003;301(5630):215–218.
106 Lin K, Dorman JB, Rodan A, Kenyon C. daf-16: An HNF-3/forkhead family member that can function to double the life-span of *Caenorhabditis elegans. Science.* 1997;278(5341):1319–1322.
107 Vicent D, Ilany J, Kondo T, et al. The role of endothelial insulin signaling in the regulation of vascular tone and insulin resistance. *J Clin Invest.* 2003;111(9):1373–1380.

94

Genetics of Coronary Artery Disease and Myocardial Infarction

The MEF2 Signaling Pathway in the Endothelium

Stephen R. Archacki*,†, Sun-Ah You*, Quansheng Xi*, and Qing Wang*,†

*Lerner Research Institute and Center for Cardiovascular Genetics, Cleveland Clinic Foundation; of Molecular Medicine, Cleveland Clinic Lerner College of Medicine of Case Western Reserve University;

†Cleveland State University, Ohio

Coronary artery disease (CAD) is the leading cause of death and disability in the United States and other developed countries (1). CAD is characterized by the formation of atherosclerotic plaques in the walls of the coronary arteries (2–5). The structure of a normal coronary artery consists of the endothelium (a single layer of endothelial cells [ECs]) on the luminal side; followed by the intima, consisting of collagens and proteoglycans; the middle layer (media) of smooth muscle cells (SMCs); and the outside layer (adventitia), consisting of connective tissues, fibroblasts, and more SMCs (Figure 94.1). Development of CAD starts with binding of blood monocytes to the endothelium through cell adhesion molecules including vascular cell adhesion molecule (VCAM)-1 and intercellular adhesion molecule (ICAM)-1, followed by infiltration of lipoproteins and monocytes into the media region, which attract oxidized lipids and form foam cells, the hallmark of an arterial lesion. Necrosis and apoptosis of foam cells lead to a necrotic core with a mass of cell debris and lipids. Macrophages and the foam cells secrete cytokines, inflammatory molecules, and growth factors that induce SMC migration, proliferation, and the production of extracellular matrix-forming plaques with fibrous caps. When the plaques are stable, the patient may not experience any symptoms of chest pain. However, plaque rupture may lead to thrombosis and secondary unstable angina, acute myocardial infarction (MI), or sudden death (6,7).

This overview of the pathophysiology of CAD and MI represents the current prevailing mechanism for the pathogenesis of CAD and MI and suggests that CAD is an inflammatory process affecting the endothelium (i.e., endothelial dysfunction) in addition to a disease of lipid metabolism, SMC proliferation and migration, and upregulation of immune responses. Recent advances in molecular genetic studies of CAD and MI have, however, provided new insights into the pathogenic mechanisms of CAD and MI. The discovery of *myocyte enhancer factor 2A (MEF2A)*, as a gene associated with CAD and MI, identifies one member of the MEF2 family in the abnormal development of the endothelium in the disease process (6,7). This chapter reviews the structure and function of MEF2 transcription factors, with emphasis on their roles in vascular development and their involvement in the development of CAD and MI.

MEF2 STRUCTURE AND FUNCTION

The MEF2 family of transcription factors has been studied most extensively in muscle cells, where they play a key role in myogenesis and morphogenesis of cardiac or skeletal muscle (8). MEF2 proteins have been implicated in neuronal cell survival, T-cell apoptosis, and cellular responses to growth factors, cytokines, and environmental stressors (8–18). Recent studies, as discussed later, suggest that MEF2 proteins also regulate vascular morphogenesis, including endothelial function and development.

The MEF2 family consists of four members: MEF2A, MEF2B, MEF2C, and MEF2D. They are nuclear phosphoproteins belonging to the MCM1 agamous deficiens serum response factor (MADS) superfamily of DNA binding proteins. The four MEF2 proteins share approximately 50% amino acid identity overall but have more than 95% similarity at the N-termini. MEF2 proteins have an identical structure of functional domains (Figure 94.2). The N-termini of the MEF2 proteins contain the MADS domain responsible for specific

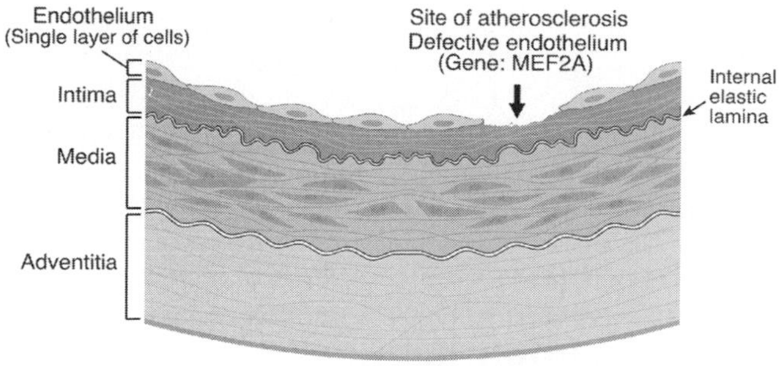

Figure 94.1. A schematic diagram of a coronary artery. It consists of the endothelium and intima, internal elastic lamina, media, and adventitia. MEF2A is expressed in the endothelium. Mutations detected in this gene lead to a dysfunctional EC layer and culminate in the generation of atherosclerosis.

DNA binding and dimerization, followed by the MEF2 domain, which increases DNA binding and mediates interactions of MEF2 proteins with other cofactors. The C-termini of MEF2 factors contain the transcriptional activation domains, which are also the targets of mitogen-activated protein kinase (MAPK) activation pathway and intracellular signaling cascades activated by calcium (19–22). The C-terminal end also contains the nuclear localization signals for transporting these transcription factors into the nucleus.

The MEF2 proteins bind as homo- and heterodimers to a *cis*-element with the consensus sequence (C/T)TA(A/T)$_4$ TA(G/A) in the promoter of many cardiac and skeletal muscle–specific genes. The three-dimensional structure of a complex between DNA and the core dimeric DNA binding domain of MEF2A (residues 1–85; containing residues 1–58 for MADS domain and residues 59–73 for MEF2 domain) has been determined. The MADS domain mediates recognition and specificity of DNA binding, and both the major and minor grooves of DNA are involved in this interaction (23).

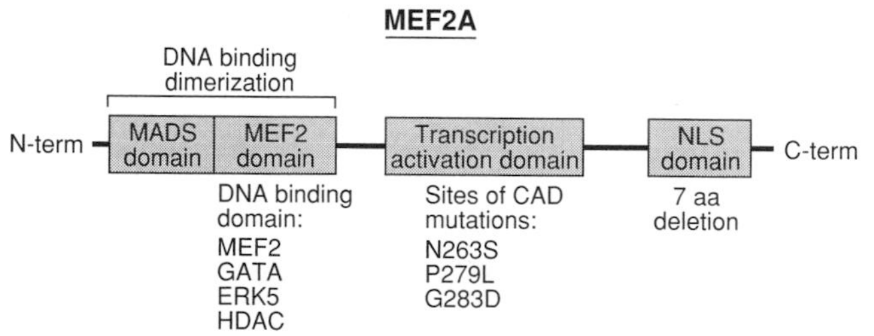

Figure 94.2. The domain structure of MEF2A and locations of CAD-associated mutations in MEF2A. The N-terminal end of the MEF2A protein contains the MADS domain responsible for specific DNA binding and dimerization, followed by the MEF2 domain that increases DNA binding and mediates interactions of MEF2 proteins with other cofactors (GATA, ERK5, HDAC, etc.). The C-terminal region of MEF2A contains the transcriptional activation domains, which are also the targets of MAP kinase activation pathways and intracellular signaling cascades activated by calcium. The C-terminal end of MEF2A contains the nuclear localization signal (NLS).

MEF2 proteins interact with many different cofactors, which may be a mechanism by which MEF2 factors control many different cellular processes. These factors, which can bind and stimulate MEF2 activity, include MyoD, GATA, nuclear factor of activated T cells (NFAT), TH receptor, p300/PCAF, 14–3–3, and extracellular signal-regulated kinase 5 (ERK5) and other proteins that suppress MEF2 function such as histone deacetylase (HDAC)4, HDAC5, HDAC7, HDAC9, interacting transcription repressor (MITR), and Cabin (23). In addition, the interaction of MEF2 with MyoD regulates skeletal muscle differentiation, and the interaction of MEF2 with GATA factors may be important for cardiovascular development (24).

MEF2 is also a key player in the transduction of signals from extracellular environment. It is involved in the induction of the *c-jun* promoter in response to signals originating from multiple cell surface receptors including G-protein–coupled receptors (24), epidermal growth factor receptor (25–27), the lipopolysaccharide receptor (28), and the CD28 costimulatory receptor in T lymphocytes (29). In response to increased intracellular calcium concentration mediated by T-cell receptor signaling, MEF2 stimulates the expression of the gene encoding the orphan nuclear hormone receptor Nur77 (30), also known as TR3, which is a potent activator of cytochrome c-mediated apoptosis (31). And finally, angiotensin II stimulates the DNA-binding activity of MEF2A and its expression at the protein level without affecting the expression level of MEF2A transcription, suggesting a post-transcriptional regulation of MEF2A expression by angiotensin II in vascular SMCs (32). The p38 MAPK-dependent pathway mediates this process. It is apparent that MEF2 is involved in several pathways triggered from the binding of ligands to cell surface receptors. Furthermore, MEF2 also targets over 30 genes that specifically bind to the promoters in nearly all skeletal and cardiac muscle (33).

MEF2A IS ASSOCIATED WITH CORONARY ARTERY DISEASE AND MYOCARDIAL INFARCTION

Association between MEF2A Mutations and Coronary Artery Disease/Myocardial Infarction

Genetic studies of a human pedigree with CAD and MI identified a seven–amino acid deletion of MEF2A that causes CAD and MI (7,34). The family studied was from Iowa (in the Midwestern United States) and has been plagued for generations by incidents of CAD and MI. Thirteen family members were affected with CAD; nine of them developed acute MI, five patients required percutaneous coronary angioplasty, and four underwent coronary artery bypass surgery. The disease in this family is inherited in an autosomal dominant fashion. The disease gene was localized to the telomeric region of the long arm of chromosome 15, designated 15q26. This is the first genetic locus mapped for autosomal dominant CAD and MI, thus the locus was designated as *adCAD1*.

The specific CAD/MI gene at this locus was identified as *MEF2A*. It contained a deletion of seven amino acids ($\Delta Q_{440}P_{441}P_{442}Q_{443}P_{444}Q_{445}P_{446}$) located close to the nuclear localization signal of MEF2A protein (Figure 94.2). All participants were screened for mutations by single-stranded conformational polymorphisms (SSCP) analysis in this gene (whether they had symptomatic heart disease or not), and the presence or absence of atherosclerosis was documented by a cardiac angiogram (6,7). The MEF2A mutation was present in every family member with detectable CAD and was absent in relatives and unrelated individuals without evidence of heart disease.

Further screening of the *MEF2A* gene with SSCP in 207 living patients in the general population with CAD/MI identified three novel missense mutations, N263S, P279L, and G283D in four patients (1.93%) (Figure 94.2) (6). None of these mutations was detected in 191 control individuals without CAD/MI as documented by direct assessment of the coronary arteries by an angiogram. The CAD cases and controls were matched for ethnicity, age, and body mass index (6). These results suggest that nearly 2% of the CAD/MI population may carry an MEF2A mutation. The P279L variant in exon 7 was independently confirmed to be associated with MI in a Spanish population (odds ratio g3.1) (34). In sum, the *MEF2A* gene has been established as the first non–lipid-related gene in which mutations have been linked to CAD and MI, connecting the MEF2 signaling pathway to an important human disease.

MEF2A Mutation Molecular Mechanisms in Coronary Artery Disease and Myocardial Infarction

The MEF2A mutations have been functionally characterized in detail. The seven–amino acid deletion disrupts the nuclear localization of MEF2A, reduces MEF2A-mediated transcription activation, and abolishes synergistic activation between MEF2A and the transcription factor GATA-1 (11). Mutant MEF2A with the seven–amino acid deletion interferes with the function of wild-type MEF2A protein, indicating that the deletion acts via a dominant negative mechanism (7). The three missense mutations have similar functional effects on MEF2A, and they reduce the transcriptional activation activity of MEF2A or synergistic transactivation by MEF2A and GATA-1 as demonstrated in HeLa cells (7). In contrast to the deletion, each of the three missense mutations acts by a loss-of-function mechanism and does not interfere with the function of wild-type MEF2A protein. Interestingly, consistent with genotype–phenotype correlation studies, the seven–amino acid deletion has a dominant negative effect in that it is associated with a much more severe clinical outcome (9 of 13 patients developed MI) compared with the missense mutations (none of the four carriers have developed MI yet) (6,7).

Because mutations in MEF2A are associated with CAD, we hypothesized that MEF2A contributes to a dysfunctional or abnormal development of the endothelium, rendering ECs then prone to inflammation and thrombosis, ultimately leading to the development of CAD and MI. Immunostaining studies with an anti-MEF2A antibody detected MEF2A protein in the endothelium of human coronary arteries (7), which co-localizes with the expression signal of platelet-endothelial adhesion molecule-(PECAM)-1/CD31, an EC marker (Figure 94.3). MEF2A protein was also detected in cultured human umbilical vascular ECs and intact human coronary arteries by Western blot analyses (Figure 94.4). And finally, MEF2A mRNA was also detected in rat SMCs in vitro (35). MEF2A protein expression was detected in proliferating SMCs but not in differentiated SMCs in a rat model of arterial injury and clinical restenosis (35). We have also detected expression of MEF2A protein in the nuclei of proliferating human SMCs

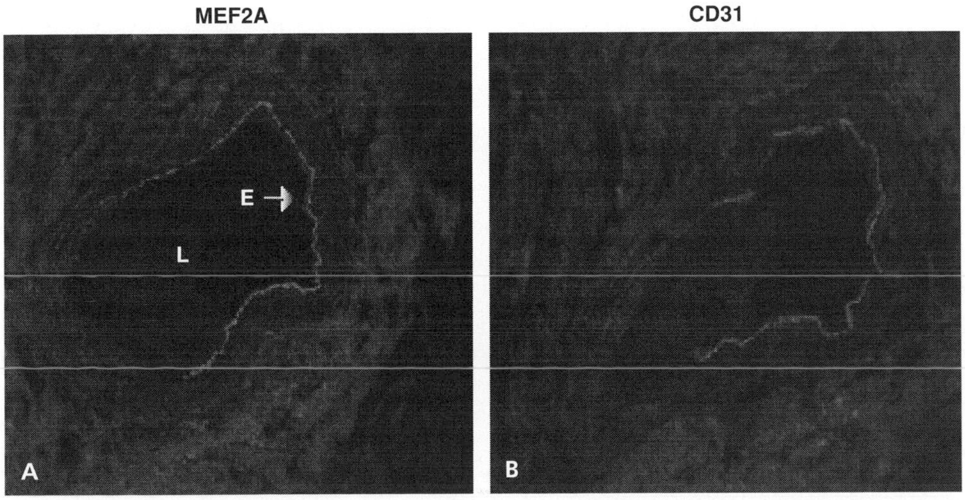

Figure 94.3. The expression of MEF2A in the coronary endothelium. (**A**) Immunostaining studies with an anti-MEF2A antibody detected a high level of expression of MEF2A protein in the endothelium of coronary arteries. (**B**) MEF2A colocalized with the expression signal of CD31, an EC marker. (Reproduced with permission from L. Wang et al. *Science* 2003;302:15780–15781.)

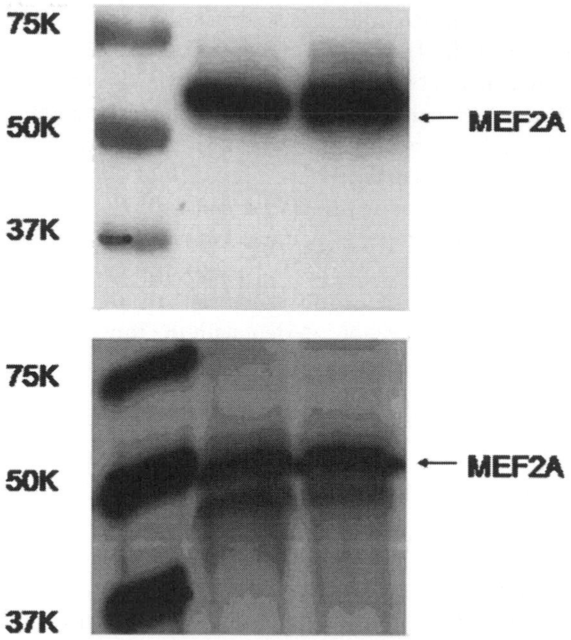

Figure 94.4. Detection of MEF2A protein in ECs and coronary arteries by Western blot analysis. (**A**) MEF2A protein was detected in protein extracted from cultured human umbilical vascular ECs. (**B**) MEF2A protein was detected from protein extracted from intact human coronary artery tissue.

(7). These results suggest that MEF2A protein is present in the target organ of CAD – namely the endothelium and SMCs of the coronary artery.

Expression of MEF2A has been previously studied in mice. *MEF2A* mRNA was detected in blood vessels during mouse early embryogenesis (36). Expression of MEF2A protein was detected as early as day 8.5 postcoitum in cells of embryonic vasculature, and the pattern is similar to that of vascular endothelial growth factor receptor 2 (VEGFR2) and von Willebrand factor (both are EC markers) (37). These expression studies suggest that MEF2A may play an important role in vascular morphogenesis.

Mice deficient in MEF2A have been created (38). Homozygous $MEF2A^{-/-}$ mice in the 129sv genetic background die suddenly within the first week of life. However, mice on mixed genetic backgrounds survive (38). No structural heart defects have been detected before death, but dilation of the right ventricle was detected in homozygous $MEF2A^{-/-}$ mice with sudden death at necropsy. The cause of the sudden death phenotype in homozygous $MEF2A^{-/-}$ mice remains largely unknown.

Based on the findings that MEF2A mutations are associated with CAD and that MEF2A is upregulated in cultured ECs and in coronary endothelium, we proposed a novel function of a transcription factor that contributes specifically to dysfunctional or abnormally developed endothelium, thus triggering the development of CAD (7). We expect that future studies will uncover a role for MEF2A in controlling the expression of many genes in the coronary endothelium. Mutations in MEF2A will alter the expression of some subsets of genes and thus reprogram the transcriptional profile in the endothelium, leading to EC dysfunction. As the endothelium forms a barrier between blood and the arterial wall, defects in this cell layer may render it more susceptible to the invasion of monocytes and macrophages and to the formation of atherosclerotic plaques. Once the arterial wall integrity is lost, a cascade of events culminates in unstable angina, MI, or sudden cardiac death. It is important to note that this hypothesis remains to be further validated.

In sum, the molecular pathway of MEF2A plays an important role in the complex pathogenic mechanisms that generate CAD and MI. It is likely that the MEF2A signaling pathway will be found to contribute to the regulation of endothelial development and function and may also be a key player involved in the complex disease process of atherosclerosis.

MEF2B, A POTENT TRANSACTIVATOR EXPRESSED IN EARLY MYOGENIC LINEAGES

Like other MEF2 members, MEF2B also binds to A/T-rich DNA sequences and regulates gene expression through interactions with other coactivators and corepressors in several tissues. In muscle cells, MEF2B is the only member of MEF2 family that regulates smooth muscle myosin heavy chain gene expression (39). In neuronal cells, MEF2B and MEF2C, together with adhesion related kinase (Ark), repress the expression of gonadotropin-releasing hormone (GnRH) (40). Although there has been no report on MEF2B in ECs, MEF2B was detected in SMCs and neointima (41).

MEF2B is the most divergent member of the vertebrate MEF2 family of transcription factors. It shares significant amino acid sequence with other MEF2 members within the MADS/MEF2 domains, whereas the C-terminal region (transactivation domain) shows relatively little similarity (42).

Members of the MEF2 family are generally kept in an inactive state when directly associated with the class II HDACs. HDACs repress transcription by deacetylating core histones, which results in chromatin condensation. In T lymphocytes, MEF2 activity is also repressed by Cabin1, which recruits class I HDAC to MEF2 target genes. Recently, there have been advances in understanding how MEF2 interacts with HDACs and Cabin1. Han and colleagues solved the crystal structure of the MADS-box/MEF2 domain of human MEF2B bound to a motif of Cabin1 and DNA (43). A stably folded MEF2 domain was shown to be on the surface of the MADS domain, and Cabin1 binds a hydrophobic groove on the MEF2 domain. The structure of the complex between MADS-box/MEF2 domain of MEF2B and the MEF2-binding motif of class II HDAC9 was also determined and is similar to the structure of the Cabin1/MEF2/DNA complex (44). Thus, MEF2 proteins recruit transcriptional corepressor Cabin1 or HDACs to specific DNA sites through the MEF2 domain.

The functional role of MEF2B remains largely unknown. Unlike other MEF2 members, MEF2B is not phosphorylated by p38 MAPK or big MAPK1 (BMK1, also called ERK-5). Thus, how MEF2B is phosphorylated still needs to be elucidated (45,46). MEF2B also binds the MEF2 consensus sequence with reduced affinity relative to the other family members, whereas MEF2A, -C and -D have similar DNA specificity (42).

MEF2C, CRITICAL FOR THE DEVELOPMENT OF THE VASCULAR SYSTEM

The *MEF2C* gene is expressed in multiple cell types, including skeletal, cardiac, and SMCs as well as the neural crest and the vascular endothelium (47–49). It plays an important role in vascular development and is required for cardiac myogenesis and morphogenesis (49–51). *MEF2C*-null mice exhibit profound cardiac defects, including a near complete loss of the right ventricle and its outflow tract, defective aortic arch looping, and a disorganized endocardium (50). Consequently, mouse embryos lacking MEF2C fail to form a functional vascular system and have major defects in vasculogenesis that are lethal by embryonic day E9.5. Although ECs do develop, they fail to organize normally into a vascular plexus, suggesting a potential role for MEF2C in EC differentiation. At the same time, SMCs do not differentiate. It is not clear whether MEF2C has a primary role for SMC differentiation or whether a defect in signaling between ECs and pericytes is responsible for this failure of SMCs to differentiate. These vascular defects are similar to that of knockout mice for VEGF or its receptor VEGFR1 (49).

Bi and colleagues, showed that MEF2C deficiency results in a complex set of vascular malformations, including the arrest of the yolk sac capillary network and the eventual disruption of the yolk sac vessels (51). The embryonic vasculature fails to remodel into a more complex vasculature of diversely sized vessels by E9.5. The vessels in the head and those caudal to the heart have enlarged lumens and a reduction in the number of capillaries, whereas the vessels located dorsally and anterior to the heart have smaller lumens compared with normal control animals. Furthermore, ECs in this region are rounded with erratic orientations with respect to the underlying myocardium (51). This suggests that a defect in EC differentiation or angiogenesis occurs and has been attributed to the decreased angiopoietin 1 (Ang1) and VEGF expression by the myocardium (51). This in turn suggests that an essential function of MEF2C in the myocardium is the regulation of proper expression of these cytokines to direct endocardial formation.

Similar embryonic abnormalities were also detected in mice null for big MAPK-1 (BMK1, also called ERK-5) and MAPK kinase kinase-3 (MEKK3) (49–54). The MAPK signaling pathways start with activation of MEKK by G-protein–coupled receptors. Activated MEKK phosphorylates a subordinate MAPK kinase (e.g., MEK5), which further activates a MAPK (e.g., BMK1 or ERK5) by phosphorylation. BMK1 can stimulate transcriptional activation of MEF2 factors by phosphorylating the transcription activation domains or by direct interactions with MEF2 factors.

Studies with *BMK1* conditional knockout mice provided direct genetic evidence that a BMK1–MEF2C pathway is critical for EC survival and blood vessel integrity (55). Cre-recombinase-mediated ablation of *BMK1* in adult mice results in the degeneration of the cardiovascular system and is accompanied by hemorrhages in multiple organs. The EC lining of blood vessels become round, irregularly aligned, and eventually apoptotic. The in vitro removal of BMK1 protein also leads to the death of ECs due in part to the deregulation of MEF2C. Additionally, endothelial-specific *BMK* knockout leads to cardiovascular defects identical to those of global *BMK* knockout mutants, whereas surprisingly, mice lacking BMK1 in cardiomyocytes develop to term without any apparent defects. Moreover, normal ECs proliferated and underwent a mesenchymal transformation in cardiac explant cultures, whereas explanted ECs from $BMK1^{-/-}$ mice failed to proliferate in vitro. Infection of *BMK1*-knockout ECs with an adenovirus encoding a constitutively active form of MEF2C fused to the VP16 activation domain provided considerable protection from apoptosis for these cells. Taken together, these results provide direct genetic evidence that the BMK1 pathway delivers survival signal through MEF2C in ECs (55).

In sum, these targeted inactivation studies of the MEF2C have demonstrated that MEF2C is an early marker of the vascular development that mediates EC organization and angiogenesis.

MEF2C Is Regulated by Ets-1

Screening for transcriptional enhancers from the mouse *MEF2C* gene identified an EC-specific, intronic enhancer that binds transcription factor v-ets erythroblastosis virus E26 oncogene homolog 1 (avian) (Ets-1), which is a transcription factor that is sufficient to drive MEF2C expression in the developing vascular endothelium. Ets-1 is active from the earliest stages in vasculogenesis in the yolk sac of the embryo and remains robustly active throughout embryonic development and into maturation to the adult form. The EC-specific enhancer of MEF2C contains four highly conserved Ets-binding sites that are efficiently bound by Ets-1 protein in vitro, and two of these sites are required for the enhancer function in embryos. The MEF2C enhancer does not drive its expression in vascular SMCs, indicating that the enhancer is specific for ECs and additional discrete elements within the *MEF2C* locus will direct expression in SMCs (56).

MEF2D, A MEMBER OF MEF2 FAMILY WITH A KEY ROLE IN APOPTOSIS

Breitbart first cloned human *MEF2D* in 1993 using a *MEF2B* cDNA probe to screen human cardiac ventricle cDNA. It was later mapped to human chromosome 1q12–23. MEF2D is ubiquitously expressed with several alternatively spliced forms (57,58). MEF2D has been shown to play a key role in T-cell receptor (TCR)-mediated apoptosis during thymic negative selection. It mediates calcium-dependent transcription of Nur77, a transcription factor involved in TCR-mediated apoptosis of thymocytes (30,59). MEF2D binds two calcium-responsive DNA elements in the *Nur77* promoter and mediates the calcium-dependent induction of Nur77 (30). It remains to be determined whether MEF2D plays any role in vascular development.

KEY POINTS

- The *MEF2* genes encode members of the MEF2 family of transcription factors. Four members in this family exist: MEF2A, MEF2B, MEF2C, and MEF2D. MEF2 proteins bind to a conserved A/T-rich sequence element bearing the consensus sequence CTA$(A/T)_4$TAG in the promoters of many skeletal and cardiac muscle genes and regulates their expression.
- MEF2 proteins share high sequence similarity in an N-terminal MADS (MCM1, agamous, deficiens, serum response factor) domain and an adjacent MEF2-specific domain. The MADS domain mediates DNA binding and dimerization and the MEF2-specific domain influences DNA-binding affinity and is responsible for interactions with transcriptional cofactors. The C-terminal regions of MEF2 factors, which are more divergent, act as transcription activation domains and contain signals that are necessary for nuclear localization.
- Although it is established that MEF2 factors play a critical role in muscle differentiation and development, subsequent studies have demonstrated that they are expressed ubiquitously and might have additional functions apart from their involvement in myogenesis.
- *MEF2A* is the first disease-associated gene identified for autosomal dominant CAD and MI. The first genetic locus for autosomal dominant CAD/MI, *adCAD1*, is in the chromosome 15q26 region.
- MEF2A protein is highly expressed in coronary endothelium, suggesting that the pathogenesis of CAD may involve deregulation of transcriptional programs in the endothelium. Dysfunctional or abnormally developed endothelium has been proposed as a triggering mechanism for development of CAD.
- MEF2B is the only member of the MEF2 family that regulates smooth muscle myosin heavy chain gene expression. It is the most divergent member of the vertebrate MEF2 family of transcription factors.
- Several studies of knockout mice of *MEF2C* have established it as a critical gene for normal development of the vascular system. An intronic enhancer of *MEF2C* is responsible for its expression in ECs. The enhancer is EC specific and a site for transcription factor Ets-1.
- MEF2D plays a role in T-cell apoptosis.

Future Goals

- To further associate MEF2A in the genesis of CAD/MI and validate these findings in the patient population at large
- To define a specific mechanism of action of MEF2A in the generation of disease
- To create the MEF2A seven–amino acid deletion in a transgenic mouse model to observe a genotype–phenotype correlation with CAD/MI

ACKNOWLEDGMENTS

This work was supported by the NIH grants P50HL77107 R01 HL73817, R01 HL65630, R01 HL66251 and an American Heart Association Established Investigator award (all to QW).

REFERENCES

1 American Heart Association. Heart disease and stroke statistics-2004 update. 2003. American Heart Association, 2003.
2 Libby P, Ridker PM, Maseri A. Inflammation and atherosclerosis. *Circulation*. 2002;105:1135–1143.
3 Lusis AJ. Atherosclerosis. *Nature*. 2000;407:233–241.
4 Shen G, Archacki SR, Wang Q. The molecular genetics of coronary artery disease and myocardial infarction. *Acute Coronary Syndrome*. 2004;6:129–141.
5 Wang Q, Chen Q. Cardiovascular disease and congenital heart defects. *Encyc Hum Genome*. 2003;1:396–411.
6 Bhagavatula MR, Fan C, Shen GQ, et al. Transcription factor MEF2A mutations in patients with coronary artery disease. *Hum Mol Genet*. 2004;13:3181–3188.
7 Wang L, Fan C, Topol SE, et al. Mutation of MEF2A in an inherited disorder with features of coronary artery disease. *Science*. 2003;302:1578–1581.
8 Lin Q, Schwarz J, Bucana C, et al. Control of mouse cardiac morphogenesis and myogenesis by transcription factor MEF2C. *Science*. 1997;276:1404–1407.
9 Quinn ZA, Yang CC, Wrana JL, et al. Smad proteins function as co-modulators for MEF2 transcriptional regulatory proteins. *Nucleic Acids Res*. 2001;29:732–742.
10 Bruneau BG, Bao ZZ, Tanaka M, et al. Cardiac expression of the ventricle-specific homeobox gene Irx4 is modulated by Nkx2–5 and dHand. *Dev Biol*. 2000;217:266–277.
11 Bi W, Drake CJ, Schwarz JJ. The transcription factor MEF2C-null mouse exhibits complex vascular malformations and reduced

cardiac expression of angiopoietin 1 and VEGF. *Dev Biol.* 1999; 211:255–267.

12 Lyons GE, Micales BK, Schwarz J, et al. Expression of mef2 genes in the mouse central nervous system suggests a role in neuronal maturation. *J Neurosci.* 1995;15:5727–5738.

13 Edmondson DG, Lyons GE, Martin JF, et al. Mef2 gene expression marks the cardiac and skeletal muscle lineages during mouse embryogenesis. *Development.* 1994;120:1251–1263.

14 Swanson BJ, Jack HM, Lyons GE. Characterization of myocyte enhancer factor 2 (MEF2) expression in B and T cells: MEF2C is a B cell-restricted transcription factor in lymphocytes. *Mol Immunol.* 1998;35:445–458.

15 Buchberger A, Arnold HH. The MADS domain containing transcription factor cMef2a is expressed in heart and skeletal muscle during embryonic chick development. *Dev Genes Evol.* 1999; 209:376–381.

16 BlackBL, Olson EN. Transcriptional control of muscle development by myocyte enhancer factor-2 (MEF2) proteins. *Annu Rev Cell Dev Biol.* 1998;14:167–196.

17 McKinsey TA, Zhang CL, Olson EN. MEF2: a calcium-dependent regulator of cell division, differentiation and death. *Trends Biochem Sci.* 2002;27:40–47.

18 Youn HD, Sun L, Prywes R, et al. Apoptosis of T cells mediated by Ca^{2+}-induced release of the transcription factor MEF2. *Science.* 1999;286:790–793.

19 Zhao M, New L, Kravchenko VV, et al. Regulation of the MEF2 family of transcription factors by p38. *Mol Cell Biol.* 1999;19: 21–30.

20 Han J, Jiang Y, Li Z, et al. Activation of the transcription factor MEF2C by the MAP kinase p38 in inflammation. *Nature.* 1997; 386:296–299.

21 Yang CC, Ornatsky OI, McDermott JC, et al. Interaction of myocyte enhancer factor 2 (MEF2) with a mitogen-activated protein kinase, ERK5/BMK1. *Nucleic Acids Res.* 1999;26: 4771–4777.

22 Ornatsky OI, Cox DM, Tangirala P, et al. Post-translational control of the MEF2A transcriptional regulatory protein. *Nucleic Acids Res.* 1999;27:646–654.

23 Black BL, Molkentin JD, Olson EN. Multiple roles for the MyoD basic region in transmission of transcriptional activation signals and interaction with MEF2. *Mol Cell Biol.* 1998;18: 69–77.

24 Buchberger A, Arnold HH. The MADS domain containing transcription factor cMef2a is expressed in heart and skeletal muscle during embryonic chick development. *Dev Genes Evol.* 1999;209: 376–381.

25 Clarke N, Arenzana N, Hai T, et al. Epidermal growth factor induction of the c-jun promoter by a Rac pathway. *Mol Cell Biol.* 1998;18:1065–1073.

26 Kato Y, Tapping RI, Huang S, et al. Bmk1/Erk5 is required for cell proliferation induced by epidermal growth factor. *Nature.* 1998;395:713–716.

27 Kato Y, Kravchenko VV, Tapping RI, et al. BMK1/ERK5 regulates serum-induced early gene expression through transcription factor MEF2C. *EMBO J.* 1997;162:7054–7066.

28 Han J, Jiang Y, Li Z, et al. Activation of the transcription factor MEF2C by the MAP kinase p38 in inflammation. *Nature.* 1997; 386:296–299.

29 Shin HM, Han TH. CD28-mediated regulation of the c-jun promoter involves the MEF2 transcription factor in Jurkat T cells. *Mol Immunol.* 1999;36:197–203.

30 Woronicz JD, Calnan B, Ngo V, Winoto A. Requirement for the orphan steroid receptor Nur77 in apoptosis of T-cell hybridomas. *Nature.* 1994;367:277–281.

31 McKinsey TA, Zhang CL, Olson EN. MEF2: a calcium-dependent regulator of cell division, differentiation and death. *Trends Biochem Sci.* 2002;27:40–47.

32 Suzuki E, Nishimatsu H, Satonaka H, et al. Angiotensin II induces myocyte enhancer factor 2- and calcineurin/nuclear factor of activated T cell-dependent transcriptional activation in vascular myocytes *Circ Res.* 2002;90:1004–1011.

33 Black BL, Olson EN. Transcriptional control of muscle development by myocyte enhancer factor-2 (MEF2) proteins. *Annu Rev Cell Dev Biol.* 1998;14:167–196.

34 Gonzalez P, Garcia-Castro M, Reguero JR, et al. The Pro279 Leu variant in the transcription factor MEF2A is associated with myocardial infarction. *J Med Genet.* 2006;43:167–169.

35 Firulli AB, Miano JM, Bi W, et al. Myocyte enhancer binding factor-2 expression and activity in vascular smooth muscle cells. Association with the activated phenotype. *Circ Res.* 1996;78:196–204.

36 Edmondson DG, Lyons GE, Martin JF, Olson EN. Mef2 gene expression marks the cardiac and skeletal muscle lineages during mouse embryogenesis. *Development.* 1994;120: 1251–1263.

37 Subramanian SV, Nadal-Ginard B. Early expression of the different isoforms of the myocyte enhancer factor-2 (MEF2) protein in myogenic as well as non-myogenic cell lineages during mouse embryogenesis. *Mech Dev.* 1996;57:103–112.

38 Naya FJ, Black BL, Wu H, et al. Mitochondrial deficiency and cardiac sudden death in mice lacking the MEF2A transcription factor. *Nat Med.* 2002;8:1303–1309.

39 Hidaka K, Morisaki T, Byun SH, et al. The MEF2B homologue differentially expressed in mouse embryonal carcinoma cells. *Biochem Biophys Res Commun.* 1995;213:555–560.

40 Allen MP, Xu M, Zeng C, et al. Myocyte enhancer factors-2B and -2C are required for adhesion related kinase repression of neuronal gonadotropin releasing hormone gene expression. *J Biol Chem.* 2000;275:39662–39670.

41 Katoh Y, Molkentin JD, Dave V, et al. MEF2B is a component of a smooth muscle-specific complex that binds an A/T-rich element important for smooth muscle myosin heavy chain gene expression. *J Biol Chem.* 1998;273:1511–1518.

42 Molkentin JD, Firulli AB, Black BL, et al. MEF2B is a potent transactivator expressed in early myogenic lineages. *Mol Cell Biol.* 1996;16:3814–3824.

43 Han A, Pan F, Stroud JC, et al. Sequence-specific recruitment of transcriptional co-repressor Cabin1 by myocyte enhancer factor-2. *Nature.* 2003;422:730–734

44 Han A, He J, Wu Y, et al. Mechanism of recruitment of class II histone deacetylases by myocyte enhancer factor-2. *J Mol Biol.* 2005;345:91–102.

45 Zhao M, New L, Kravchenko VV, et al. Regulation of the MEF2 family of transcription factors by p38. *Mol Cell Biol.* 1999;19: 21–30.

46 Kato Y, Zhao M, Morikawa A, et al. Big mitogen-activated kinase regulates multiple members of the MEF2 protein family. *J Biol Chem.* 2000;275:18534–18540.

47 Dodou E, Verzi MP, Anderson JP, et al. MEF2c is a direct transcriptional target of ISL1 and GATA factors in the anterior heart field during mouse embryonic development. *Development.* 2004;13:3931–3942.

48 Edmondson DG, Lyons GE, Martin JF, Olson EN. MEF2 gene expression marks the cardiac and skeletal muscle lineages during mouse embryogenesis. *Development*. 1994;120:1251–1263.
49 Lin Q, Lu J, Yanagisawa H, et al. Requirement of the MADS-box transcription factor MEF2C for vascular development. *Development*. 1998;125:4565–4574.
50 Lin Q, Schwarz J, Bucana C, Olson EN. Control of mouse cardiac morphogenesis and myogenesis by transcription factor MEF2C. *Science*. 1997;276:1404–1407.
51 Bi W, Drake CJ, Schwarz JJ. The transcription factor MEF2C-null mouse exhibits complex vascular malformations and reduced cardiac expression of angiopoietin 1 and VEGF. *Dev Biol.* 1999; 211:255–267.
52 Regan CP, Li W, Boucher DM, et al. Erk5 null mice display multiple extraembryonic vascular and embryonic cardiovascular defects. *Proc Natl Acad Sci USA*. 2002;99:9248–9253
53 Sohn SJ, Sarvis BK, Cado D, Winoto A. ERK5 MAPK regulates embryonic angiogenesis and acts as a hypoxia-sensitive repressor of vascular endothelial growth factor expression. *J Biol Chem.* 2002;277:43344–43351.
54 Yang J, Boerm M, McCarty M, et al. MEKK3 is essential for early embryonic cardiovascular development. *Nat Genet.* 2000; 24:309–313.
55 Hayashi M, Kim SW, Imanaka-Yoshida K, et al.Targeted deletion of BMK1/ERK5 in adult mice perturbs vascular integrity and leads to endothelial failure. *J Clin Invest*. 2004;113:1138–1148.
56 De Val S, Anderson JP, Heidt AB, et al. Mef2c is activated directly by Ets transcription factors through an evolutionarily conserved EC-specific enhancer. *Dev Biol.* 2004;275:424–434.
57 Breitbart RE, Liang CS, Smoot LB, et al. A fourth human MEF2 transcription factor, hMEF2D, is an early marker of the myogenic lineage. *Development*. 1993;118:1095–1106.
58 Hobson GM, Krahe R, Garcia E, et al. Regional chromosomal assignments for four members of the MADS domain transcription enhancer factor 2 (MEF2) gene family to human chromosomes 15q26, 19p12, 5q14, and 1q12-q23. *Genomics*. 1995;29: 704–711.
59 Liu ZG, Smith SW, McLaughlin KA, et al. A. Apoptotic signals delivered through the T-cell receptor of a T-cell hybrid require the immediate-early gene Nur77. *Nature*. 1994;367:281–284.

95

Vezf1

A Transcriptional Regulator of the Endothelium

Frank Kuhnert* and Heidi Stuhlmann†

*The Scripps Research Institute, La Jolla, California;
Stanford University Medical Center, Stanford, California;
†Weill Medical College of Cornell University, New York City, New York

This chapter summarizes our current knowledge of vascular endothelial zinc finger 1 (Vezf1) and its human orthologue DB1 and their role in vascular biology. We describe the unique features of the *Vezf1* gene and its homologues, and we will present evidence that Vezf1 acts as a transcription factor. Finally, we discuss possible roles that Vezf1 plays in the endothelium during development and in angiogenesis.

Mouse *Vezf1* originally was identified using a retroviral "gene trap" screen for early cardiovascular genes in mouse embryonic stem (ES) cells and embryos (1–4). During embryogenesis, Vezf1 expression is localized in vascular endothelium (4), as well as neuronal and mesodermal tissues. Under in vitro conditions, Vezf1 expression has been reported in cultured endothelial cells (ECs) such as human umbilical vein ECs (HUVECs), mouse EOMA and yolk sac–derived ECs, hematopoietic cell lines (T, pre-B, erythroleukemia cells, macrophages) (4), and a fetal liver–derived stromal cell line (5). The human orthologue, DB1 (also termed ZF161), was first isolated from Jurkat T cells as a protein that binds to a GT/GC-rich region in the human *interleukin (IL)-3* promoter (6). In addition, DB1 was found to be upregulated during differentiation of dendritic cells from monocytes by cDNA microarray analysis (7).

VEZF1 IS HIGHLY CONSERVED AMONG VERTEBRATES

One of the striking features of Vezf1 is the high extent of conservation of the gene structure and the nucleotide and amino acid sequences (4; F. Kuhnert and H. Stuhlmann, unpublished observations). The murine *Vezf1* gene contains six exons and five introns and spans a total length of 16.3-kb on chromosome 11. The human orthologue *DB1/ZNF16* maps to a syntenic region at 17q23.2 and shows the same gene structure with respect to the number, the position, and the size of the coding exons, as well the nucleotide sequences at the exon–intron boundaries. The mouse cDNA is 4,578-bp in length and contains 1,557 bp of coding sequences. The human *DB1* cDNA is 4,652-bp in length with an open reading frame of 1,566 bp. Alignment of mouse and human cDNAs reveals an 85% to 92% sequence identity in the coding and most of the 5′ and 3′ untranslated regions, suggesting the presence of conserved functional and structural elements in the mRNA.

The Vezf1 and DB1 open reading frames encode 518- and 521-amino-acid proteins, respectively, with a predicted molecular mass of 56 to 58 kDa. The two sequences differ only in six amino acids and in additional three glutamine residues in the human DB1 polyglutamine stretch (Figure 95.1). Amino acid alignment of putative Vezf1 orthologues from various vertebrate species illuminates a striking conservation from fish through mammals and few nonconservative changes from amphibia through mammals (Figure 95.1) (F. Kuhnert and H. Stuhlmann, unpublished observations). Thus, it is likely that crucial functional domains of Vezf1 were conserved during the evolution of vertebrate species.

FUNCTIONAL DOMAINS OF VEZF1

Vezf1 displays hallmarks of a bona fide transcription factor (Figure 95.2). For example, it contains six internal zinc fingers of the Cys2/His2 (Kruppel-like) type, a motif known to promote DNA binding. The zinc finger domain is 65% identical to mouse PUR-1, a protein that binds to purine-rich sequences and transactivates the *insulin* promoter (8,9). Its human orthologue, MAZ/SAF-1, is a zinc finger transcription regulator of c-Myc, CD4, serotonin 1α receptor, and γ-fibrinogen (10,11) and shares 65% amino acid identity in the zinc finger region with DB1/ZNF161 (Figure 95.2). No other genes with significant sequence similarities exist in the databases, defining Vezf1/DB1 and PUR-1/MAZ as the only

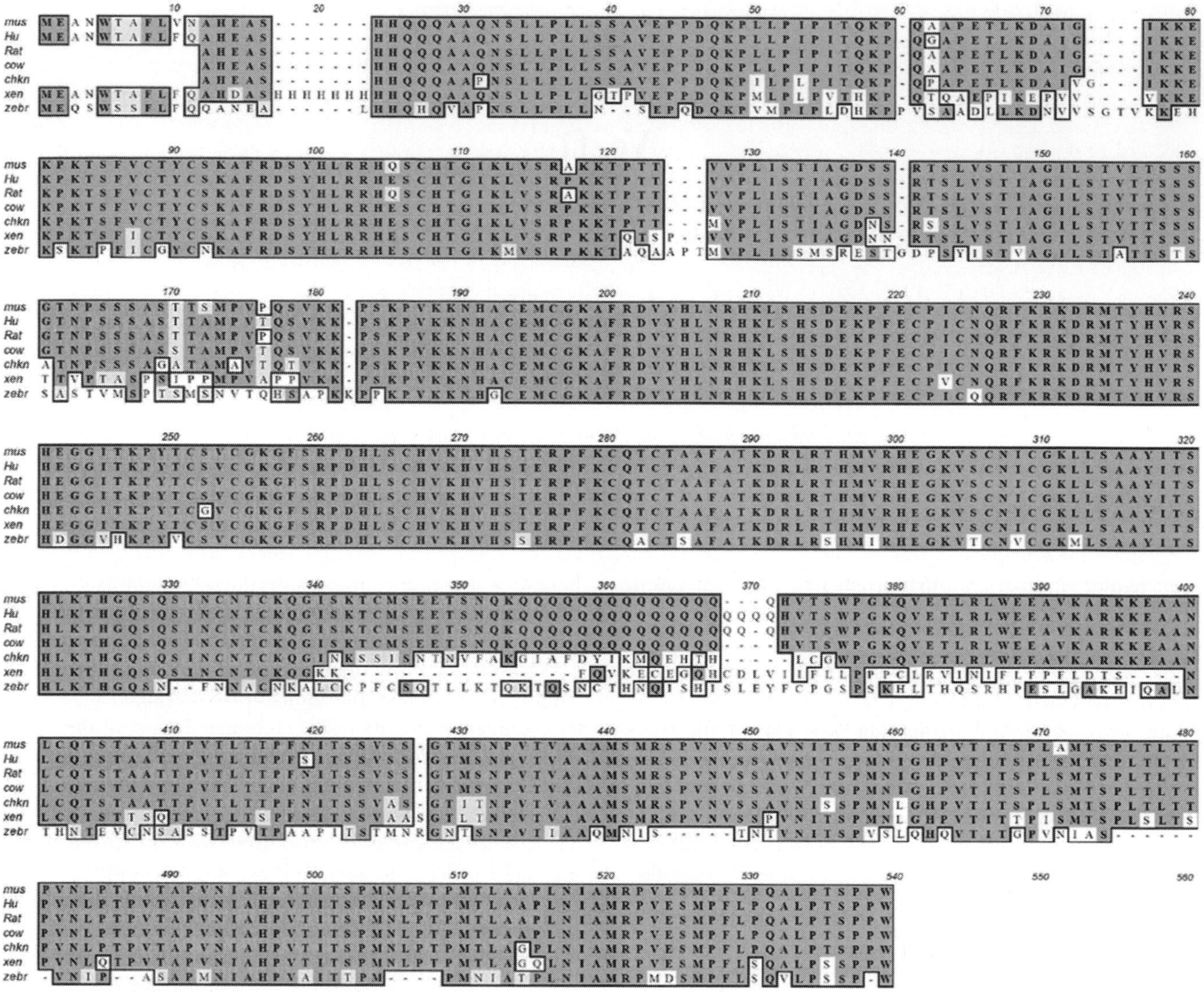

Figure 95.1. Amino acid sequence alignment of mouse Vezf1 and its orthologues from key vertebrate species.

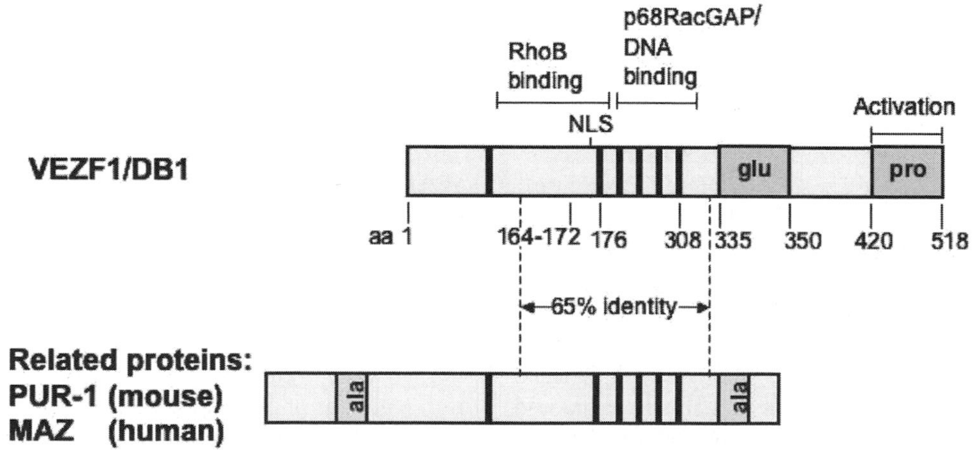

Figure 95.2. Structural and functional domains of mouse Vezf1, human DB1/ZBF161 and related proteins.

members of a small subfamily of Cys2/His2 transcription factors. In addition to the zinc finger domain, Vezf1 contains a polyglutamine stretch and a proline-rich region, known mediators of transcriptional repression and activation, respectively, at its C-terminus. The functional domains of Vezf1 were delineated by a combination of mutagenesis, transactivation, DNA-binding and subcellular localization analyses (Figure 95.2). The dissection of the Vezf1 domain structure identified a N-terminal nuclear localization sequence (NLS) KKPSKPVKK, spanning residues 164 to 172, and a strong C-terminal transcriptional transactivation domain (amino acids 421 to 518) encompassing the proline-rich region, that is functional both in endothelial and heterologous cell lines. Interestingly, this transactivation domain is activated only after deletion of the N-terminal portion of Vezf1, but is inactive in the context of the full-length protein. A similar "unmasking" process of a C-terminal transactivation domain has recently been reported for Smad proteins. The DNA binding activity was mapped to the central five-zinc finger motif. This region was found to be capable of forming specific DNA-protein complexes with oligonucleotide probes containing sequence motifs derived from the *IL-3* and the endothelial-specific *FLK-1* and *FLT-1* promoters. In addition, the central five-zinc finger domain of DB1 was shown to interact with a novel GTPase-activating protein, p68RacGAP, via yeast two-hybrid screening (12). Finally, the N-terminal region upstream of the DNA binding domain of DB1, including the first zinc finger, functions as a RhoB binding site (13).

VEZF1 TARGET GENES

Human *DB1* was cloned based on its binding activity to the CT/GC-rich region of the human *IL-3* promoter (6). In transient transfection assays DB1 augments Tax-dependent transcription without displaying any detectable effect on the transcription activity of the *IL-3* promoter by itself (6). The interaction of Vezf1 with members of the Rho GTPase machinery may represent a novel mechanism to modulate Vezf1 activity. For example, sequestration of DB1 by a prenylated, nuclear-localized form of the RhoB GTPase results in the inhibition of transcriptional potentiation of Tax-mediated transcriptional activation of the *IL-3* promoter (13). Moreover, recent studies demonstrate a RhoB interaction-dependent transactivation of the *neuropilin 1 (NRP1)* promoter by DB1 via canonical DB1 binding sites in primary ECs.

Vezf1 was shown to bind in a sequence-specific manner to the same CCCCGC core sequence in the murine promoters of *IL-3*, *FLK-1*, and *FLT-1* but failed to activate transcription of these promoters in transient transfection assays in ECs (D. Lemons, F. Kuhnert, and H. Stuhlmann; unpublished observations), suggesting that Vezf1 functions in the context of these promoters in concert with additional coactivators. Vezf1/DB1 was found, however, to transactivate the EC-restricted human *endothelin-1 (hET-1)* promoter in transient transfection assays (14). In fact, a novel Vezf1-responsive element, ACCCCC, was identified 47-bp upstream of the transcriptional start site of the *hET-1* promoter. Cotransfection of p68RacGAP potently inhibits Vezf1-dependent transcriptional activation of the *hET-1* promoter (12). *Stathmin/oncoprotein 18 (OP18)*, a microtubule-destabilizing protein, was identified as another putative Vezf1 target gene by cDNA microarray analysis of ECs treated with Vezf1 antisense oligodeoxynucleotides (AS-ODNs) (15). Downregulation of Vezf1 in proliferating ECs by AS-ODNs resulted in a 50% reduction of stathmin/OP18 expression, whereas overexpression of Vezf1 in ECs increased the expression of stathmin/OP18. Furthermore, downregulation of stathmin/OP18 phenocopied the effect of *Vezf1* knockdown on EC proliferation, migration and network formation in vitro (15). Lastly, the analysis of a large canon of potential vascular targets, including vascular endothelial growth factor (VEGF), angiopoietin, ephrin, Notch, and transforming growth factor (TGF)-β signaling pathways, in *Vezf1*$^{-/-}$ embryos by microarray and RNA in situ hybridization analysis is ongoing but thus far has not revealed any genes whose expression is regulated by Vezf1 in vivo (16).

FUNCTION OF VEZF1 DURING VASCULAR DEVELOPMENT – LOSS-OF-FUNCTION STUDIES IN MICE

Gene-targeting studies in mouse ES cells and mice established that Vezf1 function is required for the development of the blood vascular and the lymphatic system in a dose-dependent manner (16). Loss of Vezf1 function resulted in an incompletely penetrant, haploinsufficient embryonic lethal phenotype. Homozygous mutant embryos can be subdivided into two classes. Class I embryos display an early phenotype that is characterized by defective angiogenic remodeling, particularly in the developing vasculature of the aortic arch system, the head and neck, and in the intersomitic vessels, and by the loss of vascular integrity. In contrast, class II embryos display apparently normal vascular patterning but are subject to midgestation hemorrhaging (Figure 95.3A; for color reproduction, see Color Plate 95.3). Immunohistochemical and ultrastructural analysis demonstrated normal vascular smooth muscle cell/pericyte coverage in Vezf1 mutants but revealed defects in EC adhesion, tight junction formation, and extracellular matrix (ECM) organization in mutant vessels as the molecular basis for observed hemorrhaging (16).

Furthermore, haploinsufficiency was detected in about 20% of the heterozygous embryos. Affected heterozygous mutant embryos displayed dramatic lymphatic hypervascularization at midgestation, accompanied by hemorrhaging and edema in the jugular region, the first site of lymphatic sprouting from the venous system (Figure 95.3B). A low incidence of the lymphatic phenotype also was observed in class II homozygous mutant embryos. Apparently, the early lethality

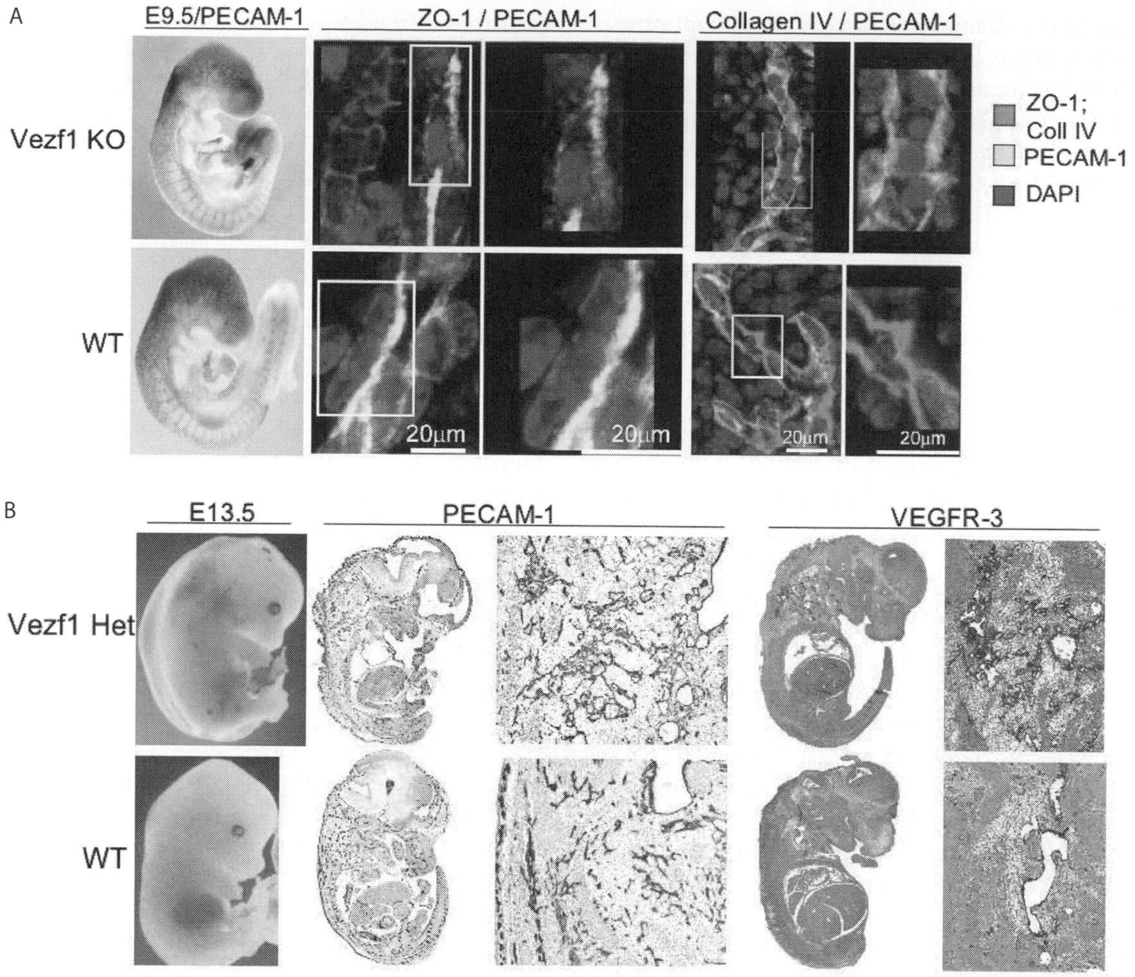

Figure 95.3. Blood vascular and lymphatic phenotypes in *Vezf1* loss-of-function embryos. (**A**) Hemorrhaging *Vezf1* KO embryos with EC defects. (**B**) Lymphatic hypervascularization, hemorrhaging, and edema in *Vezf1* heterozygous embryos. For color reproduction, see Color Plate 95.3.

in *Vezf1*$^{-/-}$ embryos precluded a more robust detection, both in frequency and severity, in homozygous mutant embryos. The lymphatic system phenotype is highly reminiscent of the human congenital malformation syndrome cystic hygroma. Furthermore, it indicates that Vezf1 functions as a negative regulator of lymphatic differentiation and/or proliferation during embryonic development. This conclusion is supported by recent studies in which knockdown of the human orthologue DB1 in primary ECs lead to an increase in lymphatic phenotype and gene expression (I. Adini and L. Benjamin, personal communication). Importantly, it raises the question of whether loss of Vezf1 function can be correlated with other known hyperplastic conditions thought to arise by excessive proliferation of lymphatic ECs, such as lymphangioma, lymphangiosarcoma, and Kaposi sarcoma.

The Vezf1 loss-of-function studies suggest that the primary defect lies in the lymphatic and blood ECs – i.e., that Vezf1 acts cell-autonomously in ECs. However, because Vezf1 expression is present but not restricted to the developing endothelium, additional defects in other Vezf1 expressing cells, such as mesenchymal cells, may contribute to the mutant phenotype. Recently, we have demonstrated that EC-specific expression of Vezf1 under the control of the *Tie-2* promoter rescues the *Vezf1*$^{-/-}$ phenotype, suggesting that endothelial Vezf1 is the most critical determinant of survival (F. Kuhnert, Z. Zou, and H. Stuhlmann; unpublished observations).

FUNCTION OF VEZF1 DURING IN VITRO DIFFERENTIATION – LOSS-OF-FUNCTION STUDIES IN EMBRYONIC STEM CELLS

In vitro ES cell differentiation into embryoid bodies recapitulates the endothelial differentiation program and maturation steps observed during vasculogenesis and angiogenesis in the embryo. Studies using ES cells in which one or both *Vezf1* alleles are genetically inactivated were consistent with the vascular phenotypes observed in mice, strongly indicating the endothelium as the primary target of Vezf1 function. Embryoid bodies derived from *Vezf1*$^{-/-}$ ES cells failed to form a well-organized and differentiated vascular network. Abnormal ECM deposition on mutant ECs was found, similar to the phenotype detected in mutant embryos. In addition, *Vezf1*$^{-/-}$ ECs showed dramatic spouting defects in a three-dimensional collagen matrix. Finally, *Vezf1*$^{-/-}$ teratocarcinomas displayed delayed differentiation and reduced and abnormal vascularization.

INTERACTION OF VEZF1 WITH RHO GTPases

Increasing evidence implicates Vezf1 interaction with Rho GTPases and associated regulatory proteins as a novel mechanism to modulate Vezf1 function. Nuclear localization of a prenylated form of RhoB and specific binding of Vezf1/DB1 to prenylated RhoB via its N-terminal region was demonstrated by Lebowitz and Prendergast (13). Sequestration of DB1 by RhoB resulted in the inhibition of transcriptional activation of the *IL-3* promoter by DB1. Recent studies demonstrate a positive effect of RhoB–Vezf1/DB1 interaction on the transactivation of certain endothelial genes such as *NRP1*. Vezf1 also was shown to bind to a novel GTPase-activating protein, p68RacGAP, in ECs (12). Rac1-specific p68RacGAP colocalizes with Vezf1 in the nucleus and the cytoplasm, inhibits Vezf1-dependent transcriptional activation of the *hET-1* promoter, and modulates EC capillary tube formation in vitro. It is unknown whether transcriptional inhibition of the *hET-1* promoter, similar to the *IL-3* promoter, is mediated by sequestration of Vezf1. The interaction between Vezf1 and components of the Rho GTPase machinery may represent an important and novel paradigm for the regulation of physiological and pathological angiogenesis, especially in light of the recent implication of RhoB in the regulation of endothelial sprouting and apoptosis (17).

CONCLUSION

Recent functional studies provided us with new insights into the biological role of Vezf1 during the development of the mammalian vascular system. These studies suggest that Vezf1 is a transcription factor that controls in a strictly dosage-specific manner normal blood vessel and lymphatic vessel development. Molecular studies suggest that interaction of Vezf1 with other cofactors, specifically Rho GTPases and Rho/Rac associated proteins are crucial to modulate its function in the endothelium.

KEY POINTS

- Vezf1 is a transcription factor that in a strictly dosage-dependent manner regulates the development of the blood and lymphatic vasculature.
- Vezf1 regulates EC adhesion, tight junction formation, and ECM organization.
- Vezf1 is a negative regulator of lymphatic EC proliferation/differentiation.
- Putative Vezf1 EC target genes include *ET-1*, *NRP1*, *FLT-1*, *FLK-1*, *IL-3*, and *stathmin/OP18*.
- The interaction between Vezf1 and components of the Rho GTPase machinery may represent a novel paradigm to modulate Vezf1 transcriptional activity.

Future Goals

- To determine if loss of Vezf1 function can be correlated with known lymphatic EC hyperplasia, such as cystic hygromas, lymphangioma, lymphangiosarcoma, and Kaposi sarcoma
- To identify genes that function in the same transcriptional regulatory circuits as Vezf1

REFERENCES

1 Kuhnert F, Stuhlmann H. Identifying early vascular genes through gene trapping in mouse embryonic stem cells. *Curr Top Dev Biol.* 2004;62:261–281.

2 Leahy A, Xiong J-W, Kuhnert F, Stuhlmann H. Use of developmental marker genes to define temporal and spatial patterns of differentiation during embryoid body formation. *J Exp Zool.* 1999;284:67–81.

3 Xiong J-W, Battaglino R, Leahy A, Stuhlmann H. Large-scale screening for developmental genes in ES cells and embryoid bodies using retroviral entrapment vectors. *Dev Dyn.* 1998;212(2):181–197.

4 Xiong J-W, Leahy A, Lee H-H, Stuhlmann H. *Vezf1*: a Zn finger transcription factor restricted to endothelial cells and their precursors. *Dev Biol.* 1999;206:123–141.

5 Hackney JA, Charbord P, Brunk BP, et al. A molecular profile of a hematopoietic stem cell niche. *Proc Natl Acad Sci USA.* 2002;99(20):13061–13066.

6 Koyano-Nakagawa N, Nishida J, Baldwin D, et al. Molecular cloning of a novel human cDNA encoding a zinc finger protein that binds to the interleukin-3 promoter. *Mol Cell Biol.* 1994;14:5099–5107.

7 Lapteva N, Ando Y, Nieda M, et al. Profiling of genes expressed in human monocytes and monocyte-derived dendritic cells using cDNA expression array. *Br J Haematol.* 2001;114(1):191–197.

8 Kennedy GC, Rutter WJ. Pur-1, a zinc finger protein that binds to purine-rich sequences, transactivates an insulin promoter in heterologous cells. *Proc Natl Acad Sci USA.* 1992;89:11498–11502.

9 Song J, Murakami H, Tsutsui H, et al. Structural organization and expression of the mouse gene for Pur-1, a highly conserved homolog of the human MAZ gene. *Eur J Biochem.* 1999;259(3):676–683.

10 Bossone SA, Asselin C, Patel AJ, Marcu KB. MAZ, a zinc finger, binds to c-myc and C2 gene sequences regulating transcriptional

initiation and termination. *Proc Natl Acad Sci USA.* 1992;89: 7452–7456.

11 Duncan DD, Stupakoff A, Hedrick SM, et al. A myc-associated zinc finger protein binding site is one of four important functional regions in the CD4 promoter. *Mol Cell Biol.* 1995;15(6): 3179–3186.

12 Aitsebaomo J, Wennerberg K, Der CJ, et al. p68RacGAP is a novel GTPase-activating protein that interacts with vascular endothelial zinc finger-1 and modulates endothelial cell capillary formation. *J Biol Chem.* 2004;279(17):17963–17972.

13 Lebowitz PF, Prendergast GC. Functional interaction between RhoB and the transcription factor DB1. *Cell Adhes Commun.* 1998;6(4):277–287.

14 Aitsebaomo J, Kingsley-Kallesen ML, Wu Y, et al. Vezf1/DB1 is an endothelial cell-specific transcription factor that regulates expression of the endothelin-1 promoter. *J Biol Chem.* 2001;14:14.

15 Miyashita H, Kanemura M, Yamazaki T, et al. Vascular endothelial zinc finger 1 is involved in the regulation of angiogenesis: possible contribution of stathmin/OP18 as a downstream target gene. *Arterioscler Thromb Vasc Biol.* 2004;24(5):878–884.

16 Kuhnert F, Campagnolo L, Xiong J-W, et al. Dosage-dependent requirement for mouse *Vezf1* in vascular system development. *Dev Biol.* 2005;283:140–156.

17 Adini I, Rabinovitz I, Sun JF, et al. RhoB controls Akt trafficking and stage-specific survival of endothelial cells during vascular development. *Genes Dev.* 2003;17(21):2721–2732.

96

Sox Genes

At the Heart of Endothelial Transcription

Neville Young and Peter Koopman

Institute for Molecular Bioscience, University of Queensland, Brisbane, Australia

The endothelium is not well understood as a fully functional organ, yet a surprising amount of data exist regarding vascular development during embryogenesis. During mouse development, endothelial cells (ECs) alone make up the vasculature until around halfway through gestation. A large number of knockout and mutant mice exist in which severe defects in the endothelial vasculature at these early stages lead to embryonic lethality. The use of these animals to dissect the genetic interactions within ECs during development has led to an increased understanding of the endothelium.

The central thread of this chapter is the transcriptional regulation of ECs by members of the *Sox* gene family, but it also will incorporate relevant information gained from other signaling hierarchies. We have amassed substantial data implicating SOX18 in vascular development (1,2). In situ analysis of *Sox18* has demonstrated expression in the condensing mesenchyme underlying the developing hair follicle, in the developing heart, and in the ECs of the vasculature. Mutations in *Sox18* lead to mice with defective hair, skin, and vasculature. The vascular defects can be particularly life-threatening, leading to generalized edema due to the leaky vasculature that can result in embryonic or postnatal lethality. Here, we review the data bearing on the molecular and cellular roles of SOX18 and related genes in regulating endothelial phenotype and function.

SOX GENES

Sox genes encode a family of transcription factors, characterized by the presence of a 79 amino acid DNA-binding high mobility group (HMG) domain, similar to that found in mammalian testis-determining factor gene (3). HMG domains can be found in a number of proteins involved in chromatin organization, general transcription and transcriptional regulation. HMGB1, for example, is a non-SOX HMG domain containing protein involved in chromatin packaging but also is secreted from necrotic cells and activated macrophages to induce angiogenesis (4). *Sry* is located on the Y chromosome and triggers the development of the testis. To date, based on sequence conservation of the HMG domain, *Sox* genes have been identified exclusively in the animal kingdom: in insects, nematodes, reptiles, birds, amphibians, and fish, with at least 20 members currently recognized in mammals (5). SOX transcription factors have both classical and architectural modes of action: they are able to bind to promoter regions to activate the transcription of target genes directly and also are able to bend DNA to an angle of between 70 and 80 degrees, thus bringing together other proteins and *cis*-acting DNA regulatory sequences to facilitate transcription.

Sox genes have diverse temporal and spatial expression patterns during embryonic development and are implicated in many cell fate decisions during early embryogenesis. Their diverse roles are best illustrated by the variety of phenotypes associated with spontaneous mutations in *Sox* genes. In humans, mutations in *Sry* cause male-to-female sex reversal and gonadal dysgenesis (6); *SOX9* mutations result in camptomelic dysplasia, a skeletal dysmorphogenesis (7); *SOX10* mutations result in Waardenburg-Shah syndrome, leading to deafness, hypopigmentation, and aganglionic megacolon, all due to defects in neural crest migration (8). Thus, SOX transcription factors seems to be involved in a variety of tissue-specific switching events that control changes in the expression of suites of genes involved in cell-type specification and differentiation.

The vast majority of SOX transcription factors appear to act as transcriptional activators; however, in a small number of cases, transcriptional repression has been ascribed to SOX proteins. Experiments to date suggest that efficient SOX protein regulation of target gene transcription requires recruitment of cofactors and partner proteins to their transactivation domains. Evidence exists in vitro suggesting that some *Sox* genes can partially activate transcription independently but that this activation is potentiated in the presence of cofactors (9). However, in vivo, it is difficult to imagine how the abundance of SOX binding motifs would be regulated efficiently in the absence of cofactor modification. Recruitment to these SOX functional domains of non-DNA binding transcription factors or other cofactors has been postulated to be a probable mechanism to generate SOX target gene specificity (10).

PHYLOGENY OF *SOX* GENES

Before addressing the structural and functional groupings of the various *Sox* genes, it is important to define exactly what a *Sox* gene is. *Sox* genes can be found scattered throughout the genome and were originally classified as encoding proteins having more than 50% amino acid sequence identity to SRY in the HMG domain; however, this definition is no longer accurate or particularly useful in distinguishing a bona fide *Sox* gene from other classes of HMG-box genes. For example, human *SOX30*, a gene with obvious structural features of a *Sox* gene, encodes a protein with only 48% identity to human *SRY* in the HMG domain. Present criteria for defining *Sox* genes use the conservation of key motifs within the HMG domain, such as the signature sequence motif RPMNAFMVW, which appears to be conserved in all non-SRY SOX proteins. The original concept of reference to SRY was a historical, and, in retrospect, poor choice for defining *Sox* genes, especially as *Sry* is present only in mammalian lineages. In addition, due to its location on the rapidly evolving Y chromosome, *Sry* is highly divergent, even between closely related species (5).

The 20 *Sox* genes found in mammals are arranged into groups A through H, based on their HMG box similarities and additional information on non-HMG domain structural motifs (5). Vertebrates typically express several members of each group, whereas invertebrate species have a single ancestral sequence for each group (the exception being Group B, where four sequences have been identified in *Drosophila*).

Relationships between members of the *Sox* family shed light on the evolutionary process as they display traits of both slow divergence and rapid change. The HMG domain, the motif that forms the core of all SOX family proteins, has accumulated relatively few sequence changes due to selection pressures to retain DNA binding and is therefore well conserved across all SOX groups. In contrast, the regions outside of the HMG domain show more rapid stochastic evolutionary changes that must have occurred via the apparent co-option of functional domains and motifs of other proteins (11). As a result of this, sequence and structure outside the HMG domain is highly divergent between SOX groups, yet conserved to a recognizable extent between members within the same SOX group. The degree of conservation correlates loosely with the degree of similarity of function between members of the same SOX group. For example, members of *Sox* group D include *Sox5* and *Sox6*, which have similar expression profiles, are thought to act redundantly in spermatogenesis and chondrogenesis, and show a high degree of structural conservation. In contrast, group E *Sox* genes are less similar in structure and have very different expression patterns and roles in development and disease.

In mammals, *Sox18* is related to two other *Sox* genes, *Sox7* and *Sox17*, which together form *Sox* group F. These three genes have similar structures and overlapping temporal and spatial expression profiles in several but not all cell types. This suggests that although the three genes are regulated differently in some tissues, they could potentially function redundantly in cells where they are coexpressed – for instance, in ECs during vascular development.

DISCOVERY OF SOX18

Sox18 was first identified as part of an effort to identify novel *Sox* genes expressed in mouse embryonic tissue. A gene with an open reading frame of 1404 base pairs, encoding a protein 468 amino acids in length, was identified and mapped to distal chromosome 2 (12,13). The encoded protein, SOX18, was found to bind to the sequence 5′-AACAAAG-3′ and to be capable of activating transcription via this sequence in GAL4 reporter assays in vitro (14) possibly via a transactivation domain mapped to a region C-terminal to the HMG domain (Figure 96.1) (5). Further study revealed that this in vitro activation could be significantly increased in the presence of known binding partners (15). It therefore seems unlikely that in vivo SOX18 activates transcription in the absence of cofactors. Much more probable is that proteins such as myocyte enhancer factor 2 (MEF2C), which are known to interact with the C-terminal region of SOX18 in ECs, will have a crucial role in regulating transcriptional specificity and efficiency. Otherwise the ability of any Sox gene to activate any gene bearing a SOX binding motif at would cause transcriptional chaos.

The sequence of *Sox18* is generally well conserved between species, suggesting a conserved and important function for the protein. However, subtle differences exist between SOX18 proteins from mouse, human, and chicken, and these may reflect different functions required of the gene product in different species. For example, murine SOX18 has a longer C-terminus than either human or chicken, whereas chicken SOX18 has a divergent *trans* activating domain when compared to the other two (Figure 96.1). It has been postulated that mouse SOX18 may recruit more DNA-binding partners in its extended C-terminus, and that chicken SOX18 may recruit a different suite of cofactors to its *trans* activation domain.

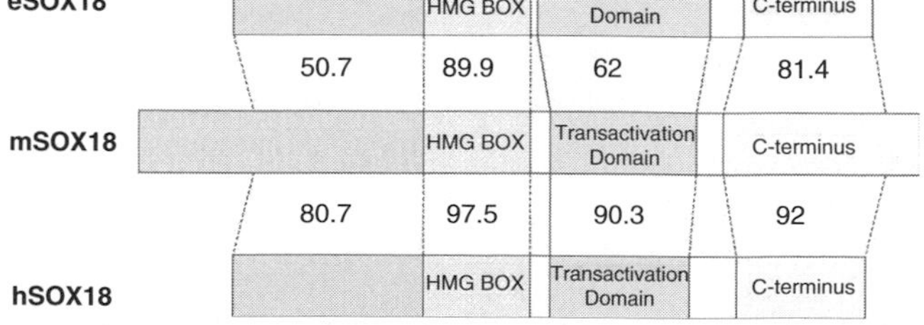

Figure 96.1. A comparison of the chicken, mouse, and human SOX18 protein structures. Known functional domains labeled and percentage identity to mouse SOX18 shown.

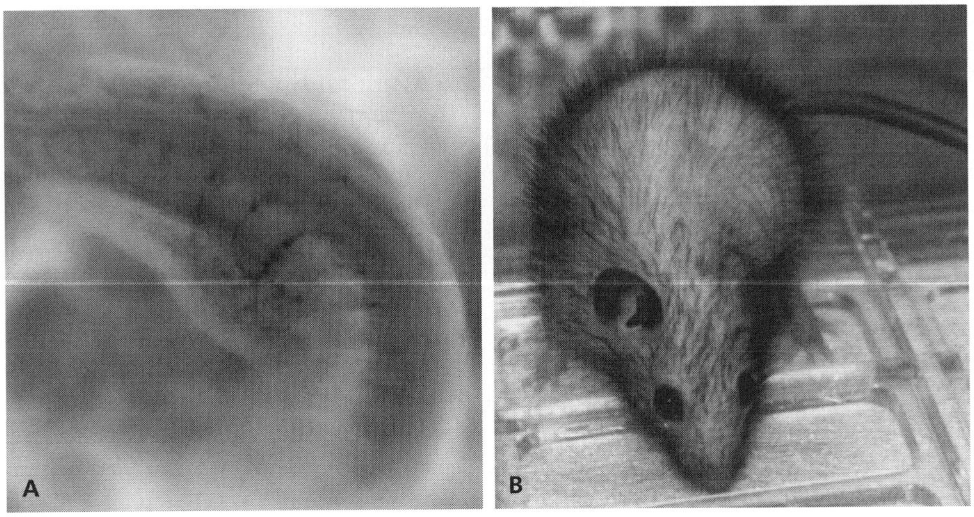

Figure 96.2. *Sox18* Expression. (**A**) *Sox18* expression in the developing intersomitic vessels and limb bud of an E9.5 mouse. (**B**) Adult mouse heterozygous for the *RaOp* dominant-negative form of SOX18.

SOX18 IS EXPRESSED IN ENDOTHELIUM OF DEVELOPING BLOOD VESSELS

Blood vessels in an embryo are initially assembled from endothelial precursors (vasculogenesis), which subsequently undergo remodeling via endothelial sprouting and branching (angiogenesis). Initially, *Sox18* expression was observed in the developing vascular endothelium and condensing mesenchyme of hair follicles in mouse embryos (1,16). Close examination revealed that at embryonic day (E)7.5, *Sox18* is expressed in the allantois and within yolk sac blood islands. At 8.0 dpc expression was observed in angioblasts of mesodermal origin (nascent ECs) around the foregut and in the presumptive endocardial cells of the cardiogenic plate. At E8.5, expression was detected throughout the developing heart and early vasculogenic network and by E9.5 could be observed in vessels forming by angiogenesis (Figure 96.2A).

The expression of *Sox18* coincides with that of other known EC markers, such as vascular endothelial growth factor (VEGF) receptor 1 (VEGFR1/FLT-1) and VEGF receptor 2 (VEGFR2/FLK-1). Interestingly, in *Flk-1*–knockout mice, in which ECs fail to differentiate, *Sox18* expression is absent, yet in *Flt-1*–knockout mice, where ECs do differentiate but fail to organize into vessels, there is *Sox18* expression.

The vasculature at these early developmental stages consists entirely of ECs prior to the appearance of other components of the vasculature such as smooth muscle cells and pericytes. Thus, the onset of expression in endothelial precursors from as early as E7.5 and the absence of expression in *Flk-1*–knockout mice indicate that SOX18 is associated with the initial events of endothelial differentiation and subsequent vascular development.

Interestingly, *Sox18* expression is transient in ECs, lasting for about 48 hours during early vasculogenic and angiogenic events before being extinguished in established vessels. This pattern of expression also is conserved between species (it is also observed in the chicken embryo) and strengthens the case for the role of SOX18 as a regulator of vascular development (16). After its initial embryonic expression SOX18 is not observed again in the vasculature except in adult tissue undergoing angiogenic remodeling – e.g., during wound healing, post myocardial ischemia, or in de novo tumor endothelium.

MUTATIONS IN SOX18 REVEAL ITS FUNCTION

Nearly all of the work relating to the role of SOX18 in EC biology has been undertaken in the developing embryo, and it is at this time that defects in SOX18 function can be seen to influence EC biology. The vascular system is one of the earliest organs to be established in the developing embryo, and its formation is critical for successful embryogenesis. Given its pattern of expression during blood vessel development, the removal of SOX18 function from the embryo would have been postulated to have grave consequences during embryogenesis. It was surprising, then, to find that *Sox18*-knockout mice are normal and viable (1). The lack of major phenotype of *Sox18*-knockout mice may be explained by the fact that other members of the *Sox* group F, namely *Sox7* and *Sox17*, have spatial and temporally overlapping expression patterns in the vasculature permitting them to act redundantly to compensate for its loss.

Somewhat fortuitously, the location of *Sox18* on mouse chromosome 2 coincided with the mapped locus of a classical, spontaneous, semidominant mouse mutant, *ragged* (*Ra*) (Figure 96.2B). *Ra* mice exhibit defects in hair and vascular development (they are bald) and although the vascular network seems to form normally, these mice suffer from edema, chylous ascites, and cyanosis. Further study revealed point mutations in *Sox18* in all four known allelic variants of *ragged*; these mutations are predicted to result in missense translation of the SOX18 C-terminus and premature truncation of the protein

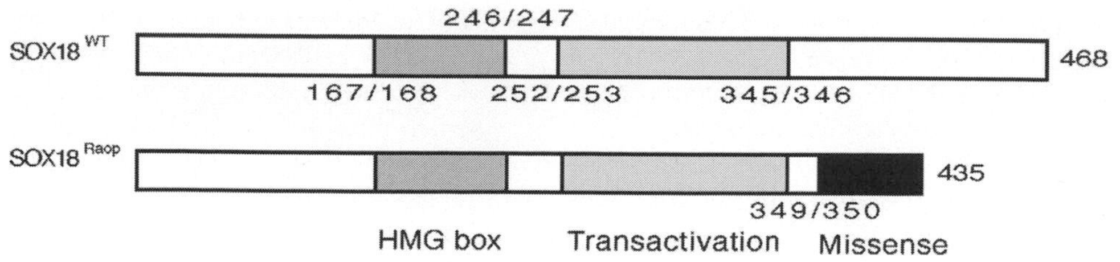

Figure 96.3. A schematic of murine SOX18 protein compared to the RaOp mutant version. Note intact HMG box and transactivation domain in RaOp mutant.

(Figure 96.3) (17). These truncated proteins were found to be unable to activate transcriptional target sequences relative to wild-type controls in GAL4 in vitro assays (14). These studies provide a ready explanation for the etiology and biochemical underpinnings of the *Ra* mutant phenotype and its semidominant mode of action: *Ra* mutant forms of SOX18 are presumably able to bind to their normal transcriptional targets but are unable to appropriately engage cofactors and partner proteins required for proper function in endothelial development. Evidence for the importance of cofactors in SOX18-targeted transcription is seen in work on the MADS box transcription factors *Mef2C*. Both *Mef2C* and *Sox18*-null mutants display overlapping phenotypic abnormalities. Further work has demonstrated that these proteins interact directly and are coexpressed in EC nuclei. MEF2C also potentiates SOX18-mediated transcription *in vivo* but this activation is reduced with *Sox18 ragged* mutants where the activation domain is truncated and the protein interactions disrupted (15).

Three of the four *Ra* mutant alleles have mutations within the defined transcriptional transactivation domain of SOX18, however, the fourth allele, termed *ragged opossum* (*RaOp*) has a mutation 3′ of the activation domain and yet, despite having this domain intact, is unable to activate transcription. This suggests that either a further functional domain exists in the C-terminus of murine SOX18, or that the activation domain extends further than previously thought, or both. The three *ragged* mutant mice with point mutations in the transactivation domain all exhibit edema, chylous ascites and cyanosis at birth in the homozygous state. However, the *RaOp* mutant mice exhibit these symptoms in the heterozygous state, whereas homozygotes die at around E13.5 due to severe hemorrhage of the vascular network. Therefore, somewhat paradoxically, the *RaOp* gene product, although apparently a more intact protein than that encoded by the other three mutant alleles, causes the most severe mutant phenotype. It has been postulated that this paradox is due to the *RaOp* gene product more efficiently sequestering cofactors and/or partner proteins required for normal regulation of endothelial development (18).

The contrast between the lack of phenotype in the *Sox18*-null mouse and the severe vascular phenotype observed in the *RaOp* mice was initially difficult to explain. However, subsequent cell transfection experiments showed that *Ra*-expressing constructs can interfere with the ability of SOX18, SOX17, and SOX7 to activate their target genes (9). Evidence also has been uncovered suggesting that all three Group F *Sox* genes are coexpressed in the vascular endothelium at the same time (19). Thus, it seems that the ragged proteins are able to act in a dominant negative fashion to interfere with the normal function of SOX18, SOX17, and SOX7, emphasizing the importance of the function of these genes in endothelial and vascular development. This also suggests that a common repertoire of cofactors exists in ECs for a short space of time when the initial vascular network is being established.

More recently, mutations in the human form of *SOX18* have been found to underlie both recessive and dominant hypotrichosis–lymphedema–telangiectasia, suggesting a role for SOX18 in the development and/or maintenance of lymphatic and blood vessels in humans (20). Interestingly the most severe of these mutations closely resembles the *ragged* mutations seen in mice.

THE SOX18 MOLECULAR PATHWAY

Signaling pathways that mediate vascular development during embryogenesis have been well characterized in recent years, and the importance of cell trophic factors such as VEGF, platelet-derived growth factor (PDGF), fibroblast growth factors (FGFs), epidermal growth factor (EGF), and their receptors have been well documented (21). However, much less is known about transcriptional regulation within ECs. SOX18 is only one of a handful of transcription factors that have been identified with a defined role in vascular development and can thus provide important clues into how endothelial phenotype is regulated.

Some insights into the function of SOX18 have been provided by analysis of *Sox18* expression in null mutants of *Flt-1* and *Flk-1* which also have severe vascular defects (22,23). FLT-1 is essential for organization of the embryonic vasculature but it is not required for EC differentiation. As mentioned earlier, normal *Sox18* expression is observed in the ECs of *Flt-1*–knockout embryos, even though they do not assemble into a normal vascular network. In contrast, no *Sox18* expression is observed in *Flk-1*–knockout mice, in which endothelial precursors do form but do not seem to differentiate into ECs. This lack of expression of *Sox18* in endothelial precursor cells (which at this stage would normally express *Sox18*) in FLK-1-deficient embryos, along with the luminal position of *Sox18*-expressing cells indicate that *Sox18* expression is associated with early vascular endothelial development. These studies suggest that *Sox18* expression is downstream of, but not necessarily directly controlled by, the VEGF–FLK-1 pathway where it may be involved in preliminary EC differentiation.

What genes are regulated by SOX18 and how is the regulation controlled? Recent findings have started to shed light on both how SOX18 mediates transcription and what genes it targets. Based on an overlapping expression pattern and the phenotypic effects observed in *Mef2C*-knockout mice, a relationship between MEF2C and SOX18 was identified (15). MEF2C is a transcription factor demonstrated to have a role in vascular development; it is expressed in ECs during development and *Mef2C*-knockout mice exhibit vascular defects similar to those observed in *VEGF* and *VEGFR1/Flt-1*–knockout mice (23) (see Chapter 32). Subsequently, SOX18 and MEF2C proteins were found to interact biochemically, and the two proteins are coexpressed in the nuclei of ECs in vivo (15). The expression of *Mef2C* enhances SOX18-mediated transcription and regulates the activity of the SOX18 activation domain, suggesting that MEF2C and SOX18 may be partners in transcriptional regulation during early vascular development.

Further work in our laboratories has revealed that SOX18 is able to activate the transcription of the *vascular cell adhesion molecule (VCAM)-1* gene (9). VCAM-1 is a member of the immunoglobulin superfamily of adhesion molecules, and is an important molecule involved in cell-cell contacts. VCAM-1 is expressed on the surface of ECs, and *VCAM-1*- knockout mice have disrupted chorioallantoic fusion and die in utero at E13.5. SOX18 activates transcription of this gene via a classical SOX binding motif in the *VCAM-1* promoter. It is therefore possible to hypothesize that the *Ragged* mutant phenotype is caused, at least in part, by an absence of VCAM-1 expression due to a lack of functional SOX18, resulting in poor cell–cell contacts in the vascular endothelium, in turn causing vascular leakage and edema. Activation of VCAM-1 by, and binding of MEF2C to, SOX18 are indeed defective in *RaOp* heterozygous mice, reaffirming the role of SOX18 in a regulative pathway governing the early development of ECs in the vasculature (9,18).

Other targets of SOX18 remain to be identified, and many candidates exist. For example, the Tie2 receptor, required for angiopoietin (Ang)-induced signaling, is expressed in ECs subsequent to *Sox18* expression. SOX binding sequences exist in a region of the *Tie2* promoter shown to be critical for endothelial expression. It also is of interest to note that the putative SOX site in the *Tie2* promoter is adjacent to an octameric binding sequence motif. OCT transcription factors previously have been demonstrated to interact with SOX factors to enhance transcription in other cell types (24).

SOX18 FUNCTION AND ENDOTHELIAL CELL PHENOTYPE

The cell type specificity, timing, and transient nature of *Sox18* expression suggests that it is involved in the initial specification of ECs. However, endothelium is clearly able to form in *RaOp* mutant mouse embryos but appears to generate vessels with impaired integrity, causing edema and hemorrhage. Further, the only transcriptional target so far identified for SOX18 is one that is implicated in regulating vascular integrity. What, then, is the true cellular function of SOX18?

Evidence from other model systems in which similar defects are observed to those seen in *Ra* mice offers clues to possible targets of, or proteins interacting with, SOX18. One important feature to note when using a candidate gene approach to identify partner proteins of SOX18 is to observe whether they are expressed at the same time in the same cells. Mice with disrupted Tie2/Ang1 signaling have phenotypes similar to *RaOp* homozygotes. Mice with disrupted Tie2 signaling initially develop a normal vasculature, but the embryos die at between E9.5 and E12.5 due to malformations of microvessel structure (25). *Tie2* is coexpressed with, and a potential target of, SOX18 in ECs and the similarities between the mutant phenotypes suggest that they may interact in the same pathway. Ultrastructural analysis of *Tie2*- and *Ang1*-knockout mice suggests that initially ECs in these mice form normal networks but that angiogenic sprouting is defective, contributing to vessel stenosis, and embryonic lethality. Interestingly, and entirely hypothetically, this may be a case where two different but important vascular signaling pathways may intersect. Earlier we suggested that the VEGF/FLK-1 pathway was upstream of SOX18, therefore the possibility exists that in the absence of SOX18 the Tie2/Ang1 pathway will not be activated, contributing further to defective EC differentiation.

Comparisons between *RaOp* homozygotes and these knockout mice show distinct phenotypic similarities, but it remains to be seen if ECs in the *RaOp* homozygotes are affected similarly to those in *Tie2*- and *Ang1*-knockout mice. However, the ultrastructural defects observed in *Tie2*-knockout mice, such as the rounding of the ECs and capillary stenosis, may in turn explain some of the gross defects observed in these as well as the *RaOp* homozygotes. If ECs do not differentiate, migrate, or proliferate normally at these early stages, then they may not interact correctly with each other or the extracellular matrix later on, preventing stabilization and maturation of the vasculature. In addition, it is known that in *PDGF*-knockout mice, a lack of recruitment of pericytes to the vasculature can lead to microaneurysms and subsequent edema similar to that observed in the *RaOp* homozygotes (26). Therefore, it also may be the case that ECs from *RaOp* homozygotes do not differentiate correctly and are unable to recruit the correct companion cells, contributing to the observed hemorrhagic phenotype.

It also may be true that SOX18 affects multiple steps in the endothelial differentiation pathway. It will be informative to study in detail the extent and patterning of the vasculature in *Ra* mutant mice, measure vascular integrity in dye leakage experiments, and examine the vasculature in *Ra* mutants at the ultrastructural level, complementing these in vivo observations with in vitro studies of the effects of SOX18 on EC migration, proliferation, and ability to generate tubes.

SOX18 AND ENDOTHELIAL DISEASE

Point mutations in SOX18 have been found to underlie the human disease hypotrichosis–lymphedema–telangiectasia (HLT) (20). The most severe, dominant forms of this disease closely resemble the phenotype of the *RaOp* mouse mutant and are caused by similar molecular lesions that would be expected to act in a dominant-negative manner. Curiously, other disease mutations can occur in the HMG-box region of SOX18, and are likely to result in a loss of function, consistent with their recessive mode of inheritance. The discrepancy between the loss-of-function phenotype in mice (mild coat color effects) and in humans (HLT) points to molecular differences in the involvement of *Sox18* and/or other group F Sox genes between the two species or differential dosage requirements between the two species. The spectrum of phenotypes in HLT further suggests that, in addition to its role in angiogenesis, SOX18 plays an important role in the development and/or maintenance of lymphatic vessels in humans.

The transient expression of *Sox18* during embryogenesis is not only a developmental phenomenon required for the initial formation of a vascular network in the embryo; *Sox18* expression also is observed transiently in adults, but only during de novo vessel formation, such as occurs during excisional wound healing (27).

Pathological neovessel formation also occurs during tumor growth. Evidence abounds that tumor growth is dependent on neovessel formation and that the molecular pathways employed by tumor cells to generate a de novo vascular network are a recapitulation of the developmental process. In 1994, a dominant-negative version of the *Flk-1* receptor was expressed in mice in which aggressive subcutaneous tumors were induced. This resulted in mice that grew tumors, but at a greatly reduced rate compared to their wild-type litter mates (28). As *Flk-1* is expressed in ECs, this work suggested that targeting the VEGF-FLK-1 pathway in EC function could be effective as an antiangiogenic tumor therapy. However, it has taken 10 years for this work to advance to successful Phase III clinical trials of a humanized monoclonal antibody, bevacizumab, directed at VEGF (29).

One of the problems with targeting ECs in antiangiogenic strategies is their physiological importance maintaining overall host homeostasis; many antiangiogenic therapies may inadvertently target existing blood vessels in the body. The specific expression pattern of *Sox18* during neovessel formation suggests that it forms the basis of a novel strategy for targeting only nascent endothelium to inhibit tumor angiogenesis. As *Sox18* expression is transient and only on in ECs participating in new growth, then targeting SOX18 function would only affect the subset of ECs participating in tumor neoangiogenesis.

Overall, the transcriptional regulation of ECs is still not well understood; however, increasing efforts from many different groups have begun to shed light on this subject. What is clear is that SOX18 plays an important role in the initial differentiation of ECs and that defects in this role compromise embryo development and survival. As work in this field advances, exciting opportunities are arising to translate knowledge of EC development during embryogenesis into treatments for patients with diseases in which pathological angiogenesis is a marker of poor prognosis.

KEY POINTS

- The transient function of SOX18 during EC differentiation in the embryo is vital for the formation of a fully functional vasculature.
- *Sox18* is one of a group of *Sox* genes involved in endothelial biology.
- Loss of function of SOX18 may be masked (in mice) by the action of the other group F *Sox* genes, *Sox7*, and *Sox17*.
- Dominant-negative mutations in SOX18 result in vascular and lymphatic dysfunction in mice and humans. In these cases, the mutant SOX18 protein is thought to interfere with the functions of wild-type SOX18 protein, and with SOX7 and SOX17 proteins.
- At least one target of SOX18, VCAM-1, is involved in determining vascular integrity.
- The activity of SOX18 is important also for adult neoangiogenesis.

Future Goals

- To identify further downstream targets of SOX18
- To study in detail the vasculature, EC phenotype, and lymphatic integrity in the *RaOp* mouse model
- To study the effects of SOX18 gain- and loss-of-function on EC migration, proliferation, and tubulogenesis in vitro
- To explore SOX18 as a novel target for pro- or antiangiogenic therapies

ACKNOWLEDGMENTS

We are grateful to Jo Bowles and Alisa Poh for critical reading of the manuscript. Our work is supported by grants from the National Health and Medical Research Council of Australia and the National Heart Foundation of Australia. Peter Koopman is a Professorial Research Fellow of the Australian Research Council.

REFERENCES

1 Pennisi D. Mice null for Sox18 are viable and display a mild coat defect. *Mol Cell Biol.* 2000;20(24):9331–9336.
2 Pennisi D, et al. Mutations in Sox18 underlie cardiovascular and hair follicle defects in ragged mice. *Nat Genet.* 2000;24(4):434–437.
3 Gubbay J, et al. A gene mapping to the sex-determining region of the mouse Y chromosome is a member of a novel family of embryonically expressed genes. *Nature.* 1990;346(6281):245–250.

4 Schlueter C, et al. Angiogenetic signaling through hypoxia: HMGB1: an angiogenetic switch molecule. *Am J Pathol.* 2005; 166(4):1259–1263.

5 Bowles J, Schepers G, Koopman P. Phylogeny of the SOX family of developmental transcription factors based on sequence and structural indicators. *Dev Biol.* 2000;227(2):239–255.

6 Berta P, et al. Genetic evidence equating SRY and the testis-determining factor. *Nature.* 1990;348(6300):448–450.

7 Wagner T, et al. Autosomal sex reversal and camptomelic dysplasia are caused by mutations in and around the SRY-related gene SOX9. *Cell.* 1994;79(6):1111–1120.

8 Bondurand N, et al. A molecular analysis of the yemenite deaf-blind hypopigmentation syndrome: SOX10 dysfunction causes different neurocristopathies. *Hum Mol Genet.* 1999;8(9):1785–1789.

9 Hosking BM, et al. The VCAM-1 gene that encodes the vascular cell adhesion molecule is a target of the Sry-related high mobility group box gene, Sox18. *J Biol Chem.* 2004;279(7):5314–5322.

10 Kamachi Y, Cheah KS, Kondoh H. Mechanism of regulatory target selection by the SOX high-mobility-group domain proteins as revealed by comparison of SOX1/2/3 and SOX9. *Mol Cell Biol.* 1999;19(1):107–120.

11 Holland PW, et al. Gene duplications and the origins of vertebrate development. *Dev Suppl.* 1994;125–133.

12 Dunn TL, et al. Sequence and expression of Sox-18 encoding a new HMG-box transcription factor. *Gene.* 1995;161(2):223–225.

13 Greenfield A, et al. The Sry-related gene Sox18 maps to distal mouse chromosome 2. *Genomics.* 1996;36(3):558–559.

14 Hosking BM, et al. Cloning and functional analysis of the Sry-related HMG box gene, Sox18. *Gene.* 2001;262(1–2):239–247.

15 Hosking BM, et al. SOX18 directly interacts with MEF2C in ECs. *Biochem Biophys Res Commun.* 2001;287(2):493–500.

16 Olsson JE, et al. Sox18 expression in blood vessels and feather buds during chicken embryogenesis. *Gene.* 2001;271(2):151–158.

17 Downes M, Koopman P. SOX18 and the transcriptional regulation of blood vessel development. *Trends Cardiovasc Med.* 2001; 11(8):318–324.

18 James K, et al. Sox18 mutations in the ragged mouse alleles ragged-like and opossum. *Genesis.* 2003;36(1):1–6.

19 Kanai-Azuma, M, et al. Depletion of definitive gut endoderm in Sox17-null mutant mice. *Development.* 2002;129(10):2367–2379.

20 Irrthum A, et al. Mutations in the transcription factor gene SOX18 underlie recessive and dominant forms of hypotrichosis-lymphedema-telangiectasia. *Am J Hum Genet.* 2003;72(6):1470–1478.

21 Carmeliet P. Mechanisms of angiogenesis and arteriogenesis. *Nat Med.* 2000;6(4):389–395.

22 Shalaby F, et al. Failure of blood-island formation and vasculogenesis in Flk-1-deficient mice. *Nature.* 1995;376(6535):62–66.

23 Fong GH, et al. Role of the Flt-1 receptor tyrosine kinase in regulating the assembly of vascular endothelium. *Nature.* 1995;376 (6535):66–70.

24 Fadel BM, Boutet SC, Quertermous T. Octamer-dependent in vivo expression of the EC-specific TIE2 gene. *J Biol Chem.* 1999; 274(29):20376–20383.

25 Sato TN, et al. Distinct roles of the receptor tyrosine kinases Tie-1 and Tie-2 in blood vessel formation. *Nature.* 1995;376(6535): 70–74.

26 Lindahl P, et al. Pericyte loss and microaneurysm formation in PDGF-B-deficient mice. *Science.* 1997;277(5323):242–245.

27 Darby IA, et al. Sox18 is transiently expressed during angiogenesis in granulation tissue of skin wounds with an identical expression pattern to Flk-1 mRNA. *Lab Invest.* 2001;81(7):937–943.

28 Millauer B, et al. Glioblastoma growth inhibited in vivo by a dominant-negative Flk-1 mutant. *Nature.* 1994;367(6463):576–579.

29 Yang JC, et al. A randomized trial of bevacizumab, an anti-vascular endothelial growth factor antibody, for metastatic renal cancer. *N Engl J Med.* 2003;349(5):427–434.

97

Id Proteins and Angiogenesis

Robert Benezra and Erik Henke

Memorial Sloan-Kettering, Cancer Center, New York City, New York

The Id proteins were the first identified naturally occurring dominant negative antagonists of a transcription factor class. They have been shown to control a wide range of cell fate decisions in embryonic and adult development in organisms ranging from fly to man (1,2). The four related members of the family (called Id1–4) all contain a helix-loop-helix (HLH) dimerization motif. The 40- to 50-amino acid HLH domain is found in many transcription factors, including the myc family of cellular oncogenes, MyoD, other E-box (CANNTG)-binding proteins, and hypoxia inducible factor (HIF)-1α. Most of these HLH-containing transcription factors have an extra basic region (hence the term *bHLH proteins*) that specifically binds to DNA. In contrast, the Id proteins lack the basic region. Consequently, Id proteins bind to and sequester bHLH proteins, preventing their homo- or heterodimerization with other bHLH factors and their association with DNA. Principal targets of the Id proteins are the ubiquitously expressed bHLH E-proteins, which cooperate with tissue-restricted bHLH proteins to regulate gene expression (1,3,4). Thus Id–E-protein interactions provide a unifying mechanism for controlling cell-type–specific gene expression in multiple cell lineages.

Loss-of-function mutations in Id1 and Id3 have been shown to lead to embryonic and adult angiogenic defects, suggesting an important role for these proteins in blood vessel formation (5). Based on a consideration of underlying transcriptional mechanisms, it seems likely that such defects are related to the liberation of otherwise Id-sequestered E-proteins and the inappropriate activation of genes under the control of bHLH transcription factors. Although the identification of E-protein partners in endothelial cells (ECs) remains elusive, certain downstream consequences of Id loss are gradually coming to light. In addition, the biological consequences of Id loss on EC mobilization and maturation, as well as the consequences of Id loss on murine models of carcinogenesis, are shedding important new insights into basic and possibly translationally meaningful questions. This chapter reviews these recent advances.

ANGIOGENESIS DEFECTS AS A RESULT OF ID LOSS

Mice that are null for *Id1* or *Id3* develop normally and display only mild phenotypic alterations. *Id1*-null mice show some defects in T-cell migration, and *Id3*-null mice have defective B-cell proliferation (Table 97-1). However, mice in which both copies of *Id1* and *Id3* are disrupted die in utero at embryonic day (E)13.5 with a characteristic hemorrhage in the forebrain. Grossly dilated vessels are apparent in the ganglionic eminence at E11.5, but by E12.5 these vessels appear to collapse upon themselves to form an aggregate of endothelium (5). Interestingly, a similar, albeit less severe phenotype than that of the *Id1/3*-double knockouts is observed in mice lacking integrin αv, an integrin implicated in angiogenic processes (6). In the case of Id loss of function, it seems reasonable to hypothesize that the specific localization of the vascular defect to the brain is due to compensation by Id2 in vessels outside the brain (7). Indeed, animals with loss-of-function mutations in Id1, -2, and -3 ($Id1^{-/-} Id2^{+/-} Id3^{-/-}$) develop more widespread hemorrhage that is not restricted to the brain. In addition, animals that lack both *Id2* copies and both copies of either *Id1* or *Id3* ($Id1^{-/-} Id2^{-/-}$ and $Id2^{-/-} Id3^{-/-}$) die at E13.5 from cardiac malformations without showing forebrain hemorrhage (8). Why the ganglionic eminence is so susceptible to the vascular perturbation in the *Id*-knockout and integrin αv-knockout mice is not understood, but the findings provide a striking example of site-specific vasculopathy (Table 97-1).

The embryonic vascular defect in *Id1/3*-double knockout mice raised the possibility that these proteins are important for postnatal angiogenesis. In contrast to the double knockout animals, mice that are missing one, two, or three copies of the *Id1/3* pair are viable, and the expected number of animals is born. When transplanted with tumor allografts, these animals demonstrate significant reductions in tumor microvessel density and loss of integrin $\alpha v\beta 3$ and matrix metalloproteinase (MMP)-2 on the surface of the remaining vessels (37). These changes are associated with a dramatic decrease in the growth

Table 97-1: Id Expression Pattern and Knockout Phenotypes

Gene	*Embryonic Expression (Refs.)*	*Adult Expression and Expression in Tumors (Refs.)*	*Angiogenic Expression (Refs.)*	*Knockout Phenotype (Refs.)*
Id1	Roofplate of neuronal tube, later in dividing neuroblasts Gut (mesenchyme), endocardiac cushion, lung, kidney, tooth, glandular structures, heart (9)	Not expressed in normal tissue, some evidence of expression in tumor cells – breast (10,11), colon carcinoma (12), thyroid cancer (13)	Most tumor endothelium; poorly differentiated, but not well differentiated, prostate tumors (14–16)	Normal development, defects in T-cell migration (17,18), animals smaller than wild-type
Id2	Roofplate of neuronal tube, later in maturating neurons Gut (endothelium), endocardiac cushion lung, kidney, tooth, glandular structures, heart (9)	Purkinje cells (cerebellum) (5,19), overexpressed in colon adenocarcinoma (12) and neuroblastoma (20)	No evidence for expression in tumor endothelium, expression in neuroectodermal tumors leads to upregulation of VEGF (21)	25% embryonic lethality, lack of lymph nodes, Peyer's patches. Reduced killer, Langerhans, and splenic cells (22,23); hydronephrosis (24) and enteric neoplastic lesions (25), immature mammary glands, lactation defects (26)
Id3	Roof and floorplate of neuronal tube, later in dividing neuroblasts, olfactory system, branchial arches, limbs, sclerotome, endocardiac cushion, outer lining of the gut (mesenchyme), lung, retina, kidneys, teeth, heart (9)	Not expressed in normal tissue; evidence of expression in colon adenocarcinomas (12)	Most tumor endothelium (14–16)	Normal development, defects in B-cell proliferation, thymocyte maturation, and humoral immunity; animals smaller than wild-type (27,28)
Id4	Absent in early neurogenesis, expressed in maturating neurons, ventral epithelium of developing stomach (9) Oligodendrocyte precursors (29)	Purkinje cells (cerebellum), testis, spleen, kidney, BM. Overexpressed in glioblastoma, oligodendrogliomas (30) and bladder cancer (31) Inactivated in highly metastatic breast tumors (32), colorectal carcinoma (33), and gastric adenocarcinoma (34) Potential tumor suppressor in leukemia (35)	No evidence for expression in tumor endothelium	Premature differentiation, neural proliferation defects, defective differentiation in oligodendrocyte lineage, malformation of different brain regions (36)

$Id1^{-/-}/Id3^{-/-}$, forebrain hemorrhage, cardiac development defective, embryonic lethal (E13.5).
$Id1^{-/-}\ Id2^{-/-}$ cardiac malformation, embryonic lethal (E13.5).
$Id2^{-/-}\ Id3^{-/-}$ cardiac malformation, embryonic lethal (E13.5).
$Id1^{-/-}/Id2^{-/-}/Id3^{-/-}$, widespread hemorrhage, cardiac development defective, embryonic lethal (E11.5).

of some tumor allografts (e.g., B6RV2 lymphoma), but surprisingly others (e.g., Lewis lung carcinoma) continue to grow, albeit more slowly, despite severe hypoxic stress, internal hemorrhage, and necrosis. Depending on the allograft model, a dose effect of Id loss also is observed. Whereas in the case of B6RV2 grafts deletion of one copy ($Id1^{+/-}\ Id3^{+/+}$) is sufficient to block tumor growth completely, Lewis lung carcinoma and murine breast carcinoma (B-CA) transplants show a stepwise stronger reduction in growth rate by subsequent deletion of *Id* alleles. In the case of Lewis lung carcinoma, despite primary tumor growth in the absence of the Ids, metastatic growth to the lung was severely perturbed (5). So, in this case at least, allograft growth in the face of reduced Id dosages and hypoxic stress was very much different from that observed at metastatic sites. Whether this is due to a global sensitivity of metastatic lesions to the type of angiogenic stress imposed by Id loss or other antiangiogenic treatment warrants further analysis.

Transplanted tumors may behave very differently from more physiologically relevant models of cancer with respect to their response to antiangiogenic intervention. A critical next step in the analysis of Id proteins as potential therapeutic targets was to determine the effect of Id loss on the vascularization and growth of spontaneous murine tumors. Three informative model systems have been employed to date and each, remarkably, provides unique insights into the effects of angiogenic stress on tumor growth and progression. The model systems include (a) mice that develop breast tumors initiated by the overexpression of the her2/neu oncogene (MMTV-her2/neu); (b) mice missing one copy of the *phosphatase and tensin homolog deleted from chromosome 10 (PTEN)* tumor suppressor (*PTEN* $^{+/-}$), which develop multiple neoplasias; and (c) mice that develop prostate tumors as a result of overexpression of the SV40 T-antigen oncoprotein in the prostate epithelium (Tramp) (14–16). Each of these strains was crossed with the *Id*-knockout mice and assayed for changes in tumor phenotype.

In contrast to the allograft models in which *Id*$^{+/+}$ tumor cells are transplanted into *Id*-null recipients, the spontaneous tumors will carry the same Id profile as the mouse bearing the tumor. Therefore, it is important to consider the potential confounding effects of Id loss in the tumor cells themselves. Careful analyses of spontaneous tumors in *Id*$^{+/+}$ mice failed to reveal evidence for Id1/3 expression in tumor cells in any of the three tumor models. In contrast, prominent Id1 expression was detected in the nuclei of the ECs infiltrating those tumors (14–16). These findings strongly suggest that any effect of Id loss is likely to be a reflection of perturbations of the vasculature.

In the case of MMTV-her2/neu animals bearing breast tumors, the loss of Id resulted in a dramatic change in the morphology of the tumors. Tumors in animals with intact *Id1* and *Id3* genes presented as dense cellular masses with little necrosis. However, when one, two, or three copies of the *Id1/3* pair were missing, a dramatic increase in cystification was observed. Fluid-filled cavities surrounded by a viable rim of cells at the periphery frequently were observed. These cells at the periphery may have "escaped" the hypoxic stress caused by the vascular defects imposed by Id loss. This hypoxic stress is manifested by the increase in HIF-1α protein present during the early stages of cyst growth. However, the volumetric growth of the cysts was equal to or greater than the dense lesions observed in the *Id* wild-type background. In an attempt to target the viable rim of cells at the periphery of the cysts, animals were treated with the Hsp90 antagonist 17-AAG. Hsp90 has been shown to act as chaperone for both her2/neu (the oncogene driving the growth of the tumor) and HIF-1α, which may be driving the "escape" from hypoxic stress by inducing very high vascular endothelial growth factor (VEGF) expression and recruitment of vessels to the viable rim. Whereas treatment with 17-AAG in the Id wild-type background led only to a modest suppression of growth, remarkably, Id loss in combination with 17-AAG administration led to the first reported complete suppression of growth of these breast lesions. Withdrawal of 17-AAG after 8 weeks led to regrowth of the tumors (despite the lesions being extremely small during treatment), making this an excellent model for further analysis of tumor dormancy (14).

Loss of one copy of *PTEN* leads to an accumulation of phosphatidyl-inositol-triphosphate and activation of the growth-promoting Akt kinase. Mice harboring this mutation develop a spectrum of tumors and hyperplastic lesions, such as lymph node hyperplasias, uterine carcinomas, prostate intraepithelial neoplasias (PIN), and pheochromocytomas (38). In situ hybridization and immunohistochemistry showed that Id1 and Id3 expression was confined to the endothelium of lymph hyperplastic lesions and uterine carcinomas, whereas the vessels of PINs, pheochromatocytomas, and non-neoplastic tissue stained negative. Id1 and Id3 were either not expressed at all or only weakly expressed in the tumor cells. *Id*-knockout mice were crossed into the *PTEN*$^{+/-}$ background, and the effect on the vascularization of multiple tumors was examined. In all tumors in which Id was expressed in the endothelium, Id loss (*Id1*$^{-/-}$ *Id3*$^{+/-}$) led to a loss of vascular integrity. The consequences of this, however, varied dramatically from lesion to lesion. The lymphoid hyperplasias, for example, showed pinpoint hemorrhages resulting from Id loss, but lesions grew at normal rates. Intrauterine carcinomas, on the other hand, showed such extensive hemorrhage and necrosis that most of the disease was eliminated by Id loss (16). Thus, this model clearly demonstrates that the effects of Id loss on tumor biology depend on tumor type, a result which may pertain to other antiangiogenic interventions.

Finally, loss of Id function in the Tramp model of prostate cancer has yielded another novel insight into the effects of antiangiogenic stress on tumor growth and progression. Studies in the Tramp model prior to crossing with *Id*-null mice revealed Id1 and Id3 expression in the nuclei of those ECs infiltrating poorly differentiated, but not well differentiated, prostate tumors – a result corroborated in human specimens. These data suggest that the expression of Id in tumor-associated ECs is not only dependent on the type of tumor, as shown in the PTEN model, but also on tumor grade. (Furthermore, as described later, only those ECs infiltrating the poorly differentiated tumors are bone marrow-derived.) As predicted from this analysis, when the Id loss-of-function mutations (*Id1*$^{-/-}$ *Id3*$^{+/-}$) were introduced into the Tramp background, only the poorly differentiated tumors displayed a dramatic increase in hemorrhage and necrosis (15). As in the case of the her2/neu cross, this disruption of vascular integrity was not followed by a dramatic suppression of tumor growth, suggesting again that combination therapy aimed at targeting hypoxia "escapees" will ultimately be the most effective therapeutic antiangiogenic strategy in human cancer.

ID PROTEINS AND PROGENITOR CELL MOBILIZATION

The inhibition of allograft tumor growth in the *Id*-knockout model allowed for a rapid determination of the functional role of bone marrow–derived endothelial progenitor cells (EPCs) in tumor vascularization. Previous studies have shown that EPCs may contribute to neoangiogenesis (39–41).

Interestingly, *Id1*$^{+/-}$ *Id3*$^{-/-}$ animals were first shown to have a defect in their ability to mobilize a CD11b^{-}/VEGF receptor (VEGFR2)$^{+}$ EPC population in response to VEGF administration (42). In these experiments, it also was shown that VEGF administration leads to an upregulation of Id1 in the bone marrow. Together with the observation that Id knockdown attenuates the growth of some tumors, these data raised the interesting possibility that Id-mediated tumor growth depends on mobilization and subsequent uptake of EPCs in tumor vasculature. To test this hypothesis, lethally irradiated *Id1*$^{+/-}$ *Id3*$^{-/-}$ and *Id*-WT animals were transplanted with bone marrow from *Id*-WT *Rosa26-LacZ*($^{+}$) mice. Four weeks after bone marrow reconstitution, the transplant recipients were injected subdermally with B6RV2 cells. In the *Id1*$^{+/-}$ *Id3*$^{-/-}$ mice, bone marrow transplantation reversed the attenuating effect of the *Id*-null background on tumor growth and vasculature (42). Indeed, nearly 100% of tumor vessels after 2 days were LacZ^{+}, indicating their derivation from the donor bone marrow. Importantly, the percentage of LacZ^{+} vessels declined over time, suggesting the replacement of the endothelium with host-derived ECs. The utilization of the bone marrow-derived cells was not simply due to the "freezing" of the peripheral vasculature in the *Id*-null animals, because bone marrow from *Rosa26-LacZ*($^{+}$) mice transplanted into irradiated wild-type animals produced similar results (42).

But is the same phenomenon observed in spontaneous tumors which, we have argued, are much more physiologically relevant with respect to angiogenesis? Here, as in the tumor growth analysis, significant differences are observed. In the *PTEN*$^{+/-}$ tumor model, the utilization of ROSA26-LacZ^{+} bone marrow–derived progenitors is clearly dependent on the tumor type: Lesions in the lymph nodes are never observed to contain bone marrow-derived ECs; however, in intrauterine carcinomas, between 10% and 20% of the ECs incorporated into the vessels are donor-derived. Importantly, even this relatively low level of bone marrow progenitor incorporation was shown to be of functional significance because, after transplanting the *Id*-null mice with wild-type ROSA26-LacZ^{+} bone marrow, the hemorrhage and necrosis observed in the uterine tumors in the *PTEN*$^{+/-}$ background was significantly reduced (16). The low percentage of progenitor incorporation, which also has been observed in human tumors marked with bone marrow from gender-mismatched transplants (43), may reflect replacement of progenitors with mature endothelium from the periphery (see next section). In the Tramp prostate model, incorporation of bone marrow–derived progenitor cells was shown to correlate with tumor grade; high-grade (poorly differentiated) tumors recruited Id-positive bone marrow–derived progenitors to about 15% of the vessels, whereas low-grade (well-differentiated) tumors utilized Id-negative host-derived endothelium (15). Here again, even low incorporation of EPCs had functional relevance, because wild-type (but not Id mutant) bone marrow significantly reduced the hemorrhage observed in the prostate tumors in the *Id*-knockout background, but had – as expected – no influence on tumor progression. In addition, Id mutant bone marrow increased the hemorrhage observed in the prostate tumors when transplanted into wild-type animals, suggesting that these cells are both necessary and sufficient to renormalize the vasculature that results from Id loss (15). Taken together, these results and recent, unpublished data obtained from treatment experiments using VEGF, support a model in which upregulation of Id in the bone marrow in response to VEGF leads to the mobilization of EPCs, which incorporate into a functionally significant subset of vessels in some tumor types (Figure 97.1). Whether the utilization and presence of these EPCs in the tumor bed varies significantly with time (as in the allografts) remains to be determined.

Recently, several other observations have provided support for the EPC hypothesis. First, Kerbel and coworkers have shown that EPC concentrations in peripheral blood can predict the angiogenic potential of various mouse strains (44). In humans, gender-mismatched bone marrow transplants yielded approximately 2% donor ECs in the skin and gut (43,45).

Several studies recently have called into question the importance of bone marrow–derived progenitor cells for the formation of tumor vessels. First, Naldini and coworkers, by performing careful confocal microscopy studies with mice transplanted with Tie2–driven green fluorescent protein (GFP)-marked bone marrow, showed quite convincingly that only rare platelet-endothelial cell adhesion molecule (PECAM)-1/CD31^{+} ECs are GFP-positive and that the contribution from the marrow is more pronounced in the hematopoietic compartment (46). This latter population of cells may contribute to angiogenesis without becoming an integral part of the vessel. In a separate study, Begley and colleagues marked the endothelium by introducing both a LacZ Cre-lox–activatable allele and a tamoxifen-inducible Cre recombinase driven by the SCL-promoter/enhancer (47). In this way, only upon tamoxifen stimulation would SCL-expressing cells be marked with β-galactosidase. These investigators found that, although these animals did not display β-galactosidase-positive cells in the bone marrow upon tamoxifen treatment (presumably because SCL is not expressed until later in the development of the lineages), they did develop β-galactosidase-positive vessels in the tumor (47). They reasoned that, if these cells were derived from bone marrow–derived progenitors, then transplantation of wild-type animals with bone marrow from the transgenics should show β-galactosidase-positive vessels upon tamoxifen treatment and tumor challenge. None, however, were observed.

How can these results be reconciled with the earlier mentioned data that showed a clear involvement of bone marrow-derived ECs in tumor angiogenesis? We would propose that all of these findings are concordant if we assume that an early EPC migrates from the bone marrow to the site of a tumor or ischemic injury and is required for the establishment of the very early vascular network. These cells, however, gradually become replaced with more mature endothelium from the periphery (Figure 97.2). This model predicts that in the

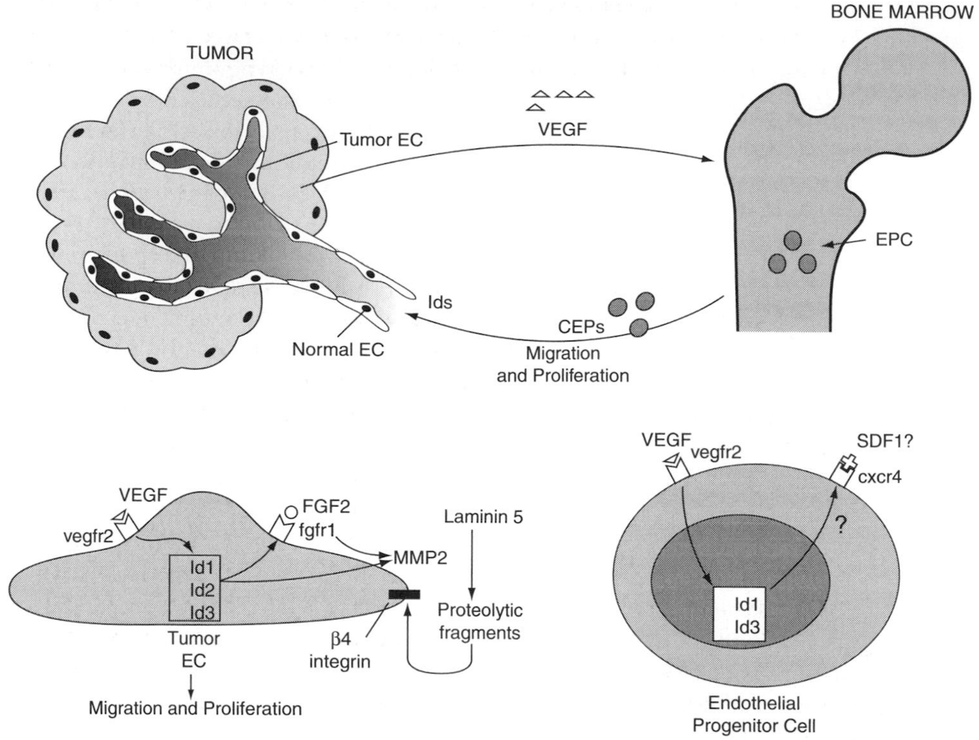

Figure 97.1. Role of Id proteins in peripheral and bone marrow-derived ECs. In the periphery, Ids are required for maintaining the expression of FGFR1, MMP-2, laminin 5, and integrin $\alpha 6\beta 45$, which form a promigratory network. This network depends on the engagement of integrin $\alpha 6\beta 4$ with an MMP-2 cleavage product of laminin 5. In the bone marrow, Ids are required for the mobilization and proliferation of EPCs, which enter the blood stream as circulating endothelial progenitors (CEPs) and migrate to the site of the tumor. Id1 and Id3 in the bone marrow are upregulated in response to VEGF, and it is speculated that this may lead to the upregulation of the chemokine receptor CXCR4. These cells may then be competent to engage stromal-derived factor (SDF)-1 from the tumor (or tumor-associated stromal cells), which facilitates their mobilization.

β-galactosidase-marked bone marrow transplant experiments of Lyden and colleagues (42), there would be a gradual drift of tumor endothelium from $LacZ^+$ to $LacZ^-$, and this trend was indeed originally reported. The Begley experiment might be interpreted as having failed to mark the early progenitors that initially migrate to the site of the tumor because these cells do not develop into mature endothelium, but rather are replaced by it from the unmarked periphery. Finally, Naldini and coworkers might have found a higher percentage of GFP-positive progenitors if an endothelial-specific marker earlier

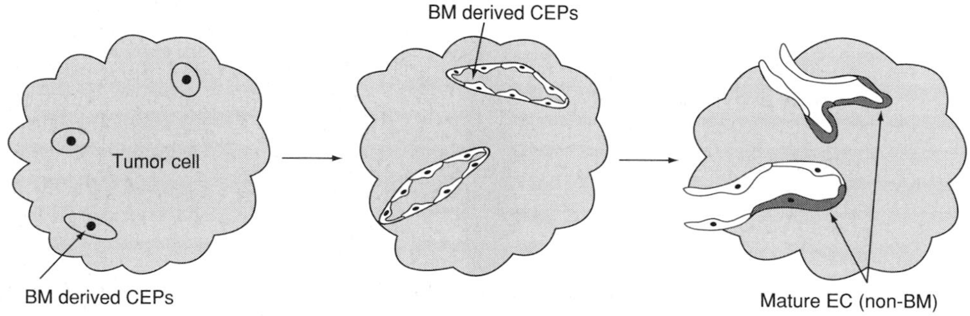

Figure 97.2. Temporal changes in bone marrow–derived contribution to the tumor vasculature. A speculative scheme is proposed whereby the early vascular network in a tumor is primarily composed of circulating endothelial progenitors (CEPs), which fuse to form blood vessels. These cells in these vessels are gradually replaced with more mature endothelium from the periphery.

than CD31 was utilized in the analysis. Although this interpretation is speculative at this point, it nonetheless provides a model compatible with most of the data published and a readily testable hypothesis.

ID1 PROTEINS AND ENDOTHELIAL CELL BIOLOGY

Because the Id proteins are themselves regulators of transcription, it seems reasonable to assume that the angiogenesis phenotype of Id loss is mediated by changes in the transcription of key mediators of EC migration and vessel assembly. Lymph hyperplastic lesions that develop in the *PTEN*$^{+/-}$ animals have a severely perturbed vascular network in the *Id1*-null background, and changes in Id-dependent gene expression profiles in the tumor endothelium were determined (16,48). Four key players in an integrin signaling pathway were all shown to be downregulated in response to Id1 loss in vivo: integrins $\alpha6$ and $\beta4$, fibroblast growth factor receptor 1 (FGFR1), MMP-2, and laminin 5 (16). When FGFR1 engages FGF2, it leads to an upregulation and enhanced secretion of MMP-2 in some cell types (49), as well as upregulation of integrins $\alpha6$ and $\beta4$ (48). MMP-2, in turn, cleaves laminin 5 to yield a promigratory proteolytic fragment that engages integrin $\alpha6\beta4$ (50). Downregulation of all four components is predicted to short-circuit this pathway and prevent proper EC migration. Inhibition of these signaling molecules, in Matrigel plug assays, was shown to phenocopy the loss of Id, demonstrating a functional significance of the array findings (16). Importantly, this model has received support from an independent genetic analysis, in which it was shown that hypomorphic mutations in integrin $\beta4$ also lead to impaired angiogenesis (51). It is of note that the lesions from which the Id-deficient ECs were derived do not recruit bone marrow–derived EPCs. It will be of interest, therefore, to determine if the same promigratory pathway is affected in the bone marrow to explain the defect in EPC mobilization observed in the *Id*-knockout animals.

The antiangiogenic protein thrombospondin (TSP)-1 has been reported to be upregulated in response to Id1 loss in murine embryonic fibroblasts, and suggests another mechanism whereby Id1 loss in the vasculature might lead to angiogenesis inhibition (52). No upregulation of TSP-1 in the vasculature of spontaneous tumors or xenografts in *Id1*-null mice has been reported, however, making such a mechanism unlikely. A genetic analysis of the *TSP-1*-knockout in the *Id1*-knockout background could be performed because both strains are viable and, if the mechanism is playing any role in tumor angiogenesis, a reversal of the *Id*-null phenotype should be observed in the *TSP-1/Id1*-double null mice.

The chemokine stromal derived factor (SDF)-1 has been shown to be upregulated by ECs in response to ischemia, which leads to the mobilization of EPCs. If the observation that the SDF-1 receptor, CXCR4, is regulated by Id1 in T cells (17) also applies to EPCs in the bone marrow, it would suggest that VEGF induction of Id1 may in turn lead to the competence of EPCs to receive an SDF1 signal from the periphery. A careful analysis of CXCR4 expression in the EPCs is clearly an important next step in providing support for this intriguing possibility.

But what controls Id expression in the neovasculature? It has been shown that VEGF in the circulation leads to a dramatic upregulation of Id1 and Id3 in the bone marrow, presumably in the progenitor cells that become mobilized (42). Activation of the mitogen-activated protein kinase (MAPK) pathway by VEGF receptor stimulation could impinge on the *Id1/3* promoters at the Egr1 site, as has been demonstrated in other cell types (53,54). Validation of this idea in ECs by transfecting appropriate *Id* promoter-reporters should be established, and other transcription factors mediating the response could be identified. Integrin engagement also has been reported to impinge on the *Id1* promoter, independent of growth factor stimulation, and, here again, analysis in ECs needs to be performed to confirm this idea. The proangiogenic statin, fluvostatin, but not VEGF also has been reported to upregulate Id1 in human dermal microvascular ECs (HDMECs) in culture (55), although VEGF does apparently upregulate Id1 expression in human umbilical vein ECs (HUVECs) (56), suggesting differential Id regulation depending on the endothelial subtype.

Another signaling cascade that impinges on the Id protein in EC cells is the transforming growth factor (TGF)-β pathway. Both bone morphogenic protein (BMP) 2 and 6 (through the BMP receptor) and low concentrations of TGF-β (through type I receptor activin-like kinase [ALK]-5) lead to an upregulation of Id1 expression via Smad1 and 5 and enhanced proliferation and migration of ECs in culture (57–59). BMP6 and TGF-β-induced stimulation of migration is inhibited by antisense Id1 treatments, demonstrating the requirement for Ids in these responses (57,58). Ectopic expression of Id1 in ECs in culture enhances cell migration and tube formation (57), demonstrating the sufficiency of Id1 expression for these effects. High concentrations of TGF-β (through the type I receptor ALK1), on the other hand, lead to an inhibition of cell proliferation and migration, the downregulation of Id1, and upregulation of the antiangiogenic plasminogen activator inhibitor (PAI)-1 (58). An important in vivo correlate of these findings is the observation that mice lacking the BMP-specific Smad1 or -5 die at midgestation with enlarged blood vessels (57,60,61) similar to those observed in the *Id1/3*-double knockout embryos. Together then, these studies indicate that the development and maturation of ECs in mice is at least in part mediated by the TGF-β/Id pathway connections.

CONCLUSION

The prevailing evidence suggests that Id proteins, and/or their upstream or downstream effectors, are potential targets for novel anticancer drug design. In particular, genetic manipulation of Id levels has validated Id as an important target in several spontaneous murine tumors that are physiologically much closer to human disease than transplanted tumors. The fact that the Ids are expressed at low levels in most normal adult tissue, but are reactivated in the tumor vasculature (and possibly in some cancers in the tumor cells themselves), suggests that targeted therapy will be associated with low toxicity. The advantage of targeting Id itself is that the transcription factor is selectively expressed in tumors. Moreover, the genetic mouse models indicate that partial loss of Id function is sufficient to inhibit tumor growth. The disadvantage is the challenge inherent in inhibiting nuclear proteins. One approach is to inhibit interactions between Id and other proteins. Over the past few years, progress has been made in targeting protein–protein interactions using small molecules. In particular, agents effective at inhibiting the Myc–Max interaction (62) or β-catenin (63) have been reported. Other approaches include the use of inhibitors targeted at mediators of Id expression or downstream effectors of Id. The unique phenotype of Id loss in these tumor models, as well as the role of the EPC as a surrogate marker, promise to facilitate drug screening.

The advantage of targeting upstream and/or downstream mediators of the Id pathway is their relative accessibility and amenability to downregulation. However, these intermediates are not in and of themselves specific to the tumor. For example, whereas MMP-2 and integrin $\alpha6\beta4$ have been identified as Id target genes, they are also expressed in Id-negative cell types, presumably through the influence of other transcriptional networks. Thus, by targeting one or another of these downstream genes, tumor specificity would be lost.

In summary, novel drug design to inhibit the activity of the Id proteins, even acknowledging the difficulty of the challenge, is a goal worthy of pursuit for the management of human cancer.

KEY POINTS

- Id proteins are negative regulators of transcription and are essential for the mobilization, migration, and proliferation of ECs to and within a tumor.
- Loss of Id in tumor ECs leads to the inactivation of key elements of a promigratory pathway involving the cleavage of laminin 5 and engagement of integrin $\alpha6\beta4$, and may account for the vascular phenotype observed in *Id*-knockout animals.
- The effects of Id loss and antiangiogenic stress depend on the tumor type and tumor grade. Identifying and characterizing the Id-dependent events in the bone marrow during tumor growth is an important next step.
- Targeting Id may be an important antiangiogenic therapeutic strategy, because it is not expressed in normal ECs and most normal adult tissue.

REFERENCES

1 Benezra R, Davis RL, Lockshon D, et al. The protein Id: a negative regulator of helix-loop-helix DNA binding proteins. *Cell.* 1990;61:49–59.

2 Ruzinova MB, Benezra R. Id proteins in development, cell cycle and cancer. *Trends Cell Biol.* 2003;13(8):410–418.

3 Sun XH, Copeland NA, Jenkins NA, Baltimore D. Id proteins Id1 and 1d2 selectively inhibit DNA binding by one class of helix-loop-helix proteins. *Mol Cell Biol.* 1991;11:5603–5611.

4 Jen Y, Weintraub H, Benezra R. Overexpression of Id protein inhibits the muscle differentiation program: in vivo association of Id with E2A proteins. *Genes Dev.* 1992;6:1466–1479.

5 Lyden D, Young AZ, Zagzag D, et al. Id1 and Id3 are required for neurogenesis, angiogenesis and vascularization of tumour xenografts. *Nature.* 1999;401(6754):670–677.

6 Bader BL, Rayburn H, Crowley D, Hynes RO. Extensive vasculogenesis, angiogenesis, and organogenesis precede lethality in mice lacking all alpha v integrins. *Cell.* 1998;95(4):507–519.

7 Jen Y, Manova K, Benezra R. Expression patterns of Id1, Id2, and Id3 are highly related but distinct from that of Id4 during mouse embryogenesis. *Dev Dyn.* 1996;207(3):235–252.

8 Fraidenraich D, Stillwell E, Romero E, et al. Rescue of cardiac defects in Id knockout embryos by injection of embryonic stem cells. *Science.* 2004;306(5694):247–252.

9 Jen Y, Manova K, Benezra R. Each member of the Id gene family exhibits a unique expression pattern in mouse gastrulation and neurogenesis. *Dev Dyn.* 1997;208:92–106.

10 Fong S, Itahana Y, Sumida T, et al. Id-1 as a molecular target in therapy for breast cancer cell invasion and metastasis. *Proc Natl Acad Sci USA.* 2003;100(23):13543–13548.

11 Lin CQ, Singh J, Murata K, et al. A role for Id-1 in the aggressive phenotype and steroid hormone response of human breast cancer cells. *Cancer Res.* 2000;60(5):1332–1340.

12 Wilson JW, Deed RW, Inoue T, et al. Expression of Id helix-loop-helix proteins in colorectal adenocarcinoma correlates with p53 expression and mitotic index. *Cancer Res.* 2001;61(24):8803–8810.

13 Kebebew E, Treseler PA, Duh QY, Clark OH. The helix-loop-helix transcription factor, Id-1, is overexpressed in medullary thyroid cancer. *Surgery.* 2000;128(6):952–957.

14 de Candia P, Solit DB, Giri D, et al. Angiogenesis impairment in Id-deficient mice cooperates with an Hsp90 inhibitor to completely suppress HER2/neu-dependent breast tumors. *Proc Natl Acad Sci USA.* 2003;100(21):12337–12342.

15 Li H, Gerald WL, Benezra R. Utilization of bone marrow-derived endothelial cell precursors in spontaneous prostate tumors varies with tumor grade. *Cancer Res.* 2004;64(17):6137–6143.

16 Ruzinova MB, Schoer RA, Gerald W, et al. Effect of angiogenesis inhibition by Id loss and the contribution of bone-marrow-derived endothelial cells in spontaneous murine tumors. *Cancer Cell.* 2003;4(4):277–289.

17 Sikder H, Huso DL, Zhang H, et al. Disruption of Id1 reveals major differences in angiogenesis between transplanted and autochthonous tumors. *Cancer Cell.* 2003;4(4):291–299.

18 Yan W, Young AZ, Soares VC, et al. High incidence of T-cell tumors in E2A-null mice and E2A/Id1 double- knockout mice. *Mol Cell Biol.* 1997;17(12):7317–7327.

19 Tzeng SF, de Vellis J. Id1, Id2, and Id3 gene expression in neural cells during development. *Glia.* 1998;24(4):372–381.

20 Lasorella A, Noseda M, Beyna M, Iavarone A. Id2 is a retinoblastoma protein target and mediates signalling by Myc oncoproteins. *Nature*. 2000;407(6804):592–598.
21 Lasorella A, Rothschild G, Yokota Y, et al. Id2 mediates tumor initiation, proliferation, and angiogenesis in Rb mutant mice. *Mol Cell Biol*. 2005;25(9):3563–3574.
22 Yokota Y, Mansouri A, Mori S, et al. Development of peripheral lymphoid organs and natural killer cells depends on the helix-loop-helix inhibitor Id2. *Nature*. 1999;397(6721):702–706.
23 Hacker C, Kirsch RD, Ju XS, et al. Transcriptional profiling identifies Id2 function in dendritic cell development. *Nat Immunol*. 2003;4(4):380–386.
24 Aoki Y, Mori S, Kitajima K, et al. Id2 haploinsufficiency in mice leads to congenital hydronephrosis resembling that in humans. *Genes Cells*. 2004;9(12):1287–1296.
25 Russell RG, Lasorella A, Dettin LE, Iavarone A. Id2 drives differentiation and suppresses tumor formation in the intestinal epithelium. *Cancer Res*. 2004;64(20):7220–7225.
26 Mori S, Nishikawa SI, Yokota Y. Lactation defect in mice lacking the helix-loop-helix inhibitor Id2. *EMBO J*. 2000;19(21):5772–5781.
27 Pan L, Sato S, Frederick JP, et al. Impaired immune responses and B-cell proliferation in mice lacking the Id3 gene. *Mol Cell Biol*. 1999;19(9):5969–5980.
28 Rivera RR, Johns CP, Quan J, et al. Thymocyte selection is regulated by the helix-loop-helix inhibitor protein, Id3. *Immunity*. 2000;12(1):17–26.
29 Kondo T, Raff M. The Id4 HLH protein and the timing of oligodendrocyte differentiation. *EMBO J*. 2000;19(9):1998–2007.
30 Liang Y, Bollen AW, Nicholas MK, Gupta N. Id4 and FABP7 are preferentially expressed in cells with astrocytic features in oligodendrogliomas and oligoastrocytomas. *BMC Clin Pathol*. 2005; 5:6.
31 Wu Q, Hoffmann MJ, Hartmann FH, Schulz WA. Amplification and overexpression of the ID4 gene at 6p22.3 in bladder cancer. *Mol Cancer*. 2005;4(1):16.
32 Umetani N, Mori T, Koyanagi K, et al. Aberrant hypermethylation of ID4 gene promoter region increases risk of lymph node metastasis in T1 breast cancer. *Oncogene*. 2005;24(29):4721–4727.
33 Umetani N, Takeuchi H, Fujimoto A, et al. Epigenetic inactivation of ID4 in colorectal carcinomas correlates with poor differentiation and unfavorable prognosis. *Clin Cancer Res*. 2004; 10(22):7475–7483.
34 Chan AS, Tsui WY, Chen X, et al. Downregulation of ID4 by promoter hypermethylation in gastric adenocarcinoma. *Oncogene*. 2003;22(44):6946–6953.
35 Yu L, Liu C, Vandeusen J, et al. Global assessment of promoter methylation in a mouse model of cancer identifies ID4 as a putative tumor-suppressor gene in human leukemia. *Nat Genet*. 2005;37(3):265–274.
36 Yun K, Mantani A, Garel S, et al. Id4 regulates neural progenitor proliferation and differentiation in vivo. *Development*. 2004; 131(21):5441–5448.
37 Lyden D, Young AZ, Zagzag D, et al. Id1 and Id3 are required for neurogenesis, angiogenesis and vascularization of tumour xenografts. *Nature*. 1999;401(6754):670–677.
38 Di Cristofano A, Pesce B, Cordon-Cardo C, Pandolfi PP. PTEN is essential for embryonic development and tumour suppression. *Nat Genet*. 1998;19(4):348–355.
39 Asahara T, Takahashi T, Masuda H, et al. VEGF contributes to postnatal neovascularization by mobilizing bone marrow-derived endothelial progenitor cells. *EMBO J*. 1999;18(14):3964–3972.
40 Takahashi T, Kalka C, Masuda H, et al. Ischemia- and cytokine-induced mobilization of bone marrow-derived endothelial progenitor cells for neovascularization. *Nat Med*. 1999;5(4):434–438.
41 Rafii S. Circulating endothelial precursors: mystery, reality, and promise. *J Clin Invest*. 2000;105(1):17–19.
42 Lyden D, Hattori K, Dias S, et al. Impaired recruitment of bone-marrow-derived endothelial and hematopoietic precursor cells blocks tumor angiogenesis and growth. *Nat Med*. 2001;7(11): 1194–1201.
43 Peters BA, Diaz LA, Polyak K, et al. Contribution of bone marrow-derived endothelial cells to human tumor vasculature. *Nat Med*. 2005;11(3):261–262.
44 Shaked Y, Bertolini F, Man S, et al. Genetic heterogeneity of the vasculogenic phenotype parallels angiogenesis; Implications for cellular surrogate marker analysis of antiangiogenesis. *Cancer Cell*. 2005;7(1):101–111.
45 Jiang S, Walker L, Afentoulis M, et al. Transplanted human bone marrow contributes to vascular endothelium. *Proc Natl Acad Sci USA*. 2004;101(48):16891–16896.
46 De Palma M, Venneri MA, Roca C, Naldini L. Targeting exogenous genes to tumor angiogenesis by transplantation of genetically modified hematopoietic stem cells. *Nat Med*. 2003;9(6): 789–795.
47 Gothert JR, Gustin SE, van Eekelen JA, et al. Genetically tagging endothelial cells in vivo: bone marrow-derived cells do not contribute to tumor endothelium. *Blood*. 2004;104(6):1769–1777.
48 Klein S, Giancotti FG, Presta M, et al. Basic fibroblast growth factor modulates integrin expression in microvascular endothelial cells. *Mol Biol Cell*. 1993;4(10):973–982.
49 Miyake H, Hara I, Yoshimura K, et al. Introduction of basic fibroblast growth factor gene into mouse renal cell carcinoma cell line enhances its metastatic potential. *Cancer Res*. 1996;56(10): 2440–2445.
50 Giannelli G, Falk-Marzillier J, Schiraldi O, et al. Induction of cell migration by matrix metalloprotease-2 cleavage of laminin-5. *Science*. 1997;277(5323):225–228.
51 Nikolopoulos SN, Blaikie P, Yoshioka T, et al. Integrin beta4 signaling promotes tumor angiogenesis. *Cancer Cell*. 2004;6(5): 471–483.
52 Volpert OV, Pili R, Sikder HA, et al. Id1 regulates angiogenesis through transcriptional repression of thrombospondin-1. *Cancer Cell*. 2002;2(6):473–483.
53 Mechtcheriakova D, Wlachos A, Holzmuller H, et al. Vascular endothelial cell growth factor-induced tissue factor expression in endothelial cells is mediated by EGR-1. *Blood*. 1999;93(11):3811–3823.
54 Mechtcheriakova D, Schabbauer G, Lucerna M, et al. Specificity, diversity, and convergence in VEGF and TNF-alpha signaling events leading to tissue factor up-regulation via EGR-1 in endothelial cells. *FASEB J*. 2001;15(1):230–242.
55 Pammer J, Reinisch C, Kaun C, et al. Inhibitors of differentiation/DNA binding proteins Id1 and Id3 are regulated by statins in endothelial cells. *Endothelium*. 2004;11(3–4):175–180.
56 Sakurai D, Tsuchiya N, Yamaguchi A, et al. Crucial role of inhibitor of DNA binding/differentiation in the vascular endothelial

growth factor-induced activation and angiogenic processes of human endothelial cells. *J Immunol.* 2004;173(9):5801–5809.

57 Valdimarsdottir G, Goumans MJ, Rosendahl A, et al. Stimulation of Id1 expression by bone morphogenetic protein is sufficient and necessary for bone morphogenetic protein-induced activation of endothelial cells. *Circulation.* 2002;106(17):2263–2270.

58 Goumans MJ, Valdimarsdottir G, Itoh S, et al. Balancing the activation state of the endothelium via two distinct TGF-beta type I receptors. *EMBO J.* 2002;21(7):1743–1753.

59 Langenfeld EM, Langenfeld J. Bone morphogenetic protein-2 stimulates angiogenesis in developing tumors. *Mol Cancer Res.* 2004;2(3):141–149.

60 Chang H, Huylebroeck D, Verschueren K, et al. Smad5 knockout mice die at mid-gestation due to multiple embryonic and extraembryonic defects. *Development.* 1999;126(8):1631–1642.

61 Lechleider RJ, Ryan JL, Garrett L, et al. Targeted mutagenesis of Smad1 reveals an essential role in chorioallantoic fusion. *Dev Biol.* 2001;240(1):157–167.

62 Berg T, Cohen SB, Desharnais J, et al. Small-molecule antagonists of Myc/Max dimerization inhibit Myc-induced transformation of chicken embryo fibroblasts. *Proc Natl Acad Sci USA.* 2002;99(6):3830–3835.

63 Emami KH, Nguyen C, Ma H, et al. A small molecule inhibitor of beta-catenin/CREB-binding protein transcription. *Proc Natl Acad Sci USA.* 2004;101(34):12682–12687.

Output

98

Introductory Essay

Endothelial Cell Output

William C. Aird

Beth Israel Deaconess Medical Center, Harvard Medical School, Boston, Massachusetts

A major theme, or organizing principle, of this book is the analogy of the endothelial cell (EC) to an input–output device. Input arises from the extracellular environment and includes numerous biochemical and biomechanical forces. These environmental cues trigger signaling in ECs through receptor-dependent and receptor-independent mechanisms. Intracellular signaling then results in post-transcriptional and/or transcriptional changes, ultimately leading to an alteration in output, or phenotype. Other sections in this book cover topics related to input and signaling. The chapters in the current section focus on the output side of the equation. An important qualification when discussing output is the scale of investigation that is being considered. For example, at the level of the single cell, ECs may undergo a change in calcium flux, cell shape, protein or mRNA expression, migration, proliferation, and/or apoptosis. Monolayers of ECs express barrier properties and may be assayed for adhesion and transmigration of leukocytes. Other (emergent) properties of the endothelium are only appreciated at the level of the whole blood vessel, organ, or organism. These include endothelial regulation of vasomotor tone and formation of new blood vessels.

In this section, we begin with novel in vivo proteomic approaches that underscore the remarkable heterogeneity of EC phenotypes at the level of the cell surface (Chapters 99 and 100). The next series of chapters focus on the role of the endothelium in hemostasis, beginning with an overview of coagulation (Chapter 101), followed by detailed discussions of specific procoagulant and anticoagulant molecules (Chapters 102–109). Another important function of the endothelium is the regulation of vasomotor tone. We have not included an overview chapter related to this topic. Rather, we focus on the pivotal role of nitric oxide (NO) in EC biology (Chapter 110). For thematic purposes, this chapter is followed by one on heme oxygenase and carbon monoxide (CO) (Chapter 111). NO and CO are produced by *and* signal in ECs. Another example of an autocrine factor that serves as both input and output are the prostanoids (Chapter 112). As reviewed in Chapter 113, the endothelium plays an important role in mediating permeability. The role for endothelium in governing leukocyte adhesion and transmigration is discussed in Chapter 114, and the component parts underlying this process are detailed in Chapters 115–118. Proliferation and migration of ECs are critical aspects of angiogenesis and are covered in a later section on health and disease. ECs display a balance between pro-survival and pro-apoptotic factors (Chapter 119). Finally, there is increasing evidence for the role of the endothelium in antigen presentation, as reviewed in Chapter 120.

The chapters in the current section offer a representative spectrum of endothelial outputs. Many of these may be considered core (or unifying) properties of the endothelium – in other words, phenotypes that are common to more than one vascular bed. For example, every EC in the body contributes to hemostatic balance (albeit through site-specific formulas of procoagulants and anticoagulants). NO is produced by many ECs, particularly those that line arterioles. All vascular beds display barrier properties. Although leukocyte adhesion occurs primarily in postcapillary venules, these blood vessel types are distributed across the body. The regulation of programmed cell death is common to all ECs.

It is well established that many (if not most) EC phenotypes are restricted in their tissue distribution. Some of these regional phenotypes are discussed in organ-based chapters in Part III. There is little question that many site-specific properties of the endothelium await discovery. Indeed, a survey of existing proteomic (see Chapters 99 and 100) and genomic data indicate a wealth of novel vascular bed–specific transmembrane and secreted proteins. Deciphering these so-called *vascular zip codes* is an important goal for the future. Finally, when considering outputs, it is important to recognize that in the human body, ECs may communicate with one another and with other tissues over large distances. For example, ECs may secrete soluble mediators or release microparticles that signal to ECs from another vascular bed. Alternatively, circulating blood cells may "pick up" information from one site of the vasculature and transfer that information to another site (for an interesting example, see Chapter 62). Future work in this area may uncover previously hidden "information highways" that further underscore the need to approach the endothelium as an integrated system.

99

Proteomic Mapping of Endothelium and Vascular Targeting in Vivo

Lucy A. Carver and Jan E. Schnitzer

Sidney Kimmel Cancer Center, San Diego, California

Endothelial cells (ECs) are highly adapted to meet the needs of local tissue and therefore acquire molecular and functional variation according to their location in the body. Although there is little question that the microenvironment of the tissue surrounding the blood vessels significantly influences EC phenotype, very little molecular information exists about vascular endothelium and the degree to which EC expression is modulated within different organs in vivo. Elucidating molecular expression and topography for ECs in multiple tissues constitutes the first step in a systems biology approach to gain a fundamental understanding of functional differences across organ systems and to frame future studies investigating how environmental inputs alter EC gene/protein expression and physiology. This information is vital also for tissue engineering, in which persists a critical lack of understanding of the properties of ECs in the context of normal tissues. By knowing EC expression in a given organ, researchers gain important topographical and functional knowledge, define the set of functional players in each endothelia, and gain key markers both to discover the factors leading to a particular phenotype and maybe to someday re-create the necessary tissue environment in culture. Finally, defining the endothelial proteome in healthy and diseased tissues may reveal further microenvironmental modulation and may prove clinically useful by identifying new disease biomarkers and, perhaps more importantly, novel targets for site-directed delivery of functional and molecular imaging agents as well as nanomedicines, gene vectors, and drugs. In this chapter, we discuss technical barriers to mapping comprehensively the vascular endothelium in vivo and how various promising discovery and validation platforms that integrate state-of-the-art global analytical techniques have overcome some of these difficulties to reveal distinct molecular signatures in normal and neoplastic tissues.

ULTRASTRUCTURAL DIVERSITY OF ENDOTHELIA IN TISSUE

Morphological Heterogeneity

The endothelium is a key gatekeeper controlling the exchange of molecules from the blood to the tissue parenchyma. It largely controls the permeability of a particular vascular bed to blood-borne molecules. The permselectivity of the EC barrier is strongly dependent on the structure and type of endothelium lining the microvasculature in different vascular beds. ECs lining the microvascular beds of different organs exhibit extensive structural differentiation that can be grouped into three primary morphologic categories: sinusoidal, fenestrated, and continuous. Sinusoidal endothelium, found in liver, spleen, and bone marrow, has large intercellular gaps (here, also termed *discontinuous endothelia*) and no basement membrane, allowing for minimally restricted transport of molecules from the capillary lumen into the tissue (or vice versa). Other endothelia are characterized by the presence of a large number of circular transcellular openings, called *fenestrae*, with a diameter of 60 to 80 nm. Fenestrated endothelia frequently have thin diaphragms and are found in organs that require rapid exchange of small molecules, such as kidney glomeruli, endocrine glands, and intestine. Continuous endothelia form a barrier with an uninterrupted cell monolayer and basement membrane with neither fenestrae within ECs nor large gaps between ECs. Most endothelia fall into this latter category.

Permeability in continuous endothelia is regulated by intercellular junctions and vesicular transcytosis. In general, water, solutes, and other small molecules are transported paracellularly (e.g., between ECs) via diffusion and convection; however, specialized proteins (aquaporins and ion channels) at the EC surface membrane may facilitate transmembrane transport into ECs and perhaps even across in select tissues (1).

Most continuous endothelia, including lung, heart, and skeletal muscle, have abundant caveolae that function for select blood proteins as vesicular carriers into and across the cell. Caveolae can constitute 50% to 70% of the cell surface plasma membrane and occupy 10% to 15% of the total cell volume (about 500–600 vesicles/μm^3) (2–4). Other continuous endothelia located in brain and testes form a very restrictive barrier by having few, if any, caveolae and by forming epithelial-like, tight junctions that prevent even small solute transport. To meet the metabolic demands of the brain tissue cells, endothelia of the blood–brain barrier must express specific transporters, even for molecules as small as glucose and amino acids (5,6).

Although originally derived from normal blood vessels, tumor vascular endothelium undergoes considerable alterations. Early tumor development is accompanied by extensive neovascularization with an increased number of intraendothelial organelles, such as vesiculo–vacuolar organelles (VVOs), and extensive distortions of its capillaries forming a chaotic network of dilated vessels that lack segmental differentiation and the strict organization of the vasculature of normal organs (7,8). Other distinctions include reduced basement membrane development with altered composition (7), increased proliferation rate of the ECs (9), and even altered cellular composition of the blood vessels themselves (7). The altered development and morphology of tumor ECs is discussed in detail in Chapter 158.

Distinct Glycocalyx and Transport Pathways

Both passive, pressure-driven transport pathways and active pathways exist for blood molecules to cross endothelium. Passive pathways include paracellular transport through highly regulated intercellular junctions for convective filtration of water, and solutes (10) as well as transcellular transport through fenestrae or transendothelial channels that may form transiently in very attenuated regions of the cell by the fusion of two or more caveolae each located on apposing plasma membranes (11). Active pathways include both coated (i.e., clathrin dependent) and noncoated (e.g., caveolae) plasmalemmal vesicles that transport plasma macromolecules bound or adsorbed from blood by shuttling their contents either into the cell (endocytosis) or from the luminal to abluminal aspect of the endothelium or vice versa (transcytosis) (12–15). The endocytic and transcellular pathways may include both fluid-phase and receptor-mediated processes. Caveolae are discussed in more detail in Chapter 73.

Capillary permeability also depends on the interaction of blood proteins such as albumin and orosomucoid with the endothelial glycocalyx (10,16–19). The luminal surface of the endothelium is exposed directly to the blood and is covered by an elaborate carbohydrate-containing, proteinaceous, polyanionic surface coat called the glycocalyx, which also includes plasmalemmal components such as the glycosylated ectodomains of integral membrane proteins, proteoglycans, glycolipids, and adsorbed blood proteins. The absence of serum proteins in vascular perfusates increases capillary permeability (10,16,18,20,21). Blood protein binding within the glycocalyx may restrict transendothelial transport via the "serum effect"[1] or may enhance specific transport of molecules by receptor- and caveolae-mediated processes (22). Serum also may reduce the coupling of macromolecular transport with hydraulic conductivity (23), thereby reducing paracellular "leakiness" in vivo and increasing the relative importance of transcellular transport. This phenomenon is not simply secondary to the effects of oncotic pressure. Protein binding is thought to form a molecular filter within transport pathways (16) that can electrostatically (24) and sterically (25) restrict the transvascular transport of water, small solutes, and macromolecules. Binding specific receptors mediating transmembrane signaling to produce transient calcium influxes, nitric oxide (NO) production, and other intracellular responses also contributes to regulating EC barrier function, probably by modulating both paracellular and transcellular transport.

This obvious and extensive morphological diversity is likely to reflect tremendous molecular heterogeneity in ECs from different organ vascular beds. Yet expression profiling has been challenging to perform, especially in vivo, where tissue microenvironmental factors influence EC phenotype. In the next sections, we discuss recent large-scale developments on establishing the molecular diversity of endothelia as well as its underlying mechanisms and possible consequences both in studying EC physiology and in the discovery of tissue-specific EC markers for site-specific imaging and drug delivery.

TECHNICAL HURDLES IN STUDYING AND MAPPING ENDOTHELIAL CELLS

Difficulty in Sampling Endothelial Cells in Vivo

Little is currently known about EC molecular expression (i.e., its proteome) and even less about differential expression among the various organ capillary beds. In part, the current paucity of data on EC proteins may relate to methodological problems in interrogating vascular endothelium in vivo. Although it appears logical to study vascular endothelia in vivo (due to morphological and molecular heterogeneity of ECs among tissues), it has been quite difficult to analyze ECs as they exist in their native state in tissue. Even in highly vascularized tissue, the endothelium represents only a very small percentage of the cells constituting each tissue. It appears critical therefore to extract EC samples from the tissue to have any chance to unmask EC molecular expression that otherwise is beyond the dynamic range of detection and is effectively "swamped out" by the much stronger signals from other cell types in the tissue and the nearly 100,000 other, more abundant, tissue proteins. ECs are not localized clumped in a region

1 The serum effect is the phenomenon by which adsorption of serum proteins within the glycocalyx of ECs renders the EC barrier impermeable to solute transport. E.g., ferritin (see Schneeberger and Hamelin. *AJP* 1984;247:206.)

of the tissue but rather exist as a very thin attenuated monolayer lining blood vessels disseminated throughout the organ, thereby making microdissection or isolation to obtain purified, uncontaminated material quite difficult. This in turn greatly hinders detection, identification, and purification of EC-specific proteins as well as any comparisons of endothelia between tissues.

Endothelial Cell Isolation

Given these realities, it is not surprising that scientists have attempted for nearly half a century to isolate ECs from the rest of the tissue/organ (26–28). As early as the 1920s, attempts were made to isolate ECs from tissue (29). With the discovery of reasonable EC marker proteins, such as angiotensin-converting enzyme (ACE) and von Willebrand factor (vWF), the first demonstrable and reproducible isolation of ECs from tissue occurred 35 years ago and constituted a major breakthrough in the field (27,30). The development of effective EC isolation techniques heralded a new era in vascular biology where, far from being a passive lining to the blood vessel, the endothelium was discovered to be an active participant in many physiological and pathological functions, including coagulation, angiogenesis, atherosclerosis, and inflammation. The ability to isolate and grow ECs in cell culture permitted many of the pathways involved in the various EC functions to be characterized in detail. ECs isolated from many different tissues showed some diversity in protein expression but yielded little tissue specificity (31).

Phenotypic Drift

Unfortunately, the successful isolation of ECs for growth in culture is not the sought-for panacea for obtaining material for mapping EC protein expression. ECs are highly plastic cells that clearly alter their phenotype according to microenvironmental cues. For example, avascular brain tissue transplanted into the coelomic cavity of chick embryos undergoes vascularization to form a competent blood–brain barrier; conversely, when avascular coelomic tissue was transplanted into the brain, the ECs that invade the mesenchymal tissue grafts form leaky capillaries and venules (32,33). ECs also change considerably when removed from their native tissue environment for isolation and growth in culture. The significant microenviromental perturbation contributes to morphologically obvious phenotypic drift and the loss of native functions as the cells de-differentiate and adapt their metabolism and protein expression to ex vivo and cell culture conditions (22,34,35). For example, the signaling molecule tumor necrosis factor (TNF)-α stimulates neovascularization in rabbit corneas whereas it inhibits proliferation of ECs grown in culture (36). Many distinctive characteristics found in vivo are lost, including expression of tissue-specific proteins and the usual abundance of caveolae, which decreases 30- to 100-fold in cultured ECs and even in ECs freshly and rapidly isolated from tissue (37–40). The rapidity of this change when isolating ECs from tissues is rather remarkable and may be indicative of the acute sensitivity and responsiveness of ECs to environmental cues in vivo.

Despite ample evidence for significant phenotypic change, the extent of the molecular chasm between ECs in vivo and in vitro is only beginning to be elucidated (37,39,40). Comparison of the cell surface proteome of ECs existing natively in vivo versus rat lung microvascular ECs (RLMVECs) grown under standard cell culture conditions has revealed significant differences in expression, including those proteins apparently induced by the tissue microenvironment (37). Of the 450 proteins identified in vivo, only 263 were found at all in vitro (58%), suggesting modulation by the unique tissue microenvironment not reproduced in cell culture (37). Out of 73 known EC marker proteins, 65 were found at the EC surface in rat lung versus only 32 in cultured RLMVECs. Forty-one known EC markers were detected only in rat lung.[2] Last, ~20 EC proteins showing tissue-restricted expression among endothelia were not detected in isolated ECs (37). Thus, the in vivo tissue microenvironment, which cannot yet be duplicated ex vivo, appears to significantly regulate distinct protein expression in ECs.

A number of biological mechanisms may account for differences in EC morphology and protein expression in vivo versus in vitro. Physical factors from hemodynamic forces can greatly influence EC phenotype, including induction and redistribution of protein expression with translocation from intracellular compartments to the cell surface (41). Cultured cells or even freshly isolated cells ex vivo lack the natural cues from the surrounding basement membrane, perivascular cells, and the tissue parenchyma to maintain expression of specific proteins. ECs interact with and respond to circulating hormones, carrier proteins, underlying perivascular cells, and the surrounding matrix of the tissue. During isolation from tissue, ECs are often subjected to harsh chemical treatment (digestion by proteolytic enzymes), mechanical disruption (mincing, mixing, centrifugation), and environmental (hypoxia) conditions. It is not surprising that this gross perturbation of ECs caused by tissue disruption can greatly alter molecular expression.

ANALYSIS OF ENDOTHELIAL CELL PROTEIN EXPRESSION AND TOPOGRAPHY

Antibodies, peptides, and ligands represent probes potentially useful for analyzing EC protein expression in cell culture and in vivo using a wide variety of techniques, including staining of tissue sections or cell monolayers, Western analysis, and whole body imaging after intravenous injection of the probe.

2 There were 73 total known EC markers. Some of these 73 proteins were identified in both the lung P and lung cultured cells; i.e. a subset of the 73 were identified in both samples; some proteins were unique to lung (not found in cultured cells) (41 of the 73 were unique to lung with 24 being identified also in cultured cells – 41 + 24 = 65) and some were unique to cultured cells (not found in the lung).

Proteins also can be analyzed indirectly through mapping of gene expression and mRNA levels using a wide variety of genomic techniques. The application of mass spectrometry (MS) to peptide sequence analysis has created a powerful means to identify directly proteins expressed by vascular ECs. Table 99-1 summarizes several direct and indirect methods to map the EC proteome (as discussed later) and describes some of their strengths and weaknesses. Analysis in vivo or in situ is greatly preferred but technically burdensome.

Cell- and Ligand-Based Approaches

Some of the first indications that EC diversity existed came from incubating cultured ECs under different conditions with a wide variety of protein ligands as well as tumor, immune, or other cells (42). Possible differences in cell surface expression were readily evident. In fact, after laborious protein purification or expression cloning, many very important EC cell adhesion molecules were identified through approaches pioneered by Gimbrone, Pober, Bevilacqua, Auerbach and others (43–47). Albumin binding proteins and important receptors to a wide variety of growth factors (e.g., vascular endothelial growth factor [VEGF], platelet-derived growth factor [PDGF], transforming growth factor [TGF]-β) ultimately have been identified by ligand-based searches of cultured ECs (18,48–53). Intravenous injection of cells[3] and ligands also has confirmed EC heterogeneity in vivo but such an approach has yielded little in the way of new molecular discoveries (54).

Monoclonal Antibody–Based Approaches

Over the last quarter century, various attempts to map EC protein expression utilized isolated ECs and their membranes or cell lysates as immunogens in mice for monoclonal antibody production (30,31,55–61). These early pioneering studies provided new antibodies to ECs and the first direct evidence for differential protein expression for endothelium from different organs. They also uncovered useful markers for comparative and functional studies of ECs but ultimately failed to yield tissue-specific antibodies or target proteins useful for targeting a single organ selectively. Such an outcome is not surprising given that organ-associated EC phenotypes are largely dependent on the tissue microenvironment (62). The hybridoma-based technology is also inherently limited by its ability to screen only a small portion of the immune repertoire and by its reduced ability to generate antibodies directed to self-antigens or to highly conserved epitopes.

Phage-Based Techniques

The concept of phage display was developed initially by the pioneering work of George Smith, who first demonstrated the ability of filamentous bacteriophage to display at their outer surface foreign polypeptides fused to a coat protein, either g3p or g8p (63). Since then, phage display has proven to be a valuable tool in the display of small polypeptides as well as large proteins. The most significant accomplishment of phage display in recent years is its success in displaying antibody fragments and manipulating large antibody libraries for the generation of monoclonal antibodies (64,65). Unlike antibodies produced from mice using hybridoma technology, phage display permits a very large repertoire of antibodies to be displayed in a library created and analyzed in vitro without necessarily a need for immunization. After successive rounds of selection and amplification of a phage library against a target, the few clones that bind to the target can be rapidly identified. With phage libraries, there is no more need for laboratory animals, human antibodies can be generated directly for medical use, and antibodies with novel specificities such as antibodies to self-antigens or to highly conserved epitopes can be generated.

Many variations of phage display screening have been used to identify EC surface proteins. Screening phage libraries by in vitro bacterial panning (66) or after intravenous injection to search for tissue-homing peptides has been reported (67–69), but so far the yield has been rather modest with few promising tissue-specific targets or validated targeting probes. Reported targeting of adipose tissue with a proapoptotic peptide that binds to prohibitin resulted in selective ablation of white fat (70). Doxorubicin conjugated to RGD- and NGR-containing peptides bind integrins and can substantially improve the therapeutic index of this common chemotherapeutic agent (68). Animal survival increased many months, consistent with tumor regression and fewer metastases seen in the treated mice. Other peptides apparently targeting prostate, brain, and kidney tissue have been identified by in vivo panning of peptide-phage display libraries (67,69). Typically, when peptides are injected intravenously, only 0.2% or less of the injected dose of organ- and tumor-specific peptides is found in the target tissue (67,68). In one study, tumor accumulation was only 2- to 10-fold more than in the control brain tissue whose highly restrictive endothelium should greatly minimize any nonspecific uptake (68). Unfortunately, most of the vascular targets recognizing these peptides remain elusive, probably because such small peptides lack the specificity and affinity necessary for purifying a single target molecule (71). The mechanism responsible for the reported beneficial responses with peptides and peptide-drug conjugates remains unclear (71), in part because, when known, the targeted proteins can be found on multiple endothelia and other cell types (38).

The in vivo panning strategy appears limited inherently by the rather poor binding capacity and specificity of short peptides as well as the difficulty in achieving intravascular conditions needed to delineate large libraries. The rapid transit of blood through tissue vessels and the rapid scavenging of intravenously injected phage by the liver and spleen (reticuloendothelial system) within seconds to minutes works against phage selection by in vivo panning. Unlike classical screening where phage are incubated with an isolated protein or even

3 For example, B16 murine melanoma cells injected IV into normal mice selectively colonize the lung suggesting that there is something present in the lung EC that permits binding of the B16 cells to the EC surface.

Table 99-1: Techniques to Study Protein Expression and Topography of ECs

Indirect Methods I

- Uses and/or generates probes to putative EC molecules that can then be used to assess specificity of expression and then to identify molecular interactions with a probe (i.e., antibody–antigen).
- Requires identification of probe interaction partner, which can be difficult for some probes such as peptides that can interact with multiple proteins.

Technique	*Advantages*	*Disadvantages*
Cell- and ligand-based approaches	1. Can identify useful EC markers	1. Laborious process to identify cell-surface binding proteins 2. Low throughput
Hybridoma technology	1. Can identify useful EC markers 2. Readily generates probes for comparative and functional EC studies	1. Target protein not readily identified 2. Cannot generate antibodies directed to self-antigens or to highly conserved epitopes 3. Can only screen small portion of immune repertoire
Phage display libraries	1. Allows rapid screening of large, randomly paired antibody light and heavy chain libraries 2. Readily generates probes for comparative and functional EC studies 3. Can screen naïve or host immune response repertoire	1. Clones often of low affinity due to loss of original light and heavy chain pairing, structural changes brought about by connecting linkers 2. Optimal conditions known for classic screening on cells and proteins in vitro
In vivo panning of phage or viral libraries (peptide or antibody)	1. Allows rapid screening of large peptide combinatorial libraries 2. Readily generates peptides for comparative and functional EC studies	1. Limited by poor binding capacity of peptides 2. Short peptides with low specificity 3. Phage rapidly removed from blood by the reticulo-endothelial system (RES) 4. Glycocalyx and hydrodynamic forces in vivo may hinder access and prevent equilibrium binding to EC surface in vivo 5. scFv have lower affinity than whole IgG

Indirect Methods II

- Identifies mRNAs to putative EC molecules. mRNA levels can be compared to other tissues to determine induction or downregulation
- Requires generation of probes for use in validating specificity of expression and targetability

Technique	*Advantages*	*Disadvantages*
Genomic analysis	1. Allows large-scale, quantitative, high-throughput molecular profiling of ECs 2. Readily detect alterations in gene expression among various tissues or under different conditions 3. Human genome largely complete for straightforward sequence identification	1. Large number of genes identified requires laborious validation 2. Provides only indirect analysis of protein expression; lacks information regarding abundance, localization, or post-translational modification

Direct Methods

- Analyzes protein expression directly
- Requires validation usually through generation of new specific probes (antibody, peptide, oligonucleotide, DNA) for testing expression in multiple cells and tissues.

Technique	*Advantages*	*Disadvantages*
MS analysis	1. Allows large-scale, high-throughput molecular profiling of ECs 2. Readily compare protein expression across tissues 3. Provides direct analysis of protein expression and post-translational modification	1. Large number of proteins identified requires laborious validation 2. Dynamic range limitations may reduce identifications 3. Quantitative analysis currently is challenging
Protein arrays (antibodies, ligands, other)	1. Can be large-scale and high-throughput 2. Rapid screening 3. Sensitive and highly differential screen	1. Requires very specific probes 2. Ultimately need probe for each distinct protein (>100,000) 3. Protein interaction may interfere to yield false-negatives 4. Depends on sample screened for EC specificity

cells in culture for sufficient time to allow equilibrium binding, in vivo most of the phage (>90%) are removed before having sufficient time "to experience" and reach equilibrium binding to the EC surface. Moreover, large carriers, such as phage or viral vectors, will experience significant streamlining to concentrate in a central coaxial distribution profile in the blood vessel tube. This effectively reduces interactions with luminal EC surfaces and probably increases the need for longer circulation times to achieve binding. In addition, the glycocalyx can function as a barrier to larger entities from binding the ECs resulting in restriction of transendothelial transport (72). Finally, validation of candidate targeting peptides has been difficult with an over-reliance on tissue staining techniques, which, for reasons discussed later, are difficult to perform in a quantitative or qualitatively rigorous manner. Further technical development is clearly needed to harness more effectively the power of phage display in creating new EC binding probes and in allowing more comprehensive mapping of endothelium in vivo.

Not surprisingly, very recent approaches from the labs pioneering this technology have begun to include several rounds of in vitro prescreening with cells or tissue components to enrich the sublibrary before in vivo panning (73,74). Another more recent phage-based strategy has been described to rapidly identify not only probes but also their cognate antigens for targeting vascular endothelia of specific organs in vivo (75). Phage libraries are differentially screened against luminal EC plasma membranes (isolated directly from tissue) by in vitro phage screening techniques using standard, well-controlled screening conditions more akin to the classic phage approaches of the past (75). Numerous phage clones reactive to EC surface determinants in lung have been identified and rapidly validated by in vivo molecular imaging. Unfortunately, rapid liver and spleen uptake has been observed for all phage tested within seconds of intravenous injection, thereby precluding validation efforts of phage targeting in vivo. To avoid this, scFv were cloned from the phage to produce minibodies that showed rapid and selective targeting after intravenous injection. Finally, the high degree of antibody specificities and affinities permit rapid and immediate identification of organ-selective endothelial target antigens by MS analysis of immunoprecipitates and Western analysis. Such novel, comprehensive strategies may provide new promise for harnessing the power of phage display for mapping vascular endothelia natively in tissue and for achieving vascular targeting of specific tissues in vivo.

Mass Spectrometry–Based Analysis

Two-dimensional (2D) gels have a long history of use in comparing protein expression across tissues but require large amounts of starting material and are time consuming and labor intensive to run (76). In addition, separation of hydrophobic, highly charged, or transmembrane proteins as well as intergel reproducibility can be problematic. High-throughput analysis can be achieved through the use of robotic gel analysis and spot-picking coupled to automated sample preparation and MS loading technology, which reduces error and radically decreases the amount of hands-on time for technical personnel. Multidimensional HPLC protein separation and identification technologies (77) coupled to database searching have been developed to analyze samples with significant diversity (>1,000 proteins). These effective MS-based "shotgun" approaches to protein identification (78) may greatly speed the discovery process by obviating the need to apply time-consuming and labor-intensive gel-based separation techniques. Various techniques, such as isotope-coded affinity tags (ICAT) (79), have been elaborated to achieve accurate quantification and sequence identification of peptides within a complex mixture.

For rapid and reproducible profiling, protein arrays are being developed that lend themselves well to high-throughput screening (80). Protein arrays are analogous to DNA arrays in that antibodies, small molecules, phage, or protein baits are immobilized on a slide that is then incubated with tagged sample proteins to estimate the concentration of specific proteins in the original sample. The success of these arrays is largely dependent on the affinity and specificity of antibody or protein probes used. In addition, the reactivity of the probe to the target protein may be decreased or abolished upon protein conformational change, protein interactions, or aggregation.

Most cell types, including ECs, are likely to express >10,000 genes and contain >100,000 distinct proteins. In this regard, shotgun MS approaches are probably better than 2D gel-based approaches, but to date no technology can effectively handle this level of molecular diversity and complexity to identify all such proteins in a cell. Multiple studies have been published describing a somewhat limited but direct proteomic analysis of ECs. Two MS analyses of whole cultured ECs (human umbilical vein ECs and a cell line designated RHE) subjected to 2D gel analysis recently identified a few proteins (<60 all together) and no cell surface enzymes or transmembrane proteins (81,82). A third study used shotgun MS approaches to identify 450 proteins (using stringent identification criteria of ≥3 peptides) in luminal EC surface membranes isolated from rat lungs using the silica-coating technique (described in Chapter 73) (37). This included ~80 known EC markers and a wide variety of protein types such as transmembrane, enzymes, signaling molecules, and cytoskeletal elements. About 60% of these proteins expressed in vivo also were detected in rat lung ECs in vitro, consistent with modulation of expression by the tissue microenvironment not reproduced ex vivo or in cell culture (37).

Various chemical techniques that enzymatically label proteins have been applied to selectively radiolabel cell surface proteins of cells in culture (83–87) or on the EC surface of blood vessels in situ (86–89). Enzymatically driven intravascular radioiodination of rat heart endothelium in situ is quite cumbersome and difficult because it requires perfusing isolated rat hearts for 15 minutes with ≥10 mCi of ^{125}I and microspheres covalently coated with lactoperoxidase and glucose peroxidase (86). Although useful in many ways, including evaluating the carbohydrate nature of cell surface glycoproteins by lectin analysis (84,86) or defining the presence of a protein or a ligand-receptor interaction by affinity chromatography using specific antibodies and ligands (e.g., albumin-binding proteins) (84,85), direct global isolation of the radiolabeled EC surface proteins away from the more abundant other tissue proteins is quite difficult.

Chemical biotinylation using lipid-impermeable hydrophilic labeling agents improved comparative mapping of the EC surface and was first done in 1987 in vitro (90) and then 10 years later in vivo (91). One approach entails direct detection of the biotinylated proteins by avidin blotting after protein separation by SDS-PAGE. Perhaps a better strategy is to first perform avidin-based chromatography before separating the proteins in the eluate by SDS-PAGE for protein staining to detect all proteins or avidin blotting to visualize only biotinylated proteins. With the development of MS techniques to identify proteins, these biotinylation techniques have been used again nearly two decades later, this time to identify many of the biotinylated proteins in cell culture (92) and by perfusion in situ (93). Unfortunately, because chemical reagents such as biotin can easily and rapidly extravasate into the tissue, this technique must be applied with caution for the mapping of endothelium in tissue and with the understanding that many proteins on non-EC tissue cells are accessible for labeling. It becomes difficult to prolong the reaction time necessary to permit uniform chemical conjugation of all EC surface proteins yet limit reagent permeation into the tissue (which occurs within seconds). Thus, such chemical labeling in situ in fact extends significantly beyond just EC surface proteins to proteins of other cells and extracellular matrix further inside the tissue. This is readily evident by avidin staining of tissue sections of in situ biotinylated organs and by some of the biotinylated proteins identified that related to non-ECs in the tissue – e.g., cadherin-16, which is selectively expressed in epithelial cells in the adult kidney (93). Such in vivo biotinylation is based on the enhanced statistical probability for labeling luminal EC surface proteins over other tissue proteins and ultimately, for EC analysis, is clearly superior to starting with all proteins in the tissue homogenate. Lastly, both radioiodination and biotinylation techniques only label a small subset of EC proteins.

Modern "omic" analysis techniques are being utilized in attempts to identify all proteins and genes expressed within a given tissue or subset of cells within a tissue often for the purpose of discovering new drug targets. Genomic and proteomic approaches can complement each other to facilitate the generation of accurate maps of total protein/gene expression of any given cell type or tissue. Proteomic analysis offers an important advantage over DNA microarray or other genome technologies because it allows the preferential identification of gene products specifically expressed in the tissue or cell of interest. Despite many strengths, including improved precision and high throughput, a major weakness of exclusively using genomic approaches for mapping protein expression is that mRNA analysis provides only an indirect analysis of proteins as gene products. Transcriptional activity does not necessarily correspond to the abundance, structure, or activity of its protein product for various reasons, including molecular half-life of mRNA versus protein product, synthesis rate, multigenic and epigenetic phenomena, as well as environmental influences, such as cell cycle, stress, and disease (94–96). Direct genomic identification seems unlikely for proteins not simply induced in expression but rather becoming new disease-induced targets primarily through post-translational modification (i.e., phosphorylation, sulfation, glycosylation, proteolytic cleavage) and/or translocation/redistribution (i.e., from poorly accessible intracellular cytosolic or membrane compartments to the outside cell surface). Thus, the importance of analyzing expression directly at the protein level is being more widely recognized, and various techniques have been developed to achieve this type of analysis. Also defining protein expression at the subcellular level is required for context, function, and targeting purposes.

VASCULAR TARGETING

A major goal of modern medical research is the discovery of new targets for detecting disease through molecular imaging (97,98) and for treating disease through directed delivery in vivo (99,100). With the completion of the human genome project, it is estimated that the human genome encodes 25,000 genes. However, because of extensive RNA processing, post-translational modification, and the formation of functional protein complexes, this gene pool expands in size to several hundred thousand to perhaps a million possible distinct proteins (101). Past genomic and proteomic analyses of samples from normal and diseased tissues have yielded thousands of candidate drug targets for diagnostic as well as potential therapeutic targeting (99,100). Yet the sheer number of candidates appears to have overwhelmed the required but time-consuming in vivo validation process (100,101) so that the dilemma in drug target discovery today has become identifying which few of these many candidates are the most meaningful and targetable in treating and imaging disease.

Unfortunately, the potential value of identifying tissue-specific targets is offset by the difficulty in gaining access to that tissue owing to the endothelial barrier (15,102). Most known tissue- and tumor-associated gene products are expressed by the cells primarily constituting the tissue (e.g., tumor cells, cardiomyocytes, hepatocytes, alveolar epithelial cells) and, as such, these potential targets remain inside tissue compartments that are poorly accessible to many agents injected into the circulation, including gene vectors, nanoparticles, and even most biological agents, such as antibodies. This lack of tissue delivery and penetration prevents effective therapy in vivo. It has become apparent that new analytical strategies are required to reduce data complexity and to focus the power of global identification technologies ("omics") to permit rapid discovery and validation of truly accessible targets in vivo.

Vascular targeting has been proposed as an alternative strategy for site-directed delivery that bypasses many of the problems associated with direct tissue/cell targeting. This approach diverts analytical focus from the poorly accessible surface of cells inside the target tissue to the luminal EC surface of blood vessels feeding the tissue/tumor. This interface is directly exposed, and thus inherently accessible, to agents circulating in the blood (103,104). The strategy is predicated on the existence of luminal EC surface proteins expressed selectively in one tissue to the extent necessary to achieve its goal. Vascular targeting of antibodies may be applied not only to the delivery of drugs for therapy but also to imaging where targeting antibodies conjugated to imaging agents may provide useful diagnostic and prognostic tools for detection and assessment of primary and metastatic lesions. But to be useful for many drug-delivery applications, this strategy still requires a means by which to cross the vascular wall for access to underlying tissue cells (15,22,71,105). A new aspect to the vascular targeting strategy (discussed in detail in Chapter 73) is the potential utility of the caveolae transcytosis pathway in endothelium for selectively overcoming this key barrier to facilitate tissue-directed delivery to underlying tissue cells (15,71,104).

To reach its potential, vascular targeting requires the discovery of tissue-specific (i.e., vascular bed–specific), accessible endothelial targets. Many attempts have been made to identify such targets and develop probes (56,67–70,106–109) but, so far, without technologies for comprehensive analysis, most directed delivery in vivo has not met theoretical expectations (71). A lung-targeting antibody has been reported (106), but the antigen was subsequently found to be thrombomodulin (110), which is expressed by many different cell types, including the EC surface of most organs. Immunotargeting the panendothelial marker, platelet-endothelial cell adhesion molecule (PECAM)-1, can improve delivery to the lung after IV injection from the usual <1% to nearly 4% of the injected dose (2- to 4-fold over liver delivery) but only when the antibody is biotinylated and complexed with streptavidin (111). Although ACE is expressed on certain epithelial cells, macrophages, and ECs, some antibodies to ACE show 10- to 20-fold greater targeting of lung than other organs and blood (112) and have been used to enhance drug delivery to the lung (113) as well as lung gene delivery and expression using adenoviral vectors (114). Tissue-homing peptides, discovered primarily via in vivo panning of phage display libraries, also have provided a modest increase in tissue delivery (67), but the levels of targeting were <1% of the injected dose and many times were just compared to delivery to brain with relative targeting indices of 6- to 20-fold. Brain uptake in vivo is notoriously low because its endothelium has very tight junctions and few caveolae, both contributing to a high resistance to even very small molecules.

VEGF and its receptors are currently being exploited as sites for tumor imaging and/or therapy (8). Antibodies to VEGF injected intravenously in tumor-bearing rodents accumulates preferentially within neoplastic tissue (115). VEGF conjugated to toxin is also effective in reducing tumor size in rodents (116). Unfortunately, because VEGF and its receptors are expressed in various normal human tissues, such as kidney, it is unclear how selective these targets will be clinically. When used with chemotherapy, Bevacizumab (Avastin; a humanized murine mAb to VEGF) has shown promise by extending median survival time by about 8 months in patients with metastatic colon cancer and 2 to 3 months in patients with non–small cell lung cancer (117,118). Thrombotic complications are evident but may be avoided through exclusion of high-risk patients.

Table 99-2 lists the tissue distribution of these and other potential tumor EC targets currently being investigated for tumor immunotherapies. They include roundabout 4 (Robo4) expressed at sites of active angiogenesis in tumor ECs, in adults (119). Robo4 is an EC-specific member of the roundabout receptor family that binds slit to inhibit EC migration in primary EC cultures (120). It has a critical role in embryonic angiogenesis, is generally absent in normal adult ECs, and is upregulated in ECs of various solid tumors (120–123). Delta 4 is a Notch ligand located in the plasma membrane that is induced by hypoxia on the endothelium at sites of active angiogenesis, including tumor ECs (124). Nucleolin is a well-known nuclear protein that can be expressed on tumor cells and on the EC surface in tumor blood vessels but not generally normal vessels (38,125). Intravenous injection of homing peptides or nucleolin antibodies accumulate in tumor vessels as well as in angiogenic vessels of implanted "matrigel" plugs (125). CD44 is a highly heterogeneous cell surface receptor with specific variants highly expressed in tumor ECs as well as in other tissues (126). A monoclonal antibody recognizing a widely distributed isoform CD44H conjugated to chemotherapeutic agents has shown promise in suppressing solid tumor growth in rats (127,128) and mice (129).

Endoglin is a RGD-containing glycoprotein (130) that is a component of the TGF-β receptor system (131). Although endoglin is expressed by a variety of different cells and on the endothelia of many organs (132), endoglin antibodies conjugated to deglycosylated ricin A chain have shown strong antitumor and anti-angiogenic efficacy when injected into mice bearing human breast tumors (133,134).

Several integrins, including $\alpha v\beta 3$, $\alpha v\beta 5$, and $\alpha 5\beta 1$, may be induced by tumors and may be a useful target site for inhibiting the growth of tumors (135,136). Integrin $\alpha 5\beta 1$ may regulate the function of $\alpha v\beta 3$ during EC migration in vitro or angiogenesis in vivo (137). Antibodies, peptides, and antagonists of $\alpha 5\beta 1$ inhibit growth factor-induced angiogenesis and cause regression of human tumors in animal models (138). Antibodies recognizing $\alpha v\beta 3$ and RGD-peptides to αv integrins cause tumor regression by preventing normal engagement which induces apoptosis selectively in proliferating ECs (68,136). Recently nanoparticles targeting $\alpha v\beta 3$ integrins in the tumor vasculature have been shown to direct gene therapy to ablate tumors (135). Selective dual $\alpha v\beta 3$ and $\alpha v\beta 5$ antagonists inhibited angiogenesis in the chick chorioallantoic membrane assay and significantly inhibited tumor growth in mice bearing subcutaneous human melanoma tumors (139).

Table 99-2: EC Surface Proteins Showing Restricted Expression

Name	*Lung*	*Heart*	*Kidney*	*Liver*	*Other Normal Tissue*	*Solid Tumor*
Highly Expressed in Lung EC						
Aminopeptidase P	+++	—	—	—	—	—
Angiotensin converting enzyme	+++	+	+	+	—	ND
Aquaporin 1	++	+	+	—	ND	ND
Carbonic anhydrase	+++	++	+	+	—	ND
Dipeptidyl peptidase IV	++	—	—	+	+	ND
Endothelin converting enzyme	++	+	+	+	ND	ND
OX-45	+++	—	—	—	—	—
PV-1	+++	+	+++	+	+	+++
Receptor for advanced glycation end products	+++	+	—	+	ND	ND
Seven transmembrane receptor	++	++	+	++	ND	ND
Thrombomodulin	++	++	++	+	+	ND
Angiogenesis/tumor-induced						
AnnA1*	—	—	—	—	—	++
AnnA2	+	+	+	+	+	++
AnnA8	+	—	—	—	ND	++
APN/CD13	++	++	—	—	+	+++
CD44H	+	+	+	+	+	+++
CEA-CAM-1	+++	—	—	+	ND	++
Del1	—	—	—	—	—	++
Delta 4	+	+	+	+	+	++
EDB fibronectin†,‡	—	—	—	—	—	++
Endoglin	++	++	—	++	+	++
EphrinA5	++	++++	+	+	ND	+++
EphrinA7	+	++++	++	+	ND	++
Integrins ($\alpha v\beta 3$, $\alpha v\alpha 5$, $\alpha 5\beta 1$)*	+	+	+	+	+	+++
Myeloperoxidase	+	+	—	—	ND	++
Neuropilin	+	++++	—	++	ND	++
Nucleolin*	+	+++	—	—	ND	+++
PSMA	—	—	+	+	+	+++
Robo4	+	—	—	—	+	++
Stanniocalcin-1	+	+	+	—	+	+++
TEM 1, 5, 7, 8	+	—	+	+/—	+	+++
Tenascin C	—	—	—	—	++	+++
Tie2	+	—	+	+	ND	++
Transferrin receptor	+	++	—	++	+++	+++
VCAM	+	—	—	—	+	++
VEGFR1	+	—	+	—	+	++
VEGFR2	+	+	+	+	+	++
Vit D BP	—	—	—	++	ND	+++

*This expression profile is only for this protein as it is expressed at the surface on the outside of the EC (not elsewhere; e.g., nucleolin is expressed fairly ubiquitously in nuclei of all cells including ECs) and integrins expressed on outside of many cell types including blood cells, fibroblasts, and epithelial cells.

†May be regulated by physiological angiogenesis in endometrium and ovaries (wound healing?)

‡Unclear if expressed at the luminal EC surface or mostly in extracellular matrix.

ND; not determined.

Prostate-specific membrane antigen (PSMA), is upregulated in neoplastic prostate epithelium (108,140–142) and neovasculature (143–146) as well as in metastatic lesions (141, 146,147). Although its expression also is observed in brain, salivary gland, and intestine (108,140,146,148,149), antibodies to PSMA (^{111}In-7E11; ProstaScint) are currently being used as an imaging agent for detecting soft-tissue metastases in prostate cancer patients at high risk of metastatic disease, and studies are underway to determine potential utility in targeted therapy (150–153).

A 165 kDa sialoglycoprotein named endosialin is found in various carcinomas, sarcomas, neuroectodermal tumors but not lymphomas (154). Monoclonal antibodies to endosialin recognize endothelium in tumors but not in normal tissues (154). Endosialin was discovered independently to be expressed in tumor ECs by SAGE analysis of normal versus colon tumor ECs, which revealed several genes upregulated in the tumor vessels (109). Tumor endothelial marker 1 (TEM1) (eventually found to be endosialin), TEM5, TEM7, and TEM8 are all transmembrane proteins whose expression is highly upregulated in tumor ECs and/or in mouse embryonic vasculature (155–157). A splice variant of TEM8 has been identified as the anthrax toxin receptor (158). Re-engineered anthrax toxin that is activated by cell-surface urokinase has been shown to cause regression of subcutaneous tumors in mice (159).

Intercellular adhesion molecule (ICAM)-1 is an adhesion molecule expressed normally on ECs, where it functions in leukocyte attachment and migration to sites of inflammation (160). During tumor angiogenesis, cytokines produced by the tumor cells activate ECs to produce increased amounts of ICAM-1 and to shed the soluble form (sICAM)-1 into the blood (161). sICAM-1 stimulates EC migration, differentiation, and vessel sprouting in vitro as well as angiogenesis and tumor growth promotion in vivo (162,163). ICAM-1 may mediate EC migration via endothelial NO synthase (eNOS) activation and actin cytoskeleton reorganization (164). Serum sICAM-1 levels are elevated in cancer, and there is a positive correlation with tumor progression (165,166). In addition, a monoclonal antibody directed against another well-known cell adhesion molecule, vascular cell adhesion molecule (VCAM)-1, when conjugated to human tissue factor, selectively induces vascular thrombosis and inhibited tumor growth in mice bearing solid, human Hodgkin tumors (167).

Lastly, there are several proteins associated with select endothelia but not necessarily located at the luminal EC surface for immediate access to circulating targeting agents. Developmentally regulated endothelial locus (Del)-1 is an extracellular matrix protein expressed in ischemic tissue that stimulates angiogenesis (168). Del-1 contains two discoidin I–like domains and may act to promote adhesion of ECs through interaction with the $\alpha\nu\beta3$ integrin receptor (169). Stanniocalcin (STC)-1 is a secreted glycoprotein that acts with hepatocyte growth factor to regulate EC maturation (170). STC-1 is highly expressed in tumor blood vessels and may be an effective tumor vascular target (171). The EDB domain of fibronectin is highly expressed by the ECs of many tumor types but undetectable in normal adult tissues (172–175). Radioiodinated EDB antibodies have recently been tested in the clinic and have shown the ability to target tumors (176). The primary location of these proteins in the extracellular matrix adjacent to the endothelium may require passive leakage of targeting agents across the vascular wall, which ultimately may slow targeting. Tenascin C is a highly upregulated extracellular matrix protein in brain and lung tumors but undetectable in normal adult tissues (177). Antibodies recognizing tenascin C have been used for many years in the clinic for imaging tumors (178–182).

Applying Proteomic Mapping to Vascular Target Discovery

Most of these vascular targets are expressed by, or proximal to, ECs in tissue and, in fact, have been identified by studying isolated ECs and tissue sections with genetic-, ligand-, and/or antibody-based probes. More direct MS-based proteomic mapping of vascular ECs in vivo is desirable and may, at least theoretically, be even more fruitful, but it is technically complex and difficult. One approach to reduce the tremendous molecular complexity of the starting analyte (the tissue with its hundreds of thousands of proteins) and to rapidly and directly identify and validate accessible vascular targets is to integrate multiple technologies – including subcellular tissue fractionation, subtractive proteomics, in silico bioinformatics, expression profiling, in vivo biodistribution assay, and molecular imaging – into a comprehensive analytical system for determining the proteins expressed at the cellular site of interest (e.g., ECs in the tissue). To focus the power of proteomic technology to a small subset of tissue proteins, namely the accessible vascular targets (i.e., the proteins expressed on the EC surface directly exposed to the blood circulation), tissue subfractionation has been applied to isolate the pertinent luminal EC plasma membranes and thereby to promote the unmasking of a small subset of proteins on the EC surface from the several hundred thousand proteins in the tissue or individual but whole ECs. This magnitude of data complexity reduction at the beginning of the analysis greatly reduces the background of nonaccessible tissue targets, thus potentially decreasing the failure rate and the overall work necessary to find valid targets.

Such an integrated approach has recently been used to discover tissue-modulated proteins showing promise in tissue-specific imaging and pharmacodelivery (37,38) (Table 99-2). A subtractive comparison of proteins expressed in rat lung ECs in vivo versus in vitro in cultured cells has been performed to discover possible tissue-restricted EC surface proteins and even identify potential lung-specific EC proteins in vivo (37). The list of proteins identified in vivo using a stringent criteria of ≥ 5 peptides versus in vitro using minimal criteria of one peptide were subtracted in silico based on the working hypothesis that tissue-dependent EC protein expression may be reduced in cell culture, which does not duplicate conditions in vivo. The list of differentially expressed candidate proteins was reduced further by bioinformatic interrogation of structure, glycosylation, and membrane orientation to yield 11 proteins likely to have domains outside the cell and therefore targetable from the bloodstream. Expression profiling with specific antibodies revealed that each protein is indeed expressed on the EC plasma membrane and that each exhibits a different restricted expression pattern among the organs tested. None of these proteins is expressed equivalently on the EC surface membranes in each organ. Endothelia in many organs other than lung lack detectable expression. A distinct molecular signature for each vascular bed is quite apparent from the expression levels of these 11 EC marker proteins in each organ. Two proteins, aminopeptidase P and OX-45, appear specific for lung endothelium. Antibodies to aminopeptidase P show lung-exclusive targeting after intravenous injection. Antibodies to OX-45 do not because leucocytes express this protein also.

A similar subtractive approach has been used to compare diseased versus normal tissue to uncover potential targetable tumor-induced EC proteins (38). Fifteen differentially expressed proteins have been identified as being enriched on the EC surface in tumor vessels (AnnexinA1, AnnexinA8, ephrinA5, ephrinA7, myeloperoxidase, nucleolin, transferrin receptor, and vitamin D-binding protein), including several proteins already implicated in tumor angiogenesis (VEGF receptor [VEGFR]-1 and -2, Tie2, aminopeptidase-N, endoglin, C-CAM-1, neuropilin 1). Although clearly expressed or even induced on the tumor EC surface in vivo, nearly all of the proteins also are found expressed restrictively in normal endothelium of select major organs. One of the identified proteins was a 34 kDa protein recognized by AnnexinA1 (AnnA1) antibodies only on tumor endothelium. In vivo imaging using radioiodinated AnnA1 antibodies shows tumor-selective imaging and targeting within one hour of intravenous injection. This unprecedented rapidity in tumor targeting probably relates to its immediate accessibility at the luminal EC surface in tumors. Targeted radioimmunotherapy is effective with survival studies showing complete remission in 80% of the animals after a single treatment with radioiodinated AnnA1 antibodies versus 100% mortality within 7 days without treatment or when using control nontargeting antibodies (38). Thus, an integrated approach to proteomic analysis is capable of identifying a potential tumor-induced vascular target capable of directing immunotargeting of tumor neovasculature to achieve significant remission even in advanced disease.

Validation of EC expression and targeting in vivo is essential. In vivo imaging techniques, such as gamma scintigraphy, SPECT, PET, and luminescence imaging are fast becoming the new "gold standard" for noninvasively obtaining information pertaining to target organ selectivity, structure and organ function, and receptors and enzymes within normal and diseased tissues (97,98). Biodistribution analyses using radioactive probes specific for the target protein are laborious but definitive in showing the tissue distribution of the probe based on the limited sample pool. Other techniques, such as immunohistochemical analyses of tissue sections, lack the dynamic range and statistical power to quantify target expression and probe targeting in vivo after intravenous injection. They tend to yield false-negative results, especially when poorly performed and controlled. However, such techniques, including immunofluorescence and electron microscopy, can provide key information about ECs and tissue processing in vivo and thus should be used only after targeting conditions are established so that more detailed information on probe transport in tissue can be obtained. Of course, probe specificity must be established for the studies to be at all meaningful.

At this time, we know very little about how drugs, gene vectors, nanoparticles, antibodies, and other agents are processed by the endothelium of each tissue and then by the cells inside the tissue. Without this fundamental knowledge, it will be difficult to achieve, in a directed fashion, the desired goals of in vivo targeting. Nanomedicine, gene therapy, and even many standard drug and biologic therapies, such as antibodies, may be greatly limited. Development of new imaging technologies as well as advanced and optimized application of current imaging modalities, such as intravital and multiphoton microscopy, will also help accelerate this vision. Better probes are also much needed, both for sensitive in vivo detection and for achieving tissue targeting in vivo.

Before site-directed therapies can approach their theoretical potential, the extent of EC heterogeneity must be determined and potential tissue-specific EC protein targets must be identified (15,71,104). Table 99-2 lists several EC surface proteins as potential targets for specific organs and solid tumors. For additional useful, accessible vascular targets to be discovered, it will be necessary to systematically and comprehensively map and compare protein expression on the endothelium of major organs and tumors, so that tissue-modulated and even organ- or tumor-specific EC surface markers may be uncovered. The development and application of more robust technologies for both discovery and validation efforts appear to be needed to achieve such goals and concretely determine the true potential for vascular targeting.

Vascular targeting is likely to go beyond application to site-directed delivery and penetration into tissue. It may someday result not only in altering the EC phenotype in vivo, for instance to address a disease, but also the effect of endothelium on the underlying tissue cells. Directing signaling modifiers to specific endothelia is likely to yield tissue-specific responses, for instance, by targeting the endothelium to change its behavior in such a way that it alters signaling to underlying tissue.

CONCLUSION

Elucidating the molecular topography of the EC surface in multiple tissues constitutes the first step in gaining a better fundamental understanding of functional differences across organ systems and forms the basis for future studies into how environmental inputs alter EC gene/protein expression and physiology. This information is vital as well for tissue engineering where there is a critical lack of understanding of the properties of ECs in the context of normal tissues. By knowing EC expression in a given organ, researchers gain markers to discover the factors leading to a particular phenotype and thus will be able to recreate the tissue environment in culture with higher fidelity. Finally, defining this vascular proteome is expected to uncover new functions and to discover clinically significant tissue-modulated and possibly even organ-specific molecules that may be useful as targets for site-directed delivery of drugs, genes, or imaging agents. We are just at our infancy in defining molecular expression of ECs in vivo. The next decade should yield very significant progress toward defining the molecular topography and diversity of endothelium in many normal and diseased tissues. The process of induced and restricted expression creates much impetus for overcoming technical barriers to map comprehensively vascular endothelium in vivo.

KEY POINTS

- ECs lining the microvascular beds of different organs exhibit extensive morphological, ultrastructural, and molecular differentiation.
- The broad molecular heterogeneity of ECs contributes to numerous technical difficulties in characterizing their physiology and expression profile but also provides the basis for discovering novel vascular targets for site-directed delivery of functional and molecular imaging agents, nanomedicines, gene vectors, and drugs.
- Numerous approaches have been used to attempt to identify proteins directly that are expressed by ECs, most recently through the use of mass spectrometric–based techniques to analyze peptide sequences exposed on the luminal EC surface.
- Vascular targeting focuses on targeting proteins on the luminal EC surface of blood vessels feeding the tissue/tumor that is directly exposed and thus inherently accessible to agents circulating in the blood and bypasses many of the problems associated with direct tissue/cell targeting.
- Proteomic analysis of luminal EC plasma membrane proteins has identified lung- and tumor-induced proteins that can be targeted by monoclonal antibodies for rapid and efficient accumulation into the tissue/tumor.

REFERENCES

1 Schnitzer JE, Oh P. Aquaporin is found on the endothelial cell surface and in caveolae: inhibition of water transport in rat lung in situ. *Microcirculation.* 1995;6:A402.
2 Bruns RR, Palade GE. Studies on blood capillaries. I. General organization of blood capillaries in muscle. *J Cell Biol.* 1968;37:244–276.
3 Johansson BR. Size and distribution of endothelial plasmalemmal vesicles in consecutive segments of the microvasculature in cat skeletal muscle. *Microvasc Res.* 1979;17:107–117.
4 Simionescu M, Simionescu N, Palade GE. Morphometric data on the endothelium of blood capillaries. *J Cell Biol.* 1974;60:128–152.
5 Maher F, Vannucci SJ, Simpson IA. Glucose transporter proteins in brain. *FASEB J.* 1994;8:1003–1011.
6 Pardridge WM. Blood-brain barrier carrier-mediated transport and brain metabolism of amino acids. *Neurochem Res.* 1998;23:635–644.
7 Blood CH, Zetter BR. Tumor interaction with the vasculature: angiogenesis and tumor metastasis. *Biochim Biophys Acta.* 1990;1032:89–118.
8 Dvorak HF, Nagy JA, Dvorak AM. Structure of solid tumors and their vasculature: implications for therapy with monoclonal antibodies. *Cancer Cells.* 1991;3:77–85.
9 Hobson B, Denekamp J. Endothelial proliferation in tumours and normal tissues: continuous labelling studies. *Br J Cancer.* 1984;49:405–413.
10 Huxley VH, Curry FE. Albumin modulation of capillary permeability: test of an adsorption mechanism. *Am J Physiol.* 1985;248:H264–H273.
11 Wagner RC, Chen S-C. Transcapillary transport of solute by the endothelial vesicular system: evidence from thin serial section analysis. *Microvasc Res.* 1991;42:139–150.
12 Ghitescu L, Bendayan M. Transendothelial transport of serum albumin: a quantitative immunocytochemical study. *J Cell Biol.* 1992;117:745–755.
13 Milici AJ, Watrous NE, Stukenbrok H, Palade GE. Transcytosis of albumin in capillary endothelium. *J Cell Biol.* 1987;105:2603–2612.
14 Palade GE. Transport in quanta across the endothelium of blood capillaries. *Anat Rec.* 1960;136:254.
15 Schnitzer JE. Caveolae: from basic trafficking mechanisms to targeting transcytosis for tissue-specific drug and gene delivery in vivo. *Adv Drug Deliv Rev.* 2001;49:265–280.
16 Curry FE. Effect of albumin on the structure of the molecular filter at the capillary wall. *Fed Proc.* 1985;44:2610–2613.
17 Haraldsson B, Rippe B. Orosomucoid as one of the serum components contributing to normal permselectivity in rat skeletal muscle. *Acta Physiol Scand.* 1987;129:127–135.
18 Schneeberger EE, Hamelin M. Interaction of serum proteins with lung endothelial glycocalyx: its effect on endothelial permeability. *Am J Physiol.* 1984;247:H206–H217.
19 Schnitzer JE. Update on the cellular and molecular basis of capillary permeability. *Trends Cardiovasc Med.* 1993;3:124–130.
20 Michel CC, Phillips ME, Turner MR. The effects of native and modified bovine serum albumin on the permeability of frog mesenteric capillaries. *J Physiol Lond.* 1985;360:333–346.
21 Levick JR, Michel CC. The effect of bovine albumin on the permeability of frog mesenteric capillaries. *Q J Exp Physiol Cogn Med Sci.* 1973;58:87–97.
22 Schnitzer JE. The endothelial cell surface and caveolae in health and disease. In: Born GVR, Schwartz CJ, eds. *Vascular Endothelium: Physiology, Pathology, and Therapeutic Opportunities.* Stuttgart: Schattauer; 1997:77–95.

23 Huxley VH, Curry FE. Differential actions of albumin and plasma on capillary solute permeability. *Am J Physiol.* 1991;260: H1645–H1654.
24 Schnitzer JE. Glycocalyx electrostatic potential profile analysis: ion, pH, steric, and charge effects. *Yale J Biol Med.* 1988;61:427–446.
25 Schnitzer JE. Analysis of steric partition behavior of molecules in membranes using statistical physics. Application to gel chromatography and electrophoresis. *Biophys J.* 1988;54:1065–1076.
26 Folkman J, Haudenschild CC, Zetter BR. Long-term culture of capillary endothelial cells. *Proc Natl Acad Sci USA.* 1979;76: 5217–5221.
27 Jaffe EA, Nachman RL, Becker CG, Minick CR. Culture of human endothelial cells derived from umbilical veins. Identification by morphologic and immunologic criteria. *J Clin Invest.* 1973;52:2745–2756.
28 Gimbrone MA Jr., Cotran RS, Folkman J. Human vascular endothelial cells in culture. Growth and DNA synthesis. *J Cell Biol.* 1974;60:673–684.
29 Lewis WH. Endothelium in tissue cultures. *Am J Anat.* 1922;30: 39–60.
30 Auerbach R, Alby L, Grieves J, et al. Monoclonal antibody against angiotensin-converting enzyme: its use as a marker for murine, bovine, and human endothelial cells. *Proc Natl Acad Sci USA.* 1982;79:7891–7895.
31 Auerbach R, Alby L, Morrissey LW, et al. Expression of organ-specific antigens on capillary endothelial cells. *Microvasc Res.* 1985;29:401–411.
32 Janzer RC, Raff MC. Astrocytes induce blood-brain barrier properties in endothelial cells. *Nature.* 1987;325:253–257.
33 Stewart PA, Wiley MJ. Developing nervous tissue induces formation of blood-brain barrier characteristics in invading endothelial cells: a study using quail-chick transplantation chimeras. *Develop Biol.* 1981;84:183–192.
34 Thum T, Haverich A, Borlak J. Cellular dedifferentiation of endothelium is linked to activation and silencing of certain nuclear transcription factors: implications for endothelial dysfunction and vascular biology. *FASEB J.* 2000;14:740–751.
35 Madri JA, Williams SK. Capillary endothelial cell culture: phenotype modulation by matrix components. *J Cell Biol.* 1983;97: 153–165.
36 Frater-Schroder M, Risau W, Hallmann R, et al. Tumor necrosis factor type alpha, a potent inhibitor of endothelial cell growth in vitro, is angiogenic in vivo. *Proc Natl Acad Sci USA.* 1987;84:5277–5281.
37 Durr E, Yu J, Krasinska KM, et al. Direct proteomic mapping of the lung microvascular endothelial cell surface in vivo and in cell culture. *Nat Biotechnol.* 2004;22:985–992.
38 Oh P, Li Y, Yu J, et al. Subtractive proteomic mapping of the endothelial surface in lung and solid tumours for tissue-specific therapy. *Nature.* 2004;429:629–635.
39 Weidner N, Folkman J, Pozza F, et al. Tumor angiogenesis: a new significant and prognostic indicator in invasive breast cancer. *J Natl Cancer Inst.* 1992;84:1875–1877.
40 Schnitzer JE. Transport functions of the glycocalyx, specific proteins, and caveolae in endothelium. In: Bassingthwaite J, Goresky CA, Linehan JH, eds. *Capillary Permeation, Cellular Transport and Reaction Kinetics.* London: Oxford Press; 1997: 31–69.
41 Rizzo V, Morton C, DePaola N, et al. Recruitment of endothelial caveolae into mechanotransduction pathways by flow conditioning in vitro. *Am J Physiol Heart Circ Physiol.* 2003;285: H1720–H1729.
42 Kramer RH, Nicolson GL. Interactions of tumor cells with vascular endothelial cell monolayers: a model for metastatic invasion. *Proc Natl Acad Sci USA.* 1979;76:5704–5708.
43 Alby L, Auerbach R. Differential adhesion of tumor cells to capillary endothelial cells in vitro. *Proc Natl Acad Sci USA.* 1984;81:5739–5743.
44 Bevilacqua MP, Stengelin S, Gimbrone MA, Seed B. Endothelial leukocyte adhesion molecule 1: an inducible receptor for neutrophils related to complement regulatory proteins and lectins. *Science.* 1989;243:1160–1164.
45 Bevilacqua MP, Pober JS, Wheeler ME, et al. Interleukin-1 activation of vascular endothelium. Effects on procoagulant activity and leukocyte adhesion. *Am J Pathol.* 1985;121:394–403.
46 Pober JS, Bevilacqua MP, Mendrick DL, et al. Two distinct monokines, interleukin 1 and tumor necrosis factor, each independently induce biosynthesis and transient expression of the same antigen on the surface of cultured human vascular endothelial cells. *J Immunol.* 1986;136:1680–1687.
47 Rice GE, Gimbrone MA Jr., Bevilacqua MP. Tumor cell-endothelial interactions. Increased adhesion of human melanoma cells to activated vascular endothelium. *Am J Pathol.* 1988; 133:204–210.
48 Schnitzer JE, Carley WW, Palade GE. Specific albumin binding to microvascular endothelium in culture. *Am J Physiol.* 1988;254:H425–H437.
49 Myoken Y, Kayada Y, Okamoto T, et al. Vascular endothelial cell growth factor (VEGF) produced by A-431 human epidermoid carcinoma cells and identification of VEGF membrane binding sites. *Proc Natl Acad Sci USA.* 1991;88:5819–5823.
50 Vaisman N, Gospodarowicz D, Neufeld G. Characterization of the receptors for vascular endothelial growth factor. *J Biol Chem.* 1990;265:19461–19466.
51 Glenn K, Bowen-Pope DF, Ross R. Platelet-derived growth factor. III. Identification of a platelet-derived growth factor receptor by affinity labeling. *J Biol Chem.* 1982;257:5172–5176.
52 Merwin JR, Newman W, Beall LD, et al. Vascular cells respond differentially to transforming growth factors beta 1 and beta 2 in vitro. *Am J Pathol.* 1991;138:37–51.
53 Hirai R, Kaji K. Transforming growth factor beta 1-specific binding proteins on human vascular endothelial cells. *Exp Cell Res.* 1992;201:119–125.
54 Nicolson GL, Custead SE. Tumor metastasis is not due to adaptation of cells to a new organ environment. *Science.* 1982;215:176–178.
55 Goerdt S, Steckel F, Schulze-Osthoff K, et al. Characterization and differential expression of an endothelial cell- specific surface antigen in continuous and sinusoidal endothelial, in skin vascular lesions and in vitro. *Exp Cell Biol.* 1989;57:185–192.
56 Hagemeier HH, Vollmer E, Goerdt S, et al. A monoclonal antibody reacting with endothelial cells of budding vessels in tumors and inflammatory tissues, and non-reactive with normal adult tissues. *Int J Cancer.* 1986;38:481–488.
57 Kaplan KL, Weber D, Cook P, et al. Monoclonal antibodies to E92, an endothelial cell surface antigen. *Arteriosclerosis.* 1983;3:403–412.
58 Koch AE, Nickoloff BJ, Holgersson J, et al. 4A11, a monoclonal antibody recognizing a novel antigen expressed on aberrant vascular endothelium. Upregulation in an in vivo model of contact dermatitis. *Am J Pathol.* 1994;144:244–259.

59 Hamburger AW, Reid YA, Pelle BA, et al. Isolation and characterization of monoclonal antibodies reactive with endothelial cells. *Tissue Cell.* 1985;17:451–459.
60 Zhu DZ, Pauli BU. Generation of monoclonal antibodies directed against organ-specific endothelial cell surface determinants. *J Histochem Cytochem.* 1991;39:1137–1142.
61 Michalak T, White FP, Gard AL, Dutton GR. A monoclonal antibody to the endothelium of rat brain microvessels. *Brain Res.* 1986;379:320–328.
62 Aird WC, Edelberg JM, Weiler-Guettler H, et al. Vascular bed-specific expression of an endothelial cell gene is programmed by the tissue microenvironment. *J Cell Biol.* 1997;138:1117–1124.
63 Smith GP. Filamentous fusion phage: novel expression vectors that display cloned antigens on the virion surface. *Science.* 1985;228:1315–1317.
64 Winter G, Griffiths AD, Hawkins RE, Hoogenboom HR. Making antibodies by phage display technology. *Annu Rev Immunol.* 1994;12:433–455.
65 Rader C, Barbas CF 3rd. Phage display of combinatorial antibody libraries. *Curr Opin Biotechnol.* 1997;8:503–508.
66 Brown CK, Modzelewski RA, Johnson CS, Wong MK. A novel approach for the identification of unique tumor vasculature binding peptides using an *E. coli* peptide display library. *Ann Surg Oncol.* 2000;7:743–749.
67 Pasqualini R, Ruoslahti E. Organ targeting *in vivo* using phage display peptide libraries. *Nature.* 1996;380:364–366.
68 Arap W, Pasqualini R, Ruoslahti E. Cancer treatment by targeted drug delivery to tumor vasculature in a mouse model. *Science.* 1998;279:377–380.
69 Arap W, Haedicke W, Bernasconi M, et al. Targeting the prostate for destruction through a vascular address. *Proc Natl Acad Sci USA.* 2002;99:1527–1531.
70 Kolonin MG, Saha PK, Chan L, et al. Reversal of obesity by targeted ablation of adipose tissue. *Nat Med.* 2004;10:625–632.
71 Schnitzer JE. Vascular targeting as a strategy for cancer therapy. *N Engl J Med.* 1998;339:472–474.
72 Vink H, Duling BR. Identification of distinct luminal domains for macromolecules, erythrocytes, and leukocytes within mammalian capillaries. *Circ Res.* 1996;79:581–589.
73 Laakkonen P, Porkka K, Hoffman JA, Ruoslahti E. A tumor-homing peptide with a targeting specificity related to lymphatic vessels. *Nat Med.* 2002;8:751–755.
74 Porkka K, Laakkonen P, Hoffman JA, et al. A fragment of the HMGN2 protein homes to the nuclei of tumor cells and tumor endothelial cells in vivo. *Proc Natl Acad Sci USA.* 2002;99:7444–7449.
75 Valadon P, Garnett JD, Testa JE, et al. Screening phage display libraries for organ-specific vascular immunotargeting in vivo. *Proc Natl Acad Sci USA.* 2006;103:407–412.
76 Ong SE, Pandey A. An evaluation of the use of two-dimensional gel electrophoresis in proteomics. *Biomol Eng.* 2001;18:195–205.
77 Washburn MP, Wolters D, Yates JR 3rd. Large-scale analysis of the yeast proteome by multidimensional protein identification technology. *Nat Biotechnol.* 2001;19:242–247.
78 Aebersold R, Mann M. Mass spectrometry-based proteomics. *Nature.* 2003;422:198–207.
79 Gygi SP, Rist B, Gerber SA, et al. Quantitative analysis of complex protein mixtures using isotope-coded affinity tags. *Nat Biotechnol.* 1999;17:994–999.
80 Templin MF, Stoll D, Schrenk M, et al. Protein microarray technology. *Drug Discov Today.* 2002;7:815–822.
81 Obermeyer N, Janson N, Bergmann J, et al. Proteome analysis of migrating versus nonmigrating rat heart endothelial cells reveals distinct expression patterns. *Endothelium.* 2003;10:167–178.
82 Bruneel A, Labas V, Mailloux A, et al. Proteomic study of human umbilical vein endothelial cells in culture. *Proteomics.* 2003;3:714–723.
83 Ghinea N, Hasu M, Popov D. Selective radioiodination of the apical and luminal cell surfaces: in vitro and in situ experiments on vascular endothelial cells with Iodogen-coated Sephadex. *Anal Biochem.* 1989;179:274–279.
84 Schnitzer JE, Carley WW, Palade GE. Albumin interacts specifically with a 60-kDa microvascular endothelial glycoprotein. *Proc Natl Acad Sci USA.* 1988;85:6773–6777.
85 Schnitzer JE, Ulmer JB, Palade GE. A major endothelial plasmalemmal sialoglycoprotein, gp60, is immunologically related to glycophorin. *Proc Natl Acad Sci USA.* 1990;87:6843–6847.
86 Schnitzer JE, Shen CP, Palade GE. Lectin analysis of common glycoproteins detected on the surface of continuous microvascular endothelium in situ and in culture: identification of sialoglycoproteins. *Eur J Cell Biol.* 1990;52:241–251.
87 Schnitzer JE. gp60 is an albumin-binding glycoprotein expressed by continuous endothelium involved in albumin transcytosis. *Am J Physiol.* 1992;262:H246–H254.
88 Merker MP, Carley WW, Gillis CN. Molecular mapping of pulmonary endothelial membrane glycoproteins of the intact rabbit lung. *FASEB J.* 1990;4:3040–3048.
89 Belloni PN, Nicolson GL. Differential expression of cell surface glycoproteins on various organ-derived microvascular endothelia and endothelial cell cultures. *J Cell Physiol.* 1988;136:398–410.
90 Cole SR, Ashman LK, Ey PL. Biotinylation: an alternative to radioiodination for the identification of cell surface antigens in immunoprecipitates. *Mol Immunol.* 1987;24:699–705.
91 De La Fuente EK, Dawson CA, Nelin LD, et al. Biotinylation of membrane proteins accessible via the pulmonary circulation in normal and hyperoxic rats. *Am J Physiol.* 1997;272:L461–L470.
92 Scheurer SB, Rybak JN, Roesli C, et al. Identification and relative quantification of membrane proteins by surface biotinylation and two-dimensional peptide mapping. *Proteomics.* 2005;5:2718–2728.
93 Rybak JN, Ettorre A, Kaissling B, et al. In vivo protein biotinylation for identification of organ-specific antigens accessible from the vasculature. *Nat Methods.* 2005;2:291–298.
94 Liotta L, Petricoin E. Molecular profiling of human cancer. *Nat Rev Genet.* 2000;1:48–56.
95 Humphery-Smith I, Cordwell SJ, Blackstock WP. Proteome research: complementarity and limitations with respect to the RNA and DNA worlds. *Electrophoresis.* 1997;18:1217–1242.
96 Gygi SP, Rochon Y, Franza BR, Aebersold R. Correlation between protein and mRNA abundance in yeast. *Mol Cell Biol.* 1999;19:1720–1730.
97 Massoud TF, Gambhir SS. Molecular imaging in living subjects: seeing fundamental biological processes in a new light. *Genes Dev.* 2003;17:545–580.
98 Rudin M, Weissleder R. Molecular imaging in drug discovery and development. *Nat Rev Drug Discov.* 2003;2:123–131.
99 Lindsay MA. Target discovery. *Nat Rev Drug Discov.* 2003;2:831–838.

100 Workman P. New drug targets for genomic cancer therapy: successes, limitations, opportunities and future challenges. *Curr Cancer Drug Targets.* 2001;1:33–47.
101 Huber LA. Is proteomics heading in the wrong direction? *Nat Rev Mol Cell Biol.* 2003;4:74–80.
102 Renkin EM. Capillary transport of macromolecules: pores and other endothelial pathways. *J Appl Physiol.* 1985;58:315–325.
103 McIntosh DP, Tan X-Y, Oh P, Schnitzer JE. Targeting endothelium and its dynamic caveolae for tissue-specific transcytosis in vivo: a pathway to overcome cell barriers to drug and gene delivery. *Proc Natl Acad Sci USA.* 2002;99:1996–2001.
104 Carver LA, Schnitzer JE. Caveolae: mining little caves for new cancer targets. *Nat Rev Cancer.* 2003;3:571–581.
105 Burrows FJ, Thorpe PE. Vascular targeting – a new approach to the therapy of solid tumors. *Pharmacol Ther.* 1994;64:155–174.
106 Hughes BJ, Kennel SK, Lee R, Huang L. Monoclonal antibody targeting of liposomes to mouse lung *in vivo. Cancer Res.* 1989;49:6214–6220.
107 Ohizumi I, Tsunoda S, Taniguchi K, et al. Identification of tumor vascular antigens by monoclonal antibodies prepared from rat-tumor-derived endothelial cells. *Int J Cancer.* 1998;77:561–566.
108 Horoszewicz JS, Kawinski E, Murphy GP. Monoclonal antibodies to a new antigenic marker in epithelial prostatic cells and serum of prostatic cancer patients. *Anticancer Res.* 1987;7:927–935.
109 St Croix B, Rago C, Velculescu V, et al. Genes expressed in human tumor endothelium. *Science.* 2000;289:1197–1202.
110 Ford VA, Stringer C, Kennel SJ. Thrombomodulin is preferentially expressed in Balb/c lung microvessels. *J Biol Chem.* 1992;267:5446–5450.
111 Muzykantov VR, Christofidou-Solomidou M, Balyasnikova I, et al. Streptavidin facilitates internalization and pulmonary targeting of an anti-endothelial cell antibody (platelet-endothelial cell adhesion molecule 1): a strategy for vascular immunotargeting of drugs. *Proc Natl Acad Sci USA.* 1999;96:2379–2384.
112 Danilov SM, Muzykantov VR, Martynov AV, et al. Lung is the target organ for a monoclonal antibody to angiotensin-converting enzyme. *Lab Invest.* 1991;64:118–124.
113 Muzykantov VR, Martynov AV, Puchnina EA, Danilov SM. In vivo administration of glucose oxidase conjugated with monoclonal antibodies to angiotensin-converting enzyme. The tissue distribution, blood clearance, and targeting into rat lungs. *Am Rev Respir Dis.* 1989;139:1464–1473.
114 Reynolds PN, Zinn KR, Gavrilyuk VD, et al. A targetable, injectable adenoviral vector for selective gene delivery to pulmonary endothelium in vivo. *Mol Ther.* 2000;2:562–578.
115 Ke L, Qu H, Nagy JA, et al. Vascular targeting of solid and ascites tumours with antibodies to vascular endothelial growth factor. *Eur J Cancer.* 1996;32A:2467–2473.
116 Olson TA, Mohanraj D, Roy S, Ramakrishnan S. Targeting the tumor vasculature: inhibition of tumor growth by a vascular endothelial growth factor-toxin conjugate. *Int J Cancer.* 1997; 73:865–870.
117 Fernando NH, Hurwitz HI. Targeted therapy of colorectal cancer: clinical experience with bevacizumab. *Oncologist.* 2004; (9Suppl1):11–8.
118 Johnson DH, Fehrenbacher L, Novotny WF, et al. Randomized phase II trial comparing bevacizumab plus carboplatin and paclitaxel with carboplatin and paclitaxel alone in previously untreated locally advanced or metastatic non-small-cell lung cancer. *J Clin Oncol.* 2004;22:2184–2191.
119 Huminiecki L, Gorn M, Suchting S, et al. Magic roundabout is a new member of the roundabout receptor family that is endothelial specific and expressed at sites of active angiogenesis. *Genomics.* 2002;79:547–552.
120 Park KW, Morrison CM, Sorensen LK, et al. Robo4 is a vascular-specific receptor that inhibits endothelial migration. *Dev Biol.* 2003;261:251–267.
121 Seth P, Lin Y, Hanai J, et al. Magic roundabout, a tumor endothelial marker: expression and signaling. *Biochem Biophys Res Commun.* 2005;332:533–541.
122 Bedell VM, Yeo SY, Park KW, et al. Roundabout4 is essential for angiogenesis in vivo. *Proc Natl Acad Sci USA.* 2005;102:6373–6378.
123 Suchting S, Heal P, Tahtis K, et al. Soluble Robo4 receptor inhibits in vivo angiogenesis and endothelial cell migration. *FASEB J.* 2005;19:121–123.
124 Mailhos C, Modlich U, Lewis J, et al. Delta4, an endothelial specific notch ligand expressed at sites of physiological and tumor angiogenesis. *Differentiation.* 2001;69:135–144.
125 Christian S, Pilch J, Akerman ME, et al. Nucleolin expressed at the cell surface is a marker of endothelial cells in angiogenic blood vessels. *J Cell Biol.* 2003;163:871–878.
126 Taniguchi K, Harada N, Ohizumi I, et al. Molecular cloning and characterization of antigens expressed on rat tumor vascular endothelial cells. *Int J Cancer.* 2000;86:799–805.
127 Ohizumi I, Taniguchi K, Saito H, et al. Suppression of solid tumor growth by a monoclonal antibody against tumor vasculature in rats: involvement of intravascular thrombosis and fibrinogenesis. *Int J Cancer.* 1999;82:853–859.
128 Makimoto H, Koizumi K, Tsunoda S, et al. Tumor vascular targeting using a tumor-tissue endothelium-specific monoclonal antibody as an effective strategy for cancer chemotherapy. *Biochem Biophys Res Commun.* 1999;260:346–350.
129 Wakai Y, Matsui J, Koizumi K, et al. Effective cancer targeting using an anti-tumor tissue vascular endothelium-specific monoclonal antibody (TES-23). *Jpn J Cancer Res.* 2000;91:1319–1325.
130 Gougos A, Letarte M. Primary structure of endoglin, an RGD-containing glycoprotein of human endothelial cells. *J Biol Chem.* 1990;265:8361–8364.
131 Cheifetz S, Bellon T, Cales C, et al. Endoglin is a component of the transforming growth factor-beta receptor system in human endothelial cells. *J Biol Chem.* 1992;267:19027–19030.
132 Gougos A, Letarte M. Identification of a human endothelial cell antigen with monoclonal antibody 44G4 produced against a pre-B leukemic cell line. *J Immunol.* 1988;141:1925–1933.
133 Seon BK, Matsuno F, Haruta Y, et al. Long-lasting complete inhibition of human solid tumors in SCID mice by targeting endothelial cells of tumor vasculature with antihuman endoglin immunotoxin. *Clin Cancer Res.* 1997;3:1031–1044.
134 Matsuno F, Haruta Y, Kondo M, et al. Induction of lasting complete regression of preformed distinct solid tumors by targeting the tumor vasculature using two new anti-endoglin monoclonal antibodies. *Clin Cancer Res.* 1999;5:371–382.
135 Hood JD, Bednarski M, Frausto R, et al. Tumor regression by targeted gene delivery to the neovasculature. *Science.* 2002; 296:2404–2407.
136 Brooks PC, Montgomery AM, Rosenfeld M, et al. Integrin alpha v beta 3 antagonists promote tumor regression by inducing apoptosis of angiogenic blood vessels. *Cell.* 1994;79:1157–1164.

137 Kim S, Harris M, Varner JA. Regulation of integrin alpha vbeta 3-mediated endothelial cell migration and angiogenesis by integrin alpha5beta1 and protein kinase A. *J Biol Chem.* 2000; 275:33920–33928.
138 Kim S, Bell K, Mousa SA, Varner JA. Regulation of angiogenesis in vivo by ligation of integrin alpha5beta1 with the central cell-binding domain of fibronectin. *Am J Pathol.* 2000;156:1345–1362.
139 Kumar CC, Malkowski M, Yin Z, et al. Inhibition of angiogenesis and tumor growth by SCH221153, a dual alpha(v)beta3 and alpha(v)beta5 integrin receptor antagonist. *Cancer Res.* 2001;61:2232–2238.
140 Israeli RS, Powell CT, Corr JG, et al. Expression of the prostate-specific membrane antigen. *Cancer Res.* 1994;54:1807–1811.
141 Sweat SD, Pacelli A, Murphy GP, Bostwick DG. Prostate-specific membrane antigen expression is greatest in prostate adenocarcinoma and lymph node metastases. *Urology.* 1998;52:637–640.
142 Bostwick DG, Iczkowski KA. Microvessel density in prostate cancer: prognostic and therapeutic utility. *Semin Urol Oncol.* 1998;16:118–123.
143 Liu H, Moy P, Kim S, et al. Monoclonal antibodies to the extracellular domain of prostate-specific membrane antigen also react with tumor vascular endothelium. *Cancer Res.* 1997; 57:3629–3634.
144 Chang SS, O'Keefe DS, Bacich DJ, et al. Prostate-specific membrane antigen is produced in tumor-associated neovasculature. *Clin Cancer Res.* 1999;5:2674–2681.
145 Chang SS, Reuter VE, Heston WD, et al. Five different anti-prostate-specific membrane antigen (PSMA) antibodies confirm PSMA expression in tumor-associated neovasculature. *Cancer Res.* 1999;59:3192–3198.
146 Silver DA, Pellicer I, Fair WR, et al. Prostate-specific membrane antigen expression in normal and malignant human tissues. *Clin Cancer Res.* 1997;3:81–85.
147 Renneberg H, Friedetzky A, Konrad L, et al. Prostate specific membrane antigen (PSM) is expressed in various human tissues: implication for the use of PSM reverse transcription polymerase chain reaction to detect hematogenous prostate cancer spread. *Urol Res.* 1999;27:23–27.
148 Troyer JK, Beckett ML, Wright GL Jr. Detection and characterization of the prostate-specific membrane antigen (PSMA) in tissue extracts and body fluids. *Int J Cancer.* 1995;62:552–558.
149 Lopes AD, Davis WL, Rosenstraus MJ, et al. Immunohistochemical and pharmacokinetic characterization of the site-specific immunoconjugate CYT-356 derived from antiprostate monoclonal antibody 7E11-C5. *Cancer Res.* 1990;50:6423–6429.
150 Tjoa BA, Simmons SJ, Elgamal A, et al. Follow-up evaluation of a phase II prostate cancer vaccine trial. *Prostate.* 1999;40:125–129.
151 Tjoa BA, Simmons SJ, Bowes VA, et al. Evaluation of phase I/II clinical trials in prostate cancer with dendritic cells and PSMA peptides. *Prostate.* 1998;36:39–44.
152 Murphy GP, Tjoa BA, Simmons SJ, et al. Phase II prostate cancer vaccine trial: report of a study involving 37 patients with disease recurrence following primary treatment. *Prostate.* 1999;39:54–59.
153 Murphy G, Tjoa B, Ragde H, et al. Phase I clinical trial: T-cell therapy for prostate cancer using autologous dendritic cells pulsed with HLA-A0201-specific peptides from prostate-specific membrane antigen. *Prostate.* 1996;29:371–380.
154 Rettig WJ, Garin-Chesa P, Healey JH, et al. Identification of endosialin, a cell surface glycoprotein of vascular endothelial cells in human cancer. *Proc Natl Acad Sci USA.* 1992;89:10832–10836.
155 Carson-Walter EB, Watkins DN, Nanda A, et al. Cell surface tumor endothelial markers are conserved in mice and humans. *Cancer Res.* 2001;61:6649–6655.
156 Walter-Yohrling J, Morgenbesser S, Rouleau C, et al. Murine endothelial cell lines as models of tumor endothelial cells. *Clin Cancer Res.* 2004;10:2179–2189.
157 Nanda A, Buckhaults P, Seaman S, et al. Identification of a binding partner for the endothelial cell surface proteins TEM7 and TEM7R. *Cancer Res.* 2004;64:8507–8511.
158 Bradley KA, Mogridge J, Mourez M, et al. Identification of the cellular receptor for anthrax toxin. *Nature.* 2001;414:225–229.
159 Liu S, Aaronson H, Mitola DJ, et al. Potent antitumor activity of a urokinase-activated engineered anthrax toxin. *Proc Natl Acad Sci USA.* 2003;100:657–662.
160 Smith CW, Rothlein R, Hughes BJ, et al. Recognition of an endothelial determinant for CD 18-dependent human neutrophil adherence and transendothelial migration. *J Clin Invest.* 1988;82:1746–1756.
161 Giavazzi R, Chirivi RG, Garofalo A, et al. Soluble intercellular adhesion molecule 1 is released by human melanoma cells and is associated with tumor growth in nude mice. *Cancer Res.* 1992;52:2628–2630.
162 Gho YS, Kim PN, Li HC, et al. Stimulation of tumor growth by human soluble intercellular adhesion molecule-1. *Cancer Res.* 2001;61:4253–4257.
163 Gho YS, Kleinman HK, Sosne G. Angiogenic activity of human soluble intercellular adhesion molecule-1. *Cancer Res.* 1999; 59:5128–5132.
164 Kevil CG, Orr AW, Langston W, et al. Intercellular adhesion molecule-1 (ICAM-1) regulates endothelial cell motility through a nitric oxide-dependent pathway. *J Biol Chem.* 2004; 279:19230–19238.
165 van de Stolpe A, van der Saag PT. Intercellular adhesion molecule-1. *J Mol Med.* 1996;74:13–33.
166 Gearing AJ, Newman W. Circulating adhesion molecules in disease. *Immunol Today.* 1993;14:506–512.
167 Ran S, Gao B, Duffy S, et al. Infarction of solid Hodgkin's tumors in mice by antibody-directed targeting of tissue factor to tumor vasculature. *Cancer Res.* 1998;58:4646–4653.
168 Ho HK, Jang JJ, Kaji S, et al. Developmental endothelial locus-1 (Del-1), a novel angiogenic protein: its role in ischemia. *Circulation.* 2004;109:1314–1319.
169 Hidai C, Zupancic T, Penta K, et al. Cloning and characterization of developmental endothelial locus-1: an embryonic endothelial cell protein that binds the alphavbeta3 integrin receptor. *Genes Dev.* 1998;12:21–33.
170 Zlot C, Ingle G, Hongo J, et al. Stanniocalcin 1 is an autocrine modulator of endothelial angiogenic responses to hepatocyte growth factor. *J Biol Chem.* 2003;278:47654–47659.
171 Gerritsen ME, Soriano R, Yang S, et al. In silico data filtering to identify new angiogenesis targets from a large in vitro gene profiling data set. *Physiol Genomics.* 2002;10:13–20.
172 Carnemolla B, Neri D, Castellani P, et al. Phage antibodies with pan-species recognition of the oncofoetal angiogenesis marker fibronectin ED-B domain. *Int J Cancer.* 1996;68:397–405.

173 Carnemolla B, Balza E, Siri A, et al. A tumor-associated fibronectin isoform generated by alternative splicing of messenger RNA precursors. *J Cell Biol*. 1989;108:1139–1148.

174 Kaczmarek J, Castellani P, Nicolo G, et al. Distribution of oncofetal fibronectin isoforms in normal, hyperplastic and neoplastic human breast tissues. *Int J Cancer*. 1994;59:11–16.

175 Castellani P, Viale G, Dorcaratto A, et al. The fibronectin isoform containing the ED-B oncofetal domain: a marker of angiogenesis. *Int J Cancer*. 1994;59:612–618.

176 Santimaria M, Moscatelli G, Viale GL, et al. Immunoscintigraphic detection of the ED-B domain of fibronectin, a marker of angiogenesis, in patients with cancer. *Clin Cancer Res*. 2003;9:571–579.

177 Carnemolla B, Castellani P, Ponassi M, et al. Identification of a glioblastoma-associated tenascin-C isoform by a high affinity recombinant antibody. *Am J Pathol*. 1999;154:1345–1352.

178 Riva P, Franceschi G, Frattarelli M, et al. Loco-regional radioimmunotherapy of high-grade malignant gliomas using specific monoclonal antibodies labeled with 90Y: a phase I study. *Clin Cancer Res*. 1999;5:S3275–S3280.

179 Riva P, Franceschi G, Frattarelli M, et al. 131I radioconjugated antibodies for the locoregional radioimmunotherapy of high-grade malignant glioma – phase I and II study. *Acta Oncol*. 1999; 38:351–359.

180 Paganelli G, Grana C, Chinol M, et al. Antibody-guided three-step therapy for high grade glioma with yttrium-90 biotin. *Eur J Nucl Med*. 1999;26:348–357.

181 Rizzieri DA, Akabani G, Zalutsky MR, et al. Phase 1 trial study of 131I-labeled chimeric 81C6 monoclonal antibody for the treatment of patients with non-Hodgkin lymphoma. *Blood*. 2004;104:642–648.

182 Reardon DA, Akabani G, Coleman RE, et al. Phase II trial of murine (131)I-labeled antitenascin monoclonal antibody 81C6 administered into surgically created resection cavities of patients with newly diagnosed malignant gliomas. *J Clin Oncol*. 2002;20:1389–1397.

100

A Phage Display Perspective

Amado J. Zurita, Wadih Arap, and Renata Pasqualini

M.D. Anderson Cancer Center, The University of Texas, Houston

The vascular system is a complex network of vessels connecting the heart with organs and tissues to maintain homeostasis in response to physiological and pathological stimuli. Blood vessels are composed of endothelial cells (ECs), mural cells (pericytes or smooth muscle cells), and basement membrane. ECs lining the inner surface of blood and lymphatic vessels play important roles in the control of vascular tone, hemostasis, tissue growth, capillary exchange, inflammation, immune response, and angiogenesis. Even though these functions, along with other anatomical and molecular features, are common to all ECs, substantial structural and functional heterogeneity is seen among them. For example, ECs in the brain form a tight continuous monolayer required for a critical barrier function (blood–brain barrier), but those in the kidney, spleen, liver, and bone marrow display fenestrations or intercellular gaps for the rapid exchange of fluids and/or cells. This diversity is clearly a result of molecular differences between EC populations, which allow complex interactions with very distinctive microenvironments. Recent studies have identified transcriptional diversity between ECs in different types of blood vessels (arteries vs. veins, vessels of different caliber) and different anatomical locations (1–4). Not surprisingly, lymphatic ECs also display distinct characteristics (5).

Blood vessels can grow from endothelial progenitors (vasculogenesis), from the sprouting and subsequent stabilization of these sprouts by mural cells (angiogenesis and arteriogenesis), or from the expansive growth of preexisting vessels (collateral growth) (6). When vessel growth is altered, in particular when excessive angiogenesis occurs, new blood vessels contribute to a long (and growing) list of disorders, including cancer, arthritis, retinopathies, obesity, atherosclerosis, and asthma (6). However, it is in solid tumors where an abnormal vasculature is a hallmark (7,8). Tumor blood vessels clearly differ both structurally and functionally from normal vasculature and from newly formed blood vessels in sites of physiological angiogenesis or chronic inflammation (9). Most tumor vessels are tortuous, leaky, and discontinuous; have an altered blood flow; and although mural cells and basement membrane are present, they are profoundly abnormal (10–12). Tumor lymphatics are also anomalous (13). These abnormalities are determined by an imbalance between levels of pro- and antiangiogenic molecules in cancer, thus activating the so-called angiogenic switch from vascular quiescence, which can occur at different stages of tumor progression depending on the tumor type and the environment (14). It is the continuous nature of the changes related to the tumor environment – forming new vessels or remodeling existing ones – that explains the tremendous heterogeneity that exists between tumors, between a primary lesion and its metastases, between different moments in single lesions, and sometimes even between different locations in single lesions (9,15). In addition, recent reports have challenged the traditional assumption that tumor-associated ECs (TECs) are genetically stable, providing further evidence for the existence of phenotypic differences between normal ECs and TECs and for heterogeneity among TECs (16–18).

Characterizing differences between diverse normal EC populations and between normal and pathological ECs is crucial for understanding vascular development, physiology, and pathophysiology, and for the development and delivery of targeted therapies. A variety of approaches have been applied to identify *molecular* markers for specific vascular beds, including the Stamper–Woodruff assay for lymphocyte homing, monoclonal antibodies, differential display, serial analysis of gene expression (SAGE), genomic microarrays, subtractive or covalent modification proteomics, and phage display. In this chapter, we review recent progress that has been made in using phage display libraries (in particular peptide libraries) as a proteomics tool to map heterogeneity of the vascular system. We focus on applications of phage display technology related to normal and tumor vasculature, mainly in vivo phage display and vascular targeting.

PHAGE DISPLAY TARGETING STRATEGIES AND THE VASCULATURE

Phage display libraries are commonly used to obtain defined peptide sequences interacting with a particular molecule of interest with high affinity (19). Typically used libraries consist of genetically encoded polypeptides or proteins (including antibodies) expressed within a coat protein, usually pIII (a minor coat protein located at one tip) or pVIII (the major coat protein forming the body), of a filamentous bacteriophage (of which the most widely employed is M13). Billions of different sequences can be displayed, allowing exposure of a vast number of individual phage clones and subsequent affinity selection of those displaying peptides that bind specifically to a given molecular target of almost any chemical nature. The typical procedure for affinity purification or (bio)panning involves incubation of a primary or amplified existing library with the target molecule, removal of unbound phage by washing or perfusion, and recovery and propagation of the bound phage by bacterial infection. Several consecutive rounds of panning, usually three to four, are typically performed to improve selection. Finally, sequencing the corresponding viral DNA in the region encoding the displayed peptide reveals the sequence displayed by a binding phage. The selected phage and cognate peptides are then used for further analysis of the ligand–target interaction.

Through different strategies, phage display has been widely used in the last decade in vascular biology, in particular in the context of angiogenesis (9,20). Originally developed to map epitope-binding sites of antibodies (21), the search for peptide fragments with new or improved specificity or affinity has led to a number of technical advances in the construction of libraries, phage screening methodologies, and sequence affinity maturation (19,22,23).

Generally speaking, phage display is carried out either in vitro (on purified molecular targets or on cells) or in vivo.

In Vitro Targeting

Cell-Free Screening on Isolated Molecular Targets

The primary application of phage display has been the isolation of high-affinity ligands for previously characterized protein targets (24–26). These procedures require the use of purified or recombinant marker molecules attached to a solid support. Phage libraries displaying cyclic random peptides or antibody fragments (single-chain Fv or Fab) have been widely used. Interestingly, it has been shown over the years that a high number of isolated ligands bind to biologically relevant sites and act as agonists or, more frequently, antagonists of the function of the target molecule (27). Quite often, the primary structures of the selected peptides resemble those within known or potential interacting proteins, a useful characteristic in mapping protein–protein interactions. Phage-borne peptides demonstrate a broad mimicking potential to linear, conformational, and nonprotein binding sites (28). In peptide libraries, it is common to observe motifs consisting of three amino acids appearing several times in difference sequence contexts and being responsible for binding. This number provides the minimal framework for the overall fold of polypeptide chains and protein–protein interactions (29). Some examples of tripeptides acting as binding motifs include RGD, LDV, and LLG (to integrins) (30,31); GFE (to membrane dipeptidase) (32,33); and NGR (to aminopeptidase N[APN]/CD13) (34).

Panning on purified blood vessel–associated molecules has led to the successful isolation of ligands (peptides or antibody fragments) to vascular targets. The vast majority of targets have been angiogenesis-related markers. For brevity's sake, we only list markers without further discussion. This includes the selection of ligands against:

- Recepters, such as vascular endothelial growth factor (VEGF)-receptor 1 (VEGFR1) (35,36), VEGFR2 (37–40), VEGFR3 (41), EphA2 (42), Tie2 (43–45), and NG2 (46)
- Adhesion molecules, such as integrins (31,47–51), galectin-3 (52), intercellular adhesion molecule (ICAM)-1 (53), P-selectin (54), and E-selectin (55)
- Growth factors or extracellular matrix molecules associated with angiogenesis, such as VEGF (56–59), angiopoietin 2 (60), membrane type-1 matrix metalloproteinase (MT1-MMP) (61), matrix metalloproteinase (MMP)-2 and MMP-9 (62), thrombospondin-1 (63), and the ED-B fragment of fibronectin (64)

Screening the Molecular Diversity of Surfaces of Cells in the Vasculature

The isolation and characterization of ligands recognizing receptors on the surface of endothelial or vascular wall–associated cells is critical for the development of vascular-targeted agents. Not only that, but targeted ligands display in many cases specific functions that can be used for diagnostic or therapeutic purposes, or to gain insight into ligand–receptor interactions. The identification of ligands binding to cell membrane–embedded receptors has several advantages over cell-free strategies. As opposed to purified receptors, those on cell surfaces are more likely to preserve their natural or biologically active conformation, which can be lost upon purification and immobilization. Also, many cell surface receptors are active as complexes whose formation may require vitality of the cell, and this sometimes provides an additional advantage for targeting specificity or ligand function. An additional advantage is that these procedures can be performed in an unbiased manner, potentially allowing the recognition of receptors of unknown distribution or previously unrecognized existence. Nevertheless, the isolation of highly specific ligands to complex cell surface molecules remains challenging in many cases (65). Moreover, ECs in culture may not truly represent the cell phenotype in vivo. Although some characteristics seem to be stable when ECs are cultured for several passages (66,67), rapid changes in gene expression programs and de-differentiation can occur ex vivo (68,69).

Various groups have reported the successful identification of specific peptide or antibody ligands after biopanning on vascular cells or even blood vessel surfaces ex vivo. In a few cases, new ligands for established cell surface receptors were sought, as in the case of VEGFR2/KDR (70), αv integrins (71), LOX-1 receptor (72), vascular adhesion molecule (VCAM)-1 (73) and ICAM-1 (74). However, most biopannings on cells have been completed with no prior information about the cell surface receptor repertoire. This is the case for those performed on human umbilical vein ECs (HUVECs) either in culture or obtained ex vivo (65,75–77), brain microvascular ECs (78), dermal microvascular ECs (79), coronary artery ECs (80), rat retinal ECs and aorta (81), and rat heart ECs and histological sections of collateral blood vessels (82).

A typical problem associated with the use of phage display on cell surfaces is the high number of nonspecific binder clones and the loss of low affinity specific binders with the washes to reduce background. Our laboratory has developed a novel technology termed *Biopanning and Rapid Analysis of Selective Interacting Ligands* (BRASIL) for the screening of cell-surface binding peptides from phage libraries (83). Using a single-step differential centrifugation to separate specific phage–cell complexes from unbound phage, this methodology allows for fast selection of peptide binders and shows an improved sensitivity and specificity over conventional methods. As a proof of principle, we used BRASIL to screen a phage display random peptide library on activated, VEGF-A_{165}-stimulated HUVECs after a library subtraction step on quiescent ECs (83). This work showed that BRASIL is well suited for characterizing vascular targets in EC membranes in vitro and recognizing functionally relevant epitopes on target molecules. BRASIL also can be used to screen phage antibody or other types of phage display libraries. Moreover, as several rounds of selection can be performed on multiple samples in a relatively short time in tandem with cell-sorting techniques, this methodology can be automated and used for high-throughput screening – for example, of circulating vascular cell populations coming from patients.

In Vivo Targeting

Vascular targeting in vivo takes advantage of blood vessel heterogeneity. In normal blood vessels, functional and structural multiplicity are linked at the organ, tissue, and single vessel level. In diseased tissues, environmental pressure also creates differences among blood vessels. Structural and functional diversity is associated with molecular differences between vascular cell populations.

The molecular diversity of the vasculature provides a rational foundation for the development of site-specific targeted bioactive molecules. Plasma-membrane antigens or receptors can be used to target diagnostic or therapeutic agents directly to normal or diseased sites. Targeted diagnostics should improve sensitivity and specificity for detection of disease. Targeted therapies should improve efficacy with reduced toxicity.

In Vivo Phage Display

In vivo phage display allows for the identification of ligand–receptor pairs specific to given sites and functional status of the vascular system (84). This technology employs the same phage selection procedures used in vitro but accomplishes it in vivo by injecting a random phage library intravenously into a living organism. After a predetermined circulation time, biopsies from the sites of interest, whole organs or tumors are taken. Phage displaying peptides or antibody fragments that home to specific organs, tissues, or vascular beds are recovered. This process can be repeated several times to improve targeting specificity and affinity. Finally, ligands are isolated, identified, and further characterized. The homing ability of individual ligands also can be validated by this methodology (32,84,85).

Although the majority of in vivo studies have been performed using random peptide libraries, some recent work has shown that it is possible to screen libraries displaying antibody fragments on phage. For example, screening a phage-scFv library led to the isolation of fragments selective for murine thymic endothelium (86). Other investigators have described biodistribution and in vivo homing ability of phage displaying a specific antibody Fab (87) and scFv (65), respectively.

One of the most important advantages of in vivo phage display over other high-throughput profiling approaches is that information on focal distribution of receptor targets is preserved up to the single cell level, which is particularly applicable in heterogeneous systems such as the vasculature. In distribution studies and using immunohistochemistry to localize phage homing, it is not uncommon to see, within a population of cells, cell surface differences that are related to target distribution and level of expression. Compared with in vitro selection, in vivo phage display provides further advantages. Cell specificity in vitro may not be stringent enough in complex systems, especially if affinity is low for an abundant cell type. Ligands isolated in vivo have shown the ability to home selectively to the site of interest and bind under the physical conditions of circulation at the target location. As a consequence, peptides that recognize plasma and ubiquitous cell surface proteins are "counterselected" from the circulating phage pool, improving homing selectivity. In addition, by adjusting the time in which the phage pool is allowed to circulate, one can favor the selection of specific peptides for the endothelium or other vascular and perivascular cells or structural components. This is particularly useful in situations in which the vascular integrity is disrupted, as in the case of cancer-associated vasculature (several examples of nonendothelial molecular targets will be discussed later in the chapter). However, despite all its powerful characteristics, in vivo phage display is technically challenging because of the difficulty in monitoring after-injection events, in particular contact time between phage and endothelial surfaces. A careful optimization of circulation times based on targeted organ and pharmacokinetic properties of distinct phage display libraries is critical with this technology (88).

Since the first description of the in vivo phage display in 1996 (84), numerous peptide ligands and corresponding vascular markers have been discovered by our laboratory and others (20,89). In many cases, selected molecular targets, although specifically expressed in a vascular bed, also were found in other locations outside the vasculature in a manner not

Table 100-1: Structure and Activity of Vascular Homing Peptides Identified by in Vivo Phage Display with Established Receptors

Peptide Ligand	*Receptor*	*Receptor Location in Vessel*	*Tissue*	*References*
Normal Tissues				
CGFECVRQCPERC	Membrane dipeptidase	EC surface	Lung	33
CPGPEGAGC	Aminopeptidase P	EC surface	Breast	95
CKGGRAKDC	Prohibitin	EC surface	Fat	97
CGRRAGGSC	Interleukin-11 receptor-α	EC surface	Prostate	94
Angiogenesis				
CDCRGDCFC (RGD-4C)	Integrins $\alpha v\beta 3$ and $\alpha v\beta 5$	EC surface	Angiogenesis	85, 106
CNGRC	Aminopeptidase N/CD13	EC/pericyte surface	Angiogenesis	34, 85
CAPGPSKSC	Glucose-regulated protein 78	EC surface	Atherosclerotic lesions	100
F3 (34 amino acids from HMGN2)	Nucleolin	EC surface	Angiogenesis	110

necessarily tissue- or cell-specific. These results illustrate an essential characteristic of in vivo phage display for target discovery: Both accessibility and target expression influence the outcome. In this respect, the vasculature provides unique accessibility to agents that are given intravenously. Another significant aspect is that, as noted earlier in this chapter, many of the peptide ligands bind to functionally important regions of their target molecules and thus play a functional role to add to their targeting capability. Some examples will be discussed in the next paragraphs.

Normal Tissue-Specific Vascular Targets

Peptide ligands targeting brain and kidney (84), skin, intestine, uterus, adrenal gland, retina (32), pancreas (32,90), lung (32,91), muscle (92), prostate (93,94), breast (95), lymph nodes (96), and fat (97) have been described. These and other results obtained with unrelated technologies are demonstrating that ECs in diverse environments exhibit different sets of cell surface markers, or similar markers displaying site-specific features.

In vivo phage display is uncovering a vascular address system of homing ligands and corresponding receptors (so-called *zip codes*) that provides unique opportunities for targeted clinical applications. It also improves our understanding of molecular interactions in vascular beds. In this regard, the identification of targeted receptors is critical (Table 100-1). Integrating the isolation of ligands from in vivo screenings with complementary biochemical approaches has yielded an array of tissue-specific vascular receptors. Two proteases have been identified by in vivo phage display as markers of the vasculature in normal tissues: membrane dipeptidase in the lung (33) and aminopeptidase P in the breast (95). However, other types of molecules also have been described. For example, we recently established prohibitin (a multifunctional membrane-associated protein) as a vascular marker of adipose tissue (97).

Organs such as the adrenal gland, kidney, pancreas, and brain contain structurally and functionally distinct regions that exhibit a unique vascular organization. To determine whether a differential expression of vascular addresses could be recognized by in vivo phage display within focal regions of the microvasculature, Yao and colleagues combined this technology with laser pressure catapult microdissection on murine pancreatic islets of Langerhans (90). Among several candidate proteins exhibiting structural homology with the peptide sequences recovered, two ephrin A-type ligand homologues, A2 and A4, were identified. Significantly, the vast majority of peptides that were isolated from the microvasculature of microdissected islet sections were not found among those from the acinar pancreas. Furthermore, sequences obtained from the acinar pancreas were also unique. This work further showed the ability of in vivo phage display to recognize molecular vascular heterogeneity.

Although the previously described studies were performed on mouse models and still allowed the identification of many putative human ligand and target homologues (34,85,98), mouse-derived probes might not always be useful for targeting applications in humans. Moreover, level of expression and distribution of targets can change among different species (94). This potential species-related diversity in ligand specificity, affinity and accessibility, and target expression and distribution should prompt careful evaluation of results obtained from animal studies before considering human applications. With this in mind, the construction of a human molecular vascular map is clearly required to obtain human-relevant data in the field of vascular proteomics and to successfully translate vascular targeting into clinical practice.

We have recently taken the first steps toward such a fundamental task (94). A brain-dead patient with an end-stage B-cell lymphoproliferative disorder received an intravenous infusion of a CX $_7$C (C = cysteine, X = any amino acid) random phage display peptide library. Fifteen minutes later, tissue biopsies were obtained from bone marrow, prostate, liver, fat, skeletal muscle, and skin to provide histopathological diagnosis and to recover phage from various organs. Phage isolated from each tissue were recovered for identification of the corresponding peptide inserts. To analyze the distribution of inserts, we designed a high-throughput pattern-recognition software for the comparison of short amino acid residue sequences (tripeptide motifs) retrieved both from phage recovered from the biopsies and from the unselected library. Analysis of the approximately 50,000 tripeptide motifs obtained from the recovered phage showed that their distribution to the different organs was not random. Some of the tripeptide motifs were recovered from a single tissue, whereas others were found in several, probably indicating binding to differentially expressed endothelial markers for the former and to ubiquitous endothelial molecules for the latter. Further analysis of shared motifs isolated from single organs (four to six amino acids long) found matches with known human proteins, potential circulating ligands (either secreted proteins or surface receptors present on circulating cells) for EC surface receptors. Among candidate human proteins potentially mimicked by selected peptide motifs, we identified interleukin (IL)-11 as a ligand mimicked by a peptide specifically enriched in prostate. The prostate-endothelium homing phage was shown in phage overlay assays to recognize the endothelium and epithelium of normal human prostate, but not other tissues, such as skin. Furthermore, it bound specifically to IL-11 receptor (IL-11α) in vitro.

These studies may form the basis of a targeted pharmacology, which could be particularly useful in the context of angiogenesis-dependent diseases.

Vascular Targets in Tissues Undergoing Angiogenesis

We have previously discussed the changes occurring in the vasculature of tissues undergoing angiogenesis. In vivo phage display is being used to investigate molecular differences between normal microvasculature and that with a dysregulated growth. Novel vascular markers have been identified, especially in the context of cancer-associated angiogenesis. Their tissue and cell distribution and accessibility to the circulation have been specifically addressed, and in some cases expression has even been correlated with disease/tumor stage. In addition, many of the vascular markers found in angiogenic blood vessels not only serve as target receptors for circulating ligands but also are functionally relevant and their activity can be modulated.

Exploring the vasculature in chronic inflammatory diseases has revealed site-specific ligands. Houston and colleagues isolated peptides that home to atherosclerotic blood vessels in low-density lipoprotein receptor-deficient mice (99). Similarly, Liu and associates probed atherosclerotic lesions in *ApoE*$^{-/-}$ mice (100). Glucose-regulated protein 78 (GRP78), a cell-surface expressed chaperone identified by our group as the molecular target for autoantibodies associated with metastatic prostate cancer (101), was recognized as the endothelial surface receptor of one selected peptide (100). In addition, another peptide with sequence similarity to tissue inhibitor of metalloproteinase (TIMP)-2 showed specific binding to atherosclerotic endothelium in vivo (100). Targeting restenotic lesions after angioplasty, using a sequential in vitro–in vivo approach, Michon and colleagues selected phage binding to proliferating vascular smooth muscle cells and, after three rounds, to physically denuded common carotid artery in *ApoE*$^{-/-}$ mice (exposing the vascular smooth muscle cell layer) (102). An interesting strategy is being pursued by a group of investigators in the United Kingdom (89). These authors performed in vivo phage display selection on human synovial tissue from patients with rheumatoid arthritis or osteoarthritis transplanted subcutaneously into SCID animals (103). Based on their previous observation of the formation of functional anastomoses between human graft and murine subdermal blood vessels (104), they claim this model allows selection on human vasculature. Further validation of homing specificity in nongrafted human tissues, functionality, and additional results on different disease models are expected. Finally, exploring the effect of aging on cardioprotection from myocardial infarction, Cai and associates found a decrease in cardiac microvascular homing of tumor necrosis factor (TNF)-α mimic peptides in vivo, which was attributed to age-related downregulation of TNF-receptor 1 in murine heart (105).

In vivo screening of phage libraries has generated a panel of diverse peptide motifs that recognize markers in tumor-associated vasculature. The first ligand motifs identified by in vivo phage display in cancer, nowadays extensively validated, were RGD (arginine-glycine-aspartic acid, embedded in a double-cyclic peptide designated RGD-4C) and NGR (asparagine-glycine-arginine) (85). The RGD-4C peptide is a selective binder of the $\alpha v\beta 3$ and $\alpha v\beta 5$ integrins (106), which are known to be selectively expressed in human tumor blood vessels and many human tumor cells (107). Similar to some of the markers described for normal tissues, peptides containing NGR bind to a peptidase/protease: APN, also known as CD13 (34). Tumor homing by RGD-4C- and NGR-containing peptides is independent of the tumor cell's origin.

APN/CD13 and aminopeptidase A (APA), another vascular protease accessible to circulating ligands with known upregulation in tumor-associated pericytes (108), are functional targets in angiogenic vasculature in that they modulate angiogenesis (34,98). Their enzymatic activities contribute to proangiogenic pathways, because their functional ablation, either through genetic or biochemical methods, significantly reduces the formation of new blood vessels in several angiogenesis-dependent conditions, including cancer and retinopathies.

By using phage libraries expressing random peptides or cDNA-encoded protein fragments, Ruoslahti and colleagues have been using an ex vivo–in vivo approach for screening of tumor angiogenesis in recent years. The phage library is first incubated on cell suspensions prepared from the target tissue, and then the enriched library is used in vivo. In one screen, phage were initially selected for binding to bone marrow cells ex vivo and subsequently in tumor-bearing mice, obtaining a ligand mimic of a fragment of the nuclear protein high mobility group (HMG) protein 2 (HMGN2) (109). A derived 31-aa synthetic peptide (F3) showed preferential accumulation in the nuclei of tumor-associated ECs and certain tumor cells. In a subsequent report, the target receptor for F3, also expressed in a nontumor angiogenesis model, was identified as cell surface nucleolin (110). Using the same strategy for screening but favoring the recognition of nonendothelial targets (by depleting the cell suspension of them), a new peptide homing to tumor cells and associated lymphatic vessels, LyP-1, was uncovered (111). LyP-1 showed no recognition of lymphatic vessels in normal tissues. Intriguingly, binding to a subset of cultured tumor cells correlated with the ability of LyP-1 to recognize the putative tumor lymphatics in orthotopic xenografts and metastatic lesions derived from them (112). Moreover, LyP-1 presented an intrinsic proapoptotic effect (112). However, the surface molecule to which LyP-1 binds, which may be implicated in the peptide's activity, has not yet been identified. As an additional proof of the molecular specialization of the vasculature and the sensitivity of in vivo phage display to detect it, two studies have shown molecular signatures associated with different stages of tumor progression, from dysplasia to invasive cancer (113,114). By using transgenic mouse models for multistage tumorigenesis (K14-HPV16 squamous [113] and RIP-Tag2 pancreatic islet [114] carcinomas), different tumor-type and tumor-stage-specific vascular homing peptides were obtained.

In vivo phage display also has made possible the identification of other peptide sequences targeting tumor vasculature, including those accompanying irradiated tumors (115), a model of rat glioblastoma (116), and human gastric cancer transplanted to the subrenal capsule of immunosuppressed mice (117).

By validating the IL-11/IL-11Rα ligand–receptor pair in human prostate cancer (94), we have recently shown that ligand–receptor pairs in the context of a normal tissue can still be present and functional in cancer, even when the cancer has metastasized (118). We found the expression of IL-11Rα in the blood vessel and tumor compartments (and its recognition by our IL-11 mimic sequence) progressively increased with prostate cancer progression. Furthermore, a proapoptotic peptide guided by the IL-11 mimetope specifically targeted and internalized in IL-11Rα-expressing prostate cancer cells, inducing apoptosis (118). Our results further illustrate the value of in vivo phage display for the identification of relevant targets in the context of human disease.

VASCULAR ADDRESSES FOR TARGETED DELIVERY

There is an obvious pharmacokinetic advantage in making use of the molecular differences in the vasculature for targeting. We have reviewed several examples of targeted ligands identified by phage display exhibiting an intrinsic functional capability. In the majority of cases, however, ligands selected for homing to blood vessels have been used to guide the delivery of coupled therapeutic and diagnostic agents. Cytotoxic drugs (85,119,120), proapoptotic peptides (93,97,121,122), thrombogenic agents (123), metalloprotease inhibitors (62), cytokines (124,125), genes (80,126–129), and liposomes (130,131) have been addressed to receptors in the vasculature. Although most of the experiments have been performed in cancer (85,93,112), other experimental models of disease also have been explored (97,132). This approach, in general, showed enhanced therapeutic efficacy and decreased side effects for targeted versus parental untargeted agents, clearly illustrating the value of the targeting technology for therapy. Tumor-targeting peptide ligands also have been used to guide and deliver imaging agents, in particular RGD (133–137).

Several factors influence the success of a targeting strategy (138). Among them, features that explain the specificity achieved for vascular targets identified by in vivo phage display have to be considered, including preferential expression of the target in an organ or tissue of interest, accessibility of the target from the circulation, and environment-dependent modifications in the target. An additional feature achieved by many targeting peptides, internalization capability, can be important for homing selectivity and intracellular delivery of functional moieties. An illustrative example is APN/CD13, which has been extensively targeted for tumor therapy in animal models (85,121,124). APN/CD13 was recognized as the principal receptor for the NGR peptide motif, a tripeptide sequence previously identified by in vivo phage display that homes strictly to the endothelium of angiogenic blood vessels (34,85). APN/CD13 is expressed in epithelial cells in the liver, intestine, bile duct, brain, kidney, prostate, and lung; in fibroblasts and smooth muscle cells; and in the normal and leukemic progeny of myeloid cells (139). Although some of these locations (in particular those in hematopoietic compartments) are accessible from the circulation, NGR-containing peptide ligands mostly recognize the endothelial and periendothelial cell compartment of angiogenic blood vessels. The basis for this restricted recognition pattern appears to be the existence of different APN/CD13 forms in normal epithelial cells, tumor-associated vessels, and myeloid cells, so that a tumor vessel-related APN/CD13 form, but not those associated with epithelial and myeloid cells, functions as an NGR receptor (139). Moreover, it has been reported that APN/CD13 has multiple enzymatic activities depending on the tissue microenvironment. Studies using antibodies and natural peptide substrates indicate that APN/CD13 undergoes conformational changes and exposure of cryptic epitopes upon binding, which regulate enzymatic activity (140). Factors present only in the tumor microenvironment might cause a differential reactivity or accessibility to different APN/CD13 ligands, such as the tumor vasculature–targeting NGR peptides. This example clearly demonstrates the importance of using a functional proteomics method, such as in vivo phage display, for the identification of ligands that target molecules with complex regulatory mechanisms and expression patterns in time and space.

KEY POINTS

- The recognition of heterogeneity in the vasculature is longstanding. However, the significance, extent, molecular diversity, and potential utility of this heterogeneity in health and disease are now being more fully appreciated and exploited thanks to the in vivo phage display technology.
- In vivo phage display provides novel tools for selective vascular targeting and identifies corresponding vascular cell surface receptors accessible to the circulation.
- In vivo phage display also improves our understanding of organ and tissue vascular specificity in health and disease and contributes to a definition of the functional role that new targets play in the pathogenesis of the many angiogenesis-related diseases.

Future Goals

- To proceed from the ligands identified by in vivo phage display to new targets in blood vessels, sophisticated state-of-the-art array proteomic technologies and precise target validation methods are needed.
- To construct, using in vivo phage display technology, a detailed map of the molecular specificities of the human vasculature that can be used to improve our understanding of human disease and to develop clinical applications

ACKNOWLEDGMENTS

Our work was funded by grants from the NIH and DOD to R. P. and W. A., and the Gillson-Longenbaugh Foundation. A.J.Z. received a fellowship from the Instituto de Salud Carlos III, Spain.

REFERENCES

1 Wang HU, Chen ZF, Anderson DJ. Molecular distinction and angiogenic interaction between embryonic arteries and veins revealed by ephrin-B2 and its receptor Eph-B4. *Cell*. 1998;93(5):741–753.

2 Lawson ND, Scheer N, Pham VN, et al. Notch signaling is required for arterial-venous differentiation during embryonic vascular development. *Development*. 2001;128(19):3675–3683.

3 Muller AM, Hermanns MI, Cronen C, Kirkpatrick CJ. Comparative study of adhesion molecule expression in cultured human macro- and microvascular endothelial cells. *Exp Mol Pathol*. 2002;73(3):171–180.

4 Chi JT, Chang HY, Haraldsen G, et al. Endothelial cell diversity revealed by global expression profiling. *Proc Natl Acad Sci USA*. 2003;100(19):10623–10628.

5 Podgrabinska S, Braun P, Velasco P, et al. Molecular characterization of lymphatic endothelial cells. *Proc Natl Acad Sci USA*. 2002;99(25):16069–16074.

6 Carmeliet P. Angiogenesis in health and disease. *Nat Med*. 2003; 9(6):653–660.

7 Hanahan D, Weinberg RA. The hallmarks of cancer. *Cell*. 2000; 100(1):57–70.

8 Carmeliet P, Jain RK. Angiogenesis in cancer and other diseases. *Nature*. 2000;407(6801):249–257.

9 Pasqualini R, Arap W, McDonald DM. Probing the structural and molecular diversity of tumor vasculature. *Trends Mol Med*. 2002;8(12):563–571.

10 Morikawa S, Baluk P, Kaidoh T, et al. Abnormalities in pericytes on blood vessels and endothelial sprouts in tumors. *Am J Pathol*. 2002;160(3):985–1000.

11 Baluk P, Morikawa S, Haskell A, et al. Abnormalities of basement membrane on blood vessels and endothelial sprouts in tumors. *Am J Pathol*. 2003;163(5):1801–1815.

12 Mollica F, Jain RK, Netti PA. A model for temporal heterogeneities of tumor blood flow. *Microvasc Res*. 2003;65(1): 56–60.

13 Jain RK, Fenton BT. Intratumoral lymphatic vessels: a case of mistaken identity or malfunction? *J Natl Cancer Inst*. 2002; 94(6):417–421.

14 Hanahan D, Folkman J. Patterns and emerging mechanisms of the angiogenic switch during tumorigenesis. *Cell*. 1996;86(3): 353–364.

15 Eberhard A, Kahlert S, Goede V, et al. Heterogeneity of angiogenesis and blood vessel maturation in human tumors: implications for antiangiogenic tumor therapies. *Cancer Res*. 2000;60(5):1388–1393.

16 Bussolati B, Deambrosis I, Russo S, et al. Altered angiogenesis and survival in human tumor-derived endothelial cells. *FASEB J*. 2003;17(9):1159–1161.

17 Streubel B, Chott A, Huber D, et al. Lymphoma-specific genetic aberrations in microvascular endothelial cells in B-cell lymphomas. *N Engl J Med*. 2004;351(3):250–259.

18 Hida K, Hida Y, Amin DN, et al. Tumor-associated endothelial cells with cytogenetic abnormalities. *Cancer Res*. 2004;64 (22):8249–8255.

19 Azzazy HM, Highsmith WE Jr. Phage display technology: clinical applications and recent innovations. *Clin Biochem*. 2002; 35(6):425–445.

20 Kolonin M, Pasqualini R, Arap W. Molecular addresses in blood vessels as targets for therapy. *Curr Opin Chem Biol*. 2001;5(3): 308–313.

21 Smith GP. Filamentous fusion phage: novel expression vectors that display cloned antigens on the virion surface. *Science*. 1985;228(4705):1315–1317.

22 Hoogenboom HR. Overview of antibody phage-display technology and its applications. *Methods Mol Biol.* 2002;178:1–37.
23 Zurita AJ, Arap W, Pasqualini R. Mapping tumor vascular diversity by screening phage display libraries. *J Control Release.* 2003; 91(1–2):183–186.
24 Scott JK, Smith GP. Searching for peptide ligands with an epitope library. *Science.* 1990;249(4967):386–390.
25 Devlin JJ, Panganiban LC, Devlin PE. Random peptide libraries: a source of specific protein binding molecules. *Science.* 1990;249(4967):404–406.
26 Cwirla SE, Peters EA, Barrett RW, Dower WJ. Peptides on phage: a vast library of peptides for identifying ligands. *Proc Natl Acad Sci USA.* 1990;87(16):6378–6382.
27 Kay BK, Hamilton PT. Identification of enzyme inhibitors from phage-displayed combinatorial peptide libraries. *Comb Chem High Throughput Screen.* 2001;4(7):535–543.
28 Smith GP. Surface presentation of protein epitopes using bacteriophage expression systems. *Curr Opin Biotechnol.* 1991;2(5): 668–673.
29 Vendruscolo M, Paci E, Dobson CM, Karplus M. Three key residues form a critical contact network in a protein folding transition state. *Nature.* 2001;409(6820):641–645.
30 Ruoslahti E. RGD and other recognition sequences for integrins. *Annu Rev Cell Dev Biol.* 1996;12:697–715.
31 Koivunen E, Ranta TM, Annila A, et al. Inhibition of beta(2) integrin-mediated leukocyte cell adhesion by leucine-leucine-glycine motif-containing peptides. *J Cell Biol.* 2001;153(5):905–916.
32 Rajotte D, Arap W, Hagedorn M, et al. Molecular heterogeneity of the vascular endothelium revealed by in vivo phage display. *J Clin Invest.* 1998;102(2):430–437.
33 Rajotte D, Ruoslahti E. Membrane dipeptidase is the receptor for a lung-targeting peptide identified by in vivo phage display. *J Biol Chem.* 1999;274(17):11593–11598.
34 Pasqualini R, Koivunen E, Kain R, et al. Aminopeptidase N is a receptor for tumor-homing peptides and a target for inhibiting angiogenesis. *Cancer Res.* 2000;60(3):722–727.
35 An P, Lei H, Zhang J, et al. Suppression of tumor growth and metastasis by a VEGFR-1 antagonizing peptide identified from a phage display library. *Int J Cancer.* 2004;111(2):165–173.
36 El-Mousawi M, Tchistiakova L, Yurchenko L, et al. A vascular endothelial growth factor high affinity receptor 1-specific peptide with antiangiogenic activity identified using a phage display peptide library. *J Biol Chem.* 2003;278(47):46681–46691.
37 Boldicke T, Tesar M, Griesel C, et al. Anti-VEGFR-2 scFvs for cell isolation. Single-chain antibodies recognizing the human vascular endothelial growth factor receptor-2 (VEGFR-2/flk-1) on the surface of primary endothelial cells and preselected CD34+ cells from cord blood. *Stem Cells.* 2001;19(1):24–36.
38 Popkov M, Jendreyko N, Gonzalez-Sapienza G, et al. Human/mouse cross-reactive anti-VEGF receptor 2 recombinant antibodies selected from an immune b9 allotype rabbit antibody library. *J Immunol Methods.* 2004;288(1–2):149–164.
39 Hetian L, Ping A, Shumei S, et al. A novel peptide isolated from a phage display library inhibits tumor growth and metastasis by blocking the binding of vascular endothelial growth factor to its kinase domain receptor. *J Biol Chem.* 2002;277(45):43137–43142.
40 Lu D, Jimenez X, Zhang H, et al. Selection of high affinity human neutralizing antibodies to VEGFR2 from a large antibody phage display library for antiangiogenesis therapy. *Int J Cancer.* 2002;97(3):393–399.
41 Persaud K, Tille JC, Liu M, et al. Involvement of the VEGF receptor 3 in tubular morphogenesis demonstrated with a human anti-human VEGFR-3 monoclonal antibody that antagonizes receptor activation by VEGF-C. *J Cell Sci.* 2004;117(Pt 13): 2745–2756.
42 Koolpe M, Dail M, Pasquale EB. An ephrin mimetic peptide that selectively targets the EphA2 receptor. *J Biol Chem.* 2002; 277(49):46974–46979.
43 Popkov M, Mage RG, Alexander CB, et al. Rabbit immune repertoires as sources for therapeutic monoclonal antibodies: the impact of kappa allotype-correlated variation in cysteine content on antibody libraries selected by phage display. *J Mol Biol.* 2003;325(2):325–335.
44 Tournaire R, Simon MP, le Noble F, et al. A short synthetic peptide inhibits signal transduction, migration and angiogenesis mediated by Tie2 receptor. *EMBO Rep.* 2004;5(3):262–267.
45 Deng SJ, Liu W, Simmons CA, et al. Identifying substrates for endothelium-specific Tie-2 receptor tyrosine kinase from phage-displayed peptide libraries for high throughput screening. *Comb Chem High Throughput Screen.* 2001;4(6):525–533.
46 Burg MA, Pasqualini R, Arap W, et al. NG2 proteoglycan-binding peptides target tumor neovasculature. *Cancer Res.* 1999; 59(12):2869–2874.
47 Koivunen E, Wang B, Ruoslahti E. Isolation of a highly specific ligand for the alpha 5 beta 1 integrin from a phage display library. *J Cell Biol.* 1994;124(3):373–380.
48 Pasqualini R, Koivunen E, Ruoslahti E. A peptide isolated from phage display libraries is a structural and functional mimic of an RGD-binding site on integrins. *J Cell Biol.* 1995;130(5):1189–1196.
49 Pasqualini R, Koivunen E, Ruoslahti E. Alpha v integrins as receptors for tumor targeting by circulating ligands. *Nat Biotechnol.* 1997;15(6):542–546.
50 Stefanidakis M, Bjorklund M, Ihanus E, et al. Identification of a negatively charged peptide motif within the catalytic domain of progelatinases that mediates binding to leukocyte beta 2 integrins. *J Biol Chem.* 2003;278(36):34674–34684.
51 Cardo-Vila M, Arap W, Pasqualini R. Alpha v beta 5 integrin-dependent programmed cell death triggered by a peptide mimic of annexin V. *Mol Cell.* 2003;11(5):1151–1162.
52 Zou J, Glinsky VV, Landon LA, et al. Peptides specific to the galectin-3 carbohydrate recognition domain inhibit metastasis-associated cancer cell adhesion. *Carcinogenesis.* 2005;26(2):309–318.
53 Shannon JP, Silva MV, Brown DC, Larson RS. Novel cyclic peptide inhibits intercellular adhesion molecule-1-mediated cell aggregation. *J Pept Res.* 2001;58(2):140–150.
54 Molenaar TJ, Appeldoorn CC, de Haas SA, et al. Specific inhibition of P-selectin-mediated cell adhesion by phage display-derived peptide antagonists. *Blood.* 2002;100(10):3570–3577.
55 Fukuda MN, Ohyama C, Lowitz K, et al. A peptide mimic of E-selectin ligand inhibits sialyl Lewis X-dependent lung colonization of tumor cells. *Cancer Res.* 2000;60(2):450–456.
56 Smith KA, Kirkpatrick N, Madden LA, et al. Isolation and characterisation of vascular endothelial growth factor-165 specific scFv fragments by phage display. *Int J Oncol.* 2003;22(2):333–338.

57 Cooke SP, Boxer GM, Lawrence L, et al. A strategy for antitumor vascular therapy by targeting the vascular endothelial growth factor: receptor complex. *Cancer Res.* 2001;61(9):3653–3659.
58 Vitaliti A, Wittmer M, Steiner R, et al. Inhibition of tumor angiogenesis by a single-chain antibody directed against vascular endothelial growth factor. *Cancer Res.* 2000;60(16):4311–4314.
59 Fairbrother WJ, Christinger HW, Cochran AG, et al. Novel peptides selected to bind vascular endothelial growth factor target the receptor-binding site. *Biochemistry.* 1998;37(51):17754–17764.
60 Cai M, Zhang H, Hui R. Single chain Fv antibody against angiopoietin-2 inhibits VEGF-induced endothelial cell proliferation and migration in vitro. *Biochem Biophys Res Commun.* 2003;309(4):946–951.
61 Ohkubo S, Miyadera K, Sugimoto Y, et al. Identification of substrate sequences for membrane type-1 matrix metalloproteinase using bacteriophage peptide display library. *Biochem Biophys Res Commun.* 1999;266(2):308–313.
62 Koivunen E, Arap W, Valtanen H, et al. Tumor targeting with a selective gelatinase inhibitor. *Nat Biotechnol.* 1999;17(8):768–774.
63 Watkins NA, Du LM, Scott JP, et al. Single-chain antibody fragments derived from a human synthetic phage-display library bind thrombospondin and inhibit sickle cell adhesion. *Blood.* 2003;102(2):718–724.
64 Pini A, Viti F, Santucci A, et al. Design and use of a phage display library. Human antibodies with subnanomolar affinity against a marker of angiogenesis eluted from a two-dimensional gel. *J Biol Chem.* 1998;273(34):21769–21776.
65 Mutuberria R, Hoogenboom HR, van der Linden E, et al. Model systems to study the parameters determining the success of phage antibody selections on complex antigens. *J Immunol Methods.* 1999;231(1–2):65–81.
66 Kriehuber E, Breiteneder-Geleff S, Groeger M, et al. Isolation and characterization of dermal lymphatic and blood endothelial cells reveal stable and functionally specialized cell lineages. *J Exp Med.* 2001;194(6):797–808.
67 Hirakawa S, Hong YK, Harvey N, et al. Identification of vascular lineage-specific genes by transcriptional profiling of isolated blood vascular and lymphatic endothelial cells. *Am J Pathol.* 2003;162(2):575–586.
68 Borsum T, Hagen I, Henriksen T, Carlander B. Alterations in the protein composition and surface structure of human endothelial cells during growth in primary culture. *Atherosclerosis.* 1982;44(3):367–378.
69 Lacorre DA, Baekkevold ES, Garrido I, et al. Plasticity of endothelial cells: rapid dedifferentiation of freshly isolated high endothelial venule endothelial cells outside the lymphoid tissue microenvironment. *Blood.* 2004;103(11):4164–4172.
70 Binetruy-Tournaire R, Demangel C, Malavaud B, et al. Identification of a peptide blocking vascular endothelial growth factor (VEGF)-mediated angiogenesis. *EMBO J.* 2000;19(7):1525–1533.
71 Holig P, Bach M, Volkel T, et al. Novel RGD lipopeptides for the targeting of liposomes to integrin-expressing endothelial and melanoma cells. *Protein Eng Des Sel.* 2004;17(5):433–441.
72 White SJ, Nicklin SA, Sawamura T, Baker AH. Identification of peptides that target the endothelial cell-specific LOX-1 receptor. *Hypertension.* 2001;37(2 Part 2):449–455.
73 Kelly KA, Allport JR, Tsourkas A, et al. Detection of vascular adhesion molecule-1 expression using a novel multimodal nanoparticle. *Circ Res.* 2005;96(3):327–336.
74 Belizaire AK, Tchistiakova L, St-Pierre Y, Alakhov V. Identification of a murine ICAM-1-specific peptide by subtractive phage library selection on cells. *Biochem Biophys Res Commun.* 2003;309(3):625–630.
75 Mutuberria R, Satijn S, Huijbers A, et al. Isolation of human antibodies to tumor-associated endothelial cell markers by in vitro human endothelial cell selection with phage display libraries. *J Immunol Methods.* 2004;287(1–2):31–47.
76 Nicklin SA, White SJ, Watkins SJ, et al. Selective targeting of gene transfer to vascular endothelial cells by use of peptides isolated by phage display. *Circulation.* 2000;102(2):231–237.
77 Maruta F, Parker AL, Fisher KD, et al. Use of a phage display library to identify oligopeptides binding to the luminal surface of polarized endothelium by ex vivo perfusion of human umbilical veins. *J Drug Target.* 2003;11(1):53–59.
78 Muruganandam A, Tanha J, Narang S, Stanimirovic D. Selection of phage-displayed llama single-domain antibodies that transmigrate across human blood-brain barrier endothelium. *FASEB J.* 2002;16(2):240–242.
79 Volkel T, Muller R, Kontermann RE. Isolation of endothelial cell-specific human antibodies from a novel fully synthetic scFv library. *Biochem Biophys Res Commun.* 2004;317(2):515–521.
80 Muller OJ, Kaul F, Weitzman MD, et al. Random peptide libraries displayed on adeno-associated virus to select for targeted gene therapy vectors. *Nat Biotechnol.* 2003;21(9):1040–1046.
81 Romeo G, Frangioni JV, Kazlauskas A. Profilin acts downstream of LDL to mediate diabetic endothelial cell dysfunction. *FASEB J.* 2004;18(6):725–727.
82 Mazur A, Deylig A, Schaper W, et al. Biopanning of single-chain antibodies expressing phages reveals distinct expression patterns of angiogenic and arteriogenic vessels. *Endothelium.* 2003;10(4–5):277–284.
83 Giordano RJ, Cardo-Vila M, Lahdenranta J, et al. Biopanning and rapid analysis of selective interactive ligands. *Nat Med.* 2001;7(11):1249–1253.
84 Pasqualini R, Ruoslahti E. Organ targeting in vivo using phage display peptide libraries. *Nature.* 1996;380(6572):364–366.
85 Arap W, Pasqualini R, Ruoslahti E. Cancer treatment by targeted drug delivery to tumor vasculature in a mouse model. *Science.* 1998;279(5349):377–380.
86 Johns M, George AJ, Ritter MA. In vivo selection of sFv from phage display libraries. *J Immunol Methods.* 2000;239(1–2):137–151.
87 Yip YL, Hawkins NJ, Smith G, Ward RL. Biodistribution of filamentous phage-Fab in nude mice. *J Immunol Methods.* 1999; 225(1–2):171–178.
88 Zou J, Dickerson MT, Owen NK, et al. Biodistribution of filamentous phage peptide libraries in mice. *Mol Biol Rep.* 2004; 31(2):121–129.
89 George AJ, Lee L, Pitzalis C. Isolating ligands specific for human vasculature using in vivo phage selection. *Trends Biotechnol.* 2003;21(5):199–203.
90 Yao VJ, Ozawa MG, Trepel M, et al. Targeting pancreatic islets with phage display assisted by laser pressure catapult microdissection. *Am J Pathol.* 2005;166(2):625–636.

91 Brown DM, Ruoslahti E. Metadherin, a cell surface protein in breast tumors that mediates lung metastasis. *Cancer Cell.* 2004;5(4):365–374.
92 Samoylova TI, Smith BF. Elucidation of muscle-binding peptides by phage display screening. *Muscle Nerve.* 1999;22(4):460–466.
93 Arap W, Haedicke W, Bernasconi M, et al. Targeting the prostate for destruction through a vascular address. *Proc Natl Acad Sci USA.* 2002;99(3):1527–1531.
94 Arap W, Kolonin MG, Trepel M, et al. Steps toward mapping the human vasculature by phage display. *Nat Med.* 2002;8(2):121–127.
95 Essler M, Ruoslahti E. Molecular specialization of breast vasculature: a breast-homing phage-displayed peptide binds to aminopeptidase P in breast vasculature. *Proc Natl Acad Sci USA.* 2002;99(4):2252–2257.
96 Trepel M, Arap W, Pasqualini R. Modulation of the immune response by systemic targeting of antigens to lymph nodes. *Cancer Res.* 2001;61(22):8110–8112.
97 Kolonin MG, Saha PK, Chan L, et al. Reversal of obesity by targeted ablation of adipose tissue. *Nat Med.* 2004;10(6):625–632.
98 Marchio S, Lahdenranta J, Schlingemann RO, et al. Aminopeptidase A is a functional target in angiogenic blood vessels. *Cancer Cell.* 2004;5(2):151–162.
99 Houston P, Goodman J, Lewis A, et al. Homing markers for atherosclerosis: applications for drug delivery, gene delivery and vascular imaging. *FEBS Lett.* 2001;492(1–2):73–77.
100 Liu C, Bhattacharjee G, Boisvert W, et al. In vivo interrogation of the molecular display of atherosclerotic lesion surfaces. *Am J Pathol.* 2003;163(5):1859–1871.
101 Mintz PJ, Kim J, Do KA, et al. Fingerprinting the circulating repertoire of antibodies from cancer patients. *Nat Biotechnol.* 2003;21(1):57–63.
102 Michon IN, Hauer AD, von der Thusen JH, et al. Targeting of peptides to restenotic vascular smooth muscle cells using phage display in vitro and in vivo. *Biochim Biophys Acta.* 2002;1591(1–3):87–97.
103 Lee L, Buckley C, Blades MC, et al. Identification of synovium-specific homing peptides by in vivo phage display selection. *Arthritis Rheum.* 2002;46(8):2109–2120.
104 Wahid S, Blades MC, De Lord D, et al. Tumour necrosis factor-alpha (TNF-alpha) enhances lymphocyte migration into rheumatoid synovial tissue transplanted into severe combined immunodeficient (SCID) mice. *Clin Exp Immunol.* 2000;122 (1):133–142.
105 Cai D, Xaymardan M, Holm JM, et al. Age-associated impairment in TNF-alpha cardioprotection from myocardial infarction. *Am J Physiol Heart Circ Physiol.* 2003;285(2):H463–H469.
106 Koivunen E, Wang B, Ruoslahti E. Phage libraries displaying cyclic peptides with different ring sizes: ligand specificities of the RGD-directed integrins. *Biotechnology (N Y).* 1995;13(3):265–270.
107 Max R, Gerritsen RR, Nooijen PT, et al. Immunohistochemical analysis of integrin alpha vbeta3 expression on tumor-associated vessels of human carcinomas. *Int J Cancer.* 1997; 71(3):320–324.
108 Schlingemann RO, Oosterwijk E, Wesseling P, et al. Aminopeptidase a is a constituent of activated pericytes in angiogenesis. *J Pathol.* 1996;179(4):436–442.
109 Porkka K, Laakkonen P, Hoffman JA, et al. A fragment of the HMGN2 protein homes to the nuclei of tumor cells and tumor endothelial cells in vivo. *Proc Natl Acad Sci USA.* 2002; 99(11):7444–7449.
110 Christian S, Pilch J, Akerman ME, et al. Nucleolin expressed at the cell surface is a marker of endothelial cells in angiogenic blood vessels. *J Cell Biol.* 2003;163(4):871–878.
111 Laakkonen P, Porkka K, Hoffman JA, Ruoslahti E. A tumor-homing peptide with a targeting specificity related to lymphatic vessels. *Nat Med.* 2002;8(7):751–755.
112 Laakkonen P, Akerman ME, Biliran H, et al. Antitumor activity of a homing peptide that targets tumor lymphatics and tumor cells. *Proc Natl Acad Sci USA.* 2004;101(25):9381–9386.
113 Hoffman JA, Giraudo E, Singh M, et al. Progressive vascular changes in a transgenic mouse model of squamous cell carcinoma. *Cancer Cell.* 2003;4(5):383–391.
114 Joyce JA, Laakkonen P, Bernasconi M, et al. Stage-specific vascular markers revealed by phage display in a mouse model of pancreatic islet tumorigenesis. *Cancer Cell.* 2003;4(5):393–403.
115 Hallahan D, Geng L, Qu S, et al. Integrin-mediated targeting of drug delivery to irradiated tumor blood vessels. *Cancer Cell.* 2003;3(1):63–74.
116 Schluesener HJ, Xianglin T. Selection of recombinant phages binding to pathological endothelial and tumor cells of rat glioblastoma by in-vivo display. *J Neurol Sci.* 2004;224(1–2):77–82.
117 Zhi M, Wu KC, Dong L, et al. Characterization of a specific phage-displayed peptide binding to vasculature of human gastric cancer. *Cancer Biol Ther.* 2004;3(12):1232–1235.
118 Zurita AJ, Troncoso P, Cardo-Vila M, et al. Combinatorial screenings in patients: the interleukin-11 receptor alpha as a candidate target in the progression of human prostate cancer. *Cancer Res.* 2004;64(2):435–439.
119 Burkhart DJ, Kalet BT, Coleman MP, et al. Doxorubicin-formaldehyde conjugates targeting alphavbeta3 integrin. *Mol Cancer Ther.* 2004;3(12):1593–1604.
120 de Groot FM, Broxterman HJ, Adams HP, et al. Design, synthesis, and biological evaluation of a dual tumor-specific motive containing integrin-targeted plasmin-cleavable doxorubicin prodrug. *Mol Cancer Ther.* 2002;1(11):901–911.
121 Ellerby HM, Arap W, Ellerby LM, et al. Anti-cancer activity of targeted pro-apoptotic peptides. *Nat Med.* 1999;5(9):1032–1038.
122 Chen Y, Xu X, Hong S, et al. RGD-Tachyplesin inhibits tumor growth. *Cancer Res.* 2001;61(6):2434–2438.
123 Hu P, Yan J, Sharifi J, et al. Comparison of three different targeted tissue factor fusion proteins for inducing tumor vessel thrombosis. *Cancer Res.* 2003;63(16):5046–5053.
124 Curnis F, Sacchi A, Borgna L, et al. Enhancement of tumor necrosis factor alpha antitumor immunotherapeutic properties by targeted delivery to aminopeptidase N (CD13). *Nat Biotechnol.* 2000;18(11):1185–1190.
125 Dickerson EB, Akhtar N, Steinberg H, et al. Enhancement of the antiangiogenic activity of interleukin-12 by peptide targeted delivery of the cytokine to alphavbeta3 integrin. *Mol Cancer Res.* 2004;2(12):663–673.
126 Trepel M, Grifman M, Weitzman MD, Pasqualini R. Molecular adaptors for vascular-targeted adenoviral gene delivery. *Hum Gene Ther.* 2000;11(14):1971–1981.

127 Jacob D, Davis J, Zhu H, et al. Suppressing orthotopic pancreatic tumor growth with a fiber-modified adenovector expressing the TRAIL gene from the human telomerase reverse transcriptase promoter. *Clin Cancer Res.* 2004;10(10):3535–3541.
128 Lamfers ML, Grill J, Dirven CM, et al. Potential of the conditionally replicative adenovirus Ad5-Delta24RGD in the treatment of malignant gliomas and its enhanced effect with radiotherapy. *Cancer Res.* 2002;62(20):5736–5742.
129 Grill J, Van Beusechem VW, Van Der Valk P, et al. Combined targeting of adenoviruses to integrins and epidermal growth factor receptors increases gene transfer into primary glioma cells and spheroids. *Clin Cancer Res.* 2001;7(3):641–650.
130 Pastorino F, Brignole C, Marimpietri D, et al. Vascular damage and anti-angiogenic effects of tumor vessel-targeted liposomal chemotherapy. *Cancer Res.* 2003;63(21):7400–7409.
131 Oku N, Asai T, Watanabe K, et al. Anti-neovascular therapy using novel peptides homing to angiogenic vessels. *Oncogene.* 2002;21(17):2662–2669.
132 Gerlag DM, Borges E, Tak PP, et al. Suppression of murine collagen-induced arthritis by targeted apoptosis of synovial neovasculature. *Arthritis Res.* 2001;3(6):357–361.
133 Haubner R, Wester HJ, Weber WA, et al. Noninvasive imaging of alpha(v)beta3 integrin expression using 18F-labeled RGD-containing glycopeptide and positron emission tomography. *Cancer Res.* 2001;61(5):1781–1785.
134 Haubner R, Weber WA, Beer AJ, et al. Noninvasive visualization of the activated alphavbeta3 integrin in cancer patients by positron emission tomography and [(18)F]Galacto-RGD. *PLoECsS Med.* 2005;2(3):E70.
135 Chen X, Tohme M, Park R, et al. Micro-PET imaging of alphavbeta3-integrin expression with 18F-labeled dimeric RGD peptide. *Mol Imaging.* 2004;3(2):96–104.
136 Janssen M, Frielink C, Dijkgraaf I, et al. Improved tumor targeting of radiolabeled RGD peptides using rapid dose fractionation. *Cancer Biother Radiopharm.* 2004;19(4):399–404.
137 Chen X, Conti PS, Moats RA. In vivo near-infrared fluorescence imaging of integrin alphavbeta3 in brain tumor xenografts. *Cancer Res.* 2004;64(21):8009–8014.
138 Allen TM. Ligand-targeted therapeutics in anticancer therapy. *Nat Rev Cancer.* 2002;2(10):750–763.
139 Curnis F, Arrigoni G, Sacchi A, et al. Differential binding of drugs containing the NGR motif to CD13 isoforms in tumor vessels, epithelia, and myeloid cells. *Cancer Res.* 2002;62(3):867–874.
140 Xu Y, Wellner D, Scheinberg DA. Cryptic and regulatory epitopes in CD13/aminopeptidase N. *Exp Hematol.* 1997;25(6):521–529.

101

Hemostasis and the Endothelium

William C. Aird

Beth Israel Deaconess Medical Center, Harvard Medical School, Boston, Massachusetts

Vascular thrombotic disorders are the leading causes of mortality in the Western world. It is interesting to note that most, if not all of these disorders, although consisting of a systemic imbalance in hemostasis, are characterized by the development of focal thrombotic lesions (Tables 101-1 and 101-2). For example, microangiopathic hemolytic anemia (e.g., hemolytic uremia syndrome and thrombotic thrombocytopenia purpura) affect every microvascular bed with the notable exception of the liver and lung (see Chapter 146). The congenital hypercoagulable states, as exemplified by the factor V Leiden mutation, are associated with systemic changes in clotting factor level and/or activity, yet generally predispose patients to an increased risk of venous, and not arterial, thrombosis (1,2). In patients with warfarin-induced skin necrosis (a rare complication of coumarin treatment), a systemic imbalance occurs in vitamin K–dependent clotting factors (circulating protein C is disproportionately low compared with circulating factors II, VII, IX, and X). Remarkably, fibrin deposition is limited to postcapillary venules of the dermis (3,4). In the final analysis, very few, if any, thrombotic diseases affect the entire vasculature (2). This observation is supported by genetic mouse models, in which the deletion of one or another natural anticoagulant mechanisms results in vascular bed–specific fibrin deposition and thrombosis (5). An important question that arises from these clinical and animal studies is how a systemic imbalance in hemostasis is ultimately manifested by local rather than diffuse vasculopathic lesions. As discussed herein, the endothelium provides an important clue to the answer.

CLASSIFICATIONS IN HEMOSTASIS

Hemostasis is strictly defined as the arrest of bleeding. *Coagulation* is the transformation of a liquid into a semisolid or solid, coherent mass. In day-to-day practice, the terms *hemostasis* and *coagulation* are used interchangeably to describe the physiological process by which blood is maintained in a fluid state within the closed circulation. Hemostasis may be classified according to several schemes, some of which are outlined here (notably absent is a discussion of congenital versus acquired hemostatic defects).

Primary versus Secondary Hemostasis

Primary hemostasis refers to the platelet response, whereas secondary hemostasis refers to the soluble circulating clotting factors that converge in a cascade of enzymatic reactions to generate the end product, fibrin. In reality, the cellular and protein components of hemostasis are highly coordinated and interdependent; they function in unison in both space and time. Nevertheless, a conceptual distinction between these two compartments is helpful when considering mechanisms, diagnosis, and therapy. Platelets arise through the membrane budding of terminally differentiated precursor cells, termed *megakaryocytes*, which are located in the bone marrow. The megakaryocyte is one of the largest cells in the mammalian body and one of the few cells in the animal kingdom with a hyperdiploid or polyploid nucleus. Once released into the blood, platelets circulate freely as small discoid-shaped anucleate cells. When activated, platelets undergo a change in shape, adhere, and aggregate. During this process, they present a newly activated cell surface for assembly of the clotting cascade – just one example of the cross-talk that occurs between primary and secondary hemostasis. (The dialogue between platelets and endothelial cells is further detailed in Chapter 64). Assays for primary hemostasis include an inspection of the peripheral blood smear, platelet count, and a small number of ancillary tests, including platelet aggregation studies and heparin-induced thrombocytopenia (HIT) antibodies.

Secondary hemostasis consists of circulating proteins that are synthesized primarily in the liver. Exceptions include factor VIII (whose source is still debated), tissue factor (TF), and some of the anticoagulant proteins (discussed later). Clotting is initiated by TF-mediated activation of factor VII (extrinsic pathway) (Figure 101.1). TF is a transmembrane protein expressed on the surface of activated monocytes, in the subendothelial layers of the blood vessel wall, and by certain subsets

Table 101-1: Hypercoagulable States Associated with Disorders of Primary Hemostasis†

	Disease/Disorder	*Site of Thrombosis*
Acquired	TTP	All organs except lung and liver
	HUS	Predominantly kidney
	MPD	Portal/hepatic veins
	PNH	Portal/hepatic veins
	DIC	Microvessels; all organs are susceptible
	HIT	Arteries, veins, often in unusual sites
	APS	Arteries and veins
	Atherosclerosis	Conduit arteries
Congenital	Sickle cell disease	Microvessels, especially joints

TTP, thrombotic thrombocytopenia purpura; HUS, hemolytic uremic syndrome; MPD, myeloproliferative disorder; PNH, paroxysmal nocturnal hemoglobinuria; DIC, disseminated intravascular coagulation; APS, antiphospholipid antibody syndrome

†Primary and secondary hemostasis are integrally linked; therefore most of these diseases also are associated with activation of the clotting cascade. This is particularly true in the case of DIC, HIT, and APS.

of endothelial cells. Interestingly, the precise source of TF that is responsible for initiating coagulation in physiological states is still an open question – one that ultimately will be addressed by studying genetic mouse models lacking TF in one or another cell lineage. The clotting cascade is amplified through the intrinsic pathway via mechanisms that involve cross-talk (factor VIIa of the extrinsic pathway activates factor IX of the intrinsic pathway) and feedback (thrombin activates factors XI and VIII of the intrinsic pathway). The clotting cascade consists of a series of linked reactions in which a serine protease, once activated, is capable of activating its downstream substrate. These enzymatic reactions are phospholipid-dependent and take place on activated cell surfaces. Some reactions are accelerated by the presence of cofactors, namely factor VIIIa (which accelerates factor IXa-mediated activation of factor X) and factor Va (which accelerates factor Xa-mediated activation of prothrombin). In the final step of the clotting reaction, the serine protease, thrombin, cleaves the structural protein, fibrinogen, to generate fibrin. Fibrin is ultimately cross-linked through the activity of factor XIIIa.

For every procoagulant reaction, a natural anticoagulant response occurs. The anticoagulant mechanisms are regulated primarily by the endothelium. The endothelium expresses tissue factor pathway inhibitor (TFPI) (6), which forms a quaternary structure with TF, VIIa, and Xa, thus inhibiting the extrinsic pathway. The endothelium synthesizes heparan, a cofactor for antithrombin III (ATIII), which neutralizes each of the serine proteases in the clotting cascade (7,8). Protein C is converted to activated protein C in the presence of endothelial membrane-bound thrombomodulin (TM) and endothelial protein C receptor (EPCR) (9). Once activated, protein C inactivates the cofactors of the clotting cascade (factors VIIIa and Va), a process that is accelerated by the cofactor protein S. Finally, the fibrinolytic system may be thought of as a natural anticoagulant mechanism, in which plasmin degrades preformed fibrin. The endothelium synthesizes several proteins involved in fibrinolytic balance, including tissue-type plasminogen activator (t-PA) and plasminogen activator (PAI)-1.

Assays for secondary hemostasis include mixing studies, partial thromboplastin time (which measures the integrity of the intrinsic pathway), prothrombin time (which measures activity of the extrinsic pathway), and specific clotting factor assays. On the anticoagulant side, assays exist for ATIII, protein C, and protein S (TFPI levels also can be measured, but provide little in the way of useful information). In addition, genetic screening is available for diagnosing the factor V Leiden mutation and the prothrombin 20210 mutation, both of which are associated with increased risk for thrombosis. Routine assays do not presently exist for those anticoagulants that are bound to the endothelium, including TM, EPCR, and heparan (for an example of a study that describes expression of TM and EPCR in human subjects, see Ref. 10).

Hemostasis as a Finely Tuned Balance

Hemostasis may be viewed as a finely tuned balance between procoagulant and anticoagulant forces (1). On the procoagulant side is the monocyte, the platelet (primary hemostasis), and the clotting cascade (secondary hemostasis). The anticoagulant side includes the four mechanisms described in the previous section (TFPI, heparan/ATIII, protein C/protein S/TM, and fibrinolysis). In addition, blood flow (or lack of stasis) facilitates clearance of activated proteases and provides the endothelium with protective levels of shear stress. The maintenance of vascular integrity (intact endothelium) and the attenuation of negatively charged membrane surfaces limit the activation of primary and secondary hemostasis. Depending on which side the scale tips toward, abnormalities in hemostasis clinically manifest either bleeding (hemorrhage) or thrombosis (clotting).

Virchow's Triad

Virchow's triad represents a time-honored approach to hypercoagulability. The triad consists of a change in the blood vessel wall (namely loss of vascular integrity), a reduction in blood flow (namely stasis) or an alteration in blood constituents. Impairment in vascular integrity and/or local alterations in blood flow – as occur, for example, following hip surgery – are sufficient to explain many causes of focal or site-specific thrombosis. It may be argued that low flow in the deep veins of the leg render this site particularly vulnerable to thrombosis in patients with congenital hypercoagulable states. Moreover, patients with congenital hypercoagulable states have an increased risk of venous thromboembolism following trauma, surgery, or immobilization (11,12). However, Virchow's triad – as it is usually interpreted – fails to explain certain thrombotic phenotypes (e.g., the lack of correlation between congenital

Table 101-2: Hypercoagulable States Associated with Disorders of Secondary Hemostasis

	Disease/Disorder	*Virchow's Triad**	*Site of Thrombosis*
Acquired	Pregnancy	S/B	DVT
	Immobilization	S	DVT
	OCP	B	DVT
	V/A catheters	T	Site of catheter
	Trauma	T	Site of trauma
	Surgery	S/T	Site of surgery, DVT
	Sepsis	S/T/B	Multiple (but not all) organs
	Congestive heart failure	S	DVT
	DIC	B	Multiple (but not all) organs
	HIT	T/B	Arteries and veins
	APS	B	Arteries and veins
	Cancer	S/T/B	DVT
	Atherosclerosis	T/B	AMI, stroke, peripheral artery
	Organ transplantation	S/T/B	Transplanted organ
	Hyperhomocysteinemia	B	Venous and arterial
	Nephrotic syndrome	B	DVT, renal vein thrombosis
Congenital	ATIII deficiency	B	DVT
	PC deficiency	B	DVT
	PS deficiency	B	DVT
	Prothrombin mutation	B	DVT
	V Leiden	B	DVT
	Fibrinolytic defects (e.g., plasminogen deficiency)	B	DVT
	Dysfibrinogenemia	B	DVT
	Sickle cell disease	S/T/B	Different organs
	Hyperhomocysteinemia	B	Arteries and veins

OCP, oral contraceptive pill; V/A, venous or arterial; PC, protein C, PS, protein S; DVT, deep venous thrombosis; AMI, acute myocardial infarction

*Listed are the major contributors in Virchow's triad (as originally defined). S, stasis; B, blood constituents; T, trauma to blood vessel wall. Virtually every disease has an endothelial component.

hypercoagulable states and arterial thrombosis). Assuming that such individuals have the same incidence of atherosclerosis and plaque rupture as the normal population, then Virchow's triad would predict a higher rate of acute myocardial infarction. Another example worth considering is warfarin-induced skin necrosis. The fundamental abnormality is a disproportionate reduction in *circulating* protein C. According to Virchow's triad, the predilection for dermal clots must be explained by local stasis and/or damage in the microvasculature of the skin. No evidence supports either of these mechanisms.

When Virchow proposed his triad in 1845, virtually no understanding existed of the biology of the vessel wall or the nature of the blood constituents. However, as argued in the sections that follow, a modified version of Virchow's triad, which takes into account the critical role of the endothelium in modulating hemostatic phenotypes, provides a conceptual framework for approaching the local nature of thrombotic disorders.

SPATIAL AND TEMPORAL DYNAMICS OF THE ENDOTHELIUM

The endothelium plays a critical role in mediating hemostasis (1,13–15). As discussed earlier, ECs express TFPI, heparan, TM, EPCR, t-PA. Other endothelial-derived anticoagulants include ecto-ADPase, prostacyclin, and nitric oxide. On the procoagulant side, ECs synthesize TF, PAI-1, von Willebrand factor (vWF), and protease-activated receptors (1). Importantly, endothelial-derived anticoagulant and procoagulant molecules are unevenly distributed throughout the vasculature (Table 101-3). For example, TFPI is expressed predominantly in capillary endothelium, EPCR in large veins and arteries, endothelial nitric oxide synthase (eNOS) on the arterial side of the circulation, vWF in veins, and TM in blood vessel types of every caliber in all organs except the brain (1,16–18). TF is not detectable in normal endothelium, whereas in a baboon model of sepsis, the gene is upregulated in a subset of endothelial cells in the marginal zone of splenic follicles (19),

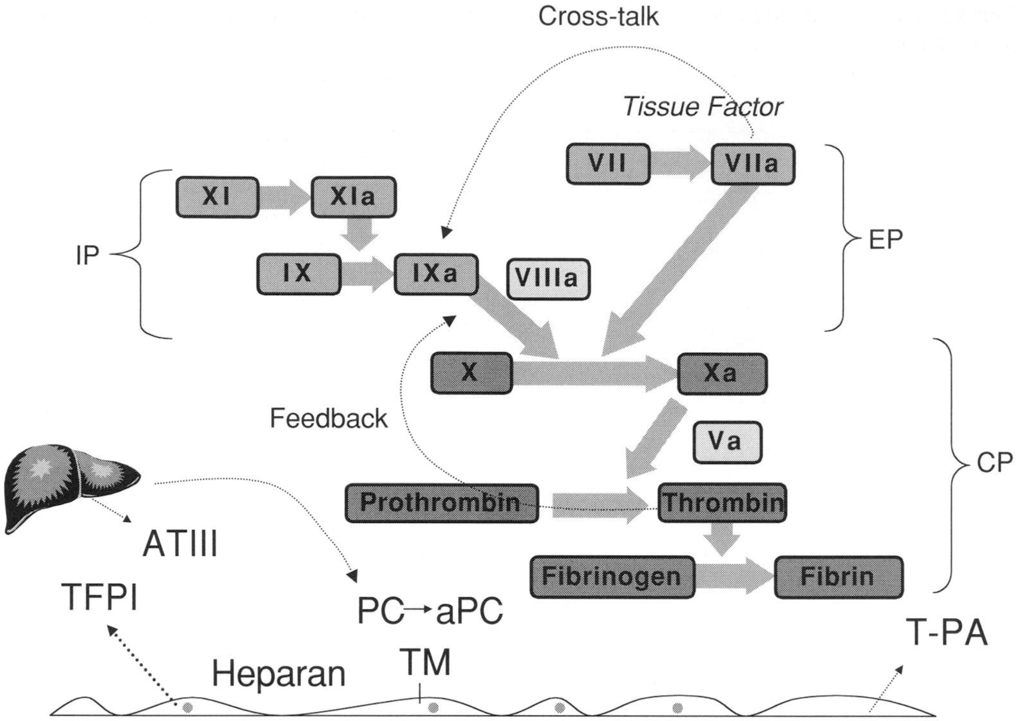

Figure 101.1. The coagulation pathway. The clotting cascade consists of an extrinsic pathway (*EP*; tissue factor, factor VII), an intrinsic pathway (*IP*; factors XI, IX, VIII), and a common pathway (*CP*; factors X, V, (pro)thrombin, and fibrinogen). Factors VII, XI, IX, X, II (thrombin) are serine proteases; factors V and VIII are cofactors; fibrinogen is a structural protein. Shown are the four major classes of natural anticoagulants: antithrombin III (ATIII)-heparan (which inhibits the serine proteases of the clotting cascade), protein C (PC)/protein S (not shown) and thrombomodulin (TM) (which inhibits the cofactors of the clotting cascade), tissue factor pathway inhibitor (TFPI) (which inhibits the extrinsic pathway), and the fibrinolytic system (A activates plasmin, which degrades fibrin). The liver and endothelium (the cell lining at bottom) both contribute to the synthesis and release of hemostatic factors. Note that factor XII is not included in the scheme. This factor, which can activate factor XI in vitro, is important to consider when interpreting results of coagulation assays. However, it does not play a critical role in mediating in vivo hemostasis. Two key links connect the extrinsic and common pathways with the intrinsic pathway. First, factor VIIa activates factor IX (cross-talk). Second, thrombin activates factors XI and VIII (feedback). t-PA; tissue-type plasminogen activator. The activated form of the serine protease is indicated by the suffix "a".

and in regions of disturbed flow (20). These data suggest that endothelial cells from different sites of the vascular tree employ site-specific "formulas" of procoagulants and anticoagulants to balance local hemostasis. These observations provide a powerful foundation for an integrated model of hemostasis (for review, see Ref. 21).

INTEGRATED MODEL OF HEMOSTASIS

The liver synthesizes a relatively constant amount of fibrinogen, serine proteases, cofactors, and anticoagulants (protein C, ATIII, protein S) (Figure 101.2). The bone marrow produces and releases into the circulation a relatively fixed number of monocytes and platelets, cells that are capable of either expressing TF or promoting clotting reactions. This net output of liver-derived proteins and marrow-derived cells is systemically distributed and integrated into the unique hemostatic balance of each and every vascular bed. Thus, when an alteration occurs in the net output of proteins (e.g., protein C or factor V Leiden deficiencies) or cells (e.g., increased activation of monocytes and/or platelets in sepsis), the changes affect the hemostatic balance in ways that differ among various sites of the vascular tree. To return to the example of warfarin-induced skin necrosis, it seems likely that the site-specific hemostatic balance of the postcapillary venular endothelium in the skin renders the dermal microvasculature particularly vulnerable to the systemic imbalance in vitamin K–dependent factors, particularly protein C. Interestingly, a similar pattern of thrombosis is seen in purpura fulminans associated with congenital homozygous protein C deficiency or meningococcemia-induced acquired protein C deficiency (22). Based on the sensitivity of dermal microvessels to protein C deficiency, it is interesting to speculate that dermal endothelium expresses lower levels of other natural anticoagulants (e.g., TFPI, heparan, and/or t-PA) compared with other

Table 101-3: Distribution of Representative Endothelial-Derived Hemostatic Factors

Factor	*Distribution*
vWF	Veins
eNOS	Arteries
TFPI	Capillaries
EPCR	Large veins and arteries
t-PA	Highest levels in brain; in lung in bronchial but not pulmonary circulation
TM	Absent in brain

vascular beds or that factors V and VIII are preferentially activated at this site.

Adding another layer of complexity (and heterogeneity) to the model is the fact that the endothelium may be differentially activated between sites of the vasculature. For example, sepsis may be associated with the local accumulation of cytokines and secondary changes in leukocyte adhesion, thrombin generation, barrier function, blood flow, and oxygenation, each of which may affect the local endothelial–derived hemostatic balance.

This revised scheme offers certain advantages over older models. First, it is more inclusive in that it recognizes the involvement of four functionally linked organ systems in mediating coagulation: namely, the liver, bone marrow, cardiovascular system, and endothelium. In this way, we are reminded of the importance of the hepatocyte in synthesizing the serine proteases, the two cofactors, fibrinogen, and the natural anticoagulants, ATIII, protein C, and protein S; the critical role of the TF-expressing monocyte in initiating coagulation; the participation of the platelet in localizing and perpetuating the coagulation response; and the importance of the EC as a manufacturer of hemostatic factors and regulator of the hemostatic balance. Second, the scheme incorporates both primary and secondary hemostasis. All too often, the cellular and soluble phases of coagulation are perceived as separate and independent entities that operate in series, when in fact they are highly integrated, parallel processes. Finally, the paradigm provides a useful conceptual framework for understanding the local nature of thrombotic diathesis. The very existence of vascular bed–specific phenotypes is enough to explain how a systemic imbalance in proteins and/or cells may be channeled into local clot formation.

It is not difficult to reconcile this model with that of Virchow. Two observations have not changed over the past 150 years. First, stasis of flow (e.g., secondary to obesity, tumor, congestive heart failure, pregnancy, or immobility) when introduced into the system, may lead to the accumulation of activated clotting factors, reduction in protective

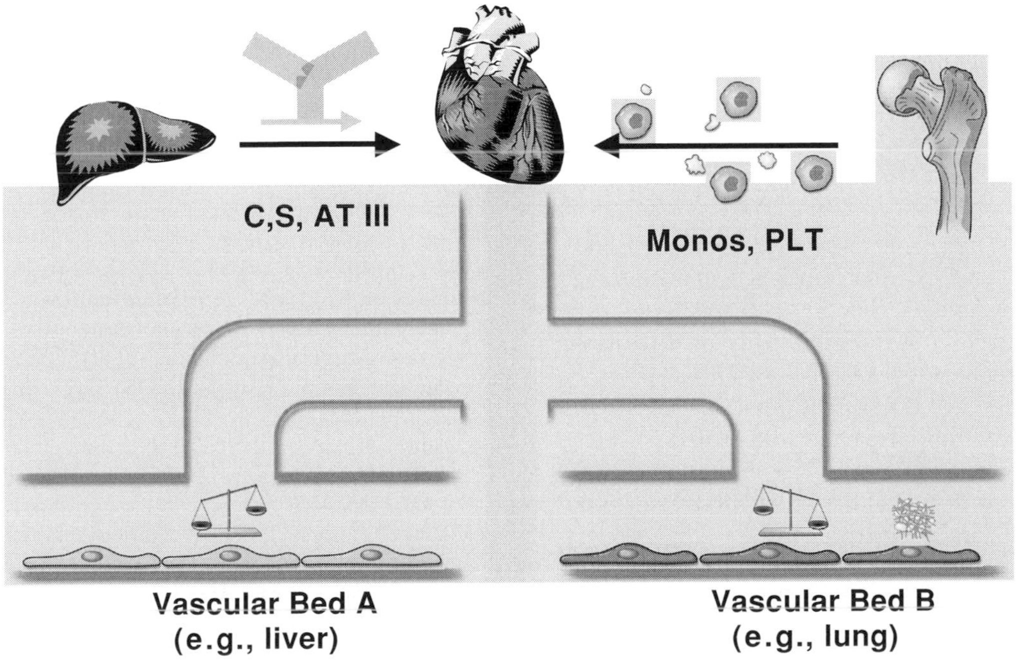

Figure 101.2. Integrated model of hemostasis. The liver (*left*) produces the serine proteases, cofactors fibrinogen of the clotting cascade (shown as Y-shape), and many of the circulating natural anticoagulants. Shown are protein C (*C*), protein S (*S*), antithrombin III (*ATIII*). The bone marrow (*right*) releases monocytes (Monos) and platelets (PLT), which are capable of expressing tissue factor and/or an activated cell surface. The liver and bone marrow–derived proteins and cells are systemically distributed and integrated into the unique hemostatic balance of each vascular bed (shown are balances in two hypothetical vascular beds). (Adapted with permission from Aird WC. *Crit Care Med.* 2001 July; 29:S28–34.)

hemodynamic forces, and downstream hypoxia – all of which may tip the balance to the procoagulant side. Second, frank disruption or denudation of the endothelium (e.g., as occurs in trauma, surgery, or catheter placement) may result in the exposure of blood to subendothelial adventitial TF and secondary thrombosis. However, two new observations warrant emphasis. First, when talking about blood constituents, we now appreciate the importance not only of the soluble clotting factors, but also of the cells – and not just the platelet, but also the monocyte, perhaps the single most important initiator of blood coagulation. Second, we now understand that the endothelium is far from a passive barrier; it is highly active, very much alive, rich in diversity, and steeped in complexity – and as a result is a critical determinant of local hemostatic balance. Based on our new knowledge, Virchow's triad may be modified – slightly – to incorporate not only of the structural but also the functional integrity of the vessel wall, particularly as it relates to the endothelium.

CONCLUSION

Virchow's triad has withstood the test of time and continues to provide a working foundation for approaching patients with thrombotic disorders. However, the strict application of nineteenth-century principles to a modern-day understanding of thrombosis neglects important advances in the fields of hemostasis and vascular biology. In this chapter, we have proposed an integrated model that honors the spirit of the original triad, while at the same time recognizing the critical role of the endothelium in mediating hemostasis.

KEY POINTS

- The endothelium is a key regulator of hemostasis and serves to channel systemic imbalances in hemostasis into focal thrombotic phenotypes.
- The endothelium synthesizes many procoagulant and anticoagulant molecules.
- Each vascular bed expresses a unique repertoire of procoagulant and anticoagulant molecules.
- The endothelium is part of an integrated model of hemostasis that includes the activity of multiple organs, cells, and soluble proteins.

Future Goals

- To determine the primary source of TF
- To elucidate/catalogue vascular bed–specific balances of procoagulant and anticoagulants
- To use that information to develop site-specific anticoagulant therapies

REFERENCES

1 Aird WC. Vascular bed-specific hemostasis: role of endothelium in sepsis pathogenesis. *Crit Care Med.* 2001;29:S28–35.
2 Rosenberg RD, Aird WC. Vascular-bed-specific hemostasis and hypercoagulable states. *N Engl J Med.* 1999;340:1555–1564.
3 Chan YC, Valenti D, Mansfield AO, Stansby G. Warfarin induced skin necrosis. *Br J Surg.* 2000;87:266–272.
4 Stewart AJ, Penman ID, Cook MK, Ludlam CA. Warfarin-induced skin necrosis. *Postgrad Med J.* 1999;75:233–235.
5 Weiler-Guettler H, Christie PD, Beeler DL, et al. A targeted point mutation in thrombomodulin generates viable mice with a prethrombotic state. *J Clin Invest.* 1998;101:1983–1991.
6 Broze GJ, Jr. Tissue factor pathway inhibitor. *Thromb Haemost.* 1995;74:90–93.
7 Bauer KA, Rosenberg RD. Role of antithrombin III as a regulator of in vivo coagulation. *Semin Hematol.* 1991;28:10–18.
8 Damus PS, Hicks M, Rosenberg RD. Anticoagulant action of heparin. *Nature.* 1973;246:355–357.
9 Esmon CT. The protein C pathway. *Chest.* 2003;124:26S–32S.
10 Faust SN, Levin M, Harrison OB, et al. Dysfunction of endothelial protein C activation in severe meningococcal sepsis. *N Engl J Med.* 2001;345:408–416.
11 Sanson BJ, Simioni P, Tormene D, et al. The incidence of venous thromboembolism in asymptomatic carriers of a deficiency of antithrombin, protein C, or protein S: a prospective cohort study. *Blood.* 1999;94:3702–3706.
12 Simioni P, Sanson BJ, Prandoni P, et al. Incidence of venous thromboembolism in families with inherited thrombophilia. *Thromb Haemost.* 1999;81:198–202.
13 Cines DB, Pollak ES, Buck CA, et al. Endothelial cells in physiology and in the pathophysiology of vascular disorders. *Blood.* 1998;91:3527–3561.
14 Bombeli T, Mueller M, Haeberli A. Anticoagulant properties of the vascular endothelium. *Thromb Haemost.* 1997;77:408–423.
15 Gross PL, Aird WC. The endothelium and thrombosis. *Semin Thromb Hemost.* 2000;26:463–478.
16 Yamamoto K, de Waard V, Fearns C, Loskutoff DJ. Tissue distribution and regulation of murine von Willebrand factor gene expression in vivo. *Blood.* 1998;92:2791–2801.
17 Osterud B, Bajaj MS, Bajaj SP. Sites of tissue factor pathway inhibitor (TFPI) and tissue factor expression under physiologic and pathologic conditions. On behalf of the Subcommittee on Tissue Factor Pathway Inhibitor (TFPI) of the Scientific and Standardization Committee of the ISTH. *Thromb Haemost.* 1995; 73:873–875.
18 Ishii H, Salem HH, Bell CE, et al. Thrombomodulin, an endothelial anticoagulant protein, is absent from the human brain. *Blood.* 1986;67:362–365.
19 Drake TA, Cheng J, Chang A, Taylor FB, Jr. Expression of tissue factor, thrombomodulin, and E-selectin in baboons with lethal *Escherichia coli* sepsis. *Am J Pathol.* 1993;142:1458–1470.
20 Lupu C, Westmuckett AD, Peer G, et al. Tissue factor-dependent coagulation is preferentially up-regulated within arterial branching areas in a baboon model of *Escherichia coli* sepsis. *Am J Pathol.* 2005;167:1161–1172.
21 Aird WC. Spatial and temporal dynamics of the endothelium. *J Thromb Haemost.* 2005;3:1392–1406.
22 Wiss K. Clotting and thrombotic disorders of the skin in children. *Curr Opin Pediatr.* 1993;5:452–457.

102

Von Willebrand Factor

Tom Diacovo

Columbia University, New York City, New York

von Willebrand factor (vWF) is a large glycoprotein of complex multimeric structure. It is synthesized exclusively by vascular endothelium and megakaryocytes, and is stored and secreted by these cell types as well as by circulating platelets. In blood, it serves not only as a carrier for factor VIII, protecting it from proteolysis, but also plays a critical role in hemostasis by promoting platelet deposition at sites of endothelial cell (EC) injury. To accomplish this task, vWF must form a "bridge" between components of the subendothelial matrix or the surface of inflamed venular endothelium and receptors expressed on circulating platelets. One remarkable feature of this plasma glycoprotein is its ability to support significant interactions with platelets only upon surface immobilization and under specific hemodynamic conditions. This avoids the disastrous consequences of platelet–vWF aggregate formation in flowing blood. The importance of vWF in hemostasis is underscored by the occurrence of clinical bleeding when plasma levels fall below 50 IU/dL or when functional defects are present in the protein due to spontaneously occurring or inherited mutations. Such abnormalities result in von Willebrand disease (vWD), the most common inherited bleeding disorder in humans, with an estimated prevalence of 1% in the general population (1,2).

HISTORICAL ASPECTS

Recognition that certain families may have a predisposition to bleeding has been described in the Talmud and other Hebrew writings as early as the second century A.D. (3). These Jewish texts decreed that offspring were exempt from circumcision if deaths due to hemorrhage occurred in other male siblings undergoing this procedure, suggesting that these individuals suffered from hemophilia. Yet, it was not until 1926 that the Finnish physician, Dr. Erik von Willebrand identified a new bleeding disorder he termed *hereditary pseudohemophilia* (4). In his initial study, he described a hematological disorder, affecting several families, which was transmitted in an autosomal dominant fashion and presented mainly as recurrent episodes of mucosal bleeding but occasionally as fatal hemorrhages. Although bleeding time was prolonged despite relatively normal platelet counts in afflicted individuals, there was a lack of a male gender bias, distinguishing it from hemophilia. Unfortunately, the etiology of this newly defined bleeding disorder remained elusive to Dr. von Willebrand throughout his career. It would eventually be named in his honor, vWD.

It was not until 45 years later when immunologic assays were devised to measure the levels of clotting factors in blood that a deficiency in a plasma protein distinct from factor VIII, termed vWF, was identified as the cause for this bleeding disorder. Subsequent cloning of the *vWF* gene, together with knowledge pertaining to its synthesis, secretion, and ligand-binding characteristics have been instrumental in elucidating the molecular mechanisms that contribute to this disorder, as well as in the development of a classification system for vWD. More recently, crystallographic analysis of specific regions contained within this plasma protein has provided important insights into potential structural mechanisms that may regulate its hemostatic properties.

VON WILLEBRAND FACTOR GENE

The gene encoding for human vWF has been cloned and located on chromosome 12p13.2, while its murine counterpart has been identified on chromosome 6. It spans 178 kb and contains 52 exons corresponding to 8.7 kb of mRNA (5). A noncoding, highly similar pseudogene has been identified on chromosome 22q11–13 in higher primates and includes a region spanning from exon 23 to 34 (6). However, it is not believed to yield functional transcripts because of the presence of a splice site and nonsense mutations. Distinct modifier genes and cell type–specific transcription factors are believed to regulate plasma levels of vWF and account for its restricted expression in megakaryocytes and ECs (7–10).

The primary translation product is a 2,813-residue precursor polypeptide, referred to as prepro-vWF, which consists of a 22-residue signal peptide, a 741-residue propeptide, and the mature subunit of 2,050 residues (molecular weight ~278 kDa) (Figure 102.1). The nascent vWF protein is highly

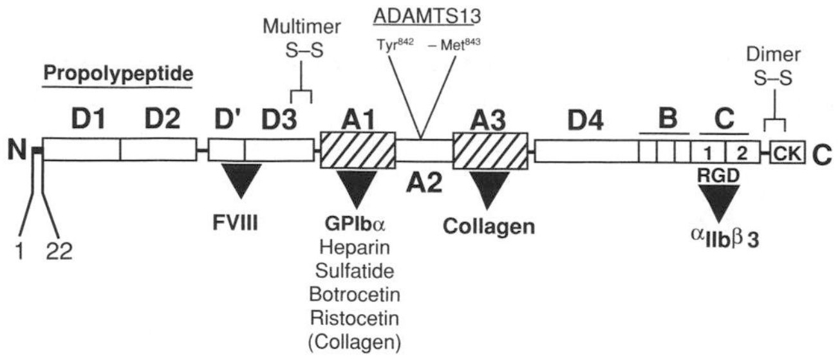

Figure 102.1. Schematic representation of prepro-vWF. The vWF precursor consists of a signal peptide (residues 1–22), a propeptide (residues 23–763), and the mature subunit (residues 764–2050). The location of the five types of conserved structural domains is indicated, as well as their respective ligands or receptors (boldface designates a major receptor or ligand). Location of intersubunit disulfide bonds yielding pro-vWF dimers or multimers of the mature protein also is indicated.

glycosylated (10%–19%), containing 22 carbohydrate side chains, 10 of which are *O*-linked to Ser or Thr residues and 12 *N*-linked to Asn residues (11). The *N*-linked oligosaccharides of vWF synthesized in endothelium are unusual in that they contain ABO blood group antigens (12). Although the presence of these blood group antigens appears to impact on vWF levels, they only account for a fraction of the total variation in the population (20%–30%). Evidence is mounting, however, that the pattern of vWF glycosylation may be important in regulating its stability and function in the circulation. This is supported by the shortened half-life of circulating vWF in animals lacking oligosaccharide branch-specific sialylation because of a deficiency in ST3Gal-IV sialyltransferase (13).

VON WILLEBRAND FACTOR PROTEIN

To provide effective hemostasis, vWF must exist in a multimeric form composed of dimers of the mature peptide. Considerable progress has been made in understanding the biochemistry of this process in endothelium, which appears to occur in an analogous manner in megakaryocytes (14). Pro-vWF subunits are assembled in the endoplasmic reticulum, "tail-to-tail" through disulfide bonds formed between their carboxy termini. Specifically, a cysteine knot (CK) domain comprised of 90 residues has been implicated in dimerization, based on a naturally occurring mutation in humans and site-directed mutagenesis studies (15). After the formation of this dimeric building block, the glycosylated, pro-vWF dimers are transported to the Golgi, whereupon multimerization occurs through the formation of "head-to-head" intersubunit disulfide bonds at the amino-termini ends of dimers. This process requires the presence of the propeptide, which is comprised of domains D1 and D2, because multimer formation in, but not transport to, the Golgi is abolished in its absence. The D1 and D2 domains posses *CXXC* motifs found in thiol:disulfide oxidoreductases, enzymes known to promote the rearrangement or the formation of disulfide bonds. Recent evidence has suggested that the propeptide may indeed function in this capacity (16). Ultimately, this process can yield vWF multimers that contain more than 40 subunits, exceeding 20,000 kDa in size; these are comparable in length to the diameter of a platelet (4 μM). The highest-molecular-weight multimers are most functionally active in hemostasis, and are ultimately stored in Weibel-Palade bodies in endothelium and in α-granules in platelets (17). In the case of the former, vWF appears to have a heterogeneous pattern of expression in blood vessels, as determined by immunohistology, with the strongest staining intensity exhibited in veins, followed by arteries, arterioles, capillaries, and venules (18).

STORAGE AND SECRETION IN ENDOTHELIUM

Two distinct pathways of secretion exist for vWF produced by endothelium: one constitutive, which is linked to synthesis and is directed toward both the apical and luminal surfaces; the other inducible, relying on storage and subsequent release from an intracellular, membrane-bound organelle known as a Weibel-Palade body, in response to physical trauma or physiological agonists (Figure 102.2). This latter pathway is most efficacious in terms of platelet recruitment. This is explained by the fact that the majority of high-molecular-weight multimers synthesized by endothelium are retained Weibel-Palade bodies. In response to agonists that raise either intracellular Ca^{2+} (i.e., thrombin) or cyclic adenosine monophosphate (cAMP) levels (i.e., epinephrine and vasopressin), Weibel-Palade bodies undergo exocytosis in a basolateral direction, resulting in an increase in the effective concentration of large multimers in the extracellular matrix (ECM) (14). By contrast, this unique polarity appears to be lost upon the biomechanical disruption of the endothelium, which has been reported to result in the bidirectional release of high-molecular-weight multimers of vWF. This alteration in secretion may prevent further

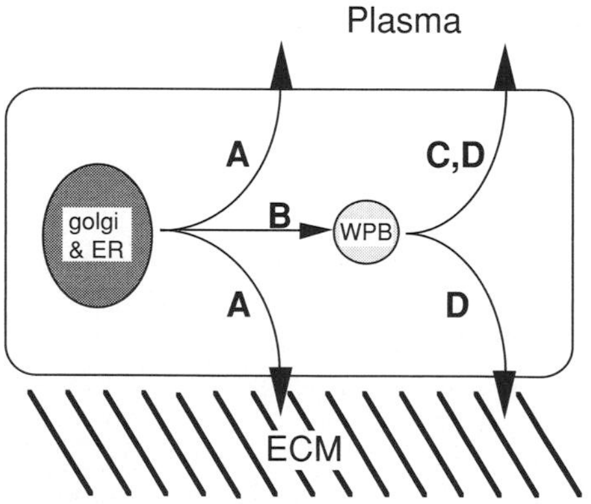

Figure 102.2. Secretion of vWF from endothelium. (**A**) Constitutive secretion to plasma and extracellular matrix (ECM). (**B**) Storage in Weibel-Palade body (WPB). (**C**) Exocytosis of WPB in response to circulating agonists. (**D**) Exocytosis of WPB in response to biomechanical disruption of the endothelium.

dislodgement of damaged endothelium by anchoring it to the ECM, while simultaneously providing an effective substrate for platelet deposition. In either case, this ensures that the resulting thrombogenic response will be maximal at the site of injury. vWF storage, however, is not limited to ECs because megakaryocytes and platelets also possess this capability (stored in organelles known as α-granules); however, this source of vWF contributes minimally to plasma levels (19). Release of α-granule contents requires platelet activation, which occurs in response to agonists generated or exposed at sites of vascular injury. Both the constitutive and inducible secretory pathways in endothelium contribute to plasma levels of mature vWF, which average approximately 100 IU/dL (range 50–200 IU/dL), with a circulating half-life of approximately 12 hours (20,21). Because the distribution of vWF in the subendothelial matrix may be variable throughout the vascular system, this circulating pool of vWF is believed to be important in ensuring rapid platelet deposition wherever injured tissue comes into contact with circulating blood.

REGULATION OF PLASMA VON WILLEBRAND FACTOR MULTIMER SIZE

Once released into the circulation, vWF is then subjected to proteolysis that converts the large multimers into a series of smaller ones. This involves cleavage of a single bond between residues Tyr842 and Met843 of its A2 domain, a process that relies on the activity of a metalloprotease known as ADAMTS13 (a disintegrin and metalloprotease, with thrombospondin-1-like domains, member 13) (22) and is enhanced in the presence of fluid shear force (23). The gene encoding this protein has been identified and localized to chromosome 9q34 in humans. The importance of vWF proteolysis is exemplified by genetic mutations or circulating autoantibodies that result in loss of protease activity, contributing to chronic thrombotic thrombocytopenic purpura (TTP). This disorder is characterized by the presence of unusually large vWF multimers (ULvWF) in plasma and the formation of thrombi in the microcirculation. It is believed that these ULvWF multimers are responsible for spontaneous platelet clumping in TTP.

STRUCTURE AND FUNCTION OF PLASMA VON WILLEBRAND FACTOR

The majority of the mature vWF protein (>90%) consists of four domains, each of which is present in multiple copies and exhibits internal homologies with one another. They are arranged in the following order: D′-D3-A1-A2-A3-D4-B1-B2-B3-C1-C2 (see Figure 102.1). In general, each domain performs a specific function, but may interact with several ligands. In regard to hemostasis, the A1 domain of vWF is solely responsible for supporting the initial capture of circulating platelets, while the primary function of its A3 domain is to anchor the vWF–platelet complex to the site of injury. Significant progress has been made in terms of elucidating the structural elements contained within these two domains that form direct contacts with or regulate their interactions with ligands.

For vWF to commence platelet recruitment to a region of vascular damage, it must first become immobilized by interacting with a reactive surface. The A3 domain (residues 910–1111 of mature vWF) is well suited for this process because it contains a functional binding site for collagen, a major component of exposed ECM (24). Evidence to support this claim is provided by an in vivo animal study demonstrating a reduction in arterial thrombosis upon administration of a function-blocking antibody to A3 (25) as well as identification of patients with a moderate bleeding disorder due to a point mutation that perturbs the binding of this domain to collagen (26). Although the A1 domain also possesses collagen binding activity (27), no overlap occurs between these domains in terms of the types of collagen that support such interactions. To date, no direct in vivo evidence exists demonstrating the importance of the A1 domain in this process.

A binding site for only two types of collagen (I and III) has been identified within the A3 domain (28). Both types are known to form fibrils composed of three helical chains, which is thought to enhance their interactions with this region of vWF (29). The multimeric structure of vWF also appears to strengthen adhesion due to the presence of numerous A3 domains, each of which can bind to exposed collagen fibrils (30). The formation of multiple A3–collagen bonds is believed to increase the overall avidity of the interaction, thus firmly anchoring vWF to the ECM.

Structurally, the A3 domain shares many common attributes with integrin I domains that also bind collagen (i.e., α-chains of integrins $\alpha 1\beta 1$, $\alpha 2\beta 1$, $\alpha 10\beta 1$, and $\alpha 11\beta 1$). This includes a "dinucleotide binding" fold, composed of a central β-sheet flanked on both sides by amphipathic α-helices, and the metal ion–dependent adhesion site (MIDAS) motif. Although present in a modified form, the MIDAS motif is not functional and, in contrast to its integrin counterparts, collagen binding is cation independent (31). Other distinctions between these two domains include unique collagen binding sites (hydrophobic and flat versus hydrophilic groove, respectively), differences in the modulation of its collagen binding affinity, and the presence of an intra-disulfide bridge linking the N- and C-termini (Cys923–Cys1109) of A3 (32,33). Although it has been speculated that the interactions of the A3 domain with collagen may regulate the binding activity of the A1 domain in terms of interactions with platelets, no direct physiological evidence exists for such a paradigm. Thus, it appears that the primary function of the A3 domain is the immobilization of vWF to components of the subendothelial matrix.

In addition to binding to components of the subendothelial matrix, vWF can also become immobilized on the surface of intact venular endothelium. Previously, it has been demonstrated that mouse mesenteric venules stimulated with the calcium ionophore A23187 can support platelet attachment and translocation in vivo, an event reliant on endothelium-bound vWF (34). Moreover, P-selectin, an adhesion molecule expressed on the surface of inflamed venular endothelium has been implicated in anchoring ultralarge vWF multimers to these cells in vitro (35). The in vivo significance of this finding, however, remains to be determined.

Once immobilized, the role of the vWF-A1 domain (residues 497–717) is to initiate the process of platelet deposition at sites of vascular injury. The critical nature of this interaction is exemplified by the bleeding disorder, termed *type 2M vWD*, which results from the incorporation of loss-of-function mutations within this domain (36). In addition, recombinant vWF multimers lacking the A1 domain cannot support platelet adhesion in flow despite retaining the ability to interact with collagen (37). In regards to platelet receptors, it is glycoprotein Ib (GPIb)-α that possesses a unique binding site for this region of vWF, which is contained within its N-terminal region (residues 1–293). GPIbα, a member of the leucine-rich repeat (LRR) family, is constitutively expressed on the surface of platelets in an active form. It exists in a complex with at least four other polypeptide chains that include GPIbβ, to which it is disulfide linked, and GPIX and GPV that associate noncovalently with GPIbα. Eight LRRs found within the binding site for the A1 domain form an elongated curve, with a region proximal to the N-terminus protruding from the protein surface (referred to as the β-finger) and a disorder loop (referred to as the β-switch) found in its C-terminus of this region (Figure 102.3) (38,39).

The structure of the A1 domain is similar to that of the A3, and includes the α/β fold, with a central β-sheet flanked by α-helices on each side as well as one intradisulfide bond (Cys509–Cys 695), but no MIDAS motif (see Figure 102.3) (40). Its overall shape is cuboid, with the top and bottom faces forming the major and minor binding sites that interact with the concave surface of GPIbα. The most extensive contact site buries approximately 1,700 Å^2 of surface area, interacting with LRRs five to eight and the C-terminal flank of the GPIbα (39). For this to occur, the β-switch region of this platelet receptor undergoes a conformational change so that it aligns itself with the central β-sheet of the A1 domain. The smaller site ($\sim$900 Å^2) accommodates the binding of the β-finger and the first LRR of GPIbα, an event that appears to require the displacement of the amino-terminal extension of the A1 domain. Based on these findings, as well as the preferential localization of mutations in humans within this region, which enhance GPIbα binding, it is speculated that the amino-terminal extension regulates the adhesive properties of this domain. Moreover, recombinant A1 proteins lacking this extension have a higher affinity for this platelet receptor (41). Despite these observations, the physiological relevance of such structural changes in this receptor–ligand pair remains to be determined, as well as the contribution of other domains to this process.

A1-MEDIATED ADHESION IN FLOW

Once immobilized on the EC surface or exposed subendothelial matrix, vWF faces a monumental task: engaging GPIbα on platelets while these cells are traveling at considerable velocities within the circulation. Moreover, hydrodynamic forces

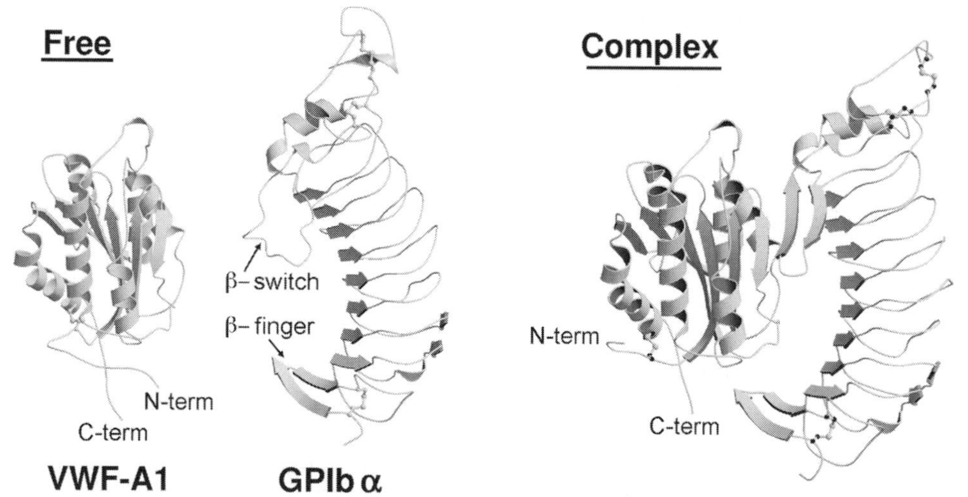

Figure 102.3. Structure of the vWF–A1 domain and the amino-terminal region of the platelet receptor GPIbα. In the uncomplexed form, the amino-terminal extension of the A1 domain lies across its globular core, whereas the β-switch region of GPIbα lacks defined secondary structure. In the binary complex, the concave surface of GPIbα formed by the LRR faces the A1 domain so that its flanking β-finger makes contacts at its base, an event that appears to require the displacement of the amino-terminal extension of the A1 domain. In addition, the β-switch region of GPIbα undergoes a conformational change that results in the formation of two β-strands that bind near the top of A1, thus facilitating interactions as well as extending the central β-sheet of this domain.

generated by flowing blood (termed *shear stress*), which are maximal at the vessel wall, tend to disrupt such interactions once formed. A remarkable feature of the A1 domain is its ability not only to maximize binding to GPIbα under conditions of high flow, but also to limit platelet–vWF aggregate formation to sites of vascular injury. Thus, it is not unexpected that this multimeric glycoprotein functions best under the prevailing hydrodynamic conditions that exist in arterioles and arteries, where injuries can result in considerable blood loss. Interestingly, the adhesive bond formed between this receptor–ligand pair is relatively labile in that platelets captured by vWF initially translocate along the injured vessel wall in the direction of flowing blood. This interaction reduces the velocity of platelets and brings them in close apposition with the exposed ECM and other domains of vWF, permitting additional adhesion molecules on platelets (such as the integrin $\alpha 2\beta 1$ and/or αIIb$\beta 3$) to engage their cognate ligands, an event that cannot precede GPIbα binding at high rates of flow (42). This permits the firm adhesion and spreading of platelets, thus retaining them at the site of vessel injury. This multistep adhesive process is analogous to that reported for leukocyte recruitment to sites of inflammation, where the selectin family of adhesion molecules performs an identical role as GPIbα (43). Despite the unique structures and ligand binding requirements of these two classes of adhesion receptors, their ability to support the rapid attachment and translocation of leukocytes or platelets, respectively, suggests that the bonds they form may possess similar kinetic properties. Indeed, previous studies evaluating the biophysical properties of the GPIbα–A1 domain bond (44) have demonstrated that flow-dependent adhesion and rapid and force-driven kinetics are the predominant features of this interaction, characteristics shared by selectin molecules (45) (Figure 102.4). Thus, the ability of these two distinct families of adhesion molecules to perform similar biological functions relies not on structural homologies, but rather on evolutionary bias toward rapid rates of association and dissociation that are paramount for cell capture and translocation in flowing blood.

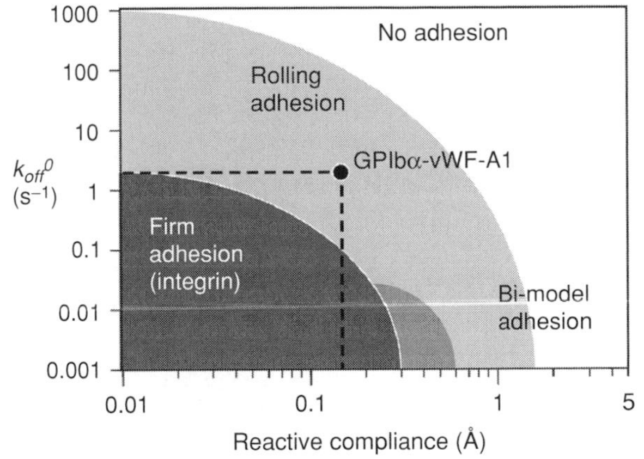

Figure 102.4. State diagram depicting the transitions in adhesion based on the kinetic properties of receptor-ligand bonds in flow. The rolling adhesion area represents the region of parameter space where rolling motion occurs over some part of shear rates range from 30 to 400 s^{-1}. The x- and y-axis denote values for the intrinsic lifetime ($k_{off}{}^{\circ}$) and reactive compliance (Å) of receptor–ligand bonds, respectively. The estimated $k_{off}{}^{\circ}$ for the GPIbα-A1 bond is $3.5 \pm 0.4\ s^{-1}$, with a reactive compliance (resistance of the bond to shear-induced dissociation) of 0.16 ± 0.002 Å, values similar to that reported for selectin–ligand interactions (45).

Based on the alteration in A1 structure that is required to accommodate GPIbα binding, it is reasonable to speculate that certain in vivo conditions may favor one conformation over another, thus limiting or enhancing adhesion. This may explain the dependency on surface-immobilization and specific flow rates to promote interactions between platelets and vWF. For example, it is conceivable that surface immobilization of vWF and the hydrodynamic forces generated by flowing blood work in concert to either directly alter the structure of the A1 domain and/or increase its effective concentration by transforming this multimeric glycoprotein from a globular to an extended-chain conformation (46). In the latter case, the increased accessibility of A1 domains would permit multiple GPIbα bonds to form, thus stabilizing platelet–vWF interactions. By contrast, the globular nature of vWF in the circulation would tend to limit the number of A1 domains able to interact with platelets. This, in combination with the rapid dissociation kinetics of the complex, may preclude additional bonds from forming, which would prevent resting platelets from spontaneously aggregating with plasma vWF in flowing blood. Although the exact mechanism responsible for regulating adhesion between this receptor–ligand pair remains to be determined, evidence is mounting that the bond lifetime of the complex must be precisely maintained. For example, function-enhancing mutations localized to either the amino-terminus extension of the A1 domain or the β-switch region of GPIbα not only share several clinical attributes (including a prolonged bleeding time, borderline thrombocytopenia, and a decrease in the high-molecular-weight multimers of vWF), but also result in an identical sixfold prolongation in bond lifetime of the mutant complex (47). Patients with these genetic disorders have increased bleeding tendencies that are believed to result from the spontaneous binding of plasma vWF to circulating platelets and the subsequent clearance of these hemostatic elements from blood. It has been speculated that prolongation in the lifetime of the mutant complex would favor multiple bond formation, thus stabilizing random interactions between platelets and vWF in the circulation. More recently, it has been demonstrated that botrocetin, a snake venom protein that promotes vWF–platelet aggregation in low-flow states, also prolongs the duration of the interaction between GPIbα and the A1 domain. By contrast to function-enhancing mutations, this occurs in the absence of any structural changes induced by botrocetin in either molecule, but rather relies on the formation of a new interface between this snake venom protein and GPIbα (48). Thus, maintaining a rapid rate of dissociation between the A1 domain and GPIbα appears to be critical for preventing inadvertent aggregation between these two hemostatic elements while in the circulation.

KEY POINTS

- ECs are the major source of vWF, a multimeric plasma glycoprotein essential for maintaining hemostasis.
- The largest vWF multimers, which are stored in Weibel-Palade bodies of endothelium and α-granules of platelets, are required for effective platelet deposition at sites of vascular injury.
- A reduction in vWF multimer size in plasma through proteolytic processing by ADAMTS13 appears to play a role in preventing inappropriate thrombus formation.
- The A1 domain of vWF is the major binding site for the platelet receptor GPIbα, and the rapid and force-driven kinetics of this interaction is well suited to promote platelet deposition at sites of vascular injury under arterial flow rates.

Future Goals

- To decipher key structural elements critical for vWF interactions with platelets, endothelium, and components of the ECM, as well as understand the biophysical properties of such interactions, as essential to designing therapies aimed at the prevention and/or amelioration of thrombotic disorders

REFERENCES

1 Werner EJ, Broxson EH, Tucker EL, et al. Prevalence of von Willebrand disease in children: a multiethnic study. *J Pediatr.* 1993;123:893–898.

2 Rodeghiero F, Castaman G, Dini E. Epidemiological investigation of the prevalence of von Willebrand's disease. *Blood.* 1987;69:454–459.

3 Rosner F. Hemophilia in the Talmud and rabbinic writings. *Ann Intern Med.* 1969;70:833–837.

4 Von Willebrand EA. Hereditar pseudohamofili. *Fin Lakaresallsk. Handl.*1926;67:7–12.

5 Ginsburg D, Handin RI, Bonthron DT, et al. Human von Willebrand factor (vWF): isolation of complementary DNA (cDNA) clones and chromosomal localization. *Science.* 1985;228:1401–1406.

6 Mancuso DJ, Tuley EA, Westfield LA, et al. Human von Willebrand factor gene and pseudogene: structural analysis and differentiation by polymerase chain reaction. *Biochemistry.* 1991;30:253–269.

7 Guan J, Guillot PV, Aird WC. Characterization of the mouse von Willebrand factor promoter. *Blood.* 1999;94:3405–3412.

8 Jahroudi N, Ardekani AM, Greenberger JS. An NF1-like protein functions as a repressor of the von Willebrand factor promoter. *J Biol Chem.* 1996;271:21413–21421.

9 Schwachtgen JL, Remacle JE, Janel N, et al. Oct-1 is involved in the transcriptional repression of the von Willebrand factor gene promoter. *Blood.* 1998;92:1247–1258.

10 Mohlke KL, Purkayastha AA, Westrick RJ, et al. Mvwf, a dominant modifier of murine von Willebrand factor, results from altered lineage-specific expression of a glycosyltransferase. *Cell.* 1999;96:111–120.

11 Varughese KI, Celikel R, Ruggeri ZM. Structure and function of the von Willebrand factor A1 domain. *Curr Protein Pept Sci.* 2002;3:301–312.

12 Matsui T, Titani K, Mizuochi T. Structures of the asparagine-linked oligosaccharide chains of human von Willebrand factor. Occurrence of blood group A, B, and H(O) structures. *J Biol Chem.* 1992;267:8723–8731.

13 Ellies LG, Ditto D, Levy GG, et al. Sialyltransferase ST3Gal-IV operates as a dominant modifier of hemostasis by concealing asialoglycoprotein receptor ligands. *Proc Natl Acad Sci USA.* 2002;99:10042–10047.

14 Wagner DD. Cell biology of von Willebrand factor. *Annu Rev Cell Biol.* 1990;6:217–246.

15 Katsumi A, Tuley EA, Bodo I, Sadler JE. Localization of disulfide bonds in the cystine knot domain of human von Willebrand factor. *J Biol Chem.* 2000;275:25585–25594.

16 Purvis AR, Sadler JE. A covalent oxidoreductase intermediate in propeptide-dependent von Willebrand factor multimerization. *J Biol Chem.* 2004;279:49982–49988.

17 Federici AB, Bader R, Pagani S, et al. Binding of von Willebrand factor to glycoproteins Ib and IIb/IIIa complex: affinity is related to multimeric size. *Br J Haematol.* 1989;73:93–99.

18 Pusztaszeri M, Seelentag W, Bosman FT. Immunohistochemical expression of endothelial markers CD31, CD34, von Willebrand factor, and Fli-1 in normal human tissues. *J Histochem Cytochem.* 2006;54(4):385–395.

19 Nichols TC, Samama CM, Bellinger DA, et al. Function of von Willebrand factor after crossed bone marrow transplantation between normal and von Willebrand disease pigs: effect on arterial thrombosis in chimeras. *Proc Natl Acad Sci USA.* 1995;92:2455–2459.

20 Gill JC, Endres-Brooks J, Bauer PJ, et al. The effect of ABO blood group on the diagnosis of von Willebrand disease. *Blood.* 1987;69:1691–1695.

21 Borchiellini A, Fijnvandraat K, ten Cate JW, et al. Quantitative analysis of von Willebrand factor propeptide release in vivo: effect of experimental endotoxemia and administration of 1-deamino-8-D-arginine vasopressin in humans. *Blood.* 1996;88:2951–2958.

22 Sadler JE, Moake JL, Miyata T, George JN. Recent advances in thrombotic thrombocytopenic purpura. *Hematology Am Soc Hematol Educ Program.* 2004;407–423.

23 Tsai HM. Physiologic cleavage of von Willebrand factor by a plasma protease is dependent on its conformation and requires calcium ion. *Blood.* 1996;87:4235–4244.

24 Cruz MA, Yuan H, Lee JR, et al. Interaction of the von Willebrand factor (vWF) with collagen. Localization of the primary collagen-binding site by analysis of recombinant vWF a domain polypeptides. *J Biol Chem.* 1995;270:10822–10827.

25 Wu D, Vanhoorelbeke K, Cauwenberghs N, et al. Inhibition of the von Willebrand (VWF)-collagen interaction by an antihuman VWF monoclonal antibody results in abolition of in vivo arterial platelet thrombus formation in baboons. *Blood.* 2002;99:3623–3628.

26 Ribba AS, Loisel I, Lavergne JM, et al. Ser968Thr mutation within the A3 domain of von Willebrand factor (VWF) in two related patients leads to a defective binding of VWF to collagen. *Thromb Haemost.* 2001;86:848–854.

27 Hoylaerts MF, Yamamoto H, Nuyts K, et al. von Willebrand factor binds to native collagen VI primarily via its A1 domain. *Biochem J*. 1997;324:185–191.

28 Lankhof H, van Hoeij M, Schiphorst ME, et al. A3 domain is essential for interaction of von Willebrand factor with collagen type III. *Thromb Haemost*. 1996;75:950–958.

29 Scott DM, Griffin B, Pepper DS, Barnes MJ. The binding of purified factor VIII/von Willebrand factor to collagens of differing type and form. *Thromb Res*. 1981;24:467–472.

30 Romijn RA, Bouma B, Wuyster W, et al. Identification of the collagen-binding site of the von Willebrand factor A3-domain. *J Biol Chem*. 2001;276:9985–9991.

31 Bienkowska J, Cruz M, Atiemo A, et al. The von Willebrand factor A3 domain does not contain a metal ion-dependent adhesion site motif. *J Biol Chem*. 1997;272:25162–25167.

32 Romijn RA, Westein E, Bouma B, et al. Mapping the collagen-binding site in the von Willebrand factor-A3 domain. *J Biol Chem*. 2003;278:15035–15039.

33 Emsley J, Knight CG, Farndale RW, et al. Structural basis of collagen recognition by integrin alpha2beta1. *Cell*. 2000;101:47–56.

34 Andre P, Denis CV, Ware J, et al. Platelets adhere to and translocate on von Willebrand factor presented by endothelium in stimulated veins. *Blood*. 2000;96:3322–3328.

35 Padilla A, Moake JL, Bernardo A, et al. P-selectin anchors newly released ultralarge von Willebrand factor multimers to the EC surface. *Blood*. 2004;103:2150–2156.

36 Sadler JE. New concepts in von Willebrand disease. *Annu Rev Med*. 2005;56:173–191.

37 Sixma JJ, Schiphorst ME, Verweij CL, Pannekoek H. Effect of deletion of the A1 domain of von Willebrand factor on its binding to heparin, collagen and platelets in the presence of ristocetin. *Eur J Biochem*. 1991;196:369–375.

38 Uff S, Clemetson JM, Harrison T, et al. Crystal structure of the platelet glycoprotein Ib(alpha) N-terminal domain reveals an unmasking mechanism for receptor activation. *J Biol Chem*. 2002; 277:35657–35663.

39 Huizinga EG, Tsuji S, Romijn RA, et al. Structures of glycoprotein Ibalpha and its complex with von Willebrand factor A1 domain. *Science*. 2002;297:1176–1179.

40 Emsley J, Cruz M, Handin R, Liddington R. Crystal structure of the von Willebrand Factor A1 domain and implications for the binding of platelet glycoprotein Ib. *J Biol Chem*. 1998;273:10396–10401.

41 Sugimoto M, Dent J, McClintock R, et al. Analysis of structure-function relationships in the platelet membrane glycoprotein Ib-binding domain of von Willebrand's factor by expression of deletion mutants. *J Biol Chem*. 1993;268:12185–12192.

42 Andrews RK, Berndt MC. Platelet physiology and thrombosis. *Thromb Res*. 2004;114:447–453.

43 Springer TA. Traffic signals for lymphocyte recirculation and leukocyte emigration: the multistep paradigm. *Cell*. 1994;76: 301–314.

44 Doggett TA, Girdhar G, Lawshe A, et al. Selectin-like kinetics and biomechanics promote rapid platelet adhesion in flow: the GPIb(alpha)-vWF tether bond. *Biophys J*. 2002;83:194–205.

45 Chang KC, Tees DF, Hammer DA. The state diagram for cell adhesion under flow: leukocyte rolling and firm adhesion. *Proc Natl Acad Sci USA*. 2000;97:11262–11267.

46 Siedlecki CA, Lestini BJ, Kottke-Marchant KK, et al. Shear-dependent changes in the three-dimensional structure of human von Willebrand factor. *Blood*. 1996;88:2939–2950.

47 Doggett TA, Girdhar G, Lawshe A, et al. Alterations in the intrinsic properties of the GPIbalpha-vWF tether bond define the kinetics of the platelet-type von Willebrand disease mutation, G233V. *Blood*. 2003;102:152–160.

48 Fukuda K, Doggett T, Laurenzi IJ, et al. The snake venom protein botrocetin acts as a biological brace to promote dysfunctional platelet aggregation. *Nat Struct Mol Biol*. 2005;12:152–159.

103

Tissue Factor Pathway Inhibitor

Alan E. Mast

Blood Research Institute, The Blood Center of Southeastern Wisconsin, Milwaukee

Tissue factor pathway inhibitor (TFPI) is a factor Xa-dependent inhibitor of the factor VIIa/tissue factor (TF) catalytic complex that initiates blood coagulation. TFPI is expressed by, and found on the surface of, endothelial cells (ECs), where it inhibits intravascular TF activity and thereby helps to prevent the development of intravascular thrombosis (1). The identification, cloning, and characterization of TFPI in the mid-1980s represented the culmination of many studies performed in the early and mid-1900s (2). These early studies demonstrated that serum contains a calcium-dependent factor that inhibits the procoagulant activity of tissue extracts and whose preincubation with serum prevents the toxicity of TF infused into mice (3). Studies by P. F. Hjort in the 1950s demonstrated that the serum factor inhibited the TF/factor VIIa/calcium complex, but not TF or factor VIIa individually, indicating that factor VIIa and TF function as a catalytic complex. In 1985, Sanders and coworkers described a factor Xa-dependent inhibitor of the factor VIIa/TF catalytic complex present in the lipoprotein fraction of plasma (4). This finding led to the purification and identification of TFPI from conditioned medium of HepG2 cells by Broze and coworkers in 1987 (5). Subsequent cloning and cDNA sequence analysis revealed TFPI to be a trivalent Kunitz-type serine protease inhibitor that inhibits factor Xa via the second Kunitz domain and factor VIIa/TF via the first Kunitz domain (6,7). Initially, TFPI was referred to as either lipoprotein-associated coagulation inhibitor (LACI) or extrinsic pathway inhibitor (EPI). For purposes of standardization, a group of interested investigators at the International Society for Thrombosis and Haemostasis, meeting in 1991, recommended the name *TF pathway inhibitor.* Since that time, two novel proteins have been identified, which has served only to renew confusion around the nomenclature. The first protein is a trivalent Kunitz-type inhibitor with structural homology to TFPI, termed TFPI-2. The name is somewhat of a misnomer in that TFPI-2 is primarily an inhibitor of plasmin within the extracellular matrix and is a poor inhibitor of factor VIIa/TF. Nevertheless, the use of this term has led some to designate TFPI as TFPI-1. The second protein is an alternatively spliced form of TFPI that is truncated prior to the third Kunitz domain. This protein has been called TFPIβ, leading some to designate TFPI as TFPIα. In this chapter, TFPI will be used to refer to the full-length form of the protein, as was originally agreed upon in 1991.

EVOLUTIONARY CONSIDERATIONS

TFPI contains three tandem Kunitz-type serine protease inhibitor domains. The prototypical Kunitz-type inhibitor is bovine pancreatic trypsin inhibitor (BPTI), also called aprotinin, an antifibrinolytic agent used clinically to prevent bleeding after heart bypass surgery. The Kunitz/BPTI protease inhibitor family is found in a wide variety of animal species but not in plants. It appears that the ancestral Kunitz-type inhibitor arose about 500 million years ago (8). Its core structure is a conserved fold of about 60 amino acids stabilized by three disulfide bridges (9). The three tandem Kunitz domains most likely assembled by gene duplication and exon shuffling (10). In addition to gene duplication, Kunitz domains also have propagated through insertion into new genes, as appears to have occurred with the Kunitz domain of the amyloid β-precursor protein that accumulates within the neuritic plaques of patients with Alzheimer disease (11,12). In addition to duplication, Kunitz domains have undergone extensive evolutionary diversification of target protease specificity. This has been accomplished by alteration of the active-site amino acids with relative conservation of the core structure of the Kunitz domain. Evolutionary selection of Kunitz domains has produced a diverse family of inhibitors with a wide variety of protease specificities (9). These include the human plasma proteins inter-α-trypsin inhibitor and pre-α-trypsin inhibitor (13), as well as other interesting proteins such as penthalaris (14), a protein found in tick saliva that contains five tandem Kunitz domains. Interestingly, diversification of the Kunitz domain in nonactive site regions has produced a homologous group of proteins that function as potassium and calcium channel blockers. These proteins have lost protease inhibitory activity and act as neurotoxins in snake venoms (15).

DEVELOPMENTAL CONSIDERATIONS

TFPI deficiency has not been described in humans, suggesting that it may be essential for development. Indeed, targeted deletion of *TFPI* in mice results in intrauterine lethality, demonstrating that TFPI is required for live birth (16). Approximately 60% of TFPI-deficient animals die between embryonic day (E)9.5 and E11.5 from apparent yolk sac hemorrhage. The 40% of mice surviving beyond E11.5 have normal organ development but have short tails and other signs of coagulopathy including diffuse fibrin(ogen) deposition in the liver and rare intravascular thrombi in the brain (16). These findings are consistent with unregulated factor VIIa/TF activity, causing a consumptive coagulopathy with subsequent bleeding. TFPI also inhibits TF activity in adult animals. This was first shown by demonstrating that wild-type mice are sensitized to TF–induced disseminated intravascular coagulation following depletion of TFPI through infusion of a polyclonal antibody directed against TFPI (17). Heterozygous mice ($TFPI^{+/-}$) have 50% of the normal plasma TFPI activity (16). These mice develop normally, without evidence of spontaneous thrombosis or consumptive coagulopathy and have the same average time to occlusive thrombosis as wild-type mice following photochemical injury of the carotid artery (18). However, in a study of responses to vascular injury, $TFPI^{+/-}$ mice produced significantly greater neointimal proliferation than wild-type littermates (19).

Similar to the TFPI-deficient mice, mice deficient in the hemostatic proteins TF (20–22), factor V (23), factor X (24), and prothrombin (25,26) also die with variable frequency between days E9.5 and E11.5. A common finding among these mice is severe hemorrhage that is possibly associated with a defect in vascular integrity. One explanation for the similar phenotype of the TFPI-deficient mice and the mice deficient in factors that TFPI regulates is the development of a consumptive coagulopathy that causes these animals to become functionally deficient in clotting factors and die from hemorrhage. However, hemorrhage cannot be the sole cause of death, because mice with a complete loss of fibrinogen- and platelet-dependent hemostatic capacity suffer from severe hemorrhage yet they survive embryogenesis (27). Alternatively, the death of TFPI-deficient embryos may result from dysregulation of coagulation factor–mediated signaling through protease-activated receptors (PARs) that are necessary for the maintenance of vascular integrity within the developing embryo. This notion is supported by the observation that 50% of PAR-1–deficient mice also die at midgestation with apparent defects in vascular integrity similar to those found in the TFPI-deficient mice (28).

GENE REGULATION AND GENE/ PROTEIN STRUCTURE

Gene Structure

The *TFPI* gene is located on chromosome 2. It contains 10 exons and encodes a 276-amino acid protein containing an acidic N-terminal region followed by three tandem Kunitz-type serine protease inhibitor domains and a highly basic C-terminal region. The first two exons are not translated. Exon 3 encodes the signal sequence that is removed during post-translational processing and the acidic N-terminal region of the mature protein. Exons 4, 6, and 9 encode the three Kunitz domains. Exons 5 and 7 encode the connecting regions between the Kunitz domains. Exon 10 encodes the highly basic C-terminal region of TFPI. Exon 8 encodes the C-terminal region of TFPIβ, the alternatively spliced form of TFPI that is truncated immediately prior to the third Kunitz domain (29,30). In TFPI, exon 8 is removed as an intron. When exon 8 is expressed, a totally new C-terminal region that contains a glycosylphosphatidylinisotol (GPI) anchor attachment signal replaces the third Kunitz domain and basic C-terminal region of TFPI, thereby producing so-called TFPIβ.

Gene Regulation

TFPI is constitutively expressed by cultured ECs, and its expression is either not altered or only slightly upregulated by endotoxin, tumor necrosis factor (TNF)-α, or interleukin (IL)-1 (1). In contrast, these inflammatory stimuli greatly increase TF expression (1). The inability of TFPI expression to "keep pace" with TF expression in ECs may contribute to the coagulation disorders associated with severe inflammation. In contrast to the lack of response to inflammatory stimuli, exposing cultured ECs to fluid flow progressively increases the amount of TFPI made by two- to threefold as the shear stress increases from minimal flow, to venous flow, to arterial flow rates (31). In cultured vascular smooth muscle cells (VSMCs), TFPI expression is increased two- to fivefold by basic fibroblast growth factor/heparin (32), epidermal growth factor, and platelet-derived growth factor B (33). Plasma levels of TFPI in normal individuals vary by threefold (34). A study examining the association of microsatellite markers throughout the genome with plasma TFPI levels in 397 individuals from 21 families found that polymorphisms in and around the *TFPI* structural gene on chromosome 2 were the major genetic determinants of variation in plasma TFPI levels (35).

Protein Structure and Function

Mature TFPI is an approximately 43-kDa protein (Figure 103.1). The second Kunitz domain of TFPI directly inhibits factor Xa, a process that is accelerated by heparin binding to the basic C-terminal region (36,37). The first Kunitz domain inhibits factor VIIa/TF in a factor Xa-dependent manner (7). The third Kunitz domain does not appear to function as a protease inhibitor, but is required for the full anticoagulant activity of TFPI in TF–initiated plasma clotting assays (38). Because rapid inhibition of factor VIIa by the first Kunitz domain depends on the presence of factor Xa, TFPI often has been described as a two-stage inhibitor of blood coagulation, first inhibiting factor Xa via the second Kunitz domain with

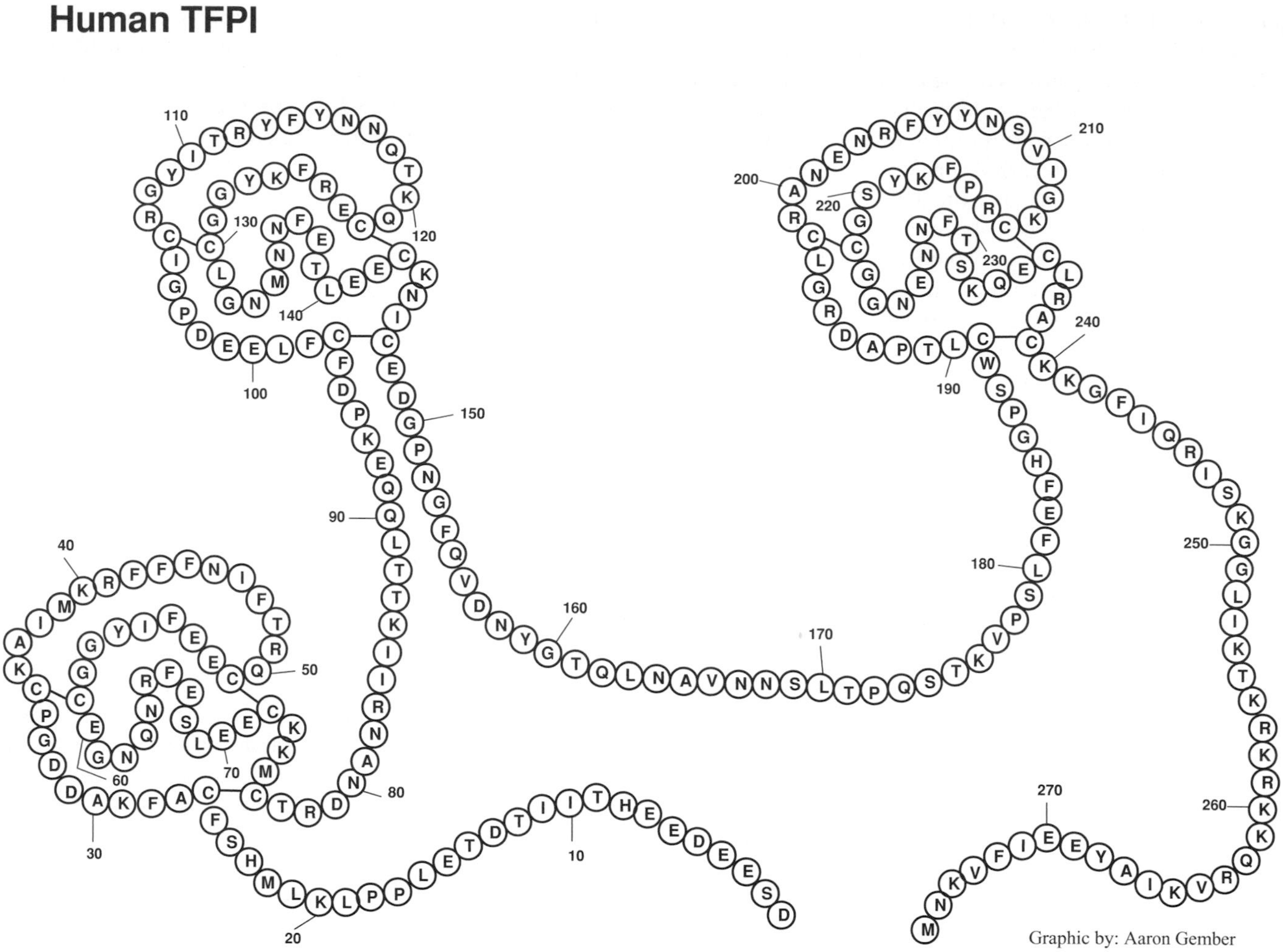

Figure 103.1. Human TFPI.

subsequent inhibition of factor VIIa/TF by the TFPI/factor Xa complex. However, recent data support a revised model of inhibition. During the initiation of blood clotting, the factor VIIa/TF complex activates factor X to factor Xa by limited proteolysis. Following activation of factor X, extended binding interactions between factor Xa and the factor VIIa/TF complex allow for factor Xa to remain transiently associated with factor VIIa/TF. Enzyme kinetic studies examining the inhibitory properties of TFPI demonstrated that the rate-limiting step for the inhibition of factor VIIa/TF by TFPI is the inhibition of the *newly* formed factor Xa (39). Therefore, it appears that it is the transient factor VIIa/TF/factor Xa ternary complex that is most efficiently inhibited by TFPI through a mechanism in which the first two Kunitz domains simultaneously bind to the active sites of factor VIIa and factor Xa. This same ternary complex has been shown to be a highly efficient activator of endothelial PARs, providing a key link between inflammation and the coagulation cascade (40–42). Blockage of endogenous TFPI activity in TNF-α-stimulated ECs increases TF–dependent PAR signaling (41). Thus, TFPI appears to play a central role in regulating this type of cellular signaling. Because factor VIIa/TF–mediated activation of both blood coagulation and inflammation occurs upon phospholipid/cellular surfaces, effective inhibition will require TFPI to be optimally localized and oriented on the cell surface.

Tissue Factor Pathway Inhibitor Location and Endothelial Association

TFPI is preferentially produced by vascular ECs regardless of the type of vessel (43,44), and it also is produced in smaller amounts by VSMCs (33,43), monocytes (45,46), megakaryocytes (44), and astrocytes (47). Its expression appears to be enhanced within arterial plaques (43,48). Approximately 85% of the intravascular TFPI is localized on the surface of the endothelium (1,49). The plasma concentration is only approximately 2.5 nM. About 50% of this plasma pool has undergone variable amounts of proteolytic degradation within its C-terminal region and is bound to lipoproteins. As a result, plasma TFPI has reduced anticoagulant activity and is unlikely to play an important role as an in vivo inhibitor of blood coagulation (50). Circulating platelets represent a third pool of TFPI, accounting for approximately 8% of the total protein in whole blood (51). Both the plasma and platelet forms of TFPI inhibit factor VIIa/TF activity when tested using in vitro assays; however, their relative activities, when compared to endothelial-bound TFPI, for inhibition of TF in vivo has not been deciphered. Most studies performed to date have focused on endothelial-associated TFPI.

The preferential location of TFPI on the endothelial surface is somewhat unexpected because, structurally, TFPI resembles a secreted protein that would be found in plasma rather than on the cell surface. The C-terminal region contains neither the membrane-spanning amino acids typical of integral membrane proteins nor the hydrophobic sequence typically found in proteins directly attached to a GPI anchor. In vivo studies indicate that the plasma concentration of TFPI increases two- to fourfold following infusion of heparin (52,53). Because the increase is rapid (<10 minutes), it appears that heparin-releasable TFPI is bound to the endothelium through nonspecific interactions between the basic C-terminal region and cell-surface glycosaminoglycans. Consistent with these data, in vitro studies have demonstrated that the binding of *Escherichia coli*-produced recombinant TFPI to cultured ECs is prevented by heparin and that C-terminally truncated forms of TFPI have very low affinity for the cell surface (54,55). However, recombinant forms of TFPI, produced either in bacteria or mammalian cells, bind to cell surfaces much differently than does endogenous TFPI produced by ECs (56). In both cultured ECs and fresh placenta, which contains endothelial and trophoblast cells not exposed to the potential artifacts of cell culture, endogenous TFPI associates with the cell surface primarily through a GPI anchor in a manner that is not altered by heparin (49,57–59). It appears that heparin-releasable TFPI represents only a small portion (between 1% and 10%) of the total TFPI on the endothelium that remains bound to the cell surface through ionic interactions between the C-terminal region and cell surface glycosaminoglycans following cleavage of the GPI anchor by endogenous enzymes on the cell surface (58). The predominance of GPI-anchored TFPI within the vasculature is important clinically because it suggests that heparin infusion does not significantly redistribute TFPI within the vasculature. Additionally, it is consistent with the wide range of plasma TFPI concentrations in normal subjects, and the general lack of correlation between the plasma concentration of TFPI and the incidence of clinically significant thrombosis (60–62).

GPI-anchored proteins have an N-terminal leader sequence that directs the polypeptide to the endoplasmic reticulum. The GPI anchor attaches near the C-terminal end of the protein and displaces a C-terminal peptide of 17 to 30 amino acids from the protein. GPI-anchored TFPI on the placental surface has an intact C-terminal polypeptide, indicating that this region was not removed within the endoplasmic reticulum (58). Therefore, it is hypothesized that TFPI is attached indirectly to the GPI anchor through binding to a GPI-anchored coreceptor. Interestingly, experiments examining the structural determinants of TFPI necessary for the association of altered recombinant forms of TFPI with the cell surface demonstrated that the active site of the third Kunitz domain and the basic C-terminal region of TFPI are required. These compelling data suggest that TFPI may interact with the endothelial surface through binding of the third Kunitz domain to a GPI-anchored serine protease (63).

An Alternatively Spliced Form of Tissue Factor Pathway Inhibitor

ECs also appear to make an alternatively spliced, truncated form of TFPI that is directly attached to a GPI anchor (29,30). This form of TFPI, called TFPIβ (Figure 103.2), contains the first two Kunitz domains present in full-length TFPI and, therefore, is theoretically able to inhibit both factor Xa and factor VIIa/TF. The alternative splice occurs following exon 7, which encodes the connecting region between the second and third Kunitz domains of full-length TFPI. In human TFPIβ, exon 8 encodes a stretch of 12 amino acids followed by a GPI anchor attachment sequence that is removed in the endoplasmic reticulum. Thus, human TFPIβ has only 12 unique amino acids that are not present in TFPI. Real-time PCR measurements of cultured ECs indicate that *TFPI*β mRNA is present at about one-tenth the level of *TFPI* mRNA (30). Purified recombinant mouse TFPIβ is a slow-binding inhibitor of factor Xa (29). In addition, enzyme kinetic studies comparing the inhibitory activity of recombinant TFPI to that of a recombinant truncated form of TFPI that contains only the first two Kunitz domains, and therefore is similar to TFPIβ, have demonstrated that loss of the third Kunitz domain and C-terminal region significantly decreases TFPI inhibitory activity in both amidolytic assays measuring antifactor Xa activity and in clot-based assays measuring anti-TF activity. Together, these data suggest that TFPIβ may be a less potent anticoagulant than TFPI (38,64). However, these assays were performed using soluble rather than cell-associated forms of TFPI and do not necessarily represent the activity of surface-associated forms of TFPI. Recently, it has been shown that when the first two Kunitz domains of TFPI are linked to annexin V to create a chimeric protein with affinity for phosphatidylserine-containing membranes, its relative potency in a TF–induced clotting assay increases 250-fold (65). Thus, TFPIβ theoretically may have significantly more anticoagulant activity when associated with the cell surface through a GPI anchor than when in solution.

Localization of Tissue Factor Pathway Inhibitor within Lipid Rafts

One consequence of the attachment of TFPI to endothelium through a GPI-anchored coreceptor is that it becomes concentrated within caveolae or lipid-raft microdomains on the cell surface along with other GPI-anchored proteins (57,59). Caveolae are 50- to 100-nm invaginations of the plasma membrane that are coated with caveolin and enriched in cholesterol and sphingomyelin. Noninvaginated domains that are biochemically similar to caveolae are referred to as lipid rafts.

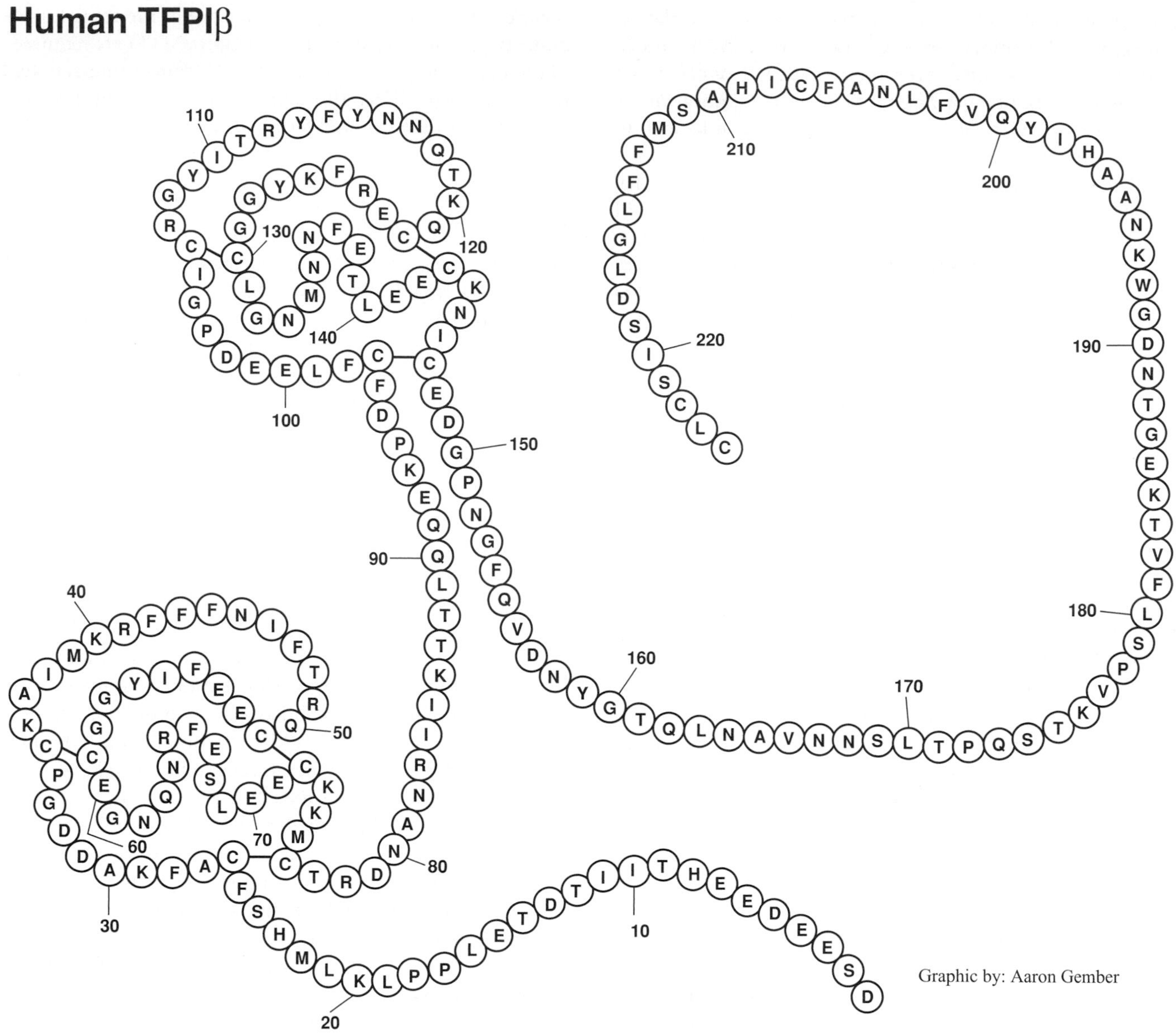

Figure 103.2. Human TFPIβ.

Caveolae and lipid rafts have been implicated as mediators of endocytosis (66), vesicular transport (67), pathogen invasion (68), and signal transduction pathways (69). Downregulation of TF procoagulant activity by TFPI coincides with the translocation of the factor VIIa/TF complex from the bulk-phase membrane into caveolae and lipid-raft microdomains (57). Because these domains inefficiently support factor VIIa/TF catalytic activity (57), the translocation of TF may act to further downregulate its procoagulant activity. However, recombinant TFPI chimeras that are directly attached to the cell surface either through a GPI anchor, which localizes the recombinant TFPI to caveolae and lipid-raft microdomains, or a hydrophobic membrane spanning domain that localizes TFPI within the bulk phase membrane, demonstrate equal ability to inhibit factor VIIa/TF (70). Thus, the physiological advantages for localization of TFPI to these domains and the role of the GPI-anchored binding protein in TFPI anticoagulant activity are not yet clear.

INTERACTIONS WITH OTHER OUTPUTS

Tissue Factor Pathway Inhibitor Directly Counterbalances Tissue Factor Activity

Inactivation of the mouse *TF* gene results in lethality of most embryos between E9.5 and E11.5 (20,21). A small percentage of *TF*-null embryos survive this period but die later during gestation. One *TF*-null mouse, delivered by cesarean section, developed normally until 4 weeks of age, before succumbing to a brain hemorrhage (71). The embryonic lethality can be rescued by expression of a human transgene that produces TF in amounts of about 1% of that found in wild-type mice (72). These TF(low) mice survive birth but develop hemostatic defects in the lung, heart, uterus, and placenta. Many of the animals die from lung hemorrhage at approximately 3 months of age. Mice living beyond 3 months suffer from cardiac fibrosis and fatal postpartum hemorrhages (73). When the TF(low) mice are bred with TFPI-deficient mice, homozygous *TFPI* $(^{-/-})$ mice are viable and reproduce (74). Moreover, the TF(low)/*TFPI*$(^{-/-})$ mice actually are healthier and live longer than the TF(low) mice containing normal amounts of TFPI. The decrease in TFPI activity increases the functional amount of TF activity in the lungs and prevents the mice from succumbing to pulmonary hemorrhage. These data provide clear in vivo evidence that TFPI directly counterbalances TF activity. The TF(low)/*TFPI*$(^{-/-})$ mice continue to develop cardiac fibrosis and postpartum hemorrhage, suggesting that the heart and placenta/uterus require higher amounts of TF activity to prevent bleeding than is provided by expression of the transgene even in the absence of TFPI (74). The variable rescue of the hemostatic abnormalities in different organs of the TF(low) mice demonstrate that an organ-specific balance of TF and TFPI is required for proper hemostasis during both embryonic development and in adult mice. In a manner similar to the TF(low) mice, factor VII deficiency also rescues *TFPI*$(^{-/-})$ mice from embryonic lethality. However, the *factor VII*$(^{-/-})$/*TFPI*$(^{-/-})$ mice suffer from fatal perinatal bleeding due to the lack of factor VII (75).

Tissue Factor Pathway Inhibitor Deficiency Exacerbates Thrombosis in Mice with Factor V Leiden

Factor V Leiden is an altered form of factor V containing a prothrombotic mutation that makes it resistant to proteolytic cleavage and inactivation by activated protein C. Mice genetically altered to produce the Factor V Leiden mutation (FVQ/Q) have a mild prothrombotic phenotype, exhibiting only variable degrees of spontaneous thrombosis depending on the genetic background of the mice (76). However, when they are bred with TFPI-deficient mice to produce a *FV(Q/Q)/TFPI*$(^{+/-})$ genotype, the mice suffer from nearly complete perinatal mortality. The mortality is associated with fibrin deposition in multiple organs, suggesting that they succumb to disseminated thrombosis (77). The thrombotic phenotype of the *FV(Q/Q)/TFPI*$(^{+/-})$ mice is rescued by breeding into a heterozygous *TF*$(^{+/-})$ background, again providing evidence that TF and TFPI directly counterbalance each other in vivo (78).

DIAGNOSTIC AND THERAPEUTIC IMPLICATIONS

Tissue Factor Pathway Inhibitor as a Risk Factor for Thrombosis

Although heterozygous *TFPI* $(^{+/-})$ deficiency is a strong risk factor for thrombosis in mice with factor V Leiden, low plasma levels of TFPI have been only weakly linked to thrombotic disease in humans. Two studies have suggested the presence of a "threshold effect" whereby patients with free (non-lipoprotein bound) plasma TFPI concentration less than 10% of the normal mean value are at increased risk (approximately twofold) for deep venous thrombosis (61) and myocardial infarction (62). However, in both studies, no difference was found in the mean free plasma TFPI level in the disease and control groups. Other published studies have reported conflicting results regarding the contribution of plasma TFPI levels and TFPI polymorphisms to the development of thrombosis. This is likely a result of the wide normal range for plasma TFPI, the various methods for measurement of plasma TFPI, and the probable poor correlation between the plasma TFPI concentration and the amount of endothelial-associated TFPI. The potential discordance between plasma and cell-associated TFPI is best illustrated in patients with abetalipoproteinemia. These individuals have greatly decreased plasma TFPI with normal amounts of endothelial-associated TFPI, yet they do not have an increased risk for thrombosis (34).

Tissue Factor Pathway Inhibitor as a Therapeutic Agent for Sepsis

The coagulation cascade is intimately connected with inflammation. As described earlier, TFPI likely inhibits blood clotting by simultaneously binding the active sites of factor VIIa and newly activated factor X within the ternary factor VIIa/TF/factor Xa complex (39). In the absence of the inhibitory effects of TFPI, factor Xa within this complex is an efficient activator of PAR-1 and -2 (42). Activation of these receptors can lead to the expression of adhesion molecules on the endothelium, with subsequent leukocyte activation and generation of an inflammatory response. In addition, TF expression is enhanced by inflammatory cytokines whereas cellular TFPI expression is largely unaffected, further contributing to the procoagulant environment associated with inflammation (1,32,33). Because of the hypothesized role for TF in propagating the inflammatory response, recombinant TFPI produced in *Escherichia coli* has been tested for use as a therapeutic agent in several animal models as well as in human trials for the treatment of sepsis and other inflammatory conditions. It improves survival in rabbit models of septic shock and lethal *Escherichia coli* peritonitis (78,79) as well as in mouse models of cecal ligation and puncture and polymicrobial gram-positive sepsis (80). In baboons infused with lethal doses of *Escherichia coli*, 70% to 100% rescue can be obtained by the infusion of recombinant TFPI beginning 30 minutes after the lethal challenge (81). The baboons receiving recombinant TFPI showed decreased activation of coagulation as evidenced by a lack of thrombosis, hemorrhage, and infarction in adrenal and renal tissues when compared with untreated controls. Increases in IL-6 and IL-8 were attenuated compared with controls, indicating that recombinant TFPI also produces a decrease in the inflammatory response within the baboons. In a subsequent study in which recombinant

TFPI was infused beginning 120 minutes after the lethal challenge, 11 of 12 treated baboons died (82). Interestingly, even though the baboons died, they again did not have evidence of thrombosis or hemorrhage in the adrenal and renal tissues. These data indicate that thrombosis and hemorrhage in these tissues is not the cause of death in this sepsis model. Additionally, they suggest that the recombinant TFPI infused at 30 minutes prevents death by altering the very early procoagulant events of sepsis, thereby preventing the development of disseminated intravascular coagulation and/or the actions of inflammatory mediators. In contrast to these data from baboons, a study of humans infused with nonlethal doses of endotoxin demonstrated that, although recombinant TFPI completely blocked the endotoxin-induced activation of coagulation, it had no effect on leukocyte activation, chemokine release, EC activation, or the acute-phase response (83). The differences between the results of these studies may reflect differences between the response of humans and baboons to inflammatory stimuli and/or differences in the response to lethal versus nonlethal challenges.

The success of recombinant TFPI infusion in preventing death from sepsis in animal models led to the initiation of clinical trials using recombinant TFPI for the treatment of sepsis. A Phase II trial demonstrated a trend toward a reduction in 28-day all-cause mortality even though the study was not powered to show efficacy (84). However, in a Phase III trial of approximately 2,000 patients with sepsis and septic shock, recombinant TFPI failed to reduce 28-day all-cause mortality (85). One potential problem with the use of recombinant TFPI as a therapeutic agent is that it appears that recombinant TFPI is much more effective at inhibiting blood coagulation than inhibiting inflammation by blocking PAR signaling when tested in an in vitro cellular model (86). This discrepancy of inhibitory action suggests that effective inhibition of inflammation by recombinant TFPI may produce excessive bleeding, and emphasizes the importance of the mechanisms through which endogenously produced TFPI associates with the endothelial surface.

KEY POINTS

- TFPI is a critical endothelial-associated anticoagulant protein that directly counterbalances TF activity in vivo.
- TFPI is a trivalent Kunitz-type serine protease inhibitor that inhibits both factor Xa and factor VIIa/TF. As such, it inhibits both TF–initiated blood coagulation and the inflammation initiated by blood coagulation protease-mediated cellular signaling.
- Although TFPI structurally resembles a soluble protein, it associates with endothelium through tight binding to a GPI-anchored coreceptor. An alternatively spliced, truncated form of TFPI, TFPIβ, directly associates with endothelium via a GPI anchor. The localization of TFPI to cellular surfaces greatly enhances its activity. The lack of, or nonphysiological, cell surface association of recombinant TFPI may limit its efficacy as a therapeutic agent.
- Low amounts of TFPI have been clearly linked to thrombotic disease in mouse models, but it has not been indisputably linked to thrombosis in humans. The poor correlation of low TFPI with thrombosis in humans may be due to a poor correlation between the plasma TFPI concentration and the amount of TFPI on the endothelial surface.

Future Goals

- To identify and characterize the GPI-anchored coreceptor that binds TFPI to the endothelial surface, to greatly enhance our understanding of how TFPI regulates TF activity
- To develop a peripheral blood assay that accurately measures the amount of TFPI associated with the endothelial surface, which will be a diagnostic advance that will allow for more accurate prediction of the contribution of low TFPI to the development of thrombotic and vascular disease in humans

REFERENCES

1 Ameri A, Kuppuswamy MN, Basu S, Bajaj SP. Expression of tissue factor pathway inhibitor by cultured endothelial cells in response to inflammatory mediators. *Blood*. 1992;79:3219–3226.

2 Broze GJ Jr. The rediscovery and isolation of TFPI. *J Thromb Haemost*. 2003;1:1671–1675.

3 Loeb L, Fleisher MS, Tuttle L. The interaction between blood serum and tissue extract in the coagulation of the blood. I. The combined action of serum and tissue extract on fluoride, hirudin, and peptone plasma; the effect of heating on the serum. *J Biol Chem*. 1922;51:461–483.

4 Sanders NL, Bajaj SP, Zivelin A, Rapaport SI. Inhibition of tissue factor/factor VIIa activity in plasma requires factor X and an additional plasma component. *Blood*. 1985;66:204–212.

5 Broze GJ Jr., Miletich JP. Isolation of the tissue factor inhibitor produced by HepG2 hepatoma cells. *Proc Natl Acad Sci USA*. 1987;84:1886–1890.

6 Wun TC, Kretzmer KK, Girard TJ, et al. Cloning and characterization of a cDNA coding for the lipoprotein- associated coagulation inhibitor shows that it consists of three tandem Kunitz-type inhibitory domains. *J Biol Chem*. 1988;263:6001–6004.

7 Girard TJ, Warren LA, Novotny WF, et al. Functional significance of the Kunitz-type inhibitory domains of lipoprotein-associated coagulation inhibitor. *Nature*. 1989;338:518–520.

8 Ikeo K, Takahashi K, Gojobori T. Evolutionary origin of a Kunitz-type trypsin inhibitor domain inserted in the amyloid beta

precursor protein of Alzheimer's disease. *J Mol Evol.* 1992;34: 536–543.

9 Pritchard L, Dufton MJ. Evolutionary trace analysis of the Kunitz/BPTI family of proteins: functional divergence may have been based on conformational adjustment. *J Mol Biol.* 1999;285: 1589–1607.

10 Girard TJ, Eddy R, Wesselschmidt RL, et al. Structure of the human lipoprotein-associated coagulation inhibitor gene. Intron/exon gene organization and localization of the gene to chromosome 2. *J Biol Chem.* 1991;266:5036–5041.

11 Tanzi RE, McClatchey AI, Lamperti ED, et al. Protease inhibitor domain encoded by an amyloid protein precursor mRNA associated with Alzheimer's disease. *Nature.* 1988;331:528–530.

12 Ponte P, DeWhitt PG, Schilling J, et al. A new A4 amyloid mRNA contains a domain homologous to serine proteinase inhibitors. *Nature.* 1988;331:525–527.

13 Enghild JJ, Thogersen IB, Pizzo SV, Salvesen G. Analysis of inter-alpha-trypsin inhibitor and a novel trypsin inhibitor, pre-alpha-trypsin inhibitor, from human plasma. Polypeptide chain stoichiometry and assembly by glycan. *J Biol Chem.* 1989;264: 15975–15981.

14 Francischetti I.M, Mather TN, Ribeiro JM. Penthalaris, a novel recombinant five-Kunitz tissue factor pathway inhibitor (TFPI) from the salivary gland of the tick vector of Lyme disease, Ixodes scapularis. *Thromb Haemost.* 2004;91:886–898.

15 Zupunski V, Kordis D, Gubensek F. Adaptive evolution in the snake venom Kunitz/BPTI protein family. *FEBS Lett.* 2003;547: 131–136.

16 Huang ZF, Broze G Jr. Consequences of tissue factor pathway inhibitor gene-disruption in mice. *Thromb Haemost.* 1997; 78:699–704.

17 Sandset PM, Warn-Cramer BJ, Maki SL, Rapaport SI. Immunodepletion of extrinsic pathway inhibitor sensitizes rabbits to endotoxin-induced intravascular coagulation and the generalized Shwartzman reaction. *Blood.* 1991;78:1496–1502.

18 Westrick RJ, Bodary PF, Xu Z, et al. Deficiency of tissue factor pathway inhibitor promotes atherosclerosis and thrombosis in mice. *Circulation.* 2001;103:3044–3046.

19 Singh R, Pan S, Mueske CS, et al. Tissue factor pathway inhibitor deficiency enhances neointimal proliferation and formation in a murine model of vascular remodelling. *Thromb Haemost.* 2003;89:747–751.

20 Toomey JR, Kratzer KE, Lasky NM, et al. Targeted disruption of the murine tissue factor gene results in embryonic lethality. *Blood.* 1996;88:1583–1587.

21 Bugge TH, Xiao Q, Kombrinck KW, et al. Fatal embryonic bleeding events in mice lacking tissue factor, the cell-associated initiator of blood coagulation. *Proc Natl Acad Sci USA.* 1996;93:6258–6263.

22 Carmeliet P, Mackman N, Moons L, et al. Role of tissue factor in embryonic blood vessel development. *Nature.* 1996;383:73–75.

23 Yang TL, Cui J, Taylor JM, et al. Rescue of fatal neonatal hemorrhage in factor V deficient mice by low level transgene expression. *Thromb Haemost.* 2000;83:70–77.

24 Dewerchin M, Liang Z, Moons L, et al. Blood coagulation factor X deficiency causes partial embryonic lethality and fatal neonatal bleeding in mice. *Thromb Haemost.* 2000;83:185–190.

25 Sun WY, Witte DP, Degen JL, et al. Prothrombin deficiency results in embryonic and neonatal lethality in mice. *Proc Natl Acad Sci USA.* 1998;95:7597–7602.

26 Xue J, Wu Q, Westfield LA, et al. Incomplete embryonic lethality and fatal neonatal hemorrhage caused by prothrombin deficiency in mice. *Proc Natl Acad Sci USA.* 1998;95:7603–7607.

27 Palumbo JS, Zogg M, Talmage KE, et al. Role of fibrinogen- and platelet-mediated hemostasis in mouse embryogenesis and reproduction. *J Thromb Haemost.* 2004;2:1368–1379.

28 Connolly AJ, Ishihara H, Kahn ML, et al. Role of the thrombin receptor in development and evidence for a second receptor. *Nature.* 1996;381:516–519.

29 Chang JY, Monroe DM, Oliver JA, Roberts HR. TFPIbeta, a second product from the mouse tissue factor pathway inhibitor (TFPI) gene. *Thromb Haemost.* 1999;81:45–49.

30 Zhang J, Piro O, Lu L, Broze GJ Jr. Glycosyl phosphatidylinositol anchorage of tissue factor pathway inhibitor. *Circulation.* 2003;108:623–627.

31 Westmuckett AD, Lupu C, Roquefeuil S, et al. Fluid flow induces upregulation of synthesis and release of tissue factor pathway inhibitor in vitro. *Arterioscler Thromb Vasc Biol.* 2000;20:2474–2482.

32 Pendurthi UR, Rao LV, Williams JT, Idell S. Regulation of tissue factor pathway inhibitor expression in smooth muscle cells. *Blood.* 1999;94:579–586.

33 Caplice NM, Mueske CS, Kleppe LS, et al. Expression of tissue factor pathway inhibitor in vascular smooth muscle cells and its regulation by growth factors. *Circ Res.* 1998;83:1264–1270.

34 Novotny WF, Brown SG, Miletich JP, et al. Plasma antigen levels of the lipoprotein-associated coagulation inhibitor in patient samples. *Blood.* 1991;78:387–393.

35 Almasy L, Soria JM, Souto JC, et al. A locus on chromosome 2 influences levels of tissue factor pathway inhibitor. Results from the GAIT study. *Arterioscler Thromb Vasc Biol.* 2005;25(7):1489–1492. Epub 2005 Apr 21.

36 Nordfang O, Bjorn SE, Valentin S, et al. The C-terminus of tissue factor pathway inhibitor is essential to its anticoagulant activity. *Biochemistry.* 1991;30:10371–10376.

37 Wesselschmidt R, Likert K, Girard T, et al. Tissue factor pathway inhibitor: the carboxy-terminus is required for optimal inhibition of factor Xa. *Blood.* 1992;79:2004–2010.

38 Lockett JM, Mast AE. Contribution of regions distal to glycine-160 to the anticoagulant activity of tissue factor pathway inhibitor. *Biochemistry.* 2002;41:4989–4997.

39 Baugh RJ, Broze GJ Jr., Krishnaswamy S. Regulation of extrinsic pathway factor Xa formation by tissue factor pathway inhibitor. *J Biol Chem.* 1998;273:4378–4386.

40 Camerer E, Huang W, Coughlin SR. Tissue factor- and factor X-dependent activation of protease-activated receptor 2 by factor VIIa. *Proc Natl Acad Sci USA.* 2000;97:5255–5260.

41 Riewald M, Ruf W. Mechanistic coupling of protease signaling and initiation of coagulation by tissue factor. *Proc Natl Acad Sci USA.* 2001;98:7742–7747.

42 Ahamed J, Belting M, Ruf W. Regulation of tissue factor-induced signaling by endogenous and recombinant tissue factor pathway inhibitor 1. *Blood.* 2005;105:2384–2391.

43 Crawley J, Lupu F, Westmuckett AD, et al. Expression, localization, and activity of tissue factor pathway inhibitor in normal and atherosclerotic human vessels. *Arterioscler Thromb Vasc Biol.* 2000;20:1362–1373.

44 Werling RW, Zacharski LR, Kisiel W, et al. Distribution of tissue factor pathway inhibitor in normal and malignant human tissues. *Thromb Haemost.* 1993;69:366–369.

45 Bajaj MS, Bajaj SP. Tissue factor pathway inhibitor: potential therapeutic applications. *Thromb Haemost.* 1997;78:471–477.

46 Ott I, Andrassy M, Zieglgansberger D, et al. Regulation of monocyte procoagulant activity in acute myocardial infarction: role of tissue factor and tissue factor pathway inhibitor-1. *Blood.* 2001;97:3721–3726.

47 Bajaj MS, Kuppuswamy MN, Manepalli AN, Bajaj SP. Transcriptional expression of tissue factor pathway inhibitor, thrombomodulin and von Willebrand factor in normal human tissues. *Thromb Haemost.* 1999;82:1047–1052.

48 Caplice NM, Mueske CS, Kleppe LS, Simari RD. Presence of tissue factor pathway inhibitor in human atherosclerotic plaques is associated with reduced tissue factor activity. *Circulation.* 1998;98:1051–1057.

49 Ott I, Miyagi Y, Miyazaki K, et al. Reversible regulation of tissue factor-induced coagulation by glycosyl phosphatidylinositol-anchored tissue factor pathway inhibitor. *Arterioscler Thromb Vasc Biol.* 2000;20:874–882.

50 Broze GJ Jr., Lange GW, Duffin KL, MacPhail L. Heterogeneity of plasma tissue factor pathway inhibitor. *Blood Coagul Fibrinolysis.* 1994;5:551–559.

51 Novotny WF, Girard TJ, Miletich JP, Broze GJ Jr. Platelets secrete a coagulation inhibitor functionally and antigenically similar to the lipoprotein associated coagulation inhibitor. *Blood.* 1988;72: 2020–2025.

52 Sandset PM, Abildgaard U, Larsen ML. Heparin induces release of extrinsic coagulation pathway inhibitor (EPI). *Thromb Res.* 1988;50:803–813.

53 Hansen JB, Sandset PM, Huseby KR, et al. Depletion of intravascular pools of tissue factor pathway inhibitor (TFPI) during repeated or continuous intravenous infusion of heparin in man. *Thromb Haemost.* 1996;76:703–709.

54 Iversen N, Sandset PM, Abildgaard U, Torjesen PA. Binding of tissue factor pathway inhibitor to cultured endothelial cells-influence of glycosaminoglycans. *Thromb Res.* 1996;84:267–278.

55 Warshawsky I, Bu G, Mast A, et al. The carboxy terminus of tissue factor pathway inhibitor is required for interacting with hepatoma cells in vitro and in vivo. *J Clin Invest.* 1995;95:1773–1781.

56 Ho G, Narita M, Broze GJ Jr., Schwartz AL. Recombinant full-length tissue factor pathway inhibitor fails to bind to the cell surface: implications for catabolism in vitro and in vivo. *Blood.* 2000;95:1973–1978.

57 Sevinsky JR, Rao LV, Ruf W. Ligand-induced protease receptor translocation into caveolae: a mechanism for regulating cell surface proteolysis of the tissue factor-dependent coagulation pathway. *J Cell Biol.* 1996;133:293–304.

58 Mast AE, Acharya N, Malecha MJ, et al. Characterization of the association of tissue factor pathway inhibitor with human placenta. *Arterioscler Thromb Vasc Biol.* 2002;22:2099–2104.

59 Lupu C, Goodwin CA, Westmuckett AD, et al. Tissue factor pathway inhibitor in endothelial cells colocalizes with glycolipid microdomains/caveolae. Regulatory mechanism(s) of the anticoagulant properties of the endothelium. *Arterioscler Thromb Vasc Biol.* 1997;17:2964–2974.

60 Adams MJ, Cardigan RA, Marchant WA, et al. Tissue factor pathway inhibitor antigen and activity in 96 patients receiving heparin for cardiopulmonary bypass. *J Cardiothorac Vasc Anesth.* 2002;16:59–63.

61 Dahm A, Van H, V, Bendz B, et al. Low levels of tissue factor pathway inhibitor (TFPI) increase the risk of venous thrombosis. *Blood.* 2003;101:4387–4392.

62 Morange PE, Simon C, Alessi MC, et al. Endothelial cell markers and the risk of coronary heart disease: the Prospective Epidemiological Study of Myocardial Infarction (PRIME) study. *Circulation.* 2004;109:1343–1348.

63 Piro O, Broze GJ Jr. Role for the Kunitz-3 domain of tissue factor pathway inhibitor-{alpha} in cell surface binding. *Circulation.* 2004;110:3567–3572.

64 Lindhout T, Willems G, Blezer R, Hemker HC. Kinetics of the inhibition of human factor Xa by full-length and truncated recombinant tissue factor pathway inhibitor. *Biochem J.* 1994;297(Pt1):131–136.

65 Chen HH, Vicente CP, He L, et al. Fusion proteins comprising annexin V and Kunitz protease inhibitors are highly potent thrombogenic site-directed anticoagulants. *Blood.* 2005;2004–2011.

66 Anderson RG, Kamen BA, Rothberg KG, Lacey SW. Potocytosis: sequestration and transport of small molecules by caveolae. *Science.* 1992;255:410–411.

67 Smart EJ, Ying YS, Conrad PA, Anderson RG. Caveolin moves from caveolae to the Golgi apparatus in response to cholesterol oxidation. *J Cell Biol.* 1994;127:1185–1197.

68 Baorto DM, Gao Z, Malaviya R, et al. Survival of FimH-expressing enterobacteria in macrophages relies on glycolipid traffic. *Nature.* 1997;389:636–639.

69 Chun M, Liyanage UK, Lisanti MP, Lodish HF. Signal transduction of a G protein-coupled receptor in caveolae: colocalization of endothelin and its receptor with caveolin. *Proc Natl Acad Sci USA.* 1994;91:11728–11732.

70 Dietzen DJ, Jack GG, Page KL, et al. Localization of tissue factor pathway inhibitor to lipid rafts is not required for inhibition of factor VIIa/tissue factor activity. *Thromb Haemost.* 2003;89:65–73.

71 Toomey JR, Kratzer KE, Lasky NM, Broze GJ Jr. Effect of tissue factor deficiency on mouse and tumor development. *Proc Natl Acad Sci USA.* 1997;94:6922–6926.

72 Parry GCN, Erlich JH, Carmeliet P, et al. Low levels of tissue factor are compatible with development and hemostasis in mice. *J Clin Invest.* 1998;101:560–569.

73 Mackman N. Role of tissue factor in hemostasis, thrombosis, and vascular development. *Arterioscler Thromb Vasc Biol.* 2004;24:1015–1022.

74 Pedersen B, Holscher T, Sato Y, et al. A balance between tissue factor and tissue factor pathway inhibitor is required for embryonic development and hemostasis in adult mice. *Blood.* 2005;105:2777–2782.

75 Chan JC, Carmeliet P, Moons L, et al. Factor VII deficiency rescues the intrauterine lethality in mice associated with a tissue factor pathway inhibitor deficit. *J Clin Invest.* 1999;103:475–482.

76 Cui J, Eitzman DT, Westrick RJ, et al. Spontaneous thrombosis in mice carrying the factor V Leiden mutation. *Blood.* 2000;96: 4222–4226.

77 Eitzman DT, Westrick RJ, Bi X, et al. Lethal perinatal thrombosis in mice resulting from the interaction of tissue factor pathway inhibitor deficiency and factor V Leiden. *Circulation.* 2002; 105:2139–2142.

78 Camerota AJ, Creasey AA, Patla V, et al. Delayed treatment with recombinant human tissue factor pathway inhibitor improves survival in rabbits with gram-negative peritonitis. *J Infect Dis.* 1998;177:668–676.

79 Matyal R, Vin Y, Delude RL, et al. Extremely low doses of tissue factor pathway inhibitor decrease mortality in a rabbit model of septic shock. *Intensive Care Med.* 2001;27:1274–1280.

80 Opal SM, Palardy JE, Parejo NA, Creasey AA. The activity of tissue factor pathway inhibitor in experimental models of superantigen-induced shock and polymicrobial intra-abdominal sepsis. *Crit Care Med*. 2001;29:13–17.
81 Creasey AA, Chang AC, Feigen L, et al. Tissue factor pathway inhibitor reduces mortality from *Escherichia coli* septic shock. *J Clin Invest*. 1993;91:2850–2856.
82 Randolph MM, White GL, Kosanke SD, et al. Attenuation of tissue thrombosis and hemorrhage by ala-TFPI does not account for its protection against E. coli – a comparative study of treated and untreated non-surviving baboons challenged with LD100 *E. coli*. *Thromb Haemost*. 1998;79:1048–1053.
83 de Jonge E, Dekkers PE, Creasey AA, et al. Tissue factor pathway inhibitor does not influence inflammatory pathways during human endotoxemia. *J Infect Dis*. 2001;183:1815–1818.
84 Abraham E, Reinhart K, Svoboda P, et al. Assessment of the safety of recombinant tissue factor pathway inhibitor in patients with severe sepsis: a multicenter, randomized, placebo-controlled, single-blind, dose escalation study. *Crit Care Med*. 2001;29:2081–2089.
85 Doshi SN, Marmur JD. Evolving role of tissue factor and its pathway inhibitor. *Crit Care Med*. 2002;30:S241–S250.
86 Ahamed J, Belting M, Ruf W. Regulation of tissue factor-induced signaling by endogenous and recombinant tissue factor pathway inhibitor-1. *Blood*. 2004;2004–2009.

104

Tissue Factor Expression by the Endothelium

Gernot Schabbauer and Nigel Mackman

The Scripps Research Institute, Department of Immunology, La Jolla, California

Tissue factor (TF) is the primary cellular initiator of blood coagulation (1). It is constitutively expressed around blood vessels and plays an essential role in hemostasis. Under normal conditions, the surface of the endothelium is antithrombotic due to the binding of the anticoagulant protein tissue factor pathway inhibitor (TFPI) and the absence of TF. In addition, "resting" endothelium expresses the receptors thrombomodulin and endothelial protein C receptor, which permits the generation of the anticoagulant protein, activated protein C. However, under pathological conditions, the surface of the endothelium becomes prothrombotic due to the induction of TF and the downregulation of anticoagulants.

Numerous studies have documented the induction of TF in cultured vascular endothelial cells (ECs) in response to a variety of agonists. Other studies have described TF expression on the endothelium in vivo, although some controversy exists about whether or not ECs express TF in vivo. This chapter reviews our current understanding of the intracellular regulatory pathways that control *TF* gene expression in cultured ECs. We also evaluate the data on TF expression by ECs in vivo.

TISSUE FACTOR EXPRESSION BY ENDOTHELIAL CELLS IN VITRO

Induction of Tissue Factor Expression in Cultured Endothelial Cells

The majority of the studies on TF expression in cultured ECs have been performed using human umbilical vein ECs (HUVECs). Quiescent cultured ECs express very low or undetectable levels of TF. The first description of TF expression by ECs was in 1977, by a group that showed a transient induction of TF activity in ECs after subculture (2). Since this time, a variety of agents have been shown to induce TF expression in ECs (Table 104-1). These agents can be broadly divided into the following classes: inflammatory mediators, growth factors, proatherogenic agents, hypertensive agents, infectious agents, and physical stimuli.

Early studies showed that the inflammatory mediators lipopolysaccharide (LPS), tumor necrosis factor (TNF)-α, and interleukin (IL)-1β induced TF expression in ECs (3–5). In addition, TF expression was upregulated by other agents, such as vascular endothelial growth factor (VEGF), CD40 ligand, thrombin, and oxygen free radicals (6–10). The proatherogenic agents, minimally modified low-density lipoprotein (MM-LDL) and homocysteine, were also shown to induce TF expression in ECs (11,12), raising the possibility that TF contributes to the development of early atherosclerotic lesions. Several infectious agents have been shown to increase TF expression in ECs (13–15), which may represent an attempt by the host to limit the spread of the infection. Finally, angiotensin II (Ang-II) was shown to induce TF expression in cultured rat aortic ECs without affecting TFPI expression (16).

We and others have elucidated the transcriptional mechanisms by which different agents induce *TF* gene expression in ECs. In the human *TF* promoter, multiple binding sites for the transcription factors AP-1, nuclear factor (NF)-κB and Egr-1 are required for maximal induction of the *TF* gene in human ECs exposed to either LPS, TNF-α, or IL-1β (17) (Figure 104.1). Mutation of the NF-κB binding site (at −179 bp relative to the start site of transcription) significantly reduces the level of induction, indicating that this binding site plays a major role in the induction of *TF* gene expression in response to inflammatory mediators. Interestingly, the NF-κB motif in the *TF* promoter differs from the consensus binding site for NF-κB because of the presence of a cytosine instead of a guanine at position 1. We found that this one nucleotide change altered the binding specificity from the classic p50/p65 heterodimer to a c-Rel/p65 heterodimer (18). Similar results were found with the porcine *TF* promoter (19). This may explain why the pattern of *TF* gene expression sometimes differs from that of genes regulated by p50/p65 heterodimers. For example, in contrast to p50/p65-regulated genes, such as *E-selectin*, TF expression is rarely detected in activated endothelium despite evidence for nuclear translocation of p65.

An early study by Clauss and colleagues (6) showed that VEGF signaling results in increased TF expression in ECs. The mechanism of induction was analyzed by two groups who found that NF-κB was not involved. One group showed that nuclear factor of activated T cells (NFAT) mediated the

Table 104-1: Induction of TF Expression in Cultured ECs

Stimulus	*References*
LPS	3, 17
TNFα	4
IL-1β	5, 17
CD40 ligand	8
Thrombin	9, 10
VEGF	6, 20
Herpes simplex virus	13
Measles	14
Rickettsia	15
Shear stress	26, 24
MM-LDL	11
oxPAPC	22
Homocysteine	12
Angiotensin II	16
Oxygen free radicals	7

effect of VEGF, whereas the second group concluded that Egr-1 was required for the induction (20,21). It appears that VEGF induction of TF expression in ECs is mediated by both NFAT and Egr-1 (see Figure 104.1). It is possible that VEGF induction of TF in ECs may contribute to tumor angiogenesis, because levels of VEGF in tumors often correlate with levels of TF.

MM-LDL was shown to induce TF expression in ECs (11). One of the oxidized lipids present on MM-LDL is oxidized 1-palmitoyl-2-arachidonoyl-sn-glycerol-3-phosphorylcholine (oxPAPC). Bochkov and colleagues (22) showed that oxPAPC also induces TF expression in ECs in an NF-κB–independent manner via activation of NFAT and Egr-1.

TF gene expression in ECs also is induced by shear stress. ECs exposed to moderate flow (15 dynes/cm^2) expressed little TF, whereas ECs exposed to low flow (0.5 dynes/cm^2) expressed significant levels of TF (23). More recently, oscillating shear stress was shown to induce TF expression in cultured ECs (24). Oscillating shear stress occurs at sites within the vasculature that are prone to the development of atherosclerotic lesions. Lin and colleagues (25) proposed that induction of TF by laminar shear stress was mediated by phosphorylation and activation of Sp1. In contrast, a later study demonstrated that shear stress induction of the *TF* gene was mediated by Egr-1 (26). It seems likely that Egr-1 is the key regulator of shear stress induction of TF in ECs.

These studies demonstrate that induction of the *TF* gene in ECs in response to different agents is mediated by different combinations of transcription factors.

Inhibition of Tissue Factor Expression in Cultured Endothelial Cells

Other studies have shown that various agents inhibit inducible TF expression (Table 104-2). For example, the anti-inflammatory cytokines IL-4 and IL-13 have been reported to suppress TF expression in ECs stimulated either with TNF-α or IL-1β (27). These studies may lead to the development of therapeutic strategies to reduce TF expression under pathological conditions, such as sepsis and atherosclerosis.

The antioxidant pyrrolidine diothiocarbamate (PDTC) has been shown to reduce LPS induction of TF expression in ECs by inhibiting NF-κB (28). A second study showed

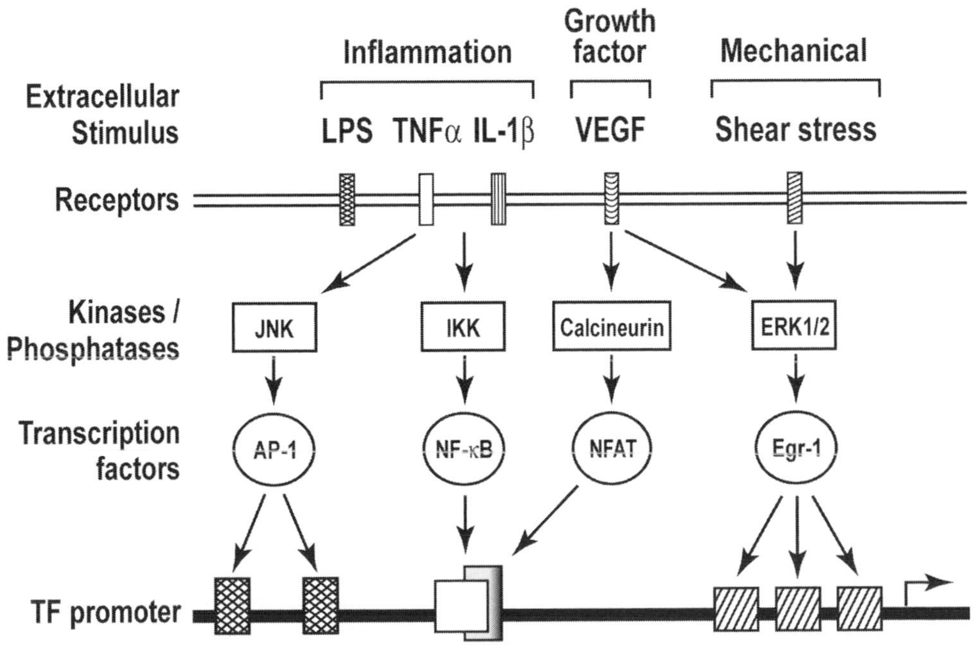

Figure 104.1. The induction of the human *TF* promoter in ECs by various agonists is mediated by a variety of intracellular signaling pathways and transcription factors.

Table 104-2: Inhibition of TF Expression in Cultured ECs

Inhibitor	*Target*	*References*
PDTC	NF-κB	28
Simvastatin	NF-κB	33
Adrenomedullin	cAMP	31
HDL	PI3K	36
Angiopoetin 1	PI3K	35

that inhibition of NF-κB by overexpression of the endogenous inhibitor, IκBα, also reduced LPS-induced TF expression in ECs (29). Surprisingly, the immunosuppressor cyclosporin inhibited *TF* gene expression in monocytes but enhanced it in ECs (30). We and others have shown that elevation of intracellular levels of cyclic adenosine monophosphate (cAMP) and activation of protein kinase A (PKA) inhibits both TNF-α and Ang-II induction of TF in ECs (31,32). The mechanism of inhibition appears to be via squelching of coactivators, such as cAMP binding protein (CBP), which are required for NF-κB–dependent transcription (32). These data indicate that the NF-κB site in the *TF* promoter is a key regulator of TF expression in ECs exposed to inflammatory mediators.

Statins, such as simvastatin, inhibit 3-hydroxy-3-methylglutaryl coenzyme A (HMG CoA) reductase and reduce the generation of geranylgeranyl pyrophosphate (GGPP). This molecule is required for efficient activation of Rho GTPase, Rho kinase, and NF-κB in ECs. Simvastatin reduced Ang-II induction of TF in ECs by inhibiting NF-κB activation (33). Therefore, statins may reduce the frequency of thrombosis by inhibiting TF expression in ECs and other cell types expressing TF within atherosclerotic plaques.

Several studies have shown that the phosphoinositide 3-kinase (PI3K)/Akt pathway suppresses TF expression in ECs. First, it was shown that inhibition of PI3K enhanced VEGF-induced TF expression in ECs (34). This enhancement appeared to be mediated by increased activation of the extracellular-signal-regulated kinase (ERK)1/2–mitogen-activated protein kinase (MAPK) pathway and expression of Egr-1. Second, Kim and colleagues (35) found that activation of the PI3K/Akt pathway by angiopoietin 1 inhibited the induction of TF in ECs exposed to VEGF or TNF-α. Third, reconstituted high-density lipoprotein (HDL) inhibited thrombin induction of TF expression in a PI3K-dependent, Akt-independent manner (36). Taken together, these studies indicated that activators of PI3K inhibit inducible TF expression in ECs.

TISSUE FACTOR EXPRESSION BY ENDOTHELIAL CELLS IN VIVO

General Considerations

Despite extensive studies showing that a variety of agents induce TF expression in cultured cells, the expression of TF in the endothelium in vivo is controversial (Table 104-3). Several issues should be addressed before discussing individual studies. First, many immunohistochemical studies conclude that TF is expressed on ECs. This can be problematic for several reasons: The quality of the anti-TF antibodies can be quite variable; a positive control tissue, such as the skin, and control antibodies often are not shown; and a strong positive signal in the vessel wall makes it difficult to distinguish between TF expression by vascular smooth muscle cells versus ECs. Second, few studies show colocalization with EC markers, such as platelet-endothelial cell adhesion molecule (PECAM)-1/CD31 and von Willebrand factor (vWF). Third, some studies measure functional TF activity in intact vessels and organs and conclude that this is due to EC TF expression because no detectable damage occurs to the endothelial barrier. Finally, the level of TF expression in stimulated cultured ECs is very low compared with other cells, such as fibroblasts. However, low levels of TF expression in the endothelium in vivo have the potential to significantly contribute to intravascular clotting due to the large number of ECs in the body and the enormous amplification of the coagulation protease cascade. Therefore, it is possible that ECs may express functional TF on their surface at levels that are below the detectable threshold of in situ hybridization and immunohistochemical techniques. Thus, researchers must be cautious in concluding that TF is not expressed by ECs. Indeed, TF expression by the endothelium may contribute to different disease processes, such as thrombosis, inflammation, and tumor angiogenesis (Figure 104.2).

Tissue Expression by Endothelial Cells in Endotoxemia and Sepsis

An early study showed that the infusion of IL-1β into rabbits induced TF expression in the aorta (37). The authors concluded that factor VIIa-dependent factor Xa formation was due to EC TF expression because scanning electron microscopy demonstrated the presence of a continuous layer of endothelium without exposure of the internal elastic lamina. No immunohistochemical studies for TF were performed to directly reveal expression of TF on the endothelium. A second study showed that LPS-treated rabbits exhibited increased levels of functional TF in the thoracic aorta (38). An alternative explanation for these results is that IL-1β and LPS disrupt the integrity of the EC barrier in the aorta, which would permit plasma clotting factors to contact the subendothelium and be activated by extravascular TF. In contrast to these in vivo studies, stimulation of human saphenous vein segments ex vivo with either LPS or thrombin failed to induce TF activity (39).

The most widely quoted paper reporting that ECs can express TF in vivo is by Drake and colleagues (40). This group analyzed TF expression in a baboon model of lethal *Escherichia coli* sepsis using a highly sensitive immunohistochemical procedure employing anti-human TF monoclonal antibodies. In septic baboons, TF was detected in ECs of the microvasculature of the marginal zone of the spleen. TF colocalized with thrombomodulin and vWF, demonstrating that the cells were

Table 104-3: TF Expression by ECs In Vivo

Disease/Agent	*Species*	*Vascular Bed*	*TF Expression*	*References*
Sepsis	Baboon	Splenic capillaries	Positive	11
Sepsis	Baboon	Renal glomerular capillaries	Positive	11
Sepsis	Baboon	Lung alveolar capillaries	Positive	11
LPS	Rabbit	Aorta	Positive	37
LPS	Rabbit	Aorta	Positive	38
LPS	Rabbit	All tissues	Negative	41
LPS	Rat	All tissues	Negative	42
Breast cancer	Human	Tumor	Positive	45
Breast cancer	Human	Tumor	Negative	47
Sickle cell disease	Human	Cec	Positive	43
Sickle cell disease	Mouse	Pulmonary veins	Positive	44
Atherosclerosis	Human	Carotid	Negative	48
Atherosclerosis	Human	Undefined	Positive	49
Cardiac vasculopathy	Rat	Cardiac	Positive	53
Tuberculosis	Human	Lung	Positive	51

ECs. A less robust TF expression in ECs was observed in the renal glomerular and lung alveolar capillaries. Importantly, EC TF expression was not observed in larger vessels in the spleen, kidney, and lung or in the microvasculature of other tissues. We analyzed EC TF expression in a rabbit endotoxemia model but failed to detect TF expression by ECs in the splenic microvasculature or in blood vessels of any other organ (41). In a rat model of LPS-induced disseminated intravascular coagulation, TF antigen was observed in monocytes but not the ECs of the microvasculature of the lung (42). These differences may be due to the different sensitivities of the immunohistochemical techniques or the use of different models and species. Interestingly, the selective expression of TF in the ECs of some tissues in septic baboons suggests the presence of different regulatory pathways in different ECs.

Tissue Factor Expression by Endothelial Cells in Sickle Cell Disease

Solovey and colleagues (43) examined TF expression in circulating ECs (CECs) from healthy human controls and patients

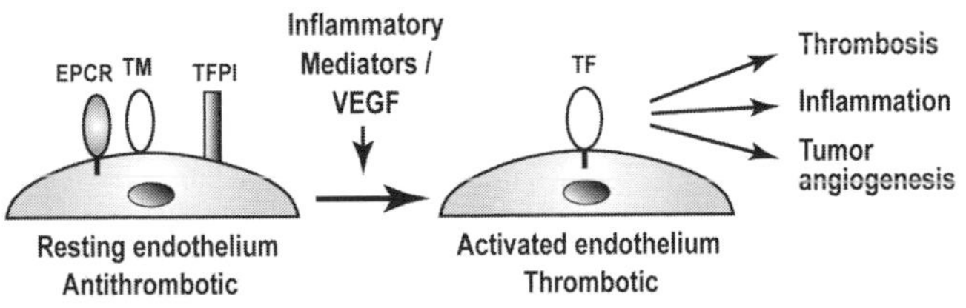

Figure 104.2. Exposure of resting ECs to inflammatory mediators or growth factors. Exposure to agonists, such as VEGF, induces TF expression and decreases the expression of the anticoagulants TFPI, thrombomodulin (TM), and endothelial protein C receptor (EPCR). TF expression by ECs in vivo may contribute to thrombosis, inflammation, and tumor angiogenesis.

with sickle cell disease as a surrogate for vessel wall endothelium. CECs from patients expressed higher levels of TF than CECs from controls. A subsequent study using a mouse model of sickle cell disease showed that TF immunostaining was significantly increased in ECs of the pulmonary veins in moderate and severe sickle cell disease (44). Hypoxia and reoxygenation also induced EC TF expression in the lung of sickle cell mice. These studies suggest that TF expression by ECs may contribute to the activation of coagulation in sickle cell disease.

Tissue Factor Expression by Tumor Endothelium

In 1996, Contrino and colleagues described TF expression by ECs in tumors of patients with invasive breast cancer but not in ECs of benign tumors (45). The authors proposed that TF may be a useful marker of the switch to the angiogenic phenotype. A subsequent study by Rickles and colleagues (46) reported TF expression by ECs in adenocarcinoma of the lung and four out of five large-cell carcinomas. In the latter study, TF expression was colocalized with CD31. In contrast, Luther and colleagues (47) failed to detect TF expression in the ECs of breast tumors. In view of the variable expression and controversy, it seems unlikely that TF can be used as a reliable marker of angiogenic ECs.

Tissue Factor Expression by Endothelial Cells in Atherosclerosis

In human atherosclerotic plaques, *TF* mRNA and protein was observed in macrophages present as foam cells and in monocytes adjacent to the cholesterol cleft, but not in the endothelium (48). In a later study, digoxigenin-labeled factor VIIa bound to vWF-positive endothelium overlying atherosclerotic plaques (49). These results suggest that TF is variably expressed by ECs overlying atherosclerotic plaques and may contribute to thrombosis in plaque erosion.

Tissue Factor Expression by Endothelial Cells in Cardiac Ischemia–Reperfusion Injury

A recent study analyzed the effect of oxygen free radicals in TF expression in coronary ECs in vitro and in vivo (7). Oxygen free radicals strongly induced TF expression in cultured rabbit coronary ECs. Similarly, TF activity was induced in the hearts of rabbits subjected to cardiac ischemia–reperfusion (I/R) injury. TF activity in the hearts was measured by perfusing them ex vivo with factor VII and factor X, and measuring factor Xa levels. The authors argued that the TF was expressed by ECs because the injury was not associated with ultrastructural vascular damage. In contrast, we failed to detect TF expression by ECs but did observe robust TF expression on cardiomyocytes in hearts of rabbits subjected to a similar I/R procedure (50). These differences indicate that researchers must be cautious in concluding that ECs express TF without direct immunohistochemical evidence.

Tissue Factor Expression by Endothelial Cells in Other Diseases

TF expression by ECs has been reported in human patients with tuberculosis and idiopathic inflammatory bowel disease (51,52). In addition, TF is expressed by ECs of cardiac vessels in rat models of Ang-II-induced cardiac vasculopathy and cardiac allograft vasculopathy (53,54). TF expression in ECs in these diseases and disease models may contribute to both coagulation and inflammation.

STUDIES USING TRANSGENIC MOUSE MODELS

We and others have used mouse models and bone marrow transplantation to determine the relative contribution of TF expression by hematopoietic and nonhematopoietic cells to LPS-induced coagulation. In these studies, we hypothesize that ECs are the major source of TF in nonhematopoietic cells, although we cannot exclude the possibility that other nonhematopoietic cells also contribute to LPS-induced coagulation. We found that mice genetically engineered to express very low levels of TF, so called low-TF mice (Table 104-4) (55), exhibited low levels of LPS-induced coagulation as measured by plasma thrombin-antithrombin (TAT) levels (56). In contrast, LPS stimulation of wild-type mice transplanted with bone marrow from low-TF mice for 6 weeks had levels of TAT that were intermediate between those of low-TF and wild-type mice. This result suggested that nonhematopoietic cells, possibly ECs, were expressing TF, and that this expression significantly contributed to LPS-induced coagulation.

In a separate study, transgenic mice were generated that expressed chimeric anticoagulant proteins under the control of the *CD31* promoter, which directed expression in ECs, platelets, and monocytes (57). LPS-induced coagulation was substantially reduced in the transgenic mice compared with wild-type mice. Bone marrow transplantation experiments showed that expression of the anticoagulants on ECs produced a greater reduction in LPS-induced coagulation than expression on platelets and monocytes. This study provides indirect evidence that TF expression by ECs contributes to LPS-induced coagulation.

Recently, we generated mice containing a *TF* gene flanked by Lox P sites. These mice will be used to determine the effect of EC-specific deletion of the *TF* gene on LPS-induced coagulation.[1]

Table 104-4: Mouse Models with Altered TF Gene Expression

Model	*Reference*
Low-TF mice	55
Floxed TF mice	Unpublished

KEY POINTS

- EC TF expression plays an important role in intravascular thrombosis and may contribute to inflammation and tumor angiogenesis.
- Induction of *TF* gene expression in ECs is primarily mediated by the transcription factors AP-1, NF-κB, and Egr-1.
- Inhibition of NF-κB and activation of PI3K reduces TF expression in ECs.
- TF is expressed by a subset of ECs in vivo in animal models of sepsis and sickle cell disease.

Future Goals

- To elucidate the contribution of EC TF in intravascular thrombosis, inflammation, and tumor angiogenesis
- To selectively target anticoagulants to the vascular endothelium to limit the pathological effects of EC TF expression

1 A recent study showed that overexpression of the Kruppel-like factor KLF2 inhibited cytokine-mediated induction of TF expression in ECs (58). This suggests that KLF2 may act as a negative regulator of TF expression in the endothelium in vivo. A second study described a detailed analysis of TF expression in various cell types in septic baboons (59). TF was found to colocalize with TFPI and CD31, indicating that it was located on the surface of EC at arterial branching areas. However, these studies cannot distinguish between TF expression by the ECs themselves versus deposition of TF positive microparticles that are generated by monocytes and other cell types.

REFERENCES

1 Mackman N. Role of tissue factor in hemostasis, thrombosis, and vascular development. *Arterioscler Thromb Vasc Biol.* 2004;24: 1015–1022.

2 Maynard JR, Dreyer BE, Stemerman MB, Pitlick FA. Tissue-factor coagulant activity of cultured human endothelial and smooth muscle cells and fibroblasts. *Blood.* 1977;50:387–396.

3 Colucci M, Balconi G, Lorenzet R, et al. Cultured human endothelial cells generate tissue factor in response to endotoxin. *J Clin Invest.* 1983;71:1893–1896.

4 Bevilacqua MP, Pober JS, Majeau GR, et al. Recombinant tumor necrosis factor induces procoagulant activity in cultured human vascular endothelium: Characterization and comparison with the actions of interleukin 1. *Proc Natl Acad Sci USA.* 1986;83: 4533–4537.

5 Bevilacqua MP, Pober JS, Majeau GR, et al. Interleukin 1 (IL-1) induces biosynthesis and cell surface expression of procoagulant activity in human vascular endothelial cells. *J Exp Med.* 1984;160:618–623.

6 Clauss M, Gerlach M, Gerlach H, et al. Vascular permeability factor: a tumor-derived polypeptide that induces endothelial cell and monocyte procoagulant activity, and promotes monocyte migration. *J Exp Med.* 1990;172:1535–1545.

7 Golino P, Ragni M, Cirillo P, et al. Effects of tissue factor induced by oxygen free radicals on coronary flow during reperfusion. *Nature Med.* 1996;2:35–40.

8 Bavendiek U, Libby P, Kilbride M, et al. Induction of tissue factor expression in human endothelial cells by CD40 ligand is mediated via activator protein 1, nuclear factor κB, and Egr-1. *J Biol Chem.* 2002;277:25032–25039.

9 Galdal KS, Lyberg T, Evensen SA, et al. Thrombin induces thromboplastin synthesis in cultured vascular endothelial cells. *Thromb Haemost.* 1985;54:373–376.

10 Takeya H, Gabazza EC, Aoki S, et al. Synergistic effect of sphingosine 1-phosphate on thrombin-induced tissue factor expression in endothelial cells. *Blood.* 2003;102:1693–1700.

11 Drake TA, Hannani K, Fei H, et al. Minimally oxidized low-density lipoprotein induces tissue factor expression in cultured human endothelial cells. *Am J Pathol.* 1991;138:601–607.

12 Fryer RH, Wilson BD, Gubler DB, et al. Homocysteine, a risk factor for premature vascular disease and thrombosis, induces tissue factor activity in endothelial cells. *Arterioscler Thromb.* 1993;13:1327–1333.

13 Key NS, Vercellotti GM, Winkelmann JC, et al. Infection of vascular endothelial cells with herpes simplex virus enhances tissue factor activity and reduces thrombomodulin expression. *Proc Natl Acad Sci USA.* 1990;87:7095–7099.

14 Mazure G, Grundy JE, Nygard G, et al. Measles virus induction of human endothelial cell tissue factor procoagulant activity *in vitro. J Gen Virol.* 1994;75:2863–2871.

15 Sporn LA, Haidaris PJ, Shi R-J, et al. Rickettsia rickettsii infection of cultured human endothelial cells induces tissue factor expression. *Blood.* 1994;83:1527–1534.

16 Nishimura H, Tsuji H, Masuda H, et al. Angiotensin II increases plasminogen activator inhibitor-1 and tissue factor mRNA expression without changing that of tissue type plasminogen activator or tissue factor pathway inhibitor in cultured rat aortic endothelial cells. *Thromb Haemost.* 1997;77:1189–1195.

17 Parry GC, Mackman N. Transcriptional regulation of tissue factor expression in human endothelial cells. *Arterioscler Thromb.* 1995;15:612–621.

18 Oeth PA, Parry GCN, Kunsch C, et al. Lipopolysaccharide induction of tissue factor gene expression in monocytic cells is mediated by binding of c-Rel/p65 heterodimers to a κB-like site. *Mol Cell Biol.* 1994;14:3772–3781.

19 Moll T, Czyz M, Holzmuller H, et al. Regulation of the tissue factor promoter in endothelial cells. *J Biol Chem.* 1995;270:3849–3857.

20 Mechtcheriakova D, Wlachos A, Holzmuller H, et al. Vascular endothelial cell growth factor-induced tissue factor expression in endothelial cells is mediated by EGR-1. *Blood.* 1999;93:3811–3823.

21 Armesilla AL, Lorenzo E, Gomez del Arco P, et al. Vascular endothelial growth factor activates nuclear factor of activated T cells in human endothelial cells: a role for tissue factor gene expression. *Mol Cell Biol.* 1999;19:2032–2043.

22 Bochkov VN, Mechtcheriakova D, Lucerna M, et al. Oxidized phospholipids stimulate tissue factor expression in human endothelial cells via activation of ERK/EGR-1 and Ca($^{++}$)/NFAT. *Blood.* 2002;99:199–206.

23 Grabowski EF, Zuckerman DB, Nemerson Y. The functional expression of tissue factor by fibroblasts and endothelial cells under flow conditions. *Blood.* 1993;81:3265–3270.

24 Mazzolai L, Silacci P, Bouzourene K, et al. Tissue factor activity is upregulated in human endothelial cells exposed to oscillatory shear stress. *Thromb Haemost.* 2002;87:1062–1068.

25 Lin M-C, Shyy JYJ, Almus-Jacobs F, et al. The shear stress-induced tissue factor gene expression is mediated through the Sp1 *Cis*-element. *J Clin Invest.* 1997;99:737–744.

26 Houston P, Dickson MC, Ludbrook V, et al. Fluid shear stress induction of the tissue factor promoter in vitro and in vivo is mediated by Egr-1. *Arterioscler Thromb Vasc Biol.* 1999;19:281–289.

27 Herbert JM, Savi P, Laplace MC, et al. IL-4 and IL-13 exhibit comparable abilities to reduce pyrogen-induced expression of procoagulant activity in endothelial cells and monocytes. *FEBS Lett.* 1993;328:268–270.

28 Orthner CL, Rodgers GM, Fitzgerald LA. Pyrrolidine dithiocarbamate abrogates tissue factor (TF) expression by endothelial cells: evidence implicating nuclear factor- kB in TF induction by diverse agonists. *Blood.* 1995;86:436–443.

29 Wrighton CJ, Hofer-Warbinek R, Moll T, et al. Inhibition of endothelial cell activation by adenovirus-mediated expression of IkBα, an inhibitor of the transcription factor NF-κB. *J Exp Med.* 1996;183:1013–1022.

30 Holschermann H, Rascher C, Oelschlager C, et al. Opposite regulation of tissue factor expression by calcineurin in monocytes and endothelial cells. *J Immunol.* 2001;166:7112–7120.

31 Sugano T, Tsuji H, Masuda H, et al. Adrenomedullin inhibits angiotensin II-induced expression of tissue factor and plasminogen activator inhibitor-1 in cultured rat aortic endothelial cells. *Arterioscler Thromb Vasc Biol.* 2001;21:1078–1083.

32 Ollivier V, Parry GCN, Cobb RR, et al. Elevated cyclic AMP inhibits NF-κB-mediated transcription in human monocytic cells and endothelial cells. *J Biol Chem.* 1996;271:20828–20835.

33 Kunieda Y, Nakagawa K, Nishimura H, et al. HMG CoA reductase inhibitor suppresses the expression of tissue factor and plasminogen activator inhibitor-1 induced by angiotensin II in cultured rat aortic cells. *Thromb Res.* 2003;110:227–234.

34 Blum S, Issbruker K, Willuweit A, et al. An inhibitory role of the phosphatidylinositol 3-kinase-signaling pathway in vascular endothelial growth factor-induced tissue factor expression. *J Biol Chem.* 2001;276:33428–33434.
35 Kim I, Oh JL, Ryu YS, et al. Angiopoietin-1 negatively regulates expression and activity of tissue factor in endothelial cells. *FASEB J*. 2002;16:126–128.
36 Viswambharan H, Ming X-F, Zhu S, et al. Reconstituted high-density lipoprotein inhibits thrombin-induced endothelial tissue factor expression through inhibition of RhoA and stimulation of phosphatidylinositol 3-kinase but not Akt/endothelial nitric oxide synthase. *Circ Res.* 2004;94:918–925.
37 Nawroth PP, Handley DA, Esmon CT, Stern DM. Interleukin 1 induces endothelial cell procoagulant while suppressing cell-surface anticoagulant activity. *Proc Natl Acad Sci USA.* 1986;83:3460–3464.
38 Semeraro N, Triggiani R, Montemurro P, et al. Enhanced endothelial tissue factor but normal thrombomodulin in endotoxin-treated rabbits. *Thromb Res.* 1993;71:479–486.
39 Solberg S, Osterud B, Larsen T, Sorlie D. Lack of ability to synthesize tissue factor by endothelial cells in intact human saphenous veins. *Blood Coagul Fibrinolysis.* 1990;1:595–600.
40 Drake TA, Cheng J, Chang A, Taylor FB Jr. Expression of tissue factor, thrombomodulin, and E-selectin in baboons with lethal *Escherichia coli* sepsis. *Am J Pathol.* 1993;142:1–13.
41 Erlich JH, Fearns C, Mathison J, et al. Lipopolysaccharide induction of tissue factor expression in rabbits. *Infect Immun.* 1999;67:2540–2546.
42 Hara S, Asada Y, Hatakeyama K, et al. Expression of tissue factor and tissue factor pathway inhibitor in rat lungs with lipopolysaccharide-induced disseminated intravascular coagulation. *Lab Invest.* 1997;77:581–589.
43 Solovey A, Gui LH, Key NS, Hebbel RP. Tissue factor expression by endothelial cells in sickle cell anemia. *J Clin Invest.* 1998;101:1899–1904.
44 Solovey A, Lollander R, Shet A, et al. Endothelial cell expression of tissue factor in sickle mice is augmented by hypoxia/reoxygenation and inhibited by lovastatin. *Blood.* 2004;104:840–846.
45 Contrino J, Hair G, Kreutzer DL, Rickles FR. In situ detection of tissue factor in vascular ECs: correlation with the malignant phenotype of human breast disease. *Nat Med.* 1996;2:209–215.
46 Shoji M, Hancock WW, Abe K, et al. Activation of coagulation and angiogenesis in cancer. Immunohistochemical localization in situ of clotting proteins and vascular endothelial growth factor in human cancer. *Am J Pathol.* 1998;152:399–411.
47 Luther T, Flossel C, Albrecht S, et al. Tissue factor expression in normal and abnormal mammary gland. *Nat Med.* 2005;2:209–215.
48 Wilcox JN, Smith KM, Schwartz SM, Gordon D. Localization of tissue factor in the normal vessel wall and in the atherosclerotic plaque. *Proc Natl Acad Sci USA.* 1989;86:2839–2843.
49 Thiruvikraman SV, Guha A, Roboz J, et al. In situ localization of tissue factor in human atherosclerotic plaques by binding of digoxigenin-labeled factors VIIa and X. *Lab Invest.* 1996;75(4): 451–461.
50 Erlich JH, Boyle EM Jr., Labriola J, et al. Inhibition of the tissue factor-thrombin pathway limits infarct size after myocardial ischemia-reperfusion injury by reducing inflammation. *Am J Pathol.* 2000;157:1849–1862.
51 Lang IM, Mackman N, Kriett JM, et al. Prothrombotic activation of pulmonary arterial endothelial cells in a patient with tuberculosis. *Hum Pathol.* 1996;27:423–427.
52 More L, Sim R, Hudson M, et al. Immunohistochemical study of tissue factor expression in normal intestine and idiopathic inflammatory bowel disease. *J Clin Pathol.* 1993;46:703–708.
53 Muller D, Mervaala E, Dechend R, et al. Angiotensin II (AT_1) receptor blockade reduces vascular tissue factor in angiotensin II-induced cardiac vasculopathy. *Am J Pathol.* 2000;157:111–122.
54 Hölschermann H, Bohle RM, Zeller H, et al. In situ detection of tissue factor within the coronary intima in rat cardiac allograft vasculopathy. *Am J Pathol.* 1999;154:211–220.
55 Parry GC, Erlich JH, Carmeliet P, et al. Low levels of tissue factor are compatible with development and hemostasis in mice. *J Clin Invest.* 1998;101:560–569.
56 Pawlinski R, Pedersen B, Schabbauer G, et al. Role of tissue factor and protease activated receptors in a mouse model of endotoxemia. *Blood.* 2004;103:1342–1347.
57 Chen D, Giannopoulos K, Shiels PG, et al. Inhibition of intravascular thrombosis in murine endotoxemia by targeted expression of hirudin and tissue factor pathway inhibitor analogs to activated endothelium. *Blood.* 2004;104:1344–1349.
58 Lin Z, Kumar A, Senbanerjee S, et al. Kruppel-like factor 2 (KLF2) regulates endothelial thrombotic function. *Circ Res.* 2005; 96:E48–E57.
59 Lupu C, Westmuckett AD, Peer G, et al. Tissue factor-dependent coagulation is preferentially up-regulated within arterial branching areas in a baboon model of *Escherichia coli* sepsis. *Am J Pathol.* 2005;167:1161–1172.

105

Thrombomodulin

Marlies Van de Wouwer and Edward M. Conway

Department for Transgene Technology and Gene Therapy, VIB, and the Center for Transgene Technology and Gene Therapy (CTG), K.U. Leuven, Belgium

Thrombin has far-reaching effects in a variety of biological systems, including coagulation, inflammation, cell survival, and cell proliferation. Modulation of the dynamic production of this key protease is essential to maintain homeostasis during development and, in adults, under a variety of pathophysiological conditions. The crucial importance of thrombin has prompted intense studies over the past few decades to characterize the molecular pathways that regulate its functional expression. The remarkable insight of several scientists led to the discovery of thrombomodulin (TM) (1), a key regulator of thrombin's activities, a predominantly vascular endothelial cell (EC) cofactor in a physiologically relevant natural anticoagulant system, and a modulator of inflammation and cell proliferation, with links to innate immunity.

THROMBOMODULIN IS A MASTER SWITCH THAT PREVENTS BLOOD CLOTTING

The discovery of TM (1,2), elegantly recounted by Esmon and Owen (1), emanated from evidence in the 1960s of the existence of a circulating thrombin-activated protein – now referred to as activated protein C (APC) – and the identification and isolation of its precursor, protein C (PC) (3). PC is a vitamin-K–dependent plasma protein, synthesized in the liver and circulating as a biologically inactive species. Activation of PC by thrombin requires TM as a cofactor (4) (Figure 105.1), which accelerates the reaction over 1,000-fold. APC, in turn, suppresses further thrombin generation through the proteolysis of coagulation factors Va and VIIIa, facilitated by the cofactor, protein S (PS). Not only does TM affect the generation of APC but, when complexed with TM, thrombin's entire substrate specificity is changed such that it no longer functions as a procoagulant molecule. For example, thrombin-TM is not able to activate platelets or factor V, or to cleave fibrinogen. Thus, TM is a biological switch for thrombin, maintaining the hemostatic balance to prevent disease and/or to respond to injury. The relevance of this pathway is supported in the clinic: Individuals with functional deficiencies of PC or PS, or with resistance to cleavage of factor V by APC (factor V Leiden polymorphism) are more prone to develop thromboembolic disease. Notably, defects in TM expression or function in humans have been difficult to document, and no reports exist of patients entirely lacking TM. This likely underlines its critical importance for survival, although it also may reflect difficulties in diagnosing TM deficiency.

THROMBOMODULIN MODULATES INFLAMMATION VIA ACTIVATED PROTEIN C GENERATION

Through activation of PC and alteration of thrombin's substrate specificity, the function of TM extends well beyond coagulation. APC has potent anti-inflammatory and vasculoprotective properties (5). This is confirmed by clinical studies showing that the administration of APC to patients with severe sepsis significantly reduces mortality (6). Several mechanisms exist by which APC suppresses inflammation. For example, APC downregulates thrombin generation, which thus yields less thrombin-induced activation of platelets, leukocytes, and ECs (7,8). Additionally, APC has direct cellular effects – mostly mediated via the cleavage of protease-activated receptor (PAR)-1 (9) – interfering with endothelial apoptosis, suppressing monocyte and macrophage cytokine signaling, preventing tumor necrosis factor (TNF)-α-induced upregulation of leukocyte adhesion molecules, and promoting endothelial barrier protection (10).

THROMBOMODULIN INTERFERES WITH FIBRINOLYSIS AND COMPLEMENT ACTIVATION VIA THROMBIN-ACTIVATABLE FIBRINOLYSIS INHIBITOR ACTIVATION

The thrombin–TM complex also modulates inflammation and fibrinolysis by activating thrombin-activatable fibrinolysis inhibitor (TAFI), a plasma procarboxypeptidase B (11).

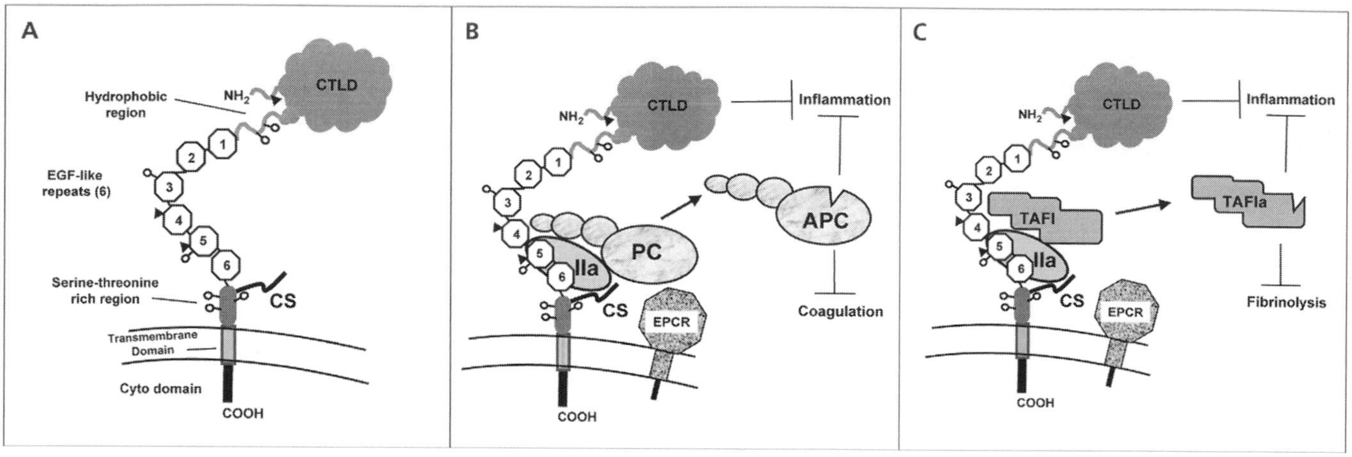

Figure 105.1. Structure of thrombomodulin and activation of protein C (PC) and thrombin activatable fibrinolysis inhibitor (TAFI). (**A**) TM is represented with its five structural domains, including the N-terminal C-type lectin-like domain (CTLD) attached via a hydrophobic region to six EGF-like repeats, the serine-threonine rich region with the attached chondroitin sulfate (CS) moiety, a transmembrane domain, and the short cytoplasmic tail. Sites of potential attachment of O-linked (*O*) and N-linked (σ) sugars are shown. (**B**) EGF-like repeats four through six of TM provide minimum cofactor function for thrombin (IIa)-mediated activation of PC, a step that is further enhanced by the EC protein C receptor (EPCR). Activated protein C (APC) cleaves coagulation cofactors Va and VIIIa (not shown), thereby downregulating thrombin generation; it also interferes with inflammation via several mechanisms. The CTLD of TM also directly suppresses inflammation. (**C**) EGF-like repeats three through six of TM provide cofactor function for thrombin-mediated activation of TAFI. The procarboxypeptidase TAFIa cleaves basic C-terminal amino acid residues of its substrates, including fibrin/ogen, and thus interferes with efficient generation of plasmin from plasminogen. TAFIa also inactivates the anaphylotoxin C3a and C5a, and the proinflammatory mediators, bradykinin and osteopontin.

Activated TAFI (TAFIa) catalyzes the removal of C-terminal basic amino acid residues from target proteins, thereby altering their functional properties. The first-described effect of TAFIa was its suppression of fibrinolysis achieved through removal of lysine residues from modified fibrinogen, which results in inefficient tissue-type plasminogen activator (t-PA)-induced conversion of plasminogen to plasmin (see Figure 105.1). Conversely, TM also may be considered profibrinolytic, because, via APC, it suppresses thrombin generation, which in turn downregulates further TAFIa production. Indeed, this may contribute to the bleeding tendency in patients with factor XI deficiency (12). TAFIa also targets and inactivates the anaphylotoxin C3a and C5a (13), and the proinflammatory mediators bradykinin and osteopontin. The physiological relevance of these biochemical pathways in health and disease remains to be resolved (14,15).

THE *TM* GENE AND ITS PRODUCT – STRUCTURE AND FUNCTIONS

Early recognition of the importance of TM in the field of coagulation led quickly, during the 1980s, to analyses of its gene structure, regulation, and expression (16–20). The single *TM* mRNA species is transcribed from an intronless gene located on chromosome 20p in humans (21). The protein is organized as a type I transmembrane glycoprotein receptor, although soluble forms exist that are generated by proteolytic cleavage of the extracellular domains (see the following section). Human TM is 557 amino acids long, organized into five structurally distinct domains (see Figure 105.1). After a short N-terminal signal peptide, it comprises a globular C-type lectin-like domain attached via a hydrophobic region to a series of six epidermal growth factor (EGF)-like repeats, a serine-threonine-rich region with sites of O-linked glycosylation that support a chondroitin sulfate moiety, a highly conserved transmembrane domain, and a short cytoplasmic tail. Interestingly, endosialin, a tumor EC marker without a known function, but with strikingly similar structural features to TM (including an N-terminal C-type lectin-like domain) also is encoded by an intronless gene (22). The potential relevance of this intriguing observation is discussed in the section, Evolutionary Considerations.

In vivo and in vitro studies have revealed the functions of several of the structural domains of TM (Table 105-1). The stretch of protein containing the EGF-like repeats is the best characterized, this being the site of interaction of TM with thrombin, PC, and TAFI (23). Thrombin binds to EGF5 and -6 via its anion binding exosite I, but activation of PC requires EGF4, -5, and -6, and activation of TAFI requires EGF3 through EGF6 (24,25). Nothing is known about the function of EGF1 and -2, although isolated EGF1 through -6 have mitogenic activity in vitro on cultured fibroblasts and vascular smooth muscle cells, implying a potential role in cellular proliferation and atherogenesis. An aspartic acid residue (Asp349) within EGF4 is required for Ca^{2+}-mediated binding of TM to the γ-carboxyglutamic acid (gla)-containing domain of PC, and thus for thrombin-mediated activation of PC. Elucidated by a combination of in vitro biochemical studies and NMR structural analyses (26), oxidation of methionine

Table 105-1: Structure–Function Correlates of TM

Structure	*Residues*	*Function*
C-type lectin-like domain (CTLD)	1–154	Anti-inflammatory (inhibits neutrophil adhesion) Mediates cell–cell interactions Antiapoptotic
Hydrophobic region	155–222	?
EGF1–2	223–304	?
EGF3–6	305–462	Activation of TAFI
EGF4–6	346–462	Activation of PC Thrombin neutralization by PCI
EGF5–6	387–462	Site for thrombin attachment Inactivation of scu-PA (+CS)
EGF1–6	223–462	Mitogenic
Chondroitin sulfate (CS) in serine-threonine rich region	463–495	Site for thrombin attachment (low affinity) Enhances neutralization of scu-PA Enhances neutralization of thrombin by PCI and heparin-AT PF4 binding, which enhances activation of PC
Transmembrane	496–521	Anchor for TM and site of cleavage by rhomboid
Cytoplasmic tail	522–557	Recognition site for rhomboid

388 in the linker between EGF4 and EGF5 abrogates thrombin's ability to activate PC, an event that likely occurs with neutrophil activation during inflammation. We can speculate that, at the onset of an inflammatory stimulus, leukocyte release of reactive oxygen species causes oxidation of methionine 388, suppressing the generation of APC and promoting local fibrin deposition. TAFI activation, which would still be able to be generated under these conditions (because it is unaffected by the oxidation of methionine 388), would interfere with fibrinolysis, enabling the clot to propagate and/or consolidate. The EGF-like repeats of TM also impede fibrinolysis by accelerating the thrombin-mediated conversion of single-chain urokinase-type plasminogen activator (scu-PA) to thrombin-cleaved two-chain urokinase-type plasminogen activator (tcu-PA), thus diminishing the generation of plasmin (27).

Adjacent to the EGF-like repeats, the chondroitin sulfate moiety of TM – in vitro – enhances thrombin-mediated activation of PC, accelerates neutralization of thrombin by the protein C inhibitor (PCI) and heparin-antithrombin, and promotes platelet factor 4 (PF4) binding to PC (28). This in turn facilitates the generation of APC by altering the conformation of PC to favor its interaction with thrombin-TM.

At the N-terminus of TM, situated furthest from the membrane, the C-type lectin-like domain is physically positioned so that it may interact with other molecules. This structure lacks anticoagulant activity, but plays a direct role in inflammation and cell survival. Transgenic mice that lack the lectin-like domain of TM respond to inflammatory stimuli with augmented neutrophil extravasation, elevated cytokines, and greater tissue destruction in murine models of myocardial ischemia–reperfusion, arthritis, and endotoxin inhalation lung injury (29). In vivo studies revealed that the lectin-like domain of TM has direct anti-inflammatory properties, protecting the vasculature by interfering with neutrophil adhesion, mediated by activation of well-conserved mitogen-activated protein kinase (MAPK) pathways. Recent evidence suggests that the lectin-like domain of TM sequesters the circulating cytokine high-mobility group box 1 (HMGB1), thereby preventing it from transmitting intracellular proinflammatory signals (30). Maintenance of the integrity of cell–cell interactions also may be mediated by the lectin-like domain via homotypic and/or heterotypic interactions (31). Overall, the lectin-like domain of TM dampens the response of the vascular endothelium to proinflammatory stimuli, interestingly via mechanisms that appear to overlap with those of APC, thus highlighting their close coordination and integration.

EVOLUTIONARY CONSIDERATIONS

As mentioned earlier, *TM* is encoded by an intronless gene (21). This is not particularly unusual, since 5% to 12% of human genes are intronless (32), including, for example, those for most heat-shock proteins, many G-protein–coupled receptors, the β_2-adrenergic receptor, type I interferons, and complement proteases C1r and C1s (33). The question of whether a lack of introns confers special properties to the gene is unclear. Because single exon genes, such as *TM*, are not susceptible to splice defects, it is reasonable to assume that they might be transcribed with greater fidelity. This does not necessarily imply that transcription rates are higher or more efficient, because introns actually may increase gene expression levels (34). Moreover, single exon genes may be disadvantaged in terms of flexibility in regulating expression levels in different tissues and under varying conditions. From an evolutionary standpoint, evidence suggests that intronless mammalian genes are more rapidly evolving (32), although this has not been evaluated for *TM*; nor has the relevance of this observation been elucidated.

TM belongs to the family of C-type lectin-like domains, which bind to protein ligands rather than to carbohydrates, do not bind calcium, and are primarily involved in immune surveillance. The domain organization of *Drosophila* proteins that contain C-type lectin-like domains is strikingly different when compared with the organization of those in mammalian or in *C. elegans* proteins. The C-type lectin-like domain of TM is unique in its composition. Whereas most C-type lectins have a calcium-binding motif, this is lacking in TM. Furthermore, the lectin-like domain of TM contains eight, rather than the usual four cysteines. Only endosialin (35), a tumor-specific endothelial marker, and C1qRp, a molecule involved in leukocyte adhesion and complement activation (36), have structurally similar N-terminal C-type lectin-like domains. Murine C1qRp and TM may be derived from a common ancestor, because their genes colocalize to the same region of chromosome 2, and because they share similar domain composition and organization.

Lectins are commonly involved in the complement system and, indeed, the mannan-binding lectin (MBL), also containing a C-type lectin, provides an alternative activation pathway. Channel-forming proteins of the complement pathway also have a cysteine-rich N-terminal region typical of the C-type lectins, and C6 shows structural similarity (31%–38%) with TM. Notably, both C6 and TM may bind thrombin, although the relevance of the C6–thrombin interaction is unknown. Preliminary data from our group suggest that the C-type lectin-like domain of TM also might be involved in the modulation of complement (unpublished observations), consistent with a report that immobilized TM interferes with complement activation.

In contrast to many key components of the coagulation cascade (e.g., factors II, V, VII, VIII, FIX, X, tissue factor, PC, and PS), little is otherwise known about TM in terms of its evolution. Biochemical studies and comparative sequence analyses support the existence of the major coagulation factors and cofactors, including TM, in all jawed vertebrates, suggesting that they evolved before the divergence of teleosts over 430 million years ago. However, no confirmatory evidence exists of the existence of TM in nonmammalian vertebrates. Although PC has been identified in *Fugu*, the closest *Fugu* orthologue for *TM* is an intronless gene with 29% identity and 46% similarity to human TM, but functional studies to confirm a relationship have not yet been performed.

ROLE OF THROMBOMODULIN IN DEVELOPMENT

During murine embryogenesis, TM first appears in the parietal endoderm at embryonic day (E)7.5, and is present in nonvascular structures at E10.5. Gene inactivation studies in mice confirmed a critical role for TM in development, unrelated to coagulation, because TM-deficient embryos die at E8.5, without clear evidence of thrombosis (37) (Table 105-2). In elegant studies, Isermann and colleagues revealed that placental trophoblasts normally express TM, which is required for their growth and survival (38,39). Embryonic lethality of TM-deficient embryos could thus be attributed to loss of TM-dependent survival of placental trophoblast cells, mediated by unregulated coagulation and/or fibrinolysis, possibly via PAR-2 or PAR-4 (39). Neither the cytoplasmic tail nor the N-terminal lectin-like domains of TM are required for normal fetal development (29,40).

Table 105-2: Transgenic Mice

Genotype	*Phenotype (Refs.)*
$TM^{-/-}$	*TM* gene inactivation results in embryonic lethality that occurs at ~E8.5 without evidence of thrombosis (37) but embryos die secondary to growth arrest of trophoblasts (38,39).
$TM^{pro/pro}$	Mice with single amino acid substitution in TM (Glu404 to Pro) causes two- to threefold decrease in TM expression and minimal PC activation. Mice are viable, but are sensitive to prothrombotic and proinflammatory stimuli (56–58).
$TM^{cyt/cyt}$	Mice lack the cytoplasmic tail of TM, are viable, and without significant defects in coagulation or inflammation (40).
$TM^{LeD/LeD}$	Mice lack the C-type lectin-like domain of TM. They have normal PC cofactor activity, without coagulation defects, but are proinflammatory and more sensitive to endotoxin and ischemia–reperfusion injury (29).
TMLox–	Endothelial-specific loss of TM causes juvenile onset of thrombosis (38).
$TAFI^{-/-}$	*TAFI* gene inactivation has no obvious thrombotic phenotype, but exhibits evidence of enhanced endogenous fibrinolysis (14,15).
$EPCR^{-/-}$	*EPCR* gene inactivation results in embryonic lethality by E10.5. This is caused by lack of trophoblast expression of EPCR, because embryos with normal expression of trophoblast EPCR, but lacking endothelial EPCR survive. As adults, however, they are prone to thrombosis in response to procoagulant stimuli (42,44).
Low EPCR mice	Mice with low EPCR (<10%) are fertile and appear normal, and are not more prone to arterial thrombosis (43).

ENDOTHELIAL PROTEIN C RECEPTOR: A PARTNER FOR THROMBOMODULIN

It is evident from the preceding sections that several proteins interact with TM to enhance and regulate the thrombin-induced activation of PC. The endothelial protein C receptor (EPCR), constitutively expressed by ECs, and similar structurally to the major histocompatibility complex class 1/CDI family of proteins (41), accelerates thrombin-mediated activation of PC 10- to 20-fold in vivo. Once generated, APC first binds to EPCR, prior to proteolyzing factors Va/VIIIa. Once bound to EPCR, APC can cleave PAR-1, initiating antiapoptotic signaling pathways (9).

The expression pattern of EPCR is distinct from that of TM, with greater prominence in the endothelium of large vessels. Similar to TM, however, trophoblast expression of EPCR of at least 1% to 10% is both sufficient and necessary for survival of the embryo (42–44). In contrast to the situation with TM, lack of EC expression of EPCR does not cause profound, spontaneous juvenile onset thrombosis (38,39) (see Table 105-2). Rather, mice lacking endothelial EPCR survive to adulthood, but suffer from thromboses only in response to procoagulant stimuli (44). These data from transgenic mouse studies support major roles for both TM and EPCR in maintaining vascular thromboresistance, although TM is clearly the more prominent player.

REGULATION OF THROMBOMODULIN EXPRESSION

In view of the central importance of TM in determining the substrate specificity of thrombin, a protease with wide biological importance, it is not surprising that expression of TM is tightly regulated via several mechanisms. Not only is TM expressed by blood ECs lining vessels of all caliber and in all organs, including the brain (45), but it also has been detected in lymphatic endothelium, astrocytes, keratinocytes, mesothelial cells, neutrophils, monocytes, platelets, VSMCs, osteoblasts, trophoblasts, and several tumor cells, in which a role in modulating thrombin might be important. The 5′-regulatory region of the human *TM* gene has been characterized – mostly through in vitro studies – dissection of which has revealed several elements that act positively or negatively. The promoter has a typical TATA box and CAAT box, with a strong transcription start signal 32 bases from the TATA box. Various potential 5′ *cis*-acting elements are present, including four Sp1 sites, a region responsive to heat shock, a series of Ets motifs, four thyroid hormone/retinoic acid response elements, a silencer element, and a shear-responsive element. The approximately 3.7-kb *TM* mRNA has a half-life of approximately 9 hours, and the 3′-untranslated region contains an A+T-rich motif that might be a site for regulation of its stability. Regions that endow tissue or cell specificity have not yet been identified. Suppression of *TM* gene transcription has been documented in response to shear stress, hypoxia, oxidized low-density lipoprotein, hyperhomocysteinemia, prostaglandins, and transforming growth factor (TGF)-β and, in vivo, to cytokines and sepsis, conditions that may be linked with endothelial damage, atherogenesis, and/or fibrin deposition (46,47). Transcriptional upregulation of TM is reported to occur with thrombin, vascular endothelial growth factor (VEGF), histamine, and statins, and in vivo in rodents after heat shock (48–51). Interestingly, inflammatory cytokines such as TNF-α or interleukin (IL)-1β that suppress TM in ECs, upregulate TM in macrophages, highlighting the complexity of its regulation.

Similar to TM, transcription of the *EPCR* gene also is suppressed by endotoxin and inflammatory cytokines. Furthermore, a soluble form of EPCR (sEPCR) is generated – particularly during sepsis – by matrix metalloproteinase-mediated cleavage of the protein from the cell surface. The function of sEPCR is not yet resolved. It may bind to PC and APC, interfering with APC anticoagulant function by altering its active site. However, sEPCR also may complex with the neutrophil elastase-like protein, proteinase 3 (PR3), whereupon the complex interferes with neutrophil integrin function, thereby diminishing neutrophil–endothelial interactions and neutrophil extravasation (5,52).

DIAGNOSTICS – IS SOLUBLE THROMBOMODULIN USEFUL?

Generated by enzymatic cleavage of the integral membrane endothelial protein, TM exists in a soluble form in the plasma and, depending on the assay system, normal levels range from 3 to 50 ng/mL. Soluble TM is comprised of multiple fragments from the extracellular portion of the molecule, some of which – those containing the intact EGF-like repeats – retain anticoagulant function (53). Soluble TM increases with vascular endothelial damage from infections, vasculitis, sepsis, and inflammation, probably due to cleavage by proteases released from activated neutrophils, and/or by an endothelial intramembranous protease, known as a rhomboid, which identifies a recognition sequence in the cytoplasmic domain of TM and cleaves it within the transmembrane domain (54). Indeed, soluble plasma TM may be a marker for disease activity in vasculitides such as systemic lupus erythematosus or arthritis, and in diabetic nephropathy. But much debate surrounds the relevance and utility of plasma TM levels, with some groups claiming that high levels are predictive for development of coronary heart disease, whereas others believe that soluble TM is vasculoprotective (55). This discordance may reflect the current lack of tools to identify which soluble fragment of TM is being measured, and which is biologically relevant for measuring. Data from our lab support the view that the lectin-like domain of TM, which is one of the circulating soluble fragments of TM, is vasculoprotective (29), although confirmation in humans is lacking.

THROMBOMODULIN MUTATIONS AND HUMAN DISEASE

Increased fibrin deposition occurs in the lungs of mice that carry a TM mutation that dramatically suppresses its capacity as a cofactor in the thrombin-mediated activation of PC (56–58). There are, however, conflicting reports as to the clinical significance of TM mutations in venous and arterial thrombotic disease in humans. Since the first report of familial thrombophilia associated with a *TM* gene mutation (59), several additional mutations and polymorphisms in the coding region and the promoter region have been described (archived in http://archive.uwcm.ac.uk/uwcm/mg/search/119613.html). Some have been evaluated in vitro to characterize effects on the functional expression of TM, but results often do not correlate with clinical findings (60). However, several studies do suggest a link. For example, in a large prospective analysis of men, the A455V and -1208–1209TTdelTT variants exhibit a small, but nonsignificant increase in risk for coronary heart disease (61). A natural mutation encoding an elongated TM protein with reduced expression was detected in a pedigree associated with myocardial ischemia. And in some segments of the population, the Ala25Thr substitution is a risk factor for ischemic heart disease (62). Overall, further large-scale studies are required to better characterize the potential association. Notably, there have been no surveys designed to evaluate the incidence of TM mutations in other diseases, such as cancer or inflammation, but these are warranted.

THERAPEUTIC CONSIDERATIONS

Preclinical trials using recombinant forms of TM that encompass EGF1 through EGF6 are effective in protecting against tissue factor- or endotoxin-induced disseminated intravascular coagulation or lung injury, and in rat models, local nonviral *TM* gene transfer prevents thrombosis of vein grafts. Phase I clinical trials suggest that recombinant soluble TM is safe (63), but its efficacy in the prevention or treatment of hypercoagulable disorders, sepsis, or inflammatory disorders remains to be proven, and at least a theoretical risk exists of bleeding complications. Support for using TM specifically to prevent inflammation comes first from studies in rabbit restenosis models, where administration of TM via adenovirus prevented restenosis and dampened the inflammatory response. Furthermore, complement activation was diminished by TM either when immobilized in a medical device, or when systemically administered in a rat model of glomerulonephritis. The in vivo anti-inflammatory effects of administering recombinant forms of the lectin-like domain of TM in mouse models of human disease (e.g., myocardial ischemia–reperfusion, arthritis) are promising, but require further confirmation. Lastly, in limited studies, recombinant TM also has antitumor effects, the mechanisms of which are under investigation.

KEY POINTS

- TM is a crucial cofactor in a major natural anticoagulant pathway, facilitating effective thrombin-induced activation of PC, switching the substrate specificity of thrombin from procoagulant and proinflammatory, to anticoagulant and anti-inflammatory.
- Critical structural domains of TM play distinct roles in inflammation, coagulation, fibrinolysis, and cell proliferation, and these are integrated to protect the organism during development and in adults in response to a variety of stresses. Delineating the structure–function correlates of TM is leading to new diagnostics, and safer, more effective disease- and organ-specific therapies.

Future Goals

- To elucidate the pathophysiological role of soluble fragments of TM that circulate in the plasma and determine if they reflect vascular disease
- To characterize genetic defects in TM (and/or its partner proteins), which may be useful in identifying individuals with a predisposition to develop hemostatic/thrombotic, proliferative, or inflammatory disorders
- To characterize TM within the context of a complex of multiple proteins – membrane-associated, soluble, and intracellular – differentially expressed in many cells and tissues, and to assess their contribution to human health and disease

ACKNOWLEDGMENTS

Support was provided by the FWO in Flanders, Belgium (Grant #G.0382.02) and the National Institute of Health, USA (Grant #HL067846–01). MVdW is supported by a fellowship from the IWT, Belgium and is an Emmanuel van der Schueren scholar.

REFERENCES

1 Esmon CT, Owen WG. The discovery of thrombomodulin. *J Thromb Haemost.* 2004;2:209–213.
2 Owen WG, Esmon CT. Functional properties of an endothelial cell cofactor for thrombin-catalyzed activation of protein C. *J Biol Chem.* 1981;256:5532–5535.
3 Stenflo J. A new vitamin K-dependent protein. Purification from bovine plasma and preliminary characterization. *J Biol Chem.* 1976;251:355–363.
4 Esmon CT, Esmon NL, Harris KW. Isolation of a membrane-bound cofactor for thrombin-catalyzed activation of protein C. *J Biol Chem.* 1982;257:7944–7947.
5 Taylor FB Jr., Stearns-Kurosawa DJ, Kurosawa S, et al. The endothelial cell protein C receptor aids in host defense against *Escherichia coli* sepsis. *Blood.* 2000;95:1680–1686.
6 Bernard GR. Drotrecogin alfa (activated) (recombinant human activated protein C) for the treatment of severe sepsis. *Crit Care Med.* 2003;31:S85–S93.
7 Minami T, Sugiyama A, Wu SQ, et al. Thrombin and phenotypic modulation of the endothelium. *Arterioscler Thromb Vasc Biol.* 2004;24:41–53.
8 Steinhoff M, Buddenkotte J, Shpacovitch V, et al. Proteinase-activated receptors: transducers of proteinase-mediated signaling in inflammation and immune response. *Endocr Rev.* 2005;26:1–43.

9 Riewald M, Petrovan RJ, Donner A, Ruf W. Activated protein C signals through the thrombin receptor PAR1 in endothelial cells. *J Endotoxin Res.* 2003;9:317–321.
10 Feistritzer C, Riewald M. Endothelial barrier protection by activated protein C through PAR1-dependent sphingosine 1-phosphate receptor-1 cross-activation. *Blood.* 2005;105:3178–3184.
11 Bajzar L, Manuel R, Nesheim M. Purification and characterization of TAFI, a thrombin-activatable fibrinolysis inhibitor. *J Biol Chem.* 1995;270:14477–14484.
12 Dai L, Mitchell M, Savidge G, Alhaq A. The profibrinolytic effect of plasma thrombomodulin in factor XI deficiency and its implications in hemostasis. *J Thromb Haemost.* 2004;2:2200–2204.
13 Campbell WD, Lazoura E, Okada N, Okada H. Inactivation of C3a and C5a octapeptides by carboxypeptidase R and carboxypeptidase N. *Microbiol Immunol.* 2002;46:131–134.
14 Nagashima M, Yin ZF, Broze GJ Jr., Morser J. Thrombin-activatable fibrinolysis inhibitor (TAFI) deficient mice. *Front Biosci.* 2002;7:D556–D568.
15 Mao SS, Holahan MA, Bailey C, et al. Demonstration of enhanced endogenous fibrinolysis in thrombin activatable fibrinolysis inhibitor-deficient mice. *Blood Coagul Fibrinolysis.* 2005;16:407–415.
16 Esmon CT, Owen WG. Identification of an endothelial cell cofactor for thrombin-catalyzed activation of protein C. *Proc Natl Acad Sci USA.* 1981;78:2249–2252.
17 Kurosawa S, Galvin J, Esmon N, Esmon C. Proteolytic formation and properties of functional domains of thrombomodulin. *J Biol Chem.* 1987;262:2206–2212.
18 Kurosawa S, Stearns DJ, Jackson KW, Esmon CT. A 10-kDa cyanogen bromide fragment from the epidermal growth factor homology domain of rabbit thrombomodulin contains the primary thrombin binding site. *J Biol Chem.* 1988;263:5993–5996.
19 Lu R, Esmon NL, Esmon CT, Johnson AE. The active site of the thrombin-thrombomodulin complex. *J Biol Chem.* 1989;264:12956–12962.
20 Ye J, Esmon C, Johnson A. The chondroitin sulfate moiety of thrombomodulin binds a second molecule of thrombin. *J Biol Chem.* 1993;268:2373–2379.
21 Jackman RW, Beeler DL, Fritze L, et al. Human thrombomodulin gene is intron depleted: nucleic acid sequences of the cDNA and gene predict protein structure and suggest sites of regulatory control. *Proc Natl Acad Sci USA.* 1987;84:6425–6429.
22 Christian S, Ahorn H, Koehler A, et al. Molecular cloning and characterization of endosialin, a C-type lectin- like cell surface receptor of tumor endothelium. *J Biol Chem.* 2001;276:7408–7414.
23 Zushi M, Gomi K, Yamamoto S, et al. The last three consecutive epidermal growth factor-like structures of human thrombomodulin comprise the minimum functional domain for protein C-activating cofactor activity and anticoagulant activity. *J Biol Chem.* 1989;264:10351–10353.
24 Kokame K, Zheng X, Sadler J. Activation of thrombin-activatable fibrinolysis inhibitor requires epidermal growth factor-like domain 3 of thrombomodulin and is inhibited competitively by protein C. *J Biol Chem.* 1998;273:12135–12139.
25 Lentz S, Chen Y, Sadler J. Sequences required for thrombomodulin cofactor activity within the fourth epidermal growth factor-like domain of human thrombomodulin. *J Biol Chem.* 1993;268:15312–15317.
26 Prieto JH, Sampoli Benitez BA, Melacini G, et al. Dynamics of the fragment of thrombomodulin containing the fourth and fifth epidermal growth factor-like domains correlate with function. *Biochemistry.* 2005;44:1225–1233.
27 DeMunk G, Groeneveld E, Rijken DC. Acceleration of the thrombin inactivation of single chain urokinase-type plasminogen activator (Pro-urokinase) by thrombomodulin. *J Clin Invest.* 1991;88:1680–1684.
28 Slungaard A, Fernandez JA, Griffin JH, et al. Platelet factor 4 enhances generation of activated protein C in vitro and in vivo. *Blood.* 2003;102:146–151.
29 Conway EM, Van de Wouwer M, Pollefeyt S, et al. The lectin-like domain of thrombomodulin confers protection from neutrophil-mediated tissue damage by suppressing adhesion molecule expression via nuclear factor kappaB and mitogen-activated protein kinase pathways. *J Exp Med.* 2002;196:565–577.
30 Abeyama K, Stern DM, Ito Y, et al. The N-terminal domain of thrombomodulin sequesters high-mobility group-B1 protein, a novel anti-inflammatory mechanism. *J Clin Invest.* 2005;115(5):1267–1274. Epub 2005 Apr 14.
31 Huang HC, Shi GY, Jiang SJ, et al. Thrombomodulin-mediated cell adhesion: involvement of its lectin-like domain. *J Biol Chem.* 2003;278(47):46750–46759. Epub 2003 Sep 1.
32 Agarwal SM. Evolutionary rate variation in eukaryotic lineage specific human intronless proteins. *Biochem Biophys Res Commun.* 2005;337(4):1192–1197.
33 Gentles AJ, Karlin S. Why are human G-protein-coupled receptors predominantly intronless? *Trends Genet.* 1999;15:47–49.
34 Le Hir H, Nott A, Moore MJ. How introns influence and enhance eukaryotic gene expression. *Trends Biochem Sci.* 2003;28:215–220.
35 Opavsky R, Haviernik P, Jurkovicova D, et al. Molecular characterization of the mouse Tem1/endosialin gene regulated by cell density in vitro and expressed in normal tissues in vivo. *J Biol Chem.* 2001;276:38795–38807.
36 McGreal E, Gasque P. Structure-function studies of the receptors for complement C1q. *Biochem Soc Trans.* 2002;30:1010–1014.
37 Healy A, Rayburn H, Rosenberg R, Weiler H. Absence of the blood-clotting regulator thrombomodulin causes embryonic lethality in mice before development of a functional cardiovascular system. *Proc Natl Acad Sci USA.* 1995;92:850–854.
38 Isermann B, Hendrickson SB, Zogg M, et al. Endothelium-specific loss of murine thrombomodulin disrupts the protein C anticoagulant pathway and causes juvenile-onset thrombosis. *J Clin Invest.* 2001;108:537–546.
39 Isermann B, Sood R, Pawlinski R, et al. The thrombomodulin-protein C system is essential for the maintenance of pregnancy. *Nat Med.* 2003;9:331–337.
40 Conway EM, Pollefeyt S, Cornelissen J, et al. Structure-function analyses of thrombomodulin by gene-targeting in mice: the cytoplasmic domain is not required for normal fetal development. *Blood.* 1999;93:3442–3450.
41 Oganesyan V, Oganesyan N, Terzyan S, et al. The crystal structure of the endothelial protein C receptor and a bound phospholipid. *J Biol Chem.* 2002;277:24851–24854.
42 Gu JM, Crawley JT, Ferrell G, et al. Disruption of the endothelial cell protein C receptor gene in mice causes placental thrombosis and early embryonic lethality. *J Biol Chem.* 2002;277:43335–43343.
43 Castellino FJ, Liang Z, Volkir SP, et al. Mice with a severe deficiency of the endothelial protein C receptor gene develop, survive,

and reproduce normally, and do not present with enhanced arterial thrombosis after challenge. *Thromb Haemost*. 2002;88:462–472.

44 Li W, Zheng X, Gu JM, et al. Extra-embryonic expression of EPCR is essential for embryonic viability. *Blood*. 2005;106(8): 2716–2722. Epub 2005 Jun 14.

45 Conway EM, Zwerts F, Van Eygen V, et al. Survivin-dependent angiogenesis in ischemic brain: molecular mechanisms of hypoxia-induced upregulation. *Am J Pathol.* 2003;163:935–946.

46 Laszik ZG, Zhou XJ, Ferrell GL, et al. Down-regulation of endothelial expression of endothelial cell protein C receptor and thrombomodulin in coronary atherosclerosis. *Am J Pathol.* 2001;159:797–802.

47 Lentz SR. Thrombosis of vein grafts: wall tension restrains thrombomodulin expression. *Circ Res.* 2003;92:12–13.

48 Hirokawa K, Aoki N. Up-regulation of thrombomodulin by activation of histamine H1-receptors in human umbilical-vein endothelial cells in vitro. *Biochem J.* 1991;276:739–743.

49 Ma SF, Garcia JG, Reuning U, et al. Thrombin induces thrombomodulin mRNA expression via the proteolytically activated thrombin receptor in cultured bovine smooth muscle cells. *J Lab Clin Med.* 1997;129:611–619.

50 Shi J, Wang J, Zheng H, et al. Statins increase thrombomodulin expression and function in human endothelial cells by a nitric oxide-dependent mechanism and counteract tumor necrosis factor alpha-induced thrombomodulin downregulation. *Blood Coagul Fibrinolysis.* 2003;14:575–585.

51 Conway E, Liu L, Nowakowski B, et al. Heat shock of vascular endothelial cells induces an upregulatory transcriptional response of the thrombomodulin gene that is delayed in onset and does not attenuate. *J Biol Chem.* 1994;269:22804–22810.

52 Esmon CT. Structure and functions of the endothelial cell protein C receptor. *Crit Care Med.* 2004;32:S298–S301.

53 Ohlin AK, Larsson K, Hansson M. Soluble thrombomodulin activity and soluble thrombomodulin antigen in plasma. *J Thromb Haemost*. 2005;3:976–982.

54 Lohi O, Urban S, Freeman M. Diverse substrate recognition mechanisms for rhomboids: thrombomodulin is cleaved by mammalian rhomboids. *Curr Biol.* 2004;14:236–241.

55 Salomaa V, Matei C, Aleksic N, et al. Soluble thrombomodulin as a predictor of incident coronary heart disease and symptomless carotid artery atherosclerosis in the Atherosclerosis Risk in Communities (ARIC) Study: a case-cohort study. *Lancet*. 1999; 353:1729–1734.

56 Weiler-Guettler H, Christie P, Beeler D, et al. A targeted point mutation in thrombomodulin generates viable mice with a prethrombotic state. *J Clin Invest.* 1998;101:1–9.

57 Weiler H, Lindner V, Kerlin B, et al. Characterization of a mouse model for thrombomodulin deficiency. *Arterioscler Thromb Vasc Biol.* 2001;21:1531–1537.

58 Rijneveld AW, Weijer S, Florquin S, et al. Thrombomodulin mutant mice with a strongly reduced capacity to generate activated protein C have an unaltered pulmonary immune response to respiratory pathogens and lipopolysaccharide. *Blood*. 2004; 103(5):1702–1709. Epub 2003 Oct 30. .

59 Ohlin A-K, Marlar R. The first mutation identified in the thrombomodulin gene in a 45 year old man presenting with thromboembolic disease. *Blood*. 1995;85:330–336.

60 Kunz G, Ohlin AK, Adami A, et al. Naturally occurring mutations in the thrombomodulin gene leading to impaired expression and function. *Blood*. 2002;99:3646–3653.

61 Aleksic N, Folsom AR, Cushman M, et al. Prospective study of the A455V polymorphism in the thrombomodulin gene, plasma thrombomodulin, and incidence of venous thromboembolism: the LITE Study. *J Thromb Haemost*. 2003;1:88–94.

62 Doggen CJ, Kunz G, Rosendaal FR, et al. A mutation in the thrombomodulin gene, 127G to A coding for Ala25Thr, and the risk of myocardial infarction in men. *Thromb Haemost*. 1998;80:743–748.

63 Moll S, Lindley C, Pescatore S, et al. Phase I study of a novel recombinant human soluble thrombomodulin, ART-123. *J Thromb Haemost.* 2004;2:1745–1751.

106

Heparan Sulfate

Nicholas W. Shworak

Angiogenesis Research Center, Dartmouth Medical School, Hanover, New Hampshire

A multitude of biological processes are regulated by heparan sulfate proteoglycans (HSPGs), which are produced by virtually all cell types. These hybrid molecules are composed of protein cores to which are attached one or more long chains of heparan sulfate (HS). Functional diversity is engendered in part from a multiplicity of core proteins and in part from the structural complexity of the HS chain. This polysaccharide is a type of glycosaminoglycan (GAG) – a long unbranched copolymer comprised of alternating amino and acid sugars. The HS sugar residues are decorated at various positions with sulfate groups, which creates a large array of short sequence motifs that bind and thereby modulate the functional properties of numerous regulatory molecules, including signaling ligands/receptors, proteases, enzymes, and lipoproteins (Table 106-1) (1,2). HS–protein interactions control cellular processes such as signaling, adhesion, migration, and vesicular trafficking. Through these cellular actions, HSPGs regulate many biological events including angiogenesis, lipoprotein metabolism, inflammation, hemostasis, developmental inductions, and axonal guidance.

This chapter focuses on only one aspect of HSPG biology – the input role of the HS chain in regulating endothelial cell (EC) signaling by "heparin-binding growth factors." For purposes of discussion, this term is defined as any signaling ligand that exhibits high affinity to heparin, a particular type of highly sulfated HS. This definition encompasses classic growth factors such as fibroblast growth factors (FGFs); vascular endothelial growth factors (VEGFs); heparin-binding epidermal growth factor (Wnts); cytokines like interleukin-3, granulocyte monocyte colony stimulating factor (GM-CSF), and interferon-γ; most chemokines; and even growth inhibitors such as endostatin. Thus, this term includes the majority of HS binding ligands (see Table 106-1). The reference to "heparin-binding" reflects the historical use of heparin chromatography to purify many of these factors. By definition, heparin is found only within mast cell granules, so it is unlikely to contribute to the in vivo roles of most of these ligands. Instead, the physiological activities of heparin-binding growth factors predominantly involve their interaction with the HS chains of cell surface and extracellular matrix (ECM) HSPGs, which are produced by virtually all cell types.

HISTORY

Although heparin is a low-abundance form of HS, it was the first form of HS discovered, and thus dominated the early history of the field. In 1916, a second-year medical student, Jay McLean, found that an extract from dog liver exhibited strong anticoagulant activity (3). This activity was designated as *heparin* to denote its isolation from liver. By 1935, clinical trials were started, heralding the longstanding use of heparin as an efficacious anticoagulant (4). The acceptance of heparin as a therapeutic agent motivated studies to discover its structure and mechanism of action.

The elucidation of heparin's mechanism of action provides a paradigm for how HS motifs convey biological activity. In 1925, William Henry Howell, McLean's mentor, determined that heparin was some form of polysaccharide and proposed that its anticoagulant activity required a plasma cofactor (5). However, almost six decades of research would be required to unravel its intricate structure and mechanism of action (4). The hypothetical plasma cofactor was ultimately identified as antithrombin (AT) in 1968, by Abildgaard (6). Soon after, Damus, Hicks, and Rosenberg found that heparin mediates its effects by binding to AT (7). Such binding was initially considered to occur through nonspecific ionic interactions; however, this notion was soon discounted. In 1976, the three independent groups of Lindahl, Rosenberg, and Sims showed that only one third of heparin molecules could bind AT, and only this population of molecules exhibited anticoagulant activity (8–10). This landmark observation suggested structural specificity. Indeed, in the early 1980s, the groups of Choay, Lindahl, and Rosenberg demonstrated that the active component of heparin was a pentasaccharide with a specific motif of

Table 106-1: A Partial Listing of Heparan Sulfate Binding Proteins. The Majority of These Ligands Can Be Defined as "Heparin-Binding Growth Factors"

Antiangiogenic	**Growth Factors**	**Morphogens**	**Proteases**
Angiostatin	Angiogenin	Activin	Cathepsin G
Endostatin	Amphiregulin	BMP-2, -4	Neutrophil elastase
TGF-β	Betacellulin	Chordin	Protease Nexin I
Interferon-γ	FGFs (1–23)	Frizzled-type peptides	**ECM/plasma components**
GCP-2	Heparin-binding EGF	Sonic hedgehog	Fibrin
IP-10	HGF	Sprouty peptides	Fibronectin
PF4	IGF-IIII	Wnts (1–13)	Interstitial collagens
Chemokines	Midkine	**Energy metabolism**	Laminins
GRO-α	Neuregulin	ApoB, ApoE	Agouti-related protein
GRP-β	Pleiotropin	Lipoprotein lipase	Pleiotropin (HB-GAM)
IL-8	PDGF-AA	Triglyceride lipases	Tenascin
GCP-2	TGF-β	**Cell adhesion**	Thrombospondin
IP-10	VEGF-$A_{165, 189}$	L-selectin	Vitronectin
PF4	**Growth factor-binding proteins**	MAC-1	**Coagulation**
Cytokines		N-CAM	AT
IL-2, -3, -4, -5, -7, -12	Follistatin	PECAM-1	Heparin cofactor II
	IGF BP-3, -5		Leuserpin
GM-CSF	TGF-β BP		Plasminogen activator inhibitor
Interferon-γ	FGF receptors		Tissue factor pathway inhibitor
TNF-α			Tissue plasminogen activator
			Thrombin

(Using data from Bernfield M, Gotte M, Park PW, et al. Functions of cell surface heparan sulfate proteoglycans. *Annu Rev Biochem.* 1999;68:729–777; and Esko JD, Selleck SB. Order out of chaos: assembly of ligand binding sites in heparan sulfate. *Annu Rev Biochem.* 2002;71:435–471.)

sulfate groups (4). It is now appreciated that distinct activities of HS are conveyed by specific motifs, with each motif binding to a distinct effector protein.

Although heparin dominated the initial HS landscape, as early as 1937, it was appreciated that heparin played only a limited role in the body, being present only in the basophilic granules of mast cells (11). (The liver capsule contains high levels of mast cells, which accounts for its high heparin content.) The ubiquitous nature of HS was not realized until the early 1970s, when it was found that virtually all cell types produce HS as a proteoglycan (4). Various core proteins were identified during the 1980s, and HS biosynthetic enzymes were cloned in the 1990s. The ubiquitous form of HS and heparin were found to be synthesized by the same family of enzymes, which led to the appreciation that heparin is simply one type of HS (1,2,12). Indeed, certain non–mast cell types produce a subpopulation of HS chains that exhibit high sulfate content indistinguishable from that of mast-cell heparin. The multifaceted nature of HS function began to emerge during the 1970s, when the first of numerous heparin/HS-binding proteins were discovered. During the early 1980s, the groups of Klagsbrun and Folkman championed the use of heparin affinity chromatography to purify EC mitogens (13). This application helped trigger investigations into HS-mediated signaling by heparin-binding growth factors, which continue to this day.

HEPARAN SULFATE PROTEOGLYCAN STRUCTURE, SYNTHESIS, AND POSTSYNTHETIC MODIFICATION

Core Proteins

The broad functional repertoire of HSPGs in part stems from a multiplicity of core proteins (Figure 106.1A). Some core proteins exclusively bear HS chains, whereas others can carry two types of GAG chains – HS and chondroitin sulfate (CS) (Table 106-2). In vertebrates, most cell-surface HS is carried by syndecans (four related transmembrane proteins extensively described in Chapter 44 and glypicans (a family of at least six homologous glycophosphatidyl-inositol [GPI]-anchored proteins). However, a variety of additional integral membrane proteins, such as betaglycan and splice variants of CD44, can occasionally bear HS chains and are thus considered "part-time" proteoglycans. The major carriers of non–surface bound HS in the ECM appear to be perlecan, agrin, and collagen XVIII (1,14). At present, 11 core proteins are known to be expressed by ECs (Table 106-2). Such complexity does not merely reflect redundancy, because specialized features of a given HSPG core protein can: (a) define localization to extracellular or intracellular compartments; (b) allow for cellular internalization, recycling, or transcellular transport; and (c) provide direct interactions with signaling and/or cytoskeletal components (1,15,16). The importance of the core protein is exemplified

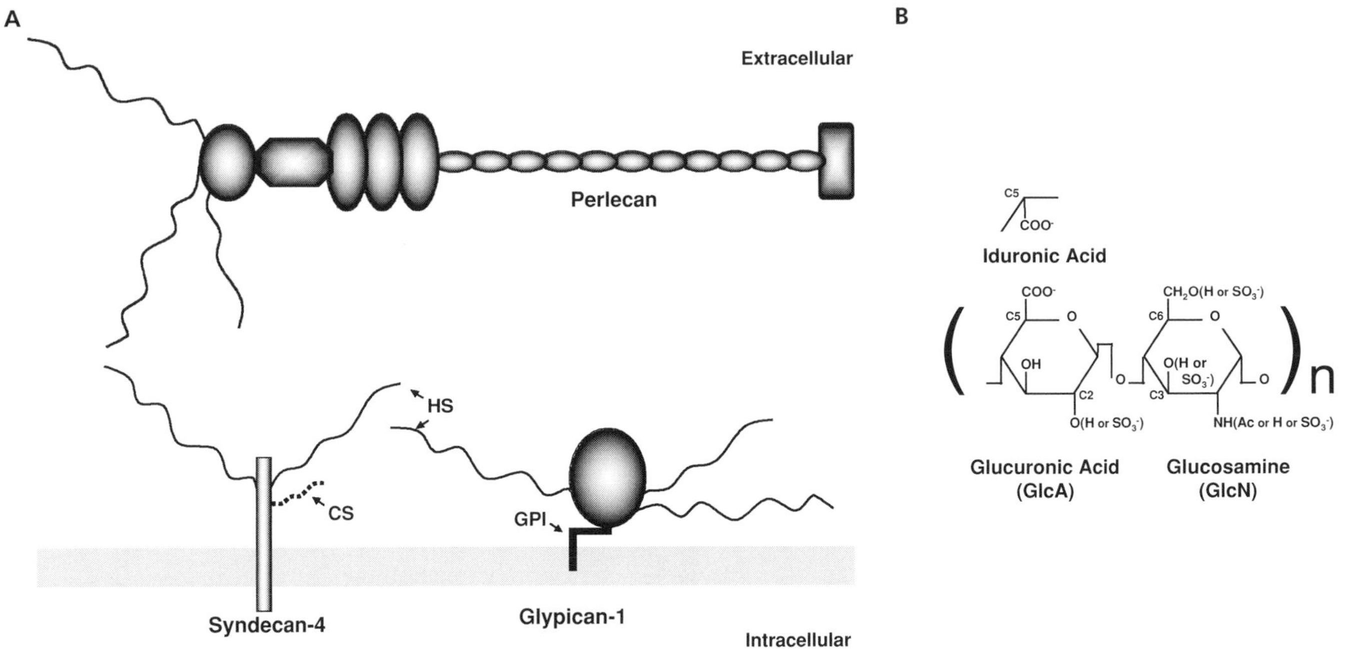

Figure 106.1. HSPG structure. (**A**) Representative HSPGs of ECs. Schematic presentation of the EC surface/basement membrane showing representative types of integral membrane (syndecan-4), GPI-linked (glypican-1), and extracellular (perlecan) HSPGs. Syndecan-4 is shown as carrying one CS (*dotted line*) and two HS (*unbroken line*) chains, but can bear multiple permutations of these GAGs. (**B**) Structure of the HS disaccharide repeat. Both HS and heparin are comprised of the same repeated disaccharide. The uronic acid can be remodeled by epimerization at C_5, which converts glucuronic acid to iduronic acid (only C_5 region shown), or by adding a sulfate at C_2. Remodeling of the glucosamine can involve replacement of the *N*-acetyl group with either a sulfate group or a proton (creating a free amino group). Sulfates can also be added at the C_3 or C_6 positions.

by Chapter 44, which elaborates on how syndecan-4 regulates EC phenotype through cell signaling events. Together, the core protein and associated machinery circumscribe a platform of such immense utility that virtually all higher eukaryotic tissues and cells employ HSPGs in multiple roles.

Heparan Sulfate Chain Structure

In addition to the multiplicity of core proteins, the complexity and functional diversity of HSPGs is further amplified by the HS component. HS is a heterogeneous linear polysaccharide consisting of a repeated disaccharide unit, hexuronic acid $\beta_{1\rightarrow4}$*N*-acetylglucosamine (GlcNAc) $\alpha_{1\rightarrow4}$ (glucuronic/iduronic acid alternating with glucosamine, that is partially decorated with *N*- and various *O*-sulfate groups (see Figure 106.1B). In large part, the specific arrangement of the sulfate moieties gives rise to distinct binding motifs that interact with an increasingly expansive list of protein effectors (2,17,18). The HS chains are quite long, ranging from 100 to 200 disaccharide units, which span approximately 40 to 80 nm, respectively. Each chain is internally repetitive, containing short blocks of highly sulfated motifs alternating with blocks containing minimal sulfation (1,15,16). Because a single HS chain can contain multiple copies of different ligand binding motifs (19), HS is ideally suited to act as a template for assembling multimolecular complexes, such as signaling complexes.

Heparan Sulfate Biosynthesis

HS synthesis involves an extensive process of post-translational modification that occurs in the Golgi apparatus (Figure 106.2). The process first employs UDP-linked sugars as substrates for polymerizing the HS chain. This backbone is then remodeled largely by sulfotransferases that employ the universal sulfate donor 3′ phosphoadensosine-5′-phosophosulfate (PAPS) to decorate the chain with sulfate groups, thereby creating distinct HS motifs (14). Although core proteins convey many functions, they do not play a substantial role in defining the types of HS motifs that occur along their HS chains (20). Instead, production of specific motifs is dictated by the sequence-specific properties of the various modification enzymes.

Synthesis initiates at certain serine residues of various core proteins by the assembly of a short-linkage tetrasaccharide (see Figure 106.2). This step is common to HS and CS synthesis, which explains why many HSPGs can also carry CS chains. Compared to HS, CS sugar chains are shorter, contain galactosamine instead of GlcNAc, and are sulfated in some distinct positions. As a result, CS exhibits biological activities different

Table 106-2: HSPG Core Proteins

Core Proteins	*Endothelial Expression Known*	*GAG Type (Number of Chains)*	*Key Features*
Cell Surface			
Syndecan family			Type I integral membrane proteins
Syndecan-1	+	HS/CS (3–5)	
Syndecan-2	+	HS/CS (3)	
Syndecan-3		HS/CS (3–5)	
Syndecan-4	+	HS/CS (3)	
Glypican family			GPI linked
Glypican-1	+	HS (3)	
Glypican-2		HS	
Glypican-3		HS	
Glypican-4		HS	
Glypican-5		HS	
Glypican-6		HS	
Betaglycan	+	HS/CS	Part-time proteoglycan
CD44	+	HS/CS	Part-time proteoglycan
FGFR2	+	HS/CS (1)	Splice variants containing the "acid box"
Extracellular matrix			
Perlecan	+	HS/CS (3)	Large multidomain protein
Agrin	+	HS (3)	
Type XVIII collagen	+	HS/CS (3)	Cleavage product is endostatin
Testican	+	HS/CS (2)	

(Using data from Bernfield M, Gotte M, Park PW, et al. Functions of cell surface heparan sulfate proteoglycans. *Annu Rev Biochem.* 1999;68:729–777; and Silbert JE, Sugumaran G. A starting place for the road to function. *Glycoconj J.* 2002; 19:227–237.)

from those of HS. Commitment of the primed structure to HS synthesis is determined by the addition of a GlcNAc residue by either EXTL2 or EXTL3, which are distinct isoforms of GlcNAc transferase I. Elongation of the HS backbone then requires the simultaneous action of enzymes from the genes *EXT1* and *EXT2* (2,12).

Distinct structures next arise through remodeling of the copolymer backbone by a semiordered series of reactions. Initially, subsets of *N*-acetylglucosamine residues are deacetylated and *N*-sulfated by HS *N*-deacetylase/*N*-sulfotransferase (NDST) (21,22). Occasional residues escape sulfation, which leaves a free amino group. Regions with a high density of *N*-sulfated glucosamine are then preferentially modified by less frequent reactions. Occasional glucuronic acid residues are converted to iduronic acid by the HS C_5 epimerase (23,24). The HS 2-sulfotransferase (HS2ST) next produces 2-*O*-sulfated iduronic acid or 2-*O*-sulfated glucuronic acid (25,26). HS 6-*O*-sulfotransferase (HS6ST) converts occasional glucosamine residues to 6-*O*-sulfated glucosamine (27). "Late" modification also includes 3-*O*-sulfation of glucosamine, which is added by HS 3-*O*-sulfotransferase (HS3ST) (20,28–30).

Multiple isozymes have been identified for NDST, HS6ST, and HS3ST. Such multiplicity serves two purposes. First, individual isoforms can exhibit distinct sequence specificities, which expands the structural diversity of HS (30–32). Second, multiple isoforms enable cell type–specific production of distinct HS motifs (2,12). These concepts are best illustrated by the seven distinct isoforms of 3-*O*-sulfotransferases (HS3ST), which comprise the largest multigene family of HS modification enzymes. Despite the large family size, 3-*O*-sulfates are the rarest HS modification (typically comprising <0.5% of total sulfate moieties) but do serve key regulatory roles (20,33). The enzymes fall into two classes (AT-type versus gD-type) based on enzymatic specificities (34). HS3ST1 is an AT-type enzyme, which preferentially modifies a specific pentasaccharide precursor structure to create HS with high-affinity binding motifs for AT (HS^{AT+}). In ECs, the level of HS3ST1 serves to regulate the level of HS^{AT+} synthesis (28). The interaction of this motif with AT is thought to convey anticoagulant and anti-inflammatory properties to the endothelium. In contrast, HS3ST2, $HS3ST3_A$, $HS3ST3_B$, HS3ST4, and HS3ST6 are gD-type enzymes, which preferentially recognize a distinct precursor structure to create a different 3-*O*-sulfated motif (HS^{gD+}), known to bind glycoprotein gD of herpes simplex virus 1. Cellular expression of gD-type enzymes enables the cellular entry of this herpes virus (35). HS^{gD+} has recently been implicated in Notch signaling (described in the following section). However, its endogenous ligand is not yet known. HS3ST5 falls into both classes because it is able to generate both motifs with equal efficiency (36). Although this system

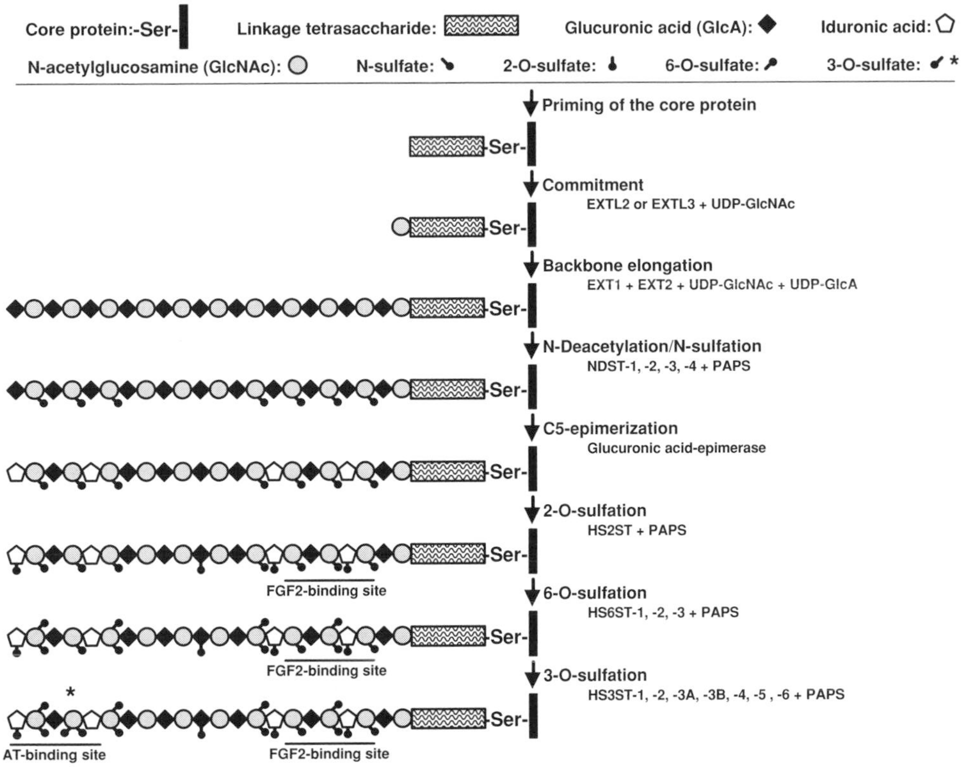

Figure 106.2. Steps of HS synthesis. This schematic indicates the major steps of HS synthesis and the involved enzymes and cofactors. This relatively ordered process creates distinct sequence motifs. Shown is the minimal sequence required to bind an FGF2 monomer. This binding site has only a few modifications, so is generated early; whereas, antithrombin (AT)-binding sites have multiple modifications and are completed later. The * indicates a residue modified by the very rare 3-*O*-sulfate moiety. Structures generated by gD-type HS3ST enzymes are not shown.

appears to show redundancy, each isoform exhibits unique cell type– and tissue-specific expression patterns. The isoforms of NDST (NDST1 to NDST4) and HS6ST (HS6ST1 to HS6ST3) also exhibit sequence specificity and cell type–specific expression (2,12). Thus, the expression of distinct combinations of HS sulfotransferases enables individual cell types to produce distinct arrays of HS motifs.

Postsynthetic Heparan Sulfate Proteoglycans Processing

The properties of HSPGs can be further modified by post-biosynthetic processing. First, cell membrane–bound proteoglycans can be shed from the cell surface. Shedding of integral membrane HSPGs involves protease clipping of the extracellular domain near the transmembrane domain, whereas a phospholipase activity liberates GPI-linked HSPGs (1). Second, HS motifs can be modified by extracellular sulfatases, which remove specific sulfate moieties. Presently known are the Sulf enzymes, which can remove 6-*O*-sulfate groups (37). Third, HS chains can be degraded by extracellular heparanases derived from leukocytes, tumor cells, or the endothelium (38). Fourth, HS chains can sustain damage by nitric oxide. Such HSPGs can undergo cellular internalization, removal of the damaged chain regions, resynthesis of fresh HS chains, and transport back to the pericellular milieu (39). Thus, multiple extracellular factors can alter HSPG chain structure. Such factors may potentially influence HS-mediated cell signaling (discussed later).

EVOLUTIONARY CONSIDERATIONS

Bona fide HS polysaccharide is not found in prokaryotes, unicellular organisms, or plants. Rigorous structural studies have found that, among multicellular animals, HS is absent only in the most primitive metazoans, the sponges. Vertebrate HS structures are clearly present, however, in the two more complex phyla *Ctenophora* and *Cnidaria*, whose ancestors represent the earliest forms of true metazoans (Eumetazoa) (40,41). The split between sponges and Eumetazoans is estimated to have occurred approximately 940 million years ago, testifying to the extreme antiquity of HS. On one hand, the evolution of HS may have conferred a selection advantage by optimizing pre-existent processes. Potential candidates include cell signaling and adhesion, which were firmly established in single-cell organisms. Additional candidates include chemokine production, innate immunity, apoptosis, and totipotent stem cells, which all occur in sponges. On the other hand, selection for the emergence of the complex HS biosynthetic pathway may have involved a unique process appearing at the divergence of Eumetazoans. In particular, this evolutionary split was enabled by the emergence of an integrated mechanism for whole-organ homeostasis (42). Sponges are not Eumetazoans because they function almost like a colony of animals. They are comprised of semiautonomous groups of cells, with each group having an independent inlet to obtain seawater that contains food and oxygen and an independent outlet for the removal of waste and carbon dioxide. In contrast, the earliest Eumetazoans have a single inlet to a common gastrovascular cavity, which is drained by a single outlet. The gastrovascular cavity is a specialized sealed compartment surrounded by a novel sheet of cells. This sheet is the original "epithelium sensu stricto," a cell layer that serves to separate and regulate the fluids in distinct extracellular compartments. Thus, all Eumetazoans are characterized by a single circulatory system that enables whole-body homeostasis (42).

This change in body plan was accompanied by an increase in cellular diversity beginning in embryogenesis. Sponges arise from a single germ layer; however, early Eumetazoans had two germ layers (endoderm and ectoderm), which enabled the formation of the new epithelium sensu stricto. Early Eumetazoans also developed neurons, which allowed for further coordination of whole-body homeostasis. Thus, the emergence of Eumetazoans featured the simultaneous development of a single "circulatory system" and a primitive nervous system. This common emergence might explain why both systems frequently employ similar signaling components – such as FGFs, ephrins, neuropilins, and Notch receptors – as elaborated later. From such distant evolutionary roots, it is not surprising that HSPGs are involved in homeostatic mechanisms, occur in multiple organ systems, serve to modulate cell type specific phenotypes, and regulate embryonic development.

The extent of the lower metazoan HS biosynthetic machinery is known for the roundworm (*Caenorhabditis elegans*) and the fruit fly (*Drosophila melanogaster*), both of which have completely sequenced genomes. These organisms possess genes encoding all the required components for HSPG production, including multiple core proteins and all the various HS biosynthetic enzymes. Numerous studies show that the gene products are functionally equivalent to their corresponding mammalian proteins (2). However, in contrast to vertebrates, these invertebrates largely lack multigene families. Thus, they exhibit only single genes for syndecan, glypican, NDST, and HS6ST. The notable exception is that *Drosophila* shows two forms of HS3ST, which appear to be the predecessors of the two major functional groupings of the large mammalian 3-*O*-sulfortransferase multigene family (34).

DEVELOPMENTAL CONSIDERATIONS

The roles for HS in vertebrate development are just beginning to be unraveled. In contrast, significant advances have been made in studies of *C. elegans* and *Drosophila*. Unfortunately, these invertebrates have a rudimentary open circulatory system, and thus are not useful for directly studying vascular development. Nevertheless, genetic investigations of these animals do reveal fundamental features of the roles of HS in development. Moreover, these studies demonstrate a role for HSPGs in several signaling pathways that are known to be operable in vertebrate ECs. Thus, the genetic analyses of such lower organisms bring to light potential ways in which HSPGs may function in mammalian ECs.

Studies of invertebrate mutants lacking various core proteins show that HSPGs are required for the development of multiple organ systems, and each core is required for signaling events that are specific in time, place, cell type, and tissue (2,43). Given that each protein conveys discrete functions, the multitude of core proteins expressed in mammalian ECs may well serve to expand endothelial functional diversity. Because lower organisms have multiple core proteins, deletion of a single core protein is not necessarily lethal. However, *Drosophila* and *C. elegans* having core proteins but completely lacking HS (due to mutations in EXT enzymes that prevent polymerization of the HS backbone) die in early embryogenesis. The lethal phenotype provides evidence for a critical role of the HS component of HSPGs during development. In *Drosophila*, such mutants show defects in pathways mediated by three HS-binding signaling ligands: hedgehog, wingless, and decapentaplegic (a form of transforming growth factor [TGF]-β) (2,43). Vertebrates have comparable ligands that are known to be involved in vasculogenesis and angiogenesis, so it is likely that mammalian ECs employ HS to regulate these pathways.

The roles of HS modification enzymes also have been revealed. In contrast to the lethality of a global HS deficiency, *C. elegans* mutants lacking C_5-epimerase, HS2ST, or HS6ST are viable. All these mutants show abnormal guidance in specific neuronal types; however, each enzyme mutation affects a specific assortment of neurons. Thus, different HS modifications convey unique properties to HS in a cell type–specific fashion. Each mutation interferes with developmental processes involving three receptors (Robo, integrin, and ephrin) (2,43). In vertebrates, signaling through all these receptors is known to involve HSPGs, with integrins and ephrin receptors controlling key endothelial functions.

Drosophila deficiencies of specific HS modification enzymes are lethal, due to the malformation of key organ systems. *Drosophila* mutants lacking a gD-type HS3ST isoform have multiple developmental defects stemming from the disruption of signaling through the Notch receptor. Because the vertebrate Notch pathway conveys an arterial cell phenotype to noncommitted ECs (44), a mammalian gD-type HS3ST isoform might control this form of EC differentiation. *Drosophila* mutants lacking HS6ST are lethal because of a malformation of the tracheal airway system (34,45). The development of the *Drosophila* trachea is analogous to vertebrate angiogenesis, because both processes employ comparable signaling components (including FGFs) to control branching morphogenesis of tubular structures. Tracheal cells coexpress the HS6ST and breathless (the *Drosophila* FGF receptor); whereas, adjacent inducing cells express branchless (the *Drosophila* FGF). Mutants lacking HS6ST activity or breathless or branchless have stunted branching of the tracheal system that is phenotypically equivalent. Moreover, HS6ST-deficient mutants have impaired FGF-dependent activation of mitogen-activated protein kinase (MAPK) (45). Thus, 6-*O*-sulfates of HS are essential for FGF-induced branching morphogenesis in *Drosophila.* These moieties also are required for mammalian FGF signaling, as elaborated later. Clearly, the identification and delineation of HS-mediated signaling mechanisms that are operable in vertebrate ECs should be facilitated by such genetic investigations into the developmental roles of HS in lower organisms.

Indeed, the roles of HS in vertebrate vessel development are beginning to come to light. HS is critical for the signaling activity of VEGF-A splice variants that contain HS-binding domains (VEGF-A_{164}, VEGF-A_{188}). Mice that exclusively express the non–HS-binding VEGF-A_{120} exhibit abnormal angiogenesis with defective vessel branching (46,47). Although individual VEGF isoforms have distinct receptor specificities, it is likely that vessel branching requires HS-binding activity, because angiogenic sprouting in zebrafish has recently been shown to require the syndecan-2 core protein (48). Moreover, FGF-stimulated angiogenesis is defective in knockin mice that express a perlecan core protein lacking HS chains (49). Mice lacking various HS modification enzymes have recently been generated. *Hs3st1*$^{-/-}$ mice are described in Chapter 107. Deficiencies in C5-epimerase, NDST1, and HS2ST greatly alter HS structure and result in multiple developmental defects that produce postnatal lethality (50–54). Given the multiple roles of HS, an evaluation of adult functions will require mice with cell type–specific deficiencies. Indeed mice selectively lacking NDST1 in ECs recently have been generated. They have reduced neutrophil infiltration into tissues, which involves reduced (a) L-selectin–mediated binding of leukocytes to ECs, (b) EC transcytosis of chemokines, and (c) EC presentation of luminal surface chemokines to leukocytes (55). Thus, future analyses of mice with global or cell type–specific deficiencies in HS modification enzymes should reveal the involvement of specific HS structures in multiple aspects of vascular development and function.

LIGAND–RECEPTOR INTERACTIONS

The Coreceptor Model of Heparan Sulfate Proteoglygan Action

HSPGs play multiple roles in signaling and can act as either positive or negative regulators (15,56). The involved mechanisms have been most extensively established for FGF signaling, which we review as a paradigm of the many ways in which the HS component of HSPGs can modulate signaling of heparin-binding growth factors.

The FGF family is comprised of more than 20 members of ligands that mediate signaling through five different receptors containing tyrosine kinase domains (FGFRs). Multiple splice variants of each FGFR exist, further expanding the complexity of this system. HSPGs modulate this system by acting as low-affinity "coreceptors" that facilitate the formation of a ternary complex in which two FGFs bind and thereby dimerize two FGFRs (57–59) (Figure 106.3A). Ligand-driven dimerization of the receptors is essential for cytoplasmic domain cross-phosphorylation, which initiates cell signaling (56).

A variety of HSPGs can act as endothelial coreceptors (including perlecan, glypican-1, and syndecans-1, -2, and -4) (60–62), which indicates a critical role for the HS chain. Solution phase studies suggest that one role of HS is to facilitate FGF dimerization into a configuration that is essential for receptor activation (63). However, the precise mechanism of ternary complex formation is disputed because two distinct HS-FGF-FGFR ternary complexes have been crystallized (56) (Figure 106.3A). One complex is asymmetric, with a single HS decasaccharide bridging two FGF1-FGFR2 dimers (64). The other complex is symmetric, with the ends of two HS fragments each stabilizing a single FGF2-FGFR1 dimer (65). One possible explanation for the structural differences is that each complex was formed with different isoforms of ligand and receptor. Given the large number of combinations of FGF receptors and ligands, perhaps multiple ways exist to generate ternary complexes. Regardless, both models show that the HS fragment(s) make similar contacts to the HS-binding residues of both the ligand and receptor. Thus, both models confirm that a major role of HS is to crosslink FGF to the FGFR (56).

Regulatory Properties of Coreceptor-Modulated Signaling

The deployment of an HSPG coreceptor engenders growth factor signaling with several novel properties and influences signaling kinetics, strength, and specificity. In part, this reflects fundamental differences between HSPGs and growth factors. For example, HSPGs are long-lived, exhibiting half-lives in the range of hours to days, compared to the rapid turnover of growth factors (15). Consequently, HSPGs can serve to stabilize growth factors and can drive sustained signaling events. Indeed, in the absence of HS, low concentrations of FGF2 can bind FGFRs and stimulate phosphorylation of MAPK (66). However, the effect is transient and does not lead to cell proliferation, whereas in the presence of HS, signaling events are sustained and mitogenesis occurs (67,68). An additional difference is that pericellular HSPGs can occur in micromolar concentrations, which greatly exceeds the nanomolar levels of growth factors (15). The influence of HSPG concentration on signaling is best addressed by considering the stoichiometry of the ligand–receptor–HS ternary complex.

Stoichiometric Considerations of Coreceptor Signaling

Because HS interacts with both FGFs and FGFRs, the successful formation of a functional HS-FGF-FGFR ternary complex is extremely dependent on the levels of HSPG present on the

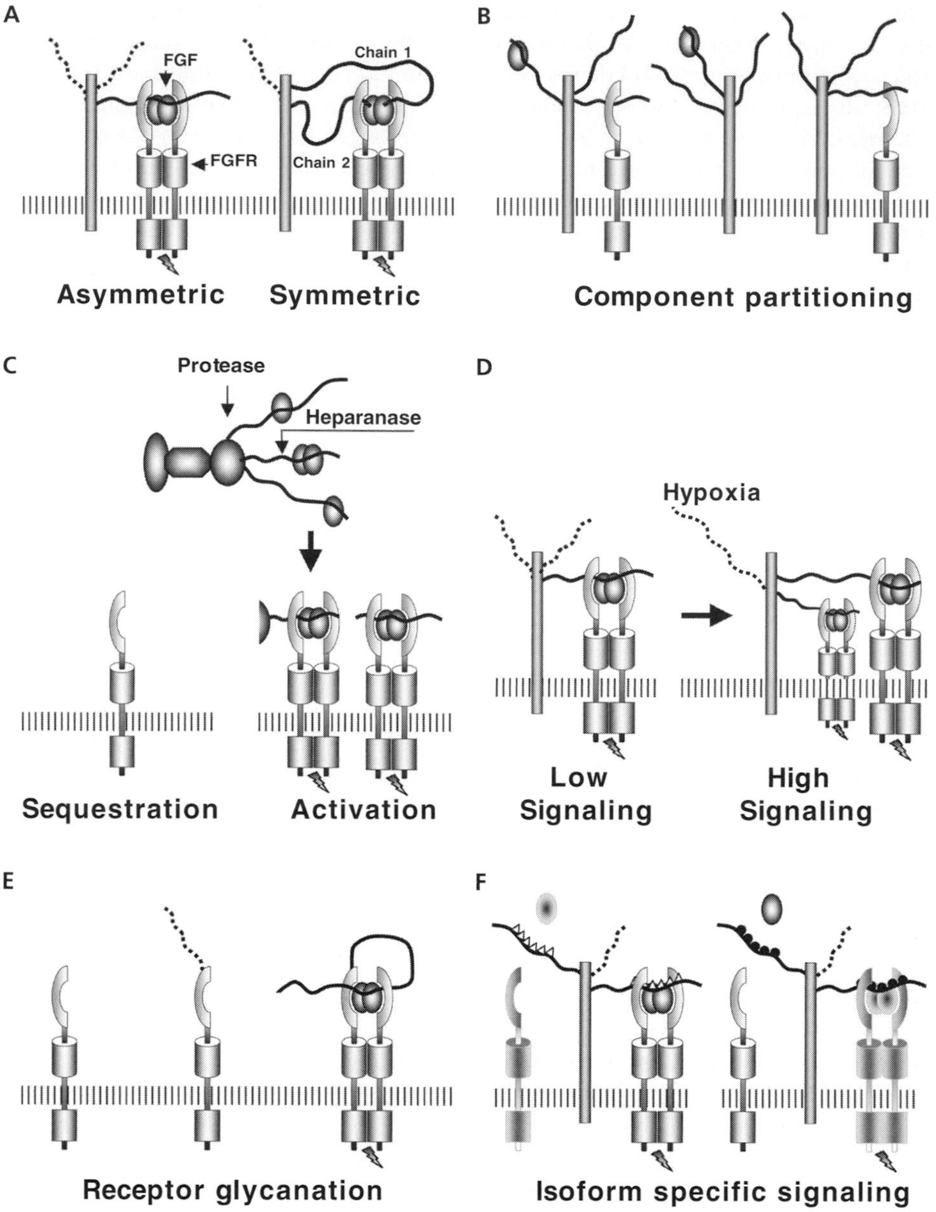

Figure 106.3. Multiple mechanisms of HSPG regulated signaling. Schematically depicted are FGFs (as indicated) and the cell membrane with FGFRs (as indicated) and usually a syndecan-4 type HSPG, which can bear variable ratios of HS (*solid lines*) and CS (*dotted lines*) chains. Formation of an HS-FGF-FGFR ternary complex results in cell signaling (*lightning bolt*). For simplicity, this figure predominantly shows asymmetric signaling complexes. (**A**) Potential ternary complexes. The asymmetric complex requires only a single HS chain, whereas the symmetric complex requires the free ends of two HS chains. (**B**) Extremely high HS levels can potentially inhibit signaling by partitioning ligands and receptors onto separate HS chains and thereby prevent ternary complex formation. (**C**) Extracellular matrix (ECM)-regulated signaling. High levels of ECM HSPGs, such as perlecan (partially shown) can sequester growth factors away from the cell membrane and inhibit signaling. Matrix-bound growth factors are liberated in conditions such as tissue injury, in which proteases and heparanase release HS–growth factor complexes that diffuse to the cell surface to initiate signaling. (**D**) Cellular responsiveness to FGF can be enhanced by hypoxia, which increases the HS content of heteroglycan-type HSPGs. (**E**) Certain FGFR splice variants can be nonglycanated or glycanated with either a CS or HS chain. The latter receptor can form a ternary complex in the absence of other HSPGs. (**F**) HS can regulate isoform-specific signaling. Presented are two cells expressing two FGFR isoforms and exposed to two FGF isoforms. HS sequence motifs (*triangles* versus *circles*) can determine which FGF:FGFR isoform combinations are functionally operable.

cell surface. This feature is readily evident for developmental morphogen gradients, in which the strength of growth factor signaling is regionally defined by the level of HSPG expression (69). Extreme concentration effects are demonstrated in HSPG-deficient cells, in which FGF signaling is restored by the addition of exogenous HS/heparin. Low concentrations of GAG stimulate FGF signaling, but high concentrations are inhibitory (60,70,71). It has been proposed that low HS/heparin concentrations promote ternary complex formation at the cell surface, whereas high GAG concentrations that exceed levels of receptors and ligands will partition each component onto separate HS chains and thereby prevent ternary-complex formation (see Figure 106.3B). Indeed at high HS/heparin concentrations, mitogenic signaling can be re-established by simply increasing the concentration of FGF (72).

The prevention of ternary complex formation by high HS levels is clearly operable in vivo, and can involve spatial segregation (see Figure 106.3C). The endothelial basement membrane contains high levels of growth factors bound to HSPGs; however, signaling does not occur because the ligands are physically separated from their cell surface–bound receptors. Thus, the basement membrane contains sequestered growth factors. However, this reservoir of growth factors can be recruited during vessel injury or remodeling. The actions of proteases or heparinase can release growth factor–HS complexes that diffuse to activate their cell surface receptors (2,15,56).

Glycanation Effects

One means of regulating HS levels is by altering the number and type of GAG chains synthesized on a core protein (glycanation). Glycanation control is germane to "heteroglycans" – in which the core proteins can be glycanated with multiple GAG types. Table 106-2 shows that ECs express several heteroglycans. Syndecan-4, for example, exhibits three GAG attachment sites that are always occupied, and each site can bear either HS or CS chains. Consequently, a single cell type can produce a variety of isoforms: pure HS-syndecan-4, pure CS-syndecan-4, and all possible permutations of heteroglycans (73). The ratio of these forms is biosynthetically controlled in response to external stimuli. For example, the exposure of ECs to hypoxia increases the cell surface HS-to-CS ratio and enhances cell responsiveness to FGF2 (74). Hypoxia does not increase core protein expression but instead enhances the expression of EXTL2, which commits GAG attachment sites to synthesis of HS chains (see Figure 106.2). Such biosynthetic control should increase the HS content on all heteroglycans expressed by a cell, and thus dramatically elevate cell surface levels of HS (see Figure 106.3D).

Glycanation also can lead to novel signaling mechanisms by placing GAG chains on certain receptors (see Figure 106.3E). An FGFR2 splice variant has recently been identified that contains a single GAG chain (75). The variant contains an alternate exon that encodes a GAG attachment site, which can bear either HS or CS chains. Compared to nonmodified or CS-modified receptors, the HS-bearing FGFR2 exhibits high affinity for FGF1. Moreover, in the absence of exogenous HS/heparin, only the HS-modified receptor responds to FGF1 stimulation with enhanced and sustained signaling events that result in mitogenesis (75). Similar forms of FGFR3 also may be possible, because this receptor also shows a similarly positioned potential GAG attachment site. Thus, the presence of an HS chain on select FGFRs appears to abrogate the need for an HSPG coreceptor. It follows that such receptors might play a dominant role in FGF signaling when cell surface levels of HSPGs are low.

Regulation of Signaling by Heparan Sulfate Sequence Motifs

Most importantly, the existence of distinct HS motifs provides a means of regulating signaling specificity. Specific HS motifs are critical for FGF signaling. For example, the HS sequence for FGF2 binding requires 2-*O*-sulfated iduronic acids (76–78). This necessity is confirmed by a cell mutant deficient in HS2ST activity, which produces HS lacking 2-*O*-sulfated iduronic acids. Consequently, the mutant cells can neither bind to nor respond to the growth factor (79). Successful signaling also requires 6-*O*-sulfates that form direct contacts with the HS binding site of the FGFR (64,65,80). This requirement is further revealed by the defective FGF-driven tracheal morphogenesis found in a HS6ST-deficient *Drosophila* mutant, as described earlier.

The dependence of FGF signaling on HS motifs allows for cell type and temporal control of ligand–receptor signaling specificity. For both FGFs and FGFRs, residues of the HS-binding regions are nonconserved across the respective families (56). Consequently, distinct FGF and FGFR isoforms recognize distinct HS motifs (81). It follows that the structural composition of cellular HS will dictate which ligand–receptor combinations are functionally operable and how efficiently they will signal (82). Because HS structure varies between cell types, two distinct cell types bearing the same FGFRs, but different HS motifs, could exhibit differential responsiveness to a given FGF isoform (see Figure 106.3F). Indeed, HS isolated from different EC phenotypes can even activate distinct FGF-FGFR combinations (83). Physiological cues also can alter the cellular expression of various HS sulfotransferases and thereby change the array of HS motifs on the cell surface (74). Such regulated HS synthesis enables the growth factor responsiveness of a single cell type to be modulated in a temporal fashion.

Apart from altering sulfotransferase expression, the array of HS motifs expressed by a cell can be modulated subsequent to HS synthesis. For example, cells can secrete endosulfatases (Sulf1 and Sulf2), which can function in the extracellular environment to remove 6-*O*-sulfates from intact HS chains. Because 6-*O*-sulfates are required for HS interaction with certain FGFRs, these enzymes could influence FGF signaling. Indeed, forced Sulf1 expression can inhibit the formation of HS-FGF-FGFR ternary complexes and thereby prevent FGF-stimulated mesoderm induction and angiogenesis (37).

Thus, the modulation of HS structure superimposes specificity control onto the myriad of FGFs and FGFRs.

DOWNSTREAM SIGNALING EFFECTS AND EFFECTS ON ENDOTHELIAL-CELL PHENOTYPE

At the cellular level, the downstream signaling events and effects on EC phenotype mediated by HS largely depend on the particular growth factor that initiates the signaling event. In large part, these effects are mediated by specific cell surface receptors; the ensuing effects are covered under individual chapters dealing with specific growth factors. However, growth factors also can induce oligomerization of HSPGs, some of which can directly function as signaling receptors. This process is best understood for the syndecan family of core proteins (see Chapter 44). At the whole-organism level, it is likely that the ability of HS to interact with a vast number of growth factors serves to coordinate certain global phenotypes or processes. Such global trends are beginning to come to light.

In the adult, the endothelium normally is comprised of highly differentiated nonproliferating cells. The endothelial basement membrane contains high levels of HS, which sequesters growth factors and thereby likely limits their function (see Figure 106.3C). Genetically engineered mice have recently revealed the importance of HS-bound growth factors stored in the basement membrane. First, knockin mice have been generated that express a form of perlecan, the major basement membrane HSPG that lacks HS chains. HS-deficient perlecan mice are viable but exhibit delayed wound healing and retarded FGF2-induced tumor growth due to defective angiogenesis (49). These phenotypes are observed in the context of tissue damage, in which basement membrane degradation normally releases HS–growth factor complexes that stimulate EC proliferation. Thus, the low levels of basement membrane HS in these mice limits the reservoir of HS-bound growth factor that can be mobilized in response to injury. Conversely, enhanced release of HS–growth factor complexes should occur in transgenic mice that overexpress the heparan cleaving enzyme, heparanase. Indeed, such mice show increased vascularization (84). Combined, these results indicate that at least one function of HS of the endothelial basement membrane is to stimulate EC proliferation in response to tissue damage.

DIAGNOSTIC AND THERAPEUTIC IMPLICATIONS

At present, HS-based therapeutics are only used to inhibit coagulation. This application has evolved from the use of unfractionated heparin, which is a very diverse mixture of molecules, to the recent deployment of a pure synthetic pentasaccharide binding site for AT, which is the active HS motif that conveys anticoagulant activity. The chemical synthesis of HS motifs is exceedingly complex and difficult to scale up for industrial production. However, the recent cloning of HS biosynthetic and degradation enzymes should enable much more facile approaches that combine enzymatic and chemical methodologies. Thus, it is likely that additional synthetic HS motifs can be exploited as therapeutics in the near future. Further studies of the sequence-specific nature of signaling by heparin-binding growth factors should identify potentially important HS motifs.

Angiogenesis is a major arena amenable to potential HS-based therapeutics. Angiogenesis is induced by the concerted action of a number of heparin-binding growth factors. Thus, HS-based therapeutics could serve to inhibit or activate this process. Inhibition is desirable in the treatment of solid tumors, which secrete heparanase and heparin-binding growth factors that stimulate tumor vascularization and hence tumor growth. Optimal treatment would involve a mixture of HS motifs capable of binding specific growth factors but incapable of activating the corresponding receptors (85,86). Such treatment could also include an HS motif that inhibits tumor-secreted heparanase (87), thereby preventing the release of HS–growth factor complexes from the endothelial basement membrane.

The activation of angiogenesis is a potential means of treating ischemic disease by generating new blood vessels to bypass obstructed vessels. Those ECs downstream of an obstruction that experience hypoxia should exhibit enhanced HS levels and therefore have enhanced sensitivity to growth factors. However, bypass vessels must originate upstream of the obstruction; these ECs are not hypoxic, so likely will show low responsiveness to growth factors. One means of circumventing the need for endogenous cell surface HS may be to treat with a growth factor containing a covalently coupled HS fragment (88). The sequence of the attached HS motifs could dictate specificity for a particular receptor isoform. Thus, HS–growth factor hybrids could exhibit far greater specificity than nonconjugated growth factors. Such specificity may be capable of exploiting endothelial phenotypic diversity. For example, HS–growth factor hybrids could potentially allow for the targeted activation of coronary versus peripheral vessels.

KEY POINTS

- HS superimposes multiple levels of control over signaling by heparin-binding growth factors.
- Changes in cell surface HS levels should influence multiple heparin-binding growth factors and thus provide a means of globally regulating multiple signaling pathways.
- Conversely, changes in HS sequence motifs provide a means of regulating the signaling capacity of individual ligand–receptor combinations.

Future Goals

- To use studies of mice deficient in various HS-sulfotransferases to clarify the aspects of EC biology regulated by specific HS-motifs; such studies may reveal novel therapeutic applications for synthetic HS motifs

REFERENCES

1 Bernfield M, Gotte M, Park PW, et al. Functions of cell surface heparan sulfate proteoglycans. *Annu Rev Biochem.* 1999;68:729–777.

2 Esko JD, Selleck SB. Order out of chaos: assembly of ligand binding sites in heparan sulfate. *Annu Rev Biochem.* 2002;71:435–471.

3 McLean J. The thromboplastic action of cephalin. *Am J Physiol.* 1916;41:250–257.

4 Röden L. Highlights in the history of heparin. In: Lane DA, Lindahl U, eds. *Heparin.* London: Edward Arnold; 1989:1–24.

5 Howell WH. The purification of heparin and its presence in blood. *Am J Physiol.* 1925;71:553–562.

6 Abildgaard U. Highly purified antithrombin III with heparin cofactor activity prepared by disc gel electrophoresis. *Scand J Clin Lab Invest.* 1968;21:89–91.

7 Damus PS, Hicks M, Rosenberg RD. Anticoagulant action of heparin. *Nature.* 1973;246:355–357.

8 Lam LH, Silbert JE, Rosenberg RD. The separation of active and inactive forms of heparin. *Biochem Biophys Res Commun.* 1976;69:570–577.

9 Hook M, Bjork I, Hopwood J, Lindahl U. Anticoagulant activity of heparin: separation of high-activity and low-activity heparin species by affinity chromatography on immobilized antithrombin. *FEBS Lett.* 1976;66:90–93.

10 Andersson LO, Barrowcliffe TW, Holmer E, et al. Anticoagulant properties of heparin fractionated by affinity chromatography on matrix-bound antithrombin III and by gel filtration. *Thromb Res.* 1976;9:575–583.

11 Jorpes JE, Holmgren H, Wilander O. Über das Vorkommen von Heparin in den Gefässwänden und in den Augen. Ein beitrag zur Physiologie der Ehrlichschen Mastzellen. *Z Mikrosk Anat Forsch.* 1937;42:279–301.

12 Rosenberg RD, Shworak NW, Liu J, et al. Heparan sulfate proteoglycans of the cardiovascular system. Specific structures emerge but how is synthesis regulated? *J Clin Invest.* 1997;99:2062–2070.

13 Shing Y, Folkman J, Sullivan R, et al. Heparin affinity: purification of a tumor-derived capillary endothelial cell growth factor. *Science.* 1984;223:1296–1299.

14 Silbert JE, Sugumaran G. A starting place for the road to function. *Glycoconj J.* 2002;19:227–237.

15 Iozzo RV, San Antonio JD. Heparan sulfate proteoglycans: heavy hitters in the angiogenesis arena. *J Clin Invest.* 2001;108:349–355.

16 Belting M. Heparan sulfate proteoglycan as a plasma membrane carrier. *Trends Biochem Sci.* 2003;28:145–151.

17 Bottaro DP. The role of extracellular matrix heparan sulfate glycosaminoglycan in the activation of growth factor signaling pathways. *Ann NY Acad Sci.* 2002;961:158.

18 Sasisekharan R, Shriver Z, Venkataraman G, Narayanasami U. Roles of heparan-sulphate glycosaminoglycans in cancer. *Nat Rev Cancer.* 2002;2:521–528.

19 Zhang L, Yoshida K, Liu J, Rosenberg RD. Anticoagulant heparan sulfate precursor structures in F9 embryonal carcinoma cells. *J Biol Chem.* 1999;274:5681–5691.

20 Shworak NW, Shirakawa M, Colliec-Jouault S, et al. Pathway-specific regulation of the synthesis of anticoagulantly active heparan sulfate. *J Biol Chem.* 1994;269:24941–24952.

21 Eriksson I, Sandbäck D, Ek B, et al. cDNA cloning and sequencing of mouse mastocytoma glucosaminyl N-deacetylase/N-sulfotransferase, an enzyme involved in the biosynthesis of heparin. *J Biol Chem.* 1994;269:10438–10443.

22 Hashimoto Y, Orellana A, Gil G, Hirschberg CB. Molecular cloning and expression of rat liver *N*-heparan sulfate sulfotransferase. *J Biol Chem.* 1992;267:15744–15750.

23 Crawford BE, Olson SK, Esko JD, Pinhal MA. Cloning, Golgi localization, and enzyme activity of the full-length heparin/heparan sulfate-glucuronic acid C5-epimerase. *J Biol Chem.* 2001;276:21538–21543.

24 Li J, Hagner-McWhirter A, Kjellen L, et al. Biosynthesis of heparin/heparan sulfate. cDNA cloning and expression of D-glucuronyl C5-epimerase from bovine lung. *J Biol Chem.* 1997;272:28158–28163.

25 Rong J, Habuchi H, Kimata K, et al. Substrate specificity of the heparan sulfate hexuronic acid 2-*O*-sulfotransferase. *Biochemistry.* 2001;40:5548–5555.

26 Kobayashi M, Habuchi H, Yoneda M, et al. Molecular cloning and expression of Chinese hamster ovary cell heparan-sulfate 2-sulfotransferase. *J Biol Chem.* 1997;272:13980–13985.

27 Habuchi H, Kobayashi M, Kimata K. Molecular characterization and expression of heparan-sulfate 6-sulfotransferase. Complete cDNA cloning in human and partial cloning in Chinese hamster ovary cells. *J Biol Chem.* 1998;273:9208–9213.

28 Shworak NW, Fritze LM, Liu J, et al. Cell-free synthesis of anticoagulant heparan sulfate reveals a limiting activity which modifies a nonlimiting precursor pool. *J Biol Chem.* 1996;271:27063–27071.

29 Shworak NW, Liu J, Fritze LMS, et al. Molecular cloning and expression of mouse and human cDNAs encoding heparan sulfate D-glucosaminyl 3-*O*-sulfotransferase. *J Biol Chem.* 1997;272:28008–28019.

30 Shworak NW, Liu J, Petros LM, et al. Multiple isoforms of heparan sulfate D-glucosaminyl 3-*O*-sulfotransferase. Isolation, characterization, and expression of human cDNAs and identification of distinct genomic loci. *J Biol Chem.* 1999;274:5170–5184.

31 Aikawa J, Grobe K, Tsujimoto M, Esko JD. Multiple isozymes of heparan sulfate/heparin GlcNAc N-deacetylase/GlcN N-sulfotransferase. Structure and activity of the fourth member, NDST4. *J Biol Chem.* 2001;276:5876–5882.

32 Habuchi H, Tanaka M, Habuchi O, et al. The occurrence of three isoforms of heparan sulfate 6-*O*-sulfotransferase having different specificities for hexuronic acid adjacent to the targeted *N*-sulfoglucosamine. *J Biol Chem.* 2000;275:2859–2868.

33 Colliec-Jouault S, Shworak NW, Liu J, et al. Characterization of a cell mutant specifically defective in the synthesis of anticoagulantly active heparan sulfate. *J Biol Chem.* 1994;269:24953–24958.

34 Kamimura K, Rhodes JM, Ueda R, et al. Regulation of Notch signaling by Drosophila heparan sulfate 3-O sulfotransferase. *J Cell Biol.* 2004;166:1069–1079.

35 Shukla D, Liu J, Blaiklock P, et al. A novel role for 3-*O*-sulfated heparan sulfate in herpes simplex virus 1 entry. *Cell.* 1999;99:13–22.
36 Xia G, Chen J, Tiwari V, et al. Heparan sulfate 3-*O*-sulfotransferase isoform 5 generates both an antithrombin-binding site and an entry receptor for herpes simplex virus, type 1. *J Biol Chem.* 2002;277:37912–37919.
37 Wang S, Ai X, Freeman SD, et al. QSulf1, a heparan sulfate 6-O-endosulfatase, inhibits fibroblast growth factor signaling in mesoderm induction and angiogenesis. *Proc Natl Acad Sci USA.* 2004;101:4833–4838.
38 Vlodavsky I, Friedmann Y. Molecular properties and involvement of heparanase in cancer metastasis and angiogenesis. *J Clin Invest.* 2001;108:341–347.
39 Fransson LA, Belting M, Cheng F, et al. Novel aspects of glypican glycobiology. *Cell Mol Life Sci.* 2004;61:1016–1024.
40 DeAngelis PL. Evolution of glycosaminoglycans and their glycosyltransferases: Implications for the extracellular matrices of animals and the capsules of pathogenic bacteria. *Anat Rec.* 2002;268:317–326.
41 Medeiros GF, Mendes A, Castro RA, et al. Distribution of sulfated glycosaminoglycans in the animal kingdom: widespread occurrence of heparin-like compounds in invertebrates. *Biochim Biophys Acta.* 2000;1475:287–294.
42 Dewel RA. Colonial origin for *Emetazoa*: major morphological transitions and the origin of bilaterian complexity. *J Morphol.* 2000;243:35–74.
43 Lee JS, Chien CB. When sugars guide axons: insights from heparan sulphate proteoglycan mutants. *Nat Rev Genet.* 2004;5:923–935.
44 Lawson ND, Vogel AM, Weinstein BM. *sonic hedgehog* and *vascular endothelial growth factor* act upstream of the Notch pathway during arterial endothelial differentiation. *Dev Cell.* 2002;3:127–136.
45 Kamimura K, Fujise M, Villa F, et al. Drosophila heparan sulfate 6-O-sulfotransferase (dHS6ST) gene. Structure, expression, and function in the formation of the tracheal system. *J Biol Chem.* 2001;276:17014–17021.
46 Carmeliet P, Ng YS, Nuyens D, et al. Impaired myocardial angiogenesis and ischemic cardiomyopathy in mice lacking the vascular endothelial growth factor isoforms VEGF164 and VEGF188. *Nat Med.* 1999;5:495–502.
47 Ruhrberg C, Gerhardt H, Golding M, et al. Spatially restricted patterning cues provided by heparin-binding VEGF-A control blood vessel branching morphogenesis. *Genes Dev.* 2002;16:2684–2698.
48 Chen E, Hermanson S, Ekker SC. Syndecan-2 is essential for angiogenic sprouting during zebrafish development. *Blood.* 2004;103:1710–1719.
49 Zhou Z, Wang J, Cao R, et al. Impaired angiogenesis, delayed wound healing and retarded tumor growth in perlecan heparan sulfate-deficient mice. *Cancer Res.* 2004;64:4699–4702.
50 Bullock SL, Fletcher JM, Beddington RS, Wilson VA. Renal agenesis in mice homozygous for a gene trap mutation in the gene encoding heparan sulfate 2-sulfotransferase. *Genes Dev.* 1998;12:1894–1906.
51 Merry CL, Bullock SL, Swan DC, et al. The molecular phenotype of heparan sulfate in the $Hs2st^{-/-}$ mutant mouse. *J Biol Chem.* 2001;276:35429–35434.
52 Li JP, Gong F, Hagner-McWhirter A, et al. Targeted disruption of a murine glucuronyl C5-epimerase gene results in heparan sulfate lacking L-iduronic acid and in neonatal lethality. *J Biol Chem.* 2003;278:28363–28366.
53 Ledin J, Staatz W, Li JP, et al. Heparan sulfate structure in mice with genetically modified heparan sulfate production. *J Biol Chem.* 2004;279:42732–42741.
54 Ringvall M, Ledin J, Holmborn K, et al. Defective heparan sulfate biosynthesis and neonatal lethality in mice lacking N-deacetylase/N-sulfotransferase-1. *J Biol Chem.* 2000;275:25926–25930.
55 Wang L, Fuster M, Sriramarao P, Esko JD. Endothelial heparan sulfate deficiency impairs L-selectin and chemokine-mediated neutrophil trafficking during inflammatory responses. *Nat Immunol.* 2005;6:902–910.
56 Pellegrini L. Role of heparan sulfate in fibroblast growth factor signalling: a structural view. *Curr Opin Struct Biol.* 2001;11:629–634.
57 Olwin BB, Rapraeger A. Repression of myogenic differentiation by aFGF, bFGF, and K-FGF is dependent on cellular heparan sulfate. *J Cell Biol.* 1992;118:631–639.
58 Ornitz DM, Leder P. Ligand specificity and heparin dependence of fibroblast growth factor receptors 1 and 3. *J Biol Chem.* 1992;267:16305–16311.
59 Zhou FY, Kan M, Owens RT, et al. Heparin-dependent fibroblast growth factor activities: effects of defined heparin oligosaccharides. *Eur J Cell Biol.* 1997;73:71–80.
60 Bonneh-Barkay D, Shlissel M, Berman B, et al. Identification of glypican as a dual modulator of the biological activity of fibroblast growth factors. *J Biol Chem.* 1997;272:12415–12421.
61 Aviezer D, Hecht D, Safran M, et al. Perlecan, basal lamina proteoglycan, promotes basic fibroblast growth factor-receptor binding, mitogenesis, and angiogenesis. *Cell.* 1994;79:1005–1013.
62 Aviezer D, Iozzo RV, Noonan DM, Yayon A. Suppression of autocrine and paracrine functions of basic fibroblast growth factor by stable expression of perlecan antisense cDNA. *Mol Cell Biol.* 1997;17:1938–1946.
63 Moy FJ, Safran M, Seddon AP, et al. Properly oriented heparin-decasaccharide-induced dimers are the biologically active form of basic fibroblast growth factor. *Biochemistry.* 1997;36:4782–4791.
64 Pellegrini L, Burke DF, von Delft F, et al. Crystal structure of fibroblast growth factor receptor ectodomain bound to ligand and heparin. *Nature.* 2000;407:1029–1034.
65 Schlessinger J, Plotnikov AN, Ibrahimi OA, et al. Crystal structure of a ternary FGF-FGFR-heparin complex reveals a dual role for heparin in FGFR binding and dimerization. *Mol Cell.* 2000;6:743–750.
66 Fannon M, Nugent MA. Basic fibroblast growth factor binds its receptors, is internalized, and stimulates DNA synthesis in Balb/c3T3 cells in the absence of heparan sulfate. *J Biol Chem.* 1996;271:17949–17956.
67 Delehedde M, Lyon M, Gallagher JT, et al. Fibroblast growth factor-2 binds to small heparin-derived oligosaccharides and stimulates a sustained phosphorylation of p42/44 mitogen-activated protein kinase and proliferation of rat mammary fibroblasts. *Biochem J.* 2002;366:235–244.
68 Delehedde M, Seve M, Sergeant N, et al. Fibroblast growth factor-2 stimulation of p42/44MAPK phosphorylation and IkappaB degradation is regulated by heparan sulfate/heparin in rat mammary fibroblasts. *J Biol Chem.* 2000;275:33905–33910.

69 Fujise M, Takeo S, Kamimura K, et al. Dally regulates Dpp morphogen gradient formation in the Drosophila wing. *Development*. 2003;130:1515–1522.

70 Jang JH, Wang F, Kan M. Heparan sulfate is required for interaction and activation of the epithelial cell fibroblast growth factor receptor-2IIIb with stromal-derived fibroblast growth factor-7. *In Vitro Cell Dev Biol Anim*. 1997;33:819–824.

71 Reich-Slotky R, Bonneh-Barkay D, Shaoul E, et al. Differential effect of cell-associated heparan sulfates on the binding of keratinocyte growth factor (KGF) and acidic fibroblast growth factor to the KGF receptor. *J Biol Chem*. 1994;269:32279–32285.

72 Padera R, Venkataraman G, Berry D, et al. FGF-2/fibroblast growth factor receptor/heparin-like glycosaminoglycan interactions: a compensation model for FGF-2 signaling. *FASEB J*. 1999;13:1677–1687.

73 Shworak NW, Shirakawa M, Mulligan RC, Rosenberg RD. Characterization of ryudocan glycosaminoglycan acceptor sites. *J Biol Chem*. 1994;269:21204–21214.

74 Li J, Shworak NW, Simons M. Increased responsiveness of hypoxic endothelial cells to FGF2 is mediated by HIF-1alpha-dependent regulation of enzymes involved in synthesis of heparan sulfate FGF2-binding sites. *J Cell Sci*. 2002;115:1951–1959.

75 Sakaguchi K, Lorenzi MV, Bottaro DP, Miki T. The acidic domain and first immunoglobulin-like loop of fibroblast growth factor receptor 2 modulate downstream signaling through glycosaminoglycan modification. *Mol Cell Biol*. 1999;19:6754–6764.

76 Habuchi H, Suzuki S, Saito T, et al. Structure of a heparan sulphate oligosaccharide that binds to basic fibroblast growth factor. *Biochem J*. 1992;285:805–813.

77 Maccarana M, Casu B, Lindahl U. Minimal sequence in heparin/heparan sulfate required for binding of basic fibroblast growth factor. *J Biol Chem*. 1993;268:23898–23905.

78 Turnbull JE, Fernig DG, Ke Y, et al. Identification of the basic fibroblast growth factor binding sequence in fibroblast heparan sulfate. *J Biol Chem*. 1992;267:10337–10341.

79 Bai X, Esko JD. Mutant defective in heparan sulfate hexuronic acid 2-*O*-sulfation. *J Biol Chem*. 1996;271:17711–17717.

80 Guimond S, Maccarana M, Olwin BB, et al. Activating and inhibitory heparin sequences for FGF-2 (basic FGF). Distinct requirements for FGF-1, FGF-2, and FGF-4. *J Biol Chem*. 1993;268:23906–23914.

81 Guimond SE, Turnbull JE. Fibroblast growth factor receptor signalling is dictated by specific heparan sulphate saccharides. *Curr Biol*. 1999;9:1343–1346.

82 Ostrovsky O, Berman B, Gallagher J, et al. Differential effects of heparin saccharides on the formation of specific fibroblast growth factor (FGF) and FGF receptor complexes. *J Biol Chem*. 2002;277:2444–2453.

83 Knox S, Merry C, Stringer S, et al. Not all perlecans are created equal: interactions with fibroblast growth factor (FGF) 2 and FGF receptors. *J Biol Chem*. 2002;277:14657–14665.

84 Zcharia E, Metzger S, Chajek-Shaul T, et al. Transgenic expression of mammalian heparanase uncovers physiological functions of heparan sulfate in tissue morphogenesis, vascularization, and feeding behavior. *FASEB J*. 2004;18:252–263.

85 Presta M, Oreste P, Zoppetti G, et al. Antiangiogenic activity of semisynthetic biotechnological heparins. Low-molecular-weight-sulfated *Escherichia coli* K5 polysaccharide derivatives as fibroblast growth factor antagonists. *Arterioscler Thromb Vasc Biol*. 2005;25(1):71–76. Epub 2004 Oct 28.

86 Miao HQ, Ornitz DM, Aingorn E, et al. Modulation of fibroblast growth factor-2 receptor binding, dimerization, signaling, and angiogenic activity by a synthetic heparin-mimicking polyanionic compound. *J Clin Invest*. 1997;99:1565–1575.

87 Ferro V, Hammond E, Fairweather JK. The development of inhibitors of heparanase, a key enzyme involved in tumour metastasis, angiogenesis and inflammation. *Mini Rev Med Chem*. 2004;4:693–702.

88 Pye DA, Gallagher JT. Monomer complexes of basic fibroblast growth factor and heparan sulfate oligosaccharides are the minimal functional unit for cell activation. *J Biol Chem*. 1999;274:13456–13461.

107

Antithrombin

Nicholas W. Shworak

Angiogenesis Research Center, Dartmouth Medical School, Hanover, New Hampshire

Antithrombin (AT) is considered to be the most important natural anticoagulant found in plasma. AT irreversibly neutralizes thrombin, factor Xa, and other activated proteases of the coagulation cascade. This neutralization reaction is dramatically enhanced by a longstanding therapeutic anticoagulant, heparin. Endothelial cells (ECs) produce HS^{AT+} – a specific form of heparan sulfate (HS) that is biochemically related to heparin. In vitro, this EC glycosaminoglycan (GAG) also can catalyze AT neutralization of coagulation proteases; consequently, HS^{AT+} has been designated as anticoagulant HS. Based on heparin's pharmacological mechanism of action, it has long been proposed that, in vivo, HS^{AT+} produced by the endothelium would activate AT and thereby regulate the antithrombotic tone of the blood vessel wall. The objective of this chapter is to evaluate whether the endothelium deploys HS^{AT+} as an output component to regulate coagulation. It is readily apparent that this issue has yet to be conclusively resolved. However, HS^{AT+} might participate in an alternative or additional function: the anti-inflammatory signaling activity of AT. Consequently, this chapter also evaluates whether the endothelium deploys HS^{AT+} as an input component that mediates cell signaling.

HISTORY

AT also has been referred to as AT-II/heparin cofactor and AT-III. Various names were proposed because AT initially was identified as a biochemical activity. At one point, four distinct antithrombin activities of plasma were described. AT-I and AT-IV reflected the thrombin inhibitory activities of fibrin and a cleavage fragment of prothrombin, respectively. AT-II/heparin cofactor and AT-III respectively reflected the heparin-dependent and -independent anticoagulant activities of AT (1). Fortunately this arcane nomenclature has been abandoned (2).

At the end of the 19th century, thrombin was known to lose activity when incubated with defibrinated plasma, thereby suggesting that blood must contain a specific inhibitor of thrombin (i.e., antithrombin activity) (3). In 1916, McLean found that extracts from ox heart and liver exhibited a distinct anticoagulant activity (4). McLean's mentor, Howell, purified a potent form of this anticoagulant from dog liver, which was designated as heparin to denote its source of isolation (5). In the early 1930s, Connaught Laboratories ushered in a new era by generating industrial quantities of heparin (6). The initial tissue source was beef liver, soon followed by beef lung; however, during the 1950s, a shift to using pig intestinal mucus/mucosa occurred (7). Industrialization enabled the therapeutic deployment of heparin in the mid 1930s, and provided the material necessary to characterize its mechanism of action (7). In 1939, Brinkhous and colleagues showed that heparin's anticoagulant action required a plasma component termed *heparin cofactor* (8). During the 1950s, several groups conducted kinetic studies indicating that plasma AT and plasma heparin cofactor activities were intimately related (7). In 1968, Abildgaard purified AT and demonstrated that it embodied both activities (9). Shortly thereafter, in 1973, Rosenberg and coworkers demonstrated that heparin catalyzed the inherent ability of AT to form covalent 1:1 inhibitory complexes with coagulation proteases (10).

The importance of AT as a natural anticoagulant was highlighted by Egeberg, who showed that a heritable deficiency was associated with thrombophilia (11). The characterization of AT deficiency was facilitated by Bock, who isolated the *AT* cDNA in 1982 (12). The next year Prochownik isolated the gene and demonstrated that multiple mutations can lead to AT deficiency (13). Today it is appreciated that more than 120 distinct mutations of this gene exist that reduce AT levels or activity and thereby convey an increased risk of thrombosis (14).

The elucidation of heparin's molecular interactions required an understanding of heparin's structure. In 1925, Howell determined that heparin was a polysaccharide (15). Shortly thereafter, it was determined that heparin was a sulfated GAG – a linear polymer of alternating acid and amino sugars. However, heparin is a complex mixture of structures; consequently, almost 60 years of research were required to determine the critical structural features that conveyed anticoagulant activity (7). The prevailing view during this period was that heparin's effects were mediated by nonspecific ionic interactions. However, structural specificity was implicated in 1976, when three separate investigations from the groups of Lindahl, Rosenberg, and Sims showed that only one third of heparin molecules could bind AT, and only this population of molecules exhibited anticoagulant activity (16–18). In the early 1980s, the groups of Casu, Choay, Lindahl, and Rosenberg demonstrated that the active component of heparin was a specific pentasaccharide motif (reviewed in 7). This motif contained a very rare substituent, a 3-*O*-sulfate group, that was essential for both high-affinity AT-binding and the enhancement of AT activity (7).

The relevance of heparin's pharmacological action to endogenous physiology was initially unclear, because heparin is not a normal constituent of blood. Since 1937, it was known that heparin was a minor body component, being present only in the basophilic granules of mast cells. (It has subsequently been shown that the liver and other sources of commercial heparin are indeed rich in mast cells [7].) In 1948, it became apparent that the body also produces HS, a structurally related compound with a lower sulfate content (19). Since the late 1960s, accumulating data have revealed that heparin is simply a minor form of HS. HS is produced by virtually all cell types as HS proteoglycans (HSPGs). Proteoglycans are hybrid molecules, comprised of a protein core to which are attached one or more GAG chains. The structure and synthesis of HSPGs are detailed in Chapter 106. During the 1970s and 1980s, it was found that ECs produce HSPGs as components of the cell surface and extracellular matrix (ECM). Moreover, a subpopulation of endothelial HSPGs were found that bear HS^{AT+}, heparin-like chains exhibiting the pentasaccharide motif that binds and enhances AT activity. Thus, it was proposed that endothelial HS^{AT+} may be an endogenous activator of AT, thereby regulating the anticoagulant tone of the blood vessel wall (7). During the 1990s, Rosenberg, Shworak, and coworkers showed that EC synthesis of HS^{AT+} was regulated by the rate-limiting action of the HS biosynthetic enzyme that adds the very rare 3-*O*-sulfate group, HS 3-*O*-sulfotransferase (OST)-1 or Hs3st1) (Figure 107.1) (20–23). They additionally cloned the mouse gene (*Hs3st1*) and generated *Hs3st1*$^{-/-}$ mice, which provided a means of evaluating the influence of endogenous HS^{AT+} on the activities of AT (24).

Since 1989, it has become apparent that AT exhibits an anti-inflammatory activity that is discrete from its anticoagulant action. The anti-inflammatory activity stems from AT-stimulated EC signaling and appears to depend on HSPGs (25,26). Thus, HS^{AT+} might regulate the anticoagulant and/or anti-inflammatory activities of AT.

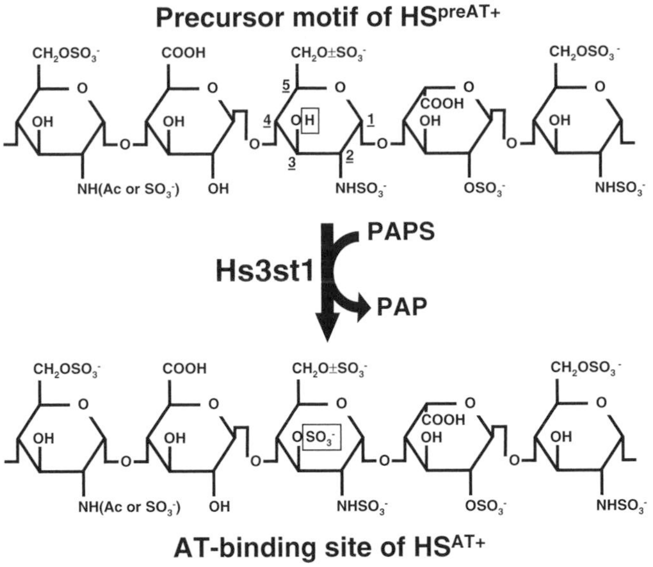

Figure 107.1. Hs3st1 regulates HS^{AT+} production. Hs3st1 acts late in HS/heparin synthesis and recognizes chain regions that conform to a specific pentasaccharide structure (*top*). AT has low affinity for this precursor structure. Hs3st1 transfers a sulfate group from the sulfate donor, PAPS, to the 3-*O*-position of the central glucosamine residue (ring positions underlined), thereby creating the high-affinity AT-binding pentasaccharide motif of HS^{AT+}/heparin (*bottom*). The sulfate donor is converted from PAPS to PAP. *Boxes indicate the altered atoms.*

THE AT PROTEIN AND GENE

Antithrombin Protein Structure and Function

To appreciate the importance of AT in regulating coagulation, one must understand the present view of the major pathways that control thrombin production (27). This process is initiated predominantly by the "extrinsic pathway." Limiting amounts tissue factor (TF) trigger a series of proteolytic activation reactions that occur on cell and platelet membranes, as detailed in Figure 107.2. Two phases of thrombin production occur. The initiation phase generates low levels of thrombin, which in turn sustains this phase through the activation of platelets and the generation of activated cofactors Va and VIIIa. These positive feedback loops also support a transition to the propagation phase, characterized by massive thrombin production. In both phases, thrombin levels are controlled by factor Xa, the limiting component of the prothrombinase complex. The predominant Xa source of the initiation phase is the *extrinsic Xase* complex (TF-VIIa). The *extrinsic Xase* complex also drives production of factor IXa and the *intrinsic Xase* complex (IXa–VIIIa), which is the major source of Xa during the propagation phase. The *intrinsic Xase* complex can also result from the factor XIa activation of factor IXa. Although a relatively minor reaction, thrombin stimulation of this portion of the intrinsic pathway can help to sustain the initiation phase when TF levels, and hence *extrinsic Xase* complex levels,

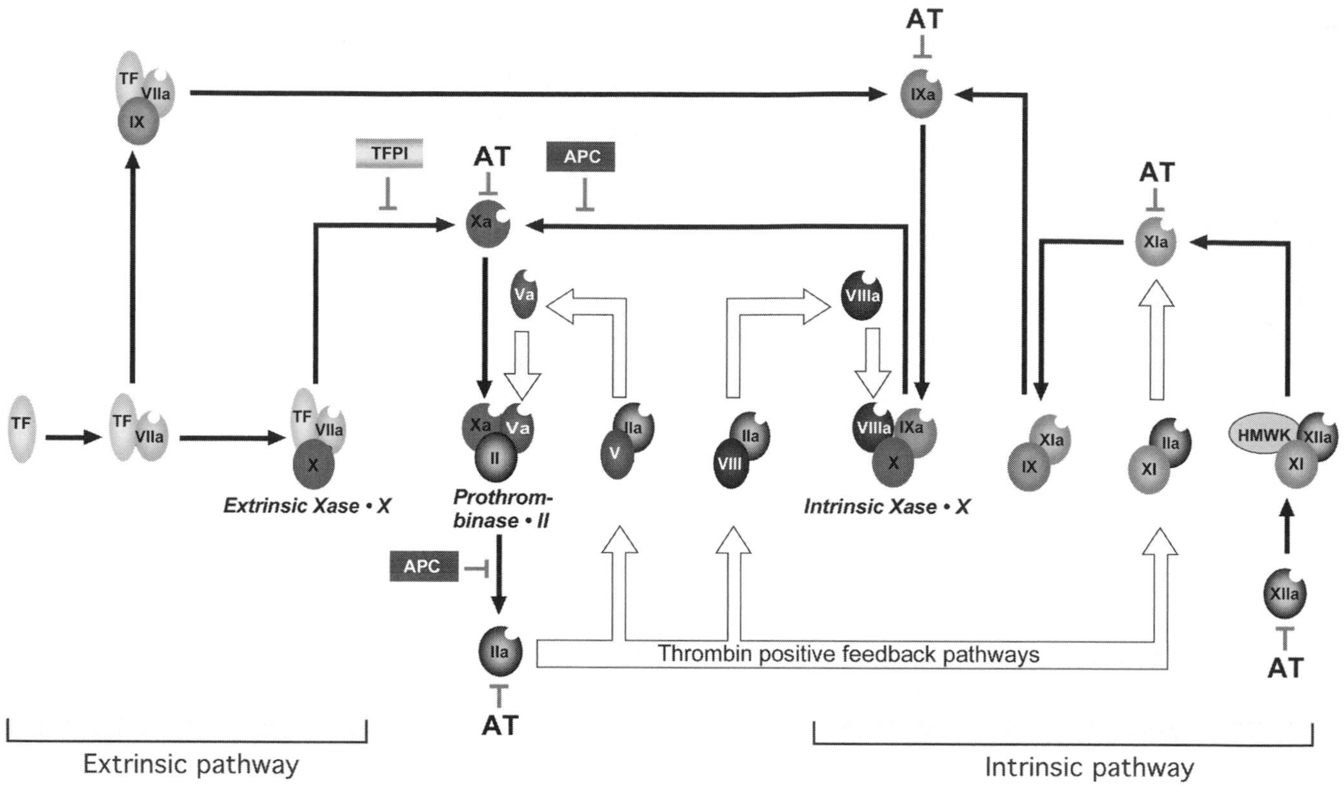

Figure 107.2. Major activators and inhibitors of thrombin generation. Thrombin generation initially is triggered (*black arrows*) by tissue factor (TF) expressed by activated ECs or revealed by endothelial denudation. TF binds and activates circulating factor VIIa. The resulting *extrinsic Xase* complex generates factor Xa. The *extrinsic Xase* complex additionally generates factor IXa, which forms the *intrinsic Xase* complex that also produces factor Xa. Factor Xa is the limiting component of the *prothrombinase* complex that cleaves prothrombin (II) to create thrombin (IIa). Thrombin then triggers positive feedback pathways (*gray arrows*) that increase levels of activated factors required for the above initiation steps. The *intrinsic Xase* complex also can result from the less understood intrinsic pathway, which starts with surface activation of factor XII. For simplicity, kallikrein/high-molecular-weight kininogen activation of Factor XII, and thrombin activation of APC are not shown. *Dark arrows* indicate components inhibited by AT, TFPI, and APC, as elaborated in the text.

are low. Activation of factor XIa by factor XII/kallikrein/high-molecular-weight kininogen has largely been thought to have dubious biologic relevance (27,28). However, recent studies of factor XIIa-deficient mice suggest this portion of the intrinsic pathway may be essential for the propagation phase of arterial thrombosis induced by major trauma (29).

The ability of AT to inhibit coagulation in the absence of GAGs has been revealed by in vitro and kinetic modeling studies. The importance of AT is highlighted by its ability to inhibit all the coagulation proteases. AT inhibits factors IIa (thrombin), IXa, Xa, XIa, XIIa, and kallikrein in their noncomplexed forms (30). Conversely, AT only inhibits factor VIIa when it is complexed with (and activated by) TF. However, AT is only a minor inhibitor of this *extrinsic Xase* complex. As predicted from the above specificities, in vitro studies show AT is only a minor regular of the initiation phase but is a major inhibitor of the propagation phase of coagulation (31). Thus, in the absence of GAGs, AT is primarily operable when there are high levels of noncomplexed factors IIa, IXa, and Xa.

The contribution of additional inhibitors of coagulation has been studied similarly. Tissue factor pathway inhibitor (TFPI) is the predominant inhibitor of the *extrinsic Xase* complex and is the major inhibitor of the initiation phase (31). The remaining *prothrombinase* and *intrinsic Xase* complexes are inhibited by activated protein C (APC), which primarily inhibits the propagation phase (27). APC is a protease that degrades factors Va and VIIIa. These molecules are cofactors, not proteases, so APC controls those components that cannot be regulated by AT. The generation of APC by thrombomodulin-bound thrombin, and the ability of protein S to enhance APC activity are covered in Chapter 108. Finally, heparin cofactor II (HCII) lacks significant anticoagulant activity in the absence of GAGs (31). However, analyses of knockout mice indicate HCII can limit arterial coagulation initiated by endothelial denudation (32). This injury exposes GAGs of the subendothelial ECM, which presumably bind HCII and activate its neutralization of thrombin. In summary, AT blocks numerous coagulation components, whereas the remaining inhibitors have more restricted and largely complementary roles.

The broad specificity and mechanism of action of AT are derived from its structure. AT is a 58,000-Da protein

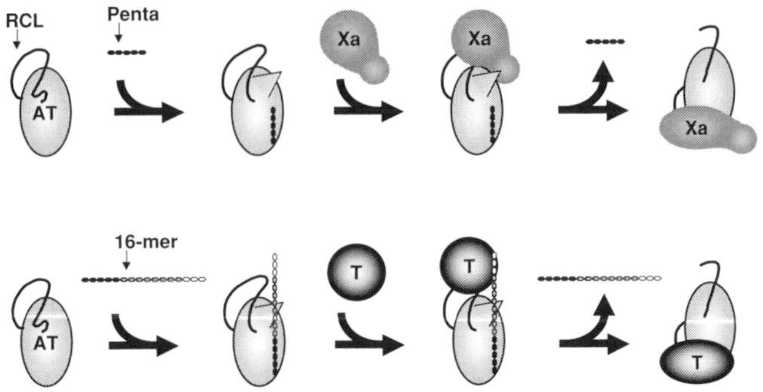

Figure 107.3. Distinct mechanisms of heparin-catalyzed protease neutralization. The speed of neutralization is limited by the rate of protease–AT binding. Heparin accelerates the protease–AT interaction by protease-specific mechanisms, as illustrated for factor Xa and thrombin (*T*). The top figure shows the minimal pentasaccharide motif (*Penta*) allosterically enhancing AT neutralization of factor Xa. Pentasaccharide binding to AT alters the conformation at one end of the reactive center loop (RCL), thereby exposing an exosite (*triangle*). Interaction of this exosite with an Xa exosite enhances the rate of Xa binding to the RCL. Cleavage of the RCL triggers a major AT conformational change that releases the pentasaccharide while simultaneously trapping and distorting the tethered protease. The bottom figure shows how a minimal 16-mer enhances thrombin binding through bridging. The 16-mer pentasaccharide motif (*black*) binds AT and produces the above conformational change; however, thrombin does not interact with the exosite for Xa (*triangle*), so that allosteric enhancement is minimal. Instead, thrombin binding is expedited by contact with the white saccharide residues. Once thrombin is bound, cleavage and entrapment occur as for Xa.

that belongs to the superfamily of serine protease inhibitors (serpins), which encompasses members such as HCII, α1-antitrypsin, angiotensinogen, and the plasminogen activator inhibitors (PAIs) 1 and 2. Serpins are "suicidal inhibitors." They have an exposed reactive center loop that serves as bait for a target protease. When a serine protease cleaves a peptide backbone, a covalent acyl–enzyme bond is created between the protease and the substrate peptide. For normal substrates, this transient bond is broken to release the protease. However, a serpin's reactive center loop functions like a mouse trap. Upon cleavage of this loop, the mousetrap springs from an open to a closed conformation. The serpin undergoes a rapid conformational change that buries the acyl–enzyme bond into the hydrophobic body of the serpin, thereby preventing bond cleavage and release of the tethered protease. This movement additionally disrupts the architecture of the protease's active site (Figure 107.3). Thus, serpins trap the protease into an inactive 1:1 complex (33,34). All serpins exhibit a similar three-dimensional structure, which reflects the common "mouse trap mechanism" of inhibition. However, AT exhibits minimal primary sequence identity to other serpins and has a very unique attribute: It is the only serpin that normally exists in a "low-activity" conformation (35). The pharmacological action of heparin exploits this unique feature.

In the absence of heparin, AT slowly neutralizes activated coagulation proteases. This heparin-independent activity is designated as the "progressive reaction." In contrast, heparin functions as a catalyst to increase the rate of protease neutralization. The precise mechanism depends on the nature of the target protease (see Figure 107.3). Neutralization of factors such as IXa and Xa require just the pentasaccharide motif of heparin/HS^{AT+}. The binding of the pentasaccharide to AT induces a conformational change that accelerates the docking of factors IXa and Xa and thereby enhances the neutralization rate by approximately 300-fold. This allosteric activation creates or exposes exosites on AT that interact with exosites of factor IXa or Xa, respectively. (Exosites are protein interaction regions that are distinct from the active site/reactive center loop.) In contrast, the pentasaccharide-induced conformational change only enhances AT neutralization of thrombin by approximately 1.6-fold. The efficient neutralization of thrombin instead requires a minimal 16-mer fragment of heparin/HS^{AT+} that functions as a bridge between AT and thrombin. One end of the fragment must contain the pentasaccharide for binding to AT, and the other end must have a highly sulfated trisaccharide for binding to thrombin. Compared to the pentasaccharide, a bridging 16-mer enhances the AT neutralization of thrombin by approximately 2,000-fold (36,37).

Although AT recognizes multiple proteases, it is not a major inhibitor of APC, an anticoagulant serine protease. Specificity against APC in large part results from the structure of AT's reactive center loop, which is poorly recognized by APC. Indeed, it appears that AT's reactive center loop has evolved residues that maximally select against recognition by APC (38). In addition, it is likely that APC also lacks exosites for AT (39). In contrast, factors IXa, Xa, and thrombin clearly have exosites that interact with AT (36,37). Thus, AT's action is selectively anticoagulant because evolution has generated multiple interactions between AT and coagulation proteases, but largely prevented interactions between AT and the anticoagulant protease, APC.

AT has different conformational and glycosylation isoforms that exhibit distinct properties. The predominant conformational isoform is latent AT (L-AT), which is a senescent form. L-AT arises when a portion of the reactive center loop spontaneously embeds into a hydrophobic region of AT, similar to the "closed mouse trap" conformation. This conformational change is irreversible and prevents L-AT from inhibiting proteases. However, L-AT may serve an alternative physiological role, because it exhibits antiangiogenic properties (40).

The glycosylation isoforms are α-AT and β-AT. α-AT is fully glycosylated, bearing four *N*-linked glycans, and it is the major form of native AT in plasma. β-AT lacks glycosylation at N135, which results in enhanced affinity for heparin (41). Given this higher affinity, and given that most HS^{AT+} is localized on the ablumenal surface, β-AT preferentially accumulates in the blood vessel wall. Consequently, β-AT may preferentially confer antithrombotic or anti-inflammatory tone to the blood vessel wall (42).

Table 107-1: Relevant Knockout and Knockin Mice

Gene	*Phenotype (Refs.)*
Serpinc1 (AT-knockout)	Homozygous knockout embryos die at embryonic day (E)15.5–16.5 of a consumptive coagulopathy characterized by extensive subcutaneous hemorrhages and fibrin(ogen) deposition in heart and liver but not in hemorrhagic regions (49). Heterozygotes are viable and, when challenged with lipopolysaccharide, show enhanced fibrin deposition in glomeruli, myocardium, and liver sinusoids (81).
Serpinc1 (type IIHBS knockin)	Homozygous knockin mice are viable but develop spontaneous life-threatening thrombosis in neonates and adults. Thrombosis in adults predominantly occurs in heart, liver, and vessels of eye, placenta, or penis. These mice show normal levels of T·AT complexes and normal venous stasis–induced thrombosis but enhanced levels of activated coagulation factors of the intrinsic pathway (67). Treatment with factor VIII blocking–antibodies can prevent thrombosis (82).
Hs3st1	Homozygous knockout mice are viable on a hybrid genetic background and show large reductions in HS^{AT+}. Knockouts show wild-type levels of fibrin accumulation and T·AT complexes, and a normal thrombotic response to arterial injury. On a pure genetic background, knockouts show intrauterine growth retardation and die ~3 days after birth. Knockout embryos and placentae do not show thrombosis or hemorrhage (24,52).
Serpind1 (HCII)	Homozygous knockout mice are viable and have normal longevity. They show a more rapid thrombotic response to arterial injury (32).

SERPINC1, THE ANTITHROMBIN GENE

SERPINC1 Structure and Function

The official designation of the *AT* gene is *SERPINC1*, which denotes AT as member 1 of serpin structural homology class C. The *SERPINC1* gene contains seven exons spanning 13.4 kbp on human chromosome 1. SERPINC1 expression is highly specific to liver hepatocytes, with only minor expression in some other organs, such as the kidneys. Consequently, plasma AT levels are predominantly controlled by the liver levels of *AT* mRNA (43).

AT behaves as a negative acute phase reactant, with plasma levels dropping after acute sepsis or surgery (44). This effect may be partially mediated by the retinoid X receptor α (RXRα) transcription factor. RXRα normally stimulates transcription of AT (45) and other negative acute phase reactants; however, hepatic nuclear levels of RXRα decline in acute sepsis (46). Proinflammatory cytokines can additionally reduce *AT* mRNA levels (47). Conversely, glucocorticoid (which is anti-inflammatory) stimulates SERPINC1 transcription (48). These alterations in gene expression are consistent with the anti-inflammatory properties of AT, as elaborated on in the next section.

SERPINC1 MUTATIONS

Numerous mutations of the human *SERPINC1* gene have been identified; these highlight the importance of AT as a natural anticoagulant. AT deficiency occurs at a frequency of about 1:3000 and is usually inherited as an autosomal dominant. Mutations are classified by the following scheme, which predicts clinical presentation:

Type I: Low functional and immunological levels of AT
Type II: AT structural variants
Subtypes:
- RS, altered reactive site function
- HBS, dysfunctional heparin binding site
- PE, pleiotropic effect

In general, type I and type II_{RS} have the most severe impact on plasma AT activity and convey the highest general risk of venous thrombosis. Afflicted individuals are virtually always heterozygotes, because complete AT deficiency is likely incompatible with human life (14). Indeed, complete AT deficiency in mice (see Table 107-1) causes intrauterine death from an extreme hypercoagulable state (49).

Increasing evidence suggests that the type II_{PE} group may exhibit the unique clinical features of severe fever-induced thrombosis (50). Several members of this group exhibit conformational instability, and have a greater tendency to convert to the inactive L-AT form. Slight increases in body temperature can dramatically accelerate this transition, so sudden-onset thrombosis can occur in response to a minor incidental fever of 39°C to 40°C (102°F to 104°F). It is important to note that L-AT can inactivate normal AT by forming inactive dimers. Such inactivation may account for the extreme severity of thrombosis in certain type II_{PE} individuals (50).

Conversely, mutations that reduce heparin binding (type II_{HBS}) have the "mildest" gene dosage effect. In contrast to the lethality of complete AT deficiency, individuals homozygous for type II_{HBS} mutations are readily found. Indeed, this category of mutations typically behaves in a recessive fashion, with thrombosis predominantly occurring in homozygotes but not heterozygotes. The exclusive viability of homozygous type II_{HBS} mutations suggests that the majority of AT's anticoagulant activity does not require a functional heparin binding site. Type II_{HBS} individuals also show an additional unique feature. Whereas other AT mutations lead to venous thrombosis, type II_{HBS} patients frequently present with arterial thrombosis (14). Thus, AT binding to endothelial HS^{AT+} may convey unique properties to the arterial wall.

THE Hs3st1 PROTEIN AND GENE

Hs3st1 Protein Structure and Function

HS biosynthesis occurs in the Golgi apparatus and is mediated predominantly by type-II integral membrane proteins. An interesting exception is Hs3st1, which exhibits the unusual structure of an intraluminal resident protein. The precursor protein gains entry into the endoplasmic reticulum via a leader sequence of 20 amino acids. Removal of these N-terminal residues by signal peptidase activity leaves an approximately 36,000-Da protein that lacks direct membrane attachment. The mature protein is modified by four *N*-linked glycans and is comprised of an approximately 30-amino acid N-terminal SPLAG-domain (rich in Ser, Pro, Leu, Ala, and Gly) and a 259-amino acid C-terminal sulfotransferase domain. It is presumed that a hypothetical membrane-bound retention partner binds to the SPLAG domain to localize Hs3st1 to a *trans*-Golgi compartment, where HS synthesis occurs. The retention mechanism is somewhat leaky. Consequently, substantial levels of Hs3st1 are secreted into the plasma. However, it is unclear if plasma Hs3st1 serves any function (22,51).

In ECs, the level of HS^{AT+} production is dictated by the rate-limiting action of Hs3st1 (see Figure 107.1). About 25% of HS chains in the Golgi contain motifs that correspond to the AT-binding site devoid of the essential 3-*O*-sulfate residue (HS^{preAT+}). Hs3st1 preferentially binds these precursors and adds the critical 3-*O*-sulfate group to create HS^{AT+} with functional AT-binding sites. This reaction employs the universal sulfate donor adenosine 3′-phosphate 5′-phosphosulphate (PAPS), which is converted into adenosine 3′-phosphate 5′-phosphate (PAP). The sulfotransferase domain contains those residues that mediate catalysis and those residues that define sequence specificity (i.e., recognizing HS^{preAT+}). HS^{preAT+} occurs in excess, and the level of Hs3st1 determines the extent of this precursor that is converted into HS^{AT+} (21). Thus, the rate-limiting action of Hs3st1 allows HS^{AT+} production to be regulated independently of total HS synthesis.

Hs3st1 Gene Structure and Function

The human gene is designated as *HS3ST1* and the mouse gene is *Hs3st1*. The gene is remarkable in that it spans more than 140 kb, but the entire coding region occurs in a single exon at the 3′ end of the gene. Thus, the vast majority of the gene encodes exons and introns of the 5′ untranslated region (UTR). This region of the mouse gene has seven exons that undergo variable splicing, to make at least 10 distinct splice variants of the 5′ UTR. Presumably, these variants enable distinct forms of translational control; however, it is not yet known if the human transcripts are equally complex (52).

Hs3st1 expression is not limited to ECs. Hs3st1 is also produced by certain neurons and by a variety of epithelial cell types such as alveolar type II cells, ovarian granulosa cells, and glomerular epithelial cells, which largely deposit HS^{AT+} in their respective basement membranes (53–55; N. Shworak, unpublished observations). Such a diversity of cell types suggests that HS^{AT+} may function in processes discrete from the coagulation cascade.

Hs3st1 expression appears to be altered in vascular inflammation, because intimal levels of both total HS and HS^{AT+} decrease with the progression of vessel inflammation and atherosclerosis (56). In particular, HS^{AT+} levels are selectively reduced by certain proinflammatory molecules such as homocysteine (57) and are selectively enhanced by the antiatherogenic apolipoprotein E (58). These alterations in HS expression are consistent with the possibility of HS^{AT+} mediating the anti-inflammatory properties of AT. In this regard, it is intriguing that type II_{HBS} AT mutants, which lack heparin binding, are the only mutant category that exhibits arterial thrombosis. As described later, the anti-inflammatory activity of AT requires direct engagement of AT with EC HSPGs. Thus, type II_{HBS} individuals may exhibit atherosclerotic lesions with an enhanced inflammatory state. Because inflammation can trigger coagulation, it is possible that type II_{HBS} arterial thrombosis is secondary to the loss of an AT/HSPG anti-inflammatory pathway.

EVOLUTIONARY CONSIDERATIONS

Evolutionary studies suggest that AT arose long after HS^{AT+}. AT is potentially very ancient, because serpins have been identified in prokaryotes, plants, animals, and viruses. However, serpins have undergone rapid evolution to fulfill highly specialized functions. Consequently, considerable primary sequence diversity exists between major phylogenetic groupings. For example, plant serpins are very distinct from animal serpins, suggesting that both lineages began with a single primordial gene that underwent extensive duplication and structural/functional divergence. Among invertebrates as well, serpin structure varies widely even between species. In contrast, orthologues of human serpins are present throughout vertebrate species, which indicates that they arose after the vertebrate-cephalochordate or the chordate-invertebrate divergence (35,59).

AT bears little resemblance to other vertebrate serpins. Its unique structure likely reflects its necessity to interact with a large number of activated coagulation proteases. Indeed, AT and most components of the coagulation cascade appear to have coevolved with the divergence of vertebrates from cephalochordates, about 450 million years ago. It is important to note that some invertebrates (e.g., the horseshoe crab, *Limulus polyphemus*) do exhibit a coagulation cascade that is driven by serine proteases and is regulated by serpins; however, the components are extremely primitive. The invertebrate coagulation proteases lack γ-carboxy glutamic acid domains, epidermal growth factor-like domains, and kringle domains. These elements enable essential interactions that are unique to the vertebrate coagulation cascade. For example, they allow the proteases to interact with lipid membranes, with other coagulation proteases, and with the protease inhibitors (60). Moreover, AT and the entire family of vertebrae serpins appear to have arisen from a common invertebrate serpin that was intracellular rather than extracellular (35). Thus, AT and a human-like coagulation cascade are unique to vertebrates.

In striking contrast, the pentasaccharide AT-binding site of HS^{AT+}/heparin has been detected in several invertebrates (61). Indeed, the entire molecular machinery for generating HSPGs arose with the emergence of Eumetazoans. In particular, *Drosophila* exhibits a gene that is clearly orthologous to human *HS3ST1* (62). Thus, HS^{AT+} emerged some 250 million years before AT, making it unlikely that HS^{AT+} initially served as an anticoagulant. It seems more likely that invertebrate HS^{AT+} plays a role in cell signaling, because this is a major primordial function of HSPGs. Moreover, *Drosophila* has a gene, paralogous to human HS3ST1, which has recently been found to regulate the Notch signaling pathway (62). This discrepancy in evolutionary appearance evokes an important unresolved question: Did AT recruit HS^{AT+} to regulate vertebrate coagulation, and/or did HS^{AT+} recruit AT to function in cell signaling?

DEVELOPMENTAL CONSIDERATIONS

The developmental expression of AT suggests that it may be multifunctional. The anticoagulant activity is highlighted by *AT*-knockout mice that die of an extreme hypercoagulable state late in gestation (49). This timing reflects the unique state of hemostatic balance during embryonic development. Levels of procoagulant and anticoagulant proteins tend to be initially low and then gradually rise to near-adult levels at term (48,63). Thus, AT-deficient embryos lack protection against the late gestational rise in procoagulant activity. However, the expression profile of AT in the embryonic mouse is dramatically different from that of the major procoagulant factors. Embryonic expression of fibrinogen and most coagulation proteases typically begins sometime after implantation and peaks in mid to late gestation, corresponding with the development of the liver, which is the major source of these plasma proteins. Although AT expression also increases in mid to late gestation, it is also highly expressed in the unfertilized egg and in preimplantation embryos (N. Shworak, unpublished observations). The expression of AT in the absence of the coagulation machinery suggests it may play an additional role that is functionally discrete from anticoagulation. HS^{AT+} also is produced by oocytes (N. Shworak, unpublished observations), raising the possibility that HS^{AT+} may be involved in very early developmental roles of AT.

ROLES IN HOMEOSTASIS

Anticoagulant Roles of Antithrombin and HS^{AT+}

The dramatic degree to which heparin enhances AT's anticoagulant activity has led some to propose that the progressive (HS-independent) activity of AT is biologically irrelevant, and that endogenous AT requires activation by endothelial HS^{AT+}. However, multiple lines of evidence suggest that AT's anticoagulant effects are mediated, at least in large part, by this progressive activity, whereas the role of endothelial HS^{AT+} is unresolved.

The vast majority of plasma AT must exist independently of HS^{AT+}, because blood normally does not contain free HS^{AT+}/heparin (64). Given the slow rate of the progressive reaction, this bulk of HS^{AT+}-free AT could potentially be relatively inert. However, in vitro analyses and kinetic modeling show that the normal plasma concentration of AT is high enough to compensate for the slow heparin-independent rate. As elaborated earlier, the rate of the HS^{AT+}-free progressive reaction is sufficient to account for AT inhibition of the propagation phase of thrombin production (31,65). This phase occurs almost exclusively on platelet membranes (28,66), which lack HS^{AT+} (N. Shworak, unpublished observations). Thus, AT's predominant anticoagulant activity does not appear to involve HS^{AT+}. This conclusion also is supported by clinical observations. Mutations affecting total AT activity cause increased risk for thrombosis in the heterozygous state, whereas hypercoagulability from mutations that specifically target heparin-dependent activity (type II_{HBS}) requires two affected AT alleles (14). Similarly, mice completely lacking AT die embryonically of a severe hypercoagulable state, whereas mice expressing a type II_{HBS} AT form are viable (although they eventually succumb to sudden-onset thrombosis) (49,67). For both of these mouse models, it is presently unclear whether thrombosis is due predominantly to the loss of AT's anticoagulant activity or its anti-inflammatory activity. However, the type II_{HBS} mice exhibited wild-type levels of thrombin·antithrombin (T·AT) complexes and also exhibit wild-type levels of thrombus induced by venous stasis (67). Combined, these results clearly indicate that the majority of AT's anticoagulant activity occurs through the progressive reaction and is independent of HS^{AT+}.

Yet the hypercoagulable state of mice and humans with type II_{HBS} mutations suggests that HS^{AT+} may contribute to the regulation of AT anticoagulant activity. To evaluate the potential contribution of HS^{AT+}, one must first consider the likely mechanism. Possibly, HS^{AT+} conveys anticoagulant tone just to the vascular EC surface, because it is "tethered" to this site by various core proteins. Such a role might initially seem unlikely, as >95% of HS^{AT+} exists in the subendothelial matrix and is not in direct contact with blood. However, a surface effect may be possible, because the perfusion of purified thrombin and AT into the hind limbs of rodents leads to an elevated rate of T•AT complex formation. HS^{AT+} and not endogenous heparin was implicated in T•AT formation, because complex levels were reduced by enzymatic degradation of HS, but were normal in mast cell (heparin) -deficient animals (52). A surface effect also is implicated by type II_{HBS} mice, which do show slight elevations in activated factors VIII, IX, XI, and XII (67). This observation suggests that HS^{AT+} might play a role in enhancing AT inhibition of the poorly understood intrinsic pathway.

Two lines of evidence suggest that the level of endothelial HS^{AT+} may define the extent of AT activation. First, the plasma AT concentration (~3.5 μM) appears saturating; that is, this level greatly exceeds the dissociation constant of AT for HS^{AT+} (~15 nM). Saturation of HS^{AT+} by AT implies a continuously acting anticoagulant system, which is consistent with the dynamic nature of hemostasis. Hemostatic tone is established actively as the net balance between ongoing antagonistic processes (procoagulant versus anticoagulant and fibrinolytic). Second, ECs have been shown to regulate HS^{AT+} levels. Only a small subpopulation of HSPGs actually bears HS^{AT+}, and this level varies between ECs isolated from different vascular beds. Moreover, analyses of mutant cell lines show that HS^{AT+} levels can be selectively altered without influencing the bulk of HS (52). Thus, the endothelium may modulate the production of HS^{AT+} as a means of fine-tuning AT activation, and hence local hemostatic balance.

This hypothesis has been tested in *Hs3st1*$^{-/-}$ mice (52). Compared with wild-types, *Hs3st1*$^{-/-}$ mice have normal tissue levels of total HS, but HS^{AT+} levels are reduced by 75% to 98%. The residual levels of HS^{AT+} were expected, because other Hs3st isoforms can contribute to HS^{AT+} production. Such large reductions in HS^{AT+} should generate a basal procoagulant state only if HS^{AT+} contributes to ongoing hemostatic tone. However, unchallenged *Hs3st1*$^{-/-}$ mice exhibit wild-type levels of tissue fibrin, arguing against a basal procoagulant state. When given a procoagulant challenge – hypoxia, which enhances lung TF expression – *Hs3st1*$^{-/-}$ and *Hs3st1*$^{+/+}$ mice both undergo comparable elevations in lung fibrin levels. Thus, a strong thrombotic challenge failed to reveal a latent procoagulant state in these animals (52). Because most of HS^{AT+} is found in the subendothelial matrix, an anticoagulant role might only occur following endothelial denudation, which would allow direct contact of the blood and subendothelial matrix. Subendothelial HS^{AT+} is almost undetectable in *Hs3st1*$^{-/-}$ mice. However, in response to chemical endothelial denudation, *Hs3st1*$^{-/-}$ and *Hs3st1*$^{+/+}$ mice have indistinguishable kinetics of occlusive thrombosis and show comparable postinjury levels of T•AT complexes (52). Hence, subendothelial HS^{AT+} influences neither occlusive thrombosis nor formation of T•AT complexes. Combined, these data suggest that hemostatic tone is not tightly linked to HS^{AT+} levels.

This conclusion is at odds with the procoagulant phenotype observed in humans and mice with type II_{HBS} (14,67). Four potential explanations exist for this discrepancy. First, low HS^{AT+} levels might be sufficient for an anticoagulant effect. Because the low residual levels of HS^{AT+} in *Hs3st1*$^{-/-}$ mice are derived from other Hs3st isoforms, multiple genes may need to be eliminated to influence hemostasis. Second, other anticoagulant mechanisms may compensate for the reduced HS^{AT+} levels. In this case, combined deficiencies in two anticoagulants may be required to detect an HS^{AT+} effect. Third, the heparin-binding site of AT may mediate its effects through GAGs other than HS^{AT+}. For example, AT neutralization of factor Xa also can be stimulated by HS that lacks an AT-binding site (HS^{AT-}). Compared to HS^{AT+}, AT has a much lower affinity for HS^{AT-}; however, substantial binding should occur under physiologic concentrations of AT (68). Because HS^{AT-} is 20 to 200 times more abundant than HS^{AT+} (23), HS^{AT-} may be predominantly responsible for AT activation. Additionally, AT neutralization of thrombin can be enhanced by a form of thrombomodulin bearing a chondroitin sulfate chain (69). In the context of these other GAGs, AT anticoagulant activity may be independent of HS^{AT+}. Indeed, overexpression of Hs3st1 in primary ECs increases HS^{AT+} levels, but fails to enhance AT activation by the cell surface (52). Fourth, the arterial thrombosis of type II_{HBS} may be secondary to an enhanced inflammatory state, as discussed earlier. Thus, it is presently unclear whether HS^{AT+} is a physiological enhancer of AT anticoagulant activity.

Although *Hs3st1*$^{-/-}$ mice do not show an obvious procoagulant state, they do exhibit other phenotypes. When generated on an inbred mouse strain, *Hs3st1*$^{-/-}$ embryos exhibit intrauterine growth retardation, and *Hs3st1*$^{-/-}$ pups die within 3 days of birth (52). The reasons for these consequences are unclear. However, the existence of these phenotypes in the absence of a procoagulant state suggests that a major role of HS^{AT+} is to regulate some process apart from coagulation. This alternate function may involve the anti-inflammatory activity of AT.

Anti-Inflammatory Activity of Antithrombin

AT exhibits an anti-inflammatory activity that can protect against the deleterious effects of systemic inflammatory response syndromes (SIRS). This acute process can lead to multiple organ dysfunction, acute respiratory distress syndrome (ARDS), and shock. SIRS can result from many clinical conditions, including trauma, disseminated intravascular

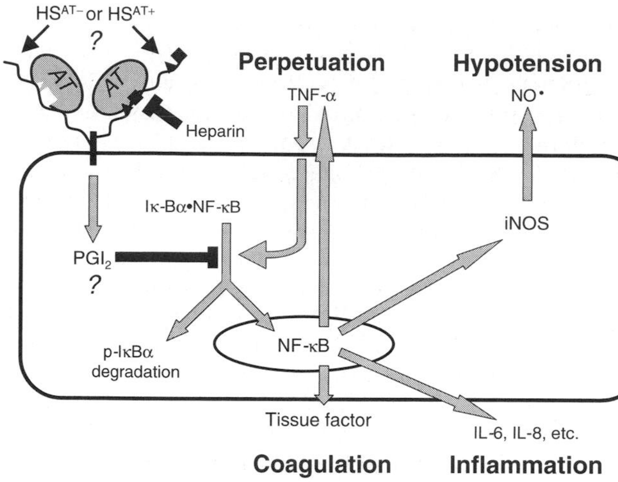

Figure 107.4. Anti-inflammatory action of AT in SIRS. Initiation of SIRS involves TNF-α activation of the transcription factor NF-κB. NF-κB normally resides in the cytoplasm in an inactive complex with its inhibitor IκBα. TNF-α induces phosphorylation of IκBα, which dissociates from the inhibitory complex and is degraded by the proteasome. The released NF-κB translocates to the nucleus to initiate the transcription of several genes, including TNF-α, which further drives EC activation. AT binding to HS^{AT+} and/or HS^{AT-} of endothelial HSPGs initiates anti-inflammatory signaling that in part enhances PGI_2 production. PGI_2, or some other factor, blocks the TNF-α–induced phosphorylation and degradation of IκBα. Consequently, AT prevents activation of NF-κB and upregulation of NF-κB–dependent genes that initiate the deleterious aspects of SIRS. Heparin blocks the pathway by competitively inhibiting AT binding to HSPGs. The pentasaccharide motif of HS^{AT+} is represented as a triangle and square.

coagulation, severe burns, acute pancreatitis, hemorrhagic shock, and sepsis. The syndrome involves a systemic overactivation of leukocytes and ECs that leads to overwhelming inflammation and coagulation. A major initiator of SIRS is tumor necrosis factor (TNF)-α (Figure 107.4). TNF-α activates the transcription factor nuclear factor (NF)-κB and thereby induces the expression of proinflammatory and procoagulant mediators such as interleukins and TF, respectively. NF-κB also induces the expression of inducible nitric oxide synthase (iNOS). In turn, iNOS causes a massive increase in levels of nitric oxide (NO) that can lead to shock. Expression of TNF-α itself is NF-κB-dependent; thus positive feedback can drive the onset of SIRS (25,26).

A large volume of data from animal and clinical trials show that intravenous therapy with AT can reduce morbidity and/or mortality in SIRS associated with sepsis, organ failure, and ischemia–reperfusion injury (25,26). A minor contribution may derive from AT-mediated neutralization of thrombin and factor Xa, which should limit further thrombin and fibrin generation, as well as proinflammatory signaling through protease-activated receptors. However, increasing evidence suggests that AT exerts its beneficial effects through direct effects on cell signaling. For example, in cultured ECs and in animals, AT stimulates the synthesis of prostacyclin (PGI_2) (70,71), a potent inhibitor of leukocyte activation, leukocyte–endothelial interactions, platelet aggregation, and NF-κB activation (Figure 107.4). AT treatment of cultured ECs also abrogates TNF-α–mediated activation of NF-κB (72), possibly through PGI_2 induction. Consequently, AT treatment prevents downstream expression of interleukin (IL)-6, IL-8, TNF-α, TF, and iNOS (25,26). Indeed, in a rat sepsis model, AT blockade of NF-κB prevented iNOS-dependent shock (73). AT also exerts signaling activities in leukocytes, which serve to inhibit leukocyte–endothelial interactions. These anti-inflammatory properties should not be surprising, because many molecular interconnections exist between coagulation and inflammation. Moreover, it is increasingly apparent that several natural anticoagulants also have anti-inflammatory activity (25,26).

In vitro and in vivo experiments show that the anti-inflammatory effects of AT (SIRS amelioration, PGI_2 production, and NF-κB inhibition) are clearly separate from its anticoagulant action. First, agents that only exhibit a pure anticoagulant activity do not induce these effects. Conversely, these effects require an AT concentration far in excess of that sufficient to resolve coagulation abnormalities. Finally, these effects are abolished by heparin, in striking contrast to heparin's enhancement of AT's anticoagulant activity (25,26). However, these experiments cannot rule out that AT anticoagulant activity at the cell surface might partially contribute to the anti-inflammatory effect.

Inhibition by heparin suggests the anti-inflammatory activity requires the binding of AT to cell surface HSPGs (see Figure 107.4). In support of this mechanism, these effects are not induced by a chemically modified form of AT that lacks heparin binding, but retains thrombin-neutralizing activity. Moreover, AT-induced PGI_2 synthesis by cultured ECs is prevented when cell surface HS chains are eliminated by enzymatic degradation or biosynthetic blockade. Additionally, the β-AT glycosylation variant exhibits enhanced affinity for HS and enhanced anti-inflammatory signaling strength. Thus, it appears that AT employs HSPGs to mediate a cell signaling event (25,26).

It is presently unclear how HSPGs mediate AT anti-inflammatory signaling. On one hand, the HS chains might facilitate the formation of a ternary complex between AT and a putative AT receptor. This mechanism would be analogous to HS-mediated signaling of growth factors. On the other hand, AT binding to the HS chain might induce proteoglycan clustering, which can directly initiate cell signaling or could influence the signaling of HS-dependent growth factors (74). It also is unclear whether AT anti-inflammatory signaling involves high-affinity interactions with HS^{AT+} and/or low-affinity interactions with HS^{AT-}. AT's anti-inflammatory effects occur only when plasma AT levels are augmented by 1.5- to threefold. Such high AT levels indicate that a low-affinity receptor is essential. Thus, HS^{AT+} cannot be the signaling "receptor," but still could play an essential role in facilitating the signaling process. The recent availability of *Hs3st1*$^{-/-}$ mice, which exhibit large reductions in vascular HS^{AT+}, will enable the testing of this possibility.

DIAGNOSTIC AND THERAPEUTIC IMPLICATIONS

AT has long been recognized as the therapeutic target of heparin. The elucidation of AT–heparin interactions has led to the development of novel heparin therapeutics with improved properties for treating venous thromboembolism. The knowledge that neutralization of factor Xa, compared to thrombin, requires a shorter heparin sequence (see Figure 107.3) has prompted the development of low-molecular-weight heparins (LMWHs), which contain shorter oligosaccharides than do unfractionated heparin. This design reduces the levels of longer fragments required for thrombin neutralization, so that LMWHs are more selective toward inhibiting factor Xa. The initial rationale was that reduced inhibition of thrombin might minimize bleeding complications. This assertion has not been borne out; however, LMWHs clearly exhibit superior pharmacokinetic properties. The smaller fragments have reduced interactions with the reticulo-endothelial system (and circulating proteins, such as platelet factor 4 [PF4]), and so have better bioavailability, a longer half-life, and reduced interindividual variation. As a result, LMWHs can be administered less frequently and do not require extensive monitoring of coagulation parameters (75). However, this drug is still a complex mixture of molecules, most of which are not mediating the drug's effects.

This limitation has been resolved with the recent release of a pure AT-binding site pentasaccharide, which exclusively targets proteases like factor Xa that are inhibited by allosteric activation of AT (see Figure 107.3). The reduced size and purity have further enhanced bioavailability, which appears to improve efficacy. Most importantly, the synthetic pentasaccharide may avoid a severe adverse reaction, heparin-induced thrombocytopenia (HIT). This autoimmune disease results from the development of antibodies against complexes of heparin with PF4. The pentasaccharide does not bind PF4 and does not cross-react with HIT antisera, so it should not elicit a HIT autoimmune response (76). Also under development is a synthetic hexadecasaccharide (16-mer) with an optimal structure for AT inhibition of thrombin by the bridging mechanism (see Figure 107.3). The 16-mer contains the AT-binding pentasaccharide at one end, so it can enhance the AT inhibition of virtually all coagulation proteases. Like the pentasaccharide, the 16-mer has good bioavailability and does not bind to PF4. However, compared to the pentasaccharide and unfractionated heparin, the 16-mer is an extremely efficient inhibitor of arterial thrombosis (77). Thus, an understanding of the anticoagulant action of AT has resulted in the development of improved therapeutics.

The anti-inflammatory activity of AT opens new doors for therapeutic application, because vascular inflammation is a common denominator in numerous disease processes. On the one hand, chronic vascular inflammation plays an important role in the development of atherogenic diseases, which are the leading killers in the Western world. Although AT intravenous therapy is not appropriate for a chronic disease, the elucidation of the AT anti-inflammatory signaling pathway might lead to the development of small molecules for selectively reducing prolonged vascular inflammation. On the other hand, many clinical conditions can trigger the acute SIRS reaction and its deleterious consequences. The impact of SIRS is exemplified by sepsis, which represents the systemic host response to infection. Severe sepsis (sepsis complicated by organ failure) is associated with 30% mortality and accounts for more than 220,000 deaths per year in the United States (78). Most patients with severe sepsis have reduced circulating levels of AT, due to suppressed hepatic synthesis and enhanced consumption by coagulation and inflammatory proteases (44). Given the acute nature of sepsis, the association of sepsis with acquired AT deficiency, and the putative anti-inflammatory effects of AT, there has been tremendous interest in exploring the therapeutic potential of AT in this syndrome.

Despite promising results in animal models and phase II clinical studies, a recent phase III trial failed to achieve its primary endpoint – a survival advantage 28 days after treatment (79). However, the trial only considered AT as an anticoagulant, and so its design ignored several key features of AT's anti-inflammatory activity. First and foremost, half of the patients in the AT treatment group received concomitant heparin treatment. A benefit should not be expected for this group, because heparin inhibits AT anti-inflammatory activity (and possibly an endothelial surface anticoagulant activity) by preventing the engagement of AT with endothelial HSPGs (see Figure 107.4). Indeed, the subgroup receiving AT without heparin showed enhanced survival as a strong trend that reached significance 90 days posttreatment. Moreover, the AT subgroup receiving heparin had an enhanced risk of hemorrhage that may have contributed to their higher mortality. Enhanced bleeding should have been anticipated as supra-physiologic levels of AT were achieved, which would increase anticoagulant tone via AT's heparin-independent activity. Thus, heparin administration is both unnecessary and contraindicated when administering supra-physiologic levels of AT. Second, plasma AT levels of the treatment group were only augmented to 1.8-fold above normal values, whereas the treatment goal was to achieve a two- to 2.5-fold increase. Because AT anti-inflammatory activity is mediated by low-affinity interactions, higher AT levels should provide a greater anti-inflammatory effect. It is important to note that maximal plasma AT levels may potentially be limited by bleeding complications arising from the heparin-independent anticoagulant activity of AT. Third, the study employed purified human plasma AT, which is suboptimal. Such preparations are predominantly composed of α-AT, whereas the β-AT form exhibits much stronger HSPG binding and anti-inflammatory activity. Deployment of a pure β-AT form could additionally reduce the need for extreme AT concentrations and thereby reduce the potential risk of bleeding complications. A suitable product appears to be available. Recombinant expression of human AT in the milk of transgenic goats results in an aberrantly glycosylated AT form with a high heparin affinity, similar to β-AT. Intriguingly, this recombinant AT has demonstrated efficacy in reducing mortality of septic baboons (80); however, its efficacy relative to plasma-derived AT presently is unknown. In summary, limitations in this trial highlight factors that may be essential for correctly deploying AT as an anti-inflammatory agent in SIRS.

KEY POINTS

- AT exhibits HS-dependent and HS-independent activities, and functions as both a major natural anticoagulant and a natural anti-inflammatory.
- The normal anticoagulant activity of AT is predominantly mediated by the slow progressive reaction, thus it is independent of endothelial HS^{AT+}.
- Endothelial HSPGs are essential for mediating the anti-inflammatory activity of AT; however, it is presently unclear whether HS^{AT+} is specifically required.
- Heparin administration enhances the anticoagulant activity but inhibits the anti-inflammatory activity.
- The elucidation of AT's anticoagulant mechanism has led to improved anticoagulants.

Future Goals

- To further the understanding of AT's anti-inflammatory activity, which should provide novel treatments for diseases involving vascular inflammation

REFERENCES

1 Seegers WH. Antithrombin III: a backward glance o'er travel'd roads. *Adv Exp Med Biol.* 1975;52:195–215.
2 Blomback M, Abildgaard U, van den Besselaar AM, et al. Nomenclature of quantities and units in thrombosis and haemostasis (recommendation 1993). A collaborative project of the Scientific and Standardization Committee of the International Society on Thrombosis and Haemostasis (ISTH/SSC) and the Commission/Committee on Quantities and Units (in Clinical Chemistry) of the International Union of Pure and Applied Chemistry-International Federation of Clinical Chemistry (IUPAC-IFCC/CQU(CC)). *Thromb Haemost.* 1994;71:375–394.
3 Contejean C. Recherches sur les injections intraveineuses de peptone et leur influence sur la coagulabilite du sang chez le chien. *Arch Physiol Norm Pathol.* 1895;7:45–53.
4 McLean J. The thromboplastic action of cephalin. *Am J Physiol.* 1916;41:250–257.
5 Howell WH, Holt E. Two new factors in blood coagulation – heparin and pro-antithrombin. *Am J Physiol.* 1918;47:328–341.
6 Scott DA, Charles AF. Studies on heparin III. The purification of heparin. *J Biol Chem.* 1933;102:431–448.
7 Röden L. Highlights in the history of heparin. In: Lane DA, Lindahl U, eds. *Heparin.* London: Edward Arnold,1989:1–24.
8 Brinkhous KM, Smith HP, Warner ED, Seegers WH. The inhibition of blood clotting: an unidentified substance which acts in conjunction with heparin to prevent the conversion of prothrombin to thrombin. *Am J Physiol.* 1939;125:683–687.
9 Abildgaard U. Highly purified antithrombin III with heparin cofactor activity prepared by disc gel electrophoresis. *Scand J Clin Lab Invest.* 1968;21:89–91.
10 Rosenberg RD, Damus PS. The purification and mechanism of action of human antithrombin-heparin cofactor. *J Biol Chem.* 1973;248:6490–6505.
11 Egeberg O. Inherited antithrombin deficiency causing thrombophilia. *Thromb Diath Haemorrh.* 1965;13:516–530.
12 Bock SC, Wion KL, Vehar GA, Lawn RM. Cloning and expression of the cDNA for human antithrombin III. *Nucleic Acids Res.* 1982; 10:8113–8125.
13 Prochownik EV, Antonarakis S, Bauer KA, et al. Molecular heterogeneity of inherited antithrombin III deficiency. *N Engl J Med.* 1983;308:1549–1552.
14 Lane DA, Bayston T, Olds RJ, et al. Antithrombin mutation database: 2nd (1997) update. For the plasma coagulation inhibitors subcommittee of the scientific and standardization committee of the international society on thrombosis and haemostasis. *Thromb Haemost.* 1997;77:197–211.
15 Howell WH. The purification of heparin and its presence in blood. *Am J Physiol.* 1925;71:553–562.
16 Lam LH, Silbert JE, Rosenberg RD. The separation of active and inactive forms of heparin. *Biochem Biophys Res Commun.* 1976; 69:570–577.
17 Hook M, Bjork I, Hopwood J, Lindahl U. Anticoagulant activity of heparin: separation of high-activity and low-activity heparin species by affinity chromatography on immobilized antithrombin. *FEBS Lett.* 1976;66:90–93.
18 Andersson LO, Barrowcliffe TW, Holmer E, et al. Anticoagulant properties of heparin fractionated by affinity chromatography on matrix-bound antithrombin III and by gel filtration. *Thromb Res.* 1976;9:575–583.
19 Jorpes JE, Gardell S. On heparin monosulfuric acid. *J Biol Chem.* 1948;176:267–276.
20 Liu J, Shworak NW, Fritze LMS, et al. Purification of heparan sulfate D-glucosaminyl 3-*O*-sulfotransferase. *J Biol Chem.* 1996;271:27072–27082.
21 Shworak NW, Fritze LM, Liu J, et al. Cell-free synthesis of anticoagulant heparan sulfate reveals a limiting activity which modifies a nonlimiting precursor pool. *J Biol Chem.* 1996;271:27063–27071.
22 Shworak NW, Liu J, Fritze LMS, et al. Molecular cloning and expression of mouse and human cDNAs encoding heparan sulfate D-glucosaminyl 3-*O*-sulfotransferase. *J Biol Chem.* 1997;272: 28008–28019.
23 Shworak NW, Shirakawa M, Colliec-Jouault S, et al. Pathway-specific regulation of the synthesis of anticoagulantly active heparan sulfate. *J Biol Chem.* 1994;269:24941–24952.
24 HajMohammadi S, Enjyoji K, Princivalle M, et al. Normal levels of anticoagulant heparan sulfate are not essential for normal hemostasis. *J Clin Invest.* 2003;111:989–999.
25 Esmon CT. Role of coagulation inhibitors in inflammation. *Thromb Haemost.* 2001;86:51–56.
26 Wiedermann Ch J, Romisch J. The anti-inflammatory actions of antithrombin – a review. *Acta Med Austriaca.* 2002;29:89–92.
27 Butenas S, Mann KG. Blood coagulation. *Biochemistry Mosc.* 2002;67:3–12.
28 Mann KG. Thrombin formation. *Chest.* 2003;124:S4–S10.
29 Pauer HU, Renne T, Hemmerlein B, et al. Targeted deletion of murine coagulation factor XII gene-a model for contact phase activation in vivo. *Thromb Haemost.* 2004;92:503–508.

30 Olson ST, Bjork I, Shore JD. Kinetic characterization of heparin-catalyzed and uncatalyzed inhibition of blood coagulation proteinases by antithrombin. *Methods Enzymol.* 1993;222:525–559.
31 van 't Veer C, Mann KG. Regulation of tissue factor initiated thrombin generation by the stoichiometric inhibitors tissue factor pathway inhibitor, antithrombin-III, and heparin cofactor-II. *J Biol Chem.* 1997;272:4367–4377.
32 He L, Vicente CP, Westrick RJ, et al. Heparin cofactor II inhibits arterial thrombosis after endothelial injury. *J Clin Invest.* 2002; 109:213–219.
33 Huntington JA, Read RJ, Carrell RW. Structure of a serpin-protease complex shows inhibition by deformation. *Nature.* 2000;407:923–926.
34 Dobo J, Gettins PG. alpha1-Proteinase inhibitor forms initial non-covalent and final covalent complexes with elastase analogously to other serpin-proteinase pairs, suggesting a common mechanism of inhibition. *J Biol Chem.* 2004;279:9264–9269.
35 Irving JA, Pike RN, Lesk AM, Whisstock JC. Phylogeny of the serpin superfamily: implications of patterns of amino acid conservation for structure and function. *Genome Res.* 2000;10:1845–1864.
36 Dementiev A, Petitou M, Herbert JM, Gettins PG. The ternary complex of antithrombin-anhydrothrombin-heparin reveals the basis of inhibitor specificity. *Nat Struct Mol Biol.* 2004;11:863–867.
37 Langdown J, Johnson DJ, Baglin TP, Huntington JA. Allosteric activation of antithrombin critically depends upon hinge region extension. *J Biol Chem.* 2004;279:47288–47297.
38 Hopkins PC, Pike RN, Stone SR. Evolution of serpin specificity: cooperative interactions in the reactive-site loop sequence of antithrombin specifically restrict the inhibition of activated protein C. *J Mol Evol.* 2000;51:507–515.
39 Olson ST, Chuang YJ. Heparin activates antithrombin anticoagulant function by generating new interaction sites (exosites) for blood clotting proteinases. *Trends Cardiovasc Med.* 2002;12:331–338.
40 O'Reilly MS, Pirie-Shepherd S, Lane WS, Folkman J. Antiangiogenic activity of the cleaved conformation of the serpin antithrombin. *Science.* 1999;285:1926–1928.
41 Turk B, Brieditis I, Bock SC, et al. The oligosaccharide side chain on Asn-135 of alpha-antithrombin, absent in beta-antithrombin, decreases the heparin affinity of the inhibitor by affecting the heparin-induced conformational change. *Biochemistry.* 1997;36:6682–6691.
42 Witmer MR, Hadcock SJ, Peltier SL, et al. Altered levels of antithrombin III and fibrinogen in the aortic wall of the alloxan-induced diabetic rabbit: evidence of a prothrombotic state. *J Lab Clin Med.* 1992;119:221–230.
43 Amrani DL, Rosenberg J, Samad F, et al. Developmental expression of chicken antithrombin III is regulated by increased RNA abundance and intracellular processing. *Biochim Biophys Acta.* 1993;1171:239–246.
44 Fourrier F, Jourdain M, Tournois A, et al. Coagulation inhibitor substitution during sepsis. *Intensive Care Med.* 1995;21(2 Suppl): S264–S268.
45 Niessen RW, Rezaee F, Reitsma PH, et al. Ligand-dependent enhancement of human antithrombin gene expression by retinoid X receptor alpha and thyroid hormone receptor beta. *Biochem J.* 1996;318(Pt 1):263–270.
46 Ghose R, Zimmerman TL, Thevananther S, Karpen SJ. Endotoxin leads to rapid subcellular re-localization of hepatic RXRalpha: a novel mechanism for reduced hepatic gene expression in inflammation. *Nucl Recept.* 2004;2:4.
47 Niessen RW, Lamping RJ, Jansen PM, et al. Antithrombin acts as a negative acute phase protein as established with studies on HepG2 cells and in baboons. *Thromb Haemost.* 1997;78:1088–1092.
48 Tejada ML, Deeley RG. Cloning of an avian antithrombin: developmental and hormonal regulation of expression. *Thromb Haemost.* 1995;73:654–661.
49 Ishiguro K, Kojima T, Kadomatsu K, et al. Complete antithrombin deficiency in mice results in embryonic lethality. *J Clin Invest.* 2000;106:873–878.
50 Carrell RW, Huntington JA, Mushunje A, Zhou A. The conformational basis of thrombosis. *Thromb Haemost.* 2001;86:14–22.
51 Yabe T, Shukla D, Spear PG, et al. Portable sulphotransferase domain determines sequence specificity of heparan sulphate 3-*O*-sulphotransferases. *Biochem J.* 2001;359:235–241.
52 Shworak NW, HajMohammadi S, de Agostini AI, Rosenberg RD. Mice deficient in heparan sulfate 3-O-sulfotransferase-1: normal hemostasis with unexpected perinatal phenotypes. *Glycoconj J.* 2002;19:355–361.
53 Girardin EP, Hajmohammadi S, Birmele B, et al. Synthesis of anticoagulantly active heparan sulfate proteoglycans by glomerular epithelial cells involves multiple 3-O-sulfotransferase isoforms and a limiting precursor pool. *J Biol Chem.* 2005;280 (45):38059–38070. Epub 2005 Aug 17.
54 Li ZY, Hirayoshi K, Suzuki Y. Expression of N-deacetylase/sulfotransferase and 3-O-sulfotransferase in rat alveolar type II cells. *Am J Physiol Lung Cell Mol Physiol.* 2000;279:L292–L301.
55 Hosseini G, Liu J, de Agostini AI. Characterization and hormonal modulation of anticoagulant heparan sulfate proteoglycans synthesized by rat ovarian granulosa cells. *J Biol Chem.* 1996; 271:22090–22099.
56 Stevens RL, Colombo M, Gonzales JJ, et al. The glycosaminoglycans of the human artery and their changes in atherosclerosis. *J Clin Invest.* 1976;58:470–481.
57 Nishinaga M, Ozawa T, Shimada K. Homocysteine, a thrombogenic agent, suppresses anticoagulant heparan sulfate expression in cultured porcine aortic endothelial cells. *J Clin Invest.* 1993; 92:1381–1386.
58 Paka L, Kako Y, Obunike JC, Pillarisetti S. Apolipoprotein E containing high density lipoprotein stimulates endothelial production of heparan sulfate rich in biologically active heparin-like domains. A potential mechanism for the anti-atherogenic actions of vascular apolipoprotein E. *J Biol Chem.* 1999;274:4816–4823.
59 Irving JA, Steenbakkers PJ, Lesk AM, et al. Serpins in prokaryotes. *Mol Biol Evol.* 2002;19:1881–1890.
60 Jiang Y, Doolittle RF. The evolution of vertebrate blood coagulation as viewed from a comparison of puffer fish and sea squirt genomes. *Proc Natl Acad Sci USA.* 2003;100:7527–7532.
61 Medeiros GF, Mendes A, Castro RA, et al. Distribution of sulfated glycosaminoglycans in the animal kingdom: widespread occurrence of heparin-like compounds in invertebrates. *Biochim Biophys Acta.* 2000;475:287–294.
62 Kamimura K, Rhodes JM, Ueda R, et al. Regulation of notch signaling by *Drosophila* heparan sulfate 3-O sulfotransferase. *J Cell Biol.* 2004;166:1069–1079.
63 Niessen RW, Lamping RJ, Peters M, et al. Fetal and neonatal development of antithrombin III plasma activity and liver messenger RNA levels in sheep. *Pediatr Res.* 1996;39:685–691.

64 Xiao H, Miller SJ, Bang NU, Faulk WP. Protein-bound heparin/heparan sulfates in human adult and umbilical cord plasma. *Haemostasis*. 1999;29:237–246.
65 Hockin MF, Jones KC, Everse SJ, Mann KG. A model for the stoichiometric regulation of blood coagulation. *J Biol Chem*. 2002;277:18322–18333.
66 Monroe DM, Roberts HR, Hoffman M. Platelet procoagulant complex assembly in a tissue factor-initiated system. *Br J Haematol*. 1994;88:364–371.
67 Dewerchin M, Herault JP, Wallays G, et al. Life-threatening thrombosis in mice with targeted Arg48-to-Cys mutation of the heparin-binding domain of antithrombin. *Circ Res*. 2003;93: 1120–1126.
68 Scully MF, Ellis V, Kakkar VV. Heparan sulphate with no affinity for antithrombin III and the control of haemostasis. *FEBS Lett*. 1988;241:11–14.
69 Bourin MC, Lundgren-Akerlund E, Lindahl U. Isolation and characterization of the glycosaminoglycan component of rabbit thrombomodulin proteoglycan. *J Biol Chem*. 1990;265:15424–15431.
70 Yamauchi T, Umeda F, Inoguchi T, Nawata H. Antithrombin III stimulates prostacyclin production by cultured aortic endothelial cells. *Biochem Biophys Res Commun*. 1989;163:1404–1411.
71 Uchiba M, Okajima K, Murakami K, et al. Effects of antithrombin III (AT III) and Trp49-modified AT III on plasma level of 6-keto-PGF1 alpha in rats. *Thromb Res*. 1995;80:201–208.
72 Oelschlager C, Romisch J, Staubitz A, et al. Antithrombin III inhibits nuclear factor kappaB activation in human monocytes and vascular endothelial cells. *Blood*. 2002;99:4015–4020.
73 Isobe H, Okajima K, Uchiba M, et al. Antithrombin prevents endotoxin-induced hypotension by inhibiting the induction of nitric oxide synthase in rats. *Blood*. 2002;99:1638–1645.
74 Tkachenko E, Rhodes JM, Simons M. Syndecans: new kids on the signaling block. *Circ Res*. 2005;96:488–500.
75 Hull RD, Pineo GF. Heparin and low-molecular-weight heparin therapy for venous thromboembolism: will unfractionated heparin survive? *Semin Thromb Hemost*. 2004;30(1 Suppl):11–23.
76 Weitz JI. New anticoagulants for treatment of venous thromboembolism. *Circulation*. 2004;110:I19–I26.
77 Bal dit Sollier C, Kang C, Berge N, et al. Activity of a synthetic hexadecasaccharide (SanOrg123781A) in a pig model of arterial thrombosis. *J Thromb Haemost*. 2004;2:925–930.
78 Balk RA. Optimum treatment of severe sepsis and septic shock: evidence in support of the recommendations. *Dis Mon*. 2004;50: 168–213.
79 Warren BL, Eid A, Singer P, et al. Caring for the critically ill patient. High-dose antithrombin III in severe sepsis: a randomized controlled trial. *JAMA*. 2001;286:1869–1878.
80 Minnema MC, Chang AC, Jansen PM, et al. Recombinant human antithrombin III improves survival and attenuates inflammatory responses in baboons lethally challenged with *Escherichia coli*. *Blood*. 2000;95:1117–1123.
81 Yanada M, Kojima T, Ishiguro K, et al. Impact of antithrombin deficiency in thrombogenesis: lipopolysaccharide and stress-induced thrombus formation in heterozygous antithrombin-deficient mice. *Blood*. 2002;99:2455–2458.
82 Dewerchin M, Van der Elst L, Singh I, et al. Inhibition of factor VIII with a partially inhibitory human recombinant monoclonal antibody prevents thrombotic events in a transgenic model of type II HBS antithrombin deficiency in mice. *J Thromb Haemost*. 2004;2:77–84.

108

Protein C

Marlies Van de Wouwer and Edward M. Conway

Department for Transgene Technology and Gene Therapy, VIB, and the Center for Transgene Technology and Gene Therapy (CTG), K.U. Leuven, Belgium

Protein C (PC) is a key component in a multiprotein, physiologically relevant pathway that regulates thrombin's procoagulant activity, but which also limits inflammatory responses and provides the vascular endothelium with protection against an array of insults. Delineation of the biochemistry and molecular biology underlying regulation of the PC pathway, complemented by preclinical and clinical studies, has led to major new insights bridging the gap between inflammation and coagulation and resulting in innovative approaches to treat thrombotic disease and sepsis.

DISCOVERY OF PROTEIN C AND A NOVEL NATURAL ANTICOAGULANT PATHWAY

In 1975, Johan Stenflo reported the discovery of a novel vitamin K–dependent factor that revolutionized our view of the molecular mechanisms by which the coagulation system is regulated and provided insights into how coagulation interfaces with other biological systems, including inflammation and cell survival. In the final steps of his strategy to isolate new vitamin K–dependent plasma proteins, Stenflo identified four peaks from a chromatography column, the third of which – "pool C" – was subsequently referred to as *protein C* (1). Shortly thereafter, Esmon and colleagues (2) recognized that PC has lipid-binding properties and is a zymogen precursor for a serine protease – so-called activated PC (APC). The importance of PC in coagulation rapidly became evident when it was determined that APC is derived from the thrombin-mediated cleavage of PC, that APC proteolytically destroys procoagulant cofactors V/Va and VIII/VIIIa, and finally that familial deficiency of PC in humans results in a hypercoagulable disorder (3–5).

STRUCTURAL FEATURES OF PROTEIN C

PC is synthesized predominantly in the liver as a 461-amino acid, long single-chain precursor, approximately 75% of which is subsequently processed prior to and during secretion to yield a two-chain species. Thus, both single- and two-chain species circulate in the plasma at a total concentration of 3 to 5 μg/mL (6). At least two other forms of PC, with differing glycosylation patterns, also circulate, but in much smaller amounts. The mature protein has an apparent molecular weight of approximately 62 kDa, and is composed of three well-defined structural domains including (a) the γ-carboxyglutamic acid (Gla) residue-containing amino-terminal domain, (b) an epidermal growth factor (EGF)-like containing domain with two EGF-like repeats, and (c) the serine protease (SP) domain (Figure 108.1). The first two domains encompass the light chain; the third makes up the heavy chain. The Gla-EGF-EGF-SP domain structure is characteristic of other serine proteases in the coagulation system, including factors VII, IX, and X.

The amino-terminus of human PC has nine Gla residues that are formed by vitamin K–dependent carboxylation of the glutamic acid residues in that region. These are crucial for calcium binding and the protein's anticoagulant function. The Gla residues mediate PC–APC interactions with negatively charged phospholipid surfaces (e.g., the surface of activated platelets and/or endothelial cells [ECs]), as well as with the endothelial PC receptor (EPCR), thereby strategically localizing the activity of PC–APC, and coordinating its interaction with thrombin and its cofactor, thrombomodulin (TM) (see the next section). The first of the two EGF-like repeats has a site for β-hydroxylation that is involved in calcium binding, whereas the second EGF-like repeat has a site for N-linked glycosylation that is important for the secretion of the molecule. The EGF-containing domain also may promote interactions with protein S (PS), factor Va, and factor VIIIa. Finally, the SP domain of PC is similar to that of other serine proteases, with a conserved catalytic triad of His^{211}, Asp^{257}, and Ser^{360}, but which uniquely has specificity for targeting factors Va and VIIIa for degradation. Also within the SP domain is a calcium-binding site that is required for calcium-dependent activation of PC by thrombin-TM.

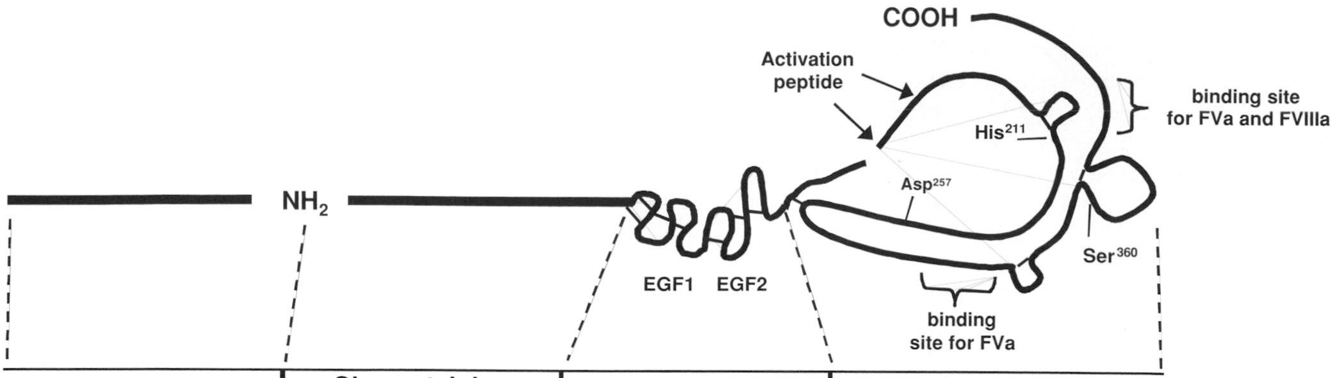

Leader-peptide	Gla-containing domain	2 EGF-like repeats	Serine protease domain
	□ 9 gla residues □ calcium-dependent interactions with cell membrane surfaces and with EPCR and TM	□ EGF1 for calcium binding—contains β-hydroxy aspartic acid residue □ EGF2 with site for N-linked glycosylation; for secretion; facilitates interactions with PS, factors Va and VIIIa	□ binding sites for factors Va and VIIIa □ calcium-dependent activation of PC by thrombin-TM □ contains His-Asp-Ser catalytic triad

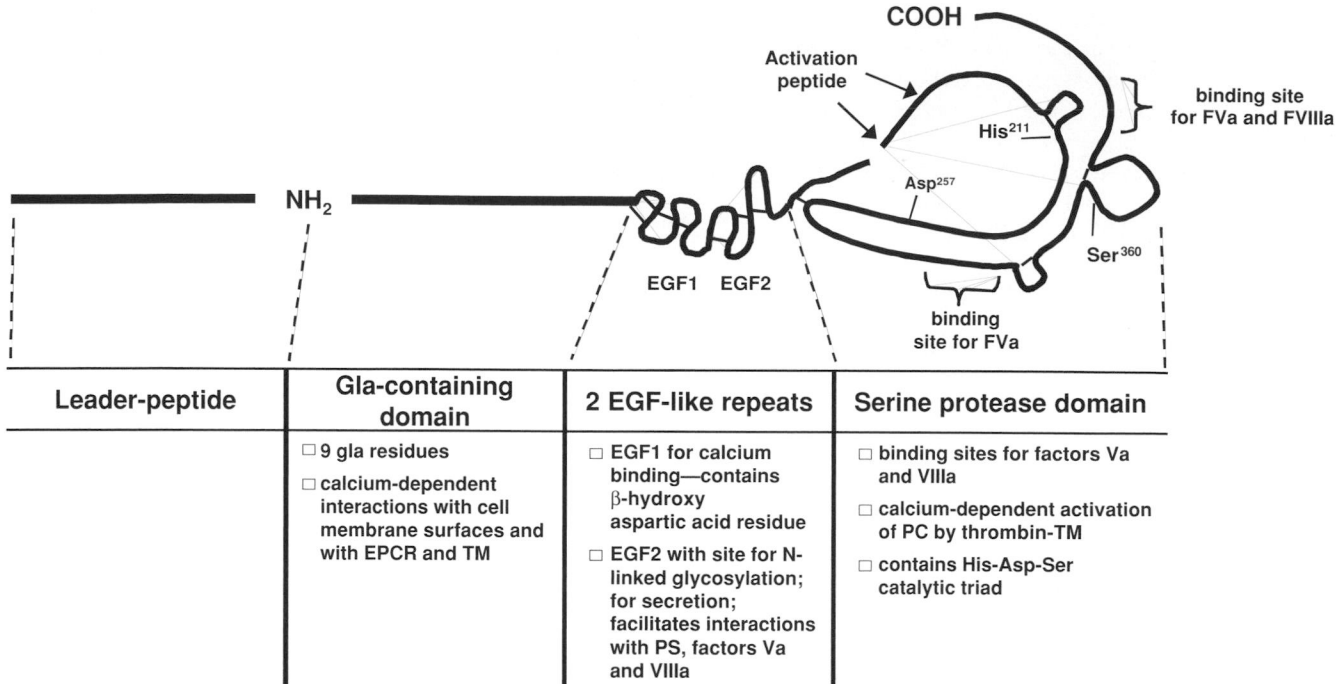

Figure 108.1. Structure of Protein C. This schematic drawing of PC shows the leader peptide, followed by three structural domains that comprise the mature protein.

PROTEIN C – THE GENE

The human and murine *PC* genes are structurally similar, each spanning approximately 15 kb, and comprising nine exons and eight introns (7,8). In humans, the gene is located on the long arm of chromosome 2. The 53-nucleotide first exon is untranslated and is followed by a 1.3-kb intron. The sequence encoding the translation initiation methionine is located in exon 2. The 5′-untranslated region contains CCAAT sequences and GC boxes, as well as several putative recognition sequences for transcription factor interactions that are important for liver-specific expression.

EVOLUTIONARY CONSIDERATIONS

Comparative studies support the notion that PC evolved prior to the divergence of teleosts over 430 million years ago, and that the vertebrate coagulation system arose during gene duplications prior to development of jawed vertebrates (9). Thus, evolutionary appearance of the Gla-EGF-EGF-SP domain structure common to factors VII, IX, X, and PC is evident in zebrafish (*Danio rerio*), fugu (*Takifugu* sp.), and chicken genomes, in which orthologues for the *PC* gene have been identified (9). It is not known whether *PC* appeared prior to or after the genes encoding the procoagulant factors. At least in the zebrafish, a functional coagulation system exists, with evidence that PC activation occurs and that the Gla domains are crucial for hemostatic integrity (10). Whether PC has additional properties in zebrafish or these other species has not yet been determined.

ACTIVATION OF PC

The major components in the PC system that promote its activation – a critical step for its physiologic function – are the transmembrane glycoprotein receptor, TM, the membrane bound EPCR, thrombin, and platelet factor 4 (PF4). TM is widely expressed by ECs in all vascular beds and in vessels of all caliber, whereas EPCR is more prominently expressed by larger-vessel ECs. When complexed with TM on the cell membrane surface, thrombin catalyzes activation of PC more than 1,000-fold (compared with activation by thrombin in the absence of TM), cleaving a single bond at the amino-terminal end of the heavy chain, with release of an activation dodecapeptide. EPCR, which binds to PC and APC with equal affinity, enhances the activation event approximately 20-fold by presenting PC to the thrombin–TM complex. Activation of PC also is accelerated up to 25-fold by PF4 (also known as CXCL4), which is secreted by activated platelets and binds to the Gla domain of PC, forming a complex with thrombin-TM and thereby altering the conformation and affinity of PC for thrombin-TM (11).

The activation of PC is exquisitely regulated. Transformation of PC to APC requires renewed sources of thrombin to be present, because once thrombin binds to TM, the serine protease inhibitors (serpins) anti-thrombin (AT) and PC inhibitor

(PCI) rapidly neutralize the thrombin – more efficiently than when thrombin is free. Thus, once thrombin generation is complete, activation of PC ceases. In contrast to the rapid-on, rapid-off nature of PC activation, neutralization of existing APC proceeds much more slowly. The half-life of APC in the circulation – approximately 20 minutes – is relatively long when compared to most serine proteases, and reflects its slow inhibition by PCI, α-2-macroglobulin and α-1-antitrypsin, a response that may provide for a more widely distributed and prolonged functional APC in the organism.

A CRITICAL ROLE FOR ACTIVATED PROTEIN C IN REGULATING COAGULATION

The earliest described and best characterized function of the PC pathway is centered around the natural anticoagulant properties of APC. As APC is generated, it competes with PC for binding to EPCR. When dissociated from EPCR, APC binds to the vitamin K–dependent plasma PS on the phospholipid surface of activated platelets and ECs (and probably other cell surfaces). There, it is positioned to proteolytically destroy and inhibit coagulation cofactors VIIIa and Va, thereby interfering with formation of the tenase and prothrombinase complexes, respectively, and suppressing further generation of factor Xa and thrombin. Degradation of the zymogens, factors V and VIII, occurs, but less efficiently. Inactivation of factor Va is enhanced by PS, whereas cleavage of factor VIIIa requires both PS and factor V (12) (Figure 108.2A). When bound to EPCR, APC is unable to inactivate factors Va/VIIIa but, in this form, transmits anti-inflammatory and vasculoprotective signals (Figure 108.2B and 108.2C; and see below). Thus, the PC-TM-EPCR system effectively switches the substrate specificity and functional properties of thrombin from a procoagulant/pro-inflammatory to an anticoagulant/anti-inflammatory molecule. Analogous to TM's switching of the substrate specificity of thrombin, EPCR may be considered a molecular switch for APC in determining whether the latter will function as an anticoagulant or an anti-inflammatory molecule. APC also promotes fibrinolysis by neutralizing plasminogen activator inhibitor (PAI)-1, an interaction that is enhanced by vitronectin, an abundant plasma and platelet glycoprotein (13) released at sites of injury. This results in local, uninhibited generation of plasmin, which may serve to regulate fibrin clot size and/or to promote clearing of cross-linked fibrin during and after injury.

ACTIVATED PROTEIN C IS A NATURAL ANTI-INFLAMMATORY MOLECULE

The ability of APC to downregulate thrombin generation and to suppress inflammation was most clearly delineated in a series of elegant in vivo studies in baboons by Taylor, Esmon, and coworkers (14,15). Treatment of baboons with APC, PC, or PS protected against the inflammatory manifestations of *Escherichia coli* endotoxin–induced sepsis; whereas blocking the activity of these proteins or EPCR exacerbated the response, promoting cytokine release, neutrophil infiltration into tissues, capillary leak syndrome, and activation of coagulation, with resulting organ failure and death (16,17). These seminal investigations were followed by many similar supportive studies. For example, heterozygous PC–deficient mice are more sensitive to polymicrobial and gram-negative bacterial sepsis, exhibiting shortened survival, higher levels of tumor necrosis factor (TNF)-α, interleukin (IL)-1β, and IL-6, with more prominent organ damage and vascular hypotension (18). In rodent models of endotoxemia, APC blunts TNF-α production and inducible nitric oxide synthase (iNOS) activity, suppresses neutrophil infiltration, and prevents hypotension and tissue damage by coagulation-independent pathways (see Figure 108.2B). Similarly, neutrophil activation and extravasation is abrogated by APC in models of ischemia–reperfusion injury, thus limiting the extent of tissue damage (19).

MECHANISMS BY WHICH ACTIVATED PROTEIN C SUPPRESSES INFLAMMATION

The profound effects of APC on the inflammatory system appear to be mediated via several mechanisms. APC inhibits the expression of tissue factor (TF) and the release of several cytokines (e.g., TNF-α, IL-1β, IL-6, IL-8, and macrophage inflammatory protein [MIP]-1α) by monocytes/macrophages (20,21) (Table 108-1). APC may promote these anti-inflammatory responses by interfering with nuclear translocation of nuclear factor (NF)-κB and activation of AP-1, which moreover prevents upregulation of cell surface leukocyte adhesion molecules and leukocyte–EC interactions. Both PC and APC inhibit neutrophil chemotaxis, but interestingly, have no apparent effects on respiratory burst or bacterial phagocytosis, suggesting specificity and limits to their actions.

EPCR is believed to play a major role in mediating many of the anti-inflammatory effects of APC, although the precise molecular mechanisms and physiologic importance is still not clear. EPCR mainly is expressed on the surface of ECs, but also is found on circulating monocytes, neutrophils, and eosinophils, and on the dendritic cells of the inflamed gut (22). Thus, it is well situated to modulate inflammation and innate immune responses. A soluble form – sEPCR – is released into the circulation during inflammation, a product of matrix metalloproteinase (MMP) cleavage of the mature protein. sEPCR binds to activated neutrophils via proteinase 3 (PR3) which in turn interacts with CD11/CD18 (Mac-1), an integrin that normally mediates leukocyte–EC interactions. sEPCR likely interferes with this interaction, preventing leukocyte adhesion and extravasation (see Figure 108.2B). Binding of APC to EPCR – membrane bound or soluble – on the leukocyte surface also may reduce intracellular calcium fluxes and prevent NF-κB translocation and the activation of downstream proinflammatory events. Moreover, EPCR–APC complexes can be

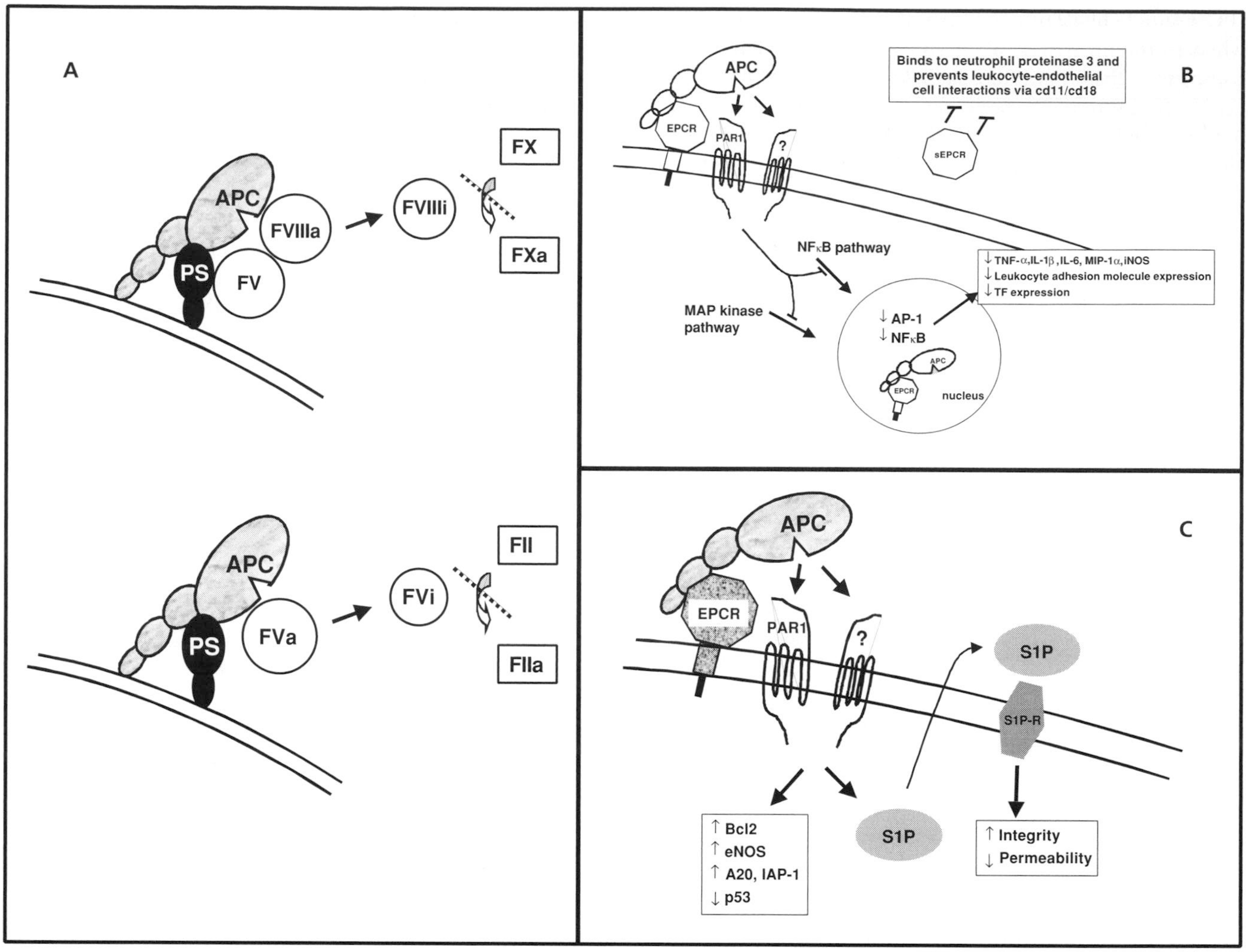

Figure 108.2. Diverse functions of APC. EPCR is a natural switch that regulates the function of APC. When free of EPCR, APC is an anticoagulant; when bound to EPCR, APC exhibits anti-inflammatory and antiapoptotic properties. (**A**) APC is an important natural anticoagulant. After a procoagulant stimulus, APC that is generated and free from EPCR localizes with PS on platelet and EC surfaces, where it cleaves and inactivates factor VIIIa with the help of factor V (*top*) and factor Va (*bottom*) to yield the inactive species FVIIIi and FVi, respectively. Lack of the active cofactors (VIIIa and Va) results in ineffective transformation of factor X to factor Xa (FXa) (*top*), and prothrombin (II) to thrombin (IIa) (*bottom*), respectively. (**B**) APC is an anti-inflammatory molecule, achieving this end via several pathways described in the text. When bound to EPCR, APC activates PAR-1 and/or other cell surface receptor(s) to induce intracellular signals that interfere with NFκB and MAPK pathways, thereby preventing transcriptional upregulation of proinflammatory mediators, such as TNF-α and IL-1β; suppressing the expression of leukocyte adhesion molecules; and downregulating tissue factor (TF) expression. APC-EPCR complexes also may be internalized and localize to the nucleus, where they may further sustain transcriptional events to prevent inflammation. Finally, soluble EPCR (sEPCR) is released during inflammation, whereupon it binds to neutrophil proteinase 3 (PR3), thereby preventing leukocyte-endothelial interactions via CD11/CD18. (**C**) When bound to EPCR, APC also has natural vasculoprotective, antiapoptotic properties. Through PAR-1 and/or other receptor(s), APC transmits intracellular signals that augment the expression of antiapoptotic proteins, decrease p53 expression, and promote release of sphingosine 1-phosphate (S1P) that then activates S1P-receptor (S1P-R), which in turn decreases vascular EC permeability and enhances vascular integrity.

internalized and translocated to the nucleus, where they may further suppress the inflammatory response, thereby sustaining the locally protective process (Figure 108.2B). EPCR expressed by ECs also may act as a cofactor in the cleavage of protease-activated receptor (PAR)-1 by APC (23). Although PAR-1 triggers events such as the release of chemokines or the upregulation of adhesion molecules that would promote inflammation, PAR-1 also induces the expression of inhibitors of apoptosis, such as inhibitor of apoptosis protein (IAP)-1 and A20, that would protect the vascular endothelium in the face of injury, ischemia, or inflammation (Figure 108.2C). Moreover, PAR-1 activation by APC may lead to sphingosine-1-phosphate (S1P) receptor-1 cross-activation that is selectively vasculoprotective (24). Several studies have concluded that, by this route, APC induces a transcriptional response in ECs that may explain many of its effects in sepsis.

Table 108-1: Diverse Functions of APC and EPCR

Functions		*Mechanism*
Anticoagulant		APC cleaves and inactivates procoagulant factors Va and VIIIa
Profibrinolytic		APC binds to and neutralizes PAI-1 – accelerated by vitronectin
Anti-inflammatory	APC	↓ TNFα, IL-1β, IL-6, IL-8, MIP-1 ↓ iNOS activity ↓ TF expression by monocytes ↓ endothelial expression of leukocyte adhesion molecules ↓ neutrophil infiltration ↓ EC permeability via S1P cross-talk
	sEPCR	Binds to neutrophil PR3 and prevents interaction with ECs APC increases Bcl2, eNOS, A20, IAP in different cell types
Antiapoptotic		APC suppresses p53 APC enhances EC integrity via S1P cross-talk

Despite these reports, experimental evidence exists that challenges the in vivo physiologic relevance of an APC–PAR-1 interaction, and whether PAR-1 is activated at all by endogenous APC (25). This raises the possibility that other PC/APC receptors exist, consistent with an early report by Hancock and colleagues (26), and implies that alternative, non–PAR-1 mediated pathways protect the vasculature and other cells from injury. Indeed, APC induces antiapoptotic signals in ECs, renal tubular cells, and neurons, partly by inhibiting p53, but also via recruitment of Bcl-2 pathways, possibly independent of PAR-1 (27).

FUNCTIONAL DEFICIENCIES OF PROTEIN C

The physiologic importance of the PC system is evident from the clinically significant hypercoagulable conditions associated with the impaired function of PC or PS (5,28). Over 100 mutations – mostly within exons, but also involving the $5'$-untranslated region of the gene – result in quantitative defects (type I) (http://www.xs4all.nl/~reitsma/Prot_C_home.htm). Over 70% of these are missense mutations. Fewer pedigrees (~20) of familial mutations yield a dysfunctional PC with normal antigen levels (type II). Deletions within the *PC* gene are uncommon. Compound heterozygotes, and simultaneous occurrence of PC mutations with other molecular defects involving the coagulation system (e.g., AT deficiency), have been reported. These tend to result in a more severe thrombotic phenotype. Heterogeneity in the development of thrombosis, even within families with the same mutation, supports the notion that the clinical expression is multifactorial, modified by environmental and genetic factors (28). Patients with heterozygous PC deficiency often have few symptoms and, when they do, may only present with thrombotic episodes as adults – most commonly with deep vein thromboses and/or pulmonary emboli – spontaneously or associated with a prothrombotic stress, such as pregnancy, preeclampsia, trauma, infection, or surgery. However, thromboses also may occur in unusual locations, not infrequently involving the arterial tree (29,30). It is interesting to note that, despite the increased sensitivity of PC-deficient mice to microbial infections, little evidence indicates that humans with PC deficiency suffer a similar enhanced risk.

Homozygous PC deficiency inevitably presents during the neonatal period, uniformly associated with life-threatening spontaneous thrombosis of the large veins (e.g., deep veins of legs, pulmonary, renal, vena cava, portal, cerebral). It may however, also involve the arterial tree, thus presenting with stroke, myocardial infarction, and/or limb ischemia (31). It also may be manifest within hours of birth by *purpura fulminans*, characterized by diffuse microvascular thrombotic lesions in the skin and rapid onset of disseminated intravascular coagulation (DIC) – a syndrome that is always fatal unless treated with PC concentrates or plasma. The rarity of homozygous PC deficiency, relative to the heterozygous state in the general population, suggests that homozygosity frequently results in nonviable fetuses and spontaneous abortions.

Pathologically similar to purpura fulminans, but usually less extensive, is warfarin-induced skin necrosis, most frequently observed at the initiation of oral warfarin-based anticoagulation, and caused by a rapid decline in PC levels relative to the procoagulant vitamin K–dependent factors IX and X. This syndrome is more prevalent in patients with underlying heterozygous PC deficiency, but also may occur with PS deficiency. Skin necrosis also is observed in some patients with severe sepsis, in whom PC levels are suppressed due to compromised liver function and increased consumption, and where the capacity to generate APC is decreased because of suppressed endothelial expression of TM and EPCR. The striking involvement of the skin microvasculature in severe PC deficiency is not explained, but is strong evidence for the importance of APC as an anticoagulant, and possibly an anti-inflammatory factor in the dermal microcirculation.

RESISTANCE TO ACTIVATED PROTEIN C

The steps leading to the exciting discovery of APC resistance, also known as Factor V_{Leiden}, were superbly related by Dahlback (32). The APC resistance phenotype is overwhelming evidence for the clinical importance of the PC system in protecting against thromboembolic disease. Factor V is a 330-kDa plasma glycoprotein that is activated through limited proteolysis at Arg306, Arg506, and Arg679 by thrombin or factor Xa, whereupon the active form – factor Va – is an essential cofactor for factor Xa mediated transformation of prothrombin to thrombin. Most factor V mutations result in quantitative factor V deficiency with abnormal bleeding. However, many patients with venous thrombosis have a missense mutation in factor V, in which Arg506 is replaced with Gln (33). This mutant factor V is resistant to APC cleavage at that site, thereby impairing not only suppression of the procoagulant activity of factor Va, but also the anticoagulant cofactor activity of factor V required for the efficient inactivation of factor VIIIa by APC-PS. The frequency of the mutation ranges from 2% to 15%, with geographical variability – more frequent in northern European latitudes – that speaks to its origin. It is thus the most common autosomally inherited hypercoagulable disorder described. Heterozygosity confers a five- to 10-fold increased risk of thrombosis, and this increases to 50- to 100-fold for homozygous patients. It is interesting to note that APC resistance may be beneficial. For example, it is associated with less bleeding in the peripartum period, as well as in hemophilia A and B. APC resistance also reduces sepsis-induced mortality, probably via the anti-inflammatory and vascular endothelial protective activities of APC, the level of which is elevated as a consequence of excess thrombin generation. Other factor V mutations have been identified that involve APC cleavage sites. For example, Arg306Thr or Arg306Gly mutations eliminate APC cleavage at Arg306, but do not apparently increase the risk of thrombosis. On the other hand, a thrombotic pedigree revealed an Ile359Thr missense mutation that resulted in diminished APC-mediated cleavage at Arg306 and lower APC cofactor-induced inactivation of factor VIIIa, similar to that seen with Factor V_{Leiden}.

REGULATING PROTEIN C EXPRESSION

Although the liver is the primary site of production of PC, extrahepatic sites have been identified, consistent with the wide expression pattern of other components of the PC system, including PS, EPCR, and TM (34). For example, PC is readily detected in tubular epithelial cells of the kidney, in the prostate, in spermatogenic cells of the testis, and in epithelial cells of the epididymis. PC also is expressed at low levels in bronchial epithelial cells, neurons, and Purkinje cells in the brain, and in cultured vascular ECs (34,35). Immediate availability of PC at these selected sites implies that it may have unique tissue-specific biologic functions, or alternatively, that sanctuaries such as the testis or brain require an additional supply over and above the normal liver-derived source.

Not surprisingly, the regulation of PC expression is, to a large extent, coordinated with that of TM and EPCR. In animal models of DIC and inflammation, PC levels are diminished partly due to consumption, but expression is further down-regulated in the liver, kidney, and testis via transcriptional mechanisms, mediated by cytokines, such TNF-α and IL-1β (34). In models of lupus nephritis or diabetic nephropathy, fibrin deposition and inflammation is exacerbated as local and systemic PC levels decrease. In patients with severe sepsis, PC levels also are reduced, and these are inversely correlated with survival.

Table 108-2: Transgenic Mice

Genotype Identification	*Phenotype (Refs.)*
$PC^{-/-}$	*PC* gene inactivation results in mice that die soon after birth from severe consumptive coagulopathy, with increased fibrin deposition as early as E12.5 (36). With decrease in maternal source of PC, total lack of PC causes embryonic lethality (37).
$PC^{+/-}$	Mice with heterozygous deficiency of PC do not exhibit spontaneous thromboses, but have enhanced sensitivity to endotoxin, responding with more severe consumptive coagulopathy, increased fibrin deposition, augmented inflammation, vascular collapse, and organ damage (18).

ROLE OF PROTEIN C IN DEVELOPMENT

Inactivation of the *PC* gene in mice resulted in a phenotype similar to that seen in humans (36), with death perinatally due to diffuse thrombosis and severe consumptive coagulopathy, particularly prominent in the brain and liver, with evidence of increased fibrin deposition in these organs evident as early as embryonic day (E)12.5 (Table 108-2). Interestingly, the *PC* "knockout" picture was strikingly different from the *TM* gene "knockout," the latter of which caused embryonic lethality by E8.5 to E9.5. This difference has indeed recently been shown to reflect a maternal transfer of PC in the *PC* knockout embryos that provided some degree of protection (37). Using novel transgenic strategies, mice with PC levels of 1% to 18% of normal have been generated; these survive beyond birth and display prothrombotic and proinflammatory phenotypes, the severity of which depends on the level of PC (37). Mice with heterozygous deficiency of PC do not develop spontaneous thromboses. But when exposed to polymicrobial sepsis, consumption of coagulation factors was more severe in the heterozygous mice, as compared with their wild-type counterparts, and they exhibited higher serum cytokine levels, more extensive tissue damage, hypotension, and earlier mortality, highlighting the protective role of PC in coagulation, inflammation, and sepsis (18,38).

THERAPEUTIC CONSIDERATIONS

Activated Protein C and Protein C in Sepsis

Targeting either the coagulation system or the inflammatory system to decrease the morbidity and mortality of patients with sepsis has been uniformly unsuccessful. However, the strategy of simultaneously interfering with both systems is supported by several preclinical trials. For unclear reasons, only APC has shown benefit in phase III clinical trials. In a large, double-blind, prospective study by Bernard and colleagues (39), the administration of recombinant human APC improved survival of critically ill patients with severe sepsis, with a 6.1% reduction in 28-day all-cause mortality, that was partly sustained at 90 days and 2.5 years (40). Response to APC was linked to a more rapid resolution of cardiovascular, respiratory, and hematopoietic system failure. The only adverse effect of APC was a statistically nonsignificant increase in serious bleeding (3.5% vs 2.0%), primarily in those with low platelet counts. Notably, the most prominent effects of APC in these trials appeared to be more related to correction of systemic coagulation parameters, rather than those of inflammation. Thus, APC reduced thrombin-antithrombin TAT complexes and F1.2 levels, but had nonsignificant effects on serum levels of TNF-α, IL-1β, IL-6, and IL-10. These markers, of course, may not reflect the local effects of APC on EC activation. Moreover, the protection by APC against sepsis-related mortality and organ dysfunction might be via antiapoptotic pathways and/or by interfering with vascular collapse, which are not directly measurable in these assays.

Can PC, rather than APC, be used to treat sepsis? This is a controversial, unresolved issue. Theoretically, administered PC may not be effectively transformed to APC during sepsis, since TM and EPCR levels are functionally and quantitatively suppressed in this setting. However, case studies indicate that even in meningococcal sepsis and purpura fulminans, in which cell-surface TM has been documented to be decreased, treatment with PC often is beneficial and associated with increased generation of APC. Overall, the efficacy of PC has not been formally tested in appropriate clinical trials and, moreover, direct comparisons with APC are lacking.

Protein C and Activated Protein C in Other Disorders

The most obvious use of PC/APC is as an anticoagulant for thromboembolic disease. Infants with homozygous PC deficiency and purpura fulminans benefit from PC infusions. Otherwise, largely due to the availability of many other effective anticoagulants, PC/APC has not been widely applied for the treatment of thromboembolic disease, nor has it been fully evaluated.

Although not approved for these uses, preclinical data suggest that APC is safe and efficacious for a variety of common, clinically important disorders. For example, APC decreases tissue damage associated with ischemia–reperfusion injury of the spinal cord, heart, gut, and kidneys, partly by modulating leukocyte trafficking, but also by interfering with fibrin deposition, promoting fibrinolysis, and stabilizing the vascular endothelium (19). In stroke models, APC protects the brain against ischemic injury reportedly by blocking p53-mediated apoptosis, acting directly on neuronal cells by mechanisms not yet fully elucidated (27,41) (see Figure 108.2C). The neuroprotection afforded by APC is particularly intriguing, and raises the possibility that APC might play a role in neurodegeneration. APC also suppresses lung inflammation and fibrosis by inhibiting bronchial epithelial cell and macrophage expression of inflammatory cytokines and platelet-derived growth factor, and by promoting fibrinolysis.

As the molecular mechanisms by which APC regulates coagulation, fibrinolysis, inflammation, cell survival, and innate immunity are being delineated, rapid progress is being made to develop mutant forms of APC that have highly specific functions, without undesirable side effects. For example, APC that lacks anticoagulant function, yet retains the capacity to abrogate proinflammatory cascades, will provide a safer and more effective anti-inflammatory agent, without increasing the risk of bleeding. Similarly, it is reasonable to consider that various "designer" forms of APC may in the foreseeable future enter the clinic for the treatment of a number of acute and chronic disorders, including organ transplant rejection, myocardial ischemia–reperfusion, hepatitis, asthma, chronic obstructive lung disease, atherosclerosis, arthritis, or multiorgan failure.

KEY POINTS

- PC is a critical component in a major "natural anticoagulant" pathway. Mediated by thrombin and accelerated by TM and EPCR, the transformation of PC to APC results in cleavage of factors Va and VIIIa and downregulation of further thrombin generation.
- Turning on and turning off PC activation is tightly regulated, in terms of time and location.
- Defects in the PC system enhance the risk of developing thromboembolic disease and, at least in rodents, also increases the inflammatory response to microbial infections.
- APC has distinct anti-inflammatory, antiapoptotic, and profibrinolytic properties that protect the organism against injurious stimuli, such as occurs during sepsis or ischemia. APC modulates these biologic processes through multiple mechanisms to achieve a protective response. EPCR is recognized to play a major role in protecting against inflammatory stimuli. PAR-1 may contribute to this process, and other receptors and intracellular pathways – many yet to be defined – are likely important. Identifying these,

and clarifying the role of EPCR and the PARs, will provide new insights for developing better diagnostic and therapeutic approaches.

- Recombinant APC is currently the only approved and clearly effective drug that decreases sepsis-related mortality in humans. Immediate challenges for clinicians and scientists are to identify the critical molecular and cellular pathways by which APC functions in these settings, the best ways to safely deliver the drug (or functional derivatives with greater specificity), how to monitor therapy, who will most benefit, and what other anti-inflammatory agents should be coadministered to improve overall outcomes.
- In view of its wide spectrum of activity, APC constitutes a novel therapy of potential benefit for many disorders, including sepsis, DIC, acute and chronic inflammatory disorders, organ ischemia–reperfusion injury, stroke, and possibly neurodegeneration.

ACKNOWLEDGMENTS

Support was provided by the FWO in Flanders, Belgium (Grant #G.0382.02) and the National Institute of Health, USA (Grant #HL067846–01). MVdW is supported by a fellowship from the IWT, Belgium, and is an Emmanuel van der Schueren scholar.

REFERENCES

1 Stenflo J. A new vitamin K-dependent protein. Purification from bovine plasma and preliminary characterization. *J Biol Chem.* 1976;251:355–363.
2 Esmon CT, Stenflo J, Suttie JW. A new vitamin K-dependent protein. A phospholipid-binding zymogen of a serine esterase. *J Biol Chem.* 1976;251:3052–3056.
3 Kisiel W, Canfield WM, Ericsson EH, Davie EW. Anticoagulant properties of bovine plasma protein C following activation of thrombin. *Biochem.* 1977;16:5824–5831.
4 Marlar RA, Kleiss AJ, Griffin JH. Human protein C: inactivation of factors V and VIII in plasma by the activated molecule. *Ann NY Acad Sci.* 1981;370:303–310.
5 Griffin JH, Evatt B, Zimmerman TS, et al. Deficiency of protein C in congenital thrombotic disease. *J Clin Invest.* 1981;68:1370–1373.
6 Stenflo J, Fernlund P. Amino acid sequence of the heavy chain of bovine protein C. *J Biol Chem.* 1982;257:12180–12190.
7 Plutzky J, Hoskins JA, Long GL, Crabtree GR. Evolution and organization of the human protein C gene. *Proc Natl Acad Sci USA.* 1986;83:546–550.
8 Jalbert LR, Rosen ED, Lissens A, et al. Nucleotide structure and characterization of the murine gene encoding anticoagulant protein C. *Thromb Haemost.* 1998;79:310–316.
9 Davidson CJ, Tuddenham EG, McVey JH. 450 million years of hemostasis. *J Thromb Haemost.* 2003;1:1487–1494.
10 Jagadeeswaran P, Sheehan JP. Analysis of blood coagulation in the zebrafish. *Blood Cells Mol Dis.* 1999;25:239–249.
11 Slungaard A, Fernandez JA, Griffin JH, et al. Platelet factor 4 enhances generation of activated protein C in vitro and in vivo. *Blood.* 2003;102:146–151.
12 Dahlback B, Villoutreix BO. Molecular recognition in the protein C anticoagulant pathway. *J Thromb Haemost.* 2003;1:1525–1534.
13 Rezaie AR. Vitronectin functions as a cofactor for rapid inhibition of activated protein C by plasminogen activator inhibitor-1. Implications for the mechanism of profibrinolytic action of activated protein C. *J Biol Chem.* 2001;276:15567–15570.
14 Taylor FJ, Chang A, Esmon CT, Hinshaw LB. Baboon model of *Escherichia coli* sepsis: description of its four stages and the role of tumor necrosis factor, tissue factors, and the protein C system in septic shock. *Curr Stud Hematol Blood Transfus.* 1991;58:8–14.
15 Taylor F, Stearns-Kurosawa D, Kurosawa S, et al. The EC protein C receptor aids in host defense against *E. coli* sepsis. *Thromb Haemost.* 1999;(Suppl):214.
16 Taylor FB. Studies on the inflammatory-coagulant axis in the baboon response to *E. coli*: regulatory roles of proteins C, S, C4bBP and of inhibitors of tissue factor. *Prog Clin Biol Res.* 1994; 388:175–194.
17 Taylor FB Jr., Peer GT, Lockhart MS, et al. Endothelial cell protein C receptor plays an important role in protein C activation in vivo. *Blood.* 2001;97:1685–1688.
18 Levi M, Dorffler-Melly J, Reitsma P, et al. Aggravation of endotoxin-induced disseminated intravascular coagulation and cytokine activation in heterozygous protein-C-deficient mice. *Blood.* 2003;101:4823–4827.
19 Levi M, Choi G, Schoots I, et al. Beyond sepsis: activated protein C and ischemia-reperfusion injury. *Crit Care Med.* 2004;32:S309–S312.
20 Grey ST, Tsuchida A, Hau H, et al. Selective inhibitory effects of the anticoagulant activated protein C on the responses of human mononuclear phagocytes to LPS, IFN-gamma, or phorbol ester. *J Immunol.* 1994;153:3664–3672.
21 Brueckmann M, Hoffmann U, Dvortsak E, et al. Drotrecogin alfa (activated) inhibits NF-kappa B activation and MIP-1-alpha release from isolated mononuclear cells of patients with severe sepsis. *Inflamm Res.* 2004;53:528–533.
22 Esmon CT. Structure and functions of the EC protein C receptor. *Crit Care Med.* 2004;32:S298–S301.
23 Riewald M, Petrovan RJ, Donner A, et al. Activation of EC protease activated receptor 1 by the protein C pathway. *Science.* 2002; 296:1880–1882.
24 Feistritzer C, Riewald M. Endothelial barrier protection by activated protein C through PAR1-dependent sphingosine 1-phosphate receptor-1 cross-activation. *Blood.* 2005;105:3178–3184.
25 Ludeman MJ, Kataoka H, Srinivasan Y, et al. PAR1 cleavage and signaling in response to activated protein C and thrombin. *J Biol Chem.* 2005;280:13122–13128.
26 Hancock WW, Grey ST, Hau L, et al. Binding of activated protein C to a specific receptor on human mononuclear phagocytes inhibits intracellular calcium signaling and monocyte-dependent proliferative responses. *Transplantation.* 1995;60: 1525–1532.
27 Cheng T, Liu D, Griffin JH, et al. Activated protein C blocks p53-mediated apoptosis in ischemic human brain endothelium and is neuroprotective. *Nat Med.* 2003;9:338–342.

28 Bovill EG, Bauer KA, Dickerman JD, et al. The clinical spectrum of heterozygous protein C deficiency in a large New England kindred. *Blood*. 1989;73:712–717.
29 Pelkonen KM, Wartiovaara-Kautto U, Nieminen MS, et al. Low normal level of protein C or of antithrombin increases risk for recurrent cardiovascular events. *Blood Coagul Fibrinolysis*. 2005; 16:275–280.
30 Haywood S, Liesner R, Pindora S, Ganesan V. Thrombophilia and first arterial ischaemic stroke: a systematic review. *Arch Dis Child*. 2005;90:402–405.
31 Seligsohn U, Berger A, Abend M, et al. Homozygous protein C deficiency manifested by massive venous thrombosis in the newborn. *N Eng J Med*. 1984;310:559–561.
32 Dahlback B. The discovery of activated protein C resistance. *J Thromb Haemost*. 2003;1:3–9.
33 Bertina RM, Koeleman BP, Koster T, et al. Mutation in blood coagulation factor V associated with resistance to activated protein C. *Nature*. 1994;369:64–67.
34 Yamamoto K, Loskutoff DJ. Extrahepatic expression and regulation of protein C in the mouse. *Am J Pathol*. 1998;153:547–555.
35 Tanabe S, Sugo T, Matsuda M. Synthesis of protein C in human umbilical vein ECs. *J Biochem (Tokyo)*. 1991;109:924–928.
36 Jalbert LR, Rosen ED, Moons L, et al. Inactivation of the gene for anticoagulant protein C causes lethal perinatal consumptive coagulopathy in mice. *J Clin Invest*. 1998;102:1481–1488.
37 Lay AJ, Liang Z, Rosen ED, Castellino FJ. Mice with a severe deficiency in protein C display prothrombotic and proinflammatory phenotypes and compromised maternal reproductive capabilities. *J Clin Invest*. 2005;115:1552–1561.
38 Gandrille S, Alach M. Polymorphism in the protein C gene detected by denaturing gradient gel electrophoresis. *Nucleic Acids Res*. 1991;19:6982.
39 Bernard GR, Vincent JL, Laterre PF, et al. Efficacy and safety of recombinant human activated protein C for severe sepsis. *N Engl J Med*. 2001;344:699–709.
40 Angus DC, Laterre PF, Helterbrand J, et al. The effect of drotrecogin alfa (activated) on long-term survival after severe sepsis. *Crit Care Med*. 2004;32:2199–2206.
41 Fernandez JA, Xu X, Liu D, et al. Recombinant murine-activated protein C is neuroprotective in a murine ischemic stroke model. *Blood Cells Mol Dis*. 2003;30:271–276.

109

Vitamin K–Dependent Anticoagulant Protein S

Björn Dahlbäck

University of Lund, Malmö, Sweden

Protein S is a vitamin K-dependent plasma glycoprotein with anticoagulant properties (1–3). It is mainly synthesized in the liver and, to a lesser degree, in several other cell types including endothelium. Protein S binds to negatively charged phospholipid membranes, where it functions as a cofactor to activated protein C (APC) in the degradation of coagulation factors Va and VIIIa, cofactors in the prothrombinase and tenase complexes, respectively. In addition, protein S has APC-independent anticoagulant activities, which directly inhibit both the prothrombinase and the tenase complexes. In vivo, endothelium presumably promotes the anticoagulant activity of protein S by providing a suitable phospholipid membrane for interaction with APC. Protein S normally circulates in human plasma at a concentration of 0.30 μM. In human plasma, 60% to 70% of protein S is bound to the high-molecular-weight C4b-binding protein (C4BP), a regulator of the classical complement pathway (3,4). The remaining 30% to 40% of protein S circulates as free protein. Thus, protein S functions as a link between the protein C anticoagulant system and the complement system.

PROTEIN S GENE AND SYNTHESIS

Protein S in plasma is mainly derived from synthesis in the liver. In addition, endothelial cells (ECs), testicular Leydig cells, and osteoblasts synthesize protein S (3). Platelets contain protein S, but it is not known whether this is derived from megakaryocytic synthesis or from uptake of plasma protein S. However, protein S synthesis has been demonstrated in a megakaryocytic cell line, suggesting megakaryocytes have the capacity to synthesize the protein. Two protein S genes (*PROS1* and *PROSP*) are found in the human genome, but only *PROS1* is expressed, whereas *PROSP* is a pseudogene. Both *PROS1* and *PROSP* are located on chromosome 3 close to but on different sides of the centromere: *PROS1* at q11.2 and *PROSP* at p21-cen. The two genes have a high degree of sequence identity between the exons (97%). The *PROS1* gene is approximately 80 kb long and contains 15 exons and 14 introns. Orangutan, rhesus monkey, and African green monkey have one *protein S* gene, but chimpanzee and gorilla have two. This suggests that the gene duplication event occurred after the branching of the orangutan from the African apes (3).

PROTEIN S STRUCTURE

Circulating human protein S is a 635-amino acid residue long, single-chain glycoprotein; the molecular weight of the apoprotein calculated from the amino acid composition is 70,690 (1–3). Protein S is extensively post-translationally modified; the mature protein contains three N-linked carbohydrate side chains and three types of modified amino acids, namely γ-carboxy glutamic acids (Gla), β-hydroxy aspartic acid (Hya), and β-hydroxy asparagines (Hyn) (3,5). The primary structures of protein S from many other species have been determined (e.g., monkey, bovine, porcine, rabbit, rat, and mouse protein S).

Mature protein S is composed of multiple domains (Figure 109.1): from the NH_2-terminus, a Gla-domain is followed by a small thrombin-sensitive region (TSR), four epidermal growth factor (EGF)-like domains, and the sex hormone binding globulin (SHBG)-like region comprising two laminin G (LamG)-type domains (1–3). The domain structure of protein S correlates with the intron/exon organization of the gene, suggesting that it has evolved through a combination of exon shuffling and gene duplication events.

The Gla-domain binds multiple Ca^{2+}-ions, and the domain has high affinity for negatively charged phospholipid membranes (3,6,7). It also interacts directly with APC and is important for the APC-cofactor activity (8). The thrombin-sensitive region contains two cysteines, forming a disulfide bridge. In human protein S, thrombin can cleave at Arg49 and Arg70, and FXa at Arg60. After cleavage, the Gla domain remains attached via the disulfide bridge, but the APC-cofactor function of protein S is lost. Several lines of experiments support the notion that the TSR is involved in the interaction between protein S and APC (1–3). Protein S contains four EGF-like domains, each encoded by a separate exon. EGF1 and EGF2 are important for expression of APC-cofactor activity,

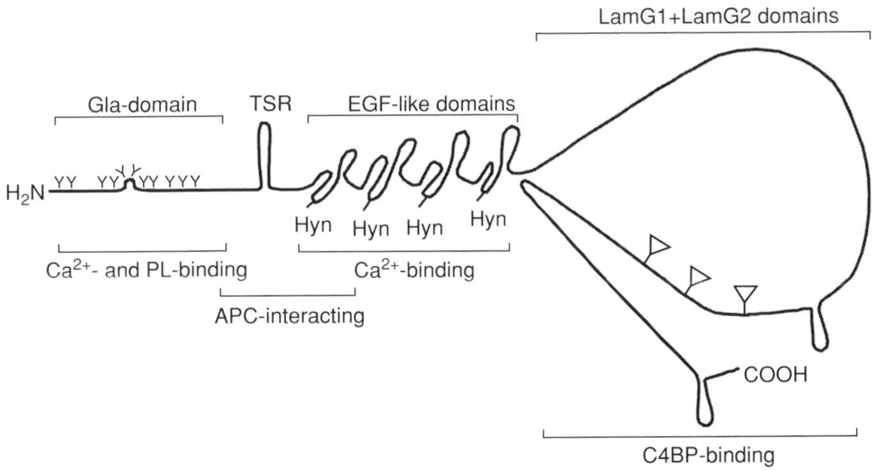

Figure 109.1. Schematic model of human protein S. Protein S is a multidomain, single-chain molecule. The Gla-domain contains eleven vitamin K–dependent Gla residues (Y). The Gla domain binds Ca^{2+} and negatively charged phospholipids. It is also involved in interaction with APC. The thrombin sensitive region (TSR) is important for interaction with APC, and cleavages by thrombin or FXa result in loss of APC-cofactor activity. The four EGF-like domains contain the modified amino acids β-hydroxy aspartic acid (Hya) or β-hydroxy asparagine (Hyn), as indicated. EGF1 and EGF2 are both important for the APC-cofactor activity of protein S. The two LamG domains constitute the C-terminal half of protein S. This region also is known as the SHBG region, and it contains the binding site for C4BP. The three N-linked carbohydrate side chains are indicated (∇).

with EGF1 interacting directly with APC (1–3). The first EGF-like domain contains a Hya, and the following three Hyn (5). The importance of the hydroxylation of Asp/Asn in protein S is not known, and recombinant human protein S synthesized under conditions that inhibit the hydroxylation still expresses full cofactor function and C4BP binding. The Hyn-containing EGF-like domains in protein S, in particular EGF4, contain very high-affinity Ca^{2+}-binding sites (K_d down to nM) (5). The Ca^{2+} binding is important for protein S to attain protease resistance and a native conformation. The carboxy-terminal SHBG-like region of protein S, comprising the two LamG domains, contains three N-linked carbohydrate side chains of unknown function. The second LamG domain has been shown to play a role in the degradation of both FVa and FVIIIa. The C4BP-binding site on protein S is fully contained in the LamG domains, with both LamG1 and LamG2 contributing to the binding. Several regions of the LamG domains have been suggested to be important for the binding, but the details of the binding site are still not known (1–3).

ACTIVATED PROTEIN C–DEPENDENT ANTICOAGULANT ACTIVITY OF PROTEIN S

Protein S functions as an APC cofactor in the degradation of both FVa and FVIIIa (3,9,10). The Gla domain of protein S plays an important role related to its ability to bind negatively charged phospholipids. Protein S and APC form a complex on this type of phospholipid membrane, with protein S increasing the affinity of APC for the membrane approximately 10-fold (Figure 109.2). However, even though the two proteins form a complex on the membrane, it has not been possible to demonstrate an interaction between the two proteins in fluid phase. The interaction between APC and protein S on the phospholipid surface affects the location of the active site of APC in relation to FVa and the membrane (11). In in vitro experiments, the presence of protein S was found to reduce the distance of the active site of APC to the phospholipid membrane by approximately 10 Å. This change in the localization of the serine protease domain of APC may be important for the cleavages in FVa and FVIIIa. The interaction between APC and protein S has mainly been characterized in in vitro systems using phospholipid vesicle preparations. In vivo, presumably ECs, platelets, and platelet microparticles provide the phospholipid surfaces for the protein S–APC interaction (12,13).

In the degradation of FVa by APC, the cleavages at Arg506 and Arg306 demonstrate different dependencies on protein S (see Figure 109.2). Thus, although protein S provides stimulation to both cleavages, the APC-mediated cleavage at Arg506 in FVa is less dependent on the presence of protein S than is the cleavage at Arg306 (10). The Arg506 cleavage, furthermore, is more efficient than that at Arg306. This is shown to be due to a positively charged exosite on the surface of the serine protease domain of APC, which interacts specifically with a negatively charged area around Arg506 (14). The cleavage at Arg506 results in partial loss of FVa activity due to reduced binding affinity for FXa, whereas the subsequent cleavage at Arg306 results in complete inactivation of FVa. FXa, when bound to FVa in the assembled prothrombinase complex, protects the Arg506 site from degradation by APC, presumably through sterical hindrance (i.e., FXa, when bound, covers the Arg506 area) (14). Protein S has been shown to abrogate the protective

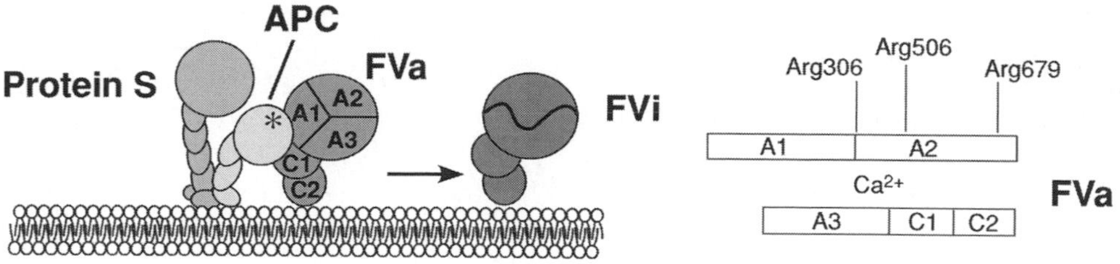

Figure 109.2. Protein S in the APC-mediated degradation of FVa. APC attacks the membrane-bound FVa at three sites: Arg306, Arg506, and Arg679. The efficiency of APC is enhanced by the presence of protein S, and the two molecules form a membrane-bound complex that efficiently degrades FVa. The Arg306 cleavage site is more dependent on the presence of protein S than is the Arg506 site. The Arg679 cleavage seems to be less important than the other two, but its physiological significance is not fully understood.

effect of FXa on the Arg506 site in FVa. Thus, protein S is important to increase the efficiency of the APC-mediated cleavages at both the Arg506 and the Arg306 sites, although the mechanisms involved in the two cleavages are slightly different.

In APC-mediated degradation of FVIIIa in the tenase complex, protein S is not the only APC cofactor of importance, because FV has been found to enhance the APC-mediated FVIIIa degradation (10). The APC-cofactor activities of FV and protein S are synergistic, and result in efficient control of the activity of the tenase complex.

ACTIVATED PROTEIN C–INDEPENDENT ANTICOAGULANT ACTIVITY OF PROTEIN S

Protein S also expresses APC-independent anticoagulant activity. The physiological importance of this activity is unclear, and its molecular mechanism has been the topic of debate (1,2). The direct anticoagulant activity has been suggested to be due to direct interactions between protein S and FVa/FVIIIa and/or FXa. The anticoagulant activity of protein S has been demonstrated in a flow system using ECs and plasma, in which protein S–depleted plasma gave considerably higher prothrombin activation than did plasma containing protein S (15). Some reports have claimed that the APC-independent anticoagulant activity of protein S is due to competition for phospholipid (16). However, such phospholipid dependence is observed only when the protein S preparations contain multimers of protein S, which are normally not present in plasma but can form during the purification of protein S (1,2).

PROTEIN S INTERACTION WITH C4B-BINDING PROTEIN

The complex in human plasma between protein S and C4BP has a 1:1 stoichiometry; the interaction is noncovalent and of high affinity (K_d approximately 10^{-10} M) (1,3,4,17). C4BP is important for the regulation of the C4bC2a complex (C3 convertase) of the classical complement pathway, functioning as a cofactor to the serine protease factor I in the degradation of C4b (3,4). C4BP is a high-molecular-weight (570 kDa) glycoprotein composed of seven identical α-chains (70 kDa) and a single β-chain (45 kDa), with the chains linked by disulfide bridges. The concentration of C4BP in plasma is approximately 0.3 μM. High-resolution electron microscopy revealed a spider- or octopus-shape for C4BP, with the different chains radiating from a central core (Figure 109.3) (18). Each α-chain contains a binding site for C4b, whereas the β-chain contains the single binding site for protein S, because the binding of protein S and C4b are independent. Molecular models of the α- and β-chains have been created, and the binding sites for C4b and protein S have been elucidated in detail (4).

Under normal conditions, at least 80% of the C4BP molecules in human plasma contain the β-chain (C4BPβ^+) and bind protein S. The concentration of C4BPβ^+ is on a molar basis approximately 30% to 40% lower than that of total protein S, and the level of free protein S is equivalent to the molar surplus of protein S over the C4BPβ^+ because the protein S–C4BP interaction is of very high affinity (1,3). C4BP is an acute-phase protein, and its concentration may increase to 400% of the normal level during inflammatory disorders. The synthesis of α-chains increases more than that of the

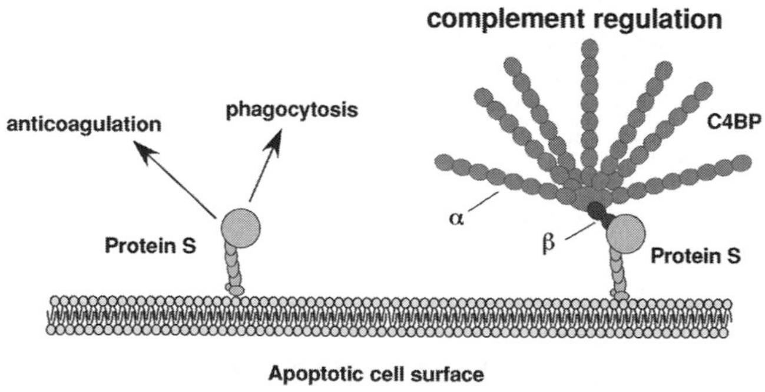

Figure 109.3. Apoptotic cells bind protein S and the protein S-C4BP complex. During apoptosis the dying cells expose the negatively charged phosphatidylserine on their surfaces. This is an "eat me" signal to neighboring cells and professional phagocytes. Both free protein S and the protein S-C4BP complex bind to the apoptotic cells via the Gla-domain of protein S. This provides the potential for regulation of both coagulation and the complement systems on the surface of apoptotic cells. Moreover, the bound protein S stimulates phagocytosis of the apoptotic cells, whereas the protein S-C4BP complexes seem to have the opposite effects. The octopus-like shape of C4BP is illustrated, the arrows indicating the α- and β-chains.

ß-chain during inflammation due to differential regulation of the α- and β-chain genes by cytokines. Therefore, the C4BP molecules that are synthesized under inflammatory stress contain only α-chains and are unable to bind protein S. The synthesis of protein S does not increase much during inflammation, and the differential regulation of α- and β-chain synthesis during inflammation ensures stable levels of free protein S also during inflammatory states, even though the plasma C4BP level may be several times higher than normal.

The interaction between protein S and C4BP provides a link between the regulation of blood coagulation and the complement system. Thus, C4BP modulates the protein S anticoagulant activity, because binding of C4BP to protein S results in the loss of the APC-cofactor activity of protein S. Likewise, protein S can be involved in affecting the C4BP-mediated regulation of the classical complement pathway by localizing the complex to the surface of negatively charged phospholipid (as on apoptotic cells) (see later discussion). Because complement activation generates local inflammation due to the release of anaphylatoxins, binding of protein S-C4BP to apoptotic cells can provide local regulation of the complement system and inhibition of inflammation in the vicinity of dying cells.

PROTEIN S BINDING TO APOPTOTIC CELLS STIMULATES PHAGOCYTOSIS

Under normal conditions, the negatively charged phosphatidylserine is not exposed on the surface of cells but located in the inner leaflet of the cell membrane. Therefore, protein S and the other vitamin K–dependent proteins do not bind to the surfaces of normal cells. Certain situations result in the exposure of phosphatidylserine on the cell surface, such as apoptosis or platelet activation. It has been shown that protein S can bind to the negatively charged phospholipids that are exposed on the surface of apoptotic cells and on platelet microparticles (19,20). The binding is calcium-dependent and mediated by the Gla domain. When bound, protein S stimulates phagocytosis of the apoptotic cells through mechanisms yet to be defined (21). The protein S–C4BP complex also binds to the apoptotic cells but this is, in contrast, found to counteract the phagocytic process (22). In addition, protein S and the protein S–C4BP complex may have other functions on the apoptotic cell surface. Thus, protein S may contribute to controlling the coagulation system on the apoptotic cell surface, whereas the protein S–C4BP complex may take part in controlling local activation of the complement pathway. In this context, it is worth mentioning that the intravenous administration of human protein S to mice subjected to a cerebral stroke model resulted in significant neuroprotection (23). Protein S reduced fibrin deposition, neutrophil infiltration, infarction size, and edema, and improved the postischemic blood flow and the motor neurological deficit. It is not clear if these beneficial effects were mainly due to the anticoagulant or the prophagocytic properties of protein S. However, it can be concluded that the beneficial effects were not exerted by protein S in complex with C4BP, because as mouse C4BP lacks the β-chain and because the mouse β-chain gene is a nonexpressed pseudogene (24).

Protein S has no known antiapoptotic activities, but it is noteworthy that protein S is homologous to the antiapoptotic protein Gas6 (product of the growth arrest specific gene number 6), with the two proteins sharing similar domain organization. Gas6 is a member of the vitamin K–dependent protein family and, as the name implies, Gas6 synthesis is induced by growth arrest, such as serum starvation in cultured cells (25–27). Several biological effects have been associated with Gas6, such as the inhibition of apoptosis, stimulation of mitogenesis, and growth stimulation of different cell types including ECs. Gas6 is widely expressed by ECs, fibroblasts, and smooth muscle cells but, unlike protein S, Gas6 is not expressed in the liver. The plasma concentration of Gas6 is 1,000-fold lower than that of protein S, and the function of Gas6 is unrelated to the protein C and coagulation systems (28). Gas6 binds to the tyrosine kinase receptors Axl, Sky, and c-Mer, and induces tyrosine phosphorylation and intracellular signaling (25–27). The Gla-domain of Gas6 binds negatively charged phospholipid membranes exposed on the surface of apoptotic cells and stimulates the phagocytosis of the apoptotic cells. Protein S also has been shown capable of functioning as a ligand for the Sky receptor. However, the physiological significance of this has been questioned because only protein S from certain species can stimulate the Sky receptor (29). Thus, bovine protein S stimulates the human Sky receptor, whereas human protein S has no or little such activity, with the structural difference between human and bovine protein S accounting for this species specificity located in the LamG1 domain of protein S (30). The three-dimensional structure of the two LamG domains of Gas6 has been determined (31). Based on this and other laminin G-type domains, models for the corresponding part of protein S have been created.

PROTEIN S DEFICIENCY AND VENOUS THROMBOSIS

Inherited protein S deficiency is a risk factor for venous thrombosis found in approximately 2% to 4% of thrombosis patients (1,3,32). The prevalence of protein S deficiency in the general population has been estimated to be between 0.03% and 0.13%. The risk of thrombosis in heterozygous carriers is about five- to 10-fold higher than in their healthy relatives. Homozygous or compound heterozygous protein S deficiency is extremely rare and incompatible with life without treatment due to extensive thrombosis. Venous thrombosis is a multifactorial disease, and the risk of thrombosis in protein S deficiency is much higher when combined with other genetic or acquired conditions predisposing for thrombosis (33). This explains the varying penetrance of thrombotic symptoms in patients with protein S deficiency.

Traditionally, inherited protein S deficiency was categorized into three different types (I–III), but convincing data have shown that there are only two types. According to the original classification, type I patients have decreased free and total protein S, type II denotes functional deficiency with normal protein levels, and type III is characterized by low free protein S but normal concentrations of total protein S. An observed coexistence of types I and III in several pedigrees suggested the two types to be phenotypic variants of the same genetic disease, and that only two distinguishable types exist, type I having low protein S, and type II having functional defects (1,3). The concentrations of protein S and C4BPβ^+ are approximately equimolar in protein S-deficient individuals, which, taken together with the high affinity of the protein S–C4BP interaction, explains the low plasma levels of free protein S that characterize type I patients. This also explains why analysis of free protein S has higher predictive value for protein S deficiency than does analysis of total protein S, and is the method of choice in evaluation of thrombosis patients (34). A majority of patients with protein S deficiency have type I, and very few have type II deficiency. Many missense and nonsense mutations have been identified in cases of protein S deficiency. However, it has been a consistent observation from many laboratories that point mutations are only found in around 50% of the protein S-deficient patients. Recently, an explanation for this was provided by genetic studies of protein S-deficient families having no identifiable point mutations. In these families, the protein S deficiency was found to be linked to the *PROS1* gene, and in many of them large deletions were identified as the genetic defect (1,35,36). Type II mutations have been found in the Gla-domain; these result in defective γ-carboxylation and presumably lead to folding problems and instability as well as poor Ca^{2+} and phospholipid binding. A Thr103 to Asn mutation in EGF1 causes type II deficiency, presumably due to poor interaction with APC. In EGF2, a Lys155 to Glu mutation results in a functional defect in protein S (protein S Tokushima). Protein S Tokushima has poor APC cofactor activity, and has been shown to interact poorly with APC, consistent with the hypothesis that EGF2 is important for the APC-cofactor activity of protein S. Inherited protein S-deficient patients have similar clinical manifestations as those with other thrombophilic conditions, mainly venous thrombosis in lower extremities (1–3,33).

The concentration of protein S in plasma is influenced by several factors (e.g., pregnancy and oral contraceptives decrease the plasma protein S concentration). During normal pregnancy, the plasma levels of both total and free protein S decrease, reaching levels found in patients with inherited deficiency (1,3,34). The plasma protein S is lower in women than in men, and higher in postmenopausal women than in premenopausal women. During treatment with oral anticoagulants (e.g., warfarin), the levels of protein S decrease, because protein S is a vitamin–K dependent protein. Acquired protein S deficiency is associated with nephrotic syndrome and disseminated intravascular coagulation. The plasma concentration of C4BP is high in patients with nephrotic syndrome, because the high molecular weight of C4BP does not allow glomerular filtration. Free protein S is lost in the urine, resulting in decreased levels of functionally active protein S, which may contribute to the thrombosis risk in nephrotic syndrome. Acquired protein S deficiency also may be the result of immunological mechanisms, as in patients with autoimmune disease or human immunodeficiency virus (HIV). Moreover, many children have been described with thrombosis and autoantibodies to protein S after varicella infections (1,3).

NOTE ADDED IN PROOF

Recently, novel insights into the mechanism of the APC-independent anticoagulant activity of protein S was gained by the report of Hackeng and colleagues that protein S serves as a cofactor to (TFPI) tissue factor pathway inhibitor in inhibition of tissue factor-induced coagulation (37).

KEY POINTS

- Vitamin K–dependent protein S binds to negatively charged phospholipid membranes and expresses both APC-dependent and APC-independent anticoagulant activities.
- In human plasma, 30% to 40% of protein S is free, whereas the rest is bound in a noncovalent high-affinity complex with the complement regulator C4b-binding protein.
- Free and C4BP-bound protein S bind to the negatively charged phospholipids exposed on the surface of apoptotic cells. Free protein S stimulates phagocytosis of apoptotic cells, whereas the protein S–C4BP complex inhibits phagocytosis.
- Inherited protein S deficiency is a risk factor for venous thrombosis, with analysis of free protein S the method of choice to diagnose the deficiency.

Future Goals

- To elucidate the importance of protein S in the phagocytosis of apoptotic cells. Which receptors does protein S interact with on the phagocytic cells?
- To demonstrate that free protein S and the protein S–C4BP complex, when bound on apoptotic cells, inhibit both blood coagulation and the complement system
- To create a protein S knockout mouse to elucidate the in vivo importance of protein S
- To determine the three-dimensional structure of protein S and the relationships between the protein S structure and its functions

REFERENCES

1 Rezende SM, Simmonds RE, Lane DA. Coagulation, inflammation, and apoptosis: different roles for protein S and the protein S-C4b binding protein complex. *Blood*. 2004;103(4):1192–1201.

2 Rigby AC, Grant MA. Protein S: a conduit between anticoagulation and inflammation. *Crit Care Med.* 2004;32(5 Suppl):S336–S341.
3 Dahlbäck B, Stenflo J. The protein C anticoagulant system. In: Stamatoyannopoulos G, Majerus PW, Perlmutter RM, Varmus H, eds. *The Molecular Basis of Blood Disease.* 3rd ed. Philadelphia: WB Saunders Company;2000:614–656.
4 Blom AM, Villoutreix BO, Dahlback B. Complement inhibitor C4b-binding protein-friend or foe in the innate immune system? *Mol Immunol.* 2004;40(18):1333–1346.
5 Stenflo J, Stenberg Y, Muranyi A. Calcium-binding EGF-like modules in coagulation proteinases: function of the calcium ion in module interactions. *Biochim Biophys Acta.* 2000;1477(1–2):51–63.
6 Dahlback B. Protein S and C4b-binding protein: components involved in the regulation of the protein C anticoagulant system. *Thromb Haemost.* 1991;66(1):49–61.
7 Nelsestuen GL, Shah AM, Harvey SB. Vitamin K-dependent proteins. *Vitam Horm.* 2000;58:355–389.
8 Saller F, Villoutreix BO, Amelot A, et al. The gamma-carboxyglutamic acid domain of anticoagulant protein S is involved in activated protein C cofactor activity, independently of phospholipid binding. *Blood.* 2005;105(1):122–130.
9 Walker FJ. Regulation of activated protein C by protein S. The role of phospholipid in factor Va inactivation. *J Biol Chem.* 1981;256(21):11128–11131.
10 Nicolaes GA, Dahlback B. Factor V and thrombotic disease: description of a janus-faced protein. *Arterioscler Thromb Vasc Biol.* 2002;22(4):530–538.
11 Yegneswaran S, Wood GM, Esmon CT, Johnson AE. Protein S alters the active site location of activated protein C above the membrane surface. A fluorescence resonance energy transfer study of topography. *J Biol Chem.* 1997;272(40):25013–25021.
12 Stern DM, Nawroth PP, Harris K, Esmon CT. Cultured bovine aortic ECs promote activated protein C-protein S-mediated inactivation of factor Va. *J Biol Chem.* 1986;261(2):713–718.
13 Hackeng TM, Hessing M, van't Veer C, et al. Protein S binding to human ECs is required for expression of cofactor activity for activated protein C. *J Biol Chem.* 1993;268(6):3993–4000.
14 Dahlback B, Villoutreix BO. Molecular recognition in the protein C anticoagulant pathway. *J Thromb Haemost.* 2003;1(7):1525–1534.
15 van't Veer C, Hackeng TM, Biesbroeck D, et al. Increased prothrombin activation in protein S-deficient plasma under flow conditions on EC matrix: an independent anticoagulant function of protein S in plasma. *Blood.* 1995;85(7):1815–1821.
16 van Wijnen M, Stam JG, van't Veer C, et al. The interaction of protein S with the phospholipid surface is essential for the activated protein C-independent activity of protein S. *Thromb Haemost.* 1996;76(3):397–403.
17 Dahlback B, Stenflo J. High molecular weight complex in human plasma between vitamin K-dependent protein S and complement component C4b-binding protein. *Proc Natl Acad Sci USA.* 1981;78(4):2512–2516.
18 Dahlback B, Smith CA, Muller-Eberhard HJ. Visualization of human C4b-binding protein and its complexes with vitamin K-dependent protein S and complement protein C4b. *Proc Natl Acad Sci USA.* 1983;80(11):3461–3465.
19 Webb JH, Blom AM, Dahlback B. Vitamin K-dependent protein S localizing complement regulator C4b-binding protein to the surface of apoptotic cells. *J Immunol.* 2002;169(5):2580–2586.
20 Dahlback B, Wiedmer T, Sims PJ. Binding of anticoagulant vitamin K-dependent protein S to platelet-derived microparticles. *Biochemistry.* 1992;31(51):12769–12777.
21 Anderson HA, Maylock CA, Williams JA, et al. Serum-derived protein S binds to phosphatidylserine and stimulates the phagocytosis of apoptotic cells. *Nat Immunol.* 2003;4(1):87–91.
22 Kask L, Trouw LA, Dahlback B, Blom AM. The C4b-binding protein-protein S complex inhibits the phagocytosis of apoptotic cells. *J Biol Chem.* 2004;279(23):23869–23873.
23 Liu D, Guo H, Griffin JH, et al. Protein S confers neuronal protection during ischemic/hypoxic injury in mice. *Circulation.* 2003;107(13):1791–1796.
24 Rodriguez de Cordoba S, Perez-Blas M, Ramos-Ruiz R, et al. The gene coding for the beta-chain of C4b-binding protein (C4BPB) has become a pseudogene in the mouse. *Genomics.* 1994;21(3):501–509.
25 Manfioletti G, Brancolini C, Avanzi G, Schneider C. The protein encoded by a growth arrest-specific gene (gas6) is a new member of the vitamin K-dependent proteins related to protein S, a negative coregulator in the blood coagulation cascade. *Mol Cell Biol.* 1993;13(8):4976–4985.
26 Melaragno MG, Fridell YW, Berk BC. The Gas6/Axl system: a novel regulator of vascular cell function. *Trends Cardiovasc Med.* 1999;9(8):250–253.
27 Crosier KE, Crosier PS. New insights into the control of cell growth; the role of the AxI family. *Pathology.* 1997;29(2):131–135.
28 Balogh I, Hafizi S, Stenhoff J, et al. Analysis of Gas6 in human platelets and plasma. *Arterioscler Thromb Vasc Biol.* 2005;25(6):1280–1286.
29 Godowski PJ, Mark MR, Chen J, et al. Reevaluation of the roles of protein S and Gas6 as ligands for the receptor tyrosine kinase Rse/Tyro 3. *Cell.* 1995;82(3):355–358.
30 Evenas P, Dahlback B, Garcia de Frutos P. The first laminin G-type domain in the SHBG-like region of protein S contains residues essential for activation of the receptor tyrosine kinase sky. *Biol Chem.* 2000;381(3):199–209.
31 Sasaki T, Knyazev PG, Cheburkin Y, et al. Crystal structure of a C-terminal fragment of growth arrest-specific protein Gas6. Receptor tyrosine kinase activation by laminin G-like domains. *J Biol Chem.* 2002;277(46):44164–44170.
32 Dahlback B. Blood coagulation. *Lancet.* 2000;355(9215):1627–1632.
33 Zoller B, Garcia de Frutos P, Hillarp A, Dahlback B. Thrombophilia as a multigenic disease. *Haematologica.* 1999;84(1):59–70.
34 Persson KE, Dahlback B, Hillarp A. Diagnosing protein S deficiency: analytical considerations. *Clin Lab.* 2003;49(3–4):103–110.
35 Lanke E, Johansson AM, Hillarp A, et al. Co-segregation of the PROS1 locus and protein S deficiency in families having no detectable mutations in PROS1. *J Thromb Haemost.* 2004;2(11):1918–1923.
36 Johansson AM, Hillarp A, Sall T, et al. Large deletions of the PROS1 gene in a large fraction of mutation-negative patients with protein S deficiency. *Thromb Haemost.* 2005;Nov;94(5):951–957.
37 Hackeng TM, Sere KM, Tans G, Rosing J. Protein S stimulates inhibition of the tissue factor pathway by tissue factor pathway inhibitor. *Proc Natl Sci Acad USA.* 2006;103:3106–3111.

110

Nitric Oxide as an Autocrine and Paracrine Regulator of Vessel Function

William C. Sessa

Boyer Center for Molecular Medicine, Yale University School of Medicine, New Haven, Connecticut

The discovery that the paramagnetic, free radical gas nitric oxide (NO) is the chemical species responsible for endothelium-dependent relaxation of large blood vessels, neurotransmission in the central and peripheral nervous systems, and activated macrophage function of the innate immune system has forged the convergence of several disciplines to study the mechanisms of NO biosynthesis in mammals and lower organisms. The global importance of NO as a highly regulated, second-messenger gas throughout the body culminated in the focused efforts of many laboratories worldwide to understand its synthesis, spectrum of activities, and metabolism. Based on the accolades ascribed to the field of NO biology, which led to a Nobel Prize in Medicine and Physiology awarded to Ignarro, Murad, and Furchgott in 1998, it could be argued that the discovery of the NO signaling pathway was a major paradigm shift in our understanding of endothelial and blood vessel function, an attribute shared by few other discoveries in vascular biology.

HISTORICAL PERSPECTIVES

The role of NO gas in the biological sciences has a long history, including its function as an environmental pollutant, as an active ingredient of gunpowder, and as the therapeutic antianginal nitroglycerin. In keeping with the focus of this volume, our historical perspective is considered from the vascular point of view.

Although the convergence of many disparate lines of scientific enquiry was responsible for the ground-swell of work in the field of NO biology, the defining moment for NO biology in the cardiovascular system and its vasoregulatory role of the endothelium in general hinged on the pioneering work of Robert Furchgott (1). A pharmacologist, Dr. Furchgott noticed that acetylcholine (Ach) was a very potent vasodilator in vivo, but was less consistent and potent in relaxing isolated blood vessels in vitro. He observed that with meticulous isolation of blood vessels, Ach always had a vasodilatory effect, whereas denudation of the inner lining of blood vessels – the endothelium – converted Ach from a vasodilator to a vasoconstrictor having direct action on the underlying smooth muscle. He then performed a classic "sandwich bioassay" experiment, in which a vessel with intact endothelium was "sandwiched" into the lumen of a denuded vessel segment to determine whether the vessel with endothelium could release a transferable factor to relax the denuded segment. The experiment was successful, and he coined the biological activity endothelium-derived relaxing factor (EDRF). The remarkable simplicity of this experiment, and the ability of students and scientists all over the world to repeat this observation, is a testimony to the fundamental nature of Furchgott's observations. This bioassay experiment also gave rise to notion that EDRF was a labile factor, inactivated by superoxide anion and oxyhemoglobin and antagonized by redox active inhibitors of soluble guanylyl cyclase (see next section).

However, a pivotal discovery from the field of tumor immunology focused researchers on finding the enzymatic mechanisms that produced EDRF. Activated mouse peritoneal macrophages inhibited tumor-cell DNA synthesis, mitochondrial respiration, and growth when L-arginine, but not D-arginine, was added to the tissue culture media (2). This cytotoxic action of activated macrophages was accompanied by the stoichiometric formation of L-citrulline and nitrite (NO_2^-) in the media. In addition, guanidino-substituted L-arginine analogs such as N^G-monomethyl-L-arginine (L-NMMA) later shown to be an inhibitor of NO synthase (NOS), prevented the L-arginine–dependent production of L-citrulline and NO_2^- and the cytotoxic actions of activated macrophages. The ability of activated macrophages to nitrosate amines, and the lack of toxicity associated with NO_2^- or nitrate (NO_3^-) in the culture medium, suggested that a nitrogen oxide more potent than NO_2^- or NO_3^- was responsible for macrophage cytotoxicity.

In an independent line of research studying the mechanisms of nitrate generation by activated macrophages (3), it was shown that macrophages metabolized isotopically labeled arginine to labeled NO_2^- and NO_3^-, thus providing evidence that they were derived from one of the two chemically equivalent guanidino nitrogens of L-arginine (4). NOS activity, for the first time, was demonstrated in the cytosol of cytokine-activated macrophages, and this crude enzyme preparation was able to metabolize L-arginine to nitrogen oxides, with NO being the intermediate in this pathway and the mediator of macrophage cytotoxicity (5). These experiments defined the substrate for NO synthesis (L-arginine), identified the first inhibitors (arginine analogs), and identified the inducible nature of NOS in macrophages.

Simultaneously, NO was found to be the principal, biologically active constituent of EDRF and an endogenous activator of soluble guanylate cyclase (6–9). Effluents from isolated perfused blood vessels were directed over preconstricted strips of endothelium-denuded vessels that were used to detect vasodilator substances released from the donor vessels. Under these conditions, NO mimicked the relaxing activity of EDRF released from the donor blood vessel. Both activities exhibited the same biological half-life, were attenuated by oxyhemoglobin and the generation of superoxide anions, and were potentiated by superoxide dismutase (9,10). The chemical identity of EDRF was suggested to be NO, based on the detection of nitrogen oxides by chemiluminescence and nitrosylhemoglobin by spectrophotometry (8,9).

However, because exogenous L-arginine did not significantly relax or potentiate the release of EDRF in freshly isolated blood vessels or cultured endothelial cells (ECs), it was very difficult to explain the mechanism of NO production in ECs. Only by culturing ECs in L-arginine–deficient media for 24 hours (at which point the intracellular L-arginine pool was depleted from approximately 100–200 μM to less than 30 μM) could the exogenous administration of L-, but not D-arginine, potentiate the release of EDRF and increase the production of nitrogen oxides (11). Similar to the studies in activated macrophages, infusion of L-[guanido-$^{15}N_2$] labeled arginine into L-arginine–deficient ECs yielded the release of ^{15}N-labeled NO. The substituted L-arginine analog, L-NMMA attenuated endothelium-dependent relaxations in various isolated vascular beds in vitro, and increased systemic blood pressure in vivo, further supporting NO as an important endogenous regulator of vascular tone (12). Soon after, L-arginine–dependent, crude NOS activity was demonstrated in cytosolic and/or membrane fractions prepared from freshly isolated or cultured ECs (13,14).

Now that L-arginine was convincingly shown to be the precursor to NO, an assay was necessary to track the enzymatic activity when isolated from cells. Work by several research teams (15–17) led to the development of two assays to track NOS activity. Bredt (15) used the conversion of radiolabeled L-arginine to L-citrulline as an assay to isolate the first NOS from brain (this method is still used as a very simple, standard assay in most laboratories). Murad's group (17) used NO activation of soluble guanylyl cyclase and cyclic guanosine monophosphate (cGMP) accumulation in a kidney cell line to track "EDRF synthase" activity and isolated the endothelial enzyme. Stuehr isolated the macrophage enzyme (18).

Bredt then used tryptic peptide sequence information to clone the first NOS isoform, neuronal NOS from rat brain (19), followed by the cloning of macrophage NOS (inducible NOS or iNOS) by Nathan (20), and endothelial NOS (eNOS) by Bloch, Michel, Harrison, and our group (21–24). Once the cDNAs were cloned, overexpression systems led to a characterization of the enzymatic mechanisms for NO synthesis and structural biology of the NOS isoforms.

EVOLUTIONARY CONSIDERATIONS

To date, many efforts are under way to define the origins of NO, NO synthesis, and NO binding proteins. Philosophically, it can be argued that NO is as ancient as life itself based on studies simulating the atmospheres of the Earth, Venus, and Mars. In these experiments, the formation of NO can be stimulated at 50% CO_2 using a simulated lightning discharge consistent with the formation of atmospheric NO under conditions during early planetary evolution (25). More relevant to cellular production of NO in biology, NOS activity can be found in protozoa, bacteria (gram-positive, but not gram-negative), plants, *Drosophila*, and fish, leading to the conclusion that NO synthesis is one of the earliest and most widespread signaling systems in living organisms (26). A recent paper identifying NOS in the plant pathogen *Streptomyces turgidiscabies* documented a role for this NOS in generating a nitrated species of the plant toxin, thaxtomin (27). Knockout of the gene or treatment of the bacteria with NOS inhibitors reduces the synthesis of thaxtomin and the virulence of the pathogenic strain, thus defining a primitive role for NOS and NO in bacteria. In mammalian NOS, the transport of electrons through the protein occurs via the modular structure of the oxygenase and reductase domains. In many of the protozoal or bacterial species exhibiting NOS activity, conserved features exist akin to mammalian NOS in the amino terminal oxygenase domain. However, bacterial NOS lacks a carboxy terminal reductase domain, suggesting that another redox protein may serve as an electron donor.

In mammals, the three NOS isoforms (nNOS, iNOS, and eNOS) share 50% amino acid identity and are considered in the class of cytochrome P450 enzymes (28,29). All the cloned NOSs contain consensus binding sites for heme, calmodulin, and L-arginine; BH4 binding in the oxygenase domain; and an electron transport component in the carboxy half-reductase domain. All NOS enzymes work as electron transfer relay stations, transferring electrons from nicotinamide adenine dinucleotide phosphate (NADPH) through the carboxy reductase domain to the heme acceptor in the amino terminus. The activated heme binds oxygen (O_2) to catalyze the oxygenation of guanidine nitrogen of the substrate L-arginine via complex chemistry. The carboxy-terminal half of mammalian NOS is similar (36% amino acid identity) to the electron transport protein, cytochrome-P450 reductase, containing consensus sequence-binding domains for the flavin nucleotides (flavin mononucleotide [FMN] and flavin adenine dinucleotide [FAD]) and NADPH. As with all cytochrome P450 enzymes, the inefficient coupling of electron transfer to substrate utilization results in an "uncoupled" enzyme producing superoxide anion instead of NO. This is an area of intense interest because it is possible that cardiovascular risk factors may reduce the synthesis of tetrahyrobiopterin (BH_4) or interfere with hsp90 binding to eNOS, thus converting eNOS from an NO generator to a superoxide generator (30).

GENE REGULATION

In general, the transcripts for *nNOS* and *eNOS* are thought to be "constitutively" expressed, whereas *iNOS* is clearly an "inducible" gene. However, many examples exist where nNOS and in particular eNOS are induced by physiological and pathophysiological stimuli (see reviews for control of gene expression, in Refs. 31–33). Most importantly for eNOS, shear stress – the mechanical force initiated by the flow of blood across an endothelial monolayer – induces eNOS mRNA and protein levels consistent with eNOS-derived NO and is important for flow-dependent dilation and remodeling of conduit vessels (24,34). Interestingly, in various species, increasing blood flow secondary to exercise training also induces eNOS, suggesting that eNOS upregulation is part of the "cardioprotective gene program" triggered by exercise training.

ROLE IN HOMEOSTASIS

Autocrine Roles of Nitric Oxide

Although NO was discovered as a diffusible gas synthesized in one cell and with actions on adjacent cells, increasing evidence suggests that NO serves as an important autocrine regulator of EC function. This discussion omits studies utilizing NO donor drugs, because these compounds do not exhibit release kinetics similar to those of enzymatic generation of NO by eNOS. Moreover, the compartmentalization of eNOS into the Golgi complex and plasma membrane is necessary for the dynamic generation of NO synthesis (35). In the context of vascular biology, treatment of ECs with inhibitors of NOS reduces the actions of several angiogenic factors in promoting NO release, proliferation, migration, tube formation, and protein secretion. In addition, blockage of endogenous NO increases mitochondrial O_2 consumption and sensitizes cells to apoptogens. The pathophysiological importance of eNOS in regulating angiogenesis has been explored in *eNOS*$^{-/-}$ mice (31,36,37). NOS inhibitors block vascular endothelial growth factor (VEGF)-induced proliferation, migration, and angiogenesis in vitro and in vivo (38,39). However, NO is not essential for the process of vasculogenesis and angiogenesis during development, because *eNOS*$^{-/-}$ mice survive and grow normally (40,41). This is in contrast to mice lacking VEGF or VEGF receptors, which die early during vascular development (42,43). However, *eNOS*$^{-/-}$ mice exposed to hindlimb ischemia do not revascularize the circulation as well as control mice and develop critical limb ischemia. Moreover, exogenous VEGF administration or *VEGF* gene therapy failed to restore angiogenesis in *eNOS*-knockout mice, supporting the notion that NO is an essential downstream element regulating VEGF-induced angiogenesis and permeability in adult mice (44–47). Recent studies have shown that *eNOS*$^{-/-}$ animals are not only refractory to VEGF, but also do not improve blood flow after ischemia, when treated with statins (48) or the chemokine stromal cell-derived factor 1 (SDF1)-α (49), suggesting that eNOS is critical for the beneficial actions of these agents.

Mechanistically, it is not clear how NO promotes angiogenic phenotypes but there are several potential mechanisms. NO activation of soluble guanylyl cyclase (sGC) and protein kinase G (PKG) activation is responsible for some of the autocrine actions of NO in cultured cells; however, many cultured ECs lose sGC or PKG, thus suggesting other targets for NO. One exciting mechanism that may explain several autocrine functions of NO is S-nitrosylation of signaling proteins (50). NO can readily form adducts with oxidized thiols to form S-nitrosothiols in proteins, and recent evidence suggests that enzymatic denitrosylation can also regulate net NO-thiol stability (51). A recent finding of the NO regulation of von Willebrand factor (vWF) secretion is an excellent example of how S-nitrosylation may regulate EC function (52). Many agonists that mobilize intracellular calcium can promote the release of prestored vWF from Weibel-Palade bodies, a unique dense secretory granule found in ECs. Blockade of NO release augments vWF secretion, suggesting that endogenous NO negatively regulates the secretion of vWF. Lowenstein and coworkers (52) identified N-ethylmaleamide–sensitive factor (NSF) as a target for S-nitrosylation. S-nitrosylation of NSF reduces adenosine triphosphate (ATP) hydrolysis, a reaction necessary for the disassembly of the proteins required for membrane fusion and secretion. Additional side-by-side studies comparing the role of the sGC/PKG pathway versus S-nitrosylation will further define the autocrine actions of NO.

Paracrine Actions of Nitric Oxide

As mentioned earlier, NO was initially discovered as an endothelial-derived factor that would diffuse across the basal lamina and relax the underlying vascular smooth muscle. The biological half-life of NO is 1 to 3 seconds in aqueous solution; it is rapidly inactivated by superoxide anion and its half-life is prolonged by superoxide dismutase (10). This ability of endothelial-derived NO or NO liberated by NO donors such as nitroglycerin is largely mediated by the activation of vascular smooth muscle -soluble guanylyl cyclase and PKG (53) and, to a lesser extent, via S-nitrosylation of calcium-activated potassium channels. Endothelial-derived NO also has effects on circulating cells, where it inhibits leukocyte and platelet adhesion and aggregation. *eNOS*$^{-/-}$ mice demonstrate increased leukocyte and platelet accumulation in certain models of inflammation consistent with endothelium-derived NO as a paracrine regulator (54).

Another example of eNOS-derived NO controlling a paracrine function is the regulation of O_2 consumption by surrounding parenchyma. Indeed eNOS-derived NO is a negative regulator of O_2 consumption in the heart and skeletal muscle by virtue of its reversible binding to components of the respiratory chain, including cytochrome oxidase (55–57). More recently, eNOS-derived NO has been implicated as a global regulator of mitochondrial biogenesis (58,59). Skeletal muscle biopsies from *eNOS*$^{-/-}$ mice contain fewer mitochondria when compared with wild-type mice. In vitro, NO can regulate mitochondrial gene expression, thus exerting a profound influence on mitochondrial number and metabolism.

THERAPEUTIC CONSIDERATIONS

Endothelial dysfunction, defined as a deficit in NO bioavailability, is a hallmark of many cardiovascular diseases. To date, data from multicenter clinical trials document that improvement in endothelium-dependent vasomotion correlates well with the beneficial actions of statin-based drugs and angiotensin-converting enzyme (ACE) inhibitors (60). This has been interpreted to suggest that maintaining a healthy eNOS/NO balance is beneficial for cardiovascular health. However, no genetic data document that mutations in the NOS/cGMP/PKG pathway cause cardiovascular disease, thus suggesting that these pathways may be genetic modifiers of polygenic disorders. In addition, other vasodilatory factors such as endothelium-derived hyperpolarizing factor (EDHF) or arachidonic acid metabolites also may be impaired and contribute to endothelial dysfunction, despite the evidence supporting impairments in NO bioactivity as the major cause of endothelial dysfunction.

Given this clinical information, what strategies are available to regulate eNOS function? Several animal studies have used *eNOS* gene transfer to improve vasomotion, blood flow, and angiogenesis; however, the clinical utility of viral approaches is still not evident. Another approach is restoration of BH_4 levels by administration of BH_4 itself or precursors that are metabolized to BH_4. Indeed, BH_4 supplementation in humans and in animal models improves endothelial dysfunction and blood flow (61). Also, as a beneficial effect of statin-based drugs, statins exert multiple functions to improve eNOS functions. Statins stabilize *eNOS* mRNA and protein levels (62) and post-translationally activate the protein kinase, Akt. Akt in turn phosphorylates eNOS (on serine 1177 in human eNOS), increasing NO release at rest and after stimulation (63). Physiologically, these mechanisms may account for the broadly accepted beneficial actions of statins that occur in the absence of cholesterol lowering. In animal models, this seems to be correct because statins do not improve blood flow in models of limb ischemia (48) or stroke (64) performed in *eNOS*$^{-/-}$ mice. Finally, a more thorough molecular understanding of how eNOS is negatively regulated by its interaction with the coat protein for caveolae formation, caveolin-1, may lead to caveolin-1 antagonists that should promote NO release (65).

KEY POINTS

- NO is the first gaseous signaling molecule discovered; it can serve as an autocrine and paracrine regulator of blood vessel function.
- In mammals, three NOS isoforms produce NO. These enzymes are a highly conserved gene family that are regulated by distinct transcriptional, post-transcriptional, and post-translational control mechanisms. The endothelial NOS (eNOS) isoform plays a major role in vascular homeostasis.
- The mechanism of action of NO on vascular cells occurs either via cGMP-dependent pathways or by protein S-nitrosylation. Genetic evidence strongly supports the cGMP/PKG pathway as important for the hemodynamic actions of NO.

Future Goals

- To determine whether maintaining a healthy eNOS/NO balance is beneficial for cardiovascular health
- To understand the transcriptional regulation and post-translational control mechanisms of eNOS, which may lead to novel modalities for enhancing eNOS function

REFERENCES

1 Furchgott RF, Zawadski JV. The obligatory role of the ECs in the relaxation of arterial smooth muscle by acetylcholine. *Nature*. 1981;288:373–376.

2 Hibbs JB Jr., Taintor RR, Vavrin Z. Macrophage cytotoxicity: role for L-arginine deiminase and imino nitrogen oxidation to nitrite. *Science*. 1987;235:473–476.

3 Stuehr D, Marletta MA. Mammalian nitrate biosynthesis: mouse macrophages produces nitrite and nitrate in response to *E. coli* lipopolysaccharide. *Proc Natl Acad Sci USA*. 1985;82:7738–7742.

4 Iyengar R, Stuehr DJ, Marletta MA. Macrophage synthesis of nitrite, nitrate, and N-nitrosamines: precursors and role of the respiratory burst. *Proc Natl Acad Sci USA*. 1985;84:6369–6373.

5 Stuehr DJ, Nathan CF. Nitric oxide: a macrophage product responsible for cytostasis and respiratory inhibition in tumor target sales. *J Exp Med*. 1989;169:1543–1545.

6 Arnold WP, Mittal CK, Katsuki S, Murad F. Nitric oxide activates guanylate cyclase and increases guanosine 3′,5′-cyclic monophosphate levels in various tissue preparations. *Proc Natl Acad Sci USA*. 1977;74:3203–3207.

7 Ignarro LJ, Lippton H, Edwards JC, et al. Mechanism of vascular smooth muscle relaxation by organic nitrates, nitrites,

nitroprusside and nitric oxide: evidence for the involvement of S-nitrosothiol as active intermediates. *J Pharmacol Exp Ther.* 1981;218:739–749.

8 Ignarro LJ, Buga GM, Wood KS, et al. Endothelium-derived relaxing factor produced and released from artery and vein is nitric oxide. *Proc Natl Acad Sci USA.* 1987;84:9265–9269.

9 Palmer RM, Ferrige AG, Moncada S. Nitric oxide release accounts for the biological activity of endothelium-derived relaxing factor. *Nature.* 1987;327:524–526.

10 Gryglewski RJ, Palmer RM, Moncada S. Superoxide anion is involved in the breakdown of endothelium-derived vascular relaxing factor. *Nature.* 1986;320:454–456.

11 Palmer RM, Ashton DS, Moncada S. Vascular endothelial cells synthesize nitric oxide from L-arginine. *Nature.* 1988;333: 664–666.

12 Rees DD, Palmer RM, Moncada S. Role of endothelium-derived nitric oxide in the regulation of blood pressure. *Proc Natl Acad Sci USA.* 1989;86:3375–3378.

13 Mulsch A. Nitric oxide synthase in native and cultured endothelial cells: calcium/calmodulin and tetrahydrobiopterin are cofactors. *J Cardiovasc Pharmacol.* 1991;17:552–556.

14 Forstermann U, Pollock JS, Schmidt HH, et al. Calmodulin-dependent endothelium-derived relaxing factor/nitric oxide synthase activity is present in the particulate and cytosolic fractions of bovine aortic endothelial cells. *Proc Natl Acad Sci USA.* 1991;88:1788–1792.

15 Bredt DS, Snyder SH. Isolation of nitric oxide synthase, a calmodulin-dependent nitric oxide synthase from porcine cerebellum. Cofactor-role, of tetrahydrobiopterin. *J Biol Chem.* 1990; 266:23790–23795.

16 Pollock JS, Forstermann U, Mitchell JA, et al. Purification and characterization of particulate endothelium-derived relaxing factor synthase from cultured and native bovine aortic endothelial cells. *Proc Natl Acad Sci USA.* 1991;88:10480–10484.

17 Schmidt HHHW, Pollock JS, Nakane M, et al. Purification of a soluble isoform of guanylyl cyclase-activating-factor synthase. *Proc Natl Acad Sci USA.* 1991;99:365–369.

18 Stuehr DJ, Cho HJ, Kwon NS, et al. Purification and characterization of the cytokine-induced macrophage nitric oxide synthase: an FAD- and FMN-containing flavoprotein. *Proc Natl Acad Sci USA.* 1991;88:7773–7777.

19 Bredt DS, Hwang PM, Glatt CE, et al. Cloned and expressed nitric oxide synthase structurally resembles cytochrome P-450 reductase. *Nature.* 1991;351:714–718.

20 Xie QW, Cho HJ, Calaycay J, et al. Cloning and characterization of inducible nitric oxide synthase from mouse macrophages. *Science.* 1992;256:225–228.

21 Janssens SP, Shimouchi A, Quertermous T, et al. Cloning and expression of a cDNA encoding human endothelium-derived relaxing factor/nitric oxide synthase [published erratum appears in *J Biol Chem.* 1992;267(31):22694]. *J Biol Chem.* 1992;267: 14519–14522.

22 Sessa WC, Harrison JK, Barber CM, et al. Molecular cloning and expression of a cDNA encoding endothelial cell nitric oxide synthase. *J Biol Chem.* 1992;267:15274–15276.

23 Lamas S, Marsden PA, Li GK, et al. Endothelial nitric oxide synthase: molecular cloning and characterization of a distinct constitutive enzyme isoform. *Proc Natl Acad Sci USA.* 1992;89: 6348–6352.

24 Nishida K, Harrison DG, Navas JP, et al. Molecular cloning and characterization of the constitutive bovine aortic endothelial cell nitric oxide synthase. *J Clin Invest.* 1992;90:2092–2096.

25 Navarro-Gonzalez R, McKay CP, Mvondo DN. A possible nitrogen crisis for Archaean life due to reduced nitrogen fixation by lightning. *Nature.* 2001;412:61–64.

26 Torreilles J. Nitric oxide: one of the more conserved and widespread signaling molecules. *Front Biosci.* 2001;6:D1161–D1172.

27 Kers JA, Wach MJ, Krasnoff SB, et al. Nitration of a peptide phytotoxin by bacterial nitric oxide synthase. *Nature.* 2004;429: 79–82.

28 Stuehr DJ. Structure-function aspects in the nitric oxide synthases. *Annu Rev Pharmacol Toxicol.* 1997;37:339–359.

29 Ghosh DK, Salerno JC. Nitric oxide synthases: domain structure and alignment in enzyme function and control. *Front Biosci.* 2003;8:D193–D209.

30 Pritchard KA Jr., Ackerman AW, Gross ER, et al. Heat shock protein 90 mediates the balance of nitric oxide and superoxide anion from endothelial nitric-oxide synthase. *J Biol Chem.* 2001; 276:17621–17624.

31 Papapetropoulos A, Rudic RD, Sessa WC. Molecular control of nitric oxide synthases in the cardiovascular system. *Cardiovasc Res.* 1999;43:509–520.

32 Wang Y, Marsden PA. Nitric oxide synthases: gene structure and regulation. *Adv Pharmacol.* 1995;34:71–90.

33 Forstermann U, Boissel JP, Kleinert H. Expressional control of the "constitutive" isoforms of nitric oxide synthase (NOS I and NOS III). *FASEB J.* 1998;12:773–790.

34 Sessa WC, Pritchard K, Seyedi N, et al. Chronic exercise in dogs increases coronary vascular nitric oxide production and endothelial cell nitric oxide synthase gene expression. *Circ Res.* 1994;74:349–353.

35 Fulton D, Gratton JP, Sessa WC. Post-translational control of endothelial nitric oxide synthase: why isn't calcium/calmodulin enough? *J Pharmacol Exp Ther.* 2001;299:818–824.

36 Huang PL. Mouse models of nitric oxide synthase deficiency. *J Am Soc Nephrol.* 2000;(11 Suppl 16):S120–S123.

37 Ortiz PA, Garvin JL. Cardiovascular and renal control in NOS-deficient mouse models. *Am J Physiol Regul Integr Comp Physiol.* 2003;284:R628–R638.

38 Ziche M, Morbidelli L, Choudhuri R, et al. Nitric oxide synthase lies downstream from vascular endothelial growth factor-induced but not basic fibroblast growth factor-induced angiogenesis. *J Clin Invest.* 1997;99:2625–2634.

39 Papapetropoulos A, Garcia-Cardena G, Madri JA, Sessa WC. Nitric oxide production contributes to the angiogenic properties of vascular endothelial growth factor in human endothelial cells. *J Clin Invest.* 1997;100:3131–3139.

40 Huang PL, Huang Z, Mashimo H, et al. Hypertension in mice lacking the gene for endothelial nitric oxide synthase [See comments]. *Nature.* 1995;377:239–242.

41 Shesely EG, Maeda N, Kim HS, et al. Elevated blood pressures in mice lacking endothelial nitric oxide synthase. *Proc Natl Acad Sci USA.* 1996;93:13176–13181.

42 Carmeliet P, Ferreira V, Breier G, et al. Abnormal blood vessel development and lethality in embryos lacking a single VEGF allele. *Nature.* 1996;380:435–439.

43 Ferrara N, Carver-Moore K, Chen H, et al. Heterozygous embryonic lethality induced by targeted inactivation of the VEGF gene. *Nature.* 1996;380:439–442.

44 Murohara T, Asahara T, Silver M, et al. Nitric oxide synthase modulates angiogenesis in response to tissue ischemia. *J Clin Invest*. 1998;101:2567–2578.

45 Fukumura D, Gohongi T, Kadambi A, et al. Predominant role of endothelial nitric oxide synthase in vascular endothelial growth factor-induced angiogenesis and vascular permeability. *Proc Natl Acad Sci USA*. 2001;98:2604–2609.

46 Aicher A, Heeschen C, Mildner-Rihm C, et al. Essential role of endothelial nitric oxide synthase for mobilization of stem and progenitor cells. *Nat Med*. 2003;9:1370–1376.

47 Gratton JP, Lin MI, Yu J, et al. Selective inhibition of tumor microvascular permeability by cavtratin blocks tumor progression in mice. *Cancer Cell*. 2003;4:31–39.

48 Sata M, Nishimatsu H, Suzuki E, et al. Endothelial nitric oxide synthase is essential for the HMG-CoA reductase inhibitor cerivastatin to promote collateral growth in response to ischemia. *FASEB J*. 2001;15:2530–2532.

49 Hiasa K, Ishibashi M, Ohtani K, et al. Gene transfer of stromal cell-derived factor-1alpha enhances ischemic vasculogenesis and angiogenesis via vascular endothelial growth factor/endothelial nitric oxide synthase-related pathway: next-generation chemokine therapy for therapeutic neovascularization. *Circulation*. 2004; 109:2454–2461.

50 Stamler JS, Lamas S, Fang FC. Nitrosylation. the prototypic redox-based signaling mechanism. *Cell*. 2001;106:675–683.

51 Liu L, Yan Y, Zeng M, et al. Essential roles of S-nitrosothiols in vascular homeostasis and endotoxic shock. *Cell*. 2004;116:617–628.

52 Matsushita K, Morrell CN, Cambien B, et al. Nitric oxide regulates exocytosis by S-nitrosylation of N-ethylmaleimide-sensitive factor. *Cell*. 2003;115:139–150.

53 Pfeifer A, Klatt P, Massberg S, et al. Defective smooth muscle regulation in cGMP kinase I-deficient mice. *EMBO J*. 1998;17:3045–3051.

54 Sanz MJ, Hickey MJ, Johnston B, et al. Neuronal nitric oxide synthase (NOS) regulates leukocyte-EC interactions in endothelial NOS deficient mice. *Br J Pharmacol*. 2001;134:305–312.

55 Shen W, Xu X, Ochoa M, et al. Role of nitric oxide in the regulation of oxygen consumption in conscious dogs. *Circ Res*. 1994; 75:1086–1095.

56 Shen W, Hintze TH, Wolin MS. Nitric oxide. An important signaling mechanism between vascular endothelium and parenchymal cells in the regulation of oxygen consumption. *Circulation*. 1995;92:3505–3512.

57 Loke KE, Messina EJ, Shesely EG, et al. Potential role of eNOS in the therapeutic control of myocardial oxygen consumption by ACE inhibitors and amlodipine. *Cardiovasc Res*. 2001;49:86–93.

58 Nisoli E, Falcone S, Tonello C, et al. Mitochondrial biogenesis by NO yields functionally active mitochondria in mammals. *Proc Natl Acad Sci USA*. 2004;101:16507–16512.

59 Nisoli E, Tonello C, Cardile A, et al. Calorie restriction promotes mitochondrial biogenesis by inducing the expression of eNOS. *Science*. 2005;310:314–317.

60 Rudic RD, Sessa WC. Nitric oxide in endothelial dysfunction and vascular remodeling: clinical correlates and experimental links. *Am J Hum Genet*. 1999;64:673–677.

61 Channon KM. Tetrahydrobiopterin: regulator of endothelial nitric oxide synthase in vascular disease. *Trends Cardiovasc Med*. 2004;14:323–327.

62 Laufs U, Liao JK. Post-transcriptional regulation of endothelial nitric oxide synthase mRNA stability by Rho GTPase. *J Biol Chem*. 1998;273:24266–24271.

63 Kureishi Y, Luo Z, Shiojima I, et al. The HMG-CoA reductase inhibitor simvastatin activates the protein kinase Akt and promotes angiogenesis in normocholesterolemic animals. *Nat Med*. 2000;6:1004–1010.

64 Endres M, Laufs U, Huang Z, et al. Stroke protection by 3-hydroxy-3-methylglutaryl (HMG)-CoA reductase inhibitors mediated by endothelial nitric oxide synthase. *Proc Natl Acad Sci USA*. 1998;95:8880–8885.

65 Drab M, Verkade P, Elger M, et al. Loss of caveolae, vascular dysfunction, and pulmonary defects in caveolin-1 gene-disrupted mice. *Science*. 2001;293:2449–2452.

111

Heme Oxygenase and Carbon Monoxide in Endothelial Cell Biology

Hong Pyo Kim, Stefan W. Ryter, and Augustine M. K. Choi

The University of Pittsburgh School of Medicine, Pennsylvania

The realization that endogenous small gas molecules can serve critical functions in vascular biology originated in the 1980s, with the identification of the endothelial cell (EC)-derived relaxing factor as equivalent to the gas nitric oxide (NO), a potent endogenous relaxant of vascular smooth muscle (1–2). NO arises from the enzymatic conversion of L-arginine to L-citrulline, by the action of constitutive and inducible NO synthase (NOS) enzymes (3). The production of NO by endothelial NOS (eNOS) provides an essential component of EC function in the regulation of vascular processes, including vascular tone, as well as the inhibition of platelet aggregation, leukocyte adherence, and smooth muscle cell (SMC) proliferation (Reviewed in Ref. 4, 5). On the other hand, the vasoactive properties of carbon monoxide (CO) have long been recognized in the context of accidental inhalation exposure. CO binds avidly to hemoglobin, with an affinity 250 times that of oxygen, and therefore causes tissue hypoxia and ultimately death at high concentration. In the mid-twentieth century, Sjostrand and later Coburn and colleagues identified the endogenous occurrence of CO in human blood as the consequence of hemoglobin turnover (6–7). The enzymatic mechanism for the endogenous production of CO from heme degradation was identified as early as 1968, with the initial characterization of microsomal heme oxygenase (8–9), which converts heme to equimolar biliverdin-IXα, CO, and free iron. The reaction requires nicotinamide adenine dinucleotide phosphate (NADPH):cytochrome p450 reductase as a source of reducing equivalents, and NADPH:biliverdin reductase (BVR) to convert biliverdin-IXα to bilirubin-IXα. Until the discovery of NO, and the elucidation of its roles in vascular signaling, endogenous CO was regarded by the scientific community as a waste product of metabolism. Increasing evidence supports the notion that, in a fashion similar to the NOS/NO axis, heme oxygenase (HO)-derived CO (HO/CO) also serves a critical role in vascular function, stimulating vasorelaxation while exerting inhibitory effects of platelet aggregation, SMC proliferation, EC apoptosis, and the macrophage-dependent release of pro-inflammatory cytokines (Figure 111.1) (5). In addition to CO generation, antioxidative or protective roles of HO activity have been proposed based on heme removal, generation of antioxidant bile pigments, and/or alteration of iron metabolism (10). This chapter reviews the functional role of the HO/CO system in the vasculature relative to the various cellular constituents, with an emphasis on the regulation of vascular tone and proliferation.

CELLULAR EFFECTS OF CARBON MONOXIDE (GENERAL MECHANISMS)

The known vascular effects of CO involve at least three general pathways, the activation of soluble guanylyl cyclase (sGC); the modulation of mitogen-activated protein kinases (MAPK) (5,11–16), including the upregulation of p38β MAPK; and/or the downregulation of extracellular-signal-regulated kinase (ERK)1/2. Additional potential mechanisms of CO action, including the activation of potassium channels and the inhibition of hemoprotein targets such as inducible NOS (iNOS) or cytochrome p450, have been proposed in the context of vasomotor tone and are discussed later. The antiapoptotic effects of low-dose exogenous CO were demonstrated in ECs and shown to depend on the activation of the p38 MAPK pathway and the upregulation of nuclear factor (NF)-κB–dependent antiapoptotic genes (16–17). The anti-inflammatory effects of HO-derived CO, demonstrated in vivo and in vitro, resulted from the p38 MAPK-dependent inhibition of pro-inflammatory cytokine production and stimulation of anti-inflammatory interleukin (IL)-10 in macrophages (12). Additional mechanisms for the anti-inflammatory effect of HO/CO include the modulation of adhesion molecule expression in ECs (18). The antiproliferative effects of CO were originally described as dependent on stimulation of cyclic guanosine 3′,5′-monophosphate (cGMP) production in vascular SMCs (19), and also as involving the upregulation of p38 MAPK,

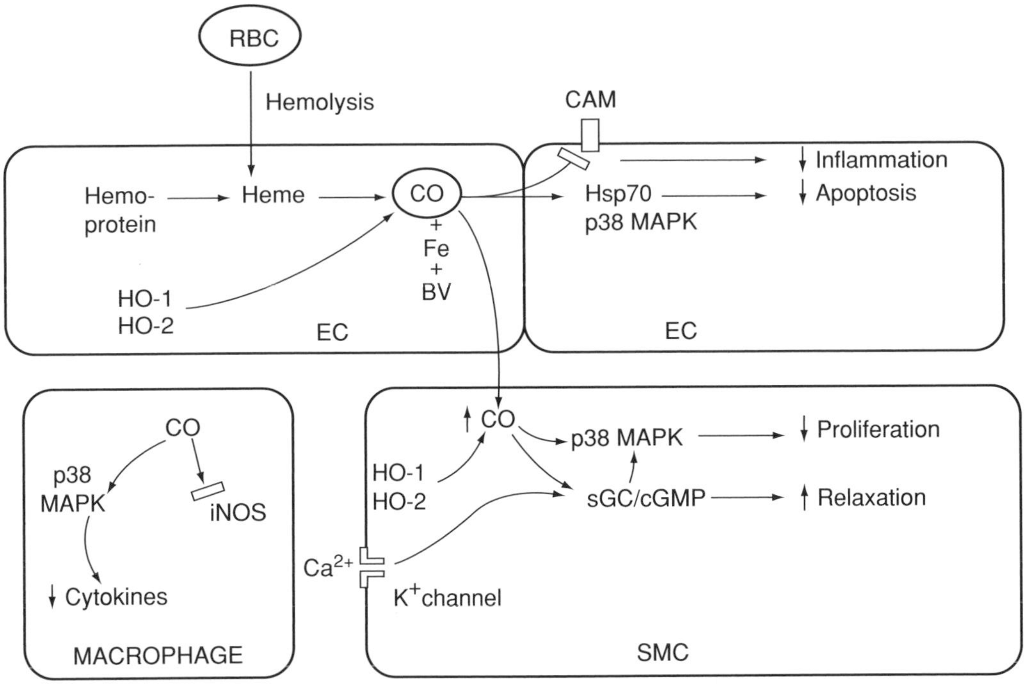

Figure 111.1. Theoretical role of HO isozymes and CO production in vascular cells. Vascular CO arises from the action of basal (HO-2) and inducible (HO-1) heme oxygenase activity. Vascular stress (from hemolysis) or toxins may cause elevation of EC HO activity, leading to EC-derived CO. Similarly, stress conditions (e.g., hypoxia, vascular injury) can cause an elevation of HO-1 in smooth muscle cells (SMCs), leading to increased SMC-derived CO. Regardless of the cellular source, CO can cause autocrine or paracrine effects on cGMP production. CO in ECs may exert antiapoptotic (by modulation of p38 MAPK and HSP70 expression) or anti-inflammatory consequences (by modulation of cellular adhesion molecule [CAM] expression) (102). CO relaxes vascular SMCs by inducing the production of cGMP and, in part, by activating calcium-dependent potassium channels. CO targeting the SMC has antiproliferative effects, also dependent on p38 MAPK activation and increased expression of p21 and caveolin-1. CO targeting macrophages has anti-inflammatory effects dependent on the downregulation of proinflammatory cytokine production and may also inhibit iNOS activation. RBC, red blood cell; BV, biliverdin.

$p21^{Waf1/Cip1}$, and the tumor suppressor protein caveolin-1 (13–15). Finally, CO may act as a neurotransmitter and modulator of the circadian clock in the central nervous system via a sGC-dependent mechanism (reviewed in 5,14).

HEME OXYGENASE ISOZYMES, PROPERTIES, AND DISTRIBUTION IN VASCULAR TISSUE

Three genetically distinct isozymes of HO have been characterized by Maines and colleagues: an inducible form, HO-1, and two constitutively expressed forms, HO-2 and HO-3, representing the products of distinct genes (20–30). Gene transcription of the inducible form HO-1 is induced in most cells and tissues to induction by exposure to various forms of chemical and physical stress (25). HO-1 contains a conserved heme-binding catalytic domain and a C-terminal hydrophobic region that allows integration with cellular membranes (31–32). HO-2, the constitutively expressed isoform, as originally characterized in testes, brain, spleen, and liver, occurs abundantly in most tissues, including the nervous and cardiovascular systems (20–28). HO-1 and HO-2 proteins differ in physical properties, including K_m for substrate, primary sequence, and molecular weight (36 kDa vs. 32 kDa, respectively, in the rat) (22). Furthermore, the expression of HO-2 does not respond to induction by stress agents, but may be regulated by glucocorticoids (33). HO-1 and HO-2 share less than 50% amino acid homology, yet share a conserved domain of 24 amino acid residues (34). HO-2 contains two additional heme-binding domains distinct from the catalytic site, with unclear functionality (35). The existence and functional relevance of a third putative isozyme, HO-3, remains controversial (29–30). As originally characterized, HO-3 demonstrated a high degree of sequence similarity to HO-2 (~90%), but it is devoid of heme catalytic activity (29). The occurrence of HO-1 and HO-2 in vascular tissues, including endothelium and vascular smooth muscle, is consistent with the proposed roles for endogenous HO/CO in vasoregulation (26,36–39). A comparative analysis of relative HO-2 content in different vascular beds is, however, currently lacking.

EVOLUTIONARY CONSIDERATIONS

The HOs display high degrees of amino acid conservation between human and rodent forms (i.e., approximately 80% amino acid homology between human and rat HO-1) (40). Proteins with functional HO activity have been identified in many species from bacteria, fungi, insects, birds, and mammals, to man. Variants of HO-1 have been described in plants such as *Arabidopsis thaliana* (41), the red algae *Cyanidium caldarium* (42), cyanobacteria (43), and in various strains of gram-positive or -negative bacteria (44–48). Recently a novel variant of HO-1, apparently with two catalytic domains, has been cloned and identified in *Escherichia coli* (48).

ENDOGENOUS PRODUCTION OF CARBON MONOXIDE

HO-dependent heme degradation accounts for at least 86% of endogenous CO production in man, with the remaining assigned to non-heme metabolic sources (e.g., cytochrome p450 reactions, lipid oxidation) (49). The non–heme-dependent CO production may increase with exposure to certain xenobiotics (e.g., methylene chloride) (50). With respect to HO activity, hemoglobin accounts for the principal source of substrate heme in tissues specialized in erythrocyte or hemoglobin turnover, such as the liver and spleen. In cell types or tissues not directly engaged in erythrocyte or hemoglobin metabolism, HO enzymes act on a heme pool derived from the turnover of endogenous cellular hemoproteins (e.g., cytochromes) (18). The ubiquitous occurrence of HO isozymes suggests that CO originates in most systemic tissue types as the product of constitutive HO-2 activity (51). This basal production of CO likely increases in parallel with the induction of HO-1 during conditions of cellular stress or injury; however, the source of additional substrate heme under stress conditions remains incompletely understood.

REGULATION OF HEME OXYGENASE-1

The induction of HO-1 represents a general transcriptional response to oxidative cellular stress in response to stimulation from diverse chemical and physical agents (52–54). The *Ho-1* gene promoter region (mouse) contains two upstream enhancer sequences (E1, E2) that occur at –4 kb and –10 kb (relative to the transcriptional start site), respectively (55–57). These enhancers mediate the transcriptional induction of the *Ho-1* gene by a majority of inducing agents, including endotoxins, heavy metal salts, phorbol esters, oxidants, and heme. E1 and E2 contain repeated stress responsive elements (StRE) consisting of recognition sequences for several transcription factors, including Cap'n'collar/basic-leucine zipper proteins, v-maf oncoprotein, and activator protein (AP)-1 (57). The binding of the NF-E2–related factor 2 (Nrf2) to StRE sites appears to represent a major mechanism for the transcriptional regulation of *Ho-1* by stress, including heme, heavy metals, and polyphenolic antioxidants (58–61). Additionally, heme can derepress transcriptional activation of *Ho-1* by binding to the transcriptional repressor Bach1, inhibiting its DNA binding activity and promoting its nuclear export (62–63).

HO-1 can be potentially induced specifically in the vasculature by a variety of stresses, including heme released from vascular hemolysis, oxidized lipoproteins, proinflammatory cytokines and bacterial lipopolysaccharide (LPS), hemodynamic changes, angiotensin, and NO (37,64). In various vascular cell types, HO-1 may respond to changes in oxygen tension (e.g., hyperoxia, hypoxia) (65). Hypoxia and hypoxia–reoxygenation treatments can promote strong induction of HO-1 protein/activity in animal ECs (66) and vascular SMCs (67).

VASCULAR RESPONSES TO INJURY IN TRANSGENIC AND KNOCKOUT MODELS

Experiments with transgenic or knockout mice have revealed the respective roles of HO-1 and/or HO-2 in various models of tissue injury (Table 111-1). For example, homozygous *Ho-1*-null mice ($Ho\text{-}1^{-/-}$) displayed increased mortality in a model of lung ischemia–reperfusion (I/R) (68), and also developed right ventricular dilation and right myocardial infarction during chronic hypoxia (10% O_2), relative to wild-type mice (69). However, $Ho\text{-}1^{-/-}$ did not differ in the development of pulmonary hypertension (69). Transgenic mice with a lung-specific HO-1 overexpressing phenotype resisted the inflammatory and hypertensive effects of hypoxia (70). *Ho-2*–null mice ($Ho\text{-}2^{-/-}$), displayed increased sensitivity to the lethal effects of hyperoxia relative to wild-type mice, despite compensatory increases in HO-1 and accumulated iron in their lungs (71). Embryo fibroblasts with the $Ho\text{-}1^{-/-}$ genotype displayed hypersensitivity to heme and H_2O_2 treatment and generated increased intracellular reactive oxygen species (ROS) production in response to these agents (72). In direct contrast, $Ho\text{-}1^{-/-}$ rats were found to be resistant rather than sensitive to the lethal effects of hyperoxia in vivo (73). The underlying mechanisms for these apparently conflicting observations remain unclear. However, collectively these studies have indicated that HO-1 and HO-2 play critical roles in the physiological distribution of cellular and tissue iron. Aberrant accumulation of tissue iron and low serum iron content have been described in both $Ho\text{-}1^{-/-}$ and $Ho\text{-}2^{-/-}$ transgenic mice (71,73–74). The $Ho\text{-}1^{-/-}$ mice were anemic, accumulated non-heme iron in the kidney and liver, but displayed reduced total iron content in the lung; whereas $Ho\text{-}2^{-/-}$ mice accumulated total lung iron without a compensatory increase in ferritin levels (71,73).

The central importance of HO-1 in tissue protection, as well as its potential clinical significance, was recently highlighted by the discovery of a child diagnosed with HO-1 deficiency, who exhibited extensive endothelial damage. The

Table 111-1: Functionality of HO-1/HO-2 Isozymes as Determined by Knockout and Transgenic Mouse Models

Model	*Observation*	*References*
Human Deficiency		
HO-1	Growth retardation, impaired iron metabolism (anemia, iron deposition)	77, 103, 104
	Chronic inflammation	72, 74
	(renal tubular injury, liver fibrosis, endothelial injury) (anemia, iron deposition)	
Knockout Mouse		
HO-1$^{-/-}$	Loss of defense mechanism against oxidative stress	82
	Increased vascular cell proliferation and hyperplasia after vascular injury	68
	Increased mortality after lung I/R injury	
	Susceptibility to chronic hypoxia (increased right ventricular dilation and right myocardial infarction)	69
HO-1$^{-/-}$ (ApoE$^{-/-}$)	Resistance to hyperoxic lung injury	73
	Accelerated atherosclerotic lesion formation	105
HO-2$^{-/-}$	Reduced breeding efficiency (ejaculatory abnormalities in male)	106
	Normal hippocampal long-term potentiation	107
	Impaired lung iron turnover	
	Susceptibility to lethal effects of hyperoxia	71
Transgenic		
Cardiac-specific	Resistance to cardiac I/R injury in normal and diabetic mice	108, 109
Lung-specific	Reduced lung inflammation, pulmonary hypertension, and vascular hypertrophy during chronic-hypoxia.	110
Neuron-specific	Resistance to brain ischemic injury	111

subject suffered from persistent hemolytic anemia characterized by marked erythrocyte fragmentation and intravascular hemolysis, with increased serum haptoglobin and low bilirubin levels. An abnormal coagulation/fibrinolysis system was described, associated with elevated circulating levels of soluble thrombomodulin and von Willebrand factor (75). Growth retardation, anemia, iron deposition, and vulnerability to stressful injury are similar characteristics, as observed in *Ho-1* gene-deleted mice (74). The underlying mechanisms for the phenotypic observations of anemia and iron deposition in animals or humans genetically altered for content of HO enzymes remain unclear. In cell culture models, Ferris and colleagues have recently described an intracellular ATP-dependent iron pump whose functioning may depend on the presence of HO-1 (76).

THE ROLE OF HEME-OXYGENASE DERIVED CARBON MONOXIDE IN VASOMOTOR CONTROL

The physiological signaling effects of NO and CO known to date involve relatively few defined mechanisms. With respect to vasodilation, sGC represents the principal molecular target of NO, such that the binding of NO to the heme moiety of the enzyme results in activation and the increased production of cGMP (77–78). In a similar fashion, CO binds to the heme iron center of sGC in an interaction that stimulates enzymatic activity. In a comparative study of the vasoactive effects of NO and CO in isolated rabbit aorta, exogenous CO produced an endothelial-independent vasorelaxant response, albeit with 1,000-fold less potency than NO under the same conditions (78). The vasodilatory properties of CO in the rabbit aorta were attributed to activation of sGC and generation of cGMP by CO, also with lesser potency than that of NO (78–79).

In cell culture experiments, endothelial-derived CO was shown to regulate cGMP production in vascular SMCs (19). The treatment of vascular SMCs directly with CO stimulated an increase in cellular cGMP levels. Exposure of these cells to hypoxia also resulted in increased levels of cGMP, an effect that was dependent on HO-1-mediated production of endogenous CO, but not NO (19). CO generated by SMCs in culture could likewise affect cGMP levels in cocultured ECs (80). Thus, CO potentially arises in all vascular cell types as the product of inducible HO activity, and it can act in an autocrine fashion at the site of generation, as well as in a paracrine fashion on other cell types distinct from the cells of origin.

The role of HO isozymes and/or CO in vasoregulation is highlighted by a series of ex vivo experiments using isolated aortic rings. The exogenous application of CO reduced contractility of isolated aortic rings in a cGMP-dependent fashion (81). Furthermore, in isolated rings, the application of metalloporphyrin inhibitors of HO activity, in the presence of NOS inhibitors, typically reduced agonist-dependent vasodilation or augmented agonist-induced vasoconstriction (26,81). Adenovirus-mediated expression of HO-1 in porcine arteries in vivo resulted in attenuated vasoconstriction in subsequently isolated aortic rings in a cGMP-dependent fashion, also in the presence of NOS inhibitors (82). Although many such studies support that exogenous CO causes endothelium-independent vasodilation, recent studies have also proposed that CO may cause a competing endothelium-dependent vasoconstriction based on direct inhibition of endothelial NOS activity by CO (83).

In addition to cGMP-dependent mechanisms of CO-induced vasodilation, as described in aortas, several cGMP-independent mechanisms have been described. Limited studies have implied a potential role for the modulation of cytochrome p-450–derived species in vascular effects of CO (84). Furthermore CO can dilate the peripheral vasculature, in part by activating calcium (Ca^{2+})-dependent potassium channels (K_{Ca}) (85–87).

For example, CO induced an endothelial-independent vasodilation in precontracted rat tail arteries dependent partially on cGMP and partially on large-conductance K_{Ca} (87). CO caused hyperpolarization of SMCs by increasing outward K^+ current, which in turn inhibited voltage-gated Ca^{2+} channels and caused smooth muscle relaxation (88). Furthermore CO increased the open probability and Ca^{2+} sensitivity of single K^+ channels in SMCs (88). In renal intralobar arteries, the inhibition of HO-2 expression, and consequently endogenous CO production, led to a decreased number of open K^+ channels in SMCs and increased vascular contractility, in a fashion reversible by exogenous CO (86). The vasodilatory action of CO observed in porcine cerebral arterioles (89) was further attributed to an increase in the effective coupling of Ca^{2+} sparks to K_{Ca} channels (85). The Ca^{2+} release of the channel blocker ryanodine, which inhibits Ca^{2+} sparks, also inhibited the CO-induced vasodilation (85). In line with this observation, the synthetic CO releasing molecule (CORM)-3 produces aortic vasodilation ex vivo and reduces blood pressure in vivo by modulation of cGMP production and K^+ channel activity (90). The relationship between the sGC/cGMP pathway and K^+ channel activity remains unclear, although recent studies in the carotid body demonstrate an association of HO-2 with the K^+ channel and imply a direct effect of HO/CO on channel activity (91).

HEME OXYGENASE-1/CARBON MONOXIDE IN VASCULAR REMODELING

Vascular remodeling, defined as changes in the size and/or structure of the adult vasculature, not only mediates physiological adaptation and healing, but also can underlie the pathogenesis of major cardiovascular diseases, such as atherosclerosis and hypertension. HO-1 induction appears to function as an adaptive response against these injurious stimuli. In support of this hypothesis, recent studies have shown that HO-1 or CO suppresses the neointima formation associated with chronic graft rejection and vascular injury (13,82,92). These effects were attributed in part to the inhibition of vascular SMC proliferation by exogenous or HO/CO (13,82). HO/CO also may exert anti-inflammatory effects (in macrophages), and derepress the fibrinolytic pathway at sites of vascular inflammation (13,68). Additionally, SMC-derived CO has been shown to exert indirect effects on SMC proliferation by downregulating the expression of endothelial-derived myogenic factors (80).

Recent reports regarding vascular remodeling also have revealed several important genes, including caveolin-1, involved in the pathogenetic process (93). Vascular injury in caveolin-1–deficient mice promotes exaggerated neointimal proliferation. Genetic ablation of caveolin-1 in mice stimulates SMC proliferation (neointimal hyperplasia), with concomitant activation of the p42/44 MAPK cascade and upregulation of cyclin D1 (93). This study prompted the investigation of the molecular relationships between HO-1 and caveolin-1. We and others showed that the activation of the p38β MAPK signaling pathway by CO led to an enhanced expression of caveolin-1 in fibroblasts and SMCs, which in turn mediated the increased expression of $p21^{Waf1/Cip1}$ and the downregulation of cyclin A, leading to growth arrest. The inhibition of intimal hyperplasia by CO treatment in a vascular injury model involved the enhanced expression of caveolin-1 in vascular SMCs (15).

HEME OXYGENASE AND CAVEOLAE

In addition to showing that HO/CO increases caveolin-1 expression, we recently demonstrated a negative feedback pathway in which caveolin-1 inhibits HO activity (Figure 11.2). Like eNOS, HO-1/HO-2 localize in part to the caveolae of ECs, and the activity of all three enzymes is modulated by caveolin-1 (94). In murine pulmonary artery ECs, HO-1, BVR, and NADPH:cytochrome p450 reductase (NPR) colocalize in plasma membrane caveolae (94). The induction of HO-1 by stress conditions or adenoviral-directed gene transfer increases the relative content of this protein in a detergent-resistant fraction containing caveolin-1. Inducible HO activity appears in plasma membrane, cytosol, and isolated caveolae. In addition, caveolae contain endogenous BVR activity, supporting the co-compartmentalization of both enzymes. Caveolin-1 physically interacts with HO-1, as shown by coimmunoprecipitation studies (94). HO activity dramatically increases in cells expressing caveolin-1 antisense transcripts, suggesting a negative regulatory role for caveolin-1 toward HO. Caveolin-1 expression attenuates LPS-inducible HO activity. Thus, caveolae may act as a platform for HO activity that potentially regulates the functional consequences of HO activity in signaling processes.

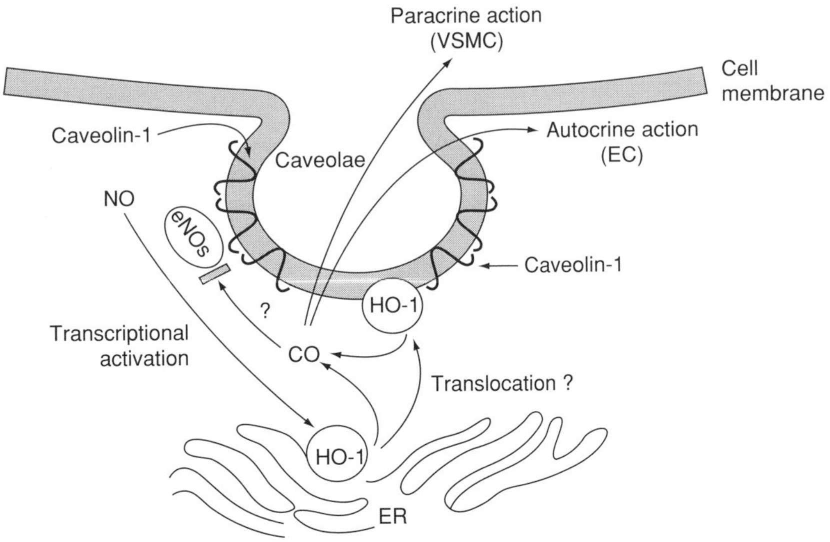

Figure 111.2. Model of HO-1 interaction with caveolae. HO-1 expressed in the endothelium by stress accumulates in the endoplasmic reticulum and also, in part, to plasma membrane caveolae. The mechanisms of translocation are not known. Caveolin-1, a resident and structural protein of the caveolae, binds to and inhibits the activity of both resident eNOS and HO-1, among other proteins. Caveolin-1 knockdown experiments show deregulation of HO activity, suggesting an inhibitory role for caveolin-1 on small gas production in the caveolae. CO (and/or NO) evolving from the caveolae may have both autocrine and paracrine effects on signaling (cGMP synthesis) and gene expression.

BILE PIGMENTS IN ENDOTHELIUM

HO-derived biliverdin is subsequently metabolized to bilirubin by the enzyme BVR. Potential protective effects of bilirubin have been determined using in vivo studies. In a rat model of hyperbilirubinemia, jaundiced Gunn rats displayed a diminished appearance of plasma biomarkers of oxidative stress compared to their nonjaundiced counterparts. These experiments suggested that hyperbilirubinemia may confer an antioxidant benefit to the circulation (95). Similar studies have been conducted in Eizai rats, which are hyperbilirubinemic due to mutations in the canalicular organic anion transporter gene. In this model, hyperbilirubinemia protected against neuronal I/R injury following occlusion of the middle cerebral artery relative to wild-type normobilirubinemic controls (96). A reduced expression of HO-1 and other damage markers was observed at the site of I/R injury in hyperbilirubinemic animals.

THERAPEUTIC APPLICATIONS

Heme Oxygenase-1 Inducers with Distinct Functions

The hydroxymethylglutaryl coenzyme A (HMG-CoA) reductase inhibitors (statins) exert cholesterol-independent, pleiotropic actions that include anti-inflammatory and antioxidative effects. Grosser and colleagues have explored the role of HO-1 as a target and mediator of statin action. In cultured human umbilical vein ECs (HUVECs), simvastatin and lovastatin increase *HO-1* mRNA levels in a concentration- and time-dependent fashion. The results show that HO-1 is a target of statins in ECs (97). Recently, potential antioxidants (e.g., vitamin E, polyphenolics) have received much attention as potential antiatherosclerotic agents. Among the polyphenols with health beneficial properties, resveratrol, a phytoalexin of grape, can protect the vascular wall from oxidation, inflammation, platelet aggregation, and thrombus formation (98). The protective properties of resveratrol have been attributed in part to HO-1 induction (98). Recently, the antiproliferative effects of rapamycin also have been associated with HO-1 induction (99).

Carbon Monoxide Therapy

Several approaches have been used to investigate the therapeutic potential of CO, ranging from direct administration of CO gas to the use of prodrugs that generate CO upon metabolism (90,100,101). A novel approach involves the use of specific CO-releasing molecules (CORMs), which will release measurable, controllable, and effective amounts of CO into biological systems. Transitional metal carbonyls, coordinated around iron, manganese, or ruthenium, recently have been developed as CORMs that, under appropriate conditions, will release CO. Such molecules have been shown to provide cardioprotection in both ex vivo and in vivo experiments (100). To date, CORMs have been largely incompatible with biological systems, because they are only soluble in organic solvents or require preactivation by physical or chemical stimuli. However, the recent development of water-soluble CORMs has provided new opportunities to investigate the pharmacological and biological features of CO without such confounding influences (101). CORM-3, a novel water-soluble CORM, recently has been used to confirm the cardioprotective actions of CO (90). These findings suggest that CORM-3 may have potential for use as a modulator of vascular function and hypertension.

KEY POINTS

- Endogenous HO/CO, in a fashion similar to NO, appears to plays an important role in maintaining EC function.
- HO-1/CO exerts vasoactive properties by activating either the sGC/cGMP pathway or enhancing K_{Ca} channel activity.
- The therapeutic potential of HO-1/CO in vascular disease originates from its antiproliferative and cytoprotective (antiapoptotic and anti-inflammatory) activity.
- CO may conceivably be delivered to humans at therapeutic concentrations by inhalation of mixed gas. Additionally, the development of safe and potent prodrugs that can act as CO donors remains an area of active investigation.

Future Goals

- To gain knowledge of the intracellular signaling mechanisms by which CO controls vital cellular processes in the treatment of human disease
- To develop sensitive methods for the detection of CO production in vitro and in vivo, and to develop specific HO inhibitors as well

REFERENCES

1 Ignarro LJ, Buga GM, Wood KS, et al. Endothelium-derived relaxing factor produced and released from artery and vein is nitric oxide. *Proc Natl Acad Sci USA*. 1987;84:9265–9269.

2 Ignarro LJ, Byrns RE, Buga GM, Wood KS. Endothelium-derived relaxing factor from pulmonary artery and vein possesses pharmacologic and chemical properties identical to those of nitric oxide radical. *Circ Res*. 1987;61:866–879.

3 Moncada S, Palmer RM, Palmer J, Higgs E. Nitric oxide: physiology, pathophysiology, and pharmacology. *Pharmacol Rev*. 1991; 43:109–142.

4 Ojita H, Liao J. Endothelial function and oxidative stress. *Endothelium*. 2004;11:123–132.

5 Ryter SW, Morse D, Choi AM. Carbon monoxide: to boldly go where NO has gone before. *Sci STKE*. 2004;230:RE6.

6 Sjostrand T. Endogenous production of carbon monoxide in man under normal and pathophysiological conditions. *Scand J Clin Lab Invest*. 1949;1:201–214.

7 Coburn RF, Blakemore WS, Forster RE. Endogenous carbon monoxide production in man. *J Clin Invest*. 1963;42:1172–1178.

8 Tenhunen R, Marver H, Schmid R. Microsomal heme oxygenase, characterization of the enzyme. *J Biol Chem*. 1969;244: 6388–6394.

9 Tenhunen R, Marver HS, Schmid R. The enzymatic conversion of heme to bilirubin by microsomal heme oxygenase. *Proc Natl Acad Sci USA*. 1968;61:748–755.

10 Ryter S, Tyrrell RM. The heme synthesis and degradation pathways, role in oxidant sensitivity. Heme oxygenase has both pro- and anti-oxidant properties. *Free Radic Biol Med*. 2000;28:289–309.

11 Furchgott RF, Jothianandan D. Endothelium-dependent and -independent vasodilation involving cyclic GMP: relaxation induced by nitric oxide, carbon monoxide and light. *Blood Vessels* 1991;28:52–61.

12 Otterbein LE, Bach FH, Alam J, et al. Carbon monoxide has anti-inflammatory effects involving the mitogen-activated protein kinase pathway. *Nat Med*. 2000;6:422–428.

13 Otterbein LE, Zuckerbraun BS, Haga M, et al. Carbon monoxide suppresses arteriosclerotic lesions associated with chronic graft rejection and with balloon injury. *Nat Med*. 2003;9: 183–190.

14 Kim HP, Ryter SW, Choi AMK. CO as a cellular signaling molecule. *Annu Rev Pharmacol Toxicol*. 2006;46:411–449.

15 Kim HP, Wang X, Nakao A, et al. Caveolin-1 expression via p38beta mitogen activated protein kinase mediates the antiproliferative effect of carbon monoxide. *Proc Natl Acad Sci USA*. 2005; 102:11319–11324.

16 Brouard S, Otterbein LE, Anrather J, et al. Carbon monoxide generated by heme oxygenase-1 suppresses endothelial cell apoptosis. *J Exp Med*. 2000;192:1015–1026.

17 Brouard S, Berberat PO, Tobiasch E, et al. Heme oxygenase-1-derived carbon monoxide requires the activation of transcription factor NF-kappa B to protect endothelial cells from tumor necrosis factor-alpha-mediated apoptosis. *J Biol Chem*. 2002;277:17950–17961.

18 Soares MP, Seldon MP, Gregoire IP, et al. Heme oxygenase-1 modulates the expression of adhesion molecules associated with endothelial cell activation. *J Immunol*. 2004;172:3553–3563.

19 Morita T, Mitsialis SA, Koike H, et al. Carbon monoxide controls the proliferation of hypoxic vascular smooth muscle cells. *J Biol Chem*. 1997;272:32804–32809.

20 Maines MD. Heme oxygenase: function, multiplicity, regulatory mechanisms, and clinical applications. *FASEB J*. 1988;2:2557–2568.

21 Maines MD, Trakshel GM, Kutty RK. Characterization of two constitutive forms of rat liver microsomal heme oxygenase. *J Biol Chem*. 1986;261:411–419.

22 Trakshel GM, Kutty RK, Maines MD. Purification and characterization of the major constitutive form of testicular heme oxygenase. *J Biol Chem*. 1986;261:11131–11137.

23 Cruse I, Maines MD. Evidence suggesting that the two forms of heme oxygenase are products of different genes. *J Biol Chem*. 1988;263:3348–3353.

24 Trakshel GM, Ewing JF, Maines MD. Heterogeneity of heme oxygenase 1 and 2 isoenzymes. *Biochem J*. 1991;275: 159–164.

25 Ryter S, Otterbein LE, Morse D, Choi AM. Heme oxygenase/carbon monoxide signaling pathways: regulation and functional significance. *Mol Cell Biochem*. 2002;234/235:249–263.

26 Zakhary R, Gaine SP, Dinerman JL, et al. Heme oxygenase 2: endothelial and neuronal localization and role in endothelium-dependent relaxation. *Proc Natl Acad Sci USA.* 1996;93:795–798.

27 Braggins PE, Trakshel GM, Kutty RK, Maines MD. Characterization of two heme oxygenase isoforms in rat spleen: comparison with the hematin-induced and constitutive isoforms of the liver. *Biochem Biophys Res Commun.* 1986;141:528–533.

28 Trakshel GM, Kutty RK, Maines MD. Resolution of rat brain heme oxygenase activity: absence of a detectable amount of the inducible form (HO-1). *Arch Biochem Biophys.* 1988;260:732–739.

29 McCoubrey WK, Huang TJ, Maines MD. Isolation and characterization of a cDNA from the rat brain that encodes hemoprotein heme oxygenase-3. *Eur J Biochem.* 1997;247:725–732.

30 Hayashi S, Omata Y, Sakamoto H, et al. Characterization of rat heme oxygenase-3 gene. Implication of processed pseudogenes derived from heme oxygenase-2 gene. *Gene.* 2004;336:241–250.

31 Shibahara S, Müller RM, Taguchi H, Yoshida T. Cloning and expression of cDNA for rat heme oxygenase. *Proc Natl Acad Sci USA.* 1985;82:7865–7869.

32 Ishikawa K, Sato M, Yoshida T. Expression of rat heme oxygenase in *Escherichia coli* as a catalytically active, full length form that binds to membranes. *Eur J Biochem.* 1991;202:161–165.

33 Raju VS, McCoubrey WK Jr., Maines MD. Regulation of heme oxygenase-2 by glucocorticoids in neonatal rat brain: characterization of a functional glucocorticoid response element. *Biochim Biophys Acta.* 1997;1351:89–104.

34 Rotenberg MO, Maines MD. Characterization of a cDNA-encoding rabbit brain heme oxygenase-2 and identification of a conserved domain among mammalian heme oxygenase isozymes: possible heme-binding site? *Arch Biochem Biophys.* 1991;290:336–344.

35 McCoubrey WK, Huang TJ, Maines MD. Heme oxygenase-2 is a hemoprotein and binds heme through heme regulatory motifs that are not involved in heme catalysis. *J Biol Chem.* 1997;272:12568–12575.

36 Christodoulides N, Durante W, Kroll MH, Schafer AI. Vascular smooth muscle cell heme oxygenases generate guanylate cyclase stimulatory carbon monoxide. *Circulation.* 1995;91:2306–2309.

37 Durante W, Schafer AI. Carbon monoxide and vascular cell function. *Int J Mol Med.* 1998;2:255–262.

38 Zhang F, Kaide J, Rodriguez-Mulero F, et al. Vasoregulatory function of the heme-heme oxygenase system. *Am J Hypertens.* 2001;14:S62-S67.

39 Zhang F, Kaide J, Wei Y, et al. Carbon monoxide produced by isolated arterioles attenuates pressure-induced vasoconstriction. *Am J Physiol Heart Circ Physiol.* 2001;281:H350-H358.

40 Yoshida T, Biro T, Cohen T, Muller R. Human heme oxygenase cDNA and induction of its mRNA by hemin. *Eur J Biochem.* 1988;171:457–461.

41 Muramoto T, Kohci T, Yakota A, et al. The *Arabidopsis* photomorphogenic mutant hyl is deficient in phytochrome chromophore biosynthesis as a result of a mutation in a plastid heme oxygenase. *Plant Cell.* 1999;11:335–348.

42 Beale SI, Cornejo J. Enzymatic heme oxygenase activity in soluble extracts of the unicellular red alga, *Cyanidium caldarium. Arch Biochem Biophys.* 1984;235:371–384.

43 Cornejo J, Willows RD, Beale SI. Phytobilin biosynthesis: cloning and expression of a gene encoding soluble ferredoxin-dependent heme oxygenase from *Synechocystis* sp. PCC 6803. *Plant J.* 1998;15:99–107.

44 Wilks A, Schmitt MP. Expression and characterization of a heme oxygenase (Hmu O) from *Corynebacterium diphtheriae.* Iron acquisition requires oxidative cleavage of the heme macrocycle. *J Biol Chem.* 1998;273:837–841.

45 Zhu W, Wilks A, Stojiljkovic I. Degradation of heme gram-negative bacteria: the product of the *hemO* Gene of *Neisseriae* is a heme oxygenase. *J Bacteriol.* 2000;182:6783–6790.

46 Ratliff M, Zhu W, Deshmukh R, et al. Homologues of neisserial heme oxygenase in gram-negative bacteria: degradation of heme by the product of the *pigA* gene of *Pseudomonas aeruginosa. J Bacteriol.* 2001;183:6394–6403.

47 Frankenberg-Dinkel N. Bacterial heme oxygenases. *Antioxid Redox Signal.* 2004;6:825–834.

48 Suits MD, Pal GP, Nakatsu K, et al. Identification of an *Escherichia coli* O157:H7 heme oxygenase with tandem functional repeats. *Proc Natl Acad Sci USA.* 2005;102:16955–16960.

49 Vreman HJ, Wong RJ, Stevenson DK. Carbon monoxide in breath, blood, and other tissues. In: Penney DG, ed. *Carbon Monoxide Toxicity.* Boca Raton: CRC Press, 2000:19–60.

50 Ryter S, Choi AMK. Signal transduction and the gasotransmitters. NO, CO, and H_2S in biology and medicine. In: Wang R, ed. *Carbon Monoxide Synthesis and Metabolism.* New York: Humana Press, 2004:187–203.

51 Maines MD. The heme oxygenase system: a regulator of second messenger gases. *Annu Rev Pharmacol Toxicol.* 1997;37:517–554.

52 Keyse SM, Applegate LA, Tromvoukis Y, Tyrrell RM. Oxidant stress leads to transcriptional activation of the human heme oxygenase gene in cultured skin fibroblasts. *Mol Cell Biol.* 1990;10:4967–4969.

53 Keyse SM, Tyrrell RM. Heme oxygenase is the major 32-kDa stress protein induced in human skin fibroblasts by UVA radiation, hydrogen peroxide, and sodium arsenite. *Proc Natl Acad Sci USA.* 1989;86:99–103.

54 Applegate LA, Luscher P, Tyrrell RM. Induction of heme oxygenase: a general response to oxidant stress in cultured mammalian cells. *Cancer Res.* 1991;51:974–978.

55 Alam J. Multiple elements within the 5′ distal enhancer of the mouse heme oxygenase-1 gene mediate induction by heavy metals. *J Biol Chem.* 1994;269:25049–25056.

56 Alam J, Camhi S, Choi AM. Identification of a second region upstream of the mouse heme oxygenase-1 gene that functions as a basal level and inducer-dependent transcription enhancer. *J Biol Chem.* 1995;270:11977–11984.

57 Choi AM, Alam J. Heme oxygenase-1: function, regulation, and implication of a novel stress-inducible protein in oxidant-induced lung injury. *Am J Respir Cell Mol Biol.* 1996;15:9–19.

58 Alam J, Stewart D, Touchard C, et al. Nrf2, a Cap'n'Collar transcription factor, regulates induction of the heme oxygenase-1 gene. *J Biol Chem.* 1999;274:26071–26078.

59 Alam J, Wicks C, Stewart D, et al. Mechanism of heme oxygenase-1 gene activation by cadmium in MCF-7 mammary epithelial cells. Role of p38 kinase and Nrf2 transcription factor. *J Biol Chem.* 2000;275:27694–27702.

60 Li N, Venkatesan MI, Miguel A, et al. Induction of heme oxygenase-1 expression in macrophages by diesel exhaust particle chemicals and quinones via the antioxidant-responsive element. *J Immunol.* 2000;165:3393–3401.

61 Balogun E, Hoque M, Gong P, et al. Curcumin activates the haem oxygenase-1 gene via regulation of Nrf2 and the antioxidant-responsive element. *Biochem J.* 2003;371:887–895.

62 Ogawa K, Sun J, Taketani S, et al. Heme mediates derepression of Maf recognition element through direct binding to transcription repressor Bach1. *EMBO J*. 2001;20:2835–2843.
63 Sun J, Hoshino H, Takaku K, et al. Hemoprotein Bach1 regulates enhancer availability of heme oxygenase-1 gene. *EMBO J*. 2002;21:5216–5224.
64 Abraham NG, Kappas A. Heme oxygenase and the cardiovascular-renal system. *Free Radic Biol Med*. 2005;39:1–25.
65 Ryter S, Choi AMK. Heme oxygenase-1: molecular mechanisms of gene expression in oxygen-related stress. *Antioxid Redox Signal*. 2002;4:625–632.
66 Ryter S, Si ML, Lai C-C, Su CY. Regulation of endothelial heme oxygenase activity during hypoxia is dependent on intracellular chelatable iron. *Am J Physiol Heart Circ Physiol*. 2000;279: H2889–H2897.
67 Morita T, Mitsialis SA, Koike H, et al. Carbon monoxide controls the proliferation of hypoxic vascular smooth muscle cells. *J Biol Chem*. 1997;272:32804–32809.
68 Fujita T, Toda K, Karimova A, et al. Paradoxical rescue from ischemic lung injury by inhaled carbon monoxide driven by derepression of fibrinolysis. *Nat Med*. 2001;7:598.
69 Yet SF, Perrella MA, Layne MD, et al. Hypoxia induces severe right ventricular dilatation and infarction in heme oxygenase-1 null mice. *J Clin Invest*. 1999;103:R23–R29.
70 Minamino T, Christou H, Hsieh CM, et al. Targeted expression of heme oxygenase-1 prevents the pulmonary inflammatory and vascular responses to hypoxia. *Proc Natl Acad Sci USA*. 2001;98: 8798–8803.
71 Dennery PA, Spitz DR, Yang G, et al. Oxygen toxicity and iron accumulation in the lungs of mice lacking heme oxygenase-2. *J Clin Invest*. 1998;101:1001–1011.
72 Poss KD, Tonegawa S. Reduced stress defense in heme oxygenase 1-deficient cells. *Proc Natl Acad Sci USA*. 1997;94:10925–10930.
73 Dennery PA, Visner G, Weng YH, et al. Resistance to hyperoxia with heme oxygenase-1 disruption: role of iron. *Free Radic Biol Med*. 2003;34:124–133.
74 Poss KD, Tonegawa S. Heme oxygenase 1 is required for mammalian iron reutilization. *Proc Natl Acad Sci USA*. 1997;94: 10919–10924.
75 Yachie A, Niida Y, Wada T, et al. Oxidative stress causes enhanced endothelial cell injury in human heme oxygenase-1 deficiency. *J Clin Invest*. 1999;103:129–135.
76 Ferris CD, Jaffrey SR, Sawa A, et al. Haem oxygenase-1 prevents cell death by regulating cellular iron. *Nat Cell Biol*. 1999;1:152–157.
77 Ignarro LJ, Degnan JN, Baricos WH, et al. Activation of purified guanylate cyclase by nitric oxide requires heme. Comparison of heme-deficient, heme-reconstituted and heme-containing forms of soluble enzyme from bovine lung. *Biochim Biophys Acta*. 1982;718:49–59.
78 Furchgott RF, Jothianandan D. Endothelium-dependent and -independent vasodilation involving cyclic GMP: relaxation induced by nitric oxide, carbon monoxide and light. *Blood Vessels*. 1991;28:52–61.
79 Hussain AS, Marks GS, Brien JF, Nakatsu K. The soluble guanylate cyclase inhibitor 1H-[1,2,4]oxadiazolo[4,3,-alpha]quinoxalin-1-one (ODQ) inhibits relaxation of rabbit aortic rings induced by carbon monoxide, nitric oxide, and glyceryl trinitrate. *Can J Physiol Pharmacol*. 1997;75: 1034–1037.
80 Morita T, Kourembanas S. Endothelial cell expression of vasoconstrictors and growth factors is regulated by smooth muscle cell-derived carbon monoxide. *J Clin Invest*. 1995;96:2676–2682.
81 Caudill TK, Resta TC, Kanagy NL, Walker BR. Role of endothelial carbon monoxide in attenuated vasoreactivity following chronic hypoxia. *Am J Physiol Regul Integr Comp Physiol*. 1998; 275:R1025–R1030.
82 Duckers HJ, Boehm M, True AL, et al. Heme oxygenase-1 protects against vascular constriction and proliferation. *Nat Med*. 2001;7:693–698.
83 Johnson FK, Johnson RA. Carbon monoxide promotes endothelium-dependent constriction of isolated gracilis muscle arterioles. *Am J Physiol Regul Integr Comp Physiol*. 2003;285: R536–R541.
84 Coceani F, Kelsey L, Seidlitz E. Carbon monoxide-induced relaxation of the ductus arteriosis in the lamb: evidence against the prime role of guanylyl cyclase. *Br J Pharmacol*. 1996;118: 1689–1696.
85 Jaggar JH, Leffler CW, Cheranov SY, et al. Carbon monoxide dilates cerebral arterioles by enhancing the coupling of $Ca2^+$ sparks to $Ca2^+$-activated K^+ channels. *Circ Res*. 2002;91:610–617.
86 Kaide JI, Zhang F, Wei Y, et al. Carbon monoxide of vascular origin attenuates the sensitivity of renal arterial vessels to vasoconstrictors. *J Clin Invest*. 2001;107: 1163–1171.
87 Wang R, Wang Z, Wu L. Carbon monoxide-induced vasodilation and the underlying mechanisms. *Br J Pharmacol*. 1997; 121:927–934.
88 Wang R, Wu L, Wang Z. The direct effect of carbon monoxide on KCa channels in vascular smooth muscle cells. *Eur J Physiol*. 1997;434:285–291.
89 Leffler CW, Nasjletti A, Yu C, et al. Carbon monoxide and cerebral microvascular tone in newborn pigs. *Am J Physiol Heart Circ Physiol*. 1999;276:H1641–H1646.
90 Foresti R, Hammad J, Clark JE, et al. Vasoactive properties of CORM-3, a novel water-soluble carbon monoxide–releasing molecule. *Br J Pharmacol*. 2004;142:453–460.
91 Williams SE, Wootton P, Mason HS, et al. Hemoxygenase-2 is an oxygen sensor for a calcium-sensitive potassium channel. *Science*. 2004;306:2093–2097.
92 Soares MP, Lin Y, Anrather J, et al. Expression of heme oxygenase-1 can determine cardiac xenograft survival. *Nat Med*. 1998;4:1073–1077.
93 Hassan GS, Jasmin JF, Schubert W, et al. Caveolin-1 deficiency stimulates neointima formation during vascular injury. *Biochemistry*. 2004;43:8312–8321.
94 Kim HP, Wang X, Galbiati F, et al. Caveolae compartmentalization of heme oxygenase-1 in endothelial cells. *FASEB J*. 2004;18: 1080–1089.
95 Dennery PA, McDonagh AF, Spitz DR, et al. Hyperbilirubinemia results in reduced oxidative injury in neonatal Gunn rats exposed to hyperoxia. *Free Radic Biol Med*. 1995;19:395–404.
96 Kitamura Y, Ishida Y, Takata K, et al. Hyperbilirubinemia protects against focal ischemia in rats. *J Neurosci Res*. 2003;71:544–550.
97 Grosser N, Hemmerle A, Berndt G, et al. The antioxidant defense protein heme oxygenase 1 is a novel target for statins in endothelial cells . *Free Radic Biol Med*. 2004;37:2064–2071.

98 Juan SH, Cheng TH, Lin HC, et al. Mechanism of concentration-dependent induction of heme oxygenase-1 by resveratrol in human aortic smooth muscle cells. *Biochem Pharmacol.* 2005; 69:41–48.

99 Visner GA, Lu F, Zhou H, et al. Rapamycin induces heme oxygenase-1 in human pulmonary vascular cells: implications in the antiproliferative response to rapamycin. *Circulation.* 2003;107:911–916.

100 Motterlini R, Sawle P, Hammad J, et al. CORM-A1: a new pharmacologically active carbon monoxide-releasing molecule. *FASEB J.* 2005;19:284–286.

101 Motterlini R, Mann BE, Foresti R. Therapeutic applications of carbon monoxide-releasing molecules. *Expert Opin Investig Drugs.* 2005;14:1305–1318.

102 Kim HP, Wang X, Zhang J, et al. Heat shock protein-70 mediates the protective effect of carbon monoxide: involvement of p38 MAPK and heat shock factor 1. *J Immunol.* 2005;175:2622–2629.

103 Ohta K, Yachie A, Fujimoto K, et al. Tubular injury as a cardinal pathologic feature in human heme oxygenase-1 deficiency. *Am J Kidney Dis.* 2000;35:863–870.

104 Kawashima A, Oda Y, Yachie A, et al. Heme oxygenase-1 deficiency: the first autopsy case. *Hum Pathol.* 2002;33:125–130.

105 Yet SF, Layne MD, Liu X, et al. Absence of heme oxygenase-1 exacerbates atherosclerotic lesion formation and vascular remodeling. *FASEB J.* 2003;17:1759–1761.

106 Burnett AL, Johns DG, Kriegsfeld LJ, et al. Ejaculatory abnormalities in mice with targeted disruption of the gene for heme oxygenase-2. *Nat Med.* 1998;4:84–87.

107 Poss KD, Thomas MJ, Ebralidze AK, et al. Hippocampal long-term potentiation is normal in heme oxygenase-2 mutant mice. *Neuron.* 1995;15:867–873.

108 Yet SF, Tian R, Layne MD, et al. Cardiac-specific expression of heme oxygenase-1 protects against ischemia and reperfusion injury in transgenic mice. *Circ Res.* 2001;89:168–73.

109 Liu X, Wei J, Peng DH, et al. Absence of heme oxygenase-1 exacerbates myocardial ischemia/reperfusion injury in diabetic mice. *Diabetes.* 2005;54:778–784.

110 Minamino T, Christou H, Hsieh CM, et al. Targeted expression of heme oxygenase-1 prevents the pulmonary inflammatory and vascular responses to hypoxia. *Proc Natl Acad Sci USA.* 2001; 98:8798–8803.

111 Panahian N, Yoshiura M, Maines MD. Overexpression of heme oxygenase-1 is neuroprotective in a model of permanent middle cerebral artery occlusion in transgenic mice. *J Neurochem.* 1999;72:1187–1203.

112

Endothelial Eicosanoids

Kenneth K. Wu

Institute of Molecular Medicine, The University of Texas Health Science Center at Houston

Eicosanoids comprise a large array of small-molecular-weight compounds that are synthesized in almost all mammalian cells from an unsaturated fatty acid, arachidonic acid (AA), chemically known as eicosatetraenoic acid. Eicosanoids are not stored in cells, but are synthesized in cells in response to stimuli. Elevated intracellular calcium and activation of protein kinase C (PKC) as the result of cell activation represent two key signaling pathways that initiate eicosanoid production. These signaling molecules activate a group of phospholipase A_2 (PLA_2), most notably the cytosolic PLA_2, that catalyze the release of AA from membrane phospholipids (1,2). AA is metabolized to form eicosanoids by many enzymes, now classified into three enzymatic pathways: (a) cyclooxygenase (COX), (b) lipoxygenase (LOX), and (c) cytochrome P450 (CYP) (Figure 112.1). The COX pathway converts AA into classic prostaglandins (PG), namely prostacyclin (also known as PGI_2) and thromboxane A_2 (TXA_2). The LOX pathway is responsible for producing leukotrienes (LT) and 5-, 12-, and 15-hydroxyeicosatetraenoic acid (HETE) (3,4). The CYP pathway generates diverse epoxy eicosatetraenoic acid (EET) and dihydroxy eicosatetraenoic acid (di-HETE) products (5).

Intracellular eicosanoids are released into extracellular milieu, where they act in an autocrine or paracrine manner. They bind a specific family of membrane eicosanoid receptors and induce diverse cellular activities via selective signaling pathways (6). Evidence is emerging that certain eicosanoids that are produced in the cell bind directly to intracellular receptors, thereby eliciting their cellular actions. Eicosanoids possess multiple, diverse biological actions that regulate and mediate the physiological functions of many organs. They play major pathophysiological roles in human diseases including cardiovascular disorders, asthma, and cancer (7–9).

The vascular endothelium occupies a pivotal position in eicosanoid biosynthesis. It is capable of synthesizing eicosanoids via all three metabolic pathways, and the eicosanoids generated are involved in mediating and regulating a variety of important physiological functions including vascular tone, blood cell reactivity, and vascular wall integrity.

HISTORY

Prostaglandins were discovered independently, in 1933–1934, by Goldblatt (10) and von Euler (11). They described an endogenous depressor substance in human seminal fluid and prostate glands. The substance was named prostaglandin by von Euler (12), and was subsequently isolated by Berström and Sjövall in 1960 (13). The purified substance comprised two chemically related compounds that were named PGE_1 and $PGF_{1\alpha}$. Additional stable prostaglandins were isolated and their structures determined during the next several years by Berström, Samuelsson, and coworkers (14,15). All the purified compounds contain 20 carbons with unsaturated bonds. It was subsequently demonstrated that prostaglandins are biosynthesized from polyunsaturated fatty acids such as AA (C20:4ω6) (16,17). Prostaglandin endoperoxides PGG_2 and PGH_2 were identified in 1973–1974 (18,19), which led to the discovery of TXA_2 (20). Prostacyclin was discovered by Vane's group in 1976, as an antiaggregatory and vasodilatory substance produced by aortic tissues (21); its chemical structure was solved in the same year (22). The actions of PGI_2 on platelets and vascular smooth muscle cells (VSMCs) via cyclic adenosine monophosphate (cAMP)/protein kinase A (PKA) signaling pathway was delineated in the late 1970s (23).

Prostaglandin H synthase (PGHS)-1 (also known as cyclooxygenase or COX-1) was purified to homogeneity during the late 1970s, and its cDNA was cloned about 10 years later (24,25). COX-2 (PGHS-2) was cloned independently during the early 1990s by two groups (26,27). PGI synthase (PGIS) was described as an enzyme of PGI_2 biosynthesis in arterial tissues during the late 1970s (21). It was purified about 10 years later (28), and its cDNA was cloned during the mid-1990s by two laboratories (29,30).

Leukotrienes and CYP metabolites were discovered during the 1980s (3,5).

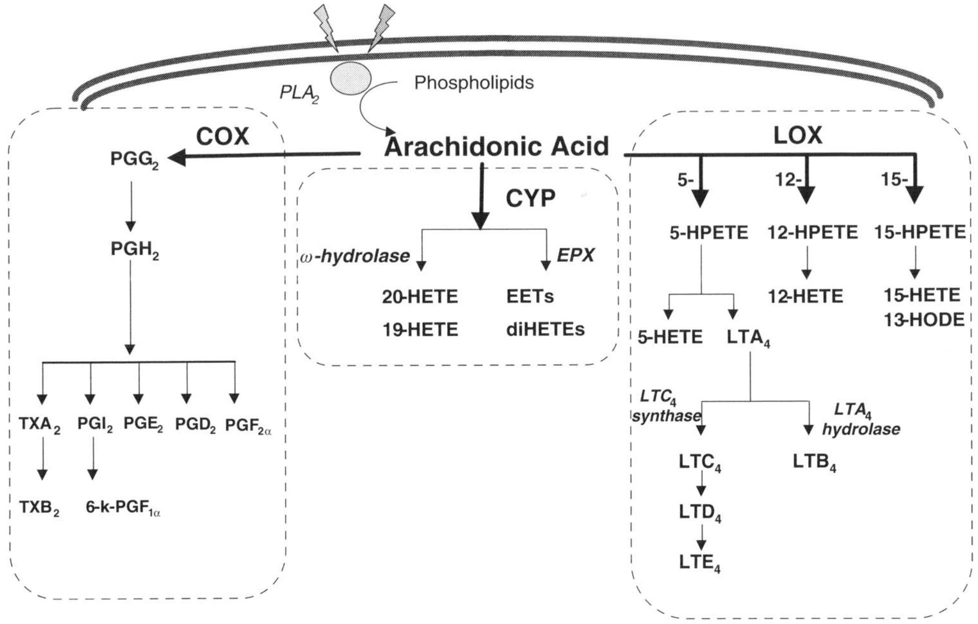

Figure 112.1. Eicosanoid biosynthesis via COX, CYP, and LOX pathways. Key eicosanoids are shown. PG, prostaglandin; TXA_2, thromboxane A_2; HETE, hydroxyeicosatetraenoic acid; EET, epoxy eicosatetraenoic acid; LT, leukotrienes.

EICOSANOID BIOSYNTHESIS IN ENDOTHELIAL CELLS

Like other properties of the intact endothelium, eicosanoid synthesis varies among different types of endothelial cells (ECs). However, the nature and mechanisms of spatial and temporal regulation of eicosanoid generation in vivo remain poorly defined. In contrast, there is a wealth of information on eicosanoid synthesis derived from studies on umbilical vein ECs (HUVECs) and bovine aortic ECs (BAECs).

Endothelial Eicosanoid Biosynthesis

Major eicosanoids produced from PLA_2-initiated AA metabolism in HUVECs and BAECs include 6-keto-$PGF_{1\alpha}$ (a stable metabolite of PGI_2) and PGE_2 derived from the COX pathway, and 15-, 11-, and 12-HETE from the LOX pathway (31,32). PGD_2, $PGF_{2\alpha}$, and 5-HETE are produced in smaller quantities. 12-hydroxyheptadecatrienoic acid (12-HHT) is invariably detected, which represents a nonenzymatic breakdown product of PG endoperoxides.

The eicosanoid profile produced by microvascular ECs is different from that produced by HUVECs or BAECs. The 6-keto-$PGF_{1\alpha}$ level is much lower, whereas EETs from the CYP pathway are more abundant. Major CYP metabolites produced by microvascular EC include 14,15 and 11,12-EET as well as 14,15- and 11,12-di-HETE (33).

The synthetic enzymes that have been clearly demonstrated to be expressed in ECs are shown in Figure 112.2. As described earlier, the synthetic enzymes are classified into

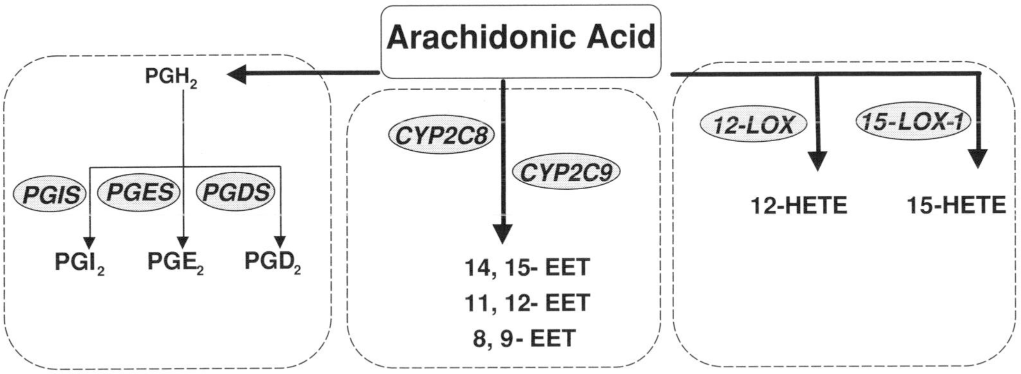

Figure 112.2. Eicosanoid synthetic enzymes expressed in ECs. Enzymes are shown in circles. COX, cyclooxygenase; PGIS, prostaglandin I_2 (PGI) synthase; PGES, PGE_2 synthase; PGDS, PGD_2 synthase; CYP2C8 and 9; cytochrome p450 family 2C8 and 2C9; 12-LOX, 12-lipoxygenase; 15-LOX-1, 15-lipoxygenase-1.

three groups according to their synthetic steps. Enzymes in the first step comprise several PLA_2 isoforms, including calcium-dependent cytosolic PLA_2 ($cPLA_2$), calcium-independent cytosolic PLA_2 ($iPLA_2$), and secretory PLA_2 ($sPLA_2$) (1). $cPLA_2$ in ECs has been characterized extensively. It is localized to the cytosol in resting ECs and, upon cell activation, is translocated to intracellular membranes where it is functionally coupled to COX, LOX, and CYP. COX-1 and COX-2 are expressed in ECs (2). COX-1 is constitutively expressed and may be induced by cytokines and tissue injury by two- to threefold (34). Analysis of its promoter function reveals that it has typical housekeeping gene features (35). By contrast, COX-2 is expressed in resting ECs at a very low level that is often not detectable by Western blots but may be detectable by flow cytometry (36). Its expression is induced by diverse agonists, including cytokines, growth factors, and shear stress (37).

Of the enzymes in the final step of the COX pathway, PGIS is constitutively expressed in ECs (29,30). Its promoter region is typical of a housekeeping gene (38). PGE synthase (PGES) comprises three isoforms: a constitutively expressed cytosolic PGES (cPGES) (39,40) and two inducible membrane-bound PGES (mPGES) (40). Expression of these isoforms in ECs has not been documented extensively. cPGES is probably constitutively expressed in ECs and is responsible for the production of basal PGE_2. mPGES was reported to be expressed in brain ECs, in which it was coexpressed with COX-2 (41). Two isoforms of PGD synthase (PGDS) have been identified: the lipocalin-type (L-PGDS) and the hematopoietic type (H-PGDS) (42,43). L-PGDS is expressed in ECs and is considered to play a role in coronary artery disease (44). L-PGDS is likely to be responsible for the small amount of PGD_2 generated from AA in ECs.

The LOX enzymes in EC are expressed in low abundance, and their regulation remains controversial. For example, a low level of 15-LOX-1 is detectable in EC, where it was reported to be induced by interleukin (IL)-4 (45), but studies from *IL-4*$^{-/-}$ mice failed to support the essential role of IL-4 for 15-LOX-1 expression (46). Other LOX, including leukocyte-type 12-LOX, platelet type 12-LOX, and 5-LOX, have been reported to be constitutively expressed in ECs, and their levels are generally very low, in accord with the synthesis of low levels of corresponding HETEs (47–49).

Human ECs express several CYP enzymes including CYP2C8, -2C9, -2J2, and -2B1 (33,50,51). The CYP2C8 and -2C9 family of enzymes is considered to be largely responsible for the formation of vasorelaxing EETs (50,51).

Endothelial Eicosanoid Synthesis by Transcellular Metabolism

Eicosanoids are produced not only from AA but also from an intermediate metabolite produced by blood cells that come in contact with ECs, a process called *transcellular metabolism.* This concept originated from work showing that PGI_2 is produced from platelet PG endoperoxides that are translocated from platelets to ECs via platelet–EC interactions (52,53). Transcellular metabolism enables ECs to synthesize certain metabolites that cannot be produced from AA. One example is the EC synthesis of LTC_4 from leukocyte-derived LTA_4. Because ECs express a very low level of 5-LOX, they do not produce detectable LTC_4. However, they are capable of synthesizing LTC_4 from LTA_4, because they express LTC_4 synthase (54).

ACTIONS OF ENDOTHELIAL EICOSANOIDS

Existing data indicate that endothelial PGI_2, EET, and 15-LOX products are physiologically and pathophysiologically important. Their actions are described in more detail in the following sections.

Prostacyclin

PGI_2 inhibits platelet aggregation induced by diverse physiological agonists by raising the platelet cAMP level in a paracrine manner (23). PGI_2 binds to a specific receptor, designated I type prostaglandin (IP) receptor, which belongs to the eicosanoid receptor superfamily (6). IP receptor is a seven-transmembrane G-coupled receptor. It is coupled to G_s protein, which is, in turn, coupled to adenyl cyclase. The cAMP generated by adenylyl cyclase binds and activates PKA and exerts inhibitory action on platelet reactivity. Enhanced PGI_2 production in ECs in response to exogenous insults probably plays an important role in suppressing platelet aggregation and maintaining a thromboresistant state. EC-produced PGI_2 relaxes VSMCs and controls the reactivity of blood leukocytes such as neutrophils, Th2 lymphocytes, and macrophages via the IP-receptor and PKA signaling pathway (55–57).

In addition to its IP receptor–mediated actions on platelets, leukocytes, and vascular cells, PGI_2 has been shown to bind peroxisome proliferator-activated receptor (PPAR)δ and activate PPARδ transcriptional activity (58). COX-2–derived PGI_2 has been shown to defend renal interstitial cells from damage by hypertonicity (59). It was suggested that the cytoprotective action of PGI_2 is mediated by PPARδ activation. However, the evidence provided in these studies is indirect (58,59). Neither study has provided clear evidence for the requirement of PPARδ for the PGI_2 cytoprotective action. It is also unclear how PPARδ activation by PGI_2 protects cells from damage. Because ECs constantly encounter environmental stresses from circulating blood, it is of major importance to determine whether PGI_2 protects endothelial survival in an autocrine fashion. It has been shown that ECs express IP receptors (60). However, no studies have reported on endothelial protection by PGI_2. Preliminary results from our laboratory reveal that PGI_2 produced by gene transfer, or synthetic PGI_2 analogs such as carbaprostacyclin, protects HUVECs from hydrogen peroxide (H_2O_2)-induced apoptosis. PPARδ is constitutively expressed in HUVECs. Overexpression of PPARδ enhances the antiapoptotic action of PGI_2, whereas suppression of PPARδ by small-interference RNA (siRNA) renders HUVECs less resistant to H_2O_2-induced apoptosis. These preliminary data suggest that PGI_2 protects ECs from oxidant-induced apoptosis in a PPARδ-dependent manner.

A major role that PGI_2 plays in blood vessels is to antagonize the action of TXA_2. The TXA_2 produced primarily by platelets and, to a lesser extent, by monocytes induces platelet aggregation and causes vasoconstriction. It plays a key role in arterial thrombosis. PGI_2 antagonizes the actions of TXA_2 on platelets and VSMCs, thereby reducing thrombosis and vasoconstriction. Recent work from IP receptor and thromboxane-type prostaglandin (TP) receptor-deficient mice supports an essential role for PGI_2 in controlling TXA_2-induced atherosclerosis and thrombosis (61).

PGI_2, thus, plays critical roles in maintaining vascular integrity, conferring thromboresistance, and inhibiting vascular inflammation. Its production is dynamically regulated to meet the challenge from frequent vascular insults.

Epoxy Eicosatetraenoic Acids

Several EETs are produced in ECs via the actions of CYP epoxygenases. CYP2C8 and -2C9 catalyze the conversion of AA to 14,15- and 11,12-EET and, in addition, CYP2C9 converts AA to 8,9-EET (62). CYP2J2 catalyzes the formation of 5,6-, 8,9-, 11,12-, and 14,15-EET (63). 11,12- and 14,15-EET possess vasodilating properties (64). The reported data indicate that they act directly on VSMCs to reduce contractility. However, 5,6- and 8,9-EETs may be converted via COX-2 pathway to form vasodilating or vasoconstrictive compounds (65). The mechanism by which EETs reduce VSMC contraction is not entirely clear. One common view is that EETs open the K_{Ca} channels of VSMCs to facilitate release of K^+ into the extracellular milieu (66,67). It remains controversial whether the actions of EETs are receptor-mediated.

Several reports show that EETs possess anti-inflammatory actions (50). EETs prevent the adhesion of activated monocytes to vascular endothelium and 11,12-EET suppresses cytokine-induced expression of adhesive molecules, such as vascular cell adhesion molecule (VCAM)-1 and E-selectin, on the endothelial surface (50). EETs have been reported to mediate the proliferation of diverse cells including VSMCs and ECs (68). EETs increase intracellular calcium concentration and activation of mitogen-activated protein kinase (MAPK) pathway in ECs and VSMCs, suggesting that the anti-inflammatory and mitogenic actions may be mediated by these pathways (69). In summary, EC-derived EETs relax VSMCs, inhibit monocyte interaction with ECs, and mediate VSMC proliferation in a paracrine manner, and they induce EC proliferation in an autocrine manner.

Lipoxygenase Products

There is considerable information regarding the role of murine 12/15-LOX (a homolog of human 15-LOX-1) in atherosclerosis. Murine 12/15-LOX is a multifunctional enzyme that catalyzes the formation of 12-HETE and 15-HETE from AA, and 13S-hydroxyoctadecatrienoic acid (13S-HODE) from linoleic acid (4). Increased 12/15-LOX eicosanoids are reported to increase monocyte–endothelial interactions in diabetic mice and accelerate atherosclerosis in *ApoE*$^{-/-}$ mice (70). The deletion of *12/15-LOX* by gene targeting reduces atherosclerosis (71). 15-LOX-1 is a human homolog of murine 12/15-LOX that catalyzes the formation of 15-HETE from AA and 13-HODE from linoleic acid. It was reported that overexpression of 15-LOX-1 increases atherosclerosis (72). These results suggest that 15-LOX-1 and 12/15-LOX play a role in atherosclerosis. However, the metabolites that are responsible for the atherosclerotic actions are not entirely clear.

It has been reported that 12-HETE derived from 12-LOX increases monocyte adhesion to human ECs (73). It is further shown that oxidized low-density lipoprotein (LDL) induces ECs to generate 12-HETE and 11,12-EET (74). The inhibition of 12-HETE suppressed LDL-induced monocyte adhesion to vascular ECs, consistent with the involvement of 12-HETE in monocyte–endothelial interactions.

STIMULATION OF PROSTACYCLIN SYNTHESIS

PGI_2 is functionally and quantitatively the most important eicosanoid produced by ECs. Resting ECs produce a basal level of PGI_2 that maintains vascular homeostasis under physiological conditions. When ECs are stimulated by diverse agents, an increased level of PGI_2 is produced. Many physiological and pathological agents have been shown to stimulate PGI_2 production. Based on kinetic differences, two types of stimulation are noted: (a) rapid stimulation independent of new COX mRNA or protein synthesis, and (b) robust PGI_2 stimulation secondary to increased COX-2 expression. The mechanisms underlying these two types of stimulation are distinct and are described separately.

Rapid Stimulation of Prostacyclin Production by Histamine, Bradykinin, and Thrombin

It was discovered in 1977 that ECs are capable of producing PGI_2 (75). Histamine was subsequently shown to stimulate the production of PGI_2 in ECs (76). It was noted that histamine induces a rapid synthesis of PGI_2, followed by a period of refractoriness to stimulation. Bradykinin, thrombin, and ionophore A23187 were also reported to stimulate EC production of PGI_2 in a similar manner (77). Kinetic analysis of 6-keto-$PGF_{1\alpha}$ production revealed a rapid response that reached a plateau at 10 minutes after the addition of histamine or thrombin (77). These physiological agonists stimulate PGI_2 production by a common signaling mechanism. They activate phospholipase C, which generates inositol triphosphate (IP3) and diacylglycerol (DAG). IP3 elicits an elevation of intracellular calcium levels, which induces cPLA translocation to the outer surface of nuclear envelope (NE) and endoplasmic reticulum (ER). There, it is functionally coupled to COX-1

(78), which is localized to the luminal membrane by hydrophobic interaction (79,80). PGIS is localized to ER and NE membrane through a single transmembrane domain, and the bulk of the enzyme is located at the cytosolic side (81). PGIS is colocalized with COX-1 in ECs (82). It is widely accepted that activated $cPLA_2$ catalyzes the release of AA from membrane phospholipids, and the released AA enters COX-1, where it is converted to PGH_2. PGH_2 enters PGIS and is converted to PGI_2. Although the functional coupling of $cPLA_2$, COX, and PGIS implies a direct transfer of substrate and intermediate metabolite from one enzyme to the next within the NE and ER membrane, the precise mechanism by which PGI_2 is synthesized from AA remains to be elucidated.

PGI_2 production declines rapidly after stimulation with histamine, thrombin, and ionophore. A plausible explanation is that COX-1 has a short half-life once it participates in the catalytic reaction. It has been estimated that COX-1 has a half-life of about 10 minutes (83). The short half-life of COX-1 catalytic activity is attributed to the inactivation and degradation of COX-1 by reactive oxygen species generated by COX-1 (84). The kinetic analysis suggests that PGIS catalytic activity also has a short lifespan (85), which is explained by autoinactivation of PGIS during catalysis (85). The so-called "suicidal" autoinactivation of COX-1 is considered to be responsible for the refractory period that ECs experience after an episode of stimulation, because it takes hours for COX-1 to be resynthesized. Thus, the extent of PGI_2 production by ECs under the stimulation by vasoactive agonists is limited and transient.

Sustained Prostacyclin Production Induced by Proinflammatory Cytokines and Growth Factors

Numerous reports have shown that PGI_2 production induced by PMA, lipopolysaccharide (LPS), cytokines (e.g., IL-1β, tumor necrosis factor [TNF]-α), and growth factors (e.g., platelet-derived growth factor) has a different time course from that produced by histamine, thrombin, and bradykinin. PGI_2 production is not detected until 30 to 60 minutes after the cytokine or growth factor stimulation, with levels reaching plateau at 4 to 6 hours and declining over a 12-hour period. In response to proinflammatory cytokines, the kinetics of PGI_2 production parallels that of COX-2, suggesting that a majority of PGI_2 is derived from COX-2. COX-2 shares a high sequence homology with COX-1; the key functional motifs, including the heme ligation site and substrate binding residue, are conserved. The x-ray crystallographic structure of COX-2 is also similar to that of COX-1 (79,80). Furthermore, a majority of COX-2 also is localized to NE and ER (82). The catalytic kinetics of COX-2 is also similar to COX-1 (2). A major difference between COX-2 and COX-1 resides in the robust expression of COX-2 induced by diverse proinflammatory and mitogenic factors (37). COX-2 proteins induced by proinflammatory mediators and mitogenic factors peak at 4 to 6 hours and return to baseline at 12 to 24 hours after stimulation, mirroring the kinetics of PGI_2 production. Consequently, the quantity of PGI_2 produced via the inducible COX-2 pathway is several orders of magnitude higher than that via COX-1 pathway. PGI_2 overproduction signaled by cytokines and mitogenic factors is likely to play an important role in maintaining vascular integrity, thromboresistance, and normal vascular tone.

Good evidence suggests COX-2–mediated in vivo PGI_2 production in humans. PGI_2 synthesis in humans has been determined by measuring the excretion rate of 2,3-dinor-6-keto $PGF_{1\alpha}$ (86). It has been shown that urinary 2,3-dinor-6-keto-$PGF_{1\alpha}$ levels in healthy human subjects are suppressed by the administration of selective COX-2 inhibitors, thus suggesting that PGI_2 production in healthy humans is catalyzed largely by COX-2 (86). The urinary 2,3-dinor-6-keto-$PGF_{1\alpha}$ levels are elevated in patients with atherosclerosis (87), suggesting a compensatory production of PGI_2 by ECs, although the exact cell types that generate PGI_2 in atherosclerosis have not been clearly defined.

A majority of PGI_2 produced in atherosclerotic patients is COX-2–dependent, but COX-1 appears also to play a role in contributing to PGI_2 production (88). Because atherosclerotic tissues are enriched with cytokines and mitogenic factors, COX-2 expression probably is induced by a multitude of proinflammatory mediators. The role of increased PGI_2 production in atherosclerosis is unclear. Circumstantial evidence suggests that it may play an important protective role.

Transcriptional Regulation of Cyclooxygenase-2

The molecular basis for transcriptional upregulation of COX-2 by PMA, LPS, proinflammatory mediators, and growth factors has been studied extensively. The promoters of human and murine *COX-2* genes are similar and contain a canonical TATA as well as a number of functionally important enhancer elements within 500 base pairs (bp) of the 5′-untranslated region (89–91). Transfection assays and DNA binding studies have led to the identification of several transactivators essential for the basal and inducible expression of COX-2 (92). Basal promoter activity in HUVECs is mediated by a cAMP response element (CRE), and a C/EBP enhancer element located at the proximal promoter region. These have been shown to bind CRE-binding protein (CREB) and C/EBPδ, respectively (93). CRE is required for promoter activation by proinflammatory mediators, and its mutation has been shown to abrogate the inducible expression of COX-2 (93,94). In addition to CRE, full induction of the *COX-2* promoter requires binding of several transcriptional activators that cooperatively recruit p300 (95–97). Interestingly, the combination of transactivators varies according to the extracellular agonist. For example, PMA stimulates binding of CREB-2, c-Jun, and C/EBPβ to their respective enhancer elements, whereas TNF-α–mediated induction of COX-2 involves the inducible binding of p50-p65 nuclear factor (NF)-κB to both κB enhancer elements of the *COX-2* promoter/enhancer region. The binding activity of p50/p65 is regulated by post-translational modification. Work from our laboratory reveals that p50 is acetylated by the action of p300 histone acetyltransferase (HAT) activity, and that the acetylated p50 exhibits an increased binding activity to *COX-2* promoter (97). C/EBPβ comprises several truncated forms as the result of using different start codons for translation (98). Two of the truncated forms have opposite actions: the liver-transcription activating protein (LAP) truncated form binds and activates gene transcription, whereas a shorter liver transcriptional inhibitory protein (LIP) inhibits gene transcription (99). Full-length as well as LAP- and LIP-truncated forms of C/EBPβ are detected in ECs. At resting state, C/EBPβ has weak DNA binding activities due to the presence of an intramolecular inhibitory element that blocks the binding domain. Stimulation of cells by PMA results in phosphorylation of C/EBPβ by P90 ribosomal S6 kinase (100). The importance of C/EBPβ binding is underscored by the inhibition of C/EBPβ binding by aspirin and sodium salicylate (100,101).

Prostacyclin Synthase

PGIS is constitutively expressed in ECs. Only a single copy of the gene has been identified (38). Its promoter region is TATA-less and GC-rich. It was reported to be inducible, but this has not been supported by solid evidence. PGIS is a cytochrome P450 protein, classified under family 8 (CYP8). It anchors to the ER and NE membrane via a single transmembrane domain, and the bulk of the enzyme is located in the cytosol. Topographic analyses reveal that its substrate entrance channel is close to the membrane surface, allowing rapid transit of the PGH_2 produced by COX to be metabolized into PGI_2 by PGIS (81). PGIS activity is suppressed by peroxynitrite generated from the interaction of nitric oxide (NO) with superoxide anion (102). Subsequently, it was shown that peroxynitrite induces the nitration of tyrosine 430 of PGIS, leading to reduced PGIS catalytic activity (103). PGIS nitration has been shown to have a pathophysiological relevance in experimental animals. It causes defective vasorelaxation of atherosclerotic bovine aortic arteries (104). It remains unclear whether PGIS is inactivated by endogenous peroxynitrite in ECs. Nevertheless, the negative regulation of PGIS activity by the cross-talk between NO, superoxide, and PGIS is likely to have important pathophysiological implications.

PROSTACYCLIN AS A THERAPEUTIC AGENT

PGI_2 and its analogs have proved to be efficacious in treating pulmonary hypertension. It was reported that urinary levels of 2,3-dinor-6-keto-$PGF_{1\alpha}$ in patients with pulmonary hypertension was reduced, consistent with defective PGI_2 production (105). Defective PGI_2 production is probably due to inactivation of PGIS (106). In a randomized clinical trial, intravenous infusion of PGI_2 produced symptomatic and hemodynamic improvement when compared with conventional therapy (107). A long-term trial using aerosolized iloprost, a stable analog of PGI_2, in patients with primary pulmonary hypertension revealed that iloprost improved exercise capacity and pulmonary hemodynamics (108).

There are reports on the beneficial effects of PGI_2 analogs in relieving the symptoms of peripheral vascular diseases. For example, oral beraprost, a PGI_2 analog, is effective in improving intermittent claudication (109). Similarly, a PGE_1 analog that acts via IP and EP receptors was reported to improve intermittent claudication from peripheral arterial disease (110).

ENGINEERING EICOSANOID BIOSYNTHESIS AND FUNCTION BY GENE TRANSFER

Eicosanoids produced by ECs are dynamically regulated. A major factor that determines the level of eicosanoid synthesis is the cellular level of the synthetic enzymes for a given eicosanoid. Because PGI_2 is the key eicosanoid produced by ECs, with a high degree of functional relevance, we have used its biosynthesis as a model to investigate eicosanoid augmentation by gene transfer. *COX-1*, *PGIS*, and a bicistronic *COX-1/PGIS* (COPI) cDNAs were constructed in an adenoviral vector. HUVECs transfected with Ad-COX-1 (50 moi) were shown to produce a twofold increase in 6-keto-$PGF_{1\alpha}$, and a marked increase in PGE_2 when compared with those transfected with control vector (111). The endoperoxide product, 12-HHT, was increased by approximately 20-fold. These results suggest that, in the presence of COX-1 overexpression, PGIS is a limiting factor in PGI_2 production. This notion was confirmed by Ad-PGIS transfer, which increased 6-keto-$PGF_{1\alpha}$ by more than fivefold without an increase in PGE_2, $PGF_{2\alpha}$, or 12-HHT (111). Results from Ad-PGIS experiments suggest that a limited COX-1 catalytic capacity exists, probably because of COX-1 autoinactivation. All the available PGH_2 generated from COX-1 is shunted into the overexpressed PGIS pathway.

To analyze the relationship between COX-1 and PGIS, HUVECs were transfected with Ad-COX-1 and Ad-PGIS at different viral titers, then assayed for eicosanoids. When Ad-COX-1 was fixed at 50 moi, increasing Ad-PGIS moi caused a selective augmentation of 6-keto-$PGF_{1\alpha}$ and a concurrent reduction in PGE_2 and $PGF_{2\alpha}$. At an equivalent moi (50:50), 6-keto-$PGI_{1\alpha}$ was the predominant product, with a level approximately 40-fold higher than Ad-PGIS alone. Surprisingly, when Ad-PGIS exceeded Ad-COX-1 (50 Ad-COX-1 to 100 Ad-PGIS) the 6-keto-$PGF_{1\alpha}$ level was reduced to less than 50% of that produced by an equivalent moi (50:50), whereas PGE_2 and $PGF_{2\alpha}$ levels remained unchanged. The results from the Ad-COX-1 and Ad-PGIS mixing experiments suggest a delicate relationship between COX-1 and PGIS in PGI_2 synthesis. Optimal PGI_2 production appears to require an appropriate molecular level of COX-1 and PGIS (111). To ensure that COX-1 and PGIS are co-overexpressed in the same cell, we transduced HUVECs with a bicistronic Ad-COPI (COX-1 and PGIS expressions are driven by identical but separate promoters). The results confirm an equivalent COX-1 and PGIS protein overexpression and a selective augmentation of PGI_2 production without a concurrent overproduction of other prostanoids.

Taken together, these results indicate that the level of eicosanoid production is determined largely by the protein levels of the synthetic enzymes. It is feasible to engineer cells for the selective augmentation of a given eicosanoid by co-overexpressing enzymes catalyzing the formation of that eicosanoid. This approach should be valuable for studying enzyme coupling and eicosanoid regulation. Furthermore, it has therapeutic potential, as demonstrated in its application to control brain ischemia–reperfusion injury, described next.

Control of Brain Ischemia–Reperfusion Injury by Cyclooxygenase-1, Prostacyclin Synthase, Cyclooxygenase/Prostacyclin Synthase Gene Transfer

It has been reported that Ad-COX-1 or Ad-COPI gene transfer directly into brain before or after ischemia reduces infarct volume (112), and that Ad-PGIS gene transfer attenuates carotid artery restenosis (113,114). Although the protective effects of these gene transfer approaches are attributed to PGI_2 augmentation, other factors such as alteration of the balance between protective and damaging eicosanoids may contribute to the protective actions. To identify the factors that protect tissue survival by COX-1–based gene transfer, we administered Ad-COX-1, Ad-COPI, and control vector to rat brain via cerebral ventricular infusion before and after ischemia–reperfusion injury. Key eicosanoids in brain tissues were measured and correlated with infarct volume. Ad-COX-1 overexpression in nonischemic brain tissues increased PGI_2, PGE_2, PGD_2, and TXA_2, with a reduction in LTB_4 and LTC_4. The eicosanoid profile in Ad-COX-1–transfected ischemic brain tissues was similar, except that TXA_2 was no longer significantly increased, and LTB_4 and LTC_4 were more significantly suppressed (112). Ad-COX-1 was effective in reducing infarct volume when administered before and even 5 hours after a 60-minute middle cerebral artery occlusion (112). The beneficial effect of *COX-1* gene transfer was attributed to: (a) augmented PGI_2; (b) augmented PGD_2, which is converted 15-deoxy-$\Delta^{12,14}$ PGJ_2 (15d-PGJ_2); and (c) reduced LTs.

The administration of Ad-COPI exerted a protective action similar to that of Ad-COX-1. However, its brain eicosanoid profile differed from that of Ad-COX-1 transfection. Consistent with the in vitro cellular data, Ad-COPI increased PGI_2 and decreased PGE_2, PGD_2, TXA_2, LTB_4, and LTC_4 (112). Thus, the beneficial effect of Ad-COPI was mediated primarily by an augmented PGI_2 production accompanied by suppression of damaging eicosanoids, notably LTB_4 and LTC_4.

Using an identical protocol, the administration of Ad-PGIS reduced infarct volume only at 72 hours before ischemia. Its infusion at 24 hours or immediately before, and at 5 or 24 hours after ischemia was no longer protective. The brain eicosanoid profile in rats infused with Ad-PGIS 72 hours before injury revealed a highly increased PGI_2 without a significant change in other eicosanoids.

These experimental data validate a fundamental principle of eicosanoid biochemistry. It is feasible to engineer eicosanoid productions through the transfer of a single or multiple synthetic enzymes. The gene transfer approach should be valuable for experimental studies and have important therapeutic implications. At present, gene therapy is limited because vectors for efficient and safe gene transfer remain unsatisfactory. However, when the vector problems are resolved, gene transfer to enrich protective and suppress proinflammatory eicosanoids should have the potential for treating diverse diseases, including ischemia–reperfusion injury.

CONCLUSION

ECs are capable of synthesizing eicosanoids via COX, LOX, and CYP pathways. Functionally important eicosanoids include PGI_2, EETs, and probably 15- and 12-HETE. Among EC eicosanoids, PGI_2 has been well characterized, and is probably the most important. It exerts diverse actions that are mediated by two receptors. The classic actions of PGI_2 on inhibiting platelet aggregation, VSMC relaxation, and leukocyte activation are mediated by a G_S-coupled IP receptor, whereas the novel actions of PGI_2 – notably cell survival and proliferation – are mediated by the nuclear factor PPARδ. PGI_2 defends ECs from apoptosis in an autocrine fashion, and controls blood and VSMC reactivity in a paracrine manner. PGI_2 in ECs is derived primarily from the COX-2/PGIS pathway (Figure 112.3). The inhibition of COX-2 abolishes PGI_2 production and results in loss of the protective mechanism and exacerbation of vascular diseases. Selective augmentation of PGI_2 by combined *COX-1*

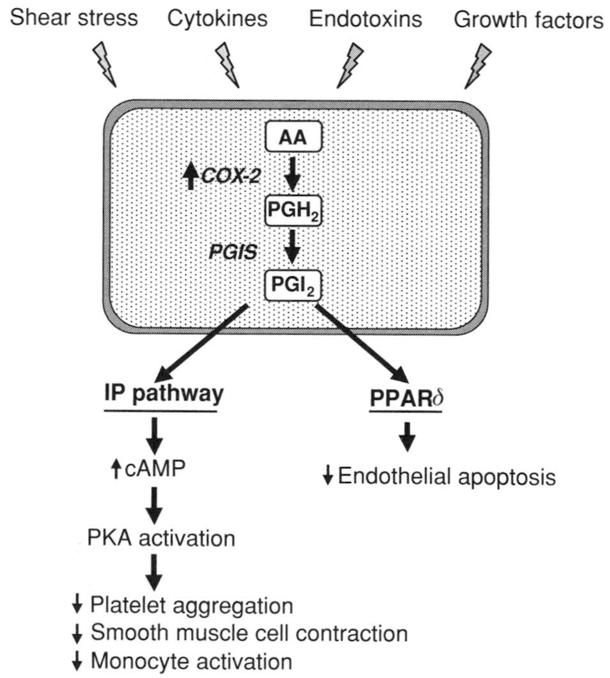

Figure 112.3. Production and action of PGI_2. PGI_2 production is catalyzed by COX-2, which is induced by physical and chemical agents shown by arrows. PGI_2 exerts multiple biological actions via PGI2 receptor (IP) pathway and peroxisome proliferator-activated receptor δ (PPARδ) pathway.

and *PGIS* gene transfer protects brain tissues from ischemia–reperfusion injury. EETs derived from CYP2C and CYP2J in ECs play an important role in relaxing VSMCs. Their mechanisms of action remain poorly understood. ECs produce 15-HETE and 12-HETE via 15-LOX-1 and 12-LOX. Their quantities are generally small relative to PGI_2 and PGE_2, and their physiological and pathophysiological roles in humans require further investigation.

KEY POINTS

- COX-2–derived PGI_2 and CYP2-derived EETs are major EC eicosanoids.
- PGI_2 controls blood cell reactivity, VSMC contractility, and EC apoptosis by binding to IP and PPARδ receptors. EETs relaxes VSMCs.

Future Goals

- To determine whether selective augmentation of PGI_2 and other protective eicosanoids by gene transfer will be therapeutically beneficial for vascular diseases and tissue infarction
- To elucidate the mechanisms by which PGI_2 exerts antiapoptotic actions and EETs inhibit vascular constriction and inflammation

ACKNOWLEDGMENTS

This work was supported by grants from the National Institutes of Health (R01-HL 50675 and P50 NS-23327).

REFERENCES

1 Dennis EA. The growing phospholipase A2 superfamily of signal transduction enzymes. *Trends Biochem Sci.* 1997;22(1):1–2.

2 Smith WL, Garavito RM, DeWitt DL. Prostaglandin endoperoxide H synthases (cyclooxygenases)-1 and -2. *J Biol Chem.* 1996;271(52):33157–33160.

3 Lewis RA, Austen KF, Soberman RJ. Leukotrienes and other products of the 5-lipoxygenase pathway. Biochemistry and relation to pathobiology in human diseases. *N Engl J Med.* 1990;323(10):645–655.

4 Brash AR. Lipoxygenases: occurrence, functions, catalysis, and acquisition of substrate. *J Biol Chem.* 1999;274(34):23679–23682.

5 Roman RJ. *P*-450 Metabolites of arachidonic acid in the control of cardiovascular function. *Physiol Rev.* 2002;82(1):131–185.

6 Narumiya S, FitzGerald GA. Genetic and pharmacological analysis of prostanoid receptor function. *J Clin Invest.* 2001;108(1):25–30.

7 Oates JA, FitzGerald GA, Branch RA, et al. Clinical implications of prostaglandin and thromboxane A2 formation (1). *N Engl J Med.* 1988;319(11):689–698.

8 McGiff JC. Prostaglandins, prostacyclin, and thromboxanes. *Annu Rev Pharmacol Toxicol.* 1981;21:479–509.

9 Levy GN. Prostaglandin H synthases, nonsteroidal anti-inflammatory drugs, and colon cancer. *FASEB J.* 1997;11(4):234–247.

10 Goldblatt MW. A depressor substance in seminal fluid. *J Soc Chem Ind.* 1933;52:1056–1057.

11 von Euler US. Zur kenntnis der pharmakologishen Wirkungen von Nativsekreten und Extrakten mannlicher accessorischer Geschlechtsdrusen. *Arch Exp Pathol Pharmakol (Naunyn-Schmeidebergs).* 1934;175:78–84.

12 von Euler US. Uber die spezifische blutdrucksen. Kende Substanz der menschlichen prostata-under samenblasekretes. *Klin Wochenschr.* 1935;14:1182–1183.

13 Bergström S, Sjovall J. The isolation of prostaglandin F from sheep prostate glands. *Acta Chem Scand.* 1960;14:1693–1701.

14 Bergström S, Ryhag R, Samuelsson B, Sjovall J. Prostaglandins and related factors 15. The structures of prostaglandin E_1, $F_{1\alpha}$ and $F_{1\beta}$. *J Biol Chem.* 1963;238:3555.

15 Samuelsson B. Prostaglandin and related factors 17. The structure of PGE_2. *J Amer Chem Soc.* 1963;85:1878–1879.

16 Bergström S, Danielson H, Samuelsson B. The enzymatic formation of prostaglandin E_2 from arachidonic acid. *Biochim Biophys Acta.* 1964;90:207–210.

17 van Dorp DA, Beerthuis RK, Nugteren DH, Vonheman H. Biosynthesis of prostaglandins. *Biochim Biophys Acta.* 1964;90:204–207.

18 Hamberg M, Samuelsson B. Detection and isolation of an endoperoxide intermediate in prostaglandin biosynthesis. *Proc Natl Acad Sci USA.* 1973;70(3):899–903.

19 Hambert M, Samuelsson B. Prostaglandin endoperoxides. Novel transformations of arachidonic acid in human platelets. *Proc Natl Acad Sci USA.* 1974;71(9):3400–3404.

20 Hamberg M, Svensson J, Samuelsson B. Thromboxanes: a new group of biologically active compounds derived from prostaglandin endoperoxides. *Proc Natl Acad Sci USA.* 1975;72(8):2994–2998.

21 Moncada S, Gryglewski R, Bunting S, Vane JR. An enzyme isolated from arteries transforms prostaglandin endoperoxides to an unstable substance that inhibits platelet aggregation. *Nature.* 1976;263(5579):663–665.

22 Whittaker N, Bunting S, Salmon J, et al. The chemical structure of prostaglandin X (prostacyclin). *Prostaglandins.* 1976;12(6):915–928.

23 Moncada S, Vane JR. Pharmacology and endogenous roles of prostaglandin endoperoxides, thromboxane A2, and prostacyclin. *Pharmacol Rev.* 1978;30(3):293–331.

24 Hemler M, Lands WE. Purification of the cyclooxygenase that forms prostaglandins. Demonstration of two forms of iron in the holoenzyme. *J Biol Chem.* 1976;251(18):5575–5579.

25 DeWitt DL, Smith WL. Primary structure of prostaglandin G/H synthase from sheep vesicular gland determined from the complementary DNA sequence. *Proc Natl Acad Sci USA.* 1988;85(5):1412–1416.

26 Xie WL, Chipman JG, Robertson DL, et al. Expression of a mitogen-responsive gene encoding prostaglandin synthase is regulated by mRNA splicing. *Proc Natl Acad Sci USA.* 1991;88(7):2692–2696.

27 Kujubu DA, Fletcher BS, Varnum BC, et al. TIS10, a phorbol ester tumor promoter-inducible mRNA from Swiss 3T3 cells,

encodes a novel prostaglandin synthase/cyclooxygenase homologue. *J Biol Chem.* 1991;266(20):12866–12872.
28 DeWitt DL, Smith WL. Purification of prostacyclin synthase from bovine aorta by immunoaffinity chromatography. Evidence that the enzyme is a hemoprotein. *J Biol Chem.* 1983; 258(5):3285–3293.
29 Miyata A, Hara S, Yokoyama C, et al. Molecular cloning and expression of human prostacyclin synthase. *Biochem Biophys Res Commun.* 1994;200(3):1728–1734.
30 Pereira B, Wu KK, Wang LH. Molecular cloning and characterization of bovine prostacyclin synthase. *Biochem Biophys Res Commun.* 1994;203(1):59–66.
31 Revtyak GE, Hughes MJ, Johnson AR, Campbell WB. Histamine stimulation of prostaglandin and HETE synthesis in human ECs. *Am J Physiol.* 1988;255(2 Pt 1):C214–C225.
32 Hopkins NK, Oglesby TD, Bundy GL, Gorman RR. Biosynthesis and metabolism of 15-hydroperoxy-5,8,11,13-eicosatetraenoic acid by human umbilical vein ECs. *J Biol Chem.* 1984;259(22): 14048–14053.
33 Bolz SS, Fisslthaler B, Pieperhoff S, et al. Antisense oligonucleotides against cytochrome P450 2C8 attenuate EDHF-mediated Ca(2+) changes and dilation in isolated resistance arteries. *FASEB J.* 2000;14(2):255–260.
34 Xu XM, Tang JL, Hajibeigi A, et al. Transcriptional regulation of endothelial constitutive PGHS-1 expression by phorbol ester. *Am J Physiol.* 1996;270(1 Pt 1):C259–C264.
35 Xu XM, Tang JL, Chen X, et al. Involvement of two Sp1 elements in basal endothelial prostaglandin H synthase-1 promoter activity. *J Biol Chem.* 1997;272(11):6943–6950.
36 Ruitenberg JJ, Waters CA. A rapid flow cytometric method for the detection of intracellular cyclooxygenases in human whole blood monocytes and a COX-2 inducible human cell line. *J Immunol Methods.* 2003;274(1–2):93–104.
37 Wu KK. Inducible cyclooxygenase and nitric oxide synthase. *Adv Pharmacol.* 1995;33:179–207.
38 Yokoyama C, Yabuki T, Inoue H, et al. Human gene encoding prostacyclin synthase (PTGIS): genomic organization, chromosomal localization, and promoter activity. *Genomics.* 1996; 36(2):296–304.
39 Tanioka T, Nakatani Y, Semmyo N, et al. Molecular identification of cytosolic prostaglandin E2 synthase that is functionally coupled with cyclooxygenase-1 in immediate prostaglandin E2 biosynthesis. *J Biol Chem.* 2000;275(42):32775–32782.
40 Jakobsson PJ, Thoren S, Morgenstern R, Samuelsson B. Identification of human prostaglandin E synthase: a microsomal, glutathione-dependent, inducible enzyme, constituting a potential novel drug target. *Proc Natl Acad Sci USA.* 1999;96(13): 7220–7225.
41 Yamagata K, Matsumura K, Inoue W, et al. Coexpression of microsomal-type prostaglandin E synthase with cyclooxygenase-2 in brain ECs of rats during endotoxin-induced fever. *J Neurosci.* 2001;21(8):2669–2677.
42 Urade Y, Hayaishi O. Biochemical, structural, genetic, physiological, and pathophysiological features of lipocalin-type prostaglandin D synthase. *Biochim Biophys Acta.* 2000;1482:259–271.
43 Urade Y, Hayaishi O. Prostaglandin D synthase: structure and function. *Vitam Horm.* 2000;58:89–120.
44 Eguchi Y, Eguchi N, Oda H, et al. Expression of lipocalin-type prostaglandin D synthase (beta-trace) in human heart and its accumulation in the coronary circulation of angina patients. *Proc Natl Acad Sci USA.* 1997;94(26):14689–14694.
45 Lee YW, Kuhn H, Kaiser S, et al. Interleukin 4 induces transcription of the 15-lipoxygenase I gene in human ECs. *J Lipid Res.* 2001;42(5):783–791.
46 Cornicelli JA, Welch K, Auerbach B, et al. Mouse peritoneal macrophages contain abundant omega-6 lipoxygenase activity that is independent of interleukin-4. *Arterioscler Thromb Vasc Biol.* 1996;16(12):1488–1494.
47 Zink MH, Oltman CL, Lu T, et al. 12-lipoxygenase in porcine coronary microcirculation: implications for coronary vasoregulation. *Am J Physiol Heart Circ Physiol.* 2001;280(2):H693–H704.
48 Nie D, Tang K, Diglio C, Honn KV. Eicosanoid regulation of angiogenesis: role of endothelial arachidonate 12-lipoxygenase. *Blood.* 2000;95(7):2304–2311.
49 Walker JL, Loscalzo J, Zhang YY. 5-Lipoxygenase and human pulmonary artery EC proliferation. *Am J Physiol Heart Circ Physiol.* 2002;282(2):H585–H593.
50 Node K, Huo Y, Ruan X, et al. Anti-inflammatory properties of cytochrome P450 epoxygenase-derived eicosanoids. *Science.* 1999;285(5431):1276–1279.
51 Fleming I, Michaelis UR, Bredenkotter D, et al. Endothelium-derived hyperpolarizing factor synthase (Cytochrome P450 2C9) is a functionally significant source of reactive oxygen species in coronary arteries. *Circ Res.* 2001;88(1):44–51.
52 Marcus AJ, Weksler BB, Jaffe EA, Broekman MJ. Synthesis of prostacyclin from platelet-derived endoperoxides by cultured human ECs. *J Clin Invest.* 1980;66(5):979–986.
53 Needleman P, Wyche A, Raz A. Platelet and blood vessel arachidonate metabolism and interactions. *J Clin Invest.* 1979; 63(2):345–349.
54 Feinmark SJ, Cannon PJ. EC leukotriene C4 synthesis results from intercellular transfer of leukotriene A4 synthesized by polymorphonuclear leukocytes. *J Biol Chem.* 1986;261(35): 16466–16472.
55 Simpson PJ, Mickelson J, Fantone JC, et al. Iloprost inhibits neutrophil function in vitro and in vivo and limits experimental infarct size in canine heart. *Circ Res.* 1987;60(5):666–673.
56 Jaffar Z, Wan KS, Roberts K. A key role for prostaglandin I2 in limiting lung mucosal Th2, but not Th1, responses to inhaled allergen. *J Immunol.* 2002;169(10):5997–6004.
57 Shinomiya S, Naraba H, Ueno A, et al. Regulation of TNFalpha and interleukin-10 production by prostaglandins I(2) and E(2): studies with prostaglandin receptor-deficient mice and prostaglandin E-receptor subtype-selective synthetic agonists. *Biochem Pharmacol.* 2001;61(9):1153–1160.
58 Forman BM, Chen J, Evans RM. Hypolipidemic drugs, polyunsaturated fatty acids, and eicosanoids are ligands for peroxisome proliferator-activated receptors alpha and delta. *Proc Natl Acad Sci USA.* 1997;94(9):4312–4317.
59 Hao CM, Redha R, Morrow J, Breyer MD. Peroxisome proliferator-activated receptor delta activation promotes cell survival following hypertonic stress. *J Biol Chem.* 2002;277(24): 21341–21345.
60 Komhoff M, Lesener B, Nakao K, et al. Localization of the prostacyclin receptor in human kidney. *Kidney Int.* 1998;54(6):1899–1908.
61 Cheng Y, Austin SC, Rocca B, et al. Role of prostacyclin in the cardiovascular response to thromboxane A2. *Science.* 2002; 296(5567):539–541.
62 Daikh BE, Lasker JM, Raucy JL, Koop DR. Regio- and stereoselective epoxidation of arachidonic acid by human cytochromes

P450 2C8 and 2C9. *J Pharmacol Exp Ther*. 1994;271(3):1427–1433.

63 Scarborough PE, Ma J, Qu W, Zeldin DC. P450 subfamily CYP2J and their role in the bioactivation of arachidonic acid in extrahepatic tissues. *Drug Metab Rev*. 1999;31(1):205–234.

64 Gebremedhin D, Ma YH, Falck JR, et al. Mechanism of action of cerebral epoxyeicosatrienoic acids on cerebral arterial smooth muscle. *Am J Physiol*. 1992;263(2 Pt 2):H519–H525.

65 Carroll MA, Balazy M, Margiotta P, et al. Renal vasodilator activity of 5,6-epoxyeicosatrienoic acid depends upon conversion by cyclooxygenase and release of prostaglandins. *J Biol Chem*. 1993;268(17):12260–12266.

66 Campbell WB, Gebremedhin D, Pratt PF, Harder DR. Identification of epoxyeicosatrienoic acids as endothelium-derived hyperpolarizing factors. *Circ Res*. 1996;78(3):415–423.

67 Dumoulin M, Salvail D, Gaudreault SB, et al. Epoxyeicosatrienoic acids relax airway smooth muscles and directly activate reconstituted KCa channels. *Am J Physiol*. 1998;275(3 Pt 1): L423–L431.

68 Munzenmaier DH, Harder DR. Cerebral microvascular EC tube formation: role of astrocytic epoxyeicosatrienoic acid release. *Am J Physiol Heart Circ Physiol*. 2000;278(4):H1163–H1167.

69 Mombouli JV, Vanhoutte PM. Endothelium-derived hyperpolarizing factor(s): updating the unknown. *Trends Pharmacol Sci*. 1997;18(7):252–256.

70 Reilly KB, Srinivasan S, Hatley ME, et al. 12/15-Lipoxygenase activity mediates inflammatory monocyte/endothelial interactions and atherosclerosis in vivo. *J Biol Chem*. 2004;279(10): 9440–9450.

71 Cyrus T, Witztum JL, Rader DJ, et al. Disruption of the 12/15-lipoxygenase gene diminishes atherosclerosis in apo E-deficient mice. *J Clin Invest*. 1999;103(11):1597–1604.

72 Harats D, Shaish A, George J, et al. Overexpression of 15-lipoxygenase in vascular endothelium accelerates early atherosclerosis in LDL receptor-deficient mice. *Arterioscler Thromb Vasc Biol*. 2000;20(9):2100–2105.

73 Patricia MK, Kim JA, Harper CM, et al. Lipoxygenase products increase monocyte adhesion to human aortic ECs. *Arterioscler Thromb Vasc Biol*. 1999;19(11):2615–2622.

74 Honda HM, Leitinger N, Frankel M, et al. Induction of monocyte binding to ECs by MM-LDL: role of lipoxygenase metabolites. *Arterioscler Thromb Vasc Biol*. 1999;19(3):680–686.

75 Weksler BB, Marcus AJ, Jaffe EA. Synthesis of prostaglandin I2 (prostacyclin) by cultured human and bovine ECs. *Proc Natl Acad Sci USA*. 1977;74(9):3922–3926.

76 Baenziger NL, Fogerty FJ, Mertz LF, Chernuta LF. Regulation of histamine-mediated prostacyclin synthesis in cultured human vascular ECs. *Cell*. 1981;24(3):915–923.

77 McIntyre TM, Zimmerman GA, Satoh K, Prescott SM. Cultured ECs synthesize both platelet-activating factor and prostacyclin in response to histamine, bradykinin, and adenosine triphosphate. *J Clin Invest*. 1985;76(1):271–280.

78 Schievella AR, Regier MK, Smith WL, Lin LL. Calcium-mediated translocation of cytosolic phospholipase A2 to the nuclear envelope and endoplasmic reticulum. *J Biol Chem*. 1995; 270(51):30749–30754.

79 Picot D, Loll PJ, Garavito RM. The X-ray crystal structure of the membrane protein prostaglandin H2 synthase-1. *Nature*. 1994; 367(6460):243–249.

80 Kurumbail RG, Stevens AM, Gierse JK, et al. Structural basis for selective inhibition of cyclooxygenase-2 by anti-inflammatory agents. *Nature*. 1996;384(6610):644–648.

81 Lin Y, Wu KK, Ruan KH. Characterization of the secondary structure and membrane interaction of the putative membrane anchor domains of prostaglandin I2 synthase and cytochrome P450 2C1. *Arch Biochem Biophys*. 1998;352(1):78–84.

82 Liou JY, Shyue SK, Tsai MJ, et al. Colocalization of prostacyclin synthase with prostaglandin H synthase-1 (PGHS-1) but not phorbol ester-induced PGHS-2 in cultured ECs. *J Biol Chem*. 2000;275(20):15314–15320.

83 Sanduja SK, Tsai AL, Matijevic-Aleksic N, Wu KK. Kinetics of prostacyclin synthesis in PGHS-1-overexpressed ECs. *Am J Physiol*. 1994;267(5 Pt 1):C1459–C1466.

84 Egan RW, Paxton J, Kuehl FA Jr. Mechanism for irreversible self-deactivation of prostaglandin synthetase. *J Biol Chem*. 1976; 251(23):7329–7335.

85 Ham EA, Egan RW, Soderman DD, et al. Peroxidase-dependent deactivation of prostacyclin synthetase. *J Biol Chem*. 1979; 254(7):2191–2194.

86 McAdam BF, Catella-Lawson F, Mardini IA, et al. Systemic biosynthesis of prostacyclin by cyclooxygenase (COX)-2: the human pharmacology of a selective inhibitor of COX-2. *Proc Natl Acad Sci USA*. 1999;96(1):272–277.

87 FitzGerald GA, Smith B, Pedersen AK, Brash AR. Increased prostacyclin biosynthesis in patients with severe atherosclerosis and platelet activation. *N Engl J Med*. 1984;310:1065–1068.

88 Belton O, Byrne D, Kearney D, et al. Cyclooxygenase-1 and -2-dependent prostacyclin formation in patients with atherosclerosis. *Circulation*. 2000;102(8):840–845.

89 Tazawa R, Xu XM, Wu KK, Wang LH. Characterization of the genomic structure, chromosomal location and promoter of human prostaglandin H synthase-2 gene. *Biochem Biophys Res Commun*. 1994;203(1):190–199.

90 Inoue H, Nanayama T, Hara S, et al. The cyclic AMP response element plays an essential role in the expression of the human prostaglandin-endoperoxide synthase 2 gene in differentiated U937 monocytic cells. *FEBS Lett*. 1994;350(1):51–54.

91 Fletcher BS, Kujubu DA, Perrin DM, Herschman HR. Structure of the mitogen-inducible TIS10 gene and demonstration that the TIS10-encoded protein is a functional prostaglandin G/H synthase. *J Biol Chem*. 1992;267(7):4338–4344.

92 Wu KK, Liou J-Y, Cieslik K. Transcriptional control of COX-2 via C/EBPβ. *Arterioscler Thromb Vasc Biol*. 2005;25:679–685.

93 Schroer K, Zhu Y, Saunders MA, et al. Obligatory role of cyclic adenosine monophosphate response element in cyclooxygenase-2 promoter induction and feedback regulation by inflammatory mediators. *Circulation*. 2002;105(23):2760–2765.

94 Inoue H, Yokoyama C, Hara S, et al. Transcriptional regulation of human prostaglandin-endoperoxide synthase-2 gene by lipopolysaccharide and phorbol ester in vascular ECs. Involvement of both nuclear factor for interleukin-6 expression site and cAMP response element. *J Biol Chem*. 1995;270(42):24965–24971.

95 Zhu Y, Saunders MA, Yeh H, et al. Dynamic regulation of cyclooxygenase-2 promoter activity by isoforms of CCAAT/enhancer-binding proteins. *J Biol Chem*. 2002;277(9):6923–6928.

96 Deng WG, Zhu Y, Wu KK. Role of p300 and PCAF in regulating cyclooxygenase-2 promoter activation by inflammatory mediators. *Blood*. 2004;103(6):2135–2142.

97 Deng WG, Zhu Y, Wu KK. Up-regulation of p300 binding and p50 acetylation in tumor necrosis factor-alpha-induced cyclooxygenase-2 promoter activation. *J Biol Chem.* 2003;278 (7):4770–4777.

98 Ossipow V, Descombes P, Schibler U. CCAAT/enhancer-binding protein mRNA is translated into multiple proteins with different transcription activation potentials. *Proc Natl Acad Sci USA.* 1993;90(17):8219–8223.

99 Descombes P, Schibler U. A liver-enriched transcriptional activator protein, LAP, and a transcriptional inhibitory protein, LIP, are translated from the same mRNA. *Cell.* 1991;67(3):569–579.

100 Cieslik KA, Zhu Y, Shtivelband M, Wu KK. Inhibition of p90 ribosomal S6 kinase-mediated CCAAT/enhancer-binding protein beta activation and cyclooxygenase-2 expression by salicylate. *J Biol Chem.* 2005;280(18):18411–18417.

101 Saunders MA, Sansores-Garcia L, Gilroy DW, Wu KK. Selective suppression of CCAAT/enhancer-binding protein beta binding and cyclooxygenase-2 promoter activity by sodium salicylate in quiescent human fibroblasts. *J Biol Chem.* 2001;276(22):18897–18904.

102 Zou MH, Ullrich V. Peroxynitrite formed by simultaneous generation of nitric oxide and superoxide selectively inhibits bovine aortic prostacyclin synthase. *FEBS Lett.* 1996;382(1–2):101–104.

103 Schmidt P, Youhnovski N, Daiber A, et al. Specific nitration at tyrosine 430 revealed by high resolution mass spectrometry as basis for redox regulation of bovine prostacyclin synthase. *J Biol Chem.* 2003;278(15):12813–12819.

104 Zou MH, Leist M, Ullrich V. Selective nitration of prostacyclin synthase and defective vasorelaxation in atherosclerotic bovine coronary arteries. *Am J Pathol.* 1999;154(5):1359–1365.

105 Christman BW, McPherson CD, Newman JH, et al. An imbalance between the excretion of thromboxane and prostacyclin metabolites in pulmonary hypertension. *N Engl J Med.* 1992; 327(2):70–75.

106 Tuder RM, Cool CD, Geraci MW, et al. Prostacyclin synthase expression is decreased in lungs from patients with severe pulmonary hypertension. *Am J Respir Crit Care Med.* 1999;159 (6):1925–1932.

107 Barst RJ, Rubin LJ, Long WA, et al. A comparison of continuous intravenous epoprostenol (prostacyclin) with conventional therapy for primary pulmonary hypertension. The Primary Pulmonary Hypertension Study Group. *N Engl J Med.* 1996; 334(5):296–302.

108 Hoeper MM, Schwarze M, Ehlerding S, et al. Long-term treatment of primary pulmonary hypertension with aerosolized iloprost, a prostacyclin analogue. *N Engl J Med.* 2000;342(25): 1866–1870.

109 Lievre M, Morand S, Besse B, et al. Oral Beraprost sodium, a prostaglandin I(2) analogue, for intermittent claudication: a double-blind, randomized, multicenter controlled trial. Beraprost et Claudication Intermittente (BERCI) Research Group. *Circulation.* 2000;102(4):426–431.

110 Belch JJ, Bell PR, Creissen D, et al. Randomized, double-blind, placebo-controlled study evaluating the efficacy and safety of AS-013, a prostaglandin E1 prodrug, in patients with intermittent claudication. *Circulation.* 1997;95(9):2298–2302.

111 Shyue SK, Tsai MJ, Liou JY, et al. Selective augmentation of prostacyclin production by combined prostacyclin synthase and cyclooxygenase-1 gene transfer. *Circulation.* 2001;103(16): 2090–2095.

112 Lin H, Lin TN, Cheung WM, et al. Cyclooxygenase-1 and bicistronic cyclooxygenase-1/prostacyclin synthase gene transfer protect against ischemic cerebral infarction. *Circulation.* 2002;105(16):1962–1969.

113 Numaguchi Y, Naruse K, Harada M, et al. Prostacyclin synthase gene transfer accelerates reendothelialization and inhibits neointimal formation in rat carotid arteries after balloon injury. *Arterioscler Thromb Vasc Biol.* 1999;19(3):727–733.

114 Todaka T, Yokoyama C, Yanamoto H, et al. Gene transfer of human prostacyclin synthase prevents neointimal formation after carotid balloon injury in rats. *Stroke.* 1999;30(2):419–426.

113

Regulation of Endothelial Barrier Responses and Permeability

Joe G.N. Garcia

Pritzker School of Medicine, University of Chicago, Illinois

The endothelial cell (EC) lining of the systemic and pulmonary vasculatures was considered for decades to exist merely as an inert and passive semipermeable cellular barrier between the blood and the interstitium of all tissues. During the early 1960s, however, Majno and Palade described the ultrastructural appearance of an actively contracted endothelium in pulmonary vessels previously exposed to the edemagenic agent histamine (1), thus igniting a controversy over whether the endothelium plays an active role in the inflammatory response. It is now well recognized that endothelial activities are critical and essential aspects of inflammation. Disruption of the integrity of the endothelial barrier results in marked increases in permeability to fluids leading to tissue edema and pain (dolor), and leukocyte infiltration into tissues (rubor and calor). When persistent and of significant intensity, this process invariably progresses to organ dysfunction. For example, in systemic inflammatory states such as sepsis, increased vascular permeability results in high morbidity and mortality via multiorgan dysfunction, including acute renal or hepatic failure, cardiac dysfunction, and respiratory insufficiency.

Despite multiple attempts to improve upon the adverse clinical outcomes associated with increased endothelial permeability, the termination of fulminant edema with restoration of endothelial integrity remains an unrealized goal. Considerable progress in the past several years, however, suggests that several barrier-enhancing agents (or their selective derivatives with greater receptor selectivity) may be on the horizon for therapeutic purposes. It also should be noted that in selective vascular cells such as the cerebral circulation, the goal may be to actually increase permeability in order to increase the access of novel therapeutics across the blood–brain barrier (Figure 113.1). Nevertheless, barrier-enhancing agents hold considerable utility not only in a variety of inflammatory disorders, in which limitation of tissue edema is a goal, but also in diverse vascular pathologies (when inflammation is not the most prominent feature) such as tumor angiogenesis, ischemia–reperfusion injury, and atherosclerosis – vascular processes that exhibit vascular leak as a fundamental phenotypic element (2,3). The overlapping involvement of vascular permeability in broad biologic processes such as inflammation and angiogenesis is underscored by recalling that the potent angiogenic agent, vascular endothelial growth factor (VEGF), was first characterized as a vascular permeability–increasing factor (VPF) and only later was linked to its capacity to profoundly promote blood vessel growth. Developmentally, the genesis of new blood vessels (vasculogenesis) or extensions of existing vessels (angiogenesis) concludes with a critical terminal maturation event: the restoration of an intact vascular barrier from the nascent and inherently leaky vessel.

Given that these temporal- and stage-dependent alterations in endothelial permeability are key to the angiogenic process, it is not surprising that other angiogenic growth factors such as hepatocyte growth factor (4), angiopoietin 1 (5), and sphingosine 1-phosphate (S1P) (6), are now recognized as angiogenic mediators with potent barrier-enhancing properties. For example, S1P proved to be the most potent EC chemotactic in serum (7) and a complete angiogenic factor (i.e., does not require additional angiogenic cofactors) (6), while demonstrating a robust and sustained duration of barrier enhancement (2,6). The ligation of specific members of the endothelial differentiation gene (Edg) family of S1P receptors results in strong barrier enhancement in vitro and in vivo (6), whereas targeted deletion of the gene encoding the barrier-enhancing $S1P_1$ (also known as Edg1) receptor resulted in embryonic lethality due to progressive edema formation and hemorrhage because of an inability to establish functional barrier integrity (8). S1P is generated by numerous cell types but is the major barrier-protective agent released by platelets (9), and significantly attenuates thrombin-induced barrier disruption (2) while rapidly restoring barrier integrity in the isolated perfused murine lung (3). Furthermore, a single intravenous infusion of S1P given after intratracheally

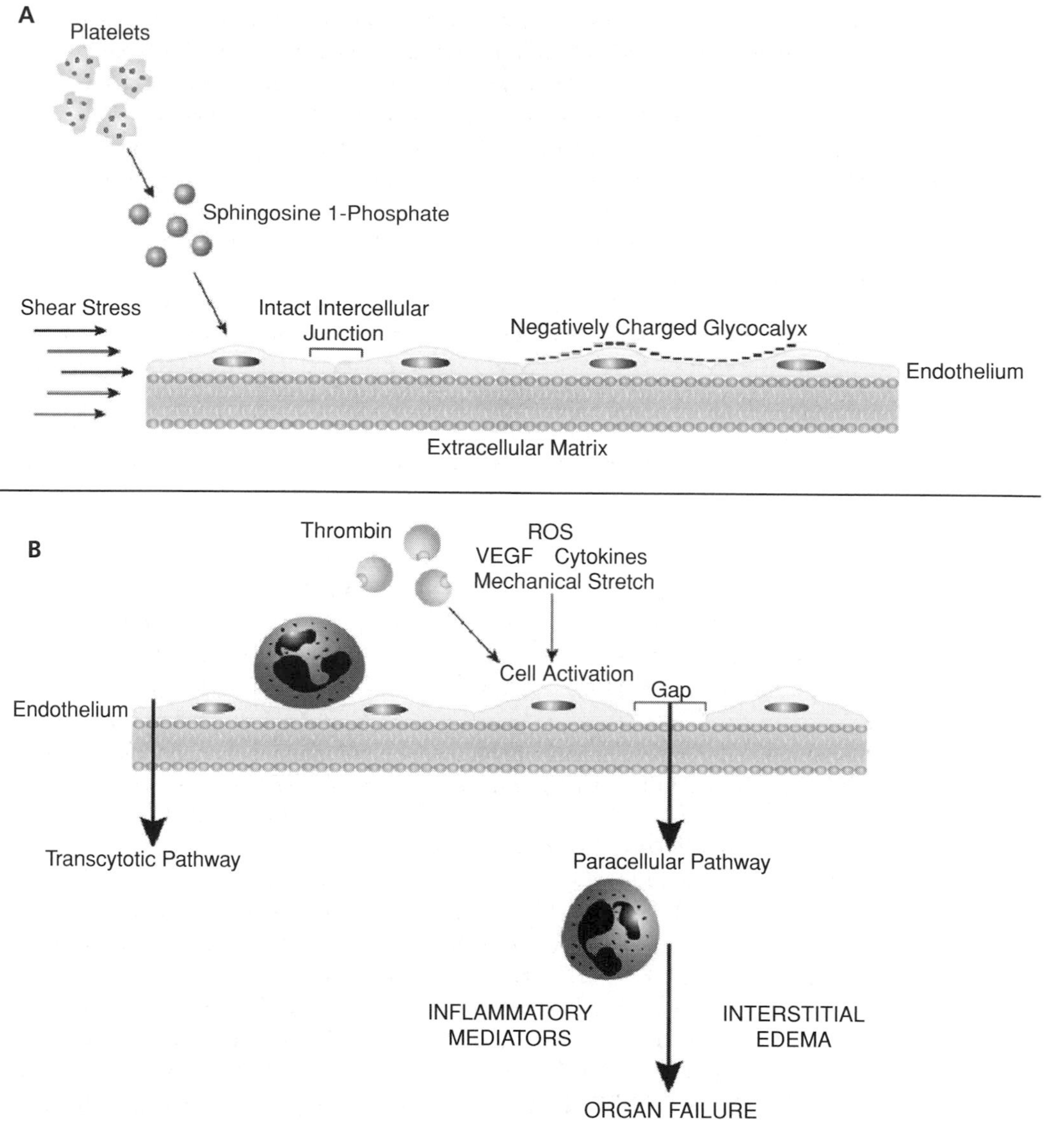

Figure 113.1. (**A**) A number of maintenance processes ensure that vascular integrity is maintained and responsive to alterations in barrier integrity by initiating signals that restore the integrity of the vascular monolayer. These include low levels of shear stress, the negatively charged glycocalyx, and barrier-protective molecules released by circulating platelets such as sphingosine 1-phosphate. (**B**) Mechanisms of increased vascular permeability via transcellular and paracellular pathways. Bioactive agonists, growth factors, cytokines, and mechanical forces (high shear stress or cyclic stretch), as well as activated leukocytes, serve to activate the vascular endothelium. This produces cellular contraction and increased passage of fluid and cells through intercellular spaces into the interstitium to initiate organ dysfunction.

administered endotoxin, produced highly significant reductions in multiple indices of inflammatory lung injury including vascular leak in both murine and canine models of inflammatory lung edema (3,10). The underlying signaling mechanisms by which S1P increases vascular integrity is discussed later in the context of G-protein–dependent signaling cascades, cytoskeletal rearrangement, and enhanced junctional integrity and focal adhesion to the extracellular matrix (ECM). The identification of novel biologically compatible agents that can restore vascular integrity offers much promise for the future management of edema formation and organ dysfunction in the critically ill.

REGIONAL DIFFERENCES IN ENDOTHELIAL CELL PERMEABILITY

ECs from different regions in the vasculature share many common features, but also exhibit site-specific properties. EC phenotypes vary between vessels of different caliber (e.g., large arteries vs. capillaries), between different organs (e.g., blood–brain barrier versus pulmonary vascular endothelium), and even between neighboring ECs (11,12). For example, pulmonary endothelium is of the continuous type, with the majority of fluid and solute exchange occurring at the level of the capillary and postcapillary microvascular endothelium because this layer has the largest surface area (relative to other vascular beds) available for both diffusion and filtration. The albumin permeability value in cultured pulmonary microvessel ECs is about one-half to one-fifth of the values of similarly cultured cells from the mainstem pulmonary artery, a feature also noted in vivo (12,13). Interestingly, cells from microvessels are significantly more restrictive to sucrose and inulin, indicating that transport occurring via paracellular pathways is reduced to a greater degree than in cells from large vessels. Additional routes of albumin transport (i.e., vesicular transport or transcytotic transport) also are involved in albumin flux across lung microvessel endothelium (14). In addition, microvascular ECs proliferate at a higher level than do macrovascular ECs and retain phenotypically distinct calcium (Ca^{2+}) and cyclic nucleotide signaling responses as well as oxidant-mediated signaling transduction (12).

Studies of ECs from different sites in the vasculature also have identified organ-specific antigens on capillary endothelium that are responsible in part for the different degrees of permeability of regional vascular beds with phenotypic differences in lectin-binding domains (15), although the correlation with barrier properties is not robust. The plasmalemmal membrane, vesicles, and paracellular channel structures also contain microdomains of anionic sites consisting of distributions of glycosaminoglycans, sialoconjugates, and monosaccharide residues. Heterogeneity in these cell surface glycoproteins, collectively known as the glycocalyx, has been described across various vascular beds and relate to the permeability characteristics of these vascular ECs in situ (16). For example, flux of albumin across glomerular capillaries is critically dependent on the negative charge of the capillary ECs and the ECM proteins constituting the basement membrane (17), with disruption of cell surface negative charge sites increasing "leakage" of albumin across the capillary–matrix complex. Because the glycocalyx in general, and heparan sulfate proteoglycans in particular, are now known to be directly linked to the endothelial cytoskeleton, these proteins may not only represent sensors for mechanical stress and cytoskeletal rearrangement, but also serve as endothelial-surface binding domains for inflammatory cationic peptides (18,19). This provides potential mechanistic insight for the recent observations that polymorphonuclear neutrophil (PMN)-derived cationic peptides are responsible for the increase in vascular permeability that evolves following EC interaction with activated PMNs (20).

VASCULAR BARRIER REGULATION: MODIFIED TENSEGRITY MODEL[1]

Vascular permeability is mediated by two well-accepted pathways: a transcellular pathway that involves either vesicular transport or the formation of channels through the endothelium, and a paracellular pathway with transport of fluid and cells occurring at the opening of intercellular junctions between neighboring ECs. The transcellular pathway utilizes a tyrosine kinase–dependent, gp60-mediated transcytotic albumin route whose regulation and function are unclear, but which may serve to uncouple protein and fluid permeability (14). However, there is general consensus that the primary mode of fluid and transendothelial leukocyte trafficking during inflammation occurs via the paracellular pathway, whose essential role in endothelial permeability has been well supported by an impressive body of research, including electron microscopy studies that demonstrate the formation of paracellular gaps at sites of active inflammation within the vasculature (1,21). The role of intercellular gaps in mediating paracellular permeability also has been determined by osmotically shrinking ECs. This process results in the formation of intercellular gaps and increased paracellular permeability, and is reversed by rehydration of the cells (22).

To study paracellular permeability, we have found it particularly useful to employ a working model of vascular barrier regulation, a modified tensegrity model, which highlights the critical importance of a delicate balance between ongoing EC contractile forces (which generate centripetal tension) and adhesive forces (generated by cell–cell and cell–matrix tethering) (Figure 113.2). Together, these competing barrier-regulatory forces govern changes in cell shape and permeability, with both contractile and tethering elements intimately linked to a complex network of endothelial cytoskeletal proteins (actin microfilaments, microtubules, intermediate filaments) that combine to regulate shape change and transduce signals within and between ECs. The EC cytoskeleton, as a result, is centrally involved in multiple endothelial biologic processes including angiogenesis, mechanotransduction, apoptosis, and critical elements of the inflammatory response, including leukocyte diapedesis and permeability to proteins. The fine tuning of these responses is governed by the over 80 actin–associated proteins (actin-binding, -capping, -nucleating, -severing proteins) identified that participate in cytoskeletal rearrangement, tensile force generation, and the regulation of cell shape, adhesion, and orchestrated cell migration, as well as endothelial junctional stability (23).

ENDOTHELIAL BARRIER REGULATION: ROLE OF THE CONTRACTILE APPARATUS

Over the past two decades, it has become increasingly clear that ECs contain an abundance of the molecular machinery necessary to generate tension via an actomyosin motor. The actin microfilament cytoskeleton, normally peripherally distributed as cortical bands in orientation to directional blood flow, is essential for the maintenance of endothelial integrity and basal barrier function. Contributing to basal barrier function and cell–cell adhesion is the presence of phosphorylated regulatory myosin light chains (MLC) within the loose subcortical actin mesh. MLC phosphorylation is catalyzed

1 The original tensegrity model was proposed by Ingber and Wang (Wang N, Butler JP, Ingber DE. *Science.* 1993;260[5111]:1080–1); also see Inger DE. Cellular tensegrity: defining new rules of biological design that govern the cytoskeleton. *J Cell Sci.* 1993;104, 613–627.

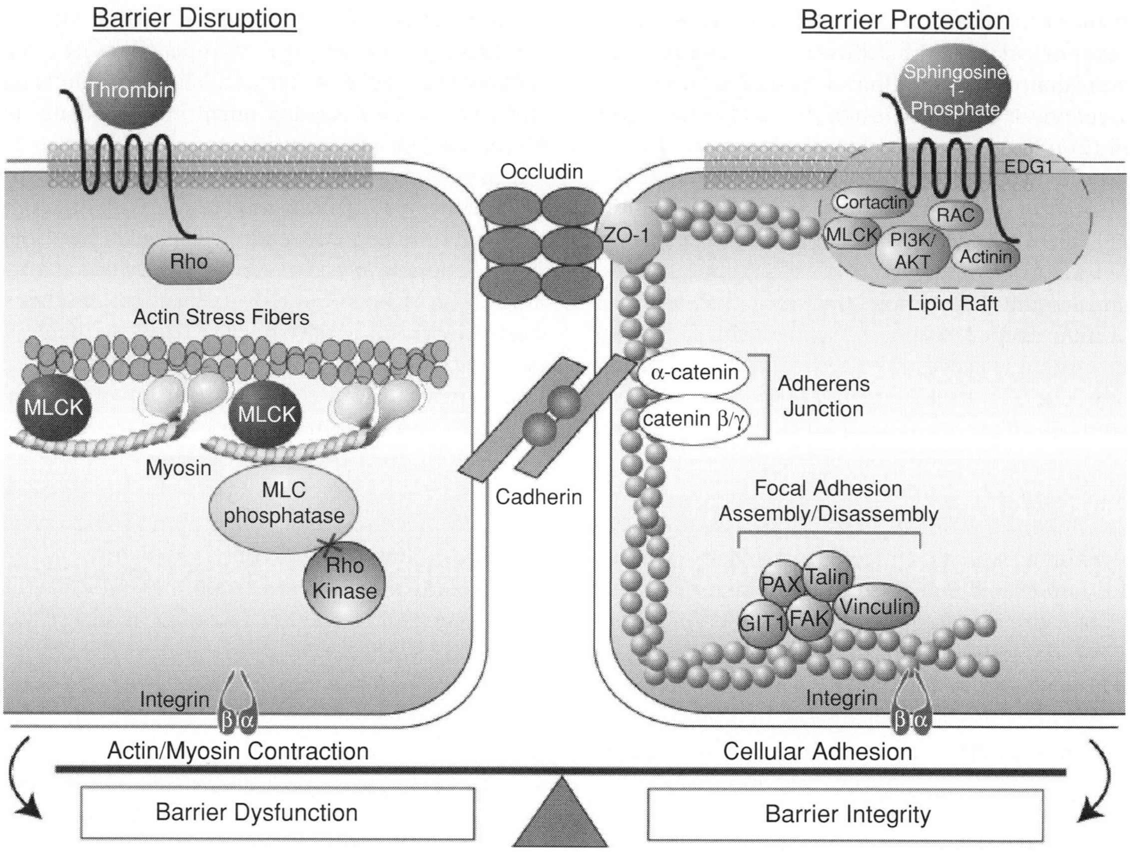

Figure 113.2. Modified model of tensegrity. In this working model of EC barrier regulation, under basal conditions, a balance exists between actomyosin contractile and cellular adhesive forces. When contractile forces predominate, as depicted in the thrombin-stimulated model in the left panel, ECs pull apart to form paracellular gaps, favoring barrier disruption. Thrombin cleavage of the labeled PAR-1 receptor on the surface of ECs, activates both heterotrimeric G-proteins (G_q, $G_{12/13}$) as well as small GTPases such as Rho. Activated Rho (Rho-GTP) induces Rho kinase, which via phosphorylation of the phosphatase regulatory subunit, inhibits MYPT. Rho kinase and MLCK activation occurs via independent pathways. Increased cytosolic Ca^{++} (via IP3 production) activates the Ca^{++}/calmodulin-dependent MLCK, with conformational changes allowing the enzyme to access the preferred substrate (MLC). Rho kinase and MLCK activation both culminate in increased MLC phosphorylation which, in turn, enables actomyosin contraction and results in increased stress fiber formation, cellular contraction, paracellular gap formation, and ultimately barrier dysfunction. In balance with these contractile forces, agents such as sphingosine 1-phosphate, a platelet-derived lipid growth factor, serve to maintain barrier function via increased cell–cell adhesion. S1P ligates $S1P_1$, resulting in the activation of the small GTPase Rac, a signaling cascade that results in cytoskeletal rearrangement and increased cortical actin with increased linkage to the adherens junction and focal adhesions. Rac activation initiates intracellular events dependent on specific PKC isotypes; PAK; LIM kinase; the actin severing protein, cofilin; and MLCK, which all contribute to increased cell–cell and cell–matrix tethering. These are spatially defined by recruitment to lipid rafts.

by the Ca^{2+}/calmodulin (CaM)-dependent serine/threonine MLC kinase (MLCK). Gene expression studies have demonstrated that the EC *MLCK* gene (*MYLK* is the gene name) is differentially expressed between micro- and macrovascular ECs (24), with more abundant expression in the microvascular endothelium. We have reported that several single nucleotide polymorphisms in *MYLK* confer increased susceptibility to sepsis and acute inflammatory lung injury (25). Direct activation of MLCK is sufficient to produce EC contraction via inducing actin stress fiber formation, barrier dysfunction, and EC permeability (25). This effect is significantly attenuated by treatment with MLCK inhibitors (26–28). In addition, MLCK inhibition prevents transforming growth factor (TGF)-β1–stimulated EC permeability (29), abolishes barrier dysfunction in rat lung models of ischemia–reperfusion injury (30), and attenuates ventilator-induced lung permeability (31). MLCK activities are tightly coordinated with the function of the small GTPase, Rho, in regulation of the contractile apparatus through its downstream effector, Rho kinase (32). Rho activation leads to phosphorylation of the myosin binding subunit of MLC phosphatase (PP1) by Rho kinase, thereby inhibiting phosphatase activity and resulting in increased levels of MLC phosphorylation, actomyosin interaction, stress fiber formation, and subsequent EC barrier dysfunction. The inhibition of either MLCK activity or Rho activation attenuates thrombin-induced EC barrier dysfunction (23). Rho/Rho kinase and MLCK may also differentially regulate MLC phosphorylation according to spatial localization.

Another critical function of the EC barrier is the regulation of PMN margination and migration into sites of acute inflammation, a complex process involving specific recognition molecules and cytokine/chemokine signaling to the cytoskeleton evoked by PMN interaction with ECs. Activation of the EC contractile apparatus by PMN adherence and diapedesis also involves MLCK activity, as indicated by the increased MLC phosphorylation after PMN–EC interaction and the observation that reduction in MLCK activity significantly attenuates leukocyte migration (33,34). Furthermore, disruption of either the actin cytoskeleton with cytochalasin B or stabilization of microtubules with paclitaxel decreases transendothelial movement, whereas disassembly of microtubules increases leukocyte migration (23). Overall, this provides a mechanistic linkage between permeability to solutes, permeability to fluids, and permeability to cells.

MODEL OF ENDOTHELIAL BARRIER REGULATION: ROLE OF JUNCTIONAL AND MATRIX ADHESION

Under homeostatic conditions, the EC cytoskeleton is focally linked to multiple membrane adhesive proteins within two intercellular complexes – adherens junctions and tight junctions. Adherens junction assembly and stabilization is essential for the maintenance of EC barrier function, with the greater complexity of the intercellular junctions (i.e., more complex involvement of adherens junction proteins) potentially accounting for the significantly lower permeabilities to small and large proteins observed in pulmonary microvessel ECs. Adherens junction complexes in ECs consist primarily of cadherin proteins bound together in a homotypic- and Ca^{2+}-dependent fashion to link adjacent ECs. The cytoplasmic tails of the cadherins are intimately linked to the actin cytoskeleton via complexes of proteins including α-, β-, and γ-catenin and p120. Together, these complexes result in decreased permeability to fluid and solute both in vitro and in vivo (35). The primary adhesive cadherin protein present in endothelial adherens junctions is vascular endothelial (VE)-cadherin, which is essential for the maintenance of EC barrier integrity, as demonstrated by increased vascular permeability induced in mice after infusion of VE-cadherin blocking antibody (35). Leukocyte binding to ECs causes a disruption of adherens junctions, with the disappearance of VE-cadherin and catenin staining in cell–cell contacts (23,35). Adherens junctions appear integral to this process because VE-cadherin blocking antibodies increase PMN diapedesis, while conversely tight junctions remain intact during this migration. Both intracellular Ca^{2+} and tyrosine phosphorylation pathways are involved, because activated PMNs increase Ca^{2+} and the phosphotyrosine content of VE-cadherin and β-catenin in association with adherens junction disruption and hyperpermeability (23).

Tyrosine phosphorylation may provide an additional regulatory link between actin cytoskeletal rearrangement and adherens junction function, because the antiadhesive protein, thrombospondin (TSP)-1 induces tyrosine phosphorylation of adherens junction proteins, actin rearrangement with stress fiber formation, and increased albumin flux across EC monolayers, whereas tyrosine kinase inhibition attenuates these effects. Thrombin alters the dissociation of the tyrosine phosphatase SHP2 from VE-cadherin complexes to produce increased levels of tyrosine phosphorylation of catenins and subsequent destabilization of adherens junction linkage to the cytoskeleton. Site-directed mutagenesis studies have tightly linked β-catenin phosphorylation on tyrosine residues with EC junctional integrity, consistent with a minimal level of basal tyrosine phosphorylation necessary for the maintenance of cell–cell contacts.

Tight junctions, consisting of a complex of proteins that include the claudins, occludin, and junctional adhesion molecule (JAM)-1, also interact with cytoskeletal actin (23). Tight junctional transmembrane proteins also include the zona occludens (ZO) family, which appear to participate in signal transduction and provide a link between occludin and the actin cytoskeleton with colocalization of polymerized actin (also referred to as F-actin) and ZO-1. Alterations in tight junctions may be signaled through the mitogen-activated protein kinase (MAPK) pathway, because both VEGF and hydrogen peroxide (H_2O_2) induce occludin dissociation from cell junctions and produce EC barrier dysfunction, which is partially blocked by MAPK inhibitors (23).

Although alterations in EC permeability depend largely on the function of intercellular adhesions, focal contacts that exist between the EC and the underlying matrix (focal adhesions) also provide additional adhesive forces in barrier regulation contributing approximately 20% of the resistance to transvascular permeability (36). Focal adhesions form functionally dynamic bridges for bidirectional signal transduction between the actin cytoskeleton and the cell–matrix interface; these bridges are comprised of ECM proteins (collagen, fibronectin, laminin, vitronectin, and proteoglycans), transmembrane α- and β integrin receptors, and cytoplasmic focal adhesion plaques (containing α-actinin, vinculin, paxillin, and talin). The core matrix proteins, because of their position and points of contact with ECs, may determine cell–substratum adhesion and thereby determine vascular permeability in normal conditions and in response to inflammatory mediators such as tumor necrosis factor (TNF)-α. Extracellular stimuli can be transmitted to the cytoskeleton through focal adhesion rearrangement linked to integrin ligation; antibodies to β_1 integrin alter EC attachment, cell spreading, and permeability. Integrin binding to the ECM induces the attachment of integrins to intracellular actin fibers, a process that stimulates tyrosine phosphorylation of multiple focal adhesion proteins (focal adhesion kinase [FAK], cortactin, and paxillin) as well as tyrosine phosphorylation–dependent Ca^{2+} influx.

In addition to cues derived from cytoskeletal integrin linkages, the ECM also can be remodeled by EC–released proteases, causing ECs layered upon this matrix to exhibit compromised monolayer integrity. Specific patterns of focal adhesion distribution and altered interactions between constituent focal adhesion proteins (including FAK, G-protein–coupled receptor kinase interacting [GIT]-1, GIT2, and paxillin) are associated with EC contraction and barrier dysfunction. In quiescent cells, FAK and paxillin are both distributed between the cytosol and focal adhesions. As discussed later, EC barrier function is increased by S1P, which induces partial disassembly of pre-existing focal adhesions coincident with the transient association of GIT1 with FAK, followed by translocation of FAK and paxillin from the cytosol to the cell periphery. FAK, paxillin and GIT2 are incorporated into new peripheral focal adhesions that are linked to the newly reinforced cortical actin ring (37,38). Integrin binding also targets activated extracellular-signal-regulated kinase (ERK) to newly formed focal adhesion sites. Reciprocally, intracellular signaling pathways that regulate cytoskeletal rearrangement also can modulate cell–matrix contacts. Rho inhibition dissociates stress fibers from focal adhesions, decreases the phosphotyrosine content of paxillin and FAK, and enhances EC barrier function (36). Similarly, v-Src–induced tyrosine phosphorylation of focal adhesion proteins is a well-established stimulus for the disassembly of these adhesive structures (36). Further studies to clarify the role of barrier-protective and barrier-disruptive tyrosine kinases in both focal adhesion and adherens junction complexes will increase our understanding of EC barrier regulation by tethering forces.

THROMBIN MODEL OF INCREASED VASCULAR PERMEABILITY

A variety of bioactive agonists, cytokines, growth factors, and mechanical forces alter pulmonary vascular barrier properties through overlapping yet often distinct mechanisms and serve to increase vascular permeability. The serine protease thrombin represents an ideal model for the examination of agonist-mediated EC activation and barrier dysfunction, because this agonist evokes numerous EC responses that regulate hemostasis, thrombosis, and vessel wall pathophysiology. The effect of thrombin on permeability is mediated by the protease-activated receptor, (PAR)-1. Thrombin ligates and proteolytically cleaves the extracellular NH_2-terminal domain of PAR-1. The cleaved NH_2-terminus functions as a tethered ligand, activating the receptor and initiating a number of downstream effects, including the activation of phospholipases A_2, C, and D; increases in cytosolic Ca^{2+}; release of von Willebrand factor (vWF), endothelin, nitric oxide (NO), and prostacyclin (PGI_2). Thrombin binding alone, without receptor cleavage, is insufficient to elicit the complex signaling cascade that results in increased endothelial permeability. The effect of thrombin on EC permeability is rapid (within 2 minutes) and reversible, and is critically dependent on G-protein–transduced signals that result in increases in cytosolic Ca^{2+}, activation of the contractile apparatus, and characteristic alterations in the cytoskeletal elements, such as marked increases in actin stress fibers. These effects have been confirmed using fluorescent microscopy and biophysical measurements (23). Critical to the increase in vascular EC permeability is a rise in intracellular Ca^{2+} concentration ($[Ca^{2+}]_i$), with significant endothelial heterogeneity in the magnitude of this response (12) (i.e., differences in spatially defined Ca^{2+} response dictate site-specific differences in permeability).

Consistent with the modified tensegrity model, EC gap formation and permeability evolves via an increase in intracellular tension, with the activated actomyosin contractile apparatus leading to thrombin-induced EC permeability, disruption in the integrity of paracellular adherens junctions, and reorganization of focal adhesion plaques. In addition to Ca^{2+}-dependent MLCK and Rho kinase activation, thrombin also induces coordinate increases in protein kinase C (PKCα, ζ) and p38 MAPK activities. These latter pathways can be evoked by edemagenic agents or mechanical signals, which does not necessarily result in an overall increase in the levels of MLC phosphorylation. PKC-mediated signaling pathways exert a prominent effect on barrier regulation (both protection and dysfunction) in a time- and species-specific manner (39). PKC is able to increase MLC phosphorylation at sites distinct from MLCK-mediated phosphorylation (Ser^{19}, Thr^{18}). This does not produce an overall increase in MLC phosphorylation or an increase in the formation of actin stress fibers. PKC activation also causes increases in the myosin phosphatase known as myosin light chain phosphatase (MYPT). These phosphatase activities are directed specifically to MLC residues Ser^{19}/Thr^{18} (39). Therefore, PKC-mediated increases in bovine EC permeability may occur through the phosphorylation of caldesmon, an actin-, myosin-, and calmodulin-binding protein, whose phosphorylation in smooth muscle alters actomyosin cross-bridges (40). Caldesmon distributes along stress fibers in a process that reflects the actin- and CaM-binding domains in caldesmon (40), and it is phosphorylated in ECs after thrombin and PMA challenge (23). Caldesmon-mediated regulation of actomyosin adenosine triphosphatase (ATPase) in smooth muscle also is modified by the actin cross-linking proteins filamin and gelsolin (23). Although filamin participates directly in barrier regulation through CaM kinase II activation, its effects on actin cytoskeletal rearrangement are regulated through Rho family GTPases, thereby providing another link with a known modulator of EC barrier function.

Unlike thrombin, TNF-α induces slow-onset barrier disruption that is independent of MLCK activity (23) but requires p38 MAPK activation; this also has been linked to contractile regulation in EC migration and lipopolysaccharide (LPS)-induced EC permeability (23,41). The mechanism through which p38 MAPK exerts these effects is unclear but may involve the actin-binding protein HSP27, a known p38 MAPK target whose actin polymerization-inhibiting activity dramatically decreases after phosphorylation in association with stress fiber development (41,42).

In the model of thrombin-induced endothelial permeability, focal adhesions remodel to withstand increased mechanical loading from contracting actomyosin stress fibers. Thrombin induces novel focal adhesion signaling, including increased association of GIT1 and GIT2 with FAK and paxillin, prominent FAK phosphorylation (at Y^{397}, Y^{576}, and Y^{925}), partial localization of GIT1 along the actin stress fibers (37,38), and rapid redistribution of focal adhesion proteins to the ends of stress fibers, with a temporal sequence similar to thrombin-induced alteration in EC permeability (37,38). As focal adhesions in adherent ECs normally undergo a constant basal rate of turnover that accelerates with cell activation, the elevated and sustained association of GIT1 with paxillin may reflect accelerated focal adhesion turnover in both barrier-promoting and barrier-disrupting models as GIT1 participates in focal adhesion disassembly. Src-kinase, which also participates in contractile regulation via phosphorylation of MLCK (43), is likely also involved in thrombin-induced focal adhesion rearrangement. Src-kinase inhibitors (PP2) blocked thrombin-induced FAK Y^{397} phosphorylation but failed to alter phosphorylation at Y^{576} and Y^{925} or thrombin-induced FAK localization and stress fiber formation, implicating other tyrosine kinases in the response to thrombin (37,38). The inhibition of FAK attenuates the permeability response to VEGF in coronary venular ECs, further validating the functional role for focal adhesion complexes in EC barrier modulation.

MECHANICAL STRESS, SIGNAL TRANSDUCTION, AND VASCULAR BARRIER REGULATION

In addition to the activation of the contractile apparatus by receptor-mediated pathways, mechanical signals also are transduced to the EC cytoskeleton and trigger complex alterations in junctional and matrix protein gene expression (36). The study of ECs in static cultures, although providing the opportunity to mechanistically evaluate the role of cytoskeletal components in physiology, has major limitations given that the endothelium in its native state is continuously exposed to mechanical forces that greatly influence cellular structure and function. Mechanical shear stress in the range of 1 to 10 dynes/cm^2 has been associated with increased monolayer integrity and barrier enhancement (44). Shear stress activates signaling pathways and leads to the upregulation of transcription factors and subsequent gene expression of various vasoactive substances, growth factors, and adhesion molecules (36). Active cytoskeletal rearrangement begins rapidly and continues to occur over several hours as ECs realign to reduce both peak shear stresses and shear stress gradients. The cellular mechanisms for sensing flow and transducing its signal are still unclear, but are believed to involve both apical actin stress fibers linked to cell–cell contact sites and integrin-mediated signal transduction.

When exposed to shear stress, ECs in static culture demonstrate altered signaling associated with cytoskeletal rearrangement, including Ca^{2+} mobilization, G-protein activation, increased tyrosine phosphorylation, and MLCK and MAPK activation (44). These pathways interact downstream to produce the complex cellular effects of flow. For example, during shear stress, the GTPases Rho and Cdc42 combine to activate MAPKs, whereas individually, Rho is necessary for flow-induced stress fiber formation and cell alignment and Cdc42 activates transcription factors. The integrated effects of these shear-induced signals on EC barrier function are variable depending on the magnitude, duration, and gradient of flow. Exposure of human pulmonary ECs to laminar shear stress (15 dyn/cm^2) results in a peripheral accumulation of focal adhesions, whereas cyclic stretch (18% elongation) induces randomly distributed focal adhesions attached to the ends of newly formed stress fibers (45,46). Shear stress activates small GTPase Rac without effects on Rho, whereas 18% cyclic stretch activates the Rho GTPase without affecting Rac. Shear stress–mediated increases in transendothelial electrical resistance was further elevated by challenge with S1P (46). Shear stress also induced FAK phosphorylation at Y^{397} and Y^{576}, whereas cyclic stretch induced FAK phosphorylation at Y^{576} (46). This suggests that mechanical and chemical factors share the signaling and cytoskeletal mechanisms involved in EC barrier regulation with differential effects of shear stress and cyclic stretch on Rho and Rac activation, focal adhesion redistribution, and site-specific FAK phosphorylation.

Shear stress maximally increases the protein expression of integrins after 12 hours of exposure, and it significantly enhances cell–matrix attachment, suggesting that flow helps maintain the EC monolayer through the augmentation of focal adhesions. However, ECs exposed to high shear gradients, or turbulent flow, develop increased permeability relative to areas of either constant, laminar flow or no flow. The majority of these studies have been performed using ECs from the systemic circulation, and any potential differences in pulmonary EC responses to flow are not well understood. Although the mechanism for this signal transduction is not well understood, the complex array of proteins that constitute the lung endothelial glycocalyx are likely involved and behave as viscoelastic anionic polymers, undergoing shear-dependent conformational changes that may function as blood-flow sensors to transduce signals into ECs. One mechanism by which shear stress may alter barrier function is through inhibiting EC apoptosis, although recent work using a TNF-α model under static conditions suggests divergent processes are involved in cytokine-induced apoptosis and permeability (23).

SPHINGOSINE 1-PHOSPHATE MODEL OF VASCULAR BARRIER RESTORATION

Despite significant progress in our understanding of the molecular and cellular events regulating increased permeability, the mechanisms of vascular barrier restoration are less

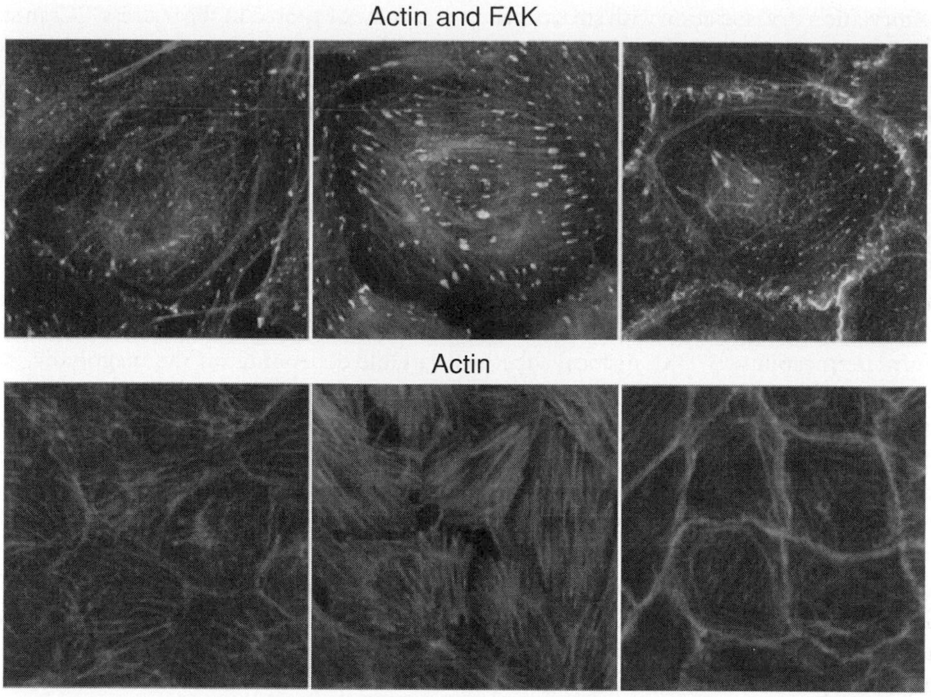

Figure 113.3. EC cytoskeletal rearrangement induced by thrombin and sphingosine 1-phosphate (S1P). These immunofluorescent images utilizing Texas red-conjugated phalloidin to identify polymerized actin filaments and focal adhesion kinase demonstrate that relative to controls, human pulmonary artery ECs stimulated with thrombin demonstrate a prominent increase in actin stress fibers associated with cell contraction and evidence of paracellular gaps (*small arrows*), with focal adhesions at the end of the actin cords. In contrast, EC stimulated with Sph-1-P reveal prominent cortical actin enhancement, a relative paucity of central stress fibers, and no paracellular gaps. Focal adhesion remodeling is associated with the cortical actin ring, as evidenced by the strong colocalization of these proteins. For color reproduction, see Color Plate 113.3.

well understood, and therapeutic agents for modulating barrier function in a clinically advantageous way remain elusive. Historically, increasing the level of cyclic nucleotides has represented the sole strategy for reducing the edema phase observed in inflammatory syndromes. Increases in intracellular cyclic adenosine monophosphate (cAMP) concentrations decrease vascular endothelial permeability and inhibit the permeability-increasing effects of thrombin and histamine in association with inhibition of F-actin reorganization (23). The reduction in paracellular permeability involves activation of 1) cAMP-dependent protein kinases that phosphorylate proteins, such as MLCK, thereby reducing MLC phosphorylation and 2) phosphatases that dephosphorylate proteins, such as MLCs, which serve to reduce stress fiber formation. Increases in cellular cAMP also prevent the increase in $[Ca^{2+}]_i$ and isotype-specific PKC activation, important downstream signals regulating permeability (12,23).

Recently, additional barrier-promoting effectors have been identified that enhance basal vascular endothelial barrier function, and either reverse or prevent agonist-mediated barrier dysfunction (2,4,5,47–50). Of these effectors, the most extensively studied is S1P (Figure 113.3; for color reproduction, see Color Plate 113.3), which is a naturally occurring sphingolipid normally found in nanomolar quantities in plasma. S1P binding to S1P receptor subtypes, also called Edg receptors (endothelial differentiation gene, although their expression is not restricted to endothelium). SIP mediates important biological functions including cell adhesion, barrier regulation, proliferation, differentiation, migration, and all survival. S1P binds to the plasma membrane heptahelical S1P receptors 1 (Edg1/S1P_1), -2 (Edg5/S1P_2), -3 (Edg3/S1P_3), -4 (Edg6/S1P_4), and -5 (Edg8/S1P_5) expressed in a variety of cell types, including endothelium, in which expression of S1P_1 and S1P_3 is high. Ligation of cell surface S1P receptors triggers a complex signaling cascade. S1P_1 signaling is coupled to the G_i pathway and activation of the small GTPase Rac1, whereas S1P_3 signaling is coupled to the G_i, $G_{q/11}$, and $G_{12/13}$ pathways, resulting in greater activation of Rho than of Rac1 (2,6,38). The importance of S1P_1 is demonstrated in Edg1 (S1P_1) receptor knockout mice, which succumb in utero to incomplete vascular maturation with progressive vascular hemorrhage (8).

Specific Rho-family guanine nucleotide exchange factors (GEFs) catalyze the exchange of Rac1-GDP (inactive) for Rac1-GTP (active). One such GEF is T-lymphoma invasion and metastasis gene 1 (Tiam 1), which is implicated in diverse functions including regulation of cell–cell adhesion (51). Selective S1P activation of Rac/S1P_1 and S1P_3/Rho depends on the S1P concentration (38) and possibly

S1P-induced recruitment of S1P receptors to specialized plasma membrane microdomains containing the scaffolding protein, caveolin-1 (51, discussed later). Rac GTPases are rapidly activated by physiologic concentrations of S1P (10 nM–2 μM) to produce barrier enhancement, whereas higher concentrations of S1P (>5 μM) activate Rho GTPases that result in active cytoskeletal rearrangement, stress fiber formation, and barrier disruption (38).

Overexpression of constitutively active Rac enhances peripheral actin polymerization in the cortical ring. Moreover, Rac activation induces lamellipodia formation, membrane ruffling, the formation of cortical actin filaments, and the spreading of ECs (2). The inhibition of Rac GTPase leads to increased EC monolayer permeability and enhances thrombin-mediated barrier dysfunction response. A variety of signaling proteins activate Rac and thus modulate molecular trafficking to and enzymatic activity at the cell periphery; this results in cortical cytoskeletal enhancement and the formation of functional adherens junction complexes. We recently reported the critical importance of Rac GTPase–dependent cortical actin rearrangement not only in the augmentation of pulmonary EC barrier function by S1P (2) but also in the barrier response to hepatocyte growth factor (HGF) (4), shear stress (44), simvastatin (52), adenosine triphosphate (ATP) (47), and oxidized phospholipids (49). Rac inactivates the actin-severing protein cofilin in the cell periphery through a signaling cascade involving p21-associated kinase (PAK)-1 and LIM kinase (2). Rac activation of the PAK family is well described and is essential for S1P-mediated cortical actin rearrangement (12).

Rac GTPase is also essential for translocation of cortactin, an F-actin binding protein that stimulates actin polymerization (but only at the cell periphery, not in transcellular actin stress fibers) and stabilizes the filamentous actin network, an event necessary for the peak barrier-enhancing response to S1P in vitro (53). Cortactin exhibits an amino acid structure that contains an N-terminal acidic region (NTA) and stimulates actin polymerization by binding the Arp2/3 complex (murine AA #1–90). Cortactin also contains a unique tandem repeat site for the direct binding of F-actin (AA #91–326), a proline and tyrosine-rich area containing sites for $p60^{src}$ phosphorylation (AA #401–495), and a C-terminal SH3 domain (#496–546), which binds junctional proteins such as ZO-1 as well as key cytoskeletal effectors such as MLCK. A blocking peptide that inhibits direct binding of EC MLCK to the SH3 domain of cortactin reduced the S1P-induced MLC phosphorylation and barrier protection (53), whereas an EC MLCK dominant-negative mutant decreased S1P-induced barrier-protective cytoskeletal rearrangement (53). A rapid increase in cortactin tyrosine phosphorylation is critical to subsequent barrier enhancement, because ECs transfected with a tyrosine-deficient mutant cortactin exhibit a blunted transepithelial electrical resistance (TER) response (53). Although cortactin has long been appreciated as a target for $p60^{src}$, we were unable to link $p60^{src}$ to cortactin peripheral translocation after S1P, despite rapid increases in the phosphotyrosine content of cortactin in ECs. PP2, a specific $p60^{src}$ inhibitor, failed to significantly attenuate S1P-induced endothelial barrier enhancement and transient overexpression of mutant tyrosine-deficient cortactin construct (site-directed mutation of three critical tyrosine residues, Y^{421}, Y^{466}, Y^{482}), accounting for more than 90% of $p60^{src}$ cortactin phosphorylation (53). This definitively demonstrated that $p60^{src}$ signaling is not required for S1P-induced cortactin translocation. Tyrosine phosphorylation of cortactin appears to be necessary, however, for peak S1P-induced endothelial barrier enhancement because ECs overexpressing the tyrosine-deficient mutant cortactin exhibit a blunted TER response. Significant inhibition of the S1P response was observed after incubation with the nonspecific tyrosine kinase inhibitors genistein, herbimycin A, or erbstatin, suggesting that tyrosine kinases other than $p60^{src}$ phosphorylate cortactin at these three critical residues during S1P-induced barrier enhancement. Dominant-negative Rac prevents the translocation of cortactin and subsequent actin polymerization in the cell periphery, and it attenuates barrier enhancement by simvastatin, S1P, ATP, and shear stress. S1P-induced barrier enhancement likely involves cortactin binding to MLCK at the site of cortical actin polymerization, thus localizing the actomyosin interaction peripherally and facilitating the stabilization of intercellular and cell–matrix adhesions. The functional role of peripheral MLC phosphorylation in promoting S1P-induced barrier enhancement remains unclear, but phosphorylating MLC at the periphery may stimulate actomyosin interactions that produce cell spreading, cell flattening, or other changes that strengthen the cortical actin ring. Alternatively, MLCK may serve as a scaffolding protein that stabilizes the cortical actin ring through multiple actin-binding sites or through other potential, yet undefined, cytoskeletal interactions. Clearly, multiple signaling proteins contribute to modulate dynamic cytoskeletal arrangements, which play a key role in the maintenance or disruption of endothelial barrier integrity.

SPHINGOSINE 1-PHOSPHATE SIGNALING IN LIPID RAFTS

Because the exact mechanisms of S1P-mediated barrier protection remain poorly understood, we examined the role of caveolin-enriched plasma membrane lipid microdomains – termed *lipid rafts* – in S1P-mediated signaling and human EC barrier regulation. Although controversial, lipid rafts have been implicated in EC migration, proliferation, adhesion, endocytosis, cholesterol and Ca^{2+} regulation, and signal transduction. We observed that S1P requires the existence of lipid rafts for phosphoinositide 3-kinase (PI3K) activation, Rac1 signaling, and EC barrier enhancement (51), observations in agreement with prior studies and the proposed role for lipid rafts as sites for receptor clustering and consequent EC signal transduction. We observed that $S1P_1$ (but not $S1P_3$) is endogenously present within lipid raft fractions, with S1P recruiting additional $S1P_1$ receptors and a fraction of cellular $S1P_3$ receptors to these EC plasma membrane microdomains. The targeted use of silent RNA (siRNA) to differentially reduce the expression of either $S1P_1$ or $S1P_3$ revealed that S1P ligation of $S1P_1$ is responsible for subsequent signaling to the EC cytoskeleton and barrier-enhancement (via PI3K activation and Rac1 signaling). Consistent with the well-appreciated barrier-disruptive role of Rho, silencing $S1P_3$ expression enhanced the initial barrier enhancement and cortical actin rearrangement induced by S1P. Although not directly tested, it is plausible that $S1P_3$ recruitment to lipid rafts after S1P challenge may serve as a negative regulator of specific $S1P_1$-mediated functions.

Endothelial lipid rafts are enriched in the phospholipid PIP_2, which serves as a substrate for the Class I PI3Ks whose activity often is modulated by regulatory subunits. S1P actively recruits to lipid rafts, PI3K catalytic subunits p110α and -β, Tiam1, and α-actinin isoforms 1 and 4 in a PI3K-dependent manner. This results in cortical actin rearrangement and EC barrier enhancement (51). The catalytic PI3K subunit, p110α, is the primary PI3K recruited to and activated within lipid rafts by S1P, an association that is consistent with increased PIP_3 production in these lipid microdomains. The exact regulatory subunit(s) that interact with p110α PI3K during S1P-mediated stimulation of ECs is currently being explored. An important downstream target of PI3K signaling is the serine/threonine kinase, protein kinase B (also known as Akt), which regulates cellular functions including survival, proliferation, and migration. Akt directly phosphorylates threonine residues within $S1P_1$ (T^{236}), an event critical for S1P-mediated EC Rac1 activation, cortical actin reorganization, and migration (6). S1P induced significant recruitment of Tiam1 into lipid rafts and increased total cellular Tiam1 activation, indicating either post-translational modifications of Tiam1 or an association between localization within lipid rafts and Tiam1 activity. Tiam1 is phosphorylated by serine/threonine kinases including Ca^{2+}/calmodulin-dependent protein kinase II and PKC-dependent mechanisms. Interestingly, we observed clear evidence for a complex formation between Tiam1 and $S1P_1$ (but not $S1P_3$) in lipid raft fractions following S1P. It remains unknown whether this interaction is direct or mediated through an adaptor protein. Tiam1 contains two pleckstrin homology (PH) domains; the N-terminal PH domain of Tiam1 is involved in Tiam1 translocation to the plasma membrane, binding to PIP_2 and/or PIP_3, and interaction with various signaling molecules. Whether S1P-induced PIP_3 production in EC lipid rafts promotes Tiam1 recruitment to lipids, in which where direct $S1P_1$ binding may occur, is unknown.

Finally, the link between lipid raft components and the EC cytoskeleton also remains poorly understood. The α-actinin family of cytoskeletal proteins regulates a variety of cellular functions including cell–cell adhesion and migration with four α-actinin isoforms identified: two nonmuscle (1 and 4) and two muscle (2 and 3). Two α-actinin monomers can form a functional dimer that can link certain cell surface adhesion receptors (i.e., α integrins, intercellular adhesion molecules [ICAMs], cadherins) to the actin cytoskeleton. Furthermore, the PI3K pathway plays an important regulatory role in α-actinin cellular localization and function. The α-actinins bind to a host of molecules including actin, vinculin, catenin, PI3K, and PIP_3 (23). Our data supports a regulatory role for the PI3K pathway in α-actinin translocation to lipid rafts. This suggests a potential involvement of adhesion receptors in S1P-induced $S1P_1$ receptor–mediated EC signaling and cytoskeletal reorganization via α-actinin. Reductions in the expression of either $S1P_1$ receptor or Tiam1 (siRNA), but not $S1P_3$, inhibited S1P-induced α-actinin-1/4 lipid raft translocation, Rac1 activation, cortical actin reorganization, and increased EC barrier function (51). These data indicate the critical importance of the PI3K pathway and α-actinin in S1P signaling to the EC cytoskeleton and barrier enhancement through the $S1P_1$ receptor.

SPHINGOSINE 1-PHOSPHATE–MEDIATED BARRIER REGULATION: ADHERENS JUNCTIONS AND FOCAL ADHESIONS

Because increased vascular permeability and secondary tissue edema occur primarily via a paracellular pathway during inflammation, there has been intense focus on the forces that regulate the integrity of the paracellular junction. Consistent with a critical role for the endothelial cytoskeleton in vascular barrier regulation, the Rho family of small GTPases – critical regulators of the nonmuscle cytoskeleton – is intimately involved in S1P-mediated cytoskeletal rearrangement and the distribution/assembly of intercellular adherens complexes and focal adhesions. Rac rapidly redistributes to areas of cell–cell contact after S1P, and dominant-negative Rac polypeptide attenuates S1P-induced VE-cadherin and β-catenin localization to areas of cell–cell contact. S1P induces the peripheral enhancement of VE-cadherin, which overlaps with F-actin and the linkage of VE-cadherin, β-, and γ-catenin to actin filaments. The localization of these proteins to areas of cell–cell contact occurs within 1 hour of S1P stimulation with VE-cadherin, α-catenin, and β-catenin partitioning into the Triton-insoluble fraction, suggesting the induction of functional adherens junction assembly with increased affinity with the endothelial cytoskeleton. The translocation of VE-cadherin to cell–cell contact regions in ECs after treatment with S1P was attenuated by microinjection of oligonucleotides designed to interrupt Edg1 and Edg3 receptor expression. Solid evidence suggests an increased linkage of the S1P-activated cortical actin cytoskeleton with newly remodeled focal adhesions (38), events that we speculate result in increased cellular tethering to adjacent cells and to the ECM. Concomitant with peripheral cytoskeletal redistribution, S1P induces $p60^{src}$-mediated phosphorylation of FAK Y^{576} and redistribution of paxillin and FAK to the cell periphery. The Src kinase inhibitor, PP2, abolished this response and the formation of the cortical actin ring, thus implicating $p60^{src}$ in the upstream signaling cascade and linking FA redistribution to the cytoskeletal alterations induced by S1P.

ENDOTHELIAL BARRIER REGULATION BY ACTIVATED PROTEIN C: SIGNALING VIA S1P RECEPTORS

Sepsis is a devastating inflammatory syndrome producing life-threatening end-organ dysfunction. It is characterized by essential and defining pathophysiological features of coagulation cascade activation and marked increases in vascular permeability. The anticoagulant serine protease, recombinant activated protein C (rhAPC; drotrecogin alfa) reduced levels of the inflammatory marker IL-6 and was the first agent to improve survival in severely septic patients. Although the mechanism underlying the prosurvival effects of APC in sepsis remains enigmatic, one mechanism of APC benefit is likely related to the recent report of the inhibition of widespread increases in vascular permeability (a characteristic of severe sepsis) and accelerated recovery from the untoward effects of edemagenic agonists. Both APC and nonactivated protein C bind to the endothelial protein C receptor (EPCR) – a crucial participant in the protein C pathway, because EPCR ligation increases the activation of protein C by approximately 20-fold. Our recent studies indicate that APC evokes human lung EC signaling paradigms that target the Rho family GTPase Rac1, with subsequent Rac1-dependent cytoskeletal rearrangement and profound EC barrier protection (Figure 113.4) (50). Finally, APC-mediated protection from the edemagenic effects of thrombin appears to occur via novel transactivation of $S1P_1$, the potent barrier-enhancing receptor for S1P. Rac1 activation, MLC phosphorylation, and cytoskeleton rearrangement are consistent with previous data concerning $S1P_1$ signaling

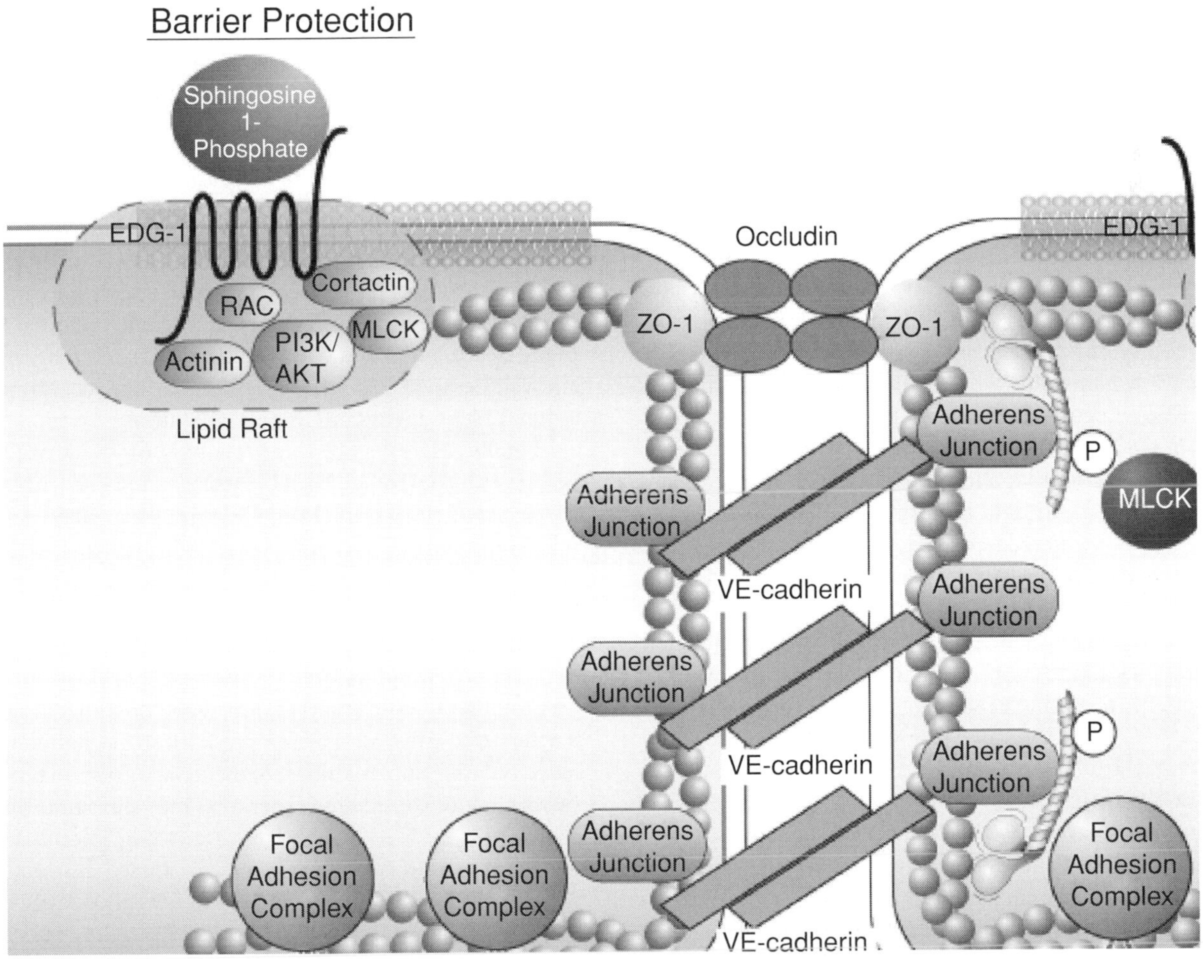

Figure 113.4. Proposed schema of APC mechanism of barrier protection against thrombin via S1P1 transactivation, lipid raft formation, and Rac signaling to the EC cytoskeleton. Prior investigation described S1P-mediated signaling involving phosphorylation of S1P1 at Thr236 via PI3K and its downstream effector Akt, which in turn activates Rac1 and leads to cortical cytoskeletal rearrangement. This cortical redistribution results in barrier protection, which is S1P1-, Rac1-, PI3K-, and Akt-dependent. We propose a model of APC-mediated barrier protection whereby APC, via EPCR and PI3K, transactivates S1P1, thus leading to EC barrier protection, which is associated with Rac1 activation and cortical actin redistribution. Prominent is the recruitment to lipid rafts of a complex including EPCR, S1P1, AKT, cortactin, and MLCK.

(50). The elevated levels of phosphorylated MLCs are spatially localized within the enhanced cortical ring, a locale in which increased tensile strength occurs with enhanced affinity between adherens junctions and focal adhesion components and the cortical cytoskeleton (53). The peripheral distribution of actin and phosphorylated MLC evoked by APC was highly reminiscent of the pattern seen with other barrier-protective agents, such as S1P, HGF, shear stress, and simvastatin, as well as with oxidized phospholipids as described earlier. Unlike S1P, APC does not directly increase baseline electrical resistance. However, both agonists exhibit a protective effect in the presence of the barrier-disruptive agent, thrombin. Utilizing $S1P_1$ phosphorylation on threonine (Thr^{236}) residues as the accepted read-out of $S1P_1$ activation, we convincingly demonstrated that APC-ligated EPCR transactivates $S1P_1$, as evidenced by robust $S1P_1$ phosphorylation, which was eliminated by EPCR blocking antisera (50). Coimmunoprecipitation studies strongly suggest an interaction between EPCR and $S1P_1$. Furthermore, use of the PI3K inhibitor LY294002 blocked APC-induced $S1P_1$ Thr^{236} phosphorylation, demonstrating that APC phosphorylates $S1P_1$ via the PI3K/Akt pathway (50). Finally, and of critical importance, $S1P_1$ silencing abrogates the APC-mediated EC barrier protection, thus indicating that $S1P_1$ expression is essential for APC-mediated barrier protection. $S1P_1$ transactivation is not a novel observation, and has been previously described in the context of ligation of growth factor receptors such as platelet-derived growth factor (PDGF) and VEGF. S1P increases tyrosine phosphorylation of the VEGF receptor, FLK-1/KDR, and S1P-mediated phosphorylation of Akt and EC NO synthase is VEGF-dependent. The exact mechanism by which growth factor receptor–$S1P_1$ transactivation occurs is unknown but has been suggested to involve increases in sphingosine kinase activity and newly generated S1P secretion that allows $S1P_1$ ligation to occur. Nevertheless, together our data indicate that $S1P_1$ is a key target for APC/EPCR-mediated barrier protection and suggests that $S1P_1$ may in fact be central to plasma membrane signaling sequences that evolve to enhance vascular integrity. The ability of APC to ameliorate the vascular leak observed in severe sepsis provides an attractive mechanism for the survival benefit of patients treated with this agent.

CONCLUSION

The integrity of the EC monolayer is a critical requirement for the preservation of organ function, and barrier disruption is recognized as a cardinal feature of inflammation angiogenesis and ischemia–reperfusion injury. Endothelial barrier properties are heterogeneous and not uniform throughout the vasculature, with capillary and postcapillary venular ECs exhibiting greater barrier properties than do macrovascular ECs (5). This reflects differences in the content of barrier regulatory components such as key cytoskeletal and adhesive elements (MLCK, FAK, cortactin), Ca^{2+} signaling pathways, and cAMP-dependent barrier protective elements. A complex interplay exists among these processes in regulating the two general pathways (transcellular and paracellular) for the movement and flow of fluid, macromolecules, and leukocytes into the interstitium. The modified tensegrity model describes paracellular gap formation as regulated by the balance of competing contractile forces. These forces generate centripetal tension and adhesive cell–cell and cell–matrix tethering forces that together regulate cell shape changes. Edemagenic agents, such as thrombin, initiate cytoskeletal rearrangement characterized by the loss of peripheral actin filaments, a concomitant increase in organized F-actin cables (stress fibers), and increased intracellular tension, via the coordinate activation of the small GTPase Rho and Ca^{++}/CaM-dependent MLCK. The resultant increase in the formation of actin stress fibers and phospho-MLC-mediated actomyosin cellular contraction destabilizes cytoskeletal-junctional linkages, culminating in increased vascular permeability.

Although significant progress has been made in understanding the molecular and cellular events that regulate permeability, the goal of modulating the EC barrier in a clinically advantageous way remains elusive. Exciting developments in the understanding of barrier restoration have led to several barrier-protective strategies that have the capacity to activate the S1P receptor, thus inducing Rac-dependent cytoskeletal cortical rearrangement; this requires Tiam1 and a strengthening of linkages of cortical actin to adherens junctions and focal adhesions. Given that structural analogs of S1P are in phase III trials, the future of barrier-enhancing therapy appears promising. These observations further confirm the potential for S1P in the treatment of acute lung injury (ALI) and may represent a novel therapeutic strategy for vascular barrier dysfunction, a signal event during inflammatory states.

Finally, identification and analysis of single nucleotide polymorphisms in candidate cytoskeletal genes with defined barrier-regulatory properties also may take us closer to understanding issues of individual variability in disease severity and therapy responses in edemagenic states. Exciting high-throughput techniques provide an opportunity to determine the critical regulatory genes responsible for complex processes such as barrier regulation and may reveal multiple novel targets for therapeutic intervention. Further understanding of barrier-restorative or barrier-preserving mechanisms likely involved after short-acting edemagenic agents may provide valuable clues about which therapies may shorten the edemagenic phase.

KEY POINTS

- EC barrier regulation is a dynamic process, with both barrier disruption and barrier restoration enhancement involving coordinated rearrangement of the endothelial cytoskeleton.

- Cytoskeleton rearrangement driven by receptor-regulated signaling occurs in spatially defined locales for barrier regulation.
- Adherens junctional proteins and focal adheren proteins are the key participants in signaling to and from the cytoskeleton.

Future Goals

- To design vascular bed–specific barrier regulatory therapies to promote the resolution of inflammatory vascular leak syndromes
- To identify key genetic variants and protein modifications that underlie the heterogeneity of susceptibility to sustained vascular permeability and result in multiorgan dysfunction

REFERENCES

1 Majno G, Palade GE. Studies on inflammation: I. The effect of histamine and serotonin on vascular permeability: an electron microscopic study. *J Biophys Biochem Cytol.* 1961;11:571–605.

2 Garcia JGN, Liu F, Verin AD, et al. Sphingosine 1-phosphate promotes endothelial cell barrier integrity by Edg-dependent cytoskeletal rearrangement. *J Clin Invest.* 2001;108(5):689–701.

3 Peng X, Hassoun PM, Sammani S, et al. Protective effects of sphingosine 1-phosphate in murine endotoxin-induced inflammatory lung injury. *Am J Respir Crit Care Med.* 2004;169(11):1245–1251.

4 Liu F, Schaphorst KL, Verin AD, et al. Hepatocyte growth factor enhances cytoskeletal rearrangement and endothelial cell barrier function and cortical cytoskeletal rearrangement: potential role of glycogen synthase kinase 3β. *FASEB J.* 2002;16:950–962.

5 Thurston G, Rudge JS, Loffe E, et al. Angiopoietin-1 protects the adult vasculature against plasma leakage. *Nat Med.* 2000;6:460–463.

6 McVerry BJ, Garcia JGN. In vitro and in vivo modulation of vascular barrier integrity by sphingosine 1-phosphate: mechanistic insights. *Cell Signal.* 2005;17(2):131–139.

7 English D, Welch Z, Kovala AT, et al. Sphingosine 1-phosphate released from platelets during clotting accounts for the potent endothelial cell chemotactic activity of blood serum and provides a novel link between hemostasis and angiogenesis. *FASEB J.* 2000;14(14):2255–2265.

8 Liu Y, Wada R, Yamashita T, et al. Edg-1, the G protein-coupled receptor for sphingosine 1 phosphate is essential for vascular maturation. *J Clin Invest.* 2000;106:951–961.

9 Schaphorst KL, Chiang E, Jacobs KN, et al. Role of sphingosine-1 phosphate in the enhancement of endothelial barrier integrity by platelet-released products. *Am J Physiol Lung Cell Mol Physiol.* 2003;285(1):L258–L267.

10 McVerry BJ, Peng X, Hassoun PM, et al. Sphingosine 1-phosphate reduces vascular leak in murine and canine models of acute lung injury. *Am J Respir Crit Care Med.* 2004;170(9):987–993.

11 Stevens T, Rosenberg R, Aird W, et al. NHLBI Workshop Report: endothelial cell phenotypes in heart, lung and blood diseases. *Am J Physiol.* 2001;281:C1422–C1433.

12 Gebb S, Stevens T. On lung endothelial cell heterogeneity. (Review). *Microvasc Circ.* 2004;68:1–12.

13 Del Vecchio PJ, Siflinger-Birnboim A, Shepard JM, et al. Endothelial monolayer permeability in macromolecules. *Fed Proc.* 1987;46:2511–2515.

14 Minshall RD, Tiruppathi C, Vogel SM, et al. Endothelial cell-surface gp60 activates vesicle formation and trafficking via G_i-coupled *Src* kinase signaling pathway. *J Cell Biol.* 2000;150:1057–1069.

15 Auerbach R, Alby L, Morrissey LW, et al. Expression of organ-specific antigens on capillary endothelial cells. *Microvas Res.* 1985;29:401–411.

16 Siflinger-Birnboim A, Cooper JA, Del Vecchio PJ, et al. Selectivity of the endothelial monolayer: effects of increased permeability. *Microvasc Res.* 1988;36:216–227.

17 Schnitzer JE, Oh P. Albondin-mediated capillary permeability to albumin. Differential role of receptors in endothelial transcytosis and endocytosis of native and modified albumins. *J Biol Chem.* 1994;269:6072–6083.

18 Dull RO, Dinavahi R, Schwartz L, et al. Lung endothelial heparan sulfates mediate cationic peptide-induced barrier dysfunction: a new role for the glycocalyx. *Am J Physiol.* 2003;285:L986–L995.

19 Dull RO, Garcia JGN. Leukocyte-induced microvascular permeability: how contractile tweaks lead to leaks. *Circ Res.* 2002;90:1143–1144.

20 Gautum N, Olofsson AM, Herwald H, et al. Heparin-binding protein (HBP/CAP37): a missing link in neutrophil-evoked alteration of endothelial permeability. *Nat Med.* 2001;7:1123–1127.

21 Hirata A, Baluk P, Fujiwara T, et al. Location of focal silver staining at endothelial gaps in inflamed venules examined by scanning electron microscopy. *Am J Physiol.* 1995;269:L403-L418.

22 Shepard JM, Goderie SK, Malik AB, et al. Effects of alterations in endothelial cell volume on albumin permeability. *J Cell Physiol.* 1987;133:389–394.

23 Dudek SM, Garcia JGN. Cytoskeletal regulation of pulmonary vascular permeability. *J Appl Physiol.* 2001;91(4):1487–1500.

24 Jen-Tsan C, Chang HY, Haraldsen G, et al. Endothelial diversity revealed by global expression profiling. *Proc Nat Acad Science USA.* 2003;100:10623–10628.

25 Gao L, Grant A, Halder I, et al. Novel polymorphisms in the myosin light chain kinase gene confer risk for acute lung injury. *Am J Respir Cell Mol Biol.* 2006;34(4):487–495.

26 Tinsley JH, de Lanerolle P, Wilson E, et al. Myosin light chain kinase transference induces myosin light chain activation and endothelial hyperpermeability. *Am J Physiol Cell Physiol.* 2000; 279:C1285–C1289.

27 Garcia JGN, Davis HW, Patterson CE. Regulation of endothelial cell gap formation and barrier dysfunction: role of myosin light chain phosphorylation. *J Cell Physiol.* 1995;163:510–522.

28 Breslin J, Sun H, Xu W, et al. Involvement of ROCK-mediated endothelial tension development in neutrophil-stimulated microvascular leakage. *Am J Physiol Heart Circ Physiol.* 2006 Feb; 290(2):H741–750. Epub 2005 Sep 19.

29 Hurst V, Goldberg PL, Minnear FL, et al. Rearrangement of adherens junctions by transforming growth factor-β1: role of contraction. *Am J Physiol.* 1999;276:L582–L595.

30 Khimenko PL, Moore TM, Wilson PS, Taylor AE. Role of calmodulin and myosin light chain kinase in lung ischemia-reperfusion injury. *Am J Physiol.* 1996;271:L121–L125.
31 Parker JC. Inhibitors of myosin light chain kinase and phosphodiesterase reduce ventilator-induced lung injury. *J Appl Physiol.* 2000;89:2241–2248.
32 Birukova A, Smurova K, Birukov KG, et al. Role of Rho GTPase in thrombin-mediated endothelial cell barrier dysfunction in pulmonary endothelium. *Microvascular Res.* 2004;67:64–77.
33 Garcia JGN, Verin AD, Herenyiova M, English D. Adherent neutrophils activate endothelial myosin light chain kinase: role in transendothelial migration. *J Appl Physiol.* 1998;84:1817–1821.
34 Hixenbaugh EA, Goeckeler ZM, Papaiya NN, et al. Stimulated neutrophils induce myosin light chain phosphorylation and isometric tension in endothelial cells. *Am J Physiol.* 1997;273:H981–H988.
35 Dejana E. Endothelial cell–cell junctions: happy together. *Nat Rev Mol Cell Biol.* 2004;5(4):261–270.
36 Romer LH, Birukov KG, Garcia JGN. The focal adhesion: paradigm for a signaling nexus. *Circ Res.* 2006;98(5):606–616.
37 Shikata Y, Birukov KG, Birukova AA, et al. Involvement of site-specific FAK phosphorylation in sphingosine-1 phosphate- and thrombin-induced focal adhesion remodeling: role of Src and GIT. *FASEB J.* 2003;17(15):2240–2249.
38 Shikata Y, Birukov KG, Garcia JGN. S1P induces FA remodeling in human pulmonary endothelial cells: role of Rac, GIT1, FAK, and paxillin. *J Appl Physiol.* 2003;94(3):1193–1203.
39 Bogatcheva NV, Verin AD, Wang P, et al. Role of species-specific MLC phosphorylation in phorbol ester-induced actin remodeling in endothelial cells. *Am J Physiol.* 2003;285:L415–L426.
40 Mirzapoiazova T, Kolosova IA, Romer L, et al. The role of caldesmon in the regulation of endothelial cytoskeleton and migration. *J Cell Physiol.* 2004;95(9):892–901.
41 Borbiev T, Birukova A, Liu F, et al. p38 MAP kinase-dependent regulation of endothelial cell permeability. *Am J Physiol Lung Cell Mol Physiol.* 2004;287(5):L911–L918.
42 Garcia JGN, Wang P, Schaphorst KL, et al. Critical involvement of p38 MAP kinase in pertussis toxin-induced cytoskeletal reorganization and lung permeability. *FASEB J.* 2002;16(9):1064–1076.
43 Birukov KG, Csortos C, Ma SF, et al. Differential regulation of alternatively spliced endothelial cell MLCK isoforms by $p60^{src}$. *J Biol Chem.* 2001;276:8567–8573.
44 Birukov KG, Birukova AA, Dudek SM, et al. Shear stress-mediated cytoskeletal remodeling and cortactin translocation in pulmonary endothelial cells. *Am J Resp Cell Mol Biol.* 2002;26:453–464.
45 Birukov KG, Jacobson JR, Flores AA, et al. Magnitude-dependent regulation of pulmonary endothelial cell barrier function by cyclic stretch. *Am J Physiol.* 2003;285:L785–L797.
46 Shikata Y, Rios A, Kawkitinarong K, et al. Differential effects of shear stress and cyclic stretch on focal adhesion remodeling, site-specific FAK phosphorylation, and small GTPases in human lung endothelial cells. *Exp Cell Res.* 2005;304(1):40–49.
47 Kolosova IA, Mirzapoiazova T, Adyshev D, et al. Signaling pathways involved in adenosine triphosphate-induced endothelial cell barrier enhancement. *Circ Res.* 2005;97(2):115–124
48 Jacobson JR, Barnard JW, Grigoryev DN, et al. Simvastatin attenuates vascular leak and inflammation in murine inflammatory lung injury. *Am J Physiol Lung Cell Mol Physiol.* 2005;288(6):L1026–L1032.
49 Birukov KG, Bochkov VN, Birukova AA, et al. Epoxycyclopentenone-containing oxidized phospholipids restore endothelial barrier function via Cdc42 and Rac. *Circ Res.* 2005;95(9):892–901.
50 Finigan JH, Dudek SM, Singleton PA, et al. Activated protein C mediates novel lung endothelial barrier enhancement: role of sphingosine 1-phosphate receptor transactivation. *J Biol Chem.* 2005;280(17):17286–17293.
51 Singleton PA, Dudek SM, Chiang ET, Garcia JGN. Regulation of sphingosine 1-phosphate-induced endothelial cytoskeletal rearrangement and barrier enhancement by S1P1 receptor, PI3 kinase, Tiam1/Rac1 and α-Actinin. *FASEB J.* 2005;19:1646–1656.
52 Jacobson JR, Birukov KG, Dudek S, et al. Cytoskeletal rearrangement and altered gene expression in endothelial cells by simvastatin. *Am J Respir Cell Mol Biol.* 2004;30(5):662–670.
53 Dudek SM, Jacobson JR, Chiang ET, et al. Pulmonary endothelial cell barrier enhancement by sphingosine 1-phosphate: roles for cortactin and myosin light chain kinase. *J Biol Chem.* 2004;279(23):24692–24700.

GLOSSARY

α-Actinin is an actin binding protein composed of two identical antiparallel peptides, with the actin binding domain close to the N-terminus, followed by four spectrin-like repeats and terminating with two EF-hands; links actin cytoskeleton to focal adhesions.

Actin-associated proteins regulate assembly and disassembly of actin filaments.

Actin microfilaments are helical protein filaments formed by the polymerization of globular actin molecules, which are major constituents of the cytoskeleton of all eucaryotic cells and a critical component of the contractile apparatus of muscles involved in cell motility, cell-substrate interactions, transport processes, cytokinesis, and the establishment and maintenance of cell morphology.

Actin polymerization is a dynamic process that results in formation of actin filaments that are derived from individual globular actin molecules. This process is regulated by many different actin-associated proteins.

Actin stress fibers (or F-actin cables) are large actin filaments that form in response to multiple signaling events and are often part of the actomyosin contractile motor.

Actomyosin motor (actomyosin cross bridges) is a contractile apparatus formed by the dynamic interaction of actin and myosin and resulting in contraction within the cell.

Arp2/3 is a seven-subunit protein that nucleates actin polymerization. When activated by various proteins, the Apr2/3 complex binds to the side of an existing actin filament and nucleates assembly of a new actin filament in a branching pattern at the cell periphery.

Caldesmon is an actin- and myosin-binding protein, which is an essential component of the cytoskeleton in smooth muscle and nonmuscle cells, and is involved in the regulation of cell contractility, division, and assembly of actin filaments.

Cofilin is a family of actin-binding proteins that disassembles actin filaments. The protein is known to sever actin filaments by creating more positive ends on filament fragments.

Cortactin is an actin-binding protein and Src target that stimulates Arp2/3 actin polymerization and interacts with multiple other cytoskeletal proteins through its C-terminal SH3 domain.

Cortical actin filaments (cortical actin ring) are peripheral actin filaments that circumferentially surround the cell and provide structure for cytoskeletal linkages within the cell.

F-actin, or filamentous actin, consists of multiple subunits of globular actin arranged in a linear fashion and provides the basic structure of the actin cytoskeleton within the cell.

Filamentous actin network (subcortical actin mesh) is a branching network of actin filaments at the cell periphery, often involved in migration.

Filamin is a V-shaped actin cross-linking protein that induces actin filaments to associate in loose networks that give some areas of the cytosol a gel-like consistency.

Gelsolin is a calcium-dependent actin-binding protein that modulates actin filament length and gelation and thus influences the structure of the cytoskeleton and plays a key role in cellular motility and differentiation. Inhibition of its activity decreases actin stress fiber–dependent contraction.

Hsp27 (heat shock protein 27) helps regulate stress fiber formation through the mitogen-activate protein kinase (MAPK) pathway.

Intermediate filaments are fibrous protein filaments (about 10 nm in diameter) that form ropelike networks within cells. One of the three prominent types of cytoskeletal filaments, including actin and microtubules.

Microtubules are polymers of α- and β-tubulin that form a lattice network of rigid hollow rods spanning the cell in a polarized fashion from nucleus to the periphery while undergoing frequent assembly and disassembly.

MLCK (myosin light chain kinase) is a $Ca2^{+}$/calmodulin-dependent enzyme that catalyzes the phosphorylation of regulatory myosin light chains (MLC), which is a key EC contractile event.

Peripheral actin filaments are structures at the periphery of the cell involved in the dynamic regulation of migration and barrier function.

Profilin is an actin-binding protein involved in cytoskeleton dynamics. It is responsible for the growth and stabilization of actin filaments and is important in the restructuring of microfilaments.

VASP (vasodilator-stimulated phosphoprotein) is a member of the Ena-VASP protein family. Ena-VASP family members contain an EHV1 N-terminal domain that binds proteins containing E/DFPPPPXD/E motifs and targets Ena-VASP proteins to focal adhesions. VASP stimulates actin nucleation and polymerization at focal adhesions.

Vinculin is located at the cytoplasmic side of focal contacts or adhesion plaques, where it binds actin and catenin.

114

Molecular Mechanisms of Leukocyte Transendothelial Cell Migration

F. William Luscinskas

Brigham and Women's Hospital and Harvard Medical School, Boston, Massachusetts

In this chapter, the focus is on three areas of interest relevant to the study of leukocyte transmigration as an output device of the vascular endothelium. The mechanisms underlying leukocyte adhesion and recruitment have been discussed in earlier chapters, and readers are referred there to obtain this information. The first section presents a brief historical perspective on the history of inflammation and, in particular, leukocyte transmigration. Second, the signaling capacity of endothelial adhesion molecules that capture leukocytes and mediate adhesion and transmigration is discussed. It is apparent that these adhesion "receptors," in particular intercellular adhesion molecule (ICAM)-1, signal during their engagement with leukocytes. Third, the process of leukocyte diapedesis at endothelial cell (EC)–cell lateral junctions and the transient and reversible alterations that occur in junctional components (e.g., transient loss of vascular endothelial [VE]-cadherin) during transmigration is addressed. This area is clearly of interest, because prolonged and/or excessive leukocyte recruitment and the inflammatory milieu together are predicted to have negative affects on the permeability barrier function of the endothelium. Finally, a discussion is included on a somewhat controversial aspect of diapedesis, namely, do leukocytes undergo diapedesis at junctional or nonjunctional sites, and what factors dictate the location? Each of these topics can be envisioned as altering or modulating the output of the endothelium.

HISTORICAL PERSPECTIVE

The study of inflammation began more that two thousand years ago with the Egyptians (see the excellent review chapter by Klaus Ley; Ref. 1). This culture described some of the signs and consequences of inflammation, notably descriptions of abscess and ulcer. It was many centuries later (during the late 1800s), however, that Elie Metchnikoff (2), a Russian embryologist and immunologist, described the process of phagocytosis by inflammatory leukocytes, and Julius Cohnheim, a German experimental histologist and pathologist, described components of the leukocyte adhesion cascade as we know it today (Figure 114.1). He articulated the process of a leukocyte "crawling" out of a blood vessel in the frog tongue and into the interstitial space, a process now termed *leukocyte transendothelial migration* (TEM), emigration, or diapedesis. In the early 20th century, Clark and Clark were among the first to describe an in vivo model to visualize the adhesion and transmigration of leukocytes in the microvessels of the rabbit ear at the light-microscopic level. They further found that leukocyte transmigration did not lead to loss of the EC barrier (3).

Another major milestone came during the early 1960s, when Marchesi, Florey, and Gowans described TEM at the ultrastructural level using fixation of tissues and electron microscopic techniques (4,5). This was the first detailed analysis of leukocyte TEM in animal models of inflammation, and this analysis and others have helped germinate the idea that leukocytes transmigrate out of the vessel wall by passing either between cell–cell lateral junctions (paracellular) or directly through the vascular ECs (transcellular). These modes of TEM are discussed later in the chapter.

Regarding the mechanisms by which leukocytes recognize a site of injury or inflammation, it was widely held, despite a paucity of evidence, that the blood leukocytes followed a gradient of chemoattractants released from sites of infection, damage, or injury (6). Little consideration was given to the idea that the vascular endothelium was involved in this process. This notion was dismissed during the mid 1980s, following reports that certain inflammatory cytokines and bacterial endotoxins could trigger vascular endothelial monolayers to become hyperadhesive for isolated blood leukocytes in vitro (7), and furthermore, that leukocytes actually transmigrated across these monolayers in the absence of exogenously added chemoattractants (8). Subsequent studies in vivo confirmed these findings (9) and began the remarkable wave of progress in unraveling the molecular and cellular mechanisms of

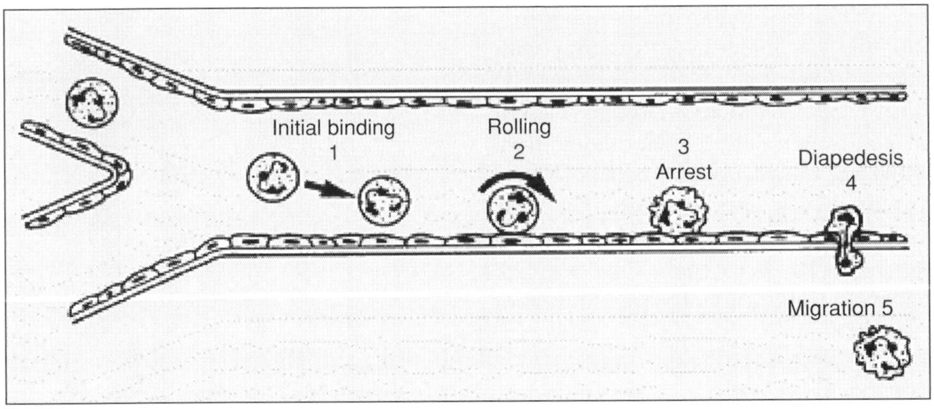

Figure 114.1. The leukocyte recruitment adhesion cascade. Model of neutrophil attachment to 6-hour TNF-α–activated HUVEC monolayers under defined laminar shear flow in vitro. This figure shows five sequential events (steps 1–5) that can be observed during neutrophil adhesive interactions with the endothelium at 1.8 dynes/cm^2.

inflammation and the central role of the vascular endothelium in this process (10). It is now well appreciated that the endothelium becomes "activated" by many types of inflammatory stimuli to induce gene transcription; protein synthesis; surface expression of adhesion molecules (vascular cell adhesion molecule [VCAM]-1, E-selectin, P-selectin, and ICAM-1); secretion of chemotactic cytokines, known as chemokines, such as interleukin (IL)-8 (also known as CXCL8) and monocyte chemoattractant protein (MCP)-1 (also known as CCR2); and chemotactic phospholipids, such as platelet-activating factor (PAF). Moreover, it is clear that general dysfunction of the endothelium is an underlying pathophysiological factor in many chronic diseases (11).

ENDOTHELIAL CELL ADHESION MOLECULE SIGNALING AND ENDOTHELIAL CELL FUNCTION

The work of many investigators has shown that peripheral blood leukocytes including neutrophils, lymphocytes, monocytes, basophils, and eosinophils use multiple adhesive steps to emigrate from the blood, with each step requiring multiple and overlapping functions of different sets of adhesion molecules. Although early studies described a two- or three-step model (12,13), the multistep adhesion cascade is conveniently divided into five steps as identified in Figure 114.1:

1. Initial attachment
2. Rolling
3. Arrest
4. Diapedesis of the endothelium
5. Migration through tissues

Figure 114.2 depicts an example of an isolated human neutrophil transmigrating across a tumor necrosis factor (TNF)-α activated human EC monolayer under physiological levels of laminar shear flow. In this system, initial attachment is primarily dependent on both E- and P-selectin, and L-selectin. Subsequently, the neutrophil rolls across the apical EC surface at approximately 10 μm/sec for variable distances before arresting, presumably due to activation of its integrins by apically presented chemokines. The arrest step is mediated by leukocyte $\beta 2$ integrins interacting with ICAM-1. For

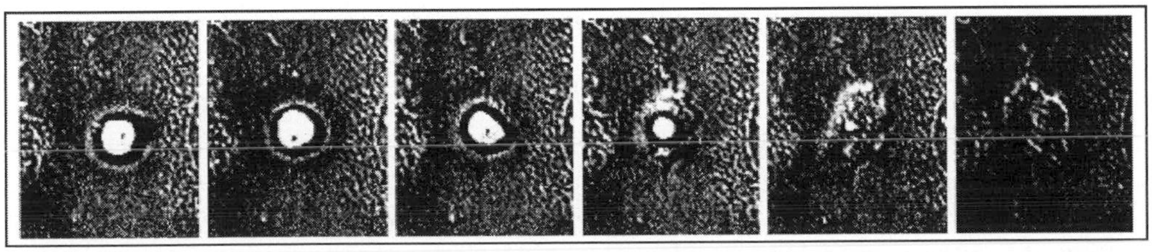

Figure 114.2. Neutrophil transmigration of cytokine-activated EC monolayer under shear flow in vitro. Neutrophil initial attachment, rolling, arrest, and transmigration across 6-hour TNF-α–treated human umbilical vein endothelium under shear flow in vitro (1.5 dynes/cm^2). Sequential frames taken by DIC microscopy depicting each event from left to right were digitized and manipulated using MetaMorph software and PhotoShop as previously described. (Reproduced with permission from Shaw SK, Ma S, Kim M, et al. Coordinated redistribution of leukocyte LFA-1 and endothelial cell ICAM-1 accompanies neutrophil transmigration. *J Exp Med.* 2004;200:1571–1580.)

other leukocyte types, including monocytes, lymphocytes, basophils, and eosinophils $\beta 1$ integrins interacting with VCAM-1 also can mediate initial attachment and arrest. The majority of neutrophils ($\sim$75%) arrest at or close to cell–cell junctions (14–16) and then crawl to intercellular cellular junctions and ultimately diapedese by squeezing between intercellular junctions. Amazingly, the entire process that leads to transmigration occurs within a few minutes and, despite the relatively high frequency of leukocyte trafficking observed, rarely is there frank damage to the EC monolayer in this system (see Figure 114.2).

Several studies have shown that the cytokine-induced EC adhesion molecules VCAM-1, ICAM-1, and E-selectin can signal upon engagement in ECs (17). E-selectin ligation by leukocytes (HL-60) triggers its association with the cytoskeleton and initiates outside-in signaling (18–20). Similar findings have been reported for leukocyte binding or mAb crosslinking to E-selectin and VCAM-1 (21). Endothelial ICAM-1 is the major receptor for $\beta 2$ integrins, and this pathway is essential for neutrophil arrest and transmigration (22). Therefore, the remainder of this discussion focuses on EC ICAM-1 as a prototype adhesion molecule that signals during leukocyte ligation.

ICAM-1 is a transmembrane glycoprotein with five extracellular IgG-like domains and a short cytoplasmic tail (28 amino acids) that appears to associate with multiple cytoskeletal linker proteins in a variety of cell types (22). Under in vitro and in vivo conditions, ECs express low levels of ICAM-1, and inflammatory stimuli can markedly increase ICAM-1 surface expression (23,24). In acute and chronic inflammatory diseases, ECs become activated and express high levels of ICAM-1, in addition to VCAM-1 and E-selectin (25–28).

The engagement of ICAM-1 in cytokine-activated ECs by adherent leukocytes or cross-linking with monoclonal antibodies triggers elevations in intracellular free calcium (Ca^{2+}) and myosin contractility (29,30), as well as the activation of small GTPases, in particular members of the Rho family (31,32) and the tyrosine kinase $p60^{Src}$ (33) (Figure 114.3). A prominent substrate for $p60^{src}$ in endothelium is the actin-binding protein, cortactin (34–37). These changes are proposed to cause cytoskeletal remodeling events that influence EC contractility and function to keep neutrophils adherent near cell–cell junctions (38). This would be a convenient mechanism to assure diapedesis occurs at a junctional location. In another study, adhesion of the leukocyte THP-1 cell line under static conditions also triggered Src kinase-dependent tyrosine phosphorylation of cortactin (39). Pharmacological inhibition of Src kinase tyrosine phosphorylation by PP2 did not prevent association of E-selectin and ICAM-1 with cortactin, but did reduce by approximately 50% THP-1 adhesion to cytokine-activated endothelial monolayers. THP-1 cells do not transmigrate in most models, so the effect of PP2 on this step is unknown. Although little is known about the function of cortactin in leukocyte transmigration, it is of interest because biochemical studies show that it binds to ICAM-1 and E-selectin (39), that it contains an actin–binding domain,

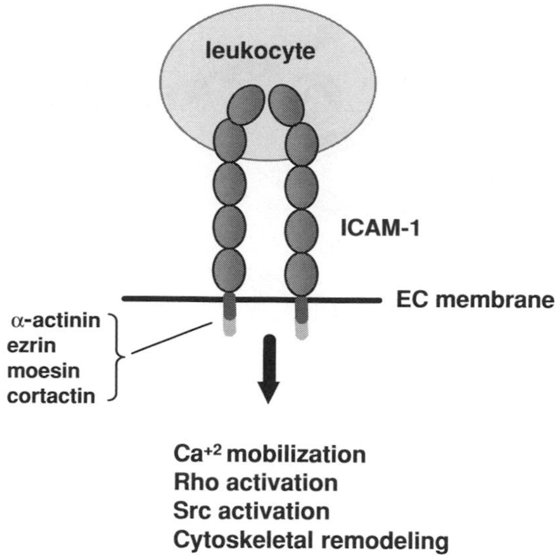

Figure 114.3. ICAM-1 engagement triggers intracellular signals in EC. ICAM-1 interacts with the adapter proteins ezrin, moesin, cortactin, and α-actinin, which link ICAM-1 to the cytoskeleton. In ECs, antibody-mediated cross-linking of ICAM-1 leads to an increase in intracellular Ca^{2+} and activation of Rho A and src kinases. These upstream signals are thought to converge with other pathways to initiate cytoskeletal remodeling (17,21,38,39). In addition, ICAM-1 clusters around those leukocytes actively transmigrating across EC (15,16,44,45), presumably through its associations with these adapter proteins.

and that a separate N-terminal domain stimulates actin polymerization (in vitro) by binding and stabilizing the Arp2/3 complex in cortical actin networks (36,37). The Arp2/3 complex is composed of actin-related proteins (Arp)-2 and -3 and five unrelated proteins (40); it is a major actin nucleating factor in cells (41). Activation of Arp2/3 is the dominant mechanism for the explosive actin polymerization characteristic of platelet shape-change on activation (42,43). Finally, recent work reported that ICAM-1 and E-selectin localize in regions of the plasma membrane that have properties consistent with "lipid rafts" (39), endowing additional properties to sites of leukocyte adhesion such as the potential for downstream signaling. Future studies are needed to ascertain the extent to which such signals alter the EC phenotype at sites of immune reactions, infection, or injury.

In addition to its adhesive and signaling capabilities, recent studies report that ICAM-1–enriched microvilli-like projections in human umbilical vein ECs (HUVECs), termed *docking structures*, can be observed during leukocyte arrest and locomotion (44) and transmigration (45), and in T-lymphoblast diapedesis (46). Pharmacological inhibition showed that maintenance of the docking structure depended, in part, on phosphoinositides [PI(4,5)P_2] and the Rho p160 kinase (ROCK) and phosphoinositide 3-kinase (PI3K) pathways. Our lab also has reported that ICAM-1 clusters at the sites of neutrophil transmigration (both junctional [16] and non-junctional sites [15]) under defined shear flow, although our data differ from those of Barreiro (46) and Carman (44,45) in that we did not observe robust microvilli-like projections containing ICAM-1 emanating from endothelium during transmigration. Rather, we observed the coordinated remodeling of both leukocyte and endothelial adhesion molecules within a narrow plane at endothelial cell–cell junctions (16). Likely explanations for these differences reside in the use of the chemoattractant PAF in the latter studies (44,45), and the use of T-lymphoblasts in the former study (46), which have a low rate of transmigration relative to neutrophils and a prolonged mobility on the apical surface that precedes transmigration. These reports, however, are consistent with the concept that leukocyte adhesion and transmigration initiates a fast assembly of specialized structures in the endothelium that contain the adhesion molecules ICAM-1, E-selectin, and VCAM-1, and actin-linking molecules that bridge these components to actin filaments. Future studies are needed to address which EC actin-binding proteins associate with ICAM-1 under resting conditions and after leukocyte ligation, and whether these associations are necessary for the observed clustering of ICAM-1 during leukocyte transmigration, both at junctional and nonjunctional sites. The in vivo counterpart is a logical target for future studies.

DIAPEDESIS AT THE JUNCTIONS

Most in vitro experimental systems have revealed that the TEM of leukocytes occurs predominantly at endothelial junctions along a paracellular pathway (47–49) (step 4 in Figure 114.1). Because leukocytes encounter multiple EC junctional molecules and molecular complexes during paracellular transmigration, the field has focused on cell surface molecules that localize at cell–cell junctions (49,50). CD99 and platelet-endothelial cell adhesion molecule (PECAM)-1/CD31 are expressed on vascular endothelium and are enriched at cell–cell lateral junctions. These molecules are expressed on most leukocyte types (49). Both molecules interact through homophilic interactions: CD99 on one EC binds to the same molecule on adjacent ECs. PECAM-1 homophilic interactions work in an analogous fashion. PECAM-1 also can function as a signaling molecule in platelets, leukocytes, and endothelium (51). During leukocyte TEM, both PECAM-1 and CD99 engage in homophilic binding of the molecule on the endothelium with the same molecule on the migrating leukocyte. Transmigrating leukocytes must cross endothelial tight junctions (TJ) and adherens junctions (AJ), which contain numerous proteins involved in selective permeability, growth control, and cell–cell adhesion (52). Endothelial AJs contain the transmembrane protein VE-cadherin, which forms a complex with cytosolic molecules α-catenin, β-catenin, plakoglobin (or γ-catenin), and p120 (53). The VE-cadherin complex then links to the actin cytoskeleton, providing stability. Recent studies have suggested the VE-cadherin complex regulates the passage of blood leukocytes (54). A recent experimental strategy directly visualized the effects of leukocyte TEM on VE-cadherin in endothelium using live-cell imaging of fluorescently tagged VE-cadherin (VE-cadherin-green fluorescent protein [GFP]) under defined laminar flow (55). Neutrophils and monocytes triggered transient and reversible displacement (4–6 μm "gaps") in VE-cadherin-GFP during transmigration. Others have reported similar alterations in VE-cadherin during neutrophil transmigration using fluorescently tagged mAb to VE-cadherin (56). The mechanism underlying displacement of VE-cadherin during leukocyte TEM is currently unknown, but may involve a localized and reversible uncoupling of the VE-cadherin from the actin cytoskeleton. These findings suggest that the transmigration of leukocytes across intercellular junctions is regulated by complex signaling events within both cell types. Burns and colleagues (57,58) also have reported that neutrophils transmigrate at tricellular or multicellular endothelial junctions that exhibit small gaps in VE-cadherin staining, although the mechanism for these small constitutive gaps is unknown.

NONJUNCTIONAL DIAPEDESIS: NOVEL MODELS OF LEUKOCYTE TRANSCELLULAR DIAPEDESIS

In vivo, the inflammatory response is a complex and extremely well orchestrated series of events that leads to dramatic changes in leukocyte–EC interactions, an increase in vessel permeability, and an accumulation of leukocytes in the extravascular space and surrounding tissues. The technique used to reliably determine the location of transmigration in tissues is ultrastructural analysis using serial sectioning and electron microscopy. This approach is not a high-throughput type analysis, and therefore frequency or statistical analysis is not performed. Nonetheless, the most detailed analysis was the publication by Feng and colleagues (59), in which multiple sets of consecutive serial electron microscopic sections were obtained from skin sites injected with the chemoattractant formyl-Met-Leu-Phe (fMLP). The images representing these serial sections were reconstructed in three dimensions using software programs. Analysis of these images revealed that most (at least 10 of the 11) openings were transendothelial pores, not inter-EC gaps. Based on these data, the authors concluded that transcellular diapedesis was not a rare event. More recent studies using intravital fluorescence video-microscopy and subsequent ultrastructural analysis also have reported evidence for lymphocyte transcellular migration across the blood–brain barrier (60,61). Additional in vivo studies using other models of inflammation have provided further evidence for transcellular transmigration (61–63).

As discussed earlier, in vivo studies have found that blood leukocytes can undergo transcellular TEM in a variety of in vivo animal models of inflammation, which suggests that this route of leukocyte transmigration is physiologically relevant. The underlying mechanisms, however, have not been identified. Historically there were few, if any, reports of leukocyte transcellular diapedesis in vitro. Two papers in 2004 reported transcellular diapedesis, but the levels ranged from a few percent to approximately 10% (45,64). Our laboratory has addressed this gap in information by utilizing in vitro live-cell fluorescence imaging to visualize the EC junction marker VE-cadherin and accurately determine the site of neutrophil arrest and TEM (paracellular or transcellular) under physiologically relevant shear flow (15). These findings confirm the recent paper by Carmen and Springer and Cinamon and colleagues (45,64) that leukocytes undergo transcellular diapedesis in vitro, and extend these findings by identifying two factors – namely high-density ICAM-1 expression and polygonal EC shape – as key mediators in neutrophil transcellular diapedesis. Elevated levels of apical ICAM-1, identical to those found on chronic (24-hour) TNF-α activated HUVECs, promote an increase in neutrophil arrest at nonjunctional locations, where high occupancy levels of ICAM-1 by the $\beta 2$ integrins lymphocyte function-associated antigen (LFA)-1 and Mac-1 trigger neutrophil-robust levels of neutrophil transcellular diapedesis. Strikingly, we found that freshly isolated peripheral blood $CD3^+$ T cells exclusively undergo paracellular diapedesis. These data indicate a clear difference in T-lymphocyte and neutrophil diapedesis behavior and indicate differential signaling downstream of ICAM-1 ligation in these two leukocyte types. Future studies are necessary to directly compare the transmigration behavior of different preparations of T cells (freshly isolated versus in vitro–generated IL-2 driven T lymphoblasts, CD4 and CD8 memory or naïve T cells) in ICAM-1 overexpressing endothelium.

KEY POINTS

- The emerging picture suggests an intimate dialog between the adherent leukocyte and the underlying endothelial monolayer that culminates in a dramatic and rapid assembly of adhesion molecules of both cell types as the leukocyte "discovers" the correct site at EC junctions.
- New in vitro models are available to study junctional and nonjunctional leukocyte transmigration pathways (15,45,64). These allow investigators to compare and contrast the cellular mechanisms underlying each pathway and to determine whether different leukocyte types utilize one or both pathways. This information can then be used to develop similar assays using in vivo models.
- EC adhesion molecules are capable of generating intracellular signals that initiate alterations in several pathways. A common output after adhesion molecule ligation is an alteration of cytoskeletal components.

Future Goals

- To determine whether ligation of EC adhesion molecules by different leukocyte types leads to common or divergent downstream signal generation
- To determine if the transcellular and paracellular transmigration pathways elicit different downstream signals in endothelium
- To elucidate the mechanisms that regulate the reversible alterations in junctional VE-cadherin complex during leukocyte transmigration

REFERENCES

1 Ley K. History of inflammation research. In: Ley K, ed. *Physiology of Inflammation*, 3rd ed. New York: Oxford University Press; 2001:1–10.

2 Metchnikoff E. *Lectures on the Comparative Pathology of Inflammation*. London: Kegan, Paul, Trench, Trubner & Co., 1893.

3 Clark ER, Clark EL. Observations on changes in blood vascular endothelium in the living animal. *Am J Anat*. 1935;57:385–438.

4 Marchesi VT. The site of leukocyte emigration during inflammation. *Quart J Exp Physiol*. 1961;46:115–118.

5 Marchesi VT, Florey HW. Electron microscopic observation on the emigration of leukocytes. *Quart J Exp Physiol*. 1960;45:343–348.

6 Leber T. Uber die entstehung der entzendung und die wirkung der entzηndungserregenden schedlichkeiten. *Fortschr Med*. 1883; 6:460–464.

7 Bevilacqua MP, Pober JS, Wheeler ME, et al. Interleukin-1 activation of vascular endothelium. Effects on procoagulant activity and leukocyte adhesion. *Am J Pathol*. 1985;121:394–403.

8 Furie MB, McHugh DD. Migration of neutrophils across endothelial monolayers is stimulated by treatment of the monolayers with interleukin-1 or tumor necrosis factor-alpha. *J Immunol*. 1989;143:3309–3317.

9 Munro JM, Pober JS, Cotran RS. Tumor necrosis factor and interferon-gamma induce distinct patterns of endothelial activation and associated leukocyte accumulation in skin of *Papio anubis*. *Am J Pathol*. 1989;135:121–133.

10 Cines DB, Pollak ES, Buck CA, et al. ECs in physiology and in the pathophysiology of vascular disorders. *Blood*. 1998;91:3527–3561.

11 Gimbrone MA. Endothelial dysfunction, hemodynamic forces, and atherosclerosis. *Thromb Haemost*. 1999;82:722–726.

12 Butcher EC. Leukocyte-endothelial cell recognition: three (or more) steps to specificity and diversity. *Cell*. 1991;67:1033–1036.

13 von Andrian UH, Chambers JD, McEvoy LM, et al. Two-step model of leukocyte-endothelial cell interaction in inflammation: distinct roles for LECAM-1 and the leukocyte beta 2 integrins in vivo. *Proc Natl Acad Sci USA*. 1991;88:7538–7542.

14 Gopalan PK, Burns AR, Simon SI, et al. Preferential sites for stationary adhesion of neutrophils to cytokine-stimulated HUVEC under flow conditions. *J Leukoc Biol*. 2000;68:47–57.

15 Yang L, Froio RM, Sciuto TE, et al. ICAM-1 regulates neutrophil adhesion and transcellular migration of TNF-α activated vascular endothelium under flow. *Blood*. 2005;106:584–592.

16 Shaw SK, Ma S, Kim M, et al. Coordinated redistribution of leukocyte LFA-1 and EC ICAM-1 accompanies neutrophil transmigration. *J Exp Med*. 2004;200:1571–1580.

17 van Buul JD, Hordijk PL. Signaling in leukocyte transendothelial migration. *Arterioscler Thromb Vasc Biol.* 2004;24:824–833.

18 Yoshida M, Westlin WF, Wang N, et al. Leukocyte adhesion to vascular endothelium induces E-selectin linkage to the actin cytoskeleton. *J Cell Biol.* 1996;133:445–455.

19 Yoshida M, Szente BE, Kiely JM, et al. Phosphorylation of the cytoplasmic domain of E-selectin is regulated during leukocyte-endothelial adhesion. *J Immunol.* 1998;161:933–941.

20 Hu Y, Szente B, Kiely JM, Gimbrone MA Jr. Molecular events in transmembrane signaling via E-selectin. SHP2 association, adaptor protein complex formation and ERK1/2 activation. *J Biol Chem.* 2001;276:48549–48553.

21 Lorenzon P, Vecile E, Nardon E, et al. EC E- and P-selectin and vascular cell adhesion molecule-1 function as signaling receptors. *J Cell Biol.* 1998;142:1381–1391.

22 Springer TA. Traffic signals for lymphocyte recirculation and leukocyte emigration: the multistep paradigm. *Cell.* 1994;76: 301–314.

23 Dustin ML, Rothlein R, Bhan AK, et al. Induction by IL-1 and interferon, tissue distribution, biochemistry, and function of a natural adherence molecule (ICAM-1). *J Immunol.* 1986;137: 245–254.

24 Dustin ML, Springer TA. Lymphocyte function associated antigen-1 (LFA-1) interaction with intercellular adhesion molecule-1 (ICAM-1) is one of at least three mechanisms for lymphocyte adhesion to cultured endothelial cells. *J Cell Biol.* 1988;107:321–331.

25 Haskard DO. Cell adhesion molecules in rheumatoid arthritis. *Curr Opin Rheumatol.* 1995;7:229–234.

26 Ross R. Atherosclerosis-an inflammatory disease. *N Engl J Med.* 1999;340:115–126.

27 Cotran RS. *Robbins Pathological Basis of Disease*, 6th ed. Philadelphia: W.B. Saunders Co., 1999.

28 Iiyama K, Hajra L, Iiyama M, et al. Patterns of vascular cell adhesion molecule-1 and intercellular adhesion molecule-1 expression in rabbit and mouse atherosclerotic lesions and at sites predisposed to lesion formation. *Circ Res.* 1999;85:199–207.

29 Adamson P, Etienne S, Couraud PO, et al. Lymphocyte migration through brain endothelial cell monolayers involves signaling through endothelial ICAM-1 via a rho-dependent pathway. *J Immunol.* 1999;162:2964–2973.

30 Greenwood J, Amos CL, Walters CE, et al. Intracellular domain of brain endothelial intercellular adhesion molecule-1 is essential for T lymphocyte-mediated signaling and migration. *J Immunol.* 2003;171:2099–2108.

31 Etienne S, Adamson P, Greenwood J, et al. ICAM-1 signaling pathways associated with Rho activation in microvascular brain endothelial cells. *J Immunol.* 1998;161:5755–5761.

32 Thompson PW, Randi AM, Ridley AJ. Intercellular adhesion molecule (ICAM)-1, but not ICAM-2, activates RhoA and stimulates c-fos and rhoA transcription in endothelial cells. *J Immunol.* 2002;169:1007–1013.

33 Durieu-Trautmann O, Chaverot N, Cazaubon S, et al. Intercellular adhesion molecule 1 activation induces tyrosine phosphorylation of the cytoskeleton-associated protein cortactin in brain microvessel endothelial cells. *J Biol Chem.* 1994;269:12536–12540.

34 Kanner SB, Reynolds AB, Parsons JT. Tyrosine phosphorylation of a 120-kilodalton pp60^{src} substrate upon epidermal growth factor and platelet-derived growth factor stimulation and in polyomavirus middle-T-antigen-transformed cells. *Mol Cell Biol.* 1991;11:713–720.

35 Webb DJ, Parsons JT, Horwitz AF. Adhesion assembly, disassembly and turnover in migrating cells – over and over and over again. *Nat Cell Biol.* 2002;4:E97–E100.

36 Weed SA, Parsons JT. Cortactin: coupling membrane dynamics to cortical actin assembly. *Oncogene.* 2001;20:6418–6434.

37 Weed SA, Karginov AV, Schafer DA, et al. Cortactin localization to sites of actin assembly in lamellipodia requires interactions with F-actin and the Arp2/3 complex. *J Cell Biol.* 2000;151:29–40.

38 Wang Q, Doerschuk CM. The p38 mitogen-activated protein kinase mediates cytoskeletal remodeling in pulmonary microvascular endothelial cells upon intracellular adhesion molecule-1 ligation. *J Immunol.* 2001;166:6877–6884.

39 Tilghman RW, Hoover RL. The Src-cortactin pathway is required for clustering of E-selectin and ICAM-1 in endothelial cells. *FASEB J.* 2002;16:1257–1259.

40 Robinson RC, Turbedsky K, Kaiser DA, et al. Crystal structure of Arp2/3 complex. *Science.* 2001;294:1679–1684.

41 Higgs H. Actin nucleation: cortactin caught in the act. *Curr Biol.* 2002;12:R593.

42 Li Z, Kim ES, Bearer EL. Arp2/3 complex is required for actin polymerization during platelet shape change. *Blood.* 2002;99: 4466–4474.

43 Bearer EL, Prakash JM, Li Z. Actin dynamics in platelets. *Int Rev Cytol.* 2002;217:137–182.

44 Carman CV, Jun CD, Salas A, Springer TA. Endothelial cells proactively form microvilli-like membrane projections upon intercellular adhesion molecule 1 engagement of leukocyte LFA-1. *J Immunol.* 2003;171:6135–6144.

45 Carman CV, Springer TA. A transmigratory cup in leukocyte diapedesis both through individual vascular endothelial cells and between them. *J Cell Biol.* 2004;167:377–388.

46 Barreiro O, Yanez-Mo M, Serrador JM, et al. Dynamic interaction of VCAM-1 and ICAM-1 with moesin and ezrin in a novel endothelial docking structure for adherent leukocytes. *J Cell Biol.* 2002;157:1233–1245.

47 Muller WA. Migration of leukocytes across endothelial junctions: some concepts and controversies. *Microcirculation.* 2001;8:181–193.

48 Allport JR, Muller WA, Luscinskas FW. Monocytes induce reversible focal changes in vascular endothelial cadherin complex during transendothelial migration under flow. *J Cell Biol.* 2001; 148:203–216.

49 Muller WA. Leukocyte-endothelial-cell interactions in leukocyte transmigration and the inflammatory response. *Trends Immunol.* 2003;24:327–334.

50 Luscinskas FW, Ma S, Nusrat A, et al. Leukocyte transendothelial migration: a junctional affair. *Semin Immunol.* 2002;14:105–113.

51 Newman PJ. Switched at birth: a new family for PECAM-1. *J Clin Invest.* 1999;103:5–9.

52 Lampugnani MG, Dejana E. Interendothelial junctions: structure, signalling and functional roles. *Curr Opin Cell Biol.* 1997;9: 674–682.

53 Lampugnani MG, Corada M, Caveda L, et al. The molecular organization of endothelial cell to cell junctions: differential association of plakoglobin, β-catenin, and α-catenin with vascular endothelial cadherin (VE-cadherin). *J Cell Biol.* 1995;129:203–218.

54 Luscinskas FW, Ma S, Nusrat A, et al. The role of endothelial cell lateral junctions during leukocyte trafficking. *Immunol Rev.* 2002;186:57–67.

55 Shaw SK, Bamba PS, Perkins BN, Luscinskas FW. Real-time imaging of vascular endothelial-cadherin during leukocyte transmigration across endothelium. *J Immunol.* 2001;167:2323–2330.

56 Su WH, Chen H-I, Jen CJ. Differential movements of VE-cadherin and PECAM-1 during transmigration of polymorphonuclear leukocytes through human umbilical vein endothelium. *Blood.* 2002;100:3597–3603.

57 Burns AR, Walker DC, Brown ES, et al. Neutrophil transendothelial migration is independent of tight junctions and occurs preferentially at tricellular corners. *J Immunol.* 1997;159:2893–2903.

58 Burns AR, Bowden RA, MacDonell SD, et al. Analysis of tight junctions during neutrophil transendothelial migration. *J Cell Sci.* 2000;113:45–57.

59 Feng D, Nagy JA, Pyne K, et al. Neutrophils emigrate from venules by a transendothelial cell pathway in response to FMLP. *J Exp Med.* 1998;187:903–915.

60 Engelhardt B, Wolburg H. Mini-review: transendothelial migration of leukocytes: through the front door or around the side of the house? *Eur J Immunol.* 2004;34:2955–2963.

61 Wolburg H, Wolburg-Buchholz K, Engelhardt B. Diapedesis of mononuclear cells across cerebral venules during experimental autoimmune encephalomyelitis leaves tight junctions intact. *Acta Neuropathol (Berl).* 2004;109:181–190.

62 Hoshi O, Ushiki T. Scanning electron microscopic studies on the route of neutrophil extravasation in the mouse after exposure to the chemotactic peptide N-formyl-methionyl-leucyl-phenylalanine (fMLP). *Arch Histol Cytol.* 1999;62:253–260.

63 Burns AR, Smith CW, Walker DC. Unique structural features that influence neutrophil emigration into the lung. *Physiol Rev.* 2003;83:309–336.

64 Cinamon G, Shinder V, Shamri R, Alon R. Chemoattractant signals and beta 2 integrin occupancy at apical endothelial contacts combine with shear stress signals to promote transendothelial neutrophil migration. *J Immunol.* 2004;173:7282–7291.

115

Functions of Platelet-Endothelial Cell Adhesion Molecule-1 in the Vascular Endothelium

Peter J. Newman and Debra K. Newman

Blood Research Institute, Blood Center of Wisconsin, Milwaukee, Wisconsin

Platelet-endothelial cell adhesion molecule (PECAM)-1 first appeared in the literature in 1985, in reports by two different groups. These reports described a 145-kDa glycoprotein of unknown function that was constitutively expressed on the surface of platelets and endothelial cells (ECs) (1); and the 120-kDa target antigen of two murine monoclonal antibodies (mAbs), termed TM2 and TM3, that were derived by immunizing mice with the monocytic leukemia cell line, THP-1 (2). The TM-2/TM-3 antigen was found on all normal peripheral blood monocytes, neutrophils, granulocytes, platelets, and mitogen-activated lymphoblasts, as well as on blasts from various leukemia patients, especially those with leukemia of myeloid origin. Clustering by the 1989 CD Workshop of seven different CD31 mAbs having similar reactivity (3), together with cloning of its cDNA by three different laboratories in 1990 (4–6), permitted a unification of these previously disparate observations. Thereafter, PECAM-1 was assigned to the cell adhesion molecule (CAM) subfamily of immunoglobulin (Ig) gene superfamily receptors.

Human PECAM-1 (Figure 115.1) is encoded by an approximately 70-kb gene (7) located at q23 of chromosome 17 (8). The gene is comprised of 16 exons, with exons 2 through 8 encoding the six extracellular Ig-like domains, exon 9 encoding the transmembrane domain, and exons 10 to 16 encoding a complex cytoplasmic domain that is subject to a rather large degree of alternative splicing (9). The cytoplasmic domain contains at least five serine and tyrosine phosphorylation sites, the most well-characterized of which are two consensus immunoreceptor tyrosine-based inhibitory motifs (ITIMs) centered on tyrosine residues 663 and 686 (10). These ITIMs, following phosphorylation during a variety of cellular activation events, serve to recruit cytosolic signaling molecules, in particular the protein tyrosine phosphatase, SHP-2 (11,12). Although the signaling properties of PECAM-1 (13) and its role in the biology of blood and vascular cells (14) have been reviewed, no compendia are devoted solely to the function of PECAM-1 in vascular endothelium. This chapter, therefore, focuses on when and where PECAM-1 becomes expressed on ECs, what happens to it as a consequence of EC activation, and the role it plays in shaping EC responses to inflammatory stimuli, flow-mediated mechanical shear stress, and chemical perturbation.

ENDOTHELIAL CELL EXPRESSION OF PLATELET-ENDOTHELIAL CELL ADHESION MOLECULE-1

PECAM-1 first appears early during mouse embryonic development in endothelial precursors within the yolk sac (9). It soon becomes detectable within the embryo itself on angioblasts (15) (mesodermally derived EC progenitors that participate in the development of blood vessels), together with other early EC markers like the basic helix-loop-helix transcription factor TAL1/SCL, the FLK-1 receptor for vascular endothelial growth factor (VEGF), and other cell adhesion and signaling molecules such as CD34 and vascular endothelial (VE)-cadherin (16). On fully mature blood vessels, PECAM-1 is a constitutively expressed, prominent component of EC junctions at all levels of the vascular tree (17,18). It also has been found on cells with significant endothelial-like properties, including trophoblasts (19), retinal pigment epithelium (20), and circulating endothelial progenitor cells (21). PECAM-1 expression on human umbilical vein ECs (HUVECs) reaches 1×10^6 molecules/cell (22). Expression of PECAM-1 within the lung vasculature may increase even further under hyperoxic conditions (23) or following irradiation (24).

As illustrated in Figure 115.2, PECAM-1 is diffusely distributed in the plasma membrane of isolated ECs, and it becomes strongly border-localized upon cell–cell contact (25). This appears to be a dynamically regulated process, because PECAM-1 has been shown to leave the junctions at the leading edge of migrating ECs during processes like wound healing or

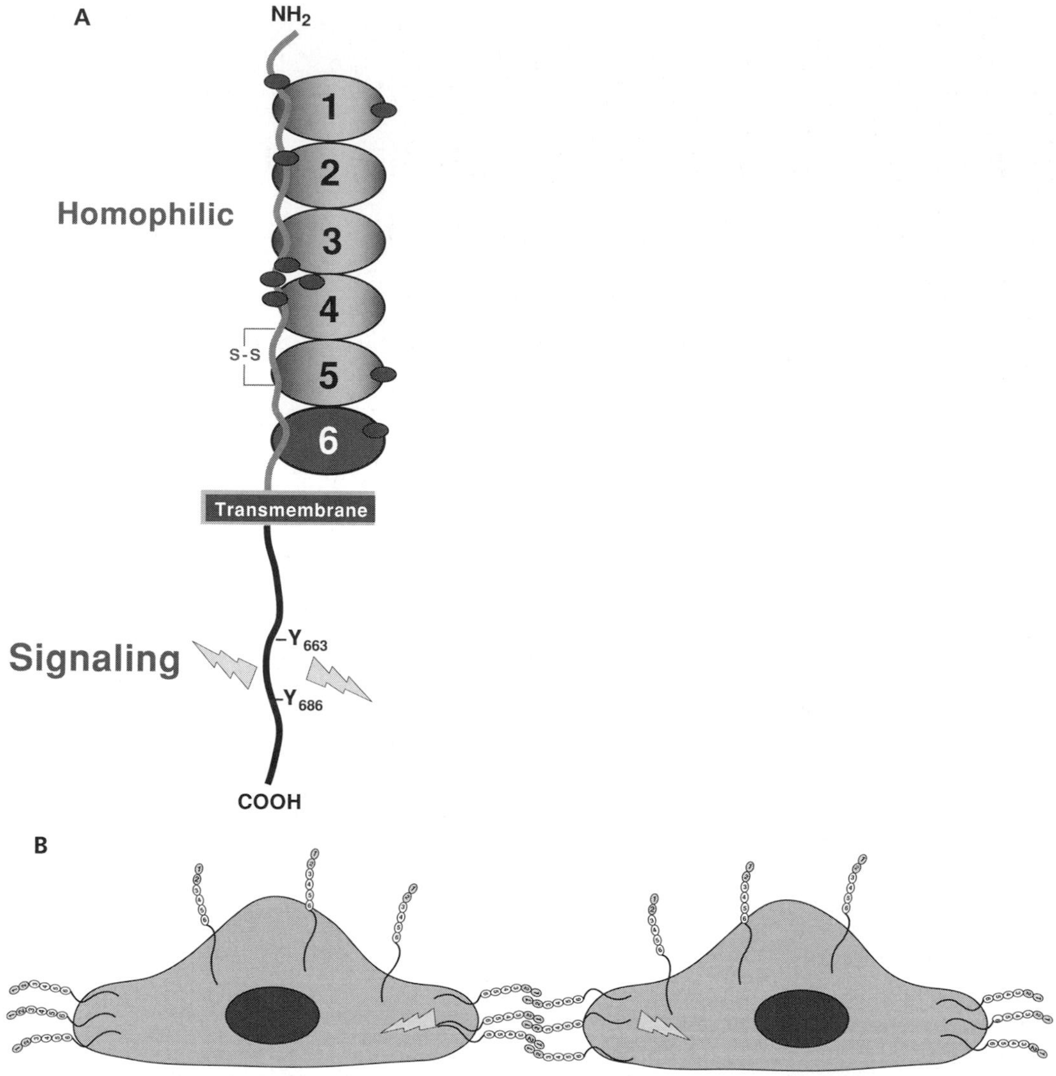

Figure 115.1. Schematic representation of PECAM-1. (**A**) PECAM-1 is a 130-kDa type I transmembrane receptor glycoprotein encoded by a 62-kb gene located on the reverse strand of the long arm of chromosome 17. The extracellular domain is comprised of six Ig Type C2 domains, with the first two mediating its homophilic binding properties. The 118-amino acid cytoplasmic domain is genetically and structurally complex, being encoded by eight different exons, some of which have been found to be alternatively spliced to yield PECAM-1 isoforms with potentially distinct signaling properties. The molecular mass of the mature protein contributed by its 711 amino acids is only approximately 80 kDa, with the remaining 50 kDa accounted for by N-linked glycosylation at nine sites, denoted by small circles. ITIMs encompassing tyrosine residues 663 and 686 contribute importantly to PECAM-1–mediated signal transduction. (**B**) Homophilic interactions of PECAM-1 at the cell–cell interface strategically position it to be able to respond to mechanical and chemical environmental signals.

angiogenesis (26). Somewhat surprisingly, border localization is mediated not by connections of the cytoplasmic domain with the underlying cytoskeleton, but rather by extracellular domain-mediated homophilic interactions, in a process that has been termed *diffusion trapping* (27). Border-enriched PECAM-1 is actually comprised of two distinct populations (28) – those molecules on the cell surface proper, and a continuously recycling pool of PECAM-1 molecules present on a surface-connected submembranous compartment that helps guide leukocytes through the junction (28) and/or affects leukocyte activation (29) during the process of transendothelial migration (TEM). Unlike its cell junctional neighbor, VE-cadherin, PECAM-1 does not appear to be enriched at adherens junctions per se, but instead is found nearer the adluminal surface (30), where it has been reported to serve as a receptor that facilitates the binding and internalization of certain strains of *Plasmodium falciparum*–infected erythrocytes (31).

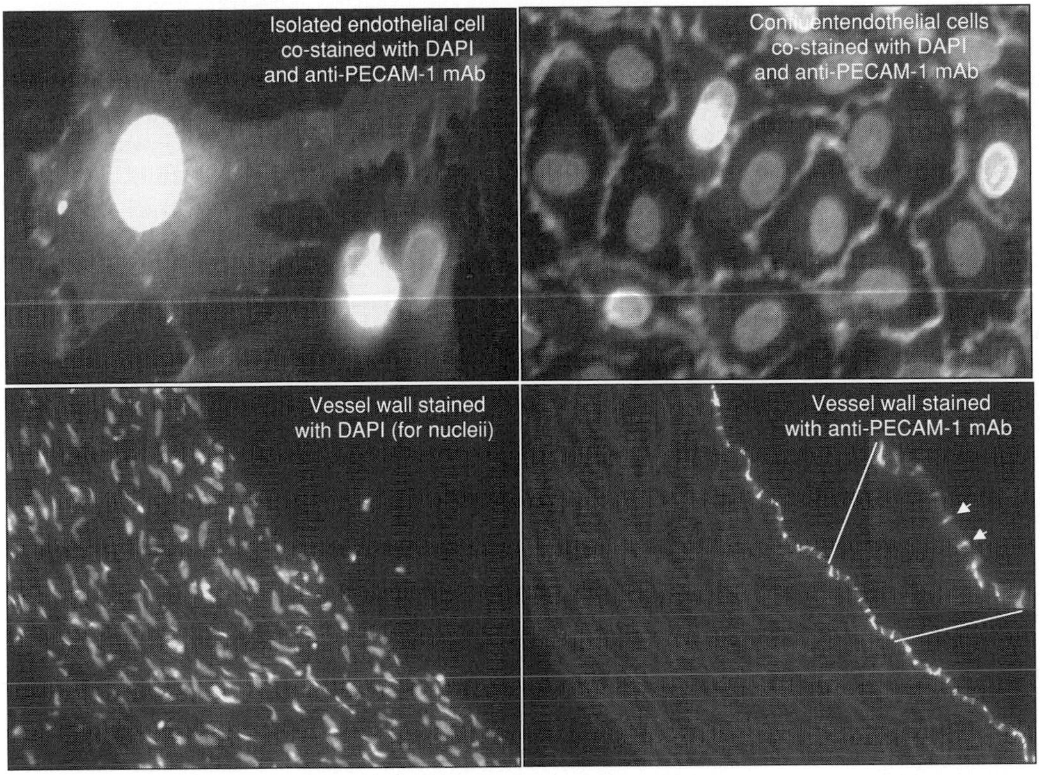

Figure 115.2. Cell contact–dependent localization of PECAM-1 at EC–cell junctions. PECAM-1 is found evenly distributed on the surface of the plasma membrane (*upper left panel*) until cells contact each other. When cells reach confluence (*upper right panel*), PECAM-1 becomes highly enriched at cell borders, both in vitro (*upper panels*) and in vivo (*lower panels*). (Photos reproduced with permission from Drs. Cunji Gao and Xia Tang, Blood Research Institute and Medical College of Wisconsin.)

PECAM-1 can be induced to change its EC junctional distribution by a number of physiological events, most notably those associated with the process of inflammation. Although neither treatment with ethyleneglycol-bis-(β-aminoethylether)-N, N, N′, N′-tetraacetic acid (EGTA) (30) nor neutrophil adhesion per se (32) causes gross redistribution of PECAM-1 from EC–cell junctions, mediators like interleukin (IL)-1 and tumor necrosis factor (TNF)-α have been found to affect both endothelial PECAM-1 expression levels as well as its subcellular distribution. For example, the treatment of intact blood vessel explants with TNF-α was found to increase the concentration of PECAM-1 at EC junctions (33), whereas herpes simplex virus infection of the cornea of live mice, which leads to locally elevated interferon (IFN)-γ levels, significantly upregulated PECAM-1 expression on corneal vascular endothelium (34). These findings are in sharp contrast to four different in vitro studies using cultured ECs, in which exposure to TNF-α and/or IFN-γ resulted in a decrease in PECAM-1 mRNA and protein levels (35,36) and a rapid redistribution out of cell–cell junctions (36–38). In either case, cytokine-mediated redistribution of PECAM-1 may affect several of its endothelial functions, including maintenance of a permeability barrier and support of leukocyte TEM.

THE ROLE OF PLATELET-ENDOTHELIAL CELL ADHESION MOLECULE-1 IN THE TRANSENDOTHELIAL MIGRATION OF LEUKOCYTES

One of the earliest described functions of EC PECAM-1 was its ability to mediate the TEM of leukocytes (39) – a property that involves homophilic interactions between the first extracellular Ig domains of leukocyte PECAM-1 and EC PECAM-1 (40). PECAM-1–deficient mice exhibit a marked defect in neutrophil and monocyte migration in response to inflammatory stimuli (41). Why this is so is not clear, but the engagement of EC PECAM-1 has been shown to initiate a number of downstream events, including: (a) tyrosine phosphorylation of its cytoplasmic ITIMs and formation of a PECAM-1/SHP-2 signaling complex (42), (b) increase in cytosolic calcium (43), (c) prostacyclin generation (43), and (d) activation of cell surface integrins (44). Of these, a recent study (29) suggests that the primary function of EC PECAM-1 in the transmigration process is to serve as a passive ligand for leukocyte PECAM-1, which, as a result of homophilic interactions with its EC counterpart, sends activation signals into the leukocyte that upregulate integrin function and facilitate the latter stages of

TEM (45). In this regard, two studies (46,47) have shown that anti-PECAM-1 mAbs inhibit leukocyte TEM by blocking their transit through the basement membrane, a collagen- and laminin-rich barrier whose transit likely requires activated leukocyte integrins. Additional evidence that PECAM-1 functions to modulate leukocyte integrin function has recently been provided by two groups who found that leukocytes became transiently delayed while traversing the perivascular basement membrane of PECAM-1–deficient mice (48,49). The nature of PECAM-1–initiated signals that activate leukocyte integrins has not been established with certainty, but may involve PECAM-1 clustering (50), activation of phosphoinositide 3-kinase (PI3K) (51), the small GTPase Rap1 (52), and/or calcium flux (53).

The ability of anti-PECAM-1 mAbs – via their ability to block leukocyte TEM – to serve as anti-inflammatory agents has been exploited numerous times. Vaporciyan and colleagues injected anti-PECAM-1 mAbs into rats, and were the first to show that such reagents in vivo could block neutrophil emigration into inflamed peritoneal and alveolar compartments, as well as across TNF-α–treated human endothelium (54). Bogen and colleagues soon thereafter observed similar effects of anti-PECAM-1 mAbs in mice (55). Beneficial effects of anti-PECAM-1 mAbs in reducing leukocyte-mediated ischemia–reperfusion injury have been demonstrated in at least six different experimental systems (56–61), providing a rationale for the eventual use of such reagents in humans.

PLATELET-ENDOTHELIAL CELL ADHESION MOLECULE-1 AS A REGULATOR OF ENDOTHELIAL CELL MIGRATION AND ANGIOGENESIS

Although individual ECs can exist for many months as stable components of the vasculature, their ability to migrate and organize into new blood vessels is important during vasculogenesis, in wound repair following vascular injury, and during the neovascularization that occurs in response to growth factors, chemokines, and solid tumor–induced angiogenesis. Many cytoskeletal elements and cell surface adhesion and signaling molecules are involved in EC migration and angiogenesis, and numerous observations suggest that PECAM-1 plays a positive role in both these processes. First, VEGF – which functions as an EC mitogen and stimulator of EC migration and new blood vessel formation – stimulates tyrosine phosphorylation of PECAM-1 (62). Second, ECs derived from PECAM-1–deficient mice show reduced migration rates in vitro (63) and form new blood vessels at a slower rate in vivo (64) – an observation similar to that found in HUVECs, in which cell migration rates were reduced by treatment with PECAM-1–specific antisense oligonucleotides (63). Third, although not universally observed (26,65,66), transfection of PECAM-1 into PECAM-1–deficient ECs (63) or PECAM-1–negative cell lines (67,68) has most often been found to enhance the cells rate of migration, as well as their ability to organize and form networks of cords on Matrigel (69). Interestingly, the PECAM-1 cytoplasmic ITIMs and their ability to recruit SHP-2 appear to be required for both these events (67). Fourth, numerous investigations have shown that anti-PECAM-1 antibodies block EC tube formation in three-dimensional collagen or laminin gels (67,69–73), and are effective inhibitors in vivo of growth factor–induced neovascularization (71) and tumor-induced angiogenesis (67,73). Taken together, these data support the notion that PECAM-1 adhesion and/or signaling contribute importantly to EC migration in wound healing and a multitude of other angiogenic responses, and it provides a rationale for the future development of anti-PECAM-1–based therapeutics to inhibit solid tumor growth.

PLATELET-ENDOTHELIAL CELL ADHESION MOLECULE-1 AS A BIOSENSOR

The observation that PECAM-1 has both adhesive as well as signaling properties, together with the fact that it is expressed at high receptor density on the surface of vascular ECs, suggested that it might function as a biosensor of local environmental changes. Evidence that this is in fact the case was first provided by Fujiwara and colleagues, who found that PECAM-1 becomes phosphorylated on ITIM tyrosine[686] within minutes following EC exposure to changes in fluid shear stress, or to mechanical changes brought about by hyper- or hypo-osmotic shock (12,74). These findings have since been confirmed (42,75,76) and extended to PECAM-1–expressing EC-like REN cells (77) and murine endothelioma cells (78). Shear stress–induced tyrosine-phosphorylation of PECAM-1 ITIMs triggers rapid translocation of SHP-2 (11,42,75) and Gab1 (42) from their cytosolic location to the inner face of the plasma membrane. Several studies have addressed the requirements for PECAM-1 tyrosine phosphorylation in response to the application of shear stress to ECs. Calcium mobilization and the activation of trimeric G-proteins, which are early responses to shear stress, are not sufficient for PECAM-1 tyrosine phosphorylation (12,74). Rather, this event requires that ECs have an intact cytoskeleton (11) and that PECAM-1 be engaged as a consequence of either cell confluence (42,78) or the adherence of nonconfluent cells to immobilized PECAM-1–specific antibodies (42). A mutant form of PECAM-1 incapable of engaging in homophilic interactions did, however, become tyrosine phosphorylated in mechanically stimulated, confluent REN cells (77), suggesting that the requirement for PECAM-1 homophilic engagement may be somewhat lax. Members of the Src (11,12,76,79) and Fps/Fes or Fer (80) families of protein tyrosine kinases have been implicated in PECAM-1 phosphorylation in response to mechanical stimulation; however, the pathways leading to the activation of these kinases in response to shear stress, and whether PECAM-1 is downstream (11,12,76,79,80) or upstream (81) of kinase activation remain important topics of investigation.

Regardless of the order of events, tyrosine phosphorylation of PECAM-1 in response to shear stress creates a scaffold for recruitment-type signaling molecules that have the potential to modulate the responses of ECs. ECs respond to shear stress by initiating a number of processes that result in biochemical, morphological, and physiological changes in the vasculature (82). PECAM-1 recently has been implicated in several of these processes. Specifically, PECAM-1–deficient ECs, derived either from PECAM-1–deficient mice or by small-interfering RNA (siRNA)-mediated knockdown of endogenous PECAM-1 expression, exhibit defective activation of Src (81), mitogen-activated protein kinase (MAPK)/extracellular-signal-regulated kinase (ERK) (42,75), and PI3K/Akt (75,81) in response to shear, although the latter two pathways were not found to be affected by PECAM-1 deficiency in a separate study (78). The upstream mechanisms by which PECAM-1 contributes to the activation of these pathways are as yet incompletely characterized, but appear to include activation of Src and integrins (81); formation of a multimolecular complex with VE-cadherin/β-catenin, VEGF receptor 2 (VEGFR2), PI3K, Tie2, SHP-2, and Gab1 (42,75,81); and a requirement for both the catalytic activity (42) and adaptor function (75) of SHP-2. PECAM-1 deficiency has predicted consequences for events downstream of these pathways, such as MAPK/ERK-dependent initiation of nuclear factor (NF)-κB–driven gene transcription (81) and activation of endothelial nitric oxide synthase (eNOS), which depends in part on activation of the PI3K/Akt pathway (75,76) and may also be regulated by colocalization of eNOS with PECAM-1 at the cell surface (76,83,84).

Although these studies provide evidence that PECAM-1 plays a role in activation of signal transduction pathways in cultured ECs exposed to shear stress, recent studies have demonstrated that the responses of PECAM-1–dependent ECs to shear stress are both observed in, and contribute significantly to, in vivo responses within the vasculature. In vessels, fluid shear stress is accomplished by increasing perfusate flow, in response to which vessels normally dilate. Flow-mediated dilation is attributable to ECs release of vasodilatory factors, including prostacyclin, nitric oxide (NO), and endothelium-derived hyperpolarizing factor (EDHF) (85). Two recent studies have reported that flow-mediated dilation is markedly impaired in the absence of PECAM-1, although a different mechanism was found to underlie impairment in each case. Thus, one study reported that PECAM-1–deficient skeletal muscle arterioles exhibited impaired dilation in response to a high temporal gradient of increased flow, in this case due to a defect in NO production (86). A second study reported that PECAM-1–deficient coronary arteries exhibited impaired dilation in response to a steady-state increase in flow; however, in this case, the impairment was due to *overproduction* of NO, superoxide anion, and peroxynitrite, which interfered with responses of vascular smooth muscle cells to EDHF (87). The molecular mechanisms by which PECAM-1 regulates flow-mediated dilation in different vascular beds have not been determined.

The physiological relevance of PECAM-1-mediated, shear stress-induced, signal transduction in ECs is just beginning to be examined in the context of cardiovascular diseases like thrombosis and atherosclerosis. Conflicting results have been obtained in murine models of these diseases; PECAM-1 deficiency has been found to have both prothrombotic (88) and atheroprotective effects (81). Although PECAM-1 deficiency has not yet been described in humans, certain allelic polymorphisms within the *PECAM-1* gene have been associated with a higher risk for developing atherosclerosis in the human population (89). Whether the ability of PECAM-1 to influence the response of vascular ECs to hemodynamic changes is mechanistically linked with its reported effects on the development of thrombotic and atherosclerotic lesions remains an intriguing avenue of future investigation.

In addition to its role as a mechanosensor, PECAM-1 also appears to modulate EC responses to peptide and lipid mediators generated during the inflammatory response. For example, when ECs are exposed to the inflammatory cytokine IL-1β, NF-κB becomes activated and translocates to the nucleus, where it upregulates the gene transcription of a number of EC surface adhesion molecules used to attract leukocytes, including intercellular adhesion molecule (ICAM)-1 and vascular cell adhesion molecule (VCAM)-1. If left unchecked, however, excessive invasion of leukocytes, particularly neutrophils, can lead to tissue injury, organ failure, and death. PECAM-1–positive transmigrating leukocytes appear to limit this untoward inflammatory response by engaging EC PECAM-1 during transit, which, via a poorly understood negative-feedback mechanism initiated by intercellular PECAM-1/PECAM-1 homophilic interactions, suppresses further NF-κB activation (90). PECAM-1 appears to exert additional anti-inflammatory effects by limiting the extent of the systemic acute phase response to bacterial lipopolysaccharide (LPS), because PECAM-1–deficient mice are highly susceptible to endotoxin-induced septic shock (91,92). This may be caused, in part, to failure of PECAM-1–deficient ECs to sustain tyrosine phosphorylation of the transcription factor, Stat3 (92), which, upon cytokine stimulation and phosphorylation by Janus kinase (Jak), dimerizes and moves to the nucleus. There, it transactivates a plethora of genes involved in controlling inflammation (93–95) and suppressing apoptosis (96). Because PECAM-1–deficient mice exhibit increased lethality following LPS challenge, it may be of future interest to determine whether people expressing low levels of EC PECAM-1 might be at greater risk for developing septicemia following bacterial infection. Finally, when lysophosphatidylcholine (Lyso PC) – a prominent phospholipid component of atherogenic lipoproteins that is also elevated in many inflammatory diseases – binds to one of its recently identified G-protein–coupled receptors on vascular endothelium, it induces tyrosine phosphorylation of PECAM-1 within minutes, and upregulates expression of ICAM-1 and VCAM-1 over a period of hours (97). The consequences of Lyso PC activation of PECAM-1 have not yet been examined.

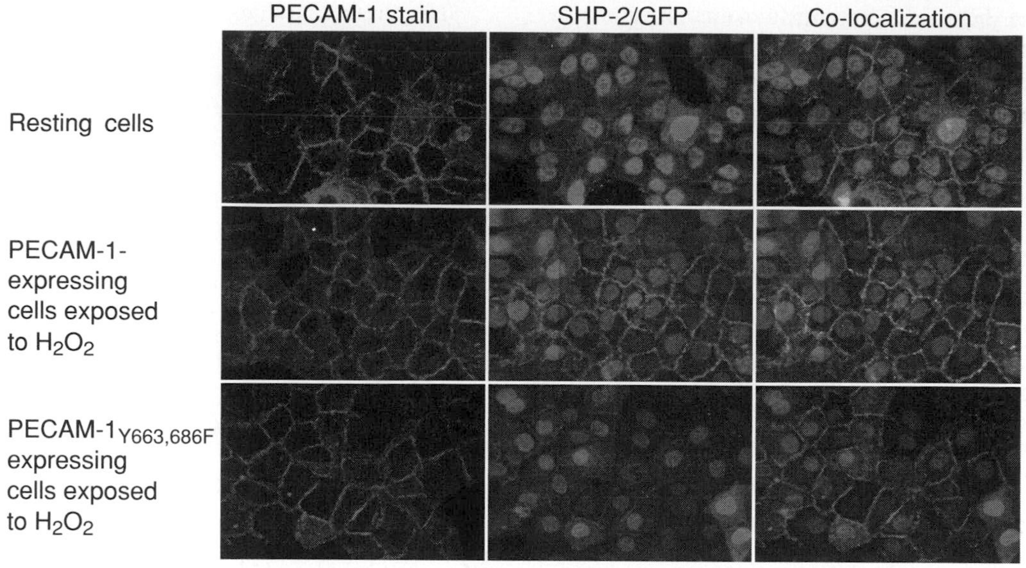

Figure 115.3. Reactive oxygen species activate PECAM-1 ITIMs and trigger cell border localization of the protein-tyrosine phosphatase, SHP-2. Endothelial-like REN cells were transfected with a green fluorescent protein (GFP) fusion protein containing the tandem SH2 domains of SHP2 together with either wild-type PECAM-1 (*top two rows*) or an ITIM-crippled ($Y_{663,686}\rightarrow F$) form of PECAM-1 (*bottom row*). In resting cells (*top row*), PECAM-1 is concentrated at cell–cell borders, whereas SHP-2 is largely cytoplasmic and nuclear. Upon exposure to oxidative stress (*middle and bottom rows*), SHP-2 is recruited to the inner face of the plasma membrane of cells expressing wild-type (*middle panels*) but not ITIM-less (*lower panels*) PECAM-1. (Reproduced with permission from Maas et al. *Am J Physiol Heart Circ Physiol.* 2003;285:H2336. Photo courtesy of Cathy Paddock, Blood Research Institute.) For color reproduction, see Color Plate 115.3.

PECAM-1 also functions as a redox sensor in vascular cells. Because intracellular tyrosine phosphatases contain a reactive site cysteine that is susceptible to oxidation, exposure of cells to even mild oxidants like hydrogen peroxide (H_2O_2) results in their inactivation. This disrupts the carefully controlled balance of intracellular tyrosine kinases and phosphatases, resulting in a marked increase in the tyrosine phosphorylation state of a number of cellular proteins, one of the more prominent of which is PECAM-1. Maas and colleagues (98) have shown that exposure of ECs to reactive oxygen species (ROS) causes rapid, reversible phosphorylation of PECAM-1 cytoplasmic domain ITIM tyrosine residues and the resultant recruitment of SHP-2 to cell–cell borders (Figure 115.3; for color reproduction, see Color Plate 115.3). Similarly, Ji and coworkers (99) found that EC exposure to H_2O_2, at concentrations expected to be achieved at sites of inflammation, resulted in SFK-dependent phosphorylation of the PECAM-1 cytoplasmic domain ITIM tyrosines and subsequent activation of a calcium-permeant, nonselective cation channel. This channel gives rise to the prolonged calcium currents that are required for ECs to open their intercellular junctions during the transendothelial passage of leukocytes. Because intact ITIMs were required for PECAM-1–dependent current activation, it is tempting to speculate that PECAM-1–mediated recruitment of SHP-2 to the inner face of the plasma membrane facilitates dephosphorylation of one or more ion channel regulatory proteins, resulting in the observed opening of plasma membrane cation channels. The identity of such proteins is unknown at present.

SUPPRESSION OF ENDOTHELIAL CELL APOPTOSIS

EC apoptosis is a normal physiological event that is thought to play an important role in branching morphogenesis during development (100), hypertension-induced remodeling of the peripheral vascular tree (101), and the development of tumor vasculature (102). In addition, microvascular EC apoptosis has been shown to be the primary cause of epithelial sloughing following radiation treatment of the gastrointestinal tract (103). Thus, identifying the cellular factors that control EC survival and death has implications for our understanding of angiogenesis (104), atherosclerosis (105), and tumor responses to radiotherapy (106).

Increasing evidence suggests that PECAM-1 is able to transmit antiapoptotic survival signals into ECs. Three different groups have found that the engagement of the PECAM-1 extracellular domain confers resistance in cultured ECs to serum-deprivation–induced cell death (107–109). The cytoprotective effects of PECAM-1 appear to operate in vivo as well. Gao and colleagues have shown that lung endothelium in PECAM-1–deficient mice is hypersensitive to radiation-induced apoptosis, compared to that derived from their wild-type counterparts (110). PECAM-1 appears to be active in suppressing only one of two major cell death pathways. Thus, whereas PECAM-1 confers no survival benefit to cells challenged with activators of the extrinsic (TNF/death receptor) pathway, it is a potent suppressor of the mitochondrial-dependent, Bax-mediated intrinsic pathway of programmed cell death (110). Because the intrinsic pathway is activated by a range of stimuli that include many genotoxic chemotherapeutic agents, ROS, and UV- and X-irradiation, targeted modulation of PECAM-1 expression might represent a novel approach to sensitizing tumor vasculature to proapoptotic anticancer agents (111).

The nature of the prosurvival signals that PECAM-1 confers to the vascular endothelium has not been firmly established. Gao and colleagues found that mutations within the homophilic binding domain or its cytoplasmic ITIMs abolish PECAM-1's cytoprotective effects, suggesting that PECAM-1–PECAM-1 interactions lead to the transmission of still-to-be-defined ITIM-dependent signals (SHP-2 binding?) that suppress apoptosis (110). An ITIM-independent mechanism by which PECAM-1 expression may confer cytoprotective properties to ECs is suggested by the recent observation that a major isoform of sphingosine kinase (SphK) is constitutively associated with part of the PECAM-1 cytoplasmic domain (that encoded by exons 9–13) in transfected HEK 293 cells overexpressing both proteins (112). How, though, might PECAM-1–SphK interactions influence the susceptibility of ECs to proapoptotic signals?

Sphingomyelin is a ubiquitous, abundant, and naturally occurring sphingolipid component of animal cell membranes that can be enzymatically catabolized to ceramide, sphingosine, and sphingosine 1-phosphate (S1P) via the sequential actions of sphingomyelinase, ceramidase, and SphK, respectively (Figure 115.4). Cells maintain a dynamic equilibrium (sometimes termed the *ceramide/S1P rheostat*) (113) among these three sphingolipid metabolites, as each can act as second messengers that activate opposing downstream signaling pathways. In particular, ceramide and sphingosine are both potent stimulators of mitochondrial-mediated EC apoptosis (114,115), while SphK serves not only to antagonize the apoptotic effects of ceramide and sphingosine by depleting them from the cellular pool, but also generates S1P, which has potent intrinsic antiapoptotic activity of its own (116).

The potential for PECAM-1 to regulate SphK/S1P signaling may be especially relevant to radiation-induced EC apoptosis, because irradiation activates sphingomyelinase (115), which is abundantly expressed in ECs (117). At the same time, it upregulates PECAM-1 expression (24,118,119). It is therefore tempting to speculate that the prosurvival effects of PECAM-1 may be due, at least in part, to its ability to recruit and concentrate SphK near sphingosine, where it can inactivate it via conversion to S1P. Manipulating cellular PECAM-1 expression levels could, at least theoretically, affect the conversion of proapoptotic sphingosine to antiapoptotic S1P, thereby exerting some control over the ceramide/S1P rheostat. For example, overexpressing PECAM-1 would be predicted to

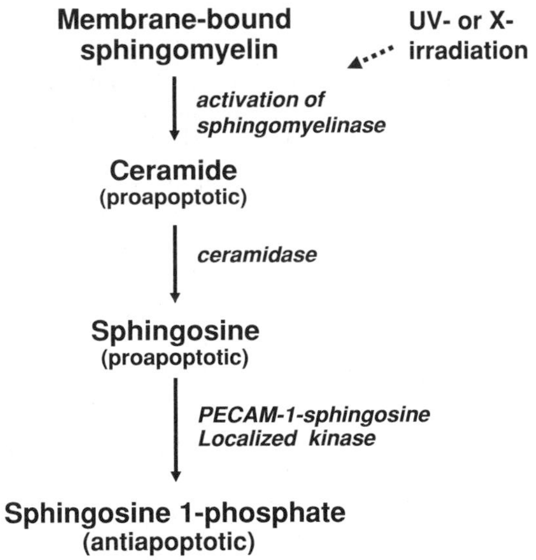

Figure 115.4. Schematic diagram showing the potential for PECAM-1 to regulate intracellular ratios of proapoptotic and antiapoptotic levels of sphingomyelin metabolites by influencing the subcellular distribution of sphingosine kinase.

increase S1P levels and thereby enhance cell survival, whereas lowering PECAM-1 would have the effect of tipping the rheostat such that cellular ceramide and sphingosine levels increase, thereby sensitizing cells to radiation-induced apoptosis. Further studies will be required to establish the relevance of this pathway to EC survival and the utility of manipulating it to promote radiation-induced injury of tumor vasculature.

BARRIER FUNCTION AND REGULATION OF VASCULAR PERMEABILITY

Ferrero and colleagues reported more than 10 years ago that PECAM-1–specific F(ab′)2 fragments are able to augment ^{125}I-labelled albumin transit across EC junctions, both in cultured cells and in mice (120). Conversely, they also found that cultured NIH/3T3 and ECV304 cells could be made less permeable to albumin if the cells were transfected with PECAM-1 cDNA. Based on their findings, they were the first to suggest that PECAM-1 contributes to EC barrier function. It was, therefore, somewhat surprising to discover that vascular development is overtly normal in PECAM-1–deficient mice (48). However, a number of studies have since shown that the stability of EC–cell junctions is more easily compromised in the blood vessels of PECAM-1–deficient mice subjected to stress and strain. For example, despite the fact that platelets are hyperreactive in PECAM-1–deficient mice (121), Mahooti and colleagues found that tail vein bleeding times are actually prolonged compared with that of wild-type mice (122). Reassuringly, these authors were able to show, using bone marrow–transplanted radiation chimeric mice, that the bleeding abnormality segregated with the phenotype of the endothelium, and not with that of circulating platelets. These data, therefore, suggest that PECAM-1 deficiency creates an EC junctional defect that becomes most obviously manifest during vascular responses to injury. Further support for an in vivo role for PECAM-1 in EC barrier function was provided by the studies of Graesser and colleagues (123), who observed that re-establishment of a vascular permeability barrier is delayed in the vessels of PECAM-1–deficient mice in skin exposed to histamine and in the brain microvasculature of mice suffering from experimental autoimmune encephalomyelitis (EAE). The most dramatic consequence of PECAM-1's absence from EC junctions, however, becomes evident during septicemia, as two different groups found that PECAM-1–deficient mice are much more susceptible to LPS-induced septic shock than are their wild-type counterparts (91,92). When challenged with LPS, the blood vessels of PECAM-1–deficient mice exhibit increased permeability and have an exaggerated loss of blood volume, with a concomitant, fatal drop in blood pressure (91). Interestingly, similar to the bleeding time phenotype, this could be corrected by expressing PECAM-1 solely on ECs. Taken together, these data support the notion that PECAM-1–mediated adhesion and/or signaling contributes appreciably to vascular integrity and the maintenance of a stable vascular permeability barrier, especially following disruptive stimuli.

The mechanism by which PECAM-1 contributes to EC barrier function is still not well understood. Several groups (70,124) found evidence for the formation of PECAM-1/β-catenin complexes, and suggested that the ability of PECAM-1 to recruit β-catenin might influence junctional stability by affecting the dynamics and assembly of VE-cadherin/β-catenin/F-actin complexes. In addition to this scaffolding function, PECAM-1 recently has been shown to regulate the ability of Jaks to phosphorylate Stat3 (92), which, in ECs, may play a protective role against LPS-induced septic shock (93,125). Whether this is related to the ability of PECAM-1 to bind SHP-2 and regulate the phosphorylation state of Jak/Stat complexes is not yet known.

KEY POINTS

- PECAM-1 is a 130-kDa member of the immunoglobulin gene superfamily, expressed at high concentrations at the intercellular junctions of confluent ECs. It contributes importantly to a variety of endothelial-specific functions, including cell migration, angiogenesis, responses to flow, and cell survival.
- The extracellular domain of PECAM-1 mediates PECAM-1/PECAM-1 homophilic interactions between cells, and is responsible for capturing PECAM-1 at cell–cell junctions. These properties are important in PECAM-1–mediated leukocyte TEM, in endothelial migration during angiogenesis, and in maintaining EC junctional integrity.
- The cytoplasmic domain of PECAM-1 contains two ITIMs that become phosphorylated and serve as a docking site for cytosolic signaling molecules, in particular the protein-tyrosine phosphatase, SHP-2. The PECAM-1/SHP-2 complex may play an important role in conferring prosurvival properties to ECs subjected to radiation injury, in the ability of PECAM-1 to mediate EC migration and blood vessel formation, and in the function of PECAM-1 as a biosensor of chemical and mechanical stress.

Future Goals

- To determine the molecular mechanisms by which PECAM-1 regulates EC responses to cytokines
- To determine how PECAM-1 protects ECs from Bax-mediated apoptosis
- To determine the mechanism by which PECAM-1 contributes vascular integrity and helps maintain the EC permeability barrier
- To determine the molecular mechanism by which PECAM-1 regulates flow-mediated dilation in different vascular beds

REFERENCES

1 van Mourik JA, Leeksma OC, Reinders JH, et al. Vascular endothelial cells synthesize a plasma membrane protein indistinguishable from platelet membrane glycoprotein IIa. *J Biol Chem.* 1985;260:11300–11306.

2 Ohto H, Maeda H, Shibata Y, et al. A novel leukocyte differentiation antigen: two monoclonal antibodies TM2 and TM3 define a 120-kd molecule present on neutrophils, monocytes, platelets, and activated lymphoblasts. *Blood.* 1985;66:873–881.

3 von dem Borne AEGK, Modderman PW. Cluster report: CD31. In: Knapp W, Dorken B, Gilks WR, et al., eds. *Leukocyte Typing IV. White Cell Differentiation Antigens.* Oxford: Oxford University Press, 1989:995.

4 Newman PJ, Berndt MC, Gorski J, et al. PECAM-1 (CD31) cloning and relation to adhesion molecules of the immunoglobulin gene superfamily. *Science.* 1990;247:1219–1222.

5 Stockinger H, Gadd SJ, Eher R, et al. Molecular characterization and functional analysis of the leukocyte surface protein CD31. *J Immunol.* 1990;145:3889–3897.

6 Simmons DL, Walker C, Power C, Pigott R. Molecular cloning of CD31, a putative intercellular adhesion molecule closely related to carcinoembryonic antigen. *J Exp Med.* 1990;171:2147–2152.

7 Kirschbaum NE, Gumina RJ, Newman PJ. Organization of the gene for human platelet/endothelial cell adhesion molecule-1 (PECAM-1) reveals alternatively spliced isoforms and a functionally complex cytoplasmic domain. *Blood.* 1994;84:4028–4037.

8 Gumina RJ, Kirschbaum N, Rao PN, et al. The human PECAM1 gene maps to 17q23. *Genomics.* 1996;34:229–232.

9 Baldwin HS, Shen HM, Yan H-C, et al. Platelet endothelial cell adhesion molecule-1 (PECAM-1/CD31): alternatively spliced, functionally distinct isoforms expressed during early mammalian cardiovascular development. *Development*. 1994;120: 2539–2553.

10 Newman PJ. Switched at birth: a new family for PECAM-1. *J Clin Invest*. 1999;103(1):5–9.

11 Masuda M, Osawa M, Shigematsu H, et al. Platelet endothelial cell adhesion molecule-1 is a major SH-PTP2 binding protein in vascular ECs. *FEBS Lett*. 1997;408:331–336.

12 Osawa M, Masuda M, Harada N, et al. Tyrosine phosphorylation of platelet endothelial cell adhesion molecule-1 (PECAM-1, CD31) in mechanically stimulated vascular ECs. *Eur J Cell Biol*. 1997;72(3):229–237.

13 Newman PJ, Newman DK. Signal transduction pathways mediated by PECAM-1. New roles for an old molecule in platelet and vascular cell biology. *Arterioscler Thromb Vasc Biol*. 2003;23:953–964.

14 Newman PJ. The biology of PECAM-1. *J Clin Invest*. 1997;99: 3–8.

15 Pinter E, Barreuther M, Lu T, et al. Platelet-endothelial cell adhesion molecule-1 (PECAM-1/CD31) tyrosine phosphorylation state changes during vasculogenesis in the murine conceptus. *Am J Pathol*. 1997;150(5):1523–1530.

16 Drake CJ, Fleming PA. Vasculogenesis in the day 6.5 to 9.5 mouse embryo. *Blood*. 2000;95(5):1671–1679.

17 Muller WA, Ratti CM, McDonnell SL, Cohn ZA. A human endothelial cell-restricted externally disposed plasmalemmal protein enriched in intercellular junctions. *J Exp Med*. 1989;170:399–414.

18 Albelda SM, Oliver PD, Romer LH, Buck CA. EndoCAM: a novel endothelial cell-cell adhesion molecule. *J Cell Biol*. 1990;110:1227–1237.

19 Blankenship TN, Enders AC. Expression of platelet-endothelial cell adhesion molecule-1 (PECAM) by macaque trophoblast cells during invasion of the spiral arteries. *Anat Rec*. 1997;247(3):413–419.

20 McKay BS, Irving PE, Skumatz CM, Burke JM. Cell-cell adhesion molecules and the development of an epithelial phenotype in cultured human retinal pigment epithelial cells. *Exp Eye Res*. 1997;65(5):661–671.

21 Bompais H, Chagraoui J, Canron X, et al. Human endothelial cells derived from circulating progenitors display specific functional properties compared with mature vessel wall ECs. *Blood*. 2004;103(7):2577–2584.

22 Newman PJ. The role of PECAM-1 in vascular cell biology. In: Fitzgerald GA, Jennings LK, Patrono C, eds. *Platelet-Dependent Vascular Occlusion*. New York: The New York Academy of Sciences; 1994:165–174.

23 Piedboeuf B, Gamache M, Frenette J, et al. Increased endothelial cell expression of platelet-EC adhesion molecule-1 during hyperoxic lung injury. *Am J Respir Cell Mol Biol*. 1998;19(4): 543–553.

24 Gaugler MH, Vereycken-Holler V, Squiban C, Aigueperse J. PECAM-1 (CD31) is required for interactions of platelets with endothelial cells after irradiation. *J Thromb Haemost*. 2004; 2(11):2020–2026.

25 Albelda SM, Muller WA, Buck CA, Newman PJ. Molecular and cellular properties of PECAM-1 (endoCAM/CD31): a novel vascular cell-cell adhesion molecule. *J Cell Biol*. 1991;114: 1059–1068.

26 Gratzinger D, Barreuther M, Madri JA. Platelet-EC adhesion molecule-1 modulates endothelial migration through its immunoreceptor tyrosine-based inhibitory motif. *Biochem Biophys Res Commun*. 2003;301(1):243–249.

27 Sun J, Paddock C, Shubert J, et al. Contributions of the extracellular and cytoplasmic domains of platelet-endothelial cell adhesion molecule-1 (PECAM-1/CD31) in regulating cell–cell localization. *J Cell Sci*. 2000;113:1459–1469.

28 Mamdouh Z, Chen X, Pierini LM, et al. Targeted recycling of PECAM from endothelial surface-connected compartments during diapedesis. *Nature*. 2003;421(6924):748–753.

29 O'Brien CD, Lim P, Sun J, Albelda SM. PECAM-1-dependent neutrophil transmigration is independent of monolayer PECAM-1 signaling or localization. *Blood*. 2003;101:2816–2825.

30 Ayalon O, Sabanai H, Lampugnani MG, et al. Spatial and temporal relationships between cadherins and PECAM-1 in cell-cell junctions of human endothelial cells. *J Cell Biol*. 1994;126:247–258.

31 Treutiger CJ, Heddini A, Fernandez V, et al. PECAM-1/CD31, an endothelial receptor for binding *Plasmodium falciparum*-infected erythrocytes. *Nat Med*. 1997;3(12):1405–1408.

32 Del Maschio A, Zanetti A, Corada M, et al. Polymorphonuclear leukocyte adhesion triggers the disorganization of endothelial cell-to-cell adherens junctions. *J Cell Biol*. 1996;135(2):497–510.

33 Ioffreda MD, Albelda SM, Elder DE, et al. TNFα induces E-selectin expression and PECAM-1 (CD31) redistribution in extracutaneous tissues. *Endothelium*. 1993;1:47–54.

34 Tang Q, Hendricks RL. Interferon γ regulates platelet endothelial cell adhesion molecule 1 expression and neutrophil infiltration into Herpes Simplex virus-infected mouse corneas. *J Exp Med*. 1996;184:1435–1447.

35 Stewart RJ, Kashour TS, Marsden PA. Vascular endothelial platelet endothelial adhesion molecule-1 (PECAM-1) expression is decreased by TNF-alpha and IFN-gamma. Evidence for cytokine-induced destabilization of messenger ribonucleic acid transcripts in bovine endothelial cells. *J Immunol*. 1996;156:1221–1228.

36 Rival Y, Maschio AD, Rabiet M-J, et al. Inhibition of platelet endothelial cell adhesion molecule-1 synthesis and leukocyte transmigration in endothelial cell by the combined action of TNF-α and IFN-gamma. *J Immunol*. 1996;157:1233–1241.

37 Romer LH, McLean NV, Yan HC, et al. IFN-gamma and TNF-alpha induce redistribution of PECAM-1 (CD31) on human ECs. *J Immunol*. 1995;154(12):6582–6592.

38 Ferrero E, Villa A, Ferrero ME, et al. Tumor necrosis factor α-induced vascular leakage involves PECAM-1 phosphorylation. *Cancer Res*. 1996;56(14):3211–3215.

39 Muller WA, Weigl SA, Deng X, Phillips DM. PECAM-1 is required for transendothelial migration of leukocytes. *J Exp Med*. 1993;178:449–460.

40 Liao F, Ali J, Greene T, Muller WA. Soluble domain 1 of platelet-endothelial cell adhesion molecule (PECAM) is sufficient to block transendothelial migration in vitro and in vivo. *J Exp Med*. 1997;185(7):1349–1357.

41 Schenkel AR, Chew TW, Muller WA. Platelet endothelial cell adhesion molecule deficiency or blockade significantly reduces leukocyte emigration in a majority of mouse strains. *J Immunol*. 2004;173(10):6403–6408.

42 Osawa M, Masuda M, Kusano K, Fujiwara K. Evidence for a role of platelet endothelial cell adhesion molecule-1 in

EC mechanosignal transduction: is it a mechanoresponsive molecule? *J Cell Biol.* 2002;158(4):773–785.

43 Gurubhagavatula I, Amrani Y, Pratico D, et al. Engagement of human PECAM-1 (CD31) on human endothelial cells increases intracellular calcium ion concentration and stimulates prostacyclin release. *J Clin Invest.* 1998;101(1):212–222.

44 Chiba R, Nakagawa N, Kurasawa K, et al. Ligation of CD31 (PECAM-1) on endothelial cells increases adhesive function of $\alpha v\beta 3$ integrin and enhances $\beta 1$ integrin-mediated adhesion of eosinophils to ECs. *Blood.* 1999;94(4):1319–1329.

45 Dangerfield J, Larbi KY, Huang MT, et al. PECAM-1 (CD31) Homophilic interaction up-regulates alpha6beta1 on transmigrated neutrophils in vivo and plays a functional role in the ability of $\alpha 6$ Integrins to mediate leukocyte migration through the perivascular basement membrane. *J Exp Med.* 2002;196(9):1201–1212.

46 Liao F, Huynh HK, Eiroa A, et al. Migration of monocytes across endothelium and passage through extracellular matrix involve separate molecular domains of PECAM-1. *J Exp Med.* 1995;182:1337–1343.

47 Thompson RD, Wakelin MW, Larbi KY, et al. Divergent effects of platelet-EC adhesion molecule-1 and beta3 integrin blockade on leukocyte transmigration in vivo. *J Immunol.* 2000;165(1):426–434.

48 Duncan GS, Andrew DP, Takimoto H, et al. Genetic evidence for functional redundancy of platelet/endothelial cell adhesion molecule-1 (PECAM-1): CD31-deficient mice reveal PECAM-1-dependent and PECAM-1-independent functions. *J Immunol.* 1999;162(5):3022–3030.

49 Thompson RD, Noble KE, Larbi KY, et al. Platelet-endothelial cell adhesion molecule-1 (PECAM-1)-deficient mice demonstrate a transient and cytokine-specific role for PECAM-1 in leukocyte migration through the perivascular basement membrane. *Blood.* 2001;97(6):1854–1860.

50 Zhao T, Newman PJ. Integrin activation by regulated dimerization and oligomerization of platelet endothelial cell adhesion molecule (PECAM)-1 from within the cell. *J Cell Biol.* 2001;152(1):65–73.

51 Pellegatta F, Chierchia SL, Zocchi MR. Functional association of platelet endothelial cell adhesion molecule-1 and phosphoinositide 3-kinase in human neutrophils. *J Biol Chem.* 1998;273(43):27768–27771.

52 Reedquist KA, Ross E, Koop EA, et al. The small GTPase, Rap1, mediates CD31-induced integrin adhesion. *J Cell Biol.* 2000;148:1151–1158.

53 Deaglio S, Dianzani U, Horenstein AL, et al. Human CD38 ligand: a 120-kDa protein predominantly expressed on endothelial cells. *J Immunol.* 1996;156:727–734.

54 Vaporciyan AA, DeLisser HM, Yan H-C, et al. Involvement of platelet endothelial cell adhesion molecule-1 in neutrophil recruitment in vivo. *Science.* 1993;262:1580–1582.

55 Bogen S, Pak J, Garifallou M, et al. Monoclonal antibody to murine PECAM-1 (CD31) blocks acute inflammation in vivo. *J Exp Med.* 1994;179:1059–1064.

56 Gumina RJ, Schultz JE, Yao Z, et al. Antibody to platelet/endothelial cell adhesion molecule-1 reduces myocardial infarct size in a rat model of ischemia-reperfusion injury. *Circulation.* 1996; 94:3327–3333.

57 Murohara T, Delyani JA, Albelda SM, Lefer AM. Blockade of platelet endothelial cell adhesion molecule-1 protects against myocardial ischemia and reperfusion injury in cats. *J Immunol.* 1996;156:3550–3557.

58 Turegun M, Gudemez E, Newman P, et al. Blockade of platelet EC adhesion molecule-1 (PECAM-1) protects against ischemia-reperfusion injury in muscle flaps at microcirculatory level. *Plast Reconstr Surg.* 1999;104(4):1033–1040.

59 Sun Z, Wang X, Lasson A, et al. Effects of inhibition of PAF, ICAM-1 and PECAM-1 on gut barrier failure caused by intestinal ischemia and reperfusion. *Scand J Gastroenterol.* 2001;36(1): 55–65.

60 Farooq MM, Serra A, Newman PJ, et al. PECAM-1/IgG attenuates peroxynitrite-mediated extremity reperfusion injury. *J Vasc Surg.* 2001;34(3):555–558.

61 Sun Z, Olanders K, Lasson A, et al. Effective treatment of gut barrier dysfunction using an antioxidant, a PAF inhibitor, and monoclonal antibodies against the adhesion molecule PECAM-1. *J Surg Res.* 2002;105(2):220–233.

62 Esser S, Lampugnani MG, Corada M, et al. Vascular endothelial growth factor induces VE-cadherin tyrosine phosphorylation in endothelial cells. *J Cell Sci.* 1998;111(Pt 13):1853–1865.

63 Gratzinger D, Canosa S, Engelhardt B, Madri JA. Platelet endothelial cell adhesion molecule-1 modulates EC motility through the small G-protein Rho. *FASEB J.* 2003;17(11):1458–1469.

64 Solowiej A, Biswas P, Graesser D, Madri JA. Lack of platelet endothelial cell adhesion molecule-1 attenuates foreign body inflammation because of decreased angiogenesis. *Am J Pathol.* 2003;162(3):953–962.

65 Schimmenti LA, Yan H-C, Madri JA, Albelda SM. Platelet endothelial cell adhesion molecule, PECAM-1, modulates cell migration. *J Cell Physiol.* 1992;153:417–428.

66 Kim CS, Wang T, Madri JA. Platelet endothelial cell adhesion molecule-1 expression modulates EC migration in vitro. *Lab Invest.* 1998;78(5):583–590.

67 Cao G, O'Brien CD, Zhou Z, et al. Involvement of human PECAM-1 in angiogenesis and in vitro EC migration. *Am J Physiol Cell Physiol.* 2002;282(5):C1181–C1190.

68 O'Brien CD, Cao G, Makrigiannakis A, DeLisser HM. Role of immunoreceptor tyrosine-based inhibitory motifs of PECAM-1 in PECAM-1-dependent cell migration. *Am J Physiol Cell Physiol.* 2004;287(4):C1103–C1113.

69 Sheibani N, Newman PJ, Frazier WA. Thrombospondin-1, a natural inhibitor of angiogenesis, regulates platelet-endothelial cell adhesion molecule-1 expression and endothelial cell morphogenesis. *Mol Biol Cell.* 1997;8:1329–1341.

70 Matsumura T, Wolff K, Petzelbauer P. Endothelial cell tube formation depends on cadherin 5 and CD31 interactions with filamentous actin. *J Immunol.* 1997;158(7):3408–3416.

71 DeLisser HM, Christofidou-Solomidou M, Strieter RM, et al. Involvement of endothelial PECAM-1/CD31 in angiogenesis. *Am J Pathol.* 1997;151:671–677.

72 Yang S, Graham J, Kahn JW, et al. Functional roles for PECAM-1 (CD31) and VE-cadherin (CD144) in tube assembly and lumen formation in three-dimensional collagen gels. *Am J Pathol.* 1999;155(3):887–895.

73 Zhou Z, Christofidou-Solomidou M, Garlanda C, DeLisser HM. Antibody against murine PECAM-1 inhibits tumor angiogenesis in mice. *Angiogenesis.* 1999;3:181–188.

74 Harada N, Masuda M, Fujiwara K. Fluid flow and osmotic stress induce tyrosine phosphorylation of an endothelial cell

128 kDa surface glycoprotein. *Biochem Biophys Res Commun.* 1995;214(1):69–74.

75 Tai LK, Zheng Q, Pan S, et al. Flow activates ERK1/2 and endothelial nitric oxide synthase via a pathway involving PECAM1, SHP2, and Tie2. *J Biol Chem.* 2005;280(33):29620–29624.

76 Fleming I, Fisslthaler B, Dixit M, Busse R. Role of PECAM-1 in the shear-stress-induced activation of Akt and the endothelial nitric oxide synthase (eNOS) in ECs. *J Cell Sci.* 2005;118(Pt 18): 4103–4111.

77 Kaufman DA, Albelda SM, Sun J, Davies PF. Role of lateral cell-cell border location and extracellular/transmembrane domains in PECAM/CD31 mechanosensation. *Biochem Biophys Res Commun.* 2004;320(4):1076–1081.

78 Sumpio BE, Yun S, Cordova AC, et al. MAPKs (ERK1/2, p38) and AKT can be phosphorylated by shear stress independently of platelet EC adhesion molecule-1 (CD31) in vascular ECs. *J Biol Chem.* 2005;280(12):11185–11191.

79 Cao MY, Huber M, Beauchemin N, et al. Regulation of mouse PECAM-1 tyrosine phosphorylation by the Src and Csk families of protein-tyrosine kinases. *J Biol Chem.* 1998;273(25):15765–15772.

80 Kogata N, Masuda M, Kamioka Y, et al. Identification of fer tyrosine kinase localized on microtubules as a platelet endothelial cell adhesion molecule-1 phosphorylating kinase in vascular endothelial cells. *Mol Biol Cell.* 2003;14(9):3553–3564.

81 Tzima E, Irani-Tehrani M, Kiosses WB, et al. A mechanosensory complex that mediates the endothelial cell response to fluid shear stress. *Nature.* 2005;437(7057):426–431.

82 Davies PF. Flow-mediated endothelial mechanotransduction. *Physiol Rev.* 1995;75(3):519–560.

83 Dusserre N, L'Heureux N, Bell KS, et al. PECAM-1 Interacts with nitric oxide synthase in human endothelial cells. Implication for flow-induced nitric oxide synthase activation. *Arterioscler Thromb Vasc Biol.* 2004;24(10):1796–1802.

84 Cheng C, van HR, de WM, et al. Shear stress affects the intracellular distribution of eNOS: direct demonstration by a novel in vivo technique. *Blood.* 2005;106(12):3691–3698.

85 Busse R, Fleming I. Regulation of endothelium-derived vasoactive autacoid production by hemodynamic forces. *Trends Pharmacol Sci.* 2003;24(1):24–29.

86 Bagi Z, Frangos JA, Yeh JC, et al. PECAM-1 mediates NO-dependent dilation of arterioles to high temporal gradients of shear stress. *Arterioscler Thromb Vasc Biol.* 2005;25(8):1590–1595.

87 Liu Y, Bubolz AB, Shi Y, et al. Peroxynitrite reduces the endothelium derived hyperpolarizing factor component of coronary flow-mediated dilation in PECAM-1-knock out mice. *Am J Physiol Regul Integr Comp Physiol.* 2006 Jan;290(1):R57–65. Epub 2005 Sep 15.

88 Falati S, Patil S, Gross PL, et al. Platelet PECAM-1 inhibits thrombus formation in vivo. *Blood.* 2006 Jan 15;107(2):535–541. Epub 2005 Sep 15.

89 Elrayess MA, Talmud PJ. Platelet endothelial cell adhesion molecule-1 (PECAM-1) & coronary heart disease. *Indian J Med Res.* 2005;121(2):77–79.

90 Cepinskas G, Savickiene J, Ionescu CV, Kvietys PR. PMN transendothelial migration decreases nuclear NFκB in IL-1β-activated ECs: role of PECAM-1. *J Cell Biol.* 2003;161(3):641–651.

91 Maas M, Stapleton M, Bergom C, et al. Endothelial cell PECAM-1 confers protection against endotoxic shock. *Am J Physiol Heart Circ Physiol.* 2005;288(1):H159–H164.

92 Carrithers M, Tandon S, Canosa S, et al. Enhanced susceptibility to endotoxic shock and impaired STAT3 signaling in CD31-deficient mice. *Am J Pathol.* 2005;166(1):185–196.

93 Benkhart EM, Siedlar M, Wedel A, et al. Role of Stat3 in lipopolysaccharide-induced IL-10 gene expression. *J Immunol.* 2000;165(3):1612–1617.

94 Takeda K, Clausen BE, Kaisho T, et al. Enhanced Th1 activity and development of chronic enterocolitis in mice devoid of Stat3 in macrophages and neutrophils. *Immunity.* 1999;10(1):39–49.

95 Welte T, Zhang SS, Wang T, et al. STAT3 deletion during hematopoiesis causes Crohn's disease-like pathogenesis and lethality: a critical role of STAT3 in innate immunity. *Proc Natl Acad Sci USA.* 2003;100(4):1879–1884.

96 Takeda K, Kaisho T, Yoshida N, et al. Stat3 activation is responsible for IL-6-dependent T cell proliferation through preventing apoptosis: generation and characterization of T cell-specific Stat3-deficient mice. *J Immunol.* 1998;161(9):4652–4660.

97 Ochi H, Kume N, Nishi E, et al. Tyrosine phosphorylation of platelet endothelial cell adhesion molecule-1 induced by lysophosphatidylcholine in cultured cells. *Biochem Biophys Res Commun.* 1998;243:862–868.

98 Maas M, Wang R, Paddock C, et al. Reactive oxygen species induce reversible PECAM-1 tyrosine phosphorylation and SHP-2 binding. *Am J Physiol Heart Circ Physiol.* 2003;285(6): H2336–H2344.

99 Ji G, O'Brien CD, Feldman M, et al. PECAM-1 (CD31) regulates a hydrogen peroxide-activated nonselective cation channel in endothelial cells. *J Cell Biol.* 2002;157(1):173–184.

100 Duval H, Harris M, Li J, et al. New insights into the function and regulation of endothelial cell apoptosis. *Angiogenesis.* 2003;6(3):171–183.

101 Intengan HD, Schiffrin EL. Vascular remodeling in hypertension: roles of apoptosis, inflammation, and fibrosis. *Hypertension.* 2001;38(3 Pt 2):581–587.

102 Liu W, Ahmad SA, Reinmuth N, et al. Endothelial cell survival and apoptosis in the tumor vasculature. *Apoptosis.* 2000; 5(4):323–328.

103 Paris F, Fuks Z, Kang A, et al. Endothelial apoptosis as the primary lesion initiating intestinal radiation damage in mice. *Science.* 2001;293(5528):293–297.

104 Chavakis E, Dimmeler S. Regulation of endothelial cell survival and apoptosis during angiogenesis. *Arterioscler Thromb Vasc Biol.* 2002;22(6):887–893.

105 Choy JC, Granville DJ, Hunt DW, McManus BM. Endothelial cell apoptosis: biochemical characteristics and potential implications for atherosclerosis. *J Mol Cell Cardiol.* 2001;33(9):1673–1690.

106 Garcia-Barros M, Paris F, Cordon-Cardo C, et al. Tumor response to radiotherapy regulated by EC apoptosis. *Science.* 2003;300(5622):1155–1159.

107 Noble KE, Wickremasinghe RG, DeCornet C, et al. Monocytes stimulate expression of the Bcl-2 family member, A1, in endothelial cells and confer protection against apoptosis. *J Immunol.* 1999;162(3):1376–1383.

108 Bird IN, Taylor V, Newton JP, et al. Homophilic PECAM-1(CD31) interactions prevent endothelial cell apoptosis but do not support cell spreading or migration. *J Cell Sci.* 1999;112(Pt 12):1989–1997.

109 Evans PC, Taylor ER, Kilshaw PJ. Signaling through CD31 protects endothelial cells from apoptosis. *Transplantation.* 2001;71(3):457–460.

110 Gao C, Sun W, Christofidou-Solomidou M, et al. PECAM-1 functions as a specific and potent inhibitor of mitochondrial-dependent apoptosis. *Blood*. 2003;102(1):169–179.

111 Bergom C, Gao C, Matsuyama S, Newman PJ. The cell adhesion and signaling molecule PECAM-1 as a molecular mediator of genotoxic chemotherapy resistance. *Cancer Research.* 2005;46:A195. [Abstract.]

112 Fukuda Y, Aoyama Y, Wada A, Igarashi Y. Identification of PECAM-1 association with sphingosine kinase 1 and its regulation by agonist-induced phosphorylation. *Biochim Biophys Acta*. 2004;1636(1):12–21.

113 Cuvillier O, Pirianov G, Kleuser B, et al. Suppression of ceramide-mediated programmed cell death by sphingosine-1-phosphate. *Nature*. 1996;381(6585):800–803.

114 Pettus BJ, Chalfant CE, Hannun YA. Ceramide in apoptosis: an overview and current perspectives. *Biochim Biophys Acta*. 2002; 1585(2–3):114–125.

115 Kolesnick R, Fuks Z. Radiation and ceramide-induced apoptosis. *Oncogene*. 2003;22(37):5897–5906.

116 Maceyka M, Payne SG, Milstien S, Spiegel S. Sphingosine kinase, sphingosine-1-phosphate, and apoptosis. *Biochim Biophys Acta*. 2002;1585(2–3):193–201.

117 Marathe S, Schissel SL, Yellin MJ, et al. Human vascular endothelial cells are a rich and regulatable source of secretory sphingomyelinase. Implications for early atherogenesis and ceramide-mediated cell signaling. *J Biol Chem*. 1998;273(7):4081–4088.

118 Quarmby S, Kumar P, Wang J, et al. Irradiation induces upregulation of CD31 in human endothelial cells. *Arterioscler Thromb Vasc Biol*. 1999;19(3):588–597.

119 Quarmby S, Hunter RD, Kumar S. Irradiation induced expression of CD31, ICAM-1 and VCAM-1 in human microvascular ECs. *Anticancer Res*. 2000;20(5B):3375–3381.

120 Ferrero E, Ferrero ME, Pardi R, Zocchi MR. The platelet endothelial cell adhesion molecule-1 (PECAM1) contributes to endothelial barrier function. *FEBS Lett*. 1995;374(3):323–326.

121 Patil S, Newman DK, Newman PJ. Platelet endothelial cell adhesion molecule-1 serves as an inhibitory receptor that modulates platelet responses to collagen. *Blood*. 2001;97:1727–1732.

122 Mahooti S, Graesser D, Patil S, et al. PECAM-1 (CD31) expression modulates bleeding time in vivo. *Am J Pathol*. 2000;157(1): 75–81.

123 Graesser D, Solowiej A, Bruckner M, et al. Altered vascular permeability and early onset of experimental autoimmune encephalomyelitis in PECAM-1-deficient mice. *J Clin Invest*. 2002;109(3):383–392.

124 Ilan N, Mahooti S, Rimm DL, Madri JA. PECAM-1 (CD31) functions as a reservoir for and a modulator of tyrosine-phosphorylated β-catenin. *J Cell Sci*. 1999;112(Pt 18):3005–3014.

125 Kano A, Wolfgang MJ, Gao Q, et al. Endothelial cells require STAT3 for protection against endotoxin-induced inflammation. *J Exp Med*. 2003;198(10):1517–1525.

116

P-Selectin

Rodger P. McEver

Oklahoma Medical Research Foundation Oklahoma City

In 1984, two groups independently described monoclonal antibodies that bound to activated but not to resting platelets (1,2). The antibodies recognized a 140-kD platelet membrane glycoprotein. Immunocytochemical analysis revealed that the protein was concentrated in the membranes of α-granules in resting platelets (3,4). Thrombin and other agonists that induced fusion of α-granules with the plasma membrane caused the protein to redistribute to the platelet surface; this finding explained why the antibodies bound only to activated platelets. The protein was originally called granule membrane protein of 140 kDa (GMP-140) or platelet activation-dependent granule-to-external-membrane protein (PADGEM). Soon after, the same groups reported that endothelial cells (ECs) also expressed this protein (5–8). In resting endothelium, the protein was concentrated in membranes of Weibel-Palade bodies. Thrombin, histamine, and other agonists that induced fusion of Weibel-Palade bodies with the plasma membrane caused the protein to redistribute to the EC surface. Therefore, GMP-140/PADGEM was a glycoprotein that could be rapidly mobilized to the cell surface from storage granules by mediators generated during injury or infection (Figure 116.1). This suggested that the protein had important functions at vascular sites where platelets and/or ECs became activated.

In 1989, one of the groups reported the primary structure of GMP-140/PADGEM derived from protein sequencing and cDNA cloning (9). This analysis revealed a modular, type I membrane glycoprotein with an N-terminal domain similar to those in calcium (Ca^{2+})-dependent (C-type) lectins, followed by an epidermal growth factor (EGF)-like module, nine short consensus repeats related to those in complement-regulatory and other proteins, a transmembrane domain, and a short cytoplasmic tail (Figure 116.2). At the same time, other groups reported the cloning of cDNAs encoding two other proteins with related modular structures (Figure 116.2). A homing receptor expressed on leukocytes mediated the adhesion of lymphocytes to lymph node endothelium by a mechanism that was believed to involve carbohydrate recognition (10,11). Endothelial leukocyte adhesion molecule (ELAM)-1, which was expressed on cytokine-activated endothelium, supported the adhesion of leukocytes in vitro (12). The structural relationship of GMP-140/PADGEM to these proteins immediately suggested a functional relationship. Shortly thereafter, leukocytes were shown to adhere to the GMP-140/PADGEM expressed on the surfaces of activated platelets and ECs (13–15). The new family of proteins received the consensus name *selectins*, to indicate that they mediated selective cell–cell interactions through the interactions of their lectin domains with putative carbohydrate ligands (16). Each family member was assigned a letter prefix to represent the cell in which it was first identified. Thus, the homing receptor was named L-selectin, ELAM-1 was named E-selectin, and GMP-140/PADGEM was named P-selectin. Since then, the properties of the selectins in vitro and in vivo have been the subject of numerous papers. The selectins have attracted great attention because of their contributions to leukocyte trafficking, and because they represent one of the earliest and best-characterized examples of a biological function for lectins in mammals. P-selectin has received particular notice as an inducible cell adhesion molecule that links the hemostatic and inflammatory responses to tissue injury.

GENE REGULATION, GENE/PROTEIN STRUCTURE, AND PROTEIN TRAFFICKING

The genes encoding all three selectins are tightly linked on syntenic regions of chromosome 1 in both humans and mice (17). They represent a classic example of the use of gene duplication and exon shuffling to generate a new gene/protein family. Almost all exons encode discrete structural domains (18). The selectin genes are located close to a group of genes encoding complement-regulatory proteins that have similar short consensus repeats. This suggests that several genes with functions during immune responses developed coordinately. The C-type lectin, EGF-like, and short consensus repeat domains are

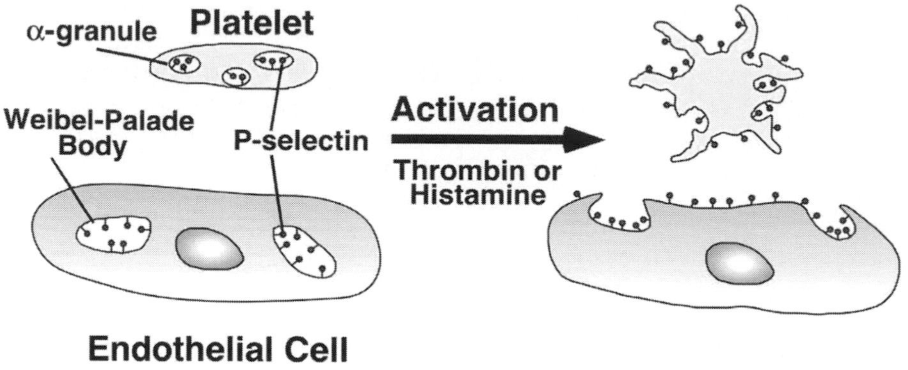

Figure 116.1. Inducible expression of P-selectin in platelets and ECs. P-selectin is constitutively synthesized by megakaryocytes and ECs, where it is sorted into the membranes of α-granules in platelets and Weibel-Palade bodies in ECs. Within seconds to minutes after cellular activation by mediators such as thrombin or histamine, P-selectin is redistributed to the plasma membrane.

derived from ancient structural modules that are also found in invertebrates. However, the specific organization of these domains in the selectin genes has been observed only in vertebrates. Expression of the selectins is restricted to hematopoietic cells and ECs, which suggests that the genes evolved to support specialized functions in the vascular system.

Expression of P-selectin is confined to megakaryocytes, from which platelets are derived, and to ECs. Recently, some macrophages, but not their parental monocytes, were also shown to express P-selectin in vivo (19). In both humans and mice, megakaryocytes and ECs synthesize P-selectin constitutively. Normally, only the ECs of postcapillary venules express P-selectin (5). Postcapillary venules, in which shear stresses are lowest, are the favored regions for leukocyte recruitment during inflammation. However, P-selectin also is expressed in the ECs of inflamed arterioles and of large arteries at sites of atheromas, usually at bifurcations having lower, oscillatory shear stresses (20). The mechanisms that confer tissue-specific expression of P-selectin are not well understood.

In mice and several other mammals, the inflammatory mediators tumor necrosis factor (TNF)-α, interleukin (IL)-1β, and lipopolysaccharide (LPS) markedly increase the transcription of the *P-selectin* gene (21). Transcripts for P-selectin peak within 4 to 6 hours, and decline to basal levels within 12 to 24 hours. The gene encoding E-selectin responds similarly, except that there is normally no basal synthesis of E-selectin in the absence of inflammatory stimuli. The mechanisms underlying inducible expression of E-selectin have been intensively studied in both humans and mice (22). Inflammatory signals mobilize the transcription factors nuclear factor (NF)-κB and activating transcription factor (ATF)-2, which migrate to the nucleus and bind to an enhanceosome in the 5′ flanking region of the *E-selectin* gene. These transcription factors cooperate with basal transcription machinery to drive gene expression.

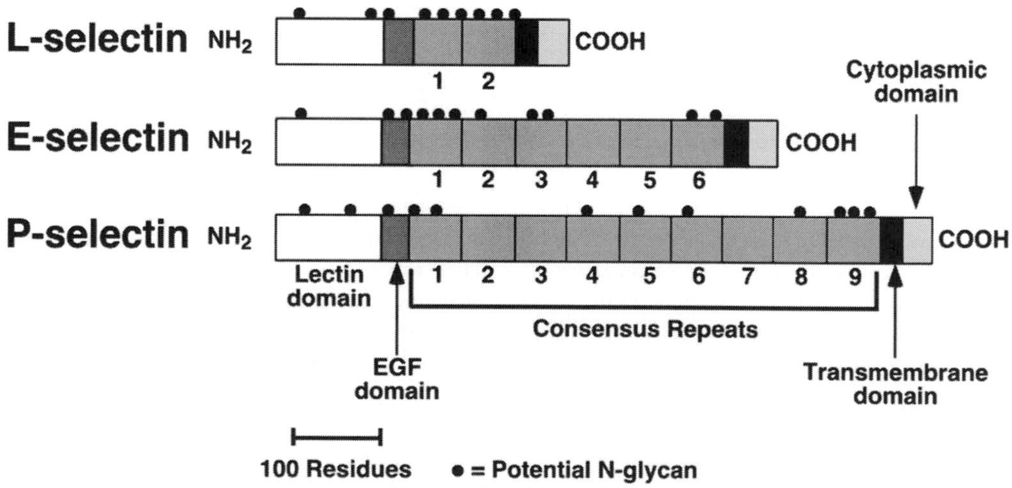

Figure 116.2. Domain organization of the selectins. Each of the selectins has an N-terminal C-type lectin domain, followed by an EGF-like domain, a variable number of consensus repeats like those in complement-regulatory proteins, a transmembrane domain, and a cytoplasmic domain.

Regulatory elements in the 5′-flanking region of the murine *P-selectin* gene are organized like those in the human and murine *E-selectin* genes (23), and site-directed mutagenesis and reporter gene experiments suggest that binding sites for NF-κB and ATF-2 in this region contribute to transcriptional upregulation in response to TNF-α and IL-1β (24). In sharp contrast, the 5′-flanking region of the human *P-selectin* gene lacks these regulatory elements. Furthermore, TNF-α, IL-1β, and LPS fail to upregulate transcription of the *P-selectin* gene in humans or baboons (25). These results suggest a divergence in the mechanisms for inducible expression of the *P-selectin* gene, but not of the *E-selectin* gene, in humans and other primates. The biological significance of this divergence is not understood. Interestingly, IL-4, oncostatin M, and complexes of IL-6 with the soluble IL-6 receptor-α subunit augment transcription of the human *P-selectin* gene in cultured human ECs (26). Whether these mediators contribute to the elevated expression of P-selectin overlying human atheromas is not known.

As mentioned, basal synthesis of P-selectin occurs in megakaryocytes/platelets and in ECs. Therefore, the most critical control step is the display of the protein at the cell surface. This is achieved initially by sorting of newly synthesized P-selectin, presumably at the *trans*-Golgi network, for delivery to regulated storage granules. Determinants in the cytoplasmic domain of P-selectin are required for sorting into Weibel-Palade bodies of ECs and into regulated storage granules of transfected endocrine cells that do not normally express P-selectin (27). These determinants presumably interact with cytosolic sorting proteins in coated transport vesicles. Unexpectedly, the cytoplasmic domain is not required to deliver P-selectin into the α-granules of platelets, which implicates an alternative sorting mechanism in these cells (28). A variety of physiological and pathological agonists can mobilize P-selectin from Weibel-Palade bodies to the cell surface. These include thrombin, histamine, complement components, hypoxia, and oxygen-derived radicals. After P-selectin moves to the surface of activated platelets and ECs, other mechanisms regulate the duration of its appearance. In vitro, P-selectin is stably expressed on the surface of activated platelets for many hours. In vivo, however, circulating activated platelets rapidly shed P-selectin from the surface by an uncharacterized proteolytic cleavage event (29). In some situations, P-selectin also may be cleaved from the surface of activated ECs. However, ECs and transfected heterologous cells also rapidly internalize P-selectin in clathrin-coated pits. Sorting determinants in the cytoplasmic tail are required for internalization (30). The protein may recycle from sorting endosomes to the cell surface or return to the *trans*-Golgi network for reincorporation into Weibel-Palade bodies. Other sorting determinants in the cytoplasmic domain direct the protein from late endosomes to lysosomes, where it is degraded (31). The net result of this intracellular trafficking is to shorten the time that P-selectin is displayed on the surface of activated ECs. This is likely an important mechanism for limiting the inflammatory response. Persistent surface expression of P-selectin probably requires a sufficiently large increase in synthesis of the protein to saturate the sorting pathway at the *trans*-Golgi network, bypass delivery to Weibel-Palade bodies, and transport newly synthesized protein directly to the cell surface. Interestingly, differential signaling by histamine and thrombin results in different rates of the internalization of P-selectin in clathrin-coated pits (32). More rapid endocytosis is associated with more efficient clustering of P-selectin in clathrin-coated pits, which enhances its adhesive function on the cell surface.

ROLE IN HOMEOSTASIS

Autocrine and Paracrine Effects

The primary function of P-selectin is to mediate the adhesion of leukocytes to activated platelets and ECs (33–35) (Figure 116.3). This can be considered both an autocrine and paracrine function in that the protein promotes adhesion of leukocytes (paracrine) to the cells that express P-selectin (autocrine). Leukocytes originally were shown to adhere to P-selectin under static conditions in vitro (13–15). During an inflammatory response in vivo, however, flowing leukocytes first tether to and roll on vascular ECs. Some of the rolling cells then decelerate until they arrest on the endothelium, and they eventually crawl between or through ECs into the underlying tissues (36). Importantly, all three selectins mediate the initial rolling adhesion of leukocytes on vascular surfaces (33–35,37). This was demonstrated in vitro by visualizing the adhesion of leukocytes to selectins in flow chambers using

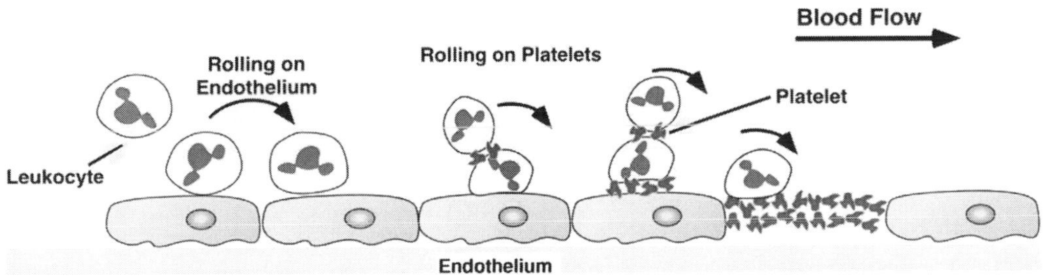

Figure 116.3. Multicellular adhesive interactions mediated by P-selectin. Leukocytes roll on P-selectin that is mobilized to the surfaces of activated ECs and activated platelets. Platelets also interact with ECs by a P-selectin–dependent mechanism that has not been fully defined.

Table 116-1: P-Selectin Knockout Mice

Genotype	*Phenotype*	*References*
P-selectin$^{-/-}$	Healthy, viable without challenge. Mild neutrophilia. Defective leukocyte rolling on P-selectin on activated platelets and ECs. Mild to moderate defects in leukocyte recruitment after inflammatory challenge.	38, 89, 90

videomicroscopy, and in vivo by observing the adhesion of leukocytes to the postcapillary venules of mice using intravital microscopy. Key tools have been monoclonal antibodies that block the adhesive functions of the selectins and gene-targeted mice that lack one or more selectins (Table 116-1). Although exceptions exist, selectin-mediated rolling usually is required before leukocytes can adhere more stably through integrins, another class of adhesion molecules. Because it can be rapidly translocated to the cell surface, P-selectin often mediates the earliest stages of leukocyte rolling on activated platelets and ECs at sites of injury (38).

Selectins enable adhesion by interactions of the membrane-distal C-type lectin domains with specific glycoconjugates on cell surfaces. Selectins primarily interact with fucose in and around a Ca^{2+}-coordination site in the lectin domain, and they secondarily interact with sialic acid (39,40). Leukocytes typically present these sugars on a determinant called sialyl Lewis X, or sLe^x (NeuAcα2–3Galβ1–4[Fucα1–3]GlcNAcβ1–), which is a terminal component of some oligosaccharides. However, P-selectin binds with very low affinity to sLe^x, and cells other than leukocytes that display sLe^x determinants adhere poorly to P-selectin under static or flow conditions (41). Indeed, P-selectin binds preferentially to a single glycoprotein on leukocytes called P-selectin glycoprotein ligand (PSGL)-1 (Figure 116.4) (34,42–44). PSGL-1 is an extended homodimeric mucin that contains many O-glycans attached to serine and threonine residues. Most of these O-glycans are not fucosylated (45), and P-selectin binds specifically to a single branched O-glycan capped with sLe^x near the membrane-distal end of PSGL-1. The site is preferred because other regions of the lectin domain of P-selectin simultaneously interact with peptide components and sulfated tyrosine residues on PSGL-1 that are near the O-glycan (40,46,47). This cooperative binding to glycan, peptide, and sulfate moieties on PSGL-1 explains why P-selectin has much higher affinity for this glycoprotein than for other proteins that have glycans capped with sLe^x. Monoclonal antibodies to the membrane-distal region of PSGL-1 block leukocyte rolling on P-selectin in vitro and in vivo (48,49), and gene-targeted mice lacking PSGL-1 exhibit little or no rolling on P-selectin in vivo (50,51). Therefore, PSGL-1 is the key physiological ligand for P-selectin on all subsets of leukocytes, including neutrophils, monocytes, lymphocytes, eosinophils, and basophils.

The flow conditions in the circulation impose significant kinetic and mechanical constraints on interactions between P-selectin and PSGL-1. To roll, cells must rapidly form and break macromolecular interactions, usually termed *bonds,* between adhesion molecules. Biochemical measurements confirm that the interaction of P-selectin with PSGL-1 involves very fast association and dissociation rates (52). Moreover, force is applied to adhesive bonds at the trailing edge of an adherent cell under flow. At the flow rates most commonly observed in postcapillary venules, the forces applied to P-selectin–PSGL-1 bonds cause them to dissociate more rapidly (53). Such bonds are called *slip bonds.* In principle, progressively higher flow rates should cause adhesive bonds to dissociate faster and faster until bond lifetimes become too short to support rolling. Yet leukocytes roll on P-selectin at nearly uniform velocities over a range of flow rates. This stabilization of rolling results from cellular alterations that dissipate force as flow rates increase (54). A major factor appears to be the dynamic extension and retraction of long membrane tethers at adhesive contacts (55,56). More tethers form at higher rates rather than at lower flow rates. At very low flow rates, a force regime paradoxically slows the dissociation of P-selectin from PSGL-1 (57). Such bonds are called *catch bonds.* Although the physiological significance of these P-selectin–PSGL-1 bonds is not clear, catch bonds may be very important for regulating L-selectin–dependent adhesion (58).

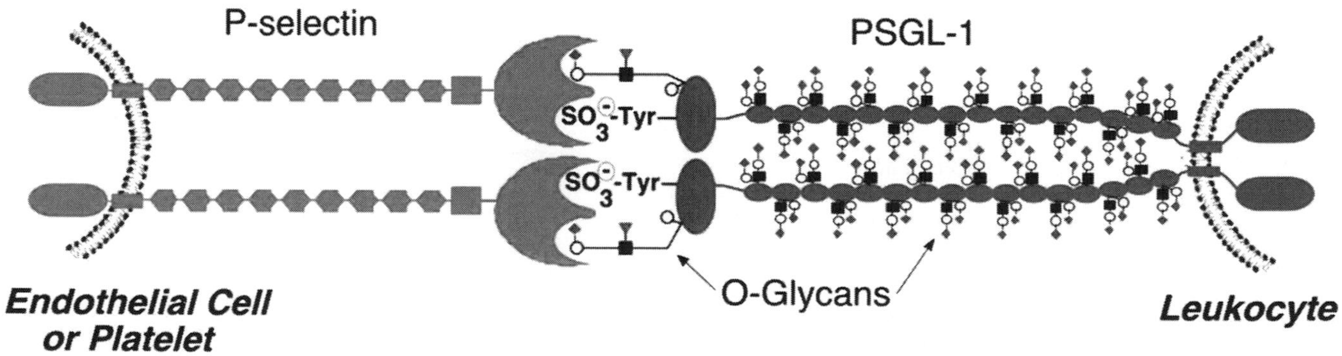

Figure 116.4. Binding of P-selectin to PSGL-1. Both P-selectin and PSGL-1 are extended homodimers that enable dimeric binding as illustrated. The lectin domain of P-selectin binds cooperatively to tyrosine sulfates, amino acids, and a specific O-glycan at the membrane-distal end of PSGL-1. Although PSGL-1 has many other O-glycans, these do not contribute significantly to the interaction with P-selectin.

Other factors control P-selectin–dependent rolling. Both P-selectin and PSGL-1 form homodimers in the membrane, which favors dimeric interactions (see Figure 116.4) (59). Dimeric interactions have longer lifetimes than monomeric interactions and distribute a given force over two bonds rather than one bond. P-selectin also clusters in the clathrin-coated pits of ECs (32,60), and PSGL-1 clusters in the tips of microvilli of leukocytes (48). These arrangements may favor bond clusters that increase the aggregate lifetime of adhesive contacts at the trailing edges of leukocytes. The extension of the binding sites of both PSGL-1 and P-selectin above the membrane increases the probability of contact (54,61). Higher molecular density on the cell surface also promotes slower and more regular rolling. In this regard, the density of P-selectin on activated platelets appears to be much higher than on activated ECs (1,7). This might allow more stable adhesion of leukocytes to platelets.

Under certain conditions, both leukocytes and platelets generate membrane fragments, or microparticles, which express PSGL-1 and P-selectin, respectively. Leukocyte microparticles can adhere to P-selectin on intact platelets, and platelet microparticles can adhere to PSGL-1 on intact leukocytes. Leukocyte microparticles also may express tissue factor, a key initiator of blood coagulation. The P-selectin–dependent accumulation of such microparticles on platelet thrombi augments thrombosis in some murine models of vascular injury (62,63). Some leukocyte microparticles can fuse with platelets (64). This suggests that leukocyte fragments, perhaps generated by the fission of long membrane tethers, might also adhere to and fuse with ECs.

Platelets have been reported to roll directly on the P-selectin of ECs (65). The platelet ligand for P-selectin has been suggested to be either PSGL-1 (66) or the glycoprotein (GP) Ib-IX complex (67), which is the major receptor for von Willebrand factor (vWF). However, it remains controversial whether platelets express PSGL-1, and the nature of the reported interaction between P-selectin and the GP Ib-IX complex requires further study. Leukocyte microparticles that adhere to ECs might also indirectly promote platelet rolling on the vessel surface. Activated ECs secrete ultralarge multimers of vWF, which may bind to P-selectin on the EC surface (68); the region of vWF that binds to P-selectin has not been identified. The ultralarge forms of vWF may enable platelets to roll on the endothelium through interactions with GP Ib-IX. In most circumstances, however, the metalloprotease ADAMTS-13 (a disintegrin and metalloprotease, with thrombospondin-1-like domains, member 13) rapidly cleaves vWF into smaller multimers, which dampens platelet adhesion to the endothelium.

In addition to its role as an adhesion receptor, P-selectin has a paracrine function as a signaling molecule when it binds to PSGL-1 on leukocytes. The cytoplasmic domain of PSGL-1 has no apparent intrinsic kinase motifs. However, engagement of PSGL-1 activates Src kinases and induces the tyrosine phosphorylation of several proteins (69–71). The enrichment of PSGL-1 in lipid rafts, where signaling molecules often cluster, may favor these responses (64). The cytoplasmic domain of PSGL-1 binds to ezrin and moesin, which link membrane proteins to the actin cytoskeleton and which can also bind to and activate the Syk kinase (72). Effector responses from PSGL-1 signaling include integrin activation (73) and both translational and transcriptional augmentation of chemokine synthesis (74,75). Cross-linking of PSGL-1 appears to be required for optimal signaling. In contrast to paracrine signaling of leukocytes through PSGL-1, autocrine signaling of ECs or platelets through P-selectin has received little attention.

Soluble forms of P-selectin circulate in the plasma of both humans and mice. In humans, some of this material represents an alternatively spliced, secreted isoform that lacks the transmembrane domain (18,76). This form has not been found in mice. Some plasma P-selectin is derived from proteolytic cleavage of the extracellular domain of P-selectin, yielding a form with a slightly lower molecular weight than the transmembrane form (28,29). Some of the material may represent P-selectin in microparticles derived from platelets and possibly from ECs. Plasma levels of P-selectin are modestly elevated in some human inflammatory and thrombotic diseases (77). Leukocytes shed microparticles expressing PSGL-1 that can also adhere to activated platelets and ECs expressing P-selectin. Circulating P-selectin has been postulated to increase the release of prothrombotic and proinflammatory microparticles from leukocytes, representing a potential paracrine function (78). However, plasma levels of P-selectin are low even in disease states. If plasma P-selectin is monomeric, its ability to signal is questionable because it would be unable to cross-link PSGL-1. Plasma concentrations are also well below the K_d for binding of monomeric P-selectin to leukocytes (52). Nevertheless, if some plasma P-selectin were oligomeric, its avidity for leukocytes and its signaling capability would increase. P-selectin in microparticles also might form multivalent interactions with PSGL-1 on leukocytes that favor signaling.

Interactions with Other Outputs

P-selectin interacts with many other molecules to promote cell adhesion and signaling. Often P-selectin cooperates with E-selectin and L-selectin to mediate leukocyte rolling. At many sites of inflammation, activated ECs express both P-selectin and E-selectin (33–35). E-selectin normally is expressed at higher densities, and it binds to more than one glycoprotein ligand on leukocytes. Interactions of P-selectin with PSGL-1 are particularly suited for mediating the initial tethering and rolling of leukocytes on ECs, whereas interactions of E-selectin with its ligands slow rolling velocities and prolong the transit times of leukocytes as they roll along venules (79). Interactions between L-selectin and PSGL-1 enable flowing leukocytes to roll on other leukocytes that have already adhered to ECs. These rolling cells may then "transfer" to the endothelium and continue to roll on P-selectin and E-selectin. Leukocytes also use L-selectin to roll on leukocyte-derived fragments expressing PSGL-1 that have been deposited on the endothelium (80). These additional interactions provide more opportunities for signaling through the engagement of E-selectin ligands or L-selectin on leukocytes. Slower rolling through cooperative selectin interactions augments encounters of leukocytes with chemokines on the apical surfaces of activated ECs. These chemokines, which generally are anchored by endothelial proteoglycans, bind to G-protein–coupled receptors on particular subsets of leukocytes. Cooperative signaling through selectin ligands and chemokine receptors enhances many effector responses in leukocytes (35,79). The most important early response is the activation of $\beta 2$- or $\alpha 4$ integrins, which further slow rolling and promote firm adhesion of leukocytes to the EC surface. The display of various combinations of selectins, chemokines, and integrin ligands on the EC surface has a major impact on which leukocyte subsets ultimately migrate into a particular inflamed tissue. Because P-selectin is expressed on both activated platelets and ECs, the possibilities for cooperation with other molecules further increase. In addition to their well-documented roles in hemostasis, platelets also secrete chemokines and other proinflammatory molecules. Cooperative signaling of P-selectin and platelet-derived chemokines might favor synthesis of tissue factor and chemokines in leukocytes (75,81).

DIAGNOSTIC AND THERAPEUTIC IMPLICATIONS

Monoclonal antibodies to P-selectin have proven very useful as probes to identify platelets that have been activated in vitro, or more specifically, to identify platelets that have fused α-granule membranes with the plasma membrane. Flow cytometry is the usual method of choice for such experiments. Initially there was great interest in using flow cytometry to identify circulating activated platelets in patients with thrombotic disorders. However, the discovery that circulating activated platelets rapidly shed P-selectin in vivo has dampened enthusiasm for such studies, because flow cytometry might only detect a subset of activated platelets that have not yet shed P-selectin from the cell surface. Flow cytometry also has identified circulating platelet–leukocyte conjugates that form through interactions of P-selectin with PSGL-1 (82). Such conjugates may be a more sensitive method for detecting platelet activation in vivo. Circulating ECs that express P-selectin have been identified in blood from patients with some inflammatory disorders. Collectively, these studies support the possibility that platelets and/or ECs are activated in thrombotic or inflammatory diseases. Whether assays that measure platelet and EC activation with antibodies to P-selectin can actually predict prognosis or response to therapy in specific diseases requires confirmation. Similarly, numerous groups have used enzyme-linked immunosorbent assays (ELISAs) to measure the levels of soluble P-selectin in plasma from patients with a variety of hematological, thrombotic, or inflammatory disorders (77). Although such levels are modestly elevated in some patients, the ability of an elevated plasma P-selectin value to predict prognosis or response to therapy has not been demonstrated with sufficient statistical power in prospective studies.

P-selectin contributes to inflammatory and thrombotic disease in a variety of experimental models in mice and other animals. The role of P-selectin has been documented with blocking monoclonal antibodies to P-selectin or with mice that lack P-selectin. Many of the early studies focused on models of ischemia–reperfusion injury, a condition that complicates treatment of hemorrhagic shock, myocardial infarction, and stroke in humans (83,84). The blockade of P-selectin function or the genetic deficiency of P-selectin markedly reduces tissue injury in these models. Such results suggested that drugs that block the function of P-selectin or other adhesion molecules might be effective therapeutics for these diseases. However, enthusiasm has declined because of the failures of antibodies to integrins or integrin ligands to improve outcome in randomized clinical trials of myocardial infarction or stroke. Small clinical trials of low-affinity selectin inhibitors also yielded negative results. These data suggest that excessive leukocyte adhesion might not be the dominating contributor to ischemia–reperfusion injury in patients, or that the appropriate subsets of patients who might respond to antiadhesion therapy have not been identified.

Despite these setbacks, anti–P-selectin therapy remains an attractive candidate to treat thrombotic and inflammatory diseases in humans. One potential disease target is sickle cell anemia, which is now widely acknowledged to be an inflammatory disease. In models of sickle cell anemia in mice, P-selectin contributes to the adhesion of leukocytes, platelets, and sickled erythrocytes to the vascular endothelium of the microcirculation, which exacerbate vaso-occlusion (85–87). A second example is deep vein thrombosis, which is an inflammatory as well as thrombotic process. P-selectin contributes to both venous inflammation and thrombosis in mouse and baboon models (88). A third example is delayed graft function in kidney and other organ transplants, which may be an example of controlled ischemia–reperfusion injury that might be more amenable to anti-selectin therapy. Again, P-selectin makes a major contribution to the pathogenesis of this disorder in animal models. Potential anti–P-selectin therapies include monoclonal antibodies to P-selectin, a recombinant soluble form of PSGL-1 fused to the Fc portion of immunoglobulin, glycosulfopeptides modeled after the P-selectin–binding domain of PSGL-1, and small molecules tailored to inhibit P-selectin function. The success of an anti–P-selectin drug candidate in a clinical trial in one of these diseases might encourage trials in other disorders.

KEY POINTS

- P-selectin is an adhesion receptor that mediates interactions among leukocytes, platelets, and ECs at sites of infection and tissue injury.

- Because P-selectin is stored in the membranes of secretory storage granules in platelets and ECs, it can be rapidly mobilized to the plasma membrane during hemostatic and inflammatory responses.
- The lectin domain of P-selectin binds cooperatively to glycan, peptide, and sulfated tyrosine moieties on its major leukocyte ligand, PSGL-1.

Future Goals

- To compare the biochemical and biophysical parameters of the interactions of each of the three selectins with their ligands on leukocytes, platelets, and ECs
- To further define the physiological and pathological functions of the selectins and their ligands in vivo

REFERENCES

1 McEver RP, Martin MN. A monoclonal antibody to a membrane glycoprotein binds only to activated platelets. *J Biol Chem.* 1984;259:9799–9804.

2 Hsu-Lin S-C, Berman CL, Furie BC, et al. A platelet membrane protein expressed during platelet activation and secretion. Studies using a monoclonal antibody specific for thrombin-activated platelets. *J Biol Chem.* 1984;259:9121–9126.

3 Stenberg PE, McEver RP, Shuman MA, et al. A platelet alpha-granule membrane protein (GMP-140) is expressed on the plasma membrane after activation. *J Cell Biol.* 1985;101:880–886.

4 Berman CL, Yeo EL, Wencel-Drake JD, et al. A platelet alpha granule membrane protein that is associated with the plasma membrane after activation. *J Clin Invest.* 1986;78:130–137.

5 McEver RP, Beckstead JH, Moore KL, et al. GMP-140, a platelet alpha-granule membrane protein, is also synthesized by vascular ECs and is localized in Weibel-Palade bodies. *J Clin Invest.* 1989; 84:92–99.

6 Bonfanti R, Furie BC, Furie B, Wagner DD. PADGEM (GMP 140) is a component of Weibel-Palade bodies of human endothelial cells. *Blood.* 1989;73:1109–1112.

7 Hattori R, Hamilton KK, Fugate RD, et al. Stimulated secretion of endothelial von Willebrand factor is accompanied by rapid redistribution to the cell surface of the intracellular granule membrane protein GMP-140. *J Biol Chem.* 1989;264:7768–7771.

8 Hattori R, Hamilton KK, McEver RP, Sims PJ. Complement proteins C5b-9 induce secretion of high molecular weight multimers of endothelial von Willebrand factor and translocation of granule membrane protein GMP-140 to the cell surface. *J Biol Chem.* 1989;264:9053–9060.

9 Johnston GI, Cook RG, McEver RP. Cloning of GMP-140, a granule membrane protein of platelets and endothelium: sequence similarity to proteins involved in cell adhesion and inflammation. *Cell.* 1989;56:1033–1044.

10 Lasky LA, Singer MS, Yednock TA, et al. Cloning of a lymphocyte homing receptor reveals a lectin domain. *Cell.* 1989;56:1045–1055.

11 Siegelman MH, van de Rijn M, Weissman IL. Mouse lymph node homing receptor cDNA clone encodes a glycoprotein revealing tandem interaction domains. *Science.* 1989;243:1165–1172.

12 Bevilacqua MP, Stengelin S, Gimbrone MA Jr., Seed B. Endothelial leukocyte adhesion molecule 1: an inducible receptor for neutrophils related to complement regulatory proteins and lectins. *Science.* 1989;243:1160–1165.

13 Larsen E, Celi A, Gilbert GE, et al. PADGEM protein: a receptor that mediates the interaction of activated platelets with neutrophils and monocytes. *Cell.* 1989;59:305–312.

14 Geng J-G, Bevilacqua MP, Moore KL, et al. Rapid neutrophil adhesion to activated endothelium mediated by GMP-140. *Nature.* 1990;343:757–760.

15 Hamburger SA, McEver RP. GMP-140 mediates adhesion of stimulated platelets to neutrophils. *Blood.* 1990;75:550–554.

16 Bevilacqua M, Butcher E, Furie B, et al. Selectins: a family of adhesion receptors. *Cell.* 1991;67:233.

17 Watson ML, Kingsmore SF, Johnston GI, et al. Genomic organization of the selectin family of leukocyte adhesion molecules on human and mouse chromosome 1. *J Exp Med.* 1990;172:263–272.

18 Johnston GI, Bliss GA, Newman PJ, McEver RP. Structure of the human gene encoding granule membrane protein-140, a member of the selectin family of adhesion receptors for leukocytes. *J Biol Chem.* 1990;265:21381–21385.

19 Tchernychev B, Furie B, Furie BC. Peritoneal macrophages express both P-selectin and PSGL-1. *J Cell Biol.* 2003;163:1145–1155.

20 Johnson-Tidey RR, McGregor JL, Taylor PR, Poston RN. Increase in the adhesion molecule P-selectin in endothelium overlying atherosclerotic plaques. Coexpression with intercellular adhesion molecule-1. *Am J Pathol.* 1994;144:952–961.

21 Gotsch U, Jager U, Dominis M, Vestweber D. Expression of P-selectin on endothelial cells is upregulated by LPS and TNF-α in vivo. *Cell Adhes Commun.* 1994;2:7–14.

22 Collins T, Read MA, Neish AS, et al. Transcriptional regulation of endothelial cell adhesion molecules: NF-κB and cytokine-inducible enhancers. *FASEB J.* 1995;9:899–909.

23 Pan J, Xia L, McEver RP. Comparison of promoters for the murine and human P-selectin genes suggests species-specific and conserved mechanisms for transcriptional regulation in endothelial cells. *J Biol Chem.* 1998;273:10058–10067.

24 Pan J, Xia L, Yao L, McEver RP. Tumor necrosis factor-α- or lipopolysaccharide-induced expression of the murine P-selectin gene in endothelial cells involves novel κB sites and a variant ATF/CRE element. *J Biol Chem.* 1998;273:10068–10077.

25 Yao L, Setiadi H, Xia L, et al. Divergent inducible expression of P-selectin and E-selectin in mice and primates. *Blood.* 1999; 94:3820–3828.

26 Yao L, Pan J, Setiadi H, et al. Interleukin 4 or oncostatin M induces a prolonged increase in P-selectin mRNA and protein in human ECs. *J Exp Med.* 1996;184:81–92.

27 Disdier M, Morrissey JH, Fugate RD, et al. Cytoplasmic domain of P-selectin (CD62) contains the signal for sorting into the regulated secretory pathway. *Mol Biol Cell.* 1992;3:309–321.

28 Hartwell DM, Mayadas TN, Berger G, et al. Role of P-selectin cytoplasmic domain in granular targeting in vivo and in early inflammatory responses. *J Cell Biol.* 1998;143:1129–1141.

29 Michelson AD, Barnard MR, Hechtman HB, et al. In vivo tracking of platelets: circulating degranulated platelets rapidly lose surface P-selectin but continue to circulate and function. *Proc Natl Acad Sci USA.* 1996;93:11877–11882.

30 Setiadi H, Disdier M, Green SA, et al. Residues throughout the cytoplasmic domain affect the internalization efficiency of P-selectin. *J Biol Chem.* 1995;270:26818–26826.
31 Green SA, Setiadi H, McEver RP, Kelly RB. The cytoplasmic domain of P-selectin contains a sorting determinant that mediates rapid degradation in lysosomes. *J Cell Biol.* 1994;124:435–448.
32 Setiadi H, McEver RP. Signal-dependent distribution of cell surface P-selectin in clathrin-coated pits affects leukocyte rolling under flow. *J Cell Biol.* 2003;163:1385–1395.
33 McEver RP, Moore KL, Cummings RD. Leukocyte trafficking mediated by selectin-carbohydrate interactions. *J Biol Chem.* 1995;270:11025–11028.
34 McEver RP. Selectins: lectins that initiate cell adhesion under flow. *Curr Opin Cell Biol.* 2002;14:581–586.
35 Ley K, Kansas GS. Selectins in T-cell recruitment to non-lymphoid tissues and sites of inflammation. *Nat Rev Immunol.* 2004;4:325–335.
36 Springer TA. Traffic signals on endothelium for lymphocyte recirculation and leukocyte emigration. *Annu Rev Physiol.* 1995; 57:827–872.
37 Lawrence MB, Springer TA. Leukocytes roll on a selectin at physiologic flow rates: distinction from and prerequisite for adhesion through integrins. *Cell.* 1991;65:859–873.
38 Mayadas TN, Johnson RC, Rayburn H, et al. Leukocyte rolling and extravasation are severely compromised in P selectin-deficient mice. *Cell.* 1993;74:541–554.
39 Graves BJ, Crowther RL, Chandran C, et al. Insight into E-selectin/ligand interaction from the crystal structure and mutagenesis of the lec/EGF domains. *Nature.* 1994;367:532–538.
40 Somers WS, Tang J, Shaw GD, Camphausen RT. Insights into the molecular basis of leukocyte tethering and rolling revealed by structures of P- and E-selectin bound to SLe(X) and PSGL-1. *Cell.* 2000;103:467–479.
41 Zhou Q, Moore KL, Smith DF, et al. The selectin GMP-140 binds to sialylated, fucosylated lactosaminoglycans on both myeloid and nonmyeloid cells. *J Cell Biol.* 1991;115:557–564.
42 Moore KL, Stults NL, Diaz S, et al. Identification of a specific glycoprotein ligand for P-selectin (CD62) on myeloid cells. *J Cell Biol.* 1992;118:445–456.
43 Sako D, Chang X-J, Barone KM, et al. Expression cloning of a functional glycoprotein ligand for P-selectin. *Cell.* 1993;75:1179–1186.
44 McEver RP, Cummings RD. Role of PSGL-1 binding to selectins in leukocyte recruitment. *J Clin Invest.* 1997;100:485–492.
45 Wilkins PP, McEver RP, Cummings RD. Structures of the O-glycans on P-selectin glycoprotein ligand-1 from HL-60 cells. *J Biol Chem.* 1996;271:18732–18742.
46 Leppänen A, Mehta P, Ouyang YB, et al. A novel glycosulfopeptide binds to P-selectin and inhibits leukocyte adhesion to P-selectin. *J Biol Chem.* 1999;274:24838–24848.
47 Leppänen A, White SP, Helin J, et al. Binding of glycosulfopeptides to P-selectin requires stereospecific contributions of individual tyrosine sulfate and sugar residues. *J Biol Chem.* 2000;275: 39569–39578.
48 Moore KL, Patel KD, Bruehl RE, et al. P-selectin glycoprotein ligand-1 mediates rolling of human neutrophils on P-selectin. *J Cell Biol.* 1995;128:661–671.
49 Norman KE, Moore KL, McEver RP, Ley K. Leukocyte rolling in vivo is mediated by P-selectin glycoprotein ligand-1. *Blood.* 1995; 86:4417–4421.
50 Yang J, Hirata T, Croce K, et al. Targeted gene disruption demonstrates that P-selectin glycoprotein ligand 1 (PSGL-1) is required for P-selectin-mediated but not E-selectin-mediated neutrophil rolling and migration. *J Exp Med.* 1999;190:1769–1782.
51 Xia L, Sperandio M, Yago T, et al. P-selectin glycoprotein ligand-1-deficient mice have impaired leukocyte tethering to E-selectin under flow. *J Clin Invest.* 2002;109:939–950.
52 Mehta P, Cummings RD, McEver RP. Affinity and kinetic analysis of P-selectin binding to P-selectin glycoprotein ligand-1. *J Biol Chem.* 1998;273:32506–32513.
53 Alon R, Hammer DA, Springer TA. Lifetime of the P-selectin: carbohydrate bond and its response to tensile force in hydrodynamic flow. *Nature.* 1995;374:539–542.
54 Yago T, Leppänen A, Qiu H, et al. Distinct molecular and cellular contributions to stabilizing selectin-mediated rolling under flow. *J Cell Biol.* 2002;158:787–799.
55 Ramachandran V, Williams M, Yago T, et al. Dynamic alterations of membrane tethers stabilize leukocyte rolling on P-selectin. *Proc Natl Acad Sci USA.* 2004;101:13519–13524.
56 Schmidtke DW, Diamond SL. Direct observation of membrane tethers formed during neutrophil attachment to platelets or P-selectin under physiological flow. *J Cell Biol.* 2000;149:719–729.
57 Marshall BT, Long M, Piper JW, et al. Direct observation of catch bonds involving cell-adhesion molecules. *Nature.* 2003;423:190–193.
58 Yago T, Wu J, Wey CD, et al. Catch bonds govern adhesion through L-selectin at threshold shear. *J Cell Biol.* 2004;166:913–923.
59 Ramachandran V, Yago T, Epperson TK, et al. Dimerization of a selectin and its ligand stabilizes cell rolling and enhances tether strength in shear flow. *Proc Natl Acad Sci USA.* 2001;98:10166–10171.
60 Setiadi H, Sedgewick G, Erlandsen SL, McEver RP. Interactions of the cytoplasmic domain of P-selectin with clathrin-coated pits enhance leukocyte adhesion under flow. *J Cell Biol.* 1998; 142:859–871.
61 Patel KD, Nollert MU, McEver RP. P-selectin must extend a sufficient length from the plasma membrane to mediate rolling of neutrophils. *J Cell Biol.* 1995;131:1893–1902.
62 Hrachovinova I, Cambien B, Hafezi-Moghadam A, et al. Interaction of P-selectin and PSGL-1 generates microparticles that correct hemostasis in a mouse model of hemophilia A. *Nat Med.* 2003;9:1020–1025.
63 Falati S, Liu Q, Gross P, et al. Accumulation of tissue factor into developing thrombi in vivo is dependent upon microparticle P-selectin glycoprotein ligand 1 and platelet P-selectin. *J Exp Med.* 2003;197:1585–1598.
64 Del Conde I, Shrimpton CN, Thiagarajan P, Lopez JA. Tissue-factor-bearing microvesicles arise from lipid rafts and fuse with activated platelets to initiate coagulation. *Blood.* 2005;106:1604–1611.
65 Frenette PS, Johnson RC, Hynes RO, Wagner DD. Platelets roll on stimulated endothelium in vivo: an interaction mediated by endothelial P-selectin. *Proc Natl Acad Sci USA.* 1995;92:7450–7454.
66 Frenette PS, Denis CV, Weiss L, et al. P-selectin glycoprotein ligand 1 (PSGL-1) is expressed on platelets and can mediate platelet-endothelial interactions in vivo. *J Exp Med.* 2000;191:1413–1422.
67 Romo GM, Dong JF, Schade AJ, et al. The glycoprotein Ib-IX-V complex is a platelet counterreceptor for P-selectin. *J Exp Med.* 1999;190:803–813.

68 Padilla A, Moake JL, Bernardo A, et al. P-selectin anchors newly released ultralarge von Willebrand factor multimers to the EC surface. *Blood*. 2004;103:2150–2156.

69 Hidari KI-PJ, Weyrich AS, Zimmerman GA, McEver RP. Engagement of P-selectin glycoprotein ligand-1 enhances tyrosine phosphorylation and activates mitogen-activated protein kinases in human neutrophils. *J Biol Chem*. 1997;272:28750–28756.

70 Evangelista V, Manarini S, Sideri R, et al. Platelet/polymorphonuclear leukocyte interaction: P-selectin triggers protein-tyrosine phosphorylation-dependent CD11b/CD18 adhesion: role of PSGL-1 as a signaling molecule. *Blood*. 1999;93:876–885.

71 Piccardoni P, Sideri R, Manarini S, et al. Platelet/polymorphonuclear leukocyte adhesion: a new role for SRC kinases in Mac-1 adhesive function triggered by P-selectin. *Blood*. 2001;98:108–116.

72 Urzainqui A, Serrador JM, Viedma F, et al. ITAM-based interaction of ERM proteins with Syk mediates signaling by the leukocyte adhesion receptor PSGL-1. *Immunity*. 2002;17:401–412.

73 Ma YQ, Plow EF, Geng JG. P-selectin binding to P-selectin glycoprotein ligand-1 induces an intermediate state of $\alpha M\beta 2$ activation and acts cooperatively with extracellular stimuli to support maximal adhesion of human neutrophils. *Blood*. 2004;104:2549–2556.

74 Weyrich AS, McIntyre TM, McEver RP, et al. Monocyte tethering by P-selectin regulates monocyte chemotactic protein-1 and tumor necrosis factor-α secretion. *J Clin Invest*. 1995;95:2297–2303.

75 Mahoney TS, Weyrich AS, Dixon DA, et al. Cell adhesion regulates gene expression at translational checkpoints in human myeloid leukocytes. *Proc Natl Acad Sci USA*. 2001;98:10284–10289.

76 Ishiwata N, Takio K, Katayama M, et al. Alternatively spliced isoform of P-selectin is present in vivo as a soluble molecule. *J Biol Chem*. 1994;269:23708–23715.

77 Blann AD, Nadar SK, Lip GY. The adhesion molecule P-selectin and cardiovascular disease. *Eur Heart J*. 2003;24:2166–2179.

78 André P, Hartwell D, Hrachovinová I, et al. Pro-coagulant state resulting from high levels of soluble P-selectin in blood. *Proc Natl Acad Sci USA*. 2000;97:13835–13840.

79 Ley K. Integration of inflammatory signals by rolling neutrophils. *Immunol Rev*. 2002;186:8–18.

80 Sperandio M, Smith ML, Forlow SB, et al. P-selectin glycoprotein ligand-1 mediates L-selectin-dependent leukocyte rolling in venules. *J Exp Med*. 2003;197:1355–1363.

81 Weyrich AS, Elstad MR, McEver RP, et al. Activated platelets signal chemokine synthesis by human monocytes. *J Clin Invest*. 1996;97:1525–1534.

82 Michelson AD, Barnard MR, Krueger LA, et al. Circulating monocyte-platelet aggregates are a more sensitive marker of in vivo platelet activation than platelet surface P-selectin: studies in baboons, human coronary intervention, and human acute myocardial infarction. *Circulation*. 2001;104:1533–1537.

83 Panés J, Perry M, Granger DN. Leukocyte EC adhesion: avenues for therapeutic intervention. *Br J Pharmacol*. 1999;126:537–550.

84 Thiagarajan RR, Winn RK, Harlan JM. The role of leukocyte and endothelial adhesion molecules in ischemia-reperfusion injury. *Thromb Haemost*. 1997;78:310–314.

85 Kaul DK, Hebbel RP. Hypoxia/reoxygenation causes inflammatory response in transgenic sickle mice but not in normal mice. *J Clin Invest*. 2000;106:411–420.

86 Wood KC, Hebbel RP, Granger DN. Endothelial cell P-selectin mediates a proinflammatory and prothrombogenic phenotype in cerebral venules of sickle cell transgenic mice. *Am J Physiol Heart Circ Physiol*. 2004;286:H1608–H1614.

87 Embury SH, Matsui NM, Ramanujam S, et al. The contribution of EC P-selectin to the microvascular flow of mouse sickle erythrocytes in vivo. *Blood*. 2004;104:3378–3385.

88 Myers D, Wrobleski S, Londy F, et al. New and effective treatment of experimentally induced venous thrombosis with anti-inflammatory rPSGL-Ig. *Thromb Haemost*. 2002;87:374–382.

89 Ley K, Bullard DC, Arbonés ML, et al. Sequential contribution of L- and P-selectin to leukocyte rolling in vivo. *J Exp Med*. 1995;181:669–675.

90 Frenette PS, Wagner DD. Insights into selectin function from knockout mice. *Thromb Haemost*. 1997;78:60–64.

117

Intercellular Adhesion Molecule-1 and Vascular Cell Adhesion Molecule-1

Silvia Muro

Institute for Environmental Medicine, University of Pennsylvania School of Medicine, Philadelphia

The vascular endothelium, positioned at the interface between the components of the blood and the interstitial milieu, plays a strategic role in the control of vascular physiology and maintenance of body homeostasis. It directs transendothelial gradients of molecules, regulates vascular tone, controls thrombosis, and is a crucial regulator of vascular redox state (1–4). Additionally, one of the most intensely studied and best characterized functions of endothelial cells (ECs) is their participation in innate and adaptive immunity, key steps in response to infection and tissue injury (5).

In a complex and elegantly coordinated process, ECs orchestrate the focal adhesion of white blood cells (WBCs) to the vessel wall and their migration into damaged tissue. This normally takes place at the level of postcapillary venules, and is governed in large part by the interaction of complementary adhesion molecules on the EC and leukocyte surfaces (Table 117-1). The generation of vasoactive compounds (e.g., prostacyclin, nitric oxide [NO]) by ECs causes vascular smooth muscle cell (VSMC) relaxation and local vasodilatation, increasing the blood flow at the site of injury and facilitating local accumulation of leukocytes (6). Flowing WBCs that collide with activated endothelium may bind to ECs through low-affinity receptors – the endothelial selectins (P- and E-selectin) – that recognize carbohydrate-based ligands on the leukocyte surface (Figure 117.1). The combined effects of blood flow and leukocyte binding to selectins result in their rolling on the EC surface. Chemotactic agents produced by ECs promote further rolling and facilitate firm adhesion via the induction of integrins on the surface of leukocytes. These integrins bind to high-affinity cell adhesion molecules (CAMs), most notably vascular cell adhesion molecule (VCAM)-1 and intercellular adhesion molecule (ICAM)-1 (7). Finally, adherent leukocytes migrate between (paracellular) or through (transcellular) ECs, a process that is mediated by a sophisticated dialogue between the two cell types, and which includes homophilic interactions between platelet-endothelial cell adhesion molecule (PECAM)-1/CD31 and CD99 (8).

Meticulous spatial and temporal regulation of VCAM-1 and ICAM-1 expression in response to specific mediators represents a complex and dynamic process permitting precise control of leukocyte anchoring to the endothelium. Additionally, VCAM-1 and ICAM-1 serve as signaling platforms that facilitate the transmigration of inflammatory cells and contribute to multiple physiological and pathological processes.

Some of the most relevant findings related to the characterization and biomedical significance of VCAM-1 and ICAM-1 are reviewed in this chapter.

HISTORICAL PERSPECTIVES

ICAM-1 was first identified in 1986, by Springer and colleagues, as a counter-receptor for the β_2 integrin, leukocyte function-associated antigen (LFA)-1 (9,10). They postulated that LFA-1 may mediate cell adhesion through a distinct molecule, based on the finding that homotypic adhesion of stimulated lymphocytes was abolished between LFA$^-$ cells but not between LFA$^-$ and LFA$^+$ cells (11). Using immunization with cells genetically deficient in LFA-1, Springer and colleagues produced monoclonal antibodies that inhibited homotypic aggregation of LFA$^+$ lymphoblastoid cells stimulated with phorbol ester, and defined ICAM-1 as a novel surface antigen involved in cell adhesion (10). ICAM-1 was soon found to be also expressed in fibroblasts, several types of epithelial cells, and ECs, where it was upregulated by cytokines and proinflammatory factors and was involved in WBC firm adhesion and transmigration (9,12). Analysis of the ICAM-1 coding sequence by two laboratories in parallel showed that this molecule did not contain RGD motifs typical of other integrin ligands but rather, it belonged to the immunoglobulin

Table 117-1: Cell Adhesion Molecules on ECs That Contribute to Inflammatory Leukocyte Infiltration

Name	*Family*	*EC Surface Expression*	*Ligands*	*Function*
P-selectin	Lectin family	Rapid and temporary induction by mediators	Fucosylated, sialylated, and sulfated ligands	Tethering rolling
E-selectin	Lectin family	Rapid and temporary induction by mediators	Fucosylated, sialylated, and sulfated ligands	Tethering rolling
VCAM-1	Ig-like family	Slow and stable induction by mediators	β_1 integrins (VLA-4)	Firm adhesion
ICAM-1	Ig-like family	Low constitutive expression. Stable upregulation by mediators	β_2 integrins (LFA-1 and Mac-1)	Firm adhesion; transmigration
PECAM-1	Ig-like family	High constitutive and stable expression	PECAM-1, β_1 and β_3 integrins	Transmigration

(Ig)-like CAM protein superfamily (13,14). Soon after this finding, two groups showed that ICAM-1 also was pathogenic, by serving as a receptor for major class rhinovirus (15,16).

While the role of ICAM-1 in leukocyte adhesion was becoming established in 1989, some evidence suggested the existence of a different inducible CAM in ECs that could mediate adhesion of both lymphocytes and nonlymphoid tumor cells to ECs in a manner that was independent of ICAM-1 and the endothelial selectins (17,18). This molecule was identified to be an immunochemically distinct, 110-kDa endothelial glycoprotein that was named inducible cell adhesion molecule

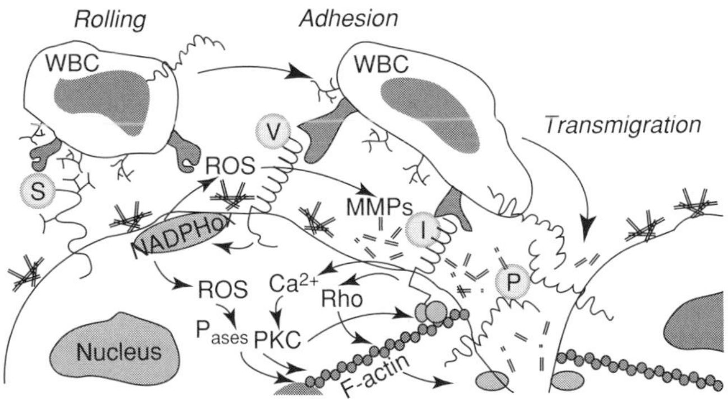

Figure 117.1. VCAM-1 and ICAM-1 in inflammation. Selectins (*S*) mediate WBC rolling on the endothelial surface, which subsequently bind to VCAM-1 (*V*) and ICAM-1 (*I*) extracellular domain via integrins, resulting in firm adhesion of leukocytes during the inflammatory process. Engagement by integrins triggers cell signal cascades through VCAM-1 and the ICAM-1 cytosolic tail, including VCAM-1–mediated activation of NADPH oxidase (NADPHox), generation of reactive oxygen species (ROS), phosphatase (Pases)-mediated cytoskeleton remodeling, and extracellular matrix degradation by MMPs. Ca^{2+}-mediated PKC signaling and Rho activation by ICAM-1 triggers formation of actin stress fibers, which may then interact with the cytosolic tail of this molecule via α-actinin and ezrin/radixin/moesin family proteins. The consequent EC retraction, together with the contribution of PECAM-1 (*P*), facilitates transmigration of leukocytes to the injured tissue.

(INCAM)-110) (17,18). *INCAM-110* cDNA was soon isolated from activated ECs, the molecule was renamed VCAM-1, and classified as an Ig-like CAM (19). VCAM-1 was observed to be additionally expressed by lymphoid cells, macrophages, and reactive mesothelial cells (20), where it served as a counter-receptor for the leukocyte β_1 integrin, very long antigen 4 (VLA)-4 (18).

EXPRESSION PATTERN, MOLECULAR STRUCTURE, AND FUNCTIONAL DOMAINS OF VASCULAR CELL ADHESION MOLECULE-1 AND INTERCELLULAR ADHESION MOLECULE-1

The Ig superfamily[1] cell adhesion molecules, VCAM-1 (CD106) and ICAM-1[2] (CD54), are type I transmembrane glycoproteins (9,10,17) (Table 117-2). They are synthesized

1 The immunoglobulin (Ig)-like superfamily of proteins is characterized by the presence of at least one 90- to 100-amino acid Ig domain consisting of two antiparallel β-sheets stabilized by a disulfide bond at its center. This structure is sheared by immunoglobulins, T-cell receptors, major histocompatibility (MHC) molecules, several lymphocyte-associated function antigens, and CAMs, including among others ICAMs, VCAM-1, and PECAM-1 (21).

2 The ICAM family comprises five type-I transmembrane glycoproteins (ICAM-1 to -5) that contain two to nine immunoglobulin-like C2-type domains. ICAM genes are clustered in chromosome region 19p13.2, except for the *ICAM-2* gene, which is located on chromosome 17q23–25 (22). ICAMs present distinct expression pattern and functions. ICAM-1 is expressed at low levels primarily on ECs but also on leukocytes, fibroblasts, and other cell types; is upregulated by cytokines; and is involved in pathological processes (e.g., adhesion and signaling during leukocyte extravasation). ICAM-2 is constitutively expressed by quiescent leukocytes, ECs, and platelets, where it is believed to contribute to physiological leukocyte transmigration and recirculation by serving as a counter-receptor for WBC LFA-1 and Mac-1 (21,23). Unlike ICAM-1, ICAM-2 expression is downregulated by inflammatory factors, hence its role in inflammation is unclear (21,23). It has been shown recently that, conversely to the luminal distribution of ICAM-1 on the EC surface, ICAM-2 is concentrated at the cell–cell borders, where it is implicated in homophilic interactions and likely contributes to vascular homeostasis and angiogenesis by regulating Rac-dependent cytoskeletal reorganizations (24). ICAM-3 (25) is constitutively and highly expressed by

as large precursor proteins, and subsequently processed to their mature forms at the rough endoplasmic reticulum and the Golgi apparatus, by proteolytic cleavage of their amino-terminal signal peptide and glycosylation at multiple sites in the region that will constitute the extracellular domain (28–30). In contrast to P-selectin, which is stored intracellularly in pre-existing Weibel-Palade bodies and is exocytosed upon endothelial activation (31), VCAM-1 and ICAM-1 rapidly traffic to the plasma membrane after their synthesis and are distributed primarily on the luminal cell surface (29,32).

VCAM-1 and ICAM-1 have been found in a variety of cell types, including both vascular and nonvascular cells. For example, in human atheroma-plaque microvessels, macrophages and VSMCs express VCAM-1 (33). VCAM-1 also is expressed by myoblasts, in which it may play a role in myogenesis. It is expressed in certain populations of dendritic cells in the lymph node and skin, bone marrow stromal cells, and synovial cells in inflamed joints (5,20,34). ICAM-1 is present in leukocytes, fibroblasts, certain epithelial cells, myoblasts, and glial and Schwann cells (35–38).

Of these various cell types, activated ECs display the highest surface density of VCAM-1 and ICAM-1. Based on studies using cultured human ECs and Scatchard analysis using specific antibodies, it is estimated that activated ECs may present approximately 10^5 to 10^6 copies of ICAM-1 (about two orders of magnitude above basal levels) and approximately 10^3 to 10^4 copies of VCAM-1 (almost undetectable in quiescent cells) at the EC luminal surface (39–41). However, ICAM-1 and VCAM-1 expression (both basal and inducible) varies among different organs and blood vessel types. For example, in skin and mesentery, VCAM-1 is most readily detected in the ECs of capillaries, postcapillary venules, and arterioles at sites of inflammation (42,43). In intact and injured rat lungs, ICAM-1 expression is discernible in capillaries and venules, but not in arterioles (44,45). Both in EC cultures and in postmortem specimens obtained from patients with acute respiratory distress syndrome, ICAM-1 is expressed at much higher levels in lung venules compared with pulmonary vein macrovessels (46,47). It is unclear to what extent these sites-specific differences in expression are intrinsic to the EC (e.g., epigenetically fixed), or are governed by the extracellular environment.

Structurally, VCAM-1 and ICAM-1 are characterized by a long glycosylated extracellular region, a single transmembrane-spanning domain, and a short cytosolic tail (Figure 117.2). The predominant form of human VCAM-1 (110 kDa) (17,18) consists of an extracellular region of 674 amino acids distributed in seven Ig-like C2-type modules, a 22-residue transmembrane region, and a 19-amino acid cytosolic domain (SwissProt P19320) (48,49). Human ICAM-1 (also a 110-kDa glycoprotein) contains a 453-residue extracellular region than comprises five Ig-like C2-type domains, a single transmembrane-spanning region of 23 amino acids, and a short cytosolic tail of 29 residues (SwissProt P05362) (14,50, 51). These distinct structural regions regulate functions attributed to VCAM-1 and ICAM-1. The Ig-like domains in the extracellular region mediate interaction with their leukocyte counter-receptors and other ligands (fibrinogen or rhinovirus, which binds to ICAM-1 Ig-domains 1 and 3, respectively), whereas the cytosolic domain mediates signal transduction events triggered by the engagement of the extracellular domains of these molecules (see Figure 117.2).

Despite similarities in the structural domains between VCAM-1 and ICAM-1, crucial conformational differences exist that dictate ligand specificity and that may influence the composition of the cellular infiltrate during inflammatory processes. In the case of VCAM-1, Ig-like domains 1 and 4 of this molecule act as ligand-binding modules for leukocyte β_1 integrins (e.g., VLA-4) (52,53). VCAM-1 binds to monocytes and lymphocytes, but not neutrophils. In contrast, domains 1 and 3 of ICAM-1 interact with β_2 integrins on all WBC types, including LFA-1 ($\alpha L\beta 2$), and the neutrophil determinant Mac-1 ($\alpha M\beta 2$) (7,32). Adhesion through Mac-1 is enhanced by binding of ICAM-1 domain 1 to fibrinogen, which acts as a bridge molecule between the EC and the leukocyte (50). Binding to fibrinogen may provide a pathway for the modulation of intercellular adhesion during inflammation and vascular injury.

Differences in ligand recognition may be explained in part by variations in the conformation of VCAM-1 and ICAM-1 Ig-like domains. Each molecule possesses an acidic residue near the c-terminus of domain 1, which has been implicated in leukocyte binding (54). However, the identity of this residue differs between receptors; ICAM-1 contains glutamic acid, which resides on a flat surface, whereas VCAM-1 contains aspartic acid, which protrudes as a loop. These structures complement the molecular surfaces of the integrins to which these CAMs bind (54). Thus, subtle conformational differences between VCAM-1 and ICAM-1 may help provide finely tuned ligand specificity.

VASCULAR CELL ADHESION MOLECULE-1 AND INTERCELLULAR ADHESION MOLECULE-1 AS SIGNAL TRANSDUCERS

VCAM-1 and ICAM-1 not only serve as adhesive surfaces for WBCs, but also function as molecular switches that activate intracellular signaling pathways that contribute to the regulation of multiple (patho)physiological processes (see Figure 117.1). Such outside-in signaling is triggered when the extracellular domain of VCAM-1 or ICAM-1 binds to adhesive leukocytes. An extensively used model to study VCAM-1– or ICAM-1–mediated signaling mimicking such

---

leukocytes, in which it mediates costimulatory signals for T-cell activation during the initiation of the immune response, but it is absent from normal ECs. ICAM-4 (26) is specifically expressed on erythrocytes and is responsible for the Landsteiner-Wiener blood group system. ICAM-5 or telencephalin (27) is expressed by subsets of neurons, but not by glial cells, exclusively within the telencephalon of mammalian brain. All ICAMs bind to the leukocyte adhesion protein, LFA-1. Additionally, ICAM-1, -2, and -4 bind to the neutrophil integrin Mac-1, and ICAM-3 binds the macrophage integrin CR4 (22).

Table 117-2: VCAM-1 and ICAM: Expression, Structure, Function, Regulation, and Animal Models

	VCAM-1	*ICAM-1*
Discovered	1989	1986
CD	CD106	CD54
Cell types	ECs, WBCs, mesothelial cells, myoblasts, dendritic cells, synovial cells	ECs, WBCs, fibroblasts, mesothelial cells, epithelial cells, myoblasts, glial and Schwann cells
Expression	Inducible	Constitutive with inducible up-regulation
Surface density in activated ECs	$\sim 10^3$–10^4	$\sim 10^5$–10^6
In vivo EC expression during inflammation	Capillaries, post-capillary venules and arterioles	Capillaries and venules
Protein	Extracellular domain: 674 aa; Transmembrane domain: 22; Cytosolic domain: 19 aa	Extracellular domain: 453 aa; Transmembrane domain: 23; Cytosolic domain: 29 aa
Ig-like domains	7	5
Molecular ligands	The WBC β_1 integrin VLA-4	Fibrinogen, rhinovirus, the WBC β_2 integrins LFA-1 and Mac-1
WBC ligands	Monocytes and lymphocytes	Monocytes, lymphocytes, and neutrophils
Signaling pathways	p42/44MAPK,ERK2,PI3K, Rho	pp60Src, p53/p56Lyn, Raf-1,ERK1,ERK2, JNK,AP1, Rho
Proliferative responses	Increased by activation of cyclin D_1	Decreased by inhibition of cdc2 cyclin-dependent kinase
ROS production	NADPH oxidase activation	Xanthine oxidase activation
Cytoskeletal remodeling	Ca^{2+} flux, Rho	Ca^{2+} flux, Rho, p60Src, PKC, PLC, HSP27, p38 MAPK
Promoter regulatory sites	Two NF-κB sites, two GATA motifs, one AP-1 site	One NF-κB site, two SP1 sequences, two AP-1 motifs
Differences in agonists	Stimulation by IL-4 No response to TGF-β	Stimulation by TGF-β No response to IL-4
Effects of shear stress	Stimulation by low shear stress Downregulation by high flow	Stimulation by low and high shear stress
Post-transcriptional modifications	Ig domains 1–2–3 GPI-anchored Ig domain 4 deletion	Deletion of Ig-like domains 2, 4, 2–3, 2–3–4.
Regulatory post-translational modifications	N/D	Binding to lectins by highly glycosylated isoforms Enhanced Mac-1 binding by defective glycosylation of Ig domain 3
Oligomerization	N/D	Dimerization through Ig domains 1–1, 4–4, and multimerization through Ig domains 4–4 between 1–1 dimers.
Internalization by ECs	Monovalent ligands via clathrin coated pits. Unknown for multivalent ligands	No uptake of monovalent ligands. Multivalent ligands via CAM-mediated endocytosis
Animal models	VCAM-1–deficient mice (*Vcam1*^{tm1Roml}; [120]) are embryonic lethal due to severe placental development and heart abnormalities.	Mice with ICAM-1 targeted mutation (*Icam1*^{tm1Bay}) (115) express ICAM-1 isoforms lacking the extracellular domain. They are viable and fertile, show mild neutrophilia, lymphocytosis, impaired lymphocyte recruitment to lungs, and suppressed contact hypersensitivity.
	Mice bearing a VCAM-1 hypomorphic mutation that results in the deletion of Ig domain 4 [121]) show impaired recruitment of lymphocytes to the lungs.	*Icam1*^{tm1Jcgr} (119) do not express ICAM-1. They are viable and fertile, and show a similar phenotype to *Icam1*^{tm1Bay} mice.

aa, amino acids; N/D, not determined.

processes consists of clustering these adhesion molecules by artificial ligands, such as cross-linking antibodies or antibody-coated beads.

Ligation of VCAM-1 or ICAM-1 activates overlapping but distinct signaling pathways (see Table 117-2). For example, binding of endothelial VCAM-1 to adherent leukocytes or cross-linking antibodies activates nicotinamide adenine dinucleotide phosphate (NADPH) oxidase in ECs, whereas WBC- and antibody-engaged ICAM-1 results in the activation of xanthine oxidase (NADPH oxidase activation has been suggested

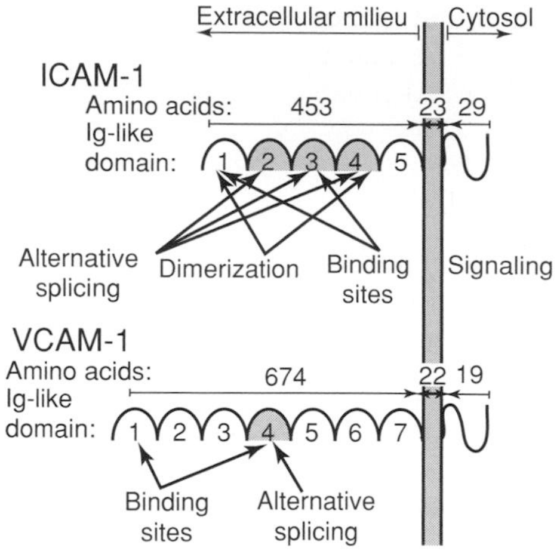

Figure 117.2. Structure of VCAM-1 and ICAM-1. Molecular architecture of VCAM-1 and ICAM-1, consisting of a large extracellular region, a single membrane-spanning domain, and a short cytosolic tail. The extracellular region contains C2-type, Ig-like domains, which are numbered starting by the domain located in the amino terminus and most distant from the plasma membrane. Distinct isoforms lacking one or several Ig-like domains can be generated from alternative splicing of the primary transcripts for these molecules. Extracellular Ig modules and intracellular domains represent docking sites for leukocyte binding and signal transduction, respectively.

but remains to be determined) and increased phosphorylation of the Src family kinases and p53/p56Lyn (55–57). Different ligands elicit distinct responses. For example, in contrast to the effects of WBCs, fibrinogen-mediated ligation of ICAM-1 activates pp60Src, but not p53/p56Lyn (58). These observations suggest that each adhesion molecule is coupled to multiple downstream signaling pathways, and that the relative activity of these pathways is determined by the nature of the ligand.

Ligand binding to the extracellular domains of VCAM-1 and ICAM-1 likely triggers signaling cascades through the cytoplasmic domain of these molecules (59), yet these pathways remain to be determined. For example, the cytosolic domain of ICAM-1 lacks a consensus tyrosine-based inhibition sequence required for recognition of molecules containing Src-homology domain 2, yet recent studies suggest that ICAM-1–mediated signaling may involve tyrosine phosphorylation of SHP-2, which may involve additional adapter proteins, such as Grb2, SOS, and Shc (59,60).

VCAM-1– and ICAM-1–mediated signaling may lead to downstream changes in gene expression (see Table 117-2). For example, binding of anti–VCAM-1 antibodies to VCAM-1 in airway smooth muscle cells leads to phosphoinositide 3-kinase (PI3K)-independent activation of extracellular-signal-regulated protein kinase (ERK)-2 and PI3K-mediated activation of p42/44 MAPK (61). In tumor necrosis factor (TNF)-α–stimulated ECs cultured under flow conditions, VCAM-1 cross-linking by multivalent anti–VCAM-1 beads induces phosphorylation of ERK2, but not ERK1, a response observed upon eosinophil binding to these cells (62). Ligation of ICAM-1 with cross-linking antibodies, mimicking leukocyte binding, also induces the activation of Raf-1, mitogen-activate protein kinases (MAPKs), ERK1 and ERK2, and/or c-jun N-terminal kinase (JNK)-dependent/AP1-mediated gene expression (59). These cascades vary according to the cell type – in brain ECs, ICAM-1 signals stimulate JNK downstream $p130^{Cas}$ tyrosine phosphorylation and recruitment of the adaptor Crk, but in astrocytes these signals operate via accumulation of cyclic adenosine monophosphate (cAMP) and activation of MAPK, ERK, and p38 MAPK (63).

Hence, the modulation of gene expression triggered by ligation of VCAM-1 or ICAM-1 may alter EC phenotype. As discussed later, outside-in signaling has been implicated in cell survival, cell growth, reactive oxygen species (ROS) generation, cytokine production, expression of certain surface molecules, and cytoskeleton rearrangements.

For example, in VSMCs VCAM-1 engagement increased expression of cyclin D1 and augmented the proliferative response to submitogenic concentrations of epidermal growth factor, suggesting that VCAM-1 may function to augment growth factor–induced responses (62). In the case of ICAM-1, fibrinogen was shown to induce apoptosis of TNF-α–activated ECs expressing the cytosolic truncated form of ICAM-1, but not wild-type ICAM-1, suggesting a protective prosurvival role for the cytosolic tail (60). In another report, ICAM-1 cross-linking by antibodies was shown to induce phosphorylation and inhibition of the cell cycle regulator, cdc2 cyclin-dependent protein kinase. This inhibition arrests or delays cell growth, because cdc2 is dephosphorylated at the G2/M interface (59).

VCAM-1 and ICAM-1 also have been demonstrated to play a role in mediating the redox state of the cell. As mentioned previously, NADPH oxidase activation and ROS generation by WBC or antibody engagement of VCAM-1 leads to changes in the redox state (56). This may in turn result in cytokine generation, upregulation of cell adhesion molecules, activation of matrix metalloproteinases (MMPs), and ultimately increased leukocyte passage. WBC- and antibody-mediated ICAM-1 cross-linking in ECs results in increased ROS production via xanthine oxidase pathway expression of the cytokines interleukin (IL)-1, IL-8, and regulated on activation, normal T-cell expressed and secreted (RANTES); the GTPase signaling molecule RhoA; surface molecules such as MHC class II (in B cells); and adhesion molecules such as VCAM-1 and ICAM-1 itself (57,59,64).

Interestingly, a number of transduction pathways mediated by VCAM-1 and ICAM-1 intersect with signaling cascades involved in cytoskeleton remodeling (see Table 117-2). For example, VCAM-1 engagement by WBCs modulates cytosolic phosphatases, stimulates calmodulin-independent calcium (Ca^{2+}) influx in ECs, and results in the regulation of several actin-binding proteins (65,66). Cross-linking of VCAM-1 on activated ECs by 10-μm anti–VCAM-1 coated beads generates a ROS-dependent remodeling of the actin cytoskeleton, in which actin coalesces immediately under the sites where beads bind to VCAM-1 (67). Upon WBC binding to activated ECs, actin reorganization mediated by VCAM-1 seems to occur independently of tyrosine kinase and PI3K signaling, but requires small GTPases (e.g., Rho) (56,65). Similarly, ICAM-1 ligation also results in cytoskeletal changes. These have been observed to be mediated by activation of p38 MAPK and heat-shock protein 27 (an actin binding protein); phosphorylation of phospholipase C (PLC)-γ1; production of inositol phosphate and Ca^{2+} release, which is responsible for protein kinase C (PKC) and p60Src activation; and tyrosine phosphorylation of cortactin (64,68,69). PKC activation by ICAM-1 induces Rho-dependent formation of actin stress fibers and activates cortactin, focal adhesion kinase (FAK), paxillin, and $p130^{Cas}$ (70,71). Thus, cytoskeletal reorganization by VCAM-1 and ICAM-1 may represent an important regulatory element of vascular permeability during pathological events.

These observations are consistent with a model of VCAM-1– and ICAM-1–mediated outside-in signaling that involves the coordinated activity of multiple binding and signaling functions (see Figure 117.1). EC changes in the redox state (e.g., due to NADPH oxidase activation upon WBC binding to VCAM-1) (56) may have a dual effect: (a) reaction of ROS on the surface of ECs and leukocytes with unpaired cysteine thiol groups that maintain inactive MMP-2 and MMP-9 propeptides may result in activation of these MMPs; and (b) diffusion of certain ROS species (e.g., hydrogen peroxide [H_2O_2]) to the cytosol and secondary activation of phosphatases that regulate actin reorganization (56). A combination of both extracellular matrix degradation and actin cytoskeleton rearrangements facilitates EC retraction at the site of leukocyte binding and, therefore, migration into the tissue. Similarly, in the case of ICAM-1, Src and PKC activation induces the Rho-dependent formation of actin stress fibers and actin-binding proteins such as cortactin, FAK, paxillin, and $p130^{Cas}$ (59). Spatial regulation of this process at sites of WBC binding may be explained by the interaction of the cytosolic domain of ligated ICAM-1 molecules with cytoskeleton-associated proteins, as in the case of actin-associated elements α-actinin, ezrin, and moesin, or the microtubule-related protein β-tubulin (72–74). These combined actions regulate vascular permeability and facilitate leukocyte transmigration.

REGULATION OF VASCULAR CELL ADHESION MOLECULE-1 AND INTERCELLULAR ADHESION MOLECULE-1 EXPRESSION AND FUNCTION

In addition to the minor structural differences between VCAM-1 and ICAM-1, other factors contribute to the specific roles for these proteins in mediating EC signaling and WBC adhesion. These include differences in the distribution of VCAM-1 and ICAM-1 across the vascular tree, as well as differences in their transcriptional regulation, and post-transcriptional and post-translational modifications.

The various cell adhesion molecules exhibit distinct patterns of gene and protein expression. VCAM-1 and endothelial selectins (P- and E-selectins) are expressed at low or undetectable levels on the surface of quiescent (nonactivated) ECs, whereas ICAM-1 is expressed at modest levels, and PECAM-1 (a related Ig-like cell adhesion molecule) at high levels (5,21). VCAM-1, P-selectin, E-selectin, and ICAM-1 are induced by agonists, both in vivo and in vitro, commonly associated with inflammation and EC activation. These include inflammatory cytokines, thrombin, lipopolysaccharide (LPS), abnormal shear stress, hypoxia, and ROS (75–77). In contrast, PECAM-1 expression in vivo and in cultured ECs is stable and does not change under pathologic conditions (75,78).

In nonvascular cells, VCAM-1 and ICAM-1 expression also is upregulated by extracellular mediators, which may contribute to adhesive events in certain settings (79). For example, ICAM-1 is present at low levels not only in ECs but also in astrocytes in normal adult brain. Its expression is upregulated during inflammatory conditions such as multiple sclerosis, or after certain pathological insults, including stroke, traumatic brain injury, and Alzheimer disease (63,80,81). Genetic and metabolic conditions, such as muscular dystrophies, diabetes, familial hypercholesterolemia, or lysosomal storage disorders (among others) lead to the pathological overexpression of adhesion molecules (including VCAM-1 and ICAM-1 and their soluble forms) in ECs, WBCs, and nonvascular cells (epithelial cells, skeletal muscle, synovial cells). This prompts inflammation, a secondary hallmark of these diseases (51,82–86).

The induction of VCAM-1 and ICAM-1 expression in such settings, as with other adhesion molecules (e.g., endothelial selectins), depends on the nature and intensity of the stimulus and the EC subtype (21,43,67,87,88). For example, phorbol ester was found to stimulate expression of VCAM-1 and ICAM-1 in human umbilical vein ECs (HUVECs), but not in dermal microvascular ECs (89,90). In vascular endothelium, interferon (INF)-γ induces the expression of ICAM-1 but not E-selectin; IL-4 stimulates VCAM-1 but not E-selectin or ICAM-1; and transforming growth factor (TGF)-β1 itself induces ICAM-1 expression but does not affect expression of VCAM-1 (21,67,87). Although certain extracellular agonists, such as IL-1 or TNF-α, induce surface expression of multiple adhesion molecules (P- and E-selectin, VCAM-1 and ICAM-1), the temporal pattern varies according to the coordinated stepwise function of these adhesion molecules (see Table 117-1). For example, in cultured ECs, P- and E-selectin (which are involved in early WBC rolling) are rapidly induced by IL-1 and TNF-α, with maximal protein expression occurring between 4 and 6 hours, followed by a rapid decline to basal levels by 24 hours. By contrast, surface VCAM-1 and ICAM-1 (which are involved in subsequent firm adhesion) increase more slowly in response to these mediators, with expression peaking at 6 to 12 hours and, in some cases, remaining elevated for several days (12).

In addition to chemical mediators, biomechanical forces also serve to regulate expression VCAM-1 and ICAM-1 (91,92). Under in vitro conditions, acute shear stress causes rapid cytoskeletal reorganization and activates the acute release of NO, activation of nuclear factor (NF)-κB, c-fos, c-jun, and SP1, and the transcriptional activation of ICAM-1, tissue factor, platelet-derived growth factor (PDGF)-B, TGF-β1, and endothelial NO synthase (eNOS) (93). Interestingly, low shear stress has been shown to increase ICAM-1 and VCAM-1 expression in cultured ECs (but not E-selectin), and increasing shear stress results in reduced VCAM-1 expression (see Table 117-2) (2,94,95).

Environmentally mediated upregulation of VCAM-1 and ICAM-1 occurs through the inducible binding of transcription factors to *cis*-regulatory elements present in the upstream promoter region of these genes (48,96,97) (see Table 117-2). The 5′ flanking region of the *VCAM-1* gene contains two NF-κB sites, as well as two functional GATA motifs and an AP-1 binding site (48,96). The *ICAM-1* promoter contains one NF-κB, two SP1, and two AP-1 regulatory motifs (97). Inducible expression of VCAM-1 and ICAM-1 is mediated either by an individual element (e.g., NF-κB), or through the coordinated action of more than one regulatory element (e.g., NF-κB plus GATA). Hence, any given agonist may modulate the expression of VCAM-1 and ICAM-1 through different mechanisms (48,63,96,98). For example, thrombin induces ICAM-1 via a PI3K-independent, PKCδ/NF-κB–dependent pathway, whereas thrombin stimulation of VCAM-1 occurs via an Akt-independent, PI3K/PKCζ/GATA-2/PKCδ/NF-κB–dependent pathway. Any given gene may respond to different agonists through unique transcriptional mechanisms. For example, thrombin-mediated induction of ICAM-1 involves a PKCδ/NF-κB–dependent pathway, whereas TNF-α activates ICAM-1 via PKCζ/NF-κB (63,98).

Multiple isoforms exist for VCAM-1 and ICAM-1 (see Figure 117.2). For example, in mice VCAM-1 exists either as a seven-domain transmembrane protein or as an alternatively spliced protein that contains the three amino-terminal Ig-domains linked to glycophosphatidylinositol (GPI) (99). In humans, two distinct isoforms of VCAM-1 have been reported: a predominant form that contains seven Ig-like domains and a molecule of six Ig modules, due to elimination of the domain 4 by alternative splicing (48,49). Although murine ICAM-1 typically contains five extracellular Ig-like domains, other isoforms may arise from alternative splicing, including proteins that lack the Ig-like domain(s) 2, 4, 2–3, or 2–3–4 (100). Importantly, different isoforms of VCAM-1 and ICAM-1 demonstrate distinct adhesive properties. For example, the mouse ICAM-1 isoform lacking Ig domains 2–3 does not bind to LFA-1 (100). Perhaps the relative expression of different isoforms represents a further level of functional control. For example, it has been observed that IFN-β1a regulates ICAM-1 RNA splicing in cultured ECs and leukocytes, leading to overexpression of certain isoforms with a relevant immunoregulatory role in multiple sclerosis (101).

VCAM-1 and ICAM-1 proteins undergo post-translational modification. The extent to which these changes influence molecular structure and/or function remains unclear. However, increasing evidence points to the importance of this process in regulating cell adhesion molecule function (see Table 117-2). For example, ICAM-2 and -3, but not ICAM-1, bind to lectins. The differential binding may be explained by the selective presence of N-linked, high-mannose glycans on ICAM-2 and -3. Interestingly, certain ICAM-1 mutants with additional N-linked glycosylations exhibit lectin ligand activity (28). Other mutations affecting residues located in the Ig domain 3 of ICAM-1 destroy the consensus sequence for glycosylation and enhance binding to purified Mac-1. Similarly, agents that interfere with ICAM-1 carbohydrate processing also affect binding to Mac-1, although not to LFA-1 (30). Therefore, the type and extent of glycosylation on ICAM-1 may regulate conformation and binding functions of this molecule. Whether this regulation operates in the case of VCAM-1 is unknown.

Additionally, ICAM-1 has been found to exist either as a monomer, as noncovalent dimers, or even as large multimers (102,103) (see Table 117-2). Dimerization occurs through either D1–D1 or D4–D4 interactions. Coordinated dimerization at the D4–D4 interface of D1–D1 dimers facilitates the formation of tetramers and larger structures (104). Oligomerization of ICAM-1 on the cell surface may regulate the conformation and orientation of this molecule, thereby controlling exposure of sites to which ligands bind. This may be particularly relevant considering that ICAM-1, as an elongated surface protein, experiences the force exerted by the blood flow. Indeed, the dimeric form of ICAM-1 is capable of binding ligands with greater affinity compared with the monomer form (102,103). Less is known about the existence of VCAM-1 oligomeric forms.

Induced synthesis of fully active, accessible VCAM-1 and ICAM-1 at the EC plasma membrane may be followed by downregulation of these molecules as a mechanism to help terminate the binding and signaling functions ascribed to them. The mechanisms underlying this phenomenon have not been fully characterized, but several hypotheses have been postulated. A soluble form of VCAM-1 has been observed to appear in vitro as a result of proteolytic cleavage by TNF-α–converting enzyme (ADAM 17) (105). In addition, ECs in culture rapidly internalize VCAM-1 upon ligation by cross-linked antibodies in an ATP-dependent manner ($t_{1/2}$ = 10–15 min), through a pathway consistent with clathrin-mediated endocytosis (106). Both membrane shedding and endocytosis may contribute to decreased VCAM-1 levels at the plasma membrane following inflammatory induction. Similarly, ECs also appear to internalize ICAM-1 in a surface density–dependent manner, but the low extent of internalization (<10% of surface ICAM-1) suggests that such a pathway plays a minor role in ICAM-1 turnover (32). Also, cleavage of ICAM-1 extracellular domain by neutrophil proteinases (e.g., elastase and cathepsin G) may account for downregulation of ICAM-1 at the EC surface (107,108). This latter process, which appears to occur independently of ICAM-1 surface density, is a potential source of soluble circulating ICAM-1. Recently, it has been shown that ECs actively internalize ICAM-1 upon ligation by anti-ICAM coated beads, which results in an acute and rapid decrease of surface ICAM-1 (80% in 30 min) (109). Internalized ICAM-1 partially (40%) recycles to the plasma membrane by 1 hour, whereas the remaining ICAM-1 fraction traffics through a

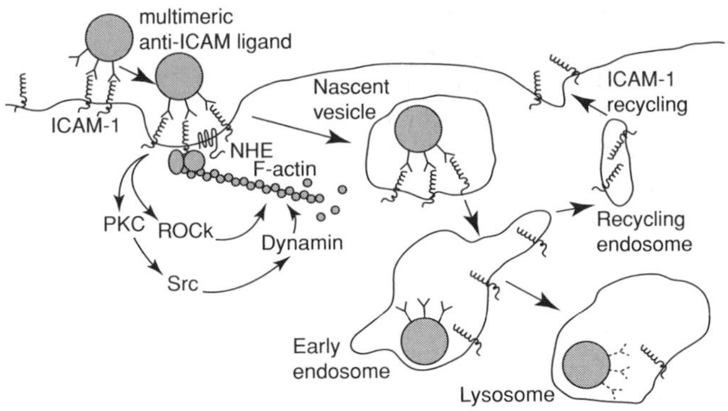

Figure 117.3. Cell adhesion molecule (CAM)-mediated endocytosis. ICAM-1 ligation by multimeric ligands (e.g., nanocarriers targeted by anti-ICAM antibodies) induces a unique endocytic pathway, CAM-mediated endocytosis, which operates through PKC, Src, and Rho-dependent kinase (ROCK) signaling. These molecules regulate activation of the large GTPase, dynamin, and plasma membrane Na^+/H^+ exchangers (NHE), resulting in actin cytoskeleton remodeling and internalization of the ligand/receptor complex. After internalization, the complex traffics to early endosomes; then ligand and receptor are differentially sorted to lysosomes and recycling compartments, respectively. Thus, the adhesive and signaling properties of ICAM-1 can be taken as a platform for designing vehicles for site-specific delivery of therapeutic agents into ECs, where ICAM-1 recycling may provide the basis for sustained targeting and prolonged effects of the delivered drugs.

degradative route (e.g., lysosomes). It is tempting to speculate that this event, which was mediated by cell adhesion molecule-mediated endocytosis (110), a novel, nonclassical endocytic pathway, also may account for regulation of ICAM-1 density at the EC surface (Figure 117.3).

VASCULAR CELL ADHESION MOLECULE-1 AND INTERCELLULAR ADHESION MOLECULE-1 AS THERAPEUTIC TARGETS IN VASCULAR PATHOLOGIES

VCAM-1 and ICAM-1 normally play an important role in homeostasis and host defense. However, under pathological conditions, dysregulation in their expression and/or function may exacerbate tissue damage and disease progression, as occurs in adult respiratory distress syndrome, acute lung injury, autoimmune disease, hypersensitivity reactions, septic shock, ischemia–reperfusion injury, host-versus-graft rejection, atherosclerosis, and diabetes, among others (39). For example, expression of VCAM-1 and ICAM-1 has been shown to be increased in ECs (and other cell types) at sites predisposed to atherosclerosis and in developing atherosclerotic lesions (111–113), suggesting a pathophysiological role for these adhesion molecules. Certain tumor cells, such as melanoma, appear to metastasize by binding to VCAM-1 on the surface of ECs (17,114). In animal models of asthma (using *ICAM-1–*null mice [115] and mice expressing a hypomorphic VCAM-1 form lacking the Ig-like domain 4 [116]), leukocyte infiltration into the lungs depended on VCAM-1 and ICAM-1 (117) (VCAM-1 and ICAM-1 animal models are shown in Table 117-2 [115,118–121]). Moreover, VCAM-1 and ICAM-1 intervene in the recruitment of T cells and lymphocytes to the central nervous system in allergic encephalomyelitis, multiple sclerosis, and meningitis (122).

Given the potential widespread importance of VCAM-1 and ICAM-1 in mediating pathology, these adhesion molecules represent valuable therapeutic targets (39,123). Blocking VCAM-1 and ICAM-1 provides protective effects in certain human disorders, such as inflammation, ischemia–reperfusion injury, acute lung injury, or arthritis (123–125). Moreover, therapeutic agents (e.g., anticancer drugs, antioxidant enzymes, thrombolytic compounds) can be coupled in a variety of formulations (e.g., liposomes, protein conjugates, targeted viruses, particulate nanocarriers, fusion proteins) with affinity carriers (e.g., full size antibodies, antibody fragments, peptides) that recognize particular endothelial adhesion molecules, thus providing site-specific delivery to the vascular areas where expression of the molecular target is prominent. For example, antibodies to VCAM-1 have been utilized as targeting ligands coupled to imaging agents or biodegradable carrier particles (immunotargeting) to provide highly selective accumulation on inflamed endothelium in vitro and in vivo (126–128). The subcellular destination of anti–VCAM-1 carriers has not been determined; therefore, the utility of this approach as a means for delivery of therapeutics either to the endothelial surface or intracellular compartments remains elusive. Nevertheless, targeted drug delivery to inflamed endothelium via anti–VCAM-1 particles may help to improve the treatment of pathologies such as cardiovascular disease, arthritis, inflammatory bowel disease, and cancer.

In the case of ICAM-1, it has been shown that monomeric antibodies to this molecule, as well as large (micron size) multimeric conjugates of biotin-streptavidin cross-linked antibodies stably bind to ICAM-1 (129,130), thus providing a useful means for site-specific drug delivery to the endothelial surface (e.g., delivery of fibrinolytics to the vascular wall) (129,131). Paradoxically, nanoscale (100–300 nm in diameter) multimeric anti–ICAM-1 conjugates and beads of similar size coated with anti–ICAM-1 antibodies are readily internalized by ECs (110). This strategy, similar to the case of a related adhesion molecule, PECAM-1, provided the basis for intracellular drug delivery of antioxidant enzymes (e.g., catalase, superoxide dismutase) for the containment of vascular oxidative stress in several pathological settings (109,132–136).

Internalization of these anti–ICAM-1 formulations occurs via a novel endocytic mechanism (see Figure 117.3), namely CAM-mediated endocytosis, which has been shown to differ from classical endothelial endocytic pathways (e.g., clathrin- and caveola-mediated endocytosis) and also from phagocytic and macropinocytic processes (137). ICAM-1 clustering by 100-nm beads coated with anti–ICAM-1 (final size = 200–250 nm diameter) signals through PKC, Src kinases, and Rho-dependent kinase. These regulate the formation of actin stress fibers and concomitant endocytosis mediated by the activity of cytosolic dynamin and membrane sodium/proton exchangers (110,136). Subsequent trafficking of anti–ICAM-1 conjugates to early endosomes and lysosomes is relatively slow (3 hours post-internalization) (134). This, in the case of antioxidant enzyme delivery, provides a sufficient therapeutic time window to cope with settings of acute oxidative stress, and it can be prolonged by supplementary drugs that influence intracellular trafficking and lysosomal degradation (132–136). Also, the partial recycling of ICAM-1 to the plasma membrane by 1 hour after conjugate internalization provides the basis for sustained delivery to ECs and prolongation of the effects of drugs via CAM-mediated endocytosis (109). Moreover, lysosomal trafficking subsequent to CAM-mediated endocytosis can be utilized as a strategy to improve classical means of enzyme replacement therapy for lysosomal storage disorders having marked endothelial contribution (138,139).

Thus, VCAM-1 and ICAM-1 represent attractive targets for the treatment of vascular pathologies, particularly those mediated through oxidative stress, thrombosis, and inflammation. Site-specific targeting of therapeutics to VCAM-1 and ICAM-1 may provide an effective means to minimize blood clearance, maximize effective delivery to disease sites, and reduce therapeutic dose, altogether increasing the efficacy of therapeutics.

KEY POINTS

- VCAM-1 and ICAM-1 are fine examples of the degree of mechanical complexity by which ECs integrate input-output signals, both in physiological and disease conditions. In response to a variety of factors, ECs regulate VCAM-1 and ICAM-1 function by controlling their expression and accessibility at the cell surface. VCAM-1 and ICAM-1 can then "sense" extracellular elements and transduce the signals intracellularly to regulate appropriate responses. Thus, the biomedical relevance of VCAM-1 and ICAM-1 may be understood from the perspective of mechanisms contributing to proper EC function and endothelial capacity to respond to environmental changes.
- VCAM-1 and ICAM-1 contribute to inflammatory cell trafficking. Their extracellular Ig-like domains provide an adhesive surface for activated leukocyte integrins, and the structural conformation of these molecules provides a fine-tune control of ligand specificity and the cellular infiltrate.
- VCAM-1 and ICAM-1 engagement by extracellular ligands serves as an outside-in transducing platform regulating EC phenotype and function (e.g., activation of kinases and transcription factors that modulate cell growth, ROS generation, cytokine production, expression of surface proteins, cytoskeletal modulation, and EC permeability).
- Regulation of VCAM-1 and ICAM-1 functions is achieved by mechanisms controlling the expression and surface accessibility of these adhesion molecules (e.g., chemical and mechanical factors affecting their transcription, and post-transcriptional and post-translational mechanisms regulating their structural conformation and oligomerization status).

Future Goals

- To understand the molecular mechanism involving VCAM-1– and ICAM-1–mediated pathogenesis, because of its direct clinical implications in diagnosis, given that these molecules can be utilized as markers for active inflammation under certain pathological conditions
- To elucidate the role of VCAM-1 and ICAM-1 as important targets for the therapy of endothelial dysfunction in the treatment of disease
- To determine if site-specific drug delivery to VCAM-1 and ICAM-1 may provide selectivity for diseased tissue and blockage of endothelial adhesion by interfering with VCAM-1 and ICAM-1 to inhibit tissue injury
- To further investigate therapy systems based on these highly versatile adhesion molecules that may ultimately lead to the discovery of novel approaches in the treatment and prevention of vascular diseases

ACKNOWLEDGMENTS

I thank Dr. Vladimir Muzykantov (Dept. Pharmacology, University of Pennsylvania School of Medicine) for his guidance on endothelial pathophysiology and vascular immunotargeting, and his enriching comments on this chapter. This work has been funded by a fellowship from Fundación Ramón Areces (Spain) and a scientist development grant from American Heart Association 0435481N to S.M.

REFERENCES

1 Cines DB, Pollak ES, Buck CA, et al. Endothelial cells in physiology and in the pathophysiology of vascular disorders. *Blood.* 1998;91(10):3527–3561.
2 Gimbrone MA Jr. Vascular endothelium, hemodynamic forces, and atherogenesis. *Am J Pathol.* 1999;155(1):1–5.
3 Simionescu M, Gafencu A, Antohe F. Transcytosis of plasma macromolecules in endothelial cells: a cell biological survey. *Microsc Res Tech.* 2002;57(5):269–288.
4 Vanhoutte PM. Endothelium-derived free radicals: for worse and for better. *J Clin Invest.* 2001;107(1):23–25.

5 Bevilacqua MP. Endothelial-leukocyte adhesion molecules. *Annu Rev Immunol.* 1993;11:767–804.
6 Zimmerman GA, Prescott SM, McIntyre TM. Endothelial cell interactions with granulocytes: tethering and signaling molecules. *Immunol Today.* 1992;13(3):93–100.
7 Springer TA. Traffic signals for lymphocyte recirculation and leukocyte emigration: the multistep paradigm. *Cell.* 1994;76 (2):301–314.
8 Worthylake RA, Burridge K. Leukocyte transendothelial migration: orchestrating the underlying molecular machinery. *Curr Opin Cell Biol.* 2001;13(5):569–577.
9 Dustin ML, Rothlein R, Bhan AK, et al. Induction by IL 1 and interferon-gamma: tissue distribution, biochemistry, and function of a natural adherence molecule (ICAM-1). *J Immunol.* 1986;137(1):245–254.
10 Rothlein R, Dustin ML, Marlin SD, Springer TA. A human intercellular adhesion molecule (ICAM-1) distinct from LFA-1. *J Immunol.* 1986;137(4):1270–1274.
11 Rothlein R, Springer TA. The requirement for lymphocyte function-associated antigen 1 in homotypic leukocyte adhesion stimulated by phorbol ester. *J Exp Med.* 1986;163(5):1132–1149.
12 Pober JS, Gimbrone MA Jr., Lapierre LA, et al. Overlapping patterns of activation of human endothelial cells by interleukin 1, tumor necrosis factor, and immune interferon. *J Immunol.* 1986; 137(6):1893–1896.
13 Simmons D, Makgoba MW, Seed B. ICAM, an adhesion ligand of LFA-1, is homologous to the neural cell adhesion molecule NCAM. *Nature.* 1988;331(6157):624–627.
14 Staunton DE, Marlin SD, Stratowa C, et al. Primary structure of ICAM-1 demonstrates interaction between members of the immunoglobulin and integrin supergene families. *Cell.* 1988; 52(6):925–933.
15 Tomassini JE, Graham D, DeWitt CM, et al. cDNA cloning reveals that the major group rhinovirus receptor on HeLa cells is intercellular adhesion molecule 1. *Proc Natl Acad Sci USA.* 1989; 86(13):4907–4911.
16 Greve JM, Davis G, Meyer AM, et al. The major human rhinovirus receptor is ICAM-1. *Cell.* 1989;56(5):839–847.
17 Rice GE, Bevilacqua MP. An inducible EC surface glycoprotein mediates melanoma adhesion. *Science.* 1989;246(4935):1303–1306.
18 Rice GE, Munro JM, Bevilacqua MP. Inducible cell adhesion molecule 110 (INCAM-110) is an endothelial receptor for lymphocytes. A CD11/CD18-independent adhesion mechanism. *J Exp Med.* 1990;171(4):1369–1374.
19 Osborn L, Hession C, Tizard R, et al. Direct expression cloning of vascular cell adhesion molecule 1, a cytokine-induced endothelial protein that binds to lymphocytes. *Cell.* 1989;59(6):1203–1211.
20 Rice GE, Munro JM, Corless C, Bevilacqua MP. Vascular and nonvascular expression of INCAM-110. A target for mononuclear leukocyte adhesion in normal and inflamed human tissues. *Am J Pathol.* 1991;138(2):385–393.
21 Springer TA. Adhesion receptors of the immune system. *Nature.* 1990;346(6283):425–434.
22 Hermand P, Huet M, Callebaut I, et al. Binding sites of leukocyte beta 2 integrins (LFA-1, Mac-1) on the human ICAM-4/LW blood group protein. *J Biol Chem.* 2000;275(34):26002–26010.
23 Staunton DE, Dustin ML, Springer TA. Functional cloning of ICAM-2, a cell adhesion ligand for LFA-1 homologous to ICAM-1. *Nature.* 1989;339(6219):61–64.
24 Huang MT, Mason JC, Birdsey GM, et al. Endothelial intercellular adhesion molecule (ICAM)-2 regulates angiogenesis. *Blood.* 2005;106(5):1636–1643.
25 Fawcett J, Holness CL, Needham LA, et al. Molecular cloning of ICAM-3, a third ligand for LFA-1, constitutively expressed on resting leukocytes. *Nature.* 1992;360(6403):481–484.
26 Bailly P, Hermand P, Callebaut I, et al. The LW blood group glycoprotein is homologous to intercellular adhesion molecules. *Proc Natl Acad Sci USA.* 1994;91(12):5306–5310.
27 Yoshihara Y, Oka S, Nemoto Y, et al. An ICAM-related neuronal glycoprotein, telencephalin, with brain segment-specific expression. *Neuron.* 1994;12(3):541–553.
28 Jimenez D, Roda P, Springer TA, Casasnovas JM. Contribution of N-linked glycans to the conformation and function of intercellular adhesion molecules (ICAMs). *J Biol Chem.* 2005 Feb 18;280(7):5854–5861. Epub 2004 Nov 15.
29 Ogawa T, Yorioka N, Ito T, et al. Ultrastructural localization of vascular cell adhesion molecule-1 in proliferative and crescentic glomerulonephritis. *Virchows Arch.* 1996;429(4–5):283–291.
30 Diamond MS, Staunton DE, Marlin SD, Springer TA. Binding of the integrin Mac-1 (CD11b/CD18) to the third immunoglobulin-like domain of ICAM-1 (CD54) and its regulation by glycosylation. *Cell.* 1991;65(6):961–971.
31 Kameda H, Morita I, Handa M, et al. Re-expression of functional P-selectin molecules on the endothelial cell surface by repeated stimulation with thrombin. *Br J Haematol.* 1997;97(2):348–355.
32 Almenar-Queralt A, Duperray A, Miles LA, et al. Apical topography and modulation of ICAM-1 expression on activated endothelium. *Am J Pathol.* 1995;147(5):1278–1288.
33 O'Brien KD, Allen MD, McDonald TO, et al. Vascular cell adhesion molecule-1 is expressed in human coronary atherosclerotic plaques. Implications for the mode of progression of advanced coronary atherosclerosis. *J Clin Invest.* 1993;92(2):945–951.
34 Rosen GD, Sanes JR, LaChance R, et al. Roles for the integrin VLA-4 and its counter receptor VCAM-1 in myogenesis. *Cell.* 1992;69(7):1107–1119.
35 Beauchamp JR, Abraham DJ, Bou-Gharios G, et al. Expression and function of heterotypic adhesion molecules during differentiation of human skeletal muscle in culture. *Am J Pathol.* 1992; 140(2):387–401.
36 Constantin G, Piccio L, Bussini S, et al. Induction of adhesion molecules on human Schwann cells by proinflammatory cytokines, an immunofluorescence study. *J Neurol Sci.* 1999; 170(2):124–130.
37 Etienne-Manneville S, Chaverot N, Strosberg AD, Couraud PO. ICAM-1-coupled signaling pathways in astrocytes converge to cyclic AMP response element-binding protein phosphorylation and TNF-alpha secretion. *J Immunol.* 1999;163(2):668–674.
38 Raju R, Dalakas MC. Gene expression profile in the muscles of patients with inflammatory myopathies: effect of therapy with IVIg and biological validation of clinically relevant genes. *Brain.* 2005;128(Pt 8):1887–1896.
39 Muro S, Muzykantov VR. Targeting of antioxidant and antithrombotic drugs to endothelial cell adhesion molecules. *Curr Pharm Des.* 2005;11(18):2383–2401.
40 Remy M, Valli N, Brethes D, et al. In vitro and in situ intercellular adhesion molecule-1 (ICAM-1) expression by endothelial cells lining a polyester fabric. *Biomaterials.* 1999;20(3):241–251.
41 Sadeghi MM, Schechner JS, Krassilnikova S, et al. Vascular cell adhesion molecule-1-targeted detection of endothelial

activation in human microvasculature. *Transplant Proc.* 2004; 36(5):1585–1591.

42 Norris P, Poston RN, Thomas DS, et al. The expression of endothelial leukocyte adhesion molecule-1 (ELAM-1), intercellular adhesion molecule-1 (ICAM-1), and vascular cell adhesion molecule-1 (VCAM-1) in experimental cutaneous inflammation: a comparison of ultraviolet B erythema and delayed hypersensitivity. *J Invest Dermatol.* 1991;96(5):763–770.

43 Pober J, Cotran RS. What can be learned from the expression of endothelial adhesion molecules in tissues? *Lab Invest.* 1991; 64(3):301–305.

44 Nishio K, Suzuki Y, Aoki T, et al. Differential contribution of various adhesion molecules to leukocyte kinetics in pulmonary microvessels of hyperoxia-exposed rat lungs. *Am J Respir Crit Care Med.* 1998;157(2):599–609.

45 Yamaguchi K, Nishio K, Sato N, et al. Leukocyte kinetics in the pulmonary microcirculation: observations using real-time confocal luminescence microscopy coupled with high-speed video analysis. *Lab Invest.* 1997;76(6):809–822.

46 Muller WA. Leukocyte-EC interactions in the inflammatory response. *Lab Invest.* 2002;82(5):521–533.

47 Muller AM, Cronen C, Muller KM, Kirkpatrick CJ. Heterogeneous expression of cell adhesion molecules by endothelial cells in ARDS. *J Pathol.* 2002;198(2):270–275.

48 Cybulsky MI, Fries JW, Williams AJ, et al. Gene structure, chromosomal location, and basis for alternative mRNA splicing of the human VCAM1 gene. *Proc Natl Acad Sci USA.* 1991; 88(17):7859–7863.

49 Osborn L, Vassallo C, Benjamin CD. Activated endothelium binds lymphocytes through a novel binding site in the alternately spliced domain of vascular cell adhesion molecule-1. *J Exp Med.* 1992;176(1):99–107.

50 D'Souza SE, Byers-Ward VJ, Gardiner EE, et al. Identification of an active sequence within the first immunoglobulin domain of intercellular cell adhesion molecule-1 (ICAM-1) that interacts with fibrinogen. *J Biol Chem.* 1996;271(39):24270–24277.

51 Hopkins AM, Baird AW, Nusrat A. ICAM-1: targeted docking for exogenous as well as endogenous ligands. *Adv Drug Deliv Rev.* 2004;56(6):763–778.

52 Chuluyan HE, Osborn L, Lobb R, Issekutz AC. Domains 1 and 4 of vascular cell adhesion molecule-1 (CD106) both support very late activation antigen-4 (CD49d/CD29)-dependent monocyte transendothelial migration. *J Immunol.* 1995;155(6):3135–3134.

53 Carlos TM, Schwartz BR, Kovach NL, et al. Vascular cell adhesion molecule-1 mediates lymphocyte adherence to cytokine-activated cultured human endothelial cells. *Blood.* 1990;76(5):965–970.

54 Wang J, Springer TA. Structural specializations of immunoglobulin superfamily members for adhesion to integrins and viruses. *Immunol Rev.* 1998;163:197–215.

55 Holland J, Owens T. Signaling through intercellular adhesion molecule 1 (ICAM-1) in a B cell lymphoma line. The activation of Lyn tyrosine kinase and the mitogen-activated protein kinase pathway. *J Biol Chem.* 1997;272(14):9108–9112.

56 Matheny HE, Deem TL, Cook-Mills JM. Lymphocyte migration through monolayers of endothelial cell lines involves VCAM-1 signaling via EC NADPH oxidase. *J Immunol.* 2000;164(12):6550–6559.

57 Wang Q, Doerschuk CM. Neutrophil-induced changes in the biomechanical properties of endothelial cells: roles of ICAM-1 and reactive oxygen species. *J Immunol.* 2000;164(12):6487–6494.

58 Gardiner EE, D'Souza SE. Sequences within fibrinogen and intercellular adhesion molecule-1 (ICAM-1) modulate signals required for mitogenesis. *J Biol Chem.* 1999;274(17):11930–11936.

59 Hubbard AK, Rothlein R. Intercellular adhesion molecule-1 (ICAM-1) expression and cell signaling cascades. *Free Radic Biol Med.* 2000;28(9):1379–1386.

60 Pluskota E, Chen Y, D'Souza SE. Src homology domain 2-containing tyrosine phosphatase 2 associates with intercellular adhesion molecule 1 to regulate cell survival. *J Biol Chem.* 2000; 275(39):30029–30036.

61 Cuvelier SL, Paul S, Shariat N, et al. Eosinophil adhesion under flow conditions activates mechanosensitive signaling pathways in human endothelial cells. *J Exp Med.* 2005;202(6):865–876.

62 Lazaar AL, Krymskaya VP, Das SK. VCAM-1 activates phosphatidylinositol 3-kinase and induces p120Cbl phosphorylation in human airway smooth muscle cells. *J Immunol.* 2001; 166(1):155–161.

63 Greenwood J, Etienne-Manneville S, Adamson P, Couraud PO. Lymphocyte migration into the central nervous system: implication of ICAM-1 signalling at the blood-brain barrier. *Vascul Pharmacol.* 2002;38(6):315–322.

64 Clayton A, Evans RA, Pettit E, et al. Cellular activation through the ligation of intercellular adhesion molecule-1. *J Cell Sci.* 1998; 111(Pt 4):443–453.

65 Alevriadou BR. CAMs and Rho small GTPases: gatekeepers for leukocyte transendothelial migration. Focus on "VCAM-1-mediated Rac signaling controls endothelial cell-cell contacts and leukocyte transmigration." *Am J Physiol Cell Physiol.* 2003;285(2):C250–C252.

66 Ricard I, Payet MD, Dupuis G. Clustering the adhesion molecules VLA-4 (CD49d/CD29) in Jurkat T cells or VCAM-1 (CD106) in endothelial (ECV 304) cells activates the phosphoinositide pathway and triggers Ca^{2+} mobilization. *Eur J Immunol.* 1997;27(6):1530–1538.

67 Tudor KS, Hess KL, Cook-Mills JM. Cytokines modulate EC intracellular signal transduction required for VCAM-1-dependent lymphocyte transendothelial migration. *Cytokine.* 2001;15(4):196–211.

68 Etienne-Manneville S, Manneville JB, Adamson P, et al. ICAM-1-coupled cytoskeletal rearrangements and transendothelial lymphocyte migration involve intracellular calcium signaling in brain endothelial cell lines. *J Immunol.* 2000;165(6):3375–3383.

69 Wang Q, Doerschuk CM. The p38 mitogen-activated protein kinase mediates cytoskeletal remodeling in pulmonary microvascular endothelial cells upon intracellular adhesion molecule-1 ligation. *J Immunol.* 2001;166(11):6877–6884.

70 Adamson P, Etienne S, Couraud PO, et al. Lymphocyte migration through brain endothelial cell monolayers involves signaling through endothelial ICAM-1 via a rho-dependent pathway. *J Immunol.* 1999;162(5):2964–2973.

71 Etienne S, Adamson P, Greenwood J, et al. ICAM-1 signaling pathways associated with Rho activation in microvascular brain endothelial cells. *J Immunol.* 1998;161(10):5755–5761.

72 Barreiro O, Yanez-Mo M, Serrador JM, et al. Dynamic interaction of VCAM-1 and ICAM-1 with moesin and ezrin in a novel endothelial docking structure for adherent leukocytes. *J Cell Biol.* 2002;157(7):1233–1245.

73 Carpen O, Pallai P, Staunton DE, Springer TA. Association of intercellular adhesion molecule-1 (ICAM-1) with actin-containing cytoskeleton and alpha-actinin. *J Cell Biol.* 1992; 118(5):1223–1234.
74 Federici C, Camoin L, Hattab M, et al. Association of the cytoplasmic domain of intercellular-adhesion molecule-1 with glyceraldehyde-3-phosphate dehydrogenase and beta-tubulin. *Eur J Biochem.* 1996;238(1):173–180.
75 Henninger DD, Panes J, Eppihimer M, et al. Cytokine-induced VCAM-1 and ICAM-1 expression in different organs of the mouse. *J Immunol.* 1997;158(4):1825–1832.
76 Schwartz BR, Wayner EA, Carlos TM, et al. Identification of surface proteins mediating adherence of CD11/CD18-deficient lymphoblastoid cells to cultured human endothelium. *J Clin Invest.* 1990;85(6):2019–2022.
77 Rothlein R, Wegner C. Role of intercellular adhesion molecule-1 in the inflammatory response. *Kidney Int.* 1992;41(3):617–619.
78 Albelda SM. Endothelial and epithelial cell adhesion molecules. *Am J Respir Cell Mol Biol.* 1991;4(3):195–203.
79 Springer TA. The sensation and regulation of interactions with the extracellular environment: the cell biology of lymphocyte adhesion receptors. *Annu Rev Cell Biol.* 1990;6:359–402.
80 Perry VH, Anthony DC, Bolton SJ, Brown HC. The blood-brain barrier and the inflammatory response. *Mol Med Today.* 1997; 3(8):335–341.
81 Sobel RA, Mitchell ME, Fondren G. Intercellular adhesion molecule-1 (ICAM-1) in cellular immune reactions in the human central nervous system. *Am J Pathol.* 1990;136(6):1309–1316.
82 Sampietro T, Tuoni M, Ferdeghini M, et al. Plasma cholesterol regulates soluble cell adhesion molecule expression in familial hypercholesterolemia. *Circulation.* 1997;96(5):1381–1385.
83 Yang PY, Rui YC. Intercellular adhesion molecule-1 and vascular endothelial growth factor expression kinetics in macrophage-derived foam cells. *Life Sci.* 2003;74(4):471–480.
84 Kowalska I, Straczkowski M, Szelachowska M, et al. Circulating E-selectin, vascular cell adhesion molecule-1, and intercellular adhesion molecule-1 in men with coronary artery disease assessed by angiography and disturbances of carbohydrate metabolism. *Metabolism.* 2002;51(6):733–736.
85 Quagliaro L, Piconi L, Assaloni R, et al. Intermittent high glucose enhances ICAM-1, VCAM-1 and E-selectin expression in human umbilical vein endothelial cells in culture: the distinct role of protein kinase C and mitochondrial superoxide production. *Atherosclerosis.* 2005;183(2):259–267.
86 Bornemann A, Anderson LV. Diagnostic protein expression in human muscle biopsies. *Brain Pathol.* 2000;10(2):193–214.
87 Suzuki Y, Tanigaki T, Heimer D, et al. TGF-beta 1 causes increased endothelial ICAM-1 expression and lung injury. *J Appl Physiol.* 1994;77(3):1281–1287.
88 Muller AM, Hermanns MI, Cronen C, Kirkpatrick CJ. Comparative study of adhesion molecule expression in cultured human macro- and microvascular endothelial cells. *Exp Mol Pathol.* 2002;73(3):171–180.
89 Lee KH, Lawley TJ, Xu YL, Swerlick RA. VCAM-1-, ELAM-1-, and ICAM-1-independent adhesion of melanoma cells to cultured human dermal microvascular endothelial cells. *J Invest Dermatol.* 1992;98(1):79–85.
90 Weber C, Negrescu E, Erl W, et al. Inhibitors of protein tyrosine kinase suppress TNF-stimulated induction of endothelial cell adhesion molecules. *J Immunol.* 1995;155(1):445–451.
91 Ando J, Kamiya A. Blood flow and vascular endothelial cell function. *Front Med Biol Eng.* 1993;5(4):245–264.
92 Malek AM, Gibbons GH, Dzau VJ, Izumo S. Fluid shear stress differentially modulates expression of genes encoding basic fibroblast growth factor and platelet-derived growth factor B chain in vascular endothelium. *J Clin Invest.* 1993;92(4): 2013–2021.
93 Ballermann BJ, Dardik A, Eng E, Liu A. Shear stress and the endothelium. *Kidney Int Suppl.* 1998;67:S100–S108.
94 Chiu JJ, Lee PL, Chen CN, et al. Shear stress increases ICAM-1 and decreases VCAM-1 and E-selectin expressions induced by tumor necrosis factor-[alpha] in endothelial cells. *Arterioscler Thromb Vasc Biol.* 2004;24(1):73–79.
95 Davies PF. Flow-mediated endothelial mechanotransduction. *Physiol Rev.* 1995;75(3):519–560.
96 Iademarco MF, McQuillan JJ, Rosen GD, Dean DC. Characterization of the promoter for vascular cell adhesion molecule-1 (VCAM-1). *J Biol Chem.* 1992;267(23):16323–16329.
97 Voraberger G, Schafer R, Stratowa C. Cloning of the human gene for intercellular adhesion molecule 1 and analysis of its 5′-regulatory region. Induction by cytokines and phorbol ester. *J Immunol.* 1991;147(8):2777–2786.
98 Minami T, Aird WC. Thrombin stimulation of the vascular cell adhesion molecule-1 promoter in endothelial cells is mediated by tandem nuclear factor-kappa B and GATA motifs. *J Biol Chem.* 2001;276(50):47632–47641.
99 Moy P, Lobb R, Tizard R, et al. Cloning of an inflammation-specific phosphatidyl inositol-linked form of murine vascular cell adhesion molecule-1. *J Biol Chem.* 1993;268(12):8835–8841.
100 King PD, Sandberg ET, Selvakumar A, et al. Novel isoforms of murine intercellular adhesion molecule-1 generated by alternative RNA splicing. *J Immunol.* 1995;154(11):6080–6093.
101 Giorelli M, De Blasi A, Defazio G, et al. Differential regulation of membrane bound and soluble ICAM 1 in human endothelium and blood mononuclear cells: effects of interferon beta-1a. *Cell Commun Adhes.* 2002;9(5–6):259–272.
102 Miller J, Knorr R, Ferrone M, et al. Intercellular adhesion molecule-1 dimerization and its consequences for adhesion mediated by lymphocyte function associated-1. *J Exp Med.* 1995;182(5):1231–1241.
103 Reilly PL, Woska JR Jr., Jeanfavre DD, et al. The native structure of intercellular adhesion molecule-1 (ICAM-1) is a dimer. Correlation with binding to LFA-1. *J Immunol.* 1995;155(2):529–532.
104 Jun CD, Shimaoka M, Carman CV, et al. Dimerization and the effectiveness of ICAM-1 in mediating LFA-1-dependent adhesion. *Proc Natl Acad Sci USA.* 2001;98(12):6830–6835.
105 Garton KJ, Gough PJ, Philalay J, et al. Stimulated shedding of vascular cell adhesion molecule 1 (VCAM-1) is mediated by tumor necrosis factor-alpha-converting enzyme (ADAM 17). *J Biol Chem.* 2003;278(39):37459–37464.
106 Ricard I, Payet MD, Dupuis G. VCAM-1 is internalized by a clathrin-related pathway in human endothelial cells but its alpha 4 beta 1 integrin counter-receptor remains associated with the plasma membrane in human T lymphocytes. *Eur J Immunol.* 1998;28(5):1708–1718.
107 Melis M, Pace E, Siena L, et al. Biologically active intercellular adhesion molecule-1 is shed as dimers by a regulated mechanism in the inflamed pleural space. *Am J Respir Crit Care Med.* 2003; 167(8):1131–1138.

108 Robledo O, Papaioannou A, Ochietti B, et al. ICAM-1 isoforms: specific activity and sensitivity to cleavage by leukocyte elastase and cathepsin G. *Eur J Immunol.* 2003;33(5):1351–1360.
109 Muro S, Gajewski C, Koval M, Muzykantov VR. ICAM-1 recycling in endothelial cells: a novel pathway for sustained intracellular delivery and prolonged effects of drugs. *Blood.* 2005;105(2):650–658.
110 Muro S, Wiewrodt R, Thomas A, et al. A novel endocytic pathway induced by clustering endothelial ICAM-1 or PECAM-1. *J Cell Sci.* 2003;116(Pt 8):1599–1609.
111 Cybulsky MI, Gimbrone MA Jr. Endothelial expression of a mononuclear leukocyte adhesion molecule during atherogenesis. *Science.* 1991;251(4995):788–791.
112 Poston RN, Haskard DO, Coucher JR, et al. Expression of intercellular adhesion molecule-1 in atherosclerotic plaques. *Am J Pathol.* 1992;140(3):665–673.
113 Iiyama K, Hajra L, Iiyama M, et al. Patterns of vascular cell adhesion molecule-1 and intercellular adhesion molecule-1 expression in rabbit and mouse atherosclerotic lesions and at sites predisposed to lesion formation. *Circ Res.* 1999;85(2):199–207.
114 Taichman DB, Cybulsky MI, Djaffar I, et al. Tumor cell surface alpha 4 beta 1 integrin mediates adhesion to vascular endothelium: demonstration of an interaction with the N-terminal domains of INCAM-110/VCAM-1. *Cell Regul.* 1991; 2(5):347–355.
115 Xu H, Gonzalo JA, St Pierre Y, et al. Leukocytosis and resistance to septic shock in intercellular adhesion molecule 1-deficient mice. *J Exp Med.* 1994;180(1):95–109.
116 Friedrich C, Cybulsky MI, Gutierrez-Ramos JC. Vascular cell adhesion molecule-1 expression by hematopoiesis-supporting stromal cells is not essential for lymphoid or myeloid differentiation in vivo or in vitro. *Eur J Immunol.* 1996;26(11):2773–2780.
117 Schleimer RP, Bochner BS. The role of adhesion molecules in allergic inflammation and their suitability as targets of antiallergic therapy. *Clin Exp Allergy.* 1998;28(Suppl 3):15–23.
118 Bullard DC, Qin L, Lorenzo I, et al. P-selectin/ICAM-1 double mutant mice: acute emigration of neutrophils into the peritoneum is completely absent but is normal into pulmonary alveoli. *J Clin Invest.* 1995;95(4):1782–1788.
119 Sligh JE Jr., Ballantyne CM, Rich SS, et al. Inflammatory and immune responses are impaired in mice deficient in intercellular adhesion molecule 1. *Proc Natl Acad Sci USA.* 1993; 90(18):8529–8533.
120 Gurtner GC, Davis V, Li H, et al. Targeted disruption of the murine VCAM1 gene: essential role of VCAM-1 in chorioallantoic fusion and placentation. *Genes Dev.* 1995;9(1):1–14.
121 Gonzalo JA, Lloyd CM, Kremer L, et al. Eosinophil recruitment to the lung in a murine model of allergic inflammation. The role of T cells, chemokines, and adhesion receptors. *J Clin Invest.* 1996;98(10):2332–2345.
122 Baron JL, Madri JA, Ruddle NH, et al. Surface expression of alpha 4 integrin by CD4 T cells is required for their entry into brain parenchyma. *J Exp Med.* 1993;177(1):57–68.
123 Muzykantov VR. *Targeting Pulmonary Endothelium.* Boston: Kluwer Academic Publishers, 2002.
124 Kumasaka T, Quinlan WM, Doyle NA, et al. Role of the intercellular adhesion molecule-1(ICAM-1) in endotoxin-induced pneumonia evaluated using ICAM-1 antisense oligonucleotides, anti-ICAM-1 monoclonal antibodies, and ICAM-1 mutant mice. *J Clin Invest.* 1996;97(10):2362–2369.
125 Rothlein R, Mainolfi EA, Kishimoto TK. Treatment of inflammation with anti-ICAM-1. *Res Immunol.* 1993;144(9):735–9; discussion 54–62.
126 Sakhalkar HS, Dalal MK, Salem AK, et al. Leukocyte-inspired biodegradable particles that selectively and avidly adhere to inflamed endothelium in vitro and in vivo. *Proc Natl Acad Sci USA.* 2003;100(26):15895–15900.
127 Tsourkas A, Shinde-Patil VR, Kelly K, et al. In vivo imaging of activated endothelium using an anti-VCAM-1 magneto-optical probe. *Bioconjug Chem.* 2005;16(3):576–581.
128 Kelly K, Allport JR, Tsourkas A, et al. Detection of vascular adhesion molecule-1 expression using a novel multimodal nanoparticle. *Circ Res.* 2005;96(3):327–336.
129 Murciano JC, Muro S, Koniaris L, et al. ICAM-directed vascular immunotargeting of antithrombotic agents to the endothelial luminal surface. *Blood.* 2003;101(10):3977–3984.
130 Wiewrodt R, Thomas AP, Cipelletti L, et al. Size-dependent intracellular immunotargeting of therapeutic cargoes into endothelial cells. *Blood.* 2002;99(3):912–922.
131 Atochina EN, Balyasnikova IV, Danilov SM, et al. Immunotargeting of catalase to ACE or ICAM-1 protects perfused rat lungs against oxidative stress. *Am J Physiol.* 1998;275(4 Pt 1):L806–L817.
132 Christofidou-Solomidou M, Scherpereel A, Wiewrodt R, et al. PECAM-directed delivery of catalase to endothelium protects against pulmonary vascular oxidative stress. *Am J Physiol Lung Cell Mol Physiol.* 2003;285(2):L283–L292.
133 Kozower BD, Christofidou-Solomidou M, Sweitzer TD, et al. Immunotargeting of catalase to the pulmonary endothelium alleviates oxidative stress and reduces acute lung transplantation injury. *Nat Biotechnol.* 2003;21(4):392–398.
134 Muro S, Cui X, Gajewski C, et al. Slow intracellular trafficking of catalase nanoparticles targeted to ICAM-1 protects endothelial cells from oxidative stress. *Am J Physiol Cell Physiol.* 2003;285(5):C1339–C1347.
135 Scherpereel A, Rome JJ, Wiewrodt R, et al. Platelet-EC adhesion molecule-1-directed immunotargeting to cardiopulmonary vasculature. *J Pharmacol Exp Ther.* 2002;300(3):777–786.
136 Muro S, Mateescu M, Gajewski G, et al. Control of intracellular trafficking of ICAM-1-targeted nanocarriers by endothelial Na^+/H^+ exchanger (NHE) proteins. *Am J Physiol Lung Cell Mol Physiol.* 2006;290(5):L809–817.
137 Muro S, Koval M, Muzykantov V. Endothelial endocytic pathways: gates for vascular drug delivery. *Curr Vasc Pharmacol.* 2004;2(3):281–299.
138 Muro S, Schuchman EH, Muzykantov VR. Lysosomal enzyme delivery by ICAM-1-targeted nanocarriers bypassing glycosylation- and clathrin-dependent endocytosis. *Mol Ther.* 2006 Jan;13(1):135–141.
139 Schuchman EH, Muro S. The development of enzyme replacement therapy for lysosomal diseases: Gaucher disease and beyond. In: Futerman AH, ed. *Gaucher Disease: Lessons Learned about Therapy of Lysosomal Disorders.* Boca Raton: CRC Press, 2006.

118

E-Selectin

David Milstone

Brigham and Women's Hospital, Harvard Medical School, Boston, Massachusetts

The discovery of E-selectin in 1987 (1) culminated a series of experiments investigating the molecular basis of endothelial activation and interactions with leukocytes during acute inflammation. This initiated an era in which several additional endothelial structures involved in leukocyte adhesion were identified, and their roles in inflammation were investigated. Exploiting these insights to develop diagnostic and therapeutic interventions for inflammatory and other diseases remains a promising area of clinical investigation.

Bevilacqua and Gimbrone (1) identified monoclonal antibodies that blocked polymorphonuclear neutrophil (PMN) adhesion to inflammatory cytokine-activated human umbilical vein endothelial cells (HUVECs) and used them to expression-clone a cell surface glycoprotein, designated endothelial-leukocyte adhesion molecule (ELAM)-1, which mediates initial PMN adhesion to inflamed endothelium. ELAM-1 possessed a then-unique, complex mosaic structure predicting, for the first time, that inflammation involved calcium-dependent interactions between endothelial cell (EC) surface lectins and carbohydrate-bearing "counter-receptors" on circulating leukocytes. At about the same time, investigators studying seemingly unrelated cell–cell interactions during inflammation – one involving megakaryocytes and platelets, the other involving leukocytes – discovered two additional lectin-like molecules with similar but distinct mosaic structures. Thus was the selectin gene family defined (2–4), consisting of E-selectin (endothelium, CD62E, previously known as ELAM)-1, P-selectin (platelet, CD62P, previously known as platelet activation-dependent granule-to-external-membrane protein [PADGEM] and granule membrane protein of 140 kDa [GMP-140]), and L-selectin (leukocyte, CD62L, previously known as MEL-14 and leukocyte adhesion molecule [LAM]-1). A large body of work has since established E-selectin not only as a crucial adhesion structure mediating leukocyte rolling, binding, and transmigration during inflammation, but also as a participant in a dialog between endothelium and leukocytes (and potentially other circulating cells as well, such as cancer cells), in which cell–cell interactions induce intracellular signaling events that result in changes in the endothelial adhesive and transmigratory phenotype.

More recently, E-selectin also has been identified as one determinant of the endothelium's ability to respond to endostatin, an antiangiogenic compound currently being evaluated as an anticancer agent. This illustrates an important guiding concept: E-selectin, and potentially other gene products with well-established importance in inflammation, may play additional yet-to-be-discovered roles either in inflammation or in other processes involving vascular endothelium. The endothelium may be particularly well suited to carry out such dual functions by virtue of its unique role as both an essential component of the supporting vasculature and an integral determinant of the specific properties of diverse tissues and organs.

EVOLUTIONARY CONSIDERATIONS

C-type lectins are common in many phyla. A protein with a selectin-like mosaic structure, but lacking an epidermal growth factor (EGF) domain, functions in peripheral nervous system development in *Drosophila* (5). The *E-*, *P-*, and *L-selectin* genes are tightly linked on chromosome 1 in both humans and mice, and show significant structural homology in their extracellular domains, suggesting this gene family arose by duplication of a common ancestral sequence. However, significantly divergent cytoplasmic domains suggest different intracellular signaling consequences of ligand engagement by each selectin. In addition, species-specific differences in transcriptional regulation at the endothelial selectin locus, presumably arising from "genetic drift," have led to potentially significant interspecies differences in E-selectin's contribution to inflammation. This has important implications for extrapolating results from experimental animals to humans (see the section "Experimental models for investigating endothelial selectin function in vivo.")

DEVELOPMENTAL CONSIDERATIONS

Fetal endothelium participates in inflammatory reactions independent of maternal endothelium and leukocytes. This occurs during chorioamnionitis (inflammation of the placenta and gestational membranes), especially with funisitis (inflammation of the umbilical cord), and during the subsequent complication of fetal pneumonia. In these cases, fetal leukocytes (including PMNs and eosinophils) transmigrate across fetal endothelium at sites of acute inflammation. However, although E-selectin is expressed and becomes cytokine-inducible during embryogenesis, the specific functional significance of this expression and its role in fetal inflammation has not been investigated. In this context, it also is worth noting that HUVECs are of fetal origin and are the cell type used in vitro to initially identify and subsequently investigate E-selectin and several other endothelial adhesion molecules important in inflammation.

Embryonic Expression

E-selectin mRNA is initially expressed at low levels at embryonic day (E)9.5 to E11.5 of murine embryogenesis, but it is not inducible by inflammatory cytokines at this stage, as determined in primary in vitro cultures of embryo-derived cells (6). Inducibility develops between E12.5 and E13.5, soon after the onset of organogenesis (lung, liver, etc.), suggesting a functional role for regulated expression in leukocyte–endothelial interactions during inflammation in the embryo. Newborn mice lacking both E- and P-selectin, but not singly deficient mice, show elevated circulating leukocytes, implying that either of the endothelial selectins can interact with embryonic leukocytes (7–9). However, expression of E-selectin protein has not been demonstrated in embryos, and the contribution of individual selectins to embryo inflammation, and its relevance to humans, has not been specifically investigated.

Despite data suggesting that E-selectin might participate in angiogenesis (10) and/or antiangiogenesis (11), growth, development, fertility, and wound healing are not affected in genetically deficient mice (8,12,13). E-selectin–deficient ECs also perform normally using in vitro angiogenesis assays (14).

Placental Expression

E-selectin mRNA and protein are developmentally regulated and constitutively expressed in the absence of inflammation in both human and murine trophoblasts during placental differentiation (Figure 118.1) (15,16). Expression occurs in murine secondary trophoblast giant cells, trophoblasts lining the central artery, and a subpopulation of labyrinthine trophoblasts located at the fetal–maternal interface. These cells line vascular channels but express a unique profile of gene products not displayed by the vascular endothelium.

Placentae lacking E-selectin show increased trophoblast glycogen cells and fewer labyrinthine PMNs, suggesting that recognition of trophoblast E-selectin by counter-receptors on other cells contributes to placental development. However, functional consequences of trophoblast expression for reproduction or embryonic development have not been reported.

GENE REGULATION

Inducible Expression in Endothelial Cells

E-selectin is one of a very few genes, including vascular endothelial (VE)-cadherin, whose expression in nonplacental tissues appears truly endothelial-specific. Most other endothelial-restricted gene products, including platelet-endothelial cell adhesion molecule (PECAM)-1, intercellular adhesion molecule (ICAM)-1, ICAM-2, P-selectin, thrombomodulin, and others, also are present in leukocytes, megakaryocytes and platelets, or various other non-ECs. As such, understanding the molecular mechanisms specifying endothelial-specific expression of E-selectin might shed light on the developmentally regulated transcriptional programs defining this lineage. However, little is known about these mechanisms.

Although highly specific, E-selectin expression is not uniform within the endothelium, either in cells cultured in vitro or throughout the normal vascular system in vivo. Most quiescent ECs in culture do not express E-selectin unless activated by one or more of a variety of inflammatory stimuli, including tumor necrosis factor (TNF)-α, interleukin (IL)-1, CD40 ligand, and bacterial endotoxin (lipopolysaccharide [LPS]). Under in vivo conditions, virtually all noninflamed murine tissues containing ECs show detectable, albeit low-level, mRNA expression, although E-selectin protein is usually undetectable. Upon activation of cultured cells, protein expression, preceded by mRNA, increases markedly within 3 to 4 hours, diminishes significantly by 6 to 8 hours, and returns to near-basal levels within 24 to 48 hours (see Figure 118.1). Expression is also dramatically upregulated during inflammation in vivo, where it varies significantly between organ beds (17,18) (our unpublished observations) and, within a given bed, is localized to physiologically important vascular structures, such as the postcapillary venules of human skin (19,20) or rabbit mesentery (21). Although both the onset and resolution of expression occur more rapidly in vivo than in vitro, expression often can be detected several hours, or even several days, after endothelial activation, depending on the inflammatory reaction studied and the tissue examined (see "Dermal EC Protein" in Figure 118.1). In addition, selected vascular beds, such as fetal and adult bone marrow, express E-selectin constitutively in vivo (9).

In contrast to E-selectin, P-selectin is present constitutively in endothelial Weibel-Palade bodies, platelet α-granules, and megakaryocytes, is rapidly mobilized to the cell surface by acute secretagogues, and regulated differently at the transcriptional level during acute inflammation in mice and humans. (See the discussion in a later section "Experimental models for investigating endothelial selectin function in vivo," which

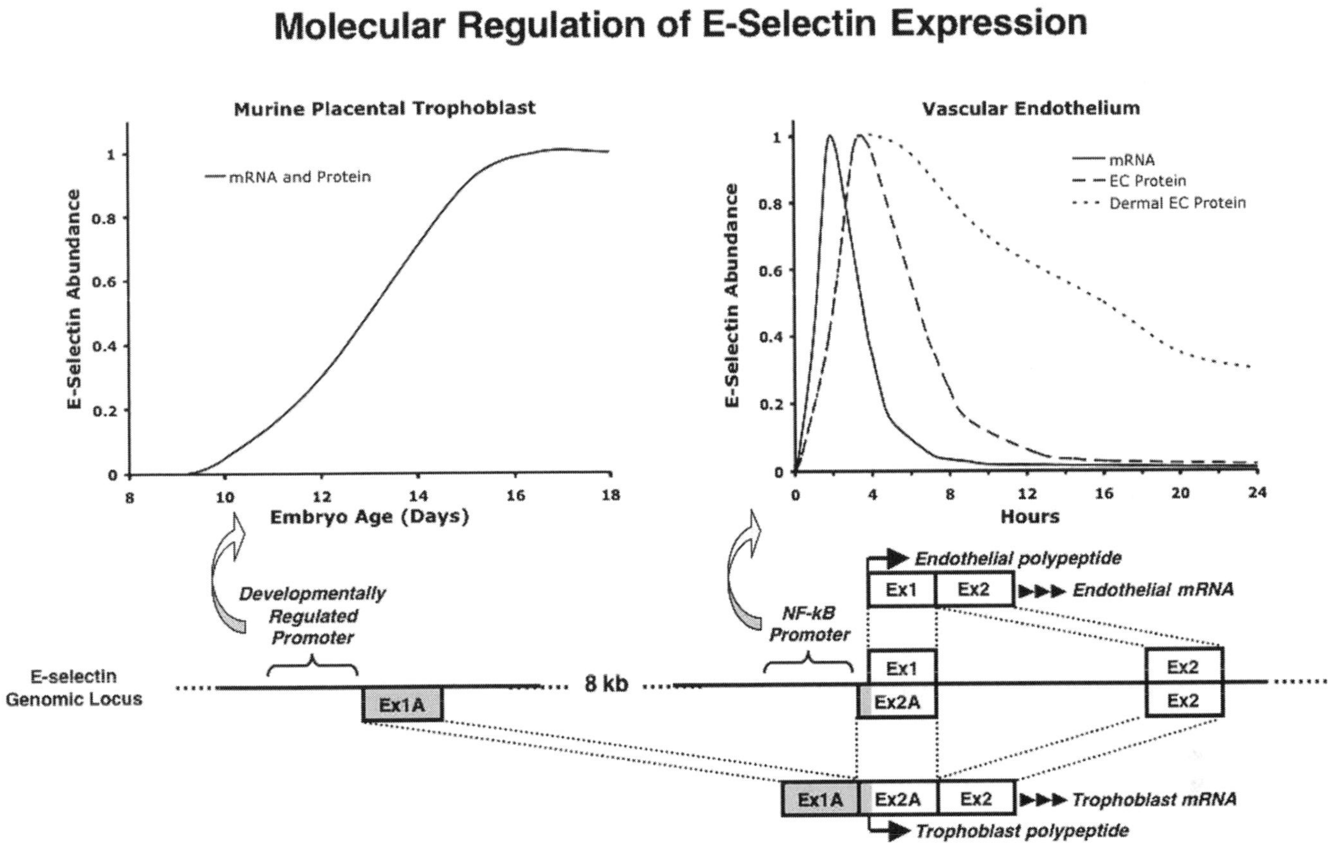

Figure 118.1. Inducible and constitutive expression of E-selectin in endothelium and trophoblast, respectively, is directed by alternative transcriptional regulatory mechanisms. *E-selectin* is transcribed in endothelium from an NF-κB–dependent promoter in response to diverse inflammatory stimuli. Developmentally regulated expression in trophoblasts, not further modulated during inflammation, utilizes a distinct transcriptional promoter and alternative first exon located several kb 5′ in the genome. Endothelium and trophoblasts are thought to produce E-selectin protein with identical primary structure. *Upper graphs*: Kinetics of E-selectin mRNA and protein expression in murine placental trophoblast (*left*) and vascular endothelium with inflammatory activation at time 0 hours (*right*). *Lower diagram*: Structural and regulatory features of the murine *E-selectin* genomic locus. The human *E-selectin* locus is organized similarly. (*Key*: Solid line with dashed ends, DNA; open boxes, translated exons; grey boxes, untranslated exons; dotted lines, mRNA splicing events leading to mature *E-selectin* mRNA; arrows, initiation of translation; arrowheads, additional translated exons not shown.)

explores the implications of these species-specific differences for the utility of various experimental models of selectin function.)

E-selectin expression is increased in specific vascular compartments under pathophysiological settings such as sepsis, vasculitis, atherosclerosis, ischemia–reperfusion injury, rheumatoid arthritis, allograft rejection, graft versus host disease, various forms of gastrointestinal inflammation, acute hepatitis, and tumor metastasis. Expression also occurs associated with endothelial proliferation in vitro and in certain in vivo settings. It is not clear if this represents a distinct activation mechanism at the molecular level or an overlap of inflammatory and growth regulatory pathways.

This temporal, spatial and physiological heterogeneity may both reflect and contribute to the integrated complexity of the in vivo milieu, and is thought to play an important role in inflammation by focusing and restricting leukocyte recruitment to appropriate temporal stages and vascular locations (18).

A nuclear factor (NF)-κB-dependent promoter regulates the rapid induction of *E-selectin* mRNA at the transcriptional level (Figure 118.2) (22,23). NF-κB interacts with additional transcriptional activators to form an enhancer complex that transmits activation signals to coactivators and basal factors, resulting in *E-selectin* gene transcription. C-jun N-terminal kinase (JNK)/stress-activated protein kinases (SAPKs), presumably acting via the phosphorylation of c-jun and activation of the transcription factor c-jun/ATF-2, have been identified as additional critical mediators of E-selectin inducibility that may, at least in part, explain the differential responsiveness of different NF-κB–dependent endothelial adhesion molecule genes in different pathological, and potentially physiological, settings (24,25).

Interestingly, stimulation of E-selectin expression by TNF-α, IL-1, or CD40 ligand renders ECs transiently nonresponsive to restimulation by the same, but not by a different, ligand. This is not due to downregulation of receptor expression. Such so-called homologous desensitization is paralleled

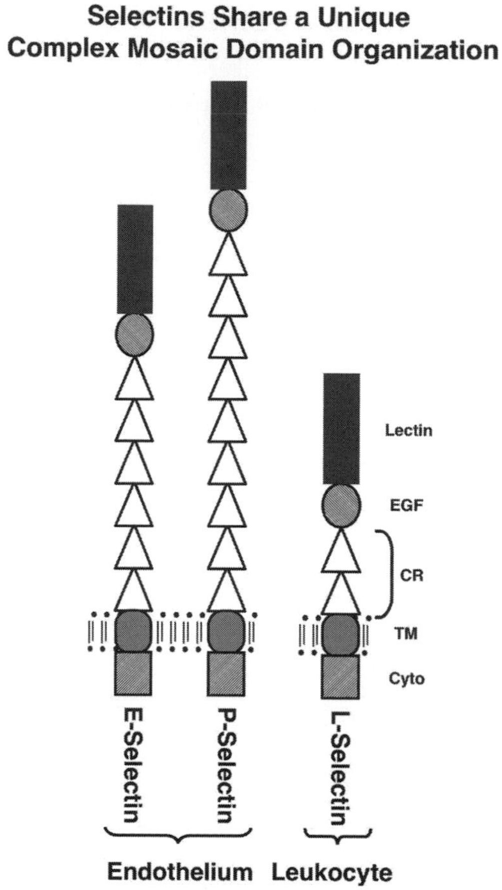

Figure 118.2. The selectins share a unique, complex, mosaic domain organization: an N-terminal C-type lectin domain, an EGF-like domain, several complement-regulatory-like domains (*CR*), a transmembrane domain (*TM*), and a short cytoplasmic domain (*Cyto*). Variable numbers of CR domains position the lectin and EGF domains of E-, P-, and L-selectin differently within the cell surface glycocalyx, contributing to selectin-specific functions in endothelial–circulating cell interactions. Balls and sticks indicate the lipid bilayer of the endothelial plasma membrane.

by changes in JNK activation, suggesting that JNK also may regulate this process (26). Homologous desensitization may limit the extent of an inflammatory response mediated predominately by a single mediator, in contrast to responses with more complicated activator profiles.

Stimulated E-selectin expression is also usually transient due to rapid degradation of *E-selectin* mRNA (27) and internalization of the mature protein from the cell surface (28), combined with inactivation of NF-κB (29). This downregulation is thought to contribute to the orderly resolution of inflammatory responses.

What emerges is a complex yet robust and self-limiting regulatory system that functions to place E-selectin protein at the EC surface rapidly and early during inflammation and to remove it equally rapidly. This constrains when and where leukocyte transmigration may occur.

"Basal" Expression in Trophoblast and Hematopoietic Endothelium

Constitutive E-selectin expression in both murine and human placentae involves complex, nested first exons, and presumed transcriptional promoters physically and functionally distinct from those involved in inducible expression in the endothelium (see Figure 118.1) (15). This implies that evolutionarily conserved and lineage-restricted transcriptional mechanisms regulate expression in homologous trophoblast populations in both species. However, additional information is lacking regarding the molecular mechanisms involved, how they are regulated, and their developmental or other functional significance. Similarly, the genetic regulatory factors and the molecular mechanisms specifying constitutive E-selectin expression in fetal and adult bone marrow endothelium are not known. This expression utilizes the same transcription start site as inducible endothelial expression (see Figure 118.1), and therefore probably uses a regulatory mechanism distinct from that specifying constitutive expression in trophoblast. How such constitutive expression initiates from a site that is also immediately adjacent to an NF-κB–dependent promoter is not clear.

PROTEIN STRUCTURE

E-selectin consists of an amino terminal, extracellular, calcium-dependent lectin domain followed by an EGF-like domain, six short near-replicate domains (short consensus repeats [SCRs]) homologous to those domains in proteins that bind complement, a transmembrane domain, and a 32-amino acid cytoplasmic domain with multiple potential phosphorylation sites and other residues subject to post-translational modification (Figures 118.2 and 118.3). This complex mosaic structure, shared with P- and L-selectin, is unique among the adhesion structures and imparts special properties to selectin-mediated interactions.

E-selectin's lectin domain interacts with carbohydrate structures bearing sialyl Lewis X, or sLe^x (NeuAcα2–3Galβ1–4[Fucα1–3]GlcNAc) in a manner similar to P- and L-selectin. However, meaningful interactions occur only with specific leukocyte counter-receptors, implying that additional structural features, most likely including the EGF domain (30), determine functionally relevant ligand specificity and, in part, differentiate specific E-, P-, and L-selectin functions. Although the EGF domain does not interact extensively with the lectin domain, absence of the EGF or SCR domains decreases ligand binding, suggesting that each also contributes to overall cell adhesion. The lectin and EGF domains in combination likely dictate ligand specificity, as has been shown for P-selectin (30). The SCR domains also may position the lectin/EGF domains in a functionally significant location within the endothelial glycocalyx and/or may mediate oligomerization or overall protein structure.

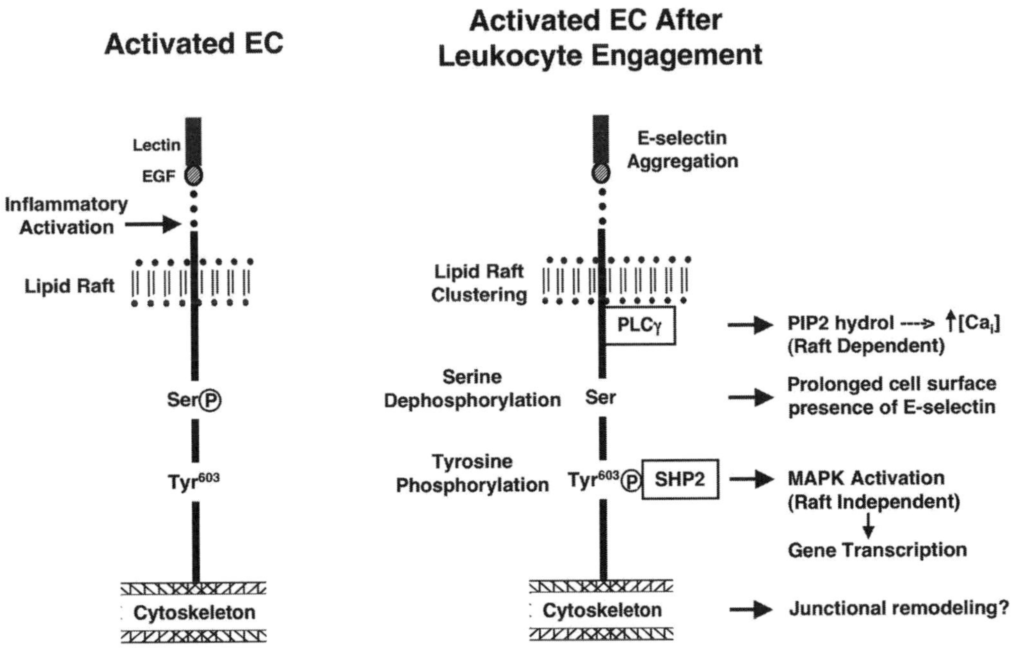

Figure 118.3. E-selectin as a signal transducing receptor. Upon leukocyte engagement with its extracellular lectin domain, E-selectin clusters in lipid rafts and its cytoplasmic domain (a) binds and activates PLC-γ and leads to PIP2 hydrolysis and increased $[Ca]_I$; (b) undergoes serine dephosphorylation that may prolong cell surface expression in certain vascular beds; (c) undergoes tyrosine^{603} phosphorylation with SHP2 recruitment, MAPK activation, and immediate-early gene transcription and; (d) interacts with the actin cytoskeleton with the potential to modulate junctional interactions with adjacent EC and influence leukocyte transmigration. Such outside-in signaling may thus modulate the phenotype of a given EC or groups of ECs during inflammation and, potentially, in other physiologic and/or pathophysiologic states as well.

More recently it has become clear that the E-selectin cytoplasmic domain, in addition to its role in endocytosis (31), interacts with intracellular components, potentially transmitting a leukocyte-dependent outside-in signaling response to the endothelium during inflammation. Upon leukocyte engagement (only PMNs have been studied to date), E-selectin, initially present in lipid rafts, associates with caveolae. This leads to phospholipase C (PLC)-γ activation, association of E-selectin with the actin cytoskeleton, cytoskeletal remodeling, specific serine residue dephosphorylation, tyrosine residue phosphorylation with mitogen-activated protein kinase (MAPK) activation, and c-fos transcription (32). These leukocyte-dependent changes, all of which require the cytoplasmic domain, occurring within the EC may contribute to leukocyte slow rolling and firm adhesion (8,33,34) and to other as yet undefined endothelial phenotypic changes resulting from leukocyte–E-selectin interaction at developing inflammatory sites.

E-SELECTIN MUTATIONS WITH CLINICAL SIGNIFICANCE

Patients with mutations of a *GDP-fucose transporter* gene, resulting in hypofucosylated glycoconjugates, suffer from leukocyte adhesion deficiency (LAD)-2 (35). This disorder is characterized by an inability to mount an inflammatory response, and is caused, in part, by defective counter-receptors for all three selectins. These patients also show developmental abnormalities, presumably reflecting other consequences of abnormal fucose metabolism.

The first report of specific, clinically evident E-selectin deficiency involved a patient with recurrent infections and an inability to mount purulent responses to infectious bacteria (36). This patient lacked E-selectin protein (assessed by immunohistochemistry) on the endothelium of inflamed skin obtained by clinical biopsy and on endothelium of biopsied noninflamed skin stimulated with TNF-α in organ culture ex vivo. This same patient showed only mildly elevated plasma soluble E-selectin (sE-selectin) levels, with a normal structural gene and normal levels of mRNA. Although the clinical presentation appears to reflect the consequences of isolated E-selectin functional deficiency, the underlying defect and molecular mechanism responsible for this unusual presentation remain unknown. This patient's *P-selectin* structural gene and expression pattern was found to be normal, suggesting that the clinical effects are due to specific abnormalities in E-selectin.

Point mutations of E-selectin associated with human disease have recently been described. E-selectin/Ser128Arg, involving a mutation within the EGF domain, is associated with altered plasma levels of sE-selectin, altered ligand affinity, increased tethering by a broader than normal range of leukocytes, increased ligation-induced phosphorylation of endothelial extracellular-signal-regulated kinase (ERK)-1/2 and p38 MAPK, and increased risk of atherosclerosis, myocardial infarction, and restenosis after angioplasty (37). E-selectin/Leu554Phe also is associated with high blood pressure (38) and atherosclerosis (39), whereas circulating concentrations of sE-selectin are associated with hypertension and obesity (40).

ROLE IN HOMEOSTASIS

E-selectin has been implicated in developmental, physiological, and pathological processes such as inflammation, hematopoiesis, tumor invasion and metastasis, and blood vessel morphogenesis.

Experimental Models for Investigating Endothelial Selectin Function In Vivo

Although genetically engineered mice have been invaluable for determining the mechanisms regulating gene expression during inflammation, their use to investigate endothelial selectin function in vivo is compromised by differences in transcriptional regulation at the human and murine selectin loci. Specifically, murine P-selectin shares with E-selectin an endotoxin- and inflammatory cytokine–inducible pattern of gene expression that is restricted to E-selectin in humans (41,42). This suggests that studies in mice may underestimate the contribution of human E-selectin to responses regulated by NF-κB, and may explain why E-selectin plays a lesser role in murine inflammation than initially anticipated. However, IL-4, oncostatin M, and IL-6/IL-6 receptor complexes can activate P-selectin expression in both species (43), suggesting that species-specific differences in inflammatory expression are complex. In fact, mice lacking both P- and E-selectin show dramatic leukocytosis, elevation of circulating cytokines, and hematopoietic abnormalities (9) that may more accurately predict the phenotype of isolated human E-selectin deficiency (although some properties may reflect lack of P-selectin–specific and/or overlapping functions). The phenotypes of mice with genetically engineered endothelial selectin alleles are summarized in the Table 118-1.

Alternative approaches involve other animals, such as rabbits or pigs, and studying the vasculature of transplanted human tissues, such as skin, in immunodeficient mice. The latter strategy, which has been accomplished by several investigators, can be combined with human bone marrow transplantation into mice to also reconstitute human leukocytes in vivo. This may most accurately represent human endothelial selectin expression and function, with cells bearing authentic counter-receptors in vivo.

Table 118-1: Mice with Genetically Engineered Endothelial Selectin Alleles

Knockout	*Phenotype In Vivo*
E-selectin (8,11,12)	Normal viability and fertility Absent slow rolling and diminished firm adhesion of WBCs on inflamed ECs Normal WBC extravasation into inflamed tissue Insensitive to antiangiogenic effects of endostatin
P-selectin (58,59)	Normal viability and fertility Nearly absent WBC rolling on inflamed ECs Delayed WBC extravasation into inflamed tissue
Combined *E-selectin* and *P-selectin* (9,13)	Opportunistic bacterial infections, including ulcerative dermatitis, leading to sepsis with death beginning at 3 months of age Absent WBC rolling on inflamed ECs Absent WBC extravasation into inflamed tissue Elevated circulating levels of hematopoietic cytokines with altered hematopoiesis

WBC, white blood cell

Cell–Cell Interactions in Inflammation

E-selectin interacts with PMNs in acute inflammation, monocytes in chronically inflamed tissues, and certain lymphocyte subtypes in immune responses. In these contexts, E-selectin is a candidate for functional importance in a vast array of human diseases and reparative processes. Studies in genetically deficient mice reveal that, in the context of inflammation, E-selectin is uniquely important in vivo for leukocyte slow rolling and subsequent firm adhesion to acutely cytokine-activated microvascular endothelium (see Table 118-1) (8,33,34). This has been demonstrated for PMN recruitment. Whether E-selectin mediates a similar specific function in monocyte and lymphocyte interactions in vivo is not known.

In general, leukocyte expression of glycosyltransferase VII is required for the production of E-selectin counter-receptors. The human disease LAD-2 described previously, in which deficient fucosylation of selectin counter-receptors leads to severe inflammatory disease, illustrates the importance of such structures.

Hematopoiesis

E-selectin expressed by the endothelium of hematopoietic organs mediates homing of hematopoietic progenitor cells from peripheral blood (44). However, E-selectin also increases apoptosis and enhances rapid differentiation of such cells, suggesting its overall effect may be to decrease long-term hematopoietic potential. In contrast, P-selectin also mediates homing but shows largely opposite effects on progenitor cell survival and differentiation. The overall effect in mice genetically deficient in both E- and P-selectin simultaneously is enhanced hematopoiesis with increased progenitor cell numbers in bone marrow and spleen (see Table 118-1), as is seen in mice genetically deficient in all selectin counter-receptors (45).

Tumor Metastasis

Studies of tumor metastasis using either specific adhesion-blocking reagents, genetically deficient mice, or mice genetically engineered to express cell surface E-selectin at ectopic sites or soluble E-selectin in the systemic circulation (see Table 118-1) (46) suggest that E-selectin can act as a recognition structure for counter-receptors on several types of circulating tumor cells, thus directly mediating transendothelial migration (TEM) during the metastatic process and providing a target for therapeutic intervention (47). Such interactions also may induce bidirectional outside-in signaling in both the tumor and the endothelium, altering cellular phenotypes in ways that further modulate the metastatic process. Circulating levels of sE-selectin also correlate with hepatic metastases, of breast and colorectal carcinoma. These levels correlate inversely with prognosis (48–50), although the mechanism and functional significance of these observations is not clear.

Blood Vessel Formation and Regression

As mentioned, genetic loss-of-function experiments show that E-selectin is not required for developmental or reproductive angiogenesis in mice. In addition, corneal angiogenesis in response to basic fibroblast growth factor (bFGF) and TNF-α is not impaired (51). However, early studies in vivo using animal models and in vitro using human cells suggested that either soluble or cell-associated E-selectin might be involved in endothelial proliferation or blood vessel formation (see Table 118-1) (10,52,53).

Recent evidence suggests E-selectin is required for the antiangiogenic action of endostatin, in vitro as well as in vivo, via a mechanism that involves the cytoplasmic domain. This implicates E-selectin–dependent intraendothelial signaling events in the molecular mechanism of endostatin action (see Table 118-1) (11). Interfering with E-selectin expression and/or signaling by recruited tumor blood vessels may thus be one mechanism by which tumor cells specify resistance to the antiangiogenic, antitumor action of endostatin.

AUTOCRINE AND PARACRINE EFFECTS

E-selectin transmits adhesion-dependent intracellular signals to ECs that must interact intimately with their neighbors to initially establish, transiently relinquish, and ultimately re-establish, in a very selective manner, a barrier to leukocyte transmigration. Newly synthesized E-selectin is localized to cholesterol-rich plasma membrane microdomains lacking caveolin. Leukocyte engagement results in the following changes:

- Larger membrane microdomains form that may contain caveolin-1 and in which PLC-γ associates with the E-selectin cytoplasmic domain, leading to phosphatidylinositol bisphosphate (PIP2) hydrolysis and increased intracellular calcium concentrations.
- The E-selectin cytoplasmic domain also becomes physically associated with the actin cytoskeleton, leading to cytoskeletal changes that may include junctional remodeling (6).
- Constitutively phosphorylated serine in the cytoplasmic domain becomes dephosphorylated (7), which can extend the life of E-selectin at the cell surface (8).
- Tyrosine^{603} becomes phosphorylated, initiating SHP2 recruitment, activation of the MAPK signaling pathway, and transcriptional activation of the immediate early response gene, c-Fos (9,10).

Interestingly, only PLC-γ hydrolysis of PIP2, but not MAPK activation, requires E-selectin localization to lipid rafts, suggesting that different signaling pathways or combinations of pathways might be recruited by activating stimuli that differentially affect the plasma membrane distribution of E-selectin.

Although the ultimate functional significance of these signaling events is not clear, they may alter the "phenotype" of the EC in which they occur. This, in turn, may influence this cell's interaction not only with the leukocyte (or potentially the hematopoietic precursor or tumor cell) initially engaging E-selectin, but also with adjacent endothelium. This presents the intriguing possibility that E-selectin may act in a cell-autonomous fashion to modulate the phenotype of the endothelium, actively supporting TEM during inflammation, hematopoiesis, or tumor cell metastasis. By comparison, the P-selectin cytoplasmic domain interacts with clathrin-coated pits (54) and is involved in intracellular trafficking, localization to EC Weibel-Palade bodies (but not platelet α-granules), turnover of the mature protein product (55), and intracellular phosphorylation events (56).

Combined with the highly regulated pattern of E-selectin expression, such "phenotypic modulation" of specific ECs also is positioned to function as part of an autocrine/paracrine regulatory system "marking" or revealing, on an integrated tissue and organ scale, exactly when and where leukocyte transmigration occurs and inflammation (or hematopoiesis or metastasis) develops. Experiments using animal models mimicking the expression and function of human endothelial selectins will help define the contribution of such effects to physiological and pathophysiological transmigration.

E-selectin–counter-receptor interactions also trigger paracrine consequences of leukocyte-endothelial interactions.

Specifically, adhesion via E-selectin can lead to activation and affinity maturation of the Mac-1 integrin receptor on leukocytes (57). This may be the mechanism underlying E-selectin's role in the transition from leukocyte rolling to firm adhesion during inflammation, as described earlier.

INTERACTIONS WITH OTHER OUTPUTS

Substantial evidence indicates that endothelial adhesion structures cooperate to orchestrate leukocyte transmigration during acute and chronic inflammation. E-selectin's importance in slow rolling and firm adhesion suggests that it interfaces, functionally if not physically, with Mac1, ICAM-1, and/or vascular cell adhesion molecule (VCAM)-1, each also implicated in these processes in various contexts, to mediate inflammatory adhesion cascades. It seems likely that different combinations of adhesion activities contributed by various adhesion structures may underlie, in part, the different leukocyte types, vascular locations, and kinetics that characterize different inflammatory states, and that perturbations of these interactions, perhaps subtle, may underlie the devastating consequences of inflammatory disease.

DIAGNOSTIC AND THERAPEUTIC IMPLICATIONS

E-selectin's unique specificity for the endothelium and strong association with cellular activation suggest it may be a useful "surrogate marker" for assessing the overall activation state of an individual's endothelium in clinical settings. This might be especially useful in patients at risk for or experiencing inflammatory or malignant/metastatic insults or disease. Therefore, the occurrence of circulating, soluble forms of E-selectin and their relation to a variety of diseases has been extensively investigated. Although E-selectin lacking the cytoplasmic domain circulates in plasma of humans and experimental animals, neither alternatively spliced isoforms lacking the transmembrane domain nor a regulated or unregulated secretion mechanism have been reported. Thus, sE-selectin probably derives from proteolytic cleavage of the mature EC surface protein, followed by release of a shortened form into the circulation.

In general, circulating levels of sE-selectin correlate well with clinical severity of sepsis and other systemic inflammatory states (37–40). Elevated sE-selectin has been observed in patients with hypertension, diabetes, and hyperlipidemia but not in patients with more localized diseases, such as atherosclerosis, in which endothelial activation also plays a prominent role. However, it is not clear that this parameter provides significant diagnostic advantages over traditional clinical assessment. It also seems unlikely that the levels of sE-selectin are adequate to block functional interactions between circulating leukocytes or tumor cells and activated endothelium, as appears possible for L-selectin. The prognostic implications of sE-selectin in malignancy were addressed earlier.

KEY POINTS

- E-selectin, a cytokine-inducible, EC surface transmembrane glycoprotein lectin, is important for leukocyte and tumor cell transmigration. It participates in adhesion-dependent outside-in signaling that may modulate the EC phenotype during inflammation and metastasis, and it may be involved in certain antiangiogenic responses.
- NF-κB–dependent E-selectin expression in inflamed vascular endothelium, in cooperation with P-selectin, leads to lectin–counter-receptor interactions, slow rolling, and firm adhesion of circulating cells, all necessary for leukocyte and tumor cell transmigration into tissues.
- When circulating cells bind E-selectin, its cytoplasmic domain interacts with the cytoskeleton and other signal transduction components in ways that may modulate the EC phenotype during inflammation.
- E-selectin's properties suggest it may play other as yet undiscovered roles in inflammation and vasculature physiology and pathophysiology, such as its recently demonstrated role in the antiangiogenic action of endostatin.

Future Goals

- To elucidate the specific contributions of E-selectin and P-selectin to leukocyte recruitment during inflammation, to tumor metastasis, to hematopoiesis, and to physiologic and pathophysiologic blood vessel formation and regression in human patients
- To determine E-selectin's prognostic and/or therapeutic utility in human patients

REFERENCES

1 Bevilacqua MP, Pober JS, Mendrick DL, et al. Identification of an inducible endothelial-leukocyte adhesion molecule. *Proc Natl Acad Sci USA*. 1987;84(24):9238–9242.

2 Johnston GI, Cook RG, McEver RP. Cloning of GMP-140, a granule membrane protein of platelets and endothelium: sequence similarity to proteins involved in cell adhesion and inflammation. *Cell*. 1989;56(6):1033–1044.

3 Siegelman MH, van de Rijn M, Weissman IL. Mouse lymph node homing receptor cDNA clone encodes a glycoprotein revealing tandem interaction domains. *Science*. 1989;243(4895):1165–1172.

4 Bevilacqua M, Butcher E, Furie B, et al. Selectins: a family of adhesion receptors. *Cell*. 1991;67(2):233.

5 Leshko-Lindsay LA, Corces VG. The role of selectins in *Drosophila* eye and bristle development. *Development*. 1997;124(1):169–180.

6 Milstone DS, O'Donnell PE, Stavrakis G, et al. E-selectin expression and stimulation by inflammatory mediators are developmentally regulated during embryogenesis. *Lab Invest*. 2000; 80(6):943–954.
7 Terry RW, Kwee L, Levine JF, Labow MA. Cytokine induction of an alternatively spliced murine vascular cell adhesion molecule (VCAM) mRNA encoding a glycosylphosphatidylinositol-anchored VCAM protein. *Proc Natl Acad Sci USA*. 1993;90(13): 5919–5923.
8 Milstone DS, Fukumura D, Padgett RC, et al. Mice lacking E-selectin show normal numbers of rolling leukocytes but reduced leukocyte stable arrest on cytokine-activated microvascular endothelium. *Microcirculation*. 1998;5(2–3):153–171.
9 Frenette PS, Mayadas TN, Rayburn H, et al. Susceptibility to infection and altered hematopoiesis in mice deficient in both P- and E-selectins. *Cell*. 1996;84:563–574.
10 Nguyen M, Strubel NA, Bischoff J. A role for sialyl Lewis-X/A glycoconjugates in capillary morphogenesis [published erratum appears in *Nature*. 1993;366(6453):368]. *Nature*. 1993;365 (6443):267–269.
11 Yu Y, Moulton KS, Khan MK, et al. E-selectin is required for the antiangiogenic activity of endostatin. *Proc Natl Acad Sci USA*. 2004;101(21):8005–8010.
12 Labow MA, Norton CR, Rumberger JM, et al. Characterization of E-selectin-deficient mice: demonstration of overlapping function of the endothelial selectins. *Immunity*. 1994;1(8):709–720.
13 Bullard DC, Kunkel EJ, Kubo H, et al. Infectious susceptibility and severe deficiency of leukocyte rolling and recruitment in E-selectin and P-selectin double mutant mice. *J Exp Med*. 1996; 183(5):2329–2336.
14 Gerritsen ME, Shen CP, Atkinson WJ, et al. Microvascular ECs from E-selectin-deficient mice form tubes in vitro. *Lab Invest*. 1996;75(2):175–184.
15 Milstone DS, Redline RW, O'Donnell PE, et al. E-selectin expression and function in a unique placental trophoblast population at the fetal-maternal interface: regulation by a trophoblast-restricted transcriptional mechanism conserved between humans and mice. *Dev Dyn*. 2000;219(1):63–76.
16 Kruse A, Merchant MJ, Hallmann R, Butcher EC. Evidence of specialized leukocyte-vascular homing interactions at the maternal/fetal interface. *Eur J Immunol*. 1999;29(4):1116–1126.
17 Eppihimer MJ, Wolitzky B, Anderson DC, et al. Heterogeneity of expression of E- and P-selectins in vivo. *Circ Res*. 1996;79(3):560–569.
18 Hickey MJ, Kanwar S, McCafferty DM, et al. Varying roles of E-selectin and P-selectin in different microvascular beds in response to antigen. *J Immunol*. 1999;162(2):1137–1143.
19 Messadi DV, Pober JS, Fiers W, et al. Induction of an activation antigen on postcapillary venular endothelium in human skin organ culture. *J Immunol*. 1987;139(5):1557–1562.
20 Petzelbauer P, Bender JR, Wilson J, Pober JS. Heterogeneity of dermal microvascular endothelial cell antigen expression and cytokine responsiveness in situ and in cell culture. *J Immunol*. 1993;151(9):5062–5072.
21 Olofsson AM, Arfors KE, Ramezani L, et al. E-selectin mediates leukocyte rolling in interleukin-1-treated rabbit mesentery venules. *Blood*. 1994;84(8):2749–2758.
22 Whelan J, Ghersa P, Hooft van Huijsduijnen R, et al. An NF kappa B-like factor is essential but not sufficient for cytokine induction of endothelial leukocyte adhesion molecule 1 (ELAM-1) gene transcription. *Nucleic Acids Res*. 1991;19(10):2645–2653.
23 Collins T, Read MA, Neish AS, et al. Transcriptional regulation of endothelial cell adhesion molecules: NF-kappa B and cytokine-inducible enhancers. *FASEB J*. 1995;9(10):899–909.
24 Min W, Pober JS. TNF initiates E-selectin transcription in human endothelial cells through parallel TRAF-NF-kappa B and TRAF-RAC/CDC42-JNK-c-Jun/ATF2 pathways. *J Immunol*. 1997;159 (7):3508–3518.
25 Read MA, Whitley MZ, Gupta S, et al. Tumor necrosis factor alpha-induced E-selectin expression is activated by the nuclear factor-kappaB and c-JUN N-terminal kinase/p38 mitogen-activated protein kinase pathways. *J Biol Chem*. 1997;272(5): 2753–2761.
26 Karmann K, Min W, Fanslow WC, Pober JS. Activation and homologous desensitization of human endothelial cells by CD40 ligand, tumor necrosis factor, and interleukin 1. *J Exp Med*. 1996;184(1):173–182.
27 Chu W, Presky DH, Swerlick RA, Burns DK. Alternatively processed human E-selectin transcripts linked to chronic expression of E-selectin in vivo. *J Immunol*. 1994;153(9):4179–4189.
28 Kluger MS, Johnson DR, Pober JS. Mechanism of sustained E-selectin expression in cultured human dermal microvascular endothelial cells. *J Immunol*. 1997;158(2):887–896.
29 Read MA, Neish AS, Gerritsen ME, Collins T. Postinduction transcriptional repression of E-selectin and vascular cell adhesion molecule-1. *J Immunol*. 1996;157(8):3472–3479.
30 Kansas GS, Saunders KB, Ley K, et al. A role for the epidermal growth factor-like domain of P-selectin in ligand recognition and cell adhesion. *J Cell Biol*. 1994;124(4):609–618.
31 Chuang PI, Young BA, Thiagarajan RR, et al. Cytoplasmic domain of E-selectin contains a non-tyrosine endocytosis signal. *J Biol Chem*. 1997;272(40):24813–24818.
32 Kiely JM, Hu Y, Garcia-Cardena G, Gimbrone MA Jr. Lipid raft localization of cell surface E-selectin is required for ligation-induced activation of phospholipase C gamma. *J Immunol*. 2003;171(6):3216–3224.
33 Kunkel EJ, Ley K. Distinct phenotype of E-selectin-deficient mice – E-selectin is required for slow leukocyte rolling in vivo. *Circ Res*. 1996;79(6):1196–1204.
34 Ley K, Allietta M, Bullard DC, Morgan S. Importance of E-selectin for firm leukocyte adhesion in vivo. *Circ Res*. 1998;83 (3):287–294.
35 Wild MK, Luhn K, Marquardt T, Vestweber D. Leukocyte adhesion deficiency II: therapy and genetic defect. *Cells Tissues Organs*. 2002;172(3):161–173.
36 DeLisser HM, Christofidou-Solomidou M, Sun J, et al. Loss of endothelial surface expression of E-selectin in a patient with recurrent infections. *Blood*. 1999;94(3):884–894.
37 Jilma B, Marsik C, Kovar F, et al. The single nucleotide polymorphism Ser128Arg in the E-selectin gene is associated with enhanced coagulation during human endotoxemia. *Blood*. 2005; 105(6):2380–2383.
38 Marteau JB, Sass C, Pfister M, et al. The Leu554Phe polymorphism in the E-selectin gene is associated with blood pressure in overweight people. *J Hypertens*. 2004;22(2):305–311.
39 Wenzel K, Stahn R, Speer A, et al. Functional characterization of atherosclerosis-associated Ser128Arg and Leu554Phe E-selectin mutations. *Biol Chem*. 1999;380(6):661–667.
40 Glowinska B, Urban M, Peczynska J, Florys B. Soluble adhesion molecules (sICAM-1, sVCAM-1) and selectins (sE selectin, sP selectin, sL selectin) levels in children and adolescents with

obesity, hypertension, and diabetes. *Metabolism.* 2005;54(8): 1020–1026.

41 Pan J, Xia L, McEver RP. Comparison of promoters for the murine and human P-selectin genes suggests species-specific and conserved mechanisms for transcriptional regulation in endothelial cells. *J Biol Chem.* 1998;273(16):10058–10067.

42 Sanders WE, Wilson RW, Ballantyne CM, Beaudet AL. Molecular cloning and analysis of in vivo expression of murine P-selectin. *Blood.* 1992;80(3):795–800.

43 Yao L, Pan J, Setiadi H, et al. Interleukin 4 or oncostatin M induces a prolonged increase in P-selectin mRNA and protein in human ECs. *J Exp Med.* 1996;184(1):81–92.

44 Frenette PS, Subbarao S, Mazo IB, et al. Endothelial selectins and vascular cell adhesion molecule-1 promote hematopoietic progenitor homing to bone marrow. *Proc Natl Acad Sci USA.* 1998;95(24):14423–14428.

45 Homeister JW, Thall AD, Petryniak B, et al. The alpha (1,3)fucosyltransferases FucT-IV and FucT-VII exert collaborative control over selectin-dependent leukocyte recruitment and lymphocyte homing. *Immunity.* 2001;15(1):115–126.

46 Biancone L, Araki M, Araki K, et al. Redirection of tumor metastasis by expression of E-selectin in vivo. *J Exp Med.* 1996;183(2): 581–587.

47 Khatib AM, Kontogiannea M, Fallavollita L, et al. Rapid induction of cytokine and E-selectin expression in the liver in response to metastatic tumor cells. *Cancer Res.* 1999;59(6):1356–1361.

48 Eichbaum MH, de Rossi TM, Kaul S, Bastert G. Serum levels of soluble E-selectin are associated with the clinical course of metastatic disease in patients with liver metastases from breast cancer. *Oncol Res.* 2004;14(11–12):603–610.

49 Uner A, Akcali Z, Unsal D. Serum levels of soluble E-selectin in colorectal cancer. *Neoplasma.* 2004;51(4):269–274.

50 Ito K, Ye CL, Hibi K, et al. Paired tumor marker of soluble E-selectin and its ligand sialyl Lewis A in colorectal cancer. *J Gastroenterol.* 2001;36(12):823–829.

51 Hartwell DW, Butterfield CE, Frenette PS, et al. Angiogenesis in P- and E-selectin-deficient mice. *Microcirculation.* 1998;5(2–3):173–178.

52 Koch AE, Halloran MM, Haskell CJ, et al. Angiogenesis mediated by soluble forms of E-selectin and vascular cell adhesion molecule-1. *Nature.* 1995;376(6540):517–519.

53 Kraling BM, Razon MJ, Boon LM, et al. E-selectin is present in proliferating endothelial cells in human hemangiomas. *Am J Pathol.* 1996;148(4):1181–1191.

54 Setiadi H, Sedgewick G, Erlandsen SL, McEver RP. Interactions of the cytoplasmic domain of P-selectin with clathrin-coated pits enhance leukocyte adhesion under flow. *J Cell Biol.* 1998; 142(3):859–871.

55 Green SA, Setiadi H, McEver RP, Kelly RB. The cytoplasmic domain of P-selectin contains a sorting determinant that mediates rapid degradation in lysosomes. *J Cell Biol.* 1994;124(4):435–448.

56 Fujimoto T, McEver RP. The cytoplasmic domain of P-selectin is phosphorylated on serine and threonine residues. *Blood.* 1993; 82(6):1758–1766.

57 Smith CW. Possible steps involved in the transition to stationary adhesion of rolling neutrophils: a brief review. *Microcirculation.* 2000;7(6 Pt 1):385–394.

58 Mayadas TN, Johnson RC, Rayburn H, et al. Leukocyte rolling and extravasation are severely compromised in P-selectin-deficient mice. *Cell.* 1993;74(3):541–554.

59 Ley K, Bullard DC, Arbones ML, et al. Sequential contribution of L- and P-selectin to leukocyte rolling in vivo. *J Exp Med.* 1995; 181(2):669–675.

119

Endothelial Cell Apoptosis

Elizabeth O. Harrington, Qing Lu, and Sharon Rounds

Providence VA Medical Center, Brown Medical School, Providence, Rhode Island

In an article published by Kerr and colleagues in 1972, the term *apoptosis*, defined as "dropping off or falling off of . . . leaves from trees," was used to describe energy-dependent cellular suicide with distinct morphologic characteristics, such as chromatin condensation, plasma membrane blebbing, and apoptotic body formation (1). The term *apoptosis,* however was first used in medicine by Hippocrates (460–370 B.C.) to describe the "falling off of the bones" in a book discussing bone fractures (2). The term was also reportedly utilized by Galen (129–201 A.D.) to describe the "dropping of the scabs" (2). Since the re-introduction of *apoptosis* to scientific and medical literature 33 years ago, close to 95,000 articles published and listed in National Library of Medicine's PubMed make use of the term.

Apoptosis or programmed cell death describes a genetically determined elimination of cells that are in excess, injured, infected, or aged. Once committed to programmed suicide, cells undergo well-ordered morphologic and molecular alterations, including cytoskeletal rearrangement, nuclear membrane collapse, chromatin condensation, cell shrinkage, plasma membrane blebbing, and the formation of disassembled membrane-enclosed vesicles, referred to as apoptotic bodies (1). DNA fragmentation also occurs by activated endonucleases, resulting in the cleavage of genomic DNA into 180- to 200-base pair fragments (1). Additionally, phosphatidylserine exposure at the cell surface acts as a chemoattractant for macrophages, thus promoting the engulfment of these apoptotic bodies and the prevention of inflammation due to limited release of intracellular contents (3,4). In contrast, cells undergoing necrosis swell and lyse, thereby releasing intracellular contents into the interstitium, leading to an inflammatory response (5). Apoptosis plays a fundamental role in normal development and tissue homeostasis, as well as in the progression of pathophysiologic diseases.

In this chapter we discuss, in general terms, the process of apoptosis, the stimuli for endothelial cell (EC) apoptosis, and evidence for EC apoptosis in normal and abnormal vasculature.

CELLULAR AND MOLECULAR BASIS OF APOPTOSIS

Various factors, such as Fas ligand (FasL), tumor necrosis factor (TNF)-α, metabolite deprivation, DNA damage, inhibition of transcription and translation, hypoxia, or interferon (IFN)-γ, can trigger programmed cell death. Although cells differentially respond to apoptotic stimuli, there are two fundamental signaling pathways by which apoptosis is mediated: the extrinsic and intrinsic pathways. The extrinsic pathway is activated when the apoptotic stimulus acts through a death receptor. Conversely, the intrinsic pathway is triggered by internal apoptotic signals involving the mitochondria. Although each apoptotic signaling pathway is activated via specific signaling mechanisms, both pathways merge and share some final common death machinery utilizing the aspartate-specific cysteinyl protease (caspase) cascades. These apoptotic pathways and their intermediate signaling molecules are delineated in Figure 119.1.

The Extrinsic Pathway

Death receptors, a family of transmembranous proteins belonging to the TNF receptor superfamily, directly bind to extracellular death ligands and transmit the signal to apoptotic-inducing intracellular machinery, ultimately leading to cell death. Death receptors have a cysteine–rich extracellular domain and a cytoplasmic death domain (DD). Through its death domain, death receptors transmit the death signals to caspases by associating with additional death domain-containing adapters. The best characterized death receptors are Fas (CD95, Apo1), TNF receptor (TNFR)-1, and death receptors 3, 4, and 5 (DR3, DR4, and DR5) (6,7).

FasL is primarily expressed in activated T cells, natural killer cells, and hepatocytes (6). FasL and Fas also are expressed in ECs (8). Upon FasL binding, Fas forms a homotrimeric complex. Additional binding of FasL promotes the formation of a hexameric Fas protein complex and the recruitment of the adapter protein, Fas-associated DD (FADD), forming

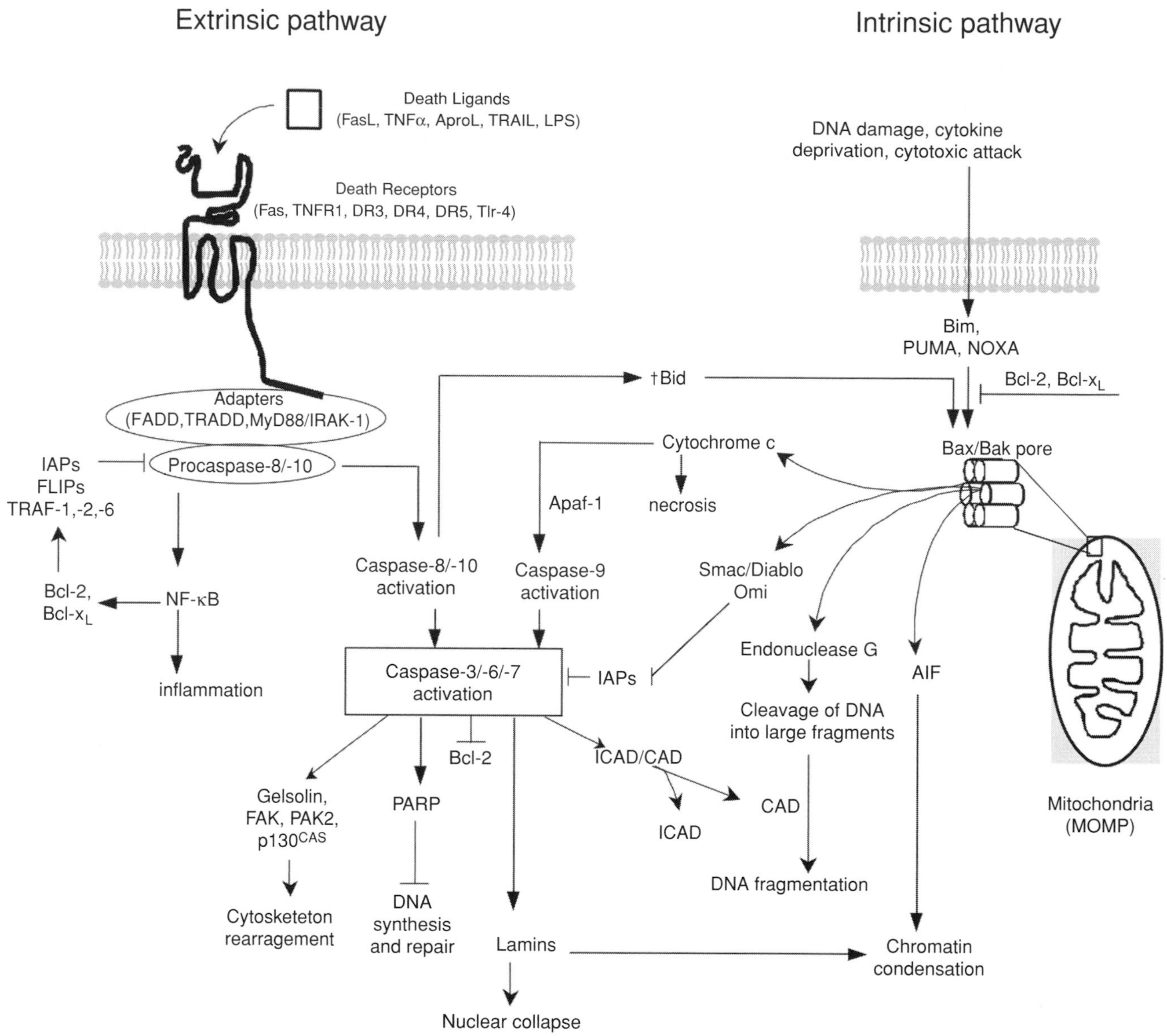

Figure 119.1. Pathways mediating the extrinsic and intrinsic apoptosis signaling response.

the death-inducing signaling complex (DISC) (see Figure 119.1). The interactions between the death domains of Fas and FADD promote recruitment of procaspase-8 via its death effector domain (DED). These protein–protein interactions cause autoproteolysis of procaspase-8, resulting in activation of caspase-8. Caspase-8 subsequently cleaves the propeptide of caspase-3, thus activating it, or it activates Bid-dependent mitochondrial pathway (9,10) and apoptosis ensues. Fas/FasL-mediated apoptosis signaling has been implicated in physiological apoptosis, such as removal of peripheral activated T cells at the end of immune response and killing of inflammatory cells at immune-privileged sites (6) and anoikis (11).

TNF-α is an inflammatory cytokine produced predominantly by activated macrophages and T cells in response to infection. TNFR1 is expressed in many tissues, including the endothelium. Similar to the Fas/FasL-mediated apoptotic pathway, ligation of TNF-α to TNFR1 causes DISC formation via its adapter molecule TNFR1-associated DD (TRADD). These protein complexes then recruit and stimulate the activation of caspase-8 and/or -10, which can ultimately induce cell death by activation of caspase-3 or by Bid-dependent mitochondrial dysfunction (9,10). The truncated form of Bid (tBid) triggers mitochondrial activation of caspase-9 by inducing Bax/Bak pore formation (9,10). However, unlike FasL, TNF-α rarely induces apoptosis. TNF-α concomitantly activates transcription factors, nuclear factor (NF)-κB and AP-1, resulting in the induction of expression of various genes that encode antiapoptotic proteins (12,13).

AproL is expressed in many tissues, yet its receptor, DR3, expressed predominantly in T cells, is upregulated during T-cell activation. Similar to TNFR1/TNF-α–induced apoptosis,

DR3/AproL also causes activation of NF-κB and AP-1 through TRADD (6). Apo2 ligand (Apo2L), also referred to as TRAIL, and its receptors, DR4 and DR5, are constitutively expressed in various tissues, as well as tumor cell lines. Apo2L binds to DR4 or DR5 and induces apoptosis through caspase-8 activation, which leads to cleavage and activation of Bid and subsequent Bid-dependent mitochondrial dysfunction. Although tumor cells are sensitive to Apo2L-triggered apoptosis, nontumorigenic cells are resistant, suggesting that a protective mechanism exists in these cells. Indeed, DcR1 and DcR2 (receptors similar to DR4 and DR5) that either lack the cytoplasmic tail or contain a truncated cytoplasmic death domain, respectively, are expressed in multiple nontumorigenic cells. These receptors act as decoys by competitively inhibiting the binding of Apo2L to DR4 or DR5, thus inhibiting apoptosis induced by this agonist (14).

The Intrinsic Pathway

Death receptor–mediated caspase activation is critical for the progression of apoptosis; however, caspase activity is not necessary to cause cell death (15). For example, the activation of Bak, a proapoptotic protein, by caspase-8–dependent Bid activation causes mitochondrial damage and cell death in the absence of caspase-3 activity (15), suggesting caspase-3–independent pathways are involved in mediating apoptosis. In fact, the mitochondria can act as stress sensors, as well as executioners of cell death. Many stress-inducing stimuli, such as cytotoxic drugs, oxidants, radiation, DNA damage, elevated intracellular calcium (Ca^{2+}), and growth factor deprivation promote mitochondrial outer membrane permeabilization (MOMP), resulting in apoptosis. This mitochondrial-dependent mechanism is referred to as an intrinsic apoptotic signaling pathway (see Figure 119.1). MOMP is regulated by the balance of the activities of antiapoptotic (such as Bcl-2, Bcl-x_L, Bfl-1, Bcl-w, Boo, Mcl-1) and proapoptotic (such as Bax, Bak, Bad, Bim, Bid, PUMA, NOXA) Bcl-2 family members (16,17). Normally, proapoptotic Bcl-2 proteins, such as Bax and Bak, are sequestered through heterodimerization in the cytoplasm by antiapoptotic Bcl-2 proteins, such as Bcl-2 and Bcl-x_L. During cellular stress, increased expression and/or activation of proapoptotic members of the Bcl-2 family causes their translocation to the mitochondrial outer membrane to form pores. These pores in turn allow some mitochondrial constituents located within the internal membrane space to escape into the cell cytoplasm; these include cytochrome-*c*, second mitochondrial-derived activator of caspase (Smac)/direct inhibitors of apoptosis protein (IAP) binding protein with low pI (Diablo), apoptosis-inducing factor (AIF), endonuclease G, and a serine protease referred to as Omi (16). The consequence of cytochrome-*c* release depends on cell type. If cells have an abundant amount of cytochrome-*c*, then the release of cytochrome-*c* does not significantly affect electron transport. The released cytochrome-*c* may rapidly bind to apoptosis protease-activating factor-1 (Apaf-1), in an adenosine triphosphate (ATP)-dependent manner. This interaction subsequently leads to the recruitment and activation of procaspase-9, with downstream activation of caspase-3, -6, or -7, culminating in apoptosis (Figure 119.1) (18,19). In contrast, in cells with high endogenous levels of caspase inhibitors, such as IAPs, the release of cytochrome-*c* is unable to effectively activate caspase-9, thus fails to induce apoptosis. Instead, the large loss of cytochrome-*c* from the mitochondrial membrane eventually disrupts the electron transport chain, driving the cells to undergo necrosis due to an increased generation of reactive oxygen species (ROS) and decreased ATP production (20). Upon release from the mitochondrial membrane, Smac/Diablo activate caspases by directly displacing IAPs from caspases, thus making the IAPs susceptible to cleavage by Omi (21,22). In addition, the release of AIF from the mitochondrial membrane allows translocation of this protein to the cell nucleus and the initiation of chromatin condensation (23). Finally, endonuclease-G contributes to DNA fragmentation by cleavage of high-molecular-weight genomic DNA, independent of caspase activity (24).

T cell receptor (TCR) ligation or cytokine deprivation has been shown to induce apoptosis by promoting mitochondrial pore formation and MOMP through the upregulation of Bim and subsequent activation of Bax/Bak (25). DNA damage also causes apoptosis through activation of p53, which in turn upregulates PUMA and NOXA expression leading to Bax activation and MOMP (26). In addition to these stimuli, in some cells, FasL-induced caspase-8 activation can promote apoptosis via a mitochondrial-dependent mechanism. In this apoptosis-inducing pathway, activated caspase-8 causes Bid truncation and activation (9). Activated Bid subsequently promotes Bax mitochondrial membrane pore formation and MOMP by displacing Bax from its heterodimeric, inactive protein complex with Bcl-2 or Bcl-x_L, thus promoting apoptosis (27).

In summary, a variety of signaling cascades may initiate apoptosis via either the extrinsic or intrinsic pathways.

Caspase Cascades

Caspases are a family of highly conserved aspartate-specific cysteinyl proteases. The caspase cascade is central to the progression of apoptosis and is utilized by both extrinsic and intrinsic pathways. All caspases have conserved amino acids and protein structure. To date, 13 mammalian caspases have been identified with distinct roles in apoptosis and inflammation. These are referred to as caspase-1 through caspase-13. All caspases are synthesized and accumulate in cells as inactive propeptides. Some caspases are activated by autoproteolysis upon the binding of adapter proteins, in response to signals initiating apoptosis. These caspases – caspase-8, -9, and -10 – are referred to as initiator caspases (28,29). Other caspases that cleave substrate proteins, culminating in cell death, are activated by the removal of bound inhibitors or by the cleavage by initiator caspases or proteases. These caspases – caspase-3, -6, and -7 – are referred to as effector caspases. Each caspase recognizes specific substrates, containing peptide motifs AXXX

(A: aspartic acid; X: any amino acid) within the N-terminus (28,29).

Caspases are executors of apoptosis (Figure 119.1). Upon receipt of the appropriate signals, the initiator caspases are activated, resulting in the immediate amplification and transmission of the signals to the effector caspases by proteolysis and activation. The activated effector caspases, in turn, cleave a select set of proteins at a single site, resulting in loss or gain of function. Over 100 proteins possessing caspase recognition motifs have been identified; these include inhibitor of caspase-activated DNase (ICAD), poly ADP-ribose polymerase (PARP), Bcl-2, lamin, and several cytoskeleton-binding proteins. In nonapoptotic cells, caspase-activated DNase (CAD) exists as an inactive complex with ICAD. Upon stimulation of a death signal, ICAD is cleaved by caspase-3, and the liberated CAD functions as an internuclease, destroying genomic DNA (28). PARP was the first substrate identified for caspase-3. Cleavage of PARP by caspase-3 inhibits DNA synthesis and repair (30). Bcl-2 prevents Bax/Bak-mediated mitochondrial pore formation by heterodimerizing with Bax or Bak, thus blocking the release of cytochrome-*c* from the mitochondria. Caspase-mediated Bcl-2 cleavage not only inactivates Bcl-2, but also produces an apoptotic-promoting proteolytic fragment (31). Lamina are cytoskeletal structures composed of lamins that underlie the nuclear envelope, with roles in chromatin organization, gene regulation, and signal transduction. Caspase cleavage of lamins disrupts the nuclear membrane and contributes to chromatin condensation (32). In addition, caspases promote the reorganization of the cytoskeleton during the progression of apoptosis by cleaving several proteins important in the formation and stabilization of the cytoskeleton, including gelsolin, focal adhesion kinase (FAK), p130cas, and p21-activated kinase 2 (PAK) (31,33–36).

REGULATORS OF APOPTOSIS

Nonpathological development and tissue homeostasis require a proper balance between survival and apoptotic signals. As expected, apoptosis is a highly regulated process, with various factors important to this control.

Bcl-2 was initially found to be overexpressed in B-cell lymphoma. This upregulation correlated with the loss of response to death signals and resistance to apoptosis (37). There are more than 15 members of the Bcl-2 family in mammalian cells. They are classified into two groups: the prosurvival or antiapoptotic subfamily, including Bcl-2 and Bcl-x_L, and the proapoptotic subfamily, including Bid, Bim, Bax, and Bak. Bcl-2 proteins, which block apoptosis, heterodimerize nonspecifically with proapoptotic proteins and sequester them in an inactive state in cytoplasm. This sequestration is protective against apoptosis by preventing MOMP. Although Bcl-2/Bcl-x_L protect against diverse cytotoxic stresses, such as irradiation and cytokine withdrawal, they are ineffective against FasL/caspase-8/caspase-3–mediated apoptosis (see Figure 119.1) (14).

As mentioned previously, upon binding of FADD with the DED domain of procaspase-8, in response to FasL ligation, procaspase-8 is activated via autoproteolytic cleavage (see Figure 119.1). FADD-like inhibitory proteins have similar DEDs, mimicking the procaspase-8 domains, but lacking caspase proteolytic activity. FLIPs regulate the progression of apoptosis by competing with procaspase-8 for binding with FADD, thus inhibiting apoptosis (38). In ECs, downregulation of c-FLIP is implicated in extracellular matrix (ECM) detachment–induced anoikis (11).

IAPs normally bind to active caspases, rendering them inactive (39). Upon stimulation of death signals, inhibitors of IAPs, such as Smac/Diablo and Omi are released from mitochondria (see Figure 119.1). Smac/Diablo displace IAPs from caspases, resulting in the activation of caspases and the promotion of apoptosis. Concomitantly, the liberated IAP is degraded by Omi, thus further enhancing apoptosis (21,22).

NF-κB is an important antiapoptotic transcription factor. NF-κB is activated by TNF-α and ApoL, as described previously (6–10,12,13). NF-κB promotes the expression of various antiapoptotic proteins, including Bcl-2, Bcl-x_L, FLIPs, IAP-1 and -2, and TNFR-associated factor-1 and -2 (12,13).

IDENTIFICATION OF APOPTOSIS

Many approaches have been developed to identify apoptotic cells based on specific morphological and biochemical cellular alterations. Frequently used assays to assess cellular apoptosis in the laboratory are listed in Table 119-1.

IMPORTANCE OF APOPTOSIS IN ENDOTHELIUM

ECs form a monolayer lining the vasculature. The functions of the endothelium are to selectively restrict the passage of solutes and macromolecules into the surrounding tissue, to maintain a nonthrombogenic surface over which the blood flows, and to regulate vascular tone. Due to the positioning of the endothelium at the interface between the blood and surrounding tissue, the endothelium is exposed to multiple stresses, mechanical and nonmechanical, which regulate vascular processes both physiologically and pathologically. One pathological consequence of these stresses in the blood vessel is the induction of apoptosis of the endothelium, which occurs in various diseases. However, EC apoptosis is also crucial in normal vascular development to regulate vasculogenesis, angiogenesis, vessel regression, and turnover of senescent EC in preexisting vessels. Thus, a careful balance of signals regulating EC survival and apoptosis is important for the maintenance of the vascular integrity.

Table 119-1: Methodologies for Detecting Apoptosis

Categories	*Features*	*Methodology*	*References*
Cell morphology	Diminished cell size and membrane blebbing	View cell morphology by electron microscopy	1
Plasma membrane integrity assay	Live cells exclude dye	Trypan blue or fluorescently labeled propidium iodide and YO-PRO-1 staining followed by light or fluorescence microscopy, respectively	11
Chromatin characterization	Chromatin condensation and DNA fragmentation	Staining with Hoechst or propidium iodide followed by fluorescence microscopy	144
DNA laddering	Endonuclease cleavage of DNA into 180–200 bp fragments	Analysis of DNA by agarose gel electrophoresis	145
Nucleosome ELISA	Endonuclease-mediated excision produces mono- and oligo-nucleosomes	DNA-affinity mediated capture of nucleosomes followed by immunodetection and quantitation of histone proteins	145
TUNEL staining	DNA strand breaks	Fluorescence labeling of DNA strand breaks using terminal deoxynucleotidyl transferase (TdT) in situ and quantitated using fluorescence microscopy or flow cytometry	144, 146
Phosphatidylserine exposure	Cell surface exposure of phosphatidylserine	Binding of fluorescently conjugated Annexin V with plasma membrane exposed phosphatidylserine and analyzed using fluorescence microscopy or flow cytometry	147
Caspase activity assay	Activation of caspases	In vitro activity assays quantitate substrate cleavage via fluorescence, colorimetric, or spectrophotometric changes in live or fixed cells and by SDS-PAGE and immunoblot analysis. In situ assays assess the ability of fluorescently labeled caspase inhibitors or antibodies directed against the mature caspase protein to bind to the caspase and quantitate via fluorescence microscopy or flow cytometry.	33, 148
Apoptotic protein translocation	Subcellular relocalization of pro- and antiapoptotic proteins	Subcellular fractionation of mitochondrial and cytoplasmic proteins and immunoblot analysis or immunofluorescent staining of cytochrome-*c*, AIF, and members of Bcl-2 family	16, 17
Mitochondrial permeabilization assay	Mitochondrial membrane depolarization	Fluorochromes sensitive to changes in the mitochondrial transmembrane potential, such as 5,5′,6,6′-tetrachloro-1,1′,3,3′-tetraethylbenzimidazolylcarbocyanine iodide (JC-1) and 1,1′,3,3,3′,3′-hexamethylindodicarbocyanine iodide ($DiIC_1$), are assessed using fluorescence microscopy or flow cytometry.	149

EC apoptosis has been shown to occur in vitro and in vivo by both a receptor-mediated, extrinsic pathway, as well as through the intrinsic pathway that results from alterations in mitochondrial function. Interestingly, just as there is a heterogeneous response to vasoactive or edematogenic stimulants by ECs derived from distinct vascular beds, it is also recognized that ECs respond differentially to apoptotic signals. For example, proliferative or angiogenic ECs express a greater number of death receptors when compared with quiescent, nonproliferating ECs; thus, they are more responsive to apoptosis-inducing agents (40). Additionally, several in vivo studies suggest that ECs undergo phenotypic changes with age, resulting in diminished vasodilatory responsiveness and exposure to increased oxidative stresses (41) that result in the ECs becoming more susceptible to apoptotic stimuli and less responsive to the apoptotic-protective effects of shear stress, as compared with younger ECs. Thus, the susceptibility of ECs to apoptosis depends on many factors, including genotypic and phenotypic characteristics of the endothelium, as well as the environmental exposures in the vascular bed from which they are derived. We will review biochemical and biomechanical stresses that regulate apoptosis of the endothelium and describe the role(s) that these pathways play in vasculopathy (summarized in Table 119-2).

Table 119-2: Stimuli Promoting EC Survival versus Apoptosis

Stimuli	*Apoptosis*	*Apoptotic Pathway*	*References*
Biomechanical			
Shear stress	↓	?	42, 43
Cyclic strain	↓	?	47
Extracellular matrix	↓	Extrinsic	11, 51, 52
Integrin	↓	Extrinsic	50
Biochemical			
Lipopolysaccharide (LPS)/endotoxin	↑	Extrinsic	59
Activated protein C	↓	Intrinsic	65, 66
NO	↓	Intrinsic	91
ROS	↑	Intrinsic	71
Antioxidants (Vitamins C and E)	↓	Intrinsic	71
oxLDL	↑	Extrinsic	104
Angiotensin II	↑	Extrinsic	68, 69
Adenosine/homocysteine	↑	Extrinsic	81
ATP/adenosine	↑	Extrinsic	81
Ceramide	↑	Intrinsic	85
Sphingosine 1-phosphate	↓	Extrinsic	86
IL-8	↓	Extrinsic	150
IL-10	↓	Extrinsic	151
Thrombospondin-1	↑	Extrinsic	152
Pigment epithelium-derived factor (PEDF)	↑	Extrinsic	153
TNF-α	↑	Extrinsic	72, 73
FGF2	↓	Extrinsic	154
VEGF	↓	Extrinsic	76, 78
Bioactive extracellular matrix proteolytic fragment			
Tumstatin	↑	Extrinsic	58
Endostatin	↑	Extrinsic	58
Arrestin	↑	Extrinsic	58
Canstatin	↑	Extrinsic	155
Angiopoietin-1	↓	Extrinsic	156
Growth factor deprivation	↑	Intrinsic	42, 52
Other			
Radiation	↑	Intrinsic	85
Temperature changes	↑	Intrinsic	157
Hypoxia	↑	Extrinsic	157

BIOMECHANICAL STIMULI INVOLVED IN ENDOTHELIAL CELL PRO- OR ANTIAPOPTOTIC SIGNALING

Shear Stress and Cyclic Strain

Mechanical, hemodynamic forces to which the endothelium is exposed include fluid shear stress from the frictional force of the blood flow and cyclic strain resulting from the rhythmic deformation of the vessels due to cardiac output forces. In addition to influencing EC protein and gene expression, proliferation, migration, and cell-cycle progression, these hemodynamic forces are important in regulating EC survival and apoptotic signaling. Early studies demonstrated that the addition of high laminar shear stress (>15 dynes/cm^2) protected ECs normally grown under static conditions from growth-factor deprivation-, TNF-α–, and hydrogen peroxide (H_2O_2)-induced apoptosis (42,43). Disruption of, or alterations in, high laminar blood flow are proapoptotic in the endothelium both in vitro (44) and in vivo, and have been shown to be important in the progression of atherosclerotic plaques (45) and vessel narrowing upon blood flow reduction (46) (Figure 119.2). Similar to shear stress, physiologic levels of cyclic strain (6%–10%), but not excessive (20%) strain, are protective against apoptotic-inducing agents (47). Not surprisingly, the two biomechanical forces promote EC survival by utilizing both overlapping and distinct signaling pathways (48).

Integrin and Extracellular Matrix Interactions

Integrins are a family of ubiquitous cell surface receptors that mediate the interactions between the cell and the ECM, thus anchoring the cells. Integrins are transmembranous proteins

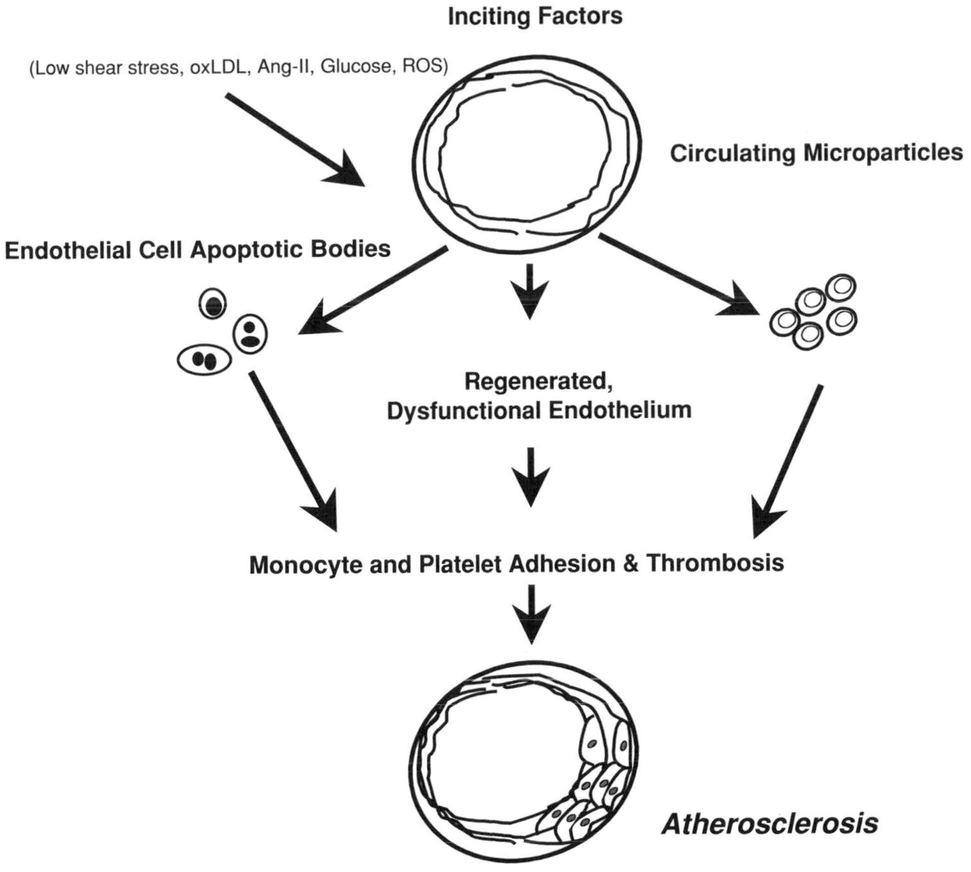

Figure 119.2. Mediators of endothelial apoptosis in atherosclerosis.

composed of α- and β- heterodimeric subunits (see Chapter 77). Currently, 18 α– and eight β-subunits have been identified, giving rise to 24 possible integrin pairs (49). Integrins interact with ECM proteins, such as fibronectin, vitronectin, collagen, heparin-sulfate proteoglycans, and laminin, to form focal adhesion complexes by organizing multiple intracellular signaling proteins with cytoskeletal structures at these focal points. These cell–ECM structures convey changes in extracellular biomechanical and biochemical environmental cues and result in various cellular responses important in EC adhesion, migration, proliferation, barrier function, and survival (50).

Loss of these cell–ECM interactions promotes apoptosis or anoikis in numerous cells, including the endothelium. ECs grown in suspension undergo apoptosis within 18 hours (51). In vitro studies demonstrate that ECs grown on ECM proteins are protected from apoptosis induced by serum deprivation or Fas-mediated cell death (11,52). Also, soluble ligands for $\alpha v\beta 3$, $\alpha 5\beta 1$, or $\alpha 2\beta 1$, which act as antagonists that disrupt the cell–ECM interactions, induce apoptosis in proliferating endothelium (53,54). Further data demonstrates that laminar shear stress upregulates the expression of EC integrins and integrin-associated proteins (55,56), suggesting that antiapoptotic signaling pathways induced by these biomechanical forces are mediated, in part, at the level of cell–ECM interactions. Finally, exposure of ECs to proangiogenic agonists promotes upregulation of $\alpha v\beta 3$ integrins (57), further suggesting a role for the integrins in survival and antiapoptotic pathways.

Angiogenesis involves EC proliferation, migration, adhesion, and tube formation. The creation of these new blood vessels, during tumor growth, requires degradation of existing ECM proteins by metalloproteinases to permit the expansion of new blood vessels. Interestingly, the proteolytic fragments of several ECM proteins, naturally found in the region of neovascularization, promote EC apoptosis (58). Endothelial apoptotic bioactive functions have been attributed to endostatin, canstatin, tumstatin, and arrestin, proteolytic fragments of collagen XVIII, and collagen IV, respectively. One point of action of these soluble matrix proteolytic products is to block integrin receptors, $\alpha v\beta 3$, $\alpha 5\beta 1$, or $\alpha 1\beta 1$, resulting in antiangiogenic signaling, in part, by promoting EC apoptosis (58). Thus, the integrity of the surrounding ECM is critically important in maintaining blood vessel survival and in regulating angiogenesis.

The roles of integrins as receptors and the ECM proteins as transmitters of biomechanical stimuli and signaling through cell survival–dependent pathways has begun to be elucidated. However, whether the function of distinct integrin molecules or ECM proteins in promoting EC survival depends on the apoptotic inducing agent and/or on cell biomechanical and biochemical cell environment remains to be determined. Moreover, if any antiapoptotic signaling pathways activated by the integrin–ECM interactions are unique to the endothelium, this would suggest the potential for vascular-specific therapies. This is an area of great research potential.

THE ROLE OF BIOCHEMICAL STIMULI IN ENDOTHELIAL CELL APOPTOTIC SIGNALING

Lipopolysaccharide

Lipopolysaccharide (LPS) or endotoxin resides in the outer membrane of gram-negative bacteria and is a potent inducer of bacterial sepsis by promoting EC activation, dysfunction, and apoptosis. Both in vitro and in vivo studies have demonstrated LPS-induced apoptosis in ECs derived from various vascular beds (59). Additionally, neutralization of the lipid A moiety of LPS attenuated EC activation by LPS (60). LPS activates the extrinsic apoptotic signaling pathway through the Toll-like receptor (TLR)-4 by forming a protein complex with LPS-binding protein (LBP) and soluble CD14 (59). This ligand/receptor interacts with the adaptor protein myeloid differentiation factor 88 (MyD88) resulting in its interaction with and activation of interleukin (IL)-1 receptor associated kinase (IRAK)-1 (59). IRAK-1 subsequently activates TNF receptor-associated (TRAF)-6, which in turn activates of NF-κB, as well as C-jun N-terminal kinase (JNK)-1 by means of a caspase cascade (59–61). The role of FADD in LPS induction of EC apoptosis is controversial (59). While it has been speculated that the purpose of NF-κB activation is to counteract apoptosis by transcriptionally upregulating several antiapoptotic genes, including IAP-1, IAP-2, and X-chromosome-linked inhibitor of apoptosis protein (XIAP) (62), suppression of NF-κB expression exacerbated TNF-α–induced apoptosis, but had no significant effect on LPS-induced EC apoptosis (63). Additionally, recent studies elucidated a role for FLICE-like inhibitory protein (FLIP) as being activated by LPS and protecting against EC apoptosis and in suppressing NF-κB activation (64), further suggesting that this transcription factor is not protective in the LPS-induced apoptotic signaling pathway.

Anticoagulant Pathway

Activated protein C (APC) is a component of the coagulation pathway with potent anticoagulant, anti-inflammatory, and antiapoptotic activities. The anticoagulation effects of APC work at the post-translational level by regulating thrombin formation, whereas the inhibitory effects of APC on inflammation and apoptosis are thought to occur at the transcriptional level. Indeed, APC enhanced the expression of prosurvival proteins including the homologue of Bcl-1, IAP-1, and endothelial nitric oxide synthase (eNOS) in human umbilical vein ECs (HUVECs) (65). Additionally, TNF-α–induced expression of NF-κB was attenuated by APC (65). These transcriptional effects by APC correlated with protection of ECs from staurosporine-induced apoptosis (65). In ECs derived from the human brain, hypoxia-induced apoptosis was blunted by pretreatment with APC; these changes correlated with modifications in signaling through p53, Bax, Bcl-2, and caspase-3 (66). Thus, while APC is protective against various apoptosis-inducing agents and proving clinically efficacious in treating sepsis, the mechanism by which APC modulates EC apoptosis remains to be more clearly defined.

Cytokines and Growth Factors

Angiotensin II is a potent vasoconstrictor implicated in various cardiac, vascular, and renal diseases. In addition to its role as a vasoconstrictor, angiotensin II has been recognized as a potent regulator of protein and gene expression and intracellular signaling pathways in various cell types, including cardiomyocytes, fibroblasts, and vascular smooth muscle cells (VSMCs) and ECs. Angiotensin II demonstrated both a proliferative and an antiproliferative effect on coronary artery ECs, depending on the angiotensin II receptor subtype stimulated, with the AT_1R stimulating and AT_2R inhibiting EC proliferation (66,67). The antiproliferative effect of angiotensin II activation through AT_2R was further shown to be proapoptotic in ECs through the activation of caspase-3 and the possible downregulation of nitric oxide (NO) production (68). Additional studies have demonstrated that angiotensin II potentiated human coronary artery EC apoptosis induced by anoxia–reoxygenation and TNF-α injury through AT_1R (69). Angiotensin II also may promote EC apoptosis through the modulation of gene expression. Indeed, angiotensin II was demonstrated to upregulate the receptor mediating oxidized low-density lipoprotein (oxLDL) uptake, lectin-like oxLDL receptor (LOX)-1, in ECs, thus potentially enhancing the uptake and subsequent injurious intracellular signaling effects of this oxidized lipoprotein (69,70). Furthermore, cytochrome-*c* release and Bcl-2 proteolysis correlated with EC apoptosis upon exposure to angiotensin II (71). Finally, the modulation of ROS levels has been suggested to be intimately involved in the progression of EC apoptosis in response to angiotensin II stimulation (71).

TNF-α, a proinflammatory cytokine produced by macrophages, has direct effects on EC secretion and surface expression of cytokines, chemokines, and growth factors important for vasoregulation and leukocyte adhesion and activation. TNF-α also promotes EC dysfunction, in part, by causing apoptosis. The varied cellular responses stem from the initial binding of TNF-α to two receptors: TNFR1 and TNFR2. This ligand–receptor interaction promotes a conformational change in the receptor that results in the recruitment of various stimulus-specific proteins. For example, in terms of the apoptotic-stimulated pathway, upon TNF-α binding to TNFR1, TRADD binds the cytosolic portion of TNFR1 and subsequently recruits FADD and procaspase-8. This complex then initiates the caspase cascade leading to EC apoptosis (72,73). TNF-α also promotes EC apoptosis by enhancing Bcl-2 proteolysis through the inactivation of mitogen-activated protein kinase (MAPK) (74). The authors speculated that the diminished levels of Bcl-2 may augment the proapoptotic signaling pathways in ECs in response to selective stimuli (74). TNF-α also upregulates various antiapoptotic proteins through NF-κB in ECs (72), thus suggesting that the sensitivity of ECs to TNF-α–induced apoptosis is regulated by the balance between proapoptotic and antiapoptotic signals.

Vascular endothelial growth factor (VEGF) is essential for angiogenesis in both physiological (e.g., pregnancy and wound healing) and pathological (e.g., tumor metastasis, diabetic retinopathy, atherosclerosis) processes. In addition, VEGF has been shown to be important in mediating "vascular protective" effects by mitigating VSMC proliferation, inflammation, and thrombosis, and by promoting EC survival (75). Hyperoxia-induced downregulation of VEGF correlated with apoptosis in ECs isolated from retinal capillaries (76). The exogenous addition of VEGF prevented experimental hyperoxia-induced EC apoptosis in vitro and in vivo (76). Additionally, VEGF rescued ECs from apoptosis induced by serum deprivation, which was dependent upon activation of phosphoinositide 3-kinase (PI3K) and Akt and inhibition of the forkhead transcription factors (77,78). Additional mechanisms by which VEGF may promote EC survival is by signaling through pathways modulated by the upregulation or stabilization of Bcl-2, NO, and prostacyclin (PGI_2) production; FAK phosphorylation and activation; and interaction with integrin receptors (75).

Endogenous Metabolites

Elevated levels of ATP or adenosine may occur in blood vessels upon the exocytotic release of nucleotides from stimulated platelet granules, during cytolytic release from necrotic cells, or from EC membrane transporters (79). Exposure of ECs to elevated levels of ATP or adenosine promotes apoptosis in vitro (80–82). EC apoptosis had been shown to be induced both by activation of type-2X purinergic (P2X) receptors and by ectonucleotidase-mediated hydrolysis of ATP (80,81). Furthermore, adenosine-induced EC apoptosis was potentiated by homocysteine and was mimicked by inhibitors of S-adenosyl-L-homocysteine hydrolase (83). Adenosine- and adenosine/homocysteine-induced apoptosis were mediated, in part, by a protein tyrosine phosphatase-dependent inhibition of p38 MAPK (84) and degradation of FAK, paxillin, and p130CAS proteins, with subsequent disruption of focal adhesion complexes (33). Additional studies have demonstrated that increased adenosine/homocysteine also causes Ras GTPase inactivation, possibly by blunting carboxylmethylation of this small GTPase through the inhibition of isoprenylcysteine carboxyl methyltransferase activity (83). Thus, it is speculated that increased levels of adenosine and homocysteine may promote EC apoptosis through multiple signaling pathways, resulting in disruption of cell–ECM interactions and anoikis.

The sphingolipid metabolites, ceramide and sphingosine 1-phosphate (S1P), have been shown to have opposing effects on EC apoptosis (see Chapter 45). Increased ceramide production in response to ionizing radiation, septic shock, and oxLDL promotes EC apoptosis (85,86). Conversely, the induction of S1P blunted EC apoptosis induced by TNF-α. Ceramide is a ubiquitous precursor synthesized de novo by the cell and utilized in the biosynthesis of sphingomyelin, a crucial component of the plasma membrane. Additional by-products of ceramide synthesis include glucosylceramide, ceramide 1-phosphate (C1P), and S1P; however, only ceramide is capable of inducing cell apoptosis (85,86). Rapid increases in ceramide production (within minutes) in response to apoptosis-inducing agents are thought to occur primarily through the degradation of sphingomyelin through the action of various sphingomyelinases, including neutral or acid sphingomyelinase (85). However, ceramide also may be increased in ECs hours later through the activation of the de novo synthetic pathway utilizing ceramide synthase (85). Ceramide is believed to ultimately produce apoptosis through the induction of the caspases, possibly through the modulation of gene expression through the stress activation pathway, JNK (85,86). Interestingly, increased production of VEGF or S1P, or the overexpression of Bcl-2, blunted ceramide-mediated EC apoptosis (86). S1P produced exogenously is thought to be protective against apoptosis in ECs by signaling through endothelial differentiation gene (EDG)-1 receptor via cell survival pathways involving Ras GTPase and extracellular-signal-regulated kinase (ERK)-1 (86).

Oxidative Stresses

Oxidative stress producing ROS is another biological mediator important in promoting EC apoptosis, occurring when the level of ROS exceeds that of cellular antioxidants. Superoxide ($O_2\cdot^-$), H_2O_2, NO, peroxynitrite ($ONOO\cdot^-$), and hydroxide radicals ($\cdot OH$) are some of the ROS produced in the endothelium. The levels of ROS may increase in the blood vessel by release from inflammatory cells or by elevated production by the endothelium or VSMCs. Intracellularly, ROS may be produced in the mitochondria or by multiple cytosolic enzymes, including nicotinamide adenine dinucleotide phosphate (NADPH) oxidase, xanthine oxidase, NOS, or myeloperoxidase. Exposure of ECs to enhanced ROS promoted apoptosis, an effect suppressed by antioxidative enzymes, such as superoxide dismutase (SOD) (87). Additionally, anoikis-induced ECs demonstrated elevated levels of ROS with subsequent caspase-3 activation (88). Incubation of the cells with antioxidants partially blunted the activation of caspase-3 resulting from anoikis, suggesting a role for the ROS in the progression of EC apoptosis (88). Additionally, increased ROS production and apoptosis occurred in ECs exposed to oxLDL (71), LPS (59), angiotensin II (71), or high glucose concentrations (89), suggesting a role for ROS in the progression of atherosclerosis, sepsis, hypertension, and diabetes. The mechanism by which ROS mediates apoptosis is believed to occur at multiple points, including caspase activation, mitochondrial dysfunction, DNA damage, and activation of other proapoptotic downstream signaling molecules (71).

NO is a vasoactive product of the pulmonary and systemic vasculature that regulates many cellular functions, including vasodilation, VSMC proliferation and migration, platelet adhesion and aggregation, and EC proliferation and apoptosis. NO and L-citrulline are produced from L-arginine and oxygen by NOS. Three isoforms of NOS have been isolated: neuronal NOS (nNOS), inducible NOS (iNOS), and eNOS, which contain significant sequence homology and share cofactor requirements for NADPH, heme, flavin adenine dinucleotide (FAD), flavin mononucleotide (FMN), tetrahydrobiopterin, and calmodulin. NO has been shown to be both protective against and an inducer of apoptosis in ECs. EC survival depends on low levels of NO when grown in the presence of angiotensin II, TNF-α, or high glucose concentrations (68,90,91). Yet, high levels of NO were itself apoptotic for ECs (91). The modes by which NO is protective against EC apoptosis are many fold, including antioxidative effects against ROS, S-nitrosylation of and inactivation of caspases, inhibition of mitochondrial cytochrome-*c* release, and upregulation of Bcl-2 expression (92). Enhanced levels of NO may promote EC apoptosis by the combining with $O_2\cdot^-$ and forming $ONOO\cdot^-$, causing DNA damage directly or indirectly by the S-nitrosylation of various proteins or lipids and resulting in the modulation of the intrinsic apoptosis signaling pathway (92).

Enhanced oxidation of LDL creates a more reactive species of the lipoprotein, which has been shown to directly correlate with the progression of atherosclerosis (93). OxLDL has numerous effects on the endothelium, including promoting apoptosis. Recent studies have demonstrated that expression of the oxLDL receptor LOX-1 was required for the apoptosis-inducing effects of oxLDL in coronary artery ECs (94). OxLDL-induced EC apoptosis correlated with enhanced FasL expression and increased caspase-3 activation (8). Additional analyses suggested that Fas-mediated apoptosis upon oxLDL exposure occurs, in part, by downregulation of the caspase inhibitor, FLIP (95). The authors speculated that this downregulation sensitizes the ECs to Fas-mediated apoptosis signaling pathway (95). Interestingly, although other proapoptotic cytokines, such as TNF-α and IFN-γ, are associated with promoting FasL expression in ECs, the signaling pathways by which these cytokines induce apoptosis in endothelium is not via the Fas-mediated pathway (8). It also has been reported that oxLDL may exert its apoptotic effects by modulating the bioavailability of NO in the endothelium (96).

ENDOTHELIAL APOPTOSIS AND VASCULAR PATHOPHYSIOLOGY

In this section, we review the evidence for endothelial apoptosis in the pathophysiology of representative disease states and the role of endothelial apoptosis in normal vascular function. In normal blood vessels, there is little evidence for EC apoptosis. Apoptosis of endothelium occurs in pathological states and appears to limit vasculogenesis and angiogenesis.

Endothelial Cell Apoptosis and Atherosclerosis

The most extensive literature regarding EC apoptosis in vivo relates to atherosclerosis. Atherosclerosis is a complex pathological process characterized by early EC activation and/or injury, accumulation of lipid-laden monocytes and inflammatory cells, VSMC proliferation, and eventual plaque rupture and vascular occlusion by thrombosis. A number of recent reviews have summarized the potential role of apoptosis in atherosclerosis (97–101).

Apoptosis in atherosclerotic plaques removed at atherectomy was demonstrated by TUNEL staining; among apoptotic cells identified included VSMCs, macrophages, and ECs (102,103). An important, and as yet unresolved, question is whether apoptosis is a cause of early EC injury and an initiating event in the process of atherosclerosis. Accumulating evidence suggests that EC apoptosis does contribute to the pathogenesis of atherosclerosis. For example, multiple risk factors associated with atherosclerosis also promote apoptosis of cultured ECs, as outlined in Table 119-2. These risk factors include oxLDL (104), hyperglycemia (105), elevated levels of homocysteine (106), and aging (107). In contrast, factors protective against vascular injury, such as laminar flow, also blunt EC apoptosis. In vivo atherosclerotic lesions are most common at sites of low or turbulent blood flow where shear stress is reduced (45), suggesting that EC apoptosis may be an important factor in the genesis of the lesion. Furthermore, in atherectomy specimens, EC apoptosis was most commonly observed downstream of stenotic areas, where shear stress is lowest (102); thus, reduced shear stress appears to predispose ECs to apoptosis in human atherosclerosis.

Interestingly, in atherosclerotic plaques and the serum of patients with heart disease, molecules that mediate cell survival and apoptosis signaling have been identified. Human atherosclerotic plaques (101) contain proapoptotic substances such as oxLDL, matrix metalloproteinases (MMPs), proinflammatory mediators (TNF-α, IFN-γ, IL-1), vasoactive substances (angiotensin II), Fas and FasL, proapoptotic Bcl-2 family members (Bax), and executioners of apoptosis (caspases). Substances protective against apoptosis also have been localized in plaques (101); these include antioxidants (heme oxygenase-1, α-tocopherol), anti-inflammatory cytokines (IL-10), growth factors (platelet-derived growth factor [PDGF], insulin-like growth factor [IGF]-1, fibroblast growth factor [FGF]-2, VEGF), and antiapoptotic Bcl family members (Bcl-x_L). In addition, serum from patients with acute coronary syndromes promoted apoptosis in cultured HUVECs, an effect that was diminished in blood drawn from the same patient a year after the clinical event (108). Proapoptotic factors Fas and Bcl-2 were increased in serum from patients with acute coronary syndromes, and the ratio of Bax/Bcl-2 was decreased 1 year later (108). In another study, proapoptotic activity in serum from patients with heart failure was an independent predictor of mortality (109). These data suggest that the balance of factors that regulate cell survival and apoptosis may determine the degree of endothelial apoptosis in patients with vascular disease.

Mediators of Endothelial Apoptosis in Atherosclerosis

Numerous causes and modulators of endothelial apoptosis have been described using cultured cells. However, the relative importance of each has not yet been clearly determined in human disease. It is possible that pro-and antiapoptotic factors vary with the temporal stage of the disease.

Significant changes in biomechanical stresses and the promotion of EC apoptosis are thought to play an important role in the initiation and progression of atherosclerosis. For example, as noted earlier, low or turbulent blood flow resulting in diminished laminar shear stress appears to predispose the endothelium to apoptosis in atherosclerotic lesions. EC anoikis is believed to occur in atherosclerosis due to the action of proteases that disrupt EC–ECM interactions (110). Indeed, activation of the inflammatory cells in atherosclerotic plaques, such as macrophages, T-lymphocytes, mast cells, and neutrophils, results in the release of enzymes capable of proteolytic degradation of ECM proteins surrounding the ECs (111). Thus, the lack of shear stress and/or increased anoikis and elevated apoptosis may be early factors in the development of EC injury in atherosclerosis.

Other mediators that promote atherosclerosis by causing apoptosis of the endothelium include multiple biochemical stimuli. OxLDL, which is known to cause apoptosis of cultured ECs, has been implicated in the pathogenesis of atherosclerosis. Indeed, its receptor, LOX-1, has been localized on ECs in atherosclerotic lesions (112). Atheromas also contain abundant glycosphingolipids and ceramide, generated by the action of sphingomyelinase (86). Ceramide, an apoptosis-inducing agent in vitro, may be an important cause of endothelial apoptosis in atherosclerosis (86).

Fas and FasL are additional proapoptotic mediators expressed in cells within atherosclerotic plaques (113). Fas–FasL interaction causes the apoptosis of monocytes; thus, it could attenuate the infiltration of these inflammatory cells into atherosclerotic lesions. Vascular ECs express FasL (8). The role of EC expression of FasL in atherosclerosis is controversial, with evidence for both exacerbation of and protection from atherosclerosis (114). Recent studies using animal models indicate that overexpression of FasL by ECs decreases atherosclerosis in apoprotein E–deficient mice (114). In addition, EC expression of soluble Fas attenuated transplant-induced arteriosclerosis in rats (115). Thus, EC expression of soluble Fas or FasL may serve to blunt, rather than exacerbate, atherosclerosis.

Mechanisms by which Endothelial Apoptosis Contributes to the Pathogenesis of Atherosclerosis

There are a number of ways in which EC apoptosis may contribute to the pathogenesis of atherosclerosis. First, studies in monkeys indicate that endothelial apoptosis is associated with impaired endothelium-dependent vasodilator function (107). Thus, endothelial apoptosis causes vascular dysfunction in vivo.

Another means by which endothelial apoptosis may contribute to the development of atherosclerosis is by stimulating the regeneration of abnormal ECs (116). Phenotypic and functional changes have been observed in ECs repopulating porcine coronary arteries after balloon denudation (117). Increased uptake of LDL and decreased cyclic guanosine monophosphate (cGMP) production were observed in ECs harvested from regenerating vascular endothelium, suggesting that regenerating endothelium may be more susceptible to atherosclerosis (117).

Thrombus formation is important in the development of acute ischemic events after plaque rupture. Apoptosis increases the expression of phosphatidylserine on cell membrane, as described earlier. Phosphatidylserine enhances tissue factor activity and thereby enhances endothelial thrombogenicity (101). Thus, apoptotic ECs are procoagulant (118). Increased tissue factor activity has been found in human atherosclerotic lesions (119). In addition, endothelial apoptosis increases platelet activation and adhesion (118), also contributing to plaque formation. In vivo evidence for the importance of endothelial apoptosis in stimulating thrombosis was reported by Durand and colleagues, using an animal model in which staurosporine was used to induce EC apoptosis and vascular thrombosis (120). Caspase inhibitors markedly decreased both endothelial apoptosis and thrombus formation, suggesting that endothelial apoptosis is important in acute ischemic events (120).

During apoptosis, cell membrane blebbing occurs, and microparticles are formed from the shed blebs (121). These microparticles strongly express procoagulant phosphatidylserine, which is redistributed to the membrane upon apoptosis. Increased circulating microparticles of endothelial origin have been described in patients with acute coronary syndromes, as compared to stable angina and noncoronary heart disease (122). Thus, both in situ and circulating fragments of apoptotic EC may contribute to thrombosis complicating plaque rupture.

In summary, good evidence suggests EC apoptosis in atherosclerotic plaques. Causes include decreased laminar shear stress, the production of oxLDL and ceramide, and the induction of anoikis. Endothelial apoptosis may contribute to the pathogenesis of atherosclerosis via decreased endothelial-dependent vasodilation and by enhancing platelet adhesion and thrombosis. It is possible that therapies to reduce EC apoptosis may ameliorate atherosclerosis. Indeed, it is possible that the beneficial effects of therapies, such as angiotensin converting enzyme (ACE) inhibitors or statins (3-hydroxy-3-methylglutaryl-CoA reductase inhibitors), may at least partially be due to inhibition of EC apoptosis (98). However, inhibition of apoptosis in other plaque cells, such as macrophages and VSMCs, could exacerbate the atherosclerotic process. Thus, any treatments aimed at inhibiting apoptosis in atherosclerosis should be specific for ECs of conductance vessels.

Endothelial Cell Apoptosis and Sepsis

Sepsis caused by gram-negative bacteremia is a cause of multisystem organ failure or sepsis syndrome, a disorder of high mortality characterized by diffuse microvascular injury and resulting in increased permeability, edema formation, and shock. A critical mediator of sepsis syndrome is LPS, a component of gram-negative bacterial cell walls. LPS stimulates the production of cytokines from inflammatory cells, including TNF-α. As noted in Table 119-2, both LPS and TNF-α cause the apoptosis of cultured ECs via the extrinsic pathway. Haimovitz-Friedman and colleagues reported that injection of LPS or TNF-α into mice caused early, diffuse EC apoptosis in intestine, lung, fat, and thymus, and this was preceded by increased tissue concentrations of ceramide (123), another inducer of apoptosis (86). Mice deficient in acid sphingomyelinase were protected against apoptosis (124), suggesting an important role for ceramide-induced endothelial apoptosis in sepsis-induced microvascular injury.

Endothelial Cell Apoptosis and Ischemia–Reperfusion Injury

Ischemia–reperfusion can cause organ injury in circumstances in which blood flow is interrupted and then restored, such as with the relief of vascular occlusion and after organ transplantation. Ischemia–reperfusion is characterized by increased ROS, inflammatory cell accumulation, and activation of neutral sphingomyelinase and ceramide accumulation (86).

Apoptosis has been demonstrated in myocardial ischemia–reperfusion injury (125,126). Using an isolated perfused rat heart model, Scarabelli and colleagues demonstrated that microvascular EC apoptosis preceded myocardial apoptosis, with radial spread from microvessels of myocyte apoptosis (127). These data suggest that apoptotic ECs release proapoptotic substances into the surrounding myocardium and cause myocyte apoptosis in a paracrine fashion. Furthermore, inhibition of apoptosis attenuated myocardial ischemia–reperfusion injury (126).

Ischemia–reperfusion injury is an early complication of organ transplantation and is related to the duration of cold storage prior to transplantation (128). Apoptosis has been reported in ischemia–reperfusion injury accompanying transplantation of lung, heart, liver, and kidney (128). Using a rat model of lung transplantation, Quadri and colleagues found that EC apoptosis was proportional to the duration of cold storage, and that caspase inhibition decreased apoptosis and enhanced the function of transplanted lungs (128). In transplanted livers, sinusoidal EC apoptosis occurred early in the course of preservation and reperfusion injury (129).

Endothelial Cell Apoptosis and Radiation-Induced Organ Injury

ECs are sensitive to radiation-induced cell injury (85). The role of radiation-induced EC apoptosis has been studied in the mechanism of radiation-induced gastrointestinal injury. Gastrointestinal injury is a key side effect of radiation, and limits the dose of treatment in cancer therapy. Using a mouse model of whole-body irradiation, Paris and colleagues demonstrated that animals deficient in acid sphingomyelinase, a protein abundantly expressed in the endothelium, did not have extensive microvascular EC apoptosis, suggesting that radiation-induced apoptosis is a result of acid sphingomyelinase–induced ceramide production (130).

Studies of tumor sensitivity to low-dose radiation also indicate a role for microvascular EC apoptosis. Tumor growth is thought to depend on angiogenesis to ensure blood supply. Because radiation commonly is used to treat cancer, the question arises as to whether the effects of radiation on tumor size are due, at least in part, to EC injury. Garcia-Barros and colleagues have reported that microvascular endothelial apoptosis occurs with low-dose radiation directed at fibrosarcomas and melanomas in mice (131). Furthermore, tumors in acid sphingomyelinase–deficient and Bax-deficient mice were resistant to radiation and displayed less radiation-induced EC apoptosis (132). Thus, both complications of and response to radiation therapy appear to be due to radiation-induced EC apoptosis.

Endothelial Cell Apoptosis in Emphysema

Emphysema is a lung disease characterized by the loss of alveolar capillary septum, resulting in the impairment of gas exchange. The imbalance of proteases and antiproteases, caused by exposure to cigarette smoke, is thought to be important in the pathogenesis of emphysema. Voelkel and colleagues have suggested that apoptosis of cells in the alveolar septum occurs in emphysema, and have suggested that epithelial and/or endothelial apoptosis may be important in the pathogenesis of the disease (133). Indeed, a role for endothelial apoptosis in emphysema is suggested by the effect of VEGF receptor antagonists that cause endothelial apoptosis in vitro and in vivo (134–136). The administration of the VEGF antagonist caused emphysema and apoptosis of alveolar wall cells in a rat model (135). In addition, injection of HUVECs caused emphysema in mice and promoted the production of antibodies directed against ECs, suggesting an immune mechanism for the disease (137). The evidence for EC apoptosis in emphysema is more extensively reviewed in Chapter 130.

Endothelial Cell Apoptosis and Thrombotic Thrombocytopenic Purpura

Thrombotic thrombocytopenic purpura (TTP) is a devastating disease characterized by thrombocytopenia, microangiopathic hemolytic anemia, fever, and neurologic and renal

Table 119-3: Pathophysiologic Conditions Characterized by Endothelial Apoptosis

Condition	*Humans (Refs.)*	*Animal Models (Refs)*
Atherosclerosis	101–103, 108, 109	114, 115
Sepsis	—	123, 124
Ischemia–reperfusion	158, 159	126–128
Radiation	—	130–132
Emphysema	133	134–137
Corpus luteum	141	—
TTP	138	—

dysfunction. Pathology shows characteristic hyalinized thrombi in microvessels. Enhanced EC apoptosis has been observed in the splenic tissues of patients with TTP and the related hemolytic-uremic syndrome (HUS) (138). In addition, plasma from patients with TTP and HUS causes the apoptosis of cultured microvascular ECs (139,140). Interestingly, cultured pulmonary and hepatic microvascular ECs were resistant to TTP plasma–induced apoptosis, and those organs are generally spared in TTP and HUS (139). This suggests that ECs from distinct vascular beds differ in susceptibility to apoptosis.

Endothelial Cell Apoptosis and the Regression of the Corpus Luteum

The ovarian follicle has the capacity for rapid microvascular development and regression during the ovulatory cycle. New blood vessels are necessary for the formation and function of the corpus luteum. On the other hand, vessel loss due to luteal EC apoptosis occurs during regression of the corpus luteum (141). Possible biochemical stimuli that promote luteal endothelial apoptosis include angiotensin II, IFN-γ, and TNF-α (see Table 119-2).

Endothelial Cell Apoptosis and the Regulation of Angiogenesis

As noted earlier, the corpus luteum is an example of a tissue that undergoes angiogenesis that is limited by apoptosis during the luteal regression phase (141). It has been suggested that EC apoptosis may limit angiogenesis involved with inflammation, development, or tumor growth (142). Thus, endothelial apoptosis may be important in limiting neovascularization, and may cause tumor regression. Integrin-mediated interactions with ECM are just one possible regulator of apoptosis limiting angiogenesis (143).

CONCLUSION

EC apoptosis is a marker of disease and limits angiogenesis (Table 119-3). EC apoptosis is believed to be an early event in atherogenesis, stimulated by low blood flow and resulting decreased laminar shear stress. In addition, endothelial apoptosis contributes to plaque rupture and resulting thrombosis. In vivo evidence suggests endothelial apoptosis in sepsis, ischemia–reperfusion and radiation injury, and emphysema. Endothelial apoptosis occurs during regression of the corpus luteum and in limiting angiogenesis in tumors.

Inhibitors of apoptosis, such as caspase inhibitors, are potential treatments to reduce vascular injury. However, use of such drugs may be problematic in that local action is needed.

KEY POINTS

- Apoptosis is important in regulating vasculogenesis, angiogenesis, and vessel regression. In addition, endothelial apoptosis regulates vascular functions.
- Extrinsic and intrinsic biochemical pathways and modulators of apoptosis have been described in detail using cultured ECs.
- Endothelial apoptosis is caused by risk factors for atherosclerosis and by other causes of vascular injury in vivo.

Future Goals

- To further study and understand biochemical pathways of endothelial apoptosis in human disease
- To determine the therapeutic potential for drugs that modulate endothelial apoptosis in a cell-specific and vascular bed-specific manner

REFERENCES

1 Kerr J, Wyllie A, Currie A. Apoptosis: a basic biological phenomenon with wide-ranging implications in tissue kinetics. *Br J Cancer*. 1972;26:239–257.

2 Esposti M. Apoptosis: who was first? *Cell Death Differ*. 1998; 5:719.

3 Fadok V, Voelker D, Campbell P, et al. Exposure of phosphatidylserine on the surface of apoptotic lymphocytes triggers specific reorganization and removal by macrophages. *J Immunol*. 1992;148:2207–2216.

4 Savill J, Fadok V. Corpse clearance defines the meaning of cell death. *Nature*. 2000;407:784–788.

5 Wyllie A, Kerr J, Currie A. Cell death: the significance of apoptosis. *Int Rev Cytol*. 1980;68:251–306.

6 Ashkenazi A, Dixit V. Death receptors: signaling and modulation. *Science*. 1998;281:1305–1308.

7 Wajant H. Death receptors. *Essays Biochem*. 2003;39:53–71.

8 Sata M, Suhara T, Walsh K. Vascular endothelial cells and smooth muscle cells differ in expression of Fas and Fas ligand and in sensitivity to Fas ligand-induced cell death. Implications for vascular disease and therapy. *Arterioscler Thromb Vasc Biol*. 2000;20:309–316.

9 Li H, Zhu H, Xu C, Yuan J. Cleavage of BID by caspase 8 mediates the mitochondrial damage in the Fas-pathway of apoptosis. *Cell.* 1998;94:491–501.
10 Boldin M, Goncharov T, Goltsev Y, Wallach D. Involvement of MACH, a novel MORT/FADD-interacting protease, in Fas/APO-1 and TNF receptor-induced cell death. *Cell.* 1996; 85:803–815.
11 Aoudjit F, Vuori K. Matrix attachment regulates Fas-induced apoptosis in endothelial cells: a role for c-FLIP and implications for anoikis. *J Cell Biol.* 2001;152:633–643.
12 Tamatani M, Che Y, Matsuzaki H, et al. Tumor necrosis factor induces Bcl-2 and Bcl-x expression through NFκB activation in primary hippocampal neurons. *J Biol Chem.* 1999;274:8531–8538.
13 Wang C, Mayo M, Korneluk R, et al. NF-κB antiapoptosis: induction of TRAF1 and TRAF2 and c-IAP1 and c-IAP2 to suppress caspase-8 activation. *Science.* 1998;281:1680–1683.
14 Marsters S, Pitti R, Sheridan J, Ashkenazi A. Control of apoptosis signaling by Apo2 ligand. *Recent Prog Horm Res.* 1999;54:225–234.
15 McCarthy N, Whyte M, Gilbert C, Evan G. Inhibition of Ced-3/ICE-related proteases does not prevent cell death induced by oncogenes, DNA damage, or the Bcl-2 homologue Bak. *J Cell Biol.* 1997;136:215–227.
16 Hung R, Chow A. Dissecting the "end game": clinical relevance, molecular mechanisms and laboratory assessment of apoptosis. *Clin Invest Med.* 2004;27:324–344.
17 Adams J, Cory S. The Bcl-2 protein family: arbiters of cell survival. *Science.* 1998;281:1322–1326.
18 Zou H, Li Y, Liu X, Wang X. An APAF-1.cytochrome c multimeric complex is a functional apoptosome that activates procaspase-9. *J Biol Chem.* 1999;274:11549–11556.
19 Slee E, Harte M, Kluck R, et al. Ordering the cytochrome c-initiated caspase cascade: hierarchical activation of caspase-2, -3, -6, -7, and -10 in a caspase-9-dependent manner. *J Cell Biol.* 1999;144:281–292.
20 Green D, Reed J. Mitochondria and apoptosis. *Science.* 1998; 281:1309–1312.
21 Du C, Fang M, Li Y, et al. Smac, a mitochondrial protein that promotes cytochrome c-dependent caspase activation by eliminating IAP inhibition. *Cell.* 2000;102:33–42.
22 Yang Q, Church-Hajduk R, Ren J, et al. Omi/HtrA2 catalytic cleavage of inhibitor of apoptosis protein (IAP) irreversibly inactivates IAPs and facilitates caspase activity in apoptosis. *Genes Dev.* 2003;17:1487–1496.
23 Susin S, Lorenzo H, Zamzami N, et al. Molecular characterization of mitochondrial apoptosis-inducing factor. *Nature.* 1999;397:441–446.
24 Li L, Luo X, Wang X. Endonuclease G is an apoptotic DNase when released from mitochondria. *Nature.* 2001;412:95–99.
25 Puthalakath H, Huang D, O'Reilly L, et al. The proapoptotic activity of the Bcl-2 family membrane Bim is regulated by interaction with the dynein motor complex. *Mol Cell.* 1999;3:287–296.
26 Nakano K, Vousden K. PUMA, a novel proapoptotic gene, is induced by p53. *Mol Cell.* 2001;7:683–694.
27 Roucou X, Rostovtseva T, Montessuit S, et al. Bid induces cytochrome c-impermeable Bax channels in liposomes. *Biochem J.* 2002;363:547–552.
28 Thornberry N, Lazebnik Y. Caspases: enemies within. *Science.* 1998;281:1312–1316.
29 Chen M, Wang J. Initiator caspases in apoptosis signaling pathways. *Apoptosis.* 2002;7:313–319.
30 Tewari M, Quan L, O'Rourke K, et al. Yama/CPP32 β, a mammalian homolog of CED-3, is a CrmA-inhibitable protease that cleaves the death substrate poly(ADP-ribose) polymerase. *Cell.* 1995;81:801–809.
31 Cheng A, Huang T, Lai C, Pan M. Induction of apoptosis by luteolin through cleavage of Bcl-2 family in human leukemia HL-60 cells. *Eur J Pharmacol.* 2005;509:1–10.
32 Taimen P, Kallajoki M. NuMA and nuclear lamins behave differently in Fas-mediated apoptosis. *J Cell Sci.* 2003;116:571–583.
33 Harrington E, Smeglin A, Newton J, et al. Protein tyrosine phosphatase-dependent proteolysis of focal adhesion complexes in endothelial cell apoptosis. *Am J Physiol.* 2001;280:L342-L353.
34 Kook S, Shim S, Choi S, et al. Caspase-mediated cleavage of p130^{cas} in etoposide-induced apoptosis Rat-1 cells. *Mol Biol Cell.* 2000;11:929–939.
35 Kothakota S, Azuma T, Reinhard C, et al. Caspase-3-generated fragment of gelsolin: effector of morphological change in apoptosis. *Science.* 1997;278:294–298.
36 Rudel T, Bokoch G. Membrane and morphological changes in apoptotic cells regulated by caspase-mediated activation of PAK2. *Science.* 1997;276:1571–1574.
37 Menendez P, Vargas A, Bueno C, et al. Quantitative analysis of bcl-2 expression in normal and leukemia human B-cell differentiation. *Leukemia.* 2004;18:491–498.
38 Thome M, Schneider P, Hofmann K, et al. Viral FLICE-inhibitory proteins (FLIPs) prevent apoptosis induced by death receptors. *Nature.* 1997;386:517–521.
39 Verhagen A, Coulson E, Vaux D. Inhibitor of apoptosis proteins and their relatives: IAPs and other BIRPs. *Genome Biol.* 2001; 2(7):Reviews3009. Epub 2001 Jul 5.
40 Volpert O, Zaichuk T, Zhou W, et al. Inducer-stimulated Fas targets activated endothelium for destruction by anti-angiogenic thrombospondin-1 and pigment epithelium-derived factor. *Nat Med.* 2002;8:349–357.
41 Brandes R, Fleming I, Busse R. Endothelial aging. *Cardiovasc Res.* 2005;66:286–294.
42 Dimmeler S, Haendeler J, Rippmann V, et al. Shear stress inhibits apoptosis of human endothelial cells. *FEBS Lett.* 1996;399:71–74.
43 Hermann C, Zeiher A, Dimmeler S. Shear stress inhibits H_2O_2-induced apoptosis of human endothelial cells by modulation of the glutathione redox cycle and nitric oxide synthase. *Arterioscler Thromb Vasc Biol.* 1997;17:3588–3592.
44 Kaiser D, Freyberg M, Friedl P. Lack of hemodynamic forces triggers apoptosis in vascular endothelial cells. *Biochem Biophys Res Commun.* 1997;231:586–590.
45 Malek A, Alper S, Izumo S. Hemodynamic shear stress and its role in atherosclerosis. *JAMA.* 1999;282:2035–2042.
46 Sho E, Sho M, Singh T, et al. Blood flow decrease induces apoptosis of endothelial cells in previously dilated arteries resulting from chronic high blood flow. *Arterioscler Thromb Vasc Biol.* 2001;21:1139–1145.
47 Liu X-M, Ensenat D, Wang H, et al. Physiologic cyclic stretch inhibits apoptosis in vascular endothelium. *FEBS Lett.* 2003; 541:52–56.
48 Haga M, Chen A, Gortler D, et al. Shear stress and cyclic strain may suppress apoptosis in endothelial cells by different pathways. *Endothelium.* 2003;10:149–157.

49 Ruegg C, Mariotti A. Vascular integrins: pleiotropic adhesion and signaling molecules in vascular homeostasis and angiogenesis. *Cell Mol Life Sci.* 2003;60:1135–1157.
50 Ingber D. Mechanical signaling and the cellular response to extracellular matrix in angiogenesis and cardiovascular physiology. *Circ Res.* 2002;91:877–887.
51 Meredith J Jr., Fazeli B, Schwartz M. The extracellular matrix as a cell survival factor. *Mol Biol Cell.* 1993;4:953–961.
52 Isik F, Gibran N, Jang Y, et al. Vitronectin decreases microvascular endothelial cell apoptosis. *J Cell Physiol.* 1998;175: 149–155.
53 Kim S, Bakre M, Yin H, Varner J. Inhibition of endothelial cell survival and angiogenesis by protein kinase A. *J Clin Invest.* 2002;110:933–941.
54 Stupack D, Puente X, Boutsaboualoy S, et al. Apoptosis of adherent cells by recruitment of caspase-8 to unligated integrins. *J Cell Biol.* 2001;155:459–470.
55 Freyberg M, Kaiser D, Graf R, et al. Proatherogenic flow conditions initiate endothelial apoptosis via thrombospondin-1 and the integrin-associated protein. *Biochem Biophys Res Commun.* 2001;286:141–149.
56 Urbich C, Walter D, Zeiher A, Dimmeler S. Laminar shear stress upregulates integrin expression: role in endothelial cell adhesion and apoptosis. *Circ Res.* 2000;87:683–689.
57 Brooks P, Clark R, Cheresh D. Requirement of vascular integrin alpha$_v$beta$_3$ for angiogenesis. *Science.* 1994;264:569–571.
58 Kalluri R. Basement membranes: structure, assembly and role in tumour angiogenesis. *Nat Rev Cancer.* 2003;3:422–433.
59 Bannerman D, Goldblum S. Mechanisms of bacterial lipopolysaccharide-induced endothelial apoptosis. *Am J Physiol.* 2003;284:L899-L914.
60 Desch C, O'Hara P, Harlan J. Antilipopolysaccharide factor from horseshoe crab, *Tachypleus tridentatus*, inhibits lipopolysaccharide activation of cultured human endothelial cells. *Infect Immun.* 1989;57:1612–1614.
61 Hull C, McLean G, Wong F, et al. Lipopolysaccharide signals an endothelial apoptosis pathway through TNF receptor-associated factor 6-mediated activation of c-Jun NH2-terminal kinase. *J Immunol.* 2002;169:2611–2618.
62 LaCasse E, Baird S, Korneluk R, MacKenzie A. The inhibitors of apoptosis (IAPs) and their emerging role in cancer. *Oncogene.* 1998;17:3247–3259.
63 Zen K, Karsan A, Stempien-Otero A, et al. NF-κB activation is required for human endothelial survival during exposure to tumor necrosis factor-alpha but not to interleukin-1β or lipopolysaccharide. *J Biol Chem.* 1999;274:28808–28815.
64 Bannerman D, Eiting K, Winn R, Harlan J. FLICE-like inhibitory protein (FLIP) protects against apoptosis and suppresses NF-κB activation induced by bacterial lipopolysaccharide. *Am J Pathol.* 2004;165:1423–1431.
65 Joyce D, Gelbert L, Ciaccia A, et al. Gene expression profile of antithrombotic protein C defines new mechanisms modulating inflammation and apoptosis. *J Biol Chem.* 2001;276:11199–11203.
66 Cheng T, Liu D, Griffin J, et al. Activated protein C blocks p53-mediated apoptosis in ischemic human brain endothelium and is neuroprotective. *Nat Med.* 2003;9:338–342.
67 Stoll M, Steckelings U, Paul M, et al. The angiotensin AT2-receptor mediates inhibition of cell proliferation in coronary ECs. *J Clin Invest.* 1995;95:651–657.
68 Dimmeler S, Rippmann V, Weiland U, et al. Angiotensin II induces apoptosis of human endothelial cells: protective effect of nitric oxide. *Circ Res.* 1997;81:970–976.
69 Li D, Yang B, Philips M, Mehta J. Proapoptotic effects of ANG II in human coronary artery endothelial cells: role of AT$_1$ receptor and PKC activation. *Am J Physiol.* 1999;276:H786-H792.
70 Li D, Zhang Y, Philips M, et al. Upregulation of endothelial receptor for oxidized low-density lipoprotein (LOX-1) in cultured human coronary artery endothelial cells by angiotensin II type 1 receptor activation. *Circ Res.* 1999:1043–1049.
71 Dimmeler S, Zeiher A. Reactive oxygen species and vascular cell apoptosis in response to angiotensin II and pro-atherosclerotic factors. *Regul Pept.* 2000;90:19–25.
72 Madge L, Pober J. TNF signaling in vascular endothelial cells. *Exp Mol Path.* 2001;70:317–325.
73 Choy J, Granvill D, Hunt D, McManus B. Endothelial cell apoptosis: biochemical characteristics and potential implications for atherosclerosis. *J Mol Cardiol.* 2001;33:1673–1690.
74 Breitschopf K, Haendeler J, Malchow P, et al. Posttranslational modification of Bcl-2 facilitates its proteasome-dependent degradation: molecular characterization of the involved signaling pathway. *Mol Biol Cell.* 2000:1886–1896.
75 Zachary I. Signaling mechanisms mediating vascular protective actions of vascular endothelial growth factor. *Am J Physiol.* 2001;280:C1375-C1386.
76 Alon T, Hemo I, Itin A, et al. Vascular endothelial growth factor acts as a survival factor for newly formed retinal vessels and has implications for retinopathy of prematurity. *Nat Med.* 1995; 1:1024–1028.
77 Abid M, Guo S, Minami T, et al. Vascular endothelial growth factor activates PI3K/Akt/Forkhead signaling in endothelial cells. *Arterioscler Thromb Vasc Biol.* 2004;24:294–300.
78 Gerber H-P, McMurtrey A, Kowalski J, et al. Vascular endothelial growth factor regulates endothelial cell survival through the phosphatidylinositol 3′-kinase/Akt signal transduction pathway. Requirement for Flk-1/KDR activation. *J Biol Chem.* 1998:30336–30343.
79 Dubyak G, el-Moatassim C. Signal transduction via P2-purinergic receptors for extracellular ATP and other nucleotides. *Am J Physiol.* 1993;265:C577-C606.
80 Goepfert C, Imai M, Brouard S, et al. CD39 modulates endothelial cell activation and apoptosis. *Mol Med.* 2000;6:591–603.
81 Dawicki D, Chatterjee D, Wyche J, Rounds S. Extracellular ATP and adenosine cause apoptosis of pulmonary artery endothelial cells. *Am J Physiol.* 1997;273:L485-L494.
82 Lelli JJ, Becks L, Dabrowska M, Hinshaw D. ATP converts necrosis to apoptosis in oxidant-injured endothelial cells. *Free Radic Biol Med.* 1998;25:964–702.
83 Kramer K, Harrington E, Lu Q, et al. Isoprenylcysteine carboxyl methyltransferase activity modulates endothelial cell apoptosis. *Mol Biol Cell.* 2003;14:848–857.
84 Harrington E, Smeglin A, Parks N, et al. Adenosine induces endothelial apoptosis by activating protein tyrosine phosphatase: A possible role of p38α. *Am J Physiol.* 2000;279: L733–L742.
85 Lin X, Fuks Z, Kolesnick R. Ceramide mediates radiation-induced death of endothelium. *Crit Care Med.* 2000;28:N87–N93.
86 Levade T, Auge N, Valdman R, et al. Sphinoglipid mediators in cardiovascular cell biology and pathology. *Circ Res.* 2001; 89:957–968.

87 Dimmeler S, Hermann C, Galle J, Zeiher A. Upregulation of superoxide dismutase and nitric oxide synthase mediates the apoptosis-suppressive effects of shear stress on endothelial cells. *Arterioscler Thromb Vasc Biol.* 1999;19:656–664.

88 Li A, Ito H, Rovira I, et al. A role for reactive oxygen species in EC anoikis. *Circ Res.* 1999;85:304–310.

89 Recchioni R, Marcheselli F, Moroni F, Pieri C. Apoptosis in human aortic endothelial cells induced by hyperglycemic condition involves mitochondrial depolarization and is prevented by N-acetyl-L-cysteine. *Metabolism.* 2002;51:1384–1388.

90 Ho F, Liu S, Liau C, et al. Nitric oxide prevents apoptosis of human endothelial cells from high glucose exposure during early stage. *J Cell Biochem.* 1999;75:258–263.

91 Shen Y, Wang X, Wilcken D. Nitric oxide induces and inhibits apoptosis through different pathways. *FEBS Lett.* 1998;433:125–131.

92 Walford G, Loscalzo J. Nitric oxide in vascular biology. *J Thromb Haemost.* 2003;1:2112–2118.

93 Steinberg D, Parthasarathy S, Carew T, et al. Beyond cholesterol. Modifications of low-density lipoprotein that increase its atherogenicity. *N Engl J Med.* 1989;320:915–924.

94 Chen J, Mehta J, Haider N, et al. Role of caspases in ox-LDL-induced apoptotic cascade in human coronary artery endothelial cells. *Circ Res.* 2004;94:370–376.

95 Sata M, Walsh K. Endothelial cell apoptosis induced by oxidized LDL is associated with the down-regulation of the cellular caspase inhibitor FLIP. *J Biol Chem.* 1998;273:33103–33106.

96 Vergnani L, Hatrik S, Ricci F, et al. Effect of native and oxidized low-density lipoprotein on endothelial nitric oxide and superoxide production: key role of L-arginine availability. *Circulation.* 2000;101:1261–1266.

97 Stefanec T. Endothelial apoptosis: could it have a role in the pathogenesis and treatment of disease? *Chest.* 2000;117:841–854.

98 Stoneman V, Bennett M. Role of apoptosis in atherosclerosis and its therapeutic implications. *Clin Sci.* 2004;107:343–354.

99 Littlewood T, Bennett M. Apoptotic cell death in atherosclerosis. *Curr Opin Lipidol.* 2003;14:469–475.

100 Dimmeler S, Haendeler J, Zeiher A. Regulation of endothelial cell apoptosis in atherothrombosis. *Curr Opin Lipidol.* 2002;13:531–536.

101 Mallat Z, Tedgui A. Apoptosis in the vasculature: mechanisms and functional importance. *Br J Pharmacol.* 2000;130:947–962.

102 Tricot O, Mallat Z, Heymes C, et al. Relation between endothelial cell apoptosis and blood flow direction in human atherosclerotic plaques. *Circulation.* 2000;101:2450–2453.

103 Isner J, Kearney M, Bortman S, Passeri J. Apoptosis in human atherosclerosis and restenosis. *Circulation.* 1995;91:2703–2711.

104 Sata M, Walsh K. Oxidized LDL activates Fas-mediated EC apoptosis. *J Clin Invest.* 1998;102:1682–1689.

105 Recchioni R, Marcheselli F, Moroni F, Pieri C. Apoptosis in human aortic endothelial cells induced by hyperglycemic condition involves mitochondrial depolarization and is prevented by N-acetyl-L-cysteine. *Metab Clin Exp.* 2002;51:1384–1388.

106 de Jong S, van den Berg M, Rauwerda J, Stehouwer C. Hyperhomocysteinemia and atherothrombotic disease. *Semin Thromb Hemost.* 1998;24:381–385.

107 Asai K, Kudej R, Shen Y-T, et al. Peripheral vascular endothelial dysfunction and apoptosis in old monkeys. *Arterioscler Thromb Vasc Biol.* 2000;20:1493–1499.

108 Valgimigli M, Agnoletti L, Curello S, et al. Serum from patients with acute coronary syndromes displays a proapoptotic effect on human endothelial cells: a possible link to pan-coronary syndromes. *Circulation.* 2003;107:264–270.

109 Rossig L, Fichtlscherer S, Heeschen C, et al. The pro-apoptotic serum activity is an independent mortality predictor of patients with heart failure. *Eur Heart J.* 2004;25:1620–1625.

110 Michel J. Anoikis in the cardiovascular system: known and unknown extracellular mediators. *Arterioscler Thromb Vasc Biol.* 2003;23:2146–2154.

111 Lindstedt K, Leskinen M, Kovanen P. Proteolysis of the pericellular matrix: a novel element determining cell survival and death in the pathogenesis of plaque erosion and rupture. *Arterioscler Thromb Vasc Biol.* 2004;24:1350–1358.

112 Kataoka H, Kume N, Miyamoto S, et al. Expression of lectin-like oxidized low-density lipoprotein receptor-1 in human atherosclerotic lesions. *Circulation.* 1999;99:3110–3117.

113 Geng Y-J, Libby P. Progression of atheroma: a struggle between death and procreation. *Arterioscler Thromb Vasc Biol.* 2002;22:1370–1380.

114 Yang J, Sato K, Aprahamian T, et al. Endothelial overexpression of Fas ligand decreases atherosclerosis in apolipoprotein E-deficient mice. *Arterioscler Thromb Vasc Biol.* 2004;24:1466–1473.

115 Wang T, Dong C, Stevenson S, et al. Overexpression of soluble Fas attenuates transplant arteriosclerosis in rat aortic allografts. *Circulation.* 2002;106:1536–1542.

116 Dimmeler S, Zeiher A. Vascular repair by circulating endothelial progenitor cells: the missing link in atherosclerosis? *J Mol Med.* 2004;82:671–677.

117 Fournet-Bourguignon M, Castedo-Delrieu M, Bidouard J, et al. Phenotypic and functional changes in regenerated porcine coronary endothelial cells: increased uptake of modified LDL and reduced production of NO. *Circ Res.* 2000;86:854–861.

118 Bombeli T, Karsan A, Tait J, Harlan J. Apoptotic vascular endothelial cells become procoagulant. *Blood.* 1997;89:2429–2442.

119 Mallat Z, Hugel B, Ohan J, et al. Shed membrane microparticles with procoagulant potential in human atherosclerotic plaques: a role for apoptosis in plaque thrombogenicity. *Circulation.* 1999;99:348–353.

120 Durand E, Scoazec A, Lafont A, et al. In vivo induction of endothelial apoptosis leads to vessel thrombosis and endothelial denudation: a clue to the understanding of the mechanisms of thrombotic plaque erosion. *Circulation.* 2004;109:2503–2506.

121 Coleman M, Sahai E, Yeo M, et al. Membrane blebbing during apoptosis results from caspase-mediated activation of ROCK I. *Nat Cell Biol.* 2001;3:339–345.

122 Mallat Z, Benamer H, Hugel B, et al. Elevated levels of shed membrane microparticles with procoagulant potential in the peripheral circulating blood of patients with acute coronary syndromes. *Circulation.* 2000;101:841–843.

123 Haimovitz-Friedman A, Cordon-Cardo C, Bayoumy S, et al. Lipopolysaccharide induces disseminated endothelial apoptosis requiring ceramide generation. *J Exp Med.* 1997;186:1831–1841.

124 Pena L, Fuks Z, Kolesnick R. Radiation-induced apoptosis of endothelial cells in the murine central nervous system: protection by fibroblast growth factor and sphingomyelinase deficiency. *Cancer Res.* 2000;60:321–327.

125 Buja L, Entman M. Modes of myocardial cell injury and cell death in ischemic heart disease. *Circulation.* 1998;98:1355–1357.
126 Yaoita H, Ogawa K, Maehara K, Maruyama Y. Attenuation of ischemia/reperfusion injury in rats by a caspase inhibitor. *Circulation.* 1998;97:276–281.
127 Scarabelli T, Stephanou A, Rayment N, et al. Apoptosis of endothelial cells precedes myocyte cell apoptosis in ischemia/reperfusion injury. *Circulation.* 2001;104:253–256.
128 Quadri S, Segall L, de Perrot M, et al. Caspase inhibition improves ischemia-reperfusion injury after lung transplantation. *Am J Transplant*. 2005;5:292–299.
129 Gao W, Bentley R, Madden J, Clavien P. Apoptosis of sinusoidal endothelial cells is a critical mechanism of preservation injury in rat liver transplantation. *Hepatology*. 1998;27:1652–1660.
130 Paris F, Fuks Z, Kang A, et al. Endothelial apoptosis as the primary lesion initiating intestinal radiation damage in mice. *Science.* 2001;293:293–297.
131 Garcia-Barros M, Paris F, Cordon-Cardo C, et al. Tumor response to radiotherapy regulated by endothelial cell apoptosis. *Science.* 2003;300:1155–1159.
132 Garcia-Barros M, Lacorazza D, Petrie H, et al. Host acid sphingomyelinase regulates microvascular function not tumor immunity. *Cancer Res.* 2004;64:8285–8291.
133 Kasahara Y, Tuder R, Cool C, et al. Endothelial cell death and decreased expression of vascular endothelial growth factor and vascular endothelial growth factor receptor 2 in emphysema. *Am J Respir Crit Care Med.* 2001;163:737–744.
134 Abdollahi A, Lipson K, Sckell A, et al. Combined therapy with direct and indirect angiogenesis inhibition results in enhanced antiangiogenic and antitumor effects. *Cancer Res.* 2003;63:8890–8898.
135 Kasahara Y, Tuder R, Taraseviciene-Stewart L, et al. Inhibition of VEGF receptors causes lung cell apoptosis and emphysema. *J Clin Invest*. 2000;106:1311–1319.
136 Shaheen R, Davis D, Liu W, et al. Antiangiogenic therapy targeting the tyrosine kinase receptor for vascular endothelial growth factor receptor inhibits the growth of colon cancer liver metastasis and induces tumor and endothelial cell apoptosis. *Cancer Res.* 1999;59:5412–5416.
137 Taraseviciene-Stewart L, Scerbavicius R, Choe K, et al. An animal model of autoimmune emphysema. *Am J Respir Crit Care Med.* 2005;171:734–742.
138 Dang C, Magid M, Weksler B, et al. Enhanced endothelial cell apoptosis in splenic tissues of patients with thrombotic thrombocytopenic purpura. *Blood.* 1999;93:1264–1270.
139 Mitra D, Jaffe E, Weksler B, et al. Thrombotic thrombocytopenic purpura and sporadic hemolytic-uremic syndrome plasmas induce apoptosis in restricted lineages of human microvascular endothelial cells. *Blood.* 1997;89:1224–1234.
140 Laurence J, Mitra D, Steiner M, et al. Plasma from patients with idiopathic and human immunodeficiency virus-associated thrombotic thrombocytopenic purpura induces apoptosis in microvascular endothelial cells. *Blood.* 1996;87:3245–3254.
141 Davis J, Rueda B, Spanel-Borowski K. Microvascular endothelial cells of the corpus luteum. *Reprod Biol Endocrinol.* 2003;1:1–15.
142 Dimmeler S, Zeiher A. Endothelial cell apoptosis in angiogenesis and vessel regression. *Circ Res.* 2000;87:434–439.
143 Stupack D, Cheresh D. Integrins and angiogenesis. *Curr Top Dev Biol.* 2004;64:207–238.
144 Bellas R, Harrington E, Sheahan K, et al. FAK blunts adenosine-homocysteine-induced endothelial cell apoptosis: Requirement for PI3-kinase. *Am J Physiol.* 2002;282:L1135-L1142.
145 Walker P, Weaver V, Lach B, et al. Endonuclease activities associated with high molecular weight and internucleosomal DNA fragmentation in apoptosis. *Exp Cell Res.* 1993;213:100–106.
146 Gorczyca W, Gong J, Darzynkiewicz Z. Detection of DNA strand breaks in individual apoptotic cells by the in situ terminal deoxynucleotidyl transferase and nick translation assays. *Cancer Res.* 1993;53:1945–1951.
147 Vermes I, Haanen C, Steffens-Nakken H, Reutelingsperger C. A novel assay for apoptosis. Flow cytometric detection of phosphatidylserine expression on earlier apoptotic cells using fluorescein labelled Annexin V. *J Immunol Methods.* 1995;184:39–51.
148 Lazebnik Y, Kaufmann S, Desnoyers S, et al. Cleavage of poly (ADP-ribose) polymerase by a proteinase with properties like ICE. *Nature.* 1994;371:346–347.
149 Reers M, Smith T, Chen L. J-aggregate formation of a carbocyanine as a quantitative fluorescent indicator of membrane potential. *Biochemistry.* 1991;30:4480–4486.
150 Li A, Dubey S, Varney M, et al. IL-8 directly enhanced endothelial cell survival, proliferation, and matrix metalloproteinases production and regulated angiogenesis. *J Immunol.* 2003;170:3369–3376.
151 Lindner H, Holler E, Ertl B, et al. Peripheral blood mononuclear cells induce programmed cell death in human endothelial cells and may prevent repair: role of cytokines. *Blood.* 1997;89:1931–1938.
152 Jimenez B, Volpert O, Crawford S, et al. Signals leading to apoptosis-dependent inhibition of neovascularization by thrombospondin-1. *Nat Med.* 2000;6:41–48.
153 Stellmach V, Crawford S, Zhou W, Bouck N. Prevention of ischemia-induced retinopathy by the natural ocular antiangiogenic agent pigment epithelium-derived factor. *Proc Natl Acad Sci USA.* 2001;98:2593–2597.
154 Karsan A, Yee E, Poirier G, et al. Fibroblast growth factor-2 inhibits endothelial cell apoptosis by Bcl-2-dependent and independent mechanisms. *Am J Pathol.* 1997;151:1775–1784.
155 Kamphaus G, Colorado P, Panka D, et al. Canstatin, a novel matrix-derived inhibitor of angiogenesis and tumor growth. *J Biol Chem.* 2000;275:1209–1215.
156 Hayes A, Huang W, Mallah J, et al. Angiopoietin-1 and its receptor Tie-2 participate in the regulation of capillary-like tubule formation and survival of endothelial cells. *Microvasc Res.* 1999;58:224–237.
157 Pohlman T, Harlan J. Adaptive responses of the endothelium to stress. *J Surg Res.* 2000;89:85–119.
158 Fischer S, Cassivi S, Xavier A, et al. Cell death in human lung transplantation: apoptosis induction in human lungs during ischemia and after transplantation. *Ann Surg.* 2000;231:424–431.
159 Fischer S, Maclean A, Liu M, et al. Dynamic changes in apoptotic and necrotic cell death correlate with severity of ischemia-reperfusion injury in lung transplantation. *Am J Respir Crit Care Med.* 2000;162:1932–1939.

120

Endothelial Antigen Presentation

Andrew H. Lichtman

Brigham and Women's Hospital, Harvard Medical School, Boston, Massachusetts

T lymphocyte–mediated cellular immunity is responsible for protection against intracellular pathogens. T cells also mediate pathological responses in autoimmune diseases, chronic inflammatory conditions, and allograft rejection. These protective and pathological roles for T cells require the recognition of antigens on the surface of other cells. This chapter reviews how endothelial cells (ECs) may play a unique role in regulating T cell–mediated immune responses by presenting antigen to circulating T cells within the lumens of blood vessels. Antigen presentation to T cells may be considered a complex output function of ECs, which is dependent on the regulated expression of several intracellular and cell surface proteins. ECs may also become susceptible to both physiological and pathological input signals as a consequence of their intimate relationships with T lymphocytes. In this chapter, a brief overview of the cell biology of antigen presentation to T cells is followed by a discussion of specific features of antigen presentation by ECs.

BACKGROUND: ANTIGEN PRESENTATION AND T-CELL HOMING

T cells are heterogeneous with respect to effector functions and history of antigen exposure. This heterogeneity is reflected in their differing requirements for activation. Regardless of the subset, T cells must recognize antigen via cell surface antigen receptors – called T-cell receptors (TCRs) – for functional activation to proceed. Most T lymphocytes recognize a complex of a proteolytically derived peptide bound to one of many possible allelic forms of a major histocompatibility complex (MHC) molecule that is expressed on the surface of another cell. These peptide–MHC complexes are the "antigens" to which the TCR binds. The $CD4^+$ subset of T cells, which includes T cells of the "helper" phenotype, recognize complexes of peptides bound to class II MHC molecules, whereas $CD8^+$ T cells, which include most cytolytic T lymphocytes (CTL), recognize peptide-class I MHC complexes. Rare subsets of T cells, such as iNKT cells, recognize lipids bound to the nonpolymorphic MHC-like CD1 molecules. The cells that display the peptide–MHC complexes for T cell recognition are called antigen presenting cells (APCs). Therefore, as a minimum requirement for a cell to perform as an effective APC for a $CD4^+$ T cell, it must be able to form peptide-class II MHC complexes and display them on the cell surface. Likewise, an APC for $CD8^+$ T cells must be able to form peptide-class I MHC complexes. The intracellular proteolytic degradation of intact proteins to form MHC-binding peptides and the assembly of the peptide-MHC complexes is called *antigen processing*. Antigen processing involves a complex series of coordinated steps in several subcellular locations, and differs in significant ways between the pathways that generate class I versus class II MHC–associated peptide complexes (1,2).

Class I Major Histocompatibility Complex Pathway of Antigen Presentation

Class I MHC expression is common to most cell types. The cell surface expression of class I MHC molecules requires assembly of a trimolecular complex that includes a short peptide (~8–11 amino acid residues long) bound within a cleft formed by the N-terminal domains of the class I MHC α-chain protein, and a noncovalently associated $\beta 2$-microglobulin molecule. Class I MHC–associated peptides are derived from cytoplasmic proteins by proteolytic degradation in the proteasome. The proteasome is a multisubunit organelle found in all nucleated cells, which degrades proteins that are tagged for destruction by ubiquitination. Proteasomal degradation of ubiquitinated proteins is an essential housekeeping function of every cell, serving to remove misfolded proteins and to regulate several signal transduction pathways. The immune system has evolved to make use of the proteasomal system to sample cytoplasmic proteins for display by class I MHC molecules. The repertoire of peptides generated by the proteasome may be altered by interferon (IFN)-γ stimulation of cells, which induces the expression of new proteasomal subunits. A subset of peptides generated by proteaosomes is delivered into the lumen of the endoplasmic reticulum (ER) by the

ATP-driven transporter associated with antigen processing (TAP). Once in the ER, chaperones facilitate the binding of the peptides to newly synthesized dimers of class I MHC α-chain and β2-microglobulin. Once peptide is bound, the stabilized complex moves through the Golgi to the cell surface. The result is the display of peptides representative of many cytoplasmic proteins, some of which may be recognized by T cells. In addition, peptides derived from ER resident proteins are presented by class I MHC molecules, and are likely generated by ER-resident proteases. Because class I MHC expression is typical of most cell types, CD8$^+$ T cells will have the opportunity to survey most cell types for the presence of microbial proteins that are in the cytoplasm, as well as mutated cytoplasmic self proteins. Because most cell types are susceptible to viral infection or potentially oncogenic mutations, the ubiquitous class I pathway is well suited for immune surveillance for viral infection and neoplasia. In addition to sampling proteins synthesized in the cytoplasm of cells, the class I pathway may also "cross-present" protein antigens that have been internalized from the extracellular milieu (3). In this case, endocytosed proteins must gain access to the cytoplasmic proteasome, or be delivered into the ER, for proteolytic degradation. Cross-presentation of exogenous protein antigens to CD8$^+$ T cells is a property best described in dendritic cells, and is an explanation for how naive microbial-specific CD8$^+$ T cells can be activated by APCs that are not directly infected by viruses (4).

Class II Major Histocompatibility Pathway of Antigen Presentation

Constitutive class II MHC expression is restricted to a small number of cell types, compared to class I MHC expression; therefore CD4$^+$ T cells have a more limited choice of APC partners with which to interact. These class II MHC–expressing cells include dendritic cells, macrophages, and B lymphocytes; these cell types also express T cell costimulatory molecules (discussed later), and are often called *professional APCs*. The dendritic cell is unique in that its primary function appears to be the transport and presentation of protein antigens. It is the only cell type that can efficiently activate naive CD4$^+$ T cells (5). The proteins represented by class II MHC–bound peptides are largely derived from the extracellular milieu. Endocytosis of these proteins into the endosome/lysosome system of organelles in APCs is the first step of the class II pathway of antigen presentation (6). The peptides are generated from the internalized proteins by proteases active in the acidic environment in these organelles. Newly synthesized class II MHC heterodimers composed of α- and β-chains, bind trimers of the 30-kD transmembrane invariant chain protein in the ER. The invariant chain blocks the peptide binding groove from occupation by peptides in the ER. In addition, via cell localization signals in its cytoplasmic tail, the invariant chain directs movement of the class II MHC/invariant trimer complex into a late endosome/lysosome compartment, which contains abundant vesicular inclusions called *multivesicular bodies*, and class II MHC–like HLA-DM molecules. The combined action of the acidic proteases and HLA-DM mediate the removal of the invariant chain and the binding of exogenously derived peptides to a cleft formed by the N-terminal domains of the α- and β-chains of the class II MHC molecules. The class II MHC peptide binding cleft is open-ended, and can accommodate a broader range of peptide lengths, typically 30 amino acid residues or longer, compared with class I MHC clefts. The final step in this pathway is delivery of the class II MHC peptide complexes to the cell surface by vesicle fusion with the plasma membrane. CD4$^+$ T cells that interact with class II MHC-expressing APCs can therefore survey a peptide sample of many extracellular proteins.

Antigen and Cell Traffic in Antigen Presentation

The expression of peptide–MHC complexes (antigens) on APCs will amount to nothing if T cells that recognize those antigens are not in the same location as the APCs. Naive T lymphocytes emerge from the thymus expressing antigen receptors, but with little ability to perform effector functions, and very few naive T cells are specific for any one particular peptide-MHC antigen. Both CD4$^+$ and CD8$^+$ naive cells must be activated by APCs in lymphoid tissues to produce large numbers of the effector cells needed to eradicate infections. The major mechanism that has evolved to ensure that rare naive T cells specific for a particular antigen encounter APCs presenting that antigen involves distinct yet overlapping trafficking patterns, which ultimately converge at the level of the parafollicular zones of lymph nodes: T cells routinely recirculate between circulating blood and lymph nodes, whereas APCs (dendritic cells), once activated by microbial products in infected tissues, become motile and move into lymphatics, eventually entering the parafollicular zones of lymph nodes. Rarely, whole microbes, microbial debris, and soluble molecules produced by microbes may enter lymphatics directly, without being associated with cells, and are brought into the lymph node via lymph flow.

In addition to presenting peptide–MHC complexes to T cells, APCs also express membrane molecules that provide signals that synergize with TCR complex–mediated signals to ensure full activation of T cells. These molecules are called *costimulators*, and they bind to receptors on those T cell that are linked to signal transduction pathways that enhance T cell proliferation and differentiation (7). The best defined costimulators are members of the B7 family, of which B7-1 (CD80) and B7-2 (CD86) are the most important. Both B7-1 and B7-2 bind to CD28 on T cells. In humans, LFA-3 (CD58) is an important costimulator that binds to CD2 on T cells. Naive CD4$^+$ T cells generally have more stringent requirements for costimulation than effector T cells. Naive CD8$^+$ T cell activation may depend on costimulation, but under conditions of strong innate responses, such as occur with highly cytopathic viral infections, the well-characterized costimulatory pathways appear not to be needed. In the absence of costimulators, recognition of antigen will not lead to activation of naive T cells, and may induce an anergic state such that the T cell will not be able to respond to antigen a second time, even if costimulators are present. Dendritic cells constitutively express low levels of costimulators, but expression is markedly enhanced by microbial products that bind to Toll-like receptors (TLRs) on the APCs. For example, the expression of CD80 and CD86 increases on dendritic cells after exposure to endotoxin as they migrate from peripheral tissues into lymph nodes. In essence, costimulator expression is a danger signal that ensures naive T cell activation will occur only when infections are present. Effector T-cell responses to antigens are not dependent on costimulation, but the responses may be enhanced by costimulatory molecules, including members of the B7 family that are distinct from B7-1 and B7-2.

Naive T cells, which are generated in the thymus, circulate throughout the body in the blood. They home to lymph nodes, where they leave the circulation. In the lymph node, they have the opportunity to encounter APCs, but if they do not recognize antigen, they move via lymphatics from one lymph node to another until they do recognize antigen or are carried back into the circulation. Naive T cell homing to lymph nodes occurs largely because of their expression of L-selectin and CCR7, which bind to peripheral lymph node addressin (PNad) and chemokines (CCL19 or CCL21), respectively – molecules that are displayed on the luminal surface of specialized high endothelial venules (HEVs) found in the parafollicular zones of lymph nodes (8). Once a naive T cell recognizes antigen on an APC in a lymph node, it becomes activated, undergoes mitotic division, and gives rise to many differentiated progeny that retain the same antigen specificity as the parent cell but gain effector functions. Many of these effector T cells re-enter the circulation, and circulate into peripheral tissues. During their differentiation, effector T cells express adhesion molecules (including E- and P-selectin ligands) and chemokine receptors (such as CCR5 and CXCR3) that favor migration into sites of inflammation. This migratory phenotype is relatively stable for the life of the effector T cell. The effector cells that get into peripheral inflammatory sites, as well as some that remain in the lymph node, perform the effector functions that aid in eliminating infections. These effector functions include macrophage and B-cell activation by $CD4^+$ effectors (so called *helper functions*) and direct killing of infected cells by $CD8^+$ effectors (CTL activity). Most effector T cells will die within days of their generation. Some of the $CD4^+$ and $CD8^+$ T cells that are derived from naive T cells have a memory phenotype, and survive in a resting state for long periods. Some of these memory T cells, called *central memory T cells*, may remain in lymph nodes. Others, called *effector memory T cells*, come to reside in peripheral mucosal sites. Memory T cells are responsible for the rapid and amplified responses upon re-exposure to antigens.

ENDOTHELIAL ANTIGEN PRESENTATION: GENERAL AND HISTORICAL CONSIDERATIONS

Even before the molecular and cell biological details of antigen processing and presentation summarized here were elucidated, investigators hypothesized that ECs may be important APCs based on the fact that they were the first cells with which T cells come into direct and intimate contact before entering lymph nodes or peripheral tissues. This unique relationship between ECs and T cells has been the basis for a general hypothesis that endothelial antigen presentation leads to the selective transendothelial migration of T cells specific for antigens found in the underlying tissue (9). In other words, ECs may sample the proteins in the surrounding tissue environment, and display them on their luminal surface as peptide–MHC complexes; circulating T cells that recognize these complexes are preferentially able to enter the tissues, compared with T cells that do not (Figure 120.1A). The relevant corollary to this hypothesis is that endothelial antigen presentation is important for the delivery of naive T cells into lymphoid tissues, where the initiation of T cell–mediated immune responses begins, and/or for the efficient delivery of effector T cells to sites where antigen is present in peripheral tissues. For the hypothesis and its corollary to be true, ECs must be able to engage in the antigen processing and peptide display functions typical of all APCs discussed earlier. ECs also would be expected to express costimulatory molecules, especially if they are involved in naive T cell activation. The history of experimental investigation of endothelial antigen presentation can be divided into two parts. The first and still active area of research is the in vitro demonstration and analysis of antigen presentation of cultured ECs to cultured T cells (10). The second active but relatively new area of research involves demonstration of endothelial antigen presentation in vivo, using experimental animal models.

MHC expression, which is the primary requirement for APC activity, was demonstrated on ECs in vitro and in vivo many years ago. Class I MHC expression is constitutive on ECs in vivo, in large and small arteries and veins, and lower levels of expression are retained on cultured ECs (10). Endothelial class I MHC expression can be upregulated by IFN-α, -β, -γ. Therefore, ECs, like most cell types, possess the minimal requirement for antigen presentation to $CD8^+$ T cells. Class II MHC expression by ECs is more variable, depending on species and cytokine exposure. Human microvessels, including postcapillary venules, constitutively express class II MHC in many tissues, but class II MHC expression in other vessels is variable (10,11). Cultured ECs from human or mouse usually lack class II MHC expression, but expression can be induced by IFN-γ treatment, regardless of the species or site of derivation (9,12). IFN-γ treatment or inflammation induces class II MHC expression on mouse microvasculature, including postcapillary venules in vivo (10). Of significance to the possible sites of endothelial antigen presentation in vivo, no compelling evidence suggests that HEVs in lymph nodes express class II MHC.

Consistent with their ability to express cell surface peptide–MHC complexes, cultured ECs from various tissue sources can present antigens to $CD4^+$ and $CD8^+$ T cells, leading to T-cell activation in vitro. Demonstration of antigen-specific activation of human T cells has been encumbered by the requirement

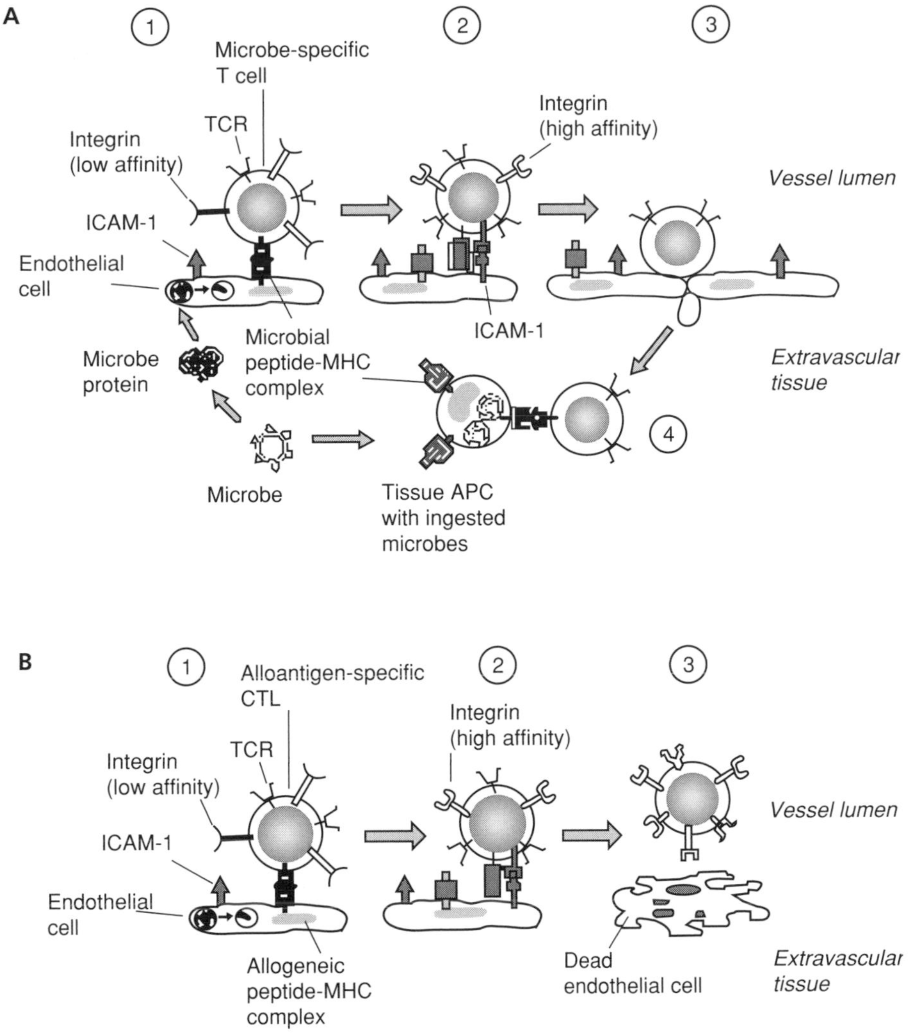

Figure 120.1. Consequences of endothelial antigen presentation. (**A**) Hypothetically, ECs may process and present antigens produced in surrounding extravascular tissue to circulating T cells that come into close contact with the vessel wall (*1*). T-cell receptor signals could then promote stable T-cell adherence by inducing integrin activation (*2*), thereby enhancing diapedesis of antigen-specific T cells (*3*), which can then interact with tissue APCs (*4*). ECs may display peptides bound to either class I or class II MHC molecules, and present them to either $CD8^+$ or $CD4^+$ T cells, respectively. There is circumstantial evidence for this physiological role of endothelial antigen presentation in immune responses, but definitive experimental proof is lacking. Selectins, chemokines, and junctional proteins, which are known to be involved in T-cell diapedesis are not shown. (**B**) EC antigen presentation to circulating T cells may result in direct T-cell–mediated damage to the EC. In the example illustrated, a circulating cytolytic T lymphocyte recognizes a peptide–MHC alloantigen presented by the EC (*1*), leading to a tight adherence (*2*), and granule release, resulting in EC death (*3*). CTL killing of ECs in allograft rejection is well substantiated. APC, antigen-presenting cell; ICAM-1, intercellular adhesion molecule-1; MHC, major histocompatibility complex; TCR, T-cell receptor.

of obtaining MHC-matched T cells and ECs, which is a requirement to model physiologic antigen presentation, and by the difficulty in obtaining T-cell populations enriched for a single antigen specificity. Nonetheless, sufficient published evidence supports the conclusion that human ECs in culture can present antigen and activate $CD4^+$ T cells. The evidence has come from studies using peripheral blood T cells from individuals previously exposed to immunogens and MHC-matched endothelium (13–16). The ECs usually are treated with IFN-γ to induce class II MHC expression. Additionally, cocultures of MHC-disparate human ECs and human T cells (either $CD4^+$ or $CD8^+$) lead to activation of the T cells, indicating that human ECs are capable of presenting alloantigens to T cells (17–19). Mouse T-cell clones or T cells from TCR-transgenic mice, which have single peptide-MHC specificity, can be simulated to proliferate and secrete cytokines in a peptide-antigen–specific manner by coculture with MHC-matched ECs and the relevant protein antigen. This is true for

CD8^{+} T cells (20,21) and for CD4^{+} T cells, if the ECs are treated with IFN-γ in order to upregulate class II MHC (22). The question of whether ECs can cross-present proteins derived from other cells to CD8^{+} T cells has been not been clearly resolved by in vitro experiments using purified endothelial cultures, but a study with pancreatic islet cultures does suggest that ECs can cross-present insulin to CD8^{+} CTL (23).

The early demonstrations of in vitro endothelial antigen presentation did not address the fundamental differences in the responsiveness of naive versus effector or memory (antigen-experienced) T cells. The migration patterns, sites of activation, and costimulatory requirements of naive versus antigen-experienced cells differ significantly; therefore, the importance of endothelial antigen presentation to these different T cells also may differ significantly.

ENDOTHELIAL ANTIGEN PRESENTATION TO NAIVE AND ANTIGEN-EXPERIENCED T CELLS: THE ROLE OF COSTIMULATORS

In vitro studies using both mouse and human cells indicate variability in the responsiveness of T cells to endothelial antigen presentation depending on the stage of T-cell differentiation and/or the activation state of the endothelium. Large numbers of naive T cells specific for a particular antigen can be isolated from TCR-transgenic mice, and their response to endothelial antigen presentation can be compared to T cells from the same mouse that have been activated in vitro with the antigen and syngeneic professional APCs. This approach has shown that naive CD4^{+} T cells do not become fully activated by culture with ECs plus antigen, but resting effector CD4^{+} T cells do (22). In contrast, naive minor histocompatibility antigen (HY antigen)-specific TCR transgenic CD8^{+} T cells can be activated by HY-expressing ECs (20). The limited published data are inconsistent with regard to the effect of IFN-γ activation of ECs on the responses of naive CD8^{+} T cells; in one case IFN-γ treatment of ECs inhibited their ability to support antigen-induced responses of naive CD8^{+} T cells (20) and, in another case, IFN-γ treatment of ECs enhanced naive CD8^{+} T-cell responses (24).

The experimental comparison of responses of naive versus antigen-experienced human T cells is difficult to achieve because naive T cells specific for any single protein antigen are very rare in human peripheral blood. However, the frequency of naive human cells reactive to allogeneic MHC antigens is high enough to support a detectable response in vitro, and CD45RA and CD45RO, which are different RNA splice variants of a cell surface glycoprotein, are reasonable, albeit not perfect, markers of naive versus memory T cells, respectively. Naive (CD45RA) human CD8^{+} T cells specific for allogenic MHC do not become activated by allo-MHC–presenting ECs, whereas memory T cells with the same allo-specificity do (25). Overall, in vitro findings are consistent with the interpretation that endothelial antigen presentation is not important for the activation of naive T cells and the initiation of primary immune responses. This interpretation also is consistent with the lack of class II MHC expression on HEVs, which are the major site of naive T cell-endothelial interaction. Furthermore, as discussed earlier, a large body of data indicate that the molecular information provided by adhesion molecules and chemokines is sufficient for homing of naive T cells into lymph nodes, in the complete absence of cognate antigen.

The failure of naive T cells to respond to endothelial antigen presentation is most likely a reflection of their stringent requirements for costimulation, and the absence of critical costimulatory molecules on ECs. B7-1 or B7-2 are the two molecules that account for the majority of costimulatory activity of APCs for naive CD4^{+} T cells in mice and a major part of the activity for naive human CD4^{+} T cells. There is very little evidence that B7-1 or B7-2 are expressed on human ECs in vivo. Expression can be induced on mouse liver sinusoidal and glomerular ECs by warm ischemia–reperfusion, which is a model for ischemia–reperfusion injury (26,27), and some expression has been reported on human hepatic sinusoidal ECs under pathological conditions (28). Expression of B7-1 and B7-2 on cultured ECs appears to vary with species and tissue source. For example, cultured pig aortic ECs express B7-1, and not B7-2 (29), whereas mouse hepatic sinusoidal endothelial express B7-2, but not B7-1. Mouse heart microvascular ECs do not express B7-1 or B7-2 (21), while mouse lung microvascular ECs express low levels of B7-1 but not B7-2 (20). With a few exceptions, most published analyses of costimulator expression on cultured human ECs have failed to detect B7-1 or B7-2 expression (30–32). In fact, naive T-cell reactivity to ECs can be induced by exogenous *B7-1* gene expression (33).

Human ECs in vivo and in vitro constitutively express CD58, which provides significant costimulatory activity for previously activated CD4^{+} and CD8^{+} T cells in vitro (10). CD58-mediated endothelial costimulation of naive human T cells in vitro can be demonstrated only when phytohemagglutinin (PHA) is used as the primary TCR stimulus (33). It is possible that endothelial costimulatory molecules, such as CD58 or members of the B7 family other than CD80 and CD86, do enhance responses of effector or memory T cells to endothelial antigen presentation in vivo. For example, inducible costimulatory molecule ICOS-ligand (B7h) is a B7-family costimulatory molecule that is expressed on ECs, and it contributes to endothelial activation of CD4^{+} T cells (34). Although strictly not a costimulatory molecule, CD40 is a tumor necrosis family (TNF)-receptor family molecule expressed on professional APCs that indirectly enhances T-cell activation by stimulating the expression of other costimulatory molecules and cytokines (35). CD154, the ligand for CD40, is expressed on T cells recently activated by antigen. Because ECs from various sources express CD40 (10), it is possible that T cells that have been activated by endothelial antigen presentation can reciprocally activate ECs through the CD154/CD40 pathway. The theoretical net result, which is consistent with in vitro experimental data, is enhanced bidirectional activation of both T cell and EC.

T CELLS ON THE MOVE: IN VITRO MODELS WITH IN VIVO RELEVANCE

The ability of cultured ECs to activate T cells in static in vitro culture does not directly address the underlying hypothesis presented at the beginning of this chapter that endothelial antigen presentation promotes selective migration of antigen-specific T cells into tissues where the antigen is present. T cells undertake two general directions of movement with respect to endothelium that can be modeled in vitro; these are movements parallel to the direction of blood flow, and movement through the endothelium.

One potential effect of endothelial antigen presentation to T cells at the apical surface of the EC is the enhancement of adhesive interactions that stop movement of the T cell parallel to flow. The result is stabilized arrest of the T cell in the face of shear stress. In the widely accepted multistep paradigm of leukocyte diapedesis, reviewed in other chapters of this volume, stable arrest is required before transendothelial migration can occur (36). Compelling in vitro and in vivo evidence suggests that chemokines displayed on the endothelial surface activate T-cell integrins via chemokine receptor signaling, and this results in higher-avidity T-cell integrin binding to ECs (37). In vitro, TCR signaling has a similar effect as chemokines in activating T-cell integrins (38). Although the affinity of a TCR for a peptide–MHC complexes is not sufficiently high to contribute to the adhesion, the coreceptor interactions that accompany antigen recognition (CD4 with class II MHC, or CD8 with class I MHC) also could serve adhesive functions. Videomicroscopic analysis of T-cell interactions with endothelium in flow chambers is a potentially informative method for determining if antigen recognition at the apical surface of ECs can promote the stable arrest of T cells in the setting of shear stress. To date, no published data have used this approach. Videomicroscopy has been used to show that T cells that adhere to endothelium in the presence of cognate antigen, but without shear stress, are moderately more resistant to being dislodged when shear flow is applied, compared with T cells that adhered in the absence of antigen (23). This is a predictable result, because it is well known that the adhesion of T cells to APCs typically increases as antigen presentation proceeds. The diameter and molecular phenotype of microvasculature in some tissues, such as postcapillary venules in the skin, suggests ample opportunity for intravascular T cells to interact with ECs (39). Typical flow chamber conditions may be less conducive to T cell–endothelial interactions than in vivo conditions. In tissue microvasculature, flow is often intermittent, the diameter of the postcapillary venules may be less than the diameter of T cells, and therefore prolonged periods occur when T cells are not moving and are closely apposed to ECs.

Studies of transmigration of T cells through endothelial monolayers under static conditions in vitro suggest that T cells that "see" antigen on ECs are more likely to transmigrate than T cells that do not see antigen (20,40). Furthermore, endothelial costimulation – for example, through the ICOS pathway – may enhance transmigration (41).

DOES ENDOTHELIAL ANTIGEN PRESENTATION OCCUR IN VIVO?

The circumstantial evidence that endothelial antigen presentation occurs in vivo is largely based on kinetic arguments, such as the early expression of class II MHC on vascular ECs prior to central nervous system (CNS) infiltration of $CD4^+$ T cells in experimental allergic encephalomyelitis (EAE) (42), and evidence of endothelial activation within 2 hours of intradermal injection of purified protein derivative (PPD) in sensitized but not naive primates (43). To provide definitive evidence that ECs actually present antigen to circulating T cells in vivo, the researcher must pick a measurable response that can only be attributable to antigen presentation, then measure that response under conditions in which no other type of APCs besides ECs could be responsible. This is a difficult task to achieve and, to date, there are a paucity of published data from studies that fulfill these requirements. The experimental approaches that have been applied to address endothelial antigen presentation in vivo vary in significant ways, and there are limitations in the interpretation of data obtained from each approach. For example, in some studies, the antigen is expressed on ECs; in others, the ECs cross-present antigen expressed in the tissue. In some studies, the endpoint is migration of antigen-specific T cells into tissues, whereas in other studies, the endpoint is intravascular interactions with the vessel wall. Nonetheless, a cumulative assessment of these different studies is that endothelial antigen presentation does occur in vivo, but the consequences are variable, depending on the vascular bed, T-cell subset, and antigen involved.

To determine if endothelial antigen presentation to T cells of a particular antigen specificity enhances preferential recruitment of those T cells, the measured endpoint is the extravascular accumulation of antigen-specific T cells in the tissues. The possibility always exists that a small number of these T cells that transmigrate out of the vessel independent of endothelial antigen presentation become activated by professional APCs in the tissue. This can lead to enhanced expression of adhesion molecules on nearby ECs, and local chemokine expression that favors migration of memory or effector T cells, regardless of specificity. Enrichment of antigen-specific T cells relative to other T cells in the transmigrated population may reflect proliferative responses of the T cells in the tissue and not preferential migration. It seems most likely that a role for endothelial antigen presentation would occur very early in a peripheral immune response to an infection, perhaps when local innate immune responses have not altered endothelial phenotype significantly. The upregulation of endothelial adhesion molecules and chemokine expression that accompanies the innate response will permit migration of effector T cells of any specificity. Likewise, endothelial antigen presentation may be important in the case of T-cell recruitment into uninflamed tissues very early in the development of tissue-specific autoimmune diseases or allograft rejection. Adoptive transfer studies in mice indicate that antigen-experienced T cells specific for protein antigens that are synthesized in the extravascular tissue of a particular organ rapidly accumulate in those tissues and cause diseases, but are not detectable in other tissues (44,45). This occurs without any prior inflammatory events in those tissues. These observations are reminiscent of the rapid development of a positive PPD reaction in sensitized individuals, and are consistent with the kinetic argument that the rapid homing of tissue antigen–specific T cells to the relevant tissue site is mediated by endothelial presentation of the antigen.

One general approach to testing whether endothelial antigen presentation does occur in vivo is to establish conditions in an experimental animal in which only ECs express a particular MHC allele, then introduce T cells with that MHC restriction into the animal and track those cells. Professional APCs are derived from bone marrow precursors. Irradiation of a mouse, followed by bone marrow transplantation using a donor with a different MHC haplotype from the host, or a gene knockout donor that does not express class I or class II MHC molecules, will result in a chimeric animal in which the radioresistant ECs express different MHC alleles than do the hematopoietically derived APCs. This approach was used to demonstrate that cardiac allografts in which only nonhematopoietically derived cells are allogeneic to the host are rejected by $CD8^+$ T cells (46), but the explicit role of ECs in activation and recruitment of T cells into the allograft could not be assessed in that model.

An alternative transgenic strategy for studying endothelial antigen presentation in vivo is the generation of mice that express a nonmouse protein exclusively in ECs, by use of endothelial-specific promoters (47). This approach suffers from the uncertainty of exclusive endothelial expression of an ectopic transgene, and the possibility that, even though the antigen is synthesized by ECs, it may be cross-presented by another APC. In theory, the transgenic approach could be very useful for intravital visualization of interactions of T cells specific for a particular protein antigen with ECs that produce, process, and present that antigen.

Perhaps the only definitive way of determining if endothelial presentation of tissue antigens to circulating T cells occurs in vivo is to inject labeled T cells specific for that antigen into animals and observe by intravital microscopy the interactions of the T cells with endothelium in vessels in the relevant tissue at early time points before any secondary effects of T-cell activation outside the vessel could occur. Quantitative comparisons can be made between the interactions of phenotypically comparable T-cell populations with or without the relevant antigen specificity, or comparisons can be made between interactions of antigen-specific T cells in tissues that do or do not express the relevant antigen. One published study that used this approach describes enhanced extravasation of male antigen (HY)/H-2^k-specific $CD8^+$ T cells across scrotal microvessels in IFN-γ treated mice that express H-2^k MHC molecules, compared to extravasations in mice that express other MHC alleles (48). The enhanced diapedesis was observed after only 10 minutes, although the number of transmigrated cells actually observed was small. Curiously, the amount of T cells adherent to the ECs did not appear to be influenced by the presence of antigen. Although the antigen-enhanced T-cell extravasation in this model may be taken as evidence for endothelial antigen presentation, the physiological relevance of the model may not be broadly applicable, because the T cells used were specific for an antigen produced by the ECs themselves. In other words, the question of whether the endothelium can promote recruitment of T cells specific for antigen present in the surrounding tissues remains unanswered. In the case of $CD8^+$ T cells, the question can be phrased, “Can ECs cross-present tissue antigens by the class I MHC pathway and thereby enhance $CD8^+$ T cell recruitment?”

Another published study using a mouse $CD8^+$ T cell clone specific for an insulin peptide suggests that EC cross-presentation of antigen does promote $CD8^+$ T-cell recruitment (23). A small number of these T cells were found in the extravascular space around some pancreatic islets 1.5 hours after injection into class I MHC–expressing mice, but not after injection into class I MHC–negative mice. This correlated with islet infiltration by the T cells at later time points in identically treated class I MHC–positive but not class I MHC–negative mice. The recruitment of the T cells into the islets depended on secretion of insulin by the islet cells. A direct demonstration of antigen-dependent T cell–endothelial interactions in vivo was not provided in that study, and therefore the conclusion that the antigen presentation by ECs enhanced migration of the T cells is based on the kinetic argument.

ENDOTHELIAL ANTIGEN PRESENTATION AS A HOMEOSTATIC OUTPUT

Although this discussion has focused on the possible role of endothelial antigen presentation in promoting T-cells recruitment and thereby enhancing adaptive immune responses, it is possible that, in some circumstances, endothelial antigen presentation may serve a homeostatic role in limiting immune responses or maintaining tolerance to self antigens. A few different types of experimental observations do suggest a regulatory role of endothelial presentation. The presence of cognate antigen appears to block transendothelial migration of T cells in some in vitro assays, especially when the T cells can be stimulated to divide independent of costimulation (40). ECs inducibly express the PD-L1, a member of the B7 family, which binds to the PD-1 receptor on T cells (49,50). PD-1 transmits negative regulatory signals that block T-cell activation. IFN-γ induces PD-L1 expression on ECs, and endothelial PD-L1 can impair both $CD4^+$ and $CD8^+$ T cell activation, including $CD8^+$ T cells killing of antigen-presenting ECs (21,50). In vivo studies in mice show that PD-L1 expression is markedly upregulated on microvascular endothelium of certain tissues, especially the heart, in response to IFN-γ (21). A potentially important physiological role for PD-L1 may be to limit T cell–mediated immune responses. Alternatively, this pathway may serve to limit damage to endothelium by T cells that recognize antigen on endothelial surfaces.

The ECs that line hepatic sinusoids appear to have a specialized role in inducing T-cell tolerance to soluble protein antigens delivered to the liver via the portal circulation (51). Unlike most ECs, liver sinusoidal ECs (LSECs) constitutively express B7-1 and B7-2, as well as class I and class II MHC (52). LSEC presentation of antigen to naive CD4$^+$ of CD8$^+$ T cells results in a loss of responsiveness to those antigens, although the mechanisms of this tolerance induction are not well defined (53,54). Thus, liver ECs may be particularly important in maintaining nonreactivity to food proteins, and delivery of alloantigens into the portal circulation may be a method for inducing allograft tolerance.

ENDOTHELIAL ANTIGEN PRESENTATION PROVIDES A TARGET FOR PATHOLOGICAL T-CELL RESPONSES

T cells mediate much of the tissue damage in autoimmune disease and allograft rejection, and this damage depends on presentation of tissue antigens to effector T cells. As discussed earlier, evidence from animal models of autoimmunity suggests that endothelial antigen presentation promotes early influx of tissue antigen-specific T cells into previously uninflamed organs. Histopathological observations suggest that ECs are a primary target for T cell–mediated allograft rejection, and CD8$^+$ CTL play a dominant role in both acute and chronic allograft rejection (55). Allogeneic CTL killing of graft vascular ECs can be considered a consequence of endothelial antigen presentation, in which the antigen is an allogeneic class I MHC-peptide complex (see Figure 120.1B). Allograft rejection has been modeled in immunodeficient mice that receive injections of human T cells and grafts of human skin or human EC-derived synthetic vessels (55,56). Although the actual site of lethal CTL–endothelial contact (e.g., luminal, interendothelial, or abluminal) is not clear from these studies, the models provide compelling data that alloreactive human CD8$^+$ T cells can be activated by and destroy ECs in vivo.

KEY POINTS

- Endothelial antigen presentation is a hypothetical mechanism by which T cells specific for particular antigens may be efficiently directed into the tissues where those antigens are located. This may influence migration of naive T cells into lymph nodes, and effector T cells into peripheral tissues.
- Extensive in vitro evidence has established that ECs are capable of displaying T-cell antigens (peptide–MHC complexes) and costimulators that are recognized by, and cause the activation of, different subsets of T cells.
- Although a small number of technically diverse studies indicate that endothelial antigen presentation occurs in experimental animals, the contribution of endothelial antigen presentation to protective immune responses in vivo remains technically difficult to demonstrate and of unknown importance.
- Endothelial antigen presentation may have different consequences in different tissue/organ contexts, and may contribute to T-cell tolerance and homeostasis.
- Endothelial antigen presentation provides a target for pathological immune responses, such as in the case of CTL-mediated killing of ECs in allograft rejection.

Future Goals

- To establish better experimental models that will permit direct visualization of antigen-dependent T cell–endothelial interactions in vivo, especially models in which the antigen is derived from surrounding tissue cells

ACKNOWLEDGMENTS

This work was supported in part by National Institutes of Health grants HL56985, AI059610, and HL56985.

REFERENCES

1 Germain RN. MHC-dependent antigen processing and peptide presentation: providing ligands for T lymphocyte activation. *Cell.* 1994;76(2):287–299.
2 York IA, Rock KL. Antigen processing and presentation by the class I major histocompatibility complex. *Annu Rev Immunol.* 1996;14:369–396.
3 Heath WR, Belz GT, Behrens GM, et al. Cross-presentation, dendritic cell subsets, and the generation of immunity to cellular antigens. *Immunol Rev.* 2004;199:9–26.
4 Yew dell JW, Haeryfar SMM. Understanding presentation of viral antigens to CD8$^+$ T cells in vivo: the key to rational vaccine design. *Ann Rev Immunol.* 2005;23(1):651–682.
5 Trombetta ES, Mellman I. Cell biology of antigen processing in vitro and in vivo. *Annu Rev Immunol.* 2005;23(1):975–1028.
6 Castellino F, Zhong G, Germain RN. Antigen presentation by MHC class II molecules: invariant chain function, protein trafficking, and the molecular basis of diverse determinant capture. *Hum Immunol.* 1997;54(2):159–169.

7 Carreno BM, Collins M. The B7 family of ligands and its receptors: new pathways for costimulation and inhibition of immune responses. *Annu Rev Immunol.* 2002;20:29–53.

8 Rot A, von Andrian UH. Chemokines in innate and adaptive host defense: basic chemokinese grammar for immune cells. *Annu Rev Immunol.* 2004;22:891–928.

9 Choi J, Enis DR, Koh KP, et al. T lymphocyte-endothelial cell interactions. *Annu Rev Immunol.* 2004;22:683–709.

10 Pober JS. Immunobiology of human vascular endothelium. *Immunol Res.* 1999;19(2–3):225–232.

11 Turner RR, Beckstead JH, Warnke RA, Wood GS. Endothelial cell phenotypic diversity. In situ demonstration of immunologic and enzymatic heterogeneity that correlates with specific morphologic subtypes. *Am J Clin Pathol.* 1987;87(5):569–575.

12 Pober JS, Gimbrone MA Jr. Expression of Ia-like antigens by human vascular endothelial cells is inducible in vitro: demonstration by monoclonal antibody binding and immunoprecipitation. *Proc Natl Acad Sci USA.* 1982;79(21):6641–6645.

13 Burger DR, Ford D, Vetto RM, et al. Endothelial cell presentation of antigen to human T cells. *Hum Immunol.* 1981;3(3):209–230.

14 Hirschberg H. Presentation of viral antigens by human vascular endothelial cells in vitro. *Hum Immunol.* 1981;2(3):235–246.

15 Pardi R, Bender JR. Signal requirements for the generation of $CD4^+$ and $CD8^+$ T-cell responses to human allogeneic microvascular endothelium. *Circ Res.* 1991;69(5):1269–1279.

16 Wagner CR, Vetto RM, Burger DR. The mechanism of antigen presentation by endothelial cells. *Immunobiology.* 1984;168(3–5):453–469.

17 Lodge PA, Haisch CE. T cell subset responses to allogeneic endothelium. Proliferation of $CD8^+$ but not $CD4^+$ lymphocytes. *Transplantation.* 1993;56(3):656–661.

18 Biedermann BC, Pober JS. Human vascular endothelial cells favor clonal expansion of unusual alloreactive CTL. *J Immunol.* 1999;162(12):7022–7030.

19 Murray AG, Libby P, Pober JS. Human vascular smooth muscle cells poorly co-stimulate and actively inhibit allogeneic $CD4^+$ T cell proliferation in vitro. *J Immunol.* 1995;154(1):151–161.

20 Marelli-Berg FM, Scott D, Bartok I, et al. Activated murine endothelial cells have reduced immunogenicity for $CD8^+$ T cells: a mechanism of immunoregulation? *J Immunol.* 2000;165(8):4182–4189.

21 Rodig N, Ryan T, Allen JA, et al. Endothelial expression of PD-L1 and PD-L2 down-regulates $CD8^+$ T cell activation and cytolysis. *Eur J Immunol.* 2003;33(11):3117–3126.

22 Perez VL, Henault L, Lichtman AH. Endothelial antigen presentation: stimulation of previously activated but not naive TCR-transgenic mouse T cells. *Cell Immunol.* 1998;189(1):31–40.

23 Savinov AY, Wong FS, Stonebraker AC, Chervonsky AV. Presentation of antigen by endothelial cells and chemoattraction are required for homing of insulin-specific $CD8^+$ T Cells. *J Exp Med.* 2003;197(5):643–656.

24 Kreisel D, Krupnick AS, Balsara KR, et al. Mouse vascular endothelium activates $CD8^+$ T lymphocytes in a B7-dependent fashion. *J Immunol.* 2002;169(11):6154–6161.

25 Dengler TJ, Pober JS. Human vascular endothelial cells stimulate memory but not naive $CD8^+$ T cells to differentiate into CTL retaining an early activation phenotype. *J Immunol.* 2000;164(10):5146–5155.

26 Satoh S, Suzuki A, Asari Y, et al. Glomerular endothelium exhibits enhanced expression of costimulatory adhesion molecules, CD80 and CD86, by warm ischemia/reperfusion injury in rats. *Lab Invest.* 2002;82(9):1209–1217.

27 Kojima N, Sato M, Suzuki A, et al. Enhanced expression of B7-1, B7-2, and intercellular adhesion molecule 1 in sinusoidal endothelial cells by warm ischemia/reperfusion injury in rat liver. *Hepatology.* 2001;34(4 Pt 1):751–757.

28 Kobayashi H, Li Z, Yamataka A, et al. Role of immunologic costimulatory factors in the pathogenesis of biliary atresia. *J Pediatr Surg.* 2003;38(6):892–896.

29 Maher SE, Karmann K, Min W, et al. Porcine endothelial CD86 is a major costimulator of xenogeneic human T cells: cloning, sequencing, and functional expression in human endothelial cells. *J Immunol.* 1996;157(9):3838–3844.

30 Marelli-Berg F, Hargreaves R, Carmichael P, et al. Major histocompatibility complex class II-expressing endothelial cells induce allospecific nonresponsiveness in naive T cells. *J Exp Med.* 1996;183(4):1603–1612.

31 Denton MD, Geehan CS, Alexander SI, et al. Endothelial cells modify the costimulatory capacity of transmigrating leukocytes and promote CD28-mediated $CD4(^+)$ T cell alloactivation. *J Exp Med.* 1999;190(4):555–566.

32 Yellin M, Brett J, Baum D, et al. Functional interactions of T cells with endothelial cells: the role of CD40L-CD40-mediated signals. *J Exp Med.* 1995;182(6):1857–1864.

33 Ma W, Pober JS. Human endothelial cells effectively costimulate cytokine production by, but not differentiation of, naive $CD4^+$ T cells. *J Immunol.* 1998;161(5):2158–2167.

34 Khayyamian S, Hutloff A, Buchner K, et al. ICOS-ligand, expressed on human endothelial cells, costimulates Th1 and Th2 cytokine secretion by memory $CD4^+$ T cells. *Proc Natl Acad Sci USA.* 2002;99(9):6198–6203.

35 Grewal IS, Flavell RA. CD40 and CD154 in cell-mediated immunity. *Annu Rev Immunol.* 1998;16:111–135.

36 Butcher EC, Picker LJ. Lymphocyte homing and homeostasis. *Science.* 1996;272(5258):60–66.

37 Ebnet K, Kaldjian EP, Anderson AO, Shaw S. Orchestrated information transfer underlying leukocyte endothelial interactions. *Annu Rev Immunol.* 1996;14:155–177.

38 Zell T, Kivens WJ, Kellermann SA, Shimizu Y. Regulation of integrin function by T cell activation: points of convergence and divergence. *Immunol Res.* 1999;20(2):127–145.

39 Pober JS, Kluger MS, Schechner JS. Human endothelial cell presentation of antigen and the homing of memory/effector T cells to skin. *Ann NY Acad Sci.* 2001;941(1):12–25.

40 Marelli-Berg FM, Frasca L, Weng L, et al. Antigen recognition influences transendothelial migration of $CD4^+$ T cells. *J Immunol.* 1999;162(2):696–703.

41 Okamoto N, Nukada Y, Tezuka K, et al. AILIM/ICOS signaling induces T-cell migration/polarization of memory/effector T-cells. *Int Immunol.* 2004;16(10):1515–1522.

42 Wilcox CE, Healey DG, Baker D, et al. Presentation of myelin basic protein by normal guinea-pig brain endothelial cells and its relevance to experimental allergic encephalomyelitis. *Immunology.* 1989;67(4):435–440.

43 Silber A, Newman W, Reimann KA, et al. Kinetic expression of endothelial adhesion molecules and relationship to leukocyte recruitment in two cutaneous models of inflammation. *Lab Invest.* 1994;70(2):163–175.

44 Kurts C, Klebba I, Davey GM, et al. Kidney protection against autoreactive $CD8(^+)$ T cells distinct from immunoprivilege and sequestration. *Kidney Int.* 2001;60(2):664–671.

45 Grabie N, Delfs MW, Westrich JR, et al. IL-12 is required for differentiation of pathogenic $CD8^+$ T cell effectors that cause myocarditis. *J Clin Invest.* 2003;111(5):671–680.
46 Kreisel D, Krupnick AS, Gelman AE, et al. Non-hematopoietic allograft cells directly activate $CD8^+$ T cells and trigger acute rejection: an alternative mechanism of allorecognition. *Nat Med.* 2002;8(3):233–239.
47 Rothermel A, Wang Y, Schechner J, et al. Endothelial cells present antigens in vivo. *BMC Immunology.* 2004;5(1):5.
48 Marelli-Berg FM, James MJ, Dangerfield J, et al. Cognate recognition of the endothelium induces HY-specific $CD8^+$ T-lymphocyte transendothelial migration (diapedesis) in vivo. *Blood.* 2004;103(8):3111–3116.
49 Eppihimer MJ, Gunn J, Freeman GJ, et al. Expression and regulation of the PD-L1 immunoinhibitory molecule on microvascular endothelial cells. *Microcirculation.* 2002;9(2):133–145.
50 Mazanet MM, Hughes CC. B7-H1 is expressed by human endothelial cells and suppresses T cell cytokine synthesis. *J Immunol.* 2002;169(7):3581–3588.
51 Knolle PA, Limmer A. Neighborhood politics: the immunoregulatory function of organ-resident liver endothelial cells. *Trends Immunol.* 2001;22(8):432–437.
52 Lohse AW, Knolle PA, Bilo K, et al. Antigen-presenting function and B7 expression of murine sinusoidal endothelial cells and Kupffer cells. *Gastroenterology.* 1996;110(4):1175–1181.
53 Limmer A, Ohl J, Kurts C, et al. Efficient presentation of exogenous antigen by liver endothelial cells to $CD8^+$ T cells results in antigen-specific T-cell tolerance. *Nat Med.* 2000;6(12):1348–1354.
54 Tokita D, Ohdan H, Onoe T, et al. Liver sinusoidal endothelial cells contribute to alloreactive T-cell tolerance induced by portal venous injection of donor splenocytes. *Transpl Int.* 2005;18(2):237–245.
55 Murray AG, Petzelbauer P, Hughes CC, et al. Human T-cell-mediated destruction of allogeneic dermal microvessels in a severe combined immunodeficient mouse. *Proc Natl Acad Sci USA.* 1994;91(19):9146–9150.
56 Zheng L, Gibson TF, Schechner JS, et al. Bcl-2 Transduction protects human endothelial cell synthetic microvessel grafts from allogeneic T cells in vivo. *J Immunol.* 2004;173(5):3020–3026.

PART III

VASCULAR BED/ORGAN STRUCTURE AND FUNCTION IN HEALTH AND DISEASE

121

Introductory Essay

The Endothelium in Health and Disease

William C. Aird

Beth Israel Deaconess Medical Center, Harvard Medical School, Boston, Massachusetts

The endothelium is an expansive organ, reaching to all recesses of the human body. The endothelium plays a critical role in homeostasis. Endothelial cells (ECs) participate in the regulation of vasomotor tone, leukocyte trafficking, coagulation, permeability, antigen presentation, innate immunity, and angiogenesis. ECs display remarkable heterogeneity in structure and function. For example, although most ECs are extremely flat (which minimizes diffusional path length), those of the high endothelial venules (HEVs) are plump and cuboidal. The structure of the intercellular junctions, the absence or presence of fenestrations, and the continuity (or lack thereof) of the abluminal basement membrane are criteria used to differentiate between continuous, fenestrated, and sinusoidal endothelium. The ECs in postcapillary venules are primarily responsible for mediating the adhesion and transmigration of leukocytes, whereas arteriolar endothelium plays a particularly important role in regulating vasomotor tone. ECs are aligned parallel to the direction of blood flow in straight segments of arteries, but not at branch points or curvatures of arteries, or in veins. Protein and mRNA expression varies among different vascular beds. Indeed, state-of-the-art proteomic approaches have uncovered a vast array of vascular bed–specific markers, or vascular "zip codes."

EC heterogeneity is mediated by two proximate ("how?") mechanisms. First, biochemical and biomechanical signals within the extracellular environment trigger post-transcriptional and/or post-translational changes in ECs. Because the net signal input varies across the vascular tree (and from one moment to the next), so too does EC output (i.e., cellular phenotype). Second, certain site-specific properties of the endothelium are epigenetically programmed and mitotically heritable. The maintenance of these latter properties are no longer dependent on or influenced by the extracellular milieu. When ECs derived from different vascular beds are cultured under identical conditions in vitro, cells tend to lose site-specific properties that are governed by the first mechanism (resulting in phenotypic drift), but retain those properties that are determined by the second mechanism. Evolutionary ("why?") explanations for the existence of EC heterogeneity are at best speculative. However, it seems likely that phenotypic heterogeneity serves at least two important purposes. First, it allows ECs to serve the varying needs of the underlying tissue. For example, the tight junctions of the blood–brain barrier protect neurons from fluctuations in the composition of blood (Chapters 123 and 124), whereas the fenestrated, discontinuous endothelium of hepatic sinusoids allows for ready access of nutrient-rich portal venous blood to the metabolic machinery of hepatocytes (Chapter 133). Second, ECs are exposed to so many different microenvironments that they must employ site-specific mechanisms to cope with these differences. For example, ECs in the inner medulla of the kidney are exposed to a profoundly hypoxic and hyperosmolar environment (Chapter 138). Although little is known about the phenotype of these particular cells, it is safe to assume that they are adapted in ways that differ from ECs in oxygen-rich environments such as the pulmonary capillary bed (Chapter 126).

In summary, far from being a giant monopoly of identical cells, the endothelium is a consortium of smaller enterprises, each uniquely adapted to the needs of the local tissue. (As with other consortiums, certain unifying or core properties apply to all constituents.) Our level understanding of the endothelium varies from one organ to another. The most intensely studied vascular beds are the blood–brain barrier (Chapters 123 and 124); the retina (Chapter 125); the coronary arteries (Chapter 132); the pulmonary and, to a lesser extent, bronchial circulation (Chapters 126 and 127); skin (Chapter 156); and HEVs (Chapter 170). Advances in these various areas have occurred largely within the context of organ-specific disciplines (a subject that we will return to in the final chapter). There has been comparatively little focus on the sinusoidal endothelium of the liver (Chapter 133) and spleen (Chapter 136), the endothelium of the large hepatic vessels (Chapter 134), and the microvascular endothelium of the gastrointestinal tract

(Chapter 135) or adipose tissue (Chapter 137). Recent studies have yielded exciting new insights into the phenotype of placental endothelium (Chapters 160 and 161) and lymphatic ECs (Chapter 169). These chapters cover a wide spectrum of vascular beds. However, the list is by no means exhaustive. Indeed, many vascular beds are not included in this volume, yet they are of tremendous interest to biomedical practitioners, including (but not limited to) ovaries, testes, and other endocrine glands; bone marrow; skeletal muscle; and thymus. These topics may be covered in a subsequent edition of the book.

The two most common terms used to describe the endothelium in disease are EC activation and EC dysfunction. Both terms were initially coined during the 1980s. EC activation describes the response of ECs to an inflammatory stimulus. The activation phenotype varies according to the nature of activation agonist, the net signal input, and the vascular bed of origin. Typically, the phenotype consists of some combination of a proadhesive and procoagulant surface, increased permeability, and release of inflammatory mediators (some might argue that proliferation also qualifies as activation). The term *activation* should not be confused with activity. The normal endothelium is highly active, constantly sensing and responding to changes in the extracellular environment. EC activation describes a phenotype, but does not address the cost of that phenotype to the host. In other words, EC activation may be adaptive or nonadaptive. In contrast, EC dysfunction is by definition nonadaptive. The latter term is used most often to describe abnormalities in endothelial-dependent vasomotor tone in atheromatous conduit arteries. However, because the endothelium is widely distributed, and because it is involved in so many different functions (over and above the regulation of vasomotor tone), the term *dysfunction* should be used more broadly to describe situations in which the EC phenotype – whether it meets some predetermined definition of activation – represents a net liability to the host.

It follows from the previous discussion that the endothelium may display a wide spectrum of phenotypes in disease. Such diversity is underscored in this section of the book. Some chapters are focused on vascular bed–restricted diseases (in reality, none are truly confined to one organ), including those in the brain (Chapters 123 and 124); retina (Chapter 125); lung (acute respiratory distress syndrome in Chapter 128, pulmonary hypertension in Chapter 129, emphysema in Chapter 130, and ischemia–reperfusion in Chapter 131); liver (Chapters 133 and 134); kidney (Chapters 138–140); thyroid (Chapter 151); skin (Chapter 156); placenta (Chapters 160 and 161); penis (Chapter 167); tumors (Chapters 158 and 159); and bone (Chapter 168).

Other chapters cover systemic illnesses that have more widespread effects on the endothelium. These include atherosclerosis (Chapter 132); infection (sepsis in Chapter 141, malaria in Chapter 142, and hemorrhagic fevers in Chapter 143); hematological syndromes (disseminated intravascular coagulation in Chapter 145, thrombotic thrombocytopenic purpura in Chapter 146, heparin-induced thrombocytopenia in Chapter 147, sickle cell disease in Chapter 148, and antiphospholipid antibody syndrome in Chapter 149); diabetes (Chapter 150); vasculitides and other chronic inflammatory disorders (Chapters 153–155); and the effects of smoking (Chapter 144) and aging (Chapter 152). These disorders involve acquired abnormalities of the endothelium, in contrast to the far less common (yet fascinating) congenital/inherited abnormalities of the endothelium, which are discussed in Chapter 122.

Increasing evidence suggests that the endothelium plays an important role in surgical illness. Thus, we have included chapters on arterial grafts (Chapter 162); burns (Chapter 163); trauma in civilian (Chapter 164) and military (Chapter 165) life; and blood transfusion (Chapter 166).

When considering the role of the endothelium in disease, an interesting and often challenging question is to what extent the endothelium is victim or perpetrator of the underlying pathophysiology. The answer is usually a combination of the two. In many diseases discussed in this section, the primary pathogenesis is nonendothelial in origin. For example, sickle cell disease is caused by a mutation in the hemoglobin gene, yet sickle red blood cells initiate a cascade of events that ultimately ensnare the endothelium. Heparin-induced thrombocytopenia (HIT) is mediated by antiplatelet factor 4-heparin antibodies that are directed primarily towards Fc receptors on the surface of platelets. Platelet activation has profound consequences on the endothelium. The pathogenesis of preeclampsia involves the release of soluble FLT-1 (sFLT-1) from the trophoblasts of the placenta. The soluble form of this receptor competitively inhibits binding of vascular endothelial growth factor (VEGF) to the endothelium. In contrast, some diseases appear to target the endothelium directly. Examples include vasculitides associated with anti-EC antibodies and sinusoidal obstruction syndrome (SOS), in which hepatic sinusoidal ECs are uniquely vulnerable to exogenous and endogenous toxins. Once the endothelium is engaged (either primarily or secondarily), it may contribute in significant ways to the underlying pathophysiology. For example, the interaction between sickle red blood cells and the endothelium results in a vicious cycle of hypercoagulability, hyperadhesiveness, and vascular occlusion. In HIT, the endothelium acquires a procoagulant phenotype that "channels" systemic changes in the hemostatic balance into local thrombus formation. In preeclampsia, sFlt-1–mediated reductions in VEGF reduce endothelial-dependent vasorelaxation. Similarly, anti-EC antibodies and SOS lead to activation and dysfunction of the endothelium.

It is uncommon to find discussions of these various diseases in the same volume. They are normally segregated into organ-specific textbooks. However, we believe that the endothelium should be viewed as an integrated system, one that transcends traditional organ-specific disciplines. Indeed, as this section underscores, the endothelium is a powerful organizing principle in human physiology and pathophysiology.

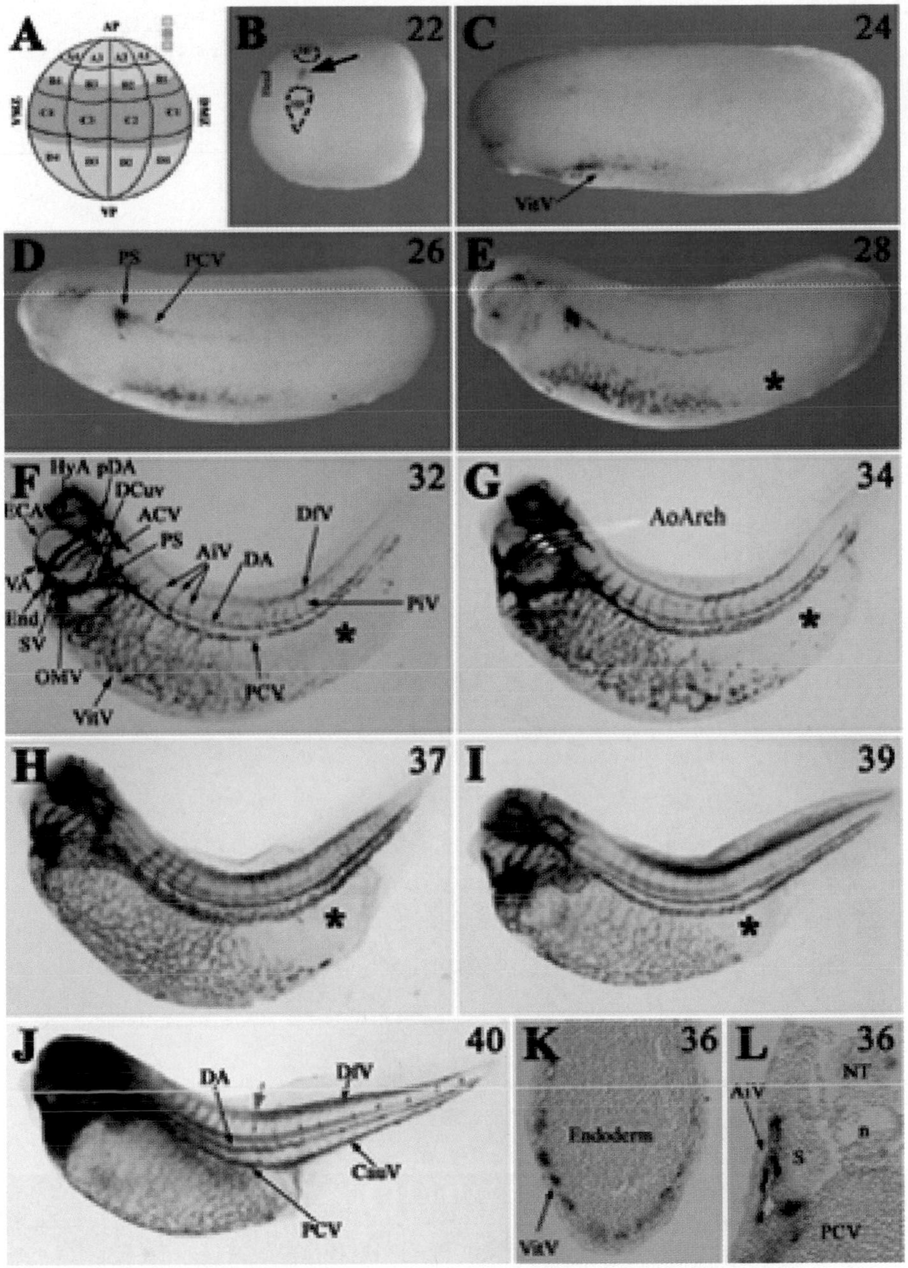

Color Plate 18.1. Early development of the vascular system in *Xenopus* embryos. (**A**) Regions of the 32-cell stage embryo contributing to the three primary germ layers. (**B–L**) Development of the vascular system as illustrated by the expression pattern of the angiopoietin tyrosine kinase receptor, Tie2; numbers at top right corner indicate developmental stage. **B,** Ventral view; **C–J,** lateral view, anterior to the left; **K–L,** transverse sections through the trunk. Asterisk (*) indicates nonvascularized tissue in the posterior. Small arrows in **J** indicate direction of circulation, whereas the large arrow indicates the anterior limit of circulation through the DfV. ACV, anterior cardinal vein; AiV, anterior intersomitic vein; AoArch, aortic arches; AP, animal pole; CauV, caudal vein; DA, dorsal aorta; DCuv, duct of Cuvier; DfV, dorsal fin vein; DMZ, dorsal marginal zone; ECA, external carotid artery; End, endocardium; HF, heart field; HyA, hyaloid artery; n, notochord; NT, neural tube; OMV, omphalomesenteric vein; PCV, posterior cardinal vein; pDA, paired dorsal aortae; PiV, posterior intersomitic veins; PS, pronephric sinus; S, somite; SV, sinus venosus; VA, ventral aorta; VitV, vitelline vessels; VMZ, ventral marginal zone; VP, vegetal pole.

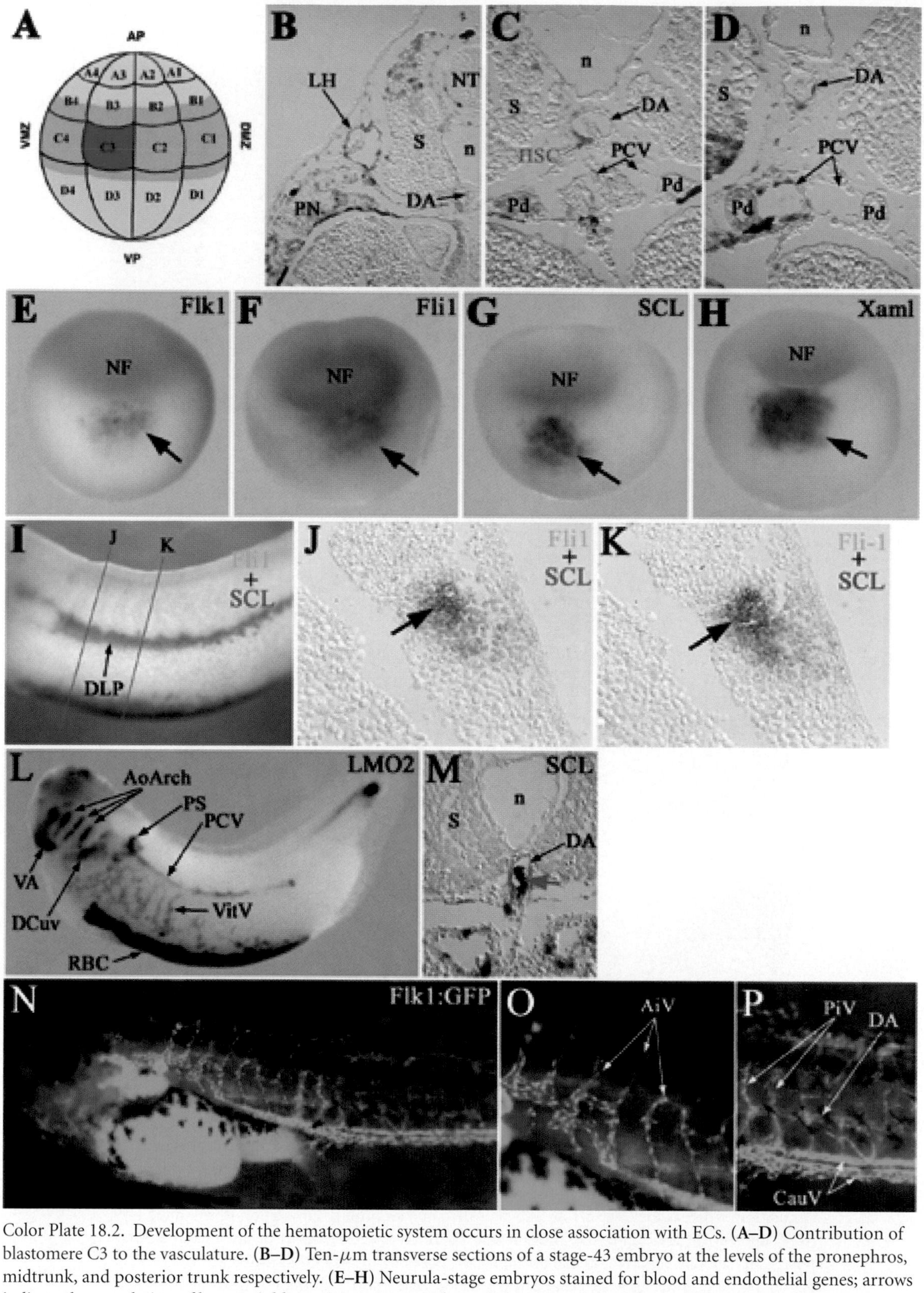

Color Plate 18.2. Development of the hematopoietic system occurs in close association with ECs. (**A–D**) Contribution of blastomere C3 to the vasculature. (**B–D**) Ten-μm transverse sections of a stage-43 embryo at the levels of the pronephros, midtrunk, and posterior trunk respectively. (**E–H**) Neurula-stage embryos stained for blood and endothelial genes; arrows indicate the population of hemangioblasts giving rise to embryonic hematopoietic cells, vitelline vessels, endocardium, and ventral aorta. (**I–K**) Tailbud-stage embryo double-stained for blood and endothelial genes showing a population of hemangioblasts located in the dorsal lateral plate (DLP, *arrows*); later, this population gives rise to the axial vessels and adult hematopoiesis. (**L**) Expression of LMO2 in blood vessels. (**M**) Hematopoietic stem cells (*arrow*) emerging in association with the ventral wall of the dorsal aorta and expressing the blood gene, SCL. (**N–P**) Transgenic embryos showing expression of GFP in blood vessels. GFP expression was under control of the *Xenopus FLK-1* promoter as indicated in reference 25. LH, lymphatic heart; NF, neural fold; Pd, pronephric duct; PN, pronephros; RBC, red blood cells.

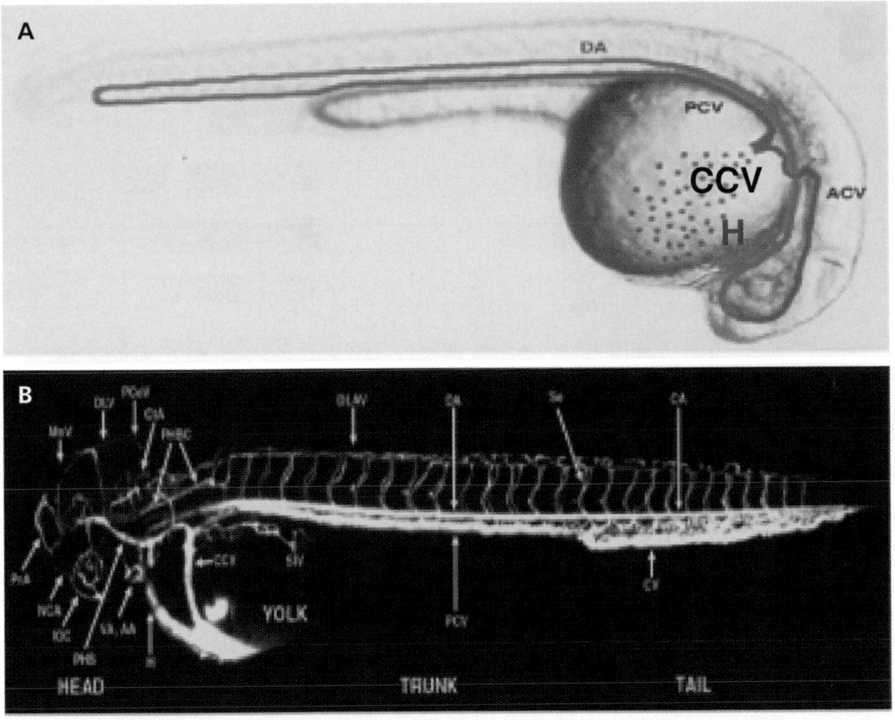

Color Plate 19.1. Anatomy of the zebrafish embryonic vasculature. (**A**) Diagram of a circulatory loop at the zebrafish embryo at 30 hpf. (**B**) Fluorescent confocal microangiogram showing the zebrafish embryonic vasculature at 60 hpf. H, Heart; VA, ventral aorta; AA, aortic arch; PHS, primary head sinus; NCA, nasal ciliary artery; PrA, prosencephalic artery; MsV, mesencephalic vein; DLV, dorsal longitudinal vein; PCeV, posterior cerebral vein; CctA, cerebellar central artery; PHBC, primary hindbrain channel; DLAV, dorsal longitudinal anastomotic vessel; DA, dorsal aorta; Se, intersegmental vessel; CA, caudal artery; CV, caudal vein; ACV, anterior cardinal vein; PCV, posterior cardinal vein; SIV, subintestinal vein; CCV, common cardinal vein (duct of Cuvier).

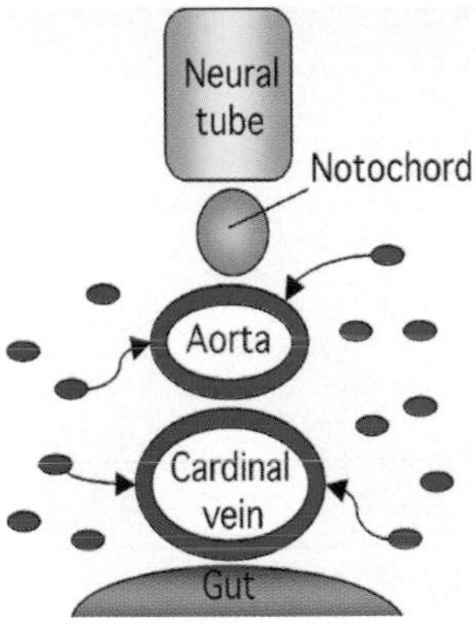

Color Plate 19.2. Assembly of the dorsal aortae and the cardinal veins in zebrafish embryos. Schematic drawing indicates the crossover section of zebrafish embryos at the trunk region, showing the neural tube, notochord, aorta, cardinal vein, and gut. Angioblasts appear to migrate and assemble into the dorsal aorta and the cardinal vein.

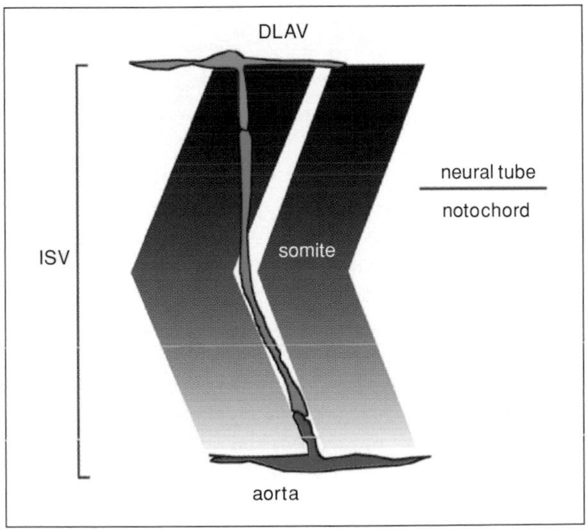

Color Plate 19.3. Model of the construction of an intersegmental vessel in zebrafish embryos. An intersegmental vessel is composed of three types of ECs: the dorsal connection to the DLAV is a T-shaped EC; the ventral connection to the aorta is an inverted T-shaped EC; the middle connection is an elongated EC. ISV, intersegmental vessel; DLAV, dorsal longitudinal anastomotic vessel.

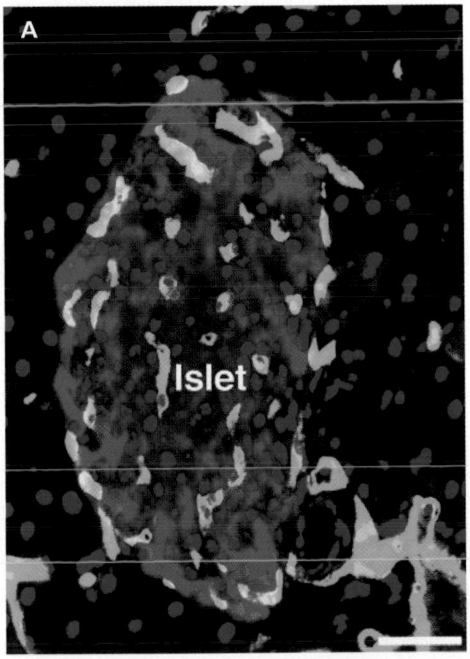

Color Plate 22.3A. The vascular system of pancreatic islets. Islets are vascularized by a dense network of fenestrated capillaries. A confocal image of a section through a mouse pancreas is shown. The β-cells are shown in red (insulin), the ECs in green (PECAM-1), and the cell nuclei in blue (DAPI). Scale bar: 50 μm.

Color Plate 51.1. Schematic of SVMP structure. P, signal sequence; Pro, latency domain; S, spacer region; Dis, disintegrin domain; Cys, cysteine-rich domain; Lec, lectin-like domain. (Reproduced with permission from Fox JW, Serrano SMT. Structural considerations of the snake venom metalloproteinases, key members of the M12 reprolysin family of metalloproteinases. *Toxicon.* 2005;45:969–985.)

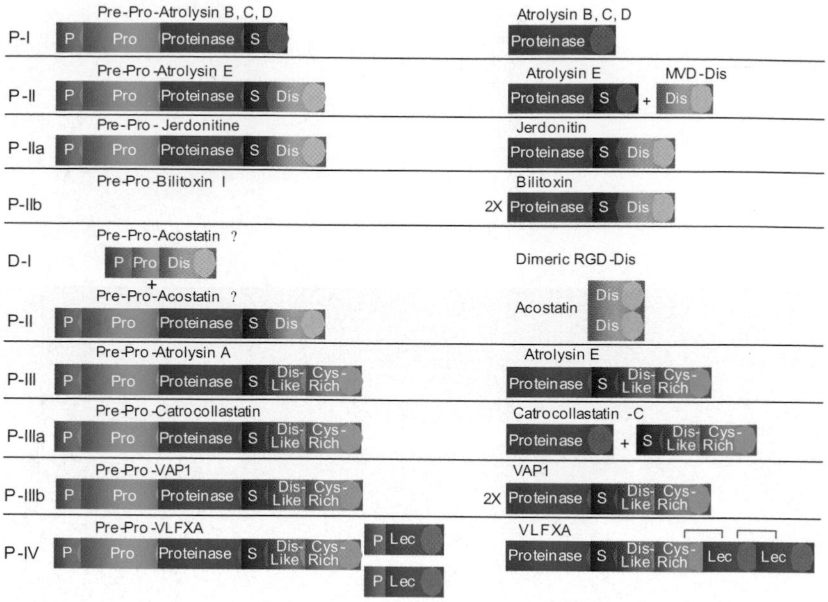

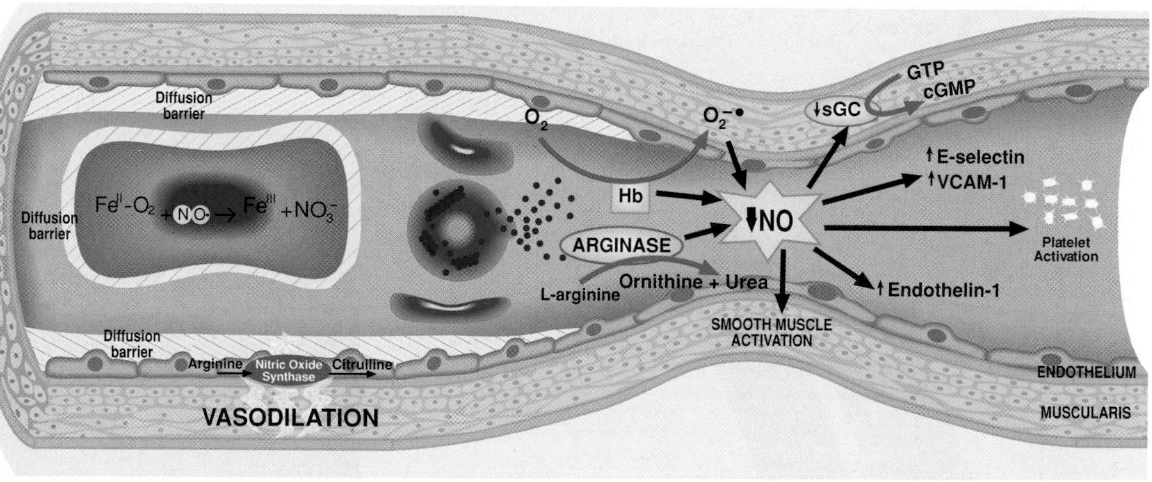

Color Plate 62.1. Disruption of NO-dependent vascular signaling by hemolysis. NO is produced by the enzyme NOS in the endothelium. Under physiological conditions, NO diffuses into the smooth muscle cells as well as into the lumen of the vessel to mediate vasodilation and a host of other responses important to maintaining vascular homeostasis. NO bioactivity is limited by its chemical reactivity with molecules such as intraerythrocytic hemoglobin and superoxide generated from xanthine oxidase. The reaction of NO with hemoglobin is limited by diffusional barriers around the red blood cell and along the endothelium. During intravascular hemolysis, cell free oxyhemoglobin in the plasma ($Fe^{II} - O_2$) rapidly scavenges NO, converting it to nitrate (NO_3^-). This decreases NO bioavailability to both smooth muscle cells and other vascular targets of NO, leading to a disruption of vascular homeostasis.

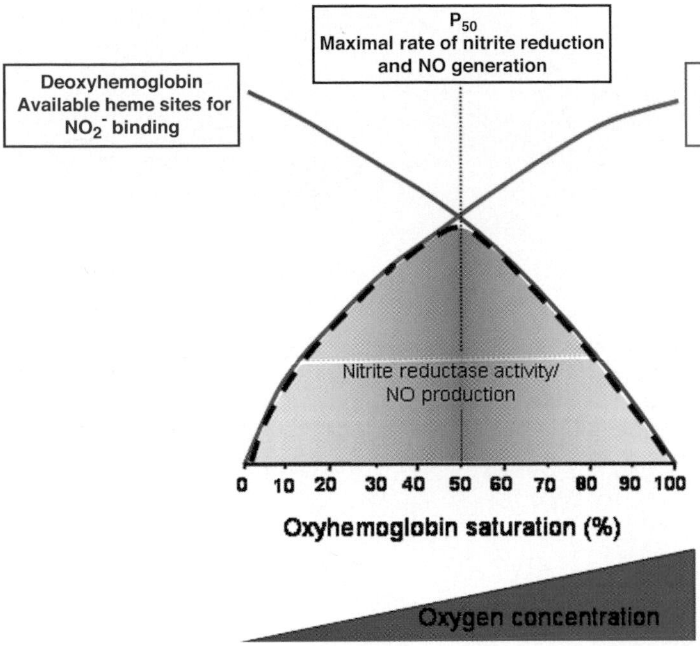

Color Plate 62.3. Allosteric regulation of nitrite reductase rate. The rate of hemoglobin-dependent nitrite reduction (*black dashed line*) depends on the number of available heme sites for reaction (*blue line*) and the amount of R state tetramers (*red line*). As O_2 saturation of hemoglobin increases, the number of available heme sites for nitrite binding decreases, decelerating the reaction. However, the R-state tetramer is increasingly stabilized, which accelerates reaction rate. The maximal reaction rate occurs at an optimal balance of these two subprocesses, which occurs when 50% of the hemoglobin is saturated with oxygen. The partial pressure at which this occurs is the P50 of hemoglobin.

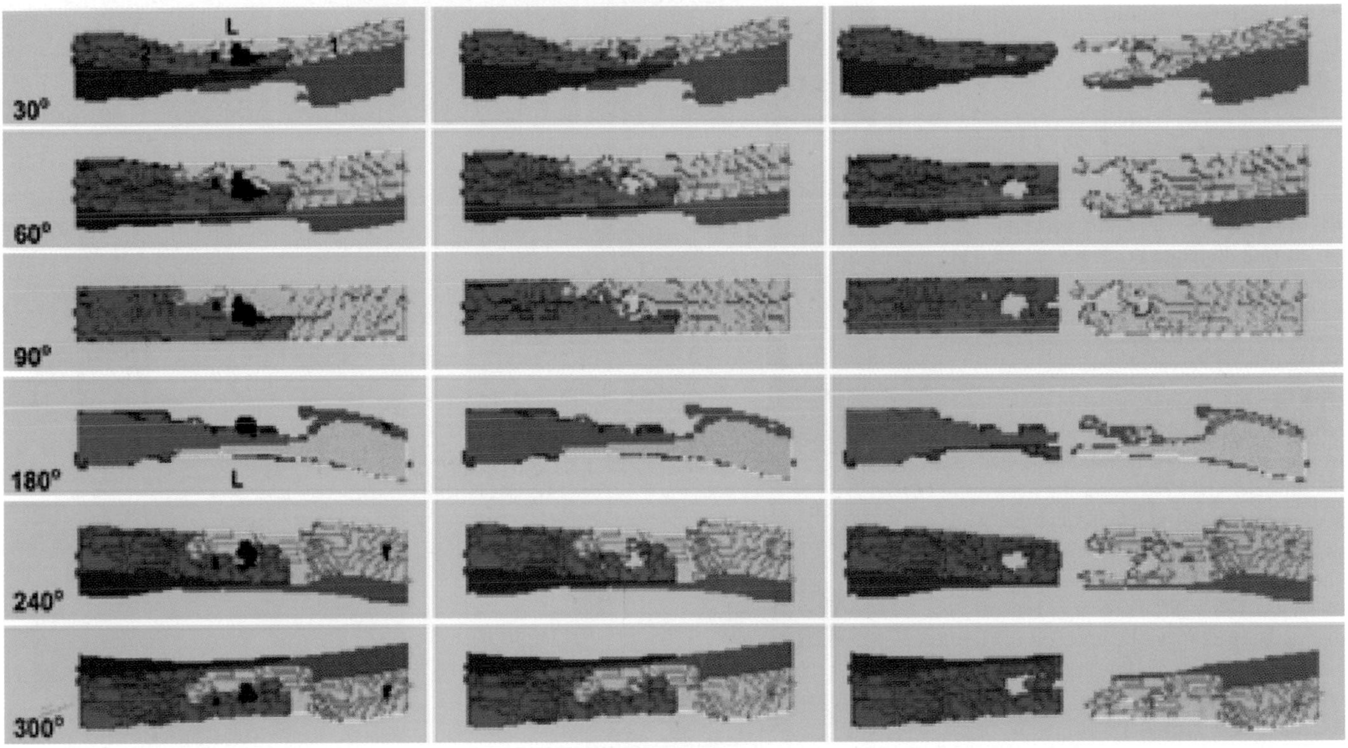

Color Plate 71.7. Computer-generated 3-D reconstruction of the pores illustrated in Figure 71.6. Panels represent successive rotations toward the viewer around a horizontal axis at the angles indicated. 0 degrees (*not shown*) corresponds to a vascular cross-section taken at right angles to the direction of blood flow and a 90-degree view looking down on the luminal surface; views $\geq$240 degrees illustrate the abluminal surface. ECs 1 (*yellow*) and 2 (*orange-brown*) represent cells correspondingly numbered in Figure 71.6. Left panel includes extravasating carbon, which is represented in black. Middle panel illustrates views corresponding to those in left panel, except that carbon was deleted from the reconstruction to facilitate visualization of the pore interior. In the right panel, the overlapping ECs are viewed separately. Two trans-EC pores are illustrated. The smaller (*left*) pore extends to the albumen by passing through only EC 2 and corresponds to the pore marked by the smaller star in Figure 71.6C; background is visible from both luminal (e.g., 90-degree) and abluminal (e.g., 240-degree) views. The larger pore passes through overlapping portions of both ECs 1 and 2 and corresponds to the pore designated by the large star in Figure 71.6C. Abluminal views (*left panel*, 240 degrees and 300 degrees) illustrate carbon present within abluminal portions of cell 1, corresponding to the carbon- containing VVO illustrated in Figure 71.6D. Note (*middle panels*, 240- and 300-degree views, here and for Figure 71.6D) that carbon does not extend through the entire thickness of cell 1. L, lumen. (Reproduced with permission from Feng D, Nagy JA, Hipp J, et al. Reinterpretation of EC gaps induced by vasoactive mediators in guinea-pig, mouse and rat: many are transcellular pores. *J Physiol.* 1997;504:747–761.)

Color Plate 71.11. Computer-generated 3-D reconstruction of portion of a venular VVO from mouse skin injected 5 min earlier with 50 ng VPF. The interior volumes of VVO vesicles and vacuoles in successive electron micrographs were traced onto transparent overlays with reference marks to retain register. Tracings were digitized at a resolution of 5.9 nm/pixel. Because section thickness (15 nm) was greater than in-plane resolution, bicubic interpolation was performed between sections, and convolutions were used to smooth surfaces. **(A–F)** Portion of aVVO (15 consecutive serial, ultra-thin EM sections illustrating 25 individual vesicles-vacuoles) reconstructed using Advanced Visual Systems (Waltham, MA) Software, here viewed in successive rotations around a horizontal axis at intervals of 30 degrees (except 15 degrees, **[C–D]**). There are two openings **(E, F)** to the vascular lumen and four to the abluminal surface **(A)**. (Reproduced with permission from Feng D, Nagy JA, Hipp J, et al. Vesiculo-vacuolar organelles and the regulation of venule permeability to macromolecules by vascular permeability factor, histamine and serotonin. *J Exp Med.* 1996;183:1981–1986.)

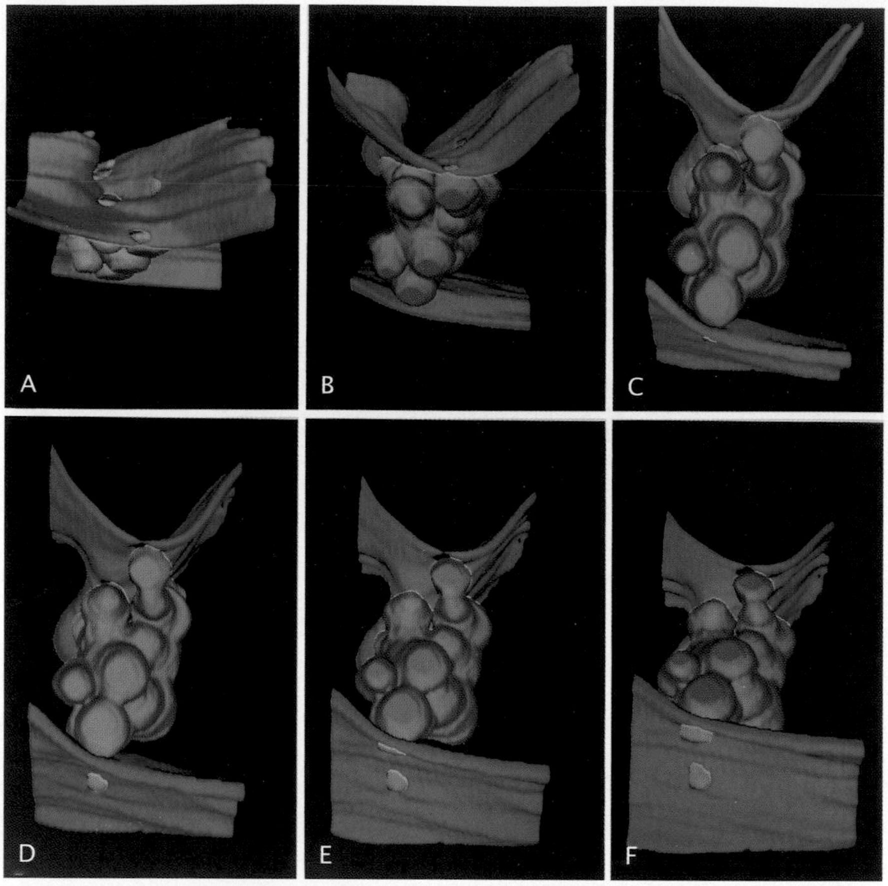

Color Plate 71.15. Computer-generated 3-D reconstruction of the transmigrating neutrophil illustrated in Figure 71.14. The panels portray successive rotations toward the viewer around a horizontal axis at angles of 60, 120, 240, and 300 degrees as indicated. 0 degrees (*not shown*) would represent a vascular cross-section at right angles to the direction of blood flow, and 90 degrees (*also not shown*) would represent a view looking directly down the luminal surface. Emigrating neutrophil (*n*) is in *upper and lower left*; in the other panels, the neutrophil was subtracted electronically to visualize the pore that passes cleanly through the cytoplasm of EC e1 (*orange-brown*) distinctly apart from the junction of e1 with e2 (*yellow*). Cytoplasmic arms of both e1 and e2 embrace the neutrophil luminally and, to a lesser extent, abluminally. L, lumen. (Reproduced with permission from Feng D, Nagy JA, Pyne K, et al. Neutrophils emigrate from venules by a trans-endothelial cell pathway in response to FMLP. *J Exp Med.* 1998;187:903–915.)

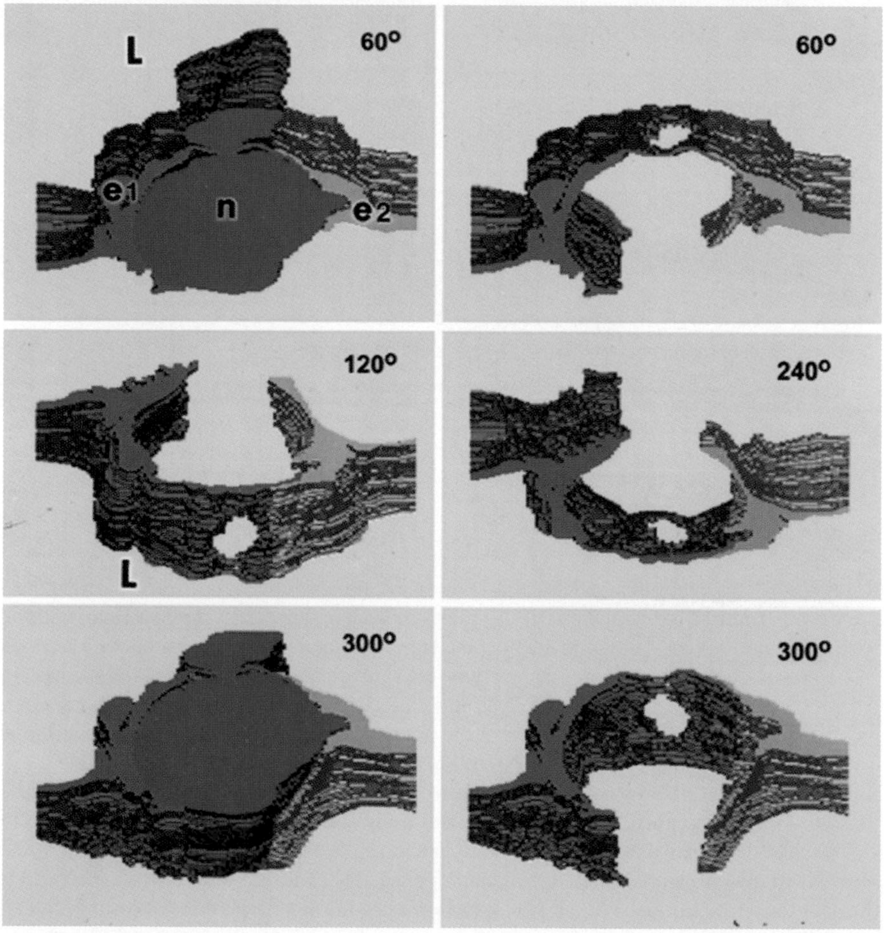

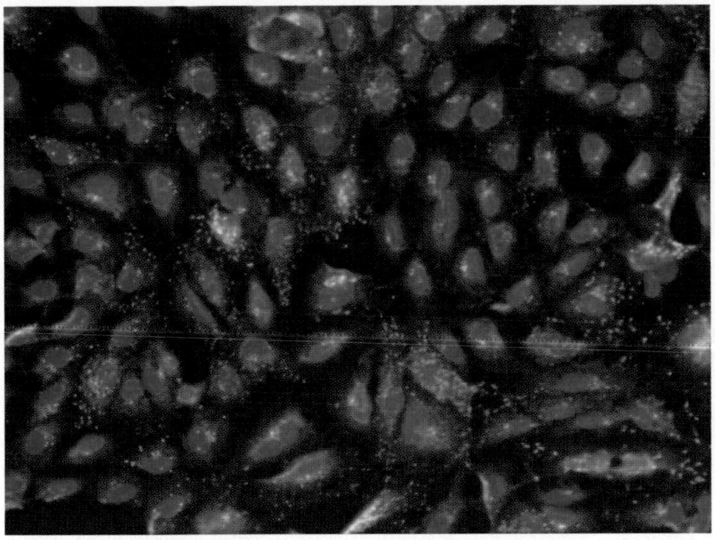

Color Plate 72.1A. Weibel-Palade bodies in ECs. Immunofluorescent images of ECs with nuclei stained blue and vWF stained green.

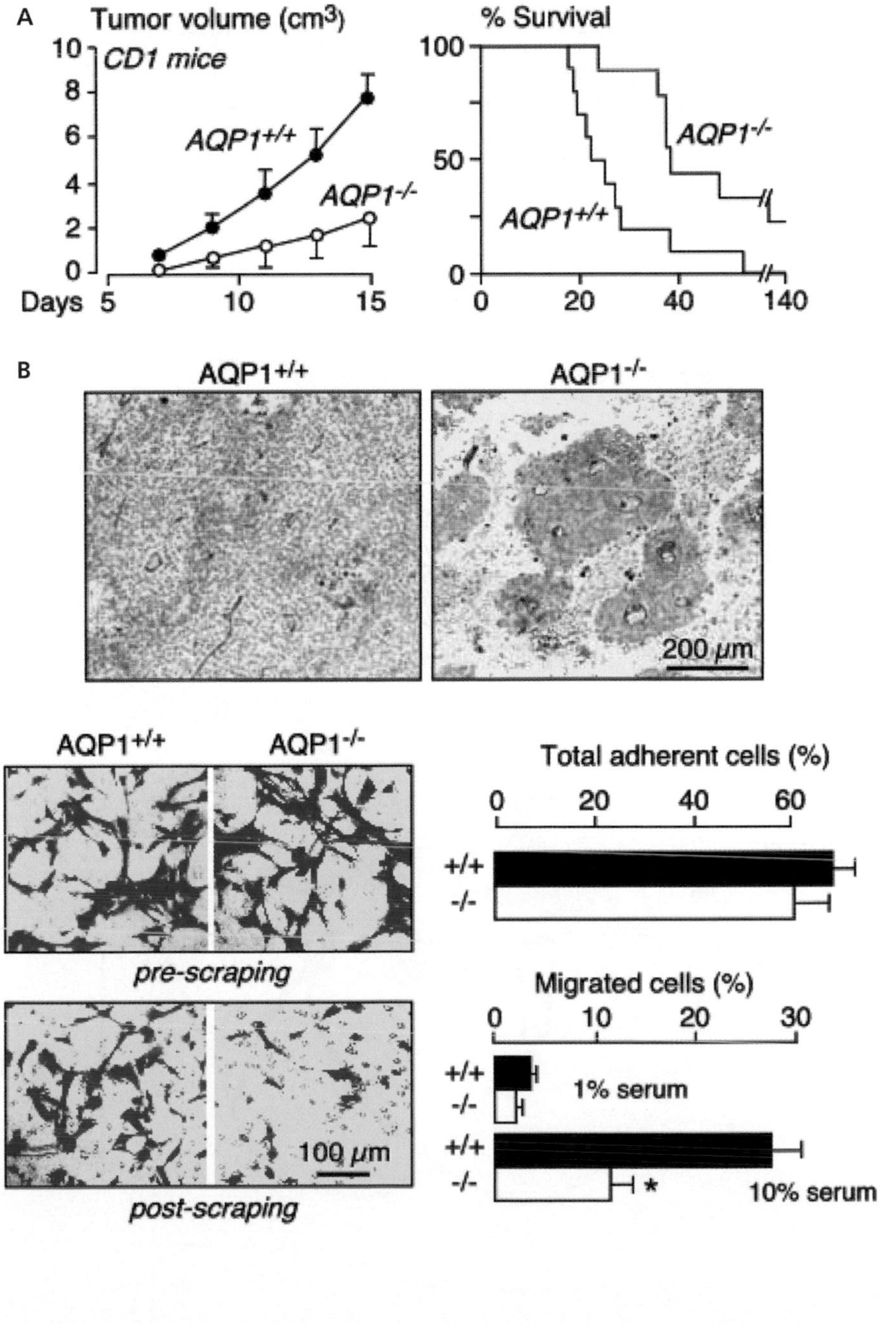

Color Plate 78.2. Impaired tumor angiogenesis and EC migration in AQP1 deficiency. (**A**) Reduced tumor growth in *AQP1*-null mice. Growth of subcutaneous melanoma in wild-type versus *AQP1*-null mice (*left*). Survival of mice with subcutaneous melanoma (*right*). (**B**) Tumor stained with isolectin-B4. Note islands of tumor surrounded by necrotic tissue in AQP1-deficient mice. (**C**) Impaired migration of ECs lacking AQP1. Adhesion and migration (*left*) of ECs quantified by counting cells in Boyden chamber assay before and after scraping. Summary of percentage adherent and migrated cells (*right*). (Reproduced with permission from Saadoun S, Papadopoulos MC, Hara-Chikuma M, Verkman AS. Targeted disruption of aquaporin-1 gene impairs angiogenesis and cell migration. *Nature.* 2005;434:786–792.)

Color Plate 88.2. Color mapping with GATA-3–selective inducible genes. Microarray (Affymetrix U133 plus) analyses were performed in duplicate using HUVECs overexpressing GATA-2 or GATA-3. Genes were filtered according to those that were induced more than twofold by GATA-3 but not GATA-2. Selected genes were analyzed using the GeneTree program. Red and green color panels represent higher and lower expression, respectively, compared with the median for that particular gene.

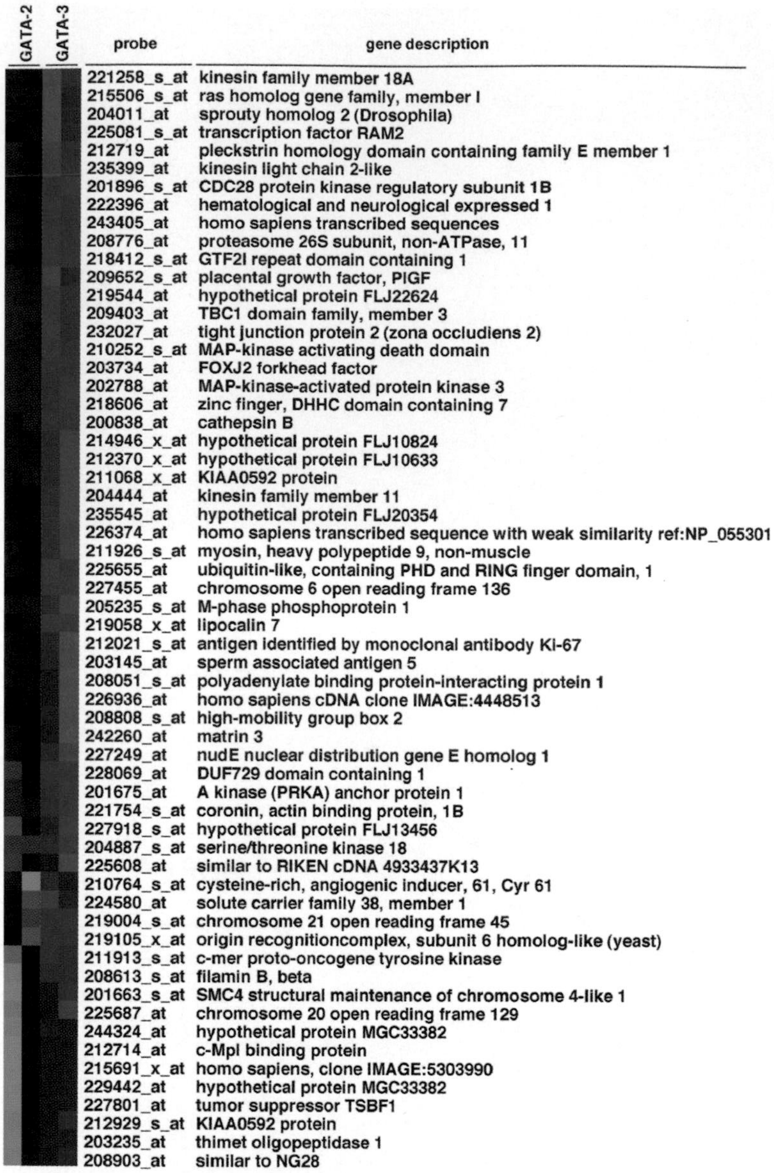

Color Plate 89.1. Ets DNA binding domain. Schematic of the highly conserved Ets DNA binding domain with a helix-loop-helix structure. Depicted is the structure of the *Ets-1* DNA binding domain. Helix 3 is shown in close proximity to the DNA, which is depicted as a brown and blue an overlapping wire-like structure. The flanking wings provide DNA binding specificity among the individual family members. A fourth helix (*top*) can function to inhibit DNA binding. (Reproduced with permission from Garvie CW, Pufall MA, Graves BJ, Wolberger C. Structural analysis of the autoinhibition of Ets-1 and its role in protein partnerships. *J Biol Chem.* 2002;277:45529–45536.)

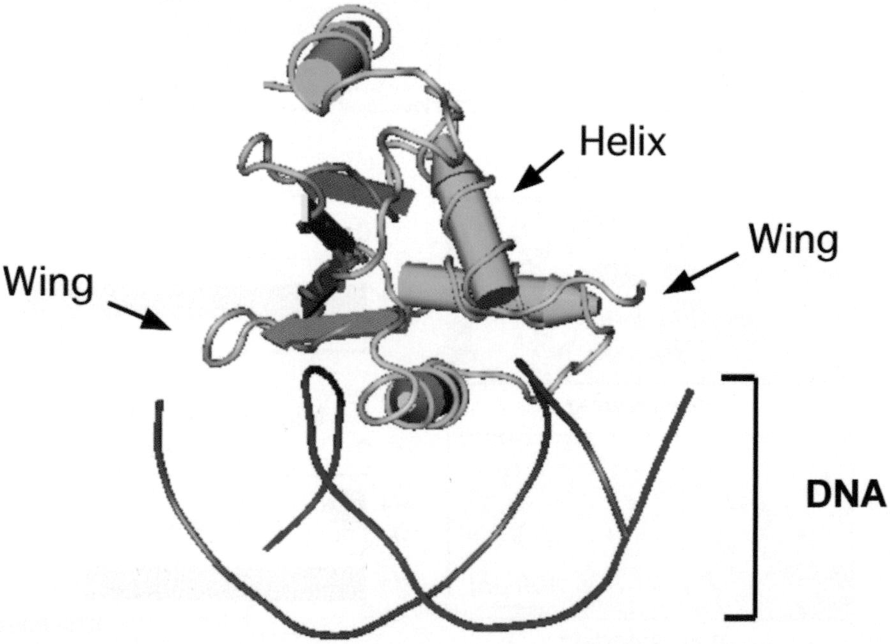

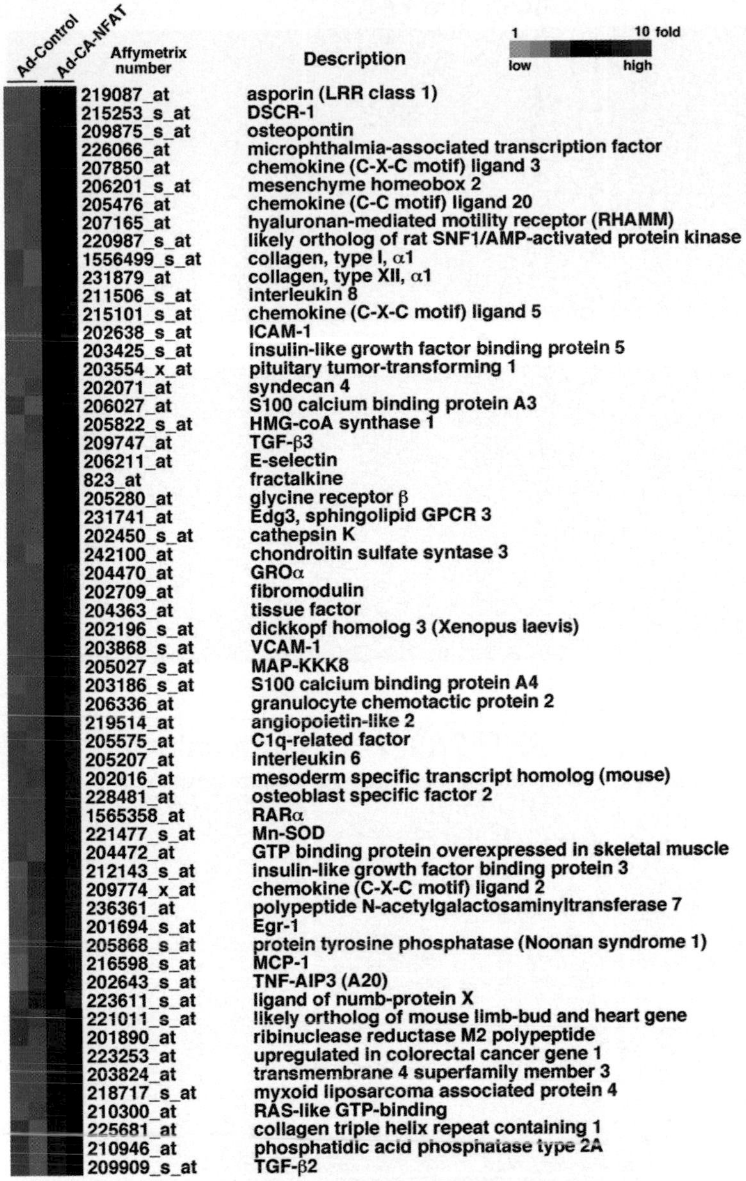
Ad-Control
Ad-CA-NFAT
Affymetrix number
Description
1
10 fold
low
high
219087_at asporin (LRR class 1)
215253_s_at DSCR-1
209875_s_at osteopontin
226066_at microphthalmia-associated transcription factor
207850_at chemokine (C-X-C motif) ligand 3
206201_s_at mesenchyme homeobox 2
205476_at chemokine (C-C motif) ligand 20
207165_at hyaluronan-mediated motility receptor (RHAMM)
220987_s_at likely ortholog of rat SNF1/AMP-activated protein kinase
1556499_s_at collagen, type I, α1
231879_at collagen, type XII, α1
211506_s_at interleukin 8
215101_s_at chemokine (C-X-C motif) ligand 5
202638_s_at ICAM-1
203425_s_at insulin-like growth factor binding protein 5
203554_x_at pituitary tumor-transforming 1
202071_at syndecan 4
206027_at S100 calcium binding protein A3
205822_s_at HMG-coA synthase 1
209747_at TGF-β3
206211_at E-selectin
823_at fractalkine
205280_at glycine receptor β
231741_at Edg3, sphingolipid GPCR 3
202450_s_at cathepsin K
242100_at chondroitin sulfate syntase 3
204470_at GROα
202709_at fibromodulin
204363_at tissue factor
202196_s_at dickkopf homolog 3 (Xenopus laevis)
203868_s_at VCAM-1
205027_s_at MAP-KKK8
203186_s_at S100 calcium binding protein A4
206336_at granulocyte chemotactic protein 2
219514_at angiopoietin-like 2
205575_at C1q-related factor
205207_at interleukin 6
202016_at mesoderm specific transcript homolog (mouse)
228481_at osteoblast specific factor 2
1565358_at RARα
221477_s_at Mn-SOD
204472_at GTP binding protein overexpressed in skeletal muscle
212143_s_at insulin-like growth factor binding protein 3
209774_x_at chemokine (C-X-C motif) ligand 2
236361_at polypeptide N-acetylgalactosaminyltransferase 7
201694_s_at Egr-1
205868_s_at protein tyrosine phosphatase (Noonan syndrome 1)
216598_s_at MCP-1
202643_s_at TNF-AIP3 (A20)
223611_s_at ligand of numb-protein X
221011_s_at likely ortholog of mouse limb-bud and heart gene
201890_at ribinuclease reductase M2 polypeptide
223253_at upregulated in colorectal cancer gene 1
203824_at transmembrane 4 superfamily member 3
218717_s_at myxoid liposarcoma associated protein 4
210300_at RAS-like GTP-binding
225681_at collagen triple helix repeat containing 1
210946_at phosphatidic acid phosphatase type 2A
209909_s_at TGF-β2

Color Plate 92.2. NFAT target genes in primary ECs. Microarray (Affymetrix U133 plus) analyses were performed in duplicate using HUVECs overexpressing constitutively active NFAT (CA-NFAT). Genes were filtered according to those induced fourfold by CA-NFAT versus control. Selected genes were analyzed by GeneTree program. Color panels with red and green represent higher and lower expression, respectively, compared with the median for that particular gene.

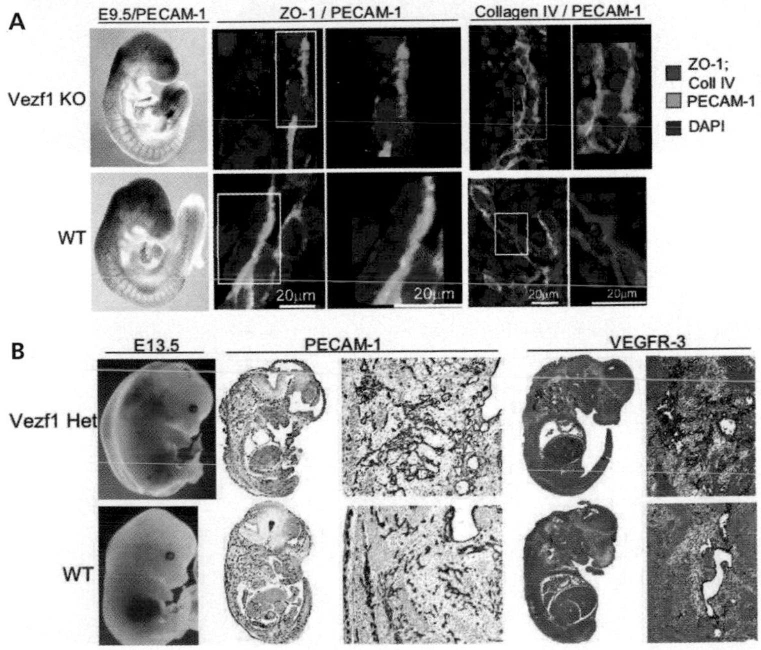

Color Plate 95.3. Blood vascular and lymphatic phenotypes in *Vezf1* loss-of-function embryos. (**A**) Hemorrhaging *Vezf1*-knockout embryos with EC defects. (**B**) Lymphatic hypervascularization, hemorrhaging, and edema in *Vezf1* heterozygous embryos.

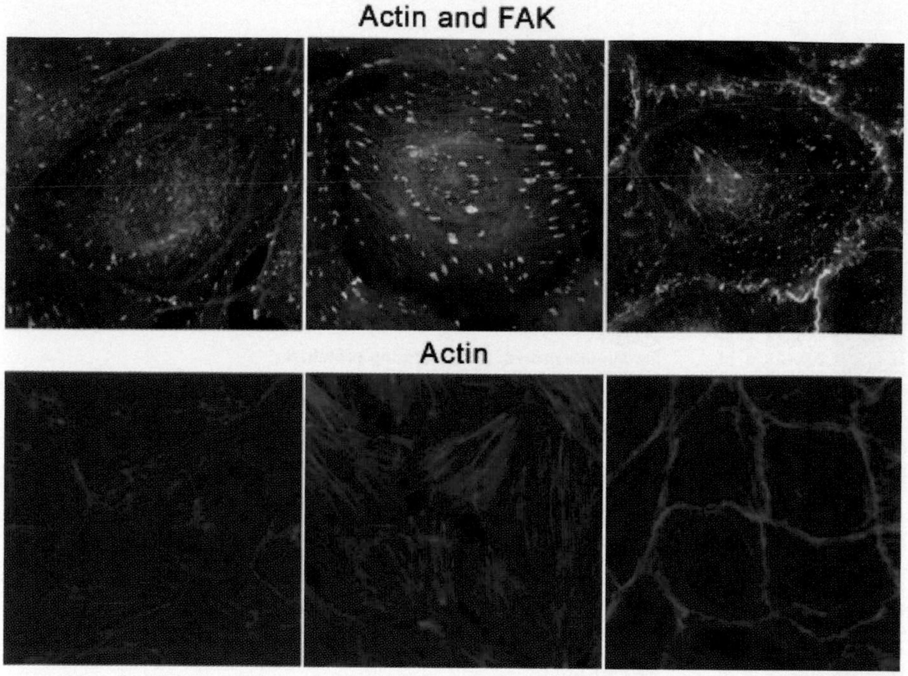

Color Plate 113.3. EC cytoskeletal rearrangement induced by thrombin and sphingosine 1-phosphate (S1P). These immunofluorescent images utilizing Texas red-conjugated phalloidin to identify polymerized actin filaments and focal adhesion kinase (*green*) demonstrate that, relative to controls, human pulmonary artery EC stimulated with thrombin demonstrate a prominent increase in actin stress fibers associated with cell contraction and evidence of paracellular gaps (*small arrows*), with focal adhesions at the end of the actin cords. In contrast, EC stimulated with Sph-1-P reveal prominent cortical actin enhancement, a relative paucity of central stress fibers, and no paracellular gaps. Focal adhesion remodeling is associated with the cortical actin ring, as evidenced by the strong colocalization of these proteins.

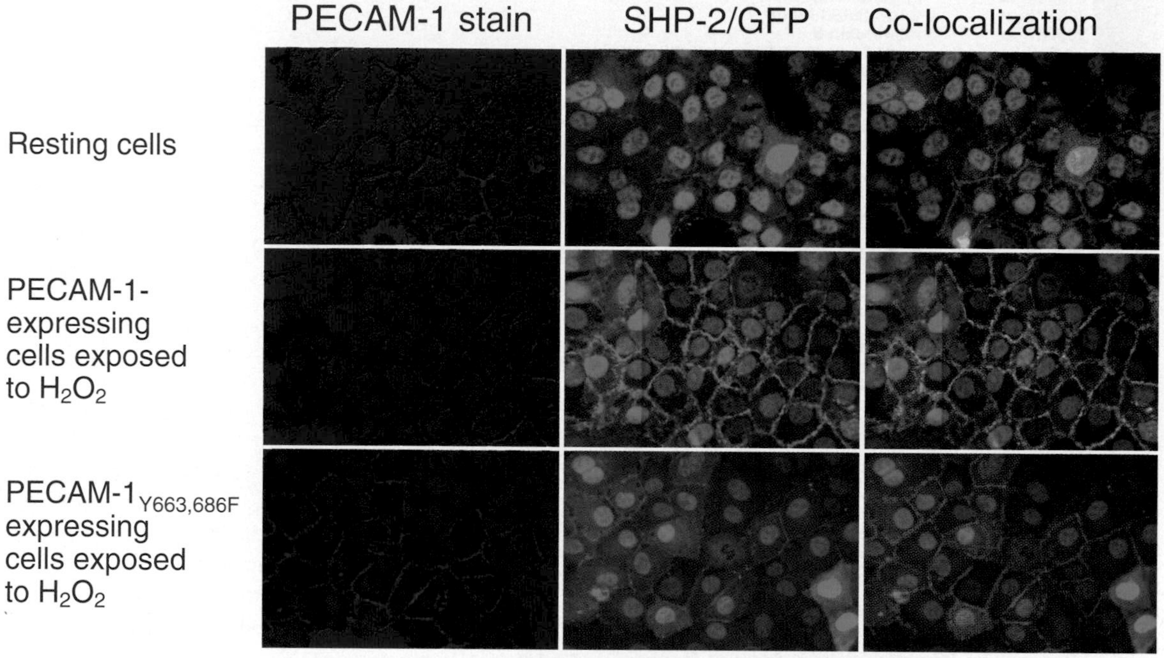

Color Plate 115.3. Reactive oxygen species activate PECAM-1 ITIMs and trigger cell border localization of the protein-tyrosine phosphatase, SHP-2. Endothelial-like REN cells were transfected with a green fluorescent protein (GFP) fusion protein containing the tandem SH2 domains of SHP2 together with either wild-type PECAM-1 (*top two rows*) or an ITIM-crippled ($Y_{663,686} \rightarrow F$) form of PECAM-1 (*bottom row*). In resting cells (*top row*), PECAM-1 is concentrated at cell–cell borders, whereas SHP-2 is largely cytoplasmic and nuclear. Upon exposure to oxidative stress (*middle and bottom rows*), SHP-2 is recruited to the inner face of the plasma membrane of cells expressing wild-type (*middle panels*) but not ITIM-less (*lower panels*) PECAM-1. (Reproduced with permission from Maas et al. *Am J Physiol Heart Circ Physiol.* 2003;285:H2336. Photo courtesy of Cathy Paddock, Blood Research Institute.)

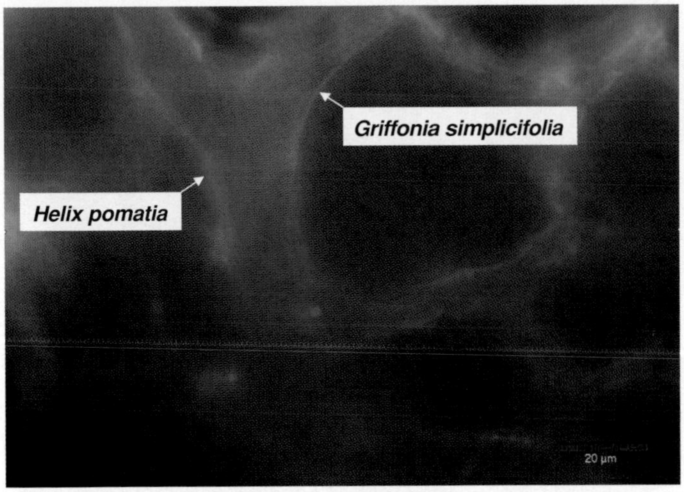

Color Plate 126.2. Infusion of TRITC-labeled *H. pomatia* and FITC-labeled *G. simplicifolia* in the intact pulmonary circulation demonstrate a clear demarcation in cell phenotype in vessels 20 to 30 μm in diameter. Note that no spectral overlap occurs, indicating ECs interact with either lectin but do not interact with both lectins. In situ epifluorescence images were tested and produced by Dr. Abu-Bakr Al-Mehdi.

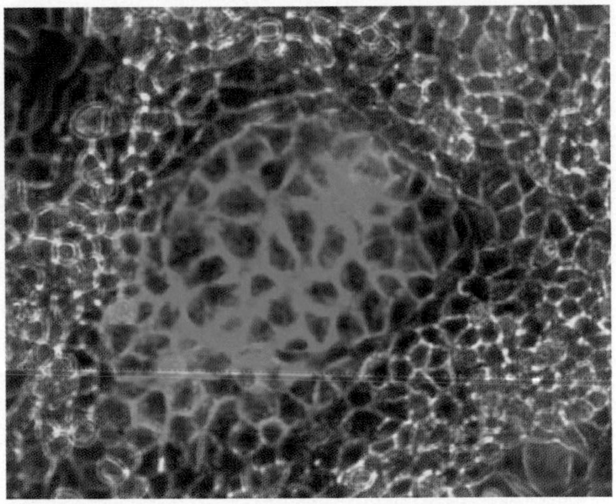

Color Plate 126.4. *G. simplicifolia* recognizes pulmonary microvascular ECs in coculture experiments. Pulmonary artery and microvascular ECs were cocultured, grown to confluence, and labeled with FITC-labeled *G. simplicifolia.* Only microvascular ECs interact with the lectin.

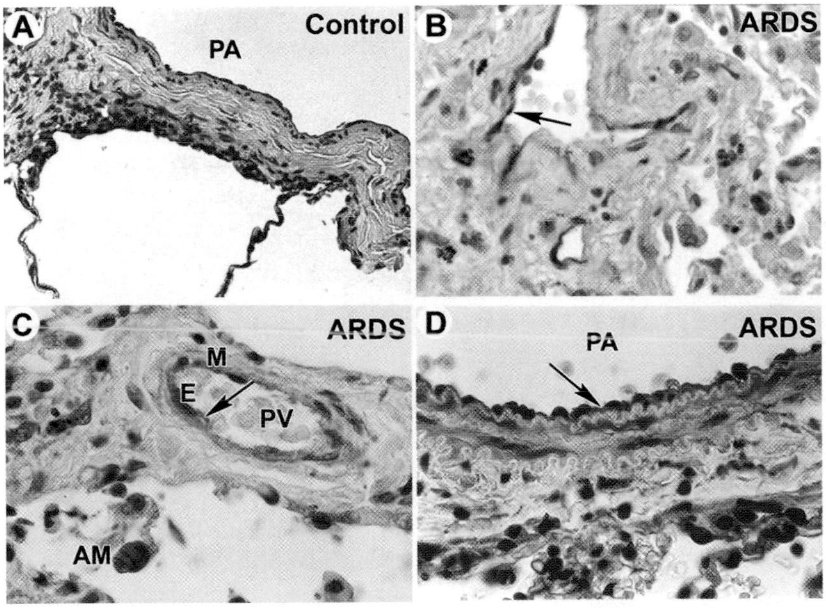

Color Plate 128.2. The endothelial phenotype is altered in lung vessels in ARDS. (**A**) Lung section from an individual without lung injury dying of acute head trauma. *PA*, pulmonary artery. (**B**) Microvessels in injured lung tissue from a patient with ARDS display surface P-selectin (*dark staining, arrow*). (**C**) Pulmonary venular (PV) ECs in a lung section from a patient dying from ARDS stain positively for ENA-78 (*arrow*). Capillary ECs were also positive for ENA-78 (*not shown*), as were alveolar macrophages (*AM*), some smooth muscular cells (*M*), and intra-alveolar PMNs (*not shown*). (**D**) ECs in the intima of a pulmonary artery (*PA*) in a lung section from a subject dying of ARDS stain intensely for COX-2. Staining for surface P-selectin (**B**), ENA-78 (**C**), and COX-2 was absent or much less evident in endothelium in vessels from subjects dying without acute lung injury or ARDS (**A**). See Ref. 5 for additional details. (Panel C reproduced with permission from Imaizumi T, et al. Human endothelial cells synthesize ENA-78: relationship to IL-8 and to signaling of PMN adhesion. *Am J Respir Cell Mol Biol.* 1997;17:181–192; panels A, B, D are used with permission from K. Albertine, G. A. Zimmerman, unpublished observations.)

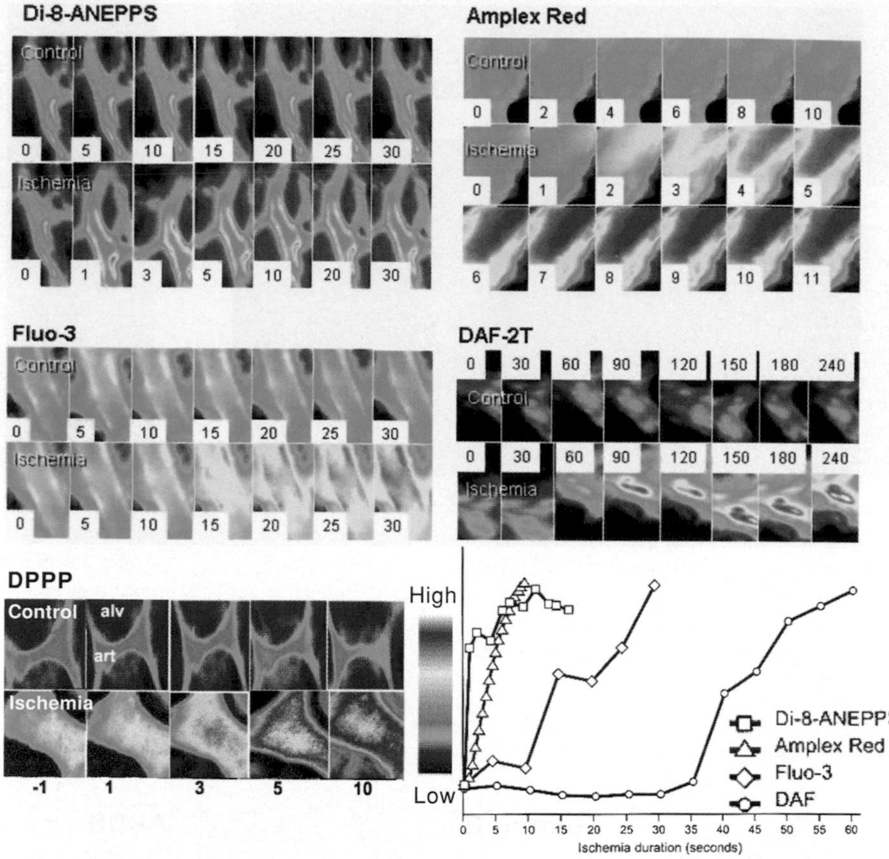

Color Plate 131.1. Fluorescence imaging of subpleural microvascular ECs in the isolated rat lung. Each set of images represents a control perfusion period followed by ischemia. Images are in pseudocolor, with red indicating higher fluorescence. The number on each panel indicates time in seconds (minutes for DPPP), either during the control observation period or after cessation of perfusion. (*Top left*): Imaging for cell membrane potential using the fluorophore di-8-ANEPPS. (*Top right*): Imaging with Amplex red for the appearance of H_2O_2 in the intravascular space. (*Middle left*): Imaging for intracellular Ca^{2+} with fluo-3. (*Middle right*): Imaging for NO with diamino fluorescein (*DAF*). (*Bottom left*): Imaging for lipid peroxidation using the hydroperoxide probe DPPP. The preset scale indicates the pseudocolor scale used in all these images. (*Bottom right*): Timing of fluorescence changes in membrane potential, ROS generation (Amplex red), intracellular Ca^{2+} (Fluo-3), and NO generation (DAF) with ischemia, as determined by imaging techniques in the isolated rat lung. The y-axis represents arbitrary units. DAF, diaminofluorescein; DPPP, diphenyl-1-pyrenylphosphine.

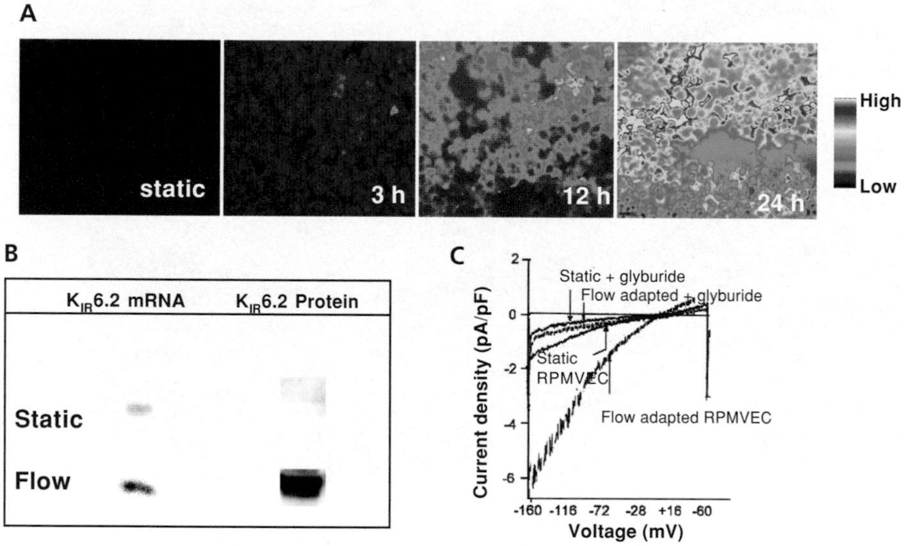

Color Plate 131.2. Induction of K_{ATP} channels during flow-adaptation. (**A**) Increase in fluorescence in cells flow-adapted at 10 dyn/cm^2 for varying periods. Cells were labeled with fluorescently labeled glyburide (BODIPY-glyburide, 50 nM). The resulting fluorescence indicating binding of glyburide to the sulfonylurea receptor (SUR) was observed with a microscope. (**B**) Representative blots of K_{IR}6.2 mRNA and protein content of RPMVECs cultured under static conditions or adapted to flow (10 dyn/cm^2 for 24 h). Total RNA was extracted, absorbed as a dot on a nitrocellulose membrane, and hybridized with 32P-labeled K_{IR}6.2 cDNA. Protein was analyzed by Western blot using polyclonal antibodies to the COOH terminus of K_{IR}6.2. (**C**) Inwardly rectifying whole cell K^+ currents (K_{IR}) in RPMVECs. Representative recordings obtained from static (no flow) cells and cells adapted to flow at a shear stress of 10 dyn/cm^2 for 24 hours. Increased current is seen for flow-adapted compared with static cells. Glyburide (K_{ATP} blocker) completely abolished the increased current.

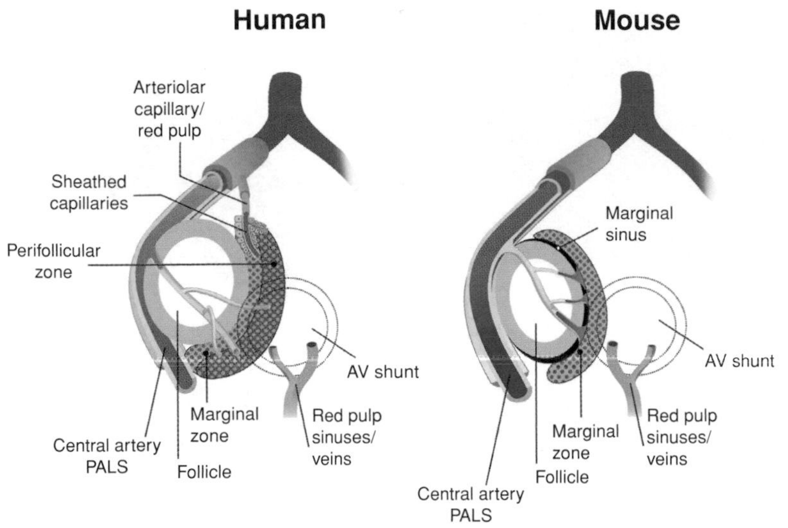

Color Plate 136.2. A hypothetical scheme of splenic circulation in man compared to rodents ("HuMouse"). After the branching of main artery, smaller arteriolar segments penetrate the white pulp. On the left side, the human structure is shown, with features characteristic for the human conditions on the left side of the drawing (SC: sheathed capillaries, PFZ: perifollicular zone). On the right side, the mouse arrangement is shown. The parts in the middle of the graph indicate the common elements in both species (MZ, marginal zone; PALS, periarteriolar lymphoid sheath).

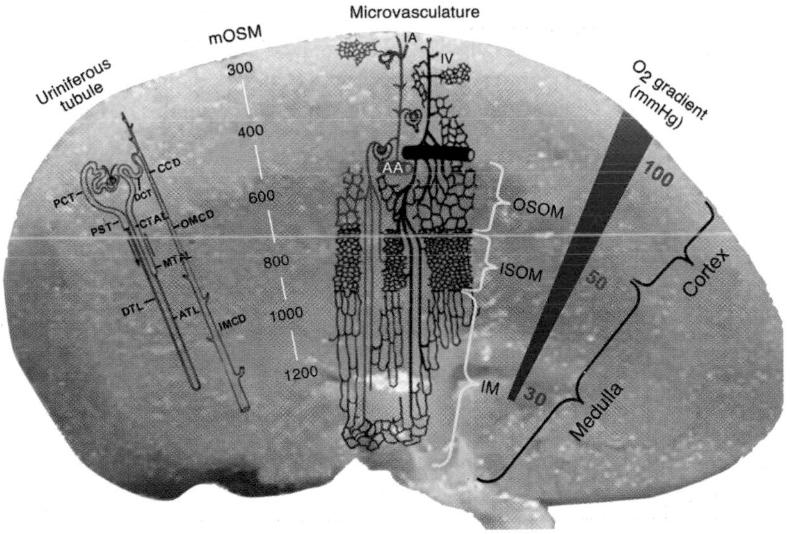

Color Plate 138.1. Schematic representation of the visional differences in the microvasculature within the kidney. The arcuate artery (*AA*) gives rise to the interlobular artery (*IA*), from which all afferent arterioles arise to perfuse the glomerular capillaries. Also shown are the osmotic and O_2 gradients from the outer cortex to the inner medulla. The medulla vascular supply arises from the juxtaglomerular efferent arterioles that branch and then descend into the medulla. As a result the outer stripe of the outer medulla (*OSOM*) has minimal arteriole capillary density. OSOM, outer stripe of the outer medulla; ISOM, inner stripe of the outer medulla; IM, inner medulla; PCT, proximal convoluted tubule; PST, proximal straight tubule; DLT, descending thin limb; ATL, ascending thin limb; MTAL, medullary thick ascending limb; CTAL, cortical thick ascending limb; DCT, distal convoluted tubule; CCD, cortical collecting duct; OMCD, outer medullary collecting duct; IMCD, inner medullary collecting duct; AA, arcuate artery; IA, interlobular artery.

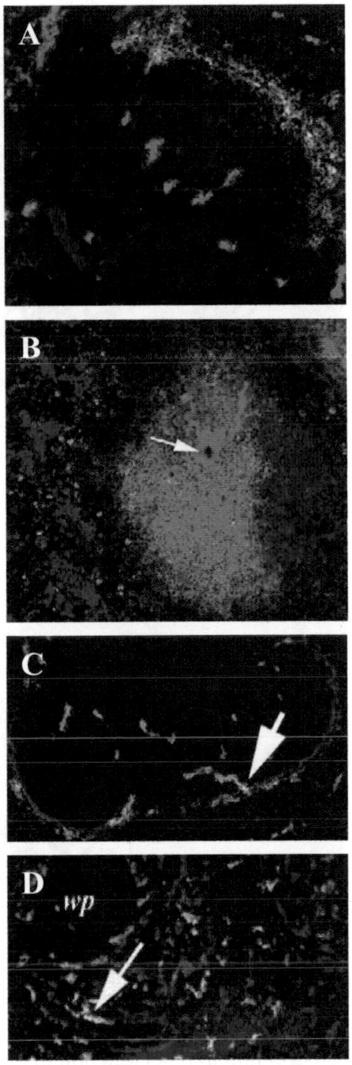

Color Plate 136.1. Endothelial labyrinth around the lymphohemopoietic compartments of the spleen. Various segments of the splenic vasculature can be identified by their different endothelial surface phenotype in separate regions of the spleen. **(A)** The main vessel types are identified as CD31-positive central artery, CD31/IBL-7/1 (*green*) double-positive arterioles in the white pulp (*turquoise*); MAdCAM-1–positive marginal (*red*) sinus coexpressing IBL-7/1 marker at some locations (*yellow*) in the marginal zone; and a few scattered IBL-7/1 positive sinuses in the red pulp (*green*). **(B)** The T cells (*green*, Thy-1) are arranged around the central arteriole (*arrow*), to which follicles containing B cells (*blue*, B220) are attached. The red pulp is outlined as the major macrophage-containing domain (*red*, F4/80). **(C)** The arrow points at a transitional area between the IBL-7/1–positive white pulp arteriole (*green*) and the MAdDCAM-1–positive segment of the marginal sinus (*red*) is shown, coexpressing both markers (*yellow*). **(D)** Sinusal heterogeneity of the red pulp area. A fraction of sinuses label only with mAb IBL-7/1 (*green*) or express IBL-7/1 and IBL- 9/2 (*red*) markers at approximately equal intensity (*arrow*), whereas the majority of red pulp sinuses and possibly venules dominantly display the IBL-9/2 antigen (62).

Color Plate 170.3. Adhesion cascades in HEVs. Examples of dedicated multistep cascade in PLNs (**A–B**) and in PPs (**C–D**). (**A**) Naïve T lymphocytes undergo tethering and rolling via L-selectin-PNAd (step 1); chemoattractant stimulation by CCR7-CCL21(step 2); and sticking (firm adhesion) via high-affinity LFA-1-ICAM-1 interactions (step 3), followed by diapedesis. The selectivity of this recruitment cascade is illustrated in (**B**); L-selectin- cells (e.g., effector memory T cells) and CCR7–leukocytes (e.g., granulocytes) fail to establish one or more of the required interactions. PPs display different "ZIP codes" in HEV segments within the interfollicular T cell area (**C**) and in upstream B follicles (**D**). MAdCAM-1 is expressed in all PP HEVs, and mediates rolling interactions by binding to either L-selectin or integrin $\alpha4\beta7$. For step 2, HEVs in different areas of the PP present distinct chemokines to rolling leukocytes: T cells are activated by CCL21, while HEVs associated with B follicles present CXCL12 and CXCL13 to stimulate B cell-expressed CXCR4 and CXCR5, respectively. Firm arrest of both B and T cells in PPs are mediated by activated $\alpha4\beta7$ and/or LFA-1 integrins.

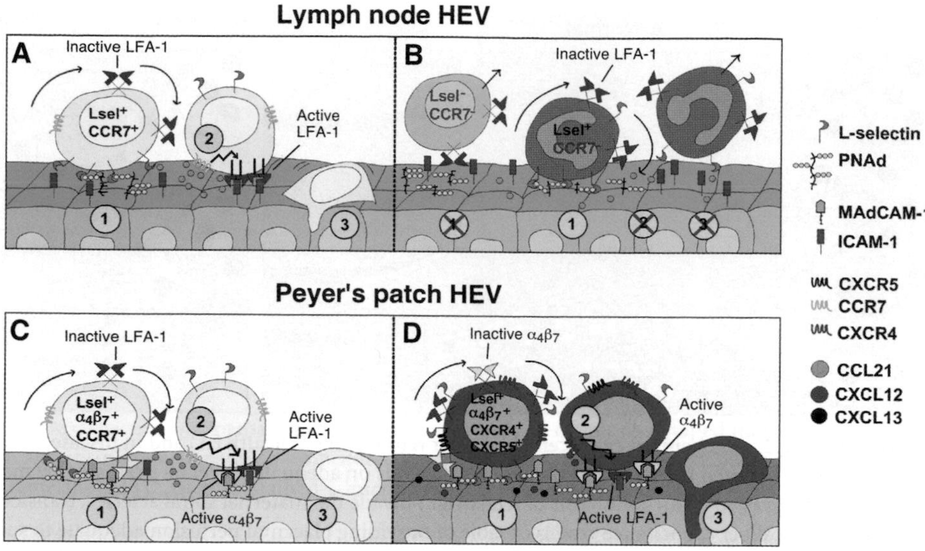

Color Plate 170.4. MECA-79 antigen expression is restricted to HEVs in PLNs. (**A**) False-color intravital micrograph showing MECA-79 coated fluorescent microspheres (*yellow*) bound to small diameter venules (order III to V) within the venular tree (*blue*) of a mouse subiliac LN. Beads do not accumulate in arterioles (*red*) or lower order venules. (**B**) Intravital micrograph showing the distribution of intravenously injected MECA-79 (*top panel*). The lower panel provides a schematic reference to help assign branching orders to the venules and to locate the superficial epigastric artery (*SEA*) and vein (*SEV*) in this preparation. (**C**) Immunohistochemistry reveals MECA-79 staining of a PLN HEV, illustrating the presence of MECA-79 on the luminal and abluminal (basement membrane) surface of the HECs. Panels in (**B**) and (**C**) were modified with permission from Stockton BM, Cheng G, Manjunath N, et al. Negative regulation of T cell homing by CD43. *Immunity.* 1998;8:373–381, and Stein JV, Rot A, Luo Y, et al. The CC chemokine thymus-derived chemotactic agent 4 (TCA-4, secondary lymphoid tissue chemokine, 6Ckine, exodus-2) triggers lymphocyte function-associated antigen 1-mediated arrest of rolling T lymphocytes in peripheral lymph node high endothelial venules. *J ExpMed.* 2000;191(1):61–76, respectively.

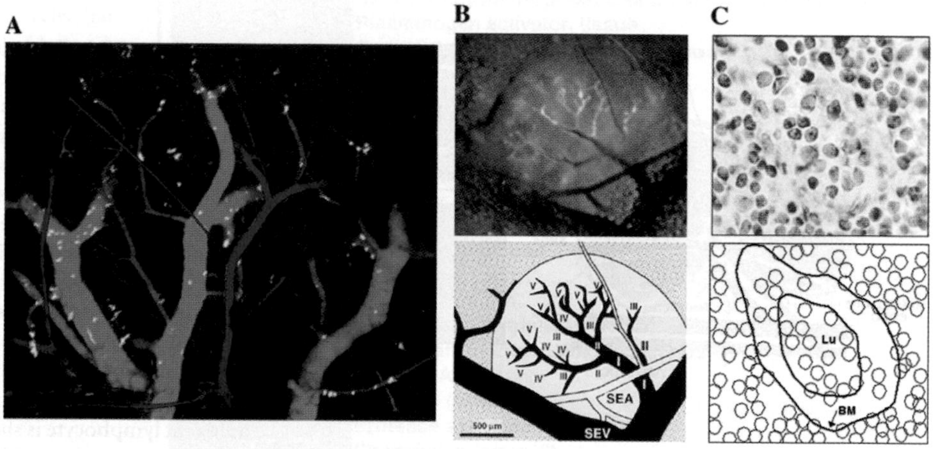

Color Plate 174.1. CECs express tissue factor. These studies of tissue factor expression by CECs illustrate the spectrum of methods that can be productively applied to these cells. A single cell is shown to be positive for endothelial marker P1H12 (**A**) and tissue factor antigen (**B**). CECs identified via P1H12 (*blue*) show a positive hybridization signal for *tissue factor* mRNA (**C**), which is not seen for control probe (**D**). Fluorescent factor VIIa binds to tissue factor-positive CECs (**E**), which is blocked by anti-tissue factor antibody (**F**). A chromogenic assay shows tissue factor function on tissue factor antigen-positive CECs (**G**), which is blocked by anti-tissue factor antibody (**H**). Tissue factor antigen-positive CECs show a positive hybridization signal (**I**) that is not seen for control probe (**J**).

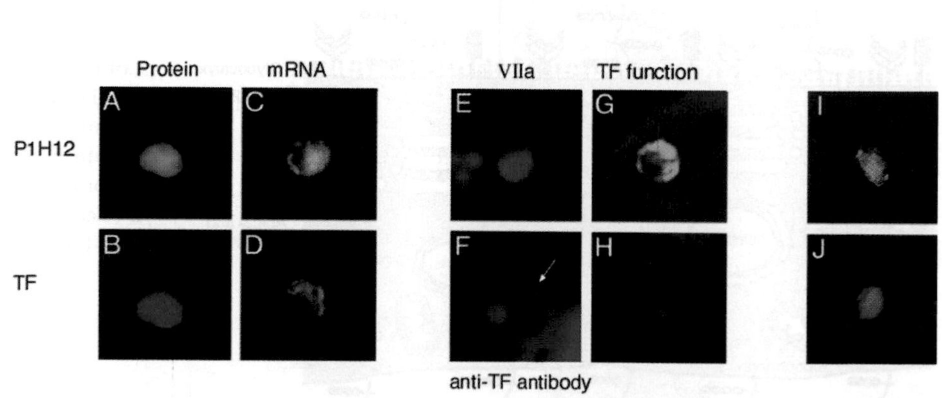

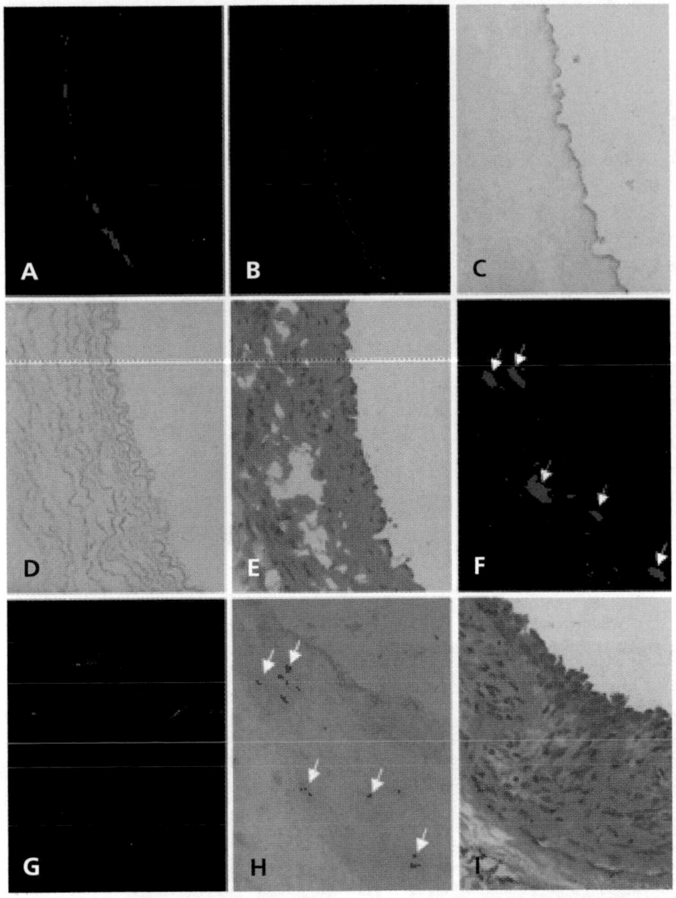

Color Plate 174.4. Administered CMMCs incorporate into vessel wall after balloon carotid injury. (**A**) Carotid section demonstrating labeled CMMCs on luminal border 4 weeks after local delivery. Colocalization staining of endothelial antigens CD31 (**B**) and BS-1 lectin (**C**) but (**D**) negative staining for macrophage marker RAM-11. (**E**) Hematoxylin and eosin staining of adjacent arterial section. (**F**) Labeled cells also detected in neointima that (**G**) do not costain for CD31 but (**H**) do stain for RAM-11, consistent with macrophage lineage. (**I**) Hematoxylin and eosin staining of adjacent arterial section. Red indicates CM-DiI fluorescence; green, CD31; arrows, colocalization of neointimal CMMCs with RAM-11 (*blue*). Magnification ×40.

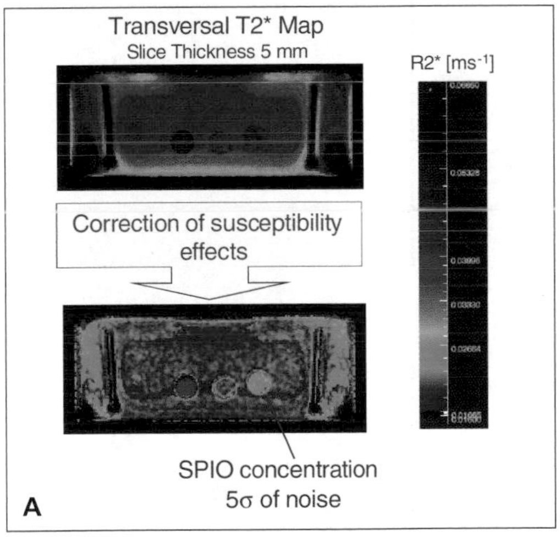

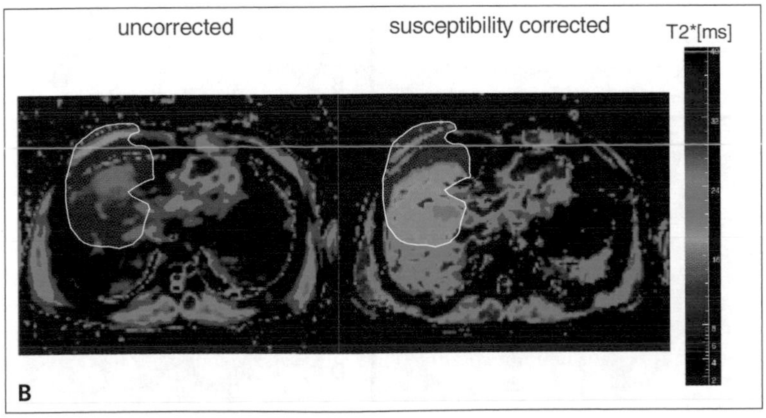

Color Plate 176.2. T2* maps of water tubes filled with SPIO contrast agents (**A**). Without susceptibility correction, no significant differences can be observed between the different tubes. After correction, SPIO concentrations as low as 5 standard deviations above the noise level could be detected. (**B**) In vivo T2* maps of the liver before and after susceptibility correction. (Reproduced with permission from Dahnke H, Schäffter T. Limits of detection of SPIO at 3.0T using T2* relaxometry. *Magn Reson Med.* 2005;53(5):1202–1206.)

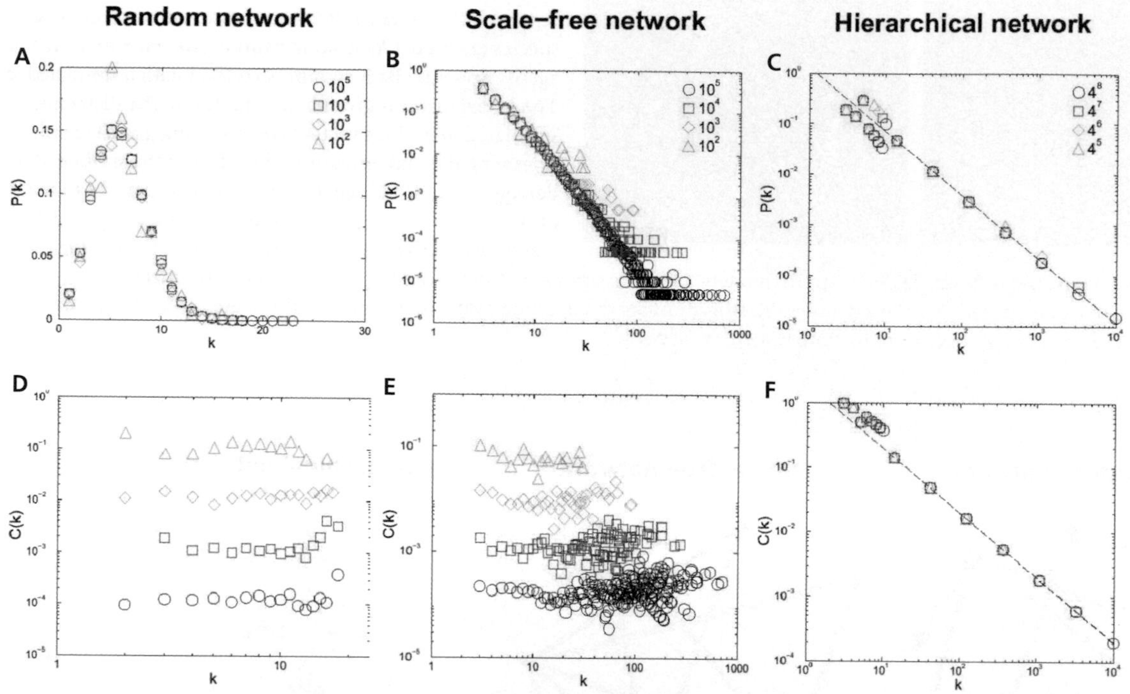

Color Plate 189.3. Properties of the three network models. (**A**) The ER model gives rise to a Poisson degree distribution $P(k)$ (the probability that a randomly selected node has exactly k links), which is strongly peaked at the average degree $\langle k \rangle$. The degree distributions for the scale-free (**B**) and the hierarchical (**C**) network models do not have a peak; they instead decay according to $P(k) \sim k^{-\gamma}$. The average clustering coefficient for nodes with exactly k neighbors, $C(k)$, is independent of k for both the ER (**D**) and the scale-free (**E**) network model. (**F**) In contrast, $C(k) \sim k^{-1}$ for the hierarchical network model.

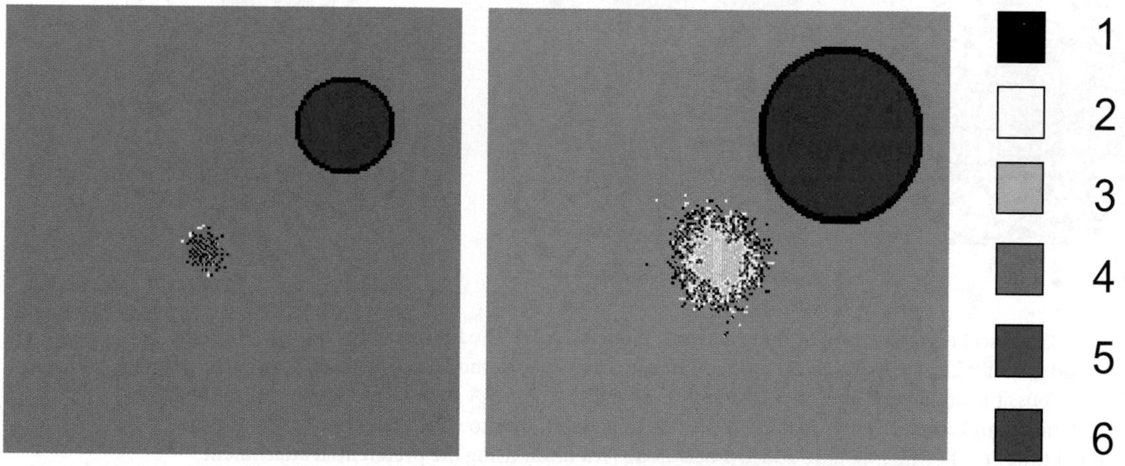

Color Plate 193.2. Depicted is a single-run, 2D visualization of the growing tumor at consecutive (scaled) time points $t = 50$ and 200 min, respectively. The key to the color coding (*right*) is as follows: (1) migrating, (2) proliferating and (3) quiescent tumor cells, (4) the extracellular matrix, (5) blood vessel outline and (6) glucose. Note the spatio-temporal expansion of the tumor toward the nutrient that diffuses from an expanding blood vessel located in the NE quadrant.

122

Hereditary Hemorrhagic Telangiectasia

A Model to Probe the Biology of the Vascular Endothelium

Mourad Toporsian and Michelle Letarte

Hospital for Sick Children, University of Toronto, Ontario, Canada

A decade ago, it was recognized that mutations in the *ENDOGLIN* (*ENG*) (1) and *ACTIVIN RECEPTOR-LIKE KINASE 1* (*ACVRL1*) (2) genes cause an autosomal dominant disorder known as hereditary hemorrhagic telangiectasia (HHT) type 1 and type 2, respectively. The predominant expression of the corresponding proteins, endoglin and ALK1, on vascular endothelial cells (ECs) and their function as transforming growth factor (TGF)-β receptors indicated that HHT is likely associated with perturbations in TGF-β signaling in ECs. Mice heterozygous for either gene develop HHT (3–6), and can be used to gain insight into the mechanisms of disease, highlighting the role of TGF-β in maintaining vascular integrity and unraveling the potential contribution of other pathways to HHT pathogenesis. Mice totally deficient in either endoglin (3–5) or ALK1 (6) do not survive past midgestation, and their analysis has revealed an essential role for these receptors in vascular development. This chapter provides an overview of HHT as a disorder of the vascular endothelium associated with the generation of abnormal structures such as telangiectases and arteriovenous malformations (AVMs), and discusses the current model of HHT pathogenesis.

HISTORICAL PERSPECTIVES

Over a century ago, HHT was recognized by the French physician Rendu as a syndrome associating cutaneous telangiectases with recurrent epistaxis (nosebleeds) (7). Following more specific clinical accounts of such manifestations by Osler (8) and the recognition of their familial nature by Weber (9), the new clinical entity was referred to as the Rendu-Osler-Weber syndrome. While the term *hereditary hemorrhagic telangiectasia* was introduced by Hanes in 1909 and is currently preferred, the early eponym is still used to designate this disorder. Until the 1940s, descriptions of clinical manifestations of HHT remained associated mostly with external visible signs, such as frequent epistaxis and small telangiectatic lesions (defined later) on the skin and in mucous membranes in oral and nasal cavities. Advances in medical imaging then revealed that AVMs –defined as abnormal direct connections between arteries and veins – were present in vital internal organs of HHT patients. AVMs are life-threatening because of their potentially lethal complications, such as hemorrhage, stroke, or brain abscess. HHT is a relatively rare disorder that is still underdiagnosed because of its varied presentation. However, the current estimated incidence is 1 in 5,000 to 8,000 individuals, with no racial or gender predilection worldwide; it can reach as high as 1 in 1,300 people in geographical isolates such as the Dutch Antilles (10).

VASCULAR LESIONS ASSOCIATED WITH HEREDITARY HEMORRHAGIC TELANGIECTASIA

Telangiectases

HHT is an autosomal dominant vascular disorder with age-related penetrance. Frequent nosebleeds are due to telangiectases in the nasal mucosa and are generally the first presenting sign of HHT. By age 21, 90% of individuals have had repeated nosebleeds. Numerous telangiectases can be present on the skin of fingers, chest, lips, eyelids, and along inner linings of mucosal cavities such as the mouth and ear. Most patients also develop telangiectases in the gastrointestinal tract with advancing age, potentially leading to hemorrhage, iron-deficiency, anemia and transfusion-dependence (11).

Telangiectases are relatively common in healthy individuals, but occur with much greater frequency in ataxia-telangiectasia; the calcinosis, Raynaud phenomenon, esophageal dysmotility, sclerodactyly, and telangiectasia (CREST) syndrome; and hereditary benign telangiectasia, and in other conditions such as pregnancy or chronic liver disease. Telangiectases coupled with recurrent epistaxis are characteristic of HHT and, in conjunction with family history and visceral AVMs, serve as the basis for HHT diagnosis. The consensus for clinical diagnosis is summarized by the Curaao criteria: (a) spontaneous and frequent epistaxis, (b) multiple telangiectases on lips, nose, oral cavity, and fingers, (c) visceral lesions and AVMs in pulmonary, cerebral, or hepatic circulations, and (d) family history of HHT. A definite diagnosis of HHT is made when three out of four criteria are satisfied, whereas a minimum of two criteria renders a diagnosis of "suspected HHT" (12). Genetic screening is available and provides a more definite diagnosis of HHT and of its type.

Telangiectases are mucosal and cutaneous lesions that represent a focal loss of capillaries; they consist of abnormally dilated venules directly connected to dilated arterioles. They contain a single endothelial layer and scant perivascular elastic tissue within the lesion. Braverman and colleagues systematically analyzed electron micrographs of dermal lesion biopsies from HHT patients and provided an overview of their development (13). The smallest clinically detectable lesions (pinpoint size) show a focally dilated postcapillary venule with prominent stress fibers in pericytes along the abluminal border. In lesions of 0.5 mm diameter, the venular lumen is expanded and shows thicker walls, with an increased number of pericytes. The upstream arteriole is dilated but remains connected to the downstream venule via short capillary segments. The 2-mm lesions display markedly dilated and convoluted venules with excessive and often uneven layers of smooth muscle cells; loss of capillaries is present, and a direct connection exists between the dilated arterioles and venules. Such lesions represent small AVMs. Figure 122.1A illustrates a telangiectatic lesion along the gastrointestinal tract.

Arteriovenous Malformations

Telangiectases or small AVMs can become much larger lesions, with direct connections between arteries and veins. The AVMs are predominantly found in HHT patients in the pulmonary, cerebral, and hepatic circulations. Because they are not externally visible, AVMs can grow to a large size before being detected clinically, and are therefore life-threatening.

Pulmonary Arteriovenous Malformations

Up to 50% of HHT patients may harbor pulmonary AVMs (PAVMs) (11) and, in fact, their presence should alert the physician to a diagnosis of HHT, because more than 80% of PAVMs are associated with this disorder (14). Rupture of a PAVM or bleeding from an endobronchial mucosal telangiectasia may result in hemoptysis. The prevalence of hemoptysis reaches 6% to 13% in HHT patients harboring PAVMs (11, 14–16). In general, HHT patients with massive pulmonary hemorrhage have a PAVM on the side of the bleeding, although rare cases present with recurrent hemoptysis without any PAVMs and have only endobronchial mucosal telangiectases (17,18). Figure 122.1B illustrates a pulmonary AVM, as revealed by angiography (19). Figure 122.1C represents schematics of a simple type of AVM, characteristic of 90% of cases, and consisting of a single segmental pulmonary artery joining a thin-walled aneurysmal sac drained by a single vein. Other PAVMs are more complex, with two or more feeding arteries and two or more draining veins (20). Embolotherapy (intentional blockage of the AVM feeding artery with coils

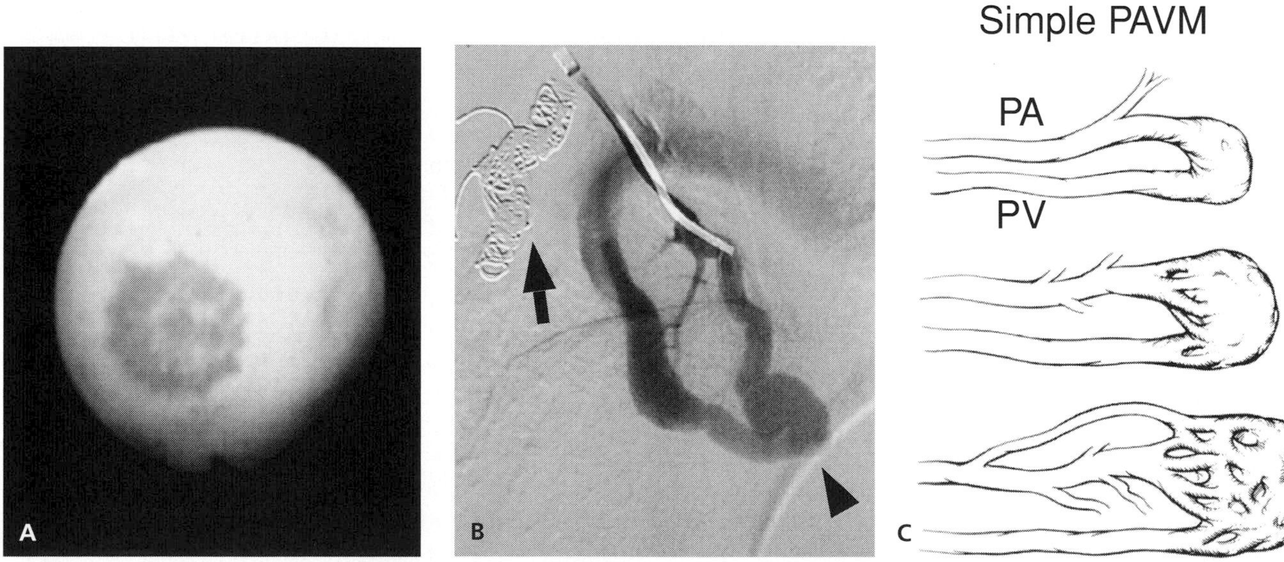

Figure 122.1. (**A**) Image of a telangiectatic lesion along the intestinal mucosa of a patient with HHT. (Courtesy of Dr. N. Marcon, St-Michael's Hospital, University of Toronto.) (**B**) Right pulmonary angiogram showing a simple PAVM in the right lower lobe (*arrowhead*). Note the presence of embolization coils (*arrow*) from prior treatment of a different AVM. (Adapted with permission from Jaskolka J, Wu L, Chan RP, Faughnan ME. Imaging of hereditary hemorrhagic telangiectasia. *Am J Roentgenol.* 2004;183:307–14.) (**C**) Schematic diagram of a simple PAVM (PA, pulmonary artery; PV, pulmonary vein). This structure is supplied by one segmental artery, which may consist of a single branch (*upper panel*), multiple branches (*middle panel*), or multiple branches with a more proximal branch arising from the same feeding artery to the aneurysm (*lower panel*). Complex PAVMs usually receive blood from many segmental arteries. (Adapted with permission from White RI, Jr., Pollak JS, Wirth JA. Pulmonary arteriovenous malformations: diagnosis and transcatheter embolotherapy. *J Vasc Interv Radiol.* 1996;7:787–804.)

or a balloon to prevent potential hemorrhage) provides an efficient means of treatment, but because most patients have multiple PAVMs that grow with time, patients must be monitored closely and the procedure repeated (16). If untreated, PAVMs can lead to neurological complications. Strokes and transient ischemic attacks occur in close to 40% of HHT patients with PAVMs due to right-to-left shunting of blood and consequently improper filtration of potential clots and bacteria (21).

Cerebral Arteriovenous Malformations

Five percent to 10% of HHT patients develop cerebral AVMs. However cerebral AVMs are relatively frequent in the general population, and only 2% of cases are associated with HHT (21). Most cerebral AVMs are believed to be congenital lesions occurring mostly in the supratentorial region but of unknown etiology; they consist of multiple arteries and veins, connecting as a fistula without an intervening normal capillary bed (22). An analysis of cases associated with HHT suggests that the appearance in the cortex of multiple micro-AVMs harboring single feeding arteries and draining veins and located in the cortex should raise suspicion of an HHT diagnosis (23).

Hepatic Arteriovenous Malformations

Liver involvement to various degrees is found in 40% of patients with HHT (24), and is characterized by malformations in hepatic vessels. Early on, microscopic telangiectases are scattered throughout the liver parenchyma, which progressively enlarge to display multiple direct abnormal arteriovenous communications. A recent three-dimensional modeling of the hepatic vasculature in HHT indicates that pathological changes include focal sinusoidal ectasia, direct communications between hepatic arteriolar branches and ectatic sinusoids (arteriovenous shunts), and large communications between the portal and central veins via ectatic sinusoids (portal vein shunts) (25). There are three distinct types of intraparenchymal right-to-left shunts: hepatic artery to hepatic vein, hepatic artery to portal vein, and portal vein to hepatic vein. Increased blood flow through even microscopic AVMs can lead to significant dilatation of the hepatic artery and its branches in the early stages, which progressively become tortuous and tangled with increasing disease severity. The progressive dilatation of the hepatic and/or portal veins and their flow abnormalities are associated with larger arteriovenous shunting. Cirrhosis and liver dysfunction as well as high cardiac output and heart failure due to increasing shunting represent end-stage manifestations of hepatic involvement (26).

GENETIC BASIS OF HEREDITARY HEMORRHAGIC TELANGIECTASIA

During the 1990s, linkage studies revealed that mutations in two distinct genes mapping to human chromosomes 9q33–34 and 12q13 and encoding endoglin (*ENG*) and activin receptor-like kinase (*ACVRL1*) cause HHT type 1 (HHT1) or HHT type 2 (HHT2), respectively (1,2). HHT1 families have a higher prevalence of pulmonary AVMs than do those with HHT2, who tend to have a generally milder phenotype and later onset (14,27). However, hepatic involvement has been observed in both groups of patients. Comprehensive phenotype/genotype correlation studies are needed to ascertain if indeed disease severity and organ specificity are related to the affected gene. About 10% to 20% of families with HHT cannot be accounted for by mutations in *ENG* or *ACVRL1* genes. A genome-wide linkage analysis recently identified an additional locus on chromosome 5 associated with HHT and referred to as HHT3 (28). It will be interesting to see if the product of this new locus is also predominantly expressed in the vascular endothelium and involved in TGF-β signaling.

Gallione and colleagues have described patients with a combined syndrome of juvenile polyposis (JP) and HHT (referred to as JPHT) and a functionally inactivating mutation in the *MADH4* gene (coding for the common Smad4), but no mutation in either *ENG* or *ACVRL1* genes (29). This suggests that HHT also can be caused by *MADH4* mutations in this combined syndrome, a further link to impaired TGF-β signaling, because the common Smad4 mediates all Smad-dependent signaling.

ENG and *ACVRL1* Mutation Analysis

ENG is comprised of 15 exons spanning 30 kb of genomic DNA. As of 2006, 156 different *ENG* mutations have been identified in patients with HHT (30). These mutations are of all types and are distributed in the first 12 exons, which encode the extracellular domain of endoglin. *ACVRL1* has 10 exons distributed over 15 kb. Currently, 123 *ACVRL1* mutations have been reported; more than half are missense substitutions, most of which are in the intracellular kinase domain (30).

Almost 80% of all reported *ENG* mutations cause frame shifts and premature stop codons that should lead to nonsense-mediated mRNA decay resulting in reduced mRNA levels and very unstable mutant proteins. Expression analysis revealed that mutant endoglin proteins are rarely detectable and, if so, are transient species that do not reach the cell surface and are destined for degradation (31). The observations that mutant proteins are unstable suggest that folding is tightly regulated, and that most mutations lead to structural instability and loss of function. Our data support the hypothesis that *ENG* haploinsufficiency leading to 50% reduction in functional protein in ECs is the underlying cause of HHT1 (31). This model implies that the disease is not due to interference by a mutant protein, indicating that the position or type of mutation does not influence disease severity. Analysis of histological sections from HHT1 patients revealed that all blood vessels showed a 50% reduction in endoglin levels, relative to platelet-endothelial cell adhesion molecule (PECAM)-1 (32). Furthermore, cerebral and pulmonary AVMs showed a lower density of endoglin due to the excessive dilatation of the vessels, whereas the endoglin:PECAM-1 ratio remained constant.

These data ruled out the loss of *ENG* heterozygosity as a potential mechanism for the generation of vascular lesions.

The quantification of ALK1 levels in ECs (33) and molecular modeling of the kinase domain mutants (34) also support haploinsufficiency as the underlying cause of HHT2. Of the 123 mutations reported for HHT2 (30), 75% occur in the intracellular kinase domain at highly conserved residues, both across species and among different type I receptors. These likely lead to the generation of misfolded and unstable proteins.

Other Diseases Associated with *ENG* and *ACVRL1* Mutations

Interestingly, a recent study has identified germline mutations in *ENG* (1711C→T, R571C and 1538 A→G, K513R) as a previously unrecognized cause of juvenile polyposis (JP) syndrome (35). These patients presented with an unusually early onset of JP (ages 3 and 5), with no clinical signs of HHT. Family history was not discussed, and one cannot rule out the possibility that these children may develop HHT later. The specific missense mutations are in conserved residues, do not appear to be normal polymorphic variants, and have not to date been identified in HHT1 patients; nor have protein levels been measured to ascertain if endoglin haploinsufficiency also is associated with these case of JP. This study therefore identifies *ENG* mutations as a possible genetic cause of JP, and it remains to be ascertained if it is due to impaired endoglin function in ECs. One cannot rule out expression of endoglin in a subset of gut epithelial cells that could be more directly implicated in JP and associated with abnormal epithelial cell cycle regulation.

Familial cases of pulmonary arterial hypertension (PAH), as well as a large proportion of somatic cases, are associated with mutations in the bone morphogenetic protein (BMP) receptor II *(BMPRII)* gene (36), which codes for the BMPRII receptor that signals through Smad1 and Smad5, a pathway shared by ALK1. Mutations in *ACVRL1* and to a lesser degree *ENG* have been recently associated with PAH (37,38). This apparent higher incidence of PAH in patients with HHT2 versus HHT1 may be related to their lower incidence of PAVMs, structures that would tend to reduce overall pulmonary vascular resistance. A recent study by Harrison and colleagues describes the case of a 3-year-old child with an *ENG* mutation who initially presented with features of PAH and, by 8 years of age, developed signs of HHT (37), at which point pulmonary vascular resistance had declined most likely due to the presence of microvascular PAVMs.

THE FUNCTIONAL PROTEINS

Endoglin

Endoglin (CD105) is an integral membrane glycoprotein initially identified on pre-B cells of childhood acute lymphocytic leukemia (39,40). It also is expressed on activated macrophages, pre-erythroblasts, developing endocardial tissue mesenchymal cells, and in cultured smooth muscle cells and fibroblasts (41). Endoglin also is present on the syncytiotrophoblast throughout pregnancy, and is transiently expressed on the extravillous cytotrophoblasts as they invade the spiral artery during the first trimester of pregnancy (41). Endoglin resides predominantly in the vascular endothelium; it was defined as an EC marker (CD105) in the Fifth International Leukocyte Typing Workshop (42). It is highly expressed on ECs of all vessel types, and considered a sensitive marker of tumor angiogenesis. Quantification of intratumor microvascular density by endoglin staining has prognostic significance in selected neoplasias (43). Indeed, antibodies to endoglin have demonstrated efficacy in animal models for tumor imaging and antiangiogenic therapy, and are promising tools for the treatment of human malignancies (44).

Endoglin is synthesized as a 68-kDa polypeptide chain and subsequently glycosylated at N- and O-linked sites, to yield a 90-kDa monomer. It contains 17 cysteine residues that generate intra- and interchain disulfide bonds, suggesting a complex and tightly regulated folding. At the surface of ECs, endoglin exists as a covalently linked 180-kDa homodimer consisting of a relatively large extracellular region (561 amino acids), a hydrophobic stretch of 25 amino acids spanning the plasma membrane, and a cytoplasmic tail of 47 residues (45), rich in phosphorylated serine and threonine (46). Endoglin is an ancillary receptor (or coreceptor) for TGF-β1 and -β3; it does not signal directly and modulates responses of these cytokines in ECs and other cell types (47). Endoglin is structurally related to betaglycan, the original TGF-β type III receptor, with the short cytoplasmic region showing the greatest sequence identity (48,49). Endoglin has therefore been referred to as a type III receptor for the TGF-β1 and -β3 isoforms; betaglycan also binds TGF-β2 and can be found in the extracellular matrix. In transfection studies, endoglin can bind other members of the TGF-β superfamily including activin, BMP-7, and BMP-2 when associated with their corresponding receptors (50). The functional implications of these interactions remain to be elucidated under physiological conditions.

Although endoglin is mostly known for its role as a TGF-β coreceptor, it is present in vast excess (400,000 molecules/EC) in comparison to TGF-β receptors (<2,000/cell), suggesting additional interactions and functions. Indeed, despite its short cytoplasmic tail and lack of obvious functional motifs, endoglin associates with the endothelial nitric oxide synthase (eNOS) activation complex (51), which includes eNOS and its allosteric regulators, heat shock protein 90 (HSP90) and caveolin-1 (Cav-1). Endoglin appears to stabilize the eNOS protein, as demonstrated by a shorter eNOS half-life in ECs from *Eng* heterozygous mice. The same studies suggested that endoglin promotes eNOS/HSP90 interaction during eNOS activation, thus favoring nitric oxide (NO) production (51). These novel findings indicate that endoglin is involved in vasoregulation mediated by its interaction with eNOS, which occurs via its intracellular short cytoplasmic tail. Interestingly,

Table 122-1: Phenotype of Mice Deficient in Selected Members of the TGF-β Superfamily

Gene Ablated	*Lethality of Null Mutant*	*Phenotype*	*References*
Tgf-β1	E10.5-P2–3 weeks	Abnormal vasculogenesis and hematopoeisis Improper EC differentiation	57
Tgf-β2	E19-P1 day	Cyanosis and respiratory distress Cranio-facial, cardiac and urogenital defects	59
Tgf-β3	P1 day	Cleft palate Abnormal and fragile lung vasculature	60
Tgf-βrII	E10.5	Defective yolk sac vasculogenesis and hematopoeisis	61
Alk1	E10–10.5	Defective yolk sac and embryo angiogenesis Hyperdilatation of large vessels	6, 62
Alk5	E9.5-E11.5	Defective yolk sac and placental vasculogenesis Defective hematopoeisis	63
Eng	E10–10.5	Defective yolk sac and embryo angiogenesis Defective heart valve formation	3, 4, 5
Smad1	E9.5–10.5	Defective yolk sac and embryo angiogenesis Defective heart valve formation	64
Smad5	E9.5–E11.5	Disorganized vessels in yolk sac Gastrointestinal and cardiac defects Chorion/allantois fusion	65

E, embryonic day

this study also demonstrates that the eNOS-endoglin association occurs in caveolar lipid rafts, subcellular compartments known to harbor many important signaling molecules, thus suggesting that the functional roles of endoglin may be even more diverse than previously recognized.

The cytoplasmic domain of endoglin also interacts with zyxin and zyxin-related protein (ZRP)-1, members of the LIM family implicated in organization of the actin cytoskeleton (52,53). This interaction occurs in the absence of exogenous TGF-β. Endoglin regulates the cellular distribution of zyxin and ZRP-1 favoring their association with actin stress fibers and possibly contributing to the structural integrity of ECs. Cytoskeletal organization is critical for angiogenesis, vascular morphology and permeability, inflammatory response, and response to shear stress, all of which are essential mechanisms in maintaining the structural and functional integrity of the vasculature.

ALK1

ALK1 is a type I serine/threonine receptor kinase of the TGF-β superfamily, specifically expressed on ECs. Interestingly, ALK1 expression is much higher in the endothelium of developing and mature arteries than in veins and capillaries (54). It is also found at epithelial-mesenchymal boundaries (54). ALK1 shares a relatively high amino-acid identity to other type I receptors including ALK2, ALK5, ALK4, and ALK3, particularly in the serine/threonine kinase subdomain. The intracellular kinase domain contains a glycine/serine-rich (GS) region that plays an important role in signaling, because its deletion abolishes the ability of TβRI to undergo phosphorylation and to mediate TGF-β–dependent responses. The extracellular domain of ALK1 contains 10 conserved cysteine residues and a potential N-linked glycosylation site. ECs possess a unique TGF-β signaling mechanism, which involves an intricate interplay between ALK1, ALK5, and endoglin. These three receptors are important modulators of the TGF-β–dependent fine tuning of the activation state of the vascular endothelium (2,55).

TGF-β SIGNALING IN ENDOTHELIAL CELLS AND ITS ROLE IN ANGIOGENESIS

Overall, TGF-β is involved in numerous cellular processes including proliferation, differentiation, apoptosis, adhesion, migration, and extracellular matrix deposition in various cell types (see Chapter 35) (56). Compelling evidence from gene ablation studies of TGF-β, its receptors, and downstream targets indicate a crucial role for TGF-β–mediated responses in cardiovascular development and the maintenance of vascular integrity. The three isoforms – TGF-β1, -β2, and -β3 are expressed in mammalian cells in tissue-specific and developmentally regulated manners. Table 122.1 summarizes the phenotypes of the *TGF-β*-null mice. In particular, TGF-β1 is expressed in endothelial, hematopoietic, and connective tissue cells, and its genetic ablation in mice causes embryonically lethal cardiovascular and hematopoietic defects (56). Some *TGF-β1*-null pups survive until weaning due to maternal TGF-β1 crossing the placenta during pregnancy and present in the milk; they then die of inflammatory disease (Figure 122.2) (57).

All three TGF-β isoforms signal through a heteromeric complex consisting of type I and type II serine/threonine

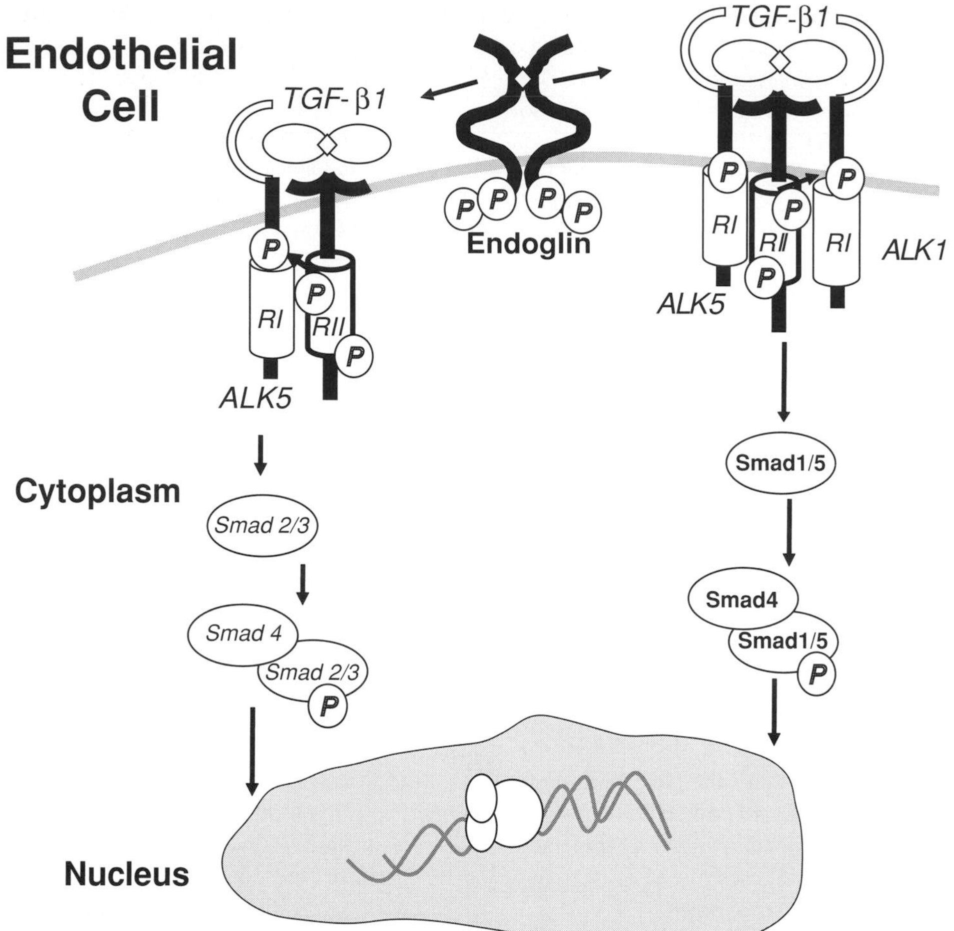

Figure 122.2. TGF-β signaling in ECs. Upon binding of the TGF-β dimer to the type II receptor (TβRII), the type I receptor is recruited and in turn signals through R-Smads. TGF-β generally signals via ALK5 leading to the phosphorylation of Smad 2/3. However, vascular ECs possess a unique TGF-β signaling mechanism whereby an additional type I receptor, ALK1, mediates TGF-β effects by activation of the Smad 1/5 pathway. The presence of this additional TGF-β signaling mechanism underscores the importance of fine tuning TGF-β–dependent responses in the vascular endothelium and the essential role for ALK1 in EC function. Although the exact role of endoglin in TGF-β pathways remains controversial, current evidence suggests that endoglin modulates TGF-β responses in ECs via its interaction with the ALK5- and ALK1-containing receptor complexes and plays a critical role in maintaining a delicate balance between these receptors.

kinase receptors (Figure 122.2). Once the TGF-β dimer binds to the type II receptor (TβRII), the type I receptor is recruited and in turn signals through receptor Smads (R-Smads). TGF-β generally signals via ALK5, leading to the phosphorylation of Smad 2/3. In ECs, an additional type I receptor, ALK1, mediates TGF-β effects by activation of the Smad 1/5 pathway (6). However, ALK5 kinase activity and the TβRII receptor are needed for optimal ALK1 activation (58). Therefore, ALK1 and ALK5 assemble into a heteromeric receptor complex with TβRII, and signal via the Smad 1/5 pathway. Thus, ALK1 is not only restricted to ECs and mutated in HHT2, but also mediates a specific TGF-β pathway that must be essential for EC function. Mice deficient in either ALK5, ALK1, TβrII, Smad1, and Smad5 all show defective cardiovascular development, substantiating the critical nature of these receptors and R-Smads in angiogenesis (see Table 122.1).

TGF-β plays an important role in determining the properties of the endothelium during angiogenesis. During the activation phase, ECs degrade the perivascular membrane and invade and migrate into the extracellular space, where they proliferate and form a lumen. In the resolution phase, cells stop migrating and proliferating and reconstitute a basement membrane to complete the generation of a new vessel. It was proposed that TGF-β signaling via the ALK1–Smad 1/5 pathway is essential for transition from the activation to the resolution phase of angiogenesis by repressing angiogenic factors and proteases and promoting recruitment of smooth muscle cells. Signaling via the ALK5–Smad 2/3 pathway induces secretion of proteases and would therefore promote the activation phase (6). However, more recent data suggest a diverging model whereby the ALK1:ALK5 ratio determines whether TGF-β induces activation (therefore proliferation and migration at a high ratio) or quiescence of the endothelium (at a low ratio) (66). The issue is still very controversial, and the simplest way to interpret current results is to conclude that TGF-β regulates angiogenesis via a fine balance between ALK5 and ALK1 signaling.

Endoglin modulates TGF-β1 and TGF-β3 responses in ECs via its interaction with the ALK5- and ALK1-containing receptor complexes (50), and plays a critical role in maintaining the fine balance between these receptors (67). However, the exact function of endoglin in the TGF-β pathway is also controversial. Lebrin and colleagues could not grow ECs from *Eng*-null embryos and have suggested that endoglin levels positively regulate growth through potentiation of TGF-β/ALK1 signaling (67). Blanco and colleagues also showed that endoglin stimulates ALK1 and interferes with ALK5 signaling (68). Our most recent data indicate that ECs derived from *Eng*-null embryos proliferate faster than do normal cells, are more responsive to growth inhibition by TGF-β1, have a normal ALK5, and an activated ALK1 pathway (69). This would imply that endoglin is not required for TGF-β–dependent activation of Smad1 per se, and we suggest that endoglin controls the levels of cell surface receptors and their binding characteristics. Future studies are needed to elucidate how endoglin might regulate TGF-β receptor expression and contribute to angiogenesis.

MURINE MODELS OF HEREDITARY HEMORRHAGIC TELANGIECTASIA

A haploinsufficiency model for HHT implies that a mutation in a single allele results in loss of function, and that *Eng* or *Alk1* heterozygous mice should develop the disease. Indeed, these heterozygous mice show signs of HHT (4,70,71), and recent studies have revealed alterations in vascular homeostasis and hemodynamic properties in the adult mice (51,72).

Strong evidence in favor of a role for endoglin and ALK1 in angiogenesis and vascular development comes from gene knockout studies in mice. *Eng*$^{-/-}$ mice die at midgestation from failure to form mature blood vessels in the yolk sac (3–5,73) and the embryo proper as do mice deficient in ALK1 (6,61) and TβRII (61) (see Table 122.1). *Eng*-null mice display severe cardiac defects, such as impaired endocardial-mesenchymal transformation, valve formation, and heart septation indicating the involvement of endoglin in cardiac development. *Alk1*-null mice also die at midgestation, and show dilated vessels and structures resembling AVMs (6,62). The angiogenic defect has been ascribed at least in part to decreased vascular smooth muscle cell differentiation and migration in both *Eng*- and *Alk1*-null mice (3,6). However, we believe that the defect originates in the endothelium. A recent study demonstrated that the *Eng*-null yolk sac endothelium secretes less TGF-β1, leading to impaired Smad2-dependent responses needed for differentiation and migration of the smooth muscle cells (73). Despite the evidence provided by the knockout models of the importance of endoglin and ALK1 in vessel development and maturation, these studies are not informative in terms of HHT pathogenesis, which is associated with the heterozygous condition. For example, arteries from *Eng*$^{+/-}$ mice display normal smooth muscle contractile function suggesting that a reduction in smooth muscle cell recruitment, as described in *Eng*$^{-/-}$ mice, is unlikely to occur.

Eng$^{+/-}$ or HHT1 Mouse

Eng$^{+/-}$ mice were first generated on the 129/Ola background strain and spontaneously developed signs of HHT such as nosebleeds, telangiectases, dilated thin-walled vessels, and AVMs, with their associated complications (4,74). Moreover, similar to the heterogeneity in clinical manifestations observed in patients with *ENG* mutations, HHT-type lesions are also highly variable between *Eng*$^{+/-}$ mice. However, later onset and reduced severity of disease have been demonstrated in *Eng*$^{+/-}$ mice with a C57BL/6 background, compared with those with a 129/Ola background. Intercrosses between these two genetic backgrounds have revealed intermediate disease severity, suggesting that both modifier genes and epigenetic factors contribute to disease heterogeneity (74).

To study the contribution of endoglin in the absence of confounding effects due to modifier genes and disease progression, we have focused on the C57BL/6 strain. We recently demonstrated that *Eng*$^{+/-}$ C57BL/6 mice display an endothelium-dependent abnormality in arterial vasomotor function, revealed in response to experimentally induced elevations in intravascular pressure (51). Resistance arteries have a reduced myogenic response or ability to contract in response to rises in perfusion pressure and may fail to limit the transmission of excessive hemodynamic stress to downstream more fragile vascular structures. This finding indicates that *Eng* haploinsufficiency is a predisposing factor, and that epigenetic or environmental triggers resulting in specific alterations in endothelial regulation of local blood flow and tone may be required to initiate clinical manifestations of HHT.

Alk1$^{+/-}$ or HHT2 Mouse

As seen in HHT2 patients, *Alk1*$^{+/-}$ mice also display age-dependent vascular lesions in the skin, oral cavity, and several internal organs (71). Interestingly, some mice have a grossly enlarged liver and high cardiac output indicative of a hyperdynamic condition from a hepatic arteriovenous shunt, usually leading to increased pulmonary pressure and/or heart failure. Unfortunately, vasomotor studies, similar to those done on *Eng*$^{+/-}$ mice (51), have not yet been performed on *Alk1*$^{+/-}$ mice. Future corroboration of similar vasomotor findings in *Alk1*$^{+/-}$ mice would not only confirm the importance of early vasoregulatory alterations in HHT pathogenesis, but would also provide new insight into novel mechanisms involved in the regulation of vascular tone by endoglin, ALK1, and their associated signaling partners.

MECHANISMS OF HEREDITARY HEMORRHAGIC TELANGIECTASIA PATHOGENESIS

HHT is a focal disease predominantly affecting the microvasculature of specific organs including the lungs, brain, liver, and skin. This can be explained by a model of pathogenesis that postulates that individuals harboring *ENG* or *Alk1* mutations have an intrinsic defect in their vascular endothelium that predisposes them to the development of vascular lesions, given the proper environmental trigger, such as sudden changes in perfusion pressure.

At first glance, the analysis of lesion development in HHT patients suggests that vascular malformations originate from inherently abnormal and/or dysfunctional postcapillary venules. However, the presence of stress fibers in pericytes in early postcapillary venular dilatations and excessive, uneven layers of smooth muscle cells in fully developed telangiectases and AVMs are suggestive of hemodynamic stress-induced venular remodeling. Such observations, along with our finding of impaired arterial regulation of local vascular tone in $Eng^{+/-}$ mice, suggest an alternative model of HHT pathogenesis that is more compatible with our current understanding of vascular physiology.

The autoregulation of local blood flow and pressure is inherent to resistance vessels. The arterial myogenic response maintains constant tissue blood flow despite potentially wide fluctuations in perfusion pressure, and thus preserves the structural integrity of the microcirculation (74). The demonstration by Toporsian and colleagues that $Eng^{+/-}$ (HHT1) resistance arteries have an endothelium-dependent

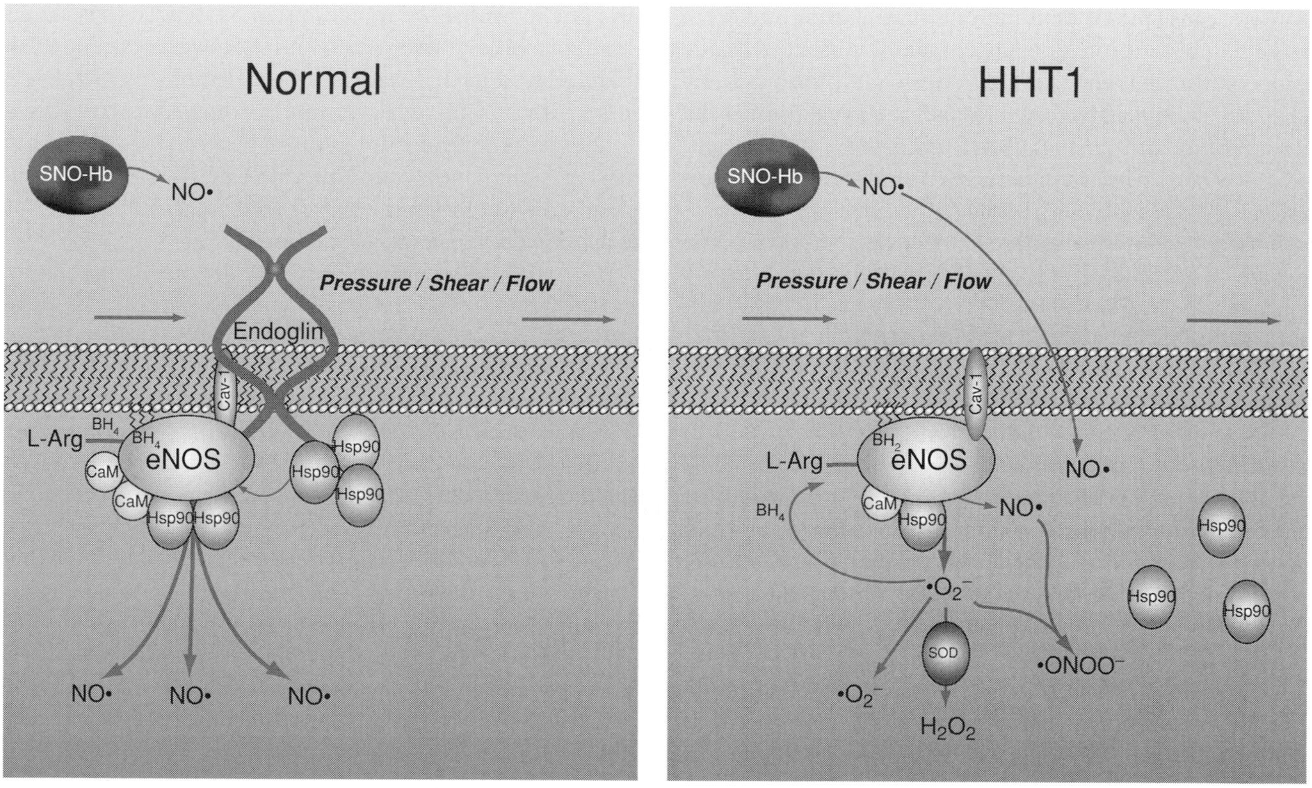

Figure 122.3. Model of HHT1 pathogenesis. A pool of endoglin resides in caveolae, where it stabilizes eNOS and acts as a molecular scaffold to bring eNOS and HSP90 in close proximity to one another, facilitating their association during EC activation. Reduced eNOS/HSP90 association in HHT1 leads to increased eNOS-derived O_2^- and the formation of H_2O_2 (via superoxide dismutase, SOD) and $ONOO^-$, which hyperpolarize smooth muscle and impair vascular contractions. Under such circumstances, resistance arteries do not adequately contract in response to sudden rises in intraluminal pressure and would thus fail to limit the transmission of excessive and potentially damaging hydrostatic pressures to downstream structures. This could contribute to the genesis of early HHT1-type vascular lesions such as capillary loss and venular dilatations. Compounds such as O_2^- or $ONOO^-$ can oxidize the essential NOS cofactor, tetrahydrobiopterin (BH_4), to dihydrobiopterin (BH_2), further uncoupling eNOS in HHT1. The preferential release of NO from *S*-nitroso-hemoglobin (SNO-Hb) in the microvasculature would also favor the local production of $ONOO^-$ and may explain the focal nature of HHT1. (Adapted with permission from Toporsian M, Gros R, Kabir MG, et al. A role for endoglin in coupling eNOS activity and regulating vascular tone revealed in hereditary hemorrhagic telangiectasia. *Circ Res.* 2005;96:684–692.)

impairment in their ability to contract in response to elevations in intravascular pressure suggests that previous reports of capillary loss and abnormal venular dilatations may indeed be consequences of a preceding functional inability of upstream resistance arteries to protect downstream, more fragile vascular structures from potentially damaging hemodynamic stress (50). The reported venular remodeling in HHT lesions is therefore probably a consequence of alterations in blood flow patterns and exposure to elevated perfusion pressures transmitted by functionally compromised upstream arterioles. This new model of HHT pathogenesis suggests that sudden increases in perfusion pressure would not be well tolerated in patients, providing a possible explanation for the frequent nosebleeds associated with the disease. It also predicts that further impairment of a compromised vasoregulatory mechanism by additional stresses such as hypoxia and inflammation will contribute to disease manifestations. We postulate that individual responsiveness to environmental stresses, related to specific polymorphisms (in putative modifier genes), may account in part for heterogeneity in age-related penetrance and clinical manifestations. The observed impairment in arterial myogenic responsiveness may lead to the generation of AVMs, as supported by a mathematical model of microcirculatory hemodynamics (76). Given the phenotypic similarities in vascular lesions observed in HHT1 and HHT2 patients, future studies demonstrating similar arterial vasomotor abnormalities in $Alk1^{+/-}$ mice would provide further support for this new model, which contends that endothelium-dependent alterations in the autoregulation of local blood flow and pressure may be the underlying mechanism of HHT pathogenesis. The predominant expression of ALK1 on arterial, compared to capillary and venular, ECs in adult mice would support this model (54).

HOW CAN ENDOGLIN AFFECT VASCULAR TONE?

The main factors involved in the minute-to-minute adjustments in vasomotor control are predominantly endothelium-derived and diffusible. They directly affect vascular smooth muscle contractility in response to sudden changes in neurohumoral and hemodynamic stimuli. If early events in the pathogenesis of HHT involve abnormal vasomotor function, endoglin (and potentially ALK1) may be critical in balancing the secretion of endothelium-derived vasoconstrictors and vasodilators, a central mechanism in the local regulation of blood flow and vascular tone. Indeed, $Eng^{+/-}$ mice display an eNOS-dependent impairment in arterial vasoreactivity, despite reduced levels of this enzyme (51). eNOS activity appears to be uncoupled in *Eng*-null cells as evidenced by severely reduced eNOS/HSP90 association and increased eNOS-derived superoxide (O_2^-). This suggests a potential scaffolding-like function for endoglin, in which its cytoplasmic tail brings HSP90 into close proximity with caveolar eNOS to ensure NO production. Furthermore, the vasomotor abnormalities in isolated mesenteric arteries are reversed by the use of antioxidants, indicating that the products of eNOS-derived O_2^- lead to the generation of endothelium-derived hyperpolarizing factors that impair smooth muscle contractions and vasomotor tone (51). Recently, Blanco and colleagues have demonstrated that both the extracellular and cytoplasmic domains of endoglin interact with ALK1 (68) suggesting the possibility that eNOS also may be part of this complex. Given the similarities in disease phenotypes between HHT1 and HHT2, future demonstration of ALK1 associating with eNOS and modulating its activation would thereby identify a common biochemical pathway in both disease subtypes and lend support for a causal role of abnormal eNOS activation in the generation of vascular malformations. Figure 122.3 illustrates the association of endoglin with eNOS/HSP90 and the postulated mechanism of HHT1 pathogenesis.

KEY POINTS

- *ENG* and *ALK1* mutations are the underlying cause of HHT, a disorder characterized by venous dilatations, focal loss of capillaries, and AVMs.
- Endoglin interacts with eNOS and modulates eNOS stability and activation, contributing to the local regulation of vascular tone.
- Early events in HHT pathogenesis may be due to abnormal vasomotor tone, subsequent to the uncoupling of eNOS activity and generation of superoxide.
- Antioxidant therapy may prevent or delay the development of vascular lesions in patients with HHT1.

Future Goals

- To define the protein interaction networks associated with endoglin, ALK1, and eNOS to elucidate the pathways critical for normal endothelial function and those defective in HHT and related disorders

REFERENCES

1 McAllister KA, Grogg KM, Johnson DW, et al. Endoglin, a TGF-beta binding protein of endothelial cells, is the gene for hereditary haemorrhagic telangiectasia type 1. *Nat Genet.* 1994;8:345–351.
2 Berg JN, Gallione CJ, Stenzel TT, et al. The activin receptor-like kinase 1 gene: genomic structure and mutations in hereditary hemorrhagic telangiectasia type 2. *Am J Hum Genet.* 1997;61:60–67.
3 Li DY, Sorensen LK, Brooke BS, et al. Defective angiogenesis in mice lacking endoglin. *Science.* 1999;284:1534–1537.
4 Bourdeau A, Dumont DJ, Letarte M. A murine model of hereditary hemorrhagic telangiectasia. *J Clin Invest.* 1999;104:1343–1351.

5 Arthur HM, Ure J, Smith AJ, et al. Endoglin, an ancillary TGF-beta receptor, is required for extraembryonic angiogenesis and plays a key role in heart development. *Dev Biol.* 2000;217:42–53.

6 Oh SP, Seki T, Goss KA, et al. Activin receptor-like kinase 1 modulates transforming growth factor-beta 1 signaling in the regulation of angiogenesis. *Proc Natl Acad Sci U S A.* 2000;97:2626–2631.

7 Rendu H. Epistaxis répétées chez un sujet porteur de petits angiomes cutanés et muqueux. *Gazette des Hôpitaux Civils et Militaires.* 1896:1322–1323.

8 Osler W. On a family form of recurring epistaxis, associated with multiple telangiectases of the skin and mucous membranes. *Bull Johns Hopkins Hosp.* 1901:333–337.

9 Weber F. Multiple hereditary developmental angiomata (telangiectases) of the skin and mucous membranes associated with recurring haemorrhages. *Lancet*. 1907:160–162.

10 Gallione CJ, Scheessele EA, Reinhardt D, et al. Two common endoglin mutations in families with hereditary hemorrhagic telangiectasia in the Netherlands Antilles: evidence for a founder effect. *Hum Genet*. 2000;107:40–44.

11 Plauchu H, de Chadarevian JP, Bideau A, Robert JM. Age-related clinical profile of hereditary hemorrhagic telangiectasia in an epidemiologically recruited population. *Am J Med Genet.* 1989;32:291–297.

12 Shovlin CL, Guttmacher AE, Buscarini E, et al. Diagnostic criteria for hereditary hemorrhagic telangiectasia (Rendu-Osler-Weber syndrome). *Am J Med Genet.* 2000;91:66–67.

13 Braverman IM, Keh A, Jacobson BS. Ultrastructure and three-dimensional organization of the telangiectases of hereditary hemorrhagic telangiectasia. *J Invest Dermatol.* 1990;95:422–427.

14 Berg JN, Guttmacher AE, Marchuk DA, Porteous ME. Clinical heterogeneity in hereditary haemorrhagic telangiectasia: are pulmonary arteriovenous malformations more common in families linked to endoglin? *J Med Genet.* 1996;33:256–257.

15 Vase P, Holm M, Arendrup H. Pulmonary arteriovenous fistulas in hereditary hemorrhagic telangiectasia. *Acta Med Scand.* 1985;218:105–109.

16 White RI, Jr., Lynch-Nyhan A, Terry P, et al. Pulmonary arteriovenous malformations: techniques and long-term outcome of embolotherapy. *Radiology.* 1988;169:663–669.

17 Ference BA, Shannon TM, White RI, Jr., et al. Life-threatening pulmonary hemorrhage with pulmonary arteriovenous malformations and hereditary hemorrhagic telangiectasia. *Chest.* 1994;106:1387–1390.

18 Lincoln MJ, Shigeoka JW. Pulmonary telangiectasia without hypoxemia. *Chest.* 1988;93:1097–1098.

19 Jaskolka J, Wu L, Chan RP, Faughnan ME. Imaging of hereditary hemorrhagic telangiectasia. *Am J Roentgenol.* 2004;183:307–314.

20 White RI, Jr., Pollak JS, Wirth JA. Pulmonary arteriovenous malformations: diagnosis and transcatheter embolotherapy. *J Vasc Interv Radiol.* 1996;7:787–804.

21 Haitjema T, Westermann CJ, Overtoom TT, et al. Hereditary hemorrhagic telangiectasia (Osler-Weber-Rendu disease): new insights in pathogenesis, complications, and treatment. *Arch Intern Med.* 1996;156:714–719.

22 Brown RD, Jr., Flemming KD, Meyer FB, et al. Natural history, evaluation, and management of intracranial vascular malformations. *Mayo Clin Proc.* 2005;80:269–281.

23 Matsubara S, Mandzia JL, ter Brugge K, et al. Angiographic and clinical characteristics of patients with cerebral arteriovenous malformations associated with hereditary hemorrhagic telangiectasia. *Am J Neuroradiol.* 2000;21:1016–1020.

24 Buscarini E, Danesino C, Olivieri C, et al. Doppler ultrasonographic grading of hepatic vascular malformations in hereditary hemorrhagic telangiectasia – results of extensive screening. *Ultraschall Med.* 2004;25:348–355.

25 Sawabe M, Arai T, Esaki Y, et al. Three-dimensional organization of the hepatic microvasculature in hereditary hemorrhagic telangiectasia. *Arch Pathol Lab Med.* 2001;125:1219–1223.

26 Buscarini E, Danesino C, Olivieri C, et al. Liver involvement in hereditary haemorrhagic telangiectasia or Rendu-Osler-Weber disease. *Dig Liver Dis.* 2005;37:635–645.

27 Letteboer TG, Mager JJ, Snijder RJ, et al. Genotype-phenotype relationship in hereditary naemorrhagic telangiectasia. *J Med Genet.* 2006;43:371–377.

28 Cole SG, Begbie ME, Wallace GM, Shovlin CL. A new locus for hereditary haemorrhagic telangiectasia (HHT3) maps to chromosome 5. *J Med Genet.* 2005;42:577–582.

29 Gallione CJ, Repetto GM, Legius E, et al. A combined syndrome of juvenile polyposis and hereditary haemorrhagic telangiectasia associated with mutations in MADH4 (SMAD4). *Lancet.* 2004;363:852–859.

30 Abdalla SA, Letarte M. Hereditary haemorrhagic telangiectasia: current views on genetics and mechanisms of disease. *J Med Genet.* 2006;43:97–110.

31 Pece N, Vera S, Cymerman U, et al. Mutant endoglin in hereditary hemorrhagic telangiectasia type 1 is transiently expressed intracellularly and is not a dominant negative. *J Clin Invest.* 1997;100:2568–2579.

32 Bourdeau A, Cymerman U, Paquet ME, et al. Endoglin expression is reduced in normal vessels but still detectable in arteriovenous malformations of patients with hereditary hemorrhagic telangiectasia type 1. *Am J Pathol.* 2000;156:911–923.

33 Abdalla SA, Pece-Barbara N, Vera S, et al. Analysis of ALK-1 and endoglin in newborns from families with hereditary hemorrhagic telangiectasia type 2. *Hum Mol Genet.* 2000;9:1227–1237.

34 Abdalla SA, Cymerman U, Johnson RM, et al. Disease-associated mutations in conserved residues of ALK-1 kinase domain. *Eur J Hum Genet.* 2003;11:279–287.

35 Sweet K, Willis J, Zhou XP, et al. Molecular classification of patients with unexplained hamartomatous and hyperplastic polyposis. *JAMA.* 2005;294:2465–2473.

36 Deng Z, Morse JH, Slager SL, et al. Familial primary pulmonary hypertension (gene PPH1) is caused by mutations in the bone morphogenetic protein receptor-II gene. *Am J Hum Genet.* 2000;67:737–744.

37 Harrison RE, Berger R, Haworth SG, et al. Transforming growth factor-beta receptor mutations and pulmonary arterial hypertension in childhood. *Circulation.* 2005;111:435–441.

38 Trembath RC, Thomson JR, Machado RD, et al. Clinical and molecular genetic features of pulmonary hypertension in patients with hereditary hemorrhagic telangiectasia. *N Engl J Med.* 2001;345:325–334.

39 Quackenbush EJ, Letarte M. Identification of several cell surface proteins of non-T, non-B acute lymphoblastic leukemia by using monoclonal antibodies. *J Immunol.* 1985;134:1276–1285.

40 Gougos A, Letarte M. Identification of a human endothelial cell antigen with monoclonal antibody 44G4 produced against a pre-B leukemic cell line. *J Immunol.* 1988;141:1925–1933.

41 St-Jacques S, Forte M, Lye SJ, Letarte M. Localization of endoglin, a transforming growth factor-beta binding protein, and of CD44

and integrins in placenta during the first trimester of pregnancy. *Biol Reprod.* 1994;51:405–413.

42 Letarte M, Greaves A. and Vera S. CD105 (endoglin) Cluster Report. In: S.F. Schlossman LB, W. Gilks, J.M. Harlan, et al. eds. *Leukocyte Typing V: white cell differentiation antigens.* New York: Oxford University Press, 1995:1756–1759.

43 Fonsatti E, Sigalotti L, Arslan P, et al. Emerging role of endoglin (CD105) as a marker of angiogenesis with clinical potential in human malignancies. *Curr Cancer Drug Targets.* 2003;3:427–432.

44 Seon BK, Matsuno F, Haruta Y, et al. Long-lasting complete inhibition of human solid tumors in SCID mice by targeting endothelial cells of tumor vasculature with antihuman endoglin immunotoxin. *Clin Cancer Res.* 1997;3:1031–1044.

45 Gougos A, Letarte M. Primary structure of endoglin, an RGD-containing glycoprotein of human endothelial cells. *J Biol Chem.* 1990;265:8361–8364.

46 Lastres P, Martin-Perez J, Langa C, Bernabeu C. Phosphorylation of the human-transforming-growth-factor-beta-binding protein endoglin. *Biochem J.* 1994;301(Pt 3):765–768.

47 Lastres P, Letamendia A, Zhang H, et al. Endoglin modulates cellular responses to TGF-beta 1. *J Cell Biol.* 1996;133:1109–1121.

48 Cheifetz S, Bellon T, Cales C, et al. Endoglin is a component of the transforming growth factor-beta receptor system in human endothelial cells. *J Biol Chem.* 1992;267:19027–19030.

49 Lopez-Casillas F, Riquelme C, Perez-Kato Y, et al. Betaglycan expression is transcriptionally up-regulated during skeletal muscle differentiation. Cloning of murine betaglycan gene promoter and its modulation by MyoD, retinoic acid, and transforming growth factor-beta. *J Biol Chem.* 2003;278:382–390.

50 Barbara NP, Wrana JL, Letarte M. Endoglin is an accessory protein that interacts with the signaling receptor complex of multiple members of the transforming growth factor-beta superfamily. *J Biol Chem.* 1999;274:584–594.

51 Toporsian M, Gros R, Kabir MG, et al. A role for endoglin in coupling eNOS activity and regulating vascular tone revealed in hereditary hemorrhagic telangiectasia. *Circ Res.* 2005;96:684–692.

52 Conley BA, Koleva R, Smith JD, et al. Endoglin controls cell migration and composition of focal adhesions: function of the cytosolic domain. *J Biol Chem.* 2004;279:27440–27449.

53 Sanz-Rodriguez F, Guerrero-Esteo M, Botella LM, et al. Endoglin regulates cytoskeletal organization through binding to ZRP-1, a member of the Lim family of proteins. *J Biol Chem.* 2004;279: 32858–32868.

54 Seki T, Yun J, Oh SP. Arterial endothelium-specific activin receptor-like kinase 1 expression suggests its role in arterialization and vascular remodeling. *Circ Res.* 2003;93:682–689.

55 ten Dijke P, Ichijo H, Franzen P, et al. Activin receptor-like kinases: a novel subclass of cell-surface receptors with predicted serine/threonine kinase activity. *Oncogene.* 1993;8:2879–2887.

56 Shi Y, Massague J. Mechanisms of TGF-beta signaling from cell membrane to the nucleus. *Cell.* 2003;113:685–700.

57 Dickson MC, Martin JS, Cousins FM, et al. Defective haematopoiesis and vasculogenesis in transforming growth factor-beta 1 knock out mice. *Development.* 1995;121:1845–1854.

58 Goumans MJ, Valdimarsdottir G, Itoh S, et al. Activin receptor-like kinase (ALK)1 is an antagonistic mediator of lateral TGF-beta/ALK5 signaling. *Mol Cell.* 2003;12:817–828.

59 Sanford LP, Ormsby I, Gittenberger-de Groot AC, et al. TGF-beta2 knockout mice have multiple developmental defects that are non-overlapping with other TGFbeta knockout phenotypes. *Development.* 1997;124:2659–2670.

60 Proetzel G, Pawlowski SA, Wiles MV, et al. Transforming growth factor-beta 3 is required for secondary palate fusion. *Nat Genet.* 1995;11:409–414.

61 Oshima M, Oshima H, Taketo MM. TGF-beta receptor type II deficiency results in defects of yolk sac hematopoiesis and vasculogenesis. *Dev Biol.* 1996;179:297–302.

62 Urness LD, Sorensen LK, Li DY. Arteriovenous malformations in mice lacking activin receptor-like kinase-1. *Nat Genet.* 2000;26:328–331.

63 Larsson J, Goumans MJ, Sjostrand LJ, et al. Abnormal angiogenesis but intact hematopoietic potential in TGF-beta type I receptor-deficient mice. *EMBO J.* 2001;20:1663–1673.

64 Lechleider RJ, Ryan JL, Garrett L, et al. Targeted mutagenesis of Smad1 reveals an essential role in chorioallantoic fusion. *Dev Biol.* 2001;240:157–167.

65 Chang H, Huylebroeck D, Verschueren K, et al. Smad5 knockout mice die at mid-gestation due to multiple embryonic and extraembryonic defects. *Development.* 1999;126:1631–1642.

66 Goumans MJ, Lebrin F, Valdimarsdottir G. Controlling the angiogenic switch: a balance between two distinct TGF-b receptor signaling pathways. *Trends Cardiovasc Med,* 2003;13:301–307.

67 Lebrin F, Goumans MJ, Jonker L, et al. Endoglin promotes endothelial cell proliferation and TGF-beta/ALK1 signal transduction. *EMBO J.* 2004;23:4018–4028.

68 Blanco FJ, Santibanez JF, Guerrero-Esteo M, et al. Interaction and functional interplay between endoglin and ALK-1, two components of the endothelial transforming growth factor-beta receptor complex. *J Cell Physiol.* 2005;204:574–584.

69 Pece-Barbara N, Vera S, Kathirkamathamby K, et al. Endoglin null endothelial cells proliferate faster and are more responsive to transforming growth factor beta1 with higher affinity receptors and an activated Alk1 pathway. *J Biol Chem.* 2005;280:27800–27808.

70 Satomi J, Mount RJ, Toporsian M, et al. Cerebral vascular abnormalities in a murine model of hereditary hemorrhagic telangiectasia. *Stroke.* 2003;34:783–789.

71 Srinivasan S, Hanes MA, Dickens T, et al. A mouse model for hereditary hemorrhagic telangiectasia (HHT) type 2. *Hum Mol Genet.* 2003;12:473–482.

72 Jerkic M, Rivas-Elena JV, Prieto M, et al. Endoglin regulates nitric oxide-dependent vasodilatation. *FASEB J.* 2004;18:609–611.

73 Carvalho RL, Jonker L, Goumans MJ, et al. Defective paracrine signalling by TGFbeta in yolk sac vasculature of endoglin mutant mice: a paradigm for hereditary haemorrhagic telangiectasia. *Development.* 2004;131:6237–6247.

74 Bourdeau A, Faughnan ME, McDonald ML, et al. Potential role of modifier genes influencing transforming growth factor-beta1 levels in the development of vascular defects in endoglin heterozygous mice with hereditary hemorrhagic telangiectasia. *Am J Pathol.* 2001;158:2011–2020.

75 Johnson PC. The myogenic response in the microcirculation and its interaction with other control systems. *J Hypertens Suppl.* 1989;7:S33–39; discussion S40.

76 Quick CM, Hashimoto T, Young WL. Lack of flow regulation may explain the development of arteriovenous malformations. *Neurol Res.* 2001;23:641–644.

123

Blood–Brain Barrier

Christian Weidenfeller and Eric V. Shusta

University of Wisconsin-Madison

The vascular endothelial cells (ECs) that separate the bloodstream from the brain interior are so impermeable that the brain vasculature is often referred to as the blood–brain barrier (BBB). As a result of its barrier properties, the BBB plays an extremely important role in central nervous system (CNS) homeostasis by protecting neurons from fluctuations in blood composition and from toxic blood-borne substances. Except for small regions of the brain, the barrier properties are found throughout the entire brain and include capillaries, arterioles, and venules as well as larger arteries and veins. As the vessels advance from the meningeal membranes that envelop the brain and into the brain matter, vascular permeability decreases rapidly and barrier-like impermeability arises. Although the BBB permeability for many substances is highly restricted, a multiplicity of molecular transport systems exist to provide the brain with necessary nutrients. The same barrier function that is critical for normal brain function also provides a formidable hurdle for delivery of therapeutics to the brain under conditions of neurological disease.

Although the endothelium provides the barrier properties of the BBB, it is the local brain microenvironment that elicits the unique phenotype. Vascular smooth muscle cells line precapillary arterioles, pericytes share a basement membrane with capillary ECs, astrocytes ensheathe the microvessels, and nerve terminals contact the endothelium. Together with the endothelium, these perivascular cell types constitute the so-called *neurovascular unit*. In this chapter, we review the microenvironmental cues that function in BBB maintenance, describe the BBB phenotype in detail, and highlight the role for the BBB in brain diseases such as stroke, multiple sclerosis (MS), and cerebral malaria. Finally, we outline the limitations that the BBB imposes on drug delivery, and introduce strategies that promise to bridge the bench-to-bedside gap for those suffering with debilitating neurological diseases.

DEVELOPMENT OF THE BLOOD–BRAIN BARRIER CONCEPT

The concept of a barrier system existing in the brain arose during the late 19th century. Bacteriologist Paul Ehrlich observed that certain aniline dyes, when administered intravenously to small animals, stained all organs except the brain. Ehrlich's interpretation of this observation was that the brain had a lower affinity for the injected dyes. A few years later, Roux and Borrel observed that tetanus toxin injected into cerebrospinal fluid (CSF) caused marked cerebral symptoms, but when administered intravenously produced no discernible cerebral effect. In 1913, Goldman injected the dye trypan blue directly into the CSF of rabbits and dogs and found that the dye readily stained the whole brain but did not enter the bloodstream and thus did not stain other organs. In this way, Goldman again showed that the CNS was separated from the blood by a barrier of some kind (1,2). The correlated findings of these investigators were attributed to special permeability properties of small brain blood vessels. However, the hypothesis that brain endothelium provides the anatomical basis of the observed barrier was not confirmed until the advent of electron microscopy allowed the high-resolution analysis of brain ultrastructure.

Using electron microscopy, Brightman and Reese demonstrated that electron-dense junctions were present between adjacent brain ECs (3). The existence of such tight junctions was used to explain the absence of horseradish peroxidase or microperoxidase in the brain after intravenous injection. The barrier to these molecules was observed regardless of vessel size or association with smooth muscle, confirming the presence of a barrier throughout the entire brain (4).

BLOOD–BRAIN BARRIER MORPHOLOGY AND ROLE OF THE MICROENVIRONMENT

When the endothelium in the brain is compared with non-brain endothelium, its general lack of permeability stands out as unique. This BBB-defining characteristic is a consequence of the local brain microenvironment. During brain development, newly formed vessels derive from the perineural plexus by angiogenesis. Thus, vessels in the brain are not formed by brain-specific precursors, but develop their BBB properties only after induction by the neural microenvironment (5).

Cross-talk between the ECs and the perivascular brain tissue is therefore essential for the development and maintenance of BBB properties. As further evidence for the role of the microenvironment, when neural tissue was grafted to peripheral sites, vessels invading the neural graft developed BBB properties (6,7). Conversely, when brain vessels invaded peripheral tissue that had been grafted into the brain, no barrier characteristics were observed (7). It is important to recognize that the cross-talk between ECs and perivascular cells is bidirectional. However, for purposes of this discussion, we focus primarily on the flow of information from the microenvironment to the endothelium.

Endothelium

ECs of the brain microvasculature have a different phenotype compared with nonbrain ECs. They are thin, measuring a mere 100 to 300 nm between luminal and abluminal compartments, and are connected by tight interendothelial junctions. Further distinguishing features of cerebral microvascular ECs include the paucity of pinocytotic activity (clathrin-mediated or caveolae-mediated) (3,8,9), high number of mitochondria (10), and the absence of fenestrations (11). The presence of tight junctions and the absence of fenestrae in brain ECs yields a high transendothelial electrical resistance (TEER) as a result of a reduced paracellular flux of ions and small charged molecules in vivo (12). These capillary features are a direct result of the local brain microenvironment. A schematic of the brain capillary and its association with the neurovascular unit is shown in Figure 123.1. The capillary lumen is lined by ECs that are connected with tight junctions at the sites of intercellular clefts. These junctions seal the clefts against even the smallest molecules and convert the EC layer into a closed interface that physically separates the blood from the brain interstitium. The phenotype of BBB endothelium is described in more detail below.

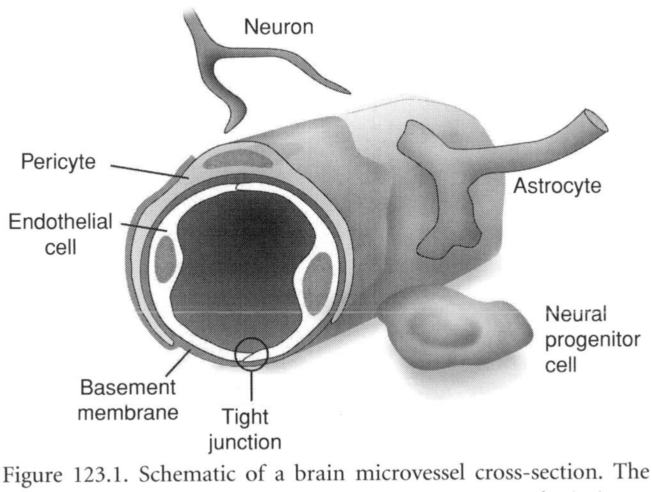

Figure 123.1. Schematic of a brain microvessel cross-section. The cellular composite forms the neurovascular unit. Note the intimate interaction between ECs and pericytes that share a basement membrane. In addition, astrocytes and neurons contact the microvessel surface. Tight junctions form between adjacent ECs. Neural stem or progenitor cells also are found in the proximity of brain microvessels.

Extracellular Matrix

The abluminal side of the BBB endothelium is surrounded by a robust basement membrane that also encloses pericytes and is contacted intimately by astrocytes. This basement membrane consists of a well-characterized composite of extracellular matrix (ECM) molecules (13). The basement membrane is composed of the ECM molecules collagen, laminin, fibronectin, entactin, heparin sulfate proteoglycans, and chondroitin sulfate proteoglycans. Brain microvascular ECs have improved growth characteristics and are more representative of the in vivo situation in terms of permeability and other BBB properties when cells are plated on collagen (14–17), fibronectin (18–21), pronectin F (22), and/or ECM secreted by astrocytes (18). The interactions between all cells of the neurovascular unit and the ECM initiate information flow that helps regulate cell proliferation, differentiation, and polarization of ECs. The ECM also anchors a variety of molecules important in normal brain function such as integrins, growth factors, laminin, and other glycoproteins (23).

Neurons

Early during embryonic development, neurons are in close contact with developing blood vessels. Although neurons appear early on in the developing brain, astrocytes encase vessels with their end-feet only later in postnatal development (24,25). It also has been observed that neuronal processes directly contact the basement membranes of ECs in the adult brain (26,27). In vitro studies provide evidence for neuronal involvement in BBB induction. Neurons influence endothelial occludin expression and sorting (28), induce biochemical activity in the form of γ-glutamyl transpeptidase (GTP) enzymatic activity that functions in mediating glucose uptake (29) and catabolism of γ-glutamyl compounds (30), and increase Na^+/K^+ ATPase activity (31). In addition, when cocultured with astrocytes and neurons, ECs were shown to support a serotoninergic phenotype in the neurons (32).

Astrocytes

Astrocytes are one type of glial cell found in the brain whose foot processes encase most of the BBB endothelial surface in vivo. Many studies have shown that astrocytes are strong inducers of BBB properties both in vivo and in vitro. Grafts of astrocytes can induce BBB-like properties in peripheral endothelium (33). In addition, when astrocytes that were positive for the intermediate filament glial fibrillary acidic protein (GFAP) are selectively ablated at the site of invasive CNS injury, the BBB remains impaired and permeable. This effect can be countered by grafting normal astrocytes that foster the repair and restoration of BBB impermeability at the injury site (34). Transgenic mice with astrocytes deficient in GFAP have an impaired BBB (35), and the GFAP-negative astrocytes are unable to induce a functional BBB in vitro, compared with normal astrocytes (36). Based on these data and the fact that GFAP is upregulated after brain injury, it has been speculated that GFAP may be an important factor for BBB induction, repair, and restoration.

Culturing with astrocyte-conditioned medium (37), astrocyte plasma membranes (31), or astrocytes either in direct contact on opposite sides of a permeable filter (15,38–40) or in indirect contact through a diffusion apparatus (37,41) is critical for the permeability, morphological, and biochemical characteristics of in vitro BBB models. Several transport systems are upregulated in brain endothelial models exposed to glial influence. Upregulation has been documented for the glucose transporter (GLUT)-1, the L- and A-amino acid carrier systems, and P-glycoprotein (25). Astrocyte influences can also induce γ-GTP (39,40,42) and alkaline phosphatase activities (38,43), stimulate growth and glucose utilization (22), and influence the polarity of P-glycoprotein (44). In addition, astrocytes increase the electrical resistance of EC monolayers (39,45), increase the level of peripheral versus distributed actin (19,37), contribute to the complexity and extent of tight junction morphology in vitro (38), and decrease the passive diffusion of impermeable molecules.

Pericytes

Pericytes are perivascular cells associated with capillaries and postcapillary venules. They are embedded in the basement membrane of the vessels and extend long cytoplasmic processes around the capillary tubes, forming so-called *peg and socket* contacts (46). Pericyte coverage of microvessels varies among vascular beds, and it has been estimated that pericytes invest 22% to 32% of rat brain microvessel surfaces (47). Three major functional roles have been ascribed to pericytes associated with CNS microvasculature. They include contractility, regulation of EC activity, and a role in inflammation. In brain and retina, the adhesive glycoprotein fibronectin has been observed at junctional sites between pericytes and ECs adjacent to adhesion plaques at the pericyte's plasma membrane (48), suggesting a mechanical linkage between the two cells. Two frequently used endothelial markers, angiopoietin and γ-GTP, are also present in pericytes. The presence of γ-GTP in brain pericytes supports the idea that, in addition to ECs, pericytes are important for BBB function (49,50). In vitro experiments using pericytes cocultured with ECs and astrocytes indicated that pericytes may be involved in capillary formation, but only when in the presence of astrocytes (51). Furthermore, pericytes have been shown to exhibit phagocytotic activity and to express macrophage markers, suggesting that pericytes are involved in the immune response at the BBB (52).

Neural Stem Cells

Yet another cell type within the brain is the neural stem cell (NSC). These cells give rise to neurons, astrocytes, and oligodendrocytes, and can be found in both the developing and adult brains (53). The regulation of NSC differentiation has not yet been completely clarified, but contact with blood vessels during development may influence their fate towards astrocytic differentiation (24), and therefore NSCs might be important for induction of BBB properties in ECs. In addition, ECs in coculture with NSCs can enhance neuronal differentiation and affect rates of self-renewal (54). Also in adult brain, NSCs are found in close apposition to the vasculature in the so-called *neural stem cell niches* of the hippocampus and subventricular zone (55). Research regarding the interactions between NSCs and ECs is in its nascent stage, but preliminary findings suggest an exciting role for the brain vasculature in the regulation of NSCs, and hence, in the regulation of brain development and repair.

Soluble Mediators

The BBB phenotype is clearly derived from physical interactions with the underlying matrix and perivascular cells; however, ECs also are responsive to soluble mediators released by the perivascular cells. Unfortunately, the identity of these components is largely unknown. A protein released from astrocytes that has been definitively implicated in the development of barrier function is glial cell line-derived neurotrophic factor (GDNF) (56). Expression of src-suppressed C-kinase substrate (SSeCKS) in astrocytes triggers angiopoietin-1 secretion that can mediate an endothelial barrier tightening in vitro, and SSeCKS positive astrocytes can be found closely interacting with brain microvessels in vivo (57). Also, the use of forskolin to elevate cellular cyclic adenosine monophosphate (cAMP) levels increases the tight junctional complexity and P-face association of the tight junction particles (58). Direct supplementation of cAMP has also been shown to increase barrier function (37) and alkaline phosphatase expression (43). Finally, when hydrocortisone is added to EC cultures, the transendothelial electrical resistance increases dramatically, and permeability to sucrose is reduced to near in vivo levels (59). Taken together, studies have shown that an intact microenvironment is essential for the proper function of the BBB. Thus, to fully understand regulation at the blood–brain interface, it is absolutely necessary to unscramble the cross-talk present between elements of the neurovascular unit.

BLOOD–BRAIN BARRIER FUNCTION

Generally speaking, the BBB performs three major functions. First, the BBB protects the brain from the blood milieu by virtue of its tight junctions. Although ions, small molecules, peptides, and proteins can pass relatively unrestricted through the endothelial lining in other vascular beds, the brain vasculature lacks fenestrae and has epithelial-like tight junctions. These tight junctions help restrict the uncontrolled passage of hydrophilic substances from the blood to the brain interstitium. In addition, the BBB displays only low levels of pinocytotic transport, which together with tight junctions help maintain brain homeostasis. Second, the BBB acts as a transport/metabolic barrier that removes metabolites and toxic substances from the brain, and also provides the brain with necessary nutrients. For this reason, BBB ECs contain a multiplicity of transport proteins that regulate the bidirectional transport of nutrients and metabolites into and out of the brain. In addition, efflux transporters remove potentially harmful compounds that enter the ECs due to their high lipophilicity. These transporters are also known as *multidrug resistance proteins*, because during disease conditions, they can dramatically lessen the brain uptake and effectiveness of drug treatments. Finally, the BBB acts as a signalling interface that functions in developmental and immune responses, among others. The origins and implications of the first two functional attributes are discussed in detail in the following sections. The role of the BBB in innate immunity is covered elsewhere in this volume (Chapter 124).

Physical Barrier

Molecular Architecture of Brain Endothelial Cell Junctions

A striking structural feature of the BBB is the tight junctional complex that provides the molecular basis for the permeability barrier. Tight junctions form a continuous belt around the ECs and seal the paracellular space (4). In electron micrographs of freeze fracture replicas of the plasma membrane, the tight junctions appear as a network of small ridges on one side of the membrane bilayer and complementary grooves on the opposite side of the bilayer (60,61). The molecular structure of the tight junctions has not been completely elucidated, but there is unequivocal evidence for both lipid and protein contributions.

The barrier phenotype is present throughout the entire brain vasculature except in small regions, known as *circumventricular organs*, that are perfused by capillaries without BBB properties and participate in neuroendocrine regulation. In a developing rodent, the brain vasculature is fenestrated until embryonic days (E)11 to E13, during which time tight junctions develop, as the fenestrations typical of peripheral capillaries disappear (62,63). During late embryonic and early postnatal development, the density of tight junction strands increases as maturation continues towards the impermeable adult BBB (64). This process restricts access to substances that normally diffuse freely into a developing embryonic brain.

The tight junctions found in brain capillaries are composed of both integral membrane proteins that link adjacent cells and peripheral membrane proteins that link the junctions to the cytoskeleton. An analysis of liver plasma membrane protein extracts led to the identification of the first transmembrane tight junction component, occludin (65). Occludin is an integral membrane protein with four transmembrane domains and two extracellular hydrophobic loops that mediate cell–cell connections. The cytoplasmic domain of occludin is highly phosphorylated when located in tight junctions (66), and this phosphorylation can regulate tight junctional permeability (67). Although the exact role of tight junction occludin has not been fully determined, existing data support a regulatory role rather than a barrier-establishing role (68). Interestingly, an occludin-deficient mouse showed no alteration in tight junction integrity, indicating that occludin has either no direct barrier function or its function is redundant and can be replaced by other tight junction components (69). The claudins make up another family of transmembrane proteins located at the tight junctional complex. In contrast to occludin, the claudins appear to play the major role in forming the seal that restricts paracellular transport. At least 24 isoforms of claudins exist, and this broad class confers junctional specificity to different cell types (70,71). Claudin-1 (72), -3 (73), -5 (73–75), and -12 (76) have been identified at the BBB. In contrast to the occludin-deficient mouse, a *claudin-5*–knockout results in an altered tight junction phenotype and increased permeability to small molecules to the brain (76). Junctional adhesion molecule (JAM) is a member of the immunoglobulin superfamily, and is also colocalized at the tight junctional complex (77). JAMs are most likely involved in occludin recruitment and tight junction assembly (78), and they have the ability to increase transcellular resistance in cells not normally forming tight junctions (77,79). The JAM proteins also are involved in the extravasation of monocytes and leukocytes in vitro and in vivo (77,80,81).

The tight junctions are linked to the cytoskeleton by a variety of accessory proteins in the cytoplasmic compartment. Important for anchoring the transmembrane proteins to the cytoskeleton are protein members of the membrane-associated guanylate kinases (MAGUKs) known as zonula occludens (ZO)-1, -2, and -3. The ZO-1 protein may act as a molecular linker bringing together claudins, ZO-2, ZO-3, cingulin, and actin (82–84). In addition to their structural functions, the MAGUKs act as binding scaffolds for signaling proteins. 7H6 antigen also is located at tight junctions and confers resistance to paracellular transport of macromolecules and metastatic cancer cells (85). The formation and maintenance of tight junctions are regulated by a variety of signaling cascades (86), indicating that tight junctions can respond to various stimuli and pathological conditions (see the section Blood–Brain Barrier Involvement in Disease).

Another important junctional component of brain ECs are the adherens junctions, which, like tight junctions, are linked to the cytoskeleton. The adherens junctions are localized to the basolateral side of the tight junctional complex and form a belt-like structure similar to that seen with the tight junctions. The major components of adherens junctions are vascular endothelial (VE)-cadherin, catenins, plakoglobulin, vinculin, and α-actinin (87). VE-cadherin is a single-transmembrane domain protein that belongs to the cell adhesion molecule superfamily, and it has been reported as being part of the adherens junctions in microvascular ECs (88,89). In epithelial cells, cadherins play a role in the formation and maintenance of the tight junctions and act as an inducer of cell polarity (90,91). Adherens junctions also form connections with the

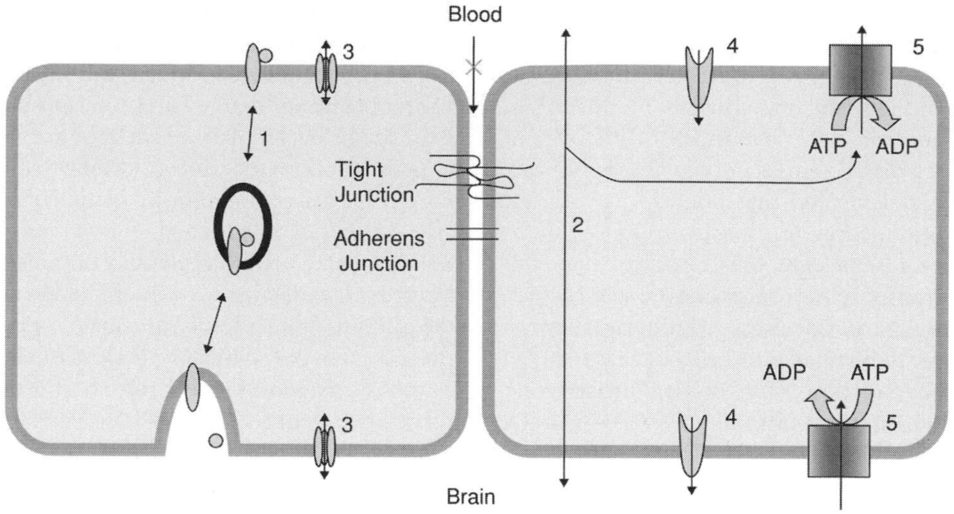

Figure 123.2. Molecular transport at the BBB. 1, receptor-mediated endocytosis/transcytosis (e.g., transferrin, insulin); 2, transcellular diffusion (e.g., oxygen, small lipophilic compounds); 3, ion channel; 4, carrier-mediated transport (e.g., GLUT1, LAT1); 5, active efflux (e.g., P-glycoprotein, MRP). For a comprehensive summary of BBB transport systems, see Zhang W, Stanimirovic DB. The transport systems of the blood–brain barrier. In: de Vries HE, ed. *The Blood–Brain Barrier and Its Microenvironment*. New York: Taylor & Francis; 2005:103–142.

ECM and play a role in signaling via the basement membrane. Therefore, the physical barrier formed by tight and adherens junctions also can function to regulate the BBB interface.

Transport/Metabolic Barrier

Molecular Transport Mechanisms at the Blood–Brain Barrier

Because the paracellular permeation of hydrophilic substances is limited by the tight junctions, a transported molecule must travel serially through both the luminal and abluminal membranes (Figure 123.2). Therefore, brain blood vessels must have additional transporter systems to meet the high energy demands of the brain. Indeed, the BBB acts as a selective barrier that modulates the influx and efflux of substances through a highly regulated composite of transport systems. A sampling of these is outlined here.

Carrier-mediated transport mediates the trans-BBB flux of small molecule energy sources, vitamins, and nutrients such as hexoses, monocarboxylic acids, amino acids, amines, and nucleosides (92,93). These carriers are highly stereospecific channels found on both endothelial membranes that mediate trans-BBB transport via the endothelial cytoplasm (see Figure 123.2). The two carrier-mediated transporters supporting the highest rate of facilitated saturable transport are the glucose (GLUT1) (94,95) and monocarboxylic acid (monocarboxylate transporter [MCT] -1) transporters (96–98). GLUT1 expression is mainly associated with the microvasculature in the brain (99,100). It is distributed asymmetrically at the BBB with an approximately threefold higher expression level on the abluminal plasma membrane compared with the luminal membrane. Its expression level is similar in the cortex, hippocampus, and cerebellum, while it is significantly lower in the olfactory bulb (101). Like GLUT1, MCT1 is expressed at both the luminal and the abluminal plasma membranes. It is crucial for the transport of lactic acid and ketone bodies to the developing brain (97). Bidirectional transport studies in vitro demonstrated that MCT1 can mediate both the uptake and efflux of monocarboxylic acids (102). To supply appropriate amounts of nutrients to the brain, some transporters not only are expressed at higher levels, but also exist as isoforms that exhibit higher affinity for their substrates. For example, the large amino acid transporter type (LAT)-1 is expressed at higher levels at the BBB than in any other tissue (103), and it has a much higher substrate affinity compared to neutral amino acid carriers localized in either the endothelium or resident cells of other tissues (104). Transport of choline as well as other molecules having quaternary ammonium groups occurs at the BBB (105). Adenosine, vitamins, and small neutral and acidic amino acid transport also have been observed at the BBB.

Receptor-mediated endocytosis also is prevalent at the BBB interface. Because nonspecific uptake of bloodstream components is limited by the lack of pinocytosis, the microvascular endothelium is equipped with receptors that mediate the selective uptake and trans-BBB transport of circulating peptides and proteins. After binding to the receptor on the luminal side of the BBB, the ligand is endocytosed and transported via the vesicular trafficking system to the abluminal side of the BBB, where the vesicle fuses with the plasma membrane and releases its cargo (see Figure 123.2). Several so-called receptor-mediated transcytosis systems have been identified in the brain microvasculature. The BBB insulin receptor has been shown to mediate the brain uptake of insulin, and results in measurable brain insulin concentrations (106). Iron is essential for all cell types and is carried in plasma and interstitial fluids by transferrin. To mediate the transport of iron-containing transferrin across the BBB, brain ECs express transferrin receptors that allow the coupled entry of iron into the brain (107,108). Leptin (109) and low-density lipoprotein (LDL) (39,110) also have been demonstrated to transport across the BBB. Ion transport and astrocyte-regulated aquaporin-based transport of water are important in maintaining homeostasis of the brain environment and have been reviewed elsewhere (111–113).

Metabolic Transformation

The combination of impermeable barriers and selective transport mechanisms is not the only means whereby the BBB helps protect the brain from uncontrolled chemical fluctuations within the interstitial fluid. Compounds that enter the ECs due to either their lipophilic nature or their affinity to a molecular transporter can be converted by metabolic processes into chemical compounds incapable of traversing the abluminal membrane. For example, L-dopa is carried into the cytoplasm of the EC by the LAT1 transporter (114), and uncontrolled passage across the abluminal membrane would be undesirable, because this compound is a precursor for dopamine. Once in the EC cytoplasm, these substances are enzymatically converted to dopamine and dihydroxyphenylacetic acid and are thus sequestered in the EC. This phenomenon is found for a broad array of compounds and is referred to as the *metabolic BBB.* Clearly the metabolic barrier must be considered when devising a brain treatment strategy, for example the use of L-dopa for Parkinson disease (115–117).

Multidrug Resistance at the Blood–Brain Barrier

As stated earlier, one of the main functions of the BBB is the protection of the brain from potentially harmful endogenous and exogenous compounds. The tight junctions seal the paracellular cleft against many solutes, but lipophilic substances can enter the cells via plasma membrane diffusion. Once in the cytoplasm, these substances can be rapidly effluxed by carrier-mediated efflux transporters. Members of the ATP-binding cassette (ABC) superfamily play an important role at the BBB. In particular, P-glycoprotein (also known as *multidrug resistance* [MDR]-1) and the multidrug-resistance associated proteins (MRP) have been demonstrated to be BBB components. P-glycoprotein was first identified in cancer cells (118), but also is expressed by normal cells in intestine, liver, and brain (119). P-glycoprotein is predominantly localized to the luminal membranes of BBB ECs and is responsible for the efflux of small-molecule substrates and therapeutics (120) from the ECs cytoplasm back into the blood vessel lumen (121). Various transporters in the MRP family have been shown to be expressed in brain endothelium on both the luminal and abluminal membranes (122,123), although the levels and species specificity are somewhat in question at this time (121). Recently, a novel brain multidrug resistance protein (BMDP) that is only distantly related to P-glycoprotein, but has high amino acid sequence similarity to the human breast cancer resistance protein (BCRP), was identified at the BBB (124).

Current genomics efforts indicate that transporters comprise between 10% to 15% of the expressed genome of the BBB (75,125,126), while transporter genes are only estimated to appear at a frequency of 3% to 5% in the human genome (127). Due to this impressive variety of transporters expressed at the BBB, a large amount of work remains to fully characterize this specialized transport interface.

BLOOD–BRAIN BARRIER VERSUS OTHER VASCULAR BARRIER SYSTEMS

Although the BBB is referred to as a unique vascular bed throughout this chapter, it is important to acknowledge that other endothelial barriers exist in the body. The blood–retinal barrier of the vascularized inner retina of the eye has many properties similar to that of the BBB, but permeability is higher (128). A peripheral blood–nerve barrier exists, although it is substantially more permeable than the BBB (129). The blood–testes barrier is endowed with impermeability characteristics by the Sertoli cells that surround a comparatively permeable endothelium. Finally, the blood–CSF barrier of the brain is clearly distinct from the BBB in both its barrier and transport characteristics. It is found at the site of the choroid plexus, and it is comprised of tightly apposed epithelial cells rather than ECs (130). Continued research focusing on these vascular and epithelial barriers promises to provide additional comparisons with the structure and function of the BBB.

BLOOD–BRAIN BARRIER INVOLVEMENT IN DISEASE

The BBB interface has been implicated in a number of diseases. This section highlights its involvement in a selected subset of these diseases from the perspective of barrier and immune function.

Human Immunodeficiency Virus

The BBB can be the target of viral, fungal, and bacterial infections. As an example, the human immunodeficiency virus (HIV) can cause neural disorders such as HIV encephalitis or acquired immunodeficiency syndrome (AIDS) dementia complex (131). Vascular changes have been noted in the brains of AIDS patients in the form of increased diameter of the cortical microvessels and thinning of the basal lamina of the blood vessels (132). In addition, HIV has been found to infect brain ECs both in vivo (133,134) and in vitro (135,136). The virus can enter the brain through the migration of infected monocytes and lymphocytes across the BBB (137–139) or the blood–CSF barrier (139,140), but evidence also suggests that the transendothelial passage of free HIV virus via adsorptive endocytosis can occur (138). Regardless of the mode of entry, HIV encephalitis and AIDS dementia correlate with the presence of monocyte infiltration and subsequent microglial activation (131).

HIV-infected monocytes elicit an inflammatory response and stimulate adhesion properties of the BBB through the release of cytokines and inflammatory mediators such as tumor necrosis factor (TNF)-α, interleukin (IL)-1β, and interferon (IFN)-γ (141–143). In vitro studies demonstrated that EC adhesion molecules E-selectin and vascular cell adhesion molecule (VCAM)-1 can then be upregulated to support the binding and migration of HIV-infected monocytes across the BBB and into the brain (138,144). The basement membrane is also a target of HIV proteins, because the transmigration process requires degradation of the basement membrane, and HIV proteins stimulate matrix metalloproteinase (MMP)-2 secretion (145). Expression of the tight junction proteins occludin and ZO-1 is decreased upon increased monocyte infiltration in AIDS patients (146,147). Based on these mechanistic observations described, the BBB appears to be intimately involved in the pathogenesis of AIDS in the central nervous system.

Cerebral Malaria

Cerebral malaria is an example of parasitic invasion of the brain. Cerebral malaria is caused by infection with *Plasmodium* parasites, with *Plasmodium falciparum* being the most common and serious pathogen. Most infected people suffer a mild febrile illness, but infection also can lead to serious complications, including multiorgan failure. In cases of cerebral malaria, CNS symptoms such as coma, seizures, and increased intracranial pressure can occur. Cerebral malaria is the most serious malarial complication, leading to about 0.5 million deaths per year (148). For patients with cerebral malaria, brain microvessels contain significantly more parasitized red blood cells than do those in patients with noncerebral malaria (149). The first step in cerebral malaria is the interaction of parasitized erythrocytes with the microvasculature in the CNS. How exactly BBB-sequestered parasitized erythrocytes cause the neurological syndrome is not well understood, because the infected erythrocytes remain in the vascular lumen. Cerebral malaria in the animal model can be associated with a complete BBB breakdown (150), but in human pathology the BBB is not affected to this extent. Studies have shown that parasitized red blood cells attach to the endothelial plasma membrane, with the EC forming pseudopodia into the capillary lumen (149). There is also evidence for adhesion receptor-mediated events (151) such as binding of parasitized red blood cells to the endothelial receptor molecules CD36, intercellular adhesion molecule-1 (ICAM-1), and CD31 as well as to thrombospondin and chondroitin sulfate A. In addition, alteration in the distribution of the junction-associated proteins occludin, ZO-1, and vinculin suggests a mechanism for brain vessel permeability (152). For further details on cerebral malaria, the reader is referred to Chapter 142.

Stroke

Stroke is the result of an injured or occluded vessel in the brain that leads to ischemic insult and edema formation. Hypoxia results in increased BBB permeability, leakage of plasma components across the BBB, and neuronal cell death, whereas reperfusion causes further damage in the affected regions because of increases in reactive oxygen species (153). Increased expression of the adhesion receptors E-selectin (154), P-selectin, and ICAM-1 may result in leukocyte recruitment to postcapillary microvessels (155). In addition, alteration of intracellular calcium concentrations and calcium-activated signaling pathways can trigger changes in junctional protein expression and distribution (156). When in vitro BBB models of rat brain ECs in coculture with astrocytes were subjected to hypoxic conditions, a 25% decrease in TEER after 4 hours of hypoxia and 95% decrease after 8 hours was observed (157). The presence of astrocytes reduced the decrease in TEER after a 4-hour hypoxic insult (157). Paracellular transport also was increased in bovine brain ECs following long duration hypoxia (24–48 hours), whether astrocytes were present (158,159). Tight junction occludin, ZO-1, and ZO-2 protein expression levels were unaltered but were relocated from cell junctions during hypoxia (159). Actin expression increases after reoxygenation and is accompanied by an increase in occludin, ZO-1, and ZO-2 protein expression and a coordinate decrease in the paracellular permeability. Finally, the basement membrane integrity is impaired in ischemic conditions (160) and may be due in part to metalloproteinase-based degradation (161). Stroke complications can be exacerbated by junctional rearrangement and breach of the BBB, and thus the BBB could prove to be an increasingly important target for therapeutic intervention.

Multiple Sclerosis

MS is a chronic inflammatory disease of the CNS manifested by BBB disturbance, local edema, and demyelination (162). Magnetic resonance imaging with contrast enhancement has demonstrated that inflammatory damage to blood vessels occurs at an early stage in the pathogenesis of MS (163,164) with local perfusion changes occurring prior to BBB breakdown (165). Experimental autoimmune encephalomyelitis (EAE) is an animal model of demyelinating disease that possesses certain pathological and clinical features of human MS. This model has provided much of the molecular basis for our understanding of MS progression and is referred to frequently here. As in the pathogenesis of aforementioned diseases like cerebral malaria or AIDS, ICAM-1 (EAE and human MS) and VCAM-1 (human MS) are upregulated at BBB endothelial sites and correlate with disease activity (166,167). These alterations may be caused by proinflammatory cytokine release (IFN-γ, TNF-α) from activated immune cells that can lead to alterations in tight junction integrity and thus BBB impermeability. As an example of the changes in tight junction integrity, abnormal levels and distributions of ZO-1 and occludin have been observed in the brains of MS patients (168). Claudin-3 is also selectively downregulated in BBB tight junctions near vascular lesions in a murine EAE model (73). Modulation of the BBB in EAE can lead to immune cell entry into the CNS, via the tight junction complex, channels or pores formed in the tight junction complex, or by emperipolesis (169). Recent genomics studies also have identified expression of myelin basic protein (MBP) at the BBB, and microvascular expression of MBP was confirmed by in situ hybridization (170,171). Taken together, these findings raise the possibility that the BBB may play an even more active role in MS pathogenesis than was previously anticipated by contributing to the T-cell recruitment process.

Brain Tumors

Significant alteration of the vascular architecture occurs in high-grade gliomas and metastatic brain tumors, whereas the vascular environment in low-grade gliomas is similar to that found in nondiseased brain (172). Whereas normal brain capillaries have diameters around 5 microns, tumor capillaries are much larger on average (up to 40 microns). In addition, the larger tumor vessels are very tortuous and irregularly spaced, unlike the angioarchitecture observed in normal brain capillaries (173–175). Higher permeability and loss of BBB function is associated with changes in the tight junctions, presence of fenestrae, and increases in vesicular trafficking. For example, studies of human brain tumor tight junction composition indicated that claudin-1 expression was lost, whereas claudin-5 and occludin expression was downregulated in hyperplastic vessels. Meanwhile, ZO-1 expression was unaffected (176,177). The net effect on the integrity of the blood–tumor barrier can be demonstrated by studies that confirmed a higher level of leakage of serum components from tumor vessels (174,178). However, a leaky blood–tumor barrier is not a universal attribute of tumor tissue in the brain. Certain tumors are resistant to fluorescein uptake (376 D) indicating a tight blood–tumor barrier (178), and size-dependent leakage kinetics govern transport across the blood–tumor barrier (179). In addition, multidrug resistance (i.e., P-glycoprotein) is still intact in glioma vessels, limiting the efficacy of chemotherapeutics (171,180). Thus, drug delivery to brain tumors still can be hampered in many instances.

Other Diseases

The BBB also is involved in a broad variety of other CNS diseases such as Parkinson disease (181), Alzheimer disease (182), hypertension (183), and the class of CNS complications resulting from bacterial, fungal, and viral infection (184). The preceding section underscores the importance of understanding the BBB regulatory mechanisms in health and disease. In many CNS diseases, the BBB may not be fully compromised, and therapeutic intervention may be desired in the presence of a still-functioning BBB. Therefore, it is critical to have brain drug targeting and delivery strategies that have the capacity to overcome the BBB.

TREATMENT OF BRAIN DISEASE: THE BLOOD–BRAIN BARRIER BOTTLENECK

Although the brain vasculature provides a large surface area (estimated to be over 21 m^2 from 400 miles of capillaries in human brain) for drug delivery, the physical, transport, and metabolic barriers presented by the brain vasculature greatly inhibit the uptake of substances from the blood. This has significant consequences for brain drug delivery and serves to limit the options available for therapy. To pass the physical barrier, candidate drugs must be both small (<500 D) and lipophilic (185). Once such a drug has entered the BBB, it may be metabolized or inactivated inside the EC or excluded by efflux transporters, and thus have limited access to underlying neural tissue. Because of these restrictions, 98% of small-molecule pharmaceuticals and almost all gene and protein medicines are excluded from the brain by the BBB (186). Therefore, only with special targeting strategies (discussed here) can these impermeable medicines cross the BBB. This highlights the challenge of delivering drugs to brain tissue. To date, only depression, schizophrenia, chronic pain, and epilepsy can be treated with small-molecule drugs that are not specifically targeted (186). Promising therapeutic candidates exist for many diseases including Parkinson disease, brain tumors, MS, and stroke, but they lack the ability to be transported into the brain parenchyma.

Invasive Strategies

One method to bypass the BBB is the use of neurosurgical techniques to introduce drugs directly into brain tissue. Strategies requiring neurosurgery are used for the implantation of polymer particles infused with a drug, but the treatment volume is limited because the extracellular fluid in brain tissue is rather static, and simple molecular diffusion only allows for a modest penetration distance of 2 to 3 mm (187). A similar method involves the direct injection of drugs into the brain ventricles. The penetration into brain tissue also is limited for intraventricular injections because this brain fluid is rapidly cleared and turned over four to five times daily by absorption across the arachnoid villi into the general circulation (93,188). To overcome these problems, convection-enhanced diffusion via intracerebral infusion of fluid can be used to increase drug concentrations over a larger region of brain tissue. Experiments using Rhesus monkeys indicate that drug distribution in the perfused region may be increased substantially (189), but bulk flow can adversely affect the brain, leading to astrogliosis (190). A similar strategy has been used to administer GDNF to the putamen of human patients as a potential treatment for Parkinson disease (191). These invasive strategies can have success if limited treatment volumes are required, such as those arising from localized, nonmetastatic brain tumors. However, for chronic conditions requiring a repetitive treatment regimen or for those diseases that affect large portions of the brain, a noninvasive drug delivery strategy is generally preferable.

BBB disruption is another invasive method that can be used to overcome the delivery problem while at the same time increasing the volume of distribution. Reversible disruption of the BBB has been investigated using hyperosmolar solutions, or vasoactive BBB disruption with histamine (192), serotonin (193), or bradykinin peptides (194,195) to reversibly disrupt the BBB and allow free paracellular passage of molecules from the blood into the brain. Although this method can be administered through the bloodstream, it is listed as an invasive approach because the process of opening the BBB can lead to toxic effects, because solutes and immune machinery normally excluded from brain gain free access (196,197).

Noninvasive Delivery via the Vasculature

A major challenge and priority is to develop novel methods that utilize the endogenous BBB molecular transport network as a conduit for transendothelial drug transport. These transport systems can be clustered into three general groups: nonspecific uptake, carrier-mediated transport, and receptor-mediated transport.

Protein Transduction Domains

Cationic protein transduction domains (PTD) fall into the realm of nonspecific carriers. One such example is the HIV TAT peptide, which is an 11-amino acid HIV-TAT sequence that mediates rapid translocation into cells. Although HIV TAT was shown to gain access to the brain interstitium after intraperitoneal injection (198), subsequent pharmacokinetic analysis indicated that the rapid clearance and broad organ uptake would necessitate prohibitively high doses (199). In general, PTDs suffer from a lack of selective targeting, which leads to widespread distribution throughout the body.

Carrier-Mediated Transport of Therapeutics

Carrier-mediated drug transport relies on the presence of endogenous transmembrane proteins that are selective and stereospecific for small-molecule solutes. For example, the Parkinson drug L-dopa gains entry to the brain by utilizing the LAT1 (114). The successful transport through the BBB is a result of L-dopa mimicking the structure of phenylalanine. However, due to the stereospecificity and steric constraints imposed by these selective membrane pores, applications are potentially limited. As another pertinent example, it is likely that the efflux of the AIDS drug zidovudine or azidodeoxythymidine (AZT) progresses in a carrier-mediated fashion (200). This exemplifies another significant problem that arises when carrier-mediated transport or membrane diffusion is used for drug delivery. The presence of efficient efflux transporters such as P-glycoprotein, MRP, and BMDP greatly limit brain uptake in many circumstances. Thus, inhibition of these efflux transporters often is presented as a strategy to enhance transport across the BBB.

Receptor-Mediated Transport of Therapeutics

As an alternative approach, receptor-mediated transport mechanisms that use the endosomal trafficking system are more compatible with the delivery of a wider range of drug cargo. Receptor-mediated transport involves the binding of an extracellular epitope of a cell surface receptor and triggering of an energy-intensive transcytosis process. This method has been successful in allowing for noninvasive transport of small-molecules (201), proteins (202–204), genes (205,206), nanoparticles (207), and liposomes up to 100 nm in size (206,208). Commonly targeted receptors are the LDL receptor (209), the transferrin receptor (202,206), and the insulin receptor (210,211). Similar, less specific processes involving absorptive-mediated transcytosis have been used with cationized proteins (212,213) that promote receptor clustering and activation of the transcytosis pathway. Such strategies often target the cell surface receptor in ways that do not disrupt the normal transport of endogenous ligand. Therefore, the impact on normal metabolic pathways is limited. In addition, this strategy is not limited by the size and shape requirements of carrier-mediated transport. Although early results derived from the receptor-mediated delivery process are promising because of the capacity to deliver drugs or drug carriers of various sizes and composition, this strategy also has some serious drawbacks that must be addressed for general clinical success. The present methods rely on receptors that are ubiquitously expressed, such as the transferrin and insulin receptors. This leads to mistargeting of potentially expensive drugs that also may have unwanted side effects in tissues other than the brain. In addition, the present methodologies generally result in a low fraction (1%–4%) of the injected dose actually reaching the brain target (201,210,211,214). However, if side effects and drug costs are not a concern, the transferrin and insulin transcytosis systems are very promising. For example, these systems result in beneficial pharmacologic responses after noninvasive delivery of brain-derived neurotrophic factor (BDNF) in rodent stroke models (203) and of tyrosine hydroxylase gene replacement therapy in Parkinson disease models (215).

Nanoparticles and Liposomes as Drug Carriers

Another exciting approach that can circumvent efflux transporters and allow for controlled release of drug in brain tissue is the use of nanoparticles or liposomes. Both nanoparticles and liposomes have been used in promising proof-of-concept studies. Their preparation is simple and scalable, and they can be designed for controlled rates of release. Nanoparticles made of polybutylcyanoacrylate (PBCA) coated with polysorbate 80 have been researched extensively (216) and were shown to be capable of delivering drugs to the brain (217). As an example of nanoparticle efficacy, intravenously injected doxorubicin-loaded, polysorbate 80-coated nanoparticles led to a 40% cure rate in rats with intracranially transplanted glioblastomas (218,219). The mechanisms for nanoparticle entry into the brain are not yet fully understood, but the polysorbate-coated nanoparticles were found to bind apolipoprotein E from blood-plasma and may mimic LDL. ECs express receptors for LDL at their luminal membrane, and this may provide a mode for brain entry. Although the mechanism of brain uptake is still unresolved, the labile nature of the particles in vivo leads to short-lived pharmacologic effects (220) and possible BBB permeabilization (221). To date, nanoparticles also suffer from a lack of brain targeting.

Antibody-targeted liposomes represent another exciting strategy for noninvasive drug delivery to the brain. Liposomes are vesicles with a phospholipid bilayer surrounding an aqueous solution. Unstabilized liposomes are removed rapidly from circulation by macrophages of the reticuloendothelial system in liver, spleen, and bone marrow (222). Incorporation of gangliosides or polyethylene glycol (PEG) derivatized lipids can prolong the half-life in vivo (223,224). Long-circulating PEG liposomes have a minimal affinity for healthy tissue in vivo and therefore provide a comparatively inert platform for drug delivery. However, the BBB permeability for unconjugated liposomes is limited (225). To overcome this problem, liposomes can be coupled with cationized proteins or antibodies against the transferrin or insulin receptors. In a recent study, liposomes were coupled with antibodies (immunoliposomes) against both the mouse transferrin receptor and the human insulin receptor. After immunoliposome injection into mice with transplanted human glial brain tumors, the transferrin receptor enabled transport across the murine tumor vasculature, whereas the insulin receptor permitted transport across the nuclear membrane of the human brain cancer cells (226). The drug payload in this case was an expression plasmid encoding antisense mRNA against the human epidermal growth factor receptor gene (EGFR). Mice treated with these liposomes had an increased lifespan compared with control mice injected with control liposomes. Phase I clinical trials with PEG-stabilized immunoliposomes are promising (227), but specific brain targeting is still limited due to the lack of availability of efficient brain targeting vectors. Similar to immunoliposomes, coupling nanoparticles with antibodies against brain microvasculature-specific receptors also could enhance target specificity. However, information about brain endothelium-specific receptors is still lacking. Genomic and proteomic studies may help to identify new possible targets for immunotargeted drug delivery (228).

KEY POINTS

- By virtue of its impermeable phenotype, the BBB plays a significant role in regulating the flow of information and material into and out of the brain.
- Understanding the participation of the BBB in disease is critical for the appropriate design of therapeutic intervention for many neurological diseases.
- The BBB presents a significant hurdle in progressing from the bench to the bedside.

Future Goals

- To further elucidate the cellular interplay of the neurovascular unit's role in health and disease; genomics, proteomics, and imaging technologies are currently being brought to bear on BBB research and will likely facilitate this process.
- To develop new strategies that allow brain drug targeting and noninvasive BBB permeability for clinical realization; such development of novel drug delivery strategies will likely require the combination of such interdisciplinary skills as antibody engineering, materials science, and pharmacokinetics.

REFERENCES

1 Davson H. History of the blood–brain barrier concept. In: Neufeldt EA, ed. *Implications of the Blood–Brain Barrier.* New York: Plenum; 1988:27–52.

2 Saunders NR, Habgood MD, Dziegielewska KM. Barrier mechanisms in the brain, I. Adult brain. *Clin Exp Pharmacol Physiol.* 1999;26(1):11–19.

3 Brightman MW, Reese TS. Junctions between intimately apposed cell membranes in the vertebrate brain. *J Cell Biol.* 1969; 40(3):648–677.

4 Reese TS, Karnovsky MJ. Fine structural localization of a blood–brain barrier to exogenous peroxidase. *J Cell Biol.* 1967; 34(1):207–217.

5 Risau W, Wolburg H. Development of the blood–brain barrier. *Trends Neurosci.* 1990;13(5):174–178.

6 Risau W, Hallmann R, Albrecht U, Henke-Fahle S. Brain induces the expression of an early cell surface marker for blood–brain barrier-specific endothelium. *EMBO J.* 1986;5(12):3179–3183.

7 Stewart PA, Wiley MJ. Developing nervous tissue induces formation of blood–brain barrier characteristics in invading endothelial cells: a study using quail-chick transplantation chimeras. *Dev Biol.* 1981;84(1):183–192.

8 Joo F. Increased production of coated vesicles in the brain capillaries during enhanced permeability of the blood–brain barrier. *Br J Exp Pathol.* 1971;52(6):646–649.

9 Bouchard P, Ghitescu LD, Bendayan M. Morpho-functional studies of the blood–brain barrier in streptozotocin-induced diabetic rats. *Diabetologia.* 2002;45(7):1017–1025.

10 Oldendorf WH, Cornford ME, Brown WJ. The large apparent work capability of the blood–brain barrier: a study of the mitochondrial content of capillary endothelial cells in brain and other tissues of the rat. *Ann Neurol.* 1977;1(5):409–417.

11 Stewart PA, Hayakawa K, Farrell CL. Quantitation of blood–brain barrier ultrastructure. *Microsc Res Tech.* 1994;27(6):516–527.

12 Crone C, Olesen SP. Electrical resistance of brain microvascular endothelium. *Brain Res.* 1982;241(1):49–55.

13 Krause D, Mischeck U, Galla HJ, Dermietzel R. Correlation of zonula occludens ZO-1 antigen expression and transendothelial resistance in porcine and rat cultured cerebral endothelial cells. *Neurosci Lett*. 1991;128(2):301–304.
14 Cecchelli R, Dehouck B, Descamps L, et al. In vitro model for evaluating drug transport across the blood–brain barrier. *Adv Drug Deliv Rev*. 1999;36(2–3):165–178.
15 Tan KH, Dobbie MS, Felix RA, et al. A comparison of the induction of immortalized endothelial cell impermeability by astrocytes. *Neuroreport*. 2001;12(7):1329–1334.
16 Brayton J, Qing Z, Hart MN, et al. Influence of adhesion molecule expression by human brain microvessel endothelium on cancer cell adhesion. *J Neuroimmunol*. 1998;89(1–2):104–112.
17 Gerhart DZ, Broderius MA, Drewes LR. Cultured human and canine endothelial cells from brain microvessels. *Brain Res Bull*. 1988;21(5):785–793.
18 Mizuguchi H, Utoguchi N, Mayumi T. Preparation of glial extracellular matrix: a novel method to analyze glial-endothelial cell interaction. *Brain Res Brain Res Protoc*. 1997;1(4):339–343.
19 Biegel D, Spencer DD, Pachter JS. Isolation and culture of human brain microvessel endothelial cells for the study of blood–brain barrier properties in vitro. *Brain Res*. 1995;692(1–2):183–189.
20 Tilling T, Engelbertz C, Decker S, et al. Expression and adhesive properties of basement membrane proteins in cerebral capillary endothelial cell cultures. *Cell Tissue Res*. 2002;310(1):19–29.
21 Tilling T, Korte D, Hoheisel D, Galla HJ. Basement membrane proteins influence brain capillary endothelial barrier function in vitro. *J Neurochem*. 1998;71(3):1151–1157.
22 Stanness KA, Guatteo E, Janigro D. A dynamic model of the blood–brain barrier "in vitro." *Neurotoxicology*. 1996;17(2): 481–496.
23 Lukes A, Mun-Bryce S, Lukes M, Rosenberg GA. Extracellular matrix degradation by metalloproteinases and central nervous system diseases. *Mol Neurobiol*. 1999;19(3):267–284.
24 Zerlin M, Goldman JE. Interactions between glial progenitors and blood vessels during early postnatal corticogenesis: blood vessel contact represents an early stage of astrocyte differentiation. *J Comp Neurol*. 1997;387(4):537–546.
25 Bauer HC, Bauer H. Neural induction of the blood–brain barrier: still an enigma. *Cell Mol Neurobiol*. 2000;20(1):13–28.
26 Cohen Z, Ehret M, Maitre M, Hamel E. Ultrastructural analysis of tryptophan hydroxylase immunoreactive nerve terminals in the rat cerebral cortex and hippocampus: their associations with local blood vessels. *Neuroscience*. 1995;66(3):555–569.
27 Estrada C, DeFelipe J. Nitric oxide-producing neurons in the neocortex:morphological and functional relationship with intraparenchymal microvasculature. *Cereb Cortex*. 1998;8(3): 193–203.
28 Cestelli A, Catania C, D'Agostino S, et al. Functional feature of a novel model of blood brain barrier: studies on permeation of test compounds. *J Control Release*. 2001;76(1–2):139–147.
29 Harik SI, Kalaria RN, Whitney PM, et al. Glucose transporters are abundant in cells with "occluding" junctions at the blood-eye barriers. *Proc Natl Acad Sci USA*. 1990;87(11):4261–4264.
30 Lieberman MW, Barrios R, Carter BZ, et al. gamma-Glutamyl transpeptidase. What does the organization and expression of a multipromoter gene tell us about its functions? *Am J Pathol*. 1995;147(5):1175–1185.
31 Tontsch U, Bauer HC. Glial cells and neurons induce blood–brain barrier related enzymes in cultured cerebral endothelial cells. *Brain Res*. 1991;539(2):247–253.
32 Stanness KA, Neumaier JF, Sexton TJ, et al. A new model of the blood–brain barrier: co-culture of neuronal, endothelial and glial cells under dynamic conditions. *Neuroreport*. 1999;10(18):3725–3731.
33 Janzer RC, Raff MC. Astrocytes induce blood–brain barrier properties in endothelial cells. *Nature*. 1987;325(6101):253–257.
34 Bush TG, Puvanachandra N, Horner CH, et al. Leukocyte infiltration, neuronal degeneration, and neurite outgrowth after ablation of scar-forming, reactive astrocytes in adult transgenic mice. *Neuron*. 1999;23(2):297–308.
35 Liedtke W, Edelmann W, Bieri PL, et al. GFAP is necessary for the integrity of CNS white matter architecture and long-term maintenance of myelination. *Neuron*. 1996;17(4):607–615.
36 Pekny M, Stanness KA, Eliasson C, et al. Impaired induction of blood–brain barrier properties in aortic endothelial cells by astrocytes from GFAP-deficient mice. *Glia*. 1998;22(4):390–400.
37 Rubin LL, Hall DE, Porter S, et al. A cell culture model of the blood–brain barrier. *J Cell Biol*. 1991;115(6):1725–1735.
38 Beck DW, Vinters HV, Hart MN, Cancilla PA. Glial cells influence polarity of the blood–brain barrier. *J Neuropathol Exp Neurol*. 1984;43(3):219–224.
39 Dehouck MP, Meresse S, Delorme P, et al. An easier, reproducible, and mass-production method to study the blood–brain barrier in vitro. *J Neurochem*. 1990;54(5):1798–1801.
40 DeBault LE. gamma-Glutamyltranspeptidase induction mediated by glial foot process-to endothelium contact in co-culture. *Brain Res*. 1981;220(2):432–435.
41 Raub TJ, Kuentzel SL, Sawada GA. Permeability of bovine brain microvessel endothelial cells in vitro: barrier tightening by a factor released from astroglioma cells. *Exp Cell Res*. 1992;199(2):330–340.
42 Roux F, Durieu-Trautmann O, Chaverot N, et al. Regulation of gamma-glutamyl transpeptidase and alkaline phosphatase activities in immortalized rat brain microvessel endothelial cells. *J Cell Physiol*. 1994;159(1):101–113.
43 Beuckmann C, Hellwig S, Galla HJ. Induction of the blood/brain-barrier-associated enzyme alkaline phosphatase in endothelial cells from cerebral capillaries is mediated via cAMP. *Eur J Biochem*. 1995;229(3):641–644.
44 Fenart L, Buee-Scherrer V, Descamps L, et al. Inhibition of P-glycoprotein: rapid assessment of its implication in blood–brain barrier integrity and drug transport to the brain by an in vitro model of the blood–brain barrier. *Pharm Res*. 1998;15(7):993–1000.
45 Stanness KA, Westrum LE, Fornaciari E, et al. Morphological and functional characterization of an in vitro blood–brain barrier model. *Brain Res*. 1997;771(2):329–342.
46 Matsusaka T. Tridimensional views of the relationship of pericytes to endothelial cells of capillaries in the human choroid and retina. *J Electron Microsc (Tokyo)*. 1975;24(1):13–18.
47 Sims DE. Recent advances in pericyte biology-implications for health and disease. *Can J Cardiol*. 1991;7(10):431–443.
48 Courtoy PJ, Boyles J. Fibronectin in the microvasculature: localization in the pericyte-endothelial interstitium. *J Ultrastruct Res*. 1983;83(3):258–273.

49 Frey A, Meckelein B, Weiler-Guttler H, et al. Pericytes of the brain microvasculature express gamma-glutamyl transpeptidase. *Eur J Biochem.* 1991;202(2):421–429.
50 Risau W. Development and differentiation of endothelium. *Kidney Int Suppl.* 1998;67:S3–S6.
51 Ramsauer M, Krause D, Dermietzel R. Angiogenesis of the blood–brain barrier in vitro and the function of cerebral pericytes. *FASEB J.* 2002;16(10):1274–1276.
52 Balabanov R, Dore-Duffy P. Role of the CNS microvascular pericyte in the blood–brain barrier. *J Neurosci Res.* 1998;53(6):637–644.
53 Gage FH. Mammalian neural stem cells. *Science.* 2000;287 (5457):1433–1438.
54 Shen Q, Goderie SK, Jin L, et al. Endothelial cells stimulate self-renewal and expand neurogenesis of neural stem cells. *Science.* 2004;304(5675):1338–1340.
55 Fuchs E, Tumbar T, Guasch G. Socializing with the neighbors: stem cells and their niche. *Cell.* 2004;116(6):769–778.
56 Igarashi Y, Utsumi H, Chiba H, et al. Glial cell line-derived neurotrophic factor induces barrier function of endothelial cells forming the blood–brain barrier. *Biochem Biophys Res Commun.* 1999;261(1):108–112.
57 Lee SW, Kim WJ, Choi YK, et al. SSeCKS regulates angiogenesis and tight junction formation in blood–brain barrier. *Nat Med.* 2003;9(7):900–906.
58 Wolburg H, Neuhaus J, Kniesel U, et al. Modulation of tight junction structure in blood–brain barrier endothelial cells. Effects of tissue culture, second messengers and cocultured astrocytes. *J Cell Sci.* 1994;107(Pt 5):1347–1357.
59 Hoheisel D, Nitz T, Franke H, et al. Hydrocortisone reinforces the blood-brain properties in a serum free cell culture system. *Biochem Biophys Res Commun.* 1998;247(2):312–315.
60 Dermietzel R. Junctions in the central nervous system of the cat. IV. Interendothelial junctions of cerebral blood vessels from selected areas of the brain. *Cell Tissue Res.* 1975;164(1):45–62.
61 Nico B, Cantino D, Bertossi M, et al. Tight endothelial junctions in the developing microvasculature: a thin section and freeze-fracture study in the chick embryo optic tectum. *J Submicrosc Cytol Pathol.* 1992;24(1):85–95.
62 Stewart PA, Hayakawa K. Early ultrastructural changes in blood–brain barrier vessels of the rat embryo. *Brain Res Dev Brain Res.* 1994;78(1):25–34.
63 Bauer HC, Bauer H, Lametschwandtner A, et al. Neovascularization and the appearance of morphological characteristics of the blood–brain barrier in the embryonic mouse central nervous system. *Brain Res Dev Brain Res.* 1993;75(2):269–278.
64 Kniesel U, Risau W, Wolburg H. Development of blood–brain barrier tight junctions in the rat cortex. *Brain Res Dev Brain Res.* 1996;96(1–2):229–240.
65 Furuse M, Hirase T, Itoh M, et al. Occludin: a novel integral membrane protein localizing at tight junctions. *J Cell Biol.* 1993;123(6 Pt 2):1777–1788.
66 Sakakibara A, Furuse M, Saitou M, et al. Possible involvement of phosphorylation of occludin in tight junction formation. *J Cell Biol.* 1997;137(6):1393–1401.
67 Hirase T, Kawashima S, Wong EY, et al. Regulation of tight junction permeability and occludin phosphorylation by Rhoa-p160ROCK-dependent and -independent mechanisms. *J Biol Chem.* 2001;276(13):10423–10431.
68 Wolburg H, Lippoldt A. Tight junctions of the blood–brain barrier: development, composition and regulation. *Vascul Pharmacol.* 2002;38(6):323–337.
69 Saitou M, Furuse M, Sasaki H, et al. Complex phenotype of mice lacking occludin, a component of tight junction strands. *Mol Biol Cell.* 2000;11(12):4131–4142.
70 Furuse M, Sasaki H, Tsukita S. Manner of interaction of heterogeneous claudin species within and between tight junction strands. *J Cell Biol.* 1999;147(4):891–903.
71 Tsukita S, Furuse M. Claudin-based barrier in simple and stratified cellular sheets. *Curr Opin Cell Biol.* 2002;14(5):531–536.
72 Huber JD, Witt KA, Hom S, et al. Inflammatory pain alters blood–brain barrier permeability and tight junctional protein expression. *Am J Physiol Heart Circ Physiol.* 2001;280(3): H1241–H1248.
73 Wolburg H, Wolburg-Buchholz K, Kraus J, et al. Localization of claudin-3 in tight junctions of the blood–brain barrier is selectively lost during experimental autoimmune encephalomyelitis and human glioblastoma multiforme. *Acta Neuropathol (Berl).* 2003;105(6):586–592.
74 Morita K, Sasaki H, Furuse M, Tsukita S. Endothelial claudin: claudin-5/TMVCF constitutes tight junction strands in endothelial cells. *J Cell Biol.* 1999;147(1):185–194.
75 Shusta EV, Boado RJ, Mathern GW, Pardridge WM. Vascular genomics of the human brain. *J Cereb Blood Flow Metab.* 2002; 22(3):245–252.
76 Nitta T, Hata M, Gotoh S, et al. Size-selective loosening of the blood–brain barrier in claudin-5-deficient mice. *J Cell Biol.* 2003;161(3):653–660.
77 Martin-Padura I, Lostaglio S, Schneemann M, et al. Junctional adhesion molecule, a novel member of the immunoglobulin superfamily that distributes at intercellular junctions and modulates monocyte transmigration. *J Cell Biol.* 1998;142(1):117–127.
78 Bazzoni G, Martinez-Estrada OM, Orsenigo F, et al. Interaction of junctional adhesion molecule with the tight junction components ZO-1, cingulin, and occludin. *J Biol Chem.* 2000; 275(27):20520–20526.
79 Aurrand-Lions M, Duncan L, Ballestrem C, Imhof BA. JAM-2, a novel immunoglobulin superfamily molecule, expressed by endothelial and lymphatic cells. *J Biol Chem.* 2001;276(4):2733–2741.
80 Del Maschio A, De Luigi A, Martin-Padura I, et al. Leukocyte recruitment in the cerebrospinal fluid of mice with experimental meningitis is inhibited by an antibody to junctional adhesion molecule (JAM). *J Exp Med.* 1999;190(9):1351–1356.
81 Palmeri D, van Zante A, Huang CC, et al. Vascular endothelial junction-associated molecule, a novel member of the immunoglobulin superfamily, is localized to intercellular boundaries of ECs. *J Biol Chem.* 2000;275(25):19139–19145.
82 Wittchen ES, Haskins J, Stevenson BR. Protein interactions at the tight junction. Actin has multiple binding partners, and ZO-1 forms independent complexes with ZO-2 and ZO-3. *J Biol Chem.* 1999;274(49):35179–35185.
83 Itoh M, Furuse M, Morita K, et al. Direct binding of three tight junction-associated MAGUKs, ZO-1, ZO-2, and ZO-3, with the COOH termini of claudins. *J Cell Biol.* 1999;147(6):1351–1363.
84 Cordenonsi M, Turco F, D'Atri F, et al. *Xenopus laevis* occludin. Identification of in vitro phosphorylation sites by protein kinase CK2 and association with cingulin. *Eur J Biochem.* 1999; 264(2):374–384.

85 Satoh H, Zhong Y, Isomura H, et al. Localization of 7H6 tight junction-associated antigen along the cell border of vascular ECs correlates with paracellular barrier function against ions, large molecules, and cancer cells. *Exp Cell Res.* 1996;222(2):269–274.
86 Matter K, Balda MS. Signalling to and from tight junctions. *Nat Rev Mol Cell Biol.* 2003;4(3):225–236.
87 Franke WW, Cowin P, Grund C, et al. The endothelial junction. The plaque and its component. In: Simionescu NS, Simionescu M, eds. *Endothelial Cell Biology in Health and Disease.* New York, London: Plenum Press, 1988.
88 Schulze C, Firth JA. Immunohistochemical localization of adherens junction components in blood–brain barrier microvessels of the rat. *J Cell Sci.* 1993;104(Pt 3):773–782.
89 Rubin LL, Barbu K, Bard F, et al. Differentiation of brain endothelial cells in cell culture. *Ann NY Acad Sci.* 1991;633:420–425.
90 McNeill H, Ozawa M, Kemler R, Nelson WJ. Novel function of the cell adhesion molecule uvomorulin as an inducer of cell surface polarity. *Cell.* 1990;62(2):309–316.
91 Gumbiner B, Simons K. The role of uvomorulin in the formation of epithelial occluding junctions. *Ciba Found Symp.* 1987; 125:168–186.
92 Drewes LR. Molecular architecture of the brain microvasculature: perspective on blood-brain transport. *J Mol Neurosci.* 2001;16(2–3):93–98; Discussion 151–157.
93 Pardridge WM. *Brain Drug Targeting; the Future of Brain Drug Development.* Cambridge, UK: Cambridge University Press, 2001.
94 Birnbaum MJ, Haspel HC, Rosen OM. Cloning and characterization of a cDNA encoding the rat brain glucose-transporter protein. *Proc Natl Acad Sci USA.* 1986;83(16):5784–5788.
95 Pardridge WM, Boado RJ, Farrell CR. Brain-type glucose transporter (GLUT-1) is selectively localized to the blood–brain barrier. Studies with quantitative western blotting and in situ hybridization. *J Biol Chem.* 1990;265(29):18035–18040.
96 Takanaga H, Tamai I, Inaba S, et al. cDNA cloning and functional characterization of rat intestinal monocarboxylate transporter. *Biochem Biophys Res Commun.* 1995;217(1):370–377.
97 Leino RL, Gerhart DZ, Drewes LR. Monocarboxylate transporter (MCT1) abundance in brains of suckling and adult rats: a quantitative electron microscopic immunogold study. *Brain Res Dev Brain Res.* 1999;113(1–2):47–54.
98 Gerhart DZ, Enerson BE, Zhdankina OY, et al. Expression of monocarboxylate transporter MCT1 by brain endothelium and glia in adult and suckling rats. *Am J Physiol.* 1997;273(1 Pt 1): E207–E213.
99 Farrell CL, Pardridge WM. Blood–brain barrier glucose transporter is asymmetrically distributed on brain capillary endothelial luminal and abluminal membranes: an electron microscopic immunogold study. *Proc Natl Acad Sci USA.* 1991;88(13):5779–5783.
100 Boado RJ, Pardridge WM. The brain-type glucose transporter mRNA is specifically expressed at the blood–brain barrier. *Biochem Biophys Res Commun.* 1990;166(1):174–179.
101 Dobrogowska DH, Vorbrodt AW. Quantitative immunocytochemical study of blood–brain barrier glucose transporter (GLUT-1) in four regions of mouse brain. *J Histochem Cytochem.* 1999;47(8):1021–1030.
102 Kido Y, Tamai I, Nakanishi T, et al. Evaluation of blood–brain barrier transporters by co-culture of brain capillary ECs with astrocytes. *Drug Metab Pharmacokinet.* 2002;17(1):34–41.
103 Boado RJ, Li JY, Nagaya M, et al. Selective expression of the large neutral amino acid transporter at the blood–brain barrier. *Proc Natl Acad Sci USA.* 1999;96(21):12079–12084.
104 Pardridge WM. Kinetics of competitive inhibition of neutral amino acid transport across the blood–brain barrier. *J Neurochem.* 1977;28(1):103–108.
105 Cornford EM, Braun LD, Oldendorf WH. Carrier mediated blood–brain barrier transport of choline and certain choline analogs. *J Neurochem.* 1978;30(2):299–308.
106 Pardridge WM, Eisenberg J, Yang J. Human blood–brain barrier insulin receptor. *J Neurochem.* 1985;44(6):1771–1778.
107 Fishman JB, Rubin JB, Handrahan JV, et al. Receptor-mediated transcytosis of transferrin across the blood–brain barrier. *J Neurosci Res.* 1987;18(2):299–304.
108 Jefferies WA, Brandon MR, Hunt SV, et al. Transferrin receptor on endothelium of brain capillaries. *Nature.* 1984;312(5990): 162–163.
109 Boado RJ, Golden PL, Levin N, Pardridge WM. Up-regulation of blood–brain barrier short-form leptin receptor gene products in rats fed a high fat diet. *J Neurochem.* 1998;71(4):1761–1764.
110 Meresse S, Delbart C, Fruchart JC, Cecchelli R. Low-density lipoprotein receptor on endothelium of brain capillaries. *J Neurochem.* 1989;53(2):340–345.
111 Keep RF, Ennis SR, Betz AL. Blood–brain barrier ion transport. In: Pardridge WM, ed. *Introduction to the Blood–Brain Barrier.* Cambridge: Cambridge University Press; 1998:207–213.
112 Badaut J, Lasbennes F, Magistretti PJ, Regli L. Aquaporins in brain: distribution, physiology, and pathophysiology. *J Cereb Blood Flow Metab.* 2002;22(4):367–378.
113 Davies DC. Blood–Brain Barrier breakdown in septic encephalopathy and brain tumours. *J Anat.* 2002;200(6): 639–646.
114 Wade LA, Katzman R. Synthetic amino acids and the nature of L-DOPA transport at the blood–brain barrier. *J Neurochem.* 1975;25(6):837–842.
115 Goldstein GW, Betz AL. The blood–brain barrier. *Sci Am.* 1986; 255(3):74–83.
116 Greig NH. Drug entry into the brain and its pharmacologic manipulation. In: Bradbury WB, ed. *Physiology and Pharmacology of the Blood–Brain Barrier.* Berlin, Heidelberg: Springer; 1992:417–437.
117 Johansson BB. The physiology of the blood–brain barrier. *Adv Exp Med Biol.* 1990;274:25–39.
118 Wong SL, Van Belle K, Sawchuk RJ. Distributional transport kinetics of zidovudine between plasma and brain extracellular fluid/cerebrospinal fluid in the rabbit: investigation of the inhibitory effect of probenecid utilizing microdialysis. *J Pharmacol Exp Ther.* 1993;264(2):899–909.
119 Beaulieu E, Demeule M, Ghitescu L, Beliveau R. P-glycoprotein is strongly expressed in the luminal membranes of the endothelium of blood vessels in the brain. *Biochem J.* 1997;326(Pt 2): 539–544.
120 Kartner N, Evernden-Porelle D, Bradley G, Ling V. Detection of P-glycoprotein in multidrug-resistant cell lines by monoclonal antibodies. *Nature.* 1985;316(6031):820–823.
121 Sun H, Dai H, Shaik N, Elmquist WF. Drug efflux transporters in the CNS. *Adv Drug Deliv Rev.* 2003;55(1):83–105.
122 Soontornmalai A, Vlaming ML, Fritschy JM. Differential, strain-specific cellular and subcellular distribution of multidrug transporters in murine choroid plexus and blood–brain barrier. *Neuroscience.* 2006;138(1):159–169. Epub 2005 Dec 19.

123 Zhang Y, Han H, Elmquist WF, Miller DW. Expression of various multidrug resistance-associated protein (MRP) homologues in brain microvessel endothelial cells. *Brain Res.* 2000;876(1–2):148–153.

124 Eisenblatter T, Galla HJ. A new multidrug resistance protein at the blood–brain barrier. *Biochem Biophys Res Commun.* 2002; 293(4):1273–1278.

125 Enerson BE, Drewes LR. The rat blood–brain barrier transcriptome. Brain03, XXIst International Symposium on Cerebral Blood Flow, Metabolism, and Function. 2003.

126 Li JY, Boado RJ, Pardridge WM. Rat blood–brain barrier genomics. II. *J Cereb Blood Flow Metab.* 2002;22(11):1319–1326.

127 Lander ES, Linton LM, Birren B, et al. Initial sequencing and analysis of the human genome. *Nature.* 2001;409(6822):860–921.

128 Steuer H, Jaworski A, Elger B, et al. Functional characterization and comparison of the outer blood-retina barrier and the blood–brain barrier. *Invest Ophthalmol Vis Sci.* 2005;46(3): 1047–1053.

129 Allt G, Lawrenson JG. The blood-nerve barrier: enzymes, transporters and receptors–a comparison with the blood–brain barrier. *Brain Res Bull.* 2000;52(1):1–12.

130 Smith DE, Johanson CE, Keep RF. Peptide and peptide analog transport systems at the blood-CSF barrier. *Adv Drug Deliv Rev.* 2004;56(12):1765–1791.

131 Glass JD, Wesselingh SL, Selnes OA, McArthur JC. Clinical-neuropathologic correlation in HIV-associated dementia. *Neurology.* 1993;43(11):2230–2237.

132 Weis S, Haug H, Budka H. Vascular changes in the cerebral cortex in HIV-1 infection: I. A morphometric investigation by light and electron microscopy. *Clin Neuropathol.* 1996;15(6):361–366.

133 Bagasra O, Lavi E, Bobroski L, et al. Cellular reservoirs of HIV-1 in the central nervous system of infected individuals: identification by the combination of in situ polymerase chain reaction and immunohistochemistry. *AIDS.* 1996;10(6):573–585.

134 An SF, Groves M, Gray F, Scaravilli F. Early entry and widespread cellular involvement of HIV-1 DNA in brains of HIV-1 positive asymptomatic individuals. *J Neuropathol Exp Neurol.* 1999;58(11):1156–1162.

135 Banks WA, Freed EO, Wolf KM, et al. Transport of human immunodeficiency virus type 1 pseudoviruses across the blood–brain barrier: role of envelope proteins and adsorptive endocytosis. *J Virol.* 2001;75(10):4681–4691.

136 Moses AV, Bloom FE, Pauza CD, Nelson JA. Human immunodeficiency virus infection of human brain capillary endothelial cells occurs via a CD4/galactosylceramide-independent mechanism. *Proc Natl Acad Sci USA.* 1993;90(22):10474–10478.

137 Price RW, Brew B, Sidtis J, et al. The brain in AIDS: central nervous system HIV-1 infection and AIDS dementia complex. *Science.* 1988;239(4840):586–592.

138 Nottet HS, Persidsky Y, Sasseville VG, et al. Mechanisms for the transendothelial migration of HIV-1-infected monocytes into brain. *J Immunol.* 1996;156(3):1284–1295.

139 Ryan G, Grimes T, Brankin B, et al. Neuropathology associated with feline immunodeficiency virus infection highlights prominent lymphocyte trafficking through both the blood-brain and blood-choroid plexus barriers. *J Neurovirol.* 2005;11(4): 337–345.

140 Petito CK, Chen H, Mastri AR, et al. HIV infection of choroid plexus in AIDS and asymptomatic HIV-infected patients suggests that the choroid plexus may be a reservoir of productive infection. *J Neurovirol.* 1999;5(6):670–677.

141 Wesselingh SL, Power C, Glass JD, et al. Intracerebral cytokine messenger RNA expression in acquired immunodeficiency syndrome dementia. *Ann Neurol.* 1993;33(6):576–582.

142 Tyor WR, Glass JD, Griffin JW, et al. Cytokine expression in the brain during the acquired immunodeficiency syndrome. *Ann Neurol.* 1992;31(4):349–360.

143 Tyor WR, Glass JD, Baumrind N, et al. Cytokine expression of macrophages in HIV-1-associated vacuolar myelopathy. *Neurology.* 1993;43(5):1002–1009.

144 Carlos T, Kovach N, Schwartz B, et al. Human monocytes bind to two cytokine-induced adhesive ligands on cultured human endothelial cells: endothelial-leukocyte adhesion molecule-1 and vascular cell adhesion molecule-1. *Blood.* 1991;77(10):2266–2271.

145 Dhawan S, Weeks BS, Soderland C, et al. HIV-1 infection alters monocyte interactions with human microvascular endothelial cells. *J Immunol.* 1995;154(1):422–432.

146 Dallasta LM, Pisarov LA, Esplen JE, et al. Blood–brain barrier tight junction disruption in human immunodeficiency virus-1 encephalitis. *Am J Pathol.* 1999;155(6):1915–1927.

147 Boven LA, Middel J, Verhoef J, et al. Monocyte infiltration is highly associated with loss of the tight junction protein zonula occludens in HIV-1-associated dementia. *Neuropathol Appl Neurobiol.* 2000;26(4):356–360.

148 Deininger MH, Kremsner PG, Meyermann R, Schluesener HJ. Differential cellular accumulation of transforming growth factor-beta1, -beta2, and -beta3 in brains of patients who died with cerebral malaria. *J Infect Dis.* 2000;181(6):2111–2115.

149 Pongponratn E, Turner GD, Day NP, et al. An ultrastructural study of the brain in fatal *Plasmodium falciparum* malaria. *Am J Trop Med Hyg.* 2003;69(4):345–359.

150 Thumwood CM, Hunt NH, Clark IA, Cowden WB. Breakdown of the blood–brain barrier in murine cerebral malaria. *Parasitology.* 1988;96(Pt 3):579–589.

151 Berendt AR, Ferguson DJ, Gardner J, et al. Molecular mechanisms of sequestration in malaria. *Parasitology.* 1994;(Suppl 108):S19–S28.

152 Brown H, Hien TT, Day N, et al. Evidence of blood–brain barrier dysfunction in human cerebral malaria. *Neuropathol Appl Neurobiol.* 1999;25(4):331–340.

153 Lievre V, Becuwe P, Bianchi A, et al. Free radical production and changes in superoxide dismutases associated with hypoxia/reoxygenation-induced apoptosis of embryonic rat forebrain neurons in culture. *Free Radic Biol Med.* 2000;29(12): 1291–1301.

154 Haring HP, Berg EL, Tsurushita N, et al. E-selectin appears in nonischemic tissue during experimental focal cerebral ischemia. *Stroke.* 1996;27(8):1386–1391; Discussion 91–92.

155 Okada Y, Copeland BR, Mori E, et al. P-selectin and intercellular adhesion molecule-1 expression after focal brain ischemia and reperfusion. *Stroke.* 1994;25(1):202–211.

156 Brown RC, Davis TP. Calcium modulation of adherens and tight junction function: a potential mechanism for blood–brain barrier disruption after stroke. *Stroke.* 2002;33(6):1706–1711.

157 Kondo T, Kinouchi H, Kawase M, Yoshimoto T. Astroglial cells inhibit the increasing permeability of brain endothelial cell monolayer following hypoxia/reoxygenation. *Neurosci Lett.* 1996;208(2):101–104.

158 Abbruscato TJ, Davis TP. Combination of hypoxia/aglycemia compromises in vitro blood–brain barrier integrity. *J Pharmacol Exp Ther*. 1999;289(2):668–675.

159 Mark KS, Davis TP. Cerebral microvascular changes in permeability and tight junctions induced by hypoxia-reoxygenation. *Am J Physiol Heart Circ Physiol*. 2002;282(4):H1485–H1494.

160 Hamann GF, Okada Y, Fitridge R, del Zoppo GJ. Microvascular basal lamina antigens disappear during cerebral ischemia and reperfusion. *Stroke*. 1995;26(11):2120–2126.

161 Petty MA, Wettstein JG. Elements of cerebral microvascular ischaemia. *Brain Res Brain Res Rev*. 2001;36(1):23–34.

162 Hemmer B, Archelos JJ, Hartung HP. New concepts in the immunopathogenesis of multiple sclerosis. *Nat Rev Neurosci*. 2002;3(4):291–301.

163 Davie CA, Hawkins CP, Barker GJ, et al. Serial proton magnetic resonance spectroscopy in acute multiple sclerosis lesions. *Brain*. 1994;117(Pt 1):49–58.

164 Miller DH, Grossman RI, Reingold SC, McFarland HF. The role of magnetic resonance techniques in understanding and managing multiple sclerosis. *Brain*. 1998;121(Pt 1):3–24.

165 Wuerfel J, Bellmann-Strobl J, Brunecker P, et al. Changes in cerebral perfusion precede plaque formation in multiple sclerosis: a longitudinal perfusion MRI study. *Brain*. 2004;127 (Pt 1):111–119.

166 Wilcox CE, Ward AM, Evans A, et al. Endothelial cell expression of the intercellular adhesion molecule-1 (ICAM-1) in the central nervous system of guinea pigs during acute and chronic relapsing experimental allergic encephalomyelitis. *J Neuroimmunol*. 1990;30(1):43–51.

167 Cannella B, Raine CS. The adhesion molecule and cytokine profile of multiple sclerosis lesions. *Ann Neurol*. 1995;37(4):424–435.

168 Plumb J, McQuaid S, Mirakhur M, Kirk J. Abnormal endothelial tight junctions in active lesions and normal-appearing white matter in multiple sclerosis. *Brain Pathol*. 2002;12(2):154–169.

169 Brosnan CF, Claudio L. *Brain Microvasculature in Multiple Sclerosis*. Cambridge: Cambridge University Press, 1998.

170 Li JY, Boado RJ, Pardridge WM. Blood–brain barrier genomics. *J Cereb Blood Flow Metab*. 2001;21(1):61–68.

171 Lesniak MS, Brem H. Targeted therapy for brain tumours. *Nat Rev Drug Discov*. 2004;3(6):499–508.

172 Plate KH, Risau W. Angiogenesis in malignant gliomas. *Glia*. 1995;15(3):339–347.

173 Vajkoczy P, Menger MD. Vascular microenvironment in gliomas. *J Neurooncol*. 2000;50(1–2):99–108.

174 Vajkoczy P, Schilling L, Ullrich A, et al. Characterization of angiogenesis and microcirculation of high-grade glioma: an intravital multifluorescence microscopic approach in the athymic nude mouse. *J Cereb Blood Flow Metab*. 1998;18(5):510–520.

175 Zama A, Tamura M, Inoue HK. Three-dimensional observations on microvascular growth in rat glioma using a vascular casting method. *J Cancer Res Clin Oncol*. 1991;117(5):396–402.

176 Liebner S, Fischmann A, Rascher G, et al. Claudin-1 and claudin-5 expression and tight junction morphology are altered in blood vessels of human glioblastoma multiforme. *Acta Neuropathol (Berl)*. 2000;100(3):323–331.

177 Papadopoulos MC, Saadoun S, Woodrow CJ, et al. Occludin expression in microvessels of neoplastic and non-neoplastic human brain. *Neuropathol Appl Neurobiol*. 2001;27(5):384–395.

178 Yuan F, Salehi HA, Boucher Y, et al. Vascular permeability and microcirculation of gliomas and mammary carcinomas transplanted in rat and mouse cranial windows. *Cancer Res*. 1994; 54(17):4564–4568.

179 Schmiedl UP, Kenney J, Maravilla KR. Dyke Award Paper. Kinetics of pathologic blood-brain-barrier permeability in an astrocytic glioma using contrast-enhanced MR. *AJNR Am J Neuroradiol*. 1992;13(1):5–14.

180 Takamiya Y, Abe Y, Tanaka Y, et al. Murine P-glycoprotein on stromal vessels mediates multidrug resistance in intracerebral human glioma xenografts. *Br J Cancer*. 1997;76(4):445–450.

181 Kortekaas R, Leenders KL, van Oostrom JC, et al. Blood–brain barrier dysfunction in parkinsonian midbrain in vivo. *Ann Neurol*. 2005;57(2):176–179.

182 Jellinger KA. Alzheimer disease and cerebrovascular pathology: an update. *J Neural Transm*. 2002;109(5–6):813–836.

183 Johansson BB. Hypertension. In: Pardridge WM, ed. *Introduction to the Blood–Brain Barrier*. Cambridge: Cambridge University Press; 1998:427–433.

184 Huang SH, Jong AY. Cellular mechanisms of microbial proteins contributing to invasion of the blood–brain barrier. *Cell Microbiol*. 2001;3(5):277–287.

185 Levin VA. Relationship of octanol/water partition coefficient and molecular weight to rat brain capillary permeability. *J Med Chem*. 1980;23(6):682–684.

186 Pardridge WM. The blood–brain barrier: bottleneck in brain drug development. *NeuroRx*. 2005;2(1):3–14.

187 Krewson CE, Klarman ML, Saltzman WM. Distribution of nerve growth factor following direct delivery to brain interstitium. *Brain Res*. 1995;680(1–2):196–206.

188 Yan Q, Matheson C, Sun J, et al. Distribution of intracerebral ventricularly administered neurotrophins in rat brain and its correlation with trk receptor expression. *Exp Neurol*. 1994;127 (1):23–36.

189 Lieberman DM, Laske DW, Morrison PF, et al. Convection-enhanced distribution of large molecules in gray matter during interstitial drug infusion. *J Neurosurg*. 1995;82(6):1021–1029.

190 Ai Y, Markesbery W, Zhang Z, et al. Intraputamenal infusion of GDNF in aged rhesus monkeys: distribution and dopaminergic effects. *J Comp Neurol*. 2003;461(2):250–261.

191 Gill SS, Patel NK, Hotton GR, et al. Direct brain infusion of glial cell line-derived neurotrophic factor in Parkinson disease. *Nat Med*. 2003;9(5):589–595.

192 Gross PM, Teasdale GM, Angerson WJ, Harper AM. H2-Receptors mediate increases in permeability of the blood–brain barrier during arterial histamine infusion. *Brain Res*. 1981;210 (1–2):396–400.

193 Winkler T, Sharma HS, Stalberg E, et al. Impairment of blood–brain barrier function by serotonin induces desynchronization of spontaneous cerebral cortical activity: experimental observations in the anaesthetized rat. *Neuroscience*. 1995;68(4):1097–1104.

194 Unterberg A, Wahl M, Baethmann A. Effects of bradykinin on permeability and diameter of pial vessels in vivo. *J Cereb Blood Flow Metab*. 1984;4(4):574–585.

195 Black KL, Cloughesy T, Huang SC, et al. Intracarotid infusion of RMP-7, a bradykinin analog, and transport of gallium-68 ethylenediamine tetraacetic acid into human gliomas. *J Neurosurg*. 1997;86(4):603–609.

196 Salahuddin TS, Johansson BB, Kalimo H, Olsson Y. Structural changes in the rat brain after carotid infusions of hyperosmolar solutions. An electron microscopic study. *Acta Neuropathol (Berl)*. 1988;77(1):5–13.

197 Lossinsky AS, Vorbrodt AW, Wisniewski HM. Scanning and transmission electron microscopic studies of microvascular pathology in the osmotically impaired blood–brain barrier. *J Neurocytol*. 1995;24(10):795–806.

198 Schwarze SR, Ho A, Vocero-Akbani A, Dowdy SF. In vivo protein transduction: delivery of a biologically active protein into the mouse. *Science*. 1999;285(5433):1569–1572.

199 Lee HJ, Pardridge WM. Pharmacokinetics and delivery of tat and tat-protein conjugates to tissues in vivo. *Bioconjug Chem*. 2001;12(6):995–999.

200 Takasawa K, Terasaki T, Suzuki H, Sugiyama Y. In vivo evidence for carrier-mediated efflux transport of 3′-azido-3′-deoxythymidine and 2′,3′-dideoxyinosine across the blood–brain barrier via a probenecid-sensitive transport system. *J Pharmacol Exp Ther*. 1997;281(1):369–375.

201 Friden PM, Walus LR, Musso GF, et al. Anti-transferrin receptor antibody and antibody-drug conjugates cross the blood–brain barrier. *Proc Natl Acad Sci USA*. 1991;88(11):4771–4775.

202 Friden PM, Walus LR, Watson P, et al. Blood–brain barrier penetration and in vivo activity of an NGF conjugate. *Science*. 1993; 259(5093):373–377.

203 Zhang Y, Pardridge WM. Neuroprotection in transient focal brain ischemia after delayed intravenous administration of brain-derived neurotrophic factor conjugated to a blood–brain barrier drug targeting system. *Stroke*. 2001;32(6):1378–1384.

204 Song BW, Vinters HV, Wu D, Pardridge WM. Enhanced neuroprotective effects of basic fibroblast growth factor in regional brain ischemia after conjugation to a blood–brain barrier delivery vector. *J Pharmacol Exp Ther*. 2002;301(2):605–610.

205 Shi N, Boado RJ, Pardridge WM. Antisense imaging of gene expression in the brain in vivo. *Proc Natl Acad Sci USA*. 2000; 97(26):14709–14714.

206 Shi N, Zhang Y, Zhu C, et al. Brain-specific expression of an exogenous gene after i.v. administration. *Proc Natl Acad Sci USA*. 2001;98(22):12754–12759.

207 Gessner A, Olbrich C, Schroder W, et al. The role of plasma proteins in brain targeting: species dependent protein adsorption patterns on brain-specific lipid drug conjugate (LDC) nanoparticles. *Int J Pharm*. 2001;214(1–2):87–91.

208 Cerletti A, Drewe J, Fricker G, et al. Endocytosis and transcytosis of an immunoliposome-based brain drug delivery system. *J Drug Target*. 2000;8(6):435–446.

209 Kreuter J. Nanoparticulate systems for brain delivery of drugs. *Adv Drug Deliv Rev*. 2001;47(1):65–81.

210 Coloma MJ, Lee HJ, Kurihara A, et al. Transport across the primate blood–brain barrier of a genetically engineered chimeric monoclonal antibody to the human insulin receptor. *Pharm Res*. 2000;17(3):266–274.

211 Pardridge WM, Kang YS, Buciak JL, Yang J. Human insulin receptor monoclonal antibody undergoes high affinity binding to human brain capillaries in vitro and rapid transcytosis through the blood–brain barrier in vivo in the primate. *Pharm Res*. 1995;12(6):807–816.

212 Wadhwani KC, Shimon-Hophy M, Rapoport SI. Enhanced permeabilities of cationized-bovine serum albumins at the blood-nerve and blood–brain barriers in awake rats. *J Neurosci Res*. 1992;32(3):407–414.

213 Bickel U, Lee VM, Trojanowski JQ, Pardridge WM. Development and in vitro characterization of a cationized monoclonal antibody against beta A4 protein: a potential probe for Alzheimer's disease. *Bioconjug Chem*. 1994;5(2): 119–125.

214 Kang YS, Bickel U, Pardridge WM. Pharmacokinetics and saturable blood–brain barrier transport of biotin bound to a conjugate of avidin and a monoclonal antibody to the transferrin receptor. *Drug Metab Dispos*. 1994;22(1):99–105.

215 Zhang Y, Schlachetzki F, Zhang YF, et al. Normalization of striatal tyrosine hydroxylase and reversal of motor impairment in experimental parkinsonism with intravenous nonviral gene therapy and a brain-specific promoter. *Hum Gene Ther*. 2004; 15(4):339–350.

216 Vauthier C, Dubernet C, Fattal E, et al. Poly(alkylcyanoacrylates) as biodegradable materials for biomedical applications. *Adv Drug Deliv Rev*. 2003;55(4):519–548.

217 Kreuter J, Alyautdin RN, Kharkevich DA, Ivanov AA. Passage of peptides through the blood–brain barrier with colloidal polymer particles (nanoparticles). *Brain Res*. 1995;674(1):171–174.

218 Gulyaev AE, Gelperina SE, Skidan IN, et al. Significant transport of doxorubicin into the brain with polysorbate 80-coated nanoparticles. *Pharm Res*. 1999;16(10):1564–1569.

219 Gelperina SE, Khalansky AS, Skidan IN, et al. Toxicological studies of doxorubicin bound to polysorbate 80-coated poly(butyl cyanoacrylate) nanoparticles in healthy rats and rats with intracranial glioblastoma. *Toxicol Lett*. 2002;126(2):131–141.

220 Friese A, Seiller E, Quack G, et al. Increase of the duration of the anticonvulsive activity of a novel NMDA receptor antagonist using poly(butylcyanoacrylate) nanoparticles as a parenteral controlled release system. *Eur J Pharm Biopharm*. 2000;49(2):103–109.

221 Olivier JC, Fenart L, Chauvet R, et al. Indirect evidence that drug brain targeting using polysorbate 80-coated polybutylcyanoacrylate nanoparticles is related to toxicity. *Pharm Res*. 1999;16(12):1836–1842.

222 Frank MM. The reticuloendothelial system and bloodstream clearance. *J Lab Clin Med*. 1993;122(5):487–488.

223 Woodle MC, Matthay KK, Newman MS, et al. Versatility in lipid compositions showing prolonged circulation with sterically stabilized liposomes. *Biochim Biophys Acta*. 1992;1105(2):193–200.

224 Uster PS, Allen TM, Daniel BE, et al. Insertion of poly(ethylene glycol) derivatized phospholipid into pre-formed liposomes results in prolonged in vivo circulation time. *FEBS Lett*. 1996;386(2–3):243–246.

225 Saito R, Bringas JR, McKnight TR, et al. Distribution of liposomes into brain and rat brain tumor models by convection-enhanced delivery monitored with magnetic resonance imaging. *Cancer Res*. 2004;64(7):2572–2579.

226 Zhang Y, Zhu C, Pardridge WM. Antisense gene therapy of brain cancer with an artificial virus gene delivery system. *Mol Ther*. 2002;6(1):67–72.

227 Matsumura Y, Hamaguchi T, Ura T, et al. Phase I clinical trial and pharmacokinetic evaluation of NK911, a micelle-encapsulated doxorubicin. *Br J Cancer*. 2004;91(10):1775–1781.

228 Shusta EV. Blood–brain barrier genomics, proteomics, and new transporter discovery. *NeuroRx*. 2005;2(1):151–161.

124

Brain Endothelial Cells Bridge Neural and Immune Networks

Kevin J. Tracey and Christine N. Metz

The Center for Patient-Oriented Research, The Feinstein Institute for Medical Research, Manhasset, New York

The brain has the highest metabolic rate of any organ structure in the human body. Although the human brain contributes only 2% to the total body weight, it uses approximately 16% of the body's cardiac output and accounts for nearly 25% of the body's oxygen consumption (1). Brain-associated endothelium includes all endothelial cells (ECs) found within the brain, including ECs of the blood–brain barrier (BBB) and circumventricular organs (CVOs) (Figure 124.1). One of the major functions of the endothelium within the brain is to direct the flow of oxygenated blood. It is estimated that approximately 600 to 700 mL of blood flows through the carotid arteries and their branches every minute to distribute oxygenated blood to the anterior portion of the brain. The capillaries, the principal vasculature within the brain, are very important components of the brain's supply system. Because of their small diameter and large number per volume of brain tissue, these capillaries provide a large surface area for delivering adequate supplies of glucose and oxygen.

BLOOD SUPPLY TO THE BRAIN

The circulatory system supplying the human brain is detailed in several references (2–4). Briefly, the brain receives its arterial supply from the carotid and vertebral arteries, which begin extracranially and run through the neck and base of the skull to reach the cranial cavity. At the base of the ear, the carotid artery bifurcates into internal and external branches. The external carotid supplies blood to the skull and meninges (membranous areas of connective tissue that envelop the brain and spinal cord consisting of the dura matter [outer layer], arachnoid [middle layer], and pia mater [inner layer closest to brain]), while the internal carotid artery provides blood to the brain (anterior portion of brain, eye, and forebrain). The internal carotid artery gives off the ophthalmic artery and then trifurcates into the posterior communicating artery, anterior cerebral artery, and middle cerebral artery. The anterior and middle cerebral arteries are terminal branches of the internal carotid artery. The anterior cerebral artery supplies the frontal lobes and the medial aspects of the parietal and occipital lobes. The middle cerebral artery (also known as the "artery of stroke") forms the largest branch of the internal carotid artery and supplies the frontoparietal somatosensory cortex. The smallest branch of the internal carotid artery, the posterior communicating artery, connects to the posterior cerebral artery, which supplies the occipital and inferior temporal lobes, including the hippocampus. The brain also receives blood via the vertebral arteries. The vertebral arteries develop from the subclavian arteries and supply or give rise to vessels that deliver blood to the spinal cord, brainstem, cerebrum, cerebellum, thalamus, hypothalamus, and portions of the posterior medial parietal lobes, and the medial, inferior temporal, and occipital lobes. Cerebral arteries emerging from the internal carotid and vertebral arteries anastomose at the base of the brain to form the Circle of Willis (also known as the cerebral arterial circle or the arterial Circle of Willis). This arrangement of arteries creates redundant vasculature that permits continuous blood flow and maintains an adequate blood supply to the brain even when any one of the smaller arteries within the Circle, or arteries feeding into the Circle, become blocked.

VENOUS DRAINAGE FROM THE BRAIN

The return flow of the cranial and spinal blood supply occurs through two venous systems. Cerebral (internal and external) and cerebellar (superior and inferior) veins drain blood from the brain (2–4). The internal cerebral veins (also known as the veins of Galen) drain the deep parts of the brain. Venous blood flows peripherally and centrally via superficial cerebral veins and deep cerebral veins, respectively, into the venous sinuses. The venous sinuses drain through the internal jugular vein.

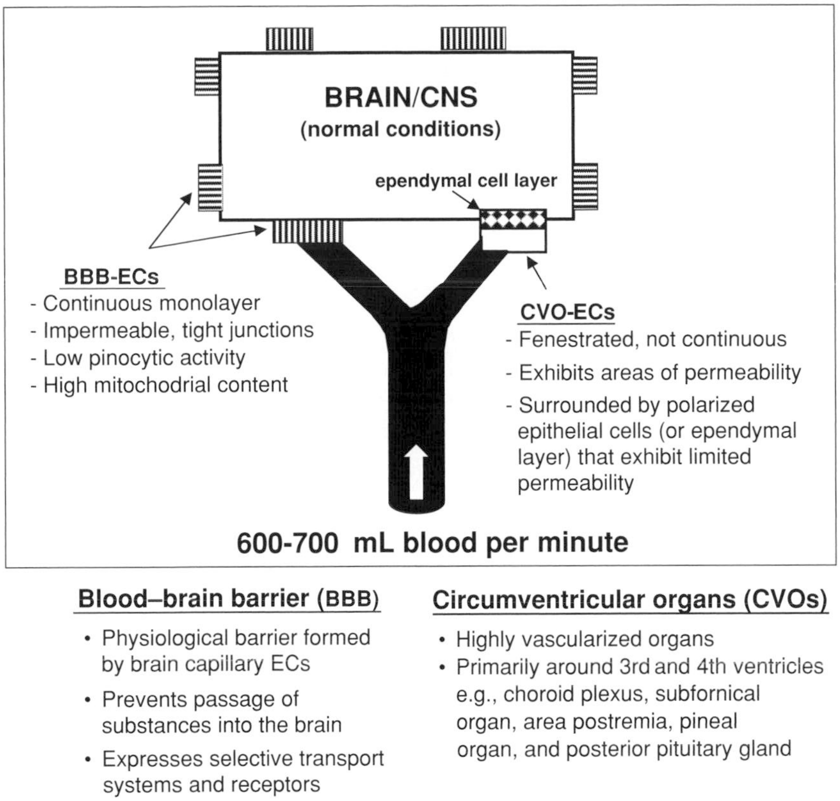
BRAIN/CNS
(normal conditions)
ependymal cell layer
BBB-ECs
- Continuous monolayer
- Impermeable, tight junctions
- Low pinocytic activity
- High mitochodrial content
CVO-ECs
- Fenestrated, not continuous
- Exhibits areas of permeability
- Surrounded by polarized epithelial cells (or ependymal layer) that exhibit limited permeability
600-700 mL blood per minute
Blood–brain barrier (BBB)
• Physiological barrier formed by brain capillary ECs
• Prevents passage of substances into the brain
• Expresses selective transport systems and receptors
Circumventricular organs (CVOs)
• Highly vascularized organs
• Primarily around 3rd and 4th ventricles e.g., choroid plexus, subfornical organ, area postremia, pineal organ, and posterior pituitary gland

Figure 124.1. The BBB protects the brain from peripheral immune system mediators. The brain/CNS receives oxygenated blood from the peripheral circulation, but it is protected from peripheral immune system mediators (cellular and noncellular) under normal physiological conditions. The BBB-EC barrier prevents the passage of hydrophilic substances and immune cells from the peripheral circulation. Unlike the continuous ECs of the BBB, ECs of the CVOs are semipermeable.

OVERVIEW OF THE STRUCTURAL PROPERTIES OF THE ENDOTHELIUM WITHIN THE BRAIN

Barrier Properties of Blood–Brain Barrier Endothelial Cells

One important difference between the circulation in the periphery (nonbrain regions of the body) and the majority of the brain is that most hydrophilic substances can easily pass from the circulation into peripheral tissues. By contrast, under normal conditions most factors present in the blood (excluding very small molecules such as amino acids, glucose, or lipid-soluble molecules [e.g., caffeine, ethanol]) cannot pass through the walls of brain ECs that form the BBB because of their location on a continuous basal lamina (5–7). As a result, the BBB protects the brain from circulating substances that could threaten brain homeostasis (see Figure 124.1). The continuous ECs of the BBB differ from fenestrated ECs (non-BBB-ECs containing small pores or fenestrae approximately 80–100 nm in diameter) typically found in most endocrine glands. BBB-ECs exhibit tight junctions of high electrical resistance that seal cell-to-cell contact sites between adjacent ECs. Unlike fenestrated ECs and sinusoidal ECs (typically found in the liver), BBB-ECs have low pinocytic activity and selective intracellular transport systems that prevent the passage of most molecules between tissues and blood, except in conditions where the integrity of the BBB is compromised (8–10). Finally, because the BBB limits the bulk flow of water across monolayers, the BBB also controls volume regulation within the central nervous system (CNS).

In contrast to the endothelial monolayer of the BBB, CVOs found within the brain (located primarily around the third and fourth ventricles) are devoid of a continuous EC barrier and exhibit areas of permeability along their endothelium (see Figure 124.1) (11–14). CVOs are highly vascularized and specialized organs. Examples of CVOs include the pineal organ (releases melatonin, neuropeptides), subfornical organ (regulates thirst), posterior pituitary gland (releases oxytocin, vasopressin), area postrema (vomiting center), and choroid plexus (produces/maintains cerebral spinal fluid). The choroid plexus is comprised of fenestrated capillaries surrounded by a layer of polarized epithelial cells (also known as ependymal cells) joined by tight junctions (15,16). Molecules that cross the vasculature of the choroid plexus must then penetrate the ependymal layer before mixing with the cerebral spinal fluid. Because very few proteins pass into the cerebral spinal fluid, it is believed that whereas the ECs of the choroid plexus are somewhat permeable, the ependymal layer is not. Like CVOs, the vasculature of brain tumors is semipermeable (17).

Cerebrovascular Innervation and the Regulation of Cerebral Blood Flow

Another important function of ECs within the brain is to regulate blood flow through interactions with the sympathetic and parasympathetic pathways. Based on the high demand for oxygenated blood within the brain, one of the major functions of brain arterial ECs is to regulate cerebral blood flow (through vasodilation and vasoconstriction). Sympathetic and parasympathetic pathways innervate most of the cerebrovasculature (e.g., larger vessels of the Circle of Willis, small pial arteries, and arterioles). Sympathetic-noradrenergic fibers secrete potent vasoconstricting substances, whereas the parasympathetic cholinergic fibers release vasodilating neurotransmitters. Vessels penetrating the brain parenchyma lack adrenergic innervation in most areas of the brain (18). Sympathetic-noradrenergic fibers arise from the superior cervical ganglion and run along the internal carotid artery as a solid nerve (internal carotid nerve). These fibers follow pial arteries and arterioles, and innervate some portions of the Circle of Willis (19). The secretion of norepinephrine and neuropeptide Y by these fibers induces vasoconstriction (20).

Cerebral blood flow is regulated primarily through cholinergic innervation of the brain vasculature. Cholinergic neurotransmitters acetylcholine (ACh) and vasoactive intestinal peptide (VIP) are potent vasodilators of human cranial arteries. The elements of a cholinergic system (acetylcholinesterase and choline acetyltransferase or ChAT) have been extensively studied using brain-derived ECs with somewhat inconsistent results. Some studies report a lack of expression of ChAT, the enzyme responsible for ACh production, by brain ECs (21,22), whereas other studies demonstrate ChAT activity/expression by these same cells (associated with capillaries, arterioles, microvessels) (23–26). The expression of ChAT by many other cell types throughout the body, including T lymphocytes, epithelial cells, mesothelial cells, smooth muscle cells, and ECs, has been established (27). Despite the controversy regarding the production of ChAT by brain ECs, both cerebral arteries and capillary ECs bind [^{3}H]-QNB (a highly specific cholinergic muscarinic antagonist), demonstrating the presence of muscarinic cholinergic receptors (mAChRs) within the brain vasculature (microvessels) (23,28). These receptors found in the cerebral microcirculation are proposed to mediate vasoconstriction, vasodilation, and activation of nitric oxide synthase (NOS) (29). It is postulated that cortical microvessels receive cholinergic input originating from basal forebrain neurons to increase cortical perfusion and dilate intracortical microvessels (30). Brain ECs receive cholinergic signals via ACh–mAChR interactions and produce nitric oxide (NO), a potent vasodilator. Similarly, other cell types found in the periphery (nonbrain regions) expressing mAChRs (e.g., keratinocytes and ECs) are believed to be responsive to cholinergic stimulation. For example, ACh induces contractions and endothelium-dependent relaxation of isolated human pulmonary arteries (that express mAChRs), however, the source of the ACh produced in vivo is not clear (31). Because ACh is degraded almost instantaneously by acetylcholinesterases, it is possible that ACh derived from ECs themselves or neighboring cells reacts with arterial ECs to regulate blood flow. Based on the high demand for oxygen by brain-associated tissues, compromised blood flow within the brain's microcirculation is detrimental to the host.

In addition to its role in the regulation of blood flow, adrenergic innervation of brain microvessels has been implicated in the regulation of BBB permeability (32,33). Brain capillary ECs express adrenergic receptors (34,35), and receptor–ligand interactions are proposed to weaken the integrity of the BBB and allow the influx of numerous peripherally derived inflammatory mediators into the brain during sepsis or infection (36).

Brain Endothelial Cells Are Targets and Producers of Inflammatory Mediators: The Neural–Immune Interface

There are numerous examples of diseases (resulting from known/unknown pathogens and non–pathogen-associated factors) that affect the brain, including sepsis, encephalomyelitis, cerebral malaria, multiple sclerosis, trauma, hypertension, ischemia, and Alzheimer disease (AD). Both exogenous agents (pathogens, toxins, and/or drugs) and endogenous factors (biological influences produced by host cells) alter BBB permeability, damage neuronal cells, and modify cerebral blood flow (37–40). While serving as the brain's barrier, BBB-ECs play a critical role in sensing inflammatory mediators produced in the periphery (nonbrain regions) and found in the circulation (outside the brain) during these diseases and conditions. Brain ECs are targets for inflammatory mediators produced by local cells found in the brain (e.g., microglial cells). In addition to sensing endogenous host mediators, BBB-ECs and non-BBB brain ECs participate in the host's innate immune response (i.e., immediate nonspecific reaction to invading pathogens) through the expression of cell surface receptors with broad specificity for pathogens and specific receptors for endogenous host-derived factors (e.g., cytokines). In response to receptor–ligand interactions, these brain ECs produce a wide range of potent substances (including cytokines, chemokines, complement factors, and vasoactive agents) that can disrupt the BBB, alter cerebral vascular circulation, and/or damage neuronal cells. In many cases, these substances increase leukocyte recruitment to the BBB and are associated with the release of additional leukocyte-derived inflammatory mediators. These substances, in turn, can further increase BBB permeability and permit the flow of harmful inflammatory mediators, cells, toxins, or pathogens into this previously protected compartment, thereby compromising brain homeostasis. Although the neurological effects of numerous infections and trauma have been recognized for many years, only recently has the direct relationship between the brain and the host innate immune response been explored.

In this chapter, we review the experimental evidence supporting the interaction between the brain – particularly brain ECs – and soluble and cellular factors of the innate immune response (both as a receiver of signals and a producer of signals) and neuroimmune interactions. In addition, we describe how the elements (both cellular and soluble) of the innate immune system (in the presence and absence of infection) alter the functions of brain ECs and how brain ECs influence neural degeneration and disease progression during two distinct conditions – sepsis and AD.

COMPARISON OF BRAIN ENDOTHELIAL CELLS AND PERIPHERAL (NONBRAIN) ENDOTHELIAL CELLS

As described in the overview of structural properties of BBB-ECs, the most prominent difference among BBB-ECs (primarily microvasculature) and non-BBB brain ECs (including ECs associated with CVOs) and nonbrain ECs is their barrier function (see Figure 124.1). BBB-ECs form a barrier because they are organized in a continuous monolayer lacking both intercellular clefts and pinocytic vessels. By contrast, most other non–BBB-ECs (including CVO-ECs) are fenestrated (or sinusoidal) and contain intercellular clefts and pinocytic vesicles (8,9). Furthermore, BBB-ECs contain more mitochondria than other ECs, enabling them to supply energy to the multiple energy-dependent transporters found in the BBB (41,42). The high metabolic capacity of BBB-ECs may be associated with carrier-mediated transport activities of the BBB-ECs.

Numerous structural differences between brain and peripheral blood vessels exist. Most of the cerebral vessels (excluding vessels that penetrate the brain parenchyma) are innervated (43). Most cerebral vessels (excluding the larger vessels of the Circle of Willis) are microvessels with internal diameters 300 microns or greater (with most $\leq$100 microns) and many of these microvessels are interconnected to form intricate bundles of tubing for efficient blood flow (44). Finally, brain arteries and arterioles (unlike nonbrain arteries and arterioles) have a single internal elastic lamella (instead of internal and outer elastic lamellae) (44). This structure is believed to contribute to brain aneurysms (or berry aneurysms) that classically occur at the point where the cerebral artery departs from the Circle of Willis.

The most comprehensive comparative studies of brain ECs versus peripheral or nonbrain ECs, at the cellular level, evaluated cerebral microvascular ECs versus human umbilical vein ECs (HUVECs). HUVECs, easily isolated from the large vein of the umbilical cord, are not microvascular (like most brain-ECs) and therefore, may not be the best comparative EC-type. ECs isolated from the brain exhibit many similarities to peripheral ECs (HUVECs); they express von Willebrand factor (vWF) antigen (a reliable cell marker for ECs), contain Weibel-Palade bodies (granules specific for ECs), and demonstrate complex junctional structures (45,46). Similarly, both human cerebral microvessel ECs (prepared from brain tissue [gray matter] after removing the pia mater and arachnoidea from the cerebrum and dissecting out the white matter) and peripheral ECs (HUVECs) express adhesion molecules, integrins, proteases, proangiogenic factors, chemokines, cytokines, growth factors, and cytokine/chemokine receptors (47).

When exposed to interferon (IFN)-γ, lipopolysaccharide (LPS), and tumor necrosis factor (TNF), primary brain capillary ECs (established from white matter) produce lower amounts of interleukin (IL)-1β and soluble intercellular adhesion molecule (ICAM)-1 compared with HUVEC (48). The reduced immunoresponsiveness of brain ECs may be due to their limited cellular interactions with elements of the host immune system. Alternatively, the observed reduced immunoresponsiveness by brain ECs may reflect ex vivo culture conditions rather than the actual activity/function of these cells within the brain in vivo.

The most detailed study comparing mRNA expression by normal human brain ECs (cerebral capillaries isolated from the gray matter) and peripheral (nonbrain) ECs (HUVECs) identified certain genes preferentially expressed exclusively by brain ECs. These include vascular endothelial growth factor (VEGF), fibroblast growth factor (FGF)-2, insulin-like growth factor (IGF) binding proteins-1, -5, and -6, follistatin (an angiogenic protein), integrin-β5 and integrin-α1, and transforming growth factor (TGF)-β_2 among others (47). By contrast, other genes were expressed by HUVECs, but not brain ECs, including platelet-endothelial cell adhesion molecule (PECAM)-1 and ICAM-2 (adhesion molecules); Tie2 and angiopoietin-2 (angiogenic factors); matrix metalloprotease (MMP)-8, -10, and -13 (proteases); CXCR4 (chemokine receptor); FLK-1 (VEGF receptor 2); and RANK (activating receptor activator of nuclear factor [NF]-κB) (47).

When analyzed for protein expression, human brain ECs (prepared from brain tissue [gray matter] after removing the pia mater and arachnoidea from the cerebrum and dissecting out the white matter) express factors associated with neuroprotective and growth supporting activities, as well as immunoregulating and angiogenic functions (47). Basal brain-derived neurotrophic factor (BDNF) and stem cell factor (SCF) expression was five times greater in brain ECs compared with HUVECs, whereas TGF-β_2 expression by brain ECs was 10 times greater than in HUVECs. Although VEGF and follistatin protein levels were undetectable in basal HUVEC cultures, human brain EC cultures contained 300 and 13 pg/mL, respectively. Furthermore, levels of IL-6 and monocyte chemoattractant protein (MCP)-1 (immunoregulatory mediators) were sixfold higher or greater in the brain EC cultures. Several recent studies suggest that IL-6 (49) and MCP-1 are neuroprotective (50). Brain ECs, but not HUVECs, expressed detectable levels of two receptors (at the protein level) involved in EC proliferation, differentiation, and angiogenesis: epidermal growth factor receptor (erbB1) and oncostatin M receptor β-subunit (47).

Together, these and other studies suggest that cerebral microvessel ECs express cell subtype–specific genes and proteins associated with neuroprotective functions, inflammatory responses, cell differentiation, blood flow regulation, and angiogenesis. These patterns are likely to reflect the unique functions of brain ECs. It is important to note that, whereas comparisons of brain-derived ECs versus peripheral ECs following ex vivo culturing are useful, the conclusions of these types of studies are limited to in vitro and ex vivo conditions; that is, they may not be useful for assessing the functions/phenotypes of these cells in vivo because the results of such studies are biased by isolation procedures, culture conditions, EC type (e.g., arteriole, venule, large vessel, capillary), and origin or tissue type/location. An important goal is to develop improved methods for assaying brain and nonbrain ECs within the context of their native environment. In addition, the isolation of specific brain EC subsets (BBB-ECs, CVO-ECs, arterial ECs, venous ECs, etc.) would allow for the comparison of these different subsets. Such tools might include the use of laser capture microdissection, rapid isolation procedures, and/or the use of in vivo phage display.

INTERACTIONS BETWEEN BRAIN ENDOTHELIUM AND IMMUNE MEDIATORS (SOLUBLE AND CELLULAR) IN DISEASE

Brain Endothelial Cells Receive Signals from the Host Immune System and Produce Inflammatory Signals

Under certain circumstances, brain ECs receive inflammatory signals generated within the brain by local immune cells (e.g., microglial cells, also known as brain macrophages). Several studies support the concept that circulating proinflammatory factors generated by cells of the peripheral immune system (i.e., the host immune system outside the brain) during infection, trauma, ischemia, or inflammation signal to the brain to evoke further neural and host inflammatory responses. ECs of the BBB form tight junctions and prevent the passage of most inflammatory mediators (produced by the peripheral immune system present in circulation) into the brain under normal conditions (see Figure 124.1). In some cases, disruption of the BBB allows the passage of mediators into the brain. However, numerous additional mechanisms have been proposed to describe how immune system–derived factors produced outside the brain reach the brain (either by physically entering the brain or by eliciting a response without entering the brain, i.e., via nerve fibers).

First, some cytokines and mediators produced in the periphery (outside the brain) can pass the intact BBB by interacting with specific BBB-EC–associated active, saturable transport systems (Figure 124.2). Examples include the putative IL-1α transport system observed in the rat BBB (51,52) and the TNF transporter found in the mouse brain (53,54). Once transported across the BBB, IL-1 is believed to induce fever, neuroendocrine activation, and behavioral changes during disease, in part, through the production of prostaglandins (PGs). These BBB-expressed transporters are regulated, in part, by the adrenergic system (55). In the case of TNF, both the p55 and p75 TNF receptors are necessary for transport across the BBB (56). Although detailed cellular and intracellular events involved in transporting TNF and IL-1 across the EC barrier are lacking, these transport systems bind their ligands in a saturable and competitive manner and then transfer the factors into the brain tissue without disrupting the BBB (56). It is not clear exactly how these transporters differ from receptors; Pan and Kastin have described "transporters" as a means to chaperone factors into the brain parenchyma from the circulation without degradation by BBB-ECs (56). Similarly, numerous specific membrane transporters involved in the passage of glucose (e.g., GLUT1), electrolytes, drugs, and other substances are localized throughout the brain.

Second, cytokines produced by immune cells (outside the brain) and found in the peripheral circulation bind to receptors present on peripheral nerves that send signals to the brain. In this case, peripherally derived cytokines induce inflammatory responses within the brain without actually entering the brain. An example of this type of messaging system is the expression of cytokine receptors by trigeminal ganglions in the periphery that activate the hypothalamus-pituitary-adrenal (HPA) axis (57). Similarly, peripheral cytokines interact with the afferent branches of the vagus nerve (58) that terminate in the nucleus tractus solitarius (NTS) of the brainstem medulla and prompt brain responses to infection and inflammation (see Figure 124.2). The afferent limb of the vagus nerve innervates many internal organs including the heart, lungs, gastrointestinal tract, and liver, and is thus well positioned to transmit host innate immune response-derived signals from peripheral tissues to the brain (59). In addition to receiving signals from the periphery, the vagus nerve responds through the release of neurotransmitters. The principal neurotransmitter released from the axon terminals of postganglionic vagal fibers following stimulation is ACh. It has been shown that ACh released by the efferent vagus nerve interacts with α-7-nicotinic ACh receptors (α7-nAChR) expressed by macrophages within the periphery to deactivate them (suppressing proinflammatory cytokine production) in vitro following LPS stimulation and in vivo during experimental sepsis (see Figure 124.2) (60–62). This novel pathway, referred to as the cholinergic anti-inflammatory pathway, represents a new function of the efferent vagus nerve: a negative feedback loop that inhibits inflammatory cytokine production and modulates both systemic and local inflammatory responses in experimental models (59,63,64). Evidence is lacking for the direct interaction between ACh produced by vagus nerve stimulation and receptors expressed on target cells.

Although previous studies have focused on the effects of cholinergic stimulation on peripheral cytokine production during infection and inflammation, the possibility remains that cholinergic stimulation also exerts anti-inflammatory

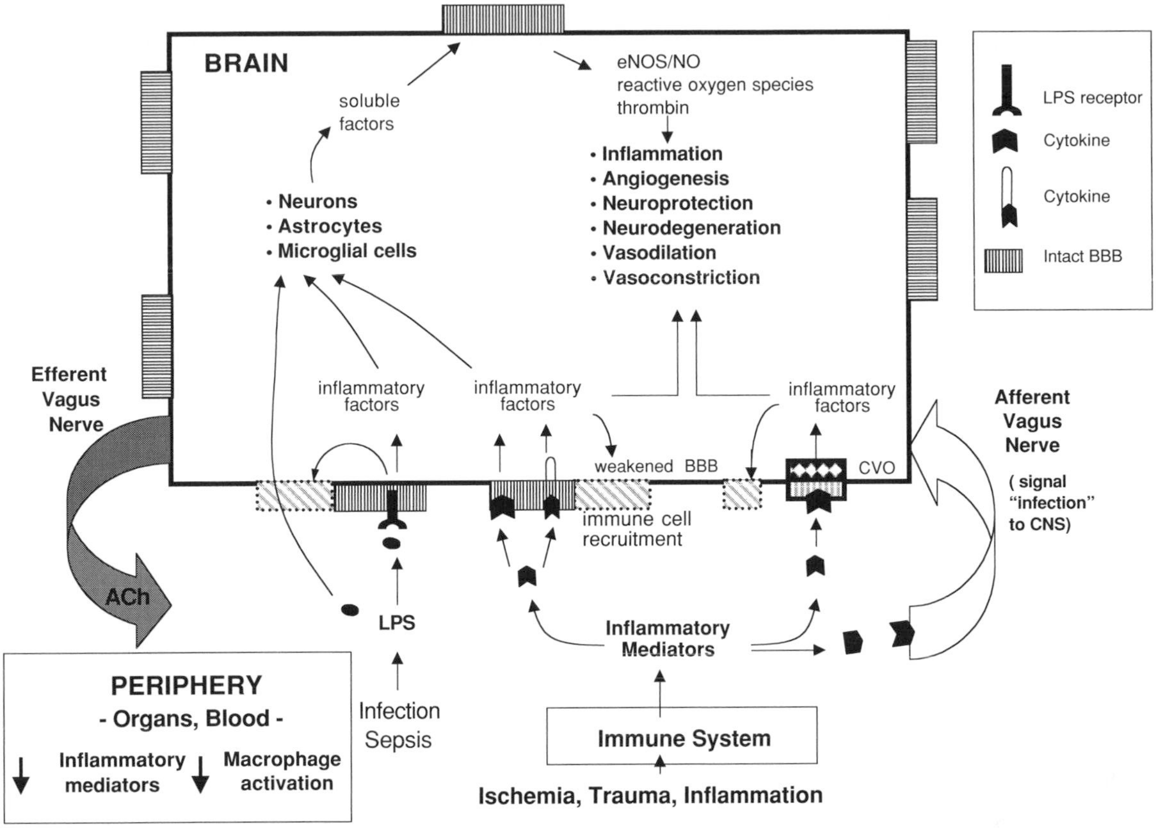

Figure 124.2. Brain EC–immune system interactions during infection and sepsis.Immune mediators produced in the periphery during inflammation and sepsis bind receptors of the afferent vagus nerve and signal to the brain that peripheral inflammatory responses have been initiated. The brain responds by releasing ACh via the efferent vagus nerve to suppress macrophage activation in the periphery. In addition, peripherally derived inflammatory mediators, pathogen products (e.g., LPS), and immune cells bind to BBB-ECs via receptors and transporters to invoke further inflammatory responses that weaken the integrity of the BBB. Peripherally derived inflammatory mediators gain access to the brain through a permeable BBB (and possibly CVOs) and elicit further inflammatory responses by local microglial cells, astrocytes, and neurons. Thus, neural EC–immune interactions regulate inflammation, angiogenesis, neurodegeneration/neuroprotection, and vasoconstriction/vasodilation within the brain during infection and sepsis.

responses within the brain (65). Interestingly, cerebral microvascular ECs and peripheral ECs express the α7-nAChR and respond to cholinergic stimulation (66–68). Although it is not clear whether ACh released by the stimulation of the vagus nerve can reach the endothelium (due to its degradation by acetylcholinesterases), cholinergic stimulation of the endothelium can be mediated via cholinergic drugs, including nicotine (which can easily enter the brain by crossing the BBB). Based on these observations and what is known about the cholinergic anti-inflammatory pathway, ECs within the brain and periphery may represent potential targets of cholinergic stimulation, leading to reduced EC inflammatory responses (e.g., adhesion molecule expression, cytokine/chemokine production) and associated leukocyte recruitment during infection. These studies are the focus of ongoing investigations in the laboratory of one of the authors of this chapter (CNM).

Third, cytokines and other pro-inflammatory mediators may gain access to the brain by passing through CVOs (see Figure 124.2). As describe earlier, CVOs, highly vascularized and specialized organs, are small barrier–deficient areas of the brain located periventricularly (e.g., the subfornical organ, choroid plexus, posterior pituitary gland, area postrema, and pineal gland). CVO-ECs are somewhat permeable and lack a continuous EC barrier found among BBB-ECs. These CVOs provide mechanisms for releasing brain-derived substances (e.g., the secretion of hormones into the systemic circulation) and potentially receiving nonbrain-derived factors into the brain. Non–BBB-ECs within the cerebral vasculature (mostly within the choroid plexus and meninges) express the IL-1 receptor type 1 (IL-1R1) (69). Although many studies suggest the entry of molecules into the brain parenchyma via CVOs, solid experimental evidence supporting this theory is lacking. It has been proposed that IL-1 and TNF pass through CVOs to activate the febrile response through activation of neurons in the anterior hypothalamus (70). Others argue that it is unlikely that molecules pass across the choroid plexus into the cerebral spinal fluid despite the fenestrated ECs due to the tight junctions between choroid epithelial (ependymal) cells surrounding the ECs (71). Evidence against the considerable permeability of CVOs to cytokines and other peripherally (nonbrain)-derived factors is provided by studies showing that the passage of IL-1α across CVOs (subfornical organ and choroid plexus) was very restricted due to tightly packed ependymal cells (72).

As discussed earlier, cerebral vascular ECs, including those associated with the BBB and CVOs, express receptors for circulating inflammatory mediators found during infection/inflammation (e.g., cytokines, leptin, and LPS) (73). Following receptor engagement, ECs can secrete secondary mediators (e.g., NO and PGE_2) that are associated with disruption of the BBB and/or induce further inflammatory responses within the brain (see Figure 124.2). For example, PGE_2 produced by ECs following receptor–ligand interactions can cross the BBB to stimulate temperature increases induced through the activity of preoptic neurons (74). Examples of cytokine receptors found on brain ECs include: IL-1R1, IL-6R, IFN-γ receptor (IFNγ-R)-1 (69,75). The IL1-R1 is expressed by ECs found mainly in the choroid plexus and meninges (69). *IL1-R1* mRNA expression found in the rat brain endothelium (both luminal and abluminal sides) of the organum vasculosum laminae terminalis (OVLT), a CVO, is constitutive and further enhanced by inflammatory mediator stimulation following intracerebral IL-1β or IL-6 injection (76). *IL-6* mRNA signals are observed in the cerebral cortex, hippocampus, and amygdala associated with blood vessels (76). Brain ECs (both CVO-ECs and BBB-ECs) also express receptors for leptin, which allows the passage of circulating leptin into the brain tissue and cerebral spinal fluid (77). In addition, brain ECs isolated from LPS-treated rats express LPS receptors (at the mRNA level) including *TLR2, TLR4,* and *CD14* (78). These LPS–EC and cytokine–EC interactions are believed to increase the permeability of the BBB and activate a series of inflammatory signals by ECs (and other cells) within the brain (78).

In Response to Inflammatory Signals, Brain Endothelial Cells Produce Numerous Proinflammatory Mediators

Brain ECs can produce numerous inflammatory factors, including IL-1 family members, IL-6, β-chemokines, and endothelial NOS (eNOS). Several brain regions lacking the BBB (the choroid plexus, meninges, median eminence, paraventricular nucleus, and arcuate nucleus) express IL-1β and IL-1 receptor antagonist (IL-1ra), with inductions observed following LPS treatment (79). *IL-1β* mRNA expression is found within the brain vasculature (BBB region), acting on IL-1 type 1 receptors localized at the interface between vasculature and perivascular glia (80). IL-1β release enhances inflammatory responses within the brain (e.g., stimulating production of PGE_2 by ECs) (81). By contrast, it has been proposed that IL-1ra expression protects the cerebral vasculature and surrounding cells during inflammation (82). Brain ECs (microvessel) constitutively express IL-6, and this basal expression can be enhanced by IL-1β stimulation (83,84). IL-6 production by brain endothelium may promote vascularization (under normal conditions and during the healing of brain injuries) (85). Moreover, IL-6 may function as a neuroprotective agent within the brain by protecting neurons from oxidative damage (49,84,86). By contrast, using an experimental model of pneumococcal meningitis, IL-6 has been shown to enhance the permeability of the BBB, suggesting that IL-6 produced by ECs within the BBB may weaken its barrier function (87).

β-Chemokines play an important role in recruiting immune cells to sites of inflammation and infection. Untreated human brain microvessel ECs express low basal levels of *MIP-1β* and regulated on activation, normal T-cell expressed and secreted (*RANTES*) mRNA and expressions are significantly upregulated by cytokine stimulation (88). LPS injection (experimental endotoxemia) induces MCP-1 expression within the brain, specifically within BBB and non-BBB vascular-associated structures (89). MCP-1 expression is thought to contribute to the early chemotactic events during cerebral inflammation. Recruitment of cytokine-secreting leukocytes to the BBB would further compromise its barrier function.

eNOS expression by brain ECs contributes to NO production within the brain. NO is the most potent vasodilator, and its production is considered to be essential for normal brain physiology. In an experimental model of sepsis using eNOS knockout mice, Connelly and colleagues showed that eNOS (via NO production) serves as an inflammatory mediator by promoting upregulation of inducible nitric oxide synthase (iNOS), and its expression significantly impacts sepsis-associated mortality (90). Constitutively expressed NO also regulates EC-associated adhesion molecule expression, leukocyte activation, and leukocyte adhesion (91). Therefore, EC-derived NO might regulate cell trafficking during infection within the brain.

Thus, brain ECs are not only targets of immune mediators, but also contribute to immune responses through the production of numerous immunomodulatory factors.

THE POTENTIAL ROLE OF BRAIN ENDOTHELIAL CELLS IN MEDIATING NEUROLOGICAL DYSFUNCTION DURING INFECTION AND SEPSIS

As discussed earlier, brain ECs have the capacity to produce numerous soluble factors during inflammation and infection. These mediators may play important roles in disrupting the BBB and regulating cell migration, neurodegeneration, brain-associated inflammatory responses, vasculogenesis, and cerebral blood flow. These processes may contribute to the neurological impairment commonly observed during sepsis. Approximately 50% of patients with progressive sepsis develop encephalopathy (or diffuse brain dysfunction) (92). Similarly, 70% of patients with bacteremia exhibit neurological symptoms ranging from lethargy to coma (93,94). In bacteremic patients, mortality directly correlates with the severity of the neurological symptoms (as measured by electroencephalograms) (94). The clinicopathological changes in sepsis-associated encephalopathy are similar to those reported in acute hemorrhagic leukoencephalitis (AHLE) (95). AHLE presents with fever, headache, stupor, confusion, delirium, and/or coma and is characterized by CNS necrosis, hemorrhage, and demyelination of the white matter (96). Although numerous studies support the hypothesis that sepsis-associated encephalopathy is due, in part, to the disruption of the BBB by proinflammatory mediators and/or EC-mediated transport of cytokines with subsequent brain inflammatory responses, additional evidence suggests that local microvascular dysfunction resulting in impaired cerebral blood flow and increased cerebrovascular resistance also contributes to sepsis encephalopathy (97). Together these observations suggest that brain-associated ECs may significantly contribute to the neurological impairment observed during sepsis (see Figure 124.2).

THE ROLE OF BRAIN-ASSOCIATED ENDOTHELIAL CELLS IN SEPSIS: THERAPEUTIC IMPLICATIONS

Based on their functional capacity, brain ECs are potential therapeutic targets for the treatment of infection/sepsis, including sepsis-associated neurological impairment. Activated protein C (APC), an anticoagulant and anti-inflammatory agent approved by the FDA for the treatment of severe sepsis in humans (98) has been shown to exert anti-inflammatory, antithrombotic, and neuroprotective effects in a murine model of stroke (99). Therefore, APC may help improve neurological responses observed during sepsis in addition to regulating peripheral coagulant/thrombotic activity. Drugs that alter NO production during sepsis also may be useful for preventing or treating encephalopathy during sepsis due to their effects on improving cerebral blood flow (97). A recent study reports the therapeutic effect of nicotine, a cholinergic agonist, in reducing mortality in experimental sepsis (62). Cholinergic agonists are postulated to interact with α7-nAChRs expressed by macrophages and possibly ECs to deactivate them and suppress proinflammatory cytokine production (61,68). These observations imply that cholinergic agonists may interact with α7-nAChR-expressing cells (present in the periphery and/or in the brain) to inhibit excessive cellular activation and inflammatory responses during sepsis. Finally, factors that protect the BBB from disruption or repair the disrupted BBB would be of interest for treating patients with severe infections and/or sepsis.

ALZHEIMER DISEASE

The Role of Brain Endothelial Cells in Alzheimer Disease: the Evidence

Despite intense investigative efforts by numerous laboratories around the world, the pathological basis for AD remains unclear. Sepsis often is associated with infectious agents and clearly involves a systemic host immune response, whereas systemic (i.e., nonbrain regions) involvement during AD has not been firmly established. Although numerous studies support the presence of an altered peripheral immune system in AD patients, the mechanistic understanding of how the peripheral immune system plays a role in AD is not understood. Several studies suggest a decline in peripheral immune responsiveness in AD patients, as evidenced by the reduced production of TNF and IL-6 by LPS-treated peripheral blood cells obtained from AD patients when compared with age-matched control blood cells (100–102). By contrast, other studies suggest increased peripheral inflammation during AD, as evidenced by increased levels of circulating IL-6, TNF, and leptin (103,104). Despite the controversy over the role of the peripheral immune system in AD pathogenesis, most studies agree that microglial cells, innate immune cells of myeloid origin present in the central nervous system, contribute to brain inflammation observed during AD (Figure 124.3) (20,105). In addition, several suggested mechanisms for AD implicate the role of the brain vasculature and local EC–immune interactions in neuronal destruction (106). In this section, we review several lines of evidence for the relationship between the brain-associated endothelium/vasculature and AD pathogenesis (107,108) and how immune mediator (cellular or soluble)-EC interactions within the brain might regulate AD progression (Figure 124.3).

AD is characterized by deficits in the production of ACh within the brain resulting in poor cerebral perfusion (Figure 124.3). Locally released ACh (from nerves and/or other ACh-releasing cell types, such as smooth muscle cells and possibly ECs, themselves) within the brain binds to heterogeneously expressed mAChRs expressed by the endothelium. ACh–mAChR interactions within the brain trigger endothelium-related NO production, which leads to cerebral vasodilation and increased blood flow. In AD, impaired ACh production and/or ACh–mAChR interactions with resultant decreased cerebral perfusion are proposed to contribute to neuronal damage. Acetylcholinesterase inhibitors (e.g., donepezil, rivastigmine, galantamine) that facilitate cholinergic transmission are a class of U.S. Food and Drug Administration (FDA)-approved drugs used for improving both cognitive and behavioral dysfunctions associated with AD (109). Recent studies using experimental models of AD show that acetylcholinesterase inhibitors also inhibit inflammatory cytokine production (IL-1) within the brain, suggesting multiple mechanisms for acetylcholinesterase inhibitors for the treatment of AD (110). Of course, other neurotransmitters may be involved in the course of AD,

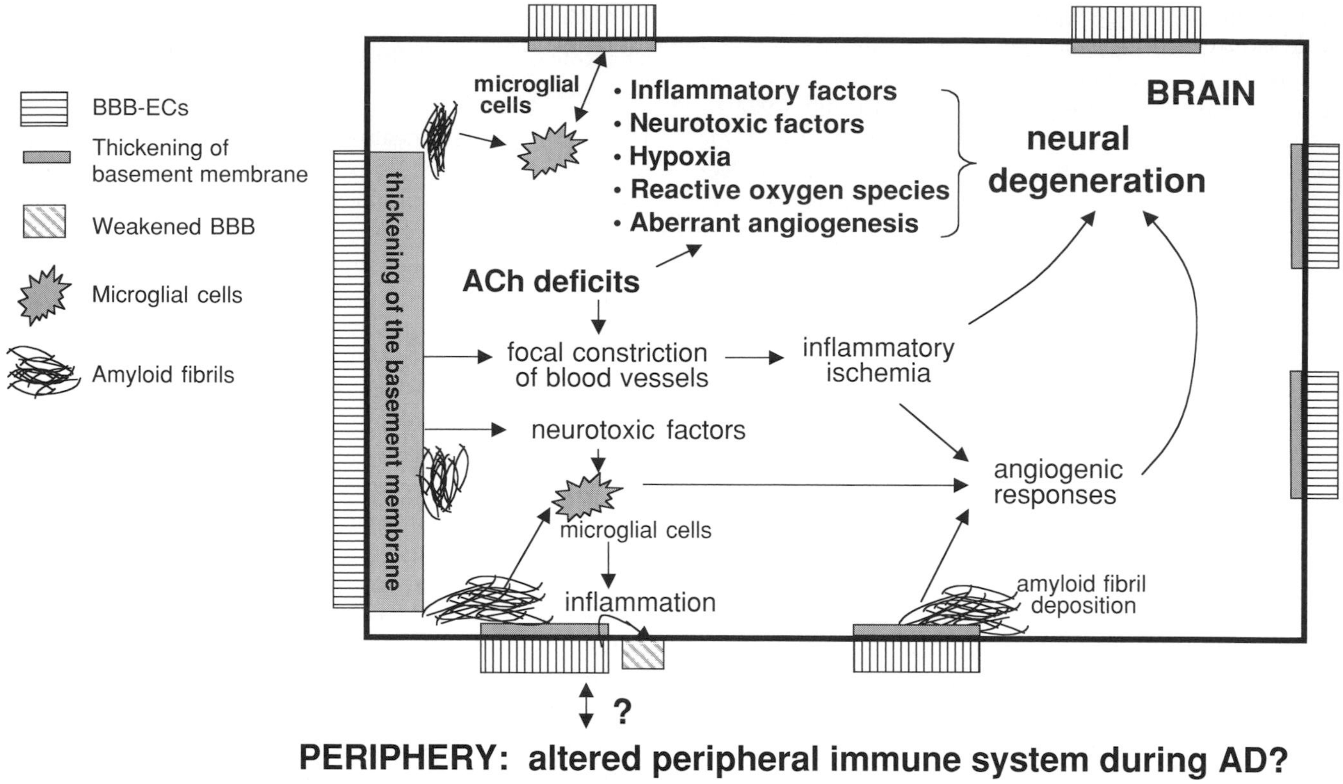

Figure 124.3. Brain EC–immune cell interactions during Alzheimer disease (AD). Unlike sepsis, the role of the peripheral immune system in the pathogenesis of AD is not known. ACh deficits within the brain lead to vasoconstriction, hypoxia, decreased cerebral perfusion, and resultant inflammatory responses by ECs and microglial (immune) cells during the pathogenesis of AD. Morphological changes in the brain's blood vessels (i.e., thickening of the basement membrane) lead to focal vasoconstriction and further hypoxia. Amyloid fibril deposition along the basement membrane of the brain's blood vessels promotes angiogenic responses by ECs and inflammation, as evidenced by microglial cell accumulation. Inflammatory mediators produced by microglial cells weaken the BBB-EC barrier. Local EC–immune cell interactions within the brain regulate numerous processes that contribute to AD pathogenesis, including hypoxia and the production of neurotoxic factors, reactive oxygen species, angiogenic factors, and inflammatory mediators.

including somatostatin, dopamine, and γ-aminobutyric acid (111).

Histopathological examination of the brain microvasculature in AD shows an increase in the basement membrane surrounding vessels and areas of vascular constriction. Early reports describe the presence of senile plaques containing amyloid fibrils that appear to be produced in the basement membranes of capillary ECs and then spread into the parenchyma (112). In addition, inflammation (demonstrated by the presence of reactive microglial cells and astrocytes) within the vicinity of amyloid plaques and surrounding the microvasculature are observed during AD (20,105). The cause of this inflammatory response is not known; however, it may be due to the presence of cytokines (IL-1, IL-6, TNF) in the CNS (secreted, in part, by activated microglial cells and/or astrocytes) and/or by the synthesis and deposition of β-amyloid (which might induce cytokine production within the brain) (20,113). This inflammatory state probably impacts microvessels surrounding regions of β-amyloid deposition and alters the integrity of the BBB (106). Most capillaries within the cerebral cortex show signs of degeneration, and microvessels exhibit sites of constriction in terminal arterioles (112). These studies examining the relationship between vascular degeneration and neural injury suggest that capillary and vascular changes occur first and are followed by neural damage and dysfunction; they support the hypothesis that aberrations in the brain microvasculature associated with perfusion deficits lead to neural degeneration in AD patients.

Evidence of altered angiogenesis is observed in the brains of AD patients. It is hypothesized that ECs within the brains of AD patients are activated to promote angiogenesis due to brain hypoxia and localized inflammation (114). Products released by activated microglial cells in areas of β-amyloid deposition may induce angiogenesis. Using a genome-wide expression profiling approach, Pogue and coworkers confirmed significant changes in the gene expression associated with stress-induced angiogenesis in brains of AD patients (115). Consistent with these findings, AD is associated with increased levels of angiogenic factors (e.g., VEGF and TGF-β) in the cerebrospinal fluid (CSF) (116). Furthermore, a recent study revealed that amyloid β-peptides, integral components of the senile plaques and vascular deposits in AD brains, promote angiogenesis by inducing EC proliferation, migration, and pseudocapillary formation in vivo (117). Although this study clearly shows the angiogenic activity of amyloid β-peptides, the results are not consistent with a previous report demonstrating the antiangiogenic effect of amyloid β-peptides in vitro and in vivo (118). Although the role of angiogenesis in AD pathogenesis remains controversial, several lines of support exist for it.

One of the most recent postulates is that the brain endothelium itself synthesizes neurotoxic factors, including reactive oxygen species that induce neuronal cell death (119) and a toxic peptide (thrombin) that selectively kills cortical neurons (120). Thrombin released from cultured brain ECs can be induced by inflammatory mediators (LPS, IL-1β, IL-6, IFN-γ, and TNF), NO releasing compounds, and by protein kinase C (PKC) inhibition. Thrombin appears to have both direct and indirect (via its cleavage of apoE4) neurotoxic effects (121). Because thrombin also promotes EC migration, proliferation, and survival (122), it may recruit additional ECs or microglial cells to the sites of damage and sustain the release of neurotoxic factors by accumulating ECs (see Figure 124.3). If the production of neurotoxic factors by brain ECs contributes to AD pathogenesis, what causes the production/release of these factors by brain ECs – activated microglial immune cells (in response to β-amyloid)?

The Role of Brain Endothelial Cells in Alzheimer Disease: Therapeutic Implications

The brain vasculature appears to play a significant role in the pathogenesis of AD through multiple mechanisms, including inflammatory, hemostatic, and angiogenic responses. Each represents a potential target for therapeutic intervention. ECs within the brain play an important role regulating blood flow in AD and mediate some of the effects of acetylcholinesterase inhibitors. Indeed, acetylcholinesterase inhibitors (e.g., donepezil, rivastigmine, galantamine) relieve, in part, the cholinergic defect observed in AD (109,123). Drugs such as nonsteroidal anti-inflammatory agents and statin-type drugs used to treat inflammatory diseases involving the endothelium (e.g., atherosclerosis) also appear to reduce the risk of AD (114,124,125). These drugs may alter the inflammatory responses of the endothelium (induced by innate immune microglial cells) to reduce neural damage and improve disease outcome in AD patients. Other drugs (including nonsteroidal anti-inflammatory drugs, statins, histamine H_2-receptor blockers, calcium-channel blockers, and other antiangiogenic agents) that modulate deviant angiogenic responses of brain ECs also may be useful in treating and preventing AD (114). Finally, microglial cells appear to regulate some EC and non-EC inflammatory responses within the brain during AD and therefore, offer another potential target for AD therapeutics.

NOVEL RESEARCH TOOLS TO STUDY THE ROLE OF BRAIN ENDOTHELIAL CELLS IN DISEASE

The functional characteristics of ECs and their associated vasculature are specific and determined, in part, by their location (organ, artery versus vein versus capillary) and the organism's overall state of health (within the brain/CNS and in the periphery). One major obstacle to improving our understanding of the role of brain vascular ECs and associated vasculature in disease is a lack of tools with which to study specific populations of brain ECs. In many studies, brain ECs are isolated and studied under ex vivo conditions. Although these studies provide important information about some of these unique populations of cells, they are limited by isolation procedures and/or culture conditions, and thus do not fully describe the functional role of ECs within the brain under normal conditions and during disease pathogenesis. In addition, methods used to isolate specific EC populations within the brain are limited. In vivo studies of the vasculature are difficult because the endothelium is one-layer thick and spread out. One recently described model system to assess the function of specific genes expressed by brain ECs employs the use of the brain-specific *vWF* gene promoter fragment (−487-+247) that targets the expression of the LacZ marker gene to brain vascular ECs in transgenic mice (126). Along these lines, transgenic mice whose brain ECs express human amyloid β-precursor protein have been generated and will be studied (126,127). Similarly, the expression of other specific genes (e.g., neuroprotective) by brain ECs, and interactions between brain ECs and microglial cells during infection/sepsis and AD, will be interesting to explore.

KEY POINTS

- Although most brain ECs are primarily microvascular, they vary depending on location within the brain. BBB-ECs exhibit tight junctions (i.e., are impermeable to most substances), low pinocytic activity, and lack fenestrations and intercellular clefts. ECs of CVOs are fenestrated and exhibit areas of permeability (see Figure 124.1). However, it is believed that ependymal cells surrounding the CVO ECs limit the transport of circulating factors into the brain.
- Brain ECs serve multiple functions in normal physiology. Tight junctions seal the contacts between continuous ECs within the BBB and prevent interactions between the brain and elements of the peripheral immune system. This barrier function of brain ECs is critical to prevent the unregulated entry of pathogens and systemic inflammatory

mediators into the brain compartment. Under normal conditions, brain ECs have limited interactions with immune cell-derived mediators. However, during infection/sepsis and AD, brain ECs serve as a bridge between neural and immune networks (see Figures 124.2 and 124.3). In some cases, peripheral immune cells or mediators interact with brain ECs and, in other cases, brain ECs interact with brain immune cells (microglial cells). During disease states, immune cell–brain EC interactions can evoke responses within the brain (inflammatory, angiogenic, neurological responses) and in the periphery (immune responses). Finally, brain ECs are not only targets of immune mediators: Brain ECs produce numerous immunomodulatory factors.

- During infection and sepsis, inflammatory mediators produced in the periphery (nonbrain regions) signal to the brain (through multiple mechanisms involving ECs, immune cells, and nerves) to elicit neurological responses (see Figure 124.2). The brain endothelium is an important cell type involved in this process: It detects and sends signals within the brain and between the brain and the periphery. Brain ECs modulate leukocyte recruitment, vasculogenesis, wound healing responses, vasodilation/vasoconstriction, coagulation/fibrinolytic activity, and both neurodegenerative and neuroprotective responses that trigger both central and peripheral effects.
- Although an altered peripheral immune system is observed during AD, the role of the peripheral immune system in mediating AD pathogenesis is not clear. The roles of brain ECs in AD pathogenesis are illustrated in Figure 124.3. Both morphological and biochemical aspects of brain endothelium are believed to be critical factors in mediating the neural damage associated with AD progression: (a) altered basement membrane structure, (b) amyloid deposition along the basement membrane, (c) aberrant angiogenesis, and (d) microvascular constriction leading to hypoxia and inflammatory responses in the brain are proposed to be the early events in AD pathogenesis. In addition, the endothelium is a potential source of neurotoxic factors implicated in neural cell damage during AD. Clearly, EC–microglial cell interactions contribute to the pathogenesis of AD.
- Based on their localization, structure, and function, brain-associated ECs represent potential targets for therapeutic strategies for the treatment of both infection/sepsis and AD.

REFERENCES

1 Magistretti P, Pellerin L, Martin J-L. Brain energy metabolism: an integrated perspective. In: *Neuropsychopharmacology: The Fifth Generation of progress.* Philadelphia: Lippincott Williams & Wilkins, 2000.

2 Bevan RD, Bevan JA. *The Human Brain Circulation: Functional Changes in Disease.* New York: Humana Press, 1994.

3 Stranding S. *Gray's Anatomy: The Anatomical Basis of Clinical Practice.* New York: Churchill Livingstone, 2004.

4 Reganchary SS, Ellenbogen RG. *Principles of Neurosurgery*, 2nd Ed. New York: Mosby, 2004.

5 Risau W, Wolburg H. Development of the blood-brain barrier. *Trends Neurosci.* 1990;13(5):174–178.

6 Staddon JM, Rubin LL. Cell adhesion, cell junctions and the blood-brain barrier. *Curr Opin Neurobiol.* 1996;6(5):622–627.

7 Joo F. Endothelial cells of the brain and other organ systems: some similarities and differences. *Prog Neurobiol.* 1996;48(3): 255–273.

8 Pollay M, Roberts PA. Blood-brain barrier: a definition of normal and altered function. *Neurosurgery.* 1980;6(6):675–685.

9 Hirano A, Kawanami T, Llena JF. Electron microscopy of the blood-brain barrier in disease. *Microsc Res Tech.* 1994;27(6): 543–556.

10 Lossinsky AS, Shivers RR. Structural pathways for macromolecular and cellular transport across the blood-brain barrier during inflammatory conditions. Review. *Histol Histopathol.* 2004; 19(2):535–564.

11 Mark MH, Farmer PM. The human subfornical organ: an anatomic and ultrastructural study. *Ann Clin Lab Sci.* 1984;14 (6):427–442.

12 Faraci FM, Choi J, Baumbach, et al. Microcirculation of the area postrema. Permeability and vascular responses. *Circ Res.* 1989; 65(2):417–425.

13 Miller AD, Leslie RA. The area postrema and vomiting. *Front Neuroendocrinol.* 1994;15(4):301–320.

14 Gross PM, Sposito NM, Pettersen SE, Fenstermacher JD. Differences in function and structure of the capillary endothelium in gray matter, white matter and a circumventricular organ of rat brain. *Blood Vessels.* 1986;23(6):261–270.

15 Groothuis DR, Levy RM. The entry of antiviral and antiretroviral drugs into the central nervous system. *J Neurovirol.* 1997; 3(6):387–400.

16 Segal MB. The choroid plexuses and the barriers between the blood and the cerebrospinal fluid. *Cell Mol Neurobiol.* 2000; 20(2):183–196.

17 Coomber BL, Stewart PA, Hayakawa K, et al. Quantitative morphology of human glioblastoma multiforme microvessels: structural basis of blood-brain barrier defect. *J Neurooncol.* 1987;5(4):299–307.

18 Peerless S, Kendall M. *The Cerebral Vessel Wall.* New York: Raven Press; 1976:175–181.

19 Hardebo J, Suzuki N. Neural pathways to the cerebral circulation. In: Bevan R, Bevan J, eds. *The Human Brain Circulation.* New York: Humana Press; 1994:37–46.

20 Pinter E, Helyes Z, Petho G, Szolcsanyi J. Noradrenergic and peptidergic sympathetic regulation of cutaneous microcirculation in the rat. *Eur J Pharmacol.* 1997;325(1):57–64.

21 Kasa P, Pakaski M, Joo F, Lajtha A. Endothelial cells from human fetal brain microvessels may be cholinoceptive, but do not synthesize acetylcholine. *J Neurochem.* 1991;56(6):2143–2146.

22 Pakaski M, Kasa P. Glial cells in coculture can increase the acetylcholinesterase activity in human brain endothelial cells. *Neurochem Int*. 1992;21(1):129–133.

23 Estrada C, Hamel E, Krause DN. Biochemical evidence for cholinergic innervation of intracerebral blood vessels. *Brain Res*. 1983;266(2):261–270.

24 Galea E, Estrada C. Periendothelial acetylcholine synthesis and release in bovine cerebral cortex capillaries. *J Cereb Blood Flow Metab*. 1991;11(5):868–874.

25 Arneric SP, Honig MA, Milner TA, et al. Neuronal and endothelial sites of acetylcholine synthesis and release associated with microvessels in rat cerebral cortex: ultrastructural and neurochemical studies. *Brain Res*. 1988;454(1–2):11–30.

26 Benagiano V, Virgintino D, Rizzi A, et al. Cholinergic nerve fibres associated with the microvessels of the human cerebral cortex: a study based on monoclonal immunocytochemistry for choline acetyltrasferase. *Eur J Histochem*. 2000;44(2):165–169.

27 Wessler I, Kirkpatrick CJ, Racke K. Non-neuronal acetylcholine, a locally acting molecule, widely distributed in biological systems: expression and function in humans. *Pharmacol Ther*. 1998;77(1):59–79.

28 Grammas P, Caspers ML. The effect of aluminum on muscarinic receptors in isolated cerebral microvessels. *Res Commun Chem Pathol Pharmacol*. 1991;72(1):69–79.

29 Linville DG, Hamel E. Pharmacological characterization of muscarinic acetylcholine binding sites in human and bovine cerebral microvessels. *Naunyn Schmiedebergs Arch Pharmacol*. 1995;352(2):179–186.

30 Hamel E. Cholinergic modulation of the cortical microvascular bed. *Prog Brain Res*. 2004;145:171–178.

31 Walch L, Taisne C, Gascard JP, et al. Cholinesterase activity in human pulmonary arteries and veins. *Br J Pharmacol*. 1997; 121(5):986–990.

32 Sarmento A, Borges N, Azevedo I. Adrenergic influences on the control of blood-brain barrier permeability. *Naunyn Schmiedebergs Arch Pharmacol*. 1991;343(6):633–637.

33 Borges N, Shi F, Azevedo I, Audus KL. Changes in brain microvessel endothelial cell monolayer permeability induced by adrenergic drugs. *Eur J Pharmacol*. 1994;269(2):243–248.

34 Durieu-Trautmann O, Foignant N, Strosberg AD, Couraud PO. Coexpression of beta 1- and beta 2-adrenergic receptors on bovine brain capillary endothelial cells in culture. *J Neurochem*. 1991;56(3):775–781.

35 Elfont RM, Sundaresan PR, Sladek CD. Adrenergic receptors on cerebral microvessels: pericyte contribution. *Am J Physiol*. 1989; 256(1 Pt2):R224–R230.

36 Tsao N, Hsu HP, Wu CM, et al. Tumour necrosis factor-alpha causes an increase in blood-brain barrier permeability during sepsis. *J Med Microbiol*. 2001;50(9):812–821.

37 Stonestreet BS, Sadowska GB, McKnight AJ, et al. Exogenous and endogenous corticosteroids modulate blood-brain barrier development in the ovine fetus. *Am J Physiol Regul Integr Comp Physiol*. 2000;279(2):R468–R477.

38 De Vries HE, Kuiper J, De Boer AG, et al. The blood-brain barrier in neuroinflammatory diseases. *Pharmacol Rev*. 1997; 49(2):143–155.

39 Dantzer R. Cytokine-induced sickness behaviour: a neuroimmune response to activation of innate immunity. *Eur J Pharmacol*. 2004;500(1–3):399–411.

40 Turrin NP, Rivest S. Unraveling the molecular details involved in the intimate link between the immune and neuroendocrine systems. *Exp Biol Med (Maywood)*. 2004;229(10):996–1006.

41 Stewart PA, Hayakawa K, Farrell CL. Quantitation of blood-brain barrier ultrastructure. *Microsc Res Tech*. 1994;27(6):516–527.

42 Fenstermacher J, Gross P, Sposito N, et al. Structural and functional variations in capillary systems within the brain. *Ann NY Acad Sci*. 1988;529:21–30.

43 Plummer HK III, Dhar M, Schuller HM. Expression of the alpha7 nicotinic acetylcholine receptor in human lung cells. *Respir Res*. 2005;6(1):29.

44 Rosenblum W. Some features of the functional anatomy. In: Bevan RD, Bevan JA, eds. *The Human Brain Circulation*. New York: Humana Press; 1994:23–35.

45 Gordon EL, Danielsson PE, Nguyen TS, Winn HR. A comparison of primary cultures of rat cerebral microvascular endothelial cells to rat aortic endothelial cells. *In Vitro Cell Dev Biol*. 1991;27A(4):312–326.

46 Thorin E, Shatos MA, Shreeve SM, et al. Human vascular endothelium heterogeneity. A comparative study of cerebral and peripheral cultured vascular endothelial cells. *Stroke*. 1997;28(2):375–381.

47 Kallmann BA, Wagner S, Hummel V, et al. Characteristic gene expression profile of primary human cerebral endothelial cells. *FASEB J*. 2002;16(6):589–591.

48 Frigerio S, Gelati M, Ciusani E, et al. Immunocompetence of human microvascular brain endothelial cells: cytokine regulation of IL-1beta, MCP-1, IL-10, sICAM-1 and sVCAM-1. *J Neurol*. 1998;245(11):727–730.

49 Juttler E, Tarabin V, Schwaninger M. Interleukin-6 (IL-6): a possible neuromodulator induced by neuronal activity. *Neuroscientist*. 2002;8(3):268–275.

50 Eugenin EA, D'Aversa TG, Lopez L, et al. MCP-1 (CCL2) protects human neurons and astrocytes from NMDA or HIV-tat-induced apoptosis. *J Neurochem*. 2003;85(5):1299–1311.

51 Banks WA, Kastin AJ. Blood to brain transport of interleukin links the immune and central nervous systems. *Life Sci*. 1991;48(25):L117–L121.

52 Banks WA, Kastin AJ, Durham DA. Bidirectional transport of interleukin-1 alpha across the blood-brain barrier. *Brain Res Bull*. 1989;23(6):433–437.

53 Gutierrez EG, Banks WA, Kastin AJ. Murine tumor necrosis factor alpha is transported from blood to brain in the mouse. *J Neuroimmunol*. 1993;47(2):169–176.

54 Lou J, Gasche Y, Zheng L, et al. Differential reactivity of brain microvascular endothelial cells to TNF reflects the genetic susceptibility to cerebral malaria. *Eur J Immunol*. 1998;28(12):3989–4000.

55 Kobayashi H, Yokoo H, Yanagita T, Wada A. Regulation of brain microvessel function. *Nippon Yakurigaku Zasshi*. 2002;119 (5):281–286.

56 Pan W, Kastin AJ. TNFalpha transport across the blood-brain barrier is abolished in receptor knockout mice. *Exp Neurol*. 2002;174(2):193–200.

57 Kobierski LA, Srivastava S, Borsook D. Systemic lipopolysaccharide and interleukin-1beta activate the interleukin 6: STAT intracellular signaling pathway in neurons of mouse trigeminal ganglion. *Neurosci Lett*. 2000;281(1):61–64.

58 Ek M, Kurosawa M, Lundeberg T, Ericsson A. Activation of vagal afferents after intravenous injection of interleukin-1beta: role

of endogenous prostaglandins. *J Neurosci.* 1998;18(22):9471–9479.
59 Tracey KJ. The inflammatory reflex. *Nature.* 2002;420(6917):853–859.
60 Borovikova LV, Ivanova S, Zhang M, et al. Vagus nerve stimulation attenuates the systemic inflammatory response to endotoxin. *Nature.* 2000;405(6785):458–462.
61 Wang H, Yu M, Ochani M, et al. Nicotinic acetylcholine receptor alpha7 subunit is an essential regulator of inflammation. *Nature.* 2003;421(6921):384–388.
62 Wang H, Liao H, Ochani M, et al. Cholinergic agonists inhibit HMGB1 release and improve survival in experimental sepsis. *Nat Med.* 2004;10(11):1216–1221.
63 Libert C. Inflammation: a nervous connection. *Nature.* 2003;421(6921):328–329.
64 Pavlov VA, Wang H, Czura CJ, et al. The cholinergic anti-inflammatory pathway: a missing link in neuroimmunomodulation. *Mol Med.* 2003;9(5–8):125–134.
65 Corcoran C, Connor TJ, O'Keane V, Garland MR. The effects of vagus nerve stimulation on pro- and anti-inflammatory cytokines in humans: a preliminary report. *Neuroimmunomodulation.* 2005;12(5):307–309.
66 Abbruscato TJ, Lopez SP, Mark KS, et al. Nicotine and cotinine modulate cerebral microvascular permeability and protein expression of ZO-1 through nicotinic acetylcholine receptors expressed on brain endothelial cells. *J Pharm Sci.* 2002;91(12):2525–2538.
67 Macklin KD, Maus AD, Pereira EF, et al. Human vascular endothelial cells express functional nicotinic acetylcholine receptors. *J Pharmacol Exp Ther.* 1998;287(1):435–439.
68 Saeed RW, Varma S, Peng-Nemeroff T, et al. Cholinergic stimulation blocks endothelial cell activation and leukocyte recruitment during inflammation. *J Exp Med.* 2005;201(7):1113–1123.
69 Konsman JP, Vigues S, Mackerlova L, et al. Rat brain vascular distribution of interleukin-1 type-1 receptor immunoreactivity: relationship to patterns of inducible cyclooxygenase expression by peripheral inflammatory stimuli. *J Comp Neurol.* 2004;472(1):113–129.
70 Li S, Goorha S, Ballou LR, Blatteis CM. Intracerebroventricular interleukin-6, macrophage inflammatory protein-1 beta and IL-18: pyrogenic and PGE(2)-mediated? *Brain Res.* 2003;992(1):76–84.
71 Ganong WF. Circumventricular organs: definition and role in the regulation of endocrine and autonomic function. *Clin Exp Pharmacol Physiol.* 2000;27(5–6):422–427.
72 Maness LM, Kastin AJ, Banks WA. Relative contributions of a CVO and the microvascular bed to delivery of blood-borne IL-1alpha to the brain. *Am J Physiol.* 1998;275(2 Pt 1):E207–E212.
73 Dinarello CA, Gatti S, Bartfai T. Fever: links with an ancient receptor. *Curr Biol.* 1999;9(4):R147–R150.
74 Engblom D, Ek M, Saha S, et al. Prostaglandins as inflammatory messengers across the blood-brain barrier. *J Mol Med.* 2002;80(1):5–15.
75 Wei YP, Kita M, Shinmura K, et al. Expression of IFN-gamma in cerebrovascular endothelial cells from aged mice. *J Interferon Cytokine Res.* 2000;20(4):403–409.
76 Cao C, Matsumura K, Shirakawa N, et al. Pyrogenic cytokines injected into the rat cerebral ventricle induce cyclooxygenase-2 in brain ECs and also upregulate their receptors. *Eur J Neurosci.* 2001;13(9):1781–1790.
77 Golden PL, Maccagnan TJ, Pardridge WM. Human blood-brain barrier leptin receptor. Binding and endocytosis in isolated human brain microvessels. *J Clin Invest.* 1997;99(1):14–18.
78 Singh AK, Jiang Y. How does peripheral lipopolysaccharide induce gene expression in the brain of rats? *Toxicology.* 2004;201(1–3):197–207.
79 Wong ML, Bongiorno PB, Gold PW, Licinio J. Localization of interleukin-1 beta converting enzyme mRNA in rat brain vasculature: evidence that the genes encoding the interleukin-1 system are constitutively expressed in brain blood vessels. Pathophysiological implications. *Neuroimmunomodulation.* 1995;2(3):141–148.
80 Wong ML, Bongiorno PB, al Shekhlee A, et al. IL-1 beta, IL-1 receptor type I and iNOS gene expression in rat brain vasculature and perivascular areas. *Neuroreport.* 1996;7(15–17):2445–2448.
81 Konsman JP, Parnet P, Dantzer R. Cytokine-induced sickness behaviour: mechanisms and implications. *Trends Neurosci.* 2002;25(3):154–159.
82 Stroemer RP, Rothwell NJ. Cortical protection by localized striatal injection of IL-1ra following cerebral ischemia in the rat. *J Cereb Blood Flow Metab.* 1997;17(6):597–604.
83 Fabry Z, Fitzsimmons KM, Herlein JA, et al. Production of the cytokines interleukin 1 and 6 by murine brain microvessel endothelium and smooth muscle pericytes. *J Neuroimmunol.* 1993;47(1):23–34.
84 Reyes TM, Fabry Z, Coe CL. Brain endothelial cell production of a neuroprotective cytokine, interleukin-6, in response to noxious stimuli. *Brain Res.* 1999;851(1–2):215–220.
85 Fee D, Grzybicki D, Dobbs M, et al. Interleukin 6 promotes vasculogenesis of murine brain microvessel endothelial cells. *Cytokine.* 2000;12(6):655–665.
86 Bissonnette CJ, Klegeris A, McGeer PL, McGeer EG. Interleukin 1alpha and interleukin 6 protect human neuronal SH-SY5Y cells from oxidative damage. *Neurosci Lett.* 2004;361(1–3):40–43.
87 Paul R, Koedel U, Winkler F, et al. Lack of IL-6 augments inflammatory response but decreases vascular permeability in bacterial meningitis. *Brain.* 2003;126(Pt 8):1873–1882.
88 Shukaliak JA, Dorovini-Zis K. Expression of the beta-chemokines RANTES and MIP-1 beta by human brain microvessel ECs in primary culture. *J Neuropathol Exp Neurol.* 2000;59(5):339–352.
89 Thibeault I, Laflamme N, Rivest S. Regulation of the gene encoding the monocyte chemoattractant protein 1 (MCP-1) in the mouse and rat brain in response to circulating LPS and proinflammatory cytokines. *J Comp Neurol.* 2001;434(4):461–477.
90 Connelly L, Madhani M, Hobbs A. Resistance to endotoxic shock in endothelial nitric oxide synthase (eNOS) knockout mice: a pro-inflammatory role for eNOS-derived NO in vivo. *J Biol Chem.* 2005;280(11):10040–10046.
91 Huang PL. Neuronal and endothelial nitric oxide synthase gene knockout mice. *Braz J Med Biol Res.* 1999;32(11):1353–1359.
92 Sprung CL, Peduzzi PN, Shatney CH, et al. Impact of encephalopathy on mortality in the sepsis syndrome. The Veterans Administration Systemic Sepsis Cooperative Study Group. *Crit Care Med.* 1990;18(8):801–806.
93 Young GB, Bolton CF, Archibald YM, et al. The electroencephalogram in sepsis-associated encephalopathy. *J Clin Neurophysiol.* 1992;9(1):145–152.

94 Young GB, Bolton CF, Austin TW, et al. The encephalopathy associated with septic illness. *Clin Invest Med.* 1990;13(6):297–304.
95 Graham DI, Behan PO, More IA. Brain damage complicating septic shock: acute haemorrhagic leucoencephalitis as a complication of the generalised Shwartzman reaction. *J Neurol Neurosurg Psychiatry.* 1979;42(1):19–28.
96 Wijdicks EF, Silbert PL, Jack CR, Parisi JE. Subcortical hemorrhage in disseminated intravascular coagulation associated with sepsis. *Am J Neuroradiol.* 1994;15(4):763–765.
97 Wilson JX, Young GB. Progress in clinical neurosciences: sepsis-associated encephalopathy: evolving concepts. *Can J Neurol Sci.* 2003;30(2):98–105.
98 Bernard GR, Vincent JL, Laterre PF, et al. Efficacy and safety of recombinant human activated protein C for severe sepsis. *N Engl J Med.* 2001;344(10):699–709.
99 Shibata M, Kumar SR, Amar A, et al. Anti-inflammatory, antithrombotic, and neuroprotective effects of activated protein C in a murine model of focal ischemic stroke. *Circulation.* 2001;103(13):1799–1805.
100 De Luigi A, Pizzimenti S, Quadri P, et al. Peripheral inflammatory response in Alzheimer's disease and multiinfarct dementia. *Neurobiol Dis.* 2002;11(2):308–314.
101 Richartz E, Stransky E, Batra A, et al. Decline of immune responsiveness: a pathogenetic factor in Alzheimer's disease? *J Psychiatr Res.* 2005;39(5):535–543.
102 Richartz E, Batra A, Simon P, et al. Diminished production of proinflammatory cytokines in patients with Alzheimer's disease. *Dement Geriatr Cogn Disord.* 2005;19(4):184–188.
103 Intebi AD, Garau L, Brusco I, et al. Alzheimer's disease patients display gender dimorphism in circulating anorectic adipokines. *Neuroimmunomodulation.* 2002;10(6):351–358.
104 Licastro F, Pedrini S, Caputo L, et al. Increased plasma levels of interleukin-1, interleukin-6 and alpha-1-antichymotrypsin in patients with Alzheimer's disease: peripheral inflammation or signals from the brain? *J Neuroimmunol.* 2000;103(1):97–102.
105 Streit WJ, Mrak RE, Griffin WS. Microglia and neuroinflammation: a pathological perspective. *J Neuroinflammation.* 2004;1(1):1–4.
106 Kalaria RN. Cerebrovascular degeneration is related to amyloid-beta protein deposition in Alzheimer's disease. *Ann NY Acad Sci.* 1997;826:263–271.
107 Budinger TF. Progenitor endothelial cell involvement in Alzheimer's disease. *Neurol Res.* 2003;25(6):617–624.
108 Grammas P, Yamada M, Zlokovic B. The cerebromicrovasculature: a key player in the pathogenesis of Alzheimer's disease. *J Alzheimers Dis.* 2002;4(3):217–223.
109 Lleo A, Greenberg SM, Growdon JH. Current pharmacotherapy for Alzheimer's disease. *Annu Rev Med.* 2006;57:513–533.
110 Pollak Y, Gilboa A, Ben Menachem O, et al. Acetylcholinesterase inhibitors reduce brain and blood interleukin-1beta production. *Ann Neurol.* 2005;57(5):741–745.
111 Mohr E, Mendis T, Rusk IN, Grimes JD. Neurotransmitter replacement therapy in Alzheimer's disease. *J Psychiatry Neurosci.* 1994;19(1):17–23.
112 Miyakawa T, Kuramoto R. Ultrastructural study of senile plaques and microvessels in the brain with Alzheimer's disease and Down's syndrome. *Ann Med.* 1989;21(2):99–102.
113 Cacquevel M, Lebeurrier N, Cheenne S, Vivien D. Cytokines in neuroinflammation and Alzheimer's disease. *Curr Drug Targets.* 2004;5(6):529–534.
114 Vagnucci AH Jr., Li WW. Alzheimer's disease and angiogenesis. *Lancet.* 2003;361(9357):605–608.
115 Pogue AI, Lukiw WJ. Angiogenic signaling in Alzheimer's disease. *Neuroreport.* 2004;15(9):1507–1510.
116 Tarkowski E, Issa R, Sjogren M, et al. Increased intrathecal levels of the angiogenic factors VEGF and TGF-beta in Alzheimer's disease and vascular dementia. *Neurobiol Aging.* 2002;23(2):237–243.
117 Cantara S, Donnini S, Morbidelli L, et al. Physiological levels of amyloid peptides stimulate the angiogenic response through FGF-2. *FASEB J.* 2004;18(15):1943–1945.
118 Paris D, Townsend K, Quadros A, et al. Inhibition of angiogenesis by Abeta peptides. *Angiogenesis.* 2004;7(1):75–85.
119 Christov A, Ottman JT, Grammas P. Vascular inflammatory, oxidative and protease-based processes: implications for neuronal cell death in Alzheimer's disease. *Neurol Res.* 2004;26(5):540–546.
120 Grammas P, Ottman T, Reimann-Philipp U, et al. Injured brain ECs release neurotoxic thrombin. *J Alzheimers Dis.* 2004;6(3):275–281.
121 Marques MA, Tolar M, Harmony JA, Crutcher KA. A thrombin cleavage fragment of apolipoprotein E exhibits isoform-specific neurotoxicity. *Neuroreport.* 1996;7(15–17):2529–2532.
122 Tsopanoglou NE, Andriopoulou P, Maragoudakis ME. On the mechanism of thrombin-induced angiogenesis: involvement of alphavbeta3-integrin. *Am J Physiol Cell Physiol.* 2002;283(5):C1501–C1510.
123 Rees TM, Brimijoin S. The role of acetylcholinesterase in the pathogenesis of Alzheimer's disease. *Drugs Today (Barc).* 2003;39(1):75–83.
124 Breitner JC, Welsh KA, Helms MJ, et al. Delayed onset of Alzheimer's disease with nonsteroidal anti-inflammatory and histamine H2 blocking drugs. *Neurobiol Aging.* 1995;16(4):523–530.
125 McGeer PL, Schulzer M, McGeer EG. Arthritis and anti-inflammatory agents as possible protective factors for Alzheimer's disease: a review of 17 epidemiologic studies. *Neurology.* 1996;47(2):425–432.
126 Jahroudi N, Schmaier A, Srikanth S, et al. Von Willebrand factor promoter targets the expression of amyloid beta protein precursor to brain vascular ECs of transgenic mice. *J Alzheimers Dis.* 2003;5(2):149–158.
127 Wirths O, Thelen K, Breyhan H, et al. Decreased plasma cholesterol levels during aging in transgenic mouse models of Alzheimer's disease. *Exp Gerontol.* 2006 Feb;41(2):220–224. Epub 2005 Nov 22.

125

The Retina and Related Hyaloid Vasculature

Developmental and Pathological Angiogenesis

Laura Benjamin

Beth Israel Deaconess Medical Center, Harvard Medical School, Boston, Massachusetts

The vasculature in the developing eye (Figure 125.1A) has been an excellent model system to study both normal angiogenesis and remodeling, as well as unique aspects of the central nervous system (CNS) vasculature. Like the rest of the CNS, microvasculature of the retina consists of endothelial cells (ECs), pericytes, and glial cells that form a barrier to tightly regulate transport from the blood into the tissue. In the retina, it is called the blood–retinal barrier (BRB). But other vascular beds in the eye are also useful for the study of developmental mechanisms (Figure 125.1B). For example, the first vasculature to form in the developing eye is the hyaloid vasculature, which feeds the early lens and spontaneously regresses as the retinal vessels form and become competent to supply the nutritional needs of the growing and metabolically demanding retina. Another specialized vasculature in the eye is the choroidal vasculature, which feeds the back of the retina through diffusion throughout the retinal pigmented epithelium. This vascular bed, like that of the retina proper, often is compromised in ocular diseases such as retinopathy and macular degeneration. This chapter, however, focuses on the study of the retinal and hyaloid vasculatures, because they have been especially useful in the elucidation of basic molecular and cellular mechanisms of vascular development.

In rodents, the hyaloid blood vessels form during embryogenesis but regress in the first week of postnatal development. During that time, the first retinal blood vessels form a primitive network of capillaries that progressively spread from the optic disc outward toward the periphery of the retina. This occurs on the surface of the retina adjacent to the vitreous, on top of the layer referred to as the ganglion cell layer (GCL). In the second week, these vessels sprout posteriorly toward the photoreceptors and then form a second primitive network parallel to the first, but just above the photoreceptor layer called the outer plexiform layer (OPL). During this time, the first layer is undergoing a remodeling toward a more mature vasculature, and all layers of the retina are increasing in cellularity. In humans, these processes are prenatal, but in most experimental smaller rodents and small mammals (such as cats and dogs), the completion of retinal development occurs in the postnatal period. Thus, in these animals, both the observation and intervention with ocular vascular development is feasible.

It has been speculated that an inhibitor of angiogenesis signals for the retinal vessels to stop just above the photoreceptor layer and prevents their invasion and disruption of this sensitive neuronal system. Some support for this theory comes from finding natural inhibitors of angiogenesis in parts of the eye, such as pigment epithelium-derived factor (PEDF) expression in the retinal pigmented epithelium (RPE) and the vitreous. The location of PEDF probably serves to inhibit retinal vitreal invasion and choroidal retinal invasion, but establishes the principle that inhibitors can be a part of natural program to limit angiogenesis into the neural portions of the retina. Thus, we might expect that an inhibitor exists that determines the limits of retinal angiogenesis at the interface of the vascular and nonvascular zones of the retina. PEDF has been used by multiple investigators to inhibit retinal neovascularization into the vitreous and choroidal neovascularization into the RPE layer (1–7). In addition, overexpression of a positive stimulator of blood vessel growth and migration such as vascular endothelial growth factor (VEGF) can override the presumed natural inhibition and allow vascular invasion of the photoreceptor layer from the retinal vessels above, albeit in a pathological manner (8). As the rodent grows larger in the third week of postnatal life, a more rudimentary vascular layer eventually forms by sprouting from the vessels that connect the two parallel vascular plexus. This intermediate vascular layer that forms between the first two is the inner plexiform layer (IPL). In humans, the entire retina matures earlier, and retinal vessels form in the third trimester before a full-term birth.

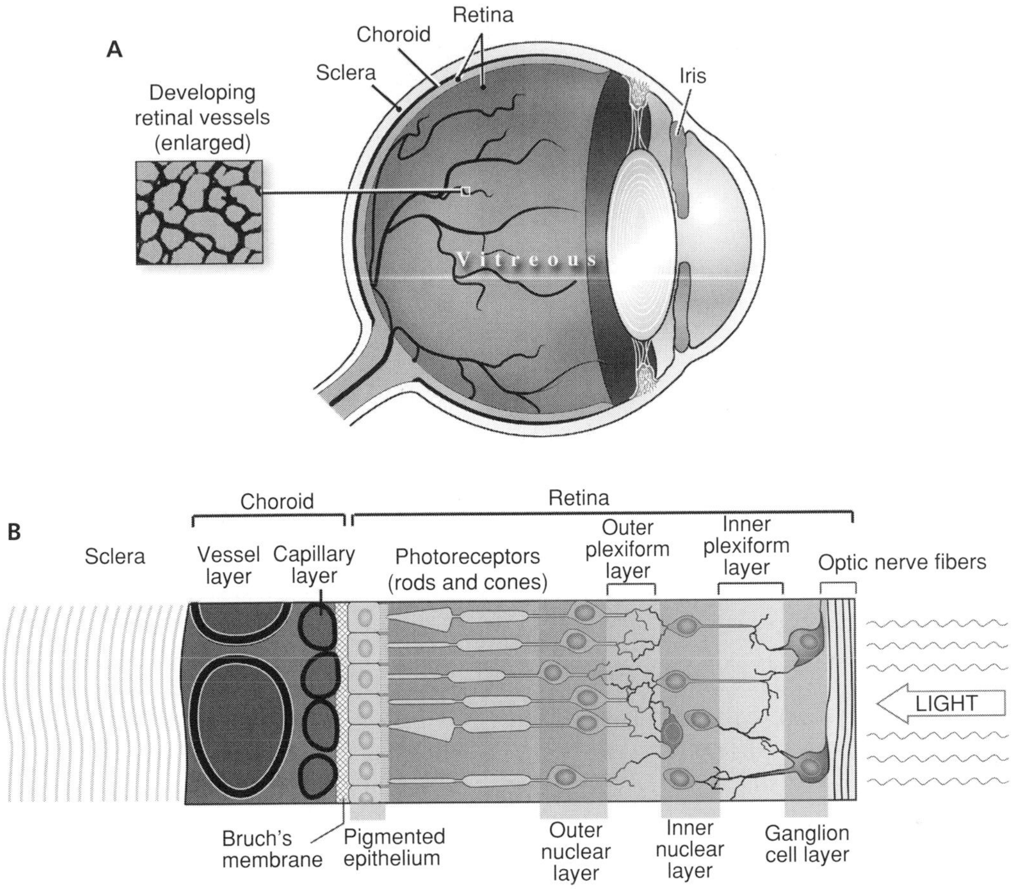

Figure 125.1. (**A**) The structures of the eye show the position of the retina, a cup-shaped organ holding the vitreous fluid. Within this cavity, the early hyaloid blood vessels are used to support the embryonic lens and developing retina. Another primitive vasculature, the pupillary membrane, covers the iris until its maturation, characterized by a programmed regression similar to the hyaloid vasculature. (**B**) The retina gets its blood supply from two sources: the deep capillary plexus and the choroid. The deep capillary plexus closest to the vitreous extends its network of vessels from the superficial surface adjacent to the vitreous, called the ganglion cell layer, through the inner nuclear layer. Most of the oxygen and nutrients supplied to the outer nuclear layer and photoreceptors in the retina come from the posterior side of the retina via the choroid and must diffuse through the Bruch's membrane and the retinal pigment epithelium to reach the retina.

ANGIOGENESIS

Angiogenesis, as opposed to vasculogenesis, relies more heavily on chaotic proliferation and primitive vessel formation than on a predetermined, patterned development. This process leads to an initial homogenous web of capillaries most easily visualized in tissues in which the architecture is planar, such as the yolk sac, retina, and perineural vascular plexus of the embryo. The initial web appears random, and organized blood flow is impossible with so many loops and bridges between vessels. However, a process called *microvascular remodeling* occurs, which rapidly transforms the initial web into an organized and heterogeneous vascular tree (9). The processes involved in remodeling are believed to be a combination of hemodynamic forces from flow, alterations in the microenvironmental levels of hypoxia as a result of oxygenated flow, and cell–cell interactions and signaling. The formation of this initial web and its subsequent remodeling can be elegantly visualized in the retina by dissecting and flat-mounting the retina after immunostaining of ECs and other cell types such as pericytes, astrocytes, and microglia (10). Perfusion with fluorescent proteins prior to resection of the eye is another way to visualize the functional vasculature and also is employed often.

Pathological angiogenesis can be analyzed using a model to simulate the disease of retinopathy of prematurity (ROP) (also called oxygen [O_2]-induced retinopathy [OIR]). This disease afflicts premature human infants born while their retinas are still immature; it is characterized by excessive and improper proliferation of blood vessels in the retina and into the vitreous (11). It can be modeled by placing the rodent in a high-O_2 environment for 5 days, during which time many immature blood vessels regress and new vessel formation is inhibited (12). Upon return to room air (normoxia), the retinal tissue senses a deficiency in oxygenation and becomes ischemic. The response to severe ischemia is responsible for the robust pathological angiogenesis that ensues.

In the rodent model, in addition to abnormal vessels (called *glomeruloid vessels*) that break through the inner limiting membrane and invade the vitreous, large regions of the retina are left unrepaired and avascular. This model is most similar to the human disease retinopathy of prematurity. However, one difference between this animal model of retinopathy and human retinopathy associated with diseases such as diabetes is that the pathophysiological insult (hyperoxia) is transient, and the healthy animal is eventually able to repair its vasculature spontaneously. Nonetheless, the study of the pathological stage in this model has been informative regarding molecules and drugs that can modulate the neovascularization process as proof-of-principles that may apply to human eye disease as well as other vascular proliferative disorders such as tumor angiogenesis.

THE RETINA IS A VALUABLE MODEL SYSTEM FOR INVESTIGATING VESSEL FORMATION

Analysis of molecular pathways that influence both the initial vessel formation and subsequent remodeling has been approached in several ways. Soluble factors have been injected into the vitreous during development and their impact visualized. Systemic drugs have been used to interrupt vascularization. Transgenic animals directing gene expression to parts of the eye that are near the retina (for example, lens-specific expression of secreted molecules is used to affect the retina via diffusion through the vitreous) or directly to the retinal vasculature have been informative. And, studies of developmental and pathological angiogenesis in genetically mutant mice have revealed subtleties in vascular function that can be difficult to observe in other organs.

Selected examples that illustrate these principles include landmark studies of VEGF function that first demonstrated the power of this molecule to act as a survival factor for newly formed blood vessels and simultaneously revealed the molecular nature of the ROP animal model. In 1995, Eli Keshet published a series of experiments that demonstrated the importance of VEGF in promoting vascular survival (13). Immediately before placing newborn rats into a hyperoxia chamber, Keshet and colleagues injected VEGF-A into one vitreous and saline into the other. In the control eye, the O_2-induced vascular regression typifies the first stage of ROP/OIR. In the VEGF-injected eye, the blood vessels were protected from this regression. Related studies on retinal VEGF expression in this model strongly implicated the O_2-sensitive properties of VEGF in mediating pathology (14,15). Glial cells in the retina play a key role in sensing hypoxia–ischemia in normal development, when they strongly express *VEGF* mRNA in the avascular zones of the retina in advance of vessel formation in those regions. Similarly, during the pathological ROP/OIR model, hyperoxia downregulates glial VEGF production, causing regression of newly formed vessels and cessation of neoangiogenesis. Return to normoxia leads to massive ischemia and VEGF production in both glial and neuronal cells. Thus, we now understand the importance of tissue oxygenation and hypoxia in the initiation of both developmental (physiological) angiogenesis and, when excessive, in pathological angiogenesis (16,17). Targeting VEGF or its receptors has been the objective of many therapeutic approaches to retinopathy (18–23). Overexpression of VEGF induces pathological angiogenesis when expressed in regions of the eye that are near retina (8,24,25).

Other growth factors, such as insulin-derived growth factor (IGF)-1, also have been implicated in the pathology of retinopathy and have been promoted as therapeutic targets (26,27). Similarly, overexpression and loss-of-function of other angiogenic cytokines such as angiopoietins (Angs) have demonstrated differential effects on vascular structure and function that reveal unique features of their action. *Ang2*-null mice are viable but, in the retina, a defect occurs in hyaloid regression that has secondary effects on retinal vascularization (28–35). These studies further demonstrate the dependence of retinal vascular development on hyaloid regression. Ang1 modulation of vascular permeability, its expression in mural cells, and its ability to restore normal patterning to retinal vessels in the absence of pericytes has also been elucidated by virtue of the unique properties of the retina (36–38).

Notably, loss-of-function studies also have been informative for a variety of genes without lethal phenotypes. Much also has been learned from the study of null mutations that showed no developmental phenotype but had profound effects on pathological angiogenesis. A striking example is the null mutation for placental growth factor (PlGF), a VEGF family member that selectively interacts with VEGF receptor 1 (VEGFR1), but not VEGFR2. PlGF-mediated activation of VEGFR1 is required for pathological angiogenesis of the retina (20,39).

The retina has been particularly useful for visualizing sprouting angiogenesis, because the planar expansion of the primary plexus allows for a clear whole-mount view of the sprouting tips as they progress across the retinal surface. A combination of immunohistochemistry, in situ hybridization, and confocal microscopy has allowed for a detailed view of the molecules involved. For example, visualization of VEGF tethered to the surface of the astrocytes, and colocalization with VEGF receptors on the cytoplasmic extensions in endothelial tip cells, beautifully demonstrate the molecular signaling of VEGF in action (40). Visualization of the sprouting tips also permitted the identification of the mild vascular developmental defects associated with the loss of RhoB. Loss of this GTPase leads to apoptotic tip cells, in part due to downregulation of Akt signaling and other molecular changes required for tip cell response to VEGF (41). Thus, the retina has proved instrumental in the analysis of gene function for a variety of different developmental stages and processes in large part because of the ease of visualizing multiple stages of vascular development and disease in this organ.

APOPTOSIS AND REMODELING IN VASCULAR DEVELOPMENT

Although many in vivo assays are available for vessel formation, few are as revealing as the hyaloid and retinal vessels for the study of remodeling and its associated vascular regression and EC apoptosis. The programmed regression of the hyaloid vessels is one of only a few examples of developmentally programmed apoptosis of entire vessels (as opposed to the vessel segments that are eliminated in microvascular remodeling). A similar process occurs in the pupillary vascular membrane that covers the iris. The hyaloid and pupillary vasculatures are transient structures. The hyaloid vasculature serves its only function to nourish the lens until the retina and its vasculature develop. These vessels emanate from the optic disc and are suspended in the vitreous. They consist of ECs thickly covered by pericytes and contacted by resident macrophages/microglia. A series of elegant experiments by Richard Lang and colleagues over the years has demonstrated an important role of macrophages in the apoptosis and regression of these two vasculatures (42,43). Early studies visualized the association of macrophages with constrictions in the vessel that preceded vessel atrophy and apoptosis. More recently, a variety of genetic models have been employed to dissect the molecular basis of regression. In mice deficient in macrophages, such as the *PU. 1*-null mice, hyaloid vessel formation is defective. In addition, wnt-7b secreted by macrophages was shown to induce proliferation of the hyaloid endothelium and to be a prerequisite for programmed cell death (44). Similarly, *Ang2*-null mice were defective in hyaloid and pupillary regression and, given that Ang2 and VEGF are expressed by pericytes, Lang attributed hyaloid regression to a loss of pericyte-derived ligands (29,45). This series of studies has helped build on the hypothesis that multiple cues are involved in regulating hyaloid development and regression, and that the proliferative signal in the presence of Ang2 leads to apoptosis. These findings not only contributed to our understanding of developmental programmed cell death of the eye vasculature, but also pointed to fundamental principles in signals that lead to apoptosis in a proliferating and developing vascular bed during microvascular remodeling. The data are consistent with the studies in tumors, in which Ang2 and VEGF determine whether co-option leads to further angiogenesis or vascular regression (46,47).

Similarly, studies in the retina itself have revealed important principles of remodeling. Early studies of pericytes, the perivascular mural cells that cover retinal capillaries, revealed a coincidence in expression of smooth muscle actin (SMA), a marker of differentiated pericytes, and remodeling during the earliest phases of microvascular remodeling. ECs produce platelet-derived growth factor (PDGF)-BB, a chemoattractant and growth factor for those pericytes having the PDGF β-receptor (PDGFRβ). Disruption of the pericyte–endothelial association by injection of PDGF-BB into the vitreous resulted in aberrant remodeling due to disruption of the normal PDGF-BB gradient, again consistent with the association of pericytes and the ability of pericytes to promote microvascular survival and normal remodeling of the microvasculature (48).

This same principle of pericyte protection of blood vessels from apoptosis has been seen in multiple tissues and settings, and with multiple molecular pathways. For example, gene targeting of EC PDGF-B expression leads to reductions in pericyte number and increased vascular fragility (49,50). In pathological retinal angiogenesis, PDGF-B inhibition also exacerbates ischemic retinopathy (51).

The ability of pericytes to protect blood vessels from VEGF loss was first described in the developing retinal vasculature, and then described in the process of tumor angiogenesis (48,52). This stabilizing function of pericytes in cancer has been further substantiated by studies showing that inhibition of pericytes by small molecule inhibitors to PDGFRβ enhances antiangiogenic therapy and vessel loss (53). As discussed previously, the expression of Ang2 interferes with pericyte–EC interactions and leaves ECs vulnerable to apoptosis in the absence of survival factors such as VEGF-A (29,33). This relationship between Ang2 and VEGF had been inferred from expression patterns of Ang2 and VEGF in tumor studies (46,47). Taken together with the phenotypes of Ang2 loss and overexpression in the eye, a hypothesis has been established that Ang2 functions to antagonize vessel stability in the absence of VEGF, but cooperates in inducing neovascularization in the presence of VEGF. VEGF-A–mediated survival function also has been reproduced with PlGF homodimers and PlGF/VEGF heterodimers, which are all similarly capable of protecting retinal blood vessel from oxygen ablation (20,54).

It has been speculated that VEGF- and Ang1–induced survival function is mediated by Akt signaling in multiple in vitro assays (55–58). In support of this, Akt signaling in the retina correlates with hyperoxia exposure in vivo. A test of this hypothesis was recently demonstrated using a transgenic model in which activated Akt can be expressed in ECs on demand via tetracycline regulation. Using this model, it was shown that, like VEGF, activation of the Akt signaling pathway can protect ECs from apoptosis in hyperoxia (59). Thus, the retina has provided a vehicle for the study of vascular remodeling and apoptosis that has elucidated several common principles of microvascular development.

THE DEVELOPMENT AND FUNCTION OF NONENDOTHELIAL CELLS IN RETINAL ANGIOGENESIS

In addition to ECs surrounded by pericytes, the retinal vasculature also contains glial cells (astrocytes and Müller cells). Astrocytes in the superficial vascular bed abutting the vitreous surround the blood vessels and contribute to the unique features of the BRB, similar to the rest of the CNS (60). Astrocytes are not only important for providing increased barrier function in the mature retina, but also play a pivotal role in the development of the retinal blood vessels (61). As the first vascular inhabitants of the newly forming retina, they lay the template on which ECs migrate and adhere, and they secrete important cytokines (e.g., VEGF) and adhesion proteins. Recent studies support a role for R-cadherin expression in mediating the effect of glial cells on vessel patterning: Antibodies to this protein inhibited branching and patterning of the inner vascular bed (62). Although astrocyte bodies are found in the superficial vascular layer abutting the vitreous, they extend processes inward to contact the endothelium in the deeper layers. Astrocytes express glial fibrillary acidic protein (GFAP) and are responsive to PDGF-A. Another glial cell in the retina is the Müller cell, which is found in the deeper layers and is associated with sprouting vessels that penetrate to form the deeper vascular beds. A recent study found that prolonged hyperoxia exposure allowed for a more normal vascular development, rather than a pathological neovascularization associated with the ROP assay described earlier. One hypothesis to explain this was that the glial cells within the deeper layers were preserved because no overwhelming ischemia was present in this protocol, and suggested that their sensitivity to extreme hypoxia might be an essential aspect of the ROP pathology (63).

Microglia are small multiprocessed cells that reside in the retina before vascularization in a regular spatial array. Microglia are phagocytotic cells that express major histocompatability complex (MHC) class I and II antigens and some macrophage markers (64,65). They can be found in association with the sprouting tips of blood vessels, but also appear to come in and out of the forming vasculature during the early stages of vessel formation and remodeling. Although they have been associates with hyaloid regression and response to ischemia in pathological settings, little attention has been paid to the role of microglia in retinal vascular development and remodeling (42,66,67).

CONCLUSION

This chapter focused primarily on lessons learned about vascular development from the study of the retinal vasculature and the associated hyaloid vessels. Other equally specialized and important vasculatures exist in the eye, such as the limbal vessels and choroid. Both of these vascular beds have been instructive in their own ways, and have relevance to important human diseases. However, for the study of normal developmental angiogenesis, in large part because of its accessibility and timing, the retina is a valuable model system. We have really just begun to unravel the molecular mechanisms that govern EC differentiation, specialization, and vascular morphogenesis. As we continue to explore new molecules and molecular interactions, no doubt the retina will continue to be a valuable tool in the analysis of developmental structure and function. Of the many interesting questions left unanswered, the study of the retina seems likely to shed light on questions of tip cell migration and guidance, mural cell interactions, and perhaps the complexity of molecular signaling in the development and function of blood vessels in general and blood vessels of the CNS in particular.

NOTE ADDED IN PROOF

A recent publication beautifully illustrates the utility of the eye as a model for novel regulation of the vasculature. The authors have shown that the cornea, which remains avascular in most animals, does so by expressing a soluble form of a VEGF receptor, sFLT, which prevents VEGF signaling. Quite interestingly, this study was able to correlate lack of sFLT expression as a change that distinguishes the cornea of the Florida manatee, one of the only animals known with blood vessels in the cornea. Vascularization of the cornea is usually associated with a pathological condition and is an obstruction for vision. However, the vessels in the manatee are small enough to only minimally interfere with the vision of this unusual animal (68).

KEY POINTS

- The rodent retina and associated hyaloid vasculature are excellent model systems to study normal and pathological angiogenesis.
- Cell–cell interactions, such those occurring among macrophages, ECs, and pericytes in the regressing hyaloid, or pericytes and ECs in the remodeling retinal vasculature, are fundamental for normal development.
- The role of VEGF-A as a survival factor, and the related signaling pathways involved in remodeling of the vasculature, has been most clearly illustrated in the retinal vasculature.
- Fundamental observations regarding EC biology as they relate to normal angiogenesis and angiogenesis in disease continue to be pioneered through the study of the rodent retina. Examples include endothelial tip cell function and regulation in sprouting angiogenesis, the role of the specialized macrophages (microglia) in microvascular development and pathology, and the study of glial and pericyte regulation of the BRB.

Future Goals

- To develop a deeper molecular understanding of how the retinal blood vessels are formed, remodeled, and functionally maintained, with attention to the cross-talk between the ECs, pericytes, astrocytes and microglia
- To develop applications to diseases such as diabetes and macular degeneration, which will become apparent with a better understanding of the normal regulation of vascular structure and function in this organ

REFERENCES

1 Matsuoka M, Ogata N, Otsuji T, et al. Expression of pigment epithelium derived factor and vascular endothelial growth factor in choroidal neovascular membranes and polypoidal choroidal vasculopathy. *Br J Ophthalmol*. 2004;88(6):809–815.

2 Auricchio A, Behling KC, Maguire AM, et al. Inhibition of retinal neovascularization by intraocular viral-mediated delivery of anti-angiogenic agents. *Mol Ther*. 2002;6(4):490–494.

3 Spranger J, Osterhoff M, Reimann M, et al. Loss of the antiangiogenic pigment epithelium-derived factor in patients with angiogenic eye disease. *Diabetes*. 2001;50(12):2641–2645.

4 Ogata N, Tombran-Tink J, Nishikawa M, et al. Pigment epithelium-derived factor in the vitreous is low in diabetic retinopathy and high in rhegmatogenous retinal detachment. *Am J Ophthalmol*. 2001;132(3):378–382.

5 Stellmach V, Crawford SE, Zhou W, Bouck N. Prevention of ischemia-induced retinopathy by the natural ocular antiangiogenic agent pigment epithelium-derived factor. *Proc Natl Acad Sci USA*. 2001;98(5):2593–2597.

6 Dawson DW, Volpert OV, Gillis P, et al. Pigment epithelium-derived factor: a potent inhibitor of angiogenesis. *Science*. 1999; 285(5425):245–248.

7 Ogata N, Wada M, Otsuji T, et al. Expression of pigment epithelium-derived factor in normal adult rat eye and experimental choroidal neovascularization. *Invest Ophthalmol Vis Sci*. 2002;43(4):1168–1175.

8 Vinores SA, Derevjanik NL, Vinores MA, et al. Sensitivity of different vascular beds in the eye to neovascularization and blood-retinal barrier breakdown in VEGF transgenic mice. *Adv Exp Med Biol*. 2000;476:129–138.

9 Risau W. Mechanisms of angiogenesis. *Nature*. 1997;386(6626): 671–674.

10 Saint-Geniez M, D'Amore PA. Development and pathology of the hyaloid, choroidal and retinal vasculature. *Int J Dev Biol*. 2004;48(8–9):1045–1058.

11 Smith LE. Pathogenesis of retinopathy of prematurity. *Semin Neonatol*. 2003;8(6):469–473.

12 Smith LE, Wesolowski E, McLellan A, et al. Oxygen-induced retinopathy in the mouse. *Invest Ophthalmol Vis Sci*. 1994;35(1):101–111.

13 Alon T, Hemo I, Itin A, et al. Vascular endothelial growth factor acts as a survival factor for newly formed retinal vessels and has implications for retinopathy of prematurity. *Nat Med*. 1995; 1(10):1024–1028.

14 Stone J, Itin A, Alon T, et al. Development of retinal vasculature is mediated by hypoxia-induced vascular endothelial growth factor (VEGF) expression by neuroglia. *J Neurosci*. 1995;15(7Pt 1): 4738–4747.

15 Pe'er J, Shweiki D, Itin A, et al. Hypoxia-induced expression of vascular endothelial growth factor by retinal cells is a common factor in neovascularizing ocular diseases. *Lab Invest*. 1995;72(6): 638–645.

16 Stone J, Chan-Ling T, Pe'er J, et al. Roles of vascular endothelial growth factor and astrocyte degeneration in the genesis of retinopathy of prematurity. *Invest Ophthalmol Vis Sci*. 1996; 37(2):290–299.

17 Pierce EA, Foley ED, Smith LE. Regulation of vascular endothelial growth factor by oxygen in a model of retinopathy of prematurity. *Arch Ophthalmol*. 1996;114(10):1219–1228.

18 Shen J, Samul R, Zimmer J, et al. Deficiency of neuropilin 2 suppresses VEGF-induced retinal neovascularization. *Mol Med*. 2004;10(1–6):12–18.

19 McLeod DS, Taomoto M, Cao J, et al. Localization of VEGF receptor-2 (KDR/Flk-1) and effects of blocking it in oxygen-induced retinopathy. *Invest Ophthalmol Vis Sci*. 2002;43(2):474–482.

20 Shih SC, Ju M, Liu N, Smith LE. Selective stimulation of VEGFR-1 prevents oxygen-induced retinal vascular degeneration in retinopathy of prematurity. *J Clin Invest*. 2003;112(1):50–57.

21 Preclinical and phase 1A clinical evaluation of an anti-VEGF pegylated aptamer (EYE001) for the treatment of exudative age-related macular degeneration. *Retina*. 2002;22(2): 143–152.

22 Rota R, Riccioni T, Zaccarini M, et al. Marked inhibition of retinal neovascularization in rats following soluble-flt-1 gene transfer. *J Gene Med*. 2004;6(9):992–1002.

23 Aiello LP, Pierce EA, Foley ED, et al. Suppression of retinal neovascularization in vivo by inhibition of vascular endothelial growth factor (VEGF) using soluble VEGF-receptor chimeric proteins. *Proc Natl Acad Sci USA*. 1995;92(23):10457–10461.

24 Ash JD, Overbeek PA. Lens-specific VEGF-A expression induces angioblast migration and proliferation and stimulates angiogenic remodeling. *Dev Biol*. 2000;223(2):383–398.

25 Lakatos L. Transgenic mice model of ocular neovascularization driven by vascular endothelial growth factor (VEGF) overexpression. *Am J Pathol*. 1998;152(5):1397–1398.

26 Smith LE. IGF-1 and retinopathy of prematurity in the preterm infant. *Biol Neonate*. 2005;88(3):237–244.

27 Poulaki V, Joussen AM, Mitsiades N, et al. Insulin-like growth factor-I plays a pathogenetic role in diabetic retinopathy. *Am J Pathol*. 2004;165(2):457–469.

28 Hackett SF, Ozaki H, Strauss RW, et al. Angiopoietin 2 expression in the retina: upregulation during physiologic and pathologic neovascularization. *J Cell Physiol*. 2000;184(3):275–284.

29 Lobov IB, Brooks PC, Lang RA. Angiopoietin-2 displays VEGF-dependent modulation of capillary structure and endothelial cell survival in vivo. *Proc Natl Acad Sci USA*. 2002;99(17):11205–11210.

30 Lip PL, Chatterjee S, Caine GJ, et al. Plasma vascular endothelial growth factor, angiopoietin-2, and soluble angiopoietin receptor tie-2 in diabetic retinopathy: effects of laser photocoagulation and angiotensin receptor blockade. *Br J Ophthalmol*. 2004; 88(12):1543–1546.

31 Patel JI, Hykin PG, Gregor ZJ, et al. Angiopoietin concentrations in diabetic retinopathy. *Br J Ophthalmol*. 2005;89(4):480–483.

32 Oshima Y, Oshima S, Nambu H, et al. Different effects of angiopoietin-2 in different vascular beds: new vessels are most sensitive. *FASEB J*. 2005;19(8):963–965.

33 Oshima Y, Deering T, Oshima S, et al. Angiopoietin-2 enhances retinal vessel sensitivity to vascular endothelial growth factor. *J Cell Physiol*. 2004;199(3):412–417.

34 Hammes HP, Lin J, Wagner P, et al. Angiopoietin-2 causes pericyte dropout in the normal retina: evidence for involvement in diabetic retinopathy. *Diabetes*. 2004;53(4):1104–1110.

35 Hackett SF, Wiegand S, Yancopoulos G, Campochiaro PA. Angiopoietin-2 plays an important role in retinal angiogenesis. *J Cell Physiol*. 2002;192(2):182–187.

36 Thurston G, Rudge JS, Ioffe E, et al. Angiopoietin-1 protects the adult vasculature against plasma leakage. *Nat Med*. 2000;6(4): 460–463.

37 Joussen AM, Poulaki V, Tsujikawa A, et al. Suppression of diabetic retinopathy with angiopoietin-1. *Am J Pathol.* 2002;160(5):1683–1693.
38 Uemura A, Ogawa M, Hirashima M, et al. Recombinant angiopoietin-1 restores higher-order architecture of growing blood vessels in mice in the absence of mural cells. *J Clin Invest.* 2002;110(11):1619–1628.
39 Luttun A, Tjwa M, Moons L, et al. Revascularization of ischemic tissues by PlGF treatment, and inhibition of tumor angiogenesis, arthritis and atherosclerosis by anti-Flt1. *Nat Med.* 2002; 8(8):831–840.
40 Gerhardt H, Golding M, Fruttiger M, et al. VEGF guides angiogenic sprouting utilizing endothelial tip cell filopodia. *J Cell Biol.* 2003;161(6):1163–1177.
41 Adini I, Rabinovitz I, Sun JF, et al. RhoB controls Akt trafficking and stage-specific survival of endothelial cells during vascular development. *Genes Dev.* 2003;17(21):2721–2732.
42 Lang RA, Bishop JM. Macrophages are required for cell death and tissue remodeling in the developing mouse eye. *Cell.* 1993; 74(3):453–462.
43 Lang R, Lustig M, Francois F, et al. Apoptosis during macrophage-dependent ocular tissue remodelling. *Development.* 1994;120(12):3395–3403.
44 Lobov IB, Rao S, Carroll TJ, et al. WNT7b mediates macrophage-induced programmed cell death in patterning of the vasculature. *Nature.* 2005;437(7057):417–421.
45 Meeson AP, Argilla M, Ko K, et al. VEGF deprivation-induced apoptosis is a component of programmed capillary regression. *Development.* 1999;126(7):1407–1415.
46 Holash J, Wiegand SJ, Yancopoulos GD. New model of tumor angiogenesis: dynamic balance between vessel regression and growth mediated by angiopoietins and VEGF. *Oncogene.* 1999; 18(38):5356–5362.
47 Holash J, Maisonpierre PC, Compton D, et al. Vessel cooption, regression, and growth in tumors mediated by angiopoietins and VEGF. *Science.* 1999;284(5422):1994–1998.
48 Benjamin LE, Hemo I, Keshet E. A plasticity window for blood vessel remodelling is defined by pericyte coverage of the preformed endothelial network and is regulated by PDGF-B and VEGF. *Development.* 1998;125(9):1591–1598.
49 Lindblom P, Gerhardt H, Liebner S, et al. Endothelial PDGF-B retention is required for proper investment of pericytes in the microvessel wall. *Genes Dev.* 2003;17(15):1835–1840.
50 Lindahl P, Johansson BR, Leveen P, Betsholtz C. Pericyte loss and microaneurysm formation in PDGF-B-deficient mice. *Science.* 1997;277(5323):242–245.
51 Wilkinson-Berka JL, Babic S, De Gooyer T, et al. Inhibition of platelet-derived growth factor promotes pericyte loss and angiogenesis in ischemic retinopathy. *Am J Pathol.* 2004;164(4):1263–1273.
52 Benjamin LE, Keshet E. Conditional switching of vascular endothelial growth factor (VEGF) expression in tumors: induction of endothelial cell shedding and regression of hemangioblastoma-like vessels by VEGF withdrawal. *Proc Natl Acad Sci USA.* 1997;94(16):8761–8766.
53 Bergers G, Song S, Meyer-Morse N, et al. Benefits of targeting both pericytes and endothelial cells in the tumor vasculature with kinase inhibitors. *J Clin Invest.* 2003;111(9):1287–1295.
54 Upalakalin JN, Hemo I, Dehio C, et al. Survival mechanisms of VEGF and PlGF during microvascular remodeling. *Cold Spring Harb Symp Quant Biol.* 2002;67:181–187.
55 Papapetropoulos A, Fulton D, Mahboubi K, et al. Angiopoietin-1 inhibits endothelial cell apoptosis via the Akt/survivin pathway. *J Biol Chem.* 2000;275(13):9102–9105.
56 Dimmeler S, Assmus B, Hermann C, et al. Fluid shear stress stimulates phosphorylation of Akt in human endothelial cells: involvement in suppression of apoptosis. *Circ Res.* 1998;83(3):334–341.
57 Morales-Ruiz M, Fulton D, Sowa G, et al. Vascular endothelial growth factor-stimulated actin reorganization and migration of endothelial cells is regulated via the serine/threonine kinase Akt. *Circ Res.* 2000;86(8):892–896.
58 Fujio Y, Walsh K. Akt mediates cytoprotection of endothelial cells by vascular endothelial growth factor in an anchorage-dependent manner. *J Biol Chem.* 1999;274(23):16349–16354.
59 Sun JF, Phung T, Shiojima I, et al. Microvascular patterning is controlled by fine-tuning the Akt signal. *Proc Natl Acad Sci USA.* 2005;102(1):128–133.
60 Provis JM. Development of the primate retinal vasculature. *Prog Retin Eye Res.* 2001;20(6):799–821.
61 Zhang Y, Stone J. Role of astrocytes in the control of developing retinal vessels. *Invest Ophthalmol Vis Sci.* 1997;38(9):1653–1666.
62 Dorrell MI, Aguilar E, Friedlander M. Retinal vascular development is mediated by endothelial filopodia, a preexisting astrocytic template and specific R-cadherin adhesion. *Invest Ophthalmol Vis Sci.* 2002;43(11):3500–3510.
63 Gu X, Samuel S, El-Shabrawey M, et al. Effects of sustained hyperoxia on revascularization in experimental retinopathy of prematurity. *Invest Ophthalmol Vis Sci.* 2002;43(2):496–502.
64 Matsubara T, Pararajasegaram G, Wu GS, Rao NA. Retinal microglia differentially express phenotypic markers of antigen-presenting cells in vitro. *Invest Ophthalmol Vis Sci.* 1999;40(13): 3186–3193.
65 Chen L, Yang P, Kijlstra A. Distribution, markers, and functions of retinal microglia. *Ocul Immunol Inflamm.* 2002;10(1):27–39.
66 Zhang C, Lam TT, Tso MO. Heterogeneous populations of microglia/macrophages in the retina and their activation after retinal ischemia and reperfusion injury. *Exp Eye Res.* 2005 Dec; 81(6):700–709. Epub 2005 Jun 20.
67 Hose S, Zigler JS Jr., Sinha D. A novel rat model to study the functions of macrophages during normal development and pathophysiology of the eye. *Immunol Lett.* 2005;96(2):299–302.
68 Ambati BK, Nozaki M, Singh H, et al. Corneal avascularity is due to soluble VEGF receptor-1. *Nature.* 2006;443:993–997.

126

Microheterogeneity of Lung Endothelium

Troy Stevens

Center for Lung Biology, University of South Alabama College of Medicine, Mobile

Endothelium forms a contiguous cell layer that separates blood from underlying tissue, in each of the body's organs. Cells comprising the endothelial system share common functions, including the ability to form a semipermeable barrier and an antiadhesive surface, produce vasoactive autocoids, move cells and chemical substances across their barrier, and transduce biophysical forces. These shared roles and the ability of endothelial cells (ECs) to adapt to localized environmental cues have led to the pervasive view that their behavior is dominantly controlled by local tissue environment.

Definitive evidence supports the idea that environmental stimuli adjust EC behavior on a moment-by-moment basis. However, heterogeneous endothelial behaviors are not only evident in cells from different organs, they are evident along the arterial-capillary-venule axis of a single organ and, indeed, they exist even among immediately adjacent cells. Such an incredible diversity brings into question whether environmental influences are sufficient to account for this phenotypic heterogeneity.

It has become clear that not all ECs arise from similar progenitors during development. Angiogenesis – the formation of new blood vessels from existing ones – occurs in large pulmonary blood vessels, whereas vasculogenesis – the formation of new blood vessels from blood islands – occurs within the lung's microcirculation. It is still debated as to whether these processes are fully distinct in nature, or whether they reflect interrelated developmental processes, although it is evident that multiple different mesenchymal precursor cells participate in endothelial biogenesis. It is likely that these different cells become uniquely imprinted as a part of their development. Such epigenetic modifications to chromatin, including methylation and acetylation, produce a stable cell memory that is vascular site–specific. The lung's circulation provides insight into the highly discrete function of ECs aligned along the arterial-capillary-venule axis. This chapter describes our current understanding of lung EC heterogeneity, considering both environmental and epigenetic contributions to phenotype specification.

ENDOTHELIAL HETEROGENEITY IN DIFFERENT PULMONARY VASCULAR SEGMENTS

Lungs receive more than 100% of the cardiac output, because they are simultaneously perfused by both the systemic (i.e., blood delivered from the left ventricle) and pulmonary (i.e., blood delivered from the right ventricle) circulations. The lung's systemic circulation provides oxygenated blood to the airways and vasovasorum to support tissue metabolic activity, whereas the lung's pulmonary circulation provides deoxygenated blood to the alveolar-capillary membrane for gas exchange. The pulmonary circulation is unique among all vascular beds, because it is the only vascular bed to receive 100% of the cardiac output, yet maintain a low blood pressure in the face of high flow rates. In contrast to the systemic circulation, blood entering the lung through the pulmonary artery possesses a "mixed venous" pO_2 ($\sim$40 mm Hg), whereas postcapillary blood possesses the highest pO_2 levels in the body following gas exchange across the alveolar-capillary membrane ($\sim$100 mm Hg) (Figure 126.1A).

By examining the pulmonary circulation's organization, it is easy to appreciate the distinctive structures of extra-alveolar and alveolar blood vessels (Figure 126.1A, B). Extra-alveolar vessels exhibit a tree-like branching pattern, from large-to-small vessels on the arterial side and from small-to-large vessels on the venous side of the circulation. Capillaries, in contrast, exhibit a web-like appearance. Although blood flow occurs continuously through extra-alveolar blood vessels, it can be discontinuous in certain capillary segments. Indeed, capillaries exhibit recruitment (of new capillary loops) and distention of vessels in response to either increased blood flow or increased venous pressure.

Pulmonary arteries and veins can be discriminated by their anatomical locations within the lung, although they are structurally more alike than are arteries and veins in the systemic circulation. Indeed, the muscle layer is smaller in a pulmonary artery than it is in a systemic artery. Pulmonary artery ECs reside on a basement membrane that separates the intima from

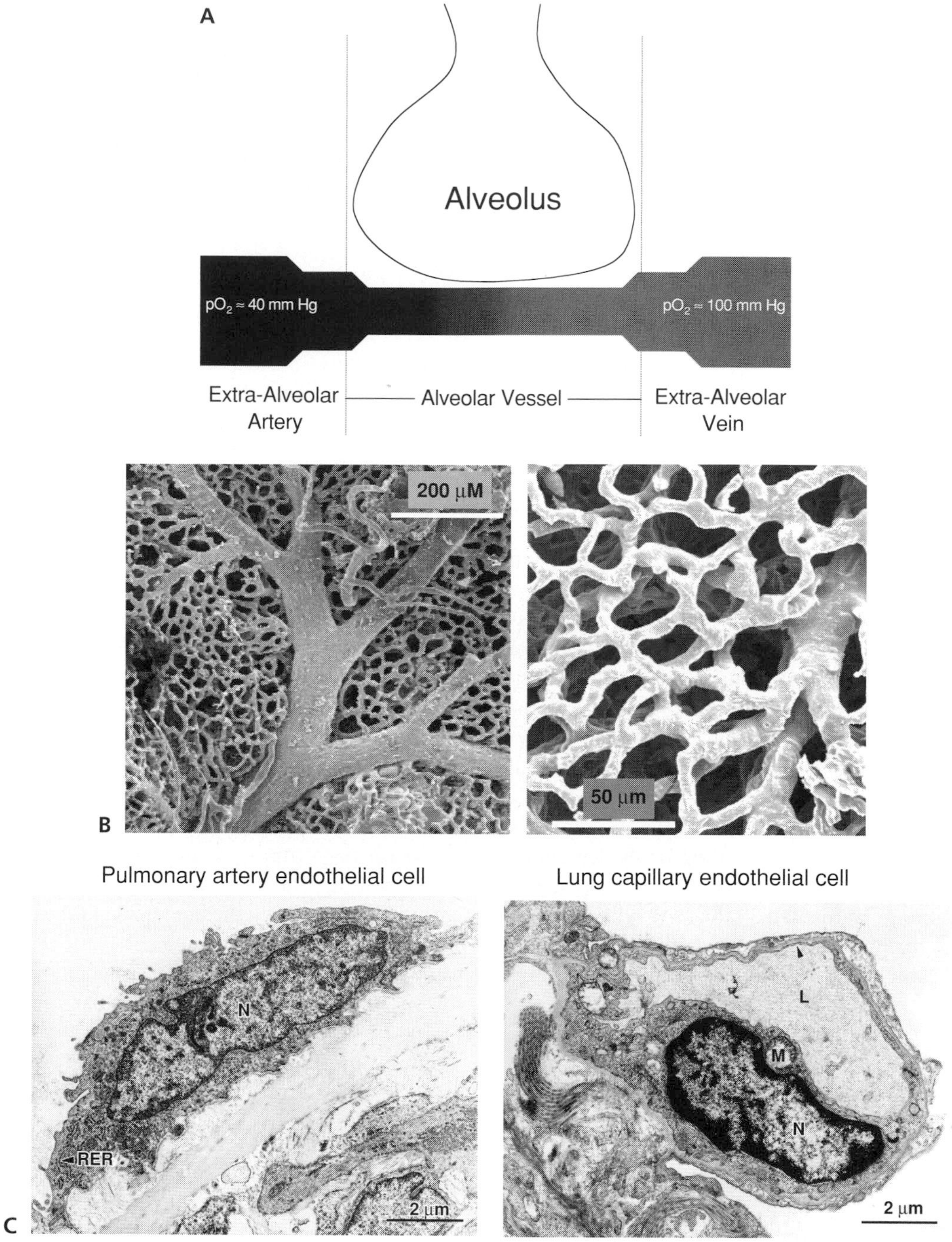

Figure 126.1. Endothelial heterogeneity exists along the arterial-capillary-venule axis in the pulmonary circulation; this structural and functional heterogeneity is apparent when extra-alveolar and alveolar ECs are compared. (**A**) The pulmonary artery arises from the right ventricle and branches into progressively smaller arterioles. Ultimately, small arterioles deliver deoxygenated ("mixed venous") blood to a dense capillary septal network. Capillaries drain blood into small venules that progressively increase in size to a large vein that returns blood to the left atria and ventricle, to provide oxygenated blood to the systemic circulation. Capillaries comprise the "alveolar" vessels, whereas conduit arteries and veins comprise "extra-alveolar" vessels. (**B**) Scanning electron micrographs of methyl methacrylate–perfused corrosion casts reveal the lung vascular architecture. The arterial segments (*left panel*) give rise to progressively smaller vessels that follow a tree-like branching pattern. Capillary segments, however, exhibit a more web-like appearance. For experimental details see Ref. 9. (**C**) Pulmonary artery and capillary ECs are shown by transmission electron microscopy. Pulmonary artery ECs reside on a thick basement membrane that separates the intima from the underlying smooth muscle layer. Capillary ECs reside on a thin matrix layer and are adjacent to underlying type I pneumonocytes. Capillary ECs possess thin cytoplasmic extensions that allow for efficient gas exchange in blood and alveoli. For experimental details see King J, Hamil T, Creighton J, et al. Structural and functional characteristics of lung macro- and microvascular EC phenotypes. *Microvasc Res.* 2004;67(2):139–151.

underlying smooth muscle cell layers (see Figure 126.1C) (1). Pulmonary artery ECs can interact with as many as six adjacent cells, which are all generally aligned in the direction of blood flow (2). The nucleus of these large vessel ECs is located in the cell's center and is surrounded by endoplasmic reticulum, Golgi apparatus, and mitochondria. In some instances, the endoplasmic reticulum can be seen extending adjacent to apical, basal, and lateral cell membranes.

Lung capillary ECs reside near type I pneumonocytes (see Figure 126.1C) (1). Only a thin basement membrane separates these two cell layers, unlike most of the body's capillaries. Multiple capillary loops exist within individual septa, so that vascular and airway surface area is optimally matched. Capillary ECs typically contact two neighboring cells and are not flow-aligned per se. The nucleus of these cells is located in the cell's center, and is also surrounded by endoplasmic reticulum, Golgi apparatus, and mitochondria. These cells possess a high number of specialized membrane invaginations – caveolae – which serve as an important site for the communication between extra- and intracellular environments. The cytoplasmic extensions of capillary ECs are extremely thin and exclude organelles from the lateral cell borders. Thus, the anatomical organization of pulmonary artery and capillary ECs is remarkably diverse.

The lung's circulation (and airway) is well-ordered to optimize gas exchange. Successful gas exchange requires that the endothelium in all the lung's vascular segments limit fluid, solute, and macromolecular access to interstitial and alveolar compartments. The earliest studies evaluating fluid flux across whole-organ vascular beds suggested that the pulmonary circulation was relatively permeable (3). These studies considered the entire pulmonary circulation as one homogeneous unit, although the permeability results were generally taken to reflect capillary barrier function, because the capillaries dominate pulmonary vascular surface area. We have learned that this idea is not entirely correct.

Several investigative groups have utilized functional measurements to divide the lung circulation into precapillary, capillary, and postcapillary segments, to assess regional differences in endothelial barrier properties (3). Although the required experimental approaches are technically challenging, these studies have all come to the same conclusion: Capillary ECs form a very restrictive barrier relative to pre- (i.e., arteriole) and postcapillary (i.e., venule) segments. The most compelling work resolved that, under basal conditions, approximately 40% of fluid flux occurs across capillary and venule segments, respectively, whereas approximately 20% of fluid flux occurs across the arterial segment (4). When these values are standardized to vascular surface area, it becomes apparent that capillary ECs exhibit a 58-fold and 26-fold more restrictive barrier function than do arterial and venule ECs, respectively. Despite the thin cytoplasmic regions and the very limited contact (i.e., junctional apposition) with immediately adjacent cells, capillary ECs exhibit a tremendous resistance to transcellular fluid flux.

Under basal conditions, some fluid, solute, and macromolecule flux occurs across the EC barrier of all lung vascular segments. A negative interstitial pressure clears this fluid into the lymphatics so that gas exchange across the alveolar-capillary membrane is not impaired by an increase in diffusion distance (3). Inflammatory conditions disrupt endothelial cell–cell junctions and result in intercellular gap formation that allows fluid exudation. If this exudative flux exceeds the rate of fluid clearance, interstitial and potentially alveolar edema occurs.

Studies assessing inflammation-induced permeability edema noted that a common first finding was perivascular cuffing around arteries and veins (3). These results were largely taken to mean that the negative interstitial pressure limits fluid accumulation at the alveolar-capillary membrane, so that fluid only accumulates at the alveolar-capillary membrane after interstitium becomes hydrated or lymphatic fluid clearance becomes compromised. However, in 1968, Whayne and Severinghaus (5) reported that hypoxia caused perivascular cuffing without concomitant disruption of the alveolar-capillary membrane, and without alveolar edema, suggesting the possibility that hypoxia selectively increased extra-alveolar EC permeability without disrupting the alveolar-capillary barrier. These results were therefore incompatible with the idea that increased lung permeability primarily reflected a loss of capillary endothelial barrier function, because the capillary segment represented the largest surface area.

Subsequent work by other labs revealed that different inflammatory agents target discrete lung vascular sites; that is, they increase EC permeability in a region-specific manner. Oxidants like hydrogen peroxide, for example, increase permeability across extra-alveolar ECs and not alveolar ECs (6,7), whereas mechanical injury, oleic acid, and 5,6- or 14,15-eicosatetranoic acid selectively increases capillary EC permeability (3,4). Taken together, these findings suggest EC heterogeneity contributes to site-specific vascular responses.

This idea was largely borne out by studies using the plant alkaloid thapsigargin, which inhibits the sarcoplasmic, endoplasmic reticulum calcium ATPase (SERCA). SERCA inhibition depletes calcium in the endoplasmic reticulum and promotes calcium entry across the plasma membrane through store-operated calcium entry pathways. By combining measurements of lung fluid filtration with immunohistochemistry and electron microscopy, it became evident that thapsigargin – e.g. the activation of store-operated calcium entry – was sufficient to increase extra-alveolar EC permeability, and not capillary EC permeability (8). Indeed, thapsigargin induces intercellular gaps between endothelium in arteries and arterioles and veins and venules, from small pre- and postcapillary ($\sim$25 μm) to extrapulmonary vessels. Pretreating the pulmonary circulation with a type 4 phosphodiesterase inhibitor, rolipram, prevented thapsigargin from increasing permeability in extra-alveolar ECs and revealed thapsigargin-induced capillary permeability (9). Thus, extra-alveolar and alveolar ECs possess different calcium channels or calcium regulatory mechanisms that are discretely activated during the process of inflammation to discriminate among vascular segments.

MARKERS OF ENDOTHELIAL CELL HETEROGENEITY IN THE INTACT PULMONARY CIRCULATION

Studies performed in the intact pulmonary circulation suggest many functional demarcations exist between extra-alveolar and alveolar ECs. Evidence that lung microvascular ECs possess an enhanced barrier function should translate into a definable difference in adhesion molecule expression. Vascular-endothelial (VE)-cadherin is the best-described molecular determinant of the EC adherens junction. ECs from all pulmonary vascular segments express VE-cadherin (2) but, interestingly, an infusion of an antibody that inhibits VE-cadherin homotypic interaction causes fulminant, hemorrhagic pulmonary edema, suggesting VE-cadherin plays an important role in capillary EC–EC adhesion (10). Some evidence suggests that capillary ECs express more VE-cadherin than do extra-alveolar ECs, although this finding is not entirely clear (2). Indeed, Bhattacharya and colleagues (11,12) have shown that capillary ECs may express E-cadherin either in addition to, or in place of, VE-cadherin; and capillary ECs also possess a high abundance of N-cadherin (T. Stevens, unpublished observations). Thus, the adherens junction in lung capillary ECs possesses a unique distribution of cadherins that may contribute to the enhanced barrier function of these cells.

Molecules in the coagulation cascade are differentially expressed among pulmonary vascular segments. The Weibel-Palade body (WPB) is a cigar-shaped organelle that contains von Willebrand factor (vWF). vWF in WPBs is a high molecular weight polymer. Its stimulated release into the circulation recruits platelets to an injury site, thereby promoting hemostasis. The membrane of WPBs contains P-selectin. Thus, in addition to releasing vWF, the stimulated secretion of WPBs increases EC surface expression of P-selectin to facilitate leukocyte recruitment (2). It is generally believed that ECs in "larger" blood vessels possess more WPBs than do capillary ECs.

Thrombin triggers the stimulated secretion of WPBs, at least in part by initiating a calcium transition. Recently, Wu and colleagues (13) have demonstrated that pulmonary capillary ECs express a voltage-gated, "T"-type calcium channel that is not expressed in ECs from extra-alveolar vessels. Thrombin activation of the T channel provides a calcium source that induces vWF secretion. Thus, the findings of Wu and colleagues (13) indicate a significant heterogeneity exists in not only in the molecular architecture of WPBs from extra-alveolar and alveolar ECs, but also in the signaling mechanisms that control their stimulated secretion.

Lung capillary ECs express a high level of angiotensin-converting enzyme (ACE). This ACE expression pattern is unique to lung, because in the systemic circulation ACE expression is most prominent in ECs of larger arteries and arterioles, with minimal capillary EC expression (14). A single pass of blood through the pulmonary circulation is sufficient to clear bradykinin from the circulation (as ACE inactivates bradykinin), and to convert angiotensin I to angiotensin II. Because angiotensin II is a vasoconstrictor that also stimulates aldosterone secretion to regulate blood volume, lung capillary ECs play a role in the control of systemic blood pressure.

LECTINS DEMARCATE ENDOTHELIAL CELL PHENOTYPES IN THE PULMONARY CIRCULATION

Given the structural and functional transition between extra-alveolar and alveolar ECs (Figure 126.2; for color reproduction, see Color Plate 126.2), investigators sought to identify approaches that could resolve the unique molecular anatomy of these cells in their in vivo environment. ECs possess a glycocalyx coat that generates a negative charge barrier and limits fluid and solute permeation across the barrier. Lectins are plant and animal proteins that interact selectively with well-defined carbohydrate moieties. Thus, lectins are useful in discriminating among EC phenotypes that differ in their glycocalyx properties.

Because lung microvascular ECs possess an enhanced barrier function relative to extra-alveolar ECs, it is reasonable to hypothesize that they also possess a distinctive glycocalyx coat. The screening of lectin binding to lung ECs in vivo and in vitro revealed that alveolar ECs preferentially interact with *Griffonia simplicifolia* and *Glycine max*, whereas pulmonary artery ECs preferentially interact with *Helix pomatia* (1). Both *G. simplicifolia* and *G. max* effectively discriminate alveolar ECs in multiple species, whereas *H. pomatia* appears to be most effective in rat. Nonetheless, it is evident using these lectins that different EC phenotypes abut one another at branch points where extra-alveolar ECs give rise to capillary segments (see Figure 126.2). Although *G. simplicifolia*–positive cells are seen

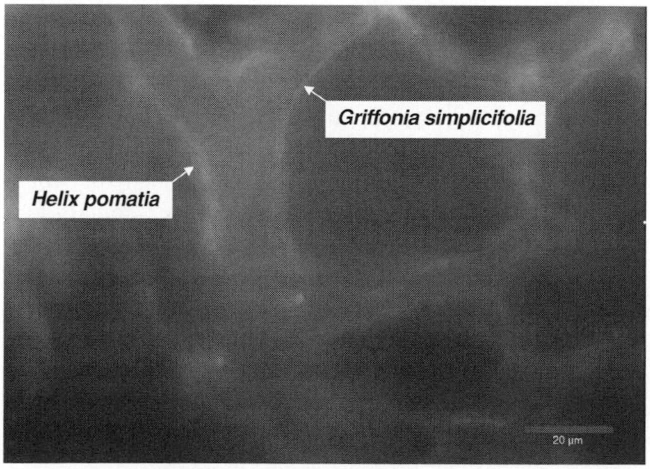

Figure 126.2. Infusion of TRITC-labeled *H. pomatia* and FITC-labeled *G. simplicifolia* in the intact pulmonary circulation demonstrate a clear demarcation in cell phenotype in vessels 20 to 30 μm in diameter. Note that no spectral overlap occurs, indicating ECs interact with either lectin but do not interact with both lectins. In situ epifluorescence images were tested and produced by Dr. Abu-Bakr Al-Mehdi. For color reproduction, see Color Plate 126.2.

immediately adjacent to *H. pomatia*–positive cells, no overlap in lectin binding occurs.

Lectins also have proven useful for isolating and purifying cultures of distinctive EC phenotypes. For example, coupling the Ryan technique for isolating lung capillary ECs from the peripheral lung with incubation of *G. simplicifolia*–coated beads allows enrichment of microvascular ECs (14). These cultures, as with pulmonary artery EC cultures, retain their lectin-specific interaction profile (Figure 126.3A). Moreover, because lectins are agglutinins, binding studies can be complemented with functional measurements to resolve different phenotypic behaviors. Confluent monolayers trypsinized into single-cell suspensions agglutinate in a cell- and lectin-specific manner, illustrating that lung macro- and microvascular ECs retain this phenotypic property in culture through multiple passages (Figure 126.3B).

It is easy to resolve macro- and microvascular ECs by visual inspection. Coculture of these cell types reveals a discrete cell–cell border at which pulmonary artery and microvascular cells grow together (see Figure 126.3C). Although they form a cell–cell junction, the molecular make-up of this junctional complex is different among the cell types. For example, surface expression of the activated leukocyte cell adhesion molecule (ALCAM; CD166) is prominent in microvascular ECs, and is absent in pulmonary artery ECs (15). It is apparent in coculture experiments that ALCAM is present between adjacent microvascular ECs but is absent at both the macro–microvascular and macro–macrovascular EC border. In addition, just as in the intact circulation, lectins retain their ability to discriminate these cell types in coculture experiments (Figure 126.4; for color reproduction see Color Plate 126.4) (15).

STABLE PHENOTYPES IN CULTURED ENDOTHELIAL CELLS

It is well recognized that passaging cells leads to phenotypic drift. For this reason, cell biologists interested in physiological applications commonly use primary or early passage cells for study. Even then, important questions remain regarding how well the behaviors of cells in culture resemble their in vivo behavior(s). Indeed, detailed comparisons between the proteomes of lung ECs in vivo and in culture reveal a significant shift in protein abundance; for example, only approximately half the integral membrane proteins and plasma membrane–associated proteins were commonly expressed when in vivo and in vitro comparisons were made (16). Thus, the process of isolating and culturing cells changes the expression (abundance) of multiple proteins, as does the act of passaging cells in culture.

Nonetheless, certain EC behaviors are retained in culture, as illustrated in the preceding section, and these behaviors accurately reflect the cell's in vivo function. Findings that lung macro- and microvascular ECs retain their differential lectin-binding patterns suggest that cells possess a memory of their vascular origin. Chi and colleagues (17) recently examined this idea by comprehensively screening mRNA expression patterns in 53 different EC types. All cells were cultured under similar environmental conditions, yet the investigators were able to discriminate among "like" and "unlike" phenotypes from their expression profiles. Indeed, global mRNA expression profiling resolved arterial, venule, and capillary fingerprints among the cells in culture.

The idea that ECs exhibit mitotically heritable, site-specific functions in vitro was borne out in studies evaluating barrier function. As in the intact circulation, cultured microvascular ECs express a unique complex of adhesion molecules that contribute to an enhanced barrier property. In addition to VE-cadherin, lung capillary ECs express N-cadherin (Troy Stevens, unpublished observations) and possibly E-cadherin (11,12). The microvascular ECs exhibit a higher resting electrical resistance than do their macrovascular counterparts, and they are more restrictive to water and macromolecular flux (15). All the above site-specific properties are maintained in tissue culture over multiple passages.

Lung microvascular ECs are also relatively insensitive to barrier disruption induced by inflammatory agonists. Thrombin, for example, maximally disrupts the pulmonary artery EC barrier in vitro within a 1 to 7 U/mL concentration range, whereas at least 10-fold higher concentrations are required to disrupt the microvascular EC barrier (18). Thrombin produces long-lasting gap formation in the pulmonary artery ECs, whereas gap formation is transient in lung microvascular ECs. This finding has led to the appreciation that microvascular cells move very rapidly, perhaps as a built-in mechanism for protecting the alveolar-capillary membrane from edema formation.

The effect of thrombin, or soluble protease-activated receptor ligands, on segment-specific permeability has not yet been examined in vivo. However, we have shown that thapsigargin induces extra-alveolar leak without directly increasing capillary permeability. Thapsigargin similarly increases permeability across pulmonary artery ECs in culture, and it is not sufficient to increase permeability across lung microvascular ECs (19). Interestingly, just as in the intact circulation, rolipram pretreatment prevents thapsigargin from inducing gaps in pulmonary artery ECs and reveals thapsigargin-induced gaps in microvascular ECs (9). Functional responses to the activation of store-operated calcium entry are therefore retained in cultured ECs, as they are in the intact pulmonary circulation. Thus, there is a growing interest in determining the signal transduction mechanisms responsible for controlling this site-specific EC behavior.

ENVIRONMENTAL CONTROL OF SEGMENT-SPECIFIC ENDOTHELIAL CELL FUNCTION

Evidence that macro- and microvascular ECs retain functionally discrete phenotypes under the same environmental condition does not detract from observations that environmental

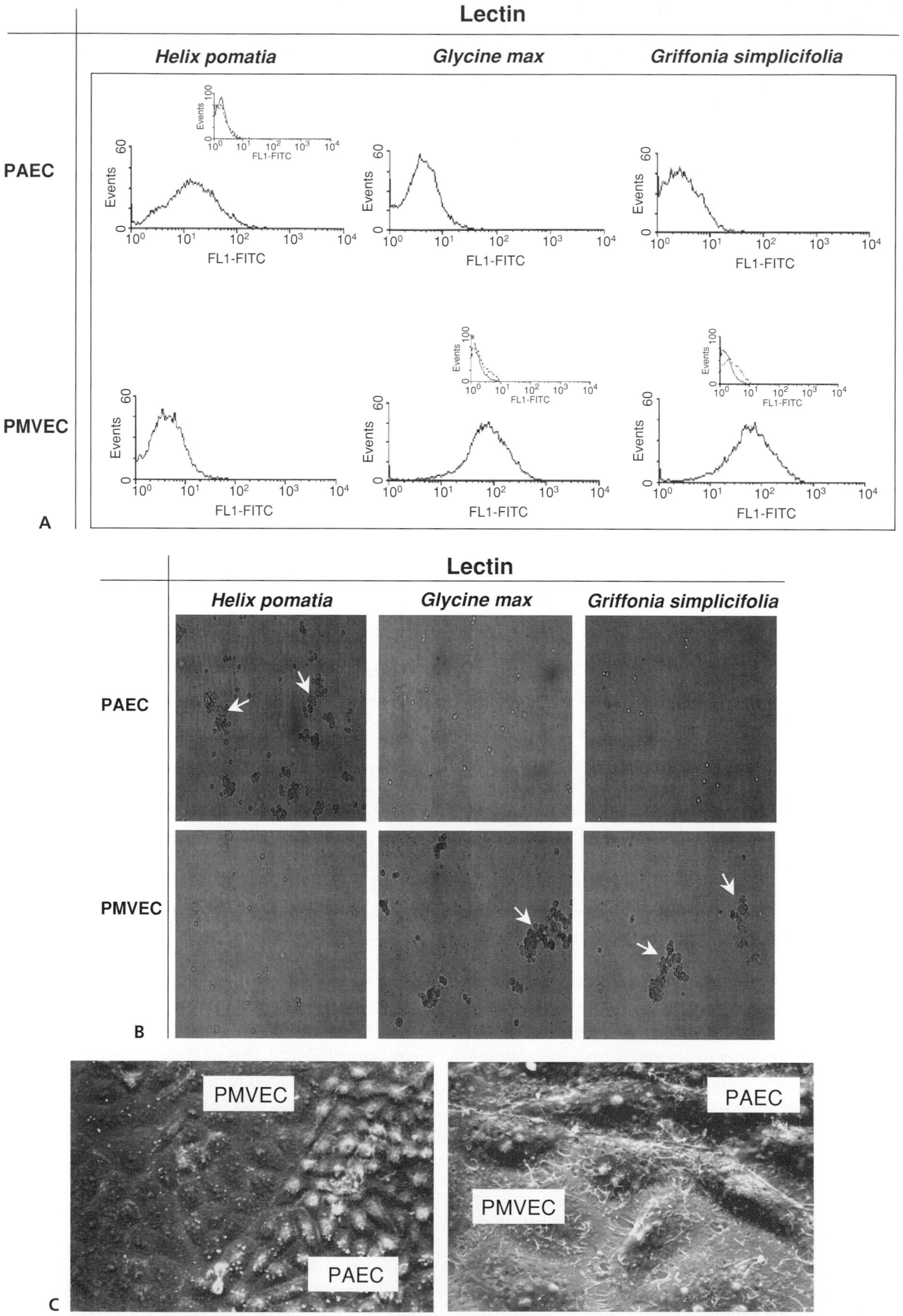

Lectin
Helix pomatia
Glycine max
Griffonia simplicifolia
PAEC
PMVEC
Events
FL1-FITC
A
B
C

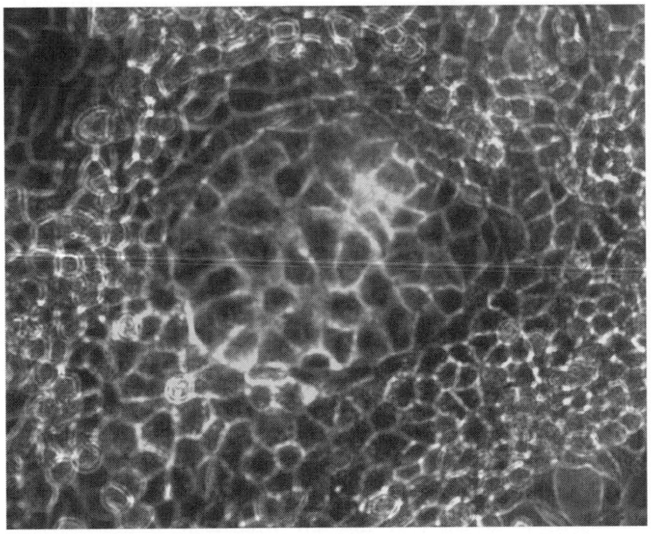

Figure 126.4. *G. simplicifolia* recognizes pulmonary microvascular ECs in coculture experiments. Pulmonary artery and microvascular ECs were cocultured, grown to confluence, and labeled with FITC-labeled *G. simplicifolia*. Only microvascular ECs interact with the lectin. For color reproduction, see Color Plate 126.4.

influences modulate EC behavior. However, it does bring into question how biophysical forces and various circulating chemical substances interact with macro- and microvascular endothelium. Very little work is available to address this important issue, although recently Al-Mehdi and colleagues (20) have assessed the EC response to flow. Although many EC culture studies are performed under static, nonflow conditions, this static environment clearly does not reflect the complex in vivo setting involving blood flow and its accompanying transmural pressure and shear stress forces. Typically, the application of perfusion to static EC cultures induces cell realignment in the direction of flow. ECs reorganize their shape from a polygonal orientation to one that is more elongated. Pulmonary artery ECs undergo this reorientation, just as do conduit cells from the systemic circulation. In contrast, lung microvascular ECs do not flow-align, even when exposed to very high levels of shear stress (20). Nonetheless, mitochondria reorganize in microvascular ECs exposed to prolonged flow, indicating that these cells do sense and respond to mechanical perturbation. Moreover, flow cessation in flow-adapted microvascular ECs induces calcium entry through T-type calcium channels and promotes oxidant production (21,22). Thus, it is clear that lung macro- and microvascular ECs each respond uniquely to environmental cues. These findings illustrate the importance of using physiologically appropriate cell sources to examine functional adaptation to environmental insults.

The complexity of environmental cues that influence EC behavior is only beginning to be appreciated. It is quite likely that unique cell–cell communication in pulmonary arteries/veins (e.g., EC to smooth muscle) and capillaries (e.g., EC to epithelial), and in matrix components within the basement membrane of extra-alveolar versus alveolar blood vessels, influence segment-specific EC function. It is also likely that different oxygen tensions in arterial (i.e., mixed venous pO_2) and venous (i.e., arterial pO_2) vessels contribute to segment-specific EC function. Detailed studies will be needed for us to better understand how environmental stimuli adjust segment-specific EC behavior.

ON ENDOTHELIAL CELL SPECIFICATION

It is not presently clear how ECs gain a stable functional memory – that is, how they establish mitotically heritable site-specific properties. However, studies of the developing pulmonary vascular circulation may shed some provocative insight into this issue. In 1997, DeMello et al. (23) used a Mercox corrosion cast model to visualize the developing mouse (20-day gestational period) pulmonary circulation by electron microscopy. Their studies revealed that the earliest evidence for pulmonary arterial development was in the pseudoglandular phase (ranging from embryonic day [E]9–16) of lung development. Only rudimentary arterial segments could be visualized at the beginning of this phase; the pulmonary circulation did not connect with capillary segments and did not return to the left ventricle. It was not until approximately E14 – near the end of the pseudoglandular phase of lung development – that arterial segments were continuous with capillary loops and ultimately with pulmonary veins. Despite the inability to visualize an "intact" circulation at E9 to 12, transmission electron microscopy demonstrated rudimentary blood lakes developing in the lung's mesenchyme. Thus, DeMello and colleagues (23) reasoned that two distinct events lead to the eventual formation of a mature pulmonary vascular bed, including angiogenesis of precapillary vessels and vasculogenesis of capillary vessels. The key event in this model is a fusion between angiogenic and vasculogenic vessels at midgestation, near the end of the pseudoglandular phase of lung development.

←

Figure 126.3. (**A**) FACS analysis reveals that macro- and microvascular ECs isolated and cultured retain their ability to uniquely interact with lectins. Insert pictures show control cells without lectin treatment, and cells treated with α-GalNAc to block *H. pomatia*, β-GalNAc to block *G. max*, and α-galactose to block *G. simplicifolia*. (**B**) Because lectins are agglutinins, their ability to agglutinate cells in the presence of trypsin was examined. *H. pomatia* agglutinates macrovascular ECs, whereas *G. max* and *G. simplicifolia* agglutinate microvascular ECs. For experimental details see King J, Hamil T, Creighton J, et al. Structural and functional characteristics of lung macro- and microvascular EC phenotypes. *Microvasc Res.* 2004;67(2):139–151. (**C**) Pulmonary artery ECs and pulmonary microvascular ECs were cocultured on gridded coverslips and observed as they grew to confluence. Pulmonary artery ECs cocultured with pulmonary microvascular ECs always generated a discernible border, illustrating unique cell recognition patterns.

There remains significant controversy as to whether this model accurately reflects the events that underlie lung vascular development. Indeed, serial sectioning of lungs from mice expressing reporter genes under the control of an endothelial-"specific" promoter has provided evidence that capillary development occurs continuously, along with the development of larger arteries and veins (24–26). These studies suggest the possibility that, in the DeMello and colleagues research (23), small capillary loops were associated with feeder vessels but could not be perfused by the Mercox solution. Thus, the exact origin of extra-alveolar and alveolar ECs remains in question.

Interestingly, work in transgenic animals has provided some mechanistic insight into what controls site-specific vascular development. In comprehensive works by multiple different labs, disruption of vascular endothelial growth factor (VEGF) signal transduction uniformly interrupts capillary development (15). In contrast, the disruption of the Notch signaling mechanism interrupts pulmonary artery development (15). Thus, these studies together suggest distinct molecular programs may be engaged as cells within different vascular locations gain endothelial specification.

Recent studies by Stewart and colleagues (27) incriminated nitric oxide (NO) in normal lung vascular development. These investigators examined endothelial NO synthase ($eNOS$)$^{-/-}$ mice, and discovered that approximately 25% of the mice died from respiratory failure within 48 hours of birth. Upon closer inspection, it was determined that capillaries developed independently of arteries and veins, suggesting the presence of artery-to-vein anastomoses, in which blood bypassed the alveolar-capillary membrane altogether. Capillaries, while present, had not simplified to form normally thin alveolar-capillary membranes. The lung vascular structures in these $eNOS^{-/-}$ mice resembled those of the alveolar-capillary dysplasia in children. Moreover, these findings of Stewart and colleagues (27) generally support the work of DeMello and colleagues (23), and suggest that NO participates in some fusion event that is essential for the eventual continuity of the lung's circulation.

At present, there is no consensus as to the independent or combined roles of angiogenesis and vasculogenesis in lung vascular development. While this issue remains an important one, a consensus point is that extra-alveolar and alveolar EC differentiation and specification occurs in different physical locations. Insight from various transgenic models supports the idea that unique gene expression programs are in place to guide appropriate EC development and to ensure that the cell functions in a manner appropriate for its vascular site. Thus, it is likely that cell-specific environments imprint highly distinctive characteristics early in the developmental course.

Although cell-specific behaviors may be imprinted early in development, there is currently no understanding of how this memory is maintained either as one ages or, indeed, when cells are placed in culture. Our group has only begun to address this issue. Because the genome is identical in cells of the body (including ECs from different vascular locations), we questioned whether proteins that control master regulatory transcriptional programs are differentially expressed among lung macro- and microvascular ECs. Histones fulfill an important role in controlling global gene expression by condensing DNA into heterochromatin and euchromatin. By controlling which DNA sequences are accessible to transcription factors, histones regulate global gene expression as a mechanism of imprinting. Interestingly, histones 2A and 2B were both overexpressed in lung microvascular ECs (when compared to pulmonary artery ECs), as was the nucleosome assembly protein. The functional significance of this observation remains to be determined, although these findings suggest that proteins organizing chromatin structure may fulfill an important role in phenotype specification, likely by controlling the activity of global transcription programs. Future studies will be required to rigorously determine the role that chromatin structure plays in endothelial phenotype specification.

IMPLICATIONS OF ENDOTHELIAL CELL HETEROGENEITY IN PULMONARY VASCULAR DISEASE

It is not often that we consider exactly why a disease process manifests itself in any specific vascular locale. However, the idea that ECs possess a site-specific phenotype suggests that imprinting may contribute to lesion site-specificity. We have already illustrated this point in our review of the developmental literature, in which $eNOS^{-/-}$ mice that died from respiratory failure early in the postnatal period exhibited an alveolar-capillary dysplasia-like anatomy (27). These findings were taken to suggest that NO fulfills an important role in the fusion of extra-alveolar and alveolar blood vessels. We have also seen that circulating inflammatory agonists can target either extra-alveolar or alveolar ECs to increase permeability in a site-specific way (3,4,6–8). Thus, there is a growing awareness that EC phenotypes contribute to the site of vascular injury.

Primary pulmonary hypertension is a rare, progressive disorder that also involves EC dysfunction. There are two prominent aspects of this disease (28). A first lesion is found in large- and intermediate-sized precapillary blood vessels, and is characterized by medial and adventitial hypertrophy/hyperplasia. Smooth muscle layers extend distally into normally nonmuscularized or poorly muscularized vascular segments, contributing significantly to the increase in pulmonary artery pressure and vascular resistance. A second lesion is found predominantly in severe forms of pulmonary hypertension, and is localized to small precapillary vessel segments, commonly originating at vessel bifurcations. This "plexiform lesion" is a lumen-occluding mass comprised of ECs that have lost the "law of the monolayer." We were particularly interested in the source of ECs found in the plexiform lesion. Histological sections show cells extending from the intima of small

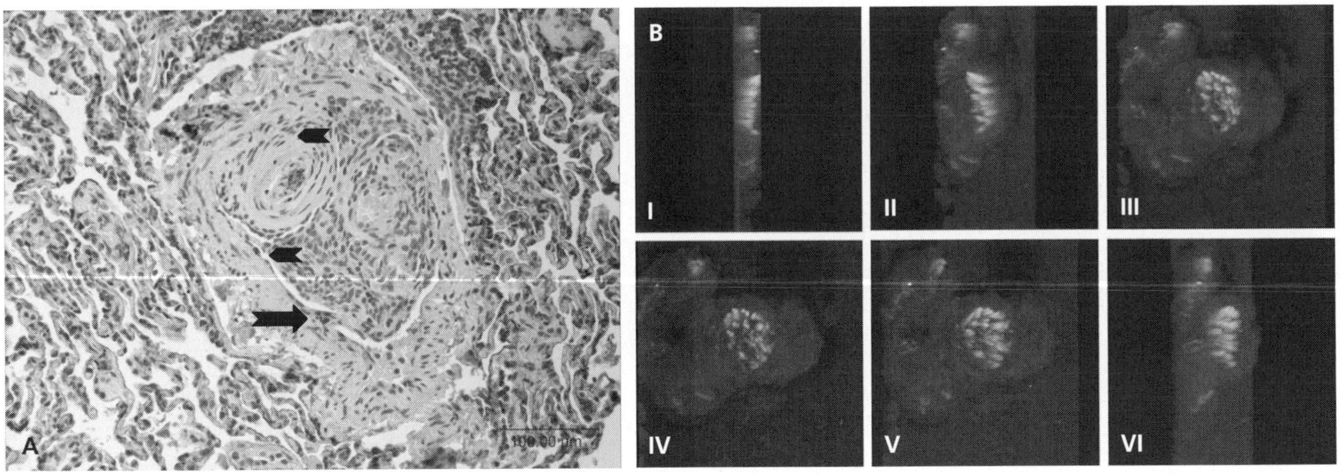

Figure 126.5. Plexogenic lesions are comprised of cells that interact with *G. simplicifolia*. (**A**) Although these lesions may extend into large blood vessels, serial sectioning reveals the initiating locus is typically a small, precapillary arteriole. This histological section of a plexogenic lesion from a patient shows EC staining for ALCAM (DAB). The arrows show sites in the vessel wall where ECs appear to have lost their intimate contact and grown into the lumen. (**B**) Human lung specimens from patients with severe pulmonary hypertension were used to analyze plexogenic lesions for their ability to interact with lectins. This 70-μm lesion interacts selectively with *G. simplicifolia* and not *H. pomatia* (not shown), suggesting cells with a microvascular EC phenotype are present in the lesion. Tissue was evaluated by confocal microscopy with 0.35-μm Z-axis sectioning and reconstructed into a 3D image. The image was then rotated 360 degrees from the lateral position I–VI and pictures captured.

precapillary vessels into the lumen (Figure 126.5A), suggesting at least some of the cells in the lumen could be of a "microvascular origin." We therefore examined whether ECs in the plexiform lesion interact with *G. simplicifolia.* Cell clusters within the lumen did, indeed, interact with this marker of microvascular ECs, suggesting that they were of a microvascular EC phenotype (Figure 126.5B). Much more definitive evidence will ultimately be required to assure that cells causing the plexiform lesions arise from microvascular ECs within the vessel wall. However, these findings provide provocative support for the idea that the EC phenotype contributes to the site of vascular injury.

CONCLUSION

Although we now recognize that ECs along a single vascular segment are highly specialized, as highlighted by the comparisons between extra-alveolar and alveolar ECs, we know very little about how such phenotype specification occurs and is maintained. We have limited information about how environmental cues interact uniquely within a given vascular segment to modify EC behavior or, in extreme circumstances, to contribute to the site-specificity of vascular disease. Indeed, the continued, rapid advance of our field will require the detailed consideration of not only how environmental cues modify EC behavior, but more specifically how these environmental influences interact with ECs within (or obtained from) a physiologically relevant vascular location.

KEY POINTS

- Extra-alveolar and alveolar ECs possess a stable memory of their vascular origin, which influences the cell's response to environmental cues and establishes a site-specific cell phenotype.
- Extra-alveolar and alveolar ECs possess distinct behaviors, including barrier function, in which alveolar ECs form a more restrictive barrier than do extra-alveolar ECs.
- Extra-alveolar and alveolar ECs retain a memory of their in vivo phenotypes, even when studied in culture under similar environmental conditions.
- EC phenotype, which is determined by the cell's vascular origin and its interaction with local environmental influences, contributes to the site-specific nature of vascular disease.

Future Goals

- To determine molecular mechanisms responsible for EC phenotype specification
- To map signaling matrices in extra-alveolar and alveolar ECs that encode their unique responses to (patho-)physiologically relevant environmental stimuli

REFERENCES

1 King J, Hamil T, Creighton J, et al. Structural and functional characteristics of lung macro- and microvascular endothelial cell phenotypes. *Microvasc Res.* 2004;67(2):139–151.

2 Thorin E, Stevens T, Patterson CE. Heterogeneity of lung endothelial cells. In: Patterson CE, ed. *Lung Endothelial Cells in Health and Disease.* Boston: Elsevier Science; 2005:277–310.

3 Parker JC, Townsley MI. Evaluation of lung injury in rats and mice. *Am J Physiol Lung Cell Mol Physiol.* 2004;286(2):L231–L246.

4 Parker JC, Yoshikawa S. Vascular segmental permeabilities at high peak inflation pressure in isolated rat lungs. *Am J Physiol Lung Cell Mol Physiol.* 2002;283(6):L1203–L1209.

5 Whayne TF Jr., Severinghaus JW. Experimental hypoxic pulmonary edema in the rat. *J Appl Physiol.* 1968;25(6):729–732.

6 Pietra GG, Johns L. Leaky intra-acinar arteries in rat lungs perfused with hydrogen peroxide. *J Appl Physiol.* 1990;69(3):1110–1116.

7 Albert RK, Kirk W, Pitts C, Butler J. Extra-alveolar vessel fluid filtration coefficients in excised and in situ canine lobes. *J Appl Physiol.* 1985;59(5):1555–1559.

8 Chetham PM, Babal P, Bridges JP, et al. Segmental regulation of pulmonary vascular permeability by store-operated Ca^{2+} entry. *Am J Physiol.* 1999;276(1 Pt 1):L41–L50.

9 Wu S, Cioffi EA, Alvarez D, et al. Essential role of a Ca^{2+}-selective, store-operated current (ISOC) in endothelial cell permeability: determinants of the vascular leak site. *Circ Res.* 2005;96(8):856–863.

10 Corada M, Mariotti M, Thurston G, et al. Vascular endothelial-cadherin is an important determinant of microvascular integrity in vivo. *Proc Natl Acad Sci USA.* 1999;96(17):9815–9820.

11 Quadri SK, Bhattacharjee M, Parthasarathi K, et al. Endothelial barrier strengthening by activation of focal adhesion kinase. *J Biol Chem.* 2003;278(15):13342–13349.

12 Safdar Z, Wang P, Ichimura H, et al. Hyperosmolarity enhances the lung capillary barrier. *J Clin Invest.* 2003;112(10):1541–1549.

13 Wu S, Haynes J Jr., Taylor JT, et al. Cav3.1 (alpha1G) T-type Ca^{2+} channels mediate vaso-occlusion of sickled erythrocytes in lung microcirculation. *Circ Res.* 2003;93(4):346–353.

14 Stevens T, Brough G, Moore TM, et al. Endothelial cells. In: Taylor SUA, ed. *Methods in Pulmonary Research.* Boston: Birkhauser Verlag; 1998:403–426.

15 Gebb S, Stevens T. On lung endothelial cell heterogeneity. *Microvasc Res.* 2004;68(1):1–12.

16 Durr E, Yu J, Krasinska KM, et al. Direct proteomic mapping of the lung microvascular endothelial cell surface in vivo and in cell culture. *Nat Biotechnol.* 2004;22(8):985–992.

17 Chi JT, Chang HY, Haraldsen G, et al. Endothelial cell diversity revealed by global expression profiling. *Proc Natl Acad Sci USA.* 2003;100(19):10623–10628.

18 Cioffi DL, Moore TM, Schaack J, et al. Dominant regulation of inter endothelial cell gap formation by calcium-inhibited type 6 adenylyl cyclase. *J Cell Biol.* 2002;157(7):1267–1278.

19 Kelly JJ, Moore TM, Babal P, et al. Pulmonary microvascular and macrovascular endothelial cells: differential regulation of Ca2+ and permeability. *Am J Physiol.* 1998;274(5 Pt 1):L810–L819.

20 Al-Mehdi AB, Schaphorst KL, Stevens T. Lung endothelial heterogeneity. In: Garcia JGN, ed. *Encyclopedia of the Microvasculature.* Boston: Elsevier Science; 2005;465–474.

21 Milovanova T, Manevich Y, Haddad A, et al. Endothelial cell proliferation associated with abrupt reduction in shear stress is dependent on reactive oxygen species. *Antioxid Redox Signal.* 2004;6(2):245–258.

22 Wei Z, Manevich Y, Al-Mehdi AB, et al. Ca^{2+} flux through voltage-gated channels with flow cessation in pulmonary microvascular endothelial cells. *Microcirculation.* 2004;11(6):517–526.

23 deMello DE, Sawyer D, Galvin N, Reid LM. Early fetal development of lung vasculature. *Am J Respir Cell Mol Biol.* 1997;16(5):568–581.

24 Schwarz MA, Zhang F, Lane JE, et al. Angiogenesis and morphogenesis of murine fetal distal lung in an allograft model. *Am J Physiol Lung Cell Mol Physiol.* 2000;278(5):L1000–L1007.

25 Schachtner SK, Wang Y, Scott Baldwin H. Qualitative and quantitative analysis of embryonic pulmonary vessel formation. *Am J Respir Cell Mol Biol.* 2000;22(2):157–165.

26 Parera MC, van Dooren M, van Kempen M, et al. Distal angiogenesis: a new concept for lung vascular morphogenesis. *Am J Physiol Lung Cell Mol Physiol.* 2005;288(1):L141–L149.

27 Han RN, Babaei S, Robb M, et al. Defective lung vascular development and fatal respiratory distress in endothelial NO synthase-deficient mice: a model of alveolar capillary dysplasia? *Circ Res.* 2004;94(8):1115–1123.

28 Stevens T, Kasper M, Cool C, Voelkel NF. Pulmonary circulation and pulmonary hypertension. In: Aird WC, ed. *Endothelial Cells in Health and Disease.* New York: Taylor & Francis; 2005:417–437.

127

Bronchial Endothelium

Elizabeth Wagner and Aigul Moldobaeva

Johns Hopkins Asthma and Allergy Center, Baltimore, Maryland

The bronchial circulation is poorly accessible and dwarfed in size by the pulmonary vasculature. Studies highlighting the unique attributes of the bronchial endothelium are few in number and represent a contemporary work-in-progress. Despite the current paucity of information regarding bronchial endothelial phenotypes, it has long been recognized that bronchial vessels undergo structural changes in pathological states. Most notable among these changes are dilation, neovascularization, and contribution to bronchial mucosal swelling.

Perhaps the first documentation of disease-associated bronchial neovascularization (hence, bronchial endothelial proliferation) is found in one of Leonardo da Vinci's anatomical drawings (circa 1513 A.D.) (Figure 127.1). Close scrutiny of a tuberculous cavity in a terminal lung unit shows bronchial vessels supplying the walls of the cavity. Whether da Vinci used artistic license in his depiction of the vasculature has been argued over the years (1).

In 1847, Rudolph Virchow demonstrated an association between chronic pulmonary artery obstruction and a proliferative bronchial phenotype (2). He interpreted these studies to suggest pulmonary vascular ischemia leads to compensatory recruitment of systemic bronchial vessels to ischemic regions. Others more recently have speculated that obstruction of even small pulmonary vessels (e.g., within a tuberculous cavity) is sufficient to induce proliferation of bronchial vessels (3).

In his original textbook (1892), Sir William Osler noted that during an asthma attack "the hyperaemia and swelling of the mucosa . . . explain well the hindrance to inspiration and expiration" (4). Although mechanisms regulating endothelial cell (EC) barrier function were not known at the time, the clinical consequences certainly were.

These and other gross morphology–based findings provided important insights into the unique features of the bronchial circulation in health and disease. Speculation about a putative role of the endothelium in mediating these properties would have to wait not only for the discovery of this cell layer, but also for the recognition that ECs are more than an inert barrier. Over the past decades, advances in ultrastructural methodology (e.g., electron microscopy), cell culture techniques, and molecular biology have paved the way for identifying and characterizing site-specific phenotypes of the bronchial endothelium that bear relevance to homeostasis and pathophysiology.

STRUCTURE

The bronchial vasculature originates from the aorta or, in some species, the intercostal arteries (5–7). The bronchial artery courses to the dorsal aspect of the carina, where it bifurcates and sends branches down the mainstem bronchi (Figure 127.2). This vascular bed perfuses the airways from the level of the carina to the terminal bronchioles. The bronchial arteries send arterioles throughout the airway adventitia that connect with capillaries that are prominent in both the adventitia as well as the mucosa of the airway wall. Thus, the bronchial vasculature forms parallel vascular plexuses, situated on either side of the airway smooth muscle. In the intraparenchymal airways, postcapillary venules collect bronchial venous drainage and drain into pulmonary venules and/or alveolar capillaries that drain into pulmonary veins and the left atrium. Additionally, the bronchial artery sends branches to large pulmonary vessels as vasa vasorum, nerves, and the visceral pleura.

The developmental origin of the bronchial circulation is not well established. Previous studies of human vascular development demonstrate that the bronchial artery arises as an outgrowth from the aorta between the ninth and twelfth week of gestation, a somewhat later time point compared with the pulmonary circulation (8,9). One or two vessels extend from the dorsal aorta and form along the cartilaginous plates of the large airways. The vessels extend longitudinally along the airways to the lung periphery as far as the terminal bronchioles. This pattern is suggestive of an angiogenic (as distinct from vasculogenic) process. However, the mechanisms underlying development of other segments of the bronchial circulation remain unclear. For example, it is not known to what extent angiogenesis or vasculogenesis (from mesoderm) contributes to the formation of fine bronchial capillaries within

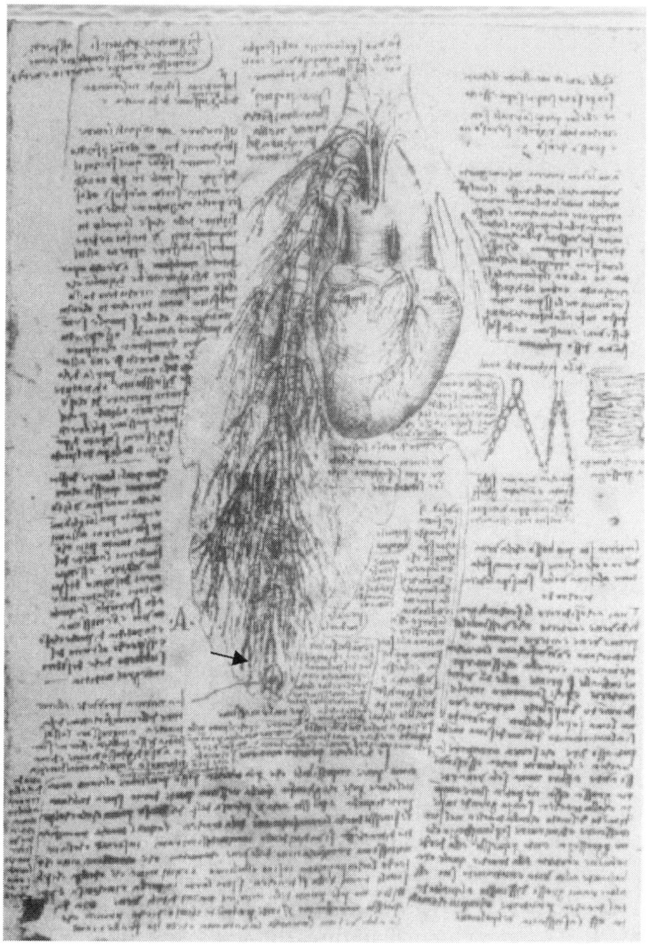

Figure 127.1. Leonardo daVinci's drawing of purported bronchial circulation. Arrow points to a tuberculous cavity in a terminal lung unit supplied by bronchial vessels. (Reproduced with permission from Mitzner W, Wagner EM. On the purported discovery of the bronchial circulation by Leonardo da Vinci. *J Appl Physiol.* 1992;73:1196–1201.)

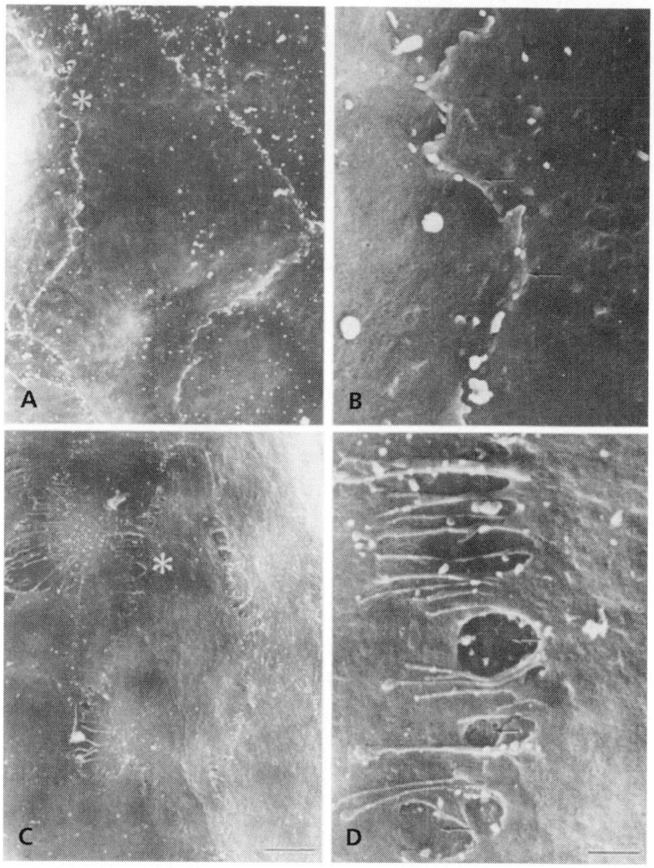

Figure 127.3. Scanning electron micrographs of the luminal surface of rat tracheal venular endothelium: **A** and **B** control, **C** and **D** after substance P. Regions marked by asterisks are enlarged in B and D. Bars, 5 μm for A and C; 1 μm for B and D. (Reproduced with permission from Baluk P, Hirata A, Thurston G, et al. Endothelial gaps: time course of formation and closure in inflamed venules of rats. *Am J Physiol.* 1997;272:L155–170.)

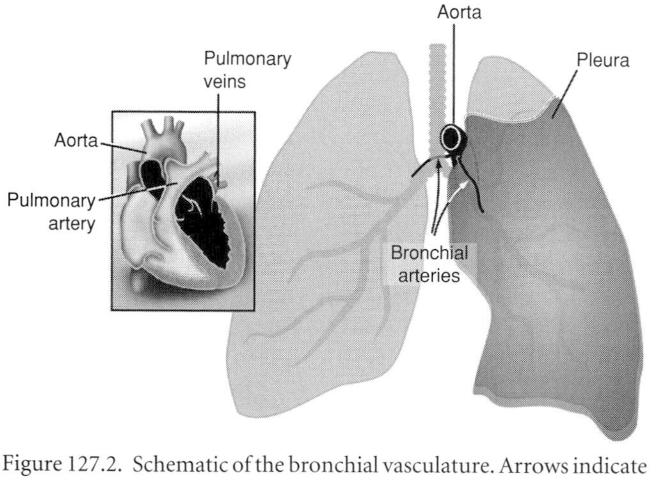

Figure 127.2. Schematic of the bronchial vasculature. Arrows indicate the site of bronchial arterial perfusion of airways, pleura and vasa vasorum of large pulmonary vessels.

the bronchial wall (namely the mucosal plexus), the vasa vasorum of large pulmonary vessel walls, or bronchial–pulmonary anastomoses.

Relatively little is known about the morphology or site-specific markers of the endothelium lining the bronchial vasculature. Capillaries are predominantly continuous and, in healthy mammals, lacking in fenestrations. Electron micrographs of the luminal surface of rat tracheal venules (Figures 127.3A/B and 127.4A) reveal uniform cell borders (*arrows*) that are narrow and closely apposed. Only recently have investigators been successful in isolating airway ECs and studying them in vitro (10,11).

BRONCHIAL ENDOTHELIUM IN HEALTH AND DISEASE

As with other vascular beds, the endothelial lining of the bronchial vasculature has been implicated in the control of vascular smooth muscle contraction/relaxation (hence blood

Table 127-1: Comparison of Bronchial and Pulmonary Vascular Beds

	Bronchial	*Pulmonary*
Surface area	Low	High
Arterial oxygen	High	Low
Arterial pressure	High	Low
Arterial resistance	High	Low
Blood flow	Low (<3% CO)	High (100% CO)
Primary function	Nutrient delivery; thermoregulation, humidification of airways	Gas exchange
Proliferative capacity	High	Low
Role for NO in vasomotor tone	Yes	Yes
Role for ACE	Yes	Yes
Leukocyte recruitment	Postcapillary	Capillary

CO, cardiac output

flow), barrier function, substrate metabolism, and leukocyte recruitment, among other common functions. Given that the bronchial and pulmonary circulations carry out distinct functions, it is not surprising that endothelium from these blood vessels express different phenotypes. For example, alveolar (pulmonary) capillaries form a tight barrier to prevent fluid accumulation and compromise in oxygen delivery, whereas bronchial capillaries play an active role in thermoregulation and humidification of ambient air. Bronchial artery ECs display a more robust proliferative capacity compared with their pulmonary counterparts, a difference that is consistent with a role for the bronchial circulation in compensating for pulmonary artery hypoperfusion. A comparison of the two lung circulations is provided (Table 127-1) and several of these features are highlighted in the next sections.

Vasomotor Tone

Previous physiological studies have demonstrated the importance of endothelial-derived substances in modulating bronchial vascular resistance. For example, systemic infusion of N-(G)-nitro-L-arginine methyl ester (L-NAME), an inhibitor of endothelial nitric oxide synthase (eNOS), resulted in decreased bronchial blood flow in sheep (13,14). Bronchial vasodilators such as bradykinin, acetylcholine, inhaled β-agonists, ionic and nonionic contrast media all function partially through the release of endothelial-derived nitric oxide (NO) (12–14). Cyclooxygenase inhibitors likewise have been shown to elicit endothelium-dependent vasoconstriction (15). Intra-arterial delivery of the potent EC-derived vasoconstrictor endothelin-1 caused a significant increase in bronchial vascular resistance in sheep (16).

These results are in keeping with observations in other systemic arteries. However, the functional significance of vasomotor regulation in the bronchial circulation is unique. For example, we and others have shown that the level of blood flow through the bronchial circulation can affect both the magnitude and time course of agonist-induced airway smooth muscle constriction by contributing to the passive wash-out of an inhaled or intra-arterial agonist (17,18). Thus, changes in airway wall perfusion due to airway EC activation may modulate airway reactivity. Additionally, changes in bronchial vascular perfusion have been shown to affect the function of vascular uptake of soluble particles deposited on airway epithelial surface (19). Thus, modulation of bronchial vascular perfusion by endothelial-derived factors can affect airway function.

Metabolism

Few studies have focused directly on the metabolic capability of the bronchial endothelium. Using an in situ perfused sheep bronchial artery preparation, Grantham and colleagues

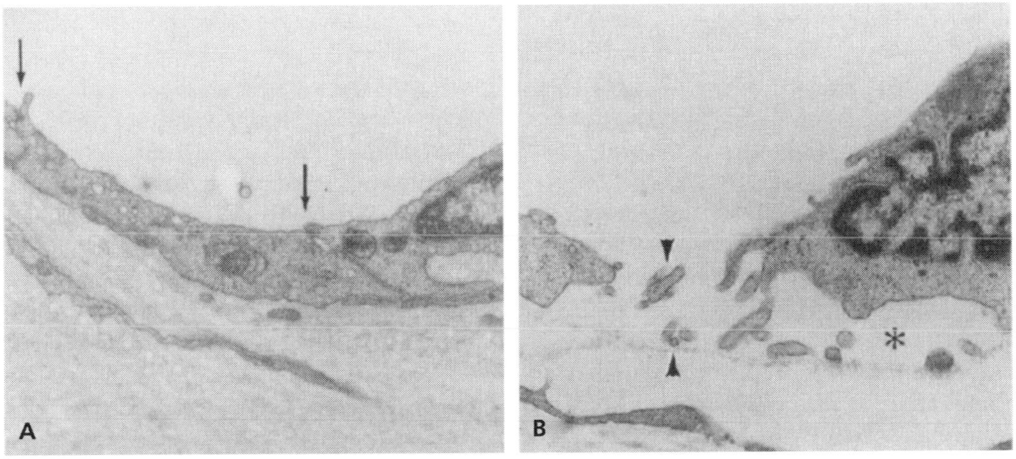

Figure 127.4. (**A**) Transmission electron micrograph showing normal EC junctions of rat tracheal venular endothelium. (**B**) ECs gaps with finger-like projections after exposure to substance P. (Reproduced with permission from Baluk P, Hirata A, Thurston G, et al. Endothelial gaps: time course of formation and closure in inflamed venules of rats. *Am J Physiol.* 1997;272:L155–170.).

demonstrated that intra-arterial administration of an angiotensin-converting enzyme (ACE) inhibitor significantly depressed metabolism of a synthetic peptide substrate for ACE (20). These authors concluded that the bronchial circulation is pharmacokinetically and metabolically active with respect to bradykinin breakdown, and that the enzyme(s) responsible for this metabolic activity is present on the surface of the vascular lumen. We confirmed in sheep the importance of ACE activity in the bronchial circulation in vivo by demonstrating the inhibition of bradykinin-mediated bronchial vasodilation after treatment with an ACE inhibitor (11). We also have demonstrated that, under in vitro conditions, cultured bronchial microvascular ECs express ACE, and that the levels and activity of this enzyme are comparable to those of pulmonary ECs. These experiments suggest an important regulatory role for bronchial endothelial ACE in the metabolism of kinin peptides known to contribute to airway pathology.

Vascular Barrier Function

Numerous studies involving animal models as well as autopsy specimens of human lungs have provided evidence for the propensity of the systemic airway circulation to contribute to abnormal fluid accumulation in both the bronchial wall and lumen. Most notably, inflammatory cytokines have been shown to induce the formation of gaps in the endothelium of postcapillary venules, which leads to exudation of plasma and protein into the interstitium (21). Figures 127.3C/D and 127.4B show electron micrographs of tracheal venular gap formation after intravenous challenge with the inflammatory mediator, substance P (21). Long, finger-like processes are shown to bridge the gaps between ECs.

To evaluate the molecular mechanisms responsible for gap formation and secondary transudation of fluid, we have recently isolated, cultured, and studied ECs from large bronchial arteries (conduit vessels) and bronchial microvessels from sheep (10). Compared with ECs from the bronchial artery, bronchial microvascular ECs displayed increased permeability both at baseline and in response to edematogenic substances, including bradykinin and thrombin. These findings run counter to those usually observed in the in vitro and in vivo studies of pulmonary endothelium, in which the microvascular endothelial barrier has been shown to be more restrictive than conduit vessel endothelium. The relative differences in permeability between bronchial and pulmonary microvascular ECs are consistent with their established roles in mediating airway humidity and alveolar dryness, respectively. Importantly, the cell culture data suggest that these differences are mitotically heritable and thus epigenetically programmed.

It has been proposed that the bronchial endothelium, by a unique mechanism of cross-talk, may contribute to the barrier function of pulmonary endothelium (22). Although the existence of bronchial artery–derived vasa vasorum in pulmonary vessels has long been recognized, the functional significance of these capillary networks has remained unexplored. Vasa vasorum have been identified in pulmonary arterioles of 200 μm diameter (23). Furthermore, in a sheep ischemia–reperfusion lung injury model, perfusion of the bronchial artery during pulmonary artery ischemia and reperfusion, or pulmonary artery reperfusion alone, significantly attenuated the increase in pulmonary endothelial permeability that occurred in the absence of bronchial artery perfusion (22). Because bronchial artery perfusion during the reperfusion phase alone was effective at reducing injury, this result excludes the possibility that mere nutrient flow through broncho-pulmonary anastomoses accounted for the preserved barrier function. These results raised the interesting hypothesis that the bronchial endothelium generates and transfers a protective substance to the pulmonary endothelium during pulmonary artery reperfusion. NO produced by bronchial endothelium is a likely, shear-induced antioxidant candidate. Subsequent preliminary studies demonstrated that perfusion of the bronchial circulation during pulmonary artery reperfusion with a NOS inhibitor attenuated the protective effect in this model. These results further highlight the potential heterogeneity of EC responses to ischemia with regard to barrier function as well as basic enzymatic function during ventilated ischemia. Furthermore, they provide evidence that the systemic circulation of the lung can have an impact on functional attributes of the pulmonary vasculature.

Leukocyte Recruitment

Leukocyte recruitment in other organs receiving systemic blood supply has been shown to involve an orchestrated series of molecular interactions between rolling leukocytes and ECs in postcapillary venules. Given that the bronchial circulation is systemic in nature, it seems likely that similar mechanisms hold true for leukocyte transendothelial migration in this vascular bed. Indeed, we recently have confirmed this hypothesis, using intravital microscopy of lipopolysaccharide (LPS)-treated airways in rats (24). Moreover, we have demonstrated that ventilatory stress imposed on bronchial endothelium (airway distension) induces leukocyte recruitment in the airway (25).

Angiogenesis

Chronic Thromboembolism

Angiogenesis in the adult lung involves the invasion of systemic vessels (bronchial, intercostal arteries) into ischemic lung parenchyma while the pulmonary vasculature remains relatively inert. In this regard, the bronchial circulation has been compared to "Mother or the Red Cross; normally accepted and unsung, but capable of giving vital help when needed" (5). Bronchial arteriograms in patients with chronic thromboembolic disease demonstrate the unique capacity of these systemic vessels to proliferate and invade the ischemic lung parenchyma. Both a dilated bronchial artery as well as a fine meshwork of vessels distal to the pulmonary occlusion can be seen (Figure 127.5) (26). Neovascularization of

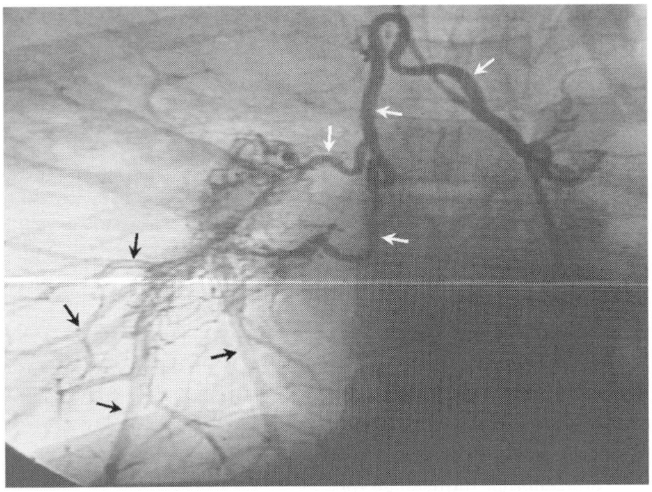

Figure 127.5. Bronchial arteriogram in a patient with thromboembolic pulmonary hypertension. The dilated bronchial artery is clearly seen, as well as a mesh of small bronchial vessels connected to the pulmonary arterial branches downstream to the embolic occlusion. White arrows point to bronchial arterial branches and black arrows point to pulmonary arterial branches distal to the occlusion. (Reproduced with permission from Endrys J, Hayat N, Cherian G. Comparison of bronchopulmonary collaterals and collateral blood flow in patients with chronic thromboembolic and primary pulmonary hypertension. *Heart.* 1997;78:171–176.)

the systemic circulation into the lung after pulmonary artery obstruction has been confirmed and studied in humans (26), sheep (27), dogs (28), pigs (29), guinea pigs (30), rats (31), and mice (32). Systemic blood flow to the lung has been shown to increase to as much as 30% of the original pulmonary blood flow after pulmonary artery occlusion (28,33). In a mouse model of chronic pulmonary thromboembolism, the proangiogenic C-X-C chemokines were shown to play a major role in vascular proliferation (34). This observation adds further support to a growing body of evidence that the Glu-Leu-Arg (ELR)-containing C-X-C chemokines and their G-protein–coupled receptor CXCR2 promote angiogenesis in several lung diseases (35).

Notably, in studies of non–small cell lung cancer (36) and idiopathic pulmonary fibrosis (37), the ELR+ C-X-C chemokines have been shown to be proangiogenic, whereas the ELR– C-X-C chemokines exhibit angiostatic properties. Based on these data, it was hypothesized that differences in proangiogenic potential of systemic versus pulmonary ECs may be explained by vascular bed–specific differences in chemokine responsiveness (38). In support of this hypothesis, macrophage inflammatory protein (MIP)-2 (an ELR-containing C-X-C) was shown to promote migration of cultured mouse aortic ECs, but not pulmonary artery ECs. Expression of the MIP-2 G-protein–coupled receptor CXCR2 did not differ between cell types. However, cathepsin S, a proteolytic enzyme important for EC motility (39), was shown to exhibit increased protein and activity in chemokine-treated aortic ECs, relative to similarly treated pulmonary ECs. These results demonstrate one potential mechanism that may contribute to lung EC subtype–specific differences in angiogenic potential.

Inflammation

Inflammation can contribute significantly to bronchial vascular remodeling and life-threatening hemoptysis. Hemoptysis in the vast majority of patients originates from the systemic, rather than pulmonary vasculature, and the bronchial vessels are involved almost universally. However, the etiology of hemoptysis requiring therapeutic embolization is variable, and the pathology of bronchial vessels leading to this acute condition is poorly understood. Several studies have focused on the chronic inflammation of asthma and airway vascular remodeling. Li and Wilson demonstrated increased vascular density in biopsy specimens of subjects with mild asthma compared with normal control volunteers (40). However, Chu and colleagues showed that only biopsy specimens from asthmatic subjects with concurrent *Mycoplasma pneumoniae* infections demonstrated significantly increased vessel numbers (41). Rodent models of inflammation have been employed to study alterations in airway vascularity, and have demonstrated increased vessel numbers, size, and permeability characteristics. Airway remodeling following *M. pulmonis* infection showed mouse strain–dependent changes in tracheal vascularity (42). Sustained alterations in the tracheal vasculature were reversed by corticosteroid treatment (43). Additionally, tracheal vessels in rats exposed to *M. pulmonis* demonstrated a significantly increased permeability in the airway vasculature when challenged with substance P after inflammation was induced (44). These results, along with the clinical features of hemoptysis, suggest that proliferating bronchial vessels express abnormal barrier function.

Hypoxia

Although the bronchial vasculature has been shown to dilate acutely during hypoxic ventilation (45,46), its proliferative capacity during chronic hypoxia has not been established. Clinically, bronchial vessels appear not to proliferate during primary pulmonary hypertension (47). However, a recent study demonstrated proliferation of the pulmonary vasa vasorum in a neonate exposed to chronic hypoxia (48). It has been suggested that blood vessels that transport venous blood, such as the systemic veins and pulmonary artery, depend more on the network of vasa vasorum than on luminal oxygen to support metabolism (49). A recent study of Davie and colleagues demonstrated a marked increase in the vascular density of the pulmonary artery adventitial vasa vasorum coincident with significant pulmonary artery hypertension (48). Furthermore, these investigators confirmed that chronic hypoxia resulted in an increase in circulating bone marrow–derived progenitor cells (c-kit^{+} cells), which also were incorporated into the new vasa vasorum within the wall of the pulmonary artery. Thus, this suggests that blood-borne progenitor cells to be delivered to the site of neovascularization may be required for this remodeling process.

KEY POINTS

- Bronchial endothelium shares the characteristic features of other systemic vascular beds, and its responses are frequently in contrast to pulmonary endothelium.
- EC-derived factors modulate vascular smooth muscle tone as they do other systemic vessels.
- Leukocyte recruitment and vascular leak occur at postcapillary sites.
- Bronchial endothelium is involved in the vigorous angiogenic response to pulmonary parenchymal ischemia.

Future Goals

- To focus studies of lung pathologies associated with the systemic circulation of the lung on the basic cellular mechanisms of bronchial ECs, despite the difficulty in accessing bronchial endothelium
- To elucidate the unique attributes of bronchial relative to pulmonary endothelial function

REFERENCES

1 Mitzner W, Wagner EM. On the purported discovery of the bronchial circulation by Leonardo da Vinci. *J Appl Physiol.* 1992; 73(3):1196–1201.

2 Virchow R. Uber die Standpunkte in den Wissenschaftlichen Medizin. *Virchow Archiv.* 1847;1:1–19.

3 Cudkowicz L. The bronchial circulation in the human. In: Butler J, ed. *The Bronchial Circulation.* Vol. 57. New York: Marcel Dekker; 1992:3–41.

4 Osler W. *The Principles and Practices of Medicine.* Vol. 1. New York: Appleton, 1892.

5 Deffebach ME, Charan NB, Lakshminarayan S, Butler J. The bronchial circulation-small, but a vital attribute of the lung. *Am Rev Respir Dis.* 1987;135:463–481.

6 Charan NB, Turk GM, Dhand R. Gross and subgross anatomy of the bronchial circulation in sheep. *J Appl Physiol.* 1984;57:658–664.

7 Wagner EM. Bronchial circulation. In: Crystal RG, West JB, Weibel ER, Barnes PJ, eds. *The Lung: Scientific Foundations.* Vol. 2. New York: Lippincott-Raven Press; 1997:1093–1105.

8 Boyden E. The time lag in the development of bronchial arteries. *Anat Rec.* 1970;166:611–614.

9 Hislop A. Airway and blood vessel interaction during lung development. *J Anat.* 2002;201:325–334.

10 Moldobaeva A, Wagner E. Heterogeneity of bronchial endothelial cell permeability. *Am J Physiol Lung Cell Mol Physiol.* 2002:26.

11 Moldobaeva A, Wagner EM. Angiotensin-converting enzyme activity in ovine bronchial vasculature. *J Appl Physiol.* 2003;95: 2278–2284.

12 Baile EM, Wang L, Verburgt L, Pare PD. Bronchial vasodilatory response to ionic and nonionic contrast media. *J Appl Physiol.* 1997;82:841–845.

13 Sasaki F, Pare P, Ernest D, et al. Endogenous nitric oxide influences acetylcholine-induced bronchovascular dilation in sheep. *J Appl Physiol.* 1995;78:539–545.

14 Carvalho P, Johnson SR, Charan NB. Non-cAMP-mediated bronchial arterial vasodilation in response to inhaled beta-agonists. *J Appl Physiol.* 1998;84:215–221.

15 Pearse DB, Dahms TE, Wagner EM. Microsphere-induced bronchial artery vasodilation: Role of adenosine, prostacyclin, and nitric oxide. *Am J Physiol Heart Circ Physiol.* 1998;274:H760–H768.

16 Wagner EM. TNF-alpha induced bronchial vasoconstriction. *Am J Physiol Heart Circ Physiol.* 2000;279:H946–H951.

17 Csete ME, Chediak AD, Abraham WM, Wanner A. Airway blood flow modifies allergic airway smooth muscle contraction. *Am Rev Respir Dis.* 1991;144:59–63.

18 Wagner EM, Mitzner WA. Bronchial circulatory reversal of methacholine-induced airway constriction. *J Appl Physiol.* 1990; 69(4):1220–1224.

19 Wagner EM, Foster WM. Interdependence of bronchial circulation and clearance of 99mTc-DTPA from the airway surface. *J Appl Physiol.* 2001;90:1275–1281.

20 Grantham CJ, Jackowski JT, Wanner A, Ryan US. Metabolic and pharmacokinetic activity of the isolated sheep bronchial circulation. *J Appl Physiol.* 1989;67:1041–1047.

21 Baluk P, Hirata A, Thurston G, et al. Endothelial gaps: time course of formation and closure in inflamed venules of rats. *Am J Physiol.* 1997;272:L155–L170.

22 Pearse DB, Wagner EM. Role of the bronchial circulation in ischemia-reperfusion lung injury. *J Appl Physiol.* 1994;76(1): 259–265.

23 Schraufnagel DE, Pearse DB, Mitzner WA, Wagner EM. Three-dimensional structure of the bronchial microcirculation in sheep. *Anat Rec.* 1995;243:357–366.

24 Lim LH, Bochner BS, Wagner EM. Leukocyte recruitment in the airways: an intravital microscopic study of rat tracheal microcirculation. *Am J Physiol Lung Cell Mol Physiol.* 2002;282:L959–L967.

25 Lim LH, Wagner EM. Airway distension promotes leukocyte recruitment in rat tracheal circulation. *Am J Respir Crit Care Med.* 2003;168:1068–1074.

26 Endrys J, Hayat N, Cherian G. Comparison of bronchopulmonary collaterals and collateral blood flow in patients with chronic thromboembolic and primary pulmonary hypertension. *Heart.* 1997;78:171–176.

27 Charan NB, Carvalho P. Angiogenesis in bronchial circulatory system after unilateral pulmonary artery obstruction. *J Appl Physiol.* 1997;82:284–291.

28 Michel RP, Hakim TS. Increased resistance in postobstructive pulmonary vasculopathy: structure-function relationships. *J Appl Physiol.* 1991;71(2):601–610.

29 Fadel E, Mazmanian GM, Chapelier A, et al. Lung reperfusion injury after chronic or acute unilateral pulmonary artery occlusion. *Am J Respir Crit Care Med.* 1998;157:1294–1300.

30 Shi W, Hu F, Kassouf W, Michel RP. Altered reactivity of pulmonary vessels in postobstructive pulmonary vasculopathy. *J Appl Physiol.* 2000;88:17–25.

31 Weibel ER. Early stages in the development of collateral circulation to the lung in the rat. *Circ Res.* 1960;8:353–376.

32 Mitzner W, Lee W, Georgakopoulos D, Wagner E. Angiogenesis in the mouse lung. *Am J Pathol.* 2000;157:93–101.

33 Michel RP, Hakim TS. Increased resistance in postobstructive

pulmonary vasculopathy: structure-function relationship. *J Appl Physiol.* 1990;71:601–610.

34 Srisuma S, Biswal SS, Mitzner WA, et al. Identification of genes promoting angiogenesis in mouse lung by transcriptional profiling. *Am J Respir Cell Mol Biol.* 2003;29:172–179.

35 Strieter RM, Polverini PJ, Kunkel SL, et al. The functional role of the ELR motif in CXC chemokine-mediated angiogenesis. *J Biol Chem.* 1995;270:27348–27357.

36 Arenberg DA, Keane MP, DiGiovine B, et al. Epithelial-neutrophil activating peptide (ENA-78) is an important angiogenic factor in non-small cell lung cancer. *J Clin Invest.* 1998; 102:465–472.

37 Strieter RM, Belperio JA, Keane MP. CXC chemokines in vascular remodeling related to pulmonary fibrosis. *Am J Respir Cell Mol Biol.* 2003;29:S67–S69.

38 Moldobaeva A, Wagner EM. Difference in proangiogenic potential of systemic and pulmonary endothelium: role of CXCR2. *Am J Physiol Lung Cell Mol Physiol.* 2005;288:L1117–L1123.

39 Shi GP, Sukhova GK, Kuzuya M, et al. Deficiency of the cysteine protease cathepsin S impairs microvessel growth. *Circ Res.* 2003;92:493–500.

40 Li X, Wilson JW. Increased vascularity of the bronchial mucosa in mild asthma. *Am J Respir Crit Care Med.* 1997;156:229–233.

41 Chu HW, Kraft M, Rex MD, Martin RJ. Evaluation of blood vessels and edema in the airways of asthma patients: regulation with clarithromycin treatment. *Chest.* 2001;120:416–422.

42 Thurston G, Murphy TJ, Baluk P, et al. Angiogenesis in mice with chronic airway inflammation: strain-dependent differences. *Am J Pathol.* 1998;153:1099–1112.

43 Thurston G, Maas K, Labarbara A, et al. Microvascular remodelling in chronic airway inflammation in mice. *Clin Exp Pharmacol Physiol.* 2000;27:836–841.

44 Dahlqvist K, Umemoto EY, Brokaw JJ, et al. Tissue macrophages associated with angiogenesis in chronic airway inflammation in rats. *Am J Respir Cell Mol Biol.* 1999;20:237–247.

45 Baile EM, Par PD. Response of the bronchial circulation to acute hypoxemia and hypercarbia in the dog. *J Appl Physiol.* 1983;55(5):1474–1479.

46 Wagner EM, Mitzner WA. Effect of hypoxia on the bronchial circulation. *J Appl Physiol.* 1988;65(4):1627–1633.

47 Remy-Jardin M, Duhamel A, Deken V, et al. Systemic collateral supply in patients with chronic thromboembolic and primary pulmonary hypertension: assessment with multidetector row helical CT angiography. *Radiology.* 2005;235:274–281.

48 Davie NJ, Crossno JT Jr., Frid MG, et al. Hypoxia-induced pulmonary artery adventitial remodeling and neovascularization: contribution of progenitor cells. *Am J Physiol Lung Cell Mol Physiol.* 2004;286:L668–L678.

49 Heistad DD, Armstrong ML. Blood flow through vasa vasorum of coronary arteries in atherosclerotic monkeys. *Arteriosclerosis.* 1986;6:326–331.

128

The Endothelium in Acute Respiratory Distress Syndrome

Mark L. Martinez and Guy A. Zimmerman

University of Utah, School of Medicine, Salt Lake City

The pulmonary endothelium, poised at the interface between air, blood, and tissue, provides both rapid and sustained responses to local and systemic perturbations. This complex vascular structure occupies a surface area of 120 m^2 and forms the intimal lining of the pulmonary arterial, venous, and capillary beds with a single continuous layer of endothelial cells (ECs) linked to each other by specialized junctions (1). The alveolar endothelium is intimately related to the alveolar epithelium both in terms of anatomic location and functions that include oxygen (O_2), carbon dioxide, water and solute transport, and barrier regulation; disruption of barrier functions of the alveolar capillary membrane is an early and critical event in the pathogenesis of acute respiratory distress syndrome (ARDS) (see later and Box 128.1). Alveolar epithelial function, which is beyond the scope of this chapter, has recently been reviewed (2).

Once thought of as passive, semipermeable conduits for nutrient and O_2 delivery – and in the lungs, contributing to separation of blood from air (1) (Box 128.1) – ECs were dismissed as structural bystanders with little or no capacity to respond to activating signals with changes in phenotype or function (3). During the 1950s, electron microscopic observations that ECs contain secretory granules, together with ongoing physiological studies of EC–leukocyte interactions, implicated the endothelium as an active participant in both physiological and pathophysiological responses to injury and inflammation (4–6). Subsequent studies clearly demonstrated that, even under normal physiological conditions, the "quiescent" endothelium is far from inactive and is involved in multiple homeostatic functions. These include, but are not limited to, cellular and nutrient trafficking, angiogenesis and vasculogenesis, regulation of vascular tone, and maintenance of blood fluidity and vascular barrier function (3–7). It is evident that ECs can respond to a host of physical and biochemical stimuli to become *activated* in both homeostatic and disease states (3–7) (Figure 128.1). EC activation in response to extracellular signals from mechanical forces, coagulation factors, cytokines, soluble growth factors, bacterial products, and toxins, and interactions with circulating platelets and leukocytes is mediated by an array of intracellular transduction pathways, leading to rapid and, in some cases, sustained functional changes. These functional alterations can lead to proinflammatory, procoagulant, and vasoconstrictive or vasodilatory responses. When these processes become dysregulated, host tissue damage and organ-specific or systemic illness with endothelial components occurs (3–5,7) (see Figure 128.1). Many of these general features of endothelial biology and function have been established using models of systemic endothelium, and have not been completely validated for pulmonary ECs (5,7). Nevertheless, these paradigms appear to be relevant to lung EC behavior under normal conditions and in ARDS; they are emphasized here. Furthermore, systemic EC function frequently is altered in ARDS (see Box 128.1).

Injury and dysregulated responses of the pulmonary endothelium are principal mechanisms in a pathologic spectrum termed *acute lung injury* (ALI) and, in its most severe form, ARDS (8,9) (see Figure 128.1). The original term, *adult respiratory distress syndrome,* evolved from a seminal report in 1967 that described 12 patients with acute onset of pulmonary injury, with characteristic radiographic and pathophysiological features that were unresponsive to traditional modes of mechanical ventilation (10). It is now known that ALI/ARDS occurs with substantial frequency in children, resulting in modification of the term (9). After early work in the field indicated that pulmonary edema is a central component of ARDS, the concept that widespread EC injury is fundamental to the pathogenesis of the syndrome emerged (9) (see Box 128.1). Subsequently, clinical guidelines established the descriptors ALI and ARDS to represent varying degrees

Text Box 128.1: Definitions, Clinical Features, and Physiology of ARDS (8–10,12,15)

ARDS is the most severe manifestation of injury to the alveolar-capillary units of the lung (alveolar-capillary membrane), a spectrum now termed *acute lung injury* (ALI). Thus ALI and ARDS are a continuum. Consensus definitions based on physiologic and radiographic abnormalities are utilized to distinguish them in clinical practice and investigation. ALI and ARDS are initiated by acute injury to the alveolar-capillary membrane. The initial insult can be delivered via the blood to the alveolar capillaries, or to the alveolar epithelial cells on the gaseous side of the alveolar-capillary units. Ultimately, both ECs and epithelial cells are involved in the pathogenesis of ARDS. Sepsis, aspiration, infectious syndromes such as influenza or SARS, and trauma with shock and/or multiple transfusions are major etiologies for ALI and ARDS in large studies. A variety of other insults can induce ALI and ARDS, including inhalation injury, pancreatitis, and drowning. Iatrogenic manipulations including hyperoxia, ventilation with high tidal volumes (ventilator-induced lung injury; VILI), and transfusion (transfusion-related lung injury; TRALI) can initiate or propagate ALI. ALI/ARDS can resolve or progress to multiple organ failure, pulmonary vascular obstruction and/or destruction, or pulmonary fibrosis.

The cardinal feature of early ARDS is flooding of the alveoli with pulmonary edema fluid rich in protein. This is termed *increased-permeability edema* (also caused by noncardiogenic pulmonary edema) to contrast with *high-pressure edema*, which results from increased pulmonary capillary hydrostatic pressure, usually due to high filling pressures in the left side of the heart. Increased-permeability edema results from altered barrier function of the alveolar capillaries, the largest capillary bed in the body, causing increased transfer of water and protein out of the vessels into the alveolar interstitium at any given microvascular hydrostatic and oncotic pressure (altered permeability coefficient in the Starling equation). Altered barrier function of the alveolar capillaries is thought to be due to acute inflammatory injury in most cases. Increased permeability of the epithelial barrier due to loosening or disruption of tight junctions between alveolar epithelial cells, which are normally tighter than those of the alveolar ECs, then results in flooding of protein-rich fluid from the interstitium into the alveolar space. Increased microvascular pressure in the alveolar capillaries due to excessive fluid administration or intercurrent cardiac dysfunction can exacerbate accumulation of edema fluid when alveolar barrier function is disrupted. Nevertheless, the utility of invasive monitoring to guide fluid management in ARDS remains controversial and is the subject of a current study in the multicenter ARDS network.

Additional features of early ALI/ARDS include alveolar epithelial cell death and denudation of the basement membrane, diffuse inflammation, and deposition of hyaline membranes. Hyaline membranes are complexes of protein and cellular debris on the alveolar surface that form in part from fibrin deposition, linking coagulation responses to alveolar-capillary membrane injury. This constellation of events results in a characteristic light and electron microscopic pattern termed *diffuse alveolar damage.*

ARDS often is considered to be isolated to the lungs, but the syndrome frequently causes systemic manifestations, and is a principal cause of multiple-organ failure. This may in part result from the effects of mediators generated in the lungs on systemic ECs. In addition, triggers of ALI/ARDS, such as sepsis, trauma, and shock, cause systemic EC activation and injury. Thus, in many cases, ARDS is a systemic syndrome, and studies that focus on systemic ECs have relevance to ALI/ARDS.

of respiratory impairment due to increased-permeability pulmonary edema, and consensus definitions based on the chest radiograph and physiological alterations were developed (8,11,12). Multiple studies indicated that this pathophysiologic spectrum results from lung cellular responses, including responses of the alveolar endothelium and epithelium, to a variety of intrinsic and extrinsic injurious agents (8,9,11–15) (see Figure 128.1 and Box 128.1). Because ALI and ARDS represent points on a spectrum of clinical severity (8,12,15), for the remainder of this chapter the term ARDS will be synonymous with ALI except where indicated.

Ongoing investigations into the functional and phenotypic changes of ECs in ARDS have resulted in altered concepts and paradigms when compared to those proposed at the time of initial clinical description of the syndrome more than 30 years ago (9). Nevertheless, most of these concepts are based on inference from studies of systemic endothelium and from studies in experimental systems, rather than on direct observations of ECs in the lungs of patients with ARDS (5). Furthermore, our ability to *translate* experimental studies of ARDS has been limited and has resulted in only a handful of moderately successful clinical interventions (8,9,14,15), indicating how much we have left to accomplish.

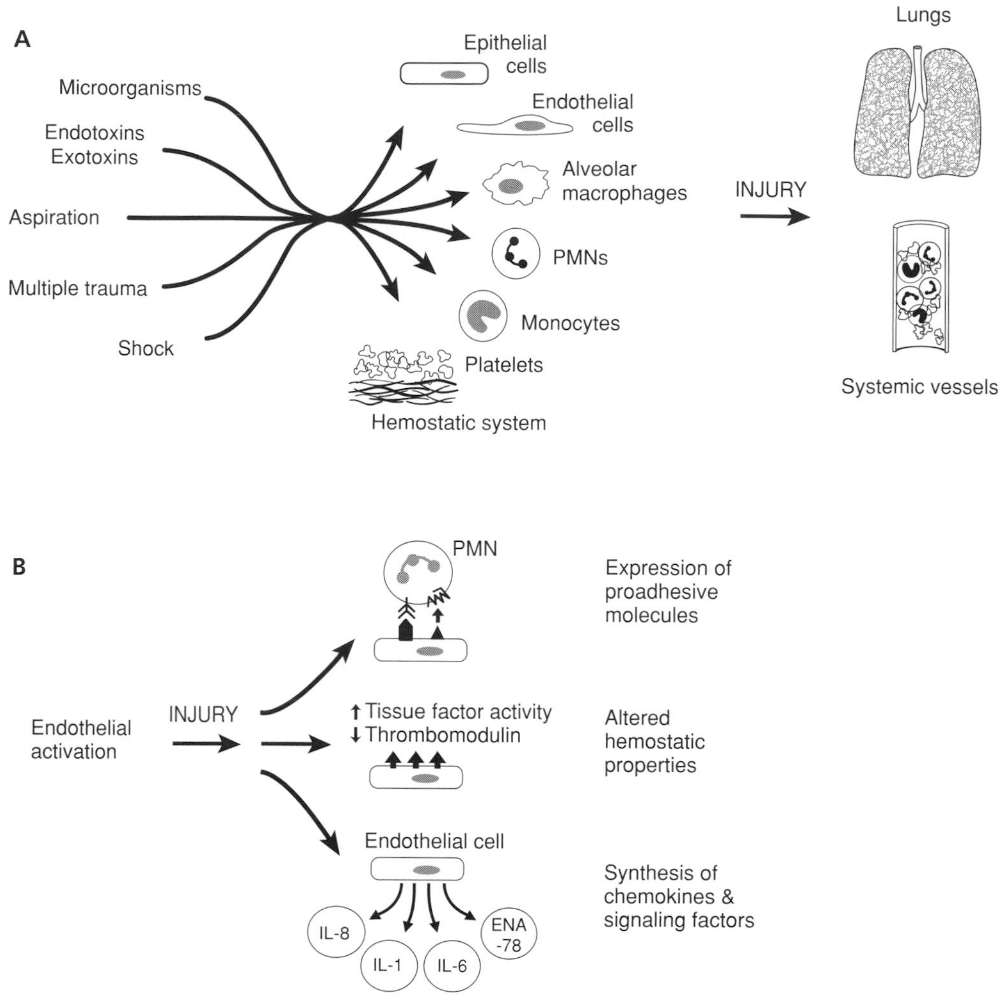

Figure 128.1. EC injury and activation are key features of ARDS. (**A**) Acute lung injury and ARDS are triggered by a variety of inciting factors that can directly or indirectly injure ECs or induce their activation. Other cellular and molecular systems participate and interact with ECs in ARDS, including leukocytes, platelets, and components of the coagulation cascade. Dysregulated endothelial responses and EC injury contribute to diffuse alveolar-capillary damage, which is a hallmark of ARDS (see Box 128.1). Evidence suggests that systemic vessels are also dysfunctional in some conditions that underlie ARDS, including sepsis. (**B**) Microbial products, such as LPS, cytokines, oxidants, and other mediators can cause endothelial activation in ARDS (**A**). Changes in EC function and phenotype in response to activation include synthesis of adhesion molecules or their translocation from subcellular stores, alterations in surface hemostatic properties, new expression of chemokines and cytokines, and synthesis of PAF, a phospholipid signaling factor. See text for details. (Adapted with permission from Zimmerman GA, McIntyre TM. Pathogenesis of sepsis and septic-induced lung injury. In: Matthay MA, ed. *Lung Biology in Health and Disease.* New York: Marcel Dekker, 2003.)

THE STUDY OF ENDOTHELIUM IN ACUTE RESPIRATORY DISTRESS SYNDROME

Alveolar-capillary membrane injury resulting in respiratory failure due to infections, trauma, or the aspiration of gastric contents has long been recognized (7,9,13,15–17) (see Figure 128.1). Observations made during the Vietnam conflict indicated that the development of protein-rich increased-permeability pulmonary edema, sometimes called “Da Nang lung,” could accompany nonpulmonary trauma, and that this fatal pulmonary edema was not, in fact, due to excessively vigorous fluid resuscitation. These findings implied that: (a) indirect injury to the pulmonary capillary endothelium had occurred, and (b) disruption of the alveolar capillary barrier was potentially mediated by ill-defined soluble factors (9,17). A significant body of evidence supporting a role for inflammatory mediators in the pathogenesis of ARDS then accumulated from biochemical studies of the sera and bronchoalveolar fluid (BALF) of ARDS patients, coupled with observations in animal models and cultured ECs (9,18–21). It is generally

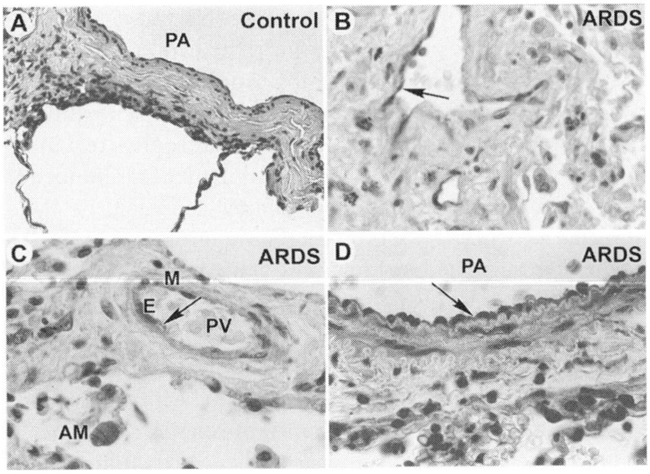

Figure 128.2. The endothelial phenotype is altered in lung vessels in ARDS. (**A**) Lung section from an individual without lung injury dying of acute head trauma. *PA*, pulmonary artery. (**B**) Microvessels in injured lung tissue from a patient with ARDS display surface P-selectin (*dark staining, arrow*). (**C**) Pulmonary venular (*PV*) ECs in a lung section from a patient dying from ARDS stain positively for ENA-78 (*arrow*). Capillary ECs were also positive for ENA-78 (*not shown*), as were alveolar macrophages (*AM*), some smooth muscular cells (*M*), and intra-alveolar PMNs (*not shown*). (**D**) EC in the intima of a pulmonary artery (*PA*) in a lung section from a subject dying of ARDS stain intensely for COX-2. Staining for surface P-selectin (**B**), ENA-78 (**C**), and COX-2 was absent or much less evident in endothelium in vessels from subjects dying without acute lung injury or ARDS (**A**). See Ref. 5 for additional details. (Panel C reproduced with permission from Imaizumi T, et al. Human endothelial cells synthesize ENA-78: relationship to IL-8 and to signaling of PMN adhesion. *Am J Respir Cell Mol Biol.* 1997;17:181–192; panels A, B, D are used with permission from K. Albertine, G. A. Zimmerman, unpublished observations.) For color reproduction, see Color Plate 128.2.

accepted that EC injury, activation, and increased permeability are key facets of ARDS, in addition to epithelial injury (5,8,9). A fundamental corollary is that inflammatory and hemostatic mediators and agents of injury can directly or indirectly activate ECs to produce dysregulated changes in their phenotype and function (5,7,9,22,23) (see Figure 128.1). A second corollary is that the pulmonary microvascular EC is the critical vascular cell involved in ARDS because of the large surface area of the lung microvascular bed and because the alveolar-capillary membrane is the principal site of injury (Box 128.1). Nevertheless, macrovascular ECs also are involved (Figure 128.2; for color reproduction, see Color Plate 128.2). A third corollary is that activated or injured ECs can release or synthesize products, such as von Willebrand factor (vWF) or other soluble components, that can be used as biomarkers of the nature and severity of endothelial injury in ARDS (8,24). Some products and factors released from activated or injured ECs can circulate and exert systemic effects (8,9).

As previously mentioned, reductionist approaches led to development of in vitro culture systems, and resulted in major advances in our understanding of EC biology (4,5,7,25). The ability to isolate, culture under defined conditions, and readily manipulate human ECs in culture has been extraordinarily informative regarding in vivo events that occur in the vessel wall in general and activation of ECs in ARDS in particular. However, it is clear that the potential exists for loss of specialized endothelial functions that are associated with native vessels or specific organ microenvironments when ECs are introduced into culture, and that there are phenotypic differences in ECs harvested from different vascular beds (26,27). Nevertheless, the study of cultured human ECs has been invaluable. For example, virtually all the adhesion and signaling molecules involved in endothelial interactions with leukocytes (6) were identified using human ECs prior to passage or in early passage culture. In addition, inferences made from these in vitro systems have been validated by in vivo models and further used as a basis for investigation of endothelial-specific activities (28–30). These surrogate systems complement clinical observations and provide a framework on which to develop more sophisticated in vivo cell biological approaches that can be translated to the bedside (5,7,31). More recently, efforts have been made to culture and study ECs derived from the lung, including macrovascular and microvascular ECs from pulmonary and bronchial circulations, and to characterize cell subtype–specific differences in phenotype (26).

EFFECTORS OF ENDOTHELIAL CELL ACTIVATION AND INJURY IN ACUTE RESPIRATORY DISTRESS SYNDROME: CYTOKINES AND BEYOND

Endothelial activation is accomplished by outside-in signaling to intracellular transduction cascades initiated by multiple surface receptors that receive signals from inflammatory, thrombotic, or mechanical stimuli (5,7). Diverse response patterns are then induced that result in physiological or pathological characteristics, depending on the nature of the stimuli and type of EC. Common clinical predispositions result in the generation of cytokines, oxidants, proteases, and physical stimuli that trigger EC activation in ARDS (Figure 128.1). Early studies focused on reactive oxygen species (ROS) and serine proteases released from activated polymorphonuclear neutrophils (PMNs) sequestered in the injured pulmonary vascular bed as key mediators of EC activation and injury (8,9). More recent work has emphasized the important role of cytokines, although it is clear that they are unlikely to be a final common pathway to EC activation in ALI/ARDS (5,8,9,31).

Cytokines are ubiquitous mediators of inflammation. These soluble polypeptides serve as chemical messengers whose generation can orchestrate either a proinflammatory state and tissue injury (e.g., tumor necrosis factor [TNF]-α and interleukin [IL]-1) or an anti-inflammatory state with tissue healing (e.g., IL-10 and IL-13) (9). Recent clinical observations indicate that using a lung-protective ventilator strategy not only decreases the mortality in patients with ALI but also reduces proinflammatory cytokine levels and the number of neutrophils in the bronchoalveolar lavage fluid of these patients (14,32).

In the lung, cytokines are produced primarily by interstitial and alveolar macrophages, which are resident in normal lung (1) and are activated in lung injury (8,9). Pulmonary ECs also may synthesize cytokines in ARDS (5,8). Cytokine receptors on ECs regulate phenotype and function in acute inflammatory responses, hemostasis, and thrombosis; acquired immunity; and cellular and vascular remodeling. Each of these variables is of clinical import in the development and resolution of ARDS (8,9). Cytokine receptors mediate intercellular communication pathways and transmit outside-in signals to intracellular effectors, including mitogen-activated protein kinase (MAPK) cascades, phosphoinositide 3-kinase (PI3K), ceramide turnover, and conserved nuclear signaling pathways that utilize the nuclear factor (NF)-κB and Janus kinase (JAK)/signal transducers and activator of transcription (STAT) families (33–39). Two of the most extensively characterized cytokines acting on human ECs are TNF-α and IL-1, which induce expression of genes that code for E-selectin and other adhesion molecules, IL-6, endogenous expression of TNF-α and IL-1 themselves, ENA-78 and other chemokines, cyclooxygenase (COX)-2, and degranulating factors for neutrophils and platelets (22,23,40–43). Thus, ECs not only respond to cytokines via at least a dozen constitutively expressed receptors, but also possess the capacity to synthesize and secrete some of the ligands recognized by these receptors, providing the basis for autocrine signaling loops (42,43). Newly discovered cytokines, such as high mobility group box (HMGB1), which mediates later events in temporal sequences in clinical and experimental lung injury and may have prognostic and therapeutic significance, also continue to be described (44,45). Local amplification involving juxtacrine and paracrine signaling between ECs and sequestered leukocytes or platelets also may contribute to the pathogenesis of ARDS, based on observations in experimental systems (Figure 128.3). Local and spatially restricted events of this nature are, however, difficult to detect in vivo and in the clinical setting using current methodology (31).

An example of complex cytokine signaling between leukocytes and ECs involves IL-6 and the soluble component of its heterotrimeric signal-transducing receptor (see Figure 128.3). IL-6, known to be increased in the sera and BALF of patients with ARDS (8,32), is incapable of activating resting human ECs due to the lack of a constitutively expressed α-subunit of the IL-6 receptor (IL-6Rα) on these cells (46,47). Human neutrophils release the IL-6Rα, which can then associate with homodimers of glycoprotein 130 (GP130) basally present on ECs (46,48). Association of soluble IL-6Rα with transmembrane GP130 homodimers constitutes a competent heterotrimeric receptor that recognizes endogenously synthesized (or exogenously generated) IL-6, inducing new expression of inflammatory or thrombotic gene products (46–48). This system has been worked out using cultured systemic ECs; it is not yet known if it operates in ARDS. Nevertheless, it provides a mechanistic paradigm by which neutrophils can signal ECs in a "retrograde" fashion (46). Additional observations indicate that it may mediate transition from an acute inflammatory response with accumulation of neutrophils to a more chronic inflammatory state with recruitment of mononuclear cells, due to the pattern of chemokines generated by IL-6-stimlated ECs (47,48). Chronic inflammation is a feature of progressive, sustained ALI that may contribute to devastating late complications in subsets of patients with ARDS, including pulmonary fibrosis and pulmonary hypertension (9,31). Interestingly, neutrophil synthesis and release of IL-6Rα can be induced by platelet activating factor (PAF), a phospholipid mediator that may have important roles in ARDS, based on both experimental and clinical studies (49). Multiple links between cytokine, chemokine, and lipid mediator pathways exist (31,49). Contributing to the complexity, effector cells in addition to leukocytes may be intermediates in signaling cascades relevant to ARDS. As an example, human platelets can release IL-1 and activate ECs via this mechanism, using previously unrecognized pathways of synthesis of the cytokine (see Figure 128.3) (49).

Sepsis is the most common cause of ARDS, and is the leading cause of mortality from this syndrome (8,9,49,50). In addition to recognizing local or systemically produced cytokines and other signaling molecules, ECs are capable of responding to bacterial products, such as lipopolysaccharide (LPS) and lipid-modified bacterial proteins with complex changes in inflammatory phenotype (49) (see Figure 128.1). LPS, alone or in concert with endogenous mediators, directly elicits the synthesis of proinflammatory cytokines and the surface expression of adhesion molecules and tissue factor activity. In some cases, it induces apoptosis of ECs (49,51–54). Identification and characterization of Toll-like receptors (TLRs), members of a family of proteins that allow host cells to recognize pathogen-associated molecular patterns (PAMPs) in an innate fashion, has generated new insight into endothelial and leukocyte signaling by LPS and other microbial products (49,52,54,55). TLRs, acting together with a complex group of adaptor and LPS recognition proteins (55), activate intracellular signal transduction cascades linked to gene regulatory pathways and to other effector mechanisms in ECs in a manner similar to that induced by cytokines (49,55). Human macrovascular and microvascular ECs express TLR2 and TLR4, and use them to differentially recognize and respond to components of gram-positive or gram-negative bacteria, fungi, and mycobacteria (52,56). The patterns of TLRs expressed on uninjured and injured animal and human ECs still are being characterized (49). LPS is specifically recognized by TLR4, resulting in outside-in signaling mediated by downstream effector cascades (54,57,58). Experimental studies in murine models demonstrate important differences in activated endothelial interactions depending on the specific microbial factor, systemic versus local challenge, and the TLR and CD14 phenotype (59). Each of these variables may be critical in sepsis-induced ARDS in humans (49).

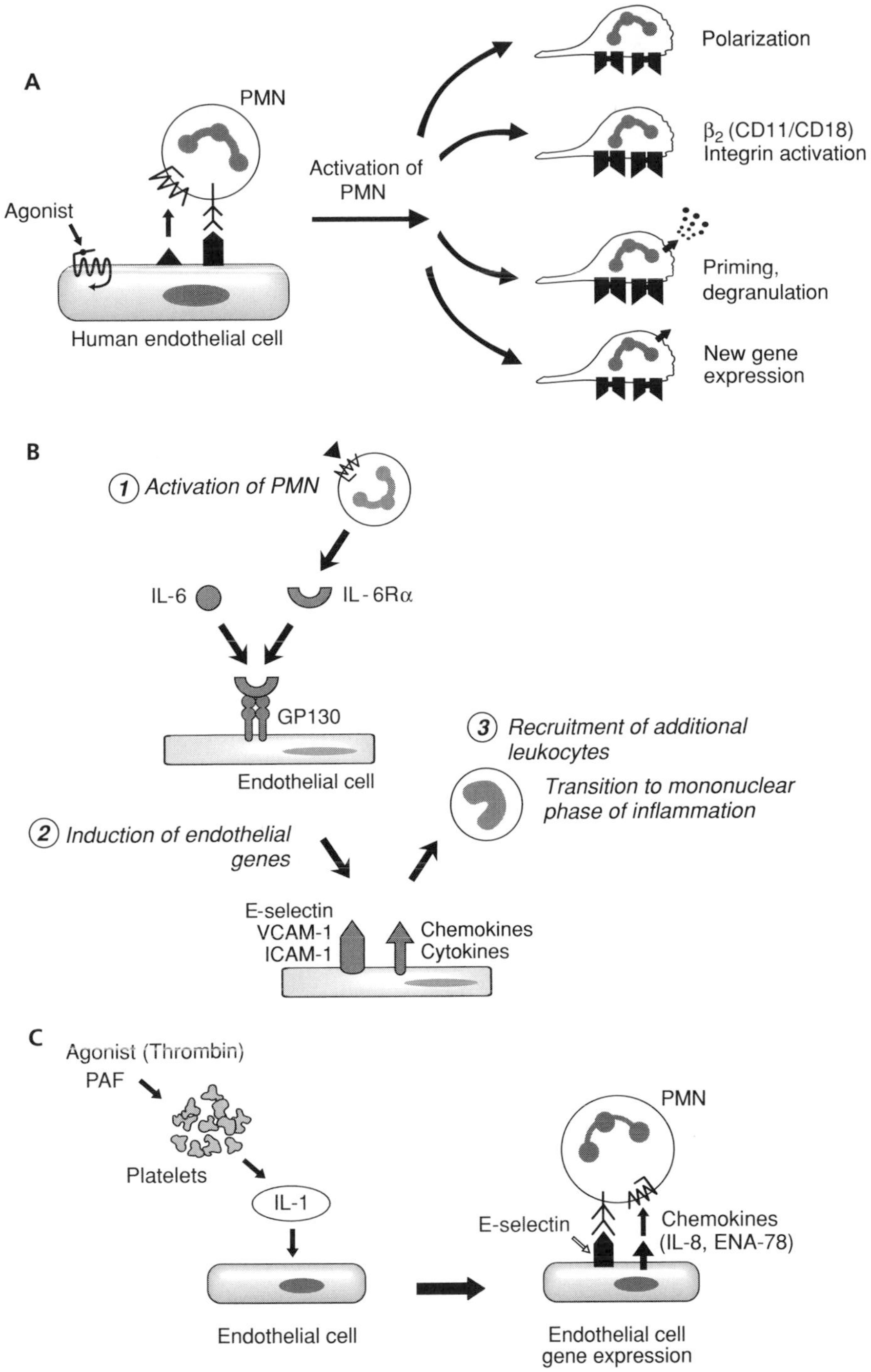

Figure 128.3. Mechanisms of endothelial–leukocyte and platelet–endothelial–leukocyte interaction that may be relevant to pulmonary vascular injury in ARDS. (**A**) Human EC activated by thrombin, cytokines, and other agents relevant to ARDS display specific patterns of adhesion and signaling molecules for PMNs. This results in local accumulation and activation of PMNs. Local activation of PMNs induces functional changes that can initiate or amplify inflammation injury. (**B**) Activation of PMNs can induce synthesis and release of IL-6Rα, which can then mediate retrograde signaling (trans-signaling) of ECs, resulting in the expression of endothelial genes that mediate the recruitment of additional leukocytes and transition to the mononuclear phase of inflammation. PAF and IL-8, which are synthesized by activated ECs, induce release of IL-6Rα from human PMNs. (**C**) Human platelets synthesize and release IL-1β when stimulated with PAF or thrombin. IL-1β can then signal the expression of genes for E-selectin and chemokines by ECs, resulting in adhesion and local activation of PMNs. (Adapted with permission from Zimmerman GA, McIntyre TM. Pathogenesis of sepsis and septic-induced lung injury. In: Matthay MA, ed. *Lung Biology in Health and Disease.* New York: Marcel Dekker, 2003.)

ALTERED ENDOTHELIAL BARRIER FUNCTION IN ACUTE RESPIRATORY DISTRESS SYNDROME

Increased microvascular EC permeability is a hallmark of ARDS (8,9) (see Box 128.1). Evidence also suggests that altered EC permeability occurs in systemic vessels in patients with ARDS, particularly those with sepsis and/or multiple organ failure (31) (Box 128.1). Highly specialized junctional structures that link adjacent cells in the interconnected monolayers of ECs that line vascular channels regulate paracellular permeability and influence the transmigration of leukocytes from the intravascular space to the extravascular milieu (60,61). Signals from activated leukocytes may alter endothelial junction and barrier characteristics in inflammatory vascular injury, including ARDS (61,62) (Figures 128.1 and 128.3). EC–matrix interactions mediated by basolateral integrins also influence permeability (4), providing both cellular attachment and outside-in signals that modify specialized intercellular junctions. As many as four types of junctions may mediate EC–EC interactions (60), although this has not been completely defined in the lung (1,62). Adherens junctions and tight junctions are critical in regulating permeability and cell polarity, and are modified by inflammatory mediators and thrombotic signals (60,61), including recently identified sphingosine-1-phosphate (S1P) (discussed later), thrombin, and other receptor-mediated agonists. Disruptions of the integrity of the EC barrier occur through changes in these multimolecular junctional structures and their interacting cytoskeletal elements, and they are likely key sites of dysregulation and increased vascular permeability during ARDS (61,62) (Box 128.1). Lung endothelial heparan sulfates recently have been identified as signaling components that link cytoskeletal reorganization with barrier dysfunction, and they also may be pulmonary endothelial sensors of shear (61).

Sphingomyelin, an abundant cell membrane component, was considered primarily a structural lipid for years. It is clear that its conversion to ceramide by sphingomyelinases and, subsequently, to S1P is important in both extra- and intracellular signaling pathways. S1P is released primarily by activated platelets, and is recognized by a family of five surface G-protein–coupled receptors, termed *S1P receptors* (S1PR) (63). This delivers outside-in signals to target cells that have S1PRs. The intracellular generation of S1P, resulting in subsequent mobilization of calcium (Ca^{2+}) and activation of NF-κB, provides a mechanism for second-messenger function in ECs (64). Initial interest in a role for S1P in increased-permeability edema in lung injury (Box 128.1) came from in vitro observations of cultured endothelium. In this model, S1P induced adherens junction assembly and cytoskeletal arrangement in an in vitro model of angiogenesis (65). In a more recent report, silencing of the $S1PR_1$ using selective knockdown by RNA interference profoundly altered EC phenotype and modified synergistic interactions in the TNF-α signaling pathway (66). Building on the observation that platelets and unidentified platelet factors can stabilize or reverse pulmonary vascular leak in animal models (67), recent studies of both small and large animal models generated evidence that exogenous S1P attenuates lung injury by decreasing endothelial permeability. These experimental systems were based on conditions relevant to ARDS, including ventilator-induced lung injury (VILI) and endotoxin challenge (68,69). The findings support a role for S1P as a barrier-enhancing agonist. In addition, they give further credibility to the utility of in vitro models in the development of new therapeutic strategies relevant to lung vascular injury and ARDS, because the original observations were made in cultured human umbilical vein ECs (HUVECs) (65).

ENDOTHELIAL CELL REGULATION OF HEMOSTASIS IN ACUTE RESPIRATORY DISTRESS SYNDROME

In situ thrombosis with fibrin microthrombi and platelet deposition on damaged endothelium was first reported in the lungs of ARDS patients at autopsy in 1976 (70). It has been demonstrated subsequently in other studies (8,31,49). Normally, the resting (nonactivated) endothelial bed is responsible for maintaining an antithrombotic and fibrinolytic surface to facilitate blood flow through the pulmonary vasculature, acting in concert with soluble regulators of coagulation in the plasma and, potentially, mechanical forces transmitted from the adjoining alveoli. In response to injurious stimuli, the endothelium changes phenotype, resulting in a prothrombotic and antifibrinolytic surface. When this process becomes dysregulated, the switch in EC phenotype initiates or amplifies pathological thrombosis and provides a mechanism that couples endothelial activation to leukocyte accumulation and other inflammatory responses (3–7,49). For example, human ECs respond to stimulation with thrombin by translocating P-selectin to their surfaces and synthesizing PAF – molecular events that mediate EC-dependent neutrophil adhesion (5,6) (see Figure 128.3). Evidence suggests that the switch to a proinflammatory and prothrombotic EC phenotype occurs in subjects with ARDS (5,8,9,49). Endothelium in pulmonary arteries and veins, in addition to microvessels in the alveolar-capillary membrane, can be involved in some patients (see Figure 128.2) (49).

The generation of tissue factor activity is a pivotal step in hemostasis and in pathological intravascular thrombosis (71–73). Tissue factor initiates the coagulation cascade by accelerating factor VIIa–dependent activation of factors IX and X, generating thrombin by sequential protease-dependent steps in the tissue factor pathway. Tissue factor is highly thrombogenic and normally found in only minute amounts in the blood. In sepsis, tissue factor expression is reported to be diffusely present on lung cells including endothelium, neutrophils, and monocyte/macrophages (71). The origin(s) of tissue factor under these conditions is not yet clear. When cells that display tissue factor are exposed to plasma, thrombin is generated and fibrin is deposited on their surfaces. Expression of tissue factor activity on ECs is induced by bacterial products and proinflammatory mediators (TNF-α, IL-1β, IL-6, C-reactive protein) in vitro (4), but factors that trigger its regional expression by endothelium in the lung remain in question (21,49,71). In addition, tissue factor–containing complexes may mediate outside-in signaling via protease-activated receptors (PARs) (discussed later). In contrast to fibroblasts and epithelial cells, the tissue factor/VIIa complex does not directly activate ECs. Rather, tissue factor-VIIa–mediated signaling in ECs requires the formation of a ternary complex involving factor X, tissue factor, and VIIa (72). Recently, blockade of the tissue factor pathway in baboons by competitive inhibition with active site-modified VIIa was reported to attenuate lung injury and decrease inflammatory cytokines in an established model of sepsis (73).

Recent experimental and clinical evidence suggests a role for signaling by coagulation proteases in endothelial responses in tissue injury. This may be important in the generation of platelet-fibrin microthrombi in ARDS (70,72). Protease-mediated signaling also may provide a molecular means for the interplay, or cross-talk, between the coagulation and inflammatory systems (49). Human ECs constitutively express surface PARs, a family of heptahelical G-protein–coupled receptors that recognize thrombin and certain other serine proteases, and transmit outside-in signals via proteolytic cleavage of a specific extracellular domain to uncover a new N-terminus that functions as a tethered ligand to activate the receptor (3,4). Thrombin, the central effector protease in the coagulation cascade, activates ECs directly via PAR-1 and -4, thereby promoting adhesion of leukocytes, synthesis of PAF and, under some conditions, the release of cytokines and alteration of barrier functions (3–7,62). Signaling via PARs provides ECs with a mechanism to regulate diverse cellular responses to local thrombin generation in physiological host defense and in vascular injury. It is unknown whether PAR expression by pulmonary or systemic ECs is altered in ARDS.

Resting endothelium has the ability to regulate hemostasis through mechanisms that include factor Xa endocytosis and a display of heparin-like glycosaminoglycans and dermatan sulfate that bind and sequester antithrombin III (ATIII) and thrombin. Mechanisms that localize and limit coagulation to sites of EC injury or activation under physiological conditions include thrombomodulin (TM), tissue factor pathway inhibitor (TFPI), ATIII, and proteins C and S (4). TFPI is a 42-kDa Kunitz-type inhibitor that is stored in a subset of EC granules and distributed to the cell surface (74). To a lesser extent, TFPI is found associated with megakaryocytes and platelets and, in soluble form, it acts by inhibiting the activity of the tissue factor/VIIa/Xa pathway (4,7,75). The role that TFPI plays in regulating coagulation cascades in ARDS is incompletely understood. In vivo, TFPI expression has been detected largely on endothelium of the microvasculature (75). LPS and cytokines have only a slight effect on TFPI expression by cultured ECs in vitro; furthermore, plasma levels of TFPI do not change in coagulopathies associated with sepsis (71). Because of the large granular stores of TFPI in endothelium, plasma levels may not reflect local activity in the lungs, and it is unclear if the increase in tissue factor relative to TFPI reported in these conditions represents a pathological imbalance between these two mediators (71,76). In a baboon model of sepsis, an infusion of TFPI was shown to increase survival and reduce lung injury and, in a phase II clinical trial in patients with severe sepsis, a trend was noted toward improved survival and lung function score with the infusion of recombinant TFPI (77). Nevertheless, a phase III clinical trial was negative.

TM, a constitutively expressed integral membrane protein found on microvascular and macrovascular ECs in vitro and in vivo, provides a further link between inflammation and coagulation. TM acts as an extracellular modifier of locally generated thrombin by binding to it and inhibiting its procoagulant activities, while enhancing its ability to activate circulating protein C, thereby generating the active form (APC) (4). Thus, this system provides a mechanism that changes thrombin to an anticoagulant molecule that activates an endogenous regulator of clotting, protein C. APC, together with the EC-derived cofactor protein S, inhibits factors Va and VIIa, thus providing a feedback inhibition loop on thrombin generation that is proportional to the magnitude of the hemostatic stimulus under regulated physiological conditions (4). TM can be cleaved from the EC surface by neutrophil elastase to produce soluble TM, which also can interact with thrombin to facilitate its activation of protein C and further enhance its profibrinolytic actions (78).

Similar to the procoagulant tissue factor pathway, the anticoagulant TM, in conjunction with APC, has complex regulatory and signaling functions (78,79). Endothelial-bound TM mediates the thrombin-dependent generation of APC. APC in turn activates the serine protease receptor, PAR-1, resulting in inhibition of NF-κB nuclear translocation, cytokine production, and expression of EC surface adhesion molecules (79,80).[1] The APC-mediated activation of PAR-1 requires binding of APC to an EC-specific coreceptor, endothelial protein C receptor (EPCR). In vitro and in vivo studies have demonstrated that inflammatory mediators that may be active in sepsis induce downregulation of both TM and EPCR on ECs, suggesting a compromise in the APC-generating capacity of the endothelium. Based on these data, as well as the observation that protein C levels are reduced in the majority of patients with severe sepsis, the therapeutic potential of recombinant human APC has been expectantly investigated in preclinical and clinical studies. In a large randomized controlled trial in patients with severe sepsis and evidence for failure of at least one organ, a 96-hour infusion with recombinant human APC was shown to decrease overall mortality by 6.1%, with a relative reduction of 19.4%. There was also a more rapid resolution of respiratory dysfunction and an increased number of ventilator-free days (81). Recent observations further indicate that APC can modify leukocyte–EC interactions in in vivo models (82), and reduce the accumulation of PMNs in the alveoli of human volunteers challenged with LPS (83).

1 As discussed elsewhere in this volume, an interesting and unresolved question is whether/how APC and thrombin mediate distinct phenotypic changes in EC by activating the same receptor (PAR-1).

A trial of recombinant APC in ALI/ARDS is in progress (M. Matthay, personal communication).

ECs may alter the regulation of fibrinolysis in ARDS through the synthesis of a complex group of factors including plasminogen, plasminogen activator inhibitors (PAI), and receptors for fibrinolytic agonists (4). ECs synthesize two forms of PAI, PAI-1 and -2 (4). PAI-1 is more active against tissue-type plasminogen activator (t-PA), which converts plasminogen to plasmin, a serine protease that degrades fibrin. In a recent in vitro study of cultured human macro- and microvascular ECs harvested from pulmonary vascular beds, it was reported that stimulation with TNF-α or LPS upregulated PAI-1 and -2 in parallel with downregulation of t-PA, a condition consistent with fibrin deposition in vivo (84). Previous studies reported that ECs obtained from the pulmonary microvascular bed of ARDS patients had a decreased "fibrinolytic potential," with decreased t-PA:PAI ratios when stimulated with LPS or TNF-α (85), although the mechanisms involved were not delineated. If present in vivo, such alterations may contribute to the pathophysiology of ARDS triggered by sepsis (49) as well as other predisposing conditions (see Figure 128.1).

Cumulative evidence suggests that the balance between anticoagulant and prothrombotic characteristics of the endothelium is altered in ARDS, and that molecular cross-talk between the inflammatory and coagulation systems is likely to be involved, although much remains to be learned. Cellular components of thrombi, including platelets and myeloid leukocytes, have the potential to initiate or amplify EC signaling and activation, adding to the complexity of these molecular interactions (see Figure 128.3) (49). Specific dissection of these interactions in ARDS may yield new approaches to therapy (31,49).

ENDOTHELIAL CELL REGULATION OF BLOOD FLOW IN ACUTE RESPIRATORY DISTRESS SYNDROME

Dysregulated pulmonary blood flow in ARDS is characterized by reduced pulmonary vasoconstriction in areas of alveolar hypoxia and vasoconstriction in areas of well-ventilated lung. This results in pronounced arterial hypoxemia due to ventilation–perfusion mismatching. Increased-permeability pulmonary edema that floods the alveoli, and the presence of hyaline membranes, further compound the gas exchange abnormalities (8,9) (see Box 128.1). The concomitant pulmonary arterial hypertension seen in many of these patients results in right ventricular dysfunction in some cases. Blood flow in the pulmonary vasculature is regulated, in large part, by vasoactive signaling factors generated by endothelium. These are thought to act primarily in a local fashion to modify vascular tone, caliber, responses to other stimuli, and hemodynamic variables (4). Nitric oxide (NO) and prostacyclin (PGI_2) are key endothelial-derived vasodilators. They also modify platelet responses and, in some cases, leukocyte–endothelial interactions (4). Thus, they have both direct and indirect vasoregulatory activities. In addition, ECs synthesize vasoconstrictor substances that are important in both acute and chronic pulmonary vascular responses. The most important of these are the endothelins (86).

ECs generate the free radical NO by converting L-citrulline to L-arginine in an enzymatic process catalyzed by NO synthases (NOS) in the presence of oxygen and nicotinamide adenine dinucleotide phosphate-oxidase (NADPH) (4). NO is a cell-permeant gas with a half-life of approximately 6 seconds. It mediates intercellular interactions in a paracrine fashion, and binds the heme-prosthetic group of guanylate cyclase in vascular smooth muscle cells, maintaining basal vascular tone and inducing vasorelaxation (4). During hypoxic vasoconstriction, it has been shown that inhaled NO (iNO) selectively dilates the pulmonary vasculature without systemic effects (87,88). NO production is catalyzed by three known NOS isoenzymes, including a "constitutive" form found primarily in ECs, eNOS, and inducible forms, iNOS and nNOS, which are upregulated by several proinflammatory mediators. The isoenzyme eNOS is involved in mediating basal vascular tone, and its expression and/or activity may be further increased by outside-in signaling via receptors for certain EC agonists (4). Therapeutically, inhaled NO has been shown to improve oxygenation in some animal models; however, clinical trials in patients with ARDS have failed to show a consistent decrease in mortality or improvement in respiratory endpoints (8,89). Characterization of the local concentrations of NO in the lungs of patients with ARDS has not been achieved.

Prostaglandins are endogenously produced by ECs through the actions of the microsomal COXs (prostaglandin H synthases). Arachidonic acid, released from membrane phospholipids in response to cellular activation (4,88), is converted to intermediates by COX-1 and -2 and to downstream products by linked eicosanoid synthases (4). PGI_2, the first EC-derived vasodilator to be described, also was shown to be a potent inhibitor of platelet activation and aggregation and to exhibit anti-inflammatory properties (4). Another COX product, thromboxane A2 (TXA2), is a potent pulmonary vasoconstrictor, increases EC permeability, and is a chief agonist for platelet aggregation (90). The COX-1 isoform is constitutively expressed in some endothelial beds, and PGI_2 generated by this enzyme is important in physiological vasoregulation and as a local antithrombotic signal (90). The inducible COX-2 is expressed in ECs in response to stimulation with IL-1, TNF-α, LPS, and other inflammatory and thrombotic mediators (90,91). Depending on the cell type and downstream prostaglandin synthases, COX-2 generates PGI_2, PGE_2, and other prostaglandins, in addition to the vasoconstrictor thromboxane. In cultured porcine pulmonary ECs cytokines and LPS, stimulation increased the concentration of TXA2 relative to PGI_2, consistent with the hypothesis that the balance between vasoconstricting and vasodilating prostaglandins is influenced by inflammatory signals (92). Prostaglandins and thromboxane have homeostatic effects on blood flow distribution to areas of better ventilation in the lung, an issue of importance in the pathobiology of ARDS. In a series of patients with ARDS, infusion of a fat emulsion that shifted the balance of prostanoids toward the vasodilator PGI_2 resulted in a worsening of oxygenation, presumably by relieving hypoxic vasoconstriction and thereby increasing pulmonary shunt physiology (93). Earlier trials of vasodilator prostaglandin analogs as therapeutic agents in ARDS also were unsuccessful (8). Although this could suggest a role for TXA2 in the management of ARDS because of its vasoconstricting properties, its prothrombotic and proinflammatory effects argue against this strategy. A confounding variable in the interpretation of clinical studies of eicosanoid administration is that the factors that regulate the expression of endogenous EC COX (see Figure 128.2) and eicosanoid synthases in ALI and ARDS are largely unknown, and alterations in their expression in the injured lung relative to endothelium in the uninflamed normal lung have not been clearly defined. In addition to regulating vascular tone and blood flow, recent in vitro evidence indicates that COX products influence EC motility and repair of endothelial EC injury (94), indicating additional roles in damaged pulmonary vessels that may affect the outcome of ARDS.

Endothelin (ET)-1 is produced primarily by ECs, and is a potent pulmonary vasoconstrictor (86,95). ET-1 also potentiates the constrictor functions of catecholamines. Preproendothelin-1 is transcriptionally induced by inflammatory cytokines, hypoxia, shear, ischemia, and ATIII, and is processed to ET-1 by ET-converting enzyme (86). ET-1 signaling occurs via two G-coupled–protein receptors. ET_A is expressed on smooth muscle cells and mediates sustained vasoconstriction, whereas ET_B is displayed primarily on ECs, where it mediates transient vasodilation by inducing NO and PGI_2 release. ET-1 previously has been shown to be increased in the lungs of patients who died with ARDS (96,97). In the serum of patients with sepsis-induced ARDS, ET-1 levels correlated both with the degree of hypoxemia and severity of organ damage (98). A subsequent small study with an inhaled ET-1 antagonist reported significant improvement in gas exchange, decreased pulmonary pressures, and decreased pulmonary shunt (95). Expression of ET-1 in parallel with NOSs has been analyzed in the lungs of patients dying of ARDS, an approach that can yield key information relevant to the patterns of vasoactive mediators that are generated (5,96). Nevertheless, changes in expression of the EC receptor systems, together with regional patterns of generation of the ET-1 and NO ligands in injured pulmonary vessels, remain incompletely characterized in ALI and ARDS.

INTERACTIONS OF ENDOTHELIAL CELLS WITH LEUKOCYTES AND PLATELETS IN ACUTE RESPIRATORY DISTRESS SYNDROME

Leukocytes and platelets traverse the circulatory system during inflammatory and immune surveillance, adhering to inflamed endothelium, exposed subendothelial matrix, and to each other (4–7,49). Interaction between leukocytes and activated pulmonary endothelium is a requisite feature of host defense against infection and repair of injury. In contrast, however, dysregulated adhesion and signaling between myeloid leukocytes and ECs are key early events in the pathological lung injury that leads to ARDS (5,8,9,31,49) (see Figure 128.1). Over the last 20 years, mechanisms that govern adhesive and signaling interactions between myeloid leukocytes and activated ECs have been extensively studied and the results extended to other classes of leukocytes (4,6). These results are reviewed elsewhere in this volume. Neutrophils, monocytes, and monocyte-derived macrophages are involved in the inflammatory responses and parenchymal damage associated with ARDS, with neutrophil adhesion and migration through the endothelium a key paradigm in this process (5,8,9) (see Figures 128.1 and 128.3).

In in vivo and in vitro models of inflammation, both leukocytes and platelets interact with activated ECs on injured vessels through a multistep cascade (4–7). Initially, leukocytes are tethered to activated ECs followed by rolling, localized signaling, tight adhesion dependent on activation of leukocytes, and subsequent transmigration between endothelial junctions in a sequence orchestrated by selectins, signaling molecules, and leukocyte integrins (4,6,7). Although the rolling of leukocytes occurs in pulmonary arterioles and venules under experimental conditions, leukocyte rolling does not occur in capillaries because of their size and, potentially, differences in expression of selectins, compared with their display by systemic microvascular beds (99). How these observations relate to injured and remodeled vessels in the lungs of patients with ARDS is largely unknown. Adhesion molecules such as P-selectin are displayed on endothelium in vessels of various sizes in the lungs of some patients with ARDS, as are signaling chemokines that can induce neutrophil accumulation and activation (5,49,100) (see Figure 128.2). Further, adhesion molecules can be displayed in a regional fashion by lung endothelium in response to clinically relevant perturbations of experimental preparations (59), and may influence leukocyte accumulation and sequestration (59,101). Leukocyte sequestration in the pulmonary microvascular bed also involves intercellular aggregation (49) and changes in deformability of the circulating cells (99). Cellular and molecular mechanisms that regulate adhesion, activation, and accumulation of myeloid and mononuclear leukocytes (Figures 128.1, 128.2, and 128.3) are key targets for future investigation aimed at interventional strategies in ALI and ARDS (6,8,9,31,49).

Mechanisms that govern the interaction of platelets with leukocytes in ARDS, and the influence of these mixed-cell aggregates on activated or injured pulmonary ECs, are incompletely characterized but ultimately may be targets for therapeutic intervention (8,9,49). Evidence suggests that platelets and leukocytes may be sequestered in pulmonary microvascular beds as a result of the initial formation of cellular aggregates in flowing blood, and that platelet–leukocyte signaling may induce fibrin thrombi in the

syndrome. Contribution of tissue factor associated with factor VII. *Am Rev Respir Dis.* 1987;136(6):1466–1474.

22 Pober JS, Gimbrone MA Jr., Lapierre LA, et al. Overlapping patterns of activation of human endothelial cells by interleukin 1, tumor necrosis factor, and immune interferon. *J Immunol.* 1986;137(6):1893–1896.

23 Bevilacqua MP, Pober JS, Majeau GR, et al. Recombinant tumor necrosis factor induces procoagulant activity in cultured human vascular endothelium: characterization and comparison with the actions of interleukin 1. *Proc Natl Acad Sci USA.* 1986;83(12):4533–4537.

24 Moss M, Gillespie MK, Ackerson L, et al. Endothelial cell activity varies in patients at risk for the adult respiratory distress syndrome. *Crit Care Med.* 1996;24(11):1782–1786.

25 Nachman RL, Jaffe EA. Endothelial cell culture: beginnings of modern vascular biology. *J Clin Invest.* 2004;114(8):1037–1040.

26 Stevens T, Rosenberg R, Aird W, et al. NHLBI workshop report: endothelial cell phenotypes in heart, lung, and blood diseases. *Am J Physiol Cell Physiol.* 2001;281(5):C1422–C1433.

27 Oh P, Li Y, Yu J, et al. Subtractive proteomic mapping of the endothelial surface in lung and solid tumours for tissue-specific therapy. *Nature.* 2004;429(6992):629–635.

28 Lucas R, Lou J, Morel DR, et al. TNF receptors in the microvascular pathology of acute respiratory distress syndrome and cerebral malaria. *J Leukoc Biol.* 1997;61(5):551–558.

29 Mehta D, Bhattacharya J, Matthay MA, Malik AB. Integrated control of lung fluid balance. *Am J Physiol Lung Cell Mol Physiol.* 2004;287(6):L1081–L1090.

30 Gutierrez-Ramos JC, Bluethmann H. Molecules and mechanisms operating in septic shock: lessons from knockout mice. *Immunol Today.* 1997;18(7):329–334.

31 Matthay MA, Zimmerman GA, Esmon C, et al. Future research directions in acute lung injury: summary of a National Heart, Lung, and Blood Institute working group. *Am J Respir Crit Care Med.* 2003;167(7):1027–1035.

32 Ranieri VM, Suter PM, Tortorella C, et al. Effect of mechanical ventilation on inflammatory mediators in patients with acute respiratory distress syndrome: a randomized controlled trial. *JAMA.* 1999;282(1):54–61.

33 Severgnini M, Takahashi S, Rozo LM, et al. Activation of the STAT pathway in acute lung injury. *Am J Physiol Lung Cell Mol Physiol.* 2004;286(6):L1282–L1292.

34 Seybold J, Thomas D, Witzenrath M, et al. Tumor necrosis factor-alpha-dependent expression of phosphodiesterase 2: role in endothelial hyperpermeability. *Blood.* 2005;105(9):3569–3576.

35 Ahmad S, Ahmad A, Gerasimovskaya E, et al. Hypoxia protects human lung microvascular endothelial and epithelial-like cells against oxygen toxicity: role of phosphatidylinositol 3-kinase. *Am J Respir Cell Mol Biol.* 2003;28(2):179–187.

36 Yost CC, Denis MM, Lindemann S, et al. Activated polymorphonuclear leukocytes rapidly synthesize retinoic acid receptor-alpha: a mechanism for translational control of transcriptional events. *J Exp Med.* 2004;200(5):671–680.

37 Modur V, Zimmerman GA, Prescott SM, McIntyre TMl. Endothelial cell inflammatory responses to tumor necrosis factor alpha. Ceramide-dependent and -independent mitogen-activated protein kinase cascades. *J Biol Chem.* 1996;271(22): 13094–13102.

38 Goggel R, Winoto-Morbach S, Vielhaber G, et al. PAF-mediated pulmonary edema: a new role for acid sphingomyelinase and ceramide. *Nat Med.* 2004;10(2):155–160.

39 Schwartz MD, Moore EE, Moore FA, et al. Nuclear factor-kappa B is activated in alveolar macrophages from patients with acute respiratory distress syndrome. *Crit Care Med.* 1996;24(8):1285–1292.

40 Pober JS, Bevilacqua MP, Mendrick DL, et al. Two distinct monokines, interleukin 1 and tumor necrosis factor, each independently induce biosynthesis and transient expression of the same antigen on the surface of cultured human vascular endothelial cells. *J Immunol.* 1986;136(5):1680–1687.

41 Bevilacqua MP, Pober JS, Wheeler ME, et al. Interleukin-1 activation of vascular endothelium. Effects on procoagulant activity and leukocyte adhesion. *Am J Pathol.* 1985;121(3):394–403.

42 Introna M, Mantovani A. Early activation signals in endothelial cells. Stimulation by cytokines. *Arterioscler Thromb Vasc Biol.* 1997;17(3):423–428.

43 Modur V, Feldhaus MJ, Weyrich AS, et al. Oncostatin M is a proinflammatory mediator. In vivo effects correlate with endothelial cell expression of inflammatory cytokines and adhesion molecules. *J Clin Invest.* 1997;100(1):158–168.

44 Ueno H, Matsuda T, Hashimoto S, et al. Contributions of high mobility group box protein in experimental and clinical acute lung injury. *Am J Respir Crit Care Med.* 2004;170(12):1310–1316.

45 Lutz W, Stetkiewicz J. High mobility group box 1 protein as a late-acting mediator of acute lung inflammation. *Int J Occup Med Environ Health.* 2004;17(2):245–254.

46 Modur V, Li Y, Zimmerman GA, et al. Retrograde inflammatory signaling from neutrophils to endothelial cells by soluble interleukin-6 receptor alpha. *J Clin Invest.* 1997;100(11):2752–2756.

47 Romano M, Sironi M, Toniatti C, et al. Role of IL-6 and its soluble receptor in induction of chemokines and leukocyte recruitment. *Immunity.* 1997;6(3):315–325.

48 Hurst SM, Wilkinson TS, McLoughlin RM, et al. Il-6 and its soluble receptor orchestrate a temporal switch in the pattern of leukocyte recruitment seen during acute inflammation. *Immunity.* 2001;14(6):705–714.

49 Zimmerman GA, McIntyre TM. Pathogenesis of sepsis and septic-induced lung injury. In: Matthay MA, ed. *Lung Biology in health and disease.* New York: Marcel Dekker; 2003:245–287.

50 Maccallum NS, Evans TW. Epidemiology of acute lung injury. *Curr Opin Crit Care.* 2005;11(1):43–49.

51 Rojas M, Woods CR, Mora AL, et al. Endotoxin-induced lung injury in mice: structural, functional, and biochemical responses. *Am J Physiol Lung Cell Mol Physiol.* 2005;288(2): L333–L341.

52 Neilsen PO, Zimmerman GA, McIntyre TM. Escherichia coli Braun lipoprotein induces a lipopolysaccharide-like endotoxic response from primary human endothelial cells. *J Immunol.* 2001;167(9):5231–5239.

53 Haimovitz-Friedman A, Cordon-Cardo C, Bayoumy S, et al. Lipopolysaccharide induces disseminated endothelial apoptosis requiring ceramide generation. *J Exp Med.* 1997;186(11):1831–1841.

54 Bannerman DD, Goldblum SE. Mechanisms of bacterial lipopolysaccharide-induced endothelial apoptosis. *Am J Physiol Lung Cell Mol Physiol.* 2003;284(6):L899–L914.

55 Basu S, Fenton MJ. Toll-like receptors: function and roles in lung disease. *Am J Physiol Lung Cell Mol Physiol.* 2004;286(5):L887–L892.

56 Faure E, Thomas L, Xu H, et al. Bacterial lipopolysaccharide and IFN-gamma induce Toll-like receptor 2 and Toll-like receptor 4 expression in human endothelial cells: role of NF-kappa B activation. *J Immunol.* 2001;166(3):2018–2024.

57 Zhang FX, Kirschning CJ, Mancinelli R, et al. Bacterial lipopolysaccharide activates nuclear factor-kappaB through interleukin-1 signaling mediators in cultured human dermal endothelial cells and mononuclear phagocytes. *J Biol Chem.* 1999;274(12):7611–7614.

58 Henneke P, Golenbock DT. TIRAP: how Toll receptors fraternize. *Nat Immunol.* 2001;2(9):828–830.

59 Andonegui G, Bonder CS, Green F, et al. Endothelium-derived Toll-like receptor-4 is the key molecule in LPS-induced neutrophil sequestration into lungs. *J Clin Invest.* 2003;111(7):1011–1020.

60 Dejana E, Corada M, Lampugnani MG. Endothelial cell-to-cell junctions. *FASEB J.* 1995;9(10):910–918.

61 Dull RO, Garcia JG. Leukocyte-induced microvascular permeability: how contractile tweaks lead to leaks. *Circ Res.* 2002;90(11):1143–1144.

62 Dudek SM, Garcia JG. Cytoskeletal regulation of pulmonary vascular permeability. *J Appl Physiol.* 2001;91(4):1487–1500.

63 McVerry BJ, Garcia JG. Endothelial cell barrier regulation by sphingosine 1-phosphate. *J Cell Biochem.* 2004;92(6):1075–1085.

64 Xia P, Wang L, Moretti PA, et al. Sphingosine kinase interacts with TRAF2 and dissects tumor necrosis factor-alpha signaling. *J Biol Chem.* 2002;277(10):7996–8003.

65 Lee MJ, Thangada S, Claffey KP, et al. Vascular endothelial cell adherens junction assembly and morphogenesis induced by sphingosine-1-phosphate. *Cell.* 1999;99(3):301–312.

66 Krump-Konvalinkova V, Yasuda S, Rubic T, et al. Stable knockdown of the sphingosine 1-phosphate receptor s1p1 influences multiple functions of human endothelial cells. *Arterioscler Thromb Vasc Biol.* 2004;25(3):546–552. Epub 2004 Dec 23.

67 Lo SK, Burhop KE, Kaplan JE, et al. Role of platelets in maintenance of pulmonary vascular permeability to protein. *Am J Physiol.* 1988;254(4 Pt 2):H763–H771.

68 McVerry BJ, Peng X, Hassoun PM, et al. Sphingosine 1-phosphate reduces vascular leak in murine and canine models of acute lung injury. *Am J Respir Crit Care Med.* 2004;170(9):987–993.

69 Peng X, Hassoun PM, Sammani S, et al. Protective effects of sphingosine 1-phosphate in murine endotoxin-induced inflammatory lung injury. *Am J Respir Crit Care Med.* 2004;169(11):1245–1251.

70 Bone RC, Francis PB, Pierce AK. Intravascular coagulation associated with the adult respiratory distress syndrome. *Am J Med.* 1976;61(5):585–589.

71 Abraham E. Coagulation abnormalities in acute lung injury and sepsis. *Am J Respir Cell Mol Biol.* 2000;22(4):401–404.

72 Riewald M, Ruf W. Mechanistic coupling of protease signaling and initiation of coagulation by tissue factor. *Proc Natl Acad Sci USA.* 2001;98(14):7742–7747.

73 Carraway MS, Welty-Wolf KE, Miller DL, et al. Blockade of tissue factor: treatment for organ injury in established sepsis. *Am J Respir Crit Care Med.* 2003;167(9):1200–1209.

74 Lupu C, Lupu F, Dennehy U, et al. Thrombin induces the redistribution and acute release of tissue factor pathway inhibitor from specific granules within human endothelial cells in culture. *Arterioscler Thromb Vasc Biol.* 1995;15(11):2055–2062.

75 Osterud B, Bajaj MS, Bajaj SP. Sites of tissue factor pathway inhibitor (TFPI) and tissue factor expression under physiologic and pathologic conditions. On behalf of the Subcommittee on Tissue factor Pathway Inhibitor (TFPI) of the Scientific and Standardization Committee of the ISTH. *Thromb Haemost.* 1995;73(5):873–875.

76 Gando S, Kameue T, Matsuda N, et al. Imbalances between the levels of tissue factor and tissue factor pathway inhibitor in ARDS patients. *Thromb Res.* 2003;109(2–3):119–124.

77 Laterre PF, Wittebole X, Dhainaut JF. Anticoagulant therapy in acute lung injury. *Crit Care Med.* 2003;31(Suppl 4):S329–S336.

78 Califano F, Giovanniello T, Pantone P, et al. Clinical importance of thrombomodulin serum levels. *Eur Rev Med Pharmacol Sci.* 2000;4(3):59–66.

79 Esmon CT. The anticoagulant and anti-inflammatory roles of the protein C anticoagulant pathway. *J Autoimmun.* 2000;15(2):113–116.

80 Sohn RH, Deming CB, Johns DC, et al. Regulation of endothelial thrombomodulin expression by inflammatory cytokines is mediated via activation of nuclear factor-kappa B. *Blood.* 2005;105(10):3910–3917. Epub 2005 Jan 27.

81 Bernard GR, Vincent JL, Laterre PF, et al. Efficacy and safety of recombinant human activated protein C for severe sepsis. *N Engl J Med.* 2001;344(10):699–709.

82 Hoffmann JN, Vollmar B, Laschke MW, et al. Microhemodynamic and cellular mechanisms of activated protein C action during endotoxemia. *Crit Care Med.* 2004;32(4):1011–1017.

83 Nick JA, Coldren CD, Geraci MW, et al. Recombinant human activated protein C reduces human endotoxin-induced pulmonary inflammation via inhibition of neutrophil chemotaxis. *Blood.* 2004;104(13):3878–3885. Erratum *Blood.* 2005;105(10):3785.

84 Muth H, Maus U, Wygrecka M, et al. Pro- and antifibrinolytic properties of human pulmonary microvascular versus artery endothelial cells: impact of endotoxin and tumor necrosis factor-alpha. *Crit Care Med.* 2004;32(1):217–226.

85 Grau GE, de Moerloose P, Bulla O, et al. Haemostatic properties of human pulmonary and cerebral microvascular endothelial cells. *Thromb Haemost.* 1997;77(3):585–590.

86 Wort SJ, Evans TW. The role of the endothelium in modulating vascular control in sepsis and related conditions. *Br Med Bull.* 1999;55(1):30–48.

87 Frostell C, Fratacci MD, Wain JC, et al. Inhaled nitric oxide. A selective pulmonary vasodilator reversing hypoxic pulmonary vasoconstriction. *Circulation.* 1991;83(6):2038–2047. Erratum: *Circulation.* 1991;84(5):2212.

88 Kaisers U, Busch T, Deja M, et al. Selective pulmonary vasodilation in acute respiratory distress syndrome. *Crit Care Med.* 2003;31(Suppl 4):S337–S342.

89 Sokol J, Jacobs SE, Bohn D. Inhaled nitric oxide for acute hypoxic respiratory failure in children and adults: a meta-analysis. *Anesth Analg.* 2003;97(4):989–998.

90 Davidge ST. Prostaglandin H synthase and vascular function. *Circ Res.* 2001;89(8):650–660.

91 Jones DA, Carlton DP, McIntyre TM, et al. Molecular cloning of human prostaglandin endoperoxide synthase type II and demonstration of expression in response to cytokines. *J Biol Chem.* 1993;268(12):9049–9054.

92 Muzaffar S, Shukla N, Lobo C, et al. Iloprost inhibits superoxide formation and gp91phox expression induced by the thromboxane A2 analogue U46619, 8-isoprostane F2alpha, prostaglandin

F2alpha, cytokines and endotoxin in the pig pulmonary artery. *Br J Pharmacol.* 2004;141(3):488–496.
93 Suchner U, Katz DP, Furst P, et al. Effects of intravenous fat emulsions on lung function in patients with acute respiratory distress syndrome or sepsis. *Crit Care Med.* 2001;29(8):1569–1574.
94 Jiang H, Weyrich AS, Zimmerman GA, McIntyre TM. Endothelial cell confluence regulates cyclooxygenase-2 and prostaglandin E2 production that modulate motility. *J Biol Chem.* 2004;279(53):55905–55913.
95 Deja M, Wolf S, Busch T, et al. The inhaled ET(A) receptor antagonist LU-135252 acts as a selective pulmonary vasodilator. *Clin Sci (Lond).* 2002;109(Suppl 48):S21–S24.
96 Albertine KH, Wang AM, Michael JR. Expression of endothelial nitric oxide synthase, inducible nitric oxide synthase, and endothelin-1 in lungs of subjects who died with ARDS. *Chest.* 1999;116(Suppl 1):S101–S102.
97 Dschietzig T, Alexiou K, Laule M, et al. Stimulation of pulmonary big endothelin-1 and endothelin-1 by antithrombin III: a rationale for combined application of antithrombin III and endothelin antagonists in sepsis-related acute respiratory distress syndrome? *Crit Care Med.* 2000;28(7):2445–2449.
98 Sanai L, Haynes WG, MacKenzie A, et al. Endothelin production in sepsis and the adult respiratory distress syndrome. *Intensive Care Med.* 1996;22(1):52–56.
99 Forlow SB, Rose CE, Ley K. Leukocyte adhesion and emigration in the lung. In: Weir ED, Reeve HL, Reeves JT, eds. *Interaction of Blood and the Pulmonary Circulation.* Armonk, NY: Futura Publishing Company;2002:255–275.
100 Imaizumi T, et al. Human endothelial cells synthesize ENA-78: relationship to IL-8 and to signaling of PMN adhesion. *Am J Respir Cell Mol Biol.* 1997;17(2):181–192.
101 Kuebler WM, Ying X, Singh B, et al. Pressure is proinflammatory in lung venular capillaries. *J Clin Invest.* 1999;104(4):495–502.
102 Sanders KA, Hueckstead T, Xu P, et al. Regulation of oxidant production in acute lung injury. *Chest.* 1999;116(Suppl 1):S56–S61.
103 Rocksen D, Ekstrand-Hammarstrom B, Johansson L, Bucht A. Vitamin E reduces transendothelial migration of neutrophils and prevents lung injury in endotoxin-induced airway inflammation. *Am J Respir Cell Mol Biol.* 2003;28(2):199–207.
104 Chow CW, Herrera Abreu MT, Suzuki T, Downey GP. Oxidative stress and acute lung injury. *Am J Respir Cell Mol Biol.* 2003;29(4):427–431.
105 Sanders SP, Zweier JL, Kuppusamy P, et al. Hyperoxic sheep pulmonary microvascular endothelial cells generate free radicals via mitochondrial electron transport. *J Clin Invest.* 1993;91(1):46–52.
106 Comhair SA, Erzurum SC. Antioxidant responses to oxidant-mediated lung diseases. *Am J Physiol Lung Cell Mol Physiol.* 2002;283(2):L246–L255.
107 Terada LS. Oxidative stress and endothelial activation. *Crit Care Med.* 2002;30(Suppl 5):S186–S191.
108 Dupont GP, Hueckstead TP, Marshall BC, et al. Regulation of xanthine dehydrogenase and xanthine oxidase activity and gene expression in cultured rat pulmonary endothelial cells. *J Clin Invest.* 1992;89(1):197–202.
109 Adachi T, Fukushima T, Usami Y, Hirano K. Binding of human xanthine oxidase to sulphated glycosaminoglycans on the endothelial-cell surface. *Biochem J.* 1993;289(Pt 2):523–527.
110 Tan S, Yokoyama Y, Dickens E, et al. Xanthine oxidase activity in the circulation of rats following hemorrhagic shock. *Free Radic Biol Med.* 1993;15(4):407–414.
111 Terada LS, Dormish JJ, Shanley PF, et al. Circulating xanthine oxidase mediates lung neutrophil sequestration after intestinal ischemia-reperfusion. *Am J Physiol.* 1992;263(3 Pt 1):L394–L401.
112 Yoshikawa T, Yoshida N, Manabe H, et al. Alpha-tocopherol protects against expression of adhesion molecules on neutrophils and endothelial cells. *Biofactors.* 1998;7(1–2):15–19.
113 Rubenfeld GD, Caldwell E, Peabody E, et al. Incidence and outcomes of acute lung injury. *N Engl J Med.* 2005;353(16):1685–1693.
114 Nagase T, Uozumi N, Ishii S, et al. Acute lung injury by sepsis and acid aspiration: a key role for cytosolic phospholipase A2. *Nat Immunol.* 2000;1(1):42–46.
115 Chinnaiyan AM, Huber-Lang M, Kumar-Sinha C, et al. Molecular signatures of sepsis: multiorgan gene expression profiles of systemic inflammation. *Am J Pathol.* 2001;159(4):1199–1209.
116 Imai Y, Kuba K, Rao S, et al. Angiotensin-converting enzyme 2 protects from severe acute lung failure. *Nature.* 2005;436(7047):112–116.
117 Nagase T, Ishii S, Kume K, et al. Platelet-activating factor mediates acid-induced lung injury in genetically engineered mice. *J Clin Invest.* 1999;104(8):1071–1076.

129

The Central Role of Endothelial Cells in Severe Angioproliferative Pulmonary Hypertension

Norbert F. Voelkel and Mark R. Nicolls

University of Colorado Health Sciences Center, Denver

Pulmonary hypertension (PH) can be transient, intermittent (caused by hypoxia or sleep apnea), or chronic. Chronic PH is most frequently irreversible because of significant pulmonary vascular alterations, a process that has been termed *remodeling* (1). The pathophysiological mechanisms of PH may be approached by considering the following two equations:

$$\Delta P = Q \times R \qquad \text{(Equation 1)}$$

$$R = 8\eta L/r^4 \qquad \text{(Equation 2)}$$

For many decades, investigators believed that intense pulmonary vasoconstriction (thus a reduction in radius or r) was responsible for the changes in pressure and vessel remodeling in many forms of severe PH (2–5). However, it is not clear whether vasoconstriction *always* occurs and is *necessary* for the development of severe pulmonary vascular hypertension. Unfortunately, acutely reversible PH (as defined by a significant reduction in the mean pulmonary artery pressure after administration of a vasodilator) is rare in adult patients (6,7). However, it occurs more frequently in children (8), where it may represent a distinct form of severe PH (9). In addition to smooth muscle proliferation, the pathology of most cases of severe, chronic, and irreversible PH consists of an abnormal proliferation of endothelial cells (ECs) within the vascular lumen (10,11). The presence of these lesions has led to the term *severe angioproliferative pulmonary hypertension* (SAPPH). In contrast, patients with PH associated with pulmonary embolism do not have angioproliferative lesions. Frequently, they display central, large-vessel clots that can be surgically removed by thromboendarterectomy.

No early diagnosis is possible for SAPPH. Moreover, the disease is not amenable to serial assessment of pulmonary vascular reactivity concomitant with histological evaluation of the lung resistance vessels. Thus, the degree to which vasoconstriction contributes to the pathogenesis of SAPPH remains uncertain.

Although atherosclerotic lesions can be documented in the main pulmonary artery and large-diameter arterial branches, SAPPH is a disease of microscopically fine precapillary arterioles. Exuberant growth of ECs leads to vascular occlusion at sites distal to bifurcations (10,11). Plexiform (or glomeruloid) proliferative lesions are the *sine qua non* signature lesions in SAPPH. These lesions occur both in primary or idiopathic arterial PH (IAPH) (12), and in secondary forms of severe PH, including left-to-right shunt, human immunodeficiency virus (HIV), or human herpesvirus (HHV)-8 infections (12–14), or a number of collagen vascular autoimmune diseases (e.g., systemic lupus erythematosus [SLE], scleroderma, or Sjögren syndrome) (15). An overlap of collagen vascular diseases with IAPH was recognized many years ago (4,16), because a significant number of patients with IAPH present with antinuclear antibodies (ANA), antiphospholipid antibodies, and Raynaud phenomenon (17). Concentric intima fibrosis, generating so-called onionskin lesions of the pulmonary precapillaries and coronary microvessels (18), frequently is observed in patients with scleroderma/calcinosis, Raynaud phenomenon, esophageal dysmotility, sclerodactyly, and telangiectasia (CREST)-associated PH. The myocardial microvessel involvement likely causes a myocardiopathy, which may account for the worse prognosis of patients with PH associated with the CREST variant of scleroderma. This chapter focuses on the contribution of ECs to the development of severe PH.

PULMONARY MICROVASCULAR ENDOTHELIAL CELLS

The impressively large metabolic capacity of the lung's microvascular endothelium was recognized during the 1970s. The first experiments demonstrating uptake and removal functions used radioactively labeled compounds such as serotonin and prostaglandin E $(PGE)_2$, which were quantitatively taken up during the first passage through the lung, after injection or infusion. Pioneers of the first hour were Alain Junod (19), Norman Gillis (20), and Christopher Dawson (21).

Using the isolated perfused rat lung preparation, it could be shown that arachidonic acid is quantitatively removed by the lung endothelium and converted to prostacyclin, its major cyclooxygenase (COX) product (22). Unfortunately, these early explorations of lung EC metabolic capacity and function did not lead to the development of a practical clinical test for EC damage in acute lung injury or severe PH.

Pulmonary microvascular ECs reside in the alveolar capillaries and extend into precapillary arterioles with diameters of 100 to 250 μm and postcapillary venules. These ECs can be histochemically and functionally distinguished from those of large pulmonary arteries and veins (23). For example, macrovascular pulmonary ECs bind the *Helix pomatia* lectin, whereas microvascular pulmonary ECs bind the *Griffonia simplicifolia* lectin, express the activated leukocyte cell adhesion molecule (ALCAM)-1, and grow faster than do large vessel ECs. Macrovascular ECs align in the direction of blood flow, whereas microvascular cells do not (23). CD34$^+$ cells seem to be more frequent among capillary ECs (24), and CD34$^+$, c-kit$^+$ cells may represent bone marrow-derived multipotent precursor cells (25). Whereas clear-cut differences exist between the phenotype of microvascular and macrovascular ECs, they share the expression of many genes, including endothelial nitric oxide synthase (eNOS), prostacyclin synthase, platelet-endothelial cell adhesion molecule (PECAM)-1, and caveolin-1. An interesting exception is vascular endothelial growth factor (VEGF), which is expressed and released in particularly high levels in microvascular ECs (at least in vitro). We postulate that this capacity to release large amounts of VEGF serves as an *autocrine survival strategy* of these cells (23).

ENDOTHELIAL CELL PROLIFERATION IN SEVERE PULMONARY HYPERTENSION

Chronic pulmonary vasoconstriction, including hypoxic vasoconstriction, leads to muscularization of the pulmonary arterioles, but not routinely to EC growth and lumen obliteration (1). The best known and well-studied model of pulmonary vascular remodeling is that of chronic hypoxia-related PH (26–29). In IAPH, hypoxic vasoconstriction is not involved, because the administration of 100% of oxygen via a breathing mask to patients with IAPH does not change the pulmonary artery pressure (it is important to remember that the term *idiopathic* is used because the cause of the PH is unknown). This fact alone – the absence of hypoxic vasoconstriction in most cases of severe PH – raises the question of what causes such "exuberant" EC proliferation in IAPH (11). Several studies, in succession, showed that the ECs in the complex vascular lesions were phenotypically abnormal (10,30–34), that EC growth in IAPH was monoclonal (35), that ECs in these lesions expressed mutated genes (36) and

Table 129-1: Immunohistochemical Characterization of the Plexiform Lesion EC Phenotype in SAPPH

	Protein Expressed or Overexpressed		*Protein Expression Diminished or Lost*
VEGF	++	Prostacyclin synthase	–
KDR	++	Prostacyclin receptor	–
5-LO	+	p27	–
FLAP	+	PPAR-γ	–
Endothelin	+	ALCAM-1	–
HIF-1α	+	N-cadherin	–
ARNT	+	TGF-βRII	–
eNOS	+	Smad	–
Survivin	+	Caveolin-1	–
P16	+	Heme oxygenase	–
c-Myc	+		
β-catenin	++		

lost expression of tumor suppressor proteins, and, finally, that the cells in the lesions were entirely devoid of apoptotic events (33).

The plexiform lesions in SAPPH are, so far, the only reported exception to the "law of the endothelial monolayer." Why this law has been broken in the precapillary arterioles in SAPPH is a matter of intense investigation. Conceptually, one may consider the "glomeruloid" plexiform lesions to represent a type of intravascular angiogenesis (37), and indeed, the lesion cells express a repertoire of angiogenesis markers, including VEGF, FLK-1/VEGF receptor 2 (VEGFR2), hypoxia-inducible factor (HIF)-1α, and hydroxycarbon nuclear translocator (ARNT) (Table 129-1). We have developed both animal (38) and cell models (discussed later) to investigate these fundamental problems of pulmonary vascular pathobiology. These endeavors have been motivated largely by the fact that these angioproliferative lesions may not be reversible, and they may form the basis of the pulmonary hypertension in IAPH.

In addition to angioproliferation, the plexiform or glomeruloid lesions also display features of inflammation. Whether the presence of inflammatory lesions is a cause or consequence of the pulmonary vascular disease has not been resolved. The presence of mast cells – which are angiogenic – in the lesions was described more than 30 years ago (39). These cells can synthesize and release histamine, leukotrienes, and VEGF and, together with T and B lymphocytes and clusters of macrophages, may provide a footprint of a localized immune response. Given that SAPPH can be associated both with autoimmunity (as in SLE) or with immune insufficiency (as in HIV infection) (40), it remains unclear what the antigenic target is in this inflammatory response. It is a striking fact that many disorders characterized by SAPPH also have significant incidences of autoantibody production. Antibody deposition in the pulmonary arteries of these patients has been already noted in SLE and Sjögren's syndrome. Additionally, patients with IAPH have been shown to have elevated plasma levels of interleukin (IL)-1 and IL-6 (41). Notably, IL-6 protects ECs from apoptosis (42) and IL-1 increases the expression of HIF-1α (43). A number of studies have demonstrated anti-EC antibodies in patients with SLE and scleroderma who also have SAPPH (44–46).

VASCULAR ENDOTHELIAL GROWTH FACTOR AND LUNG VASCULAR HOMEOSTASIS

VEGF is 50,000 times more potent in inducing vascular permeability than is histamine (47), yet the VEGF gene and protein are abundantly expressed in the adult, nongrowing lung – an organ that is required to stay "dry" if it is to serve one of its major functions: oxygen uptake and gas exchange. VEGF ligand binds preferentially to ECs, and the binding of VEGF to human umbilical vein ECs or bovine pulmonary artery ECs can be dramatically decreased by either cationic drugs such as protamine or anionic drugs like suramin. Rapid uptake and binding also can be demonstrated in the intact, integrated circulation of the isolated, perfused lung (48). We continue to be impressed by Harold Dvorak's experiments showing that high adenoviral-mediated VEGF expression resulted in angiogenesis and the formation of glomeruloid structures (49), and that angioproliferation is both VEGF dose- and time-dependent. On the other end of the VEGF expression and action spectrum are lung VEGF deficiency states that lead to a loss of lung microvessels (50) and emphysema (see Chapter 130). Taken together, these findings suggest that a homeostatic lung structure maintenance program is present (51) that significantly relies on the correct "amount" and precise timing of VEGF signaling. Indeed, it is possible that VEGF is one of the most important components of this lung structure maintenance program, because its role transcends vascular biology, stabilizes lung epithelial cell function, controls EC–lymphocyte immune interactions, and inhibits vascular smooth muscle cell apoptosis via binding to VEGF receptors (52).

VASCULAR ENDOTHELIAL GROWTH FACTOR RECEPTOR BLOCKADE CAUSES ENDOTHELIAL CELL APOPTOSIS AND CONCOMITANTLY ELEVATED SHEAR STRESS LEADS TO ENDOTHELIAL CELL PROLIFERATION

Because cause-and-effect experiments cannot be conducted in human SAPPH, we developed the first animal model of angioproliferative PH (38). The model is based on the knowledge that VEGF is an obligatory survival factor for ECs (53) and that shear stress inhibits EC apoptosis (54). To generate SAPPH, we inject adult rats with a subcutaneous dose of the combined VEGFR1 and -2 antagonist SU5416, and subject the animals to chronic hypoxia, which causes pulmonary vasoconstriction (38) and increased shear stress. Severe angioproliferative PH occurs within 3 to 5 weeks. Instead of raising shear stress by means of chronic hypoxic vasoconstriction, shear stress also can be raised by pneumonectomy, which generates the condition in which the entire cardiac output must be accommodated by the remaining lung. The combination of pneumonectomy and SU5416 VEGFR inhibition also causes pulmonary angioproliferation (55). Because VEGFR blockade induces EC apoptosis under in vitro conditions (53), we postulated that initial EC apoptosis in our model was followed by proliferation of the surviving (apoptosis-resistant) ECs, to the point of lumen obliteration (38). We tested this hypothesis directly in the CellMax system (an artificial plastic microtube system, seeded with ECs) by adding apoptosed ECs to the perfusing cell culture medium. Indeed, the addition of apoptosed ECs caused a proliferation of the original ECs (56). Based on our own data, which show that pulmonary microvascular ECs secrete in vitro large amounts of VEGF, we postulate that the maintenance of pulmonary microvascular ECs largely depends on their autocrine VEGF production (23) and signaling, and that the ECs that survive VEGFR blockade become VEGF-independent and apoptosis-resistant. This now raises the larger question: What is the nature of the VEGFR blockade-resistant cells? There can be no doubt that the plexiform lesion cells are phenotypically switched (abnormal); they display a protein expression pattern that can be interpreted as growth-promoting and anti-apoptotic (Table 129.1). The "wisdom" of this proliferating cell phenotype is to turn off antiproliferative genes, like prostacyclin synthase (32) and peroxisome proliferator-activated receptors (PPAR)-γ (33), and express antiapoptotic proteins, like Bcl-2 and survivin (N. Voelkel, unpublished observations). The phenotypic switch of ECs in SAPPH likely can be caused by multiple initiating events, including EC apoptosis (56) and viral infections (40,57). We shall return to this theme in the following section.

Sakao and colleagues (unpublished data) recently showed that commercially available human pulmonary microvascular ECs (HPMVECs) contain a small amount of CD34$^+$ EC precursor cells, which not only survive SU5416-induced EC apoptosis, but expand in number during subsequent EC passages – apparently as a consequence of the SU5416 (VEGF receptor blockade)-induced selection pressure. Thus, the "nature" of the apoptosis-surviving EC in these experiments is, in part, a c-kit$^+$ CD34$^+$ precursor cell, which, in cell culture, further transdifferentiates into α-smooth muscle actin$^+$ and tyrosine hydroxylase$^+$ cells (Sakao, unpublished data) – that is, cells that display smooth muscle and neuronal cell features. Whether c-kit$^+$ CD34$^+$ precursor cells are involved in the formation of plexiform lesions is unknown. However, the angioproliferative lumen occlusion in the animal model, combined with EC data that illustrate the emergence of apoptosis-resistant proliferating ECs, provide us with a cell biological concept for the formation of plexiform lesions.

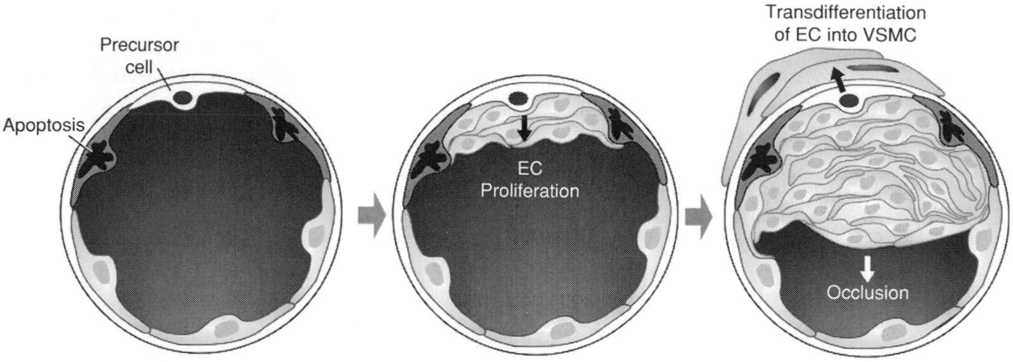

Figure 129.1. In severe angioproliferative PH, distal precapillary arterioles are the critical sites of structural changes. The model depicted in this figure postulates EC apoptosis as the initiating event. EC apoptosis can be triggered by toxins, hemodynamic stress, and cell–cell interactions. In contrast to normal vessel cell turnover, apoptosis-resistant ECs emerge and start to proliferate. At this point, the "law of the monolayer" has been broken. EC transdifferentiation into smooth muscle cells also may take place and play a role in the vessel remodeling.

ENDOTHELIAL CELL GROWTH AND VIRAL INFECTIONS

It is tempting to speculate that viral infection of ECs can alter DNA methylation and thus the expression of genes directly involved in cell growth and apoptosis control (58). Although examples in support of such a mechanism have been described (59), nothing is known about pulmonary microvascular EC methylation control of growth factor genes, or whether they are targeted by viral infection. Presently, it can only be postulated that, for example, the HHV-8 v-cyclin or G-protein–coupled receptor proteins could methylation-dependently upregulate *VEGF* gene expression in infected ECs. Although a tropism of HHV-8 for ECs has been established, it is unknown whether or how HIV infection affects growth of ECs in vivo, in patients with SAPPH, or whether it is the state of immunosuppression that permits the SAPPH to occur. VEGF protein (ligand) plasma levels are apparently not increased in HIV-infected patients (60).

FUTURE DIRECTIONS

What is needed is the thorough examination of the pathological precapillary arteriolar ECs from the lungs of patients with SAPPH ex vivo. Cell surface markers are available (Table 129-1) that allow differential sorting of phenotypically normal and abnormal cells. Cell lines must be established for the different phenotypes. These cells then can be subjected to "therapeutic trials" – as has been done for decades with various cancer cell lines. Viral infection of human pulmonary microvascular ECs – including those from patients with SAPPH – is feasible and will provide mechanistic data: How does the virus infection alter the gene expression pattern of these ECs, their growth and survival potential, and more?

THERAPEUTIC IMPLICATIONS

Presently, we have very little information whether the established treatments for SAPPH affect the angioproliferative obliterating lesions. For example, when lung tissue from patients with SAPPH who had failed continuous intravenous prostacyclin therapy were examined and compared with the tissues of patients who had not been treated with prostacyclin, no difference in the number of plexiform lesions and thickness of intima and media of nonobliterated arterioles was found (61). Why is SAPPH so difficult to treat? We believe the answer is twofold: because the patients do not come to clinical attention during the (perhaps) autoimmune initiation phase of the disease, and second, established angioproliferative lesions manifest altered EC and vascular smooth muscle phenotypes. If disobliteration is the therapeutic goal, then we must understand the biology of these "bad phenotypes." It follows that little insight can be gained by studying the behavior of normal ECs in cell culture experiments.

KEY POINTS

- The vascular pathobiology of "primary" and most forms of secondary SAPPH is primarily driven by lumen-obliterating growth of phenotypically abnormal ECs (Figure 129.1).
- The phenotypically abnormal ECs are likely cells that have survived while neighboring cells of the monolayer underwent apoptosis. The initial apoptosis could be caused by anti-EC antibodies, as a direct or indirect consequence of a viral infection or

subsequent to the interaction of an antigen presenting EC with lymphocytes.
- Effective future treatment of established SAPPH will transcend the presently used vasodilator drugs and target the abnormal EC phenotypes.

REFERENCES

1 Voelkel NF, Tuder RM. Hypoxia-induced pulmonary vascular remodeling: a model for what human disease? *J Clin Invest*. 2000; 106:733–738.

2 Wood P. Pulmonary hypertension with special reference to the vasoconstrictive factor. *Br Heart J*. 1958;20:557–570.

3 Wagenvoort CA. The pathology of primary pulmonary hypertension. *J Pathol*. 1970;101:i.

4 Voelkel NF, Reeves JT. Primary pulmonary hypertension. In: Moser K, ed. *Lung Biology in Health and Disease*. New York: Marcel Dekker; 1979:573–628.

5 Loscalzo J. Endothelial dysfunction in pulmonary hypertension. *N Engl J Med*. 1992;327:117–119.

6 Sitbon O, Humbert M, Jagot JL, et al. Inhaled nitric oxide as a screening agent for safely identifying responders to oral calcium-channel blockers in primary pulmonary hypertension. *Eur Respir J*. 1998;12:265–270.

7 Groves BM, Turkevich D, Donnellan K, et al. Current approach to treatment of primary pulmonary hypertension. *Chest*. 1988; 93:S175–S178.

8 Barst RJ, Maislin G, Fishman AP. Vasodilator therapy for primary pulmonary hypertension in children. *Circulation*. 1999;99:1197–1208.

9 Grunig E, Koehler R, Miltenberger-Miltenyi G, et al. Primary pulmonary hypertension in children may have a different genetic background than in adults. *Pediatr Res*. 2004;56:571–578.

10 Cool CD, Stewart JS, Werahera P, et al. Three-dimensional reconstruction of pulmonary arteries in plexiform pulmonary hypertension using cell-specific markers. Evidence for a dynamic and heterogeneous process of pulmonary endothelial cell growth. *Am J Pathol*. 1999;155:411–419.

11 Tuder RM, Groves B, Badesch DB, Voelkel NF. Exuberant endothelial cell growth and elements of inflammation are present in plexiform lesions of pulmonary hypertension. *Am J Pathol*. 1994;144:275–285.

12 Voelkel NF, Tuder RM. Severe pulmonary hypertensive diseases: a perspective. *Eur Respir J*. 1999;14:1246–1250.

13 Voelkel NF, Tuder RM. Cellular and molecular mechanisms in the pathogenesis of severe pulmonary hypertension. *Eur Respir J*. 1995;8:2129–2138.

14 Tuder RM, Radisavljevic Z, Shroyer KR, et al. Monoclonal endothelial cells in appetite suppressant-associated pulmonary hypertension. *Am J Respir Crit Care Med*. 1998;158:1999–2001.

15 Cool CD, Kennedy D, Voelkel NF, Tuder RM. Pathogenesis and evolution of plexiform lesions in pulmonary hypertension associated with scleroderma and human immunodeficiency virus infection. *Hum Pathol*. 1997;28:434–442.

16 Badesch DB, Tapson VF, McGoon MD, et al. Continuous intravenous epoprostenol for pulmonary hypertension due to the scleroderma spectrum of disease. A randomized, controlled trial. *Ann Intern Med*. 2000;132:425–434.

17 Karmochkine M, Cacoub P, Dorent R, et al. High prevalence of antiphospholipid antibodies in precapillary pulmonary hypertension. *J Rheumatol*. 1996;23:286–290.

18 LeRoy EC. The heart in systemic sclerosis. *N Engl J Med*. 1984;310: 188–190.

19 Junod AF. Uptake, metabolism and efflux of 14 C-5-hydroxytryptamine in isolated perfused rat lungs. *J Pharmacol Exp Ther*. 1972;183:341–355.

20 Riggs D, Havill AM, Pitt BR, Gillis CN. Pulmonary angiotensin-converting enzyme kinetics after acute lung injury in the rabbit. *J Appl Physiol*. 1988;64:2508–2516.

21 Audi SH, Dawson CA, Linehan JH, et al. Pulmonary disposition of lipophilic amine compounds in the isolated perfused rabbit lung. *J Appl Physiol*. 1998;84:516–530.

22 Voelkel NF, Gerber JG, McMurtry IF, et al. Release of vasodilator prostaglandin, PGI2, from isolated rat lung during vasoconstriction. *Circ Res*. 1981;48:207–213.

23 Stevens T, Kasper M, Cool C, Voelkel NF. Pulmonary circulation and pulmonary hypertension. In: Aird WC, ed. *Endothelial in Health and Disease*. New York: Taylor & Francis, 2005:417–437.

24 Hughes S, Yang H, Chan-Ling T. Vascularization of the human fetal retina: roles of vasculogenesis and angiogenesis. *Invest Ophthalmol Vis Sci*. 2000;41:1217–1228.

25 Heissig B, Werb Z, Rafii S, Hattori K. Role of c-kit/Kit ligand signaling in regulating vasculogenesis. *Thromb Haemost*. 2003;90: 570–576.

26 Ono S, Westcott JY, Voelkel NF. PAF antagonists inhibit pulmonary vascular remodeling induced by hypobaric hypoxia in rats. *J Appl Physiol*. 1992;73:1084–1092.

27 Eddahibi S, Hanoun N, Lanfumey L, et al. Attenuated hypoxic pulmonary hypertension in mice lacking the 5-hydroxytryptamine transporter gene. *J Clin Invest*. 2000;105:1555–1562.

28 Hoshikawa Y, Ono S, Suzuki S, et al. Generation of oxidative stress contributes to the development of pulmonary hypertension induced by hypoxia. *J Appl Physiol*. 2001;90:1299–1306.

29 Fanburg BL, Lee SL. A role for the serotonin transporter in hypoxia-induced pulmonary hypertension. *J Clin Invest*. 2000; 105:1521–1523.

30 Wright L, Tuder RM, Wang J, et al. 5-Lipoxygenase and 5-lipoxygenase activating protein (FLAP) immunoreactivity in lungs from patients with primary pulmonary hypertension. *Am J Respir Crit Care Med*. 1998;157:219–229.

31 Giaid A, Yanagisawa M, Langleben D, et al. Expression of endothelin-1 in the lungs of patients with pulmonary hypertension. *N Engl J Med*. 1993;328:1732–1739.

32 Tuder RM, Cool CD, Geraci MW, et al. Prostacyclin synthase expression is decreased in lungs from patients with severe pulmonary hypertension. *Am J Respir Crit Care Med*. 1999; 159:1925–1932.

33 Ameshima S, Golpon H, Cool CD, et al. Peroxisome proliferator-activated receptor gamma (PPARgamma) expression is decreased in pulmonary hypertension and affects endothelial cell growth. *Circ Res*. 2003;92:1162–1169.

34 Richter A, Yeager ME, Zaiman A, et al. Impaired transforming growth factor-beta signaling in idiopathic pulmonary arterial hypertension. *Am J Respir Crit Care Med*. 2004;170:1340–1348.

35 Lee SD, Shroyer KR, Markham NE, et al. Monoclonal endothelial cell proliferation is present in primary but not secondary pulmonary hypertension. *J Clin Invest.* 1998;101: 927–934.

36 Yeager ME, Halley GR, Golpon HA, et al. Microsatellite instability of endothelial cell growth and apoptosis genes within plexiform lesions in primary pulmonary hypertension. *Circ Res.* 2001;88:E2–E11.

37 Tuder RM, Chacon M, Alger L, et al. Expression of angiogenesis-related molecules in plexiform lesions in severe pulmonary hypertension: evidence for a process of disordered angiogenesis. *J Pathol.* 2001;195:367–374.

38 Taraseviciene-Stewart L, Kasahara Y, Alger L, et al. Inhibition of the VEGF receptor 2 combined with chronic hypoxia causes cell death-dependent pulmonary endothelial cell proliferation and severe pulmonary hypertension. *FASEB J.* 2001;15:427–438.

39 Heath D, Yacoub M. Lung mast cells in plexogenic pulmonary arteriopathy. *J Clin Pathol.* 1991;44:1003–1006.

40 Speich R. Pulmonary infectious complications in HIV disease. *Schweiz Rundsch Med Prax.* 1994;83:1364–1373.

41 Humbert M, Monti G, Brenot F, et al. Increased interleukin-1 and interleukin-6 serum concentrations in severe primary pulmonary hypertension. *Am J Respir Crit Care Med.* 1995;151: 1628–1631.

42 Mukaida N. Pathophysiological roles of interleukin-8/CXCL8 in pulmonary diseases. *Am J Physiol Lung Cell Mol Physiol.* 2003; 284:L566–L577.

43 Jung YJ, Isaacs JS, Lee S, et al. IL-1beta-mediated up-regulation of HIF-1alpha via an NFkappaB/COX-2 pathway identifies HIF-1 as a critical link between inflammation and oncogenesis. *FASEB J.* 2003;17:2115–2117.

44 Wusirika R, Ferri C, Marin M, et al. The assessment of anti-endothelial cell antibodies in scleroderma-associated pulmonary fibrosis. A study of indirect immunofluorescent and western blot analysis in 49 patients with scleroderma. *Am J Clin Pathol.* 2003;120:596–606.

45 Quismorio FP Jr., Sharma O, Koss M, et al. Immunopathologic and clinical studies in pulmonary hypertension associated with systemic lupus erythematosus. *Semin Arthritis Rheum.* 1984;13:349–359.

46 Nakagawa N, Osanai S, Ide H, et al. Severe pulmonary hypertension associated with primary Sjögren's syndrome. *Intern Med.* 2003;42:1248–1252.

47 Senger DR, Galli SJ, Dvorak AM, et al. Tumor cells secrete a vascular permeability factor that promotes accumulation of ascites fluid. *Science.* 1983;219:983–985.

48 Maloney J, Tuder RM, Voelkel NF. Vascular endothelial growth factor binding in the isolated perfused rat lung is decreased by protamine. ATS International Conference, May 21, 1995 (Abstract).

49 Sundberg C, Nagy JA, Brown LF, et al. Glomeruloid microvascular proliferation follows adenoviral vascular permeability factor/vascular endothelial growth factor-164 gene delivery. *Am J Pathol.* 2001;158:1145–1160.

50 Voelkel NF, Cool C, Taraceviene-Stewart L, et al. Janus face of vascular endothelial growth factor: the obligatory survival factor for lung vascular endothelium controls precapillary artery remodeling in severe pulmonary hypertension. *Crit Care Med.* 2002;30:S251–S256.

51 Voelkel N, Taraseviciene-Stewart L. Emphysema: an autoimmune vascular disease? *Proc Am Thorac Soc.* 2005;2:23–25.

52 Sakao S, Taraseviciene-Stewart L, Wood K, et al. Apoptosis of pulmonary microvascular endothelial cells stimulates vascular smooth muscle cell growth. *Am J Physiol Lung Cell Mol Physiol.* 2006;291:L362–368.

53 Gerber HP, McMurtrey A, Kowalski J, et al. Vascular endothelial growth factor regulates endothelial cell survival through the phosphatidylinositol 3′-kinase/Akt signal transduction pathway. Requirement for Flk-1/KDR activation. *J Biol Chem.* 1998;273:30336–30343.

54 Hermann C, Zeiher AM, Dimmeler S. Shear stress inhibits H_2O_2-induced apoptosis of human endothelial cells by modulation of the glutathione redox cycle and nitric oxide synthase. *Arterioscler Thromb Vasc Biol.* 1997;17:3588–3592.

55 Glenny RW, Bernard SL, Luchtel DL, et al. Inhibition of VEGF receptor-2 (VEGFR-2) combined with increased shear-stress induces plexiform lesions and pulmonary hypertension in rats. *Am J Respir Crit Care Med.* 2005;167: A711(Abstract).

56 Sakao S, Taraseviciene-Stewart L, Lee JD, et al. Initial apoptosis is followed by increased proliferation of apoptosis-resistant endothelial cells. *FASEB J.* 2005;19(9):1178–1180. Epub 2005 May 16.

57 Cool CD, Rai PR, Yeager ME, et al. Expression of human herpesvirus 8 in primary pulmonary hypertension. *N Engl J Med.* 2003;349:1113–1122.

58 Suzuki M, Toyooka S, Shivapurkar N, et al. Aberrant methylation profile of human malignant mesotheliomas and its relationship to SV40 infection. *Oncogene.* 2005;24:1302–1308.

59 Kalantari M, Calleja-Macias IE, Tewari D, et al. Conserved methylation patterns of human papillomavirus type 16 DNA in asymptomatic infection and cervical neoplasia. *J Virol.* 2004;78: 12762–12772.

60 Renwick N, Weverling GJ, Brouwer J, et al. Vascular endothelial growth factor levels in serum do not increase following HIV type 1 and HHV8 seroconversion and lack correlation with AIDS-related Kaposi's sarcoma. *AIDS Res Hum Retroviruses.* 2002;18: 695–698.

61 Achcar ROD, Demura Y, Rai PR, et al. Loss of caveolin and heme oxygenase expression in severe pulmonary hypertension. *Chest.* 2006;129(3):696–705.

130

Emphysema

An Autoimmune Vascular Disease?

Norbert F. Voelkel and Laimute Taraseviciene-Stewart

University of Colorado Health Sciences Center, Denver

Emphysema of the lungs – first described by Laennec (1) – is a manifestation of chronic obstructive lung disease, and is almost always caused by smoking (2). Rare causes of emphysema are hypersensitivity pneumonitis (3) and human immunodeficiency virus (HIV) infection (4). Histologically, emphysema is characterized by loss of alveolar septal cells, and it can be centrilobular or panlobular. In extreme cases, the lung is literally vanishing. Modern imaging technology allows both a regional and quantitative assessment of emphysema. Although traditionally thought of as chronic progressive airway disease involving proteolytic injury to lung extracellular matrix, Liebow already recognized in 1959 that the lung vessels were involved (5) as well. Recent data on human emphysema, and in animal studies, demonstrate structural alterations of small precapillary arterioles and endothelial cell (EC) dysfunction (6–9). In addition, there is evidence by Tunel staining for EC apoptosis in these small arterioles (10).

THE INVOLVEMENT OF THE ENDOTHELIUM

EC dysfunction had been demonstrated by exposing small pulmonary arteries from patients with emphysema who underwent lung resection to acetylcholine and observing impaired vasodilation when compared with normal lung artery segments of comparable diameter. This test is consistent with EC dysfunction; however, whether this ex vivo test reflects in vivo pulmonary vascular EC dysfunction remains unknown. The cause or causes of this EC dysfunction are also not understood (8,9). Subsequently, Kasahara and colleagues (10) showed increased apoptosis of ECs in lungs from patients with emphysema, as well as a decrease in the overall lung tissue expression of vascular endothelial growth factor (VEGF) and the VEGF receptor 2 (VEGFR2/FLK-1/KDR). The reduction in the expression of these proteins is likely explained by both a loss of capillary ECs and the impaired gene and protein expression of remaining VEGF-producing lung cells – as judged by immunohistochemistry of the emphysematous lungs. This begged the question of whether blockade of VEGF receptors can cause lung EC apoptosis and emphysema. Indeed, a single subcutaneous injection of the combined FLT-1/VEGFR1 and VEGFR2 inhibitor SU5416 caused alveolar septal cell apoptosis and emphysema in adult rats (11–13). SU5416 also causes impaired alveolarization in newborn rats (14). Recently, Tang and colleagues selectively ablated VEGF expression in the mouse lung by delivering intratracheal adeno-associated Cre recombinase to VEGF*loxP* mice. They demonstrated that a transient reduction in pulmonary VEGF levels results in emphysema (15).

VASCULAR ENDOTHELIAL GROWTH FACTOR IS CRITICAL FOR LUNG TISSUE HOMEOSTASIS

It appears that VEGF is important for lung health. Too much VEGF may cause EC proliferation and angioproliferative pulmonary hypertension (16) (see Chapter 129), and too little VEGF may lead to emphysema (13). In many ways, pulmonary angioproliferative hypertension and emphysema are on opposite ends of a spectrum: EC growth and survival in pulmonary hypertension, and impaired EC growth and apoptosis in emphysema. VEGF has been shown to be expressed in several cell types in the lung, including alveolar type II epithelial cells, Clara cells, fibroblasts, and macrophages. We postulate that the pulmonary microvascular ECs are particularly VEGF-dependent and sensitive to VEGF withdrawal or impaired VEGF signaling. At least in culture, rat pulmonary microvascular ECs produce about 20-fold more VEGF than do rat macrovascular (large pulmonary artery) ECs (17). Whether these ex vivo findings reflect the in vivo situation of the human lung microvascular cells is unknown.

The survival of lung microvascular ECs depends both on VEGF and an intact VEGF signaling pathway. This has been illustrated recently in a novel autoimmune emphysema model

(18) in which the immunization of adult rats with xenogeneic ECs resulted in the generation of anti-VEGFR2 antibodies. In addition, in this model, T-lymphocytes are required for emphysema development, because immunization of nude, athymic rats with xenogeneic ECs did not cause emphysema. Whether patients with cigarette smoking–induced emphysema have antiendothelial antibodies is unknown. However, this autoimmune model of emphysema provides proof of the concept that an attack of antibodies and immune-experienced T lymphocytes on the lung microvessels can cause emphysema.

The turnover rate of ECs in the lung is not well understood, nor is the contribution of resident or bone marrow–derived precursor cells. VEGF has been shown to mobilize precursor cells from the bone marrow (19), and it is, in this context, of great importance that these are patients with end-stage emphysema who can have undetectable plasma VEGF levels (20). Commercially available human pulmonary microvascular ECs contain varying numbers of $CD34^{+}$ precursor ECs, indicating perhaps that the adult human lung may rely on circulating precursor cells that have been filtered out by the lung (S. Sakao et al., unpublished observations).

Recent reports show high levels of VEGF expression in the airway mucosa of patients with asthma (21,22). High VEGF levels in these patients may be responsible for the hypervascularity of the airways (bronchial circulation) (23) and an increased number of lung parenchyma capillaries (pulmonary circulation) in patients with asthma. This could explain the greater than normal diffusing capacity frequently observed in asthmatics.

It is peculiar that VEGF protein expression is decreased in the lungs from patients with severe emphysema, because these patients are often hypoxemic, and because the lungs are infiltrated with inflammatory cells and lymphocytes known to produce VEGF (24,25). One possibility is that the *VEGF* promoter region has been modified (26) and has become relatively unresponsive to the transcription-promoting activity of hypoxia-inducible factor (HIF)-1α and cytokines like interleukin (IL)-1. Why some patients with end-stage chronic obstructive airway disease have unmeasurable levels of plasma VEGF is also unclear (20); one hypothesis is that the VEGF protein has been trapped by soluble VEGF receptors (27).

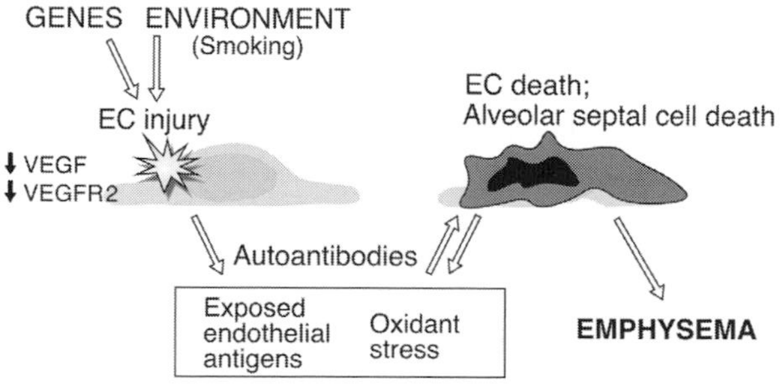

Figure 130.1. Central role of EC injury in pathogenesis of emphysema. Environmental factors and genetic mutations induce EC injury that is associated with oxidant stress and exposure of EC antigens. This results in anti-EC autoantibody production and a new attack on pulmonary microvasculature, leading to endothelial and alveolar septal cell death, lung structural damage, and the development of emphysema.

THERAPEUTIC IMPLICATIONS

Drugs that inhibit lung microvascular EC apoptosis would, theoretically, be of value in the treatment of progressive emphysema. The opposite may be true in asthma, where strategies to neutralize the effects of too much VEGF might induce lung vessel EC apoptosis and reduce the hypervascularity.

KEY POINTS

- Emphysema is also a vascular disease. Lung microvascular EC apoptosis and failure of EC regeneration may be expressions of a failure of the lung structure maintenance program.
- There may be an autoimmune component to progressive emphysema (Figure 130.1), characterized by antiendothelial antibodies and pathogenic T cells. VEGF-trapping by soluble VEGF receptor may explain undetectable plasma levels of VEGF in a subgroup of emphysema patients.

Future Goals

- To examine animal models of cigarette smoke–induced emphysema and search for anti-EC antibodies

REFERENCES

1 Laennec RT. *De auscultation mediate, on traite du diagnostic des maladies des poumons et du Coeur.* Paris: Brosson and Chaude, 1819.
2 Higgins M. Risk factors associated with chronic obstructive lung disease. *Ann NY Acad Sci.* 1991;624:7–17.
3 Malinen AP, Erkinjuntti-Pekkanen RA, Partanen PL, et al. Long-term sequelae of Farmer's lung disease in HRCT: a 14-year follow-up study of 88 patients and 83 matched control farmers. *Eur Radiol.* 2003;13:2212–2221.
4 Diaz PT, King MA, Pacht ER., et al. The pathophysiology of pulmonary diffusion impairment in human immunodeficiency virus infection. *Am J Respir Crit Care Med.* 1999;160:272–277.
5 Liebow A. Pulmonary emphysema with special reference to vascular changes. *Am Rev Respir Dis.* 1959;80:67–93.
6 Wright JL, Petty T, Thurlbeck WM. Analysis of the structure of the muscular pulmonary arteries in patients with pulmonary hypertension and COPD: National Institutes of Health nocturnal oxygen therapy trial. *Lung.* 1992;170:109–124.
7 Barbera JA, Riverola A, Roca J, et al. Pulmonary vascular abnormalities and ventilation-perfusion relationships in mild chronic

obstructive pulmonary disease. *Am J Respir Crit Care Med*. 1994; 149:423–429.
8 Dinh-Xuan AT, Higenbottam TW, Clelland CA, et al. Impairment of endothelium-dependent pulmonary-artery relaxation in chronic obstructive lung disease. *N Engl J Med*. 1991;324:1539–1547.
9 Peinado VI, Barbera JA, Ramirez J, et al. Endothelial dysfunction in pulmonary arteries of patients with mild COPD. *Am J Physiol*. 1998;274:L908–L913.
10 Kasahara Y, Tuder RM, Cool CD, et al. Endothelial cell death and decreased expression of vascular endothelial growth factor and vascular endothelial growth factor receptor 2 in emphysema. *Am J Respir Crit Care Med*. 2001;163:737–744.
11 Kasahara Y, Tuder RM, Taraseviciene-Stewart L, et al. Inhibition of VEGF receptors causes lung cell apoptosis and emphysema. *J Clin Invest*. 2000;106:1311–1319.
12 Shapiro SD. Vascular atrophy and VEGFR-2 signaling: old theories of pulmonary emphysema meet new data. *J Clin Invest*. 2000; 106:1309–1310.
13 Tuder RM, Zhen L, Cho CY, et al. Oxidative stress and apoptosis interact and cause emphysema due to vascular endothelial growth factor receptor blockade. *Am J Respir Cell Mol Biol*. 2003;29:88–97.
14 Jakkula M, Le Cras TD, Gebb S, et al. Inhibition of angiogenesis decreases alveolarization in the developing rat lung. *Am J Physiol Lung Cell Mol Physiol*. 2000;279:L600–L607.
15 Tang K, Rossiter HB, Wagner PD, Breen EC. Lung-targeted VEGF inactivation leads to an emphysema phenotype in mice. *J Appl Physiol*. 2004;97:1559–1566.
16 Tuder RM, Cool CD, Yeager M, et al. The pathobiology of pulmonary hypertension. Endothelium. *Clin Chest Med*. 2001;22: 405–418.
17 Stevens T, Kasper M, Cool C, Voelkel NF. Pulmonary circulation and pulmonary hypertension. In: Aird WC, ed. *Endothelium in Health and Disease*. New York: Taylor & Francis, 2005.
18 Taraseviciene SL, Scerbavicius R, Choe KH, et al. An animal model of autoimmune emphysema. *Am J Respir Crit Care Med*. 2005;171(7):734–742. Epub 2004 Nov 24.
19 Rabbany SY, Heissig B, Hattori K, Rafii S. Molecular pathways regulating mobilization of marrow-derived stem cells for tissue revascularization. *Trends Mol Med*. 2003;9:109–117.
20 Grazia T, Wescott J, Voelkel NF. Plasma vascular endothelial growth factor in end-stage emphysema [abstract]. *Am J Respir Crit Care Med*. 2002;165:A825.
21 Hoshino M, Takahashi M, Aoike N. Expression of vascular endothelial growth factor, basic fibroblast growth factor, and angiogenin immunoreactivity in asthmatic airways and its relationship to angiogenesis. *J Allergy Clin Immunol*. 2001;107:295–301.
22 Hoshino M, Nakamura Y, Hamid QA. Gene expression of vascular endothelial growth factor and its receptors and angiogenesis in bronchial asthma. *J Allergy Clin Immunol*. 2001;107:1034–1038.
23 Orsida BE, Li X, Hickey B, et al. Vascularity in asthmatic airways: relation to inhaled steroid dose. *Thorax*. 1999;54:289–295, 1999.
24 Mor F, Quintana FJ, Cohen IR. Angiogenesis-inflammation cross-talk: vascular endothelial growth factor is secreted by activated T cells and induces Th1 polarization. *J Immunol*. 2004;172: 4618–4623.
25 Mezquita P, Parghi SS, Brandvold KA, Ruddell A. Myc regulates VEGF production in B cells by stimulating initiation of VEGF mRNA translation. *Oncogene*. 2005;24:889–901.
26 Grishko V, Solomon M, Breit JF, et al. Hypoxia promotes oxidative base modifications in the pulmonary artery endothelial cell VEGF gene. *FASEB J*. 2001;15:1267–1269.
27 Belgore FM, Lip GY, Blann AD. Vascular endothelial growth factor and its receptor, Flt-1, in smokers and non-smokers. *Br J Biomed Sci*. 2000;57:207–213.

131

Endothelial Mechanotransduction in Lung

Ischemia in the Pulmonary Vasculature

Shampa Chatterjee and Aron B. Fisher

Institute for Environmental Medicine, University of Pennsylvania, Philadelphia

Ischemia (derived from the Greek *ischein,* to restrain, plus *haima,* blood) is defined as "local tissue anemia due to obstruction of the inflow of arterial blood" (1). *Anoxia* means the absence of oxygen. It has become common to equate ischemia with tissue anoxia. However, although the two conditions commonly coexist, anoxia does not always accompany ischemia, and ischemia has pathophysiological consequences in addition to anoxia. An example of the former is ischemia in the lung, in which continued ventilation provides normal oxygenation to lung cells in the absence of arterial blood flow. (The tissue oxygen [O_2] supply might actually increase, because O_2 is no longer transported away by the blood.) Nevertheless, anoxia is the major manifestation of ischemia in organs supplied by the systemic circulation. The consequences of ischemia unrelated to anoxia result from loss of the normal endothelial mechanical stimulation – shear stress. These effects have been described only recently. This chapter focuses primarily on events in the pulmonary circulation associated with alterations of endothelial function due to loss of shear stress with ischemia; these events represent altered mechanotransduction.

LUNG PERFUSION

The lung is perfused by the pulmonary artery, which carries the entire output of the right cardiac ventricle. Because the output of the left and right cardiac ventricles should be equal, perfusion to the lung is equal to the blood flow through the rest of the body. Furthermore, essentially the entire systemic blood flow returns through the venous system to the right side of the heart and thence to the lungs. As a consequence, the lung is a very vascular organ that accounts for approximately 30% of the vascular endothelial cells (ECs) of the body. (This discussion will not consider the bronchial circulation, which arises from systemic vessels and carries normally <0.5% of the cardiac output.) The pulmonary circulation can be interrupted – that is, made ischemic – by embolism, in situ thrombosis, compression by a mass, or surgical interruption. In these circumstances, ventilation to the affected lung segment normally would be expected to continue, at least until compromised by edema fluid or loss of surfactant followed by atelectasis (collapse of the lung). Thus, ischemia prior to the secondary responses would not result in decreased tissue oxygenation. This unique property of the lung makes it an ideal organ to separate the consequences of ischemia and reperfusion from the manifestations of anoxia and reoxygenation.

Lung Tissue Oxygenation

As discussed, cells of the alveolar area of the lung are normally protected from hypoxia because they derive their O_2 from the alveolar space.[1] Although alveolar hypoxia can markedly influence systemic organs through its effect on blood oxygenation, this has insignificant effects on lung cells because alveolar and intracellular Po_2 generally would remain far above the critical Po_2 required to maintain mitochondrial generation of adenosine triphosphate (ATP). In studies using the isolated lung, cellular metabolism was not impaired until alveolar Po_2 was decreased to less than 7 mm Hg (2). These low values may be obtained locally due to impaired blood flow in the presence of focal alveolar edema or alveolar collapse that prevents ventilation of that segment. The lung has been shown to be relatively resistant to the effects of prolonged anoxia, most likely because alternate sources of ATP generation, such as glycolysis, and low rates of energy utilization prevent acute alterations of function (3). The utilization of ATP by lung cells is primarily for routine "housekeeping" functions, without the major energy requirements for muscle activity that are seen with the

1 Although lungs would have higher Po_2 values when compared with other organs, endothelial Po_2 varies depending on the location of the pulmonary vessel. Lung vascular endothelial Po_2 ranges from 40 to100 mm in venous versus arterial vessels, whereas in other organs it shows the reverse (100–40 mm). Mean lung capillary Po_2 is normally approximately 80 mm, whereas it would be about 50 mm Hg in other organs. Whether lung ECs have greater antioxidant defenses compared to systemic ECs is not known.

heart. Although some functions such as surfactant secretion by lung epithelium and smooth muscle contractibility may be affected by focal anoxia, the effects of anoxia specifically on lung endothelial function are not well defined.

Lung anoxia was demonstrated to inhibit lung clearance of 5-hydroxytryptamine (serotonin), a function of the vascular endothelium; this effect was presumably due to depletion of cellular ATP (4). In endothelium from other organs, compromised O_2 delivery increases EC permeability (barrier function) by reducing the expression of tight junction proteins, in addition to promoting cell surface coagulant properties by suppressing the anticoagulant factor thrombomodulin (5,6). EC "activation" after hypoxia causes the production of proinflammatory mediators, adhesion molecule expression, and leukocyte and platelet adherence to endothelium, with attendant microvascular plugging (7). It is likely that similar events occur in the pulmonary endothelium in areas of focal ischemia. However, the major mechanism for cell injury due to anoxia appears to be the generation of reactive oxygen species (ROS) during the reperfusion period, as described in greater detail in the following sections (8,9).

Endothelial Mechanotransduction

A crucial insight into EC biology has been the realization that ECs represent a dynamic cell system. One of the important concepts of such dynamic systems is their adaptive response to physiological stimuli. In situ, vascular ECs normally are continuously exposed to the mechanical stimuli associated with blood flow. Thus, understanding the role that physical factors such as shear stress play in EC regulation is of utmost importance (10). Indeed, it is well established that hemodynamic forces generated by the flow of blood along the luminal surface regulate both the structure and function of ECs (11). Physical forces, such as pressure, stretch, and shear stress, act on the vascular wall and specifically on the endothelium, which transforms the mechanical stimuli into biochemical signals (mechanotransduction). Although stretch can arise from the pressure of blood on the vessel and affects both ECs and smooth muscle cells, shear results from the friction of blood against the vessel wall, and it is sensed only by ECs. Although the precise mechanisms of cellular mechanotransduction are yet to be elucidated, studies point to the involvement of stretch- and shear-responsive ion channels and cytoskeletal elements (11,12).

Most prior studies of mechanotransduction have utilized in vitro models in which increased shear elicited changes in EC cytoskeletal alignment, changes in ion conductances, activation of signal transduction pathways, and alterations in gene expression (12,13). It is predictable, then, that the endothelium would respond to the sudden reduction in flow during an ischemic episode. The studies described here, almost all of which are from our laboratory, demonstrate that pulmonary endothelium, along with the endothelium of other organs, responds to an acute decrease in shear stress. Nearly all studies have compared control versus complete absence of flow, which we have termed *oxygenated* or *nonhypoxic ischemia*, because oxygenation of lung endothelium is maintained during the ischemic period. Studies using graded reduction in flow indicate that effects of altered shear are not observed if the flow rate (i.e., shear) exceeds approximately 10% of the control value (14). Thus, the signaling paradigm presented here represents a response to marked (pathological) alterations in flow and would not be seen during normal physiological flow variations.

MODELS OF NORMOXIC LUNG ISCHEMIA

We have developed several models for investigating altered mechanotransduction in lung endothelium associated with the acute reduction in shear stress. Each of these models was developed to allow study of altered perfusate flow during adequate cellular oxygenation. All experiments evaluate the change from steady laminar flow; possible effects of oscillatory or turbulent flow have not been evaluated in these models.

The Isolated Rat or Mouse Lung

Isolated lung preparations have the advantage of studying an intact system while avoiding variables such as neuroendocrine effects, altered cardiac output, effects secondary to changes in peripheral organs, and other systemic manifestations that accompany in vivo models. A disadvantage of the isolated lung model is the development of pulmonary edema that limits the observation period of ischemia studies to a few hours. Global ischemia is produced by cessation of perfusion. Lungs are continuously ventilated with air (plus 5% carbon dioxide [CO_2] to maintain pH balance) during the ischemic period. Tissue ATP content measured after 1 hour of ischemia showed no change compared with continuously perfused lungs, confirming adequate oxygenation during the ischemic period (15,16). By contrast, the ventilation of lungs with nitrogen (N_2) resulted in a marked decrease in lung ATP content compatible with anoxia (17). The effect of ischemia is evaluated using techniques to monitor pleural surface fluorescence or to image the subpleural microvascular endothelium in the presence of various reporter fluorophores (18–20).

In Vitro Endothelial Cell Flow-Adaptation Systems

As described earlier, ECs in culture respond to an increase in flow and, within several days, reach a flow-adapted state. We reasoned that the flow-adapted state of cultured ECs would more closely approximate the in vivo condition and might be required to reproduce the changes seen with acute reduction in shear stress in the intact lung. ECs were cultured under flow either in an artificial capillary model or in a specially designed parallel plate apparatus (21). Studies were carried out using bovine pulmonary artery ECs (BPAECs) and rat or mouse pulmonary microvascular ECs (RPMVECs, MPMVECs). Full adaptation required 48 to 72 hours at a shear stress of approximately 5 dynes/cm^2, and was roughly similar for all cell types.

Several different systems that were tailored for specific situations were utilized.

Artificial Capillary System

The artificial capillary system consists of semipermeable polypropylene hollow fibers, approximately 200 μm in diameter, encased in a sealed cylinder (cartridge) with the ends forming the inlet and outlet ports to allow for perfusion (and exchange of oxygen and medium) (21,22). The system possesses both luminal and side ports that permit either luminal or abluminal flow. The latter allows by-pass of the lumen so that nutrients and oxygenation can be provided in the absence of cellular shear stress. ECs are seeded and grown on the inner surface of the capillaries and subjected to flow-adaptation by circulating cell culture medium through the end ports. For ischemia, the flow is switched to the side ports. Analysis of the medium samples obtained from these chambers using an O_2 electrode indicates that Po_2 with continuous flow is unaltered upon ischemia (23,24). Cells are removed from attachment to the artificial capillaries by trypsinization and then analyzed. The advantage of this model is the simulation of in vivo perfusion conditions and the relatively large number of cells that can be obtained for biochemical analysis.

Parallel Plate Chamber

The parallel plate chamber is designed for flow-adaptation of cells that are cultured on a glass slide or cover slip. One chamber with the dimensions of a spectrophotometer cuvette allows the monitoring of real-time fluorescence in cells using fluorescence probes and a spectrofluorometer (23), whereas another chamber can be used on the stage of a microscope for real-time microcopy (24). In these chambers, the Po_2 decreases to hypoxic values after 4 to 5 minutes of ischemia; saturation of the perfusate with O_2 or exposure of the medium to room air can maintain adequate oxygenation for longer periods (23). The advantage of these models is the ability to make measurements in real time without disturbing cells from their attachment to substrate.

ACUTE ENDOTHELIAL RESPONSE TO DECREASED SHEAR STRESS

Generation of Reactive Oxygen Species

Our initial studies with isolated rat lungs used pleural surface fluorometry to evaluate ROS production with acute cessation of flow. This method uses a light guide against the pleural surface to record light emitted from the cells of subpleural alveoli by various fluorophores (at specific wavelengths). Lungs were preperfused with dihydroethidine, a dye that reacts with superoxide anions to generate a fluorescent congener of ethidium that becomes trapped within cells by intercalation with DNA. With abrupt cessation of flow (ischemia), the pleural surface fluorescence signals showed a marked and relatively rapid (within minutes) increase in ethidium fluorescence, and a parallel decrease in hydroethidine fluorescence, indicating oxidation of the fluorophore due to generation of ROS (25). Localization of endothelium as the site of ROS generation was determined using a real-time high-resolution digital imaging system that permitted visualization of the subpleural microvasculature by either confocal fluorescence microscopy or with an epifluorescence microscope and commercially available (Metamorph) imaging software. The fluorescent probe for these studies was dihydrochlorofluorescein diacetate (H_2DCFDA), a ROS-reactive probe that is trapped within cells by deacetylation; oxidation by ROS forms fluorescent dichlorofluorescein (DCF). In ischemia, cellular DCF fluorescence increased, indicating ROS generation. The DCF signal colocalized with fluorescence from the EC marker fluorophore DiI-acetylated low-density lipoprotein (diI-AcLDL), indicating ECs as a site of ROS generation (18). The third fluorophore used to study ROS generation was amplex red, which fluoresces after reaction with hydrogen peroxide (H_2O_2) (in the presence of peroxidase). This dye is not taken up by the ECs, and remains in the pulmonary vascular space. Increased amplex red fluorescence with ischemia is compatible with ROS generation (Figure 131.1; for color reproduction, see Color Plate 131.1). The issue of intracellular versus extracellular generation of ROS is discussed later.

ROS generation as an acute response to ischemia also was evaluated in flow-adapted endothelium in vitro. Cells cultured under usual conditions failed to show a response to stop flow after a brief period of perfusion. Flow-adapted cells, on the other hand, showed a brisk and reproducible ROS generation, with ischemia as detected by H_2DCF oxidation. The results are compatible with oxidation by H_2O_2 arising from dismutation of $O_2^{-\cdot}$. For greater specificity, reduction of ferricytochrome-c (cyt-c) that is inhibitable by superoxide dismutase (SOD) was used to detect $O_2^{-\cdot}$. Reduction of cyt-c was observed within the initial 15 seconds of stop-flow, indicating a rapid generation of $O_2^{-\cdot}$. The rate of $O_2^{-\cdot}$ generation calculated from the rate of cyt-c reduction was 6.1 nmol/minutes/10^6 cells, or approximately 70% of the estimated rate of O_2 consumption (23). This surprisingly high rate of $O_2^{-\cdot}$ production indicates a respiratory burst that is similar in magnitude to that observed for polymorphonuclear neutrophils (PMNs). The cyt-c method reflects generation of $O_2^{-\cdot}$ into the extracellular space, because the protein probe was added to the medium, and $O_2^{-\cdot}$ crosses cell membranes relatively slowly. On the other hand, increased DCF fluorescence primarily reflects intracellular oxidation by H_2O_2. These results are compatible, because H_2O_2 arising from dismutation of $O_2^{-\cdot}$ can readily cross cell membranes. The dismutation of extracellular $O_2^{-\cdot}$ occurs spontaneously and can be catalyzed by extracellular superoxide dismutase.

Nicotinamide Adenine Dinucleotide Phosphate Oxidase as the Pathway for $O_2^{-\cdot}$ Generation with Ischemia

Video imaging of the pulmonary microvasculature in the intact lung detected increased ROS-associated fluorescence within 2 to 4 seconds after abrupt cessation of perfusion; this fluorescence continued to increase during the subsequent

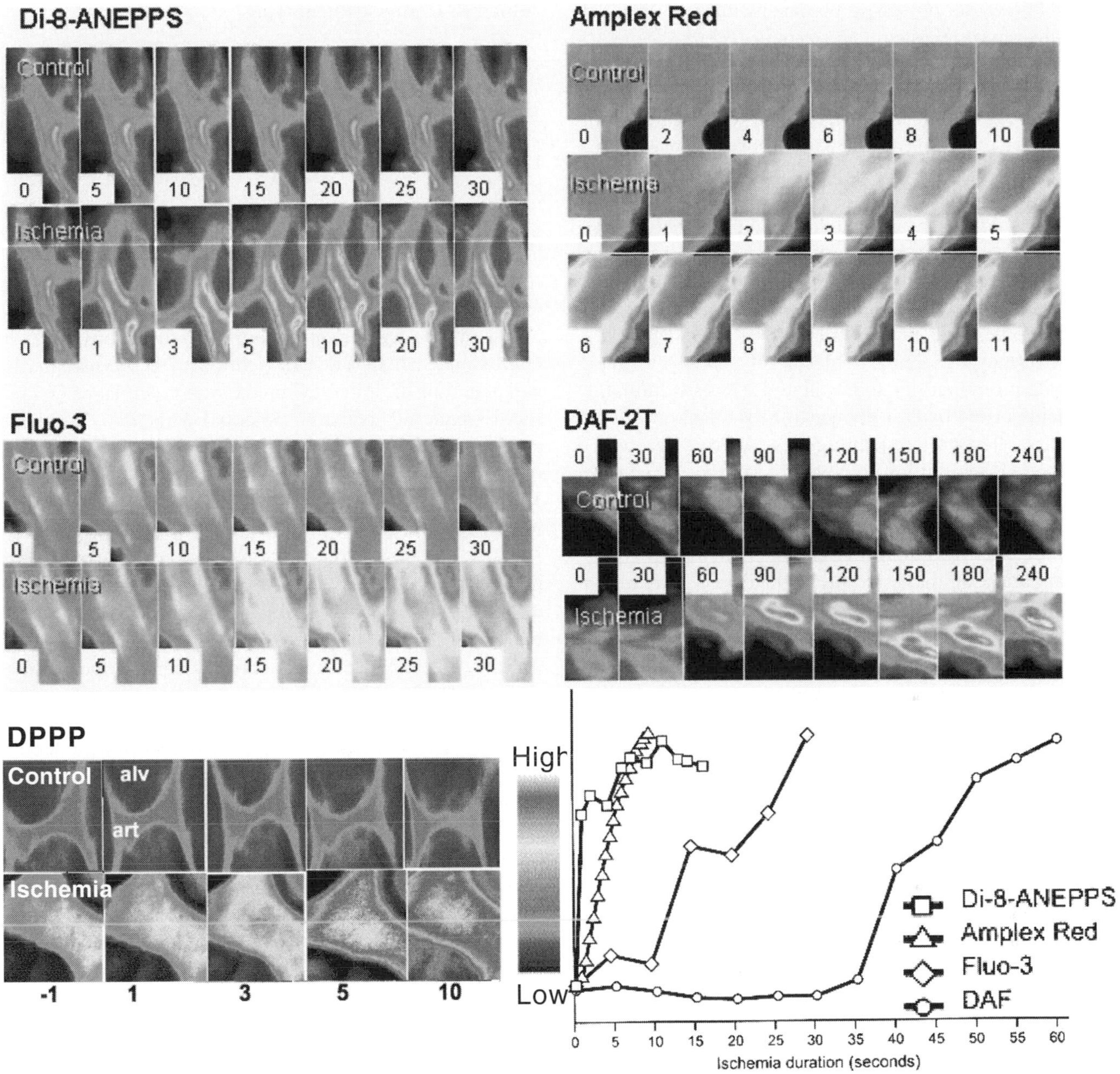

Figure 131.1. Fluorescence imaging of subpleural microvascular ECs in the isolated rat lung. Each set of images represents a control perfusion period followed by ischemia. Images are in pseudocolor. The number on each panel indicates time in seconds (minutes for DPPP), either during the control observation period or after cessation of perfusion. (*Top left*): Imaging for cell membrane potential using the fluorophore di-8-ANEPPS. (*Top right*): Imaging with Amplex red for the appearance of H_2O_2 in the intravascular space. (*Middle left*): Imaging for intracellular Ca^{2+} with fluo-3. (*Middle right*): Imaging for NO with diaminofluorescein (*DAF*). (*Bottom left*): Imaging for lipid peroxidation using the hydroperoxide probe DPPP. The preset scale indicates the pseudocolor scale used in all these images. (*Bottom right*): Timing of fluorescence changes in membrane potential, ROS generation (Amplex red), intracellular Ca^{2+} (Fluo-3), and NO generation (DAF) with ischemia, as determined by imaging techniques in the isolated rat lung. The y-axis represents arbitrary units. DAF, diaminofluorescein; DPPP, diphenyl-1-pyrenylphosphine. For color reproduction, see Color Plate 131.1.

30-minute observation period (19). Using flow-adapted cells, ROSs were produced into the extracellular space, suggesting a plasma membrane source. Allopurinol, an inhibitor of xanthine oxidase, had no effect on ROS production with ischemia, unlike its known effects in reperfusion injury. However, near total inhibition of the generation of ROS with ischemia resulted from the flavoprotein inhibitor, diphenyleneiodonium (DPI). This agent has been widely used as an inhibitor of the nicotinamide adenine dinucleotide phosphate (NADPH) oxidase pathway of ROS production in phagocytic cells.

ECs are known to possess a plasma membrane NADPH oxidase- like enzyme system that resembles the $O_2^{-\cdot}$ generating enzyme complex of PMNs (18,26–28). Mice with knockout of gp91^{phox} (*gp91*$^{phox-/-}$), the membrane flavoprotein component of NADPH oxidase in PMN, failed to generate ROS with lung ischemia, providing additional evidence that phagocyte-type NADPH oxidase is responsible for endothelial ROS generation with lung ischemia (18).

In PMNs, ROS generation requires the assembly of the NADPH oxidase complex by translocation of at least four cytosolic protein components (p47^{phox}, p67^{phox}, p40^{phox}, p21rac1) to the plasma membrane, where they associate with the cytochrome b558 heterodimer (gp91^{phox} and p22^{phox}) (29,30). Although the precise details of NADPH oxidase assembly in pulmonary ECs are still under investigation, the components of the oxidase appear to be similar to those in PMNs; other flavoprotein components (the NOX proteins) may be present in addition to gp91^{phox}, but that remains to be determined. PR-39, an inhibitor of src homology (SH)-3 groups that blocks NADPH oxidase assembly in PMNs, also inhibited ROS generation by lung endothelium in response to ischemia, indicating that assembly of the oxidase is required for maximal rates of $O_2^{-\cdot}$ production (18). In PMNs, the transmembrane protein gp91^{phox} transfers electrons across the plasma membrane to generate superoxide anions in the extracellular space. Extracellular generation of $O_2^{-\cdot}$ with ischemia also occurs with flow-adapted ECs in culture, as indicated by our observations using exogenous ferricytochrome-c.

Membrane Depolarization Is the Activating Signal for Reactive Oxygen Species Generation

Several previous observations led us to investigate changes in membrane potential as the initiating event for activation of NADPH oxidase. First, ion channel events leading to alteration in membrane potential are among the earliest responses to onset of flow. Second, local changes in membrane potential have been associated with NADPH oxidase activation in PMNs, although this issue is complicated by the response of the proton pump to extracellular transfer of electrons. To study membrane potential in intact lungs, they were preperfused with a membrane potential-sensitive dye, bis-oxonol or di-8-ANEPPS. These dyes localize in the plasma membrane where, in the depolarized state, increased rigidity of the bilayer results in better dye alignment and higher fluorescence yield. The monitoring of EC fluorescence from the pleural surface in real time revealed that cessation of flow caused membrane depolarization, as evidenced by the increase in the bis-oxonol fluorescent signal (16). For video imaging, the faster response dye, di-8-ANEPPS, was used (Figure 131.1). Rapid sampling showed a gradual increase in fluorescence within the first 1 to 2 seconds that then stabilized (19). Thus, the change in membrane potential appeared to precede activation of NADPH oxidase. Additional evidence that change in membrane potential could initiate ROS generation was obtained by perfusing lungs and treating cells with high potassium (K^+), which resulted in membrane depolarization and ROS generation. Perfusion of the intact lung or treatment of flow-adapted ECs with increasing concentrations of K^+ showed that flow cessation results in EC membrane depolarization equivalent to that observed with approximately 12 mM KCl. In accord with the Nernst equation and assuming an initial EC membrane potential of –65 mV, the decrease in membrane potential with flow cessation is approximately 20 mV (19).

Table 131-1: Mouse Phenotypes

Gene	*Survival of Null Mutant*	*Phenotype*
K_{IR}6.2	Viable	Glucose intolerance
gp91^{phox}	Viable	Abnormalities in host defense and inflammation

K_{ATP} Channels Are Responsible for Endothelial Cell Membrane Depolarization with Ischemia

Depolarization of the pulmonary microvascular endothelium with K^+ indicates that K^+ channels are, at least in part, responsible for the resting cell membrane potential. Based on inhibitor/agonist studies, we postulated that the K_{ATP} channel is an important determinant of cell membrane polarization and the subsequent ischemic response. K_{ATP} channels are composed of an inwardly rectifier pore–forming protein (K_{IR}6) and a regulatory subunit, a sulfonylurea receptor (SUR). The presence of K_{IR}6.2 and SUR1 subunits has been identified in rat and bovine pulmonary vascular ECs (Table 131-1). Pretreatment of the lungs with cromakalim (or its L-isomer, lemakalim), a K_{ATP} channel agonist, blocked membrane depolarization in ischemia (19) and prevented ROS production. Conversely, perfusing lungs with glyburide, a K_{ATP} antagonist that binds to the SUR, caused depolarization and ROS production during continued flow (i.e., in the absence of ischemia) (16). These results indicate that channel closure leads to membrane depolarization, which results in ROS generation. Lungs and ECs from mice with knockout of K_{IR}6.2 (*K_{IR}6.2*$^{-/-}$) showed a significantly diminished response to ischemia, with a lesser change in membrane potential and lesser ROS production compared with wild-type mice (31,32). We conclude that K^+ channel inactivation resulting in membrane depolarization is the earliest event following cessation of flow, and that this event is linked to ROS production through activation of membrane-localized NADPH oxidase.

Nitric Oxide Generation with Ischemia

Nitric oxide (NO) generation was demonstrated as a component of the initial response to experimental ischemia using the fluorophore DAF-2 DA. This dye is de-esterified intracellularly to DAF-2, which reacts with NO and its higher oxides to form the fluorescent DAF-2T. ECs of in situ subpleural microvessels in rat lungs showed an initial increase in DAF-2T fluorescence at about 45 seconds of ischemia, with a progressive increase over the subsequent 5 to 10 minutes (see Figure 131.1). Specific cellular localization of DAF-2T was confirmed by colocalization with the endothelial marker DiI-AcLDL (33). Abrogation of the fluorescent signal upon ischemia by pretreatment with a NO synthase (NOS) inhibitor indicated specificity of the signal for NO (33).

ECs possess at least two isoforms of NOS: endothelial NOS (eNOS) and inducible NOS (iNOS) (34). eNOS requires increased levels of calcium (Ca^{2+}) and calmodulin binding for its dissociation from caveolin and activation, whereas iNOS is Ca^{2+} independent. That eNOS is responsible for NO generation with ischemia was clear when the use of Ca^{2+}-free medium or the presence of a calmodulin inhibitor suppressed the increase in DAF-2T fluorescence with ischemia (33). Disruption of caveolae by pretreatment of cells with cyclodextrin or filipin to alter cholesterol-rich domains also inhibited endothelial NO generation, indicating that eNOS activation with ischemia requires a signal from caveolae as well as the increase in Ca^{2+}/calmodulin (22).

Intracellular Ca^{2+} Elevation with Ischemia

The effect of Ca^{2+} depletion on NO generation suggests that intracellular Ca^{2+} is increased with ischemia. Lungs were imaged in situ using fluorophores to monitor changes of intracellular Ca^{2+} in pulmonary endothelium during the ischemic episode (20,33,35). Ischemia led to increased Ca^{2+} in ECs in situ that was observed at approximately 10 to 15 seconds after the stoppage of flow, and reached a plateau at approximately 5 minutes (see Figure 131.1). Simultaneous monitoring of the Ca^{2+}- and NO-sensitive fluorophores confirmed that the Ca^{2+} increase with ischemia preceded NO generation. Temporally, the increase of intracellular Ca^{2+} followed the change in cell membrane potential, and an agent (cromakalim) that blocked membrane depolarization with ischemia also blocked Ca^{2+} increase.

Source of Intracellular Ca^{2+}

Increased intracellular Ca^{2+} with ischemia was partially prevented by perfusion with Ca^{2+}-free medium or, to a lesser extent, by pretreatment with thapsigargin, an agent that leads to the depletion of Ca^{2+} from intracellular stores through inhibition of Ca^{2+}-ATPase of endoplasmic reticulum (19,20,33). These results indicate that the ischemia-mediated increase of intracellular Ca^{2+} is due primarily to Ca^{2+} influx from both its intracellular release and Ca^{2+} influx from the extracellular medium (20,33,35).

Voltage-Gated Ca^{2+} Channels

Voltage-gated Ca^{2+} channels (VGCC) can mediate calcium influx in response to membrane depolarization (36). The presence of VGCCs has been shown in freshly isolated microvascular ECs from bovine adrenal medulla (37–39), rat brain (40), and rat lung (41), although it has been difficult to demonstrate functional channels in passaged ECs in culture. RPMVECs express both the α-1G and β3 subunits of T-type VGCC, and the expression of these proteins was significantly increased upon flow-adaptation (42). T-type Ca^{2+} channels open after a relatively small depolarization (10–20 mV), unlike L-type and other high-voltage Ca^{2+} channels, which require a larger change in membrane potential (43,44). Thus, the change in membrane potential of approximately 15 to 20 mV with ischemia in the intact lung (35) and flow-adapted ECs (42) is compatible with the involvement of T-type Ca^{2+} channels in the ischemic response. This role is supported by inhibition of Ca^{2+} influx in the presence of a T-type VGCC inhibitor (42).

Flow Adaptation and the Response to Altered Shear

Pulmonary vascular ECs in situ are constantly exposed to shear forces, and presumably are adapted to these conditions. It seems intuitive that cells cultured under static conditions, as is the usual procedure, would lose this state of adaptation. Thus, cells in culture respond to imposition of shear but would not be expected to respond to cessation of flow. We evaluated this presumption by measuring the response of ECs to ischemia as a function of duration of shear exposure in culture. Exposure of routinely (i.e., statically) cultured ECs to flow for a brief (~1/2 hour) period failed to elicit membrane depolarization or ROS generation when flow was discontinued. If cells were kept under flow for a more prolonged period (i.e., sufficiently long for flow-adaptation), both membrane depolarization and ROS generation were observed with ischemia, as described for the intact lung (21,23). A partial response to simulated ischemia was observed after 16 to 24 hours of exposure to shear, whereas a maximal response required flow-adaptation for 48 to 72 hours. Similar results for flow-adapted cells were obtained for bovine pulmonary artery ECs BPAECs, RPMVECs, and MPMVECs (21,23). K_{ATP} channel expression with flow-adaptation increased at the mRNA and protein levels (Figure 131.2; for color reproduction, see Color Plate 131.2) as did the inwardly rectified K^+-current. This induction supports the notion that K_{ATP} channels are important for the response to simulated ischemia. The failure of statically cultured cells to show membrane depolarization and ROS generation with simulated ischemia can be explained in part by their low level of K_{ATP} channel expression.

ENDOTHELIAL CELL SIGNALING IN RESPONSE TO ISCHEMIA

Recent studies have provided convincing evidence that ROS generated in low amounts during physiological perturbation act as second messengers to initiate a signaling cascade. Thus, the generation of ROS by the endothelium in response to altered shear stress suggests a signaling function. To examine the expression of signaling molecules, events subsequent to the initial response to ischemia were studied. These studies utilized primarily the in vitro artificial capillary system, which allows for a prolonged period of simulated ischemia without affecting the O_2 and nutrient supply.

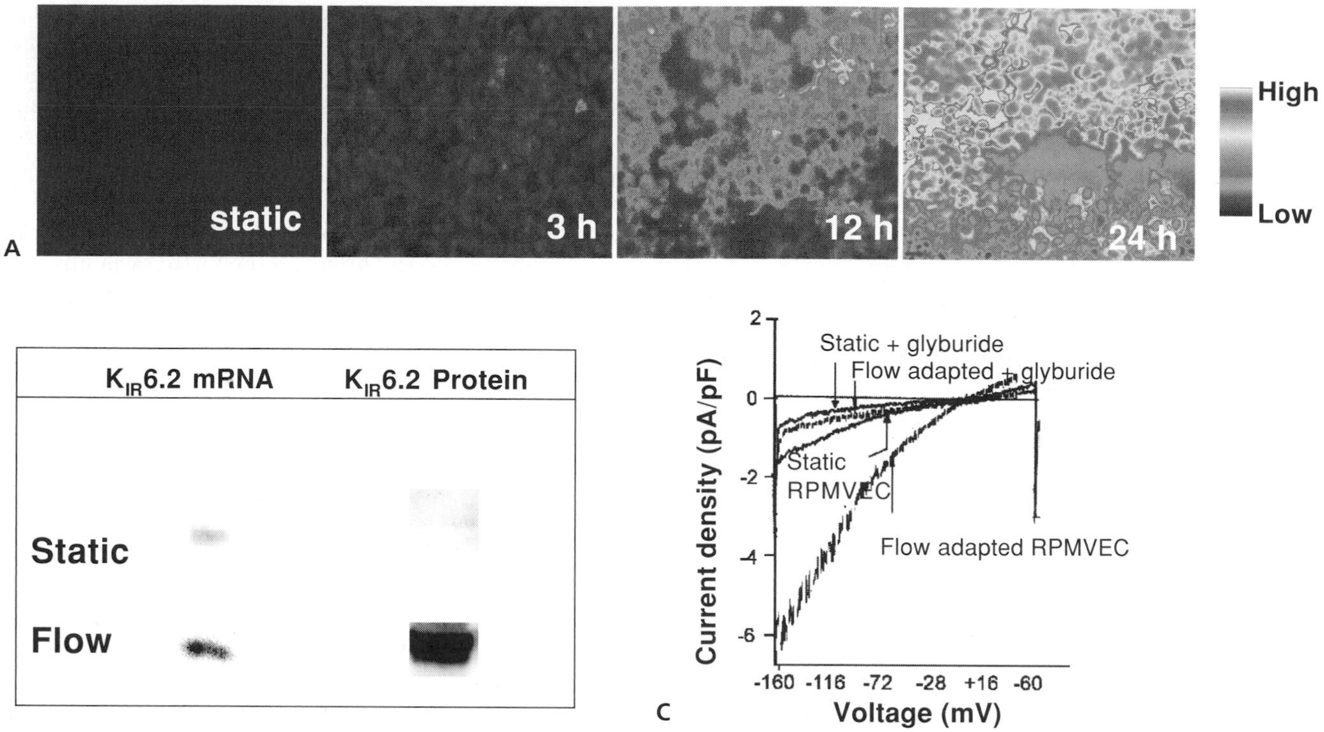

Figure 131.2. Induction of K_{ATP} channels during flow-adaptation. (**A**) Increase in fluorescence in cells flow-adapted at 10 dyn/cm^2 for varying periods. Cells were labeled with fluorescently labeled glyburide (BODIPY-glyburide, 50 nM). The resulting fluorescence indicating binding of glyburide to the sulfonylurea receptor (SUR) was observed with a microscope. (**B**) Representative blots of $K_{IR}6.2$ mRNA and protein content of RPMVECs cultured under static conditions or adapted to flow (10 dyn/cm^2 for 24 h). Total RNA was extracted, absorbed as a dot on a nitrocellulose membrane, and hybridized with ^{32}P-labeled $K_{IR}6.2$ cDNA. Protein was analyzed by Western blot using polyclonal antibodies to the COOH terminus of $K_{IR}6.2$. (**C**) Inwardly rectifying whole cell K^+ currents (K_{IR}) in RPMVECs. Representative recordings obtained from static (no flow) cells and cells adapted to flow at a shear stress of 10 dyn/cm^2 for 24 hours. Increased current is seen for flow-adapted compared with static cells. Glyburide (K_{ATP} blocker) completely abolished the increased current. For color reproduction, see Color Plate 131.2.

Sensing of Altered Shear

As defined earlier, ischemia entails a loss of blood flow and, with respect to mechanotransduction, equates to a loss of shear force. Because the cell membrane forms a boundary with the flowing blood, it seems reasonable to predict that sensing of altered shear might be elicited through membrane-anchored structures. Indeed, structures such as the specialized invaginated microdomains on EC membranes (i.e., caveolae) have been reported to be shear sensitive (45). Caveolae are enriched with cholesterol and glycosphingolipids and contain caveolin as a principal component (46).

The function of these organelles is altered by depleting membrane cholesterol through treatment with filipin or cyclodextrin (membrane-permeable and impermeable compounds, respectively). The role of membrane cholesterol in mechanotransduction may be related to the regulation of membrane fluidity, which may affect the function of membrane proteins (47,48). As described earlier, flow-adapted BPAECs depleted of cholesterol and subjected to ischemia showed decreased NO production with ischemia, indicating that NOS activation is coupled to caveolar microdomains. Our recent studies with *caveolin-1*–null mouse microvascular ECs have shown loss of the characteristic response to loss of shear. Thus, caveolae appear to serve either as a primary flow sensor or as a mechanism to integrate signals from other cellular components such as the cytoskeleton.

Activation of Signaling Cascades

Signaling by low levels of ROS is mediated through activation of mitogen-activated protein kinase (MAPK) pathways. The terminal kinases in these pathways are p38, extracellular-regulated-kinase (ERK), and c-jun N-terminal kinase (JNK). In flow-adapted ECs, ischemia resulted in phosphorylation of ERK1 and -2, indicating activation of this kinase (22). Phosphorylation was observed during the first 10 minutes of ischemia and reached a plateau at 20 to 30 minutes. Inhibitors of ROS generation suppressed ERK phosphorylation, confirming that it was a response to the oxidant signal. ERK phosphorylation in this model also was suppressed by disruption of caveolae. Thus, ERK1/2 activation in ischemia requires both ROS production and a signal from caveolae. Subsequent activation of NOS requires both ERK1/2 activation and increased Ca^{2+}/calmodulin (22). Activation of the transcription factor activator protein (AP)-1 with ischemia (see below) suggests activation of JNK, but this was not evaluated directly.

Activation of Transcription Factors and Cell Proliferation

Having established that ischemia in flow-adapted ECs activates signaling pathways, the subsequent downstream events were evaluated. Activation of eNOS has been described. Analysis of nuclear extracts from flow-adapted ECs by electrophoretic mobility shift assay indicated a nuclear translocation of nuclear factor (NF)-κB components (p65 and p50) when studied at 1 hour of ischemia. Nuclear translocation also was demonstrated for the c-Jun/c-fos subunits of AP-1 (49). Activation of both NF-κB and AP-1 was abolished by pretreatment of cells with inhibitors of ROS generation, indicating that their activation is linked to ROS-mediated signaling pathways (49).

ROS can mediate Ras-induced cell cycle progression and mammalian cell proliferation (50), an effect that may be mediated in part through activation of AP-1. After 24 hours of ischemia, flow-adapted BPAECs demonstrated an increase in ^{3}H-thymidine incorporation and an increase in the percent of cells in the proliferative phases of the cell cycle (49). In flow-adapted RPMVECs, ischemia caused a 2.5-fold increase in the cellular proliferation index and a threefold increase in the percentage of cells in the S plus G2/M phases (51). These results indicate increased DNA synthesis, cell division, and cell proliferation in response to ischemia. Proliferation after ischemia was abolished by inhibitors of cell membrane depolarization and ROS generation, indicating that proliferation results from signaling associated with ROS. Control studies eliminated the possibilities that proliferation with the ischemic model was due to loss of contact inhibition or a response to cell death (51). We postulate that ROS stimulates proliferation through a signaling cascade, as a response to cessation of flow and in an effort at neovascularization.

ISCHEMIA-INDUCED OXIDATIVE STRESS

Although ROS generation in low amounts is considered a physiological signaling mechanism, excess generation can promote cell injury due to oxidation of cellular biomolecules.

Lung Injury with Oxygenated Lung Ischemia

As described earlier, ROS are generated by pulmonary endothelium during ischemia in the presence of adequate cellular oxygenation. The primary source of ROS is the EC membrane NADPH oxidase that generates $O_2^{-\cdot}$. Dismutation produces H_2O_2, which can react with $O_2^{-\cdot}$ (especially in the presence of Fe^{2+}) to generate the hydroxyl radical ($\cdot$OH). This potent oxidant can interact rapidly with cellular biomolecules and result in the oxidation of lipids, protein, and DNA.

We have used several techniques to evaluate ischemia-induced tissue oxidative injury at the biochemical level. A recently developed imaging technique permitted the detection of lipid peroxidation with ischemia in the endothelium of the continuously ventilated isolated rat lung. The fluorophore diphenyl-2-pyrenylphosphine (DPPP) localizes in the plasma membrane and responds specifically to lipid hydroperoxides (52). Lipid peroxidation was detected at 3 minutes of ischemia using this highly sensitive method (53). Analysis of lipid peroxidation in lung homogenates by measurement of thiobarbituric acid reactive substances (TBARS) and conjugated dienes is less specific and less sensitive, in part because the background includes all cells and not just the endothelium. Using these methods, lipid peroxidation was detected at 15 to 30 minutes of ischemia, with progression at 60 minutes (21). A parallel increase in lung protein carbonyls with time of ischemia indicated protein oxidation (54). These oxidative changes were blocked by scavengers of ROS.

The presence of free Fe^{2+} greatly accelerates the rate of oxidative injury through the facilitation of $\cdot$OH generation via Fenton chemistry. Using chemical assay, free Fe^{2+} in lung homogenates subjected to ischemia increased 45% in 15 minutes and 3.5-fold (to 0.3 mmol/mg protein) in 30 minutes (55,56). Under fluorescence imaging, pulmonary ECs were the predominant site of the increase in free iron (Fe^{2+}) (56). The increase in Fe^{2+} depended on EC membrane depolarization and increased intracellular Ca^{2+}, but was independent of ROS generation. The increase in Fe^{2+} with ischemia presumably reflects its release from labile intracellular stores. The presence of Fe^{2+} chelators markedly inhibited ischemia-induced lipid peroxidation (56). Thus, oxidative injury to the pulmonary endothelium results from toxic radicals generated by the interaction of Fe^{2+} with ROS produced by NADPH oxidase. In vivo, the antioxidant capacity of the blood presumably delays (compared to isolated lungs perfused with synthetic medium) the oxidative injury associated with ischemia-mediated ROS generation.

Reperfusion-Mediated Lung Injury

Reperfusion injury has been described as the tissue damage that occurs with reperfusion following a period of ischemia. The commonly accepted mechanism is the generation of ROS associated with tissue reoxygenation (obviously, ROS cannot be produced during tissue anoxia, because some O_2 is required, by definition). This paradigm generally is called ischemia–reperfusion injury, although it would be more appropriately designated as injury due to anoxia–reoxygenation. The most widely studied biochemical pathway for ROS generation with reoxygenation is through xanthine oxidase, which arises during anoxia through proteolytic cleavage of xanthine dehydrogenase. The purine substrates for xanthine oxidase are generated during anoxia by breakdown of ATP and its metabolic products; O_2 is provided during the reperfusion phase. Anoxia–reoxygenation with continued perfusion can be studied by ventilating isolated lungs with O_2-free gas. (This experiment would be difficult to interpret in an intact animal because of the systemic effects of the resultant hypoxia.) Anoxia–reoxygenation in isolated rat lungs resulted in ROS generation, with resultant oxidation of lipid and protein components. These effects were blocked by a xanthine oxidase inhibitor (17).

The possibility of reperfusion injury associated with restoration of shear stress, as contrasted with reoxygenation, has been investigated using perfused rat lung, although studies are few and interpretation is unclear (57–60). Reperfusion of rat lung following normoxic ischemia for 30 to 60 minutes had no significant effect on lung lipid or protein oxidation beyond that observed during the ischemia period, but did result in release of lactate dehydrogenase, shedding of angiotensin-converting enzyme (ACE) from pulmonary endothelium, and lung fluid accumulation (57). These effects could have been a consequence of injury sustained during the ischemic phase of the experiments.

Ischemia–reperfusion in lungs of intact animals has been studied using various models including clamping of the lung hilus (61,62), ligation of a pulmonary artery (63,64), or lung transplantation (65). Lung injury has been observed, although the contribution of the pulmonary endothelium and the respective roles of anoxia–reoxygenation versus ischemia–reperfusion are unclear.

NORMOXIC ISCHEMIA IN SYSTEMIC VASCULAR ENDOTHELIUM

The response of pulmonary endothelium to ischemia raised the question whether similar effects occurred in other vascular beds. K^+-induced depolarization in human umbilical vein ECs (HUVECs) resulted in tyrosine kinase activation, rac translocation to the cell membrane, and $O_2^{-\cdot}$ generation, indicating an analogous response to that observed in pulmonary endothelium (66). ROS production has been demonstrated during ischemia in the heart and intestine, although this was a sidelight to the major focus (reperfusion) of the studies (67,68). ROS production in these organs occurred during the early phase of ischemia but subsequently was inhibited by lack of O_2. We have evaluated the role of ischemia in aortic endothelium using techniques similar to those described for the lung (24). Obviously, zero flow in the aorta is not compatible with life, but this preparation was used for extrapolation to other systemic vascular beds. An isolated aorta preparation consisting of a longitudinal section of freshly dissected rat aorta fixed into a flow chamber with the endothelial layer facing up was subjected to a 1-hour period of flow (to give a shear stress of 10 dynes/cm^2 for 1 hour) to reestablish flow-adaptation. Both this isolated aorta preparation and flow-adapted aortic ECs in culture showed cell membrane depolarization and ROS generation with ischemia, a response similar to that observed in pulmonary endothelium (24). As with pulmonary endothelium, membrane depolarization was associated with altered K_{ATP} channel activity, and ROS generation required the activation of NADPH oxidase. Gradual lowering of shear in the reflow-adapted aortas showed no significant change in ROS generation until shear was reduced by approximately 95% from the level used for reflow-adaptation (24). This indicates that the ischemic response is not proportional to the change in shear but rather is threshold-dependent, as observed also for the perfused rat lung (14).

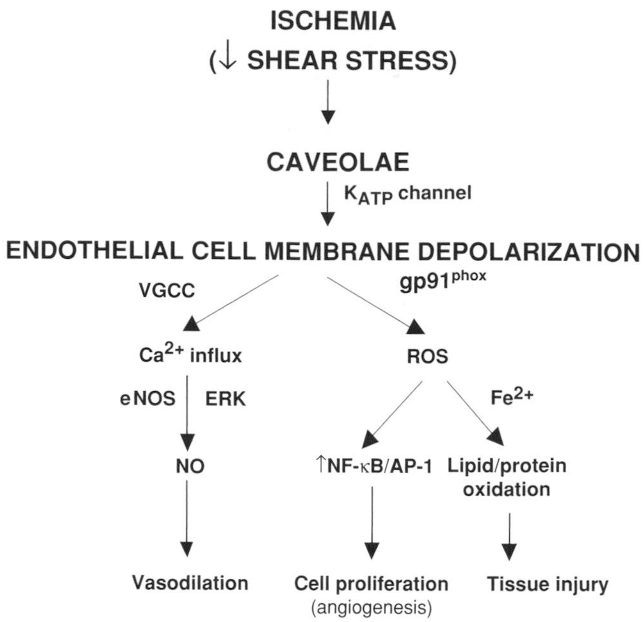

Figure 131.3. Proposed pathways for NO generation, cell proliferation, and tissue injury with ischemia. Loss of shear stress due to flow cessation is sensed by the EC, resulting in activation of NADPH oxidase, ROS generation, and elevated intracellular Ca^{2+}. ROS-dependent activation (phosphorylation) of ERK1/2 leads to Ca^{2+}-dependent activation of endothelial NO synthase and NO generation. The cell signaling cascade results in EC proliferation. NO generation, and cell proliferation might represent mechanisms to restore blood flow. Release of free Fe^{2+} in the presence of ROS results in oxidative stress with oxidation of cellular lipid and protein components.

Thus, these results indicate that the endothelial response to cessation of flow is not unique or limited to the pulmonary endothelium, and ROS generation could occur in other vascular beds as long as tissue Po_2 levels are adequate for NADPH oxidase activity. Although ischemia-mediated generation of ROS, and probably NO, requires the presence of O_2, membrane depolarization and elevated intracellular Ca^{2+} do not. Both of the latter have been observed in many tissues with ischemia, although careful study is required to dissect the respective roles of altered shear stress versus decreased tissue ATP in their genesis.

KEY POINTS

- Overall, the studies described in this chapter provide a new paradigm for ischemia, in which the loss of shear or the mechanical component of flow initiates an endothelial response (Figure 131.3).

- Our results show that cessation of flow triggers rapid depolarization (within 1–2 seconds) of the EC plasma membrane, caused by deactivation of K_{ATP} channels.
- This is followed closely by activation of membrane-localized NADPH oxidase and ROS generation. Subsequent events include release of Ca^{2+} from intracellular stores and opening of T-type voltage dependent Ca^{2+} channels, resulting in increased intracellular Ca^{2+}, caveolar-dependent phosphorylation of ERK1/2, activation of endothelial NO synthase, and NO production.
- The increased production of ROS activates several EC transcription factors (among them NF-κB and AP-1) and leads to increased EC division and proliferation.
- These responses to altered shear stress are not unique to the pulmonary circulation. They occur in aortic endothelium and possibly other vascular beds during the early phase of ischemia before O_2 is depleted. This response to ischemia may serve the physiologic purpose of promoting vasodilatation by the generation of NO and stimulating neovascularization to re-establish the impeded blood flow.

ACKNOWLEDGMENTS

We thank our many colleagues and collaborators who participated in these studies, especially A.B. Al-Mehdi, Y. Manevich, Z. Wei, and I. Matsuzaki. We especially thank T. Milovanova and J. Rossi for assistance with the manuscript.

REFERENCES

1 Merriam-Webster. ischemia (defined). In: Gove PB, ed. *Webster's Third New International Dictionary of the English Language, Unabridged*. Springfield, MA: Merriam-Webster Inc. Publishers; 1986:1197.

2 Fisher AB, Dodia C. Lung as a model for evaluation of critical intracellular PO2 and PCO. *Am J Physiol*. 1981;241: E47–E50.

3 Fisher AB, Hyde RW, Reif JS. Insensitivity of the alveolar septum to local hypoxia. *Am J Physiol*. 1972;223:770–776.

4 Steinberg H, Bassett DJ, Fisher AB. Depression of pulmonary 5-hydroxytryptamine uptake by metabolic inhibitors. *Am J Physiol*. 1975;228:1298–1303.

5 Fischer S, Wobben M, Marti HH, et al. Hypoxia-induced hyperpermeability in brain microvessel endothelial cells involves VEGF-mediated changes in the expression of zonula occludens-1. *Microvasc Res*. 2002;63:70–80.

6 Ogawa S, Gerlach H, Esposito C, et al. Hypoxia modulates the barrier and coagulant function of cultured bovine endothelium. Increased monolayer permeability and induction of procoagulant properties. *J Clin Invest*. 1990;85:1090–1098.

7 del Zoppo GJ, Mabuchi T. Cerebral microvessel responses to focal ischemia. *J Cereb Blood Flow Metab*. 2003;23:879–894.

8 Kennedy TP, Rao NV, Hopkins C, et al. Role of reactive oxygen species in reperfusion injury of the rabbit lung. *J Clin Invest*. 1989; 83:1326–1335.

9 Fisher PW, Huang YC, Kennedy TP, Piantadosi CA. PO2-dependent hydroxyl radical production during ischemia-reperfusion lung injury. *Am J Physiol*. 1993;265:L279–L285.

10 Szocs K. Endothelial dysfunction and reactive oxygen species production in ischemia/reperfusion and nitrate tolerance. *Gen Physiol Biophys*. 2004;23:265–295.

11 Barakat AI, Davies PF. Mechanisms of shear stress transmission and transduction in endothelial cells. *Chest*. 1998;114: S58–S63.

12 Davies PF, Tripathi SC. Mechanical stress mechanisms and the cell. An endothelial paradigm. *Circ Res*. 1993;72:239–245.

13 Davies PF, Barbee KA, Volin MV, et al. Spatial relationships in early signaling events of flow-mediated endothelial mechanotransduction. *Annu Rev Physiol*. 1997;59:527–549.

14 Al-Mehdi AB, Zhao G, Fisher AB. ATP-independent membrane depolarization with ischemia in the oxygen-ventilated isolated rat lung. *Am J Respir Cell Mol Biol*. 1998;18:653–661.

15 Fisher AB, Dodia C, Tan ZT, et al. Oxygen-dependent lipid peroxidation during lung ischemia. *J Clin Invest*. 1991;88: 674–679.

16 al-Mehdi AB, Shuman H, Fisher AB. Oxidant generation with $K(^{+})$-induced depolarization in the isolated perfused lung. *Free Radic Biol Med*. 1997;23:47–56.

17 Zhao G, al-Mehdi AB, Fisher AB. Anoxia-reoxygenation versus ischemia in isolated rat lungs. *Am J Physiol*. 1997;273:L1112–L1117.

18 Al-Mehdi AB, Zhao G, Dodia C, et al. Endothelial NADPH oxidase as the source of oxidants in lungs exposed to ischemia or high K^{+}. *Circ Res*. 1998;83:730–737.

19 Song C, Al-Mehdi AB, Fisher AB. An immediate endothelial cell signaling response to lung ischemia. *Am J Physiol Lung Cell Mol Physiol*. 2001;281:L993–L1000.

20 Tozawa K, al-Mehdi AB, Muzykantov V, Fisher AB. In situ imaging of intracellular calcium with ischemia in lung subpleural microvascular endothelial cells. *Antioxid Redox Signal*. 1999;1:145–154.

21 Chatterjee S, Al-Mehdi AB, Levitan I, et al. Shear stress increases expression of a KATP channel in rat and bovine pulmonary vascular endothelial cells. *Am J Physiol Cell Physiol*. 2003;285:C959–C967.

22 Wei Z, Al-Mehdi AB, Fisher AB. Signaling pathway for nitric oxide generation with simulated ischemia in flow-adapted endothelial cells. *Am J Physiol Heart Circ Physiol*. 2001;281:H2226–H2232.

23 Manevich Y, Al-Mehdi A, Muzykantov V, Fisher AB. Oxidative burst and NO generation as initial response to ischemia in flow-adapted endothelial cells. *Am J Physiol Heart Circ Physiol*. 2001;280:H2126–H2135.

24 Matsuzaki I, Chatterjee S, Debolt K, et al. Membrane depolarization and NADPH oxidase activation in aortic endothelium during ischemia reflect altered mechanotransduction. *Am J Physiol Heart Circ Physiol*. 2005;288:H336–H343.

25 Al-Mehdi AB, Shuman H, Fisher AB. Intracellular generation of reactive oxygen species during nonhypoxic lung ischemia. *Am J Physiol.* 1997;272:L294–L300.

26 Zulueta JJ, Sawhney R, Yu FS, et al. Intracellular generation of reactive oxygen species in endothelial cells exposed to anoxia-reoxygenation. *Am J Physiol.* 1997;272:L897–L902.

27 Jones SA, O'Donnell VB, Wood JD, et al. Expression of phagocyte NADPH oxidase components in human endothelial cells. *Am J Physiol.* 1996;271:H1626–H1634.

28 Babior BM. The NADPH oxidase of endothelial cells. *IUBMB Life.* 2000;50:267–269.

29 DeLeo FR, Quinn MT. Assembly of the phagocyte NADPH oxidase: molecular interaction of oxidase proteins. *J Leukoc Biol.* 1996;60:677–691.

30 Babior BM, Lambeth JD, Nauseef W. The neutrophil NADPH oxidase. *Arch Biochem Biophys.* 2002;397:342–344.

31 Chatterjee S, Levitan I, Wei Z, et al. KATP channel in the pulmonary microvasculature is an important component of the shear stress sensing mechanism. In: Marchesi VT, Veilleux CS, Goetzl EJ, et al., eds. *Experimental Biology.* Washington, DC: The Federation of American Societies for Experimental Biology; 2004:A329.

32 Zhang Q, Matsuzaki I, Chatterjee S, Fisher AB. Mice with kir6.2-knockout are resistant to endothelial cell reactive oxygen species (ROS) generation during normoxic lung ischemia. In: Marchesi VT, Veilleux CS, Goetzl EJ, et al., eds. *Experimental Biology.* Washington, DC: The Federation of American Societies for Experimental Biology;2004:A329.

33 Al-Mehdi AB, Song C, Tozawa K, Fisher AB. Ca^{2+}- and phosphatidylinositol 3-kinase-dependent nitric oxide generation in lung endothelial cells in situ with ischemia. *J Biol Chem.* 2000;275:39807–39810.

34 Ricciardolo FL, Sterk PJ, Gaston B, Folkerts G. Nitric oxide in health and disease of the respiratory system. *Physiol Rev.* 2004; 84:731–765.

35 Song C, Al-Mehdi AB, Fisher AB. An immediate endothelial cell signaling response to lung ischemia. *Am J Physiol Lung Cell Mol Physiol.* 2001;281:L993–L1000. Erratum: *Am J Physiol Lung Cell Mol Physiol* 2002;282(2):preceding L167.

36 Catterall WA. Structure and regulation of voltage-gated Ca^{2+} channels. *Annu Rev Cell Dev Biol.* 2000;16:521–555.

37 Bossu JL, De Waard M, Feltz A. Two types of calcium channels are expressed in adult bovine chromaffin cells. *J Physiol.* 1991; 437:621–634.

38 Bossu JL, Elhamdani A, Feltz A, et al. Voltage-gated Ca entry in isolated bovine capillary endothelial cells: evidence of a new type of BAY K 8644-sensitive channel. *Pflugers Arch.* 1992;420:200–207.

39 Vinet R, Vargas FF. L- and T-type voltage-gated Ca^{2+} currents in adrenal medulla endothelial cells. *Am J Physiol.* 1999;276:H1313–H1322.

40 Delpiano MA. ATP-dependent K^{+} and voltage-gated Ca^{2+} channels in endothelial cells of brain capillaries. Effect of hypoxia. *Adv Exp Med Biol.* 2000;475:435–440.

41 Wu S, Haynes J Jr., Taylor JT, et al. Cav3.1 (alpha1G) T-type Ca^{2+} channels mediate vaso-occlusion of sickled erythrocytes in lung microcirculation. *Circ Res.* 2003;93:346–353.

42 Wei Z, Manevich Y, Al-Mehdi AB, et al. Ca^{2+} flux through voltage-gated channels with flow cessation in pulmonary microvascular endothelial cells. *Microcirculation.* 2004;11:517–526.

43 Lee JH, Daud AN, Cribbs LL, et al. Cloning and expression of a novel member of the low voltage-activated T-type calcium channel family. *J Neurosci.* 1999;19:1912–1921.

44 Perez-Reyes E. Three for T: molecular analysis of the low voltage-activated calcium channel family. *Cell Mol Life Sci.* 1999;56: 660–669.

45 Rizzo V, Sung A, Oh P, Schnitzer JE. Rapid mechanotransduction in situ at the luminal cell surface of vascular endothelium and its caveolae. *J Biol Chem.* 1998;273:26323–26329.

46 Schnitzer JE, Liu J, Oh P. Endothelial caveolae have the molecular transport machinery for vesicle budding, docking, and fusion including VAMP, NSF, SNAP, annexins, and GTPases. *J Biol Chem.* 1995;270:14399–14404.

47 Gimpl G, Burger K, Fahrenholz F. Cholesterol as modulator of receptor function. *Biochemistry.* 1997;36:10959–10974.

48 Park H, Go YM, St John PL, et al. Plasma membrane cholesterol is a key molecule in shear stress-dependent activation of extracellular signal-regulated kinase. *J Biol Chem.* 1998;273:32304–32311.

49 Wei Z, Costa K, Al-Mehdi AB, et al. Simulated ischemia in flow-adapted endothelial cells leads to generation of reactive oxygen species and cell signaling. *Circ Res.* 1999;85:682–689.

50 Boonstra J, Post JA. Molecular events associated with reactive oxygen species and cell cycle progression in mammalian cells. *Gene.* 2004;337:1–13.

51 Milovanova T, Manevich Y, Haddad A, et al. Endothelial cell proliferation associated with abrupt reduction in shear stress is dependent on reactive oxygen species. *Antioxid Redox Signal.* 2004;6:245–258.

52 Takahashi M, Shibata M, Niki E. Estimation of lipid peroxidation of live cells using a fluorescent probe, diphenyl-1-pyrenylphosphine. *Free Radic Biol Med.* 2001;31:164–174.

53 Matot I, Manevich Y, Al-Mehdi AB, et al. Fluorescence imaging of lipid peroxidation in isolated rat lungs during nonhypoxic lung ischemia. *Free Radic Biol Med.* 2003;34:785–790.

54 Ayene IS, Dodia C, Fisher AB. Role of oxygen in oxidation of lipid and protein during ischemia/reperfusion in isolated perfused rat lung. *Arch Biochem Biophys.* 1992;296:183–189.

55 Zhao G, Ayene IS, Fisher AB. Role of iron in ischemia-reperfusion oxidative injury of rat lungs. *Am J Respir Cell Mol Biol.* 1997;16:293–299.

56 Al-Mehdi AB, Zhao G, Tozawa K, Fisher AB. Depolarization-associated iron release with abrupt reduction in pulmonary endothelial shear stress in situ. *Antioxid Redox Signal.* 2000;2: 335–345.

57 Eckenhoff RG, Dodia C, Tan Z, Fisher AB. Oxygen-dependent reperfusion injury in the isolated rat lung. *J Appl Physiol.* 1992;72: 1454–1460.

58 Okuda M, Furuhashi K, Nakai Y, Muneyuki M. Decrease of ischaemia-reperfusion related lung oedema by continuous ventilation and allopurinol in rat perfusion lung model. *Scand J Clin Lab Invest.* 1993;53:625–631.

59 Das KC, Misra HP. Amelioration of postischemic reperfusion injury by antiarrhythmic drugs in isolated perfused rat lung. *Environ Health Perspect.* 1994;102:117–121.

60 Imai T, Fujita T. Unilateral lung injury caused by ischemia without hypoxia in isolated rat lungs perfused with buffer solution. *J Lab Clin Med.* 1994;123:830–836.

61 Karck M, Haverich A. Nifedipine and diltiazem reduce pul-

monary edema formation during postischemic reperfusion of the rabbit lung. *Res Exp Med (Berl)*. 1992;192:137–144.
62 Friedrich I, Spillner J, Lu EX, et al. Ischemic pre-conditioning of 5 minutes but not of 10 minutes improves lung function after warm ischemia in a canine model. *J Heart Lung Transplant*. 2001;20:985–995.
63 Fadel E, Mazmanian GM, Baudet B, et al. Endothelial nitric oxide synthase function in pig lung after chronic pulmonary artery obstruction. *Am J Respir Crit Care Med*. 2000;162:1429–1434.
64 Permpikul C, Wang HY, Kriett J, et al. Reperfusion lung injury after unilateral pulmonary artery occlusion. *Respirology*. 2000; 5:133–140.
65 Divisi D, Montagna P, Jegaden O, et al. Lung transplantation by continuous perfusion in an experimental auto-transplant animal model. *Thorac Cardiovasc Surg*. 2002;50:301–305.
66 Sohn HY, Keller M, Gloe T, et al. The small G-protein Rac mediates depolarization-induced superoxide formation in human endothelial cells. *J Biol Chem*. 2000;275:18745–18750.
67 Grill HP, Zweier JL, Kuppusamy P, et al. Direct measurement of myocardial free radical generation in an in vivo model: effects of postischemic reperfusion and treatment with human recombinant superoxide dismutase. *J Am Coll Cardiol*. 1992;20:1604–1611.
68 Udassin R, Ariel I, Haskel Y, et al. Salicylate as an in vivo free radical trap: studies on ischemic insult to the rat intestine. *Free Radic Biol Med*. 1991;10:1–6.

132

Endothelium and the Initiation of Atherosclerosis

Myron I. Cybulsky

Toronto General Research Institute, University Health Network, University of Toronto, Ontario, Canada

The identification of pathogenic risk factors for atherosclerosis, such as hyperlipidemia, hypertension, and diabetes, has led to modifications in lifestyle and novel therapies that have had a significant impact on the morbidity and mortality associated with atherosclerosis-related diseases. However, despite such advances, these diseases remain highly prevalent in industrialized nations, and the recent increase in obesity and diabetes, especially in children, forecasts a resurgence in the incidence of atherosclerosis-associated morbidity. Over the last three decades, the field has rediscovered the critical role that inflammation plays in the pathogenesis of atherosclerosis. An important component of inflammation is the emigration of leukocytes from the blood into tissues. The vascular endothelium, being situated at the interface between the blood and tissues, has important regulatory functions in atherosclerosis. In normal arteries, the endothelium has homeostatic functions, including nonadhesive and antithrombotic properties, and it provides a barrier to plasma protein and lipoprotein extravasation (1). Endothelial cells (ECs) are not a passive impermeable "Teflon-like" lining, but are active cells, which respond to diverse blood- and tissue-derived stimuli. For example, in response to the repertoire of local cytokines and modified lipoprotein components, ECs synthesize and present molecules that regulate the recruitment of distinct leukocyte subpopulations to atherosclerotic lesions. Profound and diverse changes in EC phenotype in response to different cytokines have been referred to as *EC activation* (2). ECs also can respond to external mechanical forces, including hemodynamic forces of blood (3). This is probably why atherosclerotic lesions initially form at sites of arterial branches and curvatures. These regions have unique hemodynamics and EC gene expression patterns that influence signaling, the cellular composition of intimal cells, and responses to systemic risk factors for atherosclerosis.

This chapter focuses predominantly on the role of the endothelium in atherogenesis, particularly in early stages of the disease, because most mechanistic data are derived from animal models of early atherosclerosis. Insights from this field may lead to the development of novel therapeutic approaches directed at EC biology and inflammation. These therapies, in conjunction with modification of lifestyle and therapies that target lipoprotein metabolism, hypertension, and hyperglycemia, may provide synergistic benefits to populations at risk, may alleviate or delay the progression of atherosclerotic lesions, and may reduce the incidence of morbidity and mortality associated with complications of advanced lesions.

CLINICAL AND PATHOLOGICAL FEATURES OF ATHEROSCLEROSIS

Atherosclerosis is a disease of elastic and muscular arteries. It is the principal antecedent to myocardial infarction, stroke (cerebral infarction), claudication, and aneurysm formation (4). These complications generally occur in the setting of advanced atherosclerosis. Atherosclerosis develops indolently for decades prior to onset of those complications that lead to clinical manifestations. Progressive, chronic, and severe narrowing of the arterial lumen usually causes ischemia of tissues. The chronic nature of this process enables the formation of a collateral blood supply, which is promoted by tissue hypoxia. We now appreciate that infarctions usually result from sudden and complete occlusion of the vascular lumen by a thrombus associated with complications in an atherosclerotic plaque. Angiographic studies of coronary arteries have revealed that, prior to thrombosis, culprit plaques do not produce severe luminal narrowing, but may occasionally impinge into the vascular lumen (5). This is because atherosclerotic lesions are characteristically eccentric, in that they involve only part of the artery wall, which enables the uninvolved portion to undergo compensatory enlargement or outward remodeling and maintain a vascular lumen of a normal or near-normal diameter. Compensatory enlargement of atherosclerotic arteries was first described by Glagov (6), and is regulated by ECs. ECs sense changes in hemodynamic forces (shear stress) that result from alterations in the vascular lumen and institute complex biological programs that lead to normalization of the lumen size and shear stress (3). Outward arterial remodeling masks the extent of atherosclerosis that is detected by angiography, and furthermore, this technique cannot identify lesions that are predisposed to thrombosis. Therefore, investigations are under way to determine if other vascular imaging modalities, such as intravascular ultrasound, computed tomography (CT), magnetic resonance imaging (MRI) and positron emission tomography (PET) scanning, can detect the structural or metabolic features of atherosclerotic lesions that predict their clinical outcome.

Atherosclerotic lesions, also referred to as *atheromas* and *plaques*, have a range of morphological features (7,8). Initially, the intima is involved, but as lesions grow with time, the underlying media becomes atrophic, with reduced vascular smooth muscle cells (VSMCs) and extracellular matrix (ECM). In some cases, mononuclear inflammatory cells infiltrate the arterial adventitia. Atherosclerotic lesions consist of vascular and inflammatory cells, ECM, lipids, cellular debris, calcification, or even ossification. The cellular component includes ECs lining the arterial lumen and neovasculature within lesions, VSMCs derived from the media or potentially from circulating progenitors, fibroblasts or myofibroblasts, monocyte/macrophages, lymphocytes (mostly T cells), and mast cells. Monocytes and lymphocytes are recruited to the arterial intima from the blood in early as well as advanced atherosclerotic lesions. In the intima, monocytes transform into macrophages, express scavenger receptors, and engulf modified lipids to become foam cells. Macrophages can proliferate in lesions. Smooth muscle cells and fibroblasts likely synthesize the bulk of ECM and form a band of dense collagen below the arterial endothelium, termed the *fibrous cap*, which likely provides structural support to plaques. The central core contains extracellular lipid droplets, is rich in macrophages, and may contain necrotic cell debris and large cholesterol clefts. The edges or shoulder regions often contain inflammatory cells and neovasculature, indicative of angiogenesis. Lesion morphology varies from those composed predominantly of collagen with or without calcification to those with a large necrotic core and thin fibrous cap. The pathogenesis of these distinct features in lesions is poorly understood.

Atherosclerotic lesion formation can be subdivided into initiation (development of small fatty streaks), expansion (vertical and lateral growth, as well as coalescence of small fatty streaks), development of a fibro-fatty plaque (intimal VSMC recruitment, collagen deposition and formation of a fibrous cap), and growth through organization of mural thrombi. Thrombosis is associated frequently with advanced atherosclerotic lesions and, in fact, may be an important mechanism for growth of plaques (5,9). The two major causes of thrombosis are plaque rupture and plaque erosion. Plaque rupture, also referred to as plaque fissuring and fibrous cap disruption, results in the formation of a defect or gap in the fibrous cap that separates the lipid-rich core from flowing blood. Plaque erosion refers to plaque thrombosis with EC loss, but absence of rupture in the fibrous cap. Human pathological studies determined that plaque rupture occurs in most cases, especially in men. Ruptures and hemorrhages tend to occur in shoulder regions of lesions and are associated with inflammation, angiogenesis, a large necrotic core, and thin fibrous cap. The production of proteases (e.g., matrix metalloproteinases and cysteine proteases), oxygen radicals, vasoactive substances, and coagulation factors by activated immune cells, or tissue disruption resulting from hemorrhage originating from the neovasculature are potential mechanisms. Focal calcification within lesions also may provide a mechanical stimulus that predisposes to plaque rupture. Coronary artery thrombosis due to plaque erosion is less frequent than plaque rupture, tends to occur in women, and its pathogenesis is poorly understood. One possibility is that luminal ECs contribute to the initiation of thrombosis, but then migrate into the thrombus, resulting in the appearance of an erosion. Clinical data suggest that patients with unstable angina harbor multiple vulnerable plaques in more than one coronary artery (10) and that systemic factors in the setting of atherosclerosis predispose to thrombosis of multiple plaques and induce a condition termed *acute coronary syndrome* (5,9).

MURINE MODELS OF ATHEROSCLEROSIS

Significant insights into the pathogenesis of atherosclerosis have been gained from animal models, and most utilize hypercholesterolemia as the main stimulus for atherosclerotic lesion formation. Because hypercholesterolemia is only one of several major risk factors that have been identified in humans, animal models may not reflect the entire pathogenic spectrum of human atherosclerosis. Nevertheless, many important insights have been gained, especially from recently available murine models, in which embryonic stem cell and transgenic technologies have been used to alter the expression levels of various genes. The development of murine models defective in genes controlling lipid metabolism and lipoprotein expression offers valuable tools to dissect complex interactions between diet and genetics in atherosclerosis. Two models have been used predominantly – the apolipoprotein E knockout ($ApoE^{-/-}$) (11,12), and the low-density lipoprotein (LDL) receptor knockout ($LDLR^{-/-}$) (Table 132-1) (13). These hypercholesterolemic models develop atherosclerotic lesions throughout the arterial tree, and their distribution pattern and morphologic features share many similarities with human atherosclerosis, suggesting that similar pathogenic mechanisms may be involved (14–16). A significant difference in the $ApoE^{-/-}$ and $LDLR^{-/-}$ models is the degree of hypercholesterolemia and extent of atherosclerosis found when mice are fed regular chow-based diets without lipid additives. $ApoE^{-/-}$ mice fed a normal chow mouse diet develop hypercholesterolemia (predominantly elevated very low-density lipoprotein [VLDL]) and atherosclerotic lesions; however, if fed a "Western-type" diet (0.15% cholesterol, 21% fat), the hypercholesterolemia is enhanced, and lesions develop more rapidly (11,14). In contrast, $LDLR^{-/-}$ mice fed a normal chow diet have only a twofold elevation in plasma cholesterol (primarily

Table 132-1: Effects of Selective Genes on the Development of Atherosclerosis in Hypercholesterolemia Mice

Gene(s)	*Mutation*	*Atherosclerosis Model*	*Atherosclerosis Phenotype (Extent of Lesions)*	*Proposed Mechanism*	*References*
CSF-1	op/op		Reduced, markedly	Role for mono/mϕ	41–43
(M-CSF)	op/+				44
SR-A	KO	*ApoE*$^{-/-}$	Reduced/comparable	Scavenger receptors	36, 38
CD36	KO	*ApoE*$^{-/-}$ *SC* and *WD*	Reduced/comparable		37, 38
SR-B1	KO	*ApoE*$^{-/-}$	Increased (MI)	Reverse cholesterol	134
	BM deficiency	*LDLR*$^{-/-}$	Increased	transport	135
TLR-4	KO	*LDLR*$^{-/-}$	Increased		135
CD14	KO	*ApoE*$^{-/-}$	Reduced	Innate immunity –	39
MyD88	KO	*ApoE*$^{-/-}$ *WD*	Comparable	mono/mϕ signaling	40
Rag-1	KO	*ApoE*$^{-/-}$ *WD*	Reduced	and activation	39, 40
		ApoE$^{-/-}$ *SC*	Reduced	Adaptive immunity	49
		ApoE$^{-/-}$ *WD*	Comparable		50
		LDLR$^{-/-}$ *WD*	Reduced (8 wks), comparable (12 and 16 wks)		
IFN-γ R	KO	*ApoE*$^{-/-}$	Reduced	Innate and adaptive	52
IFN-γ	KO	*ApoE*$^{-/-}$ *SC* and *WD*	Reduced in males	immunity, Th1	54
		LDLR$^{-/-}$	Reduced	differentiation	55
T-bet	KO	*LDLR*$^{-/-}$	Reduced		56
TGF-β	dn TGF-βRII in T cells	*ApoE*$^{-/-}$	Increased	Innate and adaptive immunity	57
IL-10	KO	*ApoE*$^{-/-}$	Increased	Innate and adaptive	60
	Tg: T-cells (BM Tx)	*LDLR*$^{-/-}$	Reduced	immunity	59
P-selectin	KO	*LDLR*$^{-/-}$	Reduced in males only at 8–20 wks	WBC–EC interactions; and/or role for platelets	74
		ApoE$^{-/-}$	Reduced		75–77
E-selectin	KO	*ApoE*$^{-/-}$ *SC*	Smaller effect vs. P-selectin	WBC–EC interactions	75
ICAM-1	Hypomorph	*ApoE*$^{-/-}$ *SC*	Smaller effect vs. P-selectin	WBC–EC interactions	75
		LDLR$^{-/-}$	Comparable		86
VCAM-1	hypomorph	*LDLR*$^{-/-}$	Reduced	WBC–EC interactions	86
		ApoE$^{-/-}$	Reduced		87
CD11b	KO	*LDLR*$^{-/-}$	Comparable	WBC–EC interactions	90
MCP-1	KO	*LDLR*$^{-/-}$	Reduced	Chemokine	98
	KO	*Hu apoB Tg*	Reduced		97
	MMTV-MCP-1 Tg (BM Tx)	*ApoE*$^{-/-}$	Increased		96
CCR2	KO	*ApoE*$^{-/-}$	Reduced	MCP-1 receptor	99, 100
CX3CL1	KO	*ApoE*$^{-/-}$	Reduced in brachiocephalic artery, not aortic root	Chemokine (fractalkine)	121
CX3CR1	KO	*ApoE*$^{-/-}$	Reduced	Fractalkine receptor	119, 120

KO, knockout; SC, standard chow; WD, Western-type diet; mono, monocyte; mϕ, macrophage; mLDL, oxidatively-modified LDL, MMTV, mouse mammary tumor virus; Tg, transgenic; BM Tx, bone marrow transplantation; WBC, white blood cell.

intermediate-density lipoprotein [IDL]/LDL fraction) (13) and do not develop lesions in the short term (16). When fed a cholesterol-rich diet, these mice develop marked hypercholesterolemia (elevated VLDL, IDL, and LDL associated with decreased high-density lipoprotein [HDL]) and lesions throughout the aorta. Because hypercholesterolemia and aortic lesion formation in *LDLR*$^{-/-}$ mice is dependent on initiating a hypercholesterolemic diet, the onset of lesion development can be precisely controlled, and these mice provide a unique opportunity for studying early events in atherogenesis.

ENDOTHELIAL CELL ACTIVATION DURING ATHEROGENESIS

ECs form the interface between the blood and tissues and have important homeostatic functions, including maintenance of a nonadhesive and antithrombotic surface. Inflammatory cytokines, including interleukin (IL)-1 and tumor necrosis factor (TNF)-α, activate ECs and modulate their phenotype by inducing the expression of newly synthesized cell surface molecules. Inducible expression of EC adhesion molecules, such as E-selectin, P-selectin, vascular cell adhesion molecule (VCAM)-1, and intercellular adhesion molecule (ICAM)-1, and chemokines that are secreted and presented on glycosaminoglycans, actively promotes leukocyte adhesive interactions. IL-1 and TNF-α also upregulate the expression of major histocompatibility (MHC) class I and tissue factor. In contrast, interferon (IFN)-γ induces the expression of ICAM-1 and MHC class II, but not other adhesion molecules or MHC class I, and IL-4 selectively upregulates the expression of VCAM-1. These examples illustrate that different stimuli can induce unique activation of ECs, and can have profound influence on various endothelial properties, including cell adhesion, thromboresistance, cell shape, permselectivity, and immunological functions (2).

Studies of human atherosclerosis and experimental animal models have provided abundant evidence that hypercholesterolemia and associated infiltration and retention of LDL in the artery wall is an important pathogenic stimulus for the initiation and progression of atherosclerosis. LDL is retained in the intima through specific interaction with proteoglycans (17) and undergoes aggregation, oxidation, and/or enzymatic modification, which leads to the release of bioactive phospholipids that can activate ECs and incite an inflammatory response. This aspect of atherogenesis will not be reviewed here, and the reader is referred to several excellent reviews (18,19).

EC activation during atherogenesis was first observed during the early 1990s in hypercholesterolemic rabbit models (20). This study found that expression of a novel EC–leukocyte adhesion molecule, identified as VCAM-1, was upregulated selectively on ECs overlying early atherosclerotic lesions in the rabbit aorta. Expression of VCAM-1 by ECs was not observed in regions that were not involved by atherosclerotic lesions. These data suggested that the endothelium lining the arterial luminal surface actively regulates the adherence and recruitment of blood mononuclear leukocytes to the intima, which is one of the earliest events observed in atherogenesis and a necessary step for macrophage accumulation in the intima and transformation into foam cells.

Our laboratory and others compared the expression of VCAM-1, ICAM-1, and E-selectin in aortas in hypercholesterolemic rabbits, *LDLR*$^{-/-}$ and *ApoE*$^{-/-}$ mice (21,22). Northern blot analysis demonstrated increased *VCAM-1* and *ICAM-1*, but not *E-selectin*, steady-state mRNA levels in hypercholesterolemic mouse and rabbit aortas, which correlated with the extent of atherosclerotic lesion formation, as determined by staining intimal lipid deposits with oil red O (21). In small lesions, VCAM-1 and ICAM-1 were expressed predominantly by ECs, whereas in large foam cell–rich lesions, many intimal cells expressed these molecules. It is likely that expression by intimal cells accounted for increased *VCAM-1* and *ICAM-1* steady-state mRNA levels in Northern blots. VCAM-1 also was expressed by medial VSMCs adjacent to lesions. This phenotypic change may occur in activated VSMCs or cells in the process of migration to the intima.

Expression patterns of EC VCAM-1 and ICAM-1 were highly reproducible and similar in distribution among rabbits and *LDLR*$^{-/-}$ mice fed cholesterol-containing diets. EC expression was most pronounced at the periphery of both large and small lesions and extended several cells beyond the edge. VCAM-1 expression was essentially restricted to lesions, whereas ICAM-1 expression extended into the uninvolved aorta. These data suggest that adhesion molecule expression in atherosclerotic lesions of mice and rabbits may be under similar mechanistic control; however, the regulation of VCAM-1 expression is more tightly controlled than ICAM-1 by lesion-derived factors. We observed virtually identical VCAM-1 and ICAM-1 immunohistochemical staining patterns in *ApoE*$^{-/-}$ mice and Watanabe heritable hyperlipidemic rabbits fed standard laboratory chow, which indicates that hypercholesterolemia, and not other dietary factors, was responsible for upregulated VCAM-1 and ICAM-1 expression in lesions. The endothelium over central portions of rabbit, but not mouse, lesions frequently expressed VCAM-1. These observations on adhesion molecule expression patterns in atherosclerotic lesions are consistent with and extend previous observations in animal models (20,22–25).

In humans, several groups detected VCAM-1, ICAM-1, and E-selectin expression by immunohistochemical staining in advanced atherosclerotic plaques obtained at autopsy or from hearts of transplant recipients (26–31). ICAM-1 expression was found consistently in ECs over plaques and in intimal VSMCs and macrophages. As expected, the expression of E-selectin was restricted to vascular endothelium, but was variable among individuals. Caution should be exercised in interpreting these data, since the anti-E-selectin antibody used in these studies (BBA1 from British Biotechnology) has subsequently been shown to cross-react with P-selectin. Advanced human coronary artery plaques displayed focal VCAM-1 expression in luminal ECs, usually in association with inflammatory infiltrates. Focal endothelial VCAM-1 expression also was found in uninvolved vessels with diffuse intimal thickening. VCAM-1 was expressed by some smooth muscle cells and macrophages and by ECs of neovasculature at the base of plaques.

THE CONTRIBUTION OF LEUKOCYTES TO THE FORMATION TO ATHEROSCLEROTIC LESIONS

The recruitment of both blood monocytes and lymphocytes to the arterial intima is critical to the formation of an atherosclerotic lesion. This process persists even in advanced lesions (18,19,32–35). Recruited monocytes transform into macrophages, express scavenger receptors, and engulf modified lipids to become intimal foam cells. Knockout of scavenger receptor-A and CD36, a class B scavenger receptor, reduced atherosclerosis in mice (36,37), although a recent study has not supported these data (38). Monocyte recruitment, particularly at the periphery of lesions, may contribute to their lateral growth. Monocytes and macrophage foam cells also may contribute to the progression of atherosclerotic lesions by producing cytokines and growth factors. In this process, cell activation via Toll-like receptor signaling is likely an important step (39,40). Cytokines and growth factors, in turn, may amplify mononuclear leukocyte recruitment, induce migration of VSMCs into the intima, and stimulate cell replication. Because early fatty streaks are composed almost entirely of macrophage foam cells, the recruitment of monocytes to the intima may be a critical event in the initiation and expansion of these early atherosclerotic lesions.

The critical role of monocytes/macrophages in atherogenesis is highlighted by experiments using osteopetrotic (*op/op*) mice, which are deficient in macrophage colony stimulating factor (M-CSF or CSF-1) due to a point mutation in this gene. Binding of M-CSF to its receptor, c-fms, initiates many biological responses, including monocyte/macrophage chemotaxis, growth, survival, and expression of genes, including the class-A scavenger receptor. *Op/op* mice have impaired production of blood monocytes and deficiency of peritoneal and tissue macrophages. A lack of osteoclasts accounts for osteopetrosis, and an inability for teeth to erupt impairs the consumption of solid food and requires feeding of a special liquid diet. Atherosclerosis studies were carried out in *op/op* mice bred into an atherosclerosis-susceptible background, and these mice had markedly reduced atherosclerotic lesions formation (41–43). One possibility for the marked reduction in lesions may have been because *op/op* mice have reduced numbers of circulating blood monocytes. However, a subsequent study demonstrated that even heterozygous (*op/+*) mice had dramatically reduced lesion size despite only a 20% reduction in circulating monocytes and absence of osteopetrosis (44). M-CSF influences many aspects of monocyte/macrophage biology, and we do not know what aspects of atherosclerotic lesion formation were impaired in *op/+* mice.

Lymphocytes also participate in atherogenesis. Recent studies provide convincing evidence for the role of the immune system in atherogenesis (9). Lymphocytes constitute only a small fraction of cells in atherosclerotic lesions, but perhaps their most important function is to regulate the local cytokine milieu and thus either promote or suppress lesion formation. $CD4^+$ Th1 cells are the predominant T-cell subtype in atherosclerotic lesions, and they respond to antigens presented in association with MHC class II by releasing proinflammatory cytokines that activate macrophages and ECs, including IFN-γ, TNF-α, and lymphotoxin (45). $CD4^+$ T cells reactive to antigens that are present in lesions, such as oxidized LDL (oxLDL), heat shock protein 60, and *Chlamydia*, have been cloned from human atherosclerotic lesions (46–48). $ApoE^{-/-}$ mice deficient in the recombinase-activating gene 1 ($Rag\text{-}1^{-/-}$) lack adaptive immunity and develop twofold less atherosclerosis when fed standard chow, but had lesions comparable to immunocompetent $ApoE^{-/-}$ mice when fed a Western-type diet (49). In the $LDLR^{-/-}$ background, lesion development in $Rag\text{-}1^{-/-}$ mice was reduced by 54% after 8 weeks of feeding a Western-type diet; however, significant differences in lesion area gradually subsided as the diet was continued for 12 and 16 weeks (50). $ApoE^{-/-}$ mice with severe combined immunodeficiency (SCID) developed 75% less atherosclerosis, and reconstitution of these mice with $CD4^+$ T cells increased atherosclerosis (51). This proatherogenic activity may be exerted at least partly through secretion of the cytokine IFN-γ, because mice deficient in IFN-γ receptor, IFN-γ, or the transcription factor T-bet, which that is required for Th1 differentiation, develop smaller atherosclerotic lesions (52–56). In contrast to proinflammatory cytokines, transforming growth factor (TGF)-β and IL-10 are protective factors in atherogenesis (57–60) as are antibodies to oxLDL epitopes produced by B-1 B lymphocytes (61,62).

FUNCTIONS OF ADHESION MOLECULES AND CHEMOKINES IN ATHEROSCLEROSIS

The activation of ECs during atherogenesis results in the expression of adhesion molecules and chemokines in atherosclerotic lesions. Although these molecules can influence many aspects of vascular biology, one of their key functions is the recruitment of leukocytes from blood into tissues. The process of leukocyte emigration from blood into tissues has been subdivided by many investigators into tethering, rolling, arrest, stable or firm adhesion, and transendothelial migration (TEM) (63,64). Tethering is the first adhesive interaction, which captures and slows the velocity of a leukocyte as it makes contact with the endothelial monolayer. Tethering and rolling are mediated by binding of E-, P-, or L-selectin to a ligand (e.g., sialyl-Lewis^x [sLe^x] on P-selectin glycoprotein ligand [PSGL]-1, a P-selectin ligand) or by low affinity $\alpha 4\beta 1$ integrin (VLA-4, CD49d/CD29) binding to VCAM-1. Adhesive interactions that mediate rolling have rapid binding and release rates (high k_{on} and k_{off}) (65,66). Arrest, firm adhesion, and TEM (also known as diapedesis) are mediated by leukocyte integrins, which bind to members of the immunoglobulin gene superfamily (e.g., VCAM-1 and ICAM-1). Rolling leukocytes are in close proximity to endothelium and are exposed to chemokines presented on the EC surface (67,68). Chemokine binding to receptors on leukocytes initiates signals that upregulates $\beta 1$ and $\beta 2$ integrin ligand-binding capability and allows them to interact with VCAM-1 and ICAM-1 (69–71). The recruitment of specific leukocyte types to a site of inflammation is determined by the repertoire of integrins and chemokine receptors expressed by different leukocytes and by the cytokines produced by the inflammatory response. Cytokines determine what adhesion molecules and chemokines are expressed by endothelium and produced in the inflammatory site. The details of how different adhesion molecules and chemokines mediate specific steps of leukocyte–EC adhesive interactions will not be reviewed, because a number of excellent reviews on this subject are available (63,64,72,73).

During the last two decades, many lines of transgenic mice bearing a deficiency of an adhesion molecule or chemokine have been developed, and the contribution of these adhesion molecules in atherogenesis has been investigated. These mice were backcrossed into the $ApoE^{-/-}$ or $LDLR^{-/-}$ background, and the extent of atherosclerotic lesion formation was assessed by estimating the volume of lesions in the aortic root or determining the surface area of the aorta occupied by lesions. The histological features of lesions and their cellular composition also were determined.

P-selectin deficiency in the *LDLR*$^{-/-}$ background had a modest effect in male but not female mice (74), and the effect was greater in the *ApoE*$^{-/-}$ background (75–77). For example, at 4 months of age, in the *ApoE*$^{-/-}$ background, the P-selectin-deficient group had 3.5-fold smaller aortic sinus lesions than did the P-selectin wild-type group. In the P-selectin deficient group, lesions were limited to fatty streaks and contained fewer macrophages (76). By 15 months, progression to fibrous plaques was observed throughout the aorta in both groups, although lesions in the aortic sinus were smaller and less calcified in the P-selectin–deficient group. E-selectin deficiency had a relatively small effect on lesion formation in both males and females (75). Mice with combined P- and E-selectin deficiency had less lesion formation than *P-selectin*$^{-/-}$ mice alone (78). Platelets may also contribute to atherogenesis (77,79–81). P-selectin is involved in mediating the adhesion of activated platelets to ECs and leukocytes (81). The deposition of platelet chemokines onto endothelium (82) and soluble P-selectin, derived primarily from ECs, induces the formation of procoagulant microparticles (77).

Deficiency of VCAM-1 or α4 integrin results in embryonic lethality (83–85). During embryogenesis, these molecules mediate the fusion of the allantois to the chorion. This is because VCAM-1 is normally expressed on the tip of the allantois, which subsequently forms the umbilical cord and the fetal vasculature of the placenta, whereas α4 integrin is expressed on the chorion. The embryonic lethality of *VCAM-1*–null mice was circumvented by generating mice that express a mutant form of VCAM-1 at markedly reduced levels. Approximately 25% of these VCAM-1 domain 4–deficient (*VCAM-1 D4D*) mice were viable, which was sufficient for breeding them into the *LDLR*$^{-/-}$ background. Mice were fed a 1.25% cholesterol-enriched diet for 8 weeks. En face analysis of oil red O-stained aortas revealed reduced lesion area compared with *VCAM-1*$^{+/+}$ mice (86). *VCAM-1 D4D* mice were also bred into the *ApoE*$^{-/-}$ background, and lesion formation in the aortic root was quantified. These studies revealed a *VCAM-1* gene dosage effect on aortic root atherosclerotic lesions at 16 weeks of age. The aortic root lesion area was reduced by 84% and 56% in *VCAM-1*$^{D4D/D4D}$ and *VCAM-1*$^{+/D}$$4^{D}$ mice, respectively, and lesions in *VCAM-1*$^{D4D/D4D}$ mice were limited to very small nascent fatty streaks (87). Together, these studies suggest that VCAM-1 has a critical role in atherogenesis. Similarly, its ligand, the α4 integrin, should have an important function in atherosclerosis. This has been difficult to test, because mice deficient in α4 integrin are not viable. However, data supporting this have been generated using infusion of α4 integrin blocking peptide, which reduced lesion formation in mice (88). VCAM-1 and α4 integrin were also key mediators of U937 cell rolling and adhesion in an ex vivo perfusion model of the carotid artery bifurcation harvested from *ApoE*$^{-/-}$ mice (89).

In contrast to VCAM-1, ICAM-1 appears to play a minor role in the formation of early atherosclerotic lesions. In experiments carried out in parallel with *VCAM-1 D4D* mice, ICAM-1–deficient mice in the *LDLR*$^{-/-}$ background had comparable lesion formation in the aorta to wild-type littermate controls (86). In the *ApoE*$^{-/-}$ background, deficiency of ICAM-1 reduced the extent of aortic lesions, but to a lesser degree than did P-selectin deficiency (75). Leukocyte CD11b expression was not essential for the development of atherosclerosis (90).

Members of the chemokine family play important roles in atherogenesis. Although many chemokines are likely involved, the discussion here focuses on primarily monocyte chemoattractant protein (MCP)-1 (or CCL2), fractalkine (CX3CL1), and their receptors, because they illustrate most of the key concepts. Many chemokines, including MCP-1 and fractalkine, are expressed in atherosclerotic lesions, both by ECs on the lesion surface and by cells deep within the lesion (91–93). Oxidized lipids have long been implicated as important mediators of atherosclerosis and foam cell formation (94), and minimally oxLDL, but not native LDL, induces MCP-1 production in vascular wall cells such as ECs and VSMCs (95). Chemokines thus emerged as a possible molecular links between oxidized lipoproteins and foam cell recruitment to the vessel wall. In addition to being produced in lesions, certain chemokines, such as regulated on activation, normal T-cell expressed and secreted (RANTES), can be deposited on ECs by circulating blood cells, including platelets (82).

Studies in transgenic mice overexpressing MCP-1 and mice in which either MCP-1 or CCR2 (its receptor) have been genetically deleted have provided strong evidence in support of this hypothesis. In bone marrow transplantation studies in mice, overexpression of MCP-1 in the blood vessel wall macrophages leads to increased foam cell formation and increased atherosclerosis (96). Similarly, deletion of MCP-1 in mice blocked the progression of dietary-induced atherosclerosis (97,98). CCR2 is the only known receptor for MCP-1, and deletion of CCR2 in mice afforded significant protection from both macrophage accumulation and atherosclerotic lesion formation in response to a high-fat diet (99,100). Finally, studies have shown that treatment of mice with an MCP-1 antagonist resulted in a reduction in dietary-induced atherosclerosis (101). A recent study suggests that CCR2 and MCP-1 regulate the release of a key monocyte subset from the bone marrow, and mice lacking these molecules have reduced numbers of circulating monocytes (102). Collectively, these studies provide evidence that activation of CCR2 by MCP-1 plays an important role in monocyte recruitment into early atherosclerotic lesions, and suggest that interruption of the MCP-1/CCR2 axis can reduce lesion formation.

The degree to which monocytes respond to MCP-1 is influenced by a number of factors, including the number of CCR2 cell surface receptors. Studies indicate that lipoproteins contribute to the regulation of CCR2 expression. Individuals with hypercholesterolemia have higher levels of CCR2 on their monocytes; CCR2 expression correlates positively with plasma LDL cholesterol levels and negatively with plasma HDL levels

(103,104). These findings suggest that high cholesterol levels may lead to increased sensitivity of monocytes/macrophages to MCP-1, thereby increasing the movement of blood monocytes into early atherosclerotic lesions.

The subsequent downregulation of CCR2, as monocytes differentiate into macrophages, might then serve to maintain macrophages in lesions. Oxidized lipids, which accumulate in early atherosclerotic lesions not only induce MCP-1 production, but also influence the level of CCR2 expression on monocytes. Unlike native LDL, which upregulates CCR2, oxidized lipids downregulate CCR2 (103). The mechanism for this downregulation may involve the binding and internalization of oxidized lipids by CD36, a scavenger receptor, and subsequent activation of the peroxisome proliferator-activated receptor γ (PPARγ), a member of the nuclear hormone receptor family. Indeed, purified components of oxidized lipids activate PPARγ. Rosiglitazone, a synthetic ligand that activates PPARγ, decreases CCR2 expression on monocytes (105). The systemic administration of rosiglitazone inhibits the development of atherosclerosis in mice, consistent with its effects on CCR2 expression (106). Further evidence for the importance of MCP-1/CCR2 in atherosclerosis comes from the observation that individuals who are homozygous for the –2518 G/G allele in the *MCP-1* regulator region – a change that leads to increased levels of *MCP-1* mRNA – have a higher incidence of coronary artery disease (107). Although provocative, this study included relatively small groups of patients, and the clinical relevance of this polymorphism must be verified in larger studies.

Fractalkine and its receptor also have been directly linked to the development of early atherosclerotic lesions. Fractalkine is the sole member of the CX3C family, and has unique structural and functional attributes (108,109). Most chemokines are secreted peptides, but fractalkine is a type-1 transmembrane protein with an N-terminal chemokine domain anchored to the cell membrane through a contiguous extended mucin-like stalk, transmembrane, and cytoplasmic domains. TNF-α–converting enzyme (also known as a disintegrin and metalloproteinase domain 17 [ADAM17]), can cleave the mucin stalk of fractalkine near the plasma membrane and release soluble chemokine (110,111). Fractalkine binding to its seven-transmembrane domain G-protein–coupled receptor triggers signaling, but also directly mediates cell adhesion (112). Full-length transmembrane fractalkine is an efficient cell adhesion molecule, and can capture cells expressing its cognate receptor (CX3CR1) under physiologically relevant flow conditions (113,114). CX3CR1 has two common coding polymorphisms, namely V249I and T280M, which are in strong linkage disequilibrium (almost always occurring on the same allele). These have been associated with interindividual differences in susceptibility to both human immunodeficiency virus (HIV) infection and atherosclerosis (115–117). The M280 polymorphism has been linked to a decrease in the incidence of coronary artery disease in humans, and exhibits lower fractalkine binding (117,118). Mice with genetic deficiency of CX3CR1 have been crossed into the $ApoE^{-/-}$ background and showed substantially smaller lesions throughout the aorta in dietary-induced atherosclerosis (119,120). *Fractalkine*-null mice crossed into the $ApoE^{-/-}$ and $LDLR^{-/-}$ backgrounds had more dramatic reduction in lesion area at the brachiocephalic artery than in the aortic root (121).

REGIONAL DIFFERENCES IN ENDOTHELIAL CELL GENE EXPRESSION, SIGNALING, AND COMPOSITION OF THE ARTERIAL INTIMA IN THE NORMAL AORTA

Atherosclerosis is precipitated and accelerated by systemic risk factors such as elevated cholesterol, acute phase proteins, high blood pressure, oxidation products from smoking, and high blood sugar and advanced glycation endproducts in diabetes. Although the entire arterial tree is exposed to these stimuli, atherosclerotic lesions form predictably at arterial curvatures, bifurcations, and branch points, which suggests that local factors contribute to disease susceptibility, and implicates the unique and complex patterns of blood flow ("disturbed" hemodynamics) that are found in these locations (3). The prevailing hypothesis is that disturbed hemodynamic forces predispose ECs to atherosclerosis, whereas uniform laminar shear stress, found in straight segments of arteries, is atheroprotective. Hemodynamic forces may influence EC shape, proliferation, gene expression, signal transduction and permeability, transport, and retention of lipoproteins in the intima.

Abundant data from in vitro studies suggest that hemodynamic forces have profound effects on EC biology. Different shear stress profiles can induce unique repertoires of gene expression (122–124). Shear stress alters the expression of adhesion molecules in cultured ECs (125,126), and uniform laminar shear reduces cytokine-induced expression of certain adhesion molecule (127). Elucidating the molecular mechanisms of how ECs sense shear stress and of how shear modulates EC gene expression remains an area of active investigation.

In vivo studies have also provided important insights into unique biological properties of endothelium prior to the formation of atherosclerotic lesions. Unlike regions with uniform laminar flow, in which ECs are elongated and uniformly oriented in the direction of flow, ECs in regions with disturbed flow have a polygonal shape or are elongated with the cell axis oriented in random directions. In animals with normal levels of plasma lipoproteins, ECs have unique gene expression profiles in regions of arteries predisposed and protected from atherosclerosis. Low levels of VCAM-1 and ICAM-1 are expressed in regions predisposed to atherosclerosis in normal rabbits and mice, although ICAM-1 expression in normal aortas is more extensive (21). Similarly, elevated expression of VCAM-1 and ICAM-1 was found in $ApoE^{-/-}$ mice at sites predisposed to atherosclerosis prior to the formation of lesions (22). Recently, gene profiling studies of porcine aortic endothelium revealed distinct gene expression patterns in a region of the arch that is predisposed to atherosclerosis (128).

Alterations in EC gene expression may affect signaling pathways. Previously, we demonstrated striking topographic variations in EC nuclear factor (NF)-κB signal transduction. NF-κB is a dimeric transcription factor that is retained in the cell cytoplasm by an inhibitor protein (IκB). Upon stimulation, IκB kinases (IKKs) phosphorylate IκBs and target them for ubiquitination and degradation by proteasomes. NF-κB then translocates to the nucleus, where it transactivates gene expression. NF-κB induces the expression of numerous genes that are found in atherosclerotic lesions. We found that NF-κB was primed for activation in a region of the mouse ascending aorta with a high probability (HP) for developing atherosclerosis, compared to a low probability (LP) region (129). In mice with normal plasma lipoprotein levels, expression of NF-κB (p65) and IκBs was significantly higher in the cytoplasm of HP region endothelium, but NF-κB activation (determined by nuclear translocation of p65) was present in only a minority of cells. When mice were exposed to systemic endotoxin or hypercholesterolemia, NF-κB activation and upregulated expression of NF-κB–responsive genes was found preferentially in the HP region. These data illustrate how a topographic difference in endothelial signaling mediates accentuated regional gene expression in response to systemic risk factors for atherosclerosis. These findings are supported by recent gene profiling studies of porcine aortic ECs, which found upregulated expression of NF-κB elements, without significant activation (predominantly cytoplasmic localization or p65), and several broad-acting inflammatory cytokines and receptors in a region with disturbed flow (128). This study also identified enhanced antioxidant gene expression, suggesting that increased oxidative stress may occur in regions of disturbed flow. This is intriguing in light of observations from the Lusis laboratory that EC responses to oxidized lipoproteins determine the genetic susceptibility of mice to atherosclerosis (130). It is likely that regional alterations in EC gene expression will influence signaling pathways other than just NF-κB. Furthermore, hemodynamic factors also may modulate endothelial signaling molecule functions by influencing post-translational modifications, such as phosphorylation.

Low-level expression of adhesion molecules and proinflammatory genes by arterial endothelium in normal animals may induce a low-grade inflammatory response and induce the recruitment of mononuclear leukocytes. In fact, intimal macrophages have been detected predominantly at lesion-predisposed sites in normal rabbit aortae (131) as well as in humans (132). Recently, we obtained data from wild-type mice with normal levels of plasma cholesterol indicating that monocytes/macrophages are prevalent in the intima of atherosclerosis-predisposed regions in the aorta. These monocytes are recruited from the blood, and some of them express dendritic cell markers. Furthermore, the abundance of aortic intimal macrophages directly correlates with the predisposition of various mouse strains for developing atherosclerosis when placed on an appropriate genetic background (133–135). In the setting of hypercholesterolemia, lipoproteins accumulate and are modified in the arterial intima, and we propose that intimal macrophages, which are strategically located in these regions, become activated and secrete proinflammatory factors that initiate and amplify an inflammatory response. A better understanding of the milieu in which atherosclerotic lesion form and the factors that predispose certain regions to atherosclerosis may lead to novel therapeutic approaches directed at ECs and at mediators of inflammation that may alleviate or delay the progression of atherosclerosis.

KEY POINTS

- ECs regulate key aspects of atherogenesis ranging from the earliest stages to advanced lesions.
- ECs display a distinct phenotype in arterial regions predisposed to atherosclerosis.
- ECs regulate inflammation that occurs during atherogenesis by expressing adhesion molecules and presenting chemokines.
- Leukocytes play a critical role in the formation of atherosclerotic lesions and adhesion molecules, and chemokines are important mediators of leukocyte biology during atherogenesis.

Future Goals

- To elucidate the molecular mechanisms of how unique hemodynamic forces at arterial regions predisposed to atherosclerosis alter EC biology
- To develop new approaches to evaluate atherosclerotic lesion formation in murine models, such as surrogate markers and direct measurement of leukocyte recruitment into lesions, and use these to gain novel insights into the functions of adhesion molecules and chemokines

REFERENCES

1 Gimbrone MA Jr., Cybulsky MI, Kume N, et al. Vascular endothelium. An integrator of pathophysiological stimuli in atherogenesis. *Ann NY Acad Sci.* 1995;748:122–131; Discussion 31–32.

2 Pober JS, Cotran RS. Cytokines and endothelial cell biology. *Physiol Rev.* 1990;70:427–451.

3 Davies PF. Hemodynamics in the determination of endothelial phenotype and flow mechanotransduction. In: Aird W, ed., *Endothelial Biomedicine.* New York: Cambridge University Press, 2007.

4 Fuster V, Ross R, Topol EJ. *Atherosclerosis and Coronary Artery Disease.* Philadelphia: Lippincott-Raven, 1996.

5 Falk E, Shah PK. Pathogenesis of atherothrombosis – role of vulnerable, ruptured, and eroded plaques. In: Fuster V, Topol EJ, Nabel EG, eds. *Atherothrombosis and Coronary Artery Disease,* 2nd edition Philadelphia: Lippincott Williams & Wilkins; 2005;451–465.

6 Glagov S, Weisenberg E, Zarins CK, et al. Compensatory enlargement of human atherosclerotic coronary arteries. *N Engl J Med.* 1987;316:1371–1375.

7 Stary HC, Chandler AB, Glagov S, et al. A definition of initial, fatty streak, and intermediate lesions of atherosclerosis. A report from the Committee on Vascular Lesions of the Council on Arteriosclerosis, American Heart Association. *Circulation.* 1994;89: 2462–2478.

8 Stary HC, Chandler AB, Dinsmore RE, et al. A definition of advanced types of atherosclerotic lesions and a histological classification of atherosclerosis. A report from the Committee on Vascular Lesions of the Council on Arteriosclerosis, American Heart Association. *Circulation.* 1995;92:1355–1374.

9 Hansson GK. Inflammation, atherosclerosis, and coronary artery disease. *N Engl J Med.* 2005;352:1685–1695.

10 Buffon A, Biasucci LM, Liuzzo G, et al. Widespread coronary inflammation in unstable angina. *N Engl J Med.* 2002;347:5–12.

11 Plump AS, Smith JD, Hayek T, et al. Severe hypercholesterolemia and atherosclerosis in apolipoprotein E-deficient mice created by homologous recombination in ES cells. *Cell.* 1992;71:343–353.

12 Zhang SH, Reddick RL, Piedrahita JA, Maeda N. Spontaneous hypercholesterolemia and arterial lesions in mice lacking apolipoprotein E. *Science.* 1992;258:468–471.

13 Ishibashi S, Brown MS, Goldstein JL, et al. Hypercholesterolemia in low density lipoprotein receptor knockout mice and its reversal by adenovirus-mediated gene delivery. *J Clin Invest.* 1993;92:883–893.

14 Nakashima Y, Plump AS, Raines EW, et al. ApoE-deficient mice develop lesions of all phases of atherosclerosis throughout the arterial tree. *Arterioscler Thromb.* 1994;14:133–140.

15 Reddick RL, Zhang SH, Maeda N. Atherosclerosis in mice lacking apo E. Evaluation of lesional development and progression. *Arterioscler Thromb.* 1994;14:141–147.

16 Ishibashi S, Goldstein JL, Brown MS, et al. Massive xanthomatosis and atherosclerosis in cholesterol-fed low density lipoprotein receptor-negative mice. *J Clin Invest.* 1994;93:1885–1893.

17 Skalen K, Gustafsson M, Rydberg EK, et al. Subendothelial retention of atherogenic lipoproteins in early atherosclerosis. *Nature.* 2002;417:750–754.

18 Glass CK, Witztum JL. Atherosclerosis. the road ahead. *Cell.* 2001;104:503–516.

19 Steinberg D. Atherogenesis in perspective: hypercholesterolemia and inflammation as partners in crime. *Nat Med.* 2002; 8:1211–1217.

20 Cybulsky MI, Gimbrone MA Jr. Endothelial expression of a mononuclear leukocyte adhesion molecule during atherogenesis. *Science.* 1991;251:788–791.

21 Iiyama K, Hajra L, Iiyama M, et al. Patterns of vascular cell adhesion molecule-1 and intercellular adhesion molecule-1 expression in rabbit and mouse atherosclerotic lesions and at sites predisposed to lesion formation. *Circ Res.* 1999;85:199–207.

22 Nakashima Y, Raines EW, Plump AS, et al. Upregulation of VCAM-1 and ICAM-1 at atherosclerosis-prone sites on the endothelium in the ApoE-deficient mouse. *Arterioscler Thromb Vasc Biol.* 1998;18:842–851.

23 Li H, Cybulsky MI, Gimbrone MA Jr., Libby P. Inducible expression of vascular cell adhesion molecule-1 by vascular smooth muscle cells in vitro and within rabbit atheroma. *Am J Pathol.* 1993;143:1551–1559.

24 Li H, Cybulsky MI, Gimbrone MA Jr., Libby P. An atherogenic diet rapidly induces VCAM-1, a cytokine-regulatable mononuclear leukocyte adhesion molecule, in rabbit aortic endothelium. *Arterioscler Thromb.* 1993;13:197–204.

25 Sakai A, Kume N, Nishi E, et al. P-selectin and vascular cell adhesion molecule-1 are focally expressed in aortas of hypercholesterolemic rabbits before intimal accumulation of macrophages and T lymphocytes. *Arterioscler Thromb Vasc Biol.* 1997;17:310–316.

26 Poston RN, Haskard DO, Coucher JR, et al. Expression of intercellular adhesion molecule-1 in atherosclerotic plaques. *Am J Pathol.* 1992;140:665–673.

27 Printseva O, Peclo MM, Gown AM. Various cell types in human atherosclerotic lesions express ICAM-1. Further immunocytochemical and immunochemical studies employing monoclonal antibody 10F3. *Am J Pathol.* 1992;140:889–896.

28 Wood KM, Cadogan MD, Ramshaw AL, Parums DV. The distribution of adhesion molecules in human atherosclerosis. *Histopathology.* 1993;22:437–444.

29 van der Wal AC, Das PK, Tigges AJ, Becker AE. Adhesion molecules on the endothelium and mononuclear cells in human atherosclerotic lesions. *Am J Pathol.* 1992;141:1427–1433.

30 Davies MJ, Gordon JL, Gearing AJ, al e. The expression of the adhesion molecules ICAM-1, VCAM-1, PECAM, and E-selectin in human atherosclerosis. *J Pathol.* 1993;171:223–229.

31 O'Brien KD, Allen MD, McDonald TO, et al. Vascular cell adhesion molecule-1 is expressed in human coronary atherosclerotic plaques. Implications for the mode of progression of advanced coronary atherosclerosis. *J Clin Invest.* 1993;92:945–951.

32 Munro JM, Cotran RS. The pathogenesis of atherosclerosis: atherogenesis and inflammation. *Lab Invest.* 1988;58:249–261.

33 Ross R. Atherosclerosis – an inflammatory disease. *N Engl J Med.* 1999;340:115–126.

34 Lusis AJ. Atherosclerosis. *Nature.* 2000;407:233–241.

35 Libby P. Inflammation in atherosclerosis. *Nature.* 2002;420: 868–874.

36 Suzuki H, Kurihara Y, Takeya M, et al. A role for macrophage scavenger receptors in atherosclerosis and susceptibility to infection. *Nature.* 1997;386:292–296.

37 Febbraio M, Podrez EA, Smith JD, et al. Targeted disruption of the class B scavenger receptor CD36 protects against atherosclerotic lesion development in mice. *J Clin Invest.* 2000;105:1049–1056.

38 Moore KJ, Kunjathoor VV, Koehn SL, et al. Loss of receptor-mediated lipid uptake via scavenger receptor A or CD36 pathways does not ameliorate atherosclerosis in hyperlipidemic mice. *J Clin Invest.* 2005;115:2192–2201.

39 Michelsen KS, Wong MH, Shah PK, et al. Lack of Toll-like receptor 4 or myeloid differentiation factor 88 reduces atherosclerosis and alters plaque phenotype in mice deficient in apolipoprotein E. *Proc Natl Acad Sci USA.* 2004;101:10679–10684.

40 Bjorkbacka H, Kunjathoor VV, Moore KJ, et al. Reduced atherosclerosis in MyD88-null mice links elevated serum cholesterol levels to activation of innate immunity signaling pathways. *Nat Med.* 2004;10:416–421.

41 Smith JD, Trogan E, Ginsberg M, et al. Decreased atherosclerosis in mice deficient in both macrophage colony-stimulating factor (op) and apolipoprotein E. *Proc Natl Acad Sci USA.* 1995; 92:8264–8268.

42 Qiao JH, Tripathi J, Mishra NK, et al. Role of macrophage colony-stimulating factor in atherosclerosis: studies of osteopetrotic mice. *Am J Pathol.* 1997;150:1687–1699.

43 de Villiers WJ, Smith JD, Miyata M, et al. Macrophage phenotype in mice deficient in both macrophage-colony- stimulating factor (op) and apolipoprotein E. *Arterioscler Thromb Vasc Biol.* 1998;18:631–640.

44 Rajavashisth T, Qiao JH, Tripathi S, et al. Heterozygous osteopetrotic (op) mutation reduces atherosclerosis in LDL receptor- deficient mice. *J Clin Invest.* 1998;101:2702–2710.

45 Hansson GK. Immune mechanisms in atherosclerosis. *Arterioscler Thromb Vasc Biol.* 2001;21:1876–1890.

46 Xu Q. Role of heat shock proteins in atherosclerosis. *Arterioscler Thromb Vasc Biol.* 2002;22:1547–1559.

47 Stemme S, Faber B, Holm J, et al. T lymphocytes from human atherosclerotic plaques recognize oxidized low density lipoprotein. *Proc Natl Acad Sci USA.* 1995;92:3893–3897.

48 de Boer OJ, van der Wal AC, Becker AE. Atherosclerosis, inflammation, and infection. *J Pathol.* 2000;190:237–243.

49 Dansky HM, Charlton SA, Harper MM, Smith JD. T and B lymphocytes play a minor role in atherosclerotic plaque formation in the apolipoprotein E-deficient mouse. *Proc Natl Acad Sci USA.* 1997;94:4642–4646.

50 Song L, Leung C, Schindler C. Lymphocytes are important in early atherosclerosis. *J Clin Invest.* 2001;108:251–259.

51 Zhou X, Nicoletti A, Elhage R, Hansson GK. Transfer of CD4(+) T cells aggravates atherosclerosis in immunodeficient apolipoprotein E knockout mice. *Circulation.* 2000;102:2919–2922.

52 Gupta S, Pablo AM, Jiang X, et al. IFN-gamma potentiates atherosclerosis in ApoE knock-out mice. *J Clin Invest.* 1997;99: 2752–2761.

53 Whitman SC, Ravisankar P, Elam H, Daugherty A. Exogenous interferon-gamma enhances atherosclerosis in apolipoprotein $E^{-/-}$ mice. *Am J Pathol.* 2000;157:1819–1824.

54 Whitman SC, Ravisankar P, Daugherty A. IFN-gamma deficiency exerts gender-specific effects on atherogenesis in apolipoprotein $E^{-/-}$ mice. *J Interferon Cytokine Res.* 2002;22: 661–670.

55 Buono C, Come CE, Stavrakis G, et al. Influence of interferon-gamma on the extent and phenotype of diet-induced atherosclerosis in the LDLR-deficient mouse. *Arterioscler Thromb Vasc Biol.* 2003;23:454–460.

56 Buono C, Binder CJ, Stavrakis G, et al. T-bet deficiency reduces atherosclerosis and alters plaque antigen-specific immune responses. *Proc Natl Acad Sci USA.* 2005;102:1596–1601.

57 Robertson AK, Rudling M, Zhou X, et al. Disruption of TGF-beta signaling in T cells accelerates atherosclerosis. *J Clin Invest.* 2003;112:1342–1350.

58 Mallat Z, Besnard S, Duriez M, et al. Protective role of interleukin-10 in atherosclerosis. *Circ Res.* 1999;85:E17–E24.

59 Pinderski LJ, Fischbein MP, Subbanagounder G, et al. Overexpression of interleukin-10 by activated T lymphocytes inhibits atherosclerosis in LDL receptor-deficient Mice by altering lymphocyte and macrophage phenotypes. *Circ Res.* 2002;90:1064–1071.

60 Caligiuri G, Rudling M, Ollivier V, et al. Interleukin-10 deficiency increases atherosclerosis, thrombosis, and low-density lipoproteins in apolipoprotein E knockout mice. *Mol Med.* 2003; 9:10–17.

61 Caligiuri G, Nicoletti A, Poirier B, Hansson GK. Protective immunity against atherosclerosis carried by B cells of hypercholesterolemic mice. *J Clin Invest.* 2002;109:745–753.

62 Binder CJ, Horkko S, Dewan A, et al. Pneumococcal vaccination decreases atherosclerotic lesion formation: molecular mimicry between Streptococcus pneumoniae and oxidized LDL. *Nat Med.* 2003;9:736–743.

63 Springer TA. Traffic signals for lymphocyte recirculation and leukocyte emigration: the multistep paradigm. *Cell.* 1994;76: 301–314.

64 Butcher EC, Picker LJ. Lymphocyte homing and homeostasis. *Science.* 1996;272:60–66.

65 Alon R, Hammer DA, Springer TA. Lifetime of the P-selectin-carbohydrate bond and its response to tensile force in hydrodynamic flow. *Nature.* 1995;374:539–542.

66 Alon R, Chen S, Puri KD, et al. The kinetics of L-selectin tethers and the mechanics of selectin- mediated rolling. *J Cell Biol.* 1997;138:1169–1180.

67 Middleton J, Neil S, Wintle J, et al. Transcytosis and surface presentation of IL-8 by venular endothelial cells. *Cell.* 1997;91:385–395.

68 Tanaka Y, Adams DH, Hubscher S, et al. T-cell adhesion induced by proteoglycan-immobilized cytokine MIP-1 beta. *Nature.* 1993;361:79–82.

69 Lloyd AR, Oppenheim JJ, Kelvin DJ, Taub DD. Chemokines regulate T cell adherence to recombinant adhesion molecules and extracellular matrix proteins. *J Immunol.* 1996;156:932–938.

70 Campbell JJ, Qin S, Bacon KB, et al. Biology of chemokine and classical chemoattractant receptors: differential requirements for adhesion-triggering versus chemotactic responses in lymphoid cells. *J Cell Biol.* 1996;134:255–266.

71 Campbell JJ, Hedrick J, Zlotnik A, et al. Chemokines and the arrest of lymphocytes rolling under flow conditions. *Science.* 1998;279:381–384.

72 Campbell JJ, Butcher EC. Chemokines in tissue-specific and microenvironment-specific lymphocyte homing. *Curr Opin Immunol.* 2000;12:336–341.

73 Ley K. Pathways and bottlenecks in the web of inflammatory adhesion molecules and chemoattractants. *Immunol Res.* 2001; 24:87–95.

74 Johnson RC, Chapman SM, Dong ZM, et al. Absence of P-selectin delays fatty streak formation in mice. *J Clin Invest.* 1997;99:1037–1043.

75 Collins RG, Velji R, Guevara NV, et al. P-Selectin or intercellular adhesion molecule (ICAM)-1 deficiency substantially protects against atherosclerosis in apolipoprotein E- deficient mice. *J Exp Med.* 2000;191:189–194.

76 Dong ZM, Brown AA, Wagner DD. Prominent role of P-selectin in the development of advanced atherosclerosis in ApoE-deficient mice. *Circulation.* 2000;101:2290–2295.

77 Burger PC, Wagner DD. Platelet P-selectin facilitates atherosclerotic lesion development. *Blood.* 2003;101:2661–2666.

78 Dong ZM, Chapman SM, Brown AA, et al. The combined role of P- and E-selectins in atherosclerosis. *J Clin Invest.* 1998;102:145–152.

79 Methia N, Andre P, Denis CV, et al. Localized reduction of atherosclerosis in von Willebrand factor-deficient mice. *Blood.* 2001;98:1424–1428.

80 Massberg S, Brand K, Gruner S, et al. A critical role of platelet adhesion in the initiation of atherosclerotic lesion formation. *J Exp Med.* 2002;196:887–896.
81 Huo Y, Schober A, Forlow SB, et al. Circulating activated platelets exacerbate atherosclerosis in mice deficient in apolipoprotein E. *Nat Med.* 2003;9:61–67.
82 Schober A, Manka D, von Hundelshausen P, et al. Deposition of platelet RANTES triggering monocyte recruitment requires P-selectin and is involved in neointima formation after arterial injury. *Circulation.* 2002;106:1523–1529.
83 Gurtner GC, Davis V, Li H, et al. Targeted disruption of the murine VCAM1 gene: essential role of VCAM-1 in chorioallantoic fusion and placentation. *Genes Dev.* 1995;9:1–14.
84 Kwee L, Baldwin HS, Shen HM, et al. Defective development of the embryonic and extraembryonic circulatory systems in vascular cell adhesion molecule (VCAM-1) deficient mice. *Development.* 1995;121:489–503.
85 Yang JT, Rayburn H, Hynes RO. Cell adhesion events mediated by alpha 4 integrins are essential in placental and cardiac development. *Development.* 1995;121:549–560.
86 Cybulsky MI, Iiyama K, Li H, et al. A major role for VCAM-1, but not ICAM-1, in early atherosclerosis. *J Clin Invest.* 2001; 107:1255–1262.
87 Dansky HM, Barlow CB, Lominska C, et al. Adhesion of monocytes to arterial endothelium and initiation of atherosclerosis are critically dependent on vascular cell adhesion molecule-1 gene dosage. *Arterioscler Thromb Vasc Biol.* 2001;21:1662–1667.
88 Shih PT, Brennan ML, Vora DK, et al. Blocking very late antigen-4 integrin decreases leukocyte entry and fatty streak formation in mice fed an atherogenic diet. *Circ Res.* 1999;84:345–351.
89 Huo Y, Hafezi-Moghadam A, Ley K. Role of vascular cell adhesion molecule-1 and fibronectin connecting segment-1 in monocyte rolling and adhesion on early atherosclerotic lesions. *Circ Res.* 2000;87:153–159.
90 Kubo N, Boisvert WA, Ballantyne CM, Curtiss LK. Leukocyte CD11b expression is not essential for the development of atherosclerosis in mice [In process citation]. *J Lipid Res.* 2000; 41:1060–1066.
91 Nelken NA, Coughlin SR, Gordon D, Wilcox JN. Monocyte chemoattractant protein-1 in human atheromatous plaques. *J Clin Invest.* 1991;88:1121–1127.
92 Yu X, Dluz S, Graves DT, et al. Elevated expression of monocyte chemoattractant protein 1 by vascular smooth muscle cells in hypercholesterolemic primates. *Proc Natl Acad Sci USA.* 1992;89:6953–6957.
93 Wong BW, Wong D, McManus BM. Characterization of fractalkine (CX3CL1) and CX3CR1 in human coronary arteries with native atherosclerosis, diabetes mellitus, and transplant vascular disease. *Cardiovasc Pathol.* 2002;11:332–338.
94 Steinberg D. Lewis A. Conner Memorial Lecture. Oxidative modification of LDL and atherogenesis. *Circulation.* 1997;95: 1062–1071.
95 Cushing SD, Berliner JA, Valente AJ, et al. Minimally modified low density lipoprotein induces monocyte chemotactic protein 1 in human endothelial cells and smooth muscle cells. *Proc Natl Acad Sci USA.* 1990;87:5134–5138.
96 Aiello RJ, Bourassa PA, Lindsey S, et al. Monocyte chemoattractant protein-1 accelerates atherosclerosis in apolipoprotein E-deficient mice. *Arterioscler Thromb Vasc Biol.* 1999;19:1518–1525.
97 Gosling J, Slaymaker S, Gu L, et al. MCP-1 deficiency reduces susceptibility to atherosclerosis in mice that overexpress human apolipoprotein B. *J Clin Invest.* 1999;103:773–778.
98 Gu L, Okada Y, Clinton SK, et al. Absence of monocyte chemoattractant protein-1 reduces atherosclerosis in low density lipoprotein receptor-deficient mice. *Mol Cell.* 1998;2:275–281.
99 Boring L, Gosling J, Cleary M, Charo IF. Decreased lesion formation in CCR2-/- mice reveals a role for chemokines in the initiation of atherosclerosis. *Nature.* 1998;394:894–897.
100 Dawson TC, Kuziel WA, Osahar TA, Maeda N. Absence of CC chemokine receptor-2 reduces atherosclerosis in apolipoprotein E-deficient mice. *Atherosclerosis.* 1999;143:205–211.
101 Egashira K, Koyanagi M, Kitamoto S, et al. Anti-monocyte chemoattractant protein-1 gene therapy inhibits vascular remodeling in rats: blockade of MCP-1 activity after intramuscular transfer of a mutant gene inhibits vascular remodeling induced by chronic blockade of NO synthesis. *FASEB J.* 2000;14:1974–1978.
102 Serbina NV, Pamer EG. Monocyte emigration from bone marrow during bacterial infection requires signals mediated by chemokine receptor CCR2. *Nat Immunol.* 2006;7:311–317.
103 Han KH, Tangirala RK, Green SR, Quehenberger O. Chemokine receptor CCR2 expression and monocyte chemoattractant protein-1-mediated chemotaxis in human monocytes. A regulatory role for plasma LDL. *Arterioscler Thromb Vasc Biol.* 1998; 18:1983–1991.
104 Han KH, Han KO, Green SR, Quehenberger O. Expression of the monocyte chemoattractant protein-1 receptor CCR2 is increased in hypercholesterolemia. Differential effects of plasma lipoproteins on monocyte function. *J Lipid Res.* 1999;40:1053–1063.
105 Han KH, Chang MK, Boullier A, et al. Oxidized LDL reduces monocyte CCR2 expression through pathways involving peroxisome proliferator-activated receptor gamma. *J Clin Invest.* 2000;106:793–802.
106 Li M, Pascual G, Glass CK. Peroxisome proliferator-activated receptor gamma-dependent repression of the inducible nitric oxide synthase gene. *Mol Cell Biol.* 2000;20:4699–4707.
107 Szalai C, Duba J, Prohaszka Z, et al. Involvement of polymorphisms in the chemokine system in the susceptibility for coronary artery disease (CAD). Coincidence of elevated Lp(a) and MCP-1 -2518 G/G genotype in CAD patients. *Atherosclerosis.* 2001;158:233–239.
108 Bazan JF, Bacon KB, Hardiman G, et al. A new class of membrane-bound chemokine with a CX3C motif. *Nature.* 1997;385:640–644.
109 Pan Y, Lloyd C, Zhou H, et al. Neurotactin, a membrane-anchored chemokine upregulated in brain inflammation. *Nature.* 1997;387:611–617.
110 Tsou CL, Haskell CA, Charo IF. Tumor necrosis factor-alpha-converting enzyme mediates the inducible cleavage of fractalkine. *J Biol Chem.* 2001;276:44622–44626.
111 Garton KJ, Gough PJ, Blobel CP, et al. Tumor necrosis factor-alpha-converting enzyme (ADAM17) mediates the cleavage and shedding of fractalkine (CX3CL1). *J Biol Chem.* 2001;276: 37993–38001.
112 Imai T, Hieshima K, Haskell C, et al. Identification and molecular characterization of fractalkine receptor CX3CR1, which mediates both leukocyte migration and adhesion. *Cell.* 1997;91:521–530.

113 Fong AM, Robinson LA, Steeber DA, et al. Fractalkine and CX3CR1 mediate a novel mechanism of leukocyte capture, firm adhesion, and activation under physiologic flow. *J Exp Med.* 1998;188:1413–1419.
114 Haskell CA, Cleary MD, Charo IF. Molecular uncoupling of fractalkine-mediated cell adhesion and signal transduction. Rapid flow arrest of CX3CR1-expressing cells is independent of G-protein activation. *J Biol Chem.* 1999;274:10053–10058.
115 Faure S, Meyer L, Costagliola D, et al. Rapid progression to AIDS in HIV+ individuals with a structural variant of the chemokine receptor CX3CR1. *Science.* 2000;287:2274–2277.
116 McDermott DH, Halcox JP, Schenke WH, et al. Association between polymorphism in the chemokine receptor CX3CR1 and coronary vascular endothelial dysfunction and atherosclerosis. *Circ Res.* 2001;89:401–407.
117 Moatti D, Faure S, Fumeron F, et al. Polymorphism in the fractalkine receptor CX3CR1 as a genetic risk factor for coronary artery disease. *Blood.* 2001;97:1925–1928.
118 McDermott DH, Fong AM, Yang Q, et al. Chemokine receptor mutant CX3CR1-M280 has impaired adhesive function and correlates with protection from cardiovascular disease in humans. *J Clin Invest.* 2003;111:1241–1250.
119 Lesnik P, Haskell CA, Charo IF. Decreased atherosclerosis in CX(3)CR1(–/–) mice reveals a role for fractalkine in atherogenesis. *J Clin Invest.* 2003;111:333–340.
120 Combadiere C, Potteaux S, Gao J-L, et al. Decreased atherosclerotic lesion formation in CX3CR1/apolipoprotein E double knockout mice. *Circulation.* 2003;107:1009–1016.
121 Teupser D, Pavlides S, Tan M, et al. Major reduction of atherosclerosis in fractalkine (CX3CL1)-deficient mice is at the brachiocephalic artery, not the aortic root. *Proc Natl Acad Sci USA.* 2004;101:17795–17800.
122 Gimbrone MA Jr., Nagel T, Topper JN. Biomechanical activation: an emerging paradigm in endothelial adhesion biology. *J Clin Invest.* 1997;100:S61–S65.
123 Topper JN, Cai J, Falb D, Gimbrone MA Jr. Identification of vascular endothelial genes differentially responsive to fluid mechanical stimuli: cyclooxygenase-2, manganese superoxide dismutase, and endothelial cell nitric oxide synthase are selectively up-regulated by steady laminar shear stress. *Proc Natl Acad Sci USA.* 1996;93:10417–10422.
124 Dai G, Kaazempur-Mofrad MR, Natarajan S, et al. Distinct endothelial phenotypes evoked by arterial waveforms derived from atherosclerosis-susceptible and -resistant regions of human vasculature. *Proc Natl Acad Sci USA.* 2004;101:14871–14876.
125 Nagel T, Resnick N, Atkinson WJ, et al. Shear stress selectively upregulates intercellular adhesion molecule-1 expression in cultured human vascular endothelial cells. *J Clin Invest.* 1994; 94:885–891.
126 Ando J, Tsuboi H, Korenaga R, et al. Shear stress inhibits adhesion of cultured mouse endothelial cells to lymphocytes by downregulating VCAM-1 expression. *Am J Physiol.* 1994;267: C679–C687.
127 Chiu JJ, Lee PL, Chen CN, et al. Shear stress increases ICAM-1 and decreases VCAM-1 and E-selectin expressions induced by tumor necrosis factor-[alpha] in endothelial cells. *Arterioscler Thromb Vasc Biol.* 2004;24:73–79.
128 Passerini AG, Polacek DC, Shi C, et al. Coexisting proinflammatory and antioxidative endothelial transcription profiles in a disturbed flow region of the adult porcine aorta. *Proc Natl Acad Sci USA.* 2004;101:2482–2487.
129 Hajra L, Evans AI, Chen M, et al. The NF-kappa B signal transduction pathway in aortic endothelial cells is primed for activation in regions predisposed to atherosclerotic lesion formation [In process citation]. *Proc Natl Acad Sci USA.* 2000;97:9052–9057.
130 Shi W, Haberland ME, Jien ML, et al. Endothelial responses to oxidized lipoproteins determine genetic susceptibility to atherosclerosis in mice. *Circulation.* 2000;102:75–81.
131 Malinauskas RA, Herrmann RA, Truskey GA. The distribution of intimal white blood cells in the normal rabbit aorta. *Atherosclerosis.* 1995;115:147–163.
132 Bobryshev YV. Dendritic cells and their involvement in atherosclerosis. *Curr Opin Lipidol.* 2000;11:511–517.
133 Jongstra-Bilen J, Haidari M, Zhu S-N, et al. Low-grade inflammation in regions of the normal mouse arterial intima predisposed to atherosclerosis. *J Exp Med.* 2006;203:2073–2083.
134 Braun A, Trigatti BL, Post MJ, et al. Loss of SR-BI expression leads to the early onset of occlusive atherosclerotic coronary artery disease, spontaneous myocardial infarctions, severe cardiac dysfunction, and premature death in apolipoprotein E-deficient mice. *Circ Res.* 2002;90:270–276.
135 Covey SD, Krieger M, Wang W, et al. Scavenger receptor class B type I-mediated protection against atherosclerosis in LDL receptor-negative mice involves its expression in bone marrow-derived cells. *Arterioscler Thromb Vasc Biol.* 2003;23:1589–1594.

133

The Hepatic Sinusoidal Endothelial Cell

Laurie D. DeLeve

University of Southern California, Keck School of Medicine, Los Angeles

The hepatic sinusoidal endothelial cell (EC) was not recognized as a highly differentiated cell type until the sinusoid was examined by perfusion fixation combined with electron microscopy, as described in Wisse's seminal papers in 1970 and 1972 (1,2). Our understanding of hepatic sinusoidal EC characteristics took an additional step forward with the first description of a method to isolate a highly pure population of these cells (3,4).

MORPHOLOGY OF HEPATIC SINUSOIDAL ENDOTHELIAL CELL AND THE SINUSOID

The hepatic sinusoids (Figure 133.1) form the equivalent of a capillary system. The hepatic sinusoid lacks an organized basement membrane on the abluminal side of the hepatic sinusoidal ECs (SECs). The virtual space between the hepatic SECs and the hepatocytes is called the space of Disse. The space of Disse contains loosely organized extracellular matrix and resident pericytes that surround the hepatic SECs, the so-called *stellate cells.* Stellate cells are vitamin A–storing cells. In addition, they are contractile and thus capable of regulating sinusoidal diameter. On the luminal side of the hepatic sinusoidal endothelium are the resident macrophages, the Kupffer cells. When measured in vivo by light microscopy, the diameter of the hepatic sinusoid ranges from 6 to 7 μm, increasing slightly from the periportal to the centrilobular area. Sinusoids are smaller than neutrophils, which measure approximately 8.5 μm. It has been hypothesized that, because leukocytes are larger than the sinusoids, the entry of leukocytes into the sinusoid compresses the space of Disse and forces fluid out through fenestrations within hepatic SECs (discussed later), whereas with passage of the leukocyte back out of the sinusoid the space of Disse will restore the original shape and cause suction of fresh fluid back into the space (*forced sieving*). In addition, the compression of the sinusoidal cells by the leukocytes is postulated to promote mixing of the fluid within the space of Disse (*endothelial massage*) (5).

Microvascular endothelium may be classified according to whether the cell layer is continuous, discontinuous, or fenestrated; and whether it possesses an organized basement membrane on the abluminal surface. In continuous endothelium (e.g., capillaries in skin, lung, central nervous system [CNS], and muscle), the cytoplasm is continuous and the cells are connected by tight junctions (6). With the exception of the blood–brain barrier, continuous endothelium contains abundant caveolae. Discontinuous ECs have gaps between or within cells. Fenestrated ECs (e.g., capillaries of exocrine and endocrine glands, intestinal mucosa, choroid plexus, and glomeruli) form a subset of discontinuous endothelium in which ECs possess round openings or pores that transverse the body of the cell, connecting the luminal and abluminal surfaces. Hepatic SECs have fenestrations. The fenestrae are supported by a cytoskeletal ring made up of actin and myosin; they are therefore dynamic structures that can contract or dilate (7).

Endothelial fenestrae are further classified according to whether they possess a diaphragm or are open. Examples of ECs with fenestrae subtended by diaphragms are those cells in the renal peritubular capillaries and ascending vasa recta; capillaries of intestinal villi, pancreas, adrenal cortex, and endocrine glands; choriocapillaries of the brain and eye (8); and the bone marrow sinusoid (9). In mammals, only the hepatic SECs and glomerular ECs have nondiaphragmed (open) fenestrae. However, glomerular ECs have an organized basement membrane, whereas SECs do not. Thus, the phenotype of hepatic SECs is unique among all endothelium in that it is characterized by open fenestrae in the absence of an organized basement membrane. Hepatic SEC fenestrae are between 150 to 175 nm, as visualized by transmission electron microscopy or 105 to 110 nm, as ascertained by scanning electron microscopy, with somewhat larger fenestrae in the periportal region than in the centrilobular region of the liver lobule (5). As with other fenestrated ECs, the fenestrae are organized in clusters called *sieve plates.*

The term *sinusoidal* refers to capillaries that are tortuous and in intimate contact with underlying parenchymal cells. In mammals, three types of sinusoidal ECs exist: those in the bone marrow, liver, and spleen. In each case, the endothelium is discontinuous. Splenic SECs are rod-shaped cells running parallel to the longitudinal axis of the sinus and having

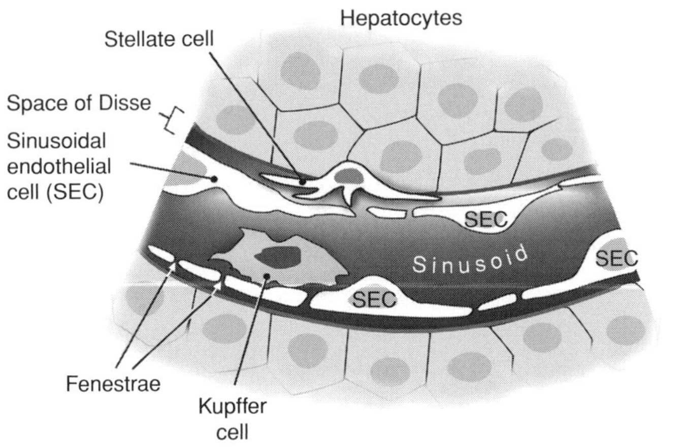

Figure 133.1. Schematic of the hepatic sinusoid.

continuous cytoplasm. The basement membrane of splenic SECs is ring-like and runs transverse to the ECs (10–12). Bone marrow SECs possess fenestrae with a diaphragm and an underlying basement membrane. The passage of red cells through the walls of the splenic and bone marrow sinus occurs through interendothelial slits that appear when cells migrate through the wall. In the remainder of the chapter, the abbreviation SEC will be used to refer to hepatic SEC.

CELL–CELL CROSS-TALK BETWEEN SINUSOIDAL ENDOTHELIAL CELLS AND OTHER CELL TYPES IN THE LIVER

The unique SEC phenotype depends on their proximity to other liver cells (13). Vascular endothelial growth factor (VEGF) is one of the major growth factors responsible for fenestration of ECs (14), a key feature of SECs. Paracrine secretion of VEGF by adjacent hepatocytes and stellate cells is essential to maintain the normal SEC phenotype (15). VEGF stimulates nitric oxide (NO) production by endothelial NO synthase (eNOS) in ECs (16–19), and autocrine production of NO is an essential regulator of SEC properties (15). However, high constitutive expression of VEGF by neighboring epithelial cells is also characteristic of other types of fenestrated endothelium, including the glomerulus and choroid plexus, which are morphologically and functionally distinct (20). Thus, in the liver, additional microenvironmentally derived paracrine factors are likely to contribute to the maintenance of this unique SEC phenotype.

Hepatocytes and stellate cells are necessary to maintain a normal SEC phenotype. Conversely, SECs play a critical role in postinjury liver regeneration and hepatic morphogenesis during development. For example, during liver regeneration, SECs are indispensable for promoting hepatocyte growth in a complex, interdependent process driven by paracrine mediators secreted by SECs, hepatocytes, and perhaps other types of liver cells (21–25). The mediators best characterized in the regulation of liver regeneration are hepatocyte growth factor (HGF) secreted by SECs (21), and VEGF secreted by hepatocytes (22,25,26), but other growth factors are also involved (23,24). Analogous to the critical role played by SECs in liver regeneration, hepatic ECs are also essential for liver organogenesis (27,28). Both vasculogenic ECs and nascent vessels are needed for the earliest stages of hepatic morphogenesis, prior to the development of functional blood vessels (27).

FUNCTIONAL CONSEQUENCES OF SINUSOIDAL ENDOTHELIAL CELL PHENOTYPE

Oxygen Delivery

The liver (like the lung) receives a dual blood supply. The hepatic artery – which branches off the abdominal aorta – delivers oxygen (O_2) to the various cells of the liver, whereas the portal vein delivers poorly oxygenated, nutrient-rich blood from the intestines. Both the hepatic artery and portal vein drain into the hepatic sinusoids. The admixture of arterial and venous blood results in a lower-than-normal capillary Po_2. According to Fick's law of diffusion, O_2 delivery is proportional to the O_2 gradient from blood to tissue and surface area, and inversely correlated with the distance. Although the Po_2 (hence O_2) gradient is reduced, the unique properties of the sinusoid serve to optimize the other variables of the equation. For example, the presence of thin cytoplasm, open fenestrae, and the lack of an organized basement membrane reduce the distance that O_2 must diffuse. Moreover, the flattened SECs cover an extremely large surface area. The importance of these adaptations is apparent in liver disease, in which the loss of SEC fenestration and the formation of an organized basement membrane (a process termed *capillarization*) lead to reduced oxygenation of hepatocytes (see the section on Capillarization and Pseudocapillarization).

Sieve Function

Because the fenestrae of SECs do not have diaphragms, and because no organized basement membrane exists, SECs are able to filter fluids, solutes and particles, chylomicron remnants and other lipoproteins, and protein-bound drugs. As a result, plasma constituents gain ready access to the space of Disse, where they may be taken up by the hepatocytes. Two examples of sieve function are discussed here: chylomicron trafficking and filtering of protein-bound drugs.

Dietary lipids are assembled with specific apolipoproteins to form chylomicrons, the largest circulating lipoproteins (>100 nm diameter). Chylomicrons are metabolized by endothelial lipoprotein lipase in the systemic circulation to form chylomicron remnants. In the liver circulation, chylomicron remnants less than 100 nm pass through the open fenestrae into the space of Disse (29–31). The role of the liver in clearance of chylomicron remnants is confirmed by the accumulation of chylomicron remnants in the blood after functional hepatectomy (32). The sieve function of SECs was demonstrated by studies using transmission electron microscopy, which showed the size differential of chylomicron remnants in sinusoidal blood versus those in the space of Disse (29–31).

Freer access of hepatocytes to plasma is also important for drug metabolism. For high-clearance drugs, unbound drug is taken up by hepatocytes, the bound and unbound fractions reequilibrate, and additional unbound drug is taken up during a single pass through the liver. Thus, fenestration allows the amount of drug taken up by the liver to exceed the unbound fraction present in the afferent circulation.

Scavenger Function and Immunity

SECs and Kupffer cells are the two hepatic members of the reticuloendothelial system. SECs play an important role in innate immunity. SECs act as scavenger cells, clearing circulating waste macromolecules and antigens. Several properties of the sinusoid suggest that they are uniquely suited to this task (33). First, SECs and Kupffer cells are in intimate contact with one another on the luminal side of the endothelium, and act as "sentinels" for gut- and portal vein–derived macromolecules and antigens. Second, SECs present a broad surface area to circulating blood. Third, sinusoidal flow is slow and intermittent, particularly in the periportal region (where the portal vein branches enter the lobule), prolonging contact time between macromolecules and SECs (34). Fourth, SECs have an active endocytotic apparatus with numerous coated pits and vesicles. The coated pits have exposed cationic residues that facilitate the specific endocytosis of negatively charged molecules taken up by receptor-mediated endocytosis. The specific activity of several lysosomal enzymes is as high or higher in SECs compared with Kupffer cells, and much higher than in hepatocytes, which permits active degradation of the endocytosed macromolecules. Finally, several receptors are present that allow SECs to eliminate an array of ligands by receptor-mediated endocytosis.

At least five types of specific receptors for endocytosis are present on mammalian SECs, and these are responsible for eliminating soluble waste macromolecules from the circulation. These include the: (a) hyaluronan receptor, (b) mannose receptor, (c) collagen α-chain receptor, (d) Fc-γ receptor, and (e) scavenger receptor (Table 133-1) (35). The macromolecules cleared by receptor-mediated endocytosis include physiological waste, such as major matrix molecules and cellular constituents, such as lysosomal enzymes; pathophysiological ligands, such as advanced glycation end products (AGEs) and oxidized low-density lipoprotein (LDL); and exogenous soluble ligands. Although SECs endocytose this colloidal and macromolecular waste, Kupffer cells phagocytose insoluble, or particulate, waste products (35). Clearance of soluble and particulate waste products by the hepatic cells of the reticuloendothelial system contributes significantly to the innate immune system.

SECs do not constitutively express E- or P-selectin; the literature is contradictory as to whether E- or P-selectin is expressed in response to activation. Intercellular adhesion molecule (ICAM)-1 (CD54) and lymphocyte function-associated (LFA)-3 (CD58) are constitutively expressed on SECs, and ICAM-1 is further upregulated upon exposure to inflammatory cytokines. Vascular cell adhesion molecule (VCAM)-1 (CD106) is not constitutively expressed (36–41), but is induced by inflammatory mediators. Although normal SECs express platelet-endothelial cell adhesion molecule (PECAM)-1 (CD31), an adhesion molecule on the cell–cell borders of ECs needed for transendothelial migration, PECAM-1 is expressed intracellularly and not on the SEC surface (15,39).

Table 133-1: Receptor-Mediated Endocytosis of SECs

Specific Receptors	*Ligands*
Hyaluronan receptor	Matrix polysaccharides and proteoglycans (e.g., hyaluronan, dermatan sulfate and chondroitin sulfates)
Collagen α-chain receptor	Collagen α-chains of several types of collagen
Scavenger receptor	Amino-terminal propeptides of types I and III procollagen AGEs Oxidized LDL Oligodeoxynucleotides Modified blood proteins
Mannose receptor	Carboxy-terminal propeptide of type I procollagen Tissue-type plasminogen activator
Fc-γ receptor	IgG-antigen immune complexes

Data used with permission from Seternes T, Sorensen K, Smedsrod B. Scavenger ECs of vertebrates: a nonperipheral leukocyte system for high-capacity elimination of waste macromolecules. *Proc Natl Acad Sci USA*. 2002;99:7594–7597.

SECs may present antigen to $CD4^+$ and $CD8^+$ T cells. In keeping with its antigen-presenting role, SECs have been shown to express major histocompatibility (MHC) class I and class II molecules, and the costimulatory molecules CD80, CD86 (42), and CD40 (43,44). Based on the anatomical and infrastructural properties outlined earlier, the sinusoids are ideally situated for eliminating gut-derived pathogenic organisms. In addition to removing pathogens from the circulation, SECs contribute to the induction of tolerance by presenting MHC class II restricted antigens to naïve $CD4^+$ T cells. This process inhibits the expansion of interferon (IFN)-γ–producing Th1 type $CD4^+$ effector cells (proinflammatory) and promotes the outgrowth of interleukin (IL)-4/IL-10–expressing Th2 type $CD4^+$ effector cells (antiinflammatory) (45). SECs also cross-present soluble antigens on MHC class I molecules, such as food antigens or self-proteins to $CD8^+$ T cells, resulting in the induction of hepatic immune tolerance (46). The ability of SECs to induce tolerance contrasts with myeloid antigen–presenting cells, which typically induce protective immunity when cross-presenting antigens to $CD8^+$ T cells. The propensity for induction of tolerance occurs under the influence of immunosuppressive mediators normally present in the hepatic microenvironment, including IL-10, prostaglandin E2, and transforming growth factor (TGF)-β (43). The induction of peripheral immune tolerance by the liver permits allogeneic liver transplantation despite MHC incompatibility, leads to immune tolerance of antigens present in the portal circulation, and results in increased acceptance of grafts when the venous drainage is routed into the portal vein (44). Hepatocytes metabolize numerous xenobiotics and are at high risk of immune attack against neoantigens formed on the cell surface by protein adducts of highly reactive metabolites formed by the hepatocyte. It has been speculated that the SEC barrier protects hepatocytes from immune attack by inducing tolerance to such neoantigens (44).

CAPILLARIZATION AND PSEUDOCAPILLARIZATION

Liver disease and aging are associated with altered SEC phenotype. In experimental animals and in humans, liver fibrosis (a consequence of many disease states) is preceded by the loss of fenestration and the formation of a fully organized basement membrane, a process termed *capillarization* (47–51). In aged experimental animals and in humans over age 40, a similar albeit less severe process occurs (termed *pseudocapillarization*), in which there is decrease or loss of fenestration and partial formation of a basement membrane (52,53).

As noted earlier, the porosity of SECs is critical for O_2 transfer to hepatocytes and clearance of fluids and solutes, including protein-bound drugs and waste products. Thus, capillarization and pseudocapillarization would be predicted to diminish hepatocyte function. Indeed, capillarization in the cirrhotic liver leads to reduced O_2 delivery to underlying hepatocytes (52), resulting in impaired O_2-dependent hepatocellular functions such as oxidative drug metabolism (54–56). Similarly, pseudocapillarization in the aged rat is associated with a decline in high-energy phosphate and other metabolites, consistent with hepatocyte hypoxia (52). It has been suggested that pseudocapillarization impairs the clearance of chylomicron remnants by the liver, and that this effect may contribute to age-related atherosclerosis (5,57,58). Pseudocapillarization also may help to explain the well-established association between aging and reduced drug metabolism, both by limiting clearance of high-clearance protein-bound drugs and by decreasing O_2 delivery required for oxidative drug metabolism (e.g., P450 metabolism).

With capillarization, a change also occurs in receptors and adhesion molecules on the surface of SECs. SECs are the major cell type responsible for removing glycosaminoglycans from the blood. One measure of cirrhosis is increased serum hyaluronan, which reflects increased production by stellate cells and decreased clearance by SECs (59,60–62). The major receptor for hyaluronan clearance is the hyaluronan receptor for endocytosis (HARE) (63,64). Given that hyaluronan clearance is decreased in cirrhosis, one might predict capillarization to be associated with reduced SEC expression of HARE. However, this has not yet been documented. Lymphatic vessel endothelial receptor (LYVE)-1, a hyaluronan receptor typically associated with lymphatic ECs, is also present on normal SECs (65). Although LYVE-1 is decreased on SECs in cirrhosis (65), LYVE-1 seems to play a minor role in hyaluronan clearance by SECs (64).

As discussed earlier, PECAM-1 is normally present in the cytoplasm of SECs, but not on the cell–cell border. Interestingly, capillarization is associated with localization of PECAM-1 at the intercellular junction (15,39,66). It remains to be established what impact this redistribution of PECAM-1 has on SEC function.

As a final example of capillarization-mediated effects on SEC receptors, the laminin receptors, $\alpha6\beta1$ and $\alpha2\beta1$, are induced with capillarization and colocalize with increased laminin deposits along the sinusoids (66).

DRUG-INDUCED TOXICITY TO SINUSOIDAL ENDOTHELIAL CELLS

Several factors render SECs particularly vulnerable to drug-induced injury. First, because orally ingested drugs gain entry to the systemic circulation via the portal vein, SECs are exposed to higher blood levels of the drug compared with other parts of the circulation. Second, SECs display significant P450 activity and can therefore readily metabolize blood-borne drugs (in some cases to more toxic intermediates) (67–69). Finally, in the process of metabolizing drugs, hepatocytes release toxic metabolites into the space of Disse, creating a concentration gradient between the hepatocyte and the lumen of the sinusoid. Given the close proximity between SECs and hepatocytes, and the lack of an intervening organized basement membrane, SECs are exposed to high levels of hepatocyte-derived toxins. This fact may explain why mild drug toxicity preferentially targets SECs, whereas more marked toxic insults are required for damage to hepatic venular ECs and other vascular beds.

Forms of toxic liver injury that may be initiated by damage to SECs include sinusoidal obstruction syndrome (SOS; also known as veno-occlusive disease), nodular regenerative hyperplasia (NRH), sinusoidal dilatation, and peliosis hepatis. All four of these diseases are characterized by damage to SECs and/or hepatic venular ECs. A number of drugs have been linked to the development of at least two of these lesions. Indeed, in some cases, a single drug has been implicated in the development of all four lesions in the same liver. Among the most common drugs that lead to SEC toxicity are azathioprine (all four lesions), 6-thioguanine (peliosis hepatis, NRH, SOS), urethane (peliosis hepatis, SOS), thorotrast (peliosis hepatis, NRH), oral contraceptives (sinusoidal dilatation, peliosis hepatis, and NRH), and anabolic steroids (peliosis hepatis, NRH) (70–73).

SINUSOIDAL OBSTRUCTION SYNDROME

Overview

This syndrome was originally called *hepatic veno-occlusive disease*, because the most obvious lesions, as detected by light microscopy, are located in the central veins (74). The first description of SOS in humans came from South Africa, in individuals who had ingested bread made from inadequately winnowed wheat contaminated by plants containing pyrrolizidine alkaloids (75). In non-Western nations, SOS is still seen in individuals who ingest pyrrolizidine alkaloids, either in the form of "bush teas" or as contaminants of the food supply. The major plant species implicated are the pyrrolizidine alkaloids *Crotalaria, Heliotropium*, and *Senecio*. In North America and Western Europe, SOS occurs sporadically as a complication of chemotherapy for malignancy or of long-term immunosuppression by azathioprine for kidney and liver transplantation (71,76–80). The chemotherapeutic drugs most clearly associated with SOS at conventional doses are gemtuzumab ozogamicin, actinomycin D, dacarbazine, cytosine arabinoside, mitramycin, 6-thioguanine, and urethane (73,81,82). Treatment of Wilms tumors, and in particular right-sided Wilms tumors, with actinomycin D plus abdominal irradiation is a greater risk factor for SOS than actinomycin D alone (83,84). The most common cause of SOS in Western nations is the myeloablative conditioning therapy used prior to bone marrow transplantation for malignancy. The incidence of SOS in patients undergoing myeloablative conditioning regimens varies from 0% to 50% (85–90). The wide range in incidence in different transplant centers is due to differences in patient selection criteria and differences in conditioning regimens. Chemotherapy regimens that include cyclophosphamide and radiation regimens using higher doses of total body irradiation carry a greater risk for developing SOS. Case fatality rates also varies widely, ranging from 0% to 67%, likely due to differences in diagnostic criteria. In regimens containing cyclophosphamide, the case fatality rate is approximately 30%, and this may be higher than for regimens without cyclophosphamide (85,89,91). The diagnosis usually is made based on the clinical features of painful hepatomegaly, weight gain, and hyperbilirubinemia (86,88). The clinical features of SOS have recently been reviewed in detail (92).

Mechanisms of Sinusoidal Obstruction Syndrome

The clinical presentation of SOS is consistent with a vascular origin of disease. In other intrinsic liver diseases, portal hypertension due to circulatory disruption is a consequence of liver function abnormalities. In contrast, in SOS, ascites due to portal hypertension precedes abnormalities in liver function. Based on the light microscopy findings of venous occlusion, it was assumed for many years that the central vein was the primary culprit in this syndrome. However clinico-pathological correlative studies have demonstrated that involvement of the central veins is not essential for the development of SOS (93). Indeed, studies in an experimental model of SOS suggest that the disease is initiated by changes within the sinusoid (discussed later). Based on the clinical and experimental studies that support the sinusoidal origin of the disease, it was proposed to change the name from hepatic veno-occlusive disease to SOS (92). This name change also serves to distinguish SOS from other forms of liver disease, characterized by occlusions of the hepatic vein, that are referred to as hepatic veno-occlusive disease. These include radiation-induced liver disease and post–liver transplantation occlusion of hepatic veins. Although changes in the sinusoid may be critical for initiating SOS and for mediating the clinical phenotype, the central veins are frequently involved in more severe disease (93). This suggests that the more marked the insult, the more extensive the involvement of distant endothelium.

In Vitro Studies

When SECs are isolated by centrifugal elutriation, yields are approximately 80×10^6 cells per rat liver and approximately 10 to 12×10^6 per mouse liver, with 98% purity or better and 99% viability or better (15). SECs isolated by this method have the characteristic features of fenestrae grouped in sieve plates at the time of isolation. A number of methods are currently in use to isolate SECs, and it remains to be determined whether these disparate methods are isolating the same populations of cells (94).

A variety of drugs and toxins implicated in SOS have been studied in SEC cultures. A common feature of these compounds is that they are more toxic to SECs compared with hepatocytes, and they are detoxified by glutathione (GSH). The following is a brief review of the factors that determine selective toxicity to SECs, as demonstrated by in vitro studies.

Monocrotaline, a pyrrolizidine alkaloid found in so-called "bush teas," is one of the best-characterized toxins that cause SOS in humans. Monocrotaline is converted by P450 to the electrophilic metabolite, monocrotaline pyrrole, and this activation occurs only in the liver. Monocrotaline pyrrole formed in SECs is detoxified by SEC GSH. Thus, when formed in excess, it may lead to profound GSH depletion. Depletion of GSH stores precedes SEC cell death in vitro. Moreover, GSH precursors that prevent monocrotaline-pyrrole–mediated depletion of GSH in SECs prevent cell death (95). Monocrotaline is more toxic to SECs than to hepatocytes, and depletion of hepatocyte GSH does not exacerbate hepatocyte toxicity (95). Taken together, these data suggest that selective toxicity to SECs is related to the formation of monocrotaline pyrrole. Monocrotaline pyrrole formed by SECs and secreted into the sinusoidal lumen is transported by red blood cells to the lung and leads to lung toxicity.

Similar to monocrotaline, dacarbazine also is activated by P450 in a liver-dependent manner. Dacarbazine is toxic to SECs, but not to hepatocytes. Dacarbazine is detoxified by GSH, but GSH depletion does not render hepatocytes susceptible to toxicity (69). Thus, as with monocrotaline, these findings suggest that the selective toxicity to SECs is related to metabolic activation of the drug.

Cyclophosphamide is employed in certain myeloablative regimens in bone marrow transplantation, and is the chemotherapeutic drug most frequently associated with SOS in this setting. Cyclophosphamide is converted by P450 to 4-hydroxycyclophosphamide, which then spontaneously tautomerizes to acrolein and phosphoramide mustard. P450-mediated activation of cyclophosphamide takes place in hepatocytes, but not SECs. Thus, cyclophosphamide is not directly toxic to primary cultures of SECs. However, in coculture studies of SECs with hepatocytes, cyclophosphamide is metabolized by hepatocytes, and the resulting metabolite (namely acrolein) exerts paracrine-mediated toxicity to SECs. Importantly, the toxicity to SECs occurs with cyclophosphamide levels in the therapeutic range, whereas much higher doses are required to damage hepatocytes (96). The differential sensitivity of SECs and hepatocytes to acrolein may be explained by differences in the GSH detoxification capacity of these two cell types (GSH is preferentially depleted in SECs). The selective susceptibility of SECs in vivo – as distinct from other ECs – is likely due to high concentrations of the metabolite in the space of Disse.

Azathioprine is a prodrug. Conjugation to glutathione, catalyzed by glutathione S-transferase, yields 6-mercaptopurine, and consumes cellular glutathione. Cells rapidly take up azathioprine, so that concentrations of orally ingested azathioprine are highest in the gut and in the liver. In vitro studies have demonstrated that profound GSH depletion precedes azathioprine-induced cell death in SECs, which suggests that GSH depletion plays a role in toxicity to SECs (95). Toxicity to SECs is greater than to hepatocytes in vitro. However, depletion of hepatocyte GSH increases toxicity and abolishes the difference in susceptibility, demonstrating that the relative resistance of hepatocytes is due to greater GSH detoxification capacity by hepatocytes. 6-Thioguanine, a metabolite downstream of 6-mercaptopurine, also causes SOS, so that the mechanism of injury might involve more than the GSH depletion that occurs when azathioprine is metabolized to 6-mercaptopurine; however it has never been examined whether 6-thioguanine might also deplete GSH through another mechanism.

In summary, in vitro studies have demonstrated common features among toxins implicated in SOS. The drugs and toxins studied in the in vitro model are selectively more toxic to SECs than to hepatocytes, they are GSH detoxified, and toxicity to SECs in vitro does not occur until SEC GSH is depleted.

In Vivo Studies

The development of a reproducible animal model of SOS has facilitated uncovering the morphological changes and the biochemical underpinnings of this syndrome (Figure 133.3). The monocrotaline rat model follows a highly reproducible course when the toxin is given by gavage in Sprague-Dawley rats, but shows tremendous variability when administered by another route or to another strain of rats. The time-course of events can be divided into pre-SOS (0–48 hours after monocrotaline), early SOS (days 3–5), and late SOS (days 6 and 7) (97).

In pre-SOS, light microscopic changes are minimal, and clinical features are absent. However, morphological changes can be detected within the first 12 hours by scanning electron microscopy, transmission electron microscopy, and in vivo microscopy. At 12 hours, there occurs a loss of SEC fenestration, formation of gaps within and between SECs, and swelling or rounding up of SECs. By 48 hours, red blood cells penetrate beneath the swollen cells into the space of Disse (98). The swollen SECs block the sinusoids, the space of Disse becomes the path of least resistance, and blood begins to flow in the space of Disse, with dissection of the sinusoidal lining. Kupffer cells, SECs, and stellate cells embolize into the lumen of the sinusoid and further block sinusoidal flow.

In early SOS, between days 3 and 5, the predominant histological features are centrilobular necrosis and hemorrhage, with loss of SECs and venular ECs (see Figure 133.2). The clinical manifestations are similar to those described in the human syndrome, namely hepatomegaly with a 24% to 80% increase in liver weight, ascites formation, and hyperbilirubinemia with bilirubin levels as high as 10 mg/dL.

In late SOS (days 6 and 7), the centrilobular necrosis resolves completely, but there is venular fibrosis and persistent damage to SECs and venular ECs with hemorrhage.

The number of perfused sinusoids reaches a nadir by day 4 and remains decreased through day 10. Kupffer cells decrease in number by day 1, but there is a progressive influx of blood-derived mononuclear cells adherent to areas denuded of SECs and venular ECs, and these aggregates of mononuclear cells contribute to the sinusoidal obstruction by day 4.

The pathophysiology of pre-SOS involves two important (and related) biochemical events: (a) altered SEC expression of matrix metalloproteinases (MMPs), and (b) changes in SEC actin cytoskeleton. Each of these mechanisms represents a potential target for preventative therapy.

MMPs are enzymes that break down extracellular matrix and allow cells to separate from the underlying matrix. Hepatic MMP-9 expression and activity is increased 12 hours after monocrotaline administration, increases markedly by 48 hours, and then continues to rise through day 4 (99). The increase in MMP-2 in the liver is delayed, and the magnitude of the elevation is much less than MMP-9. In vitro studies of SECs, hepatocytes, stellate cells, and Kupffer cells reveal that SECs are the major source of both basal and monocrotaline-induced MMP-9 and -2 activity (99). The increase in MMPs coincides with the progressive denudation of the sinusoidal lining, suggesting that increased MMP activity may cause the dehiscence of the SECs from the space of Disse. Prophylactic administration of inhibitors of MMP-9 and -2 prevents the histological changes and the clinical features of SOS, which confirms the causal role of MMPs in the disease (99).

The reactive metabolite of monocrotaline, monocrotaline pyrrole, binds covalently to F-actin in ECs, leading to its depolymerization (99,100). This process may contribute to the rounding up of SECs, which is the earliest morphological change observed by in vivo microscopy (98). Blockade of monocrotaline-induced F-actin depolymerization in SECs

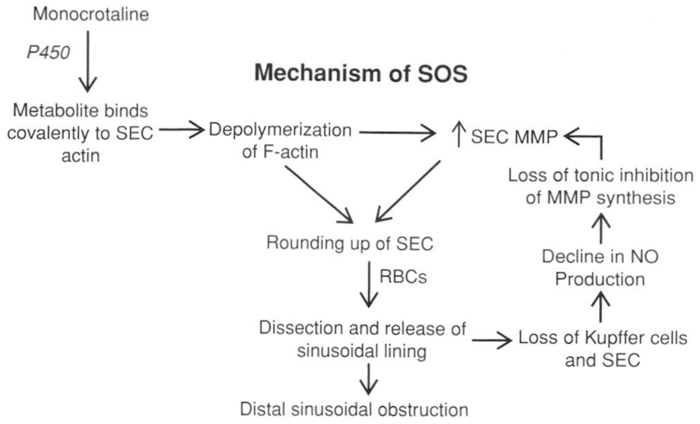

Figure 133.2. Proposed mechanisms in SOS. This scheme depicts the morphological and biochemical changes observed in the monocrotaline model of SOS. MMP, matrix metalloproteinase; NO, nitric oxide; RBC, red blood cell; SEC, sinusoidal ECs.

prevents the increase in MMP activity (99). The link between F-actin disassembly and increased MMP activity also has been observed in other models (101–103). Thus, the combined effects of F-actin depolymerization on cell shape and MMP expression, which allows SECs to be released from the underlying matrix, may lead to penetration of red blood cells beneath the endothelial barrier and dissection of endothelium off the matrix in the space of Disse.

After monocrotaline administration, hepatic vein NO levels decrease on day 1, drop by 70% on day 3, and remain low through day 8. The decline in the first 24 hours parallels the loss of Kupffer cells, whereas the subsequent reduction in NO corresponds with the loss of SECs (104). Inhibition of NO synthase (NOS) exacerbates monocrotaline toxicity. Conversely, infusion of V-PYRRO/NO, a liver-selective NO prodrug, results in dose-dependent attenuation of SOS (104). V-PYRRO/NO prevents the rounding up of SECs, thereby preserving an intact SEC lining and maintaining sinusoidal perfusion. V-PYRRO/NO prevents synthesis of MMP-9 in the monocrotaline model. Inhibition of MMP synthesis by NO is consistent with studies in other models, in which

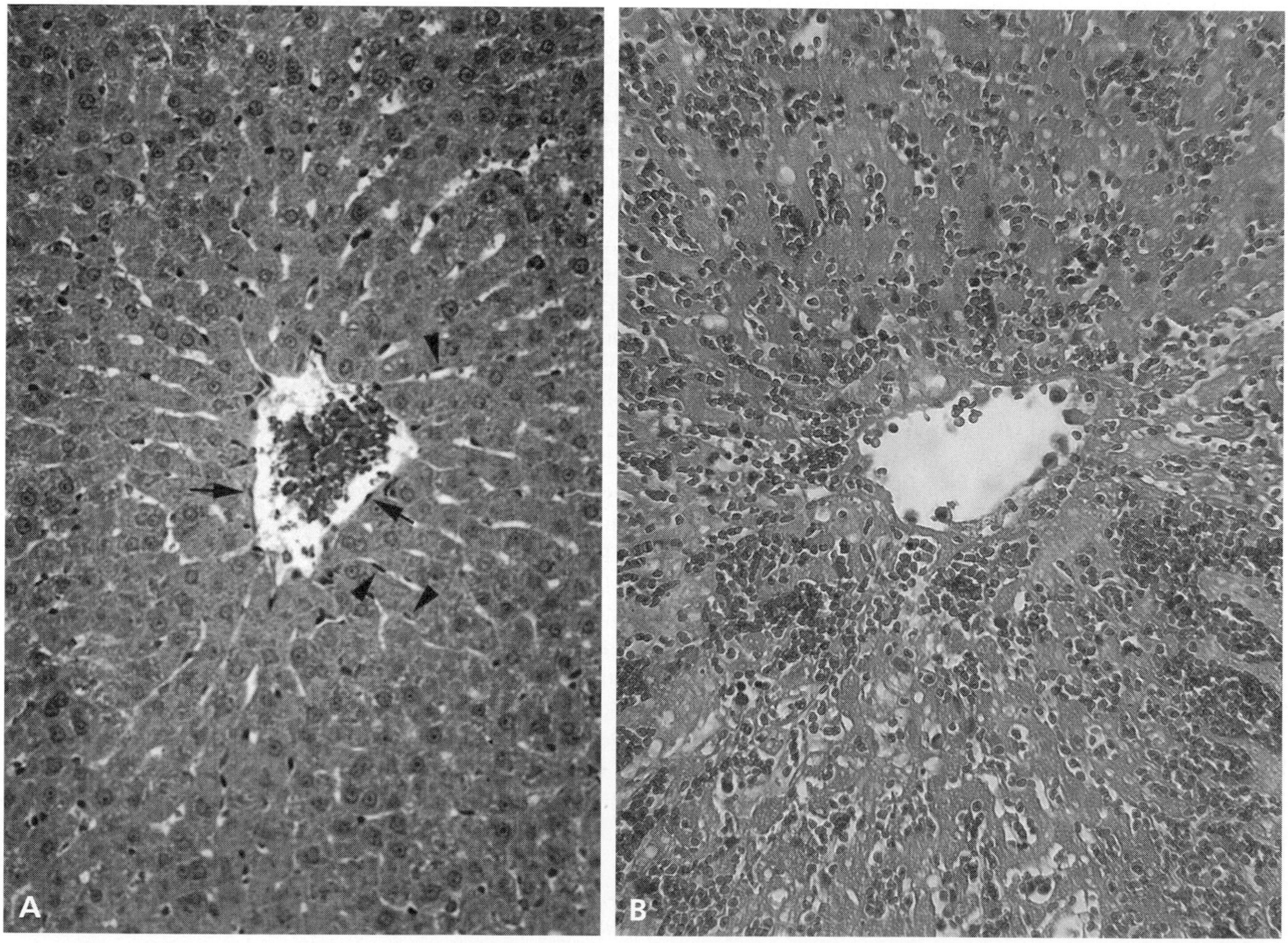

Figure 133.3. Sinusoidal obstruction syndrome. The left panel shows a normal liver. The *arrows* indicate ECs lining the central vein, and *arrowheads* indicate normal sinusoidal ECs. The right panel shows the centrilobular region of the liver in early SOS, with loss of ECs within the central vein, loss of SECs, severe centrilobular necrosis, hemorrhage, and congestion (×20 magnification).

inhibition of endogenous NO production enhances IL-1β-induced increases in synthesis of MMP 9, whereas administration of NO donors inhibits synthesis of MMPs (105–111).

This process seems to create a positive feedback loop, whereby decreased NO production leads to further reductions in NO (see Figure 133.3). Monocrotaline-mediated loss of Kupffer cells and SECs results in decreased NO production. Because NO is necessary for the tonic suppression of MMP-9, levels of this metalloproteinase increase. The change in MMP activity, in conjunction with F-actin depolymerization, leads to dissection of the space of Disse, as described earlier. The dissection causes additional SECs to slough, thus perpetuating the cycle.

The centrilobular necrosis in SOS is completely prevented by interventions that preserve the integrity of the sinusoidal lining, such as MMP inhibition and preservation of NO levels. These data strongly support the notion that hepatocyte ischemia is secondary to the sinusoidal lesion.

NODULAR REGENERATIVE HYPERPLASIA

Nodular regenerative hyperplasia (NRH) is most commonly asymptomatic, being detected incidentally at autopsy. The prevalence is around 2.5% of the population, as determined in large autopsy series (70,112). Symptomatic disease presents with signs of portal hypertension (e.g., variceal hemorrhage, ascites, and splenomegaly). However, NRH rarely leads to end-stage liver disease.

Although the etiology of NRH remains unclear, the disease appears to be associated with areas of impaired perfusion alternating with areas of reactive hyperplasia in which perfusion is maintained (113). It is believed that reduced O_2 delivery in the hypoperfused areas leads to hepatocyte apoptosis or atrophy (114). The common feature of the various risk factors for NRH is that they predispose to impaired portal venous or sinusoidal blood flow. Examples of microcirculatory toxicity that lead to NRH are long-term azathioprine immunosuppression for kidney or liver transplantation, and myeloablative chemotherapy for bone marrow transplantation. Both azathioprine (95) and myeloablative chemotherapy (96) are toxic to SECs, and have been implicated in other diseases in which the SECs are the putative target, such as peliosis hepatis and SOS (71,93,115). Thus, NRH in renal, liver, or bone marrow transplantation seems to be due to SEC toxicity, with regional impairment of the sinusoidal circulation.

PELIOSIS HEPATIS

Peliosis is a rare disease characterized by blood-filled cavities within the liver parenchyma. The peliotic lesions vary from less than 1 mm to several centimeters. Peliosis is most common in the liver, but also may involve the spleen, bone marrow, or abdominal lymph nodes. Peliosis hepatis has been linked to drugs such as azathioprine and anabolic steroids. In addition, peliosis may occur in patients with acquired immunodeficiency syndrome (AIDS), tuberculosis, leukemia, lymphoma, multiple myeloma, and myeloproliferative diseases.

Studies in patients with AIDS provide the most clear-cut evidence for the SEC origin of peliosis hepatis. In these patients, peliosis is caused by infection with *Bartonella* sp. bacilli. *Bartonella* sp. have been detected in SECs by electron microscopy (116). This leads to disruption of the SEC lining, sinusoidal dilatation, and ultimately to the formation of cavities that lack SEC lining (117,118). Later in the disease, the endothelial lining of the cavities may be partially restored. When *Bartonella* involves the organs of the reticuloendothelial system, which have a discontinuous endothelium, peliotic cavities develop, whereas *Bartonella* infection in the skin, which has a continuous endothelial lining, causes bacillary angiomatosis (118).

ACETAMINOPHEN TOXICITY

In the early stages after acetaminophen toxicity, extreme hepatic congestion occurs in both humans (119–123) and experimental animals (124–129). In the mouse, the congestion is so extreme that up to half of the red blood cell volume can accumulate in the liver (128). Ultrastructural studies have demonstrated the appearance of large pores in the SECs, accumulation of red blood cells in the space of Disse, and partial separation of SECs from the underlying hepatocytes, but without complete dehiscence (130). These circulatory changes precede hepatocyte necrosis and likely contribute to the liver toxicity (131). In vitro studies in murine liver cells have shown that acetaminophen is more toxic to SECs than to hepatocytes (132), and this may account for the microcirculatory impairment.

The early morphological changes in acetaminophen toxicity are very similar to the early features in SOS. Analogous to SOS, NO and MMPs appear to play a role in the development of the microvascular injury in acetaminophen intoxication (133–135). NO derived from eNOS protects against acetaminophen toxicity, whereas NO derived from iNOS contributes to acetaminophen injury (134). Preservation of NO levels with V-PYRRO/NO (133) or inhibition of MMP-2 and -9 attenuates acetaminophen-induced microvascular injury and parenchymal necrosis (135).

INJURY DURING LIVER TRANSPLANTATION

In liver transplantation, the donor liver is preserved and transported in a cold University of Wisconsin solution. A variable degree of injury may manifest itself when the liver is reperfused in the transplant recipient, and this ischemia–reperfusion injury can lead to graft failure. The major target for reperfusion injury is the SEC (136–139), with accompanying cell death, partial denudation of the sinusoidal lining, and deterioration of the liver microcirculation. Cold preservation leads to increased calpain protease activity in SECs (140–142), which leads to depolymerization of F-actin in SECs (143), and consequent upregulation of MMP-9 and -2 activity (144). eNOS is downregulated and iNOS is upregulated by ischemia–reperfusion, whereas preservation of NO levels and inhibition of iNOS are protective of the injury (145–147). Tonic NO production by eNOS is known to suppress new synthesis of MMPs, so that the decrease in eNOS may be a permissive factor in the increase in MMPs. Inhibition of MMPs reportedly reduces hepatocyte necrosis and aminotransferase elevations after ischemia–reperfusion (148). Upregulation of MMPs likely accounts for the dehiscence of SECs from the space of Disse in ischemia–reperfusion injury. One of the other consequences of F-actin depolymerization and increased MMP activity is increased platelet adherence to SECs (149). This accumulation of platelets and leukocytes contributes to SEC injury after cold preservation (149–151).

In addition to injury due to ischemia–reperfusion, the sinusoidal lining may be injured in live donor transplantation. When a portion of a liver is transplanted into a recipient, a mismatch occurs between the afferent circulation that was appropriate for the original liver of the recipient and the small-for-size graft. The increased blood flow is associated with transient portal hypertension and gaps in the sinusoidal lining and hepatocyte injury (152,153). The portal hypertension may damage an endothelial lining already compromised by ischemia–reperfusion.

KEY POINTS

- SECs are a highly specialized type of EC. The dedifferentiation that occurs in capillarization and pseudocapillarization has significant local and systemic consequences.
- SECs are the initial target in a number of liver diseases. Given the limited amount of research in this area, it is likely that other diseases also are initiated by injury to SECs.
- Great similarity is apparent in the early events leading to SEC damage in ischemia–reperfusion injury, SOS, and acetaminophen toxicity. It will be intriguing to determine why the damage resolves quickly in ischemia–reperfusion injury and acetaminophen intoxication, but persists in SOS.

Future Goals

- To elucidate the active endocytotic apparatus in SECs, which should lend itself to strategies to protect SEC from injury
- To examine strategies to reverse capillarization

REFERENCES

1 Wisse E. An electron microscopic study of the fenestrated endothelial lining of rat liver sinusoids. *J Ultrastruct Res.* 1970; 31:125–150.

2 Wisse E. An ultrastructural characterization of the endothelial cell in the rat liver sinusoid under normal and various experimental conditions, as a contribution to the distinction between endothelial and Kupffer cells. *J Ultrastruct Res.* 1972;38:528–562.

3 Knook DL, Sleyster EC. Separation of the Kupffer and endothelial cells of the rat liver by centrifugal elutriation. *Exp Cell Res.* 1976;99:444–449.

4 Knook DL, Blansjaar N, Sleyster EC. Isolation and characterization of Kupffer and endothelial cells from the rat liver. *Exp Cell Res.* 1977;109:317–329.

5 Wisse E, De Zanger RB, Charels K, et al. The liver sieve: considerations concerning the structure and function of endothelial fenestra, the sinusoidal wall and the space of Disse. *Hepatology.* 1985;5:683–692.

6 Risau W. Differentiation of endothelium. *FASEB J.* 1995;9:926–933.

7 Gatmaitan Z, Varticovski L, Ling L, et al. Studies on fenestral contraction in rat liver endothelial cells in culture. *Am J Physiol.* 1996;148:2027–2041.

8 Stan RV, Kubitza M, Palade GE. PV-1 is a component of the fenestral and stomatal diaphragms in fenestrated endothelia. *Proc Natl Acad Sci USA.* 1999;96:13203–13207.

9 De Bruyn PP, Michelson S. Changes in the random distribution of sialic acid at the surface of the myeloid sinusoidal endothelium resulting from the presence of diaphragmed fenestrae. *J Cell Biol.* 1979;82:708–714.

10 Weiss L, Chen LT. A scanning electron microscopic study of the spleen. *Blood.* 1974;43:665–691.

11 Blue J, Weiss L. Electron microscopy of the red pulp of the dog spleen including vascular arrangements, peri-arterial macrophage sheaths, ellipsoids, and the contractive reticular meshwork. *Am J Anat.* 1981;161:189–218.

12 Blue J, Weiss L. Vascular pathways in nonsinusal red pulp: an electron microscopic study of the cat spleen. *Am J Anat.* 1981; 161:135–168.

13 Módis L, Martinez-Hernandez A. Hepatocytes modulate the hepatic microvascular phenotype. *Lab Invest.* 1991;65:661–670.

14 Roberts WG, Palade GE. Increased microvascular permeability and endothelial fenestration induced by vascular endothelial growth factor. *J Cell Sci.* 1995;108:2369–2379.

15 DeLeve LD, Wang X, Hu L, et al. Rat liver sinusoidal endothelial cell phenotype is under paracrine and autocrine control. *Am J Physiol Gastrointest Liver Physiol.* 2004;287:G757–G763.

16 van der Zee R, Murohara T, Luo Z, et al. Vascular endothelial growth factor/vascular permeability factor augments nitric oxide release from quiescent rabbit and human vascular endothelium. *Circulation.* 1997;95:1030–1037.

17 Wu HM, Huang Q, Yuan Y, Granger HJ. VEGF induces NO-dependent hyperpermeability in coronary venules. *Am J Physiol.* 1996:271.

18 Morbidelli L, Chang CH, Douglas JG, et al. Nitric oxide mediates mitogenic effect of VEGF on coronary venular endothelium. *Am J Physiol.* 1996:270.

19 Ku DD, Zaleski JK, Liu S, Brock TA. Vascular endothelial growth factor induces EDRF-dependent relaxation in coronary arteries. *Am J Physiol.* 1993:265.

20 Risau W. Development and differentiation of endothelium. *Kidney Int Suppl.* 1998;54:S3–S6.

21 Maher JJ. Cell-specific expression of hepatocyte growth factor in liver: upregulation in sinusoidal endothelial cells after carbon tetrachloride. *J Clin Invest.* 1993;91:2244–2252.

22 Taniguchi E, Sakisaka S, Matsuo K, et al. Expression and role of vascular endothelial growth factor in liver regeneration after partial hepatectomy in rats. *J Histochem Cytochem.* 2001;49: 121–130.

23 Ross MA, Sander CM, Kleeb TB, et al. Spatiotemporal expression of angiogenesis growth factor receptors during the revascularization of regenerating rat liver. *Hepatology.* 2001;34:1135–1148.

24 Greene AK, Wiener S, Puder M, et al. Endothelial-directed hepatic regeneration after partial hepatectomy. *Ann Surg.* 2003;237: 530–535.

25 LeCouter J, Moritz DR, Li BU, et al. Angiogenesis-independent endothelial protection of liver: role of VEGFR-1. *Science.* 2003; 299:890–893.

26 Shimizu H, Miyazaki M, Wakabayashi Y, et al. Vascular endothelial growth factor secreted by replicating hepatocytes induces sinusoidal endothelial cell proliferation during regeneration after partial hepatectomy in rats. *J Hepatol.* 2001;34:683–689.

27 Matsumoto K, Yoshitomi H, Rossant J, Zaret KS. Liver organogenesis promoted by endothelial cells prior to vascular function. *Science.* 2001;294:559–563.

28 Lammert E, Cleaver O, Melton D. Role of endothelial cells in early pancreas and liver development. *Mech Dev.* 2003;120:59–64.

29 Wisse E. An electron microscopic study of the fenestrated endothelial lining of the rat liver sinusoids. *J Ultrastruct Res.* 1970;31:125–150.

30 Naito M, Wisse E. Filtration effect of endothelial fenestrations on chylomicron transport in neonatal rat liver sinusoids. *Cell Tiss Res.* 1978;190:371–382.

31 Fraser R, Bosanquet AG, Day WA. Filtration of chylomicrons by the liver may influence cholesterol metabolism and atherosclerosis. *Atherosclerosis.* 1978;29:113–123.

32 Regrave TG. Formation of cholesteryl ester-rich particulate lipid during metabolism of chylomicrons. *J Clin Invest.* 1970;49:465–471.

33 Smedsrød B, Pertoft H, Gustafson S, Laurent TC. Scavenger functions of the liver endothelial cell. *Biochem J.* 1990;266:313–327.

34 MacPhee PJ, Schmidt EE, Groom AC. Intermittence of blood flow in liver sinusoids studied by high-resolution in vivo microscopy. *Am J Physiol.* 1995;296:G692–G698.

35 Seternes T, Sorensen K, Smedsrod B. Scavenger endothelial cells of vertebrates: a nonperipheral leukocyte system for high-capacity elimination of waste macromolecules. *Proc Natl Acad Sci USA.* 2002;99:7594–7597.

36 Volpes R, Van den Oord JJ, Desmet VJ. Vascular adhesion molecules in acute and chronic liver inflammation. *Hepatology.* 1992;15:269–275.

37 Scoazec JY, Racine L, Couvelard A, et al. Endothelial cell heterogeneity in the normal human liver acinus: in situ immunohistochemical demonstration. *Liver.* 1994;14:113–123.

38 Scoazec JY, Flejou JF, D'Errico A, et al. Focal nodular hyperplasia of the liver: composition of the extracellular matrix and expression of cell-cell and cell-matrix adhesion molecules. *Hum Pathol.* 1995;26:1114–1125.

39 Xu B, Broome U, Uzunel M, et al. Capillarization of hepatic sinusoid by liver endothelial cell-reactive autoantibodies in patients with cirrhosis and chronic hepatitis. *Am J Pathol.* 2003; 163:1275–1289.

40 Arvieux C, McNab G, Adams DH. Interactions with T cells induce expression of VCAM-1 by human intrahepatic endothelial cells in vitro. *Biochem Soc Trans.* 1997:25.

41 Shimada M, Kajiyama K, Hasegawa H, et al. Role of adhesion molecule expression and soluble fractions in hepatic resection. *J Am Coll Surg.* 1998;186:534–541.

42 Lohse AW, Knolle PA, Bilo K, et al. Antigen-presenting function and B7 expression of murine sinusoidal endothelial cells and Kupffer cells. *Gastroenterology.* 1996;110:1175–1181.

43 Knolle PA, Gerken G. Local control of the immune response in the liver. *Immunol Rev.* 2000;174:21–34.

44 Knolle PA, Limmer A. Neighborhood politics: the immunoregulatory function of organ-resident liver endothelial cells. *Trends Immunol.* 2001;22:432–437.

45 Klugewitz K, Blumenthal-Barby F, Schrage A, et al. Immunomodulatory effects of the liver: deletion of activated $CD4^+$ effector cells and suppression of IFN-gamma-producing cells after intravenous protein immunization. *J Immunol.* 2002;169: 2407–2413.

46 Limmer A, Ohl J, Kurts C, et al. Efficient presentation of exogenous antigen by liver endothelial cells to $CD8^+$ T cells results in antigen-specific T-cell tolerance. *Nat Med.* 2000;6:1348–1354.

47 Schaffner F, Popper H. Capillarization of hepatic sinusoids in man. *Gastroenterology.* 1963;44:239–242.

48 Horn T, Christoffersen P, Henriksen JH. Alcoholic liver injury: defenestration in noncirrhotic livers – a scanning electron microscopic study. *Hepatology.* 1987;7:77–82.

49 Urashima S, Tsutsumi M, Nakase K, et al. Studies on capillarization of the hepatic sinusoids in alcoholic liver disease. *Alcohol Alcohol.* 1993;S1B:77–84.

50 Mori T, Okanoue T, Sawa Y, et al. Defenestration of the sinusoidal endothelial cell in a rat model of cirrhosis. *Hepatology.* 1993;17:891–897.

51 Bhunchet E, Fujieda K. Capillarization and venularization of hepatic sinusoids in porcine serum-induced rat liver fibrosis: a mechanism to maintain liver blood flow. *Hepatology.* 1993;18: 1450–1458.

52 Le Couteur DG, Cogger VC, Markus AM, et al. Pseudocapillarization and associated energy limitation in the aged rat liver. *Hepatology.* 2001;33:537–543.

53 McLean AJ, Cogger VC, Chong GC, et al. Age-related pseudocapillarization of the human liver. *J Pathol.* 2003;200:112–117.

54 Hickey PL, Angus PW, McLean AJ, Morgan DJ. Oxygen supplementation restores theophylline clearance to normal in cirrhotic rats. *Gastroenterology.* 1995;108:1504–1509.

55 Le Couteur DG, Hickey H, Harvey PJ, et al. Hepatic artery flow and propranolol metabolism in perfused cirrhotic rat liver. *J Pharmacol Exp Ther.* 1999;289:1553–1558.

56 Froomes PRA, Morgan DJ, Smallwood RA, Angus PW. Comparative effects of oxygen supplementation on theophylline and

acetaminophen clearance in human cirrhosis. *Gastroenterology*. 1999;116:915–920.
57 Le Couteur DG, Fraser R, Cogger VC, McLean AJ. Hepatic pseudocapillarisation and atherosclerosis in ageing. *Lancet*. 2002; 359:1612–1615.
58 Fraser R, Dobbs BR, Rogers GW. Lipoproteins and the liver sieve: the role of the fenestrated sinusoidal endothelium in lipoprotein metabolism, atherosclerosis, and cirrhosis. *Hepatology*. 1995;21:863–874.
59 Engstrom-Laurent A, Loof L, Nyberg A, Schroder T. Increased serum levels of hyaluronate in liver disease. *Hepatology*. 1985; 5:638–642.
60 Gressner AM, Pazen H, Greiling H. The synthesis of total and specific glycosaminoglycans during development of experimental liver cirrhosis. *Experientia*. 1977;33:1290–1292.
61 Fraser JR, Alcorn D, Laurent TC, et al. Uptake of circulating hyaluronic acid by the rat liver. Cellular localization in situ. *Cell Tissue Res.* 1985;242:505–510.
62 Gressner AM, Schafer S. Comparison of sulphated glycosaminoglycan and hyaluronate synthesis and secretion in cultured hepatocytes, fat storing cells, and Kupffer cells. *J Clin Chem Clin Biochem.* 1989;27:141–149.
63 Zhou B, Weigel JA, Fauss L, Weigel PH. Identification of the hyaluronan receptor for endocytosis (HARE). *J Biol Chem.* 2000; 275:37733–37741.
64 Weigel JA, Raymond RC, McGary C, et al. A blocking antibody to the hyaluronan receptor for endocytosis (HARE) inhibits hyaluronan clearance by perfused liver. *J Biol Chem.* 2003;278: 9808–9812.
65 Carreira CM, Nasser SM, di Tomaso E, et al. LYVE-1 is not restricted to lymph vessels: expression in normal liver blood sinusoids and down-regulation in human liver cancer and cirrhosis. *Cancer Res.* 2001;61:8079–8084.
66 Couvelard A, Scoazec JY, Feldmann G. Expression of cell-cell and cell-matrix adhesion proteins by sinusoidal endothelial cells in the normal and cirrhotic human liver. *Am J Pathol.* 1993;143:738–752.
67 Steinberg P, Lafranconi WM, Wolf CR, et al. Xenobiotic metabolizing enzymes are not restricted to parenchymal cells in rat liver. *Mol Pharmacol.* 1987;32:463–470.
68 Steinberg P, Schlemper B, Molitor E, et al. Rat liver endothelial and Kupffer cell-mediated mutagenicity of polycyclic aromatic hydrocarbons and aflatoxin B1. *Environ Health Perspect.* 1990;88:71–76.
69 DeLeve LD. Dacarbazine toxicity in murine liver cells: a novel model of hepatic endothelial injury and glutathione defense. *J Pharmacol Exp Ther.* 1994;268:1261–1270.
70 Wanless IR. Micronodular transformation (nodular regenerative hyperplasia) of the liver: a report of 64 cases among 2,500 autopsies and a new classification of benign hepatocellular nodules. *Hepatology*. 1990;11:787–797.
71 Haboubi NY, Ali HH, Whitwell HL, Ackrill P. Role of endothelial cell injury in the spectrum of azathioprine-induced liver disease after renal transplant: light microscopy and ultrastructural observations. *Am J Gastroenterol.* 1988;83:256–261.
72 Zafrani ES, Cazier A, Baudelot AM, Feldmann G. Ultrastructural lesions of the liver in human peliosis. A report of 12 cases. *Am J Pathol.* 1984;114:349–359.
73 DeLeve LD. Cancer chemotherapy. In: Kaplowitz N, DeLeve LD, eds. *Drug-induced Liver Disease*. Chapter 24. New York: Marcel Dekker; 2002:593–632.
74 Bras G, Jeliffe DB, Stuart KL. Veno-occlusive disease of the liver with non-portal type of cirrhosis occurring in Jamaica. *Arch Path.* 1954;57:285–300.
75 Willmot FC, Robertson GW. Senecio disease, or cirrhosis of the liver due to senecio poisoning. *Lancet*. 1920;2:848–849.
76 Eisenhauer T, Hartmann H, Rumpf KW, et al. Favourable outcome of hepatic veno-occlusive disease in a renal transplant patient receiving azathioprine, treated by portacaval shunt. Report of a case and review of the literature. *Digestion*. 1984;30: 185–190.
77 Katzka DA, Saul SH, Jorkasky D, et al. Azathioprine and hepatic venoocclusive disease in renal transplant patients. *Gastroenterology*. 1986;90:446–454.
78 Read AE, Wiesner RH, LaBrecque DR, et al. Hepatic veno-occlusive disease associated with renal transplantation and azathioprine therapy. *Ann Int Med.* 1986;104:651–655.
79 Liano F, Moreno A, Matesanz R, et al. Veno-occlusive hepatic disease of the liver in renal transplantation: is azathioprine the cause? [See comments]. *Nephron*. 1989;51:509–516.
80 Sterneck M, Wiesner R, Ascher N, et al. Azathioprine hepatotoxicity after liver transplantation. *Hepatology*. 1991;14:806–810.
81 Giles FJ, Kantarjian HM, Kornblau SM, et al. Mylotarg (gemtuzumab ozogamicin) therapy is associated with hepatic venoocclusive disease in patients who have not received stem cell transplantation. *Cancer*. 2001;92:406–413.
82 Rajvanshi P, Shulman HM, Sievers EL, McDonald GB. Hepatic sinusoidal obstruction following gemtuzumab ozogamicin (Mylotarg®). *Blood*. 2002;99:2310–2314.
83 Tornesello A, Piciacchia D, Mastrangelo S, et al. Veno-occlusive disease of the liver in right-sided Wilms' tumours. *Eur J Cancer.* 1998;34:1220–1223.
84 Czauderna P, Katski K, Kowalczyk J, et al. Venoocclusive liver disease (VOD) as a complication of Wilms' tumour management in the series of consecutive 206 patients. *Eur J Pediatr Surg.* 2000;10:300–303.
85 McDonald GB, Hinds MS, Fisher LD, et al. Veno-occlusive disease of the liver and multiorgan failure after bone marrow transplantation – a cohort study of 355 patients. *Ann Int Med.* 1993;118:255–267.
86 McDonald GB, Sharma P, Matthews DE, et al. Veno-occlusive disease of the liver after bone marrow transplantation: diagnosis, incidence and predisposing factors. *Hepatology*. 1984;4:116–122.
87 Ganem G, Saint-Marc Girardin MF, Kuentz M, et al. Venocclusive disease of the liver after allogeneic bone marrow transplantation in man. *Int J Radiat Biol Phys.* 1988;14:879–884.
88 Jones RJ, Lee KSK, Beschorner WE, et al. Veno-occlusive disease of the liver following bone marrow transplantation. *Transplantation.* 1987;44:778–783.
89 Carreras E, Bertz H, Arcese W, et al. Incidence and outcome of hepatic veno-occlusive disease after blood or marrow transplantation: a prospective cohort study of the European Group for Blood and Marrow Transplantation. European Group for Blood and Marrow Transplantation Chronic Leukemia Working Party. *Blood*. 1998;92:3599–3604.
90 Hasegawa S, Horibe K, Kawabe T, et al. Veno-occlusive disease of the liver after allogeneic bone marrow transplantation in children with hematologic malignancies: incidence, onset time and risk factors. *Bone Marrow Transplant*. 1998;22:1191–1197.
91 Lee JL, Gooley T, Bensinger W, et al. Veno-occlusive disease of the liver after busulfan, melphalan, and thiotepa conditioning

therapy: incidence, risk factors, and outcome. *Biol Blood Marrow Transplant*. 1999;5:306–315.

92 DeLeve LD, Shulman HM, McDonald GB. Toxic injury to hepatic sinusoids: sinusoidal obstruction syndrome (venoocclusive disease). *Sem Liver Dis.* 2002;22:623–638.

93 Shulman HM, Fisher LB, Schoch HG, et al. Venoocclusive disease of the liver after marrow transplantation: histological correlates of clinical signs and symptoms. *Hepatology*. 1994;19:1171–1180.

94 DeLeve LD, Wang X, McCuskey MK, et al. Rat liver endothelial cells isolated by anti-CD31 immunomagnetic sorting lack fenestrae and sieve plates. *Am J Physiol Gastrointest Liver Physiol.* 2006;291:G1187–G1189.

95 DeLeve LD, Wang X, Kuhlenkamp JF, Kaplowitz N. Toxicity of azathioprine and monocrotaline in murine sinusoidal endothelial cells and hepatocytes: the role of glutathione and relevance to hepatic venoocclusive disease. *Hepatology*. 1996;23:589–599.

96 DeLeve LD. Cellular target of cyclophosphamide toxicity in the murine liver: role of glutathione and site of metabolic activation. *Hepatology*. 1996;24:830–837.

97 DeLeve LD, McCuskey RS, Wang X, et al. Characterization of a reproducible rat model of hepatic veno-occlusive disease. *Hepatology*. 1999;29:1779–1791.

98 DeLeve LD, Ito I, Bethea NW, et al. Embolization by sinusoidal lining cell obstructs the microcirculation in rat sinusoidal obstruction syndrome. *Am J Physiol Gastrointest Liver Physiol.* 2003;284:G1045–G1052.

99 DeLeve LD, Wang X, Tsai J, et al. Prevention of sinusoidal obstruction syndrome (hepatic venoocclusive disease) in the rat by matrix metalloproteinase inhibitors. *Gastroenterology*. 2003; 125:882–890.

100 Lamé MW, Jones AD, Wilson DW, et al. Protein targets of monocrotaline pyrrole in pulmonary artery endothelial cells. *J Biol Chem.* 2000;275:29091–29099.

101 Werb Z, Hembry RM, Murphy G, Aggeler J. Commitment to expression of the metalloendopeptidases, collagenase and stromelysin: relationship of inducing events to changes in cytoskeletal architecture. *J Cell Biol.* 1986;102:697–702.

102 Allenberg M, Weinstein T, Li I, Silverman M. Activation of procollagenase IV by cytochalasin D and concanavalin A in cultured rat mesangial cells: linkage to cytoskeletal reorganization. *J Am Soc Nephrol.* 1994;4:1760–1770.

103 MacDougall JR, Kerbel RS. Constitutive production of 92-kDa gelatinase B can be suppressed by alterations in cell shape. *Exp Cell Res.* 1995;218:508–515.

104 DeLeve LD, Wang X, Kanel GC, et al. Decreased hepatic nitric oxide production contributes to the development of rat sinusoidal obstruction syndrome. *Hepatology*. 2003;38:900–908.

105 Eberhardt W, Beeg T, Beck KF, et al. Nitric oxide modulates expression of matrix metalloproteinase-9 in rat mesangial cells. *Kidney Int.* 2000;57:59–69.

106 Upchurch GR Jr., Ford JW, Weiss SJ, et al. Nitric oxide inhibition increases matrix metalloproteinase-9 expression by rat aortic smooth muscle cells in vitro. *J Vasc Surg.* 2001;34:76–83.

107 Matsunaga T, Weihrauch DW, Moniz MC, et al. Angiostatin inhibits coronary angiogenesis during impaired production of nitric oxide. *Circulation.* 2002;105:2185–2191.

108 Gurjar MV, DeLeon J, Sharma RV, Bhalla RC. Mechanism of inhibition of matrix metalloproteinase-9 induction by NO in vascular smooth muscle cells. *J Appl Physiol.* 2001;91:1380–1386.

109 Eagleton MJ, Peterson DA, Sullivan VV, et al. Nitric oxide inhibition increases aortic wall matrix metalloproteinase-9 expression. *J Surg Res.* 2002;104:15–21.

110 Mujumdar VS, Aru GM, Tyagi SC. Induction of oxidative stress by homocyst(e)ine impairs endothelial function. *J Cell Biochem.* 2001;82:491–500.

111 Jurasz P, Sawicki G, Duszyk M, et al. Matrix metalloproteinase 2 in tumor cell-induced platelet aggregation: regulation by nitric oxide. *Cancer Res.* 2001;61:376–382.

112 Nakanuma Y. Nodular regenerative hyperplasia of the liver: retrospective survey in autopsy series. *J Clin Gastroenterol.* 1990; 12:460–465.

113 Wanless IR, Godwin TA, Allen F, Feder A. Nodular regenerative hyperplasia of the liver in hematologic disorders: a possible response to obliterative portal venopathy. A morphometric study of nine cases with an hypothesis on the pathogenesis. *Medicine.* 1980;59:367–379.

114 Shimamatsu K, Wanless IR. Role of ischemia in causing apoptosis, atrophy, and nodular hyperplasia in human liver. *Hepatology*. 1997;26:343–350.

115 Snover DC, Weisdorf S, Bloomer J, et al. Nodular regenerative hyperplasia of the liver following bone marrow transplantation. *Hepatology*. 1989;9:443–448.

116 Leong SS, Cazen RA, Yu GS, et al. Abdominal visceral peliosis associated with bacillary angiomatosis. Ultrastructural evidence of endothelial destruction by bacilli. *Arch Pathol Lab Med.* 1992;116:866–871.

117 Scoazec JY, Marche C, Girard PM, et al. Peliosis hepatis and sinusoidal dilation during infection by the human immunodeficiency virus (HIV). An ultrastructural study. *Am J Pathol.* 1988;131:38–47.

118 Goerdt S, Sorg C. Endothelial heterogeneity and the acquired immunodeficiency syndrome: a paradigm for the pathogenesis of vascular disorders. *Clin Invest.* 1992;70:89–98.

119 Rose PG. Paracetamol overdose and liver damage. *Br Med J.* 1969;1:381–382.

120 Thompson RPH, Clark R, Wilson RA, et al. Hepatic damage from overdose of paracetamol. *Gut.* 1972;13:836.

121 Zimmerman HJ. Effects of aspirin and acetaminophen on the liver. *Arch Intern Med.* 1981;141:333–342.

122 Klatskin G, Conn HO. *Histopathology of the Liver*. Vol. 7. New York, Oxford: Oxford University Press;1993:111–142.

123 Zimmerman HJ. Syndromes of environmental hepatotoxins. In: *Hepatotoxicity – the Adverse Effect of Drugs and Other Chemicals on the Liver*. New York: Appleton-Century-Crofts;1978:279–302.

124 Dixon MF, Nimmo J, Prescott LF. Experimental paracetamol-induced hepatic necrosis: a histopathological study. *J Pathol.* 1971;103:225–229.

125 Dixon MF, Dixon B, Aparicio SR, Loney DP. Experimental paracetamol-induced hepatic necrosis: a light- and electron-microscope, and histochemical study. *J Pathol.* 1975;116:17–29.

126 Miller DJ, Pichanick GG, Fiskerstrand C, Saunders S. Hepatic erythrocyte sequestration as a cause of acute anaemia. *Am J Dig Dis.* 1977;22:1055–1059.

127 Chiu S, Bhakthan NMG. Experimental acetaminophen-induced hepatic necrosis: biochemical and electron microscopic study of cysteamine protection. *Lab Invest.* 1978;39:193–203.

128 Walker RM, Massey TE, McElligott TF, Racz WJ. Acetaminophen-induced hypothermia, hepatic congestion, and modification by N-acetylcysteine in mice. *Toxicol Appl Pharmacol.* 1981; 59:500–507.
129 Walker RM, Racz WJ, McElligott TF. Acetaminophen-induced hepatotoxic congestion in mice. *Hepatology.* 1985;5:233–240.
130 Walker RM, Racz WJ, McElligott TF. Scanning electron microscopic examination of acetaminophen-induced hepatotoxicity and congestion in mice. *Am J Physiol.* 1983;113:321–330.
131 Ito Y, Bethea NW, Abril ER, McCuskey RS. Early hepatic microvascular injury in response to acetaminophen toxicity. *Microcirculation.* 2003;10:391–400.
132 DeLeve LD, Wang X, Kaplowitz N, et al. Sinusoidal endothelial cells as a target for acetaminophen toxicity: direct action versus requirement for hepatocyte activation in different mouse strains. *Biochem Pharmacol.* 1997;53:1339–1345.
133 Liu J, Waalkes MP, Clark J, et al. The nitric oxide donor, V-PYRRO/NO, protects against acetaminophen-induced hepatotoxicity in mice. *Hepatology.* 2003;37:324–333.
134 Ito Y, Abril ER, Bethea NW, McCuskey RS. Role of nitric oxide in hepatic microvascular injury elicited by acetaminophen in mice. *Am J Physiol Gastrointest Liver Physiol.* 2004:286.
135 Ito Y, Abril ER, Bethea NW, McCuskey RS. Inhibition of matrix metalloproteinases minimizes hepatic microvascular injury in response to acetaminophen in mice. *Tox Sci.* 2005;83:190–196.
136 Caldwell-Kenkel C, Thurman RG, Lemasters JJ. Selective loss of nonparenchymal cell viability after cold ischemic storage of rat livers. *Transplantation.* 1988;45:834–837.
137 McKeown CMB, Edwards V, Phillips MJ, et al. Sinusoidal lining cell damage: the critical injury in cold preservation of liver allografts in the rat. *Transplantation.* 1988;46:178–191.
138 Caldwell-Kenkel JC, Currin RT, Tanaka Y, et al. Reperfusion injury to endothelial cells following cold ischemic storage of rat livers. *Hepatology.* 1989;10:292–299.
139 Imamura H, Brault A, Huet PM. Effects of extended cold preservation and transplantation on the rat liver microcirculation. *Hepatology.* 1997;25:664–671.
140 Aguilar HI, Steers JL, Wiesner RH, et al. Enhanced liver calpain protease activity is a risk factor for dysfunction of human liver allografts. *Transplantation.* 1997;63:612–614.
141 Kohli V, Gao W, Camargo CA Jr., Clavien PA. Calpain is a mediator of preservation-reperfusion injury in rat liver transplantation. *Proc Natl Acad Sci USA.* 1997;94:9354–9359.
142 Upadhya GA, Topp SA, Hotchkiss RS, et al. Effect of cold preservation on intracellular calcium concentration and calpain activity in rat sinusoidal endothelial cells. *Hepatology.* 2003;37:313–323.
143 Upadhya GA, Strasberg SM. Evidence that actin disassembly is a requirement for matrix metalloproteinase secretion by sinusoidal endothelial cells during cold preservation in the rat. *Hepatology.* 1999;30:169–176.
144 Upadhya GA, Harvey RP, Howard TK, et al. Evidence of a role for matrix metalloproteinases in cold preservation injury of the liver in humans and in the rat. *Hepatology.* 1997;26:922–928.
145 Cottart CH, Do L, Blanc MC, et al. Hepatoprotective effect of endogenous nitric oxide during ischemia-reperfusion in the rat. *Hepatology.* 1999;29:809–813.
146 Morisue A, Wakabayashi G, Shimazu M, et al. The role of nitric oxide after a short period of liver ischemia-reperfusion. *J Surg Res.* 2003;109:101–109.
147 Serracino-Inglott F, Virlos IT, Habib NA, et al. Differential nitric oxide synthase expression during hepatic ischemia-reperfusion. *Am J Surg.* 2003;185:589–595.
148 Curcio F, Ceriello A. Decreased cultured endothelial cell proliferation in high glucose medium is reversed by antioxidants – new insights on the pathophysiological mechanisms of diabetic vascular complications. *In Vitro Cell Dev Biol.* 1992;28A:787–790.
149 Upadhya GA, Strasberg SM. Platelet adherence to isolated rat hepatic sinusoidal endothelial cells after cold preservation. *Transplantation.* 2002;73:1764–1770.
150 Sindram D, Porte RJ, Hoffman MR, et al. Synergism between platelets and leukocytes in inducing endothelial cell apoptosis in the cold ischemic rat liver: a Kupffer cell-mediated injury. *FASEB J.* 2001;15:1230–1232.
151 Lasnier E, Blanc MC, Housset C, et al. Cytotoxic response of sinusoidal endothelial cells to polymorphonuclear leukocytes and its potential implication in hypoxia-reoxygenation injury. *Liver.* 2002;22:495–500.
152 Man K, Lo CM, Ng IO, et al. Liver transplantation in rats using small-for-size grafts: a study of hemodynamic and morphological changes. *Arch Surg.* 2001;136:280–285.
153 Man K, Fan ST, Lo CM, et al. Graft injury in relation to graft size in right lobe live donor liver transplantation: a study of hepatic sinusoidal injury in correlation with portal hemodynamics and intragraft gene expression. *Ann Surg.* 2003;237:256–264.

134

Hepatic Macrocirculation

Portal Hypertension as a Disease Paradigm of Endothelial Cell Significance and Heterogeneity

Winston Dunn and Vijay Shah

Mayo Clinic College of Medicine, Rochester, Minnesota

The hepatic circulation is a low-pressure vascular bed that accommodates large volumes of blood (1). Total hepatic blood flow in humans is approximately 1,500 mL/min and accounts for 15% to 20% of cardiac output. The liver, like the lung, receives a dual blood supply; 70% of total hepatic blood flow is from the portal vein, and 30% from the hepatic artery. Blood from the hepatic artery and portal vein supplies the liver sinusoids and ultimately drains into the hepatic vein. Any one of these segments of the hepatic circulation – the hepatic artery, portal vein, sinusoids, hepatic vein – may be inflicted by disease, with sometimes devastating consequences. This chapter examines the structure and function of the hepatic macrocirculation, with an emphasis on cirrhotic portal hypertension as a paradigm of hepatic vascular disease and endothelial cell (EC) heterogeneity.

HISTORY

The hepatic vasculature has been the subject of interest for millennia. Indeed, seminal observations of patients with portal hypertension, and its major complications of ascites and esophageal variceal hemorrhage, provided a strong foundation for our present-day knowledge of the hepatic circulation (Table 134-1). Several important themes emerge from a consideration of the history of the field. First, there is noticeable change in methodology over time from physical examination to gross anatomy, followed by studies in physiology and pathology, and culminating in methods at the interface of organ and cell physiology and molecular biology. Second, studies of the diseased liver have provided profound insights into the physiology of the normal hepatic vasculature. Last, it is instructive to reflect on the controversies and errors that arose over the centuries, including the original misconceptions of Galen regarding the vascular anatomy of the liver, and Banti's mistaken notions of the role of the spleen in the pathogenesis of cirrhosis. Often, hundreds of years were required before these errors in thought were rectified, thus highlighting the importance of challenging dogma.

STRUCTURE AND FUNCTION OF THE HEPATIC MACROCIRCULATION

Portal Venous Blood Flow

Portal venous blood is derived from the mesenteric venous circulation, which drains blood from the capillary beds of the stomach, intestines, spleen, and pancreas (Figure 134.1A). Blood flows from the portal vein into the liver sinusoids, and ultimately into the hepatic vein. This arrangement is unusual for a venous system, in that it consists of two capillary beds – the mesenteric capillaries and liver sinusoids – linked in series.

Under normal physiological conditions, the liver circulation functions as a passive reservoir accommodating venous and arterial inflow. Owing to the disproportionately large contribution of the portal vein (70%), total hepatic blood flow is largely determined by mesenteric hemodynamics. Mesenteric blood flow, in turn, is mediated primarily by mesenteric arterioles and, to a lesser extent the gut mucosa, which increases blood flow in response to nutrient absorption. Thus, portal venous inflow ultimately is controlled at the level of the mesenteric arterioles. Blood flow to the mesenteric circulation not only provides oxygen (O_2) and nutrients to the gut tissue but also plays an important role in supporting the secretion and transport of macromolecules taken up by the gut mucosa for further metabolism and processing in the liver. In keeping with these functions, total mesenteric (hence, portal venous) blood flow is markedly stimulated during the postprandial period.

Table 134-1: Timeline of Historical Landmarks Relating to Hepatic Circulation and Portal Hypertension

Time	*Investigator*	*Major Observation (Refs.)*
Paleolithic Era	Prehistoric writers in Southern France	Cave art depicts the liver as a bloody, highly vascular organ (24)
	Egyptian physicians	First recorded descriptions of the hepatic vasculature, including recognition that intestinal blood flows into the liver and that the liver is comprised of venous vascular channels (these anatomic concepts were further evaluated and debated by Hippocrates and Aristotle in the 4th to 5th century B.C., by Galen in the 2nd century A.D., by Erasistrates of Chios, Alexandria in the 3rd century, and by Herophilos of Chalcedon in the 3rd century) (25–31)
1500 B.C.	Ancient Egypt Ancient Hindus Ancient Mayans	Recognition of ascites (25,32)
Third century	Erasistrates Hippocrates Galen	Pathogenesis of ascites is actively debated (33–36)
Dark Ages, Middle Ages, 5th–10th century		Lack of substantive advances in the study of the hepatic circulation
14th century	English literature	Ascites based on Greek term for "leather water bag" appears in English language
15th century	DaVinci	Revival of the study of hepatic vascular anatomy (37)
16th century	Andrews Vesalius	(38)
16th century	Stahl of Germany	Incorrectly surmises that portal vein congestion or "abnormal plethora" is responsible for most chronic diseases (39)
1628	William Harvey	With discovery of the circulation of blood, able to correct a number of misconceptions relating to hepatic circulation that were perpetuated from the errors of Galen and Aristotle (40)
1600s	Glisson Cambridge	Validated the observations of Erasistrates and Harvey that blood flows from the intestine through hepatic channels, back to the heart, lungs, and arterial circulation, sequentially (41)
1700s – 1800s	Malpighi Wepfer Kiernan	Recognition of the hepatic acinus and portal triad (hepatic artery, portal vein, bile duct) (42–44)
1800s	Banti-Florence	Incorrectly surmises that portal hypertension is a result of splenomegaly (45,46)
1800s	Eck-Rossin	Recognition that portal hypertension could be corrected by portal caval shunting (47)
20th century	Gilbert and Villaret-France	First used the term *portal hypertension*
1950	Myers and Taylor	First demonstration of portal pressure measurement using hepatic venous pressure gradient analysis (48)
1950s	McIndoe – Mayo Clinic	Recognizes that portal hypertension is a result of hepatic vascular obstruction rather than splenomegaly (49)

Compiled from the textual recounts of the history of the hepatic circulation recently delineated by the renowned hepatologist and liver historian, Adrian Rubin. See Ref. 50. Reuben A, Groszmann R. Portal hypertension: a history. In: Sanyal A, Shah V, eds. *Portal Hypertension.* Totowa, NJ: Humana Press, 2005.

Hepatic Arterial Blood Flow

The hepatic artery, derived from the celiac plexus, provides only one-third of the total hepatic blood flow yet supplies the liver with a significant portion of its O_2 requirements. Blood from the high-pressure hepatic artery drains into hepatic sinusoids, either directly or after perfusing critical structures in the liver, including the peribiliary plexus, the vasa vasorum of hepatic veins, the portal tract interstitium, and Gleasson's capsule (2). In the sinusoids, hepatic artery blood mixes with poorly oxygenated blood from the low-pressure portal vein, providing a boost in O_2 delivery to hepatocytes. From the hepatic sinusoids, blood drains into the hepatic vein, followed by the inferior vena cava, and ultimately the right atrium. A remarkable feature of the liver circulation is the capacity of the sinusoids to "decompress" the highly pressured arterial inflow and thus maintain extremely low pressures. In the absence of such a mechanism, sinusoidal hemodynamics would be altered in

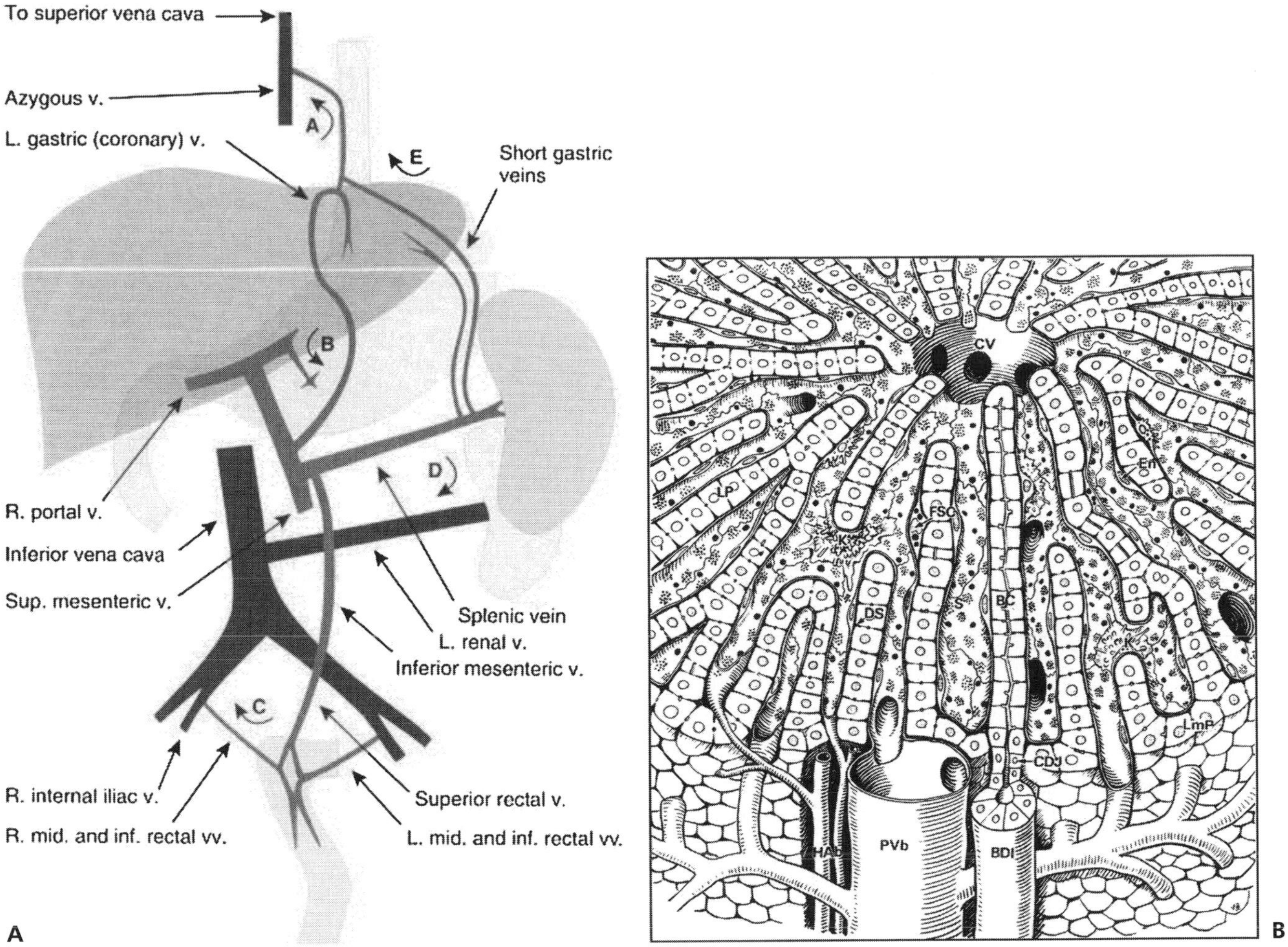

Figure 134.1. (**A**) Portal and collateral circulations. (*A*) Left gastric vein and (*E*) short gastric vein drains via the esophageal varices into the azygous vein. (*B*) Umbilical vein often is recanalized and, with the paraumbilical plexus in the abdominal wall, leads to the classical sign of caput medusae. (*C*) Superior rectal vein drains through hemorrhoid into middle rectal vein and internal iliac vein. (*D*) Abdominal viscera can drain via the retroperitoneal collaterals and drain into left renal vein. (Reprinted with permission from Feldmar A. *Sleisenger and Fordtrar's Gastro-intestinal and Liver Disease*, 7th ed. Philadelphia: Elsevier, 2002). (**B**) Vasculature architecture of the liver. Blood flow enters the liver via the portal vein (*PVb*) as well as the hepatic artery (*HAb*). Whereas portal blood enters directly into the sinusoids (*S*), hepatic arterial blood perfuses into distinct anatomic locations prior to reentering the sinusoids. Sinusoidal blood leaves the liver via the central veins (*CV*). Sinusoidal ECs (*En*) form the fenestrated sinusoidal wall. Kupffer cells (*K*) are located within the sinusoids, whereas hepatic stellate cells, also termed *fat storing cells* (*FSC*), lie within the space of Disse (*DS*), adjacent to the single layer of hepatocytes (liver plate, *LP*). Bile canaliculi (*BC*) drain bile into the interlobular bile ducts (*BDL*) via the caniliculoductar junction (*CD*) in the opposite direction to flow in the vascular channels. (Reproduced with permission from Motta P, Muto M, Fujita T. *The Liver. An Atlas of Scanning Electron Microscopy*. Tokyo: Igaku-Shoin Medical Publishers, 1978:129.)

a catastrophic way. The exact mechanism of decompression of high-pressure arterial blood within the sinusoids remains unclear.

Impairment of hepatic arterial flow normally has minimal detrimental effect on liver function since alternative portal vein-based perfusion pathways develop quickly, in part through autoregulation mechanisms described later in this chapter. However, under pathophysiological conditions such as following liver transplantation or in conditions associated with diminished portal vein flow (e.g., portal vein thrombosis and liver cirrhosis), hepatic arterial flow is of paramount importance in maintaining adequate hepatic perfusion. Significant reduction or cessation of hepatic blood flow in these settings can result in dramatic ischemic insults, not only within the biliary system, but also at the level of sinusoidally perfused hepatocytes.

Another interesting characteristic of hepatic arterial flow is its propensity to supply tumors with perfusion analogous to the role of the bronchial circulation in lung tumors. In the hepatic circulation, the unique anatomical basis of primary tumors has been exploited through the procedure of hepatic artery chemoembolization, which allows for targeted delivery of therapeutics and occlusion of the tumor vasculature with limited effects on the hepatic parenchyma and other organs.

AUTOREGULATION OF LIVER BLOOD FLOW

The total blood flow through the liver remains relatively constant over time. Such constancy, which helps to maintain normal portal pressure as well as liver clearance of xenobiotics and endogenous peptides, is mediated by autoregulation of hepatic arterial flow in response to changes in portal blood flow. This is thought to occur through a mechanism termed *the hepatic arterial buffer response* (3), a concept based on adenosine washout from the space of Mall, a thin fluid space surrounding the portal triad. Adenosine is constantly secreted around the terminal branches of the hepatic arteriole and portal venule (space of Mall) (see Figure 134.1B). When the portal blood flow increases, adenosine is washed away at a faster rate, resulting in hepatic arteriole vasoconstriction (3). Conversely, reductions in the clearance of adenosine result in adenosine accumulation and ensuing compensatory dilation of the hepatic artery. The increase in hepatic artery flow removes the excess adenosine, thereby maintaining adenosine concentration and total hepatic blood flow at a constant level. These regulatory mechanisms are vital in pathophysiological conditions associated with low portal vein flow, such as occurs with cirrhosis and portosystemic collateral shunting, hemorrhage, and surgical portocaval shunts. One would anticipate that the hepatic arterial and portal venous ECs must be playing a delicate balancing act in adenosine production and sensing; however, the endothelium-specific mechanisms responsible for this equilibrium remain largely unexplored.

ENDOTHELIUM IN DISEASE: PORTAL HYPERTENSION AS A PARADIGM FOR ENDOTHELIAL CELL HETEROGENEITY

Portal Hypertension at the Interface between Cell Biology and Plumbing

Arguably, the most important and interesting disease condition that afflicts the hepatic macrocirculation is portal hypertension. Portal hypertension results in EC-based hemodynamic derangements in nearly every vascular bed and is a prototypical disease that highlights the disparate and opposing responses of ECs in different vascular beds. The term *portal hypertension* or, more strictly, *portal venous hypertension*, refers explicitly to a pathological elevation of pressure in the veins that carry blood from the mesenteric organs (including the spleen) to the liver. Implicit in the working definition of portal hypertension is the necessary condition that the rise in portal pressure is not simply a consequence of an increase in systemic venous pressure – as might occur with congestive heart failure, for example – but rather reflects an elevated pressure gradient between the portal venous inflow to the liver and its hepatic venous outflow.

Significant advances have been made in our understanding of the cellular and molecular mechanisms of portal hypertension, including the central role of the dynamic interplay of the liver EC and the hepatic stellate cell (HSC). However,

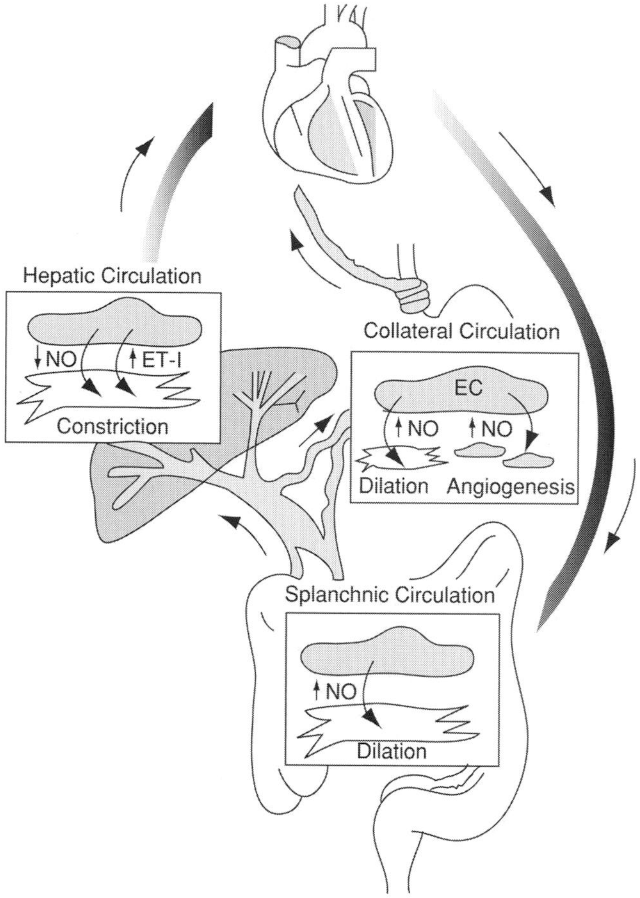

Figure 134.2. EC phenotypic changes in different vascular beds in portal hypertension. NO generation is diminished in the hepatic circulation in cirrhotic portal hypertension, thereby contributing to increased resistance, and NO generation is increased in the peripheral and mesenteric circulation, thereby contributing to increased flow into the portal circulation. The ensuing collateral circulation aims to decompress the portal circulation and is driven by NO-mediated angiogenesis as well as dilation of pre-existing collaterals. (Reprinted with permission from Shah V. Cellular and molecular basis of portal hypertension. *Clin Liver Dis.* 2001;5:629–644.)

from a pathophysiological and clinical standpoint, the problem continues to lend itself to a "plumber's approach." Blood flow through the liver is driven by the pressure gradient (ΔP) between the portal vein and hepatic vein. According to Ohm's Law, the pressure gradient ΔP is the product of hepatic resistance (R) and mesentery blood flow (Q) (Equation 1):

$$\Delta P = Q \times R \qquad \text{(Equation 1)}$$

Portal hypertension, which is defined as an elevation in the pressure gradient, may result from increased resistance to flow and/or increased inflow through the mesentery artery; both of these usually go hand-in-hand (Figure 134.2). In most cases, the initiating factor is an increase in resistance (e.g., cirrhosis, caval web, etc.). Increased resistance leads to increased inflow through a process that is not completely understood but involves porto-systemic shunting, systemic arterial vasodilation, fluid/sodium retention, and eventually increased blood flow into the portal circulation.

Increased Intrahepatic Resistance as a Form of Endothelial Dysfunction

Cirrhosis is associated with a reduction in the number and luminal diameter of sinusoids, which results in diminished cross-sectional area. According to Poiseuille's Law, the resistance (R) to flow is inversely proportional to the radius (r) of a tube to the fourth power (Equation 2):

$$R = 8\ \eta L/r^4 \qquad \text{(Equation 2)}$$

Thus, insofar as the sinusoidal surface area determines the "radius" of the hepatic vasculature, its overall reduction leads to markedly increased sinusoidal resistance. Two mechanisms contribute to the reduced surface area (hence increased resistance) in cirrhosis. First, there is mechanical obstruction of sinusoids secondary to fibrosis, anatomical derangement, and capillarization. Second, the sinusoids display elevated vascular tone. This latter process occurs through a dynamic interplay between liver ECs and HSCs. In liver injury and cirrhosis, liver ECs evidence endothelial dysfunction while the HSCs in hyperdynamic circulation syndrome become activated, expressing actin and myosin, and developing contractile properties similar to vascular smooth muscle cells (VSMCs) or pericytes. The constrictive tone of HSCs, like that of VSMCs, is modulated by a yin-and-yang balance of endothelium-derived factors, with nitric oxide (NO) and endothelin (ET)-1 seemingly most important in the hepatic circulation (4,5). The importance, in this dynamic component, of increased intrahepatic resistance lies in its susceptibility to therapeutic interventions when compared to the relatively static component of mechanical fibrosis and fixed scar. Although a sinusoidal basis for tonus changes in liver is appealing, the anatomic location of pressure drops across the liver (i.e., portal venules, sinusoids, or hepatic venules) is not fully established and is probably variable depending on the physiologic and pathobiologic circumstances.

Too Little Nitric Oxide

In cirrhosis and portal hypertension, the liver sinusoids probably experience a higher shear stress than normal, owing to the hyperdynamic circulation (6). Although shear stress is a normally a strong stimulus for release of NO in endothelium, portal hypertension is associated with an impairment in endothelium-dependent vasodilatation, a marker for NO bioactivity (7). Acetylcholine challenge into the ex vivo perfused cirrhotic liver via the portal vein results in minimal vasodilatation and even paradoxical vasoconstriction, which is akin to the endothelial dysfunction observed in other diseased vascular beds such as the atherosclerotic coronary artery (7).

Reduced intrahepatic NO production is the result of diminished endothelial NO synthase (eNOS) activity at the post-translational level. In cirrhosis, the amount of intrahepatic eNOS protein is unchanged, and overexpression of eNOS fails to improve endothelial function. Caveolin-1 protein abundance is increased in the ECs of injured and cirrhotic liver (8), and binding of caveolin to eNOS diminishes eNOS activity (6). In addition to caveolin-mediated inhibition of eNOS, cirrhosis also may be associated with diminished eNOS phosphorylation and activity. Phosphorylation at Serine 1177 by the phosphoinositide 3-kinase (PI3K)/protein kinase B (Akt) pathway increases enzyme activity. In animal models of cirrhosis, eNOS phosphorylation is diminished, and adenovirus transfection with constitutively active Akt restores intrahepatic eNOS activity and normalizes portal pressure, suggesting a defect in eNOS phosphorylation in cirrhosis (9). In support of the relevance of these post-translational changes in eNOS in human portal hypertension, statins have been shown to decrease hepatic sinusoidal resistance by increasing hepatic NO generation (although statins may have other beneficial mechanisms of action as well) (10).

NO donors can improve intrahepatic resistance but, unfortunately, these agents also may fuel a hyperdynamic circulation, characterized by NO excess in extrahepatic vascular beds. For example, low-dose isosorbide has been shown to attenuate postprandial portal pressure increases but it does so at a cost of a drop in systemic blood pressure, which can be problematic in cirrhotic patients (11). A new agent, NCX-100, is a NO donor coupled to ursodeoxycholic acid (UDCA), which has been shown in animal models to allow NO delivery specifically to the sinusoidal microcirculation without significant systemic effects (12). Clinical trials in humans are anticipated.

Too Much Endothelin-1

Another important vasoactive agent in portal hypertension and liver injury is ET-1 (5). Normally an endothelium-derived vasoconstrictor, ET-1 is mainly produced by activated HSCs in liver cirrhosis, and is probably upregulated by transforming growth factor (TGF)-β, which is increased in this disease (13). ET-1 acts in an autocrine and paracrine fashion through two distinct G-protein–coupled receptors, endothelin-A (ETA) receptor and endothelin-B (ETB) receptor. There is some controversy regarding the role of these receptors in hepatic vascular function and portal hypertension. However, it is generally believed that in ECs, ET-1 interacts with ETB to activate eNOS and mediate vasodilation (13), whereas in HSCs and VSMCs, ET binds to ETA to induce vasoconstriction in the hepatic circulation. In addition to modulating vascular tone, ET-1 also may be involved in wound healing response, which is a key mechanism of fibrosis and cirrhosis development.

Hyperdynamic Circulation

The pressure gradient across the portal bed is determined not only by resistance, but also by blood flow (see Equation 1).

Table 134-2: Systemic Vascular Changes Associated with a Rat Model of Prehepatic Portal Hypertension

	1 Hour	*1 Day*	*2 Days*	*3 Days*	*4 Days*
Total peripheral resistance	↑	↓	↓	↓	↓
Plasma volume	—	—	↑	↑	↑
Cardiac index	↓	—	—	—	↑
Mesentery artery resistance	↑	↑	—	—	↓
Mesentery inflow	↓	↓	↓	—	↑

In portal hypertension, the high pressures are often compounded by an increase in both parameters. The increase in portal venous inflow is part of a hyperdynamic circulatory state (HCS), characterized by widespread vasodilatation and increased blood flow in the mesenteric and most peripheral beds. The mechanisms underlying HCS are controversial. Although NO activity has been implicated in its pathogenesis, it is possible that NO levels are elevated as a secondary response to high flow–mediated shear stress, and that they play a role in perpetuating rather than initiating the syndrome (14). Using a prehepatic surgical model of the hyperdynamic circulation, namely portal vein ligation (PVL), Groszmann and colleagues demonstrated that NO production and NOS activation are a very early step in the development of the HCS, suggesting that NO overproduction may be a cause, rather than consequence, of HCS (15,16) (Table 134-2). However, arguing against a pathogenic role for NO in this syndrome is the demonstration – by some, but not all groups – that *eNOS*-knockout mice are susceptible to a hyperdynamic circulation in a model or prehepatic portal hypertension (17,18).

It has been hypothesized that HCS arises from bacterial translocation, endotoxemia, and tumor necrosis factor (TNF)-α, with secondary activation of endothelium and NO release. Cirrhotic patients often have bacterial translocation from the gut to mesenteric lymph nodes due to increased intestinal permeability and impaired immune clearance (19). Bacterial endotoxin triggers TNF-α production by mononuclear cells and Kupffer cells, and TNF-α then upregulates EC eNOS activity at the post-translational level via both the PI3K and Akt pathways (19). Bacterial translocation and endotoxemia are therefore targets for therapeutic intervention. Indeed, antibiotics such as norfloxacin have been shown to decrease bacterial translocation across intestinal mucosa in animal studies, and to dampen hyperdynamic circulation in human studies (20,21).

An interesting question is why portal hypertension is associated with increased eNOS/NO activity in the peripheral circulation, yet diminished activity in the liver sinusoids. Although the answer is not yet known, the observation provides a striking example of vascular bed–specific phenotypes in liver disease.

Portosystemic Collateral Circulation

Abnormally high pressures in the portal circulation may lead to portosystemic collaterals that bypass the sinusoids and thus decompress the portal bed. The collaterals – which form direct connections between the engorged portal circulation and the lower-pressure venous system – create a parallel circuit that lowers the overall resistance (R) to portal blood flow (Equation 3):

$$1/R_{\text{portal}} = 1/R_{\text{Hepatic}} + 1/R_{\text{collaterals}} \qquad \text{(Equation 3)}$$

Unfortunately, these collaterals cannot fully decompress the portal circulation. Moreover, they may result in life-threatening complications, such as the development of bleeding esophageal varices. In addition to varices, collaterals may take the form of capillary ectasias, such as portal hypertensive gastropathy, a mosaic pattern of vascular-based epithelial cell injury that occurs in patients with portal hypertension and that may present as gastrointestinal bleeding. Although physiological studies have focused on the anatomy and hemodynamics of portosystemic collaterals, more recent detailed molecular studies have begun to elucidate the growth factors and mediators responsible for vasculogenesis, neoangiogenesis, and collateralization.

Special Circulations Affected by Hyperdynamic Circulation

As alluded to earlier, portal hypertension results in EC-based hemodynamic derangements in nearly every vascular bed. Two peripheral beds that undergo clinically relevant hemodynamic changes are the kidney and lung. In the kidney, the reduced blood pressure and perceived "underfilling" of renal blood flow results in activation of endogenous vasoconstrictors and volume repletion pathways such as the renin-angiotensin, aldosterone, and antidiuretic hormone (ADH) systems (Table 134-3). These compensatory responses are not only inadequate, but they render the kidneys exquisitely sensitive to subsequent insults (e.g., small doses of nonsteroidal anti-inflammatory drugs or contrast dye), leading in some cases to renal *decompensation* (the hepatorenal syndrome). Interestingly, this form of renal dysfunction is entirely dependent on hepatic and systemic factors, and is reversed upon successful treatment of the liver disease.

The pulmonary vascular endothelium also responds in unique and interesting ways to cirrhosis and portal hypertension. In normal conditions, ET-1 mediates pulmonary artery vasoconstriction via the ETA receptor. In the so-called *hepatopulmonary syndrome,* excess ET-1 levels in the blood appear to stimulate ETB receptors in the pulmonary vasculature to promote eNOS activation, pulmonary shunting, and hypoxemia (22). Conversely in some patients, ET-1 elevation seems to act on ETA receptors to promote vascular remodeling and portopulmonary hypertension (22). It has been difficult to

Table 134-3: Systemic Vascular Changes That Occur in Human Portal Hypertension and Associated Renal Failure

	Preascitic	*Ascitic*	*HRS-II**	*HRS-I†*
Cardiac index	↑↑	↑↑	↑	↑
Heart rate/sympathetic	—	↑	↑	↑
Renin angiotensin system	—	↑	↑	↑
ET-1 plasma level	—	—	↑	↑
Mean arterial pressure	—	—	↓	↓
Renal blood flow	—	—	↓	↓
Renal natriuretic capacity	↑	↓	↓	↓
Renal prostaglandin	—	↑	↑	↓

HRS; hepatorenal syndrome; *type II HRS is a more indolent and slowly progressive form; †wild-type I HRS is a fulminant form of renal failure.

determine which patients will develop these heterogeneous and opposite responses of portal hypertension on the pulmonary vascular endothelium (23).

DIAGNOSTIC AND THERAPEUTIC IMPLICATIONS

Hepatic Venous Pressure Gradient Measurement

A variety of techniques are available for assessing volume, pressure, and flow within the mesenteric circulation. Some of these techniques are utilized predominantly for research purposes, whereas others have clear clinical utility. The identification of a technique that provides reliable, accurate, and reproducible hemodynamic information in a noninvasive manner continues to be the Holy Grail of portal circulatory investigation. One of the most mechanistically interesting methodologies is the ability to indirectly estimate portal pressure by using a wedge balloon catheter technique – hepatic venous pressure gradient (HVPG) monitoring – which is akin to the Swan Ganz measurements used to estimate systemic hemodynamics. Although the most accurate determination of portal pressure in experimental model systems can be determined by direct puncture of the portal venous vasculature, the potential complications outweigh the utility of this approach in most clinical circumstances. To perform HVPG monitoring, a balloon catheter is passed through the vena cava to the hepatic vein, where pressures are obtained with the balloon deflated within the hepatic vein and subsequently with inflation of the balloon and occlusion of the hepatic vein branch. The occluded pressure reading reflects the pressure within the hepatic sinusoids, because a column of blood is contiguous between the inflated balloon and the sinusoids. The free hepatic vein pressure is subtracted from the occluded hepatic venous pressure to provide the HVPG, an indirect assessment of pressure within the hepatic sinusoid. This technique can be useful to make a definitive diagnosis of portal hypertension, establish the site of increased portal resistance, and monitor the portal pressure response to treatments.

Diagnosing Sites of Altered Resistance According to Location

Increased resistance may be caused by structural and/or functional lesions in prehepatic, intrahepatic, and/or posthepatic locations. Within the liver, the site of increased resistance can be further identified as presinusoidal, sinusoidal, or postsinusoidal. These distinctions can often be ascertained using HVPG monitoring.

Prehepatic Resistance

Prehepatic resistance can be caused by portal vein thrombosis. In adults, it is usually related to cirrhosis, splenectomy, hypercoagulable state, and cancer. Due to the portal vein thrombus, the increased pressure in the portal vein remnant cannot be transmitted to the hepatic sinusoid, and thus HVPG-based methodology tends to underestimate hepatic venous wedge pressure (HVWP). Portal hypertension results in massive splenomegaly and varices formation. Varices may even occur along the common bile duct and occasionally cause obstructive jaundice and hemobilia. Otherwise, liver function is usually unaffected because hepatic blood flow is augmented through hepatic arterial buffer response (as discussed earlier and evidenced clinically in Figure 134.3). Over 5 to 12 weeks, the portal vein thrombus transform into a collagenous plug with tortuous venous channels; this is called *cavernous transformation.* Both cavernous transformation and the development of portal systemic collaterals are examples of venous angiogenesis or *venogenesis,* which are in contradistinction to the angiogenesis that frequently occurs in the arterial vasculature in other conditions such as coronary and peripheral arterial occlusion.

Intrahepatic Resistance

Intrahepatic presinusoidal resistance is exemplified by schistosomiasis and granulomatous disease such as sarcoidosis. In cirrhosis caused by alcoholic liver disease or viral hepatitis, the increased resistance occurs at the sinusoid. HVWP

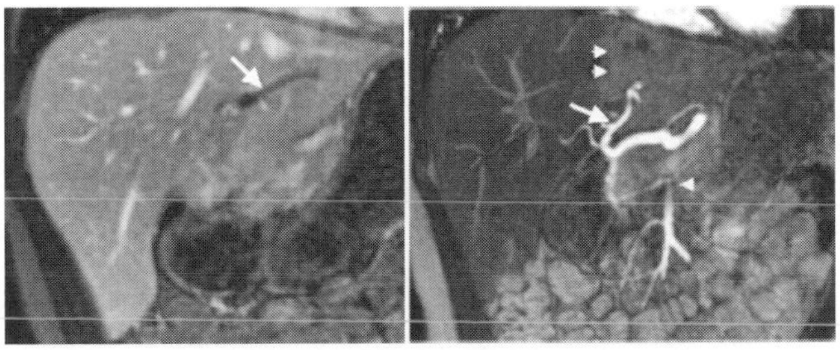

Figure 134.3. Example of hepatic blood flow autoregulation. A case of a patient with portal vein thrombosis (*left*) with compensatory marked dilatation of the hepatic artery (*right*). (Reprinted with permission from Gulberg V, Schoenberg S. Hepatic arterial buffer response: visualization by multiphasic high-resolution 3D magnetic resonance angiography. *J Hepatol.* 2004;40:181.)

accurately reflects portal venous pressure (PVP) and is elevated. Postsinusoidal intrahepatic resistance can be caused by veno-occlusive disease. Veno-occlusive disease is the occlusion of terminal hepatic venules and sinusoids as a result of endothelial damage. It is commonly seen in pyrrolizine alkaloid ingestion and aggressive chemotherapy associated with bone marrow transplant. The patient presents with weight gain, hepatomegaly, ascites, and hyperbilirubinemia.

Posthepatic Resistance

The most notable cause of posthepatic resistance in the hepatic macrocirculation is Budd-Chiari syndrome. Budd-Chiari syndrome is venous thrombosis located anywhere between the right atrium and small radicals of the hepatic vein. It is commonly seen in patients with hypercoagulable states, pregnancy, myeloproliferative disorder, and cancer. The caudate lobe of the liver drains directly into the inferior vena cava and is typically spared. Clinical presentation depends on the time course, and may include acute liver failure or insidious onset of cirrhosis. Ischemic injury and congestion are seen histologically in the liver vasculature. HVPG usually is elevated owing to the high sinusoidal pressure resulting from the occlusion.

GOALS FOR BRIDGING BENCH-TO-BEDSIDE GAP

The hepatic circulation and its involvement in portal hypertension is one example of the intense efforts being made in an attempt to bridge the bench-to-bedside gap. Endothelial research has traditionally been a physiologically oriented field. Indeed, major advances have been made using methodologies that allow researchers to translate research data into new treatment therapies for patients, including the use of β-blockers, mesenteric vasoconstrictors, nitrates, and other agents that act directly through modulating vascular phenotype, in many cases directly at the level of the EC. Today, many investigations are focused on more molecular approaches, although the emphasis on translating study data from relevant animal models to humans remains a priority for investigators in this field.

KEY POINTS

- The hepatic circulation has many unique features that distinguish it from other vascular beds; diseases of the hepatic circulation, such as portal hypertension, stimulate heterogeneous responses in ECs throughout other vascular beds.
- The hepatic circulation maintains a dual blood supply that is reciprocally and carefully regulated.
- Portal hypertension is a complex disease of the hepatic circulation that occurs from multiple different causes and affects vascular ECs in almost every organ as well as the liver.

Future Goals

- To identify noninvasive methods to estimate changes in the hepatic circulation
- To develop therapies for hepatic-circulatory diseases that can be targeted to specific vascular beds, thereby avoiding untoward complications
- To continue to translate basic advances in vascular biology to the disease models of the hepatic circulation and, in turn, translate these advances to patients with portal hypertension

REFERENCES

1 Shah V, Kamath P, de Groen P. Physiology of the splanchnic circulation. In: Topol E, Lanzer F, eds. *Theory and Practice of Vascular Diseases*. Dissertation. University of Monster, Germany, 2000.

2 Ekataksin W. The isolated artery: an intrahepatic arterial pathway that can bypass the lobular parenchyma in mammalian livers. *Hepatology*. 2000;31(2):269–279.

3 Lautt W. The 1995 Ciba-Geigy Award Lecture. Intrinsic regulation of hepatic blood flow. *Can J Pharmacol*. 1996;74:223–233.

4 Shah V. Cellular and molecular basis of portal hypertension. *Clin Liver Dis*. 2001;5:629–644.

5 Rockey DC. The cellular pathogenesis of portal hypertension: stellate cell contractility, endothelin, and nitric oxide. *Hepatology*. 1997;25:2–5.

6 Shah V, Toruner M, Haddad F, et al. Impaired endothelial nitric oxide synthase activity associated with enhanced caveolin binding in experimental liver cirrhosis. *Gastroenterology*. 1999;117:1222–1228.

7 Gupta T, Toruner M, Chung M, Groszmann R. Endothelial dysfunction and decreased production of nitric oxide in the intrahepatic microcirculation of cirrhotic rats. *Hepatology*. 1998;28:926–931.

8 Chatila R, Theise N, Shah V, et al. Caveolin-1 in normal and human cirrhotic liver. *Gastroenterology*. 2000;118:A979.

9 Morales-Ruiz M, Cejudo-Martin P, Fernandez-Varo G, et al. Transduction of the liver with activated Akt normalizes portal pressure in cirrhotic rats. *Gastroenterology*. 2003;125(2):522–531.

10 Zafra C, Abraldes J, Turnes J, et al. Simvastatin enhances hepatic nitric oxide production and decreases the hepatic vascular tone in patients with cirrhosis. *Gastroenterology*. 2004;126:749–755.

11 Bellis L, Berzigotti A, Abraldes J, et al. Low doses of isosorbide mononitrate attenuate the postprandial increase in portal pressure in patients with cirrhosis. *Hepatology*. 2003;37(2):378–384.

12 Fiorucci S, Antonelli E, Tocchetti P, Morelli A. Treatment of portal hypertension with NCX-1000, a liver-specific NO donor. A

review of its current status. *Cardiovasc Drug Rev*. 2004;22(2):135–146.
13 Rockey D, Chung J. Regulation of inducible nitric oxide synthase and nitric oxide during hepatic injury and fibrogenesis. *Am J Physiol*. 1997;273:G124–G130.
14 Hori N, Wiest R, Groszmann RJ. Enhanced release of nitric oxide in response to changes in flow and shear stress in the superior mesenteric arteries of portal hypertensive rats. *Hepatology*. 1998; 28:1467–1473.
15 Colombato L, Albillos A, Groszmann R. Temporal relationship of peripheral vasodilation, plasma volume expansion and the hyperdynamic circulatory state in portal-hypertensive rats. *Hepatology*. 1992;15:323–328.
16 Tsai M, Iwakiri Y, Cadelina G, et al. Mesenteric vasoconstriction triggers nitric oxide overproduction in the superior mesenteric artery of portal hypertensive rats. *Gastroenterology*. 2003;125(5): 1452–1461.
17 Theodorakis N, Wang Y, Skill N, et al. The role of nitric oxide synthase isoforms in extrahepatic portal hypertension: studies in gene knock-out mice. *Gastroenterology*. 2003;124(5):1500–1508.
18 Iwakiri Y, Cadeline G, Sessa W, Groszmann R. Mice with targeted deletion of eNOS develop hyperdynamic circulation associated with portal hypertension. *Am J Physiol*. 2002;283:G1074–G1081.
19 Rasaratnam B, Connelly N, Chin-Dusting J. Nitric oxide and the hyperdynamic circulation in cirrhosis: is there a role for selective intestinal decontamination? *Clin Sci (Lond)*. 2004;107(5):425–434.
20 Chin-Dustin J, Rasaratnam B, Jennings G, Dudley F. Effect of fluoroquinolone on the enhanced nitric oxide-induced peripheral vasodilation. *Ann Intern Med*. 1997;127:985–988.
21 Albillos A, de la Hera A, Gonzalez M, et al. Increased lipopolysaccharide binding protein in cirrhotic patients with marked immune and hemodynamic derangement. *Hepatology*. 2003;37 (1):208–217.
22 Zhang M, Luo B, Chen S-J, et al. Endothelin-1 stimulation of endothelial nitric oxide synthase in the pathogenesis of hepatopulmonary syndrome. *Am J Physiol*. 1999;277(5):G944–G952.
23 Benjaminov F, Prentice M, Sniderman K, et al. Portopulmonary hypertension in decompensated cirrhosis with refractor ascites. *Gut*. 2003;52(9):1355–1362.
24 Glyn D. *Lascaux and Carnac*. London: Butterworth Press, 1965.
25 Bryan C. *The Papyrus Ebers* (translated from German version). New York: D. Appleton and Company, 1931.
26 Harris C. *The Heart and Vascular System in Ancient Greek Medicine; from Alcmaen to Galen*. Oxford: Clarendon Press, 1973.
27 Singer C. *Galen on Anatomical Procedures*. Translation of surviving books with introduction and notes. London: Oxford University Press, 1956.
28 Brock A. *Galen on the Natural Faculties* (translated from the Greek). London: Loeb Classical Library, 1928.
29 Bradley S. *The Splanchnic Circulation*. New York: Oxford University Press, 1964.
30 Dobson J. Erasistratus. *Proc Roy Soc Med*. 1927;20:825–832.
31 Dobson J. Herophilus of Alexandria. *Proc Roy Soc Med*. 1925;18: 19–32.
32 Majumdar A. In: Majumdar A, ed. *Hand Book of Domestic Medicine and Common Ayurvedic Remedies*. New Delhi: Documentation and Publication Division. Central Council for Research in Ayurveda and Siddha; 1999:88–89.
33 Sprengell C. *The Aphorisms of Hippocrates, and the Sentences of Celsus: (with explanations and references to the most considerable writers in physick and physiology, both ancient and modern: to which are added, Aphorisms upon the small-pox, measles, and other distempers), not so well known to former more temperate ages*. London: Printed for R. Bonwick, et al.; 1708.
34 Dawson A. Historical notes on ascites. *Gastroenterology*. 1960;39: 790–791.
35 Runyon B. Historical aspects of treatment of patients with cirrhosis and ascites. *Semin Liver Dis*. 1997;17:163–173.
36 Reuben A. Out came copious water. *Hepatology*. 2002;36:261–264.
37 DaVinci L, Vangensteen OC, Windsor Castle RL. *Respirazione Cuore, Viscera Addominali*. Christiania, Dybwad; 1911.
38 Vesalius A. *De humani corporis Fabrica*. Basilcae: J. Oporini; 1543.
39 Stahl G. *De vena Portae, porta malorum hypochondriaco-splenectico-sufferativo, hysterico-colico hemorrhoidariorum*. Halle, 1698.
40 Harvey W. *Exercitatio anatomica du motu cordis et sanguinis in animalibus*. Frankfurt: W. Fitzer, 1628.
41 Glisson F. *Anatomia Hepatis*. London: O. Pullein, 1654.
42 Malpighi M. *De viscerum structura excercitatio anatomica*. London, 1666.
43 Wepfer J. *De dubiis anatomicis. Epistola ad Jacob Henricum Paulli*. Argentorati: Simonem Paulli, 1665.
44 Malpighi M. *Discours anatomiques sur la structure des visceres scavoir du foye, du cerveau, des reins, de la ratte, du polype du coeur et des poulmons*, 2nd Edition. Paris: L. d'Houry, 1687.
45 Banti G. Dell' anemia splenica. *Arch Scuola Anat Pat Firenze*. 1883; 124:53–122.
46 Banti G. La splenomegalia con cirrosi del fegato. *Sperimentale*. 1894;48:407–32; 47–52.
47 Eck N K. Voprosu o perevyazk e vorotno veni. Predvaritelnoye soobshtsheniye. *Voyenno-med*. 1877;130:1–2.
48 Myers J, Taylor W. An estimation of portal venous pressure by occlusive catheterization of a hepatic venule. *J Clin Invest*. 1951; 30:662–663.
49 McIndoe A. Vascular lesions of portal cirrhosis. *Arch Pathol*. 1928; 5:23–40.
50 Reuben A, Groszmann R. Portal hypertension: a history. In: Sanyal A, Shah V, eds. *Portal Hypertension*. Totowa, NJ: Humana Press, 2005.

135

Inflammatory Bowel Disease

Ossama A. Hatoum and David G. Binion

Froedtert Memorial Lutheran Hospital, Medical College of Wisconsin, Milwaukee

Human idiopathic inflammatory bowel disease (IBD), Crohn disease (CD) and ulcerative colitis (UC) are lifelong illnesses characterized by chronic inflammatory destruction of the gastrointestinal tract. IBD is estimated to affect between one to two million Americans, and is most commonly diagnosed during adolescence or early adulthood, thus representing a significant burden of disease throughout the patient's lifetime (1). The description of terminal ileitis in 1932, later commonly referred to as CD, distinguished chronic intestinal inflammation and stricture formation from intestinal tuberculosis (2), in which the resected gut tissues failed to demonstrate acid-fast bacilli on histological examination. The anatomic distribution of CD was later revised to include the large bowel (3,4). Samuel Wilks initially described UC, distinguishing it from hemorrhagic bacterial dysentery by 1859. The first description of the natural history of UC was reported in 1909. Despite decades of intense research efforts, the etiology of IBD remains obscure and, as a result, treatment options are not specific and focus on blunting the chronic inflammatory process in the gut. A majority of CD patients ultimately require surgical intervention for complications, most commonly the emergence of symptomatic strictures in the intestinal lumen. Approximately one fourth of UC patients require colectomy for the treatment of medically refractory disease or the detection of neoplastic transformation, a well-recognized complication of long-standing chronic gut inflammation. Recent evidence from genetic and immunologic investigation has identified alterations in the innate immune response and inappropriate immune activation of the mucosal immune system in response to enteric microbiologic antigens as a potential inciting influence in the development of IBD. Investigation of the molecular events in IBD also has identified a central role of the intestinal microvascular endothelium in the regulation of mucosal immunity and leukocyte trafficking into the gut, as well as its role in the regulation of intestinal perfusion, which is profoundly altered in these conditions.

THE INTESTINAL MICROCIRCULATION AND GUT-SPECIFIC ENDOTHELIUM

Perfusion of the human gut in the peritoneal cavity is supplied by the celiac axis, superior mesenteric artery (SMA), and inferior mesenteric artery (IMA), which branch from the anterior abdominal aorta, with marked variability between individuals. The celiac artery arising from the anterior aorta branches into the left gastric, common hepatic, and gastroduodenal arteries supplying the stomach, duodenum, pancreas, and liver. The SMA originates from the anterior aorta posterior to the pancreas, giving rise to four major branches – including the pancreaticoduodenal, ileocolic, right colic, and middle colic arteries – and a series of smaller jejunal and ileal branches, all of which are encased in the mesentery. These arteries provide the primary supply to the small intestine and two-thirds of the proximal large intestine. The IMA arises from the aorta just proximal to the bifurcation, branches into the left colic and multiple sigmoid arteries, and terminates in the superior rectal artery. The IMA provides the major arterial supply to the descending and sigmoid colon as well as the proximal rectum. These mesenteric intestinal arteries form a series of arcades, ultimately branching into terminal arcades, which then generate numerous straight vessels (i.e., the vasa recta) that enter the intestinal wall along the mesenteric attachment. These penetrating arterioles in the gut form a plexus in the submucosal space, which then leads to both mucosal and serosal side-branches.

The human gastrointestinal tract, particularly the small intestine, is highly vascularized, with a rich microcirculation that branches from resistance arterioles located beneath the muscularis mucosa into arcades of capillaries and venules that reach into the villous tip. The individual villi may possess a fountain-like subepithelial capillary network, which will play an essential role in physiologic function in the gut. Vascular perfusion of the gut is distinct from other organ systems, and is frequently characterized by high rates of blood flow, oxygen utilization, and transcapillary exchange of fluid and solutes. At rest, intestinal perfusion via the SMA will range from 29 to 70 mL/min/100 g intestinal tissue (5–8), whereas in the fed state, splanchnic hyperemia increases perfusion by 28% to 132% (8). Vascular perfusion of the gut is dynamic and, as expected, is determined largely on the basis of physiologic need. Work from Granger and colleagues demonstrated up to a fourfold increase in perfusion during nutritive function in the small intestine (8). Vascular perfusion of the gut is stimulated differentially in various areas of the gastrointestinal tract by partially digested food, bile, bile salts, and fats. The microcirculation in the gut, from stomach through large intestine, is involved in essentially all aspects of intestinal physiology and homeostasis. Both secretory and absorptive functions of the gut epithelium occur in a close relationship with the microcirculation. Indeed, the majority of the fluids secreted daily in the gastrointestinal tract originate from the gut's vascular supply.

In addition to the high levels of perfusion associated with the metabolic demands of digestion and absorption of nutrients, the microcirculation in the gut plays an integral role in mucosal immunity. Indeed, the gut is the largest immune organ in the body, acting at the interface between the body and the external environment of the gut lumen. Recruitment of specific circulating leukocyte populations into the gut is regulated in part by the expression of specific vascular addressins on the endothelial lining of the intestinal microcirculation, including both the lamina propria microvessels as well as the Peyer patches. These patterns of leukocyte recruitment play a key role in gut immune homeostasis, simultaneously protecting the body from exposure to deleterious pathogens, and allowing for nutrition and the absorption of digestive material without mounting an immune response to food (i.e., oral tolerance), which is integral to health. Thus, endothelial cells (ECs) lining the gut microcirculation, specifically the postcapillary venules, play a "gatekeeper" role in the selective recruitment of circulating immune cells to the antigenically rich environment of the intestine through the highly regulated expression of leukocyte addressins and chemokines (9).

THE ROLE OF THE ENDOTHELIUM IN CHRONIC INTESTINAL INFLAMMATION

The histologic pattern of tissue injury in IBD involves a spectrum of pathology. Focal intestinal inflammation is the hallmark feature of CD, which may range from crypt abscess formation to increased mononuclear inflammatory cell infiltrate accompanied by architectural distortion of intestinal crypts. Perhaps the earliest macroscopically identifiable lesion in CD is the aphthous ulcer, which occurs in the vicinity of postcapillary venules overlying lymphoid aggregates (10). The presence of noncaseating epithelioid granulomas is another hallmark feature of CD, which may be detected in from from 15% to 70% of intestinal specimens (10,11). This wide variability may reflect the methods of tissue sampling (i.e., endoscopic vs. surgical acquisition of tissues). The cellular components of granuloma formation in CD have been linked to specific cytokine and adhesion molecule expression patterns, with tumor necrosis factor (TNF)-α emerging as the key cytokine in the formation of this lesion. Late histologic features of CD are characterized morphologically by large ulcers that denude the mucosa, sinus tracts, and bowel strictures caused by transmural fibrosis.

Histologically, UC is characterized by acute and chronic inflammation and microvascular hemorrhage limited to the mucosa. The lamina propria is typically edematous, with dilation and congestion of capillaries and extravasation of red cells. In severe UC, the epithelial surface is damaged, with ulceration that may lead to more extensive bowel wall injury, including ischemic necrosis.

Early investigation into the role of ECs in IBD pathogenesis focused on histologic evaluation, characterizing the morphology of the microvasculature in chronically inflamed bowel. Using transmission electron microscopy, Dvorak and colleagues evaluated intestinal specimens from CD-resected material (12). These ultrastructural studies revealed abnormalities in the ECs lining the microcirculation in areas of the intestine affected by CD, including loss of monolayer integrity with tissue edema, extravasation of red blood cells, focal venular endothelial necrosis adjacent to areas of undamaged ECs, and EC hypertrophy.

Increased appreciation for the active role of the endothelium in the regulation of leukocyte recruitment during inflammation prompted the investigation of endothelial–leukocyte interactions in IBD. Initial investigation used histologic approaches, with various groups demonstrating increased expression of cell adhesion molecules (CAMs) in the intestinal microvasculature in both CD and UC (13). Immunolocalization studies demonstrated marked increase in E-selectin and intercellular adhesion molecule (ICAM)-1 expression in IBD intestine, whereas vascular cell adhesion molecule (VCAM)-1 expression was less clearly demonstrated (14). Subsequent investigation by Briskin and colleagues demonstrated an increase in the gut-specific homing molecule mucosal addressin cell adhesion molecule (MAdCAM)-1, which plays a major role in the recruitment of leukocytes expressing the $\alpha4$ integrin into the mucosal immune compartment (15). In normal tissues, MAdCAM-1 was shown to be constitutively expressed in the endothelium of venules of intestinal lamina propria. However, the proportion of venular endothelium within lamina propria that expresses MAdCAM-1 was increased, compared with normal tissues, at inflammatory foci associated with UC and CD (15).

In addition to these studies demonstrating increased expression of EC adhesion molecules and enhanced leukocyte recruitment (14), further investigations examined alterations in the populations of leukocytes that were recruited into the IBD gut. Alterations in leukocyte homing patterns in IBD gut were characterized by Salmi and colleagues, who demonstrated that naïve lymphocytes are preferentially recruited to the chronically inflamed intestinal microvascular endothelium, whereas normal and uninvolved intestinal microvessels preferentially bound memory lymphocytes and monocytes (16). These findings were confirmed by Burgio and colleagues, who also demonstrated an altered pattern of leukocyte binding in CD, in which monocytes and naïve T cells were again preferentially recruited to the chronically inflamed intestine through lamina propria microvessels (17). These authors also demonstrated increased expression of ICAM-1, E-selectin, and CD34 (a glycoprotein component of the peripheral node addressin) in the IBD gut microvessels. These data suggest that patterns of naïve leukocyte recruitment into the chronically inflamed IBD intestine are linked to altered expression of homing addressins on the IBD intestinal endothelium. Given the antigenically rich environment in the gut, the recruitment of naïve lymphocytes into the mucosal immune compartment may play an important role in the aberrant immune activity and perpetuation of chronic inflammation (18).

To more fully define the contribution of microvascular ECs in chronic intestinal inflammation, studies have been carried out using cultures of intestinal microvascular ECs from patients with and without IBD. Human intestinal microvascular ECs (HIMECs) have been insolated and characterized, demonstrating classic endothelial markers including Weibel-Palade bodies, expression of von Willebrand factor (vWF), platelet-endothelial cell adhesion molecule (PECAM)-1/CD31, E-selectin, ICAM-1, and VCAM-1, as well as unique patterns of leukocyte adhesion and growth, compared with human umbilical vein ECs (HUVECs) (19,20). More importantly, HIMECs also have been isolated from involved and uninvolved CD and UC intestine. HIMECs isolated from both chronically inflamed CD and UC gut demonstrated a significantly enhanced capacity to adhere leukocytes (e.g., neutrophils, as well as T-cell and monocyte-like cell lines), compared to control HIMECs (i.e., EC lines generated from "normal" margins from bowel resection for reasons other than IBD, including diverticulosis, cancer, and trauma), a phenomenon that was elicited only following activation with proinflammatory cytokines (interleukin [IL]-1β, TNF-α, and bacterial lipopolysaccharide [LPS]). Leukocyte "hyperadhesion" appears to be an acquired phenomenon, because HIMEC cultures derived from uninvolved IBD intestinal segments failed to demonstrate increased leukocyte binding (21). Interestingly, the phenotype of "hyperadhesion" following inflammatory activation was maintained in those IBD HIMECs that were passaged in vitro during the lifetime of the culture (19). The mechanisms underlying leukocyte hyperadhesion in the chronically inflamed IBD HIMECs did not appear to involve increased levels of CAM expression compared to control cultures, which prompted an investigation of possible alterations in the intracellular mechanisms that govern the downregulation of inflammatory activation in ECs. Nitric oxide (NO) plays a central role in the regulation of endothelial activation and the maintenance of vascular homeostasis, exerting a potent antiinflammatory effect and downregulating the activation of vascular ECs as well as their capacity to bind circulating leukocytes, normally an early and rate-limiting step in the inflammatory process. Control HIMECs displayed distinct patterns of NO generation through both constitutive endothelial NO synthase (eNOS; NOS3) as well as inducible NOS (iNOS; NOS2), which was expressed following inflammatory activation (22). In marked contrast, IBD HIMECs (both CD and UC) failed to express iNOS and increased levels of NO following inflammatory activation (23). Preliminary investigation of eNOS demonstrated unchanged levels of mRNA between control and IBD HIMEC lines (22). However, this investigation did not consider eNOS function, which may be altered due to activation status and cofactor association (e.g., "uncoupling" of the enzyme) (24). This loss in NO generation in IBD HIMECs was linked to enhanced leukocyte binding, because the administration of NO donors restored a normal binding pattern in the activated IBD HIMECs. Further investigation demonstrated that iNOS-derived NO in the IBD HIMECs appears to function as an endogenous antioxidant, quenching the superoxide anion, which is a central mediator of inflammatory activation in gut microvascular ECs. These data again suggest epigenetic mechanisms that lead to preservation of an altered activation status characterized by loss of NO generation in the IBD HIMECs exposed to chronic inflammatory stress in vivo. The mechanisms that lead to altered iNOS expression in the chronically inflamed IBD HIMECs have not been fully defined.

VASOREGULATION AND MICROVASCULAR DYSFUNCTION IN CHRONIC INTESTINAL INFLAMMATION

The endothelium not only provides a structural barrier between the circulation and surrounding tissue, but ECs also secrete mediators that influence vascular hemodynamics in the physiologic state. ECs contribute to the regulation of blood pressure and blood flow by releasing vasodilators such as NO and prostacyclin (PGI_2), as well as vasoconstrictors, including endothelin (ET)-1 and platelet-activating factor (PAF). The major biologic effects of these chemically diverse compounds are regulated by localization of specific receptors on vascular cells, through their rapid metabolism, or at the level of gene transcription. NO is constitutively secreted by ECs, but its production is modulated by a number of exogenous chemical and physical stimuli, whereas the other known mediators (PGI_2, ET, and PAF) are synthesized primarily in response to changes in the external environment.

An investigation of intestinal perfusion in the setting of chronic inflammation in IBD has been carried out using a variety of in vivo and in vitro techniques. Angiographic studies of the IBD intestine have demonstrated preserved anatomy in the SMA and IMA, with significant abnormalities in the vasa recta, characterized by tortuous, dilated vessels together with the loss of normal tapering and terminal coiling as they penetrate the bowel wall (25,26). In early stages of IBD, angiographic studies have demonstrated arterioles that abruptly taper as the vessels reach the bowel wall with right-angle bifurcation, bizarre distribution, and small luminal irregularities in the peripheral branches (25,26). Furthermore, advanced IBD lesions demonstrate reduced vessel diameter, decreased vascular density, and diminished blood flow in the involved segments, which may contribute to an impaired capacity to heal and resolve the chronic inflammatory "wound."

Alterations in the IBD microvascular architecture were characterized by Wakefield and colleagues (27) using scanning electron micrographs of corrosion microcasts from control and CD patients. These studies identified occlusive fibrinoid lesions in the arterioles supplying affected intestine, which were not demonstrated in uninvolved areas of bowel. Morphologically, the chronically inflamed microvessels were tapered and stenosed, compared with vessels from uninvolved and control bowel. Vascular damage was demonstrated as an early pathologic finding, which preceded the development of mucosal ulceration. Vascular damage appeared to be highest where vessels penetrated through the muscularis propria, and bursts of angiogenic vessels were seen in distal areas of the mucosal circulation. These authors concluded that multifocal gastrointestinal infarction plays a pathogenic role in the chronic inflammatory lesion in CD, with the extent of vascular damage correlating with the severity of intestinal injury.

The concept of microvascular dysfunction and relative tissue ischemia was further investigated by Funayama and colleagues, who used tissue histometry to describe remodeling in the CD intestinal microcirculation (28). Assuming that medial atrophy is an indirect measure of decreased vascular perfusion and pressure, these investigators characterized atrophy of arterial media in the submucosal CD vessels, which indirectly suggests ischemia and increased vascular resistance in deeper submucosal arteries. Taken together, these studies suggest that the microvascular anatomy in the chronically inflamed CD intestine has undergone extensive remodeling. However, these studies did not directly assay microvascular function or any potential factors that may appear to lead to these alterations in vascular architecture.

Assessment of intestinal microvascular physiology in human IBD has been characterized using direct and indirect methods to assess microvascular blood flow. Hulten and colleagues used an intraoperative isotope washout technique, as well as in vivo abdominal angiography to characterize IBD-specific alterations of intestinal blood flow, demonstrating perfusion patterns associated with distinct phases of disease (5). Early fulminant IBD with severe inflammation was characterized by increased vascular perfusion, which is typical of an acute inflammatory response. In contrast, reduced regional blood flow was seen in chronically inflamed and remodeled tissues, particularly in areas of CD stricture. These observations have been confirmed in subsequent studies that demonstrate diminished vascular perfusion associated with fibrosis and longstanding inflammation. Using endoscopic laser Doppler flowmetry (29), decreased mucosal blood flow was seen in the neoterminal ileum after ileocolectomy and in rectal mucosal perfusion in patients with longstanding UC (30).

The poorly healing, refractory inflammatory ulcerations and damage in the IBD intestine suggest that these diseases are associated with microvascular dysfunction resulting in diminished vasodilatory capacity and tissue hypoperfusion. ET-1, a potent vasoconstrictor, is known to be elevated in CD and UC patients' tissue, again suggesting an ischemic vascular supply to the chronically inflamed bowel (31). The molecular physiology underlying microvascular dysfunction in IBD was assessed directly by Hatoum and colleagues, who characterized vasodilator responses in human intestinal microvessels by measuring in vitro vasodilatory capacity in response to acetylcholine (ACh) from perfused arterioles (50–150 micron diameter) isolated from intestinal resections (32). Normal intestinal microvessels vasodilate in response to ACh via NO- and cyclooxygenase (COX)-1 and -2–dependent mechanisms, whereas chronically inflamed IBD arterioles (in both CD and UC) demonstrate a significantly diminished vasodilatory response (32,33). This decreased vasodilatory capacity in the chronically inflamed IBD microvessels was directly related to a loss of NO-dependent function. These same vessels were found to be heavily dependent on COX to maintain their vascular tone. A significantly increased expression of COX-1 and -2 occurred in the IBD arterioles compared with control (32). Microvascular endothelial dysfunction in chronically inflamed IBD tissues was associated with excess levels of oxidative stress, as measured by intravital dyes and confocal fluorescence microscopy; this was not present in vessels isolated from normal intestine or uninvolved areas of IBD bowel (32). The microvascular dysfunction identified in arterioles from chronically inflamed IBD gut was not a generalized response to inflammation, because it was not demonstrated in vessels isolated from acute inflammation (i.e., diverticulitis) and uninvolved areas of IBD. This demonstrates that medications used to treat IBD patients at the time of surgery did not contribute directly to the microvascular dysfunction, and further substantiates that intrinsic, acquired alterations in the chronically inflamed and remodeled microcirculation underlie this pathophysiology.

THE MICROVASCULAR ENDOTHELIUM AS A THERAPEUTIC TARGET IN INFLAMMATORY BOWEL DISEASE

The process of microvascular endothelial activation and CAM expression has emerged as a potential therapeutic target in the treatment of inflammatory diseases (34). Targeting endothelial–leukocyte interactions for therapeutic benefit in patients with IBD has emerged as an early area for human trials and has received intense research interest. Over the past decade, three experimental agents (antisense ICAM-1, anti-$\alpha 4$ antibody, and anti-$\alpha 4\beta$ antibody) have been extensively evaluated in placebo-controlled trials for both CD and UC. These agents have demonstrated important results regarding efficacy and safety (35).

The initial trials of antiadhesion-molecule therapy in IBD targeted ICAM-1, using an antisense oligonucleotide strategy with the compound alicaforsen (ISIS 2302, ISIS

Pharmaceuticals, Carlsbad, CA) (36). ICAM-1, a member of the immunoglobulin superfamily, is an inducible transmembrane glycoprotein that is constitutively expressed at low levels on vascular ECs, including gut microvessels, and plays a central role in the recruitment and retention of leukocytes in inflammatory foci (37–40). Following experimental success in the use of humanized antibodies targeting ICAM-1 in refractory rheumatoid arthritis patients (41), trials were carried out targeting increased ICAM-1 expression in CD patients. Alicaforsen, a 20-base phosphorothioate antisense oligodeoxynucleotide molecule designed to specifically hybridize to a sequence in the 3′ untranslated region of human *ICAM-1* messenger RNA, forms an oligodeoxynucleotide-RNA heterodimer susceptible to degradation by the ubiquitous nuclease RNase-H, and ultimately reducing cell-specific message content and ICAM-1 protein expression (42–46). In the initial trial by Yacyshyn and colleagues, a steroid-sparing effect and clinical remission was demonstrated in 40% of treated CD patients who received intravenous infusions of the agent (36). However, subsequent clinical trials of alicaforsen in CD have failed to replicate this level of success. More recent trials have focused on UC patients using topical therapy with an enema preparation of alicaforsen; this approach has demonstrated prolonged responses in the treated subjects (47,48), and further studies of the compound in the UC patient population are ongoing.

Pioneering investigation by Koizumi and colleagues defined a pivotal role for integrin–integrin ligand interaction in the pathogenesis of a spontaneous animal model of colitis that occurs in the cotton-top tamarind, a New World primate from South America (49). Using this animal model, these authors demonstrated a central role for the leukocyte $\alpha 4$ integrin interacting with specific endothelial integrin-ligands (i.e., MAdCAM-1, VCAM-1) in the pathogenesis of chronic colitis. Animals treated with an anti-$\alpha 4$ antibody demonstrated significant clinical improvement (50), and the success of these early observations led to the development of novel therapeutic strategies targeting the interaction of $\alpha 4$ integrin-positive leukocytes with their cognate endothelial ligands, including MAdCAM-1. Subsequent studies using rodent models of IBD also have demonstrated critical roles for CAMs in the initiation and perpetuation of chronic gut inflammation (51,52).

The first successful anti-$\alpha 4$ integrin trials used the humanized antibody, natalizumab, in patients with relapsing and remitting multiple sclerosis (MS) and CD. Ghosh and colleagues demonstrated remission rates of 44% and a response rate of 71% in a multicenter European CD trial. In early 2005, natalizumab was given approval by the U.S. Food and Drug Administration for inducing and maintaining remission in MS. During this time, large multicenter CD and MS trials were ongoing throughout the world, and an estimated 4,000 patients had received this compound through protocols and open-label use, when evidence linking natalizumab with severe CNS viral infection led to immediate suspension of the drug. Three cases of progressive multifocal leukoencephalopathy (PML) were identified (two MS patients and one CD patient), with two patients dying as a result of the JC viral reactivation and resultant brain damage. Retrospective analysis of stored serum samples demonstrated reactivation of JC virus with the emergence of detectable viral load in a preclinical time period in a CD patient who ultimately died from PML (53). The unexpected emergence of viral reactivation with catastrophic CNS infection suggests that the future use of widespread anti-$\alpha 4$ integrin blockade (blocking both $\alpha 4\beta 7$- and $\alpha 4\beta 1$-expressing leukocytes) would require extensive safety monitoring for any realistic potential future use.

Despite the tragic complications encountered during the early use of natalizumab, trials of selective leukocyte inhibition in the treatment of chronic inflammation and IBD have continued. A recent successful trial targeting a subset of $\alpha 4$ integrin expressing leukocytes was carried out by Feagan and colleagues (54). Using a humanized antibody targeting $\alpha 4\beta 7$ integrin (MLN02, Millenium Pharmaceuticals, Cambridge, MA), these investigators demonstrated significant response and remission in patients with active UC (response 66%, remission 33% at 6 weeks). Among the 118 patients who received the active compound, no severe infections occurred, including viral reactivation. This positive trial suggests that efforts to more selectively target leukocyte–endothelial interaction may provide a strategy to achieve both efficacy and safety in therapeutic adhesion molecule blockade for IBD.

CONCLUSION

Data from multiple lines of investigation have demonstrated that the endothelium plays an important role in the pathogenesis and clinical course of IBD. In response to the stress of chronic inflammation, endothelium in the IBD intestine demonstrates an altered physiology, with impaired vascular perfusion. This microvascular dysfunction is the result of impaired endothelial generation of NO as well as other mediators of vasorelaxation. Likewise, local endothelial populations in chronically inflamed IBD intestine demonstrate altered patterns of leukocyte recruitment, with preferential adhesion of naïve leukocytes. Finally, novel strategies targeting endothelial–leukocyte interaction are emerging as potent biologic therapies for patients with refractory IBD.

KEY POINTS

- The endothelium lining the gut microvasculature plays a critical role in mucosal immune homeostasis through expression of CAMs and vascular addressins that govern leukocyte traffic into the bowel.
- The endothelial lining of the submucosal arterioles in the human intestine regulates vascular perfusion

through NO-, COX-, and endothelium-derived hyperpolarizing factor (EDHF)-dependent mechanisms.

- The human inflammatory bowel diseases (CD, and UC) are characterized by alterations in microvascular anatomy, as well as endothelial physiology. These include impaired endothelial-dependent dilation as well as altered patterns of endothelial–leukocyte interaction. Loss of NO generation has been identified in both of these manifestations of IBD-associated endothelial dysfunction.
- Therapeutic trials of anti-CAM therapy have been initiated in patients with IBD. Molecules targeting the $\alpha 4$ integrin as well as ICAM-1 are presently undergoing investigation for both CD and UC.

REFERENCES

1 Podolsky DK. Inflammatory bowel disease. *N Engl J Med*. 2002; 347(6):417–429.

2 Crohn BB, Ginsberg L, Oppenheimer GD. Regional ilietis: a clinical and pathological entity. *JAMA*. 1932;99:1323–1329.

3 Brooke BN. Granulomatous diseases of the intestine. *Lancet*. 1959;2:745–749.

4 Lockhart-Mummery HE, Morson BC. Crohn's disease (regional enteritis) of the large intestine and its distinction from ulcerative colitis. *Gut*. 1960;1:87–105.

5 Hulten L, Lindhagen J, Lundgren O, et al. Regional intestinal blood flow in ulcerative colitis and Crohn's disease. *Gastroenterology*. 1977;72(3):388–396.

6 Hulten L, Jodal M, Lindhagen J, Lundgren O. Blood flow in the small intestine of cat and man as analyzed by an inert gas washout technique. *Gastroenterology*. 1976;70(1):45–51.

7 Hulten L, Jodal M, Lindhagen J, Lundgren O. Colonic blood flow in cat and man as analyzed by an inert gas washout technique. *Gastroenterology*. 1976;70(1):36–44.

8 Granger DN, Richardson PD, Kvietys PR, Mortillaro NA. Intestinal blood flow. *Gastroenterology*. 1980;78(4):837–863.

9 Papadakis KA, Landers C, Prehn J, et al. CC chemokine receptor 9 expression defines a subset of peripheral blood lymphocytes with mucosal T cell phenotype and Th1 or T-regulatory 1 cytokine profile. *J Immunol*. 2003;171(1):159–165.

10 Rickert RR, Carter HW. The "early" ulcerative lesion of Crohn's disease: correlative light- and scanning electron-microscopic studies. *J Clin Gastroenterol*. 1980;2(1):11–19.

11 Chambers TJ, Morson BC. The granuloma in Crohn's disease. *Gut*. 1979;20(4):269–274.

12 Dvorak AM, Monahan RA, Osage JE, Dickersin GR. Crohn's disease: transmission electron microscopic studies. II. Immunologic inflammatory response. Alterations of mast cells, basophils, eosinophils, and the microvasculature. *Hum Pathol*. 1980;11(6): 606–619.

13 Podolsky DK, Lobb R, King N, et al. Attenuation of colitis in the cotton-top tamarin by anti-alpha 4 integrin monoclonal antibody. *J Clin Invest*. 1993;92(1):372–380.

14 Jones SC, Banks RE, Haidar A, et al. Adhesion molecules in inflammatory bowel disease. *Gut*. 1995;36(5):724–730.

15 Briskin M, Winsor-Hines D, Shyjan A, et al. Human mucosal addressin cell adhesion molecule-1 is preferentially expressed in intestinal tract and associated lymphoid tissue. *Am J Pathol*. 1997; 151(1):97–110.

16 Salmi M, Jalkanen S. Human leukocyte subpopulations from inflamed gut bind to joint vasculature using distinct sets of adhesion molecules. *J Immunol*. 2001;166(7):4650–4657.

17 Burgio VL, Fais S, Boirivant M, et al. Peripheral monocyte and naive T-cell recruitment and activation in Crohn's disease. *Gastroenterology*. 1995;109(4):1029–1038.

18 Ley K, Kansas GS. Selectins in T-cell recruitment to non-lymphoid tissues and sites of inflammation. *Nat Rev Immunol*. 2004; 4(5):325–335.

19 Binion DG, West GA, Ina K, et al. Enhanced leukocyte binding by intestinal microvascular ECs in inflammatory bowel disease. *Gastroenterology*. 1997;112(6):1895–1907.

20 Haraldsen G, Rugtveit J, Kvale D, et al. Isolation and longterm culture of human intestinal microvascular ECs. *Gut*. 1995;37(2): 225–234.

21 Binion DG, West GA, Volk EE, et al. Acquired increase in leucocyte binding by intestinal microvascular endothelium in inflammatory bowel disease. *Lancet*. 1998;352(9142):1742–1746.

22 Binion DG, Fu S, Ramanujam KS, et al. iNOS expression in human intestinal microvascular ECs inhibits leukocyte adhesion. *Am J Physiol*. 1998;275(3 Pt 1):G592–G603.

23 Binion DG, Rafiee P, Ramanujam KS, et al. Deficient iNOS in inflammatory bowel disease intestinal microvascular ECs results in increased leukocyte adhesion. *Free Radic Biol Med*. 2000;29(9): 881–888.

24 Pritchard KA, Ackerman AW, Ou J, et al. Native low-density lipoprotein induces endothelial nitric oxide synthase dysfunction: role of heat shock protein 90 and caveolin-1. *Free Radic Biol Med*. 2002;33(1):52–62.

25 Mellor JA, Chandler GN, Chapman AH, Irving HC. Massive gastrointestinal bleeding in Crohn's disease: successful control by intra-arterial vasopressin infusion. *Gut*. 1982;23(10):872–874.

26 McGarrity TJ, Manasse JS, Koch KL, Weidner WA. Crohn's disease and massive lower gastrointestinal bleeding: angiographic appearance and two case reports. *Am J Gastroenterol*. 1987;82 (10):1096–1099.

27 Wakefield AJ, Sawyerr AM, Dhillon AP, et al. Pathogenesis of Crohn's disease: multifocal gastrointestinal infarction. *Lancet*. 1989;2(8671):1057–1062.

28 Funayama Y, Sasaki I, Naito H, et al. Remodeling of vascular wall in Crohn's disease. *Dig Dis Sci*. 1999;44(11):2319–2323.

29 Angerson WJ, Allison MC, Baxter JN, Russell RI. Neoterminal ileal blood flow after ileocolonic resection for Crohn's disease. *Gut*. 1993;34(11):1531–1534.

30 Tateishi S, Arima S, Futami K. Assessment of blood flow in the small intestine by laser Doppler flowmetry: comparison of healthy small intestine and small intestine in Crohn's disease. *J Gastroenterol*. 1997;32(4):457–463.

31 Murch SH, Braegger CP, Sessa WC, MacDonald TT. High endothelin-1 immunoreactivity in Crohn's disease and ulcerative colitis. *Lancet*. 1992;339(8790):381–385.

32 Hatoum OA, Binion DG, Otterson MF, Gutterman DD. Acquired microvascular dysfunction in inflammatory bowel disease: loss of nitric oxide-mediated vasodilation. *Gastroenterology*. 2003; 125(1):58–69.

33 Hatoum OA, Gauthier KM, Binion DG, et al. Novel mechanism of vasodilation in inflammatory bowel disease. *Arterioscler Thromb Vasc Biol.* 2005;25(11):2355–2361.
34 Simmons DL. Anti-adhesion therapies. *Curr Opin Pharmacol.* 2005;5(4):398–404.
35 van Assche G, Rutgeerts P. Antiadhesion molecule therapy in inflammatory bowel disease. *Inflamm Bowel Dis.* 2002;8(4):291–300.
36 Yacyshyn BR, Bowen-Yacyshyn MB, Jewell L, et al. A placebo-controlled trial of ICAM-1 antisense oligonucleotide in the treatment of Crohn's disease. *Gastroenterology.* 1998;114(6):1133–1142.
37 Dustin ML, Rothlein R, Bhan AK, et al. Induction by IL 1 and interferon-gamma: tissue distribution, biochemistry, and function of a natural adherence molecule (ICAM-1). *J Immunol.* 1986;137(1):245–254.
38 Rothlein R, Dustin ML, Marlin SD, Springer TA. A human intercellular adhesion molecule (ICAM-1) distinct from LFA-1. *J Immunol.* 1986;137(4):1270–1274.
39 Marlin SD, Springer TA. Purified intercellular adhesion molecule-1 (ICAM-1) is a ligand for lymphocyte function-associated antigen 1 (LFA-1). *Cell.* 1987;51(5):813–819.
40 Diamond MS, Staunton DE, de Fougerolles AR, et al. ICAM-1 (CD54): a counter-receptor for Mac-1 (CD11b/CD18). *J Cell Biol.* 1990;111(6 Pt 2):3129–3139.
41 Kavanaugh AF, Davis LS, Nichols LA, et al. Treatment of refractory rheumatoid arthritis with a monoclonal antibody to intercellular adhesion molecule 1. *Arthritis Rheum.* 1994;37(7):992–999.
42 Nestle FO, Mitra RS, Bennett CF, et al. Cationic lipid is not required for uptake and selective inhibitory activity of ICAM-1 phosphorothioate antisense oligonucleotides in keratinocytes. *J Invest Dermatol.* 1994;103(4):569–575.
43 Chiang MY, Chan H, Zounes MA, et al. Antisense oligonucleotides inhibit intercellular adhesion molecule 1 expression by two distinct mechanisms. *J Biol Chem.* 1991;266(27):18162–18171.
44 Bennett CF, Condon TP, Grimm S, et al. Inhibition of EC adhesion molecule expression with antisense oligonucleotides. *J Immunol.* 1994;152(7):3530–3540.
45 Hoke GD, Draper K, Freier SM, et al. Effects of phosphorothioate capping on antisense oligonucleotide stability, hybridization and antiviral efficacy versus herpes simplex virus infection. *Nucleic Acids Res.* 1991;19(20):5743–5748.
46 Wickstrom E. Oligodeoxynucleotide stability in subcellular extracts and culture media. *J Biochem Biophys Methods.* 1986;13 (2):97–102.
47 Van Deventer SJ, et al. A phase 2 dose ranging, double-blind, placebo-controlled study of allcaforsen enama in subjects with acute exacerabation of moderate left sided ulcerative colitis. *Gastroenterology.* 2005;128(4):A74.
48 Philip Bminer, et al. A phase 2 trial to assess the safety and efficacy of two dose formulation of allcaforsen enama compared with 4g enema for acute ulcerative colitis. *Gastroenterology.* 2005; 128(4):A74.
49 Koizumi M, King N, Lobb R, et al. Expression of vascular adhesion molecules in inflammatory bowel disease. *Gastroenterology.* 1992;103(3):840–847.
50 Dignass A, Lynch-Devaney K, Kindon H, et al. Trefoil peptides promote epithelial migration through a transforming growth factor beta-independent pathway. *J Clin Invest.* 1994;94(1):376–383.
51 Panes J, Granger DN. Leukocyte-endothelial cell interactions: molecular mechanisms and implications in gastrointestinal disease. *Gastroenterology.* 1998;114(5):1066–1090.
52 Connor EM, Eppihimer MJ, Morise Z, et al. Expression of mucosal addressin cell adhesion molecule-1 (MAdCAM-1) in acute and chronic inflammation. *J Leukoc Biol.* 1999;65(3):349–355.
53 Van Assche G, Van Ranst M, Sciot R, et al. Progressive multifocal leukoencephalopathy after natalizumab therapy for Crohn's disease. *N Engl J Med.* 2005;28;353(4):362–368.
54 Feagan BG, Greenberg GR, Wild G, et al. Treatment of ulcerative colitis with a humanized antibody to the alpha4beta7 integrin. *N Engl J Med.* 2005;352(24):2499–2507.

136

The Vascular Bed of Spleen in Health and Disease

Péter Balogh

Faculty of Medicine, University of Pécs, Hungary

The spleen is the largest filter of blood in the body. The major functions of the spleen are to remove aging red blood cells, to participate in innate immunity, and to promote adaptive immune response. The branching arterial system in the spleen is sheathed by lymphoid tissue (white pulp) and ends in a venous sinus system (red pulp). The red pulp, which serves to filter blood and remove senescent erythrocytes, consists of afferent arterial cords (an open blood system lacking an endothelial lining) leading to venous sinuses that are lined by discontinuous endothelium. The white pulp represents the lymphoid region of the spleen and consists of B and T lymphocyte–rich lymphoid sheaths surrounding the arterial system. Despite these common properties, important anatomical differences exist between the spleen of humans and rodents. Thus, caution is required when extrapolating results from one species to another. The goal of this chapter is to discuss the properties of endothelial cells (ECs) that line the various vascular beds of the spleen, and where necessary to point out the differences between the human and rodent architecture.

HISTORICAL BACKGROUND, OR, THE *MYSTERII PLENUM ORGANON*

For the better part of history, the spleen was attributed with the function of removing agents that would prove harmful to the individual, thus reducing contamination of the healthy blood. In contradistinction to Aristotle, Galen argued that the liver, and not the heart, was the principal organ providing the body with strength and spirit. Galen held that the spleen contained black bile and was responsible for controlling emotions by eliminating sadness and melancholy. Remarkably, this interpretation of splenic function would prevail until the mid-18th century. Even William Harvey, a physician renowned for his discovery of the circulation, believed that the sign of a properly functioning spleen was the person's ability to laugh and that its impairment would manifest in the patient's loss of cheerful spirit, replaced by sullenness (1,2).

The notion that the spleen carries out functions beyond filtering of black bile or spiritual cleansing can be traced back to the Renaissance period in Bologna and elsewhere in Italy, when it was suggested that the spleen removes harmful substances from the feces via its vascular supply. Soon thereafter, Marcello Malpighi using light microscopy was able to describe the splenic corpuscles, which were named after him and are now referred to collectively as the splenic white pulp, followed about two centuries later by Theodor Billroth, who first adequately described the microscopic structure of the red pulp (hence the term *cords of Billroth*). Based on subsequent microscopic observations, the spleen was recognized to possess three distinct regions – namely, white pulp and red pulp, separated by a marginal zone. The white pulp is further divided into follicles and periarteriolar lymphoid sheaths. Each of these regions is associated with a unique vascular bed, lined by phenotypically distinct subsets of ECs (Figure 136.1; for color reproduction, see Color Plate 136.1).

The spleen's connection with the lymphohemopoietic system was first established in the early 1770s by William Hewson, who identified lymphocytes in the spleen and thymus. Subsequent work by Rudolf Virchow revealed the similar morphology of white blood cells and splenic white pulp lymphocytes. The spleen's ability to remove blood-borne particles (including fragmented erythrocytes) was later recognized by Emil Ponfick, one of Virchow's numerous apprentices of pathology fame, and during the early decades of the 20th century the spleen's role in immunologic defense gradually came to light (2).

Morphological studies performed during the 1920s revealed considerable erythropoietic activity in the mammalian spleen during embryogenesis, a process that was subsequently localized to areas outside the sinuses. This arrangement is in contrast to what takes place in the yolk sac, where vasculogenesis and intravascular red blood cell formation occur simultaneously as a result of the parallel differentiation of hemangioblasts (3,4).

Subsequent studies during the late 1960s and early 1970s demonstrated that both the embryonic and adult spleen contain cells capable of reconstituting multilineage (lymphoid

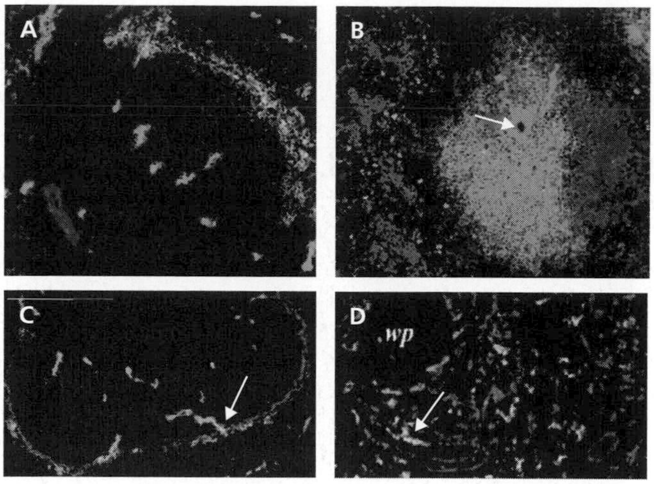

Figure 136.1. Endothelial labyrinth around the lymphohemopoietic compartments of the spleen. Various segments of the splenic vasculature can be identified by their different endothelial surface phenotype in separate regions of the spleen. (**A**) The main vessel types are identified as CD31-positive central artery, CD31/IBL-7/1 double-positive arterioles in the white pulp; MAdCAM-1–positive marginal sinus coexpressing IBL-7/1 marker at some locations in the marginal zone; and a few scattered IBL-7/1 positive sinuses in the red pulp. (**B**) The T cells (Thy-1) are arranged around the central arteriole (*arrow*), to which follicles containing B cells (B220) are attached. The red pulp is outlined as the major macrophage-containing domain (F4/80). (**C**) The arrow points at a transitional area between the IBL-7/1–positive white pulp arteriole and the MAdDCAM-1–positive segment of the marginal sinus is shown, coexpressing both markers. (**D**) Sinusal heterogeneity of the red pulp area. A fraction of sinuses label only with mAb IBL-7/1 or express IBL-7/1 and IBL-9/2 markers at approximately equal intensity (*arrow*), whereas the majority of red pulp sinuses and possibly venules dominantly display the IBL-9/2 antigen (62). For color reproduction, see Color Plate 136.1.

and myeloid) long-term hematopoiesis in irradiated animals, as well as committed precursor cells (particularly in the erythromyeloid line)(5,6).

STRUCTURE AND FUNCTION OF SPLENIC VASCULATURE

Overview of the Splenic Circulation: Structural Differences Between Human and Rodents

For its size, the spleen in both human and rodents receives a disproportionately large amount of blood flow, corresponding to approximately 5% to 10% of total cardiac output (7). This volume is pumped into the spleen through the splenic artery (a. lienalis) whose subsequent branching forms the basic scaffolding elements of the tissue. The number of arterial branches in humans is considerably higher than in rodents. In either case, small-sized arterioles (termed *central arterioles*) are surrounded by lymphatic tissue (white pulp), including a T cell–rich periarteriolar lymphoid sheath and B cell–rich follicles. These arterioles in humans are often arranged in bundles, where arterioles and two to three parallel capillaries are held together in a reticular sheath (8,9). In some individuals several vessels may originate from one larger arteriole, thus forming the penicillar arterioles, primarily in the red pulp.

The smallest branches of the central arterioles (termed *terminal arterioles*) traverse the white pulp and, depending on the species, may connect either to the marginal sinus surrounding the white pulp (in rodents) or to the red pulp sinuses in the perifollicular zone outside the marginal zone (in humans). In rats and mice, a few larger white pulp arterioles may reach the red pulp by traversing the marginal zone. Using intravital microscopy, a distinct set of white pulp terminal arterioles was described recently in mice, which continue into smaller capillaries whose bud-like terminal segment within the white pulp is arborized into loosely organized honeycomb-like structures. At these locations the blood may seep into the surrounding tissue. The spatial relationship of these complexes to the marginal sinus is currently unclear; moreover, it is also probable that the cells lining these vascular regions are of nonendothelial origin, as indicated by the lack of expression of a panendothelial *Tie2* promoter-driven reporter gene in this formation. Kinetic analysis of labeled cells has raised the possibility that these honeycomb-like regions may act as a splenic entry port for recirculating T cells into the spleen by allowing their firm adhesion to the vessel wall. Using large-molecular-weight soluble fluorophores, it appears that most of the blood flow is directed toward the veins at a relatively high velocity, with a smaller fraction of the blood extravasating into the red pulp at a slower speed. Thus, the honeycomb-like structures in the white pulp (or at the white pulp–red pulp juncture) may represent a novel "open" segment of the splenic vasculature, in addition to the red pulp sinuses (10,11).

The reticular cells and fibers constituting the circumferential reticulum enclosing the white pulp also serve as a stromal support for the marginal sinus or perifollicular space. The ambiguity as to which type of cell – endothelial or reticular/fibroblastic – forms the inner lining of lumens within this space explains the judicious use of the term *sinus-lining cells.* As an example of the above uncertainty, mucosal addressin cell adhesion molecule (MAdCAM)-1 is expressed in this segment of the vasculature in mice, whereas expression of MAdCAM-1 in the marginal zone of human spleen is limited to fibroblasts (12,13).

In contrast to that of rodents, the human spleen does not contain a marginal sinus. Instead, some arterioles terminate in a perifollicular zone (which is absent in rodents), and others reach the red pulp, in a fashion similar to that in mice and rats (12). Prior to their termination within the red pulp, these arteriolar capillaries are surrounded by a complex multilayer structure that forms the basic scaffolding of splenic cords (of Billroth), primarily composed of macrophages, dendritic cells, and a fibroblastic reticular meshwork. It is also commonly referred to as the *periarteriolar macrophage sheath* (14). In humans the dominant cell type is the sialoadhesin-positive

Table 136-1: Vascular Features of Splenic versus Lymph Node Homing of Lymphocytes

Feature	*Spleen*	*Lymph Node*[a]
Type of vascular bed functioning as exit port	Terminal arteriole/marginal sinus	Postcapillary venule
Location of the exit port	B-cell area (marginal zone or perifollicular zone)	T-cell area (paracortex)
Type of the endothelium lining the vessel	Flat (sinus endothelium)[b]	High endothelium (cuboidal dimensions)
Molecular constraints	Independent from L-selectin and its various ligands[c]	L-selectin dependent

[a] General features are listed, without detailing the differences between peripheral and MALT-associated lymph nodes.
[b] The auxiliary role of other sinus-lining cells (marginal zone macrophages) cannot be ruled out.
[c] The adhesion and migration of hemopoietic progenitors by VCAM-1/VLA-4 and of B cells by CD22 also appear to be regulated differently between the spleen and bone marrow, where the engagement of these surface molecules are involved in the BM (but not the splenic) homing (19,20).

macrophage (15). This "open" segment of the red pulp vasculature inserted between the terminal arterioles and the small venules is unique to the splenic circulation, and although it has received considerable attention in the field of splenic circulation, its mechanisms of action remain unclear.

Although the spleen is considered a peripheral lymphoid organ, it differs in fundamental ways from peripheral lymph nodes. A unique feature of the spleen is the presence of the marginal zone, a compartment composed of specialized macrophage subsets and a distinct set of B cells – the marginal zone B cells. These lymphocytes differ from their follicular counterparts in many aspects, including their phenotype, functional properties, and developmental requirements. As a result of their dominance in the marginal zone, the extravasation of circulating lymphocytes in the human and rodent spleen takes place in this B-cell rich (non-follicular) compartment, in contrast to lymph nodes and tonsils, where lymphocytes home via high endothelial venules to T-cell rich domains. The principal differences between the vascular aspects of splenic versus lymph node homing processes are summarized in Table 136-1.

In the human (sinusal) spleen the venous sinuses are lined with tightly fitting ECs, which may be pushed apart to create interendothelial slits (IES), indicating the presence of a process subject to regulation. In rodents (traditionally considered to be nonsinusal – i.e., having an absence of sinuses between arterioles and venous capillaries in their spleens), the large gaps between ECs exert a considerably lesser obstacle to blood cell passage. As a result of the open circulation, a substantial amount of blood is filtered through red pulp reticular cells and fibers before reentering the red pulp sinuses through their endothelial pores and slits. These reticular (or barrier) cells surround the arterial endings in usual pericyte-like fashion, whereas their extensions are connected to each other, thus creating an extravascular meshwork for the corpuscular filtering of extravascular blood before it reaches the venous phase of circulation (14). It is this reentry phase that serves as a checkpoint for removing those red blood cells whose physical deformability is impaired due to internal abnormalities (e.g., plasmodia in malaria infection) or are fragile (e.g., complement-damaged erythrocytes in antibody-induced anemia). The connection of blood flow between the arterial ending and the initiation of venous sinuses may be facilitated by the active movement/repositioning of reticular fibroblasts, thus bringing together two closely situated vascular segments. Moreover, these cells may even form lumens, thus complementing the endothelial gap between two vascular segments (9). The ordered arrangement of these or related cells in the splenic white pulp in mouse has been demonstrated to be involved in the transport of soluble macromolecules, acting as a molecular sieve (16).

The "open circulation" described earlier appears to be the typical form of vasculature in both human and rodents. Due to the relatively large amount of blood flow for its size, it is probable that the majority of blood volume within the spleen is likely to drain into venous sinuses without percolating through extravascular spaces for any length of time. Thus the spleen simultaneously possesses both open and closed circulatory pathways.

After their origin in the perifollicular zone and red pulp (or the equivalent marginal zone/sinus in rodents), the sinusoids converge into larger venules and finally merge into trabecular veins. The effluent blood (that has been filtered through the meshwork of red pulp chords) leaves the spleen through the v. lienalis, which is joined by the vena mesenterica superior to form the portal vein. Because the cellular content of a hematopoietic tissue is maintained by controlling simultaneously the access and egress of leukocytes, it is surprising that, in addition to the uncertainties surrounding the precise mechanism underlying the access of leukocytes and their precursors to the spleen, the control of their reverse movement from the spleen into the circulation has almost entirely been ignored.

It is generally accepted that the spleen contains lymphatic vessels, as evidenced by the presence of thin-walled vessels

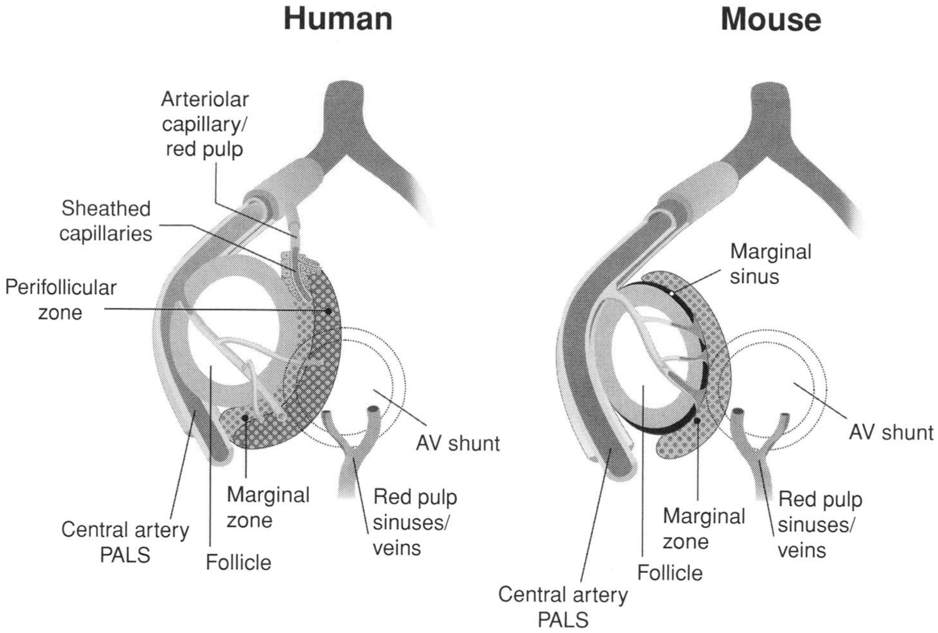

Figure 136.2. A hypothetical scheme of splenic circulation in man compared to rodents ("HuMouse"). After the branching of main artery, smaller arteriolar segments penetrate the white pulp. On the left side, the human structure is shown, with features characteristic for the human conditions on the left side of the drawing (SC: sheathed capillaries, PFZ: perifollicular zone). On the right side, the mouse arrangement is shown. The parts in the middle of the graph indicate the common elements in both species (MZ, marginal zone; PALS, periarteriolar lymphoid sheath). For color reproduction, see Color Plate 136.2.

lacking basement membrane deeply embedded in the white pulp. However, little is known about the physiological importance of these vessels (17,18).

A "minimalist" scheme of vasculature in human versus rodent spleen is shown in Figure 136.2 (for color reproduction, see Color Plate 136.2). Table 136-2 summarizes the essential differences between various splenic vascular segments in man and rodents.

Ultrastructural Characteristics of Splenic Endothelium

Although most ultrastructural studies of the splenic vasculature have focused on the unique marginal sinus (or perifollicular space) and red pulp, a small number of transmission electron microscopic studies of the white pulp have demonstrated a continuous endothelium lining the central arteriole and downstream arteriolar capillaries. The endothelium rests on a basal lamina that, like other vessels in the spleen, contains heparan sulfate, chondroitin sulfate A and/or C, chondroitin sulfate B, and sialic acid residues, substances that are also present on the luminal aspect of splenic endothelia (with the exception of red pulp sinuses) (21). The endothelial lining of the central artery is surrounded proximally by two or three layers of smooth muscle cells and distally (because the vessel terminates in the marginal sinus or red pulp sinus) by cytoplasmic processes of reticular cells (9). The transition between smooth muscle cell and reticular cell covering is marked by a progressive reduction in endothelial continuity and the appearance of fenestrated basement membrane. This structure is created by the ring fibers (or annular fibers) on the abluminal side of the endothelium, intertwined by longitudinal fibers.

The terminal segment(s) of arteriolar capillaries may form (a) simple funnel-like structures with splits and pores or cut-across, (b) small ampulla with pores, or (c) one larger ampulla covering multiple arterioles (9).

The red pulp sinuses are lined with ECs (also named *rod cells*) that run parallel to each other and to the course of blood flow. In contrast to the continuous-type vessels (central arteriole, terminal arterioles, and veins), the luminal surface of rat sinuses lacks heparan sulfate, chondroitin sulfate A and/or C, chondroitin sulfate B, and sialic acid residues, which are restricted to the fenestrated basal lamina within this compartment (21). The nuclei of sinus ECs characteristically protrude into the vessel lumen. In addition to these ovoid nuclear protrusions, the luminal aspect of the endothelium may also contain short drop-like or filiform micro-extensions. In various diseases (such as hepatic cirrhosis with portal hypertension, hereditary spherocytosis, and idiopathic thrombocytopenic purpura) the number of these extensions may increase (9). The lateral aspects of ECs are connected by two main types of intercellular connections: adherens junctions and, to a lesser extent, tight junctions. As with other vascular beds, desmosomes are absent in splenic endothelium. The actin filaments

Table 136-2: Comparison of Main Features of Some Splenic Vascular Bed Homologues in Human and Rodents

Vascular Bed Segment	*Humans*	*Rodents*
Central arteriole	Often present in non-T-area; smaller diameter indicating more numerous branching (8).	Usually runs in the central part of T-cell zone; thinner muscular layer.
Terminal arterioles	Surrounded by sialoadhesin-positive macrophages (primarily in the perifollicular zone, outside the MZ).	Sialoadhesin-positive macrophages are restricted to the marginal zone, the arterioles end in marginal sinus.
Marginal sinus	Absent	Present

(running parallel to the length of cell) are typically associated with the junctional membranes and are continuous with the actin filaments in contractible stress fibers at both forms of contact, presumably rendering the sinus wall able to adjust to volume/pressure alterations of the splenic circulation (22–24).

In addition to the outward protrusion of nuclei and microextensions, sinus endothelium in the rat spleen also displays an extensive canalicular membrane system composed of inwardly projecting channels of approximately 20-nm width. These tubular structures penetrate the entire cell, and their continuity with the plasma membrane on both sides of the EC indicates that they may be open on both ends. In addition to these tubular formations, spherical forms of membrane invaginations restricted to the membrane-proximal regions of cytoplasm (i.e., not traversing the entire cell), such as caveolae and coated pits, have also been described. The function of these compartments is currently unknown; they have been speculated to be involved in the release and storage of various substances, including calcium or endothelin, thrombomodulin or fibroblast growth factor (FGF) (25–27).

On the abluminal aspect of the sinus endothelium, the cells are connected to each other by ring fibers that overlap the above interendothelial bridges. These bands are formed by the extensions of reticular cells located nearby and are inserted into small grooves on the abluminal surface of rod cells, without causing any protrusion on their luminal aspect (9).

Developmental and Functional Characteristics of Splenic Endothelium

The heterogeneous nature of the splenic vasculature combined with its diverse functions during ontogeny has considerably hampered our understanding of EC function in this organ. In general, site-specific EC phenotypes are epigenetically predetermined and/or mediated by the local tissue microenvironment (28). In rodents, immediately after birth the tissue architecture of the spleen lacks any discernible compartmentalization, including a lack of follicle formation, separation of T-cell and B-cell zones, and the absence of marginal zone. Figure 136.1 illustrates the lymphoid/myeloid "parenchymal" and "endothelial" compartmentalization of the adult spleen. Although a substantial amount of evidence in humans indicates the heterogeneous composition of endothelium in the spleen, most of the developmental data have been compiled from animal studies. Therefore, the results obtained must be interpreted with caution.

The developing spleen represents a complex organization of hematopoietic tissue–stem cell niches, which gradually shifts in function from an overwhelmingly myelopoietic tissue (forming mainly erythroid cells and, to a lesser degree, granulocytes and megakaryocytes) to a peripheral lymphoid organ that retains its potential to support hematopoiesis. Splenic organogenesis is under the combined control of several regulators, including homeodomain transcription factors. The available evidence indicates that the spleen (already present in sharks, the most ancient group of vertebrates to have adaptive immunity) in mammals develops from the dorsal pancreatic mesenchyme, independently of pancreatic epithelium. Transgenic mice overexpressing Sonic Hedgehog factor (Shh) driven by the *Pdx-1* promoter (an essential regulator of early pancreatic development) lack pancreatic mesoderm, due to its conversion into intestinal mesenchyma, and also fail to develop a spleen (29). Increased hedgehog (Hh) signaling induced by the absence of Hh-inhibitor Hhip also blocks the formation of the spleen, for which the deficiency of FGF10 is believed to be responsible (30). The spontaneous mutation observed in dominant hemimelia (Dh) is associated with abnormal limb and visceral development affecting the splenic mesodermal plate (SMP), and it also causes asplenia (31).

Subsequent to its symmetric appearance, SMP is induced to undergo a leftward growth, a process that involves FGF9 and FGF10 under the control of Bapkx1/Nkx3.2. Asplenia is also observed in mice that are null for either Bapx1/Nkx3.2 or Wt1 or POD-1/capsulin (32). Attempts to determine genetic interaction of Hox11 with Wt1, Bapx1/Nkx3.2, and capsulin led to the conclusion that Wt1 functions downstream of Hox11 (33–40). More recently another form of abortive splenic development has been described as a result of aberrant iron metabolism and toxicity mediated by iron overload in polycythemia (*Pcm*) mutation. This condition probably affects mesenchymal stromal precursors in the red pulp area (41). Although present, the spleen in these animals is aberrantly small, its red pulp stromal architecture is atrophic, and the various vascular elements (central artery and distinct red pulp sinuses, respectively) are abnormally arranged. In mice deficient in the Nkx2.3 homeodomain transcription factor, atrophic splenic differentiation with variable severity was observed. In these animals the marginal sinus lining cells in the rudimentary spleen lacked the MAdCAM-1 glycoprotein (42,43).

The differentiation of endothelial precursors at splenic locations is paired with the induction of various transcription factors. Of these, the SCL/TAL1 encoding a basic helix-loop-helix–type transcription factor appears to be upregulated early during the hematopoietic differentiation simultaneously in both the hematopoietic cells and endothelium, including the developing spleen (44). Both in human and adult mice the expression of SCL/TAL1 protein is detected in the red pulp endothelium as well as macrophages, whereas in humans its mRNA is also expressed in the white pulp vessels, primarily at the peripheral regions of follicles (45). The deletion of the *Scl/Tal1* gene resulted in embryonic lethality preceding the formation of spleen, indicating that its function is not restricted to splenic organogenesis, but is more fundamental in vasculogenesis and endothelial differentiation (46). In addition, chromosomal rearrangement of the *SCL/TAL1* gene is frequently observed in acute T-cell leukemia. During sequential in vitro endothelial and/or hemopoietic differentiation from embryonic stem cells, BMP4 induces the differentiation of FLK-1-positive cells into FLK-1/SCL double-positive intermediates. Vascular endothelial growth factor (VEGF) interacts with FLK-1 to promote endothelial (and also hematopoietic) differentiation and expansion of these FLK-1/SCL double-positive cells (47). It is not yet known whether similar events take place in the emerging spleen. Because the VEGF receptor 3 (VEGFR3) (FLT-4) is mainly expressed on lymphatic endothelium precursors, its role in splenic vasculogenesis is likely to be only secondary, if any (48).

In addition to the VEGF family members, another important group of factors involved in the vascular modeling and angiogenesis is the angiopoietins (Ang1, Ang2, and Ang3), recognized by the receptor tyrosine kinase Tie2 (49). Ang1 elicits endothelial survival and stabilization, whereas Ang2 induces vascular destabilization and subsequent neovascularization or neoangiogenesis. Tie1 and Tie2 are also expressed by hematopoietic stem cells and are required for hematopoiesis in the bone marrow (50). The *Tie2* promoter has been used to drive EC-specific expression of several genes in vivo. In *Tie2–GFP* transgenic mice, reporter gene expression was detected in most vascular beds but was notably absent in the honeycomb-like structures at the termination of arterial capillaries in mouse (11). Similar lack of expression at the marginal sinus was observed in mice carrying other EC-specific promoters, including receptor protein tyrosine phosphatase μ, (RPTPμ), endothelial nitric oxide synthase (NOS), and von Willebrand factor (vWF) (51–53). Together, these data point to a complex ontogeny of ECs lining of splenic vessels.

In the postnatal period, the tumor necrosis factor/lymphotoxin (TNF/LT) family members are involved in directing the proper segregation of lymphoid and stromal compartments of the white pulp and marginal zone, thus rendering the spleen capable of mounting effective immune responses. These do not appear to influence the prenatal hematopoietic function (54). A striking consequence of lack of lymphotoxin-beta receptor (LTβR) signaling is the absence of MAdCAM-1 antigen from the sinus-lining cells of the marginal sinus (55). It appears to be a temporally fixed trait, as the availability of LIGHT (alternative soluble ligand for LTβR) cannot reestablish the display of MAdCAM-1 by these cells in LT-deficient animals, in contrast to the induction of follicular dendritic cell differentiation and T/B segregation (56). Interestingly, the perturbation of TNF/LT receptor signaling does not seem to affect the architecture of splenic red pulp, including its vascular elements (57).

In contrast to these conditions in which the expression of MAdCAM-1 antigen on marginal sinus-lining cells is reduced or missing, sphingosine 1-phosphate receptor $(S1P)_3$-null mice demonstrate an abnormally expanded population of MAdCAM-1–positive cells in a disorganized marginal zone region. This effect was apparently due to the nonlymphoid cells within the spleen, as established by reciprocal bone marrow transplantation and hemopoietic reconstitution (58).

Immunohistochemical studies have revealed intriguing spatial and temporal differences in EC phenotype within the spleen of humans and rodents. For example, several EC-specific cell membrane or intracellular markers are restricted to one splenic compartment, such as white pulp, marginal zone, or red pulp in both human and mice (Table 136-3). Although largely descriptive, these findings provide compelling evidence for the existence of multiple EC phenotypes within the spleen, presumably reflecting marked functional heterogeneity (59–67).

SPLENIC ENDOTHELIUM IN DISEASE

Vascular pathology of the spleen can be divided into vessel-derived and other diseases, originating from nonvascular tissue components. Although primary vascular splenic diseases are rare (and often asymptomatic), the spleen is a frequent site of secondary manifestations in a wide range of hematological, immunologic, oncologic, infectious, vascular, and systemic disorders. This last section will focus on vascular pathology due to aberrant development of vessels.

The most frequent of these disorders are splenic vascular tumors. Depending on their growth characteristics, these tumors can be divided into (a) benign (including hemangioma, hamartoma, and lymphangioma) (b) variable (littoral cell angioma, hemangioendothelioma, and hemangiopericytoma), and (c) malignant (angiosarcoma) types. On ultrasound, the primary tumors often manifest as solitary or multiplex echogenic nodules, intramural dilatations, or cysts. They

Table 136-3: Phenotypic Features of Distinct Vascular Segments of the Spleen

Antigen/Marker (Refs.)	*Region*			*Function*
	Central Arteriole and White Pulp	*Marginal Sinus*[a]	*Red Pulp*	
CD31 (8)	++	+	+/–	Modulates cell activation responses, endothelium apoptosis and cell adhesion
CD34	++	+/–	+/–	L-selectin ligand in a glycosylation-dependent manner
vWF	+	–	–	Platelet adhesion to damaged endothelium
Ulex europaeus agglutinin (UEA-I)	++	–	+	Specific for Fucα1–2gal sugar moieties
CD8 (59)	–	–	+	Coreceptor for major histocompatibility class I-associated antigen recognition on T cells
CD36 (60)	–	–	+	Class B scavenger receptor expressed by macrophages
Bandeiraea simplicifolia lectin (BS-1)	++	+	+	Specific for α-gal and α-galNAc sugar moieties
MAdCAM-1 (13)	–	++	-	Receptor for $\alpha 4\beta 7$ integrin
IBL–7/1 (61,62)	+/–[b]	++	+[c]	Unknown
IBL-9/2 (62)	–	–	++	Unknown
Stabilin-2 (63) (man and mouse)	–	–	++	Hyaluronate binding.
Bonzo/CXCR6 (64)	–	–	+	Receptor for CXCL16 involved in CD8 T cell movements in the splenic red pulp
CD104 (65) (integrin β4)	+	–	–	Noncovalently associates with integrin α6/CD49f
OX2/CD200 (66) (rat and mouse)	+	+	+	Inhibitory receptor on myeloid cells (macrophages and granulocytes)
HIS 52/RECA-1 (rat) (67)	+	+	+	Unknown

[a]Features characteristic for mice and/or rats are indicated *in italic.*
[b]The central arteriole does not express the IBL-7/1 epitope.
[c]IBL-7/1 reacts with red pulp sinus endothelium in a heterogeneous manner: IBL-7/1^{hi} vessels are IBL-9/2 negative, and the IBL-7/1^{lo} endothelia coexpress the IBL-9/2 antigen (62).

are typically located in the subcapsular region or splenic parenchymal and are often observed spreading into the surrounding tissue (68,69). Their exact origin is sometimes difficult to establish, as the proliferating lumen-forming cells may display considerable lineage infidelity (e.g., coexpression of endothelial and macrophage-related surface markers) (70). Routine histopathologic observation often reveals sinus erythrophagocytosis and cytoplasmic hemosiderin deposition, which can be correlated with the clinical symptoms of hypersplenism (anemia and thrombocytopenia).

In experimental models, several viral oncogenes have proved capable of inducing EC proliferation in the spleen. For example, middle T antigen of murine polyoma virus induces malignant proliferation of ECs in newborn mice. Interestingly, the same oncogene recruits potential tube-forming cells in the spleen of adult mice (71,72). SV7, a clone of the Moloney murine sarcoma virus 349 encoding mos protein involved in components of the mitogen-activated protein kinase transduction pathway, also induces splenic vascular tumors (73,74). Furthermore, heterozygosity of the von Hippel-Lindau (*vHL*) tumor suppressor gene appears to predispose to the development of spontaneous or carcinogen-inducible splenic vascular tumors of diverse biological behavior (75). It remains to be seen whether any of these or similar viral elements or loss of VHL tumor suppressor function leading to increased production of VEGF and angiopoietins can be responsible for the occurrence of splenic vascular tumors in humans.

In addition to the primary tumors of endothelium, vascular proliferation in the spleen may also occur as a secondary event in myelofibrosis with myeloid metaplasia (MMM). A recent study reported that in the spleen of MMM patients a characteristic microvascular shift occurred, in which the CD34-positive capillary expansion led to the substantial reduction of CD8-positive sinuses. An underlying reason for the pronounced neoangiogenesis may be the increased frequency of CD34/CD133-positive hematopoietic cell subset that would have the potential to mature into ECs. An attractive hypothesis seems to be that, in addition to the angiogenic potential of circulating hemangioblasts, the vessel formation from these local progenitors may also account for the neoangiogenesis observed in the spleen of MMM patients (76).

KEY POINTS

- The labyrinth of the spleen's vasculature is unique in possessing both open and closed circulatory pathways – still surrounded by a number of important features to solve, including its exact vascular pathways and the regulation of perfusion volume directed into the open and closed channeling systems – and precise mechanism of leukocyte homing to the spleen and their reentry into the systemic circulation.
- The splenic endothelium displays an extensive complexity in its ontogenic, developmental, and phenotypic traits related to their topographic distribution, including the possibility of heterogeneous tissue derivation from distinct ancestry. This complexity indicates that the endothelium of the spleen at different tissue compartments probably performs considerably diverse activities related to the spleen's functions in hemopoiesis and immune responses in both normal and pathological conditions.

ACKNOWLEDGMENT

Péter Balogh is a recipient of the Széchenyi István Research Fellowship of the Hungarian Academy of Sciences. Part of this work was supported by ETT grant No. 592/2003 from the Ministry of Health, Social and Family Affairs, Hungary.

REFERENCES

1 Krumbhaar EB. Functions of the spleen: (*Mysterii Plenum Organon*) Galen. *Physiol Rev.* 1926;6:160–200.
2 Lewis SM. The spleen: mysteries solved and unresolved. *Clin Haematol.* 1983;12:363–373.
3 Brodsky I, Dennis LH, Kahn SB, Brody LW. Normal mouse erythropoiesis I. The role of the spleen in mouse erythropoiesis. *Cancer Res.* 1966;26:198–201.
4 Robertson S, Kennedy M, Keller G. Hematopoietic commitment during embryogenesis. *Ann NY Acad Sci.* 1999;872:9–15.
5 Barker JE, Keenan MA, Raphals L. Development of the mouse hematopoietic system. II. Estimation of spleen and liver "stem" cell number. *J Cell Physiol.* 1969;74:51–56.
6 Medvinsky AL, Dzierzak EA. Development of the definitive hematopoietic hierarchy in the mouse. *Dev Comp Immunol.* 1998;22:289–301.
7 Ganong WF. *Review of Medical Physiology*, 21st ed. New York: Appleton and Lange, Lange Medical Publications; 2003:627–628.
8 Steiniger B, Barth P. Microanatomy and function of the spleen. *Adv Anat Embryol Cell Biol.* 2000;151:1–101.
9 Fujita T, Kashimura M, Adachi K. Scanning electron microscopy and terminal circulation. *Experientia.* 1985;41:167–179.
10 Mitchell J. Lymphocyte circulation in the spleen. Marginal zone bridging channels and their possible role in cell traffic. *Immunology.* 1973;24:93–107.
11 Grayson MH, Hotchkiss RS, Karl IE, et al. Intravital microscopy comparing T lymphocyte trafficking to the spleen and the mesenteric lymph node. *Am J Physiol Heart Circ Physiol.* 2003;284: H2213–H2226.
12 Steiniger B, Barth P, Hellinger A. The perifollicular and marginal zones of the human splenic white pulp: do fibroblasts guide lymphocyte immigration? *Am J Pathol.* 2001;159:501–512.
13 Kraal G, Schornagel K, Streeter PR, et al. Expression of the mucosal vascular addressin, MAdCAM-1, on sinus-lining cells in the spleen. *Am J Pathol.* 1995;147:763–771.
14 Weiss L. The spleen in malaria: the role of barrier cells. *Immunol Lett.* 1990;25:165–172.
15 Steiniger B, Barth P, Herbst B, et al. The species-specific structure of microanatomical compartments in the human spleen: strongly sialoadhesin-positive macrophages occur in the perifollicular zone, but not in the marginal zone. *Immunology.* 1997;92:307–316.
16 Nolte MA, Belien JA, Schadee-Eestermans I, et al. A conduit system distributes chemokines and small blood-borne molecules through the splenic white pulp. *J Exp Med.* 2003;198:505–512.
17 Saitoh K, Kamiyama R, Hatakeyama S. A scanning electron microscopic study of the boundary zone of the human spleen. *Cell Tissue Res.* 1982;222:655–665.
18 Sasou S, Sugai T. Periarterial lymphoid sheath in the rat spleen: a light, transmission, and scanning electron microscopic study. *Anat Rec.* 1992;232:15–24.
19 Papayannopoulou T, Craddock C, Nakamoto B, et al. The VLA4/VCAM-1 adhesion pathway defines contrasting mechanisms of lodgement of transplanted murine hemopoietic progenitors between bone marrow and spleen. *Proc Natl Acad Sci USA.* 1995;92:9647–9651.
20 Nitschke L, Floyd H, Ferguson DJ, Crocker PR. Identification of CD22 ligands on bone marrow sinusoidal endothelium implicated in CD22-dependent homing of recirculating B cells. *J Exp Med.* 1999;189:1513–1518.
21 Ueda H, Fujimori O, Abe M. Histochemical analysis of acidic glycoconjugates in the endothelium lining the splenic blood vessels in the rat. *Arch Histol Cytol.* 1996;59:389–397.
22 Uehara K, Miyoshi M. Tight junction of sinus endothelial cells of the rat spleen. *Tissue Cell.* 1999;31:555–560.
23 Uehara K, Miyoshi M. Stress fiber networks in sinus endothelial cells in the rat spleen. *Anat Rec.* 1999;254:22–27.
24 Uehara K, Miyoshi M. Junctions between the sinus endothelial cells of rat spleen. *Cell Tissue Res.* 1997;287:187–192.
25 Uehara K, Miyoshi M. The surface-connected canalicular system in the sinus endothelial cells of rat spleen. *Cell Tissue Res.* 1999;296:439–442.
26 Uehara K, Miyoshi M. Tubular invaginations with caveolae and coated pits in the sinus endothelial cells of the rat spleen. *Histochem Cell Biol.* 1999;112:351–358.
27 Uehara K, Miyoshi M. Localization of caveolin-3 in the sinus endothelial cells of the rat spleen. *Cell Tissue Res.* 2002;307:329–336.
28 Aird WC. Endothelium as an organ system. *Crit Care Med.* 2004; 32:S271–S279.
29 Apelqvist A, Ahlgren U, Edlund H. Sonic hedgehog directs specialised mesoderm differentiation in the intestine and pancreas. *Curr Biol.* 1997;7:801–804.
30 Kawahira H, Ma NH, Tzanakakis ES, et al. Combined activities of hedgehog signaling inhibitors regulate pancreas development. *Development.* 2003;130:4871–4879.

31 Green MC. A defect of the splanchnic mesoderm caused by the mutant gene dominant hemimelia in the mouse. *Dev Biol.* 1967; 15:62–89.

32 Hecksher-Sorensen J, Watson RP, Lettice LA, et al. The splanchnic mesodermal plate directs spleen and pancreatic laterality, and is regulated by Bapx1/Nkx3.2. *Development*. 2004;131:4665–4675.

33 Searle AG. Hereditary absence of spleen in the mouse. *Nature.* 1959;184:1419–1420.

34 Roberts CW, Shutter JR, Korsmeyer SJ. Hox11 controls the genesis of the spleen. *Nature.* 1994;368:747–749.

35 Dear TN, Colledge WH, Carlton MB, et al. The Hox11 gene is essential for cell survival during spleen development. *Development*. 1995;121:2909–2915.

36 Kanzler B, Dear TN. Hox11 acts cell autonomously in spleen development and its absence results in altered cell fate of mesenchymal spleen precursors. *Dev Biol.* 2001;234:231–243.

37 Koehler K, Franz T, Dear TN. Hox11 is required to maintain normal Wt1 mRNA levels in the developing spleen. *Dev Dyn.* 2000;218:201–206.

38 Herzer U, Crocoll A, Barton D, et al. The Wilms tumor suppressor gene wt1 is required for development of the spleen. *Curr Biol.* 1999;9:837–840.

39 Tribioli C, Lufkin T. The murine Bapx1 homeobox gene plays a critical role in embryonic development of the axial skeleton and spleen. *Development*. 1999;126:5699–5711.

40 Lu J, Chang P, Richardson JA, et al. The basic helix-loop-helix transcription factor capsulin controls spleen organogenesis. *Proc Natl Acad Sci USA.* 2000;97:9525–9530.

41 Mok H, Mendoza M, Prchal JT, et al. Dysregulation of ferroportin 1 interferes with spleen organogenesis in polycythaemia mice. *Development*. 2004;131:4871–4881.

42 Pabst O, Forster R, Lipp M, et al. NKX2.3 is required for MAdCAM-1 expression and homing of lymphocytes in spleen and mucosa-associated lymphoid tissue. *EMBO J.* 2000;19:2015–2023.

43 Wang CC, Biben C, Robb L, et al. Homeodomain factor Nkx2–3 controls regional expression of leukocyte homing coreceptor MAdCAM-1 in specialized endothelial cells of the viscera. *Dev Biol.* 2000;224:152–167.

44 Kallianpur AR, Jordan JE, Brandt SJ. The SCL/TAL-1 gene is expressed in progenitors of both the hematopoietic and vascular systems during embryogenesis. *Blood*. 1994;83:1200–1208.

45 Hwang LY, Siegelman M, Davis L, et al. Expression of the TAL1 proto-oncogene in cultured endothelial cells and blood vessels of the spleen. *Oncogene*. 1993;8:3043–3046.

46 Robb L, Elwood NJ, Elefanty AG, et al. The scl gene product is required for the generation of all hematopoietic lineages in the adult mouse. *EMBO J.* 1996;15:4123–4129.

47 Park C, Afrikanova I, Chung YS, et al. A hierarchical order of factors in the generation of FLK1- and SCL-expressing hematopoietic and endothelial progenitors from embryonic stem cells. *Development*. 2004;131:2749–2762.

48 Pajusola K, Aprelikova O, Korhonen J, et al. FLT4 receptor tyrosine kinase contains seven immunoglobulin-like loops and is expressed in multiple human tissues and cell lines. *Cancer Res.* 1992;52:5738–5743.

49 Ward NL, Dumont DJ. The angiopoietins and Tie2/Tek: adding to the complexity of cardiovascular development. *Semin Cell Dev Biol.* 2002;13:19–27.

50 Arai F, Hirao A, Ohmura M, et al. Tie2/angiopoietin-1 signaling regulates hematopoietic stem cell quiescence in the bone marrow niche. *Cell.* 2004;118:149–161.

51 Koop EA, Lopes SM, Feiken E, et al. Receptor protein tyrosine phosphatase mu expression as a marker for endothelial cell heterogeneity; analysis of RPTPmu gene expression using LacZ knock-in mice. *Int J Dev Biol.* 2003;47:345–354.

52 Guillot PV, Guan J, Liu L, et al. A vascular bed-specific pathway. *J Clin Invest.* 1999;103:799–805.

53 Aird WC, Jahroudi N, Weiler-Guettler H, et al. Human von Willebrand factor gene sequences target expression to a subpopulation of endothelial cells in transgenic mice. *Proc Natl Acad Sci USA.* 1995;92:4567–4571.

54 Fu YX, Chaplin DD. Development and maturation of secondary lymphoid tissues. *Annu Rev Immunol.* 1999;17:399–433.

55 Mackay F, Majeau GR, Lawton P, et al. Lymphotoxin but not tumor necrosis factor functions to maintain splenic architecture and humoral responsiveness in adult mice. *Eur J Immunol.* 1997;27:2033–2042.

56 Wang J, Foster A, Chin R, et al. The complementation of lymphotoxin deficiency with LIGHT, a newly discovered TNF family member, for the restoration of secondary lymphoid structure and function. *Eur J Immunol.* 2002;32:1969–1979.

57 Tumanov AV, Grivennikov SI, Shakhov AN, et al. Dissecting the role of lymphotoxin in lymphoid organs by conditional targeting. *Immunol Rev.* 2003;195:106–116.

58 Girkontaite I, Sakk V, Wagner M, et al. The sphingosine-1-phosphate (S1P) lysophospholipid receptor S1P3 regulates MAdCAM-1^+ endothelial cells in splenic marginal sinus organization. *J Exp Med.* 2004;200:1491–1501.

59 Stuart AE, Warford A. Staining of human splenic sinusoids and demonstration of unusual banded structures by monoclonal antisera. *J Clin Pathol.* 1983;36:1176–1180.

60 Buckley PJ, Dickson SA, Walker WS. Human splenic sinusoidal lining cells express antigens associated with monocytes, macrophages, endothelial cells, and T lymphocytes. *J Immunol.* 1985;134:2310–2315.

61 Balázs M, Grama L, Balogh P. Detection of phenotypic heterogeneity within the murine splenic vasculature using rat monoclonal antibodies IBL-7/1 and IBL-7/22. *Hybridoma.* 1999; 18:177–182.

62 Balázs M, Horvath G, Grama L, Balogh P. Phenotypic identification and development of distinct microvascular compartments in the postnatal mouse spleen. *Cell Immunol.* 2001;212:126–137.

63 Falkowski M, Schledzewski K, Hansen B, Goerdt S. Expression of stabilin-2, a novel fasciclin-like hyaluronan receptor protein, in murine sinusoidal endothelia, avascular tissues, and at solid/liquid interfaces. *Histochem Cell Biol.* 2003;120:361–369.

64 Matloubian M, David A, Engel S, et al. A transmembrane CXC chemokine is a ligand for HIV-coreceptor Bonzo. *Nat Immunol.* 2000;1:298–304.

65 Kennel SJ, Godfrey V, Ch'ang LY, et al. The beta 4 subunit of the integrin family is displayed on a restricted subset of endothelium in mice. *J Cell Sci.* 1992;101:145–150.

66 Hoek RM, Ruuls SR, Murphy CA, et al. Down-regulation of the macrophage lineage through interaction with OX2 (CD200). *Science.* 2000;290:1768–1771.

67 Duijvestijn AM, van Goor H, Klatter F, et al. Antibodies defining rat endothelial cells: RECA-1, a pan-endothelial cell-specific monoclonal antibody. *Lab Invest*. 1992;66:459–466.

68 Abbott RM, Levy AD, Aguilera NS, et al. From the archives of the AFIP: primary vascular neoplasms of the spleen: radiologic-pathologic correlation. *Radiographics.* 2004;24:1137–1163.
69 Kutok JL, Fletcher CD. Splenic vascular tumors. *Semin Diagn Pathol.* 2003;20:128–139.
70 Arber DA, Strickler JG, Chen YY, Weiss LM. Splenic vascular tumors: a histologic, immunophenotypic, and virologic study. *Am J Surg Pathol.* 1997;21:827–835.
71 Bautch VL, Toda S, Hassell JA, Hanahan D. Endothelial cell tumors develop in transgenic mice carrying polyoma virus middle T oncogene. *Cell.* 1987;51:529–537.
72 Williams RL, Risau W, Zerwes HG, et al. Endothelioma cells expressing the polyoma middle T oncogene induce hemangiomas by host cell recruitment. *Cell.* 1989;57:1053–1063.
73 Yuen PH, Matherne CM, Molinari-Storey LM. SV7, a molecular clone of Moloney murine sarcoma virus 349, transforms vascular endothelial cells. *Am J Pathol.* 1991;139:1449–1461.
74 Fukasawa K, Vande Woude GF. Synergy between the Mos/mitogen-activated protein kinase pathway and loss of p53 function in transformation and chromosome instability. *Mol Cell Biol.* 1997;17:506–518.
75 Kleymenova E, Everitt JI, Pluta L, et al. Susceptibility to vascular neoplasms but no increased susceptibility to renal carcinogenesis in Vhl knockout mice. *Carcinogenesis.* 2004;25:309–315.
76 Barosi G, Rosti V, Massa M, et al. Spleen neoangiogenesis in patients with myelofibrosis with myeloid metaplasia. *Br J Haematol.* 2004;124:618–625.

137

Adipose Tissue Endothelium

Gary Hausman

USDA-Agricultural Research Service, University of Georgia, Athens

Adipose tissue can be found throughout the body but is primarily distributed in several subcutaneous and visceral locations or "depots" (1). The general distribution pattern and location of adipose tissue depots and the proportion of total body fat that each depot represents varies widely between species. In humans and swine the subcutaneous depot represents a much larger proportion of total adipose tissue than in rodents. In fact, several distinct layers of subcutaneous adipose tissue are present in humans and swine. Excessive adipose tissue accumulation in humans is often associated with the adverse health consequences of the metabolic syndrome (2). An adipose tissue distribution pattern favoring visceral adipose tissue often is implicated in this syndrome.

Adipose tissue can be considered a metabolic buffer by sequestering fatty acids after a meal and releasing them when needed at a later time (3). Adipose tissue blood flow (ATBF) and its regulation is an essential component of the metabolic function of adipose tissue (3). The fact that increased ATBF leads to increased clearance of triglycerides by the tissue illustrates the physiological impact of ATBF (3).

Our understanding of the role of adipose tissue as an endocrine organ has evolved rapidly since the discovery of leptin (2). Adipose tissue is an active organ that produces and secretes a variety of factors into the circulation, including leptin and several interleukins (2), which also illustrates the physiological impact of ATBF. In contrast to ATBF, very little is known or published on adipose tissue endothelium per se. Furthermore, the widespread distribution of adipose tissue and the many location- or depot-dependent characteristics (including blood flow) (4) precludes extrapolation of results from one depot to the next.

The history of ATBF and adipose tissue endothelial biology is reviewed by Crandall and colleagues (5). Adipose tissue blood vessels were first studied in the late 1800s when investigators noted a close spatial and temporal relationship between developing adipocytes and developing blood vessels (5). In 1945, adipose tissue was shown to be significantly more vascular than expected (until that time, adipose tissue had been traditionally considered an inert tissue) (6). The first functional studies of ATBF, carried out in the 1950s, demonstrated the movement of lipid to and from adipose tissue by means of the circulation (5). Approximately 35 years ago, capillary filtration coefficient for adipose tissue was shown to be two- to threefold higher compared with resting skeletal muscle (5). Recent years have witnessed the development of sophisticated technologies to assess ATBF, including the ^{133}Xenon washout method, which permits measurement of ATBF in conscious subjects (7,8). The advent of microsphere techniques for measuring blood flow coupled with techniques to directly quantitate adipocyte cell number and size allowed a unique quantitation of ATBF per cell, which became standard for ATBF studies (4). This milestone, in turn, led to firmly establishing the remarkable influence of anatomical location or depot and metabolic state on ATBF (5) (Table 137-1).

STRUCTURE AND FUNCTION

Overview of Adipose Tissue Blood Flow

The use of new technologies to measure ATBF as well as fat cell size and number has yielded important insights into mechanisms of ATBF (3–5). ATBF is influenced by many factors including diet, hormones, innervation, exercise, and locally produced factors (3,5). ATBF fluctuation in the fed and fasted state in the rat is dependent on the depot or location, as is adipocyte hypertrophy (Table 137-1). Recent studies in humans demonstrated that basal and glucose-stimulated ATBF within the subcutaneous depot is dependent on the particular subcutaneous layer (9). It is now established that basal adipocyte blood flow per cell remains relatively constant as adipocytes undergo hypertrophy (5). Furthermore, a consistent decrease in basal blood flow per unit weight of tissue with adipose tissue growth is observed, since adipocyte number per unit weight is decreasing. Therefore, ATBF regulation is complex, involving the interactions of many physiological and anatomical influences.

Table 137-1: The Influence of Anatomical Location or Depot on Adipose Tissue Blood Flow, Growth Modality, and Innervation Density in Growing and Adult Rats

	Blood Flow		*Hypertrophy > Hyperplasia*	*Innervation Density*
	Fed	Fasted		
Mesenteric	+++	+++++	++++	++++
Epididymal	++	++++	+++	+++
Retroperitoneal	++	+++	++	++
Inguinal	+	+	+	+

Adapted with permission from Hausman DB, DiGirolamo M, Bartness TJ, et al. The biology of white adipocyte proliferation. *Obes Rev.* 2001;2:239–254. Relative levels indicated: + = lowest, +++++ = highest.

Overview of Structural Properties of Adipose Tissue Endothelium

Very little is known about the structural properties of adipose tissue endothelium in growing or aged animals and humans. Some but not all capillaries were found to be fenestrated in mouse adipose tissue (10). However, only a single depot was studied. Most information comes from studies of fetuses and young animals. Fetal adipose development is characterized by an exceptionally close spatial and temporal relationship between vascular and adipocyte development (11–16) (Table 137-2). Structural development of fetal adipose tissue arterioles and capillaries varies according to location, depot, or layer of subcutaneous tissue (5,11). In the fetus, the degree of capillary growth and elaboration relative to adipocyte lipid accretion or differentiation is predictive of the ultimate degree of vascularity and adipocyte cellularity (5,11). It is not known what regulates depot-dependent structural development of vascular beds in fetal adipose tissue. Interestingly, many functional and structural characteristics are shared by adipose tissue endothelial cells (ECs) and preadipocytes (12–16) (Table 137-2).

Lectin binding and expression of extracellular matrix components by fetal porcine adipose tissue capillary endothelium is not influenced by either depot or age (17) (Table 137-3). However, lectin binding and other qualitative cytochemical traits of adipose tissue endothelium are hormonally regulated during fetal porcine development (18). In this regard, hydrocortisone, but not thyroxine or growth hormone, influences the development of lectin binding and other cytochemical traits of adipose tissue capillary endothelium with no influence on the relative number of blood vessels (18). Lectin binding by adipose tissue endothelium during fetal development indicates that adipose tissue endothelium remains immature relative to endothelium in associated tissues like the dermis (Table 137-3).

Adipose tissue vasculature is supplied by a relatively sparse sympathetic innervation, with adrenergic nerve varicosities distributed down to the capillary level (19,20). Adrenergic innervation of adipose tissue vasculature is evident during late fetal development (19), and adrenergic innervation density and blood flow are highly correlated across several rat adipose tissue depots (Table 137-1).

Adipose tissue capillary endothelium changes markedly at the ultrastructural and structural level with adipocyte hypertrophy in the young rat (Figure 137.1). In a very short time, capillary lumen diameters are reduced by one half and numerous ultrastructural features indicate a very dynamic endothelium associated with adipocyte hypertrophy (Figure 137.1A). In contrast, adipose tissue capillary endothelium appears relatively quiescent after this period of rapid hypertrophy (Figure 137.1B). These changes in capillary endothelium are extremely rapid in the young rat and are associated with a transient and rapid rate of adipocyte hypertrophy (the subsequent rate of adipocyte hypertrophy is much slower).

Overview of Functional Properties of Adipose Tissue Endothelium and Blood Flow

Clearly the dominant function of adipose tissue endothelium and ATBF is involvement in the regulation of energy homeostasis by either sequestering or releasing energy substrates as dictated by the physiological state (3,5). In particular, adipose tissue capillary beds help to regulate blood lipid levels in two ways (20). First, very-low-density lipoproteins (VLDLs) and chylomicrons are sequestered from the blood by binding to capillary endothelial-bound lipoprotein lipase (LPL) (21). Second, the subsequent breakdown or lipolysis of VLDL and chylomicron lipid to fatty acids is catalyzed by capillary endothelial-bound LPL. Conversely, when dictated by fasting or other physiological states, another major function of ATBF is to remove the fatty acids liberated by adipocyte-derived hormone-sensitive lipase (HSL) for distribution throughout the body (3). This is demonstrated by adrenaline infusion, during which parallel increases occur in free fatty acid release from adipose tissue and ATBF (3). These functions are extremely significant to the whole animal, considering the extent of adipose tissue accumulation and distribution throughout the body. Specific influences on ATBF include glucose ingestion, mixed meal ingestion, short-term fasting, overnight fasting, and exercise, whereas fat feeding alone does not elicit an ATBF response (3–5,22). Regulation of ATBF in humans is primarily under adrenergic control (3). For example, a major proportion of postprandial enhancement of ATBF in humans is under β-adrenergic regulation (22).

The recently discovered function of adipose tissue as an endocrine organ highlights the potential significance and function of adipose tissue endothelium and ATBF in health and disease (23,24). Studies of the most chronicled of the adipose tissue "hormones" – i.e., leptin – indicate that it may be secreted in a pulsatile nature (3). Glycerol and free fatty acids are also secreted into the blood or plasma in a pulsatile manner. The pulsatility of free fatty acids and glycerol concentrations

Table 137-2: Lineage, Functional, and Structural Characteristics of Adipose Tissue ECs and Preadipocytes[a]

Studies/Experiments	*Results/Observations*	*References*
Electron microscopic studies in vivo	Capillary differentiation concurrent with early adipogenesis in perivascular cells	11
Human preadipocytes and ECs in vitro	Both cell types express and secrete PAI-1, express $\alpha v\beta 3$ integrin. Migration regulated by PAI-1	11
Cell replication rates in vitro	Depot or location dependent for rat preadipocytes and ECs	11
Adipocyte and adipose tissue SV cell conditioned media	Hypoxia-enhanced VEGF secretion: increased growth and reduced apoptosis of ECs in vitro	11, 12
Expression of Ang1 and Ang2 by adipose tissue	Adipocytes express Ang1, ECs express Ang2; Ang1 expression decreased during tissue remodeling	13
Adipose tissue induced by implanting 3T3-L1 cells into mice	VEGF signaling and PPARγ regulate angiogenesis and adipogenesis	14
Adipose tissue SV cells	Common lineage for ECs and adipocytes	15
Adipose tissue SV cells	Identification and verification of endothelial progenitor cells	16
Immunostaining in fetal and postnatal adipose tissue	Adipocyte surface antigen expressed by preadipocytes and capillary ECs	11

[a]PAI-1, plasminogen activator inhibitor-1; SV, stromal vascular cells; PPARγ, peroxisome proliferator activated receptor-γ.

may involve adipose tissue vasculature if the pulse generator is a neural-based mechanism (3).

Comparison of Adipose and Nonadipose Vascular Beds

Immunocytochemistry for laminin and type IV collagen- and lectin-binding studies demonstrate some features of adipose tissue vascular beds common to other vascular beds. For example, the distinction between arterioles and venules based on phosphatase activity develops similarly in adipose tissue and skeletal muscle but develops earlier in several other tissues (25). Furthermore, similar glycoconjugates are present on capillary endothelium in adipose tissue, muscle, and several other tissues (Table 137-2) (17). In particular, galactose residues are detected on capillary endothelium in adipose and other tissues (Table 137-3) (17).

Blood flow and hemodynamic characteristics that are similar for adipose tissue and skeletal muscle include transcapillary glucose exchange (26) and changes in blood flow/unit weight with age (27). Characteristics that distinguish adipose tissue from skeletal muscle include a greater resting blood flow/unit weight and capillary filtration coefficient for adipose tissue (3,5). The relative number and size of arterioles is greater in adipose tissue than in muscle during fetal development, which potentially provides a structural basis for greater resting ATBF (11). Finally, moderate exercise doubles ATBF while increasing skeletal muscle blood flow tenfold (28).

A number of studies demonstrate that the adipose tissue endothelium is distinct from endothelium in most other tissues in several remarkable ways. For example, the influence of antiangiogenic agents on adipose tissue accretion indicates that adipose tissue endothelium is maintained in an immature state because mature endothelium typically is not influenced by antiangiogenic agents (29,30). However, experimentally induced vascular endothelial growth factor (VEGF) inhibition in mice resulted in significant capillary regression in six other glands or organs in addition to adipose tissue (10). Nevertheless, adipose tissue is clearly distinguished by the many lineage, functional, and structural links between angiogenesis and adipogenesis (Table 137-2). In particular, a remarkable and unprecedented reciprocal regulation of angiogenesis and adipogenesis was demonstrated during adipogenesis induced in vivo (Table 137-2) (14). The quality and extent of this evidence validates the distinction of adipose tissue endothelium from other endothelium based on developmental relationships with parenchymal cells.

Table 137-3: Lectin and Extracellular Matrix Component Cytochemical Traits of Capillary Endothelium in Adipose Tissue, Dermis, and Other Tissues in the Developing Fetal Pig

	Lectins and Other Components	*Fetal Age, Days*		
		50	75	110
Dermis and other tissues*	GS-1	++	+	0
	SBA	++	+	0
	PCA	+	0	0
	UEA-1	0	0	0
	Type IV collagen	++	++	++
	Laminin	++	++	++
Adipose tissue	GS-1	++	++	++
	SBA	++	++	++
	PCA	+	+	+
	UEA-1	+	+	+
	Type IV collagen	++	++	++
	Laminin	++	++	++

Adapted with permission from Hausman GJ, Wright JT, Thomas GB. Vascular and cellular development in fetal adipose tissue: lectin binding studies and immunocytochemistry for laminin and type IV collagen. *Microvasc Res.* 1991;41:111–125.

Scoring was as follows: 0, no stain; +, intermediate staining; and ++, maximum staining.

*Included dense connective tissue around blood vessels and nerves and bands of connective tissue that separate adipose tissue layers from one another and from muscle.

Distinction between What Is Known in Vitro and in Vivo

ECs from human (31,32), rat (33), and bovine adipose tissue (34) have been isolated and characterized in vitro. However, the limited information on adipose tissue endothelium in vivo precludes a meaningful comparison of in vitro and in vivo studies on adipose tissue endothelium.

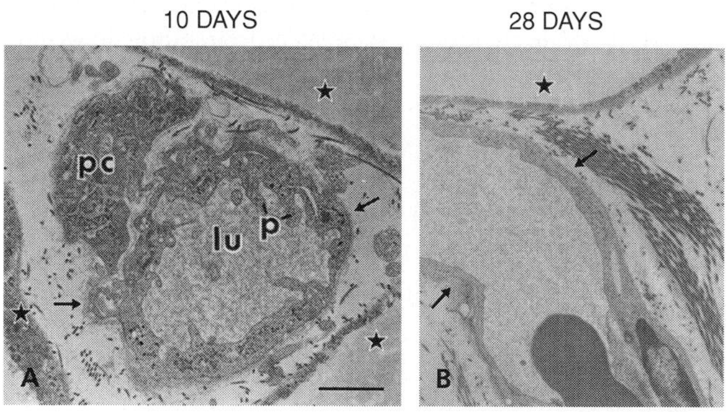

Figure 137.1. Capillaries in adipose tissue from a 10-day-old (**A**) and 28-day-old (**B**) rat. At 10 days, capillary lumens (*lu*) are very narrow and vessel walls are thin and contain many large vesicles and EC processes (*p*) extending into the lumen and the extravascular space. The highly dynamic ultrastructure of capillaries at 10 days compared to capillaries at 28 days is associated with the very rapid rate of adipocyte hypertrophy in the 10-day-old and younger rat. Capillaries in the older rats have narrow lumens and very thin walls with numerous smaller vesicles. Note the capillary external lamina (**A** and **B**, *arrows*), which is considerably thinner and more distinct at 28 days (**B**, *arrows*). Inset bar is equal to 1 μm and is also applicable to **B**. Also indicated are adipocytes (*stars*) and a perivascular cell (*pc*).

ADIPOSE TISSUE ENDOTHELIUM IN DISEASE

Obesity and type 2 diabetes are associated with general endothelial dysfunction, but very little is known about adipose tissue endothelium per se in disease (2,3). A limited number of studies include a report that obesity in humans is associated with decreased blood flow and glucose uptake in several depots (35).

DIAGNOSTIC AND THERAPEUTIC IMPLICATIONS

Various therapeutic implications and applications of understanding adipose tissue angiogenesis and ATBF have been extensively reviewed (5). Recent studies point to several novel endothelium-based means to treat or curb obesity, such as inducing overexpression of angiopoietin 1 (Ang1) in adipose tissue (Table 137-2) and targeting a proapoptotic peptide to an adipose tissue vasculature membrane protein (36). Furthermore, treatment with antiangiogenic agents can be used to selectively regress adipose tissue (29,30). Finally, evidence of endothelial progenitor cells in adipose tissue provides further support for the hypothesis that adipose tissue may serve as a source of microvascular ECs for therapeutic purposes (Table 137-2).

KEY POINTS

- Several lines of evidence indicate that adipose tissue endothelium is relatively immature, which would accommodate the very dynamic morphological changes in adipocytes and adipose tissue that occur during lipid accretion.
- Adipose tissue capillary endothelium plays a critical role in sequestering and catabolizing blood lipids for assimilation in adipocytes.
- The extent and regulation of ATBF is dependent on anatomical location and is highly correlated with adipocyte hypertrophy and innervation density.
- The nature and extent of developmental and functional relationships between adipose tissue endothelium and parenchymal cells distinguish adipose tissue from other tissues.
- Further study is necessary to elucidate the lineage, developmental, and functional relationships between adipose tissue endothelium and parenchymal cells during the development and maintenance of obesity and type 2 diabetes.
- Continued research is necessary to examine the significance and full impact of an immature adipose tissue endothelium in normal and disease states.

REFERENCES

1 Cinti S. The adipose organ. *Prostaglandins Leukot Essent Fatty Acids.* 2005;73:9–15.
2 Hauner H. Secretory factors from human adipose tissue and their functional role. *Proc Nutr Soc.* 2005;64:163–169.
3 Frayn KN. Obesity and metabolic disease: is adipose tissue the culprit? *Proc Nutr Soc.* 2005;64:7–13.
4 Hausman DB, DiGirolamo M, Bartness TJ, et al. The biology of white adipocyte proliferation. *Obes Rev.* 2001;2:239–254.
5 Crandall DL, Hausman GJ, Kral JG. A review of the microcirculation of adipose tissue: anatomic, metabolic, and angiogenic perspectives. *Microcirculation.* 1997;4:211–232.
6 Gersh IG, Still MA. Blood vessels in fat tissue: relation to problems of gas exchange. *J Exp Med.* 1945;81:219–232.
7 Moher HE, Carey GB. Technical note: adipose tissue blood flow in miniature swine (Sus scrofa) using the ^{133}xenon washout technique. *J Anim Sci.* 2002;80:1294–1298.
8 Karpe F, Fielding BA, Ilic V, et al. Monitoring adipose tissue blood flow in man: a comparison between the $^{(133)}$xenon washout method and microdialysis. *Int J Obes Relat Metab Disord.* 2002;26:1–5.
9 Ardilouze JL, Karpe F, Currie JM, et al. Subcutaneous adipose tissue blood flow varies between superior and inferior levels of the anterior abdominal wall. *Int J Obes Relat Metab Disord.* 2004; 28:228–233.
10 Kamba T, Tam BY, Hashizume H, et al. VEGF-dependent plasticity of fenestrated capillaries in the normal adult microvasculature. *Am J Physiol Heart Circ Physiol.* 2006;290:H560–576.
11 Hausman GJ, Richardson RL. Adipose tissue angiogenesis. *J Anim Sci.* 2004;82:925–934.
12 Rehman J, Traktuev D, Li J, et al. Secretion of angiogenic and antiapoptotic factors by human adipose stromal cells. *Circulation.* 2004;109:1292–1298.
13 Dallabrida SM, Zurakowski D, Shih SC, et al. Adipose tissue growth and regression are regulated by angiopoietin-1. *Biochem Biophys Res Commun.* 2003;311:563–571.
14 Fukumura D, Ushiyama A, Duda DG, et al. Paracrine regulation of angiogenesis and adipocyte differentiation during in vivo adipogenesis. *Circ Res.* 2003;93:e88–e97.
15 Planat-Benard V, Silvestre JS, Cousin B, et al. Plasticity of human adipose lineage cells toward endothelial cells: physiological and therapeutic perspectives. *Circulation.* 2004;109:656–663.
16 Miranville A, Heeschen C, Sengenes C, et al. Improvement of postnatal neovascularization by human adipose tissue-derived stem cells. *Circulation.* 2004;110:349–355.
17 Hausman GJ, Wright JT, Thomas GB. Vascular and cellular development in fetal adipose tissue: lectin binding studies and immunocytochemistry for laminin and type IV collagen. *Microvasc Res.* 1991;41:111–125.
18 Hausman GJ. The interaction of hydrocortisone and thyroxine during fetal adipose tissue differentiation: CCAAT enhancing binding protein expression and capillary cytodifferentiation. *J Anim Sci.* 1999;77:2088–2097.
19 Hausman GJ, Richardson RL. Adrenergic innervation of fetal pig adipose tissue. Histochemical and ultrastructural studies. *Acta Anat (Basel).* 1987;130:291–297.
20 Slavin BG. The morphology of adipose tissue. In: Cryer A, Van RLR, editors. *New Perspectives in Adipose Tissue: Structure Function and Development.* London: Butterworth & Co.; 1985: 23–44.
21 Mead JR, Irvine SA, Ramji DP. Lipoprotein lipase: structure, function, regulation, and role in disease. *J Mol Med.* 2002;80:753–769.
22 Ardilouze JL, Fielding BA, Currie JM et al. Nitric oxide and beta-adrenergic stimulation are major regulators of preprandial and postprandial subcutaneous adipose tissue blood flow in humans. *Circulation.* 2004;109:47–52.
23 Trayhurn P, Wood IS. Adipokines: inflammation and the pleiotropic role of white adipose tissue. *Br J Nutr.* 2004;92:347–355.
24 Fruhbeck G. The adipose tissue as a source of vasoactive factors. *Curr Med Chem Cardiovasc Hematol Agents.* 2004;2:197–208.
25 Hausman GJ. Nucleoside phosphatase and nucleotide tetrazolium reductase as markers of arteriolar differentiation in fetal pig tissue. *Histochemistry.* 1985;83:121–126.
26 Regittnig W, Ellmerer M, Fauler G, et al. Assessment of transcapillary glucose exchange in human skeletal muscle and adipose tissue. *Am J Physiol Endocrinol Metab.* 2003;285:E241–E251.
27 Delp MD, Evans MV, Duan C. Effects of aging on cardiac output, regional blood flow, and body composition in Fischer-344 rats. *J Appl Physiol.* 1998;85:1813–1822.
28 Horowitz JF, Klein S. Lipid metabolism during endurance exercise. *Am J Clin Nutr.* 2000;72:558S–63S.
29 Rupnick MA, Panigrahy D, Zhang CY, et al. Adipose tissue mass can be regulated through the vasculature. *Proc Natl Acad Sci U S A.* 2002;99:10730–10705.
30 Brakenhielm E, Cao R, Gao B, et al. Angiogenesis inhibitor, TNP-470, prevents diet-induced and genetic obesity in mice. *Circ Res.* 2004;94:1579–1588.

31 Springhorn JP, Madri JA, Squinto SP. Human capillary endothelial cells from abdominal wall adipose tissue: isolation using an anti-pecam antibody. *In Vitro Cell Dev Biol Anim.* 1995;31:473–481.

32 Hewett PW, Murray JC, Price EA, et al. Isolation and characterization of microvessel endothelial cells from human mammary adipose tissue. *In Vitro Cell Dev Biol Anim.* 1993;29A:325–331.

33 Bjorntorp P, Hansson GK, Jonasson L, et al. Isolation and characterization of endothelial cells from the epididymal fat pad of the rat. *J Lipid Res.* 1983;24:105–112.

34 Bar RS, Dolash S, Dake BL, et al. Cultured capillary endothelial cells from bovine adipose tissue: a model for insulin binding and action in microvascular endothelium. *Metabolism.* 1986;35:317–322.

35 Virtanen KA, Lonnroth P, Parkkola R, et al. Glucose uptake and perfusion in subcutaneous and visceral adipose tissue during insulin stimulation in nonobese and obese humans. *J Clin Endocrinol Metab.* 2002;87:3902–3910.

36 Kolonin MG, Saha PK, Chan L, et al. Reversal of obesity by targeted ablation of adipose tissue. *Nat Med.* 2004;10:625–632.

138

Renal Endothelium

Bruce Molitoris

Indiana Center for Biological Microscopy, Indiana University School of Medicine

The kidney, which receives approximately 20% of the cardiac output, has one of the richest and most diversified endothelial populations found within any organ. Such diversity may be explained on at least two counts. First, phenotypic diversity reflects the capacity of the endothelium to contribute to differential transport capabilities across the various segments of the nephron. Second, the endothelium must withstand unparalleled environmental extremes in oxygenation and osmolality (Figure 138.1; for color reproduction, see Color Plate 138.1). For example, endothelial cells (ECs) in the outer cortex are exposed to a normal osmolality and oxygen (O_2) tension, whereas those in the inner medullary region are exposed to an osmolality of up to 1,200 mOSM and O_2 content as low as 20 mm Hg.

From a mechanistic standpoint, phenotypic diversity of the renal endothelium must arise from site-specific epigenetic modification of ECs and/or reversible effects of the tissue/blood microenvironment. To date very little is known about the different endothelial populations within the kidney, let alone the molecular basis for their heterogeneity. Thus, the goals of this chapter are to delineate current knowledge concerning ECs in the kidney, stimulate research related to understanding the differing functions of renal ECs under physiological conditions, and explore the role of phenotypic diversity in disease processes.

STRUCTURE AND FUNCTION

The unique structure of the vascular tree within the kidney is shown in Figure 138.1. Anatomically, the kidney is divided into four zones. The cortex occupies the outermost aspect of the kidney. As one moves toward the renal pelvis from the cortex, one encounters the outer medullary region, consisting of outer and inner stripes, and finally the inner medullary region. Although the renal arteries enter the kidney via the inner renal pelvis, a series of large branching arteries deliver primary arteriolar blood into the outer cortex via the interlobular arteries terminating in a network of afferent arterioles. Although overall blood flow to the kidney averages 4 mL/g tissue/min, there is great heterogeneity within the kidney. The outer cortex receives 5–6 mL/g/min with ever deeper areas receiving less. The inner cortex, outer medulla, and inner medulla receive 2–3, 1, and 0.5 mL/g/min, respectively. Thus, over 90% of the blood flow is directed to the cortical area. Afferent arterioles have a continuous smooth muscle layer and are the major resistance vessels controlling the glomerular filtration rate (GFR). They terminate to form the glomerular capillary network where formation of the primary urine takes place via filtration. Each glomerular capillary network gives rise to a single efferent arteriole. In contrast to their afferent counterpart, efferent arterioles have a limited discontinuous layer of smooth muscle cells.

Efferent arterioles follow one of two paths. Those that arise from glomeruli within the mid and outer cortex give rise to an O_2-rich, dense network of capillaries (the so-called *peritubular plexus*), which run alongside the proximal and distal convoluted tubules of the kidney. The latter arrangement is unique in that it consists of two capillary beds (glomerulus and peritubular capillaries) separated by an arteriole. The rich cortical capillary network arises from nonanastomosing glomerular efferent arteries and joins the venous drainage system within the outer cortex as the superficial cortical veins and the interlobular veins for the mid and inner cortex. Those efferent arterioles that arise from the innermost or juxtamedullary glomeruli (at the border between the cortex and medulla) give rise to the vasa recta, which run parallel to loops of Henle and collecting tubules in the medulla (1). Descending arteriolar vasa recta branch into several vessels that penetrate the inner depths of the medulla, give rise to a dense capillary network within the outer medulla, and return to the cortex via venous ascending vasa recta (AVR). Descending vasa recta (DVR) and AVR are packed together in vascular bundles in the outer medulla like highways going up and down. The vasa recta provide the sole blood supply to the medulla.

In the inner medulla the bundles become interspersed with the nephron segments. In the inner stripe of the outer medulla vasa recta emerge from the periphery of the vascular bundles

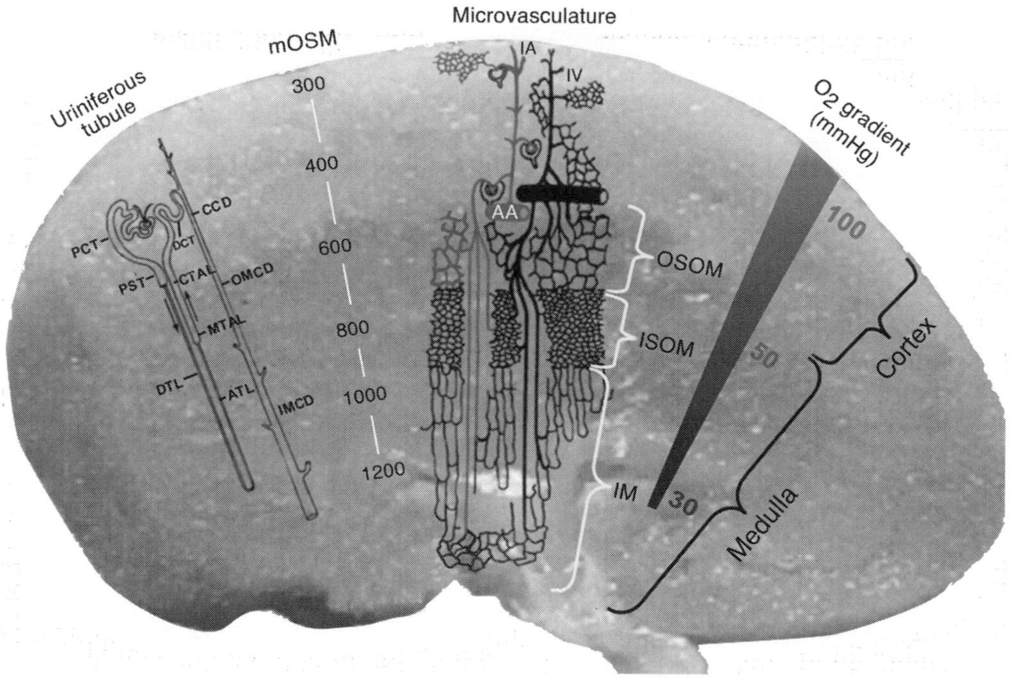

Figure 138.1. Schematic representation of the visional differences in the microvasculature within the kidney. The arcuate artery (*AA*) gives rise to the interlobular artery (*IA*), from which all afferent arterioles arise to perfuse the glomerular capillaries. Also shown are the osmotic and O_2 gradients from the outer cortex to the inner medulla. The medulla vascular supply arises from the juxtaglomerular efferent arterioles that branch and then descend into the medulla. As a result, the outer stripe of the outer medulla (*OSOM*) has minimal arteriole capillary density. OSOM, outer stripe of the outer medulla; ISOM, inner stripe of the outer medulla; IM, inner medulla; PCT, proximal convoluted tubule; PST, proximal straight tubule; DLT, descending thin limb; ATL, ascending thin limb; MTAL, medullary thick ascending limb; CTAL, cortical thick ascending limb; DCT, distal convoluted tubule; CCD, cortical collecting duct; OMCD, outer medullary collecting duct; IMCD, inner medullary collecting duct; AA, arcuate artery; IA, interlobular artery. For color reproduction, see Color Plate 138.1.

to supply blood to a dense capillary network, the so-called *interbundle plexus*, providing arterial capillary blood. In contrast, there is little in the way of an arteriole supply to the outer stripe of the outer medulla. Rather, the vessels providing primary nutrient supply and O_2 delivery to the outer stripe derive from the ascending venous vasa recta and thus this region is perfused primarily by venous capillary blood.

The unique vascular architecture of the kidney results in a gradient of decreasing oxygenation (and O_2 delivery) from the cortex to the medulla. This arrangement is essential for development of the countercurrent mechanism necessary for the recycling of osmotic compounds to maintain the marked osmotic gradient that increases as one goes from the outer cortex to the inner medullary region (Figure 138.1) (1). Importantly, the cortical medullary junction (outer stripe of the outer medulla) is a watershed zone having primarily venous capillary networks and a very low preponderance of arteriolar capillaries.

The density of the vascular space within the kidney also varies between the cortex and medullary regions. Pathologists often include the microvascular space as part of the "renal interstitium" because standard H&E stains do not identify ECs and thus one cannot differentiate the microvascular area from the true interstitial space. However, under physiological conditions the "cortical interstitial/vascular space" constitutes approximately 10% of the kidney by volume, with this percentage increasing as one ages. Within the medulla there is a gradual increase in interstitial volume from 10% to 20% in the outer medullary region to 30% to 40% in the papillary region.

ECs within this heterogeneous vascular system also vary tremendously structurally and functionally depending upon location (Table 138-1). Afferent arterioles give rise to glomerular capillaries that are lined by fine fenestrated endothelium with fenestrae ranging in size from 70 to 100 nm in diameter. Diaphragms do not extend across these fenestrae and the fenestrae are surrounded by microfilaments (2). These fenestrae are not the result of caveolae fusing, as human glomerular endothelial fenestrae are devoid of caveolin-1, and caveolin-1–deficient mice have normal glomerular endothelial fenestrae (3). Glomerular ECs are known to synthesize both nitric oxide (NO) and endothelin-1 and express vascular endothelial growth factor (VEGF) receptors that have been shown to enhance EC permeability and induce endothelial fenestrations (4). Additional evidence indicates that VEGF is important for EC survival and repair in glomerular diseases characterized by EC damage (5–7). Thus, VEGF produced by epithelial cells

Table 138-1: Properties and Functions of Renal Endothelium

Blood Vessel	*Smooth Muscle Cells*	*Endothelium*	*Unique Properties*	*Major Function*
Afferent arteriole	Continuous		Juxtaopposed to macula densa, vasoresponsive to angiotensin II, etc.	Regulate single nephron blood flow
Glomerular capillary		Continuous, fenestrated, no diaphragm	Express VEGFR, synthesize NO and endothelin-1	Selective filtration of macromolecules based on size and charge
Efferent arteriole	Discontinuous		Nonbranching in cortex, branching in juxtamedullary glomeruli (JMG)	Regulate filtration fraction
Peritubular plexus		Fenestrated facing epithelial cells	OKM5	Handle bulk of reabsorbed H_2O and ions
DVR	Discontinuous pericytes	Continuous, nonfenestrated	AQP1, urea transporter, Arise from JMG efferent arterioles	Countercurrent exchanger serving to prevent washout of osmotic gradient created by loops of Henle: enters increasingly hyperosmolar medulla, thus gains solutes and loses water
AVR	Discontinuous pericytes	Continuous, highly fenestrated, diaphragm	Extremely thin ECs	Countercurrent exchanger serving to prevent washout of osmotic gradient created by loops of Henle: leaves high osmolar environment as it moves towards cortex, thus loses in DVR and gains back water; net effect is the osmolality of blood at end of AVR approaches that of blood at beginning of DVR (slightly higher NaCl)

appears to be an important regulator of glomerular EC function and permeability. The production of VEGF is primarily by podocytes (glomerular epithelial cells), distal tubule epithelium, and collecting duct epithelium, especially in the outer medulla (see Chapter 68).

Within the cortex, microvascular ECs are fenestrated on those portions of the cell that border tubule epithelial cells. Erythropoietin-producing cells are localized in the renal cortical interstitial space adjacent to capillary endothelial and tubular epithelial cells. As efferent arterioles from juxtamedullary glomeruli penetrate into the outer stripe of the outer medulla, the outer smooth muscle layer is replaced by a discontinuous layer of pericytes (Figure 138.2). This marks the beginning of the DVR and upon further penetration into the inner medulla pericytes become less numerous. In both the inner and outer medullary regions the ascending venous vasa recta outnumber the descending arterial vasa recta. The DVR have a continuous nonfenestrated endothelium with a zona occludens limiting their permeability. They express both water channels (e.g., AQP1) and a facilitated urea transporter. In the inner medullary region, the DVR terminate and form a sparse capillary plexus that coalesces to form the AVR. Within the AVR fenestrations in ECs again occur with approximately 50% of the endothelial wall covered by fenestrae in the inner medullary region and 15% to 30% in the outer medullary region (Figure 138.2). These fenestrations have diameters of 50 to 100 nm and are bridged by a thick diaphragm. ECs within the AVR are extremely thin (Figure 138.2) and have a remarkably high hydraulic conductivity.

Little information exists regarding the nonstructural differences between renal ECs in vivo and even less is known about their characteristics or behavior in vitro. Because of the tremendous heterogeneity of ECs within the kidney, initial studies were undertaken to determine differences in expression of specific EC markers. In the human kidney, peritubular capillary ECs exhibited immunohistochemical characteristics similar to those of peripheral macrophages, including OKM5 (medullary only), von Willebrand factor (vWF), and interleukin (IL)-1 positivity. Glomerular capillary ECs expressed vWF, but not OKM5. Additional studies have identified RECA-1 and JG-12 as pan-EC-specific monoclonal antibodies for all vascular beds in the rat. Human, but not murine, tissue

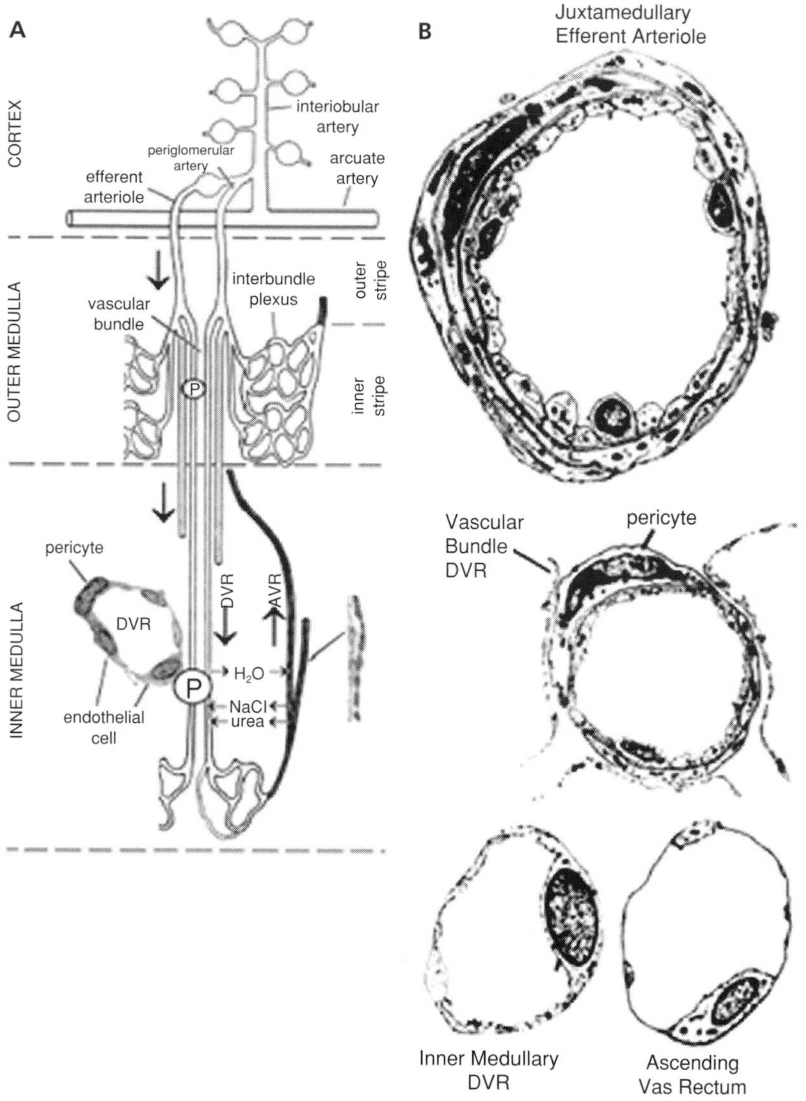

Figure 138.2. Schematic representation of the different forms of ECs found within the peritubular microvasculature of the kidney in the different zones of the kidney. (Adapted with permission from Pallone TL, Turner MR, Edwards A, Jamison RL. Countercurrent exchange is the venal medulla. *Am J Physiol Regul Integr Comp Physiol.* 2003;284:R1153–R1175.)

peritubular and glomerular ECs express HLA-DR. The presence of MHC class II proteins raises the possibility that these cells present antigen (8). The OKM5 antibody may be a specific marker for medullary ECs (9) and also reacts to macrophages.

Recently, both conditionally immortalized mouse glomerular ECs (2) and a "microvascular" EC line from mouse kidney (10) have been isolated and grown in cell culture. The glomerular ECs were isolated from H-2K-ts A58 transgenic mice carrying a gene encoding a temperature-sensitive variant of the SV 40 large tumor antigen, directing growth at 33°C and differentiation at 37°C. This immortalized mouse glomerular EC line (termed mGEnC-1) expressed vWF, podocalyxin, CD31, CD101, vascular endothelial (VE)-cadherin, GSL I-B4, and ULEX, and these cells internalized acetylated low-density lipoprotein (acLDL). Additional studies to characterize these glomerular cells have not been reported. The mouse renal microvascular EC line (REC-A4) was developed using a bead technique with anti-CD31mAb attached. Different clonal lines were propagated and one was selected for further study. These cells lacked Weibel-Palade bodies and fenestrae, but did have other characteristics highly suggestive of ECs. Moreover, the cells also showed a sensitivity towards Shiga toxin-1, demonstrated LDL uptake, formed tube-like structures on Matrigel, and demonstrated constitutive surface expression of CD31, intercellular adhesion molecule (ICAM)-1 and vascular cell adhesion molecule (VCAM)-1 (10). There is no reported evidence regarding differential lectin binding characteristics of different renal EC populations or differential binding of potentially selective antibodies to these same populations.

RENAL ENDOTHELIUM IN DISEASE: DIAGNOSTIC AND THERAPEUTIC IMPLICATIONS

The renal vascular endothelium is known to be a primary target in several disease processes including glomerular nephritis, vasculitis, lupus nephritis, preeclampsia, hemolytic uremic syndrome, ischemic acute renal failure, renal transplantation rejection, and chronic progressive kidney disease. For example, hemolytic uremic syndrome (HUS) and related thrombotic microangiopathies result from an infection with a certain verotoxin-expressing strain of *Escherichia coli* (0157:H7). The hallmark of this disease process is glomerular and renal microvascular EC injury with associated apoptosis, detachment, and intracapillary thrombus formation. The preferential involvement of glomerular and renal ECs may relate to the approximately 50-fold increase in the differential expression of the Gb3 receptor for verotoxins within these ECs. Recently a model of thrombotic microangiopathic hemolytic anemia in the rat has been developed utilizing selective renal arterial perfusion with an anti-EC antibody (6). This pathologic process, as in HUS, results in progressive glomerular and tubular interstitial damage. Notably, subcutaneously administered VEGF accelerated renal recovery in this animal model. Associated findings with VEGF therapy included an increased number of glomeruli with intact endothelium, a denser peritubular capillary network, and less interstitial fibrosis and cortical atrophy (6).

Glomerular endothelial fenestrae can also be lost within hours following VEGF removal. This results in glomerular endotheliosis, reduction/loss in GFR, and proteinuria (3). Such a mechanism may be operative in preeclampsia, where increased levels of placental-derived soluble VEGF receptor 1 inhibitor (soluble fms-like tyrosine kinase 1, sFLT-1) result in VEGF neutralization (9,11). Overexpression of VEGF leads to collapsing glomerulopathy, further demonstrating that tight regulation of VEGF is critical to glomerular development and function (11).

Microvascular endothelial damage is the salient feature associated with acute vascular rejection and, in chronic allograft nephropathy, the single most common cause of long-term graft loss (12). Peritubular capillary EC deposition of C4d was an independent predictor of graft loss. Furthermore, the number of circulating ECs in renal transplant recipients may be a marker of renal endothelial damage (13). Studying renal transplant biopsies (12) has revealed that angioregression and progressive loss of the peritubular capillary network strongly contributed to the development of interstitial fibrosis and graft dysfunction in chronic allograft nephropathy.

Ischemic acute renal failure has recently been shown to be associated with renal microvascular endothelial injury and dysfunction (14,15). During ischemic injury and/or sepsis there is an upregulation of a variety of chemokines and cytokines including IL-1, IL-6, IL-8, monocyte chemoattractant protein (MCP)-1, and tumor necrosis factor (TNF)-α, all of which are instrumental in initiating an inflammatory cascade. At the endothelial level ischemic acute renal failure was associated with an increase in vWF in the systemic circulation (16). Furthermore, utilizing the *Tie2-GFP* transgenic mouse (Jackson Laboratory), Sutton and colleagues identified and characterized EC F-actin aggregation in conjunction with alterations in the integrity of the adherens junctions within the renal microvasculature (15). Specifically, there was loss of localization of VE-cadherin immunostaining, and this was associated with a permeability defect identified using fluorescent dextrans and intravital multiphoton microscopy. Alterations have also been described in endothelial-leukocyte interactions including enhanced expression of both P- and E-selectin as well as ICAM-1 (13). The potential importance of this enhanced expression is shown by several studies utilizing knockout mice null for *P-selectin, E-selectin*, or *ICAM-1*, all of which exhibit reduced cellular injury and improved functional state of the kidney (13).

Additional studies have delineated differential inflammatory cell infiltration over time during the phases of injury and recovery following ischemia (17). The relative importance of these leukocyte populations is presently being explored. In vivo depletion of $CD4^+$ cells, but not $CD8^+$ cells, protected against ischemia-reperfusion injury (16). Furthermore, blockade of the B7–1/CD28 costimulatory pathway reduced mononuclear cell infiltration and renal functional deterioration. In the kidney, this emphasizes the importance of the AVR as B7 expression is limited to ECs of the AVR. The expression of EC adhesion molecules is not a universal phenomenon in acute renal failure (ARF), as it does not occur during toxin-induced ARF, and no leukocyte accumulation is seen in the kidney (18). Further work has demonstrated that injecting human umbilical vein ECs (HUVECs) into the systemic circulation prior to the ischemic injury results in lessening of the injury. The positive effect was greater when HUVECs were injected intra-arterially, but still present following intravenous injection. The reduction in cell epithelial injury, and renal functional deterioration, necessitated that the injected ECs express endothelial NO synthase (eNOS) expression as EC not expressing eNOS did not provide protection (19).

One particular region of the kidney seems to have progressive ischemic injury during the reperfusion period: the outer stripe, also known as the *cortical medullary junctional region*. As is shown in Figure 138.1, this is essentially a watershed zone receiving its primary and limited vascular supply from the O_2-poor ascending venous vasa recta. Following ischemic injury, this area of the kidney receives the lowest level of reperfusion, although the mechanisms underlying this lack of reperfusion have not been elucidated (14). It is generally believed that the lack of adequate reperfusion to this area leads to worsening epithelial cell injury and acute renal failure, recently described as the extension phase of ischemic cell injury within the kidney (14). This may well relate to the high expression of specific cell adhesion molecules by the venous ECs within this region resulting in white blood cell adherence (18). Multiple other factors are also involved in this lack of reperfusion, including

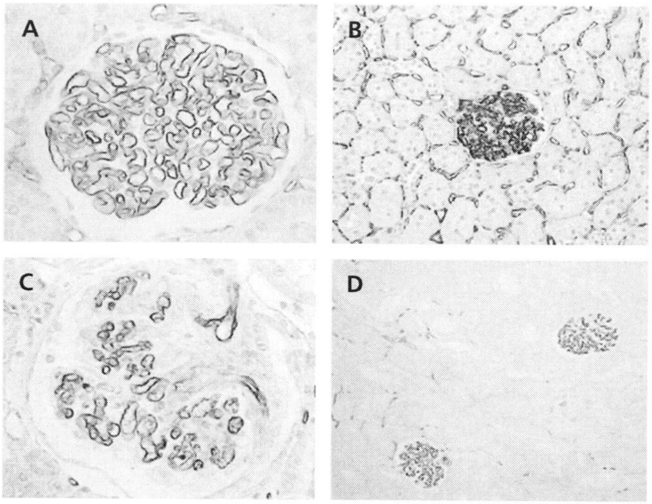

Figure 138.3. Loss of microvascular endothelium in progressive renal disease. In the contrast to normal glomerular (**A**) and peritubular (**B**) endothelium, there is a progressive loss of glomerular capillary loops (**C**) and peritubular capillaries (**D**) in the remnant kidney model of progressive renal disease. Not shown is the ability of VEGF to prevent this loss and deterioration in renal function that is associated with capillary loss. For color reproduction, see Color Plate 138.3.

coagulation abnormalities, but this lack can not be explained by reduced total RBF.

The importance of continuing microvascular inflammation and progressive microvascular dropout leading to progression of renal failure has recently been emphasized by a series of studies by Basile and associates (20). Utilizing a model of moderate to severe acute ischemic injury in rats, these investigators have demonstrated a permanent reduction in vascular density leading to reduced renal O_2 tension. Furthermore, anti-transforming growth factor (TGF)-β treatment attenuated microvascular dropout, and this was associated with a reduction in interstitial cellularity positive for the specific fibroblast marker S100A4 (20). Following severe renal ischemia, there also occurs a prolonged (6 weeks) enhanced production of IL-1β and persistent CD4$^+$-T cell infiltration (21).

The role for progressive endothelial dysfunction in chronic kidney disease has also been advanced in other models of progressive renal failure (Figure 138.3; for color reproduction, see Color Plate 138.3). An underlying theme to this work has been the essential nature of VEGF in kidney development and angiogenesis following renal injury. VEGF was first shown to play a major role in renal development by blocking VEGF activity utilizing an antibody in newborn mice. This resulted in disrupted and reduced renal cortical capillary formation, abnormal glomeruli lacking capillary tufts, and an overall reduction in number of nephrons (7). Subsequent work utilized a remnant kidney model in which one kidney is removed and the upper and lower poles of the remaining kidney are also removed. Therefore, the rat is left with one-sixth of the original kidney mass, and this remaining renal tissue undergoes a well-characterized hypertrophy followed by progressive loss of renal function associated with glomerular sclerosis, interstitial fibrosis, and proteinuria. In this model, there was an early angiogenic response with proliferation of peritubular and glomerular ECs. However, this was short-lived, and EC proliferation then decreased to below levels seen in sham operated control animals. Reduction in EC proliferation was associated with enhanced interstitial expression of the antiangiogenic factor thrombospondin (TSP)-1 and also a decrease in VEGF production (5). These changes were also correlated with progressive macrophage infiltration. Additional evidence indicated that inflammatory cytokines, including IL-1β, IL-6 and TNF-α, inhibited *VEGF* mRNA expression and protein secretion in cultured renal tubular epithelial cells. Furthermore, treatment of animals with this progressive renal disease with intravenous VEGF resulted in minimized functional deterioration and lowered mortality rates. This was associated with a marked increase in glomerular EC proliferation, peritubular capillary EC proliferation, and a twofold increase in eNOS expression. The VEGF-treated rats also exhibited less type III collagen interstitial deposition and a reduction in epithelial cell injury (22).

GOALS FOR BRIDGING BENCH-TO-BEDSIDE GAP

Clearly, understanding renal EC biology is in its infancy. Over the past 5 years the renal endothelium has been shown to be important in both acute and chronic renal disease processes involving both glomerular and peritubular capillary networks. This study will lead to enhanced investigation and adaptation of techniques and approaches developed for other vascular beds. Understanding baseline phenotypic and genotypic differences in the diverse endothelial populations within the kidney will provide insight into their role in physiological processes. This same approach following acute and chronic stimuli will pave the way for therapeutic advances. Preventing endothelial injury, enhancing its recovery, and maintaining homeostasis of the endothelium will all be important. However, there is a long way to go to improve our basic understanding before therapeutic advances will be possible.

KEY POINTS

- The renal endothelium plays a heretofore underappreciated primary role in both acute and chronic kidney diseases.
- Microvascular dropout is a primary event in chronic kidney disease progression resulting from a number of different disease processes.
- Glomerular capillary injury, GFR, and proteinuria are all related to glomerular EC alterations.

- In acute renal failure secondary to ischemia and sepsis, the microvascular endothelium plays a central role in injury and the ability of the kidney to recover.

Future Goals

- To enhance understanding of the phenotypic and genotypic differences and responses of renal ECs
- To develop techniques to visualize and manipulate renal ECs in vivo with greater sensitivity and specificity

REFERENCES

1 Pallone TL, Turner MR, Edwards A, Jamison RL. Countercurrent exchange in the renal medulla. *Am J Physiol Regul Integr Comp Physiol*. 2003; 284:R1153–R1175.

2 Rops AL, van der Vlag J, Jacobs CW, et al. Isolation and characterization of conditionally immortalized mouse glomerular endothelial cell lines. *Kidney Int*. 2004;66:2193–2201.

3 Ballermann BJ. Glomerular endothelial cell differentiation. *Kidney Int*. 2005;67:1668–1671.

4 Roberts WG, Palade GE. Increased microvascular permeability and endothelial fenestration induced by vascular endothelial growth factor. *J Cell Sci*. 1995;108(Pt 6):2369–2379.

5 Kang DH, Joly AH, Oh SW, et al. Impaired angiogenesis in the remnant kidney model: I. Potential role of vascular endothelial growth factor and thrombospondin-1. *J Am Soc Nephrol*. 2001; 12:1434–1447.

6 Kim YG, Suga SI, Kang DH, et al. Vascular endothelial growth factor accelerates renal recovery in experimental thrombotic microangiopathy. *Kidney Int*. 2000;58:2390–2399.

7 Kitamoto Y, Tokunaga H, Tomita K. Vascular endothelial growth factor is an essential molecule for mouse kidney development: glomerulogenesis and nephrogenesis. *J Clin Invest*. 1997;99: 2351–2357.

8 Muczynski KA, Ekle DM, Coder DM, Anderson SK. Normal human kidney HLA-DR-expressing renal microvascular endothelial cells: characterization, isolation, and regulation of MHC class II expression. *J Am Soc Nephrol*. 2003;14:1336–1348.

9 Maynard SE, Min JY, Merchan J, et al. Excess placental soluble fms-like tyrosine kinase 1 (sFlt1) may contribute to endothelial dysfunction, hypertension, and proteinuria in preeclampsia. *J Clin Invest*. 2003;111:649–658.

10 Gazzaniga S, Gonzalez L, Mantovani A, et al. Isolation and molecular characterization of a mouse renal microvascular endothelial cell line. *In Vitro Cell Dev Biol Anim*. 2004;40:82–88.

11 Eremina V, Sood M, Haigh J, et al. Glomerular-specific alterations of VEGF-A expression lead to distinct congenital and acquired renal diseases. *J Clin Invest*. 2003;111:707–716.

12 Ishii Y, Sawada T, Kubota K, et al. Injury and progressive loss of peritubular capillaries in the development of chronic allograft nephropathy. *Kidney Int*. 2005;67:321–332.

13 Woywodt A, Schroeder M, Gwinner W, et al. Elevated numbers of circulating endothelial cells in renal transplant recipients. *Transplantation*. 2003;76:1–4.

14 Sutton TA, Fisher CJ, Molitoris BA. Microvascular endothelial injury and dysfunction during ischemic acute renal failure. *Kidney Int*. 2002; 62:1539–1549.

15 Sutton TA, Mang HE, Campos SB, et al. Injury of the renal microvascular endothelium alters barrier function after ischemia. *Am J Physiol Renal Physiol*. 2003;285:F191–F198.

16 Ysebaert DK, De Greef KE, De Beuf A, et al. T cells as mediators in renal ischemia/reperfusion injury. *Kidney Int*. 2004;66:491–496.

17 Friedewald JJ, Rabb H. Inflammatory cells in ischemic acute renal failure. *Kidney Int*. 2004;66:486–491.

18 De Greef KE, Ysebaert DK, Persy V, et al. ICAM-1 expression and leukocyte accumulation in inner stripe of outer medulla in early phase of ischemic compared to HgCl2-induced ARF. *Kidney Int*. 2003;63:1697–1707.

19 Brodsky SV, Yamamoto T, Tada T, et al. Endothelial dysfunction in ischemic acute renal failure: rescue by transplanted ECs. *Am J Physiol Renal Physiol*. 2002;282:F1140–F1149.

20 Spurgeon KR, Donohoe DL, Basile DP. Transforming growth factor-beta in acute renal failure: receptor expression, effects on proliferation, cellularity, and vascularization after recovery from injury. *Am J Physiol Renal Physiol*. 2005;288:F568–F577.

21 Burne-Taney MJ, Yokota N, Rabb H. Persistent renal and extrarenal immune changes after severe ischemic injury. *Kidney Int*. 2005;67:1002–1009.

22 Kang DH, Hughes J, Mazzali M, et al. Impaired angiogenesis in the remnant kidney model: II. Vascular endothelial growth factor administration reduces renal fibrosis and stabilizes renal function. *J Am Soc Nephrol*. 2001;12:1448–1457.

139

Uremia

Jan T. Kielstein*,† and Danilo Fliser*

*Medical School of Hannover, Germany;
†Stanford University School of Medicine, California

The uremic syndrome can be defined as the deterioration of multiple biochemical and physiological functions in parallel with progressive renal failure. A myriad of compounds, termed *uremic toxins,* lead to a complex and variable symptomatology – the uremic syndrome. One hallmark of this syndrome is the rapid development of cardiovascular disease. A cornerstone in the complex pathogenesis of this cardio–renal interaction is the endothelium, because it controls many aspects of vascular function. The endothelium produces a wide range of regulatory molecules that, in health, provide a carefully balanced antiatherogenic environment. In contrast, endothelial dysfunction has been demonstrated repeatedly in renal failure, is present in the absence of anatomically obvious disease, and appears to be useful in the prediction of morbidity and mortality in other cardiovascular risk groups. One of the most important and intensively studied mediators of endothelial function is nitric oxide (NO). Numerous preclinical and clinical studies have shown that NO bioavailability is reduced in chronic kidney disease (CKD). NO deficiency contributes to the progressive nature of CKD, endothelial dysfunction, and associated risk for cardiovascular events. Mechanisms of NO deficiency are likely multifactorial. They include substrate limitation, competitive inhibition of NO synthase (NOS) by endogenous NOS inhibitors, and premature quenching of NO by free radicals. Recent evidence points to an important role for the NOS inhibitor, asymmetric dimethylarginine (ADMA). This compound has a wide range of actions that are consistent not only with its role in uremia pathophysiology but also as a common pathway through which cardiovascular risk factors exert their deleterious effects. Unlike other uremic toxins, it also predicts for progression of renal disease. The goal of this chapter is to review the role of uremic toxins in mediating endothelial cell (EC) dysfunction, with a particular emphasis on ADMA. Understanding the pathways involved in the accumulation of ADMA may provide a foundation for therapeutic intervention in the cardio–renal syndrome.

HISTORY

One of the first descriptions of uremia comes from Hippocrates, who wrote that "If a dropsical patient be seized with hiccup the case is hopeless" (1). Aretaeus, the Cappadocian, who worked in Alexandria in approximately the 1st century B.C. wrote, "Certain persons pass blood periodically; they are very pale, without appetite. They are languid and relaxed in their limbs, but light and agile in their head. Their eyes become dull, dim and rolling; hence may become epileptic; others are swollen, misty, dropsical; and others again are affected with melancholy and paralyses" (2). Despite this detailed description of what later turned out to be uremia, these authors did not even speculate on the underlying pathogenetic mechanisms. In his book *De Sedibus et Causis Morborum per Anatomen Indagitis,* Giovanni Battista Morgagni (1682–1771) was the first to propose that specific signs and symptoms are linked to particular anatomical changes at autopsy. In *De Sedibus,* letter XXX, Morgagni describes a patient who, in the last few months of his life, presented with symptoms of nausea, vomiting, and headache associated with episodes of loss of consciousness. The most interesting autopsy finding in this case was the description of two greatly shrunken, hard, irregularly shaped, grayish kidneys. Morgagni commented on his observations by stating that he had treated other patients with the same symptoms and that, on autopsy, they showed the same morphological picture. He therefore concluded that this renal pathology was the cause of the symptoms described (3). In 1827, Richard Bright (1789–1858) published his outstanding book *Reports of Medical Cases,* in which he addressed the same topic in considerable detail and presented 24 case reports of patients who had proteinuria and edema during life. Seventeen of them subsequently died, and at autopsy 8 of them had small "granulated" kidneys (reviewed in 4). Bright, by deriving logical scientific conclusions based on simple but well-structured clinical observations, was also the first to describe "dangerous secondary consequences which may destroy the patient at any period of the disease" in several of his patients. He classified these secondary consequences into two categories: (a) inflammation of serous or mucous membranes (such as the pericardium or pleura), and (b) cerebral hemorrhage – problems that plague patients with chronic renal disease to this day. In collaboration with William Prout (1785–1850), Bright was able to identify that the level of urea in the blood and tissues of one of his patients was particularly high, while the level in the urine was "very small." Together, Bright and Prout raised the possibility of gross chemical abnormalities occurring concurrently with the development of granulated kidneys (reviewed in 4). This suggested a chemical dimension to the clinicopathological syndrome that had so quickly gained acceptance. Jean Prévost (1790–1850) and Jean Dumas (1800–1884) in Geneva noted in 1821 that in animals the blood urea level rose markedly following bilateral nephrectomy (reviewed in 5). This marked the development of the renal function assay as it is still used today. In 1827, Friedrich Wöhler (1800–1882) in Berlin succeeded in synthesizing urea. His achievement attracted considerable attention and fortuitously occurred the same year that Richard Bright published his clinicopathological observations related to renal disease. Together, these early, seminal findings paved the way for investigating the complex syndrome of uremia.

MODERN DEFINITION OF UREMIA

The word *uremia* means "urine in the blood" and was first coined by Piorry and Lheritier in 1840 (6) to indicate the intoxication resulting from insufficient purification of the blood by the kidneys. The short and descriptive nature of the word belies the complexity of the underlying clinical syndrome (Figure 139.1). Uremia is associated with a multitude of metabolic abnormalities including fluid, electrolyte, and hormone imbalances that develop in parallel with deterioration

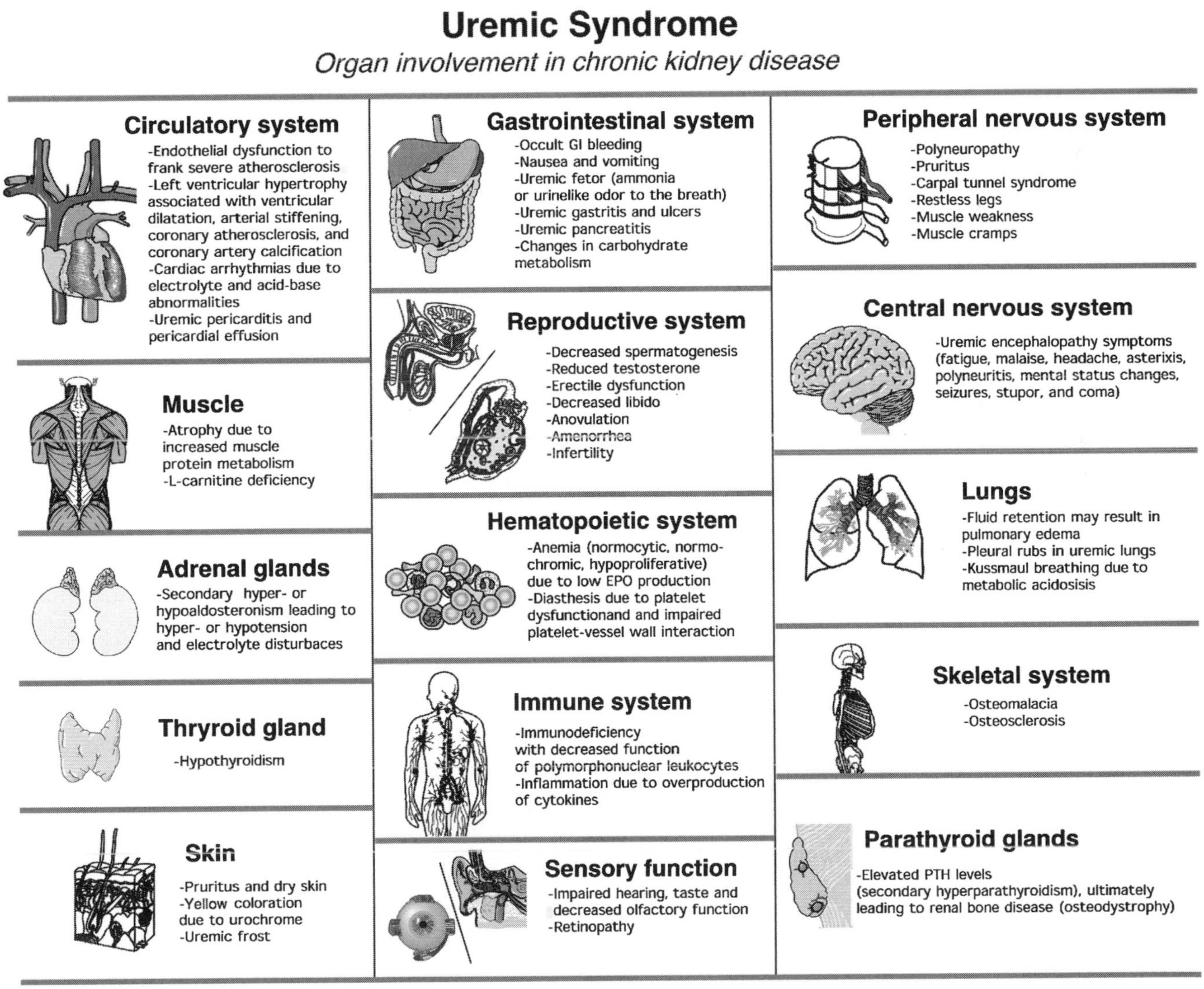
Uremic Syndrome
Organ involvement in chronic kidney disease
Circulatory system
-Endothelial dysfunction to frank severe atherosclerosis
-Left ventricular hypertrophy associated with ventricular dilatation, arterial stiffening, coronary atherosclerosis, and coronary artery calcification
-Cardiac arrhythmias due to electrolyte and acid-base abnormalities
-Uremic pericarditis and pericardial effusion
Muscle
-Atrophy due to increased muscle protein metabolism
-L-carnitine deficiency
Adrenal glands
-Secondary hyper- or hypoaldosteronism leading to hyper- or hypotension and electrolyte disturbaces
Thryroid gland
-Hypothyroidism
Skin
-Pruritus and dry skin
-Yellow coloration due to urochrome
-Uremic frost
Gastrointestinal system
-Occult GI bleeding
-Nausea and vomiting
-Uremic fetor (ammonia or urinelike odor to the breath)
-Uremic gastritis and ulcers
-Uremic pancreatitis
-Changes in carbohydrate metabolism
Reproductive system
-Decreased spermatogenesis
-Reduced testosterone
-Erectile dysfunction
-Decreased libido
-Anovulation
-Amenorrhea
-Infertility
Hematopoietic system
-Anemia (normocytic, normochromic, hypoproliferative) due to low EPO production
-Diasthesis due to platelet dysfunctionand and impaired platelet-vessel wall interaction
Immune system
-Immunodeficiency with decreased function of polymorphonuclear leukocytes
-Inflammation due to overproduction of cytokines
Sensory function
-Impaired hearing, taste and decreased olfactory function
-Retinopathy
Peripheral nervous system
-Polyneuropathy
-Pruritus
-Carpal tunnel syndrome
-Restless legs
-Muscle weakness
-Muscle cramps
Central nervous system
-Uremic encephalopathy symptoms (fatigue, malaise, headache, asterixis, polyneuritis, mental status changes, seizures, stupor, and coma)
Lungs
-Fluid retention may result in pulmonary edema
-Pleural rubs in uremic lungs
-Kussmaul breathing due to metabolic acidosisis
Skeletal system
-Osteomalacia
-Osteosclerosis
Parathyroid glands
-Elevated PTH levels (secondary hyperparathyroidism), ultimately leading to renal bone disease (osteodystrophy)

Figure 139.1. Overview of the organ involvement in chronic kidney disease/uremic syndrome. (Adapted with permission from *Dialysefibel 2.2*, 2nd edition. Günther Schönweiss, Bad Kissingen, abakiss, 1996.)

Table 139-1: Stages of Chronic Kidney Disease

Stage of Chronic Kidney Disease	*Description*	*Glomerular Filtration Rate (GFR) in mL/min/1.73 m²*
1	Kidney damage with normal or ↑ GFR	≥90
2	Kidney damage with mild ↓ GFR	60–89
3	Kidney damage with moderate ↓ GFR	30–59
4	Severe ↓ GFR	15–29
5	Kidney failure	<15 (or dialysis)

of renal function. Some of these abnormalities are reflected in the choice of synonyms that have been used for uremia (e.g., urinary poisoning, chronic Bright disease, azotemia, chronic renal insufficiency, and chronic renal disease). Uremia usually becomes clinically apparent once creatinine clearance is less than 20 mL/min, although some patients may be symptomatic at higher clearance levels. The syndrome may be heralded by the clinical onset of nausea, vomiting, fatigue, anorexia, weight loss, muscle cramps, pruritus, and change in mental status. Uremia as a clinical condition of untreated severe kidney failure is rarely seen in developed countries since the advent of dialysis treatment and renal transplantation. However, a significant segment of the world population lacks ready access to renal replacement therapy, and for individuals living in these areas uremia is a dreaded prelude to death. However, even those patients who receive optimal therapy (including renal transplantation and effective dialysis treatment) continue to demonstrate abnormal levels of so-called uremic toxins, because these therapies are aimed toward reduction and not normalization of uremic toxin levels. Thus, the terms *uremia* and *CKD* are often used interchangeably. In general, the uremic syndrome is associated with CKD, but it also may occur with acute renal failure (ARF) if loss of renal function is rapid and profound. In an attempt to define the different degrees of CKD, the National Kidney Foundation released a five-stage classification (Table 139-1) (7).

UREMIC TOXINS

Of the uremic toxins, urea is often considered the prototype. However, extensive studies addressing the possible pathophysiological role of urea in the uremic state found little evidence that urea is toxic per se. In support of this conclusion, the addition of urea to the dialysate failed to produce the typical symptoms of the uremic syndrome (8). Moreover, survival of patients on dialysis is independent of urea removal (9).

In fact, no single uremic toxin has been identified that accounts for all of the clinical manifestations of uremia. Rather an ever-expanding myriad of uremic retention solutes accumulate in the uremic patient with end-stage renal disease (ESRD), an effect that is directly or indirectly attributable to deficient renal clearance. Collectively, these retained solutes result in complex and variable symptomatology. They are referred to as uremic toxins insofar as they interact negatively with biological functions. Apart from inorganic compounds, urea, oxalic acid, parathyroid hormone (PTH), and ß2-microglobulin (ß2-M), more than 2,000 putative "uremic toxins" have been identified in uremic blood (10). In 1978 Bergstrom and Furst (11) first proposed minimal criteria for the characterization of uremic toxins: (a) they should be chemically identifiable and quantifiable (i.e., measurable) in biological fluids, (b) the concentration in uremic subjects should exceed that in nonuremic subjects, (c) the concentration should correlate with specific uremic symptoms that disappear at the normal concentration, and (d) toxic effects in a test system should be demonstrable at the concentration found in uremic subjects. All of these criteria are rarely met, however, often for technical reasons.

Recently, the concept of uremic toxins has changed, such that any substance that interferes with normal physiological functions in renal failure is recognized as a "uremic toxin." This broader definition acknowledges that concentrations in the plasma may not reflect intracellular let alone subcellular concentrations of the putative toxin. Furthermore, "normal" concentrations of various compounds may in fact be harmful to patients with CKD and lead to pathophysiology. Some compounds may play a role in the progression of renal disease and/or contribute to dialysis-related complications. Thus, the complex mosaic of clinical symptoms seen in the uremic syndrome is probably attributable to one or more of these solutes. Knowledge about the nature and kinetic behavior of the responsible compounds may provide a foundation for novel therapies.

THE PROBLEM OF RENAL FUNCTION MEASUREMENT

As we discuss later, subclinical renal disease is a risk factor for atherosclerosis, an endothelial-based disease. Thus, early diagnosis is important. Unfortunately, the reliable assessment of kidney function remains an area of uncertainty among healthcare professionals – even nephrologists. Kidney function is routinely assayed by measuring serum urea and creatinine. The relationship between the glomerular filtration rate and serum creatinine is curve-linear. Thus, serum creatinine is an insensitive marker for mild to moderate reductions in GFR. As an assay for renal function, creatinine levels are further limited by their inter-individual variability due to age, muscle mass, protein intake, and sex. Glomerular filtration rate may be estimated using formulas (e.g., Cockcroft-Gault and Modification of Diet in Renal Disease [MDRD]) of easily obtained values (plasma creatinine, serum urea nitrogen, serum albumin, age, gender, race). At present, inulin clearance is considered the gold standard for GFR measurement. Because this method is not suitable for routine use, creatinine clearance measurements are widely used. However, obtaining accurate 24-hour timed urine collections is not only labor intensive and expensive but also fraught with difficulties for patients, especially elderly ones with neurological or urological comorbidities. Thus, an important goal in the nephrology field is to identify a stable, convenient, and clinically reliable marker of GFR (1). Two markers that show promise are serum cystatin C and symmetrical dimethylarginine (SDMA). Serum cystatin C concentration is a better marker of renal dysfunction (i.e., reduced GFR) than plasma creatinine concentration, at least in elderly subjects with plasma creatinine concentrations within the normal range (12). SDMA has been found to be a reliable marker of renal function (13,14), but rather cumbersome and expensive equipment is required for its measurement.

THE EPIDEMIC OF CHRONIC KIDNEY DISEASE

Approximately 11% of the United States population suffers from some degree of CKD (7). The number of individuals with kidney failure treated by dialysis and transplantation exceeded 320,000 in 1998 and is expected to surpass 650,000 by 2010. By then, renal replacement therapy will cost up to $28 billion per year (the equivalent of the entire NIH budget for 2006). Stated another way, 0.1% of the population will encumber 2% of the national health budget.

CKD stage 5 is associated with a dramatic increase in cardiovascular morbidity and mortality (15). The cardiac mortality among dialysis patients younger than 45 years is approximately 100 times greater than in the general population (16). Even patients with incipient renal disease (CKD stage 1) have been shown to have endothelial dysfunction and an increased cardiovascular risk (17), which clearly increases with the degree of renal impairment (18). In a study of over 1,000,000 individuals enrolled in the Medicare program in the United States (19), outcomes were reported for patients with CKD alone (2.2%) and in combination with diabetes (1.6%). After two years of follow-up, the rates for requiring renal replacement therapy were 1.6 and 3.4 per 100 patient-years for patients with CKD alone and those with both disorders, respectively. More devastating, by comparison, the death rates were 17.7 and 19.9 per 100 patient-years. In addition, rates for atherosclerotic vascular disease for these two groups (per 100 patient-years) were 35.7 and 49.1, whereas heart failure rates were 30.7 and 52.3, respectively, even in the absence of fluid retention. Hence, patients with CKD are more likely to die of cardiovascular disease than to reach ESRD. Therefore, current guidelines recommend that patients with CKD be considered in the "highest risk group" for subsequent cardiovascular events and that treatment recommendations based on CVD risk stratification should take into account the highest-risk status of patients with CKD.

Despite this worrisome development, the mechanisms underlying the increased risk for cardiovascular morbidity and mortality remain obscure. Results of several cross-sectional studies have suggested that the Framingham risk factors such as older age, diabetes mellitus, systolic hypertension, left ventricular hypertrophy (LVH), and low levels of high-density lipoprotein (HDL) cholesterol are insufficient to capture the extent of CVD risk in subjects with CKD. New, nontraditional risk factors, such as the endogenous NOS inhibitor ADMA, may prove to be more specific and reliable in predicting progression of CKD (stages 1–4) to dialysis dependency as well as cardiovascular events in patients with CKD (13,20,21).

ENDOTHELIAL DYSFUNCTION IN UREMIA AND ASYMMETRIC DIMETHYLARGININE – A UREMIC TOXIN PAR EXCELLENCE

A number of conditions found in patients with uremia may lead to endothelial dysfunction, including anemia, metabolic acidosis, hyperkalemia, parathyroid and vitamin D abnormalities, endocrine abnormalities, and malnutrition. Endothelial dysfunction is characterized by reduced vasodilation capacity, a proinflammatory and/or prothrombotic activity. It is characteristic of most forms of cardiovascular disease – e.g., hypertension, coronary artery disease, chronic heart failure, peripheral artery disease, diabetes, and CKD. Mechanisms that participate in the reduced vasodilatory responses in endothelial dysfunction in general include reduced NO generation, oxidative stress, and reduced production of the hyperpolarizing factor. For purposes of this discussion, we will focus on the role of ADMA as a cause of endothelial dysfunction in renal disease.

ADMA had been known to biochemists for decades as a naturally occurring analogue of L-arginine (22). In 1992, Vallance and coworkers demonstrated elevated levels of dimethylarginines in patients with ESRD (23). In this seminal report, the authors speculated that the accumulation of ADMA, which competitively inhibits NOS, might contribute to the hypertension and predisposition to infections associated with CKD.

Thus far, over 600 reports on ADMA have been published. Collectively, these studies support a role for ADMA as a uremic toxin, a powerful marker of endothelial dysfunction and atherosclerosis, and a solid predictor of mortality in selected patient populations (20,24–26) Indeed, there is increasing evidence that ADMA represents a common pathway in mediating endothelial dysfunction associated with both traditional and nontraditional cardiovascular risk factors in uremic as well as nonuremic patients (27).

NO is the most potent biological vasodilator. It plays a critical role in inhibiting key processes associated with atherosclerosis, including monocyte adhesion, platelet aggregation, and vascular smooth muscle cell proliferation. Hence, endothelial dysfunction as a result of reduced NO activity is an early step in the course of atherosclerotic vascular disease, and evidence has accumulated that inhibition of NO synthesis by endogenous substances may be causally involved in this process (27). A family of NOS with endothelial, neuronal, and macrophage isoforms converts L-arginine to NO and citrulline by stereospecific oxidation of the terminal guanidino nitrogen

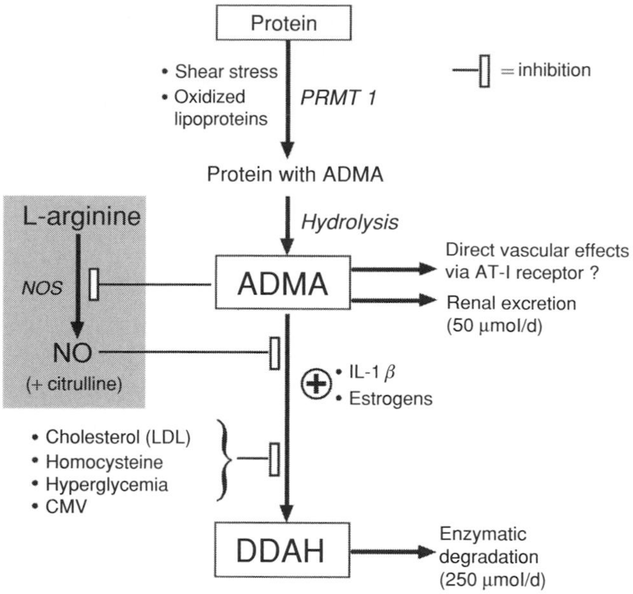

Figure 139.2. Overview of biochemical pathways for generation, elimination, and degradation of ADMA. ADMA stems from methylation of arginine residues in proteins. The reaction is catalyzed by PRMTs that transfer a methyl group to each guanidino nitrogen of an arginine residue. This reaction results in a methylated arginine derivative (protein-containing ADMA). Hydrolysis of the methylated proteins releases ADMA. All methylarginines (ADMA and SDMA) are excreted into the urine and are in part metabolized to α-keto acids by the enzyme activity of dimethylarginine pyruvate aminotransferase (DPT). The major route of ADMA elimination (250 μmol/d) occurs via degradation through the enzyme dimethylarginine dimethylaminohydrolase (DDAH). DDAH hydrolyzes ADMA to form dimethylamine and L-citrulline. Only a minor portion of ADMA (50 μmol/d) is eliminated (unchanged) in the urine. The text boxes depict factors that influence the activity of PRMT and DDAH. CMV, cytomegalovirus; IL, interleukin.

of the amino acid L-arginine (28). This process can be selectively inhibited by competitive blockade of the NOS active site with naturally occurring guanidino-substituted analogues of L-arginine such as ADMA (Figure 139.2) (29).

Humans generate approximately 300 μmol (~60 mg) ADMA per day (30). Although an analogue of L-arginine, ADMA does not appear to be directly generated from the free amino acid. Rather, ADMA is released from proteins that have been post-translationally methylated and subsequently hydrolyzed. These proteins are largely found in the nucleolus and appear to be involved in RNA processing and transcriptional control (31). Protein-arginine methyltransferases (PRMTs) catalyze the formation of methylarginine residues. A number of cell types, including human ECs, elaborate ADMA (32). To date, it is unclear whether ADMA generation is constant or whether increased PRMT activity and/or increased protein turnover might contribute to elevated ADMA levels. The latter mechanism is supported by the observation that in patients with ESRD, increased protein catabolic rate is correlated with elevated circulating levels of ADMA.

Because ADMA was found to be elevated in dialysis patients, it was believed that the compound was normally eliminated by the kidney (23). However, an early study in rabbits revealed renal excretion not to be the major route of elimination, suggesting the presence of a catabolic pathway (33). Subsequent investigations led to the identification of one such pathway, which involves dimethylarginine dimethylaminohydrolase (DDAH)–mediated hydrolysis of ADMA to dimethylamine and L-citrulline (Figure 139.2) (34). Thus far, two isoforms of DDAH have been characterized and cloned; DDAH I is predominately found in tissues that express neuronal NOS (nNOS), whereas DDAH II is localized in tissues expressing endothelial NOS (eNOS) (35).

DDAH I and II have emerged as key enzymes regulating the cellular and tissue levels of ADMA; 50 μmol ADMA per day is excreted in the urine, whereas approximately 250 μmol ADMA per day is eliminated via degradation of DDAH (30). This is in line with clinical data showing that patients with incipient renal disease and normal renal excretory function, as documented by measurement of true glomerular filtration rate using inulin clearance, have markedly elevated plasma ADMA levels (36). This could possibly be a result of decreased activity of DDAH I, which is highly expressed in brain, kidney, pancreas, and liver, or due to increased PRMT activity/protein turnover (35). In fact, a study in humans showed higher renal extraction of ADMA compared to SDMA extraction (37). Thus the kidney plays a predominant role in the elimination of ADMA, combining two mechanisms: urinary excretion as well as enzymatic degradation of ADMA.

Compelling evidence for the critical role of DDAH activity in regulating NO synthesis in vivo was demonstrated in transgenic mice overexpressing DDAH. The increased DDAH activity in these animals led to a 50% decrease of ADMA levels, which in turn increased NOS activity and thereby lowered blood pressure (38). Systemic ADMA infusion in humans resulted in a decrease in plasma cGMP and significant hemodynamic alterations, including reduced effective renal plasma flow (ERPF), sustained decrease in cardiac output, increase in systemic vascular resistance, and decrease in heart rate (39). These results are in accordance with data obtained using noninvasive techniques (30). Collectively, these preclinical and clinical findings indicate that ADMA has well-defined effects on cardiovascular and renal function in healthy subjects. It is therefore conceivable that ADMA causes sustained changes in vascular function through an intracellular action in ECs at blood concentrations found in patients with cardiovascular pathology.

ASYMMETRIC DIMETHYLARGININE AND CARDIOVASCULAR MORBIDITY AND MORTALITY IN PATIENTS WITH RENAL DISEASE

Since Lindner and coworkers first reported the association between ESRD and premature atherosclerosis/cardiovascular mortality, there has been little progress in understanding the underlying molecular mechanisms (15). Depending on age, patients on renal replacement therapy encounter a five- to 500–fold risk of dying from cardiac causes. Many studies have documented endothelial dysfunction in patients with patients undergoing dialysis (40). Given the correlation between ESRD and elevated levels of ADMA, and considering the role of this uremic toxin in inhibiting NO, it is tempting to speculate that ADMA contributes to the premature atherosclerosis/cardiovascular mortality.

Consistent with this hypothesis, a cross-sectional study revealed that ADMA is higher in dialysis patients with cardiovascular complications than in those without such complications (41). Moreover, a cohort study has demonstrated a strong and independent link between ADMA, mortality, and cardiovascular events in ESRD patients (20). In further studies, ADMA was shown (a) to be an independent risk factor for increased intima media thickness of the carotid artery (IMT) in ESRD patients (42), (b) to be an independent predictor for progression of intimal lesions (42), (c) to be strongly associated with IMT in mild to moderate chronic renal insufficiency (43), and (d) to demonstrate a strong independent link with IMT in apparently healthy individuals in the general population (44).

The Cardio–Renal Interaction

The heart and the arterial system form an integrated unit that coherently responds to hemodynamic stimuli as well as to risk factors for atherosclerosis. For example oxygen requirement is related to left ventricular geometry; coronary hypertension, with endothelial dysfunction and reduced coronary reserve; diastolic dysfunction; and structural remodeling of myocardium and vascular bed. Accordingly, in physiological and in disease states – e.g., uremia – cardiac and arterial remodeling proceed in parallel (45,46). There are several lines of evidence, which when considered collectively suggest that ADMA-induced EC dysfunction is causally related to development of LVH in dialysis patients. First, in patients with hypertension, there is a strong association between endothelial dysfunction and concentric LVH (47). Second, in patients on dialysis, the reactive hyperemic response of forearm blood flow (an outcome measure strongly influenced by endothelial function) is inversely correlated with common carotid IMT and left ventricular mass (48). Third, there is as strong independent association between ADMA levels and left ventricular mass in patients on maintenance hemodialysis (49). Fourth, ADMA is much higher in dialysis patients with concentric LVH than in those with eccentric LVH or normal left ventricular mass (Figure 139.2). Fifth, concentric LVH occurs in experimental models of chronic NO inhibition (50). Finally, severe concentric hypertrophy and vascular damage is a hallmark in the knockout mouse lacking the *eNOS* and the *ApoE* genes (51). Thus, ADMA may play a pathophysiological role in diverse cardiovascular conditions, including coronary artery disease, carotid artery disease, and heart failure.

ASYMMETRIC DIMETHYLARGININE AND MINOR RENAL DYSFUNCTION

Even minor renal dysfunction has recently been recognized as a cardiovascular risk factor (7). Interestingly, several studies found markedly elevated plasma ADMA levels not only in patients on dialysis but also in patients with progressive CDK and even in patients with incipient renal disease and normal renal function (52) Accumulation of ADMA may cause impaired acetylcholine-induced endothelium-dependent vasorelaxation in patients with incipient renal disease and normal renal function (17). In humans, as in many species, the process of aging results in deterioration of renal function even without primary renal disease. Previous studies have documented that aging affects renal hemodynamics, as manifested by increased renovascular tone and reduced dilatation of postglomerular vessels in response to stimuli such as acetylcholine or amino acids (53–56). Furthermore, in senescent individuals, reduced availability of NO is thought to be linked to the increase in blood pressure and renovascular resistance, possibly a reflection of arteriosclerosis (57,58). In comparison to concentrations in young healthy adults, ADMA concentrations have been markedly increased in the normotensive elderly and particularly in the hypertensive elderly (59). The significant relationship between high ADMA blood levels, reduced renal perfusion, and high blood pressure values is compatible with the notion that accumulation of this NOS inhibitor in senescent individuals is involved in the decrease of renal perfusion and increase of blood pressure. Recent studies indicate that ADMA may have far-reaching effects on the senescence process because in vitro it shortens telomere length by activating telomerase (60).

ASYMMETRIC DIMETHYLARGININE INHIBITS VASCULAR REPAIR

Besides leading to renal damage, ADMA might also be involved in inhibiting repair of renal lesions by endothelial progenitor cells (EPCs). EPCs have recently come into focus in cardiovascular research because they are thought to be responsible for endothelial repair (61). EPCs circulate in the cardiovascular system, where they home and incorporate into sites of active neovascularization. In fact, EPCs orchestrate re-endothelialization of damaged vessel walls, also by secreting a large number of important cytokines, which attract and govern cells that are indispensable in the process of endothelial repair. A recent study suggests that increased endothelial NO availability is required for improvement of endothelial progenitor cell mobilization, myocardial repair, and neovascularization after myocardial infarction (62). An increased ADMA level and thereby a decreased NO production may lead to a deficiency of EPCs. Indeed, a study by Thum et al. showed an inverse correlation between ADMA plasma concentration and circulating EPC levels in patients with coronary artery disease (63).

The pathophysiological relevance of ADMA may go even beyond its interference with cardiovascular function. For example, there is experimental evidence that ADMA may be implicated in dementia, endocrine alterations, osteoporosis, and fertility as well – i.e. in alterations that are also typically found in patients with advanced renal insufficiency.

KEY POINTS

- The uremic condition – i.e., CKD – is a major cardiovascular risk factor.
- Accelerated atherosclerosis due to impaired renal function is the result of the complex interaction of many compound, uremic toxins, most of which lead to reduced bioavailability of NO.
- Identifying patients at risk using reliable markers of renal function as well as better predictors of cardiovascular mortality is still an unmet need.
- Endogenous inhibition of NO synthase – e.g., via ADMA – seems to be a marker and mediator of endothelial dysfunction in uremia as well as in nonrenal patients.

Future Goals

- To foster public awareness, reliable and feasible markers of renal function, and advances in understanding the pathophysiology of injury and the repair mechanisms – e.g., by using bone marrow–derived cells – as a prerequisite to successfully meet the unique challenge of cardiovascular disease in CKD

REFERENCES

1 Adams F. *The Genuine Works of Hippocrates*. Baltimore: Williams & Wilkins, 1946.

2 Poulakou-Rebelakou E. Aretaeus on the kidney and urinary tract diseases 1. *Am J Nephrol.* 1997;17(3–4):209–213.

3 Antonello A, Calo L, Bonfante L, et al. Giovan Battista Morgagni, a pioneer of clinical nephrology. *Am J Nephrol.* 1999;19(2):222–225.

4 George CR. Development of the idea of chronic renal failure. *Am J Nephrol.* 2002;22(2–3):231–239.

5 Richet G. Early history of uremia. *Kidney Int*. 1988;33(5):1013–1015.

6 Piorry PA IHD. *Traite des alt'erations du sang*, 1st ed. Paris: Bury & JB Bailbie're, 1840.

7 Sarnak MJ, Levey AS, Schoolwerth AC, et al. Kidney disease as a risk factor for development of cardiovascular disease: a statement from the American Heart Association Councils on Kidney in Cardiovascular Disease, High Blood Pressure Research, Clinical Cardiology, and Epidemiology and Prevention. *Circulation.* 2003;108(17):2154–2169.

8 Johnson WJ, Hagge WW, Wagoner RD, et al. Effects of urea loading in patients with far-advanced renal failure. *Mayo Clin Proc*. 1972;47(1):21–29.

9 Eknoyan G, Beck GJ, Cheung AK, et al. Effect of dialysis dose and membrane flux in maintenance hemodialysis. *N Engl J Med.* 2002;347(25):2010–2019.

10 Vanholder R, De SR, Glorieux G, et al. Review on uremic toxins: classification, concentration, and interindividual variability. *Kidney Int*. 2003;63(5):1934–1943.

11 Bergstrom J, Furst P. Uremic toxins. *Kidney Int Suppl.* 1978;(8):S9–S12.

12 Fliser D, Ritz E. Serum cystatin C concentration as a marker of renal dysfunction in the elderly. *Am J Kidney Dis.* 2001;37(1):79–83.

13 Fliser D, Kronenberg F, Kielstein JT, et al. Asymmetric dimethylarginine and progression of chronic kidney disease: the mild to moderate kidney disease study. *J Am Soc Nephrol.* 2005;16(8):2456–2461.

14 Marescau B, Nagels G, Possemiers I, et al. Guanidino compounds in serum and urine of nondialyzed patients with chronic renal insufficiency. *Metabolism.* 1997;46(9):1024–1031.

15 Lindner A, Charra B, Sherrard DJ, Scribner BH. Accelerated atherosclerosis in prolonged maintenance hemodialysis. *N Engl J Med.* 1974;290(13):697–701.

16 Levey AS, Beto JA, Coronado BE, et al. Controlling the epidemic of cardiovascular disease in chronic renal disease: what do we know? What do we need to learn? Where do we go from here? National Kidney Foundation Task Force on Cardiovascular Disease. *Am J Kidney Dis.* 1998;32(5):853–906.

17 Wang D, Iversen J, Strandgaard S. Endothelium-dependent relaxation of small resistance vessels is impaired in patients with autosomal dominant polycystic kidney disease. *J Am Soc Nephrol.* 2000;11(8):1371–1376.

18 Yilmaz MI, Saglam M, Caglar K, et al. The determinants of endothelial dysfunction in CKD: oxidative stress and asymmetric dimethylarginine. *Am J Kidney Dis.* 2006;47(1):42–50.

19 Foley RN, Murray AM, Li S, et al. Chronic kidney disease and the risk for cardiovascular disease, renal replacement, and death in the United States Medicare population, 1998 to 1999. *J Am Soc Nephrol.* 2005;16(2):489–495.

20 Zoccali C, Bode-Boger S, Mallamaci F, et al. Plasma concentration of asymmetrical dimethylarginine and mortality in patients with end-stage renal disease: a prospective study. *Lancet.* 2001;358(9299):2113–2117.

21 Ravani P, Tripepi G, Malberti F, et al. Asymmetrical dimethylarginine predicts progression to dialysis and death in patients with chronic kidney disease: a competing risks modeling approach. *J Am Soc Nephrol.* 2005;16(8):2254–2256.

22 Kakimoto Y, Akazawa S. Isolation and identification of N-G, N-G- and N-G,N′-G-dimethyl-arginine, N-epsilon-mono-, di-, and trimethyllysine, and glucosylgalactosyl- and galactosyl-delta-hydroxylysine from human urine. *J Biol Chem.* 1970;245(21):5751–5758.

23 Vallance P, Leone A, Calver A, et al. Accumulation of an endogenous inhibitor of nitric oxide synthesis in chronic renal failure. *Lancet*. 1992;339(8793):572–575.

24 Kielstein JT, Bode-Boger SM, Hesse G, et al. Asymmetrical dimethylarginine in idiopathic pulmonary arterial hypertension. *Arterioscler Thromb Vasc Biol.* 2005;25(7):1414–1418.

25 Schnabel R, Blankenberg S, Lubos E, et al. Asymmetric dimethylarginine and the risk of cardiovascular events and death in

patients with coronary artery disease: results from the AtheroGene Study. *Circ Res.* 2005;97(5):E53–E59.
26 Valkonen VP, Paiva H, Salonen JT, et al. Risk of acute coronary events and serum concentration of asymmetrical dimethylarginine. *Lancet.* 2001;358(9299):2127–2128.
27 Cooke JP. Asymmetrical dimethylarginine: the Uber marker? *Circulation.* 2004;109(15):1813–1818.
28 Moncada S, Higgs A. The L-arginine-nitric oxide pathway. *N Engl J Med.* 1993;329(27):2002–2012.
29 Forstermann U, Closs EI, Pollock JS, et al. Nitric oxide synthase isozymes. Characterization, purification, molecular cloning, and functions. *Hypertension.* 1994;23(6 Pt 2):1121–1131.
30 Achan V, Broadhead M, Malaki M, et al. Asymmetric dimethylarginine causes hypertension and cardiac dysfunction in humans and is actively metabolized by dimethylarginine dimethylaminohydrolase. *Arterioscler Thromb Vasc Biol.* 2003;23(8):1455–1459.
31 Najbauer J, Johnson BA, Young AL, Aswad DW. Peptides with sequences similar to glycine, arginine-rich motifs in proteins interacting with RNA are efficiently recognized by methyltransferase(s) modifying arginine in numerous proteins. *J Biol Chem.* 1993;268(14):10501–10509.
32 MacAllister RJ, Fickling SA, Whitley GS, Vallance P. Metabolism of methylarginines by human vasculature; implications for the regulation of nitric oxide synthesis. *Br J Pharmacol.* 1994;112(1):43–48.
33 McDermott JR. Studies on the catabolism of Ng-methylarginine, Ng, Ng-dimethylarginine and Ng, Ng-dimethylarginine in the rabbit. *Biochem J.* 1976;154(1):179–184.
34 Ogawa T, Kimoto M, Sasaoka K. Purification and properties of a new enzyme, NG,NG-dimethylarginine dimethylaminohydrolase, from rat kidney. *J Biol Chem.* 1989;264(17):10205–10209.
35 Leiper JM, Santa MJ, Chubb A, et al. Identification of two human dimethylarginine dimethylaminohydrolases with distinct tissue distributions and homology with microbial arginine deiminases. *Biochem J.* 1999;343(Pt 1):209–214.
36 Kielstein JT, Boger RH, Bode-Boger SM, et al. Marked increase of asymmetric dimethylarginine in patients with incipient primary chronic renal disease. *J Am Soc Nephrol.* 2002;13(1):170–176.
37 Nijveldt RJ, Van Leeuwen PA, Van Guldener C, et al. Net renal extraction of asymmetrical (ADMA) and symmetrical (SDMA) dimethylarginine in fasting humans. *Nephrol Dial Transplant.* 2002;17(11):1999–2002.
38 Dayoub H, Achan V, Adimoolam S, et al. Dimethylarginine dimethylaminohydrolase regulates nitric oxide synthesis: genetic and physiological evidence. *Circulation.* 2003;108(24):3042–3047.
39 Kielstein JT, Impraim B, Simmel S, et al. Cardiovascular effects of systemic nitric oxide synthase inhibition with asymmetrical dimethylarginine in humans. *Circulation.* 2004;109(2):172–177.
40 Endemann DH, Schiffrin EL. Endothelial dysfunction. *J Am Soc Nephrol.* 2004;15(8):1983–1992.
41 Kielstein JT, Boger RH, Bode-Boger SM, et al. Asymmetric dimethylarginine plasma concentrations differ in patients with end-stage renal disease: relationship to treatment method and atherosclerotic disease. *J Am Soc Nephrol.* 1999;10(3):594–600.
42 Zoccali C, Benedetto FA, Maas R, et al. Asymmetric dimethylarginine, C-reactive protein, and carotid intima- media thickness in end-stage renal disease. *J Am Soc Nephrol.* 2002;13(2):490–496.
43 Nanayakkara PW, Teerlink T, Stehouwer CD, et al. Plasma asymmetric dimethylarginine (ADMA) concentration is independently associated with carotid intima-media thickness and plasma soluble vascular cell adhesion molecule-1 (sVCAM-1) concentration in patients with mild-to-moderate renal failure. *Kidney Int.* 2005;68(5):2230–2236.
44 Miyazaki H, Matsuoka H, Cooke JP, et al. Endogenous nitric oxide synthase inhibitor: a novel marker of atherosclerosis. *Circulation.* 1999;99(9):1141–1146.
45 Benedetto FA, Mallamaci F, Tripepi G, Zoccali C. Prognostic value of ultrasonographic measurement of carotid intima media thickness in dialysis patients. *J Am Soc Nephrol.* 2001;12(11):2458–2464.
46 London GM, Guerin AP, Marchais SJ, et al. Cardiac and arterial interactions in end-stage renal disease. *Kidney Int.* 1996;50(2):600–608.
47 Perticone F, Maio R, Ceravolo R, et al. Relationship between left ventricular mass and endothelium-dependent vasodilation in never-treated hypertensive patients. *Circulation.* 1999;99(15):1991–1996.
48 Pannier B, Guerin AP, Marchais SJ, et al. Postischemic vasodilation, endothelial activation, and cardiovascular remodeling in end-stage renal disease. *Kidney Int.* 2000;57(3):1091–1099.
49 Zoccali C, Mallamaci F, Maas R, et al. Left ventricular hypertrophy, cardiac remodeling and asymmetric dimethylarginine (ADMA) in hemodialysis patients. *Kidney Int.* 2002;62(1):339–345.
50 Sladek T, Gerova M, Znojil V, Devat L. Morphometric characteristics of cardiac hypertrophy induced by long-term inhibition of NO synthase. *Physiol Res.* 1996;45(4):335–338.
51 Kuhlencordt PJ, Gyurko R, Han F, et al. Accelerated atherosclerosis, aortic aneurysm formation, and ischemic heart disease in apolipoprotein E/endothelial nitric oxide synthase double-knockout mice. *Circulation.* 2001;104(4):448–454.
52 Kielstein JT, Zoccali C. Asymmetric dimethylarginine: a cardiovascular risk factor and a uremic toxin coming of age? *Am J Kidney Dis.* 2005;46(2):186–202.
53 Fliser D, Zeier M, Nowack R, Ritz E. Renal functional reserve in healthy elderly subjects. *J Am Soc Nephrol.* 1993;3(7):1371–1377.
54 Fliser D, Franek E, Joest M, et al. Renal function in the elderly: impact of hypertension and cardiac function. *Kidney Int.* 1997;51(4):1196–1204.
55 Fuiano G, Sund S, Mazza G, et al. Renal hemodynamic response to maximal vasodilating stimulus in healthy older subjects. *Kidney Int.* 2001;59(3):1052–1058.
56 Hollenberg NK, Adams DF, Solomon HS, et al. Senescence and the renal vasculature in normal man. *Circ Res.* 1974;34(3):309–316.
57 Campo C, Lahera V, Garcia-Robles R, et al. Aging abolishes the renal response to L-arginine infusion in essential hypertension. *Kidney Int Suppl.* 1996;55:S126–S128.
58 Higashi Y, Oshima T, Ozono R, et al. Aging and severity of hypertension attenuate endothelium-dependent renal vascular relaxation in humans. *Hypertension.* 1997;30(2 Pt 1):252–258.
59 Kielstein JT, Bode-Boger SM, Frolich JC, et al. Asymmetric dimethylarginine, blood pressure, and renal perfusion in elderly subjects. *Circulation.* 2003;107(14):1891–1895.
60 Scalera F, Borlak J, Beckmann B, et al. Endogenous nitric oxide synthesis inhibitor asymmetric dimethyl L-arginine accelerates endothelial cell senescence. *Arterioscler Thromb Vasc Biol.* 2004;24(10):1816–1822.

61 Kalka C, Masuda H, Takahashi T, et al. Transplantation of ex vivo expanded endothelial progenitor cells for therapeutic neovascularization. *Proc Natl Acad Sci USA*. 2000;97(7):3422–3427.
62 Landmesser U, Engberding N, Bahlmann FH, et al. Statin-induced improvement of endothelial progenitor cell mobilization, myocardial neovascularization, left ventricular function, and survival after experimental myocardial infarction requires endothelial nitric oxide synthase. *Circulation*. 2004;110(14):1933–1939.
63 Thum T, Tsikas D, Stein S, et al. Suppression of endothelial progenitor cells in human coronary artery disease by the endogenous nitric oxide synthase inhibitor asymmetric dimethylarginine. *J Am Coll Cardiol*. 2005;46(9):1693–1701.

140

The Influence of Dietary Salt Intake on Endothelial Cell Function

Paul W. Sanders

University of Alabama at Birmingham,
Department of Veterans Affairs Medical Center, Birmingham

The past two decades of endothelium-related research have altered our concepts of the function of this organ significantly. Endothelial cells (ECs) are metabolically active input–output sensors that detect and respond to mechanical and biochemical changes in the extracellular environment. Recent work has further promoted the concept that the endothelium can react to changes in dietary NaCl (also referred to as salt in this chapter) intake independently of blood pressure through changes in the extracellular milieu, particularly shear force. EC function appears to be a critical factor involved in the vascular response to changes in salt intake.

DIETARY SALT ALTERS ENDOTHELIAL CELL PRODUCTION OF TRANSFORMING GROWTH FACTOR-β AND NITRIC OXIDE

Initial studies examined the effect of salt intake on production of transforming growth factor (TGF)-β in the kidney. Expression of all three mammalian TGF-β family members – TGF-β1, β2, and β3 – increased in renal cortical tissue from rats fed a diet containing 8.0% NaCl (high salt), compared to tissue from rats maintained on a 0.3% NaCl (low salt) diet (1). Subsequent experiments focused on production of TGF-β1. Steady-state mRNA levels and production of total and active TGF-β1 were increased in glomerular preparations (1) and aortic rings (2) from rats on the high-salt diet, compared to preparations from rats on the low-salt diet. Increased TGF-β1 levels were no longer detected when aortic ring preparations were denuded of their endothelium, demonstrating that the endothelium was the source of the TGF-β1. Immunohistochemical analyses of aorta from rats on the high-salt, but not low-salt, diet revealed increased nuclear accumulation of phosphorylated Smad2/3 in the EC lining (Figure 140.1A), suggesting that TGF-β1 is signaling in ECs. Taken together, these data imply that in addition to potential paracrine effects on adjacent cells, including vascular smooth muscle, mesangial, and glomerular epithelial cells, TGF-β1 acts in an autocrine manner in ECs of the glomerulus. Although the effect of salt intake on TGF-β production was not specific to glomerular ECs – it was also observed in aortic ECs – the magnitude of the effect was greater in the glomerular preparations.

Previous studies in rats (3–6) and healthy humans (7) have demonstrated an association between increased salt intake and elevated nitric oxide (NO) production. Increased NO production permits renovascular vasorelaxation (particularly of the afferent arteriole), augments glomerular filtration rate, and improves the pressure–natriuresis curve, facilitating salt excretion (8,9). Gene transfer of the endothelial isoform of NO synthase (NOS3; also termed eNOS) into the thick ascending limb of the loop of Henle of mice with genetic deletion of NOS3 restored L-arginine-mediated NO production and inhibition of NaCl transport (10). In rats, an increase in dietary salt intake was shown to increase steady-state mRNA and protein levels of NOS3 in preparations of glomeruli and aortic rings. The increase in NOS3 protein was associated with increased baseline and calcium-ionophore–stimulated NO production (2,11). Inhibition of NO with G-(N)-nitro-L-arginine-methyl ester (L-NAME) in rats produced salt retention and salt-sensitive hypertension (12) and, if protracted, resulted in renal injury, particularly if the animals were maintained on a high-salt diet (13). In cultured nonglomerular ECs, TGF-β1 has been reported to stimulate NO production, an effect that is dependent on Smad-mediated induction of *NOS3* transcription (14,15). These latter findings raised the possibility that the autocrine effect of TGF-β1 observed in animals on high-salt diets may serve as a compensatory mechanism to induce production of NO. Indeed, studies using a neutralizing antibody to TGF-β1 confirmed a modulatory role for TGF-β1 in the EC production of NO by inhibiting the increase in NOS3

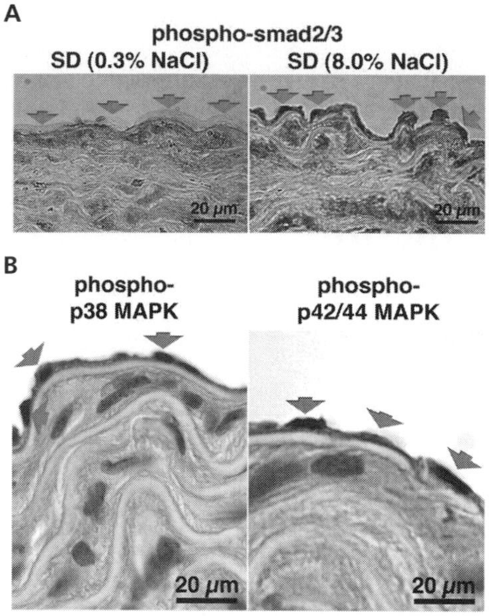

Figure 140.1. (**A**) The representative immunohistochemical analysis of phospho-Smad2/3 of aortic tissue from SD rats on low-salt (0.3% NaCl) and high-salt (8.0% NaCl) diets for four days. A polyclonal antibody that specifically recognized phosphorylated Smad2/3 (Santa Cruz) was used to show the significant activation of this pathway and nuclear localization of phospho-Smad2/3 in aortic ECs (*arrows*) of rats on the high-salt diet. EC staining of rats on the low-salt diet was faint. The vascular smooth muscle layers of both sections stained with the antibody. (**B**) Nuclear localization (*arrows*) of phospho-p38 MAPK (*left*) and phospho-p42/44 MAPK (*right*) in SD rats maintained on the 8.0% NaCl diet. (B is modified with permission from Ying W-Z, Sanders PW. Increased dietary salt activates rat aortic endothelium. *Hypertension.* 2002;39:239–244.)

protein levels in preparations from rats on the high-salt diet (2,11).

Although TGF-β1 appears to induce NOS3 expression in critical target organs that react to changes in blood pressure and blood volume, NO provides an essential inhibitory feedback mechanism. Glomerular and vascular ring preparations incubated with the NOS inhibitor, L-NAME, produced increased amounts of TGF-β1, whereas incubation with a NO donor, ($\pm$)-(E)-4-ethyl-2-[(E)-hydroxyimino]-5-nitro-3-hexenamide (NOR3), decreased production of total and active TGF-β1 (16). Thus, NO limits the production of active TGF-β1; the mechanism was independent of blood pressure and flow, because the experiments were performed in vitro under static conditions. NO has been shown to inhibit production of thrombospondin-1 in cultured mesangial cells directly in a guanylyl cyclase–dependent fashion (17). This inhibition of thrombospondin-1–mediated activation of TGF-β1 (18,19) may explain the potential benefit of NO by a mechanism of decreasing the positive feedback loop that drives expression of TGF-β in this model. The key is the development of a balance between TGF-β1 and NO. A relative excess of TGF-β1 would be detrimental if NO production does not increase, which occurs in the Dahl-Rapp salt-sensitive rat (16).

ENDOTHELIAL CELL SIGNALING MECHANISMS AND DIETARY SALT

ECs in culture respond to shear stress, the frictional force ordinarily generated by blood flow, initially through opening an inwardly rectifying potassium channel that hyperpolarizes the cell (20,21), promoting increases in intracellular calcium concentration (22). Shear stress promotes both rapid (nongenomic) and delayed (genomic) events in ECs (23). By virtue of their location and response to shear stress, ECs serve as biomechanical sensors that detect and respond to changes in blood flow. Dietary salt increases blood volume and renal blood flow in normal rats (24,25); increased flow promotes shear stress. Hemodynamic forces are strong activators of the endothelium and modulate expression of numerous genes (26). Relevant to this discussion, shear stress has been shown to induce expression of TGF-β1 in cultured ECs, and this process is inhibited by tetraethylammonium (TEA), an inhibitor of the shear-activated potassium channel (27). Similar to these studies, the initial event that stimulates TGF-β1 production by aortic ring and glomerular preparations from rats fed a high-salt diet appears to be the opening of a TEA-sensitive potassium channel, because incubation in TEA in vitro inhibited production of total and active TGF-β1 (1,2). As expected from experiments that demonstrated a direct effect of TGF-β on NOS3 expression, incubation in TEA also reduced NOS3 expression and NO production (2,11). Downstream of these cell surface events, activation of the mitogen-activated protein kinase (MAPK) pathways occurs. Dietary salt activates in a dose-dependent fashion both p38 MAPK and p42/44 MAPK pathways in aortic ring and glomerular preparations (Figures 140.2 and 140.3) (28,29). Nuclear accumulation of p38 MAPK

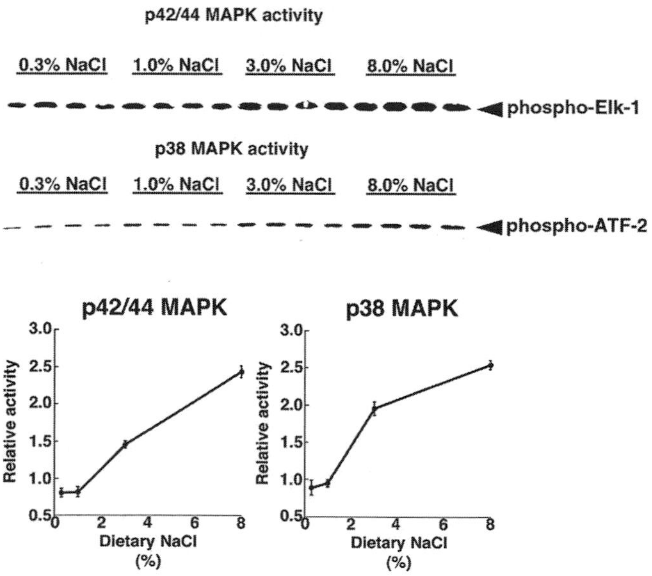

Figure 140.2. Aortic ring preparations from SD rats demonstrated a dose-dependent increase in the activities of p38 MAPK and p42/44 MAPK as the dietary intake of salt increased. (Modified with permission from Ying W-Z, Sanders PW. Increased dietary salt activates rat aortic endothelium. *Hypertension.* 2002;39:239–244.)

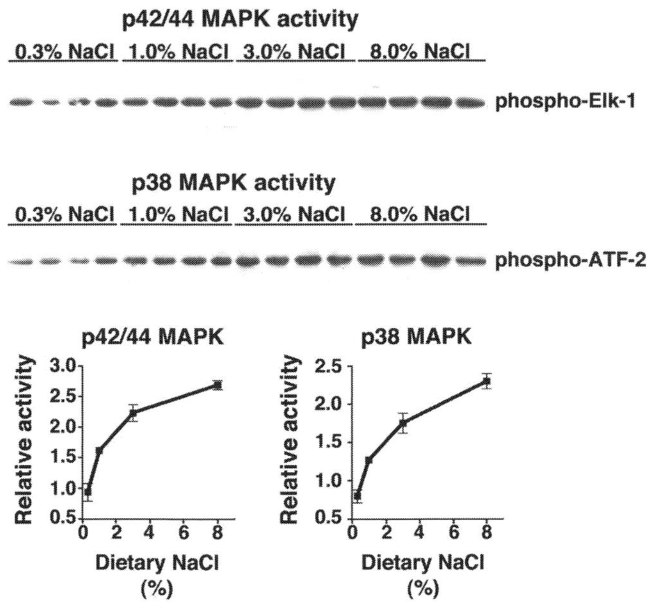

Figure 140.3. Glomerular preparations from SD rats demonstrated a dose-dependent increase in the activities of p38 MAPK and p42/44 MAPK as the dietary intake of salt increased. (Modified and reproduced with permission from Ying W-Z, Sanders PW. Dietary salt intake activates MAP kinases in the rat kidney. *FASEB J.* 2002;16:1683–1684.)

and MEK1/2 was demonstrated using immunohistochemistry (Figure 140.1B) and was associated with increased phosphorylation of the downstream transcription factors, activating transcription factor 2 (ATF2) and Elk2 (28). Using specific pharmacologic inhibitors, activation of both MAPK pathways was shown to be essential in stimulating TGF-β1 production in aorta and glomeruli of rats fed the high-salt diet (28,29). Importantly, intravenous administration of TEA just prior to harvesting the aorta inhibited the increase in glomerular MAPK activity in vivo (29). The combined data suggest that production of TGF-β1 in response to changes in salt intake is a function of multiple interdependent signaling cascades that include Smad2/3, p38 MAPK, and p42/44 MAPK. In turn, TGF-β1 has a direct impact on NOS3 expression and function. The mechanism is compatible with a shear force–mediated process, although this has not yet been demonstrated as the sole mechanism in vivo.

TGF-β1 was previously shown to induce *NOS3* mRNA expression in cultured bovine aortic and human umbilical vein ECs via the Smad2/3-signaling pathway in addition to other pathways (14,15). Simply adding excess active TGF-β1 to glomerular preparations from animals maintained on a low-salt diet did not increase NOS3 expression in the 24-hour timeframe of the study, whereas adding TGF-β1 produced a dose-dependent increase in NOS3 in glomerular preparations from rats on the high-salt diet (11). The implications are that Smad2/3-mediated signaling alone was insufficient, although activation of the Smad pathways was not examined in the latter studies and other explanations, such as sensitization of the endothelium with increased receptor numbers, are also a possibility. Increased NOS3 levels required dietary salt–induced activation of additional signaling pathways. These findings are not unique to this process of gene transcription induced by TGF-β. Activation of the Smad2/3-signaling pathway has been previously reviewed (30–35) (see Chapter 35). To summarize, binding of TGF-β1 to the type I TGF-β receptor (TβRI or activin receptor-like kinase 5[ALK5]) complex permits recruitment of the type II receptor and activation of the complex through autophosphorylation. Smad2/3 is then phosphorylated, permitting binding to Smad4 and migration of the heteromeric complex into the nucleus and binding to specific *cis*-regulatory elements in gene promoter segments. Activation of the Smad2/3 pathway is needed for transcriptional regulation of many TGF-β1–regulated genes, including the *TGF-β1* gene itself (36). The Smad complex binds DNA with low affinity and requires other binding proteins, such as ATF-2 (37,38) and activator protein (AP)-1 (39,40), to induce specific and efficient gene transcription (35,41). Thus, the combined data support activation of multiple signaling pathways that increase *TGF-β1* gene expression by ECs in response to changes in salt intake (Figure 140.4). Interestingly, these same pathways appear to be involved in the regulation of thrombospondin-1, which activates TGF-β1 (18,19) in rat proximal tubular cells and mouse fibroblasts (42).

As reviewed by Lebrin and colleagues (43), there are other effects of the Smad2/3 pathway, including inhibition of EC proliferation and migration. However, in ECs, TGF-β1 can also bind and activate ALK1, which signals through the Smad1/5 pathway. Activation of this pathway requires endoglin, which is expressed on ECs (44,45) and antagonizes the effects of ALK5-mediated cell proliferation and migration (43,46). Endoglin also appears to play an important role in determining the protein half-life of NOS3 through interactions with Hsp90 (47). Mutations in ALK1 account for one form of hereditary hemorrhagic telangiectasia, or Osler-Weber-Rendu syndrome, an autosomal dominant disease characterized by development of arteriovenous malformations and other vascular dysplasias (48,49). The potential role of ALK1 and the balance between ALK5 and ALK1 activation during increases in dietary salt intake have not been explored.

OBSERVATIONS IN PATHOPHYSIOLOGICAL STATES

Initial insight into the pathological significance of salt intake independent of blood pressure came from studies by Yu and associates (50), who showed that feeding an 8.0% NaCl diet to normotensive Wistar-Kyoto (WKY) rats and spontaneously hypertensive rats (SHR) increased TGF-β1 production and produced fibrosis in the kidney and left ventricle. The authors concluded that excessive salt intake might play a direct role in cardiovascular and renal disease. In a recent study, administration of an 8.0% NaCl diet to the Fisher-Lewis rat model of orthotopic renal transplantation accelerated the development of chronic allograft nephropathy (CAN) (51). In this model,

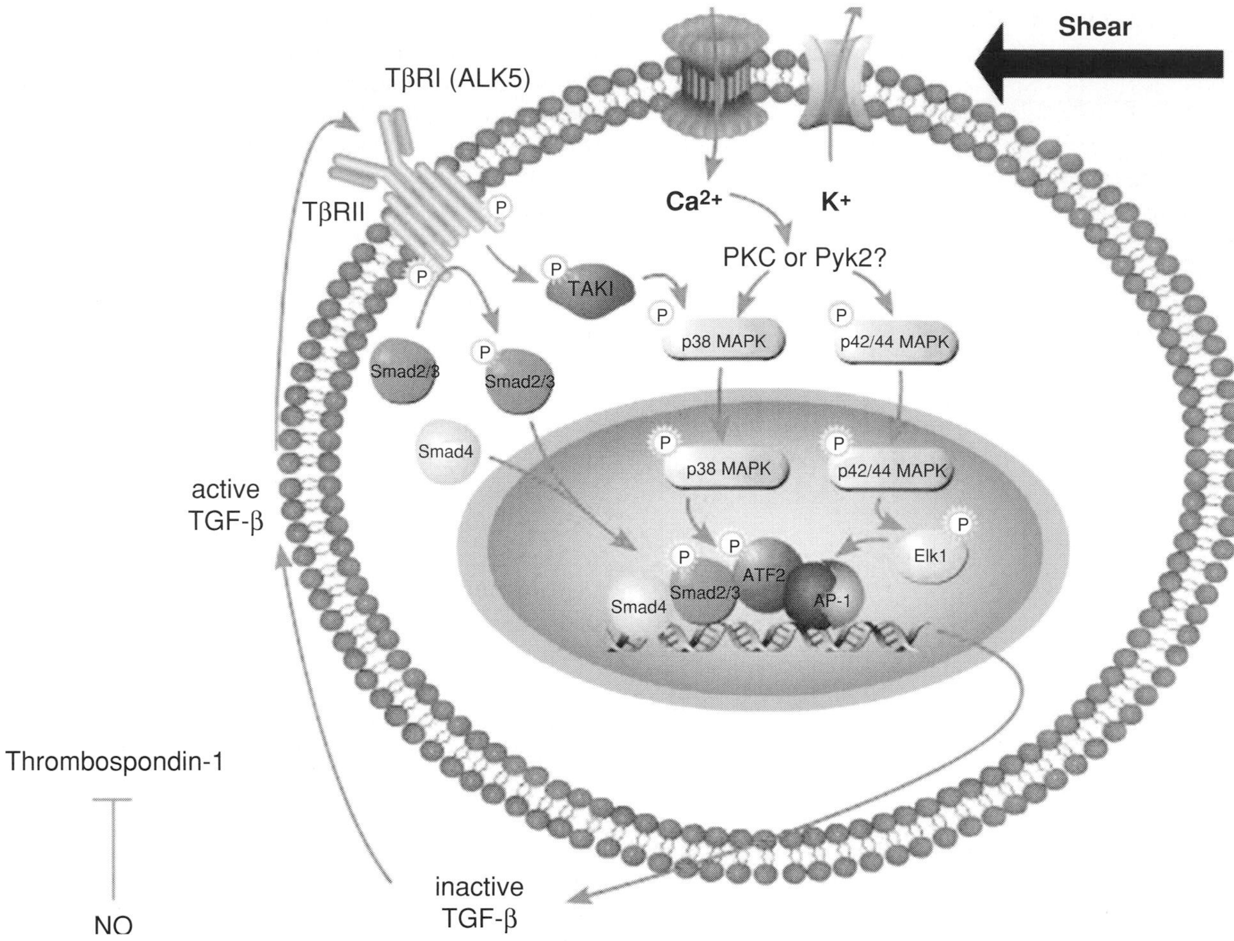

Figure 140.4. Cartoon depicting the postulated events responsible for augmented production of TGF-β by ECs in response to increases in salt intake. As shear stress increases, a TEA-sensitive potassium channel opens, permitting increases in intracellular calcium concentration. Through a process not yet clarified, both p38 MAPK and p42/44 MAPK are activated, permitting phosphorylation and activation of transcription factors that participate with the Smad2/3 signaling pathway to coordinate gene transcription. The process is amplified by increasing the production of TGF-β, which, when activated, can bind to the TβRI (ALK5) receptor. NO, perhaps by interfering with the activation of TGF-β1 or through other as yet undefined mechanisms, inhibits the positive feedback loop that drives expression of TGF-β.

the contralateral kidney was surgically excised two weeks prior to the change in salt intake. The rats did not develop hypertension over the 12 weeks of observation but demonstrated an increase in tubulointerstitial fibrosis and glomerular sclerosis in the transplanted kidneys (51), along with an increase in urinary excretion and renal cortical levels of TGF-β1 (Sanders and colleagues, unpublished observations). Control and sham rats in that study also demonstrated modest increases in albuminuria without overt renal pathology when fed the high-salt diet over the same time period (51). The association between dietary salt intake and albuminuria has been confirmed in a large cohort of human subjects in a recent study that demonstrated a significant relationship between urinary salt excretion and both left ventricular mass index and urinary albumin excretion (52). The combined studies suggest that dietary salt intake can alter cardiorenal risk factors, and further that there may be a vulnerable population that has a heightened sensitivity to the adverse cardiorenal effects of salt, such as subjects following renal transplantation.

A direct correlation between steady-state mRNA of *TGF-β1* and *NOS3* was demonstrated in normotensive Sprague-Dawley (SD) rats fed diets containing various amounts of NaCl (11,16). The relationship was also examined in the Dahl-Rapp salt-sensitive (S) rat, which is an inbred strain that develops hypertension and subsequent progressive renal failure when fed a high-salt diet. Although the relationship existed between active TGF-β1 and NO in S rats, the slope was increased such that at any level of NO, TGF-β1 levels were greater in S rats compared to SD rats. Glomerular and vascular production of TGF-β1 was increased in prehypertensive S rats and was further exaggerated with the onset of salt-sensitive hypertension (16). The mechanism accounting for the increased sensitivity of S rats is under investigation but does not appear to be related to alterations in activation of the p38 MAPK and p42/44 MAPK pathways (16). These findings again emphasize the concept of a population at risk to the adverse effects of excess salt intake. The root cause might be diminished plasticity of ECs in critical vascular beds such as the glomerulus; the application of this environmental stress in the setting of EC dysfunction ultimately creates a milieu that facilitates end-organ damage.

Reducing dietary intake of salt can slow the progression of chronic kidney disease in rats. Using the uninephrectomized SHR as a model of progressive renal failure, Benstein and associates showed that a low-sodium (0.09%) diet alone resulted in less proteinuria and glomerular sclerosis than did administration of a diuretic while continuing a standard (0.45%) sodium diet (53). In this study, systolic blood pressures determined by tail-cuff methodology did not differ among the groups over a 36-week period. These data fit well with another study that showed that administration of a diuretic to animals that continued the high salt intake did not reduce intrarenal production of TGF-β (1). Also, although the mainstays of therapy of chronic kidney disease include angiotensin-converting enzyme (ACE) inhibitors and angiotensin receptor antagonists (ARB), both of which appear to slow the progression of kidney failure in part by inhibiting the stimulation of TGF-β1 production by angiotensin II (54–57), an excess salt intake could promote TGF-β1 independently of angiotensin II. Thus, efforts to monitor and reduce salt intake through dietary restriction may slow renal disease progression through reduction in production of TGF-β1.

In a retrospective analysis of progression of chronic kidney disease, subjects with baseline creatinine clearances between 10 and 40 mL/min were grouped according to consistent urine sodium excretion rates of either less than 100 meq/d (2.3 g Na^+) or greater than 200 meq/d (4.6 g Na^+). Mean blood pressures of the groups did not differ, and both glomerular and tubulointerstitial diseases were represented in both groups. Creatinine clearance declined more rapidly in the high-salt group compared with the low-salt group (–0.51 ± 0.09 vs. –0.25 ± 0.07 mL/min/month; $p < 0.05$). Also, proteinuria increased in the high-salt group and fell in the low-salt group (58). Although this report was not a randomized, controlled trial, it presents intriguing findings that support salt restriction in patients with advanced chronic kidney disease.

Although serum TGF-β1 levels were not affected by salt intake, intrarenal production and urinary excretion of TGF-β1 increased as salt intake increased in rats (1). Ellis and colleagues (59) demonstrated an increase in urinary excretion of TGF-β1 in diabetic subjects who had either microalbuminuria or macroalbuminuria, compared to healthy controls and diabetic patients without microalbuminuria or macroalbuminuria. Although it seems unlikely that urinary TGF-β1 will have diagnostic utility, because TGF-β1 plays such an important role in progression of chronic kidney disease in a variety of disease states (60–68), serial assessment of this parameter might prove useful in determining the activity of this growth factor in the kidney and the potential benefit of an intervention. For example, decreases in urinary TGF-β1 excretion with treatment of patients with type II diabetes mellitus with an ARB has been documented (69,70). Houlihan and associates (70) also demonstrated an excellent direct correlation between urinary TGF-β1 excretion and sodium excretion in their population of patients who had type II diabetes mellitus and microalbuminuria. Although this study showed a direct correlation between salt intake and urinary TGF-β1 excretion, it did not show a decrease in urinary excretion of TGF-β1 with reduction in salt intake because the study was designed to show an effect of the ARB. As a secondary analysis, there did not appear to be sufficient numbers of patients to provide the power needed to detect an effect of reduced salt intake.

CONCLUSION

This review emphasizes the concept that the plasticity of the endothelium plays an integral role in altering vascular tone and renal function, especially in response to changes in dietary salt intake. Furthermore, with the caveat that there are limitations to these studies, the data suggest that there are populations at greater risk of developing end-organ damage from high salt intake and these individuals have lost the requisite EC flexibility to adapt to this environmental stress. Intermittent excessive salt intake probably poses no substantial risk to individuals lacking this risk factor. In contrast, those individuals with EC dysfunction from a genetic or acquired source might develop an imbalance between NOS3 and TGF-β1, permitting excess biological activity of this important fibrogenic growth factor and subsequent development or acceleration of vascular and kidney damage.

KEY POINTS

- ECs modulate production of TGF-β1 and NO in response to changes in dietary salt intake.
- Dietary salt intake promotes alterations in EC function through a process that appears to be shear mediated.
- Increased salt intake activates multiple pathways that include Smad2/3, p38 MAPK, and p42/44 MAPK pathways.
- Increased salt intake increases TGF-β1 production, which in turn increases NOS3 protein and NO production.
- NO levels adjust TGF-β1 production by ECs, supporting an important negative feedback role for NO.

Future Goals

- To define the signal transduction events involved in the upregulation of TGF-β1 by dietary salt and the mechanism by which NO inhibits EC production of TGF-β1

- To determine the functional adaptation and consequences of salt intake on EC function in the vasculature of the kidney

ACKNOWLEDGMENTS

National Institutes of Health grant (R01 DK46199) and the Office of Research and Development, Medical Research Service, Department of Veterans Affairs, supported this research.

REFERENCES

1 Ying W-Z, Sanders PW. Dietary salt modulates renal production of transforming growth factor-β in rats. *Am J Physiol.* 1998;274 (4 Pt 2):F635–F641.
2 Ying W-Z, Sanders PW. Dietary salt increases endothelial nitric oxide synthase and TGF-β1 in rat aortic endothelium. *Am J Physiol.* 1999;277(4 Pt 2):H1293–H1298.
3 Chen PY, Sanders PW. L-Arginine abrogates salt-sensitive hypertension in Dahl/Rapp rats. *J Clin Invest.* 1991;88:1559–1567.
4 Chen PY, Sanders PW. Role of nitric oxide synthesis in salt-sensitive hypertension in Dahl/Rapp rats. *Hypertension.* 1993;22: 812–818.
5 Deng X, Welch WJ, Wilcox CS. Renal vasoconstriction during inhibition of NO synthase: effects of dietary salt. *Kidney Int.* 1994;46:639–646.
6 Shultz PJ, Tolins JP. Adaptation to increased dietary salt intake in the rat: role of endogenous nitric oxide. *J Clin Invest.* 1993;91: 642–650.
7 Bech JN, Nielsen CB, Ivarsen P, et al. Dietary sodium affects systemic and renal hemodynamic response to NO inhibition in healthy humans. *Am J Physiol.* 1998;274(5 Pt 2):F914–F923.
8 Deng A, Baylis C. Locally produced EDRF controls preglomerular resistance and ultrafiltration coefficient. *Am J Physiol.* 1993;264(2 Pt 2):F212–F215.
9 Patel A, Layne S, Watts D, Kirchner KA. L-Arginine administration normalizes pressure natriuresis in the hypertensive Dahl rats. *Hypertension.* 1993;22:863–869.
10 Ortiz PA, Hong NJ, Wang D, Garvin JL. Gene transfer of eNOS to the thick ascending limb of eNOS-KO mice restores the effects of L-arginine on NaCl absorption. *Hypertension.* 2003;42(4):674–679.
11 Ying W-Z, Sanders PW. Dietary salt enhances glomerular endothelial nitric oxide synthase through TGF-β1. *Am J Physiol.* 1998; 275(1 Pt 2):F18–F24.
12 Tolins JP, Shultz PJ. Endogenous nitric oxide synthesis determines sensitivity to the pressor effect of salt. *Kidney Int.* 1994;46: 230–236.
13 Fujihara CK, Michellazzo SM, De Nucci G, Zatz R. Sodium excess aggravates hypertension and renal parenchymal injury in rats with chronic NO inhibition. *Am J Physiol.* 1994;266(5 Pt 2):F697–F705.
14 Inoue N, Venema RC, Sayegh HS, et al. Molecular regulation of the bovine endothelial cell nitric oxide synthase by transforming growth factor-β1. *Arterioscler Thromb Vasc Biol.* 1995;15:1255–1261.
15 Saura M, Zaragoza C, Cao W, et al. Smad2 mediates transforming growth factor-β induction of endothelial nitric oxide synthase expression. *Circ Res.* 2002;91:806–813.
16 Ying W-Z, Sanders PW. The interrelationship between TGF-β1 and nitric oxide is altered in salt-sensitive hypertension. *Am J Physiol Renal Physiol.* 2003;285:F902–F908.
17 Wang S, Shiva S, Poczatek MH, et al. Nitric oxide and cGMP-dependent protein kinase regulation of glucose-mediated thrombospondin 1-dependent transforming growth factor-β activation in mesangial cells. *J Biol Chem.* 2002;277:9880–9888.
18 Schultz-Cherry S, Murphy-Ullrich JE. Thrombospondin causes activation of latent transforming growth factor-beta secreted by endothelial cells by a novel mechanism. *J Cell Biol.* 1993;122(4):923–932.
19 Daniel C, Wiede J, Krutzsch HC, et al. Thrombospondin-1 is a major activator of TGF-beta in fibrotic renal disease in the rat in vivo. *Kidney Int.* 2004;65(2):459–468.
20 Olesen S-P, Clapham DE, Davies PF. Haemodynamic shear stress activates a K^+ current in vascular endothelial cells. *Nature (Lond).* 1988;331:168–170.
21 Hoger JH, Ilyin VI, Forsyth S, Hoger A. Shear stress regulates the endothelial Kir2.1 ion channel. *Proc Natl Acad Sci USA.* 2002; 99:7780–7785.
22 Shen J, Luscinskas FW, Connolly A, et al. Fluid shear stress modulates cytosolic free calcium in vascular endothelial cells. *Am J Physiol.* 1992;262(2 Pt 1):C384–C390.
23 Davies PF. Flow-mediated endothelial mechanotransduction. *Physiol Rev.* 1995;75:519–560.
24 Antonios TFT, MacGregor GA. Salt – more adverse effects. *Lancet.* 1996;348:250–251.
25 Davis JM, Häberle DA, Kawata T. The control of glomerular filtration rate and renal blood flow in chronically volume-expanded rats. *J Physiol.* 1988;402:473–495.
26 Garcia-Cardeña G, Comander J, Anderson KR, et al. Biomechanical activation of vascular endothelium as a determinant of its functional phenotype. *Proc Natl Acad Sci USA.* 2001;98:4478–4485.
27 Ohno M, Cooke JP, Dzau VJ, Gibbons GH. Fluid shear stress induces endothelial transforming growth factor beta-1 transcription and production. *J Clin Invest.* 1995;95:1363–1369.
28 Ying W-Z, Sanders PW. Increased dietary salt activates rat aortic endothelium. *Hypertension.* 2002;39:239–244.
29 Ying W-Z, Sanders PW. Dietary salt intake activates MAP kinases in the rat kidney. *FASEB J.* 2002;16:1683–1684.
30 Shi Y, Massagué J. Mechanisms of TGF-β signaling from cell membrane to the nucleus. *Cell.* 2003;113:685–700.
31 Massagué J, Chen Y-G. Controlling TGF-β signaling. *Genes Dev.* 2000;14:627–644.
32 Blobe GC, Schiemann WP, Lodish HF. Role of transforming growth factor β in human disease. *N Engl J Med.* 2000;342:1350–1358.
33 Stahl PJ, Felsen D. Transforming growth factor-β, basement membrane, and epithelial-mesenchymal transdifferentiation: implications for fibrosis in kidney disease. *Am J Pathol.* 2001;159: 1187–1192.
34 Bottinger EP, Bitzer M. TGF-beta signaling in renal disease. *J Am Soc Nephrol.* 2002;13:2600–2610.
35 Derynck R, Zhang YE. Smad-dependent and Smad-independent pathways in TGF-beta family signalling. *Nature.* 2003;425(6958): 577–584.

36 Piek E, Ju WJ, Heyer J, et al. Functional characterization of transforming growth factor beta signaling in Smad2- and Smad3-deficient fibroblasts. *J Biol Chem.* 2001;276(23):19945–19953.

37 Raingeaud J, Whitmarsh AJ, Barrett T, et al. MKK3- and MKK6-regulated gene expression is mediated by the p38 mitogen-activated protein kinase signal transduction pathway. *Mol Cell Biol.* 1996;16:1247–1255.

38 Ionescu AM, Schwarz EM, Zuscik MJ, et al. ATF-2 cooperates with Smad3 to mediate TGF-beta effects on chondrocyte maturation. *Exp Cell Res.* 2003;288(1):198–207.

39 Zhang Y, Feng X-H, Derynck R. Smad3 and Smad4 cooperate with c-Jun/c-Fos to mediate TGF-β-induced transcription. *Nature (Lond).* 1998;394:909–913.

40 Qing J, Zhang Y, Derynck R. Structural and functional characterization of the transforming growth factor-beta – induced Smad3/c-Jun transcriptional cooperativity. *J Biol Chem.* 2000; 275(49):38802–38812.

41 Shi Y, Wang Y-F, Jayaraman L, et al. Crystal structure of a Smad MH1 domain bound to DNA: Insights on DNA binding in TGF-β signaling. *Cell.* 1998;94:585–594.

42 Nakagawa T, Lan HY, Glushakova O, et al. Role of ERK1/2 and p38 mitogen-activated protein kinases in the regulation of thrombospondin-1 by TGF-β1 in rat proximal tubular cells and mouse fibroblasts. *J Am Soc Nephrol.* 2005;16:899–904.

43 Lebrin F, Deckers M, Bertolino P, Ten Dijke P. TGF-beta receptor function in the endothelium. *Cardiovasc Res.* 2005;65(3):599–608.

44 Gougos A, Letarte M. Identification of a human endothelial cell antigen with monoclonal antibody 44G4 produced against a pre-B leukemic cell line. *J Immunol.* 1988;141(6):1925–1933.

45 Gougos A, Letarte M. Primary structure of endoglin, an RGD-containing glycoprotein of human endothelial cells. *J Biol Chem.* 1990;265(15):8361–8364.

46 Goumans MJ, Valdimarsdottir G, Itoh S, et al. Activin receptor-like kinase (ALK)1 is an antagonistic mediator of lateral TGF-beta/ALK5 signaling. *Mol Cell.* 2003;12(4):817–828.

47 Toporsian M, Gros R, Kabir MG, et al. A role for endoglin in coupling eNOS activity and regulating vascular tone revealed in hereditary hemorrhagic telangiectasia. *Circ Res.* 2005:96(6):684–692. Epub 2005 Feb 17.

48 Johnson DW, Berg JN, Baldwin MA, et al. Mutations in the activin receptor-like kinase 1 gene in hereditary haemorrhagic telangiectasia type 2. *Nat Genet.* 1996;13(2):189–195.

49 Berg JN, Gallione CJ, Stenzel TT, et al. The activin receptor-like kinase 1 gene: genomic structure and mutations in hereditary hemorrhagic telangiectasia type 2. *Am J Hum Genet.* 1997;61 (1):60–67.

50 Yu HCM, Burrell LM, Black MJ, et al. Salt induces myocardial and renal fibrosis in normotensive and hypertensive rats. *Circulation.* 1998;98:2621–2628.

51 Sanders PW, Gibbs CL, Akhi KM, et al. Increased dietary salt accelerates chronic allograft nephropathy in rats. *Kidney Int.* 2001;59:1149–1157.

52 du Cailar G, Ribstein J, Mimran A. Dietary sodium and target organ damage in essential hypertension. *Am J Hypertens.* 2002; 15:222–229.

53 Benstein JA, Feiner HD, Parker M, Dworkin LD. Superiority of salt restriction over diuretics in reducing renal hypertrophy and injury in uninephrectomized SHR. *Am J Physiol.* 1990;258 (6 Pt 2):F1675–F1681.

54 Anderson PW, Zhang XY, Tian J, et al. Insulin and angiotensin II are additive in stimulating TGF-beta 1 and matrix mRNAs in mesangial cells. *Kidney Int.* 1996;50:745–753.

55 Hamaguchi A, Kim S, Izumi Y, et al. Contribution of extracellular signal-regulated kinase to angiotensin II-induced transforming growth factor-β1 expression in vascular smooth muscle cells. *Hypertension.* 1999;34:126–131.

56 Kagami S, Border WA, Miller DE, Noble NA. Angiotensin II stimulates extracellular matrix protein synthesis through induction of transforming growth factor-β expression in rat glomerular mesangial cells. *J Clin Invest.* 1994;93:2431–2437.

57 Kaneto H, Morrissey J, McCracken R, et al. Enalapril reduces collagen type IV synthesis and expansion of the interstitium in the obstructed rat kidney. *Kidney Int.* 1994;45:1637–1647.

58 Cianciaruso B, Bellizzi V, Minutolo R, et al. Salt intake and renal outcome in patients with progressive renal disease. *Miner Electrolyte Metab.* 1998;24:296–301.

59 Ellis D, Forrest KY, Erbey J, Orchard TJ. Urinary measurement of transforming growth factor-beta and type IV collagen as new markers of renal injury: application in diabetic nephropathy. *Clin Chem.* 1998;44(5):950–956.

60 Border WA, Okuda S, Languino LR, et al. Suppression of experimental glomerulonephritis by antiserum against transforming growth factor β1. *Nature.* 1990;346:371–374.

61 Akagi Y, Isaka Y, Arai M, et al. Inhibition of TGF-β1 expression by antisense oligonucleotides suppressed extracellular matrix accumulation in experimental glomerulonephritis. *Kidney Int.* 1996;50:148–155.

62 Zhu L, Herrera GA, Murphy-Ullrich JE, et al. Pathogenesis of glomerulosclerosis in light chain deposition disease: Role for transforming growth factor-β. *Am J Pathol.* 1995;147:375–385.

63 Kopp JB, Factor VM, Mozes M, et al. Transgenic mice with increased plasma levels of TGF-beta 1 develop progressive renal disease. *Lab Invest.* 1996;74(6):991–1003.

64 Shihab FS, Andoh TF, Tanner AM, et al. Role of transforming growth factor-β1 in experimental chronic cyclosporine nephropathy. *Kidney Int.* 1996;49:1141–1151.

65 Yamamoto T, Nakamura T, Noble NA, et al. Expression of transforming growth factor β is elevated in human and experimental diabetic nephropathy. *Proc Natl Acad Sci USA.* 1993;90:1814–1818.

66 Yamamoto T, Noble NA, Miller DE, Border WA. Sustained expression of TGF-β1 underlies development of progressive kidney fibrosis. *Kidney Int.* 1994;45:916–927.

67 Ziyadeh FN, Hoffman BB, Han DC, et al. Long-term prevention of renal insufficiency, excess matrix gene expression, and glomerular mesangial matrix expansion by treatment with monoclonal antitransforming growth factor-β antibody in db/db diabetic mice. *Proc Natl Acad Sci USA.* 2000;97:8015–8020.

68 Dahly AJ, Hoagland KM, Flasch AK, et al. Antihypertensive effects of chronic anti-TGF-β antibody therapy in Dahl S rats. *Am J Physiol Regul Integr Comp Physiol.* 2002;283:R757–R767.

69 Agarwal R, Siva S, Dunn SR, Sharma K. Add-on angiotensin II receptor blockade lowers urinary transforming growth factor-beta levels. *Am J Kidney Dis.* 2002;39(3):486–492.

70 Houlihan CA, Akdeniz A, Tsalamandris C, et al. Urinary transforming growth factor-beta excretion in patients with hypertension, type 2 diabetes, and elevated albumin excretion rate: effects of angiotensin receptor blockade and sodium restriction. *Diabetes Care.* 2002;25(6):1072–1077.

141

The Role of the Endothelium in Systemic Inflammatory Response Syndrome and Sepsis

Laszlo M. Hoesel and Peter A. Ward

University of Michigan Medical School, Ann Arbor

The terms *systemic inflammatory response syndrome (SIRS), sepsis, severe sepsis, septic shock and multiorgan failure (MOF)* are currently used to characterize the progressive stages of a very complex and therapeutically challenging disorder of the immune and inflammatory systems. Although SIRS is a result of a systemic activation of the innate immune system regardless of cause, sepsis, severe sepsis, and septic shock are accompanied by proven or suspected infection with or without impaired organ function. Despite tremendous research efforts for over 20 years, sepsis remains the leading cause of death in intensive care units (ICUs). SIRS and sepsis occur in approximately 750,000 patients per year in the United States, with a rising incidence of approximately 1.5% per year (1). With a mortality rate of currently 30% to 70%, sepsis and related disorders represent a major burden to the U.S. health care system, with costs estimated to be approximately $16.7 billion per year (2).

The endothelium represents the natural barrier between the intravascular and extravascular spaces. However, endothelial cells (ECs) are not inactive "border controls"; in fact, among other functions, they are involved in maintaining vasomotor tone, blood flow, local balance of various mediators, programmed cell death (apoptosis), and hemostatic balance. During infectious conditions, activated leukocytes must transmigrate to infectious sites to engulf, phagocytose, and destroy invading microbes. This is an active process not only for the migrating leukocytes but also for the ECs. During this process, numerous pro- and anti-inflammatory mediators lead to activation of the endothelium and result in the expression of various adhesion molecules, which in turn facilitate or inhibit the transmigration of activated leukocytes into tissues. In addition, ECs can generate their own inflammatory mediators and express adhesion molecules. At the same time, the body's inflammatory response leads to the activation of the coagulation (intrinsic and extrinsic pathways) as well as the fibrinolytic system. Although ECs represent an anticoagulant surface under physiologic conditions, their stimulation by a variety of proinflammatory mediators, as seen in SIRS and sepsis, results in a procoagulant phenotype. These changes clearly contribute to the development of increased microvascular permeability, edema formation, and hypotension (3). Therefore, much effort has been made to assess the interaction of ECs, leukocytes, and the coagulation system during inflammatory diseases, especially sepsis.

SYSTEMIC INFLAMMATORY RESPONSE SYNDROME AND SEPSIS

Definitions

Sepsis is not a disease but a syndrome, and the complexity and variability of symptoms observed in septic patients have resulted in difficulties in establishing an accurate, well-accepted, and reliable definition of the condition. Until only 15 years ago, sepsis was defined as the systemic host response to an infection, equated with the presence of bacteria in blood. In 1989, a simple definition of sepsis was established on the basis of clinical symptoms and a known source of infection (4). However, although many patients showed septic symptoms, detectable levels of bacteria in the blood often were not found. Therefore, the term *systemic inflammatory response syndrome* was introduced in 1992, and sepsis was redefined as SIRS associated with suspected or proven infection (5). During the following years, however, these criteria proved too nonspecific to facilitate a precise and unambiguous diagnosis or to allow staging of the disease and prognosis, although the original definition has remained. Thus, the criteria for the diagnosis of SIRS or sepsis were extended by using more clinical parameters (e.g., altered mental state, edema, hyperglycemia, hypotension, and organ dysfunction parameters) (6).

In response to a variety of infectious and noninfectious conditions, the host can react with the development of SIRS. Among the triggers of SIRS are infection, trauma, burns, ischemia and/or reperfusion injury, sterile inflammatory processes (e.g., pancreatitis), and extensive surgical procedures. The physiological changes that occur in patients with SIRS

Table 141-1: Clinical Definitions of Sepsis

Syndrome	*Clinical Findings*
SIRS	Temperature >38.5°C or <36°C Heart rate >90 beats per minute Respiratory rate >20 breaths per minute or $PaCo_2$ <32 mm Hg White blood count $>12 \times 10^9$/L or $<4 \times 10^9$/L or presence of >10% immature neutrophils
Sepsis	Two or more criteria of SIRS Documented (or suspected) infection
Severe sepsis	Sepsis associated with organ dysfunction (hypoperfusion, hypotension, lactic acidosis, oliguria, or alteration of mental state)
Septic shock	Sepsis-induced hypotension (systolic blood pressure <90 mm Hg or 40 mm Hg below baseline) despite adequate fluid resuscitation
MODS/MOF	Altered organ function such that homeostasis must be maintained with intervention

Reproduced with permission from Bone RC, Fisher CJ, Jr., Clemmer TP, et al. Sepsis syndrome: a valid clinical entity. Methylprednisolone Severe Sepsis Study Group. *Crit Care Med.* 1989;17:389–393.
$Paco_2$, arterial pressure of carbon dioxide.

are summarized in Table 141-1. In addition, possible biochemical indices of SIRS (increased levels of interleukin [IL]-6, adrenomedullin, soluble CD14, E-selectin, macrophage inflammatory protein [MIP]-1α, macrophage migration inhibitory factor [MIF], C-reactive protein [CRP] and procalcitonin) have been investigated.

SIRS triggered by infection is termed *sepsis.* Severe sepsis is defined as sepsis complicated by organ dysfunction. The most frequent cause of sepsis is bacterial (gram-positive and gram-negative) infection followed by fungal, viral, or protozoal infections. The most common sites of infections are the lungs, abdomen, and urinary tract. Interestingly, although gram-positive bacteria were previously thought to be the primary organisms causing sepsis, evidence now suggests that sepsis caused by gram-negative bacteria occurs with equal frequency (7). It should also be noted that the most common condition leading to sepsis is bacterial pneumonia.

Pathophysiological Mechanisms

Tight regulation of the immune and inflammatory systems is crucial for maintaining the balance between protective and tissue-damaging responses. SIRS and sepsis are characterized by a loss in this balance, leading to hyperactive (and often followed by hypoactive) immune and inflammatory responses. The reasons for these perturbations, however, are unknown. During the onset of sepsis, the inflammatory system becomes hyperactive, involving both cellular and humoral defense mechanisms. Endothelial and epithelial cells, as well as neutrophils, macrophages, and lymphocytes, produce proinflammatory mediators (e.g., tumor necrosis factor [TNF]-α, IL-6, IL-1, and IL-8). Although these mediators appear in the bloodstream at an early time point, another proinflammatory agent, high mobility group B1 protein (HMGB1), has been identified as a late mediator during various models of inflammation and sepsis (8). The presence of these mediators in plasma is often a harbinger of the developing SIRS and sepsis syndromes. Simultaneously, soluble elements of the immune system become activated and can be detected in the plasma/serum. The concentration of serum acute-phase proteins (such as CRP) increases, and the activated cascade of the complement system leads to the production of powerful complement split products, the anaphylatoxins C3a and C5a. Together, these processes enhance the production of proinflammatory and anti-inflammatory cytokines and chemokines. Phagocytic cells (neutrophils and macrophages) respond to these chemoattractants by releasing granular enzymes and producing reactive oxygen species (ROS), such as hydrogen peroxide (H_2O_2), and reactive nitrogen species, such as nitric oxide (NO), which are involved in the killing of ingested bacteria. ROS and NO are capable of causing tissue damage and increased vascular permeability, ultimately leading to organ injury and failure. Furthermore, the coagulation and fibrinolytic systems become activated, which can lead to disseminated intravascular coagulopathy (DIC).

At later stages of sepsis, anti-inflammatory mediators are produced, including IL-10 and IL-13, which counteract the proinflammatory effects during the early phase. The production of anti-inflammatory mediators in the setting of sepsis has been referred to as the compensatory anti-inflammatory response syndrome (CARS) (9). As sepsis progresses, neutrophils and macrophages show reduced phagocytic activity, bacterial killing, oxygen radical production, chemotaxis, and cytokine production. The mechanisms leading to this type of immunosuppression are not yet completely understood. So far, the shift from a proinflammatory to an anti-inflammatory cell response, the development of a state of immune anergy, the apoptosis-induced loss of immune cells, and the immunosuppressive effects of lymphocytes undergoing apoptosis have been described and are considered to contribute to immune suppression in patients with sepsis (10). Taken together, various functions of the immune system are compromised during the late stage of sepsis, leading to a hyporeactive host defense system and immunoparalysis.

THE ENDOTHELIUM DURING INFLAMMATION

Under physiologic conditions, ECs perform several functions to maintain homeostasis. ECs normally inhibit blood coagulation and do not stimulate the adherence of peripheral cells. Further, by producing prostacyclin, NO, and other vasoactive substances, ECs inhibit platelet aggregation and regulate the tone of arterioles and venules, thereby regulating microcirculation (11). During inflammatory states, ECs come in contact with a variety of proinflammatory mediators that change their physiologic functions profoundly. This is considered "activation of ECs." As a result, an interaction with and adhesion of leukocytes with ECs is facilitated, ECs switch from an anticoagulatory to a procoagulatory state, and an altered barrier function of ECs leads to increased permeability and impaired vasomotor tone. These changes are discussed in detail in upcoming sections.

For more than 100 years, the extravasation of leukocytes from the bloodstream into tissue has been studied and was first documented in vivo in frog mesentery and tongue during the 1800s (12) and, in 1935, in the skin of experimental animals (13). Since then, specific adhesion molecules, which are expressed on leukocytes, ECs, and platelets, have been shown to participate in this process (14). Three families of cell adhesion molecules (CAM) play a central role in leukocyte–endothelial interactions: the selectins (E, P, L), the integrins (β1, β2), and the immunoglobulin superfamily (intercellular adhesion molecule [ICAM]-1, vascular adhesion molecule [VCAM]-1, etc.). Initial contact in the blood leukocyte adhesion cascade occurs when leukocytes marginate and roll along the surface of vessels at a far slower velocity than the flow of blood. Rolling and the subsequent loose adhesion are facilitated by two interactions: that of L-selectin (on leukocytes) and its ligands (including CD34, glycosylation-dependent cell adhesion molecule [GlyCAM]-1) on ECs, and that of E-selectin (on ECs) and its corresponding ligand – for example, sialyl Lewis X compounds (sLe^x) (NeuAcα2–3Galβ1–4[Fucα1–3]GlcNAcβ1–) on leukocytes. L-selectins are constitutively expressed on leukocytes and are shed upon cell activation by the action of a membrane-associated serine protease. E-selectin is expressed on activated ECs and facilitates the adhesion of leukocytes to ECs. E-selectin is synthesized after EC stimulation with lipopolysaccharide (LPS) or proinflammatory mediators such as TNF-α and IL-1β, and binds to sLe^x compounds as one ligand among several (14). P-selectin is a third adhesion molecule that is stored intracellularly in resting platelets and within the Weibel-Palade bodies of ECs. It can be found on cell surfaces rapidly (10–15 minutes) after EC stimulation due to the fusion of Weibel-Palade granules with the cell membrane. P-selectin also can be induced in ECs after contact with the same agents that induce E-selectin (15). By binding to its ligand present on leukocytes (e.g., P-selectin glycoprotein ligand [PSGL]-1), P-selectin facilitates the adhesion of leukocytes with activated platelets and ECs. Involvement of P-selectin in inflammation processes has been demonstrated in two different models of lung injury. Intravenous infusion of cobra venom factor, leading to an acute activation of the complement system and subsequently inducing lung injury, resulted in transiently elevated expression of P-selectin as early as 30 minutes after injury (16). A second model of lung injury, the deposition of immune complexes in the lung, led to a delayed peak expression of P-selectin after 2 to 4 hours. However, the protein expression of P-selectin preceded its mRNA upregulation, suggesting release of constitutive P-selectin from EC granules followed by transcriptional upregulation. Furthermore, blockade of P-selectin with specific antibodies greatly attenuated lung injury in both models (17).

In the next step of the cascade, another set of adhesion molecules becomes engaged to facilitate the activation and firm adhesion of leukocytes to the endothelium. In addition to the role of L-selectin, neutrophils express β2 integrins, which comprise the largest group of adhesion receptors and are upregulated on the neutrophils after cell activation. β2 Integrins mediate their function by binding to specific ligands. Consisting of two noncovalent associated subunits (α and β), integrins are classified based on the β-chain and corresponding α-subunit (18). Under inflammatory conditions, the β2 integrins αLβ2 (CD11a/CD18, LFA-1) as well as αMβ2 (CD11b/CD18, Mac-1) play an important role in adhesion and transmigration of neutrophils. Another type of β integrin adhesion molecule on leukocytes, α4β1, (CD49d, VLA-4) is involved in late-stage leukocyte–EC interactions. The immunoglobulin superfamily functions as specific ligands for integrins on the surface of ECs. Important ligands for the β2 integrins are ICAM-1 through -4 (intercellular adhesion molecules), while CD11b/CD18 can also bind to complement iC3b, collagen, coagulation proteins, and fibrinogen. In contrast, very late antigen (VLA)-4 mainly binds to VCAM-1.

Once leukocytes firmly adhere to the endothelium, they transmigrate through the junctions of ECs along a chemotactic gradient. An adhesion molecule in the immunoglobulin superfamily, platelet-endothelial cell adhesion molecule (PECAM)-1, has been implicated in this process of transmigration. PECAM-1, which is expressed mostly by platelets and the endothelium, is localized at the junctions of ECs; its blockade by antibodies has been shown to inhibit leukocyte transmigration in vitro (19) and in vivo (20). Thereafter, leukocytes finally migrate into the surrounding interstitial tissue.

This paradigm of the "paracellular" route through the endothelium, in which leukocytes must pass junctional interendothelial structures on their way into the interstitium, has been challenged by several other studies. Hence, a second, "transcellular" mode of transmigration, in which lymphocytes in particular migrate through the cytoplasm of ECs, has been described. This "emperipolesis" theory was originally proposed in 1964, based on transmission electron micrographs of transmigrating leukocytes (21). This theory has recently resurfaced (22). Although molecular confirmation of emperipolesis is still lacking, such a pathway for transmigrating leukocytes, if it exists, may involve the engagement of very different molecules responsible for leukocyte transmigration. Furthermore, cascade-independent modes of leukocyte transmigration without activation of the endothelium (defined by upregulation of adhesion molecules) have been described during different types of inflammation (14). Apparently, the presence of certain chemoattractants/activators as well as the formation of intravascular neutrophil aggregates leads directly to expression of $\beta 2$ integrins and results in adhesion and transmigration through inactivated ECs (23,24).

As outlined earlier, ECs inhibit blood coagulation under physiologic conditions. This is facilitated by the expression of thrombomodulin and certain proteoglycans, activation of protein C, and the release of small amounts of tissue-type plasminogen activator (t-PA) (25–28). Under inflammatory conditions, the endothelium becomes a procoagulant surface in order to contain tissue hemorrhage. Triggered by bacterial components (e.g., LPS) and proinflammatory cytokines (e.g., TNF-α, IL-1), the expression of thrombomodulin and proteoglycans disappears and ECs start synthesizing tissue factor (29–31). In concert with decreased activation and expression of protein C, the extrinsic pathway of the coagulation pathway becomes activated. Further downstream, factors VIIa, Xa, and thrombin additionally activate the procoagulant state of ECs through a positive feedback mechanism (32). Combes and colleagues have proposed an additional mode for coagulation activation by ECs (33). Procoagulant phospholipid microparticles may be released from ECs that have undergone insertion of the complement membrane attack complex. However, to what extent these microparticles are released from ECs and contribute to a coagulation outcome remains to be determined.

In addition to cell adhesion and activation of the coagulation system, ECs are also involved in the regulation of vasomotor tone and vascular permeability. Under physiologic conditions, ECs produce and release vasodilating NO and prostacyclin as well as the vasoconstricting endothelins (28,34,35). Vasodilating mediators contribute to decreased vasomotor tone, which may result in marked hypotension if excessive. NO especially appears to play a significant role in inflammation-associated hypotension, whereas the role of prostacyclin and endothelins is not understood. ECs usually release NO in a calcium-dependent manner upon activation of constitutive endothelial NO synthase (eNOS). Under inflammatory conditions, a second, inducible NO synthase (iNOS) becomes activated in a calcium-independent manner, leading to vasodilation and hypotension (36,37). Increased permeability of the EC layer is closely associated with vasodilation and hypotension. Whereas normal ECs form a continuous, semipermeable barrier (38), activated ECs may lose their barrier function, show an increased permeability, and thereby contribute to hypotension and hypovolemia.

THE ENDOTHELIUM DURING SEPSIS

As outlined earlier, SIRS and sepsis involve the whole organism, rather than just a local inflammatory reaction. It is now widely accepted that, during sepsis, the immune system first becomes hyperactivated, leading to an uncontrolled, overwhelming immune reaction, which may harm the organism rather than support the fight against invading microorganisms. In other words, in sepsis we have "too much of a good thing." In addition to the cells of the immune system, ECs are also part of the first-line defense. During inflammatory stages involving invading microorganisms, bacterial components are among the first pathogen-associated molecular patterns (PAMPs) to activate the immune system and the endothelium.

As a consequence, ECs are involved in this septic hyperactivated state. During sepsis, the endothelium first becomes activated as a normal response to inflammation, as described earlier. However, during the further course of sepsis, with its uncontrolled immune responses, ECs become dysfunctional, presumably systemically throughout the organism. These changes can be considered an exaggeration of a normal adaptive response to inflammation/infection. In later stages of sepsis, the immune system becomes paralyzed. Eventually, these changes are associated with MOF and death. In addition to structural changes within the EC itself, functional changes comprise increased expression of cell adhesion molecules (E-selectin, ICAM-1, VCAM-1), loss of leukocyte endothelial adherence, disturbed barrier function, hemostatic imbalance, altered vasomotor tone, and increased apoptosis (39). Therefore, ECs are both victim and perpetrator during the course of sepsis. First, they become hyperactivated by contact with an overwhelming "mediator storm" of proinflammatory cytokines. Later, ECs undergo structural and functional changes, including the increased production of cytokines, which further sustains the ongoing septic process.

Lipopolysaccharide–Toll-Like Receptor Interactions

LPS, derived from the wall of gram-negative bacteria, has a high affinity for the LPS-binding protein (LBP), which is one of several acute-phase proteins in serum. LPS/LBP complexes exert their biological activity via CD14, a receptor membrane-bound (mCD14) on leukocytes. There is no conclusive evidence as to whether ECs express mCD14. However, a second, soluble form of CD14 (sCD14) has been reported to be increased in septic patients. Thus, LPS/LBP/sCD14 complexes may activate ECs without binding to mCD14 (40). Because CD14 lacks a transmembrane signaling domain, the involvement of an accessory receptor seems likely. Until recently, 10 subtypes of Toll-like receptors (TLRs) have been identified in mammals. TLR4 and, to a much lesser extent TLR2, mediate responses to LPS/LBP bound to CD14. Regulated by a variety of pro- and anti-inflammatory mediators, ECs express TLR4, which, when ligated to LPS, leads to activation of ECs during the course of sepsis. Studies using antibodies against CD14 did not completely inhibit EC activation by LPS (40). Subsequent studies have revealed that the leukocyte $\beta 2$ integrins (CD11/CD18) may also bind LPS (41). This complex, consisting of CD11/CD18-LPS, can then further activate ECs. Similarly, gram-positive bacteria are capable of inducing septic symptoms in patients. TLR-2, found to be expressed on activated ECs, mainly mediates intracellular responses to wall components of gram-positive bacteria (exotoxins, lipoteichoic acids), although conclusive evidence for TLR2-mediated signaling in the endothelium is lacking.

Activation of Complement Pathways

During sepsis and multiorgan dysfunction syndrome (MODS), plasma levels of the complement activation product C5a increase. C5a induces its inflammatory functions by interacting with specific receptors (C5aR) in the rhodopsin family of seven-transmembrane spanning, G-protein–coupled receptors (42). So far, the C5aR has been found on hemopoietic cells as well as in a variety of organs (lung, liver, kidney, brain). C5a was shown to induce rapid surface expression of P-selectin on surfaces of human umbilical vein ECs (HUVECs) (43). Recently, the ability of C5a to bind to ECs has been demonstrated (44). Furthermore, we were able to show that exposure of mouse dermal microvascular ECs (MDMECs) to LPS, IL-6, and interferon (IFN)-γ resulted in increased C5aR, and incubation with C5a and IL-6 led to increased levels of proinflammatory mediators (44). These data indicate that C5aR can be upregulated on ECs and that C5a in the copresence of additional agonists may mediate proinflammatory effects of the endothelium.

Activation of Inflammation

One of the main underlying mechanisms leading to a state of immune system hyperactivation involves the production and release of pro- and anti-inflammatory mediators (chemokines and cytokines). Chemokines are involved in attracting neutrophils to the site of inflammation, activating leukocytes after the initial rolling, and facilitating transendothelial migration of leukocytes. Endothelial and epithelial cells, as well as neutrophils, macrophages, and lymphocytes, produce large amounts of proinflammatory cytokines (e.g., TNF-α, IL-6, IL-1β, and IL-8).

TNF-α, which appears to be mainly synthesized in activated macrophages/monocytes, is a very early proinflammatory cytokine produced during SIRS/sepsis. Binding to its receptor leads to cell activation via the nuclear factor (NF)-κB transcription factor. TNF-α stimulates neutrophils and ECs to release a variety of proinflammatory mediators (45). Further, TNF-α has been shown to increase EC permeability and possibly to induce the apoptosis of ECs (39). The IL-1 family (IL-1α, IL-1β) represents another important series of cytokines produced during sepsis. IL-1 and TNF-α act synergistically, leading to further initiation of the inflammatory cascade and upregulation of vascular adhesion molecules. To counteract excessive inflammatory/immune responses, increased levels of IL-4, IL-10, IL-13 and, to some extent, IFN-γ during sepsis may be potent anti-inflammatory mediators. However, their immediate effects on ECs remain to be investigated. The chemokines IL-8 and monocyte chemoattractant protein (MCP)-1, are additional proinflammatory components during sepsis (46). Released by ECs, IL-8 plays a key role in neutrophil chemotaxis and degranulation, whereas EC-derived MCP-1 is involved in attracting monocytes.

Taken together, the mechanisms detailed here, along with the release of bacterial products, production of strong proinflammatory mediators, and increased expression of vascular adhesion molecules, lead to a state of endothelial hyperactivation. Intracellular pathways become activated, more proinflammatory cytokines and chemokines are released, and neutrophils are attracted to the site of inflammation. Neutrophils release granular enzymes, ROS (such as H_2O_2), and NO, which are capable of causing increased vascular permeability, tissue damage and, ultimately, organ injury.

Activation of Coagulation

As described earlier, the coagulation system becomes activated under septic conditions. This leads to increased levels of thrombin. As a major activator of platelets and ECs, thrombin enhances the expression by ECs of proinflammatory cytokines and chemokines as well as adhesion molecules (47). This activation of the endothelium results in intravascular thrombosis and fibrin deposition, resulting in DIC, which is associated with increasing organ dysfunction and risk of death. Recent studies in experimental sepsis (cecal ligation and puncture) in rodents indicate that in vivo blockade of C5a greatly attenuates the activation and consumptive depletion of the clotting and fibrinolytic systems (48).

Recent studies have shown that naturally occurring anticoagulants are inactivated or decreased during sepsis. In vitro experiments have shown that the addition of LPS and cytokines to ECs results in decreased levels of the anticoagulants, thrombomodulin, and t-PA, while procoagulants tissue factor, plasminogen activator inhibitor (PAI)-1, and microparticles were increased (31,33,49–51). These changes have been shown to occur in vivo as well. However, to what extent ECs are either affected by or contribute to this switch to generate an anticoagulant versus a procoagulant state remains to be seen.

Since activated protein C (APC) became the first drug approved for treatment of patients with severe sepsis, much effort has been made to investigate the APC pathway and its interaction with ECs. APC possesses profibrinolytic, anti-inflammatory, and antiapoptotic properties, and it counteracts the effects of thrombin and proinflammatory cytokines (47). Thus, the protein C pathway is currently positioned at the interface of the endothelium and the leukocyte response. Although the APC receptor on ECs and other immune cells (endothelial protein C receptor, EPCR) has been identified (52), its potential role during sepsis remains controversial. Pediatric patients with severe sepsis showed reduced levels of EPCR in skin biopsies (53). Conversely, in a murine model of sepsis, *EPCR* mRNA expression was upregulated in the liver, kidney, and lung (Table 141-2) (54). Similar results were reported after LPS injection accompanied by an increase of soluble EPCR in rodent serum, suggesting a possible shedding of EPCR from the endothelium (55). Further, recent data indicate that the microvascular anti-inflammatory effect of APC is partially mediated through the suppression of EC adhesion molecules, resulting in protection from the inflammatory insult of sepsis (47).

Table 141-2: Adhesion Molecule Knockout Models and Their Impact on Inflammation

Knockout	*Phenotype (Refs.)*
Selectins	
P-selectin	Striking leukocytosis, diminished leukocyte rolling, delayed leukocyte recruitment to inflammatory sites (61)
E-selectin	No leukocytosis, but delayed PMN recruitment to inflammatory sites (62)
L-selectin	Impaired recruitment of PMN, lymphocytes and monocytes to peritoneal cavity, impaired delayed hypersensitivity reaction, resistance to LPS-induced shock (63)
Immunoglobulins	
ICAM-1	Moderate granulocytosis, impaired PMN emigration, reduced contact hypersensitivity (64)
VCAM	Not viable; death in utero due to cardiac and other developmental abnormalities (62)
Integrins	
CD18	Impaired inflammatory response, delayed rejection of cardiac grafts (65)
CD11a	Modest delay in leukocyte recruitment to peritoneal cavity (66)
CD11b	Enhanced PMN survival (67)
$\beta 1$	Role in inflammation not yet investigated (68)
$\beta 7$	Impaired host protection in enteric helminthes infection (69)
Human deficiencies	
LAD I	Deficiencies in CD11/18 family due to mutations, impaired phagocyte emigration, chronic neutrophilia, recurrent infections without pus, normal rolling of leukocytes, but no adherence (70)
LAD II	Endogenous fucose metabolism defect, impaired selectin expression, normal leukocyte adherence but defective rolling (71,72)

PMN, polymorph neutrophil granulocytes; LAD, leukocytes adhesion deficiency

THERAPEUTIC STRATEGIES IN INFLAMMATION AND SEPSIS

Despite the current aggressive management of patients with sepsis or septic shock in ICUs, mortality remains unacceptably high. It is therefore important that more effective interventions for the treatment of sepsis are found. Recent strategies include the neutralization of bacterial products, targeting undesirable proinflammatory responses, addition of immuno-stimulatory agents to restore the immune system, and correction of abnormalities in the coagulation system. However, clinical trials using many of these promising new targets during sepsis have mostly turned out to be "dead ends," especially for pharmaceutical companies. Although the blockade of LPS, TNF-α, and interleukins resulted in improved outcome in animal models of sepsis, clinical trials failed (56). Current efforts concentrate on promising new targets during sepsis (e.g., TLRs, complement factors [especially C5a], high mobility group B1 protein, macrophage migration inhibitory factor [MIF, MIP-1α], and apoptosis). The blockade of these new proinflammatory mediators and processes has resulted in improved outcome in animal models, but clinical trials remain to be conducted. Other strategies for clinical anti-inflammatory interventions of sepsis have included inhibition or antagonism of platelet-activating factor (PAF), arachidonic acid metabolites, adhesion molecules, ROS, NO, bradykinin, phosphodiesterase, and C1 esterase, but to no avail (56). Researchers have tried to counteract the state of immunosuppression that occurs during late-stage sepsis. To restore the proinflammatory functions of phagocytic cells, IFN-γ, granulocyte colony-stimulating factor (G-CSF), and granulocyte macrophage colony-stimulating factor (GM-CSF) were tested as possible treatments in sepsis. However, despite encouraging results in animals, clinical trials in humans failed to show any beneficial effects of these molecules.

So far, only three clinical trials resulted in improved outcome of septic patients: the administration of low-dose corticosteroid in septic patients with low levels of blood cortisol (57), administration of APC in patients with severe sepsis (58), and maintenance of physiological level of blood glucose through intensive insulin therapy (59).

These strategies target soluble mediators and inflammatory cells that modulate endothelial response. Another approach is to directly block EC adhesion molecules and signaling pathways. Blockade of selectins inhibited the transmigration of neutrophils in vitro, however, in vivo experiments failed to confirm these observations, mainly due to cross-reaction of several of the antibodies (14). In contrast, the blockade of integrins (CD11/CD18) as well as ICAM-1 has been shown to inhibit neutrophil influx in almost every system tested (60). At present, antiadhesive strategies during sepsis remain at an experimental level and must be further investigated.

KEY POINTS

- ECs are victims or perpetrators of adverse events during SIRS and sepsis.
- Inflammation activates ECs, resulting in increased leukocyte adhesion, decreased vasomotor tone and permeability, as well as assumption of a procoagulant state.
- Sepsis leads to uncontrolled, overwhelming hyperactivation of the immune system, rendering ECs dysfunctional.

Future Goals

- To further determine the exact role of ECs in interaction with leukocytes
- To explore the apparent link between the inflammatory and coagulation systems
- To develop new therapeutic strategies targeting cells of the immune system and ECs

ACKNOWLEDGMENT

This work is supported by National Institute of Health (NIH) grants GM-29507, HL-31963, and GM-61656.

REFERENCES

1 Martin GS, Mannino DM, Eaton S, Moss M. The epidemiology of sepsis in the United States from 1979 through 2000. *N Engl J Med*. 2003;348(16):1546–1554.

2 Angus DC, Linde-Zwirble WT, Lidicker J, et al. Epidemiology of severe sepsis in the United States: analysis of incidence, outcome, and associated costs of care. *Crit Care Med*. 2001;29(7):1303–1310.

3 Hack CE, Zeerleder S. The endothelium in sepsis: source of and a target for inflammation. *Crit Care Med*. 2001;29(7 Suppl):S21–S27.

4 Bone RC, Fisher CJ Jr., Clemmer TP, et al. Sepsis syndrome: a valid clinical entity. Methylprednisolone Severe Sepsis Study Group. *Crit Care Med*. 1989;17(5):389–393.

5 Bone RC, Balk RA, Cerra FB, et al. Definitions for sepsis and organ failure and guidelines for the use of innovative therapies in sepsis. The ACCP/SCCM Consensus Conference Committee. American College of Chest Physicians/Society of Critical Care Medicine. *Chest*. 1992;101(6):1644–1655.

6 Levy MM, Fink MP, Marshall JC, et al. 2001 SCCM/ESICM/ACCP/ATS/SIS International Sepsis Definitions Conference. *Intensive Care Med*. 2003;29(4):530–538.

7 Abraham E, Wunderink R, Silverman H, et al. Efficacy and safety of monoclonal antibody to human tumor necrosis factor alpha in patients with sepsis syndrome. A randomized, controlled, double-blind, multicenter clinical trial. TNF-alpha MAb Sepsis Study Group. *JAMA*. 1995;273(12):934–941.

8 Wang H, Bloom O, Zhang M, et al. HMG-1 as a late mediator of endotoxin lethality in mice. *Science*. 1999;285(5425):248–251.

9 Bone RC. Sir Isaac Newton, sepsis, SIRS, and CARS. *Crit Care Med*. 1996;24(7):1125–1128.

10 Hotchkiss RS, Karl IE. The pathophysiology and treatment of sepsis. *N Engl J Med*. 2003;348(2):138–150.

11 Rees DD, Palmer RM, Moncada S. Role of endothelium-derived nitric oxide in the regulation of blood pressure. *Proc Natl Acad Sci USA*. 1989;86(9):3375–3378.

12 Cohnheim J. *Lectures in General Pathology* (Translated by McKee AD from the second German edition). Vol. 1: New Sydenham Society, 1889.

13 Clark ER, Clark EL. Observations on changes in blood vascular endothelium in the living animal. *Am J Anat*. 1935;57:385–438.

14 Albelda SM, Smith CW, Ward PA. Adhesion molecules and inflammatory injury. *FASEB J*. 1994;8(8):504–512.

15 McEver RP, Beckstead JH, Moore KL, et al. GMP-140, a platelet alpha-granule membrane protein, is also synthesized by vascular endothelial cells and is localized in Weibel-Palade bodies. *J Clin Invest*. 1989;84(1):92–99.

16 Mulligan MS, Schmid E, Till GO, et al. C5a-dependent up-regulation in vivo of lung vascular P-selectin. *J Immunol*. 1997; 158(4):1857–1861.

17 Bless NM, Tojo SJ, Kawarai H, et al. Differing patterns of P-selectin expression in lung injury. *Am J Pathol*. 1998;153(4):1113–1122.

18 Asimakopoulos G, Taylor KM. Effects of cardiopulmonary bypass on leukocyte and endothelial adhesion molecules. *Ann Thorac Surg*. 1998;66(6):2135–2144.

19 Muller WA, Weigl SA, Deng X, Phillips DM. PECAM-1 is required for transendothelial migration of leukocytes. *J Exp Med*. 1993;178(2):449–460.

20 Vaporciyan AA, DeLisser HM, Yan HC, et al. Involvement of platelet-endothelial cell adhesion molecule-1 in neutrophil recruitment in vivo. *Science*. 1993;262(5139):1580–1582.

21 Marchesi VT, Gowans JL. The migration of lymphocytes through the endothelium of venules in lymph nodes: an electron microscope study. *Proc R Soc Lond B Biol Sci*. 1964;159:283–290.

22 Engelhardt B, Wolburg H. Mini-review: transendothelial migration of leukocytes: through the front door or around the side of the house? *Eur J Immunol*. 2004;34(11):2955–2963.

23 von Andrian UH, Chambers JD, McEvoy LM, et al. Two-step model of leukocyte-endothelial cell interaction in inflammation: distinct roles for LECAM-1 and the leukocyte beta 2 integrins in vivo. *Proc Natl Acad Sci USA*. 1991;88(17):7538–7542.

24 Simon SI, Chambers JD, Sklar LA. Flow cytometric analysis and modeling of cell-cell adhesive interactions: the neutrophil as a model. *J Cell Biol*. 1990;111(6 Pt 1):2747–2756.

25 Esmon CT. Thrombomodulin as a model of molecular mechanisms that modulate protease specificity and function at the vessel surface. *FASEB J*. 1995;9(10):946–55.

26 Iversen N, Sandset PM, Abildgaard U, Torjesen PA. Binding of tissue factor pathway inhibitor to cultured endothelial cells-influence of glycosaminoglycans. *Thromb Res*. 1996;84(4):267–278.

27 Mertens G, Cassiman JJ, Van den Berghe H, et al. Cell surface heparan sulfate proteoglycans from human vascular endothelial cells. Core protein characterization and antithrombin III binding properties. *J Biol Chem*. 1992;267(28):20435–20443.

28 Weksler BB, Marcus AJ, Jaffe EA. Synthesis of prostaglandin I2 (prostacyclin) by cultured human and bovine endothelial cells. *Proc Natl Acad Sci USA*. 1977;74(9):3922–3926.

29 Ihrcke NS, Wrenshall LE, Lindman BJ, Platt JL. Role of heparan sulfate in immune system-blood vessel interactions. *Immunol Today*. 1993;14(10):500–505.

30 Moore KL, Esmon CT, Esmon NL. Tumor necrosis factor leads to the internalization and degradation of thrombomodulin from the surface of bovine aortic endothelial cells in culture. *Blood*. 1989;73(1):159–165.

31 Moore KL, Andreoli SP, Esmon NL, et al. Endotoxin enhances tissue factor and suppresses thrombomodulin expression of human vascular endothelium in vitro. *J Clin Invest*. 1987;79(1):124–130.

32 Preissner KT, Nawroth PP, Kanse SM. Vascular protease receptors: integrating haemostasis and endothelial cell functions. *J Pathol.* 2000;190(3):360–372.

33 Combes V, Simon AC, Grau GE, et al. In vitro generation of endothelial microparticles and possible prothrombotic activity in patients with lupus anticoagulant. *J Clin Invest.* 1999;104(1):93–102.

34 Furchgott RF, Cherry PD, Zawadzki JV, Jothianandan D. Endothelial cells as mediators of vasodilation of arteries. *J Cardiovasc Pharmacol.* 1984;6(Suppl 2):S336–S343.

35 Yanagisawa M, Kurihara H, Kimura S, et al. A novel peptide vasoconstrictor, endothelin, is produced by vascular endothelium and modulates smooth muscle Ca^{2+} channels. *J Hypertens Suppl.* 1988;6(4):S188–S191.

36 Marsden PA, Schappert KT, Chen HS, et al. Molecular cloning and characterization of human endothelial nitric oxide synthase. *FEBS Lett.* 1992;307(3):287–293.

37 Gross SS, Kilbourn RG, Griffith OW. NO in septic shock: good, bad or ugly? Learning from iNOS knockouts. *Trends Microbiol.* 1996;4(2):47–49.

38 Wanecek M, Weitzberg E, Rudehill A, Oldner A. The endothelin system in septic and endotoxin shock. *Eur J Pharmacol.* 2000; 407(1–2):1–15.

39 Aird WC. The role of the endothelium in severe sepsis and multiple organ dysfunction syndrome. *Blood.* 2003;101(10):3765–3777.

40 Peters K, Unger RE, Brunner J, Kirkpatrick CJ. Molecular basis of endothelial dysfunction in sepsis. *Cardiovasc Res.* 2003;60(1):49–57.

41 El-Samalouti VT, Schletter J, Brade H, et al. Detection of lipopolysaccharide (LPS)-binding membrane proteins by immuno-coprecipitation with LPS and anti-LPS antibodies. *Eur J Biochem.* 1997;250(2):418–424.

42 Gerard NP, Gerard C. The chemotactic receptor for human C5a anaphylatoxin. *Nature.* 1991;349(6310):614–617.

43 Foreman KE, Vaporciyan AA, Bonish BK, et al. C5a-induced expression of P-selectin in endothelial cells. *J Clin Invest.* 1994;94(3):1147–1155.

44 Laudes IJ, Chu JC, Huber-Lang M, et al. Expression and function of C5a receptor in mouse microvascular endothelial cells. *J Immunol.* 2002;169(10):5962–5970.

45 Varani J, Ward PA. Mechanisms of endothelial cell injury in acute inflammation. *Shock.* 1994;2(5):311–319.

46 Dinarello CA. Proinflammatory cytokines. *Chest.* 2000;118(2):503–508.

47 Joyce DE, Nelson DR, Grinnell BW. Leukocyte and endothelial cell interactions in sepsis: relevance of the protein C pathway. *Crit Care Med.* 2004;32(Suppl 5):S280–S286.

48 Laudes IJ, Chu JC, Sikranth S, et al. Anti-c5a ameliorates coagulation/fibrinolytic protein changes in a rat model of sepsis. *Am J Pathol.* 2002;160(5):1867–1875.

49 Bombeli T, Mueller M, Haeberli A. Anticoagulant properties of the vascular endothelium. *Thromb Haemost.* 1997;77(3):408–423.

50 Bevilacqua MP, Pober JS, Majeau GR, et al. Recombinant tumor necrosis factor induces procoagulant activity in cultured human vascular endothelium: characterization and comparison with the actions of interleukin 1. *Proc Natl Acad Sci USA.* 1986;83(12):4533–4537.

51 Schleef RR, Bevilacqua MP, Sawdey M, et al. Cytokine activation of vascular endothelium. Effects on tissue-type plasminogen activator and type 1 plasminogen activator inhibitor. *J Biol Chem.* 1988;263(12):5797–5803.

52 Macias WL, Yan SB, Williams MD, et al. New insights into the protein C pathway: potential implications for the biological activities of drotrecogin alfa (activated). *Crit Care.* 2005;9(Suppl 4):S38–S45.

53 Faust SN, Levin M, Harrison OB, et al. Dysfunction of endothelial protein C activation in severe meningococcal sepsis. *N Engl J Med.* 2001;345(6):408–416.

54 Ganopolsky JG, Castellino FJ. A protein C deficiency exacerbates inflammatory and hypotensive responses in mice during polymicrobial sepsis in a cecal ligation and puncture model. *Am J Pathol.* 2004;165(4):1433–1446.

55 Gu JM, Katsuura Y, Ferrell GL, Grammas P, Esmon CT. Endotoxin and thrombin elevate rodent endothelial cell protein C receptor mRNA levels and increase receptor shedding in vivo. *Blood.* 2000;95(5):1687–1693.

56 Riedemann NC, Guo RF, Ward PA. Novel strategies for the treatment of sepsis. *Nat Med.* 2003;9(5):517–524.

57 Briegel J, Forst H, Haller M, et al. Stress doses of hydrocortisone reverse hyperdynamic septic shock: a prospective, randomized, double-blind, single-center study. *Crit Care Med.* 1999;27(4):723–732.

58 Bernard GR, Vincent JL, Laterre PF, et al. Efficacy and safety of recombinant human activated protein C for severe sepsis. *N Engl J Med.* 2001;344(10):699–709.

59 van den Berghe G, Wouters P, Weekers F, et al. Intensive insulin therapy in the critically ill patients. *N Engl J Med.* 2001;345(19):1359–1367.

60 Smith CW. Endothelial adhesion molecules and their role in inflammation. *Can J Physiol Pharmacol.* 1993;71(1):76–87.

61 Mayadas TN, Johnson RC, Rayburn H, et al. Leukocyte rolling and extravasation are severely compromised in P selectin-deficient mice. *Cell.* 1993;74(3):541–554.

62 Wolitzky B, Kwee L, Terry R, et al. Targeted disruption of the murine E-selectin and VCAM-1 genes. *J Cell Biochem.* 1994;(Suppl 18A):300.

63 Tedder TF, Steeber DA, Pizcueta P. L-selectin-deficient mice have impaired leukocyte recruitment into inflammatory sites. *J Exp Med.* 1995;181(6):2259–2264.

64 Sligh JE Jr., Ballantyne CM, Rich SS, et al. Inflammatory and immune responses are impaired in mice deficient in intercellular adhesion molecule 1. *Proc Natl Acad Sci USA.* 1993;90(18):8529–8533.

65 Wilson RW, Ballantyne CM, Smith CW, et al. Gene targeting yields a CD18-mutant mouse for study of inflammation. *J Immunol.* 1993;151(3):1571–1578.

66 Schmits R, Kundig TM, Baker DM, et al. LFA-1-deficient mice show normal CTL responses to virus but fail to reject immunogenic tumor. *J Exp Med.* 1996;183(4):1415–1426.

67 Lu H, Smith CW, Perrard J, et al. LFA-1 is sufficient in mediating neutrophil emigration in Mac-1-deficient mice. *J Clin Invest.* 1997;99(6):1340–1350.

68 Mercurio AM. Lessons from the alpha2 integrin knockout mouse. *Am J Pathol.* 2002;161(1):3–6.

69 Artis D, Humphreys NE, Potten CS, et al. Beta7 integrin-deficient mice: delayed leukocyte recruitment and attenuated protective immunity in the small intestine during enteric helminth infection. *Eur J Immunol.* 2000;30(6):1656–1664.

70 Anderson DC, Schmalsteig FC, Finegold MJ, et al. The severe and moderate phenotypes of heritable Mac-1, LFA-1 deficiency: their quantitative definition and relation to leukocyte dysfunction and clinical features. *J Infect Dis*. 1985;152(4):668–689.

71 Etzioni A, Harlan JM, Pollack S, et al. Leukocyte adhesion deficiency (LAD) II: a new adhesion defect due to absence of sialyl lewis x, the ligand for selectins. *Immunodeficiency*. 1993;4(1–4):307–308.

72 von Andrian UH, Berger EM, Ramezani L, et al. In vivo behavior of neutrophils from two patients with distinct inherited leukocyte adhesion deficiency syndromes. *J Clin Invest*. 1993;91(6):2893–2897.

The Endothelium in Cerebral Malaria

Both a Target Cell and a Major Player

Valéry Combes*, Jin Ning Lou†, and Georges E. Grau*

*Bosch Institute, University of Sydney, Australia; †Institute of Clinical Medical Sciences, China-Japan Friendship Hospital, Beijing, China

On a global scale, malaria still represents a major problem of public health, principally because of its two major complications, cerebral malaria (CM) and severe malarial anemia. The disease manifestations apparent during that stage of the malaria parasite lifecycle when it is circulating within host erythrocytes are a result of complex host–parasite interactions. In man, most of the pathology is due to *Plasmodium falciparum*. In the case of CM, the microcirculation, and in particular the endothelium, appears to be pivotal in the interplay between infected erythrocytes (IEs), host immune cells, and host tissues.

We focus here on the cerebral syndrome, but the endothelium also has a central role in other aspects of malarial pathology; these have been reviewed recently elsewhere (1). For example, severe malaria may be complicated by pulmonary edema, the cause of which is still unknown. It shares many features with the adult acute respiratory distress syndrome (ARDS). Rehydration may predispose patients with severe malaria to develop pulmonary edema, although it is by no means a prerequisite. Indeed, evidence suggests that an increase in the lung capillary permeability – as distinct from simple intravascular volume overload – may be at the origin of this complication. Acute or chronic renal insufficiency is another common complication of severe malaria, and is often lethal. It occurs in adults and older children. The mechanisms leading to the acute tubular necrosis are not fully understood. A cytoadherence of IEs in glomerular capillaries sometimes can be observed, but is never as prominent as in the brain. Renal pathology in malaria often presents as a more chronic pathology, most notably with glomerulonephritis associated with deposits of immune complexes (2–5).

The histopathology of CM is dominated by *sequestration*; that is, the accumulation of mature IEs in the cerebral microvessels of patients. The description of this feature by Marchiafava and Bignami goes back at least to 1894 (6). Light and electron microscopy imaging have demonstrated electron-dense structures on the IE surface, called *knobs* (7,8), which were later shown to mediate the binding of IEs to endothelial cell (EC) membranes (9–11). Sequestration is observed in almost all organs, but particularly in the brain. However, several questions remain unanswered as to the exact involvement of IE sequestration in CM: Is sequestration the cause or the consequence of endothelial lesions? In addition to IEs, histopathological analysis of brain samples from patients who recently died of CM revealed the presence of other cell types, such as platelets and monocytes (2).

Cerebral capillaries and venules distended with IEs are the microscopic hallmark of severe falciparum malaria (3,12,13). In contrast to the peripheral blood, all mature stages of the parasite are seen within these vessels, both in adults (14–16), and in African children (17). Distended venules are more prominent in the gray matter, where they appear evenly distributed. Edema, hemorrhages, and necrosis are features of the endothelial alteration in CM.

Hemorrhages, described as "punctiform" or "ring hemorrhages," are a common pathological feature of CM and are distributed throughout the brain (including the retina and brainstem). They are more common in children who have had convulsions associated with CM than in other children (18). These ring hemorrhages contain a blocked central capillary, with an agglutinated mass of IEs surrounded by brain tissue that is necrotic and contains demyelinated fibers (14,19) or a glial reaction. It has been suggested that ring hemorrhages are a specific response to vessel blockage by IEs and reperfusion, resulting in a concentric pattern of hemorrhage surrounding a necrosed vessel (20); they are, therefore, different from the petechial hemorrhages seen after a variety of hypoxic insults, as in, for example, typhus fever. Small malarial granulomata, also called *Dürck nodules*, are a distinctive pathological feature of malaria (3). They are not found in patients who die shortly after the onset of symptoms, and probably represent a more advanced stage of repair following hemorrhage in which necrotic tissue has been replaced by neuroglial cells and microglial tissue (3,14,21). Among numerous pathological features, the retinopathy appears to be the only clinical sign distinguishing malarial from nonmalarial coma (22).

All patients with *P. falciparum* infection demonstrate IE sequestration to some extent, including immune individuals with asymptomatic parasitemia and nonimmune patients. For this reason, the mere presence of parasites in the peripheral blood or tissues is not sufficient to identify malaria as a cause of organ dysfunction or pathology. Only autopsies reveal nonmalarial causes of death in parasitemic individuals and allow the identification of features specific to malaria (22). Although postmortem analyses provide a gold standard in rigorous diagnosis, they only allow the study of the pathology of severe malaria at the terminal stage, so that informative changes early in the disease process are inaccessible (1).

REACTIVITY OF ENDOTHELIUM AS A CENTRAL DETERMINANT OF MALARIAL PATHOLOGY

Mediators Affecting the Endothelium and Produced by It

Fatal falciparum malaria is accompanied by a systemic endothelial activation (23) and a widespread induction of endothelial activation markers. In addition to the "mechanical theory" that explains CM by the IE plugging of microvessels, the "cytokine theory" suggests that an overproduction of inflammatory cytokines by host cells, in response to the presence of IEs, is responsible for the development of the neurological syndrome. A combination of both theories is more likely to be operational (24–26). In any case, cytokines, chemokines, and other mediators acting on or released by endothelium are of major importance in the pathogenesis of CM.

Cytokines and Chemokines

After the early insights of Maegraith (27,28), suggesting that host mediators could mediate the pathogenic events triggered by the parasite, several lines of evidence have suggested that tumor necrosis factor (TNF) is an important element in the pathogenesis of experimental CM (29–31). High circulating levels of TNF can be deleterious, because TNF may increase sequestration through the upregulation of host ligand molecules responsible for the cytoadherence of parasitized erythrocytes, especially intercellular adhesion molecule (ICAM)-1, vascular cell adhesion molecule (VCAM)-1, and E-selectin, thus increasing mechanical obstruction of cerebral or other blood vessels, as evidenced in man (3,32) and in the experimental mouse model for CM (33,34). Following data in experimental CM (35,36), the demonstration of high plasma TNF levels in patients with severe falciparum malaria (37–40) provided a rationale for clinical trials using agents that would prevent the release of TNF or interfere with its actions. However, attempts to prevent CM by administration of anti-TNF antibody did not reduce mortality in humans, partly because the timeframe of intervention was too narrow (41).

It is clear that high plasma TNF levels cannot be the only element responsible for the symptoms unique to cerebral and lethal *P. falciparum* malaria. Lymphotoxin (LT), a TNF-related cytokine (previously called TNF-β), might play an important role in the pathogenesis of CM, as suggested by the observation that mice null for the *LT-α* gene are protected against this complication (42). However, the possible involvement of LT-α/β in human CM still must be confirmed, although elevated circulating levels of LT-α have been reported in malaria patients (43).

Pathology in CM likely depends on an imbalance between TNF, LT-α, and other proinflammatory cytokines, on the one hand, and anti-inflammatory cytokines on the other hand (44). Among the important "anti-inflammatory" cytokines are interleukin (IL)-10 (45,46), and transforming growth factor (TGF)-β (47,48). IL-10 plays a host protective role in murine malaria (46). Treatment of CM-resistant (CM-R) mice with anti-IL-10 monoclonal antibody (mAb), in vivo, switches their phenotype toward susceptibility to CM (46). In man, correlative studies of plasma IL-10 levels and severe malaria manifestations have shown associations generally consistent with the mouse studies. Serum levels of IL-10 are elevated in acute falciparum malaria patients in Thailand prior to antimalarial therapy and decrease after treatment (49–51). It was concluded that elevated plasma cytokine levels in severe malaria are associated with systemic pathologic changes, rather than cerebral involvement, consistent with the concept of TNF proposed in mouse studies (52). In African children with malaria, plasma IL-10 concentration is elevated, including in CM patients (53). The lowest concentrations are found in severe anemia patients, and investigators suggested that the relative deficiency in IL-10 might play a role in this complication of malaria infection. IL-10 can be released by a wide range of immune and inflammatory cells. In Gabonese patients with uncomplicated *P. falciparum* infection, an increase was noted in circulating $CD4^+$ T lymphocytes expressing IL-10 (54).

As recently reviewed (55), IL-8 and macrophage inflammatory protein (MIP)-1α concentrations correlate with parasitemia and disease severity. MIP-1α is a potent inhibitor of hematopoietic cell proliferation and might be one of the agents responsible for prolonged anemia in malaria (56,57). These two molecules in the brain could be involved in the recruitment of leukocytes (neutrophils and monocytes, respectively) at the inflamed site. Similarly, a role might be postulated for the CXC chemokine GRO-α in CM; GRO-α is known to target ECs, both GRO-α and its receptor CXCR-2 are expressed by microglia, and GRO-α signaling has been implicated in the pathogenesis of brain lesions in multiple sclerosis (58). Chemokines also may be involved via their complex interactions with cytokines. In particular, IL-12 could influence pathogenesis in an unexpected way, through interaction between leukocytes ($CD4^+$, $CD8^+$, natural killer [NK]) and ECs, via factors such as interferon-inducible protein (IP)-10 and various chemokines induced by interferon (IFN)-γ (59). Also, IL-12 could upregulate cell adhesion molecules (CAMs) on ECs, arrest their cycle, and inhibit angiogenesis (59). This IL-12–induced upregulation of CAMs could take part in the homing and sequestration of leukocytes within the microvasculature.

Soluble Adhesion Molecules

During the course of CM, elevated levels of soluble ICAM-1, E-selectin and VCAM-1 were found, pointing to the existence of widespread endothelial activation (23,60–62). Moreover, elevated concentrations of soluble thrombomodulin also were found in association with the severity of the disease, indicating a substantial endothelial injury that could be of prognostic importance (61).

Endothelial Microparticles

Microparticle release is a hallmark of cell activation and apoptosis. Microparticles can originate from almost all cell types, although those released by circulating blood cells (platelets and monocytes) and ECs have been particularly well studied. The release of elevated levels of endothelial microparticles has recently been reported to occur during CM (63). These sub–micron-sized elements were found in large amounts in the plasma of Malawian children during the acute phase of CM (63). Markedly elevated levels of microparticles were specific to the neurological complication, because they did not occur in uncomplicated malaria or severe anemia. This phenomenon is another indicator of the considerable endothelial activation occurring during CM. The high plasma TNF concentrations noted in CM patients suggest that TNF plays a role in the enhancement of the endothelial microparticle release in vivo, as it does in vitro (64). In vitro, endothelial microparticles produced after TNF activation express procoagulant and proinflammatory properties, in a way similar to that of the cell from which they originate. We can thus hypothesize that, if a similar phenomenon occurs in vivo, microparticles could have a pathogenic role in the development of the cerebral syndrome and participate in the worsening of the endothelial lesion by locally activating coagulation and potentiating leukocyte adhesion. It also could be postulated that microparticles derived from brain endothelium have distant effects on other organs.

The diverse aspects of the role of brain endothelium as an input/output device – namely as a target and a player in CM pathogenesis – are summarized in Figure 142.1.

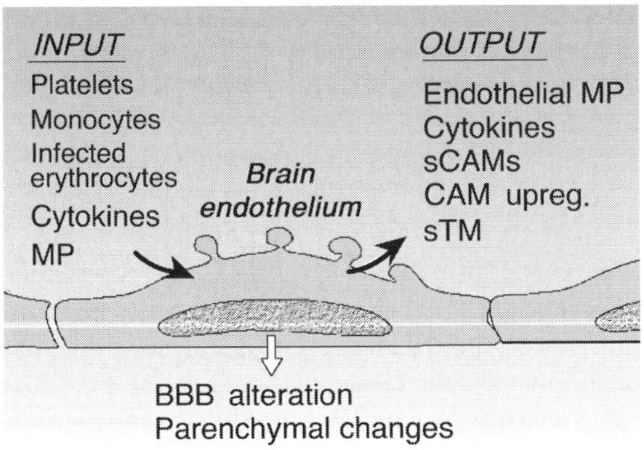

Figure 142.1. Diagrammatic representation of the involvement of brain ECs as an input/output device in the pathogenesis of cerebral malaria. MP, microparticles; sCAMs, soluble cell adhesion molecules; sTM, soluble thrombomodulin; CAM upreg., upregulation of the surface expression of cell adhesion molecules; BBB, blood–brain barrier.

Evidence for an Association between Endothelial Reactivity and Susceptibility to Disease

Because CM only occurs in a small proportion of falciparum-infected patients, and because there are no clear differences between parasites isolated from patients with neurological involvement and those isolated from uncomplicated cases, the role of the host response must be a critical determinant of CM pathophysiology. Both human and experimental murine CM are the consequence of a cascade of events involving the production of toxins by the parasite (29–31) and cytokines by the host, eventually leading to the amplification of the expression of the receptors for cytoadherence, especially ICAM-1/VCAM-1 by TNF or CD36 by IFN-γ, on brain microvascular ECs (MVECs) (1,65). This topic is discussed in detail in the section Importance of the Endothelial Compartment. Several parameters have been shown to differ in mice that are either susceptible or resistant to the neurological syndrome: Compared to their resistant counterparts, susceptible animals show higher serum levels of TNF (35), enhanced IFN-γ production capacity in response to malaria antigens, and preferential Th1 cell subset expansion in vivo (44).

Because the brain microvascular endothelium is the eventual target of proinflammatory cytokines, we postulated that the variations of its responsiveness to cytokines in different individuals could modulate this sequence of events. To address this question, we purified brain MVECs from CM-susceptible (CM-S) CBA/J mice and CM-resistant (CM-R) BALB/c mice. After showing that they equally display relevant morphological and phenotypic features of microvascular endothelium, we demonstrated that brain MVECs purified from CM-S and CM-R mice exhibit a different sensitivity to TNF. CM-S brain MVECs displayed a higher capacity to produce IL-6 and to upregulate ICAM-1 and VCAM-1 in response to TNF than did CM-R brain MVECs. In contrast, no difference was found in the induction of E-selectin after TNF challenge. CM-S brain MVECs were also significantly more sensitive to TNF-induced lysis (66). This differential reactivity to TNF was further substantiated by comparing TNF receptor expression on CM-S and CM-R brain MVECs. Although the constitutive expression of TNF receptors was comparable on cells from the two origins, TNF induced an upregulation of both p55 and p75 TNF receptors in CM-S, but not in CM-R brain MVECs. A similar regulation was found at the level of TNF receptor mRNA, but not for receptor shedding. Although a protein kinase C (PKC) inhibitor blocked the response to TNF in both CM-S and CM-R brain MVECs, an inhibitor of protein kinase A (PKA) selectively abolished the response to TNF in CM-R, but not CM-S brain MVECs, suggesting a differential protein kinase involvement in TNF-induced activation of CM-S and CM-R brain MVECs (66). These results indicate that brain MVECs purified from CM-S and CM-R mice exhibit strain-specific sensitivity to TNF, as measured by various parameters. This difference may be partly due to a differential regulation of TNF receptors and downstream signaling pathways. Moreover, a differential induction of MHC molecules on CM-S versus CM-R brain MVECs has been demonstrated (67). Altogether, these data suggest that, in addition to host T cells or monocytes, the target cell, namely the EC, also is an element accounting for the susceptibility to the neurological syndrome.

PATHOGENIC MECHANISMS LEADING TO CEREBRAL SYNDROME

Importance of the Endothelial Compartment

One of the reasons the endothelium is so important in CM is that it expresses cytoadherence receptors for IEs. This cytoadherence is mediated by a series of endothelial molecules including constitutive as well as inducible surface antigens, such as platelet-endothelial cell adhesion molecule (PECAM)-1, CD36, chondroitin sulfate A (CSA), ICAM-1, thrombospondin (TSP), $\alpha v\beta 3$ or E-selectin, P-selectin, VCAM-1, respectively (68). However, the fact that the expression of some of these antigens is not restricted to ECs, nor to brain ECs (versus other ECs), makes it difficult to decide whether they play preferential role in brain circulation and to determine the specific role of the endothelial compartment.

For example, ICAM-1 is overexpressed by both endothelial and circulating cells (33); this is the reason why experiments have been conducted in the PbA model (described in details in Refs. 2,26,29,31,32) to clarify the role of this molecule in disease pathogenesis. Whereas *ICAM-1*$^{+/+}$ mice died 6 to 8 days after infection, *ICAM-1*$^{-/-}$ mice survived 15 days. A breakdown of the blood–brain barrier was evident in infected *ICAM-1*$^{+/+}$ mice only. Interestingly, thrombocytopenia was profound in infected *ICAM-1*$^{+/+}$, but practically absent in *ICAM-1*$^{-/-}$ mice. Moreover, macrophage sequestration was evident in brain venules and lung capillaries of *ICAM-1*$^{+/+}$ mice, and was significantly less prevalent in the alveolar capillaries of infected *ICAM-1*$^{-/-}$ mice. These results indicate that, although the immune response is similar in both *ICAM-1*$^{+/+}$ and *ICAM-1*$^{-/-}$ mice, the absence of mortality in *ICAM-1*$^{-/-}$ mice correlates with a decrease of macrophage and IE trapping and a less severe thrombocytopenia. However, no information is available on the specific role of ICAM-1 of endothelial origin (69).

Nonetheless, for other molecules expressed on both endothelium and circulating cells, it has been possible to decipher the importance of the endothelial compartment using chimeric mice, generated by bone marrow transplantations, among gene knockout and wild-type mice. This has been the case for P-selectin and TNF receptor (TNFR)-2. P-selectin is a molecule expressed on activated ECs and platelets. P-selectin–deficient mice infected with PbA had a cumulative incidence of CM that was significantly reduced compared with wild-type animals, despite identical levels of parasitemia, and reduced cerebral accumulation of platelets and leukocytes. Mice deficient only in endothelial P-selectin (i.e., *P-selectin*–null mice transplanted with wild-type bone marrow) did not show any sign of CM (vascular plugging, hemorrhages, or edema), whereas mice lacking only platelet P-selectin (i.e., wild-type mice transplanted with P-selectin–deficient bone marrow) showed signs of CM similar to that seen in wild-type mice. These results suggest that endothelial P-selectin plays an important role in the pathogenesis of CM (70).

Similarly, the respective role of endothelial versus leukocyte TNFR2 has been determined using bone-marrow chimaeras. A pathogenic role of TNFR2 in CM had been demonstrated using gene knockout mice: *TNFR2*$^{-/-}$, but not *TNFR1*$^{-/-}$ mice are protected (71). This brings together the involvement of TNF and LT-α, because both cytokines bind to this receptor (31). Studies in bone marrow chimeric mice suggested that endothelial TNFR2, rather than the receptor on intravascular monocytes, is important in the pathogenesis of CM (72). Interestingly, an overexpression of TNFR2 is also evident on the cerebral microvascular endothelium in human CM in Malawian children (31).

Interactions between Endothelium and Host Cells

Red Blood Cells

Upon binding to endothelium, IEs are capable of inducing a variety of changes. Using pulmonary MVECs, it has been shown that IE adhesion upregulated several TNF-superfamily genes and apoptosis-related genes such as Bad, Bax, caspase-3, secreted apoptosis-related protein (SARP)-2, DFF45/ICAD, IFN-γ receptor2, Bcl-w, Bik, and inducible nitric oxide synthase (iNOS). Also, IEs, but not normal erythrocytes, can trigger apoptosis in pulmonary MVECs (73). A direct contact between IEs and MVECs is required, because parasite-secreted molecules did not have any effect. Along the same line, IEs induce nitric oxide (NO) release via the engagement of CD23 at the surface of MVECs (74).

Leukocytes

In murine CM, monocytes have been observed in close juxtaposition with damaged ECs (75). Although monocytes are seen in cerebral vessels in CM patients, their role remains to be established. Recently, it has been suggested that CD8^{+} T cells may be associated with endothelial disruption in experimental CM (76), whereas the CD4^{+} subset might be responsible for the initiation phase of the cerebral syndrome (2,77).

Platelets

Together with classic receptor–ligand interactions, activated platelets or fibrin also may contribute to the sequestration of IEs. In larger cerebral vessels, parasites are seen that are sequestered but not attached to the vessel wall.

Autoagglutinates (clumps of mature-stage IEs formed in the presence of nonimmune serum) have been associated with severe malaria in children (78). Autoagglutination is mediated by CD36 on platelets, although not all IEs that bind CD36 form clumps with platelets (79). Clumping in vitro with platelets may be a corollary of an in vivo role for platelets promoting interactions between IEs, ECs, and leukocytes, as suggested by studies in murine (34) and, more recently, in human (80) malaria.

In recent studies using cocultures of CD36-specific IEs (IE^{CD36}), CD36-deficient brain ECs, and platelets (81), the latter cells were shown to mediate IE binding, especially if the brain ECs had been stimulated by a proinflammatory cytokine, suggesting a novel role for platelets in IE sequestration. Although platelet bridging does not replace the direct cytoadhesion of IEs to endothelial CD36, it may represent an alternative mechanism to allow IE^{CD36} to adhere to ECs not expressing CD36, increasing sequestration in places where CD36 expression is low, such as cerebral microvessels (82,83). CD36 adhesion may be involved in the sequestration of IEs in CM by this mechanism, and these results could explain the apparent contradiction between the scanty CD36 expression in microvessels and the predominance of the CD36-binding phenotype in field isolates.

A pathogenic role of platelets on ECs can be envisaged at several levels. Platelets can alter endothelial functions directly and/or indirectly via a modulation of leukocyte functions, or even effects on normal or parasitized erythrocytes (84). The relative importance of these functions in CM pathogenesis remains to be established, but it appears that platelets can modulate IE sequestration and have substantial effects on EC functions and viability (85).

Numerous other host cells are likely to play a role in CM pathogenesis via their interaction with brain endothelium. These include neutrophils, dendritic cells, NK, and NKT cells (2).

CONCLUSION

Some of the simplest, yet most fundamental, questions remain to be answered. First, we still do not know how coma develops. Numerous parameters on the parasite side remain to be determined, in terms of adhesion phenotype, or of capacity to stimulate host responses. The complex cascades of events in the host response are far from being elucidated. Even if one remains narrowly focused on brain endothelium, a wide array of issues remain to be addressed: We need to know how the endothelium differs between sites of intense sequestration and other sites, whether platelets accumulate before or after IEs, how the coagulation and fibrinolysis cascades participate in vascular plugging, how mononuclear leukocytes accumulate in the lumen, and what the consequences are for the surrounding parenchyma. Only a more thorough understanding of the fine mechanisms of CM will enable us to propose new therapeutic approaches.

KEY POINTS

- CM is one of the deadliest complications of *P. falciparum* infection.
- Endothelium is at the center of the neurological syndrome.
- EI and host cell sequestration in deep vascular beds (brain, lungs, kidneys) activates endothelium and leads to endothelial lesions.
- The endothelium is both a target cell and a major player in CM: It produces and is modified by proinflammatory molecules.
- Cytokines play a key role in the worsening of the endothelial activation.
- Endothelial microparticles could be both a circulating hallmark of endothelial activation and also act in the worsening of the lesion, together with microparticles released by other circulating cells.
- Experiments performed in the experimental CM model show that the genetic background is reflected by a differential reactivity of the endothelium to the malarial infection. This, among other parameters, may explain why only 1% of *P. falciparum*-infected patients will develop CM.
- IE sequestration is necessary but not sufficient to the development of severe malaria.

Future Goals

- To better understand the pathogenic elements that make an uncomplicated malaria evolve into a severe malaria
- To develop antidisease molecules, rather than antiparasite molecules, which would particularly target the endothelium
- To further define the pathogenic role of microparticles in human CM
- To understand by which parameters the endothelium might differ between sites of intense sequestration and other sites

REFERENCES

1 Rogerson SJ, Grau GE, Hunt NH. The microcirculation in severe malaria. *Microcirculation.* 2004;11(7):559–576.
2 Coltel N, Combes V, Hunt NH, Grau GE. Cerebral malaria – a neurovascular pathology with many riddles still to be solved. *Curr Neurovasc Res.* 2004;1:91–110.
3 Newton CR, Krishna S. Severe falciparum malaria in children: current understanding of pathophysiology and supportive treatment. *Pharmacol Ther.* 1998;79(1):1–53.
4 Taylor WR, White NJ. Malaria and the lung. *Clin Chest Med.* 2002;23(2):457–468.

5 Eiam-Ong S. Malarial nephropathy. *Semin Nephrol.* 2003;23(1):21–33.
6 Marchiafava E, Bignami E. Malaria. In: Stedman TL, ed. *Twentieth Century Practice of Medicine: An International Encyclopedia of Modern Medical Science.* New York: William Wood & Co.; 1900:227–252.
7 Oh SS, Chishti AH, Palek J, Liu SC. Erythrocyte membrane alterations in *Plasmodium falciparum* malaria sequestration. *Curr Opin Hematol.* 1997;4(2):148–154.
8 Sharma YD. Knob proteins in falciparum malaria. *Indian J Med Res.* 1997;106:53–62.
9 Pasloske BL, Howard RJ. Malaria, the red cell, and the endothelium. *Annu Rev Med.* 1994;45:283–295.
10 Hasler T, Handunnetti SM, Aguiar JC, et al. In vitro rosetting, cytoadherence, and microagglutination properties of *Plasmodium falciparum*-infected erythrocytes from Gambian and Tanzanian patients. *Blood.* 1990;76(9):1845–1852.
11 Aikawa M, Iseki M, Barnwell JW, et al. The pathology of human cerebral malaria. *Am J Trop Med Hyg.* 1990;43(2 Pt 2):30–37.
12 Newton CR, Hien TT, White N. Cerebral malaria. *J Neurol Neurosurg Psychiatr.* 2000;69(4):433–441.
13 Newton CR, Taylor TE, Whitten RO. Pathophysiology of fatal falciparum malaria in African children. *Am J Trop Med Hyg.* 1998;58(5):673–683.
14 Spitz S. The pathology of acute falciparum malaria. *Mil Surg.* 1946;99:555–572.
15 MacPherson GG, Warrell MJ, White NJ, et al. Human cerebral malaria. A quantitative ultrastructural analysis of parasitized erythrocyte sequestration. *Am J Pathol.* 1985;119(3):385–401.
16 Aikawa M. Human cerebral malaria. *Am J Trop Med Hyg.* 1988;39(1):3–10.
17 Lemercier G, Rey M, Collomb H. Cerebral lesions of malaria in children. *Bull Soc Pathol Exot Filiales.* 1966;59(4):533–548.
18 Thomas JD. Clinical and histopathological correlation of cerebral malaria. *Trop Geogr Med.* 1971;23(3):232–238.
19 Boonpucknavig V, Boonpucknavig S, Udomsangpetch R, Nitiyanant P. An immunofluorescence study of cerebral malaria. A correlation with histopathology. *Arch Pathol Lab Med.* 1990;114(10):1028–1034.
20 Turner G. Cerebral malaria. *Brain Pathol.* 1997;7(1):569–582.
21 Edington GM. Cerebral malaria in the Gold Coast African: four autopsy reports. *Ann Trop Med Parasitol.* 1954;48(3):300–306.
22 Taylor TE, Fu WJ, Carr RA, et al. Differentiating the pathologies of cerebral malaria by postmortem parasite counts. *Nat Med.* 2004;10(2):143–145.
23 Turner GD, Ly VC, Nguyen TH, et al. Systemic endothelial activation occurs in both mild and severe malaria. Correlating dermal microvascular endothelial cell phenotype and soluble cell adhesion molecules with disease severity. *Am J Pathol.* 1998;152(6):1477–1487.
24 Berendt AR, Turner GDH, Newbold CI. Cerebral malaria: the sequestration hypothesis. *Parasitol Today.* 1994;10:412–414.
25 Clark IA, Rockett KA. The cytokine theory of human cerebral malaria. *Parasitol Today.* 1994;10:410–412.
26 Grau GE, de Kossodo S. Cerebral malaria: Mediators, mechanical obstruction or more? *Parasitol Today.* 1994;10:408–409.
27 Maegraith BG. Comments on pathophysiology. *Mil Med.* 1966;131(9):S1111–S1114.
28 Maegraith B, Fletcher A. The pathogenesis of mammalian malaria. *Adv Parasitol.* 1972;10:49–75.
29 Lou J, Lucas R, Grau GE. Pathogenesis of cerebral malaria: recent experimental data and possible applications for humans. *Clin Microbiol Rev.* 2001;14(4):810–820.
30 Clark IA, Cowden WB. The pathophysiology of falciparum malaria. *Pharmacol Ther.* 2003;99(2):221–260.
31 Hunt NH, Grau GE. Cytokines: accelerators and brakes in the pathogenesis of cerebral malaria. *Trends Immunol.* 2003;24(9):491–499.
32 Turner GD, Morrison H, Jones M, et al. An immunohistochemical study of the pathology of fatal malaria. Evidence for widespread endothelial activation and a potential role for intercellular adhesion molecule-1 in cerebral sequestration. *Am J Pathol.* 1994;145(5):1057–1069.
33 Grau GE, Pointaire P, Piguet PF, et al. Late administration of monoclonal antibody to leukocyte function-antigen 1 abrogates incipient murine cerebral malaria. *Eur J Immunol.* 1991;21(9):2265–2267.
34 Grau GE, Tacchini-Cottier F, Vesin C, et al. TNF-induced microvascular pathology: active role for platelets and importance of the LFA-1/ICAM-1 interaction. *Eur Cytokine Netw.* 1993;4(6):415–419.
35 Grau GE, Fajardo LF, Piguet PF, et al. Tumor necrosis factor (cachectin) as an essential mediator in murine cerebral malaria. *Science.* 1987;237(4819):1210–1212.
36 Clark IA, Cowden WB, Butcher GA, Hunt NH. Possible roles of tumor necrosis factor in the pathology of malaria. *Am J Pathol.* 1987;129(1):192–199.
37 Grau GE, Taylor TE, Molyneux ME, et al. Tumor necrosis factor and disease severity in children with falciparum malaria. *N Engl J Med.* 1989;320(24):1586–1591.
38 Kwiatkowski D, Hill AV, Sambou I, et al. TNF concentration in fatal cerebral, non-fatal cerebral, and uncomplicated *Plasmodium falciparum* malaria. *Lancet.* 1990;336(8725):1201–1204.
39 Kern P, Hemmer CJ, Van Damme J, et al. Elevated tumor necrosis factor alpha and interleukin-6 serum levels as markers for complicated *Plasmodium falciparum* malaria. *Am J Med.* 1989;87(2):139–143.
40 Butcher GA, Garland T, Ajdukiewicz AB, Clark IA. Serum tumor necrosis factor associated with malaria in patients in the Solomon Islands. *Trans R Soc Trop Med Hyg.* 1990;84(5):658–661.
41 van Hensbroek MB, Palmer A, Onyiorah E, et al. The effect of a monoclonal antibody to tumor necrosis factor on survival from childhood cerebral malaria. *J Infect Dis.* 1996;174(5):1091–1097.
42 Engwerda CR, Mynott TL, Sawhney S, et al. Locally up-regulated lymphotoxin alpha, not systemic tumor necrosis factor alpha, is the principle mediator of murine cerebral malaria. *J Exp Med.* 2002;195(10):1371–1377.
43 Clark IA, Gray KM, Rockett EJ, et al. Increased lymphotoxin in human malarial serum, and the ability of this cytokine to increase plasma interleukin-6 and cause hypoglycaemia in mice: implications for malarial pathology. *Trans R Soc Trop Med Hyg.* 1992;86(6):602–607.
44 de Kossodo S, Grau GE. Profiles of cytokine production in relation with susceptibility to experimental cerebral malaria. *J Immunol.* 1993;151:4811–4820.
45 Eckwalanga M, Marussig M, Tavares MD, et al. Murine AIDS protects mice against experimental cerebral malaria: down-regulation by interleukin 10 of a T-helper type 1 $CD4^+$ cell-mediated pathology. *Proc Natl Acad Sci USA.* 1994;91(17):8097–8101.

46 Kossodo S, Monso C, Juillard P, et al. Interleukin-10 modulates susceptibility in experimental cerebral malaria. *Immunology*. 1997;91(4):536–540.
47 Omer FM, Riley EM. Transforming growth factor beta production is inversely correlated with severity of murine malaria infection. *J Exp Med*. 1998;188(1):39–48.
48 Artavanis-Tsakonas K, Tongren JE, Riley EM. The war between the malaria parasite and the immune system: immunity, immunoregulation and immunopathology. *Clin Exp Immunol*. 2003; 133(2):145–152.
49 Wenisch C, Parschalk B, Narzt E, et al. Elevated serum levels of IL-10 and IFN-gamma in patients with acute *Plasmodium falciparum* malaria. *Clin Immunol Immunopathol*. 1995;74(1):115–117.
50 Ho M, Sexton MM, Tongtawe P, et al. Interleukin-10 inhibits tumor necrosis factor production but not antigen-specific lymphoproliferation in acute *Plasmodium falciparum* malaria. *J Infect Dis*. 1995;172(3):838–844.
51 Ho M, Schollaardt T, Snape S, et al. Endogenous interleukin-10 modulates proinflammatory response in *Plasmodium falciparum* malaria. *J Infect Dis*. 1998;178(2):520–525.
52 Clark IA, Ilschner S, MacMicking JD, Cowden WB. TNF and *Plasmodium berghei* ANKA-induced cerebral malaria. *Immunol Lett*. 1990;25(1–3):195–198.
53 Kurtzhals JA, Adabayeri V, Goka BQ, et al. Low plasma concentrations of interleukin 10 in severe malarial anaemia compared with cerebral and uncomplicated malaria. *Lancet*. 1998; 351(9118):1768–1772.
54 Winkler S, Willheim M, Baier K, et al. Reciprocal regulation of Th1- and Th2-cytokine-producing T cells during clearance of parasitemia in *Plasmodium falciparum* malaria. *Infect Immun*. 1998;66(12):6040–6044.
55 Brenier-Pinchart MP, Pelloux H, Derouich-Guergour D, Ambroise-Thomas P. Chemokines in host-protozoan-parasite interactions. *Trends Parasitol*. 2001;17(6):292–296.
56 Burgmann H, Hollenstein U, Wenisch C, et al. Serum concentrations of MIP-1 alpha and interleukin-8 in patients suffering from acute *Plasmodium falciparum* malaria. *Clin Immunol Immunopathol*. 1995;76(1 Pt 1):32–36.
57 Friedland JS, Ho M, Remick DG, et al. Interleukin-8 and *Plasmodium falciparum* malaria in Thailand. *Trans R Soc Trop Med Hyg*. 1993;87(1):54–55.
58 Filipovic R, Jakovcevski I, Zecevic N. GRO-alpha and CXCR2 in the human fetal brain and multiple sclerosis lesions. *Dev Neurosci*. 2003;25(2–4):279–290.
59 Strasly M, Cavallo F, Geuna M, et al. IL-12 inhibition of endothelial cell functions and angiogenesis depends on lymphocyte-endothelial cell cross-talk. *J Immunol*. 2001;166(6):3890–3899.
60 Hviid L, Theander TG, Elhassan IM, Jensen JB. Increased plasma levels of soluble ICAM-1 and ELAM-1 (E-selectin) during acute *Plasmodium falciparum* malaria. *Immunol Lett*. 1993;36(1):51–58.
61 Boehme MW, Werle E, Kommerell B, Raeth U. Serum levels of adhesion molecules and thrombomodulin as indicators of vascular injury in severe *Plasmodium falciparum* malaria. *Clin Investig*. 1994;72(8):598–603.
62 Jakobsen PH, Morris-Jones S, Ronn A, et al. Increased plasma concentrations of sICAM-1, sVCAM-1 and sELAM-1 in patients with *Plasmodium falciparum* or *P. vivax* malaria and association with disease severity. *Immunology*. 1994;83(4):665–669.
63 Combes V, Taylor TE, Juhan-Vague I, et al. Circulating endothelial microparticles in Malawian children with severe falciparum malaria complicated with coma. *JAMA*. 2004;291(21):2542–2544.
64 Combes V, Simon AC, Grau GE, et al. In vitro generation of endothelial microparticles and possible prothrombotic activity in patients with lupus anticoagulant. *J Clin Invest*. 1999;104(1):93–102.
65 Hommel M. Amplification of cytoadherence in cerebral malaria: towards a more rational explanation of disease pathophysiology. *Ann Trop Med Parasitol*. 1993;87(6):627–635.
66 Lou J, Gasche Y, Zheng L, et al. Differential reactivity of brain microvascular endothelial cells to TNF reflects the genetic susceptibility to cerebral malaria. *Eur J Immunol*. 1998;28(12):3989–4000.
67 Monso-Hinard C, Lou JN, Behr C, et al. Expression of major histocompatibility complex antigens on mouse brain microvascular endothelial cells in relation to susceptibility to cerebral malaria. *Immunology*. 1997;92(1):53–59.
68 Chen Q, Schlichtherle M, Wahlgren M. Molecular aspects of severe malaria. *Clin Microbiol Rev*. 2000;13(3):439–450.
69 Favre N, Da Laperousaz C, Ryffel B, et al. Role of ICAM-1 (CD54) in the development of murine cerebral malaria. *Microbes Infect*. 1999;1(12):961–968.
70 Combes V, Rosenkranz AR, Redard M, et al. Pathogenic role of P-selectin in experimental cerebral malaria: importance of the endothelial compartment. *Am J Pathol*. 2004;164(3):781–786.
71 Lucas R, Juillard P, Decoster E, et al. Crucial role of tumor necrosis factor (TNF) receptor 2 and membrane-bound TNF in experimental cerebral malaria. *Eur J Immunol*. 1997;27(7):1719–1725.
72 Stoelcker B, Hehlgans T, Weigl K, et al. Requirement for tumor necrosis factor receptor 2 expression on vascular cells to induce experimental cerebral malaria. *Infect Immun*. 2002;70(10):5857–5859.
73 Pino P, Vouldoukis I, Kolb JP, et al. *Plasmodium falciparum*–infected erythrocyte adhesion induces caspase activation and apoptosis in human endothelial cells. *J Infect Dis*. 2003;187(8): 1283–1290.
74 Pino P, Vouldoukis I, Dugas N, et al. Induction of the CD23/nitric oxide pathway in endothelial cells downregulates ICAM-1 expression and decreases cytoadherence of *Plasmodium falciparum*–infected erythrocytes. *Cell Microbiol*. 2004;6(9): 839–848.
75 Thumwood CM, Hunt NH, Clark IA, Cowden WB. Breakdown of the blood-brain barrier in murine cerebral malaria. *Parasitology*. 1988;96(Pt 3):579–589.
76 Belnoue E, Kayibanda M, Vigario AM, et al. On the pathogenic role of brain-sequestered alphabeta $CD8^+$ T cells in experimental cerebral malaria. *J Immunol*. 2002;169(11):6369–6375.
77 Grau GE, Piguet PF, Engers HD, et al. $L3T4^+$ T lymphocytes play a major role in the pathogenesis of murine cerebral malaria. *J Immunol*. 1986;137:2348–2354.
78 Roberts DJ, Pain A, Kai O, et al. Autoagglutination of malaria-infected red blood cells and malaria severity. *Lancet*. 2000;355 (9213):1427–1428.
79 Pain A, Ferguson DJ, Kai O, et al. Platelet-mediated clumping of *Plasmodium falciparum*–infected erythrocytes is a common adhesive phenotype and is associated with severe malaria. *Proc Natl Acad Sci USA*. 2001;98(4):1805–1810.
80 Grau GE, Mackenzie CD, Carr RA, et al. Platelet accumulation in brain microvessels in fatal pediatric cerebral malaria. *J Infect Dis*. 2003;187(3):461–466.

81 Wassmer SC, Lepolard C, Traore B, et al. Platelets reorient *Plasmodium falciparum*–infected erythrocyte cytoadhesion to activated endothelial cells. *J Infect Dis*. 2004;189(2): 180–189.
82 Silamut K, Phu NH, Whitty C, et al. A quantitative analysis of the microvascular sequestration of malaria parasites in the human brain. *Am J Pathol*. 1999;155(2):395–410.
83 Berendt AR, Ferguson DJ, Gardner J, et al. Molecular mechanisms of sequestration in malaria. *Parasitology*. 1994;108:S19–S28.
84 Lucas R, Grau GE, Matthay MA, Wendel A. Functional dissection of tumor necrosis factor in liver, brain and lungs. *Curr Trends Immunol*. 2002;4:47–59.
85 Wassmer SC, Combes V, Grau GE. Pathophysiology of cerebral malaria: role of host cells in the modulation of cytoadhesion. *Ann NY Acad Sci*. 2003;992:30–38.

143

Hemorrhagic Fevers

Endothelial Cells and Ebola-Virus Hemorrhagic Fever

Tatiana A. Afanasieva*, Victoria Wahl-Jensen†,‡, Jochen Seebach*, Herrmann Schillers§, Dessy Nikova§, Ute Ströher†,‡, Heinz Feldmann†,‡, and Hans-Joachim Schnittler*

*Institute of Physiology, Technical University, Dresden, Germany;
†National Microbiology Laboratory, Public Health, Agency of Canada, Winnipeg, Manitoba;
‡University of Manitoba, Winnipeg, Manitoba, Canada;
§Institute of Physiology, Westfalia Wilhelms University, Muenster, Germany

Viral hemorrhagic fever (VHF) is a severe multiorgan disease with strong immune involvement and diffuse vascular dysregulation, particularly of the vascular endothelium. Several families of RNA viruses are regularly associated with a VHF syndrome in humans: *Arenaviridae* (Lassa virus, Machupo virus, Junin virus, Guanarito virus, and Sabia virus), *Bunyaviridae* (Rift Valley fever virus, Crimean-Congo hemorrhagic fever virus, hantaviruses), *Flaviviridae* (Yellow fever virus, Dengue virus, Omsk hemorrhagic fever virus, and Kyasanur forest disease virus), and *Filoviridae* (Marburg virus and Ebola virus). The clinical manifestations of VHF vary and are dependent on the causative agent (see CDC homepage http://www.cdc.gov). However, some common clinical features include fever, various degrees of vascular dysregulation with bleeding tendency and shock development, and the vascular endothelium seems to be affected in most cases (1–3). Some of the VHF-causing pathogens target the endothelium directly, whereas others induce primarily indirect alterations through proinflammatory mediators released from infected target cells (e.g., monocytes/macrophages). Marburg (MARV) and Ebola viruses (EBOV) cause the most severe form of VHF and, thus, serve as important model pathogens for studying the pathogenesis and management of VHFs. Filoviruses, as well as some other hemorrhagic fever (HF) viruses, are biological safety level 4 (BSL4) agents, which somewhat complicates investigations. Filoviruses seem to target both the vascular system and the immune system, leading to the opinion that filovirus HF fever is a vascular disease as well as an immune syndrome (2–5). Although our understanding of the molecular mechanisms of VHF pathogenesis is still limited, some important scientific achievements have been made in the past decade. In this chapter we specifically focus on the role of the vascular endothelium during EBOV HF, and also address some novel methodologies that allow for investigations involving endothelial cells (ECs) outside of the BSL4 containment laboratory.

EBOLA VIRUS CAUSES THE MOST SEVERE VIRAL HEMORRHAGIC FEVER

EBOV is a filamentous, enveloped, nonsegmented negative-stranded RNA virus that causes severe hemorrhagic fever in humans and nonhuman primates. Mortality rates up to 89% have been reported for *Zaire ebolavirus* (ZEBOV) (6). Unfortunately, currently no treatments or prophylaxis for humans are available, but promising candidates are being tested in appropriate animal models (7–10). Thus, symptomatic therapy is currently the only plausible treatment. However, even this type of treatment might not be available in the developing countries where these diseases commonly occur. The host of EBOV is unknown, but primates are frequently infected and might transmit the virus to humans. Close physical contact is required for infection, and virus uptake seems to occur via skin lesions and mucous membranes. Once it enters the lymph and blood vessels, the virus spreads preferentially to organs such as lymph nodes, liver, and spleen, where it can directly access sessile monocytes/macrophages (11). This hypothetical route for EBOV infection and spread was recently supported by experiments in *Cynomolgus* monkeys.

Intramuscular injection of EBOV resulted in local infection of monocytes/macrophages and dendritic cells and virus production. The infected cells and/or the virus travel to regional lymph nodes and to the liver and spleen. The virus was then shown to infect tissue macrophages, dendritic cells, and fibroblastic reticular cells in those organs (12,13). These data strongly indicate that the monocyte-derived macrophages and dendritic cells are primary targets in EBOV HF. Macrophages and dendritic cells seem to respond heterogeneously to EBOV infection. Although monocytes/macrophages become activated and release large quantities of cytokines and proinflammatory mediators such as tumor necrosis factor (TNF)-α, interleukin (IL)-1β, IL-6, and IL-8, dendritic cells respond with an insufficient release of immunomodulatory mediators (5,14).

The clinical picture of EBOV HF is characterized by abrupt onset of nonspecific flu-like symptoms such as fever, headache, muscle and abdominal pain, and nausea. At around day 6, approximately 50% of the patients infected with ZEBOV develop maculopapular rash on the trunk and shoulders. By days 7 through 10, survivors show decreasing viremia levels, whereas increasing viremia is seen in cases with disease progression. Fatal cases develop a shock syndrome associated with edema formation, hypotension, signs of pronounced disseminated intravascular coagulation (DIC), bleeding from venopuncture sites and the gastrointestinal tract, hematuria, and melena; on average, death occurs between day 10 to 17. The clinical course of severe EBOV HF and other forms of septic shock syndrome have common features (15). It is not entirely clear how the host response is mobilized in survivors, but release of immunomodulatory mediators as well as the development of specific antibodies seems to play an important role. It has been shown that there is a clear difference in the expression pattern of cytokines between fatal cases and survivors. An initial but short-lived increase of proinflammatory cytokines such as TNF-α, IL-1β, and IL-6 in the plasma has been postulated as a marker for nonfatal infection, whereas the release of anti-inflammatory cytokines such as IL-10 and increased levels of interferon (IFN)-γ, neopterin, and IL-1 receptor antagonist (IL-1RA) in the plasma early after the onset of symptoms seems to be indicative for fatal outcome (for summary we refer the reader to Ref. 15).

ENDOTHELIAL STRUCTURES IMPORTANT FOR UNDERSTANDING OF VIRAL HEMORRHAGIC FEVER PATHOGENESIS

The symptoms developed during EBOV HF indicate that the vascular endothelium plays an important role in the pathogenesis of the disease. The vascular endothelium has multiple functions in maintaining body homeostasis. It forms the primary selective barrier between the blood and tissue, and dysregulation of EC functions can cause changes in vascular permeability that contribute to the development of hemorrhage and shock. Furthermore, blood pressure regulation is largely modulated by ECs via release of vasoactive agents such as nitric oxide (NO) and endothelin. Additionally, the vascular endothelium mediates the host immune responses through the induction of cytokines, chemokines, and cellular receptors that recruit or activate immune cells.

The balance of macromolecule distribution and exchange of water and solutes between the vasculature and the interstitial space is controlled by vascular ECs. Whereas transport of macromolecules through the endothelium (e.g., albumin) follows a transcellular pathway, the exchange of water and small solutes takes place primarily through endothelial junctions in capillaries and postcapillary venules (paracellular pathway). Under physiological conditions, the intercellular junctions are impermeable to macromolecules, with some exceptions such as in the kidneys, liver, and spleen. In general, endothelial barrier function is decreased in inflamed tissue due to reduced intercellular adhesion, leading to enhanced extravasation of small solutes and water, and finally resulting in edema formation. In severe cases of certain forms of inflammation, intercellular gap formation occurs, allowing extravasation of larger macromolecules and water into the interstitial space and further contributing to the severity of the pathological condition (16). Although the paracellular manner of increased endothelial permeability in inflammation generally is accepted, it is currently unclear whether a transcellular transport of macromolecules contributes to edema formation as well.

The composition of endothelial junctions is heterogeneous within different organs and within the microvascular bed. The common distribution of adherens junctions throughout the vasculature makes them crucial structures for maintaining the barrier function. Particularly, postcapillary venules, displaying mainly adherens-like junctions and lacking tight junction strands (17), are the target site of increased permeability and leukocyte extravasation during inflammation. This implicates the adherens junction in vascular endothelium as a main player in regulation of endothelial barrier function. The backbone of adherens junctions is formed by the integral and calcium-dependent membrane protein, vascular endothelial (VE)-cadherin, which is connected to cortical actin filaments (18,19). This connection enhances the VE-cadherin–mediated cell adhesion and barrier function dramatically. In general, a decrease in barrier function is associated with a loss of cortical actin filaments from the VE-cadherin/catenin complex, leading to stress-fiber formation and a redistribution of the VE-cadherin/catenin complex (20). Although the actin reorganization is controlled by GTPases of the Rho-family (21), tyrosine phosphorylation of VE-cadherin and the catenins, as well as activation of both actin filament dynamics and contractile mechanisms, play a fundamental role in the regulation of endothelial barrier function during inflammation (18,19,22). In filovirus infections, those phenomena (reorganization but no tyrosine phosphorylation of VE-cadherin and the catenins) are seen in cell culture models when ECs are treated with supernatants of filovirus-infected monocytes/macrophages (23–25).

After stimulation with proinflammatory agents, ECs express adhesion molecules that mediate endothelial–leukocyte interactions. The process of leukocyte migration includes three successive stages: rolling, mediated by the selectin class of adhesion molecules (e.g., E-selectin); tight adhesion, mediated by the leukocyte integrins and their EC counter-receptors (e.g., intercellular adhesion molecule [ICAM]-1, and vascular cell adhesion molecule [VCAM]-1); and transmigration, which involves platelet-endothelial cell adhesion molecule (PECAM)-1, CD99, and CD69 (26,27). In animal models of filovirus disease, white blood cells (particularly granulocytes) are adherent to postcapillary venules close to sites of virus-induced focal necrosis, but extravasation appears to be blocked for unknown reasons (28). It is assumed that the inflammatory response of the vascular system is an important component of EBOV HF pathogenesis. Therefore, uncovering the molecular mechanisms responsible for the regulation of cell junctions in filovirus infection is of central importance for the understanding of EBOV pathogenesis.

PROPOSED PATHOGENESIS

As outlined earlier, disrupted endothelial barrier and alteration of adhesion molecule expression are thought to determine disease-decisive symptoms during severe EBOV infection. They occur through a complex mechanism that involves both direct and indirect targeting of the vascular endothelium and immune competent cells (1,2,5,23,29). Several major pathological determinants are discussed as factors in this process (Figure 143.1).

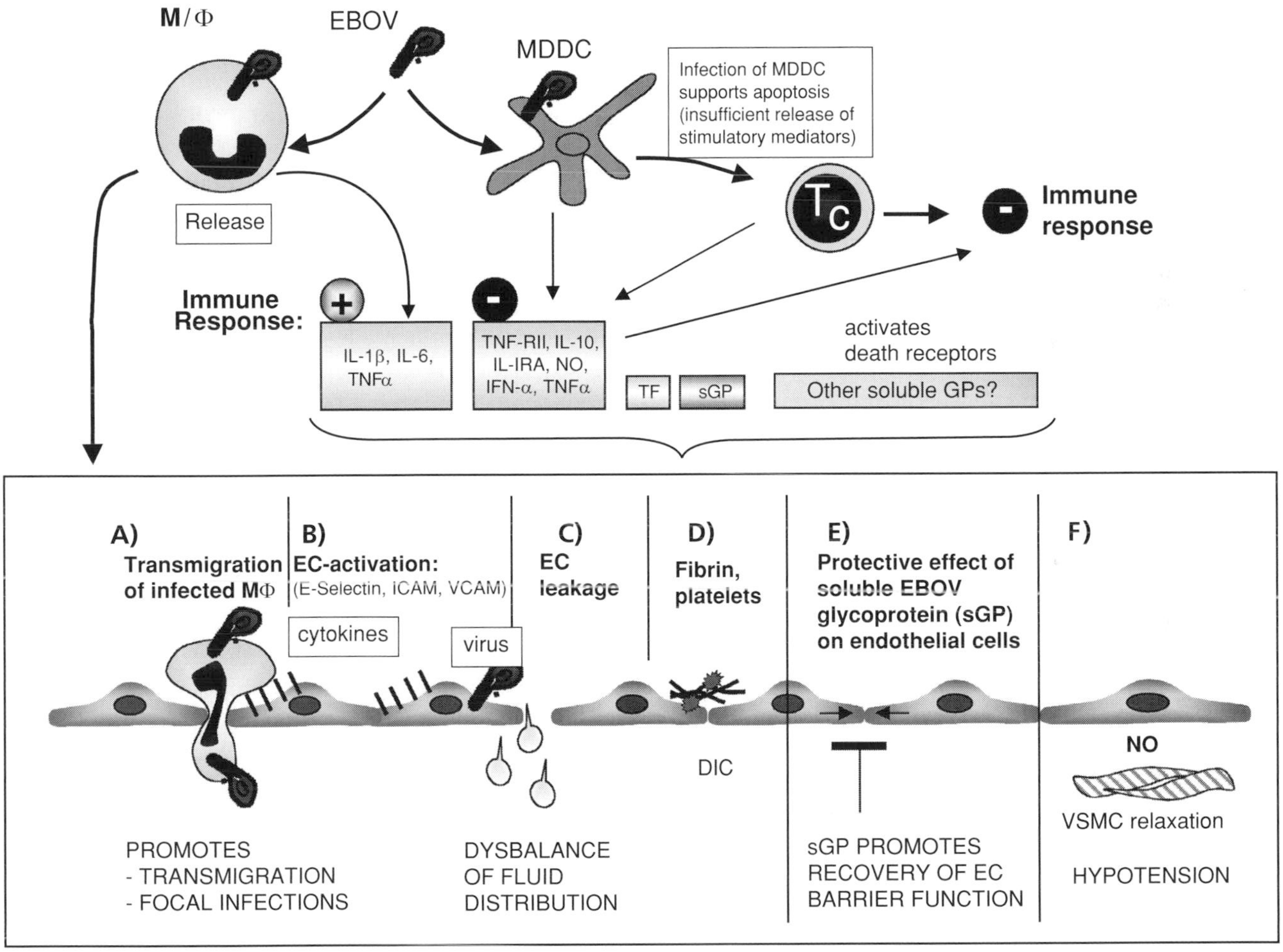

Figure 143.1. Model of Ebola virus pathogenesis. Primary target cells of EBOV include macrophages (*M*Φ) and monocyte-derived dendritic cells (MDDCs). Although infection of MΦ induces the release of cytokines and chemokines, the lack of this response by MDDCs may favor apoptosis of lymphocytes (*Tc*). In addition to release of cytokines and chemokines, infected MΦ also release tissue factor (TF) and the viral sGP. Collectively, these various mediators may exert an effect on the vascular endothelium, as shown in **A** through **F**. First, infected MΦ are activated and transmigrate through the endothelium, thereby contributing to focal infections (**A**). The endothelium also is activated and responds by increasing expression of cell adhesion molecules (**B**). This activation can occur directly via viral infection of the endothelium or indirectly as a result of mediators expressed by MΦ. Endothelial leakage (**C**) and fibrin deposition (**D**) also can be induced as an indirect effect of mediator molecules. Recent data suggest that sGP exerts an anti-inflammatory role on the endothelium by restoring the barrier function of endothelial monolayers (**E**). Finally, NO release from MΦ may contribute to vascular smooth muscle cell (VSMC) relaxation and consequently, hypotension.

SOLUBLE FACTORS PRODUCED BY PRIMARY TARGET CELLS AND THEIR IMPACT ON ENDOTHELIAL DYSFUNCTION

Mononuclear phagocytic cells are primary targets for EBOV, and infection leads to lytic replication (30). In addition to virus production, the infection of mononuclear phagocytic cells leads to their activation and subsequent release of active mediators such as proinflammatory cytokines (e.g., TNF-α, IL-1β, and IL-6) and chemokines (e.g., IL-8). The activation of monocytes/macrophages is independent of virus replication (31,32). It was recently shown that interaction of macrophages with virus-like particles, consisting of only the EBOV matrix protein (VP40) and transmembrane glycoprotein ($GP_{1,2}$) and nearly identical in morphology to infectious virus particles (Figure 143.2), is sufficient to induce expression of TNF-α, IL-6, and IL-8 similar in kinetics to activation triggered by lipopolysaccharide (LPS) (31,33). Proinflammatory cytokines and some chemokines are powerful modulators of endothelial barrier function and cause endothelial activation. Earlier studies demonstrated that the cell culture supernatants derived from filovirus-infected monocytes/macrophages were able to cause intercellular gap formation resulting from reorganization of structural components of the adherens junctions. This effect was mediated to a large extent by TNF-α and could be blocked by neutralizing monoclonal antibodies to TNF-α (24). Increased levels of cytokines were detected in vitro, in sera of infected patients, and in experimentally infected nonhuman primates (32,34,35). It is known that macrophage-derived cytokines act on multiple organs, but the vascular system and the endothelium seem to be the major targets in VHF, as also observed in septic forms of shock. Furthermore, infected macrophages release vasoactive mediators such as NO that cause loss of vascular smooth muscle tone (29). Data obtained from cell culture and in vivo models, as well as from postmortem material from humans, indicate that the endothelium responds to filovirus-mediated release of cytokines with an increase in permeability, dysregulation of vascular tone, expression of cell adhesion molecules, and a change to a procoagulant phenotype. Thus, imbalanced production of proinflammatory cytokines can be associated with negative prognosis and fatal outcome similar to septic shock.

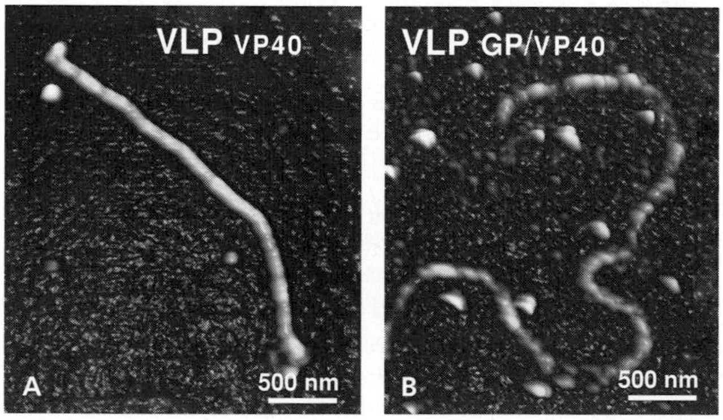

Figure 143.2. Atomic force microscopy of virus-like particles. Purified VP40-only particles (**A**) or VLPs expressing VP40 and $GP_{1,2}$ (**B**) were prepared on mica and scanned by atomic force microscopy in air mode. The particles are filamentous in morphology, and the surface of VP40-only particles (**A**) appears smooth whereas VLPs, which are studded with $GP_{1,2}$ display a rough surface (**B**).

A hallmark of filovirus infection in humans and nonhuman primates is coagulopathy, which is partly caused by thrombocytopenia, a common laboratory finding during infection. Thrombocytopenia might be the result of a dysregulation of megakaryocyte maturation in the bone marrow caused by high levels of interferon. The mechanisms triggering coagulation abnormalities in EBOV HF seem to be complex and are not well understood. However, recent studies on experimentally EBOV-infected nonhuman primates suggested that coagulation abnormalities are initiated early in the disease course through expression and release of tissue factor from infected monocytes/macrophages, a major activator of the clotting cascade (4,10–12,36). The level of tissue factor expression can be amplified further by proinflammatory cytokines (e.g., IL-6, IL-1, and TNF-α) released from infected cells during infection. In addition, shedding of tissue factor–positive monocytes/macrophage-derived microparticles, which have strong procoagulant potential, and rapid depletion of the major anticoagulant protein C were observed. This coincided with the appearance of fibrin deposition that can result in fibrin mesh networks encapsulating infected macrophages in multiple organs (e.g., liver and spleen). The coagulation disorders in EBOV infections are consistent with the syndrome of DIC. No single test is available that is sufficient to permit a definitive diagnosis of DIC. However, D-dimer, a marker for activation of both coagulation and fibrinolytic cascades (37), is dramatically increased in EBOV-infected *Cynomolgus* macaques (13). In addition, within 3 to 4 days after infection, platelets and protein C levels were decreased, and fibrin thrombi were found in vessels of several tissues, further indicating the development of DIC (13). Consequently, therapeutic strategies aimed at controlling certain coagulation and fibrinolytic parameters may be helpful in the management of EBOV HF and other VHF syndromes. Indeed, administration of recombinant nematode anticoagulant protein c2 (rNAPc2), a potent inhibitor of tissue factor-initiated blood coagulation, proved to be beneficial in EBOV-infected macaques, resulting in survival of 33% of the animals and a significant delay in death for the rest of the animals. rNAPc2-treated animals showed less prominent fibrin deposits and intravascular thromboemboli (4,10,38).

VIRAL CYTOPATHIC EFFECT

The vascular endothelium is considered a secondary target for virus replication. It was shown earlier that ECs are able to support a productive infection of filoviruses in vitro (24,39). In general, filovirus infection leads to a moderate cytopathic effect in susceptible cells and tissues. The mechanism of destruction of ECs and other target cells by EBOV is not entirely understood. Most probably, several processes play a role, including viral shutdown of cellular processes, massive production and accumulation of viral proteins in infected cells, and dramatic budding of multiple Ebola viral particles from the cell membrane. It was proposed that viral full-length glycoprotein has a specific cytotoxic potential and was therefore considered to be the major viral pathogenic determinant. Particularly, the expression of EBOV glycoprotein in EC cultures caused cell destruction (40,41), and it was concluded that this effect might be relevant in EBOV pathogenesis. However, in tissue obtained from postmortem human material and from experimentally infected nonhuman primates, endothelial destruction due to infection was less prominent (4,11,12,36,42).

EBOV-infected cells that release infectious EBOV particles produce a transmembrane glycoprotein (GP) and a number of soluble glycoproteins (Figure 143.3), and it has been postulated that they may play an important role in EBOV pathogenesis (43). All viral glycoproteins are encoded by gene 4 of the EBOV genome. The primary product of this gene is a precursor of a nonstructural soluble glycoprotein (sGP) that is post-translationally cleaved into mature sGP and Δ-peptide;

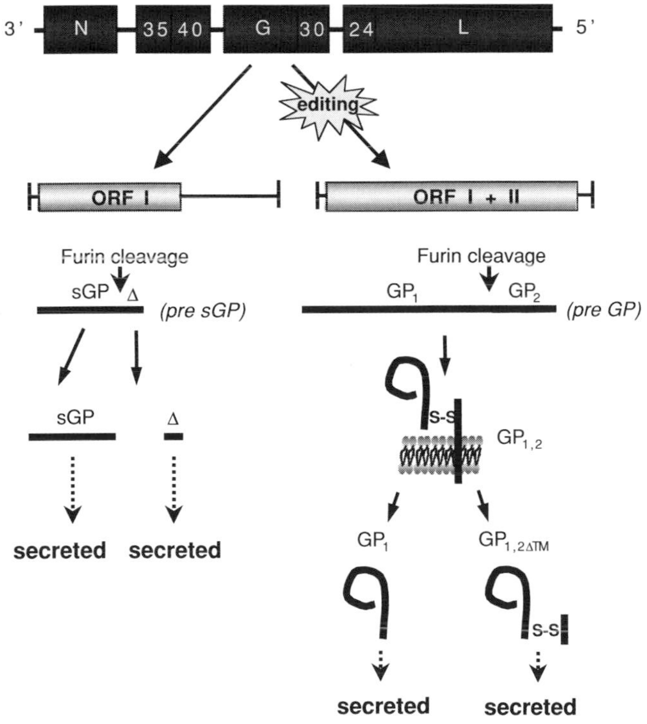

Figure 143.3. Glycoprotein expression strategy for Ebola virus. The direct transcription of viral RNA leads to the production of pre-sGP, which is post-translationally modified and cleaved into sGP and Δ-peptide (both are secreted). RNA editing (insertion of an additional/eighth adenosine at a specific sequence) leads to production of pre-GP, which is also post-translationally modified and cleaved by furin into the disulfide-linked fragments GP_1 and GP_2. The full-length $GP_{1,2}$ forms trimers on the surface of particles and mediates receptor binding and fusion to host cells.

both are secreted products, and sGP has been detected in patient serum (43,44). RNA editing is necessary to express the structural transmembrane glycoprotein GP that is also post-translationally cleaved into the disulfide-linked fragments GP_1 and GP_2 (Figure 143.3). The homotrimeric $GP_{1,2}$ forms the spikes on the virus particle, and is indispensable for receptor binding and fusion with the host cell membrane. In vitro studies have demonstrated that $GP_{1,2}$, when expressed in ECs, is cytotoxic and thus may contribute to endothelial damage during EBOV infection (40,45). In one study, the cytotoxic effect of $GP_{1,2}$ in human embryonic kidney 293T and ECs was attributed to the mucin-like domain of the GP_1 subunit (40). EBOV $GP_{1,2}$ has a coiled-coil domain that forms a trimer of helical hairpin-like loops. The hairpin structure lies adjacent to the fusion-peptide region and is inserted directly into the cell membrane of target cells, providing a potential mechanism of direct toxicity. Interestingly, a recombinant virus with a mutated editing site resulting in mainly $GP_{1,2}$ expression showed an increased cytotoxic effect but reduced virion production, indicating that RNA editing might play a role in reducing virus cytotoxicity. The pathogenic role of the soluble glycoproteins on the endothelium was postulated; however, until recently no studies were performed allowing for the dissection of the complex mechanism of virus action on the endothelium.

The use of recombinant viral technology in combination with novel biophysical methods, such as impedance spectroscopy, allows for the investigation of EBOV pathogenesis outside the BSL4 laboratory. An interesting example is the role of EBOV glycoproteins in endothelial activation and barrier function, by which virus-like particles were used in combination with impedance spectroscopy to measure the endothelial barrier function. Virus-like particles can be produced by expression of the EBOV matrix protein (VP40) and the transmembrane glycoprotein $GP_{1,2}$. Importantly, a reliable control is the expression of only VP40 that leads to virus-like particles lacking the GP. Using atomic force microscopy with a resolution of a couple of nanometers, we have compared the VLPs consisting of VP40 and GP as well as VLPs only carrying VP40 (see Figure 143.2). Particles bud from the cell membrane and display a morphology that is nearly identical to live virus but is completely free of viral genetic material and thus noninfectious.

Impedance spectroscopy is an emerging and unique biophysical method to reliably examine the endothelial barrier function using high temporal resolution and high sensitivity (45). Based on electrical impedance spectroscopy data, the transendothelial electrical resistance (TER) of a cultured EC monolayer can be determined, and it predominantly reflects the changes in paracellular permeability. EC barrier function is frequently studied in transwell filter systems by analyzing the passage of tracer substances. In contrast to impedance spectroscopy, which reliably detects changes in barrier function of about 2% (45), tracer systems are limited in temporal resolution and sensitivity and, thus, are only capable of detecting strong effects.

It was shown that infectious EBOV, as well as noninfectious virus-like particles, is able to interact with and activate ECs in vitro, as demonstrated by expression of ICAM-1, VCAM-1, and E-selectin adhesion molecules (31). Moreover, VLPs were able to increase the endothelial permeability measured by the highly sensitive impedance spectroscopy method and hydraulic conductivity. The permeability-increasing effect of VLPs might allow for extravasation of water, and thus contribute to the development of edema and shock observed during EBOV HF. These effects were not observed using VLPs lacking $GP_{1,2}$, indicating the importance of the transmembrane glycoprotein $GP_{1,2}$ in the context of a virion particle for EC activation and barrier dysfunction. The effect of VLPs on endothelial permeability corresponded to the effect of approximately 200 to 300 pg/mL of TNF-α. These changes were associated with the formation of actin stress fibers, although VE-cadherin staining was largely intact along the cell–cell junctions (31). The moderate effect of VLPs on endothelial barrier function was further enhanced in the presence of TNF-α, thus contributing to the severity of endothelial damage even in the absence of direct EC infection. Interestingly, the major soluble glycoprotein sGP failed to activate monocyte/macrophages and ECs or influence the endothelial barrier function, even when applied at high concentrations (50 μg/mL) (31). However, when sGP was administered simultaneously with TNF-α, it caused a recovery of EC barrier function, implying that sGP might be able to interfere with components of the TNF-α signaling pathway (Figure 143.4; for color reproduction, see Color Plate 143.4). This finding suggests that sGP may have an anti-inflammatory role (31). Whether sGP interferes with EBOV virus- or VLP-mediated activation remains unclear. It is noteworthy that in animal models of filovirus disease, white blood cells adhere to the endothelial lining of postcapillary venules close to virus-induced focal necrosis, but do not extravasate into the extravascular space (28). These data suggest that leukocyte adhesion is increased while transmigration is blocked or impaired, which hypothetically could be related to the observed anti-inflammatory effect of sGP.

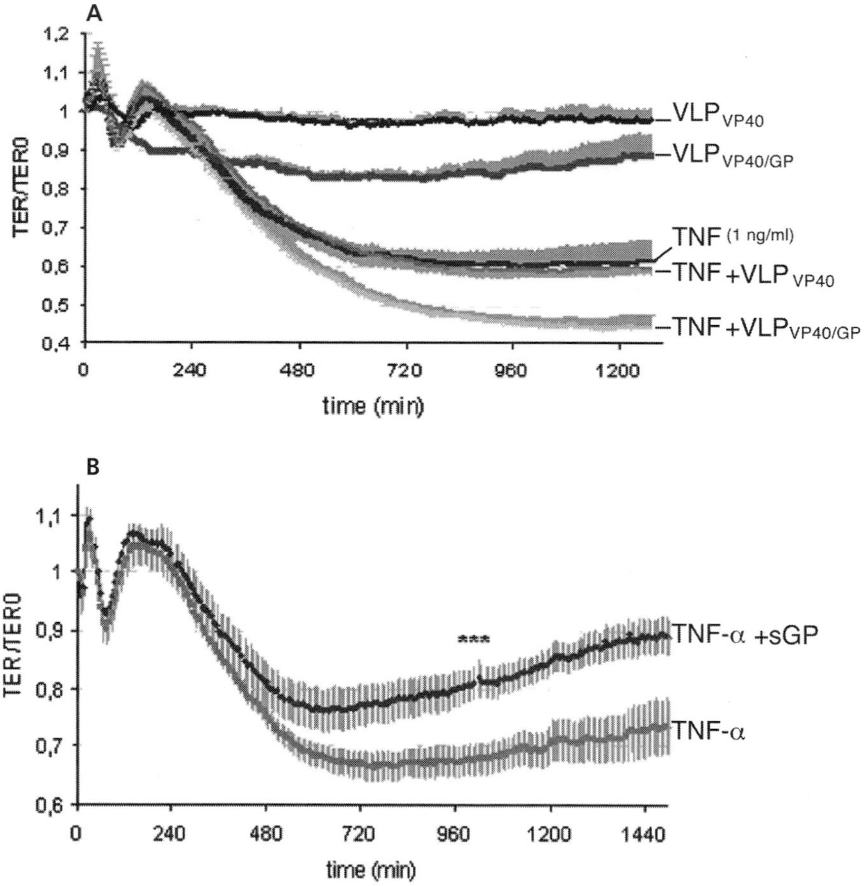

Figure 143.4. Effect of simultaneous application of TNF-α and virus-like particles or sGP on ECs. ECs were treated with TNF-α (1 ng/mL) and VLPs (1 particle/cell) or sGP (10 μg/mL). Simultaneous application of TNF-α and $VLP_{VP40/GP}$ led to further decreases in TER by approximately 17%, indicating their additive effect on reducing barrier function (**A**). Simultaneous treatment with TNF-α and sGP led to a recovery of the TNF-α–induced changes in barrier function by 15% to 20% (**B**). Changes in TER were significant ($p < 0.05$) from the time point marked by asterisks. (Reproduced with permission from Wahl-Jensen V, Afanasieva TA, Seebach J, et al. Effects of Ebola virus glycoproteins on endothelial cell activation and barrier function. *J Virol.* 2005;79:10442–10450.) For color reproduction, see Color Plate 143.4.

An additional indirect influence on vascular system dysregulation may result from infection and impairment of organs that are normally involved in the regulation of endothelial homeostasis. The filoviruses are able to infect a wide range of cells and tissues. In vitro data indicate that the heavily glycosylated surface protein of EBOV is able to bind to widely distributed cellular lectins (46). However, additional data are required to bolster this hypothesis. In addition to monocytes/macrophages and dendritic cells, hepatocytes and adrenal cortical cells are susceptible to EBOV infection. Infection of liver and the adrenal gland has been reported to occur in epidemic HF (47–49). EBOV infection results in direct damage of hepatocytes, with the formation of multiple foci of necrosis in the liver. Currently, it is not clear whether disseminated infection of the liver leads to a reduced synthesis of mediators with procoagulant activities, or to a reduced albumin synthesis that together may lead to a reduction of plasma osmotic pressure and contribute to edema formation. The adrenal cortex secretes many different hormones, including corticosteroids and mineralocorticoids that control blood pressure and the level of sodium and potassium in the body. Impaired function of the adrenal cortex will lead to hypotension and sodium loss, resulting in hypovolemia; this may contribute to shock development seen at late stages of EBOV HF and other VHFs.

CONCLUSION

The investigation of pathogenic mechanisms for several VHF-viruses, such as EBOV, MARV, and Lassa virus, is complicated by the need for a high level of biocontainment of the viruses (BSL4). This has changed lately, particularly for filoviruses, with the introduction of molecular tools and models that allow for experiments under lower biocontainment conditions. The construction of viruses such as retroviruses or vesicular stomatitis virus pseudotyped with filovirus glycoproteins, and the generation of virus-like particles that resemble the particles of infectious virus but lack the viral genetic information already have provided new information, particularly in regards to virus–cell interactions. They will become even more important tools in the future, used to study virus–cell interactions as well as various aspects of pathogenesis. Changes in EC function can be subtle and difficult to detect by morphological methods such as intercellular gap formation. To demonstrate increased paracellular permeability, highly sensitive and reliable biophysical methods such as impedance spectroscopy allow the online determination of the TER. This method is able to detect discrete effects that are not obvious by morphological or tracer techniques, such as transwell filter systems. Nevertheless, even small changes in TER of 10% to 20% increase the water permeability several-fold without an obvious redistribution of junction molecules (31). Using these more sensitive methods, in combination with genetically modified viruses, will help in the future to understand VHF pathogenesis and may provide a basis for therapy (31).

KEY POINTS

- EBOV HF pathogenesis is a complex process involving an interplay between cellular and viral factors.
- EBOV HF causes a shock syndrome that in part resembles septic shock. It is characterized by strong involvement of both the vascular ECs and immune response systems.
- Mononuclear phagocytic cells (MPCs) are primary targets for EBOV. Infection of MPC leads to the lytic replication of the virus, MPC activation, and subsequent release of active mediators such as proinflammatory cytokines, which are powerful regulators of endothelial functions and modulators of the immune system.
- Viral full-length glycoprotein has a cytotoxic potential and therefore is considered to be a major viral pathogenic determinant.
- The effect of full-length glycoprotein associated into virus-like particles on endothelial barrier function is further enhanced in the presence of proinflammatory cytokines (e.g., TNF-α), thus contributing to endothelial dysfunction even in the absence of direct EC infection.

Future Goals

- To establish reliable molecular tools and models that allow experimentation under lower biocontainment conditions
- To establish highly sensitive methods that allow the detection of minimal physiological changes in endothelial function that do not have strong morphological appearance but could have a devastating effect on organ function

ACKNOWLEDGMENTS

Work on ECs and filoviruses in Dresden was supported by grants from the German research council (DFG), grant SCHN430/2–4 and the Priority program 1130 "Infection of the endothelium" SCHN 430/3–1 and 3–2, and the MedDrive program of the Medical Faculty of the TU-Dresden. Work on filoviruses in Winnipeg was supported by Health Canada, Public Health Agency of Canada, and a grant of the Canadian Institutes of Health Research (CIHR, MOP-43921). VWJ was supported by fellowships from the Manitoba Health Research Council (MHRC) and the Department of Medical Microbiology, University of Manitoba.

REFERENCES

1 Geisbert TW, Jahrling PB. Exotic emerging viral diseases: progress and challenges. *Nat Med.* 2004;10:S110–S121.
2 Feldmann H, Steven J, Klenk HD, Schnittler HJ. Ebola virus: from discovery to vaccine. *Nat Rev Imunol.* 2003;3:77–85.

3 Schnittler H, Stroeher U, Afanasieva T, Feldmann H. The role of endothelial cells in filovirus hemorrhagic fever. In: Klenk H-D, Feldmann H, eds. *Ebola and Marburg Viruses: Molecular and Cellular Biology*. Norfolk: Horizon Biosciences; 2004:279–304.
4 Geisbert TW, Jahrling PB. Towards a vaccine against Ebola virus. *Expert Rev Vaccines*. 2003;2:777–789.
5 Mahanty S, Hutchinson K, Agarwal S, et al. Cutting edge: impairment of dendritic cells and adaptive immunity by Ebola and Lassa viruses. *J Immunol*. 2003;170:2797–2801.
6 Sanchez A, Khan AS, Zaki SR, et al. Filoviridae: Marburg and Ebola viruses. In: Fields BN, Knipe DM, eds. *Virology*. Philadelphia: Raven Press;2001:1279–1304.
7 Jones SM, Feldmann H, Stroher U, et al. Live attenuated recombinant vaccine protects nonhuman primates against Ebola and Marburg viruses. *Nat Med*. 2005;11(7):786–790. Epub 2005 Jun 5.
8 Sullivan NJ, Geisbert TW, Geisbert JB, et al. Accelerated vaccination for Ebola virus haemorrhagic fever in non-human primates. *Nature*. 2003;424:681–684.
9 Warfield KL, Bosio CM, Welcher BC, et al. Ebola virus-like particles protect from lethal Ebola virus infection. *Proc Natl Acad Sci USA*. 2003;100:15889–15894.
10 Geisbert TW, Hensley LE, Jahrling PB, et al. Treatment of Ebola virus infection with a recombinant inhibitor of factor VIIa/tissue factor: a study in rhesus monkeys. *Lancet*. 2003;362:1953–1958.
11 Schnittler HJ, Feldmann H. Marburg and Ebola hemorrhagic fevers: does the primary course of infection depend on the accessibility of organ-specific macrophages? *Clin Infect Dis*. 1998;27: 404–406.
12 Geisbert TW, Young HA, Jahrling PB, et al. Pathogenesis of Ebola hemorrhagic fever in primate models: evidence that hemorrhage is not a direct effect of virus-induced cytolysis of endothelial cells. *Am J Pathol*. 2003;163:2371–2382.
13 Geisbert TW, Young HA, Jahrling PB, et al. Mechanisms underlying coagulation abnormalities in Ebola hemorrhagic fever: overexpression of tissue factor in primate monocytes/macrophages is a key event. *J Infect Dis*. 2003;188:1618–1629.
14 Bosio CM, Aman MJ, Grogan C, et al. Ebola and Marburg viruses replicate in monocyte-derived dendritic cells without inducing the production of cytokines and full maturation. *J Infect Dis*. 2003;188:1630–1638.
15 Bray M, Mahanty S. Ebola hemorrhagic fever and septic shock. *J Infect Dis*. 2003;188:1613–1617.
16 Michel CC, Neal CR. Openings through endothelial cells associated with increased microvascular permeability. *Microcirculation*. 1999;6:45–54.
17 Simionescu M, Simionescu N, Palade GE. Characteristic endothelial junctions in different segments of the vascular system. *Thromb Res*. 1976;8:247–256.
18 Vestweber D. Regulation of endothelial cell contacts during leukocyte extravasation. *Curr Opin Cell Biol*. 2002;14:587–593.
19 Dejana E. Endothelial cell-cell junctions: happy together. *Nat Rev Mol Cell Biol*. 2004;5:261–270.
20 Bazzoni G, Dejana E. Endothelial cell-to-cell junctions: molecular organization and role in vascular homeostasis. *Physiol Rev*. 2004;84:869–901.
21 Wojciak-Stothard B, Ridley AJ. Rho GTPases and the regulation of endothelial permeability. *Vascul Pharmacol*. 2002;39:187–199.
22 Bazzoni G, Martinez Estrada O, Dejana E. Molecular structure and functional role of vascular tight junctions. *Trends Cardiovasc Med*. 1999;9:147–152.
23 Schnittler HJ, Feldmann H. Viral hemorrhagic fever – a vascular disease? *Thromb Haemost*. 2003;89:967–972.
24 Feldmann H, Bugany H, Mahner F, et al. Filovirus-induced endothelial leakage triggered by infected monocytes/macrophages. *J Virol*. 1996;70:2208–2214.
25 Bockeler M, Stroher U, Seebach J, Afanasieva T, et al. Breakdown of para-endothelial barrier function in Marburg virus infection is associated with early tyrosine phosphorylation of PECAM-1. *J Infect Dis*. 2007; In press.
26 Imhof BA, Aurrand-Lions M. Adhesion mechanisms regulating the migration of monocytes. *Nat Rev Immunol*. 2004;4:432–444.
27 Vestweber D. Lymphocyte trafficking through blood and lymphatic vessels: more than just selectins, chemokines and integrins. *Eur J Immunol*. 2003; 33:1361–1364.
28 Ryabchikova EI, Kolesnikova LV, Netesov SV. Animal pathology of filoviral infections. *Curr Top Microbiol Immunol*. 1999;235: 145–173.
29 Sanchez A, Lukwiya M, Bausch D, et al. Analysis of human peripheral blood samples from fatal and nonfatal cases of Ebola (Sudan) hemorrhagic fever: cellular responses, virus load, and NO levels. *J Virol*. 2004;78:10370–10377.
30 Schnittler HJ, Feldmann H. Molecular pathogenesis of filovirus infections: role of macrophages and endothelial cells. *Curr Top Microbiol Immunol*. 1999;235:175–204.
31 Wahl-Jensen V, Kurz SK, Hazelton PR, et al. Role of Ebola virus secreted glycoproteins and virus-like particles in activation of human macrophages. *J Virol*. 2005;79:2413–2419.
32 Stroeher U, West E, Bugany H, et al. Infection and activation of monocytes by Marburg and Ebola viruses. *J Virol*. 2001;75: 11025–11033.
33 Wahl-Jensen VM, Afanasieva TA, Seebach J, et al. Effects of Ebola virus glycoproteins on endothelial cell activation and barrier function. *J Virol*. 2005;79:10442–10450.
34 Hensley LE, Young HA, Jahrling PB, Geisbert TW. Proinflammatory response during Ebola virus infection of primate models: possible involvement of the tumor necrosis factor receptor superfamily. *Immunol Lett*. 2002;80:169–179.
35 Villinger F, Rollin PE, Brar SS, et al. Markedly elevated levels of interferon (IFN)-gamma, IFN-alpha, interleukin (IL)-2, IL-10, and tumor necrosis factor-alpha associated with fatal Ebola virus infection. *J Infect Dis*. 1999;179(Suppl 1):S188–S191.
36 Sullivan N, Yang ZY, Nabel GJ. Ebola virus pathogenesis: implications for vaccines and therapies. *J Virol*. 2003;77:9733–9737.
37 Mammen EF. Disseminated intravascular coagulation (DIC). *Clin Lab Sci*. 2000;13:239–245.
38 Geisbert TW, Hensley LE, Larsen T, et al. Pathogenesis of Ebola hemorrhagic fever in *Cynomolgus* macaques: evidence that dendritic cells are early and sustained targets of infection. *Am J Pathol*. 2003;163:2347–2370.
39 Schnittler HJ, Mahner F, Drenckhahn D, et al. Replication of Marburg virus in human endothelial cells. A possible mechanism for the development of viral hemorrhagic disease. *J Clin Invest*. 1993;91:1301–1309.
40 Chandran K, Sullivan NJ, Felbor U, et al. Endosomal proteolysis of the Ebola virus glycoprotein is necessary for infection. *Science*. 2005;308:1643–1645.
41 Sullivan NJ, Peterson M, Yang ZT, Kong WP, et al. Ebola virus glycoprotein toxicity is mediated by a dynamin-dependent protein-trafficking pathway. *J Virol*. 2005;79:547–553.

42 Zaki SR, Goldsmith CS. Pathologic features of filovirus infections in humans. *Curr Top Microbiol Immunol.* 1999;235:97–116.
43 Feldmann H, Volchkov VE, Volchkova VA, et al. Biosynthesis and role of filoviral glycoproteins. *J Gen Virol.* 2001;82: 2839–2848.
44 Sanchez A, Trappier SG, Mahy BW, et al. The virion glycoproteins of Ebola viruses are encoded in two reading frames and are expressed through transcriptional editing. *Proc Natl Acad Sci USA.* 1996;93:3602–3607.
45 Seebach J, Dieterich P, Luo F, et al. Endothelial barrier function under laminar fluid shear stress. *Lab Invest.* 2000;80:1819–1831.
46 Simmons G, Rennekamp AJ, Chai N, et al. Folate receptor alpha and caveolae are not required for Ebola virus glycoprotein-mediated viral infection. *J Virol.* 2003;77:13433–13438.
47 Yi XP, Xing SF. The distribution of epidemic hemorrhagic fever (EHF) viral antigen in tissues from EHF autopsies. *Zhonghua Bing Li Xue Za Zhi.* 1991;20:110–112.
48 Geisbert TW, Jaax NK. Marburg hemorrhagic fever: report of a case studied by immunohistochemistry and electron microscopy. *Ultrastruct Pathol.* 1998;22:3–17.
49 Davis KJ, Anderson AO, Geisbert TW, et al. Pathology of experimental Ebola virus infection in African green monkeys. Involvement of fibroblastic reticular cells. *Arch Pathol Lab Med.* 1997;121:805–819.

144

Effect of Smoking on Endothelial Function and Cardiovascular Disease

Rajat S. Barua and John A. Ambrose

University of California at San Francisco, Fresno

Cigarette smoking is the single most alterable risk factor contributing to premature morbidity and mortality in the United States. About one in five deaths from cardiovascular disease is attributable to smoking. Smoking costs Americans over $157 billion annually in health-related economic costs (1,2). Cigarette smoking predisposes to several different clinical atherosclerotic syndromes, including stable angina, acute coronary syndromes, sudden death, and stroke (1). In addition, smoking is associated with an increased incidence of aortic and peripheral atherosclerosis, which may lead to abdominal aortic aneurysms and intermittent claudication (1). Even passive smoking (environmental tobacco exposure) is reported to increase coronary artery disease (CAD) risk by 30% (1,3). Endothelial dysfunction is an important pathophysiological mechanism of atherosclerosis (4). Compelling evidence suggests that endothelial cell (EC) dysfunction is associated with (and caused by) each of the risk factors for atherosclerosis, including smoking. However, the precise mechanisms by which cigarette smoking and its components such as nicotine may cause endothelial dysfunction leading to adverse cardiovascular events remain to be fully elucidated. The present chapter summarizes the historical as well as the current perspectives on the evidence and potential mechanism(s) of cigarette smoking-related endothelial dysfunction and cardiovascular disease.

HISTORICAL PERSPECTIVES

About 500 years ago, Columbus and his crew brought tobacco (dried leaves and seeds) from the New World to Europe. By the early part of the 19th century, tobacco was rolled in paper and smoked as cigarettes (5). It was not until the late 1930s that it was initially proposed that cigarette smoking might be associated with increased cardiac mortality (5). In the 1950s, seminal prospective cohort studies on British male physicians demonstrated that cardiovascular disease is more common in smokers than in nonsmokers (5,6). Since then, innumerable epidemiological studies have confirmed an association between smoking and myocardial infarction (MI)/fatal CAD in men and women of all ages (1). In 1975, Asmussen and Kjeldsen provided the first direct evidence that cigarette smoking could result in endothelial injury by describing ultrastructural alterations in the endothelium obtained from the umbilical arteries of babies from mothers who smoked (7). In 1978, Aronow reported for the first time on the potential association between CAD and passive smoking (8).

It is now generally accepted that the cardiovascular consequences of smoking are due primarily to atherothrombotic modification of the vascular wall and increased thrombogenicity of the blood. In humans, to directly ascertain the relationship between cigarette smoking and atherosclerosis, various clinical imaging techniques such as angiography, vascular ultrasound, and transesophageal echocardiography, as well as pathologic analysis of the coronary arteries, have been employed. Utilizing these techniques, after controlling for various risk factors for atherosclerosis, multiple studies have reported that cigarette smoking is associated with an increase in the incidence and acceleration of atherosclerotic plaque formation (6).

EFFECTS OF SMOKING ON ENDOTHELIAL CELL STRUCTURE

Injury to the endothelium is believed to be an initiating event in the atherosclerotic process. It also appears to be a major determinant in causing acute cardiovascular events (4). Endothelium from the umbilical arteries of babies from mothers who smoked cigarettes during pregnancy showed an irregular appearance of the cell membrane and the formation of blebs or microvillous-like projections, indicating endothelial injury (7). In vitro, exposure to cigarette smoke extract (CSE) or a high dose of nicotine was shown to be associated with an

Table 144-1: Effect of Cigarette Smoking Exposure on the Endothelium

	Acute Exposure	*Chronic Exposure*	*Mechanisms (Based on Combination of In Vitro and In Vivo Studies)*
EC structure and survival in vivo	Increased circulating ECs as well as EC microparticles	Abnormality in ultrastructural components of ECs; increased circulating ECs	Apoptosis and necrosis in vitro, with low concentration of CSE or nicotine Primary mechanism was apoptosis; with higher concentration, necrosis was more prominent
Vasomotor function	Impaired in habitual smokers as well as in nonsmokers	Impaired in macro- as well as microvascular beds	Decreased NO availability (chronic and acute); increased ET-1 (acute); decreased PGI_2 availability
Proinflammatory effects	Increased soluble adhesion molecules (ICAM-1, P-selectin)	Increased adhesion molecules (ICAM-1, VCAM-1), circulating CRP, IL-6, TNF-α	Activation of inflammatory genes due to increased oxidative stress and reduced levels of NO
Procoagulant effects	Increased circulating tissue factor activity	Decreased stimulated t-PA release; alteration of t-PA/PAI-1 molar ratio; decreased TFPI release; increased tissue factor level as well as activity in atherosclerotic plaques	Unclear; preliminary data suggest decreased NO availability and increased oxidative stress

increase in EC death or apoptosis (6,9). It had been reported that after acute as well as chronic cigarette smoke exposure, in human blood, circulating anucleated endothelial carcasses (9,10) as well as circulating ECs were increased (11). Even in nonsmokers, acute cigarette smoke exposure was associated with an increase of these circulating cells or cell particles (10). All these findings are consistent with the concept that cigarette smoke exposure is injurious to the endothelium. Table 144-1 summarizes the effects of cigarette smoke exposure on the endothelium.

EFFECTS OF SMOKING ON ENDOTHELIAL CELL FUNCTION

It is now evident that during the atherosclerotic process, alterations in the functional characteristics of the endothelium precede changes in morphology (4). The functional changes of the endothelium associated with cigarette smoking include vasoregulatory dysfunction, and proinflammatory and prothrombotic modification.

Vasoregulatory Dysfunction

Impairment of vasoregulatory function is one of the earliest functional changes in the endothelium associated with atherosclerosis. In both animal and human models, multiple studies have demonstrated that active as well as passive cigarette smoke exposure is associated with a decrease in endothelial vasodilatory function (1,12). In humans, both acute and chronic cigarette smoke exposure impaired endothelium-dependent vasodilation (EDV) in macrovascular beds such as the coronary and brachial arteries and in microvascular beds (1,12–15). In young healthy male smokers, heavy (>1 pack/day) and light (<1 pack/week) cigarette smoking was reported to have a similar degree of impairment of brachial artery EDV (16). Recently, using high-resolution magnetic resonance imaging, it was demonstrated that chronic cigarette smoking induced global impairment of EDV in both peripheral and central conduit arteries (17). Use of nicotine (patch, spray, or gum) was also reported to be associated with partial impairment of EDV. However, the degree of impairment was less severe with nicotine use as compared to that of active cigarette smoking (18,19).

The endothelium secretes multiple vasoregulatory molecules; among them the primary vasodilator is nitric oxide (NO), and the primary vasoconstrictor is endothelin (ET)-1. As discussed later, evidence regarding the effect of smoking on these vasoregulatory molecules continues to accumulate.

Effect of Cigarette Smoke on Nitric Oxide Biology

In vitro exposure to CSE or nicotine was associated with decreased availability of EC-derived NO (6,12,20). In a more physiologically relevant in vitro experiment, Barua and colleagues demonstrated that the incubation of human umbilical vein ECs (HUVECs) and human coronary artery ECs (HCAECs) with sera from young healthy male smokers resulted in decreased NO availability (12,16,21). In these

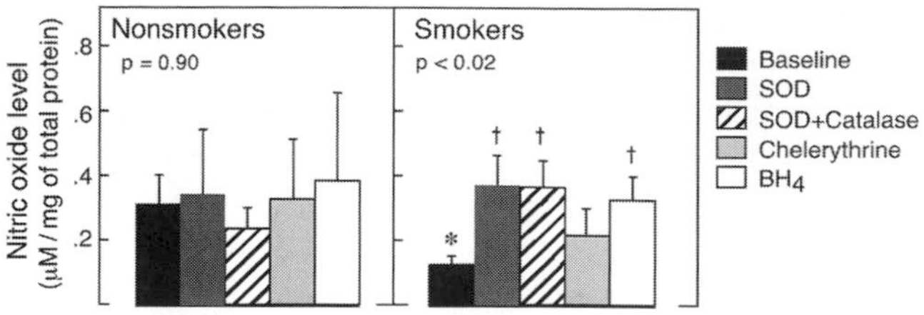

Figure 144.1. Effects of smokers' serum NO production in vitro. Confluent (~85%) HCAECs were incubated for 12 hours with equal volumes of medium and serum from nonsmokers ($n = 10$) or smokers ($n = 15$) (with or without treatment as indicated) in 12-well plates. NO production in the cell culture supernatant was determined by a chemoluminescence method. Results are presented as mean ± SEM after adjusting for the background NO and the total protein in each well. Unpaired Student's t-test: *$p < 0.02$ versus nonsmokers' baseline; paired Student's t-test: †$p < 0.04$ versus smokers' baseline. (Reproduced with permission from Barua RS, Ambrose JA, Srivastava S, et al. Reactive oxygen species are involved in smoking-induced dysfunction of nitric oxide synthase: an in vitro demonstration in human coronary artery endothelial cells. *Circulation.* 2003;107: 2342–2347.)

studies effects of sera from cigarette smokers on NO biosynthetic pathway in both cell types (HUVECs and HCAECs) were similar (12,21). Moreover, these authors found a correlation between flow-mediated brachial artery EDV and NO levels in HUVECs exposed to serum from the same individuals (12). In human subjects, it has been demonstrated indirectly in vivo by infusing N (G)-monomethyl-L-arginine (a competitive inhibitor of L-arginine, the substrate for NO biosynthesis) that the reduced EDV associated with cigarette smoking is attributable to a decreased availability of NO (1,12). However, the exact mechanisms responsible for the decrease in NO availability remain to be clearly defined.

Cigarette smoke contains a large number of free radicals and pro-oxidants. NO is also a free radical that can preferentially react with the free radicals from cigarette smoke, which would cause a decrease in NO bioavailability (1). In vitro, the addition of either free radical scavengers such as superoxide dismutase (SOD) or catalase increased NO availability in the cell culture supernatant of HCAECs exposed to smokers' serum (Figure 144.1) (21). Murohara and colleagues have shown that the CSE-induced contraction of isolated porcine coronary arteries was significantly attenuated by SOD (22). Similarly, Heitzer and colleagues reported that infusion of vitamin C, or 4 months of vitamin E oral therapy, significantly improved brachial artery EDV in human smokers (23,24). Taken together, these studies support the notion that the scavenging activity of cigarette smoke–derived free radicals leads to reduced NO bioavailability.

Cigarette smoke could also influence the availability of NO by affecting endothelial NO synthase (eNOS) expression and/or functional activity. Studies on the effect of cigarette smoking on eNOS expression have led to conflicting conclusions. In ECs isolated from the main pulmonary artery of 6- to 7-month-old pigs, CSE exposure was shown to cause a decrease in both *eNOS* mRNA and eNOS protein (20). Similarly, in lung tissue specimens from subjects who underwent lung resection for lung carcinoma, samples from smokers showed a decreased immunostaining for eNOS in ECs from all types of pulmonary vessels as compared with nonsmokers. These authors also reported that eNOS protein level was decreased in lung tissue homogenate from smokers (25). However, when serum from smokers was added to confluent HUVECs or HCAECs, eNOS protein expression was increased (12,21). Exposure of HCAECs to nicotine for 24 hours at concentrations similar to those in the blood of smokers resulted in increased *eNOS* mRNA levels (26). Recently, Beckman and colleagues, utilizing skin biopsy, demonstrated that microvessel *eNOS* mRNA levels were increased in cigarette smokers compared with healthy control subjects (27). In contrast to these conflicting studies related to smoking and eNOS expression, there is more consistency in the reported effects of cigarette smoking on eNOS activity. In various models, cigarette smoke exposure is associated with a decrease or an alteration (i.e., uncoupling) of eNOS activity that was responsible for the decreased NO, regardless of its effect on *eNOS* mRNA or protein level (20,21) (Figures 144.2 and 144.3). In vitro, the addition of free radical scavengers and/or tetrahydrobiopterin, an essential cofactor of eNOS, normalized the dysfunctional eNOS activity associated with cigarette smoking (21) (Figure 144.2). In vivo, supplementation of tetrahydrobiopterin restored the EDV in cigarette smokers (28). These findings support decreased or altered eNOS activity as a viable mechanism for the decrease in NO availability in smokers.

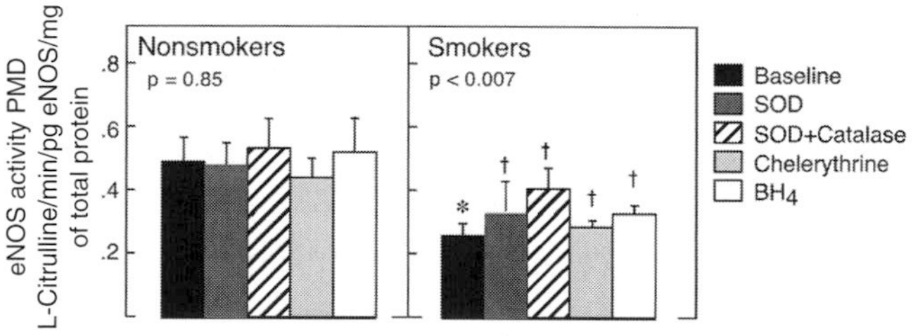

Figure 144.2. Effects of smokers' serum on eNOS activity in vitro. Confluent (~85%) HCAECs were incubated with equal volumes of medium and serum from nonsmokers ($n = 10$) or smokers ($n = 15$) (with or without treatment as indicated) in 12-well plates. After 12 hours, the HCAECs were lysed. eNOS activity of the cell lysates was determined by detecting the conversion of [^{3}H] L-arginine to [^{3}H] L-citrulline. Results are presented as the mean ± SEM after adjusting for the specific amount of eNOS protein/mg total protein. Unpaired Student's t-test: *$p < 0.02$ versus nonsmokers' baseline; paired Student's t-test: †$p < 0.04$ versus smokers' baseline; ANOVA: $p < 0.008$ for nonsmokers' baseline versus all treatment in smokers, Post hoc Fisher's PLSD: ‡$p < 0.03$ versus nonsmokers' baseline. (Reproduced with permission from Barua RS, Ambrose JA, Srivastava S, et al. Reactive oxygen species are involved in smoking-induced dysfunction of nitric oxide synthase: an in vitro demonstration in human coronary artery endothelial cells. *Circulation.* 2003;107:2342–2347.)

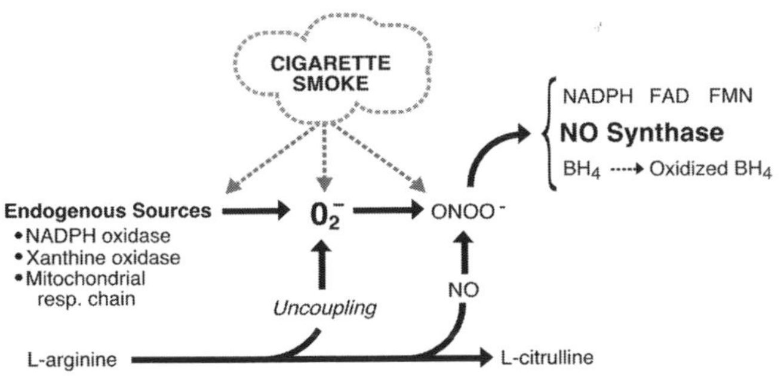

Figure 144.3. Potential mechanism(s) for decreased NO availability in smokers. *Dashed arrows* represent potential direct and indirect pathways. *Bracket* identifies eNOS and its cofactors. Normally, eNOS utilizes L-arginine to produce NO and citrulline. Free radical such as $O_2^{\cdot-}$ derived from cigarette smoke and other endogenous sources reacting with NO, not only would decrease NO availability but also would give rise to $ONOO^-$. $ONOO^-$ has been shown to oxidize BH_4. With oxidation BH_4, eNOS becomes uncoupled and produces $O_2^{\cdot-}$, which would further enhance the cellular oxidative stress. NOS, nitric oxide synthase; BH_4, tetrahydrobiopterin; NADPH, nicotinamide adenine dinucleotide phosphate reduced; FAD, flavin adenine dinucleotide; FMN, flavin adenine mononucleotide; $O_2^{\cdot-}$, superoxide; $ONOO^-$, peroxynitrite.

In addition to its vasodilatory function, NO also plays an important role in the regulation of inflammation, leukocyte adhesion, platelet activation, and thrombosis. Therefore, an alteration in endothelium-derived NO availability is likely to be one of the important steps in the initiation and progression of atherothrombotic disease (Figure 144.4).

Effect of Cigarette Smoke Exposure on Endothelium-Derived Endothelin-1

ET-1 is the most potent endogenous vasoconstrictor produced by both ECs and vascular smooth muscle cells (VSMCs). Available data regarding the effects of cigarette smoke or its components on ET-1 levels have been conflicting. In vitro, HUVECs exposed to serum from smokers or controls failed to demonstrate a difference in basal ET-1 production between nonsmokers, light smokers, and heavy smokers (16). On the other hand, HUVECs treated with medium that was conditioned with monocytes from smokers demonstrated increased release of ET-1 (29). In a rat model, acute cigarette exposure (six cigarettes for 30 minutes) was shown to increase the expression of *ET-1* mRNA in heart and lung tissues, but, with chronic cigarette exposure (six cigarettes for 30 min/day, 5 days a week for 6 months), the effect on the *ET-1* mRNA in cardiovascular tissues became insignificant (30). Using lung tissue specimens from human subjects who underwent lung resection for lung carcinoma, Barbera and colleagues found that *ET-1* mRNA and protein levels in pulmonary arteries and in lung tissue extracts were similar between smokers and nonsmokers (31). In vivo studies suggested that within the first 10 minutes of active smoking, a rise occurs in plasma or serum ET-1 levels, which is followed by a decline over time (32,33). Nicotine alone had no effect on plasma ET-1 (33). Recently, utilizing an ET A receptor (ETA) blocker (BQ-123), it was found that the vasodilatory function of the brachial artery measured by venous occlusion plethysmography did not improve in smokers, suggesting that ET-1 does not contribute substantially to vascular tone in smokers, especially via the ETA receptor (34). Thus, based on current data, the role of ET-1 in cigarette smoking–related vasodilatory dysfunction is unclear and appears to be restricted to the acute phase of smoking (10 minutes to few hours).

Effect of Cigarette Smoke Exposure on Endothelium-Derived Prostacyclin

Prostacyclin (PGI_2), a vasodilatory and antithrombogenic molecule secreted by ECs, also is affected by cigarette smoking. It was reported that aortas from rats that were chronically exposed to cigarette smoke showed a reduction in PGI_2 production (34). Similarly, Reinders and colleagues showed that incubation of HUVECs with cigarette smoke condensate (CSC) impaired the basal and stimulated (phorbol myristate acetate) production of PGI_2 (35). Ahlsten and colleagues reported that maternal smoking significantly decreased PGI_2 production in umbilical arteries from newborn infants (36). Similarly, HUVECs isolated from smoking mothers were shown to produce less PGI_2 in vitro (37). On the other hand, it was reported that incubation of nicotine with HUVECs resulted a dose-dependent increase (maximum effect with 0.05 μg/mL) in the basal level of PGI_2 (38). However, at higher doses nicotine appears to decrease the production of PGI_2 (38).

Proinflammatory Effects

The inflammatory change in the endothelial milieu leading to an increase in leukocyte and EC interaction is an essential component in the initiation and progression of atherosclerosis. In vitro, CSC was associated with a 70% to 90% increase in adherence between human monocytes and HUVECs (39). Shen and colleagues demonstrated that exposure of HUVECs to CSC resulted in increased nuclear factor (NF)-κB DNA binding activity; increased surface expression of intercellular adhesion molecule (ICAM)-1, vascular cell adhesion molecule (VCAM)-1, and platelet-endothelial cell adhesion molecule (PECAM)-1; and a twofold induction in transendothelial migration of monocytes (40). Adams and colleagues, exposing human monocytes and HUVECs to smokers' serum, found a significant increase in adhesion between these cells, which was associated with increased expression of ICAM-1 on HUVECs (41). These authors also noted that the increased adhesion between HUVECs and monocytes was reversed by L-arginine, suggesting that smoking-related impairment of NO availability was an important determinant of increased cell–cell adhesion between endothelium and leukocytes (41). Oxidized low-density lipoprotein (oxLDL) and lipid peroxidation products have been shown to be proinflammatory and, in various experimental models (including human, animal, and cell culture models), cigarette smoke exposure was associated with

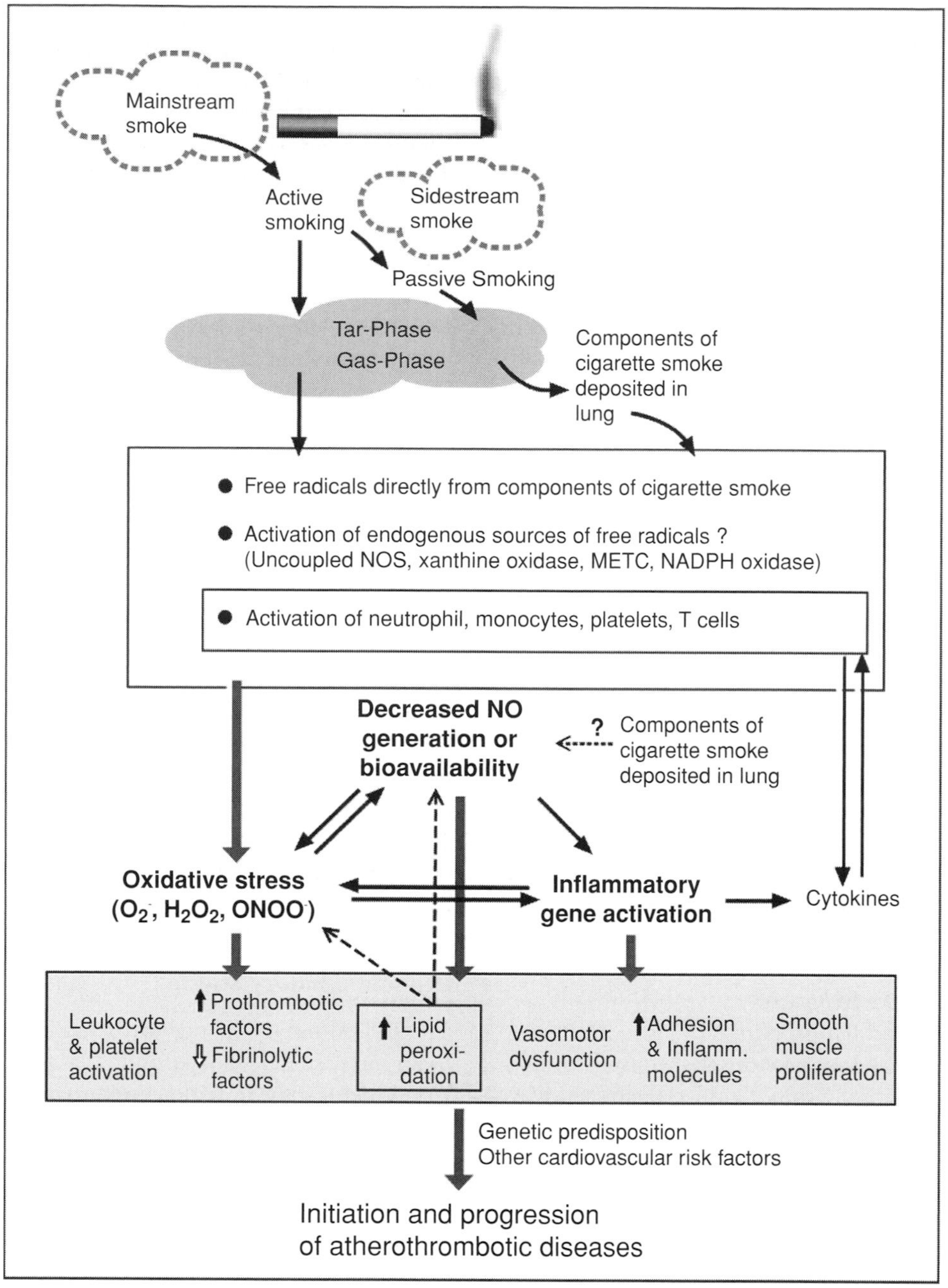

Figure 144.4. Potential pathways and mechanisms for cigarette smoking-mediated cardiovascular dysfunction. The *bold boxes* and *arrows* in the flow diagram represent the probable central mechanisms in the complex pathophysiology of cigarette smoking-mediated atherothrombotic disease. NOS, nitric oxide synthase; METC, mitochondrial electron transport chain; NADPH, nicotinamide adenine dinucleotide phosphate reduced form; $O_2^{\cdot -}$, superoxide; H_2O_2, hydrogen peroxide; $ONOO^-$, peroxynitrite. (Reproduced with permission from Ambrose JA, Barua RS. The pathophysiology of cigarette smoking and cardiovascular disease: an update. *J Am Coll Cardiol.* 2004;43:1731–1737.)

an increase in both these products (1). In humans, after controlling for multiple risk for atherosclerosis, chronic cigarette smoke exposure was found to be associated with increased levels of multiple inflammatory markers including peripheral leukocytosis, C-reactive protein (CRP), interleukin (IL)-6, and tumor necrosis factor (TNF)-α (1,42–45).

Procoagulant Effects

Epidemiological studies indicate that cigarette smoking increases the risk of acute coronary events, namely acute MI and sudden death (46). These events are mediated by thrombosis on a ruptured or eroded atherosclerotic plaque. Cigarette smokers have MI at a relatively young age, although they have fewer cardiac risk factors, and less severe underlying coronary disease on angiography, compared to nonsmokers (47,48). These data suggest that for any given degree of atherosclerosis, smokers are more hypercoagulable than are their nonsmoking counterparts.

The endothelium is a major source of both thrombotic (e.g., tissue factor) and antithrombotic (e.g., tissue factor pathway inhibitor [TFPI]-1) factors, as well as fibrinolytic (tissue-type plasminogen activator [t-PA]) and antifibrinolytic factors (plasminogen activator inhibitor [PAI]-1) (49). Thus, an imbalance in endothelial antithrombotic and fibrinolytic properties may contribute to an acute thrombotic event. In vitro, HUVECs exposed to serum from chronic smokers showed a significantly decreased TFPI-1 level, but no significant difference was found in tissue factor level (49). Atherosclerotic plaques isolated from *ApoE*$^{-/-}$ mice exposed to nonfiltered research cigarettes showed an increased tissue factor immunoreactivity and an increase in tissue factor activity (50). In smokers, 2 hours after smoking two cigarettes, an increase in circulating tissue factor activity also has been reported in human plasma (51). In vitro, HUVECs exposed to chronic smokers' serum have significant decreases in both basal and substance-P–stimulated t-PA release as well as a significant alteration in the t-PA:PAI-1 molar ratio (49). In vivo, decreased plasma t-PA antigen and activity were observed in smokers in samples obtained from brachial and coronary arteries after pharmacological (substance P or bradykinin) stimulation (1). In addition to its prothrombotic effect on the endothelium, cigarette smoking can further enhance the prothrombotic milieu by affecting platelet activation, circulating fibrinogen levels, red blood cell counts, the hematocrit, blood viscosity, and ongoing inflammatory responses (1).

CIGARETTE SMOKE COMPONENTS AND MECHANISM(S) RESPONSIBLE FOR ENDOTHELIAL DYSFUNCTION

Using a standardized automatic smoking machine designed to approximate human smoking behavior, cigarette smoke can be divided into two phases: the tar or particulate phase, and the gas phase. During the tar phase, particulate material is trapped when the smoke stream is passed through the Cambridge glass-fiber filter that retains 99.9% of all particulate material with a size >0.1 μm. The gases of the gas phase are all materials that pass through the filter. *Mainstream smoke* is the smoke drawn through the tobacco into an active smoker's mouth; it is composed of 8% tar and 92% gaseous components. *Sidestream cigarette smoke* is the smoke emitted from the burning end of a cigarette (1,52).

In 1967, the United States Federal Trade Commission adopted this method to indicate the amount of tar, nicotine, and carbon monoxide (CO) associated with existing brands of cigarettes. These machine-generated values do not reflect the actual amount of tar and nicotine that will be inhaled by a smoker from any particular cigarette, nor does it indicate the relative amount of tar and nicotine between any two brands. Because no two cigarette smokers smoke cigarettes exactly the same way, how much tar and nicotine an individual gets from a cigarette depends on how an individual smokes it (e.g., larger or more frequent puffs taken by a smoker of a low-tar/nicotine brand cigarette may deliver as much tar and nicotine as those by smoker of a high-tar/nicotine brand cigarette). Many cigarettes also have ventilation holes. Exposure to the harmful constituents in cigarette smoke substantially increases when the fingers holding the cigarette block these holes (52,53).

Cigarette smoke contains over 4,000 identified and over 100,000 unidentified components. About 500 of them have been identified in the gas phase (1,52,53). A few components, such as CO, nicotine, and some of the polycyclic aromatic hydrocarbons (PAHs), have been studied in isolation in relation to atherothrombotic dysfunction of the endothelium.

The effects of cigarette smoke–derived CO are equivocal. Earlier studies in animal models suggested that CO was associated with the induction and development of atherosclerosis. However, more recent epidemiological and animal studies have failed to demonstrate a relationship between CO exposure and EC morphology or the development of atherosclerosis (1,48). On the contrary, Thom and colleagues reported that CO-exposed bovine pulmonary artery ECs release a higher concentration of NO (54). Kourembanas and colleagues reported that CO suppressed the production of ET-1 by ECs (55).

Nicotine, a component of tar, is the only known addictive substance in cigarette smoke, and it is also its most studied component. Nicotine is vaporized from burning tobacco and carried proximally on tar droplets (mass median diameter 0.3 to 0.5 μm) that are inhaled; nicotine also is probably inhaled in gas-phase materials (56). Nicotine exerts its effect via nicotinic acetylcholine receptors (nAChRs), which are expressed in neuronal as well as non-neuronal cells (e.g., ECs and VSMCs). In neuronal cells, these receptors are thought to be involved in psychoactive action and addictive properties of tobacco. In non-neuronal cells these receptors are suggested to be involved in the regulation of mitosis, differentiation, organization of the cytoskeleton, cell–cell contact, locomotion, and migration (57). Although nicotine's role in the hemodynamic effects of smoking, such as an increase in cardiac output, heart rate, and blood pressure are well accepted, its role in endothelial dysfunction and atherothrombotic modification of the endothelium remains debatable. Lin and colleagues, utilizing en face preparations of rat thoracic aorta, showed that nicotine exposure was associated increased frequency of EC apoptosis, which resulted in enhanced transendothelial leakage of macromolecules including LDL (58). On other hand, Pittilo and colleagues reported that electron microscopic evidence of injuries in aortic endothelium from rats was minor with nicotine as compared to cigarette smoke exposure (6). In cultured ECs, low-dose nicotine ($<10^{-8}$ M) was found to increase DNA synthesis, mitogenic activity, and endothelial proliferation

(59). In reference to EDV or NO availability, in various models, nicotine exposure alone has been reported to cause no change, a decrease or an increase in these two parameters (1). Similarly, the effect of nicotine on hemostatic factors such as platelets, fibrinogen, t-PA, or PAI-1 appears to be small in the setting of smoking (1). From the current evidence, nicotine alone probably plays only a minor role in direct atherothrombotic modification of the endothelium; however, the addictive property of nicotine is likely to play a major role in perpetuating the exposure to other more detrimental components of cigarette smoke.

PAHs are found in the tar fraction of cigarette smoke. The effects of a few of them, such as benzo(a)pyrene, 7,12 dimethyl benz(a)anthracene 1-methylanthracene, phenanthrene, acenaphthylene, and naphthalene also have been studied. In vitro, HUVECs exposed to 10 μM benzo(a)pyrene for 90 minutes demonstrated increased endothelial DNA damage as assayed by single-cell gel electrophoresis (Comet assay) (60). In HCAECs, 1-methylanthracene, phenanthrene, and benzo(a)pyrene exposure caused significant apoptosis by the phospholipase A2/arachidonic acid–dependent pathway (61). In cockerels, weekly injections of benzo(a)pyrene and 7,12 dimethyl benz(a)anthracene caused an increase in atherosclerotic plaque development in the aorta (48). However, contrary data also exist. Vayssier-Taussat and colleagues, comparing tobacco smoke filtrate with benzo(a)pyrene, reported that tobacco smoke filtrate caused expression of the inducible heat shock protein 70 and heme oxygenase-1 in ECs, which produced loss of mitochondrial membrane potential and apoptosis. On the other hand, exposure to benzo(a)pyrene alone evoked neither stress proteins nor mitochondrial membrane potential (62). In isolated rat aorta, acenaphthylene and naphthalene dose-dependently relaxed the phenylephrine-induced contraction of rat aorta, and this vasorelaxation effect was diminished when endothelium was removed, suggesting that the relaxation effect of PAHs was endothelium-dependent. Pretreatment with the NO synthase inhibitor, L-N(G)-nitroarginine methyl ester, and the guanylate cyclase inhibitor, methylene blue, prevented the vasorelaxation induced by PAHs (62,63). These results suggested that the vasorelaxation effect of PAHs was mediated by activation of eNOS.

Cigarette smoke is a rich source of free radicals. The particulate (tar) phase of cigarette smoke contains $>10^{17}$ free radicals/g, and the gas phase contains $>10^{15}$ free radicals/puff. The radicals in the tar phase have a longer lifespan (hours to months), whereas the radicals in the gas phase have a shorter lifespan (seconds). In the setting of cigarette smoking, free radicals could arise not only from the gas or tar phase of cigarette smoke, but also, current evidence suggests, from endogenous sources (e.g., activated macrophages, neutrophils, and ECs) via mechanisms that include uncoupled eNOS, xanthine oxidase, nicotinamide adenine dinucleotide phosphate (NADPH) oxidase, and mitochondrial electron transport chain (see Figures 144.3 and 144.4). A reaction between free radicals such as superoxide and NO not only decreases NO availability but also generates reactive nitrogen species (peroxynitrite). Peroxynitrite has been shown to oxidize tetrahydrobiopterin, a cofactor of eNOS. With oxidized tetrahydrobiopterin, eNOS becomes uncoupled and produces superoxide anion preferentially over NO, which would further enhance the cellular oxidative stress (1) (Figure 144.3).

Current consensus suggests that free radical–mediated oxidative stress and the loss of the protective effect of NO are pivotal steps in the atherothrombotic modification of the endothelial milieu (Figure 144.4). Many of the abnormalities of the endothelium associated with cigarette smoking, including vasoregulatory dysfunction, and proinflammatory and prothrombotic effects, can largely be explained by the effects of increased oxidative stress and decreased protection from NO. Additionally, in various experimental models, antioxidants or agents that reduce the oxidative stress or increased NO availability have been shown to either improve or reverse the proatherogenic, proinflammatory, and prothrombotic attributes associated with cigarette smoking (1).

Recently, it is has been proposed that genetic predisposition also could influence intersubject variability in atherogenesis in individuals exposed to cigarette smoke. Wang and colleagues proposed that certain gene polymorphisms such as *CYP1A1 MSP* (an enzyme present in the lungs that activates smoke carcinogens) or specific *eNOS* intron-4 polymorphisms increased the susceptibility to cigarette smoke exposure–related atherosclerotic diseases including multivessel CAD and MI (1,48). Because the overall prevalence of these genetic variations has not been determined, the importance of genetic variation in the general population in relation to cigarette smoke exposure remains to be established.

EFFECT OF ENVIRONMENTAL SMOKE EXPOSURE ON THE ENDOTHELIUM

Environmental tobacco smoke results from the combination of sidestream smoke (85%) and a small fraction of exhaled mainstream smoke (15%) from smokers (1). In 1978, Aronow was the first to highlight the effect of passive smoking on CAD by showing that passive smoke exposure exacerbated angina (8). Imaging of the carotid artery with ultrasound has shown that the intimal and medial thickness (a measure of atheromatous disease) was greater in nonsmokers living with a smoker compared to that of unexposed nonsmokers, even after controlling for diet, physical activity, body mass index, alcohol intake, education, and major cardiovascular risk factors (64). In various animal models, environmental tobacco smoke exposure has been shown either to increase the area of atheromatous plaques or to increase infarct size after ligation of one coronary artery (3). In humans, active and passive smoke exposure caused a similar degree of dysfunction in EDV of the brachial artery and the coronary artery, even though the objective evidence of exposure was significantly different (serum cotinine, a stable metabolite of nicotine or carboxyhemoglobin level) (16). A similar degree of effect also was observed in platelet aggregation studies in vitro in relation to acute and passive smoke exposure (3). Meta-analyses of existing epidemiological data have suggested that the relative risk for CAD in nonsmokers exposed to environmental tobacco smoke is about 1.3 (i.e., a 30% excess risk) compared with unexposed nonsmokers (3). Interestingly, in a recent paper, Sargent and colleagues reported that a citywide smoking ban in public places over a 6-month period in Helena, Montana, reduced the incidence of acute MI by 60% during that time period (65). All these data highlight the impact of environmental tobacco smoke on atherothrombotic modification of the endothelium.

SMOKING, PREGNANCY, AND ENDOTHELIUM: EFFECT ON UTERO-PLACENTAL CIRCULATION

Cigarette smoking causes well-documented adverse effects on pregnancy and fetal health that include low birth weight, preterm delivery, perinatal morbidity, and placental complications. Even after learning that they are pregnant, 54% of women continue to smoke (66). It is estimated that smoking continues to account for 10% of all fetal mortality (67). Placental and uterine vascular dysfunction, together with thrombotic complications within these vessels, is thought to be responsible for smoking-related complications of pregnancy (68). As discussed earlier, endothelium from the umbilical arteries of babies from mothers who chronically smoked cigarettes during pregnancy showed ultrastructural evidence of endothelial injury. Utilizing Doppler ultrasonography, a significant increase of utero-placental vascular resistances was reported in chronic smokers. A direct correlation was noted between utero-placental vascular resistance and serum cotinine (a stable metabolite of nicotine often used as a surrogate marker for cigarette smoke exposure) levels in the blood from these mothers (68). Castro and colleagues reported that in habitual smokers, acute cigarette exposure significantly decreased the flow of uterine blood to the placenta (69). These observations can partly be explained by the effect of nicotine on utero-placental circulation via nicotinic receptors as well as by the decreased availability of endothelial-derived NO and PGI_2 in the utero-placental circulation of mothers who smoked.

POSITION (CHALLENGES) IN BENCH TO BEDSIDE

Although progress has been made in understanding the effect of cigarette smoking on the endothelium, the nonbinary continuous nature of the endothelium and complex chemical composition (>4,000 known components) of cigarette smoke pose a multitude of investigative and diagnostic challenges. In the in vitro studies, challenges are primarily twofold. The first major challenge is to create a dynamic in vitro endothelial culture system. It is noteworthy that most in vitro studies to date have been carried out in static culture, in which ECs, once uncoupled from their native tissue microenvironment, undergo phenotypic drift. It is unclear to what extent these findings can be translated to the in vivo setting. An important goal, which is not unique to studying the effects of smoking on the endothelium, is to develop novel cell culture systems in which the microenvironment is preserved and/or recapitulated in vitro. The second challenge is to develop an appropriate model for cigarette smoke exposure in vitro. The majority of in vitro data have involved the use of CSE or CSC. However, the concentration of CSE or CSC that simulates actual human cigarette smoke exposure is unclear, because in the circulation, cigarette smoke must first pass through the pulmonary circulation, where certain toxic components may be neutralized (e.g., by antioxidants present in the blood). Some studies have utilized serum from cigarette smokers or have isolated components of cigarette smoke as a vehicle of cigarette smoke exposure in vitro. It is unknown whether differences in the model of smoke exposure were responsible for the somewhat contradictory findings among the different in vitro models of cigarette smoking (1,6,12–20). In animal models, interspecies heterogeneity of the endothelium and differences in the behavioral pattern of cigarette smoking (habitual smoking versus control smoke exposure) must be considered when translating their relevance to humans.

In humans, various endothelium-derived biomarkers, as well as invasive and noninvasive investigative techniques, are currently being studied as research and diagnostic tools. Noninvasive evaluation of flow-mediated vasodilatory function of the brachial artery by ultrasound has been shown to be a reproducible and relatively easily applicable technique to investigate the vasoregulatory function of the endothelium in smokers. However, given the site-specific heterogeneity of the endothelium, at present, there is not enough available data to assure that this measurement of peripheral vascular endothelial vasoregulatory function would also precisely reflect the vasoregulatory function of other vascular beds, specifically the coronary circulation. New imaging techniques such as high-resolution magnetic resonance imaging and transthoracic Doppler echocardiography may improve our ability to evaluate multiple and clinically relevant vascular beds. However, to investigate the molecular mechanism(s), invasive techniques such as biopsy of the endothelium will be essential. Currently, in a living human, an ethically acceptable way to biopsy the endothelium of a clinically relevant vascular bed is not available. However, endothelium-specific evaluation of skin biopsy from smokers might provide additional insight into the potential molecular mechanism(s) underlying smoking-related endothelial dysfunction.

CLINICAL AND THERAPEUTIC IMPLICATIONS

Strategies to Reduce Oxidative Stress and Increase Nitric Oxide Availability in Smokers

From existing data, it can be hypothesized that increased oxidative stress and the loss of the protective effect of NO are the central processes responsible for cigarette smoke exposure–mediated endothelial and cardiovascular dysfunction. Therefore, increasing NO availability and reducing oxidative stress by free radical scavengers would be of potential clinical and therapeutic benefit in cigarette smoking.

As presented earlier, the in vitro addition of various antioxidants, including vitamin C and vitamin E, improved endothelial dysfunction, antagonized procoagulant effects, and reduced some inflammatory effects in various smoking models. However, human data from in vivo and prospective studies of these agents have failed to demonstrate sustained benefit. Oral vitamin C was shown to improve endothelial vasoregulatory function in healthy smokers. This effect was short-lived in smokers, and did not persist after 8 weeks of continuous vitamin C administration, despite a sustained increase of plasma ascorbate levels (70). Data on vitamin E administration in smokers are contradictory, with some studies reporting improvement in endothelial vasoregulatory function, whereas others have failed to demonstrate benefit (18,71). Similarly, prospective randomized clinical trials such as Gruppo Italiano per lo Studio della Sopravvivenza nell'Infartomiocardico (GISSI), the Heart Outcomes Prevention Evaluation (HOPE), and the Heart Protection Study failed to show any benefit of vitamin C and E in reducing cardiovascular events (72). Currently, therefore, the recommendation of antioxidants such as vitamin C and E as adjunctive therapy for primary as well as secondary prevention in smokers does not appear warranted.

In both in vitro and in vivo experimental models, several strategies have been shown to increase the NO availability in smokers, such as supplementation of L-arginine (the substrate of NO synthase) or the use of tetrahydrobiopterin, a cofactor of eNOS (1,6,26). However, long-term clinical effects of their use in smokers have not been studied. Another experimental strategy that could be useful in increasing NO bioavailability in smokers who have decreased eNOS expression in their endothelium is the utilization of viral vectors to deliver *eNOS* genes directly to the affected vasculature. Further research is required to determine their usefulness in the setting of smoking exposure.

Therapeutic Strategies for Acute Coronary Syndromes in Cigarette Smokers

It has been observed that, among cases of MI, smokers have less severe underlying coronary atherosclerosis than do nonsmokers. It also has been demonstrated that smokers with MI appear to do better after thrombolysis compared with nonsmokers, with a higher percentage experiencing complete clot lysis (25). However, it also has been shown that in smokers with MI, outcomes with coronary angioplasty were equal to or better than those with intravenous thrombolysis (25). Therefore, in smokers presenting with acute MI, coronary angioplasty may be preferable to thrombolysis. Yet, when angioplasty is not immediately available, immediate thrombolysis with an intravenous clot lysing agent may be preferable to more delayed angioplasty. Combined approaches involving clot lysis and delayed angioplasty also are being tested. Their effect in cigarette smokers is presently unknown.

Nonlinear Dose Effect of Cigarette Smoke Exposure

Cigarette smoke exposure may cause a disproportionately large effect on the endothelium at a relatively small dose. It has been demonstrated that heavy and light active smoking can cause a similar decrease in brachial artery EDV, which paralleled the in vitro observations showing equivalent alterations in NO production, eNOS protein expression, and eNOS activity in cultured HUVECs exposed to serum from the same heavy and light smokers (16). Other studies even found that passive smoke exposure is associated with a decrease in EDV, comparable to smokers (16). These observations appear to parallel the cardiovascular risk estimation from the epidemiologic studies. A meta-analysis of large cohort studies showed that in active smokers the risk for ischemic heart disease with five cigarettes per day was not a quarter, but about half that associated with 20 cigarettes per day (3). More importantly, it has been estimated that with environmental smoke exposure, even though the exposure to tobacco smoke was less than 1% of the exposure from smoking 20 cigarettes per day, the excess risk for ischemic heart disease was about 33% of that experienced by a person who smoked 20 cigarettes per day (3).

This nonlinear dosimetry of cigarette smoking raises some interesting questions: Do underlying biochemical and cellular processes become saturated with small doses of toxic components from cigarette smoke, signifying their threshold effect? Do small amounts of cigarette smoke exposure cause changes in the endogenous cellular process that multiplies the original toxin burden (i.e., free radicals) (see Figures 144.3 and 144.4) and leads to this nonlinear effect? Or, does this observation reflect how intensively cigarettes are smoked (i.e., even though the number of cigarettes smoked may be small, with more intensive cigarette smoke inhalation, the actual exposure to tobacco smoke toxins is high) (48)? Regardless of the cause, this nonlinear dosimetry stresses the important notion that there may not be a safe level for cigarette smoke exposure. Thus, strategies to eliminate smoking-related cardiovascular disease must include complete elimination of all types of cigarette smoke exposure.

Therapies for Smoking Cessation

The medications currently approved for smoking cessation are nicotine (patch and gum form) and bupropion.

As discussed earlier, nicotine exposure is the only known addictive substance of cigarettes, and it has been shown to contribute to endothelial dysfunction. Therefore, its therapeutic use raises a valid safety concern. However, nicotine alone is clearly much less hazardous compared with cigarette smoke exposure. It also has been suggested that differences in nicotine delivery between cigarette smoke exposure and pharmacotherapy also make nicotine replacement therapy less harmful. Cigarette smoke exposure delivers nicotine into the arterial circulation quite rapidly, maximizing the pharmacologic and potentially pathological effects, whereas nicotine from pharmaceutical preparations is absorbed much more slowly, resulting in much lower peak arterial concentrations that are potentially less toxic (48). Indeed, some clinical trials found no evidence that pharmacotherapy increased cardiovascular risk in patients with cardiovascular disease (48). On the contrary, it was reported that the administration of nicotine patches to smokers with known CAD reduced the size of the myocardial perfusion defect during exercise, despite higher nicotine levels while on the patch compared with when smoking alone (48). These findings are consistent with the concept that the combustion products of cigarette smoke are much more deleterious than nicotine alone.

Bupropion is an antidepressant that stimulates dopamine release and curbs the severe withdrawal symptoms of smoking cessation. The appropriate dosing of bupropion is important in the success of smoking cessation. A dose of 300 mg or 150 mg per day over 7 weeks is associated with a significant rate of smoking cessation as compared with 100 mg per day over that time (73).

Recent data from animal as well as human studies with rimonabant, a selective cannabinoid receptor antagonist, showed encouraging results in aiding smoking cessation. Currently, prospective, randomized, multicenter, double-blind, placebo-controlled clinical trials are ongoing with this new drug (74).

Smoking cessation probably is the most important intervention in preventive cardiology. However, this continues to be a major therapeutic challenge. Only approximately 10% of all smokers who try to quit are able to do so. This low rate is in part due to the strong psychological and physiological needs of dependence. However, physicians may also be at fault, because counseling for smoking cessation is often relegated to a low level of priority (26,75).

KEY POINTS

- Cigarette smoke exposure has profound and multiple deleterious effects on endothelial biology, leading to cardiac and vascular diseases.
- Smoking decreases vasodilatory function, and increases the inflammatory and thrombotic potential of the endothelium.
- From the current evidence, it appears that free radical–mediated increased oxidative stress with loss of the protective effect of NO is the central step in atherothrombotic modification of the endothelial milieu associated with cigarette smoke exposure.
- However, the precise cellular and molecular mechanism(s) associated with these observations has yet to be clearly defined. Moreover, in prospective clinical trials, some of the therapeutic strategies in smokers based on the current experimental studies have failed to show a sustained benefit. These negative results may, in part, be explained by the complexity of the endothelium – an organ that is still not completely understood. Furthermore, linear reasoning often is used to investigate or devise therapeutic strategies for endothelial dysfunction, which in itself may be too simplistic.
- The dosimetry of cigarette smoke exposure is nonlinear, which results in a disproportionately large effect with relatively small exposure. Therefore, any level of cigarette smoke exposure appears to be unsafe for health. There is a need for physicians, communities as well as legislative bodies to actively support measures that significantly reduce or eliminate any form of cigarette smoke exposure.

Future Goals

- To improve protocols used to isolate and investigate the endothelium
- To integrate the advances in proteomics, such as protein-based chips and molecular imaging techniques, to further strengthen our ability to investigate and diagnose smoking-related endothelial dysfunction

REFERENCES

1 Ambrose JA, Barua RS. The pathophysiology of cigarette smoking and cardiovascular disease: an update. *J Am Coll Cardiol.* 2004;43:1731–1737
2 American Heart Association. *2004 Heart and Stroke Statistical Update.* Dallas, TX: American Heart Association, 2003.
3 Law MR, Wald NJ. Environmental tobacco smoke and ischemic heart disease. *Prog Cardiovasc Dis.* 2003;46:31–38.
4 Ross R. The pathogenesis of atherosclerosis: a perspective for 1990s. *Nature.* 1993;362:801–809.
5 Black HR. Smoking and cardiovascular disease. In: Laragh JH, Brenner BM, eds. *Hypertension: Pathophysiology, Diagnosis And Management*, 2nd ed. New York, NY: Raven Press Ltd.; 1995: 2621–2647.
6 Pittilo RM. Cigarette smoking, endothelial injury and cardiovascular disease. *Int J Exp Pathol.* 2000;81:219–230.
7 Asmussen I, Kjeldsen K. Intimal ultrastructure of human umbilical arteries. Observations on arteries from newborn children of smoking and nonsmoking mothers. *Circ Res.* 1975;36:579–589.
8 Aronow WS. Effect of passive smoking on angina pectoris. *N Engl J Med.* 1978;299:21–24.
9 Davis JW, Shelton L, Eigenberg, DA, et al. Effect of tobacco and non-tobacco cigarette smoking on endothelium and platelets. *Clin Pharm Ther.* 1985;35:529–533.

10 Davis JW, Shelton L, Eigenberg DA, Hignite CE. Lack of effect of aspirin on cigarette smoke-induced increase in circulating endothelial cells. *Haemostasis.* 1987;17:66–69.

11 Blache D, Bouthillier D, Davignon J. Acute influence of smoking on platelet behaviour, endothelium and plasma lipids and normalization by aspirin. *Atherosclerosis.* 1992;93:179–188

12 Barua RS, Ambrose JA, Eales-Reynolds LJ, et al. Dysfunctional endothelial nitric oxide biosynthesis in healthy smokers with impaired endothelium-dependent vasodilatation. *Circulation.* 2001;104:1905–1910.

13 Lekakis J, Papamichael C, Vemmos C, et al. Effect of acute cigarette smoking on endothelium-dependent brachial artery dilatation in healthy individuals. *Am J Cardiol.* 1997;79:529–531

14 Kugiyama K, Yasue H, Ohgushi M, et al. Deficiency in nitric oxide bioactivity in epicardial coronary arteries of cigarette smokers. *J Am Coll Cardiol.* 1996;28:1161–1167.

15 Ijzerman RG, Serne EH, van Weissenbruch MM, et al. Cigarette smoking is associated with an acute impairment of microvascular function in humans. *Clin Sci (Lond).* 2003;104:247–252.

16 Barua RS, Ambrose JA, Eales-Reynolds LJ, et al. Heavy and light cigarette smokers have similar dysfunction of endothelial vasoregulatory activity: an in vivo and in vitro correlation. *J Am Coll Cardiol.* 2002;39:1758–1763.

17 Wiesmann F, Petersen SE, Leeson PM, et al. Global impairment of brachial, carotid, and aortic vascular function in young smokers: direct quantification by high-resolution magnetic resonance imaging. *J Am Coll Cardiol.* 2004;16;44:2056–2064.

18 Neunteufl T, Heher S, Kostner K, et al. Contribution of nicotine to acute endothelial dysfunction in long-term smokers. *J Am Coll Cardiol.* 2002;39:251–256.

19 Sarabi M, Lind L. Short-term effects of smoking and nicotine chewing gum on endothelium-dependent vasodilation in young healthy habitual smokers. *J Cardiovasc Pharmacol.* 2000;35:451–456.

20 Su Y, Han W, Giraldo C, et al. Effect of cigarette smoke extract on nitric oxide synthase in pulmonary artery endothelial cells. *Am J Respir Cell Mol Biol.* 1998;19:819–825.

21 Barua RS, Ambrose JA, Srivastava S, et al. Reactive oxygen species are involved in smoking-induced dysfunction of nitric oxide biosynthesis and upregulation of endothelial nitric oxide synthase: an in vitro demonstration in human coronary artery endothelial cells. *Circulation.* 2003;107:2342–2347.

22 Murohara T, Kugiyama K, Ohgushi M, et al. Cigarette smoke extract contracts isolated porcine coronary arteries by superoxide anion-mediated degradation of EDRF. *Am J Physiol.* 1994;266: H874–H880.

23 Heitzer T, Just H, Munzel T. Antioxidant vitamin C improves endothelial dysfunction in chronic smokers. *Circulation.* 1996; 94:6–9.

24 Heitzer T, Yla Herttuala S, Wild E, et al. Effect of vitamin E on endothelial vasodilator function in patients with hypercholesterolemia, chronic smoking or both. *J Am Coll Cardiol.* 1999; 33:499–505.

25 Barbera JA, Peinado VI, Santos S, et al. Reduced expression of endothelial nitric oxide synthase in pulmonary arteries of smokers. *Am J Respir Crit Care Med.* 2001;164:709–713.

26 Zhang S, Day I, Ye S. Nicotine induced changes in gene expression by human coronary artery endothelial cells. *Atherosclerosis.* 2001;54:277–283.

27 Beckman JA, Liao JK, Hurley S, et. al. Atorvastatin restores endothelial function in normocholesterolemic smokers independent of changes in low-density lipoprotein. *Circ Res.* 2004;95:217–223.

28 Heitzer T, Brockhoff C, Mayer B, et al. Tetrahydrobiopterin improves endothelium-dependent vasodilation in chronic smokers: evidence for a dysfunctional nitric oxide synthase. *Circ Res.* 2000; 86:E36–E41.

29 Fennessy FM, Moneley DS, Wang JH, et al. Taurine and vitamin C modify monocyte and endothelial dysfunction in young smokers. *Circulation.* 2003;107:410–415.

30 Adachi C, Naruse M, Ishihara Y, et al. Effects of acute and chronic cigarette smoking on the expression ofendothelin-1 mRNA of the cardiovascular tissues in rats. *J Cardiovasc Pharmacol.* 2000;36:S198–S200.

31 Haak T, Jungmann E, Raab C, Usadel KH. Elevated endothelin-1 levels after cigarette smoking. *Metabolism.* 1994;43:267–269.

32 Goerre S, Staehli C, Shaw S, Luscher TF. Effect of cigarette smoking and nicotine on plasma endothelin-1 levels. *J Cardiovasc Pharmacol.* 1995;26:S236–S238.

33 Nohria A, Garrett L, Johnson W, et al. Endothelin-1 and vascular tone in subjects with atherogenic risk factors. *Hypertension.* 2003;42:43–48.

34 Pittilo RM, Mackie IJ, Rowles PM, et al. Effects of cigarette smoking on the ultrastructure of rat thoracic aorta and its ability to produce prostacyclin. *Thromb Haemost.* 1982;48:173–176.

35 Reinders JH, Brinkman HJ, van Mourik JA, de Groot PG. Cigarette smoke impairs endothelial cell prostacyclin production. *Arteriosclerosis.* 1986;6:15–23.

36 Ahlsten G, Ewald U, Tuvemo T. Maternal smoking reduces prostacyclin formation in human umbilical arteries. A study on strictly selected pregnancies. *Acta Obstet Gynecol Scand.* 1986;65: 645–649.

37 Busacca M, Balconi G, Pietra A, et al. Maternal smoking and prostacyclin production by cultured endothelial cells from umbilical arteries. *Am J Obstet Gynecol.* 1984;148:1127–1130.

38 Boutherin-Falson O, Blaes N. Nicotine increases basal prostacyclin production and DNA synthesis of human endothelial cells in primary cultures. *Nouv Rev Fr Hematol.* 1990;32:253–258.

39 Kalra VK, Ying Y, Deemer K, et al. Mechanism of cigarette smoke condensate induced adhesion of human monocytes to cultured endothelial cells. *J Cell Physiol.* 1994;160:154–162.

40 Shen Y, Rattan V, Sultana C, Kalra VK. Cigarette smoke condensate-induced adhesion molecule expression and transendothelial migration of monocytes. *Am J Physiol.* 1996;270: H1624–H1633.

41 Adams MR, Jessup W, Celermajer DS. Cigarette smoking is associated with increased human monocyte adhesion to endothelial cells: reversibility with oral L-arginine but not vitamin C. *J Am Coll Cardiol.* 1997;29:491–497.

42 Tracy RP, Psaty BM, Macy E, et al. Lifetime smoking exposure affects the association of C-reactive protein with cardiovascular disease risk factors and subclinical disease in healthy elderly subjects. *Arterioscler Thromb Vasc Biol.* 1997;17:2167–2176.

43 Bermudez EA, Rifai N, Buring JE, et al. Relation between markers of systemic vascular inflammation and smoking in women. *Am J Cardiol.* 2002;89:1117–1119.

44 Mendall MA, Patel P, Asante M, et al. Relation of serum cytokine concentrations to cardiovascular risk factors and coronary heart disease. *Heart.* 1997;78:273–277.

45 Tappia PS, Troughton KL, Langley-Evans SC, Grimble RF. Cigarette smoking influences cytokine production and antioxidant defenses. *Clin Sci (Lond).* 1995;88:485–489.

46 Burns DM. Epidemiology of smoking-induced cardiovascular disease. *Prog Cardiovasc Dis.* 2003; 46:11–29.
47 Metz L, Waters DD. Implications of cigarette smoking for the management of patients with acute coronary syndromes. *Prog Cardiovasc Dis.* 2003;46:1–9.
48 Benowitz NL. Cigarette smoking and cardiovascular disease: pathophysiology and implications for treatment. *Prog Cardiovasc Dis.* 2003;46:91–111.
49 Barua RS, Ambrose JA, Saha DC, Eales-Reynolds LJ. Smoking is associated with altered endothelial-derived fibrinolytic and antithrombotic factors: an in vitro demonstration. *Circulation.* 2002;106:905–908.
50 Matetzky S, Tani S, Kangavari S, et al. Smoking increases tissue factor expression in atherosclerotic plaques: implications for plaque thrombogenicity. *Circulation.* 2000;102:602–604.
51 Sambola A, Osende J, Hathcock J, et al. Role of risk factors in the modulation of tissue factor activity and blood thrombogenicity. *Circulation.* 2003;107:973–977.
52 Pryor WA, Stone K. Oxidants in cigarette smoke. Radicals, hydrogen peroxide, peroxynitrate, and peroxynitrite. *Ann NY Acad Sci.* 1993;686:12–27.
53 Hoffmann D, Hoffmann I. The changing cigarette, 1950–1995. J *Toxicol Environ Health.* 1997;50:307–364.
54 Thom SR, Xu YA, Ischiropoulos H. Vascular endothelial cells generate peroxynitrite in response to carbon monoxide exposure. *Chem Res Toxicol.* 1997;10:1023–1031.
55 Kourembanas S, Morita T, Liu YX, Christou H. Mechanisms by which oxygen regulates gene expression and cell ± cell interaction in the vasculature. *Kidney Int.* 1997;51:438–443.
56 United States Department of Health and Human Services, Public Health Service. *The Health Consequences of Smoking: Nicotine Addiction. A Report of the Surgeon General.* Washington DC: Department of Health and Human Services, 1988:CDC Publication No. 88–8406.
57 Cooke JP, Bitterman H. Nicotine and angiogenesis: a new paradigm for tobacco-related diseases. *Ann Med.* 2004;36:33–40.
58 Lin SJ, Hong CY, Chang MS, et al. Long-term nicotine exposure increases aortic endothelial cell death and enhances transendothelial macromolecular transport in rats. *Arterioscler Thromb.* 1992;12:1305–1312.
59 Villablanca AC. Nicotine stimulates DNA synthesis and proliferation in vascular endothelial cells in vitro. *J Appl Physiol.* 1998;84:2089–2098.
60 Annas A, Brittebo E, Hellman B. Evaluation of benzo(a)pyrene-induced DNA damage in human endothelial cells using alkaline single cell gel electrophoresis. *Mutat Res.* 2000;471:145–155.
61 Tithof PK, Elgayyar M, Cho Y, et al. Polycyclic aromatic hydrocarbons present in cigarette smoke cause endothelial cell apoptosis by a phospholipase A2-dependent mechanism. *FASEB J.* 2002; 16:1463–1464.
62 Vayssier-Taussat M, Camilli T, Aron Y, et al. Effects of tobacco smoke and benzo[a]pyrene on human endothelial cell and monocyte stress responses. *Am J Physiol Heart Circ Physiol.* 2001;280:H1293–H1300.
63 Kang JJ, Cheng YW. Polycyclic aromatic hydrocarbons-induced vasorelaxation through activation of nitric oxide synthase in endothelium of rat aorta. *Toxicol Lett.* 1997;93:39–45.
64 Howard G, Burke GL, Szklo M, et al. Active and passive smoking are associated with increased carotid wall thickness. The Atherosclerosis Risk in Communities Study. *Arch Intern Med.* 1994;154:1277–1282.
65 Sargent RP, Shepard RM, Glantz SA. Reduced incidence of admissions for myocardial infarction associated with public smoking ban: before and after study. *BMJ.* 2004;328:977–980.
66 Ebrahim SH, Floyd RL, Merritt RK2nd, et al. Trends in pregnancy-related smoking rates in the United States, 1987–1996. *JAMA.* 2000;283:361–366.
67 Kleinman JC, Pierre MB Jr, Madans JH, et al. The effects of maternal smoking on fetal and infant mortality. *Am J Epidemiol.* 1988;127:274–282.
68 Salafia C, Shiverick K. Cigarette smoking and pregnancy II: vascular effects. *Placenta.* 1999;20:273–279.
69 Castro LC, Allen R, Ogunyemi D, et al. Cigarette smoking during pregnancy: acute effects on uterine flow velocity waveforms. *Obstet Gynecol.* 1993;81:551–555.
70 Raitakari OT, Adams MR, McCredie RJ, et al. Oral vitamin C and endothelial function in smokers: short-term improvement but no sustained beneficial effect. *J Am Coll Cardiol.* 2000;25:1616–1621.
71 Neunteufl T, Priglinger U, Heher S, et al. Effects of vitamin E on chronic and acute endothelial dysfunction. *J Am Coll Cardiol.* 2000;35:277–283.
72 Hamilton CA, Miller WH, Al-Benna S, et al. Strategies to reduce oxidative stress in cardiovascular disease. *Clin Sci (Lond).* 2004;106:219–234.
73 Hurt RD, Sachs DP, Glover ED, et al. A comparison of sustained-release bupropion and placebo for smoking cessation. *N Engl J Med.* 1997;337:1195–1202.
74 Boyd ST, Fremming BA. Rimonabant – a selective CB1 antagonist. *Ann Pharmacother.* 2005;39:684–690.
75 Lu JT, Creager MA. The relationship of cigarette smoking to peripheral arterial disease. *Rev Cardiovasc Med.* 2004;5:189–193.

145

Disseminated Intravascular Coagulation

Marcel Levi

University of Amsterdam, The Netherlands

Disseminated intravascular coagulation (DIC) is a syndrome characterized by the systemic activation of coagulation, leading to the intravascular deposition of fibrin in the (micro)vasculature and the simultaneous consumption of coagulation factors and platelets (1). DIC may result in the formation of microvascular thrombi and thereby compromise an adequate blood supply to various organs, which may contribute to organ failure. Simultaneous exhaustion of coagulation factors, protease inhibitors, and platelets, due to ongoing coagulation activation, may lead to serious bleeding. It is important to emphasize that DIC is not a disease in itself but is always secondary to an underlying disorder, for example sepsis, severe trauma, some forms of cancer, or serious immunologic or toxic reactions. A diagnosis of DIC can only be made when one of these underlying disorders is present and is further based on a scoring system, using laboratory results (platelet count, prothrombin time, and fibrin degradation products). Although the pathophysiology of DIC may vary according to the underlying clinical disorder, certain pathogenic mechanisms are common to most if not all cases, including the mediatory role of inflammatory cytokines, the tissue factor–dependent initiation of intravascular coagulation, and the attenuation of natural anticoagulants and fibrinolysis.

The endothelium plays a central role in all major pathways involved in the pathogenesis of DIC. Endothelial cells (ECs) release and respond to myriad proinflammatory mediators, resulting in the induction of tissue factor expression and downregulation of important physiological anticoagulant pathways (2). However, rather than being a unidirectional relationship, the interaction between inflammation and coagulation involves significant cross-talk in which the endothelium appears to play a pivotal role (3). For example, activated coagulation proteases, such as the tissue factor–factor VIIa complex, factor Xa, and thrombin not only promote fibrin deposition, but also bind to protease-activated receptors (PARs) on various cell types, including ECs, thus promoting additional proinflammatory and procoagulant responses. Activated protein C (APC) can bind to the endothelial protein C receptor (EPCR), thereby affecting nuclear factor (NF)-κB nuclear translocation and subsequently influencing inflammatory gene expression and inhibition of tissue factor expression on ECs (as well as monocytes). The intricate relationship between inflammation and coagulation on the EC surface has major consequences for the pathogenesis of microvascular failure and subsequent multiple organ failure in the setting of DIC.

INFLAMMATION AND DISSEMINATED INTRAVASCULAR COAGULATION

Acute inflammation, as occurs in response to severe infection or trauma, results in a systemic activation of the coagulation system (1). It was initially thought that this systemic activation of coagulation was a result of direct activation of the contact system by microorganisms or endotoxin. However, during the 1990s, it became apparent that the principal initiator of inflammation-induced thrombin generation is tissue factor, and that the contact system is not involved. Because tissue factor is expressed on cytokine-activated mononuclear cells, which play a pivotal role in the host response to infection, it was hypothesized that microorganism-induced activation of mononuclear cells resulted in tissue factor–mediated activation of coagulation. However, other than in severe meningococcemia (4), it has proved difficult to demonstrate ex vivo tissue factor expression on the monocytes of septic patients or experimental animals systemically exposed to microorganisms.

It has become apparent that cytokines mediate many of the responses triggered by severe inflammation, and cells other than circulating mononuclear cells may thereby play an important role (2). Evidence has accumulated to suggest that more complex mechanisms might be involved in the relationship between inflammation and the activation of coagulation (5). ECs respond to the cytokines expressed and released by activated leukocytes but can also release cytokines themselves. Furthermore, ECs are able to express adhesion molecules and growth factors that may not only promote the inflammatory response further but also affect the coagulation response. However, it has recently become clear that, in addition to these mostly indirect effects of the endothelium, ECs interfere directly with the initiation and regulation of fibrin formation and removal during severe infection. In fact, ECs play a prominent role in all three major pathogenetic pathways associated with coagulopathy in sepsis: tissue factor–mediated thrombin generation, dysfunctional anticoagulant pathways, and decreased fibrinolysis. Interestingly, rather than being a one-way direction of inflammation leading to coagulation, both systems intensely interact, so that coagulation also can importantly modulate inflammatory activity. Binding of coagulation proteases (like thrombin), tissue factor, or anticoagulant proteins (like APC) to specific cell receptors on mononuclear cells or ECs may affect cytokine production or inflammatory cell apoptosis. This complex cross-talk between the two systems is relevant for many disease states associated with DIC, including the systemic inflammatory response syndrome (SIRS).

INITIATION OF COAGULATION IN DISSEMINATED INTRAVASCULAR COAGULATION

The initiation of coagulation in DIC is primarily mediated by the tissue factor–factor VIIa pathway. Tissue factor is a transmembrane protein that is constitutively expressed on a number of cells throughout the body (6), predominantly in tissues not in direct contact with blood, such as the adventitial layer of larger blood vessels and subcutaneous tissue. Tissue factor expressed at the cell surface can interact with factor VII, either in its zymogen or activated form (factor VIIa). Upon complexing in its zymogen form, factor VII is activated, and the tissue factor–factor VIIa complex catalyzes the conversion of factors IX and X (7). Factors IXa and Xa enhance the activation of factor X and prothrombin, respectively. In cells in contact with blood, tissue factor expression is induced by the action of several compounds including cytokines, C-reactive protein, and advanced glycosylated end products (8). Tissue factor expression at the surface of activated mononuclear cells can be induced in vitro by a number of proinflammatory mediators and is also detectable ex vivo in patients with severe sepsis and in experimental models of endotoxemia in humans (4,9). However, evidence is emerging that ECs also play an important role in the generation of tissue factor in DIC. Under in vitro conditions, various cytokines (such as tumor necrosis factor [TNF]-α and interleukin [IL]-1) have been found to induce tissue factor expression in vascular ECs, and various observations support the notion of EC involvement in tissue factor–mediated activation of coagulation during, for example, severe infection (8,10).

REGULATION OF COAGULATION IN DISSEMINATED INTRAVASCULAR COAGULATION

Thrombin generation is limited by antithrombin (AT), the protein C system, and tissue factor pathway inhibitor (TFPI). In DIC, all three regulatory systems are defective – primarily as a result of endothelial dysfunction.

AT is the main inhibitor of thrombin and factor Xa. In subjects with DIC, AT levels are very low due to consumption, impaired synthesis, and degradation by elastase from activated neutrophils (11). A reduction in glycosaminoglycan availability at the perturbed endothelial surface (due to the influence of cytokines on its synthesis by ECs) also may lead to reduced AT function, because glycosaminoglycans act as a physiological heparin-like cofactor of AT (12).

The involvement of endothelial dysfunction in the impaired function of the protein C system is even more apparent (13). Under physiological conditions, protein C is activated by thrombin bound to the EC membrane–associated molecule, thrombomodulin. ECs, primarily of large blood vessels, express EPCR, which augments the activation of protein C at the cell surface (14). APC then decelerates the coagulation cascade by inactivating factors Va and VIIIa through proteolytic cleavage. However, in DIC, in addition to the already low levels of protein C (as a result of impaired synthesis and consumption), a major cause of protein C system dysfunction is the downregulation of thrombomodulin at the endothelial surface. Proinflammatory cytokines, such as TNF-α and IL-1, may significantly downregulate the expression of thrombomodulin, as suggested by cell culture experiments (15). These data were corroborated by observations in patients with severe gram-negative septicemia, in whom decreased thrombomodulin and EPCR expression at the EC surface and impaired activation of protein C were seen (16).

A third inhibitory mechanism of thrombin generation involves tissue TFPI, which exists in several pools – either EC-associated or lipoprotein-bound in plasma. This molecule inhibits the tissue factor–factor VIIa complex by forming a quaternary complex in which factor Xa is the fourth component. Clinical studies in patients with DIC have not provided clues as to its importance because, in the majority of patients, the levels of tissue TFPI are not diminished compared with normals. The relevance of TFPI in DIC is illustrated by the fact that (a) TFPI depletion sensitizes rabbits to DIC induced by tissue factor infusion (17), and (b) TFPI infusion protects against the harmful effects of *Escherichia coli* in primates (18). TFPI not only blocked DIC, but all baboons challenged with lethal doses of *E. coli* showed a marked improvement in vital functions compared with controls and survived the experiment without apparent complications. The beneficial effects of TFPI may be due, in part, to its attenuating effect on IL-6 generation, and to its capacity to bind endotoxin and interfere with its interaction with CD14 (3). A study in healthy human volunteers confirmed the potential of tissue TFPI to block the procoagulant pathway triggered by endotoxin (19).

FIBRINOLYSIS IN DISSEMINATED INTRAVASCULAR COAGULATION

Inhibition of the fibrinolytic system is another key element in the pathogenesis of fibrin deposition during DIC. Major fibrinolytic activators and inhibitors are synthesized and stored in ECs. Although the initial response in bacteremia and endotoxemia is an increase in fibrinolytic activation (mediated by the almost immediate release of plasminogen activators), this is only short-lived and is rapidly shut off by a sustained increase in the main inhibitor of fibrinolysis, plasminogen activator inhibitor (PAI)-1) (20). TNF-α and IL-1 increase PAI-1 synthesis or release from ECs and also decrease plasminogen activator synthesis. TNF-α and endotoxin stimulate PAI-1 production in the liver, kidney, lung, and adrenal glands of mice (21). Furthermore, *PAI-1*–null mice challenged with endotoxin do not develop thrombi in the kidney (22). These experiments demonstrate that fibrinolytic action is required to reduce the extent of intravascular fibrin accumulation.

CROSS-TALK BETWEEN COAGULATION AND INFLAMMATION

Coagulation activation yields proteases that not only interact with coagulation protein zymogens, but also with specific cell receptors to induce signaling pathways that mediate inflammatory responses. Coagulation proteins, such as factor Xa, factor VIIa, thrombin, and fibrin, can activate ECs, eliciting the synthesis of IL-6 and/or IL-8 (2). The most important mechanisms by which coagulation proteases influence inflammation is by binding to PARs, of which four types (PAR 1–4) have been identified, all belonging to the family of transmembrane domain, G-protein–coupled receptors (23). PARs are localized in the vasculature on ECs, mononuclear cells, platelets, fibroblasts, and smooth muscle cells (23). The exact role of PARs in the pathogenesis of DIC remains to be established. In vivo evidence for a role of coagulation-protease stimulation of inflammation comes from experiments showing that the administration of recombinant factor VIIa to healthy human subjects causes a small but significant three- to fourfold rise in plasma levels of IL-6 and IL-8 (24).

All three physiological anticoagulant pathways are capable of influencing inflammatory activity. This is most prominently shown for the protein C pathway, and is fully dependent on EC function (25). Indeed, APC has been found to inhibit endotoxin-induced production of TNF-α, IL-1β, IL-6, and IL-8 by cultured monocytes/macrophages (26). Further, APC abrogates endotoxin-induced cytokine release and leukocyte activation in rats in vivo (27). Blocking the protein C pathway by a monoclonal antibody in septic baboons exacerbates the inflammatory response, as evidenced by increased levels of proinflammatory cytokines and increased leukocyte infiltration and tissue destruction at histological analysis (28). Conversely, administration of APC ameliorates the inflammatory activation in various models of DIC (29,30). An infusion of APC abrogates inflammatory activity and improves organ function and survival in an experimental *E. coli* sepsis model in baboons (29). Furthermore, in models of DIC in rats, the administration of APC resulted in a significant improvement of organ function, associated with lower levels of inflammatory cytokines and less leukocyte infiltration (30). Mice with a one-allele targeted disruption of the *protein C* gene (resulting in heterozygous protein C deficiency) have not only a more severe DIC when challenged with endotoxin but also demonstrate significant differences in inflammatory responses, as shown by higher levels of circulating proinflammatory cytokines (31). Finally, septic patients receiving recombinant human APC have lower IL-6 levels than placebo-treated controls (32). It is likely that the effects of APC on inflammation are mediated by the EPCR, which may modulate downstream inflammatory processes (25). Binding of APC to the EPCR was shown to affect gene expression profiles of cells by inhibiting endotoxin-induced calcium fluxes in the cell and by blocking NF-κB nuclear translocation, which is a prerequisite for increases in proinflammatory cytokines and adhesion molecules (33). The EPCR–APC complex itself can translocate from the plasma membrane into the cell nucleus, which may be another mechanism for the modulation of gene expression, although the relative contribution of nuclear translocation and cell surface signaling is unclear at present (34). The role of EPCR in the pathogenesis of DIC is being investigated in ongoing experimental and clinical studies.

CLINICAL RELEVANCE OF COAGULATION AND INFLAMMATION IN DISSEMINATED INTRAVASCULAR COAGULATION

The relevance of the interaction between coagulation and inflammation in severe DIC is becoming increasingly clear, and this knowledge has indeed been demonstrated to be applicable to improve the clinical management of these patients. Based on the assumption that defective physiological anticoagulant mechanisms play a pivotal role in the pathogenesis of DIC, the restoration of these pathways or the stimulation of these systems by administration of supraphysiological doses of their pivotal components may be a logical approach in the (supportive) treatment of patients. The administration of recombinant human APC has been most successful in that respect. A beneficial effect of recombinant human APC was demonstrated in two randomized controlled trials, a dose-ranging clinical trial in patients with sepsis and a large multicenter efficacy trial (35). Interestingly, the beneficial effect was most distinct in patients who had overt DIC (36). Trials using AT and TFPI were less successful, although confounding factors (such as trial design and concomitant use of anticoagulant treatment) may have influenced the outcome. Also, strategies aimed at the inhibition of tissue factor may hypothetically be effective in suppressing the activation of coagulation in DIC. This approach has been shown to be successful in experimental DIC models but this strategy has not yet been evaluated in clinical trials.

Further unraveling of the intricate relationship between coagulation and inflammation and the pivotal role of ECs in DIC will probably lead to novel potential points of impact for preventive and therapeutic strategies for DIC and other diseases in which (micro)vascular thrombosis and inflammation are key events.

KEY POINTS

- DIC is a syndrome characterized by a systemic activation of coagulation, leading to microvascular thrombosis and the simultaneous consumption of coagulation factors and platelets.
- Pathogenetic pathways that play a role in the development of DIC include tissue factor–dependent activation of coagulation, defective physiological anticoagulant pathways (such as the AT system and the protein C system), and impaired fibrinolysis, due to elevated levels of PAI-1. ECs occupy a prominent position in the occurrence of each of these mechanisms.
- Considerable cross-talk occurs between inflammation and coagulation in the pathogenesis of DIC, and also occurs for the major part at the EC surface.

Future Goals

- To delineate more precisely the role of bidirectional relationship between coagulation and inflammation in DIC in vivo
- To explore the EC as a target to modulate the derangement of inflammation and coagulation leading to DIC

REFERENCES

1 Levi M, ten Cate H. Disseminated intravascular coagulation. *N Engl J Med*. 1999;341(8):586–592.

2 van der Poll T, de Jonge E, Levi M. Regulatory role of cytokines in disseminated intravascular coagulation. *Semin Thromb Hemosct*. 2001;27(6):639–651.

3 Levi M, van der Poll T, Buller HR. Bidirectional relation between inflammation and coagulation. *Circulation*. 2004;109(22):2698–2704.

4 Osterud B, Flaegstad T. Increased tissue thromboplastin activity in monocytes of patients with meningococcal infection: related to an unfavourable prognosis. *Thromb Haemost*. 1983;49(1):5–7.

5 Marshall JC. Inflammation, coagulopathy, and the pathogenesis of multiple organ dysfunction syndrome. *Crit Care Med*. 2001;29 (Suppl 7):S99–S106.

6 Mann KG, van't Veer C, Cawthern K, Butenas S. The role of the tissue factor pathway in initiation of coagulation. *Blood Coagul Fibrinolysis*. 1998;1(Suppl 9):S3–S7.

7 ten Cate H, Bauer KA, Levi M, et al. The activation of factor X and prothrombin by recombinant factor VIIa in vivo is mediated by tissue factor. *J Clin Invest*. 1993;92(3):1207–1212.

8 Camerer E, Kolsto AB, Prydz H. Cell biology of tissue factor, the principal initiator of blood coagulation. *Thromb Res*. 1996;81(1): 1–41.

9 Franco RF, de Jonge E, Dekkers PE, et al. The in vivo kinetics of tissue factor messenger RNA expression during human endotoxemia: relationship with activation of coagulation. *Blood*. 2000;96(2):554–559.

10 Edgington TS, Mackman N, Fan ST, Ruf W. Cellular immune and cytokine pathways resulting in tissue factor expression and relevance to septic shock. *Nouv Rev Fr Hematol*. 1992;34:S15–S27.

11 Levi M, ten Cate H, van der Poll T. Disseminated intravascular coagulation: state of the art. *Thromb Haemost*. 1999;82:695–705.

12 Bourin MC, Lindahl U. Glycosaminoglycans and the regulation of blood coagulation. *Biochem J*. 1993;289(Pt 2):313–330.

13 Levi M, de Jonge E, van der Poll T. Rationale for restoration of physiological anticoagulant pathways in patients with sepsis and disseminated intravascular coagulation. *Crit Care Med*. 2001;29(Suppl 7):S90–S94.

14 Laszik Z, Mitro A, Taylor FBJ, et al. Human protein C receptor is present primarily on endothelium of large blood vessels: implications for the control of the protein C pathway. *Circulation*. 1997; 96(10):3633–3640.

15 Nawroth PP, Handley DA, Esmon CT, Stern DM. Interleukin 1 induces endothelial cell procoagulant while suppressing cell-surface anticoagulant activity. *Proc Natl Acad Sci USA*. 1986; 83(10):3460–3464.

16 Faust SN, Levin M, Harrison OB, et al. Dysfunction of endothelial protein C activation in severe meningococcal sepsis. *N Engl J Med*. 2001;345(6):408–416.

17 Sandset PM, Warn-Cramer BJ, Rao LV, et al. Depletion of extrinsic pathway inhibitor (EPI) sensitizes rabbits to disseminated intravascular coagulation induced with tissue factor: evidence supporting a physiologic role for EPI as a natural anticoagulant. *Proc Natl Acad Sci USA*. 1991;88(3):708–712.

18 Creasey AA, Chang AC, Feigen L, et al. Tissue factor pathway inhibitor reduces mortality from *Escherichia coli* septic shock. *J Clin Invest*. 1993;91(6):2850–2856.

19 de Jonge E, Dekkers PE, Creasey AA, et al. Tissue factor pathway inhibitor (TFPI) dose-dependently inhibits coagulation activation without influencing the fibrinolytic and cytokine response during human endotoxemia. *Blood*. 2000;95:1124–1129.

20 Biemond BJ, Levi M, ten Cate H, et al. Plasminogen activator and plasminogen activator inhibitor I release during experimental endotoxaemia in chimpanzees: effect of interventions in the cytokine and coagulation cascades. *Clin Sci (Colch)*. 1995;88(5): 587–594.

21 Sawdey MS, Loskutoff DJ. Regulation of murine type 1 plasminogen activator inhibitor gene expression in vivo. Tissue specificity and induction by lipopolysaccharide, tumor necrosis factor-alpha, and transforming growth factor-beta. *J Clin Invest*. 1991; 88(4):1346–1353.

22 Yamamoto K, Loskutoff DJ. Fibrin deposition in tissues from endotoxin-treated mice correlates with decreases in the expression of urokinase-type but not tissue-type plasminogen activator. *J Clin Invest*. 1996;97(11):2440–2451.

23 Coughlin SR. Thrombin signalling and protease-activated receptors. *Nature*. 2000;407(6801):258–264.

24 de Jonge E, Friederich PW, Levi M, van der Poll T. Activation of coagulation by administration of recombinant factor VIIa elicits interleukin-6 and interleukin-8 release in healthy human subjects. *Clin Diagn Lab Immunol*. 2003;10:495–497.

25 Esmon CT. New mechanisms for vascular control of inflammation mediated by natural anticoagulant proteins. *J Exp Med*. 2002;196(5):561–564.

26 Yuksel M, Okajima K, Uchiba M, et al. Activated protein C inhibits lipopolysaccharide-induced tumor necrosis factor-alpha production by inhibiting activation of both nuclear factor-kappa B and activator protein-1 in human monocytes. *Thromb Haemost*. 2002;88(2):267–273.

27 Murakami K, Okajima K, Uchiba M, et al. Activated protein C attenuates endotoxin-induced pulmonary vascular injury by inhibiting activated leukocytes in rats. *Blood*. 1996;87(2):642–647.

28 Taylor FBJ, Stearns-Kurosawa DJ, Kurosawa S, et al. The endothelial cell protein C receptor aids in host defense against *Escherichia coli* sepsis. *Blood*. 2000;95(5):1680–1686.

29 Taylor FBJ, Dahlback B, Chang AC, et al. Role of free protein S and C4b binding protein in regulating the coagulant response to *Escherichia coli*. *Blood*. 1995;86(7):2642–2652.

30 Okajima K. Regulation of inflammatory responses by natural anticoagulants. *Immunol Rev*. 2001;184:258–274.

31 Levi M, Dorffler-Melly J, Reitsma PH, et al. Aggravation of endotoxin-induced disseminated intravascular coagulation and cytokine activation in heterozygous protein C deficient mice. *Blood*. 2003;101:4823–4827.

32 Bernard GR, Vincent JL, Laterre PF, et al. Efficacy and safety of recombinant human activated protein C for severe sepsis. *N Engl J Med*. 2001;344(10):699–709.

33 Hancock WW, Grey ST, Hau L, et al. Binding of activated protein C to a specific receptor on human mononuclear phagocytes inhibits intracellular calcium signaling and monocyte-dependent proliferative responses. *Transplantation*. 1995;60(12):1525–1532.

34 Esmon CT. Role of coagulation inhibitors in inflammation. *Thromb Haemost*. 2001;86(1):51–56.

35 Bernard GR, Ely EW, Wright TJ, et al. Safety and dose relationship of recombinant human activated protein C for coagulopathy in severe sepsis. *Crit Care Med*. 2001;29(11):2051–2059.

36 Dhainaut JF, Yan SB, Joyce DE, et al. Treatment effects of drotrecogin alfa (activated) in patients with severe sepsis with or without overt disseminated intravascular coagulation. *J Thromb Haemost*. 2004;2:1924–1933.

146

Thrombotic Microangiopathy

Jeffrey Laurence

Weill Medical College of Cornell University, New York

The fundamental lesion in idiopathic thrombotic thrombocytopenic purpura (TTP), sporadic hemolytic-uremic syndrome (HUS), and some forms of familial HUS/TTP is a thrombotic microangiopathy (1–3). Hyaline thrombi, composed predominantly of platelets and accompanied by localized endothelial cell (EC) activation, proliferation, and detachment, are found exclusively in the microvasculature of involved tissues (1). These thrombi subserve the clinical manifestations of both disorders. Initial treatment consists of plasmapheresis and plasma infusions, often requiring over 20 liters of plasma. Failure to institute prompt plasma exchange invariably results in death from renal failure, stroke, myocardial infarction, or other thrombotic events.

Although TTP is recognized classically by the pentad of thrombocytopenia, microangiopathic hemolytic anemia, fluctuating neurological signs, renal abnormalities, and fever, and HUS by a triad of signs – the hemolytic anemia and fever seen in TTP plus prominent renal disease (4) – pathological changes overlap significantly. Sporadic HUS patients often have multiple organ involvement, including dermal, ocular, cardiac, gastrointestinal, and neurological changes, similar to that seen in TTP (1–3). In one study, cerebral microthrombi were found in 13 of 26 HUS patients at autopsy; reciprocally, renal dysfunction occurs in 40% to 80% of TTP cases (3). Only two organs are absolutely or relatively spared in classic TTP/sporadic HUS: The lung is virtually never involved (5), and the liver is an infrequent site of pathology (2).

It is critical – in terms of pathophysiology and selection of appropriate therapeutic interventions – to distinguish between two clinical syndromes commonly labeled HUS. The epidemic type, occurring primarily in children younger than 5 years of age and typically preceded by bloody diarrhea, is linked to the shiga-like verocytotoxins (4). In contrast to classic TTP/sporadic HUS, the pulmonary microvasculature can be a major site of thrombosis in this form of HUS, and also in those disorders with the clinical features of either HUS or TTP, but arising in the setting of malignancy, cancer chemotherapy, and bone marrow or organ transplantation (6). Larger vessels of all tissue lineages also may be involved. There is no clear evidence for benefit of plasma exchange in these forms of HUS or TTP-like disorders (7,8). In contrast, most cases of sporadic HUS respond, at least initially, to plasma infusions or plasma exchange in a manner identical to idiopathic TTP (3). These distinctions are summarized in Table 146-1.

Renal microangiopathy accompanies both sporadic and epidemic, cancer- and transplant-related forms of HUS, but distinct histological patterns are found in the kidney. In epidemic (shigatoxin-associated) HUS, as in cancer, cancer chemotherapy, and transplant-linked disease, fibrin thrombi predominate, larger vessel thrombosis with cortical necrosis and inflammation is common, and no EC proliferation occurs within involved vessels (9). In sporadic HUS, as in classic idiopathic TTP, platelet thrombi predominate, with EC activation characterized by EC proliferation and expression of activation markers in vitro and in vivo, including Fas, CD54, and CD62E (10,11); severe intimal proliferation leading to luminal stenosis; and absence of inflammatory changes (5,9). The latter findings, together with EC detachment from affected microvasculature and their appearance in the peripheral blood (12), are consistent with an apoptotic process. Indeed, a high frequency of apoptotic ECs from splenic sinusoidal lumina has been described in classic TTP in vivo (10).

Microvascular EC (MVEC) abnormalities seen in idiopathic TTP/sporadic HUS in vivo (5,10) parallel findings from our laboratory with an experimental model distinguishing among types of TTP/HUS. Apoptosis could be induced in vitro by low levels of plasma (0.1–1%) from idiopathic TTP patients in primary human MVECs of renal, cerebral, cardiac, tonsillar, hepatic, and dermal origin, but not in pulmonary MVECs nor in large vessel ECs, such as those from human umbilical vein (13,14). This reflects the pathology and distribution of platelet thrombi in these disorders. The effects were paralleled by the induction of procoagulant features in ECs susceptible to TTP plasma-mediated apoptosis (13,14), findings that are briefly summarized later. Procoagulant properties are typical of ECs undergoing apoptosis: They become proadhesive for nonactivated platelets (15), a phenomenon enhanced by microparticle release, which is also characteristic of TTP (11).

In contrast, necrosis and inflammation are hallmarks of epidemic, cancer-, and transplant-linked HUS and TTP-like

Table 146-1: Clinical, Pathologic, and Pathophysiological Distinctions among Thrombotic Microangiopathies

Disorder	*Vessels Involved*	*Vessel Pathology*	*Mechanism of EC Injury*	*Response to Plasma Infusion*	*ADAMTS133 Deficiency*
Idiopathic and some forms of familial TTP; HIV-1–, ticlopidine-, and pregnancy-linked TTP	Microvessels; all except lung; hepatic rare	Platelets > fibrin thrombi; EC activation and proliferation, absence of inflammation; ultrastructural changes consistent with EC apoptosis	Apoptosis of susceptible vascular beds	Yes	Yes (subsets)
Sporadic HUS	Microvessels; all except lung; predominant renal, hepatic rare	Platelets > fibrin thrombi; EC activation and proliferation, absence of inflammation; ultrastructural changes consistent with EC apoptosis	Apoptosis of susceptible vascular beds	Yes	Yes (subsets)
Epidemic HUS (shigatoxin-related)	All microvessels, including lung, also larger vessels	In kidney: fibrin predominant; no EC proliferation; apoptosis, necrosis and inflammation present	Apoptosis, necrosis, and inflammation	No	No
Malignancy, chemotherapy, transplantation	All microvessels, including lung, also larger vessels	Fibrin	Necrosis and inflammation	Rare	No

disorders (5). This is reminiscent of injury by EC toxins or autoantibodies, and is consistent with the necrotic cell death observed following incubation of large-vessel or microvascular ECs with plasma from patients with shigatoxin-related HUS (9,13).

In summary, as illustrated in Table 146-1, two major forms of thrombotic microangiopathy exist: (a) plasma exchange–responsive acquired and familial TTP/sporadic HUS linked to MVEC apoptosis and platelet thrombi, and (b) HUS/TTP-like disease, including epidemic forms of HUS, resistant to plasma therapy and characterized by fibrin thrombi, inflammation, and necrosis of blood vessels of all sizes and tissue lineages.

A recently identified enzymatic abnormality may assist in discriminating between these two forms of thrombotic microangiopathy. In 2001, mutations leading to severe deficiency (<5% of control values) of this enzyme (ADAMTS13 – a disintegrin-like and metalloprotease with thrombospondin type 1 motif – a metalloprotease that cleaves ultra-high molecular weight von Willebrand factor [vWF] multimers) was shown to be the cause of some forms of familial TTP (16). Three years earlier, presence of an IgG autoantibody to ADAMTS13, with resulting severe decrease in enzymatic activity, was shown to be specific for at least a subset of classic TTP/sporadic HUS cases (17), although the magnitude of that subset is uncertain (18). These changes were not seen in other forms of TTP/HUS, save for those linked to human immunodeficiency virus (HIV)-1 infection and ticlopidine treatment (discussed later), and pregnancy. This suggests that at least a subset of idiopathic TTP/sporadic HUS cases represents an autoimmune disorder. Indeed, for the 10% to 20% of patients who fail initial plasma therapy, and the 40% with disease recurrence, immune suppression with glucocorticoids, vincristine, anti-CD20 monoclonal antibody, or splenectomy may be effective.

The remainder of this chapter reviews evidence that EC injury is a fundamental event in classic TTP and sporadic HUS, along with studies of the prevalence of ADAMTS13 deficiency in TTP and HUS. This work suggests that development of classic TTP/sporadic HUS may require "two hits" on the endothelium: MVEC apoptosis, related to known and unknown agents, and altered activity of vWF cleaving protease(s) such as ADAMTS13 or some other platelet-specific event.

CLUES TO PATHOPHYSIOLOGY OF THROMBOTIC THROMBOCYTOPENIA PURPURA

Initiation of Microvascular Endothelial Cell Damage in Thrombotic Thrombocytopenic Purpura

We had hypothesized that MVEC injury is one primary event in classic TTP/sporadic HUS, and that it is induced by plasma factors that initiate a cascade leading to MVEC apoptosis, thereby obviating the inflammatory reactions typical of endemic, cancer-, and transplant-associated forms of thrombopathy. The exact nature of that apoptotic process, and the inducing agents involved, are active areas of investigation.

Large-vessel ECs respond differently than do MVECs to a variety of injuries both in vitro and in vivo, reflected by differences in EC proliferation, migration, cytokine production, and phenotype (19–21). Substantial evidence suggests EC activation in the absence of inflammation, restricted to the microvasculature, in classic TTP/sporadic HUS. Histopathology reveals increased vacuolization and aggregation of rounded ECs, with mitochondrial swelling (3). These ultrastructural changes are typical of apoptotic cells in vitro, indicating that our system for inducing MVEC apoptosis by TTP plasmas in vitro mimics at least one prominent characteristic of the disease in vitro. But it does not establish cause and effect.

Indeed, studies of typical TTP lesions have been unable to establish whether EC damage precedes the deposition of platelet thrombi within a vessel lumen, or whether the extensive EC activation and proliferation surrounding these thrombi is simply a response to prior platelet aggregation with vascular occlusion. Clues to the importance of initial MVEC damage may come from a study of situations in which the incidence of classic TTP/sporadic HUS is dramatically increased.

Thrombotic Thrombocytopenic Purpura, Human Immunodeficiency Virus Infection, and Ticlopidine

Agents (biological, chemical, or both) responsible for initiating classic TTP/sporadic HUS, and the MVEC damage associated with it, remain generally unknown, save for two intriguing exceptions. Viruses have long been suspected as inciting agents. For example, TTP and HUS have occurred sporadically in association with human T cell-lymphotropic virus type 1 (22) and cytomegalovirus infections (3). But the clearest pathophysiological clues may derive from analysis of two settings associated with a marked increase in incidence of TTP clinically and pathologically indistinguishable from the idiopathic form: (a) HIV-1 infection, with TTP occurring in 1.4% of a cohort of 1,070 consecutively diagnosed patients with acquired immune deficiency syndrome (AIDS) (23) versus 0.3 per 100,000 in the general population (3); and (b) ticlopidine treatment, with 0.06% of patients taking this thienopyridine ADP receptor antagonist and antiplatelet agent developing TTP within 1.5 to 6 weeks of initiating drug therapy (24). Successful suppression of HIV viremia with highly active antiretroviral therapies has virtually eliminated the association of thrombotic microangiopathy with HIV (25). Although ticlopidine is now rarely used clinically, its associated risk for TTP was accelerated by prior EC damage, as may occur during vascular stenting (24). Clopidogrel, a thienopyridine related structurally and functionally to ticlopidine, is a commonly prescribed antiplatelet agent and also has been linked to TTP, although with a much lower incidence compared with ticlopidine (26).

Thrombotic microangiopathies occurring in the setting of active HIV infection or ticlopidine use share three major features with idiopathic TTP and sporadic HUS: (a) they respond clinically to plasma exchange, (b) low levels of patient plasmas induce apoptosis in restricted lineages of MVECs in vitro (i.e., sparing pulmonary MVECs) (13,27), and (c) severe deficiency of plasma ADAMTS13 activity has been noted in these forms (17). TTP occurring in the setting of pregnancy also has been linked to ADAMTS13 deficiency, and typically responds to plasma exchange (28), but the occurrence of EC apoptosis has not been assessed in vitro or in vivo.

The underlying cause of MVEC injury in these clinically and pathologically similar disorders – idiopathic TTP/sporadic HUS, and HIV- and ticlopidine-linked thrombopathies – is important to discern. One potential shared factor is disruption of MVEC interactions with the extracellular matrix (ECM). This leads to EC apoptosis in vitro through a variety of mechanisms (27), including altered ECM signaling pathways linked to stress-induced mitogen-activated protein kinases (MAPKs); EC detachment from ECM-supported substrate, a form of apoptosis known as anoikis; and the expression of unligated integrin adhesion molecules. Suppression of ECM proteins linked to EC apoptosis in vitro has been described in HIV disease in vivo and ticlopidine-exposed ECs in vitro (27). Finally, our lab has documented similar effects of ticlopidine and plasmas from idiopathic and ticlopidine-linked TTP patients on MVEC activation, apoptosis, ECM proteins, and induction of MAPK pathways linked to ECM-based signaling (27).

In general, sustained EC activation through phosphorylation of extracellular-signal-regulated protein kinase (ERK)-1/2 is associated with cell survival, whereas prolonged stimulation through p38 MAPK leads to apoptosis. However, the outcome of activation of a particular pathway often depends on the nature of the apoptotic signal. Our lab reported that pharmacological levels of ticlopidine initiated apoptosis in primary human MVECs with size and lineage restrictions identical to that of plasmas from idiopathic and HIV-infected TTP patients. These changes were accompanied by prolonged induction of ERK1/2 and p38 MAPK only in TTP-susceptible MVECs; apoptosis initiated by ticlopidine or plasma from ticlopidine-treated patients who developed TTP was abrogated by inhibitors of these kinases (27).

In an attempt to identify the etiological agent(s) of EC activation and apoptosis in idiopathic TTP patient plasmas, our lab has undertaken genomic and proteomic approaches. Initial clues were derived from a broad search for changes in mRNA expression in dermal and renal versus pulmonary

MVECs following TTP plasma exposure. A preliminary screen using a Hu6800 GeneChip Array (Affymetrix) indicated that pulmonary MVECs had greater basal expression of three genes encoding antiapoptotic proteins for ECs: osteoprotegerin, vascular endothelial growth factor (VEGF), and VEGF receptor 2 (VEGFR2) (29), and one related to a membrane protein protective against complement-mediated EC apoptosis, CD55 (J. Laurence, unpublished data). In addition, expression of the antiapoptotic genes Bcl-xL and Bcl-2 was increased in pulmonary but not dermal MVECs on exposure to TTP plasma (29). Retrovirus-mediated gene transfer confirmed that overexpression of Bcl-2 and Bcl-xL could protect dermal MVECs against TTP plasma-mediated damage (30). In addition, our group and others have shown that interleukin (IL)-1 and tumor necrosis factor (TNF)-α, cytokines linked to MVEC apoptosis, ECM changes, and activation of p38 MAPK are elevated in idiopathic TTP/sporadic HUS plasmas (11,31). Although the concentrations of IL-1 and TNF-α found in those plasmas are much lower than are required to induce MVEC apoptosis in vitro, they could synergize with each other and with additional TNF superfamily members, such as TNF-related apoptosis-inducing ligand (TRAIL), also upregulated in TTP (28,29), to initiate apoptosis. In association with the Proteomics Program at Cornell-Ithaca we are utilizing matrix-assisted laser desorption ionization-time of flight (MALDI-TOF/TOF) to identify protein differences between idiopathic TTP and control plasmas, directed by findings from the gene array analyses.

ADAMTS13 Deficiency and Endothelial Cell Damage: the "Two-Hit" Hypothesis

It has been proposed that *ADAMTS13* polymorphisms might influence the expression of TTP as a phenotype in ticlopidine-treated patients (32). This could also hold true for HIV-1 infection, pregnancy, or other potential inciting agents in classic, plasma exchange–responsive TTP, and might underlie the paradox of why the incidence of TTP in the setting of HIV infection and ticlopidine use, albeit high, is not greater still. Concerns include the relatively low prevalence of severe ADAMTS13 deficiency in some large series of patients with classic, acquired TTP (18), its lack of predictive value for response to plasma exchange (33), and the lack of a phenotype in common strains of *ADAMTS13*-null mice (34). These issues are addressed next. An excellent overview of the regulation of plasma vWF multimer size is provided in Chapter 102. Here the physiology of ADAMTS13 is reviewed with respect to TTP.

Apart from serving as a carrier for coagulation factor VIII, vWF mediates the rapid adhesion of platelets to sites of vascular injury. A pool of ultra-large vWF multimers is released from ECs in response to EC activation; these multimers have significantly greater capacity to bind platelets than do smaller vWF units. Once immobilized, the affinity of vWF for the gpIb-IX–V receptor complex on the surface of circulating platelets is further enhanced, leading to tethering of platelets at sites of vascular damage and formation of platelet plugs (35). Normally, vWF multimers circulate in a globular form; when exposed to increased shear forces, present normally in the microvasculature (predominantly the arterioles, and not venules or capillaries) and in damaged vessels, they unravel into string-like conformations, facilitating their platelet-tethering potential (35).

ADAMTS13 is a zinc-dependent metalloproteinase produced by hepatic stellate cells and ECs and secreted into the blood as an active enzyme with a plasma concentration of 1 μg/mL and a half-life of 2 to 3 days (35). Although it is unclear how this protein is regulated and localized, it cleaves vWF and its multimers very inefficiently in the absence of shear forces. In their presence, it serves to limit the build-up of platelet thrombi. It was thus logical to postulate that ADAMTS13 deficiency, acquired via an autoantibody inhibitor, might underlie classic TTP/sporadic HUS and, indeed, initial reports found an absolute association between deficiency of this enzyme and idiopathic, HIV-, and ticlopidine-linked TTP/HUS (36). Severe decreases in ADAMTS13 activity were not found in other forms of HUS or TTP-like disorders (36).

However, subsequent reports showed a much lower incidence of severe enzyme deficiency, regardless of the presence of anti-ADAMTS13 autoantibody. In one large survey, severe deficiency of this vWF-cleaving protease (CP) activity was found in only 16 of 48 (33%) patients with idiopathic TTP (18). ADAMTS13 suppression was not associated with the presence of neutralizing antibodies in the majority of patients (37). It was also recognized that partial deficiency of ADAMTS13 (activity between 10–49% of normal) occurs in normal pregnancies and myriad disease states, and may itself be a consequence of thrombosis, not its cause. Reductions in ADAMTS13 arise as a result of increased thrombin-dependent inactivation during the course of a thrombotic event (35).

The lack of phenotype in some ADAMTS13-deficient mice is also troubling. Homozygous null C57BL mice lacking all vWF-CP activity do not show microangiopathic hemolytic anemia, thrombocytopenia, or vWF-rich platelet thrombi in the microvasculature, despite the fact that circulating ultra-large vWF multimers are present in plasmas of these mice, as in acute TTP in humans (38). Challenge of the null mice with endotoxin failed to induce disease. Spontaneous thrombocytopenia does occur in a subset of CASA/Rk (a strain with elevated plasma vWF) *ADAMTS13*-null mice, and administration of the endemic HUS toxin, verotoxin-2, did cause an HUS-like syndrome in these animals (34). However, as noted earlier, shigatoxin-linked thrombotic microangiopathy is not associated with ADAMTS13 deficiency in humans. In addition, several familial disorders cause HUS/TTP without affecting ADAMTS13; all relate to mutations in anticomplement factors, the most prominent being factor H and CD46 (39,40). Indeed, incubation of MVECs with plasmas of idiopathic TTP patients, all of whom had ADAMTS13 deficiency, led to the deposition of the complement components C3 and membrane attack complex, and cell death, which was blocked by anticomplement agents (41). More research is needed to define the role of ADAMTS13, as well as other mutations and polymorphisms that might contribute to a predisposition to development of TTP.

Procoagulant Phenotype of Endothelial Cells in Thrombotic Thrombocytopenic Purpura

Virtually all properties of normal microvascular endothelium are altered in TTP. ECs synthesize many substances involved in coagulation and fibrinolysis, including vWF, thrombomodulin, tissue-type plasminogen activator (t-PA), plasminogen activator inhibitor, protein S, prostacyclin (PGI_2), and nitric oxide. Alterations in levels of these substances have been reported in TTP, and correlate with disease activity, although that may simply reflect EC injury (3). Our lab examined three in vitro phenomena related to a potential procoagulant EC phenotype: PGI_2 production, membrane expression of annexin II, and tissue factor production (14).

Tissue factor levels were unchanged in MVEC cultures exposed to TTP plasma. Membrane expression of annexin II was suppressed by exposure of MVECs susceptible to TTP plasma-mediated apoptosis to TTP plasmas, but the consequences of altering EC annexin II binding sites – receptors for plasminogen and t-PA – in vivo are unclear. By translocating annexin II to its external surface, ECs may actively modulate plasmin generation, blocking formation of a fibrin clot. Its relative loss thus could encourage thrombus formation, but fibrin is only a small component of the TTP clot. Basal levels of PGI_2 (assessed by measuring its stable metabolite, 6-keto-prostaglandin F1α) were significantly lower in TTP plasma–exposed dermal versus pulmonary MVECs, with very low levels of sodium arachidonate-induced PGI_2 in the former (14). Failure to increase production of PGI_2 by apoptotic ECs may be of pathological importance, because PGI_2 normally suppresses platelet aggregation induced by cytokines, shear stress, and other means. In addition, as suggested earlier, apoptotic ECs are procoagulant, and may shed procoagulant membrane blebs (42).

DESIGN OF EXPERIMENTAL THERAPEUTICS FOR CLASSIC THROMBOTIC THROMBOCYTOPENIC PURPURA/SPORADIC HEMOLYTIC-UREMIC SYNDROME

Options are limited for the treatment of TTP in patients who do not respond initially to plasma exchange, or who relapse frequently. The production of recombinant ADAMTS13 and demonstration of its efficacy in normalizing vWF-CP activity in vitro in plasmas of patients with TTP secondary to congenital absence of ADAMTS13 (43), in conjunction with methods to suppress autoantibodies to this enzyme, could aid in defining its role and potential therapeutic value in TTP. Defining the interactions among pro- and antiapoptotic factors for MVECs, their regulation by circulating triggers of EC apoptosis, and the involvement of ECM components and their MAPK-related signaling cascades may suggest the trial of novel ECM-directed treatments and help define the nature of the apoptotic stimuli in TTP/sporadic HUS.

KEY POINTS

- MVEC injury is a fundamental event in idiopathic as well as HIV- and ticlopidine-associated thrombotic microangiopathies. We postulate that this injury involves MVEC activation and apoptosis related to an as yet unidentified plasma factor(s). Development of clinical disease may require "two hits:" MVEC perturbation and deficiency of vWF-CP activities such as ADAMTS13.
- Idiopathic TTP, sporadic HUS, and TTP linked to HIV and ticlopidine share several features: involvement of all MVEC lineages except for the pulmonary system, sparing larger vessels; development of platelet thrombi with minimal fibrin content; EC lesions lacking necrosis or inflammation; severe deficiency of ADAMTS13 in about half of cases; plasmas capable of inducing apoptosis in MVECs in vitro; and clinical response to plasma exchange.
- Shigatoxin-associated HUS and HUS/TTP-like syndromes occurring in the setting of malignancy, chemotherapy, and bone marrow and organ transplantation have pathological and clinical features distinct from familial, acquired idiopathic TTP/sporadic HUS, and HIV-1 and ticlopidine-linked disease: involvement of vasculature of all sizes and tissues, including the lung; predominant fibrin thrombi; inflamed and necrotic EC lesions; ADAMTS13 plasma levels not severely depressed; failure of patient plasmas to induce MVEC apoptosis in vitro; and atypical clinical response to plasma exchange.

Future Goals

- To identify apoptosis inducing factor(s) in TTP plasmas
- To define the interactions among ADAMTS13 deficiency, microvascular EC injury, and perhaps genetic polymorphisms in establishing the TTP phenotype

REFERENCES

1 Amorosi EL, Ultmann JE. Thrombotic thrombocytopenic purpura: report of 16 cases and review of the literature. *Medicine.* 1966;45:139.

2 Ridolfi RL, Bell WR. Thrombotic thrombocytopenic purpura. Report of 25 cases and review of the literature. *Medicine.* 1981;60: 413–428.

3 Ruggenenti P, Remuzzi G. The pathophysiology and management of thrombotic thrombocytopenic purpura. *Eur J Hematol.* 1996;56:191–207.

4 Neild GH. Hemolytic-uremic syndrome in practice. *Lancet*. 1994; 343:398–401.
5 Asada Y, Sumiyoshi A, Hayashi T, et al. Immunohistochemistry of vascular lesions in thrombotic thrombocytopenic purpura, with special reference to factor VIII related antigen. *Throm Res.* 1985;38:469–479.
6 Upadhyaya K, Barwick K, Fishaut M, et al. The importance of nonrenal involvement in hemolytic-uremic syndrome. *Pediatrics.* 1980;165:115–120.
7 Dundas S, Murphy J, Soutar RL, et al. Effectiveness of therapeutic plasma exchange in the 1996 Lanarkshire *Escherichia coli* O157:H7 outbreak. *Lancet*. 1999;354:1327–1330.
8 Moake JL, Byrnes JJ. Thrombotic microangiopathies associated with drugs and bone marrow transplantation. *Hematol Oncol Clin North Am.* 1996;10:485–497.
9 Kaplan BS, Meyers KE, Schulman SL. The pathogenesis and treatment of hemolytic-uremic syndrome. *J Am Soc Nephrol.* 1998; 8:1126–1133.
10 Dang CT, Magid MS, Weksler B, et al. Enhanced endothelial cell apoptosis in splenic tissues of patients with thrombotic thrombocytopenic purpura. *Blood.* 1999;93:1264–1270.
11 Jimenez JJ, Wenche JY, Mauro LM, et al. Endothelial microparticles released in thrombotic thrombocytopenic purpura express von Willebrand factor and markers of endothelial activation. *Brit J Hematol.* 2003;123:896–902.
12 Lefevre P, George F, Durand JM, Sampol J. Detection of circulating endothelial cells in thrombotic thrombocytopenic purpura. *Thromb Hemost*. 1993;69:522.
13 Laurence J, Mitra D, Steiner M, et al. Plasma from patients with idiopathic and human immunodeficiency virus-associated thrombotic thrombocytopenic purpura induces apoptosis in microvascular endothelial cells. *Blood.* 1996;87:3245–3254.
14 Mitra D, Jaffe EA, Weksler B, et al. Thrombotic thrombocytopenic purpura and sporadic hemolytic-uremic syndrome plasmas induce apoptosis in restricted lineages of human microvascular endothelial cells. *Blood.* 1997;89:1224–1234.
15 Bombeli T, Schwartz BR, Harlan JM. Endothelial cells undergoing apoptosis become proadhesive for nonactivated platelets. *Blood.* 1999;93:3831–3838.
16 Levy GG, Nichols WC, Lian EC, et al. Mutations in a member of the ADAMTS gene family cause thrombotic thrombocytopenic purpura. *Nature.* 2001;413:488–494.
17 Tsai H-M. Advances in the pathogenesis, diagnosis, and treatment of thrombotic thrombocytopenic purpura. *J Amer Soc Nephrol.* 2003;14:1072–1081.
18 Vesely SR, George JN, Lammle B, et al. ADAMTS13 activity in thrombotic thrombocytopenic purpura-hemolytic uremic syndrome: relation to presenting features and clinical outcomes in a prospective cohort of 142 patients. *Blood.* 2003;102:60–68.
19 Keusch GT, Acheson DWK, Aaldering L, et al. Comparison of the effects of shiga-like toxin on cytokine- and butyrate-treated human umbilical and saphenous vein endothelial cells. *J Infect Dis.* 1996;173:1164.
20 Madri JA, Bell L, Marx M, et al. Effects of soluble factors and extracellular matrix components on vascular cell behavior in vitro and in vivo: models of de-endothelialization and repair. *J Cell Biochem.* 1991;45:123.
21 Cines DB, Pollak ES, Buck CA, et al. Endothelial cells in physiology and in the pathophysiology of vascular disorders. *Blood.* 1998;91:3527–3561.
22 Ucar A, Fernandez HF, Byrnes JJ, et al. Thrombotic microangiopathy and retroviral infections: A 13-year experience. *Am J Hematol.* 1994;45:304–309.
23 Gervasoni C, Ridolfo AL, Bibi T, et al. Thrombotic thrombocytopenic purpura and hemolytic-uremic syndrome in HIV-infected patients. XIth International Conference on AIDS, Vancouver, July 7–12, 1996, Abst Mo. B. 1299.
24 Bennett CL, Davidons CJ, Raisch DW, et al. Thrombotic thrombocytopenic purpura associated with ticlopidine in the setting of coronary artery stenting and stroke prevention. *Ann Intern Med.* 1999;159:2524–2528.
25 Gervasoni C, Ridolfo AL, Vaccarezza M, et al. Thrombotic microangiopathy in patients with acquired immunodeficiency syndrome before and during the era of introduction of highly active antiretroviral therapy. *Clin Infect Dis.* 2002;35:1534–1540.
26 Bennett CL, Connors JM, Carwile JM, et al. Thrombotic thrombocytopenic purpura associated with clopidogrel. *N Engl J Med.* 2000;342:1773–1777.
27 Mauro M, Zlatopolskiy A, Raife TJ, Laurence J. Thienopyridine-linked thrombotic microangiopathy: association with endothelial cell apoptosis and activation of MAP kinase signalling cascades. *Brit J Hematol.* 2004;124:200–210.
28 Ezra Y, Rose M, Eldor A. Therapy and prevention of thrombotic thrombocytopenic purpura during pregnancy: a clinical study of 16 pregnancies. *Am J Hematol.* 1996;51:1–6.
29 Kim J, Wu H, Hawthorne L, et al. Endothelial cell apoptosis genes associated with the pathogenesis of thrombotic microangiopathies: an application of oligonucleotide genechip technology. *Microvasc Res.* 2001;62:83–93.
30 Mitra D, Kim J, MacLow C, et al. Role of caspases 1 and 2 and Bcl-2-related molecules in endothelial cell apoptosis associated with thrombotic microangiopathies. *Am J Hematol.* 1998;59:279–287.
31 Mauro M, Kim J, Costello C, Laurence J. Role of transforming growth factor 1 in microvascular endothelial cell apoptosis associated with thrombotic thrombocytopenic purpura and hemolytic-uremic syndrome. *Am J Hematol.* 2001;66:12–22.
32 Sadler JE. A new name in thrombosis, ADAMTS13. *Proc Natl Acad Sci USA.* 2002;99:11552–11554.
33 George JN. Thrombotic thrombocytopenic purpura: a syndrome that keeps evolving. *J Clin Apheresis.* 2004;19:63–65.
34 Motto DG, Chauhan AK, Zhu G, et al. Shigatoxin triggers thrombotic thrombocytopenic purpura in genetically susceptible ADAMTS13-deficient mice. *J Clin Invest*. 2005;115:2752–2761.
35 Crawley JTB, Lam JK, Rance JB, et al. Proteolytic inactivation of ADAMTS13 by thrombin and plasmin. *Blood.* 2005;105:1085–1093.
36 Tsai H-M, Lian EC-Y. Antibodies to von Willebrand factor-cleaving protease in acute thrombotic thrombocytopenic purpura. *N Engl J Med.* 1998;339:1585–1594.
37 Peyvandi F, Ferrari S, Lavoretano S, et al. von Willebrand factor cleaving protease (ADAMTS-13) and ADAMTS-13 neutralizing autoantibodies in 100 patients with thrombotic thrombocytopenic purpura. *Brit J Hematol.* 2004;127:433–439.
38 Motto D, Zhang W, Zhu G, et al. Additional experimental and/or genetic factors are required to trigger TTP in ADAMTS13-deficient mice. *Blood.* 2004;104:A77.
39 Noris M, Ruggementi P, Perna A, et al. Hypocomplementemia discloses genetic predisposition to hemolytic uremic syndrome and thrombotic thrombocytopenic purpura: role of factor H abnormalities. *J Am Soc Nephrol.* 1999;10:281–293.

40 Richards A, Kemp EJ, Liszewski MK, et al. Mutations in human complement regulator, membrane cofactor protein (CD46), predispose to development of familial hemolytic uremic syndrome. *Proc Natl Acad Sci USA*. 2003;100:12966–12971.

41 Ruiz-Torres M, Casiraghi FM, Galbusera M, et al. Complement activation: the missing link between ADAMTS-13 deficiency and microvascular thrombosis of thrombotic microangiopathies. *Thromb Hemost*. 2005;93:443–452.

42 Casciola-Rosen L, Rosen A, Petri M, Schlissel M. Surface blebs on apoptotic cells are sites of enhanced procoagulant activity: implications for coagulation events and antigenic spread in systemic lupus erythematosus. *Proc Natl Acad Sci USA*. 1996;93:1624–1629.

43 Antoine G, Zimmermann K, Plaimauer B, et al. ADAMTS13 gene defects in two brothers with constitutional thrombotic thrombocytopenic purpura and normalization of von Willebrand factor-cleaving protease activity by recombinant human ADAMTS13. *Brit J Hematol*. 2003;120:821–824.

147

Heparin-Induced Thrombocytopenia

Andreas Greinacher and Theodore E. Warkentin

Institute of Immunology and Transfusion Medicine, Ernst-Moritz-Arndt-University-Greifswald, Germany; McMaster University, Hamilton, Canada

Heparin-induced thrombocytopenia (HIT) is an immune-mediated adverse drug reaction that results when pathogenic immunoglobulin (Ig)G antibodies are formed that recognize neoepitopes on a "self" protein – platelet factor 4 (PF4) – after PF4 has formed complexes with heparin (1). Multimolecular complexes of heparin, PF4, and IgG form on platelet surfaces (2). Occupancy of the platelet FcγIIa receptors by the resulting immune complexes induces platelet activation, with concomitant activation of coagulation (3), perhaps via the generation of procoagulant platelet microparticles (MPs). Once triggered, the prothrombotic risk persists for a time, even after stopping heparin. This scheme of pathogenesis is fundamentally unique.

Unlike HIT, other immune-mediated thrombocytopenic disorders are caused by antibodies that recognize one or more platelet surface glycoproteins. The Fc moieties of the platelet-bound antibodies are recognized by Fc receptors of the mononuclear phagocytic system; this results in the clearance (not activation) of the antibody-coated platelets and often produces severe thrombocytopenia and mucocutaneous bleeding. This is also the mechanism of all other (non-HIT) drug-induced immune thrombocytopenias, except that the antibodies bind to a drug (or drug metabolite)/glycoprotein complex.

The clinical importance of HIT derives from its high frequency (3–5%) in certain patient populations, particularly postoperative patients receiving unfractionated heparin (UFH) antithrombotic prophylaxis for 1 to 2 weeks, and because of its paradoxical strong association with thrombosis (odds ratio, 20–40) (1). The frequency of HIT is less with low-molecular-weight heparin (LMWH), compared with UFH. Venous thrombosis predominates, despite HIT being a primary platelet activation disorder. Understanding the pathophysiological basis of venous thrombosis predominance in HIT might help unravel the role that platelets play in contributing to venous thrombosis. This concept has been highlighted recently by a report (4) that venous thrombosis occurs more frequently than expected by chance in patients with atherosclerosis, thus potentially linking arteriopathy and resulting platelet activation with systemic hypercoagulability.

In addition to their platelet-activating activities, it has been proposed that HIT antibodies induce a procoagulant shift within the endothelial cell (EC) (5). This concept is supported by in vitro evidence that HIT antibodies bind to ECs (2,6) and induce procoagulant changes on microvascular ECs (MVECs) (7). Further, HIT antibodies can activate monocytes (8) and neutrophils (9) and lead to the formation of platelet-monocyte-neutrophil aggregates (9). Thus, pancellular activation could underlie the unusual prothrombotic nature of HIT.

An important unresolved issue is whether the prothrombotic risk of HIT is explained primarily by in vivo platelet activation per se, or whether it results from additional activation of ECs or leukocytes. Table 147-1 lists arguments both pro and con. Our thesis maintains that platelets activated by HIT antibodies or their products (e.g., MPs) are critically important in mediating EC and leukocyte activation in HIT and, in turn, that EC and leukocyte activation is a key determinant of the prothrombotic and proinflammatory clinical phenotype. Thus, although EC and leukocyte alterations are important in the clinical expression of HIT (thrombosis, systemic inflammatory response-like syndrome), these are likely the result of platelet-derived MPs and other mediators produced by activated platelets, and not directly through HIT antibody binding to ECs and leukocytes. As more platelets become activated, more pronounced activation of these other cells occurs. Such a model would be consistent with the well-known association in HIT between presence of platelet-activating IgG antibodies (rather than nonactivating antibodies of IgM and IgA class) and clinical manifestations of disease, including the correlation between the magnitude of platelet count fall and risk of thrombosis. In addition, this model also could account for the unusually severe prothrombotic phenotype of HIT, because pancellular interactions augment the pathogenicity of HIT antibodies beyond mere platelet activation.

Table 147-1: Evidence For and Against a Major Role for Endothelium in the Pathogenesis of Heparin-Induced Thrombocytopenia

For	*Against*
1. Venous thrombosis (which is common in HIT) is not typical for a primary platelet activation disorder.	1. The magnitude of the platelet count fall in HIT correlates well with the risk and severity of thrombotic events.
2. Hyperplastic and proliferating ECs plus immunoglobulin deposition identified in thrombi obtained from patients with HIT.	2. Platelet-activating anti-PF4/heparin IgG antibodies strongly correlate with HIT, but not anti-PF4/heparin IgM and IgA (that presumably would bind to ECs).
3. Acute pain in vessels can be associated with intravenous heparin bolus use in a patient with HIT antibodies.	3. Activation of ECs in vitro by HIT antibodies has been shown for microvascular, not macrovascular, ECs, whereas HIT usually causes macrovascular thrombosis.
4. There is a unique distribution of arterial thrombosis (limb artery occlusion > thrombotic stroke > myocardial infarction) that differs from typical atherothrombosis.	4. Vasculitides with activating anti-EC antibodies usually evinces small/medium-sized vessels (cf. HIT).
5. HIT has higher risk of arterial thrombosis ("white clots") than do other hypercoagulability states.	5. Renal failure is not a feature of HIT (cf. vasculitides).

CLINICAL PICTURE

HIT is often called the *white clot syndrome* based on the typical appearance of artery-occluding, platelet-rich thrombi. Early clinicians described these as "pale, soft, salmon-colored clots" that "histologically . . . were comprised mostly of fibrin, platelets and leukocytes; red cells were rare" (1). One immunohistologic study (10) of ischemic tissues obtained from patients with HIT noted hyperplastic ECs either covering, or infiltrating into, hyaline (platelet-containing) thrombi within small arteries; in some lesions, a marked intraluminal hyperplasia of ECs with less hyaline material was seen. Immunoperoxidase staining revealed presence of immunoglobulin deposition within these platelet- and EC-containing microthrombi. Another histopathologic study (11) of HIT thrombosis reported layered ("onion skin") accumulation of platelets with minimal fibrin content on ECs, without associated ulcerative or arteriosclerotic vessel abnormalities.

It remains uncertain whether "white clots" in HIT are pathologically distinct – reflecting a unique pathogenesis – or whether they represent an extreme degree of hypercoagulability not necessarily unique to HIT. For example, adenocarcinoma-associated disseminated intravascular coagulation (DIC) can mirror HIT in its extreme prothrombotic nature (e.g., high risk of bilateral deep vein thrombosis [DVT], pulmonary embolism; coumarin-associated venous gangrene).

Overall, venous thrombosis is more common than arterial thrombosis in HIT (ratio, ~4:1). However, in arteriopathic patient populations (e.g., postcardiac surgery), the ratio is about 1:1. Most thrombotic events are macrovascular, involving large veins or arteries. However, microvascular thrombosis can occur, particularly in association with coumarin therapy (3). The syndrome of venous limb gangrene complicating HIT is associated with warfarin treatment, because the vitamin K antagonist induces a protein C depletion that predisposes to thrombotic occlusion of small venules, causing acral (distal extremity) necrosis, usually in a limb already affected by large vein thrombosis.

Venous thrombosis in HIT most often involves proximal lower limb veins, with about half of these patients evincing symptomatic pulmonary embolism. Rarely, adrenal vein thrombosis (with potential for unilateral or bilateral adrenal hemorrhagic necrosis) or cerebral venous (dural sinus) thrombosis occurs. Arterial thrombosis in HIT most often involves lower limb arteries, such as the distal aortic, aortoiliac, or iliofemoral arteries. Less common arterial thrombi are thrombotic stroke and myocardial infarction, followed by mesenteric artery thrombosis, upper limb artery thrombosis, and renal artery thrombosis. The relative frequency of HIT-associated arterial thrombosis (aortic/iliofemoral > cerebral > myocardial) is the converse of that observed in typical atherothrombosis, arguing at least indirectly that the prothrombotic nature of HIT is special.

Vascular injury resulting from the use of intravascular catheters directly influences risk of thrombosis in HIT. About 10% of HIT patients who have current (or recent use of) central venous catheters develop upper-limb DVT, which invariably occurs in the same limb as the central line (12). Also, femoral artery puncture or catheter use predisposes to arterial thrombosis in patients with HIT.

A minority (about 10–20%) of patients who develop HIT while receiving subcutaneous UFH or LMWH develop skin lesions at the heparin injection sites. These lesions range in appearance from erythematous plaques to frank necrosis (13). The former lesions exhibit "eczema with spongiosis and exocytosis, without microthrombi or signs of vasculitis," with "mild superficial dermal perivascular infiltrate" sometimes observed (14). In contrast, biopsies from necrotic dermal lesions show "epidermal and dermal necrosis with numerous intravascular thrombi in the superficial dermis, accompanied sometimes by signs of leukocytoclastic vasculitis" (14). Often, the platelet count fall is minimal (<30%) or begins a few days after the heparin is discontinued. We have observed that thrombi of large arteries, rather than of veins, are more likely to occur in patients with HIT who develop skin lesions (particularly, necrosis) at heparin injection sites, particularly when a marked platelet count fall occurs.

The pathophysiology of heparin injection site skin lesions in HIT remains poorly understood. Perhaps the presence of FcγIIa receptors within the subset of ECs found in the superficial (but not deep) dermal vascular plexus (15) plays a role. In humans, only platelets and dermal MVECs bear FcγIIa receptors as their only class of Fcγ receptors (15).

Occasionally, patients develop intense limb pain (corresponding to a concurrent or imminent site of DVT), or pain at the catheter infusion site, within minutes of receiving intravenous heparin.

THE HIT ANTIGEN

HIT-mediated cell activation depends on antibody binding to cells. Thus, we will review the interaction of PF4 with glycosaminoglycans (GAGs), including those found on ECs.

Structure of Heparin

UFH is a polydisperse mixture of GAGs with molecular weights (MW) ranging from 5 to 40 kDa (mean, 13 kDa, or about 40 saccharide units). It is comprised of alternating D-glucosamine residues linked 1→4 to either L-iduronic acid or D-glucuronic acid. Its principal repeating unit is the trisulfated disaccharide,(1→4)-*O*-∀-L-iduronic acid-2-sulfate (1→4)-*O*-∀-D-glucosamine-2,6-disulfate (1→, which represents 75% to 90% of the heparin chain. The remaining 10% to 25% disaccharide units differ in the number and positions of sulfate groups.

Heparin is present naturally in the liver and mast cells. Its biologic function is unknown, but it is not primarily involved in control of hemostasis. In pharmacologic concentrations, the major effect of heparin is the catalysis of antithrombin (AT)-mediated inactivation of certain activated coagulation factors. The binding of heparin to AT is mediated by a well-defined pentasaccharide sequence within heparin with a central ∀-D-glucosamine-2,3,6-trisulfate unit.

Heparin is the GAG with the highest charge density (SO_3^-/COO^- ratio, 2.0–2.5). The high density of negative charge accounts for several limitations of heparin, including osteoporosis (binding to osteoclasts), suboptimal dose-response effect (binding to von Willebrand factor [vWF] and other acute phase reactants), and also risk of HIT (charge interaction with PF4).

Heparan Sulfate

Heparan sulfate (HS) is widely distributed throughout the vasculature. It is less-sulfated (by about 50%) than heparin, and has a major role in regulating hemostasis. It is found on EC surfaces as a proteoglycan; it consists of a protein core to which sulfated oligosaccharides are attached. Although only a minority of HS molecules have AT-catalyzing properties, most are probably capable of binding PF4, although with a lower affinity than heparin. It is unclear if, in a situation of intense platelet activation and PF4 release (as in HIT), whether PF4 binds in large quantities to endothelial HS to localize immunoinjury to endothelium. In vitro, the addition of HS inhibits the platelet-activating effects of HIT antibodies in the presence of heparin (16).

Structure of Platelet Factor 4

PF4 is a member of the CXC subfamily of chemokines (designated CXCL4). It is released from α-granules of activated platelets and, under physiologic conditions, attaches to the endothelial surface by binding to GAGs; it also binds to specific receptors, CXCR3A and CXCR3B (17). PF4 is a compact homotetrameric globular protein with a subunit MW of 7,780 Da (70 amino acids per each of the four subunits), with net positive charge. The C-terminal α-helices, which contain four lysine residues each that are involved in binding heparin, are arranged as antiparallel pairs on the surface of each extended β-sheet. These lysine residues are predominantly on one side, resulting in a ring of strong, positive charge that runs perpendicularly across the helices.

PF4 was the first chemokine described to inhibit neovascularization. In addition to its antiangiogenic properties, PF4 inhibits chemotaxis and hematopoiesis. It also interferes with the binding of fibroblast growth factor 2 (FGF2) and vascular endothelial growth factor (VEGF) to their receptors by competing with HS proteoglycans on the cell surface and by impairing the growth factor's ability to form homodimers (18). PF4 also can antagonize epidermal growth factor (EGF)- and VEGF-mediated EC stimulation (19).

A nonallelic variant gene of PF4/CXCL4, $PF4_{var1}/PF4_{alt}$, designated CXCL4L1, is expressed in ECs (20,21) and was recently also isolated from human platelets (22). PF4 and $PF4_{var1}$ differ in only three amino acids at the C-terminus. However $PF4_{var1}$ seems to be more potent in inhibiting chemotaxis of human MVECs toward interleukin (IL)-8/CXCL8 or FGF2 (22). In vivo, $PF4_{var1}$ was also more effective than PF4 in inhibiting bFGF-induced angiogenesis in rat corneas (22). However, no information is available whether PF4 and $PF4_{var1}$ differ in tertiary structure. It is possible that $PF4_{var1}$ interacts differently with ECs than does PF4, because the differences in amino acids reside in the heparin-binding region, with implications for modulating the pathogenesis of HIT.

Platelet Factor 4-Sulfated Polysaccharide Complexes

PF4 has the highest affinity for heparin among proteins stored within platelet α-granules. At low concentrations (0.1–1.0 IU/mL) of shorter heparin chains, and high concentrations of PF4, the binding capacity of a PF4 tetramer for heparin is not saturated by one heparin molecule, thus permitting simultaneous binding to more than one heparin chain. If a heparin molecule is longer than 16 monosaccharides, it is able to bind simultaneously to, and thereby bridge, two PF4 tetramers. Thus, at certain concentrations of heparin and PF4, there results the formation of large, multimolecular PF4/heparin complexes that can be subsequently dissociated by adding very high concentrations of heparin (2,23,24). The ability of a polysaccharide to form complexes with PF4 depends on its degree of sulfation (DS); this occurs in the order: UFH > LMWH > HS > dermatan sulfate > chondroitin-6-sulfate > chondroitin-4-sulfate. In addition to the DS, other structural parameters that influence the affinity of a polysaccharide to PF4 include type of uronic acid and the location of the sulfates on the amino sugar. Although HS interacts with PF4, it does not form the HIT antigen in vitro, but rather inhibits its formation with heparin.

Platelet-Activating Immune Complexes

Heparin binding to PF4 exposes one or more neoepitopes on PF4. The majority of HIT antibodies recognize a noncontiguous conformational epitope on PF4. At least two distinct antigen sites have been identified. Combining the techniques of atomic force microscopy, photon correlation spectroscopy, and quantitative immunoassays performed using three presentations of PF4 (PF4 alone, PF4/heparin, and PF4/heparin/PF4 "sandwiches"), we found evidence that the HIT antigen is formed when at least two PF4 tetramers are brought into close approximation by polysaccharide-induced charge neutralization (25).

HIT antibodies bind to PF4/heparin complexes via their $F(ab')_2$ domains, with IgG as the predominant immunoglobulin class. Thus, divalent IgG binding to multimolecular PF4/heparin complexes produces large IgG-containing immune complexes. On the platelet surface, the subsequent interaction of the HIT-IgG Fc moieties with the platelet FcγIIa receptors leads to cross-linking of these receptors and, consequently, strong platelet activation. However, by incubating ECs with HIT antibodies, we did not find EC activation (26).

Platelet Factor 4 Interactions with Low-Molecular-Weight Heparin

HIT antigen generation depends not only on the concentration but also on the chain length of heparin. LMWH has reduced affinity for platelets, ECs, and plasma proteins such as PF4. Accordingly, LMWH is less likely to form multimolecular complexes with PF4 (27), and the complexes formed are smaller (24), possibly explaining the lower frequency of HIT with LMWH.

Whereas homogeneous heparin fragments containing 12 or more saccharide units form multimolecular complexes with PF4 that are recognized by HIT antibodies, fragments containing 10 residues induce antigen formation only weakly, and fragments containing 8 or fewer residues bind weakly (23) or not at all in vitro (28). However, we recently showed that the smallest heparin-like molecule used for anticoagulation, the pentasaccharide (fondaparinux), was able to induce anti-PF4/heparin antibodies in vivo as frequently as did LMWH (although, interestingly, these antibodies did not recognize PF4/fondaparinux in vitro, although they bound well to PF4/UFH and PF4/LMWH) (29).

Platelet Factor 4 Interactions with Nonheparin Polysaccharides

Changes in PF4 structure that lead to induction of HIT antibodies in vivo can be induced by polyanions other than heparin, such as pentosan polysulfate, hypersulfated chondroitin sulfate, and PI-88 (an antiangiogenic agent) (16,27). Cross-reactivity of HIT antibodies for these PF4-polyanion complexes depends on certain features of polyanion structure, particularly the DS and/or MW (16,23). We synthesized structurally well-defined sulfated polysaccharides (23) with a variety of β-1,3-glucan sulfate DS, MW, and sulfation patterns, as well as differences in glycosidic side chains. Increases in DS or in MW (or both), or branching of the polysaccharide (by glycosidic substitution) resulted in increased formation of antigen sites on PF4.

In theory, pathological conditions that increase sulfation of endothelial HS could promote PF4 binding in a way that allows the epitopes relevant for HIT to form even in the absence of pharmacologic heparin. Such a scenario could produce a prothrombotic autoimmune disorder. Indeed, in some individuals, HIT antibodies can be detected even though they had never previously been exposed to heparin.

HIT ANTIBODIES AND ENDOTHELIAL CELLS

HIT antibodies could lead to EC activation by both direct and indirect mechanism (Figure 147.1) (30).

The first evidence that HIT antibodies might interact directly with ECs was provided in 1987, by Cines and colleagues (5). HIT serum deposited IgG, IgM, IgA, and complement to human umbilical vein ECs (HUVECs) in amounts greater than control. The addition of heparin was not necessary. PF4 most likely bound to endothelial HS, as shown by loss of IgG binding following treatment of HUVECs by heparinase. It has been suggested (6) that binding of HIT antibodies to ECs in these early studies without addition of PF4 may be explained by use of serum (rather than plasma), which contains significant levels of PF4 released from platelets during clotting.

ECs incubated with HIT serum expressed increased tissue factor–associated procoagulant activity. The authors noted that "additional increases in procoagulant activity were expressed when ECs were incubated with patients' serum in the presence of gel-filtered platelets" (5). This is relevant to work by later investigators who noted the important role of platelets in generating endothelial procoagulant activity by HIT serum/plasma (discussed subsequently). Indeed, expression of CD40 ligand on activated platelets produces certain changes in ECs, including an inflammatory response (31).

In 1994, two studies were reported investigating the interaction of HIT antibodies, PF4, and endothelium (2,6).

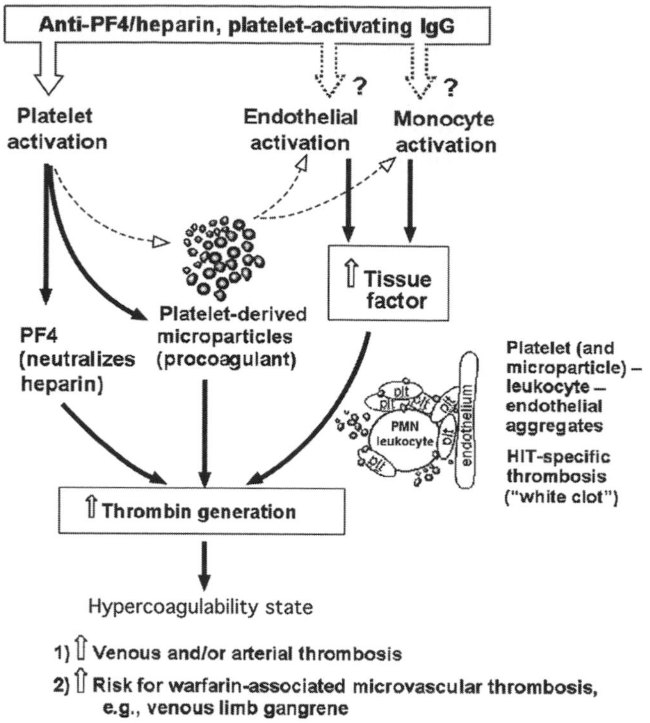

Figure 147.1. Two models for thrombosis in HIT. Activation of platelets (*Plt*) by anti–platelet factor 4 (PF4)/heparin IgG antibodies (HIT antibodies) leads to formation of procoagulant, platelet-derived microparticles, and neutralization of heparin by PF4 released from activated platelets. This leads to marked increase in thrombin ("hypercoagulability state") characterized by an increased risk of venous and arterial thrombosis, as well as increased risk for coumarin-induced venous limb gangrene. In this model, platelet activation per se is the primary factor explaining prothrombotic risk. In contrast, unique pathogenetic mechanisms operative in HIT could explain unusual thromboses, such as arterial "white clots." For example, HIT antibodies have been shown to activate endothelium and monocytes (leading to cell surface tissue factor expression), although this stimulation may be largely "indirect" through poorly defined mechanisms involving platelet activation and, possibly, the formation of platelet-derived microparticles. Further, aggregates of platelets and leukocytes (monocytes, neutrophils) have been described in HIT. To what extent these cooperative interactions between platelets, platelet-derived microparticles, neutrophils, monocytes, and endothelium lead to arterial (or venous) thrombotic events in HIT, either in large or small vessels, remains unclear. (Reproduced, with modifications, with permission from Warkentin TE. An overview of the heparin-induced thrombocytopenia syndrome. *Semin Thromb Hemost.* 2004;30:273–283.)

Visentin and coworkers (6) showed that IgG and, sometimes, IgM from HIT plasmas bound to HUVECs in a PF4-dependent, but heparin-independent, fashion. Further, antibody binding to HUVECs could be inhibited by excess heparin, but not by an Fc receptor-blocking monoclonal antibody. These investigators hypothesized that in acute HIT, PF4 released from activated platelets binds to HS on ECs, creating the neoepitopes recognized by HIT antibodies and leading to endothelial immunoinjury and thrombosis, for example by complement activation. In such a model, HIT antibodies of IgM class alone could be pathogenic, despite their inability to activate platelets via FcγIIa receptors.

We used classic adsorption/elution experiments to study the reactivity of HIT antibodies against HUVECs (2). Purified HIT-IgG tested positive in both platelet activation and PF4/heparin immunoassays. This IgG fraction was then adsorbed by HUVECs and, after washing, the antibodies were eluted by low pH. The eluate again tested positive in both activation and antigen assays. Thus, these experiments showed that the HIT antibodies recognize the same epitope(s) on platelets, ECs, and PF4/heparin complexes coated onto a microtiter plate. However, as we added low concentrations of UFH (0.1 IU/mL) to facilitate antibody binding to endothelium, we did not address whether endothelial HS itself permits binding of PF4-reactive HIT antibodies. In contrast to the hypothesis of Visentin and colleagues, in these experiments, no PF4 was added to the HUVEC culture. Because purified IgG fractions were also used in these experiments, this indicates that PF4 is directly expressed by ECs. A likely (though as yet untested) explanation is that the HIT antibodies bound to (EC-derived) $PF4_{var1}$.

Herbert and coworkers (32) found that HIT serum, together with UFH (0.5 IU/mL) and in the presence of washed platelets, activated HUVECs, as shown by increased expression of E-selectin, intracellular adhesion molecule (ICAM)-1, vascular cell adhesion molecule (VCAM)-1, and tissue factor. In addition, the activated HUVECs also synthesized and released soluble IL-1β, IL-6, plasminogen activator inhibitor (PAI)-1, and tumor necrosis factor (TNF)-α. EC lysis did not occur. A platelet glycoprotein IIb/IIIa receptor antagonist completely inhibited HUVEC activation under these conditions, as did apyrase (an adenosine diphosphate [ADP]-degrading enzyme) and ATPγS (inhibitor of ADP-induced platelet activation). Adhesion of platelets to HUVECs activated by HIT serum also was inhibited by the glycoprotein IIb/IIIa receptor antagonist. However, these studies did not address why platelets appeared necessary for the (HIT antibody-induced) endothelial activation response, nor whether the platelets were themselves activated under the experimental conditions used.

Blank and colleagues (7) reported that MVECs (derived from human bone marrow) responded differently to HIT antibodies than did HUVECs. Specifically, only MVECs could be directly activated by affinity-purified PF4-reactive HIT antibodies, thus leading to endothelial expression of E-selectin, P-selectin, and VCAM-1 (but not ICAM-1), as well as release of IL-6, vWF, and thrombomodulin (TM). In contrast, HUVECs could be activated by HIT antibodies in the presence of PF4 only if washed platelets or TNF-α had been added. Pretreatment of ECs with heparinase (which degrades endothelial-associated GAG) abolished binding of the HIT antibodies. Additional experiments showed that adhesion of a monocyte-macrophage cell line (U937) to MVECs (from human bone marrow) could be enhanced by prior treatment of the endothelium by HIT antibodies and PF4, an effect abrogated by pretreatment with a

Table 147-2: Summary of Methodologies Used in Studies of Heparin-Induced Thrombocytopenia Involving Endothelial Cells

	EC Culture Medium				*Effect on HUVECs*	
References	*Heparin*	*Human Sera*	*Source of HIT Antibodies*	*Dilution*	*Binding of Antibodies*	*Activation*
5	No	No	Sera*	1:2 to 1:4	Yes	Tissue factor ↑
6	Yes	No	Sera*	1:10	Yes†	Not done
2	Yes	Yes	IgG	1:50	Yes	Not done
32	Yes	No	Sera IgG	1:205 μg/mL	–	Adhesion molecules and cytokines ↑§
7	No	No	IgG	10 μg/mL	Yes‡	Not assessable due to preactivation
33	?	?	MAbKKO	100 μg/mL	–	TM ↓§
26	Yes	Yes	Sera IgG	1:10 to 1:132 μg/mL	Yes	No activation

§ Heparin was adsorbed from the sera
*Only after addition of PF4
†Only after preactivation of HUVECs with TNF
‡Only after addition of washed platelets

cocktail of antibodies blocking E-selectin, P-selectin, and VCAM-1 (7). This study supports the concept that EC heterogeneity may have implications for the pathogenesis of HIT. However, whereas in these in vitro studies HIT antibodies had their most prominent effect on MVECs, macrovascular (not microvascular) thrombosis is the major clinical effect of HIT.

Asmis and colleagues (33) observed that the addition of the murine HIT antibody-mimicking monoclonal antibody, named KKO (with anti-PF4/heparin specificity), to HUVECs in the presence of heparin did not result in endothelial changes unless platelets were added that had been activated by the HIT antibodies. Under these conditions, HUVECs downregulated TM, in line with the procoagulant effects on HUVECs in the presence of platelets (32).

In our own investigations (26), HUVECs were used to study a wide variety of effects of HIT antibodies on macrovascular ECs, including procoagulant (e.g., release of vWF, cleavage of TM, inhibition of protein C generation, tissue factor upregulation), proinflammatory (upregulation of VCAM-1), and cytotoxic (complement-dependent ^{51}Cr-release) properties. Although HIT antibodies bound to HUVECs, no in vitro EC-activating effect was found. However, when platelets or platelet-derived MPs were added, these led to HUVEC activation, as shown by VCAM-1 upregulation. Ex vivo investigation found that HIT patients showed elevated serum TM levels, suggesting EC activation might occur in vivo. However, TM levels did not differ from those patients with non-HIT thrombocytopenia.

In summary, these studies of HIT antibody–endothelial interaction show important discrepancies (Table 147-2). In particular, there is disagreement as to whether HIT antibodies directly activate ECs in the absence of platelets or platelet-derived MPs. These studies are difficult to compare, however, because they employed different methods and experimental conditions (e.g., use of serum versus plasma, addition of heparin, etc.).

ANIMAL MODELS OF HEPARIN-INDUCED THROMBOCYTOPENIA

Several challenges arise when developing animal models of HIT (34–38). For example, animals do not generate HIT antibodies with heparin treatment. Accordingly, Blank and colleagues immunized mice with affinity-purified anti-PF4/heparin antibodies obtained from patients with HIT (34,35). These mice developed specific anti-idiotypic antibodies against the human HIT antibodies, followed 2 months later by formation of anti–anti-idiotypic antibodies. Unfortunately, although these mice apparently become thrombocytopenic when treated with UFH, they do not develop thrombosis.

Other problems with a murine model of HIT include lack of platelet Fcγ receptors and nonreactivity of human HIT antibodies against murine PF4/heparin complexes. Accordingly, Reilly and colleagues (36,37) developed a novel murine model employing double-transgenic FcγRIIA/hPF4 mice – mice that express human FcγRIIa on their platelets as well as human PF4. When these mice were treated with the HIT-mimicking murine monoclonal antibody, KKO, that recognizes human PF4/heparin, and then given heparin, the mice developed severe thrombocytopenia and fibrin-rich thrombi in multiple organs, including the pulmonary vasculature. Thus, several salient features of human clinical HIT were recapitulated in this model. Interestingly, mice transgenic only for human PF4 developed neither thrombocytopenia nor thrombosis when treated with the HIT-mimicking monoclonal antibody. This is compatible with our concept that, in HIT, ECs are activated *indirectly* via mediators released from activated platelets.

Platelets obtained from nonhuman primates (*Macaca mulatta*; the Rhesus monkey) can be activated by HIT serum, suggesting that a primate animal model for HIT might be feasible (38).

ENDOTHELIAL HETEROGENEITY

The unusually high rate of thrombosis in HIT hints toward involvement of ECs. However, the thrombotic complications manifest predominantly in large vessels, where the EC surface area to blood volume ratio is the lowest. Shedding of TM from the EC surface is a specific marker for in vivo activation of ECs (39). TM levels are increased in HIT patients, compared with normal controls. However, comparison with non-HIT thrombocytopenic patients shows no major difference in TM levels. Indeed, non-HIT patients have a trend to even higher TM levels, despite absence of clinical thrombosis.

$PF4_{var1}$ also may be important to explain the discrepant in vitro findings, as well as the different thrombotic manifestations in HIT. Potentially, the amounts of PF4 and $PF4_{var1}$ bound to the surface of ECs could differ, depending on the intravascular location of the ECs. Furthermore, inflammatory cytokines may influence PF4 binding and expression on the EC surface. Because platelet granules contain several chemokines (e.g., regulated on activation, normal T-cell expressed and secreted [RANTES]), platelets might be necessary for a change of the activation status of those ECs required for neoepitope expression on PF4.

MULTIPLE CELLULAR ACTIVATION

Studies by Khairy and colleagues (9) suggest that HIT antibodies induce leukocyte–platelet aggregate formation. The leukocytes are comprised of monocytes and neutrophils, and are activated best together with activated platelets. High plasma myeloperoxidase levels suggest that leukocyte activation could occur in vivo in patients with HIT. However, histomorphology studies suggest a marked predominance of platelets and a paucity of leukocytes within the white thrombi of HIT (11).

Basic studies (not involving HIT sera) have shown that procoagulant platelet MPs lead to increased binding of monocytes to HUVECs by upregulating cell adhesion molecules, such as ICAM-1 on HUVECs and CD11a/CD18 on monocytes (40). Platelet MPs themselves, via transcellular lipid metabolism, can effect the activation of adjacent platelets, monocytes, and ECs (41). To what extent such cellular "crosstalk" pathways exist or contribute to thrombotic events in HIT remains unknown.

KEY POINTS

- Platelets activated by HIT antibodies, or their products (e.g., MPs), are critically important in mediating EC and leukocyte activation in HIT. In turn, EC and leukocyte activation is a key determinant of the prothrombotic and proinflammatory clinical phenotype.
- Platelet activation by HIT antibodies occurs via platelet $Fc\gamma$IIa receptors; however, ECs (except dermal MVECs) lack $Fc\gamma$ receptors.
- Increasing in vitro evidence indicates that pancellular activation (platelets, ECs, monocytes, neutrophils) occurs in HIT.

Future Goals

- To establish that ECs are activated in vivo in patients with HIT
- To determine the mechanisms by which HIT antibodies activate ECs directly or indirectly
- To clarify the role of $PF4_{var1}$ for HIT antibody–EC interaction

REFERENCES

1 Warkentin TE, Greinacher A, eds. *Heparin-induced Thrombocytopenia*, 3rd ed. New York: Marcel Dekker, 2004.
2 Greinacher A, Pötzsch B, Amiral J, et al. Heparin-associated thrombocytopenia: isolation of the antibody and characterization of a multimolecular PF4–heparin complex as the major antigen. *Thromb Haemost*. 1994;71:247–251.
3 Warkentin TE, Elavathil LJ, Hayward CPM, et al. The pathogenesis of venous limb gangrene associated with heparin-induced thrombocytopenia. *Ann Intern Med.* 1997;127:804–812.
4 Prandoni P, Bilora F, Marchiori A, et al. An association between atherosclerosis and venous thrombosis. *N Engl J Med.* 2003;348:1435–1441.
5 Cines DB, Tomaski A, Tanenbaum S. Immune endothelial-cell injury in heparin-associated thrombocytopenia. *N Engl J Med.* 1987;316:581–589.
6 Visentin GP, Ford SE, Scott JP, Aster RH. Antibodies from patients with heparin-induced thrombocytopenia/thrombosis are specific for platelet factor 4 complexed with heparin or bound to ECs. *J Clin Invest*. 1994;93:81–88.
7 Blank M, Shoenfeld Y, Tavor S, et al. Anti-platelet factor 4/heparin antibodies from patients with heparin-induced thrombocytopenia provoke direct activation of microvascular ECs. *Int Immunol.* 2002;14:121–129.
8 Arepally GM, Mayer IM. Antibodies from patients with heparin-induced thrombocytopenia stimulate monocytic cells to express cells to express tissue factor and secrete interleukin-8. *Blood.* 2001;98:1252–1254.
9 Khairy M, Lasne D, Amelot A, et al. Polymorphonuclear leukocyte and monocyte activation induced by plasma from

patients with heparin-induced thrombocytopenia in whole blood. *Thromb Haemost*. 2004;92:1411–1419.
10 Kwaan HC, Sakurai S. Endothelial cell hyperplasia contributes to thrombosis in heparin-induced thrombocytopenia. *Semin Thromb Hemost*. 1999;25(Suppl 1):23–27.
11 Hermanns B, Janssens U, Handt S, Fuzesi L. Pathomorphological aspects of heparin-induced thrombocytopenia II (HIT-II syndrome). *Virchows Arch*. 1998;432:541–546.
12 Hong AP, Cook DJ, Sigouin CS, Warkentin TE. Central venous catheters and upper-extremity deep-vein thrombosis complicating immune heparin-induced thrombocytopenia. *Blood*. 2003; 101:3049–3051.
13 Warkentin TE. Heparin-induced skin lesions. *Br J Haematol*. 1996;92:494–497.
14 Wutschert R, Piletta P, Bounameaux H. Adverse skin reactions to low molecular weight heparins: frequency, management and prevention. *Drug Saf*. 1999;20:515–525.
15 Gröger M, Sarmay G, Fiebiger E, et al. Dermal microvascular ECs express CD32 receptors in vivo and in vitro. *J Immunol*. 1996;156: 1549–1556.
16 Greinacher A, Michels I, Mueller-Eckhardt C. Heparin-associated thrombocytopenia: The antibody is not heparin specific. *Thromb Haemost*. 1992;67:545–549.
17 Lasagni L, Francalanci M, Annunziato F, et al. An alternatively spliced variant of CXCR3 mediates the inhibition of EC growth induced by IP-10, Mig, and I-TAC, and acts as functional receptor for platelet factor 4. *J Exp Med*. 2003;197:1537–1549.
18 Perollet C, Han ZC, Savona C, et al. Platelet factor 4 modulates fibroblast growth factor 2 (FGF-2) activity and inhibits FGF-2 dimerization. *Blood*. 1998;91:3289–3299.
19 Sulpice E, Bryckaert M, Lacour J, et al. Platelet factor 4 inhibits FGF2-induced EC proliferation via the extracellular signal-regulated kinase pathway but not by the phosphatidylinositol 3-kinase pathway. *Blood*. 2002;100:3087–3094.
20 Green CJ, Charles RS, Edwards BF, Johnson PH. Identification and characterization of PF4var1, a human gene variant of platelet factor 4. *Mol Cell Biol*. 1989;9:1445–1451.
21 Eisman R, Surrey S, Ramachandran B, et al. Structural and functional comparison of the genes for human platelet factor 4 and PF4alt. *Blood*. 1990;76:336–344.
22 Struyf S, Burdick MD, Proost P, et al. Platelets release CXCL4L1, a nonallelic variant of the chemokine platelet factor-4/CXCL4 and potent inhibitor of angiogenesis. *Circ Res*. 2004;95:855–857.
23 Greinacher A, Alban S, Dummel V, et al. Characterization of the structural requirements for a carbohydrate based anticoagulant with a reduced risk of inducing the immunological type of heparin-associated thrombocytopenia. *Thromb Haemost*. 1995; 74:886–892.
24 Sachais BS, Higazi AA, Cines DB, et al. Interactions of platelet factor 4 with the vessel wall. *Semin Thromb Hemost*. 2004;30:351–358.
25 Greinacher A, Gopinadhan M, Günther JU, et al. Close approximation of two platelet factor 4 tetramers by charge neutralization forms the antigen recognized by HIT antibodies. *Arterioscler Thromb Vasc Biol*. 2006;26:2386–2393.
26 Hartmann W, Greinacher A, Lubenow L, et al. Heparin-induced thrombocytopenia: in vitro studies on the interaction of HIT-antibodies with ECs. *Platelets* 2004;15:245. [Abstract].
27 Greinacher A, Michels I, Liebenhoff U, et al. Heparin-associated thrombocytopenia: immune complexes are attached to the platelet membrane by the negative charge of highly sulphated oligosaccharides. *Br J Haematol*. 1993;84:711–716.
28 Visentin GP, Moghaddam M, Beery SE, et al. Heparin is not required for detection of antibodies associated with heparin-induced thrombocytopenia/thrombosis. *J Lab Clin Med*. 2001; 138:22–31.
29 Warkentin TE, Cook RJ, Marder VJ, et al. Anti-platelet factor 4/heparin antibodies in orthopedic surgery patients receiving antithrombotic prophylaxis with fondaparinux or enoxaparin. *Blood*. 2005;106(12):3791–3796. Epub 2005 Aug 18.
30 Warkentin TE. An overview of the heparin-induced thrombocytopenia syndrome. *Semin Thromb Hemost*. 2004;30:273–283.
31 Henn V, Slupsky JR, Grafe M. CD40 ligand on activated platelets triggers an inflammatory reaction of ECs. *Nature*. 1998;391;591–594.
32 Herbert JM, Savi P, Jeske WP, Walenga JM. Effect of SR121566A, a potent GP IIb-IIIa antagonist, on the HIT serum/heparin-induced platelet mediated activation of human ECs. *Thromb Haemost*. 1998;80:326–331.
33 Asmis LM, Demming C, Arepally GM, Rade JJ. The interaction of platelets and ECs modulate the procoagulant/anticoagulant balance in heparin-induced thrombocytopenia. *Blood*. 2003:102. [Abstract].
34 Blank M, Cines DB, Arepally G, et al. Pathogenicity of human anti-platelet factor 4 (PF4)/heparin in vivo: generation of mouse anti-PF4/heparin and induction of thrombocytopenia by heparin. *Clin Exp Immunol*. 1997;108:333–339.
35 Blank M, Eldor A, Tavor S, et al. A mouse model for heparin-induced thrombocytopenia. *Semin Hematol*. 1999;36: 12–16.
36 Reilly MP, Taylor SM, Hartman NK, et al. Heparin-induced thrombocytopenia/thrombosis in a transgenic mouse model requires human platelet factor 4 and platelet activation through FcγRIIA. *Blood*. 2001;98:2442–2447.
37 McKenzie SE, Reilly MP. Heparin-induced thrombocytopenia and other immune thrombocytopenias: lessons from mouse models. *Semin Thromb Hemost*. 2004;30:559–568.
38 Ahmad S, Jeske WP, Walenga JM, et al. Human anti-heparin-platelet factor 4 antibodies are capable of activating primate platelets: towards the development of a HIT model in primates. *Thromb Res*. 2000;100:47–54.
39 Blann AD, Seigneur M, Steiner M, et al. Circulating endothelial cell markers in peripheral vascular disease: relationship to the location and extent of atherosclerotic disease. *Eur J Clin Invest*. 1997;27:916–921.
40 Barry OP, Pratico D, Savani RC, FitzGerald GA. Modulation of monocyte-endothelial cell interactions by platelet microparticles. *J Clin Invest*. 1998;102:136–144.
41 Barry OP, FitzGerald GA. Mechanisms of cellular activation by platelet microparticles. *Thromb Haemost*. 1999;82:794–800.

148

Sickle Cell Disease Endothelial Activation and Dysfunction

Robert P. Hebbel

Vascular Biology Center, University of Minnesota, Minneapolis

Sickle cell anemia is characterized by hemolytic anemia, acute painful episodes that are believed to be caused by vaso-occlusion, and various organ-specific complications (1). Although the fundamental molecular basis for this disease is the inherited presence of the mutant sickle hemoglobin, the pathogenesis of clinical sickle disease is exceedingly complex. The studies described here, both bedside-to-bench and bench-to-bedside, have suggested that the endothelium contributes substantially to the vascular pathobiology of sickle disease and helps govern clinical phenotype (2). Indeed, we regard sickle cell anemia as an "endothelial disease" in which the endothelium is abnormally activated and dysfunctional. It is likely that the vascular pathobiology of this disease is affected by the endothelial cell (EC)'s functions as a space-defining physical barrier, an adhesive surface, a regulator of vascular tone, a balancer of the anticoagulant and procoagulant properties of blood and vessel wall, a participant in the inflammatory response, and a responsive surface that is dynamically alterable.

THE BEGINNING: FROM BEDSIDE TO BENCH

After the first medical literature report of a patient with sickle disease in 1910 (3), early studies identified the relationship between deoxygenation and cytoplasmic changes and red cell sickling (4), and ultimately the molecular character of the disease as being caused by an abnormal hemoglobin molecule (5). Thereafter, the traditional view of sickle disease explained pathophysiology simply via the sickling phenomenon, a view we now recognize as being highly oversimplified (6). The bedside observation of marked heterogeneity in clinical phenotype, both from person-to-person and from time-to-time for any given patient, stimulated our original interest in the endothelial biology of this disease. Specifically, we were struck by the apparent inability of traditional explanations for sickle disease genesis to explain what was observed at the bedside. This led us to pose and investigate a series of hypotheses regarding the endothelial biology of this disease (2). The first of these, regarding red cell adhesion to endothelium, began a fundamental shift in the understanding of sickle disease pathobiology.

Red Blood Cell Adhesion to Endothelium

The seminal endothelial biology experiments (7,8) for sickle disease demonstrated that sickle red blood cell (RBCs), even when oxygenated, are abnormally adhesive to cultured vascular ECs. This has since been verified by using various in vitro, ex vivo, and in vivo systems (2), including the sickle transgenic mouse itself (Table 148-1) (9). Patient-to-patient variability in red cell adhesiveness to EC (using human umbilical vein ECs [HUVECs]) in vitro correlates with clinical vaso-occlusive severity (Figure 148.1). Elaborate studies in rodents have indicated that sickle vaso-occlusion probably involves two steps: initiation due to endothelial adhesion of deformable red cells, followed by a propagation phase caused by log-jamming of poorly deformable RBCs (10). Flow chamber and in vivo observations indicate that sickle red cell adhesion occurs under low shear stresses characteristic of the postcapillary venule (2). However, endothelial adhesion of red cells increases dramatically as vessel size approaches the limiting diameter (11), so its role in capillaries requires further study. The importance of this phenomenon derives from its ability to slow microvascular transit time, fulfilling the physical-chemical requirement to allow for the delay time between deoxygenation and substantial polymer formation (12).

The mechanisms involved in the abnormal adhesion of sickle red cells to ECs have been identified largely through studies using simplified buffer systems, and relatively little is known about this event in the more physiologic, complex plasma environment. The mechanisms currently supported by the literature are depicted in Figure 148.2. These involve

Table 148-1: Murine Models of Sickle Cell Disease

Sickle Mouse	*Murine-α Globin*	*Murine-β Globin*	*Trans-Gene*	*Phenotype*
NY1DD	+/+	Del/del	$\alpha^H\beta^S$	Mild sickle state, normal white count, elevated monocyte count, elevated serum amyloid P, low endothelial tissue factor converts to high tissue factor after H/R.
S+S-Antilles	+/+	Del/del	$\alpha^H\beta^S$ and $\alpha^H\beta^{S\text{-Antilles}}$	Moderate sickle state, elevated monocyte count, elevated serum amyloid P, moderate endothelial tissue factor.
BERK	ko/ko	ko/ko	$\alpha^H\beta^S$	Severe sickle state, elevated white count, elevated monocyte count, elevated serum amyloid P, high endothelial tissue factor

Abbreviations: del, deleted; ko, knocked out.

various adhesion molecules on the EC (e.g., CD36, $\alpha v\beta 3$, GPIb, P-selectin, vascular cell adhesion molecule [VCAM]-1), on the red cell (e.g., CD36, $\alpha 4\beta 1$, sulfated lipids), and/or involving plasma bridging molecules (e.g., thrombospondin, von Willebrand factor [vWF], fibronectin, fibrinogen). Some

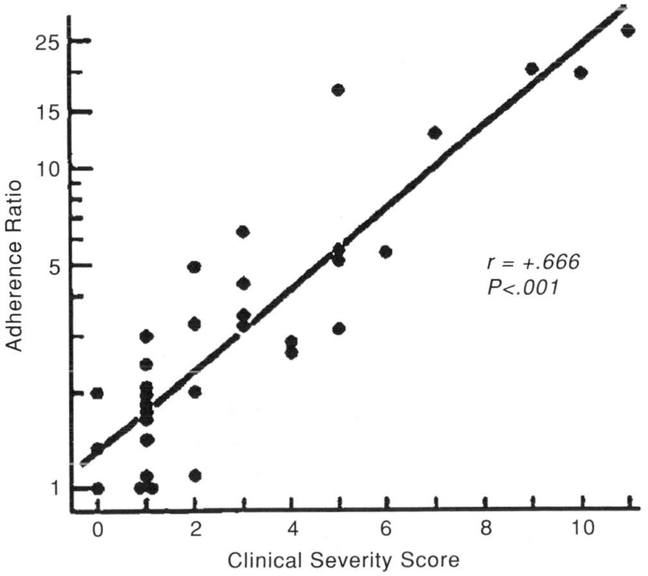

Figure 148.1. Clinical vaso-occlusive severity correlates with erythrocyte adhesiveness to endothelium. (Reproduced with permission from Hebbel RP, Boogaerts MAB, Eaton JW, Steinberg MH. Erythrocyte adherence to endothelium in sickle cell anemia: possible determinant of disease severity. *N Engl J Med.* 1980;302:992–995.)

of the identified mechanisms could utilize structures displayed by the quiescent EC, whereas others would be relevant only to activated ECs. Recently, a possible role for multicellular interactions (e.g., leukocyte/EC/RBC) in this phenomenon has received some attention (13). The many mechanisms are discussed in detail elsewhere (2). It seems unlikely that any single "most important" mechanism exists. Rather, mechanisms may differ from patient to patient, or from organ to organ, or time to time for a given patient.

Flow model experiments generally have identified adhesion mechanisms that employ defined ligand–receptor interactions (2) that are high affinity and are most likely determined by on-rate of adhesion events (14). Consequently, it is often assumed that only sickle adhesion employing such high-affinity interactions is of physiological relevance. However, there could be a role for lower-affinity mechanisms that do not occur rapidly but, rather, benefit from a high number of receptors. In fact, the demonstration of a correlation between sickle red cell adhesivity and clinical severity (Figure 148.1) was probably reflective of a low-affinity adhesion mechanism. It seems probable that sickle red cell adhesion to endothelium in pathobiology involves a variety of affinities (14). It should be noted that virtually everything known about sickle red cell adhesion to endothelium is derived from observation in a two-dimensional, flat universe (i.e., flow chambers or relatively large caliber vessels). The microcirculation, however, includes small vessels of constraining diameter. Adhesion science has not yet been able to evaluate, for example, the presumably critical interplay between adhesiveness and deformability in such a demanding, three-dimensional environment.

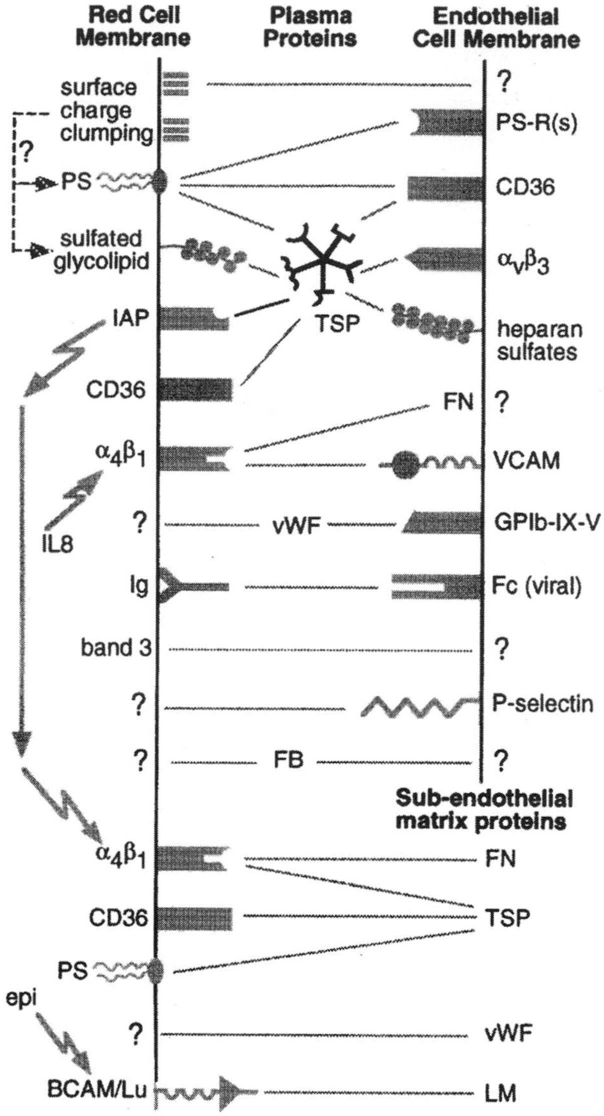

Figure 148.2. Identified mechanisms of red cell adhesion to endothelium. (Reproduced with permission from Hebbel RP, Osarogiagbon R, Kaul D. The endothelial biology of sickle cell disease: inflammation and a chronic vasculopathy. *Microcirculation.* 2004;11:129–151.) PS, phosphatidylserine; IAP, integrin-associated protein; epi, ephinephrine; FN, fibronectin; LM, laminin.

THE ENDOTHELIUM: BEDSIDE, BENCH, AND MOUSE

The apparent relevance of EC activation to sickle red cell adhesion to endothelium in vitro led our group to focus on the endothelium itself.

Circulating Endothelial Cells

Seeking to understand the activation state of the endothelium in the living subject, we developed the approach of examining the phenotype of circulating ECs (CECs). We found that the CECs in sickle subjects are abnormally activated (even remote from any acute clinical events), being proadhesive (expressing VCAM-1, intercellular adhesion molecule [ICAM]-1, $\alpha V\beta 3$, selectins [15]), procoagulant (expressing tissue factor [16]), and oxidatively stressed (expressing heme oxygenase-1 [17]). Notably, these adhesion molecules are directly relevant to leukocyte–EC and/or erythrocyte–EC interactions. These observations are presumably instructive because CEC phenotype would be governed by the same milieu that determines the phenotype of in situ vessel wall endothelium. In support of this hypothesis, we did note a correspondence between the activated phenotype of CECs and vessel wall endothelium in the sickle transgenic mouse (18). Notably, however, the most striking feature was the observation of a marked heterogeneity in vessel wall endothelial phenotype, varying from organ to organ and between small and large vessels (18). We believe that such heterogeneity in endothelial phenotype and biology extends to humans and is highly relevant to understanding sickle disease pathobiology.

Although it has not been proven that the phenotype of CECs corresponds to that of vessel wall endothelium in humans with sickle disease, our studies of endothelial tissue

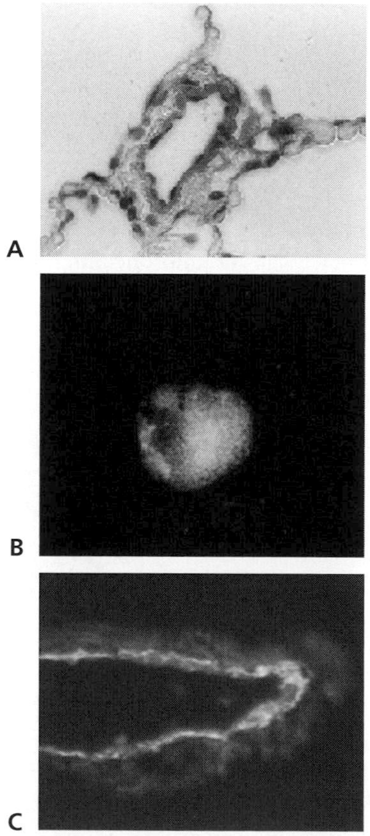

Figure 148.3. Evidence for endothelial tissue factor expression in sickle disease. (**A**) Tissue factor–positive endothelium in the lung of a sickle cell anemia patient. (**B**) A circulating EC staining positive for tissue factor antigen and tissue factor mRNA. (**C**) Tissue factor–positive vein in lung of a sickle transgenic mouse. A, reproduced with permission from Solovey A, Hebbel RP. *J Lab Clin Med.* 2001;137(6); Cover photo. B, Solovey A, Gui L, Key NS, Hebbel RP. *J Clin Invest.* 1998;101:1899–1904. C, Solovey A, Kollander R, Shet A, et al. *Blood.* 2004;104:840–846. For color reproduction, see Color Plate 148.3.

factor expression are consistent with this notion (Figure 148.3; for color reproduction, see Color Plate 148.3). We observed abnormal tissue factor expression on pulmonary endothelium in a sickle patient who had tissue factor–positive CECs and died of thromboembolism. We found that CECs from sickle patients are abnormally tissue factor positive (16). And we documented that the pulmonary venous endothelium in the sickle transgenic mouse is abnormally tissue factor positive (19).

Inflammation

The activated CECs in sickle patients are indicative of an abnormal inflammatory state. Other evidence for this includes chronic elevation of leukocyte counts, shortened leukocyte half-life, abnormal activation of granulocytes and monocytes, activation of the coagulation system, and elevations of soluble VCAM-1 and platelet counts (2). Patients also have elevated numbers of microvesicles derived from monocytes and ECs (20), a probable sign of cellular activation. The blood of sickle patients reveals variably but chronically elevated levels of EC perturbants (1). Sickle mice share this inflammatory phenotype with humans (2).

Clinically, leukocytosis is a correlative risk factor for mortality (21) and several major complications of sickle disease. Notably, leukocytosis predicts which babies will subsequently develop a severe clinical course (22). The large hydroxyurea trial showed correlation between subject leukocyte counts and pain crisis rates (23). Administration of granulocytosis-inducing growth factors to sickle patients has precipitated acute vaso-occlusive crises. In acute chest syndrome and acute painful crisis, clinical benefit is derived from corticosteroids. Intriguingly, it has been noted that a recent febrile illness is an independent risk factor for stroke, implicating proximate inflammation in its genesis. Thus, it seems likely that the inflammatory state of sickle disease comprises a vicious cycle, with inflammation causing occlusion leading to further inflammation. The chronic activation state of CECs from patients with sickle cell anemia is consistent with this notion.

Two unique aspects of sickle disease may underlie this chronic inflammatory state.

Adherent Cells and Sickled Cells

The adhesion of oxygenated sickle red cells to the endothelium triggers an injury response (not seen after normal RBC adhesion) that includes release of peroxidation byproducts, prostacyclin, 15-HETE, and endothelin-1; upregulation of adhesion molecules; inhibition of vasorelaxation; and activation of nuclear factor (NF)-κB (2). It is likely that interaction of sickled (i.e., deoxygenated) sickle red cells with ECs has effects as well. Interestingly, it has been observed that nocturnal blood oxygen saturation correlates inversely with a marker of EC activation, soluble VCAM-1 (24). Sickled red cells stimulate cultured ECs to produce endothelin-1 (25). Thus, it would seem that the relentless irritation of endothelium by red cells in the sickle patient may contribute to the endothelial biology of this disease. Indeed, we specifically posit that this process results in long-range endothelial signaling. For example, the pulmonary arterial circulation is flooded with presickled red cells. The resultant "conditioned" blood would proceed to bathe the pulmonary venous endothelium and other critical downstream endothelium, such as in the Circle of Willis.

Reperfusion Injury

We have proposed that the predominant inflammation-inciting event in sickle cell disease consists of repeated cycles of ischemia and reperfusion (2). This "reperfusion injury" physiology is a well-known pathobiological paradigm, in which events that accompany resolution of an occlusion, during reperfusion, lead to the generation of reactive oxygen species and an intense inflammatory response (26). To test this idea in sickle transgenic mice, we examined animals both at ambient air and after transient exposure to hypoxia. Although this would comprise a modest hypoxic stress for normal animals, for sickle mice it increases the occurrence of sickling and, therefore, risk for vaso-occlusion and ischemia.

In our study examining biochemical footprints of oxidative biology (27), we found evidence for abnormal oxidant generation by sickle mice even at ambient air, and this was exaggerated by hypoxia–reoxygenation (Figure 148.4). In our study using intravital microscopy to evaluate leukocyte–EC interactions in the cremaster circulation in vivo (28), we similarly observed abnormal leukocyte adhesion and rolling and emigration, and this was exaggerated for sickle mice by hypoxia–reoxygenation (Figure 148.4). In a third study, we found that sickle mice abnormally express tissue factor on endothelium of pulmonary veins (see Figure 148.3), and that this too was greatly exaggerated in sickle mice after hypoxia–reoxygenation (19). We believe that these observations in sickle mice are relevant to human biology and suggest that reperfusion injury physiology is a part of the vascular pathobiology of sickle disease.

Vasoregulation and Nitric Oxide Biology

It appears that sickle disease is a state of diminished nitric oxide (NO) bioavailability, although females seem to be protected (29). In sickle disease, a deficiency of NO could result from several processes, including consumption by plasma hemoglobin (30), consumption from excessive superoxide generation due to reperfusion injury or uncoupling of endothelial NO synthase (eNOS) or activation of vessel wall xanthine oxidase, or myeloperoxidase activity from inflammatory cells. Additional roles of arginase or L-arginine availability may be relevant. There may be important regional differences (such as between resistance and conduit vessels) in vasoregulation. In general, data seem to indicate a state of vascular instability, with upregulation of both vasodilatory (possibly mediated by heme oxygenase and carbon monoxide [CO]) and vasoconstricting (possibly mediated by endothelin-1 or peroxides) processes (31). Intriguingly, an apolipoprotein A-1 mimetic has been reported to improve the vasodilatory capacity of arterial rings from sickle mice (32).

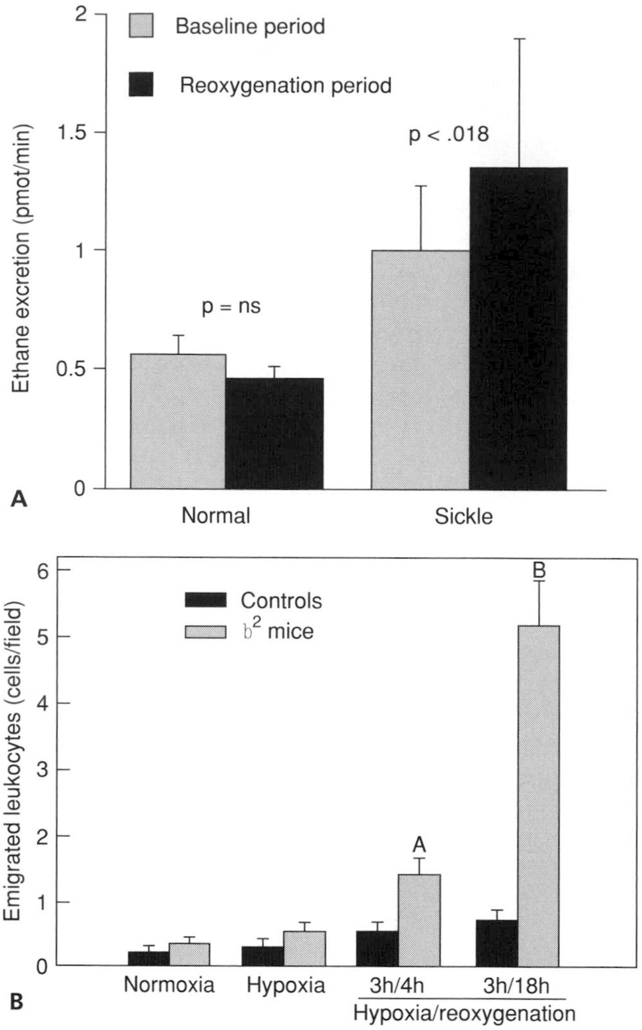

Figure 148.4. Reperfusion injury physiology in sickle transgenic mice. (**A**) Ethane, a marker of lipid peroxidation, in expired gas is at higher level in sickle animals at ambient air; this is exaggerated after exposure to hypoxia/reoxygenation, compared with normal animals. (Reproduced with permission from Osarogiagbon UR, Choong S, Belcher JD, et al. Reperfusion injury pathophysiology in sickle transgenic mice. *Blood.* 2006;96:314–320.) (**B**) Similarly, sickle mice show abnormal emigration of inflammatory cells, which is exaggerated after hypoxia/reoxygenation. (Reproduced with permission from Kaul DK, Hebbel RP. Hypoxia/reoxygenation causes inflammatory response in transgenic sickle mice but not in normal mice. *J Clin Invest.* 2000; 106:411–420.)

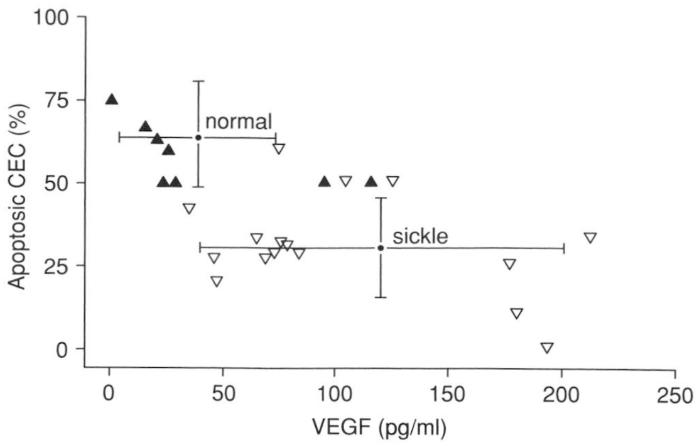

Figure 148.5. Circulating EC apoptosis correlates inversely with plasma VEGF level. (Reproduced with permission from Solovey A, Gui L, Ramakrishnan S, et al. Sickle cell anemia as a possible state of enhanced anti-apoptotic tone: survival effect of vascular endothelial growth factor on circulating and unanchored ECs. *Blood.* 1999;93:3824–3830.)

A deficiency of NO could have a significant impact on the endothelial biology of sickle disease. For example, NO inhibits endothelial expression of VCAM-1, interleukin (IL)-6, and tissue factor (33), and has multiple other salubrious actions.

Angiogenic Tone

Sickle disease seems to comprise a state of heightened angiogenic "tone." Plasma levels of vascular endothelial growth factor (VEGF), placental growth factor (PlGF), and erythropoietin are elevated (2); all are promotive of angiogenesis. Studies of CECs suggest that the balance of pro- versus antiangiogenic factors in sickle blood is tipped toward angiogenesis, because CECs in sickle blood exhibit much less apoptosis than do CECs in normals, and this correlates significantly with VEGF levels (34) (Figure 148.5).

Significantly, some clinical aspects of sickle disease are consistent with heightened angiogenic tone: (a) some children who develop occlusive disease at the Circle of Willis also develop neovascular formation (moyamoya) (35); (b) sickle cell disease entails risk for neovascularizing retinopathy, and the greater risk for this in hemoglobin SC disease versus sickle cell anemia (1) is consistent with our observations of VEGF (proangiogenic) versus thrombospondin (TSP) (antiangiogenic) levels in patient plasmas (VEGF is elevated in both, but thrombospondin is concurrently elevated only in sickle cell anemia subjects) (2); and (c) pulmonary edema is likely part of the acute chest syndrome. This syndrome can result from disparate proximate events that initiate a pathobiology in which the potentially deleterious effects of hypoxia are augmented by NO deficiency and inflammatory mediators (1,2,36). We suspect that the genesis of this syndrome is promoted by pulmonary vascular permeabilization that could derive from multiple factors, including liberation of permeabilizing fatty acids, thrombin, and elevated levels of VEGF (also known as *vascular permeability factor*).

Coagulation Activation

The clinical spectrum of sickle disease manifestations includes thrombosis. This plays a role in triggering acute stroke, as a result of underlying vessel wall changes at the Circle of Willis (35), and it accompanies pulmonary arterial wall disease (37). Underlying such complications is an abnormal stimulation of the coagulation system in sickle patients, which tends to be activated at all times, with greater disturbance evident in conjunction with acute vaso-occlusive crisis (38). Evidence for this includes excessive generation of thrombin, alteration of plasma clotting factor levels, activation of platelets and fibrinolysis, and consumption of antithrombotic substances.

This coagulation activation presumably is related to the inflammatory state of sickle disease, reflecting the fact that the EC resides at the interface of the inflammatory and coagulation systems. Such activation implies abnormal expression of tissue factor – the trigger of the clotting cascade. In sickle blood, this is abnormally expressed by monocytes (39) and CECs (16), the latter suggesting that there may be abnormal tissue factor expression by the vessel wall endothelium. Supporting observations in human and mouse were noted earlier. A portion of the microparticles in sickle blood that are derived from ECs are also tissue factor positive (20). Thus, we suspect that abnormal endothelial expression of tissue factor may occur in human sickle disease. A number of biological modifiers elevated in sickle blood (e.g., IL-1, tumor necrosis factor (TNF)-α, endotoxin, and activated monocytes) are able to trigger endothelial expression of tissue factor in vitro.

The coagulopathy story in sickle disease is not all tissue factor–based, because the abnormally externalized phosphatidylserine of the sickle red cell membrane plays a substantial role in accelerating the coagulation (40) after it has been triggered by tissue factor.

Chronic Vasculopathy

Sickle disease includes a component of chronic vasculopathy, with similar changes noted in multiple organs. Correspondingly, these organs show clinical disease: stroke, pulmonary hypertension, fetal growth retardation and mortality, autosplenectomy, priapism, and renal failure (1,2). Most thoroughly described is disease of large/medium vessels at the Circle of Willis, which tend to develop intimal hyperplasia, proliferative changes, and disruption of the internal elastic lamina (2,35). Small vessel disease also occurs in the sickle brain (35).

It is notable that this large-vessel vasculopathy shares a nonrandomness of location (albeit in a different distribution) and similar key pathological changes to atherosclerotic disease. However, the sickle lesions are not described as having a fatty streak (sickle patients have low blood lipids). Nonetheless, it is notable that inflammation is a key etiologic factor in both sickle disease (1,2) and atherosclerosis (41). We have speculated that the relevant process in sickle vessels is abnormal stimulation of ECs by inflammatory mediators, such as IL-1β, that causes the formation of the same oxidized lipid species implicated as proximate triggers of the early inflammatory process in atherosclerosis (2).

Importantly, it is not known if monocytes and inflammatory changes might be involved in the vasculopathy of sickle disease. In this regard, the current understanding of sickle vasculopathy is analogous to the level of understanding of atherosclerosis during the 1960s, before the exhaustive use of animal models revealed the primacy of inflammation in atherogenesis (41). In sickle disease, additional factors can contribute, such as chronic hypoxia, effects of adherent blood cells, stimulation by thrombin and platelet release products, antiapoptotic effects of elevated levels of VEGF, diminished NO bioavailability, and perpetual elevation of shear rates. Nonetheless, we believe that chronic inflammation, stemming from reperfusion injury and adverse effects of adherent red cells, is probably the key etiologic factor.

THERAPEUTIC IMPLICATIONS: FROM BENCH TO BEDSIDE

The laboratory studies of endothelial biology in sickle disease have underpinned our bench-to-bedside program of developing novel therapeutics for this disease.

One approach would be to use anti-inflammatory drugs of sufficient potency to impair endothelial transcriptional control of inflammatory surface molecules. A pilot study using sulfasalazine, a powerful (albeit nonspecific) NF-κB inhibitor, demonstrated the efficacy of this, by monitoring the phenotype of CECs as a surrogate marker for vessel wall endothelium (18). In addition, this drug is demonstrated to offer substantial ameliorative effects on leukocyte–endothelial interactions in the sickle mouse, thereby improving microvascular flow (42). Sulfasalazine, however, was ineffective as an inhibitor of tissue factor expression on CECs.

In studies of endothelial tissue factor expression in the sickle transgenic mouse in vivo, we have identified lovastatin as an effective inhibitor (19). Aside from their lipid-lowering effects, statins are anti-inflammatory, improve NO bioavailability, and have beneficial antiproliferative effects (43). Ongoing studies looking for additional effective tissue factor inhibitors in the live mouse have identified several additional small molecules for potential use (R. P. Hebbel, unpublished observations). Bedside pilot studies of some of these approaches are underway or planned.

Other strategies can be envisioned. The demonstrated involvement of reperfusion-injury physiology as an inflammation-inciting event in sickle disease raises the possibility that therapeutics aimed at this process might have benefit; possibilities would include allopurinol, antioxidants, and P-selectin blockers (27,28,42). Anticoagulation intervention might be considered, but this has not been adequately studied in the context of sickle cell disease. The role for red cell adhesion to endothelium in sickle pathobiology has stimulated interest in antiadhesive therapeutics. Given the multiplicity of adhesion mechanisms (2), we believe that to have efficacy, an antiadhesive agent will have to exhibit broad inhibitor activity. For example, nonspecifically antiadhesive "vascular lubricants" (44) are more likely to work than are agents that specifically interfere with a single adhesion mechanism.

PHENOTYPIC DIVERSITY AND ENDOTHELIAL GENOMICS

Subjects with sickle disease exhibit marked phenotypic diversity, most of which is unexplained (1). We expect that the application of new genomics methods to sickle disease will reveal that genetically determined, person-to-person differences in endothelial biology contribute substantially to this. Using the case of Circle of Willis disease, for example, we expect that participants could include aspects of endothelial biology that help regulate coagulation biology, shear-response genes, NO biology, and inflammation signaling (among others). Preliminary data consistent with this concept have been presented using global gene expression profiling of ECs (45), whole-genome single nucleotide polymorphism (SNP) analysis (46), and quantitative trait loci (QTL) identification (47).

KEY POINTS

- Sickle cell anemia is an "endothelial disease" with EC activation.
- All aspects of endothelial function are relevant.
- Clinical correlates are consistent with laboratory and murine data.
- The abnormal state of inflammation suggests novel targets for therapeutics.
- Therapies targeting endothelial transcriptional control may hold promise.

REFERENCES

1 Hebbel RP. Pathobiology of sickle cell disease. In: Hoffman R, Benz EJ, Shattil SJ, et al., eds. *Hematology, Basic Principles and Practice.* Philadelphia: Elsevier; 2005:591–604.

2 Hebbel RP, Osarogiagbon R, Kaul D. The endothelial biology of sickle cell disease: inflammation and a chronic vasculopathy. *Microcirculation.* 2004;11:129–151.

3 Herrick JB. Peculiar elongated and sickle-shaped red blood corpuscles in a severe case of anemia. *Arch Intern Med.* 1910;6:517–521.

4 Sherman IJ. The sickling phenomenon, with special reference tot the differentiation of sickle cell anemia from sickle cell trait. *Bull Johns Hopkins Hosp.* 1940;67:309–324.

5 Pauling L, Itano H, Singer SJ, Wells IC. Sickle cell anemia: a molecular disease. *Science.* 1949;110:543–548.

6 Embury SH. The not-so-simple process of sickle cell vasoocclusion. *Microcirculation.* 2004;11:101–113.

7 Hoover R, Rubin R, Wise G, Warren R. Adhesion of normal and sickle erythrocytes to endothelial monolayer cultures. *Blood.* 1979;54:872–876.

8 Hebbel RP, Yamada O, Moldow CF, et al. Abnormal adherence of sickle erythrocytes to cultured vascular endothelium. Possible mechanism for microvascular occlusion in sickle cell disease. *J Clin Invest.* 1980;65:154–160.

9 Kaul DK, Fabry ME, Constantini F, et al. In vivo demonstration of red cell-endothelial interaction, sickling and altered microvascular response to oxygen in the sickle transgenic mouse. *J Clin Invest.* 1995;96:2845–2853.

10 Fabry ME, Fine E, Rajanayagam V, et al. Demonstration of endothelial adhesion of sickle cells in vivo: a distinct role for deformable sickle cell discocytes. *Blood.* 1992;79:1602–1611.

11 Kaul DK, Fabry ME, Nagel RL. Microvascular sites and characteristics of sickle cell adhesion to vascular endothelium in shear flow conditions: pathophysiological implications. *Proc Natl Acad Sci USA.* 1989;86:3356–3360.

12 Ferrone FA. Polymerization and sickle cell disease: a molecular view. *Microcirculation.* 2004;11:115–128.

13 Turhan A, Weiss LA, Mohandas N, et al. Primary role for adherent leukocytes in sickle cell vascular occlusion: a new paradigm. *Proc Natl Acad Sci USA.* 2002;99:3047–3051.

14 Hebbel RP. Adhesive interactions of sickle erythrocytes with endothelium. *J Clin Invest.* 1997;99:2561–2564.

15 Solovey A, Lin Y, Browne P, et al. Circulating activated endothelial cells in sickle cell anemia. *N Engl J Med.* 1997;337:1584–1590.

16 Solovey A, Gui L, Keys NS, Hebbel RP. Tissue factor expression by endothelial cells in sickle cell anemia. *J Clin Invest.* 1998;101:1899–1904.

17 Nath KA, Grande JP, Haggard JJ, et al. Oxidative stress and induction of heme oxygenase-1 in the kidney in sickle cell disease. *Am J Pathol.* 2001;158:893–903.

18 Solovey AA, Solovey AN, Harkness J, Hebbel RP. Modulation of endothelial cell activation in sickle cell disease: a pilot study. *Blood.* 2001;97:1937–1941.

19 Solovey A, Kollander R, Shet A, et al. Endothelial cell expression of tissue factor in sickle mice is augmented by hypoxia/reoxygenation and inhibited by lovastatin. *Blood.* 2004;104:840–846.

20 Shet A, Aras O, Gupta K, et al. Sickle blood contains tissue factor positive microparticles derived from endothelial cells and monocytes. *Blood.* 2003;102:2678–2683.

21 Platt OS, Brambilla DJ, Rosse WF, et al. Mortality in sickle cell disease: life expectancy and risk factors for early death. *N Engl J Med.* 1994;330:1639–1644.

22 Miller ST, Sleeper LA, Pegelow CH, et al. Prediction of adverse outcomes in children with sickle cell disease. *N Engl J Med.* 2000;342:1612–1613.

23 Charache S, Barton FB, Moore RD, et al. Anemia. Hydroxyurea and sickle cell anemia. Clinical utility of a myelosuppressive "switching" agent. *Medicine.* 1996;75:300–326.

24 Setty BN, Stuart MJ, Dampier C, et al. Hypoxemia in sickle cell disease: biomarker modulation and relevance to pathophysiology. *Lancet.* 2003;362(9394):1450–1457.

25 Phelan M, Perrine SP, Brauer M, Faller DV. Sickle erythrocytes, after sickling, regulate the expression of the endothelin-1 gene in human endothelial cells in culture. *J Clin Invest.* 1995;96:1145–1151.

26 Granger DN, Korthuis RJ. Physiologic mechanisms of postischemic tissue injury. *Annu Rev Physiol.* 1998;1995:311–332.

27 Osarogiagbon UR, Choong S, Belcher JD, et al. Reperfusion injury pathophysiology in sickle transgenic mice. *Blood.* 2000;96:314–320.

28 Kaul DK, Hebbel RP. Hypoxia/reoxygenation causes inflammatory response in transgenic sickle mice but not in normal mice. *J Clin Invest.* 2000;106:411–420.

29 Gladwin MT, Schechter AN, Ognibene FP, et al. Divergent nitric oxide bioavailability in men and women with sickle cell disease. *Circulation.* 2003;107:271–278.
30 Reiter CD, Wang X, Tanus-Santos JE, et al. Cell-free hemoglobin limits nitric oxide bioavailability in sickle-cell disease. *Nat Med.* 2002;8:1383–1389.
31 Nath KA, Katusic ZS, Gladwin MT. The perfusion paradox and vascular instability in sickle cell disease. *Microcirculation.* 2004; 11:179–193.
32 Ou J, Ou Z, Jones DW, et al. L-4F, an apolipoprotein A-1 mimetic, dramatically improves vasodilation in hypercholesterolemia and sickle cell disease. *Circulation.* 2003;107:2337–2341.
33 Yang Y, Loscalzo J. Regulation of tissue factor expression in human microvascular endothelial cells by nitric oxide. *Circulation.* 2000;101:2144–2148.
34 Solovey A, Gui L, Ramakrishnan S, et al. Sickle cell anemia as a possible state of enhanced anti-apoptotic tone: survival effect of vascular endothelial growth factor on circulating and unanchored endothelial cells. *Blood.* 1999;93:3824–3830.
35 Prengler M, Pavlakis SG, Prohovnik I, Adams RJ. Sickle cell disease: the neurological complications. *Ann Neurol.* 2002;51: 543–552.
36 Stuart MJ, Setty BN. Sickle cell acute chest syndrome: pathogenesis and rationale for treatment. *Blood.* 1999;94:1555–1560.
37 Haque AK, Gokhale S, Rampy BA, et al. Pulmonary hypertension in sickle cell hemoglobinopathy: a clinicopathologic study of 20 cases. *Hum Pathol.* 2002;33:1037–1043.
38 Francis RB, Hebbel RP. Hemostasis. In: Embury SH, Hebbel RP, Mohandas N, Steinberg MH, eds. *Sickle Cell Disease: Basic Principles and Clinical Practice.* New York: Raven Press; 1994: 299–310.
39 Key NS, Slungaard A, Dandelet L, et al. Whole blood tissue factor procoagulant activity is elevated in patients with sickle cell disease. *Blood.* 1998;91:4216–4223.
40 Setty BN, Rao AK, Stuart MJ. Thrombophilia in sickle cell disease and the red cell connection. *Blood.* 2001;98:3228–3233.
41 Libby P. Inflammation in atherosclerosis. *Nature.* 2002;420:868–874.
42 Kaul DK, Liu X-D, Choong S, et al. Anti-inflammatory therapy ameliorates leukocyte adhesion and microvascular flow abnormalities in transgenic sickle mice. *Am J Physiol.* 2004;287:H293–H301.
43 Undas A, Brozek J, Musial J. Anti-inflammatory and antithrombotic effects of statins in the management of coronary artery disease. *Clin Lab.* 2002;48:287–296.
44 Smith CM, Hebbel RP, Tukey, DP, et al. Pluronic F-68 reduces the endothelial adherence and improves the rheology of liganded sickle erythrocytes. *Blood.* 1987;69:1631–1636.
45 Hebbel RP, Jiang A, Hillery C, et al. Genetic influence on the systems biology of sickle stroke risk detected by endothelial gene expression. American Society of Hematology Meeting. 2005 [Abstract].
46 Sebastiani P, Ramoni MR, Nolan V, et al. Multigenic dissection and prognostic modeling of overt stroke in sickle cell anemia. *Blood.* 2004;104:A460. [Abstract]
47 Peters LL, Zhang W, Lambert AJ, et al. Quantitative trait loci for baseline white blood count, platelet count, and mean platelet volume. *Mamm Genome.* 2005;16(10):749–763. Epub 2005 Oct 29.

149

The Role of Endothelial Cells in the Antiphospholipid Syndrome

Jacob H. Rand and Xiao-Xuan Wu

Montefiore Medical Center, Bronx, New York

The antiphospholipid antibody syndrome (APS) is an enigmatic autoimmune disorder in which vascular thrombosis and/or recurrent spontaneous pregnancy losses occur in patients who have positive blood tests for antibodies that recognize phospholipid-binding proteins, predominantly β_2-glycoprotein I (β2GPI). The clinical tests that are currently available to detect the presence of APS antibodies are surrogate assays that were empirically derived from the biologic false-positive syphilis test and from unusual coagulation assays. The latter, paradoxically, report the *inhibition* of phospholipid-dependent coagulation reactions (referred to as the lupus anticoagulant [LA] phenomenon [1]). The clinical manifestations of the syndrome include venous and arterial thrombosis and embolism, disseminated large and small vessel thrombosis with accompanying multiorgan ischemia and infarction, stroke, premature coronary artery disease, and recurrent spontaneous pregnancy losses. Although the mechanism(s) of the syndrome have not yet been definitively established, a growing body of evidence supports a role for vascular endothelium in the disease process.

Formal criteria – known as the *Sapporo Investigational Criteria* – have been developed to define the diagnosis of APS (2). These criteria have been updated recently as the Sydney Investigational Criteria (3). These include a clinical history of one or more episodes of vascular thrombosis (involving any site) or history for pregnancy morbidity, together with laboratory evidence for a positive LA test or a medium- or high-titer anticardiolipin (aCL) immunoglobulin (Ig)G and/or IgM antibody measured by a standard assay for β2GPI-dependent aCL antibodies. The laboratory abnormalities must be persistent – that is, present on two or more occasions, at least 6 weeks apart.

In retrospect, the first serologic evidence for the existence of this disorder had been the description of a "biological false-positive" serological test for syphilis (BFP-syphilis test) in 1952 (4). This laboratory anomaly was associated with systemic lupus erythematosus (SLE) (5) and with an anticoagulant phenomenon (6). Remarkably, it was learned that this anticoagulant activity was not associated with bleeding manifestations in vivo, but rather with thrombosis, embolism (7), and recurrent pregnancy losses (8). A major step toward the recognition of this syndrome was the development, in 1983, of a quantitative test that detected antibodies against the anionic phospholipid, cardiolipin (diphosphatidylglycerol), and the major antigen in the syphilis test reagent (9).

This syndrome was proposed to be a distinct entity, the *anticardiolipin syndrome*, in 1985 (10) and was later renamed the *antiphospholipid antibody syndrome* (11). Additional terms used for the disorder are *Hughes syndrome*, the *autoimmune thrombosis syndrome*, and *the syndrome of the black swan*. APS is considered "primary" in the absence of SLE, and "secondary" in its presence.

ETIOLOGY

The trigger that incites autoimmunity in this disorder has not yet been established. Because cells undergoing apoptosis (programmed cell death) expose phosphatidylserine on their cytoplasmic membranes (12), it has been suggested that this membrane alteration may trigger an autoimmune response in susceptible patients (13). Consistent with this idea, immunization of mice with apoptotic cells coated with β2GPI can result in the generation of autoantibodies against β2GPI, the major antigen recognized by APS antibodies (13). It also has been proposed that these autoantibodies arise in response to microbial infections, particularly cytomegalovirus (CMV), because β2GPI shares sequence similarities with microbial pathogens (14,15). There have also been case reports of children who developed fulminant APS with thrombosis following varicella infections (16).

Antibodies against anionic phospholipids arise in response to infections such as syphilis and Lyme disease; however, these differ from those present in APS in that the former generally recognize phospholipid epitopes directly, whereas those

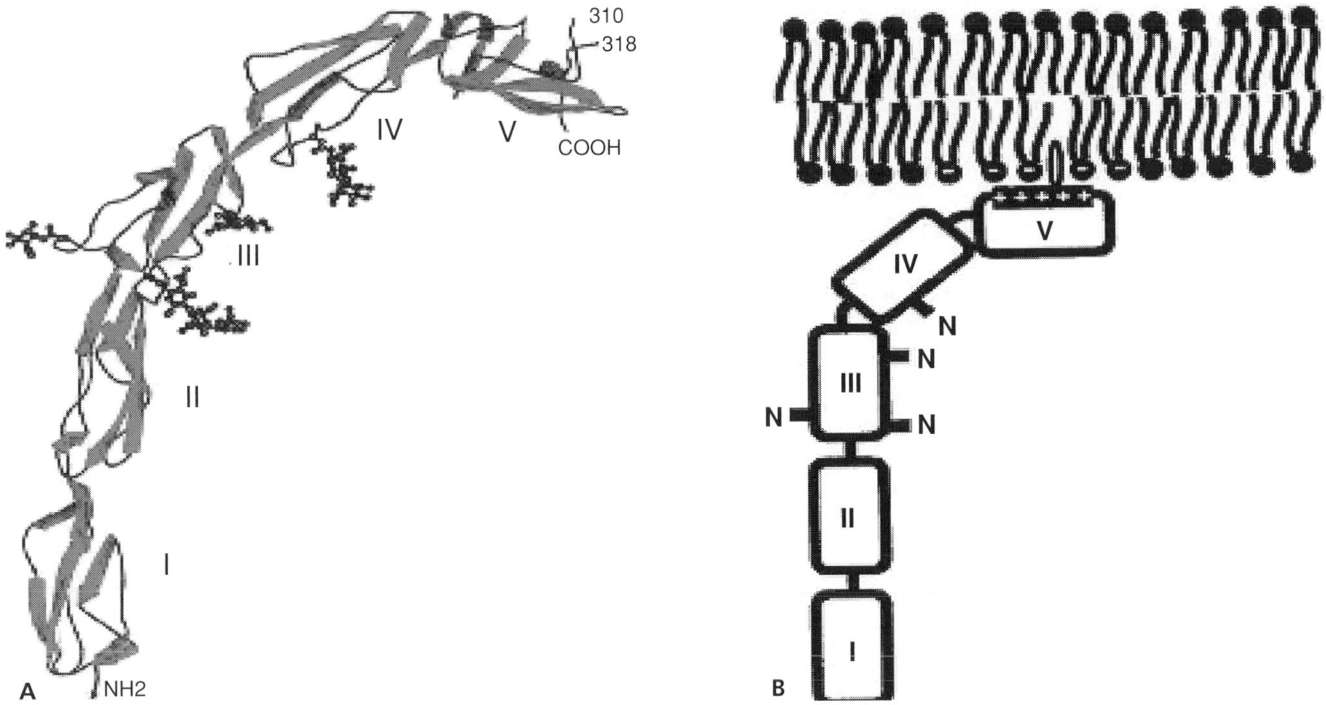

Figure 149.1. Structure of human β2GPI. (**A**) Ribbon model of β2GPI based on crystal structure: The protein is composed of an extended chain of five SCR domains having a "fish-hook" appearance. The structure of SCR domain V deviates from the standard fold of the four other domains and forms the putative phospholipid-binding site. β-strands are shown in light gray and helices in dark gray. (**B**) The structural data suggest a simple membrane-binding mechanism in which the cationic patch of domain V has an affinity for anionic phospholipid. The stretch of Ser 311 to Lys317 forms a hydrophobic loop that inserts into the lipid bilayer and positions Trp316 at the interface region between the acyl chains and the phosphate head groups of the lipids, thereby anchoring the β2GPI in the membrane. Current data support the hypothesis that APS antibodies reactive against β2GPI mainly recognize epitopes on domains I and II, and that antibody-mediated dimerization of β2GPI markedly increases the affinity of β2GPI for phospholipid. (Reproduced with permission from Bouma B, et al. *EMBO J* 1999;18:5166–5174.) For color reproduction, see Color Plate 149.1A.

generated in APS appear to mainly recognize epitopes on phospholipid-binding protein cofactors – primarily β2GPI. However, exceptions exist to this generalization, because anti-β2GPI antibodies have been detected in some patients with syphilis, leptospirosis, leishmaniasis (17), and Hansen disease (18). The antiphospholipid (aPL) antibodies generated in response to infections are not associated with thrombosis, embolism, or recurrent pregnancy losses. Although it is clear that patients with the syndrome can have IgG, IgM, and IgA antibodies, the significance of isolated IgA aPL antibodies is unclear – anticardiolipin IgA antibodies appear to be associated with clinical manifestations, whereas anti-β2GPI IgA antibodies do not (19). There appears to be a genetic component to APS, because familial clustering of raised APS antibody levels (20) and human leukocyte antigen (HLA) linkages (21–25) have been reported. aPL antibodies can arise in response to drugs – chlorpromazine is the major culprit. Although there have been case reports associating drug-induced aPL antibodies with thrombosis (26), a statistically significant association has not to our knowledge been demonstrated.

ANTIGENIC SPECIFICITIES

β2-Glycoprotein I

The major antigenic target that is recognized by APS antibodies is β2GPI – also referred to as *apolipoprotein H*. β2GPI is a highly glycosylated single-chain plasma protein composed of 326 amino acids with a molecular weight of 50 kDa (27) (Figure 149.1; for color reproduction, see Color Plate 149.1A). The protein is synthesized in the liver and enterocytes (28), and also in endothelial cells (ECs), astrocytes, neurons, and lymphocytes (29). β2GPI appears to behave as an acute-phase reactant (28). β2GPI is a member of the complement control protein or short consensus repeat (SCR) superfamily, and it has five SCR domains (also known as "sushi domains"), each with a set of 16 conserved residues and two fully conserved disulfide bonds. The crystal structure of the protein (30,31) indicates that it is shaped like a fishhook, with the end of the hook located at the carboxy-terminus. The structure suggests that the protein inserts into phospholipid membranes via its fifth SCR domain (Figure 149.1), via a "barb" that includes a positively charged cluster at amino acids 282 through 287 (32) and hydrophobic sequences at 313 through 316 (33), which are present near the carboxy-terminus. Binding of β2GPI to phospholipid membrane may require annexin A2, and is increased by APS antibodies.

Proposed Functions of β2GPI

The biological functions of the serum protein β2GPI are still unknown. Although human β2GPI deficiency was first described over 30 years ago (34), it is not associated with any known disease or thrombotic tendency. Heterozygosity for β2GPI deficiency was found in about 6% of 812 Japanese subjects and was not associated with thrombosis (35). One patient with thrombosis and homozygous β2GPI deficiency was described; however, his homozygous deficient brother and several heterozygous family members were asymptomatic (36).

β2GPI may play a role in the clearance of apoptotic cells by binding to them and thereby providing an "eat me" signal to macrophages (37). However, no evidence for manifestations of deficient apoptotic clearance has yet been described in humans or in animal models. *β2GPI*-null mice appear anatomically and histologically normal (38). Impairment of thrombin generation has been detected in their plasmas in vitro. A role for the protein in the reproductive process was suggested by the finding that less than the expected percentage of offspring of the *β2GPI* null heterozygotes were homozygous for the disrupted alleles (38).

β2GPI has multiple effects on coagulation, both anticoagulant and procoagulant (39). It exerts anticoagulant activity by inhibiting phospholipid-dependent coagulation reactions such as prothrombinase, tenase, and factor XII in vitro. β2GPI inhibits factor Xa on activated platelets – an effect that is counteracted by APS antibodies (40) – and also binds factor XI and inhibits its activation (39). The anticoagulant effects of the protein are abolished when it is "nicked" by plasmin at Lys317–Thr318 (41), an effect that has a has a negative-feedback effect on fibrinolysis, because the nicked form of β2GPI binds and inactivates plasminogen (39). The procoagulant properties of the protein promotes the inhibition of activated protein C (APC) (39). β2GPI can also increase coagulation by inhibiting the protein Z–dependent protease inhibitor (PZI), an inhibitor of factor Xa (42); reduced inhibition of factor Xa could lead to increased thrombin generation and greater susceptibility to thrombosis.

Remarkably, aPL antibodies appear to induce a gain-of-function abnormality in β2GPI by binding to nonfunctional epitopes on the molecule and thereby dimerizing it, increasing its affinity (or more formally, its *avidity*) for phospholipids bilayers. Most APS antibodies appear to recognize domain I of β2GPI (43,44), and APS antibody binding to domains I and II has been shown to increase the binding of the protein to membrane phospholipid (30). It recently has been demonstrated that IgG antibodies from APS patients that recognize a specific epitope consisting of Gly40-Arg43 are at increased risk for thrombosis (45). The increased binding of IgG–β2GPI complexes has been demonstrated to result from the increased affinity of the divalent complexes for the phospholipids bilayer (46). Consistent with this concept, the binding of divalent β2GPI, as opposed to monovalent β2GPI, also is required for endothelial activation (47).

β2GPI in Endothelial Cells

There appear to be three binding sites for β2GPI on the surface of ECs – the phospholipid bilayer itself, annexin A2, and heparan sulfate (HS). β2GPI has been shown to bind to the apical membranes of ECs via annexin A2 (47), which also serves as a receptor for plasminogen and tissue-type plasminogen activator (t-PA) (48). It has been proposed that β2GPI binding to the endothelial surface may also occur via attachment to HS, because heparitinase reduces β2GPI binding (49).

Antiphospholipid Antibody Syndrome Cofactors Other Than β2GPI

Additional cofactors and antigenic targets of APS antibodies include prothrombin (coagulation factor II), coagulation factor V, protein C, protein S, annexin A5, high-molecular-weight kininogen, and low-molecular-weight kininogen. Interestingly, protein C can be a target of aCL in the presence of cardiolipin and β2GPI, leading to protein C dysfunction (50). Also, antibodies of some APS patients have been found to recognize heparin and inhibit the formation of antithrombin III–thrombin complexes (51).

The remarkable heterogeneity of aPL antibodies, even within a single patient (52), adds to the difficulties in defining the pathogenic antiphospholipid response(s). Oxidation of phospholipids may be necessary for APS antibody recognition (53). The epitopes for some APS antibodies appear to be adducts of oxidized phospholipid and protein such as β2GPI (54). Some affinity-purified cardiolipin-binding antibodies in sera from patients with SLE appear to cross-react with oxidized low-density lipoprotein (LDL) (55). Elevated levels of these antibodies have been proposed to be markers for arterial thrombosis (56).

PATHOGENIC MECHANISMS INVOLVING ENDOTHELIUM

Overview

APS antibodies have been shown to have several effects on vascular endothelium that convert the endothelial–blood interface from a phenotype that promotes blood fluidity to one that is procoagulant and proadhesive (Table 149-1). The procoagulant phenotype is believed to result from a remarkable array of effects including the increased expression of tissue factor and anionic phospholipids on the apical cell membrane, the reduction of endogenous anticoagulant mechanisms such as annexin A5 shielding and the thrombomodulin/protein C/protein S system, and increased expression of cell surface adhesion molecules that result from the antibody-mediated triggering of signaling pathways.

Table 149-1: Endothelial Mechanisms Proposed in APS

Procoagulant Effects of APS Antibodies
Induction of tissue factor expression by ECs
APS antibody-mediated disruption of the annexin A5 shield
Interference with the protein C pathway
Inhibition of thrombin formation
Decreased activation of protein C by thrombomodulin-thrombin
Inhibition of assembly of the protein C complex
Inhibition of protein C activity
Protecting factors Va and VIIIa from proteolysis by APC
Antibodies against EPCR
Protein S deficiency
Inhibition of antithrombin-III activity via cross-reactivity with heparins
Inhibition of fibrinolysis via:
Elevation of PAI-1
Anti–β2GPI-mediated inhibition of the autoactivation of factor XII
Proadhesive Effects of APS Antibodies
Induction of receptors for cell adhesion molecules on endothelium
Vascular injury
Apoptosis
Microparticle formationv
Increased endothelin-1
Altered balance of eicosanoid synthesis

Procoagulant Effects of Antiphospholipid Antibody Syndrome Antibodies

Induction of Tissue Factor Expression by Endothelial Cells

APS antibodies have been shown to increase the expression of tissue factor by cultured ECs (57,58). The upregulation of tissue factor expression in cultured ECs was inhibited by fluvastatin (59). Anti-β2GPI antibodies increased the expression of tissue factor in cultured ECs, a response that was associated with the redistribution of nuclear factor (NF)-κB. However, the time course of activation suggested that NF-κB is an intermediate in the process of EC activation and that its response to anti-β2GPI antibodies is indirect (60). APS antibodies also increase tissue factor expression in monocytes (61).

Antiphospholipid Antibody Syndrome Antibody–Mediated Disruption of the Annexin A5 Shield

Annexin A5 has potent anticoagulant activity in vitro, which is based on its high affinity for anionic phospholipids and its capacity to displace coagulation factors from phospholipid surfaces (62). The protein has the shape of a concave disk, with the phospholipid and calcium binding domains present on the convex surface. Annexin A5 clusters on exposed membrane phospholipid (62) and forms two-dimensional crystalline arrays (Figure 149.2) (63,64). The potent anticoagulant properties of the protein are a consequence of this crystallization of annexin A5, which shields the bilayer from availability for coagulation reactions. The affinity of annexin A5 for heparin oligosaccharides suggests that cell surface HS also may play a role in the binding of annexin A5 to the bilayer (65).

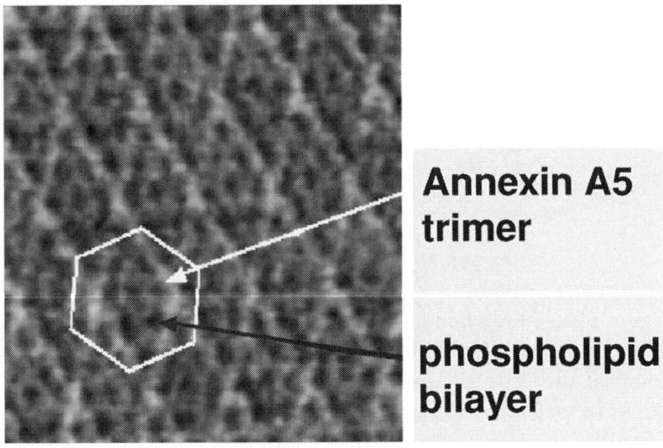

Figure 149.2. Crystal structure of annexin A5. A high-resolution atomic force image of annexin A5 that has crystallized over a phospholipid bilayer that is composed of 30% phosphatidylserine and 70% phosphatidylcholine. The highly ordered two-dimensional crystal structure is apparent. The crystals assemble via initial trimerization of annexin A5, followed by the progressive trimerization of the trimers on the bilayer.

Significant data support the concept that annexin A5 has a thrombomodulatory function in the circulation. Annexin A5 is highly expressed by vascular ECs (66). The removal of annexin A5 from the EC surface by treatment with ethylene glycol tetraacetic acid (EGTA) exposes the apical membrane to circulating blood and accelerates the coagulation of plasma exposed to the cells (67). Plasma from SLE patients decreased the binding of annexin A5 to ECs, and this was associated with atherothrombosis (68). Infusion of pregnant mice with polyclonal anti–annexin A5 antibodies resulted in placental infarction and pregnancy wastage, indicating that annexin A5 is necessary for maintenance of placental integrity (69). IgG fractions from APS patients reduced the quantity of annexin A5 on cultured ECs and accelerated the coagulation of plasma exposed to these cells (67). Antibodies against annexin A5 have been observed in the majority of patients with Takayasu arteritis (70).

The APS antibody-mediated reduction of annexin A5 occurs on noncellular phospholipid surfaces via displacement by APS antibodies in the presence of β2GPI (71), and results in acceleration of phospholipid-dependent coagulation reactions (71,72). This procoagulant property of the antibodies offers a plausible explanation for the LA paradox – that antibodies that can inhibit phospholipids-dependent coagulation reactions are associated with thrombotic, rather than bleeding, manifestations (Figure 149.3; for color reproduction, see Color Plate 149.3). This model was validated through experiments

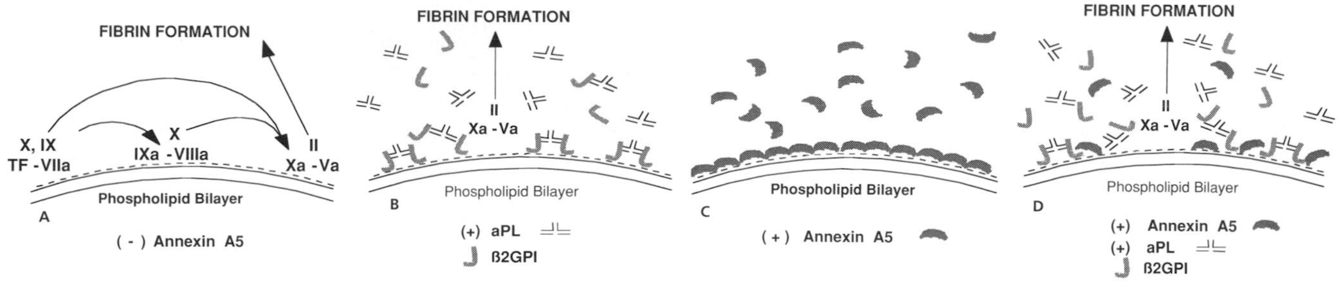

Figure 149.3. Proposed mechanisms for the LA and the APS antibody procoagulant effect. (**A**) Anionic phospholipids, predominantly phosphatidylserine (negative charges), serve as potent cofactors for the assembly of three different coagulation complexes: the tissue factor–VIIa complex, the IXa–VIIIa complex, and the Xa–Va complex, and thereby accelerate blood coagulation. The tissue factor complexes yield factors IXa or factor Xa, the IXa complex yields factor Xa, and the Xa formed from both of these reactions is the active enzyme in the prothrombinase complex that yields factor IIa (thrombin), which in turn cleaves fibrinogen to form fibrin. (**B**) LA effect. aPL antibody–β2GPI complexes can prolong coagulation times, compared with control antibodies, when limiting quantities of anionic phospholipid are available. This occurs via antibody recognition of domains I or II on the β2GPI, which results in dimers and pentamers of antibody–β2GPI complexes having high affinity for phospholipid via domain V. These high-affinity complexes reduce the access of coagulation factors to anionic phospholipids, thereby resulting in a LA effect in those conditions in which the antibody-cofactor has a sufficiently high affinity. (**C**) Annexin A5, in the absence of aPL antibodies, serves as a potent anticoagulant by crystallizing over the anionic phospholipid surface, shielding it from availability to bind coagulation proteins. (**D**) aPL antibody-mediated disruption of annexin A5 shield. aPL-β2GPI complexes with high affinity for phospholipid membranes disrupt the ability of annexin A5 to form ordered crystals on the phospholipid surface. This results in a net increase of the amount of anionic phospholipid available for promoting coagulation reactions. For color reproduction, see Color Plate 149.3.

using atomic force microscopic imaging (Figure 149.4) (73), and these findings were recently translated into clinical type assays with the demonstration that antibodies from APS patients induce resistance to annexin A5 anticoagulant activity (Figure 149.5) (74). APS antibody-mediated interference with the binding of annexin A5 also has been demonstrated using flow cytometry (75).

The available data support the hypothesis that annexin A5 has a thrombomodulatory role in vivo and that interference with its crystallization on the surfaces of vascular ECs may provide a thrombogenic mechanism for APS. This potent anticoagulant protein may play a thromboregulatory role at the vascular–blood interface by shielding anionic phospholipids, which would otherwise serve as efficient cofactors for the assembly of coagulation factor complexes, from participating in coagulation reactions.

Interference with Protein C Pathway

The protein C pathway, one of the major endogenous antithrombotic mechanisms, is initiated when thrombin binds to thrombomodulin on ECs (76). This binding modifies the substrate specificity of thrombin; the enzyme loses its procoagulant specificities and cleaves protein C – concentrated near the cell surface by the endothelial protein C receptor (EPCR) – to APC. APC complexes with the free form of protein S and proteolyses coagulation factors Va and VIIIa. APS antibodies can interfere with the protein C system by: (a) inhibiting the formation of thrombin; (b) decreasing the activation of protein C by the thrombomodulin-thrombin complex; (c) inhibiting the assembly of the protein C complex; (d) inhibiting the activity of protein C, directly or via its cofactor protein S; and (e) binding to factors Va and VIIIa in a manner that protects

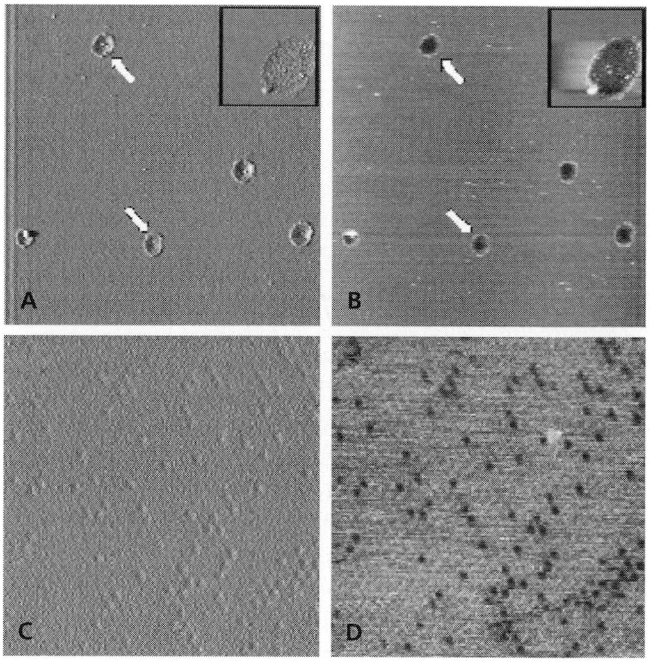

Figure 149.4. Atomic force imaging of effect of APS antibodies on annexin A5 crystallization. When aPL mAb and (2GPI were added to the annexin A5 crystal lattice formed on the bilayer, circular pits appeared (*arrows* in **A** and **B**), indicating disruptions in the crystal lattice. A representative pit is shown at higher magnification in the insets. Moreover, at higher resolution (**C, D**) more vacancy defects (small round dark holes) in the crystalline lattice are apparent. Amplitude images (**A, C**) processed with a $\times 1$ convolution, and height images (**B, D**) processed by "zero-order flatten." Magnifications 10 μm $\times$ 10 μm scan (**A, B**); 500 nm $\times$ 500 nm (**C, D**). (Reproduced with permission from Rand JH, et al. *Am J Pathol.* 2003;163:1193–1200.)

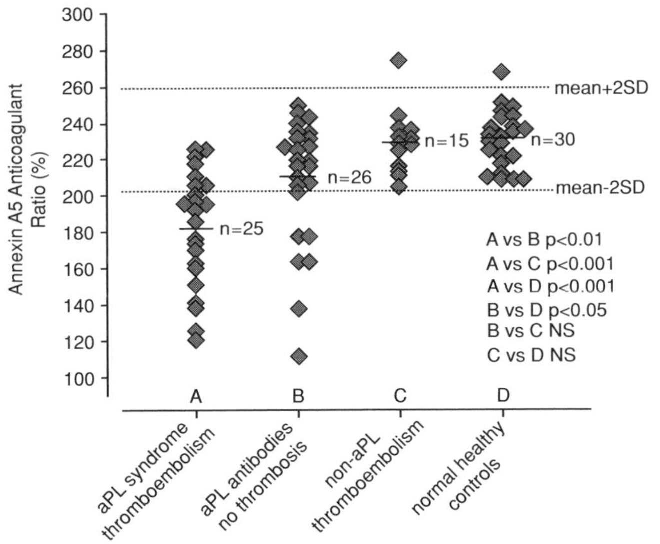

Figure 149.5. Resistance to annexin A5 anticoagulant activity in APS. Annexin A5 anticoagulant ratios were performed as described (74) and defined as the coagulation time using normal pooled plasma of phospholipid that was exposed to patient plasma in the presence of annexin A5, divided by the coagulation time in the absence of annexin A5 × 100%. The annexin A5 anticoagulant ratios for the antiphospholipid (aPL) syndrome with thromboembolism group (**A**) was significantly decreased (mean ± SD: 182 ± 31%, n = (25) as compared to the aPL antibodies without thrombosis history group (**B**) (210 ± 35%, n = (26, $p < 0.01$), the non-aPL thromboembolism group (**C**) (229 ± 16%, n = (15, $p < 0.001$), and the normal healthy control group (**D**) (231 ± 14%, n = (30, $p < 0.001$). The ratio for the plasmas from the aPL antibodies without thrombosis history group (**B**) also was reduced significantly compared with the normal healthy control group (**D**) ($p < 0.05$). There were no significant differences in annexin A5 anticoagulant ratio for the non-aPL thromboembolism group (**C**) compared with the aPL antibodies without thrombosis history group (**B**) and the normal healthy control group (**D**). Error bars are shown for mean + 2SD of normal healthy controls (**D**). (Reproduced with permission from Rand JH, et al. *Blood.* 2004;104:2783–2790.)

them from proteolysis by APC (77). Interestingly, oxidation of phosphatidylethanolamine may enhance the anticoagulant activity of APC; inhibition of this process by APS antibodies may thereby promote thrombin generation (78). It recently has been shown that autoantibodies against EPCR can be detected in APS patients, and that these are an independent risk factor for fetal death (79); however, a correlation with vascular thrombosis has not yet been demonstrated. Patients with APS also frequently have protein S deficiency (80,81).

Additional Procoagulant Effects

Antibodies from patients with APS can inhibit tissue factor pathway inhibitor (TFPI) activity, thereby further increasing tissue factor activity and consequently factor Xa generation (82,83). The recent finding that low TFPI levels are a risk factor for deep vein thrombosis (84) supports the idea that antibody-mediated reduction of this protein may be clinically significant.

It has been demonstrated that at least some APS antibodies cross-react with heparin and heparinoid molecules (which are highly polyanionic) and inhibit the acceleration of antithrombin III activity (51). As mentioned earlier, the nicked form of β2GPI inhibits fibrinolysis, which may be even further impaired in APS via the elevation of plasminogen activator inhibitor (PAI)-1 levels (80). The reason for elevated PAI-1 in APS is unclear, and may be related to the wider prevalence of the 4G/4G polymorphism (85). Additional impairment of fibrinolysis may occur via anti-β2GPI–mediated inhibition of the autoactivation of factor XII (86), and the ensuing reductions of kallikrein and urokinase.

Proadhesive Effects of Antiphospholipid Antibody Syndrome Antibodies

The binding of APS antibodies to β2GPI on the EC surface through annexin A2 is associated with the activation of ECs (47). As mentioned earlier, this activation requires the antibody-mediated dimerization of β2GPI, because monomeric Fab1 anti-β2GPI fragments inhibit this effect (47). Annexin A2 has other roles, including its serving as a platform on vascular endothelium for promoting fibrinolysis, by binding to both t-PA and plasminogen (87).

The exposure of cultured ECs to APS antibodies promoted monocyte adhesion to the cells (88), an effect mediated by β2GPI (89). The proadhesive effects of APS antibodies have been shown to be mediated by intercellular cell adhesion molecule (ICAM)-1, vascular cell adhesion molecule (VCAM)-1, and P-selectin (90). It has been demonstrated that both the APS antibody-mediated enhancement of leukocyte adhesion and increased thrombosis in the pinch injury model are reduced in transgenic ICAM-1–deficient mice, ICAM-1/P-selectin–deficient mice, and in mice infused with anti-VCAM–1 antibodies (91). In a study of VCAM-1 expression in cultured human umbilical vein ECs (HUVECs), IgG fractions from eight of 14 APS patients increased VCAM-1 expression, compared with controls; aspirin abrogated these increases (92). Treatment with APS antibodies also increased the surface expression of E-selectin on HUVECs over 400-fold and induced activation of NF-κB (93). A role for E-selectin was confirmed in experiments in which APS antibodies significantly increased leukocyte adhesion to ECs and increased thrombus size in C57BL/6 J mice, but not in E-selectin–deficient mice (93). The incubation of cultured ECs with APS antibodies also stimulated chemokine synthesis and secretion (94).

The signaling pathways for the increased expression of adhesion molecules have been explored. Activation of ECs by anti-β2GPI occurs via NF-κB activation (60). In one study, incubation of ECs with anti-β2GPI antibodies resulted in a redistribution of NF-κB from the cytoplasm to the nucleus after several hours (60), which contrasted with the more rapid redistribution that took place after tumor necrosis factor (TNF) incubation, suggesting that, although NF-κB is a critical intermediate in the process of EC activation, the response to anti-β2GPI antibodies is indirect (60). Upstream signaling events were investigated using human microvascular ECs in studies that demonstrated that the signaling cascade in response to human anti-β2GPI IgM mAbs and polyclonal affinity-purified anti-β2GPI IgG was comparable to that activated by lipopolysaccharide (LPS) or interleukin (IL)-1. Because anti-β2GPI antibodies and LPS followed the same time kinetics as IL-1 receptor-activated kinase (IRAK) phosphorylation, it was suggested that the Toll-like receptor (TLR) family might be involved in this signaling (95). It has been hypothesized that cross-linking may occur between microbe-like epitopes on β2GPI bound to annexin A2 and TLR4, and that this cross-linking may trigger signaling (96).

Proapoptotic Effects of Antiphospholipid Antibody Syndrome Antibodies

APS antibodies have been reported to recognize and injure cultured vascular ECs (97–99). A subset of APS antibodies that recognize annexin A5 induce apoptosis in ECs (100). It also has been shown that procoagulant endothelial microparticles are generated in APS patients (101).

Other Effects of Antiphospholipid Antibody Syndrome Antibodies on Endothelium

Plasma levels of endothelin-1, which is thought to play a role in arterial tone, vasospasm, and thrombotic arterial occlusion, were significantly increased in APS patients with arterial thrombosis (102). APS antibodies also may alter the balance of eicosanoid synthesis toward prothrombotic moieties, as indicated by the presence of an increased quantity of thromboxane metabolites in the urine of APS patients compared with controls (103). However, other studies have disputed this (104). APS antibodies may show cross-reactivity against oxidized LDL (53,105), and may thereby be associated with an increased risk of atherosclerosis (106).

CONCLUSION

In summary, although the pathophysiological mechanisms for thrombosis in APS have not been firmly established, the accumulating evidence suggests a central role for APS antibody-mediated prothrombotic and proadhesive effects on endothelium. The adverse effects of APS antibodies are mediated by their recognition of "cofactor" proteins, primarily β2GPI. The effects appear to be due to antibody-mediated dimerization of β2GPI, which increases the affinity of the antibody–antigen complex for the vascular surface and thereby disrupts endogenous anticoagulant mechanisms and stimulates EC activation. The major task that lies ahead is to move the understanding of the disorder from that of a phenomenological syndrome for which the diagnosis is based on a constellation of clinical history and empirically derived surrogate tests, to a disease whose pathophysiology is understood. This will lead to more accurate mechanistically based tests and, hopefully, to targeted treatments.

KEY POINTS

- APS is an autoimmune thrombotic condition for which the mechanism has not yet been established.
- Current diagnosis is mainly based on empirically derived surrogate tests – the LA" phenomenon and enzyme-linked immunosorbent assays (ELISAs) for antibodies against proteins that bind anionic phospholipids.
- β2GPI appears to be a biologically significant antigenic target in this condition. Although its precise role is unknown, it appears likely that antibody-mediated dimerization and pentamerization of the protein cause gain-of-function abnormalities that affect vascular endothelium through the inhibition of endogenous antithrombotic mechanisms or the activation of proadhesive and procoagulant pathways.

Future Goals

- To establish the actual mechanisms for thrombosis
- To develop mechanistically based diagnostic tests
- To develop treatments that directly target the disease mechanisms

ACKNOWLEDGMENTS

This work was supported in part by the National Institutes of Health/National Heart Lung and Blood Institute grant #HL-61331. We appreciate the collaboration of Dr. Douglas Taatjes and Mr. Anthony Quinn of the University of Vermont Core Imaging Facility in providing the atomic force images for the figures.

REFERENCES

1 Feinstein DI, Rapaport SI. Acquired inhibitors of blood coagulation. In: Spaet TH, ed. *Progress in Hemostasis and Thrombosis.* New York, NY: Grune & Stratton;1972:75–95.
2 Wilson WA, Gharavi AE, Koike T, et al. International consensus statement on preliminary classification criteria for definite antiphospholipid syndrome: report of an international workshop. *Arthritis Rheum.* 1999;42:1309–1311.
3 Miyakis S, Lockshin MD, Atsumi T, et al. International consensus statement on an update of the classification criteria for definite antiphospholipid syndrome (APS). *J Thromb Haemost.* 2006;4:295–306.
4 Moore JE, Mohr CF. Biologically false positive serological tests for syphilis: type, incidence, and cause. *J Am Med Assoc.* 1952; 150:467–473.

5 Moore JE, Lutz WB. Natural history of systemic lupus erythematosus: approach to its study through chronic biologic false positive reactors. *J Chronic Dis*. 1955;1:297–316.

6 Conley CL, Hartmann RC. A hemorrhagic disorder caused by circulating anticoagulant in patients with disseminated lupus erythematosus. *J Clin Invest*. 1952;31:621.

7 Bowie WEJ, Thompson JH, Pascuzzi CA, Owen GA. Thrombosis in systemic erythematosus despite circulating anticoagulants. *J Clin Invest*. 1963;62:416–430.

8 Laurell A, Nilsson I. Hypergammaglobulinemia, circulating anticoagulant and biologic false-positive Wassermann reaction. *J Lab Clin Med*. 1957;49:694–707.

9 Harris EN, Gharavi AE, Boey ML, et al. Anticardiolipin antibodies: detection by radioimmunoassay and association with thrombosis in systemic lupus erythematosus. *Lancet*. 1983; 2:1211–1214.

10 Hughes GR. The anticardiolipin syndrome. *Clin Exp Rheumatol*. 1985;3:285–286.

11 Harris EN, Hughes GRV, Gharavi AE. The antiphospholipid antibody syndrome. *J Rheumatol*. 1987;(Suppl 13): 210.

12 Martin SJ, Reutelingsperger CP, McGahon AJ, et al. Early redistribution of plasma membrane phosphatidylserine is a general feature of apoptosis regardless of the initiating stimulus: inhibition by overexpression of Bcl-2 and Abl. *J Exp Med*. 1995; 182:1545–1556.

13 Levine JS, Subang R, Koh JS, Rauch J. Induction of antiphospholipid autoantibodies by beta2-glycoprotein I bound to apoptotic thymocytes. *J Autoimmun*. 1998;11:413–424.

14 Blank M, Krause I, Fridkin M, et al. Bacterial induction of autoantibodies to beta2-glycoprotein-I accounts for the infectious etiology of antiphospholipid syndrome. *J Clin Invest*. 2002; 109:797–804.

15 Gharavi AE, Pierangeli SS, Harris EN. Viral origin of antiphospholipid antibodies: endothelial cell activation and thrombus enhancement by CMV peptide-induced APL antibodies. *Immunobiology*. 2003;207:37–42.

16 Padmakumar B, Sun J, Satchithananthan G, et al. Deep venous thrombosis and pulmonary embolism following chickenpox. *Ann Trop Paediatr*. 2004;24:271–274.

17 Santiago M, Martinelli R, Ko A, et al. Anti-beta2 glycoprotein I and anticardiolipin antibodies in leptospirosis, syphilis and Kala-azar. *Clin Exp Rheumatol*. 2001;19:425–430.

18 Loizou S, Singh S, Wypkema E, Asherson RA. Anticardiolipin, anti-beta(2)-glycoprotein I and antiprothrombin antibodies in black South African patients with infectious disease. *Ann Rheum Dis*. 2003;62:1106–1111.

19 Samarkos M, Davies KA, Gordon C, Loizou S. Clinical significance of IgA anticardiolipin and anti-beta(2)-GP1 antibodies in patients with systemic lupus erythematosus and primary antiphospholipid syndrome. *Clin Rheumatol*. 2005;25(2):199–204. Epub 2005 Aug 10.

20 Hellan M, Kuhnel E, Speiser W, et al. Familial lupus anticoagulant: a case report and review of the literature. *Blood Coagul Fibrinolysis*. 1998;9:195–200.

21 Sebastiani GD, Galeazzi M, Morozzi G, Marcolongo R. The immunogenetics of the antiphospholipid syndrome, anticardiolipin antibodies, and lupus anticoagulant. *Semin Arthritis Rheum*. 1996;25:414–420.

22 Goldstein R, Moulds JM, Smith CD, Sengar DP. MHC studies of the primary antiphospholipid antibody syndrome and of antiphospholipid antibodies in systemic lupus erythematosus. *J Rheumatol*. 1996;23:1173–1179.

23 Granados J, Vargas AG, Drenkard C, et al. Relationship of anticardiolipin antibodies and antiphospholipid syndrome to HLA-DR7 in Mexican patients with systemic lupus erythematosus (SLE). *Lupus*. 1997;6:57–62.

24 Wilson WA, Gharavi AE. Genetic risk factors for aPL syndrome. *Lupus*. 1996;5:398–403.

25 Sanchez ML, Katsumata K, Atsumi T, et al. Association of HLA-DM polymorphism with the production of antiphospholipid antibodies. *Ann Rheum Dis*. 2004;63:1645–1648.

26 Lillicrap MS, Wright G, Jones AC. Symptomatic antiphospholipid syndrome induced by chlorpromazine. *Br J Rheumatol*. 1998;37:346–347.

27 Hughes GR, Khamashta MA, Gharavi AE, Wilson WA, eds. *Lupus* (Special Issue, *7th International Symposium on Antiphospholipid Antibodies*, Vol. 5). New Orleans: Stockton Press, 1996.

28 Averna M, Paravizzini G, Marino G, et al. Liver is not the unique site of synthesis of beta 2-glycoprotein I (apolipoprotein H): evidence for an intestinal localization. *Int J Clin Lab Res*. 1997;27:207–212.

29 Caronti B, Calderaro C, Alessandri C, et al. Beta2-glycoprotein I (beta2-GPI) mRNA is expressed by several cell types involved in anti-phospholipid syndrome-related tissue damage. *Clin Exp Immunol*. 1999;115:214–219.

30 Bouma B, de Groot PG, van den Elsen JM, et al. Adhesion mechanism of human beta(2)-glycoprotein I to phospholipids based on its crystal structure. *EMBO J*. 1999;18:5166–5174.

31 Schwarzenbacher R, Zeth K, Diederichs K, et al. Crystal structure of human beta2-glycoprotein I: implications for phospholipid binding and the antiphospholipid syndrome. *EMBO J*. 1999;18:6228–6239.

32 Del Papa N, Sheng YH, Raschi E, et al. Human beta 2-glycoprotein I binds to endothelial cells through a cluster of lysine residues that are critical for anionic phospholipid binding and offers epitopes for anti-beta 2-glycoprotein I antibodies. *J Immunol*. 1998;160:5572–5578.

33 Mehdi H, Naqvi A, Kamboh MI. A hydrophobic sequence at position 313–316 (Leu-Ala-Phe-Trp) in the fifth domain of apolipoprotein H (beta2-glycoprotein I) is crucial for cardiolipin binding. *Eur J Biochem*. 2000;267:1770–1776.

34 Cleve H. Genetic studies on the deficiency of ß2 glycoprotein I of human serum. *Humangenetik*. 1968;5:294–304.

35 Yasuda S, Tsutsumi A, Chiba H, et al. Beta(2)-glycoprotein I deficiency: prevalence, genetic background and effects on plasma lipoprotein metabolism and hemostasis. *Atherosclerosis*. 2000;152:337–346.

36 Bancsi LF, van dL, I, Bertina RM. Beta 2-glycoprotein I deficiency and the risk of thrombosis. *Thromb Haemost*. 1992;67: 649–653.

37 Balasubramanian K, Schroit AJ. Characterization of phosphatidylserine-dependent beta2-glycoprotein I macrophage interactions. Implications for apoptotic cell clearance by phagocytes. *J Biol Chem*. 1998;273:29272–29277.

38 Sheng Y, Reddel SW, Herzog H, et al. Impaired thrombin generation in beta 2-glycoprotein I null mice. *J Biol Chem*. 2001; 276:13817–13821.

39 Yasuda S, Atsumi T, Ieko M, Koike T. Beta2-glycoprotein I, anti-beta2-glycoprotein I, and fibrinolysis. *Thromb Res*. 2004; 114:461–465.

40 Shi W, Chong BH, Hogg PJ, Chesterman CN. Anticardiolipin antibodies block the inhibition by beta 2-glycoprotein I of the factor Xa generating activity of platelets. *Thromb Haemost.* 1993;70:342–345.

41 Ohkura N, Hagihara Y, Yoshimura T, et al. Plasmin can reduce the function of human beta2 glycoprotein I by cleaving domain V into a nicked form. *Blood.* 1998;91:4173–4179.

42 Forastiero RR, Martinuzzo ME, Lu L, Broze GJ. Autoimmune antiphospholipid antibodies impair the inhibition of activated factor X by protein Z/protein Z-dependent protease inhibitor. *J Thromb Haemost.* 2003;1:1764–1770.

43 Reddel SW, Wang YX, Sheng YH, Krilis SA. Epitope studies with anti-beta 2-glycoprotein I antibodies from autoantibody and immunized sources. *J Autoimmun.* 2000;15: 91–96.

44 McNeeley PA, Dlott JS, Furie RA, et al. Beta2-glycoprotein I-dependent anticardiolipin antibodies preferentially bind the amino terminal domain of beta2-glycoprotein I. *Thromb Haemost.* 2001;86:590–595.

45 de Laat B, Derksen RH, Urbanus RT, de Groot PG. IgG antibodies that recognize epitope Gly40-Arg43 in domain I of beta 2-glycoprotein I cause LAC, and their presence correlates strongly with thrombosis. *Blood.* 2005;105:1540–1545.

46 Willems GM, Janssen MP, Pelsers MM, et al. Role of divalency in the high-affinity binding of anticardiolipin antibody-beta 2 glycoprotein I complexes to lipid membranes. Biochemistry. 1996; 35:13833–13842.

47 Zhang J, McCrae KR. Annexin A2 mediates endothelial cell activation by antiphospholipid/anti-beta2 glycoprotein I antibodies. *Blood.* 2005;105:1964–1969.

48 Porter TF. Antiphospholipid antibodies and infertility. *Clin Obstet Gynecol.* 2001;44:29–35.

49 Meroni PL, Tincani A, Sepp N, et al. Endothelium and the brain in CNS lupus. *Lupus.* 2003;12:919–928.

50 Atsumi T, Khamashta MA, Amengual O, et al. Binding of anticardiolipin antibodies to protein C via beta2-glycoprotein I (beta2-GPI): a possible mechanism in the inhibitory effect of antiphospholipid antibodies on the protein C system. *Clin Exp Immunol.* 1998;112:325–333.

51 Shibata S, Harpel PC, Gharavi A, et al. Autoantibodies to heparin from patients with antiphospholipid antibody syndrome inhibit formation of antithrombin III-thrombin complexes. *Blood.* 1994;83:2532–2540.

52 Lieby P, Soley A, Levallois H, et al. The clonal analysis of anticardiolipin antibodies in a single patient with primary antiphospholipid syndrome reveals an extreme antibody heterogeneity. *Blood.* 2001;97:3820–3828.

53 Witztum JL, Horkko S. The role of oxidized LDL in atherogenesis: immunological response and anti-phospholipid antibodies. *Ann NY Acad Sci.* 1997;811:88–96.

54 Horkko S, Miller E, Branch DW, et al. The epitopes for some antiphospholipid antibodies are adducts of oxidized phospholipid and beta2 glycoprotein 1 (and other proteins). *Proc Natl Acad Sci USA.* 1997;94:10356–10361.

55 Vaarala O, Puurunen M, Lukka M, et al. Affinity-purified cardiolipin-binding antibodies show heterogeneity in their binding to oxidized low-density lipoprotein. *Clin Exp Immunol.* 1996;104:269–274.

56 Amengual O, Atsumi T, Khamashta MA, et al. Autoantibodies against oxidized low-density lipoprotein in antiphospholipid syndrome. *Br J Rheumatol.* 1997;36:964–968.

57 Rustin MH, Bull HA, Machin SJ, et al. Effects of the lupus anticoagulant in patients with systemic lupus erythematosus on endothelial cell prostacyclin release and procoagulant activity. *J Invest Dermatol.* 1988;90:744–748.

58 Branch DW, Rodgers GM. Induction of endothelial cell tissue factor activity by sera from patients with antiphospholipid syndrome: a possible mechanism of thrombosis. *Am J Obstet Gynecol.* 1993;168:206–210.

59 Ferrara DE, Swerlick R, Casper K, et al. Fluvastatin inhibits upregulation of tissue factor expression by antiphospholipid antibodies on endothelial cells. *J Thromb Haemost.* 2004;2:1558–1563.

60 Dunoyer-Geindre S, de Moerloose P, Galve-de RochemonteixB, et al. NFkappaB is an essential intermediate in the activation of endothelial cells by anti-beta(2)-glycoprotein 1 antibodies. *Thromb Haemost.* 2002;88:851–857.

61 Wolberg AS, Roubey RA. Mechanisms of autoantibody-induced monocyte tissue factor expression. *Thromb Res.* 2004;114:391–396.

62 Andree HAM, Hermens WT, Hemker HC, Willems GM. Displacement of factor Va by annexin V. In: Andree HAM, ed. *Phospholipid Binding and Anticoagulant Action of Annexin V.* Maastricht, The Netherlands: Universitaire Pers Maastricht; 1992:73–85.

63 Mosser G, Ravanat C, Freyssinet JM, Brisson A. Sub-domain structure of lipid-bound annexin-V resolved by electron image analysis. *J Mol Biol.* 1991;217:241–245.

64 Voges D, Berendes R, Burger A, et al. Three-dimensional structure of membrane-bound annexin V. A correlative electron microscopy-X-ray crystallography study. *J Mol Biol.* 1994;238: 199–213.

65 Capila I, Hernaiz MJ, Mo YD, et al. Annexin V-heparin oligosaccharide complex suggests heparan sulfate-mediated assembly on cell surfaces. *Structure.* 2001;9:57–64.

66 Flaherty MJ, West S, Heimark RL, et al. Placental anticoagulant protein-I: measurement in extracellular fluids and cells of the hemostatic system. *J Lab Clin Med.* 1990;115:174–181.

67 Rand JH, Wu XX, Andree HA, et al. Pregnancy loss in the antiphospholipid-antibody syndrome–a possible thrombogenic mechanism. *N Engl J Med.* 1997;337:154–160.

68 Cederholm A, Svenungsson E, Jensen-Urstad K, et al. Decreased binding of annexin v to endothelial cells: a potential mechanism in atherothrombosis of patients with systemic lupus erythematosus. *Arterioscler Thromb Vasc Biol.* 2005;25:198–203.

69 Wang X, Campos B, Kaetzel MA, Dedman JR. Annexin V is critical in the maintenance of murine placental integrity. *Am J Obstet Gynecol.* 1999;180:1008–1016.

70 Tripathy NK, Sinha N, Nityanand S. Anti-annexin V antibodies in Takayasu's arteritis: prevalence and relationship with disease activity. Clin Exp Immunol. 2003;134:360–364.

71 Rand JH, Wu XX, Andree HAM, et al. Antiphospholipid antibodies accelerate plasma coagulation by inhibiting annexin-V binding to phospholipids: a "lupus procoagulant" phenomenon. *Blood.* 1998;92:1652–1660.

72 Rand JH, Wu XX, Giesen P. A possible solution to the paradox of the "lupus anticoagulant": antiphospholipid antibodies accelerate thrombin generation by inhibiting annexin-V. *Thromb Haemost.* 1999;82:1376–1377.

73 Rand JH, Wu XX, Quinn AS, et al. Human monoclonal antiphospholipid antibodies disrupt the annexin A5 anticoagulant crystal shield on phospholipid bilayers: evidence from

atomic force microscopy and functional assay. *Am J Pathol.* 2003;163:1193–1200.
74 Rand JH, Wu XX, Lapinski R, et al. Detection of antibody-mediated reduction of annexin A5 anticoagulant activity in plasmas of patients with the antiphospholipid syndrome. *Blood.* 2004;104:2783–2790.
75 Tomer A. Antiphospholipid antibody syndrome: rapid, sensitive, and specific flow cytometric assay for determination of anti-platelet phospholipid autoantibodies. *J Lab Clin Med.* 2002;139:147–154.
76 Esmon CT. The anticoagulant and anti-inflammatory roles of the protein C anticoagulant pathway. *J Autoimmun.* 2000;15: 113–116.
77 de-Groot PG, Horbach DA, Derksen RH. Protein C and other cofactors involved in the binding of antiphospholipid antibodies: relation to the pathogenesis of thrombosis. *Lupus.* 1996;5:488–493.
78 Esmon NL, Safa O, Smirnov MD, Esmon CT. Antiphospholipid antibodies and the protein C pathway. *J Autoimmun.* 2000;15:221–225.
79 Hurtado V, Montes R, Gris JC, et al. Autoantibodies against EPCR are found in antiphospholipid syndrome and are a risk factor for fetal death. *Blood.* 2004;104:1369–1374.
80 Ames PR, Tommasino C, Iannaccone L, et al. Coagulation activation and fibrinolytic imbalance in subjects with idiopathic antiphospholipid antibodies – a crucial role for acquired free protein S deficiency. *Thromb Haemost.* 1996;76:190–194.
81 Crowther MA, Johnston M, Weitz J, Ginsberg JS. Free protein S deficiency may be found in patients with antiphospholipid antibodies who do not have systemic lupus erythematosus. *Thromb Haemost.* 1996;76:689–691.
82 Salemink I, Blezer R, Willems GM, et al. Antibodies to beta2-glycoprotein I associated with antiphospholipid syndrome suppress the inhibitory activity of tissue factor pathway inhibitor. *Thromb Haemost.* 2000;84:653–656.
83 Adams MJ, Donohoe S, Mackie IJ, Machin SJ. Anti-tissue factor pathway inhibitor activity in patients with primary antiphospholipid syndrome. *Br J Haematol.* 2001;114:375–379.
84 Dahm A, Van H, V, Bendz B, et al. Low levels of tissue factor pathway inhibitor (TFPI) increase the risk of venous thrombosis. *Blood.* 2003;101:4387–4392.
85 Forastiero R, Martinuzzo M, Adamczuk Y, et al. The combination of thrombophilic genotypes is associated with definite antiphospholipid syndrome. *Haematologica.* 2001;86:735–741.
86 Schousboe I, Rasmussen MS. Synchronized inhibition of the phospholipid mediated autoactivation of factor XII in plasma by beta 2-glycoprotein I and anti-beta 2- glycoprotein I. *Thromb Haemost.* 1995;73:798–804.
87 Kim J, Hajjar KA. Annexin II: a plasminogen-plasminogen activator co-receptor. *Front Biosci.* 2002;7:D341–D348.
88 Simantov R, LaSala JM, Lo SK, et al. Activation of cultured vascular endothelial cells by antiphospholipid antibodies. *J Clin Invest.* 1995;96:2211–2219.
89 Meroni PL, Raschi E, Camera M, et al. Endothelial activation by aPL: a potential pathogenetic mechanism for the clinical manifestations of the syndrome. *J Autoimmun.* 2000;15:237–240.
90 Pierangeli SS, Espinola RG, Liu X, Harris EN. Thrombogenic effects of antiphospholipid antibodies are mediated by intercellular cell adhesion molecule-1, vascular cell adhesion molecule-1, and P-selectin. *Circ Res.* 2001;88:245–250.
91 Pierangeli SS, Liu X, Espinola R, et al. Functional analyses of patient-derived IgG monoclonal anticardiolipin antibodies using in vivo thrombosis and in vivo microcirculation models. *Thromb Haemost.* 2000;84:388–395.
92 Dunoyer-Geindre S, Kruithof EK, Boehlen F, et al. Aspirin inhibits endothelial cell activation induced by antiphospholipid antibodies. *J Thromb Haemost.* 2004;2:1176–1181.
93 Espinola RG, Liu X, Colden-Stanfield M, et al. E-Selectin mediates pathogenic effects of antiphospholipid antibodies. *J Thromb Haemost.* 2003;1:843–848.
94 Cho CS, Cho ML, Chen PP, et al. Antiphospholipid antibodies induce monocyte chemoattractant protein-1 in endothelial cells. *J Immunol.* 2002;168:4209–4215.
95 Raschi E, Testoni C, Bosisio D, et al. Role of the MyD88 transduction signaling pathway in endothelial activation by antiphospholipid antibodies. *Blood.* 2003;101:3495–3500.
96 Meroni PL, Raschi E, Testoni C, et al. Innate immunity in the antiphospholipid syndrome: role of toll-like receptors in endothelial cell activation by antiphospholipid antibodies. *Autoimmun Rev.* 2004;3:510–515.
97 Dueymes M, Levy Y, Ziporen L, et al. Do some antiphospholipid antibodies target endothelial cells? *Ann Med Interne Paris.* 1996;147(Suppl 1):22–23.
98 Matsuda J, Gotoh M, Gohchi K, et al. Anti-endothelial cell antibodies to the endothelial hybridoma cell line (EAhy926) in systemic lupus erythematosus patients with antiphospholipid antibodies. *Br J Haematol.* 1997;97:227–232.
99 Navarro M, Cervera R, Teixido M, et al. Antibodies to endothelial cells and to beta 2-glycoprotein I in the antiphospholipid syndrome: prevalence and isotype distribution. *Br J Rheumatol.* 1996;35:523–528.
100 Nakamura N, Ban T, Yamaji K, et al. Localization of the apoptosis-inducing activity of lupus anticoagulant in an annexin V-binding antibody subset. *J Clin Invest.* 1998;101: 1951–1959.
101 Dignat-George F, Camoin-Jau L, Sabatier F, et al. Endothelial microparticles: a potential contribution to the thrombotic complications of the antiphospholipid syndrome. *Thromb Haemost.* 2004;91:667–673.
102 Atsumi T, Khamashta MA, Haworth RS, et al. Arterial disease and thrombosis in the antiphospholipid syndrome: a pathogenic role for endothelin 1. *Arthritis Rheum.* 1998;41:800–807.
103 Lellouche F, Martinuzzo M, Said P, et al. Imbalance of thromboxane/prostacyclin biosynthesis in patients with lupus anticoagulant. *Blood.* 1991;78:2894–2899.
104 Hasselaar P, Derksen RH, Blokzijl L, de Groot PG. Thrombosis associated with antiphospholipid antibodies cannot be explained by effects on endothelial and platelet prostanoid synthesis. *Thromb Haemost.* 1988;59:80–85.
105 Horkko S, Miller E, Dudl E, et al. Antiphospholipid antibodies are directed against epitopes of oxidized phospholipids. Recognition of cardiolipin by monoclonal antibodies to epitopes of oxidized low density lipoprotein. *J Clin Invest.* 1996;98:815–825.
106 Vaarala O. Antiphospholipid antibodies and atherosclerosis. *Lupus.* 1996;5:442–447.

150

Diabetes

Angelika Bierhaus*, Hans-Peter Hammes†, and Peter P. Nawroth*

*University of Heidelberg, Germany; †University of Mannheim, Germany

Patients with diabetes mellitus suffer from increased cardiovascular morbidity and mortality. The vascular endothelium forms the inner lining of blood vessels and serves as an important autocrine and paracrine regulator of vascular function. Loss of its modulatory function is thought to play a central role in the pathogenesis of late diabetic vascular complications, in both type 1 and type 2 diabetes. The increased flux of glucose under diabetic conditions results in the activation of various pathways that metabolize the excess of glucose, yet promote endothelial dysfunction. Clinical studies have consistently shown that intensive glycemic control delays the onset and progression of vascular complications in type 1 and, to a lesser extent, type 2 diabetes. Thus, hyperglycemia is an important mediator of endothelial dysfunction. However, additional factors appear to contribute to endothelial dysfunction in type 2 diabetes, and in certain patients with type 1 diabetes (e.g., those with diabetic nephropathy or obesity). In this chapter, we focus primarily on the effect of glucose on endothelial cells (ECs), but also refer to glucose-independent alterations of EC function in type 1 and type 2 diabetes.

EFFECTS OF HIGH GLUCOSE ON ENDOTHELIAL CELLS

Hyperglycemia-induced cellular changes influence and amplify each other; therefore, activation of ECs and subsequent endothelial dysfunction is the result of interacting cellular pathways rather than a single cellular event. Accumulating data support the hypothesis that oxidative stress is a common link, because several pathways seem to contribute to the oxidative stress associated with acute or chronic hyperglycemia (1–5). The polyol pathway is one of the possible biochemical mechanisms by which hyperglycemia could impair the function and structure of ECs (6). Increased prostanoid synthesis and glucose auto-oxidation also generates reactive oxygen species (ROS). Oxidation accompanying protein glycation is another major source of oxygen free radicals (1–5). A close relationship between oxidative stress and glycemic control has been demonstrated by showing a significant positive correlation between malonaldehyde (MDA) and fasting blood glucose, glycated hemoglobin, and activation of the redox-sensitive transcription factor nuclear factor (NF)-κB. Moreover, the hypothesis that free radicals and subsequent oxidative stress might mediate the effects of hyperglycemia is supported by the observation that antioxidants counteract many of the injurious effects of hyperglycemia, such as excessive activation of the polyol and hexosamine pathways, induction of protein kinase C (PKC), advanced glycation end product (AGE) formation, and NF-κB activation (3).

ECs, however, do not only react to ROS, but also actively release ROS, thereby activating surrounding cells of the vascular wall. Hyperglycemic conditions and/or the presence of free fatty acids further induce ECs to produce free oxygen radicals via a PKC-dependent activation of nicotinamide adenine dinucleotide phosphate (NADPH)-oxidase (7). Binding of AGEs to the receptor for AGE (RAGE) also results in increased endothelial NADPH activity. In addition to production of oxygen free radicals, depletion of antioxidative capacities may play an important role in the pathogenesis of diabetes-associated endothelial dysfunction. Consistently, high glucose and AGEs have been shown to reduce intracellular glutathione levels in cultured ECs (8,9).

MITOCHONDRIAL SUPEROXIDE OVERPRODUCTION AS AN INTEGRATOR OF ENDOTHELIAL CELL DYSFUNCTION

Recently, Brownlee and coworkers demonstrated hyperglycemia-induced superoxide overproduction by the mitochondrial electron transport chain as one integrator of the various metabolic changes contributing to the development of endothelial dysfunction (3,4,10) (Figure 150.1). Intracellular glucose oxidation during glycolysis results in nicotinamide adenine dinucleotide (NADH) production, which donates reductive equivalents to the mitochondrial respiratory chain. In addition, pyruvate transported into the mitochondria is

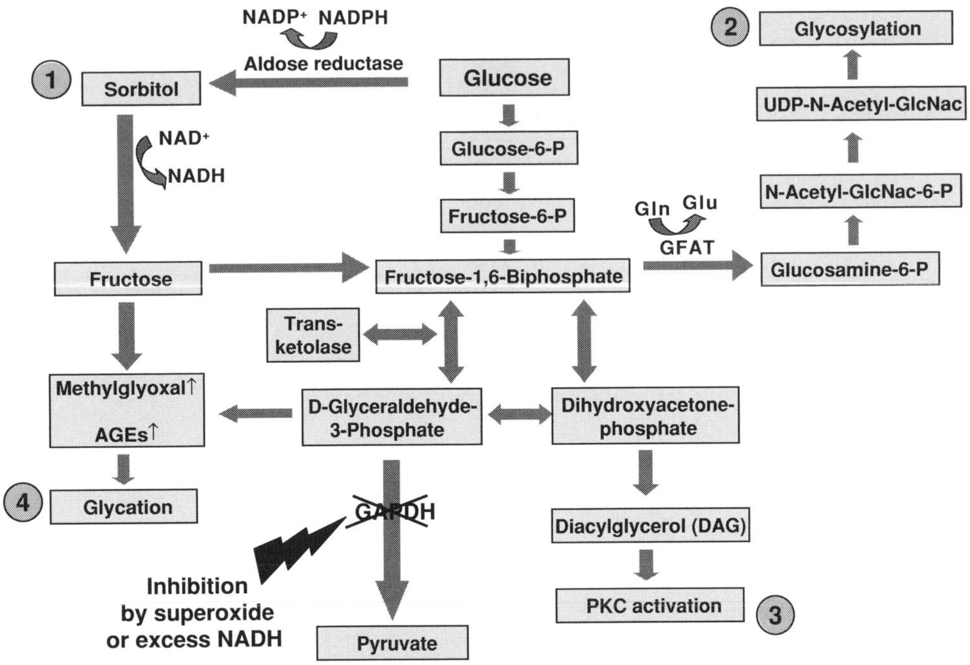

Figure 150.1. Four main molecular mechanisms have been implicated in glucose-mediated vascular damage: The polyol pathway (1), the hexosamine pathway (2), activation of PKC (3), and AGE formation (4). Hyperglycemia-induced superoxide overproduction by the mitochondrial electron transport chain acts as integrator of these various metabolic changes by inhibition of GAPDH. AGEs, advanced glycation end-products; GAPDH, glyceraldehyde phosphate dehydrogenase; GFAT, glutamine:fructose-6-phosphate amidotransferase; P, phosphate.

oxidized in the tricarboxylic cycle to form CO_2, H_2O, NADH, and flavin adenine dinucleotide ($FADH_2$). Mitochondrial NADH and $FADH_2$ subsequently provide energy for the production of adenosine triphosphate (ATP) through oxidative phosphorylation by the mitochondrial respiratory chain. The electron flow through the mitochondrial respiratory chain is triggered by four inner membrane-associated multienzyme complexes (complex I–IV), cytochrome c, and the mobile electron carrier ubiquinone. Electron transport finally results in a difference in potential at the inner mitochondrial membrane that provides energy for ATP production. Consequently, a high mitochondrial membrane potential leads to increased half-life of the superoxide (O_2^-)-producing electron transport systems and results in the increased generation of ROS. An increased availability of glucose therefore results in increased O_2^- release and NADH formation.

Subsequently, superoxide anions can induce metabolic changes by inhibition of glyceraldehyde-3-phosphate dehydrogenase (GAPDH). During glycolysis, glucose is normally converted via glucose-6-phosphate, fructose-6-phosphate, and fructose-1,6-biphosphate to form the triose glyceraldehyde-3-phosphate. The latter is oxidized by GAPDH to form 1,3-diphosphoglycerate using NAD^+ as substrate. An excess of superoxide reduces GAPDH by more than 60%. Consumption of NAD^+ due to an excess of NADH also reduces GAPDH activity. Inhibition of GAPDH results in the accumulation of upstream intermediates that are metabolized in pathways of glucose overutilization. These are: (a) the conversion of glucose to fructose by the polyol pathway; (b) the increased flux of fructose-6-phosphate through the hexosamine pathway; (c) the increased de novo synthesis of diacylglycerol (DAG), a cofactor of PKC, from dihydroxyacetone phosphate; and (d) the increased formation of reactive intracellular AGEs from accumulating trioses (see Figure 150.1). Furthermore, oxidative stress could be exacerbated by impaired endogenous antioxidant protection mechanisms. Because an excessive activation of the polyol pathway, increased PKC activation, and increased AGE formation are thought to contribute to endothelial dysfunction (discussed in a later section), blocking increased oxidative stress, including unchecked superoxide production, might at least in part restore the metabolic and vascular imbalance observed in diabetes.

THE POLYOL PATHWAY AS ONE CAUSE OF ENDOTHELIAL DYSFUNCTION

One pathway used in the presence of glucose overutilization is the polyol pathway, in which the enzyme aldose reductase catalyses the NADPH/H^+-dependent reduction of sorbitol, before the latter is oxidized to fructose (3,4,6,10) (see Figure 150.1). This leads to depletion of NADPH and generation of reactive oxidants. Because NADPH is required for generation of nitric oxide (NO) from L-arginine, the depletion of NADPH leads to reduced NO formation. Furthermore, the generation of NADH results in an increase in the $NADH/NAD^+$ ratio and in a pronounced cellular redox imbalance. This might be of pivotal importance, particularly for the neuronal vasculature. Increased polyol pathway flux has been linked to neuronal complications in experimental diabetes; these complications are attenuated by aldose reductase inhibitors, with defective NO-mediated vasodilation being a particular target. Another consequence is activation of the nuclear protein poly (ADP-ribose)-polymerase (PARP), that in turn synthesizes the polymer (ADP)-ribose, thereby further lowering the availability of NAD^+. Because NADPH also is needed to regenerate glutathione, a hyperactive polyol pathway reduces the antioxidative defense capacities.

INCREASED FLUX THROUGH THE HEXOSAMINE PATHWAY AS ANOTHER CAUSE OF ENDOTHELIAL DYSFUNCTION

Inhibition of GAPDH results in the accumulation of fructose-6-phosphate (10) that represents a substrate for the enzyme glutamine:fructose-6-phosphate amidotransferase (GFAT). GFAT converts fructose-6-phosphate into glucosamine-6-phosphate and thereby provides the substrate for proteoglycan synthesis and the formation of O-linked proteins (see Figure 150.1). Thus, hyperglycemia impairs the functional properties of cytosolic and nuclear proteins via O-linked glycosylation modification (O-GlcNAcylation). Consistently, O-GlcNAcylation immunoreactivity is barely detectable in control ECs from nonatherosclerotic small vessels, whereas the endothelial layer of carotid plaques from diabetic patients shows an intense signal in both cytoplasm and nuclear compartments (11).

Recently, hyperglycemia has been shown to induce a 2.4-fold increase in hexosamine pathway activity in cultured ECs and to significantly increase the O-glycosylation of the transcription factor SP1. Because the glycosylated form of SP1 seems to be more transcriptionally active than the deglycosylated form, modification of endothelial SP1 results in increased activation of SP1-regulated genes such as plasminogen activator inhibitor (PAI)-1 or NF-κB p65 (12). Furthermore, O-GlcNAcylation directly affects the metabolic branch of insulin signaling, such as insulin receptor (IR)-mediated activation of the IR substrate (IRS)/ phosphoinositide 3-kinase (PI3K)/Akt pathway (11). In addition, posttranslational O-acetylglucosaminylation of the Akt-binding site in endothelial NO synthase (eNOS) results in inhibition of eNOS activity. NO produced by eNOS is known to inhibit matrix metalloprotease (MMP)-2 and -9 activity. Reduction of eNOS activity by activation of the hexosamine pathway in human aortic ECs was associated with increased expression of MMP-2 and MMP-9 and decreased expression of tissue inhibitor of metalloproteinases (TIMP)-3, which might contribute to the development of macrovascular disease (11).

ACTIVATION OF PROTEIN KINASE C AS MEDIATOR OF ENDOTHELIAL DYSFUNCTION

The conversion of triose-phosphates into dihydroxyacetone phosphate and subsequently into the lipid second messenger DAG provides a cofactor for at least nine members of the PKC family (13) (see Figure 150.1). Binding of DAG in the presence of calcium induces conformational changes in the respective PKC molecules and results in opening of the active center and subsequent interaction with the substrate. Several PKC isoforms also can be activated in the absence of calcium (Ca^{2+}). The hyperglycemia-dependent activation of PKC is enhanced in the presence of insulin, because the IR induces phosphorylation of phosphatidylinositol-(4,5)-biphosphate through PI3K, which results in the release of DAG and phosphatidylinositol-3,4,5-triphosphate (PIP3), which in turn activates the PKC-isoform PKCζ. Excessive PKC activation induces phosphorylation, translocation, and activation of cytosolic phospholipase A2, resulting in the liberation of arachidonic acid and increased production of prostaglandin (PG)E_2, which are known inhibitors of Na^+,$K(^+)$-ATPase. Furthermore, excessive PKC activation impairs phospholipase-C-mediated Ca^{2+} signaling and reduces NO generation in ECs. Through induction of endothelin (ET)-1 activity, inhibition of the actin-linked regulatory protein caldesmon, and increased vascular leakage and loss of endothelial barrier function, PKCs enhance contractility and permeability of ECs. In addition, hyperglycemia inhibits thymidine incorporation and cell growth of cultured human umbilical vein ECs (HUVECs) via a PKC-dependent mechanism (14). The consequences are manifold, the most important for the retina being changes in blood flow and vascular permeability due to vasoactive molecules such as vascular endothelial growth factor (VEGF) and ET-1 (15,16). PKCβ overexpression in mice results in excess retinal neovascularization in a model of proliferative retinopathy, whereas mice with a genetic ablation of the gene develop less proliferation (17). Pharmacological inhibition of PKCβ normalizes hyperglycemia-induced impaired blood flow through the retina of hyperglycemic animals (15). PKC inhibition also reduces laser-induced retinal neovascularization, indicating that PKC activation is a possible therapeutic target in diabetic patients with differing levels of retinopathy (18).

In addition to these effects, high glucose levels induce ROS production via PKC-dependent activation of NADPH oxidase, which results in increased EC apoptosis (19). Expression of NADPH oxidase components not only has been shown in vitro, but also in vascular tissues in experimental diabetes models, suggesting that PKC-dependent activation of NADPH oxidase may be an essential mechanism responsible for increased oxidative stress in diabetes (7). Thus, activation of cytoplasmic oxidative stress via NADPH oxidase is a possible and likely additional stress effector in the chronic hyperglycemia-exposed endothelium. In vitro data, however, suggest that high glucose exposure of retinal ECs and Müller cells produces ROS predominantly by the mitochondria, via inhibition of complex II of the respiratory chain, but not by inhibition of NADPH oxidase–reduced oxidative stress (20). Although long-term data on retinal outcome of the mitochondrial inhibition of ROS are not yet available, the effect of catalytic antioxidants has been tested. These compounds readily distribute to the mitochondria; have a low redox potential, which enables them to recycle virtually all redox systems; and are regenerated by hyperglycemia-induced production of NADH. They can inhibit all critical downstream biochemical pathways involved in diabetic retinopathy, and they normalize the formation of acellular capillaries and pericyte loss almost completely in diabetic rats.

In addition, PKC has been shown to activate the proinflammatory transcription factor NF-κB and subsequent proadhesive and prothrombotic gene expression; to induce the fibrinolysis inhibitor PAI-1, apparently by boosting the binding activity of SP1 transcription factors in the *PAI-1* promoter; and to promote synthesis of fibronectin and other proteins involved in matrix expansion.

THE ROLE OF ADVANCED GLYCATION END-PRODUCTS IN DIABETIC ENDOTHELIAL DYSFUNCTION

Intracellular precursors of AGEs affect EC function through different mechanisms (21,22). They modify intracellular proteins, including growth factors, and alter cellular function. In addition, AGEs precursors become available outside cells and modify extracellular matrix molecule proteins, thereby affecting cell–matrix signaling through interference with adhesion molecules and integrins. Proteins modified by AGEs can further bind to cell surface receptors such as RAGE, AGE-R1 (OST-48), AGE-R2 (80K-H phosphoprotein), AGE-R3 (galectin-3), and CD36, but also to other newly identified AGE-binding proteins such as N-glycans, ezrin, and megalin, and thereby induce cellular activation (23–25). Experimental proof of the relevance of AGE accumulation for vascular damage in diabetes has been provided by chemical inhibition using structurally unrelated AGE inhibitors. Long-term treatment of diabetic rodents with AGE inhibitors limited the structural damage of macrovascular and microvascular vessels in these animals to a large extent.

POLY(ADP-RIBOSE)-POLYMERASE ACTIVITY AS CAUSE FOR ENDOTHELIAL DYSFUNCTION

The simultaneous generation of O_2^- and reactive NO results in the formation of peroxynitrite, which is a potent stimulus of DNA single-strand breakage. Recent studies have demonstrated that hyperglycemia-associated oxidative stress increases DNA damage, as evidenced by increasing amounts of 8-hydroxyguanine and 8-hydroxyguanosine in plasma and tissue of diabetic rats. DNA single strands activate the nuclear enzyme PARP-1, which functions as a DNA-damage sensor and promotes inflammatory signal transduction processes (26). PARP activation results in depletion of its substrate NAD^+, slowing of electron transport, and reduction of ATP, and finally leads to cellular dysfunction. In ECs exposed to high glucose, PARP activation appears to be involved in both promoting nitrosative and oxidative stress and upregulating adhesion molecules and inflammation (27). Most remarkably, PARP-dependent poly(ADP-ribosyl)ation of GAPDH was identified as one cause of hyperglycemia-induced GAPDH inhibition in ECs (28). Both the hyperglycemia-induced decrease in GAPDH activity and its poly(ADP-ribosyl)ation were prevented by decreasing hyperglycemia-induced superoxide through overexpression of either uncoupling protein (UCP)-1 or manganese superoxide dismutase (MnSOD).

Although PARP activity is increased in ECs of large and small vessels in diabetes, PARP protein is upregulated in the diabetic retina mostly in Müller cells. Both PARP protein expression and PARP activity are normalized by the specific PARP inhibitor PJ34. PARP protein interacts with both the p50 and the p65 subunit of NF-κB, affecting EC survival (but not that of pericytes) in vitro. In vivo, inhibition of PARP activity (and protein upregulation) protects ECs and pericytes from loss (29). Furthermore, PARP inhibition can affect angiotensin (Ang)-II expression, suggesting interference with several pathways that lead to retinal capillary regression in vivo.

SUBNORMAL TRANSKETOLASE ACTIVITY AS A CAUSE FOR ENDOTHELIAL DYSFUNCTION

As summarized earlier, hyperglycemia-induced overproduction of superoxide by the mitochondrial electron transport chain and subsequent inhibition of GAPDH results in the accumulation of the upstream metabolites fructose-6-phosphate and glyceraldehyde-3-phosphate, which are also end-products of the nonoxidative branch of the pentose phosphate pathway. In the pentose phosphate pathway, the thiamine-dependent enzyme transketolase diverts these metabolites into pentose-5-phosphates and erythrose-4-phosphate (Figure 150.2). Because diabetic patients have a subnormal level of erythrocyte transketolase activity, inadequate transketolase activity could be considered as one reason for triose accumulation under hyperglycemic conditions (30). Consistently, supplementation with the transketolase cofactor benfotiamine, a lipid-soluble thiamine derivate with greater bioavailability than thiamine, resulted in activation of the pentose-phosphate pathway and thereby prevented activation of pathways involved in hyperglycemia-induced vascular damage in both cultured ECs and the microvasculature of diabetic rats (31).

INCREASED SYNTHESIS OF ENDOTHELIN-1 IN DIABETES

One of the most important functions of the endothelium is the regulation of blood flow. This is achieved by tight control of pro- and anticoagulant properties, but also by regulating

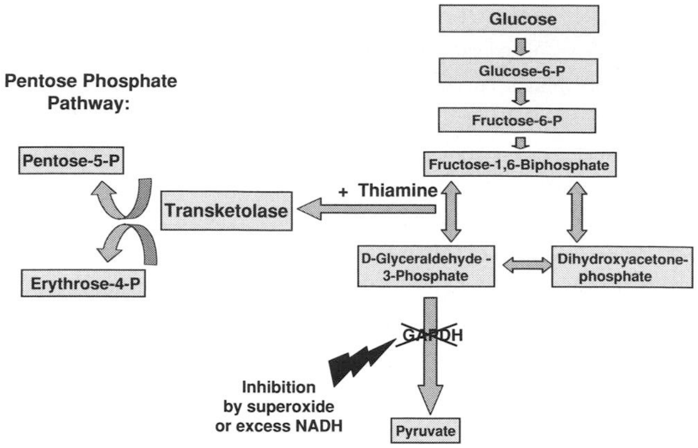

Figure 150.2. Fructose-6-phosphate and glyceraldehyde-3-phosphate are also end-products of the nonoxidative branch of the pentose phosphate pathway, in which the thiamine-dependent enzyme transketolase diverts upstream metabolites from glycolysis into pentose-5-phosphate and erythrose-4-phosphate. GAPDH, glyceraldehyde phosphate dehydrogenase; P, phosphate.

cell–cell interaction and vascular tone. ET-1 is both a potent vasoconstrictor and mitogenic factor that has been implicated as a cause of micro- and macrovascular complications in diabetes mellitus. Upregulation of ET-1 appears to be a consequence of the NO–Ang-II imbalance that contributes to end-organ injury (32). In addition, in vitro data have shown that oxidative stress, high glucose levels, and AGEs induce ET-1 expression via NF-κB activation (33). However, the pathway by which the high-glucose environment of diabetes mediates increased levels of ET-1 has still not been completely elucidated. Recent data indicate that endothelin-converting enzyme (ECE)-1, which converts inactive big ET-1 to active ET-1 peptide may play an important role, because high glucose induces PKC-mediated induction of ECE in cultured ECs (34). Thus, ECE may be another factor contributing to the elevated ET-1 peptide levels observed in diabetes. Increased ET-1 levels have been described in type 2 diabetic patients with increased intima media thickness (IMT) (35), providing further support for the role of ET-1 in late diabetic vascular disease. In addition, elevated circulating levels of ET-1 have been found in patients with microvascular disease. Consistently, ET receptor antagonists restore blood flow and nerve conduction velocity in experimental models of diabetes (36).

REDUCED AVAILABILITY OF VASCULAR NITRIC OXIDE

One of the most important mediators of vascular homeostasis leading to vascular relaxation is NO, a small, short-lived, highly reactive radical and the metabolic product of L-arginine by the enzyme NO synthase (NOS). Its complex interaction with the autocrine and paracrine systems, particularly Ang-II, modulates vasoconstriction and vasodilatation as well as the architectural remodeling of the vascular bed (37).

To date, three forms of NOS (eNOS, inducible NOS [iNOS], and neuronal NOS [nNOS]) have been described, of which eNOS is constitutively expressed in ECs. eNOS, which is located in membrane caveolae and the cytoplasm, requires tetrahydrobiopterin (BH4), NADPH, $FADH_2$, and flavin mononucleotide (FMN) as additional cofactors for its activity. The L-arginine required for NO synthesis is derived from a membrane-associated compartment distinct from the bulk intracellular amino acid pool, such as "lipid rafts" or caveolae (38). Under physiological conditions, the vascular endothelium constitutively releases NO and mediates relaxation of smooth muscle cells and regulation of the vascular tone (Figure 150.3). Consistently, loss of NO bioavailability due to reduced synthesis and increased scavenging by ROS is a feature of endothelial dysfunction in vascular disease. Endothelial dysfunction in diabetes is characterized by decreased NO bioactivity, increased superoxide production, and oxidation of BH4 (39). Particularly, ROS generated by NADPH oxidases have been shown to promote oxidation of BH4. Oxidized BH4 uncouples eNOS, resulting in decreased production of NO and generation of superoxide instead. Thus, uncoupling of eNOS in diabetes results in ROS production instead of NO-dependent vascular protection (Figure 150.3). Consistently, diabetic mice transgenic for guanosine triphosphate-cyclohydrolase (GTPCH)-1, the rate limiting enzyme for BH4-synthesis, present with selective augmentation of endothelial BH4 levels, reduced endothelial superoxide production, and preservation of NO-mediated vasodilatation (40). This indicates that BH4 is an important mediator of eNOS regulation in diabetes and that alteration of the balance of bioactive radicals in addition to a defective antioxidative defense (8) further promotes oxidative stress and vascular complications.

Although high concentrations of perioxynitrite ($ONOO^-$ >100 μM) also uncouple eNOS and inactivate prostacyclin synthase in diabetes (41), low concentrations of $ONOO^-$ (1–10 μM) activates c-Src-PI3K to generate an unknown molecule or a second messenger, which recruits AMP-activated protein kinase (AMPK) to form an active complex, resulting in AMPK activation and increasing AMPK-dependent eNOS-Ser^{1177} phosphorylation. $ONOO^-$-concentrations needed for AMPK activation are much lower than those conferring eNOS uncoupling, indicating that $ONOO^-$ will preferentially activate AMPK (42). Notably, the antidiabetic drug metformin also activates AMPK and thereby prevents oxidative stress in cultured ECs exposed to high glucose. This implies that low concentrations of $ONOO^-$, as generated by clinically relevant concentrations of metformin, activate AMPK and thereby precondition the endothelium to reduce excessive oxidant stress induced by hyperglycemia (42).

FLOW-MEDIATED DILATATION IN INSULIN RESISTANCE AND DIABETES

A noninvasive method to determine endothelial dysfunction is the determination of postischemic vasodilatation in the

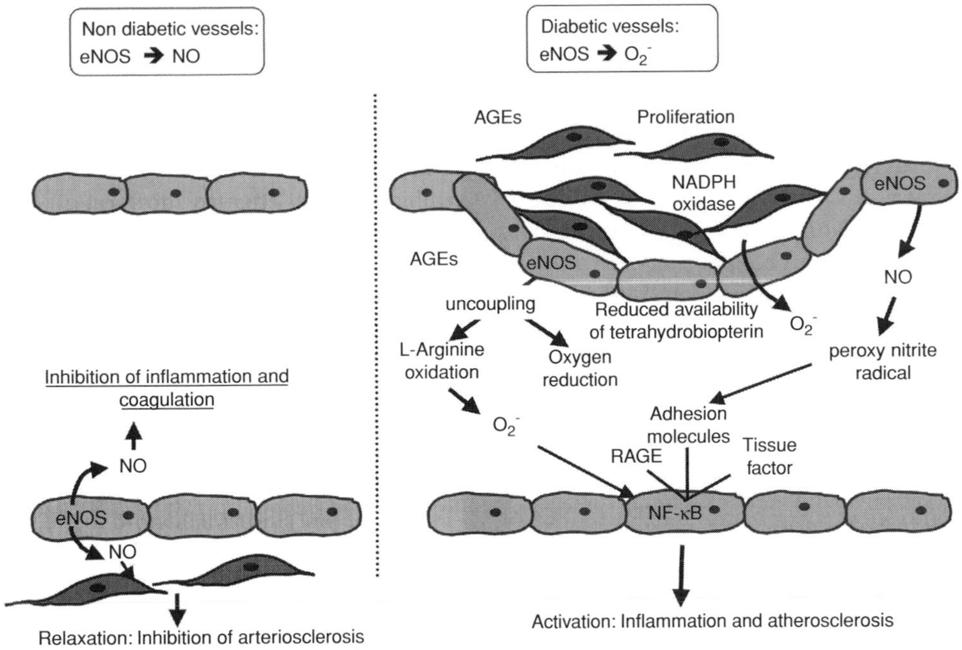

Figure 150.3. Under physiological conditions, the vascular endothelium constitutively releases NO and mediates relaxation of smooth muscle cells and regulation of the vascular tone. Endothelial dysfunction in diabetes caused by uncoupling of endothelial NO results in decreased NO bioactivity and increased superoxide (O_2^-) production, and thus promotes inflammation and atherosclerosis. AGEs, advanced glycation end products; eNOS, endothelial nitric oxide synthase; NO, nitric oxide; RAGE, receptor for AGEs.

forearm, which results in enhanced flow in the brachial artery and shear stress–induced vasodilatation. This flow-mediated (endothelial-dependent) dilatation (FMD) depends on the ability of the endothelium to produce vasodilator-endowed substances, mainly NO. Patients with a lower level of FMD are thought to have a lower production of NO, indicating endothelial dysfunction (43). Consistently, FMD decreases with increasing age and in patients with diabetes mellitus. Hyperglycemia is sufficient to change FAD, because acute hyperglycemia attenuates endothelium-dependent vasodilation in healthy adults without diabetes (44). In healthy postmenopausal women, a decrease in FMD is associated with a significantly increased risk for diabetes (45). Furthermore, nonobese women with polycystic ovary syndrome (PCOS) demonstrated an association of reduced FMD with both insulin resistance and mild chronic inflammation (46).

Insulin resistance also impairs endothelial function in young nondiabetic men. Although these young subjects exhibit a marked reduction in peripheral endothelial function early in life, no differences in IMT are observed (47). This indicates that endothelial dysfunction already is evident at an early stage, when morphological changes of the vessel wall are not apparent. In contrast, however, obesity, increased waist-to-hip ratio, and a higher percentage of body fat content were associated with increased IMT early in life and were paralleled by chronic subclinical inflammation and dysfunctional fibrinolysis independent of impaired glucose and lipid metabolism (48). Remarkably, impaired FMD response is also a common manifestation in children with type 1 diabetes, and is associated with increased carotid artery IMT, implying that endothelial dysfunction in children with type 1 diabetes may predispose for early atherosclerosis (49).

It is not known whether insulin resistance, hyperglycemia, and genetic factors affecting endothelial dysfunction are pathogenetically related or different entities. Because healthy relatives of patients with type 2 diabetes exhibit impaired FMD (43), however, there might be an independent genetic component in endothelial dysfunction. Therefore, insulin resistance could be a cause, rather than a consequence, of endothelial dysfunction.

ENDOTHELIAL DYSFUNCTION INFLUENCES INSULIN ACTION

It is not clear to what extent endothelial dysfunction is a cause or consequence of diabetes. A recent prospective study of 840 apparently healthy, nonobese, postmenopausal women demonstrated that endothelial function significantly influences the future development of diabetes, independent of age and several other risk factors for diabetes (45). This is consistent with the hypothesis of Pinkney, who defined the endothelium as the principal controlling factor of insulin concentrations in the interstitium and the amount of insulin effectively reaching the target organ (50). Because IRs are expressed on ECs of large and small blood vessels, they are thought to participate actively in insulin-regulated glucose homeostasis. It is hypothesized that the endothelium is a barrier controlling the contact between insulin and insulin-sensitive cells. Consistently, dysfunctional endothelium would be accompanied by progressive insulin resistance. Transendothelial insulin transport is believed to be a determinant of insulin action on glucose disposal. Although whether transendothelial insulin transport occurs in vivo is controversial, in vitro studies have demonstrated an IR-mediated insulin transport across the endothelial barrier (51).

When mice with a vascular EC-specific knockout of the IR (VENIRKO) were studied, glucose homeostasis under basal conditions was comparable with wild-type mice (52). *eNOS* and *ET-1* transcription, however, was significantly reduced in ECs. VENIRKO mice presented with insulin resistance, indicating that inactivation of the IR on ECs affects maintenance of the vascular tone and regulation of insulin activity. Furthermore, studies in VENIRKO mice and mice with vascular endothelial insulin-like growth factor (IGF) receptor knockout (VENIFARKO) indicate a role of endothelial insulin and IGF-1 in retinal neovascularization (53). Notably, however, these mice exhibit a normal blood–brain barrier (BBB) and a normal blood–retinal barrier (BRB) function, indicating that neither insulin nor IGF-1 signaling in vascular ECs is required for development and maintenance of the BBB or BRB (54).

Insulin-dependent stimulation of NO production is mediated by PI3K signaling, which activates Akt, which in turn phosphorylates eNOS on serine^{1177} and induces *eNOS* transcription (55). The impairment of those signaling pathways mediating potentially protective insulin effects by insulin-resistant states might contribute to the accelerated atherosclerosis associated with impaired insulin sensitivity. In addition, inhibition of insulin-dependent PI3K either by the hexosamine pathway (55) or by inhibitors such as wortmannin (56) enhances insulin-induced mitogenic effects by induction of mitogen-activated protein kinase (MAPK). Activation of the MAPK pathway, in turn, mediates activation of proinflammatory, proadhesive, prothrombotic, and profibrotic mechanisms.

ENDOTHELIAL DYSFUNCTION THROUGH VESICULATION AND INCREASED PERMEABILITY

An increase in vessel wall permeability is a common occurrence in hyperglycemia, in particular in diabetic nephropathy and retinopathy. Breakdown of the BRB is a frequent and characteristic lesion in the diabetic retina. From fluorescein angiograms in human diabetic retinas, microaneurysms can appear leaky, and have venous beading and retinal neovascularizations (57). Macular edema, the most characteristic indicator of BRB breakdown, shows several features of increased transition of intravascular elements into retinal tissues. The underlying mechanisms that contribute to increased vascular permeability are complex, and they comprise activation of permeability-enhancing factors, transendothelial transport mechanisms, and junctional protein changes. Among the many growth factors involved during the evolution of diabetic retinopathy, VEGF plays a predominant role (58). Its vasopermeation-enhancing function is well-characterized. VEGF is upregulated early during the course of nonproliferative diabetic retinopathy, both in humans and in experimental animals, and it has been correlated with increased vessel permeability. VEGF-mediated induction of vascular permeability occurs through endothelial fenestration, vesicular transport, and transport through intercellular clefts. Hyperglycemia, AGEs, ROS, and hypoxia are all well documented stimulators of VEGF in the diabetic retina. Diabetes and VEGF also share the propensity to affect paracellular permeability in association with changes in occludin distribution. Several other molecules of junctional families are likely involved in affecting permeability, such as zona occludens (ZO)-1 and claudin-5.

The tightness of the BRB also depends on the presence of perivascular cells, in particular pericytes. It has been proposed that the exceptionally dense capillary coverage with pericytes is the cause for the tightness of the BRB. As an indirect indication, mice with a homozygous deletion of platelet-derived growth factor (PDGF)-B, a pericyte recruiting factor, die near the time of birth from cerebral hemorrhages and edema, indicating severely altered BBB function. The absence of pericytes is associated with an increase in VEGF levels. In turn, intravitreal injection of angiostatin reduces VEGF expression and corrects increased permeability in early diabetic rat retina (59).

Alterations in the glomerular basement membrane are likely to affect the permeability characteristics of the glomerular capillaries and may contribute to proteinuria, glycosuria, and hypertension. At later stages of disease, mesangial and glomerular expansion results in collapse of the glomerular capillaries and subsequent glomerulosclerosis. The pathways leading to disturbed endothelial permeability and enhanced leakiness seem to be related to the pathways involved in endothelial activation and dysfunction. ROS and AGEs seem to play a major role, since blockade of RAGE interaction with soluble RAGE reduced hyperpermeability in diabetic mice. Consistently reducing AGE formation by using aminoguanidine reduced albumin excretion in experimental and clinical diabetes.

OXIDIZED LOW-DENSITY LIPOPROTEIN AND ENDOTHELIAL DYSFUNCTION

Binding of oxidized low-density lipoproteins (LDL) to the lectin-like oxidized LDL receptor (LOX)-1, which is upregulated in response to oxidative stress, induces ROS production and decreases NO released from ECs (60). Engagement of LOX-1 further induces the expression of ET-1, AT(1) receptor, and cell adhesion molecules. In addition, LOX-1 works as an adhesion molecule for activated platelets and neutrophils. Activated platelets and oxidized LDL in turn induce the generation of endothelial-derived microparticles (EDMPs), which are elevated in diabetic patients, in particular in those with diabetic nephropathy. Thus, elevated EDMPs may be a sign of vascular complications in type 2 diabetic patients. Statins have been shown to inhibit upregulation of LOX-1 and to increase eNOS expression in human coronary artery ECs.

THE ROLE ADIPOCYTOKINES IN REGULATION OF VASCULAR ENDOTHELIAL FUNCTION

Cytokines known to play a pivotal role in the immune response and inflammation possess additional endocrine and metabolic functions. Consistently, several of these cytokines are regarded as independent risk factors for vascular disease. Because visceral and subcutaneous adipose tissues are the major sources of cytokines (adipokines), increased adipose tissue mass is associated with alteration in adipokine production, such as overexpression of tumor necrosis factor (TNF)-α, interleukin (IL)-6, and PAI-1, and reduction in the plasma protein adiponectin (61). Recent experimental and clinical studies provide evidence that reduced circulating levels of adiponectin in visceral adiposity not only contribute to insulin resistance and dysglycemia, but also promote endothelial dysfunction. Although forearm blood flow in human subjects during reactive ahyperemia (a model for endothelium-dependent vasodilation) is negatively correlated with adiponectin, circulating adiponectin levels are positively associated with arterial vasodilation in response to nitroglycerin (a measure of endothelium-independent vasodilation). The proinflammatory status associated with these changes provides a potential link between insulin resistance and endothelial dysfunction in patients with type 2 diabetes. Several clinical studies have proven that weight reduction and physical activity followed by adipose tissue reduction reduces adipokines, increases adiponectin, reduces inflammation, and improves insulin sensitivity and endothelial function. Thus, adiponectin might link obesity, endothelial dysfunction, and diabetic vascular complications.

THE INFLAMMATORY COMPONENT OF ENDOTHELIAL DYSFUNCTION

ECs participating in the inflammatory response undergo a number of changes that finally result in loss of vascular integrity, activation of proinflammatory transcription factors such as NF-κB, cytokine expression, increased expression of leukocyte adhesion molecules, and a change from an antithrombotic to a prothrombotic phenotype (61–64) (Figure 150.4). In diabetes, initial inflammatory activation by increased oxidative stress, high glucose, oxidized LDLs, and PKC activation is perpetuated by engagement of RAGE, which in turn upregulates the receptor and initiates a cycle of sustained cellular perturbation. One unique feature of RAGE is sustained NF-κB activation due to de novo synthesis of *p65* mRNA, which results in a constantly growing pool of excess transcriptionally active NF-κB p65 (65). Increased *NF-κB* transcription, expression, and activity has been shown in the microvascular endothelium of patients with type 1 diabetes

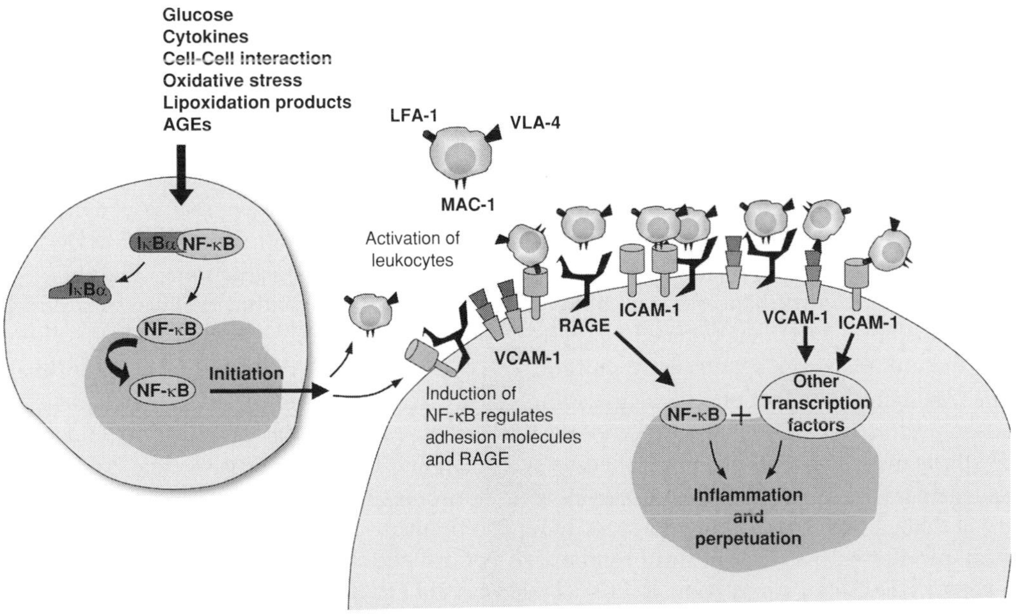

Figure 150.4. Initial inflammatory stimuli such as oxidative stress, high glucose, or cytokines induce activation of the proinflammatory transcription factor NF-κB. Subsequently, enhanced expression of NF-κB-regulated adhesion molecules on ECs promotes leukocyte adhesion, which in turn results in further NF-κB activation. NF-κB activation is perpetuated by engagement of RAGE, which upregulates the receptor and initiates sustained cellular perturbation and chronic inflammation.

(65). One of the cytokines induced by NF-κB is IL-6, which in turn activates C reactive protein (CRP). Low-grade inflammation determined by serum hs-CRP already is observed in children with type 1 diabetes and increased IMT of the common carotid arteries, implying that the proatherogenic effects of CRP were potentiated in the presence of hyperglycemia (66).

Platelet-activating factor (PAF), a phospholipid mediator of inflammation released by activated ECs, is significantly increased in patients with diabetes (67). In addition to changing the permeability of the glomerular basement membrane (GBM), PAF induces and enhances NF-κB activation. Enhanced expression of NF-κB–regulated adhesion molecules by the endothelium may further account for vascular damage in diabetics. Neutrophils from diabetic animals have been shown to exhibit higher levels of surface integrin expression and integrin-mediated adhesion. Blocking the surface integrin subunit CD18 has been shown to lead to decreased leukostasis in the diabetic retinal microvasculature, implying a central role of β2 integrins and their counter-receptors – intercellular adhesion molecule (ICAM)-1 and RAGE – in mediating endothelial dysfunction. The expression of ICAM-1, RAGE, and the very late activation antigen (VLA)-4 (integrin alpha4b) receptor vascular cell adhesion molecule (VCAM)-1 is significantly increased in animal models of diabetes. In accordance with the hyperglycemic induction of adhesion molecules in experimental diabetes, E-selectin, VCAM-1, and ICAM-1 levels are significantly higher in patients with type 1 (68) and type 2 (69) diabetes. Moreover, sera from diabetic patients significantly enhance leukocyte adhesion to cultured ECs, despite normal levels of IL-1β and TNF-α (70). Monocytes from patients with diabetes mellitus also demonstrate increased adhesion to ECs in culture, further indicating that hyperglycemia promotes leukocyte adhesion to the endothelium.

DISTURBANCES OF THE COAGULATION EQUILIBRIUM OF BLOOD IN DIABETES MELLITUS

Thrombogenesis depends on the balance between coagulation and fibrinolysis in the vasculature. Diabetes mellitus is characterized in part by major disturbances of the blood's fluid-coagulating equilibrium (63,64). The reduced synthesis of factors protecting the endothelium, such as prostacyclin and NO, is accompanied by increased adhesion and aggregation growth of platelets on dysfunctional and/or injured endothelium. Furthermore, the activated platelets promote other steps in hemostasis, in particular coagulation, and thereby promote hypercoagulability in diabetes mellitus. While quiescent ECs under physiological conditions provide an antithrombotic surface not expressing tissue factor, stimulation of ECs with glucose, AGEs, or cytokines results in tissue factor expression and generation of a prothrombotic surface. Although excessive tissue factor expression in patients with type 1 diabetes is most likely the result of hyperglycemia, tissue factor expression in patients with type 2 diabetes might further be promoted by inflammatory mediators. Consistent with the tissue factor–mediated hypercoagulability, coagulation activation markers, such as prothrombin activation fragment 1+2 and thrombin–anti-thrombin complexes, are elevated in diabetes. The plasma levels of many clotting factors including fibrinogen, factor VII, factor VIII, factor XI, factor XII, kallikrein, and von Willebrand factor (vWF) are also augmented in patients with diabetes mellitus (71). Conversely, the levels of the anticoagulant molecules such as thrombomodulin and protein C (PC) are decreased. Although vascular ECs normally synthesize activators for fibrinolysis, such as tissue-type plasminogen activator (t-PA) and urokinase-type plasminogen activator (u-PA), the fibrinolytic system and thereby the primary means of removing clots is relatively inhibited in diabetes (72). This is due to both abnormal clot structures that are more resistant to degradation and to the increase in PAI-1 synthesis, which is at least in part mediated by excessive PKC and/or hexosamine pathway activation (12).

ENDOTHELIUM, ANGIOGENESIS, AND VASCULAR REPAIR

VEGF is the most prominent member of the class of growth factors involved in repair or response-to-injury type of neovascularization (73). It is present at low levels in the normal retina. Several cells in the eye are able to constantly produce VEGF, such as ECs, pericytes, Müller cells, and ganglion cells (74). Local VEGF production in the vasculature has been found to be associated with a fenestrated capillary phenotype, such as in the glomerulus of the kidney or the choroid. In the inner retina, the local production of VEGF may have different functions, as indicated in the upcoming discussion (75).

VEGF is upregulated several-fold in the vitreous and aqueous of patients with active ocular neovascularization, irrespective of the cause of retinopathy. Several animal models of hypoxia-induced proliferative retinopathy exhibit high retinal VEGF levels. Many studies have confirmed and extended these findings, indicating that retinal diseases associated with ischemia of a certain degree are associated with hypoxia-induced VEGF upregulation. VEGF is also a permeability-enhancing factor within the eye, thus possibly contributing to the pathogenesis of macular edema. Intravitreal injection of VEGF into the eyes of rodents and primates mimics some early features of nonproliferative diabetic retinopathy, such as increased vascular permeability and microaneurysm formations. Thus, it is possible that VEGF may be responsible for some of the changes observed during the earliest stages of retinopathy (76). However, from retinal digest preparations of diabetic rats, it is known that a focal increase in the numbers of ECs and an increase in the deposition of PAS-positive material occurs within the vessel wall, which is suggestive of increased vascular permeability and proliferative activity of the retina in the absence of occlusive vascular changes. The expression of VEGF was therefore studied in these models that do not develop proliferative retinal changes, and it was found that VEGF is upregulated in association with its receptors FLK-1 and FLT-1 (77). Two important questions arise from this and the concomitant studies: What is the stimulus for VEGF upregulation? And, why is VEGF upregulated? With the available data, it appears that VEGF is induced by hyperglycemia-induced ROS production. Other likely inducers during the early course of retinopathy, in which ischemia due to capillary dropout may not yet be a significant contributor, are AGEs, which are also known to induce intraocular VEGF production and ROS. For the second question, it is more difficult to find a definitive answer. VEGF, among other growth factors, may promote endothelial resistance to oxidative stress. As mentioned earlier, VEGF is also a survival factor for ECs (78). Chronic hyperglycemia induces a comprehensive stress, not one limited to vascular ECs in the diabetic retina. An imbalance between survival factor demand and provision can induce programmed cell death. In fact, apoptosis has been observed in both vascular and neuroglial structures in diabetic retinas, which can be reduced either by the application of growth factors or by inhibitors of oxidative stress. The upregulation of VEGF in the early diabetic retina may represent a stress-induced cell response.

The recent discovery of circulating endothelial progenitor cells (EPCs) has substantially altered our understanding of new blood vessel formation. EPCs can differentiate from both bone marrow and circulating cells carrying hematopoietic stem cell markers. In vivo, they take part in vasculogenesis after ischemia in different animal models of limb ischemia, myocardial infarction, and wound healing. In metabolic disease, the outgrowth and function of EPCs in vitro is defective, and numbers of EPCs correlate with classical risk factors for vascular complications. EPCs from patients with diabetes are reduced in number and function (79). Similarly, EPCs from type 2 diabetes exhibit impaired proliferation, adhesion, and incorporation into vascular structures (80). As a consequence, EPC dysfunction may reduce the vascular regenerative potential of diabetic patients and contribute to the pathogenesis of vascular complications. Remarkably, Ang-II subtype 1-receptor blockade has been shown to increase the number of EPCs in patients with type 2 diabetes mellitus (81), demonstrating potential options for regenerating impaired repair mechanisms, although the mechanisms resulting in impaired EPC function under hyperglycemic condition are still poorly understood.

THE ANTIPHOSPHOLIPID ANTIBODY SYNDROME AND ENDOTHELIAL DYSFUNCTION

Endothelial function and vascular inflammation play major roles in the pathogenesis and clinical outcome of patients with systemic lupus erythematosus (SLE) and antiphospholipid antibody syndrome (APS), characterized by premature atherosclerosis of the legs as a first symptom. Recent studies indicate that abnormal ankle-brachial indexes (ABIs) are more common in primary APS than in healthy subjects. Consistently, patients with primary APS have a high prevalence of carotid artery IMT and a decreased lumen diameter, which might be promoted in the presence of diabetes. However, a linkage of APS to type 1 diabetes mellitus has not yet been established, and the association with type 2 diabetes is rare. Therefore, further studies are needed to clarify the relationship between diabetes and APS (82).

ENDOTHELIAL DYSFUNCTION IN DIABETIC MACROVASCULAR DISEASE

The term *microangiopathy* summarizes the fact that large-vessel disease in diabetes involves not only atherosclerosis, but also a form of diabetic angiopathy (82,83). Although the histology of lesions in the arterial wall of diabetic subjects and the classical risk factors for these alterations resemble those observed in the general population, diabetes seems to enhance the impact of the various risk factors. A small prospective study measuring carotid IMT in pancreas transplant recipients recently has shown an IMT improvement after transplantation within 2 years (84), paralleled by normalization of HbA1c and improved renal function. Because amelioration of IMT was independent of changes in body mass index, smoking, lipid levels, and use of hypolipidemic drugs, these data imply that glucose improvement might be an important factor in diabetic macrovascular disease. However, a variety of studies has shown that the correlation of macroangiopathy with the duration of diabetes and HbA1c is rather weak. This indicates that the different mechanisms described earlier, such as increased oxidative stress, disturbed blood flow, inflammation, activated hemostasis, AGE formation, and the generation of abnormal lipoproteins, all contribute to diabetic macroangiopathy. A central role for ECs is suggested due to the observation of elevated plasma levels of vWF, a marker of vascular endothelial dysfunction, in type 2 diabetes patients with macrovascular disease (85).

ENDOTHELIAL DYSFUNCTION IN DIABETIC NEPHROPATHY

Diabetic nephropathy is characterized by progressive excretion of urinary protein and loss of filtration, resulting in reduced clearance of metabolic waste products and finally leading to uremia (86). Mesangial expansion due to cytokine-induced proliferation of cells and matrix overproduction correlates with albuminuria and seems to contrast with another focus of microvascular damage, namely retinopathy.

However, clear evidence suggests that the diabetic renal ECs, both in the glomeruli and in the peritubular space, are responsive to chronic hyperglycemia (86). Earlier studies have indicated intraglomerular neovascularization in fenestrated vessels. Microvascular endothelial loss both in glomerular and peritubular capillaries of progressive renal disease is directly linked to impaired blood flow, the development of renal ischemia, and scarring. Important effectors of progressive glomerulosclerosis are transforming growth factor (TGF)-β, induced by transcriptional changes involving the hexosamine pathway, and the renin-angiotensin-aldosterone-system (RAAS). The RAAS is stimulated by hyperglycemia and also by autocrine mechanisms, involving growth factors, such as VEGF. VEGF is constitutively expressed in glomerular ECs to maintain fenestration as one hallmark of this vascular bed. Acute and chronic hyperglycemia both induce VEGF upregulation in a variety of glomerular cells, including the endothelium. Podocytes covering the filtration barrier from the outside, and which are heavily involved in the maintenance of the barrier function, are an important source of VEGF. As in the retina, the loss of VEGF from one cell source (such as podocytes) impairs the development and function of glomerular capillaries.

ENDOTHELIAL DYSFUNCTION IN DIABETIC NEUROPATHY

Because vasoregulation of nerve microvessels is regulated at the arteriolar, venular, and EC level, the etiology of diabetic neuropathy seems to have a large vascular component. Therefore, the pathophysiology of nerve dysfunction in diabetes can at least in part be explained by reduced endoneurial microcirculation, impaired blood flow due to perturbed NO levels, prostanoid metabolism, and alterations in endoneurial metabolism (87,88) (Figure 150.5).

Experimental diabetic models have demonstrated that the infusion of glucose or induction of diabetes results in abnormal vasodilatation by NO and prostacyclin. Because the reduced NO release can be improved in the presence of antioxidants, it is likely that increased oxidative stress is one major factor inhibiting NO release. However, it is not known whether antioxidants primarily improve nerve blood flow or normalize endoneurial oxidative metabolism. An overactive polyol pathway and the presence of AGEs able to quench NO may further promote the depletion of NO. Elevated plasma homocysteine levels, which are strong predictors for diabetic neuropathy in patients with type 2 diabetes, increase asymmetrical dimethylarginine (ADMA), an endogenous inhibitor of NO synthase, and thereby also decrease NO generation by ECs.

In addition to impaired NO-mediated vasodilation, upregulation of NF-κB is likely to support endoneurial ischemia by inducing NF-κB–regulated proinflammatory and procoagulant gene products, including tissue factor and ET-1. Recent studies demonstrate that ligation of AGEs to the surface receptor RAGE perpetuates activation of NF-κB in cultured ECs and neuronal cells. In sural nerve biopsies, ligands of RAGE, RAGE, activated NF-κB p65, and IL-6 colocalized in the microvasculature of patients with diabetic neuropathy, but not in controls (89). Activation of NF-κB and NF-κB–dependent gene expression was upregulated in peripheral nerves of diabetic mice, induced by AGEs and prevented by RAGE blockade. NF-κB activation was blunted in diabetic $RAGE^{-/-}$ mice compared with robust enhancement in strain-matched diabetic controls. Loss of pain perception was largely

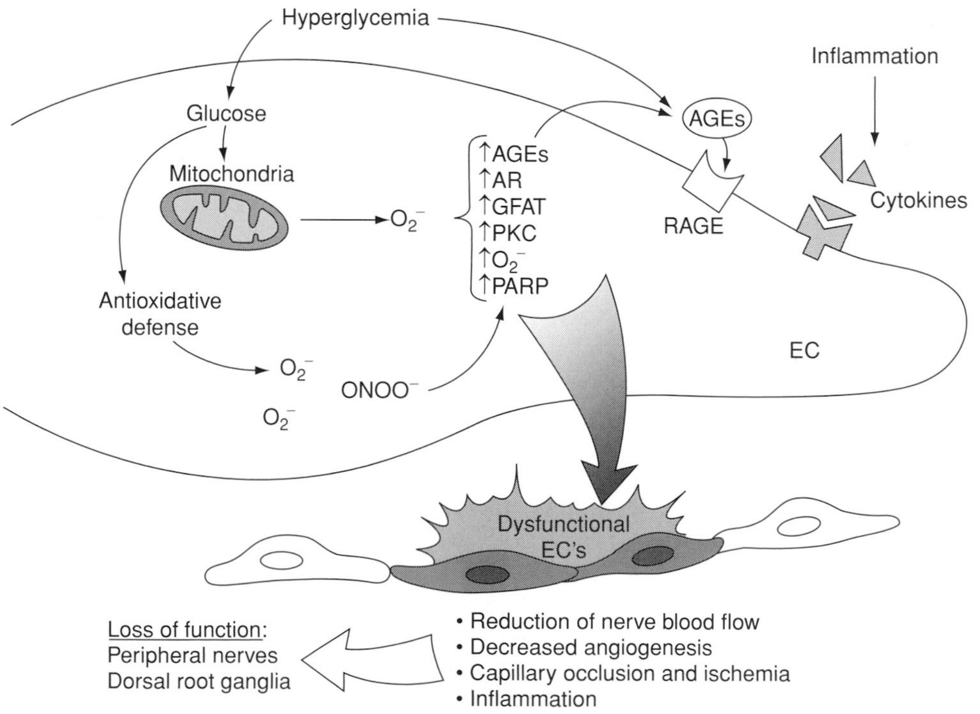

Figure 150.5. Pathways contributing to vascular dysfunction in diabetic neuropathy. AGEs, advanced glycation end-products; AR, aldose reductase; GFAT, glutamine:fructose-6-phosphate amidotransferase; PARP, poly(ADP-ribose)-polymerase; PKC, protein kinase C; RAGE, receptor for AGE.

prevented in $RAGE^{-/-}$ mice, although $RAGE^{-/-}$ mice were not protected from diabetes-induced loss of PGP9.5 positive plantar nerve fibers. In addition, diabetic neuropathy–associated loss of pain perception was reversed in wild-type mice treated with sRAGE (89). Although RAGE is known to induce NF-κB activation, it is not clear if this pathway underlies RAGE-dependent neuronal dysfunction in diabetes. It is speculated that NF-κB activation in cells of the vessel wall of the vasa nervorum results in vascular dysfunction and leads to subsequent neuronal dysfunction. An association between microangiopathy and neuropathy has long been considered, and the AGE–RAGE axis could be one mechanism by which vascular changes ultimately lead to neuronal damage. Thus, chronic vascular dysfunction might directly mediate sensory deficits.

The vasa nervorum expresses ET-1 receptors and thus is susceptible to ET-1–dependent vasoconstriction, which results in prolonged nerve ischemia and infarction. Microvascular ischemia might be further promoted by a reduction in vascular endothelial thrombomodulin expression recently observed in the nerve microvasculature of diabetic patients. Restoring vascularization of the vasa nervorum through VEGF-induced angiogenesis in mice results in reversal of experimental diabetic neuropathy, thereby supporting the concept that microvascular ischemia in the vasa nervorum might promote diabetic neuropathy (90). When intravascular oxygen saturation and blood flow was determined in human sural nerves using microlight-guide spectrophotometry, a reduced nerve oxygenation and impaired blood flow in patients with mild-moderate sensory motor diabetic neuropathy was observed. Remarkably, intravascular oxygen saturation was higher in patients with painful neuropathy, compared with those without pain, indicating that hemodynamic factors also could play an important role in the pathogenesis of neuropathic pain (91).

ENDOTHELIAL DYSFUNCTION IN DIABETIC RETINOPATHY

Small-vessel damage in the eye, termed *retinopathy*, affects every patient with longstanding diabetes, provided that he has had an extended period of insufficient glycemic control. The disease is divided into an early nonproliferative stage, during which progressive vascular occlusion occurs, and an advanced stage, during which new vessels are formed by angiogenesis. Predominantly in the elderly, a diabetic maculopathy is distinguished; this condition also arises from the stimuli following progressive capillary dropout and reflects an increase in retinal permeability around the macula. The term *nonproliferative diabetic retinopathy* is misleading from a vessel-oriented standpoint, because there is already significant formation of new blood vessels, although they are not growing beyond the retina borders. The clinical stages of diabetic retinopathy are shown in Table 150-1.

The two most characteristic features of incipient diabetic retinopathy are increased vascular permeability and progressive vascular occlusion (92). The time course of hyperglycemic retinal damage has been studied in diabetic animal models (92). The very first sign of hyperglycemic damage is a transient breakdown of the BRB. Structural defects start with the loss of intramural pericytes. With progressive hyperglycemic exposure, ECs disappear focally, leaving completely acellular capillaries behind (Figure 150.6). In areas unaffected by capillary dropout, EC numbers increase over time, suggesting a dual response to hyperglycemic injury. Acellular capillaries are the most significant lesions in the diabetic retina, because they represent the universal phenotype of hyperglycemia-induced vascular cell damage and are the triggers of subsequent vascular changes (Figure 150.6). Earlier studies in humans correlating in vivo perfusion (by fluorescein angiography) with postmortem retinal vascular digest preparations demonstrated that acellular capillaries represent capillary nonperfusion, and that microaneurysms (which are the earliest clinically detectable signs of diabetic retinopathy in a patient with diabetes), cluster around areas of capillary dropout (94). Leaky microaneurysms, focal hard exudates, and dot-blot

Table 150-1: Clinical Stages of Diabetic Retinopathy

Nonproliferative Diabetic Retinopathy	
Stage	*Fundus characteristic*
Mild	Microaneurysm, retinal hemorrhage
Moderate	+ venous beading + IRMAs present
Severe	(4–2–1 rule)
	Microaneurysms + retinal hemorrhages in four quadrants
	Venous beading in two (or more) quadrants
	IRMAs in one (or more) quadrant(s)
Proliferative Diabetic Retinopathy	
Neovascularization from the disc (NVD)	
Neovascularization elsewhere (NVE)	
Preretinal/vitreous hemorrhage	
Retinal detachment	

IRMA, intraretinal microvascular abnormalities

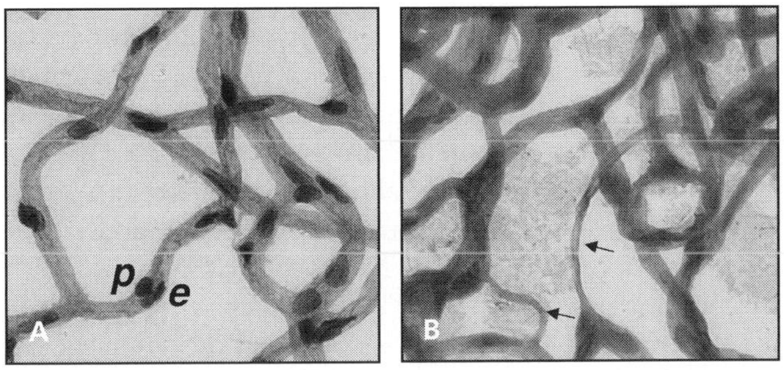

Figure 150.6. Retinal digest preparation of a normal (**A**), and a diabetic (**B**) rat retina, showing acellular capillaries upon exposure to hyperglycemia of <25 mmol/L glucose over a period of 30 weeks. Arrows indicate acellular capillaries. In the normal rat retina, a homogenous distribution of ECs (*e*) and pericytes (*p*) is present.

hemorrhages are direct indicators of a breakdown of the BRB. With increasing disease duration, and greater formation of nonperfused retinal areas, the retina responds to progressive capillary dropout with the formation of new blood vessels that penetrate through the inner limiting membrane into the vitreous body. In general, the number of newly formed vessels often correlates with the extent of nonperfusion in the retina. These neovascularizations tend to spread within the preretinal vitreous and are prone to rhexis and preretinal bleeding.

TARGETING DIABETIC VASCULAR COMPLICATIONS – REVERSIBILITY OF ENDOTHELIAL DYSFUNCTION

A number of drugs that improve endothelial function have been shown to significantly lower the risk for developing late diabetic vascular complications (95,96). Insulin-sensitizing drugs that improve endothelium-dependent FMD also reduce the risk for vascular disease (45,95,97,98). Other promising strategies include inhibition of ROS by antioxidative drugs (88,95,99), blockade of receptors that initiate the cellular response to hyperglycemia (100), and targeting "supporting enzymes" such as transketolase, cofactors such as BH4 (95), and detoxifying mechanisms such as glyoxalase-1. For example, activating transketolase has been shown to reduce AGEs, lower elevated hexosamine pathway metabolites, and activate PKC in ECs kept in high-glucose conditions. Furthermore, in vivo, long-term administration of benfotiamine, the lipid-soluble prodrug of vitamin B_1, reduced retinal AGEs, NF-κB activation, hexosamine pathway metabolites, and PKC activation to near normal levels. The result of these biochemical improvements was a major reduction in retinal capillary damage. The numbers of acellular capillaries were reduced by 80% in diabetic animals treated for 9 months.

KEY POINTS

- Although type 1 and type 2 diabetes are diseases of a different etiopathogenesis, both are characterized by disorders of glucose metabolism, alterations in insulin action (even patients with type 1 diabetes demonstrate decreased glucose disposal rates), and multifaceted endothelial responses to injury. The underlying biochemical abnormality, referred to as the *unifying concept*, seems to be the mitochondrial overproduction of ROS; however, the mechanisms leading to tissue-specific damage, such as vessel wall inflammation in large vessels, fibrosis in kidneys, thickening of endoneurial vessel basal lamina in neurons, and proliferation in the eye are not completely understood.
- In light of the considerable success of modern therapies for lowering blood glucose, lipid levels, and hypertension, it is even more fascinating to see that nonpharmaceutical interventions such as lifestyle changes and psychosocial stress relief management also improve endothelial dysfunction in diabetic patients. The latter implies that we are still far from understanding endothelial dysfunction in diabetes. In particular, results from pancreas transplantation but also from initial experiments using stem cell therapy approaches imply that ECs have a high potential to regenerate from hyperglycemic damage.
- Thus, diabetes research, with a focus on EC biology is beginning to recognize that the future will involve not only a better understanding of mechanisms underlying vascular complications, but also a more detailed knowledge of endogenous protective and repair mechanisms. This will provide the basis for a pathogenesis-oriented treatment that not only targets the underlying damaging mechanism, but also supports the cellular defense.

REFERENCES

1 Son SM, Whalin MK, Harrison DG, et al. Oxidative stress and diabetic vascular complications. *Curr Diab Rep*. 2004;4:247–252.

2 Lipinski B. Pathophysiology of oxidative stress in diabetes mellitus. *J Diabetes Complicat*. 2001;15:203–210.

3 Brownlee M. Biochemistry and molecular cell biology of diabetic complications. *Nature*. 2001;414:813–820.

4 Nishikawa T, Edelstein D, Brownlee M. The missing link: a single unifying mechanism for diabetic complications. *Kidney Int Suppl*. 2000;77:S26–S30.

5 Evans JL, Goldfine ID, Maddux BA, Grodsky GM. Oxidative stress and stress-activated signaling pathways: a unifying hypothesis of Type 2 diabetes. *Endocr Rev*. 2002;23:599–622.

6 Ramasamy R, Oates O. Aldose reductase and vascular stress. In: Marso SP, Stern DM, eds. *Diabetes and Cardiovascular Disease*. Philadelphia: Lippincott Williams & Wilkins; 2004:55–74.

7 Inoguchi T, Sonta T, Tsubouchi H, et al. Protein kinase C-dependent increase in reactive oxygen species (ROS) production in vascular tissues of diabetes: role of vascular NAD(P)H oxidase. *J Am Soc Nephrol*. 2003;14(8 Suppl 3):S227–S232.

8 Weidig P, McMaster D, Bayraktutan U. High glucose mediates pro-oxidant and antioxidant enzyme activities in coronary endothelial cells. *Diabetes Obes Metab*. 2004;6:432–441.

9 Bierhaus A, Chevion S, Chevion M, et al. Advanced glycation end product (AGEs) induced activation of NF-κB is suppressed by α-lipoic acid in cultured endothelial cells. *Diabetes*. 1997;46:1481–1490.

10 Nishikawa T, Edelstein D, Du XL, et al. Normalizing mitochondrial superoxide production blocks three pathways of hyperglycaemic damage. *Nature*. 2000;404:787–790.

11 Federici M, Menghini R, Mauriello A, et al. Insulin-dependent activation of endothelial nitric oxide synthase is impaired by O-linked glycosylation modification of signaling proteins in human coronary endothelial cells. *Circulation.* 2002;106:466–472.

12 Du XL, Edelstein D, Rossetti L, et al. Hyperglycemia-induced mitochondrial superoxide overproduction activates the hexosamine pathway and induces plasminogen activator inhibitor-1 expression by increasing SP1 glycosylation. *Proc Natl Acad Sci USA.* 2000;97:12222–12226.

13 Way KJ, Katai N, King GL. Protein kinase C and the development of diabetic vascular complications. *Diabet Med.* 2001;18:945–959.

14 Rojas S, Rojas R, Lamperti L, et al. Hyperglycaemia inhibits thymidine incorporation and cell growth via protein kinase C, mitogen-activated protein kinases and nitric oxide in human umbilical vein endothelium. *Exp Physiol.* 2003;88:209–219.

15 Ishii H, Jirousek MR, Koya D, et al. Amelioration of vascular dysfunctions in diabetic rats by an oral PKC beta inhibitor. *Science.* 1996;272:728–731.

16 Park JY, Takahara N, Gabriele A, et al. Induction of endothelin-1 expression by glucose: an effect of protein kinase C activation. *Diabetes.* 2000;49:1239–1248.

17 Suzuma K, Takahara N, Suzuma I, et al. Characterization of protein kinase C beta isoform's action on retinoblastoma protein phosphorylation, vascular endothelial growth factor-induced endothelial cell proliferation, and retinal neovascularization. *Proc Natl Acad Sci USA.* 2002;99:721–726.

18 Danis RP, Bingaman DP, Jirousek M, Yang Y. Inhibition of intraocular neovascularization caused by retinal ischemia in pigs by PKCbeta inhibition with LY333531. *Invest Ophthalmol Vis Sci.* 1998;39:171–179.

19 Quagliaro L, Piconi L, Assaloni R, et al. Intermittent high glucose enhances apoptosis related to oxidative stress in human umbilical vein endothelial cells: the role of protein kinase C and NAD(P)H-oxidase activation. *Diabetes.* 2003;52:2795–2804.

20 Du Y, Miller CM, Kern TS. Hyperglycemia increases mitochondrial superoxide in retina and retinal cells. *Free Radic Biol Med.* 2003;35:1491–1499.

21 Baynes J. Glycation and advanced glycation reactions. In: Gries FA, Cameron NE, Low PA, Ziegler D, eds. *Textbook of Diabetic Neuropathy.* Stuttgart: Georg Thieme-Verlag; 2003:96–105.

22 Bierhaus A, Hofmann MA, Ziegler R, Nawroth PP. The AGE/RAGE pathway in vascular disease and diabetes mellitus. Part I: the AGE-concept. *Cardiovasc Res.* 1998;37:586–600.

23 Thornalley PJ. Cell activation by glycated proteins. AGE receptors, receptor recognition factors and functional classification of AGEs. *Cell Mol Biol.* 1998;44:1013–1023.

24 Schmidt AM, Stern DM. RAGE: a new target for the prevention and treatment of the vascular and inflammatory complications of diabetes. *Trends Endocrinol Metab.* 2000;11:368–375.

25 Lu C, He JC, Cai W, et al. Advanced glycation endproduct (AGE) receptor 1 is a negative regulator of the inflammatory response to AGE in mesangial cells. *Proc Natl Acad Sci USA.* 2004;101:11767–11772.

26 Garcia Soriano F, Virag L, Jagtap P, et al. Diabetic endothelial dysfunction: the role of poly(ADP-ribose) polymerase activation. *Nat Med.* 2001;7:108–113.

27 Piconi L, Quagliaro L, Da Ros R, et al. Intermittent high glucose enhances ICAM-1, VCAM-1, E-selectin and interleukin-6 expression in human umbilical endothelial cells in culture: the role of poly(ADP-ribose) polymerase. *J Thromb Haemost.* 2004;2:1453–1459.

28 Du X, Matsumura T, Edelstein D, et al. Inhibition of GAPDH activity by poly(ADP-ribose) polymerase activates three major pathways of hyperglycemic damage in endothelial cells. *J Clin Invest.* 2003;112:1049–1057.

29 Zheng L, Szabo C, Kern TS. Poly(ADP-ribose) polymerase is involved in the development of diabetic retinopathy via regulation of nuclear factor-kappaB. *Diabetes.* 2004;53:2960–2967.

30 Babaei-Jadidi R, Karachalias N, Ahmed N, et al. Prevention of incipient diabetic nephropathy by high-dose thiamine and benfotiamine. *Diabetes.* 2003;52:2110–2120.

31 Hammes HP, Du X, Edelstein D, et al. Benfotiamine blocks three major pathways of hyperglycemic damage and prevents experimental diabetic retinopathy. *Nat Med.* 2003;9:294–299.

32 Lam HC. Role of endothelin in diabetic vascular complications. *Endocrine.* 2001;14:277–284.

33 Quehenberger P, Bierhaus A, Fasching P, et al. Endothelin-1 transcription is under control of nuclear factor-κB in AGE-stimulated cultured endothelial cells. *Diabetes.* 2000;49:1561–1670.

34 Keynan S, Khamaisi M, Dahan R, et al. Increased expression of endothelin-converting enzyme-1c isoform in response to high glucose levels in endothelial cells. *J Vasc Res.* 2004;41:131–140.

35 Migdalis IN, Kalogeropoulou K, Karmaniolas KD, et al. Plasma levels of endothelin and early carotid atherosclerosis in diabetic patients. *Res Commun Mol Pathol Pharmacol.* 2000;108:15–25.

36 Cameron NE, Cotter MA. Effects of a nonpeptide endothelin-1 ETA antagonist on neurovascular function in diabetic rats: interaction with the renin-angiotensin system. *J Pharmacol Exp Ther.* 1996;278:1262–1268.

37 Huang PL. Endothelial nitric oxide synthase and endothelial dysfunction. *Curr Hypertens Rep.* 2003;5:473–480.

38 Williams TM, Lisanti MP. The Caveolin genes: from cell biology to medicine. *Ann Med.* 2004;36:584–595.

39 Alp NJ, Channon KM. Regulation of endothelial nitric oxide synthase by tetrahydrobiopterin in vascular disease. *Arterioscler Thromb Vasc Biol.* 2004;24:413–420.

40 Alp NJ, Mussa S, Khoo J, et al. Tetrahydrobiopterin-dependent preservation of nitric oxide-mediated endothelial function in diabetes by targeted transgenic GTP-cyclohydrolase I overexpression. *J Clin Invest.* 2003;112:725–735.

41 Zou MH, Cohen R, Ullrich V. Peroxynitrite and vascular endothelial dysfunction in diabetes mellitus. *Endothelium.* 2004;11:89–97.

42 Zou MH, Kirkpatrick SS, Davis BJ, et al. Activation of the AMP-activated protein kinase by the anti-diabetic drug metformin in vivo. Role of mitochondrial reactive nitrogen species. *J Biol Chem.* 2004;279:43940–43951.

43 Schiekofer S, Balletshofer B, Andrassy M, et al. Endothelial dysfunction in diabetes mellitus. *Semin Thromb Hemost.* 2000;26:503–511.

44 Title LM, Cummings PM, Giddens K, Nassar BA. Oral glucose loading acutely attenuates endothelium-dependent vasodilation in healthy adults without diabetes: an effect prevented by vitamins C and E. *J Am Coll Cardiol.* 2000;36:2185–2191.

45 Rossi R, Cioni E, Nuzzo A, et al. Endothelial-dependent vasodilation and incidence of type 2 diabetes in a population of healthy postmenopausal women. *Diabetes Care.* 2005;28:702–707.

46 Tarkun I, Arslan BC, Canturk Z, et al. Endothelial dysfunction in young women with polycystic ovary syndrome: relationship

with insulin resistance and low-grade chronic inflammation. *Obstet Gynecol Surv.* 2005;60:180–181.
47 Balletshofer BM, Rittig K, Stock J, et al. Insulin resistant young subjects at risk of accelerated atherosclerosis exhibit a marked reduction in peripheral endothelial function early in life but not differences in intima-media thickness. *Atherosclerosis.* 2003;171:303–309.
48 Balletshofer BM, Rittig K, Stock J, et al. Zusammenhang zwischen Fibrinolyseaktivität und Intima-Media-Dicke bei Übergewicht und Insulinresistenz. *Diabetes und Stoffwechsel.* 2005;4:57–62.
49 Jarvisalo M, Raitakari M, Toikka J, et al. Endothelial dysfunction and increased arterial intima-media thickness in children with type 1 diabetes. *Circulation.* 2004;109:1750–1755.
50 Pinkney JH, Stehouwer CD, Coppack SW, Yudkin JS. Endothelial dysfunction: cause of the insulin resistance syndrome. *Diabetes.* 1997;46(Suppl 2):S9–S13.
51 King GL, Johnson SM. Receptor-mediated transport of insulin across endothelial cells. *Science.* 1985;227:1583–1586.
52 Vicent D, Ilany J, Kondo T, et al. The role of endothelial insulin signaling in the regulation of vascular tone and insulin resistance. *J Clin Invest.* 2003;111:1373–1380.
53 Kondo T, Vicent D, Suzuma K, et al. Knockout of insulin and IGF-1 receptors on vascular endothelial cells protects against retinal neovascularization. *J Clin Invest.* 2003;111:1835–1842.
54 Kondo T, Hafezi-Moghadam A, Thomas K, et al. Mice lacking insulin or insulin-like growth factor 1 receptors in vascular endothelial cells maintain normal blood-brain barrier. *Biochem Biophys Res Commun.* 2004;317:315–320.
55 Federici M, Pandolfi A, De Filippis EA, et al. G972R IRS-1 variant impairs insulin regulation of endothelial nitric oxide synthase in cultured human endothelial cells. *Circulation.* 2004;109:399–405.
56 Montagnani M, Golovchenko I, Kim I, et al. Inhibition of phosphatidylinositol 3-kinase enhances mitogenic actions of insulin in endothelial cells. *J Biol Chem.* 2002;277:1794–1799.
57 Frank RN. Diabetic retinopathy. *N Engl J Med.* 2004;350:48–58.
58 Duh E, Aiello LP. Vascular endothelial growth factor and diabetes: the agonist versus antagonist paradox. *Diabetes.* 1999;48: 1899–1906.
59 Hellström M, Gerhardt H, Kalén M, et al. Lack of pericytes leads to endothelial hyperplasia and abnormal vascular morphogenesis. *J Cell Biol.* 2001;153:543–553.
60 Sakurai K, Sawamura T. Stress and vascular responses: endothelial dysfunction via lectin-like oxidized low-density lipoprotein receptor-1: close relationships with oxidative stress. *J Pharmacol Sci.* 2003;91:182–186.
61 Ritchie SA, Ewart MA, Perry CG, et al. The role of insulin and the adipocytokines in regulation of vascular endothelial function. *Clin Sci (Lond).* 2004;107:519–532.
62 Sjoholm A, Nystrom T. Endothelial inflammation in insulin resistance. *Lancet.* 2005;12(365):610–612.
63 Dandona P, Aljada A, Chaudhuri A, Mohanty P. Endothelial dysfunction, inflammation and diabetes. *Rev Endocr Metab Disord.* 2004;5(1):89–97.
64 Grant PJ. Inflammatory, atherothrombotic aspects of type 2 diabetes. *Curr Med Res Opin.* 2005;1(Suppl 21):5–12.
65 Bierhaus A, Schiekofer S, Schwaninger M, et al. Diabetes-associated sustained activation of the transcription factor NF-κB. *Diabetes.* 2001;50:2792–2809.
66 Mangge H, Schauenstein K, Stroedter L, et al. Low grade inflammation in juvenile obesity and type 1 diabetes associated with early signs of atherosclerosis. *Exp Clin Endocrinol Diabetes.* 2004;112:378–382.
67 Koltai M, Hosford D, Braqut P. Role of PAF and cytokines in microvascular tissue injury. *J Lab Clin Med.* 1992;119:461–466.
68 Clausen P, Jacobsen P, Rossing K, et al. Plasma concentrations of VCAM-1 and ICAM-1 are elevated in patients with Type 1 diabetes mellitus with microalbuminuria and overt nephropathy. *Diabet Med.* 2000;17:644–649.
69 Sampson MJ, Davies IR, Brown JC, et al. Monocyte and neutrophil adhesion molecule expression during acute hyperglycemia and after antioxidant treatment in type 2 diabetes and control patients. *Arterioscler Thromb Vasc Biol.* 2002;22:1187–1193.
70 Morigi M, Angioletti S, Imberti B, et al. Leukocyte-endothelial interaction is augmented by high glucose concentrations and hyperglycemia in a NF-κB-dependent fashion. *J Clin Invest.* 1998;101:1905–1915.
71 Carr ME. Diabetes mellitus: a hypercoagulable state. *J Diabetes Complications.* 2001;15:44–54.
72 Sobel BE. Fibrinolysis and diabetes. *Front Biosci.* 2003;8:D1085–D1092.
73 Bates DO, Hillman NJ, Williams B, et al. Regulation of microvascular permeability by vascular endothelial growth factors. *J Anat.* 2002;200:81–97.
74 Hammes HP, Lin J, Bretzel RG, et al. Upregulation of the vascular endothelial growth factor/vascular endothelial growth factor receptor system in experimental background diabetic retinopathy of the rat. *Diabetes.* 1998;47:401–406.
75 Aiello LP, Avery RL, Arrigg PG, et al. Vascular endothelial growth factor in ocular fluid of patients with diabetic retinopathy and other retinal disorders. *N Engl J Med.* 1994;331:1480–1487.
76 Pierce EA, Avery RL, Foley ED, et al. Vascular endothelial growth factor/vascular permeability factor expression in a mouse model of retinal neovascularization. *Proc Natl Acad Sci USA.* 1995;92:905–909.
77 Grant MB, Afzal A, Spoerri P, et al. The role of growth factors in the pathogenesis of diabetic retinopathy. *Expert Opin Investig Drugs.* 2004;13:1275–1293.
78 Alon T, Hemo I, Itin A, et al. Vascular endothelial growth factor acts as a survival factor for newly formed retinal vessels and has implications for retinopathy of prematurity. *Nat Med.* 1995; 1:1024–1028.
79 Loomans CJ, de Koning EJ, Staal FJ, et al. Endothelial progenitor cell dysfunction: a novel concept in the pathogenesis of vascular complications of type 1 diabetes. *Diabetes.* 2004;53:195–199.
80 Tepper OM, Galiano RD, Capla JM, et al. Human endothelial progenitor cells from type II diabetics exhibit impaired proliferation, adhesion, and incorporation into vascular structures. *Circulation.* 2002;106:2781–2786.
81 Bahlmann FH, de Groot K, Mueller O, et al. Stimulation of endothelial progenitor cells: a new putative therapeutic effect of angiotensin II receptor antagonists. *Hypertension.* 2005;45:526–529.
82 Hunt BJ. The endothelium in atherogenesis. *Lupus.* 2000;9:189–193.
83 Vinik A, Flemmer M. Diabetes and macrovascular disease. *J Diabetes Complications.* 2002;16:235–245.

84 Larsen JL, Colling CW, Ratanasuwan T, et al. Pancreas transplantation improves vascular disease in patients with type 1 diabetes. *Diabetes Care*. 2004;27:1706–1711.
85 Kessler L, Wiesel ML, Attali P, et al. Von Willebrand factor in diabetic angiopathy. *Diabetes Metab*. 1998;24:327–336.
86 Stehouwer CD. Endothelial dysfunction in diabetic nephropathy: state of the art and potential significance for non-diabetic renal disease. *Nephrol Dial Transplant*. 2004;19:778–781.
87 Cameron NE, Eaton SE, Cotter MA, Tesfaye S. Vascular factors and metabolic interactions in the pathogenesis of diabetic neuropathy. *Diabetologia*. 2001;44:1973–1988.
88 Bierhaus A, Humpert PM, Rudofsky G, et al. New treatments for diabetic neuropathy – pathogenetically oriented treatment. *Curr Diab Rep*. 2003;3:452–458.
89 Bierhaus A, Haslbeck KM, Humpert PM, et al. Loss of pain perception in diabetic neuropathy is dependent on a receptor of the immune globulin superfamily. *J Clin Invest*. 2004;114:1741–1751.
90 Schratzberger P, Schratzberger G, Silver M, et al. Favorable effect of VEGF gene transfer on ischemic peripheral neuropathy. *Nat Med*. 2000;6:405–413.
91 Eaton SE, Harris ND, Ibrahim S, et al.: Increased sural nerve epineurial blood flow in human subjects with painful diabetic neuropathy. *Diabetologia*. 2003;46:934–939.
92 Ashton N. Pathogenesis of diabetic retinopathy. In: Little HL, Jack RL, Patz AP, Forsham PH, eds. *Diabetic Retinopathy*. New York: Thieme-Stratton Inc.; 1983:85–106.
93 Engerman RL. Animal models of diabetic retinopathy. *Trans Am Acad Ophthalmol Otolaryngol*. 1976;81:OP710–OP715.
94 Bresnick GH, Davis MD, Myers FL, de Venecia G. Clinicopathologic correlations in diabetic retinopathy. II. Clinical and histologic appearances of retinal capillary microaneurysms. *Arch Ophthalmol*. 1977;95:1215–1220.
95 Puddu P, Puddu GM, Cravero E, Muscari A. Different effects of antihypertensive drugs on endothelial dysfunction. *Acta Cardiol*. 2004;59:555–564.
96 Lonn E. Antiatherosclerotic effects of ACE inhibitors: where are we now? *Am J Cardiovasc Drugs*. 2001;1:315–320.
97 Knowler WC, Barrett-Connor E, Fowler SE, et al. Reduction in the incidence of type 2 diabetes with lifestyle intervention or metformin. *N Engl J Med*. 2002;346:393–403.
98 Caballero AE, Saouaf R, Lim SC, et al. The effects of troglitazone, an insulin-sensitizing agent, on the endothelial function in early and late type 2 diabetes: a placebo-controlled randomized clinical trial. *Metabolism*. 2003;52:173–180.
99 Rosen P, Du X, Tschope D. Role of oxygen derived radicals for vascular dysfunction in the diabetic heart: prevention by alphatocopherol? *Mol Cell Biochem*. 1998;188:103–111.
100 Schmidt AM, Stern DM. RAGE: a new target for the prevention and treatment of the vascular and inflammatory complications of diabetes. *Trends Endocrinol Metab*. 2000;11:368–375.

151

The Role of the Endothelium in Normal and Pathologic Thyroid Function

Jamie Mitchell, Anthony Hollenberg, and Sareh Parangi

Beth Israel Deaconess Medical Center, Harvard Medical School, Boston, Massachusetts

Angiogenesis, the formation of new blood vessels from existing vessels, has been recognized as playing an important role in the pathophysiology of many benign and malignant diseases (1,2). This chapter focuses on the role of the endothelium in the thyroid gland and explores the advances that have been made in our understanding of how the process of angiogenesis is involved in the pathophysiology of benign and malignant diseases of this gland.

HISTORY OF ENDOTHELIAL BIOLOGY IN THE THYROID

The structure and function of the thyroid gland has been the intense focus of study by physiologists, anatomists, surgeons, and basic scientists. Most attention has focused on the thyroid follicular cell, one of the most thoroughly studied cells in the human body. By comparison, the thyroid endothelium has received little attention. In the mid 1970s, there was increasing recognition that the endothelium of endocrine organs, including the thyroid, played an important role in homeostasis. At first, microscopy was used to study the structural detail of the fenestrated endothelium present in normal rat thyroid (3) as well as benign and malignant human thyroid tissue (4). The importance of angiogenesis in the progression of malignant disease was recognized during the early 1970s, with the realization by Folkman and colleagues that, for malignant tumor growth to progress beyond approximately 2 mm in diameter, the development of a vascular supply must occur (5). The role of endothelial cell (EC) proliferation in thyroid disorders was appreciated as early as 1978, with work performed by Wollman and colleagues (6). Using an experimental model of goiter induction by thiouracil and low-iodine diets in rats, they showed that capillaries within the thyroid enlarged by the third treatment day, with an increase in EC mitoses and fusion of capillaries, as seen by electron microscopy. These changes in blood vessel morphology were not observed in extrathyroid tissue, suggesting that they were directly related to goiter formation. Additionally, in this model, EC proliferation preceded that of the thyroid epithelial cells, suggesting that angiogenesis may be an important initiating event in goitrogenesis, rather than a response to an increased metabolic demand caused by epithelial cell proliferation.

Additional insights into the importance of EC regulation in thyroid disease came during the subsequent two decades, but nearly all were related to angiogenesis in the thyroid gland. Little research was undertaken into the structure and function of the normal thyroid endothelium. One study performed by Goodman and colleagues showed that epithelial follicular cells in vitro released a growth factor into the medium that promoted EC growth and migration (7). High endothelial venules in the thyroid gland were studied in humans and rats, and thought to be the site of immunological responses in some autoimmune diseases of the thyroid (8). In 1998, Ishiwata and colleagues made the observation that increased microvessel density in papillary thyroid cancers was predictive of a worse prognosis in the form of disease-free survival (9), and Segal and colleagues determined that the degree of angiogenesis occurring in follicular tumors of the thyroid predicted frequency of intrathyroid tumor spread (10). As these and other studies began to reveal a direct relationship between angiogenesis and thyroid disease, it became increasingly clear that this process played an important role in the pathogenesis of benign and malignant disorders of the thyroid gland.

As interest in angiogenesis and its role in the malignant disease intensified, a great deal of effort was directed at characterizing the molecular mechanisms involved in the regulation of this process. One of the primary driving forces behind this research was the premise that therapeutic inhibition of blood vessel formation would selectively interfere with tumor growth. It is now well established that angiogenesis is tightly regulated by both stimulatory and inhibitory factors. Under normal conditions, these factors are well balanced, maintaining an equilibrium that prevents angiogenesis from occurring. Once this balance becomes disturbed, either via an increase in stimulatory signals, and/or a decrease in

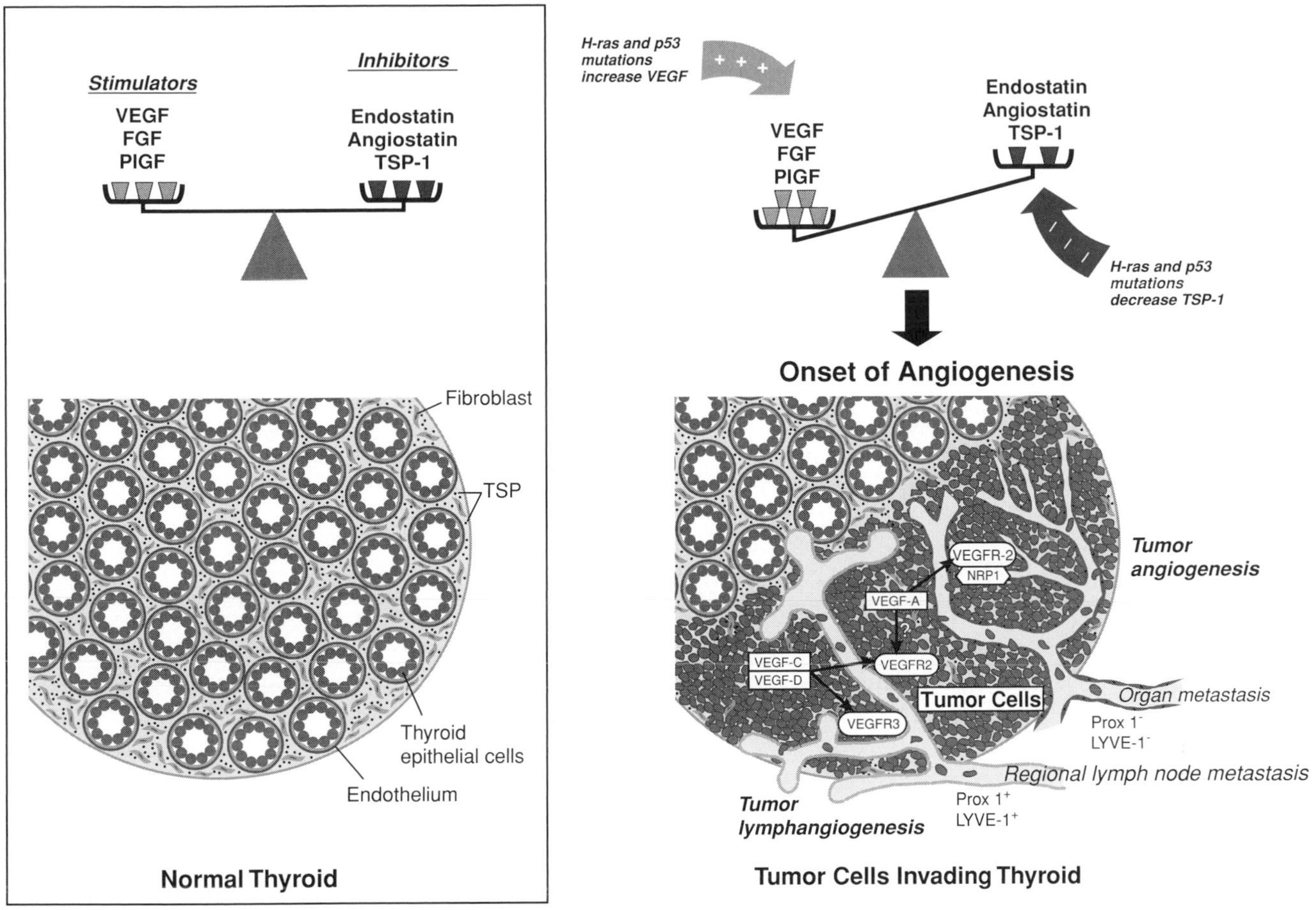

Figure 151.1. The balance of stimulators and inhibitors of angiogenesis is important in the normal thyroid gland. Disturbance of this balance, for example by Ras and p53 mutations, can lead to increases in vascular endothelial growth factor (VEGF), decreases in thrombospondin (TSP-1), and the onset of angiogenesis. Thyroid tumor cells and stromal cells provide pro- and antiangiogenic signals that are important for thyroid angiogenesis and lymphangiogenesis.

inhibitory signals, angiogenesis occurs (Figure 151.1). This has been referred to as the *angiogenic switch,* and is considered critical in the pathogenesis of malignant tumors (1). Subsequent studies demonstrated that angiogenesis is regulated by tissue-specific balances of pro and antiangiogenic factors. These findings provide a powerful rationale for developing tumor-type–specific antiangiogenesis therapies.

STRUCTURE AND FUNCTION

The thyroid gland has a rich blood supply, with an estimated flow rate of 5 mL/g tissue per minute. The entire blood volume (approximately 5 liters), which moves through the lungs approximately once every minute and through the kidneys once every 5 minutes, moves through the thyroid approximately once every hour. While the thyroid represents only 0.4% of total body weight, it accounts for 2% of total blood flow. The principal arterial supply of the thyroid is derived from the paired superior and inferior thyroid arteries (Figure 151.2). Additional arterial supply to the gland is provided by the thyroidea ima, and small branches from laryngeal and tracheoesophageal arteries. The inferior thyroid arteries arise from the thyrocervical trunk off the subclavian artery in most cases. They then run superiorly behind the common carotid artery before turning medially and entering the thyroid gland. The superior thyroid arteries arise as the first branch of the external carotid artery. They course inferiorly to enter the superior poles of the thyroid gland. Frequently, the artery branches at this point, with the main branch coursing over the anterior surface of the superior pole, and the smaller branch running posterior to the superior pole before entering the thyroid gland. The thyroidea ima artery is encountered in less than 10% of patients but, when present, arises from the aorta or brachiocephalic artery and courses superiorly along the anterior aspect of the trachea, finally entering the isthmus or one of the lower poles. Branching of the large paired arteries occurs on the surface of the thyroid, where several orders of branching results in a large vascular network. Only after a significant degree of branching do small arteries penetrate deep into the thyroid parenchyma. The penetrating vessels arborize among the follicles, ultimately sending a follicular artery to each follicle.

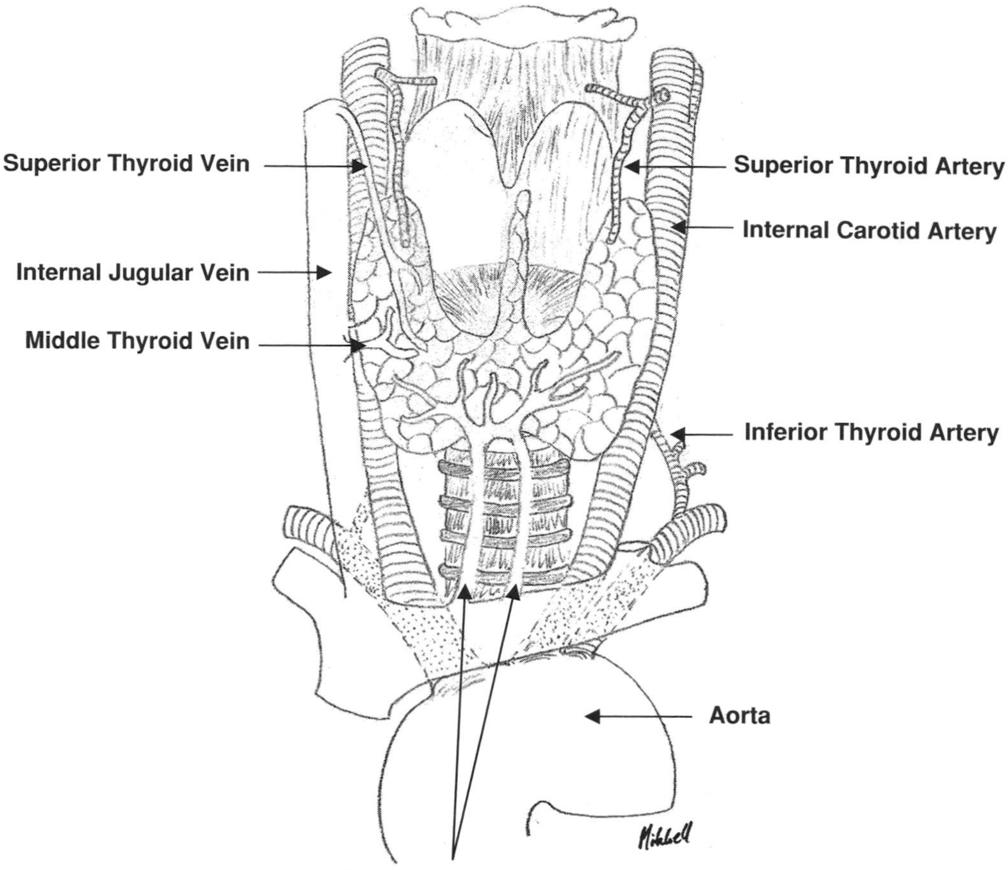

Figure 151.2. Vascular anatomy of the thyroid gland.

Venous drainage of the thyroid occurs through the superior, middle, and inferior thyroid veins (see Figure 151.2). The superior thyroid veins are formed from a confluence of vessels from the upper pole and course superiorly, crossing the carotid artery high in the neck to drain into the internal jugular veins. The middle thyroid veins drain into the internal jugular veins as well, after crossing the carotid artery anterior to the inferior thyroid artery. The inferior thyroid veins descend from the inferior poles and isthmus and drain into the brachiocephalic veins.

With the advent of transmission electron microscopy, a great deal was revealed about the structural characteristics of the ECs in the thyroid gland. The basement membrane of the follicular cells is separated from the basement membrane of adjacent ECs by a space of only a few Å (3,11). The membrane of the ECs is interrupted frequently by fenestrae of approximately 450Å in diameter. At these points, the lumen of the capillary appears to be in direct contact with the basement membrane of the thyroid follicular cells. This structural arrangement, which is thought to be similar to other endocrine organ endothelium, is believed to allow free diffusion of materials (e.g., thyroid hormone) out of the cell and into the intravascular space (Figure 151.3).

Although much of what is known about thyroid endothelial structure and function is similar to that seen in other vascular beds, recent discoveries have hinted at the fact that endothelial structure and/or function may be more tissue- or organ-specific than previously believed. A new member of the vascular endothelial growth factor (VEGF) family of angiogenic proteins has been discovered that is specific for endocrine organs, called endocrine gland VEGF (EG-VEGF). Although thus far this protein is believed to be primarily expressed in steroidogenic endocrine glands (such as the adrenal gland), these findings suggest that there may be other endothelial factors as yet undiscovered that are specific for different organ systems in the body (12,13).

Recent studies have begun to clarify the role of thyroid hormone (T_4 and T_3) in angiogenesis in the thyroid gland. For example, Davis and colleagues demonstrated that physiologic concentrations of T_4 promote angiogenesis. Their results suggest that this effect is initiated at the plasma membrane of ECs. According to their model, T_4 is converted to T_3 by cellular 5′-deiodinase activity; T_3 enters the cell and cell nucleus, forms a complex with nuclear thyroid hormone receptor (TRβ1), and initiates transcription of one or more genes relevant to angiogenesis. The presence of propylthiouracil (PTU) in the assay system did not diminish the action of T_4, however, indicating that production of T_3 via deiodination of T_4 did not contribute to the action of T_4 and occurs via activation of the extracellular-signal-regulated kinase 1 (ERK1) and

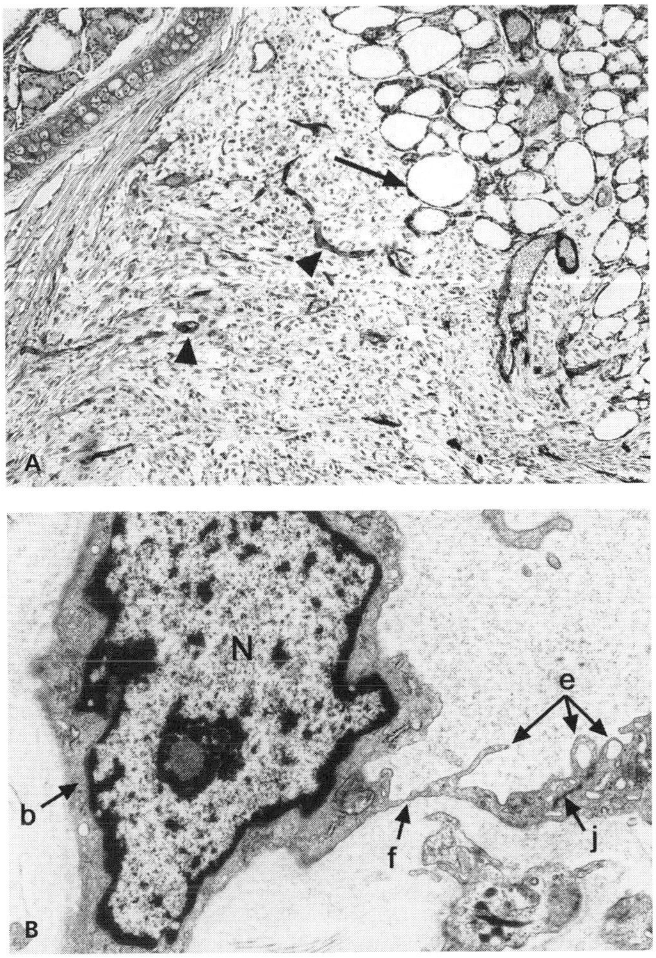

Figure 151.3. (**A**) CD31 staining of mouse ECs in normal mouse thyroid and orthotopically implanted follicular carcinoma of the thyroid CD31. Note the regular pattern of ECs in the remaining normal thyroid macrofollicules (*black arrow*) versus irregular CD31 pattern of the neovessels in the tumor (*black arrowhead*). (**B**) Medium-magnification electron microscopic image of human thyroid tumor microvessel shows the EC nucleus (*N*), basal lamina (*b*), cytoplasmic folds (*e*), junction (*j*), and fenestra (*f*). Final magnification ×18,600. (Picture courtesy of Dr. Ann Dvorak and Rita Monahan.)

-2/mitogen-activated protein kinase (MAPK) signal transduction pathway. This ultimately leads to increased expression of the proangiogenic factor fibroblast growth factor 2 (FGF2) at the EC membrane (14) with EC release of FGF2 and enhancement of the autocrine effect of released FGF2 on angiogenesis. This pathway remains to be proven in vivo.

ANGIOGENIC BALANCE IN THYROID

Studies investigating angiogenesis in the thyroid gland began with attempts to determine which of the known angiogenic stimulators and inhibitors were present in follicular cells and ECs of this organ. Several groups of researchers have shown that many of the proangiogenic factors and angiogenic inhibitors listed in Table 151-1 also exist in the thyroid

Table 151-1: The Effect of Benign and Malignant Diseases of the Thyroid on Expression of Angiogenic Factors

Normal Thyroid	*Benign Thyroid Disease*	*Malignant Thyroid Disease*
Stimulators	**Upregulated**	**Upregulated**
FGF1/FGF2/FGFR1 VEGF	FGF1/FGF2/FGFR1 VEGF/VEGFR1/ VEGFR2	VEGF/VEGFR2 Ang2
MMP-2	PlGF	VEGFC/VEGFR3
Ang2/Tie2	Ang1/Tie2 EGF TGF-β	bFGF MMP-1, -2, -3, -9
Inhibitors	**Downregulated**	**Downregulated**
Angiostatin TSP-1	TSP-1	TSP-1

gland. Studies using human thyroid tissue have determined that FGF1 and -2, as well as fibroblast growth factor receptor (FGFR)1, the receptor for these factors, are all expressed by follicular cells (15). Sato and colleagues demonstrated that normal rat thyroid follicular cells express VEGF and the VEGF receptors VEGFR1 (also known as FLT-1) and VEGFR (also known as FLK-1). Exposure to elevated levels of thyroid stimulating hormone (TSH) resulted in increased *VEGF* mRNA expression as well as increase in the mRNA expression of the *VEGFR1* and *VEGFR2*. These in vitro and in vivo findings suggest that VEGF is produced by thyroid follicles in response to stimulators of TSH receptors, via the protein kinase A and C pathways. VEGF, a secretable angiogenesis factor, subsequently stimulates VEGFR1 on ECs in a paracrine manner, leading to their proliferation and producing hypervascularity of the thyroid gland, as seen in patients with Graves disease (16). Follicular cells from the thyroid gland of rats have been shown to express angiopoietin 1 (Ang1) (17), whereas ECs purified from human thyroids have been reported to express Tie1, the receptor for Ang1 (18). Members of the matrix metalloproteinase (MMP) family of proangiogenic proteins also have been characterized in human primary thyroid follicular cell cultures (19). These proteins have been found to be secreted by fibroblasts present in the thyroid (20).

In addition to these proangiogenic factors, several angiogenic inhibitors have been found to be expressed in the thyroid. Ramsden and colleagues determined that primary cultured human follicular cells secrete angiostatin (19), whereas other studies have shown that thrombospondin (TSP)-1 is secreted from the basolateral pole of thyroid follicular cells (21). Collectively, the data from these and other studies provide compelling evidence for a functional role of many of the known angiogenic stimulators and inhibitors in regulating thyroid gland angiogenesis. Moreover, they lay a strong foundation for investigating the nature and mechanisms of the imbalance that underlies the development of benign and malignant diseases of the thyroid.

THYROID ENDOTHELIUM IN DISEASE

Benign Disease

The thyroid gland is one of the most common sites of benign epithelial hyperplasia, affecting nearly 15% of the adult population. Although these conditions, unlike malignancies, are generally not life-threatening, they nevertheless frequently require surgical intervention to relieve compression of the trachea or laryngeal nerves (22). Therefore, the ability to prevent these conditions or better treat them could potentially be of significant clinical benefit. Increasing evidence suggests an important role of angiogenesis in mediating benign thyroid disease. Thus, an understanding of the mechanisms underlying new blood vessel formation in these diseases may have therapeutic implications. Two benign diseases of the thyroid gland are associated with an increase in vascularity: hyperplastic goiter (multinodular goiter) and thyroiditis. As previously mentioned, the study by Wollman and colleagues in 1978 provided some of the first evidence that angiogenesis was important in the pathogenesis of benign diseases of the thyroid (6). Since then, many more studies have been conducted to examine the role of angiogenic factors in these conditions.

Hyperplastic Goiter

Both in vivo and in vitro studies in rats and humans have shown that the FGFs are important in goitrogenesis. For example, treatment of rats with systemic FGF1 has been shown to induce goiter formation (23). It also has been shown that follicular cells from human nontoxic multinodular goiters display increased expression of FGF1 and -2 compared with follicular cells from normal thyroid glands. The receptor for these factors, FGFR1, was also shown to be upregulated in these follicular cells (15). Additional evidence for the importance of FGF in goitrogenesis came from a study performed by Logan and colleagues, which showed that rats treated with methimazole and a low-iodine diet (a goitrogenic treatment) displayed upregulation of FGF2 expression in the follicular cell nuclei as well as the stroma of the thyroid, and that neovascularization accompanied this upregulation (24). The data from these and other studies provide strong evidence for the concept that FGF1 and -2 play an important role in goitrogenesis.

Evidence also exists for the other proangiogenic factors, with VEGF and the angiopoietins playing an important role in the pathogenesis of benign thyroid epithelial cell hyperplasia. Studies performed in rats have shown that, after chemical induction of goiter with methimazole, expression of VEGF, placental growth factor (PlGF; another member of the VEGF family of proteins), and its receptors VEGFR1 and -2 were upregulated in ECs (16,25,26). Similar studies have shown that Ang1 and its receptor Tie2 are also upregulated in the ECs of experimentally induced goiters in rats (27).

In addition to the upregulation of proangiogenic factors, the angiogenic inhibitor TSP-1 has been shown to be downregulated after the induction of goiters with methimazole and a low-iodine diet in rats, and this downregulation is concurrent with the stimulation of angiogenesis (28). These studies strongly suggest that angiogenesis is important in the pathophysiology of goitrogenesis, involving a disruption in the normal balance of angiogenic stimulators and inhibitors in the thyroid follicular cells and ECs. The mechanisms by which the normal balance of angiogenic stimulators and inhibitors is altered leading to angiogenesis and goitrogenesis remain unclear.

Autoimmune Thyroid Disease

As previously mentioned, angiogenesis is implicated in the development of several pathologic conditions, including inflammatory disorders of the thyroid such as thyroiditis (29,30). Several studies have demonstrated an important role for angiogenic factors in the pathogenesis of benign inflammatory diseases of the thyroid, including Graves disease and Hashimoto thyroiditis. For example, studies comparing healthy subjects and patients with Hashimoto thyroiditis showed a significant upregulation of *VEGF* gene expression in the serum from Hashimoto subjects, which correlated with increased vascular flow on ultrasound vascular flow studies, as well as an increase in VEGF protein levels in the thyroid gland (31,32). Another study found that primary human thyroid follicular cells in vitro showed increased expression of *VEGF* mRNA after incubation with Grave immunoglobulin(Ig)G (16). Serum levels of VEGF have been shown to be significantly elevated in patients with Graves disease and Hashimoto thyroiditis compared with healthy controls (33). In this study, the thyroids of those patients with thyroiditis were shown to have significantly increased vascularity compared with the thyroids of healthy subjects (31).

Another study examining the role of various angiogenic growth factors in thyroids from patients with subacute thyroiditis (often caused by an inflammatory reaction to a viral infection) suggested a complex process whereby increased expression of VEGF, FGF2, transforming growth factor (TGF)-β, and epidermal growth factor (EGF) in infiltrating monocytes and macrophages and thyroid follicular cells regulate endothelial and follicular proliferation (as measured by increased bromodeoxyuridine (BrdU) uptake in thyroid follicular cells after treatment with EGF) (34). As with goitrogenesis, regulation of angiogenesis appears to play an important role in the pathophysiology of inflammatory conditions of the thyroid.

In summary, a growing body of evidence supports a role for angiogenesis in the pathogenesis of hyperplastic goiter formation and inflammatory disorders of the thyroid (see Table 151-1). The extent to which the imbalance in pro- and antiangiogenic factors is an initiating event that drives follicular cell growth or is a secondary response to paracrine-mediated follicular cell-derived signals remains to be determined. Whereas the early studies by Wollman and colleagues showed that neovascularization preceded follicular cell proliferation in an animal model of goitrogenesis, suggesting that angiogenesis may be the initiating event rather than a response to follicular cell growth (6), additional studies are needed to further elucidate the mechanisms at work here. Whatever the primary mechanism, it seems clear that angiogenesis is necessary for sustaining the follicular cell hyperplasia in animal models, thus providing a rationale for testing antiangiogenesis therapy in animal models. In fact, very recently Ramsden and colleagues (35) have demonstrated that inhibition of angiogenesis in rats with chemically induced goiters was possible with use of three combined antiangiogenic gene-therapy adenoviral vectors (RAd-sTie2, RAdDN-FGFR1, and RAd-sVEGFR1). In thyroids from animals treated with all three antiangiogenic vectors, there was marked, significant inhibition of Tie2, FGFR1, VEGFR1, FGF2, and VEGF expression compared with control goiters. When used individually, RAdDN-FGFR1 partially prevented goiter and RAd-sVEGFR1 partially reduced vascular volume. Their effects were not additive. RAd-sTie2 did not reduce goiter mass or vascular volume when used alone, but was essential for complete goiter inhibition. Therefore, inhibitors of angiogenesis may prove to be useful clinical tools in the future treatment of these benign disorders of the thyroid gland.

Malignant Disease

Although therapeutic manipulation of the angiogenic balance holds promise for controlling benign thyroid conditions, its greatest potential lies in the treatment of thyroid cancer. Malignancies of the thyroid encompass many different histotypes, with varying phenotypic characteristics and clinical behaviors (36). Papillary carcinomas are thyroid malignancies (derived from follicular thyroid cells) that most commonly metastasize via lymphatics to cervical lymph nodes. Follicular carcinomas of the thyroid are generally more aggressive tumors that may give rise to hematogenous metastases. Anaplastic cancers of the thyroid are extremely aggressive tumors that possess a significant potential to become invasive, and are almost invariably fatal. Each of these tumors of the thyroid arises through malignant transformation of thyroid follicular epithelial cells. Another malignant tumor, medullary thyroid carcinoma, arises from neural crest–derived C cells present in small numbers in the thyroid. These cancers generally have a worse prognosis if not diagnosed when the primary tumor is small (<1 cm) and metastasize via both lymphatic and hematogenous routes (37).

As mentioned previously, it has been recognized that, for malignant tumors to progress, they must establish a new vascular supply through angiogenesis (5). A great deal of energy has focused on attempting to discern which angiogenic factors are involved in the pathogenesis of the various thyroid malignancies, how these differences may explain their unique clinical behavior, and how this knowledge can be utilized for the development of new, potentially curative therapies for these malignancies.

Studies analyzing human thyroid malignancies have shown that expression of several isoforms of VEGF is significantly upregulated in papillary, follicular, medullary, and undifferentiated carcinomas compared with normal thyroid tissue (37). Expression of the VEGF receptor VEGFR2 also is upregulated in the ECs of all tumor types. Other in vitro studies using cultured thyroid cancer cells have provided some evidence that, in some thyroid cancers, insulin-like growth factor (IGF)-1 induced upregulation of VEGF production, associated with activation of activator protein (AP)-1 and hypoxia-inducible factor (HIF)-1α via signaling through the phosphoinositide 3-kinase (PI3K)/Akt kinase cascade is important (38). These data indicate that IGF-1 stimulates VEGF synthesis in thyroid carcinomas by an Akt-AP-1/HIF-1α–dependent pathway, thus providing the framework for clinical use of small-molecule inhibitors, including geldanamycin analogs, to abrogate proangiogenic cascades in thyroid cancer. VEGF levels have been correlated with pathologic stage and progression, suggesting their potential prognostic significance.

Ang2 also has been shown to be upregulated in all thyroid tumor types, except microscopic papillary carcinomas (37). The fact that neither VEGF nor Ang2 was upregulated in microscopic papillary carcinomas, a hypovascular stage of thyroid neoplasia, reinforces the notion that these proangiogenic factors are critical for the progression of tumors from early to more advanced stages. Ang2 was originally believed to antagonize the proangiogenic function of Ang1 (39), and these findings speak to the complex nature of interactions between various angiogenic factors and their regulation.

Because members of the VEGF and Ang family are upregulated in both benign and malignant diseases of the thyroid, these factors are clearly not sufficient for the angiogenic switch to a malignant phenotype. The hallmark of malignancy is the ability to metastasize via blood or lymphatics. This requires the ability of the malignant cells not only to stimulate neovascularization, but also to invade existing blood and lymphatic vessels. In addition to the obvious molecular and genomic changes in the follicular cells themselves, other groups of proteins also may be involved in angiogenesis. For example, one member of the VEGF family of proteins, VEGF-C, is believed to be involved specifically in lymphangiogenesis, because its receptor, VEGFR3, is expressed only by lymphatic endothelium (40–42). Studies of human thyroid malignancies have provided evidence that VEGF-C and VEGFR3 are strongly upregulated in the majority of tumors with a propensity for lymphatic dissemination, including papillary, medullary, and undifferentiated thyroid cancers (37,43).

The MMP family is another group of proteins involved in angiogenesis that may be particularly important in the ability of tumors to invade lymphatic and blood vessels. These proteins include more than 26 zinc-dependent endopeptidases that are thought to be tightly regulated by a group of naturally occurring inhibitors (tissue inhibitor of metalloproteinase

[TIMP]) and by interactions with other MMPs (44,45). These proteins play an important role in angiogenesis through their ability to degrade the basement membrane of ECs, allowing remodeling to occur for new vessel formation. These same properties make them important in the pathologic processes of tumor invasion into lymphatics and blood vessels. Certain MMPs, including MMP-2 and -9, which are known to be important in vascular invasion, have been shown to be upregulated in certain thyroid tumors, including papillary, medullary, follicular, and anaplastic carcinomas, compared with normal thyroid tissue (46,47). Although all studies are not in agreement, mounting evidence suggests that these factors may be the key to developing a malignant phenotype for some thyroid neoplasms. This is reinforced by the fact that upregulation of these factors has not been described in benign thyroid disease. Certainly, additional studies are required for further elucidation of the exceedingly complex regulatory mechanisms involved in the process of angiogenesis, as well as those disturbances in this regulation that lead to pathologic conditions in the thyroid. It is clear that angiogenesis and lymphangiogenesis are key processes involved in the pathogenesis of benign and malignant diseases of the thyroid. Although a great deal is still to be discovered, what is currently known is already being utilized to develop new therapies for thyroid diseases.

DIAGNOSTIC AND THERAPEUTIC IMPLICATIONS

Routine assays for the thyroid endothelium do not exist. Lymphovascular or capsular invasion (as observed by the pathologist) is considered a sign of malignant diseases of the thyroid and is often used to differentiate between benign and malignant conditions. Thyroid malignancies with a high incidence of lymphovascular invasion seen on routine histochemical staining are considered more clinically aggressive. However, the use of endothelial markers (e.g., CD31) to measure vascular density or vascular invasion is not standard.

The endothelium of the thyroid represents a potential therapeutic target in both benign and malignant thyroid conditions. At one time, Lugol's iodine solution was used to treat patients with large goiters. Although this therapy reduces thyroid vascularity, there is no evidence that it functions at the level of the endothelium (48). Ramsden and colleagues showed that adenoviral-mediated dominant negative inhibition of Tie2, VEGFR1, and FGFR1 resulted in complete inhibition of goiter formation in a mouse model of goiter (35). When used alone, recombinant FGFR1 adenoviral vector prevented goiter, recombinant VEGFR1 adenovirus partially reduced vascular volume, and Tie2 recombinant adenoviral vector failed to reduce goiter mass or vascular volume.

More attention has been paid to the potential of endothelial-directed therapies in malignant diseases of the thyroid. Recent exciting studies have shown that papillary carcinomas of the thyroid often are initiated by a set of transforming events with rearrangements of the rearranged during transfection (RET) receptor or activating mutations of the BRAF or RAS oncogenes that target proteins that act along a linear signaling cascade, resulting in upregulation of chemokines and their receptors (49). It is interesting that these genes seem to act cooperatively in autocrine and paracrine loops that are relevant of the establishment of the neoplastic phenotype, including autonomous proliferation, motility, and angiogenesis. The therapeutic potential of interfering with the process of angiogenesis in malignant tumors has been recognized for some time. Even though research has focused on the discovery and characterization of the effectors of angiogenesis, it is the endothelium itself and its response to these signals that is central to that potential. Three primary strategies exist for accomplishing this: (a) use of vascular targeting agents to destroy existing vessels, (b) stimulating endogenous inhibitors of angiogenesis or administering angiogenic inhibitors (antiangiogenic agents), and (c) inhibiting the production of factors that activate or promote angiogenesis. These strategies can of course also be combined, especially when compelling clinical or molecular evidence suggests that such an approach may be synergistic.

Vascular Targeting Agents

One strategy that exploits the endothelium to treat malignancies involves the use of vascular targeting agents that cause directed destruction of tumor vessels. Vascular targeting agents differ conceptually from antiangiogenic agents in that, whereas the goal of the latter is to stop the formation of new blood vessels, the goal of the former is to destroy existing tumor vessels. Vascular targeting agents exploit pathophysiological differences between normal and tumor ECs and result in a characteristic central tumor necrosis (50).

One example of a vascular targeting agent is combrestatin A4 prodrug (CA4P), which aims to destroy the tubulin cytoskeleton required for rapidly proliferating and immature tumor ECs, resulting in cell death. Studies have shown that this agent was directly cytotoxic against human anaplastic thyroid cancer cell lines in vitro. Additional in vivo studies using CA4P in a xenografted nude mouse model of anaplastic thyroid cancer resulted in significant reductions in tumor volumes (likely via both direct tumor and vascular toxicity), but also showed significant toxicity to the animals (51).

Additional strategies using vascular targeting agents include those that selectively bind to components in the endothelium unique to tumor blood vessels. The targeting moiety is usually an antibody or peptide directed against a marker that is selectively upregulated in the endothelium of tumors compared with that of normal tissues. Proteins involved in angiogenesis or vascular remodeling, such as VEGF receptors and $\alpha V\beta 3$ integrins, and cellular adhesion molecules, such as vascular cell adhesion molecule (VCAM)-1 and E-selectin, are examples of endothelial components targeted in this fashion (52). Such strategies have been shown to result in thrombosis of tumor vessels, EC death, alterations in EC shape, and the redirection of host immune defenses to attack tumor endothelium (53,54). Other investigators have attempted to utilize vascular targeting agents to deliver toxins to tumor endothelium. For example, studies have been performed using diphtheria toxin and gelonin conjugated to VEGF-A to induce regression of tumors in mouse models (55,56). One limitation of these vascular targeting strategies is that many of the targeted markers upregulated in tumor endothelium also are upregulated in the endothelium of vessels at sites of inflammation, tissue remodeling, and physiologic angiogenesis.

Antiangiogenic Agents

Many studies using animal models are being performed examining the efficacy of antiangiogenic agents in the treatment of thyroid cancers, with encouraging results. Many of these have used models of anaplastic thyroid cancer, given the poor prognosis of this disease and the lack of alternative treatments available to this patient population. Bauer and colleagues administered an anti-VEGF monoclonal antibody (mAb) recognizing VEGF isoforms 121, 165, and 189 in a subcutaneous nude mouse model of thyroid cancer. They showed a 60% reduction in tumor size after treatment compared with untreated controls (57).

Another study performed by Xu and colleagues examined the use of manumycin, a farnestyl transferase inhibitor, which also is known to inhibit angiogenesis by reducing VEGF production, and paclitaxel, which inhibits endothelial migration, in the treatment of anaplastic thyroid cancer in a mouse model. They found that, when used in combination, manumycin and paclitaxel significantly inhibited tumor growth compared with controls (58). Pietras and colleagues showed that treatment of a xenograft mouse model of anaplastic thyroid cancer with STI 571 (Gleevec), which selectively blocks the platelet-derived growth factor (PDGF) receptor kinase, the c-Kit receptor kinase, and the Abl and Arg nonreceptor kinases, in combination with tamoxifen (Taxol) resulted in a significantly increased antitumor effect compared with Taxol alone (59).

Other studies examining the efficacy of inhibition of endogenous angiogenic factors have been performed using animal models of follicular thyroid cancer. One study used PTK 787/ZK222584 (PTK/ZK, Novartis and Schering), an inhibitor designed to block the VEGF receptor tyrosine kinase. Nude mice implanted subcutaneously with follicular thyroid cancer cells treated with PTK/ZK had 41% less tumor growth compared with controls, as well as reduced microvascular staining (60). Another study by Soh and colleagues showed that treatment of follicular thyroid cancer in a xenograft mouse model with mAb to VEGF resulted in significant reductions in both tumor volumes and blood vessel density compared to controls (61). Subcutaneous delivery of the angiogenic inhibitor endostastin in a mouse xenograft model of human follicular thyroid cancer was shown to result in an 84% reduction in tumor volume compared with saline controls (62).

In an in vitro and in vivo model of medullary cancer in severe combined immunodeficiency mice, treatment with STI571 resulted in inhibition of RET phosphorylation, cell proliferation, and inhibition using the small-molecule FGFR inhibitor. PD173074 also was tested and resulted in abrogation of FGF1–mediated FGFR4 phosphorylation in TT cells, an effect that was accompanied by significant arrest of cell proliferation and tumor growth in vivo. Moreover, the combination of STI571 and PD173074 resulted in greater suppression of cell proliferation in vitro and tumor control in vivo than that achieved using either agent alone. These data highlight RET and FGFR4 as therapeutic targets and suggest a potential role for the combined use of tyrosine kinase inhibitors in the management of inoperable medullary thyroid cancers (63).

Current Clinical Trials

All these preclinical studies have been encouraging as to the potential of some of these agents to contribute significantly to the future treatment of thyroid cancers. In a thorough search of the National Institute of Health PDQ clinical trials database (May 2005), approximately one dozen current phase II clinical trial are using strategies that are partially or entirely antiangiogenic in their scope against any form of thyroid cancer. They cannot all be listed here due to page constraints but some agents will be reviewed.

One trial involves the use of combrestatin A4 phosphate (CA4P), a microtubule depolymerizing agent that leads to endothelial cytoskeletal changes by binding to tubulin and, in animal studies, can lead to extensive ischemic necrosis in tumors resistant to conventional anticancer treatments. In the first phase I clinical trial using CA4P (64) three patients with anaplastic thyroid cancer and two with medullary cancer were treated. One patient with anaplastic thyroid cancer with pretreatment-persistent disease in the thyroid bed and a 3-cm cervical lymph node with metastatic anaplastic thyroid cancer achieved a complete response and was disease free 30 months post-therapy (dose level 3). Surgical exploration of the neck in this patient post CA4P therapy revealed a complete pathologic response. Of note, acute coronary syndrome, with EKG changes that resolved without negative sequelae, was seen in a number of patients undergoing treatment with CA4P, including one patient with anaplastic thyroid cancer (65). Currently, a phase II trial of CA4 P is accruing patients with anaplastic thyroid cancer.

Another phase II trial is examining the role of ZD6474, a low-molecular-weight tyrosine kinase inhibitor (an anilinoquinazoline compound) originally identified as a chemical inhibitor of VEGFR2, with additional activity against the EGF receptor (EGFR) (66). This compound also has been found to inhibit RET/PTC transformed fibroblasts injected into nude mice (67). This combination might prove to be effective, especially in medullary carcinoma of the thyroid in which both angiogenesis and RET/PTC mutations play an important role in the development and spread of this malignancy. In a recent study by Vidal and colleagues (68), RET-mediated eye phenotypes in a *Drosophila* model of multiple endocrine neoplasias type 2A and 2B were suppressed by use of oral ZD6474, with high efficacy and low toxicity.

Another agent under current study is bortezomib (PS-341, Velcade), the first proteasome inhibitor evaluated in human clinical trials. It has been approved by the U.S. Food and Drug Administration for use in patients with refractory or relapsed multiple myeloma. Preclinical study results show that bortezomib suppresses tumor cell growth, induces apoptosis, overcomes resistance to standard chemotherapy agents and radiation therapy, and inhibits angiogenesis (69).

BAY-43–9006, is a novel bi-aryl-urea that is a potent inhibitor of Raf-1, a member of the RAF/MEK/ERK signaling pathway. It also has been shown to be effective against several tyrosine kinases involved in neovascularization, such as VEGFR2 and VEGFR3, PDGF receptor B (PDGFRB), FLT-3, and c-kit (70). Given the recent developments indicating the importance of both RAF signaling and angiogenesis in thyroid cancer, it is logical that this agent is being rapidly tested in phase II clinical trials for patients with locally advanced, metastatic, or locally recurrent thyroid cancers.

CONCLUSION

This chapter has focused on the role of the endothelium in the thyroid gland and explored the advances that have been made in our understanding of how the process of angiogenesis is involved in the pathophysiology of benign and malignant diseases of this gland. The thyroid follicular cell has been one of the most studied cells in the human body, and the thyroid endothelium has been clearly the focus of less intense study than it deserves. However, every day, new advances are being made in understanding the cross-talk between thyroid follicular cells, stroma, and thyroid endothelium; these advances will undoubtedly lead to logically combined therapeutic agents to treat both benign and malignant thyroid conditions.

KEY POINTS

- Angiogenesis and ECs play an important role in the pathogenesis of benign and malignant disorders of the thyroid gland, but they are understudied.
- Proangiogenic factors such as FGF1, FGF2, VEGF, and the angiopoietins play an important role in the pathogenesis of benign thyroid epithelial cell hyperplasia and goitrogenesis.
- Regulation of angiogenesis appears to play an important role in the pathophysiology of inflammatory conditions of the thyroid.
- Studies analyzing human thyroid malignancies have shown that the expression of several isoforms of VEGF is significantly upregulated in papillary, follicular, medullary, and undifferentiated carcinomas compared with normal thyroid tissue.
- Studies of human thyroid malignancies have provided evidence that VEGF-C and VEGFR3 are strongly upregulated in the majority of tumors with a propensity for lymphatic dissemination, including papillary, medullary, and undifferentiated thyroid cancers.
- The endothelium of the thyroid represents a potential therapeutic target in both benign and malignant thyroid conditions.
- Many phase II clinical trials directed against thyroid cancer are currently using strategies that are partially or entirely antiangiogenic in scope.

REFERENCES

1 Hanahan D, Folkman J. Patterns and emerging mechanisms of the angiogenic switch during tumorigenesis. *Cell.* Aug 9 1996; 86(3):353–364.
2 Folkman J. Angiogenesis in cancer, vascular, rheumatoid and other disease. *Nat Med.* Jan 1995;1(1):27–31.
3 Hansen J, Skaaring P. Scanning electron microscopy of normal rat thyroid. *Anat Anz.* 1973;134(3):177–185.
4 Ludatscher RM, Gellei B, Barzilai D. Ultrastructural observations on the capillaries of human thyroid tumours. *J Pathol.* Jun 1979;128(2):57–62.
5 Folkman J. Tumor angiogenesis: therapeutic implications. *N Engl J Med.* Nov 18 1971;285(21):1182–1186.
6 Wollman SH, Herveg JP, Zeligs JD, Ericson LE. Blood capillary enlargement during the development of thyroid hyperplasia in the rat. *Endocrinology*. Dec 1978;103(6):2306–2314.
7 Goodman AL, Rone JD. Thyroid angiogenesis: endotheliotropic chemoattractant activity from rat thyroid cells in culture. *Endocrinology*. Dec 1987;121(6):2131–2140.
8 Kabel PJ, Voorbij HA, de Haan-Meulman M, Pals ST, Drexhage HA. High endothelial venules present in lymphoid cell accumulations in thyroids affected by autoimmune disease: a study in men and BB rats of functional activity and development. *J Clin Endocrinol Metab.* Apr 1989;68(4):744–751.
9 Ishiwata T, Iino Y, Takei H, Oyama T, Morishita Y. Tumor angiogenesis as an independent prognostic indicator in human papillary thyroid carcinoma. *Oncol Rep.* Nov-Dec 1998;5(6):1343–1348.
10 Segal K, Shpitzer T, Feinmesser M, Stern Y, Feinmesser R. Angiogenesis in follicular tumors of the thyroid. *J Surg Oncol.* Oct 1996; 63(2):95–98.
11 Ishimura K, Okamoto H, Fujita H. Freeze-etching images of capillary endothelial pores in the liver, thyroid and adrenal of the mouse. *Arch Histol Jpn.* Apr 1978;41(2):187–193.
12 LeCouter J, Kowalski J, Foster J, et al. Identification of an angiogenic mitogen selective for endocrine gland endothelium. *Nature.* Aug 30 2001;412(6850):877–884.
13 Samson M, Peale FV, Jr., Frantz G, Rioux-Leclercq N, Rajpert-De Meyts E, Ferrara N. Human endocrine gland-derived vascular endothelial growth factor: expression early in development and in Leydig cell tumors suggests roles in normal and pathological

testis angiogenesis. *J Clin Endocrinol Metab.* Aug 2004;89(8): 4078–4088.

14 Davis FB, Mousa SA, O'Connor L, et al. Proangiogenic action of thyroid hormone is fibroblast growth factor-dependent and is initiated at the cell surface. *Circ Res.* Jun 11 2004;94(11):1500–1506.

15 Thompson SD, Franklyn JA, Watkinson JC, Verhaeg JM, Sheppard MC, Eggo MC. Fibroblast growth factors 1 and 2 and fibroblast growth factor receptor 1 are elevated in thyroid hyperplasia. *J Clin Endocrinol Metab.* Apr 1998;83(4):1336–1341.

16 Sato K, Yamazaki K, Shizume K, et al. Stimulation by thyroid-stimulating hormone and Grave's immunoglobulin G of vascular endothelial growth factor mRNA expression in human thyroid follicles in vitro and flt mRNA expression in the rat thyroid in vivo. *J Clin Invest.* Sep 1995;96(3):1295–1302.

17 Cocks HC, Ramsden JD, Watkinson JC, Eggo MC. Thyroid stimulating hormone increases angiogenic growth factor expression in rat thyrocytes. *Clin Otolaryngol.* Dec 2000;25(6):570–576.

18 Patel VA, Logan A, Watkinson JC, et al. Isolation and characterization of human thyroid endothelial cells. *Am J Physiol Endocrinol Metab.* Jan 2003;284(1):E168–176.

19 Ramsden JD, Yarram S, Mathews E, Watkinson JC, Eggo MC. Thyroid follicular cells secrete plasminogen activators and can form angiostatin from plasminogen. *J Endocrinol.* Jun 2002;173 (3):475–481.

20 Hofmann A, Laue S, Rost AK, Scherbaum WA, Aust G. mRNA levels of membrane-type 1 matrix metalloproteinase (MT1-MMP), MMP-2, and MMP-9 and of their inhibitors TIMP-2 and TIMP-3 in normal thyrocytes and thyroid carcinoma cell lines. *Thyroid.* Mar 1998;8(3):203–214.

21 Prabakaran D, Kim P, Kim KR, Arvan P. Polarized secretion of thrombospondin is opposite to thyroglobulin in thyroid epithelial cells. *J Biol Chem.* Apr 25 1993;268(12):9041–9048.

22 Bidey SP, Hill DJ, Eggo MC. Growth factors and goitrogenesis. *J Endocrinol.* Mar 1999;160(3):321–332.

23 Chanoine JP, Stein GS, Braverman LE, et al. Acidic fibroblast growth factor modulates gene expression in the rat thyroid in vivo. *J Cell Biochem.* Dec 1992;50(4):392–399.

24 Logan A, Black EG, Gonzalez AM, Buscaglia M, Sheppard MC. Basic fibroblast growth factor: an autocrine mitogen of rat thyroid follicular cells? *Endocrinology.* Apr 1992;130(4):2363–2372.

25 Viglietto G, Romano A, Manzo G, et al. Upregulation of the angiogenic factors PlGF, VEGF and their receptors (Flt-1, Flk-1/KDR) by TSH in cultured thyrocytes and in the thyroid gland of thiouracil-fed rats suggest a TSH-dependent paracrine mechanism for goiter hypervascularization. *Oncogene.* Nov 27 1997; 15(22):2687–2698.

26 Wang JF, Milosveski V, Schramek C, Fong GH, Becks GP, Hill DJ. Presence and possible role of vascular endothelial growth factor in thyroid cell growth and function. *J Endocrinol.* Apr 1998; 157(1):5–12.

27 Ramsden JD, Cocks HC, Shams M, et al. Tie-2 is expressed on thyroid follicular cells, is increased in goiter, and is regulated by thyrotropin through cyclic adenosine 3′,5′-monophosphate. *J Clin Endocrinol Metab.* Jun 2001;86(6):2709–2716.

28 Patel VA, Hill DJ, Eggo MC, Sheppard MC, Becks GP, Logan A. Changes in the immunohistochemical localisation of fibroblast growth factor-2, transforming growth factor-beta 1 and thrombospondin-1 are associated with early angiogenic events in the hyperplastic rat thyroid. *J Endocrinol.* Mar 1996;148(3):485–499.

29 Folkman J. Angiogenesis and angiogenesis inhibition: an overview. *Exs.* 1997;79:1–8.

30 Ferrara N, Keyt B. Vascular endothelial growth factor: basic biology and clinical implications. *Exs.* 1997;79:209–232.

31 Ramsden JD. Angiogenesis in the thyroid gland. *J Endocrinol.* Sep 2000;166(3):475–480.

32 Klein M, Picard E, Vignaud JM, et al. Vascular endothelial growth factor gene and protein: strong expression in thyroiditis and thyroid carcinoma. *J Endocrinol.* Apr 1999;161(1):41–49.

33 Iitaka M, Miura S, Yamanaka K, et al. Increased serum vascular endothelial growth factor levels and intrathyroidal vascular area in patients with Graves' disease and Hashimoto's thyroiditis. *J Clin Endocrinol Metab.* Nov 1998;83(11):3908–3912.

34 Toda S, Nishimura T, Yamada S, et al. Immunohistochemical expression of growth factors in subacute thyroiditis and their effects on thyroid folliculogenesis and angiogenesis in collagen gel matrix culture. *J Pathol.* Aug 1999;188(4):415–422.

35 Ramsden JD, Buchanan MA, Egginton S, Watkinson JC, Mautner V, Eggo MC. Complete inhibition of goiter in mice requires combined gene therapy modification of angiopoietin, VEGF and FGF signaling. *Endocrinology.* Apr 7 2005.

36 Pierotti MA, Bongarzone I, Borello MG, Greco A, Pilotti S, Sozzi G. Cytogenetics and molecular genetics of carcinomas arising from thyroid epithelial follicular cells. *Genes Chromosomes Cancer.* May 1996;16(1):1–14.

37 Bunone G, Vigneri P, Mariani L, et al. Expression of angiogenesis stimulators and inhibitors in human thyroid tumors and correlation with clinical pathological features. *Am J Pathol.* Dec 1999;155(6):1967–1976.

38 Poulaki V, Mitsiades CS, McMullan C, et al. Regulation of vascular endothelial growth factor I expression by insulin-like growth factor I in thyroid carcinomas. *J Clin Endocrinol Metab.* Nov 2003; 88(11):5392–5398.

39 Maisonpierre PC, Suri C, Jones PF, et al. Angiopoietin-2, a natural antagonist for Tie2 that disrupts in vivo angiogenesis. *Science.* Jul 4 1997;277(5322):55–60.

40 Joukov V, Pajusola K, Kaipainen A, et al. A novel vascular endothelial growth factor, VEGF-C, is a ligand for the Flt4 (VEGFR-3) and KDR (VEGFR-2) receptor tyrosine kinases. *Embo J.* Apr 1 1996;15(7):1751.

41 Kaipainen A, Korhonen J, Mustonen T, et al. Expression of the fms-like tyrosine kinase 4 gene becomes restricted to lymphatic endothelium during development. *Proc Natl Acad Sci U S A.* Apr 11 1995;92(8):3566–3570.

42 Kukk E, Lymboussaki A, Taira S, et al. VEGF-C receptor binding and pattern of expression with VEGFR-3 suggests a role in lymphatic vascular development. *Development.* Dec 1996;122(12): 3829–3837.

43 Hung CJ, Ginzinger DG, Zarnegar R, et al. Expression of vascular endothelial growth factor-C in benign and malignant thyroid tumors. *J Clin Endocrinol Metab.* Aug 2003;88(8):3694–3699.

44 Murphy G, Knauper V, Cowell S, et al. Evaluation of some newer matrix metalloproteinases. *Ann N Y Acad Sci.* Jun 30 1999;878: 25–39.

45 Denis LJ, Verweij J. Matrix metalloproteinase inhibitors: present achievements and future prospects. *Invest New Drugs.* 1997;15 (3):175–185.

46 Komorowski J, Pasieka Z, Jankiewicz-Wika J, Stepien H. Matrix metalloproteinases, tissue inhibitors of matrix metalloproteinases and angiogenic cytokines in peripheral blood of patients with thyroid cancer. *Thyroid.* Aug 2002;12(8):655–662.

47 Frigugliett IC, Mello ES, Castro IV, Filho GB, Alves VA. Metalloproteinase-9 immunoexpression and angiogenesis in thyroid follicular neoplasms: relation to clinical and histopathologic features. *Head Neck.* Jul 2000;22(4):373–379.

48 Robert J, Mariethoz S, Pache JC, et al. Short- and long-term results of total vs subtotal thyroidectomies in the surgical treatment of Graves' disease. *Swiss Surg.* 2001;7(1):20–24.

49 Melillo RM, Castellone MD, Guarino V, et al. The RET/PTC-RAS-BRAF linear signaling cascade mediates the motile and mitogenic phenotype of thyroid cancer cells. *J Clin Invest.* Apr 2005;115(4):1068–1081.

50 Thorpe PE. Vascular targeting agents as cancer therapeutics. *Clin Cancer Res.* Jan 15 2004;10(2):415–427.

51 Dziba JM, Marcinek R, Venkataraman G, Robinson JA, Ain KB. Combretastatin A4 phosphate has primary antineoplastic activity against human anaplastic thyroid carcinoma cell lines and xenograft tumors. *Thyroid.* Dec 2002;12(12):1063–1070.

52 Brooks PC, Montgomery AM, Rosenfeld M, et al. Integrin alpha v beta 3 antagonists promote tumor regression by inducing apoptosis of angiogenic blood vessels. *Cell.* Dec 30 1994;79(7):1157–1164.

53 Burrows FJ, Derbyshire EJ, Tazzari PL, et al. Up-regulation of endoglin on vascular endothelial cells in human solid tumors: implications for diagnosis and therapy. *Clin Cancer Res.* Dec 1995;1(12):1623–1634.

54 Thorpe PE, Burrows FJ. Antibody-directed targeting of the vasculature of solid tumors. *Breast Cancer Res Treat.* 1995;36(2): 237–251.

55 Ramakrishnan S, Olson TA, Bautch VL, Mohanraj D. Vascular endothelial growth factor-toxin conjugate specifically inhibits KDR/flk-1-positive endothelial cell proliferation in vitro and angiogenesis in vivo. *Cancer Res.* Mar 15 1996;56(6):1324–1330.

56 Veenendaal LM, Jin H, Ran S, et al. In vitro and in vivo studies of a VEGF121/rGelonin chimeric fusion toxin targeting the neovasculature of solid tumors. *Proc Natl Acad Sci U S A.* Jun 11 2002;99(12):7866–7871.

57 Bauer AJ, Terrell R, Doniparthi NK, et al. Vascular endothelial growth factor monoclonal antibody inhibits growth of anaplastic thyroid cancer xenografts in nude mice. *Thyroid.* Nov 2002;12(11):953–961.

58 Xu G, Pan J, Martin C, Yeung SC. Angiogenesis inhibition in the in vivo antineoplastic effect of manumycin and paclitaxel against anaplastic thyroid carcinoma. *J Clin Endocrinol Metab.* Apr 2001;86(4):1769–1777.

59 Pietras K, Rubin K, Sjoblom T, et al. Inhibition of PDGF receptor signaling in tumor stroma enhances antitumor effect of chemotherapy. *Cancer Res.* Oct 1 2002;62(19):5476–5484.

60 Schoenberger J, Grimm D, Kossmehl P, Infanger M, Kurth E, Eilles C. Effects of PTK787/ZK222584, a tyrosine kinase inhibitor, on the growth of a poorly differentiated thyroid carcinoma: an animal study. *Endocrinology.* Mar 2004;145(3):1031–1038.

61 Soh EY, Eigelberger MS, Kim KJ, et al. Neutralizing vascular endothelial growth factor activity inhibits thyroid cancer growth in vivo. *Surgery.* Dec 2000;128(6):1059–1065;discussion 1065–1056.

62 Ye C, Feng C, Wang S, Liu X, Lin Y, Li M. Antiangiogenic and antitumor effects of endostatin on follicular thyroid carcinoma. *Endocrinology.* Sep 2002;143(9):3522–3528.

63 St Bernard R, Zheng L, Liu W, Winer D, Asa SL, Ezzat S. Fibroblast growth factor receptors as molecular targets in thyroid carcinoma. *Endocrinology.* Mar 2005;146(3):1145–1153.

64 Dowlati A, Robertson K, Cooney M, et al. A phase I pharmacokinetic and translational study of the novel vascular targeting agent combretastatin a-4 phosphate on a single-dose intravenous schedule in patients with advanced cancer. *Cancer Res.* Jun 15 2002;62(12):3408–3416.

65 Cooney MM, Radivoyevitch T, Dowlati A, et al. Cardiovascular safety profile of combretastatin a4 phosphate in a single-dose phase I study in patients with advanced cancer. *Clin Cancer Res.* Jan 1 2004;10(1 Pt 1):96–100.

66 Ciardiello F, Caputo R, Damiano V, et al. Antitumor effects of ZD6474, a small molecule vascular endothelial growth factor receptor tyrosine kinase inhibitor, with additional activity against epidermal growth factor receptor tyrosine kinase. *Clin Cancer Res.* Apr 2003;9(4):1546–1556.

67 Carlomagno F, Vitagliano D, Guida T, et al. Efficient inhibition of RET/papillary thyroid carcinoma oncogenic kinases by 4-amino-5-(4-chloro-phenyl)-7-(t-butyl)pyrazolo[3,4-d]pyrimidine (PP2). *J Clin Endocrinol Metab.* Apr 2003;88(4):1897–1902.

68 Vidal M, Wells S, Ryan A, Cagan R. ZD6474 suppresses oncogenic RET isoforms in a Drosophila model for type 2 multiple endocrine neoplasia syndromes and papillary thyroid carcinoma. *Cancer Res.* May 1 2005;65(9):3538–3541.

69 Jung L, Holle L, Dalton WS. Discovery, Development, and clinical applications of bortezomib. *Oncology (Huntingt).* Dec 2004;18(14 Suppl 11):4–13.

70 Wilhelm SM, Carter C, Tang L, et al. BAY 43–9006 exhibits broad spectrum oral antitumor activity and targets the RAF/MEK/ERK pathway and receptor tyrosine kinases involved in tumor progression and angiogenesis. *Cancer Res.* Oct 1 2004;64(19):7099–7109.

152

Endothelial Dysfunction and the Link to Age-Related Vascular Disease

Jay M. Edelberg* and May J. Reed†

*Weill Medical College of Cornell University, New York;
†University of Washington, Seattle

Vascular pathology and its associated clinical entities have a broad impact on the health of older persons. Specifically, cardiovascular disease and stroke account for the majority of morbidity and mortality in individuals over the age of 65 (1). Other conditions related to vascular pathology, such as erectile dysfunction (ED) and impaired wound healing, also have a significant effect on the lives of older persons. Understanding the changes in the vasculature that occur with aging may provide insights into many disorders that affect the elderly.

Aging is associated with multiple changes in endothelial cell (EC) function that limit the capacity of older blood vessels to dynamically regulate and restore blood flow (Table 152-1). Although older persons have higher vascular risk factor profiles as a result of cumulative environmental exposures (e.g., smoking) and an increased prevalence of endogenous pathological conditions (e.g., diabetes and hyperlipidemia), inherent age-related changes in endothelial function contribute significantly to subsequent vascular dysfunction. Accordingly, it is important to emphasize that, although many factors have a role in the development of EC dysfunction, the normal aging process is itself an independent risk factor for developing vascular diseases. The overall goal of this chapter is to focus on changes in endothelial function related to "normal" aging in the general population, and to discuss the interaction between aging and other cardiovascular risk factors in promoting vascular disease in the elderly.

THE AGED ENDOTHELIAL CELL

Little is known about how "normal aging" affects ECs in the absence of disease and/or age-related physiologic changes impacting vascular and endothelial function. Many of the observations are extrapolated from experiments using fibroblasts that have incurred numerous population doublings in vitro (2,3). When ECs are examined, they are usually late-passage human umbilical vein ECs (HUVECs) or other lines that grow easily in culture (4,5). It is important to note that in vitro studies usually result in the loss of features that reflect differences in the function of large- versus small-vessel ECs.

Spontaneous changes that have been observed in cells aged in culture, and that are relevant to aging in vivo, include attrition of telomeres leading to unstable chromosomes and subsequent inhibition of cell cycle progression (6). The resultant senescent EC phenotype is defined by decreased proliferation, expression of β-galactosidase at pH 6, accumulation of cellular/chromatin debris, and increased expression of cell cycle inhibitors (7). Senescent cells also have stress-induced alterations, including stabilization of p53 and hypophosphorylation of Rb with subsequent activation of cell cycle inhibitors (8,9). In addition, whereas senescent cells produce less nitric oxide (NO), they accumulate reactive nitrogen and other toxic free radical species that can activate the apoptotic pathway (10,11). To this end, the endothelial regulatory pathways governing apoptosis and senescence are not mutually exclusive. The choice of one pathway over the other is a complex balance determined by the type and concentration of exogenous stress, as well as the presence or absence of proteins, such as caspase inhibitors, that control cell survival (12).

ENDOTHELIAL FUNCTIONS/DYSFUNCTIONS

In vivo ECs orchestrate multiple functions, which collectively serve to regulate vascular supply to tissues. For example, the finely tuned balance between coagulation and fibrinolytic pathways maintains optimal hemostasis/blood fluidity within the intravascular space. Endothelial-mediated vasodilation and vasoconstriction serve to adjust flow from one moment to the next according to the needs of the local tissue, and to rapidly respond to major disruptions in supply or demand brought about by thrombosis and/or end-organ damage. The endothelium also plays a central role in angiogenesis, which supports additional blood flow requirements by ischemic tissues.

Table 152-1: Endothelial Dysfunction and Aging

Disease/ Pathophysiology	*Thrombosis*	*Vasodilation/ Constriction*	*Angiogenesis*
Coronary artery disease	+	+	+
Stroke	+		
Wound healing			+
Erectile dysfunction		+	

Age-related endothelial dysfunctions correlated with vascular diseases affecting older persons.

In addition, ECs modulate immune and inflammatory activities, impacting on overall host defense mechanisms and vascular homeostasis. In the aging blood vessel, these and other endothelial functions are altered, contributing to an increase in the severity of vascular pathology in older persons.

Thrombotic Regulation

Hemostatic regulation is one of the central properties of the vascular endothelium. The normal endothelial lining provides a dynamic antithrombotic surface that suppresses procoagulant enzyme activity and platelet aggregation in the local circulation. The anticoagulant properties of unactivated, quiescent ECs are maintained to a large degree by the expression of the glycosaminoglycan, heparan, and the enzymatic cofactor, thrombomodulin (TM) as well as tissue factor pathway inhibitor (TFPI) and the endothelial protein C receptor (EPCR). One of the significant actions of endothelial heparan is to promote the inactivation of thrombin and factor Xa by antithrombin III (ATIII). TM acts as a substrate-modulating cofactor for activated protein C (APC) to direct the inactivation of factors Va and VIIIa.

Vascular injury induced through acute stimuli (such as vessel trauma) and chronic conditions (such as atherosclerosis) reverse the hemostatic regulatory pathways and induce a procoagulant milieu. Disruption of the endothelial lining also exposes tissue factor and the collagen of the subendothelial extracellular matrix to activate thrombin and platelet aggregation. Activation of the intact endothelium may result in increased adhesion of platelets and leukocytes, secretion of von Willebrand factor (vWF), expression of tissue factor, and downregulation of natural anticoagulant mechanisms such as TM.

Quiescent ECs also contribute to the resolution of thrombi, secreting enzymes (tissue-type plasminogen activator [t-PA] and urokinase-type plasminogen activator [u-PA]) that activate plasminogen to plasmin, an enzyme that degrades fibrin. In addition, ECs release inhibitors that quench activating enzymes as well as plasmin itself (plasminogen activator inhibitors [PAI]-1, and α2-antiplasmin, respectively). The EC membrane also provides a surface for the regulation of the enzymes that bind to annexin-2 (plasminogen receptor) and the urokinase receptor.

Previous studies have shown that aging is associated with increased coagulability and decreased fibrinolysis due to changes in endothelial- and nonendothelial-mediated functions, suggesting a potential basis for the predisposition of older individuals to thrombotic coronary and cerebrovascular events. Large-scale population-based studies have found a correlation between advanced age and increased levels of plasma coagulation factors and D-dimers (13,14). Moreover, older persons demonstrate longer thrombolysis time compared with younger individuals (15,16). Mechanistically, this may be related to increases in PAI-1, which is found to be upregulated in vitro in aged human ECs (4) as well as in the in vivo endothelium of "Klotho" mice, a genetic murine model of aging (17). Notably, however, in studies of lipopolysaccharide (LPS)-induced thrombosis, older mice revealed an increase in PAI-1 that was restricted to perivascular cells, including hepatocytes and renal glomerular cells, but not ECs (18). This increase in PAI-1 was linked to elevations in tumor necrosis factor (TNF)-α (18), which suggests that age-associated changes in TNF receptor-1 pathways may contribute to the increase in PAI-1 and thrombosis in the aging vasculature (19). In addition, the production of coagulation inhibitors, such as APC and ATIII, has been shown to be altered in some groups of elderly (20,21). Similarly, heparan, the endogenous template for ATIII, has been shown to be structurally altered in aging (22). Although age-related changes have been described only with respect to its interactions with vascular growth factors, changes in heparan could impact endothelial hemostatic regulation either directly or indirectly through modulation of coagulation enzymes and growth factors.

The clinical impact of senescent shifts in coagulation and fibrinolytic pathways is complex in the aging population. An increase in prothrombotic function is associated with an increased risk of acute coronary events, which is the major cause of morbidity and mortality in the elderly (1). Conversely, however, the serum of centenarians shows increased expression of biomarkers for hypercoagulability (14), including high levels of PAI-1 (23). These data suggest that longevity-related genes may counter and/or redirect the coagulant pathways to promote healthy aging in long-lived individuals.

Vasomotor Tone

ECs govern blood flow by regulating the constriction and dilation of blood vessels. Changes in local oxygen consumption or disruption of end-organ blood flow due to vascular thrombosis triggers EC pathways aimed at acutely compensating for the dynamic needs of tissue beds. Through the balance of local factors, including NO and endothelin (ET)-1, ECs can govern smooth muscle tone in the expansion and constriction of blood vessels to adjust the delivery of oxygenated blood to specific organs.

Age-associated changes in ECs as well as smooth muscle cell function can limit the capacity of older blood vessels to regulate local blood flow. The altered regulation of vascular tone impairs the vasodilation of both coronary and peripheral blood vessels in older persons. Clinical studies often have employed forearm brachial blood flow as a surrogate index for coronary vascular dynamics. In this model, older persons have decreased brachial endothelium-dependent vasodilation, as compared with younger individuals with otherwise matched clinical characteristics and risk factors (24). These data correlate with impaired acetylcholine-induced endothelium-dependent coronary vasodilation in vivo (25), but are not uniform throughout the aged vasculature (26). For example, aging may have a greater impact on smaller blood vessels of the mesenteric tree relative to larger conduits such as the aorta (27). The larger arterial beds, however, are more affected by age-related decreases in vasoelastic properties, causing increases in arterial stiffness (28). Such findings highlight that vascular changes in the aged often reflect events that are specific to the local vasculature.

Mechanistically, the changes in endothelium-mediated vasoactive function have been linked to alterations in two integral mediators of these respective functions: NO and ET-1. Coronary basal and stimulated NO production is decreased with aging (29). Moreover, studies in rodent models reveal that NO-dependent aortic vasodilation is diminished (30). ET-1-induced coronary vasoconstriction is increased in the aged (31,32). Functionally, aging rodent hearts have less resistance to myocardial ischemia and reperfusion injury in association with decreased coronary NO production (33). Age-related deficits in hypoxia-inducible factor (HIF)-1–mediated expression of vascular endothelial growth factor (VEGF) may impair the codependent vasodilatory actions of NO and VEGF (34). The compromise in early compensatory actions and regulation of vascular tone contributes to impaired angiogenesis and can increase the subsequent risk of apoptosis and necrosis of dependent tissues.

In addition to changes in vasoactivity, large arteries in the aged demonstrate increased stiffness due to protein cross-linking and deposition of advanced glycation end-products (AGEs) in the extracellular matrix. The decreases in elasticity limit the dynamic recoil of the arterial system (28). The resultant elevation in blood pressure required to compensate for the increase in conduit stiffness could, in turn, lead to further endothelial dysfunction in the aging vasculature. Novel agents developed as AGE cross-link "breakers" improve cardiovascular function in aging animal models and offer the potential for countering the effects of aging, as well as diabetes, on the arterial system (35).

Angiogenesis

Disruption of blood flow due to injury, thrombosis, or vasoconstriction induces a cascade of growth factors that enhance cell proliferation and migration. Specifically, the resulting hypoxia promotes the expression of a set of genes regulated, in part, by HIF. HIF stimulates the expression of VEGF, which in turn promotes EC proliferation and migration as well as local vasodilation. Other factors induced in the hypoxic state include fibroblast growth factor 2 (FGF2) and platelet-derived growth factor (PDGF), which function synergistically with VEGF to promote angiogenesis. Endothelial precursor cells (EPCs) also are recruited from the circulation and contribute to angiogenesis and the expression of growth factors.

Physiological and pathological angiogenesis is depressed in most older tissues including the heart, peripheral vascular beds (36), and select disease-matched tumor beds (37,38). Indeed, cardiac angiogenesis induced by myocardial hypertrophy is significantly depressed in the aging heart, as evidenced by 20% less capillary growth (39,40) and subsequent impaired myocardial blood flow (41,42). Angiographic studies of risk factor–matched subsets of patients with coronary artery disease reveal decreased collateral circulation in older individuals (43).

The age-related impairment of angiogenesis is due to deficits in growth factor synthesis and dysregulation of angiogenic cytokine pathways, both in peripheral ECs and EPCs. Experimental models have demonstrated impairments in the expression and function of a number of angiogenic factors, including VEGF (34,36), FGF2 (44,45), transforming growth factor (TGF)-β (46), and PDGF (47). Age-associated decreases in the concentration and function of EPCs compound this dysfunction (48,49). Functionally, the replacement of deficient angiogenic growth factors, such as FGF, VEGF, and PDGF, improves the migration and proliferation of aged, multipassaged ECs and ECs isolated from aging mice (50,51). Moreover, treatment of aged tissues (both in the basal state and after ischemic injury) with angiogenic growth factors (36,52) or EPCs from transplanted young bone marrow cells increases subsequent capillary density and improves vascular function.

It is important to note that age-related macular degeneration, a common cause of blindness in the elderly, is an exception to the rule regarding a lack of angiogenesis in aged tissues. In the vascularized form of this condition, high levels of VEGF and alterations in other growth factors from the retinal pigmented epithelium contribute to excessive neovascularization and subsequent increased risk of blindness (53). To this end, the recent use of VEGF inhibitors for the treatment of this disease reveals the importance of tissue-specific regulation of angiogenesis in the clinical manifestations of endothelial angiogenic function and dysfunction in the elderly (54).

AGE-ASSOCIATED DISEASES – THE LINK TO ENDOTHELIAL DYSFUNCTION

The combined impairments in hemostasis, vasodilation, and angiogenesis increase the risk of associated vascular diseases in older persons. For example, the prothrombotic tendency may lead to stroke and myocardial infarction (MI), alterations in vascular tone predispose to endothelial dysfunction, and depressed angiogenesis might contribute to delayed tissue repair in older persons.

Cardiovascular Disease: Link to Multiple Endothelial Dysfunctions

Ischemic cardiovascular disease is the most common cause of morbidity and mortality in the U.S. population over age 65 (1). MIs in the elderly result in poorer outcomes and are associated with higher rates of infarction-related mortality, compared with younger individuals. In addition, those surviving initial cardiac events are more likely to develop congestive heart failure, suggesting that age-related changes in the vasculature predispose older individuals to an increased risk of clinically significant pathology.

Endothelial dysfunction is one of the most important predictive factors for cardiovascular disease in the general population (55), and its impact is heightened with age. For example, age-associated increases in antifibrinolytic parameters, such as endothelial-derived PAI-1 (56), could contribute to the increased incidence of cardiovascular disease in the aging population. Mechanistically, however, PAI-1's role as an acute phase reactant, like C-reactive protein, also could be the basis of the association between cardiovascular disease and increased inflammatory mediators.

Physiologically, the response of the aging heart to coronary occlusions is impaired, leading to increased myocardial injury, when compared with similar arterial compromise in younger hearts. Previous studies have demonstrated that episodes of cardiac tissue hypoxia caused by transient coronary occlusion induce vasodilatory actions that result in cardioprotection in the young rodent heart (57). In the aged heart, this ischemic preconditioning can be significantly depressed (58–60). This deficit, which reflects inherent endothelial dysfunction and impaired vasodilation, contributes to worse clinical outcomes for older persons after MI.

Age-related decreases in angiogenesis may also contribute to increased pathology following myocardial injury in older persons. Angiographic studies have shown that subgroups of older patients with diabetes have decreased collateral circulation relative to younger counterparts otherwise matched for clinical characteristics (43). Experimental studies aimed at reversing these age-related changes have demonstrated that restoration of PDGF-based pathways at the time of acute coronary occlusion can significantly reduce the extent of myocardial injury in the aging rodent heart (61). Interestingly, the PDGF-induced angiogenic factors (including VEGF and angiopoietin 2 [Ang2]) rapidly suppress myocardial apoptosis. These data suggest that restoring angiogenic pathways in the aging heart may have clinical utility in conjunction with present reperfusion therapies (e.g., thrombolytics and primary angioplasty) for acute MI.

Stroke: Link to Prothrombotic Predisposition

Advanced age has been shown to be a risk factor for stroke and its complications (62,63). Indeed, stroke is the single greatest source of major morbidity in older persons. The neuropathology resulting from stroke can contribute to cognitive impairment in older persons. Often seen in the setting of systolic hypertension, "subclinical" strokes are a major cause of dementia and functional decline in aging (64). Thromboembolic occlusion of a vessel is the primary pathophysiology of stroke and, similar to cardiovascular disease, is related in part to endothelial dysfunction. Consequently, age-associated changes in endothelial pathways may contribute directly to the increased risk of stroke in the geriatric population.

In addition to hypertension and aging itself, atrial fibrillation and diabetes are the greatest risk factors for stroke in older persons. Each of these conditions is more prevalent in older persons and is associated with impaired endothelial function. Hypertension has been linked to impaired vasodilation. Atrial fibrillation has been associated with increased endothelial activation (65). Diabetes results in greater degrees of oxidative stress and release of inflammatory mediators, which are detrimental to endothelial function and increase the risk of vascular damage (66). Thus, due to the coexistence of multiple risk factors in many elders, optimal prevention of stroke in the aged requires control of hypertension, minimization of thrombus formation in the large vessels and atria, and improved management of diabetes.

Aging and Delayed Wound Healing: Link to Impaired Angiogenesis

Aging is associated with delayed wound healing in the skin and other organs. In the basal state, there are fewer capillaries in aged skin and other tissues (67), and the rate of capillary growth in wounds is decreased in older animals (68). The decline in neovascularization has been attributed to decreases in both the proliferation and migration of ECs in the wound bed (69). This deficit is due, in part, to reduced levels and function of endothelial angiogenic factors, such as the growth factors TGF-β and VEGF (36,70,71). Other age-related defects in EC behavior during injury include increased adhesion to leukocytes, enhanced response to TNF-α, and greater interleukin (IL)-1 production (72–75).

It has been shown that the replacement of deficient angiogenic growth factors can increase the growth of capillaries in aged tissues (36,52). However, the extrapolation to enhanced wound repair (as a direct result of increased neovascularization, as opposed to the numerous other changes induced by the application of these angiogenic factors) remains to be proved. Notably, in animal models, treatment with inhibitors of angiogenesis had inconsistent effects on cutaneous wound repair and has been shown to modulate functions other than the growth of new vessels (76–78). The importance of angiogenesis to wound healing may be tissue-specific: A lack of newly formed vessels has been shown to be detrimental to repair of intestinal anastomoses (79). It is important to note that none of the studies using inhibitors of angiogenesis were performed on aged animals. In this context, it is likely that a limitation on capillary growth that is not detrimental to healing in a young animal may have significant consequences for an aged animal, wherein multiple other impairments coexist.

Erectile Dysfunction: Link to Impaired Vasodilation

ED, which has received an unprecedented amount of public attention in recent years, is a primary defect of endothelial-mediated vasodilation. Aging, diabetes, hypertension, and smoking are independent risk factors for ED. Accordingly, men with ED also have an increased risk of developing cardiovascular diseases (80,81), demonstrating that the manifestations of endothelial dysfunction are broad with respect to clinical presentations of vascular pathology in older persons.

One of the underlying mechanisms of ED is a lack of NO, in conjunction with increased ET-1, in the aged penile endothelium. Acetylcholine-mediated endothelial vasodilation is impaired in the aged penile vasculature, similar to brachial arterial studies that correlate with an increased risk of impaired endothelial function (24). Mechanistically, this decrease in NO may be related to an age-related suppression in penile expression of VEGF (82). Experimental approaches to restore these pathways through VEGF gene therapy have shown increased endothelial NO synthase (eNOS) expression and improved penile function in an aging rat model (83). Moreover, diabetic models suggest that VEGF may act through the suppression of apoptosis in the penile cavernosa (84), a role potentially analogous to the antiapoptotic actions of exogenous angiogenic growth factors on the ischemic myocardium (61).

Notably, it is the pharmacological treatment of ED that represents the most significant advance in the development of therapies aimed at reversing the age-associated impairment in endothelial/vascular function. Specifically, the role of sildenafil, an inhibitor of phosphodiesterase (PDE)-5 activity in human corpora cavernosa (85), restores vasoactive erectile function and has revolutionized the treatment of ED. Moreover, the benefits of PDE-5 inhibition extend beyond ED to improve vascular function in other beds. Early trials have shown beneficial actions of sildenafil in the pulmonary arterial system to treat pulmonary hypertension (86,87). Similarly, PDE-5 inhibition improves cutaneous microcirculation in patients with coronary artery disease (88). Mechanistically, the actions of PDE-5 inhibition extend beyond vasodilation, potentially improving angiogenesis (89) and inducing cardiac preconditioning (86,90). The expanded understanding gained from the targeting of PDE-5 suggests that clinical pharmacology developed to counter age-associated endothelial vascular impairment in a specific vascular bed may have broader applications for reversing senescent endothelial dysfunction. Deficits in function in different regions of the vasculature may share common pathways in endothelial dysfunction and that development of interventions aimed at improving EC function in one organ may have utility in the treatment of other tissues in older persons.

KEY POINTS

- Those endothelial functions that are altered with aging – vasodilation, hemostasis, and angiogenesis – have a significant influence on many clinical conditions that affect older persons.
- In aging ECs, thrombosis and activation are increased, and vasodilation and angiogenesis are decreased.
- By improving the understanding of the mechanisms that relate endothelial dysfunction with age-related pathology, future therapies may be developed to specifically reverse the predisposition to vascular diseases in the elderly.

REFERENCES

1 American Heart Association. *Americans and Cardiovascular Disease Biostatistical Fact Sheet – Older Americans and Cardiovascular Disease.* Dallas: American Heart Association, 2001.

2 Kulju KS, Lehman JM. Increased p53 protein associated with aging in human diploid fibroblasts. *Exp Cell Res.* 1995;217(2):336–345.

3 Brookes S, Rowe J, Gutierrez Del Arroyo A, et al. Contribution of p16(INK4a) to replicative senescence of human fibroblasts. *Exp Cell Res.* 2004;298(2):549–559.

4 Comi P, Chiaramonte R, Maier JA. Senescence-dependent regulation of type 1 plasminogen activator inhibitor in human vascular endothelial cells. *Exp Cell Res.* 1995;219(1):304–308.

5 Reed MJ, Corsa AC, Kudravi SA, et al. A deficit in collagenase activity contributes to impaired migration of aged microvascular endothelial cells. *J Cell Biochem.* 2000;77(1):116–126.

6 Herbig U, Jobling WA, Chen BP, et al. Telomere shortening triggers senescence of human cells through a pathway involving ATM, p53, and p21(CIP1), but not p16(INK4a). *Mol Cell.* 2004;14(4):501–513.

7 Chen J, Brodsky SV, Goligorsky DM, et al. Glycated collagen I induces premature senescence-like phenotypic changes in endothelial cells. *Circ Res.* 2002;90(12):1290–1298.

8 Breitschopf K, Zeiher AM, Dimmeler S. Pro-atherogenic factors induce telomerase inactivation in endothelial cells through an Akt-dependent mechanism. *FEBS Lett.* 2001;493(1):21–25.

9 Kurz DJ, Decary S, Hong Y, et al. Chronic oxidative stress compromises telomere integrity and accelerates the onset of senescence in human endothelial cells. *J Cell Sci.* 2004;117(Pt 11):2417–2426.

10 Hoffmann J, Haendeler J, Zeiher AM, Dimmeler S. TNFalpha and oxLDL reduce protein S-nitrosylation in endothelial cells. *J Biol Chem.* 2001;276(44):41383–41387.

11 Unterluggauer H, Hampel B, Zwerschke W, Jansen-Durr P. Senescence-associated cell death of human endothelial cells: the role of oxidative stress. *Exp Gerontol.* 2003;38(10):1149–1160.

12 Rebbaa A, Zheng X, Chou PM, Mirkin BL. Caspase inhibition switches doxorubicin-induced apoptosis to senescence. *Oncogene.* 2003;22(18):2805–2811.

13 Mari D, Mannucci PM, Coppola R, et al. Hypercoagulability in centenarians: the paradox of successful aging. *Blood*. 1995;85 (11):3144–3149.

14 Sagripanti A, Carpi A. Natural anticoagulants, aging, and thromboembolism. *Exp Gerontol*. 1998;33(7–8):891–896.

15 Chong AY, Blann AD, Patel J, et al. Endothelial dysfunction and damage in congestive heart failure: relation of flow-mediated dilation to circulating endothelial cells, plasma indexes of endothelial damage, and brain natriuretic peptide. *Circulation*. 2004;110(13):1794–1798.

16 Ikarugi H, Yamashita T, Aoki R, et al. Impaired spontaneous thrombolytic activity in elderly and in habitual smokers, as measured by a new global thrombosis test. *Blood Coagul Fibrinolysis*. 2003;14(8):781–784.

17 Takeshita K, Yamamoto K, Ito M, et al. Increased expression of plasminogen activator inhibitor-1 with fibrin deposition in a murine model of aging, "Klotho" mouse. *Semin Thromb Hemost*. 2002;28(6):545–554.

18 Yamamoto K, Shimokawa T, Yi H, et al. Aging accelerates endotoxin-induced thrombosis: increased responses of plasminogen activator inhibitor-1 and lipopolysaccharide signaling with aging. *Am J Pathol*. 2002;161(5):1805–1814.

19 Cai D, Xaymardan M, Holm JM, et al. Age-associated impairment in TNF-alpha cardioprotection from myocardial infarction. *Am J Physiol Heart Circ Physiol*. 2003;285(2):H463–H469.

20 Bauer KA, Weiss LM, Sparrow D, et al. Aging-associated changes in indices of thrombin generation and protein C activation in humans. Normative Aging Study. *J Clin Invest*. 1987;80(6):1527–1534.

21 Hager K, Setzer J, Vogl T, et al. Blood coagulation factors in the elderly. *Arch Gerontol Geriatr*. 1989;9(3):277–282.

22 Feyzi E, Saldeen T, Larsson E, et al. Age-dependent modulation of heparan sulfate structure and function. *J Biol Chem*. 1998; 273(22):13395–13398.

23 Mannucci PM, Mari D, Merati G, et al. Gene polymorphisms predicting high plasma levels of coagulation and fibrinolysis proteins. A study in centenarians. *Arterioscler Thromb Vasc Biol*. 1997;17(4):755–759.

24 Taddei S, Virdis A, Mattei P, et al. Aging and endothelial function in normotensive subjects and patients with essential hypertension. *Circulation*. 1995;91(7):1981–1987.

25 Egashira K, Inou T, Hirooka Y, et al. Effects of age on endothelium-dependent vasodilation of resistance coronary artery by acetylcholine in humans. *Circulation*. 1993;88(1):77–81.

26 Tominaga M, Fujii K, Abe I, et al. Hypertension and ageing impair acetylcholine-induced vasodilation in rats. *J Hypertens*. 1994;12(3):259–268.

27 Matz RL, de Sotomayor MA, Schott C, et al. Vascular bed heterogeneity in age-related endothelial dysfunction with respect to NO and eicosanoids. *Br J Pharmacol*. 2000;131(2):303–311.

28 Lakatta EG, Levy D. Arterial and cardiac aging: major shareholders in cardiovascular disease enterprises. Part I: aging arteries: a "set up" for vascular disease. *Circulation*. 2003;107(1):139–146.

29 Amrani M, Goodwin AT, Gray CC, Yacoub MH. Ageing is associated with reduced basal and stimulated release of nitric oxide by the coronary endothelium. *Acta Physiol Scand*. 1996;157(1):79–84.

30 Imaoka Y, Osanai T, Kamada T, et al. Nitric oxide-dependent vasodilator mechanism is not impaired by hypertension but is diminished with aging in the rat aorta. *J Cardiovasc Pharmacol*. 1999;33(5):756–761.

31 Tschudi MR, Luscher TF. Age and hypertension differently affect coronary contractions to endothelin-1, serotonin, and angiotensins. *Circulation*. 1995;91(9):2415–2422.

32 Goodwin AT, Amrani M, Marchbank AJ, et al. Coronary vasoconstriction to endothelin-1 increases with age before and after ischaemia and reperfusion. *Cardiovasc Res*. 1999;41(3):554–562.

33 Gao F, Christopher TA, Lopez BL, et al. Mechanism of decreased adenosine protection in reperfusion injury of aging rats. *Am J Physiol Heart Circ Physiol*. 2000;279(1):H329–H338.

34 Rivard A, Berthou-Soulie L, Principe N, et al. Age-dependent defect in vascular endothelial growth factor expression is associated with reduced hypoxia-inducible factor 1 activity. *J Biol Chem*. 2002;275:29643–29647.

35 Vaitkevicius PV, Lane M, Spurgeon H, et al. A cross-link breaker has sustained effects on arterial and ventricular properties in older rhesus monkeys. *Proc Natl Acad Sci USA*. 2001;98(3):1171–1175.

36 Rivard A, Fabre JE, Silver M, et al. Age-dependent impairment of angiogenesis. *Circulation*. 1999;99(1):111–120.

37 Marinho A, Soares R, Ferro J, et al. Angiogenesis in breast cancer is related to age but not to other prognostic parameters. *Pathol Res Pract*. 1997;193:267–273.

38 Pili R, Guo Y, Chang J, et al. Altered angiogenesis underlying age-dependent changes in tumor growth. *J Natl Cancer Inst*. 1994; 86:1303–1314.

39 Tomanek RJ. Response of the coronary vasculature to myocardial hypertrophy. *J Am Coll Cardiol*. 1990;15(3):528–533.

40 Tomanek RJ, Aydelotte MR, Torry RJ. Remodeling of coronary vessels during aging in purebred beagles. *Circ Res*. 1991;69(4): 1068–1074.

41 Hachamovitch R, Wicker P, Capasso JM, Anversa P. Alterations of coronary blood flow and reserve with aging in Fischer 344 rats. *Am J Physiol*. 1989;256(1 Pt 2):H66–H73.

42 Czernin J, Muller P, Chan S, et al. Influence of age and hemodynamics on myocardial blood flow and flow reserve. *Circulation*. 1993;88(1):62–69.

43 Melidonis A, Tournis S, Kouvaras G, et al. Comparison of coronary collateral circulation in diabetic and non-diabetic patients suffering from coronary artery disease. *Clin Cardiol*. 1999;22: 465–471.

44 Garfinkel S, Hu X, Prudovsky IA, et al. FGF-1-dependent proliferative and migratory responses are impaired in senescent human umbilical vein endothelial cells and correlate with the inability to signal tyrosine phosphorylation of fibroblast growth factor receptor-1 substrates. *J Cell Biol*. 1996;134(3):783–791.

45 Augustin-Voss HG, Voss AK, Pauli BU. Senescence of aortic endothelial cells in culture: effects of basic fibroblast growth factor expression on cell phenotype, migration, and proliferation. *J Cell Physiol*. 1993;157(2):279–288.

46 Reed MJ. Wound repair in older patients: preventing problems and managing the healing. Interview by Marc E. Weksler. *Geriatrics*. 1998;53(5):88–94.

47 Sarzani R, Arnaldi G, Takasaki I, et al. Effects of hypertension and aging on platelet-derived growth factor and platelet-derived growth factor receptor expression in rat aorta and heart. *Hypertension*. 1991;18(Suppl 5):III93–III99.

48 Hill JM, Zalos G, Halcox JP, et al. Circulating endothelial progenitor cells, vascular function, and cardiovascular risk. *N Engl J Med*. 2003;348(7):593–600.

49 Edelberg JM, Tang L, Hattori K, et al. Young adult bone marrow-derived endothelial precursor cells restore aging-impaired cardiac angiogenic function. *Circ Res*. 2002;90:E89–E93.

50 Watanabe Y, Lee SW, Detmar M, et al. Vascular permeability factor/vascular endothelial growth factor (VPF/VEGF) delays and induces escape from senescence in human dermal microvascular endothelial cells. *Oncogene*. 1997;14(17):2025–2032.

51 Arthur WT, Vernon RB, Sage EH, Reed MJ. Growth factors reverse the impaired sprouting of microvessels from aged mice. *Microvasc Res*. 1998;55(3):260–270.

52 Edelberg JM, Lee SH, Kaur M, et al. Platelet-derived growth factor-AB limits the extent of myocardial infarction in a rat model: feasibility of restoring impaired angiogenic capacity in the aging heart. *Circulation*. 2002;105(5):608–613.

53 Witmer AN, Vrensen GF, Van Noorden CJ, Schlingemann RO. Vascular endothelial growth factors and angiogenesis in eye disease. *Prog Retin Eye Res*. 2003;22(1):1–29.

54 Gragoudas ES, Adamis AP, Cunningham ET Jr., et al. Pegaptanib for neovascular age-related macular degeneration. *N Engl J Med*. 2004;351(27):2805–2816.

55 Widlansky ME, Gokce N, Keaney JF Jr., Vita JA. The clinical implications of endothelial dysfunction. *J Am Coll Cardiol*. 2003;42(7):1149–1160.

56 Fay WP, Murphy JG, Owen WG. High concentrations of active plasminogen activator inhibitor-1 in porcine coronary artery thrombi. *Arterioscler Thromb Vasc Biol*. 1996;16(10):1277–1284.

57 Rochetaing A, Kreher P. Reactive hyperemia during early reperfusion as a determinant of improved functional recovery in ischemic preconditioned rat hearts. *J Thorac Cardiovasc Surg*. 2003;125(6):1516–1525.

58 Tani M, Honma Y, Takayama M, et al. Loss of protection by hypoxic preconditioning in aging Fischer 344 rat hearts related to myocardial glycogen content and Na^+ imbalance. *Cardiovasc Res*. 1999;41(3):594–602.

59 Abete P, Ferrara N, Cacciatore F, et al. Angina-induced protection against myocardial infarction in adult and elderly patients: a loss of preconditioning mechanism in the aging heart? *J Am Coll Cardiol*. 1997;30(4):947–954.

60 Fenton RA, Dickson EW, Meyer TE, Dobson JG Jr. Aging reduces the cardioprotective effect of ischemic preconditioning in the rat heart. *J Mol Cell Cardiol*. 2000;32(7):1371–1375.

61 Xaymardan M, Zheng J, Duignan I, et al. Senescent impairment in synergistic cytokine pathways that provide rapid cardioprotection in the rat heart. *J Exp Med*. 2004;199(6):797–804.

62 Simons L, McCallum J, Friedlander Y, Simons J. Risk factors for ischemic stroke – Dubbo study of the elderly. *Stroke*. 1998;29:1341–1346.

63 Petty GW, Brown RD Jr., Whisnant JP, et al. Survival and recurrence after first cerebral infarction: a population-based study in Rochester, Minnesota, 1975 through 1989. *Neurology*. 1998;50(1):208–216.

64 Hazzard WR, John P, Blass JP, et al. *Principles of Geriatric Medicine and Gerontology*, 5th ed. New York: McGraw-Hill, 2003.

65 Marin F, Roldan V, Climent V, et al. Is thrombogenesis in atrial fibrillation related to matrix metalloproteinase-1 and its inhibitor, TIMP-1? *Stroke*. 2003;34(5):1181–1186.

66 Basta G, Schmidt AM, De Caterina R. Advanced glycation end products and vascular inflammation: implications for accelerated atherosclerosis in diabetes. *Cardiovasc Res*. 2004;63(4):582–592.

67 Gilchrest B. Aging of the Skin. In: Hazzard W, Blass H, Ettinger W, et al., eds. *Principles of Geriatric Medicine and Gerontology*. New York: McGraw-Hill, 1998.

68 Holm-Pedersen P, Zederfeldt B. Strength development of skin incisions in young and old rats. *Scand J Plast Reconstr Surg*. 1971;5(1):7–12.

69 Reed MJ, Penn PE, Li Y, et al. Enhanced cell proliferation and biosynthesis mediate improved wound repair in refed, caloric-restricted mice. *Mech Ageing Dev*. 1996;89(1):21–43.

70 Beck LS, DeGuzman L, Lee WP, et al. One systemic administration of transforming growth factor-beta 1 reverses age- or glucocorticoid-impaired wound healing. *J Clin Invest*. 1993;92(6):2841–2849.

71 Reed MJ, Corsa A, Pendergrass W, et al. Neovascularization in aged mice: delayed angiogenesis is coincident with decreased levels of transforming growth factor beta1 and type I collagen. *Am J Pathol*. 1998;152(1):113–123.

72 Molenaar R, Visser WJ, Verkerk A, et al. Peroxidative stress and in vitro ageing of endothelial cells increases the monocyte-endothelial cell adherence in a human in vitro system. *Atherosclerosis*. 1989;76(2–3):193–202.

73 Ashcroft GS, Horan MA, Ferguson MW. The effects of ageing on cutaneous wound healing in mammals. *J Anat*. 1995;187(Pt 1):1–26.

74 Ashcroft GS, Horan MA, Ferguson MW. The effects of ageing on wound healing: immunolocalisation of growth factors and their receptors in a murine incisional model. *J Anat*. 1997;190(Pt 3):351–365.

75 Swift ME, Kleinman HK, DiPietro LA. Impaired wound repair and delayed angiogenesis in aged mice. *Lab Invest*. 1999;79(12):1479–1487.

76 Berger A, Feldman A, Gnant M, et al. The angiogenesis inhibitor, endostatin, does not affect murine cutaneous wound healing. *J Surg Res*. 2000(91):26–31.

77 Bloch W, Huggel K, Sasaki T, et al. The angiogenesis inhibitor endostatin impairs blood vessel maturation during wound healing. *FASEB J*. 2000;14:2373–2376.

78 Chavakis E, Aicher A, Heeschen C, et al. Role of beta2-integrins for homing and neovascularization capacity of endothelial progenitor cells. *J Exp Med*. 2005;201(1):63–72.

79 te Velde EA, Voest EE, van Gorp JM, et al. Adverse effects of the antiangiogenic agent angiostatin on the healing of experimental colonic anastomoses. *Ann Surg Oncol*. 2002;9(3):303–309.

80 Deanfield J, Donald A, Ferri C, et al. Endothelial function and dysfunction. Part I: methodological issues for assessment in the different vascular beds: a statement by the Working Group on Endothelin and Endothelial Factors of the European Society of Hypertension. *J Hypertens*. 2005;23(1):7–17.

81 Vlachopoulos C, Rokkas K, Ioakeimidis N, et al. Prevalence of asymptomatic coronary artery disease in men with vasculogenic erectile dysfunction: a prospective angiographic study. *Eur Urol*. 2005;48(6):996–1003.

82 Rajasekaran M, Kasyan A, Jain A, et al. Altered growth factor expression in the aging penis: the Brown-Norway rat model. *J Androl*. 2002;23(3):393–399.

83 Park K, Ahn KY, Kim MK, et al. Intracavernosal injection of vascular endothelial growth factor improves erectile function in aged rats. *Eur Urol*. 2004;46(3):403–407.

84 Yamanaka M, Shirai M, Shiina H, et al. Vascular endothelial growth factor restores erectile function through inhibition of apoptosis in diabetic rat penile crura. *J Urol*. 2005;173(1):318–323.

85 Boolell M, Allen MJ, Ballard SA, et al. Sildenafil: an orally active type 5 cyclic GMP-specific phosphodiesterase inhibitor for the

treatment of penile erectile dysfunction. *Int J Impot Res.* 1996; 8(2):47–52.

86 Kukreja RC, Ockaili R, Salloum F, et al. Cardioprotection with phosphodiesterase-5 inhibition – a novel preconditioning strategy. *J Mol Cell Cardiol.* 2004;36(2):165–173.

87 Schulze-Neick I, Hartenstein P, Li J, et al. Intravenous sildenafil is a potent pulmonary vasodilator in children with congenital heart disease. *Circulation.* 2003;108(Suppl 1):II167–II173.

88 Park JW, Mrowietz C, Chung N, Jung F. Sildenafil improves cutaneous microcirculation in patients with coronary artery disease: a monocentric, prospective, double-blind, placebo-controlled, randomized cross-over study. *Clin Hemorheol Microcirc.* 2004;31 (3):173–183.

89 Zhang L, Zhang RL, Wang Y, et al. Functional recovery in aged and young rats after embolic stroke: treatment with a phosphodiesterase type 5 inhibitor. *Stroke.* 2005;36(4):847–852.

90 Kukreja RC, Salloum F, Das A, et al. Pharmacological preconditioning with sildenafil: Basic mechanisms and clinical implications. *Vascul Pharmacol.* 2005;42(5–6):219–232.

153

Kawasaki Disease

Jane C. Burns

University of California, San Diego School of Medicine

Kawasaki disease (KD), an acute vasculitis of infancy and early childhood, affords a unique opportunity to study acute endothelial cell (EC) damage in the setting of previously healthy arteries unaffected by underlying disease processes such as atherosclerosis, hypertension, or diabetes. KD is now the most common form of acquired heart disease in children in the United States and Japan (1,2). The vasculitis presents with clinical signs that include fever, rash, conjunctival injection, edema and erythema of the extremities, and mucosal erythema (3). The vasculitis is self-limited, but without treatment, one in four children will develop coronary artery aneurysms occasionally accompanied by aneurysms of other medium-sized, muscular, extraparenchymal arteries (4). Echocardiography during the first 2 months after onset of fever is used to classify patients as having normal, dilated, or aneurismal coronary arteries. Long-term sequelae of the coronary artery aneurysms include ischemic heart disease and myocardial infarction. The acute inflammation can be abrogated in the majority of patients with a single dose (2 g/kg) of intravenous immunoglobulin (IVIG) and aspirin (80–100 mg/kg/day), which reduces the aneurysm rate to 3% to 5% (5).

The etiology of KD remains unknown, although an infectious cause is suspected based on seasonality and clustering of cases (2) and the similarity of clinical signs to other infectious diseases. In addition, the peak incidence in infants and children younger than 5 years, coupled with the rare occurrence of KD in adults and infants younger than 3 months of age, is consistent with infection with a widely disseminated agent that causes asymptomatic infection in most hosts and the acquisition of protective immunity and passage of transplacental antibodies. The finding of immunoglobulin (Ig)A-secreting plasma cells infiltrating into the trachea and small airways, coupled with molecular evidence of an oligoclonal IgA response, suggests a pathogen with a respiratory portal of entry (6,7). Genetic factors likely influence disease susceptibility because the disease is over-represented in children of Asian descent and among siblings and other family members of an index case (8,9). Study of the pathophysiology of KD is hampered by lack of a convincing animal model, inability to sample affected tissues during life, and the paucity of autopsy material due to the rare occurrence of fatal cases. In this chapter, we review the evidence for acute EC injury associated with KD and discuss studies that address whether EC dysfunction is also a late complication of KD in children and young adults.

PATHOLOGY OF THE VESSEL WALL IN KAWASAKI DISEASE

Our understanding of the pathology of KD is based largely on autopsy data due to the inability to biopsy affected arteries during life. Initial autopsy studies focused on similarities between KD and infantile polyarteritis nodosa, a systemic vasculitis of young infants recognized only at autopsy (10,11). The hallmark of the arteritis observed in autopsy tissues in both disease entities is the presence of lesions within the same patient at different stages in the evolution of the vasculitis, as outlined in Table 153-1 (12,13). The earliest changes are observed in the endothelium of the musculo-elastic arteries and include swelling, proliferation, enlarged nuclei, and frank degeneration with adherent platelets entrapped in fibrin (14). The endothelium of the vasa vasorum of larger arteries can be similarly involved. More advanced lesions show edema and inflammatory cell infiltrate in the subendothelial space. This progresses to destruction of the media, with necrosis associated with infiltration of monocytes and lymphocytes of the memory T cell ($CD45RO^+$) and cytotoxic/suppressor T cell ($CD8^+$) phenotypes extending from lumen to adventitia, which results in a transmural vasculitis (15). Replacement of the intima and media with fibrous connective tissue, thinning of the media with aneurysm formation, scarring, and stenosis complete the progression of the vascular lesion. Coronary artery rupture is extremely rare but can occur during the subacute phase. Stenosis occurs over a period of months to years, and these lesions may remain silent until the moment of acute thrombotic occlusion, often decades after the initial acute illness (16,17).

Remodeling of the arterial wall may be followed by gradual thrombotic occlusion and recanalization of the vessel. In a

Table 153-1: Progression of Arterial Lesions in Acute KD as Deduced from Autopsy Data

Intimal mononuclear cell accumulation with intimal thickening
Mixed inflammatory cell infiltration of outer adventitial sites
Lymphocytic infiltration of luminal side of intima
Progressive medial disruption with a mixed inflammatory infiltrate and edema
Intimal fibrinoid necrosis
Luminal thrombosis and aneurismal dilatation

Adapted with permission from Mandell GL, ed. *Atlas of infectious diseases*, Vol. XI, Pediatric Infectious Diseases. Philadelphia: Current Medicine; 1999: 10.1–10.14.

study of arterial lesions after KD, high levels of growth factors, including transforming growth factor (TGF)-β1, platelet-derived growth factor (PDGF), and basic fibroblast growth factor (bFGF) were noted in the muscular layers, whereas vascular endothelial growth factor (VEGF) was strongly expressed by the ECs of these neomicrovessels (18). Presumably, the artery recanulates by proteolysis of the thrombus with seeding of endothelium via migration and proliferation of endothelial progenitor cells. Thus, KD provides not only a model for acute EC destruction but also neoangiogenesis and EC proliferation.

The state of the endothelium in the microvasculature during acute and convalescent KD in living patients has been studied in three different vascular beds: skin, conjunctiva, and right ventricular myocardium. In the skin, nonspecific degenerative changes were noted in the endothelium of both subcutaneous venules and arterioles (19). Direct evidence of EC activation during acute KD comes from examination of skin biopsies that demonstrated high levels of intercellular adhesion molecule (ICAM)-1, endothelial adhesion molecule (ELAM)-1 (also known as E-selectin), and human leukocyte antigen (HLA) DQ expression on capillary ECs (20). Study of conjunctival biopsies by electron microscopy revealed areas of vacuolated endothelial cytoplasm with adherent platelet clumps, suggesting focal EC damage (21) (Figure 153.1).

In a study of right ventricular biopsies in 201 KD patients studied 1 week to 11 years after onset of acute KD, all patients had some degree of EC degeneration detected by light microscopy (22). In addition, all patients had evidence of myocarditis with lymphocytic infiltrates and varying degrees of fibrosis. Disarray, abnormal branching, and hypertrophy of cardiac myocytes also was noted. A second study of right ventricular biopsies obtained 2 months to 23 years after disease onset in 54 KD patients revealed changes in the endothelium by electron microscopy that included atrophy, degeneration, and fresh thrombus attached to damaged ECs of the microvessels (23). ECs in the arterioles and small arteries showed proliferation, swelling, vacuolization, and projection into the lumen. These findings were noted in six (20%) of 30 age-matched controls with nonvasculitic heart disease as compared with 26 (48.1%) of 54 KD patients. Disturbingly, these changes also were noted in KD patients who had normal coronary arteries as assessed by echocardiography during their acute and subacute stages.

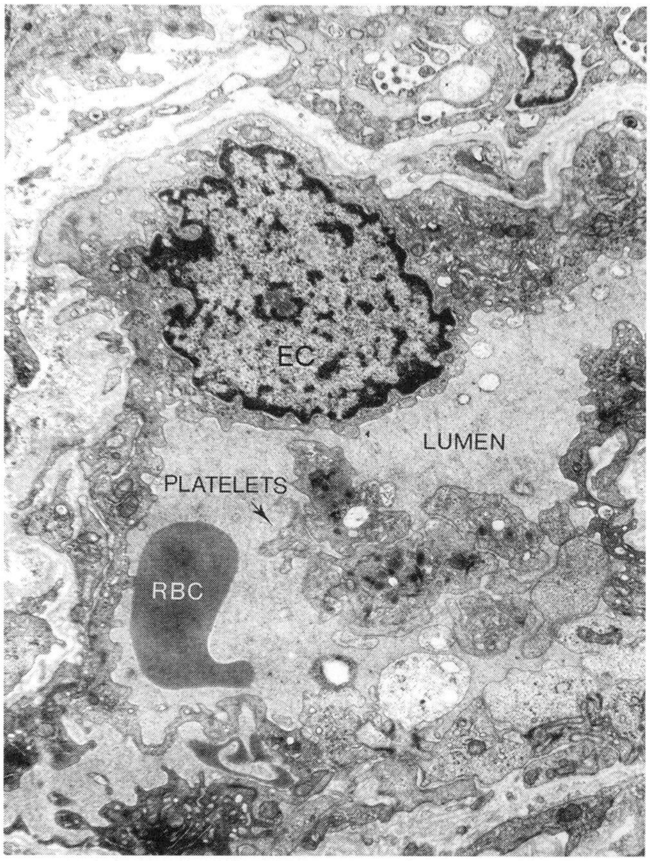

Figure 153.1. Electron photomicrograph of conjunctival biopsy obtained on illness day 4 from a 3-month-old patient with KD prior to intravenous gamma globulin treatment. ECs have areas of vacuolated cytoplasm with adherent platelet clumps suggesting focal EC damage (21). Magnification ×14,280. (Reproduced with permission from Burns JC, Wright JD, Newburger JW, et al. Conjunctival biopsy in patients with Kawasaki disease. *Pediatr Pathol Lab Med.* 1995; 15:547–553.)

In a histologic study of coronary arteries in a child with normal echocardiograms during the acute and subacute phase of KD who died 13 months later, expression of PDGF-A and inducible nitric oxide synthase (iNOS) was elevated in the regenerating intima. This suggests that persistent inflammation and repair of EC injury continues for months after the acute illness has subsided (24).

ROLE OF ANTIENDOTHELIAL CELL ANTIBODIES

Early observations by Leung and coworkers suggested that circulating anti-EC antibodies (AECAs) might be involved

Table 153-2: Noninvasive Assessment of Endothelial Cell Function in Children Following Acute Kawasaki Disease

Method	*Population of KD Subjects*	*Results**	*Conclusion*	*References*
BAR	20 British subjects age 11–19 yrs	Reduced flow-mediated dilation; normal response to NTG	Abnormal EC function in KD subjects	44
BAR	39 Chinese subjects age 3–14 yrs	Reduced flow-mediated dilation reversible with IV infusion of vitamin C; normal response to NTG	Abnormal EC function in KD subjects	46
BAR	24 Canadian subjects, mean age 14.3 ± 1.8 yrs	No difference in flow-mediated dilation between KD subjects and controls ($n = 11$)	Normal EC function in KD subjects	55
BAR	24 Japanese subjects, mean age 8.3 ± 4.1 yrs	Reduced flow-mediated dilation with greatest decrease in KD subjects with aneurysms	Abnormal EC function in KD subjects	52
PET	27 Japanese subjects age, 17.2 ± 3.2 yrs	Decreased myocardial flow reserve; reduced response to cold pressor challenge; normal response to adenosine triphosphate	Abnormal EC function in KD subjects	56
LDF	27 U.S. subjects age 0.33–16 yrs	Normal hyperemic response to acetylcholine iontophoresis	Normal EC function in micro-vasculature in convalescent KD subjects	53

*Compared to age-matched normal controls
BAR, brachial artery reactivity; NTG, nitroglycerin; LDF, laser Doppler fluximetry. Studies assessed both children with aneurysms, transient dilation, and normal echocardiograms during the acute illness.

in the pathogenesis of the vasculitis of KD. Patient sera incubated with interleukin (IL)-1– or tumor necrosis factor (TNF)-α–stimulated human umbilical vein ECs (HUVECs) caused cell lysis in the presence of complement, whereas incubation with unstimulated HUVECs had no effect (25). This suggested that, in acute KD, cytokine-mediated activation of ECs leads to the expression of neoantigens that are targeted by cytotoxic antibodies. However, subsequent studies of AECAs in KD yielded conflicting results (26–28). Some investigators detected AECAs during the acute phase and demonstrated cytotoxicity in the absence of HUVEC stimulation with cytokines (28,29). The current consensus is that anti-EC antibodies also are present in the serum of children suffering from a variety of different febrile illnesses and that such antibodies are a nonspecific epiphenomenon associated with generalized immune activation rather than important mediators of the vasculitis in KD (30,31).

ENDOTHELIAL CELL BIOMARKERS IN ACUTE KAWASAKI DISEASE

Several studies have measured circulating biomarkers of EC activation and injury during the acute phase of KD, and these conclude that both processes occur. Studies have documented increased plasma levels of von Willebrand factor (vWF) during the acute phase of KD (32–35). Many studies have examined levels of circulating soluble adhesion molecules, selectins, and integrins as markers of EC activation. Biomarkers of activation including ICAM-1, vascular cell adhesion molecule (VCAM)-1, E-selectin, and P-selectin all have elevated levels during acute KD (36,37). Elevated circulating levels of ICAM-1 also have been correlated with increased TNF-α levels and increased severity of coronary artery lesions resulting from KD (37). The finding during acute KD of elevated levels of S100 proteins, specifically S100A8, S100A9, and S100A12, which interact with the receptor for advanced glycation end-products (RAGE) on ECs and are potent chemotactic agents, may be one mechanism by which increased expression of adhesion molecules and activation of ECs occurs (38–40).

The finding of circulating ECs during the acute and subacute phase of KD, coupled with the detection of endothelial progenitor cells in highest numbers during the subacute phase, is further evidence of EC injury and repair (18,41). Levels of circulating progenitor cells were highest in patients who developed coronary artery dilatation or aneurysms, thus suggesting that the number of circulating cells may correlate with the severity of vessel wall injury (42). Assays of EC microparticles in plasma as a measure of EC activation have not been performed.

ENDOTHELIAL CELL FUNCTION LATE AFTER KAWASAKI DISEASE

Classic methods for studying EC function in vivo, including response to acetylcholine infusion in the coronary arteries, brachial artery flow-mediated dilatation, myocardial blood flow following cold challenge, and laser Doppler fluximetry have yielded an overall impression of subclinical EC dysfunction even in individuals without coronary artery damage detected by echocardiogram during the acute illness (43–56). The most reproducible of these studies has been the direct infusion of acetylcholine into the coronary arteries at the time of angiography. Results from four studies demonstrated failure to vasodilate in response to acetylcholine in coronary artery segments with either persistent or regressed aneurysms, whereas coronary artery segments without previous aneurysms vasodilated normally in these same subjects (47–49,51). A case report of a 21-year-old woman who had suffered KD at age 19 months described coronary artery spasm in response to acetylcholine infusion despite a history of normal echocardiograms during the acute illness and angiographically normal coronary arteries prior to infusion of acetylcholine (54). Whether this isolated case is an anomaly or whether EC dysfunction occurs in a subset of patients with normal echocardiograms during the acute phase of KD must be studied in a larger population of these patients as young adults.

Although invasive methods of assessing EC function in KD patients have yielded consistent results, studies using noninvasive methods have yielded conflicting results, perhaps due to technical issues (44,46,52,53,55,56) (Table 153-2). None of the four studies reporting assessments of flow-mediated dilation of the brachial artery used high-resolution transducers (>10-MHz linear-array transducer) nor did they provide assessments of inter- and intraobserver variability in a comparable population of children, as recommended by an international task force (57). Studies in Chinese, Japanese, and British children found reduced brachial artery flow-mediated dilation (EC-dependent response) in patients with a history of KD regardless of their coronary artery status during the acute disease and normal responses to nitroglycerin (EC-independent response) when compared with age-matched control subjects (44,46, 52). This suggested diffuse EC dysfunction in children and young adults with a history of KD 5 to 17 years prior to testing. In contrast, a third study in Canadian children found no evidence for EC dysfunction using essentially the same study design (55).

Further studies using noninvasive techniques have included ^{15}O-water positron emission tomography (PET) following adenosine triphosphate administration (EC-independent response) and cold pressor challenge (EC-dependent). Based on reduced myocardial blood flow following cold challenge, the authors concluded that KD patients had impaired EC function regardless of their coronary artery status during their acute disease (56) (see Table 153-2). A single study using laser Doppler fluximetry before and after acetylcholine iontophoresis found normal EC responses in the microcirculation in the skin 0.9 to 11 years after KD (53). If confirmed by further studies, these data suggest that EC dysfunction late after KD may be restricted to medium-sized muscular arteries.

CONCLUSION

KD is characterized by acute EC activation and injury, which in some cases is associated with transmural inflammation and destruction of the vessel wall architecture. This leads to the formation of coronary artery aneurysms, the hallmark of this disease. Emerging evidence suggests that EC injury may continue for years following the acute illness in children who develop aneurysms. Thus, KD provides a model to study both acute and chronic EC injury and recovery in a population of otherwise healthy young subjects.

KEY POINTS

- KD provides a model to study both acute and chronic EC injury and recovery in a population of otherwise healthy young subjects.
- Acute EC injury occurs early in the vasculitic process.
- Recovery from acute EC damage may take months to years.
- Abnormal EC function may occur in a subset of patients years after recovery from the acute illness.

Future Goals

- To understand the environmental (probably infectious) trigger for acute KD and the mechanism by which this causes acute EC injury
- To understand the complex genetics that influence susceptibility to disease and disease outcome

REFERENCES

1 Taubert KA, Rowley AH, Shulman ST. Nationwide survey of Kawasaki disease and acute rheumatic fever. *J Pediatr*. 1991;119 (2):279–282.

2 Burns JC, Cayan DR, Tong G, et al. Seasonality and temporal clustering of Kawasaki Syndrome in Japan, 1987–2000. *Epidemiology*. 2005;16:220–225.

3 Kawasaki T, Kosaki F, Okawa S, et al. A new infantile acute febrile mucocutaneous lymph node syndrome (MLNS) prevailing in Japan. *Pediatrics*. 1974;54(3):271–276.

4 Tanaka N. Pathological study of Kawasaki disease (MCLS): with special reference to sequelae. *Jpn J Med Sci Biol*. 1979;32(4):245–246.

5 Newburger JW, Takahashi M, Beiser AS, et al. A single intravenous infusion of gamma globulin as compared with four infusions in the treatment of acute Kawasaki syndrome. *N Engl J Med*. 1991;324(23):1633–1639.

6 Rowley AH, Shulman ST, Spike BT, et al. Oligoclonal IgA response in the vascular wall in acute Kawasaki disease. *J Immunol.* 2001;166(2):1334–1343.
7 Rowley AH, Shulman ST, Mask CA, et al. IgA plasma cell infiltration of proximal respiratory tract, pancreas, kidney, and coronary artery in acute Kawasaki disease. *J Infect Dis.* 2000;182(4):1183–1191.
8 Dergun M, Kao A, Hauger SB, et al. Familial occurrence of Kawasaki syndrome in North America. *Arch Pediatr Adolesc Med.* 2005;159(9):876–881.
9 Uehara R, Yashiro M, Nakamura Y, Yanagawa H. Kawasaki disease in parents and children. *Acta Paediatr.* 2003;92(6):694–697.
10 Tanaka N, Sekimoto K, Naoe S. Kawasaki disease. Relationship with infantile periarteritis nodosa. *Arch Pathol Lab Med.* 1976; 100(2):81–86.
11 Landing BH, Larson EJ. Are infantile periarteritis nodosa with coronary artery involvement and fatal mucocutaneous lymph node syndrome the same? Comparison of 20 patients from North America with patients from Hawaii and Japan. *Pediatrics.* 1977;59(5):651–662.
12 Burns JC, Felsburg PJ, Wilson H, et al. Canine pain syndrome is a model for the study of Kawasaki disease. *Perspect Biol Med.* 1991;35(1):68–73.
13 Naoe S, Takahashi K, Masuda H, Tanaka N. Kawasaki disease. With particular emphasis on arterial lesions. *Acta Pathol Jpn.* 1991;41(11):785–797.
14 Amano S, Hazama F, Hamashima Y. Pathology of Kawasaki disease: I. Pathology and morphogenesis of the vascular changes. *Jpn Circ J.* 1979;43(7):633–643.
15 Brown TJ, Crawford SE, Cornwall ML, et al. CD8 T lymphocytes and macrophages infiltrate coronary artery aneurysms in acute Kawasaki disease. *J Infect Dis.* 2001;184(7):940–943.
16 Burns JC, Shike H, Gordon JB, et al. Sequelae of Kawasaki disease in adolescents and young adults. *J Am Coll Cardiol.* 1996; 28(1):253–257.
17 Kato H, Inoue O, Kawasaki T, et al. Adult coronary artery disease probably due to childhood Kawasaki disease. *Lancet.* 1992; 340(8828):1127–1129.
18 Suzuki A, Miyagawa-Tomita S, Komatsu K, et al. Active remodeling of the coronary arterial lesions in the late phase of Kawasaki disease: immunohistochemical study. *Circulation.* 2000;101(25): 2935–2941.
19 Hirose S, Hamashima Y. Morphological observation on the vasculitis in the mucocutaneous lymph node syndrome. *Eur J Pediatr.* 1978;129:21–27.
20 Leung DY, Cotran RS, Kurt-Jones E, et al. Endothelial cell activation and high interleukin-1 secretion in the pathogenesis of acute Kawasaki disease. *Lancet.* 1989;2(8675):1298–1302.
21 Burns JC, Wright JD, Newburger JW, et al. Conjunctival biopsy in patients with Kawasaki disease. *Pediatr Pathol Lab Med.* 1995; 15(4):547–553.
22 Yutani C, Go S, Kamiya T, et al. Cardiac biopsy of Kawasaki disease. *Arch Pathol Lab Med.* 1981;105(9):470–473.
23 Liu AM, Ghazizadeh M, Onouchi Z, Asano G. Ultrastructural characteristics of myocardial and coronary microvascular lesions in Kawasaki disease. *Microvasc Res.* 1999;58(1):10–27.
24 Suzuki A, Miyagawa-Tomita S, Komatsu K, et al. Immunohistochemical study of apparently intact coronary artery in a child after Kawasaki disease. *Pediatr Int.* 2004;46(5):590–596.
25 Leung DYM, Geha RS, Newburger JW, et al. Two monokines, interleukin 1 and tumor necrosis factor, render cultured vascular endothelial cells susceptible to lysis by antibodies circulating during Kawasaki syndrome. *J Exp Med.* 1986;164:1958–1972.
26 Guzman J, Fung M, Petty RE. Diagnostic value of anti-neutrophil cytoplasmic and anti-endothelial cell antibodies in early Kawasaki disease. *J Pediatr.* 1994;124(6):917–920.
27 Tizard EJ, Baguley E, Hughes GR, Dillon MJ. Antiendothelial cell antibodies detected by a cellular based ELISA in Kawasaki disease. *Arch Dis Child.* 1991;66(2):189–192.
28 Kaneko K, Savage CO, Pottinger BE, et al. Antiendothelial cell antibodies can be cytotoxic to endothelial cells without cytokine pre-stimulation and correlate with ELISA antibody measurement in Kawasaki disease. *Clin Exp Immunol.* 1994;98(2):264–269.
29 Fujieda M, Oishi N, Kurashige T. Antibodies to endothelial cells in Kawasaki disease lyse endothelial cells without cytokine pretreatment. *Clin Exp Immunol.* 1997;107(1):120–126.
30 Nash MC, Shah V, Reader JA, Dillon MJ. Anti-neutrophil cytoplasmic antibodies and anti-endothelial cell antibodies are not increased in Kawasaki disease. *Br J Rheumatol.* 1995;34(9):882–887.
31 Grunebaum E, Blank M, Cohen S, et al. The role of anti-endothelial cell antibodies in Kawasaki disease – in vitro and in vivo studies. *Clin Exp Immunol.* 2002;130(2):233–240.
32 Burns JC, Glode MP, Clarke SH, et al. Coagulopathy and platelet activation in Kawasaki syndrome: identification of patients at high risk for development of coronary artery aneurysms. *J Pediatr.* 1984;105(2):206–211.
33 Irazuzta JE, Elbl F, Rees AR. Factor VIII related antigen (von Willebrand's factor) in Kawasaki disease. *Clin Pediatr (Phila).* 1990;29(6):347–348.
34 Nash MC, Shah V, Dillon MJ. Soluble cell adhesion molecules and von Willebrand factor in children with Kawasaki disease. *Clin Exp Immunol.* 1995;101(1):13–17.
35 Falcini F, Generini S, Pignone A, et al. Are angiotensin converting enzyme and von Willebrand factor circulating levels useful surrogate parameters to monitor disease activity in Kawasaki disease? *Endothelium.* 1999;6(3):209–215.
36 Takeshita S, Dobashi H, Nakatani K, et al. Circulating soluble selectins in Kawasaki disease. *Clin Exp Immunol.* 1997;108(3): 446–450.
37 Furukawa S, Imai K, Matsubara T, et al. Increased levels of circulating intercellular adhesion molecule 1 in Kawasaki disease. *Arthritis Rheum.* 1992;35(6):672–677.
38 Foell D, Ichida F, Vogl T, et al. S100A12 (EN-RAGE) in monitoring Kawasaki disease. *Lancet.* 2003;361(9365):1270–1272.
39 Ye F, Foell D, Hirono KI, et al. Neutrophil-derived S100A12 is profoundly upregulated in the early stage of acute Kawasaki disease. *Am J Cardiol.* 2004;94(6):840–844.
40 Ebihara T, Endo R, Kikuta H, et al. Differential gene expression of S100 protein family in leukocytes from patients with Kawasaki disease. *Eur J Pediatr.* 2005;164(7):427–431.
41 Yu X, Hirono KI, Ichida F, et al. Enhanced iNOS expression in leukocytes and circulating endothelial cells is associated with the progression of coronary artery lesions in acute Kawasaki disease. *Pediatr Res.* 2004;55(4):688–694.
42 Nakatani K, Takeshita S, Tsujimoto H, et al. Circulating endothelial cells in Kawasaki disease. *Clin Exp Immunol.* 2003;131(3): 536–540.
43 Cheung YF, Yung TC, Tam SC, et al. Novel and traditional cardiovascular risk factors in children after Kawasaki disease:

implications for premature atherosclerosis. *J Am Coll Cardiol.* 2004;43(1):120–124.

44 Dhillon R, Clarkson P, Donald AE, et al. Endothelial dysfunction late after Kawasaki disease. *Circulation.* 1996;94(9):2103–2106.

45 Noto N, Okada T, Yamasuge M, et al. Noninvasive assessment of the early progression of atherosclerosis in adolescents with Kawasaki disease and coronary artery lesions. *Pediatrics.* 2001; 107(5):1095–1099.

46 Deng YB, Li TL, Xiang HJ, et al. Impaired endothelial function in the brachial artery after Kawasaki disease and the effects of intravenous administration of vitamin C. *Pediatr Infect Dis J.* 2003; 22(1):34–39.

47 Yamakawa R, Ishii M, Sugimura T, et al. Coronary endothelial dysfunction after Kawasaki disease: evaluation by intracoronary injection of acetylcholine. *J Am Coll Cardiol.* 1998;31(5):1074–1080.

48 Mitani Y, Okuda Y, Shimpo H, et al. Impaired endothelial function in epicardial coronary arteries after Kawasaki disease. *Circulation.* 1997;96(2):454–461.

49 Iemura M, Ishii M, Sugimura T, et al. Long term consequences of regressed coronary aneurysms after Kawasaki disease: vascular wall morphology and function. *Heart.* 2000;83(3):307–311.

50 Muzik O, Paridon SM, Singh TP, et al. Quantification of myocardial blood flow and flow reserve in children with a history of Kawasaki disease and normal coronary arteries using positron emission tomography. *J Am Coll Cardiol.* 1996;28(3):757–762.

51 Sugimura T, Kato H, Inoue O, et al. Vasodilatory response of the coronary arteries after Kawasaki disease: evaluation by intracoronary injection of isosorbide dinitrate. *J Pediatr.* 1992;121(5 Pt 1):684–688.

52 Kadono T, Sugiyama H, Hoshiai M, et al. Endothelial function evaluated by flow-mediated dilatation in pediatric vascular disease. *Pediatr Cardiol.* 2005;26(4)385–390.

53 Kurio G, Zhiroff KA, Jih LJ, et al. Non-invasive determination of endothelial cell function in Kawasaki syndrome. Submitted.

54 Murakami H, Hirokami M, Hanawa N, et al. Acetylcholine-induced coronary spasm with a history of Kawasaki disease: case report. *Circ J.* 2003;67(3):273–274.

55 Silva AA, Maeno Y, Hashmi A, et al. Cardiovascular risk factors after Kawasaki disease: a case-control study. *J Pediatr.* 2001;138 (3):400–405.

56 Furuyama H, Odagawa Y, Katoh C, et al. Altered myocardial flow reserve and endothelial function late after Kawasaki disease. *J Pediatr.* 2003;142(2):149–154.

57 Corretti MC, Anderson TJ, Benjamin EJ, et al. Guidelines for the ultrasound assessment of endothelial-dependent flow-mediated vasodilation of the brachial artery: a report of the International Brachial Artery Reactivity Task Force. *J Am Coll Cardiol.* 2002; 39(2):257–265.

154

Systemic Vasculitis

Autoantibodies Targeting Endothelial Cells

Miri Blank*, Sonja Praprotnik†, and Yehuda Shoenfeld*,‡

*Sheba Medical Center, Tel-Hashomer, Tel-Aviv University, Israel;
†University Clinical Center, Ljubljana, Slovenia;
‡Tel-Aviv University, Israel

The vasculitides are defined by the presence of leukocytes in the vessel wall, producing reactive damage to mural structures. Compromise of the lumen leads to tissue ischemia and necrosis. In general, affected vessels vary in size, type, and location in association with the specific vasculitic disorder. In recent years, substantial progress has been made in identifying the attributes of specific types of vasculitis, thus allowing for accurate diagnosis. One approach to classifying vasculitides categorizes them, in part, on the basis of the predominant type of vessel affected (Table 154-1). Accordingly, vasculitides are classified into large-vessel vasculitis, medium-sized vessel vasculitis, and small-vessel vasculitis. Large-vessel vasculitis primarily affects the aorta and its primary branches, but medium-sized arteries may also be affected. Medium-sized vessel vasculitis is a disease of small and medium-sized muscular arteries. Small-vessel vasculitis affects vessels smaller than arteries, such as arterioles, venules, and capillaries. It is important to note that small-vessel vasculitis sometimes also affects arteries, and thus the vascular distribution overlaps.

The exact mechanisms underlying these disorders are unclear. Three different pathogenic models of disease have been advanced to help explain why the lesions of a particular vasculitic syndrome are found only in specific vessels. First, the distribution of the antigen responsible for the vasculitis may be differentially distributed. Second, the recruitment and accumulation of the inflammatory infiltrate is likely to be determined by site-specific properties of the endothelial cell (EC), including the expression of adhesion molecules, the secretion of mediators, peptides and hormones, and the specific interaction with inflammatory cells. Finally, nonendothelial structures of the vessel wall may be involved in controlling regional inflammatory processes. All three mechanisms are likely to contribute to the site-specific nature of endothelial injury and vasculitis.

Several causes exist for EC stimulation and injury associated with vasculitis. For example, giant-cell arteritis is caused by pathogenic immune responses that are ultimately controlled by T cells (1). The typical lesions, granulomas in the vessel wall, are formed by $CD4^+$ T cells and macrophages. $CD4^+$ T cells undergo in situ activation in the adventitia, where they interact with indigenous dendritic cells. Tissue injury is mediated by several distinct sets of macrophages. Wegener granulomatosis (WG) is characterized by an early lesion involving neutrophils and ECs as both targets and active participants, and antineutrophil cytoplasmic antibodies (ANCAs), which may enhance immunoinflammatory events that contribute to disease (2). Another group of autoantibodies, anti-EC antibodies (AECAs), were found in many autoimmune diseases showing vasculitis.

ANTIENDOTHELIAL CELL ANTIBODIES

Antibodies that react with EC structures were first reported during the early 1970s. Sera from patients with various rheumatic diseases, including systemic lupus erythematosus (SLE) and scleroderma, were reactive with capillaries in the mouse kidney specimens (3). Since then, several methodological approaches have demonstrated the existence and potential pathogenic role of AECAs in a wide variety of inflammatory diseases (4). AECAs bind to different structures on endothelial membranes, mainly through the $F(ab)_2$ portion of the immunoglobulin (Ig). IgG, IgM, and IgA isotypes of those antibodies have been reported. AECAs target antigens have not yet been defined, but there are likely to be many of them. AECA-positive sera has been shown to display broad reactivity against ECs from different species (mouse, bovine, and human) and multiple vascular beds, including aorta, umbilical

Table 154-1: Classification of Primary Systemic Vasculitis

Vasculitic Type	*Classification*	*Antigens Targeted by AECA*
Large vessel	Giant cell (temporal) arteritis	
	Takayasu arteritis	70-kD protein; proteins range of 60–65 kD
Medium-sized vessel	Classic polyarteritis nodosa	
	Kawasaki disease	
Small vessel	Wegener granulomatosis	PR3, human lysosomal associated membrane protein 2 (h-lamp-2); range of 25-, 68-, 125-, 155-, 180-kD proteins
	Churg-Strauss syndrome	
	Microscopic polyangiitis (microscopic polyarteritis)	Myeloperoxidase
	Henoch-Schönlein purpura	
	Essential cryoglobulinemic vasculitis	
	Cutaneous leukocytoclastic angiitis	

cord vein, saphenous vein, and small vessels from kidney, skin, omenta, and brain (4,5). One interpretation of these findings is that AECAs are nonspecific and target commonly expressed endothelial antigens. However, increasing evidence suggests that AECAs demonstrate specificity for microvascular or macrovascular ECs, and that this specificity correlates with the clinical pattern of vasculopathy.[1]

DISEASE ASSOCIATION AND PREVALENCE OF ANTIENDOTHELIAL CELL ANTIBODIES

AECAs can be detected in various clinically distinct, immunologically mediated diseases in which the common factor appears to be immune/inflammatory-mediated damage to the vessel wall. The largest disease group in which AECAs have been detected includes patients with systemic vasculitides. AECAs also are associated with other systemic autoimmune diseases. In the next sections, we discuss the occurrence of AECAs in these various syndromes and, where possible, emphasize the correlation between site-specific pathology and in vitro AECA specificity.

Small-Vessel Vasculitides

Wegener Granulomatosis and Microscopic Polyangiitis

WG and microscopic polyangiitis (MPA) are systemic disorders in which inflammation and fibrinoid necrosis of small-vessel walls may affect several organs. WG and MPA share many common pathological features including pulmonary capillaritis and pauci-immune focal necrotizing and/or crescentic glomerulonephritis. However, an important difference between the two diseases is the presence of granulomatous inflammation in WG, which is not observed in MPA. Several investigative groups have reported the occurrence of AECAs in both WG and MPA sera, which are distinct from antineutrophil cytoplasmic antibodies (ANCAs), the other serologic marker present in the majority of these patients.

The presence of AECAs in WG has been reported to correlate with disease activity and risk for relapse (6). For example, a prospective analysis of patients with ANCA-negative WG showed that persistence of AECAs after remission was associated with a high risk of relapse, and that a rise in AECA titers preceded the development of relapse (6).

Several in vitro studies have demonstrated that AECAs from the sera of patients with WG are able to activate ECs (7–9). Incubation of human umbilical vein ECs (HUVECs) with AECA IgG from WG patients resulted in increased expression of E-selectin, intercellular adhesion molecule (ICAM)-1, and vascular cell adhesion molecule (VCAM)-1 (7). The upregulation of adhesion molecules was followed by increased leukocyte adhesion. Once bound to ECs, AECAs also induced a parallel increase in interleukin (IL)-1, IL-6, and chemokines (IL-8 and monocyte chemotactic protein [MCP]-1). Each of these studies employed similar methods to detect AECAs and to evaluate EC phenotype, and thus collectively provide compelling evidence for AECA-mediated activation of HUVECs.

More recently, it was demonstrated that AECAs from patients with WG and MPA react more strongly with human microvascular ECs (MVECs) cultivated from unaffected target organs such as the nose, kidney, and lung, compared with HUVECs. Thus, HUVECs may not always be the suitable target in AECA assays (10). Vascular bed–specific AECAs did not result in EC lysis. Interestingly, the same profile of AECAs persisted despite clinical remission of the disease, albeit at lower titers. Binding of WG-AECAs to kidney and nasal ECs was lost upon treatment with interferon (IFN)-γ and tumor necrosis

1 In these assays, HUVECs (until passage 4) or microvascular cells are grown in 96-well ELISA plates for about 48 hours until confluent. Half the plate of monolayer cells is fixed with 0.4% glutaraldeyde; the other half undergoes no fixation. The plates are blocked with 3% BSA at 37C. The antibodies are added at different concentrations in triplicate for 2 hours of incubation and probed with antihuman IgG/IgM conjugated to alkaline phosphatase and appropriate substrate. The data are read at 405 nm.

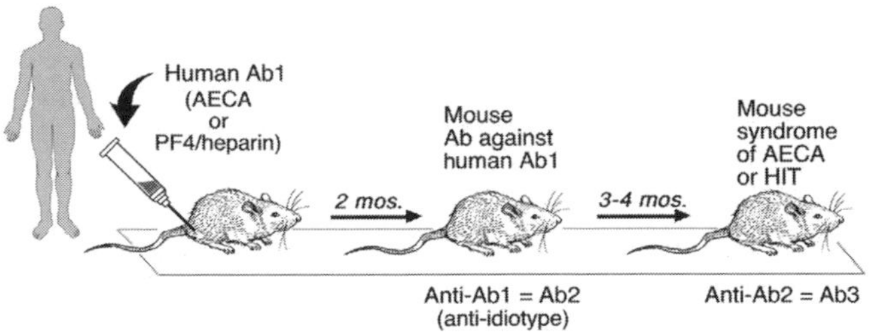

Figure 154.1. Generation of mouse AECAs following immunization with human AECAs. Based on Jerne theory, mice immunized with antibody-1 (Ab1), will develop anti–anti-Ab1 (Ab2) and later anti-Ab2 (Ab3), which will resemble the binding properties of Ab1. Naïve mice such as BALB/c or C3H were immunized with human AECAs or human anti-PF4/heparin complex emulsified in adjuvant and booster 3 weeks later. Eight weeks later, the immunized mice developed anti-human-AECA (Ab2, anti-idiotypic Abs), and 3 to 4 months later anti–anti-idiotypic Abs (Ab3), which had the biological activity of Ab1 (human AECAs or human anti-PF4/heparin). The generation of Ab3 was followed by the development of clinical manifestation similar to vasculitis or HIT respectively.

factor (TNF)-α. However, these two cytokines were cytotoxic (30%) to nasal and lung ECs, an effect that was accentuated (60%) by addition of WG and MPA serum. This increased sera effect may be a result of combined cytokine effect and complement-dependent cytotoxicity (10). These findings suggest that AECAs are not directly involved in destruction of the vessel wall but may instead sensitize ECs to the effects of other host factors involved in the initiation of the disease.

A previous study demonstrated the capacity of human AECAs to induce vasculitis-like lesions in mice (11). Naïve BALB/c mice, injected with IgG fractions from patients with WG, developed histopathologic signs of renal and pulmonary vasculitis. These IgG fractions were shown to contain both anti-proteinase 3 (PR3; the target antigen of cytoplasmic ANCA) and AECA activity. To demonstrate an independent pathogenic role for AECAs, naïve mice were immunized with human WG IgG depleted of anti-PR3 antibodies but still displaying AECAs. These animals developed anti–human-IgG (Ab2) and later anti–antihuman-IgG (Ab3) (characterized as murine AECA; Figure 154.1), which was followed by glomerular vascular inflammation (12). This experimental model, therefore, supports a direct connection between the occurrence of autoimmune AECAs and the appearance of vasculitic histologic lesions.

Together with the Churg-Strauss syndrome (CSS), WG and MPA represent a triad of small-vessel vasculitides characterized by the presence of ANCAs in patient serum. ANCAs are a family of antibodies directed against enzymes that are stored in primary granules of polymorphonuclear neutrophils and lysosomes of monocytes. Based on immunofluorescence patterns, a distinction is made between cytoplasmic ANCAs (c-ANCAs), which target PR3, and perinuclear ANCAs (p-ANCAs), which target myeloperoxidase (MPO). The immunofluorescence pattern in MPA and CSS is usually perinuclear with specificity for MPO. The presence of c-ANCAs, with antibodies to PR3 has a nearly 95% specificity for WG, and the titer correlates with disease activity. In addition to being a serological marker for WG, substantial evidence points to a pathogenic role of ANCAs. In vitro, sera or purified IgG from ANCA-positive patients, as well as monoclonal antibodies directed against MPO or PR3, have been found to induce an oxidative burst in healthy human neutrophils that are pretreated with inflammatory cytokines, such as TNF-α, or bacterial lipopolysaccharide (LPS) (13). Furthermore, it has been shown that ANCAs stimulate neutrophils to damage human ECs. For example, neutrophil-derived PR3 and elastase released at sites of neutrophil and/or monocyte infiltration may signal in a paracrine manner to induce apoptosis of ECs (13). Proinflammatory cytokines also can induce PR3 expression on the surface of ECs, thereby converting these cells into ANCA targets (14).

Antineutrophil Cytoplasmic Antibodies–Positive Necrotizing and Crescentic Glomerulonephritis

Among the first reports of vascular bed–specific antibodies were studies of patients with ANCA-positive necrotizing and crescentic glomerulonephritis (NCGN) (15). ANCA-positive sera from patients with NCGN reacted with glycoproteins from a membrane preparation of neutrophils, designated as gp170/80–110 and verified to be identical with human lysosomal-associated membrane protein 2 (h-lamp-2). Unexpectedly, these sera also cross-reacted with a 130-kDa EC membrane glycoprotein (gp130) of the renal microvasculature. This latter glycoprotein is also present on the surface of ECs of intestinal capillaries and placental capillaries, but not on ECs of arteries and arterioles. Monoclonal antibodies against gp170/80–110 as well as rabbit anti-gp130 failed to bind unstimulated or IL-1–treated HUVECs. Interestingly, 14 of 16 patients with NCGN had IgG that was specific for gp130 and gp170/80–110. The relationship between gp130 and h-lamp-2 has not been established yet. A possible explanation is that the 130-kd antigen shares one or several epitopes with h-lamp-2. It is also possible that the 130-kd membrane protein is an isoform of h-lamp-2, which differs primarily in its carbohydrate side chain (15).

Large-Vessel Vasculitides

Kawasaki disease (KD) is an arteritis of large, medium, and small arteries, particularly the coronary arteries (see Chapter 153). The disease usually occurs in children, and is often associated with a mucocutaneous lymph node syndrome. IgM AECAs were detected in up to 72% of patients with KD, whereas the prevalence of IgG ranged from 12% to 42% (16,17). Antibody titers, measured by cyto-enzyme-linked immunosorbent assay (ELISA) with nonfixed HUVECs, correlated with disease activity and, in some instances, disappeared in patients with convalescent KD (18). AECAs in KD are cytotoxic to HUVECs and human saphenous vein ECs in the absence or presence of prestimulation with IFN-γ, IL-1, and TNF-α (16). IFN-inducible neoautoantigens expressed on

ECs appear to be distinct from IL-1– and TNF-α– inducible neoautoantigens. We have shown recently that mice immunized with AECAs from a patient with KD develop murine AECAs (see Figure 154.1) (17). The murine AECAs increased monocyte adhesion to ECs in vitro, similar to the AECAs used for immunization. Furthermore, all mice that developed AECAs had proteinuria and IgG deposition in the renal mesangium. No histological or immunofluorescence evidence of cardiac vasculitis was detected. These findings suggest that AECAs may play a pathophysiological role in certain KD manifestations.

AECAs have been detected in other large-vessel diseases such as giant-cell arteritis and thromboangiitis obliterans. TA primarily affects the aorta and its primary branches. The inflammation may be localized to a portion of the thoracic or abdominal aorta and branches, or may involve the entire vessel. Giant-cell arteritis and temporal arteritis (TA) are chronic vasculitides of large- and medium-sized vessels. Although it may be generalized, vessel inflammation most prominently involves the cranial branches of the arteries originating from the aortic arch. Thromboangiitis obliterans is a nonatherosclerotic, segmental, inflammatory disease that most commonly affects the small and medium-sized arteries, veins, and nerves of the extremities. AECAs derived from patients with TA have been shown to activate ECs (discussed in the next section) (19). In another study, TA-derived AECAs were reported to mediate complement-dependent cytotoxicity of ECs (20). Patients with active thromboangiitis obliterans had significantly higher AECA titers than did those with inactive disease (21). Taken together, these studies support a pathogenic role for AECAs in most large-vessel vasculopathies. AECAs also were detected in a large proportion of patient with giant-cell arteritis, but the association with disease activity was not evaluated (22).

We have shown that AECAs from TA patients preferentially bind and activate macrovascular, compared with microvascular ECs (19). Monoclonal antibodies were selected, the mixture of which produced 100% inhibition of binding of the original IgG from the patient with TA to HUVECs. All monoclonal antibodies possessed high activity against macrovascular ECs (i.e., anti-HUVEC activity), but none had significant anti-MVEC activity. Four of the six monoclonal antibodies activated ECs, as evidenced by increased IL-6 and von Willebrand factor (vWF) secretion. The four monoclonal antibodies induced EC expression of adhesion molecules and increased adhesion of monocytes to ECs. In addition, these monoclonal antibodies stimulated the nuclear translocation of the nuclear factor (NF)-κB transcription factor (23). Moreover, immunohistochemical studies demonstrated considerable antihuman aortic EC activity of the monoclonal antibodies, whereas antihuman MVEC antibodies (from patients with heparin-induced thrombocytopenia [HIT]) or normal human IgG did not react with human aorta. Again, the distinct predilection of TA-derived AECA monoclonal antibodies for macrovascular antigens corresponds with the distribution of pathological lesions in this large artery vasculitis.

Systemic Lupus Erythematosus and Antiphospholipid Syndrome

A high prevalence of AECAs has been described in patients with clinically active SLE (up to 80%) (22). However, AECA binding was not specific to ECs, since binding reactivity was partially absorbed by both fibroblast and peripheral blood mononuclear cell membrane suspensions (24). The high prevalence of nonspecific AECAs in that study may have been due to the method utilized for AECA detection, namely ELISA with HUVEC membranes. It is difficult to completely rule out any contamination of cytoplasmic structures in the cell membrane preparation. Furthermore, the high percentage of AECA-positive patients may reflect a referral bias, with a greater representation of severe clinical cases. In other studies, an association between the AECA titer, SLE disease activity, and lupus nephritis (25) or SLE-associated psychiatric manifestations (26) was reported, but the prevalence of AECAs in SLE patients (as detected by cyto-ELISA with HUVECs) in the last quoted study was lower and ranged from 29.4% in patients without psychiatric manifestation to 64.7% in patients with psychiatric manifestation. No association was found between anti-DNA antibodies and AECAs in SLE patients (27).

Binding of AECAs to ECs was reported to be unrelated to the presence of circulating immune complexes and antinuclear antibodies in the serum (5,24). Antigens recognized by AECAs in SLE sera have been partially characterized by immunoblotting techniques that have demonstrated large numbers of bands (24,28). Moreover, analysis of the antigens on EC surface membrane immunoprecipitated by AECA-positive SLE sera identified several constitutive proteins with molecular weights ranging from 25 to 200 kDa. Although most of the endothelial antigens were also recognized by AECAs from patients with WG or MPA, some were precipitated by SLE sera alone (29). These data point to the existence of both disease-specific and disease-nonspecific antibodies. Other studies have implicated constitutively expressed sulfate proteoglycan or heparin-like molecules as antigenic determinants for AECAs in a subset of SLE patients (24). SLE-derived AECAs also may recognize antigens in the subendothelial matrix, including collagen (30).

Antibodies directed against constitutive endothelial antigens, both anti-DNA and anti-β_2-glycoprotein I (anti-β_2GPI) antibodies, can display antiendothelial activity due to their ability to recognize their respective molecules adhered on the EC surface (25,31). Recently, binding of antiphospholipid IgG or both polyclonal and monoclonal anti-β_2GPI antibodies has been shown to induce phenotypic changes in cultured HUVECs (32), including upregulation of E-selectin, ICAM-1, and VCAM-1; increased secretion of proinflammatory cytokines such as IL-1 and IL-6; and increased procoagulant activity. In addition, mononuclear leukocytes, adhering to activated endothelium, can be activated by endothelial inflammatory cytokines and induced to display a procoagulant phenotype, further contributing to the thrombophilic state in antiphospholipid antibody syndrome (APS). Similar induction of a proinflammatory phenotype in vitro was also shown in the anti-EC IgG fraction from SLE patients, which did not react with either double-stranded DNA or β_2GPI (8,31). Indeed, true AECAs directed against constitutive endothelial antigens and distinct from anti-β_2GPI antibodies also have been found in the sera of patients with both APS and SLE-associated APS (31).

An analysis of AECA specificity in primary APS and SLE revealed differences in both the pattern of antibody binding and band intensity between membrane antigens on HUVECs and those on human MVECs (33). Of 17 primary APS sera, antibody binding to HUVEC membranes was found in nine sera and to human MVEC membranes in seven sera. Binding at 72 to 79 kDa was confined to HUVECs. In 32 SLE sera, binding to HUVECs and human MVEC membranes was detected in 17 and 22 sera, respectively, with binding at 135 to 155 kDa being confined to human MVECs. As already mentioned, some anti-EC reactivity in APS may be directed to epitopes on phospholipid-binding proteins, especially β_2GPI.

Antiendothelial Cell Antibodies in Transplantation

Progressive narrowing and ultimate luminal occlusion of the arteries and arterioles of a transplanted organ are the hallmark of chronic rejection. Immunologically mediated injury of ECs may be the initiating event. Indeed, increasing evidence suggests that AECAs are associated with transplant rejection (34). AECAs can recognize major histocompatibility complex (MHC) antigens in MHC-mismatched allografts, but even in well-matched allografts, antibodies can be targeted to antigens of the EC/monocyte system or to EC-specific determinants (34). It was shown that serum AECA titers increase significantly in renal transplant recipients with acute rejection, but decrease after formal rejection (immunosuppressive) therapy. In recipients with acute rejection, the prerenal transplant AECA titer was higher than in those without acute rejection, leading to the conclusion that pre- and postrenal transplant AECA titers may be useful for predicting and monitoring acute rejection, respectively (35). Recent studies of cardiac allograft recipients support an association between AECAs and transplant-induced coronary artery disease (36). In one study, these antibodies were shown to be directed against vimentin, a cytoskeletal protein with a molecular mass of 56 to 58 kDa (37), whereas another report implicated triose phosphate isomerase (molecular weight of 28 kDa), the 85-kd human CD36 antigen (a cell surface glycoprotein), and a 50-kDa keratin-like protein (a member of the intermediate filament protein expressed in epithelial cells) as target antigens (37). Although these antibodies may prove useful as diagnostic markers for post-transplantation coronary artery disease, which, if any, play a direct pathogenic role in mediating this complication remains unknown (36).

Cytomegalovirus (CMV) infection is associated with an increased incidence of acute and chronic allograft rejection in renal and cardiac transplant recipients. There are reports suggesting that AECA levels, as determined by cyto-ELISA with HUVECs, are elevated during CMV infections in transplant recipients. These observations support the concept that CMV infections are associated with an increased humoral immune response to EC antigens, which may in turn increase the risk for vascular rejection. However, further investigations have shown that AECAs induced by CMV infections are not specific to ECs and react with multiple cell types such as fibroblasts, keratinocytes, platelets, peripheral blood mononuclear cells, and Burkett lymphoma Raji B cells (38). Consequently, elevated AECA levels seen in CMV-infected transplant recipients may reflect a polyclonal activation of the humoral immune response, which is of uncertain pathogenic significance.

Other Diseases

AECAs have been found in other autoimmune disease in addition to systemic vasculitis and SLE. For example, antibodies are often detected in the sera of patients with thrombotic thrombocytopenic purpura (TTP) and hemolytic uremic syndrome (HUS). CD36 has been proposed as a target antigen in TTP (39). In four consecutive patients with TTP, affinity-purified AECA F(ab)$_2$ bound and activated human bone marrow ECs (HBMECs), but not HUVECs (40). The activation was manifested by enhanced thrombomodulin, IL-6, and vWF release from the EC, followed by raised levels of adhesion molecules (E-selectin, P-selectin, VCAM-1) and CD36 expression as well as monocyte adhesion to ECs. It is noteworthy that TTP is characterized by small-vessel thrombosis only (in the brain, kidney, or skin), whereas thrombosis of large blood vessels is never observed.

AECAs also have been described in patients with HIT (41). HIT is a complex clinical syndrome in which individuals sensitized to heparin develop thrombocytopenia, which may be associated with severe thrombotic events. HIT is mediated by antibodies to epitopes formed by complexes between heparin or other glycosaminoglycans and platelet factor 4 (PF4) (see Chapter 147). In addition, sera from patients with HIT and thrombosis contain antibodies that react with ECs (42). Using a model of active immunity, we studied the effect of IgG anti-PF4/heparin antibody in vivo (43). Naïve BALB/c mice actively immunized with human anti-PF4/heparin antibodies subsequently developed thrombocytopenia after administration of unfractionated heparin (see Figure 154.1). The results of this study support the importance of anti-PF4/heparin antibodies in the pathogenesis of HIT. Because the thrombotic phenomena in HIT occur in large vessels (arteries and veins) as well as in capillaries, we investigated the interaction of affinity-purified anti-PF4/heparin antibodies in cultured microvascular HBMECs and macrovascular HUVECs from six patients with HIT (19). HIT antibodies directly activated unstimulated MVECs, causing release of IL-6, vWF, and soluble thrombomodulin; increased expression of the adhesion molecules P-selectin, E-selectin, and VCAM-1; and enhanced monocyte adhesion. In contrast, the HIT antibodies reacted with macrovascular ECs only after pretreatment of the cells with TNF-α.

Behçet disease affects primarily small vessels, although large-vessel involvement is observed in 15% to 35% of patients. However, when sera from the same patients were exposed to microvascular (omental) or macrovascular (human umbilical vein) ECs with the use of a cyto-ELISA, the binding to MVECs was seen in 43% of patient sera versus 26% of patient sera recognizing HUVECs (44,45).

Naturally Occurring Antiendothelial Cell Antibodies

The occurrence of IgG AECAs in the circulation of healthy individuals and in pooled normal human Ig has been demonstrated by ELISA using fixed HUVECs (46). Further characterization by immunoblotting revealed that natural AECAs recognize a restricted and conserved set of endothelial antigens, with almost no differences between healthy individuals (46). Analysis of AECAs in the serum of SLE patients in the same experiment revealed a higher amount of activity and wider spectrum of antigenic specificities. Recent observations support the concept that natural AECAs increase the anti-inflammatory properties of ECs through selective inhibition of thromboxane A_2, endothelin-1, and metalloproteinase (MMP)-9 secretion, as well as through inhibition of the EC proinflammatory response to TNF-α (36).

IDENTIFICATION OF ANTIENDOTHELIAL CELL ANTIBODY TARGET ANTIGENS

A major challenge in elucidating the mechanisms of AECA-mediated EC activation is to identify the culprit antigen(s). As mentioned earlier, many candidate antigens have been identified, including those that are constitutively expressed (e.g., sulfate proteoglycan- or heparin-like molecules), upregulated by inflammatory mediators, or adherent to ECs (e.g., DNA or DNA-histone complexes, β_2GPI, PF4-heparan complexes, PR3 and/or myeloperoxidase) (4,35).

In a recent study, a molecular cloning strategy was used to isolate a panel of candidate EC antigens in patients with SLE (47). The candidate autoantigens identified included EC-specific plasminogen activator inhibitor, the classic lupus antigen, ribosomal P protein PO, and proteins that have never been described as putative antigens in SLE, such as ribosomal protein L6, elongation factor 1α, adenyl cyclase-associated protein, profilin II, and two novel proteins. Furthermore, it was shown that a patient with active disease and nephritis had autoantibodies that recognized a spectrum of EC autoantigens different from those recognized by the autoantibodies of another patient who had active disease without nephritis. The authors pointed out that these different autoantigen profiles may help to explain the wide clinical spectrum seen in SLE (47). In a recent study, heat-shock protein 60 (Hsp60) was recognized as a new antigen on ECs. Namely, most of the SLE tested sera (73%) bound to a 60-kDa EC surface polypeptide that was identified as human Hsp60. Incubation of ECs with these anti-Hsp60 antibodies induced apoptosis in a time- and dose-dependent manner (48). It also has been observed that IgM AECAs from patients with Behçet disease recognized human α-enolase on human dermal MVECs (49).

The observation that certain antibodies preferentially target microvascular or macrovascular ECs (and that such patterns mirror the clinical phenotype) suggests that AECAs may recognize vascular bed–specific antigens (50,51). It is well established that ECs display marked heterogeneity in structure, mRNA and protein expression, and function. Antigenic heterogeneity is evident between vessels of different sizes, between same-sized vessels from different organs, and even between neighboring ECs. An important goal for the future is to map the site-specific antigenic determinants for AECAs, and to determine which – if any – of these antigens are involved in the pathophysiology of vasculitis.

CIRCULATING ENDOTHELIAL CELLS AND VASCULITIS

Several investigators have reported that circulating ECs were detected in peripheral blood in various conditions with vascular injury (52). These cells are probably detached from the vascular wall, but the mechanism of detachment remains enigmatic. It is well established that the number of circulating ECs increases in small-vessel vasculitis and KD, reflecting the extent of the disease activity (53). In coronary artery disease, circulating ECs were detected in patients with acute myocardial infarction and unstable angina. These findings were well in line with current concepts of plaque vulnerability, and patients with stable angina did not exhibit elevated cell number (54). It is not known whether circulating ECs might be capable of causing an inflammatory response on their own. In addition to circulating ECs, endothelial microparticles (EMPs) have recently emerged as markers of vascular activation. They result from an exocytotic budding process that translocates phosphatidylserine from the inner to the outer leaflet of the cell membrane. This inversion of membrane polarity is followed by blebbing of the membrane surface, leading to the formation of microparticles and their release in the extracellular environment. Even if there is no real consensus on microparticle definition, it is accepted that they are submicron elements (from 0.1 to 1 μm) that express at their surfaces phosphatidylserine and membrane antigens characteristic of their cell of origin. It was demonstrated in a recent in vitro study that EMPs were able to express adhesive receptors similar to those expressed by activated ECs. In the interaction with human monocytes, EMPs also were able to stimulate tissue factor–mediated procoagulant activity, suggesting a novel mechanism by which EMPs may participate in the dissemination and amplification of procoagulant cellular response. EMPs may be involved in target-organ damage in severe hypertension, diabetes, the procoagulant state in pregnancy, and systemic vasculitis (55,56).

CONCLUSION

Increasing evidence points to a pathogenic role for AECAs (and ANCAs) in the inflammatory process that leads to vasculitis. The differences found between ECs from large and small vessels, and the consequent reactivity of AECAs in small and large vessel diseases and ANCA in small-vessel vasculitis, underscore the importance of using cells derived from vessels of appropriate size when studying macro- or microangiopathies. The wide range of AECA frequencies in a defined disease may therefore be attributed to the various sources of ECs used for their detection. Some sera apparently negative for AECAs may be reactive if ECs of appropriate types are used. Consequently, it is rational to classify AECAs into one of two groups of antibodies, against either microvascular or macrovascular ECs.

KEY POINTS

- EC injuries in vasculitis can be phenotyped into microvascular and macrovascular involvements, based on targeting specific EC molecules by anti-EC antibodies.
- These polyclonal populations of antibodies cause activation of the ECs into procoagulation or other pathophysiological states.

Future Goals

- To better understand the mechanisms that cause involvement of macrovascular ECs in some diseases, and microvascular endothelium in others
- To use gene array and proteomics approaches to better understand the signaling cascades provoked by the diverse subclasses of anti-EC antibodies on the different phenotypes of ECs, with the goal of developing more site-specific therapies

REFERENCES

1 Weyand CM, Ma-Krupa W, Goronzy JJ. Imunopathways in giant cell arteritis and polymialgia rheumatica. *Autoimmun Rev.* 2004;3(1):46–53.

2 Csernok E, Müller A, Gross WL. Immunopathology of ANCA-associated vasculitis. *Int Med.* 1999;38(10):759–765.

3 Tan EM, Pearson CM. Rheumatic disease sera reactive with capillares in the mouse kidney. *Arthritis Rheum.* 1972;15:23–28.

4 Praprotnik S, Blank M, Meroni PL, et al. Classification of anti-endothelial cell antibodies into antibodies against microvascular and macrovascular endothelial cells: the pathogenic and diagnostic implications. *Arthritis Rheum.* 2001;44(7):1484–1494.

5 Del Papa N, Gambini D, Meroni PL. Anti-endothelial cell antibodies and autoimmune diseases. *Clin Rev Allergy.* 1994;12(3):275–286.

6 Gobel U, Eichhorn J, Kettritz R, et al. Disease activity and autoantibodies to endothelial cells in patients with Wegener's granulomatosis. *Am J Kidney Dis.* 1996;28(2):186–194.

7 Del Papa N, Guidali L, Sironi M, et al. Anti-EC IgG antibodies from patients with Wegener's granulomatosis bind to human endothelial cells in vitro and induce adhesion molecule expression and cytokine secretion. *Arthritis Rheum.* 1996;39(5):758–766.

8 Carvalho D, Savage CO, Isenberg D, Pearson JD. IgG anti-endothelial cell autoantibodies from patients with systemic lupus erythematosus or systemic vasculitis stimulate the release of two endothelial cell-derived mediators, which enhance adhesion molecule expression and leukocyte adhesion in an autocrine manner. *Arthritis Rheum.* 1999;42(4):631–640.

9 Muller Kobold AC, van Wijk RT, Franssen CF, et al. In vitro up-regulation of E-selectin and induction of interleukin-6 in endothelial cells by autoantibodies in Wegener's granulomatosis and microscopic polyangiitis. *Clin Exp Rheumatol.* 1999;17(4):433–440.

10 Holmen C, Christensson M, Pettersson E, et al. Wegener's granulomatosis is associated with organ-specific anti-endothelial cell antibodies. *Kidney Int.* 2004;66(3):1049–1060.

11 Blank M, Tomer Y, Stein M, et al. Immunization with anti-neutrophil cytoplasmic antibody (ANCA) induces the production of mouse ANCA and perivascular lymphocyte infiltration. *Clin Exp Immunol.* 1995;102(1):120–130.

12 Damianovich M, Gilburd B, George J, et al. Pathogenic role of anti-endothelial cell antibodies in vasculitis. An idiotypic experimental model. *J Immunol.* 1996;156(12):4946–4951.

13 Rarok AA, Limburg PC, Kallenberg CG. Neutrophil-activating potential of antinetrophil cytoplasm autoantibodies. *J Leukoc Biol.* 2003;74(1):3–15.

14 Sibelius U, Hattar K, Schenkel A, et al. Wegener's granulomatosis: anti-proteinase 3 antibodies are potent inductors of human endothelial cell signaling and response. *J Exp Med.* 1998;187(4):497–503.

15 Kain R, Matsui K, Exner M, et al. A novel class of autoantigens of anti-neutrophil cytoplasmic antibodies in necrotizing and crescentic glomerulonephritis: the lysosomal membrane glycoprotein h-lamp-2 in neutrophil granulocytes and a related membrane protein in glomerular endothelial cells. *J Exp Med.* 1995;181(2):585–597.

16 Fujieda M, Oishi N, Kurashige T. Antibodies to endothelial cells in Kawasaki disease lyse ECs without cytokine pretreatment. *Clin Exp Immunol.* 1997;107(1):120–126.

17 Grunebaum E, Blank M, Cohen S, et al. The role of anti-endothelial cell antibodies in Kawasaki disease – in vitro and in vivo studies. *Clin Exp Immunol.* 2002;130(2):233–240.

18 Tizard EJ, Baguley E, Hughes GR, Dillon MJ. Antiendothelial cell antibodies detected by a cellular based ELISA in Kawasaki disease. *Arch Dis Child.* 1991;66:189–192.

19 Blank M, Krause I, Goldkorn T, et al. Monoclonal anti-endothelial cell antibodies from a patient with Takayasu arteritis activate endothelial cells from large vessels. *Arthritis Rheum.* 1999;42(7):1421–1432.

20 Tripathy NK, Upadhyaya S, Sinha N, Nityanand S. Complement and cell mediated cytotoxicity by antiendothelial cell antibodies in Takayasu's arteritis. *J Rheumatol.* 2001;28(4):805–808.

21 Eichhorn J, Sima D, Lindschau C, et al. Anti-endothelial cell antibodies in thromboangiitis obliterans. *Am J Med Sci.* 1998;315(1):17–23.

22 Navarro M, Cervera R, Font J, et al. Anti-endothelial cell antibodies in systemic autoimmune diseases: prevalence and clinical significance. *Lupus.* 1997;6:521–526.
23 Preston G, Zarella CS, Pendergraft WF, et al. Novel effect of neutrophil-derived proteinase 3 and elastase on the vascular endothelium involve in vivo cleavage of NF-κB and proapoptotic changes in JNK, ERK, and p38 MAPK signaling signaling pathways. *J Am Soc Nephrol.* 2002;13(12):2840–2849.
24 van der Zee JM, Siegert CE, de Vreede TA, et al. Characterization of anti-EC antibodies in systemic lupus erythematosus (SLE). *Clin Exp Immunol.* 1991;84(2):238–244.
25 Chan TM, Yu PM, Tsang KL, Cheng IK. Endothelial cell binding by human polyclonal anti-DNA antibodies: relationship to disease activity and endothelial functional alterations. *Clin Exp Immunol.* 1995;100(3):506–513.
26 Conti F, Alessandri C, Bompane D, et al. Autoantibody profile in systemic lupus erythematosus with psychiatric manifestations: a role for anti-endothelial-cell antibodies. *Arthritis Res Ther.* 2004; 6(4):R366–R372.
27 Moscato S, Pratesi F, Bongiorni F, et al. Endothelial cell binding by systemic lupus antibodies: functional properties and relationship with anti-DNA activity. *J Autoimmun.* 2002;18(3): 231–238.
28 Li JS, Liu MF, Lei HY. Characterization of anti-endothelial cell antibodies in the patients with systemic lupus erythematosus: a potential marker for disease activity. *Clin Immunol Immunopathol.* 1996;79(3):211–216.
29 Del Papa N, Conforti G, Gambini D, et al. Characterization of the endothelial surface proteins recognized by anti-endothelial antibodies in primary and secondary autoimmune vasculitis. *Clin Immunol Immunopathol.* 1994;70(3):211–216.
30 Yasici AZ, Behrendt M, Cooper D, et al. The identification of endothelial cell autoantigens. *J Autoimmun.* 2000;15(1):41–49.
31 Papa ND, Raschi E, Moroni G, et al. Anti-EC IgG fractions from systemic lupus erythematosus patients bind to human endothelial cells and induce a pro-adhesive and a pro-inflammatory phenotype in vitro. *Lupus.* 1999;8(6):423–429.
32 George J, Blank M, Levy Y, et al. Differential effects of anti-beta2-glycoprotein 1 antibodies on endothelial cells and on the manifestations of experimental antiphospholipid syndrome. *Circulation.* 1998;97:900–906.
33 Hill MB, Phipps JL, Hughes P, Greaves M. Anti-EC antibodies in primary antiphospholipid syndrome and SLE: patterns of reactivity with membrane antigens on microvascular and umbilical venous cell membranes. *Br J Heamatol.* 1998;103(2): 416–421.
34 Perrey C, Brenchley PE, Johnson RW, Martin S. An association between antibodies specific for endothelial cells and renal transplant failure. *Transpl Immunol.* 1998;6(2):101–106.
35 Shin YS, Yang CW, Ahn HJ, et al. Clinical significance of anti-endothelial cell antibody in renal transplant recipients. *Korean J Intern Med.* 2001;16(1):24–29.
36 Faulk WP, Rose M, Meroni PL, et al. Antibodies to endothelial cells identify myocardial damage and predict development of coronary artery disease in patients with transplanted hearts. *Hum Immunol.* 1999;60(9):826–832.
37 Wheeler CH, Collins A, Dunn MJ, et al. Characterization of endothelial antigens associated with transplant-associated coronary artery disease. *J Heart Lung Transplant.* 1995;14(6 Pt 2): S188–S197.
38 Toyoda M, Petrosian A, Jordan SC. Immunological characterization of anti-endothelial cell antibodies induced by cytomegalovirus infection. *Transplantation.* 1999;68(9):1311–1318.
39 Schultz DR, Arnold PI, Jy W, et al. Anti-CD36 autoantibodies in thrombotic thrombocytopenic purpura and other thrombotic disorders: identification of an 85 kD form of CD36 as a target antigen. *Br J Haematol.* 1998;103(3):849–857.
40 Praprotnik S, Blank M, Levy Y, et al. Anti-endothelial cell antibodies from patients with thrombotic thrombocytopenic purpura specifically activate small vessel ECs. *Int Immunol.* 2001;13(2):203–210.
41 Cines DB, Tomaski A, Tannenbaum S. Immune endothelial-cell injury in heparin-associated thrombocytopenia. *N Engl J Med.* 1987;316(10):581–589.
42 Blank M, Shoenfeld Y, Tavor S, et al. Anti-platelet factor 4/heparin antibodies from patients with heparin induced thrombocytopenia provoke direct activation of microvascular endothelial cell. *Int Immunol.* 2002;14(2):121–129.
43 Blank M, Cines DB, Arepally G, et al. Pathogenicity of human anti-platelet factor 4 (PF4)/heparin in vivo: generation of mouse anti-PF4/heparin and induction of thrombocytopenia by heparin. *Clin Exp Immunol.* 1997;108(2):333–339.
44 Lee KH, Cho HJ, Kim HS, et al. Activation of extracellular signal regulated kinase 1/2 in human dermal microvascular endothelial cell. *J Dermatol Sci.* 2002;30(1):63–72.
45 Schirmer M, Calamia KT, O'Duffy JD. Is there a place for large vessel disease in the diagnostic criteria of Behcet's disease? *J Rheumatol.* 1999;26(12):2511–2512.
46 Ronda N, Leonardi S, Orlandini G, et al. Natural anti-endothelial cell antibodies (AECA). *J Autoimmun.* 1999;13(1):121–127.
47 Frampton G, Moriya S, Pearson JD, et al. Identification of candidate endothelial cell autoantigens in systemic lupus erythematosus using a molecular cloning strategy: a role for ribosomal P protein P0 as an endothelial cell autoantigen. *Rheumatology (Oxford).* 2000;39(10):1114–1120.
48 Dieudé M, Senécal JL, Raymond Y. Induction of endothelial cell apoptosis by heat-shock protein 60-reactive antibodies from anti-endothelial cell autoantibodies positive systemic lupus erythematosus patients. *Arthritis Rheum.* 2004;50(10):3221–3231.
49 Lee KH, Chung HS, Kim HS, et al. Human alpha-enolase from endothelial cells as a target antigen of anti-EC antibody in Behcet's disease. *Arthritis Rheum.* 2003;48(7): 2025–2035.
50 Ribatti D, Nico B, Vacca A, et al. Endothelial cell heterogeneity and organ specificity. *J Haematother Stem Cell Res.* 2002;11(1): 81–90.
51 Rosenberg RD, Aird WC. Vascular-bed – specific hemostasis and hypercoagulable states. *N Engl J Med.* 1999;340(20):1555–1564.
52 Haubitz M, Woyvodt A. Circulating endothelial cell and vasculitis. *Internal Med.* 2004;43(8):660–667.
53 Nakatani K, Takeshita S, Tsujimoto H, et al. Circulating endothelial cells in Kawasaki disease. *Clin Exp Immunol.* 2003;131(3): 536–540.
54 Mutin M, Canavy I, Blann A, et al. Direct evidence of endothelial injury in acute myocardial infarction und unstable angina by demonstration of circulating endothelial cells. *Blood.* 1999;93(9):2951–2958.
55 Sabatier F, Roux V, Anfosso F, et al. Interaction of endothelial microparticles with monocytic cells in vitro induces tissue factor-dependent procoagulant activity. *Blood.* 2002;99(11):3962–3970.
56 Pendergraft WF 3rd, Pressler BM, Jennette JC, et al. Autoantigen complementarity: a new theory implicating complementary proteins as initiators of autoimmune disease. *J Mol Biol.* 2005; 83(1):12–25.

155

High Endothelial Venule-Like Vessels in Human Chronic Inflammatory Diseases

Jean-Philippe Girard

Institute of Pharmacology and Structural Biology, Toulouse, France

HISTORY OF HIGH ENDOTHELIAL VENULE-LIKE VESSELS IN RHEUMATOID ARTHRITIS AND OTHER HUMAN CHRONIC INFLAMMATORY DISEASES

High endothelial venules (HEVs) are anatomically distinct postcapillary venules found in lymphoid organs. They support high levels of lymphocyte extravasation from the blood (1–3). The precise relationship between the structure and function of these specialized blood vessels has intrigued and fascinated many investigators over the past 100 years. The peculiar structure of HEVs in lymph nodes from *Macacus cynomolgus* was first observed by Thome in 1898 (4). He described these vessels as being composed of plump, cuboidal endothelial cells (ECs) whose surface strongly bulges into the vascular lumen. These ECs looked like columnar epithelial cells and Thome wrote that, at first notice, one is more inclined to think of the duct of a gland rather than that of a blood vessel. It is this cuboidal or columnar appearance that has given rise to the names *high ECs* and *high endothelial venules.* The observations of Thome were confirmed in other species, including human, by von Schumacher, who was the first to note the presence of numerous lymphocytes within the wall of HEVs (5). Vessels of the HEV type, characterized by a plump endothelial lining with many adherent and infiltrating lymphocytes, were later observed in Peyer patches and tonsils. Although the occurrence of lymphocytes within HEV walls was first noted in 1899 (5), the direction and physiological significance of lymphocyte migration through HEVs was not fully appreciated until the classical autoradiographic experiments of Gowans and Knight (6). In their landmark paper, published in 1964, these authors demonstrated that radioactively labeled small lymphocytes that were transfused into the blood of rats migrated rapidly into lymph nodes and Peyer patches by crossing the walls of HEVs. These experiments conclusively showed that HEVs are the site of a large-scale migration of lymphocytes from the blood into secondary lymphoid organs. Evidence accumulated over the past 25 years indicates that HEV-like vessels develop in chronically inflamed tissues and appear to share with HEVs this unique capacity to recruit large numbers of circulating lymphocytes (1). The structure, phenotype, and functional role of these HEV-like vessels in human chronic inflammatory diseases is the focus of this chapter.

In rheumatoid arthritis (RA), a chronic systemic inflammatory disease affecting the joints, small blood vessels with HEV characteristics were first described during the 1980s by Antony J. Freemont (7) and Morris Ziff (8). Freemont noticed that areas of lymphocyte aggregation in the diseased synovia (>150 lymphocytes/mm^2) contained small blood vessels easily distinguished by the plump, cuboidal appearance of their ECs (Figure 155.1), the unusual thickness of their perivascular sheaths, and the presence of numerous lymphocytes within their walls. These HEV-like vessels were never observed outside areas of lymphocytic aggregation in the diseased synovia or in normal tissues (7). Although the most striking feature of the ECs from the HEV-like vessels found in RA was their unusual height, ultrastructural analysis also revealed certain additional features, including a prominent Golgi complex, abundant mitochondria closely associated with rough endoplasmic reticulum, numerous ribosomes, and a large oval nucleus contrasting with the flattened nuclei from the usual squamous ECs. This indicated an intense biosynthetic activity generally not observed in ECs from other vessels, which typically have a poorly developed Golgi apparatus. In addition, several different membrane-bound vesicular structures were seen in the cytoplasm, including secretory vesicles, Weibel-Palade bodies, and multivesicular bodies that are also characteristic of lymph-node HEV ECs. Altogether, these ultrastructural findings showed that the plump ECs of HEV-like vessels in RA exhibit features that allow them to be regarded as metabolically active, with a predominantly biosynthetic and secretory function.

In his pioneering study, Freemont observed that the ECs from the HEV-like vessels in diseased synovia were less plump

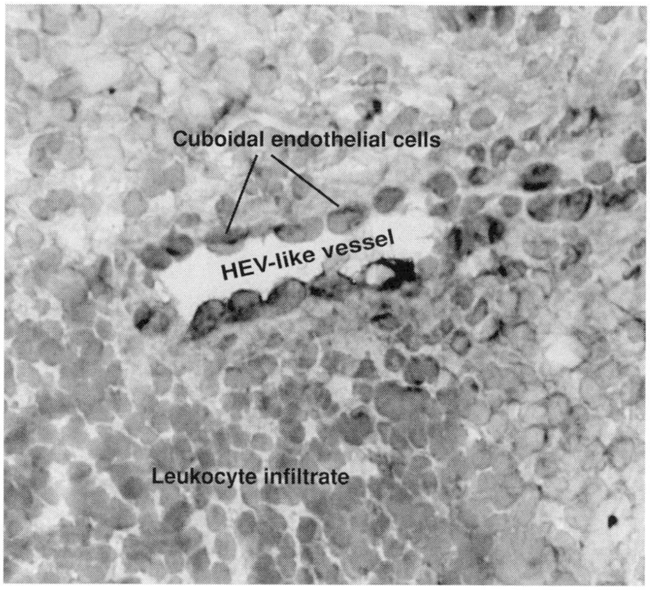

Figure 155.1. HEV-like vessel in RA synovium. HEV-like vessels in the inflamed synovium always are found in association with dense accumulations of lymphocytes. The plump, cuboidal morphology of the ECs can be discerned easily. Staining of the cuboidal ECs with HEV marker MECA-79 is more intense on the side of the leukocyte infiltrate.

when the lymphocyte count was less than 150/mm^2 (7). A few years later, in 1986, Iguchi and Ziff used electron microscopy to perform morphometric analysis of blood vessels in RA synovial membrane and found a strong correlation between the "tallness" or "plumpness" of the ECs and the number and percentage of perivascular lymphocytes (8). This confirmed the intimate relationship between plump endothelium and lymphoid aggregation in RA and strongly suggested that lymphocytes emigrated through the HEV-like vessels to form perivascular lymphoid aggregates in the RA synovium. This latter possibility was further supported by functional studies published the same year (1986) by Sirpa Jalkanen and Eugene Butcher (9), who demonstrated, using the Stamper-Woodruff assay of lymphocyte adhesion to frozen sections in vitro (see Chapter 170 for a description of the assay), the capacity of HEV-like vessels in the RA synovium to support lymphocyte adhesion. Freemont extended these studies by showing that the capacity of HEV-like vessels in frozen sections of RA synovium to bind lymphocytes in vitro was strongly correlated with the plumpness of the ECs and their capacity to incorporate large amounts of radioactive sulfate, a unique metabolic property shared with HEV ECs from lymphoid tissues (10). He proposed that HEV-like vessels may be responsible for the selective accumulation of lymphocytes in the RA synovium and that these observations could be exploited therapeutically "by using antibodies to inhibit selective lymphocyte adhesion to synovial blood vessels or by interfering with synthesis of the sulfated macromolecule implicated in endothelial control of lymphocyte traffic it may be possible to reduce lymphocyte accumulation within diseased synovium" (10).

As early as 1983, Freemont extended the observations made in RA to other diseases and described the presence of HEV-like vessels exhibiting plump ECs and numerous lymphocytes within their walls in many human chronic inflammatory diseases (11). In several completely unrelated disease states affecting distinct tissues (stomach, dermis/subcutaneous tissue, thyroid, liver, kidney, muscle), he found that infiltrates containing significant numbers of lymphocytes (>150 lymphocytes/mm^2) always contained HEV-like vessels with plump endothelium that mediated sulfate uptake and in vitro lymphocyte adhesion (12). These observations strongly suggested that changes in the local venous microvasculature to a type capable of promoting lymphocyte traffic into the tissue were not disease- or organ-specific but were probably a universal finding in sites of lymphocyte accumulation in chronically inflamed tissue. Freemont also made several interesting observations about the mechanisms that govern the formation of HEV-like vessels in diseased tissues. First, recognizable HEVs developed only after lymphocyte migration had been initiated, suggesting that their formation was not the primary event in lymphocyte migration. Second, HEVs developed in sites that, under normal conditions, contained no such vessels. Third, plump ECs from HEV-like vessels did not show mitotic activity. He concluded from these observations that HEV-like vessels developed from remodeling of existing vessels, and their development was a response to, or a consequence of, lymphocyte accumulation in the tissue. However, once developed, these HEV-like vessels were likely to participate in a positive feedback mechanism for increasing lymphocyte entry into the diseased tissue, thus contributing to the amplification and maintenance of chronic inflammation.

PHENOTYPE AND FUNCTION OF HIGH ENDOTHELIAL VENULE-LIKE VESSELS IN HUMAN CHRONIC INFLAMMATORY DISEASES

High Endothelial Venule Markers MECA-79, HECA-452, and 6-Sulfo Sialyl Lewis x Ligands for Lymphocyte L-Selectin

Many studies performed over the past 15 years have revealed that HEV-like vessels induced by chronic inflammation in extralymphoid sites are phenotypically similar to HEVs from lymphoid tissues (1). This is illustrated by the results obtained using monoclonal antibodies MECA-79 and HECA-452, two of the best HEV markers currently available (Table 155-1). These two antibodies, which recognize carbohydrate epitopes found on the 6-sulfo sialyl Lewis X (sLe^x; NeuAcα2–3Galβ1–4[Fucα1–3]GlcNAcβ1–) ligands for the lymphocyte homing receptor L-selectin (Figure 155.2), were initially produced in the laboratory of Eugene C. Butcher in 1988 (13,14) and, since that date, have been widely used for the identification and characterization of HEVs and HEV-like vessels in human tissues and chronic inflammatory diseases.

Table 155-1: MECA-79$^+$ HEV-Like Vessels in Human Chronic Inflammatory Diseases

Affected Organs	*Diseases*	*HEV Markers*	*HEV-Associated Chemokines*	*Other Features*	*References*
Synovium	Rheumatoid arthritis	MECA-79$^+$ HECA-452$^+$ 2F3$^+$	CCL21$^+$ CXCL13$^+$ CXCL12$^+$ CCL19$^+$	Cuboidal endothelium Sulfate uptake	7, 10, 15, 41, 50–54
Gut	Inflammatory bowel diseases (Crohn disease, ulcerative colitis)	MECA-79$^+$ HECA-452$^+$ 2F3$^+$ MAdCAM-1$^+$	CCL21$^+$ CXCL13$^+$	Cuboidal endothelium	11, 14, 25, 31, 43, 50, 60
Thyroid	Autoimmune thyroiditis (Grave disease, Hashimoto disease)	MECA-79$^+$ HECA-452$^+$ 2F3$^+$	CCL21$^+$	Cuboidal endothelium Sulfate uptake	12, 14, 15, 25, 42, 58
Skin	Allergic contact dermatitis, psoriasis, *Lichen planus*, cutaneous lupus erythematosus, cutaneous lymphomas	MECA-79$^+$ HECA-452$^+$	CCL21$^+$	Cuboidal endothelium	15, 25, 36, 50, 55
Stomach	*Helicobacter pylori* gastritis	MECA-79$^+$ HECA-452$^+$ MAdCAM-1$^+$	CCL21$^+$	Cuboidal endothelium	29, 44, 56
Lung	Asthma	MECA-79$^+$ HECA-452$^+$ 2F3$^+$			27
Lung	Bronchiectasis	MECA-79$^+$		Cuboidal endothelium	34
Heart	Acute allograft rejection	MECA-79$^+$ HECA-452$^+$ 2F3$^+$		Fuc-TVII$^+$	26
Kidney	Acute allograft rejection	MECA-79$^+$ HECA-452$^+$ 2F3$^+$			28
Kidney	Glomerulonephritis	MECA-79$^+$ 2F3$^+$		Cuboidal endothelium	32
Muscle	Inflammatory myopathies	MECA-79$^+$		Cuboidal endothelium	33
Nasal mucosa	Experimentally induced allergic rhinitis	MECA-79$^+$		Cuboidal endothelium	35
Conjunctiva	Conjunctival inflammation after cataract surgery	MECA-79$^+$ 2F3$^+$			37

High Endothelial Venule Marker MECA-79

MECA-79 stains all HEVs within lymphoid tissues (both luminal and abluminal surfaces of HEV ECs) and does not react with human umbilical vein ECs (HUVECs), nor with ECs of postcapillary venules in spleen, thymus, or nonlymphoid tissues (13,15). MECA-79 antibody inhibits both lymphocyte migration through HEVs in vivo and lymphocyte adhesion to human lymph node and tonsil HEVs in vitro. Although initially produced against mouse HEVs, MECA-79 shows a wide species cross-reactivity and, in both mouse and

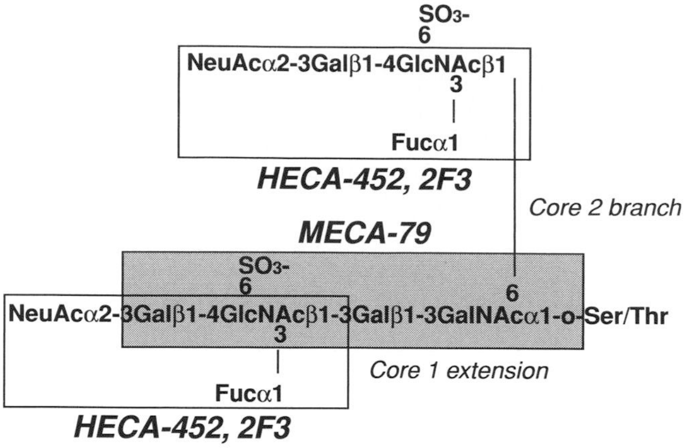

Figure 155.2. Structure of the carbohydrate epitopes recognized by HEV markers MECA-79, HECA-452, and 2F3. The three antibodies react with 6-sulfo sLe^x ligands for lymphocyte L-selectin, displayed in HEVs from lymphoid tissues and HEV-like vessels from chronically inflamed human tissues. The 6-sulfo sLe^x epitope, which is a major capping group of L-selectin ligands, is shown both on extended core 1 and core 2 branched structures, formed respectively by core 1 extending β1,3-N-acetylglucosaminyltransferase and core 2 branching β1,6-N-acetylglucosaminyltransferase. HECA-452 and 2F3 antibodies recognize the sialic acid (NeuAc) and fucose residues of the 6-sulfo sLe^x structure (*white boxes*). MECA-79 reacts with a sulfate-dependent epitope on the extended core 1 structure (*gray box*).

human, recognizes a sulfate-dependent carbohydrate epitope (see Figure 155.2) decorating a set of heavily O-glycosylated sialomucin-type HEV counter-receptors for lymphocyte L-selectin, referred to as peripheral node addressins (PNAds, see Chapter 170) (13,15,16). Recent work from the groups of Minoru Fukuda and John B. Lowe has revealed that the minimal carbohydrate epitope recognized by MECA-79 corresponds to an extended core 1 O-glycan structure modified with 6-sulfo N-acetyllactosamine (17). Moreover, MECA-79 also can bind to the sialylated and fucosylated derivative of this structure, which constitutes the 6-sulfo sialyl Lewis X epitope displayed on PNAd (see Figure 155.2) and is recognized by lymphocyte L-selectin (16–18). An HEV-specific sulfotransferase enzyme designated LSST/HEC-GlcNAc6ST (GlcNAc6ST-2/CHST4) has independently been characterized by the groups of Minoru Fukuda and Steven D. Rosen, and shown to be essential for the elaboration of the MECA-79 epitope and the generation of functional L-selectin ligands within the lumen of HEVs (16,19–21). In addition, a second HEV-expressed GlcNAc-6-O-sulfotransferase identified by Takashi Muramatsu and colleagues and designated GlcNAc6ST-1 (CHST2) also has been found to contribute to the generation of the MECA-79 epitope in HEVs (particularly the abluminal one) and the synthesis of 6-sulfo sialyl Lewis ligands for lymphocyte L-selectin (22). Mice lacking both GlcNAc6ST-1 and GlcNAc6ST-2 demonstrated complete elimination of MECA-79 and 6-sulfo sialyl Lewis X epitopes in HEVs and considerably reduced L-selectin mediated lymphocyte adhesion to HEVs (23,24).

Induction of MECA-79 in HEV-like vessels from RA and other human chronic inflammatory diseases was first reported by Butcher and colleagues in 1993 (15). They observed MECA-79$^+$ vessels with plump endothelium in many cutaneous sites of chronic inflammation, including allergic contact dermatitis, psoriasis, lichen planus, and cutaneous lymphomas. Most importantly, the venules that reacted with MECA-79 were nearly always associated with lymphoid infiltrates, and the degree of MECA-79 expression correlated with the extent of mononuclear cell infiltration present in each lesion. Although MECA-79 reacted with entire vessel segments in most instances, occasional examples of partial reactivity were observed in which only one side of the vessel or even a single EC was MECA-79$^+$. Similar observations were made by Butcher and colleagues in extracutaneous sites of chronic inflammation, particularly synovium in RA and thyroid in lymphocytic thyroiditis, in which vascular MECA-79 reactivity was most pronounced in association with extensive lymphoid infiltrates (Figure 155.3). These studies have been extended by Risto Renkonen and coworkers, who have carried out systematic

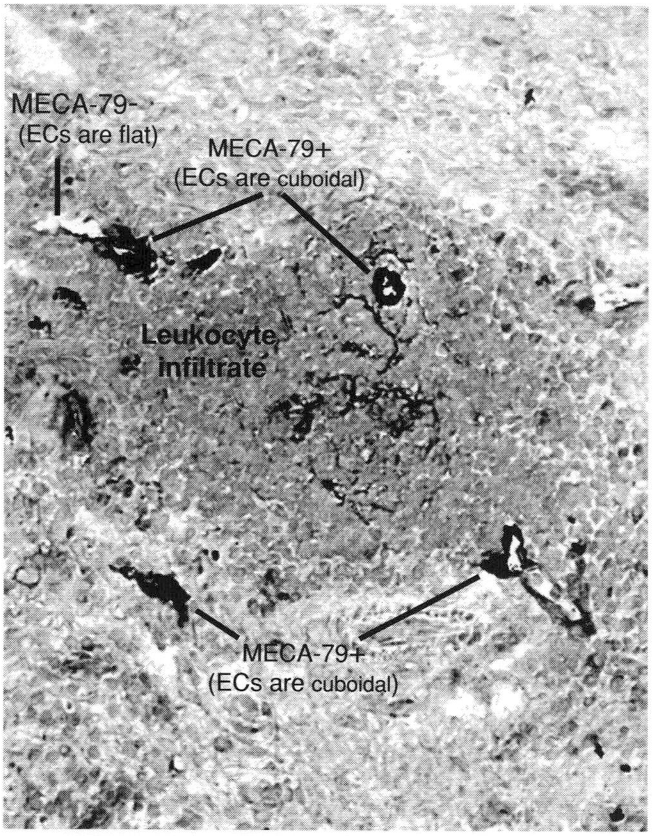

Figure 155.3. Selective induction of HEV marker MECA-79 on HEV-like vessels in RA. Although MECA-79 reacts with the entire vessel segment in the center of the leukocyte infiltrate, partial reactivity is observed at the periphery. Very often, only one side of the vessel (the one close to the infiltrate) is MECA-79$^+$, and a single vessel contains both MECA-79$^+$ cuboidal ECs and MECA-79$^-$ flat ECs. As previously described in lymphoid tissues, a reticular pattern of MECA-79 staining is seen around the HEV-like vessel in the leukocyte infiltrate.

surveys in humans, involving large numbers of independent samples. Immunohistochemical staining with MECA-79 was performed on more than 400 specimens from chronically inflamed tissues and their corresponding noninflamed tissues (25). The endothelial expression of the MECA-79 epitope in postcapillary venules was absent in control tissues but induced de novo in inflamed tissues from patients with thyroiditis, psoriasis, ulcerative colitis, bronchial asthma, and heart and kidney allograft rejection (25–28). Induction of MECA-79 was found both on HEV-like and flat-walled vessels, and the intensity of staining and percentage of MECA-79$^+$ vessels were strongly correlated with the degree of lymphocytic infiltration in the inflamed tissues (25,26). Moreover, differences were observed between the inflammations associated with lymphocytic (MECA-79$^+$ vessels) or neutrophil (MECA-79$^-$ vessels) infiltrations in the same organs, such as between psoriasis and vasculitis in the skin (25), or bronchial asthma and adult respiratory distress syndrome in the lungs (27), further supporting the possibility that MECA-79 induction is a hallmark of lymphocyte infiltration in inflamed human tissues.

Interestingly, expression of MECA-79 has been shown to correlate with progression of inflammation in several diseases. During heart and kidney allograft rejection, MECA-79 expression was induced at the onset and, during acute rejection episodes, correlated with the severity of rejection; it decreased to background levels as the rejection resolved (26,28). Similar observations were recently reported by Fukuda and colleagues, who studied a large series (>140) of chronic inflammatory gastritis human specimens associated with *Helicobacter pylori* infection of the gastric mucosa (29). The number of MECA-79$^+$ HEV-like vessels and the number of patients exhibiting MECA-79$^+$ HEV-like vessels were both highly correlated with the progression of inflammation. In addition, analysis of gastric biopsies obtained from 17 patients with chronic active gastritis before and after eradication of *H. pylori* infection revealed that eradication of *H. pylori* is associated with the disappearance of MECA-79$^+$ HEV-like vessels and lymphocyte infiltration in the gastric mucosa.

The molecular mechanisms involved in the induction of the MECA-79 epitope in human chronically inflamed tissues remain poorly characterized. However, recent data from Rosen and colleagues suggest that the HEV-specific sulfotransferase HEC-GlcNAc6ST is likely to be key. Double staining of human RA samples with MECA-79 and an antibody to the HEV sulfotransferase revealed that all the MECA-79$^+$ vessels were also positive for the sulfotransferase (30). In contrast, MECA-79 staining and the HEV sulfotransferase were not observed in synovial vessels from osteoarthritis patients, used as controls for RA.

A complete list of human inflammatory diseases for which MECA-79$^+$ HEV-like or flat-walled vessels have been observed can be found in Table 155-1. In addition to the aforementioned diseases, these include Crohn disease (31), glomerulonephritis (32), inflammatory myopathies (33), lung bronchiectasis (34), experimentally induced allergic rhinitis (35), cutaneous lupus erythematosus (36), and conjunctival inflammation induced by cataract surgery (37). The diversity of diseases and affected organs suggest that MECA-79$^+$ HEV-like vessels may appear in most (if not all) human nonlymphoid tissues after long-standing chronic inflammation.

High Endothelial Venule Marker HECA-452

Another monoclonal antibody recognizing a carbohydrate epitope expressed on human HEV and HEV-like vessels but not on HUVECs or ECs from other flat-walled vessels was described by Butcher and colleagues in 1988. Elegant studies by Reiji Kannagi and coworkers have revealed that this antibody, HECA-452, belongs to a group of variant-type anti-sLe^x antibodies (HECA-452, 2F3, 2H5) which, unlike the classical type anti-sLe^x antibodies (CSLEX-1, FH-6), recognize the 6-sulfo sLe^x ligands for L-selectin on human HEVs (18,38). HECA-452 recognizes both the sialic acid and fucose residues of the 6-sulfo sLe^x ligand and, similar to the other anti-sLe^x antibodies (including 2F3), does not cross-react with rodent HEVs, likely because of species-specific modifications of sialic acid residues. Expression of HECA-452 and related sLe^x epitope 2F3 in ECs has been shown to be induced by the α13 fucosyltransferase Fuc-TVII (38), which is specifically expressed by HEVs in secondary lymphoid organs (39) and known to play a key role in the synthesis of HEV-specific L-selectin ligands in vivo (40). It should be noted that the α13 fucosyltransferase Fuc-TVII also controls fucosylation of selectin ligands in leukocytes (40), which may explain the cross-reaction of HECA-452 with monocytic cells, dendritic cells, and the cutaneous lymphocyte antigen (CLA) subset of skin-homing memory lymphocytes (14,41,42).

Induction of the HECA-452 epitope on HEV-like vessels was first described in Crohn disease (14), autoimmune thyroiditis (14,42) and RA (41). HECA-452$^+$ vessels in these chronically inflamed tissues resembled HEVs seen in lymphoid organs; they had, however, a more variable appearance, with irregularity in the morphology of the endothelium, from flat to cuboidal. In all these diseases, HECA-452$^+$ vessels always were found in association with dense accumulations of lymphocytes, and positive vessels were never observed in acutely inflamed tissues lacking intense local infiltration of lymphocytes. The presence of diffusely distributed lymphocytes in these acutely inflamed conditions indicated that induction of the HECA-452 epitope at the EC surface was not an absolute requirement for low levels of lymphocyte extravasation. In RA, the morphological and phenotypic changes of the endothelium from HECA-452$^-$ flat-walled vessels to HECA-452$^+$ HEV-like vessels were found to follow the influx of dendritic cells in the synovium (41).

Induction of HECA-452 and related sLe^x epitope 2F3 in human inflammatory diseases has been further characterized through immunohistochemical analyses of large series of patient specimens. Similarly to MECA-79, HECA-452 and/or 2F3 were significantly upregulated on small blood vessels (HEV-like and/or flat-walled) in inflamed tissues from patients with thyroiditis, ulcerative colitis, bronchial asthma, conjunctival inflammation induced by cataract surgery, glomerulonephritis, and heart and kidney allograft rejection (25–28,32,37), and their expression was found to correlate with progression of inflammation during acute heart and kidney rejection episodes (26,28). The staining patterns observed with the two anti- sLe^x antibodies in the diseased tissues were similar but not identical because, for example, HECA-452 was induced in 25% of vessels from psoriatic lesions, whereas 2F3 remained completely negative in skin (25). Interestingly, induction of the two epitopes during heart allograft rejection was associated with the upregulation of fucosyltransferase Fuc-TVII (26).

In *H. pylori*–associated chronic inflammatory gastritis, HECA-452 was found to be coexpressed with MECA-79 on the same HEV-like vessels, and its expression on these vessels was dependent on continuous *H. pylori* infection and highly correlated with progression of inflammation (29). Moreover, the HEV-like vessels detected by MECA-79 and HECA-452 in *H. pylori*–associated gastritis appeared to be functional for recruitment of L-selectin–expressing lymphocytes, because they bound a soluble recombinant form of L-selectin (L-selectin/IgM chimera). Indeed, a large number of B and T lymphocytes were recruited in the gastric mucosa infected with *H. pylori*. Together with those other studies showing simultaneous induction in many diseases of the sulfate-dependent epitope MECA-79 and the sialic acid and fucose-dependent epitopes HECA-452 and/or 2F3 (see Table 155-1), these later results suggest that functional 6-sulfo sLe^x ligands for L-selectin (see Figure 155.2) are induced on HEV-like vessels in many human chronic inflammatory diseases. Therefore, HEV-like vessels are likely to be fully competent for L-selectin-mediated lymphocyte tethering and rolling, the first essential steps of the multistep lymphocyte recruitment cascade (see Chapter 170).

The mucosal HEV-specific EC adhesion molecule mucosal addressin cell adhesion molecule (MAdCAM)-1, a ligand for both L-selectin (when decorated with the MECA-79 epitope) and lymphocyte $\alpha 4\beta 7$ integrin, known to mediate the initial interaction of lymphocytes with mucosal HEVs in Peyer patches, also has been shown to be induced on HEV-like vessels in chronic inflammatory bowel diseases (Crohn disease, ulcerative colitis) and *H. pylori* gastritis (43,44). This suggests that MAdCAM-1 is likely to cooperate with the 6-sulfo sLe^x ligands for L-selectin in the initial capture and rolling interactions of lymphocytes in chronic inflammatory diseases associated with the gastrointestinal tract.

Chemokines and Lymphocyte Migration Through High Endothelial Venule-Like Vessels

Chemokines and chemokine receptors support lymphocyte migration through HEVs by triggering the integrin-mediated arrest of rolling lymphocytes (2,3). Circulating B and T lymphocytes express multiple chemokine receptors, including CCR7, the receptor for chemokines CCL21 (SLC) and CCL19 (ELC); CXCR4, the receptor for CXCL12 (SDF1); and, in the case of B lymphocytes, CXCR5, the receptor for CXCL13 (BLC/BCA-1). The so-called *lymphoid chemokines* (or HEV-associated chemokines) CCL21, CCL19, CXCL12, and CXCL13 have all been found to be associated with the luminal surface of HEVs in lymphoid tissues (45–48). However, only CCL21 is produced by the HEV ECs themselves (49). The three other chemokines are produced by non-ECs in the lymphoid tissues and subsequently transported to the HEVs, transcytosed, and displayed on the HEV lumen (46–48). A similar scenario seems to apply to the RA synovium, where the four lymphoid chemokines have been detected on HEV-like and flat-walled vessels (50–54). The ECs in chronically inflamed synovium acquire the ability to synthesize CCL21, a property that is normally restricted to HEV ECs from lymphoid tissues (50). In contrast, although CXCL12 was displayed on RA synovial endothelium (54), in situ hybridization revealed that ECs did not express the corresponding mRNA, which was mainly produced by fibroblast-like synoviocytes (52). Similarly, CXCL13 was found on ECs but was thought to be produced by follicular dendritic cells and synovial fibroblasts (51).

Among the four lymphoid chemokines induced on endothelium in chronic inflammatory diseases (see Table 155-1), CCL21 and CXCL13 are likely to be the most important for recruitment of T and B lymphocytes, respectively. In RA synovium, Uli von Andrian and colleagues observed induction of CCL21 both on MECA-79$^+$ HEV-like vessels and MECA-79$^-$ flat-walled vessels (50). Indeed, 85% of blood vessels in the inflamed synovium were found to express CCL21. Similarly, most blood vessels (66%) in tissue samples from patients with ulcerative colitis expressed CCL21. CCL21 is highly potent for attracting naive T cells, and CCL21 induction in blood vessels correlated positively with the influx of CD45Ra$^+$ naïve T cells in the inflamed tissues from patients with RA or ulcerative colitis (50). Together, these findings suggest that induction of CCL21 on HEV-like vessels and flat-walled vessels is an important determinant for naive T-cell migration into chronically inflamed human tissues. In addition, CCL21 induction on endothelium during chronic inflammation also may be critical for recruitment of memory lymphocytes. For example, in T-cell autoimmune infiltrative skin diseases (atopic dermatitis, lichen planus, etc.), expression of CCL21 was highly induced in ECs from venules, and the receptor for CCL21, CCR7, also was found to be highly expressed on the infiltrating T-cells, the majority of which expressed the memory CD45Ro$^+$ phenotype (55). CXCL12 also may participate in the recruitment of memory lymphocytes during chronic inflammation, particularly in the RA synovium, where endothelial expression of CXCL12 was associated with the presence of large numbers of perivascular CXCR4$^+$ CD45Ro$^+$ lymphocytes (54).

In a number of other chronic inflammatory diseases, the induced expression of CCL21 in ECs may cause abnormal recruitment of T cells from the circulation to sites of pathologic inflammation. These include chronic *H. pylori* gastritis (56), chronic inflammatory liver disease (57), and autoimmune thyroiditis (58), in which CCL21 was shown to be induced on HECA-452$^+$ HEV-like vessels. In addition, CCL21 also was found to be induced on blood–brain barrier venules in experimental autoimmune encephalomyelitis (EAE), an animal model for multiple sclerosis and, together with CCL19, appeared to be important for aberrant T-cell infiltration into the inflamed brain (59).

CXCL13 has been found to be induced on endothelium both in RA and ulcerative colitis (51,60). In vitro, CXCL13 is a strong chemoattractant for B lymphocytes. Endothelial induction of CXCL13 in chronically inflamed tissues is therefore likely to mediate recruitment and transendothelial migration of large numbers of $CXCR5^+$ B lymphocytes. Indeed, in ulcerative colitis, Per Brandtzaeg and coworkers have found that induction of CXCL13 on HEV-like vessels was associated with infiltration of numerous $CXCR5^+$ B lymphocytes in irregular lymphoid aggregates (60).

In addition to mediating lymphocyte recruitment into chronically inflamed tissues, CXCL13 and CCL21 also may participate in the development of HEV-like vessels and ectopic lymphoid tissue in these inflamed sites, in collaboration with the tumor necrosis factor (TNF)/lymphotoxin pathways.

MECHANISMS GOVERNING THE DEVELOPMENT OF HIGH ENDOTHELIAL VENULE-LIKE VESSELS IN CHRONIC INFLAMMATION

The development of specific protocols for the isolation of HEV ECs from fresh human tissues (61) has opened the way to the molecular characterization of these specialized ECs, allowing for example, the identification of genes involved in HEV sulfation pathways (62) or genes encoding nuclear and secreted factors preferentially expressed in human HEVs (61,63). Most importantly, the use of HEV ECs freshly purified from human tonsils has demonstrated that HEV ECs exhibit a remarkable plasticity and rapidly lose their specialized characteristics outside the lymphoid tissue microenvironment (64). Although the mechanisms governing the development of HEV-like vessels in chronically inflamed human tissues are not yet completely understood, the latter study strongly suggested that lymphoid cells surrounding HEVs, and the soluble cytokines/chemokines produced by these cells, were likely to be key for the induction of HEV-like vessels during chronic inflammation.

Because HEV-like vessels are always associated with dense lymphocytic infiltration in sites of chronic inflammation, lymphocytes are likely to play an important role in the induction of the specialized HEV phenotype. However, the loss of HEV characteristics in mouse lymph nodes deprived of afferent lymphatics, despite normal numbers of recirculating lymphocytes in the blood, clearly shows that recirculating lymphocytes are not sufficient to induce the HEV phenotype (1). This also is supported by the fact that generalized induction of HEV-like vessels has not been reported in human diseases characterized by uncontrolled lymphocyte overpopulation (e.g., chronic lymphocytic leukemia, acute lymphoblastic leukemia, etc.). Therefore, cooperation between lymphocytes and other cell types found in the infiltrates may be required for the induction and maintenance of HEV-like vessels. Dendritic cells are likely to be key in this respect, because studies of HEV-like vessel development in RA or autoimmune thyroiditis, and HEV induction during development of the human immune system, have revealed that the appearance of dendritic cells always precedes the development of HEVs (1).

A key role for soluble mediators (cytokines, chemokines) in HEV-like vessel development during human chronic inflammatory diseases is supported by the results obtained in transgenic and knockout mice models, which indicate that lymphoid chemokines cooperate with TNF/lymphotoxin cytokines for the induction of HEV-like vessels in chronically inflamed nonlymphoid tissues. Transgenic expression in pancreatic islets of cytokines TNF-α; lymphotoxin-α (TNF-β); membrane-bound lymphotoxin $\alpha 1\beta 2$ (65–67); or lymphoid HEV-associated chemokines CCL21 (68), CXCL13 (69) and, to a lesser extent, CCL19 or CXCL12 (70), induced chronic inflammatory infiltrates containing HEV-like vessels expressing HEV markers MECA-79 and/or MAdCAM-1. These vessels closely resembled lymph nodes with respect to both cellular composition and organization. This process of de novo formation of organized lymphoid tissue in chronic inflammation was named *lymphoid neogenesis* by Ruddle and colleagues in 1996 (65–67,71,72). Over the last 10 years, lymphoid neogenesis has been described in many human chronic inflammatory diseases (72), including RA (50,51,53), inflammatory bowel diseases (60), autoimmune thyroiditis (58), *H. pylori*-induced gastritis (56), and chronic inflammatory liver disease (57). Although not observed in all patients, lymphoid neogenesis is believed to be a key determinant in the pathogenesis of these diseases by allowing local production of autoantibodies and direct priming of autoreactive naive T and B lymphocytes at the site of inflammation (50,71,72).

The membrane-bound lymphotoxin $\alpha 1\beta 2$ and the chemokines CCL21 and CXCL13 are the most potent inducers of lymphoid neogenesis (66,69,70), and both CCL21 and CXCL13 have been proposed to act through recruitment and upregulation of lymphotoxin $\alpha 1\beta 2$ on T and B lymphocytes, respectively (69,70). Ectopic expression of lymphotoxin $\alpha 1\beta 2$ in pancreatic islets has been shown to be sufficient for induction of HEV-like vessels expressing the HEV sulfotransferase HEC-GlcNAc6ST and luminal $MECA\text{-}79^+$ PNAd ligands for L-selectin, as well as MAdCAM-1 and chemokine CCL21 (66). In addition, lymphotoxin $\alpha 1\beta 2$ was recently proposed to bind to the lymphotoxin receptor (LTβR) expressed on HEV ECs, and to regulate HEV-specific gene expression through the alternative nuclear factor (NF)-κB pathway (73). An intriguing possibility is that continuous stimulation of LTβR on HEV ECs by lymphotoxin $\alpha 1\beta 2$ expressed at the surface of recirculating lymphocytes and/or dendritic cells may be required for induction and/or maintenance of HEVs in lymphoid tissues and HEV-like vessels in chronically inflamed tissues. This would explain why HEV-like vessels are always found in association with extensive lymphoid infiltrates in human chronic inflammatory diseases, and why HEV markers such as

MECA-79 are only induced in ECs that are in direct contact with the leukocyte infiltrate (Figure 155.3).

Nuclear transcription factors acting downstream of the LTβR in ECs are likely to be key for the induction and maintenance of HEV-like vessels. Studies in mice indicate that binding of lymphotoxin-$\alpha\beta$ to the LTβR may activate expression of MECA-79$^+$ PNAd and other HEV-specific genes (sulfotransferase HEC-GlcNAc-6ST, chemokine CCL21) through the alternative NF-κB pathway, which is known to regulate NF-κB–inducing kinase (NIK) and IκB kinase complex-α kinase (IKKα) activities and the subsequent processing of NF-κB2 p100 to p52, resulting in nuclear translocation of RelB:p52 heterodimers (66,73,74). NF-κB2 and RelB, two members of the NF-κB transcription factor family, may therefore play a key role in HEV-specific gene expression (73,74). Because NF-κB transcription factors regulated by the TNF/lymphotoxin pathway also can be activated in a variety of other cell types, other EC- or microenvironment-specific cues must exist that work together with the aforementioned signaling to elicit a HEV-specific phenotype. For example, nuclear transcription factors regulated by other pathways and/or encoded by HEV master control genes might also exist and participate in the development of HEV-like vessels. An attractive candidate is NF-HEV, a homeodomain-like nuclear factor preferentially expressed in human HEVs, which may regulate HEV-specific genes (63). In addition, signals originating from the extracellular matrix (ECM) also may play important roles in the induction and/or maintenance of HEVs (1). Possible candidates include the matricellular proteins Hevin and Mac-25, which are preferentially and abundantly expressed in HEVs (61,75).

HIGH ENDOTHELIAL VENULE-LIKE VESSELS IN DISEASE: THERAPEUTIC IMPLICATIONS

The development of HEV-like vessels in chronic inflammatory diseases is tightly associated with the recruitment of large numbers of lymphocytes into the diseased tissues, thus favoring the expansion and perpetuation of an autoimmune response. Therefore, interfering with the development and/or maintenance of these HEV-like vessels or with the function of HEV-associated molecules that control lymphocyte recruitment (e.g., HEV-specific adhesion molecules PNAd and MAdCAM-1, HEV-associated chemokines CCL21, CXCL13), is likely to provide therapeutic benefits in many distinct human chronic inflammatory diseases (see Table 155-1).

As previously discussed, the presence of HEV-like vessels and/or associated markers (MECA-79, HECA-452, 2F3) in diseased tissues is strongly correlated to the progression of inflammation. The example of *H. pylori* gastritis is particularly interesting in this respect, because the disappearance of MECA-79$^+$HECA-452$^+$ HEV-like vessels after treatment was strikingly correlated with the disappearance of lymphocyte infiltration in the gastric mucosa (29). Similar clinical observations have been reported in heart or kidney allograft rejection episodes (26,28), and in RA, in which HEV-like vessels are observed in the rheumatoid synovium of patients with active untreated disease, whereas tissue samples from patients whose disease has been modified by treatment exhibit a flatter endothelium (76). Altogether, these clinical observations indicate that the formation of HEV-like vessels – and thus recruitment of lymphocytes to chronic inflammatory sites – can be reversed.

In view of the critical role played by the TNF/lymphotoxin cytokines in the induction and/or maintenance of HEVs, specific inhibitors of these cytokines are likely to be useful to inhibit the formation and/or maintenance of HEV-like vessels in chronic inflammatory human diseases. Indeed, the soluble TNF receptor Etanercept, which blocks both TNF-α and lymphotoxin-α and has notable efficacy in the treatment of RA, may act, at least in part, at the level of the endothelium by inhibiting the expression of HEV-associated adhesion molecules and chemokines. Antagonists of membrane lymphotoxin-$\alpha\beta$ currently in development are likely to be even more efficient and specific for the inhibition of HEV-like vessel development in clinical settings. Interestingly, one of these lymphotoxin-$\alpha\beta$ antagonists – the soluble LTβR-immunoglobulin fusion protein, which blocks development of HEV-like vessels induced by transgenic expression of chemokines CXCL13 (69) or CCL21 (70) – has shown therapeutic efficacy in various preclinical models of human chronic inflammatory diseases (77).

Lymphoid chemokines, such as CCL21 and CXCL13, are potent inducers of HEV-like vessels and ectopic lymphoid tissue in transgenic mice. Therefore, induction of these lymphoid chemokines at sites of inflammation in human diseases could convert the lesion from an acute to a chronic state, and blocking their activity or receptors on lymphocytes might have significant therapeutic value in human chronic inflammatory diseases. Small-molecule inhibitors of CCR7 and CXCR5, currently in development, may be useful for that purpose. However, the four HEV-associated lymphoid chemokines (see Table 155-1), as well as several inflammatory chemokines, are induced on HEV-like and flat-walled vessels in RA synovium, and it is unclear at present whether blocking the activity of a single chemokine (such as CCL21 or CXCL13) will be sufficient for therapeutic efficacy. The use of broad spectrum or promiscuous chemokine inhibitors may avoid this problem of chemokine redundancy.

HEV-specific adhesion molecules MECA-79$^+$PNAd and MAdCAM-1 constitute other interesting therapeutic targets to block lymphocyte infiltration into chronically inflamed human tissues. Indeed, the mucosal HEV-specific MAdCAM-1 is a validated target for inflammatory bowel diseases. A therapeutic antibody targeting lymphocyte $\alpha4\beta7$ integrin (natalizumab), which disrupts $\alpha4\beta7$MAdCAM-1–mediated lymphocyte/endothelium interactions, has shown therapeutic efficacy in phase I clinical trials for Crohn disease (78). Disruption of L-selectin/MECA-79$^+$PNAd interactions also may provide significant therapeutic benefits. Indeed, targeting MECA-79$^+$PNAd recently has been shown to have therapeutic efficacy in a sheep model of human asthma (79).

Although detrimental in chronic inflammatory diseases, the development of HEV-like vessels and lymphoid neogenesis could be beneficial in other diseases such as cancer, in setting up a local site for antigen presentation. Indeed, targeting of lymphotoxin to a tumor in mouse has been shown to induce functional HEV-like vessels within the tumor, thus allowing naive lymphocyte recruitment, priming, and expansion in the ectopic lymphoid tissue induced at the tumor site (80). Interestingly, tumor venules with HEV-characteristics have been observed in areas of lymphocytic infiltration into human tumors (81). ECs from these venules exhibited a plump, cuboidal morphology, and expressed the HEV-specific marker MECA-79 (81,82). HEV-like vessels therefore may play an important role not only in human chronic inflammatory diseases but also in human cancer.

CONCLUSION

The HEV endothelium is unique among vascular endothelium due to its capacity to recruit large numbers of lymphocytes. HEV-like vessels induced by chronic inflammation in non-lymphoid tissues appear to be structurally, phenotypically, and functionally similar to HEVs from lymphoid tissues. A better understanding of the mechanisms controlling development and maintenance of HEV endothelium could therefore provide the basis of novel therapeutic approaches for the treatment of human chronic inflammatory diseases (RA, inflammatory bowel diseases, autoimmune thyroiditis, *H. pylori* gastritis, asthma, atopic dermatitis, psoriasis, etc.) in which HEV-like vessels facilitate large-scale influx of lymphocytes, leading to amplification and maintenance of chronic inflammation.

KEY POINTS

- HEV-like vessels in human chronic inflammatory diseases affecting distinct tissues (synovium, colon, thyroid, stomach, skin, lung, etc.) have been shown to express HEV-specific molecules (MECA-79, MAdCAM-1, HECA-452, and 6-sulfo sLe^x ligands for lymphocyte L-selectin), which mediate the initial capture and rolling interactions of lymphocytes.
- These HEV-like vessels also express HEV-associated lymphoid chemokines (CCL21, CCL19, CXCL13, and CXCL12), which induce integrin-mediated arrest, transendothelial migration, and recruitment of lymphocytes into the inflamed tissues.
- HEV ECs exhibit a remarkable plasticity, and rapidly lose their specialized characteristics outside the lymphoid tissue microenvironment. Studies in transgenic and knockout mice models indicate that lymphoid chemokines CCL21 and CXCL13 may cooperate with cytokines of the TNF/lymphotoxin family, particularly membrane-bound lymphotoxin $\alpha 1\beta 2$, for the regulation of HEV-specific gene expression through the alternative NF-κB pathway. Therefore, cytokines and chemokines produced by lymphoid cells are likely to be key for induction and maintenance of HEV-like vessels in human chronic inflammatory diseases.

Future Goals

- Because interfering with the development and/or maintenance of HEV-like vessels, or with the function of HEV-associated molecules that control lymphocyte recruitment, is likely to provide therapeutic benefits in many distinct human chronic inflammatory diseases, future studies should aim at developing biological therapeutic agents (antibodies, soluble receptors) targeting membrane lymphotoxin and HEV-associated lymphoid chemokines, which appear to control development of HEV-like vessels and associated ectopic lymphoid tissues, or HEV-specific adhesion molecules such as MECA-79^+PNAd and MAdCAM-1, a therapeutic target already validated in Phase I clinical trials for Crohn disease.
- To further characterize the ECs from HEV-like vessels at the molecular level: Application of high-throughput genomic and proteomic approaches to these ECs, freshly isolated from RA synovium or other human chronically inflamed tissues, is likely to reveal a treasury of as-yet-unknown genes and molecules with potential clinical importance. These may include HEV master control genes that would have considerable applications to control the development of HEV-like vessels in clinical situations, permitting therapies that either inhibit formation of HEV-like vessels in human chronic inflammatory diseases (in which these vessels are detrimental), or stimulate HEV-like vessel development in other clinical settings, such as human cancer (in which these vessels could be beneficial by allowing massive lymphocyte recruitment within the tumors).

REFERENCES

1 Girard JP, Springer TA. High endothelial venules (HEVs): specialized endothelium for lymphocyte migration. *Immunol Today.* 1995;16(9):449–457.

2 von Andrian UH, Mempel TR. Homing and cellular traffic in lymph nodes. *Nat Rev Immunol.* 2003;3(11):867–878.

3 Miyasaka M, Tanaka T. Lymphocyte trafficking across high endothelial venules: dogmas and enigmas. *Nat Rev Immunol.* 2004; 4(5):360–370.

4 Thome R. Endothelien als Phagocyten. *Arch Mikrosk Anat.* 1898; 52:820–842.

5 von Schumacher S. Ueber Phagocytose und die Abfuhrwege de Leucocyten in den Lymphdrusen. *Arch Mikrosk Anat.* 1899;54: 311–328.

6 Gowans JL, Knight EJ. The route of recirculation of lymphocytes in the rat. *Proc R Soc Lond B Biol Sci.* 1964;159:257–282.

7 Freemont AJ, Jones CJ, Bromley M, Andrews P. Changes in vascular endothelium related to lymphocyte collections in diseased synovia. *Arthritis Rheum.* 1983;26(12):1427–1433.

8 Iguchi T, Ziff M. Electron microscopic study of rheumatoid synovial vasculature. Intimate relationship between tall endothelium and lymphoid aggregation. *J Clin Invest.* 1986;77(2):355–361.

9 Jalkanen S, Steere AC, Fox RI, Butcher EC. A distinct endothelial cell recognition system that controls lymphocyte traffic into inflamed synovium. *Science.* 1986;233(4763):556–558.

10 Freemont AJ. Molecules controlling lymphocyte-endothelial interactions in lymph nodes are produced in vessels of inflamed synovium. *Ann Rheum Dis.* 1987;46(12):924–928.

11 Freemont AJ. A possible route for lymphocyte migration into diseased tissues. *J Clin Pathol.* 1983;36(2):161–166.

12 Freemont AJ. Functional and biosynthetic changes in endothelial cells of vessels in chronically inflamed tissues: evidence for endothelial control of lymphocyte entry into diseased tissues. *J Pathol.* 1988;155(3):225–230.

13 Streeter PR, Rouse BT, Butcher EC. Immunohistologic and functional characterization of a vascular addressin involved in lymphocyte homing into peripheral lymph nodes. *J Cell Biol.* 1988;107(5):1853–1862.

14 Duijvestijn AM, Horst E, Pals ST, et al. High endothelial differentiation in human lymphoid and inflammatory tissues defined by monoclonal antibody HECA-452. *Am J Pathol.* 1988;130(1):147–155.

15 Michie SA, Streeter PR, Bolt PA, et al. The human peripheral lymph node vascular addressin. An inducible endothelial antigen involved in lymphocyte homing. *Am J Pathol.* 1993;143(6):1688–1698.

16 Rosen SD. Ligands for L-selectin: homing, inflammation, and beyond. *Annu Rev Immunol.* 2004;22:129–156.

17 Yeh JC, Hiraoka N, Petryniak B, et al. Novel sulfated lymphocyte homing receptors and their control by a Core1 extension beta 1,3-N-acetylglucosaminyltransferase. *Cell.* 2001;105(7):957–969.

18 Mitsuoka C, Sawada-Kasugai M, Ando-Furui K, et al. Identification of a major carbohydrate capping group of the L-selectin ligand on high endothelial venules in human lymph nodes as 6-sulfo sialyl Lewis X. *J Biol Chem.* 1998;273(18):11225–11233.

19 Hiraoka N, Petryniak B, Nakayama J, et al. A novel, high endothelial venule-specific sulfotransferase expresses 6- sulfo sialyl Lewis(x), an L-selectin ligand displayed by CD34. *Immunity.* 1999;11(1):79–89.

20 Bistrup A, Bhakta S, Lee JK, et al. Sulfotransferases of two specificities function in the reconstitution of high endothelial cell ligands for L-selectin. 1999;145(4):899–910.

21 Hemmerich S, Bistrup A, Singer MS, et al. Sulfation of L-selectin ligands by an HEV-restricted sulfotransferase regulates lymphocyte homing to lymph nodes. *Immunity.* 2001;15(2):237–247.

22 Uchimura K, Kadomatsu K, El-Fasakhany FM, et al. N-acetylglucosamine 6-O-sulfotransferase-1 regulates expression of L-selectin ligands and lymphocyte homing. *J Biol Chem.* 2004; 279(33):35001–35008.

23 Uchimura K, Gauguet JM, Singer MS, et al. A major class of L-selectin ligands is eliminated in mice deficient in two sulfotransferases expressed in high endothelial venules. *Nat Immunol.* 2005;6(11):1105–1113.

24 Kawashima H, Petryniak B, Hiraoka N, et al. N-acetylglucosamine-6-O-sulfotransferases 1 and 2 cooperatively control lymphocyte homing through L-selectin ligand biosynthesis in high endothelial venules. *Nat Immunol.* 2005;6(11): 1096–1104.

25 Renkonen J, Tynninen O, Hayry P, et al. Glycosylation might provide endothelial zip codes for organ-specific leukocyte traffic into inflammatory sites. *Am J Pathol.* 2002;161(2):543–550.

26 Toppila S, Paavonen T, Nieminen MS, et al. Endothelial L-selectin ligands are likely to recruit lymphocytes into rejecting human heart transplants. *Am J Pathol.* 1999;155(4):1303–1310.

27 Toppila S, Paavonen T, Laitinen A, et al. Endothelial sulfated sialyl Lewis x glycans, putative L-selectin ligands, are preferentially expressed in bronchial asthma but not in other chronic inflammatory lung diseases. *Am J Respir Cell Mol Biol.* 2000;23(4):492–498.

28 Kirveskari J, Paavonen T, Hayry P, Renkonen R. De novo induction of endothelial L-selectin ligands during kidney allograft rejection. *J Am Soc Nephrol.* 2000;11(12):2358–2365.

29 Kobayashi M, Mitoma J, Nakamura N, et al. Induction of peripheral lymph node addressin in human gastric mucosa infected by *Helicobacter pylori*. *Proc Natl Acad Sci USA.* 2004;101(51): 17807–17812.

30 Pablos JL, Santiago B, Tsay D, et al. A HEV-restricted sulfotransferase is expressed in rheumatoid arthritis synovium and is induced by lymphotoxin-alpha/beta and TNF-alpha in cultured endothelial cells. *BMC Immunol.* 2005;6(1):6.

31 Salmi M, Granfors K, MacDermott R, Jalkanen S. Aberrant binding of lamina propria lymphocytes to vascular endothelium in inflammatory bowel diseases. *Gastroenterology.* 1994;106(3): 596–605.

32 Takaeda M, Yokoyama H, Segawa-Takaeda C, et al. High endothelial venule-like vessels in the interstitial lesions of human glomerulonephritis. *Am J Nephrol.* 2002;22(1):48–57.

33 De Bleecker JL, Engel AG, Butcher EC. Peripheral lymphoid tissue-like adhesion molecule expression in nodular infiltrates in inflammatory myopathies. *Neuromuscul Disord.* 1996;6(4):255–260.

34 Collett C, Munro JM. Selective induction of endothelial L-selectin ligand in human lung inflammation. *Histochem J.* 1999; 31(4):213–219.

35 Jahnsen FL, Lund-Johansen F, Dunne JF, et al. Experimentally induced recruitment of plasmacytoid (CD123high) dendritic cells in human nasal allergy. *J Immunol.* 2000;165(7):4062–4068.

36 Farkas L, Beiske K, Lund-Johansen F, et al. Plasmacytoid dendritic cells (natural interferon- alpha/beta-producing cells) accumulate in cutaneous lupus erythematosus lesions. *Am J Pathol.* 2001;159(1):237–243.

37 Kirveskari J, Helinto M, Moilanen JA, et al. Hydrocortisone reduced in vivo, inflammation-induced slow rolling of leukocytes and their extravasation into human conjunctiva. *Blood.* 2002;100(6):2203–2207.

38 Mitsuoka C, Kawakami-Kimura N, Kasugai-Sawada M, et al. Sulfated sialyl Lewis X, the putative L-selectin ligand, detected on endothelial cells of high endothelial venules by a distinct set of

anti- sialyl Lewis X antibodies [published erratum appears in *Biochem Biophys Res Commun.* 1997;233(2):576.] *Biochem Biophys Res Commun.* 1997;230(3):546–551.

39 Smith PL, Gersten KM, Petryniak B, et al. Expression of the alpha(1,3)fucosyltransferase Fuc-TV in lymphoid aggregate high endothelial venules correlates with expression of L- selectin ligands. *J Biol Chem.* 1996;271(14):8250–8259.

40 Maly P, Thall A, Petryniak B, et al. The alpha(1,3)fucosyltransferase Fuc-TV controls leukocyte trafficking through an essential role in L-, E-, and P-selectin ligand biosynthesis. *Cell.* 1996;86(4):643–653.

41 van Dinther-Janssen AC, Pals ST, Scheper R, et al. Dendritic cells and high endothelial venules in the rheumatoid synovial membrane. *J Rheumatol.* 1990;17(1):11–17.

42 Kabel PJ, Voorbij HA, de Haan-Meulman M, et al. High endothelial venules present in lymphoid cell accumulations in thyroids affected by autoimmune disease: a study in men and BB rats of functional activity and development. *J Clin Endocrinol Metab.* 1989;68(4):744–751.

43 Briskin M, Winsor-Hines D, Shyjan A, et al. Human mucosal addressin cell adhesion molecule-1 is preferentially expressed in intestinal tract and associated lymphoid tissue. *Am J Pathol.* 1997;151(1):97–110.

44 Dogan A, Du M, Koulis A, et al. Expression of lymphocyte homing receptors and vascular addressins in low-grade gastric B-cell lymphomas of mucosa-associated lymphoid tissue. *Am J Pathol.* 1997;151(5):1361–1369.

45 Stein JV, Rot A, Luo Y, et al. The CC chemokine thymus-derived chemotactic agent 4 (TCA-4, secondary lymphoid tissue chemokine, 6Ckine, exodus-2) triggers lymphocyte function-associated antigen 1-mediated arrest of rolling T lymphocytes in peripheral lymph node high endothelial venules. *J Exp Med.* 2000;191(1):61–76.

46 Baekkevold ES, Yamanaka T, Palframan RT, et al. The CCR7 ligand elc (CCL19) is transcytosed in high endothelial venules and mediates T cell recruitment. *J Exp Med.* 2001;193(9):1105–1112.

47 Okada T, Ngo VN, Ekland EH, et al. Chemokine requirements for B cell entry to lymph nodes and Peyer's patches. *J Exp Med.* 2002;196(1):65–75.

48 Ebisuno Y, Tanaka T, Kanemitsu N, et al. Cutting edge: the B cell chemokine CXC chemokine ligand 13/B lymphocyte chemoattractant is expressed in the high endothelial venules of lymph nodes and Peyer's patches and affects B cell trafficking across high endothelial venules. *J Immunol.* 2003;171(4):1642–1646.

49 Gunn MD, Tangemann K, Tam C, et al. A chemokine expressed in lymphoid high endothelial venules promotes the adhesion and chemotaxis of naive T lymphocytes. *Proc Natl Acad Sci USA.* 1998;95(1):258–263.

50 Weninger W, Carlsen HS, Goodarzi M, et al. Naive T cell recruitment to nonlymphoid tissues: a role for endothelium-expressed CC chemokine ligand 21 in autoimmune disease and lymphoid neogenesis. *J Immunol.* 2003;170(9):4638–4648.

51 Takemura S, Braun A, Crowson C, et al. Lymphoid neogenesis in rheumatoid synovitis. *J Immunol.* 2001;167(2):1072–1080.

52 Pablos JL, Santiago B, Galindo M, et al. Synoviocyte-derived CXCL12 is displayed on endothelium and induces angiogenesis in rheumatoid arthritis. *J Immunol.* 2003;170(4):2147–2152.

53 Page G, Lebecque S, Miossec P. Anatomic localization of immature and mature dendritic cells in an ectopic lymphoid organ: correlation with selective chemokine expression in rheumatoid synovium. *J Immunol.* 2002;168(10):5333–5341.

54 Buckley CD, Amft N, Bradfield PF, et al. Persistent induction of the chemokine receptor CXCR4 by TGF-beta 1 on synovial T cells contributes to their accumulation within the rheumatoid synovium. *J Immunol.* 2000;165(6):3423–3429.

55 Christopherson KW 2nd, Hood AF, Travers JB, et al. Endothelial induction of the T-cell chemokine CCL21 in T-cell autoimmune diseases. *Blood.* 2003;101(3):801–806.

56 Mazzucchelli L, Blaser A, Kappeler A, et al. BCA-1 is highly expressed in *Helicobacter pylori*-induced mucosa-associated lymphoid tissue and gastric lymphoma. *J Clin Invest.* 1999;104(10): R49–R54.

57 Grant AJ, Goddard S, Ahmed-Choudhury J, et al. Hepatic expression of secondary lymphoid chemokine (CCL21) promotes the development of portal-associated lymphoid tissue in chronic inflammatory liver disease. *Am J Pathol.* 2002;160(4):1445–1455.

58 Armengol MP, Cardoso-Schmidt CB, Fernandez M, et al. Chemokines determine local lymphoneogenesis and a reduction of circulating $CXCR4^+$ T and CCR7 B and T lymphocytes in thyroid autoimmune diseases. *J Immunol.* 2003;170(12):6320–6328.

59 Alt C, Laschinger M, Engelhardt B. Functional expression of the lymphoid chemokines CCL19 (ELC) and CCL 21 (SLC) at the blood-brain barrier suggests their involvement in G-protein-dependent lymphocyte recruitment into the central nervous system during experimental autoimmune encephalomyelitis. *Eur J Immunol.* 2002;32(8):2133–2144.

60 Carlsen HS, Baekkevold ES, Johansen FE, et al. B cell attracting chemokine 1 (CXCL13) and its receptor CXCR5 are expressed in normal and aberrant gut associated lymphoid tissue. *Gut.* 2002;51(3):364–371.

61 Girard JP, Springer TA. Cloning from purified high endothelial venule cells of hevin, a close relative of the antiadhesive extracellular matrix protein SPARC. *Immunity.* 1995;2(1):113–123.

62 Girard JP, Baekkevold ES, Feliu J, et al. Molecular cloning and functional analysis of SUT-1, a sulfate transporter from human high endothelial venules. *Proc Natl Acad Sci USA.* 1999; 96(22):12772–12777.

63 Baekkevold ES, Roussigne M, Yamanaka T, et al. Molecular characterization of NF-HEV, a nuclear factor preferentially expressed in human high endothelial venules. *Am J Pathol.* 2003;163(1):69–79.

64 Lacorre DA, Baekkevold ES, Garrido I, et al. Plasticity of endothelial cells: rapid dedifferentiation of freshly isolated high endothelial venule endothelial cells outside the lymphoid tissue microenvironment. *Blood.* 2004;103(11):4164–4172.

65 Kratz A, Campos-Neto A, Hanson MS, Ruddle NH. Chronic inflammation caused by lymphotoxin is lymphoid neogenesis. *J Exp Med.* 1996;183:1461–1472.

66 Drayton DL, Ying X, Lee J, et al. Ectopic LT alpha beta directs lymphoid organ neogenesis with concomitant expression of peripheral node addressin and a HEV-restricted sulfotransferase. *J Exp Med.* 2003;197(9):1153–1163.

67 Ruddle NH. Lymphoid neo-organogenesis: lymphotoxin's role in inflammation and development. *Immunol Res.* 1999;19(2–3): 119–125.

68 Fan L, Reilly CR, Luo Y, et al. Ectopic expression of the chemokine TCA4/SLC is sufficient to trigger lymphoid neogenesis. *J Immunol.* 2000;164:3955–3959.

69 Luther SA, Lopez T, Bai W, et al. BLC expression in pancreatic islets causes B cell recruitment and lymphotoxin-dependent lymphoid neogenesis. *Immunity*. 2000;12:471–481.

70 Luther SA, Bidgol A, Hargreaves DC, et al. Differing activities of homeostatic chemokines CCL19, CCL21, and CXCL12 in lymphocyte and dendritic cell recruitment and lymphoid neogenesis. *J Immunol*. 2002;169(1):424–433.

71 Weyand CM, Kurtin PJ, Goronzy JJ. Ectopic lymphoid organogenesis: a fast track for autoimmunity. *Am J Pathol*. 2001;159(3):787–793.

72 Hjelmstrom P. Lymphoid neogenesis: de novo formation of lymphoid tissue in chronic inflammation through expression of homing chemokines. *J Leukoc Biol*. 2001;69(3):331–339.

73 Drayton DL, Bonizzi G, Ying X, et al. I kappa B kinase complex alpha kinase activity controls chemokine and high endothelial venule gene expression in lymph nodes and nasal-associated lymphoid tissue. *J Immunol*. 2004;173(10):6161–6168.

74 Carragher D, Johal R, Button A, et al. A stroma-derived defect in NF-kappaB2-/- mice causes impaired lymph node development and lymphocyte recruitment. *J Immunol*. 2004;173(4):2271–2279.

75 Girard JP, Baekkevold ES, Yamanaka T, et al. Heterogeneity of endothelial cells: the specialized phenotype of human high endothelial venules characterized by suppression subtractive hybridization. *Am J Pathol*. 1999;155(6):2043–2055.

76 Middleton J, Americh L, Gayon R, et al. Endothelial cell phenotypes in the rheumatoid synovium: activated, angiogenic, apoptotic and leaky. *Arthritis Res Ther*. 2004;6(2):60–72.

77 Gommerman JL, Browning JL. Lymphotoxin/light, lymphoid microenvironments and autoimmune disease. *Nat Rev Immunol*. 2003;3(8):642–655.

78 Ghosh S, Goldin E, Gordon FH, et al. Natalizumab for active Crohn's disease. *N Engl J Med*. 2003;348(1):24–32.

79 Rosen SD, Tsay D, Singer MS, et al. Therapeutic targeting of endothelial ligands for L-selectin (PNAd) in a sheep model of asthma. *Am J Pathol*. 2005;166(3):935–944.

80 Schrama D, thor Straten P, Fischer WH, et al. Targeting of lymphotoxin-alpha to the tumor elicits an efficient immune response associated with induction of peripheral lymphoid-like tissue. *Immunity*. 2001;14(2):111–121.

81 Freemont AJ. The small blood vessels in areas of lymphocytic infiltration around malignant neoplasms. *Br J Cancer*. 1982;46(2):283–288.

82 Salmi M, Grenman R, Grenman S, et al. Tumor endothelium selectively supports binding of IL-2-propagated tumor-infiltrating lymphocytes. *J Immunol*. 1995;154(11):6002–6012.

156

Endothelium and Skin

Peter Petzelbauer, Marion Gröger, Robert Loewe, and Rainer Kunstfeld

Medical University Vienna, Austria

The skin surface sets the boundary between the body and the external environment and serves three main functions. It protects against outside damage from physical, chemical, or infection-related stress, and it regulates body temperature and fluid balance. To fulfill these diverse functions, the skin has three anatomic layers: the epidermis, dermis, and subcutis, which serve independent and joint functions.

The epidermis, the outmost coating of the skin, is formed by a multilayer of keratinocytes. The epidermis is nonvascularized and separated by a basement membrane from the underlying zone of highly vascularized connective tissue, called the dermis. The dermo-epithelial zone is not linear, but rather is an interdigitated interface. Keratinocytes form the rete ridges, which protrude into the dermis. Concurrently, between these rete ridges, the dermis extends upward and forms the papillae. Morphologically and functionally distinct from this so-called papillary dermis is the deeper portion of the dermis, the reticular dermis, which harbors eccrine and apocrine glands (for sweat and scent production, respectively). The subcutaneous fat layer (subcutis) below resides on the muscle fascia. Terminal hairs root in the fat layer; the sebaceous glands pertinent to the hair follicle reside within the dermis (Figure 156.1; for color reproduction, see Color Plate 156.1).

VASCULAR ARCHITECTURE

The knowledge about the structure and organization of the cutaneous vasculature has been developed over the past 25 years. Before that, it was assumed that cutaneous vessels form a random network. Irwin Braverman was the pioneer in setting up a model for the complex vascular network of the skin (1). This chapter focuses on skin blood vessels; the characteristics of the lymphatic vascular system are described in Chapter 169.

The structure and organization of the microvasculature in developing human skin was analyzed by light and electron microscopy in conjunction with computer reconstructions. At embryonic day (E)35 to 40, simple, capillary-like vessels are morphologically identifiable in the presumptive dermis. A single plane of vessels and discontinuous vascular segments are seen in specimens at E40 to E50, whereas two planes and a capillary network emerge at E50 to E75. By electron microscopy, endothelial junctions appear similar to those in adult endothelial cells (ECs), but little or no basal lamina surrounds the vessel. By the end of the first trimester, the major vascular organization of the dermis appears defined, but the basal lamina still consists of amorphous deposits and eventually thickened segments (2).

In adult skin, blood vessels follow anatomic structures. Within the papillary dermis, paired arterioles and venules form an interconnected network of vessels. They run on a plane parallel to and just beneath the epidermis and are called *superficial vascular plexus* (SVP) (3). Capillaries arise from the arterioles, extend upward within the papillary dermis between the epidermal rete ridges, and then loop back down to the venules, forming arcade-like structures. In normal skin, most of this capillary loop is invested with a homogeneous basement membrane that resembles the segment of the arteriole. The capillaries acquire a venular-like investiture (multilayered basement membrane) only at the level of the deepest rete, just proximal to the anastomosis with the venule of the SVP (see Figure 156.1). Arterioles and venules within the SVP do not possess an internal or external elastic membrane.

Vessels of the SVP are connected by short, straight connections to the arterioles and venules of a deeper planar network of anastomosing vessels called the *deep vascular plexus* (DVP), which resides within the reticular dermis (see Figure 156.1). The plane of the DVP is parallel to that of the SVP and runs above the boundary between the reticular dermis and the underlying subcutis. The DVP is fed through penetrating vessels from the subcutis and drained by valve-containing veins into the subcutaneous fat. Valves are positioned at sites where the small vessels connect to larger ones and imply a mechanism involved in forward propulsion of blood. The DVP contains arterioles and venules of larger caliber consisting of three layers: an intima, media, and adventitia. Capillary networks connecting the arterioles and venules of the DVP

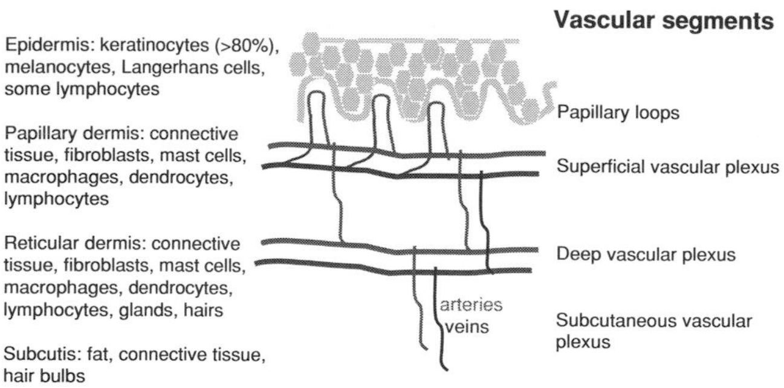

Figure 156.1. Schematic drawing of skin architecture. For color reproduction, see Color Plate 156.1.

provide nourishment for the adnexial structures within the reticular dermis.

Dermal microvessels are surrounded by flat adventitial cells, ultrastructurally most closely resembling a fibroblast, called *veil cells.* They are found in the dermis and infrequently around capillaries of the subcutis. Veil cells are probably specific for skin vessels. These cells are negative for T- or B-cell markers and negative for human leukocyte antigen (HLA)-DR and CD1a expression. They express GPIbα, von Willebrand factor (vWF), and factor XIIIa, a coagulation transglutaminase; these cells are called *XIIIa-positive dendrocytes* (4,5). Interestingly, factor XIIIa-positive dendrocyte rarefaction is found in the classic type of Ehlers-Danlos syndrome, but the significance of this finding is not clear.

Unlike veil cells, pericytes are an integral component of the vessel wall. They extend long cytoplasmic processes over the surface of ECs and form interdigitating contacts. They cover small arterioles, capillaries, and venules on their abluminal surfaces. In the microvasculature, ECs and pericytes are anatomical and functional neighbors. They are separated by the EC basal lamina, which is not completely dense and allows punctuate direct contact and interdigitation. Pericytes are best identified by immunohistochemistry using antibodies to the cell surface 3G5 ganglioside antigen, which reacts with pericytes of several different tissues (6). A subpopulation of 3G5-positive pericytes coexpress smooth muscle antigen (SMA) (6). The coverage of ECs by pericytes varies considerably between different microvessel types, and the location of pericytes on the microvessel is not random but appears to be correlated with function. At points of pericyte–EC contact, communicating gap junctions, tight junctions, and adhesion plaques are present. In large venules, the pericytes only partially encircle the EC cell tube and are more randomly placed. In postcapillary venules, pericytes completely encircle and grip ECs through multiple contact points from their lateral processes, with one pericyte making contact to two to four ECs (7).

Pericytes are rich in contractile proteins. Based on their complex interdigitation with ECs of postcapillary venules, it can be assumed that they are responsible, at least in part, for gap formation in response to factors including histamine or bradykinin (although cultured ECs are capable of contraction in the absence of pericytes). Moreover, interactions between pericytes and endothelium are important for the maturation, remodeling, and maintenance of the vascular system via the secretion of growth factors or modulation of the extracellular matrix.

PHENOTYPIC FEATURES OF SKIN ENDOTHELIAL CELLS

ECs of the skin share most biochemical/molecular features common for all ECs in the body. They express vWF, the single most widely used marker for vascular endothelium, which is found in Weibel-Palade bodies or in the biosynthetic organelles. Differences in levels of vWF expression in different vascular segments of the skin have not been reported. Skin ECs uniformly express the pan-endothelial marker CD144 (vascular endothelial [VE]-cadherin). CD31 (platelet-endothelial cell adhesion molecule [PECAM]-1) expression is shared by endothelium and platelets, but also is found on circulating leukocytes. As in all endothelium, skin ECs can be recognized by their ability to take up acetylated low-density lipoprotein (acLDL) via the "scavenger cell pathway" of LDL metabolism. At least four different types of scavenger receptors exist; some are shared with or are unique to macrophages, whereas some are probably unique to endothelium. Like ECs of other organs, skin ECs secrete CD143 (angiotensin-converting enzyme [ACE]); this characteristic is shared with macrophages. Moreover, normal skin ECs express the anticoagulant protein CD141 (thrombomodulin). This feature is shared with trophoblast cells of the placenta, which, like endothelium, form a blood-compatible surface that prevents coagulation. Interestingly, thrombomodulin expression by skin ECs is significantly downregulated in patients with meningococcal sepsis, which correlates with low endothelial protein C receptor expression and with low plasma protein C levels (8).

Human vascular ECs express blood group H antigens recognized by the lectin *Ulex europaeus* agglutinin (UEA)-1. This antigen also is expressed on erythrocytes and on certain specialized epithelia; for example, in the skin it is expressed on keratinocytes.

In the skin, as in other organs, blood capillaries are the terminal segment of the arterial tree, and consist of tubes formed by a continuous string of elongated ECs, each curved to form a segment of a hollow cylinder. The arterial capillary connects with the venular capillary, which empties out into the postcapillary venule. Electron microscopic examinations of skin sections distinguish between these vascular segments. The arterial capillary is surrounded by a dense basement membrane, which develops lamellae in the venular capillary. The postcapillary venule has an increased external diameter, as compared with the capillary segments. Moreover, pericytes form two to three layers around postcapillary venules, as opposed to only one layer in the venular capillary (1). Phenotypic evaluations of these particular vascular segments of the skin commonly use the term *microvessel* without a precise assignment of a certain molecular phenotype to one of these three microvascular beds. In keeping with this convention, we use the term *microvascular EC* (MVEC) to refer to all three vascular segments.

In 1985, Schlingemann and colleagues described an antibody called *Pathologische Anatomie Leiden-endothelium* (PAL-E), which bound to virtually all capillary ECs of the body and to most venules, but not to arterioles, arteries, and large veins (9). The molecular identity of PAL-E appears to be a glycoprotein associated to caveolae, which has homology to human fenestrated endothelial-linked structure protein (FELS) and to rat plasmalemmal vesicle (PV)-1 (10). The capillaries of most tissues, including those of the skin, are positive for CD34, a molecule that also is found on immature hematopoietic cells. Interestingly, monoclonal antibodies to different CD34 epitopes variably recognize hematopoietic cells, but uniformly bind to blood capillaries. Skin blood capillaries (and capillaries from most other organs) express vascular endothelial growth factor receptor (VEGFR)-1, -2, and -3 proteins, whereas arterioles, arteries, venules, and veins do not (at least as tested by immunohistochemistry) (11). Skin MVECs constitutively express class II major histocompatibility (MHC) molecules, implying that they function in antigen presentation (12). So far, in addition to skin ECs, constitutive HLA class II expression has been described only in renal peritubular and glomerular capillaries. CD36, a receptor for collagen, thrombospondin, and *Plasmodium falciparum*, is found on virtually all capillaries. The situation is different in the skin; within the SVP, there exists a small subpopulation of microvessels that is CD36 negative, but the functional relevance of this finding is unknown (13). The SVP is unique also with regard to expression of the low-affinity Fcγ receptor CD32 (14). The CD32$^+$ population appears to reside within the CD36$^-$ EC population of the papillary dermis. CD32 expression also has been described on liver MVECs. Blood capillaries of the SVP are typically negative for mannose macrophage receptor (CD206) expression. At this site, CD206 is found on lymphatic capillaries. In contrast, in the DVP of the skin, a fraction of medium-sized blood vessels are CD206$^+$ (15).

MVECs in different organs may express a unique profile of sialomucins. Sialomucins are a heterogeneous class of highly O-glycosylated sialic acid–rich glycoproteins that are easily accessible on the cell surface. Depending on the mode of glycosylation, they can either prevent or support adhesion, or even accomplish both functions. During the adhesive steps between inflammatory cells and endothelial surfaces, they largely mediate tethering and rolling. As an example of an antiadhesive sialomucin, skin ECs may express podocalyxin. Proadhesive sialomucins include the L-selectin ligands glycosylation-dependent cell adhesion molecule (GlyCAM)-1, mucosal addressin cell adhesion molecule (MAdCAM)-1, and CD34, as well as endomucin, endoglycan, and other members of the CD34 family of proteins (16,17). Among these L-selectin ligands, skin ECs express CD34, endomucin, and endoglycan, but expression of these proteins is certainly not restricted to the skin (18–20). All sialomucins acquire their adhesive function only when decorated with appropriate carbohydrate epitopes. Two main epitopes have been described, namely sialyl Lewis X (sLe^x) (NeuAcα2–3Galβ1–4[Fucα1–3]GlcNAcβ1–) epitopes, and sulfo-sLe^x epitopes.

Table 156-1: Phenotype of Postcapillary Venules of the Superficial Vascular Plexus

Normal Skin	*Acute Inflammation*	*Chronic Inflammation*
vWF	E-selectin (CD62E) high	E-selectin (CD62E)
PECAM-1 (CD31)	P-selectin(CD62P)	P-selectin (CD62P)
CD34	VCAM-1 (CD106) low	VCAM-1 (CD106) high
VE-cadherin (CD144)	CXCR4 downregulated	MECA-79–reactive epitopes
PAL-E	CCR10 high	HECA-452–reactive epitopes
CXCR4		
CD32		
CCR10 low		
CD36 negative		

sLe^x epitopes are carbohydrates decorated with sialylated and fucosylated lactosamines, which can be detected by antibody HECA-452. Under normal conditions, skin endothelium shows only minimal reactivity with antibody HECA-452 (21). The sialomucin carrying this epitope in skin ECs is not known. On lymphocytes, the HECA-452-reactive epitope has been described as a sLe^x-like carbohydrate displayed on P-selectin glycoprotein ligand (PSGL)-1; this epitope is also called *cutaneous leukocyte-associated antigen* (CLA), and is typically expressed on skin-homing T cells.

Sulfo-sLe^x epitopes are carbohydrates decorated with sialylated, fucosylated, and sulfated lactosamines, which can be detected with the antibody MECA-79. They are collectively called *peripheral node addressins* and function as L-selectin ligands. In the skin, under normal conditions, MECA-79–reactive vessels are found only in proximity to hair follicles. Here, the majority of MECA-79 reactivity is displayed by CD34 molecules (22). Interestingly, a direct association exists between MECA-79 reactivity of skin endothelium and the appearance of naïve T cells within the draining lymphatic vessel, indicating that MECA-79 plays a role in the recirculation of naïve T cells through the skin (23). Whether this indicates that the skin may function as a secondary lymphoid organ remains to be clarified. In situations of skin inflammation, expression of HECA-452 as well as MECA-79 epitopes on ECs can be dramatically induced (21,22,24) (Table 156-1).

MVECs of the skin can express E- and P-selectin. E-selectin is absent in normal skin, but is significantly induced in inflammatory reactions by cytokines like tumor necrosis factor (TNF)-α or interleukin (IL)-1. When expressed, it is typically found on the venule-like portions of the capillaries (13). This is the site where inflammatory cells enter the tissue. On activated skin endothelium in cell culture, E-selectin expression is more robust and persistent, compared to large-vessel endothelium. This appears to be a general feature of MVECs and not a skin-specific reaction pattern (25). P-selectin also is found on endothelium. In the endothelium of normal skin, analogous to the endothelium of other organs, P-selectin is prestored in Weibel-Palade bodies. In response to mediators such as histamine or thrombin, these storage organelles are degranulated, and P-selectin is released and transported to the cell surface.

In inflammation, ECs show induced expression of intercellular adhesion molecule (ICAM)-1 (CD54) and de novo expression of vascular cell adhesion molecule (VCAM)-1 (CD106). VCAM-1 plays a role in rolling and – more importantly – in tight adhesion of lymphocytes as well as monocytes, eosinophils, and basophils. In contrast to E-selectin, TNF-α–induced VCAM-1 expression on skin endothelium is relatively weak in cell culture as well as in tissue organ culture, compared with large vessel–derived ECs (13). Glomerular ECs behave similarly (26). Robust VCAM-1 induction in skin ECs requires the presence of TNF-α plus interferon (IFN)-α or IFN-γ (24). It is not clear whether this is a skin-specific phenomenon. Because the skin is naturally exposed to high amounts of exogenous toxins or antigens, one could speculate that the requirement for IFN plus TNF-α for efficient VCAM-1 expression by skin ECs might be a protective mechanism to raise the threshold for chronic inflammation.

Other endothelial molecules involved in tight adhesion of inflammatory cells are ligands for β2 integrins. They appear pivotal for lymphocyte homing to the skin. In mice lacking the β2 integrin subunit, CD18, allergic contact dermatitis and delayed-type hypersensitivity reactions were impaired (27). These experiments did not specify the type of β2 integrin responsible for impaired skin inflammation. The β2 integrin family is composed of four members: CD11a/CD18 (lymphocyte function-associated antigen-1 [LFA-1]), CD11b/CD18 (Mac-1), CD11c/CD18 (p150,95), and CD11d/CD18 (28). Ligands for LFA-1 and Mac-1 expressed on ECs include ICAM-1, -2, and -3, and junctional adhesion molecule (JAM)-A and JAM-C. For CD11c/CD18, no direct endothelial ligands are yet described. CD11c/CD18 is a fibrinogen receptor expressed on neutrophils, monocytes, and macrophages, and on subsets of lymphocytes. CD11c indirectly interacts with VE-cadherin expressed on ECs through interposed fibrin fragments (29). Ligands for CD11d/CD18 expressed on ECs include VCAM-1 and ICAM-3.

To visualize chemokine-binding sites on skin endothelium, in situ binding assays for the association of radioiodinated chemokines with structures in normal human skin have been performed. These assays demonstrated the presence of specific and saturable binding sites for IL-8 (CXCL8), regulated on activation, normal T-cell expressed and secreted (RANTES; CCL5), monocyte chemoattractant protein (MCP)-1 (CCL2), and MCP-3 (CCL7) on venules and small veins but not on arterioles, arteries, or capillaries (30). Constitutively expressed chemokine receptors on normal skin endothelium include CCR2, -3, -4, and -8, and CXCR1, -2, -3, and -4 (31,32). Expression of D6, a nonsignaling CC chemokine receptor, was found on afferent lymphatics in skin and not on blood vessels (33).

For transmigration, inflammatory cells must cross the multilayered molecular zipper of interendothelial junctions. In skin microvessels (as with vessels of other organs), molecules configuring this zipper include JAMs, PECAM-1 (CD31), CD99, and VE-cadherin (CD144). Each of these molecules is constitutively expressed and participates in zipper formation by homophilic adhesion with neighboring ECs. PECAM-1 and CD99 contribute to leukocyte transmigration by building homophilic and/or heterophilic adhesion to leukocytes (34,35). VE-cadherin interacts with leukocytes through an interposed fragment of fibrin, which is formed through thrombin-induced fibrinogen activation followed by plasmin or urokinase digestion (29).

Finally, it should be noted that in normal skin in vivo, ECs investing blood capillaries can be differentiated from ECs investing lymphatic vessels. The former express markers like N-cadherin or CXCR4; the latter are positive for Prox-1, lymphatic vessel endothelial hyaluronan receptor (LYVE-1), podoplanin, and VEGFR3. However, at least in cell culture, expression of the lymphatic markers podoplanin and Prox-1 is induced by IL-3 even on blood vessel ECs. Conversely, antibodies blocking IL-3 function downregulate podoplanin and Prox-1 on lymphatics, indicating that transdifferentiation between blood and lymphatic endothelium may occur (36).

SKIN ENDOTHELIUM IN THERMOREGULATION

The regulation of the core temperature is vital to maintain normal physiological body functions and is thus adjusted within a very narrow range. Because the skin forms the strategic boundary between the body and the thermal environment, it plays a key role in thermoregulation and serves as both a source of thermal information and an effector organ for controlling thermal homeostasis. Vasoconstriction of cutaneous vessels allows conservation of body heat, whereas vasodilation and secondary elevation of blood flow to the skin promotes convective heat transfer from the core to the periphery of the body.

In glabrous skin (i.e., the palms, soles, and lips) the dermis contains arteriovenous anastomoses, which are absent in nonglabrous skin. These anastomoses shunt blood from the arterial directly into the venous beds, thereby bypassing capillary loops. This anatomic difference is the basis for the vigorous ability of glabrous skin to autoregulate blood flow, which is not a feature of nonglabrous skin.

Nonglabrous skin is subject to reflex thermoregulation of blood flow. This process is mediated by two branches of the sympathetic nervous system: a noradrenergic vasoconstrictor system and a cholinergic vasodilatory system. The noradrenergic vasoconstrictor system is tonically active, is further induced by cold exposure, and can mediate subtle modulations in vascular tone under normothermic conditions. In response to body heating, an initial vasodilation is achieved by reflex removal of active vasoconstrictor tone. With further heating, and usually coincident with the onset of sweating, a profound active vasodilatation is mediated through the release of vasodilator transmitters (predominantly acetylcholine) from the parasympathetic cholinergic nerves. It should be noted that additional neuromediators are secreted by autonomic nerve fibers as well as by sensory nerve fibers of the skin, such as substance P, calcitonin gene–related peptide, vasoactive intestinal peptide, and pituitary adenylate cyclase activating polypeptide, which directly or indirectly induce vasodilation (37). In addition to neural control, local factors such as venous congestion, increased transmural pressure, or local alterations in temperature are capable of modulating skin blood flow. The response to thermal stress may be profound. Under thermoneutral conditions, skin blood flow is approximately 250 mL/min. Heat exposure may result in cutaneous blood flows of up to 6 L/min (~60% of the cardiac output) (38). This change requires increased cardiac output and redistribution of blood flow from other areas such as the splanchnic vasculature.

Dysregulation in the skin thermoregulatory system is characteristic of several dermatological conditions. For example, postmenopausal hot flashes appear directly related to changing estrogen levels. In these individuals, estrogen replacement relieves hot flashes and also reduces resting body temperature by 0.5°C (39). The loss of body temperature by estrogen is thought to be based on the active vasodilator function of estrogens (38). Because the precise mechanism underlying postmenopausal hot flashes is not known, it is unclear how estrogen relieves these flashes (38).

Evidence suggests that diabetic patients have an impaired thermoregulatory response of skin vessels, which correlates with an increased risk of heat illness (38).

Two cutaneous microvascular disorders that may be related to pathologic reflex or local thermoregulation are Raynaud phenomenon and erythromelalgia. Neither disorder is well understood with regard to etiology or pathophysiology. Patients with Raynaud disease suffer from an exaggerated vasospastic response to cold or emotion resulting in ischemia of digits. Abnormalities in Raynaud include impaired vascular function, impaired release of vasoconstrictors and dilators, and impaired neuronal function. This has been recently reviewed by Herrick (40). The sympathetic nervous system is a major player in thermoregulation. The α_{2c}-adrenoreceptor is most important in thermoregulation (41). Indeed, isolated dermal arterioles from patients with Raynaud exhibited abnormal contraction in response to α_2-adrenergic agonist UK 14,304 at 31°C, compared with arterioles derived from controls (42). Another study attributed this increased α_2-adrenoreceptor reactivity to the smooth muscle cell component and not to ECs of the dermal arterioles (43). Another dysregulation potentially contributing to the impaired thermoregulatory response in patients with Raynaud has been related to calcitonin gene–related peptide, a very potent vasodilator and the predominant neurotransmitter in capsaicin-sensitive sensory nerves. Reduced numbers of calcitonin gene–related peptide-immunoreactive nerve fibers have been found in digital skin from patients with Raynaud, and intravenous application of calcitonin gene–related peptide has been reported to improve blood flow in a small number of these patients (44). Finally, elevated circulating levels of asymmetrical dimethylarginine (an endogenous inhibitor of endothelial nitric oxide synthase) and of the vasoconstrictor endothelin-1 have been found in patients with Raynaud phenomenon (45).

Erythromelalgia is a condition of intermittent erythema of peripheral acral skin associated with the sensation of burning pain. The redness and the warmer temperature of the affected limb is accompanied by a massive increase in blood flow. It has been assumed that symptoms are due to a distal small-fiber neuropathy resulting in sudomotor abnormalities (i.e., absent or markedly reduced sweat production) and in reduced vasoconstriction in response to reflex stimulation of the adrenergic vasoconstrictor system (38).

ANGIOGENESIS IN THE SKIN

Cutaneous angiogenesis occurs in wound healing, inflammation, and tumorigenesis. New vessels form by the sprouting of ECs from pre-existing vascular structures. The term *vasculogenesis* refers to the formation of blood vessels in embryos by the recruitment of endothelial progenitor cells to sites of new vessel formation. Endothelial progenitor cells also exist in the adult and contribute to new vessel formation, a process that has been termed *postnatal vasculogenesis*. These phenomena are not specific to skin, and are thus not discussed in the chapter. The general role of growth factors in orchestrating angiogenesis is discussed in Chapter 157; those of chemokines and their receptors have been reviewed recently (46).

Skin angiogenesis differs in interesting ways from that occurring in other organs. Skin harbors hair follicles, which, in addition to their role in forming hair shafts, appear to play a role in new vessel formation. The follicle contains a distinct population of presumptive follicular stem cells that express nestin, also a marker for neural stem cells. These nestin-expressing follicle cells are located in the follicular bulge region. During the follicular growth cycle, they differentiate into the various cell types of the hair follicle and, in addition, can form a variety of epidermal cells (47). Recently, it has been shown that these nestin-positive cells also can develop into cutaneous blood vessels that originate from hair follicles and form a follicle-linked vascular network (47). The quantitative contribution of "hair follicle-derived" angiogenesis to "conventional" angiogenesis through local sprouting and/or hematogenic seeding of endothelial progenitors is still unknown.

Skin endothelium is also unique by virtue of its close proximity to keratinocytes, which form the epidermal coating. Keratinocytes have been shown to express a wide range of angiogenic factors, including members of the fibroblast or transforming growth factor (TGF) protein family, platelet-derived growth factor, or vascular endothelial growth factor

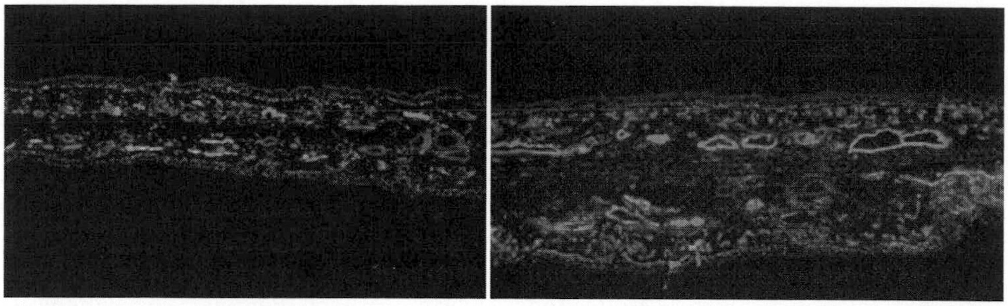

Figure 156.2. In wild-type (*left image*) or in K14/VEGF-A transgenic mice (*right image*), delayed-type hypersensitivity reactions (DTH) were induced by oxazolone (animals and methods are described by Kunstfeld et al. [85]). Seven days later, ears were subjected to immunofluorescence labeling. DTH reactions were strongly enhanced and prolonged in VEGF-A transgenics. Note the increased numbers of blood vessels and the dilated lymphatics in transgenic animals as compared with controls. All nuclei are labeled using Hoechst stain. (Photos courtesy of M. Detmar, unpublished.) For color reproduction, see Color Plate 156.2.

(VEGF) (48). The pathogenic potential of aberrant growth factor expression by epidermal cells is exemplified in psoriasis. This disease is characterized by a distinct pattern of vascular remodeling. Psoriatic vessels become tortuous and elongated. The well-defined boundary between arteriolar endothelium resting on a homogenous basement membrane and venular endothelium resting on a laminated basement membrane becomes vague. Specifically, the intrapapillary portions of capillaries elongate and acquire a venular phenotype. Light microscopic investigations of eruptive psoriatic lesions reveal that vascular alterations are the first ultrastructurally detectable changes within the dermal compartment. These alterations consist of gap formation within postcapillary venules, EC hypertrophy (they acquire a cuboidal shape similar to that seen in high endothelial venules), and compression of the vascular lumen, all of which precede the invasion of inflammatory cells into the tissue. Gene expression studies have demonstrated that VEGF-A is typically overexpressed in lesional psoriatic keratinocytes (49). This situation can be mimicked in a transgenic animal model, in which VEGF-A is overexpressed within the epidermal compartment. In this model, the dermis shows enhanced vascularity (Figure 156.2; for color reproduction, see Color Plate 156.2). Subsequently, inflammation occurs in a very similar manner to that seen in true psoriasis (50). In humans, a linkage between psoriasis and a single nucleotide polymorphism of VEGF-A has been described (51). Specifically, the G to C polymorphism at position +405, which promotes the translation of large VEGF isoforms, is frequently found in psoriatic patients.

Transgenic mice overexpressing signal transducer and activation of transcription 3 (STAT3) within the epidermis also display a psoriasis-like phenotype, which includes increased numbers of dermal blood vessels (52). Although STAT3 overexpression results in enhanced VEGF-A expression in transgenic keratinocytes, it is not clear whether the proliferation of dermal vessels in these animals is mediated by VEGF or by another angiogenic pathway, for example, induction by fibroblast growth factors (FGFs) or by metalloproteinases.

In addition to proangiogenic factors, keratinocytes have been shown to express inhibitors of angiogenesis, such as thrombospondin (TSP)-1 (53). In a transgenic mouse model, targeted overexpression of TSP-1 within the basal layer of the epidermis, and in the outer root sheath of the hair follicle, resulted in impaired granulation tissue formation and angiogenesis following wounding. In these animals, TSP-1 overexpression inhibited wound angiogenesis even in the presence of high VEGF-A expression in vivo and in vitro (54). The molecular mechanisms of TSP-1–mediated inhibition of angiogenesis remain to be established.

These examples show that skin angiogenesis is under the control of keratinocyte-derived factors. Thus, skin angiogenesis may be indirectly influenced by environmental factors, such as ultraviolet (UV) irradiation, which leads to pronounced downregulation of TSP-1 and an upregulation of VEGF in keratinocytes. Increased expression of VEGF-A has been shown to increase the cutaneous damage that occurs after UVB exposure, suggesting that the VEGF signaling pathway might serve as a novel target for the prevention of UVB-induced photodamage (55).

SKIN ENDOTHELIUM IN DISEASE

The dermatologist classifies skin inflammation by eye and describes inflammatory skin diseases according to body distribution, color, and shape of the lesions. The clinical pattern of the individual lesion is mainly determined by the type of inflammatory cell involved (e.g., neutrophils or lymphocytes) and the layer(s) of skin affected (epidermis, dermis, and/or subcutis). For example, pronounced edema and sparse leukocyte recruitment in the superficial dermis presents as a wheal (e.g., the primary lesion seen in urticaria). By contrast, pronounced lymphocyte accumulation around vessels of the SVP produces a flat, red papule (e.g., the primary lesion seen in eczema or psoriasis). Lymphocyte accumulation in the superficial and deep dermis manifests as a nodule (e.g., the primary lesion seen in lupus erythematosus). These

Table 156-2: Compartmentalization of Inflammatory Skin Diseases

	Papillary Dermis (SVP)				
	Tips of Papillae	*Papillary Dermis below the Rete*	*Reticular Dermis (DVP)*	*Subcutis*	*Composition of Infiltrate*
Psoriasis	++	+			Ly, PMN
Pityriasis rosea Viral rashes Lichen ruber planus Graft versus host disease	+	++			Ly (Eo, PMN)
Eczema Herpes simplex Urticaria Secondary lues	+	++ ++	+		Ly (Eo) mixed
Leukocytoclastic vasculitis Erythema elevatum et diutinum Erysipelas/Cellulitis Sweet syndrome		++	+		PMN
Perniones Polymorphic light eruption Lupus erythematodes		++	++		Ly
Rheumatoid vasculitis Scleroderma Necrobiosis lipoidica		++	++	+	Ly (Mp, Pl)
Livedoid vasculitis Churg-Strauss syndrome Wegener disease Polyarteritis nodosa			++	++	mixed
Erythema nodosum Erythema nodosum leprosum Nodular vasculitis (erythema induratum)				++	Ly, Mp
Septic vasculitis	++	++	++	++	PMN

PMN, neutrophil; Ly, lymphocyte; Eo, eosinophil; Mp, macrophage; Pl, plasma cell. +, segment(s) of skin vasculature that may also participate in the inflammatory reaction; ++, segment(s) of skin vasculature that is dominantly involved by the inflammatory reaction.

different clinical patterns of skin lesions correspond to distinct microscopic features seen by light microscopy in biopsy specimens. The dermatopathologist, like the clinician, applies an algorithmic method based on pattern analysis to establish a diagnosis. Pattern analysis involves correlating the type of inflammatory cell and the location of the inflammatory infiltrate within the skin.

Site and Cell Type-Specific Patterns of Inflammation

As shown in Table 156-2, many skin diseases are associated with a signature pattern. Such "compartmentalization" of inflammation strongly favors the concept that distinct segments of the skin vascular system respond differently to pathogenic stimuli, thereby providing specific "ZIP codes" for inflammatory cells.

Eczema and Psoriasis

In eczema and in psoriasis, inflammatory cells exit postcapillary venules of the superficial vessels and migrate upward into the epidermis, but fundamental differences exist in terms of the anatomic location of inflammatory cells in relation to the vascular system (Figure 156.3; for color reproduction, see Color Plate 156.3). In eczema, postcapillary venules of the SVP exhibit EC hypertrophy, mitosis, and basal lamina alterations. At this site, lymphocytes exit the vascular system and migrate upward into the epidermis without any preference to rete ridges. Vessels within the tips of the papillary dermis typically are not the site of lymphocyte evasion (56). The epidermal involvement appears random. Endothelial adhesion molecule expression (e.g., E-selectin) usually is confined to the SVP and absent from vessels within the tips of the papillae.

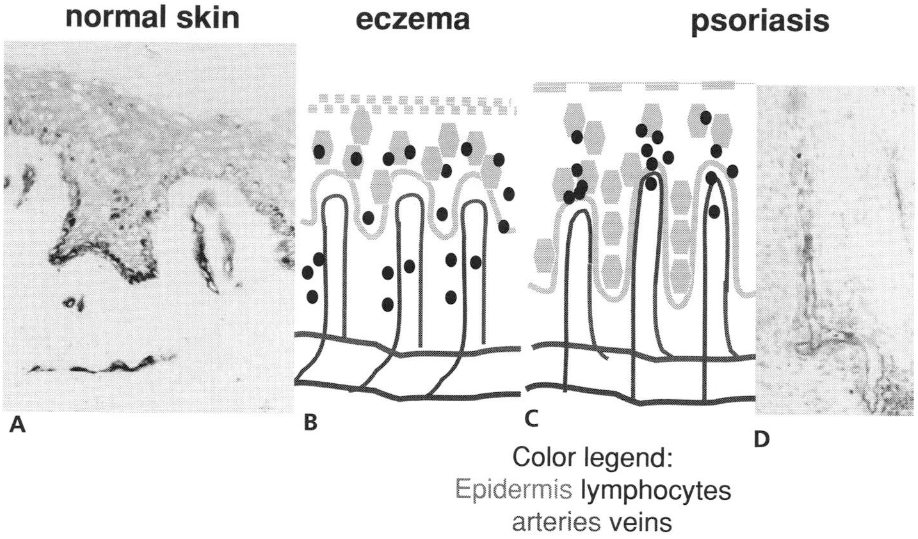

Figure 156.3. (**A**) A normal skin specimen treated ex vivo with TNF for 4 hours. As a result, postcapillary venules of the superficial vascular plexus express E-selectin. All vessels are labeled using *Ulex europaeus* agglutinin. The capillary loop within the tip of the papillae does not express E-selectin and is arteriolar by morphology. (**B**) A schematic diagram of the skin in eczema. The morphology of papillary loops is comparable to that of normal skin; lymphocytes escape the postcapillary venules of the superficial vascular plexus and randomly migrate upward into the epidermis. (**C**) Psoriatic skin. The papillary loops are venular, and express E-selectin and VCAM-1. Lymphocytes and neutrophils evade the capillaries within the tips of the papillae (where the blood vessel "kisses" the epidermis) and produce the phenomenon of "squirting papillae." (**D**) An immunohistochemistry of psoriatic skin using an anti–VCAM-1 antibody. The vessel within the papillae shows VCAM-1 reactivity along its entire length. For color reproduction, see Color Plate 156.3.

In psoriasis, the pathology typically is found within the tips of the papillae. ECs of the venular portion of the capillary loop proliferate (57), and the venules become elongated and dilated. At the turnaround point, capillaries touch the epidermal compartment (called the *kissing* phenomenon). At these sites, lymphocytes and neutrophils exit the vessel lumen and migrate up into the epidermis. Dermatopathologists have termed this impressive image "squirting papillae," which are overlaid by mounds of parakeratosis and permeated by neutrophils at their summits. As the morphologic substrate of this phenomenon, the capillary loops become venular, and express adhesion molecules like E-selectin and VCAM-1 (58). The pathogenic role of the aberrant adhesion molecule expression in vessels within the tips of the papillae is supported by immunohistochemistry studies of lesional skin in cyclosporine-treated psoriatic plaques. In these lesions, the first sign of regression is loss of EC adhesion molecule expression within the tips of the papillae (59).

Leukocytoclastic Vasculitis

In leukocytoclastic vasculitis, vessel destruction occurs, predominantly involving postcapillary venules of the superficial, but not the DVP. This disease has striking similarities to the Arthus reaction. The Arthus reaction is severely suppressed in *Fcγ receptor*–knockout mice. Based on that analogy, it has been proposed that Fcγ receptors play a crucial role in initiating vasculitis. Indeed, skin MVECs express Fcγ receptor IIa (CD32), which is largely confined to vessels of the SVP. This is the same site at which vessel destruction occurs in leukocytoclastic vasculitis (14,60).

Experimental Models for Skin Inflammation

Compartmentalization of inflammatory reactions may be modeled in severe combined immune-deficient (SCID) mice. In one study, these animals were grafted either with superficial or deep portions of human skin, and then injected intravenously with human T cells (61). Although circulating human T cells alloreactive to the human skin migrated equally into both human grafts, they were phenotypically distinct. T cells entering skin grafts containing the epidermis and the SVP were CLA^+ positive, whereas T cells entering skin grafts containing the DVP were CLA^- negative (61). This indicates that vessels of the SVP have a certain address for CLA^+ cells that is not present in deep vessels. CLA has been shown to interact with endothelial E-selectin (62), but E-selectin was expressed equally on ECs of grafts containing the superficial as well as the DVP. Thus, additional regulatory pathways for the recruitment of CLA^{hi+} T cells to the SVP of the skin must exist.

A potential candidate is the chemokine cutaneous T cell–attracting chemokine (TARC or CCL17), which is produced by cutaneous venules, induced in inflamed skin, and absent on intestinal vessels (63). Its receptor, CCR4, is expressed on the CLA^+ memory T cells. The interaction of TARC and CCR4 triggers E-selectin–mediated adhesion of CLA^+ T cells (63). However, skin infiltration in *CCR4*$^{-/-}$ mice is not impaired, indicating that additional regulatory pathways exist (63).

Another candidate for SVP-specific recruitment of CLA^{hi+} T cells is the chemokine thymus and activation-regulated chemokine (CTACK or CCL27). In immunohistochemical studies, CTACK is constitutively expressed by basal keratinocytes, and is detectable on ECs and fibroblasts of the papillary, but not of the reticular dermis (64). In contrast, cultured dermal MVECs (derived from SVP) and fibroblasts do not express CTACK. One explanation for the discrepancy between the in vivo and in vitro findings is that CTACK is expressed and secreted by keratinocytes in the epidermis, and then captured and presented by dermal cells through glycosaminoglycans or through its receptor, CCR10.

The precise role for CTACK in mediating site-specific homing of lymphocytes is controversial. In an animal model of contact hypersensitivity, skin homing of $CD4/CLA/CCR10^+$ T cells is blocked by neutralizing anti-CTACK antibodies (64). This experiment points toward CTACK as the major component for conferring skin specificity for CLA^+ cells. However, in another model of contact hypersensitivity, involving adoptive transfer of T lymphocytes, anti-CTACK antibodies impaired skin homing of *CCR4*$^{-/-}$ T cells only but had no effect on homing of wild-type cells (63). The situation is further complicated when diverse chemokine receptor expression profiles of T_H1 or T_H2 subsets of T cells are taken into consideration (65).

Cell Adhesion Molecules in Acute and Chronic Inflammation

EC adhesion molecule expression profiles differ not only among diseases, but they also show a striking diverse pattern in acute compared with chronic disease. In acute dermatitis, ECs express high levels of E-selectin, P-selectin, moderate amounts of VCAM-1, and are largely nonreactive with antibody MECA-79 (i.e., they are negative for sialyted, fucosylated, and sulfated carbohydrate moieties corresponding to peripheral node addressins). The inflammatory infiltrate consists mainly of $CD45RO^+$ memory T cells with some neutrophils and eosinophils intermingled. In line with this observation are experiments with *L-selectin*$^{-/-}$ mice, in which allergen-induced skin inflammation was not impaired during the first 20 hours after challenge (66). This indicates that L-selectin ligands do not play an important role during the early phase of the inflammatory reaction. In more chronic lesions, such as those of psoriasis, ECs express E-selectin and high amounts of VCAM-1 (13). Moderate amounts of ECs react with antibody HECA-452 and MECA-79 (21). In longstanding lesions, such as those in chronic lupus erythematosus or mycosis fungoides (cutaneous T-cell lymphoma), MECA-79 reactivity on ECs plays a dominant role (22,67). The role of L-selectin and its ligands in chronic inflammation is also supported by experiments in *L-selectin*–knockout mice. In a model for chronic graft rejection, skin inflammation is delayed and attenuated in *L-selectin*$^{-/-}$ mice (68).

Epidermal–Dermal Communication

The role of keratinocytes in eliciting skin inflammation is highlighted by experiments using genetically modified animals. It should be noted that the interpretation of data derived from mouse skin must take into consideration that the mouse dermis is less complex than that of the human. No clear distinction exists between the superficial and DVP, and the capillary loops within the tips of the papillae are absent. Thus, distinct features of certain skin diseases do not exist in mouse skin. Moreover, the composition of inflammatory infiltrates differs. For example, in allergic contact dermatitis in humans, the infiltrate is predominantly composed of lymphocytes, whereas in mice the majority of cells are neutrophils. Within the mouse epidermis even accumulations of neutrophils may occur (spongiotic neutrophilic abscesses).

Results from mouse models shown in Table 156-3 highlight the fact that the epidermal compartment has profound effects on the function of dermal ECs. Overexpression of genes targeted to keratinocytes may result in vessel alterations and spontaneous skin inflammation as shown, for example, in VEGF-A or STAT3 transgenic animals (50,52). On the other hand, deletion of genes within the epidermis also may cause vessel alterations and spontaneous skin inflammation, as seen for example, in mice with inducible epidermal deletion of JunB and its functional companion c-Jun. These animals also suffer from arthritis (69). Other overexpression systems also may show skin disease. Conditional overexpression of Tie2 from its minimal promoter drives expression in ECs, keratinocytes, and epithelial cells of the hair follicle. These mice show spontaneous skin inflammation, which can be "cured" by treating animals with cyclosporine (70). Genes listed in Table 156-3 are more or less functionally unrelated, although endothelial and inflammatory responses show overlapping patterns. This indicates that several different genes, when aberrantly expressed within the epidermis, can be the driving force for dermal skin inflammation. Moreover, these models show that, once inflammation has been established, a gene expression analysis may reveal many up- and downregulated factors and the starting point of the inflammatory reaction may be obscured. Finally, these models indicate that chronic skin inflammation must not necessarily be antigen-driven.

These findings mirror the situation in chronic skin diseases in humans, in which many genes are found activated or inhibited, although the primary cause of the disease remains obscure. In fact, for psoriasis, a scenario could be envisioned in which there is a genetic aberrant expression or function of one or more genes in the epidermal or vascular compartment of the skin, which does not lead to pathology under normal situations. However, cofactors like trauma, toxic irritation, or altered bacterial skin flora may induce persistent activation or insufficient counter-regulation and precipitate disease. Such a possibility is exemplified in the chemokine

Table 156-3: Animals with Aberrant Gene Expression Causing Spontaneous Skin Inflammation

	Angiogenesis	*Neutrophils*	*Micro-Abscess*	*Lymphocytes*	*References*
Gene overexpression targeted to the epidermis					
VEGF	Yes	Yes	Yes	Yes	50
Amphiregulin	Yes	Yes	Yes	Yes	72
Transforming growth factor-β1	Yes	Yes	Yes	Yes	73
Bone morphogenetic protein-6	Yes	Yes	Yes	Yes	74
β1, α2β1, α5β1 integrin		Yes	Yes	Yes	75
MEK1		Yes		Yes	76
Bone marrow kinase gene in chromosome X	Yes	Yes			77
Interleukin-18		Yes		Yes	78
Interleukin-1α		Yes		Yes	79
Interleukin-1α and IL-1 receptor		Yes		Yes	80
IFN-γ	Yes	Yes	Yes	Yes	75
STAT3	Yes	Yes	Yes	Yes	52
Gene deletion targeted to the epidermis					
JunB, c-Jun	Yes	Yes	Yes	Yes	69
Other transgenics					
HLA-B27 and β_2-microglobulin		Yes		Yes	81
Tie2	Yes	Yes	Yes	Yes	70
Kallikrein 7		Yes		Yes	82
Knockout mice					
CD18 (strain-dependent)		Yes	Yes	Yes	83
Epiregulin*					84

*Inflammatory response, not further characterized.

receptor *D6*-knockout mouse (71). D6, a nonsignaling decoy receptor, internalizes and degrades β-chemokines. Expression is restricted to nonhematopoietic cells such as ECs lining lymphatics of skin. The knockout mouse is born and develops normally, but following topical application of phorbol esters, persistent inflammation occurs with similarities to psoriasis. In humans, mutations have been found that, by themselves, do not cause disease, but are pathogenic in the context of cofactors. This has been shown for some point mutations in proteins of the Ras oncogene family, which in context with other hits serve as cofactors in oncogenesis. For psoriasis, TNF or IL-10 promoter polymorphism and VEGF-A point mutations have been found associated with psoriasis (51), and psoriasis susceptibility loci have been identified; it can be expected that the list will extend.

KEY POINTS

- The skin vasculature is unique in several aspects. The skin has several functionally distinct vascular units, the loops within the tips of the papillae, the SVP, the DVP, and the subcutaneous plexus. These vascular segments may respond to exogenous or endogenous triggers individually or conjointly, thereby "modeling" clinical skin disease expression.
- Due to the close proximity to the epidermis, keratinocytes can shape the reaction patterns of skin vessels; ECs of the SVP may respond to keratinocyte-derived factors like CTACK or VEGF, resulting in an altered vascular architecture, phenotype, and function.
- Due to their proximity to the epidermis, skin vessels are indirectly exposed to and may react to environmental factors that then may induce and/or maintain EC pathology (e.g., UV irradiation or infection).
- Skin ECs have certain unique features that may account for skin-specific pathologies. Examples are expression of CD32 and peripheral node addressins (recognized by antibody MECA-79), or the presentation of keratinocyte-derived CTACK. These skin-specific features may represent novel targets for the treatment of skin diseases.

Future Goals

- To develop therapies that inhibit keratinocyte-derived CTACK or VEGF production, or prevent sialylation or sulfation of endothelial sialomucins in chronic skin diseases such as atopic dermatitis, psoriasis, or cutaneous lupus erythematosus

ACKNOWLEDGMENTS

We are indebted to our teacher Dr. Klaus Wolff, chairman emeritus of the Department of Dermatology / Medical University of Vienna to whom we wish to devote this chapter. This work was -in part-supported by the Austrian science foundation (grant #S94).

REFERENCES

1 Braverman IM. The cutaneous microcirculation. *J Investig Dermatol Symp Proc*. 2000;5(1):3–9.
2 Johnson CL, Holbrook KA. Development of human embryonic and fetal dermal vasculature. *J Invest Dermatol*. 1989;93(Suppl 2): S10–S17.
3 Braverman IM, Yen A. Ultrastructure of the human dermal microcirculation. II. The capillary loops of the dermal papillae. *J Invest Dermatol*. 1977;68(1):44–52.
4 Sueki H, Telegan B, Murphy GF. Computer-assisted three-dimensional reconstruction of human dermal dendrocytes. *J Invest Dermatol*. 1995;105(5):704–708.
5 Gibran NS, Nickoloff BJ, Holbrook KA. Ontogeny and characterization of factor XIIIa+ cells in developing human skin. *Anat Embryol (Berl)*. 1996;193(1):35–41.
6 Helmbold P, Wohlrab J, Marsch WC, Nayak RC. Human dermal pericytes express 3G5 ganglioside – a new approach for microvessel histology in the skin. *J Cutan Pathol*. 2001;28(4): 206–210.
7 Braverman IM, Sibley J. Ultrastructural and three-dimensional analysis of the contractile cells of the cutaneous microvasculature. *J Invest Dermatol*. 1990;95(1):90–96.
8 Faust SN, Levin M, Harrison OB, et al. Dysfunction of endothelial protein C activation in severe meningococcal sepsis. *N Engl J Med*. 2001;345(6):408–416.
9 Schlingemann RO, Dingjan GM, Emeis JJ, et al. Monoclonal antibody PAL-E specific for endothelium. *Lab Invest*. 1985;52(1):71–76.
10 Niemela H, Elima K, Henttinen T, et al. Molecular identification of PAL-E, a widely used endothelial-cell marker. *Blood*. 2005;106 (10):3405–3409.
11 Witmer AN, Dai J, Weich HA, et al. Expression of vascular endothelial growth factor receptors 1, 2, and 3 in quiescent endothelia. *J Histochem Cytochem*. 2002;50(6):767–777.
12 Pober JS, Kluger MS, Schechner JS. Human endothelial cell presentation of antigen and the homing of memory/effector T cells to skin. *Ann NY Acad Sci*. 2001;941:12–25.
13 Petzelbauer P, Bender JR, Wilson J, Pober JS. Heterogeneity of dermal microvascular endothelial cell antigen expression and cytokine responsiveness in situ and in cell culture. *J Immunol*. 1993;151(9):5062–5072.
14 Gröger M, Sarmay G, Fiebiger E, et al. Dermal microvascular endothelial cells express CD32 receptors in vivo and in vitro. *J Immunol*. 1996;156(4):1549–1556.
15 Gröger M, Holnthoner W, Maurer D, et al. Dermal microvascular endothelial cells express the 180-kDa macrophage mannose receptor in situ and in vitro. *J Immunol*. 2000;165(10):5428–5434.
16 Kanda H, Tanaka T, Matsumoto M, et al. Endomucin, a sialomucin expressed in high endothelial venules, supports L-selectin-mediated rolling. *Int Immunol*. 2004;16(9):1265–1274.
17 Fieger CB, Sassetti CM, Rosen SD. Endoglycan, a member of the CD34 family, functions as an L-selectin ligand through modification with tyrosine sulfation and sialyl Lewis x. *J Biol Chem*. 2003;278(30):27390–27398.
18 Samulowitz U, Kuhn A, Brachtendorf G, et al. Human endomucin: distribution pattern, expression on high endothelial venules, and decoration with the MECA-79 epitope. *Am J Pathol*. 2002;160(5):1669–1681.
19 Kuhn A, Brachtendorf G, Kurth F, et al. Expression of endomucin, a novel endothelial sialomucin, in normal and diseased human skin. *J Invest Dermatol*. 2002;119(6):1388–1393.
20 Sassetti C, Van ZA, Rosen SD. Identification of endoglycan, a member of the CD34/podocalyxin family of sialomucins. *J Biol Chem*. 2000;275(12):9001–9010.
21 Renkonen J, Tynninen O, Hayry P, et al. Glycosylation might provide endothelial zip codes for organ-specific leukocyte traffic into inflammatory sites. *Am J Pathol*. 2002;161(2):543–550.
22 Lechleitner S, Kunstfeld R, Messeritsch-Fanta C, et al. Peripheral lymph node addressins are expressed on skin endothelial cells. *J Invest Dermatol*. 1999;113(3):410–414.
23 Mackay CR, Marston W, Dudler L. Altered patterns of T cell migration through lymph nodes and skin following antigen challenge. *Eur J Immunol*. 1992;22(9):2205–2210.
24 Lechleitner S, Gille J, Johnson DR, Petzelbauer P. Interferon enhances tumor necrosis factor-induced vascular cell adhesion molecule 1 (CD106) expression in human endothelial cells by an interferon-related factor 1-dependent pathway. *J Exp Med*. 1998;187(12):2023–2030.
25 Kluger MS, Johnson DR, Pober JS. Mechanism of sustained E-selectin expression in cultured human dermal microvascular endothelial cells. *J Immunol*. 1997;158(2):887–896.
26 Murakami S, Morioka T, Nakagawa Y, et al. Expression of adhesion molecules by cultured human glomerular endothelial cells in response to cytokines: comparison to human umbilical vein and dermal microvascular endothelial cells. *Microvasc Res*. 2001;62(3):383–391.
27 Grabbe S, Varga G, Beissert S, et al. Beta2 integrins are required for skin homing of primed T cells but not for priming naive T cells. *J Clin Invest*. 2002;109(2):183–192.
28 Mayadas TN, Cullere X. Neutrophil beta2 integrins: moderators of life or death decisions. *Trends Immunol*. 2005;26(7):388–395.
29 Petzelbauer P, Zacharowski PA, Miyazaki Y, et al. The fibrin-derived peptide Bbeta15–42 protects the myocardium against ischemia-reperfusion injury. *Nat Med*. 2005;11(3):298–304.
30 Rot A. In situ binding assay for studying chemokine interactions with endothelial cells. *J Immunol Methods*. 2003;273(1–2):63–71.
31 Salcedo R, Resau JH, Halverson D, et al. Differential expression and responsiveness of chemokine receptors (CXCR1–3) by human microvascular endothelial cells and umbilical vein endothelial cells. *FASEB J*. 2000;14(13):2055–2064.
32 Cheng SS, Lukacs NW, Kunkel SL. Eotaxin/CCL11 suppresses IL-8/CXCL8 secretion from human dermal microvascular endothelial cells. *J Immunol*. 2002;168(6):2887–2894.
33 Nibbs RJ, Kriehuber E, Ponath PD, et al. The beta-chemokine receptor D6 is expressed by lymphatic endothelium and a subset of vascular tumors. *Am J Pathol*. 2001;158(3):867–877.
34 Liao F, Huynh HK, Eiroa A, et al. Migration of monocytes across endothelium and passage through extracellular matrix involve separate molecular domains of PECAM-1. *J Exp Med*. 1995;182(5):1337–1343.
35 Schenkel AR, Mamdouh Z, Chen X, et al. CD99 plays a major role in the migration of monocytes through endothelial junctions. *Nat Immunol*. 2002;3(2):143–150.

36 Gröger M, Loewe R, Holnthoner W, et al. IL-3 induces expression of lymphatic markers Prox-1 and podoplanin in human endothelial cells. *J Immunol.* 2004;173(12):7161–7169.
37 Steinhoff M, Stander S, Seeliger S, et al. Modern aspects of cutaneous neurogenic inflammation. *Arch Dermatol.* 2003;139(11): 1479–1488.
38 Charkoudian N. Skin blood flow in adult human thermoregulation: how it works, when it does not, and why. *Mayo Clin Proc.* 2003;78(5):603–612.
39 Brooks EM, Morgan AL, Pierzga JM, et al. Chronic hormone replacement therapy alters thermoregulatory and vasomotor function in postmenopausal women. *J Appl Physiol.* 1997;83(2): 477–484.
40 Herrick AL. Pathogenesis of Raynaud's phenomenon. *Rheumatology (Oxford).* 2005;44(5):587–596.
41 Chotani MA, Flavahan S, Mitra S, et al. Silent alpha(2C)-adrenergic receptors enable cold-induced vasoconstriction in cutaneous arteries. *Am J Physiol Heart Circ Physiol.* 2000;278(4): H1075–H1083.
42 Furspan PB, Chatterjee S, Freedman RR. Increased tyrosine phosphorylation mediates the cooling-induced contraction and increased vascular reactivity of Raynaud's disease. *Arthritis Rheum.* 2004;50(5):1578–1585.
43 Flavahan NA, Flavahan S, Liu Q, et al. Increased alpha2-adrenergic constriction of isolated arterioles in diffuse scleroderma. *Arthritis Rheum.* 2000;43(8):1886–1890.
44 Bunker CB, Reavley C, O'Shaughnessy DJ, Dowd PM. Calcitonin gene-related peptide in treatment of severe peripheral vascular insufficiency in Raynaud's phenomenon. *Lancet.* 1993; 342(8863):80–83.
45 Rajagopalan S, Pfenninger D, Kehrer C, et al. Increased asymmetric dimethylarginine and endothelin 1 levels in secondary Raynaud's phenomenon: implications for vascular dysfunction and progression of disease. *Arthritis Rheum.* 2003;48(7):1992–2000.
46 Romagnani P, Lasagni L, Annunziato F, et al. CXC chemokines: the regulatory link between inflammation and angiogenesis. *Trends Immunol.* 2004;25(4):201–209.
47 Amoh Y, Li L, Yang M, et al. Nascent blood vessels in the skin arise from nestin-expressing hair-follicle cells. *Proc Natl Acad Sci USA.* 2004;101(36):13291–13295.
48 Uchi H, Terao H, Koga T, Furue M. Cytokines and chemokines in the epidermis. *J Dermatol Sci.* 2000;24(Suppl 1):S29–S38.
49 Detmar M, Brown LF, Claffey KP, et al. Overexpression of vascular permeability factor/vascular endothelial growth factor and its receptors in psoriasis. *J Exp Med.* 1994;180(3):1141–1146.
50 Xia YP, Li B, Hylton D, et al. Transgenic delivery of VEGF to mouse skin leads to an inflammatory condition resembling human psoriasis. *Blood.* 2003;102(1):161–168.
51 Young HS, Summers AM, Bhushan M, et al. Single-nucleotide polymorphisms of vascular endothelial growth factor in psoriasis of early onset. *J Invest Dermatol.* 2004;122(1):209–215.
52 Sano S, Chan KS, Carbajal S, et al. Stat3 links activated keratinocytes and immunocytes required for development of psoriasis in a novel transgenic mouse model. *Nat Med.* 2005;11(1):43–49.
53 Wight TN, Raugi GJ, Mumby SM, Bornstein P. Light microscopic immunolocation of thrombospondin in human tissues. *J Histochem Cytochem.* 1985;33(4):295–302.
54 Streit M, Velasco P, Riccardi L, et al. Thrombospondin-1 suppresses wound healing and granulation tissue formation in the skin of transgenic mice. *EMBO J.* 2000;19(13):3272–3282.
55 Hirakawa S, Fujii S, Kajiya K, et al. Vascular endothelial growth factor promotes sensitivity to ultraviolet B-induced cutaneous photodamage. *Blood.* 2005;105(6):2392–2399.
56 Dvorak AM, Mihm MC Jr., Dvorak HF. Morphology of delayed-type hypersensitivity reactions in man. II. Ultrastructural alterations affecting the microvasculature and the tissue mast cells. *Lab Invest.* 1976;34(2):179–191.
57 Creamer D, Allen MH, Sousa A, et al. Localization of endothelial proliferation and microvascular expansion in active plaque psoriasis. *Br J Dermatol.* 1997;136(6):859–865.
58 Petzelbauer P, Pober JS, Keh A, Braverman IM. Inducibility and expression of microvascular endothelial adhesion molecules in lesional, perilesional, and uninvolved skin of psoriatic patients. *J Invest Dermatol.* 1994;103(3):300–305.
59 Petzelbauer P, Stingl G, Wolff K, Volc-Platzer B. Cyclosporin Asuppresses ICAM-1 expression by papillary endothelium in healing psoriatic plaques. *J Invest Dermatol.* 1991;96(3):362–369.
60 Gröger M, Fischer GF, Wolff K, Petzelbauer P. Immune complexes from vasculitis patients bind to endothelial Fc receptors independent of the allelic polymorphism of FcgammaRIIa. *J Invest Dermatol.* 1999;113(1):56–60.
61 Kunstfeld R, Lechleitner S, Gröger M, et al. HECA-452+ T cells migrate through superficial vascular plexus but not through deep vascular plexus endothelium. *J Invest Dermatol.* 1997;108 (3):343–348.
62 Rossiter H, van RF, Mudde GC, et al. Skin disease-related T cells bind to endothelial selectins: expression of cutaneous lymphocyte antigen (CLA) predicts E-selectin but not P-selectin binding. *Eur J Immunol.* 1994;24(1):205–210.
63 Reiss Y, Proudfoot AE, Power CA, et al. CC chemokine receptor (CCR)4 and the CCR10 ligand cutaneous T cell-attracting chemokine (CTACK) in lymphocyte trafficking to inflamed skin. *J Exp Med.* 2001;194(10):1541–1547.
64 Homey B, Alenius H, Muller A, et al. CCL27-CCR10 interactions regulate T cell-mediated skin inflammation. *Nat Med.* 2002;8(2):157–165.
65 Bonecchi R, Bianchi G, Bordignon PP, et al. Differential expression of chemokine receptors and chemotactic responsiveness of type 1 T helper cells (Th1s) and Th2s. *J Exp Med.* 1998;187(1): 129–134.
66 Catalina MD, Estess P, Siegelman MH. Selective requirements for leukocyte adhesion molecules in models of acute and chronic cutaneous inflammation: participation of E- and P- but not L-selectin. *Blood.* 1999;93(2):580–589.
67 Farkas L, Beiske K, Lund-Johansen F, et al. Plasmacytoid dendritic cells (natural interferon- alpha/beta-producing cells) accumulate in cutaneous lupus erythematosus lesions. *Am J Pathol.* 2001;159(1):237–243.
68 Tang ML, Hale LP, Steeber DA, Tedder TF. L-selectin is involved in lymphocyte migration to sites of inflammation in the skin: delayed rejection of allografts in L-selectin-deficient mice. *J Immunol.* 1997;158(11):5191–5199.
69 Zenz R, Eferl R, Kenner L, et al. Psoriasis-like skin disease and arthritis caused by inducible epidermal deletion of Jun proteins. *Nature.* 2005;437(7057):369–375.
70 Voskas D, Jones N, Van SP, et al. A cyclosporine-sensitive psoriasis-like disease produced in Tie2 transgenic mice. *Am J Pathol.* 2005;166(3):843–855.
71 Jamieson T, Cook DN, Nibbs RJ, et al. The chemokine receptor D6 limits the inflammatory response in vivo. *Nat Immunol.* 2005; 6(4):403–411.

72 Cook PW, Piepkorn M, Clegg CH, et al. Transgenic expression of the human amphiregulin gene induces a psoriasis-like phenotype. *J Clin Invest*. 1997;100(9):2286–2294.
73 Li AG, Wang D, Feng XH, Wang XJ. Latent TGFbeta1 overexpression in keratinocytes results in a severe psoriasis-like skin disorder. *EMBO J*. 2004;23(8):1770–1781.
74 Blessing M, Schirmacher P, Kaiser S. Overexpression of bone morphogenetic protein-6 (BMP-6) in the epidermis of transgenic mice: inhibition or stimulation of proliferation depending on the pattern of transgene expression and formation of psoriatic lesions. *J Cell Biol*. 1996;135(1):227–239.
75 Carroll JM, Crompton T, Seery JP, Watt FM. Transgenic mice expressing IFN-gamma in the epidermis have eczema, hair hypopigmentation, and hair loss. *J Invest Dermatol*. 1997;108(4):412–422.
76 Hobbs RM, Silva-Vargas V, Groves R, Watt FM. Expression of activated MEK1 in differentiating epidermal cells is sufficient to generate hyperproliferative and inflammatory skin lesions. *J Invest Dermatol*. 2004;123(3):503–515.
77 Paavonen K, Ekman N, Wirzenius M, et al. Bmx tyrosine kinase transgene induces skin hyperplasia, inflammatory angiogenesis, and accelerated wound healing. *Mol Biol Cell*. 2004;15(9):4226–4233.
78 Konishi H, Tsutsui H, Murakami T, et al. IL-18 contributes to the spontaneous development of atopic dermatitis-like inflammatory skin lesion independently of IgE/stat6 under specific pathogen-free conditions. *Proc Natl Acad Sci USA*. 2002;99(17):11340–11345.
79 Groves RW, Mizutani H, Kieffer JD, Kupper TS. Inflammatory skin disease in transgenic mice that express high levels of interleukin 1 alpha in basal epidermis. *Proc Natl Acad Sci USA*. 1995;92(25):11874–11878.
80 Groves RW, Rauschmayr T, Nakamura K, et al. Inflammatory and hyperproliferative skin disease in mice that express elevated levels of the IL-1 receptor (type I) on epidermal keratinocytes. Evidence that IL-1-inducible secondary cytokines produced by keratinocytes in vivo can cause skin disease. *J Clin Invest*. 1996;98(2):336–344.
81 Hammer RE, Maika SD, Richardson JA, et al. Spontaneous inflammatory disease in transgenic rats expressing HLA-B27 and human beta 2m: an animal model of HLA-B27-associated human disorders. *Cell*. 1990;63(5):1099–1112.
82 Hansson L, Backman A, Ny A, et al. Epidermal overexpression of stratum corneum chymotryptic enzyme in mice: a model for chronic itchy dermatitis. *J Invest Dermatol*. 2002;118(3):444–449.
83 Bullard DC, Scharffetter-Kochanek K, McArthur MJ, et al. A polygenic mouse model of psoriasiform skin disease in CD18-deficient mice. *Proc Natl Acad Sci USA*. 1996;93(5):2116–2121.
84 Shirasawa S, Sugiyama S, Baba I, et al. Dermatitis due to epiregulin deficiency and a critical role of epiregulin in immune-related responses of keratinocyte and macrophage. *Proc Natl Acad Sci USA*. 2004;101(38):13921–13926.
85 Kunstfeld R, Hirakawa S, Hong YK, et al. Induction of cutaneous delayed-type hypersensitivity reactions in VEGF-A transgenic mice results in chronic skin inflammation associated with persistent lymphatic hyperplasia. *Blood*. 2004;104(4):1048–1057.

157

Angiogenesis

Helmut G. Augustin

University of Heidelberg, and the German Cancer Research Center (DKFZ)

The formation of blood vessels, or vascular morphogenesis, is a fundamental biological process (Figure 157.1). In fact, the cardiovascular system is the first organ that develops in the mammalian embryo. Consequently, genetic manipulation of key regulators of the angiogenic cascade often yields dramatic phenotypes that are not compatible with life and lead to early embryonic lethality. This in turn implies that the process of blood vessel formation is under strong evolutionary pressure, and that every living animal with a closed circulation that has successfully made it through embryonic development is capable of executing the angiogenic program. As such, the formation of blood vessels is a conserved, hierarchically structured process with a high degree of similarity, both molecularly as well as functionally, in such diverse species as zebrafish (1), frog (2), mouse (3,4), and man (3,4). The goals of this chapter are to discuss: (a) the history of vascular morphogenesis research, (b) the basic principles of physiological and pathological growth of blood vessels, (c) concepts of the induction and inhibition of blood vessel growth (angiogenic switch, angiogenic balance), (d) techniques and strategies to quantitatively and qualitatively assess the structural and functional status of neovasculature, and (e) established and emerging avenues for therapeutically interfering with the growth of blood vessels (antiangiogenesis, therapeutic angiogenesis). The topic of lymphangiogenesis is reviewed separately in Chapter 169.

DEFINITIONS

The term a*ngiogenesis* is widely used when referring to the process of blood vessel formation (Table 157-1). Yet, more specifically, angiogenesis denotes the formation of new capillaries from pre-existing ones (see Figure 157.1). This occurs primarily through classical sprouting angiogenesis. Alternatively, nonsprouting angiogenesis (also called *intussusception* or *intussusceptive microvascular growth* [IMG]) (5) describes the formation of a vascular network from an endothelial cell (EC)-lined vessel by focally inserting a tissue pillar or by longitudinal fold-like splitting of a vessel (6,7) (Figure 157.1).

In contrast to the formation of capillary networks via mechanisms of sprouting and nonsprouting angiogenesis, the term *arteriogenesis* describes the mechanisms of adaptive vessel growth whereby larger arteries are formed from a pre-existing network of collateral arterioles in response to changes in blood flow that lead to increased biomechanical forces (shear stress), for example, as a consequence of a local ischemic event (7,8). Arteriogenesis refers primarily to the pathology-associated remodeling of the adult vasculature associated with an intense inflammatory response (9). This type of postnatal adaptive response should be distinguished from developmental arteriogenesis, in which the neovascular capillary network is physiologically remodeled to form large arteries. A combination of genetic determinants (10) and biomechanical forces (11) is believed to be responsible for the physiological program that leads to the formation of larger arteries and veins.

The third basic mechanism of vascular morphogenesis is called *vasculogenesis.* Vasculogenesis describes the primary in situ differentiation of ECs from mesodermal precursor cells (angioblasts) and their subsequent organization into a primary capillary plexus (see Figure 157.1) (12). Angioblasts share a common lineage with hematopoietic cells (hemangioblasts), reflecting the close relationship between resident vessel wall cells and the mobile cells in the circulation (13). Vasculogenesis is primarily a developmental process that precedes and occurs concomitantly with the angiogenic phase of vessel growth. Yet, the identification of endothelial progenitor cells (EPCs) in the adult circulation in 1997 (14) has opened a whole new field of research aimed at exploring the mechanisms and contribution of vascular stem and progenitor cells to vascular morphogenic processes in the adult, as well as their therapeutic potential (15). Although the quantitative contribution of *adult vasculogenesis* to vascular morphogenic processes in the adult is the focus of intense research and the subject of some controversy, it is widely recognized that cells originating in stem cell niches within the bone marrow and

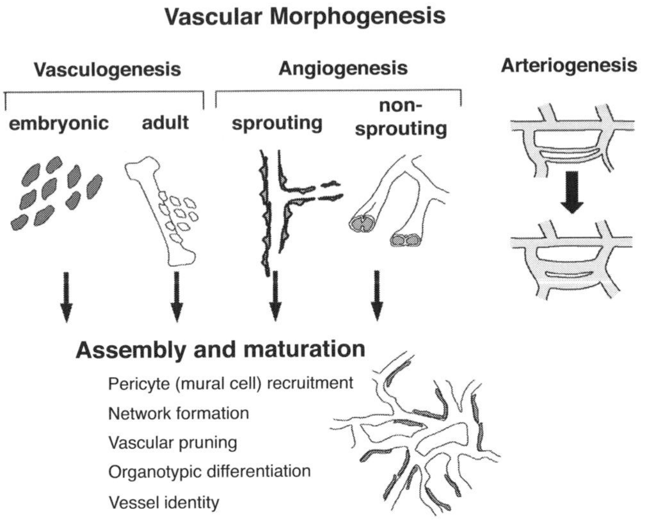

Figure 157.1. Principles of vascular morphogenesis. Vasculogenesis, angiogenesis, and arteriogenesis contribute to the formation of a structured network of blood vessels. Vasculogenesis occurs during development by the coalescence of in situ differentiating angioblastic cells that form the first primitive vascular plexus. Adult vasculogenesis refers to the recruitment of endothelial stem and progenitor cells from the bone marrow and other angioblastic niches. Angiogenesis can occur as classical sprouting angiogenesis or as nonsprouting angiogenesis (intussusception). Arteriogenesis is characterized by adaptive growth of arterioles in response to increased, flow-mediated biomechanical forces.

other parts of the body can, in principle, contribute to the growth of blood vessels in the adult.

HISTORY OF VASCULAR MORPHOGENESIS RESEARCH

In 1787, the British surgeon John Hunter first used the term *angiogenesis* to describe the growth of blood vessels in the reindeer antler (16). One hundred and fifty years later, Arthur Herting described the growth of new blood vessels in the placenta of pregnant monkeys (17). Modern angiogenesis research dates back to 1971, when Judah Folkman of Harvard Medical School introduced the hypothesis that the growth of tumors is angiogenesis-dependent (18). In fact, a PubMed search employing the key term *angiogenesis* yields less than 10 published papers prior to Dr. Folkman's seminal article in the *New England Journal of Medicine*, in contrast to the almost 30,000 angiogenesis-related publications that have been published in the 35 years since. Of these, 18 original publications have so far been cited more than 1,000 times, qualifying them as angiogenesis citation classics (Table 157-2).

Dr. Folkman's original hypothesis was slow to gain acceptance. Except for cartilage and cornea, blood vessels are found in all tissues of the body and almost any cell of the body is within 100 to 150 μm of the next capillary (maximal effective diffusion distance for oxygen). It was therefore widely believed that tumors grow along pre-existing blood vessels, a concept that has recently been revived and named *vessel cooption*[1] (20). Although the angiogenesis dependency of tumor growth is solidly established, the relative contribution of vessel cooption to tumor vascularization remains to be determined.

A major research goal during the 1980s was to identify and therapeutically block a putative tumor angiogenesis factor (TAF) (18). We now know that TAF is not a single factor. Indeed, a long list of angio-stimulatory and angio-inhibitory molecules has been identified (Table 157-3). The nature of the "angiogenic cocktail" that drives the angiogenic cascade during the physiological and pathological growth of blood vessels is far from being understood.

Several angiogenesis factors were identified during the 1980s. These include the fibroblast growth factors (FGFs), the transforming growth factors (TGFs), platelet-derived EC growth factor (PD-ECGF, thymidine phosphorylase), and tumor necrosis factor (TNF)-α (21). The search for angiogenesis-inducing growth factors during these years was strongly stimulated by the discovery that many growth factors bind heparin, a property that facilitated their purification (22). All the angiogenic growth factors identified during these years are capable of inducing angiogenesis in specific in vitro and in vivo assays. Yet, they are not vascular-specific in their activity. Instead, they act pleiotropically to regulate the function of many cell types. This lack of specificity limited their potential as therapeutic targets.

The identification of the first specific vascular growth factor, vascular endothelial growth factor (VEGF) in 1989 (23–25) revolutionized the field of angiogenesis research. Whereas a total of <1,000 angiogenesis-related articles were published prior to 1990, over 3,500 papers are now published every year. Indeed, angiogenesis research has been (and continues to be) one of the fastest growing fields in all of biomedicine (Figure 157.2).

The EC mitogen VEGF (26) had been identified 6 years earlier as vascular permeability factor (VPF) (27). Yet, its association with the angiogenic process was not noted at the time. The molecular mechanisms underlying VEGF/VPF-mediated permeability and mitogenicity remain poorly understood to this day.

VEGF exerts its vascular-specific functions through the almost EC-specific expression of its receptors, VEGFR1 (28) and VEGFR2 (29). The importance of the ligand and receptors was established during the mid-1990s through genetic inactivation experiments that identified VEGF and its receptors as strictly rate-limiting for the developmental vascular

1 Vessel cooption describes the process whereby growing tumors prey on the pre-existing vasculature to grow along these vessels and to remodel pre-existing vessels for their vascularization purposes. Although it is generally accepted that angiogenesis is the primary mode of tumor vascularization, vessel cooption has been described to occur in at least 20% of lung tumors (19). Likewise, vessel cooption may be one of the reasons for limited efficacy of antiangiogenic therapies.

Table 157-1: Vascular Morphogenesis

Biological Process	*Definition*	*Embryonic*	*Adult – Physiology*	*Adult – Pathology*
Vasculogenesis	In situ differentiation of ECs from mesodermal precursor cells	Generation of primary capillary plexus	Bone marrow–derived EPCs may contribute to physiological vascular morphogenesis	Bone marrow–derived EPCs may contribute to the formation of new vessels during pathology (tumors, ischemic events)
Angiogenesis	Formation of new capillaries from pre-existing capillaries	Expansion of vasculogenesis-derived vascular plexus	Female reproductive system	Tumors, wound healing, inflammation
Arteriogenesis	Formation of larger arteries from a pre-existing network of collateral arterioles	Remodeling of neovascular capillary network	Not described	Adaptive remodeling of adult vasculature following local ischemia (e.g., heart)
Vessel wall assembly	3D assembly of growing vasculature	Formation of anastomosing vascular network; recruitment of mural cells	n.s.	n.s.
Acquisition of vessel identity	Formation of different types of vessel structures	Formation of different caliber blood vessels and development of arteriovenous asymmetry	n.s.	n.s.
Organotypic differentiation	Molecular and anatomical differentiation of assembled vasculature	Formation of continuous and discontinuous endothelia; expression of organotypic molecular phenotype	n.s.	n.s.

n.s., not studied in much molecular and mechanistic detail. Yet, it can be implied that the later steps of the angiogenic cascade are similarly regulated as during developmental vascular morphogenic processes. However, the differences between the formation of a regular and fully differentiated vascular network in normal organs and the chaotropic vascular network associated with pathological neovascularization are mechanistically poorly understood.

morphogenesis (28–31). Their genetic inactivation consequently leads to early embryonic lethality. In fact, the heterozygous early embryonic lethal phenotype of VEGF-deficient mice is to this day the most dramatic phenotype of the many thousands of knockout mice that have been generated during the last 15 years (30,31).

VEGF may be viewed as a hierarchically high master switch of the angiogenic cascade. Indeed, 11 of the 18 all-time angiogenesis citation classics focus on VEGF (23,24,27–34) or its regulation (35) (see Figure 157.1). The finding that VEGF expression is induced by hypoxia (35) may be conceptually among the most important angiogenesis-related discoveries, because it linked a pleiotropic milieu factor (hypoxia) to the expression of a specific angiogenesis growth factor (i.e., VEGF). Moreover, the discovery paved the way for significant advances in hypoxia research, culminating in the identification of prolyl hydroxylases as a molecular candidate for the long-sought-after oxygen sensor (36).

Despite its requisite role in angiogenesis, VEGF acts in concert with other growth factors. The angiopoietins (Ang1 and Ang2) were discovered as ligands of the EC receptor tyrosine kinase Tie2 less than 10 years ago (37). Gene ablation experiments in mice suggested a role for Ang1/Tie2 signaling in remodeling and stabilization of the growing neovasculature (38). Ang1 acts as an agonistic Tie2 ligand and is involved in regulating vessel maturation and remodeling by acting as a survival factor for ECs and controlling the association of ECs with mural cells (pericytes and smooth muscle cells) (37). Low-level Ang1-dependent constitutive Tie2 phosphorylation appears to be involved in maintaining the quiescent phenotype of the resting vasculature (39). Ang2 functions are more complex. Ang2 acts as a functional antagonist of Ang1 by binding to Tie2 without exerting signal transducing functions (40). Intriguingly, Ang2 is produced almost exclusively by ECs and acts as an autocrine regulator of vessel stabilization and destabilization (41,42). As such, Ang2 may facilitate angiogenesis in the presence of angiogenic activity (e.g., VEGF stimulation) or induce vessel regression in the absence of angiogenic activity (e.g., during regression of the cyclic ovarian corpus luteum) (43).

The Eph receptor tyrosine kinases comprise the largest known family of tyrosine kinase receptors and interact in a

Table 157-2: Milestones of Angiogenesis Research According to Frequency of Citation

Discovery	*First Author*	*Senior Author*	*Year*	*Citation**	*References*
Prognostic value of intratumoral microvessel density counts	Noel Weidner	Judah Folkman	1991	2,266	81
Identification of VEGF	David W. Leung	Napoleone Ferrara	1989	2,271	24
Hypoxia regulation of VEGF expression	Dorit Shweiki	Eli Keshet	1992	2,165	35
Induction of angiogenesis by TGF-β	Anita S. Roberts	Anthony S. Fauci	1986	2,048	99
Discovery of endogenous angiogenesis inhibitor endostatin	Michael O'Reilly	Judah Folkman	1997	2005	59
Discovery of endogenous angiogenesis inhibitor angiostatin	Michael O'Reilly	Judah Folkman	1994	1,938	58
Inhibition of tumor angiogenesis by anti-VEGF antibody	K. Jin Kim	Napoleone Ferrara	1993	1,760	33
Embryonic lethal phenotype of Flk-1 (VEGFR2)-deficient mice	Fouad Shalaby	Andre C. Schuh	1995	1,684	29
Discovery of tumor-derived vascular permeability factor (VPF)	Donald R. Senger	Harold F. Dvorak	1983	1,512	27
Embryonic lethal phenotype of	Peter Carmeliet	Andras Nagy	1996	1,496	30
heterozygous VEGF-deficient mice	Napoleone Ferrara	Mark W. Moore	1996	1,403	31
Identification of VEGF as tumor angiogenesis factor	Karl H. Plate	Werner Risau	1992	1,384	32
Role of integrin $\alpha v\beta 3$ during tumor angiogenesis	Peter C. Brooks	David A. Cheresh	1994	1,334/1,312	49/100
Embryonic lethal phenotype of VEGFR1-deficient mice	Guo-Hua Fong	Martin L. Breitman	1995	1,230	28
Cyclooxygenase (COX) regulation of angiogenesis	Masahiko Tsujii	Raymond N. DuBois	1998	1,228	101
Regulation of vasculogenesis and angiogenesis by VEGF/VEGFR2	Birgit Millauer	Axel Ullrich	1993	1,176	34
Identification of VEGF, identity with VPF	Pamela J. Keck	Daniel T. Connolly	1989	1,163	23
VHL regulation of HIF-dependent angiogenesis factor	Patrick Maxwell	Peter Ratcliffe	1999	1,120	136
Discovery of angiopoietin 2	Peter Maisonpierre	George Yancopoulos	1997	1,085	40
Identification of interleukin-8 as angiogenic factor	Alisa E. Koch	Robert M. Strieter	1992	1,056	137
Demonstration of angiogenic switch in transgenic RipTag model	Judah Folkman	Douglas Hanahan	1989	1,054	138
Discovery of angiopoietin 1	Chitra Suri	George Yancopoulos	1996	1,050	38
Detection of VEGF in ocular fluid	Lloyd Paul Aiello	George King	1994	1,025	139

*This list includes all angiogenesis-related original publications that have been cited more than 1,000 times (out of approximately 30,000 original publications [based on Science Citation Index – ISI Web of Science as of February 2007]).

specific, yet somewhat promiscuous manner with their corresponding ephrin ligands (44). Although initially characterized as regulators of axonal outgrowth in the nervous system, gene inactivation experiments in mice have revealed key roles for ephrinB2 and its EphB2, EphB3, and EphB4 receptors during vascular development (44). Mouse embryos lacking ephrinB2 or EphB4 (or the combination of EphB2 and EphB3) suffer dramatic defects in early angiogenic remodeling that are similar to those seen in mice lacking Ang1 or Tie2 (45). Moreover, ephrinB2 and EphB4 display remarkable asymmetric arteriovenous expression patterns, with ephrinB2 selectively marking arterial vessels and EphB4 preferentially being expressed by venous ECs (45). This asymmetric expression pattern suggests a critical role of ephrinB2 and EphB4 in establishing arterial versus venous identity of the growing vasculature. The identification of vascular morphogenic functions of ephrinB2 and EphB4 has led to the development of two rapidly growing novel fields of vascular biology research, namely the molecular analysis of the structural and functional mechanisms of arteriovenous differentiation (46), and the molecular study of the commonalities between vascular and neuronal guidance and patterning mechanisms (47).

Angiogenic activation induces a complex gene expression program in ECs, enabling the cells to execute the molecular

Table 157-3: Positive and Negative Regulators of Angiogenesis

Stimulators (Refs.)	*Inhibitors (Refs.)*
Peptide growth factors	**Peptide inhibitors (proteolytic fragments)**
VEGF-A, -B, -C, -D, -E (102)	Angiostatin (plasminogen fragment) (60, 77)
Placenta growth factor (PlGF) -1, -2 (74)	Endostatin (collagen XVIII fragment) (59, 60)
Angiopoietin (Ang)-1 (38)	Vasostatin (calreticulin fragment) (60, 110)
Fibroblast growth factor (FGF) -1, -2 (103)	Tumstatin (collagen IV fragment) (60)
Platelet-derived growth factor (PDGF-BB) (104)	Canstatin (collagen IV fragment) (60, 111)
Endocrine gland-VEGF (EG-VEGF) (105)	Arresten (collagen IV fragment) (60)
Transforming growth factor (TGF)-α, -β (106, 107)	Vasostatin (chromogranin A fragment) (60)
Hepatocyte growth factor (HGF, Scatter factor) (108)	Endorepellin (perlecan fragment) (60, 112)
Insulin-like growth factor (IGF)-1 (109)	Antithrombin III fragment (60, 113)
	Alphastatin (fibrinogen fragment) (114)
Hemostatic regulators	**Extracellular matrix molecules**
Tissue factor (115)	Thrombospondin (117)
Thrombin (116)	
Semaphorins	**Semaphorins**
Semaphorin-4D (SEMA4D) (118)	Semaphorin-3A (SMEA3A) (119)
	Semaphorin-3F (SEMA3F) (120)
Multifunctional cytokine/immune mediators	**Multifunctional cytokines/immune mediators**
Tumor necrosis factor (TNF)-α (low dose) (121)	Tumor necrosis factor (TNF)-α (high dose) (121)
CXC-Chemokines	**CXC-Chemokines**
Interleukin (IL)-8; CXCL8	Platelet factor 4 (PF4) (122)
Stromal-derived factor (SDF)-1; CXCL12 (64)	IP-10 (CXCL10) (123)
	Gro-β (CXCL2) (124)
Enzymes	**Inhibitors of enzymatic activity**
Cyclooxygenase (COX)-2 (101)	TIMP-1, -2, -3, -4 (127)
PD-ECGF, thymidine phosphorylase (125)	Plasminogen activator inhibitor (PAI)-1, -2 (53)
Angiogenin (ribonuclease A homologue) (126)	
Hormones	**Hormones/metabolites**
Estrogens (128)	2-Methoxyestradiol (2-ME) (131)
Prostaglandin-E_1, -E_2 (129)	Proliferin-related protein (130)
Follistatin (67)	
Proliferin (130)	
Oligosaccharides	**Oligosaccharides**
Hyalorunan oligosaccharides (132)	Hyaluronan, high molecular weight species (132)
Gangliosides (133)	
Hematopoietic growth factors	
Erythropoietin (134)	
G-CSF (135)	
GM-CSF (135)	

VEGF, vascular endothelial growth factor; HB-EGF, heparin-binding epidermal growth factor-like growth factor; PD-ECGF, platelet-derived EC growth factor; G-CSF, granulocyte colony stimulating factor; GM-CSF, granulocyte macrophage colony stimulating factor; TIMP, tissue metalloproteinase inhibitor; IP-10, interferon-γ-inducible protein-10.

tasks required for new blood vessel formation (48). The invasive ingrowth of angiogenic ECs involves the upregulation of a distinct set of adhesion molecules including the integrin heterodimers $\alpha v\beta 3$ and $\alpha v\beta 5$ (49,50) as well as homotypic interactions involving vascular endothelial (VE)-cadherin and other junctional molecules (51). Likewise, invading ECs (also called *tip cells*, in contrast to following cells, which are called *stalk cells*) (52) display a shift of their proteolytic balance toward a proinvasive phenotype with upregulated expression of plasminogen activators (tissue-type plasminogen activator [t-PA] and urokinase-type plasminogen activator [u-PA]) and downregulation of plasminogen activator inhibitor (PAI)-1

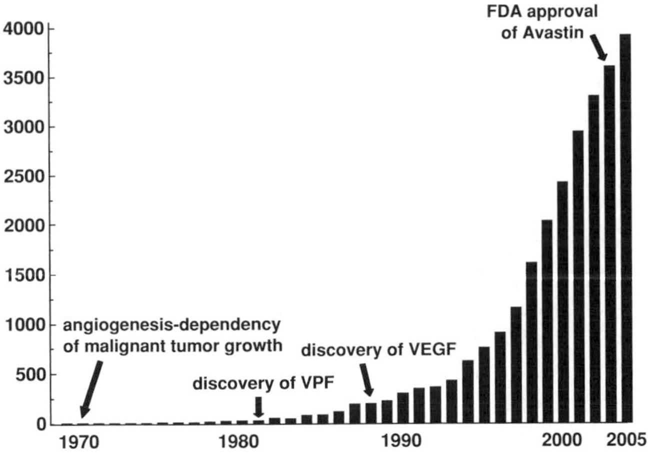

Figure 157.2. Number of angiogenesis-related articles published in *Medline*. The annual number of angiogenesis-related publications has increased by more than a factor of 10 in the last 10 years.

(53). Angiogenic ECs deposit their own extracellular matrix, rearrange their cytoskeleton, and activate their proliferative machinery. Importantly, each of these steps represents a potential molecular target for antiangiogenesis therapy (54).

Although much remains be learned about the angiogenic process, it is reasonable to conclude that the primary molecular building blocks for angiogenesis have been identified. In contrast, the mechanisms underlying physiological angioinhibition and angio-regression are less well understood. The first endogenous antiangiogenic activity from cartilage was isolated in 1975 (55). Thrombospondin (TSP)-1 was identified in 1978 (56), but its antiangiogenic activity was not recognized until 1990 (57). In the mid-1990s, the discovery of the polypeptides angiostatin (58) and endostatin (59) provided new insights into the endogenous balance of angiogenic factors (see Table 157-2). Angiostatin and endostatin are cryptic fragments of plasminogen and type XVIII collagen, respectively, that are liberated through the activities of cleaving proteases. These prototypic endogenous inhibitors of angiogenesis have stimulated the systematic search for other endogenous inhibitors of angiogenesis, many of which are cryptic fragments of larger molecules with different functions (mostly molecules of the extracellular matrix and coagulation system) (60). Although the molecular mode of action of most of these molecules remains to be elucidated, endogenous angiogenesis inhibitors are being increasingly validated in complex in vivo model systems, as has been demonstrated for TSP-1, endostatin, and tumstatin in gene ablation experiments (61,62).

Today, the field of vascular morphogenesis with its three primary avenues of investigation – vasculogenesis, angiogenesis, and arteriogenesis – is solidly established and still growing (see Figure 157.2). Future studies will likely lead to the identification of additional specific regulators of the angiogenic cascade. For example, the cytokine endocrine gland–derived (EG)-VEGF and the chemokine stromal cell–derived factor (SDF)-1 control vascular morphogenesis in only selected vascular beds (63,64). Additional organ- and caliber-specific vascular growth factors are likely to be discovered. Likewise, the study of the molecular basis of vascular assembly, network formation, organotypic differentiation, control of directional flow, and vascular maturation is in its nascence.

PHYSIOLOGICAL AND PATHOLOGICAL GROWTH OF BLOOD VESSELS

The growth of blood vessels is primarily a physiological developmental process. As such, the first primitive vasculogenesis-derived capillary plexus expands by angiogenesis and intussusceptive microvascular growth (65). Sprouting and nonsprouting angiogenesis contribute to an increasing complexity of the growing vascular network. The network assembles and matures by the recruitment of smooth muscle cells and pericytes, eventually allowing directional flow of blood. The morphogenic events leading to a mature vascular network involve several additional steps, including vessel wall assembly, maturation, acquisition of vessel identity and organotypic differentiation. Vascular remodeling is a distinct process that describes the adaptive reorganization of an existing, mature vasculature in a physiologically or pathologically challenged microvascular bed.

Angiogenesis in the adult is normally downregulated (an exception is the female reproductive system). ECs cells in quiescent[2] blood vessels have very low turnover rates and divide only once every few months or years. Yet, they are not postmitotic cells; they may be rapidly recruited into the cell cycle in response to angiogenic stimuli (66). In fact, there may be few cell types in the human body with such a pronounced capacity to switch reversibly between resting and proliferating states. For example, many cell populations such as intestinal epithelial cells or keratinocytes are programmed for continuous turnover. In contrast, postmitotic cells such as neurons and cardiomyocytes have a very limited regenerative capacity. The fact that most adult ECs are quiescent, yet have the potential to rapidly proliferate, points to the existence of tonically active, although reversible antiangiogenic mechanisms. Although these mechanisms remain poorly understood, evidence suggests an autocrine role of the endothelium in this process. For example, the activin/follistatin system has been shown to control EC activity in an autocrine manner (67). Similarly, the inhibitor of the vascular maintenance and quiescence regulating Ang1/Tie2 system, Ang2, is almost exclusively expressed by ECs, suggesting a primarily autocrine mode of action (41).

Unlike other organs in the healthy adult, the female reproductive system displays angiogenic activity. The ovary and uterus undergo distinct endocrine-mediated morphological changes during the reproductive cycle; these are associated with discrete phases of tissue growth, maturation, and

2 Here we use the term *quiescent* narrowly to describe nonproliferating ECs.

regression. Concomitantly, blood vessel growth, maturation, and regression accompany and to some extent even precede the reproductive cycle, making angiogenesis a critical and rate-limiting prerequisite for normal reproductive function (68). Indeed, inhibition of key regulators of the angiogenic cascade leads to disruption of the reproductive cycle and female infertility (69).

The female reproductive system notwithstanding, angiogenesis in the adult is almost always associated with pathological processes. According to PubMed, approximately 90% of research papers related to pathological angiogenesis focus on tumor angiogenesis. However, a growing appreciation exists for the role of angiogenesis in other pathological processes, including disorders of the eye (e.g., diabetic retinopathy, retinopathy of prematurity [ROP], age-dependent macular degeneration [AMD]), acute and chronic inflammatory reactions (e.g., osteoarthritis, atherosclerosis, hypertensive vascular remodeling), and wound healing. Adaptation to changes in oxygen tension may be a primary molecular determinant underlying each of these processes (70).

Pathological angiogenesis is widely considered to result from the activation of the same molecular program that governs physiological angiogenesis during embryonic development. As such, tumor angiogenesis is considered to be an *oncofetal mechanism*, in which a normal developmental process is reactivated in the course of adult disease. Oncofetal mechanisms are attractive therapeutic targets because their inhibition – at least in nonpregnant patients – should be associated with tumor-specific effects. Indeed, extensive experience in clinical trials indicates that antiangiogenic therapies are well tolerated and do not appear to cause major unwanted side effects. For example, although prolonged anti-VEGF treatment may be associated with an increase in blood pressure, other side effects are rare (71).

The concept of an oncofetal mechanism, although conceptually helpful, belies important differences between mechanisms of developmental and tumor angiogenesis. Unlike physiological angiogenesis, pathological angiogenesis often is associated with an inflammatory response, and cells recruited during inflammatory reactions are well known to exert angiogenesis-modulating activities (72). Likewise, tumor growth involves numerous interactions with the body's immune system that may similarly modulate the angiogenic program. Physiological angiogenesis follows a well-coordinated, hierarchically structured program that leads to the formation of a stable, anatomically structured vascular network whose pattern is optimized in an organ-specific manner to meet the demands of the underlying tissue. In contrast, tumor angiogenesis is an uncontrolled and unidirectional process (i.e., the tumor has no built-in program for vascular stabilization or vascular regression), and leads to a highly dense, chaotic, and poorly organized vasculature. These considerations have led to the notion that a pruned, *normalized* vasculature with fewer blood vessels (e.g., resulting from an antiangiogenic therapy) may in fact be functionally more effective (73).

Genetic studies have provided further evidence for differences in regulatory mechanisms underlying developmental and tumor angiogenesis. Genetic inactivation of many angiogenic growth factors and their corresponding receptors (e.g., VEGF, VEGFR1, VEGFR2, Ang1, Tie1, Tie2, ephrinB2, EphB4) leads to embryonic lethality, pointing to the rate-limiting importance of these molecules for developmental angiogenesis. Knockouts of other candidate angiogenic molecules have failed to demonstrate embryonic lethality and were thus widely ignored as angiogenesis factors throughout much of the 1990s. However, it is now clear that many such factors may play a selective role in pathological angiogenesis in the adult. For example, although placenta growth factor (PlGF)-deficient mice develop normally and are viable and fertile, adult $PlGF^{-/-}$ mice have impaired angiogenesis, plasma extravasation, and collateral growth during ischemia, inflammation, wound healing, and cancer (74). Similar discrepancies between developmental and adult angiogenesis have been established for regulatory molecules of the proteolytic balance during angiogenesis, for example, plasminogen activators (t-PA and u-PA) and PAI-I (53). It is widely recognized that specific (e.g., PlGF) and pleiotropic (e.g., fibrinolytic mediators, FGF) angiogenic growth factors that are not rate-limiting for developmental angiogenesis may nevertheless be important and even rate-limiting regulators of adult pathological angiogenesis.

INDUCTION AND INHIBITION OF BLOOD VESSEL GROWTH

The identification of molecules with angio-stimulatory and angio-inhibitory activity (see Table 157-3) has supported the concept that angiogenesis – like most biological processes – is controlled by a balance of positive and negative regulatory factors. The angiogenic balance dictates the net angiogenic phenotype, namely induction of angiogenesis or vessel regression (75). Most angiogenesis stimulators have a short half-life and act locally. In contrast, many angiogenesis inhibitors are long-lived and appear to act systemically. The concept of an angiogenic balance is attractive and plausible. Yet, few experiments to systematically manipulate this balance locally or systemically have been performed. Moreover, the molecular mechanisms of most angio-inhibitory molecules are still poorly understood.

Inhibitors of angiogenesis fall under two classes. The first includes a small number of molecules that act by inhibiting proangiogenic factors. For example, transmembrane EC surface receptors of angiogenic cytokines (e.g., VEGFRs and Tie2) transduce signals that allow angiogenically activated ECs to execute the angiogenic program. In contrast, soluble receptor derivatives (sVEGFR1 [sFLT-1] and sTie2) act as ligand traps that bind the corresponding growth factor and thereby inhibit angiogenesis (76). Similarly, Ang1 acts as a stimulating ligand of Tie2, whereas Ang2 acts as an inhibitor of Ang1/Tie2 signaling (37). The second, larger group of molecules comprises those endogenous angiogenesis inhibitors (see Table 157-3) that act independently of the angiogenesis-inducing molecules through poorly defined mechanisms. Examples of the latter include the collagen XVIII and plasminogen fragments endostatin and angiostatin, respectively (59,77).

The best example of a dynamically regulated angiogenic balance in physiology is found in the female reproductive system. Angiogenesis in the cyclic corpus luteum is initiated following ovulation by massive induction of VEGF, which shifts the balance in favor of proangiogenic activity (68). Conversely, angio-regression during luteolysis is initiated by a combination of VEGF downregulation and induction of Ang2 expression (78). Thus, physiological vessel regression in the corpus luteum is initiated by concomitant downregulation of the stimulator (VEGF) and an upregulation of the inhibitor (Ang2). The role of other endogenous angiogenesis inhibitors during reproductive angiogenesis has yet to be studied.

A shift of the angiogenic balance toward the induction of angiogenesis also occurs during tumor growth and progression. This *angiogenic switch* occurs during the transition from an avascular tumor toward an angiogenic tumor (79). The induction of angiogenesis becomes a prerequisite for the growth of the tumor if its size exceeds a diameter that allows nourishment of the tumor by diffusion. A malignant primary tumor or a small micrometastasis can persist for months or even years as dormant tumor as long as the tumor has not activated the angiogenic switch (80). The switch can result as a consequence of the increasing hypoxia within the growing avascular tumor or as a consequence of a specific angiogenic gene program activated by the tumor cells during tumor progression; for example, by specific mutations that affect angiogenesis-regulating oncogenes or tumor suppressor genes. The molecular mechanisms of the angiogenic switch during tumor progression are still poorly understood (75,79). This is largely due to the limited availability of suitable experimental models to study the angiogenic switch. The angiogenic switch does occur in *natural* tumor models (i.e., spontaneous carcinogenesis animal models, chemical carcinogenesis or genetic carcinogenesis models [e.g., RipTag]). Transplantation models with injected tumor cells are poorly suited for studying the angiogenic switch, because most established tumor cell lines are angiogenic and therefore have a committed angiogenesis phenotype.

QUANTITATIVE AND QUALITATIVE ASSESSMENT OF A NEOVASCULATURE

The demonstration of a correlation between an increase in microvessel density (MVD) and tumor progression by Weidner and colleagues in 1991 is the most frequently cited publication in the field of angiogenesis research (see Table 157-2) (81). This publication has inspired much research, and MVD counting studies have been performed for many different types of malignant and benign tumors (82). To date, more than 1,500 MVD studies have been published. Collectively, these studies have firmly established the relationship between angiogenesis and tumor progression. Yet, they also have led to the widely held misconception that MVD counts are equated with angiogenesis. MVD counts, which are typically based on immunohistochemical analyses of pan-endothelial markers (CD31, CD34, von Willebrand factor [vWF]), describe the average intercapillary distance in tumors, and are thus reflective of the angio-architectural properties of the tumor vasculature. This is an important parameter, because it is the goal of an antiangiogenic tumor therapy to reduce the intercapillary distance such that it becomes rate-limiting for the growth of the tumor.

In 1972, Brem and colleagues proposed a microscopic angiogenesis grading system (MAGS) to quantify the angiogenic status of tumors (83). The score, which incorporates measures of vascular density, number of EC nuclei, and cytological properties of tumor-associated ECs, was used to establish an angiogenic rank order in different human brain tumors (83). As an alternative approach, MVD counting of marker molecules that are specific to proliferative ECs (e.g., CD105 or VEGFR2) has been used to assess the angiogenic status of tumors. However, it is not clear whether the latter approach is an improvement over MVD counting based on pan-endothelial marker molecules. A number of investigators have measured EC proliferation by double-labeling immunohistochemical techniques as a kinetic parameter of the tumor neovasculature. In fact, given the extremely low turnover of quiescent ECs in normal tissues (months to years), EC proliferation may be the most reliable marker of active angiogenesis in a tumor (66). Other advanced microscopic techniques to study angiogenic blood vessels (e.g., fluorescence, confocal, multiphoton and electron microscopy, and casting techniques) have been extensively studied in experimental models for high resolution two-dimensional and three-dimensional assessment of angiogenic vascular networks, but have been used only sparingly in the analysis of human tumor neovasculature (84).

An alternative to pathology-based, invasive examination of the tumor vasculature is the measurement of circulating soluble or cellular biomarkers (*surrogate markers*) of angiogenesis (85). A number of activation-associated endothelial-derived molecules can be detected in the circulation either as specific soluble splice forms of corresponding EC surface receptors and adhesion molecules or as a consequence of proteolytic shedding of EC surface molecules (86). These include the soluble forms the angiogenic EC surface receptors VEGFR1 (sFLT-1), VEGFR2 (sKDR) and Tie2 (sTie2); and soluble forms of inflammatory markers, such as sE-selectin, soluble intercellular adhesion molecule (ICAM)-1, and soluble vascular adhesion molecule (VCAM)-1 (87). Similarly, circulating concentrations of the angiogenic factors VEGF and FGF2 are upregulated in many cancer patients (88).

An emerging novel biomarker of angiogenesis is the circulating EPC. Quantitative analysis of circulating EPCs may prove to be a useful readout for the intensity of local angiogenesis (89). Similarly, the transcriptome of peripheral blood cells – assayed by DNA microarrays – has been proposed as a surrogate biomarker for cancer, and may be useful for following the response to antiangiogenic therapies (90).

The widespread use of pathological specimens and the emerging use of surrogate marker assays to assess tumor angiogenesis are complemented by rapid developments in the field of noninvasive imaging (84). Positron emission tomography, x-ray computed tomography, magnetic resonance imaging, ultrasound, and optical imaging all are being extensively explored as imaging modalities in tumor angiogenesis. These techniques are capable not only of imaging the neovasculature, but also of functionally assessing the neovasculature (e.g., determination of blood flow, blood volume, empirical semiquantitative hemodynamic parameters, and quantitative hemodynamic parameters from pharmacokinetic modeling) (91).

THERAPEUTIC STRATEGIES TO INTERFERE WITH THE GROWTH OF BLOOD VESSELS

Translational prospects in the field of angiogenesis revolve around the therapeutic inhibition of abnormal vessel growth in several disease states, most notably cancer. The clinical approval of the anti-VEGF antibody Avastin (bevacizumab) for use in combination with chemotherapy for the first-line treatment of colorectal cancers in 2004 has established antiangiogenesis as the first therapy directed against the tumor stroma that complements established surgical, chemotherapeutic, and radiotherapeutic tumor approaches (92).

Tumor angiogenesis is a particularly attractive therapeutic target for a number of reasons: (a) angiogenesis is downregulated in normal healthy tissues, thus therapy is associated with minimal side effects even after prolonged treatment; (b) the tumor vasculature is not clonogenic, but rather is derived from normal blood vessels, and thus its pharmacological inhibition should not lead to the development of resistance; (c) each tumor capillary may supply hundreds of tumor cells, a ratio that provides for amplification of drug effect; and (d) in contrast to the interstitial location of tumor cells, direct contact of the vasculature to the circulation allows efficient access of therapeutic agents.

A number of approaches have been employed to inhibit pathological angiogenesis (93). Pharmacological inhibition of angiogenesis is aimed at interfering with the angiogenic cascade or the immature neovasculature. Pharmacological agents may be synthetic or semisynthetic substances, endogenous inhibitors of angiogenesis, or biological antagonists of the angiogenic cascade. In contrast to the direct inhibition of angiogenesis, vascular targeting is aimed at utilizing specific molecular determinants of the neovasculature for the delivery of a biological, chemical, or physical activity that will then locally act angiocidal or tumoricidal (94). A comprehensive website summarizing the status of tumor antiangiogenic compounds in various stages of clinical trials is maintained by the National Cancer Institute at: www.cancer.gov/clinicaltrials/developments/antiangio-table.

Clinical experience has established that antiangiogenic tumor therapies work most effectively in combination with chemotherapy or radiotherapy. Conceptually, this is somewhat surprising. An antiangiogenic therapy is, in principle, aimed at starving the tumor, which should lead to less effective delivery of a concomitantly given chemotherapeutic drug. The mechanisms for the synergy between antiangiogenesis and chemotherapy are not well understood. Yet, increasing evidence suggests that antiangiogenic tumor therapy *prunes* the immature and chaotic tumor vasculature to lead to less, but more normal appearing and functioning tumor blood vessels. This *normalization* effect is believed to allow more effective access of chemotherapeutic drugs to their respective therapeutic target (73). As surprising as the synergy between antiangiogenesis and chemotherapy is the observed synergy between antiangiogenesis and radiotherapy (95,96). Antiangiogenesis should, in principle, be a prohypoxic therapy. Yet, hypoxia is the primary cause for radioresistance of tumors. Thus, a tumor vessel *normalization* effect must similarly be responsible for the synergy between antiangiogenesis and radiotherapy.

Antiangiogenesis research has driven the field. Yet, a number of indications may benefit from an *induction* of angiogenesis, including wound healing, cardiac ischemia, and peripheral limb ischemia (97). Various approaches have been taken to therapeutically deliver angiogenic cytokines such as VEGF and FGF2. These include the local administration of recombinant proteins and gene therapeutic delivery of angiogenic cytokines. Individual cytokine therapy may have limitations, because it may be capable of inducing a neovascular response, but it may not induce the growth of a patent neovascular network that is stable for prolonged periods of time. This notion has led to alternative strategies aimed at inducing the complex endogenous angiogenic program and not just a single cytokine. For example, experiments are under way to locally induce hypoxia-inducible factor (HIF)-1, a key regulator of the hypoxia response program, which is able to control the complex endogenous program of angiogenesis induction (98).

KEY POINTS

- Angiogenesis is a fundamental developmental biological process that is critically associated with the growth of tumors and other important human diseases. Manipulation of angiogenesis (antiangiogenesis, therapeutic angiogenesis) holds the promise of novel therapeutic avenues for some of the most devastating human diseases.

- Angiogenesis is a physiological process of embryonic and postnatal development largely downregulated in the adult. As an oncofetal mechanism, it is associated with the growth of tumors, and can be considered one of the most important tumor progression and metastasis-enabling mechanism.
- The VEGF/VEGFR system acts as a hierarchically high master switch of the angiogenic cascade.
- As an intricately regulated process, angiogenesis is controlled by a balance of angio-stimulatory and angio-inhibitory molecules.
- Clinical translation of angiogenesis research is limited by the lack of appropriate angio-diagnostic tools and procedures and the limited availability of reliable biomarkers.
- The first antiangiogenic tumor therapy has received clinical approval. Others are to follow. Non-neoplastic indications of antiangiogenic and proangiogenic therapies are intensely explored in as diverse disciplines such as ophthalmology, dermatology, gynecology, and cardiology.

REFERENCES

1 Kidd KR, Weinstein BM. Fishing for novel angiogenic therapies. *Br J Pharmacol*. 2003;140(4):585–594.

2 Cleaver O, Tonissen KF, Saha MS, Krieg PA. Neovascularization of the Xenopus embryo. *Dev Dyn*. 1997;210(1):66–77.

3 Carmeliet P. Angiogenesis in health and disease. *Nat Med*. 2003; 9(6):653–660.

4 Folkman J. Angiogenesis in cancer, vascular, rheumatoid and other disease. *Nat Med*. 1995;1(1):27–31.

5 Djonov V, Baum O, Burri PH. Vascular remodeling by intussusceptive angiogenesis. *Cell Tissue Res*. 2003;314(1):107–117.

6 Risau W. Mechanisms of angiogenesis. *Nature*. 1997;386(6626): 671–674.

7 Carmeliet P. Mechanisms of angiogenesis and arteriogenesis. *Nat Med*. 2000;6(4):389–395.

8 Buschmann I, Schaper W. The pathophysiology of the collateral circulation (arteriogenesis). *J Pathol*. 2000;190(3):338–342.

9 Hoefer IE, Grundmann S, van Royen N, et al. Leukocyte subpopulations and arteriogenesis: specific role of monocytes, lymphocytes and granulocytes. *Atherosclerosis*. 2005;181(2):285–293.

10 Zhong TP, Childs S, Leu JP, Fishman MC. Gridlock signalling pathway fashions the first embryonic artery. *Nature*. 2001; 414(6860):216–220.

11 le Noble F, Moyon D, Pardanaud L, et al. Flow regulates arterial-venous differentiation in the chick embryo yolk sac. *Development*. 2004;131(2):361–375.

12 Risau W, Flamme I. Vasculogenesis. *Ann Rev Cell Dev Biol*. 1995; 11:73–91.

13 Choi K, Kennedy M, Kazarov A, et al. A common precursor for hematopoietic and endothelial cells. *Development*. 1998; 125(4):725–732.

14 Asahara T, Murohara T, Sullivan A, et al. Isolation of putative progenitor endothelial cells for angiogenesis. *Science*. 1997; 275(5302):964–967.

15 Rafii S, Lyden D. Therapeutic stem and progenitor cell transplantation for organ vascularization and regeneration. *Nat Med*. 2003;9:702–712.

16 Hunter J. Lectures on the principles of surgery. In: Palmer J, ed. *The Works of John Hunter*. London, 1835.

17 Herting AT. Angiogenesis in the early human chorio and in primary placenta of the macaque monkey. *Contrib Embryol*. 1935; 25:37–91.

18 Folkman J. Tumor angiogenesis: therapeutic implications. *N Engl J Med*. 1971;285:1182–1186.

19 Pezzella F, Pastorino U, Tagliabue E, et al. Non-small-cell lung carcinoma tumor growth without morphological evidence of neo-angiogenesis. *Am J Pathol*. 1997;151:1417–1423.

20 Holash J, Maisonpierre PC, Compton D, et al. Vessel cooption, regression, and growth in tumors mediated by angiopoietins and VEGF. *Science*. 1999;284:1994–1998.

21 Klagsbrun M, D'Amore PA. Regulators of angiogenesis. *Annu Rev Physiol*. 1991;53:217–239.

22 Shing Y, Folkman J, Sullivan R, et al. Heparin affinity: purification of a tumor-derived capillary endothelial cell growth factor. *Science*. 1984;223:1296–1299.

23 Keck PJ, Hauser SD, Krivi G, et al. Vascular permeability factor, an endothelial cell mitogen related to PDGF. *Science*. 1989;246: 1309–1312.

24 Leung DW, Cachianes G, Kuang WJ, et al. Vascular endothelial growth factor is a secreted angiogenic mitogen. *Science*. 1989; 246:1306–1309.

25 Plouet J, Schilling J, Gospodarowicz D. Isolation and characterization of a newly identified endothelial cell mitogen produced by AtT-20 cells. *EMBO J*. 1989;8:3801–3806.

26 Ferrara N, Davis Smyth T. The biology of vascular endothelial growth factor. *Endocr Rev*. 1997;18:4–25.

27 Senger DR, Galli SJ, Dvorak AM, et al. Tumor cells secrete a vascular permeability factor that promotes accumulation of ascites fluid. *Science*. 1983;219:983–985.

28 Fong GH, Rossant J, Gertsenstein M, Breitman ML. Role of the Flt-1 receptor tyrosine kinase in regulating the assembly of vascular endothelium. *Nature*. 1995;376:66–70.

29 Shalaby F, Rossant J, Yamaguchi TP, et al. Failure of blood-island formation and vasculogenesis in Flk-1-deficient mice. *Nature*. 1995;376:62–66.

30 Carmeliet P, Ferreira V, Breier G, et al. Abnormal blood vessel development and lethality in embryos lacking a single VEGF allele. *Nature*. 1996;380:435–439.

31 Ferrara N, Carver Moore K, Chen H, et al. Heterozygous embryonic lethality induced by targeted inactivation of the VEGF gene. *Nature*. 1996;380:439–442.

32 Plate KH, Breier G, Weich HA, Risau W. Vascular endothelial growth factor is a potential tumour angiogenesis factor in human gliomas in vivo. *Nature*. 1992;359:845–848.

33 Kim KJ, Li B, Winer J, et al. Inhibition of vascular endothelial growth factor-induced angiogenesis suppresses tumour growth in vivo. *Nature*. 1993;362:841–844.

34 Millauer B, Wizigmann Voos S, Schnurch H, et al. High affinity VEGF binding and developmental expression suggest Flk-1 as a major regulator of vasculogenesis and angiogenesis. *Cell*. 1993;72:835–846.

35 Shweiki D, Itin A, Soffer D, Keshet E. Vascular endothelial growth factor induced by hypoxia may mediate hypoxia-initiated angiogenesis. *Nature*. 1992;359:843–845.
36 Epstein AC, Gleadle JM, McNeill LA, et al. C. elegans EGL-9 and mammalian homologs define a family of dioxygenases that regulate HIF by prolyl hydroxylation. *Cell*. 2001;107:43–54.
37 Thurston G. Role of Angiopoietins and Tie receptor tyrosine kinases in angiogenesis and lymphangiogenesis. *Cell Tissue Res*. 2003;314:61–68.
38 Suri C, Jones PF, Patan S, et al. Requisite role of angiopoietin-1, a ligand for the TIE2 receptor, during embryonic angiogenesis. *Cell*. 1996;87:1171–1180.
39 Wong AL, Haroon ZA, Werner S, et al. Tie2 expression and phosphorylation in angiogenic and quiescent adult tissues. *Circ Res*. 1997;81:567–574.
40 Maisonpierre PC, Suri C, Jones PF, et al. Angiopoietin-2, a natural antagonist for Tie2 that disrupts in vivo angiogenesis. *Science*. 1997;277:55–60.
41 Fiedler U, Scharpfenecker M, Koidl S, et al. The Tie-2 ligand Angiopoietin-2 is stored in and rapidly released upon stimulation from endothelial cell Weibel-Palade bodies. *Blood*. 2004;103:4150–4156.
42 Scharpfenecker M, Fiedler U, Reiss Y, Augustin HG. The Tie-2 ligand Angiopoietin-2 destabilizes quiescent endothelium through an internal autocrine loop mechanism. *J Cell Sci*. 2005;118:771–780.
43 Hanahan D. Signaling vascular morphogenesis and maintenance. *Science*. 1997;277:48–50.
44 Augustin HG, Reiss Y. EphB receptors and ephrinB ligands: regulators of vascular assembly and homeostasis. *Cell Tissue Res*. 2003;314:25–31.
45 Adams RH. Vascular patterning by Eph receptor tyrosine kinases and ephrins. *Semin Cell Dev Biol*. 2002;13:55–60.
46 Lawson ND, Weinstein BM. Arteries and veins: making a difference with zebrafish. *Nat Rev Genet*. 2002;3:674–682.
47 Carmeliet P, Tessier-Lavigne M. Common mechanisms of nerve and blood vessel wiring. *Nature*. 2005;436:193–200.
48 St Croix B, Rago C, Velculescu V, et al. Genes expressed in human tumor endothelium. *Science*. 2000;289:1197–1202.
49 Brooks PC, Clark RA, Cheresh DA. Requirement of vascular integrin alpha v beta 3 for angiogenesis. *Science*. 1994;264:569–571.
50 Hynes RO. A reevaluation of integrins as regulators of angiogenesis. *Nat Med*. 2002;8:918–921.
51 Bazzoni G, Dejana E. Endothelial cell-to-cell junctions: molecular organization and role in vascular homeostasis. *Physiol Rev*. 2004;84:869–901.
52 Gerhardt H, Golding M, Fruttiger M, et al. VEGF guides angiogenic sprouting utilizing endothelial tip cell filopodia. *J Cell Biol*. 2003;161:1163–1177.
53 Noel A, Maillard C, Rocks N, et al. Membrane associated proteases and their inhibitors in tumour angiogenesis. *J Clin Pathol*. 2004;57:577–584.
54 Ziche M, Donnini S, Morbidelli L. Development of new drugs in angiogenesis. *Curr Drug Targets*. 2004;5:485–493.
55 Brem H, Folkman J. Inhibition of tumor angiogenesis mediated by cartilage. *J Exp Med*. 1975;141:427–439.
56 Lawler JW, Slayter HS, Coligan JE. Isolation and characterization of a high molecular weight glycoprotein from human blood platelets. *J Biol Chem*. 1978;253:8609–8616.
57 Good DJ, Polverini PJ, Rastinejad F, et al. A tumor suppressor-dependent inhibitor of angiogenesis is immunologically and functionally indistinguishable from a fragment of thrombospondin. *Proc Natl Acad Sci USA*. 1990;87:6624–6628.
58 O'Reilly MS, Holmgren L, Shing Y, et al. Angiostatin: a novel angiogenesis inhibitor that mediates the suppression of metastases by a Lewis lung carcinoma. *Cell*. 1994;79:315–328.
59 O'Reilly MS, Boehm T, Shing Y, et al. Endostatin: an endogenous inhibitor of angiogenesis and tumor growth. *Cell*. 1997;88:277–285.
60 Nyberg P, Xie L, Kalluri R. Endogenous inhibitors of angiogenesis. *Cancer Res*. 2005;65:3967–3979.
61 Sund M, Hamano Y, Sugimoto H, et al. Function of endogenous inhibitors of angiogenesis as endothelium-specific tumor suppressors. *Proc Natl Acad Sci USA*. 2005;102:2934–2939.
62 Adams JC, Lawler J. The thrombospondins. *Int J Biochem Cell Biol*. 2004;36(6):961–968.
63 LeCouter J, Lin R, Ferrara N. Endocrine gland-derived VEGF and the emerging hypothesis of organ-specific regulation of angiogenesis. *Nat Med*. 2002;8:913–917.
64 Tachibana K, Hirota S, Iizasa H, et al. The chemokine receptor CXCR4 is essential for vascularization of the gastrointestinal tract. *Nature*. 1998;393:591–594.
65 Risau W. Differentiation of endothelium. *FASEB J*. 1995;9:926–933.
66 Eberhard A, Kahlert S, Goede V, et al. Heterogeneity of angiogenesis and blood vessel maturation in human tumors: implications for antiangiogenic tumor therapies. *Cancer Res*. 2000;60:1388–1393.
67 Kozian DH, Ziche M, Augustin HG. The activin-binding protein follistatin regulates autocrine endothelial cell activity and induces angiogenesis. *Lab Invest*. 1997;76:267–276.
68 Augustin HG. Vascular morphogenesis in the ovary. *Baillieres Best Pract Res Clin Obstet Gynaecol*. 2000;14:867–882.
69 Ferrara N, Chen H, Davis Smyth T, et al. Vascular endothelial growth factor is essential for corpus luteum angiogenesis. *Nat Med*. 1998;4:336–340.
70 Carmeliet P, Jain RK. Angiogenesis in cancer and other diseases. *Nature*. 2000;407:249–257.
71 Mulcahy MF, Benson AB 3rd. Bevacizumab in the treatment of colorectal cancer. *Expert Opin Biol Ther*. 2005;5:997–1005.
72 Jackson JR, Seed MP, Kircher CH, et al. The codependence of angiogenesis and chronic inflammation. *FASEB J*. 1997;11:457–465.
73 Jain RK. Normalization of tumor vasculature: an emerging concept in antiangiogenic therapy. *Science*. 2005;307:58–62.
74 Carmeliet P, Moons L, Luttun A, et al. Synergism between vascular endothelial growth factor and placental growth factor contributes to angiogenesis and plasma extravasation in pathological conditions. *Nat Med*. 2001;7:575–583.
75 Hanahan D, Folkman J. Patterns and emerging mechanisms of the angiogenic switch during tumorigenesis. *Cell*. 1996;86:353–364.
76 Aiello LP, Pierce EA, Foley ED, et al. Suppression of retinal neovascularization in vivo by inhibition of vascular endothelial growth factor (VEGF) using soluble VEGF-receptor chimeric proteins. *Proc Natl Acad Sci USA*. 1995;92:10457–10461.
77 MS OR, Holmgren L, Shing Y, et al. Angiostatin: a novel angiogenesis inhibitor that mediates the suppression of metastases by a Lewis lung carcinoma. *Cell*. 1994;79:315–328.

78 Goede V, Schmidt T, Kimmina S, et al. Analysis of blood vessel maturation processes during cyclic ovarian angiogenesis. *Lab Invest*. 1998;78:1385–1394.

79 Bergers G, Benjamin LE. Tumorigenesis and the angiogenic switch. *Nat Rev Cancer*. 2003;3:401–410.

80 Udagawa T, Fernandez A, Achilles EG, et al. Persistence of microscopic human cancers in mice: alterations in the angiogenic balance accompanies loss of tumor dormancy. *FASEB J*. 2002;16:1361–1370.

81 Weidner N, Semple JP, Welch WR, Folkman J. Tumor angiogenesis and metastasis – correlation in invasive breast carcinoma. *N Engl J Med*. 1991;324:1–8.

82 Hlatky L, Hahnfeldt P, Folkman J. Clinical application of antiangiogenic therapy: microvessel density, what it does and doesn't tell us. *J Natl Cancer Inst*. 2002;94:883–893.

83 Brem S, Cotran R, Folkman J. Tumor angiogenesis: a quantitative method for histologic grading. *J Natl Cancer Inst*. 1972;48: 347–356.

84 McDonald DM, Choyke PL. Imaging of angiogenesis: from microscope to clinic. *Nat Med*. 2003;9:713–725.

85 Davis DW, McConkey DJ, Abbruzzese JL, Herbst RS. Surrogate markers in antiangiogenesis clinical trials. *Br J Cancer*. 2003; 89:8–14.

86 Drevs J. Soluble markers for the detection of hypoxia under antiangiogenic treatment. *Anticancer Res*. 2003;23:1159–1161.

87 Byrne GJ, Ghellal A, Iddon J, et al. Serum soluble vascular cell adhesion molecule-1: role as a surrogate marker of angiogenesis. *J Natl Cancer Inst*. 2000;92:1329–1336.

88 Gasparini G. Clinical significance of determination of surrogate markers of angiogenesis in breast cancer. *Crit Rev Oncol Hematol*. 2001;37:47–114.

89 Monestiroli S, Mancuso P, Burlini A, et al. Kinetics and viability of circulating endothelial cells as surrogate angiogenesis marker in an animal model of human lymphoma. *Cancer Res*. 2001;61:4341–4344.

90 DePrimo SE, Wong LM, Khatry DB, et al. Expression profiling of blood samples from an SU5416 Phase III metastatic colorectal cancer clinical trial: a novel strategy for biomarker identification. *BMC Cancer*. 2003;3:3.

91 Miller JC, Pien HH, Sahani D, et al. Imaging angiogenesis: applications and potential for drug development. *J Natl Cancer Inst*. 2005;97:172–187.

92 Ferrara N, Hillan KJ, Gerber HP, Novotny W. Discovery and development of bevacizumab, an anti-VEGF antibody for treating cancer. *Nat Rev Drug Discov*. 2004;3:391–400.

93 Scappaticci FA. Mechanisms and future directions for angiogenesis-based cancer therapies. *J Clin Oncol*. 2002;20: 3906–3927.

94 Neri D, Bicknell R. Tumour vascular targeting. *Nat Rev Cancer*. 2005;5:436–446.

95 Zips D, Eicheler W, Geyer P, et al. Enhanced susceptibility of irradiated tumor vessels to vascular endothelial growth factor receptor tyrosine kinase inhibition. *Cancer Res*. 2005;65:5374–5379.

96 Garcia-Barros M, Paris F, Cordon-Cardo C, et al. Tumor response to radiotherapy regulated by endothelial cell apoptosis. *Science*. 2003;300:1155–1159.

97 Simons M, Ware JA. Therapeutic angiogenesis in cardiovascular disease. *Nat Rev Drug Discov*. 2003;2:863–871.

98 Pugh CW, Ratcliffe PJ. Regulation of angiogenesis by hypoxia: role of the HIF system. *Nat Med*. 2003;9:677–684.

99 Roberts AB, Sporn MB, Assoian RK, et al. Transforming growth factor type beta: rapid induction of fibrosis and angiogenesis in vivo and stimulation of collagen formation in vitro. *Proc Natl Acad Sci USA*. 1986;83:4167–4171.

100 Brooks PC, Montgomery AM, Rosenfeld M, et al. Integrin alpha v beta 3 antagonists promote tumor regression by inducing apoptosis of angiogenic blood vessels. *Cell*. 1994;79:1157–1164.

101 Tsujii M, Kawano S, Tsuji S, et al. Cyclooxygenase regulates angiogenesis induced by colon cancer cells. *Cell*. 1998;93:705–716.

102 Ferrara N, Gerber HP, LeCouter J. The biology of VEGF and its receptors. *Nat Med*. 2003;9:669–676.

103 Javerzat S, Auguste P, Bikfalvi A. The role of fibroblast growth factors in vascular development. *Trends Mol Med*. 2002;8: 483.

104 Hellstrom M, Kalen M, Lindahl P, et al. Role of PDGF-B and PDGFR-beta in recruitment of vascular smooth muscle cells and pericytes during embryonic blood vessel formation in the mouse. *Development*. 1999;126:3047–3055.

105 Lecouter J, Lin R, Ferrara N. EG-VEGF: a novel mediator of endocrine-specific angiogenesis, endothelial phenotype, and function. *Ann NY Acad Sci*. 2004;1014:50–57.

106 Larsson J, Goumans MJ, Sjostrand LJ, et al. Abnormal angiogenesis but intact hematopoietic potential in TGF-beta type I receptor-deficient mice. *EMBO J*. 2001;20:1663–1673.

107 Vinals F, Pouyssegur J. Transforming growth factor beta1 (TGF-beta1) promotes endothelial cell survival during in vitro angiogenesis via an autocrine mechanism implicating TGF-alpha signaling. *Mol Cell Biol*. 2001;21:7218–7230.

108 Rosen EM, Goldberg ID. Scatter factor and angiogenesis. *Adv Cancer Res*. 1995;67:257–279.

109 Kondo T, Vicent D, Suzuma K, et al. Knockout of insulin and IGF-1 receptors on vascular endothelial cells protects against retinal neovascularization. *J Clin Invest*. 2003;111:1835–1842.

110 Pike SE, Yao L, Jones KD, et al. Vasostatin, a calreticulin fragment, inhibits angiogenesis and suppresses tumor growth. *J Exp Med*. 1998;188:2349–2356.

111 Kamphaus GD, Colorado PC, Panka DJ, et al. Canstatin, a novel matrix-derived inhibitor of angiogenesis and tumor growth. *J Biol Chem*. 2000;275:1209–1215.

112 Mongiat M, Sweeney SM, San Antonio JD, et al. Endorepellin, a novel inhibitor of angiogenesis derived from the C terminus of perlecan. *J Biol Chem*. 2003;278:4238–4249.

113 O'Reilly MS, Pirie-Shepherd S, Lane WS, Folkman J. Antiangiogenic activity of the cleaved conformation of the serpin antithrombin. *Science*. 1999;285:1926–1928.

114 Staton CA, Brown NJ, Rodgers GR, et al. Alphastatin, a 24-amino acid fragment of human fibrinogen, is a potent new inhibitor of activated endothelial cells in vitro and in vivo. *Blood*. 2004;103:601–606.

115 Carmeliet P, Mackman N, Moons L, et al. Role of tissue factor in embryonic blood vessel development. *Nature*. 1996;383: 73–75.

116 Tsopanoglou NE, Maragoudakis ME. Role of thrombin in angiogenesis and tumor progression. *Semin Thromb Hemost*. 2004;30:63–69.

117 Armstrong LC, Bornstein P. Thrombospondins 1 and 2 function as inhibitors of angiogenesis. *Matrix Biol.* 2003;22:63–71.
118 Conrotto P, Valdembri D, Corso S, et al. Sema4D induces angiogenesis through Met recruitment by Plexin B1. *Blood.* 2005;105:4321–4329.
119 Serini G, Valdembri D, Zanivan S, et al. Class 3 semaphorins control vascular morphogenesis by inhibiting integrin function. *Nature.* 2003;424:391–397.
120 Bielenberg DR, Hida Y, Shimizu A, et al. Semaphorin 3F, a chemorepulsant for endothelial cells, induces a poorly vascularized, encapsulated, nonmetastatic tumor phenotype. *J Clin Invest.* 2004;114:1260–1271.
121 Fajardo LF, Kwan HH, Kowalski J, et al. Dual role of tumor necrosis factor-alpha in angiogenesis. *Am J Pathol.* 1992;140: 539–544.
122 Bikfalvi A. Platelet factor 4: an inhibitor of angiogenesis. *Semin Thromb Hemost.* 2004;30:379–385.
123 Angiolillo AL, Sgadari C, Taub DD, et al. Human interferon-inducible protein 10 is a potent inhibitor of angiogenesis in vivo. *J Exp Med.* 1995;182:155–162.
124 Cao Y, Chen C, Weatherbee JA, et al. gro-beta, a -C-X-C-chemokine, is an angiogenesis inhibitor that suppresses the growth of Lewis lung carcinoma in mice. *J Exp Med.* 1995;182: 2069–2077.
125 Brown NS, Bicknell R. Thymidine phosphorylase, 2-deoxy-D-ribose and angiogenesis. *Biochem J.* 1998;334:1–8.
126 Kishimoto K, Liu S, Tsuji T, et al. Endogenous angiogenin in endothelial cells is a general requirement for cell proliferation and angiogenesis. *Oncogene.* 2005;24:445–456.
127 Stetler-Stevenson WG. Matrix metalloproteinases in angiogenesis: a moving target for therapeutic intervention. *J Clin Invest.* 1999;103:1237–1241.
128 Rubanyi GM, Johns A, Kauser K. Effect of estrogen on endothelial function and angiogenesis. *Vascul Pharmacol.* 2002;38:89–98.
129 Amano H, Hayashi I, Endo H, et al. Host prostaglandin E(2)-EP3 signaling regulates tumor-associated angiogenesis and tumor growth. *J Exp Med.* 2003;197:221–232.
130 Jackson D, Volpert OV, Bouck N, Linzer DI. Stimulation and inhibition of angiogenesis by placental proliferin and proliferin-related protein. *Science.* 1994;266:1581–1584.
131 Klauber N, Parangi S, Flynn E, et al. Inhibition of angiogenesis and breast cancer in mice by the microtubule inhibitors 2-methoxyestradiol and taxol. *Cancer Res.* 1997;57:81–86.
132 Rooney P, Kumar S, Ponting J, Wang M. The role of hyaluronan in tumour neovascularization. *Int J Cancer.* 1995;60:632–636.
133 Ziche M, Alessandri G, Gullino PM. Gangliosides promote the angiogenic response. *Lab Invest.* 1989;61:629–634.
134 Ribatti D, Presta M, Vacca A, et al. Human erythropoietin induces a pro-angiogenic phenotype in cultured endothelial cells and stimulates neovascularization in vivo. *Blood.* 1999;93: 2627–2636.
135 Minamino K, Adachi Y, Okigaki M, et al. Macrophage colony-stimulating factor (M-CSF), as well as granulocyte colony-stimulating factor (G-CSF), accelerates neovascularization. *Stem Cells.* 2005;23:347–354.
136 Maxwell PH, Wiesener MS, Chang GW, et al. The tumor suppressor protein VHL targets hypoxia-inducible factors for oxygen-dependent proteolysis. *Nature.* 1999;399:271–275.
137 Koch AE, Polyerini PJ, Kunkel SL, et al. Interleukin-8 as a macrophage-derived mediator of angiogenesis. *Science.* 1992; 258:1798–1801.
138 Folkman J, Watson K, Ingber, et al. Induction of angiogenesis during the transition from hyperplasia to neoplasia. *Nature.* 1989;339:58–61.
139 Aiello LP, Avery RL, Arrigg PG, et al. Vascular endothelial growth factor in ocular fluid of patient with diabetic retinopathy and other retinal disorders. *N Engl J Med.* 1994;331:1480–1487.

158

Tumor Blood Vessels

Harold F. Dvorak

Beth Israel Deaconess Medical Center, Harvard Medical School, Boston, Massachusetts

It has been known for more than a century that tumors have their own blood supply and, for the better part of that time, that the tumor vasculature is highly abnormal, differing from that of normal tissues with respect to organization, structure, and function (1). At one time it was thought that tumors had a vasculature that was superior to that of normal tissues; this misconception arose because tumor vessels are often of large size and were therefore more readily visualized by angiography and macroscopy than the smaller but more numerous and functionally superior capillaries of normal tissues. By the late 1970s, it also was clear that tumor vessels were heterogeneous, hyperpermeable to plasma and plasma proteins, and that tumor blood flow was unevenly distributed and, overall, significantly lower than that in normal tissues. Further, it was known that tumor vessels were induced by tumor-secreted products, although the tumor angiogenic factor(s) responsible were just beginning to be investigated. In the years that followed, much has been learned about the molecules responsible for angiogenesis, particularly vascular endothelial growth factor (VEGF)-A, and about the steps and mechanisms by which tumors induce the new vascular supply that they require to grow beyond minimal size. Recent successes with blocking agents against VEGF-A suggest that antiangiogenesis may provide a valuable new approach to tumor therapy (2). This chapter reviews the properties of tumor blood vessels, their differences from normal vessels, and the steps and mechanisms by which they form.

THE NORMAL MICROVASCULATURE AND STROMA

Before discussing tumors, it is important to review the structure of normal tissues to provide a standard of comparison. In general, normal tissues are comprised of two discrete but interdependent compartments, the parenchyma (generally epithelium) and stroma (vascular connective tissue). Stroma includes the microvasculature but is also comprised of extracellular fluid, fibroblasts, and fibroblast products that include interstitial collagens, proteoglycans, hyaluronan, and other compounds. Stroma is required to maintain tissue structure and the microvascular component to provide gas exchange, nutrition, waste disposal, and other functions.

In most normal tissues, the microvasculature consists of three types of blood vessel: arterioles, capillaries, and venules, all linked in series (3). Arterial blood enters arterioles, then capillaries, which flow into postcapillary venules and, subsequently, into veins. In some tissues (e.g., skin), blood can bypass capillaries by way of arteriovenous shunts. Although part of a continuum with some degree of overlap, each of the three major types of normal microvessel has a characteristic structure and function. Arterioles are 10- to 20-μm diameter vessels whose lining endothelium is covered with one or more layers of smooth muscle cells. Arteriolar tone is regulated by autonomic, generally sympathetic, nerves that modulate vascular smooth muscle cell contraction and, in this way, regulate blood pressure and flow. Smooth muscle relaxation is regulated in part by nitrogen oxide secreted by endothelium.

Capillaries are smaller vessels, 4 to 9 μm in diameter, which are lined by a thin, flattened but continuous layer of endothelium and are enveloped by basement membrane and a variable coating of pericytes. In some tissues (e.g., kidneys, endocrine glands) the endothelium is fenestrated. Fenestrae are 50- to 150-nm diameter zones of extreme endothelial cell (EC) thinning that, in most cases, are closed by diaphragms. Capillaries are normally spaced at intervals of no greater than 100 to 200 μm – distances that correspond to the diffusion range of oxygen (O_2). This is important because capillaries are the principal "exchange vessels" and are engaged in tissue nutrition and waste disposal. Plasma solute and its low-molecular-weight constituents (i.e., O_2, glucose, and salts) pass freely from proximal portions of capillaries, largely, it is thought, by processes of convection and diffusion. Distally, these same processes lead to reabsorption of interstitial fluid and solute, including metabolic waste products such as carbon dioxide (CO_2). Capillary endothelium is the principal barrier to molecular exchange between blood and underlying tissue, and two pathways have been implicated: paracellular (through inter-EC junctions) and transcellular (via caveolae and fenestrae). Small molecules below the size of plasma proteins can use the former pathway; however, plasma proteins are too large to pass through EC junctions in normal tissues and extravasate almost exclusively by caveolae (4). Caveolae are approximately 70-nm diameter membrane-bound vesicles that shuttle across EC cytoplasm from the vascular lumen to the ablumen where they discharge their cargo of plasma protein-rich solute into the tissues.

Venules are larger than capillaries, typically 20 μm or greater, and are lined by a taller endothelium, basement membrane, and smooth muscle cells. They play a lesser role in metabolite transport but are the key segment of the microvasculature that is activated in both humoral and cellular inflammation. Majno demonstrated many years ago that venules are the primary site of the solute and plasma protein leakage induced by inflammatory mediators such as histamine and serotonin (5). He also proposed a mechanism for inflammatory venular hyperpermeability, namely, that inflammatory mediators caused venular ECs to pull apart, generating inter-EC gaps through which fluid and macromolecules could extravasate (6). However, more recent studies have provided strong evidence that many of the endothelial openings induced by inflammatory mediators in endothelium are transcellular, not intercellular (7). Also, a structure has been identified in venular endothelium, the vesiculo-vacuolar organelle or VVO, that likely accounts for much of the transendothelial plasma flux that occurs in response to VEGF-A and other mediators (8,9) (Chapters 71 and 74). VVOs are grape-like clusters of uncoated, largely parajunctional cytoplasmic vesicles and vacuoles that together form an entity that traverses endothelial cytoplasm from lumen to ablumen (10). The individual vesicles and vacuoles that comprise VVOs share many properties with caveolae and are linked to each other and to the luminal and abluminal plasma membranes by stomata that are normally closed by thin diaphragms resembling those that close fenestrae. VEGF-A, histamine, and other inflammatory mediators are thought to cause these diaphragms to open, providing a transcellular pathway for plasma and plasma protein extravasation.

Leukocytes also leave the vasculature by traversing venules. As with the passage of fluid, it was originally thought that leukocytes crossed venules through opened inter-EC junctions. However, more recent structural studies, making use of serial sections and three-dimensional reconstructions, have shown that granulocytes and monocytes can pass through ECs, possibly by way of VVOs, independent of intercellular junctions (11,12). More work is required to sort out the relative importance of paracellular versus transcellular pathways for both fluid and inflammatory cell extravasation. The problem is a difficult one because inter-EC junctions are complicated, irregular structures. Also, VVOs are concentrated parajunctionally, and individual VVO vesicles may open to the intercellular cleft above and below specialized junctions, as well as to the luminal and abluminal surfaces (Chapter 71).

THE TUMOR VASCULATURE AND STROMA

As in normal tissues, solid tumors are comprised of parenchyma (the malignant cells) and a vascularized stroma, which they induce and in which they are dispersed. All tumors, regardless of their site of origin, require stroma if they are to grow beyond minimal size, and for the same reasons that normal tissues require stroma – maintenance of structure, nourishment, gas exchange, and waste disposal. The numbers of new blood vessels and the amount of connective tissue stroma vary considerably among different tumors, and, in general, are not closely linked to the degree of tumor malignancy. Highly malignant tumors may have much or little vascular stroma, but all growing tumors require at least some. Even so-called "liquid" tumors have stroma. Thus, ascites tumors induce a vascularized connective tissue stroma resembling that of solid tumors in the tissues lining the peritoneal cavity, and the tumor cells themselves are suspended in ascites fluid, a plasma protein–rich exudate resulting from the increased permeability of peritoneal lining vessels (13–15). In the case of leukemias, the circulating blood itself serves as stroma but, in addition, vascularized stroma also may be induced in tumor-involved bone marrow (16–18).

Superficially, tumor stroma resembles normal connective tissue and includes, in addition to blood vessels, fibroblasts and their products such as collagen and proteoglycans. However, upon closer inspection, the composition of tumor stroma differs strikingly from that of normal tissues (19,20). Differences include a significantly higher content of plasma protein–rich interstitial fluid, deposits of extravascular fibrin and its breakdown products, matrix proteins such as alternatively spliced fibronectins and tenascin that are normally found only in fetal development, abnormal proteoglycans (21) and, the subject of this chapter, highly abnormal blood vessels (19,22).

Organization of Tumor Blood Vessels

Tumors, like normal tissues, are supplied by what are believed to be normal arteries and are drained by apparently normal veins. The new blood vessels that tumors induce, therefore, correspond primarily to that portion of the vascular tree that is situated between arteries and veins – that portion that, in normal tissues, corresponds to the microvasculature. The types of new blood vessels induced by individual tumors, and the patterns they form, differ strikingly from one another. However, the vascular pattern of a given tumor remains relatively constant even after repeated transplantation and metastatic spread. Thus, as Warren observed, the vascular morphology of a specific tumor is characteristic for that tumor but may not be unique to that tumor (1).

Whereas normal microvessels are arranged in a hierarchy of evenly spaced, well-differentiated arterioles, capillaries, and venules, tumor microvessels follow a chaotic pattern and are hierarchically disorganized (1). A characteristic feature is spatial heterogeneity (uneven distribution), and the frequency of localized zones of increased vascular density has been used as a predictor of clinical outcome (23). One thought is that these microvascular "hot spots" provide a point of entry of cancer cells into the blood to form metastases; by contrast, tumor cells only rarely invade muscle-coated arteries or arterioles.

Tumor vessels often exhibit a serpentine course, branch irregularly, and form arteriovenous shunts. In many cases, vessels are most abundant at the tumor–host interface, where they are often distributed circumferentially, forming a prominent mantle around nodules of tumor cells. Internal portions of tumors are typically less well vascularized. Blood flow through the tumor vasculature does not follow a consistent, unidirectional path. Rather, the tumor vasculature is organized as a maze of interconnected vessels through which blood flows haphazardly and irregularly. Not all open vessels are perfused continuously, and, over intervals of even a few minutes, blood flow may follow different paths and even proceed in alternating directions through the same vessel (24).

The tumor vasculature is also heterogeneous over longer periods of time (1). Morphometric studies have shown that vascular volume, length, and surface area all increase during early tumor growth. As a result, small tumors tend to be relatively well vascularized. Later, however, vascular growth slows and the tumor blood supply becomes progressively deficient, leading to central ischemia and necrosis. Thus, tumors are said to outgrow their blood supply or, put in other words, the developing vasculature and other stromal elements fail to keep pace with tumor cell growth.

Structure of Tumor Blood Vessels

Tumor blood vessels are a caricature of their normal counterparts. They do not conform to the hierarchical pattern of normal vascular beds and are not easily classified. Writing in the late 1970s, Warren (1) described nine distinct types of tumor vessel (Table 158-1). Although this classification has been superseded, it is of historical interest because it represents the first serious attempt to classify tumor vessels. It is also worth presenting here because Warren's classification was only published in monographs that are now long out of print. Not all the vessel types that Warren described were found in the same tumor; rather, his list represents a composite of vessel types found in a survey of a large number of different human and animal tumor types. Warren concluded that the main difference between the tumor and normal vasculature was that, in the former, capillaries and veins became tortuous and dilated. In fact, however, the commonly expressed idea that tumor vessels are "dilated" is a misconception. Correctly understood, dilatation results from relaxation of vascular smooth muscle cells, as occurs, for example, in normal arterioles in response to nitric oxide. Tumor vessels characteristically have a paucity of smooth muscle cells, and their average enlarged size reflects intrinsic properties of vascular development, not muscle relaxation.

Table 158-1: The Tumor Microcirculation According to Warren

Vessel Type	*Vessel Properties*
Arteries, arterioles	Resemble those of normal tissues but may have reduced innervation and vasomotion
Capillaries	Resemble normal capillaries as far as is known, i.e., have intact and complete EC lining, intercellular junctions, basement membrane, etc.
Capillary sprouts	Blind-ended, growing branches of existing capillaries
Sinusoidal vessels	Large vessels with nonparallel EC lining, occasional pericytes
Blood channels with no or incomplete EC lining	Blood percolates in channels lined wholly or in part by tumor cells. Found in melanomas and some sarcomas
Giant capillaries	Large (up to 50 μm diameter) vessels lined only by ECs with some fibrous supporting tissue; commonly tortuous and describe arcs around the expanding tumor
Capillaries with fenestrated endothelium	Common in endocrine tumors but also found in many other types of tumor
Arteriovenous anastomoses	Similar to those found in normal tissues where arterioles flow directly into venules or small veins

Adapted from Warren B. The vascular morphology of tumors. In Peterson H-I, ed. *Tumor Blood Circulation: Angiogenesis, Vascular morphology and Blood Flow of Experimental and Human Tumors.* Boca Raton: CRC Press; 1979:1–47.

Tumor Blood Flow

For some years it was thought that tumors had a more extensive blood supply than that of normal tissues (25). However, in now classic studies, Gullino (20) demonstrated that blood perfusion rates in animal tumors were generally much lower than those of the normal tissues from which they arose. Moreover, as tumors grew in size, their average perfusion rate decreased further as an increasingly inadequate blood supply led to focal necrosis (25). Subsequent studies by many investigators have confirmed and generalized these observations, although, as with other properties of tumor vessels, extreme heterogeneity exists.

What accounts for the relatively reduced blood flow found in tumors? Although all the reasons responsible are not as yet fully understood, measurements by many investigators and particularly by the Jain laboratory have shown that all the variables affecting blood flow are altered in experimental tumors (26). Blood flow (Q) in any vascular system is proportional to microvascular pressure drop (ΔP) and inversely proportional to blood viscosity (η), and to a factor referred to as extrinsic geometric resistance (z) (Equation 1).

$$Q = \Delta P / \eta z \qquad \text{(Equation 1)}$$

Whereas pressures in the arteries supplying normal and tumor vessels are quite similar, microvascular pressures within tumors are elevated, and venous pressures are significantly reduced (27). Also, tumor blood vessels exhibit greater resistance to flow than do the vessels supplying normal tissues. Because of their serpentine course, thin walls, exposure to increased interstitial pressure, and other local factors, extrinsic geometric resistance is increased, frequently in experimental tumors by more than an order of magnitude. Finally, the viscosity of the blood within tumor vessels is increased for at least two reasons. First, the hematocrit is increased because tumor vessels are hyperpermeable and leak plasma, resulting in sluggish blood flow and rouleaux formation. Second, blood viscosity is increased as the result of the Fahraeus-Lindqvist effect (viscosity increases with increasing vessel size; tumor vessels are enlarged, on average, compared with their normal counterparts) (28).

Attempts have been made to increase or decrease tumor blood flow using a variety of vasoactive drugs but the response has been variable and inconsistent (25). In general, the vascular beds of most tumors behave as rigid tubes that are in a state of close to maximal diameter; little capacity is available to increase flow in response to higher vascular pressure, probably because of a paucity of muscularized blood vessels that can be induced to dilate. The net result is low maximum perfusion capacity (i.e., high vascular resistance) compared with most normal tissues.

Tumor Vessel Function

Not unexpectedly, abnormal organization and structure and reduced blood flow lead to deficient tumor vessel function. The uneven distribution of vessels results in an unequal distribution of O_2 and nutrients and uneven clearance of waste products, resulting in zones of metabolic deficiency, ischemia, and necrosis. Also, the increased average diameter of tumor vessels results in an altered surface area–to-volume ratio that further contributes to inefficient tissue nutrition. As a result, and also because of arteriovenous shunts, nutrients are not taken up efficiently by tumors, as is manifest by the increased oxygenation of the venous blood draining tumors. Poor clearance of CO_2 and other metabolites, coupled with high tumor cell glyocolytic activity, results in a tumor microenvironment that is acidic compared with that of normal tissues (pH $\sim$7.2 versus pH $\sim$7.4) (29).

Tumor Vessel Hyperpermeability

Another general abnormality of the tumor vasculature is increased permeability to plasma – particularly to plasma proteins, large molecules that escape only very slowly from normal blood vessels. As early as 1959, investigators from several different laboratories reported increased clearance of plasma proteins from tumor blood vessels (30). In parallel studies, Gullino (20) found that tumor interstitial fluid was increased both in quantity and in plasma protein content, as compared with normal tissues, resulting in an increase in interstitial colloid osmotic pressure. He and others suggested that vascular hyperpermeability, coupled with a lack of functional lymphatics, accounted for the edematous state of tumors.

Subsequent studies have demonstrated that hyperpermeability is indeed a general property of the blood vessels supplying malignant tumors. This finding has led to a parallel literature, summarized in the next sections, that describes tumor vessel hyperpermeability from structural and functional (physiological) perspectives.

Structural Basis of Tumor Vessel Hyperpermeability

As noted above plasma components can extravasate across vascular endothelium by paracellular (through inter-EC junctions) and transcellular (caveolae, fenestrae, and VVOs) routes (see Chapter 71). Inter-EC junctions allow the passage of water and other small molecules but normally exclude plasma proteins and similarly sized molecules. Shuttling across capillary endothelium, caveolae can transport solutes of both large and small size; they also may join together to form continuous, transcapillary pores. Fenestrae are specialized structures confined to specialized endothelia; they normally restrict the passage of plasma proteins, while permitting the passage of peptide and steroid hormones. Finally, VVOs, like caveolae, provide a system for the passage of both small and large molecules such as plasma proteins.

Which of these pathways are responsible for the passage of plasma and plasma proteins in tumor vessels? There is as yet disagreement in the literature. Some investigators have observed gaps in tumor endothelium and have argued that these represent the opening of interendothelial junctions (6,31). We, however, are impressed with the sturdiness of inter-EC junctions, and are persuaded that the bulk of tumor vessel leakiness occurs through VVOs and the trans-EC pores that may result from VVOs activated by VEGF-A or other tumor-secreted vasoactive mediators (32).

Physiological Measurements of Tumor Vessel Permeability

As usually understood in the cancer literature, the term *permeability* refers to the increased extravasation of solutes, particularly plasma proteins or other macromolecules, from tumor blood vessels. Physiologists, however, have a stricter definition of permeability, confining the use of this term to changes in the intrinsic properties of the vascular wall. Because endothelium is the limiting barrier to the extravasation of plasma solutes from blood vessels, any changes in permeability must take place at the level of the vascular endothelium. Physiologists regard the endothelium as a semipermeable membrane, permeated by pores of varying size, and they attribute the passage of molecules across that membrane to two driving forces: convection and diffusion. Based on these considerations, Starling, Renkin, Michel, Curry, Rippe, and others (33–35) have developed equations that describe in mathematical detail the passage of solutes across blood vessels (Equations 2–4):

$$J_{s-convection} = J_{v-convection}[1 - \sigma]C_p \qquad \text{(Equation 2)}$$

where $J_{s-convection}$ is the rate of flux of a solute such as plasma albumin; $J_{v\text{-}convection}$ is the rate of flux of the solvent (plasma water); σ is the reflection coefficient of the vascular wall to the solute as compared to that of water; and C_p is the concentration of solute in plasma.

$$J_{v-convection} = L_p A[p_v - p_i] - \sigma[\pi_v - \pi] \qquad \text{(Equation 3)}$$

where $J_{v-convection}$ is the rate of flux of solvent; L_p is the hydraulic conductivity of the vessel wall; A is the surface area of the vascular endothelium across which flux takes place; P_c and P_i represent hydrostatic pressures within the circulation and interstitium; σ is the osmotic reflection coefficient and π_v and π_i are the osmotic pressures within the vessel and interstitium, respectively.

$$J_{s-diffusion} = PA[C_v - C_i] \qquad \text{(Equation 4)}$$

where $J_{s-diffusion}$ is the diffusive flux of the solute across the vessel wall under conditions of zero flow; P is the diffusive permeability of the vessel wall; A is the surface area available for exchange; and C_v and C_i are the concentrations of the solute in plasma and the interstitium, respectively.

From these equations, it is clear that the intrinsic properties of the vascular endothelium and the particular solute under investigation have central roles in regulating transvascular passage; these intrinsic properties are reflected in the variables L_p, σ, and P (33–35). In addition, however, a number of other factors also can have important effects in regulating the passage of solutes across the EC barrier. These include the surface area available for plasma exchange, hydrostatic and osmotic pressures within the blood and interstitial tissues, solute concentrations within and outside the vasculature, and the ratio of convective to diffusive flux. It also must be said that these equations explain solute extravasation in terms of convection and diffusion and therefore apply if solutes are passing through trans-EC pores or interendothelial gaps. However, the endothelium is more than a semipermeable membrane, and solutes may cross ECs not only through open pores but also by shuttling caveolae and through VVOs, where passage from vesicle to vesicle is regulated by the opening and closing of diaphragms.

So, which of these factors account for the hyperpermeability of tumor blood vessels? The increased interstitial pressure, both hydrostatic and osmotic, that results from the increased volume and protein content of tumor interstitial fluid would be expected to retard extravasation of molecules of all sizes, including proteins. Similarly, tumors have reduced overall vascular surface area for solute flux. Nonetheless, plasma and plasma proteins extravasate from tumor blood vessels to a much greater extent than from normal vessels, and this results, in large part, from changes in the intrinsic properties of the vascular endothelium lining tumor blood vessels. These changes are all the more impressive in that they involve only a subset of tumor blood vessels, particularly mother vessels localized at the tumor–host interface.

Other Properties of Tumor Vascular Endothelial Cells

Another property shared almost universally by tumor vascular endothelium is an increased rate of proliferation. In a comprehensive study and literature review, Denekamp and Hobson (36) reported that the vascular endothelium labeling index (LI) for tumors was significantly higher than that of the vasculature of normal tissues (0.22%). However, as with other properties of tumor vessels, great variability was noted; studies of 131 individual tumors demonstrated an EC LI that varied from 3.6% to 32.3%. EC LI did not correlate closely with tumor growth rate, but was generally significantly lower than that of the tumor cells themselves (range of tumor cell LI: 7.1–60.5%). Thus, the rate of tumor cell proliferation in individual tumors is not directly related to the rate of EC proliferation in the tumor vessels responsible for supplying them with nutrients.

Although increased permeability and proliferation are properties shared by the vascular endothelium supplying nearly all tumors, other EC properties, although irregularly expressed, often differentiate tumor vessels from those of normal tissues. Thus, in different tumors, tumor vascular ECs are reported to express on their surfaces increased anionic phospholipids, the $\alpha v\beta 3$ integrin, fibronectins, H/T cadherin, platelet-derived growth factor receptor (PDGFR)-β, NG2/ human melanoma proteoglycan (HMP), Tie, and other molecules, and increased or decreased leukocyte adhesion receptors (37–46). Recently, Hida and colleagues (47) have reported that blood vessels in several animal and human tumors express cytogenetic abnormalities.

TUMOR ANGIOGENESIS

Very early in their growth, and, interestingly, also at very late stages of malignant progression, tumors may satisfy their nutritional and waste removal needs by coopting the normal vasculature (48). In some cases, this is dramatically manifest as tumor cells grow in cuffs around pre-existing normal blood vessels (49,50). However, for the most part, growing tumors must induce the formation of new blood vessels and do so even before they invade the surrounding tissues (19,51,52). Tumor angiogenesis was recognized early in the 20th century by Goldman (53), Algire (54), and others. In elegant experiments that made use of filters permeable to large and small molecules but not to cells, Greenblatt and Shubi (55) and Ehrman and Knoth (56) demonstrated that tumors induced angiogenesis by producing a diffusible substance that could pass through filters into the surrounding tissues where angiogenesis actually took place. Folkman greatly extended this work during the early 1970s (57), proposing that tumor growth and metastasis depended on the induction of new blood vessels (angiogenesis), and that interfering with angiogenesis could provide a novel therapeutic approach. There followed a spirited search to identify the tumor factor(s) responsible for inducing angiogenesis, and a large number of candidate molecules have now been identified that can induce new blood vessel formation both in vitro and in vivo (58–62). These include proteins such as VEGF-A, other members of the vascular permeability factor (VPF)/VEGF family such as placenta growth factor (PlGF), cytokines such as fibroblast growth factor (FGF), transforming growth factor (TGF)-β, PDGF, interleukin (IL)-8, ephrins, and angiopoietins, as well as a variety of small molecules. VEGF-A is thought to be the most important of these angiogenic factors for several reason: (a) it is expressed by nearly all malignant tumors; (b) its receptors are selectively (although not exclusively) expressed on vascular endothelium and are consistently upregulated on tumor vascular endothelium; (c) many other angiogenic factors act by upregulating VEGF-A expression; (d) blocking VEGF-A or its receptors can inhibit the growth of animal tumors and, more recently, human tumors; and, finally, (e) because as is shown later, it can induce many of the types of blood vessels found in tumors (61,62).

Vascular Endothelial Growth Factor-A

VEGF-A was discovered as the result of microscopic studies that followed the growth pattern of tumor cells at early stages after their transplantation into syngeneic animal tissues. The critical observation was that such tumors grew as clumps of neoplastic cells dispersed in an edematous fibrin gel (63,64). The Irish pathologist O'Meara (65) had described what he thought to be fibrin strands on tumor cells during the late 1950s, but the methodologies available at that time were not sufficient to provide a definitive answer. Only rare tumors have the capacity to produce fibrinogen, the fibrin precursor. Rather, the fibrin found in tumors results from plasma fibrinogen. Therefore, two essential requirements exist for fibrin deposition in tumors (or indeed in any tissue outside of the vasculature): (a) permeability of tumor blood vessels to fibrinogen and other necessary plasma clotting proteins, and (b) extravascular activation of the clotting system to transform extravasated fibrinogen to fibrin. The second step is not unique to tumors in that vascular leakage in normal tissues also results in activation of the clotting system because tissue factor is expressed by many tissue cells (66). Thus, the hyperpermeability of the tumor vasculature distinguishes it from normal, and led us to search for a tumor product that could induce vascular permeability. Initial studies revealed that a vascular permeability factor was indeed present in the culture supernatants of a wide variety of tumors and also in vivo in tumor ascites fluid, and that it was a secreted protein that acted directly on vascular endothelium (64). Using the Miles assay to screen column fractions for their ability to induce vascular permeability, Donald Senger purified VPF to homogeneity (67). Sometime later, our colleagues at the Monsanto Company made use of Senger's N-terminal amino acid sequencing data to clone VPF, and they showed that it induced angiogenesis in vivo and promoted the proliferation of cultured ECs (68). Independently and in a back-to-back publication, Napoleone Ferrara and his colleagues at Genentech cloned the same protein as an EC mitogen and called it vascular endothelial growth factor (69). More recently, because other related growth factors have been found, VPF/VEGF is best referred to as VEGF-A.

VEGF-A is an alternatively spliced protein that exists in several different isoforms, each with a similar although not identical range of activities (70,71). VEGF-A_{165} (the mouse equivalent, VEGF-A_{164}, is one amino acid shorter), the most widely expressed and best studied isoform, is a multifunctional cytokine with a vascular permeabilizing activity some 50,000 times that of histamine. It also stimulates EC migration and is mitogenic for ECs, although only under low serum conditions that are unlikely to occur in vivo. VEGF-A is also an EC survival factor, able to avert both EC apoptosis and senescence (72,73). Aside from its vascular permeabilizing activities, VEGF-A's activities in angiogenesis depend to a large extent on its ability to reprogram the pattern of EC gene expression. All of VEGF-A's activities are thought to be mediated through two receptor tyrosine kinases, VEGFR1 (also known as FLT-1) and VEGFR2 (also known as KDR), and through a more recently described nontyrosine kinase receptor, neuropilin (74–76). Although expressed on normal endothelium, these receptors typically are overexpressed by tumor blood vessel ECs (61,62).

Normal versus Tumor Angiogenesis

The microvasculature of normal tissues arises in the embryo by the twin processes of vasculogenesis (formation of vessels from primitive stem cell precursors) and angiogenesis (formation of vessels from pre-existing vessels). Although incompletely understood, these processes are highly regulated and result from the balanced secretion, in appropriate amounts and sequence, of many different cytokines and inhibitors, including VEGF-A (60,77,78). The result is a hierarchy of the intricate and highly ordered and differentiated microvascular channels of normal tissues described earlier, each with well-defined anatomic and physiologic properties.

In contrast, the pathological angiogenesis found in tumors (and also in wound healing and inflammation) is a much cruder process that involves the unbalanced secretion of a smaller subset of cytokines and particularly VEGF-A (61,62). If this general hypothesis is correct, then it should be possible to produce surrogate tumor blood vessels by expressing VEGF-A (or other cytokines) in normal tissues – that is, it should be possible to generate tumor-like vessels in the absence of tumor cells. In fact, this approach has been successful. Nonreplicating adenoviral vectors engineered to express VEGF-A can be injected into normal mouse tissues where they infect host cells that serve for a time as factories producing VEGF-A. The result is an angiogenic response that mimics that of many tumors (62,79). Also, as in tumors (80), there is debate as to whether, and to what extent, the ECs that line the newly formed vessels induced by VEGF-A develop from pre-existing vascular endothelium or from bone marrow–derived circulating endothelial progenitor cells.

Adenoviral vectors offer several advantages for generating tumor-like blood vessels. They allow angiogenesis to be induced in any tissue accessible to vector injection; cytokine expression levels can be readily varied by adjusting viral dose; different cytokines can be combined, together or in sequence; and no foreign matrix need be introduced. A further advantage of this approach is that it allows insight into the steps and mechanisms by which tumor vessels of different types form. Finally, adenoviral vectors are not integrated into the genome, and therefore they are expressed for only a limited period of time (several weeks); as a result, it is possible to determine the effects of exogenous cytokine expression on the new blood vessels that have been induced. One disadvantage exists: namely, that experiments must be performed in immunodeficient mice to avoid an immune response directed against the vector.

TUMOR-SURROGATE VESSELS INDUCED BY AD-VEGF-A_{164}

Using an adenoviral vector engineered to express VEGF-A_{164}, it is possible to induce the formation of at least six different types of blood vessels (Table 158-2), all of which are found in human and animal tumors. These include mother vessels, bridged mother vessels, capillaries, glomeruloid vascular proliferations (glomeruloid bodies), and arteriovenous malformations (stabilized mother vessels). In addition, Ad-VEGF-A_{164} induces arteriogenesis and a highly abnormal form of lymphangiogenesis (Figure 158.1; for color reproduction, see Color Plate 158.1) (81,82). The extent to which lymphatics are induced in tumors remains to be determined.

Table 158-2: Proposed Classification of Tumor Blood Vessels

Vessel Type	*Vessel Properties*
Feeder vessels	Large, often tortuous arteries and veins whose properties have not been well studied
Mother vessels	Large, thin-walled, hyperpermeable, lightly fenestrated pericyte-poor sinusoids that are engorged with red blood cells
Bridged mother vessels	Mother vessels in which EC processes extend into and across the lumen
Capillaries	Resemble normal capillaries as far as is known
Glomeruloid microvascular proliferations	Poorly organized vascular structures that macroscopically resemble renal glomeruli; they are comprised of ECs and pericytes with minimal vascular lumens and reduplicated basement membranes
Vascular malformations (stabilized mother vessels)	Mother vessels that have acquired an asymmetric coat of smooth muscle cells or, alternatively, fibrous connective tissue; resemble arteriovenous malformations found in other settings
Vascular mimicry	Blood-filled spaces lined by tumor cells that may contribute to tumor circulation

All these vessel types except for vascular mimicry can be induced by VEGF-A_{164}.

Mother Vessels

The term *mother* vessels was coined by Paku and Paweletz (83) to refer to the first type of new blood vessel to form in experimental tumors. These are also the first vessel type to appear in response to VEGF-A; hence, the other types of vessels that evolve from mother vessels may be regarded as *daughter* vessels (62,79,81). Mother vessels would seem to correspond to several of the types of blood vessel categorized by Warren (see Table 158-2) as *giant capillaries* and *sinusoidal vessels*; in addition, although not capillaries, they are lightly fenestrated. We have defined mother vessels as enlarged, thin-walled, hyperpermeable, pericyte-poor sinusoids (79). They arise from pre-existing venules and capillaries through a three-step process of basement membrane degradation, pericyte detachment, and extensive increase in cross-sectional area. This process begins within hours of Ad-VEGF-A_{164} injection and is similar in all normal tissues studied, including brain (84). Initially (at least for the first 48 hours) mother vessel formation proceeds without significant EC division. Basement membrane degradation is an essential step because basement membranes are noncompliant (nonelastic) structures that do not allow microvessels to increase by more than approximately 30% in cross-sectional area (85), far less than the fivefold expansion characteristic of many mother vessels. The specific proteases responsible for basement membrane degradation have not as yet been identified. Whether pericytes detach from mother vessels by an active process or simply fall off as the result of basement membrane degradation is not known. Vascular expansion of capillaries and venules is accommodated by membrane stored in VVOs and caveolae. VVOs provide a particularly abundant source of stored intracellular membrane, corresponding to more than

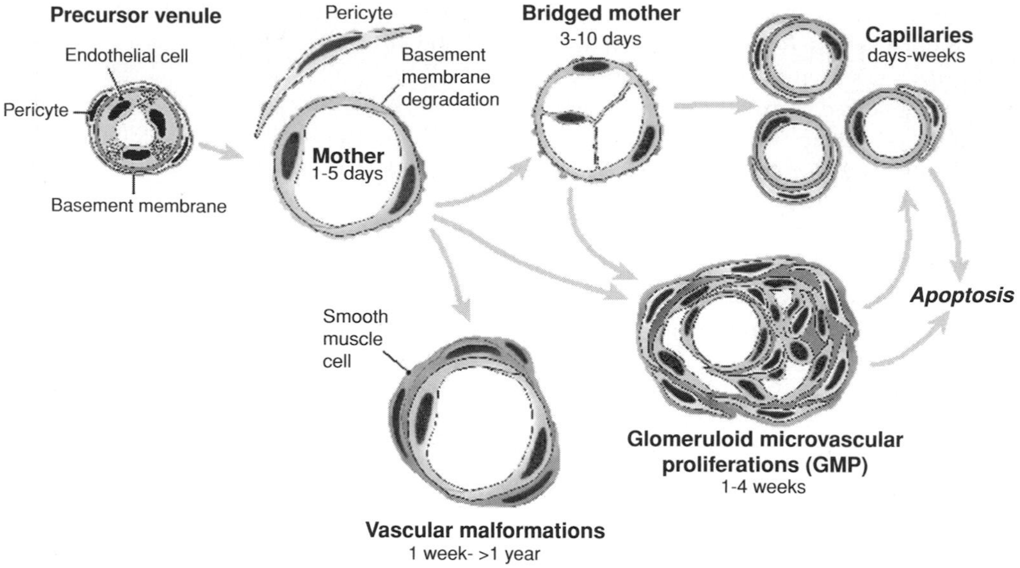

Figure 158.1. Schematic diagram depicting the types and sequence of new vessel formation in response to Ad-VEGF-A_{164}. (Reproduced with modification, with permission from Pettersson A, Nagy JA, Brown LF, et al. Heterogeneity of the angiogenic response induced in different normal adult tissues by vascular permeability factor/vascular endothelial growth factor. *Lab Invest.* 2000;80:99–115.) For color reproduction, see Color Plate 158.1.

twice that of the plasma membrane of pre-existing venular ECs (10,30). As mother vessels develop, the EC cytoplasm becomes greatly thinned and VVOs decrease in both number and complexity as they contribute their membranes to the greatly expanded plasma membrane. Vessels of this description are common in both animal and human tumors, and a similar process likely occurs in the formation of mother vessels in healing wounds (e.g., skin wounds, myocardial infarcts) (86,87).

Mother vessels are highly permeable to circulating macromolecules, such as plasma proteins and fluorsceinated macromolecular dextrans, and account to a large extent for the overall hyperpermeability of the vasculature in animal tumors. As in normal venules exposed to VEGF-A or other permeabilizing mediators, circulating macromolecules such as ferritin extravasate through VVOs. This may seem surprising in that, as noted earlier, mother vessel ECs have many fewer VVOs than are found in normal venular endothelium. The answer likely lies in the fact that, although fewer in number, the VVOs remaining are less complex (have fewer vesicles and vacuoles), and the vascular endothelium is greatly thinned; as a result, the transcellular pathway is greatly shortened.

Macromolecules also extravasate from mother vessels by way of fenestrations, although these cover less than 1% of the endothelial surface; this is of interest because the fenestrae of normal vessels are not thought to be permeable to large molecules such as plasma proteins.

Mother vessels are transitional forms, dependent on exogenous VEGF-A, and we postulate that they are the vessel type most susceptible to anti–VEGF-A therapy. However, mother vessels evolve into several different types of daughter vessels, as listed in Table 158-2. Although the steps involved are understood in part, the molecular mechanisms have not been elucidated nor is it known why different mother vessels evolve into different types of daughter vessel.

Bridged Mother Vessels and Capillaries

One process by which mother vessels evolve into capillaries was originally discovered in tumor vessels (15) and involves intraluminal bridging by vascular ECs. These endothelial bridges divide mother vessels into several smaller channels that are then thought to separate from each other to form independent capillaries (Figure 158.2; for color reproduction, see Color Plate 158.2) (79). A similar bridging process has been found in the new blood vessels that supply healing wounds (e.g., myocardial infarcts) (87) and in skeletal muscle exposed to chronic vascular dilatation (88). ECs extend cytoplasmic processes into and across mother vessel lumens, forming transluminal bridges that divide blood flow into multiple smaller-sized channels. This is unexpected, because mammalian cells normally migrate on surface matrices and would not be expected to extend processes into a flowing stream of blood. In fact, they may not be doing so. Because they are highly permeable to plasma, mother vessel lumens are packed with red blood cells and have little retained plasma; this results in increased viscosity and diminished blood flow, both of which favor intravascular clotting. Clotting is further favored in that VEGF-A upregulates EC expression of tissue factor (89), the procoagulant that initiates the

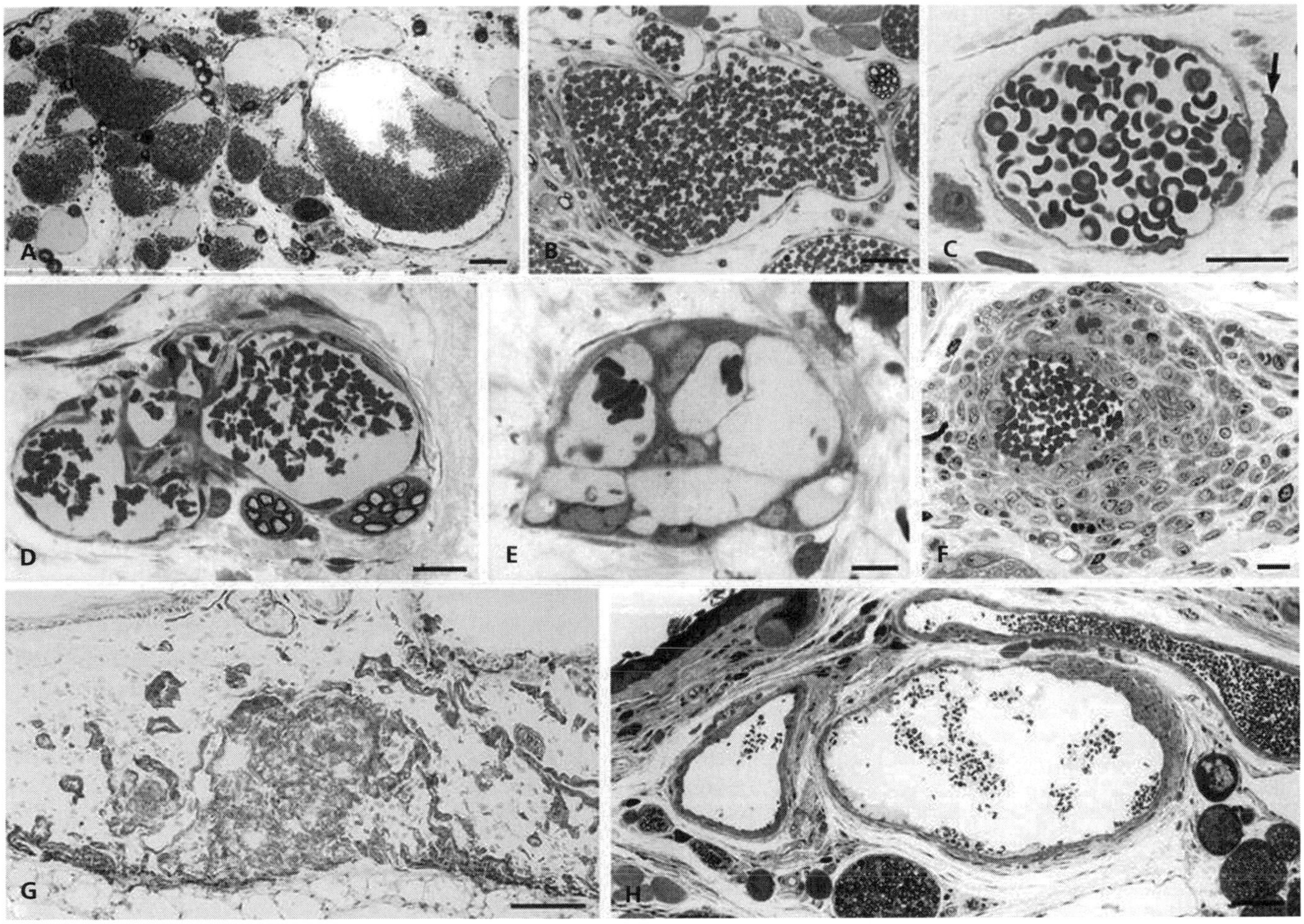

Figure 158.2. Illustration of the various types of new, tumor-surrogate blood vessels induced by Ad-VEGF-A_{164} in Giemsa-stained, 1-micron Epon sections (**A–F, H**) and following immunohistochemical staining for CD31 (**G**). (**A–C**) Typical mother vessels, enlarged, thin-walled sinusoids with high hematocrit (as the result of vascular permeability to plasma) and few pericytes. In **C**, arrow indicates a detached pericyte. (**D, E**) Bridging of mother vessels by extension of EC processes that divide the lumen into multiple smaller channels. (**F, G**) Typical glomeruloid microvascular proliferations. (**H**) Vascular malformations; mother vessels that have become stabilized by acquiring an asymmetric coating of smooth muscle cells. Magnification bars: **A, H**, 50 μm; **B–F**, 15 μm; **G**, 25 μm. (**F** reproduced with modification, with permission from Pettersson A, Nagy JA, Brown LF, et al. Heterogeneity of the angiogenic response induced in different normal adult tissues by vascular permeability factor/vascular endothelial growth factor. *Lab Invest.* 2000;80:99–115; **H** reproduced with modification, with permission from *Cold Spring Harbor Symp Quant Biol.* 2002;67:227–237.) For color reproduction, see Color Plate 158.2.

extrinsic clotting pathway. In fact, mother vessels in tumors and those induced by Ad-VEGF-A_{164} commonly develop intravascular clots and may thrombose. Thus, although not proven, it is possible that the EC processes that project into mother vessel lumens are attached to a fibrin matrix. Whatever the mechanism, the smaller channels separate from each other to form individual, smaller-caliber capillaries that are, as far as is known, normal in structure.

Glomeruloid Microvascular Proliferations (Glomeruloid Bodies)

Glomeruloid microvascular proliferations (GMPs) are poorly organized vascular structures that resemble renal glomeruli (hence the name). They are commonly found in a number of different human tumors including glioblastoma multiforme and cancers of the stomach, breast, and other locations (79,90,91). GMPs are hyperpermeable to plasma and plasma proteins but, because they are poorly perfused, likely account for much less plasma extravasation than mother vessels. Nascent GMPs first appear as focal accumulations of large, poorly differentiated, CD31- and VEGFR2-positive cells in the endothelial lining of mother vessels (79,90). The source of these cells, whether locally from mother vessel endothelium or systemically from circulating endothelial progenitor cells, is not known. Whatever their source, these cells proliferate rapidly, extending into mother vessel lumens, and also outward into the surrounding extravascular matrix. In this manner, they encroach on and compress the mother vessels from which they had arisen, eventually dividing single large mother vessel lumens into much smaller channels that are often too small to admit the passage of red blood cells. The great majority of cells comprising GMPs continue to express EC markers. However, as GMPs grow, cells expressing pericyte markers (NG2, PDGFR-β) and ultrastructural characteristics also appeared. Electron microscopy showed that pericytes were situated peripherally to small clusters of ECs that formed small or no evident vascular lumens. At least in this experimental setting, macrophages, identified by markers (F4/80, HLA DR2) and characteristic ultrastructure, also accumulated peripherally. An additional prominent feature is deposition of abnormal multilayered basal lamina.

The GMPs so generated were found to be dependent on the continued presence of exogenous VEGF-A_{164} for their maintenance as well as for their generation. As adenoviral vector-derived VEGF-A_{164} expression declined, GMPs underwent apoptosis and progressively devolved into smaller, normal-appearing capillaries (90). Similarly, all the human tumors known to form GMP also express VEGF-A, and tumors such as glioblastoma multiforme, which makes unusually large amounts of VEGF-A, are among those that most commonly have GMPs. Recent evidence suggests that the presence of GMPs in human tumors is a negative prognostic sign (91,92).

Fidler (93) has proposed a minor variation of this model of GMP induction, postulating that ECs lining large tumor vessels (presumably mother vessels) project inwardly to form transcellular bridges, as originally described by Nagy and colleagues (15). On the other hand, Dome and colleagues (94) have proposed a third, very different model in which tumor cells form GMPs by attaching to the basement membranes of pre-existing cerebral microvessels and pulling entire capillaries (ECs and pericytes) into loops and coils. In this model, the vessels play only a passive role and exhibit very little in the way of cell proliferation. This model obviously cannot explain the formation of GMPs induced by Ad-VEGF-A_{164} in the absence of tumor cells (90).

Vascular Malformations (Stabilized Mother Vessels)

Thin-walled mother vessels lacking adequate pericyte and basement membrane support are subject to thrombosis or collapse. Although some mother vessels avoid these fates by evolving into capillaries or GMPs, others maintain their large size by acquiring a coat of smooth muscle cells and/or a rim of collagen-expressing fibroblasts. Such vessels are readily distinguished from normal arteries and veins by their inappropriately large size (for their tissue location), by their thinner and often asymmetric muscular coat, and by the presence of perivascular fibrosis. Vessels of this description closely resemble the vascular malformations found independent of malignant tumors – for example in skin and brain – thus suggesting a mechanism by which such malformations may form (95). As their structure would suggest, these vessels are not thought to be hyperpermeable; also, unlike mother vessels and glomeruloid bodies, they persist indefinitely, long after adenoviral vector-induced VEGF-A_{164} expression has ceased (Figure 158.3; for color reproduction, see Color Plate 158.3). Thus, stabilized mother vessels have acquired independence from exogenous VEGF-A, although it is possible that they are maintained by VEGF-A secreted by the smooth muscle cells

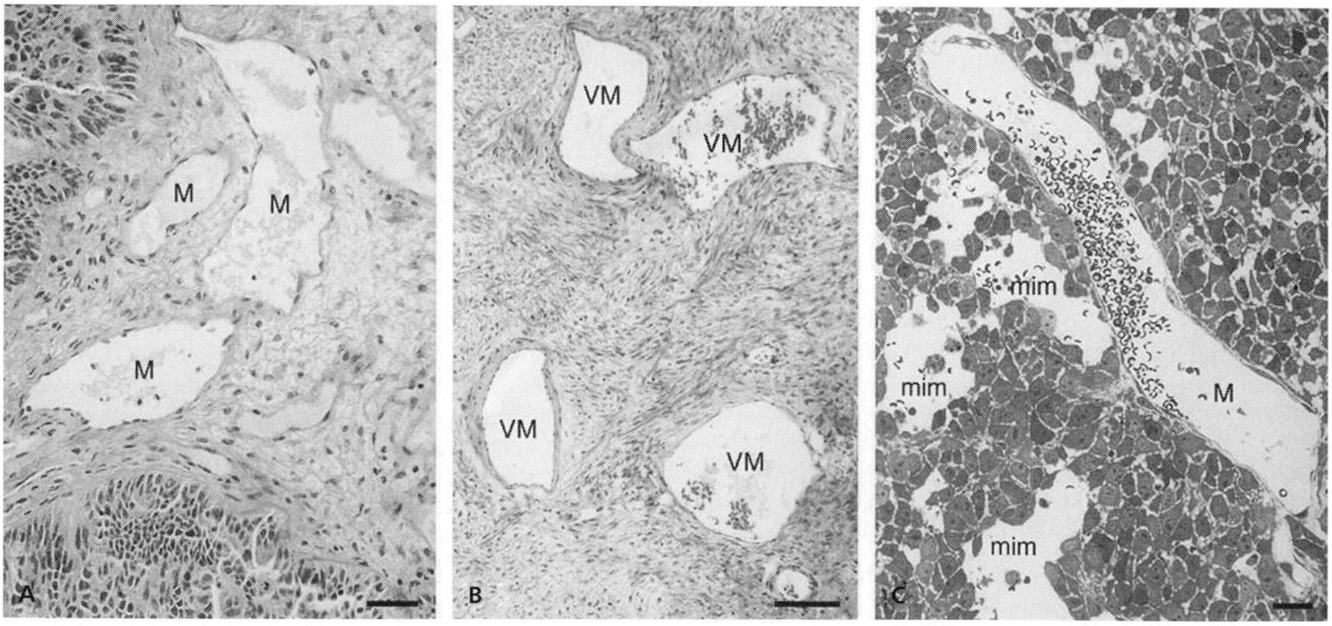

Figure 158.3. Illustration of mother vessels (*M*), vascular malformations (*VM*) in human ovarian carcinomas (**A**, **B**, hematoxylin and eosin-stained, paraffin-embedded sections) and vascular mimicry in a mouse B16 melanoma (**C**, Giemsa stained 1-micron plastic section). Magnification bars: 50 μm. For color reproduction, see Color Plate 158.3.

that closely envelop them. This finding has important implications, because these vessels would not be expected to be susceptible to anti–VEGF-A therapy.

Vascular Mimicry and Mosaic Vessels

In some tumors, particularly melanomas, red blood cell–filled spaces have been found that are apparently lined by tumor cells instead of ECs. Such vessels have been referred to as examples of vascular mimicry – tumor cells that have acquired some of the structural and molecular properties of ECs and form channels that mimic those of blood vessels (96). That such spaces exist is not in question, but their significance is much debated (97). The issue hangs on whether these spaces are a part of a functional vascular system or whether they simply reflect hemorrhage into tumors from leaky but otherwise conventional vessels. Very recent studies have investigated this question further and report the finding of "mosaic" vessels in some experimental tumors; that is, vessels that are lined by an attenuated endothelium with sites that do not immunostain for EC markers, gaps marked by an apparently incomplete basement membrane (98).

TUMOR VASCULAR ENDOTHELIUM AS A THERAPEUTIC TARGET

Because tumors, like normal tissues, depend on their vascular supply for survival and growth, attempts have been made to prevent angiogenesis as a strategy for inhibiting tumor growth. A variety of agents that block VEGF-A or its receptors have proved efficacious in animal studies and, recently, an antibody to VEGF-A has been effective in inhibiting the progression of certain types of cancer in patients (2). As noted earlier, certain tumor vessels require exogenous VEGF-A for their survival; therefore, inhibition of VEGF-A activity also may serve to destroy pre-existing vessels as well as preventing angiogenesis (72). These findings, coupled with the structural and functional differences that distinguish tumor from normal endothelium, have suggested that tumor vessels might express unique markers that could serve as therapeutic targets. A number of investigators are pursuing this line of investigation, using a variety of approaches that include brute-force biochemistry (99,100), selective isolation of the luminal plasma membranes of tumor vessel endothelium (101), and phage-display technology (102–105). Some promising targets have been identified, but it is as yet too soon to determine whether any will be useful in cancer therapy.

A final related point worth mentioning is that of tumor vessel normalization. Recent reports have called attention to an improvement in the quality of tumor blood vessels following antiangiogenic and other types of therapy (106). Vascular normalization could improve tumor access to standard chemotherapy and, by providing better oxygenation, render tumors more susceptible to radiation therapy.

KEY POINTS

- Tumor blood vessels are heterogeneous and highly abnormal with respect to organization, structure, and function.
- Physiological angiogenesis, although as yet incompletely understood, results from the balanced secretion, in appropriate amounts and sequence, of many different cytokines and inhibitors, including VEGF-A.
- Tumor angiogenesis results from the unbalanced secretion of a small number of mediators and growth factors, particularly VEGF-A.
- The hyperpermeability of tumor blood vessels results in extravascular deposition of a fibrin provisional stroma analogous to that occurring in wound healing. Thus, tumors and healing wounds generate vascularized stroma by similar mechanisms.

Future Goals

- To define the molecular mechanisms by which mother and the various types of daughter vessel form
- To determine which types of tumor vessels are affected by anti–VEGF-A therapy and whether additional types must be targeted to make antiangiogenesis therapy more effective

REFERENCES

1 Warren B. The vascular morphology of tumors. In: Peterson H-I, ed. *Tumor Blood Circulation: Angiogenesis, Vascular Morphology and Blood Flow of Experimental and Human Tumors.* Boca Raton: CRC Press; 1979:1–47.

2 Hurwitz H, Fehrenbacher L, Novotny W, et al. Bevacizumab plus irinotecan, fluorouracil, and leucovorin for metastatic colorectal cancer. *N Engl J Med.* 2004;350(23):2335–2342.

3 Guyton A, Hall J. *Textbook of Medical Physiology*, 10th ed. Philadelphia: Saunders, 2000.

4 Palade G. The microvascular endothelium revisited. In: Simionescu N, Simionescu M, eds. *Endothelial Cell Biology in Health and Disease.* New York: Plenum Press, 1988.

5 Majno G, Palade GE, Schoefl GI. Studies on inflammation. The site of action of histamine and serotonin along the vascular tree: a topographic study. *J Biophys Biochem Cytol.* 1961;11:607–626.

6 Majno G, Shea SM, Leventhal M. Endothelial contraction induced by histamine-type mediators: an electron microscopic study. *J Cell Biol.* 1969;42(3):647–672.

7 Neal CR, Michel CC. Transcellular gaps in microvascular walls of frog and rat when permeability is increased by perfusion with the ionophore A23187. *J Physiol.* 1995;488(Pt 2):427–437.

8 Feng D, Nagy JA, Hipp J, et al. Vesiculo-vacuolar organelles and the regulation of venule permeability to macromolecules by vascular permeability factor, histamine, and serotonin. *J Exp Med.* 1996;183(5):1981–1986.

9 Feng D, Nagy JA, Hipp J, et al. Reinterpretation of endothelial cell gaps induced by vasoactive mediators in guinea-pig, mouse and rat: many are transcellular pores. *J Physiol*. 1997;504(Pt 3): 747–761.
10 Dvorak AM, Kohn S, Morgan ES, et al. The vesiculo-vacuolar organelle (VVO): a distinct endothelial cell structure that provides a transcellular pathway for macromolecular extravasation. *J Leukoc Biol*. 1996;59(1):100–115.
11 Carman CV, Springer TA. A transmigratory cup in leukocyte diapedesis both through individual vascular endothelial cells and between them. *J Cell Biol*. 2004;167(2):377–388.
12 Feng D, Nagy JA, Pyne K, et al. Neutrophils emigrate from venules by a transendothelial cell pathway in response to FMLP. *J Exp Med*. 1998;187(6):903–915.
13 Nagy JA, Masse EM, Herzberg KT, et al. Pathogenesis of ascites tumor growth: vascular permeability factor, vascular hyperpermeability, and ascites fluid accumulation. *Cancer Res*. 1995; 55(2):360–368.
14 Nagy JA, Meyers MS, Masse EM, et al. Pathogenesis of ascites tumor growth: fibrinogen influx and fibrin accumulation in tissues lining the peritoneal cavity. *Cancer Res*. 1995;55(2):369–375.
15 Nagy JA, Morgan ES, Herzberg KT, et al. Pathogenesis of ascites tumor growth: angiogenesis, vascular remodeling, and stroma formation in the peritoneal lining. *Cancer Res*. 1995;55(2):376–385.
16 Kini AR. Angiogenesis in leukemia and lymphoma. *Cancer Treat Res*. 2004;121:221–238.
17 Molica S, Vacca A, Levato D, et al. Angiogenesis in acute and chronic lymphocytic leukemia. *Leuk Res*. 2004;28(4):321–324.
18 Ribatti D, Scavelli C, Roccaro AM, et al. Hematopoietic cancer and angiogenesis. *Stem Cells Dev*. 2004;13(5):484–495.
19 Brown LF, Guidi AJ, Schnitt SJ, et al. Vascular stroma formation in carcinoma in situ, invasive carcinoma, and metastatic carcinoma of the breast. *Clin Cancer Res*. 1999;5(5):1041–1056.
20 Gullino P. Extracellular compartments of solid tumors. In: Becker F, ed. *Cancer: A Comprehensive Treatise*. New York: Plenum Press; 1975:327–354.
21 Yeo TK, Brown L, Dvorak HF. Alterations in proteoglycan synthesis common to healing wounds and tumors. *Am J Pathol*. 1991;138(6):1437–1450.
22 Dvorak H. Tumor architecture and targeted delivery. In: Abrams P, Fritzberg A, eds. *Radioimmunotherapy of Cancer*. New York: Marcel Dekker, Inc.; 2000:107–135.
23 Weidner N. Angiogenesis as a predictor of clinical outcome in cancer patients. *Hum Pathol*. 2000;31(4):403–405.
24 Chaplin DJ, Olive PL, Durand RE. Intermittent blood flow in a murine tumor: radiobiological effects. *Cancer Res*. 1987;47(2): 597–601.
25 Peterson H-I. The microcirculation of tumors. In: Orr F, Buchanan M, Weiss L, eds. *Microcirculation in Cancer Metastasis*. Boca Raton: CRC Press; 1991:277–298.
26 Jain RK. Determinants of tumor blood flow: a review. *Cancer Res*. 1988;48(10):2641–2658.
27 Sevick EM, Jain RK. Geometric resistance to blood flow in solid tumors perfused ex vivo: effects of tumor size and perfusion pressure. *Cancer Res*. 1989;49(13):3506–3512.
28 Sevick EM, Jain RK. Viscous resistance to blood flow in solid tumors: effect of hematocrit on intratumor blood viscosity. *Cancer Res*. 1989;49(13):3513–3519.
29 Stubbs M, McSheehy PM, Griffiths JR, Bashford CL. Causes and consequences of tumour acidity and implications for treatment. *Mol Med Today*. 2000;6(1):15–19.
30 Dvorak HF, Nagy JA, Feng D, et al. Vascular permeability factor/vascular endothelial growth factor and the significance of microvascular hyperpermeability in angiogenesis. *Curr Top Microbiol Immunol*. 1999;237:97–132.
31 McDonald DM, Thurston G, Baluk P. Endothelial gaps as sites for plasma leakage in inflammation. *Microcirculation*. 1999; 6(1):7–22.
32 Feng D, Nagy JA, Pyne K, et al. Pathways of macromolecular extravasation across microvascular endothelium in response to VPF/VEGF and other vasoactive mediators. *Microcirculation*. 1999;6(1):23–44.
33 Bates DO, Lodwick D, Williams B. Vascular endothelial growth factor and microvascular permeability. *Microcirculation*. 1999; 6(2):83–96.
34 Michel CC, Curry FE. Microvascular permeability. *Physiol Rev*. 1999;79(3):703–761.
35 Rippe B, Haraldsson B. Transport of macromolecules across microvascular walls: the two-pore theory. *Physiol Rev*. 1994;74 (1):163–219.
36 Denekamp J, Hobson B. Endothelial-cell proliferation in experimental tumours. *Br J Cancer*. 1982;46(5):711–720.
37 Burg MA, Pasqualini R, Arap W, et al. NG2 proteoglycan-binding peptides target tumor neovasculature. *Cancer Res*. 1999; 59(12):2869–2874.
38 Kaipainen A, Vlaykova T, Hatva E, et al. Enhanced expression of the tie receptor tyrosine kinase messenger RNA in the vascular endothelium of metastatic melanomas. *Cancer Res*. 1994;54(24):6571–6577.
39 Lauri D, Needham L, Martin-Padura I, Dejana E. Tumor cell adhesion to endothelial cells: endothelial leukocyte adhesion molecule-1 as an inducible adhesive receptor specific for colon carcinoma cells. *J Natl Cancer Inst*. 1991;83(18):1321–1324.
40 Plate KH, Breier G, Farrell CL, Risau W. Platelet-derived growth factor receptor-beta is induced during tumor development and upregulated during tumor progression in endothelial cells in human gliomas. *Lab Invest*. 1992;67(4):529–534.
41 Ran S, Downes A, Thorpe PE. Increased exposure of anionic phospholipids on the surface of tumor blood vessels. *Cancer Res*. 2002;62(21):6132–6140.
42 Renkonen R, Paavonen T, Nortamo P, Gahmberg CG. Expression of endothelial adhesion molecules in vivo. Increased endothelial ICAM-2 expression in lymphoid malignancies. *Am J Pathol*. 1992;140(4):763–767.
43 Salmi M, Grenman R, Grenman S, et al. Tumor endothelium selectively supports binding of IL-2-propagated tumor-infiltrating lymphocytes. *J Immunol*. 1995;154(11):6002–6012.
44 Wu NZ, Klitzman B, Dodge R, Dewhirst MW. Diminished leukocyte-endothelium interaction in tumor microvessels. *Cancer Res*. 1992;52(15):4265–4268.
45 Wyder L, Vitaliti A, Schneider H, et al. Increased expression of H/T-cadherin in tumor-penetrating blood vessels. *Cancer Res*. 2000;60(17):4682–4688.
46 Zocchi MR, Poggi A. Lymphocyte-endothelial cell adhesion molecules at the primary tumor site in human lung and renal cell carcinomas. *J Natl Cancer Inst*. 1993;85(3):246–247.
47 Hida K, Hida Y, Amin DN, et al. Tumor-associated endothelial cells with cytogenetic abnormalities. *Cancer Res*. 2004;64(22): 8249–8255.

48 Clark WH. Tumour progression and the nature of cancer. *Br J Cancer*. 1991;64(4):631–644.

49 Holash J, Maisonpierre PC, Compton D, et al. Vessel cooption, regression, and growth in tumors mediated by angiopoietins and VEGF. *Science*. 1999;284(5422):1994–1998.

50 Leenders WP, Kusters B, de Waal RM. Vessel co-option: how tumors obtain blood supply in the absence of sprouting angiogenesis. *Endothelium*. 2002;9(2):83–87.

51 Brem SS, Jensen HM, Gullino PM. Angiogenesis as a marker of preneoplastic lesions of the human breast. *Cancer*. 1978;41(1):239–244.

52 Heffelfinger SC, Miller MA, Yassin R, Gear R. Angiogenic growth factors in preinvasive breast disease. *Clin Cancer Res*. 1999;5(10):2867–2876.

53 Goldman E. The growth of malignant disease in man and the lower animals with special reference to the vascular system. *Lancet North Am Ed*. 1907;2:1236–1240.

54 Algire G. Microscopic studies of the early growth of a transplantable melanoma of the mouse using the transparent chamber technique. *J Natl Cancer Inst*. 1943;4:1.

55 Greenblatt M, Shubi P. Tumor angiogenesis: transfilter diffusion studies in the hamster by the transparent chamber technique. *J Natl Cancer Inst*. 1968;41(1):111–124.

56 Ehrmann RL, Knoth M. Choriocarcinoma. Transfilter stimulation of vasoproliferation in the hamster cheek pouch. Studied by light and electron microscopy. *J Natl Cancer Inst*. 1968;41(6):1329–1341.

57 Folkman J. Tumor angiogenesis: therapeutic implications. *N Engl J Med*. 1971;285(21):1182–1186.

58 Carmeliet P, Jain RK. Angiogenesis in cancer and other diseases. *Nature*. 2000;407(6801):249–257.

59 Folkman J, Klagsbrun M. Angiogenic factors. *Science*. 1987;235(4787):442–447.

60 Yancopoulos GD, Davis S, Gale NW, et al. Vascular-specific growth factors and blood vessel formation. *Nature*. 2000;407(6801):242–248.

61 Dvorak HF. Vascular permeability factor/vascular endothelial growth factor: a critical cytokine in tumor angiogenesis and a potential target for diagnosis and therapy. *J Clin Oncol*. 2002;20(21):4368–4380.

62 Dvorak HF. Rous-Whipple Award Lecture. How tumors make bad blood vessels and stroma. *Am J Pathol*. 2003;162(6):1747–1757.

63 Dvorak HF, Dvorak AM, Manseau EJ, et al. Fibrin gel investment associated with line 1 and line 10 solid tumor growth, angiogenesis, and fibroplasia in guinea pigs. Role of cellular immunity, myofibroblasts, microvascular damage, and infarction in line 1 tumor regression. *J Natl Cancer Inst*. 1979;62(6):1459–1472.

64 Dvorak HF, Orenstein NS, Carvalho AC, et al. Induction of a fibrin-gel investment: an early event in line 10 hepatocarcinoma growth mediated by tumor-secreted products. *J Immunol*. 1979;122(1):166–174.

65 O'Meara RA, Jackson RD. Cytological observations on carcinoma. *Ir J Med Sci*. 1958;171(391):327–328.

66 Dvorak HN, Senger DR, Dvorak AM, et al. Regulation of extravascular coagulation by microvascular permeability. *Science*. 1985;227(4690):1059–1061.

67 Senger DR, Galli SJ, Dvorak AM, et al. Tumor cells secrete a vascular permeability factor that promotes accumulation of ascites fluid. *Science*. 983;219(4587):983–985.

68 Keck PJ, Hauser SD, Krivi G, et al. Vascular permeability factor, an endothelial cell mitogen related to PDGF. *Science*. 1989;246(4935):1309–1312.

69 Leung DW, Cachianes G, Kuang WJ, et al. Vascular endothelial growth factor is a secreted angiogenic mitogen. *Science*. 1989;246(4935):1306–1309.

70 Brown LF, Detmar M, Claffey K, et al. Vascular permeability factor/vascular endothelial growth factor: a multifunctional angiogenic cytokine. *EXS*. 1997;79:233–269.

71 Ferrara N, Keyt B. Vascular endothelial growth factor: basic biology and clinical implications. *EXS*. 1997;79:209–232.

72 Benjamin LE, Golijanin D, Itin A, et al. Selective ablation of immature blood vessels in established human tumors follows vascular endothelial growth factor withdrawal. *J Clin Invest*. 1999;103(2):159–165.

73 Watanabe Y, Lee SW, Detmar M, et al. Vascular permeability factor/vascular endothelial growth factor (VPF/VEGF) delays and induces escape from senescence in human dermal microvascular endothelial cells. *Oncogene*. 1997;14(17):2025–2032.

74 Miao HQ, Klagsbrun M. Neuropilin is a mediator of angiogenesis. *Cancer Metastasis Rev*. 2000;19(1–2):29–37.

75 Veikkola T, Alitalo K. VEGFs, receptors and angiogenesis. *Semin Cancer Biol*. 1999;9(3):211–220.

76 Cross MJ, Dixelius J, Matsumoto T, Claesson-Welsh L. VEGF-receptor signal transduction. *Trends Biochem Sci*. 2003;28(9):488–494.

77 Beck L Jr., D'Amore PA. Vascular development: cellular and molecular regulation. *FASEB J*. 1997;11(5):365–373.

78 Carmeliet P. Mechanisms of angiogenesis and arteriogenesis. *Nat Med*. 2000;6(4):389–395.

79 Pettersson A, Nagy JA, Brown LF, et al. Heterogeneity of the angiogenic response induced in different normal adult tissues by vascular permeability factor/vascular endothelial growth factor. *Lab Invest*. 2000;80(1):99–115.

80 Lyden D, Hattori K, Dias S, et al. Impaired recruitment of bone-marrow-derived endothelial and hematopoietic precursor cells blocks tumor angiogenesis and growth. *Nat Med*. 2001;7(11):1194–1201.

81 Nagy JA, Dvorak AM, Dvorak HF. VEGF-A(164/165) and PlGF: roles in angiogenesis and arteriogenesis. *Trends Cardiovasc Med*. 2003;13(5):169–175.

82 Nagy JA, Vasile E, Feng D, et al. Vascular permeability factor/vascular endothelial growth factor induces lymphangiogenesis as well as angiogenesis. *J Exp Med*. 2002;196(11):1497–1506.

83 Paku S, Paweletz N. First steps of tumor-related angiogenesis. *Lab Invest*. 1991;65(3):334–346.

84 Stiver SI, Tan X, Brown LF, et al. VEGF-A angiogenesis induces a stable neovasculature in adult murine brain. *J Neuropathol Exp Neurol*. 2004;63(8):841–855.

85 Swayne GT, Smaje LH, Bergel DH. Distensibility of single capillaries and venules in the rat and frog mesentery. *Int J Microcirc Clin Exp*. 1989;8(1):25–42.

86 Dvorak HF. Tumors: wounds that do not heal. Similarities between tumor stroma generation and wound healing. *N Engl J Med*. 1986;315(26):1650–1659.

87 Ren G, Michael LH, Entman ML, Frangogiannis NG. Morphological characteristics of the microvasculature in healing myocardial infarcts. *J Histochem Cytochem*. 2002;50(1):71–79.

88 Egginton S, Zhou AL, Brown MD, Hudlicka O. Unorthodox angiogenesis in skeletal muscle. *Cardiovasc Res*. 2001;49(3):634–646.
89 Clauss M, Grell M, Fangmann C, et al. Synergistic induction of endothelial tissue factor by tumor necrosis factor and vascular endothelial growth factor: functional analysis of the tumor necrosis factor receptors. *FEBS Lett*. 1996;390(3):334–338.
90 Sundberg C, Nagy JA, Brown LF, et al. Glomeruloid microvascular proliferation follows adenoviral vascular permeability factor/vascular endothelial growth factor-164 gene delivery. *Am J Pathol*. 2001;158(3):1145–1160.
91 Straume O, Chappuis PO, Salvesen HB, et al. Prognostic importance of glomeruloid microvascular proliferation indicates an aggressive angiogenic phenotype in human cancers. *Cancer Res*. 2002;62(23):6808–6811.
92 Goffin JR, Straume O, Chappuis PO, et al. Glomeruloid microvascular proliferation is associated with p53 expression, germline BRCA1 mutations and an adverse outcome following breast cancer. *Br J Cancer*. 2003;89(6):1031–1034.
93 Fidler IJ, Yano S, Zhang RD, et al. The seed and soil hypothesis: vascularisation and brain metastases. *Lancet Oncol*. 2002;3(1): 53–57.
94 Dome B, Timar J, Paku S. A novel concept of glomeruloid body formation in experimental cerebral metastases. *J Neuropathol Exp Neurol*. 2003;62(6):655–661.
95 McKee P. *Pathology of the Skin with Clinical Correlations*. London: Mosby International, 1996.
96 Folberg R, Hendrix MJ, Maniotis AJ. Vasculogenic mimicry and tumor angiogenesis. *Am J Pathol*. 2000;156(2):361–381.
97 McDonald DM, Munn L, Jain RK. Vasculogenic mimicry: how convincing, how novel, and how significant? *Am J Pathol*. 2000; 156(2):383–388.
98 di Tomaso E, Capen D, Haskell A, et al. Mosaic tumor vessels: cellular basis and ultrastructure of focal regions lacking endothelial cell markers. *Cancer Res*. 2005;65(13):5740–5749.
99 Nanda A, St Croix B. Tumor endothelial markers: new targets for cancer therapy. *Curr Opin Oncol*. 2004;16(1):44–49.
100 St Croix B, Rago C, Velculescu V, et al. Genes expressed in human tumor endothelium. *Science*. 2000;289(5482):1197–1202.
101 Oh P, Li Y, Yu J, et al. Subtractive proteomic mapping of the endothelial surface in lung and solid tumours for tissue-specific therapy. *Nature*. 2004;429(6992):629–635.
102 Arap W, Pasqualini R, Ruoslahti E. Cancer treatment by targeted drug delivery to tumor vasculature in a mouse model. *Science*. 1998;279(5349):377–380.
103 Brown DM, Ruoslahti E. Metadherin, a cell surface protein in breast tumors that mediates lung metastasis. *Cancer Cell*. 2004;5(4):365–374.
104 Koivunen E, Arap W, Valtanen H, et al. Tumor targeting with a selective gelatinase inhibitor. *Nat Biotechnol*. 1999;17(8):768–774.
105 Yao VJ, Ozawa MG, Trepel M, et al. Targeting pancreatic islets with phage display assisted by laser pressure catapult microdissection. *Am J Pathol*. 2005;166(2):625–636.
106 Jain RK. Normalization of tumor vasculature: an emerging concept in antiangiogenic therapy. *Science*. 2005;307(5706):58–62.

159

Kaposi's Sarcoma

Kimberly E. Foreman

Loyola University Medical Center, Maywood, Illinois

In 1872, Moritz Kaposi (1837–1902), a Hungarian dermatologist, described five male patients presenting with deep red to brown-violet plaques or nodules on the skin of the lower extremities (Figure 159.1; for color reproduction, see Color Plate 159.1) (1). This disease, later designated Kaposi's Sarcoma (KS), was uniformly fatal for his patients. Kaposi stated: "The disease must, from our present experience, be considered from the onset not only as incurable, but also as deadly."

Today, KS is recognized as the most common tumor in human immunodeficiency virus (HIV)-1 infected individuals. The tumors are multifocal, highly vascularized neoplasms characterized histologically by spindle-shaped tumor cells that form slit-like vascular spaces often filled with erythrocytes (see Figure 159.1). The lesions also contain fibroblasts, endothelial cells (ECs, indicating angiogenesis), and an inflammatory infiltrate consisting of lymphocytes, eosinophils, and plasma cells. Early lesions resemble granulation tissue and, as the disease progresses, the tumor cells coalesce to form large masses. In these late-stage lesions, a vast majority of the tumor cells, as well as some surrounding ECs, are infected with the KS-associated herpesvirus (KSHV, also known as human herpesvirus [HHV]-8). This virus, which is believed to be the etiologic agent responsible for KS, is discussed in this chapter. As noted by Kaposi, the extremities are initially involved in most cases; however, the disease can progress to involve virtually any internal organ, particularly the lungs and gastrointestinal tract. Curative therapy has not been developed for KS, and current treatments are only able to temporarily relieve its symptoms (2).

THE KAPOSI'S SARCOMA–ASSOCIATED HERPESVIRUS

Research in KS changed dramatically in 1994, when Chang, Moore, and colleagues identified novel herpesvirus-like DNA sequences in over 90% of KS lesions (3). Subsequent work has confirmed these findings, and KSHV has since been consistently associated with KS and two lymphoproliferative disorders: primary effusion lymphoma and a subset of multicentric Castleman disease (4,5). KSHV is a lymphotropic virus (i.e., it persistently infects B-lymphocytes) that shares several features with other herpesviruses including a large, double-stranded DNA genome, replication within the nucleus, and the ability to establish a latent infection. Latency permits the life-long persistent infection of an individual, because only a small subset of viral proteins are produced, and the virus can effectively hide from the immune system. When appropriately stimulated, the virus can switch from the latent state to lytic reactivation. During the lytic cycle, viral replication occurs. A majority of the viral genes are expressed, progeny virions are produced, and the host cell is ultimately destroyed. In KS, the majority of tumor cells are latently infected, with only 1% to 3% of the cells undergoing reactivation (6).

Strong evidence indicates KSHV is the etiologic agent of KS. KSHV DNA has been detected in all four forms of KS and in virtually all tested KS lesions, seroepidemiologic studies found KSHV antibodies in most KS patients, and KSHV infection can predict the future development of KS (7–9). It is, however, currently unknown how KSHV causes KS. Clues can be gained from analysis of the KSHV genome, which has been sequenced. In addition to viral structural proteins and those necessary for viral replication, the KSHV genome contains a large number of unique genes encoding for homologues of human cellular proteins. Recent studies have identified many of these viral homologues as biologically active proteins that can induce proliferation, alter cell cycle regulation, and inhibit apoptosis, indicating they may play an important role in KS pathogenesis. In addition, KSHV alters expression of a variety of host cellular proteins to evade attack by the immune system.

EPIDEMIOLOGICAL FORMS OF KAPOSI'S SARCOMA

Four clinical-epidemiological forms of KS are currently recognized: (a) classical, (b) endemic (African), (c) iatrogenic (transplant-associated), and (d) acquired immunodeficiency syndrome (AIDS)-related (epidemic). Although these forms are histologically indistinguishable, each possesses unique

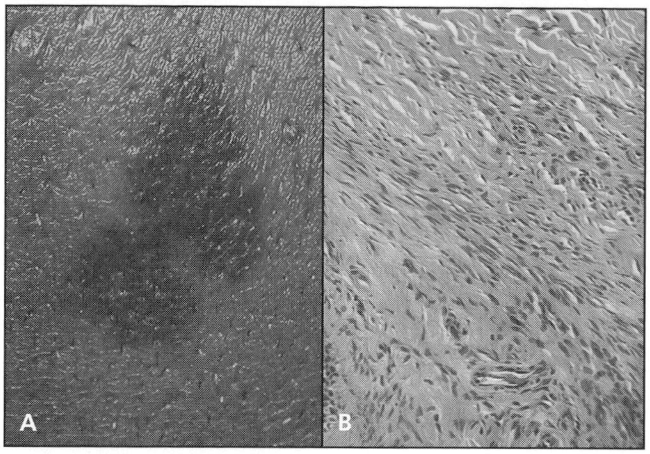

Figure 159.1. (**A**) Kaposi's sarcoma lesion. (**B**) Hematoxylin and eosin (H&E) stain of a Kaposi's sarcoma lesion. Note the fascicles of spindle-shaped tumor cells in the center of the picture, which are characteristic of these lesions. For color reproduction, see Color Plate 159.1.

characteristics with respect to anatomic involvement, prognosis/aggressive nature of the tumors, age, sex, and geographical origin of the patient.

Classical KS occurs sporadically in older men (over 50 years) of Mediterranean, Eastern European, or Jewish descent. Men are preferentially afflicted at a 15:1 male:female ratio. Interestingly, individuals born in these regions (or whose families originate from these regions) are at an increased risk for developing classic KS long after their emigration to Europe or North America (10). The tumors in classic KS generally are confined to the skin, progress slowly, and are frequently thought of as a relatively benign, cosmetic nuisance. Visceral or mucosal involvement occurs in up to 10% of patients, and arises only after a period of years (11).

Endemic KS (also known as African KS) occurs in children and young adults in sub-Saharan Africa. Prior to the AIDS epidemic, KS accounted for 9% of all reported cancers in Uganda; now, in combination with the AIDS-related form, KS accounts for almost half of all cancers in Ugandan men (12,13). Endemic KS in young adults tends to have extensive involvement of the skin. Frequently, these patients suffer from painful edema, and their tumors can ulcerate and infiltrate the surrounding tissue. Endemic KS in children is surprisingly different from that found in young adults. The neoplasm generally begins as a lymphadenopathy rather than a cutaneous lesion and rapidly progresses to the visceral organs. The prognosis is poor, with virtually 100% fatality within 3 years (14).

Iatrogenic KS occurs in approximately 0.4% of transplant patients in the United States, but in countries where endemic or classical KS occurs, it can be the most common post-transplant tumor, affecting up to 5% of transplant patients. The tumors tend to occur 2 to 5 years post-transplantation, and are most commonly found in patients with renal transplants, although it is unclear if this is related to the large number of renal transplants that are performed, or if it is due to some other unidentified factor. Iatrogenic KS can be aggressive, yet the lesions tend to regress if immunosuppressive therapy is reduced or discontinued. Several studies have attempted to determine if KSHV infection in these patients is the result of reactivation of pre-existing KSHV infection or if the recipient acquired KSHV from the organ donor, presumably through free virus or KSHV-infected B cells within the organ. Early reports indicated that both mechanisms were likely involved; however, an elegant study by Luppi and colleagues demonstrated that the KS tumor cells in five of eight (63%) renal transplant patients harbored either genetic or antigenic markers of the donor (15–17). The most striking finding in this report was the identification of the Y chromosome in KS tumor cells from female transplant patients who received an organ from a male donor (17). The simplest explanation for these findings is that individuals with no history of KS can harbor precursor tumor cells that are suppressed or killed in an immunocompetent host but, when introduced into the immunosuppressed transplant patient, the cells can circumvent normal growth control and result in tumor formation.

AIDS-related KS was initially identified in 1981, by dermatologists in San Francisco and New York, who noted the unusual diagnosis of clinically aggressive KS in otherwise healthy homosexual males (18). This observation, together with an increase in *Pneumocystis carinii* pneumonia, heralded the beginning of the AIDS epidemic in the United States. In this population, KS can be extremely aggressive. Patients can suffer from local progression as well as rapid dissemination to visceral organs, leading to death. At the beginning of the AIDS epidemic, approximately 20% of HIV-1 infected individuals were afflicted with this disease. With the introduction of highly active antiretroviral therapy (HAART) to treat HIV infection, the incidence of AIDS-KS has dramatically decreased, although it continues to be the most common tumor in HIV-1 positive individuals. This is particularly true in developing countries where state-of-the-art antiretroviral therapies are not readily available. Although it is a common misconception that KS tumor cells are infected with HIV-1, there is no evidence that HIV-1 can infect these cells. It is thought that HIV-1 contributes to KS pathogenesis through a dysregulation of the immune system that allows KSHV to escape normal immune surveillance and promotes the growth and proliferation of the tumor cells.

THE ORIGIN OF THE KAPOSI'S SARCOMA TUMOR CELL

The origin of the KS tumor cell has been debated since the time of Kaposi. Today, most experts favor the EC (or an EC precursor) as the likely progenitor of the spindle-shaped tumor cell, although compelling arguments still can be made for other cell types, including the dermal dendrocyte and pluripotent mesenchymal precursors. Most data supporting the endothelial origin of KS come from immunohistochemical staining studies identifying specific EC markers on the tumor cells. Using factor VIII-related antigen (von Willebrand factor [vWF]) and binding of *Ulex europaeus* lectin I as markers for vascular endothelium, several early studies concluded that the proliferating spindle cells as well as the cells lining the vascular channels within the tumor were of endothelial origin (19,20). Many of these studies went further to try to identify the type of endothelium (vascular or lymphatic) based on the staining pattern. Not all studies, however, confirmed these early results. Some investigators failed to detect endothelial-specific markers on the tumor cells although the proliferating ECs in the vasculature were positive (21,22). Additional studies attempted to confirm or refute the immunostaining-based results using alternative techniques, such as the identification of Weibel-Palade bodies (endothelial-specific storage vesicles for vWF) by electron microscopy; yet contradictory results continued to plague these investigations (23–25).

As additional endothelial-specific antigens have been identified over the years, studies aimed at proving or disproving the endothelial origin of KS have been regularly published (summarized in Table 159-1). Data demonstrating the combined expression of phenotypic markers such as CD31, CD34, E-selectin, and vascular endothelial growth factor receptor 2 (VEGFR2) (also known as KDR/FLK-1) have strengthened support for an EC origin and, more recently, the detection of markers restricted to lymphatic endothelium (VEGFR3 [FLT-4] and podoplanin) have led many investigators to identify lymphatic ECs as the precursor of KS. Perhaps the most compelling evidence for an endothelial origin comes from recent studies demonstrating that ECs (derived from either the umbilical vein or dermal microvasculature) can be infected with KSHV in vitro resulting in conversion of the cells from their traditional cobblestone morphology to spindle cells and, in some cases, resulting in transformation (26). This infection induced the expression of several lymphatic-specific proteins such as Prox1 and lymphatic vessel endothelial receptor (LYVE)-1, effectively "re-programming" these blood vessel–derived cells to express a lymphatic phenotype (27).

Although the data linking KS cells with ECs (either vascular or lymphatic) are abundant, these findings cannot reconcile the coexpression of EC markers with macrophage-, smooth muscle–, and dendritic cell–specific markers that are also found on KS tumor cells (see Table 159-1). As mentioned earlier, investigators have argued that infection of normal ECs results in cellular reprogramming that promotes the expression of an unusual combination of antigenic markers. Other investigators have interpreted these findings to mean the spindle cells are actually a mixture of different cell types and, as the lesions progress, a single cell type may become more prominent (28). Still others have suggested that the KS tumor cell is derived from a pluripotent progenitor cell. Infection of the cell with KSHV and/or stimulation from the cytokine-rich microenvironment (discussed later) may promote differentiation of the cell down a unique pathway. Clearly, additional studies are needed to definitively identify the precursor of the KS tumor cell.

CYTOKINES AND KAPOSI'S SARCOMA PATHOGENESIS

It has been hypothesized that KS begins as a polyclonal hyperplasia triggered by immune activation/dysregulation that may develop into a true monoclonal malignancy as the disease progresses (29). In early lesions, only 10% of the tumor cells are infected with KSHV, and this increases to virtually all of the tumor cells in late-stage lesions. These infected cells are believed to produce and release viral proteins as well as a unique combination of cytokines, chemokines, and angiogenic factors that act through autocrine and paracrine mechanisms to promote the growth, proliferation, and survival of KSHV-infected cells and uninfected cells. The importance of cytokines in KS was originally identified in studies using cultured KS tumor cells that showed the cells were highly dependent on cytokines and growth factors for growth and proliferation (30). These cells produce a large number of biologically active cytokines including interleukin (IL)-1β, IL-6, IL-8, platelet-derived growth factor (PDGF), hepatocyte growth factor (HGF), granulocyte macrophage colony stimulating factor (GM-CSF), transforming growth factor (TGF)-β, oncostatin M (OSM), basic fibroblast growth factor (bFGF), and vascular endothelial growth factor (VEGF). Subsequent studies confirmed the expression of these factors in KS lesions using in situ hybridization and immunohistochemical staining.

The role of KSHV in producing these factors is currently unclear, because cultured KS tumor cells do not appear to be infected with KSHV (except during the earliest passages) (31). This loss of KSHV is not understood, and may be due to inappropriate culture conditions that are unable to support KSHV maintenance in the tumor cells or, perhaps, noninfected cells within the culture have a growth advantage. In either case, studies of ECs infected with KSHV or expressing individual KSHV proteins have begun to provide insight into the link between cytokines and KSHV. Hayward and colleagues showed that KSHV infection of microvascular ECs induced IL-8 and TGF-β production, whereas expression of the constitutively active KSHV-encoded G-protein–coupled receptor (vGPCR; a homologue of the IL-8 receptor, CXCR2) or viral IL-6 induced VEGF expression (32–34). In addition, mediators produced by KS tumor cells may work in concert with other factors associated with immune dysfunction to create the cytokine-rich microenvironment characteristic of KS. For example, HIV-1 infected cells (including T cells, monocytes/macrophages, and dendritic cells) have been shown to secrete inflammatory cytokines such as tumor necrosis factor (TNF)-α and IL-1β, as well as induce the expression of cytokines by neighboring cells. HIV-1–infected patients are known to have higher circulating levels of IL-1, IL-3, interferon (IFN)-γ, TNF-α, TNF-β, TGF-β, OSM, and bFGF, which is probably a result of chronic viral infection and stimulation of the immune system. Taken together, the complex "cytokine storm" within AIDS-KS lesions may be partly responsible for the aggressive nature of AIDS-related KS.

Table 159-1: Antigen Expression by Kaposi's Sarcoma Tumor Cells in Vivo

Antigen or Respective Antibody	*Normal Distribution*	*KS Tumor Cell Expression*	*References*
Endothelial cells			
CD31 (PECAM)	Endothelium	Positive	49
CD34	Vascular endothelium Hematopoietic progenitor cells	Positive	49–52
CD62E (E-selectin)	Activated endothelium	Positive	53
VEGFR2	Endothelium Hematopoietic progenitor cells	Positive	54
VE-cadherin	Vascular endothelium	Positive	55
VEGFR3	Lymphatic endothelium	Positive	56–58
Podoplanin	Lymphatic endothelium	Positive	58
Factor VIII related antigen (von Willebrand factor)	Vascular endothelium	Positive	19, 20, 59–61
		Equivocal	22, 62–64
		Negative	21, 65
Ulex europaeus lectin I	Vascular endothelium	Positive	61, 62, 64
	Lymphatic endothelium*	Equivocal	22
		Negative	66
PAL-E	Vascular endothelium	Negative	64, 67
EN4	Vascular endothelium Lymphatic endothelium	Positive	64, 67
OKM5 (CD36)	Vascular endothelium	Positive	68
		Equivocal	69
HCl-1	Vascular endothelium	Equivocal	63, 68
HLA-DR/Ia	Vascular endothelium	Positive	67, 68
		Equivocal	63
		Negative	62
E92	Vascular endothelium	Positive	68
E431	Vascular endothelium	Positive	63
Thrombomodulin	Endothelium	Positive	53
B721	Vascular endothelium Vascular smooth muscle Activated lymphocytes	Equivocal	63
Weibel-Palade bodies	Endothelium	Positive	24, 69
		Negative	23, 70
Dendritic cells/monocytes/macrophages			
CD14	Monocytes/macrophages Dendritic cells (some)	Equivocal	55, 69, 71, 72
CD68	Monocytes/macrophages Dendritic cells Hematopoietic progenitor cells	Positive	55
CD206 (Mannose receptor)	Macrophages Dendritic cells	Positive	55
Factor XIIIa	Dermal dendritic cells	Positive	69, 71, 72
Fibroblasts/monocytes			
IB10	Fibroblast, monocyte/macrophage	Positive	73
Smooth muscle cells			
Smooth muscle cell α-actin	Vascular smooth muscle cells Pericytes	Positive	74

*Binding of *Ulex europaeus* lectin I to lymphatic ECs appears controversial.

ANGIOGENESIS

Angiogenesis is a complex process that is regulated by the precise balance of proangiogenic and antiangiogenic modulators. Not only are growth factors (such as VEGF and bFGF) necessary to stimulate EC proliferation and migration, but angiogenesis also requires the breakdown of the extracellular matrix by matrix metalloproteinases (MMPs), attraction of pericytes and macrophages, stimulation of smooth muscle cell proliferation and migration, and deposition of a new extracellular matrix. Establishment of adequate vascularization is essential for the development and growth of all tumors, and KS is no exception. In fact, KS is characterized by prominent vascularization, and key histologic features of the lesions include extravasated erythrocytes around new vessels and slit-like vascular spaces running alongside normal dermal vessels. These slit-like spaces are abortive vascular structures formed by the KS tumor cells. Interestingly, because the tumor cells may be of endothelial origin, mediators of angiogenesis also may function as growth factors for these cells. Indeed, many of the cytokines known to induce KS cell proliferation (including HGF, VEGF, bFGF, IL-8) are also key mediators of angiogenesis. KS cells also express the necessary receptors for these proteins including the HGF receptor c-Met, VEGFR2, and VEGFR3. Expression of VEGFR1 (FLT-1) has been reported by some groups, but was not confirmed by others (35,36).

VEGF and bFGF have been the focus of many studies in KS, because they appear to be particularly important in KS initiation and progression. High levels of both VEGF and bFGF are found in KS lesions, although circulating levels of only bFGF, not VEGF, are increased in AIDS-KS patients. Elevated levels of these angiogenic factors within the lesions may be related to the presence of inflammatory cytokines in the KS microenvironment that are known to induce both bFGF and VEGF expression (including TNF-α, IL-1β, PDGF, and IFN-γ). Elegant studies by Ensoli and colleagues found that simple overexpression of VEGF and bFGF synergistically induced angiogenic KS-like lesions in nude mice and vascular permeability/edema in guinea pigs (37). Interestingly, VEGF not only induces EC proliferation, but also induces expression of Bcl-2, an important antiapoptotic protein, which may enhance survival of the cells forming the growing vasculature (38). VEGF also stimulates production of MMPs, urokinase-type plasminogen activator, and collagenases by ECs, which are all involved in the breakdown of extracellular matrix. More recently, studies have implicated VEGF in promoting the entry of KSHV into target cells. The Raf oncoprotein, a member of the mitogen-activated protein kinase (MAPK) signaling cascade, induces VEGF expression, which subsequently enhances KSHV infection (39). It is currently unclear how VEGF promotes KSHV infection, although use of pharmacologic agents to block VEGFR1 and VEGFR2 signaling effectively blocked this enhancement.

As already mentioned, AIDS-associated KS is a particularly aggressive form of the disease. Although the immunosuppression caused by HIV-1 infection is clearly involved, studies implicate a role for factors other than immune dysregulation. One potentially important factor is the HIV-1 encoded Tat protein. Extracellular Tat has been implicated in promoting angiogenesis, and it functions as a growth factor for KS tumor cells. Tat synergizes with bFGF, but not VEGF, to induce angiogenic KS-like lesions characterized by EC proliferation and migration in vivo (40). Studies indicate this interaction involves engagement of the integrin receptors $\alpha 5\beta 1$ and $\alpha v\beta 3$ by Tat, as the blockade of the RGD (arginine-glycine-aspartic acid)-binding site in $\alpha v\beta 3$ inhibits lesion formation in mice in response to bFGF and Tat (41).

Recent studies have focused on the role of KSHV and KSHV-encoded proteins in promoting angiogenesis. KSHV-infected ECs express higher levels of VEGF, VEGFR1, VEGFR2, and VEGFR3 than do uninfected controls. Blockade of VEGF or VEGFR with antibodies or antisense oligonucleotides inhibited proliferation of KSHV-infected ECs (42). As already mentioned, KSHV-vGPCR has been shown to activate the MAPK signaling cascade resulting in VEGF expression. KSHV-vGPCR also induces constitutive activation of nuclear factor (NF)-κB and signaling activator protein (AP)-1 transcription factors. This was associated with expression of NF-κB–dependent cytokines such as IL-1β, IL-6, TNF-α, IL-8, and the AP-1-dependent bFGF. Moreover, transgenic mice expressing vGPCR under control of the *CD2* promoter developed angioproliferative lesions in the ears, tail, nose, and paws (43). These lesions strongly resemble human KS on gross and microscopic examination. Despite these exciting findings, many questions remain. The human *CD2* promoter targets gene expression to hematopoietic cells, and these transgenic mice express vGPCR only in natural killer cells and T-lymphocytes. These cells are not infected with KSHV in human disease, which complicates interpretation of the results. Furthermore, vGPCR is expressed only during lytic reactivation of the KSHV lifecycle and, therefore, can only be detected in less than 5% of cells within the lesion. It is currently unknown if expression in this small subset of cells can have a significant paracrine effect on latently infected neighboring tumor cells; however, some studies indicate that vGPCR can promote the tumorigenic potential of viral latent genes.

Several other KSHV-encoded proteins also appear to induce VEGF. Expression of vIL-6 in NIH-3T3 cells resulted in tumor formation in mice that was more vascularized than in control tumors. Enhanced VEGF expression could also be detected in these animals (34). KSHV also encodes for three chemokine homologues vMIP-I, vMIP-II, and vMIP-III. All three proteins have been shown to stimulate angiogenesis (44).

ANTIANGIOGENIC AGENTS AS A THERAPEUTIC OPTION FOR KAPOSI'S SARCOMA

Currently, no cure exists for KS. Radiation, cryotherapy, and chemotherapy are generally used for localized lesions, whereas systemic chemotherapy is necessary for disseminated disease. A need clearly exists for more effective, less toxic therapies, and antiangiogenic agents represent an attractive therapeutic target for patients with KS (as well as other cancers).

Around the time HAART therapy (including protease inhibitors) became widely used in the treatment of HIV-1 disease, the incidence of KS decreased dramatically in the United States. Although protease inhibitors were designed to specifically block the HIV-1 lifecycle and prevent virion production, studies have shown that protease inhibitors also block bFGF- and VEGF-induced angiogenesis by inhibiting MMP-2 proteolytic activity (45). MMP-2 is critical in degrading the basement membrane, and inhibition of this enzyme appears to block endothelial and KS cell invasion and prevent angiogenesis. Given this effect on neovascularization as well as the overall improvement in immunocompetence, it is understandable why HAART therapy/protease inhibitors have become a first-line therapeutic option for AIDS-related KS.

Similarly, studies have been initiated or recently completed to identify other therapeutic agents that target various aspects of angiogenesis. Clinical trials are currently underway for thalidomide, an agent that likely inhibits angiogenesis through several different mechanisms. Studies have shown that thalidomide blocks bFGF and VEGF effects, inhibits TNF-α synthesis, blocks NF-κB activation, downregulates adhesion molecule expression, stimulates T cells, and inhibits IL-6 and IL-12 production. In a phase II clinical study, 35% of AIDS-KS patients had a partial response, with 60% of those patients showing significant reductions in KSHV viral load (46). In another phase II trial, eight of 17 of patients (47%) had a partial response, and an additional two patients had stable disease (47). Unfortunately, not all antiangiogenesis agents have shown such promising results. Despite exciting preclinical and phase I results, TNP-70 (fumagillin, an inhibitor of methionine aminopeptidase, which is critically important for EC proliferation and angiogenesis), SU5416 (a VEGFR2 inhibitor), and IM862 (a dipeptide with antiangiogenic properties) were determined ineffective in phase II and phase III trials. It has been suggested that the problem may be that VEGF is more important during early tumor growth, when vessels are developing, than in tumors with established vasculature. Additional agents continue to be tested in preclinical studies with KS tumor cells (the prominent angiogenesis in these models is ideal for initial tests), while other agents are in clinical trials for other cancers. The most promising of these agents is bevacizumab, a humanized monoclonal antibody against VEGF, which in combination with standard chemotherapy has recently been approved by the U.S. Food and Drug Administration for colorectal cancer (48).

KEY POINTS

- KS is a potentially life-threatening neoplasm characterized by spindle-shaped tumor cells and prominent angiogenesis. The tumor cells are believed to be derived from either vascular or lymphatic endothelium.
- The tumor cells are somewhat unique in that the cytokine-rich microenvironment of KS provides growth factors that not only promote tumor cell proliferation, but also induce angiogenesis.
- KSHV, a human herpesvirus, is believed to be the etiologic agent responsible for KS. Virally encoded homologues of human proteins appear to play a key role in promoting expression of cytokines that function in an autocrine and paracrine fashion to promote tumor and EC growth and proliferation.
- Although angiogenesis is necessary for all tumors to grow and metastasize, KS is characterized by prominent vascularization, making antiangiogenic therapies particularly attractive targets for treating this neoplasm. Clinical trials are in progress to test this possibility.

ACKNOWLEDGMENT

Due to page limitations, the author was unable to cite all of the important primary literature that is discussed in this chapter. This work was supported in part by a grant from the National Institutes of Health/National Cancer Institute CA108450.

REFERENCES

1 Kaposi M. Idiopatisches multiples pigmentsarkom der haut. *Arch Dermatol Syphilis*. 1872;4:265–273.
2 von Roenn JH. Clinical presentations and standard therapy of AIDS-associated Kaposi's sarcoma. *Hematol Oncol Clin North Am*. 2003;17:747–762.
3 Chang Y, Cesarman E, Pessin MS, et al. Identification of herpesvirus-like DNA sequences in AIDS-associated Kaposi's sarcoma. *Science*. 1994;266:1865–1869.
4 Cesarman E, Chang Y, Moore PS, et al. Kaposi's sarcoma-associated herpesvirus-like DNA sequences in AIDS-related body-cavity-based lymphomas. *N Engl J Med*. 1995;332:1186–1191.
5 Soulier J, Grollet L, Oksenhendler E, et al. Kaposi's sarcoma-associated herpesvirus-like DNA sequences in multicentric Castleman's disease. *Blood*. 1995;86:1276–1280.
6 Sun R, Lin SF, Staskus K, et al. Kinetics of Kaposi's sarcoma-associated herpesvirus gene expression. *J Virol*. 1999;73:2232–2242.
7 Huang YQ, Li JJ, Kaplan MH, et al. Human herpesvirus-like nucleic acid in various forms of Kaposi's sarcoma. *Lancet*. 1995; 345:759–761.
8 Kedes DH, Operskalski E, Busch M, et al. The seroepidemiology of human herpesvirus 8 (Kaposi's sarcoma-associated herpesvirus) – distribution of infection in KS risk groups and evidence for sexual transmission. *Nat Med*. 1996;2:918–924.
9 Whitby D, Howard MR, Tenant-Flowers M, et al. Detection of Kaposi's Sarcoma associated herpesvirus in peripheral blood of

HIV-infected individuals and progression to Kaposi's sarcoma [see comments]. *Lancet*. 1995;346:799–802.

10 Ross RK, Casagrande JT, Dworsky RL, et al. Kaposi's sarcoma in Los Angeles, California. *J Natl Cancer Inst*. 1985;75:1011–1015.

11 Cottoni F, Masala MV, Piras P, et al. Mucosal involvement in classic Kaposi's sarcoma. *Br J Dermatol*. 2003;148:1273–1274.

12 Cook-Mozaffari P, Newton R, Beral V, Burkitt DP. The geographical distribution of Kaposi's sarcoma and of lymphomas in Africa before the AIDS epidemic. *Br J Cancer*. 1998;78:1521–1528.

13 Wabinga HR, Parkin DM, Wabwire-Mangen F, Nambooze S. Trends in cancer incidence in Kyadondo County, Uganda, 1960–1997. *Br J Cancer*. 2000;82:1585–1592.

14 Templeton AC, Bhana D. Prognosis in Kaposi's sarcoma. *J Natl Cancer Inst*. 1975;55:1301–1304.

15 Cattani P, Capuano M, Graffeo R, et al. Kaposi's sarcoma associated with previous human herpesvirus 8 infection in kidney transplant recipients. *J Clin Microbiol*. 2001;39:506–508.

16 Luppi M, Barozzi P, Santagostino G, et al. Molecular evidence of organ-related transmission of Kaposi's Sarcoma-associated herpesvirus or human herpesvirus-8 in transplant patients. *Blood*. 2000;96:3279–3281.

17 Barozzi P, Luppi M, Facchetti F, et al. Post-transplant Kaposi's Sarcoma originates from the seeding of donor-derived progenitors. *Nat Med*. 2003;9:554–561.

18 Friedman-Kien AE, Lauberstein L, Marmor M, et al. Kaposi's sarcoma and pneumocystis pneumonia among homosexual men: New York City and California. *MMWR*. 1981;30:305–308.

19 Guarda LG, Silva EG, Ordonez NG, Smith JL Jr. Factor VIII in Kaposi's sarcoma. *Am J Clin Pathol*. 1981;76:197–200.

20 Nadji M, Morales AR, Ziegles-Weissman J, Penneys NS. Kaposi's sarcoma: immunohistologic evidence for an endothelial origin. *Arch Pathol Lab Med*. 1981;105:274–275.

21 Burgdorf WH, Mukai K, Rosai J. Immunohistochemical identification of factor VIII-related antigen in endothelial cells of cutaneous lesions of alleged vascular nature. *Am J Clin Pathol*. 1981;75:167–171.

22 Miettinen M, Holthofer H, Lehto VP, et al. Ulex europaeus I lectin as a marker for tumors derived from endothelial cells. *Am J Clin Pathol*. 1983;79:32–36.

23 Mcnutt NS, Fletcher V, Conant MA. Early lesions of Kaposi's sarcoma in homosexual men. An ultrastructural comparison with other vascular proliferations in skin. *Am J Pathol*. 1983;111:62–77.

24 Leu HJ, Schneider J, Hardmeier T, et al. Kaposi's sarcoma and malignant lymphoma in AIDS. *Virchows Arch A Pathol Anat Histopathol*. 1984;403:205–212.

25 Marquart KH. Weibel-Palade bodies in Kaposi's sarcoma cells. *J Clin Pathol*. 1987;40:933.

26 Ciufo DM, Cannon JS, Poole LJ, et al. Spindle cell conversion by Kaposi's sarcoma-associated herpesvirus: formation of colonies and plaques with mixed lytic and latent gene expression in infected primary dermal microvascular endothelial cell cultures. *J Virol*. 2001;75:5614–5626.

27 Hong YK, Foreman K, Shin JW, et al. Lymphatic reprogramming of blood vascular endothelium by Kaposi's Sarcoma-associated herpesvirus. *Nat Genet*. 2004;36:683–685.

28 Roth WK, Brandstetter H, Sturzl M. Cellular and molecular features of HIV-associated Kaposi's sarcoma. *AIDS*. 1992;6:895–913.

29 Gallo RC. Some aspects of the pathogenesis of HIV-1 associated Kaposi's sarcoma. *J Natl Cancer Inst Monogr*. 1998;23:55–57.

30 Ensoli B, Nakamura S, Salahuddin SZ, et al. AIDS-Kaposi's sarcoma-derived cells express cytokines with autocrine and paracrine growth effects. *Science*. 1989;243:223–226.

31 Foreman KE, Friborg J, Kong W, et al. Propagation of a human herpesvirus from AIDS-associated Kaposi's sarcoma. *N Engl J Med*. 1997;336:163–171.

32 Poole LJ, Yu Y, Kim PS, et al. Altered patterns of cellular gene expression in dermal microvascular endothelial cells infected with Kaposi's sarcoma-associated herpesvirus. *J Virol*. 2002;76:3395–3420.

33 Sodhi A, Montaner S, Patel V, et al. The Kaposi's sarcoma-associated herpes virus G protein-coupled receptor up-regulates vascular endothelial growth factor expression and secretion through mitogen-activated protein kinase and p38 pathways acting on hypoxia-inducible factor 1alpha. *Cancer Res*. 2000;60:4873–4880.

34 Aoki Y, Jaffe ES, Chang Y, et al. Angiogenesis and hematopoiesis induced by Kaposi's sarcoma-associated herpesvirus-encoded interleukin-6. *Blood*. 1999;93:4034–4043.

35 Masood R, Cai J, Zheng T, et al. Vascular endothelial growth factor/vascular permeability factor is an autocrine growth factor for AIDS-Kaposi's Sarcoma. *Proc Natl Acad Sci USA*. 1997;94:979–984.

36 Brown LF, Tognazzi K, Dvorak HF, Harrist TJ. Strong expression of kinase insert domain-containing receptor, a vascular permeability factor/vascular endothelial growth factor receptor in AIDS-associated Kaposi's sarcoma and cutaneous angiosarcoma. *Am J Pathol*. 1996;148:1065–1074.

37 Samaniego F, Markham PD, Gendelman R, et al. Vascular endothelial growth factor and basic fibroblast growth factor present in Kaposi's sarcoma (KS) are induced by inflammatory cytokines and synergize to promote vascular permeability and KS lesion development. *Am J Pathol*. 1998;152:1433–1443.

38 Nor JE, Christensen J, Liu J, et al. Up-regulation of Bcl-2 in microvascular endothelial cells enhances intratumoral angiogenesis and accelerates tumor growth. *Cancer Res*. 2001;61:2183–2188.

39 Hamden KE, Ford PW, Whitman AG, et al. Raf-induced vascular endothelial growth factor augments Kaposi's sarcoma-associated herpesvirus infection. *J Virol*. 2004;78:13381–13390.

40 Ensoli B, Gendelman R, Markham P, et al. Synergy between basic fibroblast growth factor and HIV-1 Tat protein in induction of Kaposi's sarcoma. *Nature*. 1994;371:674–680.

41 Barillari G, Sgadari C, Palladino C, et al. Inflammatory cytokines synergize with the HIV-1 Tat protein to promote angiogenesis and Kaposi's sarcoma via induction of basic fibroblast growth factor and the alpha v beta 3 integrin. *J Immunol*. 1999;163:1929–1935.

42 Masood R, Cesarman E, Smith DL, et al. Human herpesvirus-8-transformed endothelial cells have functionally activated vascular endothelial growth factor/vascular endothelial growth factor receptor. *Am J Pathol*. 2002;160:23–29.

43 Yang TY, Chen SC, Leach MW, et al. Transgenic expression of the chemokine receptor encoded by human herpesvirus 8 induces an angioproliferative disease resembling Kaposi's sarcoma. *J Exp Med*. 2000;191:445–453.

44 Stine JT, Wood C, Hill M, et al. KSHV-encoded CC chemokine vMIP-III is a CCR4 agonist, stimulates angiogenesis, and selectively chemoattracts TH2 cells. *Blood*. 2000;95:1151–1157.

45 Sgadari C, Barillari G, Toschi E, et al. HIV protease inhibitors are potent anti-angiogenic molecules and promote regression of Kaposi's Sarcoma. *Nat Med*. 2002;8:225–232.

46 Fife K, Howard MR, Gracie F, et al. Activity of thalidomide in AIDS-related Kaposi's sarcoma and correlation with HHV8 titre. *Int J STD AIDS*. 1998;9:751–755.

47 Little RF, Wyvill KM, Pluda JM, et al. Activity of thalidomide in AIDS-related Kaposi's sarcoma. *J Clin Oncol*. 2000;18:2593–2602.

48 Willett CG, Boucher Y, di Tomaso E, et al. Direct evidence that the VEGF-specific antibody bevacizumab has antivascular effects in human rectal cancer. *Nat Med*. 2004;10:145–147.

49 Nickoloff BJ. PECAM-1 (CD31) is expressed on proliferating endothelial cells, stromal spindle-shaped cells, and dermal dendrocytes in Kaposi's sarcoma. *Arch Dermatol*. 1993;129:250–251.

50 Nickoloff BJ. The human progenitor cell antigen (CD34) is localized on endothelial cells, dermal dendritic cells, and perifollicular cells in formalin-fixed normal skin, and on proliferating endothelial cells and stromal spindle-shaped cells in Kaposi's sarcoma. *Arch Dermatol*. 1991;127:523–529.

51 Kraffert C, Planus L, Penneys NS. Kaposi's sarcoma: further immunohistologic evidence of a vascular endothelial origin. *Arch Dermatol*. 1991;127:1734–1735.

52 Kanitakis J, Narvaez D, Claudy A. Expression of the CD34 antigen distinguishes Kaposi's sarcoma from pseudo-Kaposi's sarcoma (acroangiodermatitis). *Br J Dermatol*. 1996;134:44–46.

53 Zhang YM, Bachmann S, Hemmer C, et al. Vascular origin of Kaposi's sarcoma. Expression of leukocyte adhesion molecule-1, thrombomodulin, and tissue factor. *Am J Pathol*. 1994;144:51–59.

54 Brown LF, Tognazzi K, Dvorak HF, Harrist TJ. Strong expression of kinase insert domain-containing receptor, a vascular permeability factor/vascular endothelial growth factor receptor in AIDS-associated Kaposi's sarcoma and cutaneous angiosarcoma. *Am J Pathol*. 1996;48:1065–1074.

55 Uccini S, Sirianni MC, Vincenzi L, et al. Kaposi's sarcoma cells express the macrophage-associated antigen mannose receptor and develop in peripheral blood cultures of Kaposi's sarcoma patients. *Am J Pathol*. 1997;150:929–938.

56 Jussila L, Valtola R, Partanen TA, et al. Lymphatic endothelium and Kaposi's sarcoma spindle cells detected by antibodies against the vascular endothelial growth factor receptor-3. *Cancer Res*. 1998;58:1599–1604.

57 Partanen TA, Alitalo K, Miettinen M. Lack of lymphatic vascular specificity of vascular endothelial growth factor receptor 3 in 185 vascular tumors. *Cancer*. 1999;86:2406–2412.

58 Weninger W, Partanen TA, Breiteneder-Geleff S, et al. Expression of vascular endothelial growth factor receptor-3 and podoplanin suggests a lymphatic endothelial cell origin of Kaposi's sarcoma tumor cells. *Lab Invest*. 1999;79:243–251.

59 Flotte TJ, Hatcher VA, Friedman-Kien AE. Factor VIII-related antigen in Kaposi's sarcoma in young homosexual men. *Arch Dermatol*. 1984;120:180–182.

60 Facchetti F, Lucini L, Gavazzoni R, Callea F. Immunomorphological analysis of the role of blood vessel endothelium in the morphogenesis of cutaneous Kaposi's sarcoma: a study of 57 cases. *Histopathology*. 1988;12:581–593.

61 Little D, Said JW, Siegel RJ, et al. Endothelial cell markers in vascular neoplasms: an immunohistochemical study comparing factor VIII-related antigen, blood group specific antigens, 6-keto-PGF1 alpha, and Ulex europaeus 1 lectin. *J Pathol*. 1986;149: 89–95.

62 Beckstead JH, Wood GS, Fletcher V. Evidence for the origin of Kaposi's sarcoma from lymphatic endothelium. *Am J Pathol*. 1985;119:294–300.

63 Scully PA, Steinman HK, Kennedy C, et al. AIDS-related Kaposi's sarcoma displays differential expression of endothelial surface antigens. *Am J Pathol*. 1988;130:244–251.

64 Jones RR, Spaull J, Spry C, Jones EW. Histogenesis of Kaposi's sarcoma in patients with and without acquired immune deficiency syndrome (AIDS). *J Clin Pathol*. 1986;39:742–749.

65 Millard PR, Heryet AR. An immunohistological study of factor VIII related antigen and Kaposi's sarcoma using polyclonal and monoclonal antibodies. *J Pathol*. 1985;146:31–38.

66 Walker RA. Ulex europeus I – peroxidase as a marker of vascular endothelium: its application in routine histopathology. *J Pathol*. 1985;146:123–127.

67 Nadimi H, Saatee S, Armin A, Toto PD. Expression of endothelial cell markers PAL-E and EN-4 and Ia-antigens in Kaposi's sarcoma. *J Oral Pathol*. 1988;17:416–420.

68 Rutgers JL, Wieczorek R, Bonetti F, et al. The expression of endothelial cell surface antigens by AIDS-associated Kaposi's sarcoma. Evidence for a vascular endothelial cell origin. *Am J Pathol*. 1986;122:493–499.

69 Rappersberger K, Tschachler E, Zonzits E, et al. Endemic Kaposi's sarcoma in human immunodeficiency virus type 1-seronegative persons: demonstration of retrovirus-like particles in cutaneous lesions. *J Invest Dermatol*. 1990;95:371–381.

70 Dictor M, Carlen B, Bendsoe N, Flamholc L. Ultrastructural development of Kaposi's sarcoma in relation to the dermal microvasculature. *Virchows Arch A Pathol Anat Histopathol*. 1991; 419:35–43.

71 Nickoloff BJ, Griffiths CE. The spindle-shaped cells in cutaneous Kaposi's sarcoma. Histologic simulators include factor XIIIa dermal dendrocytes. *Am J Pathol*. 1989;135:793–800.

72 Nickoloff BJ, Griffiths CE. Factor XIIIa-expressing dermal dendrocytes in AIDS-associated cutaneous Kaposi's sarcomas. *Science*. 1989;243:1736–1737.

73 Simonart T, Hermans P, Schandene L, Van Vooren JP. Phenotypic characteristics of Kaposi's sarcoma tumour cells derived from patch-, plaque- and nodular-stage lesions: analysis of cell cultures isolated from AIDS and non-AIDS patients and review of the literature. *Br J Dermatol*. 2000;143:557–563.

74 Weich HA, Salahuddin SZ, Gill P, et al. AIDS-associated Kaposi's sarcoma-derived cells in long-term culture express and synthesize smooth muscle alpha-actin. *Am J Pathol*. 1991;139:1251–1258.

160

Endothelial Mimicry of Placental Trophoblast Cells

Hartmut Weiler and Rashmi Sood

Blood Center of Wisconsin, Blood Research Institute, Milwaukee

The placenta has evolved as a specialized organ in a subgroup of mammals to enable the prolonged intrauterine development of the fetus. It is a gender-specific and transitory organ that is essential for the exchange of nutrients and metabolites between the circulatory systems of the mother and fetus. The vascular bed of the placenta is established de novo at the onset of pregnancy, and regresses after delivery. The tissue mass of the placenta comprises both maternal and fetal elements. As a result of the so-called *hemochorial* type of placentation observed in humans, the vascular space containing maternal blood in the placenta is lined by two different types of cells: endothelial cells (ECs) of maternal origin and trophoblast cells of fetal origin. On the other hand, fetal blood vessels emanating from the embryo/fetus into the placental tissue remain intact, and fetal and maternal blood therefore remains segregated. Thus, whereas in all other vascular beds the blood vessel endothelium is the principal gatekeeper between tissue and blood, in the placenta a different cell type, the trophoblast cell, occupies part of the interface between (maternal) blood and (fetal) tissue. By their particular anatomical position at the blood–tissue boundary, trophoblast cells are equivalent to ECs in other organs, and fulfill EC-like functions. The concept of *endothelial mimicry* by trophoblast cells was initially coined to describe a remarkable process of remodeling of the maternal arteries supplying blood to the placenta, during which so-called *endovascular* fetal trophoblast cells replace the maternal endothelium in these arteries and induce an erosion of smooth muscle cells regulating the vascular tone of these arteries. It is now evident that not only endovascular trophoblast cells, but other fetal trophoblast cells in the placenta, adopt EC-like properties that allow them to function in a similar fashion to ECs in other segments of the vascular tree. Thus, in the vascular bed of the placenta, one encounters cellular and genetic heterogeneity of the vascular lining that is (under normal physiological conditions) not found anywhere else in the body. This unique architecture may underlie highly vascular bed–specific pathologies that emerge only during pregnancy and are caused by the failure of trophoblast cells to fulfill EC-like functions.

ENDOTHELIAL MIMICRY BY ANATOMICAL LOCATION

Successful implantation of the blastocyst-stage embryo into the uterine wall triggers proliferation and differentiation of maternal stromal cells in the wall of the uterus. This results in a thickening of the uterine wall, in which the embryo then becomes embedded. This process of decidualization is accompanied by the growth of new blood vessels that extend from existing uterine arteries toward the site of implantation to deliver maternal blood to the embryo. A characteristic feature of placentation in humans is that maternal blood vessels supplying the decidua lose or fail to maintain their integrity, resulting in an "open-ended" circulation exposing the outermost layer of the implanted blastocyst to direct contact with maternal blood. This type of placentation is termed *hemochorial*. The outermost fetal cell population is represented by trophoblast cells derived from the fetal trophectoderm (TE). This cell population emerges as the result of the very first developmental lineage allocation in embryogenesis. This allocation specifies three cell populations in the blastocyst around the time of implantation: TE, primitive endoderm (PE), and the epiblast (EPI). The latter two constitute the inner cell mass (ICM), which contains the pluripotent embryonic stem cell population that gives rise to the embryo itself; PE gives rise to extraembryonic membranes (amnion, chorion), and TE is the source of all the trophoblast derivatives that establish the fetal aspect of the future placenta. For reviews of early postimplantation development, see Refs. 1–4.

Around the second week of pregnancy, the conceptus is surrounded by an outer layer of trophoblast cells that have fused into a multinucleated syncytium to form the syncytiotrophoblast. The syncytiotrophoblast layer is the first to be in contact with maternal blood, and it remains the principal fetal cell type at the blood–tissue interface. Between the second and third week of gestation, cytotrophoblast cells underlying the syncytium organize into columns (villi) that expand

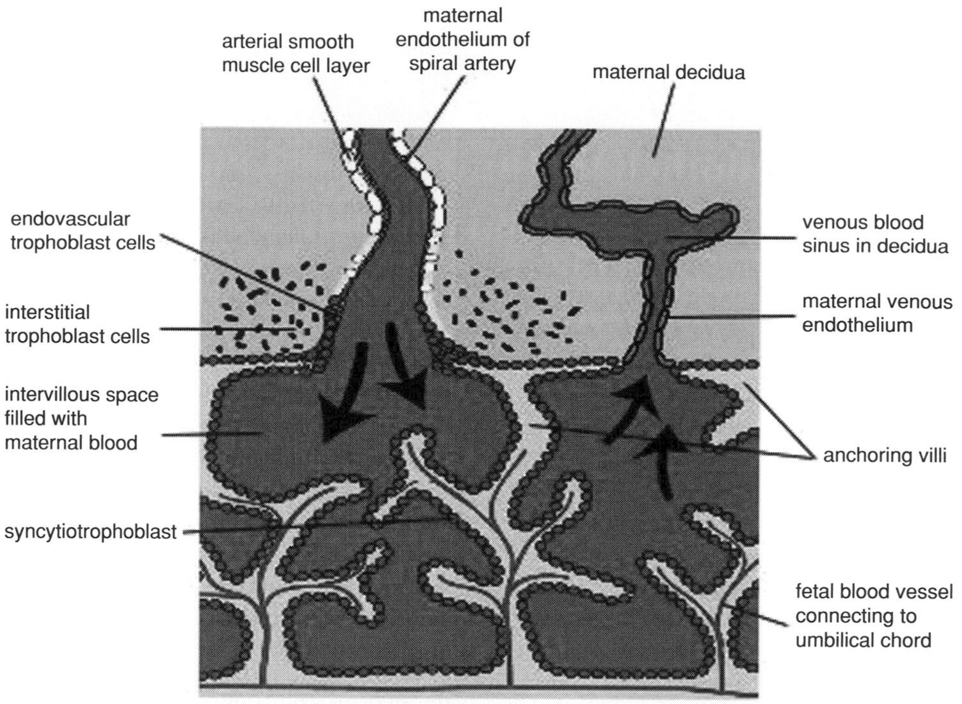

Figure 160.1. Schematic representation of the vascular bed in the placenta. Maternal blood reaches the intervillous space through spiral arteries (*arrows* indicating the principal flow of blood). These vessels are remodeled through interaction with trophoblast cells (endovascular trophoblasts) that migrate out from the tip of anchoring villi and eventually line the terminal segments of the arteries. The surface of the fetal villi exposed to maternal blood is covered by trophoblast cells. Fetal blood vessels are always lined by an intact fetal endothelium. Blood exits through maternal veins into the cavernous sinus within the decidua. As a result, the lining of the intervillous space filled with maternal blood is made up almost exclusively of fetal trophoblast cells.

into the decidua. Initially, the columns consist entirely of cytotrophoblast (primary villi). Extraembryonic mesenchyme penetrates the primary villi and forms the core of the cytotrophoblast columns (secondary villi). By the third week, the villi expand into the spaces filled with maternal blood (lacunae), and the mesenchymal core of the columns is vascularized by fetal blood capillaries (tertiary villi), which are contiguous with the umbilical vessels. Villi that make direct contact with decidual tissue of the mother are termed *anchoring villi* (Figure 160.1).

As a consequence, the peculiar anatomy of hemochorial placentation creates a vascular bed in which the lining of the maternal blood space is comprised of ECs lining the maternal blood vessels leading into the placenta and of fetal trophoblast cells covering the surface of the placental villi. This heterogeneity with respect to cellular and genetic composition is unique to the vascular bed of the placenta.

ENDOVASCULAR TROPHOBLAST AND REMODELING OF MATERNAL ARTERIES

Maternal blood is supplied to the placenta through radial arteries that emanate from the arcuate blood vessels of the uterine wall. As they enter the decidua, the former become coiled spiral arteries from which blood flows into blood lacunae into which the trophoblastic villi protrude. During the course of pregnancy, the morphology and functional properties of these supplying arteries change dramatically. These alterations are brought about in part by the interaction of spiral arteries with a specialized subpopulation of trophoblast cells, the invasive extravillous trophoblast. These trophoblast cells invade the maternal decidua from the tip of anchoring villi. Some of the extravillous trophoblast cells remain in the interstitium of the decidua, while others invade the distal portions of the spiral arterioles closest to the site of implantation. Initially, the latter form clusters of "endovascular" trophoblasts, which plug the outlets of the spiral arteries and, in this way, suppress flow through these arteries. It is thought that this "clogging" of maternal blood vessels sustains a state of relative hypoxia within the deep placental bed, which could in theory support or sustain an initial, hypoxia-driven burst of vascularization of the villi by fetal capillaries. Gradual resolution of endovascular trophoblast clusters, which may involve apoptosis, coincides with the onset of measurable blood flow in the placental bed at around 3 months into gestation.

After the onset of flow, spiral arteries undergo a remodeling process that results in the dilation of these vessels, loss of elasticity and vasomotor control, loss of vascular wall smooth muscle cells, and invasive replacement of maternal ECs by fetal trophoblasts. The dilation and erosion of smooth muscle cells does not appear to depend on direct trophoblast–vessel interactions, but could be triggered by nitric oxide (NO) and/or other mediators released from remote trophoblast cells or by uterine lymphocytes, in particular a specific subset of so-called uterine natural killer cells. The mechanisms governing the evident tropism of endovascular trophoblasts toward arteries, as opposed to the venous segment of the maternal vessels, are largely unknown, and might involve chemokine signals released from platelets adhering to the terminal segments of the arteries (5), and/or cell–cell interactions mediated by the ephrin/ephrin-receptor system expressed by trophoblast cells and the vascular endothelium (6,7). The population of the spiral artery lumen and wall with endovascular trophoblasts is accompanied by a switch in the expression of adhesion receptors. Cytotrophoblast cells giving rise to of endovascular trophoblast cells produce typical epithelial adhesion markers, such as E-cadherin and $\alpha 6\beta 4$ integrin. In contrast, the trophoblast cells replacing the endothelium of spiral arteries downregulate these receptors, and switch to expressing typical EC adhesion molecules, such as vascular endothelial (VE)-cadherin, platelet-endothelial cell adhesion molecule (PECAM)-1, and $\alpha V\beta 3$ integrin, as well as sialyl Lewis X (sLe^x) (NeuAcα2–3Galβ1–4[Fucα1–3]GlcNAcβ1–) carbohydrate moieties, which might engage E- and P-selectins present on resident endothelium (8). The adoption of an EC-like adhesion receptor phenotype has been termed *endothelial mimicry*, or *epithelial–endothelial transformation* (9,10). It is thought that the sum of these alterations in the structure of the spiral arteries enhances maternal blood flow to the placenta, and – more importantly – uncouples the regulation of blood flow into the placental vascular bed from maternal vasomotor control.

Much of the interest in the molecular basis of the remodeling of uterine arteries and the consequences for placental perfusion has been fueled by the single most devastating disease of pregnancy, preeclampsia. In general, it is thought that the disease is the result of defective feto-maternal interactions regulating the perfusion of the placenta, including those summarized here. Abnormally low perfusion may trigger the release of trophoblast-derived signals into the systemic vasculature, which then cause the various pathologies characteristic of preeclampsia. The current knowledge about this disease is summarized in Chapter 161.

TROPHOBLAST CELLS REGULATE HEMOSTASIS AT THE FETO-MATERNAL INTERFACE IN AN ENDOTHELIAL-LIKE MANNER

The anatomical location of trophoblast cells places them in direct contact with maternal blood. In virtually all nonplacental organ systems, a major function of the blood vessel endothelium is to maintain a noncoagulant surface. Normal endothelium expresses a wide spectrum of molecules that proactively suppress the activity of the coagulation system and various aspects of platelet function. A survey of the literature provides ample evidence that trophoblast cells maintain an anticoagulant expression profile similar to that of ECs (Table 160-1). This comprises gene products that mediate intravascular hemostasis, including regulators of blood coagulation thrombomodulin (TM); endothelial protein C receptor (EPCR), the inhibitor of coagulation initiation tissue factor pathway inhibitor (TFPI)-1; the ectoenzyme ecto-ADPase/CD39, which suppresses ADP-dependent platelet activation; the thrombin inhibitor protease nexin (PN)-1; and the anticoagulant annexin V; fibrinolytic regulators such as urokinase-type plasminogen activator (u-PA), tissue-type plasminogen activator (t-PA), and plasminogen activator inhibitor (PAI)-1; and regulators of vascular tone and leukocyte interactions (including inducible NO synthase [iNOS], prostacyclin synthase, cyclooxygenase [COX]-2, and E-selectin).

A systematic analysis of the gene expression program of differentiating mouse trophoblast cells shows that this endothelial-like phenotype is conserved between mice and humans, and that trophoblast cells can acquire the EC-like anticoagulant phenotype in a differentiation-dependent manner in vitro. The differentiation-associated expression of a subgroup of these hemostatic regulators occurs in a concerted fashion, suggestive of a common regulatory mechanism that coordinates their expression during trophoblast differentiation (Figure 160.2; for color reproduction, see Color Plate 160.2).

Trophoblast cells are able not only to regulate the initiation and extent of the blood coagulation reaction, but also sense the presence of activated coagulation factors. Several reports describe the expression of the thrombin receptor protease-activated receptor (PAR)-1 in trophoblast cells of the human and mouse placenta, and demonstrate that – at least in vitro – the principal coagulation protease thrombin can modulate the invasiveness and proliferation of human and mouse trophoblast cells (11–14). Likewise, activation of the platelet arm of hemostasis is known to modulate placental function, and soluble mediators released by activated platelets stimulate trophoblast migration and invasion, and induce expression of $\alpha 1$ integrin (5).

In summary, these observations suggest that the ability to control and sense the activation state of coagulation in the maternal blood space of the placenta is a defining and intrinsic feature of trophoblast cells, akin to that of the endothelium in other vascular compartments.

TROPHOBLAST CELLS AND ACTIVATED ENDOTHELIUM

As outlined earlier, the anatomical location of trophoblast cells places them in direct contact with maternal blood. Akin to blood vessel endothelium in other organs, the trophoblastic

Table 160-1: Hemostasis-Related Genes Expressed by Mouse and Human Trophoblasts

		In Situ Expression (Refs.)	
Gene	*Increased during in Vitro Differentiation of Mouse Trophoblast*[auo]	*In Embryonic Day 9.5 Mouse Placenta*	*In Preterm Human Placenta*
Tissue factor	—	TGC (14, 44)	CT (48, 49)
TM	145-fold	TGC (14, 44, 50, AUO)	SYN (51, 52, AUO)
TFPI	10-fold	TGC (AUO)	SYN, CT (53, AUO)
EPCR	4-fold	TGC (45, AUO)	SYN, CT (AUO)
PAR-1	—	Lab (AUO)	SYN, CT (11–13)
PAR-2	2-fold	—	SYN, CT (13)
PAR-3	—	—	SYN, CT (12, 13)
PAR-4	—	TGC (AUO)	
PN-1	5-fold	ST (AUO)	CT (54, AUO)
HS 3-O-sulfotransferase (3-OST-1)	18-fold	TGC (AUO)	—
Annexin (A2)	2-fold		SYN (55)
Annexin (A5)	—		Tr Memb (38, 56, 57)
u-PA-receptor (uPAR)	—		I, endo, peri tr (58, 59)
t-PA	8-fold		I, endo, peri tr (59)
u-PA	—		Term chorion (60)
PAI-1	14-fold		Invading tr (61)
Ecto-ADPase (CD39)	140-fold	TGC	SYN, CT (62, AUO)
Ecto-5′-nucleotidase	4-fold		
Nitric oxide synthase-1 (nNOS)	—		
iNOS	11-fold		
Prostacyclin synthase	2-fold		(63)
PGE synthase 2	2-fold		
COX-2	8-fold		

Fold increases in expression are averages of four independent expression profiling experiments, each on trophoblast stem cells and their in vitro differentiated derivatives (AUO; authors' unpublished observations). Sites of in situ gene expression in developing mouse and human placenta are indicated: TGC, trophoblast giant cells; Lab, labyrinth; ST, spongiotrophoblasts; CT, cytotrophoblasts; SYN, synctiotrophoblasts; Tr Memb, trophoblast membranes; I tr, invading trophoblasts; endo tr, endovascular trophoblasts; peri tr, perivascular trophoblasts.

lining of the maternal blood space in the placenta produces a battery of gene products that suppress the activation of coagulation and platelets. On the other hand, the placenta is also a rich source of the initiator of coagulation, tissue factor. Surprisingly, tissue factor antigen and, more importantly, procoagulant activity is also detected in syncytiotrophoblast membranes contacting maternal blood (15). Such an expression of tissue factor on cells at the blood–tissue interface is in striking contrast to all other vascular beds, in which tissue factor expression is strictly excluded from ECs. The sequestration of tissue factor from inadvertent contact with blood in a normal, unperturbed blood vessel is essential to prevent initiation of the blood coagulation pathways. However, tissue factor expression in ECs may be induced in response to a variety of stimuli usually associated with the presence of inflammatory challenges. Proinflammatory cytokines, ligands for Toll-like receptors, and the principal coagulation protease, thrombin, not only induce tissue factor expression, but also evoke increased production of adhesion receptors that regulate endothelial–leukocyte interactions. Moreover, these mediators simultaneously suppress the transcription of anticoagulant gene products, most notably endothelial thrombomodulin. This transition of the endothelial phenotype from a noncoagulant and antiadhesive phenotype to a state of enhanced coagulation and leukocyte interactions has been termed *endothelial activation*, and appears to reflect a principal switch in a concerted gene expression program that is under the control of defined transcription factors (16,17). Thus, trophoblast cells exhibit, even under normal conditions, at least one hallmark of activated endothelium – expression of tissue factor. This procoagulant feature of trophoblast cells has functional implications in the manifestation of organ-specific pathologies of dysregulated hemostasis in the placenta (see following sections).

The paradigm of trophoblast cells resembling "activated" endothelium extends to a critical regulator of endothelial–leukocyte interactions, E-selectin. This selectin is found on a subpopulation of trophoblast cells that line, or are in close apposition to, vascular channels of the maternal blood compartment in the decidua, and therefore are likely identical to endovascular trophoblast cells (18–20). E-selectin is an EC-specific lectin whose expression is – akin to tissue factor – induced by proinflammatory cytokines and

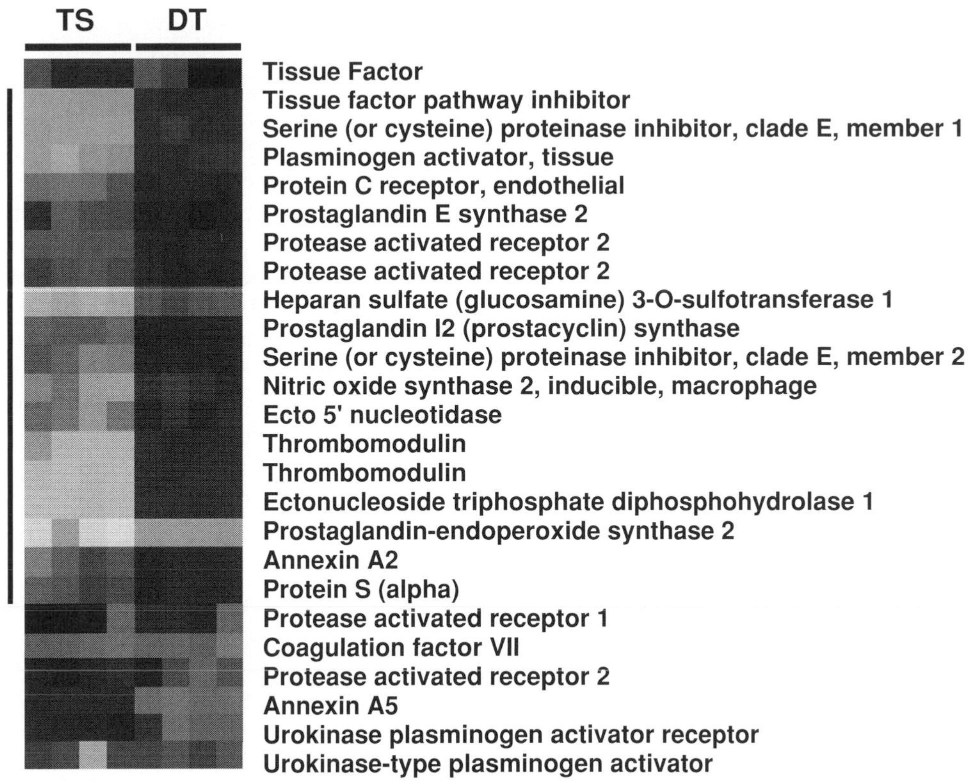

Figure 160.2. Expression of hemostasis-related genes during the in vitro differentiation of mouse trophoblast cells. The fetal trophoblast component of the placenta originates from a pool of trophoblast stem (TS) cells that proliferate and give rise to differentiated trophoblast derivates (DT). This differentiation process of TS cells is recapitulated in vitro. The figure shows the result of a gene chip-based analysis of the changes in mRNA expression of hemostasis-related genes that occur during in vitro differentiation of TS cells. Each gene is represented by a single row of colored boxes; each experiment by a single column. The color of the box represents expression levels in comparison to median expression for that gene in all experiments. Gray represents downregulated (lower than median) expression; dark gray represents upregulated (higher than median) expression; genes unchanged in expression level during differentiation are colored black. Sixteen of the 22 hemostasis-related genes expressed in mouse trophoblasts are upregulated during differentiation, essentially following an "off-on" pattern. For color reproduction, see Color Plate 160.2.

lipopolysaccharide (LPS); E-selectin binds counter-receptors on neutrophils and monocytes and regulates leukocyte recruitment to, and rolling on, activated endothelium. The maternal endothelium lining vascular channels also displays a highly unusual profile of adhesion receptor expression. These maternal vessels display a combination of P-selectin and mucosal addressin cell adhesion molecule (MAdCAM)-1 that appears unique to the placenta (21). E-selectin expression on trophoblast cells, and the peculiar distribution of other leukocyte interaction receptors, may play an important role in allowing the antigenically "foreign" embryo to escape immune rejection by the mother. Theoretically, these adhesion receptors could mediate site-specific homing of leukocyte populations to defined anatomical regions in the placenta. Such an anatomical sequestration, recruitment of specific leukocyte subpopulations, or exclusion of populations with a high potential for allogeneic rejection may, in concert with other well-characterized mechanisms, help to maintain an immunologically permissive environment for the fetal allograft.

"ENDOTHELIAL DYSFUNCTION" OF TROPHOBLAST CELLS: A CAUSE OF THROMBOPHILIA-ASSOCIATED PREGNANCY COMPLICATIONS?

Pregnancy is an independent risk factor for the occurrence of venous thrombosis in the mother. Possibly as a response to the unique physiological challenges to the mother during pregnancy and childbirth, the production of coagulation factors is augmented and the activity of the protein C system declines over gestation. These changes lead to heightened hemostatic system activity and produce a physiological procoagulant state in pregnant females (22). Increased activation of the blood coagulation system also may have adverse effects on the health of the fetus: A body of evidence suggests that inherited and acquired thrombophilia of the mother, including defects in the function of the natural protein C anticoagulant pathway, contribute to the pathogenesis of fetal loss at various stages of gestation. A meta-analysis of 31 quality-controlled epidemiological studies shows that the factor V (fV) Leiden mutation is associated with early and late recurrent fetal loss, and late, nonrecurrent fetal loss; protein S deficiency is associated with recurrent fetal loss; and activated protein C (APC) resistance unrelated to the fV Leiden mutation increases the odds of early recurrent fetal loss by more than threefold (23–25). Thrombophilia also has been associated with obstetric complications other than fetal loss – preeclampsia, intrauterine growth restriction (IUGR), placental abruption, and stillbirth (22).

It is noteworthy that none of these studies assessed the genotype of the fetus. Yet, one would expect that the incidence of fetal loss is increased if the mother's fV Leiden carrier status (rendering the maternal blood resistant to the anticoagulant of APC) coincided with reduced expression or function of thrombomodulin or the protein C receptor EPCR by fetal trophoblast cells (resulting in diminished activation of the mother's protein C at the villus surface). Such a failure of fetal trophoblast cells to sustain the natural protein C anticoagulant mechanism would lead to a vascular bed–specific epistasis of maternal and fetal risk factors that could result in localized coagulation activation in the placenta. Conversely, because of the aforementioned segregation of fetal and maternal circulatory systems, the fV Leiden status of the fetus would only affect the fetal intravascular compartment, but remain without consequence for hemostasis in the maternal blood contacting the placental villi. This concept may be an important issue to consider when evaluating epidemiological data on the association of maternal thrombophilia and fetal loss, because fetal defects in trophoblast function may be a critical modifying risk factor for pregnancy complications in mothers with pre-existing coagulation defects.

Dysfunction of trophoblast cells may also be the mechanism underlying fetal loss associated with the antiphospholipid antibody syndrome (APS). APS is an acquired autoimmune disorder that occurs in patients having elevated levels of antibodies against anionic phospholipid–protein complexes. It is characterized by clinical episodes of venous, arterial, or small-vessel thrombosis. In pregnant women, this syndrome is associated with recurrent spontaneous abortions, premature birth, intrauterine fetal death, preeclampsia, and intrauterine growth retardation (26,27). Up to 20% of women with recurrent miscarriage have laboratory evidence of antiphospholipid (aPL) antibodies (28). Of women with APS, approximately 15% to 20% have obstetric complications, and up to 50% to 75% of their pregnancies have a poor outcome (29,30). Antiphospholipid antibodies are a heterogeneous family of immunoglobulins that recognize a rather wide range of plasma proteins bound to anionic surfaces. β_2-glycoprotein I (β_2GPI), a phospholipid-binding protein, is the major antigenic target for aPL antibodies. Several proteins of the coagulation pathway, including prothrombin, protein C, protein S, EPCR, and annexin V are also targets of aPL antibodies (31–33). Thus, more than one biochemical mechanism may be involved in this syndrome, and both maternal systemic and local fetal components of the placenta are equally likely targets of the dysregulation and fetal loss induced by aPL antibodies. aPL antibodies inhibit the inactivation of activated factor V (fVa) by aPC on a phospholipid surface (34,35). The term *acquired resistance to APC* has been coined to identify this condition, which is phenotypically very similar to the resistance to aPC seen in fV Leiden patients. aPL antibodies also have been proposed to disrupt the annexin V anticoagulant shield on placental trophoblasts. IgG fraction from APS patient serum reduces annexin V on cultured trophoblasts in vitro, whereas a marked reduction occurs in annexin V expression on synctiotrophoblasts lining the placental villi in APS patients in vivo (36–38). In vitro studies show that removal of annexin V from the surface of mouse trophoblasts by treatment with ethylene glycol tetraacetic acid (EGTA) accelerates the coagulation of plasma exposed to these cells (39). Infusion of pregnant mice with polyclonal antiannexin V antibodies results in placental infarction and pregnancy wastage (40). The binding of aPL antibodies to β_2GPI leads to activation of cultured ECs, manifested by upregulation of cell surface adhesion molecules and increased secretion of interleukin-6 and prostaglandins. Annexin II, the molecule mediating high-affinity binding of β_2GPI to the endothelial surface, is highly expressed on mouse trophoblast cells and upregulated in differentiated cells characteristic of the chorioallantoic placenta (41). Although the contributions of thrombosis and trophoblast abnormalities in the pathogenesis of APS-associated fetal loss still need experimental demonstration, it is clear that complement activation is an absolute requirement in this process (42). The latter finding was based on the surprising observation that the beneficial effect of heparin treatment in an animal model of aPL-induced fetal loss was not due to heparin's anticoagulant effect, but rather to the intrinsic ability of heparin to inhibit complement activation (43).

Like humans, mice exhibit the hemochorial type of placentation. Despite anatomical differences, the basic architecture of the placental vascular bed and the relation of trophoblast cells to the maternal blood space are identical in both species (2). Analysis of mice with greatly reduced tissue factor activity, and of mice lacking the membrane receptor components of the protein C pathway (e.g., thrombomodulin and EPCR), has provided insight into the role of these coagulation regulators in placental development and function.

First, it was shown that constitutive expression of tissue factor by placental trophoblast cells is essential for normal placental function (44). In mice with greatly diminished tissue factor function, the placenta of tissue factor–deficient mothers carrying embryos with normal tissue factor function contains numerous blood pools in the maternal vascular compartment. If the maternal tissue factor defect coincides with reduced tissue factor function on the fetal aspects of the placenta (as occurs in tissue factor–deficient mothers carrying tissue factor–deficient embryos), pregnancy fails due to fatal hemorrhage of the mother. This clearly shows that normal placental function requires normal maternal *and* fetal tissue factor function.

Second, the analysis of mice lacking thrombomodulin or EPCR showed that the absence of either one protein C pathway component from trophoblast cells resulted in fetal loss at a time point roughly equivalent to the end of the first trimester in human gestation (14,45–47). These observations document that the ability of trophoblast cells to suppress the coagulation reaction in the maternal blood space are critical for placental function.

Third, in a mouse model of maternal thrombophilia secondary to the fV Leiden mutation, fetal loss occurs only if maternal thrombophilia coincides with a fetal defect in the trophoblasts' anticoagulant capacity (H. Weiler, unpublished observations). Mice that are homozygous for the fV Leiden polymorphism only rarely develop spontaneous thrombosis, and they appear able to sustain multiple pregnancies without complications. Likewise, in contrast to mice completely lacking thrombomodulin, mice with reduced thrombomodulin function are viable and show no reproductive abnormalities. However, a fetus that is homozygous for the thrombomodulin defect (which results in reduced ability of trophoblast cells to control coagulation) cannot survive in a mother that is a homozygous fV Leiden carrier (but has nearly normal thrombomodulin function). On the other hand, a fetus that is homozygous for both the fV Leiden mutation and the thrombomodulin defect can develop normally when carried by a mother who is only heterozygous for the Leiden allele. Moreover, the resulting mice are viable and do not appear to develop spontaneous thrombosis. Thus, in the placenta, the epistasis of a maternal fV Leiden defect and reduced fetal thrombomodulin function results in fetal loss, whereas the same gene–gene interaction does not disrupt vascular function in vascular beds other than the placenta. In part, the particular sensitivity of the placental vascular bed may be secondary to the constitutive presence of the blood clotting initiator tissue factor in trophoblast cells.

KEY POINTS

- The vascular bed of the placenta exhibits a unique cellular and genetic heterogeneity, in which the blood–tissue interface is comprised of both maternal endothelium and fetal trophoblast cells. These trophoblasts resemble ECs by anatomical localization, and by an endothelial-like ability to regulate hemostasis.
- Maternal arteries supplying blood to the placenta undergo a unique remodeling process that involves loss of smooth muscle cells from the arterial wall and replacement of maternal ECs by invasive "endovascular trophoblast" cells. This remodeling process is essential for the adequate perfusion of the placental vascular bed.
- The hemostatic balance in the placenta is regulated by coagulation factors and platelets in maternal blood, by maternal ECs, and by fetal trophoblast cells. Defects in the trophoblast cells' ability to regulate hemostasis in an EC-like manner are associated with placental pathology and reproductive failure.

Future Goals

- To determine the role of trophoblast cells in severe pregnancy complications such as preeclampsia and APS-associated pregnancy loss
- To undertake epidemiological studies required to determine the contribution of fetal trophoblast cell defects in determining the odds for the occurrence of pregnancy complications in pregnant women, especially in women with thrombophilic risk factors

REFERENCES

1 Chaddha V, Viero S, Huppertz B, Kingdom J. Developmental biology of the placenta and the origins of placental insufficiency. *Semin Fetal Neonatal Med*. 2004;9:357–369.
2 Georgiades P, Ferguson-Smith AC, Burton GJ. Comparative developmental anatomy of the murine and human definitive placentae. *Placenta*. 2002;23:3–19.
3 Red-Horse K, Zhou Y, Genbacev O, et al. Trophoblast differentiation during embryo implantation and formation of the maternal-fetal interface. *J Clin Invest*. 2004;114:744–754.
4 Rossant J. Lineage development and polar asymmetries in the peri-implantation mouse blastocyst. *Semin Cell Dev Biol*. 2004; 15:573–581.
5 Sato Y, Fujiwara H, Zeng BX, et al. Platelet-derived soluble factors induce human extravillous trophoblast migration and differentiation: platelets are a possible regulator of trophoblast infiltration into maternal spiral arteries. *Blood*. 2005;106(2):428–435.
6 Goldman-Wohl D, Greenfield C, Haimov-Kochman R, et al. Eph and ephrin expression in normal placental development and preeclampsia. *Placenta*. 2004;25:623–630.
7 Sapin V, Bouillet P, Oulad-Abdelghani M, et al. Differential expression of retinoic acid-inducible (Stra) genes during mouse placentation. *Mech Dev*. 2000;92:295–299.
8 Kaufmann P, Black S, Huppertz B. Endovascular trophoblast invasion: implications for the pathogenesis of intrauterine growth retardation and preeclampsia. *Biol Reprod*. 2003;69: 1–7.
9 Damsky CH, Fisher SJ. Trophoblast pseudo-vasculogenesis: faking it with endothelial adhesion receptors. *Curr Opin Cell Biol*. 1998;10:660–666.
10 Iruela-Arispe ML. Normal placentation: a tale that requires an epithelial-to-endothelial conversion. *J Clin Invest*. 1997;99:2057–2058.
11 Even-Ram S, Uziely B, Cohen P, et al. Thrombin receptor overexpression in malignant and physiological invasion processes. *Nat Med*. 1998;4:909–914.

12 Even-Ram SC, Grisaru-Granovsky S, Pruss D, et al. The pattern of expression of protease-activated receptors (PARs) during early trophoblast development. *J Pathol*. 2003;200:47–52.
13 O'Brien PJ, Koi H, Parry S, et al. Thrombin receptors and protease-activated receptor-2 in human placentation: receptor activation mediates extravillous trophoblast invasion in vitro. *Am J Pathol*. 2003;163:1245–1254.
14 Isermann B, Sood R, Pawlinski R, et al. The thrombomodulin-protein C system is essential for the maintenance of pregnancy. *Nat Med*. 2003;9:331–337.
15 Aharon A, Brenner B, Katz T, et al. Tissue factor and tissue factor pathway inhibitor levels in trophoblast cells: implications for placental hemostasis. *Thromb Haemost*. 2004;92:776–786.
16 Suzuki T, Aizawa K, Matsumura T, Nagai R. Vascular implications of the Kruppel-like family of transcription factors. *Arterioscler Thromb Vasc Biol*. 2005;25:1135–1141.
17 Lin Z, Kumar A, SenBanerjee S, et al. Kruppel-like factor 2 (KLF2) regulates endothelial thrombotic function. *Circ Res*. 2005; 96:E48–E57.
18 Milstone DS, Redline RW, O'Donnell PE, et al. E-selectin expression and function in a unique placental trophoblast population at the fetal-maternal interface: regulation by a trophoblast-restricted transcriptional mechanism conserved between humans and mice. *Dev Dyn*. 2000;219:63–76.
19 Kruse A, Hallmann R, Butcher EC. Specialized patterns of vascular differentiation antigens in the pregnant mouse uterus and the placenta. *Biol Reprod*. 1999;61:1393–1401.
20 Zhou Y, Fisher SJ, Janatpour M, et al. Human cytotrophoblasts adopt a vascular phenotype as they differentiate. A strategy for successful endovascular invasion? *J Clin Invest*. 1997;99:2139–2151.
21 Kruse A, Merchant MJ, Hallmann R, Butcher EC. Evidence of specialized leukocyte-vascular homing interactions at the maternal/fetal interface. *Eur J Immunol*. 1999;29:1116–1126.
22 Hoffman R, Brenner B. Thrombophilia related issues in women and children. *Semin Thromb Hemost*. 2005;31:97–103.
23 Lin J, August P. Genetic thrombophilias and preeclampsia: a meta-analysis. *Obstet Gynecol*. 2005;105:182–192.
24 Kovalevsky G, Gracia CR, Berlin JA, et al. Evaluation of the association between hereditary thrombophilias and recurrent pregnancy loss: a meta-analysis. *Arch Intern Med*. 2004;164:558–563.
25 Rey E, Kahn SR, David M, Shrier I. Thrombophilic disorders and fetal loss: a meta-analysis. *Lancet*. 2003;361:901–908.
26 Wilson WA, Gharavi AE, Koike T, et al. International consensus statement on preliminary classification criteria for definite antiphospholipid syndrome: report of an international workshop. *Arthritis Rheum*. 1999;42:1309–1311.
27 Lockshin MD. Pregnancy loss and antiphospholipid antibodies. *Lupus*. 1998;7(Suppl 2):S86–S89.
28 Stephenson MD. Frequency of factors associated with habitual abortion in 197 couples. *Fertil Steril*. 1996;66:24–29.
29 Lynch A, Marlar R, Murphy J, et al. Antiphospholipid antibodies in predicting adverse pregnancy outcome. A prospective study. *Ann Intern Med*. 1994;120:470–475.
30 Lockwood CJ, Romero R, Feinberg RF, et al. The prevalence and biologic significance of lupus anticoagulant and anticardiolipin antibodies in a general obstetric population. *Am J Obstet Gynecol*. 1989;161:369–373.
31 de Groot PG, Horbach DA, Derksen RH. Protein C and other cofactors involved in the binding of antiphospholipid antibodies: relation to the pathogenesis of thrombosis. *Lupus*. 1996;5:488–493.
32 Galli M. Non beta 2-glycoprotein I cofactors for antiphospholipid antibodies. *Lupus*. 1996;5:388–392.
33 Esmon NL, Safa O, Smirnov MD, Esmon CT. Antiphospholipid antibodies and the protein C pathway. *J Autoimmun*. 2000; 15:221–225.
34 Marciniak E, Romond EH. Impaired catalytic function of activated protein C: a new in vitro manifestation of lupus anticoagulant. *Blood*. 1989;74:2426–2432.
35 Freyssinet JM, Cazenave JP. Lupus-like anticoagulants, modulation of the protein C pathway and thrombosis. *Thromb Haemost*. 1987;58:679–681.
36 Rand JH, Wu XX, Andree HA, et al. Pregnancy loss in the antiphospholipid-antibody syndrome – a possible thrombogenic mechanism. *N Engl J Med*. 1997;337:154–160.
37 Rand JH, Wu XX, Guller S, et al. Reduction of annexin-V (placental anticoagulant protein-I) on placental villi of women with antiphospholipid antibodies and recurrent spontaneous abortion. *Am J Obstet Gynecol*. 1994;171:1566–1572.
38 Krikun G, Lockwood CJ, Wu XX, et al. The expression of the placental anticoagulant protein, annexin V, by villous trophoblasts: immunolocalization and in vitro regulation. *Placenta*. 1994;15:601–612.
39 Vogt E, Ng AK, Rote NS. Antiphosphatidylserine antibody removes annexin-V and facilitates the binding of prothrombin at the surface of a choriocarcinoma model of trophoblast differentiation. *Am J Obstet Gynecol*. 1997;177:964–972.
40 Wang X, Campos B, Kaetzel MA, Dedman JR. Annexin V is critical in the maintenance of murine placental integrity. *Am J Obstet Gynecol*. 1999;180:1008–1016.
41 Sood R, Kalloway S, Mast AE, et al. Fetomaternal cross talk in the placental vascular bed: control of coagulatin by trophoblast cells. *Blood*. 2006;107(8):3173–3180
42 Holers VM, Girardi G, Mo L, et al. Complement C3 activation is required for antiphospholipid antibody-induced fetal loss. *J Exp Med*. 2002;195:211–220.
43 Girardi G, Redecha P, Salmon JE. Heparin prevents antiphospholipid antibody-induced fetal loss by inhibiting complement activation. *Nat Med*. 2004;10:1222–1226.
44 Erlich J, Parry GC, Fearns C, et al. Tissue factor is required for uterine hemostasis and maintenance of the placental labyrinth during gestation. *Proc Natl Acad Sci USA*. 1999;96:8138–8143.
45 Crawley JT, Gu JM, Ferrell G, Esmon CT. Distribution of endothelial cell protein C/activated protein C receptor (EPCR) during mouse embryo development. *Thromb Haemost*. 2002;88:259–266.
46 Isermann B, Hendrickson SB, Hutley K, et al. Tissue-restricted expression of thrombomodulin in the placenta rescues thrombomodulin-deficient mice from early lethality and reveals a secondary developmental block. *Development*. 2001;128:827–838.
47 Li W, Zheng X, Gu JM, et al. Extra-embryonic expression of EPCR is essential for embryonic viability. *Blood*. 2005;106(8):2716–2222.
48 Estelles A, Gilabert J, Grancha S, et al. Abnormal expression of type 1 plasminogen activator inhibitor and tissue factor in severe preeclampsia. *Thromb Haemost*. 1998;79:500–508.

49 Lanir N, Aharon A, Brenner B. Procoagulant and anticoagulant mechanisms in human placenta. *Semin Thromb Hemost*. 2003;29: 175–184.

50 Weiler-Guettler H, Aird WC, Rayburn H, et al. Developmentally regulated gene expression of thrombomodulin in postimplantation mouse embryos. *Development*. 1996;122:2271–2281.

51 Fazel A, Vincenot A, Malassine A, et al. Increase in expression and activity of thrombomodulin in term human syncytiotrophoblast microvilli. *Placenta*. 1998;19:261–268.

52 Maruyama I, Bell CE, Majerus PW. Thrombomodulin is found on endothelium of arteries, veins, capillaries, and lymphatics, and on syncytiotrophoblast of human placenta. *J Cell Biol*. 1985; 101:363–371.

53 Edstrom CS, Calhoun DA, Christensen RD. Expression of tissue factor pathway inhibitor in human fetal and placental tissues. *Early Hum Dev*. 2000;59:77–84.

54 White EA, Baker JB, McGrogan M, Kitos PA. Protease nexin 1 is expressed in the human placenta. *Thromb Haemost*. 1993;69: 119–123.

55 Kaczan-Bourgois D, Salles JP, Hullin F, et al. Increased content of annexin II (p36) and p11 in human placenta brush-border membrane vesicles during syncytiotrophoblast maturation and differentiation. *Placenta*. 1996;17:669–676.

56 Rambotti MG, Spreca A, Donato R. Immunocytochemical localization of annexins V and VI in human placentae of different gestational ages. *Cell Mol Biol Res*. 1993;39:579–588.

57 Donohoe S, Kingdom JC, Mackie IJ, et al. Ontogeny of beta 2 glycoprotein I and annexin V in villous placenta of normal and antiphospholipid syndrome pregnancies. *Thromb Haemost*. 2000;84:32–38.

58 Pierleoni C, Samuelsen GB, Graem N, et al. Immunohistochemical identification of the receptor for urokinase plasminogen activator associated with fibrin deposition in normal and ectopic human placenta. *Placenta*. 1998;19:501–508.

59 Floridon C, et al. Localization and significance of urokinase plasminogen activator and its receptor in placental tissue from intrauterine, ectopic and molar pregnancies. *Placenta*. 1999;20: 711–721.

60 Liu YX, Hu ZY, Liu K, et al. Localization and distribution of tissue type and urokinase type plasminogen activators and their inhibitors Type 1 and 2 in human and rhesus monkey fetal membranes. *Placenta*. 1998;19:171–180.

61 Feinberg RF, Kao LC, Haimowitz JE, et al. Plasminogen activator inhibitor types 1 and 2 in human trophoblasts. PAI-1 is an immunocytochemical marker of invading trophoblasts. *Lab Invest*. 1989;61:20–26.

62 Makita K, Shimoyama T, Sakurai Y, et al. Placental ecto-ATP diphosphohydrolase: its structural feature distinct from CD39, localization and inhibition on shear-induced platelet aggregation. *Int J Hematol*. 1998;68:297–310.

63 Premyslova M, Li W, Alfaidy N, et al. Differential expression and regulation of microsomal prostaglandin E(2) synthase in human fetal membranes and placenta with infection and in cultured trophoblast cells. *J Clin Endocrinol Metab*. 2003;88:6040–6047.

161

Placental Vasculature in Health and Disease

S. Ananth Karumanchi and Hai-Tao Yuan

Beth Israel Deaconess Medical Center, Harvard Medical School, Boston, Massachusetts

Vascular development occurs through the processes of angiogenesis and vasculogenesis (1,2). Angiogenesis is the process of neovascularization from pre-existing blood vessels, whereas vasculogenesis is the process of blood vessel generation from angioblast precursor cells. The human placenta undergoes high levels of both angiogenesis and vasculogenesis during fetal development (3). Additionally, the placenta undergoes a process of vascular mimicry (also referred to as *pseudovasculogenesis*), in which the placental cytotrophoblasts convert from an epithelial to an endothelial phenotype during normal placental development (4). When placental vascular development is deranged, serious complications such as intrauterine growth restriction (IUGR) and preeclampsia can occur. This chapter discusses placental vascular development during health and in disease, with an emphasis on the role of placental angiogenic factors in these processes.

PLACENTAL VASCULAR DEVELOPMENT IN HEALTH

Implantation

Placentation, which is characteristic of and specific to mammals, allows for the development of the fetus within a protective maternal environment (5). The placenta provides oxygen (O_2) and nutrients to, and transfers wastes from, the developing fetus. The placenta is a highly vascular organ, containing both embryonic and maternal blood vessels (Figure 161.1; for color reproduction, see Color Plate 161.1).

During embryonic development, the blastocyst separates into two cell populations: an outer polarized cell layer (the trophoectoderm) and the nonpolarized inner cell mass. The trophoectoderm gives rise to extraembryonic trophoblasts, whereas the inner cell mass is destined to form the embryo proper.

Implantation of the blastocyst into the uterine wall, which occurs approximately 1 week after fertilization, involves a complex interplay between the trophoblasts and maternal cells. The process involves initial adhesion of the blastocyst to the uterine wall, stable adhesion between blastocyst and uterine epithelium, and invasion of trophoblasts through the epithelial wall. In some ways, these steps are analogous to those associated with leukocyte transmigration in the postcapillary venules. For example, both processes are dependent on the selectin receptor–counter-receptor interactions (6). The highly proliferative trophoblasts ultimately invade the entire endometrium, the outer one-third of the myometrium, and the maternal circulation. Hypoxia is an important driving force for trophoblast proliferation (7). As trophoblasts invade the uterus, they are exposed to increasing concentrations of O_2, at which time they exit cell cycle and differentiate.

Successful implantation requires a receptive uterus. Maternal receptivity is orchestrated by a combination of pituitary and ovarian hormones. During this process, the endometrium becomes more vascular. Implantation is associated with increased permeability of these blood vessels. Much of our understanding of these early events during implantation has been inferred from studies in nonhuman primates and rodents (8).

Invading trophoblasts form placental cords or villi. The villi contain an inner (relative to embryo) layer of mononuclear cytotrophoblasts and an outer layer of continuous multinucleated cytoplasm, the syncytiotrophoblast. An important function of trophoblast tissue is to establish a functional circulation as a means of promoting maternal–fetal exchange. The fundamental steps necessary for successful placentation include: (a) trophoblast invasion, (b) vascularization of the trophoblast tissue to establish and maintain a fetoplacental vasculature, and (c) subsequent maternal vascular remodeling to gain uteroplacental circulation. Both vasculogenesis and angiogenesis are involved in placental vascular formation.

Placental Vasculogenesis

Syncytiotrophoblasts invade maternal tissue, with the depth of invasion depending on the species, the deepest being in humans. Further development leads to the penetration of cytotrophoblastic cones into the syncytiotrophoblastic mass and the development of lacunae, which eventually becomes the intervillous space. Continuing growth and differentiation

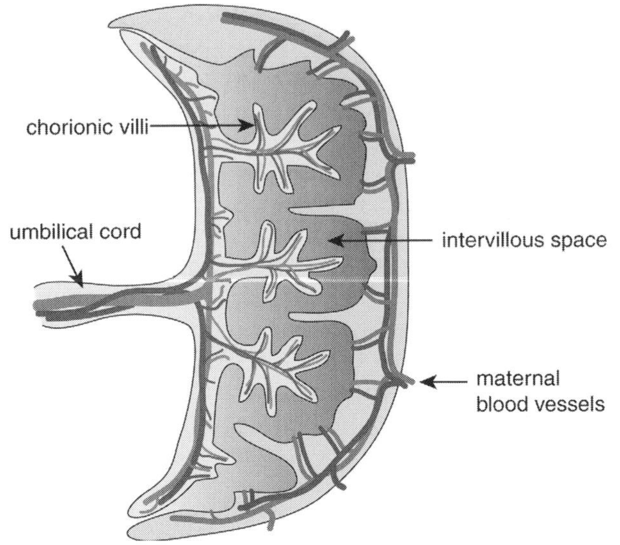

Figure 161.1. A schematic of a normal human placenta depicting the fetal and maternal circulations. For color reproduction, see Color Plate 161.1.

of the trophoblasts leads to branching of the trophoblast villi and the shaping of a placental labyrinth. Extraembryonic mesenchyme gives rise to mesenchymal cores that differentiate and grow into the center of the cytotrophoblast cores. These mesenchymal cells differentiate into endothelial cells (ECs), forming the first capillaries of the fetal placental vasculature. In humans, placental vasculogenesis is evident by approximately embryonic day (E)21 to 22 (3). At this stage, cords of hemangiogenetic cells are present, and some demonstrate primitive lumen formation. These cords further develop, so that by approximately E32, most villi show the presence of capillary structures (3).

Maternal Vascular Remodeling and Placental Pseudovasculogenesis

The formation of the maternal–placental circulation requires remodeling of maternal blood vessels (namely, the spiral arteries). In humans, during the mid to late first trimester, the trophoblasts invade deep into the endometrium and part of the myometrium of the uterus, causing complete remodeling of the upper parts of maternal spiral arterioles. Through the open endings of maternal vessels that are created by trophoblast invasion, maternal blood is released into the placental labyrinth, flows around the trophoblast villi, and is drained by spiral veins (Figure 161.2; for color reproduction, see Color Plate 161.2). When cytotrophoblasts invade maternal spiral arteries, they replace ECs of the maternal spiral arterioles, a process common in all hemochorial placentation. During this process, the endovascular cytotrophoblasts convert from an epithelial to endothelial phenotype, a process termed *pseudovasculogenesis* or *vascular mimicry* (Figure 161.2) (see Chapter 160) (4). This process involves cytotrophoblast stem cells losing epithelial markers, such as E-cadherin and $\alpha 6\beta 3$ integrin, and gaining expression of endothelial markers, including vascular endothelial (VE)-cadherin and $\alpha v\beta 3$ integrin (4,9).

In addition to replacing maternal ECs, cytotrophoblasts also remodel the highly muscular tunica media of spiral arterioles, a process that requires matrix metalloproteinase (MMP)-9 (10,11), which then transforms the highly resistant maternal vessels into larger capacitance vessels with low resistance (see Figure 161.2) (12). Uterine blood flow during pregnancy increases more than 20-fold; the functional consequence of spiral arteriole remodeling is to maximize the capacity of the materno-placental circulation and to provide sufficient blood supply for placenta and fetus at low blood pressure (13). During intrauterine pregnancies, spiral arteries from both the implantation and nonimplantation regions display these physiological changes, although the mechanisms for these changes remain unknown (14). Placental O_2 tension has been suggested to be one of the major regulators of cytotrophoblast migration and differentiation (7). It also has been hypothesized that decidual natural killer (NK) cells and/or activated macrophages may play a role in the vascular remodeling noted during pregnancy, but definitive evidence for this theory is lacking.

Fetal Circulation and Placental Villous Angiogenesis

The fetal circulation enters the placenta via the umbilical vessels embedded within the umbilical cord. Inside the placenta, fetal circulation branches successively into units within the cotyledons and then into capillary loops within the chorionic villi. From E32 until the end of the first trimester, the endothelial tube segments formed by vasculogenesis in the placental villi are transformed into primitive capillary networks by the balanced interaction of two parallel mechanisms: (a) elongation of pre-existing tubes by nonbranching angiogenesis, and (b) ramification of these tubes by lateral sprouting (sprouting angiogenesis). A third process, termed *intussusceptive microvascular growth*, rarely contributes. In the third month of pregnancy, some of the centrally located endothelial tubes of immature intermediate villi achieve larger diameters of 100 μm and more. Within a few weeks, they establish thin media- and adventitia-like structures by concentric fibrosis in the surrounding stroma and by differentiation of precursor smooth muscle cells expressing α- and γ-smooth muscle actins in addition to vimentin and desmin. This is followed soon after by the expression of smooth muscle myosin (3,15). These vessels are forerunners of villous arteries and veins.

From week 25 postconception until term, patterns of villous vascular growth switch from prevailing branching angiogenesis to a prevalence of nonbranching angiogenesis. Analysis of proliferation markers at this stage reveals a relative reduction of trophoblast proliferation and an increase in endothelial proliferation along the entire length of these structures, resulting in nonsprouting angiogenesis by proliferative elongation. The final length of these peripheral capillary loops exceeds 4,000 μm (16,17). They grow at a rate that exceeds that of

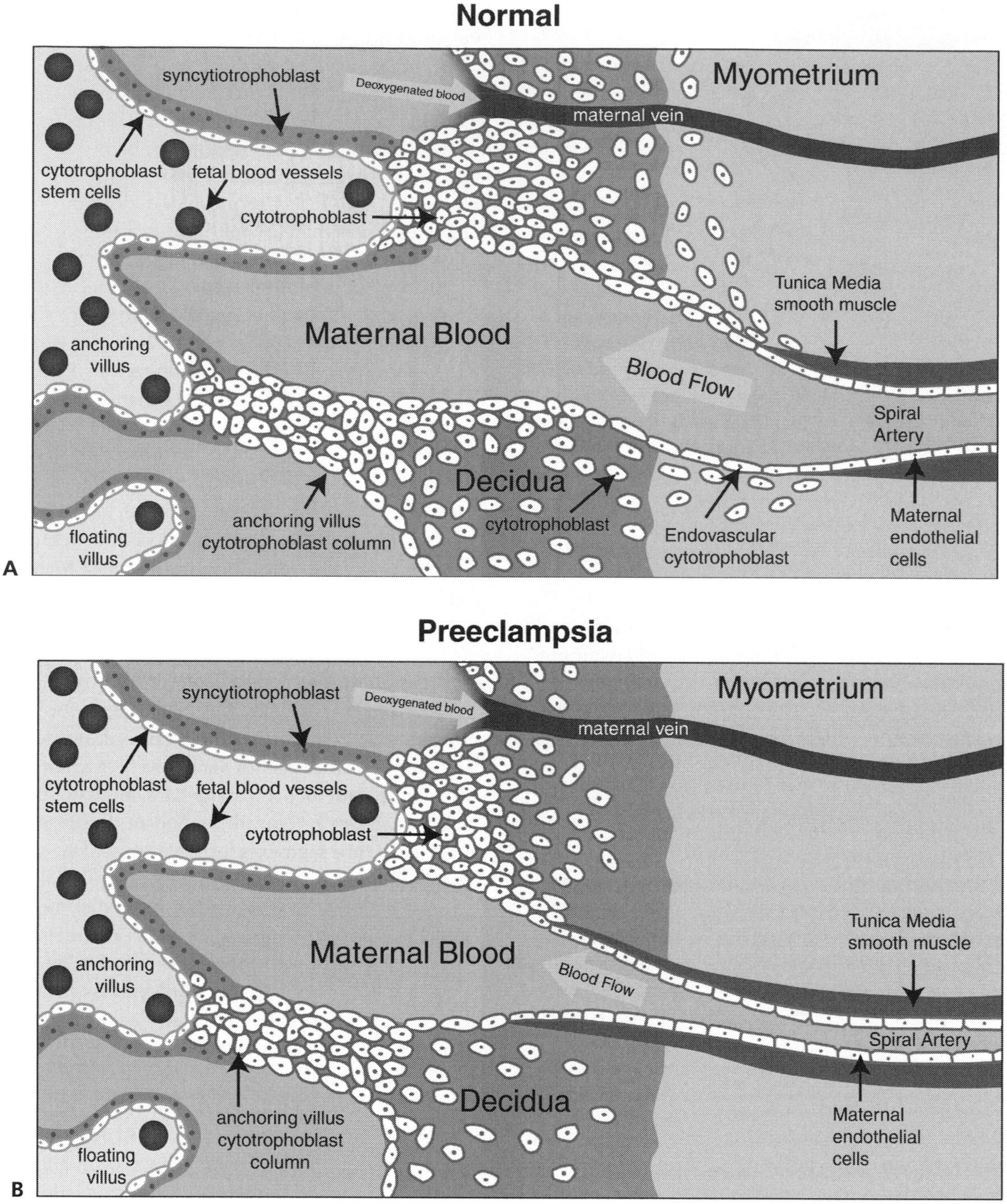

Figure 161.2. A schematic of placental vascular remodeling in health (*upper panel*) and in disease, preeclampsia (*lower panel*). Exchange of O_2, nutrients, and waste products between the fetus and the mother depends on adequate placental perfusion by maternal vessels. In normal placental development, invasive cytotrophoblasts of fetal origin invade the maternal spiral arteries, transforming them from small-caliber resistance vessels to high-caliber capacitance vessels capable of providing placental perfusion adequate to sustain the growing fetus. During the process of vascular invasion, the cytotrophoblasts differentiate from an epithelial phenotype to an endothelial phenotype, a process referred to as *pseudovasculogenesis* (*upper panel*). In preeclampsia, cytotrophoblasts fail to adopt an invasive endothelial phenotype. Instead, invasion of the spiral arteries is shallow, and they remain small-caliber, resistance vessels (*lower panel*). This is thought to lead to placental ischemia and secretion of antiangiogenic factors. (Reproduced with permission from Lam C, Lim KH, Karumanchi SA. Circulating angiogenic factors in the pathogenesis and prediction of preeclampsia. *Hypertension.* 2005;46:1077–1085.) For color reproduction, see Color Plate 161.2.

the villi themselves, resulting in coiling of the capillaries. The looping capillaries bulge toward, and obtrude into, the trophoblastic surface and thereby contribute to formation of the terminal villi. Each of the latter is supplied by one or two capillary coils and is covered by an extremely thin (<2 μm) layer of trophoblast that contributes to the so-called *vasculosyncytial membranes.* These are the principal sites of diffusional exchange of gases between mother and fetus. Normally, the capillary loops of five to ten such terminal villi are connected to each other in series by the slender, elongated capillaries of the central mature intermediate villus.

The fetal vessels (chorionic vessels) from the individual cotyledons of the placenta unite at the placental surface to form the umbilical vessels that then traverse in the umbilical cord. The umbilical cord consists of one vein and two arteries (see Figure 161.1). The connective tissue surrounding these vessels in the umbilical cord is referred to as *Wharton jelly.* Most umbilical cords are twisted at birth, probably related to fetal activity in utero. The umbilical vein carries the oxygenated blood from the placenta to the fetus and the umbilical arteries carry the deoxygenated blood back to the placenta. The umbilical vein was the first tissue used to derive ECs for cell culture studies (18), and these cells continue to date as the most common primary ECs used in vascular biology research. The ECs of the umbilical veins and arteries are unusually rich in organelles, which are not usually noted in vessels elsewhere in the body. Furthermore, the umbilical veins lack vasa vasorum, a finding probably related to the oxygenated blood that it carries.

Vascular Growth Factors and Inhibitors During Normal Placental Development

Placental vascularization involves a complex interaction of several regulatory factors. The list of pro- and antiangiogenic molecules thought to play a role in placental vascular development is expanding exponentially (Table 161-1) and has been reviewed in detail elsewhere (19), but the families of vascular endothelial growth factor (VEGF) and angiopoietins (Angs) gene products have been most extensively studied.

As discussed in Chapter 32, VEGF-A was initially defined, characterized, and purified for its ability to induce vascular leak and permeability, as well as for its ability to promote vascular EC proliferation (20,21). The members of this family include VEGF-A, VEGF-B, VEGF-C, VEGF-D, and placental growth factor (PlGF). VEGF-A is an endothelial-specific mitogen and a survival factor that exists in four isoforms: VEGF-A_{121}, VEGF-A_{165}, VEGF-A_{189}, and VEGF-A_{206}. Inactivation of a single *VEGF* gene resulted in embryonic lethality in heterozygous embryos at E11 to 12, and significant defects in placental vasculature were observed, implicating VEGF in placental vascular development (22). The high-affinity receptor tyrosine kinases for VEGF-A include FLT-1 (also referred to as VEGFR1) and kinase-insert domain-containing receptor (KDR, human/FLK1, murine), also known as VEGFR2 (23). KDR mediates the major growth and permeability actions of VEGF, whereas FLT-1 may have a negative role, either by acting as a decoy receptor or by suppressing signaling through KDR. PlGF was the first VEGF relative identified that was found to be abundantly expressed in the placenta, and it acts by binding to FLT-1 but not KDR (24). Although PlGF can potentiate the angiogenic activity of VEGF, reproduction was not affected in mice with an isolated *PlGF*$^{-/-}$ genotype, and no placental or embryonic angiogenesis defects were reported (25,26). VEGF-C and VEGF-D, based on their ability to bind the lymphatic-specific FLT-3 receptor (also known as VEGFR3), seem to be important for lymphatic development (27,28). Alternative splicing of *FLT-1* results in the production of an endogenously secreted antiangiogenic protein referred to as soluble FLT-1 (sFLT-1), which lacks the cytoplasmic and transmembrane domain but retains the ligand-binding domain (Figure 161.3) (29). Thus, sFLT-1 can antagonize VEGF and PlGF by binding to them and preventing interaction with their endogenous full-length receptors (29,30). Increased production of sFLT-1 has been shown to play a pathogenic role in the severe endothelial dysfunction in preeclampsia (discussed later).

The Tie2 receptor binds a family of ligands termed *angiopoietins,* of which there are four members (Ang1 to Ang4). Ang1–mediated activation of Tie2 promotes endothelial survival and capillary sprouting (31–33). The effects of

Table 161-1: List of Pro- and Antiangiogenic Factors That Govern Placental Development

Proangiogenic Factors
1. VEGF-A, -B, -C, and -D
2. PlGF
3. Ang1, -2
4. TGFβ1, -β3
5. Hepatocyte growth factor
6. Granulocyte colony stimulating factor (CSF) and granulocyte macrophage CSF
7. Proliferin
8. PDGF
9. Estrogens
10. Leptin
11. IGF-I, IGF-II
12. Interleukin-8
13. Ephrins
14. Fibroblast growth factor 2
Antiangiogenic Factors
1. sFLT-1
2. sFLK-1
3. Endostatin
4. Tumor necrosis factor-α
5. Prolactin
6. Endostatin
7. Thrombospondin-1

Data used with permission from Zygmunt M, Herr F, Munstedt K, et al. Angiogenesis and vasculogenesis in pregnancy. *Eur J Obstet Gynecol Reprod Biol.* 2003;110(Suppl 1):S10–8.

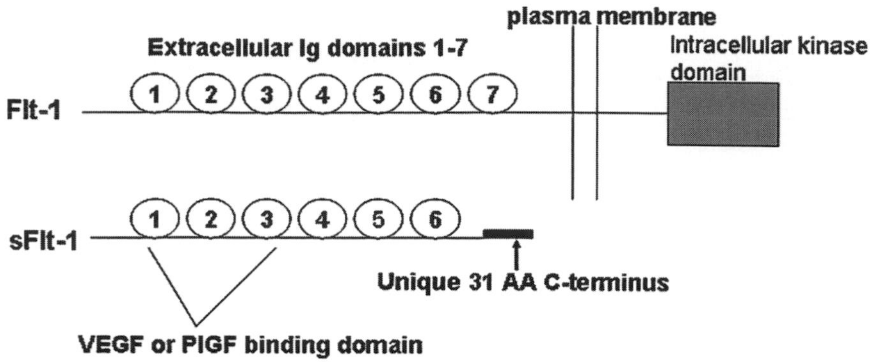

Figure 161.3. The protein structures of FLT-1 and sFLT-1 are illustrated. FLT-1 has seven immunoglobulin domains (IgG), which are thought to mediate ligand binding to VEGF and PlGF. sFLT-1 protein has a unique 31-AA C-terminus region derived from alternative splicing, and lacks the transmembrane and cytoplasmic domains. (Reproduced with permission from Karumanchi SA, Lim KH, Sukhatme VP, August P, eds. *Pathogenesis of preeclampsia.* Wellesley, MA: UpToDate; 2005.)

Ang1 and VEGF on sprouting are synergistic (31). In addition to its direct effects on ECs, Tie2 activation has been reported to induce maturation of adjacent smooth muscle and pericyte precursors via paracrine action of endothelial-derived factors such as platelet-derived growth factor (PDGF)-B (34). Ang1 also has been shown to inhibit capillary permeability (35,36), preventing plasma leakage in response to VEGF. In the presence of abundant VEGF, Ang2 is thought to destabilize vascular networks and facilitate sprouting, as during tumor growth (37,38). Conversely, with low ambient VEGF levels, Ang2 may cause vessel regression, as in corpus luteum involution (39).

Although few in vivo functional studies in humans demonstrate how placentation is regulated on a molecular basis by vascular growth factors, current knowledge about the expression patterns of angiogenic factors provides insights into potential molecular regulatory processes. In all primates studied to date, including humans, VEGF and its receptors have been localized in the placenta during early pregnancy, at the time when the syncytiotrophoblast penetrates maternal tissue, before vascular development and remodeling takes place. VEGF is expressed in maternal decidual cells (40–42) and in invading cytotrophoblasts (41,43). Early in gestation, of the two VEGF receptors, only FLT-1 expression is localized in the invading extravillous trophoblast (40,41,44). These observations raise the possibility that VEGF–FLT-1 interactions contribute to early trophoblast invasion. Later in gestation, within the trophoblast tree or columns (see Figure 161.2), VEGF is localized to cytotrophoblast cells (40,43,45), and FLT-1 to extravillous trophoblasts (40,43,45,46). This pattern suggests that, in addition to regulating trophoblast invasion, the VEGF–FLT-1 axis may play an important role in the coordination of trophoblast differentiation and migration. VEGF-A has been demonstrated to induce cytotrophoblast invasion in vitro, an effect that is blocked by exogenous sFLT-1 (45). Although homozygous knockout mice for both FLT1 and KDR have been shown to have defective fetal and placental vasculogenesis and angiogenesis (47,48), definitive in vivo evidence of impaired trophoblast invasion is still lacking, because the normal invasion of trophoblasts is relatively shallow even at baseline in rodents (49), in contrast to human placentas, in which robust invasion occurs. Recent studies using in vitro trophoblast cultures suggest that PlGF and VEGF-C (two related growth factors) also are expressed by the invading cytotrophoblasts and may contribute to cytotrophoblast invasion and differentiation via FLT1 and FLT3 signaling, respectively (50).

The development of the trophoblast tree occurs simultaneously with formation of the fetoplacental vascular system. This latter process involves differentiation and proliferation of fetal ECs, tubule formation, and vessel stabilization. In the majority of studies, the KDR receptor was found to be exclusively expressed on ECs or mesenchymal cells from which ECs differentiate (40,43,44,46,51), whereas VEGF expression was localized in the trophoblast, indicating that EC differentiation, migration, and proliferation, as the essential steps for building the primary vascular network, are mediated by a VEGF/KDR ligand–receptor pair in a paracrine system.

The angiopoietins also are expressed during the very early placentation period in primates (52), indicating their involvement in the regulation of trophoblast growth. In primates, Ang1 is highly expressed in the syncytiotrophoblast, whereas its receptor, Tie2, is located in the cytotrophoblast (40,43). Similar observations have been made in the human placenta (43,53). Ang1 was shown to stimulate trophoblast growth and migration in vitro (53). *Ang1* gene expression has been found to increase with gestation (40). Thus, Ang1 expression may trigger the in-growth of the cytotrophoblast cones in the syncytiotrophoblast, whereas the relatively higher expression later during placentation may be required for branching of the villi and shaping the labyrinth.

Fetal capillaries within the placental villi are thin-walled, to allow O_2 diffusion, whereas the chorion vessels are stabilized by a thick wall of pericytes, smooth muscle cells, or both to fulfill their task of collecting and draining all fetal placental blood. Tie2 has been shown to be expressed at high levels in the endothelium of chorion vessels, and at low levels in the fetal capillaries of the villi (40). These results suggest that the

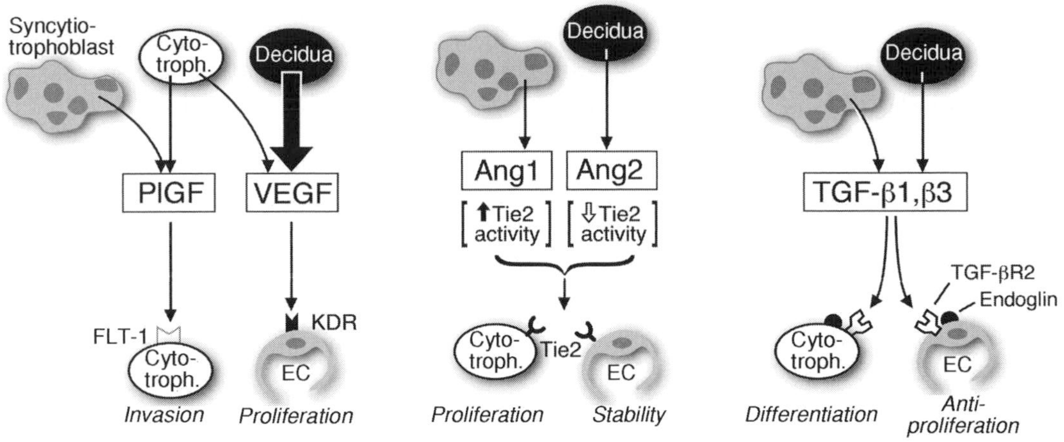

Figure 161.4. Summary of the role of angiogenic factors in placental trophoblast differentiation/invasion and angiogenesis. Cytotroph., cytotrophoblast.

Ang1/Tie2 ligand–receptor pair acts on the fetal vasculature, especially on chorion vessels, to induce stability. In humans, Ang1 expression was found in the media of stem villous vessels at term (53), which is also consistent with its reported role in vessel maturation and stabilization.

Maternal vessels must be remodeled to attain uteroplacental circulation. In humans, the invasion process of the trophoblast is so deep that upper parts of maternal spiral arterioles become completely digested, and the placental labyrinth is filled by the open endings of the spiral arterioles. In humans, Ang2 is localized in the syncytiotrophoblast, whereas Tie2 is expressed in the maternal endothelium (53,54), indicating a paracrine mechanism during maternal remodeling (55). In some mammals, such as rats, invasion is more superficial, and the integrity of maternal vessels is preserved throughout pregnancy. In these animals, maternal vessels form gaps through which maternal blood is released into the labyrinth.

In primates, high expression of *Ang2* mRNA is detected during early gestation, when maternal remodeling mainly takes place. Ang2 is a possible candidate for induction of maternal vascular transformation because it destabilizes the vasculature (39). This destabilization process is a local event, which may be driven by Ang2/Tie2. In humans, although the remodeling process of maternal vessels may be supported by Ang2/Tie2, it is mainly driven by the aggressive trophoblast invasion. Finally, transforming growth factor (TGF)-β1 and -3, two angiogenic proteins made by the maternal decidua and the syncytiotrophoblasts, have been shown more recently to also affect cytotrophoblast invasion and differentiation through endoglin signaling (56,57).

In summary, abundant evidence supports a role for angiogenic growth factors in hemochorial placentation in human and nonhuman primates. The most comprehensive study investigating both growth factor families, VEGF and angiopoietins, and their receptors during early, mid, and late pregnancy has been undertaken in marmoset monkeys, and the study indicates a tight spatial and temporal regulation of diverse developmental processes during placentation by angiogenic inducers (40,52). In accordance with these reports, it has been specifically hypothesized that VEGF/FLT-1 and Ang1/Tie2 may be involved in trophoblast differentiation and invasion; VEGF/KDR and Ang1/Tie2 may trigger fetoplacental vascular development, whereas Ang2/Tie2 may support the remodeling processes of the maternal vasculature (Figure 161.4) (40).

In contrast to proangiogenic proteins, little attention has been paid to the characterization of antiangiogenic proteins in the placenta. The first major endogenous inhibitor of angiogenesis that has been known to be made in abundant quantities in the placenta is sFLT-1, a potent circulating antiangiogenic molecule (58). This potent antiangiogenic molecule is abundantly expressed by the cytotrophoblasts, and goes up in expression with increasing gestation (29,30) (see Figure 161.3). Other antiangiogenic proteins that have been found to be expressed in the placenta include thrombospondin-1, endostatin, and prolactin. However, the precise role of these proteins during normal placentation is unclear. In general, early on in pregnancy, the proangiogenic proteins are overexpressed and probably account for placental angiogenesis and the increase in placental mass that accompanies fetal development (59). Toward the end of pregnancy, antiangiogenic factors increase in expression, possibly in preparation for delivery (59). In addition to the gestational age-dependent distribution of these proteins, spatial differences also occur in the expression of the angiogenic proteins in the human placenta. For example, at term, placental extracts derived from the decidual side have been found to have stronger antiangiogenic activity, as compared with chorionic villus extracts (60).

PLACENTAL VASCULAR DEVELOPMENT IN DISEASE

As discussed earlier, during the first two trimesters of pregnancy, the spiral arteries undergo extensive remodeling. This is thought to reduce maternal blood flow resistance and increase uteroplacental perfusion. Abnormal spiral arterial remodeling is a key pathological feature seen in preeclampsia and in IUGR. In addition, impaired villous angiogenesis is noted in IUGR placentas. Other pregnancy complications, such as placental abruption and intrauterine fetal death (IUFD), also have been hypothesized to be due to defective placental vascular development and coagulation abnormalities; however, the mechanisms mediating these disorders are less well understood.

Preeclampsia

Preeclampsia occurs in 5% of all human pregnancies (61), and is the leading cause of maternal death in the industrialized world. This condition is characterized by the new onset of hypertension and proteinuria, developing after 20 weeks of gestational age. Variant forms of preeclampsia include eclampsia, which is characterized by grand mal seizure; and HELLP syndrome, which is manifested by *h*emolysis, *e*levated *l*iver enzymes and *l*ow *p*latelet count. These latter conditions are associated with life-threatening complications, including cerebral hemorrhage, lung edema, liver hemorrhage, and disseminated intravascular coagulation. Preeclampsia is associated with IUGR and iatrogenic preterm birth of the fetus, resulting in a high risk of early neonatal death and infant mortality or morbidity (61). Current treatment consists of symptom relief and, ultimately, delivery.

A pathophysiological role for the placenta in preeclampsia is supported by the observation that this syndrome occurs solely in the presence of a placenta or a hydatidiform mole, and remits dramatically in the postpartum period after delivery of the placenta (62). The major pathological abnormality in the preeclamptic placenta is insufficient maternal spiral artery remodeling. The decrease in remodeling of the uterine spiral arterioles is thought to be due to defective endovascular invasion by cytotrophoblasts and impaired pseudovasculogenesis (see Figure 161.2) (63,64). The hypothesis that defective trophoblastic invasion with accompanying uteroplacental hypoperfusion may lead to preeclampsia is supported by both animal and human studies. Pathologic examination of placentas from pregnancies with advanced preeclampsia often reveals numerous placental infarctions and sclerotic narrowing of arterioles (65). In one-third to one-half of preeclamptic women however, these changes are not pronounced. Using ultrasound estimation, the uteroplacental blood flow usually is diminished, and uterine vascular resistance usually is increased in preeclamptic women (66). Placental ischemia induced by mechanical constriction of the uterine arteries or the aorta produces hypertension, proteinuria and, variably, glomerular endotheliosis in several animal species (66). However, placental ischemia alone, as seen in many cases of IUGR, does not appear to be sufficient to produce preeclampsia. Thus, although uteroplacental ischemia is an important trigger of preeclampsia, it may be absent in some cases, and the maternal response to placental ischemia is variable.

Studies during the 1980s implicated maternal endothelial dysfunction as a key contributor to the clinical symptoms in preeclamptic patients (67,68). Roberts and colleagues proposed that EC injury caused increased sensitivity to pressor agents, vasoconstriction, and activation of the coagulation cascade as the basis of preeclampsia (69). Evidence of EC injury was provided by the classic renal lesion of preeclampsia – glomerular endotheliosis – in which glomerular capillary ECs are engorged with intracellular inclusions (70). Blood from preeclamptic patients demonstrates markers of endothelial injury, including von Willebrand Factor (71–73), cellular fibronectin (74–76), and thrombomodulin (77,78). Levels of circulating prostacyclins (normally made by healthy ECs) also are decreased in preeclamptic patients (79). Blood vessels from preeclamptic patients reveal decreased endothelial-mediated vasodilator function (80). Data from many studies support the notion that the maternal serum in preeclampsia has soluble factors that mediate endothelial dysfunction. For example, serum or plasma from preeclampsia patients has been shown to alter endothelial phenotype in vitro, including altered expression of vascular cell adhesion molecules and nitric oxide (NO), and changes in prostaglandin balance (81,82). Interestingly, preeclamptic patients also have been noted to have lower skin capillary density compared with healthy pregnant patients, a finding that suggests that defective angiogenesis may be implicated in the pathogenesis of preeclampsia (83). Finally, women with a prior history of preeclampsia have an increased incidence of ischemic heart disease, stroke, and hypertension (84). The association between preeclampsia and subsequent development of cardiovascular disease points to the existence of systemic endothelial dysfunction in these individuals.

To identify the soluble factors that mediate maternal endothelial dysfunction, we performed gene expression profiling of placental tissue from women with and without preeclampsia by using microarray chips. We found messenger RNA for *sFLT-1* to be upregulated dramatically in preeclamptic placentas (85). We also showed that increased systemic levels of sFLT-1 were present in patients with preeclampsia that decreased to baseline 48 hours after delivery. Hence, we hypothesized that excess circulating sFLT-1 levels may lead to an antiangiogenic state and cause endothelial dysfunction and the clinical syndrome of preeclampsia. Increased circulating sFLT-1 concentrations in patients with preeclampsia were associated with decreased circulating levels of free VEGF and free PlGF, a finding that also has been reported by a number of groups (85–88). Furthermore, in a recent study, uterine vein sFLT-1 concentrations were almost four- to five-fold higher than the peripheral venous concentrations, suggesting that the predominant source of maternal sFLT-1 was the placenta (89). By using endothelial tube formation assay, an established in vitro model of angiogenesis, we found that serum from patients with preeclampsia inhibited endothelial tube formation. Notably, 48 hours postpartum, this antiangiogenic effect had disappeared from the serum, suggesting that the inhibition of tubes noted with serum from preeclamptic patients was caused by a circulating factor released by the placenta. When sFLT-1 was added to normotensive serum at concentrations noted in patients with preeclampsia, tube formation did not occur, mimicking the effects seen with serum from preeclamptic patients. This antiangiogenic effect could be restored by adding exogenous VEGF and PlGF (85). These results suggest that the antiangiogenic properties of serum from preeclamptic patients were caused by blockade of VEGF and PlGF by excess circulating sFLT-1. Finally, exogenous gene transfer of sFLT-1 using an adenoviral vector into pregnant rats produced hypertension, proteinuria, and glomerular endotheliosis, the classic pathologic renal lesion of preeclampsia (85). This effect also was seen in nonpregnant animals, suggesting that the effects of sFLT-1 on the maternal vasculature were direct and not dependent on the placenta. Furthermore, a soluble form of VEGF receptor-2 antagonist (sFLK-1; which does not antagonize PlGF), when given exogenously, did not induce a preeclamptic phenotype in pregnant rats, suggesting that antagonism of both VEGF and PlGF is necessary to induce the maternal syndrome. Hence, we concluded that excess sFLT-1 made by preeclamptic placentas may be responsible for the hypertension and proteinuria of preeclampsia by inducing a deficiency of VEGF and PlGF. It is still unknown whether the excess sFLT-1 made in preeclampsia is a primary phenomenon or secondary to placental ischemia.

Alterations in both circulating angiogenic proteins have been shown to antedate clinical preeclampsia, suggesting important diagnostic and predictive implications for this syndrome (86,87,90,91). Alterations in sFLT-1 were useful to predict proximity to the clinical disease, whereas alterations in PlGF were found to be very useful to predict preterm preeclampsia (86). Alterations in the levels of sFLT-1 and free PlGF were greater in women with an earlier onset of preeclampsia and in women in whom preeclampsia was associated with a small-for-gestational-age infant (86,92). The free (unbound) PlGF also is filtered freely into the urine and therefore may also serve to predict the subsequent development of preeclampsia (93). More recently, circulating levels of yet another placental antiangiogenic protein, soluble endoglin, have been reported to be elevated in patients with preeclampsia. Soluble endoglin amplifies the vascular damage mediated by sFLT-1 in pregnant rats, inducing a severe preeclampsia-like syndrome with features of the HELLP syndrome (94). The precise role of soluble endoglin and its relationship with sFLT-1 currently is being explored.

Circumstantial evidence suggests that antagonism of VEGF and PlGF may have a role in hypertension and proteinuria. VEGF induces NO and vasodilatory prostacyclins in ECs, suggesting a role in decreasing vascular tone and blood pressure (95,96). Exogenous VEGF has been found to accelerate renal recovery in rat models of glomerulonephritis and experimental thrombotic microangiopathy (97,98). More recently, exogenous VEGF was shown to ameliorate post–cyclosporine-mediated hypertension, endothelial dysfunction, and nephropathy (99). Additionally, in recent antiangiogenic clinical trials, VEGF signaling inhibitors have resulted in hypertension and proteinuria (100). The organs targeted in preeclampsia (such as the glomerulus or the hepatic sinusoids) have fenestrated endothelia. It has been shown that VEGF induces endothelial fenestrae in vitro (101), and the loss of 50% of VEGF production in the mouse glomerulus leads not only to glomerular endotheliosis but also to loss of glomerular endothelial fenestrae (102). Collectively, these data suggest that VEGF is important not only in blood pressure regulation, but also in maintaining the integrity of the glomerular filtration barrier, and that antagonism of VEGF signaling, as occurs with excess sFLT-1, might lead to endothelial dysfunction, proteinuria, and hypertension. Finally, the excess sFLT-1 theory in preeclampsia is consistent with the evolutionary explanation for preeclampsia discussed by Haig and colleagues (Chapter 17). However, there are limitations and several unanswered questions to the sFLT-1 story. Although placental hypoxia has been suggested as the cause of excess sFLT-1, the precise mechanisms of excess sFLT-1 production by the placenta are not known. The exact role of sFLT-1 in normal placental development and in placental pseudovasculogenesis also is not clear. No brain abnormalities (eclampsia) were reported in sFLT-1–treated animals. Although sFLT-1 was increased in most patients with preeclampsia, it was not increased in some patients with mild preeclampsia (103). Moreover, the relationship of sFLT-1 with known risk factors for preeclampsia (such as obesity, pre-existing hypertension) is unclear; one hypothesis is that a threshold for sFLT-1 to cause disease exists, below which a normal pregnancy proceeds and above which preeclampsia results, and that women with risk factors may represent a group whose threshold has been effectively lowered, rendering them more susceptible to sFLT-1 and resulting in the maternal syndrome at levels that match those of normal pregnancy. Additional synergistic factors that are elaborated by the placenta, such as circulating soluble endoglin, may play a role in the pathogenesis of the generalized endothelial dysfunction noted in preeclampsia.

More recently, endostatin (a circulating fragment of collagen XVIII and an endogenous inhibitor of angiogenesis) also was reported to be increased modestly in patients with preeclampsia (104). However, it is unclear at present if these alterations are functionally relevant in the pathogenesis of preeclampsia or whether this reflects an increase of endothelial basement membrane turnover.

It also has been hypothesized by Wallukat and colleagues that agonistic angiotensin 1 (AT1) receptor autoantibodies noted in women with preeclampsia may play a pathogenic role (105). They hypothesized that these antibodies, which activate the AT1 receptor, may account for the increased angiotensin (Ang)-II sensitivity of preeclampsia. The same investigators later showed that these AT1 autoantibodies, like Ang-II itself, stimulate ECs to produce tissue factor, an early marker of endothelial dysfunction (106). Xia and coworkers found AT1 autoantibodies decreased the invasiveness of immortalized human trophoblasts in an in vitro invasion assay, suggesting that these autoantibodies might contribute to defective placental pseudovasculogenesis as well (107). AT1 autoantibodies are not limited to pregnancy; they also appear to be increased in malignant renovascular hypertension and vascular rejection in nonpregnancy (108,109). In addition, these antibodies have been identified in women with abnormal second-trimester uterine artery Doppler studies who did not develop preeclampsia, suggesting this antibody may be a nonspecific response to placental hypoperfusion (110).

Intrauterine Growth Restriction

Normal fetal growth is a complex interplay of fetal, maternal, and placental health. An abnormality in any of these three systems leads to small-for-gestational age babies. Such babies who arise as a result of placental vascular insufficiency are referred to as IUGR. Although IUGR frequently is associated with severe premature preeclampsia, IUGR also can occur in the absence of any evidence of preeclampsia. Low birth weight in IUGR babies also has been suggested as a predisposing factor for increased hypertension and cardiovascular diseases later in life (111). Normal pregnancy results in a significant increase in diastolic blood flow velocity as a result of physiological maternal vascular remodeling. In pregnancies complicated by IUGR, severe placental vascular insufficiency occurs, and the Doppler waveforms are similar to the nonpregnant state (66). This can be demonstrated by several methods, including Doppler ultrasound and magnetic resonance imaging of the uterine vessels. Similar to preeclampsia, placental bed biopsies in human IUGR show impaired migration of the extravillous trophoblasts and remodeling of the maternal spiral arterioles (12). It is still unclear why some patients with IUGR have the full-blown maternal syndrome of preeclampsia, while others do not. Human IUGR placentas, but not preeclamptic placentas, also have been found to have defective villous capillarization (112). The morphology of the trophoblast covering these diseased villi are also abnormal, with reduced amounts of proliferating villous cytotrophoblasts and an apoptotic syncytiotrophoblast. Recent work suggests that decreased syncytial fusion may lead to the increased syncytial apoptosis noted. Placental hypoxia is thought to preserve the syncytiotrophoblast integrity, and higher levels of O_2 are thought to inhibit cytotrophoblast proliferation and accelerate syncytial apoptosis. There is less O_2 extraction and consumption by the IUGR fetus, and the resulting higher levels of Po_2 have been hypothesized to inhibit the fetoplacental angiogenesis (113). However, studies documenting lack of relative hypoxia in IUGR placentas (without preeclampsia) are still lacking.

Impairment of insulin-like growth factor (IGF)-1 and -2 signaling pathways in experimental animals leads to IUGR (114–116). Recently, two babies with IUGR were reported to have mutations of the *IGF-1R* gene (117). Although abnormalities in IGF signaling are a relatively uncommon cause of IUGR, impaired IGF-2 signaling through IGF1R is the first molecular abnormality reported in this complex disease. IGFs are potent proangiogenic molecules both in vitro and in vivo (118–120). Decreased levels of pregnancy-associated plasma protein (PAPP)-A also have been associated with IUGR; however, its role in the pathogenesis is not well understood (121,122). Abnormalities in other angiogenic proteins have been shown in small clinical studies, but definitive pathogenic role for these factors have not been demonstrated so far. Serum concentrations of sFLT-1 have been found to be increased modestly in patients with IUGR without preeclampsia (88), a finding that has not been confirmed by others (123). More recently, endostatin concentrations in the fetus and neonate were reported to be lower in pregnancies complicated by IUGR (124). Leptin and asymmetric dimethylarginine also have been reported to be elevated during placental insufficiency (125,126), but the cause-and-effect relationship of these molecules has not yet been established.

KEY POINTS

- Normal placental development is characterized by angiogenesis, vasculogenesis, and vascular mimicry, and abnormality in these physiological processes may lead to disorders such as preeclampsia and IUGR.
- A normal pregnancy requires a balance between pro- and antiangiogenic factors that are expressed in the placenta.
- Abnormalities in the signaling of these angiogenic factors may lead to defective trophoblast differentiation, spiral arterial vascular remodeling, and villous angiogenesis of the placenta.
- Increased placental production of circulating sFLT-1, with accompanying defective VEGF signaling in the placenta and in the maternal vasculature, leads to preeclampsia.
- As our understanding continues to advance based on molecular and genetic techniques, we are hopeful that new interventions may improve our management of placental vascular disorders in the near future.

Future Goals

- To undertake studies, especially in animals (nonhuman primates that resemble human placentation) lacking angiogenic factor signaling components specifically in the placenta, to better define the role of angiogenic proteins and vascularization during placental development and maternal health
- To identify and characterize novel angiogenic factors that control trophoblast differentiation and vascular remodeling, to help not only in understanding the pathogenesis of these placental vascular disorders, but also to suggest novel clinical diagnostic and therapeutic options for patients with these disorders

- To continue large-scale multicenter trials (such as the Genetics of Preeclampsia [GOPEC]) trial to shed light on the susceptibility genes for the various placental vascular diseases (127)

REFERENCES

1 Risau W, Flamme I. Vasculogenesis. *Annu Rev Cell Dev Biol.* 1995;11:73–91.

2 Risau W. Mechanisms of angiogenesis. *Nature* 1997;386(6626): 671–4.

3 Demir R, Kaufmann P, Castellucci M, et al. Fetal vasculogenesis and angiogenesis in human placental villi. *Acta Anat (Basel).* 1989;136(3):190–203.

4 Zhou Y, Fisher SJ, Janatpour M, et al. Human cytotrophoblasts adopt a vascular phenotype as they differentiate. A strategy for successful endovascular invasion? *J Clin Invest.* 1997;99(9): 2139–2151.

5 Cross JC, Werb Z, Fisher SJ. Implantation and the placenta: key pieces of the development puzzle. *Science.* 1994;266(5190): 1508–1518.

6 Genbacev OD, Prakobphol A, Foulk RA, et al. Trophoblast L-selectin-mediated adhesion at the maternal-fetal interface. *Science.* 2003;299(5605):405–408.

7 Genbacev O, Zhou Y, Ludlow JW, Fisher SJ. Regulation of human placental development by oxygen tension. *Science.* 1997; 277(5332):1669–1672.

8 Norwitz ER, Schust DJ, Fisher SJ. Implantation and the survival of early pregnancy. *N Engl J Med.* 2001;345(19):1400–1408.

9 Jaffe R. First trimester utero-placental circulation: maternal-fetal interaction. *J Perinat Med.* 1998;26(3):168–174.

10 Roth I, Fisher SJ. IL-10 is an autocrine inhibitor of human placental cytotrophoblast MMP-9 production and invasion. *Dev Biol.* 1999;205(1):194–204.

11 Lim KH, Zhou Y, Janatpour M, et al. Human cytotrophoblast differentiation/invasion is abnormal in pre- eclampsia. *Am J Pathol.* 1997;151(6):1809–1818.

12 Brosens IA, Robertson WB, Dixon HG. The role of the spiral arteries in the pathogenesis of preeclampsia. *Obstet Gynecol Annu.* 1972;1:177–191.

13 Khong TY, De Wolf F, Robertson WB, Brosens I. Inadequate maternal vascular response to placentation in pregnancies complicated by pre-eclampsia and by small-for-gestational age infants. *Br J Obstet Gynaecol.* 1986;93(10):1049–1059.

14 Craven CM, Morgan T, Ward K. Decidual spiral artery remodelling begins before cellular interaction with cytotrophoblasts. *Placenta.* 1998;19(4):241–252.

15 Kohnen G, Kertschanska S, Demir R, Kaufmann P. Placental villous stroma as a model system for myofibroblast differentiation. *Histochem Cell Biol.* 1996;105(6):415–429.

16 Kaufmann P, Bruns U, Leiser R, et al. The fetal vascularisation of term human placental villi. II. Intermediate and terminal villi. *Anat Embryol (Berl).* 1985;173(2):203–214.

17 Kaufmann P, Luckhardt M, Schweikhart G, Cantle SJ. Cross-sectional features and three-dimensional structure of human placental villi. *Placenta.* 1987;8(3):235–247.

18 Gimbrone MA Jr., Cotran RS, Folkman J. Human vascular endothelial cells in culture. Growth and DNA synthesis. *J Cell Biol.* 1974;60(3):673–684.

19 Zygmunt M, Herr F, Munstedt K, et al. Angiogenesis and vasculogenesis in pregnancy. *Eur J Obstet Gynecol Reprod Biol.* 2003; 110(Suppl 1):S10–S18.

20 Dvorak HF. Vascular permeability factor/vascular endothelial growth factor: a critical cytokine in tumor angiogenesis and a potential target for diagnosis and therapy. *J Clin Oncol.* 2002; 20(21):4368–4380.

21 Ferrara N, Gerber HP. The role of vascular endothelial growth factor in angiogenesis. *Acta Haematol.* 2001;106(4):148–156.

22 Ferrara N, Davis-Smyth T. The biology of vascular endothelial growth factor. *Endocr Rev.* 1997;18(1):4–25.

23 Shibuya M. Structure and function of VEGF/VEGF-receptor system involved in angiogenesis. *Cell Struct Funct.* 2001;26(1): 25–35.

24 Persico MG, Vincenti V, DiPalma T. Structure, expression and receptor-binding properties of placenta growth factor (PlGF). *Curr Top Microbiol Immunol.* 1999;237:31–40.

25 Carmeliet P, Moons L, Luttun A, et al. Synergism between vascular endothelial growth factor and placental growth factor contributes to angiogenesis and plasma extravasation in pathological conditions. *Nat Med.* 2001;7(5):575–583.

26 Charnock-Jones DS, Burton GJ. Placental vascular morphogenesis. *Baillieres Best Pract Res Clin Obstet Gynaecol.* 2000; 14(6):953–968.

27 Jussila L, Alitalo K. Vascular growth factors and lymphangiogenesis. *Physiol Rev.* 2002;82(3):673–700.

28 Olofsson B, Jeltsch M, Eriksson U, Alitalo K. Current biology of VEGF-B and VEGF-C. *Curr Opin Biotechnol.* 1999;10(6):528–535.

29 Kendall RL, Thomas KA. Inhibition of vascular endothelial cell growth factor activity by an endogenously encoded soluble receptor. *Proc Natl Acad Sci USA.* 1993;90(22):10705–10709.

30 He Y, Smith SK, Day KA, et al. Alternative splicing of vascular endothelial growth factor (VEGF)-R1 (FLT-1) pre-mRNA is important for the regulation of VEGF activity. *Mol Endocrinol.* 1999;13(4):537–545.

31 Papapetropoulos A, Garcia-Cardena G, Dengler TJ, et al. Direct actions of angiopoietin-1 on human endothelium: evidence for network stabilization, cell survival, and interaction with other angiogenic growth factors. *Lab Invest.* 1999;79(2):213–223.

32 Koblizek TI, Weiss C, Yancopoulos GD, et al. Angiopoietin-1 induces sprouting angiogenesis in vitro. *Curr Biol.* 1998;8(9): 529–532.

33 Davis S, Aldrich TH, Jones PF, et al. Isolation of angiopoietin-1, a ligand for the TIE2 receptor, by secretion-trap expression cloning. *Cell.* 1996;87(7):1161–1169.

34 Carmeliet P. Mechanisms of angiogenesis and arteriogenesis. *Nat Med.* 2000;6(4):389–395.

35 Thurston G, Rudge JS, Ioffe E, et al. Angiopoietin-1 protects the adult vasculature against plasma leakage. *Nat Med.* 2000;6(4): 460–463.

36 Thurston G, Suri C, Smith K, et al. Leakage-resistant blood vessels in mice transgenically overexpressing angiopoietin-1. *Science.* 1999;286(5449):2511–2514.

37 Stratmann A, Acker T, Burger AM, et al. Differential inhibition of tumor angiogenesis by tie2 and vascular endothelial growth factor receptor-2 dominant-negative receptor mutants. *Int J Cancer.* 2001;91(3):273–282.

38 Tanaka S, Mori M, Sakamoto Y, et al. Biologic significance of angiopoietin-2 expression in human hepatocellular carcinoma. *J Clin Invest*. 1999;103(3):341–345.

39 Maisonpierre PC, Suri C, Jones PF, et al. Angiopoietin-2, a natural antagonist for Tie2 that disrupts in vivo angiogenesis. *Science*. 1997;277(5322):55–60.

40 Wulff C, Wilson H, Dickson SE, et al. Hemochorial placentation in the primate: expression of vascular endothelial growth factor, angiopoietins, and their receptors throughout pregnancy. *Biol Reprod*. 2002;66(3):802–812.

41 Charnock-Jones DS, Sharkey AM, Boocock CA, et al. Vascular endothelial growth factor receptor localization and activation in human trophoblast and choriocarcinoma cells. *Biol Reprod*. 1994;51(3):524–530.

42 Cooper JC, Sharkey AM, McLaren J, et al. Localization of vascular endothelial growth factor and its receptor, flt, in human placenta and decidua by immunohistochemistry. *J Reprod Fertil*. 1995;105(2):205–213.

43 Geva E, Ginzinger DG, Zaloudek CJ, et al. Human placental vascular development: vasculogenic and angiogenic (branching and nonbranching) transformation is regulated by vascular endothelial growth factor-A, angiopoietin-1, and angiopoietin-2. *J Clin Endocrinol Metab*. 2002;87(9):4213–4224.

44 Clark DE, Smith SK, Sharkey AM, Charnock-Jones DS. Localization of VEGF and expression of its receptors flt and KDR in human placenta throughout pregnancy. *Hum Reprod*. 1996; 11(5):1090–1098.

45 Zhou Y, McMaster M, Woo K, et al. Vascular endothelial growth factor ligands and receptors that regulate human cytotrophoblast survival are dysregulated in severe preeclampsia and hemolysis, elevated liver enzymes, and low platelets syndrome. *Am J Pathol*. 2002;160(4):1405–1423.

46 Hildebrandt VA, Babischkin JS, Koos RD, et al. Developmental regulation of vascular endothelial growth/permeability factor messenger ribonucleic acid levels in and vascularization of the villous placenta during baboon pregnancy. *Endocrinology*. 2001;142(5):2050–2057.

47 Shalaby F, Rossant J, Yamaguchi TP, et al. Failure of blood-island formation and vasculogenesis in Flk-1-deficient mice. *Nature*. 1995;376(6535):62–66.

48 Fong G, Rassant J, Gertenstein M, et al. Role of Flt-1 receptor tyrosine kinase in regulation of assembly of vascular endothelium. *Nature*. 1995;376:66–67.

49 Ain R, Canham LN, Soares MJ. Gestation stage-dependent intrauterine trophoblast cell invasion in the rat and mouse: novel endocrine phenotype and regulation. *Dev Biol*. 2003;260 (1):176–190.

50 Zhou Y, Bellingard V, Feng KT, et al. Human cytotrophoblasts promote endothelial survival and vascular remodeling through secretion of Ang2, PlGF, and VEGF-C. *Dev Biol*. 2003;263(1):114–125.

51 Helske S, Vuorela P, Carpen O, et al. Expression of vascular endothelial growth factor receptors 1, 2 and 3 in placentas from normal and complicated pregnancies. *Mol Hum Reprod*. 2001; 7(2):205–210.

52 Rowe AJ, Wulff C, Fraser HM. Localization of mRNA for vascular endothelial growth factor (VEGF), angiopoietins and their receptors during the peri-implantation period and early pregnancy in marmosets (*Callithrix jacchus*). *Reproduction*. 2003; 126(2):227–238.

53 Dunk C, Shams M, Nijjar S, et al. Angiopoietin-1 and angiopoietin-2 activate trophoblast Tie-2 to promote growth and migration during placental development. *Am J Pathol*. 2000; 156(6):2185–2199.

54 Zhang EG, Smith SK, Baker PN, Charnock-Jones DS. The regulation and localization of angiopoietin-1, -2, and their receptor Tie2 in normal and pathologic human placentae. *Mol Med*. 2001;7(9):624–635.

55 Goldman-Wohl DS, Ariel I, Greenfield C, et al. Tie-2 and angiopoietin-2 expression at the fetal-maternal interface: a receptor ligand model for vascular remodelling. *Mol Hum Reprod*. 2000;6(1):81–87.

56 Caniggia I, Taylor CV, Ritchie JW, et al. Endoglin regulates trophoblast differentiation along the invasive pathway in human placental villous explants. *Endocrinology*. 1997;138(11):4977–4988.

57 Caniggia I, Grisaru-Gravnosky S, Kuliszewsky M, et al. Inhibition of TGF-beta 3 restores the invasive capability of extravillous trophoblasts in preeclamptic pregnancies. *J Clin Invest*. 1999;103(12):1641–1650.

58 Clark DE, Smith SK, He Y, et al. A vascular endothelial growth factor antagonist is produced by the human placenta and released into the maternal circulation. *Biol Reprod*. 1998;59(6): 1540–1548.

59 Bdolah Y, Sukhatme VP, Karumanchi SA. Angiogenic imbalance in the pathophysiology of preeclampsia: newer insights. *Semin Nephrol*. 2004;24(6):548–556.

60 Stallmach T, Duc C, van Praag E, et al. Feto-maternal interface of human placenta inhibits angiogenesis in the chick chorioallantoic membrane (CAM) assay. *Angiogenesis*. 2001;4(1):79–84.

61 Walker JJ. Pre-eclampsia. *Lancet*. 2000;356(9237):1260–1265.

62 Page EW. The relation between hydatid moles, relative ischemia of the gravid uterus and the placental origin of eclampsia. *Am J Obstet Gynecol*. 1939;37:291–293.

63 Red-Horse K, Zhou Y, Genbacev O, et al. Trophoblast differentiation during embryo implantation and formation of the maternal-fetal interface. *J Clin Invest*. 2004;114(6):744–754.

64 Zhou Y, Damsky CH, Fisher SJ. Preeclampsia is associated with failure of human cytotrophoblasts to mimic a vascular adhesion phenotype. One cause of defective endovascular invasion in this syndrome? *J Clin Invest*. 1997;99(9):2152–2164.

65 Moldenhauer JS, Stanek J, Warshak C, et al. The frequency and severity of placental findings in women with preeclampsia are gestational age dependent. *Am J Obstet Gynecol*. 2003;189(4): 1173–1177.

66 Harrington K, Cooper D, Lees C, et al. Doppler ultrasound of the uterine arteries: the importance of bilateral notching in the prediction of pre-eclampsia, placental abruption or delivery of a small-for-gestational-age baby. *Ultrasound Obstet Gynecol*. 1996;7(3):182–188.

67 Roberts JM, Taylor RN, Musci TJ, et al. Preeclampsia: an endothelial cell disorder. *Am J Obstet Gynecol*. 1989;161(5): 1200–1204.

68 Ferris TF. Pregnancy, preeclampsia, and the endothelial cell. *N Engl J Med*. 1991;325(20):1439–1440.

69 Roberts JM. Endothelial dysfunction in preeclampsia. *Semin Reprod Endocrinol*. 1998;16(1):5–15.

70 Fisher KA, Luger A, Spargo BH, Lindheimer MD. Hypertension in pregnancy: clinical-pathological correlations and remote prognosis. *Medicine (Baltimore)*. 1981;60(4):267–276.

71 Thorp JM Jr., White GC 2nd, Moake JL, Bowes WA Jr. von Willebrand factor multimeric levels and patterns in patients with severe preeclampsia. *Obstet Gynecol*. 1990;75(2):163–167.

72 Redman S, Sargent. Preeclampsia: an excessive maternal inflammatory response to pregnancy. *Am J Obstet Gynecol*. 1999;180 (2):499–506.

73 Calvin S, Corrigan J, Weinstein L, Jeter M. Factor VIII: von Willebrand factor patterns in the plasma of patients with preeclampsia. *Am J Perinatol*. 1988;5(1):29–32.

74 Lockwood CJ, Peters JH. Increased plasma levels of ED1+ cellular fibronectin precede the clinical signs of preeclampsia. *Am J Obstet Gynecol*. 1990;162(2):358–362.

75 Friedman SA, de Groot CJ, Taylor RN, et al. Plasma cellular fibronectin as a measure of endothelial involvement in preeclampsia and intrauterine growth retardation. *Am J Obstet Gynecol*. 1994;170(3):838–841.

76 Taylor RN, Crombleholme WR, Friedman SA, et al. High plasma cellular fibronectin levels correlate with biochemical and clinical features of preeclampsia but cannot be attributed to hypertension alone. *Am J Obstet Gynecol*. 1991;165(4 Pt 1):895–901.

77 Minakami H, Takahashi T, Izumi A, Tamada T. Increased levels of plasma thrombomodulin in preeclampsia. *Gynecol Obstet Invest*. 1993;36(4):208–210.

78 Boffa MC, Valsecchi L, Fausto A, et al. Predictive value of plasma thrombomodulin in preeclampsia and gestational hypertension. *Thromb Haemost*. 1998;79(6):1092–1095.

79 Mills JL, DerSimonian R, Raymond E, et al. Prostacyclin and thromboxane changes predating clinical onset of preeclampsia: a multicenter prospective study. *JAMA*. 1999;282(4):356–362.

80 Ashworth JR, Warren AY, Baker PN, Johnson IR. A comparison of endothelium-dependent relaxation in omental and myometrial resistance arteries in pregnant and nonpregnant women. *Am J Obstet Gynecol*. 1996;175(5):1307–1312.

81 Roberts JM, Edep ME, Goldfien A, Taylor RN. Sera from preeclamptic women specifically activate human umbilical vein endothelial cells in vitro: morphological and biochemical evidence. *Am J Reprod Immunol*. 1992;27(3–4):101–108.

82 Roberts JM. Preeclampsia: what we know and what we do not know. *Semin Perinatol*. 2000;24(1):24–28.

83 Hasan KM, Manyonda IT, Ng FS, et al. Skin capillary density changes in normal pregnancy and pre-eclampsia. *J Hypertens*. 2002;20(12):2439–2443.

84 Jonsdottir LS, Arngrimsson R, Geirsson RT, et al. Death rates from ischemic heart disease in women with a history of hypertension in pregnancy. *Acta Obstet Gynecol Scand*. 1995;74(10): 772–776.

85 Maynard SE, Min JY, Merchan J, et al. Excess placental soluble fms-like tyrosine kinase 1 (sFLT1) may contribute to endothelial dysfunction, hypertension, and proteinuria in preeclampsia. *J Clin Invest*. 2003;111(5):649–658.

86 Levine RJ, Maynard SE, Qian C, et al. Circulating angiogenic factors and the risk of preeclampsia. *N Engl J Med*. 2004;350(7): 672–683.

87 Polliotti BM, Fry AG, Saller DN, et al. Second-trimester maternal serum placental growth factor and vascular endothelial growth factor for predicting severe, early-onset preeclampsia. *Obstet Gynecol*. 2003;101(6):1266–1274.

88 Tsatsaris V, Goffin F, Munaut C, et al. Overexpression of the soluble vascular endothelial growth factor receptor in preeclamptic patients: pathophysiological consequences. *J Clin Endocrinol Metab*. 2003;88(11):5555–5563.

89 Bujold E, Romero R, Chaiworapongsa T, et al. Evidence supporting that the excess of the sVEGFR-1 concentration in maternal plasma in preeclampsia has a uterine origin. *J Matern Fetal Neonatal Med*. 2005;18(1):9–16.

90 Taylor RN, Grimwood J, Taylor RS, et al. Longitudinal serum concentrations of placental growth factor: evidence for abnormal placental angiogenesis in pathologic pregnancies. *Am J Obstet Gynecol*. 2003;188(1):177–182.

91 Hertig A, Berkane N, Lefevre G, et al. Maternal serum sFLT1 concentration is an early and reliable predictive marker of preeclampsia. *Clin Chem*. 2004;50(9):1702–1703.

92 Chaiworapongsa T, Romero R, Espinoza J, et al. Evidence supporting a role for blockade of the vascular endothelial growth factor system in the pathophysiology of preeclampsia. Young Investigator Award. *Am J Obstet Gynecol*. 2004;190(6):1541–1547; discussion 7–50.

93 Levine RJ, Thadhani R, Qian C, et al. Urinary placental growth factor and risk of preeclampsia. *JAMA*. 2005;293(1):77–85.

94 Venkatesha S, Toporsian M, Lam C, et al. Soluble endoglin contributes to the pathogenesis of preeclampsia. *Nat Med*. 2006.

95 Morbidelli L, Chang CH, Douglas JG, et al. Nitric oxide mediates mitogenic effect of VEGF on coronary venular endothelium. *Am J Physiol*. 1996;270(1 Pt 2):H411–H415.

96 He H, Venema VJ, Gu X, et al. Vascular endothelial growth factor signals endothelial cell production of nitric oxide and prostacyclin through flk-1/KDR activation of c-Src. *J Biol Chem*. 1999;274(35):25130–25135.

97 Kim YG, Suga SI, Kang DH, et al. Vascular endothelial growth factor accelerates renal recovery in experimental thrombotic microangiopathy. *Kidney Int*. 2000;58(6):2390–2399.

98 Masuda Y, Shimizu A, Mori T, et al. Vascular endothelial growth factor enhances glomerular capillary repair and accelerates resolution of experimentally induced glomerulonephritis. *Am J Pathol*. 2001;159(2):599–608.

99 Kang DH, Kim YG, Andoh TF, et al. Post-cyclosporine-mediated hypertension and nephropathy: amelioration by vascular endothelial growth factor. *Am J Physiol Renal Physiol*. 2001; 280(4):F727–F736.

100 Yang JC, Haworth L, Sherry RM, et al. A randomized trial of bevacizumab, an anti-vascular endothelial growth factor antibody, for metastatic renal cancer. *N Engl J Med*. 2003;349(5): 427–434.

101 Esser S, Wolburg K, Wolburg H, et al. Vascular endothelial growth factor induces endothelial fenestrations in vitro. *J Cell Biol*. 1998;140(4):947–959.

102 Eremina V, Sood M, Haigh J, et al. Glomerular-specific alterations of VEGF-A expression lead to distinct congenital and acquired renal diseases. *J Clin Invest*. 2003;111(5):707–716.

103 Powers RW, Roberts JM, Cooper KM, et al. Maternal serum soluble fms-like tyrosine kinase 1 concentrations are not increased in early pregnancy and decrease more slowly postpartum in women who develop preeclampsia. *Am J Obstet Gynecol*. 2005; 193(1):185–191.

104 Hirtenlehner K, Pollheimer J, Lichtenberger C, et al. Elevated serum concentrations of the angiogenesis inhibitor endostatin in preeclamptic women. *J Soc Gynecol Investig*. 2003;10(7):412–417.

105 Wallukat G, Homuth V, Fischer T, et al. Patients with preeclampsia develop agonistic autoantibodies against the angiotensin AT1 receptor. *J Clin Invest*. 1999;103(7):945–952.

106 Dechend R, Homuth V, Wallukat G, et al. AT(1) receptor agonistic antibodies from preeclamptic patients cause vascular cells to express tissue factor. *Circulation.* 2000;101(20):2382–2387.
107 Xia Y, Wen H, Bobst S, et al. Maternal autoantibodies from preeclamptic patients activate angiotensin receptors on human trophoblast cells. *J Soc Gynecol Investig*. 2003;10(2):82–93.
108 Dechend R, Muller DN, Wallukat G, et al. AT1 receptor agonistic antibodies, hypertension, and preeclampsia. *Semin Nephrol.* 2004;24(6):571–579.
109 Dragun D, Muller DN, Brasen JH, et al. Angiotensin II type 1-receptor activating antibodies in renal-allograft rejection. *N Engl J Med*. 2005;352(6):558–569.
110 Walther T, Wallukat G, Jank A, et al. Angiotensin II type 1 receptor agonistic antibodies reflect fundamental alterations in the uteroplacental vasculature. *Hypertension.* 2005;46(6):1275–1279.
111 Baschat AA, Hecher K. Fetal growth restriction due to placental disease. *Semin Perinatol*. 2004;28(1):67–80.
112 Mayhew TM, Wijesekara J, Baker PN, Ong SS. Morphometric evidence that villous development and fetoplacental angiogenesis are compromised by intrauterine growth restriction but not by pre-eclampsia. *Placenta.* 2004;25(10):829–833.
113 Chaddha V, Viero S, Huppertz B, Kingdom J. Developmental biology of the placenta and the origins of placental insufficiency. *Semin Fetal Neonatal Med*. 2004;9(5):357–369.
114 DeChiara TM, Efstratiadis A, Robertson EJ. A growth-deficiency phenotype in heterozygous mice carrying an insulin-like growth factor II gene disrupted by targeting. *Nature*. 1990; 345(6270):78–80.
115 Crossey PA, Pillai CC, Miell JP. Altered placental development and intrauterine growth restriction in IGF binding protein-1 transgenic mice. *J Clin Invest*. 2002;110(3):411–418.
116 Baker J, Liu JP, Robertson EJ, Efstratiadis A. Role of insulin-like growth factors in embryonic and postnatal growth. *Cell*. 1993; 75(1):73–82.
117 Abuzzahab MJ, Schneider A, Goddard A, et al. IGF-I receptor mutations resulting in intrauterine and postnatal growth retardation. *N Engl J Med*. 2003;349(23):2211–2222.
118 Lopez-Lopez C, LeRoith D, Torres-Aleman I. Insulin-like growth factor I is required for vessel remodeling in the adult brain. *Proc Natl Acad Sci USA*. 2004;101(26):9833–9838.
119 Herr F, Liang OD, Herrero J, et al. Possible angiogenic roles of insulin-like growth factor II and its receptors in uterine vascular adaptation to pregnancy. *J Clin Endocrinol Metab*. 2003; 88(10):4811–4817.
120 Rabinovsky ED, Draghia-Akli R. Insulin-like growth factor I plasmid therapy promotes in vivo angiogenesis. *Mol Ther*. 2004; 9(1):46–55.
121 Krantz D, Goetzl L, Simpson JL, et al. Association of extreme first-trimester free human chorionic gonadotropin-beta, pregnancy-associated plasma protein A, and nuchal translucency with intrauterine growth restriction and other adverse pregnancy outcomes. *Am J Obstet Gynecol*. 2004;191(4):1452–1458.
122 Smith GC, Stenhouse EJ, Crossley JA, et al. Early pregnancy levels of pregnancy-associated plasma protein a and the risk of intrauterine growth restriction, premature birth, preeclampsia, and stillbirth. *J Clin Endocrinol Metab*. 2002;87(4):1762–1767.
123 Shibata E, Rajakumar A, Powers RW, et al. Soluble fms-like tyrosine kinase 1 is increased in preeclampsia but not in normotensive pregnancies with small-for-gestational-age neonates: relationship to circulating placental growth factor. *J Clin Endocrinol Metab*. 2005;90(8):4895–4903.
124 Malamitsi-Puchner A, Boutsikou T, Economou E, et al. The role of the anti-angiogenic factor endostatin in intrauterine growth restriction. *J Soc Gynecol Investig*. 2005;12(3):195–197.
125 Lepercq J, Guerre-Millo M, Andre J, et al. Leptin: a potential marker of placental insufficiency. *Gynecol Obstet Invest*. 2003; 55(3):151–155.
126 Savvidou MD, Hingorani AD, Tsikas D, et al. Endothelial dysfunction and raised plasma concentrations of asymmetric dimethylarginine in pregnant women who subsequently develop pre-eclampsia. *Lancet*. 2003;361(9368):1511–1517.
127 Lachmeijer AM, Dekker GA, Pals G, et al. Searching for preeclampsia genes: the current position. *Eur J Obstet Gynecol Reprod Biol*. 2002;105(2):94–113.

GLOSSARY

Chorionic villus is a finger-like projection of the placenta that contains fetal blood vessels and that grows into the intervillous space.

Decidua is hormonally altered endometrium of the uterus, present during pregnancy or during the end of the luteal phase of menstruation.

Extravillous trophoblasts are a type of cytotrophoblast that invades into the maternal decidua and the myometrium. When they invade into the maternal vessels present in the decidua, they are referred to as *endovascular trophoblasts.* The counterpart of these cells in rodents is referred to as *trophoblast giant cells.*

Glomerular endotheliosis is the signature lesion noted in the kidneys of patients with preeclampsia. It is characterized by glomerular endothelial swelling with loss of endothelial fenestrae and relatively preserved epithelial foot processes.

Hemochorial placentation is a type of placenta (as in humans) in which the placental tissue is in direct contact with maternal blood supply. This makes it easy for placentally derived molecules to enter the maternal bloodstream.

Placental labyrinth represents that part of the placenta in which the fetal/maternal exchange of oxygen (O_2) and nutrients occur. It is a system of channels, half of which contain maternal blood lacunae (intervillous spaces) lined by syncytiotrophoblasts; the remaining half contains fetal capillaries, lined by fetal endothelium.

Spiral artery is a branch of the uterine artery that traverses the myometrium and the decidua to deliver blood to the placenta.

Trophoblasts are the epithelial cells that constitute the bulk of the placenta. The outer layer of trophoblasts form a shell that is in contact with the decidua; it fuses fuse to form multinucleated cells referred to as *syncytiotrophoblasts.* The inner layer of cells remain unfused and are referred to as *cytotrophoblasts.* The cytotrophoblasts can perform specialized functions, such as invasion and differentiation.

Vascular mimicry or **pseudovasculogenesis** occurs when the cytotrophoblasts invade and line the maternal blood vessels; they convert from an epithelial to endothelia phenotype. This process is analogous to the vascular mimicry noted in metastatic cancers, in which epithelial cells line the vascular spaces.

162

Endothelialization of Prosthetic Vascular Grafts

Thomas S. Monahan and Frank W. LoGerfo

Beth Israel Deaconess Medical Center, Harvard Medical School, Boston, Massachusetts

Each year in the United States, over 100,000 patients will undergo lower-extremity arterial reconstruction for peripheral vascular disease, and an additional 400,000 patients will have a coronary artery bypass graft (CABG) (1). Conduits for arterial reconstruction or bypass may be separated into two broad categories: autologous and prosthetic. Autologous conduits are derived from the patient's own tissue and include saphenous vein grafts, arm vein grafts, and arterial grafts such as radial artery or hypogastric artery grafts. Prosthetic conduits include fabric or plastic grafts such as expanded polytetrafluoroethylene (ePTFE or Gore-Tex) and polyester or Dacron. Other prosthetic grafts are fabricated from denatured, cross-linked tissue such as human umbilical cord vein grafts.

Regardless of the prosthetic graft used, the clinical experience with small diameter (<6 mm) prosthetic arterial grafts has been poor. The surface of all prosthetic grafts is thrombogenic, and rapidly becomes covered with a fibrin coating or pseudointima (Figure 162.1). In humans, this surface remains thrombogenic and acellular for the lifetime of the conduit. These grafts are unable to tolerate low-flow conditions. In contrast, autologous conduits retain a viable endothelium that actively inhibits thrombosis, making such grafts suitable for the low-flow conditions associated with CABG procedures or long bypass grafts to small arteries. One solution to the prosthetic graft problem is to try to create and maintain a healthy autologous endothelial surface on prosthetic arterial grafts.

For the same reason, it is desirable to minimize injury to the existing endothelium when implanting autologous bypass grafts. If the endothelium is injured, highly thrombogenic subendothelial collagen is exposed to circulating blood, leading to platelet adhesion (2). Platelets release factors that cause direct thrombosis or clot formation and cause adjacent endothelium to release growth factors that stimulate hyperplasia of the underlying vascular smooth muscle cells (VSMCs); this is known as intimal hyperplasia (IH). IH is initiated by the pathologic migration and proliferation of VSMCs from the media to the intima and a subsequent change in phenotype from a native contractile phenotype to a pathologic secretory phenotype (3). The resulting lesion consists of these secretory VSMCs and extracellular matrix. The absence of an intact endothelium, and the resulting specific genetic signaling events that occur, trigger the changes associated with the pathogenesis of IH (3–5). Evidence also suggests that adventitial fibroblasts contribute to the pathogenesis of IH (6). Irrespective of the origin of the offending cells, preservation of an intact, uninjured endothelium at the time of surgical bypass is important in preventing the formation of IH in autologous bypass grafts. The end result of IH is a narrowing in the bypass graft, which increases the incidence of graft thrombosis.

Because the body of a prosthetic conduit is acellular, IH cannot occur except at the junction of the graft and the host artery (the anastomosis), where it is referred to as neointimal anastomotic hyperplasia (NIH). NIH tends to be greater at the outflow end of prosthetic grafts, where it can lead to a gradual compromise in flow through the graft. Because the surface of the graft remains thrombogenic, delayed thrombosis of the graft occurs as flow rate decreases. A natural approach to limiting the consequences of anastomotic IH is to provide a conduit with either an endothelium or endothelial-like structure to maintain an actively antithrombogenic surface (7,8).

HISTORY OF BIOENGINEERED PROSTHETIC GRAFTS

It is well established that autogenous grafts (especially small-caliber grafts ≤6 mm) have superior long-term patency compared to prosthetic grafts. However, up to 30% of patients requiring a small-caliber vascular graft do not have autogenous tissue of sufficient length or quality to use for arterial reconstruction (9). Unacceptable patency rates for small-caliber expanded polytetrafluoroethylene (ePTFE) and polyester grafts have driven researchers to attempt to develop a synthetic graft that confers similar long-term patency rates to those achieved with autologous tissue.

The Holy Grail of prosthetic vascular grafts would embody several essential characteristics. The most important characteristic of a prosthetic graft is mechanical strength. It must be capable of withstanding the constant strain of arterial

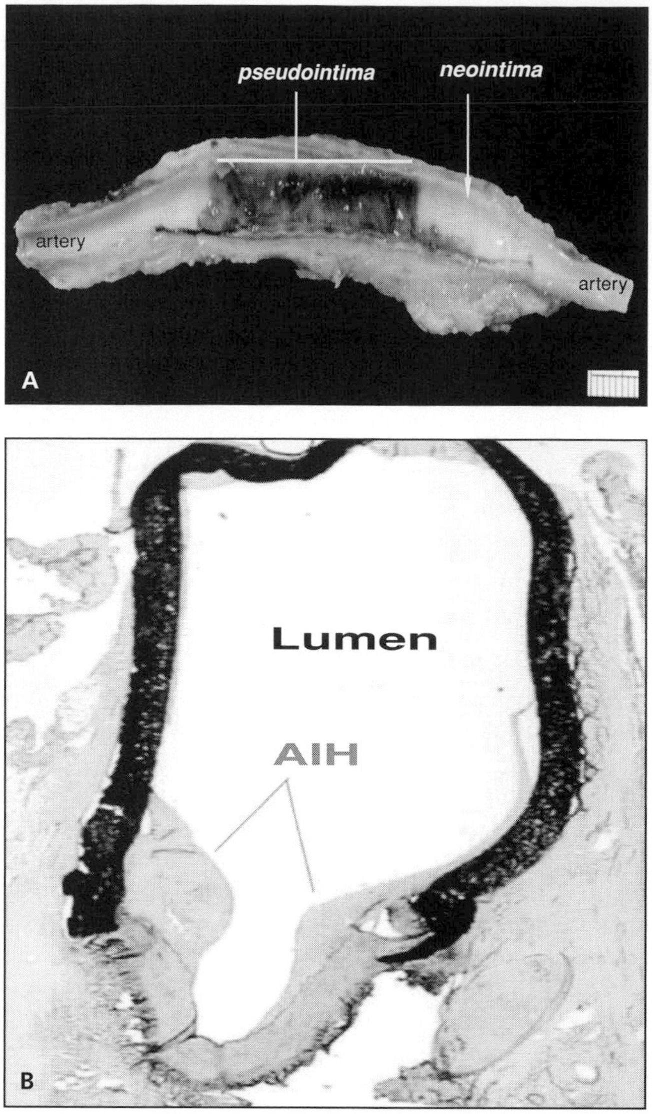

Figure 162.1. Neointima formation on a 6-mm Dacron vascular interposition graft implanted in an experimental canine model. A glistening neointima (**A**) is readily observed as early as 30 days after grafting into the arterial circulation. At this relatively early time point, there is already a readily identifiable neointima (**B**) that has formed at the anastomosis.

pressures for the life of the graft without fatigue; failure to do so would result in disruption of the graft and hemorrhage. In addition to mechanical strength, the graft must have favorable suturability and handling characteristics, comparable to those of presently available conduits. Finally, the ideal prosthetic graft would be resistant to both thrombosis and infection (10). An additional desirable characteristic of a synthetic graft is the ability to grow over time, as would be required in the special case of pediatric cardiovascular surgery (11).

Since the first clinical demonstration that synthetic materials could be employed as conduits in the arterial circulation over 50 years ago (12), surgeons have been attempting to improve on available prosthetics to embody the characteristics of the ideal vascular graft. Over the years, polyester and ePTFE have demonstrated the desired mechanical durability, ease of handling, and availability required for a successful prosthetic, but lack the antithrombogenic properties essential for long-term patency in narrow-caliber grafts.

Histological analysis of implanted vascular grafts has revealed a cellular ingrowth that occurs with time. Experimental studies have demonstrated that in nonhuman primates, endothelial cells (ECs) have the capability of populating the mid-portion of ePTFE grafts (13); this suggests that the same events are possible in humans and could lead to a fully endothelialized surface. However, significant endothelial ingrowth has not been observed in humans. Despite speculation that the endothelial ingrowth observed in nonhuman primates is attributed to differences in porosity of the prosthetic grafts, there is no evidence of in vivo endothelialization of prosthetic grafts in the clinical setting, regardless of porosity. In polyester, ePTFE, and all other prosthetic grafts in humans, the graft develops a fibrin clot pseudointima that lacks the antithrombogenic qualities of a true endothelium-lined intima (see Figure 162.1). At the junction or anastomosis with the host artery, multiple factors come into play that lead to the development of localized IH. In this case the ECs grow from the host artery into the graft for a distance of only 5 to 10 mm. At this location, IH occurs with prosthetic arterial grafts, and the lesion can become flow limiting. It appears that those ECs that grow into the prosthetic at these locations are activated, and release cytokines that initiate hyperplasia.

The creation of bioengineered vascular prosthetics was fostered by the demonstration that human ECs could be grown successfully in tissue culture (14). With the technology to isolate and grow ECs, the logical extension was to use these cells to coat the lumen of a prosthetic graft with autologous ECs ex vivo and then to use the endothelial-seeded graft for conduit. The first attempts to seed the lumen of grafts involved incubating the graft with cell suspensions in culture medium, a "one-step" procedure. However, incubating the cells on the surface of the graft resulted in fewer than 10% of cells attaching and remaining on the graft under flow conditions (15). In addition to poor retention rates, the ECs failed to demonstrate the flattened phenotype of native endothelium. On electron microscopic examination of the luminal surface of both ePTFE and Dacron grafts, it was evident that these prosthetics represent a hostile environment that largely precludes cellular adhesion and the growth of a confluent endothelium.

After consideration of the failures of the early prosthetic grafts, second-generation or "two-step" grafts were created. These ePTFE prosthetic grafts were first precoated with fibrin glue, then lined with cultured autogenous ECs. This approach demonstrated improved retention of cells; however, a large number of the autogenous cells were still lost under flow conditions. Nonetheless, with the improved substrate for cellular attachment, the remaining ECs survived and, after 4 weeks in an experimental animal model, demonstrated high cellular density and a mature, well-differentiated appearance. In contrast, the control, untreated grafts were lined by a fibrin pseudointima rich in adherent platelets, similar to our own experience with these grafts (16).

Other investigators attempted to create an endothelialized vascular graft using alternative scaffoldings that more closely resembled the surface of a native vessel. One of the first reports described using a collagen matrix to first culture bovine VSMCs, then apply ECs and finally fibroblasts. This graft was not strong enough to tolerate even very low (<10 mm Hg) intraluminal pressures. When the graft was bolstered with a Dacron mesh, the intraluminal burst pressure was only raised to the physiological level (17).

Various subsequent attempts were made to strengthen the structural component of synthetic grafts to accommodate higher intraluminal burst pressures. Biodegradable polymers such as polyglycolic acid (PGA) exhibit porosity and the ease of handling desired of a synthetic graft. However, this polymer degrades in just 2 months, preventing the newly forming cellular graft from acquiring the strength necessary to tolerate arterial pressures. When PGA is combined with other copolymers, such as polyhydroxyalkanoate (PHA), the mechanical strength of the graft is increased to levels that would allow implantation in the arterial circulation. PGA-PHA grafts have been successfully implanted in the infrarenal aortic position in lambs with excellent results. These cellular grafts demonstrated superior patency to acellular, control grafts and, on retrieval after 5 months, demonstrated stress-strain curves similar to native vessels.

PGA-PHA grafts begin to take on the appearance of native vessels after 5 months of maturation in the arterial circulation. In addition to exhibiting the strength and thrombosis-resistance of native vessels, histologically they have a distinct intima and media confirmed by immunohistochemical analysis (18). To date, this represents one of the most promising approaches to the creation of a prosthetic vascular graft. The time required to create such a graft, the cost associated with the graft, and the inability to size a graft appropriate for the required reconstruction all limit clinical utility.

CLINICAL EXPERIENCE WITH ENDOTHELIALIZED VASCULAR GRAFTS

These improvements in the endothelialization of prosthetic arterial grafts have led to several clinical trials. In a phase I trial, 49 patients with no available autogenous conduit requiring femoropopliteal bypass grafting were randomized into two treatment groups. The first included 16 patients with either critical limb ischemia or disabling claudication who received a standard bypass graft with an untreated 6-mm ePTFE graft. The second treatment group was composed of 33 patients with similar indications for operations who received the previously described 6-mm ePTFE graft treated with the "two-stage" autogenous endothelial seeding. The patients receiving the treated graft enjoyed significantly improved graft patency compared with the untreated group at almost 3 years follow-up (19).

With promising phase I data, this procedure was adopted for routine operations in 153 patients lacking adequate autogenous conduit. In this series, both 6-mm and 7-mm ePTFE grafts were used as conduit to both the above-knee and below-knee popliteal arteries, with a preponderance of grafts to the above-knee popliteal artery. These grafts had a 7-year patency rate of 62% by Kaplan-Meier analysis (20), a rate that compares favorably to other published series evaluating patency of ePTFE grafts (21). This did not, however, represent an improvement over well-established outcomes with standard nonendothelialized grafts.

One obvious disadvantage to evaluating the long-term morphologic changes in prosthetic grafts in humans as opposed to animals is the inability to have access to grafts for study. In the large study published by Meinhart and colleagues (20), failures occurred in the endothelialized grafts that allowed both radiographic and histologic assessment of the "two-stage" graft after long-term implantation in the human arterial system. One patient who received bilateral engineered grafts experienced a unilateral graft failure. Interestingly, the regions of stenosis were in the mid-portion of the graft, in sharp contrast to the typical stenosis of anastomotic IH (Figure 162.2). This finding suggests a significantly different pathologic mechanism of graft failure in these engineered grafts (22). In a "bare" graft, the acellular pseudointima cannot develop a hyperplastic cellular lesion. Apparently, however, the endothelialized surface is capable of forming the typical lesion of IH seen in autogenous grafts. Thus, the antithrombogenic advantages of an endothelialized surface may be offset by the development of flow-limiting stenosis in the body of the graft.

Scanning electron microscopy demonstrated an intact EC layer over the entire length of the graft, consistent with the animal data for the "two-stage" graft. Immunohistochemical analysis confirmed that the endothelium lining the graft was functional and displayed all the expected markers of healthy endothelium, including the EC marker CD34, von Willebrand factor, and Weibel-Palade bodies. The endothelium also rested on a basement membrane consisting of type IV collagen (22).

The development of a small-caliber prosthetic graft with acceptable long-term patency is of particular interest for surgery of the coronary arteries. The small caliber of coronary arteries necessitates conduit much smaller than the smallest clinically approved 6-mm ePTFE graft. The previously described studies all employed endothelial-coated grafts with diameters of at least 6 mm. Using a similar approach, cardiac surgeons have begun to engineer endothelialized vascular grafts for patients with no other available conduit. In place of ePTFE, Lamm and colleagues used cryopreserved allograft veins (CAV) as the scaffolding for their endothelium. The CAVs were mechanically de-endothelialized, precoated with autogenous serum (analogous to the previously described pretreatment with fibrin glue), and then treated with autogenous ECs from unusable conduits supplied from the bypass graft recipient. In their series, they used 15 autologous endothelialized vein allografts in 12 patients requiring CABG surgery and who lacked adequate autogenous conduit. Only two grafts

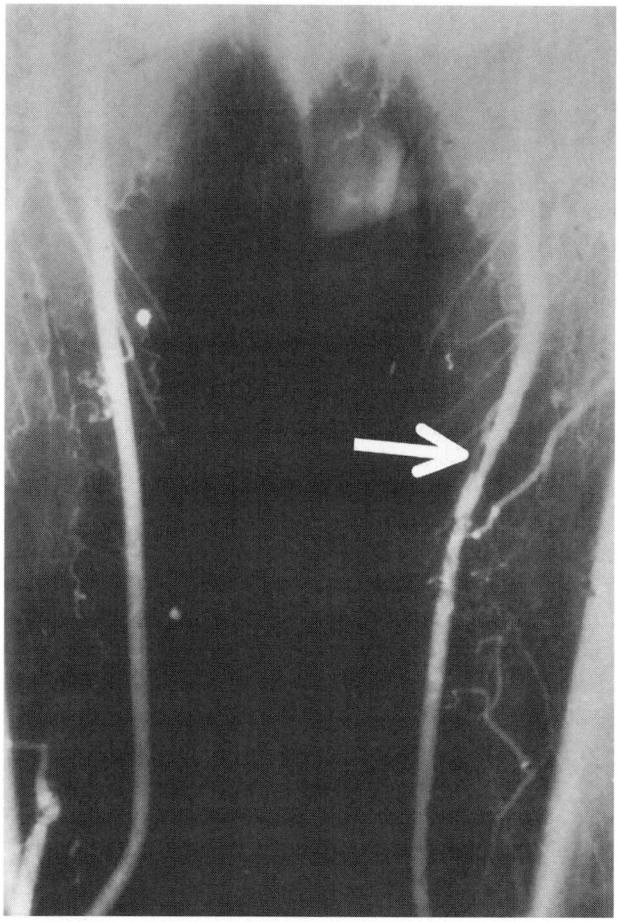

Figure 162.2. Angiogram of failing endothelialized ePTFE graft. Note that the graft on the patient's right is patent and smooth from its origin from the femoral artery to its insertion on the popliteal artery. In contrast there is a 75% stenosis (*arrow*) of the graft with similar construction on the contralateral side. This pattern of occlusion is dramatically different from that more commonly observed, which is caused by anastomotic IH and follows the pattern observed in vein bypass grafting. (Adapted with permission from Deutsch M, et al. In vitro endothelialization of expanded polytetrafluoroethylene grafts: a clinical case report after 41 months of implantation. *J Vasc Surg.* 1997;25:757–763.)

occluded on angiographic follow-up, and one patient expired due to causes unrelated to the graft, which was patent on postmortem exam (23). These studies are very encouraging, showing that, over the last 20 years, significant strides have been made to create an artificial graft with a functional endothelium. In at least 12 patients, these grafts have proved life-saving.

Lamm and colleagues used cryopreserved veins for their scaffolding to grow a cellular prosthetic arterial graft. Other matrices have been used clinically in the creation of a prosthetic graft. The first reported case of using an entirely synthetic biodegradable scaffold was presented in 2001. Peripherally harvested cells seeded onto a PGA copolymer scaffold were used to create a prosthetic graft to bypass a completely occluded pulmonary artery in a 4-year-old girl. At 7 months' follow-up, the patient was doing well, and there was no evidence of aneurysm formation or graft occlusion (11,24). A larger series of synthetic vascular grafts used to treat over 40 children with congenital heart disease from the same group of investigators reports a patency of over 95% for the first year (25). These findings suggest that grafts created in this manner might potentially grow with the child, thus obviating the need for later surgical procedures.

FUTURE DIRECTIONS AND INSIGHTS

Although much progress has been made in the creation of an endothelial-lined prosthetic graft, much work needs to be done for this technology to reach mainstream cardiovascular surgery practice. As noted, few trials demonstrate adequate patency of grafts of less than 6 mm in diameter. Additionally, because it takes a significant amount of time to produce these grafts, it is not feasible to prepare them at the time of surgery. The present technology is not available for patients needing emergent bypass grafting, and no selection of graft sizes is available to the surgeon at the time of operation. Thus, today's grafts fail to meet the Holy Grail criteria.

Alternative strategies for achieving the thrombosis-resistant surface required for small-caliber grafts are being investigated. One strategy is to bind an antithrombogenic molecule such as heparin to the luminal surface of the graft, thus preventing thrombosis. Early studies suggest that this treatment increases patency rates in an animal model (26). In that study, significant activity of the bound heparin occurred after 12 weeks of implantation in the arterial system. Our own group has developed the technology to bind both the anticoagulant hirudin and vascular endothelial growth factor (VEGF) to the luminal surface of Dacron. The purpose of this model is to both prevent early thrombosis and to promote the ingrowth of native vascular ECs. Early in vivo data in animals are encouraging.

Investigators also are modifying the surfaces of polymers used as structural scaffolding for prosthetic vascular grafts. Various peptides can be applied to the surface of the graft to induce VSMC ingrowth and attachment, in the same manner that VEGF might induce EC ingrowth in Dacron grafts. Growth factors such as transforming growth factor (TGF)-β1 have been attached to the surface of PGA to stimulate matrix production and increase the strength of the graft (27). Ultimately, the strength provided by a robust vascular smooth muscle ingrowth will need to be balanced against a narrowed graft lumen, as observed in IH.

An obvious need exists for a thrombosis- and infection-resistant synthetic conduit. The previously described studies highlight some of the many advances that have been made over the past 20 years in the bioengineering of prosthetic arterial grafts. These studies should serve to provide future engineers with direction in the construction of vascular grafts, but also to draw attention to the limitations and requirements dictated by vascular biology.

KEY POINTS

- An intact, functional endothelium is essential for vein-graft patency; without a functional endothelium, prosthetic grafts less than 6 mm in diameter have unacceptable patency rates.
- The "Holy Grail" of prosthetic grafts would combine strength, suturability, ready availability, and resistance to thrombosis.
- Prosthetic vascular grafts are available, but lack a thrombosis-resistant lining, thus requiring a diameter of at least 6 mm.
- Synthetic vascular grafts composed of cultured ECs on artificial scaffolding have been used successfully in small trials, but are not practical for widespread use in patients.

Future Goals

- To create a quiescent endothelium: At this time, prosthetic grafts seeded with autogenous endothelium are of limited utility due to cost, availability, time needed to fabricate the graft, and IH. Perhaps this can be accomplished using different substrates, subendothelium, or through genetic manipulation. Promising future directions in the development of prosthetic arterial grafts include binding of antithrombogenic proteins to create an actively antithrombotic surface.

REFERENCES

1 Rutherford RB. *Vascular Surgery*, 5th ed. Philadelphia: W. B. Saunders, 2000.
2 Gingrich RD, Hoak JC. Platelet-endothelial cell interactions. *Semin Hematol.* 1979;16(3):208–220.
3 Conte MS, Mann MJ, Simosa HF, et al. Genetic interventions for vein bypass graft disease: a review. *J Vasc Surg.* 2002;36(5):1040–1052.
4 Kalish JA, Willis DJ, Li C, et al. Temporal genomics of vein bypass grafting through oligonucleotide microarray analysis. *J Vasc Surg.* 2004;39(3):645–654.
5 Willis DJ, Kalish JA, Li C, et al. Temporal gene expression following prosthetic arterial grafting. *J Surg Res.* 2004;120(1):27–36.
6 Sartore S, Chiavegato A, Faggin E, et al. Contribution of adventitial fibroblasts to neointima formation and vascular remodeling: from innocent bystander to active participant. *Circ Res.* 2001; 89(12):1111–1121.
7 Graham LM, Brothers TE, Vincent CK, et al. The role of an endothelial cell lining in limiting distal anastomotic intimal hyperplasia of 4-mm-I. D. Dacron grafts in a canine model. *J Biomed Mater Res.* 1991;25(4):525–533.
8 Bordenave L, Remy-Zolghadri M, Fernandez P, et al. Clinical performance of vascular grafts lined with ECs. *Endothelium.* 1999; 6(4):267–275.
9 Huynh T, Abraham G, Murray J, et al. Remodeling of an acellular collagen graft into a physiologically responsive neovessel. *Nat Biotechnol.* 1999;17(11):1083–1086.
10 Conte MS. The ideal small arterial substitute: a search for the Holy Grail? *FASEB J.* 1998;12(1):43–45.
11 Kakisis JD, Liapis CD, Breuer C, et al. Artificial blood vessel: the Holy Grail of peripheral vascular surgery. *J Vasc Surg.* 2005; 41(2):349–354.
12 Blakemore AH, Voorhees AB Jr. The use of tubes constructed from vinyon N cloth in bridging arterial defects; experimental and clinical. *Ann Surg.* 1954;140(3):324–334.
13 Clowes AW, Zacharias RK, Kirkman TR. Early endothelial coverage of synthetic arterial grafts: porosity revisited. *Am J Surg.* 1987; 153(5):501–504.
14 Jarrell B, Levine E, Shapiro S, et al. Human adult endothelial cell growth in culture. *J Vasc Surg.* 1984;1(6):757–764.
15 Vohra R, Thomson GJ, Carr HM, et al. Comparison of different vascular prostheses and matrices in relation to endothelial seeding. *Br J Surg.* 1991;78(4):417–420.
16 Zilla P, Preiss P, Groscurth P, et al. In vitro-lined endothelium: initial integrity and ultrastructural events. *Surgery.* 1994; 116(3):524–534.
17 Weinberg CB, Bell E. A blood vessel model constructed from collagen and cultured vascular cells. *Science.* 1986;231(4736):397–400.
18 Shum-Tim D, Stock U, Hrkach J, et al. Tissue engineering of autologous aorta using a new biodegradable polymer. *Ann Thorac Surg.* 1999;68(6):2298–2304; discussion: 2305.
19 Zilla P, Deutsch M, Meinhart J, et al. Clinical in vitro endothelialization of femoropopliteal bypass grafts: an actuarial follow-up over three years. *J Vasc Surg.* 1994;19(3):540–548.
20 Meinhart JG, Deutsch M, Fischlein T, et al. Clinical autologous in vitro endothelialization of 153 infrainguinal ePTFE grafts. *Ann Thorac Surg.* 2001;71(Suppl 5):S327–S331.
21 Veith FJ, Gupta SK, Ascer E, et al. Six-year prospective multicenter randomized comparison of autologous saphenous vein and expanded polytetrafluoroethylene grafts in infrainguinal arterial reconstructions. *J Vasc Surg.* 1986;3(1):104–114.
22 Deutsch M, Meinhart J, Vesely M, et al. In vitro endothelialization of expanded polytetrafluoroethylene grafts: a clinical case report after 41 months of implantation. *J Vasc Surg.* 1997;25(4):757–763.
23 Lamm P, Juchem G, Milz S, et al. Autologous endothelialized vein allograft: a solution in the search for small-caliber grafts in coronary artery bypass graft operations. *Circulation.* 2001;104(12 Suppl 1):I108–I114.
24 Shin'oka T, Imai Y, Ikada Y. Transplantation of a tissue-engineered pulmonary artery. *N Engl J Med.* 2001;344(7):532–533.
25 Matsumura G, Hibino N, Ikada Y, et al. Successful application of tissue engineered vascular autografts: clinical experience. *Biomaterials.* 2003;24(13):2303–2308.
26 Begovac PC, Thomson RC, Fisher JL, et al. Improvements in GORE-TEX vascular graft performance by Carmeda BioActive surface heparin immobilization. *Eur J Vasc Endovasc Surg.* 2003; 25(5):432–437.
27 Mann BK, Schmedlen RH, West JL. Tethered-TGF-beta increases extracellular matrix production of vascular smooth muscle cells. *Biomaterials.* 2001;22(5):439–444.

163

The Endothelium's Diverse Roles Following Acute Burn Injury

Rob Cartotto

Sunnybrook and Women's College Health Sciences Centre, University of Toronto, Canada

It is estimated that each year in the United States, 1.25 million burn injuries occur, and it is known that between 60,000 and 80,000 people require in-hospital care for their burns, and that 1.4 people per 100,000 of the population will die as a result of a burn (1). Morbidity from a burn also may be considerable and includes disfigurement as well as the possibility of permanent impairment of functional abilities. The cost of hospital care for a patient with flame burns and/or smoke inhalation injury ranges from $29,560 to $117,506 (USD) per patient, whereas the cost of a single fire-related death, including loss of future earning potential, is estimated at between $250,000 and $1 million (1). It should be readily apparent then, that burn injuries are a major source of morbidity, mortality, and financial loss.

The endothelium, which at the outset might seem to be a relatively minor player in the complex overall picture of a burn injury, in fact has a major role in the pathophysiology that follows a burn. The thesis of this chapter is that the burn wound should be conceptualized as a dynamic "organ," and that the endothelium within this organ, in concert with the coagulation, fibrinolytic, and inflammatory pathways, is responsible for many of the local as well as systemic derangements that follow a burn.

Thermal injury to the skin and soft tissues acutely produces local damage at the site of the injury and systemic derangements distant from the injury. Although heat itself is the initial insult at the injury site, the host response to the heat-induced injury is responsible for both ongoing local tissue damage, as well as numerous remote pathophysiological alterations. The effect of heat on a cell depends on an inverse relationship between the temperature to which the cell is heated and the duration of exposure to that temperature (2). For example, exposure of pig skin to 60°C for 3 seconds produces irreversible epidermal necrosis, whereas exposure of the same skin to 43°C for several hours will result only in mild cellular protein denaturation, and exposure below 40°C to 42°C indefinitely does not produce direct cellular injury (2).

Historically, the burn wound was considered as a relatively inert and passive source of fluid leak and infection. However, it is now recognized that the microvascular endothelium of the burn wound, through its interactions with the coagulation and fibrinolysis pathways, and the acute inflammatory cascade, plays a critical role in integrating and mediating the host response – both locally and systemically (Figure 163.1). Viewed from this perspective, the burn wound may be considered a specialized "organ" (3), complete with its own endothelial-lined blood supply and characterized by discrete physiological and pathophysiological properties. The notion that the burn wound contains blood vessels with an active endothelium suggests that this cell layer may be not only central in mediating the host response, but also may be leveraged for therapeutic gain.

CONCEPTUALIZATION OF THE BURN WOUND

When heat is applied to the skin, the local vasculature responds with immediate vasodilatation in an effort to dissipate heat. If the amount of heat energy delivered to the skin, which is, as noted, based on an inverse relationship between temperature and duration of exposure (2), is greater than the capacity of the local circulation to dissipate heat, then direct heat injury ensues. Heat-induced injury includes malfunction of intracellular enzymes, cell swelling, progressive protein denaturation, and ultimately coagulation necrosis of tissue (2,4). (In this sense, "coagulation" refers both to the complete denaturation of cellular proteins and structure such that normal cellular architecture is completely obliterated, as well as to total occlusion of the microcirculation, with coagulated blood and cells including erythrocytes, neutrophils, and platelets.) This process leads to a three-dimensional zone of injury (termed the *zone of coagulation*) at the site of maximal heat energy (5) (see Figure 163.1). This region is characterized by complete cell death, obliteration of the micro- and macrocirculation,

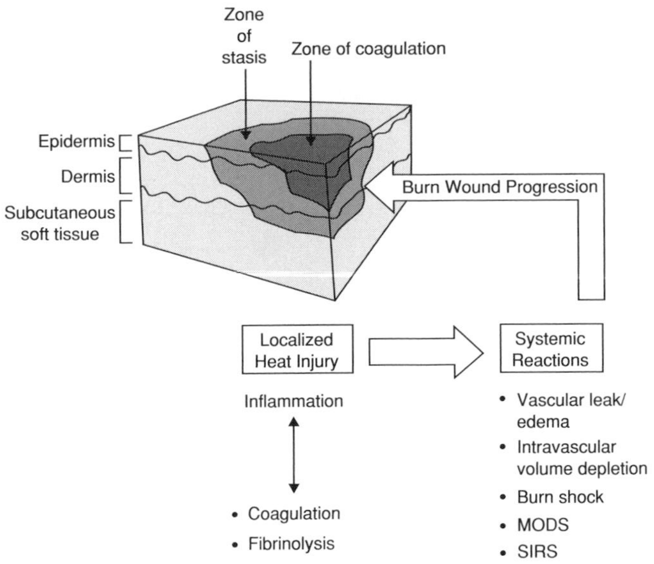

Figure 163.1. Schematic view of the concentric zones of the burn wound demonstrating the dynamic zone of stasis, in which deregulation of the normal homeostatic balance between the endothelium, coagulation, fibrinolysis, and inflammation results in local burn wound progression and systemic manifestations such as the systemic inflammatory response syndrome (SIRS), multiple organ dysfunction syndrome (MODS), and burn shock.

and loss of blood flow (5,6). The protective potential of heat shock proteins (HSPs), which are proteins generated by cells upon heating, is not entirely understood. Some experimental models suggest that induction of the heat shock response with subsequent production of HSPs improves outcome following experimental burns (7).

Concentrically surrounding the zone of coagulation is a much larger three-dimensional zone of injury, termed the *zone of stasis*, in which the microcirculation is packed with red blood cells, neutrophils, and platelet microthrombi and microemboli. Cells in this zone have sustained a partial heat injury and have been rendered ischemic by the impaired blood flow. The net result is an intense inflammatory response, characterized by a complex interplay between resident and infiltrating cells (endothelial cells [ECs], platelets, leukocytes), and soluble mediators. Locally, these reactions lead to the accumulation of interstitial fluid and further tissue ischemia. Systemically, they may produce distant effects including the systemic inflammatory response syndrome (SIRS), the acute respiratory distress syndrome (ARDS), burn shock, and multiple organ dysfunction syndrome (MODS) (8). The zone of stasis is dynamically unstable. Normally, it "falls victim" to its own intensity, and undergoes coagulation necrosis, creating the appearance of concentric expansion of the zone of coagulation (6). Clinically, this is appreciated as an apparent "extension" or "deepening" of the burn wound, which is observed over the first 72 hours following a burn. In exceptional circumstances, for example with experimental delivery of anti-inflammatory agents to the burn wound, the zone of stasis may partially recover (4). Indeed this zone represents a vital target for future therapies in burn patients.

Finally, in a third concentric zone peripheral and deep to the zone of stasis, is the *zone of hyperemia*, in which heat dissipation was sufficient to avoid cellular injury and in which temporary vasodilatation will be followed by complete recovery.

Burns are classified based on the depth of injury into the skin. First-degree burns involve only an epidermal injury. Second-degree (or partial-thickness) burns have sustained injury to the epidermis and some part of the upper dermis, while leaving some amount of the deeper dermis and the underlying subdermal vascular plexus unscathed, the exception being a deep second-degree burn in which injury extends well down into the dermis, with potential compromise of the subdermal vascular plexus. A third-degree (or full-thickness) burn includes injury to the epidermis and the entire dermal thickness, and typically features injury to or complete obliteration of the subdermal vascular plexus.

THE ENDOTHELIUM AND ACUTE BURN WOUND INTERSTITIAL EDEMA FORMATION

Rapid and profound edema formation is a fundamental characteristic of the acute burn wound and is directly related to activation and dysfunction of ECs within the zone of stasis. The immediate effect of heat is the disruption of numerous cellular enzyme systems. One important enzyme system that becomes impaired is the Na,K-ATPase pump (3,9). ECs with dysfunctional Na,K-ATPase pumps become "flooded" with sodium and water, and their transmembrane potential diminishes from –90 mv to –60 mv. Electron microscopic studies have directly demonstrated that ECs, particularly those lining postcapillary venules, swell and begin to pull away from each other, creating large gaps that allow the efflux of fluid and proteins from the circulation into the interstitial space (3,10–12) (Figure 163.2). Compounding the effects of swelling-induced endothelial gap formation are the effects of numerous acute inflammatory mediators, such as histamine, bradykinin, complement, prostaglandins, thrombin, and leukotrienes, which have inundated the zone of stasis in response to heat-induced tissue damage and which act to increase EC permeability (Figure 163.2). The resulting loss of barrier function leads to burn wound edema and, if the wound is large enough, intravascular volume depletion. Although other forces contribute to loss of intravascular volume (e.g., changes in oncotic pressure and hydrostatic pressures within the capillary) (13), the breakdown of normal EC–EC contacts and loss of barrier function is largely responsible for the generation of the edema and fluid shifts that are so characteristic of a major burn injury. As ECs swell, and as fluid accumulates in the interstitium, the vessel lumen diameter narrows and blood flow within the zone of stasis is further impeded, promoting tissue ischemia and wound progression.

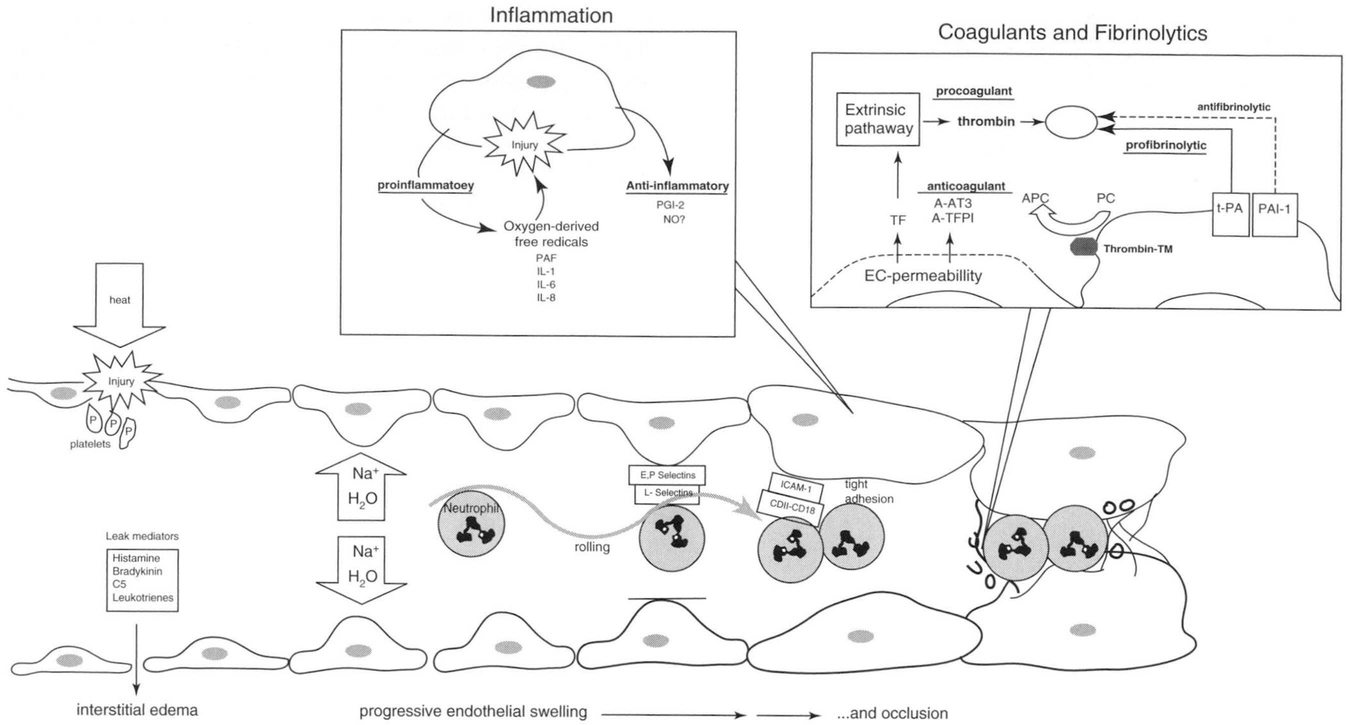

Figure 163.2. Schematic diagram demonstrating multiple roles of the endothelium in the pathophysiology of the burn wound. Progressive EC swelling and microvascular lumen occlusion are shown from left to right. P, platelets; TF, tissue factor; A-AT3, activated antithrombin 3; A-TFPI, activated tissue factor pathway inhibitor; PC, protein C; APC, activated protein C; TM, thrombomodulin; t-PA, tissue plasminogen activator; PAI-1, plasminogen activator inhibitor; OFR, oxygen-derived free radicals.

ENDOTHELIUM IS A TARGET FOR AND A SOURCE OF ACUTE INFLAMMATORY MEDIATORS IN THE ZONE OF STASIS

Numerous inflammatory mediators are liberated within the zone of stasis following thermal tissue injury. These mediators may act upon the endothelium and, in many cases are produced by the endothelium itself. They may exert both local and systemic effects.

Histamine

Within the zone of stasis, resident mast cells, stimulated by burn injury and activation of the complement cascade, release histamine, which causes vasodilatation and acts on the endothelium to increase microvascular permeability. Histamine's ability to induce endothelial microvascular leakage appears to be responsible for the early rapid phase of edema formation in the burn wound (3,8,14). Experimentally, this response can be greatly inhibited by pretreatment with H_1-histamine receptor antagonists (15). Furthermore, within the burn zone of stasis histamine appears to stimulate xanthine oxidase activity within ECs, resulting in oxidization of hypoxanthine and liberation of the oxygen free radical superoxide (O_2^-) (4,16,17), which in turn mediates ongoing microvascular injury. In an experimental burn model in rats, allopurinol, a xanthine oxidase inhibitor, was found to reduce edema formation in the zone of stasis 16. Thus, within the zone of stasis, the endothelium itself may become a source of oxygen free radicals (see Figure 163.2).

Kinins

The kinin system is activated following thermal injury by both activated Hageman factor (XIIa) and by damaged tissue cells to produce bradykinin and lysyl bradykinin (18,19). Bradykinin, in particular, causes an increase in permeability across the microvascular endothelium within the zone of stasis (8).

Complement

The complement system becomes activated mainly through the alternative pathway following burn injury (8,20,21). Although complement plays a particularly important role in chemoattraction, phagocyte activation, bacterial opsonization, and target cell lysis, certain complement components (e.g., C5a) act on the microvascular endothelium to increase permeability (8).

Platelet-Activating Factor

ECs (along with platelets, neutrophils, and monocytes) are an important source of platelet-activating factor (PAF) following thermal injury. A major effect of PAF is to cause increased microvascular permeability across the endothelium within the zone of stasis, resulting in prolonged ongoing edema formation in the burn wound (8).

Cytokines

Cytokines are intercellular messenger proteins that are produced not only by cells of the immune system such as lymphocytes, monocytes, and macrophages, but also by the endothelium. Within the zone of stasis, ECs are both a source of, and target for, cytokines. For example, ECs express interleukin (IL)-1, IL-6, and IL-8 (see Figure 163.2). Following burn injury, increased levels of IL-1, IL-6, and IL-8 have been observed both systemically (8,22) as well as in burn blister fluid (the exudative fluid contained within the epidermal-domed blisters and bullae that are the classic clinical finding in second degree [partial thickness] burns of the skin) (8,23). These interleukins have a variety of effects including generation of fever, promotion of clot formation through activation of the extrinsic coagulation pathway, and stimulation of the cascade of neutrophil chemotaxis, neutrophil adhesion to the endothelial surface, and release of injurious reactive oxygen metabolites and hydrolytic enzymes by activated, adherent neutrophils (8,24).

The Arachidonic Acid Cascade

Activation of the arachidonic acid cascade, resulting in increased production of the end-products (prostaglandins, thromboxanes, lipoxins, and leukotrienes) in burned tissue, was recognized almost 30 years ago (3,8,25). ECs, leukocytes, platelets, and mast cells all generate leukotrienes (C4, D4, and E4), which probably all act on the microvascular endothelium within the zone of stasis to increase permeability and cause edema formation (3,8,26) (see Figure 163.2). During the acute inflammatory phase of burn injury, burn-injured endothelium fails to secrete normal amounts of prostacyclin (PGI_2), favoring platelet and leukocyte aggregation and adhesion to the endothelium (3). However, in the latter, prolonged post-traumatic phase, increased endothelial production of PGI_2 appears to play a role in mediating prolonged vasodilatation and inhibiting platelet aggregation within and around the burn wound.

PLATELETS ARE ACTIVATED AND ADHERE TO THE ENDOTHELIUM IN THE BURN WOUND

In vivo models of burn injury in the mouse ear demonstrate that, within the zone of stasis, platelets rapidly adhere to endothelial surfaces that have been injured by heat (6,27). Upon adhesion, the platelets become activated and release thromboxane A_2, which promotes vasoconstriction and platelet aggregation within the already compromised microcirculation. Surface membrane phospholipids of the activated platelets (e.g., formerly platelet factor [PF-3]) become exposed in order to bind coagulation factors during thrombin formation (21). The activated platelets also secrete numerous cytokines (transforming growth factor-β, insulin-like growth factor, platelet-derived growth factor, and epidermal growth factor), which collectively serve to activate and attract macrophages and fibroblasts, cells that are necessary in the later phases of wound repair and regeneration (8).

ENHANCED LEUKOCYTE–ENDOTHELIAL ADHESION FOLLOWING BURN INJURY

It is recognized that activated neutrophils, attracted to the zone of stasis by numerous chemotactic factors, adhere to the microvascular endothelium and in so doing release reactive oxygen metabolites and hydrolytic enzymes that cause ongoing tissue injury as well as further microvascular leakage (8). The toxic inflammatory "soup" within the zone of stasis consists of many chemotactic agents for neutrophils, including complement component C5a, leukotriene B4, thrombin, PAF, IL-1, IL-8, and tumor necrosis factor (TNF). Chemoattractant activity, along with sluggish blood flow in the congested microcirculation, helps to bring neutrophils into proximity with the endothelium (28,29). In the next step, leukocytes "roll" along the endothelial surface, a process that is mediated by an interaction between L-selectin, a glycoprotein on the neutrophil surface, and E- selectin and P-selectin, which are molecules expressed on the EC surface in response to various stimuli such as cytokines, thrombin, complement, histamine, and leukotrienes (28). Rolling is followed by tight adhesion between leukocyte and endothelium. This latter process is mediated by CD11/CD18 and intercellular adhesion molecule (ICAM)-1. CD11/CD18, a member of the integrin family of glycoproteins, is expressed both constitutively and in response to inflammatory stimuli on the surface of leukocytes. ICAM-1 is a member of immunoglobulin supergene family, which is largely restricted to ECs. ICAM-1 is normally expressed at low levels, but is significantly upregulated on the surface of activated ECs (28). Following experimental burn injury in a rabbit model, administration of anti-ICAM antibodies improved blood flow in the zone of stasis, suggesting an important role for this adhesion molecule there (30).The CD11/CD18–ICAM-1 interaction results in a tight adhesion of the now stationary leukocyte to the EC surface (28) (see Figure 163.2). The adherent neutrophils then begin the process of transendothelial migration and, ultimately, take on a phagocytic role as they enter the wound. During adhesion/transmigration and in the process of phagocytosis, neutrophils release oxygen-derived free radicals and proteolytic enzymes into the local microenvironment (Figure 163.2). Neutrophils are considered an important source of oxygen free radicals after burn injury (17). These oxygen free radicals participate in cell membrane damage through lipid peroxidation, and are believed to be key elements in progression of the burn wound as well as in the development of burn shock and remote organ injury following burns (17,31).

Following burn injury, CD11/CD18 expression is increased on the surface of circulating leukocytes (28,32). Administration of antibodies to either CD18 or ICAM-1 both prior to and 30 minutes following burn injury in a rabbit model resulted in improved microvascular perfusion as measured by laser Doppler and reduced wound progression in the burn wound zone of stasis. These findings support an important role for neutrophil–endothelium adhesion in mediating burn wound progression (30). Importantly, they raise the possibility that this process may be amenable to therapeutic targeting.

ENDOTHELIUM AND NITRIC OXIDE PRODUCTION FOLLOWING BURN INJURY

The endothelium is a major source of nitric oxide (NO). The endothelium produces NO from the oxidation of L-arginine, catalyzed by the constitutively expressed enzyme endothelial NO synthase (eNOS), as well as by inducible NO synthase (iNOS), which is induced by pathological stimuli (33). NO regulates vasomotor tone, inhibits platelet aggregation, and has both proinflammatory as well as anti-inflammatory effects. Within the burn wound zone of stasis, upregulation of both eNOS and iNOS activity results in increased NO production (33,34), which may in turn contribute to the development of local edema through potentiation of substance P release, the recruitment and adhesion of neutrophils, and the dilatation of blood vessels (33,35,36). On the other hand, NO also may act locally to attenuate inflammation by inhibiting platelet aggregation and by downregulating endothelial expression of certain adhesion molecules, including selectins and ICAM-1 (33,37). Human and animal studies showing elevated levels of the stable NO metabolite NO_3^- in both plasma and urine post-burn support the notion that increased systemic production of NO occurs following a burn injury (33,34,38). In one study, administration of the nonspecific NOS inhibitor N^G–monomethyl-L-arginine, (L-NNMA), attenuated this effect (39). This increased systemic production of NO is believed to contribute to increased gastrointestinal bacterial translocation, pulmonary microvascular leakage and leukocyte infiltration, and myocardial dysfunction following major burn injury (33,40,41). However, these effects are highly variable and largely depend on the time point post injury at which NO is acting.

ENDOTHELIUM AND ALTERATIONS IN COAGULATION AND FIBRINOLYSIS FOLLOWING BURN INJURY

The endothelium plays a central role in regulating hemostasis. Through a finely tuned balance of procoagulant activity, anticoagulant forces, and fibrinolytic components, the endothelium normally presents a nonthrombogenic surface to circulating blood (21,24). On the procoagulant side, ECs express von Willebrand factor (vWF), tissue factor, as well as inflammation-related proadhesive molecules such as selectins and ICAM-1. On the anticoagulant side, ECs produce heparan, which serves as a cofactor for antithrombin III (ATIII); tissue factor pathway inhibitor (TFPI), which inhibits factor VIIa-TF; and thrombomodulin (TM), which catalyses conversion of protein C to activated protein C (APC), a powerful anticoagulant, profibrinolytic, and anti-inflammatory agent (42,43). In addition, ECs also produce NO and PGI_2, which inhibit platelet adhesion and aggregation. The endothelium is intimately involved in the regulation of fibrinolysis through the production of tissue-type plasminogen activator (t-PA) and plasminogen activator inhibitor (PAI)-1. t-PA converts plasminogen to plasmin. Plasmin in turn lyses fibrin (and fibrinogen), producing D-dimers. Increased circulating levels of D-dimers are diagnostic of ongoing coagulation and fibrinolysis (21). PAI-1 is a natural inhibitor of t-PA (21) (see Figure 163.2). Thrombin represents an important link between coagulation and inflammation. In addition to its role in cleaving fibrinogen to generate fibrin, thrombin has proinflammatory effects that include neutrophil chemotaxis, endothelial adhesion molecule expression (selectins), barrier dysfunction, and promotion of PAF expression (44).

When thrombin binds to TM on the surface of ECs, it acquires anticoagulant and anti-inflammatory properties. TM-bound thrombin activates protein C. APC is a central modulator of coagulation, fibrinolysis, and inflammation (43). The anticoagulant effects of APC include inhibition of thrombin generation by inactivation of clotting factors Va and VIIIa, and inhibition of tissue factor release from monocytes and endothelium. Profibrinolytic effects of APC include inactivation of PAI-1 and thrombin-activated fibrinolysis inhibitor (TAFI). Finally, APC's anti-inflammatory effects include the suppression of inflammatory cytokines production, such as IL-1 and TNF, and inhibition of selectin-mediated neutrophil adhesion to the endothelium (43).

It is widely believed that the endothelium plays a pathophysiological role in severe sepsis. Under these conditions, the endothelium acquires a procoagulant surface, attenuates fibrinolysis, expresses proadhesion molecules and inflammatory mediators, and undergoes apoptosis (24,43,44). The end result is progressive occlusion of the microcirculation and cell death, which may contribute to organ failure and mortality. Although less well characterized than in sepsis, a similar cascade of events appears to occur following burn injury. Two studies involving a total of 103 burn patients with moderate to severe burns (45–47) demonstrated that, in the initial days following burn injury, intense burn-size–dependent activation of coagulation occurs. The hypercoaguable state was characterized by increased levels of clotting factor VIIa and thrombin-antithrombin (TAT) levels, combined with depression of the anticoagulants ATIII and protein C. Presumably, burns result in rapid activation of the extrinsic pathway through upregulation of tissue factor on the surface of monocytes (and possibly ECs). Theoretically, collagen exposure in the burn wound can activate the intrinsic limb of coagulation, although there is little evidence for the existence of this pathway in vivo (47).

Simultaneously, there also appears to be burn-size–dependent activation of fibrinolysis, characterized by increased levels of t-PA, D-dimers, and reduced levels of plasminogen (46,47). However, the activation of fibrinolysis appears to yield over time to a general inhibition of fibrinolysis, as characterized by progressive elevation of the antifibrinolytic PAI-1 over the first 7 days post burn (47). This is a particularly interesting finding because it was recognized nearly 30 years ago that inhibition of fibrinolysis existed in the plasma of thermal injury patients (48), and 10 years ago that burn blister fluid was antifibrinolytic (49).

This state of hypercoagulation with relatively ineffective fibrinolysis could well play a role in the progressive occlusion of the microcirculation within the burn wound zone of stasis, and be partly responsible for the phenomenon of wound progression. Furthermore, these derangements in coagulation and fibrinolysis are associated with development of organ dysfunction remote from the burn seen after burn injury. In the study by Garcia, organ failure scores were significantly correlated with ATIII, plasminogen, and protein C levels (47). Thus, based on the limited data available from burn patients, there appear to be some important similarities between the procoagulant/antifibrinolytic/inflammatory state recognized during sepsis, and the local burn wound and systemic state following a major burn injury.

KEY POINTS

- The burn wound is best conceptualized as a dynamic, specialized organ, complete with its own endothelium-lined blood supply. In concert with the coagulation, fibrinolysis, and inflammatory pathways, the endothelium plays a central and multifaceted role in directing local progression of the burn wound, as well as many of the systemic derangements seen following a major thermal injury.
- Endothelial gap formation and increased microvascular endothelial permeability are the basis of the profound edema that characterizes the burn wound, as well as the intravascular volume depletion that follows a major burn injury.
- Recognition of the role of enhanced endothelium–leukocyte adhesion following burn injury is a relatively new and unexplored area. Inhibition of this adhesion process by antiadhesion molecule antibodies appears to improve perfusion of the burn wound. However, the effects of this approach in clinical burn care, and the possible infectious risks of this approach, remain unknown.
- The development of a hypercoaguable state and an activated endothelium with relative suppression of fibrinolysis at the microcirculatory level following an acute burn injury has been recognized, but is not as well characterized as it is during sepsis. However, there appear to be similarities between the two processes, and this hypercoaguable/hypofibrinolytic state could well be the basis of burn wound progression and distant organ dysfunction following a major burn injury.

Future Goals

- To further characterize the coagulation and fibrinolysis state at a microvascular level following acute burn injury is necessary; the role of agents such as recombinant APC, which appears to have a profound effect during sepsis, offer theoretically ideal benefits with respect to control of burn wound progression and the systemic alterations following major burns.

REFERENCES

1 Pruitt BA, Goodwin CW, Mason AD. Epidemiological, demographic, and outcome characteristics of burn injury. In: Herndon DN, ed. *Total Burn Care*, 2nd ed. London, England: Saunders Co.; 2002:16–30.

2 Moritz AR, Henriquez FC. Studies of thermal injury II. The relative importance of time and surface temperature in the causation of cutaneous burns. *Am J Pathol*. 1947;23:695–720.

3 Arturson G. Local effects. In: Settle JAD, ed. *Burns Management*, 1st ed. London, England: Churchill Livingstone; 1996:83–95.

4 Williams WG, Phillips LG. Pathophysiology of the burn wound. In: Herndon DN, ed. *Total Burn Care*, 1st ed. London, England: WB Saunders Co. Ltd.; 1996:63–71.

5 Jackson DMG. The diagnosis of the depth of burning. *Br J Surg*. 1953;40:588–596.

6 Boykin JV, Eriksson E, Pittman RN. In vivo microcirculation of scald burn and the progression of postburn dermal ischemia. *Plast Reconstr Surg*. 1980;66:191–198.

7 Meyer TN, da Silva AL, Viera EC, et al. Heat shock response reduces mortality after severe experimental burns. *Burns*. 2000; 26:233–238.

8 Arturson G. Pathophysiology of the burn wound and pharmacologic treatment. The Rudi Hermans Lecture 1995. *Burns*. 1996;22:255–268.

9 Monafo WW, Deitz F, Halikiopoulos H. Water and electrolyte fluxes in muscle, skin, and tendon after thermal injury and ischemia. *Burns*. 1976;3:80–86.

10 Arturson G. Microvascular permeability to macromolecules in thermal injury. *Acta Physiologica Scand (Suppl)*. 1979;463:111–122.

11 Nozaki M, Guest MM, Bond TP, Larson DL. Permeability of blood vessel after thermal injury. *Burns*. 1980;6:213–221.

12 Cotran RS. The delayed and prolonged vascular leakage in inflammation II: an electron microscope study of the vascular response after thermal injury. *Am J Pathol*. 1965;46:589–620.

13 Lund T, Onarheim G, Wiig H, Reed RK. Mechanisms behind the increased dermal imbibition pressure in acute burn edema. *Am J Physiol.* 1989;256:H940.
14 Friedl HP, Till GO, Trentz O, Ward PA. Role of histamine, complement, and xanthine oxidase in thermal injury of the skin. *Am J Pathol.* 1989;135:203–217.
15 Wilhelm DL, Mason B. Vascular permeability changes in inflammation: the role of endogenous permeability factors in mild thermal injury. *Br J Exp Pathol.* 1960;41:487–506.
16 Till GO, Guilds LS, Mahrougui M, et al. Role of xanthine oxidase in thermal injury of the skin. *Am J Pathol.* 1989;135:195–202.
17 Latha B, Babu M. The involvement of free radicals in burn injury: a review. *Burns.* 2001;27:309–317.
18 Zachariae E, Henningsen S. Plasma kinins in inflammation-relation to other mediators and leukocytes. *Scand J Clin Lab Investig.* 1969;107(Suppl):85.
19 Holder A, Neely AN. Hageman factor dependent kinin activation in burns and its theoretical relationship to post burn immunosuppression syndrome and infection. *J Burn Care Rehabil.* 1990;11:496–503.
20 Gelfand JA, Donelan M, Burke JF. Preferential activation and depletion of the alternative complement pathway by burn injury. *Ann Surg.* 1983;198:58–62.
21 Dries DJ. Activation of the clotting system and complement after trauma. *New Horizons in Critical Care Med.* 1996;4:276–288.
22 Cannon JG, Friedburg JS, Gelfand JA, et al. Circulating interleukin 1ß and tumor necrosis factor α concentrations after burn injury in humans. *Crit Care Med.* 1992;20:1414–1419.
23 Ono I, Gunji H, Zhang JZ, et al. A study of cytokines in burn blister fluid related to wound healing. *Burns.* 1995;21:352–355.
24 Hack CE, Zeerleder S. The endothelium in sepsis: source of and a target for inflammation. *Crit Care Med.* 2001;29:S21–S27.
25 Arturson G. Prostaglandins in human burn wound secretion. *Burns.* 1977;3:112.
26 Dobke MK, Hayes EC, Baxter CR. Leukotrienes LTB4 and LTC4 in thermally injured patients' plasma and burn blister fluid. *J Burn Care Rehabil.* 1987;8: 189–191.
27 Boykin JV, Eriksson E, Pittman RN. Microcirculation of a scald burn: an in vivo experimental study of the hairless mouse ear. *Burns.* 1980;7:335–338.
28 Korthius RJ, Anderson DC, Granger DN. Role of neutrophil-endothelial cell adhesion in inflammatory disorders. *J Crit Care.* 1994;9:47–71.
29 Mayers I, Johnson D. The nonspecific inflammatory response to injury. *Can J Anaesth.* 1998;45:871–879.
30 Mileski W, Borgstrom D, Lightfoot E, et al. Inhibition of leukocyte-endothelial adherence following thermal injury. *J Surg Res.* 1992;52:334–339.
31 Saez J, Ward P, Ganther B, Vivaldi E. Superoxide radical in the pathogenesis of burn shock. *Circ Shock.* 1984;12:229–239.
32 Nelson R, Hasslen S, Ahrenholz D, et al. Influence of minor thermal injury on expression of complement receptor CF3 on human neutrophils. *Am J Pathol.* 1986;125:563–571.
33 Rawlingson A. Nitric oxide, inflammation and acute burn injury. *Burns.* 2003;29:631–640.
34 Gamelli RL, George M, Sharp-Pucci M, et al. Burn induced nitric oxide release in humans. *J Trauma.* 1995;39:869–877.
35 Yonehara N, Yoshimura M. Interaction between nitric oxide and substance P on heat induced inflammation in the rat paw. *Neurosci Res.* 2000;36:35–43.
36 Rawlingson A, Greenacre SA, Brain SD. Functional significance of iNOS induction, neutrophil accumulation and protein nitration in thermally injured rat cutaneous microvasculature. *Am J Pathol.* 2003;162:1373–1380.
37 Lindemann S, Sharafi M, Spiecker M, et al. NO reduces PMN adhesion to human vascular endothelial cells due to downregulation of ICAM 1 mRNA and surface expression. *Thromb Res.* 2000;97:113–123.
38 Preiser JC, Reper P, Vlasselaer D, et al. Nitric oxide production is increased in patients after burn injury. *J Trauma.* 1996;40:368–371.
39 Becker WK, Shippee RL, McManus AT, et al. Kinetics of NO production following experimental thermal injury in rats. *J Trauma.* 1993;34:855–862.
40 Chen LW, Hsu CM, Cha MC, et al. Changes in gut mucosal nitric oxide synthase (NOS) activity after thermal injury and its relation to barrier failure. *Shock.* 1999;11:104–110.
41 Soejima K, Traber LD, Schmalsteig FC, et al. Role of nitric oxide in vascular permeability after combined burns and smoke inhalation injury. *Am J Respir Crit Care Med.* 2001;163:745–752.
42 Esmon CT. Protein C anticoagulation pathway and its role in controlling microvascular thrombosis and inflammation. *Crit Care Med.* 2001;29:S48–S52.
43 Grinnell BW, Joyce D. Recombinant human activated protein C: a system modulator of vascular function for severe sepsis. *Crit Care Med.* 2001;29:S53–S61.
44 Faust SN, Heyderman RS, Levin M. Coagulation in severe sepsis: a central role for thrombomodulin and activated protein C. *Crit Care Med.* 2001;29:S62–S67.
45 Aird WC. Vascular bed specific hemostasis: role of endothelium in sepsis pathogenesis. *Crit Care Med.* 2001;29:S28–S35.
46 Kowal-Vern A, Gamelli RL, Walenga JM, et al. The effect of burn wound size on hemostasis: a correlation of the hemostatic changes to the clinical state. *J Trauma.* 1992;33:50–57.
47 Garcia-Avello A, Lorente JA, Cesar-Perez LJ, et al. Degree of hypercoagulability and hyperfibrinolysis is related to organ failure and prognosis after burn trauma. *Thromb Res.* 1998;89:59–64.
48 Arturson G, Rammer L. Endogenous inhibition of fibrinolysis in patients with severe burns. *Acta Chir Scand.* 1974;140:181–184.
49 Rockwell WB, Ehrlich HP. Fibrinolysis inhibition in burn blister fluid. *J Burn Care Rehabil.* 1990;11:1–6.

164

Trauma-Hemorrhage and Its Effects on the Endothelium

Yukihiro Yokoyama and Irshad H. Chaudry

Center for Surgical Research and Department of Surgery, University of Alabama at Birmingham

Despite improvements in critical care medicine, trauma continues to be one of the leading causes of death in developed countries. Severe trauma induces an elevation of circulating inflammatory cytokines (1), upregulation of vasoconstrictors (2), and alteration in coagulation cascade (3). It also leads to an impairment of endothelial function and circulatory alterations in major organs, which can lead to multiple organ dysfunction syndrome (MODS) (4) and eventually to multiple organ failure (MOF). In this regard, numerous studies have focused on examining the mechanisms of endothelial dysfunction following trauma so that effective treatment strategies may be developed.

In the literature, a tendency exists to group trauma and massive hemorrhage under the same heading. Indeed, massive hemorrhage rarely occurs (with the exception of aortic aneurysm) in the absence of trauma (experimental models notwithstanding), and severe trauma usually is accompanied by hemorrhage. However, it is important to recognize that simple hemorrhage per se or tissue trauma without massive blood loss may trigger distinct pathways and lead to differences in host response. Severe hemorrhage (with and without tissue trauma) or severe tissue trauma (with or without massive blood loss) eventually leads to the activation of the same final pathways that lead to organ dysfunction and eventual failure, if the process is not controlled and corrected in due course. Furthermore, the site of injury (e.g., brain, thoracic, abdominal, and extremity) and the modality of resuscitation (e.g., massive blood transfusion versus resuscitation by Ringer's lactate) also may result in different host response. In this regard, allogeneic blood transfusion clearly will elicit immunosuppression and produce an entirely different response than resuscitation with Ringer's lactate. Furthermore, additional complicating issues include whether the injury is internal or external, penetrating or blunt. Although these confounding factors can influence the initial responses, many similarities have been observed with respect to the process of organ dysfunction and failure including endothelial dysfunction among these insults.

Not only is endothelium vulnerable to the effects of trauma, it also plays a key pathophysiological role in mediating the trauma phenotype. Trauma induces inflammatory and coagulation responses, which in turn result in endothelial activation. Even with trauma on an extremity, endothelial dysfunction is affected in the remote major organs (5,6). Once activated, endothelial cells (ECs) may contribute to organ injury (Figure 164.1). Data from extensive basic and clinical research in trauma support four potentially important mechanisms of EC dysfunction and/or endothelial-mediated organ dysfunction: (a) enhanced inflammatory processes, (b) cellular damage by excessively produced oxygen-derived free radicals, (c) impaired tissue circulation due to an imbalance between vasodilators and vasoconstrictors, and (d) the damaging effect of gut-derived mesenteric lymph. Finally, there is increasing evidence for the role of gender dimorphism in dictating clinical outcomes in trauma.

MECHANISMS OF ENDOTHELIAL DYSFUNCTION AND ENDOTHELIAL-MEDIATED ORGAN INJURY FOLLOWING TRAUMA-HEMORRHAGE

Enhanced Inflammatory Processes

Hemorrhagic shock is associated with increased leukocyte adhesion and extravasation (7). Leukocyte trafficking is mediated, in large part, by the inducible expression of adhesion molecules (e.g., selectins, integrins, and members of the immunoglobulin superfamily) on the surface of the endothelium. The various rodent models of simple hemorrhage, soft-tissue trauma with or without hemorrhage, and soft-tissue trauma plus bone fracture and hemorrhage are listed in Table 164-1 (8–28). Although various species of animals such as dogs, pigs, monkeys, and horses have been utilized to study trauma-hemorrhage, we will restrict discussion to trauma-hemorrhage models in rodents.

Table 164-1: Rat and Mouse Models of Hemorrhage and Trauma-Hemorrhage

Models	*References*
Anesthetized fixed-volume bleed-out model of hemorrhage	9–14
Anesthetized fixed-pressure model of hemorrhagic shock	15–18
Unanesthetized fixed-volume hemorrhage model	19, 20
Unanesthetized fixed-pressure model of hemorrhage	21–26
Unanesthetized nonheparinized model of trauma-hemorrhagic shock	27, 28

In a mouse model of hemorrhagic shock, P-selectin was shown to increase in the lungs, liver, heart, kidney, intestinal mesentery, stomach, small bowel, and colon (29). In another study, hemorrhagic shock resulted in early induction of intercellular adhesion molecule (ICAM)-1 and vascular cell adhesion molecule (VCAM)-1 in the pulmonary microvessels and in the marginal and trabecular areas of the spleen (30), as evidenced by immunostaining. Furthermore, it remains unclear whether these organ-specific patterns of expression/induction correlate with the degree of leukocyte extravasation. Although the precise mechanism(s) by which hemorrhage leads to upregulation of adhesion molecules is unclear, it likely involves activation of ECs by some combination of proinflammatory cytokines, chemokines, hypoxia, low flow, and/or reactive oxygen species (ROS) (31).

These and other similar observations suggest that therapeutic targeting of leukocyte–endothelial interactions may attenuate hemorrhagic shock–induced organ damage (Table 164-2). For example, monoclonal antibodies targeted against CD18 (human leukocyte adherence glycoprotein) substantially decreased injury to the lung, liver, and gastrointestinal mucosa in hemorrhage models (32). A recombinant type I tumor necrosis factor (TNF)-binding protein also has been reported to reduce leukocyte adhesion in the liver sinusoids as well as mean adhesion time of leukocytes in the hepatic central vein (33). Resuscitation with 25% albumin after hemorrhagic shock significantly attenuated the increase in leukocyte CD11b, L-selectin, and endothelial ICAM-1 expression (34). Furthermore, high-density lipoproteins (HDLs) have been shown to inhibit the cytokine-induced expression of EC adhesion molecules both in vitro and in vivo, and to reduce inflammatory infiltration of neutrophils and development of MODS following hemorrhagic shock in rats (35). Other therapies that may prove useful in reducing leukocyte adhesion in hemorrhagic shock include adrenocorticotropin (36), hypertonic saline (37), pentoxifylline (38), low-molecular-weight heparin (39), and nonanticoagulant heparin (40). Any benefit of these agents is hypothetical and/or anecdotal and will require confirmation in clinical studies. Certainly none has become a gold standard for the treatment of hemorrhagic shock in patients.

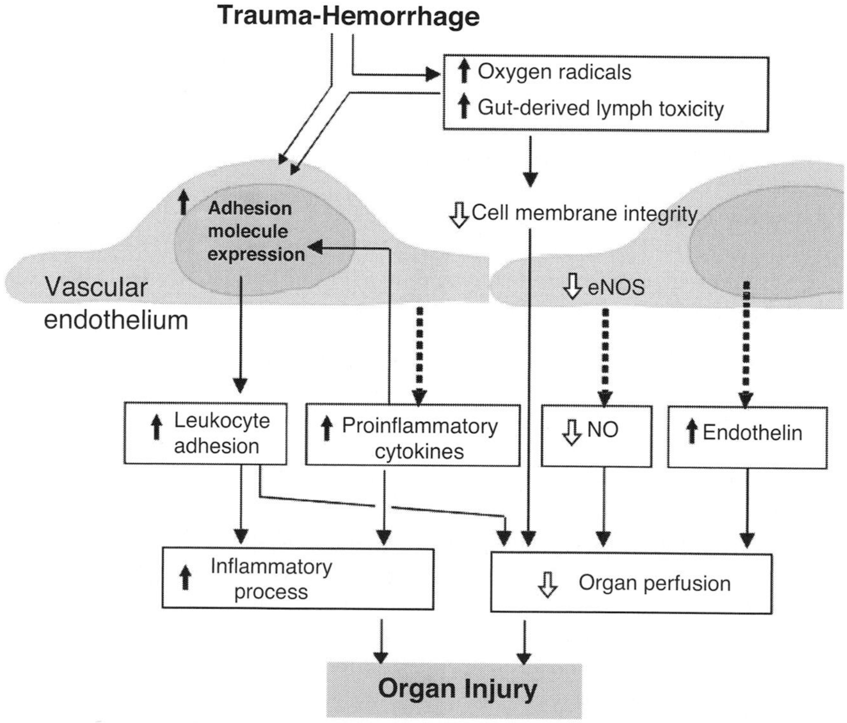

Figure 164.1. Possible mechanisms involved in endothelium-related organ injury following trauma and hemorrhage. NO, nitric oxide; eNOS, endothelial nitric oxide synthase.

Table 164-2: Mechanisms of Complications in Trauma-Hemorrhage

Conditions	*Complications*	*Mechanisms*	*Models*	*References*
Trauma	Hemorrhage; massive transfusion; hypotension; tissue lysis	Ischemia, I/R, activation of innate immunity (danger response via TLR); fat emboli, increased endothelin, cytokines, free radicals	H/R femur fracture	1, 3, 7 5, 6, 30
Hemorrhage	Massive transfusion; hypotension	Hypoxia/reoxygenation, ischemia, I/R, oxygen-derived free radicals, increased proinflammatory cytokines, clotting cascades, P-selectin	I/R, H/R	1, 7, 29, 30–33, 35, 36, 38, 39, 42, 43, 58, 60–62, 64
Massive transfusion	TRALI	Proinflammatory activation	Two-hit model (sepsis and transfusion)	29, 30 (see also Chapter 166)

I/R, ischemia/reperfusion; H/R, hemorrhagic shock/resuscitation; TRALI, transfusion-related acute lung injury; TLR, Toll-like receptor.

Increased Oxygen-Derived Free Radical Formation

Severe hypoxic stress induced by hemorrhagic shock and subsequent resuscitation results in the formation of oxygen-derived free radicals. These products interact directly with ECs, resulting in lipid peroxidation and cellular damage, increased adhesion of leukocytes to endothelium, and cytokine release (41). In a mouse model of hemorrhage, ROS, derived from either endothelial xanthine oxidase or neutrophil nicotinamide adenine dinucleotide phosphate (NADPH) oxidase, have been shown to mediate increased expression of P-selectin in the vascular beds of the lung, heart, liver, kidney, stomach, small bowel, and colon (42). Furthermore, the production of prostacyclin (PGI_2), a vasodilator and inhibitor of platelet aggregation, was inhibited by oxygen radicals in the splanchnic circulation, thus leading to impaired vascular relaxation and decreased mesenteric perfusion (43). Oxygen radicals also may be associated with impaired acetylcholine-induced relaxation of blood vessels after hemorrhagic shock (44).

Impaired Endothelium-Dependent Vasodilator Response

Marked hemodynamic alterations due to severe hemorrhage markedly affect EC function. Furthermore, shock without blood loss, such as septic shock, also showed similar EC dysfunction as observed after hemorrhagic shock (45,46). Studies have also shown that hypoxemia per se without any blood loss produces immunosuppression (47,48). In view of this, it might be expected that EC functions would also be depressed following hypoxemia. Studies in fact indicate this to be the case (49). In one study, the response to the endothelium-dependent vasodilator, acetylcholine, was significantly depressed in the thoracic aorta of rats subjected to hemorrhagic shock (50). A similar defect was observed in renal (51), cerebral (52), pulmonary (53), and coronary arteries. (54) In contrast, endothelium-independent vasodilation induced by adenosine remained unchanged in the shock state. These results suggest that hemorrhagic stress selectively depresses endothelial function in the arterial system. This hypothesis is further supported by studies showing decreased endothelial nitric oxide synthase (*eNOS*) mRNA expression in the cerebral arterial endothelium after subarachnoid hemorrhage (55). Moreover, administration of L-arginine, the precursor for the endothelium-derived relaxing factor, nitric oxide (NO), was shown to restore vascular reactivity in a feline model of hemorrhage (56).

One potentially negative effect of NO is the formation of peroxynitrite, which is formed when excessive NO is produced, and acts as a highly toxic oxygen radical. A moderate amount of NO is useful in maintaining tissue perfusion and inhibiting leukocyte adhesion to endothelium. However, excessive amounts of NO are detrimental, because they form harmful oxygen radicals, which not only directly damage the endothelium, but also upregulate leukocyte adhesion molecules. This hypothesis is supported by several lines of investigation. First, in a rat model of hemorrhagic shock, the administration of exogenous NO in the liver resulted in significantly reduced adhesion of leukocytes to sinusoidal endothelium (57), indicating a suppressive effect of NO on the expression of adhesion molecules. Sodium nitroprusside, an NO donor, also significantly decreased neutrophil adhesion and migration in rat mesentery after hemorrhage (58). In contrast, inhibition of inducible NO synthase (iNOS), which is upregulated in the gut following hemorrhagic shock, leads to significant downregulation of gut ICAM-1 expression, indicating that excessive production of NO could serve as an inducer of intestinal ICAM-1 expression (59). NO scavenger reduces *ICAM-1* mRNA expression in the lung and causes a subsidence of inflammatory response after hemorrhagic shock (60). The amount of NO production may differentially affect the gut function. Gut mucosa barrier function is impaired by hemorrhagic shock and leads to an enhanced bacterial translocation. Under such conditions, NO maintains intestinal integrity, thus preventing bacterial translocation by maintaining microcirculation through its vasodilating effects. On the other hand, excessive amounts of NO damage gut barrier function (61). These controversial issues need further clarification by additional studies.

Increased Vasoconstrictor Production and Increased Endothelial Sensitivity to Vasoconstrictors

Endothelin, a potent vasoconstrictor, is primarily produced by vascular ECs. Increased production of endothelin, or increased vascular sensitivity to endothelin, induces persistent vasoconstriction and impairment in organ microcirculation. Plasma endothelin levels are increased during stress conditions, including hemorrhagic shock, and the elevation of plasma endothelin linearly correlates with the amount of blood loss (62). Vasoconstrictive response to exogenous endothelin also is enhanced after hemorrhage or trauma-hemorrhage, which is evident in the hepatic portal system (63). In this regard, blockade of endothelin receptor prevents systemic and portal hyperemia in the liver (64). Circulating levels of other vasoconstrictors, such as thromboxane A_2, also are increased following hemorrhage, and these also could be potential therapeutic targets.

The Effect of Gut-Derived Mesenteric Lymph on Posthemorrhage Endothelial Dysfunction

The splanchnic organs are subjected to severe hypoxia during hemorrhagic shock, and blood flow to the organs is persistently reduced even 24 hours after resuscitation. In response to hemorrhage, the gut produces a variety of inflammatory factors including TNF, interleukins (ILs), platelet activating factor, and oxygen radicals. Recent studies have indicated an important role of splanchnic lymph rather than portal venous blood as a causative factor in producing EC dysfunction and subsequent organ injury after hemorrhagic shock (65). Mesenteric lymph collected from posthemorrhage animals increases the permeability and even cell death of human umbilical vein ECs (HUVECs). It also activates neutrophil superoxide formation to a greater extent than does portal venous plasma (65). Given that mesenteric lymph drains into the subclavian vein via the thoracic duct, it is interesting to speculate that this fluid contributes to hemorrhagic shock–induced lung injury. In support of this notion, mesenteric lymph collected from rats subjected to trauma-hemorrhagic shock has been shown to increase permeability of cultured pulmonary microvascular ECs (66). Moreover, ligation of the mesenteric lymphatic ducts in rat model of hemorrhagic shock prevented upregulation of P-selectin and acute lung injury (67). At present, the identification of factor(s) in the mesenteric lymph responsible for producing tissue injury remains elusive, although size fractionation studies suggests that one putative mediator is larger than 100 kDa (68). Elucidation of this factor could provide useful information in establishing a treatment strategy for the patients subjected to severe hemorrhage and/or trauma.

GENDER DIMORPHISM IN THE OUTCOME OF TRAUMA-HEMORRHAGE WITH SPECIAL FOCUS ON ENDOTHELIAL FUNCTIONS

Extensive studies have shown gender dimorphism in cardiovascular and immune response to stressful conditions. In this regard, proestrus female rats are better able to tolerate the deleterious effects of trauma-hemorrhage on cardiovascular and immune functions, compared with males (69). Additionally, male rats subjected to hemorrhagic shock had increased gut injury and lung permeability compared with female rats (70). Furthermore, mesenteric lymph from male rats after hemorrhagic shock was more cytotoxic to HUVECs compared with that from female rats (70). Interestingly, the effects of trauma-hemorrhage in precastrated males were similar to female animals, whereas those of ovariectomized female animals were similar to intact males. These results indicate that the prevailing sex hormone milieu at the time of injury is an important determinant in maintaining or inhibiting organ and immune functions following trauma-hemorrhage.

Vascular endothelial function and associated organ perfusion also show gender dimorphism. Although cardiac and splanchnic perfusion was significantly decreased following trauma-hemorrhage in male rats, such alterations were attenuated in proestrus females (71). Moreover, males, but not females, demonstrated endothelial dysfunction in the intestinal vasculature, as evidenced by impaired response to acetylcholine (71). An understanding of these gender-specific responses may provide important insights into therapy. Indeed, as is discussed later, studies in animal models of trauma-hemorrhage have established a role for several sex hormones and hormone receptor inhibitors in improving EC function.

Estrogen

Estrogen (E_2), the primary female sex hormone, acts not only as a gonadal hormone, but also as a vasomodulator. In a rat model of trauma-hemorrhage, administration of a single subcutaneous dose of E_2 to male animals restored the depressed cardiac, hepatic, and immune functions (72–74). Furthermore, E_2 administration improved splanchnic perfusion and oxygen consumption in the small intestine when trauma-hemorrhage was followed by a "second hit," such as sepsis (75). These results suggest that E_2 increases organ perfusion, perhaps through its stimulation of vasodilatory factors (e.g., PGI_2 and NO synthesis) and inhibition of vasoconstricting mediators (e.g., angiotensin II and endothelin) (Figure 164.2). Indeed, E_2 induces the expression of eNOS in cultured ECs (76). This effect is independent of cytosolic calcium (Ca^{2+}) mobilization, and may be associated with rapid activation of the phosphoinositide 3-kinase (PI3K)/Akt pathway (76,77). Moreover, E_2 inhibits endothelin release from

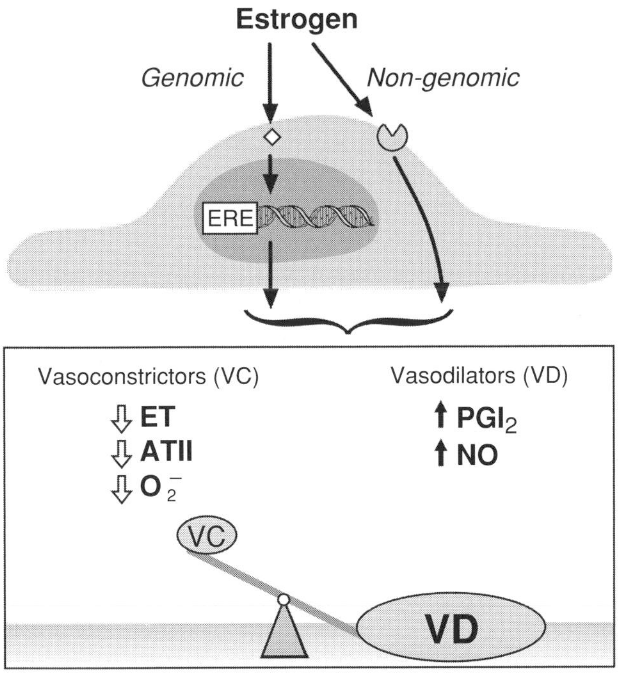

Figure 164.2. Estrogen-mediated regulation in the release of vasoactive agents from vascular ECs. Actions may be mediated either by the activation of ERE through genomic pathways or by direct activation of intracellular signal transductions through nongenomic pathways. ERE, estrogen-response element; ET, endothelin; O_2^-, superoxide anion; ATII, angiotensin II; PGI_2, prostacyclin; NO, nitric oxide.

cultured ECs by inhibiting mitogen-activated protein kinase (MAPK) activity (78). Finally, there is evidence that E_2 attenuates endothelin-mediated vasoconstriction of the hepatic portal system following trauma-hemorrhage (63).

Estrogens, like other steroid hormones, cross the cell membrane and form complexes with cytoplasmic receptors. These ligand–receptor complexes enter the nucleus and bind to specific DNA sequences called E_2-responsive elements. This process, which occurs over hours, leads to increased expression of target genes and alteration in cellular function. In addition to the latter "genomic" effects of estrogen, E_2 exerts "nongenomic" effects through receptor-mediated or receptor-independent pathways (79). These activities do not require new transcription or translation, and occur very rapidly (on the order of minutes). The nongenomic pathway activates MAPK, cyclic guanosine monophosphate (cGMP) production, and rapid release of NO from ECs (see Figure 164.2) (80). A recent report indicates that the nongenomic pathway contributes to the vasodilatory effects of E_2 through the activation of endothelial E_2 receptors located on the cell surface (79).

Two different types of E_2 receptors (ERs) have been identified in mammalian tissues: ER-α and ER-β. These receptors differ in their function, ligand affinity, and tissue distribution. ER-α is believed to play a more prominent role than ER-β in regulating endothelial function. ER-α correlates functionally with E_2-mediated rapid activation of eNOS, production of cGMP, release of NO, and activation of MAPK in cultured ECs (79), and increased production of NO in the thoracic aorta (81).

Other mechanisms of E_2-mediated salutary effects on endothelium may involve inhibition of leukocyte adhesion and reduced production of oxygen-derived free radicals. Inflammatory cytokine-induced expression of E-selectin is suppressed by E_2, through a process that requires the DNA-binding domain of ER-α (82). Estrogen causes a time- and concentration-dependent decrease in oxygen radical formation in HUVECs (83). Estrogen also increases the activities of superoxide dismutase, catalase, glutathione peroxidase, and glutathione reductase in female bovine aortic ECs, which leads to decreased oxygen-derived free radical production, compared to those in male (84).

Dehydroepiandrosterone

Dehydroepiandrosterone (DHEA) is the most abundant steroid hormone in plasma and, depending on the hormonal milieu, serves as an intermediate in the synthesis of testosterone and E_2. DHEA has both androgenic and estrogenic effects. The administration of a single dose of DHEA following trauma-hemorrhage restored the depressed splenocyte and macrophage functions under those conditions (85). When sepsis was induced by cecal ligation and puncture at 48 hours after trauma-hemorrhage, DHEA administration following trauma-hemorrhage significantly increased the survival rate of animals (80). The impaired cardiac output and hepatocellular function following trauma-hemorrhage were also restored by DHEA administration (86). These results indicate the presence of a mechanism similar to estrogen with regard to the regulation of vascular tone. However, since DHEA is a precursor for both androgen and E_2, it remained unclear whether the administered DHEA acted through androgenic or estrogenic pathway in those animal models. This question was resolved by administering DHEA along with an ER antagonist (ICI 182,780), which abolished the salutary effects of DHEA (62). These findings indicate that the effects of DHEA on cardiac and hepatocellular functions after trauma-hemorrhage are mediated via the ER. Nonetheless, it is still unclear whether the ERs are activated by direct binding of DHEA, or by E_2 synthesized through the metabolism of DHEA (87).

Androstenediol

Androstenediol is one of the metabolites of DHEA. Studies have shown that administration of androstenediol following trauma-hemorrhage and resuscitation attenuated the reduced portal blood flow, bile production, and serum albumin levels under those conditions (87). This was associated with upregulation of hepatic eNOS expression in the liver, which might produce appropriate amounts of NO in the liver and maintain hepatic microcirculation. As stated earlier, upregulation of iNOS and greater amounts of NO production are thought to be among the major contributors to hepatic injury after severe hemorrhagic shock. Additionally, endothelin is another mediator responsible for deteriorating liver circulation after trauma-hemorrhage (63). Interestingly, in vivo administration of androstenediol significantly inhibited the upregulation of these factors at both mRNA and protein levels (87). This may at least partially explain the mechanism of androstenediol-mediated salutary effects in the hepatic circulation following trauma-hemorrhage. Moreover, androstenediol showed salutary effects on cardiac function and splanchnic circulation, both of which are also decreased following trauma-hemorrhage (88).

Flutamide

Flutamide, a nonsteroidal androgen receptor antagonist, has protective effects on immune function in male mice following trauma-hemorrhage (89). Although IL-2 and IL-3 release by splenocytes, splenocyte proliferative capacity, as well as splenic and peritoneal macrophage IL-1 release were significantly reduced following trauma-hemorrhage, these alterations were minimized by administration of flutamide following trauma-hemorrhage. The survival rate following trauma-hemorrhage and induction of subsequent sepsis was also significantly increased in the flutamide-treated group (90). In addition, the decreased left ventricular performance, cardiac output, hepatic microvascular blood flow, and hepatocellular function following trauma-hemorrhage were significantly improved by flutamide treatment (91). Androgens also appear to play an inhibitory role in small intestinal endothelial function. In rats with low plasma androgen levels (castrated males, proestrus females, and estrogen-treated non-castrated males), a positive correlation was noted between plasma androgen levels and acetylcholine-induced perfusion pressure reduction in the small intestine (92). Intestinal tissue NO levels were also significantly higher in these animals, indicating an inhibitory role of androgens on endothelium-derived NO production. In this regard, flutamide administration following trauma-hemorrhage attenuated the depressed endothelium-dependent vascular relaxation (i.e., the production of vascular endothelium-derived NO) (93). Organ blood flow and O_2 consumption also were improved in the liver, small intestine, and kidneys in males. Interestingly, the expression of thromboxane A_2 receptor and the action of thromboxane A_2 on the artery have been shown to be enhanced by testosterone (94). In contrast, the synthesis of PGI_2 has been shown to be inhibited by testosterone (95). Although the precise mechanism responsible for the beneficial effects of flutamide on endothelial function remains unknown, the findings described here may indicate one of the possible mechanisms.

FUTURE DIRECTIONS OF ENDOTHELIUM RESEARCH FOLLOWING TRAUMA-HEMORRHAGIC SHOCK

Extensive studies using animal models have helped to elucidate mechanisms of endothelial dysfunction following trauma-hemorrhage. Collectively, these data have provided a strong foundation for therapeutic intervention. However, despite promising results in animals, clinical trials with these agents have yielded disappointing results. For example, leukocyte adhesion blockade with a recombinant human monoclonal antibody against CD18 failed to demonstrate efficacy in a phase II clinical study of trauma patients with hypotension from hemorrhagic shock (96). Although resuscitation with hypertonic saline plus dextran was reported to benefit animals with trauma (97), the effect was limited to hemodynamic improvement in human patients (98). Although ligation of the mesenteric lymphatic duct protects against lung injury in animal models of trauma-hemorrhage, this procedure is not practical in the clinical setting. Thus, effective treatment strategies for patients with major traumatic injuries still must be developed.

Based on the findings with estrogen, DHEA, androstenediol, and flutamide, these hormones/receptor antagonists may prove useful as a therapeutic adjunct in trauma-hemorrhage. These hormones/receptor antagonists may not only prevent organ dysfunction by preserving the integrity of endothelial function, but may also improve and restore immune and cardiovascular depression following trauma-hemorrhage. Another advantage of these agents is that, in contrast to anticytokine or antichemokine therapies, they target a broad spectrum of homeostatic functions, and thus restore integrity of the entire organism (including endothelial function). In addition, they are endogenous mediators and pose little risk to patients, even when used chronically. Nonetheless, further studies are required to elucidate the precise nature and mechanisms by which sex steroids influence the endothelium, particularly following trauma-hemorrhage and other low-flow conditions.

KEY POINTS

- Endothelial dysfunction occurs following trauma-hemorrhage and contributes to the resulting organ damage.
- The major mechanisms involved in endothelium-mediated organ injury are: (a) enhanced inflammatory processes; (b) cellular damage by excessive production of oxygen-derived free radicals; and (c) impaired tissue circulation due to an imbalance between vasodilators and vasoconstrictors.
- Female sex hormones, especially estrogen, have been shown to inhibit these responses, thereby suggesting that the use of estrogen following trauma-hemorrhage could be a useful therapeutic adjunct for preventing multiple organ dysfunctions following trauma-hemorrhage.

- Although several mediators and agents have been shown to be useful in preventing endothelial dysfunction following trauma-hemorrhage in animal models, many of them have not been shown to be effective in the clinical arena following trauma.
- There appears to be a hormonal regulation of endothelial functions under stressful conditions such as trauma-hemorrhage. In this regard, sex hormones such as estrogen, DHEA, androstenediol, or the hormone receptor antagonist, flutamide, have been shown to be effective in restoring trauma-hemorrhage-induced endothelial function.

Future Goals

- To examine the use of sex hormones or their receptor antagonists in the clinical arena to determine whether such adjuncts improve the outcome following injury; the side effects of these agents may be minimal because most of the proposed agents are administered only once after injury, and all these agents are widely used clinically without major complications.

ACKNOWLEDGMENTS

This work was supported by NIH grant R37 GM39519.

REFERENCES

1 Ayala A, Wang P, Ba ZF, et al. Differential alterations in plasma IL-6 and TNF levels following trauma and hemorrhage. *Am J Physiol.* 1991;260:R167–R171.
2 Boldt J, Menges T, Kuhn D, et al. Alterations in circulating vasoactive substances in the critically ill – a comparison between survivors and non-survivors. *Intensive Care Med.* 1995;21(3):218–225.
3 Harrigan C, Lucas CE, Ledgerwood AM. The effect of hemorrhagic shock on the clotting cascade in injured patients. *J Trauma.* 1989;29(10):1416–1421; discussion: 1421–1422.
4 Aird WC. The role of the endothelium in severe sepsis and multiple organ dysfunction syndrome. *Blood.* 2003;101(10):3765–3777.
5 Williams JJ, Moalli R, Calista C, Herndon JH. Pulmonary endothelial injury and altered fibrinolysis after femur fracture in rabbits. *J Orthop Trauma.* 1990;4(3):303–308.
6 Ashburn JH, Baveja R, Kresge N, et al. Remote trauma sensitizes hepatic microcirculation to endothelin via caveolin inhibition of eNOS activity. *Shock.* 2004;22(2):120–130.
7 Davis JM, Stevens JM, Peitzman A, et al. Neutrophil migratory activity in severe hemorrhagic shock. *Circ Shock.* 1983;10:199–204.
8 Bitterman H, Smith BA, Lefer AM. Beneficial actions of antagonism of peptide leukotrienes in hemorrhagic shock. *Circ Shock.* 1988;24:159–168.
9 Ikai I, Ozaki N, Shimahara Y, et al. Significance of hepatic mitochondrial redox potential on the concentrations of plasma amino acids following hemorrhagic shock in rats. *Circ Shock.* 1989;27:63–72.
10 Schumer W. Localization of the energy pathway block in shock. *Surgery.* 1968;64:55–59.
11 Campion DS, Lynch LJ, Rector FC, et al. Effect of hemorrhagic shock on transmembrane potential. *Surgery.* 1969;66:1051–1059.
12 Drucker WR, DeKiewiet JC. Glucose uptake by diaphragms from rats subjected to hemorrhagic shock. *Am J Physiol.* 1964;206:317–320.
13 Weil MH, Afifi AA. Experimental and clinical studies on lactate and pyruvate as indicators of the severity of acute circulatory failure. *Circulation.* 1970;41:989–1001.
14 van der Meer C, Valkenburg PW, Snijders PM, et al. A method for hemorrhagic shock in the rat. *J Pharmacol Methods.* 1987;17:75–82.
15 Chaudry IH, Wang P, Singh G, et al. Rat and mouse models of hypovolemic-traumatic shock. In: Schlag G, Redl H, eds. *Pathophysiology of Shock, Sepsis and Organ Failure.* Berlin: Springer-Verlag; 1993:371–383.
16 Schwacha MG, Wang P, Chaudry IH. Trauma models for studying the influence of gender and aging. In: Souba W, Wilmore D, eds. *Surgical Research.* New York: Academic Press; 2001:357–366.
17 Holden WD, Depalma RG, Drucker WR, McKalen A. Ultrastructural changes in hemorrhagic shock: electron microscopic study of liver, kidney and striated muscle cell in rats. *Ann Surg.* 1965;162:517–536.
18 Mela L, Bacalzo LV Jr, Miller LD. Defective oxidative metabolism of rat liver mitochondria in hemorrhagic and endotoxin shock. *Am J Physiol.* 1971;220:571–577.
19 LePage GA. The effects of hemorrhage on tissue metabolites. *Am J Physiol.* 1946;147:446–453.
20 Abraham E, Richmond JN, Chang YH. Effects of hemorrhage on interleukin-1 production. *Circ Shock.* 1988;25:33–40.
21 Sayeed MM, Baue AE. Mitochondrial metabolism of succinate, β-hydroxybutyrate, and α-ketoglutarate in hemorrhagic shock. *Am J Physiol.* 1971;220:1275–1281.
22 Chaudry IH, Sayeed MM, Baue AE. Effect of hemorrhagic shock in tissue adenine nucleotides in conscious rats. *Can J Physiol Pharmacol.* 1974;52:131–137.
23 Machiedo GW, Ghuman S, Rush BF, et al. The effect of ATP-$MgCl_2$ infusion on hepatic cell permeability and metabolism after hemorrhagic shock. *Surgery.* 1981;90:328–335.
24 Rush BF, Redan JA, Flanagan JJ, et al. Does the bacteremia observed in hemorrhagic shock have clinical significance? A study in germ-free animals. *Ann Surg.* 1989;210:342–347.
25 Stephan RN, Kupper TS, Geha AS, et al. Hemorrhage without tissue trauma produces immunosuppression and enhances susceptibility to sepsis. *Arch Surg.* 1987;122:62–68.
26 Ertel W, Morrison MH, Ayala A, et al. Blockade of prostaglandin production increases cachectin synthesis and prevents depression of macrophage functions following hemorrhagic shock. *Ann Surg.* 1991;213:265–271.

27 Wang P, Singh G, Rana MW, et al. Preheparinization improves organ function following hemorrhage and resuscitation. *Am J Physiol.* 1990;259:R645–R650.
28 Crowell JW, Read WI. In vivo coagulation – a probable cause of irreversible shock. *Am J Physiol.* 1955;183:565–569.
29 Akgur FM, Zibari GB, McDonald JC, et al. Kinetics of P-selectin expression in regional vascular beds after resuscitation of hemorrhagic shock: a clue to the mechanism of multiple system organ failure. *Shock.* 2000;13(2):140–144.
30 Sun LL, Ruff P, Austin B, et al. Early up-regulation of intercellular adhesion molecule-1 and vascular cell adhesion molecule-1 expression in rats with hemorrhagic shock and resuscitation. *Shock.* 1999;11(6):416–422.
31 Hawrylowicz CM, Howells GL, Feldmann M. Platelet-derived interleukin 1 induces human endothelial adhesion molecule expression and cytokine production. *J Exp Med.* 1991;174(4): 785–790.
32 Vedder NB, Winn RK, Rice CL, et al. A monoclonal antibody to the adherence-promoting leukocyte glycoprotein CD18 reduces organ injury and improves survival from hemorrhagic shock and resuscitation in rabbits. *J Clin Invest.* 1988;31:939–944.
33 Maier M, Strobele H, Voges J, et al. Attenuation of leukocyte adhesion by recombinant TNF-binding protein after hemorrhagic shock in the rat. *Shock.* 2003;19(5):457–461.
34 Powers KA, Kapus A, Khadaroo RG, et al. 25% Albumin modulates adhesive interactions between neutrophils and the endothelium following shock/resuscitation. *Surgery.* 2002;132(2):391–398.
35 Cockerill GW, McDonald MC, Mota-Filipe H, et al. High density lipoproteins reduce organ injury and organ dysfunction in a rat model of hemorrhagic shock. *FASEBJ.* 2001;15(11):1941–1952.
36 Bertuglia S, Giusti A. Influence of ACTH-(1–24) and plasma hyperviscosity on free radical production and capillary perfusion after hemorrhagic shock. *Microcirculation.* 2004;11(3):227–238.
37 Angle N, Hoyt DB, Cabello-Passini R, et al. Hypertonic saline resuscitation reduces neutrophil margination by suppressing neutrophil L selectin expression. *J Trauma.* 1998;45(1):7–12; discussion: 12–13.
38 Barroso-Aranda J, Schmid-Schonbein GW. Pentoxifylline pretreatment decreases the pool of circulating activated neutrophils, in-vivo adhesion to endothelium, and improves survival from hemorrhagic shock. *Biorheology.* 1990;27(3–4):401–418.
39 Balogh Z, Wolfard A, Szalay L, et al. Dalteparin sodium treatment during resuscitation inhibits hemorrhagic shock-induced leukocyte rolling and adhesion in the mesenteric microcirculation. *J Trauma.* 2002;52(6):1062–9; discussion: 1070.
40 Wang P, Ba ZF, Reich SS, et al. Effects of a non-anticoagulant heparin on cardiovascular and hepatocellular function after hemorrhagic shock. *Am J Physiol.* 1996;270:H1294–H1302.
41 Myers SI, Hernandez R. Oxygen free radical regulation of rat splanchnic blood flow. *Surgery.* 1992;112(2):347–354.
42 Akgur FM, Brown MF, Zibari GB, et al. Role of superoxide in hemorrhagic shock-induced P-selectin expression. *Am J Physiol Heart Circ Physiol.* 2000;279(2):H791–H797.
43 Myers SI, Hernandez R. Role of oxygen-derived free radicals on rat splanchnic eicosanoid production during acute hemorrhage. *Prostaglandins.* 1992;44(1):25–36.
44 Zakaria el R, Garrison RN, Spain DA, Harris PD. Impairment of endothelium-dependent dilation response after resuscitation from hemorrhagic shock involved postreceptor mechanisms. *Shock.* 2004;21(2):175–181.
45 Wang P, Ba ZF, Chaudry IH. Endothelium-dependent relaxation is depressed at the macro- and microcirculatory levels during sepsis. *Am J Physiol.* 1995;269:R988–R994.
46 Wang P, Wood TJ, Zhou M, et al. Inhibition of the biologic activity of tumor necrosis factor maintains vascular endothelial cell function during hyperdynamic sepsis. *J Trauma.* 1996;40:694–701.
47 Ertel W, Morrison MH, Ayala A, Chaudry IH. Hypoxemia in the absence of blood loss or significant hypotension causes inflammatory cytokine release. *Am J Physiol.* 1995;269:R160–R166.
48 Knöferl M, Jarrar D, Schwacha MG, et al. Severe hypoxemia in the absence of blood loss causes a gender dimorphic immune response. *Am J Physiol.* 2000;279:C2004–C2010.
49 Williams JM, Pearce WJ. Age-dependent modulation of endothelium-dependent vasodilatation by chronic hypoxia in ovine cranial arteries. *J Appl Physiol.* 2006;100(1):22–232.
50 Wang P, Ba ZF, Chaudry IH. Endothelial cell dysfunction occurs after hemorrhage in nonheparinized but not in preheparinized models. *J Surg Res.* 1993;54:499–506.
51 Szabo C, Farago M, Horvath I, et al. Hemorrhagic hypotension impairs endothelium-dependent relaxations in the renal artery of the cat. *Circ Shock.* 1992;36:238–241.
52 Quan L, Sobey CG. Selective effects of subarachnoid hemorrhage on cerebral vascular responses to 4-aminopyridine in rats. *Stroke.* 2000;31(10):2460–2465.
53 Isotani E, Azuma H, Suzuki R, et al. Impaired endothelium-dependent relaxation in rabbit pulmonary artery after subarachnoid hemorrhage. *J Cardiovasc Pharmacol.* 1996;28(5):639–644.
54 Dignan RJ, Wechsler AS, DeMaria EJ. Coronary vasomotor dysfunction following hemorrhagic shock. *J Surg Res.* 1992;52:382–388.
55 Hino A, Tokuyama Y, Weir B, et al. Changes in endothelial nitric oxide synthase mRNA during vasospasm after subarachnoid hemorrhage in monkeys. *Neurosurgery.* 1996;39(3):562–567; discussion: 567–568.
56 Szabo C, Csaki C, Benyo Z, et al. Role of the L-arginine-nitric oxide pathway in the changes in cerebrovascular reactivity following hemorrhagic hypotension and retransfusion. *Circ Shock.* 1992;37(4):307–316.
57 Bauer C, Kuntz W, Ohnsmann F, et al. The attenuation of hepatic microcirculatory alterations by exogenous substitution of nitric oxide by s-nitroso-human albumin after hemorrhagic shock in the rat. *Shock.* 2004;21(2):165–169.
58 Miyabe M, Yanagi K, Ohshima N, et al. Sodium nitroprusside decreases leukocyte adhesion and emigration after hemorrhagic shock. *Anesth Analg.* 2002;94(2):296–301.
59 Hierholzer C, Kalff JC, Billiar TR, et al. Induced nitric oxide promotes intestinal inflammation following hemorrhagic shock. *Am J Physiol Gastrointest Liver Physiol.* 2004;286(2):G225–G233.
60 Hierholzer C, Menezes JM, Ungeheuer A, et al. A nitric oxide scavenger protects against pulmonary inflammation following hemorrhagic shock. *Shock.* 2002;17(2):98–103.
61 Palmer RM, Bridge L, Foxwell NA, Moncada S. The role of nitric oxide in endothelial cell damage and its inhibition by glucocorticoids. *Br J Pharmacol.* 1992;105(1):11–12.
62 Chang H, Wu GJ, Wang SM, Hung CR. Plasma endothelin level changes during hemorrhagic shock. *J Trauma.* 1993;35(6):825–833.

63 Yokoyama Y, Toth B, Kitchens WC, et al. Estradiol's effect on portal response to endothelin-1 after trauma-hemorrhage. *J Surg Res.* 2004;121(1):25–30.

64 Pannen BH, Bauer M, Noldge-Schomburg GF, et al. Regulation of hepatic blood flow during resuscitation from hemorrhagic shock: role of NO and endothelins. *Am J Physiol.* 1997;272(6 Pt 2):H2736–H2745.

65 Upperman JS, Deitch EA, Guo W, et al. Post-hemorrhagic shock mesenteric lymph is cytotoxic to endothelial cells and activates neutrophils. *Shock.* 1998;10(6):407–414.

66 Deitch EA, Adams CA, Lu Q, Xu DZ. Mesenteric lymph from rats subjected to trauma-hemorrhagic shock are injurious to rat pulmonary microvascular endothelial cells as well as human umbilical vein endothelial cells. *Shock.* 2001;16(4):290–293.

67 Adams CA Jr., Sambol JT, Xu DZ, et al. Hemorrhagic shock induced up-regulation of P-selectin expression is mediated by factors in mesenteric lymph and blunted by mesenteric lymph duct interruption. *J Trauma.* 2001;51(4):625–631; discussion: 631–632.

68 Adams CA Jr., Xu DZ, Lu Q, Deitch EA. Factors larger than 100 kd in post-hemorrhagic shock mesenteric lymph are toxic for endothelial cells. *Surgery.* 2001;129(3):351–363.

69 Jarrar D, Wang P, Cioffi WG, et al. The female reproductive cycle is an important variable in the response to trauma-hemorrhage. *Am J Physiol Heart Circ Physiol.* 2000;279(3): H1015–H1021.

70 Adams CA Jr, Magnotti LJ, Xu DZ, et al. Acute lung injury after hemorrhagic shock is dependent on gut injury and sex. *Am Surg.* 2000;66(10):905–912.

71 Ba ZF, Kuebler JF, Rue LW III, et al. Gender dimorphic tissue perfusion response after acute hemorrhage and resuscitation: role of vascular endothelial cell function. *Am J Physiol Heart Circ Physiol.* 2003;284(6):H2162–H2169.

72 Mizushima Y, Wang P, Jarrar D, et al. Estradiol administration after trauma-hemorrhage improves cardiovascular and hepatocellular functions in male animals. *Ann Surg.* 2000;232(5):673–679.

73 Knöferl M, Jarrar D, Angele MK, et al. 17β-estradiol normalizes immune responses in ovariectomized females after trauma-hemorrhage. *Am J Physiol Cell Physiol.* 2001;281:C1131–C1138.

74 Knöferl M, Angele MK, Diodato MD, et al. Female sex hormones regulate macrophage function after trauma-hemorrhage and prevent increased death rate from subsequent sepsis. *Ann Surg.* 2002;235:105–112.

75 Kuebler JF, Jarrar D, Toth B, et al. Estradiol administration improves splanchnic perfusion following trauma-hemorrhage and sepsis. *Arch Surg.* 2002;137(1):74–79.

76 Caulin-Glaser T, Garcia-Cardena G, Sarrel P, et al. 17 beta-estradiol regulation of human endothelial cell basal nitric oxide release, independent of cytosolic Ca^{2+} mobilization. *Circ Res.* 1997;81(5):885–892.

77 Haynes MP, Sinha D, Russell KS, et al. Membrane estrogen receptor engagement activates endothelial nitric oxide synthase via the PI3-kinase-Akt pathway in human endothelial cells. *Circ Res.* 2000;87(8):677–682.

78 Dubey RK, Jackson EK, Keller PJ, et al. Estradiol metabolites inhibit endothelin synthesis by an estrogen receptor-independent mechanism. *Hypertension.* 2001;37(2 Part 2):640–644.

79 Russell KS, Haynes MP, Sinha D, et al. Human vascular endothelial cells contain membrane binding sites for estradiol, which mediate rapid intracellular signaling. *Proc Natl Acad Sci USA.* 2000;97(11):5930–5395.

80 Simoncini T, Fornari L, Mannella P, et al. Novel non-transcriptional mechanisms for estrogen receptor signaling in the cardiovascular system. Interaction of estrogen receptor alpha with phosphatidylinositol 3-OH kinase. *Steroids.* 2002;67(12): 935–939.

81 Darblade B, Pendaries C, Krust A, et al. Estradiol alters nitric oxide production in the mouse aorta through the alpha-, but not beta-, estrogen receptor. *Circ Res.* 2002;90(4):413–419.

82 Tyree CM, Zou A, Allegretto EA. 17beta-Estradiol inhibits cytokine induction of the human E-selectin promoter. *J Steroid Biochem Mol Biol.* 2002;80(3):291–297.

83 Wagner AH, Schroeter MR, Hecker M. 17beta-estradiol inhibition of NADPH oxidase expression in human endothelial cells. *FASEBJ.* 2001;15(12):2121–2130.

84 Si ML, Al-Sharafi B, Lai CC, et al. Gender difference in cytoprotection induced by estrogen on female and male bovine aortic endothelial cells. *Endocrine.* 2001;15(3):255–262.

85 Angele MK, Catania RA, Ayala A, et al. Dehydroepiandrosterone: an inexpensive steroid hormone that decreases the mortality due to sepsis following trauma-induced hemorrhage. *Arch Surg.* 1998;133(12):1281–1288.

86 Jarrar D, Wang P, Cioffi WG, et al. Mechanisms of the salutary effects of dehydroepiandrosterone after trauma-hemorrhage: direct or indirect effects on cardiac and hepatocellular functions? *Arch Surg.* 2000;135(4):416–422; discussion: 422–423.

87 Shimizu T, Szalay L, Choudhry MA, et al. Mechanism of salutary effects of androstenediol on hepatic function after trauma-hemorrhage: role of endothelial and inducible nitric oxide synthase. *Am J Physiol Gastrointest Liver Physiol.* 2005;288(2):G244–G250.

88 Shimizu T, Choudhry MA, Szalay L, et al. Salutary effects of androstenediol on cardiac function and splanchnic perfusion after trauma-hemorrhage. *Am J Physiol Regul Integr Comp Physiol.* 2004;287(2):R386–R390.

89 Wichmann MW, Angele MK, Ayala A, et al. Flutamide: a novel agent for restoring the depressed cell-mediated immunity following soft-tissue trauma and hemorrhagic shock. *Shock.* 1997;8(4):242–248.

90 Angele MK, Wichmann MW, Ayala A, et al. Testosterone receptor blockade after hemorrhage in males. Restoration of the depressed immune functions and improved survival following subsequent sepsis. *Arch Surg.* 1997;132(11):1207–1214.

91 Remmers DE, Wang P, Cioffi WG, et al. Testosterone receptor blockade after trauma-hemorrhage improves cardiac and hepatic functions in males. *Am J Physiol.* 1997;273(6 Pt 2):H2919–H2925.

92 Ba ZF, Yokoyama Y, Toth B, et al. Gender differences in small intestinal endothelial function: inhibitory role of androgens. *Am J Physiol Gastrointest Liver Physiol.* 2004;286(3):G452–G457.

93 Ba ZF, Wang P, Koo DJ, et al. Attenuation of vascular endothelial dysfunction by testosterone receptor blockade after trauma and hemorrhagic shock. *Arch Surg.* 2001;136(10):1158–1163.

94 Karanian JW, Ramwell PW. Effect of gender and sex steroids on the contractile response of canine coronary and renal blood vessels. *J Cardiovasc Pharmacol.* 1996;27(3):312–319.

95 Nakao J, Change WC, Murota SI, Orimo H. Testosterone inhibits prostacyclin production by rat aortic smooth muscle cells in culture. *Atherosclerosis.* 1981;39(2):203–209.
96 Rhee P, Morris J, Durham R, et al. Recombinant humanized monoclonal antibody against CD18 (rhuMAb CD18) in traumatic hemorrhagic shock: results of a phase II clinical trial. Traumatic Shock Group. *J Trauma.* 2000;49(4):611–619; discussion: 619–620.
97 Corso CO, Okamoto S, Ruttinger D, Messmer K. Hypertonic saline dextran attenuates leukocyte accumulation in the liver after hemorrhagic shock and resuscitation. *J Trauma.* 1999;46(3):417–423.
98 Younes RN, Aun F, Accioly CQ, et al. Hypertonic solutions in the treatment of hypovolemic shock: a prospective, randomized study in patients admitted to the emergency room. *Surgery.* 1992; 111(4):380–385.

165

Coagulopathy of Trauma

Implications for Battlefield Hemostasis

Anthony E. Pusateri and John B. Holcomb

U.S. Army Institute of Surgical Research, Fort Sam Houston, Texas

There is increasing appreciation for the role of the endothelium in orchestrating the host response to infection and injury, as occurs in sepsis, trauma, or surgery. An urgent need exists to identify novel therapies that attenuate or modify the host response and thus improve patient outcome. This chapter focuses on the nature and mechanisms of coagulopathy associated with trauma, particularly that related to combat injury. Trauma-related coagulopathy is qualitatively different from that seen in sepsis and surgery. Trauma suffered in combat, as distinct from civilian life, introduces additional variables that may impact on the bleeding phenotype. Although at present we have little understanding of how the endothelium is influenced by and/or mediates the host response in trauma, it seems likely that this organ will ultimately serve as an important diagnostic tool and therapeutic target.

NATURE OF COMBAT WOUNDS

Causes of Death

Approximately 50% of those who die from combat injury die from exsanguinating hemorrhage, which is the leading cause of death from combat trauma (1,2). Historically, a small percentage of these deaths have been from sites where the bleeding might be controlled by compression. This population of casualties will be reduced through the use of improved training for self care, medics, and other first-responder personnel, coupled with effective equipment like improved tourniquets or hemostatic dressings. The military has aggressively and successfully pursued these measures. The largest percentage of potentially preventable hemorrhagic combat deaths, however, is due to noncompressible (internal) hemorrhage (47%). The challenge is to develop means with which medics can initiate control of truncal hemorrhage through nonsurgical means. This represents a significant research hurdle. Many of these combat deaths occur from almost immediate exsanguination from wounds that are not amenable to surgical repair. However, there exists a population of potentially preventable deaths, comprising approximately 6% of all combat deaths and 47% of the potentially preventable deaths, that may be salvageable with improved hemorrhage-control methods for internal bleeding (1,2). These casualties often die during transport, and also are represented in the deployed hospitals where hemorrhagic death early after admission is not uncommon. This discussion is primarily focused on this population.

Wounding Patterns

Combat trauma differs from civilian trauma in several important ways. In combat trauma, the distribution of mechanisms of injury is strongly dependent on the branch of service and the tactics of the enemy. For example, burns are extremely common among those wounded in armored fighting vehicles, but not so among dismounted infantry (1). Military personnel in downed aircraft suffer burns and massive blunt force injury with a lethality rate approaching 80%. Among dismounted troops or those in open vehicles, penetrating trauma dominates (>80%) and, in the current Global War on Terrorism, explosions account for 77% of all injuries, resulting in casualties with significant soft tissue injury and multiple penetrating wounds. Small-arms fire accounts for 17%, whereas significant burns account for 5% of all evacuated casualties (2). A common theme in military trauma, one that differentiates it from civilian trauma, is the predominance of multiple massive soft-tissue injuries, the number of penetrating wounds in each patient casualty, and the sheer number of those patients who arrive daily.

Evacuation Time

Due to tactical and environmental factors, the time required for evacuating combat casualties for surgical care is often prolonged far beyond civilian urban transport times. Under combat conditions, medics and others often must stabilize and maintain a severely injured patient for several hours before handing them off to others, who must maintain the casualty patient during transport. Environmental factors, austere field conditions, and prolonged evacuation times make the combat casualty patient particularly prone to developing life-threatening, trauma-related coagulopathy.

MECHANISMS OF TRAUMA COAGULOPATHY

Incidence and Importance

In severe trauma, blood loss, tissue damage, inflammatory response, immobilization, and environmental factors can set the stage for the development of coagulopathy. The derangements in the blood coagulation system occur rapidly and have been observed within the first hour and even minutes following trauma. By the time of arrival at the emergency department, 28% (2,994 of 10,790) of trauma patients had a detectable coagulopathy that was associated with poor outcome (3,4), and the incidence of coagulopathy was increased with increasing injury severity scores (Figure 165.1). Patients who have sustained traumatic injuries and hemorrhage are at risk for coagulopathy due to factors such as hypothermia, acidosis, and hemodilution (5–9), as well as the actions of immune cells and inflammatory mediators (10–14). Consumptive processes also play a role (15). Patients with traumatic brain injury are prone to develop coagulopathy, which may be either hypercoagulable or hyperfibrinolytic, as rapidly as 1 to 4 hours after injury (16–24). Release of tissue factor from injured brain tissue has been proposed as a likely causative factor of the coagulopathy associated with head trauma (11,25); however, it does not appear that this fully explains the phenomenon (16,24). The remainder of this discussion focuses on those trauma patients without head injury who are at risk of bleeding to death.

In the normal state, a balance exists among procoagulant, anticoagulant, fibrinolytic, and antifibrinolytic mechanisms, and platelet function also is regulated. The result is a system that allows circulating blood to remain in a fluid state yet respond within seconds to a vascular insult to provide a compartmentalized procoagulant response. This system allows hemostasis to occur without appreciably altering the systemic anticoagulant balance. Hypothermic, acidotic, hemodilutional, and inflammatory changes occur rapidly following severe trauma, and often are present and compounded by standard treatment by the time of admission to the emergency department. Trauma-related coagulopathy is multifactorial and characterized by dysregulation of balanced mechanisms. Given the compressed time span during which the conditions associated with coagulopathy develop, it is apparent that multiple factors contribute to coagulopathy simultaneously, especially in the most severely injured patients. At a given point during the posttrauma time course of a casualty, some coagulopathic processes may dominate, while others are masked by dominant processes or are not yet fully developed. It becomes apparent that resuscitation interventions can contribute either favorably or unfavorably to this process. The following discussion addresses in some detail hypothermia, acidosis, hemodilution, systemic inflammatory response, and consumptive processes as they pertain to trauma coagulopathy during the first 12 hours after trauma. This is the critical period following severe trauma, largely determining if the patient will live or die (26,27). It is also a period when the care of a combat casualty patient may significantly differ from that of a civilian casualty patient.

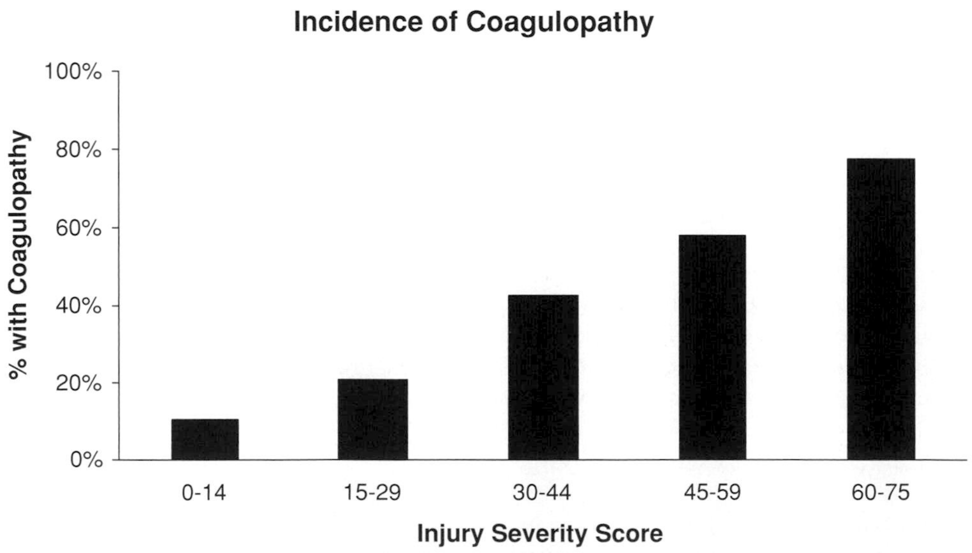

Figure 165.1. Coagulopathy in acute trauma. (Reproduced with permission from Brohi K, Singh J, Heron M, Coats T. Acute traumatic coagulopathy. *J Trauma*. 2003;5:1127–1130.)

Hypothermia

Combat casualty patients are at particular risk for developing hypothermia due to such factors as the field environment, use of cold prehospital fluid resuscitation, wounding mechanisms that result in loss of clothing and tissue, and extended evacuation times. It is well known that hypothermia slows the coagulation process. Extended plasma clotting times are detectable below 35°C (28,29). Enzymatic rates for the tissue factor/factor VIIa and the prothrombinase complexes decrease with temperature, consistent with the expected twofold increase in enzymatic rate for each 10°C increase in reaction temperature for most enzymes (30). The inhibitory activity of antithrombin III (ATIII) declines with temperature to a lesser degree than do coagulation enzyme rates (31), suggesting that changes in both pro- and anticoagulant components of the balance may contribute to the effects of hypothermia on coagulation. The rate of whole-blood clot formation is slowed, but final clot strength is not reduced by hypothermia (28,32,33). Hypothermia reversibly impairs platelet function, resulting in defects in primary hemostasis and possibly other functions. Template bleeding times are extended by hypothermia (34,35). Platelet aggregation in response to thrombin is greatly reduced, as is shear-induced platelet surface adhesion when the temperature declines to 33°C (31). Platelet aggregation in response to shear increases during hypothermia, but this is not associated with functional activation, and may represent a mechanism that reduces the number of platelets available to respond physiologically to vascular insult (36). Platelet thromboxane production is decreased during hypothermia (34,35). In whole blood, hypothermia slows thromboxane production, serotonin release, and thrombin generation (37). Taken together, alterations in pro- and anticoagulant processes, and changes in platelet function, combine to impair hemostasis during hypothermia.

Acidosis

Plasma clotting times are prolonged as pH is reduced (38). The activity of the tissue factor/factor VIIa complex decreases by 55%, and activity of the prothrombinase complex declines by 70% as pH levels decline from 7.4 to 7.0 (30). There also appears to be a platelet defect resulting from acidosis that involves both platelet depletion and altered function. Following induced acidosis at pH 7.1, capillary bleeding time was prolonged in swine (32). This was accompanied by a reduced whole blood clotting response to collagen, a platelet agonist. Acidosis reduces the ability of platelets to aggregate in response to the physiological agonists epinephrine, adenosine diphosphate (ADP), and particularly collagen (39,40). Reduced responsiveness of platelets to physiological agonists under acidotic conditions may involve inhibition of extracellular calcium influx (41,42). In apparent contrast with reduced platelet responsiveness in acidosis, a number of early studies demonstrated that prolonged acidosis resulted in platelet depletion. In a recent study, acidosis increased platelet activation status and the number of circulating aggregates, while decreasing platelet counts (43). It is not clear how this phenomenon occurs. This in vivo activation may be a shear-related phenomenon or the result of a nonfunctional partial activation of platelets in response to acidosis. The end result appears to be that a population of platelets aggregate in response to acidosis and are rapidly removed from the circulation, whereas remaining platelets may be less able to form an initial platelet plug in response to physiological agonists at the site of a vascular injury. The impairments in coagulation enzyme and platelet function induced by acidosis can be significant in terms of hemostasis. The combined effects of hypothermia and acidosis on certain hemostatic parameters appears to be greater than the effects of either condition alone (Figure 165.2) (32,45).

Hemodilution

Hypovolemic shock resulting from hemorrhage leads to a hemodilution due to the altered balance between blood pressure and tissue oncotic pressures. The remaining coagulation factors and platelets are diluted as the body shifts fluid from the extravascular to intravascular compartment to restore circulating volume (46). Further dilution occurs with the standard crystalloid resuscitation practices. This dilution of coagulation proteins and functional platelets is compounded by massive transfusion of standard blood products that have a lower functionality than native blood (46). Mild hemodilution (up to 20% to 30%) with saline or lactated Ringer's solution results in a slightly hypercoagulable state, as measured by some, but not all, laboratory tests (47), but this hypercoagulability is reversed at higher levels of dilution (48). Low platelet counts are associated with poor hemostasis and reduced clot strength. Red cells contribute to hemostasis by concentrating platelets near the endothelial surface, metabolically enhancing platelet responsiveness, and possibly by contributing procoagulant surfaces at the site of thrombus formation. Therefore, reduced red cell count from hemodilution can also contribute to coagulopathy (49). Colloidal solutions can impair blood clotting to a degree beyond that explained by simple dilutional effects, depending on the nature of the colloid molecule. As an example, we have observed that fluids containing high-molecular-weight hetastarch reduced the velocity of clot formation by 48% and final clot strength by 24% compared to normal saline or lactated Ringer's solution, over a range of hemodilutions in vitro (A.E. Pusateri, J.B. Holcomb, unpublished observations). Hetastarches, dextrans, and gelatin induce a von Willebrand disease-like condition that impairs hemostasis by effectively reducing factor VIII and von Willebrand factor (vWF) concentrations. Among hetastarch formulations, those that employ molecules with higher molecular weights and higher degrees of hydroxyethyl substitution have the greater effect. Albumin appears to have little, if any, negative effect on coagulation (50). A degree of hemodilution occurs naturally as part of the body's attempt to restore circulating blood volume. This by itself may induce changes in hemostatic function. Much more

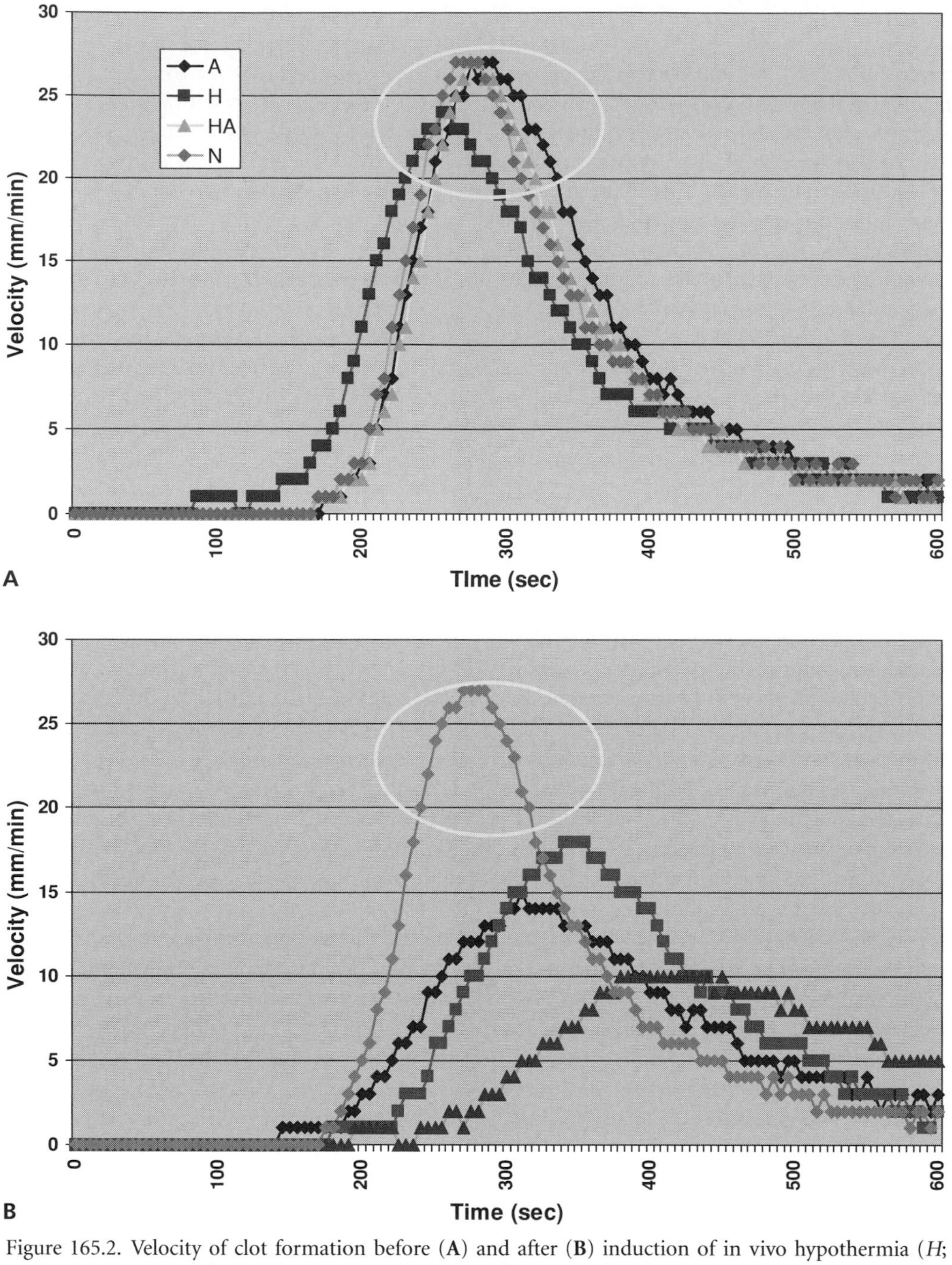

Figure 165.2. Velocity of clot formation before (**A**) and after (**B**) induction of in vivo hypothermia (*H*; core temperature 32°C), acidosis (*A*; pH 7.1), or hypothermia and acidosis (*HA*; core temperature 32°C and pH 7.1). Velocity of clot formation measured using thromboelastography, as previously described (44). Coagulopathic conditions decrease the peak velocity of clot formation and prolong the time required to reach peak velocity (32).

significant are the dilutional effects of common fluid resuscitation regimens.

Inflammatory Response and Consumptive Processes

Historically, the term disseminate intravascular coagulation (DIC) has been used to describe the consumptive processes that may occur following trauma. It is now understood (ISTH DIC subcommittee), and we agree that the consumptive process with regard to trauma occurs along a continuum and is not limited to "classical" DIC, as occurs in classical sepsis. This is especially true during the early posttrauma period (15). This is an important distinction, in terms of pathophysiology, nomenclature, and treatment. Nonetheless, the term DIC will be used here for continuity with the majority of the literature. A combination of mechanisms, including release into the circulation of phospholipids and tissue factor from damaged tissues, mechanical damage to the endothelium, and hemolysis may contribute to DIC (11). Additionally, it is clear that immune cells and inflammatory mediators play a major role in the development of DIC (10–13). It appears that inflammation and DIC synergistically contribute to poor outcome in trauma patients (51,52). The severity of the systemic inflammatory response syndrome (SIRS) following trauma corresponds with the development of organ failure (53,54) and is an independent predictor of length of stay and mortality (55,56). In patients with trauma, the duration of SIRS is a key determinant for the development of multiple-organ dysfunction syndrome (MODS) (56). Further, persistent systemic thrombin activity associated with DIC is closely linked to SIRS (14,57). Following severe trauma, immune cells and inflammatory mediators impinge upon hemostatic regulatory mechanisms at multiple levels, including effects on the procoagulant, anticoagulant, and fibrinolytic mechanisms, as well as on platelet function.

Procoagulant Response

Tissue factor is elevated within hours of severe trauma, both as plasma tissue factor (15,57–60) and as tissue factor expressed on circulating monocytes (23). In a series of trauma patients with a mean injury severity score of 24, monocyte tissue factor expression was elevated to 25% to 30% of monocytes at the time of presentation to the emergency department (compared to an average of 2% expression for healthy controls). These patients also had more platelet–monocyte aggregates (23). Rapid elevations in plasma free tissue factor have been observed in patients with severe trauma without DIC (58), but tissue factor levels have been most highly elevated in trauma patients who developed SIRS (57), DIC (60), and adult respiratory distress syndrome (ARDS) (58,61). The source of this early rise in plasma tissue factor is not clear. It may be the result of cleavage/shedding from tissue factor–bearing monocytes (23). It also may be due to release of tissue factor from endothelial cells (ECs) in response to cytokines (62).

Suppression of Physiological Anticoagulant Mechanism

The impaired function of natural regulatory pathways of coagulation activation also contributes to the procoagulant state and intravascular fibrin deposition. Plasma levels of ATIII, the most important inhibitor of thrombin, are reduced in trauma patients with DIC (20,53,59), and the degree of reduction has been linked to the severity of DIC as well as morbidity and mortality in trauma patients. In 1996, Owings and colleagues demonstrated that reduced ATIII levels in trauma patients were associated with DIC, ARDS, and deep vein thrombosis (DVT) (20). Lower ATIII levels were associated with organ failure in another study (53). This reduction in ATIII levels appears to be the result of a combination of consumption due to continued thrombin generation, degradation by neutrophil elastase, and impaired ATIII synthesis (11). Other anticoagulant systems also appear to be involved. Low protein C (59), an imbalance between the levels of tissue factor and tissue factor pathway inhibitor (58,63), and loss of thrombomodulin from ECs (64) also have been associated with coagulopathy after trauma.

Changes in Fibrinolysis

A consistent finding in trauma-related DIC is that the fibrinolytic system is impaired. An increased level of plasminogen activator inhibitor (PAI)-1 has been a consistent finding in posttraumatic DIC (51,59,65), and this elevation is observed very early in the posttrauma period. In another study, tissue-type plasminogen activator (t-PA) and PAI-1 were elevated in trauma patients with DIC (66). This may reflect an initial fibrinolytic activity due to endothelial release of t-PA that is offset by a sustained large increase in PAI-1, as has been observed in response to experimental ischemia–reperfusion injury (67,68). Gando and colleagues reported reduced α2-antiplasmin concentrations in trauma patients, perhaps reflecting reduced requirements resulting from suppression of plasminogen activation by PAI-1 (15).

Platelet–Leukocyte Interaction

Shortly following trauma, platelet activation can be detected in the circulation. In trauma patients, the percentage of circulating activated platelets (as indicated by P-selectin and GPIIb-IIIa expression) was increased approximately 10-fold over the levels in normal controls at the time of arrival at the emergency department (69). In addition to exposing circulating procoagulant phospholipid surfaces, this activation results in enhanced platelet–leukocyte interactions, which can promote leukocyte adhesion to endothelium (70).

Role of Cytokines

Inflammatory cytokines and other products released by immune cells play a significant role in the derangement of coagulation and fibrinolysis. Elevations in interleukin (IL)-1 (65), IL-6 (53,66), IL-8 (53), and tumor necrosis factor (TNF)-α (65) have been documented within 12 hours after trauma in patients who developed DIC. These cytokines are related to posttraumatic inflammatory response and organ failure, as well as DIC (53,65,66).

IL-1 increases endothelial procoagulant activity (71,72). Additionally, IL-1 decreases endothelial t-PA and increases PAI-1 (73), leading to reduced fibrinolysis. IL-6 is procoagulant (74), through actions on monocytes (75) and endothelium (62), but does not appear to affect the fibrinolytic system (74). IL-8 may exert procoagulant activity by increasing monocyte tissue factor expression (75). TNF-α is procoagulant (76,77), enhancing tissue factor expression by ECs (62,76) while suppressing the protein C pathway (76,78–80). TNF-α also stimulates plasminogen activation and PAI-1 activity, yielding a net antifibrinolysis (73,77,81). The actions of cytokines during the early posttrauma period are consistent with the overall procoagulant and antifibrinolytic state that favors the deposition of fibrin within the microvasculature and the systemic consumption of coagulation proteins and platelets.

Elastase

Neutrophil-derived elastase is elevated systemically within hours of trauma (51,53), and has been related to poor outcome. Neutrophil activation is associated with endothelial injury, which is important in the development of MODS in posttrauma DIC patients (52,53,64). Elastase cleaves thrombomodulin, impairing the protein C system of anticoagulation (82–84), and may act synergistically with oxygen-free radicals in damaging the endothelium (85).

Oxidation and Nitration

Peroxynitrite ($ONOO^-$) is a potent and a relatively long-lived cytotoxic oxidant formed in vivo in the reaction between superoxide anion (O_2^-) and nitric oxide (NO). Neutrophils and activated macrophages that produce significant quantities of O_2^- and NO during inflammatory responses may be the major sources of $ONOO^-$. When $ONOO^-$ is generated in excess, it may damage cells and functional proteins by oxidation and nitration, potentially resulting in altered function of platelets and proteins of the coagulofibrinolytic system. Under these conditions, NO is scavenged in the production of $ONOO^-$, diminishing the normal regulatory influence of NO and potentially leading to further dysregulation of the local hemostatic balance. At low concentrations of $ONOO^-$, platelet response to physiological agonists is inhibited in a reversible manner (86,87). However, at higher concentrations, $ONOO^-$ results in partial activation of platelets and damages platelets such that their ability to respond to normal physiological agonists, such as collagen (86) and thrombin (88–90), is reduced. $ONOO^-$ also reduced cyclooxygenase activity (91). Incubation of normal plasma with peroxynitrite resulted in dose-dependent increases in prothrombin time (PT) and activated partial thromboplastin time (aPTT), and in prolonged clot formation time and reduced clot strength as determined by thromboelastography (92). Components of the procoagulant, anticoagulant, and fibrinolytic mechanisms can be impaired by $ONOO^-$. Exposure of fibrinogen to $ONOO^-$ results in oxidation of fibrinogen and a resultant dose-dependent decrease in the ability of fibrinogen to form a fibrin clot (93). Peroxynitrite incubation decreases the activities of tissue factor (94–96), factor VII, factor VIII complex, and factor X (92). ATIII and protein C activity also are impaired by peroxynitrite (92), as is t-PA (97). Oxidants other than $ONOO^-$ may have similar effects (98). Oxidation and nitration of a number of cellular- and plasma-phase molecules can result in the dysregulation of the hemostatic mechanism at multiple levels.

Summary of Coagulopathy Mechanisms

Within the first hours following severe trauma, blood loss, tissue damage, immobilization, and environmental factors can set the stage for the development of coagulopathy. Multiple factors, including hypothermia, acidosis, hemodilution (including both "natural" and that due to resuscitation measures), inflammatory response, and consumptive processes impair hemostasis and potentially predispose to later complications by deranging the normal hemostatic mechanism. The system is affected at multiple levels of regulation, including pro- and anticoagulant mechanisms, pro- and antifibrinolytic mechanisms, and platelet function. Involved components include the plasma, platelets, red cells, immune cells, and endothelium.

NEED AND OPPORTUNITY FOR INITIAL HEMOSTATIC RESUSCITATION AND INFLAMMATORY MODULATION

Present initial resuscitation regimens provide intravascular volume to enhance stroke volume and cellular perfusion but do nothing hemostatically, other than contribute to hemodilution. For the majority of trauma patients, this is adequate, because they are minimally injured and in reality do not require resuscitation. In the urban trauma environment, this is probably adequate for 75% of the patients who arrive at a trauma center, but even in an arena of very rapid evacuation, fully 25% (who are the most seriously injured) arrive already coagulopathic. In this environment, advanced care, including blood products and other resources, will be rapidly available after a relatively short evacuation. In contrast, the military casualty patient and many rural transport patients often must be stabilized for several hours prior to receiving surgical care or blood products. Conventional strategies for these casualties serve to promote the coagulopathy in the severely injured patient. A new hemostatic resuscitation approach offers the opportunity to deliver drugs or other compounds that could enhance survival by mitigating factors associated with coagulopathy. This could result in a decrease or cessation of bleeding, which could reduce overall blood loss, the period or severity of shock, lessen acidosis and hypothermia, and improve survival. Delivery of an immunomodulatory drug might dampen the initial inflammatory response to trauma which, by extension, could reduce the consumptive processes and related longer-term sequelae, such as SIRS, MODS, and multiple-organ failure (MOF).

Although currently no approved hemostatic resuscitation regimen exists, some promising drugs may prove useful. Recombinant activated factor VII has been used to improve hemostasis in hemophilia patients for years, and use in other coagulopathies, including those resulting from trauma, is gaining attention around the world. The theoretic risk for this drug is that it might induce DIC or thrombosis in trauma patients who have systemic exposure of tissue factor and procoagulant phospholipids. However, data from a recent prospective, randomized, blinded, multicenter trial suggest that the safety fears related to this theoretic risk are unfounded in this population (99). As the safety and efficacy of this drug for trauma becomes more certain, it may be possible to employ this drug in the prehospital setting as part of a resuscitation regimen. Resuscitation solutions that contain hypertonic saline inhibit immune cell function. Evidence suggests that early resuscitation with hypertonic solutions may minimize the initial inflammatory response, mitigating the drastic shifts in hemostatic regulatory mechanisms due to the activation of immune cells and release of inflammatory mediators. Freeze-dried platelets, plasma, and fibrinogen also are becoming available (100). Combination therapy using these and other products could form the basis for a hemostatic resuscitation strategy, which may then be incorporated into a global damage control resuscitation paradigm that would address more than just intravascular volume and tissue oxygenation. Focused research examining these and other approaches is needed to develop a hemostatic resuscitation regimen for the treatment of combat casualty patients and potentially other trauma patients.

KEY POINTS

- Trauma-related coagulopathy is multifactorial, involving the nearly simultaneous effects of hypothermia, acidosis, hemodilution, inflammatory responses, and consumptive processes in various combinations that result in dysregulation at multiple levels of the hemostatic mechanism.
- Combat casualty patients differ from civilian trauma patients in both the severity and pattern of wounds, as well as in the expected time required for evacuation to surgical care. Further advances in prehospital stabilization and treatment are essential to improve battlefield hemostasis.
- Current prehospital and hospital resuscitation strategies using crystalloids, colloids, and blood products can worsen trauma-induced coagulopathy in severely injured casualties

Future Goals

- To undertake research aimed at furthering our understanding of the interaction between the coagulation and inflammatory processes, thus allowing us to define the optimal hemostatic resuscitation strategy

REFERENCES

1 Champion HR, Bellamy RF, Roberts CP, Leppaniemi A. A profile of combat injury. *J Trauma.* 2003;54(Suppl 5):S13–S19.

2 Holcomb JB. *Ongoing review of combat casualty data for Operations Enduring Freedom and Iraqi Freedom.* United States Department of Defense Joint Trauma Database, 2005.

3 MacLeod JB, Lynn M, McKenney MG, et al. Early coagulopathy predicts mortality in trauma. *J Trauma.* 2003;55(1):39–44.

4 Brohi K, Singh J, Heron M, Coats T. Acute traumatic coagulopathy. *J Trauma.* 2003;54(6):1127–1130.

5 Bernabei AF, Levison MA, Bender JS. The effects of hypothermia and injury severity on blood loss during trauma laparotomy. *J Trauma.* 1992;33(6):835–839.

6 Cosgriff N, Moore EE, Sauaia A, et al. Predicting life-threatening coagulopathy in the massively transfused trauma patient: hypothermia and acidoses revisited. *J Trauma.* 1997;42(5):857–861; Discussion 61–62.

7 Gentilello LM, Jurkovich GJ, Stark MS, et al. Is hypothermia in the victim of major trauma protective or harmful? A randomized, prospective study. *Ann Surg.* 1997;226(4):439–447; Discussion 47–49.

8 Jurkovich GJ, Greiser WB, Luterman A, Curreri PW. Hypothermia in trauma victims: an ominous predictor of survival. *J Trauma.* 1987;27(9):1019–1024.

9 Luna GK, Maier RV, Pavlin EG, et al. Incidence and effect of hypothermia in seriously injured patients. *J Trauma.* 1987;27(9):1014–1018.

10 Esmon CT. Crosstalk between inflammation and thrombosis. *Maturitas.* 2004;47(4):305–314.

11 Levi M, de Jonge E, van der Poll T, ten Cate H. Disseminated intravascular coagulation. *Thromb Haemost.* 1999;82(2):695–705.

12 Levi M, Keller TT, van Gorp E, ten Cate H. Infection and inflammation and the coagulation system. *Cardiovasc Res.* 2003;60(1):26–39.

13 Levi M, ten Cate H, van der Poll T. Endothelium: interface between coagulation and inflammation. *Crit Care Med.* 2002;30(Suppl 5):S220–S224.

14 Gando S. Disseminated intravascular coagulation in trauma patients. *Semin Thromb Hemost.* 2001;27(6):585–592.

15 Martinowitz U, Michaelson M. Guidelines for the use of recombinant activated factor VII (rFVIIa) in uncontrolled bleeding: a report by the Israeli Multidisciplinary rFVIIa Task Force. *J Thromb Haemost.* 2005;3(4):640–648.

16 Gando S, Nanzaki S, Kemmotsu O. Coagulofibrinolytic changes after isolated head injury are not different from those in trauma patients without head injury. *J Trauma.* 1999;46(6):1070–1076; Discussion 6–7.

17 Hulka F, Mullins RJ, Frank EH. Blunt brain injury activates the coagulation process. *Arch Surg.* 1996;131(9):923–927; Discussion 7–8.

18 Kearney TJ, Bentt L, Grode M, et al. Coagulopathy and catecholamines in severe head injury. *J Trauma.* 1992;32(5):608–611; Discussion 11–12.

19 Kuo JR, Chou TJ, Chio CC. Coagulopathy as a parameter to predict the outcome in head injury patients – analysis of 61 cases. *J Clin Neurosci.* 2004;11(7):710–714.

20 Kushimoto S, Yamamoto Y, Shibata Y, et al. Implications of excessive fibrinolysis and alpha(2)-plasmin inhibitor deficiency in patients with severe head injury. *Neurosurgery.* 2001;49(5):1084–1089; Discussion 9–90.

21 Owings JT, Bagley M, Gosselin R, et al. Effect of critical injury on plasma antithrombin activity: low antithrombin levels are associated with thromboembolic complications. *J Trauma.* 1996;41(3):396–405; Discussion 6.

22 Scherer RU, Spangenberg P. Procoagulant activity in patients with isolated severe head trauma. *Crit Care Med.* 1998;26(1):149–156.

23 Sorensen JV, Jensen HP, Rahr HB, et al. Haemostatic activation in patients with head injury with and without simultaneous multiple trauma. *Scand J Clin Lab Invest.* 1993;53(7):659–665.

24 Utter GH, Owings JT, Jacoby RC, et al. Injury induces increased monocyte expression of tissue factor: factors associated with head injury attenuate the injury-related monocyte expression of tissue factor. *J Trauma*. 2002;52(6):1071–1077; Discussion 7.
25 Penner JA. Disseminated intravascular coagulation in patients with multiple organ failure of non-septic origin. *Semin Thromb Hemost*. 1998;24(1):45–52.
26 Acosta JA, Yang JC, Winchell RJ, et al. Lethal injuries and time to death in a level I trauma center. *J Am Coll Surg*. 1998;186(5):528–533.
27 Hoyt DB, Bulger EM, Knudson MM, et al. Death in the operating room: an analysis of a multi-center experience. *J Trauma*. 1994; 37(3):426–432.
28 Kettner SC, Sitzwohl C, Zimpfer M, et al. The effect of graded hypothermia (36 degrees C-32 degrees C) on hemostasis in anesthetized patients without surgical trauma. *Anesth Analg*. 2003; 96(6):1772–1776.
29 Reed RL 2nd, Bracey AW Jr., Hudson JD, et al. Hypothermia and blood coagulation: dissociation between enzyme activity and clotting factor levels. *Circ Shock*. 1990;32(2):141–152.
30 Meng ZH, Wolberg AS, Monroe DM 3rd, Hoffman M. The effect of temperature and pH on the activity of factor VIIa: implications for the efficacy of high-dose factor VIIa in hypothermic and acidotic patients. *J Trauma*. 2003;55(5):886–891.
31 Wolberg AS, Meng ZH, Monroe DM 3rd, Hoffman M. A systematic evaluation of the effect of temperature on coagulation enzyme activity and platelet function. *J Trauma*. 2004;56(6): 1221–1228.
32 Pusateri AE, Delgado AV, Martini WZ, Uscilowicz JM. Differential effects of acidosis and hypothermia on clot development in swine. *J Thromb Haemost*. 2005;3(Suppl 1):P0803.
33 Watts DD, Trask A, Soeken K, et al. Hypothermic coagulopathy in trauma: effect of varying levels of hypothermia on enzyme speed, platelet function, and fibrinolytic activity. *J Trauma*. 1998;44(5):846–854.
34 Valeri CR, Feingold H, Cassidy G, et al. Hypothermia-induced reversible platelet dysfunction. *Ann Surg*. 1987;205(2):175–181.
35 Valeri CR, Khabbaz K, Khuri SF, et al. Effect of skin temperature on platelet function in patients undergoing extracorporeal bypass. *J Thorac Cardiovasc Surg*. 1992;104(1):108–116.
36 Zhang JN, Wood J, Bergeron AL, et al. Effects of low temperature on shear-induced platelet aggregation and activation. *J Trauma*. 2004;57(2):216–223.
37 Pusateri AE, Delgado AV, Martini WZ, et al. Effects of acidosis and hypothermia on whole blood clotting function in swine. *J Thromb Haemost*. 2005;3(Suppl 1):P1402.
38 Chaimoff C, Creter D, Djaldetti M. The effect of pH on platelet and coagulation factor activities. *Am J Surg*. 1978;136(2):257–259.
39 Lamberth EL Jr., Warriner RA 3rd, Batchelor ED, Des Prez RM. Effect of metabolic acidosis and alkalosis on human platelet aggregation induced by epinephrine and ADP. *Proc Soc Exp Biol Med*. 1974;145(2):743–746.
40 oude Egbrink MG, Tangelder GJ, Slaaf DW, et al. Influence of hypercapnia and hypoxia on rabbit platelet aggregation. *Thromb Res*. 1990;57(6):863–875.
41 Gende OA. Capacitative calcium influx and intracellular pH cross-talk in human platelets. *Platelets*. 2003;14(1):9–14.
42 Marumo M, Suehiro A, Kakishita E, et al. Extracellular pH affects platelet aggregation associated with modulation of store-operated $Ca^{(2+)}$ entry. *Thromb Res*. 2001;104(5):353–360.
43 Delgado AV, Dong JF, Chambers JP, Pusateri AE. Effects of hypothermia and acidosis on swine platelet activation and aggregation in vivo. *J Thromb Haemost*. 2005;3(Suppl 1):P2084.
44 Pusateri AE, Ryan KL, Delgado AV, et al. Effects of increasing doses of activated recombinant factor VII on haemostatic parameters in swine. *Thromb Haemost*. 2005;93(2):275–283.
45 Martini WZ, Pusateri AE, Uscilowicz JM, et al. Independent contributions of hypothermia and acidosis to coagulopathy in Swine. *J Trauma*. 2005;58(5):1002–1010.
46 Armand R, Hess JR. Treating coagulopathy in trauma patients. *Transfus Med Rev*. 2003;17(3):223–231.
47 Ruttmann TG, James MF, Aronson I. In vivo investigation into the effects of haemodilution with hydroxyethyl starch (200/0.5) and normal saline on coagulation. *Br J Anaesth*. 1998;80(5):612–616.
48 Fenger-Eriksen C, Anker-Moller E, Heslop J, et al. Thrombelastographic whole blood clot formation after ex vivo addition of plasma substitutes: improvements of the induced coagulopathy with fibrinogen concentrate. *Br J Anaesth*. 2005;94(3): 324–329.
49 Valeri CR, Cassidy G, Pivacek LE, et al. Anemia-induced increase in the bleeding time: implications for treatment of nonsurgical blood loss. *Transfusion*. 2001;41(8):977–983.
50 de Jonge E, Levi M. Effects of different plasma substitutes on blood coagulation: a comparative review. *Crit Care Med*. 2001; 29(6):1261–1267.
51 Gando S, Kameue T, Nanzaki S, et al. Increased neutrophil elastase, persistent intravascular coagulation, and decreased fibrinolytic activity in patients with posttraumatic acute respiratory distress syndrome. *J Trauma*. 1997;42(6):1068–1072.
52 Gando S, Kameue T, Matsuda N, et al. Combined activation of coagulation and inflammation has an important role in multiple organ dysfunction and poor outcome after severe trauma. *Thromb Haemost*. 2002;88(6):943–949.
53 Nast-Kolb D, Waydhas C, Gippner-Steppert C, et al. Indicators of the posttraumatic inflammatory response correlate with organ failure in patients with multiple injuries. *J Trauma*. 1997; 42(3):446–454; Discussion 54–55.
54 Roumen RM, Hendriks T, van der Ven-Jongekrijg J, et al. Cytokine patterns in patients after major vascular surgery, hemorrhagic shock, and severe blunt trauma. Relation with subsequent adult respiratory distress syndrome and multiple organ failure. *Ann Surg*. 1993;218(6):769–776.
55 Malone DL, Kuhls D, Napolitano LM, et al. Back to basics: validation of the admission systemic inflammatory response syndrome score in predicting outcome in trauma. *J Trauma*. 2001; 51(3):458–463.
56 Napolitano LM, Ferrer T, McCarter RJ Jr., Scalea TM. Systemic inflammatory response syndrome score at admission independently predicts mortality and length of stay in trauma patients. *J Trauma*. 2000;49(4):647–652; Discussion 52–53.
57 Gando S, Kameue T, Nanzaki S, et al. Participation of tissue factor and thrombin in posttraumatic systemic inflammatory syndrome. *Crit Care Med*. 1997;25(11):1820–1826.
58 Gando S, Kameue T, Matsuda N, et al. Imbalances between the levels of tissue factor and tissue factor pathway inhibitor in ARDS patients. *Thromb Res*. 2003;109(2–3):119–124.
59 Gando S, Kameue T, Nanzaki S, Nakanishi Y. Disseminated intravascular coagulation is a frequent complication of systemic inflammatory response syndrome. *Thromb Haemost*. 1996;75 (2):224–228.

60 Gando S, Nanzaki S, Sasaki S, Kemmotsu O. Significant correlations between tissue factor and thrombin markers in trauma and septic patients with disseminated intravascular coagulation. *Thromb Haemost*. 1998;79(6):1111–1115.

61 Gando S, Nanzaki S, Morimoto Y, et al. Systemic activation of tissue-factor dependent coagulation pathway in evolving acute respiratory distress syndrome in patients with trauma and sepsis. *J Trauma*. 1999;47(4):719–723.

62 Szotowski B, Antoniak S, Poller W, et al. Procoagulant soluble tissue factor is released from endothelial cells in response to inflammatory cytokines. *Circ Res*. 2005;96(12):1233–1239.

63 Gando S, Nanzaki S, Morimoto Y, et al. Tissue factor pathway inhibitor response does not correlate with tissue factor-induced disseminated intravascular coagulation and multiple organ dysfunction syndrome in trauma patients. *Crit Care Med*. 2001; 29(2):262–266.

64 Gando S, Nakanishi Y, Kameue T, Nanzaki S. Soluble thrombomodulin increases in patients with disseminated intravascular coagulation and in those with multiple organ dysfunction syndrome after trauma: role of neutrophil elastase. *J Trauma*. 1995; 39(4):660–664.

65 Gando S, Nakanishi Y, Tedo I. Cytokines and plasminogen activator inhibitor-1 in posttrauma disseminated intravascular coagulation: relationship to multiple organ dysfunction syndrome. *Crit Care Med*. 1995;23(11):1835–1842.

66 Kowal-Vern A, Sharp-Pucci MM, Walenga JM, et al. Trauma and thermal injury: comparison of hemostatic and cytokine changes in the acute phase of injury. *J Trauma*. 1998;44(2):325–329.

67 Schoots IG, Levi M, Roossink EH, et al. Local intravascular coagulation and fibrin deposition on intestinal ischemia-reperfusion in rats. *Surgery*. 2003;133(4):411–419.

68 Schoots IG, Levi M, van Vliet AK, et al. Inhibition of coagulation and inflammation by activated protein C or antithrombin reduces intestinal ischemia/reperfusion injury in rats. *Crit Care Med*. 2004;32(6):1375–1383.

69 Jacoby RC, Owings JT, Holmes J, et al. Platelet activation and function after trauma. *J Trauma*. 2001;51(4):639–647.

70 Ogura H, Kawasaki T, Tanaka H, et al. Activated platelets enhance microparticle formation and platelet-leukocyte interaction in severe trauma and sepsis. *J Trauma*. 2001;50(5):801–809.

71 Bevilacqua MP, Pober JS, Majeau GR, et al. Interleukin 1 (IL-1) induces biosynthesis and cell surface expression of procoagulant activity in human vascular endothelial cells. *J Exp Med*. 1984;160(2):618–623.

72 Bevilacqua MP, Pober JS, Wheeler ME, et al. Interleukin-1 activation of vascular endothelium. Effects on procoagulant activity and leukocyte adhesion. *Am J Pathol*. 1985;121(3):394–403.

73 Schleef RR, Bevilacqua MP, Sawdey M, et al. Cytokine activation of vascular endothelium. Effects on tissue-type plasminogen activator and type 1 plasminogen activator inhibitor. *J Biol Chem*. 1988;263(12):5797–5803.

74 Stouthard JM, Levi M, Hack CE, et al. Interleukin-6 stimulates coagulation, not fibrinolysis, in humans. *Thromb Haemost*. 1996;76(5):738–742.

75 Neumann FJ, Ott I, Marx N, et al. Effect of human recombinant interleukin-6 and interleukin-8 on monocyte procoagulant activity. *Arterioscler Thromb Vasc Biol*. 1997;17(12):3399–3405.

76 Nawroth PP, Stern DM. Modulation of endothelial cell hemostatic properties by tumor necrosis factor. *J Exp Med*. 1986; 163(3):740–745.

77 van der Poll T, Jansen PM, Van Zee KJ, et al. Tumor necrosis factor-alpha induces activation of coagulation and fibrinolysis in baboons through an exclusive effect on the p55 receptor. *Blood*. 1996;88(3):922–927.

78 Conway EM, Rosenberg RD. Tumor necrosis factor suppresses transcription of the thrombomodulin gene in endothelial cells. *Mol Cell Biol*. 1988;8(12):5588–5592.

79 Fukudome K, Esmon CT. Identification, cloning, and regulation of a novel endothelial cell protein C/activated protein C receptor. *J Biol Chem*. 1994;269(42):26486–26491.

80 Moore KL, Esmon CT, Esmon NL. Tumor necrosis factor leads to the internalization and degradation of thrombomodulin from the surface of bovine aortic endothelial cells in culture. *Blood*. 1989;73(1):159–165.

81 Biemond BJ, Levi M, Ten Cate H, et al. Plasminogen activator and plasminogen activator inhibitor I release during experimental endotoxaemia in chimpanzees: effect of interventions in the cytokine and coagulation cascades. *Clin Sci (Lond)*. 1995;88 (5):587–594.

82 Kobayashi H, Sadakata H, Suzuki K, et al. Thrombomodulin release from umbilical endothelial cells initiated by preeclampsia plasma-induced neutrophil activation. *Obstet Gynecol*. 1998;92(3):425–430.

83 MacGregor IR, Perrie AM, Donnelly SC, Haslett C. Modulation of human endothelial thrombomodulin by neutrophils and their release products. *Am J Respir Crit Care Med*. 1997;155(1): 47–52.

84 Miyazaki Y, Inoue T, Kyi M, et al. Effects of a neutrophil elastase inhibitor (ONO-5046) on acute pulmonary injury induced by tumor necrosis factor alpha (TNFalpha) and activated neutrophils in isolated perfused rabbit lungs. *Am J Respir Crit Care Med*. 1998;157(1):89–94.

85 Fujita H, Morita I, Ishikawa K, Murota S. The synergistic effect of elastase and hydrogen peroxide on vascular endothelial cell injury is due to the production of hydroxylradical in the endothelial cells. *J Atheroscler Thromb*. 1996;3(1):32–38.

86 Brown AS, Moro MA, Masse JM, et al. Nitric oxide-dependent and independent effects on human platelets treated with peroxynitrite. *Cardiovasc Res*. 1998;40(2):380–388.

87 Low SY, Sabetkar M, Bruckdorfer KR, Naseem KM. The role of protein nitration in the inhibition of platelet activation by peroxynitrite. *FEBS Lett*. 2002;511(1–3):59–64.

88 Mondoro TH, Shafer BC, Vostal JG. Peroxynitrite-induced tyrosine nitration and phosphorylation in human platelets. *Free Radic Biol Med*. 1997;22(6):1055–1063.

89 Nowak P, Wachowicz B. The effects of peroxynitrite on pig platelet lipid peroxidation and the secretory process induced by thrombin. *Cytobios*. 2001;106(Suppl 2):179–187.

90 Nowak P, Wachowicz B. Studies on pig blood platelet responses to peroxynitrite action. *Platelets*. 2001;12(6):376–381.

91 Boulos C, Jiang H, Balazy M. Diffusion of peroxynitrite into the human platelet inhibits cyclooxygenase via nitration of tyrosine residues. *J Pharmacol Exp Ther*. 2000;293(1): 222–229.

92 Nielsen VG, Crow JP, Mogal A, et al. Peroxynitrite decreases hemostasis in human plasma in vitro. *Anesth Analg*. 2004;99(1): 21–26.

93 Lupidi G, Angeletti M, Eleuteri AM, et al. Peroxynitrite-mediated oxidation of fibrinogen inhibits clot formation. *FEBS Lett*. 1999;462(3):236–240.
94 Adam JM, Ettelaie C, Naseem KM, et al. Modification of tissue factor by peroxynitrite influences its procoagulant activity. *FEBS Lett*. 1998;429(3):347–350.
95 Gerlach M, Keh D, Bezold G, et al. Nitric oxide inhibits tissue factor synthesis, expression and activity in human monocytes by prior formation of peroxynitrite. *Intensive Care Med*. 1998; 24(11):1199–1208.
96 Nielsen VG, Crow JP. Peroxynitrite decreases rabbit tissue factor activity in vitro. *Anesth Analg*. 2004;98(3):668–671.
97 Nielsen VG, Crow JP, Zhou F, Parks DA. Peroxynitrite inactivates tissue plasminogen activator. *Anesth Analg*. 2004;98(5): 1312–1317.
98 Stief TW. Regulation of hemostasis by singlet-oxygen (1DeltaO2*). *Curr Vasc Pharmacol*. 2004;2(4):357–362.
99 Boffard KD, Riou B, Warren B, et al. Recombinant factor VIIa as adjunctive therapy for bleeding control in severely injured trauma patients: two parallel randomized, placebo-controlled, double-blind clinical trials. *J Trauma*. 2005;59(1):8–15; Discussion 8.
100 Fries D, Krismer A, Klingler A, et al. Effect of fibrinogen on reversal of dilutional coagulopathy: a porcine model. *Br J Anaesth*. 2005;95(2):172–177.

166

The Effects of Blood Transfusion on Vascular Endothelium

Christopher C. Silliman

Bonfils Blood Center and the Departments of Pediatrics and Surgery, University of Colorado School of Medicine, Denver

Blood transfusions reduce the mortality associated with such conditions as major trauma, major surgery, and myelotoxic chemotherapy and, as such, have revolutionized medical care. With the exception of acute hemolytic transfusion reactions, the practice of blood transfusion was once considered to be relatively risk-free. It is now clear that blood transfusions are associated with short- and long-term side effects, including the transmission of infectious agents, transfusion-related immunomodulation (TRIM), and proinflammatory stimulation of the innate immune system (Table 166-1) (1–3). As an interface between circulating blood and underlying tissue, the endothelium is preferentially exposed to cellular and noncellular mediators contained within blood products. Thus, transfusion may result in phenotypic alteration of the endothelium.

INFECTIOUS AGENTS

Owing to marked improvements in screening assays (particularly nucleic acid amplification–based protocols), transfusion-transmitted viral infections rarely occur today. However, it is important to recognize that pathogenic viruses from other parts of the world may undergo a change in their natural range of distribution, and thus become introduced into the United States, as illustrated by the recent reports of West Nile Virus (4). Many viruses and viral proteins, most notably cytomegalovirus (CMV) and the Tat proteins from human immunodeficiency virus (HIV)-1, have been shown to interact directly with the endothelial cell (EC) membrane, resulting in increased surface expression of adhesion molecules and the synthesis and release of chemokines (5,6). The resulting proinflammatory phenotype may lead to the firm adhesion of granulocytes, polymorphonuclear neutrophil (PMN)-mediated acute lung injury (ALI), or other tissue damage (5,6). Some viruses, such as HIV-1 or Bunyaviridae, efficiently enter and replicate in the vascular endothelium and cause massive endothelial damage, whereas other viruses, including Hanta virus or Arenaviridae, do not actually infect the cell (7). The ultimate impact of viral infection on the endothelium depends on many factors, including the type and titer of virus, mode of cellular entry, nature of the host response, and vascular bed involved (7).

In contrast to advances in eliminating viral pathogen transmission, bacterial contamination remains a leading cause of transfusion-related morbidity and mortality. Indeed, continued efforts to prolong the storage time of platelets will likely result in even higher rates of bacterial infection (8). The skin flora of the donor is the usual source of bacterial contamination. However, there are exceptions. For example, donors with reptile pets are at risk for transmitting more virulent organisms, including *Salmonella* species (8). Once infused into the recipient, bacteria may directly infect ECs. More commonly, components of the bacterial wall (e.g., lipopolysaccharide) activate pattern recognition receptors (e.g., Toll-like receptors) on the surface of monocytes and ECs, resulting in a proinflammatory and procoagulant phenotype (9). Bacteria also may damage the endothelium through the release of exotoxins, which induce pores in the endothelium and result in vascular leak (7,10). Finally, parasites also may be transmitted by transfusion and may have similar effects on endothelium. Malaria infection is the most recognized parasitic complication of blood transfusion and, although rare in the United States (0.3/1,000,000), represents the most common transfusion-related parasitic infection worldwide (11). *Trypanosoma cruzi*, the etiologic agent in Chagas disease and endemic to Central and South America, has been transmitted by transfusion to four immunocompromised recipients in the past 50 years (11).

Transmission of prion disease (e.g., variant Creutzfeldt-Jakob disease [vCJD]) is a rare complication (12). Prions may infect the vascular endothelium. Infected ECs are highly efficient at replicating these proteins, and thus may serve as a primary and sustained reservoir of blood-borne prions (13). Although prions have been implicated in relatively few cases

Table 166-1: Complications of Transfusion

Complication	*Most Common Product Associated with Complication*	*Incidence*	*Mechanism*	*Mechanism(s) of EC Dysfunction*	*Reduced with Leukoreduction*	*Reduced with Washing*	*Reduced with Fresher Blood*
Infection							
HIV	Plasma-containing or plasma-derived components	1/2,400,000–1/1,400,000	Presence in plasma or plasma-derived products	Direct interaction with EC membrane; direct entry	No	No	No
CMV	Cellular components with leukocytes	Rarely due to transfusion	Infected blood donor leukocytes	Direct interaction with EC membrane; direct entry	Yes	No	No
Hepatitis C	Plasma-containing or plasma-derived components	1/1,700,000–1/872,000	Presence in plasma or plasma-derived products	Direct interaction with EC membrane; direct entry	No	No	No
Bacteria	PCs, rarely PRBCs		Contamination of donor blood	Activation of TLR	No	Not known	No
Prion: Variant Creutzfeldt-Jakob disease (vCJD)	???	Rare (2 cases)	Long incubation periods	Direct infection	In experimental models only	No	No
TRIM	PRBCs, WB, and PCs	Not known	Passenger leukocytes and/or soluble biologic response modifiers (BRMs)	Not known	Variable results; Two randomized, controlled trials (RCTs) demonstrate efficacy	Limited data, one RCT shows efficacy	Not known
Massive Transfusion (MODS/MOF)	PRBCs	Not known	Presence of soluble proinflammatory BRMs	Proinflammatory action of BRMs in stored blood	No	Not Known	Yes
TRALI	Platelets, FFP, PRBCs, IVIG	1/5,000–1/3,23 USA1/769, 230–1/7,900 EU	Antibody-mediated (single insult) vs. two insults (clinical status 1st; antibodies or BRMs 2nd)	???	No	Likely, but no RCTs	Possibly if due to BRMs, no RCTs

WB, whole blood; TLR, Toll-like receptors; IVIG, intravenous immunoglobulin; MODS, multiorgan dysfunction syndrome.

of spongiform encephalopathy, the potential for transfusion-related infection may be enormous due to our inability to detect these agents, the long incubation period between exposure and the clinical signs of infection, and our relative inexperience with these pathogens that may lead to a delayed diagnosis (12).

TRANSFUSION-RELATED IMMUNOMODULATION

TRIM was first recognized in 1978 by Opelz and colleagues, who demonstrated enhanced survival of renal allografts in multiply transfused hosts as compared with nontransfused organ recipients (14). Other manifestations of TRIM include increased postoperative infections, death from multiple organ failure, and increased mortality related to the age of the transfused product (2). Hypothetical concerns include the accelerated growth or spread of malignancies, and reactivation of latent viruses (2). The mechanisms of TRIM are poorly defined, but presumably involve soluble mediators and/or donor cells in the transfusion product. Prestorage leukoreduction has been shown to decrease the incidence of TRIM, supporting a pathophysiological role for donor leukocytes (2,15). Interestingly, transfusion of fresher, compared with stored, blood components leads to a decreased incidence of postoperative pneumonia, suggesting that biologic response modifiers either of leukocyte or noncellular origin may be the causative agents (16). Randomized, controlled clinical trials are needed before recommending the routine use of fresher blood products. Washing of cellular components removes biologic response modifiers that accumulate in the plasma during routine storage (3,17). However, such a procedure has not been shown to prevent TRIM, is expensive and time consuming, and is thus reserved for patients with severe allergic reactions to blood components.

Little information is available on whether or how TRIM alters the phenotype of the endothelium or to what extent the endothelium contributes to the TRIM phenotype. Extrapolating from the observed immunomodulatory effects of viruses (especially HIV-1), drugs, and ethanol on the endothelium, TRIM may be predicted to result in endothelial damage and/or barrier dysfunction and an increased risk for secondary invasion by viral pathogens or toxic substances into discrete organs, particularly the liver (18).

TRANSFUSION AND INFLAMMATION

Massive Transfusion

Massive transfusion, defined as the replacement of one blood volume per day or the transfusion of 10 units of packed red blood cells (PRBCs), has long been recognized as a predisposing factor for ALI and the acute respiratory distress syndrome (ARDS) (19–22). Many investigators have assumed that transfusions are not pathogenic per se, but rather represent a surrogate marker for shock and its effects on organ dysfunction. However, PRBC transfusion has been shown to be an independent risk factor for the development of multiple organ failure (MOF) in traumatically injured patients (23). In that study, transfusion of more than six units of PRBCs within the first 12 hours increased the risk for MOF in patients with intermediate injury severity scores (23). Furthermore, the risk for MOF directly correlated with the storage age of the PRBCs transfused, such that patients who received PRBC units of greater than 21 days of storage developed MOF, whereas those transfused with units that were banked for shorter periods did not (24).

Recent studies have shown that the plasma fraction of stored, but not fresh, PRBCs causes proinflammatory activation of cultured pulmonary human microvascular ECs (HMVECs) (25,26). In these experiments, the plasma fraction from PRBCs on day 42 (the upper limit of storage time for this blood product) resulted in increased surface expression of intercellular cellular adhesion molecule (ICAM)-1; increased synthesis and release of chemokines, namely interleukin (IL)-8, growth-related oncogene (GRO)-α, and epithelial neutrophil activating peptide (ENA)-78; and increased adherence of PMNs (25,26). In contrast to these results, plasma fractions taken from the identical PRBC unit on day 1 failed to elicit a proinflammatory phenotype. The day-42 plasma activity was localized to the chloroform-soluble fraction, implicating a role for lipids that accumulate during routine PRBC storage (25,26).

To rule out a confounding effect of leukocytes during PRBC storage, we recently collected units of blood from healthy donors. The samples were divided into two fractions: One was leukoreduced by filtration; the other was left unmodified (control). Plasma derived from the PRBCs (stored for 1 or 42 days) was then added to pulmonary HMVECs. Plasma from day-42 PRBCs, whether prestorage leukoreduced or not, caused increased surface expression of ICAM-1, whereas day-1 plasma from the identical units had no such effect (C. Silliman, unpublished observations). Consistent with the results of previous studies, the lipid fraction was shown to contain the responsible activity (C. Silliman, unpublished observations). Together, these data suggest that, during routine storage of PRBCs, a leukocyte-independent, biologically active lipid accumulates in the plasma fraction that causes proinflammatory activation of HMVECs and may predispose massively transfused patients to ALI or MOF. Further work is required to identify the nature and source of the biologically active lipid.

Similar to these findings, plasma from whole blood–derived platelet concentrates demonstrate storage-dependent activation of HMVECs (27). In contrast, platelet concentrates collected by apheresis techniques did not activate HMVECs, irrespective of their storage age (27). Because whole blood–derived platelet concentrates are not widely used in the developed world, these effects may be of little concern. However, in developing countries, in remote areas, or on the battlefield in which apheresis platelet concentrates are not available, one must account for such potential untoward effects from transfusions of whole blood–derived platelets.

Transfusion-Related Acute Lung Injury

Transfusion-related acute lung injury (TRALI) is the leading cause of transfusion-related mortality over the past 2 reporting years (28). TRALI was first reported in 1951, but was not recognized as a distinct clinical entity until 1985 (29,30). TRALI remains underdiagnosed and under-reported (31–34). The syndrome is clinically indistinguishable from ALI, and is manifested by tachypnea, cyanosis, dyspnea, fever, hypoxia, and pulmonary edema (21,30,32). The differential diagnosis of patients who have pulmonary insufficiency following transfusion includes transfusion-related circulatory overload, anaphylactic transfusion reactions, and transfusions contaminated with bacteria (33). Treatment consists of aggressive respiratory support. The mortality rate from TRALI ranges from 5% to 25% (33).

Antigranulocyte Antibodies Directed against Human Leukocyte Antigen Class I or Granulocyte-Specific Antigens

The pathophysiology of TRALI is complex, and a number of different models have been proposed. The first postulates that TRALI is due solely to the infusion of antigranulocyte antibodies directed against human leukocyte antigen (HLA) class I or granulocyte-specific antigens. Infusion of these antibodies into a patient with the cognate antigen present on their granulocytes results in activation of complement, pulmonary sequestration of PMNs, and PMN-mediated damage (30). In addition, infusion of viable granulocytes into a patient with antileukocyte antibodies also may cause TRALI. Although this mechanism is thought to account for less than 5% of all cases, it has particular importance for patients requiring granulocyte transfusions (30,35). An animal model was developed that reinforced this hypothesis using infused human neutrophil antigen (HNA)-3a antibodies, HNA-3a$^+$ human PMNs, and human plasma as a complement source that were infused into isolated, perfused rabbit lungs (36). If any of these components were deleted, including substitution of nonspecific immunoglobulins, lung injury did not occur. However, one must take into account the effect of the tubing, because such surfaces may act to promote PMN priming and the transition from a nonadhesive to an adhesive phenotype (33). This model has been updated and appears to require at least two events to precipitate ALI (37).

Antibodies Directed Against Human Leukocyte Antigen Class II Antigens

A second antibody-mediated pathogenesis involves the infusion of HLA class II antibodies directed against specific HLA class II antigens in the recipient (38). Upon engagement of these antibodies to their cognate antigens on monocytes, cytokines are produced that may cause or exacerbate TRALI; moreover, following prolonged incubation with cytokines in vitro, PMNs may express HLA class II antigens, and such antibodies may have the capacity to directly affect the microbicidal arsenal of the PMN (39–41). The exact details of PMN-mediated ALI secondary to HLA class II antibodies requires further elucidation (33,42).

Two-Event Model

A two-event model also has been proposed for TRALI: The first event involves proinflammatory activation of the pulmonary endothelium, and the second event is the transfusion itself (22,43). Many different clinical conditions result in activation of pulmonary endothelium (the first event), including sepsis, trauma, and surgery (21,32,33). Transfusion (the second event) results in the delivery of donor antibodies directed against antigens on the surface of PMNs adhering to activated lung endothelium, or against other biological response modifiers that accumulate during routine storage of the cellular product, resulting in accentuated PMN-induced endothelial damage and capillary leak (33). Although the plasma component of blood products is believed to mediate the second event, whole blood–derived platelets are most frequently implicated in this syndrome, followed by fresh frozen plasma (FFP) and PRBCs (33).

TRALI may be studied using in vitro and/or in vivo models. Under in vitro conditions, pulmonary microvascular ECs treated with endotoxin demonstrate increased expression of ICAM-1 and chemokines, as well as firm adhesion and priming of PMNs (first event) (44). Subsequent addition of lysophosphatidylcholine (lyso-PC), the lipid product that accumulates during storage of cellular blood products, promoted nicotinamide adenine dinucleotide phosphate (NADPH) oxidase activity in the adherent/primed PMNs (second event) (44). The resulting burst of reactive oxygen species resulted in cytokine-dependent EC damage (44). These findings mimic the proposed two-event model of TRALI and suggest that the cell culture system may prove useful for dissecting the molecular basis for this phenomenon.

To generate an in vivo model of TRALI, endotoxin was systemically administered to Sprague Dawley rats via intraperitoneal injection and resulted in active adhesion (as distinct from physical sequestration) of PMNs to pulmonary endothelium for up to 12 hours (first event) (45). The second event involved the transfusion of stored autologous blood components. In this two-event model, plasma and lipids from stored but not fresh PRBCs and platelet concentrates resulted in TRALI (45,46). The histology of the rat lungs demonstrated the presence of firmly adherent PMNs in the pulmonary vasculature, PMN-mediated damage of the endothelium, and infiltration of alveoli with PMNs and fluid (45,46). These findings are identical to a those reported in a recent autopsy of fatal TRALI (45,47).

Among the transfusion-transmitted mediators that have been implicated in TRALI are antibodies directed against specific antigens on the surface of granulocytes, for example, antibodies to HNA-3a (5b), including those against HLA class I,

HLA class II, and granulocyte-specific antigens; soluble CD40 ligand (sCD40L), which accumulates during the storage of platelet concentrates; complement, which may be caused by the interaction of cells with tubing employed for collection; and biologically active lipids that accumulate during routine storage of all cellular components (44,45,48). Recent in vitro data have determined that both granulocyte-specific antibodies and sCD40L have the capacity to prime the PMN NADPH oxidase and cause PMN-mediated HMVEC damage in this vitro model of TRALI (48,49).

TRALI also occurs in neutropenic hosts, albeit rarely (21,22). These cases appear to defy both the two-event and antibody-mediated models, because fewer leukocytes are present to cause pulmonary injury. In these patients, platelet-mediated release of vascular endothelial growth factor (VEGF) may cause fenestration of the pulmonary endothelium and result in mild capillary leak identical to the mild pulmonary leak caused by VEGF in isolated, perfused rat lungs (50). Although preliminary results have implicated VEGF in a case of TRALI in a neutropenic patient, more data are required to confirm the role of this permeability and growth factor in TRALI (50).

OTHER EFFECTS OF TRANSFUSION ON ENDOTHELIUM

Storage of PRBCs for longer than 15 days yields many red cells that are stiff and unable to navigate the tortuous capillary beds in various organs (51,52). Investigators have postulated that these stored PRBCs may cause vaso-occlusion by adhering to the vascular endothelium. Consistent with this hypothesis, older (hence stiffer) RBCs demonstrate increased adherence to human umbilical vein ECs (HUVECs) or bovine ECs (51,53). In addition, older RBCs have reduced intracellular levels of 2,3-DPG and increased oxygen (O_2) affinity, which renders them less efficient in O_2 delivery. Transfusion of older RBCs may lead to increased mortality in cardiac bypass patients, microcirculatory hypoperfusion in the critically ill, acute chest syndrome in sickle cell anemia, and certain clinical manifestations in patients with hereditary stomatocytosis postsplenectomy, including dyspnea, chest pain, and abdominal pain (15,51,53–55). Recent in vitro data suggest that prestorage leukoreduction of PRBCs and treatment of ECs with nitric oxide (NO) may restore membrane fluidity and attenuate the widespread adherence of older RBCs to endothelium (56). However, further studies are required before advocating widespread administration of younger RBCs, prestorage leukoreduction, and/or NO treatment in these patient populations (56,57).

The effects of FFP on PMN–endothelial interactions have been investigated because FFP is commonly employed in the injured or surgical patient for the treatment of severe bleeding. Despite the potential for FFP to contain bioactive substances derived from leukocytes that may cause proinflammatory activation of the vascular endothelium, in vitro FFP "treatment" of HUVECs in cocultures with human PMNs actually decreased the interactions between PMNs and HUVECs (58). In contrast, albumin, which has been employed for decades for colloid resuscitation of patients, causes proinflammatory activation of endothelial monolayers and vascular endothelium in animal models, that results in increased surface expression of cellular adhesion molecules and PMN adhesion (59). These proinflammatory effects of albumin may be due to its role as a carrier of bioactive lipids and other proinflammatory agents, especially if the source is outdated human blood.

The development of hemoglobin substitutes as O_2 carriers has been hampered by their effects on the vascular endothelium, particularly their interference with NO- and endothelin-1-mediated vascular tone (60–63). In fact, the vasoconstrictive properties of one hemoglobin-based O_2 carrier has been cited as the reason for the unexpected mortality in a recent randomized control trial (64–67). Polymerized hemoglobin solutions appear not to cause significant vasoconstriction and, as opposed to other hemoglobin substitutes that cause a proinflammatory activation of ECs, appear to have an anti-inflammatory capacity (64,67). Further work is required to discern the effects of all of these compounds on human vascular endothelium in vivo.

Finally, although improved renal allograft survival was initially attributed to transfusions, it has become increasingly clear that the multiple transfusion of an allograft recipient may lead to rejection of the allograft (68,69). This transfusion-related rejection is not relegated to any one type of transplant and, in the case of corneal allograft rejection, has been related to endothelial rejection (68).

AVOIDANCE

Many of the effects of transfusion may be avoided by washing and resuspending the cells in saline (3,17). The proinflammatory potential of stored blood on PMNs and ECs is abrogated by this procedure (3) (C. Silliman, unpublished observations). Clinical data in cancer patients suggest that washing of all ABO-matched components imparts better survival, possibly due to eradication of soluble mediators or plasma proteins that may be etiologic in immunomodulation of the host (70). Plasma itself may potentially impart detrimental effects on the transfused host, especially if collection is through some of the newer machinery that causes extensive cellular exposure to plastic tubing, identical to cardiopulmonary bypass or dialysis equipment, resulting in complement activation and subsequent transfusion of activated complement into an ill patient (71). Further work is required to identify the real risk of adverse events due to FFP infusion in such patients. In addition, more restrictive transfusion practices have decreased the number of adverse events and improved survival in injured patients, despite increases in both patient age and injury severity score, two indicators of worse outcome (D. Ciesla, unpublished observations). Thus, if a transfusion is not clinically indicated, it should be avoided.

KEY POINTS

- Transfusion of cellular components and plasma is vital to restore O_2-carrying capacity and maintain hemostasis, but it is important to recognize that transfusions may have a number of deleterious effects, including the transmission of infectious agents, host immunomodulation, and proinflammatory activation of the innate immune system. The vascular endothelium plays an important role in all these processes, and strategies to decrease the effects of transfusion on host endothelium must be implemented as therapeutic options become available.
- More restrictive transfusion policies, the use of fresher or washed components, and transfusion of prestorage leukoreduced PRBCs are strategies that may decrease the effects of transfusion on the vascular endothelium, but experimentation in vitro and in vivo as well as multicenter clinical trials must be completed to possibly decrease the adverse events associated with transfusion.
- For example, the role of stored platelet transfusions, which may contain large amounts of VEGF, in the metastatic or recurrent potential of solid tumors has never been explored and may provide a key element in the supportive care of patients with malignancies.

Future Goals

- To emphasize that the clinician must be wary of using in vitro or animal data to make clinical decisions, because many of these experiments may have little to do with the clinical condition of human patients, despite the best attempts of the investigators to approximate human disease

REFERENCES

1 Dodd RY, Notari EP, Stramer SL. Current prevalence and incidence of infectious disease markers and estimated window-period risk in the American Red Cross blood donor population. *Transfusion.* 2002;42(8):975–979.
2 Kao KJ. Mechanisms and new approaches for the allogeneic blood transfusion-induced immunomodulatory effects. *Transfus Med Rev.* 2000;14(1):12–22.
3 Silliman CC, Moore EE, Johnson JL, et al. Transfusion of the injured patient: proceed with caution. *Shock.* 2004;21(4):291–299.
4 Fiebig EW, Busch MP. Emerging infections in transfusion medicine. *Clin Lab Med.* 2004;24(3):797–823, viii.
5 Bussolino F, Mitola S, Serini G, et al. Interactions between endothelial cells and HIV-1. *Int J Biochem Cell Biol.* 2001;33(4):371–390.
6 Grundy JE, Lawson KM, MacCormac LP, et al. Cytomegalovirus-infected endothelial cells recruit neutrophils by the secretion of C-X-C chemokines and transmit virus by direct neutrophil-endothelial cell contact and during neutrophil transendothelial migration. *J Infect Dis.* 1998;177(6):1465–1474.
7 Hippenstiel S, Suttorp N. Interaction of pathogens with the endothelium. *Thromb Haemost.* 2003;89(1):18–24.
8 Blajchman MA. Incidence and significance of the bacterial contamination of blood components. *Dev Biol (Basel).* 2002;108:59–67.
9 Beutler B. Innate immunity: an overview. *Mol Immunol.* 2004;40(12):845–859.
10 Bhakdi S, Grimminger F, Suttorp N, et al. Proteinaceous bacterial toxins and pathogenesis of sepsis syndrome and septic shock: the unknown connection. *Med Microbiol Immunol (Berl).* 1994;183(3):119–144.
11 *Technical Manual*, 15th ed. Bethesda: AABB, 2005.
12 Goodnough LT, Hewitt PE, Silliman CC. Transfusion medicine: Joint ASH and AABB educational session. *Hematology (Am Soc Hematol Educ Program)* 2004;457–472.
13 Simak J, Holada K, D'Agnillo F, et al. Cellular prion protein is expressed on endothelial cells and is released during apoptosis on membrane microparticles found in human plasma. *Transfusion.* 2002;42(3):334–342.
14 Opelz G, Terasaki PI. Improvement of kidney-graft survival with increased numbers of blood transfusions. *N Engl J Med.* 1978;299(15):799–803.
15 van de Watering LM, Hermans J, Houbiers JG, et al. Beneficial effects of leukocyte depletion of transfused blood on postoperative complications in patients undergoing cardiac surgery: a randomized clinical trial. *Circulation.* 1998;97(6):562–568.
16 Vamvakas EC, Carven JH. Transfusion and postoperative pneumonia in coronary artery bypass graft surgery: effect of the length of storage of transfused red cells. *Transfusion.* 1999;39(7):701–710.
17 Goldfinger D, Lowe C. Prevention of adverse reactions to blood transfusion by the administration of saline-washed red blood cells. *Transfusion.* 1981;21(3):277–280.
18 Witte MH, Borgs P, Way DL, et al. AIDS, alcohol, endothelium, and immunity. *Alcohol.* 1994;11(2):91–97.
19 Bernard GR, Artigas A, Brigham KL, et al. Report of the American-European Consensus conference on acute respiratory distress syndrome: definitions, mechanisms, relevant outcomes, and clinical trial coordination. Consensus Committee. *J Crit Care.* 1994;9(1):72–81.
20 Roffey P, Thangathurai D, Mikhail M, et al. TRALI and massive transfusion. *Resuscitation.* 2003;58(1):121.
21 Silliman CC, Paterson AJ, Dickey WO, et al. The association of biologically active lipids with the development of transfusion-related acute lung injury: a retrospective study. *Transfusion.* 1997;37(7):719–726.
22 Silliman CC, Ambruso DR, Boshkov LK. Transfusion-related acute lung injury (TRALI). *Blood.* 2004;105(6):2266–2273. Epub 2004 Nov 30.
23 Sauaia A, Moore FA, Moore EE, et al. Early predictors of postinjury multiple organ failure. *Arch Surg.* 1994;129(1):39–45.
24 Zallen G, Offner PJ, Moore EE, et al. Age of transfused blood is an independent risk factor for postinjury multiple organ failure. *Am J Surg.* 1999;178(6):570–572.
25 Silliman CC, Elzi DJ, Hiester AA. Plasma from stored platelets and red cells activate human pulmonary endothelium. *Transfusion.* 1998;38:S96S.

26 Silliman CC, Wyman TH. Stored packed red blood cells cause release of chemokines: interleukin-8 (IL-8), epithelial neutrophil activating peptide-78 (ENA-78), and growth-related oncogene-alpha from human pulmonary microvascular endothelial cells. *Blood*. 2001;98:A828.
27 Silliman CC, Elzi DJ, Hiester AA. Plasma from whole blood derived platelets but not apheresis platelets activates pulmonary endothelial cells. *Blood*. 1998;92:A563.
28 Holness L, Knippen MA, Simmons L, Lachenbruch PA. Fatalities caused by TRALI. *Transfus Med Rev*. 2004;18(3):184–188.
29 Barnard RD. Indiscriminate transfusion: a critique of case reports illustrating hypersensitivity reactions. *NY State J Med*. 1951; 51(20):2399–2402.
30 Popovsky MA, Moore SB. Diagnostic and pathogenetic considerations in transfusion-related acute lung injury. *Transfusion*. 1985;25(6):573–577.
31 Popovsky MA, Chaplin HC Jr., Moore SB. Transfusion-related acute lung injury: a neglected, serious complication of hemotherapy. *Transfusion*. 1992;32(6):589–592.
32 Silliman CC, Boshkov LK, Mehdizadehkashi Z, et al. Transfusion-related acute lung injury: epidemiology and a prospective analysis of etiologic factors. *Blood*. 2003;101(2):454–462.
33 Silliman CC, Ambruso DR, Boshkov LK. Transfusion-related acute lung injury. *Blood*. 2005;105(6):2266–2273.
34 Wallis JP. Transfusion-related acute lung injury (TRALI) – under-diagnosed and under-reported. *Br J Anaesth*. 2003;90(5): 573–576.
35 Bux J, Becker F, Seeger W, et al. Transfusion-related acute lung injury due to HLA-A2-specific antibodies in recipient and NB1-specific antibodies in donor blood. *Br J Haematol*. 1996;93(3): 707–713.
36 Seeger W, Schneider U, Kreusler B, et al. Reproduction of transfusion-related acute lung injury in an ex vivo lung model. *Blood*. 1990;76(7):1438–1444.
37 Bux J, Hardt O, Kohstall M, et al. Reproduction of granulocyte antibody-mediated TRALI in an ex-vivo rat lung model. *Blood*. 2003;102:A94.
38 Kopko PM, Popovsky MA, MacKenzie MR, et al. HLA class II antibodies in transfusion-related acute lung injury. *Transfusion*. 2001;41(10):1244–1248.
39 Kopko PM, Paglieroni TG, Popovsky MA, et al. TRALI: correlation of antigen-antibody and monocyte activation in donor-recipient pairs. *Transfusion*. 2003;43(2):177–184.
40 Riesbeck K, Billstrom A, Tordsson J, et al. Endothelial cells expressing an inflammatory phenotype are lysed by superantigen-targeted cytotoxic T cells. *Clin Diagn Lab Immunol*. 1998;5(5): 675–682.
41 Waldman WJ, Knight DA, Adams PW, et al. In vitro induction of endothelial HLA class II antigen expression by cytomegalovirus-activated $CD4^+$ T cells. *Transplantation*. 1993;56(6):1504–1512.
42 Kopko PM, Popovsky MA. Pulmonary injury from transfusion-related acute lung injury. *Clin Chest Med*. 2004;25(1):105–111.
43 Kopko PM, Marshall CS, MacKenzie MR, et al. Transfusion-related acute lung injury: report of a clinical look-back investigation. *JAMA*. 2002;287(15):1968–1971.
44 Wyman TH, Bjornsen AJ, Elzi DJ, et al. A two-insult in vitro model of PMN-mediated pulmonary endothelial damage: requirements for adherence and chemokine release. *Am J Physiol Cell Physiol*. 2002;283(6):C1592–C1603.
45 Silliman CC, Voelkel NF, Allard JD, et al. Plasma and lipids from stored packed red blood cells cause acute lung injury in an animal model. *J Clin Invest*. 1998;101(7):1458–1467.
46 Silliman CC, Bjornsen AJ, Wyman TH, et al. Plasma and lipids from stored platelets cause acute lung injury in an animal model. *Transfusion*. 2003;43(5):633–640.
47 Dry SM, Bechard KM, Milford EL, et al. The pathology of transfusion-related acute lung injury. *Am J Clin Pathol*. 1999;112 (2):216–221.
48 Blumberg N, Boshkov LK, Silliman CC, et al. CD40 ligand (CD154) as a cofactor in the development of transfusion-related acute lung injury (TRALI). *Blood*. 2004;104(11):A237–A238.
49 Curtis BR, Kelher M, McLaughlin N, et al. The two-event model of transfusion-related acute lung injury: antibodies to HNA-3a cause PMN cytotoxicity. *Blood*. 2004;104:A237.
50 Boshkov LK, Maloney J, Bieber S, Silliman CC. Two cases of TRALI from the same platelet unit: implications for pathophysiology and the role of PMNs and VEGF. *Blood*. 2000;96(11):A655.
51 Luk CS, Gray-Statchuk LA, Cepinkas G, Chin-Yee IH. WBC reduction reduces storage-associated RBC adhesion to human vascular endothelial cells under conditions of continuous flow in vitro. *Transfusion*. 2003;43(2):151–156.
52 Tinmouth A, Chin-Yee I. The clinical consequences of the red cell storage lesion. *Transfus Med Rev*. 2001;15(2):91–107.
53 Eichelbronner O, Sielenkamper A, Cepinskas G, et al. Endotoxin promotes adhesion of human erythrocytes to human vascular endothelial cells under conditions of flow. *Crit Care Med*. 2000;28(6):1865–1870.
54 Tissot Van Patot MC, MacKenzie S, Tucker A, Voelkel NF. Endotoxin-induced adhesion of red blood cells to pulmonary artery endothelial cells. *Am J Physiol*. 1996;270(1 Pt 1):L28–L36.
55 Smith BD, Segel GB. Abnormal erythrocyte endothelial adherence in hereditary stomatocytosis. *Blood*. 1997;89(9):3451–3456.
56 Space SL, Lane PA, Pickett CK, Weil JV. Nitric oxide attenuates normal and sickle red blood cell adherence to pulmonary endothelium. *Am J Hematol*. 2000;63(4):200–204.
57 Weiner DL, Hibberd PL, Betit P, et al. Preliminary assessment of inhaled nitric oxide for acute vaso-occlusive crisis in pediatric patients with sickle cell disease. *JAMA*. 2003;289(9):1136–1142.
58 Nohe B, Kiefer RT, Ploppa A, et al. The effects of fresh frozen plasma on neutrophil-endothelial interactions. *Anesth Analg*. 2003;97(1):216–221.
59 Isbister JP, Fisher MM. Adverse effects of plasma volume expanders. *Anaesth Intensive Care*. 1980;8(2):145–151.
60 Abassi Z, Kotob S, Pieruzzi F, et al. Effects of polymerization on the hypertensive action of diaspirin cross-linked hemoglobin in rats. *J Lab Clin Med*. 1997;129(6):603–610.
61 Faivre-Fiorina B, Caron A, Fassot C, et al. Presence of hemoglobin inside aortic endothelial cells after cell-free hemoglobin administration in guinea pigs. *Am J Physiol*. 1999;276(2 Pt 2): H766–H770.
62 Gould SA, Moss GS. Clinical development of human polymerized hemoglobin as a blood substitute. *World J Surg*. 1996;20(9): 1200–1207.
63 Kim HW, Tai J, Greenburg AG. Alpha adrenergic activation and hemoglobin mediated contraction in the isolated rat thoracic aorta. *Artif Cells Blood Substit Immobil Biotechnol*. 2001;29(5): 367–380.
64 Cheng AM, Moore EE, Johnson JL, et al. Polymerized hemoglobin induces heme oxygenase-1 protein expression and inhibits intercellular adhesion molecule-1 protein expression in

human lung microvascular endothelial cells. *J Am Coll Surg.* 2005;201(4):579–584.

65 Sloan EP, Koenigsberg M, Gens D, et al. Diaspirin cross-linked hemoglobin (DCLHb) in the treatment of severe traumatic hemorrhagic shock: a randomized controlled efficacy trial. *JAMA.* 1999;282(19):1857–1864.

66 Sloan EP, Koenigsberg M, Brunett PH, et al. Post hoc mortality analysis of the efficacy trial of diaspirin cross-linked hemoglobin in the treatment of severe traumatic hemorrhagic shock. *J Trauma.* 2002;52(5):887–895.

67 Toussaint-Hacquard M, Devaux Y, Longrois D, et al. Biological response of human aortic endothelial cells exposed to acellular hemoglobin solutions developed as potential blood substitutes. *Life Sci.* 2003;72(10):1143–1157.

68 Hwang DG, Kramer SG. Corneal allograft rejection after multiple blood transfusions. *Am J Ophthalmol.* 1993;116(4):451–455.

69 Kupiec-Weglinski JW. Graft rejection in sensitized recipients. *Ann Transplant.* 1996;1(1):34–40.

70 Blumberg N, Heal JM, Rowe JM. A randomized trial of washed red blood cell and platelet transfusions in adult acute leukemia [ISRCTN76536440]. *BMC Blood Disord.* 2004;4(1):6.

71 Ambruso DR, Giclas P, Silliman CC, Kelher M, Geier S. Complement activation associated with transfusion-related acute lung injury. *Blood.* 2004;104:A971–A972.

167

The Role of Endothelium in Erectile Function and Dysfunction

Muammer Kendirci and Wayne J.G. Hellstrom

Tulane University Health Sciences Center, New Orleans, Louisiana

Penile erection is a neurovascular event that depends on neural integrity, a functional vascular system, and healthy cavernosal tissues (1). Physiological erectile function involves three synergistic and simultaneous processes: (a) a neurogenically mediated increase in penile arterial inflow, (b) relaxation of cavernosal smooth muscle, and (c) restriction of venous outflow from the corpora cavernosa. The corpus cavernosum of the penis is composed of a meshwork of interconnected smooth muscle cells lined by vascular endothelium. In addition, endothelial cells (ECs) and underlying smooth muscle line the small-resistance helicine arteries that supply blood to the corpora cavernosa during penile tumescence. Structural and functional alterations in the vascular tree of the corpus cavernosum or impairment of any combination of neurovascular processes can result in erectile dysfunction (ED). ED has traditionally been classified as psychogenic, organic, or a combination of these two entities. More recent data show that more than 80% of ED cases have an organic basis, with vascular disease being the most common etiology (2). Although ED is a natural consequence of aging, its severity is directly related to the number and degree of vascular risk factors, such as hypertension, cigarette smoking, atherosclerosis, hypercholesterolemia, and diabetes mellitus (3) (Table 167-1). Hence, endothelial dysfunction in the penile vascular bed can lead to ED (4).

ED is defined as the consistent inability to attain or maintain an erection sufficient for satisfactory sexual intercourse (5). The Massachusetts Male Aging Study (MMAS), a substantial epidemiological survey that quantified the prevalence of ED in a noninstitutionalized population of men in the Boston suburbs (3), revealed that 52% of 1,290 men aged 40 to 70 years had some degree of ED; with almost 10% exhibiting a total absence of erectile function. It has been estimated that 25 to 30 million men in the United States have partial to complete ED (5). Extrapolation of this data estimates that the worldwide incidence of ED will increase from 152 million men in 1995 to 322 million men by the year 2025 (6). Ongoing calculations from the participants in the MMAS reveal the overall incidence of ED after an average follow-up of 8.8 years to increase to 26 cases per 1,000 man-years (7).

VASCULAR ENDOTHELIUM AND PENILE ERECTION

The vascular endothelium of the penis plays a pivotal role in modulating vascular tone and blood flow into the penis in response to humoral, neural, and mechanical stimuli. The endothelium releases various factors that affect the contractile and relaxatory activity of the underlying vascular smooth muscle. In addition, physical hemodynamic changes caused by alterations in penile blood flow and shear stress release various mediators that modulate the underlying smooth muscle tone. In endothelial dysfunction, the regulatory role of the endothelium is hindered, resulting in decreased responsiveness to vasodilatory mediators and/or increased sensitivity to various vasoconstricting agents. The term *endothelial dysfunction* implies a decrease in endothelium-dependent corpora cavernosal smooth muscle relaxation, for the most part secondary to increased destruction or total loss of nitric oxide (NO) bioactivity in the vascular tree.

NO serves many vital physiological functions, including regulation of vascular tone, immunomodulation, neurotransmission and, for our purposes, penile erection (8). Furthermore, NO inhibits smooth muscle cell proliferation and collagen synthesis, and can have a direct toxic effect on these tissues. The main or constitutive NO synthase (NOS) isoforms (endothelial NOS [eNOS] and neuronal NOS [nNOS]) are involved in the initiation and maintenance of penile erection and are coupled to calcium (Ca^{2+}) and calmodulin (9). NO involved in signaling events that regulate neurotransmission and penile vascular tone is produced by (a) eNOS from ECs lining the cavernosal smooth muscle cells; (b) agonist-induced activation via acetylcholine released from cholinergic nerves, and (c) nNOS activity initiated by nonadrenergic noncholinergic (NANC) neurons. NO diffuses

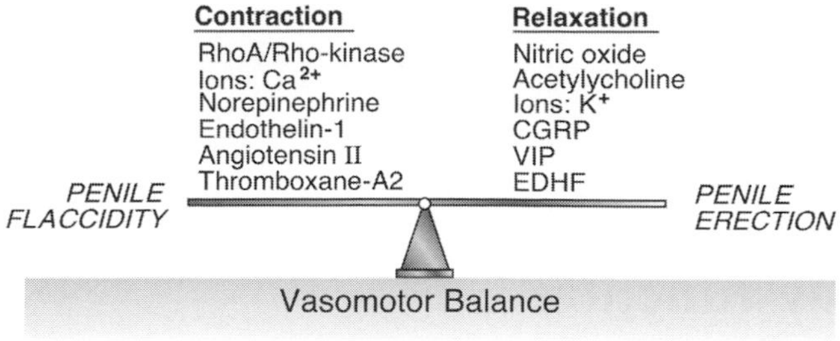

Figure 167.1. Endothelial, neural, and humoral factors mediating vasoconstriction and vasodilation in penile erection and flaccidity.

into the adjacent smooth muscle cells and stimulates the enzyme guanylate cyclase (GC). GC increases the synthesis of cyclic guanosine monophosphate (cGMP), which in turn, through a cascade of events, decreases intracellular Ca^{2+} concentrations and causes smooth muscle relaxation and penile erection.

Table 167-1: Common Risk Factors for Erectile Dysfunction

Vascular diseases
Coronary artery disease
Peripheral vascular disease
Atherosclerosis
Hypercholesterolemia
Hypertension
Smoking
Diabetes mellitus
Aging
Neurological disorders
Hormonal disorders
Psychological conditions
Radical prostatectomy
Oxidative stress
Drug-induced
Lifestyle issues

PHYSIOLOGY AND PHARMACOLOGY OF PENILE ERECTION

The balance between the contractile (α-adrenergic, endothelin (ET), angiotensin (Ang), RhoA/Rho-kinase, thromboxane A_2) and vasodilatory second-messenger systems (GC/cGMP and adenylate cyclase/cyclic guanosine monophosphate [AC/cAMP]) determines the overall tone of the corpora cavernosal smooth muscle of the penis (10) (Figures 167.1 and 167.2). This balance is controlled by both central and peripheral mechanisms and involves a plethora of neurotransmitters acting through various signal transduction pathways. In the corpora cavernosa, the vascular endothelium and cavernosal arteries are a source of both vasorelaxing (NO, prostacyclin [PGI_2], endothelium-derived hyperpolarizing factor [EDHF]) and vasoconstricting (Ang-II, ET-1, and Rho-kinase) factors.

The vasodilators prostaglandin E (PGE) and PGI_2 are produced by ECs in the vasculature. Prostaglandin synthesis is activated by NO through shear stress from blood flow in the penis. Activation of specific PGE receptors in the penis increases intracellular concentrations of cAMP via activation of AC, causing a reduction in intracellular Ca^{2+} and cavernosal smooth muscle relaxation (11). Both PGE_2 and its derivative PGE_1 are potent vasorelaxing agents in human corpora cavernosa smooth muscle. PGE_1 has been employed clinically because of its highly efficacious local activity in the medical treatment of ED.

EDHF is released by endothelium-dependent agonists (acetylcholine or bradykinin) or by shear stress that triggers the synthesis of a cytochrome P450 metabolite and acts by hyperpolarizing the underlying cavernosal smooth muscle (12,13). Recent studies have shown that EDHF mediates relaxation of human penile resistance arteries, which are normally resistant to NOS and cyclooxygenase (COX) inhibition (14). The authors postulate that EDHF plays more of a precise role in the endothelium-dependent relaxation of the penile vascular bed. Studies from our laboratory have determined that the pharmacological inhibitor of cytochrome P450 2C, sulfaphenazole, can attenuate the cavernosal nerve–mediated erectile responses in the rat, suggesting that the cytochrome P450 metabolite may mediate an EDHF-dependent smooth muscle effect in the penis (15).

ET-1, a member of the ET family, is generated by the vascular endothelium and is a very potent vasoconstrictor of the penile vasculature (16). Two ET receptors modulate the effects of ET-1, namely ET_A, located on the underlying smooth muscle, and ET_B, located on both the smooth muscle and vascular endothelium. ET_A receptors mediate contraction and promote smooth muscle growth, whereas ET_B receptors

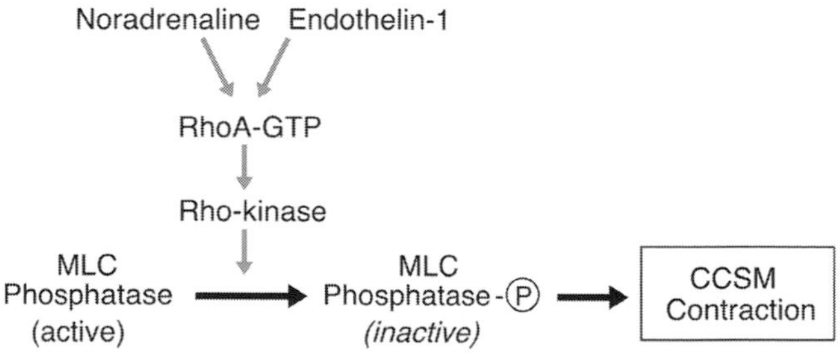

Figure 167.2. Rho-kinase pathway. The detumescent state of the corpus cavernosum smooth muscle (CCSM) is considered to be mediated by release of contractile neurotransmitters or modulators including noradrenaline and ET-1. GTP-RhoA activates Rho-kinase and inhibits MLC phosphatase, increasing MLC20 phosphorylation by basal level activity of MLC kinase. The resulting myosin phosphorylation and subsequent CCSM contraction occur without a change in sarcoplasmic Ca^{2+} concentration. (Adapted with permission from Andersson KE. Erectile physiological and pathophysiological pathways involved in erectile dysfunction. *J Urol.* 2003;170:S6–13.)

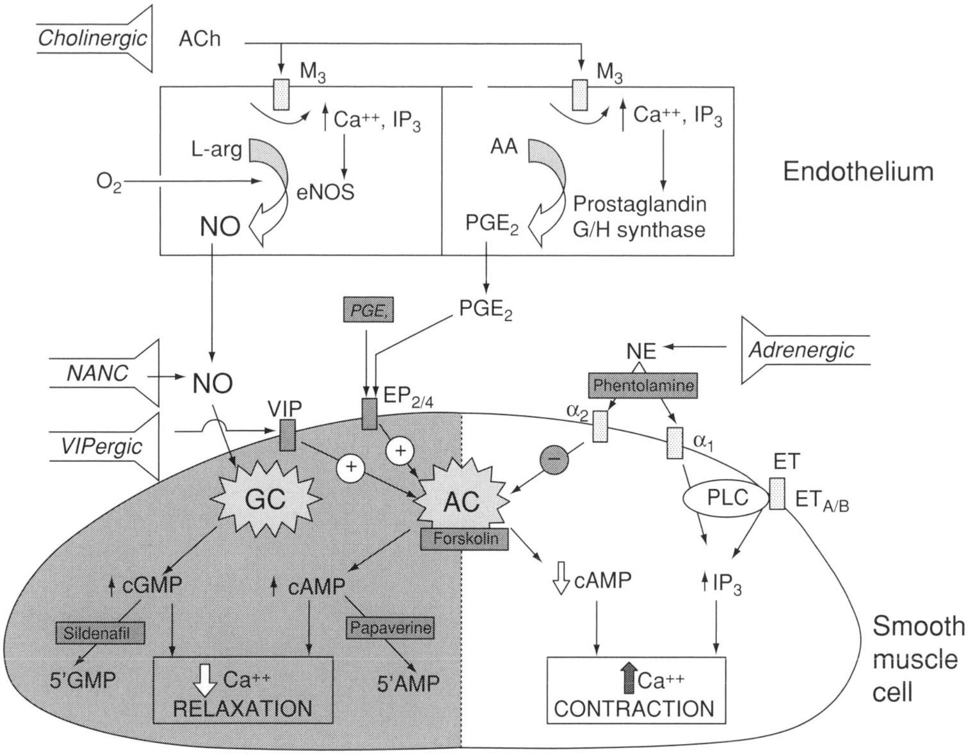

Figure 167.3. The interaction between vascular endothelium and cavernosal smooth muscle cells. (Adapted with permission from Moreland RB, Hsieh G, Nakane M, Brioni JD. The biochemical and neurologic basis for the treatment of male erectile dysfunction. *J Pharmacol Exp Ther.* 2001;296:225–234.)

on the endothelium promote NO- and prostacyclin-mediated vasorelaxation and contractions of the smooth muscle. In vitro cavernosal tissue strip experiments reveal that ET-1 functions as a vasodilator at low doses via ET_B receptor activation and as a vasoconstrictor via ET_A receptor activation at high doses (16). Despite the presence of ET receptors in the corpora cavernosa, current animal and clinical data do not support a central role for ET-1 in regulating the erectile response (17,18). However, studies in diabetic animal and human corpus cavernosal strips have demonstrated upregulation of ET_A receptors and downregulation of ET_B receptors, suggesting that ET-1 may be more involved in the pathophysiology of diabetic ED (19,20).

Ang-II, another potent vasoconstrictor, has been hypothesized to have a local functional role in the regulation of the vascular tissue in the penis (21). ECs in the corpora cavernosa appear to contain a local Ang-II paracrine system that modulates vascular tone and sustains cavernosal smooth muscle cell contraction (22). In vitro studies have confirmed that Ang-II causes a dose-dependent contraction of cavernosal tissues (23). In human studies, men with organic ED have been observed to have elevated Ang-II levels, supporting a role for Ang-II in the pathogenesis of organic ED (24).

Contraction of cavernosal smooth muscle is primarily mediated by Ca^{2+}-sensitization (Figure 167.3). This mechanism involves either Ca^{2+}-independent promotion of myosin light chain kinase (MLCK) or attenuation of myosin light chain phosphatase (MLCP) activity (25). A principal regulator of MLCP is the serine/threonine kinase, Rho-kinase. RhoA, a member of the Ras low-molecular-weight GTP-binding proteins, mediates agonist-induced activation of Rho kinase and contraction of smooth muscle cells (26). Norepinephrine (NE) and ET-1 both stimulate the GTPase RhoA in vascular smooth muscle cells via a RhoA/Rho-kinase-dependent pathway (Figure 167.2). The role of the RhoA/Rho-kinase pathway in the regulation of cavernous smooth muscle tone in both human ECs and human corpora cavernosa smooth muscle cells grown in culture has recently been described (27,28).

As a corollary, direct administration of the Rho-kinase inhibitor Y-27632 into the corpora cavernosa causes a dose-dependent increase in intracavernosal pressure and erection in rats (29). Treatment with Y-27632 also potentiates NO-mediated increase in erectile function (30). Moreover, adeno-associated viral gene transfer of the RhoA mutant to the rat corpora cavernosa enhanced erectile function, suggesting that inhibition of RhoA/Rho-kinase expression in the penis can augment cavernosal smooth muscle relaxation and erectile function (31). Given that NO partly mediates the erectile response via inhibition of RhoA/Rho-kinase Ca^{2+}-sensitizing pathway, it has recently been postulated that endogenous NO-mediated vasodilation may occur through the inhibition of Rho-kinase vasoconstrictor activity. Recent evidence suggests that the RhoA/Rho-kinase signal transduction pathway is an important signal mediator of penile vascular EC function

(32). Further exploration of this exciting area of study is anticipated.

DIABETES AND PENILE ENDOTHELIAL DYSFUNCTION

Diabetes mellitus is associated with a high prevalence of ED, because at least 50% of diabetic men are afflicted (33). The major underlying factors contributing to diabetic ED in cavernosal tissues are a decrease in the number of nitrergic NOS-containing nerve fibers, reduction in nNOS and eNOS activity, and impaired endothelial- and neurogenic-mediated smooth muscle relaxation (34,35). Diabetic animal studies have documented significant reductions in cavernosal eNOS and nNOS protein and gene expression, and decreased cavernosal cGMP levels in the penile tissues (36–38). Furthermore, endothelium-independent cavernosal smooth muscle relaxation is impaired in diabetic ED (39). These and other observations reveal that diabetes attenuates endothelial- and neurogenic-NO neurotransmission and decreases soluble GC, cGMP, and protein kinase G (PKG) (39). Diabetic ED is the result of a dysfunctional NO/cGMP system and reduced NO production, most likely because of decreased eNOS and nNOS in the penile vasculature.

In diabetic vascular dysfunction, the RhoA/Rho-kinase pathway plays an important role in suppressing eNOS gene expression and enzyme activity in ECs by inhibiting Akt (40). In alloxan-induced diabetic rabbits, researchers have demonstrated that Rho-kinase levels are increased in corpora cavernosal smooth muscle, showing that cavernosal smooth muscle tone is modulated in part via Rho-kinase sensitization of the endothelin-mediated contractile pathway. Hence, Rho-kinase may be a key component of the molecular mechanism in the origin of diabetes-induced ED (20).

In streptozotocin-induced diabetic rats, our laboratory also noted an increase in RhoA and Rho-kinase levels in the corpora cavernosa and a reduction in eNOS protein and enzyme activity (32). This increased expression of RhoA/Rho-kinase was countered by the Rho-kinase inhibitor, Y-27632, which caused an enhanced erectile response. Adeno-associated viral vector encoding for a dominant negative RhoA mutant was directly injected into the corpora cavernosa to reduce RhoA-Rho-kinase expression (32). After gene transfer, the RhoA/Rho-kinase expression was significantly reduced, eNOS protein and activity were increased, and erectile responses to cavernosal nerve stimulation were restored. These data reveal that there is an upregulation of cavernosal Rho-kinase expression when endothelial-derived NO production in the penis is reduced.

An interesting concept in deciphering the actions of these pathways is the study of arginase. Arginase competes with eNOS for the substrate L-arginine in ECs throughout the vascular system. Thus, increased expression of arginase in diabetic corpora cavernosa may diminish endothelial-derived NO biosynthesis by competing for the substrate L-arginine. Studies have demonstrated an increase in arginase activity, arginase II protein, and gene expression in diabetic human corpora cavernosa compared with nondiabetic cavernosal tissue (41). Furthermore, diabetic cavernosal tissues exhibit a significantly lower conversion of L-arginine to L-citrulline, suggesting that the constitutive Ca^{2+}-dependent eNOS/NO pathway is impaired in diabetes (41). In vitro studies in diabetic rabbits treated with long-term oral L-arginine have reported increased endothelium-dependent cavernosal smooth muscle relaxation. In theory, L-arginine transport or availability to interact with eNOS in the endothelium of diabetics is reduced, thereby contributing to endothelial dysfunction (42). Preliminary clinical observations have not supported oral ingestion of L-arginine to ameliorate ED.

Another interesting observation is the presence of advanced glycation end-products (AGEs) in diabetes (43). A common consequence of AGE formation is the pathologic cross-linking of collagen, leading to vascular thickening with a loss of elasticity, endothelial dysfunction, and subsequently, atherosclerosis of the vascular tree. AGE accumulation in endothelial and cavernosal smooth muscle cells have been documented to reduce eNOS levels via intracellular glycation and alteration of eNOS activity (44). These data suggest that AGE formation may in some manner be another pathway that contributes to the pathogenesis of diabetic endothelial dysfunction and ED. Our laboratory has demonstrated that aminoguanidine, an AGE inhibitor, halts the time-dependent progression of impaired erectile responses in the streptozotocin-induced diabetic rat (45,46). Aminoguanidine may prevent the diabetes-induced changes in the connective tissue composition of the microvascular wall of the arterioles supplying the penis. Another interesting observation is that delayed administration of ALT-711, an AGE cross-link breaker, improved erectile function and decreased AGE levels, suggesting that AGE cross-link breakers may have a significant therapeutic potential for reversing the consequences of established diabetic ED (47).

AGING AND PENILE ENDOTHELIAL DYSFUNCTION

Aging is known to be associated with diminished vascular endothelial function and reduced erectile function. Various pathophysiological mechanisms have been postulated for age-related ED, including decreased NANC fibers in the corpora cavernosa, reduction in constitutive NOS activity, altered NO bioavailability, impaired endothelium-dependent smooth muscle relaxation, and increased degradation or scavenging of NO (48–51). Although normal or elevated levels of eNOS protein have been reported in the corpora cavernosa of aged rats by some researchers, eNOS activity and endothelial-derived NO biosynthesis generally are reduced with aging (48). This may, in part, be explained by increased production of reactive oxygen species (ROS) and reduced enzyme activity of eNOS in the aged penile vasculature.

Researchers have demonstrated a significant increase of eNOS protein and mRNA, constitutive NOS activity, endothelial-derived NO, and cavernosal cGMP levels in the penis of aged rats after transfection with adenovirus encoded with the eNOS isoform (48,52). There was a corresponding increase in erectile function as measured by cavernosal nerve stimulation and after intracavernous acetylcholine administration. Hence, age-related diminished endothelial function (reduced NO biosynthesis) can be reversed in the penile vascular endothelium using gene-based therapy (50).

As mentioned, arginase has a theoretical role in regulating eNOS activity in the penile vascular endothelium and is considered a possible contributing factor in age-related endothelial dysfunction and ED (53). Decreased constitutive NOS activity and NO bioavailability in the aged penis in animals may be due to a lack of L-arginine substrate available to produce eNOS. Studies have shown that there is an increase in arginase activity and arginase I mRNA and protein expression in the corpora cavernosa of aged mice (15). Recent experiments have shown that a combination of an arginase inhibitor and adeno-associated virus encoding an antisense sequence to arginase I has restored eNOS activity and improved endothelium-dependent erectile responses to acetylcholine and cavernosal nerve stimulation (15).

HYPERCHOLESTEROLEMIA AND PENILE ENDOTHELIAL DYSFUNCTION

Impaired endothelium-dependent relaxation in the vascular beds of men with hypercholesterolemia has been well established (54,55). The association between hypercholesterolemia and ED was originally attributed to atherosclerosis in the hypogastric-cavernosal arterial vascular bed, thereby causing a reduction in penile blood flow (56). Experimental studies in rabbits revealed that hypercholesterolemia produces a marked reduction in cavernosal smooth muscle cells, ECs, and elastic fibers, and increases in collagen III and IV in the cavernosal tissues (57). Histological evaluation of rabbits placed on high-cholesterol diets demonstrated significant cavernosal smooth muscle cell degeneration with a loss of intercellular contacts (58).

Gholami and colleagues revealed that penises of hypercholesterolemic rats had lower nerve content, fewer ECs, and higher smooth muscle content than those of rats with normal cholesterol levels (59). In this study, rats placed on high-cholesterol diets demonstrated hypermyelination, severe atrophy of axons, a decrease in the number and size of nonmyelinated axons, disarray of the smooth muscle cells with scant myofilaments and foamy cytoplasm, and denuded endothelial lining of the sinusoids covered by numerous platelets, suggesting that a high-fat diet caused ED via both neurogenic and vascular alterations. Vascular endothelial growth factor (VEGF) has been used to upregulate eNOS in the corpus cavernosum and increase the production of NO (60). Intravenous and intracavernosal administration of VEGF in hypercholesterolemic rats increases EC content and function, and improves endothelium-dependent smooth muscle relaxation in the corpus cavernosum. These findings have led researchers to postulate that VEGF may in some manner protect the corporal endothelium from hypercholesterolemia-induced injury, thus preserving endothelium-dependent corporal smooth muscle relaxation (59,61,62).

OXIDATIVE STRESS AND PENILE ENDOTHELIAL DYSFUNCTION

ECs produce ROS in response to shear stress, endothelium-derived agonists, including acetylcholine and bradykinin, and also in various vascular disease states. Potential sources of ROS in ECs include nicotinamide adenine dinucleotide phosphate (NADPH) oxidase (generates superoxide anion), lipooxygenase, cyclooxygenase, peroxidases, cytochrome P450s, xanthine oxidase, and eNOS (63). The reaction of superoxide anions and NO in the vascular endothelium or smooth muscle cells triggers the formation of the highly toxic molecule, peroxynitrite (64). Due to its toxic effects, peroxynitrite can cause direct tissue injury, alterations in vascular tone, oxidation of vascular proteins and lipids, and overall organ dysfunction (65). The antioxidants superoxide dismutase (SOD), catalase, glutathione peroxidase and reductase play a pivotal role at the cellular level in protecting against ROS (66). Among the three types of SOD isoforms in the human body, copper- and zinc-containing SOD (CuZn-SOD) and extracellular-SOD (EC-SOD) have been identified in the penis, predominantly in the endothelial and cavernosal smooth muscle cells (67). Increased levels of superoxide anions in the endothelium and cavernosal smooth muscle cells contribute to ED by causing endothelial dysfunction and reducing cavernosal NO biosynthesis. Apart from reducing NO biosynthesis, superoxide anions cause Ca^{2+} mobilization, which results in reduced Ca^{2+} levels in the cavernosal ECs. Overall, increased oxidative stress and superoxide anion production alter the penile vasculature and impairs endothelial-derived NO in the erectile tissues, resulting in ED (15,68).

Oxidative stress is quite prominent in certain chronic disease states, including diabetes, hypercholesterolemia, and aging, and is associated with significant changes in the endothelium and smooth muscle cells in the penis (69,70). As noted, superoxide anions react with NO and form peroxynitrite. In aging, increased production of peroxynitrite accelerates the degeneration of nerves and ECs involved in the erectile process (67). An imbalance in superoxide anion generation and inactivation in the penile vasculature causes impaired endothelium-dependent smooth muscle relaxation and ED (67). In aged rats experimentally transfected with adeno cytomegalovirus (CMV)-EC-SOD, there was a significant increase in EC-SOD mRNA and protein expression and a significant reduction of superoxide anions. Moreover, EC-SOD gene therapy increased cGMP levels in the corpus cavernosum and was found to enhance the in vivo erectile response to cavernosal nerve stimulation. These observations implicate EC-SOD as beneficial in limiting superoxide anion production and preventing some of the outcomes of age-related ED (67).

The effects of diabetes on endothelium-dependent and NANC-mediated cavernosal smooth muscle relaxation are well established in diabetic animal models and in vitro studies (34,37,71,72). Oxidative stress has been suggested to also play an important role in diabetic endothelial dysfunction. An in vitro study by Khan and colleagues demonstrated that SOD treatment restored endothelial- and NANC-mediated cavernosal smooth muscle relaxation, inferring the significance of the production of superoxide anions in diabetic endothelial dysfunction (73). Another related study in streptozotocin-induced diabetic rats, malondialdehyde, an oxidative stress marker, and superoxide anions were increased in the corpus cavernosum (74). Adenoviral gene transfer of EC-SOD to the diabetic rat penis reduced cavernosal malondialdehyde and superoxide anion levels and restored cavernosal nerve-mediated erectile function in vivo (75).

Oxidative stress is similarly involved in the endothelial dysfunction observed in hypercholesterolemia. In hypercholesterolemic rabbits, endothelium-dependent corpus cavernosal smooth muscle relaxation is impaired and superoxide levels are correspondingly elevated (76). In response to this oxidative insult, a significant increase in cavernosal CuZn-SOD and manganese SOD (MnSOD) levels occurs in order to scavenge the increased superoxide anions. Despite the elevated SOD levels and increased scavenging activities, there remain significant impairments in endothelium-dependent smooth muscle relaxation. Other studies have documented elevated low-density lipoprotein (LDL) peroxidation and the early development of atherosclerosis (77). An oxidative stress-derived pathogen, ox-LDL, formed by superoxide anions and peroxynitrite, also has been shown to impair endothelium-dependent relaxation in the penis in hypercholesterolemia (78).

A recent study assessed the effect of sildenafil on SOD formation and $p47^{phox}$ (the active subunit of NADPH oxidase) expression in cultured corpus cavernosal smooth muscle cells (79). Cavernosal smooth muscle cells derived from rabbit penis were incubated with U46619 (thromboxane A_2 analogue) with and without sildenafil. This study demonstrated that sildenafil inhibited the SOD formation effect of U46619. The investigators concluded that sildenafil is a potent inhibitor of SOD formation in cavernosal smooth muscle cells, suggesting the inhibitory role of sildenafil in therapy of ED related to intrapenile oxidative stress.

CONCLUSION

Endothelial and erectile dysfunctions have many origins, yet share similar vascular risk factors and pathophysiological mechanisms. A healthy vascular endothelium plays an important role in normal erectile function. Vascular disease states, such as diabetes, hypertension, atherosclerosis, advanced age, hypercholesterolemia, and coronary artery disease, are associated with impaired endothelial function. Alterations in the vascular endothelium of the penis decrease the responsiveness to vasodilatory mediators and increase the sensitivity to vasoconstrictors, resulting in ED. Vascular risk factors in these disease states also cause decreased NO bioavailability and impaired eNOS enzyme activity, which is crucial for cavernosal smooth muscle relaxation and penile erection. Because normal penile erection necessitates coordinated arterial endothelium-dependent vasorelaxation and cavernosal endothelium-dependent smooth muscle relaxation, insights into the pathophysiological mechanisms involved in endothelial dysfunction and their relation to various vascular comorbidities will lead researchers and clinicians to innovations in prevention and EC-based strategies to cure ED.

KEY POINTS

- Approximately 52% of men aged 40 to 70 years have some degree of ED. Normal penile erection requires a coordinated arterial endothelium-dependent vasorelaxation and cavernosal endothelium-dependent smooth muscle relaxation. Endothelial and erectile dysfunctions have a multifactorial origin and share similar risk factors and pathophysiologies. Vascular risk factors, such as diabetes, hypertension, atherosclerosis, advanced age, hypercholesterolemia, and coronary artery disease, are associated with impaired vascular endothelial and erectile function. As a corollary, the onset of ED should direct clinicians to search for underlying vascular risk factors. Erectile function mirrors endothelial and vascular health.
- Penile erection is a neurovascular event that relies on neural integrity, a functional vascular endothelial system, and healthy cavernosal smooth muscle.
- A healthy vascular endothelium plays an important role in erectile function. Any alteration in the vascular endothelium of the penis may result in a decreased responsiveness to vasodilator mediators or an increased sensitivity to vasoconstrictors, with the net effect being ED.
- Common consequences of vascular risk factors on the endothelium are impairment in NO bioavailability and decreased eNOS enzyme activity.

Future Goals

- To advance our understanding of the pathophysiology of endothelial dysfunction, thus leading to novel prevention strategies and EC-based pharmacological and gene-related therapies to correct ED

- To promulgate to the public those lifestyle changes that will preserve and/or improve vascular endothelial and erectile function

REFERENCES

1 Giuliano FA, Rampin O, Benoit G, Jardin A. Neural control of penile erection. *Urol Clin North Am.* 1995;22(4):747–766.

2 Meuleman EJ. Prevalence of erectile dysfunction: need for treatment? *Int J Impot Res.* 2002;14(Suppl 1):S22–S28.

3 Feldman HA, Goldstein I, Hatzichristou DG, et al. Impotence and its medical and psychosocial correlates: results of the Massachusetts Male Aging Study. *J Urol.* 1994;151(1):54–61.

4 Maas R, Schwedhelm E, Albsmeier J, Boger RH. The pathophysiology of erectile dysfunction related to endothelial dysfunction and mediators of vascular function. *Vasc Med.* 2002;7(3):213–225.

5 NIH Consensus Conference. Impotence. NIH Consensus Development Panel on Impotence. *JAMA.* 1993;270(1):83–90.

6 Ayta IA, McKinlay JB, Krane RJ. The likely worldwide increase in erectile dysfunction between 1995 and 2025 and some possible policy consequences. *BJU Int.* 1999;84(1):50–56.

7 Johannes CB, Araujo AB, Feldman HA, et al. Incidence of erectile dysfunction in men 40 to 69 years old: longitudinal results from the Massachusetts male aging study. *J Urol.* 2000;163(2):460–463.

8 Ignarro LJ, Cirino G, Casini A, Napoli C. Nitric oxide as a signaling molecule in the vascular system: an overview. *J Cardiovasc Pharmacol.* 1999;34(6):879–886.

9 Ignarro LJ, Bush PA, Buga GM, et al. Nitric oxide and cyclic GMP formation upon electrical field stimulation cause relaxation of corpus cavernosum smooth muscle. *Biochem Biophys Res Commun.* 1990;170(2):843–850.

10 Andersson KE. Neurophysiology/pharmacology of erection. *Int J Impot Res.* 2001;13(Suppl 3):S8–S17.

11 Moreland RB, Albadawi H, Bratton C, et al. O2-dependent prostanoid synthesis activates functional PGE receptors on corpus cavernosum smooth muscle. *Am J Physiol Heart Circ Physiol.* 2001;281(2):H552–H558.

12 Busse R, Edwards G, Feletou M, et al. EDHF: bringing the concepts together. *Trends Pharmacol Sci.* 2002;23(8):374–380.

13 Fleming I. Cytochrome p450 and vascular homeostasis. *Circ Res.* 2001;89(9):753–762.

14 Angulo J, Cuevas P, Fernandez A, et al. Calcium dobesilate potentiates endothelium-derived hyperpolarizing factor-mediated relaxation of human penile resistance arteries. *Br J Pharmacol.* 2003;139(4):854–862.

15 Bivalacqua TJ, Usta MF, Champion HC, et al. Endothelial dysfunction in erectile dysfunction: role of the endothelium in erectile physiology and disease. *J Androl.* 2003;24(Suppl 6):S17–S37.

16 Christ GJ, Lerner SE, Kim DC, Melman A. Endothelin-1 as a putative modulator of erectile dysfunction: I. Characteristics of contraction of isolated corporal tissue strips. *J Urol.* 1995;153(6):1998–2003.

17 Becker AJ, Uckert S, Stief CG, et al. Systemic and cavernous plasma levels of endothelin 1 in healthy males during different functional conditions of the penis. *World J Urol.* 2000;18(3):227–231.

18 Dai Y, Pollock DM, Lewis RL, et al. Receptor-specific influence of endothelin-1 in the erectile response of the rat. *Am J Physiol Regul Integr Comp Physiol.* 2000;279(1):R25–R30.

19 Sullivan ME, Dashwood MR, Thompson CS, et al. Alterations in endothelin B receptor sites in cavernosal tissue of diabetic rabbits: potential relevance to the pathogenesis of erectile dysfunction. *J Urol.* 1997;158(5):1966–1972.

20 Chang S, Hypolite JA, Changolkar A, et al. Increased contractility of diabetic rabbit corpora smooth muscle in response to endothelin is mediated via Rho-kinase beta. *Int J Impot Res.* 2003;15(1):53–62.

21 Kifor I, Williams GH, Vickers MA, et al. Tissue angiotensin II as a modulator of erectile function. I. Angiotensin peptide content, secretion and effects in the corpus cavernosum. *J Urol.* 1997;157(5):1920–1925.

22 Park JK, Kim SZ, Kim SH, et al. Renin angiotensin system in rabbit corpus cavernosum: functional characterization of angiotensin II receptors. *J Urol.* 1997;158(2):653–658.

23 Becker AJ, Uckert S, Stief CG, et al. Possible role of bradykinin and angiotensin II in the regulation of penile erection and detumescence. *Urology.* 2001;57(1):193–198.

24 Becker AJ, Uckert S, Stief CG, et al. Plasma levels of angiotensin II during different penile conditions in the cavernous and systemic blood of healthy men and patients with erectile dysfunction. *Urology.* 2001;58(5):805–810.

25 Chitaley K, Webb RC, Mills TM. The ups and downs of Rho-kinase and penile erection: upstream regulators and downstream substrates of rho-kinase and their potential role in the erectile response. *Int J Impot Res.* 2003;15(2):105–109.

26 Sward K, Mita M, Wilson DP, et al. The role of RhoA and Rho-associated kinase in vascular smooth muscle contraction. *Curr Hypertens Rep.* 2003;5(1):66–72.

27 Essler M, Amano M, Kruse HJ, et al. Thrombin inactivates myosin light chain phosphatase via Rho and its target Rho kinase in human endothelial cells. *J Biol Chem.* 1998;273(34):21867–21874.

28 Rees RW, Ziessen T, Ralph DJ, et al. Human and rabbit cavernosal smooth muscle cells express Rho-kinase. *Int J Impot Res.* 2002;14(1):1–7.

29 Chitaley K, Wingard CJ, Clinton Webb R, et al. Antagonism of Rho-kinase stimulates rat penile erection via a nitric oxide-independent pathway. *Nat Med.* 2001;7(1):119–122.

30 Mills TM, Chitaley K, Lewis RW, Webb RC. Nitric oxide inhibits RhoA/Rho-kinase signaling to cause penile erection. *Eur J Pharmacol.* 2002;439(1–3):173–174.

31 Chitaley K, Bivalacqua TJ, Champion HC, et al. Adeno-associated viral gene transfer of dominant negative RhoA enhances erectile function in rats. *Biochem Biophys Res Commun.* 2002;298(3):427–432.

32 Bivalacqua TJ, Champion HC, Usta MF, et al. RhoA/Rho-kinase suppresses endothelial nitric oxide synthase in the penis: a mechanism for diabetes-associated erectile dysfunction. *Proc Natl Acad Sci USA.* 2004;101(24):9121–9126.

33 Hakim LS, Goldstein I. Diabetic sexual dysfunction. *Endocrinol Metab Clin North Am.* 1996;25(2):379–400.

34 Saenz de Tejada I, Goldstein I, Azadzoi K, et al. Impaired neurogenic and endothelium-mediated relaxation of penile smooth

muscle from diabetic men with impotence. *N Engl J Med.* 1989; 320(16):1025–1030.

35 Cellek S, Rodrigo J, Lobos E, et al. Selective nitrergic neurodegeneration in diabetes mellitus – a nitric oxide-dependent phenomenon. *Br J Pharmacol.* 1999;128(8):1804–1812.

36 Podlasek CA, Zelner DJ, Bervig TR, et al. Characterization and localization of nitric oxide synthase isoforms in the BB/WOR diabetic rat. *J Urol.* 2001;166(2):746–755.

37 Bivalacqua TJ, Usta MF, Champion HC, et al. Gene transfer of endothelial nitric oxide synthase partially restores nitric oxide synthesis and erectile function in streptozotocin diabetic rats. *J Urol.* 2003;169(5):1911–1917.

38 Escrig A, Marin R, Abreu P, et al. Changes in mating behavior, erectile function, and nitric oxide levels in penile corpora cavernosa in streptozotocin-diabetic rats. *Biol Reprod.* 2002;66(1): 185–189.

39 Way KJ, Reid JJ. The effects of diabetes on nitric oxide-mediated responses in rat corpus cavernosum. *Eur J Pharmacol.* 1999;376(1–2):73–82.

40 Ming XF, Viswambharan H, Barandier C, et al. Rho GTPase/Rho kinase negatively regulates endothelial nitric oxide synthase phosphorylation through the inhibition of protein kinase B/Akt in human endothelial cells. *Mol Cell Biol.* 2002;22(24): 8467–8477.

41 Bivalacqua TJ, Hellstrom WJ, Kadowitz PJ, Champion HC. Increased expression of arginase II in human diabetic corpus cavernosum: in diabetic-associated erectile dysfunction. *Biochem Biophys Res Commun.* 2001;283(4):923–927.

42 Yildirim S, Ayan S, Sarioglu Y, et al. The effects of long-term oral administration of L-arginine on the erectile response of rabbits with alloxan-induced diabetes. *BJU Int.* 1999;83(6):679–685.

43 Singh R, Barden A, Mori T, Beilin L. Advanced glycation end-products: a review. *Diabetologia.* 2001;44(2):129–146.

44 Seftel AD, Vaziri ND, Ni Z, et al. Advanced glycation end products in human penis: elevation in diabetic tissue, site of deposition, and possible effect through iNOS or eNOS. *Urology.* 1997; 50(6):1016–1026.

45 Usta MF, Bivalacqua TJ, Yang DY, et al. The protective effect of aminoguanidine on erectile function in streptozotocin diabetic rats. *J Urol.* 2003;170(4 Pt 1):1437–1442.

46 Usta MF, Bivalacqua TJ, Koksal IT, et al. The protective effect of aminoguanidine on erectile function in diabetic rats is not related to the timing of treatment. *BJU Int.* 2004;94(3):429–432.

47 Usta MF, Kendirci M, Gur S, Foxwell NA, Bivalacqua TJ, Cellek S, Hellstrom WJ. The breakdown of preformed advanced glycation end products reverses erectile dysfunction in streptozotocin-induced diabetic rats: preventive versus curative treatment. *J Sex Med.* 2006;3(2):242–50; discussion 250–252.

48 Bivalacqua TJ, Champion HC, Mehta YS, et al. Adenoviral gene transfer of endothelial nitric oxide synthase (eNOS) to the penis improves age-related erectile dysfunction in the rat. *Int J Impot Res.* 2000;12(Suppl 3):S8–S17.

49 Cartledge JJ, Eardley I, Morrison JF. Nitric oxide-mediated corpus cavernosal smooth muscle relaxation is impaired in ageing and diabetes. *BJU Int.* 2001;87(4):394–401.

50 Carrier S, Nagaraju P, Morgan DM, et al. Age decreases nitric oxide synthase-containing nerve fibers in the rat penis. *J Urol.* 1997;157(3):1088–1092.

51 Gonzalez-Cadavid NF, Rajfer J. Molecular pathophysiology and gene therapy of aging-related erectile dysfunction. *Exp Gerontol.* 2004;39(11–12):1705–1712.

52 Champion HC, Bivalacqua TJ, Hyman AL, et al. Gene transfer of endothelial nitric oxide synthase to the penis augments erectile responses in the aged rat. *Proc Natl Acad Sci USA.* 1999;96(20): 11648–11652.

53 Mori M, Gotoh T. Regulation of nitric oxide production by arginine metabolic enzymes. *Biochem Biophys Res Commun.* 2000;275(3):715–719.

54 Rosenfeld ME. Oxidized LDL affects multiple atherogenic cellular responses. *Circulation.* 1991;83(6):2137–2140.

55 Kugiyama K, Kerns SA, Morrisett JD, et al. Impairment of endothelium-dependent arterial relaxation by lysolecithin in modified low-density lipoproteins. *Nature.* 1990;344(6262):160–162.

56 Azadzoi KM, Goldstein I. Erectile dysfunction due to atherosclerotic vascular disease: the development of an animal model. *J Urol.* 1992;147(6):1675–1681.

57 Yesilli C, Yaman O, Anafarta K. Effect of experimental hypercholesterolemia on cavernosal structures. *Urology.* 2001;57(6): 1184–1188.

58 Junemann KP, Aufenanger J, Konrad T, et al. The effect of impaired lipid metabolism on the smooth muscle cells of rabbits. *Urol Res.* 1991;19(5):271–275.

59 Gholami SS, Rogers R, Chang J, et al. The effect of vascular endothelial growth factor and adeno-associated virus mediated brain derived neurotrophic factor on neurogenic and vasculogenic erectile dysfunction induced by hyperlipidemia. *J Urol.* 2003;169(4):1577–1581.

60 Lin CS, Ho HC, Chen KC, et al. Intracavernosal injection of vascular endothelial growth factor induces nitric oxide synthase isoforms. *BJU Int.* 2002;89(9):955–960.

61 Henry GD, Byrne R, Hunyh TT, et al. Intracavernosal injections of vascular endothelial growth factor protects endothelial dependent corpora cavernosal smooth muscle relaxation in the hypercholesterolemic rabbit: a preliminary study. *Int J Impot Res.* 2000;12(6):334–339.

62 Byrne RR, Henry GD, Rao DS, et al. Vascular endothelial growth factor restores corporeal smooth muscle function in vitro. *J Urol.* 2001;165(4):1310–1315.

63 Munzel T, Hink U, Heitzer T, Meinertz T. Role for NADPH/NADH oxidase in the modulation of vascular tone. *Ann NY Acad Sci.* 1999;874:386–400.

64 Wolin MS. Interactions of oxidants with vascular signaling systems. *Arterioscler Thromb Vasc Biol.* 2000;20(6):1430–1442.

65 Beckman JS, Koppenol WH. Nitric oxide, superoxide, and peroxynitrite: the good, the bad, and ugly. *Am J Physiol.* 1996;271 (5 Pt 1):C1424–C1437.

66 Kunsch C, Medford RM. Oxidative stress as a regulator of gene expression in the vasculature. *Circ Res.* 1999;85(8):753–766.

67 Bivalacqua TJ, Armstrong JS, Biggerstaff J, et al. Gene transfer of extracellular SOD to the penis reduces O2-* and improves erectile function in aged rats. *Am J Physiol Heart Circ Physiol.* 2003;284(4):H1408–H1421.

68 Cartledge JJ, Eardley I, Morrison JF. Impairment of corpus cavernosal smooth muscle relaxation by glycosylated human haemoglobin. *BJU Int.* 2000;85(6):735–741.

69 Drew B, Leeuwenburgh C. Aging and the role of reactive nitrogen species. *Ann NY Acad Sci.* 2002;959:66–81.

70 Taddei S, Virdis A, Ghiadoni L, et al. Age-related reduction of NO availability and oxidative stress in humans. *Hypertension.* 2001;38(2):274–279.

71 Cartledge JJ, Eardley I, Morrison JF. Advanced glycation end-products are responsible for the impairment of corpus cavernosal smooth muscle relaxation seen in diabetes. *BJU Int*. 2001;87 (4):402–407.
72 Gur S, Ozturk B, Karahan ST. Impaired endothelium-dependent and neurogenic relaxation of corpus cavernosum from diabetic rats: improvement with L-arginine. *Urol Res*. 2000;28(1): 14–19.
73 Khan MA, Thompson CS, Jeremy JY, et al. The effect of superoxide dismutase on nitric oxide-mediated and electrical field-stimulated diabetic rabbit cavernosal smooth muscle relaxation. *BJU Int*. 2001;87(1):98–103.
74 Ryu JK, Kim DJ, Lee T, et al. The role of free radical in the pathogenesis of impotence in streptozotocin-induced diabetic rats. *Yonsei Med J*. 2003;44(2):236–241.
75 Bivalacqua TJ, Usta MF, Kendirci M, et al. Superoxide anion production in the rat penis impairs erectile function in diabetes: influence of extracellular superoxide dismutase gene therapy. *J Sex Med*. 2005:2187–2198.
76 Kim SC, Kim IK, Seo KK, et al. Involvement of superoxide radical in the impaired endothelium-dependent relaxation of cavernous smooth muscle in hypercholesterolemic rabbits. *Urol Res*. 1997;25(5):341–346.
77 Rubbo H, Trostchansky A, Botti H, Batthyany C. Interactions of nitric oxide and peroxynitrite with low-density lipoprotein. *Biol Chem*. 2002;383(3–4):547–552.
78 Ahn TY, Gomez-Coronado D, Martinez V, et al. Enhanced contractility of rabbit corpus cavernosum smooth muscle by oxidized low density lipoproteins. *Int J Impot Res*. 1999;11(1):9–14.
79 Koupparis AJ, Jeremy JY, Muzaffar S, et al. Sildenafil inhibits the formation of superoxide and the expression of gp47 NAD[P]H oxidase induced by the thromboxane A2 mimetic, U46619, in corpus cavernosal smooth muscle cells. *BJU Int*. 2005;96(3):423–427.
80 Lue TF. Erectile dysfunction. *N Engl J Med*. 2000;342(24):1802–1813.
81 Andersson KE. Erectile physiological and pathophysiological pathways involved in erectile dysfunction. *J Urol*. 2003;170 (2 Pt 2):S6–S13. Discussion S-4.
82 Moreland RB, Hsieh G, Nakane M, Brioni JD. The biochemical and neurologic basis for the treatment of male erectile dysfunction. *J Pharmacol Exp Ther*. 2001;296(2):225–234.

168

Avascular Necrosis

Vascular Bed/Organ Structure and Function in Health and Disease

Chantal Séguin

Montreal General Hospital, McGill University, Canada

Avascular necrosis (AVN) of the femoral head is a clinical entity in which bone death occurs as a result of interruption of blood at the level of the microcirculation. The etiology and pathogenesis of nontraumatic AVN have not been fully elucidated. The understanding of this disease progression is important for several reasons. First, AVN is a devastating musculoskeletal condition that strongly impacts those affected. Second, it tends to occur in young people. Third, current treatments are suboptimal since they are, for the most part, palliative rather than curative. Finally, we are currently unable to identify individuals who will develop AVN, even in the group considered at high risk; consequently, practitioners are not able to stop the progression of the disease or reverse the process once AVN has developed.

Although the actual prevalence of the disease is unknown, an estimated 10,000 to 20,000 new patients with AVN are diagnosed each year in the United States (1). AVN is the underlying diagnosis in 5% to 18% of the more than 500,000 total hip arthroplasties performed yearly in the United States and Western Europe (1).

AVN has been associated with a variety of risk factors, and classified as either secondary or idiopathic. Environmental risk factors include hyperlipidemia, steroid use, alcohol, various blood dyscrasias (e.g., hemoglobinopathies, coagulopathies), pregnancy, hyperbaric exposure, use of chemotherapeutic agents, systemic lupus erythematosus, inflammatory bowel disease, lipid storage diseases (Gaucher disease), and familial thrombophilias (1).

The condition may occur in up to 30% of patients with lupus erythematosus. In a large U.S. study of 2,590 patients with sickle cell disease, Milner and associates reported a prevalence of 9.8% (1). There is also a high incidence of AVN among organ transplant recipients. Allogenic bone marrow transplantation carries the highest incidence of AVN, reported to be 10% in the United States. Likewise, a 42.8% incidence of AVN of the femoral head has been found in a review reported by Arlet and colleagues of 138 patients with lower extremity vascular occlusive disease demonstrated by arteriography (2). Many recent studies also have described AVN in association with human immunodeficiency virus (HIV) infections (3). More recently, some studies have shown an association of AVN of the mandible with the chronic use of bisphosphonates, although no cases were seen in hip AVN (4). AVN also has been described after septic shock associated with multiorgan failure and after renal transplantation (5).

Several studies suggest that alcohol and steroid use may account for 90% to 100% of all nontraumatic "secondary" cases of AVN (6). The effects of both alcohol and corticosteroids have been related to alterations in lipid metabolism through intravascular coagulation and fat embolism phenomena. However, of those people fulfilling the risk criteria for AVN development, most never develop the disease. This finding suggests that a second event is necessary to initiate the development of AVN in a subset of patients with an acquired or inherited predisposition for this disorder (6).

HISTOLOGY OF THE FEMORAL HEAD IN AVASCULAR NECROSIS

The immunochemistry of 14 femoral heads with late-stage nontraumatic AVN in a study conducted in Denmark in 1995 by Starklint and colleagues (7) identified five histologically distinct zones in most of the affected heads. In the transitional zone, several areas with intravascular aggregations of newly formed and older fibrin clots were noticed, mainly on the venous side of the vascular system. Small vessels were collapsed, with a few endothelial cells (ECs) clumped together in the center of a concentric fibrous tissue. These and other findings suggest that obstruction of the venous outflow due to intravascular thrombosis is important in the pathogenesis of nontraumatic AVN of the femoral head.

ROLE OF ENDOTHELIUM IN AVASCULAR NECROSIS

Accumulating evidence suggests that the endothelium is involved in the development of AVN. Glucocorticoids have been shown to alter EC phenotypes (see Chapter 180). Nishimura and colleagues, in their histopathologic study of veins in steroid-treated rabbits, were able to show damage to ECs and smooth muscle cells via electron microscopy (8). Borcsok and colleagues reported that glucocorticoids induced the expression of endothelin-1 in ECs via an endothelin-A receptor-dependent mechanism (9). EC dysfunction under glucocorticoid influence also was demonstrated by Kitajima (10). We have shown that the addition of dexamethasone 1,000 mM to human umblicial vein ECs (HUVECs) results in increased expression of vascular cell adhesion molecule (VCAM)-1. We are currently studying the effect of plasma from a cohort of AVN patients versus healthy controls on the induction of activation markers in HUVECs (C. Séguin et al., unpublished observations).

Other studies have demonstrated endothelial dysfunction and subsequent platelet thrombosis following intravascular bubble formation in dysbaric osteonecrosis (11) and a possible role for lipid peroxide on ECs by suppressing their proliferation (12).

AVN is a frequent bone complication seen in patients with sickle cell disease. Bone infarcts can occur at multiple sites. The endothelium plays a critical role in the pathophysiology of sickle cell anemia. In sickle cell anemia, the endothelium acquires an activation phenotype, promoting increased leukocyte adhesion and displaying a procoagulant surface. These changes contribute to reduced blood flow and ischemia. Indirect evidence that the vascular endothelium is abnormally activated in sickle cell disease comes from the study of circulating ECs (CECs). CECs from sickle cell patients have abnormally increased expression of VCAM-1, intercellular adhesion molecule (ICAM)-1, E-selectin, and tissue factor, reflecting a proadhesive and procoagulant state (13). Sickle cell anemia also has been associated with reduced nitric oxide (NO) bioactivity in resistance vessels, with resultant impaired vasodilation (14). EC dysfunction has been implicated in such complications as stroke and pulmonary hypertension. It is interesting to speculate that phenotypic changes in the bone endothelium contributes to the initiation and/or progression of AVN.

The identification of vascular bed–specific markers that are associated with an increased risk for AVN and/or contribute to the pathogenesis of this disease might provide a foundation for molecular diagnostic tests aimed at detecting individuals at high risk before the disease occurs or progresses to irreversible stages. Also, such information might lead to the development of novel therapies, with the goal of reducing repeated surgical treatments, the resultant morbidity, and consequently the associated costs to our health system.

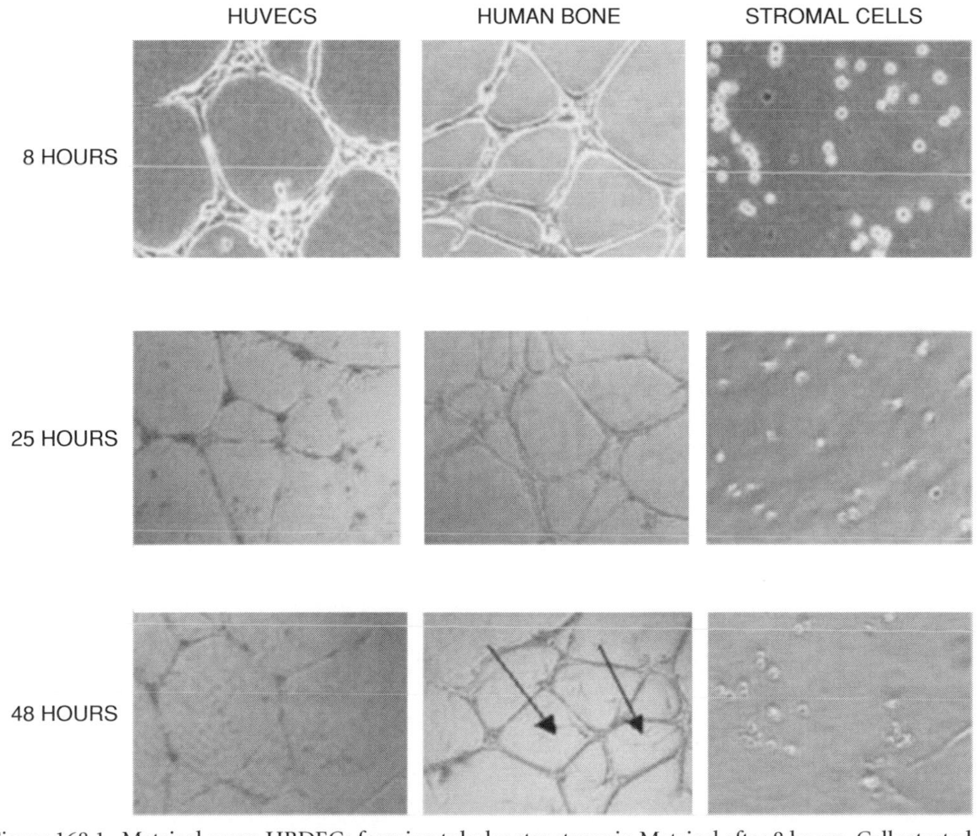

Figure 168.1. Matrigel assay. HBDECs forming tubular structures in Matrigel after 8 hours. Cells started to migrate to the inside of the lumens after 48 hours (*arrows*).

ISOLATION OF HUMAN BONE-DERIVED ENDOTHELIAL CELLS

The first successful isolation of bone-derived ECs was reported by Zhang in a mouse study model using a 0.5% collagenase (15). We have successfully isolated human bone-derived ECs (HBDECs) from unaffected (control) and affected (AVN) femoral heads from human subjects undergoing surgery (C. Séguin, unpublished observations). These studies demonstrate the presence of a population of cells that are positive for the EC surface markers, platelet-endothelial cell adhesion molecule (PECAM)-1/CD31, CD105 (endoglin), and von Willebrand factor. Furthermore, when assayed in Matrigel, these cells formed tubular structures, suggesting their capability to undergo angiogenesis (Figure 168.1). These preliminary results hold great promise. Although the cells were isolated from the cancellous bone of the femoral head, this technique can be expanded to isolate ECs from other human bone tissues. The cells isolated could provide a primary population of ECs for the biological analysis of human vascular bone diseases.

KEY POINTS

- AVN is a devastating musculoskeletal disorder affecting mainly young people; it is associated with significant morbidity in those affected, and it impacts highly on the health care system.
- Intravascular thrombosis is manifested on histopathological bone studies, implying the contribution of the vascular system in the pathogenesis of AVN.
- Evidence suggests that EC dysfunction or activation is a key mechanism underlying the development of microvascular thrombus and as such, supports the hypothesis of bone-derived endothelium in the development of vascular bone diseases.

Future Goals

- To develop a suitable animal model based on clinical disease state, to appreciate better the contribution of bone-derived endothelium in the pathogenesis of vascular bone diseases

REFERENCES

1 Urbaniak JR, Jones JP Jr. *Osteonecrosis: Etiology, Diagnosis, and Treatment.* Rosemont, IL: American Orthopaedic Association, 1997.

2 Arlet J. Nontraumatic avascular necrosis of the femoral head. Past, present and future. *Clin Orthop.* 1992;(277):12–21.

3 Calza L, Manfredi R, Chiodo F. Osteonecrosis in HIV-infected patients and its correlation with highly active antiretroviral therapy (HAART). *Presse Med.* 2003;32(13 Pt 1):595–598.

4 Ruggiero S, Mehrotra B, Rosenberg T, Engroff S. Osteonecrosis of the jaws associated with the use of biphosphonates: a review of 63 cases. *J Oral Maxillofac Surg.* 2004;62:527–534.

5 Bolland MJ, Hood G, Bastin ST, et al. Bilateral femoral head osteonecrosis after septic shock and multiorgan failure. *J Bone Miner Res.* 2004;19(3):517–520.

6 Urbaniak JR, Harvey EJ. Revascularization of the femoral head in osteonecrosis. *J Am Acad Orthop Surg.* 1998;6:44–54.

7 Starklint H, Lausten GS, Arnoldi CC. Microvascular obstruction in avascular necrosis. Immunochemistry of 14 femoral heads. *Acta Orthop Scand.* 1995;66(1): 9–12.

8 Nishimura T, Matsumoto T, Nishino M, Tomita K. Histopathologic study of veins in steroid treated rabbits. *Clin Orthop Relat Res.* 1997;(334):37–42.

9 Borcsok I, Schairer HU, Sommer U, et al. Glucocorticoids regulate the expression of the human osteoblastic endothelin A receptor gene. *J Exp Med.* 1998;188(9):1563–1573.

10 Kitajima I. Steroid-induced osteoporosis. *Clin Calcium.* 2001;11 (5):582–588.

11 Slichter SJ, Stegall P, Smith K, et al. Dysbaric osteonecrosis: a consequence of intravascular bubble formation, endothelial damage, and platelet thrombosis. *J Lab Clin Med.* 1981;98(4):568–590.

12 Iio H, Ake Y, Saegusa Y, Mizuno K. The effect of lipid peroxide on osteoblasts and vascular endothelial cells – the possible role of ischemia-reperfusionin the progression of avascular necrosis of the femoral head. *Kobe J Med Sci.* 1996;42(6):361–373.

13 Solovey A, Kollander R, Shet A, et al. Endothelial cell expression of tissue factor in sickle mice is augmented by hypoxia/reoxygenation and inhibited by lovastatin. *Blood.* 2004;104(3):840–846. Epub Apr 08 2004.

14 Eberhardt RT, McMahon L, Duffy SJ, et al. Sickle cell anemia is associated with reduced nitric oxide bioactivity in peripheral conduit and resistance vessels. *Am J Hematol.* 2003;74:104–111.

15 Zhang Y, Fujita N, Oh-hara T, et al. Production of interleukin-11 in bone-derived endothelial cells and its role in the formation of osteolytic bone metastasis. *Oncogene.* 1998;16:693–703.

169

Molecular Control of Lymphatic System Development

Darren Kafka and Young-Kwon Hong

Keck School of Medicine, University of Southern California, Los Angeles

The human body has two major circulatory systems: the blood and lymphatic vascular systems. Both systems share functional and anatomical similarities and play complementary roles in tissue perfusion and fluid circulation. Nonetheless, the two systems have attracted different levels of scientific and clinical attention. Although the blood vascular system has been extensively studied, in contrast, the lymphatic system has been neglected, despite the essential roles it plays in tissue fluid homeostasis, immune response, and lipid absorption. This is largely due to the lack of lymphatic-specific molecular tools to visualize and characterize lymphatic vessels. However, a number of recent discoveries involving the characterization of lymphatic-specific molecules and lymphatic endothelial cells (ECs) have begun to unravel the mystery of the lymphatic system and have opened a new era of lymphatic research. Moreover, studies of genetic mouse models have provided important insights into the molecular mechanisms underlying lymphatic development, as well as the genetic basis of human diseases associated with the lymphatic system. In this chapter, we review some of the landmark findings that have significantly advanced our current understanding of the lymphatic system.

STRUCTURE AND FUNCTION OF THE LYMPHATIC VASCULAR SYSTEM

The blood vascular system is a closed circular network in which blood leaves and returns to the heart after flowing through the arteries, tissue capillaries, and veins. In comparison, the lymphatic system is a blunt-ended linear system: The initial lymphatic vessels begin at the interstitial spaces of peripheral tissues and organs, and they are connected to the thicker collecting lymphatic vessels, which are eventually connected to the inferior vena cava through the thoracic duct (Figure 169.1, Table 169-1). From peripheral tissues and organs, lymphatic capillaries drain protein-rich interstitial fluid, extravasated cells, and large extracellular molecules (collectively called lymph), eventually transporting it back to the blood vascular system for recirculation. When this draining function of the lymphatic system is compromised by surgery, radiation, or infections, the lymph fluid accumulates and cause swellings (lymphedema) in the affected tissues or organs. In fact, surgical lymph node removal for the purpose of tumor staging is one of the most common causes of tissue swelling. Therefore, the major function of the lymphatic system is to control tissue-fluid homeostasis. In addition, the lymphatic system serves as a conduit for the absorption of large molecules and lipids in the digestive systems. Lymphatics also play a critical role in the trafficking of lymphocytes and antigen-presenting cells to regional lymph nodes. Unfortunately, many malignant tumors take advantage of the lymphatic system for their dissemination.

Despite their overlapping functional and anatomical features, the blood vascular and lymphatic systems display distinct phenotypes. Blood capillaries possess a compact, three-layered structure in which the blood vascular ECs (BECs) are surrounded by a continuous basement membrane (the liver sinusoids are an exception), which in turn is covered to a variable degree by smooth muscle cells/pericytes. ECs of the blood capillaries tend to be tightly connected to each other through adherens, gap junctions, and tight junctions. By comparison, the lymphatic capillaries consist of a single layer of lymphatic ECs (LECs) that partially overlap one another. They tend to have large, irregularly shaped lumens. In contrast to BECs, LECs lack tight junctions. With the exception of the larger collecting lymphatic vessels that have luminal valves, LECs in lymphatic capillaries are rarely surrounded by a continuous basement membrane or smooth muscle cells/pericytes. Instead, they are directly connected to the surrounding extracellular matrix through anchoring filaments (Figure 169.2). Under normal conditions, most lymphatic capillaries remain collapsed. When interstitial pressure increases, the anchoring filaments exert tension on the LECs, pulling open the intercellular junctions, and allowing interstitial fluid to enter the lumen of lymphatic capillaries.

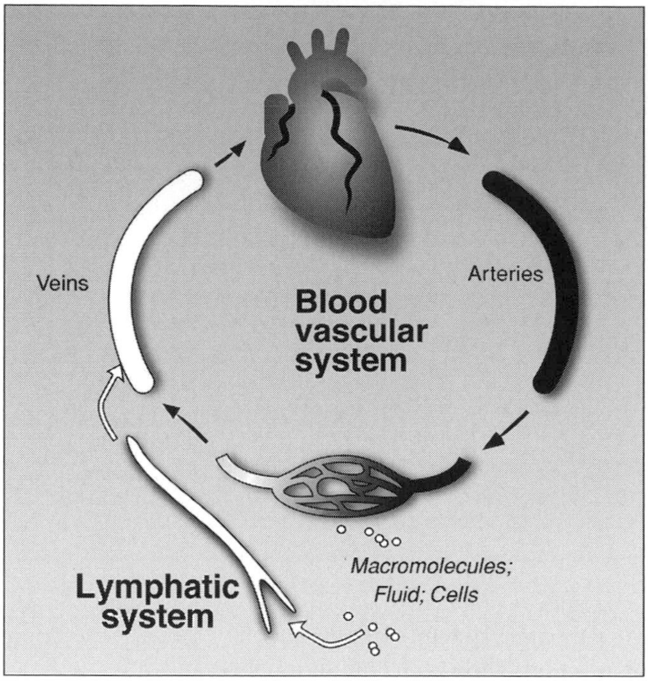

Figure 169.1. Schematic illustration of the blood vascular and lymphatic systems. The blood vascular system is a circular and closed system, whereas the lymphatic system is a blunt-ended and linear system. Fluid, macromolecules, and cells extravasated from blood capillaries flow into lymphatic capillaries in peripheral tissues and are then transported back to the blood vascular system for recirculation.

In addition to lymphatic vessels, the lymphatic system includes lymphatic organs such as the lymph nodes, tonsils, Peyer patches, spleen, and thymus. These so-called secondary lymphoid organs control the quality of immune responses by serving as a fluid filter in which pathogens and other foreign agents encounter immune cells. Most vascularized tissues and organs have lymphatic vessels, with the exception of the brain, spinal cord, and retina. Although lymphatic vessels do not exist in the brain and spinal cord, studies found that tracers injected into the cerebrospinal fluid (CSF) or brain parenchyma were cleared into lymphatic vessels external to the cranium and into lymph nodes in the head and neck (1). This finding established a physiological relationship between the CSF and extracranial lymph compartments. Although the normal cornea lacks both blood and lymphatic vessels, various corneal diseases and surgical manipulation can induce lymphangiogenesis as well as angiogenesis (1).

DEVELOPMENT OF THE LYMPHATIC SYSTEM: HISTORICAL AND PHYLOGENETIC PERSPECTIVES

Lymphatics were initially described as "white blood vessels" by Hippocrates. After that first description, the lymphatic system was virtually ignored until 1627, when the Italian anatomist Gasper Asellius rediscovered lymphatic vessels as "milky veins" in the mesentery of a dog (2). These cursory observations contrasted with William Harvey's seminal and thorough description of the closed blood vascular system in the following year (1628) in his book, *An Anatomical Study of the Motion of the Heart and Blood in Animals* (3). Following Asellius's finding of the mesenteric lymphatic vessels, the collecting lymphatic vessels and the thoracic duct were identified and their fluid-draining function was further characterized. At the beginning of the 20th century, two competing hypotheses were proposed regarding the genesis of the lymphatic system. One hypothesis (centrifugal model) claimed that lymphatic endothelium derives from blood vascular endothelium and that lymphatic vessels are formed through the sprouting of preexisting blood vessels. The other hypothesis (centripetal model), which was more widely held then, posited that LECs are independently differentiated from mesenchymal cell–derived lymphangioblasts, and that the lymphatic connection to the blood vascular system is acquired only later. Using ink-injection experiments in pig embryos, Florence Sabin showed, in 1902, that the initial lymphatic sacs were derived from embryonic veins, and that these primitive lymphatic plexuses gradually spread throughout the body to form the lymphatic network (4,5). Although Sabin's experiment strongly supported the centrifugal model, it would take another 100 years to resolve this debate.

About four decades ago, Leak and Burke pioneered a comparative morphological study of blood and lymphatic vessels and – for the first time – attempted to establish a defined culture of LECs (6). However, these efforts were hampered by the absence of available lymphatic-specific markers. During the past decade, new lymphatic-specific molecular markers, such as vascular endothelial growth factor receptor 3 (VEGFR3), lymphatic vessel endothelial receptor (LYVE)-1, and podoplanin have been identified and characterized. One of the landmark (and unexpected) discoveries in lymphatic research was made in 1999, when *Prox1*-knockout mice were shown to lack a lymphatic system (7). This and subsequent studies showed that a subset of embryonic cardinal vein ECs express Prox1 during early embryogenesis, and that the Prox1^{+} cells bud off and migrate out to form the initial lymphatic vessels (7,8). This finding was consistent with Sabin's data and largely confirmed the blood vascular origin of mammalian lymphatic system. In contrast to mammals, lymphatic vessels in *Xenopus* tadpoles have been reported to develop from lymphangioblasts or, through transdifferentiation, from venous ECs (9). Moreover, several studies of avian development revealed that independent lymphangioblasts contributed to the formation of the lymphatic system in early wing buds, limb buds, and the chorioallantoic membrane (10–12). Whether lymphangioblasts exist in mammals and contribute to mammalian lymphangiogenesis remains unclear.

From an evolutionary standpoint, the lymphatic system first appeared in vertebrates (13). Although zebrafish (*Danio*

Table 169-1: Schematic Comparisons between the Blood Vascular and Lymphatic Systems

	Blood Vessels/BEC	*Lymphatics/LEC*
Structure	Closed, circular	Open, linear
Division	Arteries, arterioles, capillaries, venules, veins	Capillaries, collecting vessels, lymph nodes, thoracic duct
Constituents	Blood	Lymph (protein-rich interstitial fluid, extravasated cells and large extracellular molecules)
Functions	Hemostasis, inflammation, leukocyte trafficking, barrier function	Tissue fluid homeostasis; absorption of large molecules and lipids in the digestive systems; trafficking of lymphocytes and antigen-presenting cells to regional lymph nodes
Vessel wall	Depending on vessel type, vascular bed: adherens and tight junctions, continuous basement membrane (except in liver sinusoids), pericytes or vascular smooth muscle cells	Overlapping ECs, no tight junctions, discontinuous basement membrane; few pericytes (except large collecting vessels); collecting vessels contain interluminal valves to prevent reverse flow
Development	Vasculogenesis and angiogenesis	Lymphangiogenesis (budding from cardinal vein)
Origin	Mesoderm	Mesoderm
Examples of cell-type specific markers	CD34, PAL-E	LYVE-1, VEGFR3, Prox1, and podoplanin
Absence	Cartilage, cornea	Cartilage, brain, spinal cord, and the retina, cornea
Activation	Response to inflammatory mediators, usually consists of combination of procoagulant and proadhesive properties and increased permeability	Response to inflammatory mediators; HGFR is expressed in inflammation or wound-associated lymphatic vessels
Heterogeneity	Phenotypic heterogeneity is well established. EC phenotypes vary between tissues, within a given organ and even between neighboring ECs	The extent of EC heterogeneity in lymphatics remains to be established

rerio) provide a valuable experimental model system for the study of angiogenesis, they do not have a lymphatic system.[1] Later, amphibians and reptiles developed specialized lymph hearts that drain and transport the interstitial fluid to the blood vascular system for recirculation. The primitive lymphatic system further evolved in birds and mammals to include lymph nodes, which play important roles in immune functions. The lymphatic system of *Xenopus* (frogs) has been used as a simple experimental tool for studies of the development and function of the lymphatic system (15–20).

1 It has since been reported that zebrafish have a functional lymphatic system (14).

MOLECULAR PLAYERS IN LYMPHATIC DEVELOPMENT

One of the landmark advances in lymphatic research was the establishment of defined cultures of LECs (21–24). Cultured ECs isolated from dermal microvascular lymphatic and blood vessels have been shown to retain many, but not all, of their lineage-specific phenotypes. Furthermore, cell morphology and growth characteristics of these two different types of ECs are largely indistinguishable. Several microarray analyses revealed that approximately 96% to 98% of transcripts were comparably expressed between the two cell types, with only 2% to 3% of genes (corresponding to approximately

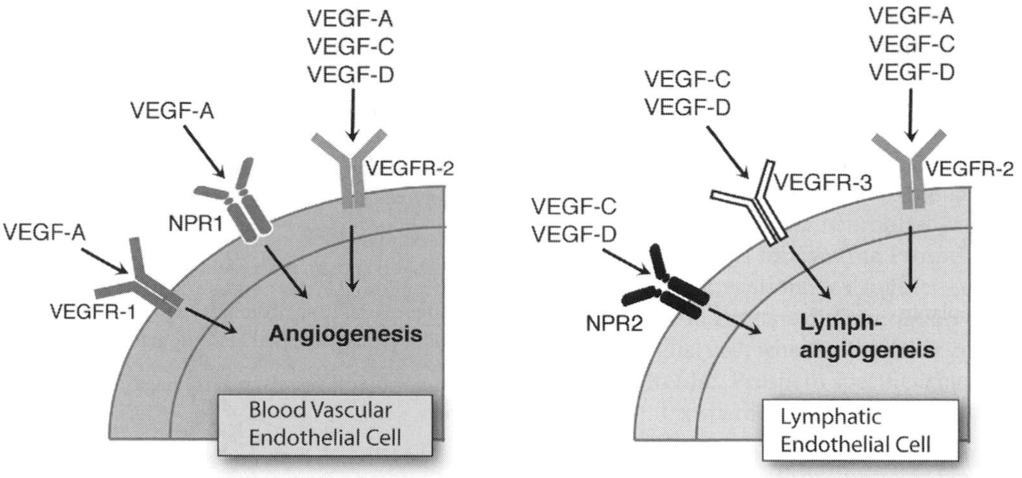

Figure 169.4. Schematic illustration of specific growth factors and their receptors expressed by blood vascular and lymphatic ECs. Blood vascular ECs express VEGFR1, VEGFR2, and NPR1, whereas lymphatic ECs express VEGFR2, VEGFR3, and NPR2. VEGF-A, -C, and -D can promote angiogenesis and lymphangiogenesis.

transcriptional profiles of embryonic venous ECs. Indeed, we and others have shown that ectopic expression of Prox1 in differentiated primary BECs isolated from human foreskin was sufficient to reprogram BECs to adopt an LEC phenotype (25,34). When expressed in BECs, Prox1 downregulated several BEC-specific genes such as E-selectin, interleukin (IL)-8, laminin, collagens, NRP1, and vascular endothelial growth factor (VEGF)-C, and upregulated certain lymphatic-specific genes including VEGFR3, podoplanin, and desmoplakin I and II (25,34). Furthermore, knockout and overexpression studies have shown that Prox1 directs the progenitor cells of the mammalian retina and the eye lens to exit from the cell-cycle and to induce postmitotic differentiation programs (35–37). However, Prox1 does not appear to regulate the cell cycle during lymphatic development, suggesting that the mechanism of cellular differentiation induced by Prox1 may be cell-type specific (7,8). A recent study of *Prox1* mutant mice provided new insights into the role of lymphatics in fat accumulation (38). The latter study showed that the leakage of the lymph from structurally compromised lymphatic vessels in Prox1-deficient mice causes adult-onset obesity. Indeed, lymphedema resulting from lymphatic dysfunction due to the surgical removal of lymph nodes often is associated with fat accumulation and tissue fibrosis. Therefore, this study provided a long-suspected, yet hitherto unproven, link between adipogenesis and the lymphatic system.

Prox1 has been also implicated in Kaposis's sarcoma (KS), the most common neoplasm in human immunodeficiency virus (HIV)-infected patients (39–41). The causative agent in KS is KS-associated herpes virus (KSHV, also known as HHV-8). When KSHV infects BECs in vitro, the virus reprograms the cell type to resemble LECs by upregulating Prox1. KSHV is also capable of infecting cultured LEC s and, interestingly, the virus downregulates Prox1 expression to respecify the cell type toward that of BECs (39). The pathological significance of this Prox1-mediated cell type reprogramming remains to be further studied.

Vascular Endothelial Growth Factor Receptor 3

VEGFR3, also known as FLT-4, is a member of the FMS-like tyrosine kinase family that comprises seven extracellular immunoglobulin (Ig) homology domains, a single transmembrane domain, and a split intracellular tyrosine kinase domain. It is structurally related to VEGFR1/FLT-1 and VEGFR2/KDR/FLK-1, the principle receptors for VEGF-A (42,43). VEGFR3 serves as a major receptor for VEGF-C and VEGF-D, two important lymphangiogenesis factors and, unlike VEGFR1 and VEGFR2, does not bind to VEGF-A (Figure 169.4). After biosynthesis, the fifth Ig domain is proteolytically processed, and the resulting polypeptide chains remain linked by a disulfide bond. Binding of VEGFR3 to its ligands induces dimerization of the receptor and transphosphorylation of the intracellular domains. This creates docking sites for the adaptor proteins Shc and Grb-2, and leads to the activation of Akt, extracellular-signal-regulated protein kinase (ERK)-1/2, related adhesion focal tyrosine kinase (RAFTK), and nuclear factor (NF)-κB, resulting in altered gene expression, cell survival, and proliferation (44).

VEGFR3 was the first gene whose expression was found to be specific for LECs in adults (43). Although VEGFR3 is also expressed in ECs of the cardinal vein and in angioblasts of the head mesenchyme during early development, its expression becomes restricted to lymphatic endothelium at a later developmental stage and in adults (43). In contrast, VEGFR3 expression in arterial ECs is tightly repressed by Notch signaling, an essential signaling pathway that determines arterial versus venous cell fates (45,46). *VEGFR3*-knockout mice die at E9.5 and display abnormal blood vascular organization, defective lumens in large blood vessels, and fluid accumulation in the pericardial cavity (47). However, initial vasculogenesis and angiogenesis appear to occur normally.

Because $VEGFR3^{-/-}$ embryos die prior to formation of the lymphatic system, the knockout model could not be used to ascertain the role of VEGFR3 in lymphatic development. Overexpression of soluble VEGFR3 (a competitive inhibitor of membrane-bound VEGFR3) in transgenic mice inhibited fetal lymphangiogenesis and induced regression of preformed lymphatic vessels without affecting blood vessels, resulting in tissue lymphedema and fibrosis (48). Moreover, separate studies showed that neutralizing VEGF-C/VEGFR3 signaling suppresses tubular morphogenesis of lymphatic vessels and inhibits tumor cell entry and spread through lymphatic vessels (49–51). Notably, some tumor- or wound-associated blood vessels are reported to express VEGFR3 (52). Studies of transgenic mice expressing a mutant form of VEGF-C (VEGF-C156S) that selectively binds and activates VEGFR3, but not VEGFR2, revealed that activation of VEGFR3 alone is sufficient to promote lymphangiogenesis (53).

Vascular Endothelial Growth Factor-C and -D

VEGF-C and VEGF-D, originally cloned by virtue of their affinity to VEGFR3, bind to both VEGFR2 and VEGFR3. Their interaction with VEGFR3 strongly promotes lymphangiogenesis (54–58). VEGF-C is widely expressed by a various cell types such as mesenchymal cells around embryonic veins, activated macrophages, skeletal muscle cells, and smooth muscle cells surrounding large arteries (54,59–61). Interestingly, VEGF-C is also expressed by BECs (21), implying that the blood vascular system may control growth and/or maintenance of lymphatic vessels. Both VEGF-C and -D are synthesized as precursor proteins with the N- and C-terminal propeptides and, when secreted, undergo proteolytic processing by secretory pro-protein convertases such as furin, PC5 and PC7, or plasmin. When fully processed, the isoforms acquire an increased affinity to VEGFR2 and therefore induce angiogenesis in vivo (53,62–65). A mutant form of VEGF-C that contains a cysteine to serine point mutation at residue 156 was shown to selectively activate VEGFR3, but not VEGFR2 (66). Activation of VEGFR3 by VEGF-C and/or VEGF-D promotes proliferation, survival, and migration of cultured human LECs and the VEGF-C/VEGFR3 pathway delivers a major lymphangiogenic signal in vivo (24,53,67). Importantly, tumor cells overexpressing VEGF-C and VEGF-D were shown to promote tumor lymphangiogenesis and metastasis in mice (68,69). These studies demonstrated – for the first time – that the expression of lymphangiogenic factors by tumor cells in a mouse model significantly promote tumor lymphangiogenesis and that the enhanced lymphangiogenic activity results in an increased incidence of tumor metastasis to regional lymph nodes and distant organs. A strong positive correlation between tumor metastasis and the density of tumor-associated lymphatic vessels has been consistently demonstrated in many different types of cancers, a list that continues to grow (70–75). Moreover, the tumor-associated lymphatic vessel density was shown to be a novel prognostic factor for malignant melanoma in human (75).

VEGF-C–deficient mice display normal blood vessel development, but absent lymphatic vessel formation (61). These data indicate that VEGF-C is essential for embryonic lymphangiogenesis, but dispensable for vasculogenesis and angiogenesis (61). Although the mutant mice exhibited Prox1 expression, $Prox1^{+}$ venous ECs failed to migrate out to form the initial lymph sacs. Notably, the sprouting defect was rescued by exogenous administration to embryos of VEGF-C or VEGF-D, but not VEGF-A (61). These results suggest that the lymphatic specification of $Prox1^{+}$ venous ECs and subsequent migration of committed LECs independently regulated processes.

Interestingly, heterozygous VEGF-C mutant mice demonstrated cutaneous lymphatic hypoplasia and lymphedema, showing a haploinsufficiency effect of VEGF-C for the normal development and function of the lymphatic system. Surprisingly, VEGF-D–deficient mice seem to develop a normal lymphatic system despite the putative role of this growth factor in physiological and pathological lymphangiogenesis (76). It will be interesting to investigate if *VEGF-C* and *VEGF-D*–double knockout mice display more severe lymphatic phenotypes than do the VEGF-C single mutant mice.

Vascular Endothelial Growth Factor Receptor 2

VEGFR2, also known as KDR or FLK-1, is a receptor for VEGF-A, VEGF-C, and VEGF-D. VEGFR2 is expressed by both BECs and LECs in vitro and in situ (21,23,77). Although the role of VEGF-A/VEGFR2 signaling in angiogenesis has been well established (see Chapter 32), its contribution to lymphangiogenesis is currently a matter of controversy. VEGF-A can potently enhance proliferation and tube formation of cultured LECs (21,77). Also, a local injection of adenovirus expressing murine VEGF-A_{164} into mouse ears resulted in pronounced and persistent in vivo lymphangiogenesis (78). Similarly, transgenic mice that specifically overexpress murine VEGF-A_{164} in the skin showed enhanced lymphangiogenesis – in addition to angiogenesis – during tissue repair and in skin inflammation (77,79). Blockade of VEGFR2 signaling using a VEGFR2 blocking antibody inhibited both angiogenesis and lymphangiogenesis in healing wounds, indicating the importance of VEGFR2 for repair-associated lymphangiogenesis (78). In contrast, adenoviral-mediated overexpression of the human VEGF-A_{165} isoform did not show distinct lymphangiogenic activity in mouse models (80,81). This discrepancy may be due to a possible species-specificity of VEGF-A. VEGF-A/VEGFR2 signaling also induces lymphangiogenesis in the sentinel lymph node as well as in tumors, and significantly promotes tumor metastasis (82).

Lymphatic Vessel Endothelial Receptor-1

LYVE-1 is the lymph vessel endothelial hyaluronan (HA) receptor-1 and, along with VEGFR3 and podoplanin (discussed in the next section), was the first to be identified as a lymphatic vessel marker. HA, a large mucopolysaccharide polymer (10^2–10^4 kDa), is a prolific component of the extracellular matrix in various tissues, and plays an important role in maintaining structural integrity of connective tissues by serving as a ubiquitous substratum for cell adhesion, migration, and differentiation (83,84). Extracellular HA undergoes a constant turnover under normal conditions and HA released from the tissue matrix is transported through lymphatic vessels to the lymph nodes, where it is partially degraded and subsequently transported to the liver for complete hydrolysis. LYVE-1 was originally identified by its high sequence homology to the inflammatory leukocyte homing receptor CD44 that is predominantly expressed by blood vascular endothelium (85). Although LYVE-1 is absent from BECs, it is expressed by activated tissue macrophages and sinusoidal endothelium of the liver and the spleen, where high-molecular-weight HA is absorbed and degraded (83,84). LYVE-1 also is implicated in the trafficking of cells within lymphatic vessels and lymph nodes. Despite the important role of LYVE-1 in HA homeostasis and its specific expression in lymphatic vessels, LYVE-1-deficient mice appear normal – no obvious lymphatic vascular malfunctions or morphological abnormalities have been detected thus far (86).

Podoplanin

Podoplanin, a mucin-type transmembrane sialoglycoprotein, is expressed by LECs, but not BECs in vivo and in vitro (21,23,25,87,88). The expression of podoplanin is also detected in a wide range of other cell types, including lung type I alveolar cells, choroid plexus cells, ciliary epithelial cells of the eye, osteocytes, and kidney podocytes. It has been independently cloned by a number of groups, and variably named OTS8, E11 antigen, RTI40, murine gp38, canine gp40, human gp36, aggrus, and murine PA2.26 (87–92). During mouse development, podoplanin is expressed at around E9 in the central nervous system and the foregut, but not yet in the vascular system (88,89,93). Around E11.5 to E12.5, its expression becomes apparent in all ECs in the cardinal vein, including the budding Prox1$^+$ cells (88). Similar to VEGFR3 and LYVE-1, the expression of podoplanin is maintained in the budding cells that are committed to the lymphatic lineage, but is progressively downregulated in venous ECs (88). Podoplanin-deficient mice die at birth due to lung failure (88,89). Importantly, newborn mutant mice display lymphedema in the limbs due to impaired cutaneous lymphatic transport (88). The podoplanin mutant mice develop a normal blood vasculature, but the lymphatic networks of their intestine and skin are disorganized. In vitro functional analyses showed that podoplanin is essential for the migration, adhesion, and tube formation of LECs (89). Because *podoplanin*-null mice die at birth, the generation of tissue-specific knockout mice will be essential to further study the role of podoplanin in adult lymphatics.

Angiopoietins

Angiopoietin 1 (Ang1) and Ang2 are competing ligands for the EC-specific tyrosine kinase receptor Tie2 (94,95). Whereas Ang1 recruits perivascular cells and stabilizes mature blood vessels, Ang2 interferes with this stabilizing signal, resulting in disturbed interactions between BECs and surrounding pericytes and dismantling of the EC–basement membrane–pericyte structure. Ang2 induces angiogenic sprouting when VEGF is present, but promotes vessel regression in the absence of VEGF (96,97). Ang2 is expressed by the smooth muscle cells of large arteries, large veins, and venules, and its expression is enhanced by VEGF-A and hypoxia in microvascular ECs at sites of vascular remodeling (97–99). A study of *Ang2*-knockout mice revealed that Ang2 is dispensable for embryonic vascular development but is required for subsequent angiogenic remodeling (100). *Ang2*-knockout mice also display major lymphatic vessel defects and develop chylous ascites and lymphedema due to lymphatic dysfunction (100). Their large lymphatic vessels are structurally irregular and leaky, and the smaller lymphatic vessels display abnormal patterning. Interestingly, overexpression of Ang1 rescued the defects of the lymphatic system, but not those of the blood vascular system in Ang2-deficient mice (100). This suggests that Ang2 acts as a Tie2 agonist for lymphatic vessels but as an antagonist for blood vessels, adding additional complexity to the exquisite interplays between Ang1/2 and other angiogenic factors (100). Although LECs express Tie2, the lymphatic phenotype of Tie2-deficient mice was not studied due to their embryonic lethality at E10.5, when the lymphatic system is about to appear (101,102). Recently, Ang1 was shown to promote lymphatic sprouting and hyperplasia in a mouse model (103).

Syk and SLP-76

Spleen tyrosine kinase (Syk) and SH2 domain-containing leukocyte protein of 76 kDa (SLP-76), are two adaptor molecules involved in signaling in hematopoietic cells. The protein tyrosine kinase Syk is widely expressed in hematopoietic cells and transfers signals from activated immunoreceptors that result in diverse cellular responses such as cell proliferation, differentiation, and phagocytosis (104). SLP-76, also known as lymphocyte cytosolic protein 2 (LCP2), is a substrate of Syk and serves as an adaptor protein for the signal transduction (105). Syk- and SLP-76–mediated hematopoietic signaling was shown to be required for normal development of the lymphatic system because Syk- or SLP-76–deficient mutant mice displayed incomplete separation of developing lymphatic vessels from embryonic vein during embryogenesis (106). This failure of separation gave rise to an arterio-venous-lymphatic shunting and, as the result, the mutant mice exhibited cardiomegaly, elevated cardiac output, and admixture of blood with lymph (106). The Syk- or SLP-76–deficient mouse embryos develop extensive hemorrhage into the developing cutaneous lymphatic vessels at around E12. Interestingly, chimeric vessels that are composed of both LYVE-1^+ lymphatic- and LYVE-1^- BECs were found in the mutant mice (106). Because neither EC type expresses Syk or SLP-76, these findings suggest that hematopoietic cells might be involved in the separation of the two vascular systems (106). Consistent with this notion, transplantation of bone marrow cells from SLP-76–deficient mice into lethally irradiated wild-type mice reproduced the same lymph–vascular phenotype as that of SLP-76–deficient mice (106). Collectively, these findings raise the possibility that hematopoietic cells contribute to the development of the lymphatic system. However, it remains to be answered how Syk and SLP-76 influence lymphatic development.

CC Chemokine Ligand 21

CC chemokine ligand 21 (CCL21) plays an important role in inflammatory responses. CCL21 is also known as secondary lymphoid-tissue chemokine (SLC), Exodus-2, or 6Ckine. CCL21 is predominantly expressed by lymphatic endothelium, high endothelial venules (HEVs), and the T-cell area cells in lymph nodes, and Peyer patches. CCL21 serves as a ligand of chemokine receptor 7 (CCR7) and promotes adhesion and stimulates migration of thymocytes, T lymphocytes, macrophages, and neutrophils via high-affinity binding to the receptor. It is the first chemokine identified that mediates homing of lymphocytes and migration of antigen-stimulated dendritic cells into secondary lymphoid organs (107,108).

Platelet-Derived Growth Factor

Platelet-derived growth factor (PDGF) was first isolated by its sequence homology to a viral oncogene, and was subsequently shown to promote the proliferation of mesenchymal cells. PDGF consists of two chains and five members (PDGF-AA, -BB, -AB, -CC, and -DD) constitute the PDGF family. PDGF receptors (PDGFRs) comprise two subunits (α and β), forming PDGFR-$\alpha\alpha$, -$\alpha\beta$, and -$\beta\beta$. Ligand binding to a receptor subunit induces homo- or heterodimerization of the receptors, resulting in their autophosphorylation. PDGF has been implicated in pathological angiogenesis. Recently, the PDGF family also was found to promote lymphangiogenesis (109). PDGF-BB stimulated mitogen-activated protein kinase (MAPK) activity and the cell motility of isolated LECs. PDGF-BB potently induced growth of lymphatic vessels in vivo and promoted tumor metastasis, independent of the VEGFR3 signaling pathway. Consistently, PDFGR-β was shown to be expressed in LECs (109).

Hepatocyte Growth Factor

Hepatocyte growth factor (HGF) was found to promote lymphangiogenesis both in vitro and in vivo (110). Activated LECs in wounds or inflamed skin display significantly higher levels of HGF receptor (HGFR), whereas little or no HGFR expression was detected in lymphatic vessels from normal tissues. Notably, HGF-induced proliferation of LECs did not require activation of VEGFR3, and HGF-induced cell migration was partially mediated by integrin $\alpha 9$ (110).

Insulin-Like Growth Factors

Insulin-like growth factor (IGF)-1 and -2 were recently reported to promote lymphangiogenesis in a mouse cornea assay (111). IGF-1–induced lymphangiogenesis was not blocked by administration of soluble VEGFR3, suggesting that the VEGFR3 signaling pathway is not required for IGF-1–mediated lymphangiogenesis. IGF-1 and IGF-2 stimulate phosphorylation of intracellular signaling components, such as Akt, Src, and ERK in LECs.

Neuropilin 2

NRP2 is a nonkinase type I transmembrane receptor and functions as a neuronal guidance molecule. NRP2 binds to the class III semaphorins and several VEGF family members including VEGF-A_{165}, placental growth factor (PlGF)-2, and VEGF-C (112). Although NRP1 is predominantly expressed in arterial ECs, NRP2 is mainly expressed in veins. Whereas NRP2 is strongly expressed in visceral lymphatic vessels, its expression is weak in cutaneous lymphatic capillaries (34,113,114). Interestingly, NRP2 mutant mice display a severe reduction of small lymphatic vessels in all tissues including the skin, gut, heart, diaphragm, and lung, but develop normal arteries, veins, as well as larger, collecting lymphatic vessels (113). This mutant phenotype is consistent with the finding that NRP2 binds VEGF-C and that VEGF-C signaling through VEGFR3 may be enhanced by NRP2, resembling the NRP1–mediated promotion of VEGF-A signaling through VEGFR2 (61,114). Therefore, NRPs are essential not only for only axon guidance, but also for angiogenesis and lymphangiogenesis by serving as multifunctional semaphorin and VEGF family receptors.

Fibroblast Growth Factor 2

Fibroblast growth factor 2 (FGF2) potently stimulates angiogenesis. Similarly, FGF2 was also shown to strongly activate proliferation of cultured LECs and in vivo lymphatic vessel growth in the mouse cornea assay. However, it was proposed that the lymphangiogenic activity of FGF2 is indirectly mediated through the VEGF-C/VEGFR2 signal pathway based on a mouse cornea assay in which neutralizing VEGFR3 blocked FGF2-induced lymphangiogenesis. However, FGF receptors were shown to be expressed in LECs both in vitro and in vivo (115). Interestingly, a low dose of FGF2 induces lymphangiogenesis without accompanying angiogenesis, suggesting that lymphatic vessel growth does not require preceding growth of blood vessels.

LYMPHEDEMA AND THE DEVELOPMENT OF MOLECULAR THERAPIES

Lymphedema refers to chronic tissue swelling (most commonly of the extremities) that is caused by the accumulation of interstitial fluid due to lymphatic dysfunction resulting from lymphatic hypoplasia, lymphatic vessel malformation, impaired lymphatic transport, and/or obstruction of lymph flow. Lymphedema can be divided into primary (genetic) and secondary (acquired) forms, based on underlying etiology (116–119). Primary lymphedema arises from a developmental abnormality of the lymphatic system caused by genetic defects. Depending on the age of onset, primary lymphedema is further subdivided into three groups: congenital lymphedema (at birth), lymphedema praecox (early onset), and lymphedema tarda (late onset).

Congenital lymphedema represents all forms of lymphedema that are clinically evident at birth, and accounts for 10% to 25% of all primary lymphedema. Females are more commonly affected than males (by a factor of 2), and lower extremity swelling is threefold more common than is upper extremity swelling. A subgroup of congenital lymphedema (Milroy disease) has been recently linked to the *VEGFR3* locus on distal chromosome 5q, and various missense mutations in the *VEGFR3* gene have been identified in patients with this disorder (120–122).

Lymphedema praecox (Meige disease) is the most common form (60%–80%) of primary lymphedema and becomes clinically apparent in childhood and around puberty. Females are affected four times more than males, and patients display hypoplastic lymphatic networks with reduced caliber and number of lymphatic vessels. Lymphedema-distichiasis, a subset of lymphedema praecox, is a rare autosomal dominant disease characterized by double rows of eyelashes in addition to swelling of the limbs. Lymphedema-distichiasis has been genetically linked to the *FOXC2* gene, which encodes a member of the forkhead/winged-helix family of transcription factors that are involved in diverse developmental pathways (123–125). Foxc2-deficient mice display lymphatic dysfunction, associated with defective valves of the collecting lymphatic vessels, irregular lymphatic patterning, and abnormal pericyte coverage on lymphatic capillaries (126). Hypotrichosis-lymphedema-telangiectasia, another subset of lymphedema praecox, has been recently described to be unusually associated with hypotrichosis and telangiectasia (127,128). Mutations of the SRY-related transcription factor SOX18 were found to be responsible for both recessive and dominant forms of this disease.

The third form of primary lymphedema is called lymphedema tarda, which is defined as a late onset lymphedema (after age 35), and constitutes about 10% of all primary lymphedema. This form of lymphedema displays a tortuous, hyperplastic pattern of lymphatic vessels characterized by an increase in diameter and number. Notably, patients with this form of lymphedema often lack functional lymph valves.

Secondary (acquired) lymphedema is mainly caused by functionally compromised lymphatics due to surgery, radiation, infection (filariasis), or compression. Lymph dissection (lymphadenectomy) is a common procedure for staging tumors of breast, gynecological, head and neck, testicular, prostate, bladder, or colon cancers, rendering patients at an increased risk for lymphedema. To date, the highest incidence of secondary lymphedema is observed among breast cancer patients who undergo radiation therapy following auxiliary lymphadenectomy. Several studies show that 25% to 56% of breast cancer patients develop mild-to-severe lymphedema after cancer treatment (129–132). Worldwide, the most common cause of lymphedema is filariasis, the direct infestation of lymph nodes by the parasite *Wuchereria bancrofti*. Other causes of lymphedema include vein stripping, vascular surgery, lipectomy, and burns. Lymphedema presents considerable social, economic, and psychological hardship to patients and their families. Unfortunately, there is no currently available cure for this disfiguring disease.

Lymphedema also has been described in a number of genetic mouse models (Table 169-3). The *Chy* mouse, originally generated by ethylnitrosourea-induced mutagenesis, displays chronic lymphedema that is attributed to an amino acid substitution mutation in the highly conserved VEGFR3 tyrosine kinase domain (similar to human Milroy disease) (114). Targeted overexpression of a soluble form of VEGFR3 in the skin resulted in functional blockade of endogenous VEGF-C and VEGF-D signaling, and a lymphedema-like phenotype (48). Furthermore, similar to human lymphedema-distichiasis, the heterozygote Foxc2 mutant mice developed generalized hypoplasia of lymphatic vessels and lymph nodes associated with occasional hindlimb swelling (133). Finally, mouse embryos with targeted deletion of the *Net* gene, an Ets ternary complex transcription factor, showed an accumulation of chyle in the thoracic cage (chylothorax) and dilated lymphatic vessels (lymphangiectasis) (134).

To date, a number of lymphangiogenic factors have been reported, including VEGF-A, VEGF-C, VEGF-D, FGF2, PDGF, IGF-1, IGF-2, Ang1, and HGF, and the list is rapidly growing. Among them, VEGF-C has been best characterized and provides a hope to treat lymphedema at least in animal models. Application of a VEGF-C–expressing adenovirus promoted generation of functional new lymphatic vessels in the *Chy* mutant mice (114). Likewise, adenovirus-mediated overexpression of VEGF-C in normal skin or at the edge of epigastric skin flaps in mice has been shown to induce lymphangiogenesis (80,115). Furthermore, when human recombinant VEGF-C was applied to rabbit ears, of which lymphatics were previously surgically ablated to induce lymphedema, a single dose of VEGF-C was sufficient to increase lymphangiogenesis and to reverse the tissue hypercellularity of the lymphedema (136). Similarly, VEGF-C gene therapy augmented postnatal lymphangiogenesis and ameliorated secondary lymphedema in the rabbit ear and mouse tail lymphedema models (137). Notably, however, lymphatic vessels induced by VEGF-C in a tumor model exhibit a retrograde draining pattern possibly

Table 169-3: Mutant Mouse Models Showing Abnormalities of the Lymphatic Vascular System

Genes	*Models*	*References*	*Phenotypes*
Ang2	KO	98	Chylous ascites, subcutaneous edema, and poor interstitial fluid uptake; abnormal lymphatic patterning
FOXC2	KO	103, 124	Abnormal lymphatic vascular patterning, increased pericyte investment of lymphatic vessels, agenesis of valves and lymphatic dysfunction
HGF	TG	108, 139	Increased lymphangiogenesis
Integrin $\alpha 9$	KO	140	Chylothorax, excessive pleural fluid, edema
NRP2	KO	111	Reduction of small lymphatic vessels during development
Podoplanin	KO	86	Lymphedema, dilation of lymphatic vessels, and diminished lymphatic transport
Prox1	KO	7	No lymphatic vasculature developed
SLP-79 and Syk	KO	104	Failure of blood–lymphatic separation
VEGF-C	TG	66	Hyperplastic lymphatic vessels
VEGFR3	KO	46	Cardiovascular failure and defect remodeling of vascular networks
VEGFR3	*Chy* mice	112	Chylous ascites, hypoplastic cutaneous lymphatic vessels, swelling of limbs

KO, knockout; TG, transgenic

due to defective intraluminal valves of the vessels (138). Therefore, although VEGF-C provides promising data to be used as a new therapy against lymphedema, additional studies remain to be performed on the functionality of VEGF-C–induced lymphatic vessels and on the long-term toxicity and efficacy of VEGF-C.

CONCLUSION

Since its early description by Hippocrates, the lymphatic system has received far less attention than the blood vascular system. This relative anonymity of the lymphatics is, at least in part, due to its elusive molecular, anatomical, and functional features. However, in recent years, the discovery of novel lymphatic-specific molecular markers and a defined LEC culture has significantly improved our understanding of the genesis and function of the system. Despite remarkable advances during the past decade, many unanswered questions remain. What are the ultimate signaling mechanisms that induce the differentiation program of LECs during embryogenesis? What are the molecular mechanisms underlying maturation and regeneration of lymphatic vessels in adults? To what extent is the lymphatic endothelium heterogeneous in structure and function in the adult, and how do the spatial and temporal dynamics of the system contribute to health and disease? What are the molecular interactions between tumor cells and the LECs of tumor-associated lymphatic vessels, and how do these interactions mediate tumor metastases? The answer to these and related questions should provide a strong foundation for developing new therapies for patients with lymphedema and/or cancer.

KEY POINTS

- The lymphatic system is derived from the blood vascular system during development.
- Prox1, LYVE-1, podoplanin, and VEGFR3 are expressed in LECs but not in BECs.
- A number of molecular players are involved in development of the lymphatic system.
- An obstruction of the lymphatic system results in lymphedema.

Future Goals

- To understand the inductive signal and molecular mechanism underlying lymphatic differentiation of venous BECs during development

- To identify and characterize additional BEC- or LEC-specific molecules essential for lineage-specific differentiation
- To identify and characterize lymphatic endothelial precursor cells that are essential for physiological and pathological lymphangiogenesis
- To develop novel therapies to treat primary and secondary lymphedema

REFERENCES

1 Cursiefen C, Chen L, Dana MR, Streilein JW. Corneal lymphangiogenesis: evidence, mechanisms, and implications for corneal transplant immunology. *Cornea*. 2003;22(3):273–281.

2 Asellius G. *De Lactibus Sive Lacteis Venis*. Milan: Mediolani, 1627.

3 Harvey W. *An Anatomical Study of the Motion of the Heart and Blood in Animals*. London, 1628.

4 Sabin FR. On the origin of the lymphatic system from the veins and the development of the lymph hearts and thoracic duct in the pig. *Am J Anat*. 1902;1:367–391.

5 Sabin FR. On the development of the superficial lymphatics in the skin of the pig. *Am J Anat*. 1904;3:183–195.

6 Leak LV, Burke JF. Fine structure of the lymphatic capillary and the adjoining connective tissue area. *Am J Anat*. 1966;118(3):785–809.

7 Wigle JT, Oliver G. Prox1 function is required for the development of the murine lymphatic system. *Cell*. 1999;98(6):769–778.

8 Wigle JT, Harvey N, Detmar M, et al. An essential role for Prox1 in the induction of the lymphatic endothelial cell phenotype. *EMBO J*. 2002;21(7):1505–1513.

9 Ny A, Koch M, Schneider M, et al. A genetic *Xenopus laevis* tadpole model to study lymphangiogenesis. *Nat Med*. 2005;11(9):998–1004.

10 Schneider M, Othman-Hassan K, Christ B, Wilting J. Lymphangioblasts in the avian wing bud. *Dev Dyn*. 1999;216(4–5):311–319.

11 Wilting J, Papoutsi M, Othman-Hassan K, et al. Development of the avian lymphatic system. *Microsc Res Tech*. 2001;55(2):81–91.

12 He L, Papoutsi M, Huang R, et al. Three different fates of cells migrating from somites into the limb bud. *Anat Embryol (Berl)*. 2003;207(1):29–34.

13 Rusznyak I, Foeldi M, Szabo G. *Lymphatics and Lymph Circulation*. Oxford: Pergamon Press, 1967.

14 Yaniv K, Isogai S, Castranova D, et al. Live imaging of lymphatic development in the zebrafish. *Nat Med*. 2006;12(6):711–716.

15 Ohtani O, Ohtani Y, Li RX. Phylogeny and ontogeny of the lymphatic stomata connecting the pleural and peritoneal cavities with the lymphatic system – a review. *Ital J Anat Embryol*. 2001;106(2 Suppl 1):251–259.

16 Cerra MC, Amelio D, Tavolaro P, et al. Pericardium of the frog, Rana esculenta, is morphologically designed as a lymphatic space. *J Morphol*. 2003;257(1):72–77.

17 Baldwin AL, Ferrer P, Rozum JS, Gore RW. Regulation of water balance between blood and lymph in the frog, Rana pipiens. *Lymphology*. 1993;26(1):4–18.

18 Greber K, Schipp R. Early development and myogenesis of the posterior anuran lymph hearts. *Anat Embryol (Berl)*. 1990;181(1):75–82.

19 Koyama T, Horimoto M. Lymphatic microcirculation in frog lung. *Biorheology*. 1988;25(1–2):219–226.

20 Day JB, Rech RH, Robb JS. Pharmacological and microelectrode studies on the frog lymph heart. *J Cell Physiol*. 1963;62:33–41.

21 Hirakawa S, Hong YK, Harvey N, et al. Identification of vascular lineage-specific genes by transcriptional profiling of isolated blood vascular and lymphatic endothelial cells. *Am J Pathol*. 2003;162(2):575–586.

22 Podgrabinska S, Braun P, Velasco P, et al. Molecular characterization of lymphatic endothelial cells. *Proc Natl Acad Sci USA*. 2002;99(25):16069–16074.

23 Kriehuber E, Breiteneder GS, Groeger M, et al. Isolation and characterization of dermal lymphatic and blood endothelial cells reveal stable and functionally specialized cell lineages. *J Exp Med*. 2001;194:797–808.

24 Makinen T, Veikkola T, Mustjoki S, et al. Isolated lymphatic endothelial cells transduce growth, survival and migratory signals via the VEGF-C/D receptor VEGFR-3. *EMBO J*. 2001;20:4762–4773.

25 Petrova TV, Makinen T, Makela TP, et al. Lymphatic endothelial reprogramming of vascular endothelial cells by the Prox-1 homeobox transcription factor. *EMBO J*. 2002;21(17):4593–4599.

26 Oliver G, Sosa-Pineda B, Geisendorf S, et al. Prox 1, a prospero-related homeobox gene expressed during mouse development. *Mech Dev*. 1993;44(1):3–16.

27 Tomarev SI, Sundin O, Banerjee-Basu S, et al. Chicken homeobox gene Prox 1 related to *Drosophila Prospero* is expressed in the developing lens and retina. *Dev Dyn*. 1996;206(4):354–367.

28 Hong YK, Detmar M. Prox1, master regulator of the lymphatic vasculature phenotype. *Cell Tissue Res*. 2003;314(1):85–92.

29 Doe CQ, Chu-LaGraff Q, Wright DM, Scott MP. The prospero gene specifies cell fates in the *Drosophila* central nervous system. *Cell*. 1991;65(3):451–464.

30 Vaessin H, Grell E, Wolff E, et al. Prospero is expressed in neuronal precursors and encodes a nuclear protein that is involved in the control of axonal outgrowth in *Drosophila*. *Cell*. 1991;67(5):941–953.

31 Jacobs JR. Perturbed glial scaffold formation precedes axon tract malformation in *Drosophila* mutants. *J Neurobiol*. 1993;24(5):611–626.

32 Kauffmann RC, Li S, Gallagher PA, et al. Ras1 signaling and transcriptional competence in the R7 cell of *Drosophila*. *Genes Dev*. 1996;10(17):2167–2178.

33 Schaefer JJ, Oliver G, Henry JJ. Conservation of gene expression during embryonic lens formation and cornea-lens transdifferentiation in *Xenopus laevis*. *Dev Dyn*. 1999;215(4):308–318.

34 Hong YK, Harvey N, Noh YH, et al. Prox1 is a master control gene in the program specifying lymphatic endothelial cell fate. *Dev Dyn*. 2002;225(3):351–357.

35 Wigle JT, Chowdhury K, Gruss P, Oliver G. Prox1 function is crucial for mouse lens-fibre elongation. *Nat Genet*. 1999;21(3):318–322.

36 Dyer MA, Livesey FJ, Cepko CL, Oliver G. Prox1 function controls progenitor cell proliferation and horizontal cell genesis in the mammalian retina. *Nat Genet*. 2003;34(1):53–58.
37 Edqvist PH, Hallbook F. Newborn horizontal cells migrate bi-directionally across the neuroepithelium during retinal development. *Development*. 2004;131(6):1343–1351.
38 Harvey NL, Srinivasan RS, Dillard ME, et al. Lymphatic vascular defects promoted by Prox1 haploinsufficiency cause adult-onset obesity. *Nat Genet*. 2005;37(10):1072–1081.
39 Wang HW, Trotter MW, Lagos D, et al. Kaposi sarcoma herpesvirus-induced cellular reprogramming contributes to the lymphatic endothelial gene expression in Kaposi sarcoma. *Nat Genet*. 2004;36(7):687–693.
40 Hong YK, Foreman K, Shin JW, et al. Lymphatic reprogramming of blood vascular endothelium by Kaposi sarcoma-associated herpesvirus. *Nat Genet*. 2004;36(7):683–685.
41 Carroll PA, Brazeau E, Lagunoff M. Kaposi's sarcoma-associated herpesvirus infection of blood endothelial cells induces lymphatic differentiation. *Virology*. 2004;328(1):7–18.
42 Kaipainen A, Korhonen J, Pajusola K, et al. The related FLT4, FLT1, and KDR receptor tyrosine kinases show distinct expression patterns in human fetal endothelial cells. *J Exp Med*. 1993; 178(6):2077–2088.
43 Kaipainen A, Korhonen J, Mustonen T, et al. Expression of the fms-like tyrosine kinase 4 gene becomes restricted to lymphatic endothelium during development. *Proc Natl Acad Sci USA*. 1995;92(8):3566–3570.
44 Tammela T, Petrova TV, Alitalo K. Molecular lymphangiogenesis: new players. *Trends Cell Biol*. 2005;15(8):434–441.
45 Lawson ND, Scheer N, Pham VN, et al. Notch signaling is required for arterial-venous differentiation during embryonic vascular development. *Development*. 2001;128(19):3675–3683.
46 Torres-Vazquez J, Kamei M, Weinstein BM. Molecular distinction between arteries and veins. *Cell Tissue Res*. 2003;314(1):43–59.
47 Dumont DJ, Jussila L, Taipale J, et al. Cardiovascular failure in mouse embryos deficient in VEGF receptor-3. *Science*. 1998;282(5390):946–949.
48 Makinen T, Jussila L, Veikkola T, et al. Inhibition of lymphangiogenesis with resulting lymphedema in transgenic mice expressing soluble VEGF receptor-3. *Nat Med*. 2001;7(2):199–205.
49 Lin J, Lalani AS, Harding TC, et al. Inhibition of lymphogenous metastasis using adeno-associated. *Cancer Res*. JID – 2984705R 2005;65(15):6901–6909.
50 He Y, Rajantie I, Pajusola K, et al. Vascular endothelial cell growth factor receptor 3-mediated activation of. *Cancer Res*. JID – 2984705R 2005;65(11):4739–4746.
51 Persaud K, Tille J-C, Liu M, et al. Involvement of the VEGF receptor 3 in tubular morphogenesis demonstrated. *J Cell Sci*. JID – 0052457 2004;117(Pt 13):2745–2756.
52 Kubo H, Fujiwara T, Jussila L, et al. Involvement of vascular endothelial growth factor receptor-3 in maintenance of integrity of endothelial cell lining during tumor angiogenesis. *Blood*. 2000;96(2):546–553.
53 Veikkola T, Jussila L, Makinen T, et al. Signalling via vascular endothelial growth factor receptor-3 is sufficient for lymphangiogenesis in transgenic mice. *EMBO J*. 2001;20:1223–1231.
54 Joukov V, Pajusola K, Kaipainen A, et al. A novel vascular endothelial growth factor, VEGF-C, is a ligand for the Flt4 (VEGFR-3) and KDR (VEGFR-2) receptor tyrosine kinases. *EMBO J*. 1996;15(7):1751.
55 Lee J, Gray A, Yuan J, et al. Vascular endothelial growth factor-related protein: a ligand and specific activator of the tyrosine kinase receptor Flt4. *Proc Natl Acad Sci USA*. 1996;93(5):1988–1992.
56 Yamada Y, Nezu J, Shimane M, Hirata Y. Molecular cloning of a novel vascular endothelial growth factor, VEGF-D. *Genomics*. 1997;42(3):483–488.
57 Achen MG, Jeltsch M, Kukk E, et al. Vascular endothelial growth factor D (VEGF-D) is a ligand for the tyrosine kinases VEGF receptor 2 (Flk1) and VEGF receptor 3 (Flt4). *Proc Natl Acad Sci USA*. 1998;95(2):548–553.
58 Orlandini M, Marconcini L, Ferruzzi R, Oliviero S. Identification of a c-fos-induced gene that is related to the platelet-derived growth factor/vascular endothelial growth factor family. *Proc Natl Acad Sci USA*. 1996;93(21):11675–11680.
59 Eichmann A, Corbel C, Jaffredo T, et al. Avian VEGF-C: cloning, embryonic expression pattern and stimulation of the differentiation of VEGFR2-expressing endothelial cell precursors. *Development*. 1998;125(4):743–752.
60 Kukk E, Lymboussaki A, Taira S, et al. VEGF-C receptor binding and pattern of expression with VEGFR-3 suggests a role in lymphatic vascular development. *Development*. 1996;122 (12):3829–3837.
61 Karkkainen MJ, Haiko P, Sainio K, et al. Vascular endothelial growth factor C is required for sprouting of the first lymphatic vessels from embryonic veins. *Nat Immunol*. 2004;5(1):74–80.
62 Joukov V, Sorsa T, Kumar V, et al. Proteolytic processing regulates receptor specificity and activity of VEGF-C. *EMBO J*. 1997; 16(13):3898–3911.
63 Stacker SA, Stenvers K, Caesar C, et al. Biosynthesis of vascular endothelial growth factor-D involves proteolytic processing which generates non-covalent homodimers. *J Biol Chem*. 1999; 274(45):32127–32136.
64 Cao Y, Linden P, Farnebo J, et al. Vascular endothelial growth factor C induces angiogenesis in vivo. *Proc Natl Acad Sci USA*. 1998;95(24):14389–14394.
65 Siegfried G, Basak A, Cromlish JA, et al. The secretory proprotein convertases furin, PC5, and PC7 activate VEGF-C to induce tumorigenesis. *J Clin Invest*. 2003;111(11):1723–1732.
66 Joukov V, Kumar V, Sorsa T, et al. A recombinant mutant vascular endothelial growth factor-C that has lost vascular endothelial growth factor receptor-2 binding, activation, and vascular permeability activities. *J Biol Chem*. 1998;273(12):6599–6602.
67 Jeltsch M, Kaipainen A, Joukov V, et al. Hyperplasia of lymphatic vessels in VEGF-C transgenic mice. *Science*. 1997;276:1423–1425.
68 Skobe M, Hawighorst T, Jackson DG, et al. Induction of tumor lymphangiogenesis by VEGF-C promotes breast cancer metastasis. *Nat Med*. 2001;7(2):192–198.
69 Stacker SA, Caesar C, Baldwin ME, et al. VEGF-D promotes the metastatic spread of tumor cells via the lymphatics. *Nat Med*. 2001;7(2):186–191.
70 Tille JC, Nisato R, Pepper MS. Lymphangiogenesis and tumour metastasis. *Novartis Found Symp*. JID – 9807767 2004;256:112–131; Discussion 32–36, 259–269.
71 He Y, Karpanen T, Alitalo K. Role of lymphangiogenic factors in tumor metastasis. *Biochim Biophys Acta*. JID – 0217513 2004; 1654(1):3–12.
72 Lohela M, Saaristo A, Veikkola T, Alitalo K. Lymphangiogenic growth factors, receptors and therapies. *Thromb Haemost*. JID – 7608063 2003;90(2):167–184.

73 Duff SE, Li C, Jeziorska M, et al. Vascular endothelial growth factors C and D and lymphangiogenesis in. *Br J Cancer*. JID – 0370635 2003;89(3):426–430.

74 Pepper MS, Tille J-C, Nisato R, Skobe M. Lymphangiogenesis and tumor metastasis. *Cell Tissue Res*. JID – 0417625 2003; 314(1):167–177.

75 Dadras SS, Paul T, Bertoncini J, et al. Tumor lymphangiogenesis: a novel prognostic indicator for cutaneous melanoma metastasis and survival. *Am J Pathol*. 2003;162(6):1951–1960.

76 Baldwin ME, Halford MM, Roufail S, et al. Vascular endothelial growth factor D is dispensable for development of the lymphatic system. *Mol Cell Biol*. 2005;25(6):2441–2449.

77 Hong Y-K, Lange-Asschenfeldt B, Velasco P, et al. VEGF-A promotes tissue repair-associated lymphatic vessel formation via VEGFR-2 and the a1b1 and a2b1 integrins. *FASEB J*. 2004;18 (10):1111–1113.

78 Nagy JA, Vasile E, Feng D, et al. Vascular permeability factor/vascular endothelial growth factor induces lymphangiogenesis as well as angiogenesis. *J Exp Med*. 2002;196(11):1497–1506.

79 Kunstfeld R, Hirakawa S, Hong YK, et al. Induction of cutaneous delayed-type hypersensitivity reactions in VEGF-A transgenic mice results in chronic skin inflammation associated with persistent lymphatic hyperplasia. *Blood*. 2004;104(4):1048–1057.

80 Enholm B, Karpanen T, Jeltsch M, et al. Adenoviral expression of vascular endothelial growth factor-C induces lymphangiogenesis in the skin. *Circ Res*. 2001;88(6):623–629.

81 Byzova TV, Goldman CK, Jankau J, et al. Adenovirus encoding vascular endothelial growth factor-D induces tissue-specific vascular patterns in vivo. *Blood*. 2002;99(12):4434–4442.

82 Hirakawa S, Kodama S, Kunstfeld R, et al. VEGF-A induces tumor and sentinel lymph node lymphangiogenesis and promotes lymphatic metastasis. *J Exp Med*. 2005;201(7):1089–1099.

83 Jackson DG. The lymphatics revisited: new perspectives from the hyaluronan receptor LYVE-1. *Trends Cardiovasc Med*. 2003; 13(1):1–7.

84 Jackson DG. Biology of the lymphatic marker LYVE-1 and applications in research into lymphatic trafficking and lymphangiogenesis. *APMIS*. 2004;112(7–8):526–538.

85 Banerji S, Ni J, Wang SX, et al. LYVE-1, a new homologue of the CD44 glycoprotein, is a lymph-specific receptor for hyaluronan. *J Cell Biol*. 1999;144(4):789–801.

86 Gale NW, Prevo R, Espinosa J, et al. Normal lymphatic development and function in mice deficient for the lymphatic nyaluronan receptor LYVE-1. *Mol Cell Biol*. 2007;27(2):595–604.

87 Wetterwald A, Hoffstetter W, Cecchini MG, et al. Characterization and cloning of the E11 antigen, a marker expressed by rat osteoblasts and osteocytes. *Bone*. 1996;18(2):125–132.

88 Schacht V, Ramirez MI, Hong YK, et al. T1alpha/podoplanin deficiency disrupts normal lymphatic vasculature formation and causes lymphedema. *EMBO J*. 2003;22(14):3546–3556.

89 Ramirez MI, Millien G, Hinds A, et al. T1alpha, a lung type I cell differentiation gene, is required for normal lung cell proliferation and alveolus formation at birth. *Dev Biol*. 2003;256(1):61–72.

90 Kato Y, Fujita N, Kunita A, et al. Molecular identification of Aggrus/T1alpha as a platelet aggregation-inducing factor expressed in colorectal tumors. *J Biol Chem*. 2003;278(51): 51599–51605.

91 Zimmer G, Oeffner F, Von Messling V, et al. Cloning and characterization of gp36, a human mucin-type glycoprotein preferentially expressed in vascular endothelium. *Biochem J*. 1999; 341(Pt 2):277–284.

92 Nose K, Saito H, Kuroki T. Isolation of a gene sequence induced later by tumor-promoting 12-O-tetradecanoylphorbol-13-acetate in mouse osteoblastic cells (MC3T3-E1) and expressed constitutively in ras-transformed cells. *Cell Growth Differ*. 1990; 1(11):511–518.

93 Rishi AK, Joyce-Brady M, Fisher J, et al. Cloning, characterization, and developmental expression of a rat lung alveolar type I cell gene in embryonic endodermal and neural derivatives. *Dev Biol*. 1995;167:294–306.

94 Maisonpierre PC, Suri C, Jones PF, et al. Angiopoietin-2, a natural antagonist for Tie2 that disrupts in vivo angiogenesis. *Science*. 1997;277(5322):55–60.

95 Suri C, Jones PF, Patan S, et al. Requisite role of angiopoietin-1, a ligand for the TIE2 receptor, during embryonic angiogenesis. *Cell*. 1996;87(7):1171–1180.

96 Holash J, Wiegand SJ, Yancopoulos GD. New model of tumor angiogenesis: dynamic balance between vessel regression and growth mediated by angiopoietins and VEGF. *Oncogene*. 1999;18(38):5356–5362.

97 Holash J, Maisonpierre PC, Compton D, et al. Vessel cooption, regression, and growth in tumors mediated by angiopoietins and VEGF. *Science*. 1999;284(5422):1994–1998.

98 Mandriota SJ, Pepper MS. Regulation of angiopoietin-2 mRNA levels in bovine microvascular endothelial cells by cytokines and hypoxia. *Circ Res*. 1998;83(8):852–859.

99 Mandriota SJ, Pyke C, Di Sanza C, et al. Hypoxia-inducible angiopoietin-2 expression is mimicked by iodonium compounds and occurs in the rat brain and skin in response to systemic hypoxia and tissue ischemia. *Am J Pathol*. 2000;156 (6):2077–2089.

100 Gale NW, Thurston G, Hackett SF, et al. Angiopoietin-2 is required for postnatal angiogenesis and lymphatic patterning, and only the latter role is rescued by angiopoietin-1. *Dev Cell*. 2002;3:411.

101 Dumont DJ, Gradwohl G, Fong GH, et al. Dominant-negative and targeted null mutations in the endothelial receptor tyrosine kinase, tek, reveal a critical role in vasculogenesis of the embryo. *Genes Dev*. 1994;8(16):1897–1909.

102 Sato TN, Tozawa Y, Deutsch U, et al. Distinct roles of the receptor tyrosine kinases Tie-1 and Tie-2 in blood vessel formation. *Nature*. 1995;376(6535):70–74.

103 Tammela T, Saaristo A, Lohela M, et al. Angiopoietin-1 promotes lymphatic sprouting and hyperplasia. *Blood*. 2005;105 (12):4642–4648.

104 Yanagi S, Inatome R, Takano T, Yamamura H. Syk expression and novel function in a wide variety of tissues. *Biochem Biophys Res Commun*. 2001;288(3):495–498.

105 Clements JL. Known and potential functions for the SLP-76 adapter protein in regulating T-cell activation and development. *Immunol Rev*. 2003;191:211–219.

106 Abtahian F, Guerriero A, Sebzda E, et al. Regulation of blood and lymphatic vascular separation by signaling proteins SLP-76 and Syk. *Science*. 2003;299(5604):247–251.

107 Tangemann K, Gunn MD, Giblin P, Rosen SD. A high endothelial cell-derived chemokine induces rapid, efficient, and subset-selective arrest of rolling T lymphocytes on a reconstituted endothelial substrate. *J Immunol*. 1998;161(11):6330–6337.

108 Gunn MD, Tangemann K, Tam C, et al. A chemokine expressed in lymphoid high endothelial venules promotes the adhesion

and chemotaxis of naive T lymphocytes. *Proc Natl Acad Sci USA*. 1998;95(1):258–263.

109 Cao R, Bjorndahl MA, Religa P, et al. PDGF-BB induces intratumoral lymphangiogenesis and promotes lymphatic metastasis. *Cancer Cell*. 2004;6(4):333–345.

110 Kajiya K, Hirakawa S, Ma B, et al. Hepatocyte growth factor promotes lymphatic vessel formation and function. *EMBO J*. 2005;24(16):2885–2895.

111 Bjorndahl M, Cao R, Nissen LJ, et al. Insulin-like growth factors 1 and 2 induce lymphangiogenesis in vivo. *Proc Natl Acad Sci USA*. 2005;102(43):15593–15598.

112 Neufeld G, Cohen T, Shraga N, et al. The neuropilins: multifunctional semaphorin and VEGF receptors that modulate axon guidance and angiogenesis. *Trends Cardiovasc Med*. 2002; 12(1):13–19.

113 Yuan L, Moyon D, Pardanaud L, et al. Abnormal lymphatic vessel development in neuropilin 2 mutant mice. *Development*. 2002;129(20):4797–4806.

114 Karkkainen MJ, Saaristo A, Jussila L, et al. A model for gene therapy of human hereditary lymphedema. *Proc Natl Acad Sci USA*. 2001;98:12677–12682.

115 Shin JW, Min M, Larrieu-Lahargue F, et al. Prox1 promotes lineage-specific expression of fibroblast growth factor (FGF) receptor-3 in lymphatic endothelium: a role for FGF signaling in lymphangiogenesis. *Mol Biol Cell*. 2006;17(2):576–584.

116 Child AH, Beninson J, Sarfarazi M. Cause of primary congenital lymphedema. *Angiology*. JID – 0203706 1999;50(4):325–326.

117 Fonkalsrud EW. Congenital malformations of the lymphatic system. *Semin Pediatr Surg*. JID – 9216162 1994;3(2):62–69.

118 Mortimer PS. The pathophysiology of lymphedema. *Cancer*. JID – 0374236 1998;83(Suppl 12):2798–2802.

119 Rockson SG. Precipitating factors in lymphedema: myths and realities. *Cancer*. JID – 0374236 1998;83(Suppl 12):2814–2816.

120 Evans AL, Brice G, Sotirova V, et al. Mapping of primary congenital lymphedema to the 5q35.3 region. *Am J Hum Genet*. 1999; 64(2):547–555.

121 Karkkainen MJ, Ferrell RE, Lawrence EC, et al. Missense mutations interfere with VEGFR-3 signalling in primary lymphoedema. *Nat Genet*. 2000;25:153–159.

122 Irrthum A, Karkkainen MJ, Devriendt K, et al. Congenital hereditary lymphedema caused by a mutation that inactivates VEGFR3 tyrosine kinase. *Am J Hum Genet*. 2000;67(2):295–301.

123 Bell R, Brice G, Child AH, et al. Analysis of lymphoedema-distichiasis families for FOXC2 mutations reveals small insertions and deletions throughout the gene. *Hum Genet*. 2001; 108(6):546–551.

124 Fang J, Dagenais SL, Erickson RP, et al. Mutations in FOXC2 (MFH-1), a forkhead family transcription factor, are responsible for the hereditary lymphedema-distichiasis syndrome. *Am J Hum Genet*. 2000;67(6):1382–1388.

125 Finegold DN, Kimak MA, Lawrence EC, et al. Truncating mutations in FOXC2 cause multiple lymphedema syndromes. *Hum Mol Genet*. 2001;10(11):1185–1189.

126 Petrova TV, Karpanen T, Norrmen C, et al. Defective valves and abnormal mural cell recruitment underlie lymphatic vascular failure in lymphedema distichiasis. *Nat Med*. 2004;10(9):974–981.

127 Irrthum A, Devriendt K, Chitayat D, et al. Mutations in the transcription factor gene SOX18 underlie recessive and dominant forms of hypotrichosis-lymphedema-telangiectasia. *Am J Hum Genet*. 2003;72(6):1470–1478.

128 Hosking BM, Wang SCM, Downes M, et al. . The VCAM-1 gene that encodes the vascular cell adhesion molecule is a. *J Biol Chem*. JID – 2985121R 2004;279(7):5314–5322.

129 Pezner RD, Patterson MP, Hill LR, et al. Arm lymphedema in patients treated conservatively for breast cancer: relationship to patient age and axillary node dissection technique. *Int J Radiat Oncol Biol Phys*. 1986;12(12):2079–2083.

130 Kiel KD, Rademacker AW. Early-stage breast cancer: arm edema after wide excision and breast irradiation. *Radiology*. 1996; 198(1):279–283.

131 Hinrichs CS, Watroba NL, Rezaishiraz H, et al. Lymphedema secondary to postmastectomy radiation: incidence and risk factors. *Ann Surg Oncol*. 2004;11(6):573–580.

132 Ozaslan C, Kuru B. Lymphedema after treatment of breast cancer. *Am J Surg*. 2004;187(1):69–72.

133 Kriederman BM, Myloyde TL, Witte MH, et al. FOXC2 haploinsufficient mice are a model for human autosomal dominant lymphedema-distichiasis syndrome. *Hum Mol Genet*. 2003;12(10):1179–1185.

134 Ayadi A, Zheng H, Sobieszczuk P, et al. Net-targeted mutant mice develop a vascular phenotype and up-regulate egr-1. *EMBO J*. 2001;20(18):5139–5152.

135 Saaristo A, Tammela T, Timonen J, et al. Vascular endothelial growth factor-C gene therapy restores lymphatic flow across incision wounds. *FASEB J*. 2004;18(14):1707–1709.

136 Szuba A, Skobe M, Karkkainen MJ, et al. Therapeutic lymphangiogenesis with human recombinant VEGF-C. *FASEB J*. 2002;16(14):1985–1987.

137 Yoon YS, Murayama T, Gravereaux E, et al. VEGF-C gene therapy augments postnatal lymphangiogenesis and ameliorates secondary lymphedema. *J Clin Invest*. 2003;111(5):717–725.

138 Isaka N, Padera TP, Hagendoorn J, et al. Peritumor lymphatics induced by vascular endothelial growth factor-C exhibit abnormal function. *Cancer Res*. 2004;64(13):4400–4404.

139 Cao R, Bjorndahl MA, Gallego MI, et al. Hepatocyte growth factor is a lymphangiogenic factor with an indirect mechanism of action. *Blood*. 2006;107(9):3531–3536.

140 Huang XZ, Wu JF, Ferrando R, et al. Fatal bilateral chylothorax in mice lacking the integrin alpha9beta1. *Mol Cell Biol*. 2000;20(14):5208–5214.

170

High Endothelial Venules

Jean-Marc Gauguet, Roberto Bonasio, and Ulrich H. von Andrian

The CBR Institute for Biomedical Research and Department of Pathology, Harvard Medical School, Boston, Massachusetts

Leukocytes, the cellular component of the immune system, are generated in primary lymphoid organs (bone marrow [BM] and thymus), but it is in the secondary lymphoid organs (SLOs) – which include peripheral lymph nodes (PLNs), mesenteric lymph nodes (MLNs), Peyer patches (PPs), spleen, appendix, and tonsils – that the immune response is orchestrated. These highly specialized organs collect antigen (Ag) and Ag-presenting cells (APCs) from distinct anatomical regions and serve as filters for Ag arriving from the periphery. Their extravascular environment is designed to optimize lymphocyte recognition of, and subsequent responses to, cognate Ag. A remarkable property of SLOs is their ability to recruit vast numbers of blood-borne B and T cells, which function as the backbone of the adaptive immune response. With the exception of the spleen, all SLOs contain specialized post-capillary and small collecting venules, called high endothelial venules (HEVs). These serve as the principal site of lymphocyte entry from the blood (1,2). HEVs express organ-specific patterns of traffic molecules, which define a unique vascular address that is not found in other microvascular beds. These molecules coordinate the recruitment of circulating lymphocytes by promoting multistep adhesion cascades involving selectins, chemokines, integrins, and their respective ligands or counter-receptors (3,4). Selectins recognize ligands with extensive post-translational carbohydrate modifications. Indeed, HEVs express abundantly glycosylated adhesion molecule ligands, which serve to recruit circulating lymphocytes (5). Although selectins and their ligands are constitutively active, integrins require activation to bind to their ligands (6), and activation typically comes via signals delivered when chemokines or other chemoattractants presented on endothelial cells (ECs) in HEVs (which will be referred to throughout the chapter as high ECs, or HECs) bind their G-protein–coupled receptors (7). In this chapter, we discuss our current knowledge about the functional, structural, and molecular characteristics of HEVs that allow them to serve as the portal to SLOs. We examine the development of HEVs in normal and pathological settings, and explore how these unique microvessels respond to environmental cues. In addition, we discuss techniques to study these structures as well as emerging technologies that may pave the way for future discoveries.

HISTORICAL AND COMPARATIVE PERSPECTIVES

HEVs have been the subject of scientific scrutiny since as early as the 19th century. Histological examination of SLO endothelium revealed at least three types of vascular beds: (a) prototypical flat-walled vessels that comprise arteries, most venules, and capillaries; (b) fenestrated ECs similar to those found in kidney, liver, spleen, and BM; and (c) a unique postcapillary vascular bed lined by cuboidal (or high) ECs – the HEVs. Another interesting feature of HEVs was that numerous lymphocytes could be found in close association with them, and some lymphocytes could be seen traversing them. These findings led to a great deal of controversy, for even after it was firmly established that HEVs are part of the circulatory system, the prevalent initial interpretation was that lymphocytes were generated in lymphoid organs and secreted via HEVs into the blood (8,9). A much clearer understanding emerged only during the early 1960s, when Gowans and Knight demonstrated that small (i.e., naïve) lymphocytes recirculate from the blood into lymph nodes (LNs) and via the thoracic duct back into the blood (10). These authors injected ^{3}H-labeled lymphocytes intravenously into rats and recorded the recovery of labeled cells in thoracic duct lymph. The recirculating cells first appeared 2 to 4 hours after injection and peaked in the lymph after 24 hours. More than 90% of the injected cells were recovered from chronic thoracic duct fistulas within 4 days, indicating that virtually all lymphocytes migrate actively and continuously (10).

The critical link between lymphocyte recirculation and HEVs was revealed when Marchesi and Gowans used electron microscopy (EM) to demonstrate that these vessels serve as the entry site for circulating lymphocytes into SLOs (1). In this landmark study, intravenously injected lymphocytes could be "seen penetrating the endothelium of the venules" in LNs; lymphocytes were found within the vessel lumen, between ECs, and in the region of the subendothelial basement membrane. Labeled cells in the LN parenchyma were in close proximity to HEVs, suggesting that the new arrivals had recently crossed the microvascular barrier.

In most mammals, the chief direction of lymphocyte migration is from the blood across HEVs into lymphoid tissues, although some cells also may migrate in the opposite direction. In sheep and rat PLNs (11,12), lymphocytes use HEVs predominantly to migrate into lymphoid organs, whereas in pig MLNs, lymphocytes migrate from the lymphoid tissue into the bloodstream (13). Despite these apparent species differences, HEVs remain essential vascular components of lymphocyte traffic in all mammals.

STUDYING HEVS: IN VITRO AND IN VIVO MODELS

In Vitro Microscopy

The light microscope first allowed investigators to visualize HEVs, and our understanding of these structures has deepened with advances in microscopy technology. EM allowed investigators to peer into HECs to examine their intracellular components and architecture (discussed in the next section) (14). The first description of how lymphocytes use HEVs to gain entry into lymphoid organs utilized EM (1). In the years since, there has been a multitude of EM studies of SLOs, most recently to examine the transport of soluble lymph-borne factors across HEVs into the vessel lumen (15).

The first in vitro tool to study lymphocyte interaction with HEVs was described by Stamper and Woodruff (16). In that assay, frozen sections of lymphoid organs were incubated with a lymphocyte suspension, gently washed, and fixed with glutaraldehyde. Using dark-field microscopy, it was then possible to visualize and enumerate lymphocytes that interacted with HEVs. A later modification of the Stamper-Woodruff assay reduced the number of lymphocytes needed (17). Because lymphoid organs rapidly recruit and transiently sequester circulating lymphocytes, an in vivo homing assay can be used to measure the number of transfused lymphocytes (labeled with radioisotopes, fluorescent dyes, or genetically encoded markers) that enter into a lymphoid organ within a certain period (typically 1–24 hours) (17). Investigators also used these assays to test the role of specific traffic molecules, for example, by treating lymphocytes, tissue sections, or animals with antibodies or with an enzyme such as neuraminidase (18–21).

Immunohistochemistry

Immunohistochemistry and affinity chromatography with HEV-specific monoclonal antibodies (mAbs), and in situ hybridization have been among the key technologies used to identify and characterize molecules produced by HECs that are essential for lymphocyte trafficking. For example, the mAb MECA-79 was critical for the identification of peripheral lymph node addressin (PNAd), a group of sulfated glycans that are expressed in both human and mouse PLN HEVs (20); Mel-14 permitted identification and characterization of L-selectin (18); and the mAb ER-TR7 has been instrumental in helping investigators visualize the fibroblastic reticular architecture of the LN (22). In situ hybridization was instrumental, for example, in identifying the chemokine, CCL21 (SLC/TCA-4/6Ckine/exodus 2) (23).

Culturing High Endothelial Cells

Several attempts have been made to grow and study primary HECs ex vivo (24,25). After in vitro culture, these cells preserve some of the properties of bona fide HECs, including a certain ability to bind lymphocytes (26), and the expression of some proteins and antigens characteristic of freshly isolated HECs (27,28). However, no culture conditions used so far have been able to produce HECs that fully recapitulate the phenotype observed in vivo (27,28). The major obstacle to maintaining these cells in vitro appears to be the difficulty in reproducing the unique microenvironment found in the PLN, for it has been shown that 48 hours after HECs are removed from their in vivo setting, they rapidly lose expression of several important HEC-specific genes (29). Likely, this loss of HEV gene expression is due to the absence of afferent lymph-derived factors, which are necessary for maintenance of the HEV phenotype.

Genomics

The specialized phenotype of HEVs must originate from a unique repertoire of expressed genes. Because bona fide HECs cannot be grown long-term in vitro, mRNA for expression profiling requires the cumbersome purification of relatively few HECs from single-cell suspensions of mammalian lymphoid tissues. Quantitative 3′-cDNA libraries have been generated using this approach for both murine PLN (30) and PP (31) HEVs, leading to the identification of several known and unknown genes expressed in HEVs, but not in flat ECs. Human tonsil HEVs have been compared with other human ECs by subtractive hybridization and similar molecular techniques (32–34). These studies have identified several HEV-specific surface proteins, enzymes, molecules involved in sulfate metabolism, and a nuclear transcription factor, NF-HEV, which is much more abundantly expressed in HEVs compared with other vascular beds (35). It remains to be determined whether and to what extent NF-HEV is necessary and/or sufficient to induce or regulate HEV function or phenotype. Genomic analysis of HECs also has demonstrated that the PLN microenvironment exerts control on HEC gene expression. When Girard and colleagues compared mRNA extracted from freshly isolated human tonsil HECs with mRNA purified from HECs following 2 days in vitro culture, they observed a dramatic loss in mRNA expression of several important HEV genes, including *FT-VII*, Duffy antigen-related chemokine receptor (*DARC*) and newly identified HEV-specific genes such as collagen XV (29). Using reverse transcription-polymerase chain reaction (RT-PCR), it was shown that HECs and medullary venular ECs in PLNs differentially express FucT-IV and FucT-VII, which leads to high levels of L-selectin ligand activity in HECs and lower levels in medullary venules (36). Data from these genomic analyses may allow researchers to determine with greater confidence whether attempts to culture HECs in vitro produce cells that recapitulate their in vivo origin.

Intravital Microscopy

The ability to visualize and record leukocyte dynamics in the vessels of a living animal has revealed fascinating complexity. Advances in microscopic, fluorescence, and biological technology have dramatically aided in this endeavor (37). Successful recordings require a surgical preparation that sufficiently exposes the specimen (typically a vascular bed) of interest without disrupting blood supply or inducing excessive surgical trauma, achieves an adequate degree of immobilization to dampen motion resulting from normal respiration or peristalsis, and prevents sample desiccation by superfusion with physiological buffer. Animals also must be anesthetized during the procedure, which adds to the level of sophistication required compared to in vitro studies. Despite these challenges, the earliest studies date back to the 19th century, when scientists examined vascular beds in the translucent tissues of frogs (38,39). Recent intravital microscopy (IVM) experiments have continued to use bright-field, but increasingly epifluorescence, microscopy to examine leukocyte migration. During the early 1990s, studies of leukocyte migration initially examined the effect of inflammation on leukocyte adhesion to various non-LN vascular beds, including those of the cremaster and mesentery (40–42). More recent studies have investigated the interaction of lymphocytes with HEVs in a living animal, including the murine PP (43,44) and subiliac LN (also called inguinal LN) (45,46). IVM in solid organs such as the PPs and PLNs requires leukocyte visualization by epifluorescence. Leukocytes can be stained with intravital fluorophores through the intravenous injection of dyes such as rhodamine 6G, which labels nucleated cells (leaving red blood cells unstained). The drawback to this technique is that the observer cannot unambiguously determine what kind of leukocyte is being studied, although mononuclear cells can sometimes be distinguished from polymorphonuclear cells (46). This obstacle can be overcome by purifying and labeling a cell population of interest ex vivo, or by using cells from a transgenic mouse that expresses enhanced green fluorescent protein (EGFP) or another fluorescent protein) in a specific leukocyte subset (37,47–49).

IVM has been a powerful tool to dissect the multistep adhesion cascades for lymphocyte homing via HEVs (50). However, traditional epifluorescence-based IVM cannot generate three-dimensional images, which negates its usefulness in studies of extravascular events, such as diapedesis or migration within the parenchyma. Multiphoton microscopy (MPM) (51) has revolutionized this field by allowing investigators to generate optical sections deep within intact lymphoid and nonlymphoid tissues (52). Recent studies have analyzed the behavior of lymphocytes in intact, freshly excised LNs (53–55) and, more recently, in intravital settings using anesthetized mice (56–58). MP-IVM also has been used to evaluate the localization and interaction of dendritic cells (DCs) with the reticular fiber network of PLNs (59).

MORPHOLOGY AND ANATOMY OF HIGH ENDOTHELIAL VENULES

Functional Anatomy of Peripheral Lymph Nodes and Peyer Patches

PLNs are strategically located along lymphatics, which drain virtually all peripheral tissues. The prototypical PLN is a bean-shaped parenchymatous organ, composed of lymphoid and myeloid leukocytes and stromal cells, surrounded by a fibrous capsule that confers structural integrity. PLNs are subdivided into two main areas: cortex and medulla (Figure 170.1; for color reproduction, see Color Plate 170.1). The cortex comprises the superficial B-cell area that contains B follicles and germinal centers and the deeper T-cell area, or paracortex. The structural unit of the deep cortex is the paracortical cord, a column of T cells and antigen-presenting cells delimited by lymph-draining sinuses (60). At the center of each cord is a HEV, which merges with other HEVs into incrementally larger collecting venules while draining blood toward the medulla. The medulla contains a highly developed system of lymphatic sinuses that receive lymph fluid from the subcapsular and cortical sinuses and coalesce at the PLN's hilus into the efferent lymphatic vessel (61). The paracortical cords extend into the medulla, where they are referred to as *medullary cords*, and contain many plasma cells and macrophages (62).

PPs are organized lymphoid organs in the wall of the small intestine. In contrast to PLNs, PPs do not receive afferent lymph; instead, antigen influx into PPs occurs via M cells, a specialized type of epithelial cell that interfaces with the gut lumen and continuously transcytoses antigenic material from the intestinal cavity into the subepithelial dome, an area rich in DCs (63,64). The PP parenchyma is dominated by large B follicles, whereas T cells are confined to the smaller interfollicular area.

Microvascular Organization of Peripheral Lymph Nodes and Peyer Patches

The microvascular organization varies considerably between different PLNs, but has a number of shared features. We distinguish a main feeding artery and a main collecting vein, both of which access the organ at the hilum (see Figure 170.1A). The arterial endothelium is not specialized to support interactions with circulating leukocytes; on the other hand, the degree of organization on the venular side is remarkable. In mouse inguinal LNs, starting from the main collecting venule,

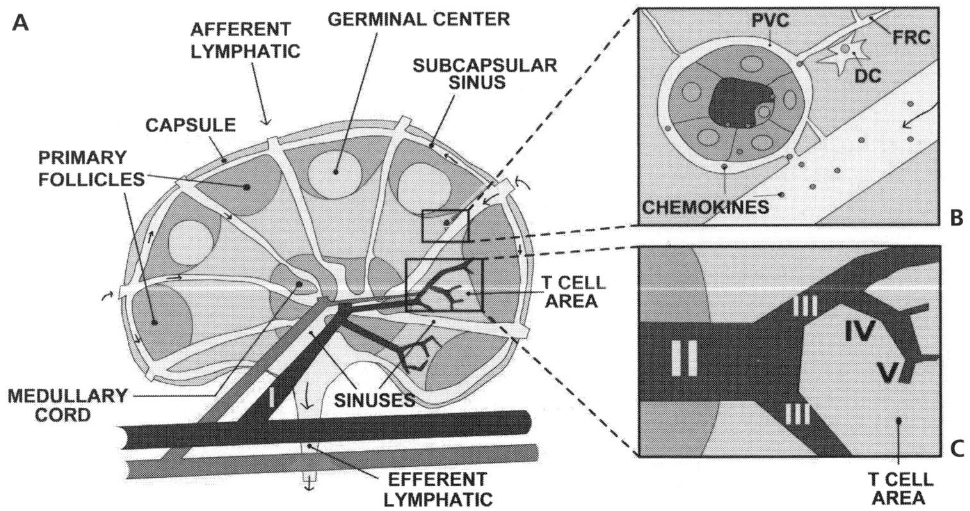

Figure 170.1. Functional anatomy of lymph nodes. (**A**) A schematic representation of a mouse LN is provided. The venous microcirculation as well as the feeding artery are shown in black and gray, respectively. The collecting venule that drains blood from the medulla is marked as order I. For better clarity, only a segment of the intranodal venular tree is shown. Lymph-draining conduits are shown in light gray. Arrows indicate the direction of lymph flow. (**B**) Cross-section of an HEV with fibroblastic reticular cell conduits (FRC) that drain lymph fluid and lymph-borne small molecules such chemokines (shown as small circles) from lymphatic sinuses toward the HEV. EM studies suggest that the FRC conduits drain lymph into a narrow perivenular channel (PVC) surrounding each HEV. Upon delivery to HEVs, chemokines are transcytosed for presentation in the lumen. An adherent lymphocyte is shown inside the HEV. Also shown is a resident DC in close association with the FRC. (**C**) Architecture of a venular tree showing the branching orders characteristic of mouse subiliac LNs. For color reproduction, see Color Plate 170.1.

a distinct hierarchy of branches can usually be identified (65) (Figure 170.1C). The endothelium of the first and second branching orders in the medulla has a flat appearance similar to endothelium in nonlymphoid tissues. In contrast, endothelium in higher-order venules in the paracortex and deep cortex has a cuboidal, cobblestone-like shape. Thus, only microvessels with a higher branching order are HEVs. Ultrastructural studies indicate that ECs in the transition region between HEVs and flat-walled venules often are arranged in an overlapping plate pattern (66).

Whereas HEVs in PLNs are restricted to the T-cell area, HEVs in PPs originate in B follicles. From here, they drain blood toward the interfollicular T-cell zone, where they merge into larger collecting venules. Although no apparent structural differences exist between HEVs in PLNs and PPs (67), biochemical, immunohistochemical, and functional studies have uncovered a number of tissue-specific differences, which manifest in the selective recruitment of distinct lymphocyte subsets.

Lymph Flow in Peripheral Lymph Nodes and Fibroblastic Reticular Cells

Lymph is drained to PLNs by a system of blind-ending vessels lined by specialized lymphatic ECs. Afferent lymphatic vessels perforate the capsule of the PLN on its convex aspect, distal to the hilum, and drain into the subcapsular sinus (see Figure 170.1A). Lymph-borne cells, particulate material, and soluble molecules have different fates after entering a PLN. DCs and effector memory lymphocytes invade the subcapsular sinus floor and migrate to their proper location in the cortex (68,69). Particulate material and large molecules are excluded from the parenchyma (70) and are phagocytosed by subcapsular phagocytes or drain to the efferent lymphatic vessel through the sinus system. Low-molecular-weight molecules with a radius of less than 4 nm are channeled into the parenchyma along a network of collagen fibers (71,72) that are ensleeved by fibroblastic reticular cells (FRCs). This sizing, column-like drainage system is called the *FRC conduit* (50,60). A monoclonal antibody, ER-TR7, has been used recently to visualize the FRC network within PLNs. FRCs also encircle HEVs, and EM studies have suggested the existence of a narrow perivascular gap between them, the perivenular channel (PVC), that is thought to receive lymph fluid via the FRC conduit (Figure 170.1B).

FRCs directly contact some of the hematopoietic components of the lymph node, and recent studies have demonstrated a significant functional consequence of these interactions. In vitro, FRCs recruit lymphocytes, and these interactions induce FRCs to generate a stromal network that supports cell migration and immune cell contacts (73). In addition to the paracortical cords surrounding HEVs, FRCs are particularly dense within the cortical ridge, which is the boundary between T- and B-cell follicles (74). Tissue-derived DCs are found near the cortical ridge (74), and also adjacent to HEVs (69), where they can interact with newly arrived naïve T cells. In fact, some DCs form intimate contacts with FRCs (see Figure 170.1B),

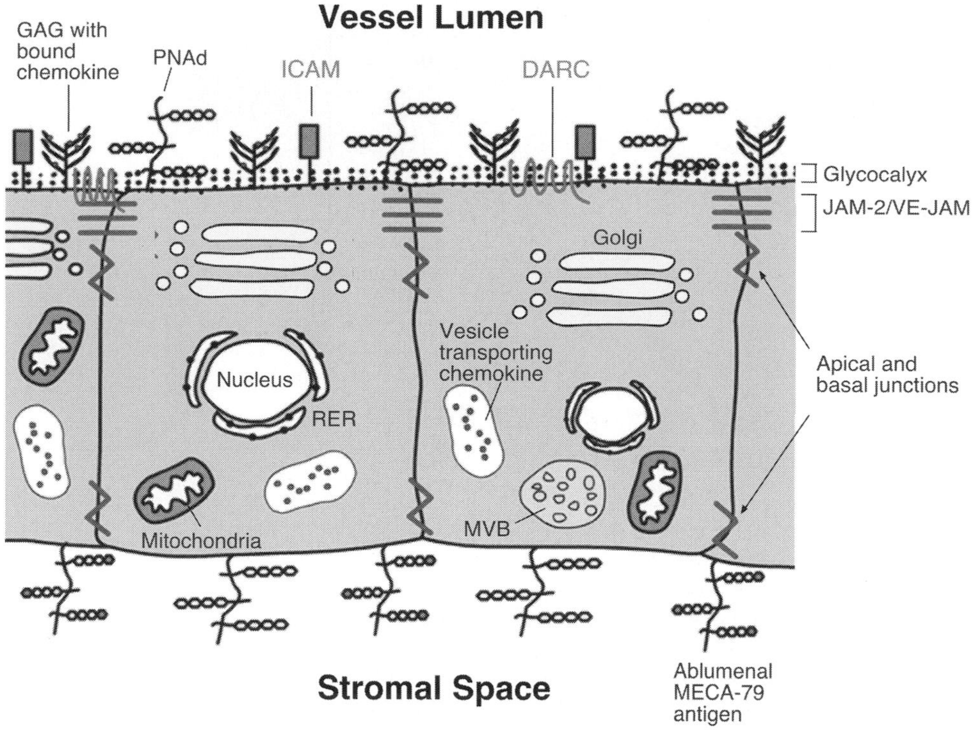

Figure 170.2. Characteristic ultrastructural features of HECs in PLN. Specialized subcellular and molecular features that distinguish HECs from other type of endothelium are indicated in the figure. Most importantly, the prominent Golgi apparatus supports the generation of L-selectin ligands and the MECA-79 antigen by post-translational modifications of sialomucins. For color reproduction, see Color Plate 170.2.

which allow them to sample and present to naïve T cells antigen drained through the FRC from the periphery (75). These in vitro and in vivo findings suggest that FRCs generate structures that can help organize and coordinate the various hematopoietic components of a PLN and optimize antigen sampling and T-cell activation.

Ultrastructure of High Endothelial Cells

The ultrastructure of HECs is dominated by a prominent Golgi apparatus, which is often oriented toward the luminal aspect of the cell (14) (Figure 170.2; for color reproduction, see Color Plate 170.2). In addition, HECs have abundant mitochondria and ribosomes, sparse rough endoplasmic reticulum, and multivesicular bodies (14,76,77). These ultrastructural characteristics, and the ability of PLN HECs to incorporate substantial amounts of sulfate (78), are atypical for conventional ECs and were interpreted as indicative of a cellular machinery with high metabolic activity geared toward the biosynthesis of glycoproteins (14).

Shallow pits (12) and a glycocalyx that appears thicker than the glycocalyx on non-HEV endothelium (79) line the apical surface of HECs. In addition, microvillous-like protrusions project into the luminal space, possibly in order to facilitate tethering of circulating lymphocytes (34). Another important feature of HEVs is the existence of discontinuous "spot-welded" junctions between HECs (79). These discontinuous junctions differ from the tight junctions that characterize capillary and arterial endothelium, but are similar to the "nonoccluding" junctions found in other postcapillary venules, and likely facilitate the passage of lymphocytes between adjacent HECs (2).

Chemokine Presentation on High Endothelial Cells

Chemokines produced by HECs or surrounding stromal cells are presented on the endothelial surface, where they induce integrin activation and firm arrest of rolling leukocytes (47). The immobilization and display of chemokines is thought to be mediated by glycosylaminoglycans (GAGs) (see Figure 170.2). GAGs are proteoglycans that carry negatively charged sulfate and carboxyl groups, which permit electrostatic interactions with basic peptide motifs present in most chemokines (80,81). Heparan sulfate are highly expressed on ECs and are likely candidates to bind and present chemokines. Versican is a sulfated proteoglycan that has been demonstrated to block the functional activity of some chemokines, including CCL21, without affecting the interaction of CCL21 with its receptor, CCR7 (82). This function of versican has only been demonstrated in vitro; its expression pattern and function in SLO HECs remains to be investigated.

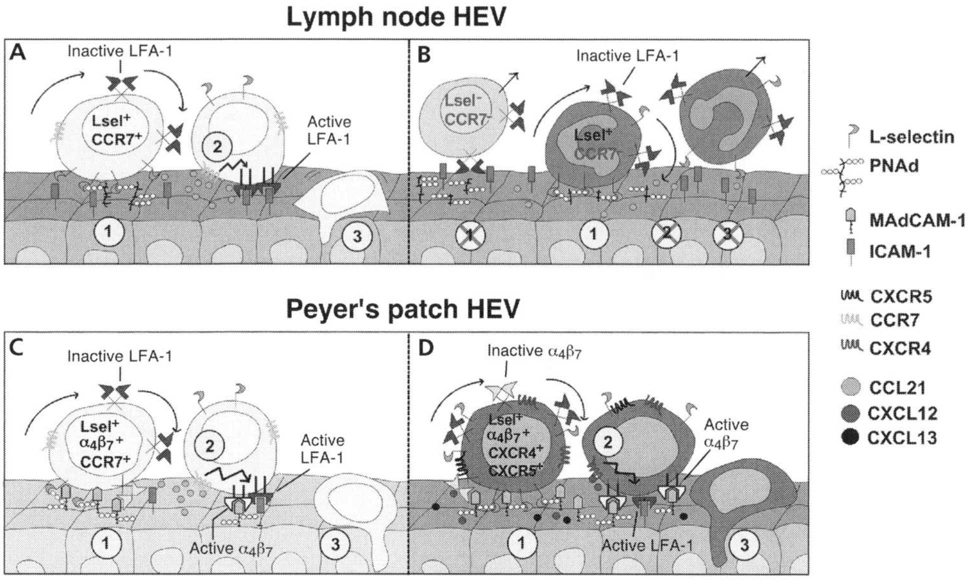

Figure 170.3. Adhesion cascades in HEVs. Examples of dedicated multistep cascade in PLNs (**A–B**) and in PPs (**C–D**). (**A**) Naïve T lymphocytes undergo tethering and rolling via L-selectin-PNAd (step 1); chemoattractant stimulation by CCR7–CCL21 (step 2); and sticking (firm adhesion) via high-affinity LFA-1–ICAM-1 interactions (step 3), followed by diapedesis. The selectivity of this recruitment cascade is illustrated in (**B**); L-selectin$^-$ cells (e.g., effector memory T cells) and CCR7$^-$ leukocytes (e.g., granulocytes) fail to establish one or more of the required interactions. PPs display different "ZIP codes" in HEV segments within the interfollicular T cell area (**C**) and in upstream B follicles (**D**). MadCAM-1 is expressed in all PP HEVs, and mediates rolling interactions by binding to either L-selectin or integrin $\alpha4\beta7$. For step 2, HEVs in different areas of the PP present distinct chemokines to rolling leukocytes: T cells are activated by CCL21, while HEVs associated with B follicles present CXCL12 and CXCL13 to stimulate B cell–expressed CXCR4 and CXCR5, respectively. Firm arrest of both B and T cells in PPs are mediated by activated $\alpha4\beta7$ and/or LFA-1 integrins. For color reproduction, see Color Plate 170.3.

DARC is a nonsignaling seven-transmembrane receptor that binds both inflammatory CC and CXC chemokines. DARC is highly expressed in HEVs (34), and has been suggested to play a role in chemokine transport, because it is found in endothelial caveolae (83), which, in flat endothelium, have been shown to function as conduits for chemokine translocation from the abluminal to the luminal surface (84). Studies in knockout mice suggest that DARC does not play a role in regulating constitutive lymphocyte trafficking to PLNs (85). However, DARC can scavenge inflammatory chemokines, including CCL2, which suggests a role for DARC in regulating inflammation, possibly during "remote control" of monocyte recruitment to PLNs (49).

Leukocyte Recruitment via High Endothelial Venules

It is widely accepted that tissue- and subset-specific leukocyte migration is governed by a sequence of molecularly distinct adhesion and signaling events (3,4,86). Adhesion cascades are initiated by a tethering step that allows leukocytes to bind loosely to ECs (step 1A). Once attachment has occurred, marginated cells are pushed forward by the bloodstream, resulting in a slow rolling motion along the vascular wall (step 1B). Subsequently, rolling cells encounter chemotactic stimuli on or near the EC surface that can bind to specific leukocyte receptors (step 2). Chemoattractant binding, in turn, induces rapid intracellular signaling and triggers activation-dependent adhesion steps that allow leukocytes to stick firmly (step 3) and, eventually, to emigrate through the vessel wall.

Although microvessels in most nonlymphoid tissues can only support substantial leukocyte traffic upon exposure to inflammatory mediators, HEVs must recruit large numbers of lymphocytes in the absence of inflammation. Thus, HEVs constitutively express a unique pattern of traffic molecules that are fundamental for the normal function of the immune system.

The Multistep Adhesion Cascade in Peripheral Lymph Nodes

Detailed IVM analyses have defined the specific adhesion cascades that mediate T- and B-cell homing to LNs (Figure 170.3A; for color reproduction, see Color Plate 170.3) (46,47,87–89). Tethering and rolling of both subsets is mediated by L-selectin (CD62L). The endothelial L-selectin ligand in PLNs is PNAd, an *O*-linked carbohydrate moiety, the major components of which are recognized in humans and mice by the mAb MECA-79 (20,90). The structure of the MECA-79 antigen and HEV-expressed L-selectin ligands is discussed in the section "PLN Ligands and MECA-79."

Firm arrest of rolling T and B cells in LN HEVs is mediated by lymphocyte function-associated antigen (LFA)-1 (46,91,92). LFA-1–deficient lymphocytes home poorly to LNs, although the observed magnitude of the homing defect was somewhat variable between two independently generated knockout strains (92,93), possibly due to the different backgrounds of the animals used. In one strain, residual LFA-1–independent homing was mediated by the family of $\alpha 4$ integrins, $\alpha 4\beta 7$ and $\alpha 4\beta 1$, which interact with vascular cell adhesion molecule (VCAM)-1 in LN HEVs (93).

HEC ligands for LFA-1 include intercellular adhesion molecule (ICAM)-1 and -2. ICAM-2 contains two Ig-like domains, whereas ICAM-1 contains five; these Ig-like domains function as the binding site for LFA-1 (94,95). Although both ICAM-1 and -2 are expressed on HECs, deficiency in either ICAM-1 or -2 does not result in any defect in lymphocyte homing to SLOs (96–98). Rather, a defect in homing to PLNs is only observed in ICAM-1–deficient mice treated with ICAM-2 blocking antibodies or wild-type mice treated with both ICAM-1 and -2 blocking antibodies (99), suggesting a redundant function of ICAM-1 and -2.

CCL21 (also called SLC, TCA-4, 6-C-kine, or exodus 2) activates integrins on naïve and central memory T cells (47,87). CCL21 is constitutively expressed by HECs and binds to CCR7 (100–102). A second CCR7 agonist, CCL19 (also called ELC or MIP-3β), is expressed by lymphatic endothelium and interstitial cells within LNs, but not by HEVs. However, lymph-borne CCL19 can be transported to the luminal surface of HEVs and induces integrin activation on rolling T cells (15). The physiological role of CCL19 in lymphocyte homing remains to be determined.

Mice carry two isoforms of CCL21: CCL21ser is expressed in HECs, whereas CCL21leu is generated in lymphatic endothelium (103). A spontaneous mutant strain called *plt/plt* (*p*aucity in *l*ymph node *T* cells) is deficient in CCL21ser and CCL19 (103,104). LNs of *plt/plt* mice and CCR7-deficient mice contain few naïve T cells, but the B-cell compartment is much less affected. LNs in these mutant animals also contain substantial numbers of central memory T cells (105,106). This indicates that B cells and central memory T cells may respond not only to CCR7 agonists, but also to another integrin-activating signal in HEVs. Indeed, CXCL12 (SDF-1α), the ligand for CXCR4, can trigger B-cell and central memory T-cell arrest in HEVs (88,107). CCR7 and CXCR4 can independently maintain B-cell homing (albeit at somewhat lower levels than in wild-type animals, where both function simultaneously). Interestingly, although CXCL12 potently induces integrin activation on rolling naïve T cells in vitro (101,108), it does so very poorly in vivo (109). Thus, despite low-level expression of CXCL12 in murine HEVs, naïve T cells require CCR7 signals, at least in the Balb/c and DDD/1 genetic background (47,88). On the other hand, CXCL12 and CXCR4 have a modest role in T-cell homing in C57BL/6 mice, indicating that some chemokine pathways may be regulated in a strain-dependent manner (88). CXCL13 has been detected on the surface of PLN HECs (110); however, its contribution to B-cell homing is likely small because B-cell homing to PLNs is almost completely abrogated in the absence of signaling through CCR7 and CXCR4 (88).

Although B cells can respond to two (or more, depending upon the tissue) distinct integrin activation signals in HEV, B-cell homing to LNs is significantly less efficient than that of T cells. This is because the L-selectin levels on B cells are approximately 50% lower than on T cells (111). T cells from *L-selectin*$^{+/-}$ heterozygous mice express similar L-selectin levels as *L-selectin*$^{+/+}$ B cells; however, *L-selectin*$^{+/-}$ T cells still traffic to PLNs better than do wild-type B cells, suggesting that factors other than absolute expression of L-selectin regulate lymphocyte trafficking to PLNs (111). In experiments with pre–B-cell clones stably transfected with human L-selectin, we observed that at least 50,000 L-selectin molecules per cell were needed for efficient rolling in LN HEVs (45,112, and unpublished data). Thus, it is important to keep in mind that the mere presence of L-selectin on a leukocyte does not necessarily predict its potential to home to PLNs, even when this cell expresses all other prerequisite traffic molecules (i.e., CCR7 and LFA-1). A leukocyte may be deemed L-selectin^{+} based on flow cytometric criteria, but this would be of little consequence if the expressed copy number is substantially lower than that on naïve T cells ($\sim$70,000–100,000 molecules/cell).

The fact that a sequence of three distinct molecular steps must be successfully engaged in HEVs explains why some leukocytes home to PLNs, whereas others do not. For example, granulocytes express L-selectin and LFA-1, but not CCR7 or CXCR4; mature myeloid DCs express CCR7 and LFA-1, but not L-selectin; and effector CTLs lose both L-selectin and CCR7. Consequently, granulocytes roll, but fail to arrest, whereas DCs and effector cells cannot tether or roll in LN HEVs (46,87,113) (see Figure 170.3B).

L-selectin–independent homing via HEVs can be induced through intravenous injections of activated platelets (114,115). Circulating activated platelets express P-selectin on their surface, which mediates platelet binding to PNAd on HEVs and, simultaneously, to P-selectin glycoprotein ligand (PSGL)-1 on circulating leukocytes (114). This platelet bridge can transiently restore lymphocyte homing in L-selectin–deficient mice. Indeed, L-selectin–deficient animals mount poor cutaneous hypersensitivity (CHS) responses when they are sensitized with a cutaneous hapten antigen, because naïve T cells do not migrate into skin-draining LNs (116). When the homing cascade is restored during the sensitization phase by infusing activated platelets for several hours, the CHS response is restored (115). Of note, a small fraction of leukocytes roll in a P-selectin–dependent manner in PLN HEVs of L-selectin–deficient mice, even when no platelets are being infused (115). However, it is not yet clear if this minor rolling pathway contributes to physiologic L-selectin–independent homing and whether the relevant sources of P-selectin in this setting are circulating activated platelets or ECs or both.

The Multistep Adhesion Cascade in Peyer Patches and Mesenteric Lymph Nodes

The homing cascade just discussed applies primarily to skin-draining LNs. HEVs in mucosa-associated SLOs, such as MLNs and PPs, express not only PNAd, but also the mucosal addressin cell adhesion molecule (MAdCAM)-1, a ligand for the $\alpha4\beta7$ integrin (19,117). The $\alpha4\beta7$/MAdCAM-1 pathway can mediate selectin-independent lymphocyte tethering and rolling in HEVs (44). Thus, L-selectin deficiency compromises lymphocyte homing to PLNs more severely than to MLNs (118), whereas $\beta7$ integrin deficiency results in reduced homing to MLNs, but has no effect in PLNs (119). Deficiency for both L-selectin and $\alpha4\beta7$ results in the profound inability of lymphocytes to adhere to PP HEVs or enter into PPs (120), although a nominal role for P-selectin was reported when surgically traumatized PPs were examined by IVM (121).

PP HEVs express MAdCAM-1, but not PNAd on their luminal surface (although MECA-79–reactive material is present at the abluminal side of PP HEVs [20]). MAdCAM-1 is a bifunctional adhesion molecule. The first Ig domain of MAdCAM-1 contains the binding site for $\alpha4\beta7$ (122), whereas MAdCAM-1's mucin domain can be decorated with L-selectin ligands. Indeed, affinity-purified MAdCAM-1 from MLNs is recognized by the mAb MECA-79 (123). By contrast, MAdCAM-1 in PP HEVs is not detected by MECA-79, but nevertheless supports lymphocyte tethering and rolling via both L-selectin and $\alpha4\beta7$ integrin (44). In its active conformation, $\alpha4\beta7$ can mediate firm arrest (124); however, naïve lymphocytes express relatively little $\alpha4\beta7$ (125) and require additional engagement of LFA-1 for firm arrest, whereas $\alpha4\beta7$ is primarily critical to slow down the rolling cells (44) (see Figure 170.3C).

IVM experiments have shown that the integrin-activating chemokines in PP HEVs are segmentally presented (88,110,126). In B follicles, HEVs present CXCL12 and CXCL13, the ligands for CXCR4 and CXCR5, respectively (see Figure 170.3D). Signals through these two receptors induce integrin activation on rolling B cells, but not on T cells (88,110). Conversely, as soon as HEVs enter into the interfollicular T-cell area, they express CCL21 and stimulate preferential T-cell arrest (126). Thus, HEVs are not only distinct between different lymphoid tissues, but segmental specialization even occurs within individual microvessels. The factors that orchestrate this remarkable endothelial subspecialization remain to be identified.

Remote Control

Unique mechanisms exist to modulate leukocyte recruitment to PLNs during inflammation (47,49). Chemokines, such as CCL2 (monocyte chemoattractant protein [MCP]-1), produced at a peripheral site of inflammation, can drain via the lymph to the subcapsular sinus where they enter the FRC conduit and are channeled toward HEVs in the cortex (72) (see Figure 170.2B). Chemokines reaching the abluminal surface of HEVs are then transcytosed within vesicles to the luminal side and presented to rolling leukocytes (15,49). The mechanism of chemokine transport across HEVs has not been determined, but might involve caveolae, which have been shown to mediate chemokine transcytosis in dermal microvessels (84).

This presentation of cutaneous-derived chemokines on HEVs in draining LNs, termed *remote control,* allows the rapid recruitment of circulating monocytes, which express CCR2, the receptor for CCL2 (49). Monocytes do not express CCR7 or CXCR4, and are therefore excluded from resting PLNs. However, by discharging CCL2 into the lymph, inflamed peripheral tissues project a potent monocyte chemoattractant signal onto HEVs in draining LNs, which triggers integrin activation (49). Within the LN, monocytes can differentiate into macrophages or DCs and participate in the ensuing immune response. In addition, inflammation induces mRNA for CXCL9 (monokine induced by gamma-interferon [MIG]) in draining LNs and presentation of this chemokine on a subset of HEVs that supports monocyte adhesion in vitro (127). It is not known whether CXCL9 is produced by the inflamed HEVs themselves or by other intra- (or extra-) nodal cells from where the chemokine might have been transported to the HEVs.

The ability to remotely modulate the multistep adhesion cascade in HEVs by discharging chemokines into the lymph enables peripheral tissues to control the composition and function of leukocytes in draining LNs. However, recent work suggests that there also may be counter-regulatory mechanisms. Lymphatic ECs express the nonsignaling serpentine receptor D6, which binds promiscuously to a number of chemokines, including CCL2 (128). Upon ligation, D6 triggers the rapid internalization and degradation of its cargo (129), and recent studies have shown that this ability of D6 to clear chemokines is critical for helping to resolve continuous leukocyte influx into inflammatory sites (130). The expression of D6 on lymphatic endothelium suggests that D6 also may function to prevent uncontrolled seepage of chemokines through the lymph to draining PLNs.

Transendothelial Migration

The final step of the adhesion cascade, transmigration, is also the most poorly understood at the molecular level. The EC has solved the apparent paradox of letting a leukocyte pass to the abluminal side to access the tissue without compromising vascular integrity (131), and HEVs have further evolved to allow this process to occur constitutively and continually. Adhesion molecules including platelet-endothelial cell adhesion molecule (PECAM)-1 (CD31) (132) and CD99 (133) play roles in transendothelial migration of inflammatory cells in nonlymphoid tissues; however, their contribution to lymphocyte diapedesis across HEVs has yet to be demonstrated. One candidate are the junctional adhesion molecules (JAMs), which are found at intercellular boundaries in HEVs and normal endothelium and contribute to leukocyte transmigration in vitro and in vivo (134–136). It remains unknown whether the JAM family members contribute to lymphocyte transmigration through HEVs. Integrins function during transmigration (137), and a recent study has demonstrated that transmigrating leukocytes form a *transmigratory cup* composed of LFA-1–ICAM-1 interactions (138). Recently, ephrinA1 has been detected on HEVs and, in vitro, ephrinA1 enhances CD4 T-cell migration to CCL19 and CXCL12 and increases actin polymerization (139). These findings suggest that ephrinA1 may promote CD4 transmigration across HEVs, although further investigation is needed.

One interesting question is whether leukocytes transmigrate between adjacent ECs or directly through an individual EC. Recent evidence suggests that a significant fraction (5%–10%) of migrating neutrophils, monocytes, and lymphocytes take a transcellular route through a human umbilical vein EC monolayer in vitro (138). In vivo, a transcellular route of migration of lymphocytes through HEVs has been suggested by EM studies (1,140); however, definitive evidence is still lacking.

Additional High Endothelial Venule Adhesion Molecules

In addition to the selectins and integrins, other classes of adhesion molecules have been implicated in lymphocyte trafficking to SLOs. Vascular adhesion protein (VAP)-1 is an amine oxidase that is expressed on HECs of PLNs as well as on ECs from vessels supplying inflamed sites (141,142). Antibodies against VAP-1 and VAP-1–deficient mice have demonstrated a clear role for VAP-1 in regulating lymphocyte trafficking to PLNs and PPs, as well as at sites of inflammation (143,144). In particular, the absence of VAP-1 results in an increased leukocyte rolling velocity, indicating that VAP-1 strengthens leukocyte–endothelial interactions. All reports suggest that the enzymatic activity of VAP-1 is critical to its function, although it remains unclear what molecule(s) undergo modification to strengthen leukocyte rolling interactions.

Common lymphatic endothelial and vascular endothelial receptor (CLEVER)-1 is a large (270–300-kDa) glycoprotein that was identified on lymphatic and high ECs (145). Although it is not known whether CLEVER-1 presents MECA-79–reactive antigen or can bind L-selectin, antibodies against CLEVER-1 block lymphocyte adhesion to HEVs in addition to rolling and transmigration of leukocytes on vascular endothelium in vitro (145,146). The exact function of CLEVER-1, however, is unclear, because other reports have suggested that CLEVER-1 functions as a scavenger receptor (147).

Although all adhesion molecules previously described promote lymphocyte adhesion, CD43 negatively regulates lymphocyte homing to SLOs (148). The large and negatively charged CD43 is thought to decrease the ability of circulating lymphocytes to interact with endothelial adhesion molecules.

HIGH ENDOTHELIAL VENULE–SPECIFIC LYMPHOCYTE TRAFFIC MOLECULES

Peripheral Lymph Node Ligands and MECA-79

The mAb MECA-79 immunoprecipitates a set of glycans on the surface of HECs in both murine and human lymphoid tissues (20,90,149). Affinity-purified MECA-79–reactive glycoproteins support L-selectin binding in vitro (90), and a relatively high dose of MECA-79 blocks short-term lymphocyte recruitment to PLNs (20,36), as well as rolling in HEVs in vivo (45,150). Tissue sections stained with MECA-79 reveal reactive material on both the luminal and abluminal surface of PLN HECs (Figure 170.4C; for color reproduction, see Color Plate 170.4), but only on the abluminal side of PP HECs (20,151). The physiological role of abluminal MECA-79 antigen is not understood, but it is clear that it is biosynthetically distinct from the luminal material (151,152). IVM studies of the localization of fluorescent MECA-79 and MECA-79–conjugated fluorescent latex beads have shown that MECA-79 antigen is highly restricted to cortical HEVs within the PLN vasculature (36,148) (Figure 170.4A,B). In the HEV lumen, MECA-79 localizes to microvillous-like protrusions, which may facilitate L-selectin–dependent tethering of circulating lymphocytes (34). The transition from MECA-79$^+$ to MECA-79$^-$ ECs is typically very abrupt, from one EC to the next (36). This transition in immunoreactivity is located at the corticomedullary boundary, and can be so accurate that a vessel running parallel to the juncture can be immunoreactive on the side that faces the cortex, while the endothelium on the medullary side remains negative for MECA-79 (153).

The MECA-79 epitope decorates several sialomucins identified in PLNs. These include GlyCAM-1 (Sgp50), a secreted glycoprotein (154,155); CD34 (Sgp90) (156); podocalyxin (157); and one (or more) large molecular specie(s) termed Sgp^{200}, which remain(s) to be identified (158). Two additional proteins that may support lymphocyte recruitment to PLNs are endomucin (159,160) and endoglycan (161,162). Endomucin is expressed on HEVs, can carry the MECA-79 epitope, and can bind L-selectin (163), whereas endoglycan functions as an L-selectin ligand requiring tyrosine sulfation and glycosylation, but is not recognized by MECA-79. Given our current knowledge, these glycoproteins probably serve as interchangeable scaffolds that are functionally redundant, at least in the context of L-selectin ligand presentation. For example, CD34 is a major component detected by MECA-79 in human tonsils (164), but CD34-deficient mice have normal lymphocyte migration to lymphoid tissues (Table 170-1) (165).

Glycosylation-dependent cell adhesion molecule (GlyCAM)-1 expression is restricted to HEVs and mammary epithelial cells (166), podocalyxin was originally described in the kidney, and CD34 is expressed on many different cell types, including non-HEV endothelium and cells of hematopoietic origin (167). However, on non-HEV cell types, none of these molecules react with MECA-79 or support L-selectin binding. Indeed, L-selectin ligand presenting sialomucins undergo

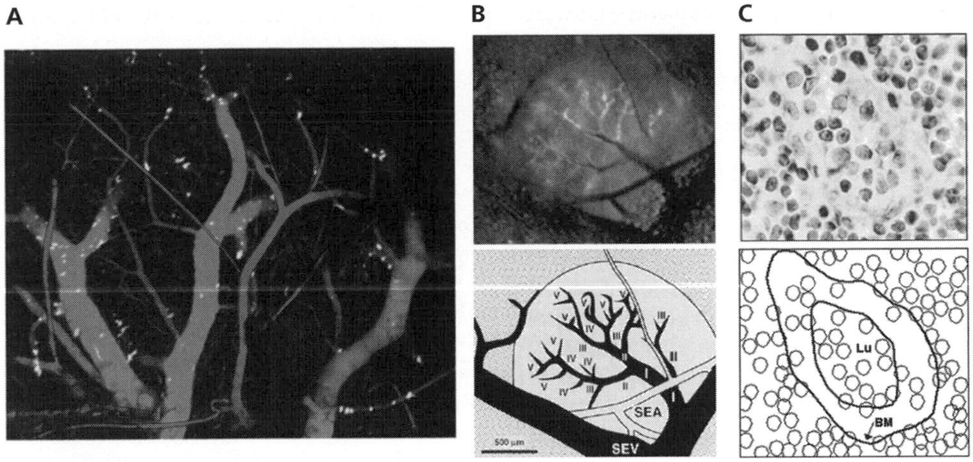

Figure 170.4. MECA-79 antigen expression is restricted to HEVs in PLNs. (**A**) False-color intravital micrograph showing MECA-79 coated fluorescent microspheres bound to small diameter venules (order III to V) within the venular tree of a mouse subiliac LN. Beads do not accumulate in arterioles or lower order venules. (**B**) Intravital micrograph showing the distribution of intravenously injected MECA-79 (*top panel*). The lower panel provides a schematic reference to help assign branching orders to the venules and to locate the superficial epigastric artery (*SEA*) and vein (*SEV*) in this preparation. (**C**) Immunohistochemistry reveals MECA-79 staining of a PLN HEV, illustrating the presence of MECA-79 on the luminal and abluminal (basement membrane) surface of the HECs. Panels in (**B**) and (**C**) were modified with permission from Stockton BM, Cheng G, Manjunath N, et al. Negative regulation of T cell homing by CD43. *Immunity.* 1998;8:373–381, and Stein JV, Rot A, Luo Y, et al. The CC chemokine thymus-derived chemotactic agent 4 (TCA-4, secondary lymphoid tissue chemokine, 6Ckine, exodus-2) triggers lymphocyte function-associated antigen 1-mediated arrest of rolling T lymphocytes in peripheral lymph node high endothelial venules. *J Exp Med.* 2000;191(1):61–76, respectively. For color reproduction, see Color Plate 170.4.

extensive HEV-specific post-translational modifications that are essential for their function in lymphocyte migration.

Recent studies have demonstrated that not all L-selectin ligands in PLNs are reactive with MECA-79. Indeed, medullary PLN venules support L-selectin–dependent rolling, yet they do not stain with MECA-79 (36); the biochemical nature of this medullary venule-expressed L-selectin ligand is not currently understood, although it does require α (1,3)-fucosylation (discussed in the next section). Additionally, HEC-*GlcNAc6ST*$^{-/-}$ mice completely lack MECA-79 antigen on the luminal surface of PLN HEVs, although L-selectin ligands are still present (150).

Post-Translational Modification of Peripheral Lymph Node High Endothelial Venule Proteins

The importance of HEV-specific carbohydrate modifications of HEV proteins for lymphocyte trafficking was first suggested by observations that specific mono- and polysaccharides inhibit lymphocyte binding to LN HEVs (168–170). Detailed biochemical studies of affinity-purified MECA-79 and L-selectin–reactive material (170–173) have shown that most HEV-expressed L-selectin ligands are sialylated, fucosylated, and sulfated *O*-linked carbohydrates that decorate a select group of sialomucins in HEVs (Table 170-2), and studies in knockout mice have revealed the contributions of enzymes responsible for these modifications in vivo (Table 170-3).

Sialic Acid

Rosen and colleagues demonstrated that removal of sialic acid from LN sections with sialidase blocked lymphocyte binding to HEVs (174), and intravenous injection of sialidase inhibited lymphocyte homing to PLNs (21). Interestingly, sialic acid modification of HEV glycoproteins is not necessary for MECA-79 reactivity (175), but sialylation is essential for L-selectin binding (176). Indeed, the major capping group that binds L-selectin is a sulfated form of sialyl Lewis X, or sLe^x (NeuAcα2–3Galβ1–4[Fucα1–3]GlcNAcβ1–) (172,173). Sialic acid is not essential for lymphocyte migration to PPs, since treatment of PP HEVs with sialidase does not affect lymphocyte binding or homing to PPs (21,174). However, treatment of purified MAdCAM-1 with sialidase blocked in vitro L-selectin binding to MAdCAM-1 (123), suggesting that the interaction of $\alpha 4\beta 7$ integrins with MAdCAM-1 does not depend on sialylation, whereas L-selectin binding to MAdCAM-1 requires it. The family of α2,3 sialyltransferases is large, and the sialyltransferase(s) responsible for the sialylation of L-selectin ligands in PLN HEVs remain(s) to be determined.

α(1,3)-Fucosylation

α(1,3) Fucosylation is essential for the activity of virtually all physiological selectin ligands, including those in HEVs. Fucosyltransferases-IV (FucT-IV) and FucT-VII are essential for the generation of selectin ligands on leukocytes and ECs

Table 170-1: Effect of Disrupting Genes Involved in Lymphocyte Recruitment to Lymphoid Organs

Gene	*Leukocyte Trafficking in Knockout*	*References*
L-selectin (CD62L)	Hypocellular PLNs and impaired lymphocyte trafficking Impaired DTH responses	118, 216
CD34	Defects in early hematopoietic progenitor development No defects in lymphocyte trafficking	165
CD43	Increased lymphocyte migration to secondary lymphoid organs	148
CD44	Delayed or subtle defect in entry of CD44-deficient lymphocytes to PLNs	217, 218
LFA-1 (CD11aCD18/$_{\alpha L}\beta 2$)	Impaired lymphocyte trafficking to PLNs, MLNs and PPs	93, 92
β_7 Integrin	Impaired lymphocyte trafficking to MLNs and PPs and impaired adhesion to PP HEVs	119
ICAM-1 (CD54)	Defects in DTH response and neutrophil recruitment to inflammatory sites. Defective lymphocyte recruitment to SLOs only when anti-ICAM-2 antibodies used	96, 97, 99
ICAM-2 (CD102)	No defect in lymphocyte recruitment to lymphoid organs	98
CCR7	Impaired lymphocyte and DC trafficking to PLNs and PPs	105
CCL21ser/CCL19 (*plt/plt* mice)	Impaired T cell trafficking to PLNs and PPs	219, 47, 126, 88
CXCR5, CXCL13 (BLC)	Impaired B-cell homing and migration to B follicles in PPs and spleen	220, 88
CXCR4, CXCL12 (SDF-1α)	Impaired naïve T and B cell and central memory T cell migration to lymphoid organs in the absence of CCR7 ligands	88, 107
D6	Hyperactive inflammatory response	130
DARC	Potential defect in inflammatory chemokine transport or presentation on endothelium	221, 85
VAP-1	Defect in lymphocyte recruitment to PLNs and inflammatory sites	144

(see Tables 170-2 and 170-3) (177–180). Neutrophil recruitment in *FucT-VII*$^{-/-}$ mice is severely impaired due to an inability of cells to generate functional E- and P-selectin ligands (178). Lymphocyte trafficking to PLNs is reduced by approximately 80% in FucT-VII–deficient mice (178). *FucT-IV/VII* double knockout mice have a significantly more complete (>95% reduction) defect in lymphocyte homing to PLNs compared with FucT-VII–deficient mice, indicating that both FucTs can generate L-selectin ligands in HEVs (180). *FucT-VII*$^{-/-}$ and *FucT-IV/VII*$^{-/-}$ mice also have a decreased inflammatory response, related to decreased P- and E-selectin ligand activity (178,180). Surprisingly, FucT-IV–deficient mice have moderately increased lymphocyte homing to PLNs and MLNs (36).

To explain this counterintuitive observation in *FucT-IV*$^{-/-}$ PLNs, one must consider the acceptor preferences of FucT-IV and FucT-VII (181,182). FucT-VII requires an α (2,3)-sialylated terminal lactosamine to generate sLe^x-like selectin ligands. FucT-IV also can use the sialylated lactosamine substrate, albeit less efficiently than FucT-VII. However, unlike FucT-VII, FucT-IV additionally fucosylates internal GlcNAc residues in polylactosamine (181) and uses nonsialylated lactosamine to generate Lewis^x-like structures (180), which do not interact with selectins. Thus, FucT-IV may compete with α(2,3)sialyltransferase(s) and/or FucT-VII for terminal lactosamine as a shared precursor substrate on *O*-linked glycans. Consequently, FucT-IV may divert some terminal lactosamine acceptor moieties toward the Lewis^x pathway that does not yield L-selectin ligands, and thus away from the α(2,3)sialyltransferase/FucT-VII synthetic route. In wild-type mice, FucT-IV thus attenuates FucT-VII–dependent production of sLe^x-based L-selectin ligands. Without FucT-IV, increased acceptor availability for α(2,3)sialyltransferase/FucT-VII permits increased selectin ligand

Table 170-2: Sialyl Lewis^x Structures and HEV Carbohydrate Synthesis Enzymes

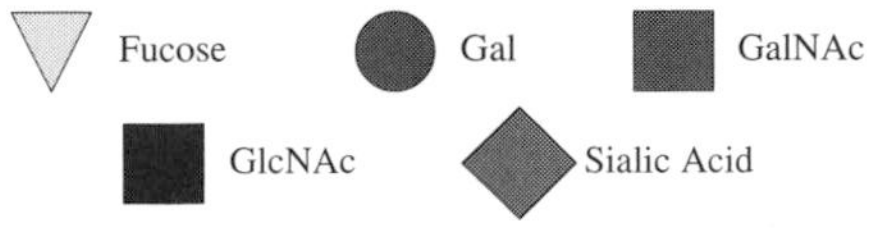

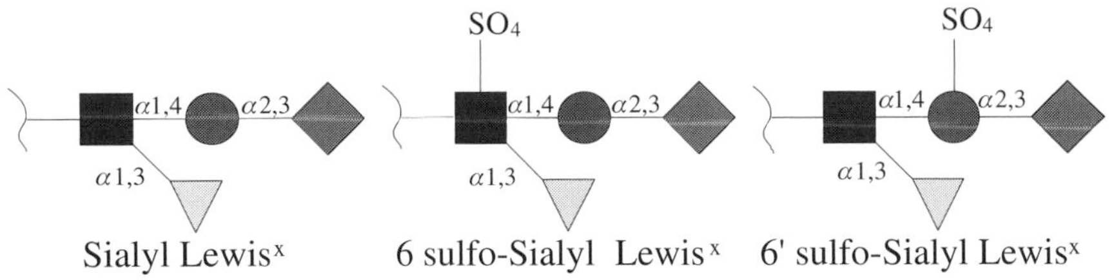

Enzyme	Enzyme Reaction
C1GlcNAcT	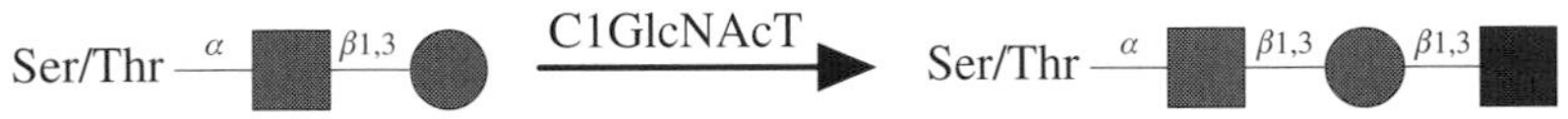
C2GlcNAcT-1	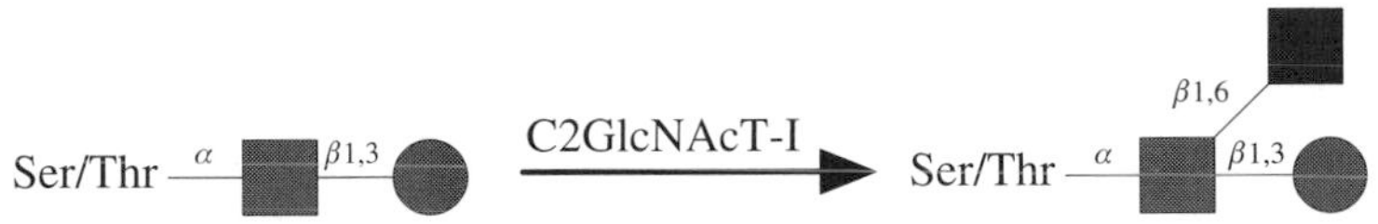
HEC-GlcNAc6ST (GlcNAc6ST-2)[1], GlcNAc6ST-1[2]	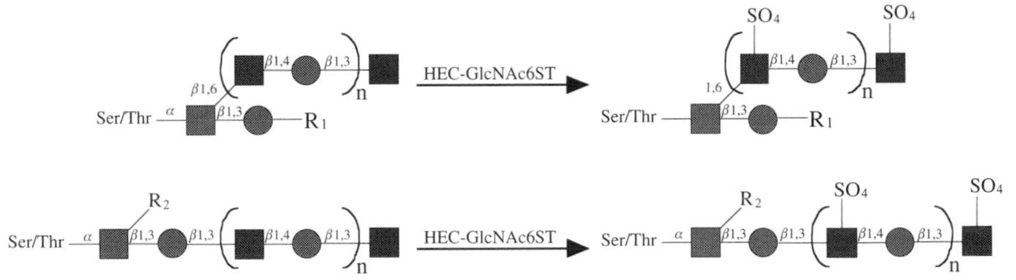 Internal sulfations may or may not be present.
FucT-VII (FucT-IV)	These fucosylated, sialylated, sulfated structures represent *bona fide* L-selectin ligands. 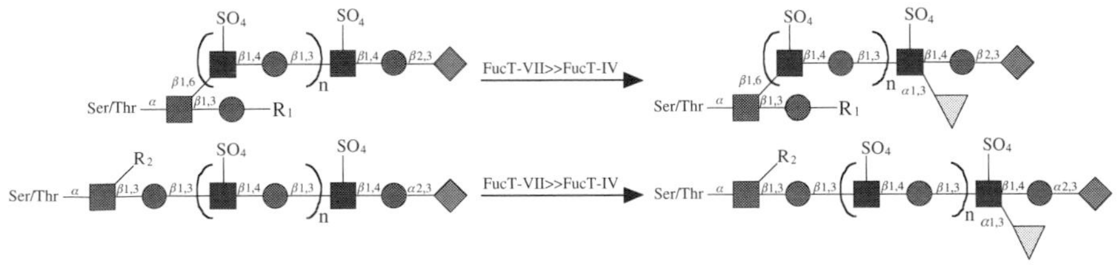
FucT-IV[3]	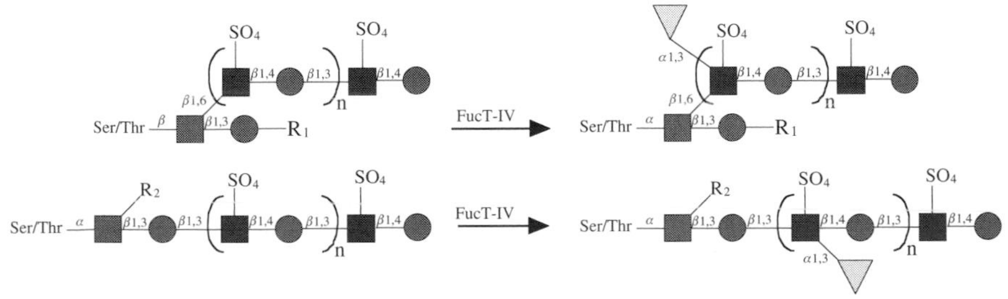

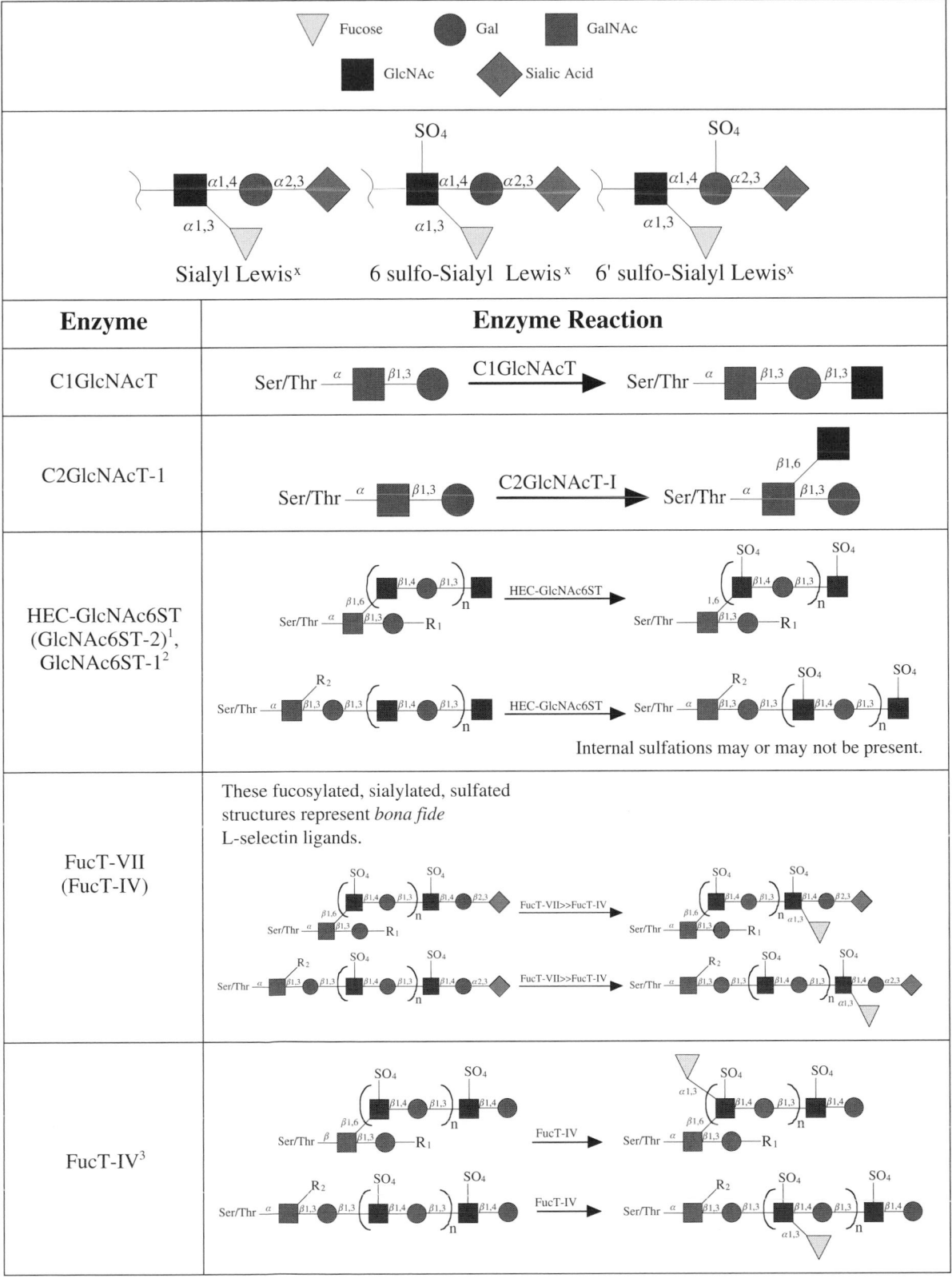

[1]HEC-GlcNAc6ST generates luminal MECA-79 antigen on LN HECs.
[2]GlcNAc6ST-1 generates abluminal MECA-79 antigen on LN and PP HECs.
[3]Unlike FucT-VII, FucT-IV fucosylates nonsialylated carbohydrates and internal GlcNAc residues on extended lactosamines; however, fucosylation of terminal GlcNAc residues can also occur. In addition, FucT-1V fucosylates glycolipids (not depicted) (222,223).
Biantennary glycans arise from a core1 branch at R_1 or a core2 branch at R_2.

Table 170-3: Phenotype of HEC-Modifying Enzyme Knockouts

Enzymes	*Knockout Phenotype*			*References*
	Lymphocyte Homing to PLNs	*MECA-79 Staining of HEVs*	*L-selectin IgM Staining of HEVs*	
C2GlcNAcT-1	Normal/Decreased	Normal	Absent	193, 190, 89
HEC-GlcNAc6ST (GlcNAc6ST-2)	50% Reduction	Only abluminal staining	Only abluminal staining	151, 150, 89
GlcNAc6ST-1	20% Reduction	Only luminal staining present	NA[1]	152
FucT-VII	80% Reduction	Normal	Absent	178
FucT-IV	25% Increased	NA	Normal	180, 36
FucT-VII/FucT-IV	>95% Reduction	NA	NA	180

C2GlcNAcT-1, core2 β-1,6-*N*-acetylglucosaminyltransferase-1; HEC-GlcNAc6ST, high EC GlcNAc-6-sulfotransferase; FucT-VII, α(1,3)fucosyltransferase-VII; FucT-IV, α(1,3)fucosyltransferase-IV.
[1]NA, Data not available.

production, whereas FucT-VII deficiency has a lesser impact, since sLe^x-based L-selectin ligands could still be generated if FucT-IV is abundantly expressed.

Experimental evidence supports this scenario and has revealed additional complexity, because FucT-IV and FucT-VII are not uniformly expressed within the venular tree in PLNs (36). For example FucT-IV is expressed at higher levels in the medullary collecting venules of mouse PLNs than in para- and subcortical HEVs. In medullary venules, FucT-IV is required to generate an additional L-selectin ligand distinct from PNAd (36). Although MECA-79 staining highlights para- and subcortical HEVs and requires primarily FucT-VII (148), the medullary L-selectin ligand is not recognized by MECA-79, but still supports rolling of L-selectin$^+$ lymphocytes (36).

Sulfation

The critical role of sulfation of HEV proteins was first suggested by the unique ability of HEVs among vascular endothelium to incorporate large amounts of sulfate (78). Indeed, metabolic inhibition of sulfation reduces L-selectin ligand activity in PLNs (171). There are two sulfated sLe^x moieties in PLNs, one containing Gal-6-SO_4 (6′-sulfo Lewis^x) and the other containing GlcNAc-6-SO_4 (6-sulfo Lewis^x) (see Table 170-2) (173), but only 6-sulfation of sLe^x is thought to enhance L-selectin binding (183,184), although some in vitro evidence suggests that the 6′-sulfo modification can also promote L-selectin binding (185). The GlcNAc-6-O-sulfotransferases responsible for generation of L-selectin ligands in HEVs are GlcNAc6ST-1 (186) and HEC-GlcNAc6ST/LSST/GlcNAc6ST-2 (Table 170-2) (187,188). In reconstitution experiments, GlyCAM-1–IgG coexpressed with HEC-GlcNAC6ST and FucT-VII supports greater rolling of L-selectin$^+$ cells than does GlyCAM-1–IgG expressed with FucT-VII alone (189). Similar results were obtained with CD34 (188). Mice deficient in HEC-GlcNAc6ST have hypocellular PLNs, impaired (~50%) lymphocyte homing to PLNs, reduced L-selectin binding to HEVs, and a near absence of MECA-79–reactive material in the lumen of HEVs (151,190). The latter is not surprising, since the epitope recognized by MECA-79 requires 6-sulfation of sLe^x (158,175). GlcNAc6ST-1 is the sulfotransferase responsible for the generation of abluminal MECA-79 antigen in PLN and PP HEVs (152), and mice deficient in GlcNAc6ST-1 have a 20% to 30% reduction in lymphocyte homing to PLNs and PPs, suggesting that GlcNAc6ST-1 is also responsible for the biosynthesis of some luminal L-selectin ligands on HEVs (see Table 170-3) (152).

O-Linked Carbohydrates

O-linked glycans in core2 (i.e., GlcNAcβ1,6[Galβ1,3]-GalNAc) and core1 (i.e., GlcNAcβ1,3[Galβ1,3]-GalNAc) linkages represent the major components of L-selectin ligands in HEVs (191) (see Table 170-2). Two enzymes, core2 β-1,6-*N*-acetylglucosaminyltransferase-1 (C2GlcNAcT-1) (192) and core1-β1,3-*N*-acetylglucosaminyltransferase (C1GlcNAcT) (175) have been identified as being expressed in PLN HECs and are suggested to contribute significantly to the synthesis of L-selectin ligands in PLNs. Although C2GlcNAcT-I is critical for the generation of P- and E-selectin ligands, no defect was observed in lymphocyte trafficking to peripheral lymphoid organs of mice deficient for this enzyme (193), although more recent data indeed suggests that these animals have defects in lymphocyte trafficking to PLNs (89,190) (see Table 170-3). Residual L-selectin ligand activity in C2GlcNAcT-1–deficient mice may be due to additional core2 β-1,6-*N*-acetylglucosaminyltransferases and/or the activity of C1GlcNAcT, which has been shown in vitro to be critical in the generation of MECA-79–reactive glycans and to contribute to the synthesis of L-selectin ligands (175) (see Table 170-2).

Mucosal Addressin Cell Adhesion Molecule-1

MAdCAM-1 in MLN and PP HEVs can support L-selectin interactions if appropriately glycosylated. Interestingly, MAdCAM-1 contains Ig domains capable of binding to integrins, including $\alpha4\beta7$ and, more weakly, to $\alpha4\beta1$, but it also has a mucin-like region rich in serine and threonine, which bears *O*-linked carbohydrate modifications that can support L-selectin binding (194). MECA-79 can immunoprecipitate MAdCAM-1 from MLNs, but not PPs, and MAdCAM-1 from MLNs supports L-selectin binding in vitro (123). However, MAdCAM-1 in PP HEVs can also present L-selectin ligands (which are not recognized by MECA-79) and support L-selectin tethering and rolling in vivo (44).

PLASTICITY OF ENDOTHELIUM AND HIGH ENDOTHELIAL VENULES

Perinatal Switch

In the developing mouse, the main addressin expressed in both LN and PP HEVs is MAdCAM-1 (195). It is thought that a population of $CD45^{+}CD4^{+}CD3^{-}IL7R\alpha^{hi}$ cells is required to provide the signals necessary for the further development of LNs and PPs (196). These rare cells (1%–2% of total peripheral blood lymphocytes in the mouse fetus) express $\alpha4\beta7$ integrin and must bind to MAdCAM-1 to migrate to LNs (195). On the other hand, studies in rats found almost no HEVs in LNs prenatally (197). ECs in cortical vessels acquired a cuboid morphology only gradually after birth, indicating that HEV commitment may be uniquely regulated in different species.

Lymph-Borne High Endothelial Venule Differentiation Factors

Once developed, the mature HEV phenotype requires constant maintenance signals. Hendriks and colleagues reported that HEVs from PLNs that had undergone occlusion of their afferent lymphatics became flat and lost their characteristic morphology (198). Subsequent studies showed that lymphocytes do not adhere to the lymph-deprived HEVs (199), and genes that contribute to the generation of L-selectin ligands are turned off (200,201). The nature of the HEV-sustaining lymph-borne signal is still mysterious. Lymph flow per se is probably not essential, since HEVs are also found in PPs, which do not possess afferent lymphatics.

Ectopic High Endothelial Venules

Normal endothelium can differentiate into HEVs when given the appropriate stimuli (see Chapter 155). Chronic inflammation in the setting of certain infections and autoimmune diseases can promote the organization of cellular infiltrates into lymph node–like structures that contain T and B follicles, DCs, and HEVs in a process termed *lymphoid neogenesis* (202–204). Postcapillary venules in these aggregates assume a HEV-like morphology, express PNAd and/or MAdCAM-1, and recruit naïve T cells from the blood (205). These de novo HEVs express the repertoire of carbohydrate-modifying enzymes including core1-β3GlcNAcT, core2 GlcNAcT-1, FucT-VII, and HEC-GlcNAc6ST, typically found in bona fide HECs (190,206). Induced expression of CCL19, CCL21 or CXCL13 is sufficient to cause lymphoid neogenesis, at least in certain organs such as the pancreas (207–209).

How the recruitment of naïve lymphocytes induces the differentiation of ectopic HEVs is not yet clear. A likely player is lymphotoxin (LTα), a member of the TNF family that is expressed by activated CD4 and CD8 T cells (210), and also by the above mentioned $CD4^{+}CD3^{-}IL7R\alpha^{hi}$ cells that seed nascent LNs. Indeed, LT$\alpha\beta$, its receptor (LTβR), and downstream signaling components including IκB kinase complex-α coordinate the genesis of SLOs and several important genes including GlyCAM-1, HEC-GlcNAc6ST, and the chemokines CCL19, CCL21, and CXCL13 (211–214). Moreover, expression of LT$\alpha\beta$ is sufficient and necessary to induce lymphoid neogenesis (204,208,214).

Recent studies have demonstrated a significant role for stromal cells in the induction of SLOs. When a thymus-derived stromal cell line expressing LTβR, VCAM-1, and LTα was injected into the renal subcapsular space, the transplanted cells could induce the formation of SLO-like tissue containing organized T- and B-cell areas and HEVs (215). Aggregate formation was enhanced when activated DCs were coinjected with the stromal cells. MLN cell suspensions from newborn mice injected intradermally into the skin of newborn or adult mice generate ectopic lymphoid aggregates (205). These newborn MLN suspensions contained stromal cells that expressed VCAM-1, ICAM-1, and LTβR, as well as $CD45^{+}CD4^{+}CD3^{-}$ hematopoietic cells; however, if the MLN cells were derived from $LT\alpha^{-/-}$ mice, ectopic aggregates could not form. Interestingly, in this system, the donor MLN cells gave rise to the lymphatic endothelium, HEVs, stromal cells, and FRCs, but the hematopoietic components were of host origin. These two systems could prove to be very powerful tools for further dissecting the molecular events necessary for normal and ectopic lymphoid organogenesis.

CONCLUSION

Although the past two decades have seen considerable progress in our understanding of the functions and molecular and biochemical underpinnings of HEVs, many fundamental questions remain to be answered. What does it take to make an HEV? As we have seen, environmental input from constitutive lymph flow or factors arising from chronic inflammation are required to maintain HEVs and to turn regular ECs into HECs, respectively. What are the unique signaling and transcriptional events that induce and sustain this dramatic phenotypic alteration? Can we interfere with these signals for therapeutic purposes? Can we use them to grow and study HECs in vitro? What makes a PLN HEV different from a PP HEV, and are these different from ectopic HEVs? Another gray area is how lymphocytes transmigrate across HEVs. Is leukocyte recruitment regulated at this last step of the adhesion cascade, or is diapedesis a default process? To answer these questions, novel technologies have begun to be employed together with established techniques.

KEY POINTS

- HEVs are highly specialized postcapillary venules that have evolved to function as the portal for leukocyte access to most SLOs.
- ECs forming HEVs display a distinct pattern of gene expression aimed at maximizing their efficiency in recruiting lymphocytes.
- HEVs in PLNs have the unique capability of efficiently recruiting naïve T cells through a three-step cascade mediated by sequential interactions between L-selectin: PNAd, CCR7:CCL21, and LFA-1:ICAM-1.
- Although HEVs are usually found in secondary lymphoid organs, any venular endothelium can acquire an HEV-like phenotype during chronic inflammation; conversely HEVs in SLOs can revert to normal flat endothelium when lymph drainage is interrupted.

Future Goals

- To determine the molecular mechanism of transmigration of lymphocytes across HEVs
- To identify molecular signals that convert flat to HEV-like endothelium in inflamed tissues, in order to develop novel approaches to control chronic inflammation and autoimmunity

REFERENCES

1 Marchesi VT, Gowans JL. The migration of lymphocytes through the endothelium of venules in lymph nodes: an electron microscope study. *Proc R Soc*. 1964;B 159:283–290.
2 Girard J-P, Springer TA. High endothelial venules (HEVs): specialized endothelium for lymphocyte migration. *Immunol Today*. 1995;16:449–457.
3 Springer TA. Traffic signals for lymphocyte recirculation and leukocyte emigration: the multi-step paradigm. *Cell*. 1994; 76:301–314.
4 von Andrian UH, Mackay CR. T-cell function and migration. Two sides of the same coin. *N Engl J Med*. 2000;343(14):1020–1034.
5 Rosen SD. Ligands for L-selectin: homing, inflammation, and beyond. *Annu Rev Immunol*. 2004;22:129–156.
6 Takagi J, Springer TA. Integrin activation and structural rearrangement. *Immunol Rev*. 2002;186:141–163.
7 Laudanna C, Kim JY, Constantin G, Butcher E. Rapid leukocyte integrin activation by chemokines. *Immunol Rev*. 2002;186:37–46.
8 Thomé R. Endothelien als Phagocyten (aus den Lymphdrüsen von Macacus cynomolgous). *Arch Mikroskop Anat*. 1898;52: 820–842.
9 von Schumacher S. Über Phagocytose und die Abfuhrwege der Leukocyten in den Lymphdrüsen. *Arch Mikroskop Anat*. 1899:54.
10 Gowans JL, Knight EJ. The route of re-circulation of lymphocytes in the rat. *Proc R Soc Lond B Biol Sci*. 1964;159:257–282.
11 Hall JG, Morris B. The origin of the cells in the efferent lymph from a single lymph node. *J Exp Med*. 1965;121:901–911.
12 Anderson AO, Anderson ND. Lymphocyte emigration from high endothelial venules in rat lymph nodes. *Immunology*. 1976; 31:731–748.
13 Pabst R, Binns RM. Heterogeneity of lymphocyte homing physiology: several mechanisms operate in the control of migration to lymphoid and non-lymphoid organs in vivo. *Immunol Rev*. 1989;108:83–109.
14 Claesson MH, Jorgensen O, Ropke C. Light and electron microscopic studies of the paracortical post-capillary high-endothelial venules. *Z Zellforsch Mikrosk Anat*. 1971;119(2):195–207.
15 Baekkevold ES, Yamanaka T, Palframan RT, et al. The CCR7 ligand ELC (CCL19) is transcytosed in high endothelial venules and mediates T cell recruitment. *J Exp Med*. 2001;193(9):1105–1112.
16 Stamper HB Jr., Woodruff JJ. Lymphocyte homing into lymph nodes: in vitro demonstration of the selective affinity of recirculating lymphocytes for high-endothelial venules. *J Exp Med*. 1976;144:828.
17 Butcher EC, Scollay RG, Weissman IL. Organ specificity of lymphocyte migration: mediation by highly selective lymphocyte interaction with organ-specific determinants on high endothelial venules. *Eur J Immunol*. 1980;10:556–561.
18 Gallatin WM, Weissman IL, Butcher EC. A cell-surface molecule involved in organ-specific homing of lymphocytes. *Nature*. 1983;304:30–34.
19 Streeter PR, Lakey-Berg E, Rouse BTN, et al. A tissue-specific endothelial cell molecule involved in lymphocyte homing. *Nature*. 1988;331:41–46.
20 Streeter PR, Rouse BTN, Butcher EC. Immunohistologic and functional characterization of a vascular addressin involved in lymphocyte homing into peripheral lymph nodes. *J Cell Biol*. 1988;107:1853–1862.
21 Rosen SD, Chi SI, True DD, et al. Intravenously injected sialidase inactivates attachment sites for lymphocytes on high endothelial venules. *J Immunol*. 1989;142:1895–1902.
22 Van Vliet E, Melis M, Foidart JM, Van Ewijk W. Reticular fibroblasts in peripheral lymphoid organs identified by a monoclonal antibody. *J Histochem Cytochem*. 1986;34(7):883–890.
23 Nagira M, Imai T, Yoshida R, et al. A lymphocyte-specific CC chemokine, secondary lymphoid tissue chemokine (SLC), is a highly efficient chemoattractant for B cells and activated T cells. *Eur J Immunol*. 1998;28:1516–1523.
24 Ager A. Isolation and culture of high endothelial cells from rat lymph nodes. *J Cell Sci*. 1987;87:133–144.
25 Ise Y, Yamaguchi K, Sato K, et al. Molecular mechanisms underlying lymphocyte recirculation: I. Functional, phenotypical and morphological characterization of high endothelial cells cultured in vitro. *Eur J Immunol*. 1988;18:1235–1244.
26 Ager A, Mistry S. Interaction between lymphocytes and cultured high endothelial cells: An in vitro model of lymphocyte

migration across high endothelial venule endothelium. *Eur J Immunol.* 1990;18:1265–1274.
27 Derry CJ, Faveeuw C, Mordsley KR, Ager A. Novel chondroitin sulfate-modified ligands for L-selectin on lymph node high endothelial venules. *Eur J Immunol.* 1999;29(2):419–430.
28 Baekkevold ES, Jahnsen FL, Johansen FE, et al. Culture characterization of differentiated high endothelial venule cells from human tonsils. *Lab Invest.* 1999;79(3):327–336.
29 Lacorre DA, Baekkevold ES, Garrido I, et al. Plasticity of endothelial cells: rapid dedifferentiation of freshly isolated high endothelial venule endothelial cells outside the lymphoid tissue microenvironment. *Blood.* 2004;103(11):4164–4172.
30 Izawa D, Tanaka T, Saito K, et al. Expression profile of active genes in mouse lymph node high endothelial cells. *Int Immunol.* 1999;11(12):1989–1998.
31 Saito K, Tanaka T, Kanda H, et al. Gene expression profiling of mucosal addressin cell adhesion molecule-1^+ high endothelial venule cells (HEV) and identification of a leucine-rich HEV glycoprotein as a HEV marker. *J Immunol.* 2002;168(3):1050–1059.
32 Girard J-P, Springer TA. Cloning from purified high endothelial venule cells of hevin, a close relative of the antiadhesive extracellular matrix protein SPARC. *Immunity.* 1995;2:113–122.
33 Girard JP, Baekkevold ES, Amalric F. Sulfation in high endothelial venules: cloning and expression of the human PAPS synthetase. *FASEB J.* 1998;12(7):603–612.
34 Girard JP, Baekkevold ES, Yamanaka T, et al. Heterogeneity of endothelial cells: the specialized phenotype of human high endothelial venules characterized by suppression subtractive hybridization. *Am J Pathol.* 1999;155(6):2043–2055.
35 Baekkevold ES, Roussigne M, Yamanaka T, et al. Molecular characterization of NF-HEV, a nuclear factor preferentially expressed in human high endothelial venules. *Am J Pathol.* 2003;163(1):69–79.
36 M'Rini C, Cheng G, Schweitzer C, et al. A novel endothelial L-selectin ligand activity in lymph node medulla that is regulated by a(1,3)-fucosyltransferase-IV. *J Exp Med.* 2003;198(9):1301–1312.
37 Mempel TR, Scimone ML, Mora JR, von Andrian UH. In vivo imaging of leukocyte trafficking in blood vessels and tissues. *Curr Opin Immunol.* 2004;16(4):406–417.
38 Wagner R. *Erläuterungstafeln zur Physiologie und Entwicklungsgeschichte.* Leipzig: Leopold Voss, 1839.
39 Cohnheim J. *Lectures on General Pathology: A Handbook for Practitioners and Students.* London: The New Sydenham Society, 1889.
40 von Andrian UH, Chambers JD, McEvoy LM, et al. Two-step model of leukocyte-endothelial cell interaction in inflammation: distinct roles for LECAM-1 and the leukocyte β_2 integrins *in vivo. Proc Natl Acad Sci USA.* 1991;88:7538–7542.
41 Ley K, Gaehtgens P, Fennie C, et al. Lectin-like cell adhesion molecule 1 mediates leukocyte rolling in mesenteric venules in vivo. *Blood.* 1991;77:2553–2555.
42 Mayadas TN, Johnson RC, Rayburn H, et al. Leukocyte rolling and extravasation are severely compromised in P-selectin-deficient mice. *Cell.* 1993;74:541–554.
43 Bjerknes M, Cheng H, Ottaway CA. Dynamics of lymphocyte-endothelial interactions in vivo. *Science.* 1986;231:402–405.
44 Bargatze RF, Jutila MA, Butcher EC. Distinct roles of L-selectin and integrins $\alpha4\beta7$ and LFA-1 in lymphocyte homing to Peyer's patch-HEV in situ: the multistep model confirmed and refined. *Immunity.* 1995;3:99–108.
45 von Andrian UH. Intravital microscopy of the peripheral lymph node microcirculation in mice. *Microcirculation.* 1996;3:287–300.
46 Warnock RA, Askari S, Butcher EC, von Andrian UH. Molecular mechanisms of lymphocyte homing to peripheral lymph nodes. *J Exp Med.* 1998;187:205–216.
47 Stein JV, Rot A, Luo Y, et al. The CC chemokine thymus-derived chemotactic agent 4 (TCA-4, secondary lymphoid tissue chemokine, 6Ckine, exodus-2) triggers lymphocyte function-associated antigen 1-mediated arrest of rolling T lymphocytes in peripheral lymph node high endothelial venules. *J Exp Med.* 2000;191(1):61–76.
48 Singbartl K, Thatte J, Smith ML, et al. A CD2-green fluorescence protein-transgenic mouse reveals very late antigen-4-dependent $CD8^+$ lymphocyte rolling in inflamed venules. *J Immunol.* 2001;166(12):7520–7526.
49 Palframan RT, Jung S, Cheng G, et al. Inflammatory chemokine transport and presentation in HEV: a remote control mechanism for monocyte recruitment to lymph nodes in inflamed tissues. *J Exp Med.* 2001;194(9):1361–1374.
50 von Andrian UH, Mempel TR. Homing and cellular traffic in lymph nodes. *Nat Rev Immunol.* 2003;3:867–878.
51 Denk W, Strickler JH, Webb WW. Two-photon laser scanning fluorescence microscopy. *Science.* 1990;248(4951):73–76.
52 Cahalan MD, Parker I, Wei SH, Miller MJ. Two-photon tissue imaging: seeing the immune system in a fresh light. *Nat Rev Immunol.* 2002;2(11):872–880.
53 Miller MJ, Wei SH, Parker I, Cahalan MD. Two-photon imaging of lymphocyte motility and antigen response in intact lymph node. *Science.* 2002;296(5574):1869–1873.
54 Stoll S, Delon J, Brotz TM, Germain RN. Dynamic imaging of T cell-dendritic cell interactions in lymph nodes. *Science.* 2002;296(5574):1873–1876.
55 Bousso P, Robey E. Dynamics of $CD8^{(+)}$ T cell priming by dendritic cells in intact lymph nodes. *Nat Immunol.* 2003;4(6):579–585.
56 Miller MJ, Wei SH, Cahalan MD, Parker I. Autonomous T cell trafficking examined in vivo with intravital two-photon microscopy. *Proc Natl Acad Sci USA.* 2003;100(5):2604–2609.
57 Mempel TR, Henrickson SE, von Andrian UH. T cell priming by dendritic cells in lymph nodes occurs in three distinct phases. *Nature.* 2004;427(6970):154–159.
58 Mazo IB, Honczarenko M, Leung H, et al. Bone marrow is a major reservoir and site of recruitment for central memory $CD8^+$ T cells. *Immunity.* 2005;22(2):259–270.
59 Lindquist RL, Shakhar G, Dudziak D, et al. Visualizing dendritic cell networks in vivo. *Nat Immunol.* 2004;5(12):1243–1250.
60 Gretz JE, Anderson AO, Shaw S. Cords, channels, corridors and conduits: critical architectural elements facilitating cell interactions in the lymph node cortex. *Immunol Rev.* 1997;156:11–24.
61 Witmer MD, Steinman RM. The anatomy of peripheral lymphoid organs with emphasis on accessory cells: light-microscopic immunocytochemical studies of mouse spleen, lymph node, and Peyer's patch. *Am J Anat.* 1984;170(3):465–481.
62 Krall WJ, Braun J. In situ lacZ retrovirus-marked lymphocytes define a B cell microenvironment in the lymph node medulla. *New Biol.* 1992;4(5):581–590.
63 Bockman DE, Stevens W. Gut-associated lymphoepithelial tissue: bidirectional transport of tracer by specialized epithelial

cells associated with lymphoid follicles. *J Reticuloendothel Soc.* 1977;21(4):245–254.
64 Mowat AM. Anatomical basis of tolerance and immunity to intestinal antigens. *Nat Rev Immunol.* 2003;3(4):331–341.
65 von Andrian UH, M'Rini C. In situ analysis of lymphocyte migration to lymph nodes. *Cell Adhes Commun.* 1998;6 (2–3):85–96.
66 De Bruyn PPH, Cho Y. Structure and function of high endothelial postcapillary venules in lymphocyte recirculation. In: Grundman E, Vollmer T, eds. *Reaction Patterns of the Lymph Node.* Berlin: Springer-Verlag;1990:85–103.
67 Yamaguchi K, Schoefl GI. Blood vessels of the Peyer's patch in the mouse: I. Topographic studies. *Anat Rec.* 1983;206(4):391–401.
68 Farr AG, Cho Y, De Bruyn PP. The structure of the sinus wall of the lymph node relative to its endocytic properties and transmural cell passage. *Am J Anat.* 1980;157(3):265–284.
69 Bajenoff M, Granjeaud S, Guerder S. The strategy of T cell antigen-presenting cell encounter in antigen-draining lymph nodes revealed by imaging of initial T cell activation. *J Exp Med.* 2003;198(5):715–724.
70 Anderson AO, Anderson ND. Studies on the structure and permeability of the microvasculature in normal rat lymph nodes. *Am J Pathol.* 1975;30:387–418.
71 Clark SL Jr. The reticulum of lymph nodes in mice studied with the electron microscope. *Am J Anat.* 1962;110:217–258.
72 Gretz JE, Norbury CC, Anderson AO, et al. Lymph-borne chemokines and other low molecular weight molecules reach high endothelial venules via specialized conduits while a functional barrier limits access to the lymphocyte microenvironments in lymph node cortex. *J Exp Med.* 2000;192(10):1425–1440.
73 Katakai T, Hara T, Sugai M, et al. Lymph node fibroblastic reticular cells construct the stromal reticulum via contact with lymphocytes. *J Exp Med.* 2004;200(6):783–795.
74 Katakai T, Hara T, Lee JH, et al. A novel reticular stromal structure in lymph node cortex: an immuno-platform for interactions among dendritic cells, T cells and B cells. *Int Immunol.* 2004;16(8):1133–1142.
75 Sixt M, Kanazawa N, Selg M, et al. The conduit system transports soluble antigens from the afferent lymph to resident dendritic cells in the T cell area of the lymph node. *Immunity.* 2005; 22(1):19–29.
76 Anderson ND, Anderson AO, Wyllie RG. Specialized structure and metabolic activities of high endothelial venules in rat lymphatic tissues. *Immunology.* 1976;31:455–473.
77 Freemont AJ, Jones CJP. Light microscopic, histochemical and ultrastructural studies of human lymph node paracortical venules. *J Anat.* 1983;136:349–362.
78 Andrews P, Milsom DW, Ford WL. Migration of lymphocytes across specialized vascular endothelium V. Production of a sulphated macromolecule by high endothelial cells in lymph nodes. *J Cell Sci.* 1982;57:277–292.
79 Anderson AO, Shaw S. T cell adhesion to endothelium: the FRC conduit system and other anatomic and molecular features which facilitate the adhesion cascade in lymph node. *Semin Immunol.* 1993;5:271–282.
80 Kuschert GS, Coulin F, Power CA, et al. Glycosaminoglycans interact selectively with chemokines and modulate receptor binding and cellular responses. *Biochemistry.* 1999;38(39): 12959–12968.
81 Middleton J, Patterson AM, Gardner L, et al. Leukocyte extravasation: chemokine transport and presentation by the endothelium. *Blood.* 2002;100(12):3853–3860.
82 Hirose J, Kawashima H, Yoshie O, et al. Versican interacts with chemokines and modulates cellular responses. *J Biol Chem.* 2001;276(7):5228–5234.
83 Chaudhuri A, Nielsen S, Elkjaer ML, et al. Detection of Duffy antigen in the plasma membranes and caveolae of vascular endothelial and epithelial cells of nonerythroid organs. *Blood.* 1997;89(2):701–712.
84 Middleton J, Neil S, Wintle J, et al. Transcytosis and surface presentation of IL-8 by venular endothelial cells. *Cell.* 1997; 91:1001–1011.
85 Kashiwazaki M, Tanaka T, Kanda H, et al. A high endothelial venule-expressing promiscuous chemokine receptor DARC can bind inflammatory, but not lymphoid, chemokines and is dispensable for lymphocyte homing under physiological conditions. *Int Immunol.* 2003;15(10):1219–1227.
86 Butcher EC, Picker LJ. Lymphocyte homing and homeostasis. *Science.* 1996;272:60–66.
87 Weninger W, Crowley MA, Manjunath N, von Andrian UH. Migratory properties of naive, effector, and memory CD8($^{+}$) T cells. *J Exp Med.* 2001;194(7):953–966.
88 Okada T, Ngo VN, Ekland EH, et al. Chemokine requirements for B cell entry to lymph nodes and Peyer's patches. *J Exp Med.* 2002;196(1):65–75.
89 Gauguet JM, Rosen SD, Marth JD, von Andrian UH. Core 2 branching beta1,6-N-acetylglucosaminyltransferase and high endothelial cell N-acetylglucosamine-6-sulfotransferase exert differential control over B- and T-lymphocyte homing to peripheral lymph nodes. *Blood.* 2004;104(13): 4104–4112.
90 Berg EL, Robinson MK, Warnock RA, Butcher EC. The human peripheral lymph node vascular addressin is a ligand for LECAM-1, the peripheral lymph node homing receptor. *J Cell Biol.* 1991;114:343–349.
91 Hamann A, Westrich DJ, Duijevstijn A, et al. Evidence for an accessory role of LFA-1 in lymphocyte-high endothelium interaction during homing. *J Immunol.* 1988;140:693–699.
92 Andrew DP, Spellberg JP, Takimoto H, et al. Transendothelial migration and trafficking of leukocytes in LFA-1- deficient mice. *Eur J Immunol.* 1998;28(6):1959–1969.
93 Berlin-Rufenach C, Otto F, Mathies M, et al. Lymphocyte migration in lymphocyte function-associated antigen (LFA)-1- deficient mice. *J Exp Med.* 1999;189(9):1467–1478.
94 Staunton DE, Marlin SD, Stratowa C, et al. Primary structure of intercellular adhesion molecule 1 (ICAM-1) demonstrates interaction between members of the immunoglobulin and integrin supergene families. *Cell.* 1988;52:925–933.
95 Staunton DE, Dustin ML, Springer TA. Functional cloning of ICAM-2, a cell adhesion ligand for LFA-1 homologous to ICAM-1. *Nature.* 1989;339:61–64.
96 Xu H, Gonzalo JA, St. Pierre Y, et al. Leukocytosis and resistance to septic shock in intercellular adhesion molecule 1-deficient mice. *J Exp Med.* 1994;180:95–109.
97 Sligh JE Jr, Ballantyne CM, Rich S, et al. Inflammatory and immune responses are impaired in mice deficient in intercellular adhesion molecule 1. *Proc Natl Acad Sci USA.* 1993;90:8529–8533.
98 Gerwin N, Gonzalo JA, Lloyd C, et al. Prolonged eosinophil accumulation in allergic lung interstitium of ICAM-2

deficient mice results in extended hyper-responsiveness. *Immunity*. 1999;10:9–19.
99 Lehmann JC, Jablonski-Westrich D, Haubold U, et al. Overlapping and selective roles of endothelial intercellular adhesion molecule-1 (ICAM-1) and ICAM-2 in lymphocyte trafficking. *J Immunol*. 2003;171(5):2588–2593.
100 Gunn MD, Tangemann K, Tam C, et al. A chemokine expressed in lymphoid high endothelial venules promotes the adhesion and chemotaxis of naive T lymphocytes. *Proc Natl Acad Sci USA*. 1998;95:258–263.
101 Campbell JJ, Hedrick J, Zlotnik A, et al. Chemokines and the arrest of lymphocytes rolling under flow conditions. *Science*. 1998;279:381–384.
102 Campbell JJ, Bowman EP, Murphy K, et al. 6-C-kine (SLC), a lymphocyte adhesion-triggering chemokine expressed by high endothelium, is an agonist for the MIP-3beta receptor CCR7. *J Cell Biol*. 1998;141:1053–1059.
103 Vassileva G, Soto H, Zlotnik A, et al. The reduced expression of 6Ckine in the plt mouse results from the deletion of one of two 6Ckine genes. *J Exp Med*. 1999;190(8):1183–1188.
104 Nakano H, Gunn MD. Gene duplications at the chemokine locus on mouse chromosome 4: multiple strain-specific haplotypes and the deletion of secondary lymphoid-organ chemokine and EBI-1 ligand chemokine genes in the plt mutation. *J Immunol*. 2001;166(1):361–369.
105 Forster R, Schubel A, Breitfeld D, et al. CCR7 coordinates the primary immune response by establishing functional microenvironments in secondary lymphoid organs. *Cell*. 1999;99(1):23–33.
106 Gunn MD, Kyuwa S, Tam C, et al. Mice lacking expression of secondary lymphoid organ chemokine have defects in lymphocyte homing and dendritic cell localization. *J Exp Med*. 1999; 189:451–460.
107 Scimone ML, Felbinger TW, Mazo IB, et al. CXCL12 mediates CCR7-independent homing of central memory cells, but not naive T cells, in peripheral lymph nodes. *J Exp Med*. 2004; 199(8):1113–1120.
108 Cinamon G, Shinder V, Alon R. Shear forces promote lymphocyte migration across vascular endothelium bearing apical chemokines. *Nat Immunol*. 2001;2(6):515–522.
109 Weninger W, Carlsen HS, Goodarzi M, et al. Naïve T cell recruitment to non-lymphoid tissues: a role for endothelium-expressed CCL21 in autoimmune disease and lymphoid neogenesis. *J Immunol*. 2003;170:4638–4648.
110 Ebisuno Y, Tanaka T, Kanemitsu N, et al. Cutting edge: the B cell chemokine CXC chemokine ligand 13/B lymphocyte chemoattractant is expressed in the high endothelial venules of lymph nodes and Peyer's patches and affects B cell trafficking across high endothelial venules. *J Immunol*. 2003;171(4):1642–1646.
111 Tang ML, Steeber DA, Zhang XQ, Tedder TF. Intrinsic differences in L-selectin expression levels affect T and B lymphocyte subset-specific recirculation pathways. *J Immunol*. 1998; 160(10):5113–5121.
112 Stein JV, Cheng G, Stockton BM, et al. L-selectin-mediated leukocyte adhesion in vivo: Microvillous distribution determines tethering efficiency, but not rolling velocity. *J Exp Med*. 1999;189:37–50.
113 Robert C, Klein C, Cheng G, et al. Gene therapy to target dendritic cells from blood to lymph nodes. *Gene Ther*. 2003; 10(17):1479–1486.
114 Diacovo TG, Puri KD, Warnock RA, et al. Platelet-mediated lymphocyte delivery to high endothelial venules. *Science*. 1996; 273:252–255.
115 Diacovo TG, Catalina MD, Siegelman MH, von Andrian UH. Circulating activated platelets reconstitute lymphocyte homing and immunity in L-selectin-deficient mice. *J Exp Med*. 1998;187(2):197–204.
116 Catalina MD, Carroll MC, Arizpe H, et al. The route of antigen entry determines the requirement for L-selectin during immune responses. *J Exp Med*. 1996;184:2341–2351.
117 Berlin C, Berg EL, Briskin MJ, et al. $\alpha 4 \beta 7$ integrin mediates lymphocyte binding to the mucosal vascular addressin MAdCAM-1. *Cell*. 1993;74:185–195.
118 Arbones ML, Ord DC, Ley K, et al. Lymphocyte homing and leukocyte rolling and migration are impaired in L-selectin-deficient mice. *Immunity*. 1994;1:247–260.
119 Wagner N, Lohler J, Kunkel EJ, et al. Critical role for $\beta 7$ integrins in formation of the gut-associated lymphoid tissue. *Nature*. 1996;382:366–370.
120 Steeber DA, Tang ML, Zhang XQ, et al. Efficient lymphocyte migration across high endothelial venules of mouse Peyer's patches requires overlapping expression of L-selectin and beta7 integrin. *J Immunol*. 1998;161(12):6638–6647.
121 Kunkel EJ, Ramos CL, Steeber DA, et al. The roles of L-selectin, beta 7 integrins, and P-selectin in leukocyte rolling and adhesion in high endothelial venules of Peyer's patches. *J Immunol*. 1998;161(5):2449–2456.
122 Tan K, Casasnovas JM, Liu J-h, et al. The structure of immunoglobulin superfamily domains 1 and 2 of MAdCAM-1 reveals novel features important for integrin recognition. *Structure*. 1998;6:793–801.
123 Berg EL, McEvoy LM, Berlin C, et al. L-selectin-mediated lymphocyte rolling on MAdCAM-1. *Nature*. 1993;366:695–698.
124 Berlin C, Bargatze RF, von Andrian UH, et al. $\alpha 4$ integrins mediate lymphocyte attachment and rolling under physiologic flow. *Cell*. 1995;80:413–422.
125 Hamann A, Andrew DP, Jablonski-Westrich D, et al. Role of α_4-integrins in lymphocyte homing to mucosal tissues in vivo. *J Immunol*. 1994;152:3282–3293.
126 Warnock RA, Campbell JJ, Dorf ME, et al. The role of chemokines in the microenvironmental control of T versus B cell arrest in Peyer's patch high endothelial venules. *J Exp Med*. 2000;191(1):77–88.
127 Janatpour MJ, Hudak S, Sathe M, et al. Tumor necrosis factor-dependent segmental control of MIG expression by high endothelial venules in inflamed lymph nodes regulates monocyte recruitment. *J Exp Med*. 2001;194(9):1375–1384.
128 Nibbs RJ, Kriehuber E, Ponath PD, et al. The beta-chemokine receptor D6 is expressed by lymphatic endothelium and a subset of vascular tumors. *Am J Pathol*. 2001;158(3):867–877.
129 Fra AM, Locati M, Otero K, et al. Cutting edge: scavenging of inflammatory CC chemokines by the promiscuous putatively silent chemokine receptor D6. *J Immunol*. 2003;170(5):2279–2282.
130 Jamieson T, Cook DN, Nibbs RJ, et al. The chemokine receptor D6 limits the inflammatory response in vivo. *Nat Immunol*. 2005;6(4):403–11. Epub 2005 Mar 6.
131 Huang AJ, Furie MB, Nicholson SC, et al. Effects of human neutrophil chemotaxis across human endothelial cell monolayers on the permeability of these monolayers to ions and macromolecules. *J Cell Physiol*. 1988;135(3):355–366.

132 Muller WA, Weigl SA, Deng X, Phillips DM. PECAM-1 is required for transendothelial migration of leukocytes. *J Exp Med.* 1993;178:449–460.

133 Schenkel AR, Mamdouh Z, Chen X, et al. CD99 plays a major role in the migration of monocytes through endothelial junctions. *Nat Immunol.* 2002;3(2):143–150.

134 Muller WA. Leukocyte-endothelial-cell interactions in leukocyte transmigration and the inflammatory response. *Trends Immunol.* 2003;24(6):327–334.

135 Worthylake RA, Burridge K. Leukocyte transendothelial migration: orchestrating the underlying molecular machinery. *Curr Opin Cell Biol.* 2001;13(5):569–577.

136 Johnson-Leger C, Aurrand-Lions M, Imhof BA. The parting of the endothelium: miracle, or simply a junctional affair? *J Cell Sci.* 2000;113(Pt 6):921–933.

137 Weber C, Lu C-F, Casasnovas JM, Springer TA. Role of $\alpha_L\beta_2$ integrin avidity in transendothelial chemotaxis of mononuclear cells. *J Immunol.* 1997;159:3968–3975.

138 Carman CV, Springer TA. A transmigratory cup in leukocyte diapedesis both through individual vascular endothelial cells and between them. *J Cell Biol.* 2004;167(2):377–388.

139 Aasheim HC, Delabie J, Finne EF. Ephrin-A1 binding to $CD4^+$ T lymphocytes stimulates migration and induces tyrosine phosphorylation of PYK2. *Blood.* 2004;105(7):2869–2876. Epub 2004 Dec 7.

140 Farr AG, De Bruyn PP. The mode of lymphocyte migration through postcapillary venule endothelium in lymph node. *Am J Anat.* 1975;143(1):59–92.

141 Salmi M, Jalkanen S. A 90-kilodalton endothelial cell molecule mediating lymphocyte binding in humans. *Science.* 1992;257: 1407–1409.

142 Salmi M, Jalkanen S. VAP-1: an adhesin and an enzyme. *Trends Immunol.* 2001;22(4):211–216.

143 Salmi M, Yegutkin GG, Lehvonen R, et al. A cell surface amine oxidase directly controls lymphocyte migration. *Immunity.* 2001;14(3):265–276.

144 Stolen CM, Marttila-Ichihara F, Koskinen K, et al. Absence of the endothelial oxidase AOC3 leads to abnormal leukocyte traffic in vivo. *Immunity.* 2005;22(1):105–115.

145 Irjala H, Elima K, Johansson EL, et al. The same endothelial receptor controls lymphocyte traffic both in vascular and lymphatic vessels. *Eur J Immunol.* 2003;33(3):815–824.

146 Salmi M, Koskinen K, Henttinen T, et al. CLEVER-1 mediates lymphocyte transmigration through vascular and lymphatic endothelium. *Blood.* 2004;104(13):3849–3857.

147 Prevo R, Banerji S, Ni J, Jackson DG. Rapid plasma membrane-endosomal trafficking of the lymph node sinus and high endothelial venule scavenger receptor/homing receptor stabilin-1 (FEEL-1/CLEVER-1). *J Biol Chem.* 2004;279(50): 52580–52592.

148 Stockton BM, Cheng G, Manjunath N, et al. Negative regulation of T cell homing by CD43. *Immunity.* 1998;8:373–381.

149 Imai Y, True DD, Singer MS, Rosen SD. Direct demonstration of the lectin activity of gp90 mel, a lymphocyte homing receptor. *J Cell Biol.* 1990;111:1225–1232.

150 van Zante A, Gauguet J-M, Bistrup A, et al. Lymphocyte-HEV interactions in lymph nodes of a sulfotransferase-deficient mouse. *J Exp Med.* 2003;198(9):1289–1300.

151 Hemmerich S, Bistrup A, Singer MS, et al. Sulfation of L-selectin ligands by an HEV-restricted sulfotransferase regulates lymphocyte homing to lymph nodes. *Immunity.* 2001;15(2):237–247.

152 Uchimura K, Kadomatsu K, El-Fasakhany FM, et al. N-acetylglucosamine 6-O-sulfotransferase-1 regulates expression of L-selectin ligands and lymphocyte homing. *J Biol Chem.* 2004; 279(33):35001–35008.

153 Sainte-Marie G, Peng FS. High endothelial venules of the rat lymph node. A review and a question: is their activity antigen specific? *Anat Rec.* 1996;245(4):593–620.

154 Lasky LA, Singer MS, Dowbenko D, et al. An endothelial ligand for L-selectin is a novel mucin-like molecule. *Cell.* 1992;69:927–938.

155 Brustein M, Kraal G, Mebius RE, Watson SR. Identification of a soluble form of a ligand for the lymphocyte homing receptor. *J Exp Med.* 1992;176:1415–1419.

156 Baumhueter S, Singer MS, Henzel W, et al. Binding of L-selectin to the vascular sialomucin, CD34. *Science.* 1993;262:436–438.

157 Sassetti C, Tangemann K, Singer MS, et al. Identification of podocalyxin-like protein as a high endothelial venule ligand for L-selectin: parallels to CD34. *J Exp Med.* 1998;187(12):1965–1975.

158 Hemmerich S, Butcher EC, Rosen SD. Sulfation-dependent recognition of high endothelial venules (HEV)-ligands by L-selectin and MECA 79. *J Exp Med.* 1994;180:2219–2226.

159 Morgan SM, Samulowitz U, Darley L, et al. Biochemical characterization and molecular cloning of a novel endothelial-specific sialomucin. *Blood.* 1999;93(1):165–175.

160 Samulowitz U, Kuhn A, Brachtendorf G, et al. Human endomucin: distribution pattern, expression on high endothelial venules, and decoration with the MECA-79 epitope. *Am J Pathol.* 2002;160(5):1669–1681.

161 Sassetti C, Van Zante A, Rosen SD. Identification of endoglycan, a member of the CD34/podocalyxin family of sialomucins. *J Biol Chem.* 2000;275(12):9001–9010.

162 Fieger CB, Sassetti CM, Rosen SD. Endoglycan, a member of the CD34 family, functions as a L-selectin ligand through modification with tyrosine sulfation and sialyl Lewis x. *J Biol Chem.* 2003;278:27390–27398.

163 Kanda H, Tanaka T, Matsumoto M, et al. Endomucin, a sialomucin expressed in high endothelial venules, supports L-selectin-mediated rolling. *Int Immunol.* 2004;16(9):1265–1274.

164 Puri KD, Finger EB, Gaudernack G, Springer TA. Sialomucin CD34 is the major L-selectin ligand in human tonsil high endothelial venules. *J Cell Biol.* 1995;131:261–270.

165 Cheng J, Baumhueter S, Cacalano G, et al. Hematopoietic defects in mice lacking the sialomucin CD34. *Blood.* 1996;87(2): 479–490.

166 Dowbenko D, Kikuta A, Fennie C, et al. Glycosylation-dependent cell adhesion molecule 1 (GlyCAM-1) mucin is expressed by lactating mammary gland epithelial cells and is present in milk. *J Clin Invest.* 1993;92:952–960.

167 Baumhueter S, Dybdal N, Kyle C, Lasky LA. Global vascular expression of murine CD34, a sialomucin-like endothelial ligand for L-selectin. *Blood.* 1994;84:2554–2565.

168 Stoolman LM, Rosen SD. Possible role for cell-surface carbohydrate-binding molecules in lymphocyte recirculation. *J Cell Biol.* 1983;96(3):722–729.

169 Stoolman LM, Tenforde TS, Rosen SD. Phosphomannosyl receptors may participate in the adhesive interaction between lymphocytes and high endothelial venules. *J Cell Biol.* 1984;99(4 Pt 1):1535–1540.

170 Yednock TA, Butcher EC, Stoolman LM, Rosen SD. Receptors involved in lymphocyte homing: relationship between a carbohydrate-binding receptor and the MEL-14 antigen. *J Cell Biol.* 1987;104:725–731.

171 Imai Y, Lasky LA, Rosen SD. Sulphation requirement for GlyCAM-1, an endothelial ligand for L-selectin. *Nature.* 1993; 361:555–557.

172 Hemmerich S, Rosen SD. 6′-Sulfated sialyl Lewis x is a major capping group of GlyCAM-1. *Biochemistry.* 1994;33:4830–4835.

173 Hemmerich S, Bertozzi CR, Leffler H, Rosen SD. Identification of the sulfated monosaccharides of GlyCAM-1, an endothelial-derived ligand for L-selectin. *Biochemistry.* 1994;33:4820–4829.

174 Rosen SD, Singer MS, Yednock TA, Stoolman LM. Involvement of sialic acid on endothelial cells in organ-specific lymphocyte recirculation. *Science.* 1985;228:1005–1007.

175 Yeh JC, Hiraoka N, Petryniak B, et al. Novel sulfated lymphocyte homing receptors and their control by a Core1 extension beta 1,3-N-acetylglucosaminyltransferase. *Cell.* 2001;105(7):957–969.

176 Imai Y, Singer MS, Fennie C, et al. Identification of a carbohydrate based endothelial ligand for a lymphocyte homing receptor. *J Cell Biol.* 1991;113:1213–1221.

177 Smith PL, Gersten KM, Petryniak B, et al. Expression of the $\alpha(1,3)$ fucosyltransferase Fuc-TVII in lymphoid aggregate high endothelial venules correlates with expression of L-selectin ligands. *J Biol Chem.* 1996;271:8250–8259.

178 Maly P, Thall AD, Petryniak B, et al. The $\alpha(1,3)$ fucosyltransferase Fuc-TVII controls leukocyte trafficking through an essential role in L-, E-, and P-selectin ligand biosynthesis. *Cell.* 1996; 86:643–653.

179 Weninger W, Ulfman LH, Cheng G, et al. Specialized contributions by alpha(1,3)-fucosyltransferase-IV and FucT-VII during leukocyte rolling in dermal microvessels. *Immunity.* 2000;12(6):665–676.

180 Homeister JW, Thall AD, Petryniak B, et al. The alpha(1,3) fucosyltransferases FucT-IV and FucT-VII exert collaborative control over selectin-dependent leukocyte recruitment and lymphocyte homing. *Immunity.* 2001;15(1):115–126.

181 Niemela R, Natunen J, Majuri ML, et al. Complementary acceptor and site specificities of Fuc-TIV and Fuc-TVII allow effective biosynthesis of sialyl-Trelex and related polylactosamines present on glycoprotein counterreceptors of selections. *J Biol Chem.* 1998;273:4021–4026.

182 Lowe JB. Glycosylation, immunity, and autoimmunity. *Cell.* 2001;104(6):809–812.

183 Sanders WJ, Katsumoto TR, Bertozzi CR, et al. L-selectin-carbohydrate interactions: Relevant modifications of the Lewis x trisaccharide. *Biochemistry.* 1996;35:14862–14867.

184 Galustian C, Lawson AM, Komba S, et al. Sialyl-Lewis^x sequence 6-*O* -sulfated at N-Acetylglucosamine rather than at galactose is the preferred ligand for L-selectin and De-N-acetylation of the sialic acid enhances the binding strength. *Biochem Biophys Res Commun.* 1997;240:748–751.

185 Tsuboi S, Isogai Y, Hada N, et al. 6′-Sulfo sialyl Lex but not 6-sulfo sialyl Lex expressed on the cell surface supports L-selectin-mediated adhesion. *J Biol Chem.* 1996;271(44):27213–27216.

186 Uchimura K, Muramatsu H, Kadomatsu K, et al. Molecular cloning and characterization of an N-acetylglucosamine-6-O-sulfotransferase. *J Biol Chem.* 1998;273(35):22577–22583.

187 Bistrup A, Bhakta S, Lee JK, et al. Sulfotransferases of two specificities function in the reconstitution of high endothelial cell ligands for L-selectin. *J Cell Biol.* 1999;145:899–910.

188 Hiraoka N, Petryniak B, Nakayama J, et al. A novel, high endothelial venule-specific sulfotransferase expresses 6- sulfo sialyl Lewis(x), an L-selectin ligand displayed by CD34. *Immunity.* 1999;11(1):79–89.

189 Tangemann K, Bistrup A, Hemmerich S, Rosen SD. Sulfation of a high endothelial venule-expressed ligand for L-selectin. Effects on tethering and rolling of lymphocytes. *J Exp Med.* 1999;190(7):935–942.

190 Hiraoka N, Kawashima H, Petryniak B, et al. Core 2 branching {beta}1,6-N-acetylglucosaminyl transferase and high endothelial venule-restricted sulfotransferase collaboratively control lymphocyte homing. *J Biol Chem.* 2004;279(4): 3058–3067.

191 Hemmerich S, Leffler H, Rosen SD. Structure of the O-glycans in glyCAM-1, an endothelial-derived ligand for L-selectin. *J Biol Chem.* 1995;270:12035–12047.

192 Williams D, Schachter H. Mucin synthesis. I. Detection in canine submaxillary glands of an N-acetylglucosaminyltransferase which acts on mucin substrates. *J Biol Chem.* 1980;255 (23):11247–11252.

193 Ellies LG, Tsuboi S, Petryniak B, et al. Core 2 oligosaccharide biosynthesis distinguishes between selectin ligands essential for leukocyte homing and inflammation. *Immunity.* 1998;9:881–890.

194 Briskin MJ, McEvoy LM, Butcher EC. MAdCAM-1 has homology to immunoglobulin and mucin-like adhesion receptors and to IgA1. *Nature.* 1993;363:461–464.

195 Mebius RE, Streeter PR, Michie S, et al. A developmental switch in lymphocyte homing receptor and endothelial vascular addressin expression regulates lymphocyte homing and permits $CD4^+CD3^-$ cells to colonize lymph nodes. *Proc Natl Acad Sci USA.* 1996;93:11019–11024.

196 Fu YX, Chaplin DD. Development and maturation of secondary lymphoid tissues. *Annu Rev Immunol.* 1999;17:399–433.

197 Belisle C, Sainte-Marie G. Tridimensional study of the deep cortex of the rat lymph node. V: Postnatal development of the deep cortex units. *Anat Rec.* 1981;200(2):207–220.

198 Hendriks HR, Eestermans IL, Hoefsmit EC. Depletion of macrophages and disappearance of postcapillary high endothelial venules in lymph nodes deprived of afferent lymphatic vessels. *Cell Tissue Res.* 1980;211(3):375–389.

199 Hendriks HR, Duijvestijn AM, Kraal G. Rapid decrease in lymphocyte adherence to high endothelial venules in lymph nodes deprived of afferent lymphatic vessels. *Eur J Immunol.* 1987; 17:1691–1695.

200 Mebius RE, Dowbenko D, Williams A, et al. Expression of GlyCAM-1, an endothelial ligand for L-selectin, is affected by afferent lymphatic flow. *J Immunol.* 1993;151: 6769–6776.

201 Swarte VV, Joziasse DH, Van den Eijnden DH, et al. Regulation of fucosyltransferase-VII expression in peripheral lymph node high endothelial venules. *Eur J Immunol.* 1998;28(10):3040–3047.

202 Mebius RE. Organogenesis of lymphoid tissues. *Nat Rev Immunol.* 2003;3(4):292–303.

203 Ruddle NH. Lymphoid neo-organogenesis: lymphotoxin's role in inflammation and development. *Immunol Res.* 1999;19(2–3): 119–125.

204 Kratz A, Campos-Neto A, Hanson MS, Ruddle NH. Chronic inflammation caused by lymphotoxin is lymphoid neogenesis. *J Exp Med*. 1996;183:1461–1472.
205 Cupedo T, Jansen W, Kraal G, Mebius RE. Induction of secondary and tertiary lymphoid structures in the skin. *Immunity*. 2004;21(5):655–667.
206 Bistrup A, Tsay D, Shenoy P, et al. Detection of a sulfotransferase (HEC-GlcNAc6ST) in high endothelial venules of lymph nodes and in high endothelial venule-like vessels within ectopic lymphoid aggregates: relationship to the MECA-79 epitope. *Am J Pathol*. 2004;164(5):1635–1644.
207 Fan L, Reilly CR, Luo Y, et al. Cutting edge: ectopic expression of the chemokine TCA4/SLC is sufficient to trigger lymphoid neogenesis. *J Immunol*. 2000;164(8):3955–3959.
208 Luther SA, Lopez T, Bai W, et al. BLC expression in pancreatic islets causes B cell recruitment and lymphotoxin-dependent lymphoid neogenesis. *Immunity*. 2000;12(5):471–481.
209 Chen SC, Vassileva G, Kinsley D, et al. Ectopic expression of the murine chemokines CCL21a and CCL21b induces the formation of lymph node-like structures in pancreas, but not skin, of transgenic mice. *J Immunol*. 2002;168(3):1001–1008.
210 Millet I, Ruddle NH. Differential regulation of lymphotoxin (LT), lymphotoxin-β (LT-β), and TNF-α in murine T cell clones activated through the TCR. *J Immunol*. 1994;152:4336–4346.
211 De Togni P, Goellner J, Ruddle NH, et al. Abnormal development of peripheral lymphoid organs in mice deficient in lymphotoxin. *Science*. 1994;264:703–707.
212 Koni PA, Sacca R, Lawton P, et al. Distinct roles in lymphoid organogenesis for lymphotoxins alpha and beta revealed in lymphotoxin beta-deficient mice. *Immunity*. 1997;6(4):491–500.
213 Drayton DL, Bonizzi G, Ying X, et al. I kappa B kinase complex alpha kinase activity controls chemokine and high endothelial venule gene expression in lymph nodes and nasal-associated lymphoid tissue. *J Immunol*. 2004;173(10):6161–6168.
214 Drayton DL, Ying X, Lee J, et al. Ectopic LT alpha beta directs lymphoid organ neogenesis with concomitant expression of peripheral node addressin and a HEV-restricted sulfotransferase. *J Exp Med*. 2003;197(9):1153–1163.
215 Suematsu S, Watanabe T. Generation of a synthetic lymphoid tissue-like organoid in mice. *Nat Biotechnol*. 2004;22(12):1539–1545.
216 Tedder TF, Steeber DA, Pizcueta P. L-selectin-deficient mice have impaired leukocyte recruitment into inflammatory sites. *J Exp Med*. 1995;181:2259–2264.
217 Protin U, Schweighoffer T, Jochum W, Hilberg F. CD44-deficient mice develop normally with changes in subpopulations and recirculation of lymphocyte subsets. *J Immunol*. 1999;163(9):4917–4923.
218 Stoop R, Gal I, Glant TT, et al. Trafficking of CD44-deficient murine lymphocytes under normal and inflammatory conditions. *Eur J Immunol*. 2002;32(9):2532–2542.
219 Nakano H, Tamura T, Yoshimoto T, et al. Genetic defect in T lymphocyte-specific homing into peripheral lymph nodes. *Eur J Immunol*. 1997;27:215–221.
220 Forster R, Mattis AE, Kremmer E, et al. A putative chemokine receptor, BLRI, directs B cell migration to defined lymphoid organs and specific anatomic compartments of the spleen. *Cell*. 1996;87:1037–1047.
221 Dawson TC, Lentsch AB, Wang Z, et al. Exaggerated response to endotoxin in mice lacking the Duffy antigen/receptor for chemokines (DARC). *Blood*. 2000;96(5):1681–1684.
222 Huang MC, Laskowska A, Vestweber D, Wild MK. The alpha (1,3)-fucosyltransferase Fuc-TIV, but not Fuc-TVII, generates sialyl Lewis X-like epitopes preferentially on glycolipids. *J Biol Chem*. 2002;277:47786–47795.
223 Alon R, Feizi T, Yuen C-T, et al. Glycolipid ligands for selectins support leukocyte tethering and rolling under physiologic flow conditions. *J Immunol*. 1995;154:5356–5366.

171

Hierarchy of Circulating and Vessel Wall–Derived Endothelial Progenitor Cells

David A. Ingram and Mervin C. Yoder

Herman B. Wells Center for Pediatric Research,
Indiana University School of Medicine, Indianapolis

The level of endothelial cell (EC) proliferation in normal, mature vessels in most mammals remains poorly defined but in general is reported to be extremely low, if not nonexistent. In fact, until approximately 50 years ago, the predominant view held that ECs lining vessels do not undergo mitosis. However, the advent of tritiated thymidine labeling studies and modifications of the Hautchen preparation permitted direct analysis of EC mitosis in vessels recovered after labeling in vivo (1). In some experimental animals, such as rats, guinea pigs, pigs, and dogs, the tritiated thymidine labeling studies demonstrated 0.1% to 3.0% EC turnover daily (2,3). Endothelial proliferation rates were correlated with the age of the subject and appeared to decline rapidly after birth with most adult vessel endothelium displaying mitosis in <1% of the cells daily (4). Furthermore, the sites of endothelial replication were not homogenously distributed but appeared to occur in clustered areas nearest vessel bifurcations where flow was disturbed and often turbulent (2). Whether these dividing ECs were unique and possessed proliferative potential that was lacking in other mature endothelium or these focal areas of replicating cells merely represented the sites of greatest vessel injury and endothelial turnover has not yet been determined. It has been well documented that EC division may reach 50% of the cells in the thoracic aorta following experimentally induced hypertension, re-endothelialization of organized clots or injured vessels after arterial denudation, or following experimentally induced vascular constriction (5).

In marked contrast to the slow turnover of ECs in normal vessels, in vitro plating of ECs derived from human or animal vessels is associated with brisk EC proliferation. For example, human umbilical vein ECs (HUVECs) and bovine aortic ECs (BAECs) are two commonly studied models for in vitro analysis of EC functions. Both HUVECs and BAECs proliferate well initially, but cell division wanes with time and cells eventually develop replicative senescence and fail to divide after approximately 40 population doublings. It is unknown if each EC derived from the vessels possesses similar proliferative potential or if only some of the cells can divide.

Angiogenesis is the process of new vessel formation from preexisting vessels; this is the process reported to give rise to new vessels in adult subjects. Recently, bone marrow–derived circulating endothelial progenitor cells (EPCs) have been identified and have been reported to play a vasculogenic role in new vessel formation, at least in some experimental murine ischemic or tumor models (6,7). Contrasting evidence indicates that bone marrow–derived EPCs do not contribute significantly to the endothelial lining of normal arterial, venous, and capillary vessels during development and play only a minor role in murine tumor neoangiogenesis (8–10). A comparison between the proliferative potential of circulating EPCs and the ECs that reside in normal vessels has not been previously addressed.

Emerging evidence to support the use of EPCs for angiogenic therapies or as biomarkers to assess a patient's cardiovascular disease risk and progression is accumulating and is gathering enthusiasm. However, there is no uniform definition of an EPC, which makes interpretation of these studies problematic. Although a hallmark of stem and progenitor cells (e.g., hematopoietic, intestinal, neuronal) is their ability to proliferate and give rise to functional progeny, EPCs are primarily defined by the expression of selected cell surface antigens (determined using monoclonal antibodies and flow cytometry), growth as transient colonies in vitro, incorporation into host vessels or new vessel formation, and release of angiogenic molecules in animal models of vascular disease. Reliance on cell surface expression of molecules as a definitive means for progenitor cell identification can be problematic because the expression may vary with the physiological state of the cell. Until this point, there has been no assay developed to assess the proliferative potential of individual ECs or EPCs and thus no comparative analysis completed.

PROGENITOR CELL PARADIGMS FOR OTHER CELL LINEAGES

Assays to measure the clonal growth of hematopoietic stem and progenitor cells are relatively new tools available to the researcher. Some of the earliest methods of identifying hematopoietic clones arose while investigating the mechanisms used by the host to recover from radiation injury. Whole-body irradiation of animals was noted to depress hematopoiesis and cause life-threatening deficiencies in all circulating blood cells (pancytopenia) in exposed animals (11). In 1956, Ford and colleagues provided definitive evidence that hematopoietic cells and not plasma or subcellular molecules conferred protection to the host from death by pancytopenia following irradiation. Using donor cells from a strain of mice carrying a balanced (but morphologically identifiable) chromosome translocation, blood cells of a lethally irradiated mouse were completely replaced by those of the donor cells bearing the chromosomal marker after transplantation (12). However, evidence that single multipotent hematopoietic progenitor cells (cells giving rise to more than one lineage of blood cells) could be identified in vivo was not reported until 1961. By injecting donor marrow cells into a lethally irradiated recipient animal and examining the recipient spleen for hematopoietic colonies 8 to 12 days later, Till and McCulloch had devised the first in vivo clonal hematopoietic assay. Each colony of hematopoietic cells in the spleen was demonstrated to arise from a single precursor cell, the colony forming unit in spleen cell (CFU-S). Other assays to identify hematopoietic clones followed (13).

Pluznik and Sachs (14) and Bradley and Metcalf (15) reported that murine hematopoietic cells could be cultured in vitro and that addition of soluble fluid (i.e., urine or pregnant uterine extract) from several murine organs resulted in the in vitro formation of myeloid colonies. Furthermore, these investigators reported that each myeloid colony developing in vitro arose from a single precursor cell called the colony-forming unit in culture, or CFU-C. Use of this assay permitted identification of numerous hematopoietic growth factors and cytokines important for erythroid, myeloid, and multipotent progenitor cell proliferation and differentiation.

Plating of single hematopoietic cells in special double-layer agar cultures with multiple recombinant cytokines also permitted the identification of clonal hematopoietic progenitors that were highly proliferative (high proliferative potential-colony forming cells, or HPP-CFCs). HPP-CFC colonies contained >50,000 cells and were visible in the culture dishes without need for magnification. HPP-CFC clones could be plucked from the agar medium, dispersed into a single cell suspension, and replated in CFC assays with emergence of secondary HPP-CFCs, as well as committed erythroid and myeloid and multipotent progenitors (16). Due to these properties, HPP-CFCs are still considered the most primitive hematopoietic progenitor cell that can be cultured in vitro without the presence of hematopoietic stromal cells in co-culture.

Thus, the hematopoietic system is organized as a hierarchy of cells (Figure 171.1) that progress from the stem cell, possessing the most proliferative potential, through successive progenitor cell stages that sequentially lose proliferative potential but display increasing evidence of lineage commitment, to the final mature cells of each lineage that are highly differentiated and for the most part, devoid of proliferative potential. Having established this robust hierarchical order of progenitor cell proliferative potential (at a clonal level) using the in vitro assays, further studies of cell surface antigen expression and in vivo functional analysis permitted prospective identification and further characterization of these hematopoietic progenitors.

IDENTIFICATION OF ENDOTHELIAL PROGENITOR CELLS FROM HUMAN ADULT PERIPHERAL AND UMBILICAL CORD BLOOD

Hematopoietic progenitor cells and EPCs share a number of cell surface markers in the developing yolk sac and embryo, and genetic disruption of numerous genes affects both hematopoietic and EC development, suggesting they originate from a common precursor, the hemangioblast. Although a hierarchy of hematopoietic stem and progenitor cells has been well established (as discussed earlier), evidence to support a similar hierarchy of stem and progenitor cells (based on differences in proliferative potential) for the endothelial lineage has not been postulated. We hypothesized that a hierarchy of EPCs exists for the endothelial lineage.

To isolate EPCs to address this question, we harvested mononuclear cells (MNCs) from peripheral blood of healthy adults and umbilical cords of normal term infants and observed for endothelial colony formation (17). Interestingly, the number of EC colonies per equivalent blood volume was increased 15-fold in cord blood compared to adult samples, and the cord blood–derived EC colonies emerged in culture one week earlier than adult colonies. Further, cord blood EPC–derived EC colonies consistently appeared larger compared with adult colonies. Thus, in the process of isolating EPCs to address our experimental question, we detected distinct differences in the size, frequency, and time of appearance between adult and cord blood EPC–derived EC colonies. These observations suggested that cord blood EPCs may be derived from a different population of progenitors than those previously identified as adult EPCs.

Before testing the proliferative and clonogenic potential of cord blood and adult EPC–derived ECs, we verified that the cell progeny derived from cord blood and adult EPCs were not contaminated with other cell types (17). This is important because prior studies have shown that the cell progeny derived from some EPC-derived colonies isolated from adult and cord blood MNCs contain cells, which express the hematopoietic specific cell surface antigen CD45 or other markers of the monocyte macrophage and dendritic cell lineages. We determined that after initial EPC-derived EC passage, the cells formed monolayers of spindle shaped cells with "cobblestone" morphology.

Immunophenotyping revealed that both cord blood– and adult blood–derived ECs expressed the EC surface antigens

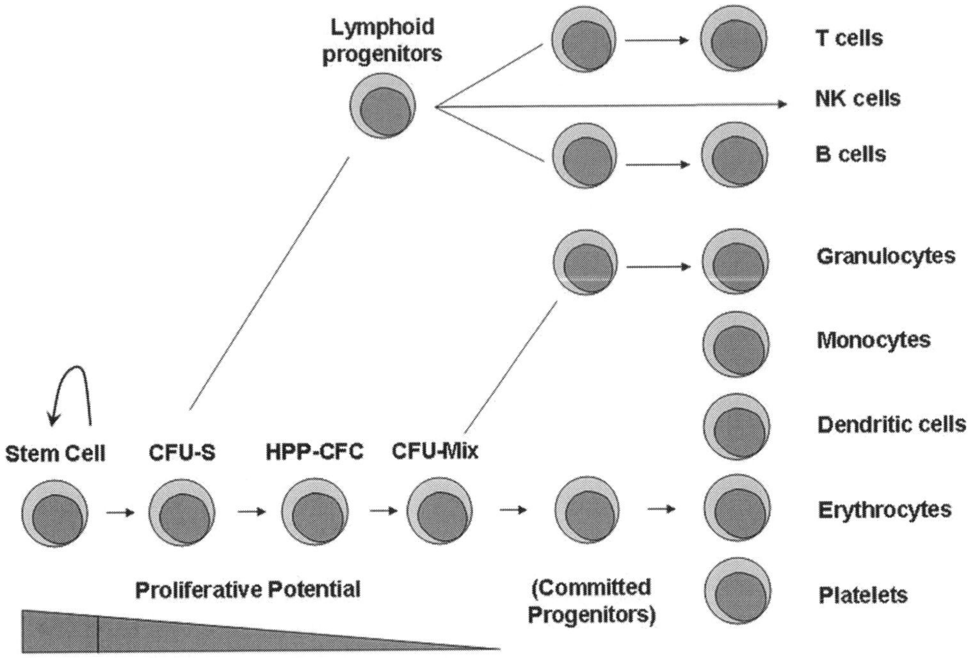

Figure 171.1. Hierarchical organization of the hematopoietic system stressing the progressive loss of proliferative potential associated with discrete steps in progenitor cell differentiation. Stem cells self-renew and demonstrate the highest proliferative potential, whereas mature myeloid cells are generally nonreplicative (only lymphocytes and macrophages retain replicative potential as mature cells). Colony forming unit-spleen, CFU-S; high proliferative potential-colony forming cell, HPP-CFC; colony forming unit-mixed lineage, CFU-Mix.

CD31, CD141 (thrombomodulin), CD105 (endoglin), CD146 (S-endo1), CD144 (vascular endothelial [VE]-cadherin), von Willebrand factor (vWF), and FLK-1. Importantly, both cell populations were devoid of cell surface expression of the mature hematopoietic cell proteins CD45 and CD14. However, a small percentage of both cord blood and adult cells did express CD34, AC133, and CD117 (c-kit), which are cell surface antigens previously identified on both endothelial and hematopoietic progenitor cells. Nevertheless, plating the EPC-derived ECs that expressed CD34, AC133, or CD117 (c-kit) failed to give rise to hematopoietic colony forming cells in methylcellulose assays demonstrating that these cells do not possess hematopoietic activity.

We next tested whether the EPC-derived cells would incorporate DiI-acetylated-low-density lipoprotein (DiI-Ac-LDL), form capillary-like structures in Matrigel, and upregulate vascular cell adhesion molecule (VCAM)-1 in response to cytokine stimulation, which are some characteristics of ECs. To better visualize whether cord blood or adult ECs form capillary-like structures in Matrigel, early passage (1–2) cord and adult cells derived from the initial EPCs were transduced with a retrovirus encoding the enhanced green fluorescent protein (EGFP) and selected for EGFP expression. Both cord blood and adult cells subcultured from the adherent EC colonies uniformly incorporated DiI-Ac-LDL, formed capillary-like structures in Matrigel, and upregulated VCAM-1 in response to recombinant human interleukin (IL)-1 stimulation. Thus, these studies confirm that the cell progeny derived from cord blood– and adult blood–derived EPCs were endothelial in origin and were not contaminated with or derived from hematopoietic cells.

ASSAYS TO QUANTITATE THE PROLIFERATIVE POTENTIAL OF ENDOTHELIAL PROGENITOR CELLS

As a first estimate of the proliferative potential of EPC-derived cord blood and adult ECs, cells were passaged in vitro. Remarkably, cord blood EPC–derived ECs could be expanded for at least 100 population doublings without obvious signs of replicative senescence. In contrast, adult EPC–derived ECs could not be cultured for much beyond 30 population doublings (17). To quantitate and compare the proliferative kinetics of cord blood and adult EPC–derived ECs, we calculated the population doubling times (PDTs) and cumulative population doubling levels (CPDLs) during a defined time in culture (60 days). There was a 2.5-fold decrease in the PDTs and a 1.5-fold increase in the CPDLs of cord blood EPC–derived ECs compared to adult EPC–derived cells. Similar PDTs and CPDLs of adult peripheral blood–derived ECs has been reported by others, who tested the proliferative kinetics of EPC-derived cells isolated from healthy adult donors (17).

As a second strategy to compare the growth potential of the EPC-derived ECs, we replated cord blood and adult blood EPC-derived ECs at 1 to 2 passages in limiting dilutions in culture medium defined for optimal EC growth in vitro. At cell concentrations of $\leq 100/cm^2$, one could readily discern colonies of cells varying in size and cell morphology in addition to many single adherent ECs. Cord blood EPC–derived ECs formed more colonies that were larger in size than those EC colonies emerging from the replated adult blood EPC–derived ECs. These data suggested that EPC-derived ECs demonstrate a wide variability in proliferative potential and that some of the EPC-derived ECs from cord blood display much higher proliferative potential than any of the ECs derived from adult blood EPCs.

To quantitatively and stringently interrogate the proliferative potential of individual EPCs, we devised a single cell assay for the clonal analysis of EPC colony forming activity (Figure 171.2) (17). Cord blood and adult blood EPC–derived ECs at 1 to 2 passages were transduced with a retrovirus encoding EGFP and 24 hours later, $EGFP^+$ cells were removed from the plates and sorted as single cells into 96-well plates using a FACS Vantage sorter. Wells were examined the following morning to assure that at least one cell was present in each well and any wells with more than one cell were eliminated from consideration. After two weeks of culture, wells were examined to determine the percentage of wells that contained more than one cell and the total EC number of each colony formed.

Remarkably, the percentage of single cells undergoing at least one cell division was increased fivefold for cord blood EPC–derived ECs compared to ECs from adult EPCs. Further, the average number of cell progeny derived from a single-plated cord blood EPC–derived ECs was 100-fold greater compared to the number of cells derived from an individual adult EPC-derived ECs. Greater than 80% of the single adult ECs that divided gave rise to small colonies or clusters of cells ranging in number from 2 to 50 cells. A small population of single adult ECs did form colonies containing greater than 500 cells. In contrast, at least 60% of the single-plated cord blood EPC–derived ECs that divided formed well-circumscribed colonies containing between 2,000 and 10,000 cells in the 14-day culture period. A representative of the photomicrograph of the colonies derived from single cord or adult cultured cells is shown in Figure 171.3. These single-cell studies demonstrated that there are different types of endothelial colony forming cells which could be discriminated by their cell autonomous proliferative potential, and that EPC-derived ECs display a hierarchy of proliferative potentials similar to the hematopoietic progenitor cell hierarchy.

In the hematopoietic cell system, the most proliferative progenitor cell type that can be cultured in vitro in the absence of a stromal cell monolayer is termed the *HPP-CFC* (discussed earlier). We tested whether a cell with similar proliferative potential is present within adult and cord blood EPC–derived ECs to further define and classify the EPC hierarchy. The clonal progeny derived from single-plated cord blood or adult

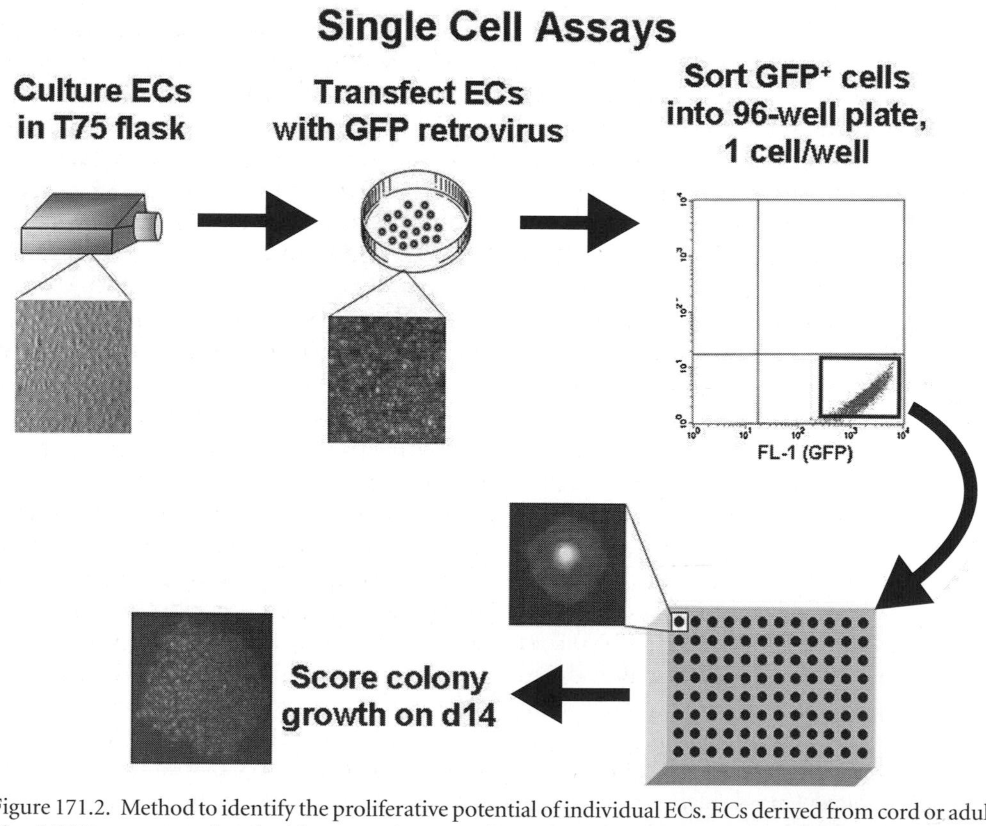

Figure 171.2. Method to identify the proliferative potential of individual ECs. ECs derived from cord or adult blood EPCs were transduced with a retrovirus encoding green fluorescent protein and the expressing cells identified and single cells plated using a flow cytometric sorter. The green fluorescing cells were visualized through a fluorescence microscope and the number of cells in each well directly counted.

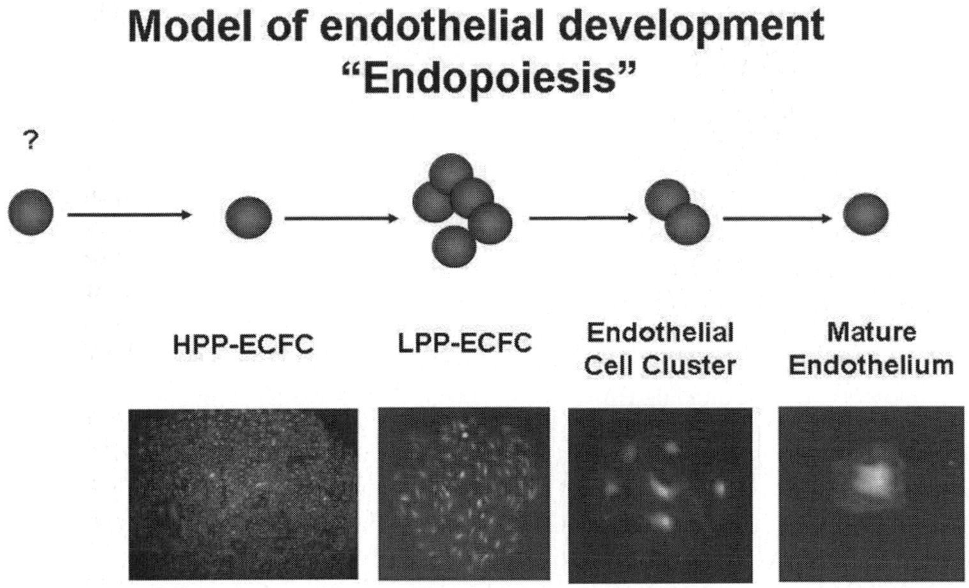

Figure 171.3. Endopoiesis: the process of EC development. We propose that high proliferative potential-endothelial colony forming cells (HPP-ECFCs) are the most proliferative endothelial progenitor cells (EPCs) and that these cells demonstrate high replating potential. HPP-ECFCs can give rise to at least secondary HPP-ECFCs and all other EPCs and mature ECs. Low proliferative potential-endothelial colony forming cells (LPP-ECFCs) do not replate into secondary LPP-ECFCs but do form EC clusters and mature ECs. Endothelial cluster forming cells give rise only to mature nondividing ECs.

EPC–derived ECs were trypsinized, replated, and cultured into 24-well tissue culture plates for seven days. After replating the clonal progeny of over 1000 single adult EPC-derived cells (those primary wells containing >50 cells), we detected only one secondary colony in the wells (1 of 1,000 plated) after 14 days of culture. In contrast, approximately one-half (205 of 421 plated) of the clonal progeny of single-plated cord blood EPC–derived ECs (those primary wells >50 ECs) formed secondary colonies or rapidly grew to confluence in 24-well plates. Secondary colonies or confluent cell monolayers derived from single cord blood ECs were serially passaged into progressively larger tissue culture plates. Strikingly, some single-plated cord blood EPC–derived ECs yielded at least 10^{12} cells in long-term culture and all (11 independent clones tested) produced at least 10 million progeny. These data suggested that the cord blood EPCs contained ECs with additional EPC activity that was similar to the HPP-CFC activity displayed by highly proliferative hematopoietic progenitor clones (cells that can give rise to high numbers of progeny and that can be replated into at least secondary HPP-CFCs).

Given that murine genetic studies clearly show that the origin of ECs is closely linked to hematopoietic cell development, it is not surprising that our data showed that EPCs can be discriminated by similar clonogenic and growth characteristics. We have proposed that EPCs can be identified using similar terminology to that utilized for defining hematopoietic cell progenitors (Figure 171.3) (17). High proliferative potential-endothelial colony forming cells (HPP-ECFCs) give rise to macroscopic colonies that form secondary and tertiary colonies upon replating. We have provided evidence that HPP-ECFCs give rise to all subsequent stages of endothelial progenitors in addition to replating into secondary HPP-ECFCs. Low proliferative potential-endothelial colony forming cells (LPP-ECFCs) form colonies, which contain greater than 50 cells, but do not form secondary LPP-ECFC colonies upon replating. This EPC stage represents the most proliferative population of EPCs that can be isolated from normal adult peripheral blood. They do give rise to EC clusters. EC clusters are composed of groups of fewer than 50 cells and do not replate into colonies or clusters. Our data have also indicated that mature differentiated ECs can be resolved in the single cell assay and are defined as cells that are nonproliferative. Thus, the circulating EPCs emerging from cord and adult blood represent EPCs possessing different proliferative potentials and at different stages of EPC development. The fact that EPC-derived ECs further display EPC activity raises the question of whether vessel-derived ECs also possess EPC activity.

VESSEL-DERIVED ENDOTHELIAL CELLS AND ENDOTHELIAL PROGENITOR CELLS

To compare the clonogenic capacity of single ECs derived from vessel walls, we established monolayers of ECs derived from cord blood circulating EPCs and from ECs isolated from human umbilical veins and human aorta (18). Cord blood EPC–derived ECs, HUVECs, and human aortic ECs (HAECs) expressed the EC surface antigens CD31, CD141, CD105, CD146, CD144, vWF, and FLK-1 but not the hematopoietic cell surface antigens CD45 and CD14. Thus, cord blood

EPC–derived ECs, HUVECs, and HAECs expressed a similar profile of EC-specific antigens.

Utilizing the single cell assay for EPC colony-forming activity, we compared the percentage of single-plated ECs that divided at least once and the distribution of HPP-ECFCs, LPP-ECFCs, and endothelial clusters derived from cord blood EPC–derived ECs, HUVECs, and HAECs. The percentage of single cord blood ECs undergoing at least one cell division was 55%. Similarly, 52% of single HUVECs and 53% of single HAECs divided after 14 days. Nearly half (47%) of the single cord blood ECs that divided possessed HPP-ECFC activity and gave rise to colonies of >2000 ECs. Strikingly, 28% of single dividing HUVECs and 27% of single dividing HAECs also possessed HPP-ECFC activity. In addition, single cord blood ECs, HUVECs, and HAECs possessed LPP-ECFC activity and formed smaller colonies that contained between 51 and 500 cells. Thus, similar to the circulating EPCs, the ECs lining umbilical veins and the human aorta comprise resident EPCs at different stages of maturation possessing different levels of proliferative potential.

These data suggest a new conceptual framework for determining both the origin and function of EPCs in maintaining vessel integrity. Whereas HUVECs and HAECs have long been considered to be fully differentiated mature ECs, we have now shown that a complete hierarchy of EPCs can be identified in HUVEC and HAEC monolayers and discriminated by their clonogenic potential (18). These data explain why HUVECs and HAECs can be passaged extensively in vitro given that HPP-ECFCs and LPP-ECFCs are present at nearly 50% frequency in HUVEC and HAEC populations. Future studies may clarify whether the sites of EC proliferation previously identified in the tritiated thymidine labeling studies can be confirmed to be the sites of EPC residence in the endothelium lining vessels in vivo.

ENDOTHELIAL PROGENITOR CELLS IN DISEASE AND DIAGNOSTIC IMPLICATIONS

Great emphasis has recently been placed on defining the role of circulating marrow–derived EPCs in repair of damaged vascular endothelium or in the process of tumor angiogenesis. Data obtained in murine studies appear to confirm a role of circulating EPCs in vessel repair following a variety of experimentally induced vascular injuries or following implantation of certain tumors. Likewise, marrow-derived EPCs appear to contribute to sites of vascular turnover in both animal and human transplantation studies (6,7). However, other studies employing highly sophisticated marking strategies indicate that marrow-derived EPCs play no role or a minimal role in neovascularization, vessel repair, or normal vessel growth and development (8–10). The controversies surrounding this question may be clarified by utilizing methods of detection that clearly discriminate between EPCs with high proliferative potential identified using the single cell assay and the more numerous marrow-derived EPCs, which can be cultured short-term in vitro into small colonies of cells that co-express certain monocyte and macrophage antigens, as well as more typical EC surface proteins, but fail to give rise to EC progeny that can be significantly passaged ex vivo (7,19–22). The latter human EPCs cannot be cultured at a single cell level and thus clonal analysis of proliferative potential cannot be determined. As suggested by some authors, these EPCs may be better defined as angiogenic progenitors or vascular progenitors than as EPCs (20).

The newly identified HPP-ECFC and associated progenitors have not been enumerated in the blood or blood vessels of many patients with diseases. One recent preliminary report suggests that the concentration of the HPP-ECFCs in patients with cardiovascular ischemic disease correlates with patient status such that the higher the concentration of the HPP-ECFCs, the higher the risk that the patient has ongoing severe ischemic disease (22). Further comparative studies with circulating marrow–derived EPCs will be required to determine the interrelationship of these cells with the HPP-ECFCs and associated progenitors in the pathogenesis of cardiovascular ischemic disorders.

Likewise, novel strategies will need to be developed to assay for the number and distribution of HPP-ECFCs and associated EPCs that reside in large and small arterial and venous vessels and capillaries. One method would be to biopsy readily accessible vessels with an intravascular catheter and then enumerate the number and distribution of the EPCs using the single cell EPC colony forming assay. A more invasive procedure would be to use open biopsy of superficial arterial and venous vessels to obtain ECs for enumeration of the various HPP-ECFCs and associated EPCs. Alternatively, some unique cell surface markers for the HPP-ECFCs and associated EPCs may be soon isolated that will permit sampling of the blood for cells that are mobilized from the vessels and can be detected by flow cytometric methods.

THERAPEUTIC IMPLICATIONS OF HIGH PROLIFERATIVE POTENTIAL-ENDOTHELIAL COLONY FORMING CELLS

The finding that vascular ECs comprise HPP-ECFCs and associated EPCs provides a new context to reexamine the process of angiogenesis. As mature ECs derived from umbilical veins and human aortas fail to divide in vitro, it is unlikely that these cells play a significant role in the migration, proliferation, and neocapillary development at sites of new vessel formation. Rather, the HPP-ECFCs and associated EPCs are more likely to be the targets of recruitment by the growth factors and molecules secreted to recruit the cells with proliferative potential out of the existing endothelial intima and into the tissues via sprouting.

The availability of an in vitro assay to identify these cells may now permit screening for additional molecules that may play a role in neoangiogenesis and, alternatively, provide a potentially new strategy to screen for antiangiogenic molecules. These approaches may identify new families of molecules that may be tested and applied to treat the vast number of patients with cardiovascular disease.

GOALS FOR BRIDGING BENCH-TO-BEDSIDE GAP

Identification of this novel hierarchy of EPCs that both circulate and reside in the endothelial intima of arterial and venous vessels will provide new pathways for correlating the biology of these EPCs with specific disease states and may permit opportunities for consideration of cell replacement therapies. Although numerous clinical studies are contemplating the use of whole bone marrow cells or a particular subset of marrow-derived EPCs with certain cell surface markers as a therapy for patients with critical ischemic limb disease or following myocardial ischemic disease, it is unclear if the cells under consideration will be long-lasting contributors to formation of a replenished endothelial pool of HPP-ECFCs and associated EPCs in the new vessels. The recent identification of HPP-ECFCs in several large animal species (bovine, porcine, canine, nonhuman primate; M. C. Yoder, unpublished observations) now opens the possibility for comparative analysis of marrow-derived EPCs and vessel-derived ECs with EPC activity to determine the specific roles that each of these cells plays in new vessel formation in animal models that will be relevant to human disease.

KEY POINTS

- The vascular system represents a stem and progenitor cell niche for the endothelial lineage.
- Circulating EPCs represent a hierarchy of progenitors with different proliferative potentials.
- The proliferative potentials of EPCs can be quantitated.
- The complete hierarchy of endothelial stem/progenitor cells is present in the endothelium of arterial, venous, and capillary vessels.

Future Goals

- To pursue future studies that will permit prospective isolation of these stem/progenitor cells, determine when vascular endothelium becomes seeded with these cells, and determine the role these cells play in vascular repair and regeneration

REFERENCES

1 Schwartz SM, Benditt EP. Cell replication in the aortic endothelium: a new method for study of the problem. *Lab Invest*. 1973; 28(6):699–707.
2 Caplan BA, Schwartz CJ. Increased endothelial cell turnover in areas of in vivo Evans Blue uptake in the pig aorta. *Atherosclerosis*. 1973;17(3):401–417.
3 Schwartz S, Benditt E. Clustering of replicating cells in aortic endothelium. *Proc Natl Acad Sci USA*. 1976;73:651–653.
4 Schwartz SM, Benditt EP. Aortic endothelial cell replication. I. Effects of age and hypertension in the rat. *Circ Res*. 1977;41(2):248–255.
5 Schwartz SM, Stemerman MB, Benditt EP. The aortic intima. II. Repair of the aortic lining after mechanical denudation. *Am J Pathol*. 1975;81(1):15–42.
6 Rafii S, Lyden D. Therapeutic stem and progenitor cell transplantation for organ vascularization and regeneration. *Nat Med*. 2003;9:702–712.
7 Urbich C, Dimmeler S. Endothelial progenitor cells: characterization and role in vascular biology. *Circ Res*. 2004;95:343–353.
8 Gothert JR, Gustin SE, van Eekelen JA, et al. Genetically tagging endothelial cells in vivo: bone marrow-derived cells do not contribute to tumor endothelium. *Blood*. 2004;104(6):1769–1777.
9 Peters BA, Diaz LA, Polyak K, et al. Contribution of bone marrow-derived endothelial cells to human tumor vasculature. *Nat Med*. 2005;11(3):261–262.
10 Stadtfeld M, Graf T. Assessing the role of hematopoietic plasticity for endothelial and hepatocyte development by non-invasive lineage tracing. *Development*. 2005;132(1):203–213.
11 Jacobson L, Marks E, Robson M, et al. The effect of spleen protection on mortality following X-irradiation. *J Lab Clin Med*. 1949; 34:1538–1542.
12 Ford C, Hamerton J, Barnes W, Loutit J. Cytological identification of radiation-chimaeras. *Nature*. 1956;177:452–454.
13 Till J, McCulloch E. A direct measurement of the radiation sensitivity of normal mouse bone marrow cells. *Radiat Res*. 1961; 14:213–222.
14 Pluznik D, Sachs L. The cloning of normal "mast" cells in tissue culture. *J Cell Comp Physiol*. 1965;66:319–327.
15 Bradley T, Metcalf D. The growth of mouse bone marrow cells in vitro. *Aust J Exp Biol Med Sci*. 1966;44:287–300.
16 Kriegler AB, Verschoor SM, Bernardo D, Bertoncello I. The relationship between different high proliferative potential colony-forming cells in mouse bone marrow. *Exp Hematol*. 1994;22(5): 432–440.
17 Ingram D, Mead L, Tanaka H, et al. Identification of a novel hierarchy of endothelial progenitor cells utilizing human peripheral and umbilical cord blood. *Blood*. 2004;104:2752–2760.
18 Ingram DA, Mead LE, Moore DB, et al. Vessel wall derived endothelial cells rapidly proliferate because they contain a complete hierarchy of endothelial progenitor cells. *Blood*. 2004;105(7):2783–2786.
19 Conejo-Garcia JR, Buckanovich RJ, Benencia F, et al. Vascular leukocytes contribute to tumor vascularization. *Blood*. 2005;105 (2):679–681.
20 Rehman J, Li J, Orschell C, March K. Peripheral blood "endothelial progenitor cells" are derived from monocyte/macrophages and secrete angiogenic growth factors. *Circulation*. 2003;107: 1164–1172.
21 Schatteman GC, Awad O. Hemangioblasts, angioblasts, and adult endothelial cell progenitors. *Anat Rec A Discov Mol Cell Evol Biol*. 2004;276(1):13–21.
22 Urbich C, Heeschen C, Aicher A, et al. Relevance of monocytic features for neovascularization capacity of circulating endothelial progenitor cells. *Circulation*. 2003;108(20):2511–2516.

PART IV

DIAGNOSIS AND TREATMENT

172

Introductory Essay

Diagnosis and Treatment

Mansoor Husain

Physiology and Laboratory Medicine & Pathobiology, University of Toronto; Division of Cell & Molecular Biology, Toronto General Hospital Research Institute, Ontario, Canada

In previous sections of this textbook, chapters have been assembled to convey the molecular and cellular basis of organ pathophysiology, as viewed through the lens of vascular biology in general and the endothelium in specific. How the endothelium interacts with adjacent vascular smooth muscle cells, circulating blood cells, and the underlying parenchyma of each organ have been exampled, and specific chapters have focused on particular organs and their disease states. These sections provided the knowledge upon which the ensuing chapters are built, but which focus now on diagnosis and treatment.

The endothelium has remarkable, yet largely untapped diagnostic potential. From a bedside physician's viewpoint, the endothelium is difficult to assay. It is hidden from view and cannot be inspected, palpated, percussed, or auscultated. Moreover, the endothelium is poorly circumscribed and therefore not readily amenable to conventional radiological studies. However, the endothelium has several properties that make it an ideal target for therapy. First, it is widely distributed and therefore provides a window into every organ in the body. Second, it is rapidly and preferentially exposed to systemically delivered agents. Third, as an input–output device, the endothelium is highly malleable and responsive to exogenously administered compounds.

In Part IV, the authors provide their perspective on how current knowledge may affect real or potential advances in the diagnosis and treatment of endothelial-based diseases. Advances in diagnosis include the measurement of circulating soluble mediators (Chapter 173), and circulating blood endothelial cells (Chapter 174); or circulating microparticles (Chapter 175). Other diagnostic assays include molecular imaging (Chapter 176), real-time imaging (Chapter 177), and flow studies (Chapter 178). Chapters on therapy focus on statins (Chapter 179), steroids (Chapter 180), nitrates (Chapter 181), alteration of flow (Chapter 182), drug-coated stents (Chapter 183), tissue engineering (Chapter 184) and drug/viral targeting (Chapters 185 and 186).

In this introductory essay, I will posit how information such as this may enable a conceptual shift in the classification and approach to therapy of cardiovascular diseases. It is my hope that readers of this section may likewise imagine and eventually actualize the clinical value of the collected writings to other areas of medicine.

The World Health Organization has determined that cardiac and vascular diseases are now the leading causes of mortality in our world, accounting for one of every three deaths.[1] This holds true for both the developed and developing worlds, where aging populations and rapid industrialization are ever increasing the incidence of high blood pressure, stroke, heart attack, and heart failure. Indeed, it can be stated that diseases of the heart and blood vessels are an enormous health care burden poised only to worsen in the coming decades. As an organ crucially involved in the maintenance of normal cardiac and vascular structure and function, the endothelium has emerged as a pivotal target in the fight against this global pandemic. Moreover, an appreciation for how primary or secondary abnormalities in function of this cell type may affect and inform on cardiovascular disease progression has begin to change the way in which we think about this challenge.

Much excitement has been raised about the potential to harvest and therapeutically employ circulating endothelial progenitor cells or bone marrow–derived progenitors of other cell lineages. Although still in its infancy, the opportunity at hand for truly regenerative medicine has never been greater. However, current proposals for cell-based treatments of end-stage heart failure resemble on some level the indiscriminate treatment of anemia with a blood transfusion. Would not knowing the basis of the problem help us better design these trials? It would seem obvious that dramatic improvements in our *diagnostic* capabilities will be as important and

1 *Atlas of Heart Disease and Stroke.* Judith Mackay and George Mensah. 2004. Non-Serial Publication of the World Health Organization. ISBN 92-4-156276-5.

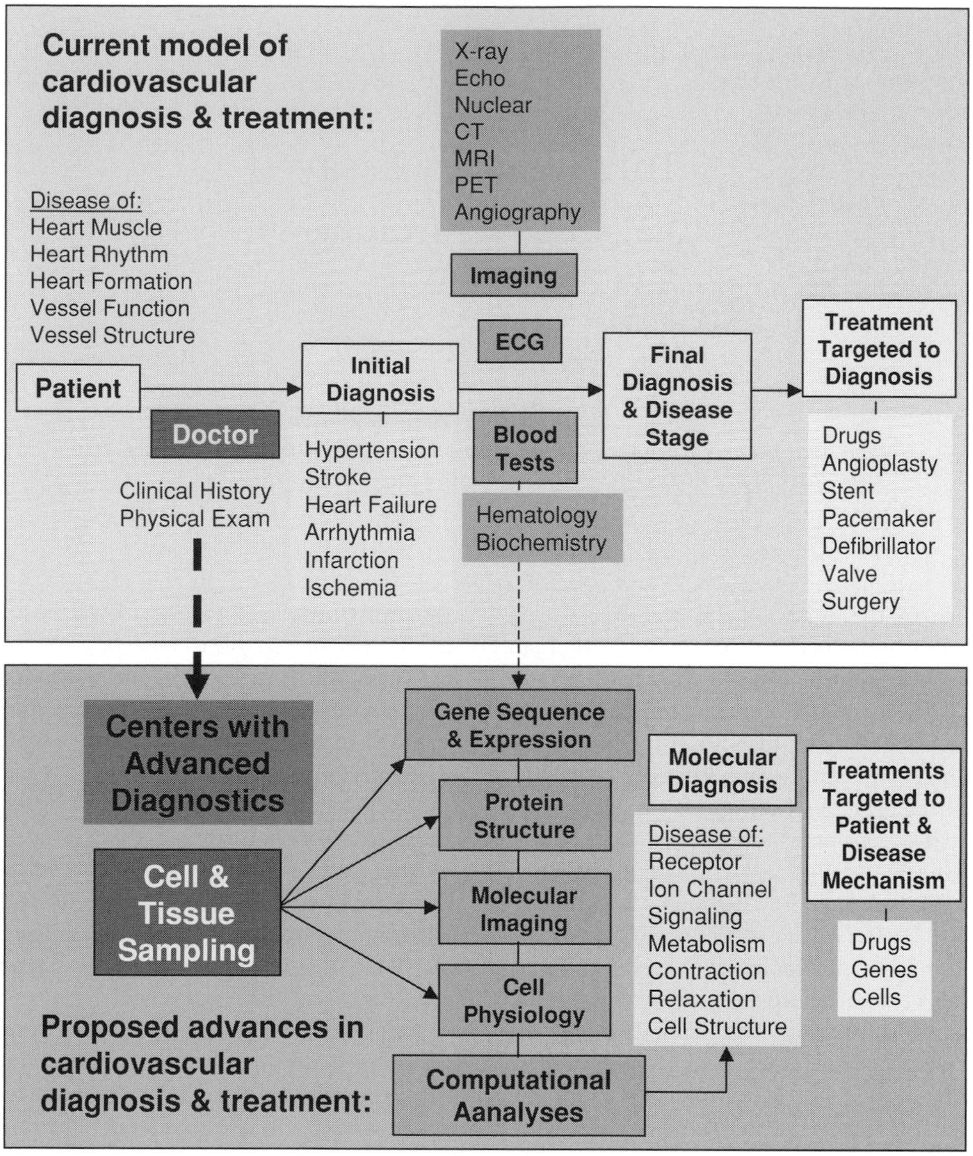

Figure 172.1. Staging of cardiovascular disease.

are as imperative as our need to implement advanced therapies.

Currently, "modern" approaches to the diagnosis and treatment of cardiovascular (and other) diseases are based on a physician's ability to obtain a history of illness from the patient and to integrate that information with findings from a physical examination and tests such as electrocardiogram, imaging, and blood chemistry. Together, these data are used to classify a patient as having a particular disease or not, stage its degree and extent, and to begin treatment based on this categorization (Figure 172.1). Unfortunately, existing diagnostic methods remain superficial in their level of interrogation, with many different genetic, environmental, and complex molecular contributors to disease remaining unrecognized. Although advanced assessments of endothelial function are being developed, few centers in the world have widely implemented noninvasive or ex vivo assays of the molecular physiology of endothelial, cardiac, or vascular muscle cells or applied them to the rationale design of treatment regimens. As such, our patients are lumped together in relatively crude diagnostic groupings and are treated as if they were identical. Recognition of the limits and hazards of this paradigm is increasing, however until we individualize treatments on the basis of a deeper understanding of the fundamental mechanisms underlying disease, only some of our patients benefit, others do not, and some may even be harmed by current strategies.

Taking, for example, a widespread disease as simple in pathophysiology and yet as complex in cause as hypertension, we recognize that far too many patients go unrecognized, require multiple classes of antihypertensive mediations to normalize their blood pressure, or remain poorly controlled despite polypharmacy. For this and many other diseases, it seems logical that the more precise a "match" between mechanism and treatment, the better the efficacy of treatments will be, with fewer unnecessary side effects.

We should seek to identify people at risk earlier in the development of their disease and to individualize care based on mechanisms that may be unique to a given patient and their particular genetic–environment interactions. To enable this, a novel approach is proposed, which seeks to acquire and integrate diagnostic data from more fundamental levels of interrogation, yielding a more informed "molecular" classification of disease. To accomplish this, healthcare institutions must first develop an infrastructure that increases the levels of interrogation currently available for cardiovascular disease. Furthermore, to realize the full potential of a corresponding revolution in genomic and proteomic analyses, such an infrastructure must also be able to integrate this information in comprehensive databases allowing computational analyses of the "systems biology" of each disease, correlating these with clinical outcomes. Such an infrastructure is critical to the formation of new models of cardiovascular disease and to the development of the next generation of more targeted therapeutic strategies (Figure 172.1).

Linking and harnessing the knowledge from this and other textbooks with that of public and private repositories of genomic, proteomic, physiomic, and clinical data will determine future advances in care. Bioinformatic approaches designed to interrogate such data sets are where connections and inter-relationships currently invisible to healthcare researchers and practitioner may emerge. Training programs will be equally critical. The education of medical personnel will need to include a thorough understanding of the relevant novel diagnostic techniques and methods of complex data acquisition, management, and analysis. Where there is greatest need, and where centers that initiate such an infrastructure may excel, is in the opportunity to develop truly *translational* individuals, who can interpret the language and needs of both fundamental and clinical scientists and act as synergistic intermediaries to the benefit of both. Close collaboration and regular exchanges with like-minded institutions and infrastructure will accelerate significantly the integration of this conceptual shift in the principles and practices of medicine.

173

Circulating Markers of Endothelial Function

Andrew D. Blann, William Foster, and Gregory Y.H. Lip

City Hospital, Birmingham, United Kingdom

The vascular endothelium is a dynamic and metabolically active organ with numerous functions, including regulation of the coagulation cascade and local blood flow, control of fluid passage into the media, and the regulation of cell migration through the vascular intima (1,2). These many functions are reflected in the variety of molecules expressed at the cell surface and secreted by endothelial cells (ECs). The appearance of such molecules in measurable quantities in the blood allows indirect measurement of the status of the endothelium, which can exist in various different states: resting, activated, and/or dysfunctional. Stimulation of resting endothelium (e.g., by inflammatory cytokines) results in an activation state, characterized by the upregulation of various cell adhesion molecules and procoagulant molecules. Dysfunction can be thought of as an inability of the endothelium to carry out its normal physiological functions (2,3).

The ideal test of endothelial function would be simple, quick, and inexpensive to perform, sensitive, endothelial-specific, correlated with a known disease process, and deranged in a predictable way in disease processes. In other words, it would be a reliable surrogate marker for a specific endothelial disease process. In addition, the marker would be stable enough to measure reliably in the laboratory, yet also have sufficiently rapid turnover or short half-life to provide a meaningful measure of current endothelial status. These criteria effectively remove nitric oxide (NO) from a list of useful markers because it cannot be measured directly or in a simple manner in the laboratory. Other well-established molecules, such as plasminogen activator inhibitor (PAI)-1, soluble intercellular adhesion molecule (ICAM)-1, and soluble vascular cell adhesion molecule (VCAM)-1, suffer from the disadvantage that they are also expressed by non-ECs (e.g., platelets, adipocytes, activated leukocytes, and smooth muscle cells) and are thus nonspecific.

Those plasma markers that fulfill these criteria include von Willebrand factor (vWF), soluble E-selectin, endothelin (ET)-1, soluble thrombomodulin (sTM), and tissue-type plasminogen activator (t-PA) (4). In addition to these soluble markers, there is increasing evidence that cellular markers, including endothelial microparticles (EMPs) and circulating ECs (CECs), provide important insights into the status of the endothelium (5,6). In this chapter, we first review molecular plasma markers of endothelial integrity and then discuss the present status of cellular markers.

CIRCULATING SOLUBLE MARKERS

von Willebrand Factor

vWF, the prototype marker of endothelial damage, was first described as such 30 years ago (7). vWF is a multimeric procoagulant protein encoded on chromosome 12, whose expression is limited to ECs and megakaryocytes. The *vWF* gene is highly complex, consisting of 52 exons and encoding a 250 kDa protein that forms the basic monomer. The mature molecule comprises numerous monomers forming polymers or multimers and is subject to post-translational modification before it becomes functional, reaching a size of up to 1 MDa (8). The importance of vWF in promoting hemostasis is clear from the clinical consequences of its deficiency – namely, the hemophilia-like disorder, von Willebrand disease. vWF binds and stabilizes factor VIII and mediates platelet adhesion to the site of vessel injury and aggregation by binding platelets and collagen. It is also the most extensively studied circulating marker of endothelial function.

vWF is constitutively secreted by the endothelium in low levels in healthy individuals, but further release can occur from preformed intracellular stores (i.e., Weibel-Palade bodies) when ECs are stimulated or damaged in vitro by a variety of stimuli, including inflammatory cytokines (e.g., interleukin [IL]-1, tumor necrosis factor [TNF]-α), histamine, elastase, and vasopressin (7–9). Whether the same processes occur in vivo is unclear but is widely presumed.

Levels of vWF rise in a number of nonpathological situations, notably in normal pregnancy and following exercise; levels are also determined by the ABO blood group (9–11). However, circulating levels are also elevated in many pathological conditions, including atherosclerosis; inflammatory connective tissue diseases, such as rheumatoid arthritis and systemic lupus erythematosus; and in many cancers (9). Increased levels are associated with each of the major risk factors for atherosclerosis (cigarette smoking, hypertension, diabetes mellitus, obesity, and elevated serum cholesterol) (12,13), as well as several minor risk factors including impaired glucose tolerance (14), chronic renal impairment (15), and rheumatoid arthritis (16). Effective treatment of cardiovascular risk factors results in a lowering of vWF levels (17).

High vWF levels have been shown to have prognostic significance, conferring, for example, an increased risk of first and subsequent myocardial infarction, stroke, thromboembolism, complications from peripheral vascular disease, and death (9,18). As a short-term prognostic marker, elevated levels of vWF following admission for acute coronary syndromes were found to predict later adverse events (19). This also applied to patients with preexisting myocardial disease (20). Even in apparently healthy, asymptomatic individuals, elevated levels of vWF predict an adverse cardiovascular outcome (21).

Thus, vWF is unequivocally raised in patients with major risk factors for atherosclerosis and in those with frank disease, and raised levels carry a poor prognosis. Furthermore, given the established role of vWF as a procoagulant, the increased levels may contribute to pathogenesis, rather than simply serving as a marker for EC dysfunction (9).

Soluble Thrombomodulin

TM is an endothelial transmembrane protein that binds thrombin. This binding causes a conformational change in the structure of thrombin, increasing its affinity for protein C. Thus thrombomodulin has anticoagulant properties and is partly responsible for the prevention of intravascular thrombus formation (22). sTM can be measured in plasma and serum and presumably reflects the loss of the molecule from the endothelial surface into the blood (23–25). Elevated serum levels of sTM may thus reflect increased surface expression with normal surface loss, increased loss of the molecule with normal expression, or a combination of the two. The antithrombotic effects of the soluble form of TM have been shown experimentally by injecting mice with thrombin; those mice pretreated with TM were protected against widespread, lethal thrombus seen in the other mice (26).

Increased levels of sTM have been found in prothrombotic conditions, including acute and chronic atrial fibrillation, advanced heart failure, disseminated intravascular coagulation, and normal pregnancy (27–31). Levels of sTM have been found to correlate with the presence and severity of atheromatous disease – e.g., carotid intima-media thickness (32) – and extent of peripheral vascular disease (33). sTM is also elevated in patients with ischemic heart disease (34).

In terms of predicting future cardiovascular events, sTM is elevated in patients with many of the classical cardiovascular risk factors, including hypertension, type I and type II diabetes mellitus, and obesity (35–37). Smoking has been associated with both increased (38) and decreased (39) levels. sTM has also been found to be predictive of future cardiovascular events in patients with established atherosclerosis (40,41). However, interestingly, patients not known to have atherosclerotic disease do not have this association; the presence of elevated sTM levels actually predicted a *lower* risk of incident coronary artery disease in subjects investigated for the ARIC study (42).

Following peripheral angioplasty, elevated levels predict late restenosis (43). However, there is controversy as Gerdes and colleagues (44) reported that raised sTM predicts stroke, but not myocardial infarction, in a group of patients who had had a recent myocardial infarction or ischemic stroke or who had peripheral arterial disease. Low sTM levels occurred in smokers, whereas elevated levels were associated with diabetes, although there was no relationship between sTM and mild peripheral (lower limb) artery disease (45). More recently, the same group showed an interaction between low levels of sTM and the nonspecific inflammatory marker sICAM-1 in predicting the development of coronary heart disease, with sICAM-1 only being associated with worse outcomes when sTM levels were low (46).

Thus, levels of sTM, produced as the consequence of shedding or cleavage of anticoagulant membrane thrombomodulin, are increased in association with most risk factors for atherosclerosis. However, there is debate as to the power of such data in predicting adverse cardiovascular outcome.

Endothelin

ET is a 21-amino acid peptide that results from cleavage of so-called big ET by the action of ET converting enzyme (ECE), which is located on the EC membrane. Although three isoforms of ET (ET-1, ET-2, and ET-3) exist, ET-1 appears to be the predominant form in the cardiovascular system. Constitutive release of ET-1 by the endothelium is important for the maintenance of vascular tone (47). However, ET-1 stored in Weibel-Palade bodies may be released upon endothelial activation, resulting in further vasoconstriction (48). In fact, ET-1 is one of the most potent vasoconstrictors known.

ET binds to two receptor types, ETA and ETB. Although vascular smooth muscle cells express both types of receptor, ET-1–mediated activation of the ETA receptor is primarily responsible for mediating vasoconstriction (49). In contrast to ETA, which is predominantly expressed by vascular smooth muscle cells, ETB is found in a variety of tissues including endothelium, macrophages, and vascular adventitia (50,51). Experimentally, ET-1 produces powerful vasoconstriction (particularly at higher doses; lower doses may initially cause vasodilatation) in isolated human lung and pulmonary artery samples (52). Other data have shown that in vivo infusions of ET-1 produce endothelial dysfunction in healthy volunteers (defined as impaired forearm blood flow increase in response to infused acetylcholine) (53). Moreover, blockade of ETA receptors by specific antagonists improved endothelial function in people with endothelial dysfunction in the context of atheroma. This fits well with other studies of the internal mammary artery (harvested at the time of coronary artery bypass grafting), where ETA receptor blockade improved endothelial function in vitro (54).

Increased circulating levels of ET-1 are found in the plasma of patients who have endothelial dysfunction as measured by other means. This is clearly demonstrable following local endothelial injury caused by coronary artery angioplasty and stenting (55), where levels are elevated immediately post procedure. Other data from patients with coronary endothelial dysfunction (based on abnormal vasoconstrictor response to intracoronary acetylcholine injection) support this, with ET-1 levels higher in these subjects (56). Physiologically, ET-1 levels rise in normal pregnancy, but diabetic women and women who develop preeclampsia have particularly high levels (57,58).

Many of the classical cardiovascular risk factors are associated with increased ET-1 levels, and in most cases the level normalizes with effective treatment of the risk factor. In the case of hypertension, blood pressure treatment with an angiotensin-converting enzyme inhibitor (ACEI) lowered ET-1 levels. It is not clear whether this is a blood pressure lowering effect or a direct effect of ACEIs on ET release, as previous work has suggested that in vitro release of ET from cultured human ECs is inhibited by ACEIs (59). In hypercholesterolemic patients, ET-1 production is also increased, and treatment with the statin Pravastatin has been found to reduce ET-1 levels (60,61). Levels of ET-1 are elevated in patients with type I and type II diabetes mellitus, even when the diabetes is well controlled and without complications, but levels are elevated further in the presence of diabetic vasculopathy (62–64). Some clarification of the possible mechanisms underlying this observation comes from experimental work involving infusion of insulin into the forearm circulation of healthy volunteers; ET-1 activity was found to increase in response to the insulin infusion (65). Other experimental work using human umbilical vein ECs (HUVECs) has shown that stimulating these cells with TNF-α resulted in increased secretion of ET-1 (66). This not only provides an intriguing link between inflammation and endothelial dysfunction, but also fits with work in obesity, where ET-1 levels have been found to correlate with TNF-α levels (67).

Perhaps surprisingly, whereas ET-1 levels are higher in hypertensive patients and hypercholesterolemic patients, increased levels do not independently predict future cardiovascular events in these groups (68). Cigarette smoking is associated with increased ET-1 levels, and levels rise in the immediate time after smoking a cigarette (69,70).

Thus, increased levels of ET-1 are associated with in most of the risk factors for cardiovascular disease and generally fall with treatment of the risk factor (with the notable exception of diabetes); levels are also elevated following vascular injury and in patients with endothelial dysfunction. However (in contrast to vWF and sTM), levels are not directly related to prognosis.

Soluble E-Selectin

E-selectin, an endothelial-specific surface ligand that facilitates leukocyte rolling, is a marker of endothelial activation. The membrane-bound form is not found on normal, resting endothelium (71) but is expressed by ECs when activated by various stimuli (72). Soluble E-selectin, presumed to be shed or digested from the membrane, is found in the plasma of healthy persons and is increased in various patient groups. Lack of the cytoplasmic domain suggests proteolytic cleavage as the mechanism of release (73).

Consistent with the experimental observation that HUVECs express E-selectin in response to activation by inflammatory cytokines is the observation that soluble E-selectin is elevated in a number of inflammatory conditions such as Wegener and systemic vasculitis (74,75). In hyperlipidemic patients, levels are higher than in controls; lipid-lowering treatment has been shown to reduce the levels (76,77). Higher soluble E-selectin levels are also associated with a Western diet, smoking, hypertension, and diabetes mellitus (78–81), where it correlates with hemoglobin AIC (HbAIc), and increased levels predict future development of diabetes (82–84). Furthermore, treatment of diabetic patients (and healthy controls) with an ACEI has been shown to lower soluble E-selectin levels, although other studies have failed to correlate soluble E-selectin with HbA1c (85,86).

Finally, soluble E-selectin levels rise in the hours following myocardial infarction but return to baseline by five days; levels are also elevated in acute ischemic stroke (87–89). In addition, plasma levels of soluble E-selectin predict restenosis in patients who have undergone peripheral arterial balloon angioplasty (90). However, although one follow-up study found soluble E-selectin to be significantly related to future death, another found that measurement of this adhesion molecule is unlikely to add much predictive information compared to that obtained from established risk factors (91,92). For a recent review on soluble E-selectin, see reference 93.

Soluble E-selectin is therefore a marker of endothelial (inflammatory) activation and is elevated in a variety of situations in which the endothelium is stimulated or dysfunctional. Although levels are increased in many high-risk patients, and elevated levels appear to predict the onset of diabetes in patients at risk, debate continues.

Tissue-Type Plasminogen Activator

Thrombogenesis and thrombolysis occur in vivo as a dynamic balance that maintains vascular integrity and prevents excessive bleeding. This balance is brought about partly by the complex relationship between prothrombotic and anticoagulant factors. However, once thrombosis has started there is a balance between production and breakdown of the particular thrombus. In vivo, plasmin is produced by cleavage of the proenzyme plasminogen, by the action of t-PA. However, in vivo, free plasmin (i.e., plasmin not bound to fibrin) is rapidly inactivated, and t-PA activity is potently inhibited by the action of PAI-1 and PAI-2, the two endogenous inhibitors of t-PA. Both t-PA and PAI are endothelial products, although the latter may also arise from platelets and adipose tissue (4,94,95).

Table 173-1: Follow-up Studies Allowing Comparisons between Plasma Markers of Endothelial Function

Patients Recruited (Refs.)	*Years of Follow-up*	*Markers Studied*	*Conclusion*
Post-MI, age < 70 years (100)	4.9	vWF, t-PA	vWF predicts MACE
Rheumatoid arthritis (101)	2	vWF, t-PA	vWF predicts thromboembolism
On oral anticoagulation (102)	3.8	vWFt-PA	vWF predicts all cause and vascular mortality
Angina pectoris (20)	2	vWF, t-PA	t-PA ($p = 0.02$) > vWF ($p < 0.05$) predicts MACE
Hypertension (103)	3	vWF, sE-selectin	vWF predicts MACE
Hypercholesterolemia (104)	4	vWF, sE-selectin	vWF predicts MACE
Post-MI (41)	4	sTM, sE-selectin	sTM predicts MACE
Post MI, age < 70 years (105)	10	vWF, t-PA	vWF ($p = 0.006$) > t-PA ($p = 0.025$) predicts MACE
Unstable CAD (107)	14 days and 30 days	vWF, ET	Acute rise in vWF predicts MACE
Mixed CAD and PAD (107)	3.8	vWF, sTM	vWF ($p < 0.001$) > sTM ($p < 0.01$) predicts MACE
Free of PAD but some CAD (108)	5	vWF, t-PA	vWF predicts development of new PAD
Peripheral angioplasty (109)	0.5	vWF, sTM sE-selectin	None predict restenosis
Essentially healthy (110)	2–2.5	vWF, sTM	Neither predicted stroke
Diabetics (111)	9	vWF, t-PA sE-selectin	vWF and t-PA both predict mortality
Stable CAD (112)	13	vWF, t-PA	t-PA predicts coronary events
Free of CAD (113)	5	vWF, sTM	vWF predicts "hard" CAD

MACE, major adverse cardiovascular event (myocardial infarction [MI], stroke; coronary artery bypass grafting; etc. – see individual references for details); CAD, coronary artery disease; PAD, peripheral artery disease.

Levels of t-PA are frequently described as abnormal following cardiovascular events (96,97). However, the pathophysiological significance of elevated (or suppressed) levels of this molecule (and PAI-1) is very difficult to ascertain, not least because the actual level of any one molecule may well be less important than the biological activity of that molecule, which is of course a complex interaction involving the amount of molecule, its location, the amount and location of substrate, and the activity of its inhibitors (98,99). Nevertheless, in patients with atherosclerosis, levels are altered in those who go on to develop coronary artery disease, and it is therefore predictive of primary cardiovascular events in at-risk patients. (4,20,100–102).

Thus, whereas the clinical value of t-PA and its analogues in treating acute thrombotic events is now established, its value as an endothelial marker and/or predictor of adverse cardiovascular events is not as well accepted, possibly for technical reasons.

Which Plasma Marker Is Most Useful?

The previous discussion has indicated various examples of studies promoting vWF, sTM, ET, soluble E-selectin, and t-PA as endothelial markers predictive of adverse outcome, generally thrombosis precipitating myocardial or cerebrovascular infarction. However, few have placed these molecules side-by-side to allow a comparison (Table 173-1). Because of their (direct) involvement in hemostasis, the most frequently compared are vWF and t-PA, and of these, most studies find vWF to be the better predictor of adverse outcomes (100–113). Although most studies have focused on cardiovascular disease, both vWF and soluble E-selectin predict outcome in patients acutely ill with multisystem organ failure and/or sepsis (114).

CELLULAR MARKERS

Circulating Endothelial Cells

Recent research has recognized new populations of nonhematopoietic cells in the blood. One of these cell types, the CEC, may be defined by the expression of membrane glycoprotein CD146 (115). Small numbers of CECs are found in health, but raised numbers are present in a wide variety of human conditions, including inflammatory, immune, infectious, neoplastic, and cardiovascular disease and seem likely to be evidence of profound vascular insult (5,116). They are thought to originate when ECs are sloughed from the vessel wall following some form of pathological insult. Precise phenotypic characterization is required under the microscope or

Table 173-2: Summary of Soluble Plasma Endothelial Markers in the Risk Factors of Atherosclerosis, in Actual Disease, and as Predictors of Disease

Marker	*Association Smoking*	*Association Diabetes*	*Association Hypertension*	*Association Hypercholesterolemia*	*Association Atherosclerosis*	*Association Others*	*Predict Primary Events*	*Predict Secondary Events*
vWF	√ (1)	√ (1)	√ (1)	√ (1)	√	Pregnancy, ABO blood group, RA, SLE, many cancers	√, AMI	√, AMI, stroke
sTM	+/–	√	√	??	√	Thrombotic conditions, AF, DIC	??	√ (stroke < AMI)
ET	?	√ (But does not reverse with good diabetic control)	√ (Reverses with therapy)	√ (Reverses with therapy)	√ (Coronary artery EC dysfunction)	Postangioplasty, stenting, pregnancy.	No	No
sE-selectin	+/–	√ (+/– correlates with HbA1c, reverses with therapy)	√	√ (Reverse with therapy)	√ (Weakly)	Inflammatory disease, cancer, sepsis	No	No
t-PA	? √	√	√ (Also predicts hypertension)	??	√	Some cancers, e.g., raised in glioma, not in multiple myeloma.	√ (AMI, stroke)	√ (AMI)

Abbreviations as in Table 173-1, plus ET, endothelin; AMI, acute myocardial infarction; RA, rheumatoid arthritis; SLE, systemic lupus erythematosus. (1) reduced as the risk factor recedes. √ common agreement.

flow cytometer by the expression of endothelial markers (e.g., vWF, vascular endothelial [VE]-cadherin) together with the absence of the expression of leukocyte (CD45) and immaturity markers (CD133) (5).

Although increased numbers of CECs correlate with other markers of vascular disease (e.g., vWF, soluble E-selectin) (117,118), questions regarding the precise definition, cell biology, and origin of CECs remain. For example, CECs may be alive, apoptotic, damaged, or necrotic, and they may possess procoagulant and/or proinflammatory properties. However, because these cells seem to be representative of in situ endothelium, their phenotype may provide useful information about their vascular bed of origin. Indeed, whatever their phenotype, there is growing evidence that CECs may well be a novel biomarker, the measurement of which will have utility in various clinical settings related to vascular injury. Despite this promise, progress is impeded by the diversity of methodologies used to detect these cells. Accordingly, results are sometimes inconclusive and even conflicting. Nevertheless, increased CECs in coronary artery disease (119,120) predict future adverse cardiovascular events (121) suggesting they may move from being simply a research index to having a role in the clinic.

Endothelial Microparticles

Cellular activation, damage, and apoptosis may lead to the release of EMPs (6,122,123). Although the significance of EMPs remains unclear, there is a growing view that EMPs can function as important diffusible mediators of cytokines and adhesins, thus promoting cellular signaling and activity (124). In vitro stimulation of HUVECs by angiogenic growth factors (vascular endothelial growth factor [VEGF], fibroblast growth factor 2 [FGF2]) resulted in EMP formation rich in matrix metalloproteinases and capillary cordlike structures, suggesting a potential in vivo role in angiogenesis (125).

There may be a role for EMPs in thrombotic disease as they display procoagulant activity, defined by platelet factor 3 activity and expression of tissue factor (126–129). In prothrombotic states such as sickle cell disease, raised tissue factor-positive EMPs in patients compared with controls correlated strongly with procoagulant activity (129). Of note, Boulanger and colleagues (130) reported severe endothelial dysfunction in rat aorta after incubation with human EMPs, affecting the endothelial NO transduction pathway, but not NO synthase expression. EMPs may also be proinflammatory as they tend to bind and activate monocytes, resulting in cytokine release (e.g., TNF-α, IL-1β) which causes further paracrine and/or autocrine activation of monocytes and endothelium (6).

These studies raise the possibility of EMPs being mediators of vascular insult and inflammation in disease, rather than just markers of endothelial dysfunction. More data are required to answer these issues, particularly in its relationship with cancer, of which there are currently no reports to the best of our knowledge.

CONCLUSION

Recent reviews have underlined the importance of the endothelium in a variety of human diseases (1–3,131). The endothelium releases/sheds circulating cells and molecular markers (of which the most promising and most frequently cited is vWF) (9,132) that reflect its integrity and activity. In most cases markers are released in health as well as disease, with the levels varying according to the general particular disease process (e.g., hypertension); in others the release of the marker is only in response to a particular (acute or precipitating) event or events (e.g., septicemia) (Table 173-2). Nevertheless, a knowledge of patterns of release of such cells and molecules allows a greater understanding of relationship between the vascular tree and cardiovascular physiology and pathology and thus the targeting (e.g., by ACEIs, statins) of particular disease processes.

KEY POINTS

- Endothelial physiology can be determined by certain cells and molecules in the plasma.
- Plasma makers include vWF, sTM, soluble E-selectin, ET, and t-PA.
- vWF is the most reliable plasma marker of endothelial perturbation: raised levels consistently predict adverse cardiovascular events.
- CECs are present in the blood of subjects with a variety of cardiovascular diseases and in cancer.

REFERENCES

1 Cines DB, Pollak ES, Buck CA, et al. Endothelial cells in physiology and in the pathophysiology of vascular disorders. *Blood.* 1998;91:3527–3561.
2 Aird WC. Spatial and temporal dynamics of the endothelium. *J Thromb Haemost*. 2005;3:1392–1406.
3 Felmeden DC, Lip GY. Endothelial function and its assessment. *Expert Opin Investig Drugs*. 2005;14(11):1319–1336.
4 Blann AD, Taberner DA. A reliable marker of endothelial cell dysfunction: does it exist? *Br J Haematol*. 1995;90:244–248.
5 Blann AD, Woywodt A, Bertolini F, et al. Circulating endothelial cells: biomarker of vascular disease. *Thromb Haemost*. 2005;93:228–235.
6 Jimenez JJ, Jy W, Mauro LM, et al. Endothelial microparticles (EMP) as vascular disease markers. *Adv Clin Chem*. 2005;39: 131–157.
7 Boneu B, Abbal M, Plante J, Bierme R. Factor-VIII complex and endothelial damage. *Lancet*. 1975;i1430.
8 Zimmerman TS. Factor VIII/von Willebrand factor. Structure and function. *Ann NY Acad Sci*. 1987;509:53–59.

9 Blann AD. Plasma von Willebrand factor, thrombosis and the endothelium: the first 30 years. *Thromb Haemost*. 2006;95(1): 49–55.

10 Stirling Y, Woolf L, North WR, et al. Haemostasis in normal pregnancy. *Thromb Haemost*. 1984;52:176–182.

11 Caekebeke KMJ, Koster T, Briet E. Bleeding times, blood groups and von Willebrand factor. *Br J Haem*. 1989;73:217–220.

12 Lufkin EG, Fass DN, O'Fallon WM, Bowie EJ. Increased von Willebrand factor in diabetes mellitus. *Metabolism*. 1979;28: 63–66.

13 Blann AD, McCollum CN, Lip GY. Relationship between plasma markers of endothelial cell integrity and the Framingham cardiovascular disease risk-factor scores in apparently healthy individuals. *Blood Coagul Fibrinolysis*. 2002;13:513–518.

14 Leurs PB, Stolk RP, Hamulyak K, et al. Tissue factor pathway inhibitor and other endothelium-dependent hemostatic factors in elderly individuals with normal or impaired glucose tolerance and type 2 diabetes. *Diabetes Care*. 2002;25:1340–1345.

15 Stam F, van Guldener C, Schalkwijk CG, et al. Impaired renal function is associated with markers of endothelial dysfunction and increased inflammatory activity. *Nephrol Dial Transplant*. 2003;18:892–898.

16 McEntegart A, Capell HA, Creran D, et al. Cardiovascular risk factors, including thrombotic variables, in a population with rheumatoid arthritis. *Rheumatology*. 2001;40:640–644.

17 Felmeden DC, Blann AD, Spencer CG, et al. A comparison of flow-mediated dilatation and von Willebrand factor as markers of endothelial cell function in health and in hypertension: relationship to cardiovascular risk and effects of treatment: a substudy of the Anglo-Scandinavian Cardiac Outcomes Trial. *Blood Coagul Fibrinolysis*. 2003;14:425–431.

18 Jager A, van Hinsbergh VW, Kostense PJ, et al. von Willebrand factor, C-reactive protein, and 5-year mortality in diabetic and nondiabetic subjects: the Hoorn Study. *Arterioscler Thromb Vasc Biol*. 1999;19:3071–3078.

19 Montalescot G, Collett JP, Lisojn L, et al. Effects of various anticoagulant treatments on von Willebrand factor release in unstable angina. *J Am Coll Cardiol*. 2000;36:110–114.

20 Thompson SG, Kienast, Pyke SDM, et al. Hemostatic factors and the risk of myocardial infarction or sudden death in patients with angina pectoris. *N Engl J Med*. 1995;332:635–641.

21 Meade TW, Cooper JA, Stirling Y, et al. Factor VIII, ABO blood groups and the incidence of ischaemic heart disease. *Br J Haem*. 1994;88:601–607.

22 Esmon CT. The roles of protein C and thrombomodulin in the regulation of blood coagulation. *J Biol Chem*. 1989;264:4743–4749.

23 Ishii H, Uchiyama H, Kazama M. Soluble thrombomodulin antigen in conditioned medium is increased by damage of endothelial cells. *Thromb Haemost*. 1991;65:618–623.

24 Sawada K, Yamamoto H, Yago H, Suehiro S. A simple assay to detect endothelial cell injury: measurement of released thrombomodulin from cells. *Exp Mol Pathol*. 1992;57:116–123.

25 Seigneur M, Dufourcq P, Conri CI, et al. Plasma thrombomodulin: new approach of endothelium damage. *Int Angiol*. 1993;12:355–359.

26 Gomi K, Zushi M, Honda G, et al. Antithrombotic effect of recombinant human thrombomodulin on thrombin-induced thromboembolism in mice. *Blood*. 1990;75:1396–1399.

27 Marin F, Roldan V, Climent VE, et al. Plasma von Willebrand factor, soluble thrombomodulin, and fibrin D-dimer concentrations in acute onset non-rheumatic atrial fibrillation. *Heart*. 2004;90:1162–1166.

28 Li-Saw-Hee FL, Blann AD, Lip GY. A cross-sectional and diurnal study of thrombogenesis among patients with chronic atrial fibrillation. *J Am Coll Cardiol*. 2000;35:1926–1931.

29 Cugno M, Mari D, Meroni PL, et al. Haemostatic and inflammatory biomarkers in advanced chronic heart failure: role of oral anticoagulants and successful heart transplantation. *Br J Haematol*. 2004;126:85–92.

30 Asakura H, Jokaji H, Saito M, et al. Plasma levels of soluble thrombomodulin increase in cases of disseminated intravascular coagulation with organ failure. *Am J Hematol*. 1991;38:281–287.

31 de Moerloose P, Mermillod N, Amiral J, Reber G. Thrombomodulin levels during normal pregnancy, at delivery and in the post-partum: comparison with tissue-type plasminogen activator and plasminogen activator inhibitor-1. *Thromb Haemost*. 1998;79:554–556.

32 Petit L, van Oort FV, Le Gal G, et al. Association of postmenopausal hormone replacement therapy with carotid atherosclerosis and soluble thrombomodulin: the vascular aging (EVA) study. *Thromb Res*. 2002;105:291–297.

33 Blann AD, Seigneur M, Steiner M, et al. Circulating endothelial cell markers in peripheral vascular disease: relationship to the location and extent of atherosclerotic disease. *Eur J Clin Invest*. 1997;27:916–921.

34 Blann AD, Amiral J, McCollum CN. Circulating endothelial cell/leucocyte adhesion molecules in ischaemic heart disease. *Br J Haematol*. 1996;95:263–265.

35 Dohi Y, Ohashi M, Sugiyama M, et al. Circulating thrombomodulin levels are related to latent progression of atherosclerosis in hypertensive patients. *Hypertens Res*. 2003;26(6):479–483.

36 Rigla M, Fontcuberta J, Mateo J, et al. Physical training decreases plasma thrombomodulin in type I and type II diabetic patients. *Diabetologia*. 2001;44:693–699.

37 Porreca E, Di Febbo C, Fusco L, et al. Soluble thrombomodulin and vascular adhesion molecule-1 are associated to leptin plasma levels in obese women. *Atherosclerosis*. 2004;172:175–180.

38 Markuljak I, Ivankova J, Kubisz P. Thrombomodulin and von Willebrand factor in smokers and during smoking. *Nouv Rev Fr Hematol*. 1995;37:137–139.

39 Blann AD, Steele C, McCollum CN. Influence of smoking on soluble adhesion molecules and endothelial cell markers. *Thromb Res*. 1997;85:433–438.

40 Blann AD, McCollum CN. von Willebrand factor and soluble thrombomodulin as predictors of adverse events among subjects with peripheral or coronary atherosclerosis. *Blood Coagul Fibrinolysis*. 1999;10:375–380.

41 Blann AD, Amiral J, McCollum CN. Prognostic value of increased soluble thrombomodulin and increased soluble E-selectin in ischaemic heart disease. *Eur J Haematol*. 1997;59: 115–120.

42 Salomaa V, Matei C, Aleksic N, et al. Soluble thrombomodulin as a predictor of incident coronary heart disease and symptomless carotid artery atherosclerosis in the Atherosclerosis Risk in Communities (ARIC) Study: a case-cohort study. *Lancet*. 1999;353:1729–1734.

43 Tsakiris DA, Tschopl M, Jager K, et al. Circulating cell adhesion molecules and endothelial markers before and after transluminal angioplasty in peripheral arterial occlusive disease. *Atherosclerosis*. 1999;142:193–200.

44 Gerdes VE, Kremer Hovinga JA, Ten Cate H, et al. Soluble thrombomodulin in patients with established atherosclerosis. *J Thromb Haemost*. 2004;2:200–201.

45 Salomaa V, Matei C, Aleksic N, et al. Cross-sectional association of soluble thrombomodulin with mild peripheral artery disease; the ARIC study. Atherosclerosis Risk in Communities. *Atherosclerosis*. 2001;157:309–314.

46 Wu KK, Aleksic N, Ballantyne CM, et al. Interaction between soluble thrombomodulin and intercellular adhesion molecule-1 in predicting risk of coronary heart disease. *Circulation*. 2003; 107:1729–1732.

47 Haynes WG, Webb DJ. Contribution of endogenous generation of endothelin-1 to basal vascular tone. *Lancet*. 1994;344:852–854.

48 Russell FD, Skepper JN, Davenport AP. Human endothelial cell storage granules: a novel intracellular site for isoforms of the endothelial converting enzyme. *Circ Res*. 1998;83:314–321.

49 Yang Z, Krasnici N, Luscher TF. Endothelin-1 potentiates human smooth muscle cell growth to PDGF: effects of ETA and ETB receptor blockade. *Circulation*. 1999;100:5–8.

50 McEwan PE, Valdenaire O, Sutherland L, et al. A nonradioactive method for localization of endothelin receptor mRNA in situ. *J Cardiovasc Pharmacol*. 1998;31(Suppl 1):S443–S446.

51 Fan J, Unoki H, Iwasa S, Watanabe T. Role of endothelin-1 in atherosclerosis. *Ann NY Acad Sci*. 2000;902:84–93.

52 Bennett RT, Jones RD, Morice AH, et al. Vasoconstrictive effects of endothelin-1, endothelin-3, and urotensin II in isolated perfused human lungs and isolated human pulmonary arteries. *Thorax*. 2004;59:401–407.

53 Bohm F, Ahlborg G, Pernow J. Endothelin-1 inhibits endothelium-dependent vasodilatation in the human forearm: reversal by ETA receptor blockade in patients with atherosclerosis. *Clin Sci*. 2002;102:321–327.

54 Verma S, Lovren F, Dumont AS, et al. Endothelin receptor blockade improves endothelial function in human internal mammary arteries. *Cardiovasc Res*. 2001;49:146–151.

55 Wainstein MV, Goncalves SC, Zago AJ, et al. Plasma endothelin-1 levels after coronary stenting in humans. *Am J Cardiol*. 2003; 92:1211–1214.

56 Lerman A, Holmes DR Jr, Bell MR, et al. Endothelin in coronary endothelial dysfunction and early atherosclerosis in humans. *Circulation*. 1995;92:2426–2431.

57 Wolff K, Carlstrom K, Fyhrquist F, et al. Plasma endothelin in normal and diabetic pregnancy. *Diabetes Care*. 1997;20:653–656.

58 Ajne G, Wolff K, Fyhrquist F, et al. Endothelin converting enzyme (ECE) activity in normal pregnancy and preeclampsia. *Hypertens Pregnancy*. 2003;22:215–224.

59 Yoshida H, Nakamura M. Inhibition by angiotensin converting enzyme inhibitors of endothelin secretion from cultured human endothelial cells. *Life Sci*. 1992;50:PL195–PL200.

60 Cardillo C, Kilcoyne CM, Cannon RO 3rd, Panza JA. Increased activity of endogenous endothelin in patients with hypercholesterolemia. *J Am Coll Cardiol*. 2000;36:1483–1488.

61 Glorioso N, Troffa C, Filigheddu F, et al. Effect of the HMG-CoA reductase inhibitors on blood pressure in patients with essential hypertension and primary hypercholesterolemia. *Hypertension*. 1999;34:1281–1286.

62 Ciarla MV, Bocciarelli A, Di Gregorio S, et al. Autoantibodies and endothelial dysfunction in well-controlled, uncomplicated insulin-dependent diabetes mellitus patients. *Atherosclerosis*. 2001;158:241–246.

63 Donatelli M, Hoffmann E, Colletti I, et al. Circulating endothelin-1 levels in type 2 diabetic patients with ischaemic heart disease. *Acta Diabetol*. 1996;33:246–248.

64 Perfetto F, Tarquini R, de Leonardis V, et al. Angiopathy affects circulating endothelin-1 levels in type 2 diabetic patients. *Acta Diabetol*. 1995;32:263–267.

65 Cardillo C, Nambi SS, Kilcoyne CM, et al. Insulin stimulates both endothelin and nitric oxide activity in the human forearm. *Circulation*. 1999;100:820–825.

66 Scalera F. Intracellular glutathione and lipid peroxide availability and the secretion of vasoactive substances by human umbilical vein endothelial cells after incubation with TNF-alpha. *Eur J Clin Invest*. 2003;33:176–182.

67 Winkler G, Lakatos P, Salamon F, et al. Raised serum TNF-alpha level as a link between endothelial dysfunction and insulin resistance in normotensive obese patients. *Diabet Med*. 1999;16:207–211.

68 Schneider JG, Tilly N, Hierl T, et al. Raised plasma endothelin-1 levels in diabetes mellitus. *Am J Hypertens*. 2002;15:967–972.

69 Haak T, Jungmann E, Raab C, Usadel KH. Elevated endothelin-1 levels after cigarette smoking. *Metabolism*. 1994;43:267–269.

70 Hirai Y, Adachi H, Fujiura Y, et al. Plasma endothelin-1 level is related to renal function and smoking status but not to blood pressure: an epidemiological study. *J Hypertens*. 2004;22:713–718.

71 Cotran RS, Gimbrone MA Jr, Bevilacqua MP, et al. Induction and detection of a human endothelial activation antigen in vivo. *J Exp Med*. 1986;164:661–666.

72 Newman W, Beall LD, Carson CW, et al. Soluble E-selectin is found in supernatants of activated endothelial cells and is raised in the serum of patients with septic shock. *J Immunol*. 1993;150:644–654.

73 Cowley HC, Heney D, Gearing AJH, et al. Increased circulating adhesion molecule concentrations in patients with the systemic inflammatory response syndrome: a prospective cohort study. *Crit Care Med*. 1994;22:651–657.

74 Boehme MWJ, Schmitt WH, Youinou P, et al. Clinical relevance of Raised serum thrombomodulin and soluble E-selectin in patients with Wegener's granulomatosis and other systemic vasculitides. *Am J Med*. 1996;101:387–394.

75 Hackman A, Abe Y, Insull W, et al. Levels of soluble cell adhesion molecules in patients with dyslipidemia. *Circulation*. 1996;93:1334–1338.

76 Wang Y, Blessing F, Walli AK, et al. Effects of heparin-mediated extracorporeal low-density lipoprotein precipitation beyond lowering proatherogenic lipoproteins – reduction of circulating proinflammatory and procoagulatory markers. *Atherosclerosis*. 2004;175:145–150.

77 Lopez-Garcia E, Schulze MB, Fung TT, et al. Major dietary patterns are related to plasma concentrations of markers of inflammation and endothelial dysfunction. *Am J Clin Nutr*. 2004;80:1029–1035.

78 Winkelmann BR, Boehm BO, Nauck M, et al. Cigarette smoking is independently associated with markers of endothelial dysfunction and hyperinsulinaemia in nondiabetic individuals with coronary artery disease. *Curr Med Res Opin*. 2001;17:132–141.

79 Miller MA, Kerry SM, Cook DG, Cappuccio FP. Cellular adhesion molecules and blood pressure: interaction with sex in a multi-ethnic population. *J Hypertens*. 2004;22:705–711.

80 Leinonen E, Hurt-Camejo E, Wiklund O, et al. Insulin resistance and adiposity correlate with acute-phase reaction and

soluble cell adhesion molecules in type 2 diabetes. *Atherosclerosis*. 2003;166:387–394.
81 Smulders RA, Stehouwer CD, Schalkwijk CG, et al. Distinct associations of HbA1c and the urinary excretion of pentosidine, an advanced glycosylation end-product, with markers of endothelial function in insulin-dependent diabetes mellitus. *Thromb Haemost*. 1998;80:52–57.
82 Bluher M, Unger R, Rassoul F, et al. Relation between glycaemic control, hyperinsulinaemia and plasma concentrations of soluble adhesion molecules in patients with impaired glucose tolerance or Type II diabetes. *Diabetologia*. 2002;45:210–216.
83 Meigs JB, Hu FB, Rifai N, Manson JE. Biomarkers of endothelial dysfunction and risk of type 2 diabetes mellitus. *JAMA*. 2004; 291:1978–1986.
84 Schalkwijk CG, Smulders RA, Lambert J, et al. ACE-inhibition modulates some endothelial functions in healthy subjects and in normotensive type 1 diabetic patients. *Eur J Clin Invest*. 2000; 30:853–860.
85 Steiner M, Reinhardt KM, Krammer B, et al. Increased levels of soluble adhesion molecules in type (non-insulin) diabetes mellitus are independent of glycaemic control. *Thromb Haemost*. 1994;72:979–984.
86 Fasching P, Veitl M, Rohac M, et al. Elevated concentrations of circulating adhesion molecules and their association with microvascular complications in insulin-dependent diabetes mellitus. *J Clin Endocrinol Metab*. 1996;81:4313–4317.
87 Siminiak T, Dye JF, Egdell RM, et al. The release of soluble adhesion molecules ICAM-1 and E-selectin after acute myocardial infarction and following coronary angioplasty. *Int J Cardiol*. 1997;61:113–118.
88 Squadrito F, Altavilla D, Ioculano M, et al. Soluble E-selectin levels in acute human myocardial infarction. *Int J Microcirc Clin Exp*. 1995;15:80–84.
89 Cherian P, Hankey GJ, Eikelboom JW, et al. Endothelial and platelet activation in acute ischemic stroke and its etiological subtypes. *Stroke*. 2003;34:2132–2137.
90 Belch JJ, Shaw JW, Kirk G, et al. The white blood cell adhesion molecule E-selectin predicts restenosis in patients with intermittent claudication undergoing percutaneous transluminal angioplasty. *Circulation*. 1997;95:2027–2031.
91 Blankenberg S, Rupprecht HJ, Bickel C, et al. Circulating cell adhesion molecules and death in patients with coronary artery disease. *Circulation*. 2001;104:1336–1342.
92 Malik I, Danesh J, Whincup P, et al. Soluble adhesion molecules and prediction of coronary heart disease: a prospective study and meta-analysis. *Lancet*. 2001;358:971–975.
93 Roldan V, Marin F, Lip GYH, Blann AD. Soluble E selectin in cardiovascular disease and its risk factors. *Thromb Haemost*. 2003;90:1007–1020.
94 Torr-Brown SR, Sobel BE. Attenuation of thrombolysis by release of plasminogen activator inhibitor type-1 from platelets. *Thromb Res*. 1993;72:413–421.
95 Lundgren CH, Brown SL, Nordt TK, et al. Elaboration of type-1 plasminogen activator inhibitor from adipocytes. A potential pathogenetic link between obesity and cardiovascular disease. *Circulation*. 1996;93:106–110.
96 Blann AD, Dobrotova M, Kubisz P, McCollum CN. The relationship between soluble P-selectin, von Willebrand factor, tissue plasminogen activator and plasminogen activator inhibitor in atherosclerosis. *Thromb Haemost*. 1995;74:626–630.
97 Lowe GDO, Danesh J, Lewington S, et al. Tissue plasminogen activator antigen and coronary heart disease. *Eur Heart J*. 2004; 25:252–259.
98 de Bono D. Significance of raised plasma concentrations of tissue-type plasminogen activator and plasminogen activator inhibitor in patients at risk from ischaemic heart disease. *Br Heart J*. 1994;71:504–507.
99 Jeanneau C, Sultan Y. Tissue plasminogen activator in human megakaryocytes and platelets: immunocytochemical localization, immunoblotting and zymographic analysis. *Thromb Haemost*. 1988;59:529–534.
100 Jansson JH, Nilsson TK, Johnson O. von Willebrand factor in plasma: a novel risk factor for recurrent myocardial infarction and death. *Br Heart J*. 1991;66:351–355.
101 Brannstrom M, Jansson JH, Boman K, Nilsson TK. Endothelial haemostatic factors may be associated with mortality in patients on long-term anticoagulant treatment. *Thromb Haemost*. 1995; 74:612–615.
102 Wallberg-Jonsson S, Dahlen GH, Nilsson TK, et al. Tissue plasminogen activator, plasminogen activator inhibitor-1 and von Willebrand factor in rheumatoid arthritis. *Clin Rheumatol*. 1993;12:318–324.
103 Blann AD, Waite MA. von Willebrand factor and soluble E-selectin in hypertension: influence of treatment and value in predicting the progression of atherosclerosis. *Coron Artery Dis*. 1996;7:143–147.
104 Blann AD, Miller JP, McCollum CN. von Willebrand factor and soluble E-selectin in the prediction of cardiovascular disease progression in hyperlipidaemia. *Atherosclerosis*. 1997;132:151–156.
105 Jansson JH, Nilsson TK, Johnson O. von Willebrand factor, tissue plasminogen activator, and dehydroepiandrosterone sulphate predict cardiovascular death in a 10 year follow up of survivors of acute myocardial infarction. *Heart*. 1998;80:334–337.
106 Montalescot G, Philippe F, Ankri A, et al. Early increase of von Willebrand factor predicts adverse outcome in unstable coronary artery disease: beneficial effects of enoxaparin. French Investigators of the ESSENCE Trial. *Circulation*. 1998;98:294–299.
107 Blann AD, McCollum CN. von Willebrand factor and soluble thrombomodulin as predictors of adverse events among subjects with peripheral or coronary atherosclerosis. *Blood Coagul Fibrinolysis*. 1999;10:375–380.
108 Smith FB, Lee AJ, Hau CM, et al. Plasma fibrinogen, haemostatic factors and prediction of peripheral arterial disease in the Edinburgh Artery Study. *Blood Coagul Fibrinolysis*. 2000;11:43–50.
109 Tsakiris DA, Tschopl M, Wolf F, et al. Platelets and cytokines in concert with endothelial activation in patients with peripheral arterial occlusive disease. *Blood Coagul Fibrinolysis*. 2000; 11:165–173.
110 Johansson L, Jansson JH, Boman K, et al. Prospective study on soluble thrombomodulin and von Willebrand factor and the risk of ischemic and hemorrhagic stroke. *Thromb Haemost*. 2002;87:211–217.
111 Stehouwer CD, Gall MA, Twisk JW, et al. Increased urinary albumin excretion, endothelial dysfunction, and chronic low-grade inflammation in type 2 diabetes: progressive, interrelated, and independently associated with risk of death. *Diabetes*. 2002;51:1157–1165.

112 Niessner A, Graf S, Nikfardjam M, et al. Circulating t-PA antigen predicts major adverse coronary events in patients with stable coronary artery disease – a 13-year follow-up. *Thromb Haemost.* 2003;90:344–350.
113 Morange PE, Simon C, Alessi MC, et al. Endothelial cell markers and the risk of coronary heart disease: the Prospective Epidemiological Study of Myocardial Infarction (PRIME) study. *Circulation.* 2004;109:1343–1348.
114 Reinhart K, Bayer O, Brunkhorst F, Meisner M. Markers of endothelial damage in organ dysfunction and sepsis. *Crit Care Med.* 2002;30(Suppl 5):S302–S312.
115 George F, Brisson C, Poncelat P, et al. Rapid isolation of human endothelial cells from whole blood using S-Endo 1 monoclonal antibody coupled to immuno-magnetic beads: demonstration of endothelial injury after angioplasty. *Thromb Haemost.* 1992;67:147–153.
116 Mancuso P, Burlini A, Pruneri G, et al. Resting and activated endothelial cells are increased in the peripheral blood of cancer patients. *Blood.* 2001;97:3658–3661.
117 Makin A, Chung NAY, Silverman SH, et al. Assessment of endothelial damage in atherosclerotic vascular disease by quantification of circulating endothelial cells. *Eur Heart J.* 2004;25:371–376.
118 Chong AY, Blann AD, Patel J, et al. Endothelial dysfunction and damage in congestive heart failure: relation of flow-mediated dilation to circulating endothelial cells, plasma indexes of endothelial damage, and brain natriuretic peptide. *Circulation.* 2004;110:1794–1798.
119 Mutin M, Canavy I, Blann A, et al. Direct evidence of endothelial injury in acute myocardial infarction and unstable angina by demonstration of circulating endothelial cells. *Blood.* 1999;93:2951–2958.
120 Quilici J, Banzet N, Paule P, et al. Circulating endothelial cell count as a diagnostic marker for non-ST elevation acute coronary syndromes *Circulation.* 2004;110:1586–1591.
121 Lee KW, Lip GY, Tayebjee M, et al. Circulating endothelial cells, Von Willebrand Factor, interleukin-6 and prognosis in patients with acute coronary syndromes. *Blood.* 2005;105:526–532.
122 Jimenez JJ, Wenche J, Mauro LM, et al. Endothelial cells release phenotypically and quantitatively distinct microparticles in activation and apoptosis. *Thomb Res.* 2003;109:175–180.
123 Laurence J, Mitra D, Steiner M, et al. Plasma from patients with idiopathic and human immunodeficiency virus-associated thrombotic thrombocytopenic purpura induces apoptosis in microvascular endothelial cells. *Blood.* 1996:3245–3254.
124 Freyssinet JM. Cellular microparticles: what are they bad or good for? *J Thromb Haemost.* 2003;1:1655–1662.
125 Taraboletti G, D'Ascenzo S, Borsotti P, et al. Shedding of the matrix metalloproteinases MMP-2, MMP-9 and MT1-MMP as membrane vesicle-associated components by endothelial cells. *Am J Pathol.* 2002;160:673–680.
126 Mallat Z, Benamer H, Hugel B, et al. Elevated levels of shed membrane microparticles with procoagulant potential in the peripheral circulating blood of patients with acute coronary syndromes. *Circulation.* 2000;101:841–843.
127 Casciola-Rosen L, Rosen A, Petri M, et al. Surface blebs on apoptotic cells are sites of enhanced procoagulant activity: implications for coagulation events and antigenic spread in systemic lupus erythematosus. *Proc Nat Acad Sci USA.* 1996;93:1624–1629.
128 Jimenez JJ, Wenche J, Mauro LM, et al. Endothelial microparticles released in thrombotic thrombocytopaenic purpura express von Willebrand factor and markers of endothelial activation. *Br J Haematol.* 2003;123:896–902.
129 Shet AS, Aras O, Gupta K, et al. Sickle blood contains tissue factor-positive microparticles derived from endothelial cells and monocytes. *Blood.* 2003;102:2678–2683.
130 Boulanger CM, Scoazec A, Ebrahimian T, et al. Circulating microparticles from patients with myocardial infarction cause endothelial dysfunction. *Circulation.* 2001;104:2649–2652.
131 Hwa C, Sebastian A, Aird WC. Endothelial biomedicine: its status as an interdisciplinary field, its progress as a basic science, and its translational bench-to-bedside gap. *Endothelium.* 2005; 12:139–151.
132 Yarnell J, McCrum E, Rumley A, et al. Association of European population levels of thrombotic and inflammatory factors with risk of coronary heart disease: the MONICA Optional Haemostasis Study. *Eur Heart J.* 2005;26:332–342.

174

Blood Endothelial Cells

Robert D. Simari*, Rajiv Gulati†, and Robert P. Hebbel‡

*Mayo Clinic College of Medicine, Rochester, Minnesota; † University of Birmingham, United Kingdom; ‡ University of Minnesota Medical School, Minneapolis

In 1963, in an attempt to define the source of endothelium on vascular grafts, Stump and colleagues suspended a Dacron patch within the lumen of a prosthetic vascular graft in the aorta of a juvenile pig (1). As early as 14 days following placement, islands of endothelial cells (ECs) were identified on the patch surface. Because the patch had been isolated from contact with both the prosthesis and native vascular tissue, these findings implicated circulating blood as the source of ECs. These observations were not actively pursued from an experimental or clinical context until recently.

This original observation of a vascular source of endothelium has been corroborated in chimeric transplantation models that have enabled the discrimination of host- and donor-derived cells by genetic markers. These studies have revealed bone marrow–derived circulating progenitors to contribute to both endothelial and intimal smooth muscle cell formation in multiple models of vascular injury as reviewed by Sata (2). Moreover, treatment with 3-hydroxy-3-methyl-glutaryl-CoA (HMG-CoA) reductase inhibitors (statins) appears to accelerate the incorporation of bone marrow–derived ECs following arterial denudation in rodent models (3,4). These studies confirm the presence of cells with either an endothelial phenotype or endothelial potential within human blood. Furthermore, strategies have been developed to utilize these cells for the prevention and treatment of vascular disease. In this chapter, identification, classification, and potential translational uses of these cells will be discussed.

CIRCULATING ENDOTHELIAL CELLS

Definition and Phenotype

The blood of normal individuals contains circulating ECs (CECs) as well as monocytic cells with the potential to develop endothelial features in culture (culture-modified mononuclear cells – CMMCs) and progenitors capable of differentiation into ECs (so-called *true endothelial progenitor cells* [EPCs]). Investigators studying CECs have recently produced a consensus report (5). Interpretation of this literature is dependent, of course, on methods used to define CECs. Older studies, using identification methods limited to histochemical staining and morphology, suggested a frequency of around 1 CEC per 1 million nucleated cells in normal blood. Modern studies, using monoclonal antibodies to identify CECs, have confirmed this frequency (5). A variety of surface markers have been used to identify CECs. Most commonly this has been carried out using monoclonal antibodies to CD146 plus another endothelial marker (e.g., vascular endothelial [VE]-cadherin or von Willebrand factor [vWF]). Insofar as *true* EPCs exhibit some phenotypic overlap with mature ECs in terms of markers (6), it may be that CEC counting includes some progenitors. However, available data on the latter cell type suggest they are represented in blood at much lower frequency than are CECs.

Based on the literature (5) and our own experience (7), we believe that application of rigorous and appropriate methods should identify <10 CECs per mL of blood from normal individuals. Then, there are outlier studies, such as one which applied fluorescence-activated cell sorting (FACS) technology and detected up to 7,000 "CECs" per mL in normals (8). It is probable that the latter study was identifying microparticles, if not outright artifacts, rather than true CECs. We, therefore, recommend a definition that includes the CECs actually being shown to be nucleated (7), and we have used this in our own studies described in the next section. FACS has been successfully used to enumerate CECs (9).

Origin of Circulating Endothelial Cells

It is believed that CECs are derived from the vessel wall. Evidence for this was provided by examination of sex chromosomes in CECs of currently healthy humans who had previously undergone a gender-mismatched bone marrow transplant (10). Virtually all CECs exhibited the recipient genotype. This presumably indicates a natural, low-level turnover of endothelium, in which injured or apoptotic cells depart from the vessel wall. The residence time of CECs in blood has not been directly measured, but the rapid changes in CEC counts we have observed in sickle cell anemia suggest rapid turnover. Elevated CEC counts have been reported in a broad spectrum of inflammatory, cardiovascular, infectious, and blood diseases. Examples include septic shock, lupus, vasculitis, scleroderma, coronary syndromes and manipulation, venous insufficiency, arteritis, pulmonary hypertension, peripheral vascular disease, cytomegalovirus (CMV), Rickettsial infection, sickle cell anemia, and thalassemia. A common thread in the pathophysiology of these conditions is the existence of vascular damage, which may account for the increased CEC counts (5). However, in the absence of genetic marking data for CECs in these various disease states, the true origin of these cells remains speculative.

Usefulness of Circulating Endothelial Cell Enumeration

Individual CEC counts should be interpreted cautiously. For example, in sickle cell anemia we found that CEC counts fluctuated chaotically for individual subjects followed longitudinally (7). In fact, sickle cell anemia is a disease in which vascular damage is believed to be chronic and ongoing (and to fluctuate in intensity), so it perhaps is not surprising to see significant changes in CEC counts in that disease. However, if this observation holds true for other disease situations (and this is not known), this indicates that single time-point CEC counts would be of limited value for an individual patient.

On the other hand, examination of groups of subjects reveals instructive trends. For example, CEC counts tend to be higher in rough proportion to disease activity or severity in sickle cell anemia, coronary syndromes, peripheral artery disease, Mediterranean spotted fever, lupus, and inflammatory vasculitis, among others (5). Encouragingly, CEC counts have been observed to correlate with plasma vWF level in patients with cardiovascular disease (11). If confirmed and extended, such associations may reveal that CEC level is a measure of degree of vascular injury. In studies of lupus patients, number of CECs was observed to correlate with brachial artery endothelial dysfunction (9). However, before CEC counts can be used diagnostically, it will be important to establish the temporal stability or lack thereof in any given disease situation.

Clinical Usefulness of Circulating Endothelial Cell Phenotype

We have demonstrated that CEC phenotype can be used as a surrogate marker of vessel wall endothelium activation status (7,12). We find that CEC phenotype is the same for live versus dead, or for apoptotic versus nonapoptotic cells. So we believe that phenotype is acquired before cell death and has been governed by the same milieu as has the vessel wall endothelium. Such phenotyping possibly can be instructive in a variety of situations.

Two-thirds of CECs in normals exhibit evidence of apoptosis (TUNEL positivity and fragmented nuclei). Undoubtedly this reflects the EC's well-known intolerance of being unanchored, so that once dislodged from vessel wall it rapidly undergoes apoptosis. In sickle disease patients, only one-third of CECs are apoptotic (13). Although this could reflect the presence of a younger population of CECs (e.g., due to ongoing vascular insult), it is interesting that such a degree of CEC apoptosis correlated inversely with plasma levels of vascular endothelial growth factor (VEGF), a major survival and apoptosis-inhibiting factor for ECs. We suggest that monitoring CEC apoptosis could be a surrogate indicator for "apoptotic tone" of the vessel wall, possibly useful for following disease activity or antiangiogenic therapeutics.

Analysis of CECs might provide insights into the pathogenesis of disease. For instance, CECs express tissue factor in sickle cell anemia, but not in normal subjects (14) (Figure 174.1; for color reproduction, see Color Plate 174.1). Studies of CECs in sickle cells anemia reveal the spectrum of phenotypic features that can be evaluated on CECs – e.g., antigen expression, mRNA presence, binding function, and biochemical activity of tissue factor (Figure 174.1). Endothelial adhesion molecules (intercellular adhesion molecule [ICAM]-1, vascular cell adhesion molecule [VCAM]-1, selectins) are abnormally expressed by CECs in sickle cell anemia (7) and scleroderma (9). In lupus, CECs are described as having lowered levels of endothelial protein C receptor (5). Such findings are consistent with the procoagulant states of these diseases.

Biological Significance of Circulating Endothelial Cells

CECs are present in such small numbers that it seems highly unlikely that they would directly have an impact on disease pathobiology – e.g., as procoagulant cells. On the other hand, we believe that CECs, as surrogate markers, provide insight into the biology of endothelium remaining in the vessel wall. Direct evidence for this is limited to sickle transgenic mice, where CECs had an activated/pro-adhesive phenotype and so did tissue endothelium (12).

One wonders about the state of the vessel wall from which CECs have been liberated. Although there are no data that address this, the CEC counts achieved in the sickle cell disease model (up to 70/mL) would represent perhaps 20 cm^2 of denuded vessel wall, which is not a trivial amount. We will not be surprised if CEC counts are eventually found to correlate with thrombotic risk because of the concomitant denuded surface.

ENDOTHELIAL-LIKE CELLS GROWN FROM BLOOD: BLOOD OUTGROWTH ENDOTHELIAL CELLS AND ENDOTHELIAL PROGENITOR CELLS

Endothelial-like cells can be generated from continued culture of peripheral blood samples from human (10), sheep (15), mice (16), rabbit (17), and dog (R. Hebbel, unpublished observations) in endothelial growth conditions. For clarity, we use the term *blood outgrowth ECs* (BOECs) for homogenous populations of cells generated in culture having endothelial morphology (typically cobblestone) and expressing markers

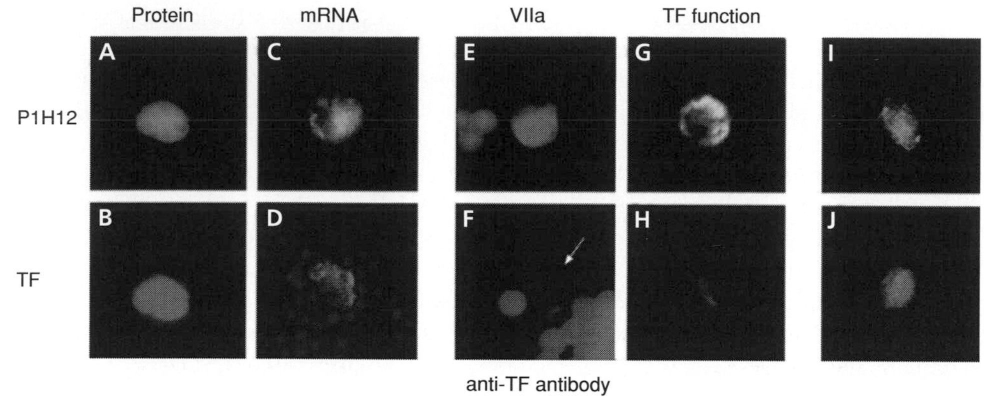

Figure 174.1. CECs express tissue factor. These studies of tissue factor expression by CECs illustrate the spectrum of methods that can be productively applied to these cells. A single cell is shown to be positive for endothelial marker P1H12 (**A**) and tissue factor antigen (**B**). CECs identified via P1H12 show a positive hybridization signal for *tissue factor* mRNA (**C**), which is not seen for control probe (**D**). Fluorescent factor VIIa binds to tissue factor-positive CECs (**E**), which is blocked by anti-tissue factor antibody (**F**). A chromogenic assay shows tissue factor function on tissue factor antigen–positive CECs (**G**) which is blocked by anti-tissue factor antibody (**H**). Tissue factor antigen–positive CECs show a positive hybridization signal (**I**) that is not seen for control probe (**J**). For color reproduction, see Color Plate 174.1.

of mature differentiated ECs, but not myelomonocytic markers such as CD14 or CD45, nor the stem cell marker AC133. In contrast, shorter-term cultures of peripheral blood can also yield a heterogeneous population of spindle-shaped cells that share endothelial and myelomonocytic markers. This heterogeneous population of spindle-shaped cells (18), often mistakenly referred to as EPCs, are also known as CMMCs. BOECs are most likely derived from true endothelial progenitors, whereas CMMCs arise from monocytes (18). Additionally, although CECs have limited proliferative potential, they cannot be formally excluded as a source of endothelial-like cells obtained in culture.

Blood Outgrowth Endothelial Cells

Lin and colleagues first demonstrated robust true endothelial outgrowth from human blood, yielding 10^{19} BOECs (10). Examination of the genotype of BOECs grown from blood in humans who had previously received a gender-mismatched marrow transplant allowed for identification of the bone marrow as the source of the outgrowth cells. The strategy for generating BOECs in vitro is highly reproducible. Using unfractionated peripheral blood, mononuclear cells are plated onto collagen I and grown with 2% fetal bovine serum and endothelial growth factors. Nonadherent cells are discarded on the second day of culture. Endothelial growth exhibits two phases. The early outgrowth phase is relatively slow and includes limited expansion of recipient-derived CECs. The second, delayed phase consists of robust outgrowth from a transplantable marrow-derived cell in peripheral blood, a putative bona fide EPC (Figure 174.2). During this second phase, cells acquire cobblestone morphology and express distinct endothelial markers, including vWF, thrombomodulin, CD34, CD31, CD36, VE-cadherin, α_v integrin, FLK-1, and P1H12. Moreover, they contain Weibel-Palade bodies. BOECs are negative for CD14 (a monocyte marker), CD45 (a marker for hematopoietic cells), and AC133 (a stem cell marker) (Figure 174.3). Thus, BOECs share markers of differentiated ECs, do not possess myelomonocytic markers (typical of loosely defined EPCs), and have a cobblestone morphology.

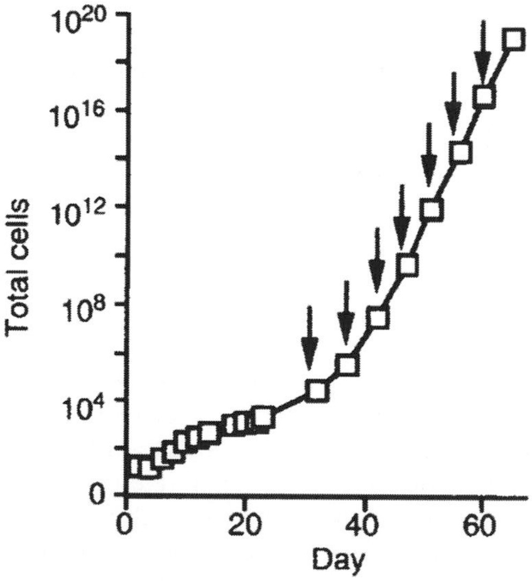

Figure 174.2. Endothelial expansion from buffy coat mononuclear cells of normal blood. On day 2 and for subsequent passages (*arrows*), the number of ECs was confirmed by staining for P1H12 and vWF, and that number was consistent with the cell count by morphology. All data points plotted as mean $\pm$ SD (n = 5 for culture up to passage 6; n = 4 for subsequent passages).

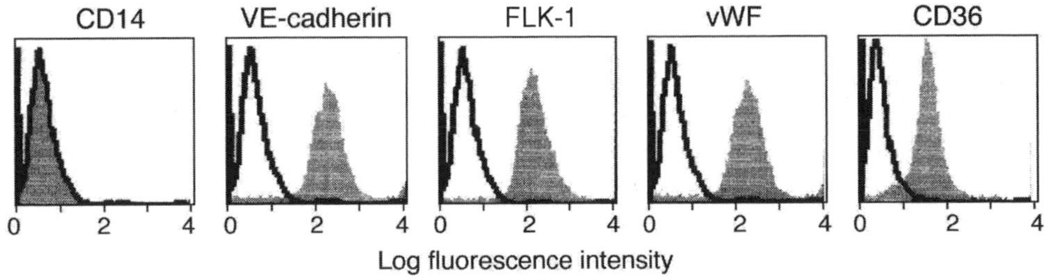

Figure 174.3. Flow cytometry analysis of outgrowth EC phenotype. In each graph, the black line outlines the region of fluorescent intensity for cells labeled with negative control antibody. The filled region identifies cells labeled with antibody for the expression marker indicated above each graph. Outgrowth cells are negative for CD14 (monocyte marker) and positive for FLK-1, vWF, CD36, and the endothelial-specific marker VE-cadherin.

Endothelial Progenitor Cells

For purposes of clarity, we propose that the term *EPCs* be restricted to putative endothelial progenitors located in either peripheral blood or marrow. The putative circulating EPCs and marrow EPCs may well be the same, as there is precedent for spillage of marrow stem cells into the peripheral circulation. Unfortunately, the literature on EPCs is greatly confused due to application of inappropriate terminology and unproven assumptions.

"True" Endothelial Progenitor Cells

There is strong in vitro and in vivo evidence – including the BOEC findings described – for presence of "true" EPCs in blood and bone marrow (6,10,19–22). Phenotypic data need to take into account that few markers are endothelial-specific. In aggregate, data suggest that the true EPC has the phenotype CD34$^+$, FLK-1$^+$, FGFR1$^+$, AC133$^+$. The population of blood and marrow cells that carries these markers tends to also express c-kit, Tie2, P1H12, and VE-cadherin, but to be negative for vWF and uptake of acetylated low-density lipoprotein (acLDL). The relevant cell may reside in the c-kit bright subpopulation (21). It should be noted that all studies of EPCs to date have been performed on bulk populations of cells and have not yet been critically tested by studies using single-cell culture.

A stem cell should be self-renewing, multipotent, and transplantable. True endothelial progenitors can be transplanted in humans (10,23) and animals (24), but whether the true EPC has self-renewing capacity has not been defined. EPCs may be committed progenitors rather than a self-renewing or pluripotent progenitor cell. It has been suggested that the source of EPCs is a hemangioblast, analogous to that seen in early development (25,26). Evidence suggests that adult mice do have hemangioblasts, able to be transplanted as a single cell and able to generate both endothelial and hematopoietic lineage cells (24). A report from human studies will require confirmation (27). Other possibilities exist. The enormous plasticity of some marrow cells has been noted (28), and multipotent adult progenitor cells (MAPCs) have recently been generated in culture and are capable of generating ECs (29).

Culture Modified Mononuclear Cells

Asahara and colleagues established cultures of CD34$^+$ cells (but only at 16% purity in a population of mononuclear cells) from blood and obtained modified cells they referred to as *EPCs* (16). Unfortunately, no evidence was provided to support the designation of these cells as progenitors. Moreover, unlike BOECs, the morphology of these cells was spindle shaped, and the cells expressed rather low levels of endothelial markers and CD34 and very high levels of myelomonocytic markers – e.g., 27% CD45 positive and approximately 90% CD14 positive (16,30,31). It would seem that these "EPCs" are more likely to have been CMMCs, absent evidence to the contrary.

In our opinion, the term *EPC* has been inappropriately applied in a number of recent publications to cells that are actually CMMCs (e.g., 16,30–36) or even BOECs (15). The term *EPC* has been repeatedly applied to cells within short-term cultures of peripheral blood mononuclear cells that exhibit dual-positivity for uptake of acLDL and *Ulex europaeus* lectin binding. In fact, Rehman and colleagues used the Asahara method but found that the majority of "EPCs" expressed monocyte/macrophage markers and that the dual labeled acLDL/*U. europaeus* positive cells are derived from monocytes/macrophages (37). Taken together, we regard this as insufficient evidence for an EPC, because neither of these markers is specific for ECs and, in fact, both also identify monocytes (37,38), the cell population within which the so-called EPCs reside. In our experience, short-term cultures of peripheral blood mononuclear cells are dominated by cells of monocytic origin.

Other spindle-shaped cells are characteristic of Kaposi sarcoma and may be obtained by culturing blood (39). The origin of these cells is thought to be from cells lining splenic sinuses. The culture of these cells is promoted by cytokines (39) and simulated by high concentrations of growth factors used by some to culture "EPCs" – e.g., high serum concentrations. CMMCs grown from blood are elongated, and they exhibit both endothelial and myelomonocytic markers (e.g., CD14) (39). Such cells at best can be regarded only as "endothelial-like" cells.

Fernandez-Pujol and colleagues reported culture of CD14^{-} positive peripheral blood monocytes on fibronectin with endothelial growth factors. The resulting cells were of mixed morphologies, including spindle shapes, and were called "endothelial-like" because they exhibited not only endothelial markers but also CD14 and CD68 (40). Similarly, culture of monocytes on fibronectin and with endothelial growth factors (41) or bovine brain extract (42) yields spindle-shaped cells that share endothelial and macrophagocytic lineage markers. Interestingly, CMMCs derived from CD34^{-} cells can be incorporated into vessel walls as endothelial-like cells, as long as there is help from cocultured CD34^{+} positive cells (42). It is also important to note that CMMCs exhibit endothelial functions such as tube formation in vitro, as well as uptake of acLDL and binding of *U. europaeus* lectin, markers sometimes used with the unwarranted assumption that they are endothelial specific.

THERAPEUTIC USES OF BLOOD OUTGROWTH ENDOTHELIAL CELLS AND ENDOTHELIAL-LIKE CELLS OBTAINED FROM BLOOD

Therapeutic vascular interventions are limited by sequelae initiated by mechanical damage to the vessel wall. Disruption of the endothelial layer exposes the subendothelial surface and induces a cascade of events, resulting in impaired vasorelaxation, thrombus formation, and intimal hyperplasia. Restoration of endothelial integrity has been shown to modify these events in multiple animal models, resulting in improved vascular remodeling.

Re-endothelization of injured vascular segments is associated with attenuation of neointimal formation. Indeed, early restoration of endothelial integrity through local delivery of VEGF was associated with reduced intimal formation, providing evidence to support a role for re-endothelization in suppressing intimal hyperplasia (43). This approach is undergoing phase 1 clinical evaluation following both coronary and peripheral intervention (44,45).

In addition to indirect approaches to enhance endothelization such as growth factor delivery, direct strategies have been used to enhance re-endothelization at the site of injury. The ability to generate BOECs and endothelial-like cells from peripheral blood supports a strategy that obviates surgical harvest of mature endothelium. In a rabbit model, we generated CMMCs from peripheral blood, demonstrating this population to contain true precursors to BOECs (46). Balloon injury to a clamp-isolated segment of carotid artery was performed under direct vision, followed by intraluminal administration of autologous CMMCs (cultured for seven days from peripheral blood). CMMCs were then allowed to dwell for 20 minutes prior to restoring normal arterial flow (Figure 174.4; for color reproduction, see Color Plate 174.4). By labeling CMMCs with a fluorescent membrane dye prior to administration, we were able to identify delivered cells in the vessel wall up to 4 weeks after injury. Labeled cells were seen lining the lumen expressing markers of endothelial – but not monocyte/macrophage – lineage. It could not be determined whether these cells arose from monocytic transdifferentiation or from true, rare endothelial precursors. Labeled cells were also detected in deeper layers expressing macrophage but not endothelial markers, indicating either defined adoption of macrophage phenotype or phagocytosis of delivered (labeled) cells by resident macrophages.

Encouraged by the detection of delivered cells up to 4 weeks following delivery, we proceeded to evaluate the effect of CMMCs on parameters of vascular structure and function, comparing local delivery of a saline-cell suspension immediately after balloon injury with saline alone as a control. To evaluate the effect of CMMC delivery on vascular reactivity we examined excised carotid rings in an organ chamber 4 weeks after balloon injury and cell delivery. CMMC delivery was associated with a marked improvement in endothelium-dependent vasorelaxation compared with saline, although vasorelaxation responses did not achieve those of normal, uninjured vessels (Figure 174.5). Interestingly, we also demonstrated that CMMC delivery was associated with a 55% reduction in neointimal thickening. Finally, by exploiting the ability of endothelium to block extravasation of Evans blue dye, we determined that CMMC delivery was associated with significantly enhanced arterial re-endothelization (46).

BOECs have also been tested for their vasculoprotective effects. Kaushal and colleagues demonstrated that seeding of decellularized vascular grafts with autologous ovine BOECs dramatically improved graft patency (15). We and others have evaluated the effect of local BOEC delivery following balloon injury of a native artery in a rabbit model (47,48). We hypothesized that BOEC delivery after arterial injury may modify the vascular response in an endothelium-dependent manner. Importantly, we also elected to compare the effect of delivering peripheral blood mononuclear cells (PBMCs) as a control for cultured cell therapy. Our results, together with those of Griese and colleagues, indicated that BOEC delivery was associated with an overall reduction in neointimal formation compared to delivery of both saline (47,48) and PBMCs (47).

It is interesting to note that the functional and structural effects of BOEC delivery are almost identical with those of CMMC delivery (46,47). Acknowledging that this comparison has limitations, it remains somewhat surprising that BOECs, with their more distinct endothelial properties and with their delivery in approximately 10-fold greater numbers than CMMCs, were not clearly superior in their vascular protection, as measured by vasorelaxation and neointimal formation.

Other investigators have tested the strategy of vasoprotection with blood-derived ECs using complementary approaches. Fujiyama transplanted subpopulations of human mononuclear cells (MNCs) into balloon-injured arteries of athymic rats (49) following pretreatment with gene transfer of monocyte chemoattractant protein (MCP)-1. The authors demonstrated that local delivery of CD14^{+} monocytes was associated with accelerated endothelization and reduced

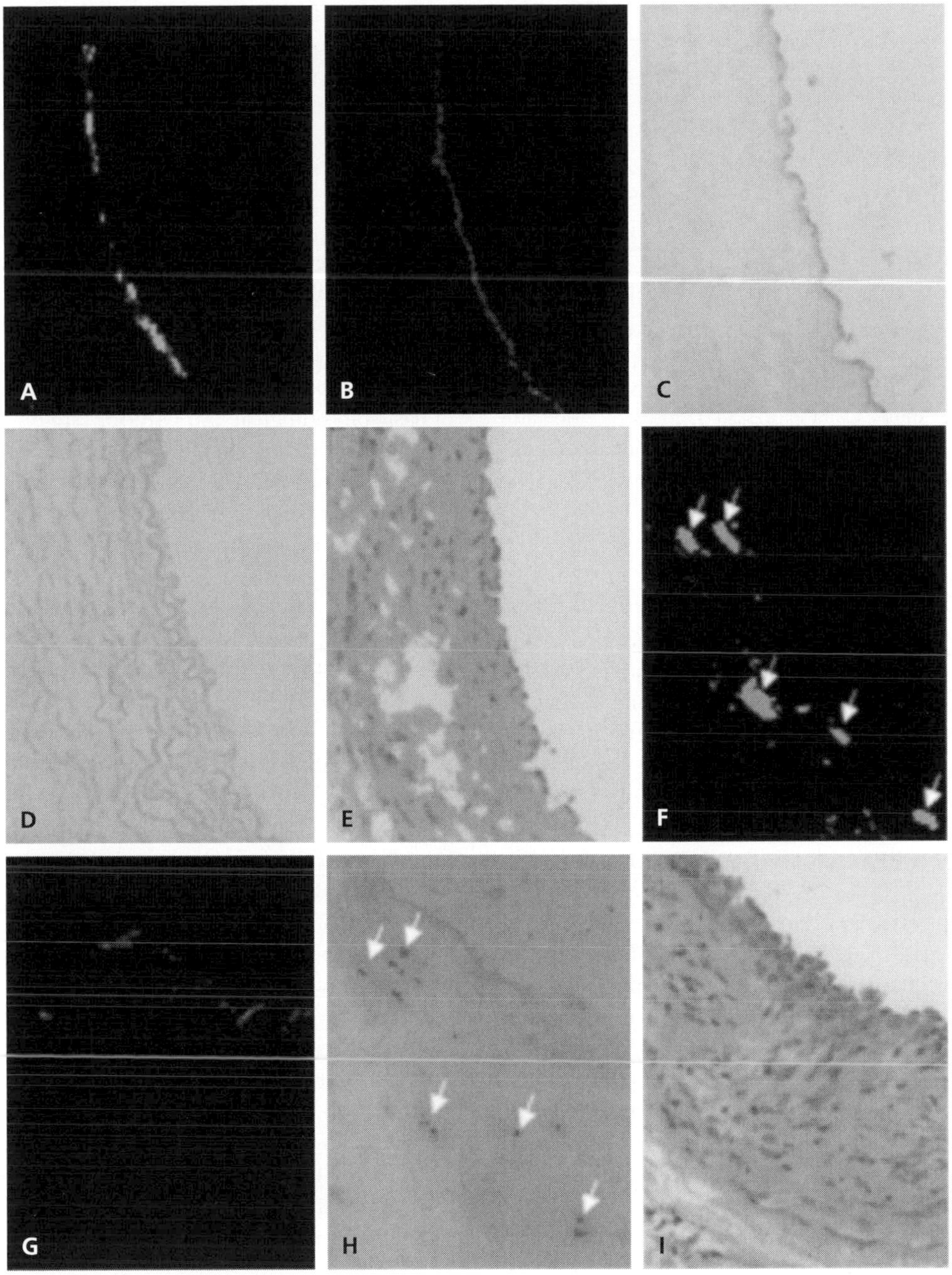

Figure 174.4. Administered CMMCs incorporate into vessel wall after balloon carotid injury. (**A**) Carotid section demonstrating labeled CMMCs on luminal border 4 weeks after local delivery. Colocalization staining of endothelial antigens CD31 (**B**) and BS-1 lectin (**C**) but (**D**) negative staining for macrophage marker RAM-11. (**E**) Hematoxylin, and eosin staining of adjacent arterial section. (**F**) Labeled cells also detected in neointima that (**G**) do not costain for CD31 but (**H**) do stain for RAM-11, consistent with macrophage lineage. (**I**) Hematoxylin and eosin staining of adjacent arterial section. Gray indicates CM-DiI fluorescence; light gray, CD31; arrows, colocalization of neointimal CMMCs with RAM-11. Magnification ×40. For color reproduction, see Color Plate 174.4.

neointimal formation compared to delivery of $CD34^+$ cells or unselected MNCs. It was suggested that the delivered monocytes were able to "transdifferentiate" into mature ECs in vivo. This was concluded from the coexpression of endothelial antigens and lack of monocytic markers on prelabeled delivered cells. However, is not possible to exclude an effect of MCP-1 in explaining these observations. Moreover, delivered monocytes may have fused with resident endothelium, as has been proposed in other studies (50). In a splenectomized mouse model, Werner determined that systemic delivery of cultured spleen-derived mononuclear cells homed to mechanically injured vessel segments. Cell administration accelerated

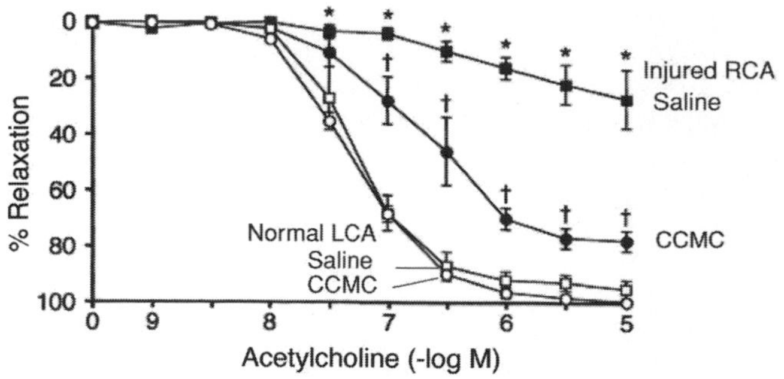

Figure 174.5. Enhanced endothelium-dependent vasoreactivity in carotid rings from CMMC-treated rabbits. Vessels from CMMC-treated rabbits show markedly enhanced vasoreactivity to acetylcholine 4 weeks after injury. Data expressed as mean % relaxation ± SEM (*p <0.05 for CMMC- vs. saline-treated injured arteries). Uninjured left carotid arteries retained largest responses ($p < 0.05$ for maximal relaxation and EC_{50} compared with CMMC rings). RCA indicates right carotid artery; LCA, left carotid artery.

re-endothelization and reduced neointimal formation at the site of injury (51). These recent findings underscore the need for additional studies aimed at determining whether monocyte–endothelial transdifferentiation is a genuine pathophysiological process.

MECHANISM OF EFFECTS OF CELL DELIVERY

The modification of vascular responses by cell therapy is likely to be at least in part related to restoration of endothelial integrity and the resultant barrier properties conferred. Additionally, ECs may modulate vascular structure and function through secreted factors (52). Given this, it is entirely possible that nonluminally located ECs (47) may also contribute to the vascular protection conferred by cell delivery. Furthermore, it is not clear how much endothelial resurfacing relates directly to incorporation of delivered cells. For example, Griese (48) used retroviral labeling to enable the identification of delivered BOECs following balloon injury and demonstrated at 4 weeks no labeled cells were apparent in the same location. Despite this apparent loss of delivered cells, overall arterial re-endothelization was enhanced. Similarly, Conte noted loss of genetic label following delivery of cultured venous ECs in a rabbit model (53). Similarly, we found CMMC-derived cells to be detectable on a minority of the luminal arterial surface 4 weeks after balloon injury and cell delivery, despite markedly increased endothelization (46). Although loss of label or genetic marker may account for some of these findings, it is also likely that a potential mechanism of the benefit seen with cell delivery occurs through paracrine mechanisms. Cultured monocyte-lineage cells have been shown to secrete growth factors and cytokines in vitro (37). These factors include a variety of growth factors and cytokines that may promote vascular healing (54). These findings demonstrate the need for more studies to define the relative contribution of direct re-endothelization and associated paracrine mechanisms involved in delivery of BOECs and endothelial-like cells in models of vascular disease.

CONCLUSION

The dogma that vascular cells are limited to the vasculature and have vascular precursors has been proven incorrect. Normal human blood contains ECs and cells with the potential to develop an endothelial phenotype in culture. Our understanding of blood ECs and cells capable of developing an endothelial phenotype will increase our fundamental understanding of basic physiological and pathophysiological mechanisms of hematologic and vascular disease. In addition, the understanding of the origin and function of these cells has led to novel diagnostic and therapeutic opportunities. Yet, the hopes for new therapies would be best served by a shared accuracy of definitions and lack of hyperbolic claims of potency.

KEY POINTS

- Circulating blood contains mature ECs from vascular origins.
- Evidence suggests that circulating blood contains "bona fide" endothelial progenitors that have not been fully characterized.
- Circulating monocytic cells can develop endothelial features in culture and these cells have been mislabeled "endothelial progenitor cells."
- Autologous delivery of cells with endothelial features has been shown to modify vascular form and function with translational potential.

Future Goals

- To identify "bona fide" bone marrow and circulating endothelial precursors and effect the translation of strategies that use these cells for autologous vascular repair

REFERENCES

1 Stump MM, Jordan GL, DeBakey ME, Halpert B. Endothelium grown from circulating blood on isolated intravascular Dacron hub. *Am J Pathol.* 1963;43:361–363.
2 Sata M, Walsh K. Circulating vascular progenitor cells contribute to vascular repair, remodeling, and lesion formation. *Trends Cardiovasc Med.* 2003;13:249–253.
3 Walter DH, Rittig K, Bahlmann FH, et al. Statin therapy accelerates reendothelialization: a novel effect involving mobilization and incorporation of bone marrow-derived endothelial progenitor cells. *Circulation.* 2002;105:3017–3024.

4 Werner N, Priller J, Laufs U, et al. Bone marrow-derived progenitor cells modulate vascular reendothelialization and neointimal formation: effect of 3-hydroxy-3-methylglutaryl coenzyme A reductase inhibition. *Arteriolscler Thromb Vasc Biol.* 2002;22: 1567–1572.
5 Blann AD, Woywodt A, Bertolini F, et al. Circulating endothelial cells. Biomarker of vascular disease. *Thromb Haemost.* 2005; 93:228–235.
6 Burger PE, Coetzee S, McKeehan WL, et al. Fibroblast growth factor receptor-1 is expressed by endothelial progenitor cells. *Blood.* 2002;100:3527–3535.
7 Solovey A, Lin Y, Browne P, et al. Circulating activated endothelial cells in sickle cell anemia. *N Engl J Med.* 1997;337:1584–1590.
8 Mancuso P, Burlini A, Pruneri G, et al. Resting and activated endothelial cells are increased in the peripheral blood of cancer patients. *Blood.* 2001;97:3658–3661.
9 Rajagopalan S, Somers EC, Brook RD, et al. Endothelial cell apoptosis in systemic lupus erythematosus: a common pathway for abnormal vascular function and thrombosis propensity. *Blood.* 2004;103:3677–3683.
10 Lin Y, Weisdorf DJ, Solovey A, Hebbel RP. Origins of circulating endothelial cells and endothelial outgrowth from blood. *J Clin Invest.* 2000;105:71–77.
11 Makin AJ, Blann AD, Chung NA, et al. Assessment of endothelial damage in atherosclerotic vascular disease by quantification of circulating endothelial cells. Relationship with von Willebrand factor and tissue factor. *Eur Heart J.* 2004;25:371–376.
12 Solovey AA, Solovey AN, Harkness J, Hebbel RP. Modulation of endothelial cell activation in sickle cell disease: a pilot study. *Blood.* 2001;97:1937–1941.
13 Solovey A, Gui L, Ramakrishnan S, et al. Sickle cell anemia as a possible state of enhanced anti-apoptotic tone: survival effect of vascular endothelial growth factor on circulating and unanchored endothelial cells. *Blood.* 1999;93:3824–3830.
14 Solovey A, Gui L, Key N, Hebbel R. Tissue factor expression by endothelial cells in sickle cell anemia. *J Clin Invest.* 1998;101: 1899–1904.
15 Kaushal S, Amiel GE, Guleserian KJ, et al. Functional small-diameter neovessels created using endothelial progenitor cells expanded ex vivo. *Nat Med.* 2001;7:1035–1040.
16 Asahara T, Murohara T, Sullivan A, et al. Isolation of putative progenitor endothelial cells for angiogenesis. *Science.* 1997;275:964–967.
17 Shintani S, Toyoaki M, Hisao I, et al. Augmentation of postnatal neovascularization with autologous bone marrow transplantation. *Circulation.* 2001;103:897–903.
18 Gulati R, Jevremovic D, Peterson TE, et al. Diverse origin and function of cells with endothelial phenotype obtained from adult human blood. *Circ Res.* 2003;93:1023–1025.
19 Gehling U, Ergun S, Schumacher U, et al. *In vitro* differentiation of endothelial cells from AC133-positive progenitor cells. *Blood.* 2000;95:3106–3112.
20 Peichev M, Naiyer A, Pereira D, et al. Expression of VEGFR-2 and AC133 by circulating human CD34[1] cells identifies a population of functional endothelial precursors. *Blood.* 2000;95:952–958.
21 Kocher AA, Schuster MD, Szabolcs MJ, et al. Neovascularization of ischemic myocardium by human bone-marrow-derived angioblasts prevents cardiomyocyte apoptosis, reduces remodeling and improves cardiac function. *Nat Med.* 2001;7:412–430.
22 Quirici N, Soligo D, Caneva L, et al. Differentiation and expansion of endothelial cells from human bone marrow CD133($^{+}$) cells. *Br J Haematol.* 2001;115:186–194.
23 Ikpeazu C, Davidson MK, Halteman D, et al. Donor origin of circulating endothelial progenitors after allogeneic bone marrow transplantation. *Biol Blood Marrow Transplant.* 2000;6:301–308.
24 Grant MB, May WS, Caballero S, et al. Adult hematopoietic stem cells provide functional hemangioblast activity during retinal neovascularization. *Nat Med.* 2002;8:607–612.
25 Choi K, Kennedy M, Kazarov A, et al. A common precursor for hematopoietic and endothelial cells. *Development.* 1998;125: 725–732.
26 Nishikawa SI, Nishikawa S, Hirashima M, et al. Progressive lineage analysis by cell sorting and culture identifies FLK1^{+}VE-cadherin^{+} cells at a diverging point of endothelial and hemopoietic lineages. *Development.* 1998;125:1747–1757.
27 Pelosi E, Valtieri M, Coppola S, et al. Identification of the hemangioblast in postnatal life. *Blood.* 2002;100:3203–3208.
28 Krause DS. Plasticity of marrow-derived stem cells. *Gene Ther.* 2002;9:754–758.
29 Reyes M, Dudek A, Jahagirdar B, et al. Origin of endothelial progenitors in human postnatal bone marrow. *J Clin Invest.* 2002; 109:337–346.
30 Asahara T, Takahashi T, Masuda H, et al. VEGF contributes to postnatal neovascularization by mobilizing bone marrow-derived endothelial progenitor cells. *EMBO J.* 1999;18:3964–3972.
31 Kalka C, Masuda H, Takahashi T, et al. Transplantation of ex vivo expanded endothelial progenitor cells for therapeutic neovascularization. *Proc Natl Acad Sci USA.* 2000;97:3422–3427.
32 Kalka C, Masuda H, Takahashi T, et al. Vascular endothelial growth factor (165) gene transfer augments circulating endothelial progenitor cells in human subjects. *Circ Res.* 2000;86:1198–1202.
33 Llevadot J, Murasawa S, Kureishi Y, et al. HMG-CoA reductase inhibitor mobilizes bone marrow-derived endothelial progenitor cells. *J Clin Invest.* 2001;108:399–405.
34 Dimmeler S, Aicher A, Vasa M, et al. HMG-CoA reductase inhibitors (statins) increase endothelial progenitor cells via the PI 3-kinase/Akt pathway. *J Clin Invest.* 2001;108:391–397.
35 Ito H, Rovira, II, Bloom ML, et al. Endothelial progenitor cells as putative targets for angiostatin. *Cancer Res.* 1999;59:5875–5877.
36 Badorff C, Brandes RP, Popp R, et al. Transdifferentiation of blood-derived human adult endothelial progenitor cells into functionally active cardiomyocytes. *Circulation.* 2003;107:1024–1032.
37 Rehman J, Li J, Orschell CM, March KL. Peripheral blood "endothelial progenitor cells" are derived from monocyte/macrophages and secrete angiogenic growth factors. *Circulation.* 2003;107:1164–1169.
38 Voyta JC, Via DP, Butterfield CE, Zetter BR. Identification and isolation of endothelial cells based on their increased uptake of acetylated-low density lipoprotein. *J Cell Biol.* 1984;99:2034–2040.
39 Browning PJ, Sechler JM, Kaplan M, et al. Identification and culture of Kaposi's sarcoma-like spindle cells from the peripheral blood of human immunodeficiency virus-1-infected individuals and normal controls. *Blood.* 1994;84:2711–2720.
40 Fernandez Pujol B, Lucibello FC, Gehling UM, et al. Endothelial-like cells derived from human CD14 positive monocytes. *Differentiation.* 2000;65:287–300.

41 Schmeisser A, Garlichs CD, Zhang H, et al. Monocytes coexpress endothelial and macrophagocytic lineage markers and form cord-like structures in Matrigel and angiogenic conditions. *Cardiovasc Res.* 2001;49:671–680.
42 Harraz M, Jiao C, Hanlon HD, et al. $CD34^-$ blood-derived human endothelial progenitors. *Stem Cells.* 2001;19:304–312.
43 Asahara T, Chen D, Tsurumi Y, et al. Accelerated restitution of endothelial integrity and endothelium-dependent function after phVEGF165 gene transfer. *Circulation.* 1996;94:3291–3302.
44 Hedman M, Hartikainen J, Syvanne M, et al. Safety and feasibility of catheter-based local intracoronary vascular endothelial growth factor gene transfer in the prevention of postangioplasty and in-stent restenosis and in the treatment of chronic myocardial ischemia: phase II results of the Kuopio Angiogenesis Trial (KAT). *Circulation.* 2003;107:2677–2683.
45 Losordo DW, Isner JM, Diaz-Sandoval LJ. Endothelial recovery: the next target in restenosis prevention. *Circulation.* 2003;107:2635–2637.
46 Gulati R, Jevremovic D, Peterson TE, et al. Autologous culture-modified mononuclear cells confer vascular protection after arterial injury. *Circulation.* 2003;108:1520–1526.
47 Gulati R, Jevremovic D, Witt TA, et al. Modulation of the vascular response to injury by autologous blood-derived outgrowth endothelial cells. *Am J Physiol Heart Circ Physiol.* 2004;287:H512–H517.
48 Griese DP, Ehsan A, Melo LG, et al. Isolation and transplantation of autologous circulating endothelial cells into denuded vessels and prosthetic grafts: implications for cell-based vascular therapy. *Circulation.* 2003;108:2710–2715.
49 Fujiyama S, Amano K, Uehira K, et al. Bone marrow monocyte lineage cells adhere on injured endothelium in a monocyte chemoattractant protein-1-dependent manner and accelerate reendothelialization as endothelial progenitor cells. *Circ Res.* 2003;93:980–989.
50 Alvarez-Dolado M, Pardal R, Garcia-Verdugo JM, et al. Fusion of bone-marrow-derived cells with Purkinje neurons, cardiomyocytes and hepatocytes. *Nature.* 2003;425:968–973.
51 Werner N, Junk S, Laufs U, et al. Intravenous transfusion of endothelial progenitor cells reduces neointima formation after vascular injury. *Circ Res.* 2003;93:E17–E24.
52 Nugent HM, Rogers C, Edelman ER. Endothelial implants inhibit intimal hyperplasia after porcine angioplasty. *Circ Res.* 1999;84:384–391.
53 Conte M, Birinyi L, Miyata T, et al. Efficient repopulation of denuded rabbit arteries with autologous genetically modified endothelial cells. *Circulation.* 1994;23:2161–2169.
54 He T, Smith LA, Harrington S, et al. Transplantation of circulating endothelial progenitor cells restores endothelial function of denuded rabbit carotid arteries. *Stroke.* 2004;35:2378–2384.
55 Tevaearai HT, Ekhart AD, Walton GB, et al. Myocardial gene transfer and overexpression of B_2-Adrenergic receptors potentiates the functional recovery of unloaded failing hearts. *Circulation.* 2002;106:124–129.
56 Jackson KA, Majka SM, Wang H, et al. Regeneration of ischemic cardiac muscle and vascular endothelium by adult stem cells. *J Clin Invest.* 2001;107:1395–1402.

175

Endothelial Microparticles

Biology, Function, Assay, and Clinical Application

Yeon S. Ahn, Lawrence Horstman, Eugene Ahn, Wenche Jy, and Joaquin Jimenez

University of Miami School of Medicine, Florida

Endothelial cells (ECs) have long been implicated in the pathogenesis of various clinical disorders such as thrombosis, inflammation, and vascular diseases (1,2). However our knowledge of precisely how ECs contribute to the pathogenesis of human disease remains limited. A major obstacle to the study of endothelial disturbances is the difficulty in sampling ECs from patients for investigation. Clinical tests to assess or monitor EC perturbations are not widely available, yet there is an urgent need for diagnostic assays that are clinically useful, simple, and economical. The availability of such methods promises new insights into the role of ECs in the pathogenesis of vascular disorders and will assist clinicians in the early diagnosis and prevention of common atherothrombotic conditions, the leading cause of mortality in our hemisphere.

This situation has improved in the last few years, as it has become evident that ECs, like other blood cells, shed membranous vesicles, termed *endothelial microparticles* (EMPs), in amounts and phenotypes that reflect the health of the parent cell. EMP assays offer a new and sensitive means of assessing endothelial status (3).

ECs normally present a nonadhesive and anticoagulant surface to flowing blood, but EC injury can provoke cell adhesion and initiation of coagulation with consequent thrombosis and inflammation (1,4). EMPs consist of somewhat heterogeneous species of microparticles (<1 μm diameter) carrying many biomarkers of the parent cells, the analysis of which can therefore provide useful information about EC status (3). For example, it has been shown that EMPs released upon activation exhibit distinctive antigenic phenotypes and functional properties compared to those released upon apoptosis (5).

Elevated EMPs have been reported in large vessel (macrovascular) diseases such as acute myocardial infarctions (MI) and coronary artery disease (CAD), as well as in small vessel (microvascular) diseases such as thrombotic thrombocytopenic purpura (TTP), preeclampsia, and diabetes mellitus (DM) (3). Subsequently, elevation of EMPs was also documented in several autoimmune and inflammatory disorders, most notably systemic lupus erythematosus (SLE) and multiple sclerosis (MS). The list of diseases associated with elevated EMPs continues to grow.

The goal of this chapter is to highlight the emerging role for EMPs as a diagnostic marker in clinical disease. The first part discusses their mechanisms of generation, phenotypic heterogeneity, assay methods, and putative functions. The second part reviews clinical studies of EMPs in various human disorders. Future prospects for EMP research are considered at the end of the chapter.

WHAT ARE ENDOTHELIAL MICROPARTICLES?

As their name implies, EMPs are submicroscopic membranous particles (vesicles) of size range below about 1.0 μm, which are shed from the parent ECs upon stimulation by activating agents such as tumor necrosis factor (TNF)-α or during apoptosis. The process of MP release is termed *vesiculation* and appears to follow an earlier stage of membrane budding or blebbing before pinching off as free particles. They carry with them many of the membrane proteins and phospholipids (PLs) of the parent cell (3).

In many respects, EMPs resemble the MPs shed into plasma from the circulating blood cells – i.e., platelet MPs (PMPs), leukocyte MPs (LMPs), and red blood cell MPs (RMPs). A 1999 review article on PMPs covers many topics relevant also to EMPs (6).

It is important to note that our use of the term *MP* refers strictly to plasma membrane–derived microparticles and not to recently described "exosomes," which are generated from intracellular membranes (late endosomes), and have specialized functions such as antigen presentation (7). Exosomes, which are smaller at 60 to 80 nm than most plasma-membrane derived MPs, are not included in this chapter although there may be some overlap because of certain shared antigens and uncertainty about relevant size ranges of exosomes and limit of size detection of EMPs (8).

HISTORICAL CONSIDERATIONS

Wolf first described MPs derived from platelets (PMPs) in 1967, referring to them as *platelet dust* (9). In 1977, employing electron microscopy, Zucker-Franklin observed PMPs together with RMPs in the blood of patients with immune thrombocytopenic purpura (ITP) (10). EMPs were first explicitly studied by Hamilton and colleagues, who in 1990 used flow cytometry to detect their release from human umbilical vein ECs (HUVECs) in tissue culture following stimulation by complement (11). These methods had been developed by members of the same group in their earlier pioneering studies of PMPs. EMPs were detected in part by fluorescent antibody against the stimulus, complement, leading to the observation that much or most of the added complement was shed off from the HUVECs (attached to EMPs), depleting the parent cell of added complement.

However, reliable measurement of EMPs in patient blood is more challenging than studies in cell culture because blood contains many different kinds of MPs. A further complication is that at least some types of EMPs may adhere to other blood cells, suppressing their detection. As a result, clinical applications of EMP assays went unexploited for almost a decade.

Since 1990, our lab has been investigating PMPs, particularly in certain subgroups of ITP patients. ITP is an autoimmune disorder in which auto-antibodies bind to platelets, inducing their immune-mediated destruction by different mechanisms (12,13). Patients with ITP are usually at risk for bleeding due to low platelet counts. However, we identified a subgroup of patients who rarely bleed despite severe thrombocytopenia and showed that their blood contained high levels of procoagulant PMPs, apparently functioning to protect them against hemorrhage (13,14). In related studies, we found that PMPs bind to and activate neutrophils to form grape-like cell aggregates, suggesting a role for PMPs in inflammation as well as hemostasis (15). In 1998, our interest shifted from PMPs to EMPs due to increasing awareness of the need for novel diagnostic assays in endothelial-based diseases.

Our first studies of EMPs appeared in early 2000 (16), followed later that year by findings in vivo (17) and other publications soon thereafter (18,19). The first reported clinical application of EMP assay appeared in 1999 by Combes and colleagues (20). Since then, an exponentially growing number of publications on EMPs has appeared, and steady progress has been made in measurement methodologies and understanding their clinical significance.

ENDOTHELIAL CELL PERTURBATION

Under resting conditions, ECs provide a nonthrombogenic, nonadhesive surface and regulate vascular permeability and tone by release of nitric oxide (NO) and other agents to maintain smooth blood flow in the vasculature (21). The hallmark of perturbed ECs is a phenotypic change toward a procoagulant, pro-adhesive state, promoting adhesion of blood cells, coagulation, and inflammation. EC "perturbation" encompasses a wide range of pathologies but for present purposes may be divided into endothelial *activation* and *apoptosis*.

Endothelial Cell Activation

As used here, EC activation refers to the reversible response of ECs to pro-inflammatory cytokines such as TNF-α, interleukin (IL)-1 or interferon (IFN)-γ (22). These are genetic responses requiring 4 to 24 hours, characterized by profound morphological and phenotypic changes (22,23). Activation of nuclear factor (NF)-κB and ensuing induction of pro-survival genes of the iap and xiap family promote EC survival (23,24). The same sequence of events induces surface expression of pro-adhesive molecules such as vascular cell adhesion molecule (VCAM)-1, intercellular adhesion molecule (ICAM)-1, and E-selectin (22–24), facilitating leukocyte adhesion and migration, and release of IL-6, IL-8, and monocyte chemoattractant protein (MPC)-1, which further augment the inflammatory response (25). TNF-α also elicits release of large multimers of von Willebrand factor (vWF) and P-selectin from Weibel-Palade bodies. The large vWF multimers promote platelet adhesion to the endothelial surface, leading to their activation and aggregation (26). In addition, activated ECs express higher levels of plasminogen activator inhibitor (PAI)-1 and lower levels of tissue-type plasminogen activator (t-PA) (4,21). Activated ECs may express tissue factor and promote blood coagulation, at least in vitro (27).

Endothelial Cell Apoptosis

Apoptotic ECs exhibit distinctive procoagulant properties. Loss of membrane symmetry exposes anionic PL such as phosphatidyl serine (PS), together with release of PS-rich membrane blebs that support thrombin generation (28,29). EC detachment resulting from apoptosis exposes the extracellular matrix, which potently activates platelet aggregation (30). Despite many features shared between activated and apoptotic ECs, the intracellular mechanisms are distinct. Known inducers of EC apoptosis include chemotherapeutic agents, high glucose and advanced glycation end-products, hypoxia, hyperoxia, and growth factor deprivation (31,32).

Markers of Endothelial Cell Dysfunction

A variety of soluble molecules released from ECs have been assayed by enzyme-linked immunosorbent assay (ELISA) to

Table 175-1: Responses of EC and EMP Release during EC Activation or Apoptosis

		Activation (TNF-α)		*Apoptosis (GFD)*		*Untreated (Control)*	
Marker	*Cell Line*	*Whole EC (FL Units)*	*EMP (Counts)*	*Whole EC (FL Units)*	*EMP (Counts)*	*Whole EC (FL Units)*	*EMP (Counts)*
CD31	RmVEC	↓	↑↑	↓↓↓	↑↑↑	230 ± 25	0.80 ± 0.04
	BmVEC	↓↓	↑↑	↓↓	↑↑↑	225 ± 30	0.7 ± 0.1
	CMVEC	↓	↑	↓↓↓	↑↑	10 ± 2	0.1 ± 0.01
CD105	RmVEC	↓	↑	↓↓↓	↑↑	330 ± 25	4.0 ± 0.6
	BmVEC	↓↓	↑↑	↓↓↓	↑↑↑	255 ± 25	4.0 ± 1.0
	CMVEC	↓↓	↑↑	↓↓↓	↑↑	3.0 ± 0.5	1.5 ± 0.2
CD51	RmVEC	↑	↑↑	no change	↑↑	17 ± 3	1.5 ± 0.3
	BmVEC	↑↑	↑↑	no change	↑↑	15.5 ± 4.0	2.0 ± 0.5
	CMVEC	↑	↑↑	↓	no change	20.0 ± 4.5	2.0 ± 0.5
CD62E	RmVEC	↑↑↑	↑↑↑	no change	no change	3.0 ± 0.6	3.5 ± 0.7
	BmVEC	↑↑↑	↑↑	no change	no change	2.5 ± 1.0	4.0 ± 1.5
	CMVEC	↑	↑	↓	↓↓	2.0 ± 0.5	4.0 ± 0.5
CD54	RmVEC	↑↑↑	↑↑↑	no change	no change	6.0 ± 1.5	3.0 ± 1.0
	BmVEC	↑↑↑	↑↑	no change	no change	7.5 ± 2.5	3.5 ± 1.0
	CMVEC	↑↑	↑	no change	↓	3.0 ± 1.0	3.5 ± 0.5
CD106	RmVEC	↑↑↑	↑	no change	no change	1.5 ± 0.2	1.0 ± 0.5
	BmVEC	↑↑↑	↑↑	no change	no change	1.5 ± 0.1	1.0 ± 0.5
	CMVEC	↑↑↑	↑↑	no change	no change	2.0 ± 0.5	0.5 ± 0.1
Annexin	RmVEC	↑↑	↑↑	↑↑	↑↑↑	3.0 ± 0.7	0.20 ± 0.04
	BmVEC	↑	↑↑	↑↑	↑↑↑	4.0 ± 1.0	0.5 ± 0.1
	CMVEC	↑	↑	↑↑	↑↑	3.0 ± 0.2	0.5 ± 0.1

The table shows EC responses and EMP release as judged by fluorescent-labeled antibodies to markers indicated at left, on whole ECs and EMPs after 24 hours of treatment to induce either activation (TNF-α) or apoptosis (growth factor deprivation, GFD). Whole ECs were measured by flow cytometry after detachment by trypsin in terms of mean fluorescent intensity; EMPs are relative numbers detected per mL of media supernatant. Results are shown for three cell lines: renal microvascular ECs (RmVEC), brain microvascular ECs (BmVEC), and coronary artery macrovascular ECs (CMVEC). The arrows indicate increase (*up*) or decrease (*down*) relative to untreated control values at right: ↑ = 15% to 2-fold; ↑↑ = 2- to 10-fold; ↑↑↑ = >10-fold; ↓ = 15% to 25% reduction; ↓↓ = 25%–50%; ↓↓↓ = >50% reduction. EMP results at right are in terms of numbers detected $\times 10^6$/mL. All experiments were repeated at least three times and reproducibility was good as shown by $\pm\sigma$. The upper three markers are constitutive; the lower four (*beneath heavy line*) are inducible – i.e. respond to stimulation. Apoptosis was evaluated by TUNEL assay and checked for viability by trypan blue dye exclusion: only 0.5% to 2% of TNF-α-treated cells were positive for apoptosis; whereas 50% ± 10% and 65% ± 10% of RmVEC and BmVEC, respectively, were positive following deprivation of growth factor. However, CMVEC resisted apoptosis, only 5% ± 3% being positive after this treatment.

assess endothelial dysfunction (33). These include VCAM-1, ICAM-1, E-selectin, thrombomodulin (TM) (34), and vWF (3,35). However, none of them is yet widely accepted for clinical testing. A major drawback of these methods is lack of specificity. ICAM-1, for example, is expressed by activated cells of many lineages. Similarly, vWF is not specific for ECs, because it is also released by megakaryocytes and platelets (35). Some of these soluble markers have been shown to be partly bound to MPs (3), and therefore their measurement may overlap with EMP assay.

Circulating ECs (CECs) have been described in patients with cardiovascular diseases (36,37). Because CECs are present in quite small numbers, a large volume of blood is required for their measurement. In recent years, the EMP assay is emerging as a promising new marker of endothelial dysfunction, as reviewed in the following sections. However, there is a limitation of EMP study as well. EMPs may not carry certain markers of parent cells, and biologic markers of parent cells are not equally expressed on EMPs, as seen in Table 175-1. Accordingly, combined analysis of both CECs and EMPs should improve overall evaluation of ECs.

MECHANISMS OF ENDOTHELIAL MICROPARTICLE GENERATION

EMPs are released from the parent cells under many conditions, including: (a) activation or apoptosis induced by numerous agents, (b) partial or complete lysis such as by complement, (c) oxidative injury, and (d) other insults such as high shear stress (38). The detailed mechanisms of MP release remain obscure. However, a rise in cytosolic calcium concentration, either from internal stores or from plasma, appears to be a necessary triggering event or common pathway for vesiculation in ECs and other cells. Elevated calcium can induce cytoskeletal contraction, thought to be the driving force for apoptotic membrane blebbing (39). Shedding of membrane blebs requires partial breakdown of the cytoskeleton. It was shown that calcium-dependent proteases such as calpain and caspase (40) are capable of breaking down cytoskeleton to facilitate MP release.

The mechanism of EMP release is probably similar to that for other cells such as platelets (6). More recently, Jy and colleagues have shown by fluorescent microscopy that release of CD31 (platelet-endothelial cell adhesion molecule [PECAM]-1)$^+$ EMPs is preceded by localized accretions of CD31 on the cell surface of renal microvascular ECs, a phenomenon known as capping (41). Specifically, we showed that EMP release correlated with the percentage of fully capped cells, measured at different time intervals (0, 2, 4, 8, 24 hours from addition of agonist TNF-α). Fluorescent images of caps resolved numerous MPs apparently in the act of being shed from microvilli (or pseudopodia) protruding from the caps. These observations suggest that capping of membrane proteins (or other focal adhesions) is a necessary prelude to the shedding of at least some species of EMPs (Figure 175.1). It is now well known that membrane surface proteins are quite mobile and can agglomerate by virtue of lipid rafts (42,43), areas of distinctive lipid and protein composition, studied by differential solubility in detergents, a topic now overlapping with phenomena such as patching and capping, focal adhesions, and receptor clustering. A variety of enzymes, known or hypothetical, have been implicated in membrane blebbing and vesiculation, including scramblase, floppase, calpain, and PL translocase (6). Many lines of evidence suggest that EMP generation is not a random or unregulated response but is a programmed shedding, varying with specific stimuli to result, for example, in EMPs expressing characteristic phenotypes or functions in defined proportions, in vitro and in vivo.

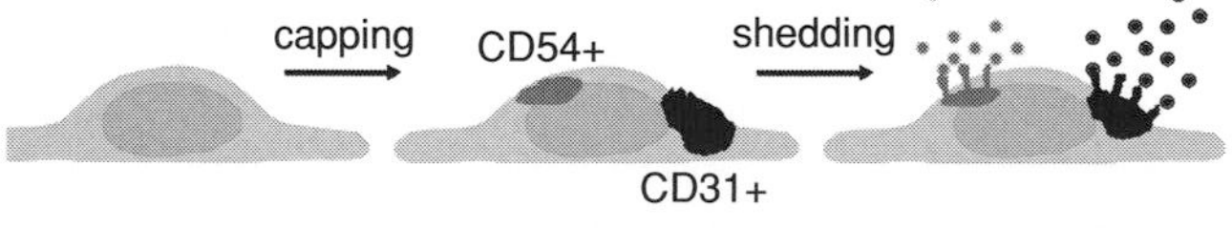

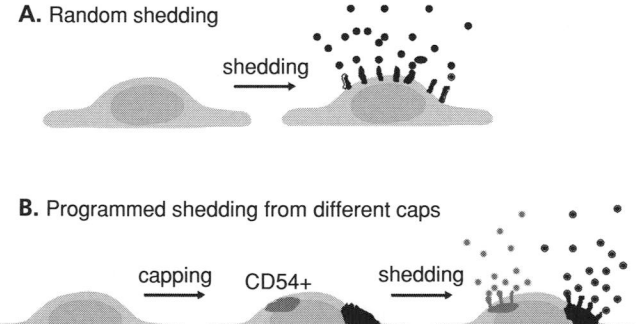

Figure 175.1. Models of microparticle shedding from endothelium. (**A**) Model of random shedding: endothelial microparticles (EMPs) were shed randomly from membrane surface of endothelium. Each microparticle carried all or most surface antigens of endothelium. (**B**) Programmed shedding from specific caps: cell stimulation induces surface antigens clustering/capping, followed by shedding of EMPs from different caps. This process will result in different species (phenotypes) of EMPs.

HETEROGENEITY OF ENDOTHELIAL MICROPARTICLES

EMPs display heterogeneity in size, density, and antigenic profile. These differences may occur in response to the same agonist, owing to variation in timing, incubation conditions, or the vascular origin of the ECs. Phenotypic heterogeneity of EMPs is also seen with different agonists. For example, Jimenez and colleagues employed flow cytometry of seven markers in three different EC types to show that TNF-α and growth factor deprivation result in different phenotypes of ECs and EMPs (Table 175-1) (5). In these studies, it was noted that the ratio of EMPs counted by two markers, CD62E (E-selectin)$^+$ EMPs/CD31$^+$ EMPs, was always more than 4.0 in activation but less than 4.0 in apoptosis. CD54 (ICAM-1)$^+$ EMPs/CD31$^+$ EMPs gave similar discrimination. Whether these criteria can be applied to interpret the EC status in clinical settings has not been established. The answer to this could influence treatment modalities or preventive strategies.

When applied to patients with TTP, the above criterion (ratio of CD62 [E-selectin]$^+$ EMPs/CD31$^+$ EMPs) led to the conclusion that EMPs in TTP are activated, not apoptotic (5). These data are consistent with our findings that neither TNF-α nor plasma from TTP patients induces EC apoptosis in vitro (5,18). Others have shown that TTP plasma does induce apoptosis of certain EC subtypes in cell culture (44). One likely explanation for these discrepancies is that other investigators have used "proapoptotic" conditions prior to adding the plasma of TTP patients (44). These conditions include decreased levels of growth factors in contrast to culture conditions in which ECs are maintained in optimal growth conditions at all times. Thus, it is conceivable that ECs cultivated in the presence of suboptimal levels of growth factors and fetal calf serum may be more susceptible to apoptosis (45).

This method may be applicable to other diseases to resolve the same question, such as whether apoptotic ECs are prominent in specific coronary conditions, a continuing controversy. Further potential applications of EMP phenotype analysis are under investigation.

Electron Micrographic Evidence of Endothelial Microparticle Heterogeneity

Representative electron micrographs of EMPs from TNF-α–activated or serum-deprived apoptotic renal microvascular ECs are shown in Figure 175.2. The principal

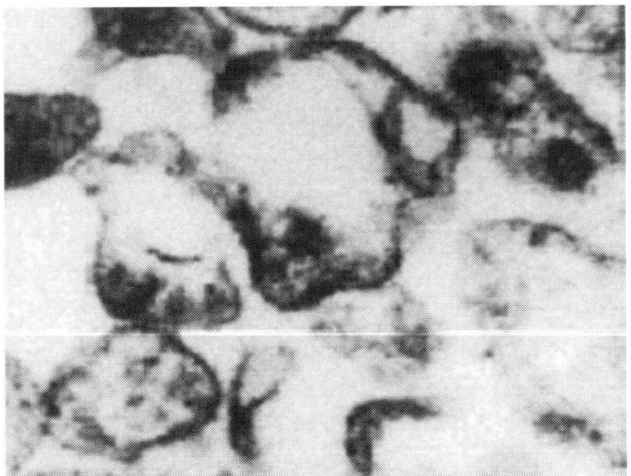

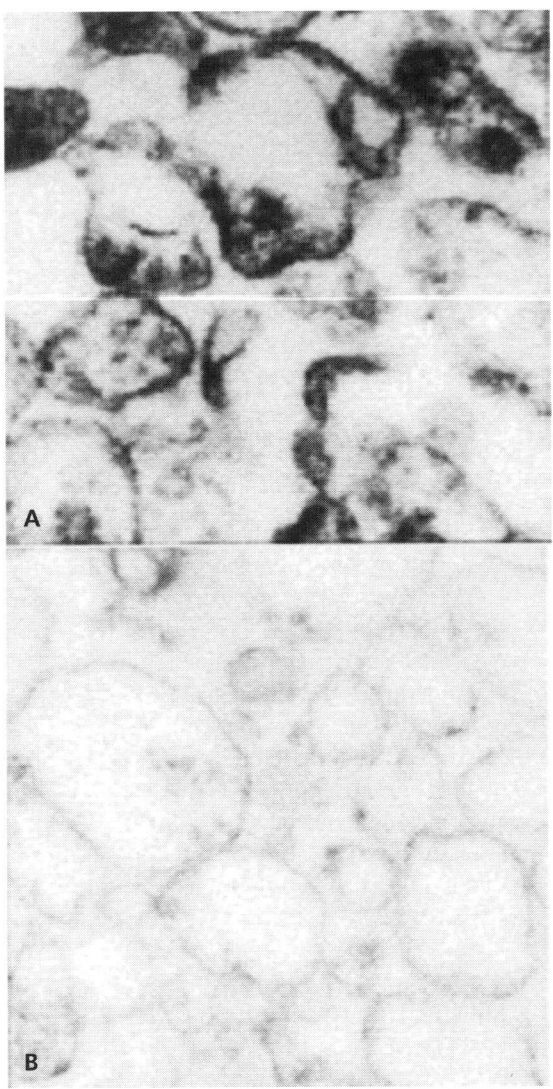

Figure 175.2. Electron microscopy of EMPs induced by apoptosis or activation. Morphology of EMPs released from (**A**) apoptotic or (**B**) activated renal ECs.

observations are that (a) EMPs exist in a range of sizes and densities; (b) EMPs derived from activated ECs mainly comprise plasma membrane vesicles, lacking significant granule or electron-dense material (Figure 175.2A); and (c) EMPs resulting from EC apoptosis consist mainly of electron-dense vesicles and/or fragments of granules (Figure 175.2B). For other examples of electron micrographs of EMPs, the reader is referred to (20,46).

Clinical Evidence of Endothelial Microparticle Heterogeneity

In studies of MS (19) and CAD (47), it was found that the diagnostic significance of EMP levels depended on which marker was used to identify and count them. For example, when CD51 (vitronectin receptor) positivity was the criterion, patients with MS showed chronic elevation irrespective of exacerbation or remission; but when CD31 positivity was the criterion, significant elevation during exacerbation compared to remission was clear (19). This was also observed in a study of EMPs in CAD: $CD31^{+}$ EMPs identified acute disease whereas $CD51^{+}$ EMPs identified chronic stage (47) (discussed later). Consistent with the link between CD51 and chronic stage disease, Simak and colleagues reported that in patients with paroxysmal nocturnal hemoglobinuria (PNH), CD105 (endoglin)/CD54 (ICAM-1) EMPs were elevated during chronic activation (and failed to correlate with hemolysis) whereas CD105 (endoglin)/CD144 (vascular endothelial [VE]-cadherin) EMPs were detected in acute states (48). This is consistent with our findings in MS and CAD.

COMPOSITION OF ENDOTHELIAL MICROPARTICLES (AND OTHER MICROPARTICLES)

MPs consist mainly of PLs and proteins, but their composition depends on cell origin and the stimulating agent or condition inducing their release. Some MPs may carry cytoplasmic materials. The lipid bilayers of the plasma membrane have been shown to be disturbed prior to release of MPs (29), leading to expression of negatively charged PS on the MPs, which has been shown to play an important role in blood coagulation (29).

Weerheim and colleagues analyzed the PL composition of MPs in normal blood and found that they include 60% phosphatidylcholine (PC) with the remainder being sphingomyelin, phosphatidylethanolamine (PE), and PS (49). In contrast, Fourcade and colleagues reported that MPs obtained from synovial fluids from the inflamed joints of arthritis patients contained more evenly distributed PC, PE, sphingomyelin, and lysophospholipids (50). MPs in the synovial fluids are mainly derived from leukocytes whereas MPs in blood are mainly released from platelets (51,52). It is possible that differences in lipid composition of MPs may be due to different cell origins or types of stimulation.

Profiles of surface antigens on MPs may provide information about the status of the parent cell. We have found in vitro that EMPs from activated ECs are enriched in CD62E (E-selectin) and CD54 (ICAM-1) antigens. In contrast, EMPs derived from apoptotic ECs are enriched in CD31 antigens (5). However, not all surface antigens of the parent cells are necessarily expressed on MPs. For example, T-cell–derived MPs lack CD28 and CD45 antigens, although they are among the most abundant antigens on the parent T-cells (53). We also observed that CD51 antigens are highly expressed on ECs but weakly on EMPs in EC tissue culture stimulated with TNF-α. It therefore appears that MP shedding is a highly selective and variable process (Figure 175.2). It may involve selective variations in membrane antigen clustering (or capping) associated with distinctive types of lipid rafts (41,54), depending on particular activation pathways.

MEASUREMENT OF ENDOTHELIAL MICROPARTICLES

Methods of Endothelial Microparticle Assay

Flow Cytometry

The majority of studies to date have employed flow cytometry to assay EMPs. Most commonly, platelet-poor plasma (PPP) is incubated with fluorescence-labeled, endothelium-specific antibodies (or annexin V [AnV]) for 20 to 30 min and then sorted by FACS (3). Results are usually expressed in counts per L or mL. However, detailed procedures vary significantly in different laboratories. Among the most importance inter-lab differences are speed of centrifugation to prepare PPP and the labeling of antibodies used for identification. At this time, no consensus or standardized protocols for EMP assay by flow cytometry exist.

A preferred method in our laboratory is the use of 2-color combination of PE-labeled anti-CD31 and FITC-labeled anti-CD42 (55,56) in a single run. Because CD31 occurs also on platelets, platelet-specific CD42 allows for distinction between PMPs ($CD31^{+}$, $CD42^{+}$) and EMPs ($CD31^{+}$, $CD42^{-}$). The CD31 marker has the further advantage of being quite bright (abundant), as shown in the EMP assay, and therefore is sensitive. Other investigators report favorably on use of CD144 VE-cadherin).

This basic method is simple with fast turnaround time and is thus well suited for clinical testing. In addition, by using whole blood samples and multiple markers, interaction of EMPs with other populations of cells can be investigated; new studies exploring the interaction of EMPs with leukocytes and platelets have been reported in several laboratories including ours (57,58). Similar studies were performed with PMPs or LMPs, delineating cell–cell interactions mediated by various MPs (e.g., PMP–leukocyte interactions) (6,15).

EMPs can be further characterized using additional markers. These can be tentatively classified into two groups based on patterns of EC expression: (a) inducible markers reflecting EC activation including CD54 (ICAM-1), CD62E (E-selectin), CD106 (VCAM-1), tissue factor, and MCP-1; and (b) constitutive markers, expressed in the resting state, including CD51/CD61 (vitronectin receptor, $\alpha_V\beta_3$ integrin complex), CD31, CD105 (endoglin), CD144 (VE-cadherin), and CD146 (S-endo1). In several studies, inducible antigens have been reported to be more elevated during the acute phase of some diseases relative to constitutive antigens. However, some of these markers, notably CD54 (ICAM-1) and tissue factor, are not EC-specific, but rather are widely expressed by other activated cells. However, MPs from nonendothelial origin can be excluded by using a second antibody specific to the non-ECs.

Annexin V/Prothrombinase Assay

In this widely used method (59–63), MPs are captured by their binding to AnV in the presence of calcium, then assayed for prothrombinase activity. This involves incubation of immobilized (captured) MPs with factors Xa and Va as well as prothrombin and starting the reaction by addition of calcium. Thrombin generation is monitored using a chromogenic substrate. Quantity of PL expressed on MPs is inferred from a standard curve obtained with defined liposomes. Secondary identification of MP types (PMPs, EMPs, LMPs) is performed by ELISA with cell-specific antigens.

The most serious objection to this method is that only a fraction of EMPs are positive for AnV (small fraction for activated MPs, larger fraction for apoptotic MPs). AnV binding depends on anionic PLs, chiefly PS, which is largely responsible for prothrombinase activity; thus, this assay will miss all MPs that are negative for AnV binding and selectively favors EMPs from apoptotic ECs.

Limitations Associated with Endothelial Microparticles Assays

Although EMPs are now regarded as highly promising new markers of EC perturbation, many problems remain to be resolved before EMP assays reach the clinical mainstream (64,65). A major limiting factor has been the lack of consensus or standardization of methods, making it difficult or impossible to compare data between labs. As a first step toward addressing this problem, a committee was organized under auspices of the International Society of Thrombosis and Haemostasis. The recent forum detailed methods of detection from six labs (including our own) (55) (Table 175-2). Future workshops will address the need for and implementation of a uniform method, or at minimum the use of a reference standard sample for MP assay.

POTENTIAL FUNCTIONS OF ENDOTHELIAL MICROPARTICLES (AND OTHER MICROPARTICLES)

As of yet, little is definitely established about EMP functions. However, circumstantial evidence has led to a growing consensus that EMPs can act as diffusible messengers, carrying molecules to specific targets and interacting with blood cells to modulate inflammation and thrombosis.

Figure 175.3 summarizes recent concepts of EMPs in blood coagulation, inflammation, and thrombosis.

Coagulation

Coagulation is usually triggered by tissue factor but completing the coagulation cascade requires a membrane surface exposing anionic PLs such as PS to facilitate assembly of the tenase and prothrombinase complexes. Most (but not all) MPs do in fact carry anionic PLs, especially PS, as well as tissue factor (66). Both of these procoagulant activities are known to occur on EMPs, LMPs, and PMPs (18,46,59,61,67). In addition, some MPs also carry tissue factor pathway inhibitor (TFPI) (68). However, the relative magnitude of tissue factor and TFPI activities on different types of MPs is not yet clear, nor is the importance of these activities on MPs

Table 175-2: Techniques for Measuring of Endothelial Microparticles by Different Laboratories

Authors (Refs.)	*Principal Technique*	*Quantitation*	*Anticoagulant*	*Prepare PPP*	*Isolation MP Pellet*	*Generic MP Detection*	*Endothelial Markers*
Biro et al. (112)	Flow cytometry	Counts	Citrate	1,550 g, 20 min	18,000 g 30 min	Annexin V	CD31,CD62E or CD144
Dignat-George et al. (113)	Flow cytometry	Counts	Citrate	1,500 g, 15 min 13,000 g, 2 min	–	Annexin V	CD51,CD144 or CD146
Hugel et al. (60)	Solid phase capture	Prothrombinase activity	Citrate	1,500 g, 15 min 13,000 g, 2 min	–	Annexin V; tissue factor	CD31 or CD62E
Jimenez et al. (56)	Flow cytometry	Counts	Citrate	200 g, 10 min 1,500 g, 7 min	–	–	$CD31^{+}$/CD42- or CD62E
Shet et al. (62)	Flow cytometry	Counts	Citrate	13,000 g, 10 min	100,000 g 60 min	Annexin V	CD144

known in relation to the procoagulant properties of whole cells.

Tissue factor activity in particular has been difficult to quantitate, with conflicting reports, apparently because of variable expression of cryptic tissue factor and masking by TFPI (69,70); we have attributed variable results partly to sensitivity to sulfhydryl redox state, masking by TFPI and to variably prolonged lag phase (71). Although the majority of MPs are procoagulant, some reports have demonstrated that MPs may be anticoagulant under certain conditions. Tans and colleagues suggested that PMPs could serve an anticoagulant function by supporting the protein C/S/TM pathway (72). Gris and colleagues demonstrated that a large fraction of protein S is MP-associated and that clinical assays using PEG precipitation cause underestimation of protein S because much of it is precipitated with MPs (73). More recently, Berckmans and colleagues have shown that low levels of thrombin generation by MPs in normal controls is via the contact pathway (independent of tissue factor) and may serve an anticoagulant function, on the grounds that protein C could thereby be activated by the trace of thrombin (51). It is now appreciated that a low but steady state of tissue factor activity is essential to normal hemostasis, but held in delicate balance, possibly regulated by MPs.

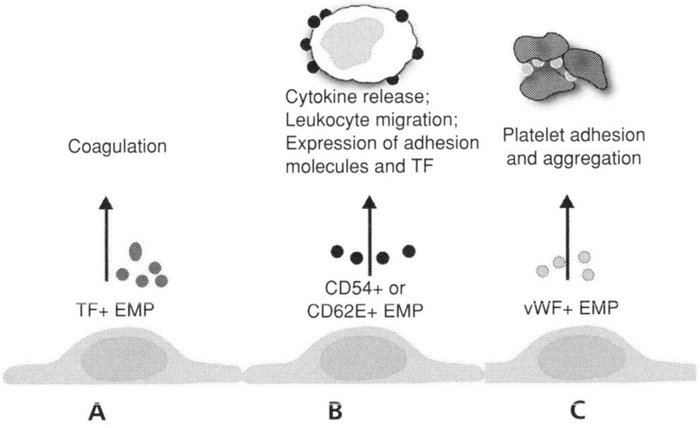

Figure 175.3. Potential functions of EMPs. Several subspecies of EMPs and their potential functions have been identified: (**A**) tissue factor positive (TF^{+}) EMPs have been demonstrated to initiate and promote blood coagulation; (**B**) $CD54^{+}$ or $CD62E^{+}$ EMPs are capable of binding to leukocytes and inducing leukocyte activation; (**C**) von Willebrand factor positive (vWF^{+}) EMPs may cause platelet adhesion and aggregation.

Leukocyte Adhesion and Transmigration

Jy and colleagues first demonstrated that PMPs interact with leukocytes (15). They showed that PMPs can bind to leukocytes via a P-selectin–dependent pathway. The binding of PMPs to leukocytes leads to leukocyte activation and aggregation, with formation of large grape-like clusters of leukocyte-PMP complexes (15). Barry and colleagues showed that PMPs interact with ECs to result in upregulation of CD54 (ICAM-1) on ECs and the subsequent adhesion of monocytes to ECs (74). Mesri and Altieri also showed that LMPs can interact with ECs to induce IL-6 release (75).

Recently Sabatier and colleagues demonstrated that EMPs can bind to and activate cultured THP-1 monocytic cells, as judged by induced expression of tissue factor antigen, and that this effect could be largely inhibited by anti-CD54 (ICAM-1) antibodies (76). Similarly, Jy and colleagues reported that EMPs induce expression of the activation marker CD11b in whole blood–derived leukocytes via a CD54 (ICAM-1)-dependent mechanism (58). In the latter study, EMPs interacted most strongly with monocytes, less so with neutrophils, and hardly at all with lymphocytes.

Interestingly, EMPs derived from TNF-α–activated ECs had a greater effect on leukocytes compared with those derived from ECs undergoing apoptosis resulting from serum deprivation (58). In binding studies, EMPs prelabeled with CD54 (ICAM-1) exhibited greatest binding to U937 monocytic cells, compared with EMPs prelabeled with PECAM-1 (CD31), E-selectin (CD62E), or vitronectin receptor (CD51). Moreover, the CD54 (ICAM-1)-labeled EMPs were largely depleted in the cell-free supernatant, consistent with the majority having bound to the U937 cells (58). Based on these observations, it was proposed that the unexpectedly low concentration of CD54 (ICAM-1)$^+$ EMPs found in blood in various disease states may be explained by their preferential binding to leukocytes, reducing their free concentration.

Finally, Jy and colleagues also demonstrated that monocytes with adherent EMPs were facilitated in their passage through an endothelial monolayer (58). Taken together, these findings suggest that at least one function of EMPs may be to modulate inflammation via leukocyte activation and transendothelial migration. The complex interactions among EMPs, platelets, and leukocytes is summarized in Figure 175.3.

Endothelial Cell Function

It has been reported that MPs released from platelets or leukocytes can influence endothelial function. For example, Boulanger and colleagues reported that MPs from patients with acute MI impaired endothelium-dependent relaxation in isolated arteries (77). In contrast, MPs isolated from patients with nonischemic chest pain had no such effect. High concentrations of MPs from MI patients affected neither endothelium-independent relaxation to sodium nitroprusside nor expression of the endothelial NO synthase (eNOS). The origin of the MPs isolated from MI patients was not identified. In another study, VanWijk and colleagues demonstrated that MPs isolated from preeclampsia patients impaired endothelium-dependent relaxation in isolated myometrial arteries (78,79). It was suggested that these MPs may contain oxidized PLs, which are potent inhibitors of endothelium-dependent arterial relaxation (80). However, some MPs may have a beneficial effect on endothelium. Barry and colleagues showed that PMPs can transfer arachidonic acid to ECs and result in prostacyclin production, which will induce vascular relaxation (81). Recently, it was reported that PMPs promote proliferation, survival, migration, and tube formation in HUVECs (82). These results suggest that the lipid component(s) of the PMPs may be major active factor(s) and that protein component(s) may be minor contributor(s). Other pro- and antiangiogenic factors associated with PMPs were recently reviewed (8).

Endothelial Microparticles as Diffusible Biomessengers

EMPs and other MPs are now considered as important biomessengers, delivering adhesins, peptides, cytokines, and pro- and anticoagulant factors to modulate thrombosis and inflammation, and perhaps other functions – e.g., antigen presentation (8) and angiogenesis (82). However little is known of the details of such putative roles.

Recent studies demonstrated that EMPs carry unusually large vWF (ULvWF) multimers and participate in thrombosis and hemostasis (57,83). EMP-associated vWF promotes formation of platelet aggregates that are more stable (less dissociable, more irreversible) than aggregates formed by free vWF (57).

A multitude of adhesion molecules (adhesins) are associated with EMPs, including VCAM-1, E-selectin, CD54 (ICAM-1), CD144 (VE-cadherin), and vitronectin receptor, as well as cytokines and pro- and antiangiogenic factors such as vascular endothelial growth factor (VEGF). In addition, conformationally distinctive prion proteins, believed to aid in the propagation of spongiform encephalopathies (SEs), have been detected on EMPs released by infected ECs (84). This finding suggests that EMPs may be active players in the infectious process of SEs.

A recent report from Hussein and colleagues documented caspase-3 inside of EMPs and other MPs (85). They speculate that caspase-3 may be involved in membrane vesiculation and that shedding of caspase-3 in MPs is one way ECs minimize or escape from apoptotic commitment. They also propose the possibility that caspase-3 in MPs may be delivered to other cells, inducing apoptosis in the target cell. Consistent with this hypothesis are studies demonstrating that incubation of MPs from acute coronary syndrome (ACS) patients with ECs in vitro induces EC dysfunction (77,86). The functional role of caspases in MPs, if any, remains unclear. However, it is tempting to speculate caspase-positive MPs deliver death signals to other cells. It is easy to imagine them being phagocytosed, thus liberating active caspase-3 in phagocytes. (It is known that PS signals phagocytosis via a PS receptor (87,88), although the specificity of this is recently questioned (89).

CLINICAL APPLICATIONS

Coronary Artery Disease

Mallat and colleagues first reported elevated cell-derived MPs in patients with ACS (67). They used the AnV/prothrombinase (ELISA) assay to measure procoagulant activity of cell-derived MPs and specific antibodies (anti-CD146 (S-endo1), anti-CD31, anti-CD3, anti-CD11a, and anti-GP1b) to determine their origin. The circulating procoagulant MPs were significantly higher in patients with acute MI and unstable angina (UA) than in those with stable angina (SA) or controls who had negative angiographic findings for CAD but were not different between SA and controls. In ACS, MPs did not discriminate acute MI from UA. ACS was associated with MPs positive for CD146 (S-endo1) or CD31, but not for GP1b (platelet marker), CD3 (lymphocyte marker), CD11a (marker for activated leukocytes). They concluded that CD31$^+$/CD146 (S-endo1)$^+$ MP are of endothelial origin and contribute to thrombin formation in the injured endothelium.

Using flow cytometry for EMP assay, Bernal-Mizrachi and colleagues reported elevated EMPs in ACS, contributing a number of new insights (47). Two species of EMPs

($CD51^+$ and $CD31^+/CD42b^-$) as well as PMPs were measured in patients with MI, UA, SA, and healthy controls. $CD31^+/CD42b^-$ EMPs and PMPs were elevated in patients with ACS (acute MI plus UA) relative to SA and healthy controls. However, $CD51^+$ EMPs were similarly elevated in both ACS and SA, suggesting that these two species of EMPs reflect different aspects of the pathology (89). Thus, $CD31^+$ EMPs mark acute ischemic injury whereas $CD51^+$ EMPs mark chronic injury. Also of interest was the observation that patients with a first episode of ACS had significantly higher $CD31^+$ EMPs than in those with chronic recurrent episodes, suggesting that the capacity for generation of EMPs is reduced in those with recurring ischemia. In those having a first episode, EMPs were significantly higher in acute MI than UA, suggesting that EMP levels reflect severity of vascular lesions. In a recent study, EMP levels were compared between various angiographic lesions in patients with ACS, revealing that EMP levels were higher in those with high risk lesions compared to low risk lesions by Ambrose classification (90).

In a prospective, randomized study, Morel and colleagues (91) evaluated the effect of the antioxidant vitamin C on $AnV^+/CD31^+$ EMP generation in patients with acute MI. Administration of vitamin C significantly decreased EMP generation in those with dyslipidemia and other high-risk factors compared to placebo group. This may suggest that oxidation underlies endothelial injury in hyperlipidemia and other risk conditions.

A recent study by Boulanger and colleagues demonstrated that circulating MPs from patients with acute MI caused impaired relaxation of rat aortic rings in vitro and caused severe endothelial dysfunction in healthy blood vessels by affecting endothelial NO transduction pathway, whereas MPs isolated from patients with nonischemic diseases had no such effect (77). Heloire and colleagues (92) observed that EMPs released from ECs in culture bind to platelets to form EMP-platelet aggregates and found that circulating EMP–platelet aggregates varied among stable CAD, acute MI, and healthy controls. They reported that EMP–platelet aggregates were significantly increased in patients with stable CAD compared with healthy controls. However the levels were lower in the first hours following acute MI, suggesting that EMPs released from activated ECs on atherosclerotic plaques were consumed locally and contributed to recruitment of platelets and formation of thrombus in injured coronary arteries. These findings indicate that EMPs are not only the markers of EC injury but also play a role in the pathogenesis of occlusive vascular disorders.

Metabolic Syndromes and Postprandial Hyperlipidemia

Metabolic Syndrome

Metabolic syndrome is a combined disturbance of insulin and glucose metabolism, obesity, dyslipidemia, and hypertension, which affects some 10% of middle-aged males in the United States. This condition presages atherosclerosis and CAD and is thus a serious challenge to public health (93,94). We observed elevated EMPs ($CD31^+$ or CD54 $[ICAM\text{-}1]^+$), PMPs ($CD62P^+$), and platelet–leukocyte conjugates in patients with metabolic syndrome, suggesting underlying activation of the endothelium, platelets, and white blood cells (95). These findings indicate that early prophylactic intervention is called for with this syndrome, and that EMP assay can assist by monitoring the status of metabolic syndrome.

Postprandial Hyperlipidemia

Ferreira and colleagues documented a rapid rise of EMPs in parallel with postprandial hyperlipidemia, at 1 and 3 hours following ingestion of a high-fat meal in young healthy volunteers, free of known risk factors (96). Correlation analysis revealed a positive association between levels of $CD31^+/CD42^-$ EMPs and triglyceride ($p = 0.003$) (96). However, EMPs measured by CD62E (E-selectin) showed no such relation, providing further evidence for the existence of distinct EMP species. There was no significant increase in EMPs or triglyceride levels following an isocaloric low-fat meal, indicating that neither the caloric intake nor the postprandial state alone affected EMP release.

In sum, these studies indicate that EMPs are sensitive markers of endothelial disturbance in different stages of atherosclerosis. EMPs were released from normal healthy endothelium upon transient injury brought on by hyperlipidemia following fatty meals, in early chronic stages of atherosclerosis (metabolic syndrome) and in late advanced states of atherosclerosis (acute MI) in CAD.

Thrombotic Thrombocytopenic Purpura

In TTP, EC injury elicits platelet adhesion and aggregation in the microvasculature, leading to consumptive thrombocytopenia, microangiopathic hemolytic anemia, and fluctuating neurological dysfunction. Unusually large multimers of vWF (ULvWF) released from injured ECs are normally rapidly proteolyzed by ADAMTS13 in plasma. In TTP, a deficiency of ADAMTS13 results in impaired proteolysis of unusually large vWF. These multimers, in turn, play a pivotal role in initiating platelet adhesion and aggregation within the microvasculature. The presence of procoagulant EMPs in patients with TTP was first reported by Jimenez and colleagues (18). Compared to normal controls or ITP patients, both $CD31^+/CD42b^-$ and $CD51^+$ EMPs were significantly elevated in TTP (18,97); levels rose in acute phases, returning toward normal with approach to remission (18). In a sequential study in patients with recurring TTP, we observed that EMP elevation appeared to precede other markers of disease activity such as thrombocytopenia and LDH. In more recent studies, CD62E $(E\text{-selectin})^+$ EMPs were shown to be a more sensitive marker of TTP compared with $CD31^+$ EMPs (5,83). CD62E $(E\text{-selectin})^+$ EMPs have other properties of interest, as discussed in the following section.

Recently, it was shown that many of the EMPs (particularly the CD62 $[E\text{-selectin}]^+$ population) in TTP are positive for vWF (83). This observation is significant, given the putative pathophysiological role for aberrant ADAMTS13-mediated processing/cleavage of ULvWF in TTP (98). It has recently been shown by a ristocetin-based assay that platelet aggregates induced by vWF-positive EMPs are significantly more stable (resistant to dissociation) than aggregates formed by free soluble vWF (57). This effect is expected to enhance formation of platelet-rich thrombi in vivo at sites of EC injury. These data raise the interesting possibility that vWF$^+$ EMP-platelet interactions contribute to TTP pathogenesis.

Preeclampsia

Gonzalez-Quintero and colleagues conducted a prospective study investigating CD31$^+$/CD42b$^-$ and CD51$^+$ EMPs in pregnant women with preeclampsia and found that both types of EMPs were elevated in preeclampsia compared to normal pregnancy (99). EMPs in normal pregnancy were also higher than nonpregnant normal controls. More recently, the same group detected the presence of CD31$^+$/CD42b$^-$ or CD62E (E-selectin)$^+$ EMPs in a larger number of patients with preeclampsia and gestational hypertension (100). Both EMP populations were significantly elevated in preeclampsia (> gestational hypertension > normal pregnant controls). They also observed a correlation between mean arterial pressure and both subsets of EMPs. Moreover, EMP levels paralleled the degree of proteinuria.

VanWijk and colleagues (78,101) investigated various subsets of MPs from ECs, T-helper and suppressor cells, B cells, platelets, granulocytes, monocytes, and erythrocytes in patients with preeclampsia, normal pregnancy, and nonpregnant healthy women. They observed that total numbers of MPs were similar in all three groups and that PMPs constituted the largest proportion of MPs in all groups. In patients with preeclampsia, MPs derived from T-helper and suppressor cells and granulocytes were significantly increased compared to normal pregnant women. Elastase concentration was also significantly increased and it correlated with granulocyte MPs and with blood pressure. Although EMP levels (CD62E [E-selectin]$^+$ and CD144 [VE-cadherin]$^+$ EMPs) were higher in preeclampsia, the difference was not statistically significant in this small number of patients. They suggested that activations of immune system, inflammatory cells, and endothelium are present in preeclampsia and could play a central role in the pathogenesis.

Bretelle and colleagues (102) investigated total MPs (defined as AnV -positive), EMPs (CD51$^+$), and PMPs (CD41$^+$) in normal pregnancy and pregnancy complicated by preeclampsia and intrauterine growth restriction (IUGR). They reported increased total MPs (with increased procoagulant activity), EMPs, and PMPs in normal pregnancy compared with normal nonpregnant women. Unexpectedly in pathologic pregnancy, total MPs, PMPs, or EMPs were lower or not different compared to normal pregnancy. It was speculated that MPs might be sequestered in placenta or bound to other cells (102).

Conflicting reports on MPs in preeclampsia among these different studies remain to be explained. However these investigations employed different assays, measured different species of MPs and included a relatively small number of patients. Nevertheless, the discordance exemplifies the problems associated with MP assays and underscores the need for standardization of the MP assay.

Diabetes Mellitus

Sabatier and colleagues measured CD51$^+$ EMPs, CD41$^+$ PMPs, CD45$^+$ or CD66b$^+$ LMPs, and total AnV$^+$ MPs in patients with type I and 2 DM and age matched healthy controls (103). In type 1 DM, EMPs, PMPs, and total AnV$^+$ MPs were significantly elevated compared to the healthy controls. Total MPs were associated with increased procoagulant activity, which in turn was correlated with HbA1c. The levels of EMPs were correlated with albuminuria in type 1 DM and highest among those with microvascular complications. In contrast, in type 2 DM, total AnV$^+$ MPs were significantly elevated without concomitant increase of their procoagulant activity. These latter patients demonstrated increased levels of LMPs but not EMPs or PMPs. Nomura and colleagues documented elevations of EMPs, PMPs, soluble P-selectin, and CD40L as well as antioxidized LDL antibodies in DM (104). EMP levels correlated with nephropathy and antioxidized LDL antibody titers. Taken together, these studies suggest that procoagulant cell-derived MPs may play a role in the pathogenesis of diabetic vasculopathy, and that EMPs can serve as a marker for endothelial dysfunction in this disease.

Hypertension

Preston and colleagues studied CD31$^+$/CD42b$^-$ EMPs and CD41$^+$ PMPs in patients with severe malignant hypertension (diastolic blood pressure >120 mm Hg), moderate hypertension (95–100 mm Hg), and normotensive controls and found that CD31$^+$/CD42b$^-$ EMPs were most significantly elevated in malignant and moderate hypertension compared to normotensives (105). EMPs correlated with systemic and diastolic blood pressure. PMPs were increased in malignant hypertension but not in moderate hypertension. EMPs were highest in malignant hypertension followed by moderate hypertension, whereas elevated PMPs were seen only in malignant hypertension. They also measured conventional soluble markers of EC disturbance, finding that sVCAM-1, slCAM-1, and vWF did not correlate with blood pressure. Among soluble markers of EC activation, only sVCAM-1 was significantly correlated with EMP levels. These findings indicate that EMPs are a more sensitive marker of endothelial injury associated with hypertension than more widely used soluble markers.

Antiphospholipid Antibody Syndrome and Systemic Lupus Erythematosus

Combes and colleagues first reported elevated EMPs in 30 patients with positive lupus anticoagulant (LA) (20). CD31$^+$/CD51$^+$ EMPs in patients with LA were significantly elevated relative to controls. EMP levels were about twofold higher in LA patients with a history of thrombotic events compared to those without a history of thrombosis. No association was observed between EMPs and age, sex, or anticoagulation.

Recently, Dignat-George and colleagues compared EMPs in five groups of patients: group 1 with antiphospholipid antibody syndrome (APS); group 2 with SLE without thrombosis but with positive antiphospholipid antibodies (APLA); group 3 with SLE without thrombosis and negative APLA; group 4 with thrombosis without SLE or APLA or autoimmune disease; group 5 normal controls matched for sex and age (63). They found elevated CD51$^+$ EMPs in groups 1 and 2 but not in groups 3 or 4. In parallel studies, plasma from different groups of patients was tested for its ability to generate EMPs in EC culture. They found that only APS plasma induced release of EMPs in vitro. They concluded that autoimmune processes associated with APLA induce release of EMPs, predisposing to thrombosis in APS.

Multiple Sclerosis

Minagar and colleagues first documented elevated levels of EMPs in patients with MS (19). Two EMP phenotypes were assayed: CD31$^+$/CD42b$^-$ and CD51$^+$. CD31$^+$ EMPs were significantly higher in patients with exacerbation compared to those in clinical remission and healthy controls and correlated well with gadolinium enhancement in brain MRI and clinical disease activity. On the other hand, CD51$^+$ EMPs did not differ between patients in exacerbation and in remission yet were significantly higher than in healthy controls. These data are consistent with those observed in patients with CAD, providing further evidence that CD31$^+$ EMPs are a marker of acute endothelial injury, whereas CD51$^+$ EMPs reflect chronic endothelial dysfunction.

In a later study, Jy and colleagues demonstrated increased EMP–monocyte conjugates (EMP-mo-C) in patients with MS (58). EMP-mo-C were measured by two-color flow cytometry using anti-CD45 with anti-CD54 (ICAM-1) or anti-CD62E (E-selectin). EMP-mo-C were significantly elevated in patients with MS in exacerbation compared to those in remission or healthy controls. Moreover, EMP62E-mo-C were positively correlated with free CD62E (E-selectin)$^+$ EMP levels and the gadolinium enhancement in MRI lesions. However, CD54 (ICAM-1)$^+$ EMPs were much lower than CD62E (E-selectin)$^+$ EMPs and EMP54-mo-C were not correlated with free CD54 (ICAM-1)$^+$ EMPs. Under in vitro conditions, EMPs were shown to bind preferentially to monocytes (compared with neutrophils or lymphocytes) and to induce monocyte activation and transendothelial migration (58). CD54 (ICAM-1)$^+$ EMPs were most avid in binding, and this would explain very low CD54 (ICAM-1)$^+$ EMPs in peripheral blood of patients with MS (i.e., it exists but in bound form, not free). This study provided clues to the functional role of EMPs in MS, an inflammatory disorder. A more recent study confirmed these findings and extended them to gain insight on therapy with interferon (106).

Systemic Vasculitis

Brogan and colleagues studied CD62E (E-selectin)$^+$ and CD105 (endoglin)$^+$ EMPs, along with PMPs, in pediatric systemic vasculitis (107). An elevation in CD62E (E-selectin)$^+$ EMPs and to a lesser extent CD105 (endoglin)$^+$ EMPs was observed in this disorder, relative to healthy or febrile controls. EMPs correlated with the activity of vasculitis and acute phase reaction products and were significantly higher than in patients who underwent remission. The elevation in EMPs in this disorder provides further evidence of EC injury (activation) in vasculitis.

Combes and colleagues reported increased CD51$^+$ EMPs in pediatric patients with uncomplicated malaria, cerebral malaria, and severe malaria with coma (108). EMPs were elevated upon admission in cerebral malaria compared to uncomplicated malaria or normal controls. Those with severe malaria presenting with coma had the highest EMP levels at admission. This finding indicates EC injury in severe and cerebral malaria and supports previous evidence for EC involvement in *P. falciparum* infection (108).

In a study by Ogura and colleagues, CD31$^+$/CD54 (ICAM-1)$^+$ EMPs and EMP–neutrophil conjugates (EMP-N-C; CD31$^+$/CD62E [E-selectin]$^+$) were evaluated in patients with severe systemic inflammatory response syndrome (SIRS) (109). Patients exhibited elevated platelet and leukocyte counts, high C-reactive protein, and increased soluble markers such as thrombin-antithrombin III complexes, PAI-1, and D-dimer. Also elevated in SIRS were soluble CD62E (E-selectin), TM, PAI-1, and leukocyte oxidative activity. Circulating EMPs were significantly increased in SIRS patients, however, no significant correlation was observed between EMPs and soluble CD62E (E-selectin), TM or PAI-1. EMPs from SIRS were found to bind preferentially to neutrophils to form EMP–neutrophil complexes that were significantly elevated in SIRS. Based on these findings, it was hypothesized that EMPs bind to neutrophils to promote inflammation and thus may play an active role in SIRS-associated inflammation.

Sepsis

Sepsis is associated with high mortality, and endothelial activation and apoptosis have been implicated in systemic multiorgan failure. The high mortality in sepsis was attributed in the past to uncontrolled hyperinflammation but recent findings challenge this, indicating that immunosuppression and impaired inflammatory reaction contributes to mortality (110). Soriano and colleagues studied EMP profiles in sepsis, along with intracellular NO (111), and investigated whether EMP profiles can predict mortality. Thirty-five patients with severe sepsis were recruited and EMPs (CD31$^+$/CD41$^-$ EMPs, CD62 [E-selectin]$^+$ EMPs) and platelet–leukocyte conjugates were assayed along with other markers on days 1, 2, and 3 of admission. EMP levels were compared between survivors and nonsurvivors. Overall mortality was 51% at 28 days. CD31$^+$ EMPs and platelet–leukocyte conjugates were significantly

Table 175-3: Summary of EMP Markers Used in Various Clinical Settings

	CD146+/ *CD31*+	*CD51*+	*CD31*+/ *CD51*+	*CD31*+/*CD42*−	*CD54*+/ *CD31*+	*CD62E*+	*vWF*+	*CD144*+	*CD105*+	*CD105*+ *CD54*+ *or CD144*+
ACS	√	√ (=SA)		√ (>SA)						
SA		√		√						
Metabolic syndrome					√					
PP hyperlipidemia				√		No				
TTP		√		√		√	√			
Preeclampsia		√		√		√		√		
Type 1 DM		√								
HTN				√						
LA			√							
MS remission		√		√						
MS exacerbation		No		√						
Vasculitis						√			√	
CM		√								
SIRS					√					
Sepsis				Reduced						
SCD									√	√
PNH									√	√
APS				√						

CM, cerebral malaria.

depressed, whereas NO in leukocytes was increased in nonsurvivors compared to survivors. However CD62E (E-selectin)$^+$ EMPs did not differ between two groups. EMP profiles measured at diagnosis of sepsis appear to be useful in predicting mortality in sepsis and support the hypothesis that reduced EC activation and impaired inflammatory response is related to mortality in sepsis. Therapeutic measures that suppress inflammatory responses should be carefully reassessed in the treatment of sepsis.

Sickle Cell Disease

Shet and colleagues investigated circulating MPs in patients with sickle cell disease (SCD) crises, those with steady-state SCD, and normal controls (66). MPs were assayed by flow cytometry. Tissue factor expression and procoagulant activity of MPs was also evaluated. Multiple kinds of MPs were detected in patients with SCD, with MPs derived from erythrocytes (RMPs) being most abundant, followed by EMPs, PMPs, and LMPs. Total MPs were elevated in crisis compared to steady-state. EMPs and monocyte-derived MPs were positive for tissue factor and had procoagulant activity. Total MPs and tissue factor-positive MPs were highest in crisis, lowest in healthy controls, and intermediate in steady-state SCD. In functional studies, MPs from SCD but not from controls exhibited increased procoagulant activity. These data are consistent with the established role for inflammation and endothelial dysfunction in SCD and raise the possibility that procoagulant MPs may contribute to the thrombotic phenotype in this disease (66).

Paroxysmal Nocturnal Hemoglobinuria

Simak and colleagues measured EMPs and MPs derived from platelets, leukocytes, and erythrocytes in patients with PNH, SCD, aplastic anemia (AA), and normal healthy controls (48). Employing three-color flow cytometry, EMPs were defined as CD105 (endoglin)$^+$/CD41$^-$, CD45$^-$; PMPs as CD41$^+$, CD105 (endoglin)$^-$, CD45$^-$; LMPs as CD45$^+$, CD105 (endoglin)$^-$, CD41$^-$; and RMPs as CD235a$^+$, CD105 (endoglin)$^-$, CD45$^-$. PS-positive MPs were detected by AnV-binding. They observed that CD105 (endoglin)$^+$ EMPs expressing AnV were significantly higher in SCD and PNH than in AA or healthy controls; highest in SCD followed by PNH. An even more striking difference was seen when EMPs expressing inflammatory markers (CD105 (endoglin)$^+$, CD54 [ICAM-1]$^+$ EMPs) were compared. MPs derived from platelets, leukocytes, and erythrocytes were also investigated and it was found that AnV$^+$ MPs and CD41$^+$ PMPs were significantly higher in SCD and PNH, whereas MPs derived from erythrocytes and leukocytes were significantly higher in SCD compared to PNH, AA, and healthy controls. Other species of EMPs (CD144 [VE-cadherin]$^+$ CD105 [endoglin]$^+$ MPs) were also increased in SCD and PNH over AA and controls. They concluded that MP profiles suggested inflammatory status of activated endothelium in SCD and PNH, but not in AA and that analysis of EMPs appears promising to provide useful information on the status vascular endothelium.

KEY POINTS

- EMPs are membranous vesicles released upon perturbation of ECs such as activation or apoptosis. They are heterogeneous in size, density, phenotypes, and likely functions and carry membrane proteins and PLs of the parent cells. Accordingly, EMP analysis can provide valuable information regarding the status of ECs.
- EMP assays is emerging as a useful biomarker reflecting endothelial disturbances in various thrombotic, inflammatory, and vascular disorders. Main types of clinical disorders to which EMP analysis are now applied include (a) thrombotic disorders such as ACS, hypertension, DM; (b) small-vessel diseases such as TTP, preeclampsia, and some ischemic dementias; (c) autoimmune and inflammatory disorders such as MS, SLE, APS, systemic vasculitis, and sepsis; and (d) hemolytic disorders such as SCD and PNH. By employing extended panels of markers for phenotype analysis, the assay can detect multiple species of EMPs, and their interaction with other cells that have been applied in clinical studies (Table 175-3).

Future Goals

- To achieve standardization of methods, which is a major challenge that must be surmounted before EMP assays can enter the mainstream of clinical testing (55)
- To understand the mechanisms of release, which are important to relating type of injury to molecular pathways responsible for a specific phenotype of EMPs. Our finding that capping seems to precede release of at least some species of EMPs (40) is also a step in that direction. Some workers are developing methods that specifically inhibit MP release, a potentially valuable research tool.
- To understand the functions of EMPs. There is growing consensus that EMPs and other MPs can act as diffusible messengers, transporting bioactive agents to initiate and mediate coagulation, inflammation, and cell-to-cell interactions (64). EMPs are also known to express tissue factor, promote coagulation, carry off complement, activate leukocytes, and enhance leukocyte transmigration (57,74). The role of EMPs in binding large-multimer vWF, promoting unusually stable platelet aggregates is recently documented in TTP (57). New roles of EMPs remain to be elucidated in the future.
- To search for greater specificity. Clinical studies showed that EMP assay can yield specific information such as activation versus apoptosis and acute versus chronic injury and even mild or transient or advanced EC perturbations. However, greater specificity of the assay is needed to distinguish various endothelial perturbation in clinical disorders. It is expected that more refined phenotypic profiling of EMPs along with functional assay will lead to further improvements in discriminating specific disease states or kinds of endothelial injury. Ideally, future research could lead to the ability to identify the site of vascular injury (organ or vascular bed), because it is already known that each organ or vascular bed contains unique markers ("addresses"), which might be carried on EMPs. This will further improve the sensitivity and specificity of the assay.

REFERENCES

1 Aird WC. Endothelial cell dynamics and complexity theory. *Crit Care Med.* 2002;30:S180–S185.

2 Szekanecz Z, Koch AE. Vascular endothelium and immune responses: implications for inflammation and angiogenesis. *Rheum Dis Clin North Am.* 2004;30:97–114.

3 Horstman LL, Jy W, Jimenez JJ, Ahn YS. Endothelial microparticles as markers of endothelial dysfunction (Review). *Front Biosci.* 2004;9:1118–1135.

4 Becker BF, Heindl B, Kupatt C, Zahler S. Endothelial function and hemostasis. *Z Kardiol.* 2000;89:160–167.

5 Jimenez JJ, Jy W, Mauro L, et al. Endothelial cells release phenotypically and quantitatively distinct microparticles in activation and apoptosis. *Thromb Res.* 2003;109:175–180.

6 Horstman LL, Ahn YS. Platelet microparticles: a wide-angle perspective (review). *Crit Rev Oncol Hematol.* 1999;30:111–142.

7 Denzer K, Kleijmeer MJ, Heijnen HFG, et al. Exosome: from internal vesicle of the multivesicular body to intercellular signaling device (review). *J Cell Sci.* 2000;113:3365–3374.

8 Horstman LL, Jy W, Jimenez J, et al. New horizons in the analysis of circulating cell-derived microparticles (review). *Keio J Med.* 2004;53:210–230.

9 Wolf P. The nature and significance of platelet products in human plasma. *Br J Haematol.* 1967;13:269–288.

10 Zucker-Franklin D, Karpatkin S. Red cell and platelet fragmentation in idiopathic thrombocytopenic purpura. *N Engl J Med.* 1977;297:517–523.

11 Hamilton KK, Hattori R, Esmon CT, Sims PJ. Complement proteins C5b-9 induce vesiculation of the endothelial plasma membrane and expose catalytic surface for assembly of the prothrombinase enzyme complex. *J Biol Chem.* 1990;265:3809–3814.

12 Ahn YS, Horstman LL. Idiopathic thrombocytopenic purpura: pathophysiology and management. *Int J Hematol.* 2002;76:123–131.
13 Jy W, Horstman LL, Arce M, Ahn YS. Clinical significance of platelet microparticles in autoimmune thrombocytopenias. *J Lab Clin Med.* 1992;119:334–345.
14 Ahn YS, Horstman LL, Jy W, et al. Vascular dementia in patients with immune thrombocytopenic purpura (ITP). *Thromb Res.* 2002;107:337–344.
15 Jy W, Mao WW, Horstman LL, et al. Platelet microparticles bind, activate and aggregate neutrophils in vitro. *Blood Cells Mol Dis.* 1995;21:217–231
16 Jimenez J, Jy W, Horstman LL, et al. Microvascular endothelial cell microparticles (EMP): markers of endothelial cell activation/damage. *J Invest Med.* 2000;48(160):191A.
17 Minagar A, Jy W, Jimenez JJ, et al. Increased endothelial microparticles (EMP) and platelet activation in multiple sclerosis (MS) in exacerbation and presence in plasma of a factor inducing EMP generation in vitro. *Blood.* 2000;96:A43.
18 Jimenez J, Jy W, Mauro L, et al. Elevated endothelial microparticles in thrombotic thrombocytopenic purpura (TTP): findings from brain and renal microvascular cell culture and patients with active disease. *Br J Haematol.* 2001;112:81–90.
19 Minagar A, Jy W, Jimenez JJ, et al. Elevated plasma endothelial microparticles in multiple sclerosis. *Neurology.* 2001;56:1319–1324.
20 Combes V, Simon AC, Grau GE, et al. In vitro generation of endothelial microparticles and possible prothrombotic activity in patients with lupus anticoagulant. *J Clin Invest.* 1999;104:93–102.
21 Bombeli T, Mueller M, Haeberli A. Anticoagulant properties of the vascular endothelium. *Thromb Haemost.* 1997;77:408–423.
22 Pober JS. Effects of tumor necrosis factor and related cytokines on vascular endothelial cells. *Ciba Symp.* 1987;131:17–184.
23 Madge LA, Pober JS. TNF signaling in vascular endothelial cells. *Exp Mol Pathol.* 2001;70:317–325.
24 Stehlik C, de Martin R, Kumabashiri I, et al. Nuclear factor (NF)-kappaB-regulated X-chromosome-linked iap gene expression protects endothelial cells from tumor necrosis factor alpha-induced apoptosis. *J Exp Med.* 1998;188:211–216.
25 Yeh ETH, Willerson JT. Coming of age of CRP: inflammatory markers in cardiology. *Circulation.* 2003;107:370–372.
26 Moake JL, Rudy CK, Troll JH, et al. Unusually large plasma factor VIII: von Willebrand factor multimers in chronic relapsing TTP. *N Engl J Med.* 1982;307:1432–1435.
27 Mulder AB, Hegge-Paping KS, Magielse CP, et al. Tumor necrosis factor alpha-induced endothelial tissue factor is located on the cell surface rather than in the subendothelial matrix. *Blood.* 1994;84:1559–1566.
28 Freyssinet JM, Toti F, Hugel B, et al. Apoptosis in vascular disease. *Thromb Haemost.* 1999;82:727–735.
29 Zwaal RFA, Schroit AJ. Pathophysiologic implications of membrane phospholipid asymmetry in blood cells. *Blood.* 1997;89:1121–1132.
30 Ruf A, Morgenstern E. Ultrastructural aspects of platelet adhesion on subendothelial structures. *Sem Thromb Haemost.* 1995;21:119–122.
31 Min C, Kang E, Yu SH, et al. Advanced glycation end products induce apoptosis and procoagulant activity in cultured human umbilical vein endothelial cells. *Diabetes Res Clin Pract.* 1999;46:197–202.
32 Hogg N, Browning J, Howard T, et al. Apoptosis in vascular endothelial cells caused by serum deprivation, oxidative stress and transforming growth factor-beta. *Endothelium.* 1999;7:35–39.
33 Andre AJH, W WN. Circulating adhesion molecules in disease. *Immunol Today.* 1993;14:506–512.
34 Cella G, Randi ML. Soluble thrombomodulin as predictor of incident coronary heart disease (One of 2 letters responding to Salomaa). *Lancet.* 1999;354:425–426.
35 Mannucci PM. Von Willebrand factor: a marker of endothelial damage? *Arterioscler Thromb Vasc Biol.* 1998;18:1359–1362.
36 Hladovec J, Prerovsky I, Stanek V, Fabian J. Circulating endothelial cells in acute myocardial infarction and angina pectoris. *Klin Wochenschr.* 1978;56:1033–1036.
37 Woywodt A, Bahlmann F, Groot KD, et al. Circulating endothelial cells: life, death, detachment and repair of the endothelial cell layer. *Nephrol Dial Transplant.* 2002;17:1728–1730.
38 Miyazaki Y, Nomura S, Miyake T, et al. High shear stress can initiate both platelet aggregation and shedding of procoagulant containing microparticles. *Blood.* 1996;88:3456–3464.
39 Mills JC, Stone NL, Erhardt J, Pittman RN. Apoptotic membrane blebbing is regulated by myosin light chain phosphorylation. *J Cell Biol.* 1998;140:627–636.
40 Rohn TT, Cusack SM, Kessinger SR, Oxford JT. Caspase activation independent of cell death is required for proper cell dispersal and correct morphology in PC12 cells. *Exp Cell Res.* 2004;295:215–225.
41 Jy W, Jimenez JJ, Mauro LM, et al. Agonist-induced capping of adhesion proteins and microparticle shedding in culture of human renal vascular endothelial cells. *Endothelium.* 2002;9:179–189.
42 Millan J, Montoya MC, Sancho D, et al. Lipid rafts mediate biosynthetic transport to the T lymphocyte uropod subdomain and are necessary for uropod integrity and function. *Blood.* 2002;99:978–984.
43 Dietzen DJ, Jack GG, Page KL, et al. Localization of tissue factor pathway inhibitor to lipid rafts is not required for inhibition of factor VIIa / tissue factor activity. *Thromb Haemost.* 2003;89:65–73.
44 Mitra D, Jaffe E, Weksler B, et al. Thrombotic thrombocytopenic purpura and sporadic hemolytic uremic syndrome plasmas induce apoptosis in restricted lineages of human microvascular endothelial cells. *Blood.* 1997;89(4):1224–1234.
45 Jimenez JJ, Jy W, Mauro LM, et al. Response to letter-to-editor by Laurence (Invited). *Br J Haematol.* 2004;125:416.
46 Kagawa H, Komiyama Y, Nakamura S, et al. Expression of functional tissue factor on small vesicles of lipopolysaccharide-stimulated human vascular endothelial cells. *Thromb Res.* 1998;91:297–304.
47 Bernal-Mizrachi L, Jy W, Jimenez JJ, et al. High levels of circulating endothelial microparticles in patients with acute coronary syndromes.. *Am Heart J.* 2003;145:962–970.
48 Simak J, Holada K, Risitano AM, et al. Elevated circulating endothelial membrane microparticles in paroxysmal nocturnal hemoglobinuria. *Br J Haematol.* 2004;125:804–813.
49 Weerheim AM, Kolb AM, Sturk A, Nieuwland R. Phospholipid composition of cell-derived microparticles determined by one-dimensional high-performance thin-layer chromatography. *Anal Biochem.* 2002;302:191–198.
50 Fourcade O, Simon M-F, Viode C, et al. Secretory phospholipase A2 generates the novel lipid mediator lysophosphatidic

acid in membrane microvesicles shed from activated cells. *Cell.* 1995;80:919–927.

51 Berckmans RJ, Neiwland R, Boing AN, Sturk A. Cell-derived microparticles circulate in healthy humans and support low grade thrombin generation. *Thromb Haemost.* 2001;85:639–646.

52 Berckmans RJ, Nieuwland R, Tak PP, et al. Cell-derived microparticles in synovial fluid from inflamed arthritic joints support coagulation exclusively via a factor VII-dependent mechanism. *Arthritis Rheum.* 2002;46:2857–2866.

53 Blanchard N, Lankar D, Faure F, et al. TCR activation of human T cells induces the production of exosomes bearing the TCR/CD3/zeta complex. *J Immunol.* 2002;168:3235–3241.

54 Salzer U, Prohaska R. Stomatin, flotillin-1, and flotillin-2 are major integral proteins of erythrocyte lipid rafts. *Blood.* 2001;97: 1141–1143.

55 Jy W, Horstman LL, Jimenez JJ, Ahn YS. Measuring circulating cell-derived microparticles (forum). *J Thromb Haemost.* 2004; 2:1842–1851.

56 Jimenez JJ, Jy W, Horstman LL, et al. Measuring circulating cell-derived microparticles. *J Thromb Haemost.* 2004;2:1850–1851.

57 Jy W, Jimenez JJ, Mauro LM, et al. Endothelial microparticles (EMP) interact with platelets via a vWF dependent pathway to form platelet aggregates, more resistant to dissociation than those induced by soluble vWF. *J Thromb Haemost.* 2005; 3(6):1301–1308.

58 Jy W, Minagar A, Jimenez JJ, et al. Endothelial microparticles (EMP) bind to monocytes to activate and enhance transmigration: elevated circulating EMP-monocyte conjugates in multiple sclerosis. *Front Biosci.* 2004;9:3137–3144.

59 Mallat Z, Hugel B, Ohan J, et al. Shed membrane microparticles with procoagulant potential in human atherosclerotic plaques: a role for apoptosis in plaque thrombogenicity. *Circulation.* 1999; 99:348–353.

60 Hugel B, Zobairi F, Freyssinet JM. Measuring circulating cell-derived microparticles. *J Thromb Haemost.* 2004;2:1846–1847.

61 Biro E, Sturk-Maquelin N, Vogel GMT, et al. Human cell-derived microparticles promote thrombus formation in vivo in a tissue factor-dependent manner. *J Thromb Haemost.* 2003;1: 2561–2568.

62 Shet AS, Key NS, Hebbel RP. Measuring circulating cell-derived microparticles. *J Thromb Haemost.* 2004;2:1848–1850.

63 Dignat-George F, Camoin-Jau L, Sabatier F, et al. Endothelial microparticles: a potential contribution to the thrombotic complications of the antiphospholipid syndrome (with editorial, pg 636–638). *Thromb Haemost.* 2004;91:667–673.

64 Freyssinet JM. Cellular microparticles: what are they bad or good for? *J Thromb Haemost.* 2003;1:1655–1662.

65 Ahn YS, Jy W, Jimenez JJ, Horstman LL. More on: cellular microparticles: what are they bad or good for? (Letter, invited). *J Thromb Haemost.* 2004;2:1215–1216.

66 Shet AS, Aras O, Gupta K, et al. Sickle blood contains tissue factor-positive microparticles derived from endothelial cells and monocytes. *Blood.* 2003;102:2678–2683.

67 Mallat Z, Benamer H, Hugel B, et al. Elevated levels of shed membrane microparticles with procoagulant potential in the peripheral circulating blood of patients with acute coronary symptoms. *Circulation.* 2000;101:841–843.

68 Steppich B, Mattisek C, Sobczyk D, et al. Tissue factor pathway inhibitor on circulating microparticles in acute myocardial infarction. *Thromb Haemost.* 2005;93:35–39.

69 Camerer E, Kolsta AB, Prydz H. Cell biology of tissue factor, the principal initiator of blood coagulation. *Thromb Res.* 1996;81:1–41

70 Bajaj MS, Birktoft JS, Steer SA, Bajaj SP. Structure and biology of tissue factor pathway inhibitor. *Thromb Haemost.* 2001; 86:959–972.

71 Horstman LL, Jy W, Bidot CJ, et al. Tissue factor (TF) activity in trauma patients is suppressed on cell-derived microparticles (MP). *Blood.* 2004;104:A297.

72 Tans G, Rosing J, Christella M, et al. Comparison of anticoagulant and procoagulant activities of stimulated platelets and platelet-derived microparticles. *Blood.* 1991;77:2641–2648.

73 Gris JC, Toulon P, Brun S, et al. The relationship between plasma microparticles, protein S, and anticardiolipid antibodies in patients with human immunodeficiency virus infection. *Thromb Haemost.* 1996;76(1):38–45.

74 Barry OP, Pratico D, Savani RC, FitzGerald GA. Modulation of monocyte-endothelial cell interactions by platelet microparticles. *J Clin Invest.* 1998;102:136–144.

75 Mesri M, Altieri DC. Leukocyte microparticles stimulate endothelial cell cytokine release and tissue factor production in a JNK1 signalling pathway. *J Biol Chem.* 1999;274:23111–23118.

76 Sabatier F, Roux V, Anfosso F, et al. Interaction of endothelial microparticles with monocytic cells in vitro induces tissue factor-dependent procoagulant activity. *Blood.* 2002;99:3962–3970.

77 Boulanger CM, Scoazec A, Ebrahimian T, et al. Circulating microparticles from patients with myocardial infarctions cause endothelial dysfunction. *Circulation.* 2001;104:2649–2652.

78 VanWijk MJ, Nieuwland R, Boer K, et al. Microparticle subpopulations are increased in preeclampsia: Possible involvement in vascular dysfunction? *Am J Obstet Gynecol.* 2002;187:450–456.

79 VanWijk MJ, Svedas E, Boer K, et al. Isolated microparticles, but not whole plasma, from women with preeclampsia impair endothelium-dependent relaxation in isolated myometrial arteries from healthy pregnant women. *Am J Obstet Gynecol.* 2002;187:1686–1693.

80 Rikitake Y, Hirata K, Kawashima S, et al. Inhibition of endothelium-dependent arterial relaxation by oxidized phosphatidylcholine. *Atherosclerosis.* 2000;152:79–87.

81 Barry OP, Pratico D, FitzGerald G. Platelet microparticles enhance adhesive interactions between monocytes and endothelial cells. *J Clin Invest.* 1997;45:A271.

82 Kim HK, Song KS, Chung JH, et al. Platelet microparticles induce angiogenesis in vitro. *Br J Haematol.* 2004;124:376–384.

83 Jimenez JJ, Jy W, Mauro LM, et al. Endothelial microparticles released in thrombotic thrombocytopenic purpura express von Willebrand factor and markers of endothelial activation. *Br J Haematol.* 2003;123:896–902.

84 Simak J, Holada K, D'Agnillo F, et al. Cellular prion protein is expressed on endothelial cells and is released during apoptosis on membrane microparticles found in human plasma. *Transfusion.* 2002;42:334–342.

85 Abid Hussein MN, Nieuwland R, Hau CM, et al. Cell-derived microparticles contain caspase-3 in vitro and in vivo. *J Thromb Haemost.* 2005; 3(5):888–896.

86 Brodsky SV, Zhang F, Nasjletti A, Goligorsky MS. Endothelium-derived microparticles impair endothelial function in vitro. *Am J Physiol-Heart Circ Physiol.* 2004;286:H1910–H1915.

87 Fadok VA, Bratton DL, Rose DM, et al. A receptor for phosphatidylserine-specific clearance of apoptotic cells. *Nature.* 2000;405:85–90.

88 Li MO, Sarkisian MR, Mehal WZ, et al. Phosphatidylserine receptor is required for clearance of apoptotic cells. *Science.* 2003;302:1560–1563.

89 Bose J, Gruber AD, Helming L, et al. The phosphatidylserine receptor has essential functions during embryogenesis but not in apoptotic cell removal. *J Biol.* 2004;3:15.

90 Bernal-Mizrachi L, Jy W, Fiero C, et al. Endothelial microparticles correlate with high-risk angiographic lesions in acute coronary syndromes. *Int J Cardiol.* 2004;97:439–446.

91 Morel O, Jesel L, Hugel B, et al. Protective effects of vitamin C on endothelium damage and platelet activation during myocardial infarction in patients with sustained generation of circulating microparticles. *J Thromb Haemost*. 2003;1:171–177.

92 Heloire F, Weill B, Weber S, Batteux. F. Aggregates of endothelial microparticles and platelets circulate in peripheral blood. Variations during stable coronary disease and acute myocardial infarction. *Thromb Res.* 2003;110:173–180.

93 Ito MK. The metabolic syndrome: pathophysiology, clinical relevance, and use of niacin. *Ann Pharmacother*. 2004;38:277–285.

94 Steinbaum SR. The metabolic syndrome: an emerging health epidemic in women. *Prog Cardiovasc Dis.* 2004;46:321–336.

95 Arteaga R, Soriano AO, Chirinos JA, et al. Elevated endothelial microparticles (EMP) and platelet activation in patients with the metabolic syndrome (MTS) at low to intermediate risk for cardiovascular diseases. *Diab Vasc Dis Res.* 1:S34–S36.

96 Ferreira AC, Peter AA, Mendez AJ, et al. Postprandial hypertriglyceridemia increases circulating levels of endothelial cell microparticles. *Circulation.* 2004;110:3599–3603.

97 Jimenez JJ, Jy W, Horstman LL, et al. TTP plasma induces release of procoagulant EMP (endothelial microparticles) from endothelial cells in culture; and EMP is markedly elevated in patients. *Blood.* 2000;96:A531.

98 Levy GG, Nichols WC, Lian EC, et al. Mutations in a member of the ADAMTS gene family cause thrombotic thrombocytopenic purpura. *Nature.* 2001;413:488–494.

99 Gonzalez-Quintero V, Jimenez JJ, Jy W, et al. Elevated plasma endothelial microparticles in preeclampsia. *Am J Obstet Gynecol.* 2003;189:589–593.

100 Gonzalez-Quintero VH, Smarkusky L, Jimenez JJ, et al. Elevated endothelial microparticles in preeclampsia correlate with proteinuria and hypertension. *Am J Obstet Gynecol.* 2004;191:1418–1424.

101 VanWijk MJ, Boer K, Berckmans RJ, et al. Enhanced coagulation activation in preeclampsia: the role of APC resistance, microparticles and other plasma constituents. *Thromb Haemost*. 2002;88:415–420.

102 Bretelle F, Sabatier F, Desprez D, et al. Circulating microparticles: a marker of procoagulant state in normal pregnancy and pregnancy complicated by preeclampsia or intrauterine growth restriction. *Thromb Haemost*. 2003;89:486–492.

103 Sabatier F, Darmon P, Hugel P, et al. Type 1 and type 2 diabetic patients display different patterns of cellular microparticles. *Diabetes.* 2002;51:2840–2845.

104 Nomura S, Shouzu A, Omoto S, et al. Activated platelet and oxidized LDL induce endothelial membrane vesiculation: clinical significance of endothelial cell-derived microparticles in patients with type 2 diabetes. *Clin Appl Thromb Hemost*. 2004;10:205–215.

105 Preston RA, Jy W, Jimenez JJ, et al. Effect of severe hypertension on endothelial and platelet microparticles. *Hypertension.* 2003;41:211–217.

106 Jimenez JJ, Jy W, Mauro LM, et al. Elevated endothelial microparticle-monocyte complexes induced by multiple sclerosis plasma and the inhibitory effects of interferon-beta 1b on release of endothelial microparticles, formation and transendothelial migration of monocyte-endothelial microparticle complexes. *Mult Scler*. 2005;11(3):310–315.

107 Brogan RA, Shah V, Brachet C, et al. Endothelial microparticles in vasculitis of the young. *Arthritis Rheum.* 2004;50:927–936.

108 Combes V, Taylor TE, Juhan-Vague I, et al. Circulating endothelial microparticles in Malawaian children with sever falciparum malaria complicated with coma. *JAMA.* 2004;291:2542–2544.

109 Ogura H, Tanka H, Koh T, et al. Enhanced production of endothelial microparticles with increased binding to leukocytes in patients with severe systemic inflammatory response. *J Trauma.* 2004;56:823–830.

110 Hotchkiss RS, Karl IE. The pathophysiology and treatment of sepsis. *N Engl J Med.* 2003;348:138–150.

111 Soriano AO, Jy W, Valdivia MA, et al. Levels of endothelial and platelet microparticles and their interactions with leukocytes correlate with organ dysfunction and predict mortality in severe sepsis. *Crit Care Med.* 2005;33:2540–2546.

112 Biro E, Nieuwland R, Sturk A. Measuring circulating cell-derived microparticles. *J Thromb Haemost.* 2004;2(10):1843–1844.

113 Dignat-George F, Sabatier F, Camoin-Jau L, et al. Measuring circulating cell-derived microparticles. *J Thromb Haemost.* 2004;2(10): 1844–1846.

176

Molecular Magnetic Resonance Imaging

Susan B. Yeon*, Andrea J. Wiethoff†, Warren J. Manning*, Elmar Spuentrup‡, and Rene M. Botnar*,§

*Beth Israel Deaconess Medical Center, Harvard Medical School, Boston, Massachusetts; †EPIX Pharmaceuticals, Cambridge, Massachusetts; ‡Technical University of Aachen, Germany; §Technical University Munich, Germany

Clinical assessment of the endothelium and vasculature by magnetic resonance imaging (MRI) has conventionally focused on assessment of lumen integrity using contrast and noncontrast angiographic techniques and assessment of macroscopic alterations in vessel wall structure (1–5). Phase-contrast MRI may be employed to measure intravascular flow velocity, and cine MRI may be used to assess vascular distensibility (6). These techniques can be applied to detect endothelial dysfunction as measured by flow-mediated dilation (as has been studied more extensively using ultrasound techniques) (7). Thus, MRI offers a range of applications for the assessment of the macroscopic structure and function of the vascular system. In addition, MRI may be used to assess microvascular characteristics and effects, including microvascular density and vascular permeability (8,9), regional tissue perfusion (10), and microvascular obstruction (11). Building on these MRI capabilities, the development of molecular MRI targeted to detect alterations in the endothelial cell (EC) and its environment may allow integration of novel information about the state of the endothelium into conventional MRI vascular assessment. The endothelium is a particularly appealing site for targeting by molecular probes because of its functional importance and because it is bathed by the bloodstream into which such probes are conventionally administered.

The imaging of molecular targets is a developing method for improving the characterization and detection of normal and disease states. MRI can provide high spatial resolution and structural definition, which is useful for imaging processes at the molecular and cellular level (12). Furthermore, the availability of a wide range of MR scanners, ranging from small-bore animal scanners to whole-body clinical systems, provides a means to bridge the gap between experimental models and clinical application. The availability of high-field animal scanners is especially important because many basic molecular imaging experiments are first performed in small animals for which particularly high spatial resolution is required for meaningful detection and localization. In contrast to some of the other imaging modalities (e.g., single photon emission computed tomography [SPECT], positron emission tomography [PET]), however, MRI has inherently lower sensitivity for probe detection, which complicates its use for imaging low quantities of molecular markers.

The goals of this chapter are to provide an overview of the MR physics important for molecular imaging and to review the current state of molecular MRI of the endothelium. The initial sections first review the basic principles of MR, with an emphasis on signal and contrast manipulation and instrumentation requirements and then discuss how the properties of instrumentation have an impact on the development of contrast agents. The later sections focus on the basic properties of targeted contrast agents, including target identification, targeting approaches, and signal amplification strategies. Finally, current and potential applications of cardiovascular molecular MRI will be described.

PRINCIPLES OF MAGNETIC RESONANCE

MRI is based on the principle of nuclear magnetic resonance. Nuclei consisting of an odd number of protons and/or neutrons have a magnetic moment. When placed inside a strong magnetic field, some of the nuclei align with the magnetic field, establishing a net longitudinal magnetization with the proton nuclei of water (hydrogen nuclei from water are the most abundant in the body). The nuclei precess at a frequency directly proportional to the strength of the main magnetic field. Recovery of magnetization in the longitudinal direction (T1) and decay of magnetization in the transverse plane (T2, T2*) are the basis of soft-tissue contrast in MRI.

Magnetic Resonance Signal Intensity

Signal intensity in MRI primarily depends on the local values of the longitudinal (1/T1) and transverse (1/T2) relaxation rate of water protons. Depending on the pulse sequence, signal usually tends to increase with shorter T1 (higher 1/T1) and decrease with shorter T2 (higher 1/T2) relaxation times. The environment in which the nuclei are located determines the MR signals created. Therefore, by manipulating the chemical environment around the protons, the signal can be altered. MR contrast agents have been developed as a way to modulate the chemical environment inside an organism. The relaxivities r1 and r2, which are commonly expressed in $(\text{mM} \times \text{s})^{-1}$s indicate the increase in 1/T1 and 1/T2 per concentration of contrast agent:

$$1/\text{T1} = 1/\text{T1}_0 + \text{r1}\,[\text{contrast agent}] \qquad \text{(Equation 1)}$$

$$1/\text{T2} = 1/\text{T2}_0 + \text{r2}\,[\text{contrast agent}] \qquad \text{(Equation 2)}$$

with T1_0 and T2_0 being the relaxation times of native tissue (i.e., tissue devoid of exogenous contrast agent).

Gadolinium (Gd)-based contrast agents usually increase 1/T1 and 1/T2 in similar amounts ($\text{r2/r1} \cong 1\text{–}2$) (13–15) whereas iron particle–based contrast agents have a much stronger effect on increasing 1/T2 ($\text{r2/r1} > 10$) (16). Gadolinium-based contrast agents therefore lead to a positive contrast effect (detected as an increase in signal intensity or brightness), whereas iron particle–based contrast agents usually cause a negative contrast effect (detected as a decrease in signal intensity or darkness). MR pulse sequences that emphasize differences in T1 and T2 are commonly referred to as *T1-* and *T2-weighted* sequences. Apart from their effect in increasing 1/T2, iron particles also increase $1/\text{T2}^*$ due to their effect on the local magnetic field B_0, thus causing local field inhomogeneities ΔB_0. This additional effect leads to even more severe signal decay.

$$1/\text{T2}^* = 1/\text{T2} + \gamma \Delta B_0 \qquad \text{(Equation 3)}$$

Iron-based contrast agents are therefore best imaged using T2*-weighted imaging sequences. For signal quantification, T2-weighted multiecho spin echo sequences or T2*-weighted multiecho gradient echosequences can be used to generate T2 or T2* maps. Typical r1 and r2 values of currently approved Gd-based contrast agents are in the range of $\text{r1} = 3\text{–}5\ (\text{mM} \times \text{s})^{-1}$ and $\text{r2} = 5\text{–}6\ (\text{mM} \times \text{s})^{-1}$. The relaxivities of iron-based contrast agents are significantly higher $\text{r1} = 20\text{–}25\ (\text{mM} \times \text{s})^{-1}$ and $\text{r2} = 100\text{–}200\ (\text{mM} \times \text{s})^{-1}$. Due to the low concentrations at which molecular imaging targets are generally found, relaxivity is important in the design of molecular contrast agents and will be discussed further in the second part of this chapter.

Magnetic Resonance Image Acquisition

For cardiac imaging, cardiac as well as respiratory motion must be effectively handled. Motion compensation requirements for noncoronary vessels such as the aorta or the carotid arteries are less stringent.

Cardiac and Respiratory Motion Compensation

MRI signal generation and detection are governed by nuclear magnetic resonance physics and subject to instrumental constraints. Because magnetic resonance processes are time-dependent, sufficient time is required for data collection and image generation. Often, the time required for data acquisition is long enough to encompass several cardiac or respiratory cycles. Motion occurring during the acquisition period can dramatically diminish image quality. Therefore, significant work has gone into developing "gated" imaging protocols to eliminate motion artifacts during data acquisition. When data are acquired over multiple cardiac and respiratory cycles (segmented data acquisition), synchronization with the electrocardiogram (ECG) and/or the position of the diaphragm is mandatory. Because the ECG is distorted (elevated T-wave) when recorded from a patient in a high magnetic field, state-of-the-art MR scanners use four or more leads for R-wave detection ($\geq$2 ECG traces) to differentiate between the R-wave and the so-called T-wave artifact (17). The T-wave artifact is caused by the magneto-hydrodynamic effect (MHD). The deflection of rapidly moving ions by the main magnetic field produces additional voltage that is superimposed on the ECG signal. The MHD artifact is strongest during maximal flow in systole and increases with increasing field strengths.

Small-bore animal scanners are commonly equipped with less sophisticated ECG gating hardware and software. In most cases, R-wave detection is performed by simple threshold algorithms. In addition, both clinical and small-animal systems are equipped with respiratory sensors (bellows) that enable respiratory motion gating. However, simple gating mechanisms may cause an interruption in MR data acquisition and thus signal variations due to altered MR steady-state conditions. This can lead to artifacts, especially if inversion recovery sequences are used in concert with contrast agents. Gating schemes that acquire data at a near-constant TR and subsequently label data as accepted or rejected based on the respiratory position of the diaphragm help overcome this limitation. A drawback of all gating schemes is the increased scanning time they entail.

Molecular Magnetic Resonance Imaging Sequences

MRI sequences can be divided into spin echo (SE) (typically 2D) and gradient echo (GRE) (typically 3D) sequences. Several types of SE or GRE imaging sequences are used for optimal contrast depending on the type of agent, location, motion, and other variables. Table 176-1 outlines the general parameters for the most common classes of sequences.

Spin Echo Sequences

ECG-triggered and nontriggered T1- and T2-weighted spin echo (SE) sequences belong to the standard sequence repertoire of every MR scanner. These sequences are used extensively for neurological, body, and musculoskeletal imaging

Table 176-1: Magnetic Resonance Imaging Sequences

	Spin Echo		*Steady-State GRE*	*IR GRE*	*GRE*
Weighting	T1w	T2w	T1w	T1w	T2*
Prepulses	(DIR)	(DIR)	(REST)	IR	–
ECG gating	Yes	Yes	No	Yes	(Yes)
Respiratory gating	Yes	Yes	(Yes)	Yes	(Yes)
TR	TE	>5 msec	Shortest	~5–10 msec	>2,000 msec
TR effective	300–700 msec	>2,000 msec	–	$TR_{IR} \geq 1$ HB	–
TE	5–15 msec	50–150 msec	Shortest	< 5 ms	4–10 msec
Flip angle	90	90	30–50	30–50	5–20

DIR, double inversion prepulse; REST, regional saturation band; IR, inversion recovery prepulse; GRE, gradient echo; HB, heart beat; (), optional

since they provide excellent image quality and can provide variable T1 or T2 weighting by adjusting the echo (TE) and repetition time (TR). T1-weighted SE sequences are characterized by short TEs (5–15 msec) and TRs (300–700 msec), whereas T2-weighted SE sequences have long TEs (50–150 msec) and TRs (>2,000 msec). In the presence of a T1-lowering contrast agent, high-resolution images with excellent soft-tissue contrast with concomitant T1 weighting can be achieved. In applications for which morphological details or hypointense blood (*black blood*) appearance in concert with visualization of contrast uptake are required, spin echo approaches are often the method of choice. A disadvantage of fast spin echo (FSE) sequences is their inability to demonstrate increasing contrast effect from higher contrast agent concentrations (15) and frequently observed suboptimal contrast-induced signal enhancement due to the relatively high signal from surrounding tissues.

The maximum MR signal is reached at Gd concentrations of approximately 1 mM for a typical contrast agent with a relaxivity r1 of 4 $(\text{mM} \times \text{s})^{-1}$. For higher concentrations, the T2 effect begins decreasing the maximal achievable signal due to the finite achievable TE, whereas T2 weighting increases with increasing contrast agent concentrations (18).

T1-Weighted Three-Dimensional Gradient Echo Sequences

Non–ECG-triggered fast radio frequency (RF) spoiled three-dimensional (3D) gradient echo sequences (TE < 5 msec, TR < 10 msec, flip angle = 30°–50°) are heavily T1 weighted and exhibit a near-linear relationship between contrast agent concentration and MR signal intensity (15).

These sequences are therefore especially well suited for higher contrast agent concentrations. Due to their short scan times (5–60 sec) and excellent background suppression, these sequences are the workhorse in first-pass contrast-enhanced angiography of the large vessels and in molecular imaging of nonmoving tissues and organs. A disadvantage of this approach is that it generally produces a hyperintense (bright) appearance of blood, which makes it a suboptimal candidate for molecular imaging of the vessel wall. The use of saturation pulses can help minimize the inflow (blood signal enhancing) effect, as demonstrated in a study using molecular MRI on fibrin (19).

T1-Weighted Inversion Recovery Three-Dimensional Gradient Echo Sequences

T1-weighted inversion recovery sequences are particularly useful if ECG triggering or respiratory gating is required for suppression of cardiac or respiratory motion artifacts. Typical scan parameters include TE < 5 msec, TR = 5 to 10 msec, flip angle = 30° to 50°, 10 to 30 RF excitations per heart beat, and bandwidth = 100 to 300 Hz/pixel. The choice of the inversion repetition time (TR_{IR}) (≥1 heart beat) determines the optimum inversion delay TI:

$$TI = \ln 2 \times T1 - T1 \times \ln(\exp(-TR_{IR}/T1) + 1) \qquad \text{(Equation 4)}$$

and thus the maximum achievable signal intensity of the administered contrast agent. T1 is the longitudinal relaxation time of the suppressed tissues. Longer inversion repetition times TR_{IR} (>1 cardiac cycle) lead to longer optimal inversion delays TI and thus to higher signal intensities at the site of contrast uptake. The drawback of this approach is the increased scanning time that it requires. An example of this approach was provided by early proof-of-concept work (20) in which an IR gradient echo sequence was used in the detection of coronary thrombi that had been labeled with Gd-DTPA fibrinogen in vitro (Figure 176.1).

Advantages of inversion recovery sequences are their excellent background suppression and flow insensitivity. Due to little signal contamination from surrounding tissues, these sequences are particularly useful for the visualization of small amounts of contrast uptake at a specific target site. A limitation is the lack of morphological information provided, although it is possible to acquire morphological information in a separate scan and overlay the images to obtain anatomical localization. Unlike images produced by nontriggered 3D-gradient echo sequences, this approach provides images in which blood signal is well suppressed, and thus allows for targeted imaging of the endothelium, vessel wall, or intraluminal thrombus.

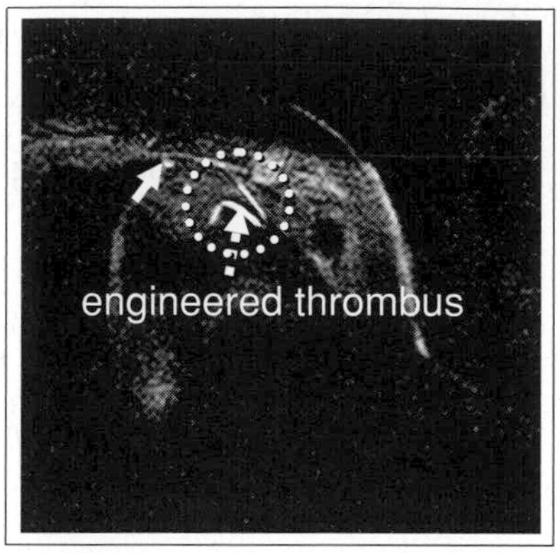

Figure 176.1. Example of an ECG-gated inversion recovery (IR) sequence for in vivo imaging of Gd-labeled coronary thrombus. Note the high contrast-to-noise ratio due to the excellent suppression of surrounding tissues and blood. This approach is well suited for determining whether the chosen target is present or absent. (Reproduced with permission from Botnar RM, Buecker A, Wiethoff AJ, et al. In vivo magnetic resonance imaging of coronary thrombosis using a fibrin-binding molecular magnetic resonance contrast agent. *Circulation.* 2004;110:1463–1466.)

T2-Weighted Gradient Echo Sequences*

Due to the prominent T2* effect of iron-based contrast agents, imaging using these agents is done using predominantly T2*-weighted gradient echo sequences. Typical parameters include TE = 4 to 10 msec, TR $\geq$2,000 msec heart beat, and flip angle = 5° to 20°. Due to the overall low signal in T2*-weighted images, morphological information is limited and sometimes requires additional scans to allow for colocalization between the site of contrast uptake and the corresponding morphological image.

Spatial Resolution

Due to the small size of the molecules and cells to be investigated by molecular MRI, spatial resolution is an important limiting factor. The spatial resolution necessary to identify molecules or cells that underlie biological processes depends heavily on the imaging sequence employed and the question to be answered. If the task is to identify the presence or absence of biological markers, low-resolution images ($\sim$1–2 mm) may be sufficient, and imaging is best done with T1-weighted IR sequences in conjunction with the use of Gd-based contrast agent or with T2*-weighted sequences in conjunction with the use of iron oxide–based contrast agent. If identification and localization of a molecular marker with respect to small anatomical structures is desired, spatial resolution and partial volume effects become crucial. Because limited data are currently available, general recommendations cannot be made. However, for minimization of partial volume effects, IR sequences appear very promising, because native tissues can be quite effectively nulled over a wide T1 range (e.g., myocardium = 800 msec; blood = 1,200 msec) thereby creating high contrast between the agent and the surrounding tissues (see Figure 176.1).

Magnetic Resonance Imaging Scanner

Field Strength

The field strength of most clinical scanners is 1.5 T (64 MHz). Recently introduced high-field systems operate at 3 T (128 MHz) and have been shown to be advantageous for neurological applications because of the twofold gain in the signal-to-noise ratio (SNR). Body imaging at 3 T has not yet been as successful due to greater technical hurdles, which are primarily related to B_0 field inhomogeneities (suboptimal shimming), limited RF penetration, and increased RF heating. Nevertheless, the potential increase in SNR that a higher field strength can provide may prove useful for imaging at the molecular and cellular level using targeted contrast agents.

For development of T1-lowering contrast agents, the field strength dependency of the longitudinal relaxivity r1 plays a critical role. This field strength dependency also is referred to as nuclear magnetic resonance dispersion (NMRD) (15,21). If the product between the Larmor frequency ω (which equals field strength) and the correlation time (fluctuation of local magnetic field induced by contrast agent) τc exceeds 1, r1 begins to decrease (15,21). Most contrast agents in clinical use today were optimized for 1.5 T ($\tau c \cong 2.5$ nsec). When developing contrast agents for 3 T or higher field systems, lower correlation times should be sought to achieve maximal longitudinal relaxivities r1. Conversely, the transverse relaxivity r2 has a different behavior and may even increase at higher field strengths. Thus, at higher field strengths, the ratio r2/r1 usually increases. T2* agents, such as iron oxide particle–based agents, therefore should be well suited for use in 3 T and higher field MR systems. On the other hand, T1 agents, such as Gd-based agents, demonstrate less contrast effect at higher field strengths.

Physiological Monitoring and Motion Compensation

Most clinical MR scanners are equipped with ECG sensors and peripheral pulse units (PPU) to allow for scan synchronization with the patient's heartbeat. In addition, respiratory bellows or MR navigators are used to monitor and correct for respiratory motion. The newest scanner generations are also equipped with blood pressure cuffs and blood oxygen saturation sensors to allow constant monitoring of vital signs in critically ill patients.

Receiver Coils

In cardiac MRI, phased-array coils (four to six coil elements) are required to meet current standards for imaging. This is because of the need for submillimeter resolution, especially when coronary artery or vessel wall imaging is included. Imaging of processes at the molecular level is likely to require even higher spatial resolution. Cardiovascular molecular imaging will benefit from improvements in phased-array coil technology, which has recently advanced to 16- and 32-channel receiver technology. Analogous to developments in multislice computed tomography (CT) technology, such improvements can be expected to reduce imaging time, which should reduce motion and other image artifacts. Furthermore, evidence from recent work (22) suggests that phased-array coil technology is likely to increase SNR, which would be particularly beneficial for imaging small molecular and cellular targets.

Data Analysis

T1 Measurements

Measurements of the T1 relaxation time are usually performed using inversion recovery sequences. By changing the inversion delay, TI, between the nonselective inversion prepulse and data acquisition, signal from tissue A ($T1 = T1_A$) will be nulled if the inversion delay TI fulfills the condition $TI = T1_A \times \ln 2$. Most T1 measurement approaches are based on the Look and Locker sequence, which acquires multiple images along the T1 relaxation curve after an initial inversion prepulse (23). Several new approaches have been proposed to reduce imaging time (24) and to enable T1 measurements in moving organs such as the heart (25).

T2 Maps*

T2* maps are acquired by sampling the signal along the free induction decay (FID) curve using multiple echo times (TE) at a constant repetition time (TR). The most common approaches are based on gradient echo sequences with signal sampling along Cartesian trajectories (26,27). The drawback is the relatively long scanning time when using this approach. In a recent study, Schaeffter and colleagues proposed a faster approach by taking advantage of the undersampling properties of radial imaging (28). Undersampled radial subimages with differing echo times (TE) were reconstructed from a complete radial data set that was acquired using multiple TEs. An exponential pixel-by-pixel fit of the FID as derived from the undersampled subimages then allows generation of T2* maps (Figure 176.2; for color reproduction, see Color Plate 176.2) (29).

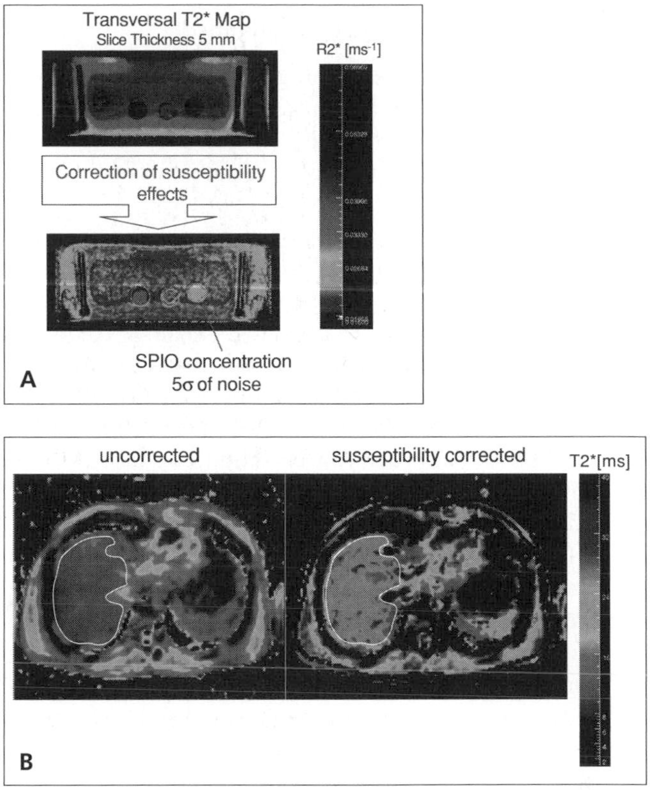

Figure 176.2. T2* maps of water tubes filled with SPIO contrast agents (**A**). Without susceptibility correction, no significant differences can be observed between the different tubes. After correction, SPIO concentrations as low as 5 standard deviations above the noise level could be detected. (**B**) In vivo T2* maps of the liver before and after susceptibility correction. (Reproduced with permission from Dahnke H, Schäffter T. Limits of detection of SPIO at 3.0T using T2* relaxometry. *Magn Reson Med.* 2005;53(5):1202–1206.) For color reproduction, see Color Plate 176.2.

ENDOGENOUS CONTRAST

Recently, groups have taken advantage of the intrinsic T1 and T2 differences of diseased tissues to image both thrombus (30,31) and vulnerable plaque (32–34). The presence of methemoglobin in thrombi produces T1 shortening, leading to high signal intensity on T1-weighted images of thrombi, such as in deep vein thrombosis (35) and pulmonary embolism (36). However, this effect declines over time as the thrombus becomes organized.

Toussaint and Yuan have characterized plaques using four types of contrast weightings (T1, T2, proton-density, and 3D time-of-flight) and shown good correlation with histopathological examination in carotid endarterectomy specimens (32,37). Fayad and colleagues have performed similar studies in the aorta using T1, T2, and proton density–weighted sequences (38).

TARGETED CONTRAST AGENTS

Although advances in MRI hardware are important to improved molecular imaging, advances in molecular imaging with MRI are increasingly dependent on the development of exogenous probes. Successful molecular probe development requires the selection of appropriate biologically and clinically relevant targets and effective strategies for meeting the challenges of sensitivity, specificity, spatial localization, and safety required for accurate diagnosis. Methods that provide for significant signal amplification are needed to detect molecular markers effectively.

Probe Detection by Magnetic Resonance

The sensitivity of a molecular imaging probe depends on the *distribution of the probe*, the *strength of the probe signal*, and the *means of detection of the probe.* The distribution of a probe is determined by the mode of its administration, its compartmentalization (degree of passage into various intravascular, extracellular, and intracellular spaces), and its mechanisms and speed of clearance. Probe distribution, together with probe-target affinity, determines the concentration of probe at the target for detection. For magnetic resonance, the strength of the achievable probe signal is determined by the concentration of contrast agent and the relaxivity of the agent. However, probe signal may not increase proportionately to increases in probe concentration (see next section). In addition, the distribution of probe in nontargeted regions is an important consideration for timing of imaging following contrast administration and for methods of probe detection to optimize the target-to-background signal ratio. The plasma half-life of an agent must be sufficient to expose target receptors to the agent. On the other hand, it may be necessary to wait for plasma concentrations of agent to fall sufficiently to distinguish luminal contrast content from vessel wall contrast uptake, although methods may be employed to reduce this requirement (39).

Magnetic Resonance Signal Amplification – General Considerations

MR contrast agents are not detected directly, but by their effect on water protons. Molar concentrations of water protons are required to provide a sufficient effect on signal intensity, so high local concentrations of contrast agents (for example, typically μM to mM for those agents having a r1 in the 4–20 $mM^{-1}s^{-1}$ range) are needed to alter the chemical environment sufficiently for detectable signal effects. Because many molecular targets are found in the nM range in the body, new methods of signal generation are being developed to enable the MRI of such molecular targets.

T1 Effects

Most MR contrast agents are based on Gd complexes (14,15,21) (Figure 176.3A) or, less commonly, iron oxide particles (Figure 176.3B) (16). Gd(III) is ideally suited for use as an MRI contrast agent not only because it has seven unpaired electrons, but also the symmetry of its electronic states produces an electron spin relaxation time slow enough to interact significantly with neighboring water protons (15,21).

Relaxivity is affected by a number of contrast agent properties including *hydration number*, the *distance between the ion and the solvent proton*, *solvent exchange rate*, *electronic relaxation time*, and *rotational correlation time.* The hydration number (number of water molecule coordination sites) for Gd chelates is generally 1 or greater (4,21). Obtaining the minimal distance between the ion and bound protons (r) is important because the relaxation rate is inversely proportional to r (6). The ability of the bound protons to rapidly exchange with free water allows for distribution of the relaxation effects throughout the bulk water (Figure 176.4). A longer electronic relaxation time also leads to higher relaxivity.

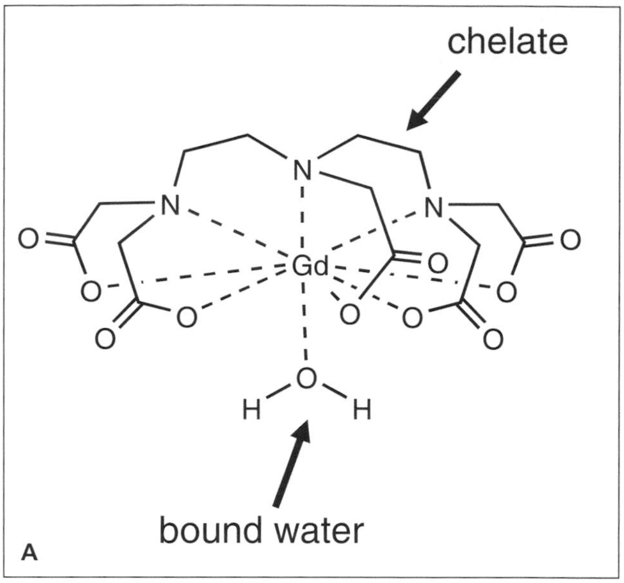

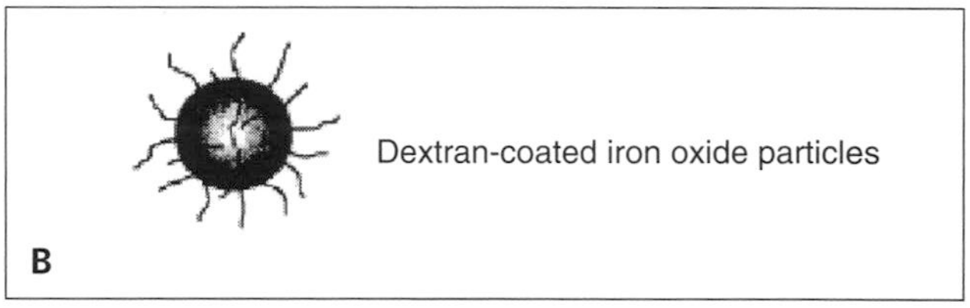

Figure 176.3. (**A**) Example of a Gd complex demonstrating the chelation of the otherwise toxic Gd^{3+} metal. The free water interacts with the surrounding protons, thereby locally changing the T1 and thus increasing the MR signal strength on T1-weighted images. Reproduced with permission from Caravan P, Ellison JJ, McMurry TJ, Lauffer RB. Gd(III) chelates as MRI contrast agents: structure, dynamics, and applications. *Chem Rev.* 1999;99:2293–352. (**B**) Example of a dextran-coated iron oxide particle.

The rotational correlation time (τR) is approximately 0.1 ns for approved agents. Because increases in rotational correlation time (τR) enhance relaxivity, various efforts in contrast agent design have focused on increasing this parameter (40). τR is lengthened by the formation of conjugates between the metal ion complex and slowly moving structures such as proteins, polymers, or dendrimers. However, the increase in relaxivity resulting from an increase in molecular weight of the contrast complex may be less than expected due to internal flexibility of the molecular structure (21).

Molecular MRI probes frequently involve attachment of Gd complexes to small ligands (e.g., small molecules, peptides) that in turn attach to slowly moving targets (e.g., proteins); thus, lengthening of the rotational correlation time is accomplished, providing a convenient means of amplifying the detection of contrast agents positioned at molecular targets. Because the unbound fraction of molecular probe will retain a lower

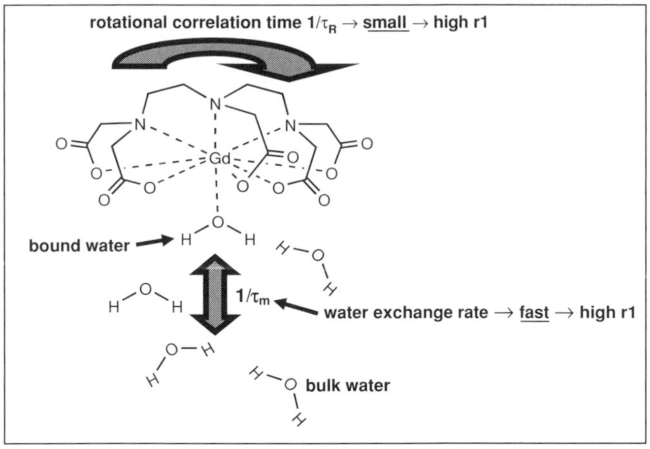

Figure 176.4. Schematic of water exchange between bound and bulk water. Fast water exchange leads to a more efficient energy transfer between Gd and bulk water, thereby creating a higher relaxivity r1. Reduction of the tumbling motion (rotational relaxation rate) of the Gd-chelate increases the relaxivity r1 as well. Binding to a target or construction of macromolecular agents (i.e., dendrimers) is a means of slowing down the rotational motion of Gd compounds.

r1, a good target-to-background signal ratio is achieved. This has been termed *receptor-induced magnetization enhancement* (RIME) (41,42). An early example of this type of contrast agent is MS-325 (EPIX Pharmaceuticals, Cambridge, MA). MS-325 is an intravascular contrast agent that reversibly binds to albumin in plasma. When bound to albumin, the relaxivity increases to approximately 20 to 23 $mM^{-1}s^{-1}$ from the 6 $mM^{-1}s^{-1}$ observed in buffer at 65 MHz (15).

A different approach to improve T1 effects without improvement in the relaxivity per Gd atom is by increasing the number of Gd per targeting complex (i.e., ligand, nanoparticle, micelle, or liposome), thus producing multivalency of Gd attachment to ligand (43). Such multivalency within the targeting complex may be sufficient to offset the flat or even diminished relaxivity effect of individual Gd atoms. This effect can be exploited for both Gd chelate-polymer (consisting of Gd chelate covalently or noncovalently bound to peptide, protein, or antibody) probes as well as for iron-oxide or Gd particulate probes, although the potential magnitude of this effect is much greater for particles (e.g., Gd/antibody ratio of ~20 Gd/antibody molecule vs. ~100 Gd/20 nm nanoparticle vs. ~300,000 Gd/300 nm nanoparticle). Thus, particulate probes offer the potential for greatly improved sensitivity over conventional chelated complexes because of the much higher achievable concentration of Gd (or iron oxide).

Estimates for the minimal Gd concentration required for detection depend on the relaxivity of the given Gd complex, which varies with the field strength. As noted earlier, for T1 agents, the maximum relaxivity attainable decreases with increasing field strength. Aime and colleagues found that, for an agent with high relaxivity (~80 $mM^{-1} \times s^{-1}$), the threshold for detection was $4 \pm 1 \times 10^7$ complexes/cell or approximately 15 μM (44). In an animal study of coronary

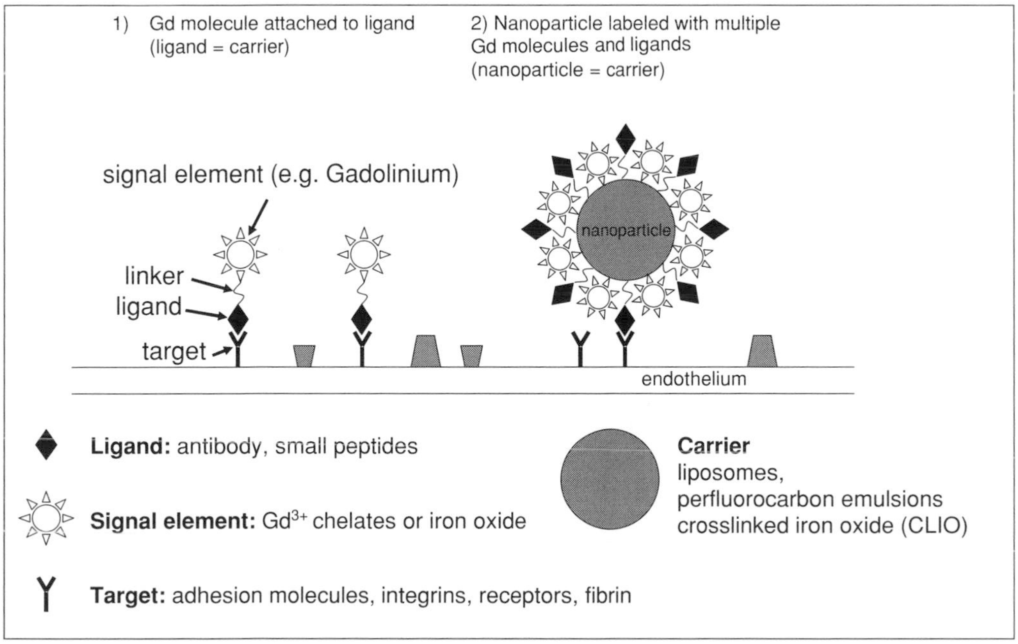

Figure 176.5. Schematic of molecular contrast agents targeted against endothelial activation. The basic components of each molecular contrast agent consist of a ligand that binds to a specific target and a signal element, which, in case of MR, is made of a Gd^{3+} chelate or an iron oxide. These two basic components can be directly (1) linked to each other or may be attached to or incorporated within a larger nanoparticle (or carrier) as demonstrated in (2).

Table 176-2: Examples of Magnetic Resonance Molecular Imaging Probes

Biological Processes	*Targets*	*Ligand*	*Carrier*	*Signal-Generating Component*	*Size/Weight*	*Relaxivities* $(mM \times s)^{-1}$	*Disease*	*References*
Thrombosis	Fibrin	Antifibrin F(ab)' fragment	Perfluoro-carbon nanoparticle	10,000–50,000 Gd3+	~250 nm	r1 = 0.18–0.54 mL × s−1 × pmol−1/ nanoparticle	CVD	72
	Fibrin	Peptide	Peptide	4 Gd^{3+}	~4,000 kDa	r1 = 21/Gd^{3}; r1 = 84/molecule	CVD	20
	Platelets	RGD-peptide	USPIO nanoparticle	USPIO			CVD	87
Angiogenesis	$\alpha v\beta 3$	Peptidomimetic vitronectin antagonist	Nanoparticle	~90,000 Gd^{3+}	~270 nm	18/25 (/Gd^{3+}) $1.7{*}10^{6}/2.4{*}10^{6}$ (/nanoparticle)	CVD, cancer	84, 85
	E-selectin	Antihuman E-selectin F(ab')2 fragment	CLIO nanoparticle	CLIO	~40 nm	0.3–0.6 mg Fe/mL T2 = 29–40 msec bound T2 ≅ 1,500 msec unbound	CVD, cancer	80
Apoptosis	Phosphat-idylserine	Annexin-V	CLIO nanoparticle	CLIO	~40 nm	0.3–0.6 mg Fe/mL T2 = 29–40 msec bound T2 ≅ 1,500 msec unbound	CVD, cancer	88
Vascular inflamma-tion	E-selectin	Antihuman E-selectin F(ab')2 fragment	CLIO nanoparticle	CLIO	~40 nm	0.3–0.6 mg Fe/mL T2 = 29–40 msec bound T2 ≅ 1,500 msec unbound	CVD, cancer	80
Neoplasia	Macrophage		USPIO nanoparticle	USPIO	~20–30 nm	r1 = 7 r2 = 81	CVD, CNS	89, 90

CNS, central nervous system; CVD, cardiovascular disease.

thrombosis, we found that Gd concentrations between 100 and 150 μM translated into an SNR of approximately 11 ± 2, which allowed for target detection (20).

As alluded to earlier, the effects of MR contrast agent concentration are nonlinear. Although contrast agent distribution has an impact on contrast concentration, contrast agent compartmentalization and local phenomena may alter local agent relaxivity. As Gd concentration increases, T1 falls. However, if T1 < TR/2 (TR = repetition time for imaging), tissue will recover nearly fully before the subsequent RF pulse. Also, at high concentrations, Gd will reduce T2 to the order of the TE (echo time) so increasing concentrations will decrease MR signal intensity. As noted earlier and demonstrated in Figures 176.2 and 176.3, fast T1-weighted gradient-echo sequences (especially 3D ones) typically have a larger scalable range than do spin-echo sequences (15).

*T2**

The synthesis and use of stable, nanosized iron oxide particles for use as MR contrast agent have been extensively described (16,45). Iron oxide particles have differential effects on 1/T1 and 1/T2, depending on their size. Superparamagnetic iron oxide (SPIO) particles produce much larger increases in 1/T2 than in 1/T1, so they are best imaged with T2-weighted scans, which reveal signal decrease (45). SPIO particles produce a marked disturbance in surrounding magnetic field homogeneity, especially apparent when a nonhomogeneous distribution produces a T2* susceptibility effect. On the other hand, ultra-small superparamagnetic particles of iron oxide (USPIOs) have a greater effect on 1/T1 than do SPIO particles, so they can be used for T1-weighted imaging (46).

Although iron oxide–based agents have greater relaxivity per metal atom than do Gd-based agents, Gd-based agents provide positive T1 signal enhancement, which is more readily distinguished from artifact than is negative signal effects and has a larger potential scalable range for detection.

Safety

Although $GdCl_3$ is very toxic, chelating Gd(III) (as with diethylenetriamine pentaacetate [DTPA]) largely reduces its toxic profile (47–49). With a high thermodynamic stability

Modality							
MRI	+	+	+	+	+	+	+
Ultrasound	+	+	+	−	±	+	+
Fluorescence	−	+	+	+	−	+	+
PET	−	+	+	+	−	+	+
Target	Flow-mediated vasodilatation	Adhesion molecules	Macrophages	MMPs Cathepsin	Lipid core Fibrous cap	$\alpha_V\beta_3$ integrin	Fibrin Platelets $\alpha_{IIb}\beta_3$ integrin Tissue factor
Process	Endothelial dysfunction	Endothelial activation	Inflammation	Proteolysis Apoptosis		Angiogenesis	Thrombosis

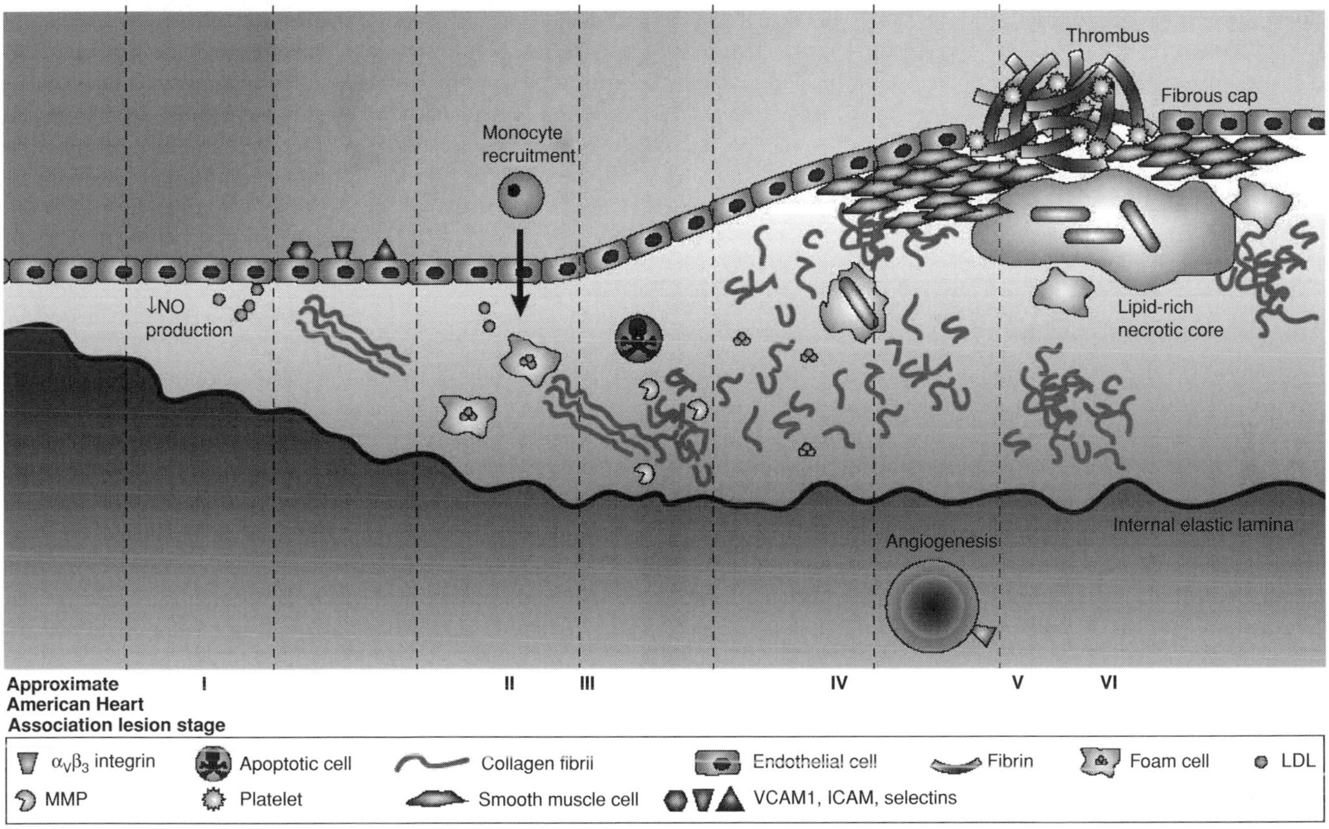

Figure 176.6. Imaging targets in atherothrombosis. Illustration of processes of atherogenesis ranging from prelesional endothelial dysfunction (*left*) through monocyte recruitment, to the development of advanced plaque complicated by thrombosis (*right*). The mechanisms are grossly simplified but focus on components (for example, cell adhesion molecules, macrophages, connective tissue elements, lipid core, and fibrin) and processes (for example, apoptosis, proteolysis, angiogenesis, and thrombosis) in plaques that have been imaged or that present useful potential imaging targets. Symbols indicate the feasibility (+ or –) of imaging using each of the modalities listed (see text for details). ICAM, intercellular cell adhesion molecule; LDL, low-density lipoprotein; MMP, matrix metalloproteinase; MRI, magnetic resonance imaging; NO, nitric oxide; PET, positron emission tomography; VCAM, vascular cell adhesion molecule. (Adapted with permission from Choudhury RP, Fuster V, Fayad ZA. Molecular, cellular and functional imaging of atherothrombosis. *Nat Rev Drug Discov.* 2004;3:913–925.)

constant ($Kgd\text{-}L = 10^{26}$ for Gd-DTPA), the chelated metal complex is essentially inert and is cleared at a significantly faster rate than the rate of dissociation of the metal ion from the complex. However, the plasma clearance rate of chelated Gd is still sufficiently slow to provide ample contrast for MRI applications. Thus, the chelation process transforms an otherwise highly toxic salt into a useful diagnostic reagent (18,50,51).

Although conventional Gd chelates are not hydrophobic and therefore generally remain extracellular, some novel molecular contrast agents are more likely to enter cells, raising concern for greater potential for toxicity. Other imaging probes, such as macromolecular intravascular agents, also may have longer elimination half-times than conventional contrast agents, with potential increased risk of in vivo Gd chelate dissociation (51). Additional concerns include antigenicity and biological interaction. When selecting a suitable target for imaging, the selected target would preferably not be involved in biological processes that might be inhibited by the binding of a targeting molecular imaging probe. Additionally, for probes designed for clinical use, it is important to ascertain whether binding of the molecular probe to its target or to other sites, such as serum proteins, affects binding of therapeutic agents that might have an impact either on drug efficacy or pharmacokinetics.

Target Identification

Molecular probe target selection criteria include substantiality of biological and clinical relevance and feasibility of target identification, which depends on factors such as adequacy of target density and accessibility of targets to probes.

Targeting

A specific disease process or cell type may be targeted by a contrast agent using either passive or active means. Although the term *molecular imaging* may be most accurately applied to probes that bind to specific molecular markers (*active targeting*), similar targeting effects may be obtained using probes having distribution characteristics that favor concentration in particular cell types or in regions of disease activity (*passive targeting*).

Passive Targeting

Examples of passive targeting agents are iron oxide particles that may be used to label components of the reticuloendothelial system (RES). SPIO nanoparticles (diameter ~200 nm) are recognized by the RES and are rapidly removed from the bloodstream (52). SPIO uptake detectable by MRI and histology has been found in the atherosclerotic plaques of hyperlipidemic rabbits in regions of high macrophage content (53). USPIOs (diameter ~18 nm) are not immediately scavenged by the hepatic and splenic RES, so they have a longer intravascular half-life, and are small enough to transmigrate the capillary wall via vesicular transport and interendothelial junctions (16). USPIOs are also phagocytosed by macrophages in plaque in hyperlipidemic rabbits (54) and humans (55).

Active Targeting

A particular cell type or disease process may be identified by targeting specific molecular markers. Potential targets may be selected from a variety of molecular markers associated with various cell types or diseases. The targeting of cells has generally focused on binding to cell surface receptors through the use of ligands such as antibodies or low-molecular-weight targeting molecules (56). MRI-targeted probes are constructed by linking such ligands to contrast agent with paramagnetic or superparamagnetic properties.

Cell surface markers (such as CD4, CD8, Mac1) on immune cells may be targeted by labeled antibodies to image areas of immune response, such as murine encephalitis (57).

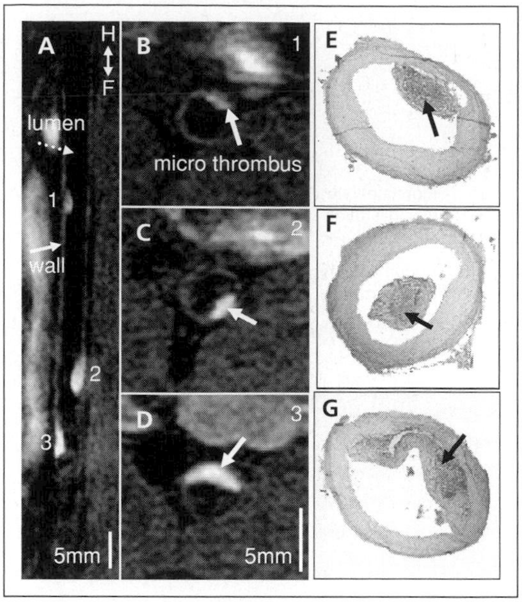

Figure 176.7. (**A**) Reformatted view of a coronal 3D data set showing the subrenal aorta approximately 20 hours post EP-1873 administration. Three well-delineated mural thrombi (*arrows*) can be observed, with good contrast between thrombus (*numbered*), arterial blood (*dotted arrow*), and the vessel wall (*dashed arrow*). The in-plane view of the aorta allows simultaneous display of all thrombi, showing head, tail, length, and relative location. (**B–D**) Corresponding cross-sectional views show good agreement with histopathology (**E–G**). (Reproduced with permission from Botnar RM, Perez AS, Witte S, et al. In vivo molecular imaging of acute and subacute thrombosis using a fibrin-binding magnetic resonance imaging contrast agent. *Circulation.* 2004;109:2023–2029.)

Monoclonal antibodies against tumor-associated antigens (e.g., Ra96, HER-2/neu receptor) may serve as ligands to identify the presence of tumor cells (58,59). Aime and colleagues have proposed an alternative approach in which negatively charged contrast agent noncovalently binds to positively charged polyamino acids that preferentially accumulate on the surface of tumor cells (60).

However, even if contrast enhancement with a targeted probe is observed in or around the specified target, the results must be initially interpreted with caution because nontargeted contrast agents may accumulate in areas of interest; for example, the accumulation of nontargeted Gd-based agents in areas of atherosclerotic plaque (39,61–63). That is, contrast may accumulate at sites of interest due to compartmental kinetics rather than molecular targeting. Therefore, careful use of controls and competition experiments are necessary to help establish the basis of contrast effect.

Following the contact of probe with the cell surface, the internalization of probe within cells by endocytosis or phagocytosis may serve as a useful mechanism to increase local contrast agent concentration (56). Non–receptor-mediated endocytosis may be exploited in tumor cells and other dividing cells (56,64). Receptor-mediated endocytosis may be a useful

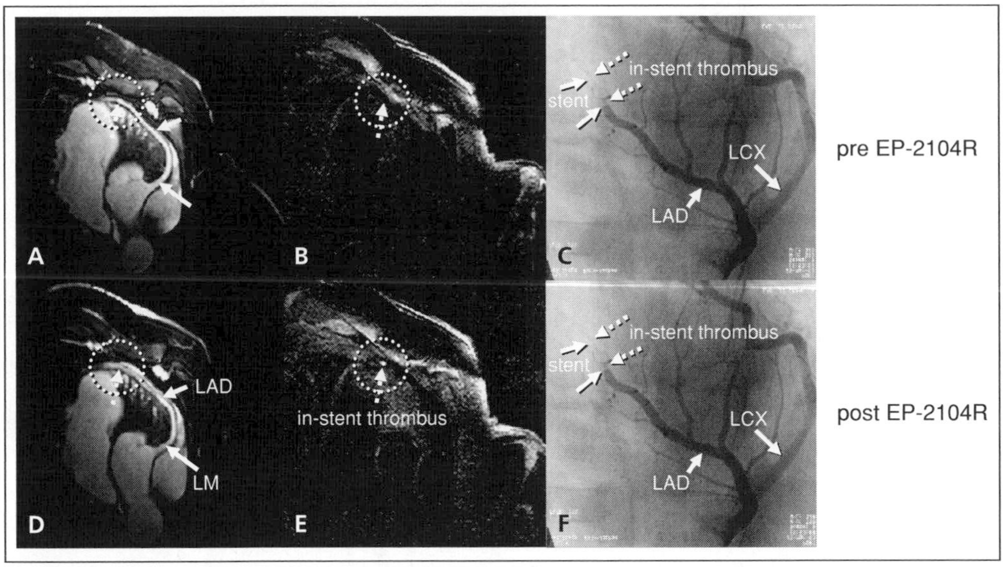

Figure 176.8. In vivo MR molecular imaging of coronary in-stent thrombosis. Bright blood images of the left main and left anterior descending coronary arteries before (**A**) and after (**D**) stent placement and injection of a fibrin-binding Gd-labeled contrast agent, EP-2104R. No apparent thrombus and no stent artifacts are visible on the post-stent placement and post-EP-2104R bright blood images (**D**). Black blood IR images before (**B**) and after stent placement and EP-2104R (**E**). A bright spot (*arrow*) is visible after intracoronary injection of EP-2104R (**E**). Thrombus was subsequently confirmed by x-ray angiography (**C, F**). (Reproduced with permission from Botnar RM, Buecker A, Wiethoff AJ, et al. In vivo magnetic resonance imaging of coronary thrombosis using a fibrin-binding molecular magnetic resonance contrast agent. *Circulation.* 2004;110:1463–1466.)

mechanism for internalization for other types of cells, including ECs (65). Phagocytosis of particulate imaging probes may serve as a useful means for contrast enhancement, as observed for identification of atherosclerotic plaque through uptake of SPIOs by macrophages within plaque. However, phagocytosis of particulate probes by the RES, particularly the Kupffer cells in the liver, may limit the availability of such probes to other targets. Therefore, mechanisms for evading the RES, such as surface coverage by poly(ethylene glycol) (PEG), an amphiphilic polymeric surfactant, are under investigation (66).

Ligand Identification

Once an appropriate molecular target has been identified, appropriate candidate ligands must be identified and selected. Methods for creating and screening libraries of peptides, such as phage display technology, are useful in identifying optimal peptide ligands for various proteins, such as receptors and antibodies. Using such an approach, a ligand with high affinity (low Kd value) for the target can be identified.

Imaging Probe Construction

Following the identification of an appropriate targeting ligand, a molecular imaging probe may be constructed by attaching the ligand to a suitable signal element (typically Gd[III] or Fe oxide). The agent carrier may be the ligand itself or a larger construct to which the ligand and contrast agent is attached (Figure 176.5). For MRI, the imaging probe is generally in the form of a metal chelate bound to a ligand (such as a peptide) or a carrier such as a nanometer-sized particle or liposome with associated signal elements (Gd or Fe oxide) and attached targeting ligand.

The avidity of the molecular probe to target may be optimized by conjugating multiple ligands to the contrast agent carrier (multivalency of ligands). An important consideration in such attachment is the preservation of specificity and affinity of binding of the ligand to the target molecule following carrier and signal element attachment. In vitro binding studies are useful to assess whether avid ligand–target binding has been retained.

On the other hand, the impact on signal of the molecular probe may be enhanced by attaching multiple signal-enhancing molecules (such as Gd chelate or iron oxide) per ligand or carrier (as discussed earlier under MR Signal Amplification).

To verify that the T1 shortening effect of a molecular probe is due to the effect of paramagnetic Gd molecules bound to its chosen specific ligand, a nonparamagnetic analog of the probe can be synthesized (for example, Gd replaced by yttrium [Y] or lanthanum [La]). Y and La have negligible MR

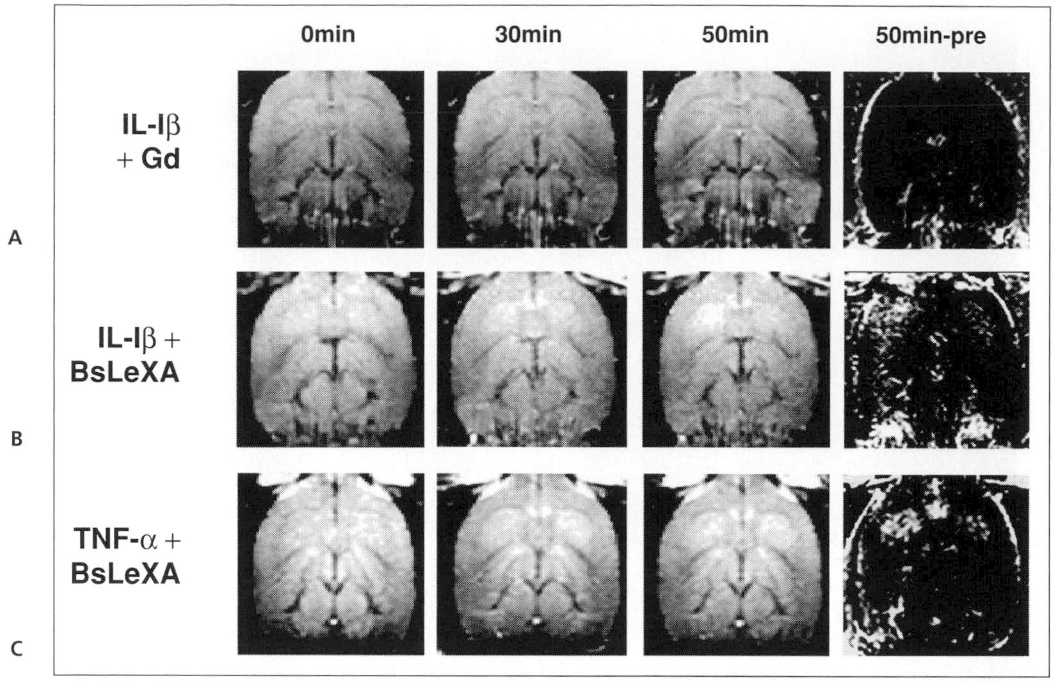

Figure 176.9. (**A**) T1-weighted images from an animal injected intracerebrally with IL-1β and intravenously with Gd-DTPA-BMA, acquired 0, 30, and 50 minutes after contrast agent administration. No changes in signal intensity were observed at any time (as shown by the difference [50 min-pre] image), indicating that there is no blood–brain barrier breakdown and entry of the contrast agent into the brain tissue. (**B**) T1-weighted images from an animal injected intracerebrally with IL-1β and intravenously with Gd-DTPA-B(sLe^x)A acquired at 0, 30, and 50 minutes after contrast agent administration. A region of increased signal intensity can be seen in the injected striatum (*left*). The degree of signal change appeared to be maximal approximately 50 minutes after contrast agent injection, and is more apparent in the difference (50 min-pre) image. This finding was consistent in the three animals injected with IL-1β and Gd-DTPA-B(sLe^x)A. (**C**) T1-weighted images from an animal injected intracerebrally with TNF-α and intravenously with Gd-DTPA-B(sLe^x)A, acquired 0, 30, and 50 minutes after contrast agent administration. A region of increased signal intensity can be seen in the injected striatum (*left*), and again the degree of signal change appeared to be maximal at approximately 50 minutes. (Reproduced with permission from Sibson NR, Blamire AM, Bernades-Silva M, et al. MRI detection of early endothelial activation in brain inflammation. *Magn Reson Med.* 2004;51:248–252.)

contrast properties. Displacement of Gd-labeled ligands by nonparamagnetic labeled ligands can then be demonstrated by MR imaging in cell cultures or animal models using an excess (i.e., 10-fold dose) of the nonparamagnetic labeled agents.

SOME POTENTIAL APPLICATIONS FOR ENDOTHELIAL ASSESSMENT BY MOLECULAR MAGNETIC RESONANCE

Targeted Imaging

A variety of MR probes have been developed to study various biological processes (e.g., thrombosis, angiogenesis, inflammation, neoplasia) and diseases (e.g., cancer, cardiovascular disease, stroke, diabetes) by targeting a spectrum of molecular markers, as summarized in Table 176-2. Molecular probes generally target either changes in receptor expression or alterations in metabolic processes (67). Most of these agents are in the preclinical stage, with only a few approved for clinical use. Table 176-2 provides examples of contrast agents under development for investigational and diagnostic purposes, with potential application for monitoring of cardiovascular therapies. Of note, probes being developed for use in one context may well have applicability for other disease processes. For example, markers of angiogenesis may be important for the study of neoplasia and atherosclerotic lesions, as well as in the study of therapies aimed to stimulate new vessel growth (Figure 176.6). Figure 176.6 demonstrates the role of the endothelium in the initiation, progression, and complication of atherosclerosis (68). The course of disease can be divided into multiple sequential steps, characterized by distinct biological processes associated with various molecular targets (68).

As noted earlier, the endothelium is a convenient site for targeting since it is directly adjacent to the bloodstream into which contrast agents are conventionally administered.

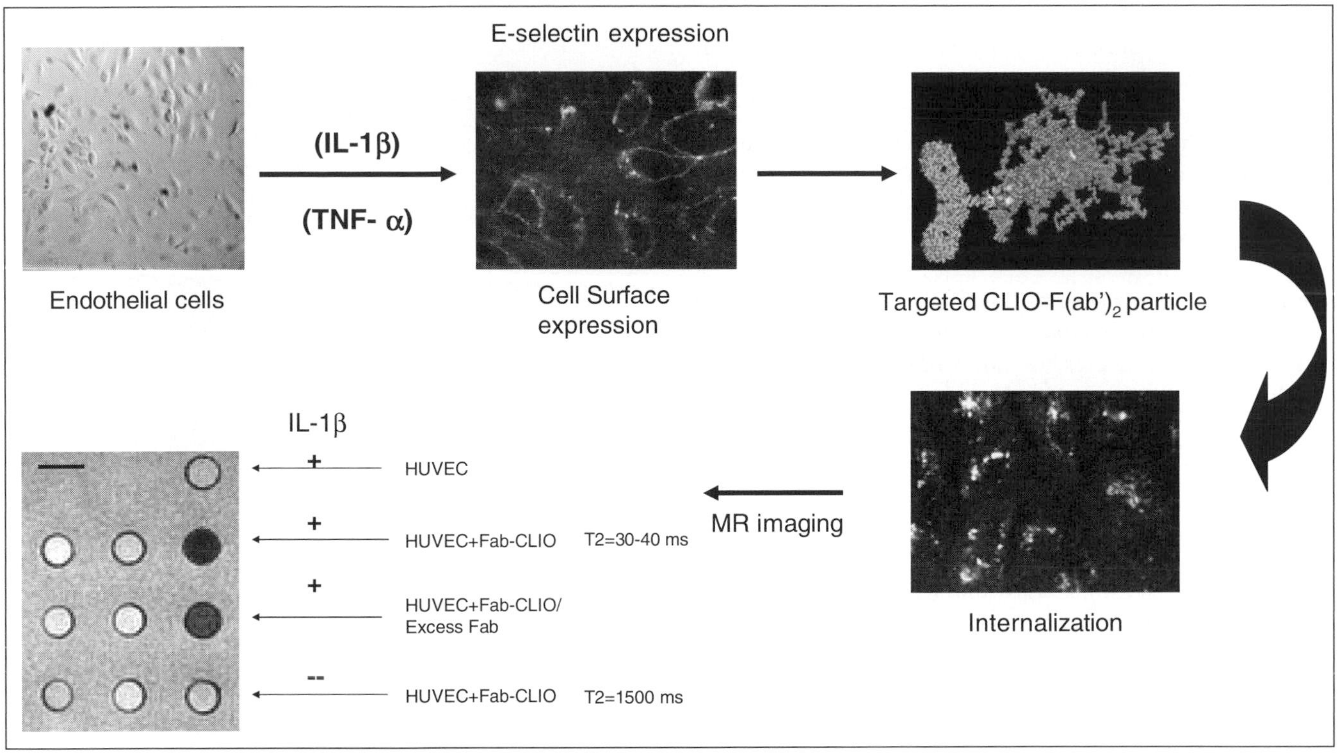

Figure 176.10. EC activation after treatment of cells with IL-1β or TNF-α. Subsequently, an E-selectin targeted CLIO-F(ab')2 particle was administered. After internalization of the contrast agent, MR imaging was performed. In the unbound state (before IL-1β and TNF-α injection), the T2 relaxation time was approximately 1,500 msec (*bright image*). Conversely, T2 was reduced to approximately 30 to 40 msec when CLIO-F(ab')2 was bound to E-selectin (*dark image*). The binding of CLIO-F(ab')2 to IL-1β–activated cells was approximately 100 to 200 times higher than for nonactivated ECs. (Reproduced with permission from Kang HW, Josephson L, Petrovsky A, et al. Magnetic resonance imaging of inducible E-selectin expression in human EC culture. *Bioconjug Chem.* 2002;13:122–127.)

Thus, the barriers to probe delivery encountered when targeting parenchymal cells (69) may be avoided. However, probe access to potential targets for investigation within the *sub*endothelium may be limited by probe size (70).

Magnetic Resonance Imaging of Thrombosis

Imaging of fibrin has potential clinical applications for the diagnosis of several significant medical conditions, including acute coronary syndromes, deep venous thrombosis, and pulmonary emboli. In recent studies by Yu (71), Flacke (72), and our group (19,20), Gd-labeled fibrin-avid nanoparticles and small peptides have been successfully used to image thrombus in the jugular vein (72), aorta (19), and the pulmonary (73) and coronary arteries (20,73). Figure 176.7 demonstrates the imaging of acute thrombus in an animal model of plaque rupture using EP-1873. Gd concentrations as low as approximately 50 μM ($r1 \cong 21\ mM^{-1} \times s^{-1}$ per Gd) were sufficient for ready visualization of mural and lumen-encroaching thrombus. A similar compound, EP-2104R, enabled the imaging of coronary in-stent thrombosis (Figure 176.8) (20) and pulmonary embolism (73) in an experimental animal model. The administered dose was 4 to 7.5 μmol/kg, much lower than that for conventional nontargeted Gd-based contrast agents (typically ~0.1 mmol/kg).

Magnetic Resonance Imaging of Atherosclerotic Plaque Using HDL-Like Nanoparticles

Frias and coworkers investigated the use of a high-density lipoprotein (HDL)-like nanoparticle loaded with Gd-DTPA-DMPE phospholipid, as well as fluorescent phospholipid, to target atherosclerotic plaque (70). Potential advantages of such a particle include its small size (~9 nm diameter), its localization to atherosclerotic plaque, its composition of mainly endogenous, protein and lipid components, and its contrast agent content (15–20 Gd-DTPA-DMPE per particle), which yields a relaxivity (r1) of 10.4 $mM^{-1}s^{-1}$. MRI performed 24 hours following administration of this particle to ApoE-knockout mice revealed a mean normalized enhancement ratio for the atherosclerotic aortic wall of 35%. Confocal fluorescence microscopy revealed some intracellular macrophage localization.

Magnetic Resonance Imaging of Endothelial Activation and Injury

Vascular inflammation and associated endothelial activation are believed to play an integral role in the initiation and progression of atherosclerosis. Endothelial activation is characterized by the upregulation of leukocyte adhesion molecules (intercellular adhesion molecule [ICAM]-1 and vascular adhesion molecule [VCAM]-1, E-selectin, and P-selectin) on the EC surface (74). The adhesion molecules facilitate tethering, adhesion, and transendothelial migration of leukocytes, including monocytes. Differentiation of monocytes into macrophages and subsequent digestion of lipoproteins by macrophages occurs in a later stage and eventually leads to the accumulation of lipid-filled macrophages, which are believed to be a precursor of rupture-prone vulnerable plaque. Although the specific mechanisms by which the adhesion molecules contribute to this process have not yet been determined, studies of atherogenic mice with various adhesion molecule deficiencies has indicated some role for each of the inducible adhesion molecules, particularly VCAM-1 and P-selectin (74–76).

In recent work by Sibson and colleagues (77) and Barber and coworkers (78), early endothelial activation was observed in focal ischemia in mice brains (Figure 176.9) (77) and in brain inflammation in rats (after interleukin [IL]-1β and tumor necrosis factor [TNF]-α–induced E- and P-selectin upregulation) using a novel MR contrast agent. This novel Gd-labeled contrast agent, Gd-DTPA-B(sLe^x)A (79) consists of the sialyl Lewis^x (sLe^x) carbohydrate, which interacts with both E- and P-selectin. The relaxivity was measured as 3.5 $mM^{-1} \times s^{-1}$ at 1.5T and thus is similar to Gd-DTPA.

Kang and colleagues have studied an antihuman E-selectin antibody, F(ab'), attached to a cross-linked iron oxide (CLIO) nanoparticle (80). This novel targeted contrast agent was tested in human umbilical vein ECs (HUVECs) culture before and after IL-1β– and TNF-α–induced E-selectin expression. In the unbound state (before IL-1β or TNF-α injection), the T2 relaxation time was approximately 1,500 msec. T2 was reduced to approximately 30 to 40 msec when CLIO-F(ab')2 was bound to E-selectin (Figure 176.10). The binding of CLIO-F(ab')2 to IL-1β activated cells was approximately 100 to 200 times higher than for nonactivated ECs. An antibody-based approach was also taken by Sipkins and colleagues, who developed antibody conjugated paramagnetic liposomes (ACPLs) targeted to ICAM-1 to identify autoimmune encephalitis (81).

Kelly and coworkers developed a novel VCAM-1–targeted MRI and fluorescence imaging agent (65). They identified several VCAM-1–mediated cell-internalizing peptides through the use of iterative phage display performed on endothelium under physiological flow conditions. The selected peptide was found to have 12-fold higher target-to-background ratios compared to VCAM-1 monoclonal antibody. Studies using the VCAM-1–targeted peptide attached to a magnetofluorescent nanoparticle demonstrated the successful identification of ECs activated by TNF-α as well as VCAM-1–expressing cells in atherosclerotic lesions in *ApoE*$^{-/-}$ mice.

An alternative approach for the identification of vascular injury has been studied by Yamamoto and coworkers, who developed a novel contrast agent, EB-DTPA-Gd, based on the chemical structure of Evans Blue dye (82). Using this contrast agent, significant signal enhancement was observed at the site

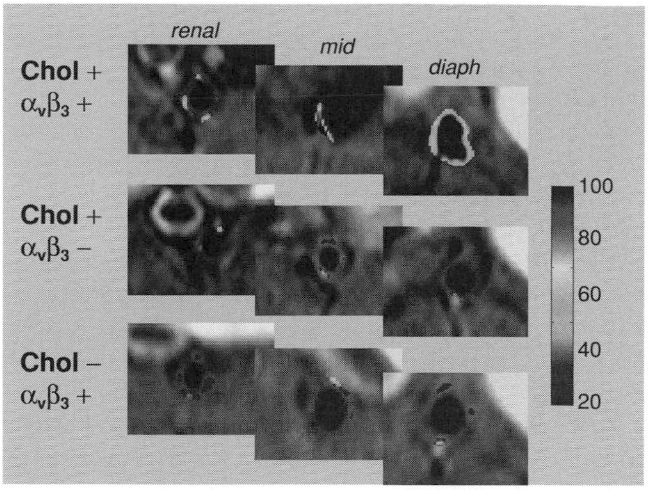

Figure 176.11. Percent enhancement maps from individual aortic segments of cholesterol-fed rabbit at renal artery, mid-aorta, and diaphragm 2 hours after treatment in cholesterol-fed rabbit given $\alpha v\beta 3$-targeted nanoparticles (*upper row*). No to minimal enhancement in cholesterol-fed animal after administration of nonspecific contrast agent (*middle row*). Similarly, no enhancement in a control animal after injection of $\alpha v\beta 3$-targeted nanoparticles (*lower row*). (Reproduced with permission from Winter PM, Morawski AM, Caruthers SD, et al. Molecular imaging of angiogenesis in early-stage atherosclerosis with alpha(v)beta3-integrin-targeted nanoparticles. *Circulation.* 2003;108:2270–2274.) For color reproduction, see Color Plate 176.11.

of common carotid balloon injury in the rat. These promising methods merit further study, with potential applications for various models of endothelial activation.

Magnetic Resonance Imaging of Angiogenic Endothelium

As noted earlier, the identification of angiogenesis may be useful in studying tumors and atherosclerotic plaque as well as for assessing therapies aimed at stimulating new vessel growth. Integrins, such as $\alpha v\beta 3$, are overexpressed in activated neovascular ECs, which are believed to play an integral role in tumor growth and the initiation and development of atherosclerosis. Wickline, Lanza, and coworkers have developed perfluoro-nanoparticles that can carry as many as 90,000 paramagnetic Gd chelates per particle and can be targeted against various biomarkers by attaching appropriate ligands. They have developed methods for identifying angiogenic vessels by targeting the $\alpha v\beta 3$ integrin (83). A probe consisting of Gd-perfluorocarbon nanoparticle linked to anti-$\alpha v\beta 3$ monoclonal antibody produced MRI signal enhancement of the capillary bed in a corneal micropocket model of angiogenesis (83). A similar approach was taken by Guciione, Li, and Bednarski, who attached anti-$\alpha v\beta 3$ antibody to a lipid-based polymerized vesicle to image an angiogenic tumor (67). A non–antibody-based approach was also taken by Wickline and coworkers, who attached a peptidomimetic vitronectin antagonist to the nanoparticles to again target $\alpha v\beta 3$ integrin. Using this approach, they were able to image angiogenic vessels in nascent Vx-2 rabbit tumors (84) and in early-stage atherosclerosis (Figure 176.11; for color reproduction, see Color Plate 176.11) (85).

Endothelial-Targeted Therapy

Targeted imaging probes could potentially serve as vehicles for targeted cell therapy. An example of such an approach was provided by Hood and colleagues, who modified polymerized nanoparticles targeted with anti-$\alpha v \beta 3$ antibody to contain a plasmid with a mutant form of the gene encoding Raf-1 (86). This mutant form was known to promote apoptosis of ECs and suppress angiogenesis in vivo. Following injection of the modified nanoparticles, significant reductions in tumor burden were observed in mice bearing melanoma and metastatic colorectal cancer (86).

CONCLUSION

Molecular MRI has great promise as a tool to improve our understanding of biological processes as well as to aid in clinical diagnosis. These methods may assist not only in monitoring response to treatment but also may provide vehicles for targeted therapies. Although progress in the field must overcome numerous technical challenges to optimize imaging methods and to develop novel targeted agents, a multidisciplinary approach to these problems will lead to further advances in this field.

ACKNOWLEDGMENTS

We thank Alexei Bogdanov PhD, Peter Caravan PhD, Hannes Dahnke PhD, Tobias Schaeffter PhD, Nicola Sibson PhD, and Sam Wickline MD for providing some of the material for the illustrations. Drs. Yeon, Spuentrup, and Botnar have received research grant support from EPIX Pharmaceuticals, Inc. Dr. Manning receives research grant support from Philips Medical Systems and EPIX Pharmaceuticals, Inc. He is also a consultant to EPIX Pharmaceuticals, Inc.

REFERENCES

1 Rofsky NM, Adelman MA. MR angiography in the evaluation of atherosclerotic peripheral vascular disease. *Radiology*. 2000; 214:325–338.
2 Joarder R, Gedroyc WM. Magnetic resonance angiography: the state of the art. *Eur Radiol.* 2001;11:446–453.
3 Meaney JF. Magnetic resonance angiography of the peripheral arteries: current status. *Eur Radiol.* 2003;13:836–852.
4 Goyen M, Debatin JF. Gadopentetate dimeglumine-enhanced three-dimensional MR-angiography: dosing, safety, and efficacy. *J Magn Reson Imaging*. 2004;19:261–273.
5 Olin JW, Kaufman JA, Bluemke DA, et al. Atherosclerotic Vascular Disease Conference: Writing Group IV: imaging. *Circulation.* 2004;109:2626–2633.
6 Oliver JJ, Webb DJ. Noninvasive assessment of arterial stiffness and risk of atherosclerotic events. *Arterioscler Thromb Vasc Biol.* 2003;23:554–566.
7 Wiesmann F, Petersen SE, Leeson PM, et al. Global impairment of brachial, carotid, and aortic vascular function in young smokers: direct quantification by high-resolution magnetic resonance imaging. *J Am Coll Cardiol.* 2004;44:2056–2064.
8 Brasch R, Turetschek K. MRI characterization of tumors and grading angiogenesis using macromolecular contrast media: status report. *Eur J Radiol.* 2000;34:148–155.
9 Neeman M, Dafni H. Structural, functional, and molecular MR imaging of the microvasculature. *Annu Rev Biomed Eng*. 2003; 5:29–56.
10 Barkhausen J, Hunold P, Jochims M, Debatin JF. Imaging of myocardial perfusion with magnetic resonance. *J Magn Reson Imaging*. 2004;19:750–757.
11 Wu KC, Zerhouni EA, Judd RM, et al. Prognostic significance of microvascular obstruction by magnetic resonance imaging in patients with acute myocardial infarction. *Circulation.* 1998;97:765–772.
12 Johnson GA, Benveniste H, Black RD, et al. Histology by magnetic resonance microscopy. *Magn Reson Q.* 1993;9:1–30.
13 Weinmann HJ, Brasch RC, Press WR, Wesbey GE. Characteristics of gadolinium-DTPA complex: a potential NMR contrast agent. *AJR Am J Roentgenol.* 1984;142:619–624.
14 Laniado M, Weinmann HJ, Schorner W, et al. First use of GdDTPA/dimeglumine in man. *Physiol Chem Phys Med NMR.* 1984;16:157–165.
15 Caravan P, Ellison JJ, McMurry TJ, Lauffer RB. Gadolinium(III) chelates as MRI contrast agents: structure, dynamics, and applications. *Chem Rev.* 1999;99:2293–2352.
16 Weissleder R, Elizondo G, Wittenberg J, et al. Ultrasmall superparamagnetic iron oxide: characterization of a new class of contrast agents for MR imaging. *Radiology.* 1990;175:489–493.
17 Fischer SE, Wickline SA, Lorenz CH. Novel real-time R-wave detection algorithm-based on the vectorcardiogram for accurate gated magnetic resonance acquisitions. *Magn Reson Med.* 1999;42:361–370.
18 Gries H. Extracellular MRI contrast agents based on gadolinium. In: *Topics in Current Chemistry.* Berlin Heidelberg: Springer-Verlag; 2002:2–24.
19 Botnar RM, Perez AS, Witte S, et al. In vivo molecular imaging of acute and subacute thrombosis using a fibrin-binding magnetic resonance imaging contrast agent. *Circulation.* 2004;109:2023–2029.
20 Botnar RM, Buecker A, Wiethoff AJ, et al. In vivo magnetic resonance imaging of coronary thrombosis using a fibrin-binding molecular magnetic resonance contrast agent. *Circulation.* 2004; 110:1463–1466.
21 Tóth E HL, Merback AE. Relaxivity of gadolinium(III) complexes: theory and mechanism. In: Merbach AE, ed. *The Chemistry of Contrast Agents in Medical Magnetic Resonance Imaging.* Chichester: John Wiley & Sons; 2001:45–119.
22 Niendorf T, Hardy CJ, Giaquinto RO, et al. Toward a single breath-hold whole-heart coverage coronary MRA using highly accelerated parallel imaging with a 32-channel MR system. *Magn Reson Med.* 2006;56(1):167–176.
23 Look DC, Locker DR. Time saving in measurement of NMR and EPR relaxation times. *Rev Sci Instrum.* 1970;41:250–251.

24 Henderson E, McKinnon G, Lee TY, Rutt BK. A fast 3D look-locker method for volumetric T1 mapping. *Magn Reson Imaging*. 1999;17:1163–1171.

25 Messroghli DR, Radjenovic A, Kozerke S, et al. Modified Look-Locker inversion recovery (MOLLI) for high-resolution T1 mapping of the heart. *Magn Reson Med.* 2004;52:141–146.

26 Reeder SB, Faranesh AZ, Boxerman JL, McVeigh ER. In vivo measurement of T2* and field inhomogeneity maps in the human heart at 1.5 T. *Magn Reson Med.* 1998;39:988–998.

27 Clare S, Francis S, Morris PG, Bowtell R. Single-shot T2(*) measurement to establish optimum echo time for fMRI: studies of the visual, motor, and auditory cortices at 3.0 T. *Magn Reson Med.* 2001;45:930–933.

28 Dahnke H, Weiss S, Schaeffter T. Simultaneous T2* mapping and anatomical imaging using a fast radial multi-gradient-echo acquisition. In: *ISMRM.* Kyoto: International Society for Magnetic Resonance Imaging; 2004:1745.

29 Dahnke H, Schäffter T. Limits of detection of SPIO at 3.0T using T2* relaxometry. *Magn Reson Med.* 2005; 53(5):1202–1206.

30 Moody AR, Murphy RE, Morgan PS, et al. Characterization of complicated carotid plaque with magnetic resonance direct thrombus imaging in patients with cerebral ischemia. *Circulation.* 2003;107:3047–3052.

31 Murphy RE, Moody AR, Morgan PS, et al. Prevalence of complicated carotid atheroma as detected by magnetic resonance direct thrombus imaging in patients with suspected carotid artery stenosis and previous acute cerebral ischemia. *Circulation.* 2003;107:3053–3058.

32 Toussaint JF, LaMuraglia GM, Southern JF, et al. Magnetic resonance images lipid, fibrous, calcified, hemorrhagic, and thrombotic components of human atherosclerosis in vivo. *Circulation.* 1996;94:932–938.

33 Yuan C, Petty C, O'Brien KD, et al. In vitro and in situ magnetic resonance imaging signal features of atherosclerotic plaque-associated lipids. *Arterioscler Thromb Vasc Biol.* 1997;17:1496–1503.

34 Fayad ZA, Fallon JT, Shinnar M, et al. Noninvasive in vivo high-resolution magnetic resonance imaging of atherosclerotic lesions in genetically engineered mice. *Circulation.* 1998;98:1541–1547.

35 Moody AR. Direct imaging of deep-vein thrombosis with magnetic resonance imaging. *Lancet*. 1997;350:1073.

36 van Beek EJ, Wild JM, Fink C, et al. MRI for the diagnosis of pulmonary embolism. *J Magn Reson Imaging*. 2003;18:627–640.

37 Yuan C, Mitsumori LM, Ferguson MS, et al. In vivo accuracy of multispectral magnetic resonance imaging for identifying lipid-rich necrotic cores and intraplaque hemorrhage in advanced human carotid plaques. *Circulation.* 2001;104:2051–2056.

38 Fayad ZA, Nahar T, Fallon JT, et al. In vivo magnetic resonance evaluation of atherosclerotic plaques in the human thoracic aorta: a comparison with transesophageal echocardiography. *Circulation.* 2000;101:2503–2509.

39 Sirol M, Itskovich VV, Mani V, et al. Lipid-rich atherosclerotic plaques detected by gadofluorine-enhanced in vivo magnetic resonance imaging. *Circulation.* 2004;109:2890–2896.

40 Aime S FM, Terreno E, Botta M. Protein-Bound Metal Chelates. In: Merbach AE, ed. *The Chemistry of Contrast Agents in Medical Magnetic Resonance Imaging.* Chichester: John Wiley & Sons; 2001:193–241.

41 Bach-Gansmo T. Ferrimagnetic susceptibility contrast agents. *Acta Radiol Suppl.* 1993;387:1–30.

42 Nivorozhkin AL, Kolodziej AF, Caravan P, et al. Enzyme-activated Gd(3^+) magnetic resonance imaging contrast agents with a prominent receptor-induced magnetization enhancement. *Angew Chem Int Ed Engl.* 2001;40:2903–2906.

43 Sieving PF, Watson AD, Rocklage SM. Preparation and characterization of paramagnetic polychelates and their protein conjugates. *Bioconjug Chem.* 1990;1:65–71.

44 Aime S, Cabella C, Colombatto S, et al. Insights into the use of paramagnetic Gd(III) complexes in MR-molecular imaging investigations. *J Magn Reson Imaging*. 2002;16:394–406.

45 Ferrucci JT, Stark DD. Iron oxide-enhanced MR imaging of the liver and spleen: review of the first 5 years. *AJR Am J Roentgenol.* 1990;155:943–950.

46 Small WC, Nelson RC, Bernardino ME. Dual contrast enhancement of both T1- and T2-weighted sequences using ultrasmall superparamagnetic iron oxide. *Magn Reson Imaging*. 1993;11: 645–654.

47 Bousquet JC, Saini S, Stark DD, et al. Gd-DOTA: characterization of a new paramagnetic complex. *Radiology*. 1988;166:693–698.

48 Weinmann HJ, Press WR, Gries H. Tolerance of extracellular contrast agents for magnetic resonance imaging. *Invest Radiol.* 1990;25(Suppl 1):S49–S50.

49 Bartolini ME, Pekar J, Chettle DR, et al. An investigation of the toxicity of gadolinium-based MRI contrast agents using neutron activation analysis. *Magn Reson Imaging*. 2003;21:541–544.

50 Niendorf HP, Haustein J, Cornelius I, et al. Safety of gadolinium-DTPA: extended clinical experience. *Magn Reson Med.* 1991;22:222–228; Discussion 229–232.

51 Brücher E. Kinetic stabilities of gadolinium (III) chelates used as MRI contrast agents. In: *Topics in Current Chemistry.* Berlin Heidelberg: Springer-Verlag; 2002:103–122.

52 Pouliquen D, Le Jeune JJ, Perdrisot R, et al. Iron oxide nanoparticles for use as an MRI contrast agent: pharmacokinetics and metabolism. *Magn Reson Imaging*. 1991;9:275–283.

53 Schmitz SA, Coupland SE, Gust R, et al. Superparamagnetic iron oxide-enhanced MRI of atherosclerotic plaques in Watanabe hereditable hyperlipidemic rabbits. *Invest Radiol.* 2000;35:460–471.

54 Ruehm SG, Corot C, Vogt P, et al. Magnetic resonance imaging of atherosclerotic plaque with ultrasmall superparamagnetic particles of iron oxide in hyperlipidemic rabbits. *Circulation.* 2001;103:415–422.

55 Kooi ME, Cappendijk VC, Cleutjens KB, et al. Accumulation of ultrasmall superparamagnetic particles of iron oxide in human atherosclerotic plaques can be detected by in vivo magnetic resonance imaging. *Circulation.* 2003;107:2453–2458.

56 Jacques V, Desreux, JF. Kinetic stabilities of gadolinium (III) chelates used as MRI contrast agents. In: *Topics in Current Chemistry.* Berlin Heidelberg: Springer-Verlag; 2002:125–164.

57 Pirko I, Johnson A, Ciric B, et al. In vivo magnetic resonance imaging of immune cells in the central nervous system with superparamagnetic antibodies. *FASEB J*. 2004;18:179–182.

58 Gohr-Rosenthal S, Schmitt-Willich H, Ebert W, Conrad J. The demonstration of human tumors on nude mice using gadolinium-labelled monoclonal antibodies for magnetic resonance imaging. *Invest Radiol.* 1993;28:789–795.

59 Artemov D. Molecular magnetic resonance imaging with targeted contrast agents. *J Cell BioChem.* 2003;90:518–524.

60 Aime S, Botta M, Garino E, et al. Non-covalent conjugates between cationic polyamino acids and GdIII chelates: a route

for seeking accumulation of MRI-contrast agents at tumor targeting sites. *Chemistry*. 2000;6:2609–2617.
61 Yuan C, Kerwin WS, Ferguson MS, et al. Contrast-enhanced high resolution MRI for atherosclerotic carotid artery tissue characterization. *J Magn Reson Imaging*. 2002;15:62–67.
62 Kramer CM, Cerilli LA, Hagspiel K, et al. Magnetic resonance imaging identifies the fibrous cap in atherosclerotic abdominal aortic aneurysm. *Circulation*. 2004;109:1016–1021.
63 Weinmann HJ, Ebert W, Misselwitz B, Schmitt-Willich H. Tissue-specific MR contrast agents. *Eur J Radiol*. 2003;46:33–44.
64 Yeh TC, Zhang W, Ildstad ST, Ho C. Intracellular labeling of T-cells with superparamagnetic contrast agents. *Magn Reson Med*. 1993;30:617–625.
65 Kelly KA, Allport JR, Tsourkas A, et al. Detection of vascular adhesion molecule-1 expression using a novel multimodal nanoparticle. *Circ Res*. 2005;96:327–336.
66 Gupta AK, Wells S. Surface-modified superparamagnetic nanoparticles for drug delivery: preparation, characterization, and cytotoxicity studies. *IEEE Trans Nanobioscience*. 2004;3:66–73.
67 Guccione S, Li KC, Bednarski MD. Vascular-targeted nanoparticles for molecular imaging and therapy. *Methods Enzymol*. 2004;386:219–236.
68 Choudhury RP, Fuster V, Fayad ZA. Molecular, cellular and functional imaging of atherothrombosis. *Nat Rev Drug Discov*. 2004;3:913–925.
69 Zhu H, Baxter LT, Jain RK. Potential and limitations of radioimmunodetection and radioimmunotherapy with monoclonal antibodies. *J Nucl Med*. 1997;38:731–741.
70 Frias JC, Williams KJ, Fisher EA, Fayad ZA. Recombinant HDL-like nanoparticles: a specific contrast agent for MRI of atherosclerotic plaques. *J Am Chem Soc*. 2004;126:16316–16317.
71 Yu X, Song SK, Chen J, et al. High-resolution MRI characterization of human thrombus using a novel fibrin-targeted paramagnetic nanoparticle contrast agent. *Magn Reson Med*. 2000;44:867–872.
72 Flacke S, Fischer S, Scott MJ, et al. Novel MRI contrast agent for molecular imaging of fibrin: implications for detecting vulnerable plaques. *Circulation*. 2001;104:1280–1285.
73 Spuentrup E, Buecker A, Katoh M, et al. Molecular magnetic resonance imaging of coronary thrombosis and pulmonary emboli with a novel fibrin-targeted contrast agent. *Circulation*. 2005;111(11):1377–1382. Epub 2005 Feb 28.
74 Cybulsky MI, Charo I.F. Leukocytes, adhesion molecules, and chemokines. In: Fuster V, ed. *Atherothrombosis and Coronary Artery Disease*. Philadelphia: Lippincott Williams & Wilkins; 2005:489–503.
75 Dansky HM, Barlow CB, Lominska C, et al. Adhesion of monocytes to arterial endothelium and initiation of atherosclerosis are critically dependent on vascular cell adhesion molecule-1 gene dosage. *Arterioscler Thromb Vasc Biol*. 2001;21:1662–1667.
76 Dong ZM, Brown AA, Wagner DD. Prominent role of P-selectin in the development of advanced atherosclerosis in ApoE-deficient mice. *Circulation*. 2000;101:2290–2295.
77 Sibson NR, Blamire AM, Bernades-Silva M, et al. MRI detection of early endothelial activation in brain inflammation. *Magn Reson Med*. 2004;51:248–252.
78 Barber PA, Foniok T, Kirk D, et al. MR molecular imaging of early endothelial activation in focal ischemia. *Ann Neurol*. 2004;56:116–120.
79 Laurent S, Vander Elst L, Fu Y, Muller RN. Synthesis and physicochemical characterization of Gd-DTPA-B(sLex)A, a new MRI contrast agent targeted to inflammation. *Bioconjug Chem*. 2004;15:99–103.
80 Kang HW, Josephson L, Petrovsky A, et al. Magnetic resonance imaging of inducible E-selectin expression in human endothelial cell culture. *Bioconjug Chem*. 2002;13:122–127.
81 Sipkins DA, Gijbels K, Tropper FD, et al. ICAM-1 expression in autoimmune encephalitis visualized using magnetic resonance imaging. *J Neuroimmunol*. 2000;104:1–9.
82 Yamamoto T, Ikuta K, Oi K, et al. In vivo MR detection of vascular endothelial injury using a new class of MRI contrast agent. *Bioorg Med Chem Lett*. 2004;14:2787–2790.
83 Anderson SA, Rader RK, Westlin WF, et al. Magnetic resonance contrast enhancement of neovasculature with alpha(v)beta(3)-targeted nanoparticles. *Magn Reson Med*. 2000;44:433–439.
84 Winter PM, Caruthers SD, Kassner A, et al. Molecular imaging of angiogenesis in nascent Vx-2 rabbit tumors using a novel alpha(nu)beta3-targeted nanoparticle and 1.5 tesla magnetic resonance imaging. *Cancer Res*. 2003;63:5838–5843.
85 Winter PM, Morawski AM, Caruthers SD, et al. Molecular imaging of angiogenesis in early-stage atherosclerosis with alpha(v)beta3-integrin-targeted nanoparticles. *Circulation*. 2003; 108:2270–2274.
86 Hood JD, Bednarski M, Frausto R, et al. Tumor regression by targeted gene delivery to the neovasculature. *Science*. 2002; 296:2404–2407.
87 Johansson LO, Bjornerud A, Ahlstrom HK, et al. A targeted contrast agent for magnetic resonance imaging of thrombus: implications of spatial resolution. *J Magn Reson Imaging*. 2001;13:615–618.
88 Schellenberger EA, Hogemann D, Josephson L, Weissleder R. Annexin V-CLIO: a nanoparticle for detecting apoptosis by MRI. *Acad Radiol*. 2002;9(Suppl 2):S310–S311.
89 Rogers J, Lewis J, Josephson L. Use of AMI-227 as an oral MR contrast agent. *Magn Reson Imaging*. 1994;12:631–639.
90 Dousset V, Delalande C, Ballarino L, et al. In vivo macrophage activity imaging in the central nervous system detected by magnetic resonance. *Magn Reson Med*. 1999;41:329–333.

177

Real-Time Imaging of the Endothelium

Peter L. Gross

St. Michael's Hospital, University of Toronto, Ontario, Canada

Real-time imaging offers a powerful diagnostic tool to evaluate the many endothelial functions in an organism. Although the clinical use of real-time imaging of the endothelium is in its infancy, the use of this tool in diagnosing endothelial dysfunction in animal models is widespread and, in many cases, state-of-the-art.

The initial attraction of real-time imaging was the "wow factor." A picture tells a thousand words – and a movie is even better! Seeing images that put in place concepts that previously were only imagined is a powerful tool.

But the real attraction of real-time imaging of the endothelium is that it allows for the spatial and temporal evaluation of experimental systems that are at a higher order of complexity (1) compared with traditional static assays. Historically, ignoring such complexity has impeded progress in endothelial research. Although studying protein structure in a crystal is complex and no doubt yields useful information, the function of that protein in a membrane, let alone in a cultured cell, is not always predictable. Similarly, the response to a stimulant of endothelial cells (ECs) in culture does not predict the response of ECs in vivo. Real-time imaging provides a window into functioning endothelium in the context of its native microenvironment (Table 177-1).

The aim of advancing technology is to allow observations in the natural, undisturbed environment. Until this is optimal, imaging of the endothelium has tended to suffer "the observer's paradox," in which the observation affects the outcome. To obfuscate this has often meant that real-time imaging of the endothelium is limited to the microcirculation of organ surfaces or thin tissues, with a few exceptions.

The goals of this chapter are to (a) catalogue established windows on the endothelium, (b) outline which endothelial functions are evaluable using real-time techniques, (c) provide a technical update on novel microscopic techniques that allow for real-time imaging of the endothelium, and (d) discuss translational and potential clinical applications of these tools.

WINDOWS ON THE ENDOTHELIUM – FROM BENCH TO BEDSIDE

The inefficiency of light penetration into tissue (reflective- or epi-illumination) or through tissue (transillumination) has limited the types of organs (and locations within an organ) where the endothelium can be evaluated both in real-time and in vivo. Clinically, only the retinal vascular bed is visualized easily and routinely. This is underscored by the importance of funduscopy as a clinical skill. Although basic scientists have designed methods to visualize vessels in myriad organs, most of these require skill to be reproducible.

BENCH

Retinal vessels are evaluated routinely by basic scientists (2). The most notable examples are antiangiogenic studies in proliferative retinopathy (3). The ear vessel of a mouse or rat is visible by transilluminated light. The autofluorescence of hair has, however, limited the broad use of this window using fluorescence in vivo microscopy; shaving introduces inflammatory artifacts but in some reports hairless mice have been used (4).

Since Atherton and Born (5) first described the technique of surgically exposing the mesentery through an abdominal wall incision, the mesenteric circulation is easily visualized (6). This technique has limitations; because of the need to stabilize the tissue only single vascular branches are visualized and very small capillaries are not seen.

The dorsal skin fold implant on mice has been the basis of a plethora of reports describing the process of angiogenesis. A glass window within a metal holder is surgically implanted on the dorsum of the animal such that the window allows observation of blood vessels of the skin from below. Arteries, veins, and capillaries can be seen. The window can be observed for as long as the animal is immobilized; the implants have been reported to last for at least 14 days (7). Vessels in tissue

Table 177-1: Windows on the Endothelium

Window	*Depth Limited*	*Surgery Required*	*Applicable to Clinic*
Skull bone marrow	Yes, by microscopy	Yes	No
Brain	Yes, by microscopy	Yes	No
Ear vessels	Yes, by microscopy	No	No
Tongue microvasculature	Yes	No	Yes
Carotid artery	No	Yes	No
Lymph Nodes	Yes, by microscopy	Yes	No
Liver	Yes, by microscopy	Yes	No
Mesentery	No, can see whole tissue	Yes	No
Skin	Yes, by microscopy	No	Yes
Skin Fold	No	Yes, but it heals	No
Cremaster muscle	No, can see whole tissue	Yes	No
Tendons	No	Yes	No

Depth limited refers to whether visualization of the whole tissue can be achieved using standard microscopic techniques, and whether depth penetration techniques are required (multiphoton or near-infrared).

allografts (e.g., lung or tumor tissue) can be studied after surgically placing allografts in dorsal skin fold implants (8).

Surface vessels of the liver (9), lymph nodes (10), skull bone marrow (11), and brain (12) have been evaluated; all also require surgical exposure. These windows are limited in that only surface vessels can be studied. New techniques such as two-photon microscopy and near-infrared (IR) fluorescence allow the visualization of deeper vessels. The recent description of the T-cell activation in six dimensions that occurs in lymph nodes represents state-of-the-art in real-time imaging (13,14).

The cremaster muscle allows for visualization of capillaries, arteries, and veins. Branches can be followed over long distances, allowing the study of events on interactions later in the circulation (15). Limitations include the requisite use of male mice, the requirement for surgical exposure, motion artifacts from arterial pulsations, and an upper limit on the size of vessels (rarely larger than 100 μm in diameter).

Larger vessels, such as the carotid artery of experimental rodents, have been visualized using real-time imaging. To accomplish this, the artery must be surgically exposed so that a special light device can be inserted under the vessel to transilluminate it from below. This technique has been used to evaluate thrombotic occlusion models (16).

Bedside

Retinal vessels are evaluated routinely by clinicians. Direct visualization allows for the surveillance of many diseases of the retinal circulation, such as proliferative retinopathy in diabetes and arteriolar sclerosis.

Capillaroscopy is an established procedure for the evaluation of dermal vessels. Traditionally, this has been used to visualize vessels in the nail-bed but now is used to visualize vessels at other skin sites (17). Capillary density and shape are evaluated and typical changes occur with psoriasis, vascular insufficiency, diabetes, and aging.

Submucosal vessels have been evaluated using orthogonal polarization spectral imaging (OPS) (18). Taking advantage of the properties of polarized light, the device allows a probe, generally placed under the tongue, to visualize the submucosal microcirculation. This technique allows for visualization of undisturbed human vessels. Because the technique relies on polarized light, fluorescence techniques are not possible. Blood vessel diameter, flow, and leukocyte–endothelial interactions have been evaluated in the context of diabetes, hypertension, anesthesia, solid organ transplantation, and sepsis (19). Small studies have reported that parameters derived from OPS can predict patient survival in sepsis (20) and predict rejection of intestinal transplants (21). The system is available commercially as a product called Cytoscan from Cytometrics, of Philadelphia, Pennsylvania (www.cytometrics.com).

ENDOTHELIAL FUNCTIONS EXPLORED BY REAL-TIME IMAGING

Permeability

Tagged-albumin has been used extensively to measure vascular permeability. This approach has been adapted using fluorescent-tagged albumin to measure vascular leakage in the microvasculature in real-time (22).

Cell Adhesion Molecule Expression

The dynamic adhesion of leukocytes to ECs is the oldest and most studied endothelial function explored by real-time imaging. For example, P- and E-selectin expression in endothelium is routinely evaluated by quantifying leukocyte rolling in these vessels. This is an indirect method to study a process involved in inflammatory cell recruitment but, by using gene-deleted animals, it can be very specific for EC adhesion molecule expression in the specific condition studied (23). Because many cell adhesion molecule interactions have shear dependency, vessels with similar hemodynamic properties are usually compared. The state-of-the-art in this method involves using dynamic tracking of a transfused, fluorescent dye–labeled leukocyte to describe the required length of the endothelium and cell adhesion molecules required for leukocyte arrest on the endothelium(15). Fluorescent-tagged antibodies also can be used to determine the luminal expression of molecules on the endothelium; this is most profoundly demonstrated by the spatial separation of two L-selectin ligands using the monoclonal antibody MECA-79 and a MECA-79–independent, $\alpha(1,3)$-fucosyltransferase-IV–dependent molecule (24).

Similar techniques have been used for evaluating the EC adhesion molecules involved in the transmigration of cells through the endothelial barrier. Multiphoton imaging has been used to visualize the movement of breast cancer cells through the extracellular matrix toward vessels and then their intravasation into the vessel (25). This technique is ripe for evaluating the role of EC adhesion molecules in mediating this "reverse" transport.

Finally, real-time imaging has proved useful in understanding the pathogenic role for the endothelium in sickle cell occlusion. It seems that the endothelium is a central player, linking inflammation and occlusion. Leukocytes adhere to inflamed endothelium in postcapillary venules via P-selectin and E-selectin. Then sickled red blood cells adhere to the leukocytes, and this results in vasoocclusion (26,27).

Coagulant and Thrombotic Activity

Recently, the real-time imaging of thrombosis in animal models has advanced knowledge in this field. Thrombotic clots, previously induced by chemical endothelial denudation, can now be induced by subtle injury to a single EC (28). The roles of platelet-endothelial cell adhesion molecule (PECAM)-1 (29) and endothelial P-selectin (29,30) in thrombus formation have been elucidated using these techniques. Soon, the roles of other endothelial molecules on thrombus formation in these models will be reported.

Vasomotor Tone

Techniques exist to study blood velocity and vessel diameter, specifically in small vessels and even capillaries using real-time imaging. Given these parameters, shear can be measured continuously and in real-time (31). The effect of nitric oxide (NO) inhibitors, vasoactive drugs, and the effects of gene deletions in this pathway have been studied.

Regeneration

Some aspects of our burgeoning knowledge of angiogenesis and vessel repair have been tested using real-time imaging. Angiogenesis is quantified by measuring vessel density. The expression of genes involved in angiogenesis, the source of the new endothelium, and physical properties of the young vessels have been evaluated with real-time imaging (7). Currently, most reports have involved the dorsal skin fold model, which is the only model ideal for evaluation of processes over days. The early processes of angiogenesis have not been tested with real-time imaging.

MICROSCOPE METHODOLOGIES

Bright-Field Microscopy

Traditional bright-field microscopy has been used to image vascular beds in real-time. The limitations of this approach include poor penetration of light through the tissue, limited accessibility of vascular beds, and difficulty in identifying cellular elements without specific markers. Nonetheless, because of the low cost and the lack of interfering factors, this technique remains powerful and ubiquitous. Polarization techniques offer enhanced contrast between objects based on their different refractive indices. Properly implemented, polarization allows clearer images that make cellular identification easier.

Fluorescence Microscopy

Fluorescence microscopy adds specificity to the identification of objects, cellular or molecular. Historically, limitations have been photobleaching of the fluorophore, photodamage by the excitation light, and the slow speed required to image multiple fluorescent colors. Recent advances in microscopy have lessened each of these limitations.

More affordable, sensitive imaging devices (driven by military and consumer demand) and better methods to switch excitation and/or emission filters have allowed for an increase in the speed with which multiple-color fluorescent images can be obtained (32). This also has lessened the intensity of excitation light, thus limiting tissue damage.

Confocal microscopes offer two advantages over traditional systems. Confocal images allow for better quantification of the amount of fluor using optical slice techniques. Tomograms of closely spaced images allow for quantification of fluor. Confocal images also allow for the evaluation of true optical resolution colocalization, and they can be used to determine if two fluors are optically colocalized.

Costly multiphoton systems offer all the advantages of confocal systems but also allow for deeper penetration into tissue. The longer wavelengths used in multiphoton systems scatter less at refractive-index changes. Multiphoton systems also have better colocalization than standard multiprobe confocal systems because, in the latter, the excitation beams are not the same single-point source. Finally, second-harmonic generation allows for imaging of α-helical proteins, such as collagen and laminin in the extracellular matrix, without the need for a fluorescent label (25).

Fluorescent Tags

The progress in imaging fluorescence warrants a review of the techniques that couple fluorophores to biologically relevant molecules.

Traditional fluorophores had variable permeability into the cell. Now fluorophores are conjugated to lipophilic substances through bonds that are broken once the molecule enters the cell (33), thus making it almost impossible for the fluorophore to leave the cell.

Newer fluorophores (Alexa series from Molecular Probes, probes.com) are brighter and more bleach-resistant are than traditional fluorescent dyes such as fluorescein. Quantum dots (34) (available at qdots.com) are semiconductor nanocrystals that are brightly fluorescent, bleach-resistant, and excited by white light, which limits tissue damage. The technologies used to attach quantum dots to biological molecules are improving, as are efforts to decrease in vivo toxicity (35). Quantum dots that emit in the near IR have been described (36).

Traditional and newer fluors can be linked to even tiny amounts of protein, especially antibodies. Peptides can be synthesized with single amino acids prelabeled with fluor. Flour/quencher combinations allow the evaluation of protease activity through the process of fluorescence resonance energy quenching (FREQ). This has been explored to evaluate the activities of thrombin (37) and cathepsin B (38).

The cloning of the green fluorescent protein (GFP) from the bioluminescent jellyfish *Aequorea victoria* and the generation of gain-of-function mutants, most of which are available in expression vectors, have allowed gene products to be visualized using real-time imaging. GFP is a 27-kDa monomeric polypeptide that needs no substrate or cofactor to fluoresce. It is specific, relatively bright, and physiologically compatible. The advantages of GFP over luciferase in similar gene-based assays are that it has at least a 100-fold increased signal to noise ratio, no need for substrate, and the ability to image directly with multicolors using GFP family molecules (YFP, etc.) (39,40). GFP has been linked to a growing list of gene promoters, including Tie2 (41) and vascular endothelial growth factor (VEGF) (42).

The development of technology to use near-IR fluorescent probes as markers offers the promise of extending real-time imaging to larger and deeper vessels. The scatter of light is inversely proportional to wavelength and directly proportional to refractive index changes. Thus, longer wavelengths can penetrate tissue (where multiple refractive index changes occur) more deeply using the same tolerable amount of scatter. Fluorescent probes in the near IR have imaged cathepsin B activity to delineate atherosclerosis in living mice (38). Similar probes have the potential to allow visualization of probes as deep as 2 cm below the surface.

CONCLUSION

Real-time imaging of endothelial function is currently limited largely to visible vessels, most commonly in animal models. In the future, advances in imaging modalities such as multiphoton microscopy and near-IR probes will allow real-time imaging to move to clinical use. Until then, lessons learned in animal models using real-time imaging will allow advances in ultrasound and magnetic resonance imaging (see Chapter 176) to be applied in the clinic (43).

KEY POINTS

The limitation of real-time imaging of the endothelium is light penetration into the tissue. Strategies to overcome this limitation include (a) the use of thin tissue, (b) the observation of vessels only at the surface of organs, (c) the use of brighter and more biocompatible fluorophores, (d) the use of near-IR fluorophores, and (e) multiphoton microscopy.

REFERENCES

1 Cahalan MD, Parker I, Wei SH, Miller MJ. Real-time imaging of lymphocytes in vivo. *Curr Opin Immunol.* 2003;15(4):372–377.

2 Gotte M, Joussen AM, Klein C, et al. Role of syndecan-1 in leukocyte-endothelial interactions in the ocular vasculature. *Invest Ophthalmol Vis Sci.* 2002;43(4):1135–1141.

3 Ishida S, Yamashiro K, Usui T, et al. Leukocytes mediate retinal vascular remodeling during development and vaso-obliteration in disease. *Nat Med.* 2003;9(6):781–788.

4 Rosen ED, Raymond S, Zollman A, et al. Laser-induced noninvasive vascular injury models in mice generate platelet- and coagulation-dependent thrombi. *Am J Pathol.* 2001;158(5):1613–1622.

5 Atherton A, Born GV. Quantitative investigations of the adhesiveness of circulating polymorphonuclear leucocytes to blood vessel walls. *J Physiol.* 1972;222(2):447–474.

6 Ni H, Denis CV, Subbarao S, et al. Persistence of platelet thrombus formation in arterioles of mice lacking both von Willebrand factor and fibrinogen. *J Clin Invest.* 2000;106(3):385–392.

7 Brown EB, Campbell RB, Tsuzuki Y, et al. In vivo measurement of gene expression, angiogenesis and physiological function in tumors using multiphoton laser scanning microscopy. *Nat Med.* 2001;7(7):864–868.

8 Sikora L, Johansson AC, Rao SP, et al. A murine model to study leukocyte rolling and intravascular trafficking in lung microvessels. *Am J Pathol.* 2003;162(6):2019–2028.

9 Oda M, Yokomori H, Han JY. Regulatory mechanisms of hepatic microcirculation. *Clin Hemorheol Microcirc.* 2003;29(3–4):167–182.

10 Diacovo TG, Puri KD, Warnock RA, et al. Platelet-mediated lymphocyte delivery to high endothelial venules. *Science.* 1996; 273(5272):252–255.

11 Mazo IB, Gutierrez-Ramos JC, Frenette PS, et al. Hematopoietic progenitor cell rolling in bone marrow microvessels: parallel contributions by endothelial selectins and vascular cell adhesion molecule 1. *J Exp Med.* 1998;188(3):465–474.

12 Kerfoot SM, Long EM, Hickey MJ, et al. TLR4 contributes to disease-inducing mechanisms resulting in central nervous

system autoimmune disease. *J Immunol.* 2004;173(11):7070–7077.

13 von Andrian UH. Immunology. T cell activation in six dimensions. *Science.* 2002;296(5574):1815–1817.

14 Miller MJ, Wei SH, Parker I, Cahalan MD. Two-photon imaging of lymphocyte motility and antigen response in intact lymph node. *Science.* 2002;296(5574):1869–1873.

15 Kunkel EJ, Dunne JL, Ley K. Leukocyte arrest during cytokine-dependent inflammation in vivo. *J Immunol.* 2000;164(6):3301–3308.

16 Jirouskova M, Chereshnev I, Vaananen H, et al. Antibody blockade or mutation of the fibrinogen gamma-chain C-terminus is more effective in inhibiting murine arterial thrombus formation than complete absence of fibrinogen. *Blood.* 2004;103(6):1995–2002.

17 Hern S, Mortimer PS. Visualization of dermal blood vessels – capillaroscopy. *Clin Exp Dermatol.* 1999;24(6):473–478.

18 Lindert J, Werner J, Redlin M, et al. OPS imaging of human microcirculation: a short technical report. *J Vasc Res.* 2002;39 (4):368–372.

19 Ince C. The microcirculation unveiled. *Am J Respir Crit Care Med.* 2002;166(1):1–2.

20 Sakr Y, Dubois MJ, De Backer D, et al. Persistent microcirculatory alterations are associated with organ failure and death in patients with septic shock. *Crit Care Med.* 2004;32(9):1825–1831.

21 Masetti M, Cautero N, Lauro A, et al. Three-year experience in clinical intestinal transplantation. *Transplant Proc.* 2004;36(2): 309–311.

22 von Dobschuetz E, Pahernik S, Hoffmann T, et al. Dynamic intravital fluorescence microscopy – a novel method for the assessment of microvascular permeability in acute pancreatitis. *Microvasc Res.* 2004;67(1):55–63.

23 Frenette PS, Mayadas TN, Rayburn H, et al. Double knockout highlights value of endothelial selectins. *Immunol Today.* 1996; 17(5):205.

24 M'Rini C, Cheng G, Schweitzer C, et al. A novel endothelial L-selectin ligand activity in lymph node medulla that is regulated by alpha(1,3)-fucosyltransferase-IV. *J Exp Med.* 2003;198(9):1301–1312.

25 Condeelis J, Segall JE. Intravital imaging of cell movement in tumours. *Nat Rev Cancer.* 2003;3(12):921–930.

26 Frenette PS. Sickle cell vasoocclusion: heterotypic, multicellular aggregations driven by leukocyte adhesion. *Microcirculation.* 2004;11(2):167–177.

27 Hebbel RP, Osarogiagbon R, Kaul D. The endothelial biology of sickle cell disease: inflammation and a chronic vasculopathy. *Microcirculation.* 2004;11(2):129–151.

28 Falati S, Gross P, Merrill-Skoloff G, et al. Real-time in vivo imaging of platelets, tissue factor and fibrin during arterial thrombus formation in the mouse. *Nat Med.* 2002;8(10):1175–1181.

29 Falati S, Patil S, Gibbins J, et al. Opposing effects of P-selectin and PECAM-1 on arterial thrombus growth and stability in vivo. *J Thromb Haemost.* 2003.

30 Falati S, Liu Q, Gross P, et al. Accumulation of tissue factor into developing thrombi in vivo is dependent upon microparticle P-selectin glycoprotein ligand 1 and platelet P-selectin. *J Exp Med.* 2003;197(11):1585–1598.

31 Hungerford JE, Sessa WC, Segal SS. Vasomotor control in arterioles of the mouse cremaster muscle. *FASEB J.* 2000;14(1):197–207.

32 Celi A, Merrill-Skoloff G, Gross P, et al. Thrombus formation: direct real-time observation and digital analysis of thrombus assembly in a living mouse by confocal and widefield intravital microscopy. *J Thromb Haemost.* 2003;1(1):60–68.

33 Baker GR, Sullam PM, Levin J. A simple, fluorescent method to internally label platelets suitable for physiological measurements. *Am J Hematol.* 1997;56(1):17–25.

34 Michalet X, Pinaud FF, Bentolila LA, et al. Quantum dots for live cells, in vivo imaging, and diagnostics. *Science.* 2005;307(5709): 538–544.

35 Seydel C. Quantum dots get wet. *Science.* 2003;300(5616):80–81.

36 Zheng J, Zhang C, Dickson RM. Highly fluorescent, water-soluble, size-tunable gold quantum dots. *Phys Rev Lett.* 2004;93 (7):077402.

37 Jaffer FA, Tung CH, Gerszten RE, Weissleder R. In vivo imaging of thrombin activity in experimental thrombi with thrombin-sensitive near-infrared molecular probe. *Arterioscler Thromb Vasc Biol.* 2002;22(11):1929–1935.

38 Chen J, Tung CH, Mahmood U, et al. In vivo imaging of proteolytic activity in atherosclerosis. *Circulation.* 2002;105(23):2766–2771.

39 Hoffman R. Green fluorescent protein imaging of tumour growth, metastasis, and angiogenesis in mouse models. *Lancet Oncol.* 2002;3(9):546–556.

40 Yang M, Li L, Jiang P, et al. Dual-color fluorescence imaging distinguishes tumor cells from induced host angiogenic vessels and stromal cells. *Proc Natl Acad Sci USA.* 2003;100(24):14259–14262.

41 Motoike T, Loughna S, Perens E, et al. Universal GFP reporter for the study of vascular development. *Genesis.* 2000;28(2):75–81.

42 Fukumura D, Xavier R, Sugiura T, et al. Tumor induction of VEGF promoter activity in stromal cells. *Cell.* 1998;94(6):715–725.

43 Choudhury RP, Fuster V, Fayad ZA. Molecular, cellular and functional imaging of atherothrombosis. *Nat Rev Drug Discov.* 2004;3(11):913–925.

178

Diagnosing Endothelial Cell Dysfunction

Aristides Veves and Roy Freeman

Beth Israel Deaconess Medical Center, Harvard Medical School, Boston, Massachusetts

One of the major developments in medicine during the last two decades is the delineation of the role of endothelium in the development of cardiovascular disease. Soon after the first observations in animals, methods were devised that could evaluate endothelial function in humans. The initial methods were invasive, but subsequently noninvasive methods were established. During the last decade, these noninvasive methods have been standardized and widely used for the clinical research, while the possibility of using them in standard clinical practice is also raised. In this chapter we review the most commonly employed methods for assaying endothelium-dependent vasodilation in both the macro- and microcirculation.

MACROCIRCULATION

Endothelium-Dependent versus Endothelium-Independent Vasodilation

Although the endothelium regulates the vascular tone by the balanced secretion of vasoconstrictors and vasodilators, it should be remembered that this is mainly achieved by the action of these vasomodulators on the vascular smooth muscle cell (VSMC). This can have serious implications in the interpretation of the results of various tests in humans. Thus, in the theoretical case in which VSMC vasodilatory capacity is impaired, those tests that assess the ability of endothelial cells (ECs) to produce vasodilators such as nitric oxide (NO) – endothelium-dependent vasodilation – will be abnormal. Therefore, correct interpretation of the data requires the consideration of both the endothelium-dependent and -independent vasodilation measurements that are collectively referred to as *vascular reactivity measurements.*

Venous Plethysmography

Venous plethysmography, or venous occlusive plethysmography, was the first technique used to measure vascular reactivity (1). This technique employs a mercury in-silastic strain gauge coupled to a plethysmograph. The gauge is placed at the upper third of the forearm. An arm blood pressure cuff with rapid inflators that applies a 40 mm Hg pressure is used to achieve forearm venous occlusion. The average forearm blood flow (FBF) from at least five measurements performed at 10- to 15-second intervals is usually recorded. In some laboratories, the hand blood flow is excluded from the analysis. To achieve this, the blood flow to the hand is blocked at the wrist level by a cuff that is inflated to suprasystolic pressure levels. Once the baseline FBF has been measured, it is usually followed by brachial intra-arterial infusion of substances that produce either endothelium-dependent vasodilation (typically acetylcholine [ACh] or methacholine) or endothelium-independent vasodilation (typically sodium nitroprusside). The change in FBF usually is expressed as milliliter (mL) per minute per 100 mL of tissue. In addition, the ratio of the mean blood pressure to FBF represents the vascular resistance. In most laboratories, the opposite hand is employed as a control by injecting normal saline and performing the same FBF measurements. In a laboratory with appropriate training and experience, this technique has a satisfactory reproducibility, especially after the infusion of vasodilators, ranging from 0.90 to 0.97 over a 3-week period (2).

Venous plethysmography has been used extensively in human studies that examined the effect of various conditions (such as diabetes and hypertension) on vascular reactivity and the possibility of reversing endothelial dysfunction after various therapeutic interventions (3–5).

Invasive Blood Flow Measurements Using the Thermodilution Technique

The thermodilution technique was developed by Ganz and colleagues and employs a double-lumen, 5-French thermodilution catheter that can evaluate the instantaneous mixing of an infused indicator (cold saline) in the vein and measure blood flow in this vessel (6). This technique has been used mainly to measure blood flow in the leg through the catheterization of the femoral vein and the infusion of vasodilators in the femoral artery (7–8).

Noninvasive Assessment of Brachial Artery Vascular Reactivity

The noninvasive measurement of brachial artery vascular reactivity using standard ultrasound techniques is currently the most commonly used method to evaluate the function of the endothelium and vascular smooth muscle. The technique was first described by Celermajer and colleagues in 1992 (9). The main principle of the technique is to use high-resolution ultrasound to follow changes in the brachial artery diameter after employing methods that either directly stimulate the endothelium and cause endothelium-dependent vasodilation or act directly on the VSMC and result in endothelium-independent dilation. Detailed guidelines regarding subject preparation, image acquisition, data analysis, and quality control have been published in a technical report by the International Brachial Artery Reactivity Task Force (10).

Of the noninvasive measurements, flow-mediated vasodilation (FMD) is the main technique used to measure endothelium-dependent vasodilation. In brief, the subject is placed in a supine position, and the brachial artery is imaged above the antecubital fossa in the longitudinal plane (Figure 178.1). A 7-MHz or greater linear array transducer is used and two-dimensional grayscale scans are taken at rest and after the stimulation that leads to artery dilation. To measure the endothelium-dependent vasodilation, FMD stimulation is employed. This method involves inducing upper limb ischemia by occluding the brachial artery with a pneumatic tourniquet that is inflated 50 mmHg higher than the systolic blood pressure. The cuff can be placed either at the upper arm, an area proximal to the area where ultrasound imaging of the artery is performed, or on the upper forearm, an area distal to the ultrasound probe. The artery typically is occluded for 5 minutes. This leads to tissue hypoxia and pH changes in the tissues distal to the occlusion, mainly the muscles. As a result, immediately after the removal of the occlusion, a vasodilation occurs in the muscle and skin microcirculation that is directly related to hypoxia, changes in the tissue pH, and accumulation of deoxyhemoglobin. This causes a brief period of high blood flow in the brachial artery (reactive hyperemia), which creates increased shear stress that leads to the stimulation of the endothelium and causes vasodilatation. The brachial artery dilation typically peaks around 60 seconds after the removal of the occlusion, and the ultrasound images are taken 30 to 120 seconds after removal of occlusion. The baseline and highest postinflation brachial artery diameter are measured using sophisticated computer programs. Vessel diameter measurements are taken at end diastole, incident with the R-wave on the electrocardiogram (ECG). The percent of the postinflation increase of the artery diameter over the baseline indicates the FMD.

Initial studies employed forearm occlusion, whereas subsequent studies used upper arm occlusion (11). Upper arm occlusion leads to higher FMD values, but forearm occlusion has an advantage in that the brachial artery does not collapse during the occlusion period, making the image acquisition easier. Previous studies in our unit have shown a very satisfactory correlation in measurements taken with either upper arm or forearm occlusion in the same subjects (12). Therefore, both techniques are equally reliable and can be employed in clinical studies.

The most important factor that determines the reproducibility of the technique is the training of the personnel who are involved in the data acquisition and analysis. Because this method is technically challenging, only well-trained examiners should perform the tests. Various manual or semiautomatic techniques also are available for the image analysis, but they also require examiners with considerable experience. In well-organized laboratories, the interobserver difference has been found to be around 1.7%, and the coefficient of variation 1.4% (9). In addition, satisfactory reproducibility also has been demonstrated in longitudinal studies in both normal subjects and subjects with reduced FMD. Thus, the measurement error (1%–3%) was significantly less than the difference between normal and abnormal FMD responses (7%–10%) (9,10).

FMD can be measured in other peripheral arteries, such as the radial, axillary, and superficial femoral arteries (13,14). However, the reproducibility of these measurements is less than that of the brachial artery, and they are not widely accepted.

Endothelial Function Assessment of Coronary Arteries

Coronary artery vascular tone is maintained by the same mechanisms that are involved in the maintenance of the vascular tone in the peripheral arteries. Because the coronary arteries are not easily accessible, the measurement of diameter changes is more difficult to quantify. The invasive method is most commonly used, and it involves infusion of ACh into the left coronary artery. The changes in the coronary artery diameter are assessed by quantitative angiography. In healthy vessels, ACh produces vasodilation, whereas in vessels with endothelial dysfunction, it causes vasoconstriction (15). The functional state of the myocardial microcirculation can be assessed using coronary intravascular Doppler that can measure changes in the blood flow in resistant vessels during the infusion of substances that produce endothelium-dependent or -independent vasodilation (16). Both these measurements can be performed in the same patient during the same session.

The correlation between coronary and peripheral artery functional measurements is very comparable. This is not surprising given the systemic nature of mechanisms that cause endothelial dysfunction and indicates that the changes in the endothelial dysfunction are comparable (17). As a result, peripheral artery measurements have been proposed as a

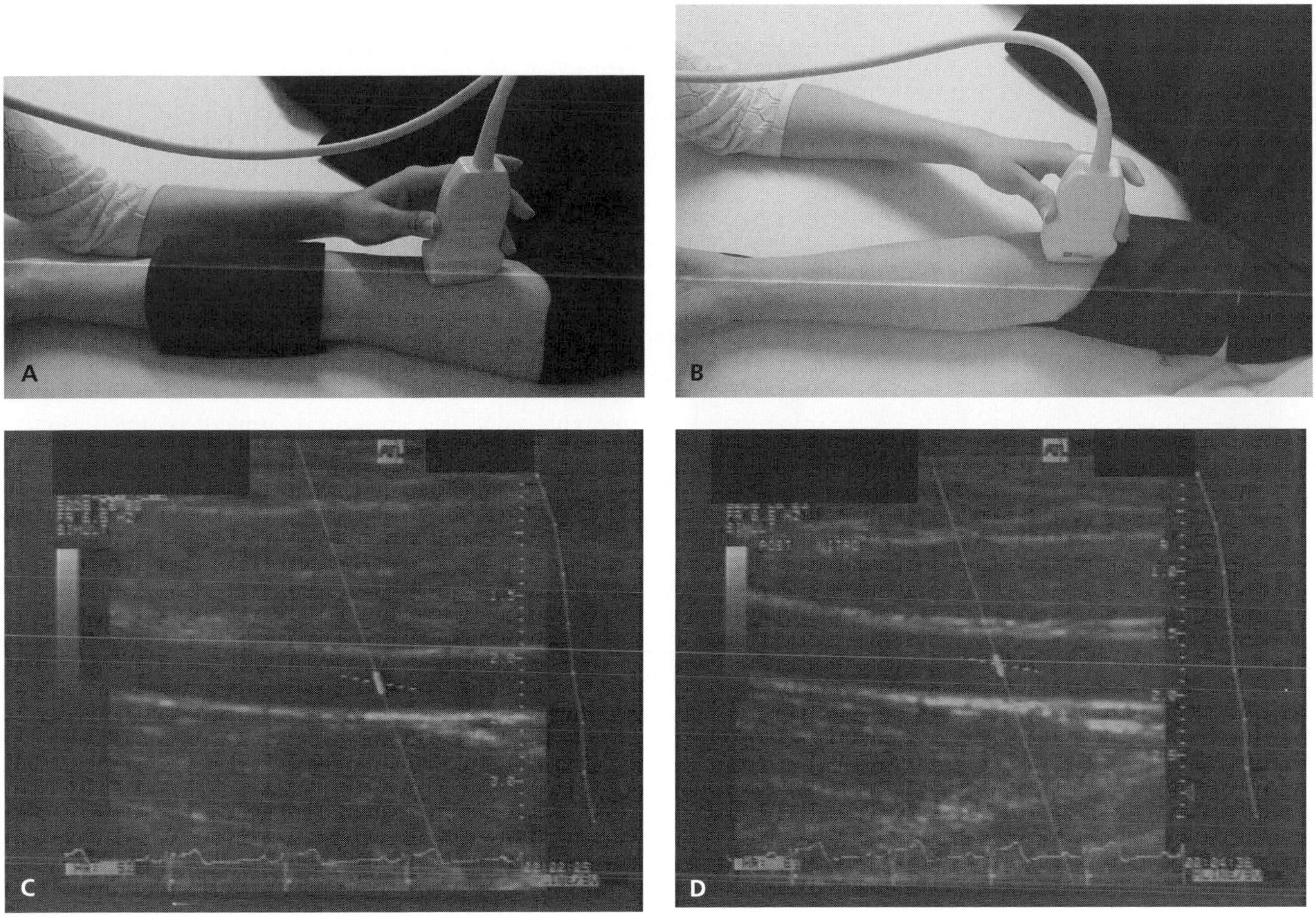

Figure 178.1. (**A**) The assessment of flow-mediated vasodilation in the brachial artery. A 7.0-MHz or greater linear array transducer is used to image the brachial artery above the antecubital fossa in the longitudinal plane. A regular sphygmomanometer is employed to occlude the artery blood flow. The sphygmomanometer can be placed either at the forearm (**A**) or the upper arm level (**B**). Two-dimensional grayscale scans are taken, one at rest at rest, before the cuff inflation (**C**) and 1 minute after the cuff deflation that leads to artery dilation (**D**). The percentage of the postocclusive artery diameter increase over the baseline represents the FMD.

surrogate means of identifying subjects at risk of cardiovascular events (18).

Anatomic Imaging of the Vessels

Common Carotid Artery Intimal-Medial Thickness

An increase in the intimal-medial thickness (IMT) of the carotid artery is one of the first morphological changes that occurs early in the course of atherosclerotic development. IMT changes can be easily detected and quantified using high-resolution ultrasound B-mode imaging. This noninvasive technique allows visualization of early atherosclerotic plaque formation prior to the luminal narrowing caused by the late stages of atherosclerosis. Imaging usually is performed with high-resolution real-time scanners equipped with 7- to 8-MHz probes. The ultrasound pattern is characterized by two parallel echogenic lines separated by a hypoechoic or anechoic space. This scan pattern is defined as the "double line pattern." The inner (luminal) line, which is generally more regular, smooth, and thin than the outer one represents the intima, the hypoechoic line the media, and the outer echogenic line the adventitia. The distance from the inner echogenic line to the interface between the hypoechoic line and the second echogenic line represents the IMT of the vascular wall (Figure 178.2) (19).

The technique is easy to perform and has a satisfactory reproducibility, with an interobserver coefficient of variation around 10% and an intraobserver coefficient of variation of approximately 5% (20). Furthermore, numerous studies have shown that IMT is associated with conventional risks for the development of atherosclerosis, and that it can predict the development of cardiovascular events (21,22). As a result, at present, IMT measurements are widely employed in clinical trials.

Pulse Wave Velocity

This technique uses Doppler probes to measure the speed of propagation of the forward-going pulse wave in the absence of reflections from the periphery, and is a surrogate measurement of the compliance of the studied artery. Pulse wave

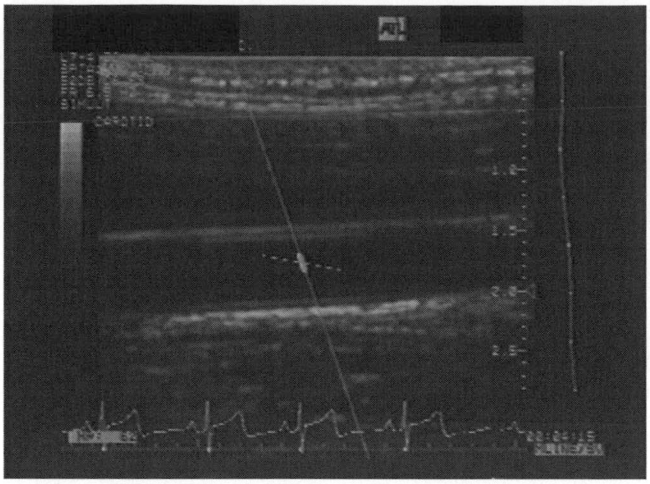

A

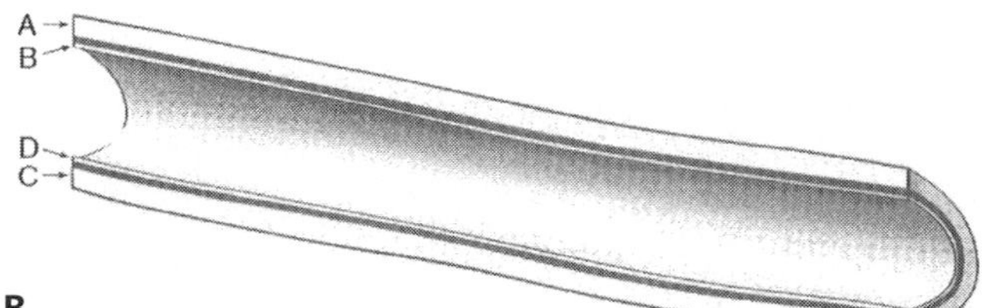

B

Figure 178.2. (**A**) Image of the common carotid artery. A 7.5-MHz linear array transducer and high-resolution ultrasound were used. (**B**) Simplified diagram of the arterial wall boundaries indicating the adventitia-media (*A*) of the near wall, intima-blood boundaries (*B*) for the near wall, and adventitia-media (*C*) and intima-blood boundaries (*D*) for the far wall. (Reproduced with permission from Danias P, Saouaf R. Noninvasive Methods to Assess Vascular Function and Pathophysiology. In: Johnstone MT, Veves A. *Diabetes and Cardiovascular Disease*, 2nd ed. Totowa, NJ: Humana Press; 2005:431–450.)

velocity (PWV) depends on the biophysical properties of the artery, and is increased in conditions that lead to stiff arteries, such as diabetes mellitus, hypercholesterolemia, coronary artery disease, and renal failure (23,24). Alternatively, the elastic properties of the aorta can be measured by employing cardiovascular magnetic resonance imaging (MRI). The major advantage of this technique is its accuracy, which allows for the detection of small differences and can be very useful for carrying out clinical trials with limited numbers of participants (25). Despite initial encouraging results, neither PWV nor MRI has gained universal acceptance, and these techniques have not been widely used in clinical trials.

MICROCIRCULATION

Measurements of Microcirculatory Blood Flow Using Laser Doppler Flowmetry

Laser Doppler flowmetry is a well-established and widely used technique for evaluating blood flow in the skin microcirculation. This technique measures the microcirculatory flux, which is determined by a combination of velocity and the number of moving red blood cells (RBCs) within the arterioles, capillaries, and postcapillary venules.

Laser Doppler flowmetry utilizes the coherent properties of laser light to measure the velocity of a collection of moving RBCs within the illuminated area. A laser beam is delivered to the skin via a fiber optic light guide. The reflected light is gathered by a second set of light guides (photodetectors) and the Doppler-shifted fraction of the light signal and the mean Doppler frequency shift is calculated. A value, measured in mV, is obtained that is proportional to the quantity of moving RBCs and the mean velocity of those cells within the sample volume of cutaneous tissue. This value is a measure of superficial microvascular perfusion. This technology does not permit the assessment of absolute perfusion values (e.g., mL/min/100 gram tissue). The technique has been validated in a study comparing direct measurements of the capillary flow velocity using dynamic capillaroscopy with laser Doppler flowmetry (26).

Either a single-point laser probe, which evaluates the microvascular blood flow at one point of the skin, or a laser perfusion scanner, which evaluates the blood flow in an area of skin, can be used (Figure 178.3). In contrast to the single-point laser probe, which assesses blood flow at a single point across time, the perfusion scanner sequentially scans up to several hundred points in an area of skin, producing a local image of cutaneous microvascular perfusion. Cutaneous blood flow has temporal and spatial heterogeneity. The single-point Doppler overcomes the deficiencies in temporal heterogeneity, whereas the laser Doppler perfusion scanner overcomes the spatial heterogeneity. Integrating flow probes may help increase the cutaneous sampling area compared to single-point flow probes. The measuring area of the probe has several "probe tips." Each probe tip has its own delivery fiber and photodetecting fiber. The values from each tip are optically integrated into one value. This method allows measurement over a larger area and thereby minimizes the effects of spatial heterogeneity and improves reproducibility of the technique.

Provocative Techniques

Vasoactive Drug Delivery

Ionophoresis and microdialysis are techniques in which agents can be delivered to local area with precision while minimizing the systemic effects associated with oral ingestion or parenteral infusions. Iontophoresis is a noninvasive method of introducing charged substances into the skin. A small charge allows the transcutaneous delivery of substances in high concentration without or with minimal trauma or pain. The number of molecules transported is a function of the length of stimulation, the current used, and the size of the stimulator. ACh, a positively charged ion, is delivered using anodal current, whereas the anion sodium nitroprusside is delivered using cathodal current.

The minimally invasive technique of microdialysis permits the delivery of larger, water-soluble molecules that lack a

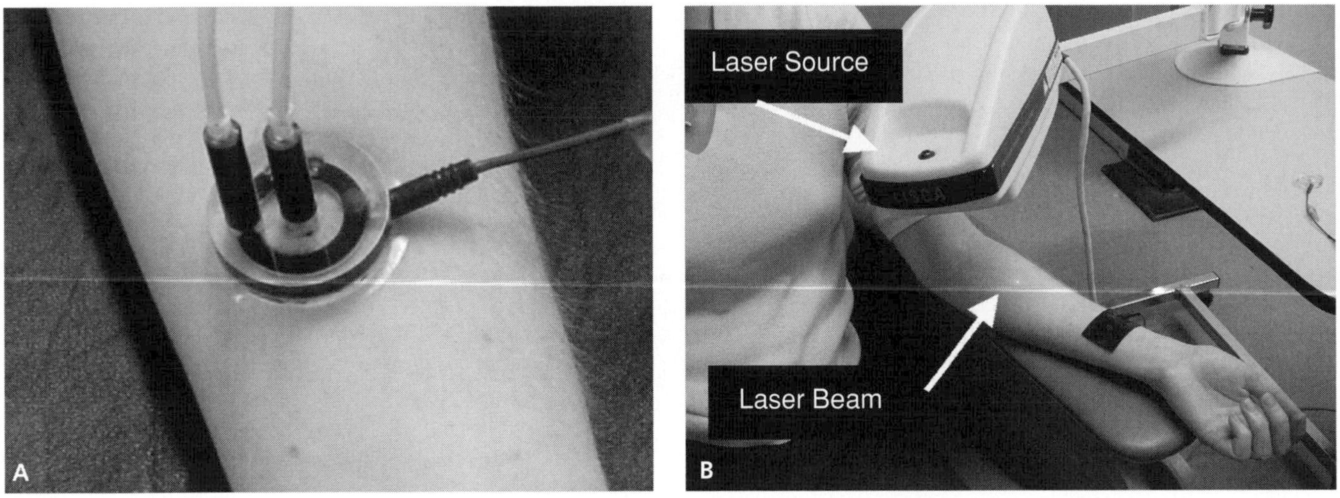

Figure 178.3. (**A**) Measurements of direct and indirect effect of vasoactive substance using single-point laser probes: One probe is used in direct contact with the iontophoresis solution chamber (*shaded ring*) and measures the direct response. The center probe measures the indirect response (nerve axon-related effect). A small quantity (<1 ml) of 1% Ach chloride solution or 1% sodium nitroprusside solution is placed in the iontophoresis. A constant current of 200 mA is applied for 60 seconds achieving a dose of 6 mC/cm^2 between the iontophoresis chamber and a second nonactive electrode placed 10 to 15 cm proximal to the chamber (black strap around the wrist). This current causes a movement of solution to be iontophoresed toward the skin. (**B**) Laser Doppler flowmetry: A helium-neon laser beam is emitted from the laser source to sequentially scan the circular hyperemic area (seen surrounding the laser beam) produced by the iontophoresed vasoactive substance to a small area on the volar surface of the forearm.

charge. Microdialysis probes are inserted into the upper dermis under needle guidance and perfused with saline or Ringer's solution at a rate of 2 to 5 μL/min. This technique is well tolerated and also can be used to sample water-soluble molecules within the extracellular space.

Vasoactive Agents

In studies, ACh is typically used to provoke endothelial-dependent vasodilation, whereas the NO donor, sodium nitroprusside, is used to provoke endothelial-independent vasodilation. ACh can be delivered by iontophoresis or microdialysis.

ACh binds to M_2 muscarinic receptors on the endothelial surface and elicits an increase in cutaneous blood flow by a direct receptor-mediated effect. Acting via a G-protein–coupled receptor and catalyzed by constitutively expressed endothelial NO synthase (eNOS), ACh stimulates the production of NO, which is a potent mediator of endothelium-dependent vasodilation. Other factors released by the microvascular endothelium in response to ACh include vasoactive prostanoids and endothelium-derived hyperpolarizing factors (EDHF). Thus, intradermal delivery of exogenous ACh produces vasodilation via these nitridergic and non-nitridergic mechanisms.

Although NO plays a prominent role in vasodilation, the relative contribution of these factors, and most importantly NO and vasoactive prostanoids, to ACh-mediated cutaneous vasodilation is not definitively resolved. We and others (27–29) have reported that iontophoresed ACh-mediated vasodilation of the forearm microcirculation is not substantially mediated by a prostanoid-dependent mechanism (Figure 178.4), whereas others have suggested that a prostanoid-dependent mechanism may play a prominent role in ACh-mediated vasodilation (30,31). The different results obtained in these studies may relate to the use of different acetylsalicylic acid (ASA) and other cyclooxygenase (COX) inhibitor formulations and doses, different routes of administration, varying iontophoretic protocols, and differing amounts of anodal current, which may cause vasodilation via a prostanoid-dependent pathway.

Microdialysis allows the use of NOS inhibitors such as *N*(G)-nitro-L-arginine methyl ester (L-NAME) to inhibit NOS production, and COX inhibitors, such as the nonselective COX inhibitor, ketorolac, to inhibit generation of prostanoid and thromboxane products, thereby permitting the elucidation of the relative contributions of NO and vasoactive prostanoids to ACh-mediated vasodilation. Nevertheless, microdialysis studies have not definitively resolved this issue. Kellogg and colleagues (32) reported that both NO and vasoactive prostanoids play a role in ACh-provoked vasodilation, whereas Holowatz and colleagues (33) suggested that vasoactive prostanoids play a prominent role in ACh-mediated vasodilation. Different microdialysis protocols may be responsible for the different results.

Endothelium-independent vasodilation is provoked by iontophoresis or microdialysis of the NO-donor, sodium nitroprusside. Vasodilation evoked by sodium nitroprusside is typically one-third less than that evoked by iontophoresis of Ach (34,35) and is equivalent to that achieved by increasing the local skin temperature to 42°C (36,37).

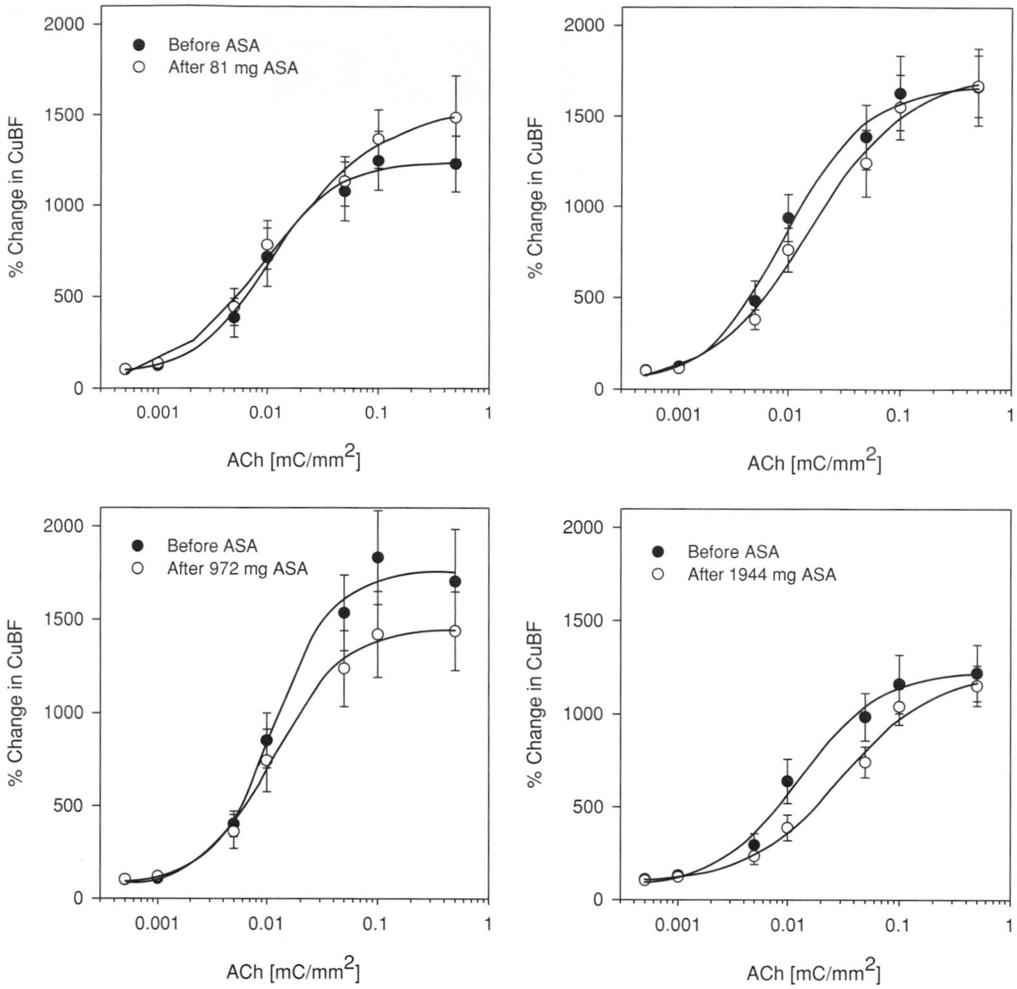

Figure 178.4. Iontophoresis of the endothelium-dependent vasodilator Ach resulted in a significant increase in cutaneous blood flow both before and after the administration of ASA (ANOVA, $p < 0.0001$). This response was not changed by any of the four doses of ASA. Data are expressed as mean ± SEM.

Local Heat

Local heating of the skin also provokes vasodilation. Although the mechanisms underlying the vasodilatory response to local heating are still incompletely understood, it is recognized that vasodilation is mediated by neurogenic reflexes and locally released neurotransmitters and neuromodulators. Although there is some variability in the response to local heating – in part related to the peak and starting temperature, the duration of heat application, and the rate of heat increase – there is a stereotyped response that appears generalizable across experimental paradigms. An initial brisk vasodilatory response peaks within minutes. This is followed by a short nadir that is, in turn, followed by a secondary dilation to a sustained plateau.

The initial vasodilatory response to local warming is evoked by local axon reflex–mediated antidromic neurotransmitter release from sensory afferents; this response is reduced by topical anesthesia in the region of local heating and is not attenuated by more proximal nerve blockade. The vasodilatory response primarily is mediated by the release of calcitonin gene-related peptide (CGRP) and substance P from C-fiber nociceptors. The sustained plateau phase of local heating-induced vasodilation is mediated by local generation of NO. This response is attenuated by pretreatment of the locally warmed area of skin with the NOS inhibitor, L-NAME. The interaction between the mediators of these two responses to local heating is not fully elucidated.

Clinical Utility of Microvascular Assessment

Laser Doppler flowmetry has been widely used to evaluate endothelial function in the limbs of diabetic patients (34,38,39). Impairment of endothelium-dependent vasodilation alone and in combination with endothelium-independent microvascular blood flow has been shown in different studies. The differences between studies depend on the duration of diabetes, the severity of the vascular complications of diabetes, and the site of study: Endothelium-dependent

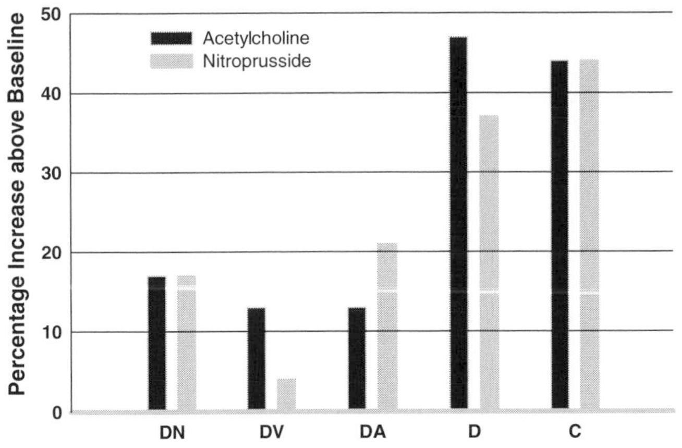

Figure 178.5. The endothelium-dependent and -independent vasomotor response to ACh and nitroprusside iontophoresis (DN, diabetic neuropathy; DV, diabetic vascular diseases; DA, Charcot arthropathy; D, Diabetic controls; C, Controls).

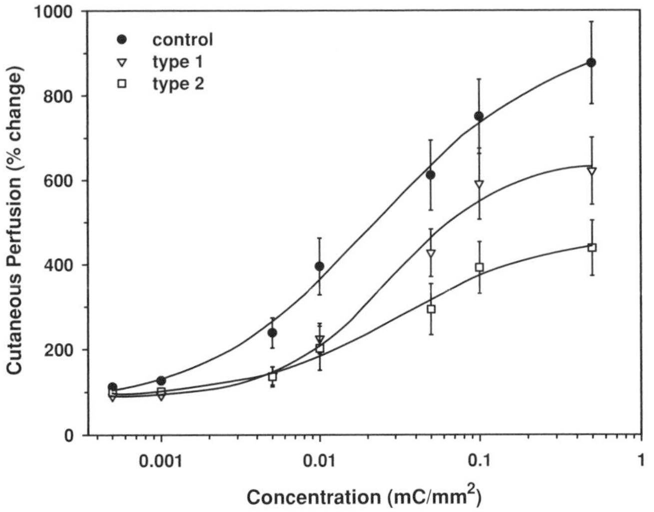

Figure 178.6. The endothelium-dependent vasomotor response to ACh iontophoresis. Application of ACh led to an increase in endothelium-dependent blood flow that is impaired in diabetic patients ($p < 0.005$). The maximum blood flow increase in type 2 diabetes is reduced compared to controls ($p = 0.001$). A trends is noted toward impaired endothelial-mediated blood flow in type 1 diabetic patients compared with controls. A rightward shift of the dose-response curves occurs in the diabetic subjects that did not reach significance. Data are expressed as mean ± SEM.

and -independent vasodilatation are significantly lower at the foot level when compared to the forearm level in both healthy controls and in diabetic patients (40). The drug delivery protocol also has an important influence on study results.

For example, we studied endothelial-mediated microvascular blood flow in neuropathic diabetic patients with a predisposition to ulceration. Vasodilation on the dorsal foot in response to heating and iontophoresis of a single, maximal dose of Ach (endothelium-dependent response) and sodium nitroprusside (endothelium-independent response) were measured using single-point laser Doppler and laser Doppler imaging in diabetic patients with neuropathy, diabetic patients with neuropathy and peripheral macrovascular disease, diabetic patients with Charcot arthropathy, diabetic patients without complications, and in healthy control subjects. The endothelial-mediated response to Ach was reduced in the patients with diabetic neuropathy, diabetic patients with Charcot arthropathy, and diabetic patients with peripheral macrovascular disease compared with control patients with diabetes and healthy control subjects. The nonendothelial-mediated response to sodium nitroprusside also was reduced in the patients with diabetic neuropathy, diabetic patients with Charcot arthropathy, and diabetic patients with vascular disease compared with control patients with diabetes and healthy control subject groups. A significant reduction in vasodilation occurred in the vascular disease group compared with all other groups (39) (Figure 178.5). Using the same iontophoretic protocol but assessing microvascular function in the forearm with laser perfusion imaging, endothelial-dependent and -independent vasodilation was observed in patients with impaired glucose tolerance and in relatives of type 2 diabetic patients (41).

In a follow-up study, we examined less severely afflicted patients at the forearm, a site that is less susceptible to peripheral neuropathy and vascular disease. In addition, we modified the iontophoretic protocol, using incremental doses of vasoactive agents to reveal subtle changes in the reactivity of the microcirculation. Using this dose-response technique, endothelium-dependent vasodilation of the forearm cutaneous microcirculation was attenuated in both diabetic subject groups. The response was significantly less in type 2 diabetic subjects than in type 1 subjects. There was no significant abnormality in endothelium-independent vasodilation in either diabetic group.

The duration of diabetes in type 2 diabetics was a significant predictor of the maximum endothelium-mediated vasodilation. These data suggest that, in type 2 diabetic patients, endothelial dysfunction precedes VSMC dysfunction (34). There was evidence of mild or early small fiber impairment in the forearm in both patient groups; both type 1 and type 2 diabetic subjects had impaired axon reflex (C-fiber)-mediated vasodilation. (Figure 178.6). In contrast, patients with primary (AL; immunoglobulin light-chain associated) amyloidosis had impaired forearm endothelial-dependent vasodilation with normal endothelium-independent and C-fiber–mediated vasodilation (35).

Clinicopathological Correlations

Skin biopsy has been used to examine the relationship been microvascular function and pathological changes in the skin. One possible mechanism for impairment of endothelial function in diabetic neuropathy is the reduction in the expression of the eNOS) activity. We have tested this hypothesis by evaluating the immunohistochemistry staining for eNOS of foot skin biopsies taken from diabetic neuropathic patients with or without peripheral vascular disease and healthy subjects. The results showed reduced staining for the diabetic patients when compared with the healthy subjects (39). Similar results were reported by other investigators using immunohistochemistry and Western blotting techniques (42).

Recently, data has emerged showing poly (ADP-ribose) polymerase (PARP) to be involved in endothelial dysfunction as well. PARP is a nuclear enzyme that responds to oxidative DNA damage by activating an inefficient cellular metabolic cycle, leading to cell necrosis. PARP activation was higher in patients with type 2 diabetes, patients with glucose intolerance, and individuals with a family history of type 2 diabetes groups than in the healthy controls (43). The activation of PARP was associated with changes in the vascular reactivity of the skin microcirculation in forearm biopsies taken from these subjects, supporting the hypothesis that PARP activation contributes to changes in microvascular reactivity. Further study is required to prove this association.

KEY POINTS

- The function of both the EC and VSMC can be reliably evaluated in both the microcirculation and macrocirculation. This has contributed to the understanding of the pathophysiology of atherosclerotic development in humans.
- Although the first described techniques were invasive, these assessments can be currently performed in the peripheral arteries and skin microcirculation using noninvasive methods.
- Noninvasive techniques are robust enough to be employed for the conduction of interventional studies.
- The assessment of endothelial function in the coronary arteries still requires invasive techniques.

Future Goals

- To standardize techniques so that collaboration between various centers and the conduction of multicenter studies will be facilitated
- To acquire more information regarding the correlation between changes in the micro- and macrocirculation and changes in the microcirculation of various organs

REFERENCES

1 Hokanson DE, Sumner DS, Strandness DE Jr. An electrically calibrated plethysmograph for direct measurement of limb blood flow. *IEEE Trans Biomed Eng*. 1975;22:25–29.

2 Lind L, Sarabi M, Millgard J. Methodological aspects of the evaluation of endothelium-dependent vasodilatation in the human forearm. *Clin Physiol*. 1998;18:81–87.

3 Cockcroft JR, Chowienczyk PJ, Benjamin N, Ritter JM. Preserved endothelium-dependent vasodilatation in patients with essential hypertension. *N Engl J Med*. 1994;330:1036–1040.

4 Johnstone MT, Creager SJ, Scales KM, et al. Impaired endothelium-dependent vasodilation in patients with insulin-dependent diabetes mellitus. *Circulation*. 1993;88:2510–2516.

5 Ting HH, Timimi FK, Boles KS, et al. Vitamin C improves endothelium-dependent vasodilation in patients with non-insulin-dependent diabetes mellitus. *J Clin Invest*. 1996;97:22–28.

6 Ganz V, Hlavova A, Fronek A, et al. Measurement of blood flow in the femoral artery in man at rest and during exercise by local thermodilution. *Circulation*. 1964;30:86–89.

7 Baron AD, Brechtel G, Wallace P, Edelman SV. Rates and tissue sites of noninsulin and insulin mediated glucose uptake in humans. *Am J Physiol*. 1988;255:E769–E774.

8 Steinberg HO, Brechtel G, Johnson A, et al. Insulin mediated skeletal muscle vasodilation is nitric oxide dependent: a novel action of insulin to increase nitric oxide release. *J Clin Invest*. 1994;94:1172–1179.

9 Celermajer DS, Sorensen KE, Gooch VM, et al. Non-invasive detection of endothelial dysfunction in children and adults at risk of atherosclerosis. *Lancet*. 1992;340:1111–1115.

10 Corretti MC, Anderson TJ, Benjamin EJ, et al. Guidelines for the ultrasound assessment of endothelial-dependent flow-mediated vasodilation of the brachial artery: a report of the International Brachial Artery Reactivity Task Force. *J Am Coll Cardiol*. 2002;39:257–265.

11 Caballero AE, Arora S, Saouaf R, et al. Micro- and macro-vascular reactivity is impaired in subjects at risk for Type 2 diabetes. *Diabetes*. 1999:48:1863–1867.

12 Saouaf R, Arora S, Smakowski P, et al. Reactive hyperemic response of the brachial artery to proximal and distal occlusion. *Acad Radiol*. 1998;5:556–560.

13 Agewall S, Doughty RN, Bagg W, et al. Comparison of ultrasound assessment of flow-mediated dilatation in the radial and brachial artery with upper and forearm cuff positions. *Clin Physiol*. 2001;21(1):9–14.

14 Gaenzer H, Neumayr G, Marschang P, et al. Flow-mediated vasodilation of the femoral and brachial artery induced by exercise in healthy nonsmoking and smoking men. *J Am Coll Cardiol*. 2001;38(5):1313–1319.

15 Ludmer PL, Selwyn AP, Shook TL, et al. Paradoxical vasoconstriction induced by acetylcholine in atherosclerotic coronary arteries. *N Engl J Med*. 1986;315(17):1046–1051.

16 Barbato E, Piscione F, Bartunek J, et al. Role of beta2 adrenergic receptors in human atherosclerotic coronary arteries. *Circulation*. 2005;111:288–294.

17 Anderson TJ, Uehata A, Gerhard MD, et all. Close relation of endothelial function in the human coronary and peripheral circulations. *J Am Coll Cardiol*. 1995;26(5):1235–1241.

18 Gokce N, Keaney JF Jr, Hunter LM, et al. Risk stratification for postoperative cardiovascular events via noninvasive assessment of endothelial function: a prospective study. *Circulation*. 2002;105(13):1567–1572.

19 Danias P, Saouaf R. Noninvasive methods to assess vascular function and pathophysiology. In: Johnstone MT, Veves A, eds. *Diabetes and Cardiovascular Disease*, 2nd ed. Totowa, NJ: Humana Press; 2005:431–450.

20 Salonen R, Haapanen A, Salonen JT. Measurement of intima-media thickness of common carotid arteries with high-resolution B-mode ultrasonography: inter- and intra-observer variability. *Ultrasound Med Biol.* 1991;17:225–230.

21 Heiss G, Sharrett AR, Barnes R, et al. Carotid atherosclerosis measured by B-mode ultrasound in populations: associations with cardiovascular risk factors in the ARIC study. *Am J Epidemiol.* 1991;134:250–256.

22 O'Leary DH, Polak JF, Kronmal RA, et al. Carotid-artery intima and media thickness as a risk factor for myocardial infarction and stroke in older adults. Cardiovascular Health Study Collaborative Research Group. *N Engl J Med.* 1999;340:14–22.

23 Cameron JD, Bulpitt CJ, Pinto ES, Rajkumar C. The aging of elastic and muscular arteries: a comparison of diabetic and non-diabetic subjects. *Diabetes Care.* 2003;26:2133–2138.

24 Guerin AP, Blacher J, Pannier B, et al. Impact of aortic stiffness attenuation on survival of patients in end-stage renal failure. *Circulation.* 2001;103:987–992.

25 Danias PG, Tritos NA, Stuber M, et al. Comparison of aortic elasticity determined by cardiovascular magnetic resonance imaging in obese versus lean adults. *Am J Cardiol.* 2003;91:195–199.

26 Tooke JE, Ostergren J, Fagrell B. Synchronous assessment of human skin microcirculation by laser Doppler flowmetry and dynamic capillaroscopy. *Int J Microcirc Clin Exp.* 1983;2:277–284.

27 Bergh off M, Kathpal M, Kilo S, et al. Vascular and neural mechanisms of ACh-mediated vasodilation in the forearm cutaneous microcirculation. *J Appl Physiol.* 2002;92(2):780–788.

28 Morris SJ, Shore AC. Skin blood flow responses to the iontophoresis of acetylcholine and sodium nitroprusside in man: possible mechanisms. *J Physiol (Lond).* 1996;496(Pt 2):531–542.

29 Dalle-Ave A, Kubli S, Golay S, et al. Acetylcholine-induced vasodilation and reactive hyperemia are not affected by acute cyclooxygenase inhibition in human skin. *Microcirculation.* 2004; 11(4):327–336.

30 Khan F, Davidson NC, Littleford RC, et al. Cutaneous vascular responses to acetylcholine are mediated by a prostanoid-dependent mechanism in man. *Vasc Med.* 1997;2(2):82–86.

31 Noon JP, Walker BR, Hand MF, Webb DJ. Studies with iontophoretic administration of drugs to human dermal vessels in vivo: cholinergic vasodilatation is mediated by dilator prostanoids rather than nitric oxide. *Br J Clin Pharmacol.* 1998; 45(6):545–550.

32 Kellogg DL Jr, Zhao JL, Coey U, Green JV. Acetylcholine-induced vasodilation is mediated by nitric oxide and prostaglandins in human skin. *J Appl Physiol.* 2005;98(2):629–632.

33 Holowatz LA, Thompson CS, Minson CT, Kenney WL. Mechanisms of acetylcholine-mediated vasodilatation in young and aged human skin. *J Physiol.* 2005;563(Pt 3):965–973.

34 Kilo S, Bergh off M, Hilz M, Freeman R. Neural and endothelial control of the microcirculation in diabetic peripheral neuropathy. *Neurology.* 2000;54(6):1246–1252.

35 Berghoff M, Kathpal M, Khan F, et al. Endothelial dysfunction precedes C-fiber abnormalities in primary (AL) amyloidosis. *Ann Neurol.* 2003;53(6):725–730.

36 Kellogg DL, Crandall CG, Liu Y, et al. Nitric oxide and cutaneous active vasodilation during heat stress in humans. *J Appl Physiol.* 1998;85(3):824–829.

37 Kellogg DL, Liu Y, Kosiba IF, O'Donnell D. Role of nitric oxide in the vascular effects of local warming of the skin in humans. *J Appl Physiol.* 1999;86(4):1185–1190.

38 Morris SJ, Shore AC, Tooke JE. Responses of the skin microcirculation to acetylcholine and sodium nitroprusside in patients with NIDDM. *Diabetologia.* 1995;38(11):1337–1344.

39 Veves A, Akbari CM, Primavera J, et al. Endothelial dysfunction and the expression of endothelial nitric oxide synthetase in diabetic neuropathy, vascular disease, and foot ulceration. *Diabetes.* 1998;47(3):457–463.

40 Arora S, Smakowski P, Frykberg RG, et al. Differences in foot and forearm skin microcirculation in diabetic patients with and without neuropathy. *Diabetes Care.* 1998;21(8):1339–1344.

41 Caballero AE, Arora S, Saouaf R, et al. Microvascular and macrovascular reactivity is reduced in subjects at risk for type 2 diabetes. *Diabetes.* 1999;48(9):1856–1862.

42 Jude EB, Boulton AJ, Ferguson MW, Appleton I. The role of nitric oxide synthase isoforms and arginase in the pathogenesis of diabetic foot ulcers: possible modulatory effects by transforming growth factor beta 1. *Diabetologia.* 1999;42(6):748–757.

43 Szabo C, Zanchi A, Komjati K, et al. Poly(ADP-Ribose) polymerase is activated in subjects at risk of developing Type 2 diabetes and is associated with impaired vascular reactivity. *Circulation.* 2002;106(21):2680–2686.

179

Statins

James K. Liao

Brigham and Women's Hospital, Harvard Medical School, Boston, Massachusetts

Risk factors for cardiovascular disease, such as cigarette smoking, hypertension, and elevated serum lipid levels impair endothelial function and lead to the development of atherosclerosis. Recent studies suggest that 3-hydroxy-3-methylglutaryl-CoA (HMG-CoA) reductase inhibitors or statins reduce cardiovascular events, in part, by improving endothelial function. Statins reduce plasma cholesterol levels, thereby decreasing the uptake of modified lipoproteins by endothelial cells (ECs). There is increasing evidence, however, that statins may also exert effects beyond cholesterol lowering. Many of these cholesterol-independent or "pleiotropic" vascular effects of statins appear to involve restoring or improving endothelial function by increasing the bioavailability of nitric oxide (NO), promoting re-endothelization, reducing oxidative stress, inhibiting inflammatory responses, and increasing fibrinolysis. Thus, the endothelium-dependent effects of statins may contribute to many of the beneficial effects of statin therapy in cardiovascular disease.

STATINS AND ENDOTHELIAL FUNCTION

Statins inhibit an early rate-limiting step in cholesterol biosynthesis (Figure 179.1). This leads to increased hepatic low-density lipoprotein (LDL) receptors and enhanced uptake of cholesterol by the liver. Indeed, therapeutic doses of statins potently reduce serum cholesterol levels in humans, and several large clinical trials have demonstrated that inhibition of HMG-CoA reductase by statins markedly decreases the incidence of cardiovascular events in hypercholesterolemic individuals (1,2). Because of the strong association between elevated serum cholesterol levels and coronary atherosclerotic disease, the reduction of serum cholesterol levels by statins has been proposed to be the predominant mechanism underlying the beneficial effects of these drugs. Indeed, acute plasma LDL aphaeresis improves endothelium-dependent vasodilatation (3), suggesting that statins could restore endothelial function, in part, by lowering serum cholesterol levels.

In some studies with statins, however, restoration of endothelial function occurs before significant reduction in serum cholesterol levels, suggesting that there are additional effects on endothelial function beyond those of cholesterol reduction. Some of the early benefits of statin therapy include improvement in vasomotor response to endothelium-dependent agonists (4), enhancement in coronary blood flow (5), and reduction of adhesion molecule expression in vivo (6) (Table 179-1). The mechanism is due, in part, to statin's ability to increase endothelial NO production by stimulating and upregulating endothelial NO synthase (eNOS) (7,8). Furthermore, statins have been shown to restore eNOS activity in the presence of hypoxia (9) and oxidized low-density lipoprotein (ox-LDL) (8), conditions which lead to endothelial dysfunction. Statins also increase the expression of tissue-type plasminogen activator (t-PA) (10) and inhibit the expression of endothelin (ET)-1, a potent vasoconstrictor and mitogen (11). Statins, therefore, exert many favorable effects on the endothelium and attenuate endothelial dysfunction in the presence of atherosclerotic risk factors.

MECHANISMS UNDERLYING THE PLEIOTROPIC EFFECTS OF STATINS

Changes in Endothelial Nitric Oxide Synthase and Nitric Oxide – a Common Mechanistic Thread

There is growing appreciation that statins exert many of their pleiotropic effects via eNOS/NO. Statins decrease the extent of cerebral and myocardial ischemia–reperfusion injury in rodents without changes in serum cholesterol levels (12,13). The vascular protective effects of statins are associated with increased blood flow, attenuated P-selectin expression, and decreased leukocyte adherence (14). Pretreatment with statins increases *eNOS* mRNA expression as well as eNOS activity, and vascular protection by statins is completely abolished in *eNOS*-knockout mice or after cotreatment with the NOS inhibitor, N(G)-nitro-L-arginine methyl ester L-NAME (12,13).

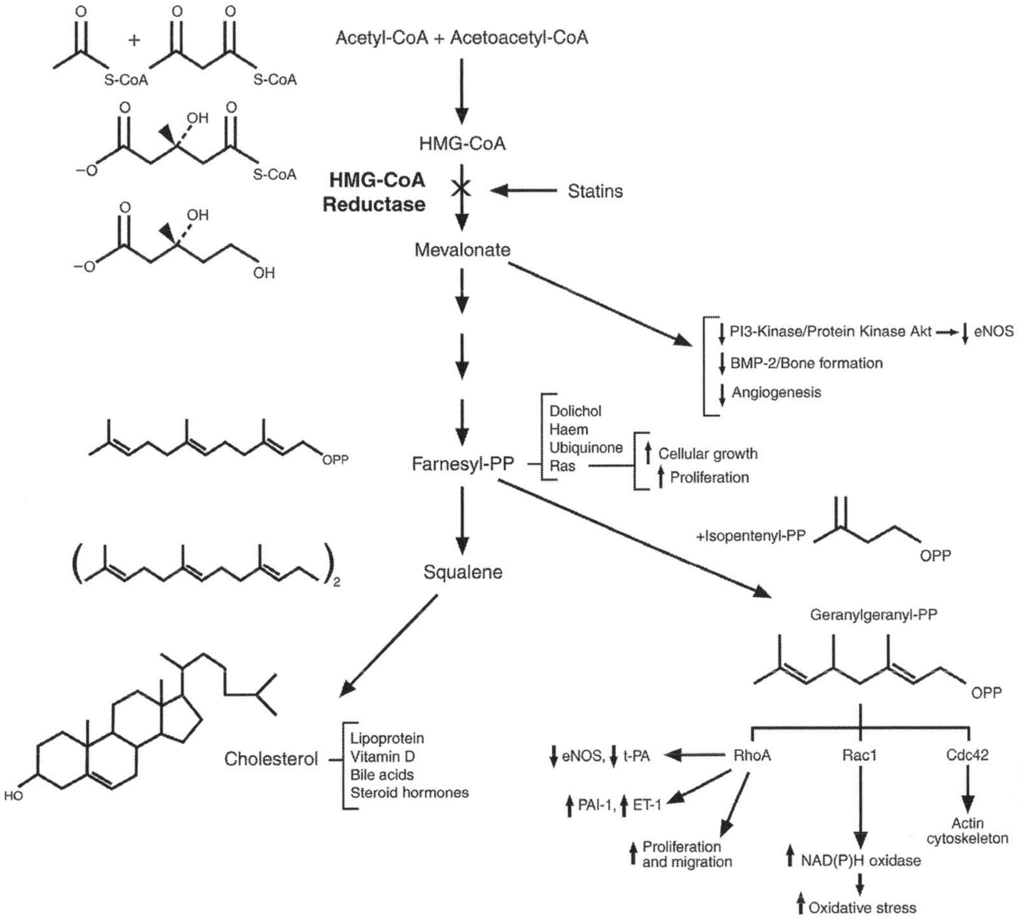

Figure 179.1. Inhibition of HMG-CoA reductase by statins decreases the synthesis of isoprenoids and cholesterol. The isoprenoid, geranylgeranyl (GG), is an important lipid attachment for Rho, which permits the subsequent membrane translocation and activation of Rho and Rho kinase. Inhibition of Rho/Rho-kinase activity by statins leads to an increase in eNOS expression and activity. Furthermore, statins activate eNOS by stimulating the PI3K/Akt pathway by a yet unidentified mechanism.

A clue to the mechanisms by which statins alter eNOS/NO activity (and exert other pleiotropic effects) is found in the biochemistry of the HMG-CoA reductase pathway. Inhibition of the HMG-CoA reductase pathway not only reduces cholesterol production, but also prevents the formation of various isoprenoid intermediates (see Figure 179.1). For example, farnesylpyrophosphate (FPP) and geranylgeranylpyrophosphate (GGPP) serve as important lipid attachments for the post-translational modification of a variety of proteins, including the subunit of heterotrimeric G-proteins and small guanosine triphosphate (GTP)-binding protein Ras, and Ras-like proteins, such as Rho, Rab, Rac, Ral, or Rap. Protein isoprenylation allows the covalent attachment, subcellular localization, and intracellular trafficking of several membrane-associated proteins. Although the effects of statins on Ras and Rho isoprenylation are reversed in the presence of FPP and GGPP, respectively, the effects of statins on eNOS expression are only reversed with GGPP and not by FPP or LDL cholesterol (15). Indeed, direct inhibition of GGPP or Rho leads to increases in eNOS expression (15,16). These findings are consistent with a non–cholesterol-lowering effect of statins and suggest that inhibition of Rho by statins mediates the increase in eNOS expression. Statins upregulate eNOS expression by prolonging *eNOS* mRNA half-life but not *eNOS* gene transcription (15). Because hypoxia, oxidized LDL, and cytokines such as tumor necrosis factor (TNF)-α decrease eNOS expression by reducing *eNOS* mRNA stability, the ability of statins to prolong eNOS half-life may make them effective agents in counteracting conditions that downregulate eNOS expression.

Furthermore, it has been recently reported by Kureishi and colleagues that statins can activate protein kinase Akt (7). The serine-threonine kinase Akt is an important regulator of various cellular processes including cell metabolism and apoptosis. Stimulation of receptor tyrosine kinases and G-protein–coupled receptors leads to the activation of phosphoinositide 3-kinase (PI3K), the products of which, namely 3′ phospholipids, promote the phosphorylation and activation of Akt. Inhibitors of PI3K, such as wortmannin, block the effects of statins on Akt activation (7). Akt has been shown to modulate

Table 179-1: Statin Mediators, Effects, and Benefits

Effect	*Mediator*	*Benefit*
Reduction in activity of NADPH oxidase	Rac1	Reduction of oxidative Stress
Decrease in synthesis of ET-1	Rho	Improvement of endothelial function
Decrease in the expression of AT_1-receptor	Rho	Improvement of endothelial function
Decrease in the expression of t-PA	Rho	Reduction in thrombosis
Increase in the expression of PAI-1	Rho	Reduction in thrombosis
Decrease in the expression of adhesion molecules	LFA-1, Rho	Reduction of inflammation
Increase in eNOS activity	Rho, Akt	Improvement of endothelial function
Increase in number and differentiation of circulating ECs	Akt	Increase in neovascularization and re-endothelization
Inhibition of apoptosis	Akt	Increase in cell survival

several targets, such as caspase-9 and eNOS, by phosphorylation. Consequently, activation of Akt by statins inhibits apoptosis and increases NO production in cultured ECs. Therefore, in addition to stabilizing *eNOS* mRNA by inhibition of Rho, there is increasing evidence that activation of the PI3K/Akt pathway also may contribute to the endothelium-dependent effects of statins, although the precise mechanisms by which PI3K is activated by statins are not yet identified.

Vasomotor Tone

In addition to their effects on eNOS/NO statins may have other effects on the balance of vasomotor mediators. Several vasoconstrictors, including ET-1 or angiotensin (Ang)-II, counteract the vasodilating effect of NO and thus contribute to endothelial dysfunction and the development of atherosclerosis. Circulating concentrations and tissue immunoreactivity of ET-1 are increased in patients with severe atherosclerosis (17–19). ET-1 has also been shown to function as a mitogenic agent. Exposure to ox-LDL leads to an increased production and release of ET-1 (20), which promotes neointimal proliferation of atherosclerotic lesions (20). Statins have been shown to inhibit *pre–proET-1* mRNA expression in a concentration-dependent manner and to reduce immunoreactive ET-1 in bovine ECs, a phenomenon which has been suggested to be mediated by Rho proteins (11,21). Furthermore, statins modulate the renin-angiotensin-system by downregulating the expression of angiotensin receptor subtype 1 (AT_1) in a Rho A–dependent manner and attenuating the biological function of Ang-II in vascular smooth muscle cells (VSMCs) (22).

Redox State

Another potential mechanism by which statins may improve endothelial function is through their antioxidant effects. For example, statins attenuate Ang-II–induced free radical production in cultured VSMCs by inhibiting Rac1-mediated nicotinamide adenine dinucleotide phosphate (NADPH) oxidase activity and downregulating angiotensin AT_1-receptor expression (23). More recently, Wassmann and colleagues reported that atorvastatin reduced VSMC mRNA expression of essential NADPH oxidase subunits $p22^{phox}$ and *NOX1* by a mechanism that might involve the translocation of Rac1 from the cytosol to the cell membrane (23,24). Because NO is scavenged by reactive oxygen species (ROS), these findings indicate that the antioxidant properties of statins also may contribute to their ability to improve endothelial function. Furthermore, withdrawal of statin treatment in mice has been shown to impair endothelium-dependent relaxation by increasing vascular superoxide anion generation via a pathway involving the Rac-dependent activation of the $gp91^{phox}$-containing vascular NADPH oxidase (25). ROS directly affects the endothelial function, and the endothelium itself has also been shown to generate ROS (26).

Inflammation

Atherosclerosis is a complex inflammatory process that is characterized by the presence of monocytes or macrophages and T lymphocytes in the atheroma. Inflammatory cytokines secreted by these macrophages and T lymphocytes can modify endothelial function, VSMC proliferation, collagen degradation, and thrombosis. An early step in atherogenesis involves monocyte adhesion to the endothelium and penetration into the subendothelial space. Statins have been shown to reduce the number of inflammatory cells in atherosclerotic plaques and therefore possess anti-inflammatory properties. The mechanisms have yet to be fully elucidated, but may involve inhibition of adhesion molecules such as intercellular adhesion molecule (ICAM)-1, interleukin (IL)-6, and IL-8, which are involved in the recruitment of inflammatory cells. In addition, a recent study has shown that statins can suppress the inflammatory response independent of HMG-CoA reductase inhibition by binding directly to a novel regulatory site of the β_2 integrin, leukocyte function antigen (LFA)-1 (27). This regulatory site serves as a major counter-receptor for ICAM-1 on leukocytes. The non-NO mechanism of the anti-inflammatory properties of statins was further elucidated by Yoshida and colleagues, who recently demonstrated that cerivastatin reduced monocyte adhesion to vascular endothelium by decreasing expression of integrins such as vascular endothelial adhesion molecule (VCAM)-1 and actin polymerization through the inactivation of RhoA (28).

A clinical marker of inflammation is high-sensitivity C-reactive protein (hs-CRP). hs-CRP is an acute phase reactant that is produced by the liver in response to proinflammatory cytokines such as IL-6, and reflects low-grade systemic inflammation. Elevated levels of hs-CRP have been shown to be predictive of increased risk for coronary artery disease (CAD) (29). CRP could contribute to the development of atherosclerosis by binding to modified LDL within atherosclerotic plaques (30) and inhibiting eNOS (31,32). However, further studies are needed to fully elucidate the role CRP plays in atherosclerosis.

Statin therapy lowers hs-CRP levels in hypercholesterolemic patients. In the Cholesterol and Recurrent Events (CARE) trial, statins significantly decreased plasma hs-CRP levels over a 5-year period in patients who did not experience recurrent coronary events (33). Similarly, an analysis of baseline and 1-year follow-up from the Air Force/Texas Coronary Atherosclerosis Prevention Study (AFCAPS/TexCAPS) demonstrated that hs-CRP levels were reduced in statin-treated patients who were free of acute major coronary events (34). These studies, therefore, indicate that statins are effective in decreasing systemic and vascular inflammation. However, any potential clinical benefits conferred by lowering hs-CRP are difficult to separate from those of the lipid-lowering effects of statins without performing further clinical studies.

Fibrinolysis

Plasminogen activator inhibitor type (PAI)-1 is the major endogenous inhibitor of t-PA, and it also plays a pivotal role in the regulation of fibrinolysis. High PAI-1 plasma levels and decreased levels of t-PA activity have been shown to be associated with coronary heart disease (35). *PAI-1* mRNA has also been found in human atherosclerotic lesions, underscoring its potential role in the development of these disorders. Increasing evidence from in vitro studies suggest that statins positively affect the fibrinolytic system of cultured ECs. In these studies, a decrease in PAI-1 and an increase in t-PA was observed after treatment with statins in ECs (10,36). Statins may therefore interfere with the progression of the atherosclerotic plaque as well as with thrombotic events in hyperlipidemic patients, independently of their ability to reduce plasma cholesterol, but further studies are needed to delineate its physiological significance.

STATINS AND RE-ENDOTHELIZATION

Stimulation of re-endothelization or neovascularization is a therapeutic aim to reduce ischemia-induced tissue injury. Postnatal neovascularization was mainly attributed to angiogenesis, that is, the proliferation, migration, and remodeling of preexisting ECs. However, some studies have recently demonstrated that bone marrow–derived endothelial precursor cells (EPCs) also are involved in this process. EPCs can be grown out of isolated $CD133^+$ or $CD34^+$ cells. Transplantation of these cells leads to postnatal neovascularization in the ischemic hindlimb, augments ischemia-induced neovascularization in vivo (37), and even improves postischemic cardiac function (38).

Recent studies revealed that statins also promote vasculogenesis. Llevadot and colleagues demonstrated in vitro that simvastatin evokes proliferation, migration, and cell survival of EPCs (39). The signal pathway for this effect includes activation of protein kinase Akt, which was confirmed by functional blocking with dominant negative Akt overexpression. Dimmeler and colleagues showed in vitro and in vivo that statins not only increase the number of circulating ECs, but also induce their differentiation (40). This might be of clinical relevance, because it has been recently reported by Walter and coworkers that induction of these cells with statins is associated with an accelerated re-endothelization after carotid balloon injury (41).

In contrast, some studies report an antiangiogenic effect of statins, which might be mediated by RhoA (42). These conflicting effects of statins may be related to the dose used. Low doses of a statin may activate endothelial Ras and promote Akt and eNOS phosphorylation and lead to an angiogenic effect, whereas higher statin doses are antiangiogenic although they promote an increase in eNOS protein expression (43). This hypothesis remains controversial because high doses of statins also have been shown to be angiogenic (44), and further studies are necessary to clarify this topic.

KEY POINTS

- Although the reduction of plasma cholesterol levels by statins improves endothelial function and limits atherosclerosis, statins may exert cholesterol-independent effects on the endothelium
- The endothelial-protective effects of statins could lead to further improvements in vascular function.
- Most of noncholesterol effects of statins are mediated by inhibition of isoprenoid synthesis and increases in NO release or bioavailability.

Future Goals

- To determine which of the statin effects are more predominant, in terms of clinical outcome, in patients with low or average cholesterol levels, although both direct and indirect endothelium-dependent effects of statins play important roles in vascular protection

ACKNOWLEDGMENTS

The work described in this article was supported in part by the National Institutes of Health (HL-52233 and NS-10828) and the American Heart Association Bugher Foundation Award.

REFERENCES

1 Randomised trial of cholesterol lowering in 4444 patients with coronary heart disease: the Scandinavian Simvastatin Survival Study (4S). *Lancet*. 1994;344(8934):1383–1389.

2 The Long-Term Intervention with Pravastatin in Ischaemic Disease (LIPID) Study Group. Prevention of cardiovascular events and death with pravastatin in patients with coronary heart disease and a broad range of initial cholesterol levels. *N Engl J Med.* 1998;339(19):1349–1357.
3 Tamai O, Matsuoka H, Itabe H, et al. Single LDL apheresis improves endothelium-dependent vasodilatation in hypercholesterolemic humans. *Circulation.* 1997;95(1):76–82.
4 Dupuis J. Mechanisms of acute coronary syndromes and the potential role of statins. *Atheroscler Suppl.* 2001;2(1):9–14.
5 Baller D, Notohamiprodjo G, Gleichmann U, et al. Improvement in coronary flow reserve determined by positron emission tomography after 6 months of cholesterol-lowering therapy in patients with early stages of coronary atherosclerosis. *Circulation.* 1999;99(22):2871–2875.
6 Seljeflot I, Tonstad S, Hjermann I, Arnesen H. Reduced expression of endothelial cell markers after 1 year treatment with simvastatin and atorvastatin in patients with coronary heart disease. *Atherosclerosis.* 2002;162(1):179–185.
7 Kureishi Y, Luo Z, Shiojima I, et al. The HMG-CoA reductase inhibitor simvastatin activates the protein kinase Akt and promotes angiogenesis in normocholesterolemic animals. *Nat Med.* 2000;6(9):1004–1010.
8 Laufs U, La Fata V, Plutzky J, Liao JK. Upregulation of endothelial nitric oxide synthase by HMG CoA reductase inhibitors. *Circulation.* 1998;97(12):1129–1135.
9 Laufs U, Fata VL, Liao JK. Inhibition of 3-hydroxy-3-methylglutaryl (HMG)-CoA reductase blocks hypoxia-mediated downregulation of endothelial nitric oxide synthase. *J Biol Chem.* 1997;272(50):31725–31729.
10 Essig M, Nguyen G, Prie D, et al. 3-Hydroxy-3-methylglutaryl coenzyme A reductase inhibitors increase fibrinolytic activity in rat aortic endothelial cells. Role of geranylgeranylation and Rho proteins. *Circ Res.* 1998;83(7):683–690.
11 Hernandez-Perera O, Perez-Sala D, Navarro-Antolin J, et al. Effects of the 3-hydroxy-3-methylglutaryl-CoA reductase inhibitors, atorvastatin and simvastatin, on the expression of endothelin-1 and endothelial nitric oxide synthase in vascular endothelial cells. *J Clin Invest.* 1998;101(12):2711–2719.
12 Endres M, Laufs U, Huang Z, et al. Stroke protection by 3-hydroxy-3-methylglutaryl (HMG)-CoA reductase inhibitors mediated by endothelial nitric oxide synthase. *Proc Natl Acad Sci USA.* 1998;95(15):8880–8885.
13 Lefer AM, Campbell B, Shin YK, et al. Simvastatin preserves the ischemic-reperfused myocardium in normocholesterolemic rat hearts. *Circulation.* 1999;100(2):178–184.
14 Scalia R, Gooszen ME, Jones SP, et al. Simvastatin exerts both anti-inflammatory and cardioprotective effects in apolipoprotein e-deficient mice. *Circulation.* 2001;103(21):2598–2603.
15 Laufs U, Liao JK. Post-transcriptional regulation of endothelial nitric oxide synthase mRNA stability by Rho GTPase. *J Biol Chem.* 1998;273(37):24266–24271.
16 Laufs U, Endres M, Stagliano N, et al. Neuroprotection mediated by changes in the endothelial actin cytoskeleton. *J Clin Invest.* 2000;106(1):15–24.
17 Lerman A, Edwards BS, Hallett JW, et al. Circulating and tissue endothelin immunoreactivity in advanced atherosclerosis. *N Engl J Med.* 1991;325(14):997–1001.
18 Lerman A, Holmes DR Jr, Bell MR, et al. Endothelin in coronary endothelial dysfunction and early atherosclerosis in humans. *Circulation.* 1995;92(9):2426–2431.
19 Zeiher AM, Goebel H, Schachinger V, Ihling C. Tissue endothelin-1 immunoreactivity in the active coronary atherosclerotic plaque. A clue to the mechanism of increased vasoreactivity of the culprit lesion in unstable angina. *Circulation.* 1995;91(4):941–947.
20 Martin-Nizard F, Houssaini HS, Lestavel-Delattre S, et al. Modified low density lipoproteins activate human macrophages to secrete immunoreactive endothelin. *FEBS Lett.* 1991;293(1–2):127–130.
21 Hernandez-Perera O, Perez-Sala D, Soria E, Lamas S. Involvement of rho GTPases in the transcriptional inhibition of preproendothelin-1 gene expression by simvastatin in vascular endothelial cells. *Circ Res.* 2000;87(7):616–622.
22 Ichiki T, Takeda K, Tokunou T, et al. Downregulation of angiotensin II type 1 receptor by hydrophobic 3- hydroxy-3–methylglutaryl coenzyme A reductase inhibitors in vascular smooth muscle cells. *Arterioscler Thromb Vasc Biol.* 2001;21(12):1896–1901.
23 Wassmann S, Laufs U, Baumer AT, et al. Inhibition of geranylgeranylation reduces angiotensin II-mediated free radical production in vascular smooth muscle cells: involvement of angiotensin AT1 receptor expression and Rac1 GTPase. *Mol Pharmacol.* 2001;59(3):646–654.
24 Wassmann S, Laufs U, Muller K, et al. Cellular antioxidant effects of atorvastatin in vitro and in vivo. *Arterioscler Thromb Vasc Biol.* 2002;22(2):300–305.
25 Vecchione C, Brandes RP. Withdrawal of 3-hydroxy-3-methylglutaryl coenzyme A reductase inhibitors elicits oxidative stress and induces endothelial dysfunction in mice. *Circ Res.* 2002;91(2):173–179.
26 Matsubara T, Ziff M. Increased superoxide anion release from human endothelial cells in response to cytokines. *J Immunol.* 1986;137(10):3295–3298.
27 Weitz-Schmidt G, Welzenbach K, Brinkmann V, et al. Statins selectively inhibit leukocyte function antigen-1 by binding to a novel regulatory integrin site. *Nat Med.* 2001;7(6):687–692.
28 Yoshida M, Sawada T, Ishii H, et al. HMG-CoA reductase inhibitor modulates monocyte-endothelial cell interaction under physiological flow conditions in vitro: involvement of Rho GTPase-dependent mechanism. *Arterioscler Thromb Vasc Biol.* 2001;21(7):1165–1171.
29 Ridker PM, Cushman M, Stampfer MJ, et al. Inflammation, aspirin, and the risk of cardiovascular disease in apparently healthy men. *N Engl J Med.* 1997;336(14):973–979.
30 Bhakdi S, Torzewski M, Klouche M, Hemmes M. Complement and atherogenesis: binding of CRP to degraded, nonoxidized LDL enhances complement activation. *Arterioscler Thromb Vasc Biol.* 1999;19(10):2348–2354.
31 Venugopal SK, Devaraj S, Yuhanna I, et al. Demonstration that C-reactive protein decreases eNOS expression and bioactivity in human aortic endothelial cells. *Circulation.* 2002;106(12):1439–1441.
32 Verma S, Wang CH, Li SH, et al. A self-fulfilling prophecy: C-reactive protein attenuates nitric oxide production and inhibits angiogenesis. *Circulation.* 2002;106(8):913–919.
33 Ridker PM, Rifai N, Pfeffer MA, et al. Long-term effects of pravastatin on plasma concentration of C-reactive protein. The Cholesterol and Recurrent Events (CARE) Investigators. *Circulation.* 1999;100(3):230–235.
34 Ridker PM, Rifai N, Clearfield M, et al. Measurement of C-reactive protein for the targeting of statin therapy in the

primary prevention of acute coronary events. *N Engl J Med.* 2001; 344(26):1959–1965.

35 Aznar J, Estelles A. Role of plasminogen activator inhibitor type 1 in the pathogenesis of coronary artery diseases. *Haemostasis.* 1994;24(4):243–251.

36 Bourcier T, Libby P. HMG CoA reductase inhibitors reduce plasminogen activator inhibitor-1 expression by human vascular smooth muscle and endothelial cells. *Arterioscler Thromb Vasc Biol.* 2000;20(2):556–562.

37 Murohara T, Ikeda H, Duan J, et al. Transplanted cord blood-derived endothelial precursor cells augment postnatal neovascularization. *J Clin Invest.* 2000;105(11):1527–1536.

38 Kawamoto A, Gwon HC, Iwaguro H, et al. Therapeutic potential of ex vivo expanded endothelial progenitor cells for myocardial ischemia. *Circulation.* 2001;103(5):634–637.

39 Llevadot J, Murasawa S, Kureishi Y, et al. HMG-CoA reductase inhibitor mobilizes bone marrow – derived endothelial progenitor cells. *J Clin Invest.* 2001;108(3):399–405.

40 Dimmeler S, Aicher A, Vasa M, et al. HMG-CoA reductase inhibitors (statins) increase endothelial progenitor cells via the PI 3-kinase/Akt pathway. *J Clin Invest.* 2001;108(3):391–397.

41 Walter DH, Rittig K, Bahlmann FH, et al. Statin therapy accelerates reendothelialization: a novel effect involving mobilization and incorporation of bone marrow-derived endothelial progenitor cells. *Circulation.* 2002;105(25):3017–3024.

42 Park HJ, Kong D, Iruela-Arispe L, et al. 3-hydroxy-3-methylglutaryl coenzyme A reductase inhibitors interfere with angiogenesis by inhibiting the geranylgeranylation of RhoA. *Circ Res.* 2002;91(2):143–150.

43 Urbich C, Dernbach E, Zeiher AM, Dimmeler S. Double-edged role of statins in angiogenesis signaling. *Circ Res.* 2002;90(6):737–744.

44 Sata M, Nishimatsu H, Suzuki E, et al. Endothelial nitric oxide synthase is essential for the HMG-CoA reductase inhibitor cerivastatin to promote collateral growth in response to ischemia. *FASEB J.* 2001;15(13):2530–2532.

180

Steroid Hormones

James K. Liao

Brigham and Women's Hospital, Harvard Medical School, Boston, Massachusetts

Many cellular responses to steroid hormones involve the transcriptional modulation of target genes by the prototypical nuclear receptor, the steroid hormone receptor. In the classic model of steroid hormone action, steroid hormones such as estrogen and glucocorticoids function essentially as ligand-dependent transcription factors by either activating or repressing gene expression through direct interactions with DNA or other transcription factors. However, recent evidence suggests an important role for the nontranscriptional effects of steroid hormones in the vascular system, particularly in endothelial cells (ECs), where they mediate the rapid activation of endothelial nitric oxide synthase (eNOS).

For example, the activated estrogen receptor (ER) mediates signaling cascades that culminate in direct protective effects such as vasodilation, inhibition of response to vessel injury, limiting myocardial injury after infarction, and attenuating cardiac hypertrophy. These effects of ER are mediated by rapid signaling pathways at the membrane and in the cytoplasm via various second messengers including protein kinases. Similarly, the nontranscriptional actions of the glucocorticoid receptor (GR) involve the rapid activation of protein kinases, such as phosphoinositide 3-kinase (PI3K) and Akt, leading to the activation of eNOS. These rapid, nongenomic pathways of estrogen and glucocorticoids may provide the pharmacological basis for future therapeutic approaches to cardiovascular diseases (Table 180-1).

SIGNALING PATHWAYS MEDIATED BY STEROID HORMONES

The study of steroid hormone action has provided many important insights into the regulation of cellular functions by nuclear receptors and, at the same time, has revealed surprising levels of biological complexity. Upon binding to steroid hormones, steroid hormone receptors such as ER and GR homodimerize (3,33), recruit coactivators and corepressors (34–36), and act as ligand-dependent transcription factors, which regulate gene expression through interaction with their corresponding DNA enhancer sequences called *steroid response elements* (SREs) (37). In the case of ER and GR, these SREs are ERE and GRE, respectively (2,38). However, recent studies suggest that not all the actions of estrogen and glucocorticoids are mediated via this classical genomic pathway. For example, in the heart and vasculature, some of the rapid, nonnuclear effects of ER and GR include vasodilation (6,11,12), inhibition of response to vessel injury (6,11), and reduction in myocardial and cerebral injury after infarction (11,12). In contrast to the hours required for transcription-dependent effects to become evident, nonnuclear steroid actions (alternatively referred to as *nontranscriptional* or *nongenomic*) in multiple cell types peaks minutes after stimulation. Other characteristics include immunity to inhibitors of RNA synthesis and recruitment of membrane or cytosol-localized signaling components. These include second messengers (e.g., calcium and nitric oxide [NO]), receptor tyrosine kinases (e.g., epidermal growth factor receptor [EGFR] and insulin-like growth factor-1 receptor [IGFR]), G-protein–coupled receptors (GPCRs), and protein kinases (e.g., PI3K, serine-threonine kinase [Akt], mitogen-activated protein kinase [MAPK] family members, nonreceptor tyrosine kinase [Src], and protein kinases A and C).

ESTROGEN SIGNALING IN ENDOTHELIAL CELLS

Nongenomic Pathways

The PI3K/Akt signaling cascade is one downstream target of nonnuclear estrogenic signaling (6,7,39). In the vasculature, short-term exposure to estrogen leads to vasodilation via NO-dependent pathways (40). The secretion of NO by healthy vessels relaxes vascular smooth muscle cells (VSMCs) and inhibits platelet activation in a cyclic guanosine 3′,5′-monophosphate (cGMP)-dependent mechanism. In cultured ECs, estrogen enhances NO release within minutes without altering the expression of eNOS (6–8,41). ER activates PI3K activity by association with the p85α regulatory subunit of PI3K in a ligand-dependent manner (6) (Figure 180.1). Furthermore, the chaperone heat shock protein 90 (HSP90) interacts with both eNOS and Akt, and modulates

Table 180-1: Steroid Hormone Transcription Factors

	ER	*GR*	*References*
Transcription			
Transcriptional activation	√ (via ERE)	√ (via GRE)	1–3
Transrepression		√	4, 5
Nongenomic pathways			
NO	√	√	6–12
Fibroblast growth factor receptor (FGFR)	√		13
Insulin-like growth factor receptor (IGFR)	√		14–16
GPCR	√		9,15,17
PI3K	√	√	6,11
MAPK	√	√	8,18–24
p38	√		25
cPLA2 inhibition		√	26
JNK-AP1 inhibition		√	5
Phenotypes			
Myocardial and vascular protection during I/R	√	√	11,27
Antiapoptotic	√		28–31
EC migration	√		28
Proangiogenic	√		28,29
Membrane integrity	√		28,32
Reduced post-MI and stroke vascular inflammation and ischemia	√	√	6,11,12

eNOS activity by acting as a scaffold to regulate Akt-dependent phosphorylation of eNOS (42).

MAPKs are a common target of non-nuclear estrogenic signaling. Induction of eNOS and inducible NOS (iNOS) in cardiac myocytes is blocked by the MAPK inhibitor PD98059 (43), which may have clinical relevance because NO inhibits caspase activation and prevents the development of congestive heart failure (44). Estrogen also activates extracellular-signal-regulated kinase (ERK1/2) in ECs, cardiomyocytes, colon cancer, breast cancer, and bone; and inhibits ERK1/2 in VSMCs and lung myofibroblasts (20,21,45,46). In the heart, ER also selectively activates the p38 isoform of MAPK to modulate the development of pressure-overload hypertrophy, which is consistent with recruitment of p38 MAPK in other models of cardiac hypertrophy (20). In cultured ECs, estrogen preserves the actin cytoskeleton during ischemia, prevents cell death, and enhances angiogenesis after injury by rapidly and selectively activating antiapoptotic p38β MAPK and inhibiting proapoptotic p38α MAPK, leading to upregulation of MAPK-activated protein kinase (MAPKAPK)-2 and phosphorylation of HSP27 (28). Downstream effects include preservation of stress fiber formation and membrane integrity, prevention of hypoxia-induced apoptosis, and induction of both EC migration and the formation of primitive capillary tubes (9,21,22,47).

Membrane Estrogen Receptors

Membrane binding sites for estrogen have been implicated in downstream signaling. Additional indirect evidence for a membrane ER comes from immunohistochemistry and studies with membrane-impermeable ligands or overexpressed nuclear receptor. The trafficking of classical ER to different cellular compartments may be regulated by the nature of stimulation. For example, in VSMCs transfected with ER, MAPK activation mediates nuclear translocation of ERα from the membrane fraction by both estrogen-dependent and -independent mechanisms (48). Indeed, recent studies suggest that ER and an alternatively-spliced form of ER, ER46, are targeted to the membrane by palmitoylation (49,50). ER appears to associate with membrane caveolae through palmitoylation of Cys^{447} (50). For example, in fractionated EC plasma membranes, ER has been localized to caveolae, and estrogen stimulates eNOS in isolated caveolae in an ER- and calcium-dependent manner (9,51,52). Within the caveolae, evidence suggests that HSP90, eNOS and Cav-1 exist in a heterotrimeric complex in ECs that modulates eNOS activity depending on intracellular calcium levels (53,54).

Non-nuclear ER signaling also involves membrane heterotrimeric G proteins. In Chinese hamster ovary cells transfected with *ERα* cDNA, ERα in the membrane fractions activated Gα_q and Gα_s and rapidly stimulated inositol phosphate production and adenylyl cyclase activity, respectively (55). The G-protein activation also has been shown in ECs, where estrogen activation of eNOS can be inhibited with ICI 182,780, RGS-4 (a regulator of G-protein signaling specific for Gα_i and Gα_q) and pertussis toxin (specific for Gα_i) (17).

SELECTIVE ESTROGEN RECEPTOR MODULATORS

An increasingly detailed understanding of the ER signaling network and its pleiotropic cellular effects has made the receptor an attractive pharmacological target. Selective ER modulators (SERMs) are ER ligands, which can have varying degrees of agonist or antagonist activities depending on the cell, promoter, and coregulator context.

Tamoxifen, the prototypical SERM, has indirect cardiovascular protective effects via reduction of serum total cholesterol and low-density lipoprotein (LDL) levels (56). Unfortunately, its strong agonist activity in the endometrium leads to endometrial hyperplasia and low-grade cancers. Raloxifene, a nonsteroidal compound, is similar to tamoxifen but it is less agonistic in the endometrium (57). Although administered

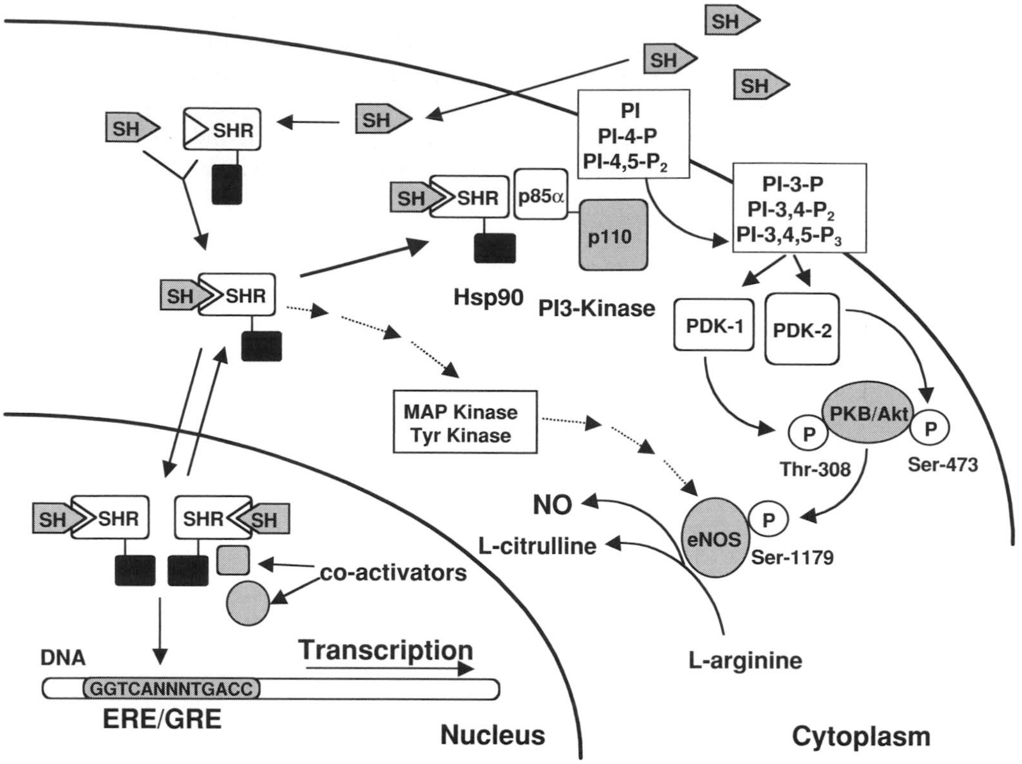

Figure 180.1. Nuclear and non-nuclear actions of the estrogen and glucocorticoid receptor. Association of steroid hormone receptor (SHR) with the regulatory p85 subunit of PI3K stimulates formation of 3'-phosphorylated phosphatidylinositols (e.g., phosphatidylinositol 3,4,5-trisphosphate, PIP_3) from phosphatidylinositol phosphate precursors (e.g., phosphatidylinositol 4,5-trisphosphate, PIP_2). This leads to the subsequent recruitment and activation of protein kinase Akt, which enhances NO release through phosphorylation of eNOS. In the nucleus, SHR binds either directly to estrogen or glucocorticoid response elements (ERE or GRE) or modulates the function of other transcription factors.

primarily for bone preservation, raloxifene also reduces serum triglycerides and serum fibrinogen levels. Like estrogen, raloxifene and its analog LY117018 (58) stimulate eNOS activity in ECs via PI3K- and ERK-dependent pathways (59), suggesting a possible mechanism for coronary artery relaxation (60). Raloxifene also improves endothelium-dependent vasorelaxation in hypertensive rats by enhancing the expression and activity of nitric oxide synthase (NOS) (61).

EM-800, a nonsteroidal compound, has higher affinity for ER than any other SERM (62). In addition to demonstrating potent antitumor activity in the uterus and breast, EM-800 may also prevent bone loss and lower serum cholesterol and triglyceride levels (63). Furthermore, in vitro studies in ECs suggest that EM-800, like estrogen, enhances NO release by sequential activation of MAPKs and PI3K/Akt, implicating a direct vascular protective effect (64).

GLUCOCORTICOIDS AND VASCULAR DISEASE

Corticosteroids are essential for normal development and the stress responses through the regulation of intermediary metabolism and the immune system (65,66). Produced by the adrenal glands under the regulation of the hypothalamus-pituitary gland axis, corticosteroids were among the first hormones to be identified and were later found to exert their effects through a specific DNA-binding protein, the GR (33). Molecular analysis of the anti-inflammatory actions of GR, however, later revealed a second important mechanism of GR function in the nucleus. Activated GR could negatively regulate expression of inflammatory genes through direct protein–protein interaction with proinflammatory transcription factors without DNA binding (termed *transrepression*) (67,68). Recent findings suggest that there is yet another level of GR action with particular importance for the vascular system. Similar to ER, the rapid, nontranscriptional effect of GR was found to mediate tissue protection in myocardial infarction (MI) and stroke through activation of eNOS, which was mediated by the PI3K/Akt pathway (11,12).

Corticosteroids have been used in the treatment of cardiovascular disease, such as acute MI and stroke, with conflicting results (69). Several features make steroids attractive candidates for the treatment of acute ischemia. The degree of ischemic tissue damage is related to the extent of the inflammatory response, which is coordinated through the interaction of leukocytes with activated vascular endothelium.

Corticosteroids inhibit EC activation and leukocyte–endothelial interaction, thereby exerting prominent, and very rapid, anti-inflammatory effects on the vasculature via transrepression mechanisms and eNOS activation (70). Steroid treatment improves the survival of patients with acute MI and protects the myocardium from experimental ischemic injury (71–73). In ischemic stroke, steroids can improve clinical outcome in a subset of patients with severe stroke, and administration of high doses of corticosteroids substantially reduces tissue damage in experimental focal cerebral ischemia (74–77). In addition, high-dose steroid treatment decreases vascular inflammation and ischemic tissue damage after myocardial infarction and stroke through direct vascular effects involving the nontranscriptional activation of eNOS (11,12).

The prolonged use of glucocorticoids, however, is often limited by their adverse effects, attributed to a delayed genomic response initiated by GR. This is exemplified by the clinical syndrome of hyperglycemia, dyslipidemia, hypertension, osteoporosis, and impaired wound healing in patients with chronic corticosteroid excess (i.e., Cushing syndrome) (78). In contrast to short-term high-dose steroid treatment with increases eNOS activity, the downregulation of eNOS expression by chronic steroid treatment is thought to contribute to steroid-induced hypertension (79). Indeed, continued low-dose steroid administration increases ischemic injury after global ischemia of the brain (80). In MI, impaired wound healing and cardiac remodeling can lead to cardiac rupture within 2 weeks of MI (81,82), and this has lead to discontinuation of steroid treatment for acute ischemia. Therefore, defining the exact molecular mechanisms involved in the beneficial and detrimental effects of GR could have important therapeutic implications.

Rapid Actions of GR

Similar to estrogen, corticosteroids appear to exert very rapid "nontranscriptional" effects (83). However, it should be noted that this concept of nongenomic effects of steroid hormones does not exclude the possibility of indirect downstream transcriptional modulation secondary to the initiated signaling pathways. Corticosteroids alter amphibian behavior and increase inositol trisphosphate in VSMCs within minutes of administration (84,85). In mice, the antianaphylactic effect of high-dose corticosteroids occurs acutely and is unaltered by the transcriptional inhibitor actinomycin D (86). Furthermore, in patients with active rheumatoid arthritis, steroids rapidly inhibit leukocyte recruitment into inflamed joints (87). In MI and ischemic stroke, high-dose steroids cause a transient decrease in blood pressure and systemic vascular resistance accompanied by an increase in coronary and cerebral blood flow within minutes of administration (12,72,88).

There is growing evidence for other rapid nontranscriptional actions of GR, especially with regard to inflammation (89,90). The synthetic corticosteroid dexamethasone (Dex) rapidly inhibits cytosolic phospholipase A2 ($cPLA_2$) activity and release of arachidonic acid in immune cells, which are important mediators of inflammation. Inhibition was reversed by the GR antagonist RU486 (mifepristone) and by pharmacological inhibition of Src, suggesting that rapid inhibition of $cPLA_2$ by GR is mediated by Src (24). Whether this pathway is important in ECs remains to be determined. Interestingly, a nongenomic effect of GR on rapid kinase signaling is also involved in the inhibition of AP-1 by steroids through transrepression mechanisms. The transcriptional activity of the AP-1 subunit, c-Jun, is enhanced through phosphorylation on Ser63/73 by members of the c-Jun-N-terminal kinase (JNK) subfamily in response to inflammatory cytokines (91). Indeed, activated GR prevented phosphorylation of c-Jun by blocking the JNK signaling cascade (92). Furthermore, acute Dex treatment interfered with JNK activation in the cytoplasm and nucleus without altering JNK subcellular distribution. These effects did not involve direct interaction between JNK and GR and were independent of DNA binding, since a dimerization-defective GR mutant, which lacks transcriptional capacity, is still able to suppress JNK activation in response to Dex (93).

Glucocorticoid and Endothelial Nitric Oxide Synthase

Many beneficial actions of corticosteroids in vivo involve the vascular endothelium. An important endogenous mediator of vascular integrity is endothelium-derived NO (94). NO produced by eNOS possesses anti-inflammatory, antiatherogenic, and anti-ischemic properties (95,96). Enhanced NO production by administration of the eNOS substrate, L-arginine, or upregulation of eNOS by statins, confers stroke protection (97–99), and transgenic mice overexpressing eNOS show decreased leukocyte accumulation and reduced vascular lesion formation following vascular injury (100). Conversely, mice with targeted disruption of eNOS (*eNOS*$^{-/-}$) exhibit increased vascular inflammation and larger cerebral infarctions following experimental ischemia (101,102), whereas inhibition of NOS activity decreases cerebral blood flow (CBF) and increases infarct size after ischemia (103).

In two different mouse models of ischemic injury, transient myocardial ischemia and transient focal cerebral ischemia, the protective effects of steroids were mediated by the nontranscriptional activation of eNOS by GR (11,12). High-dose corticosteroids significantly reduced ischemic tissue damage to the heart and the brain, whereas low doses were ineffective. Cotreatment with the GR antagonist RU486 completely reversed the effects of Dex, suggesting that GR mediated tissue protection. Furthermore, protection from ischemia was mediated by NO, since the beneficial actions of steroids were absent in *eNOS*$^{-/-}$ mice or blocked by cotreatment with the NOS inhibitor N(G)-nitro-L-arginine methyl ester (L-NAME) (11,12). Steroid treatment acutely increased eNOS activity in vitro and in vivo, which led to decreased vascular inflammation and increased vasorelaxation and regional cerebral blood flow in a NO-dependent manner (12).

Corticosteroids have recently been reported to activate protein kinase Akt (6,11,12,104), which is an important regulator of cell cycle progression and mediator of cellular survival downstream of PI3K (105). In a ligand-dependent manner, corticosteroids activate PI3K and Akt in ECs and reconstituted COS cells, which is blocked by cotreatment with RU486 or the PI3K inhibitor, LY294002, but not by transcriptional inhibitors (11,12). Although induction of nontranscriptional actions requires high doses of steroids (in contrast to estrogen, which occurs at physiological concentrations), these effects are nevertheless specifically mediated by GR. This was demonstrated in transfection studies in COS7 cells, which lack endogenous GR. Whereas Dex did not activate PI3K or Akt in the absence of GR, it readily induced Akt kinase activation after transfection with GR, or the dimerization defective GR mutant (A458T), which is unable to transactivate target genes (12). This suggests a DNA binding–independent mechanism of activation. The fact that PI3K activation can be suppressed by the GR antagonist RU486 further argues against the involvement of transrepression in this process, because the RU486 compound still induces transrepression of AP-1–dependent gene expression in reporter assays (106).

Recently, we found that the binding of thyroid hormone to the thyroid hormone receptor-$\alpha 1$ (TR$\alpha 1$) activates PI3k/Akt and eNOS, leading to the rapid nongenomic neuro-protective effects of thyroid hormone in stroke (107).

KEY POINTS

- Steroid hormones, such as estrogen and glucocorticoids, can exert vascular protective effects through nongenomic mechanisms.
- A prominent feature of both ER and GR is their ability to rapidly activate the PI3K/Akt pathway in ECs, leading to the activation of eNOS.
- By enhancing endothelial NO production, steroid hormones decrease postischemic vascular inflammation and increase regional blood flow.
- The therapeutic use of steroid hormones is limited by their side effects, which are typically associated with their genomic actions.

Future Goals

- To explore the rapid, nontranscriptional actions of steroid hormones in ECs, so that the biological and pharmacological actions of these steroid hormones could be considerably broadened, thus leading to the development of a novel class of drugs that selectively activates their nontranscriptional actions

ACKNOWLEDGMENTS

The work described in this article was supported in part by the National Institutes of Health (DK062729, HL070274, and HL080187) and the American Heart Association Bugher Foundation Award.

REFERENCES

1 Krust A, Green S, Argos P, et al. The chicken oestrogen receptor sequence: homology with v-erbA and the human oestrogen and glucocorticoid receptors. *EMBO J*. 1986;5(5):891–897.

2 Klein-Hitpass L, Tsai SY, Greene GL, et al. Specific binding of estrogen receptor to the estrogen response element. *Mol Cell Biol*. 1989;9(1):43–49.

3 Tsai SY, Carlstedt-Duke J, Weigel NL, et al. Molecular interactions of steroid hormone receptor with its enhancer element: evidence for receptor dimer formation. *Cell*. 1988;55(2):361–369.

4 Reichardt HM, Tuckermann JP, Gottlicher M, et al. Repression of inflammatory responses in the absence of DNA binding by the glucocorticoid receptor. *EMBO J*. 2001;20(24):7168–7173.

5 Tuckermann JP, Reichardt HM, Arribas R, et al. The DNA binding-independent function of the glucocorticoid receptor mediates repression of AP-1-dependent genes in skin. *J Cell Biol*. 1999;147(7):1365–1370.

6 Simoncini T, Hafezi-Moghadam A, Brazil DP, et al. Interaction of oestrogen receptor with the regulatory subunit of phosphatidylinositol-3-OH kinase. *Nature*. 2000;407(6803): 538–541.

7 Hisamoto K, Ohmichi M, Kurachi H, et al. Estrogen induces the Akt-dependent activation of endothelial nitric-oxide synthase in vascular endothelial cells. *J Biol Chem*. 2001;276(5):3459–3467.

8 Chen Z, Yuhanna IS, Galcheva-Gargova Z, et al. Estrogen receptor alpha mediates the nongenomic activation of endothelial nitric oxide synthase by estrogen. *J Clin Invest*. 1999;103(3):401–406.

9 Chambliss KL, Shaul PW. Rapid activation of endothelial NO synthase by estrogen: evidence for a steroid receptor fast-action complex (SRFC) in caveolae. *Steroids*. 2002;67(6):413–419.

10 Haynes MP, Sinha D, Russell KS, et al. Membrane estrogen receptor engagement activates endothelial nitric oxide synthase via the PI3-kinase-Akt pathway in human endothelial cells. *Circ Res*. 2000;87(8):677–682.

11 Hafezi-Moghadam A, Simoncini T, Yang E, et al. Acute cardiovascular protective effects of corticosteroids are mediated by non-transcriptional activation of endothelial nitric oxide synthase. *Nat Med*. 2002;8(5):473–479.

12 Limbourg FP, Huang Z, Plumier JC, et al. Rapid nontranscriptional activation of endothelial nitric oxide synthase mediates increased cerebral blood flow and stroke protection by corticosteroids. *J Clin Invest*. 2002;110(11):1729–1738.

13 Albuquerque ML, Akiyama SK, Schnaper HW. Basic fibroblast growth factor release by human coronary artery endothelial cells is enhanced by matrix proteins, 17beta-estradiol, and a PKC signaling pathway. *Exp Cell Res*. 1998;245(1):163–169.

14 Chen DB, Bird IM, Zheng J, Magness RR. Membrane estrogen receptor-dependent extracellular signal-regulated kinase pathway mediates acute activation of endothelial nitric oxide synthase by estrogen in uterine artery endothelial cells. *Endocrinology*. 2004;145(1):113–125.

15 Razandi M, Pedram A, Park ST, Levin ER. Proximal events in signaling by plasma membrane estrogen receptors. *J Biol Chem*. 2003;278(4):2701–2712.

16 Happerfield LC, Miles DW, Barnes DM, et al. The localization of the insulin-like growth factor receptor 1 (IGFR-1) in benign and malignant breast tissue. *J Pathol*. 1997;183(4):412–417.

17 Wyckoff MH, Chambliss KL, Mineo C, et al. Plasma membrane estrogen receptors are coupled to endothelial nitric-oxide synthase through Galpha(i). *J Biol Chem.* 2001;276(29):27071–27076.

18 van Eickels M, Grohe C, Cleutjens JP, et al. 17beta-estradiol attenuates the development of pressure-overload hypertrophy. *Circulation.* 2001;104(12):1419–1423.

19 Endoh H, Sasaki H, Maruyama K, et al. Rapid activation of MAP kinase by estrogen in the bone cell line. *Biochem Biophys Res Commun.* 1997;235(1):99–102.

20 de Jager T, Pelzer T, Muller-Botz S, et al. Mechanisms of estrogen receptor action in the myocardium. Rapid gene activation via the ERK1/2 pathway and serum response elements. *J Biol Chem.* 2001;276(30):27873–27880.

21 Hwang KC, Lee KH, Jang Y. Inhibition of MEK1,2/ERK mitogenic pathway by estrogen with antiproliferative properties in rat aortic smooth muscle cells. *J Steroid Biochem Mol Biol.* 2002; 80(1):85–90.

22 Karas RH, Gauer EA, Bieber HE, et al. Growth factor activation of the estrogen receptor in vascular cells occurs via a mitogen-activated protein kinase-independent pathway. *J Clin Invest.* 1998;101(12):2851–2861.

23 Migliaccio A, Piccolo D, Castoria G, et al. Activation of the Src/p21ras/Erk pathway by progesterone receptor via cross-talk with estrogen receptor. *EMBO J.* 1998;17(7):2008–2018.

24 Croxtall JD, Choudhury Q, Flower RJ. Glucocorticoids act within minutes to inhibit recruitment of signalling factors to activated EGF receptors through a receptor-dependent, transcription-independent mechanism. *Br J Pharmacol.* 2000; 130(2):289–298.

25 Anter E, Chen K, Shapira OM, et al. p38 mitogen-activated protein kinase activates eNOS in endothelial cells by an estrogen receptor alpha-dependent pathway in response to black tea polyphenols. *Circ Res.* 2005;96(10):1072–1078.

26 Wallner BP, Mattaliano RJ, Hession C, et al. Cloning and expression of human lipocortin, a phospholipase A2 inhibitor with potential anti-inflammatory activity. *Nature.* 1986; 320(6057):77–81.

27 Pare G, Krust A, Karas RH, et al. Estrogen receptor-alpha mediates the protective effects of estrogen against vascular injury. *Circ Res.* 2002;90(10):1087–1092.

28 Razandi M, Pedram A, Levin ER. Estrogen signals to the preservation of endothelial cell form and function. *J Biol Chem.* 2000;275(49):38540–38546.

29 Spyridopoulos I, Sullivan AB, Kearney M, et al. Estrogen-receptor-mediated inhibition of human endothelial cell apoptosis. Estradiol as a survival factor. *Circulation.* 1997;95(6):1505–1514.

30 Razandi M, Pedram A, Levin ER. Plasma membrane estrogen receptors signal to antiapoptosis in breast cancer. *Mol Endocrinol.* 2000;14(9):1434–1447.

31 Lee EJ, Jakacka M, Duan WR, et al. Adenovirus-directed expression of dominant negative estrogen receptor induces apoptosis in breast cancer cells and regression of tumors in nude mice. *Mol Med.* 2001;7(11):773–782.

32 Groten T, Pierce AA, Huen AC, Schnaper HW. 17 beta-estradiol transiently disrupts adherens junctions in endothelial cells. *FASEB J.* 2005;19(10):1368–1370.

33 Hollenberg SM, Weinberger C, Ong ES, et al. Primary structure and expression of a functional human glucocorticoid receptor cDNA. *Nature.* 1985;318(6047):635–641.

34 Tsai SY, Tsai MJ, O'Malley BW. Cooperative binding of steroid hormone receptors contributes to transcriptional synergism at target enhancer elements. *Cell.* 1989;57(3):443–448.

35 Hollenberg SM, Evans RM. Multiple and cooperative trans-activation domains of the human glucocorticoid receptor. *Cell.* 1988;55(5):899–906.

36 Umesono K, Evans RM. Determinants of target gene specificity for steroid/thyroid hormone receptors. *Cell.* 1989;57(7):1139–1146.

37 Mangelsdorf DJ, Thummel C, Beato M, et al. The nuclear receptor superfamily: the second decade. *Cell.* 1995;83(6):835–839.

38 Giguere V, Hollenberg SM, Rosenfeld MG, Evans RM. Functional domains of the human glucocorticoid receptor. *Cell.* 1986;46(5):645–652.

39 Honda K, Sawada H, Kihara T, et al. Phosphatidylinositol 3-kinase mediates neuroprotection by estrogen in cultured cortical neurons. *J Neurosci Res.* 2000;60(3):321–327.

40 Denninger JW, Marletta MA. Guanylate cyclase and the NO/cGMP signaling pathway. *Biochim Biophys Acta.* 1999;1411 (2–3):334–350.

41 Lantin-Hermoso RL, Rosenfeld CR, Yuhanna IS, et al. Estrogen acutely stimulates nitric oxide synthase activity in fetal pulmonary artery endothelium. *Am J Physiol.* 1997;273(1 Pt 1):L119–L126.

42 Fontana J, Fulton D, Chen Y, et al. Domain mapping studies reveal that the M domain of hsp90 serves as a molecular scaffold to regulate Akt-dependent phosphorylation of endothelial nitric oxide synthase and NO release. *Circ Res.* 2002;90(8):866–873.

43 Nuedling S, Kahlert S, Loebbert K, et al. Differential effects of 17beta-estradiol on mitogen-activated protein kinase pathways in rat cardiomyocytes. *FEBS Lett.* 1999;454(3):271–276.

44 Mital S, Barbone A, Addonizio LJ, et al. Endogenous endothelium-derived nitric oxide inhibits myocardial caspase activity: implications for treatment of end-stage heart failure. *J Heart Lung Transplant.* 2002;21(5):576–585.

45 Jessop HL, Sjoberg M, Cheng MZ, et al. Mechanical strain and estrogen activate estrogen receptor alpha in bone cells. *J Bone Miner Res.* 2001;16(6):1045–1055.

46 Keshamouni VG, Mattingly RR, Reddy KB. Mechanism of 17-beta-estradiol-induced Erk1/2 activation in breast cancer cells: A role for HER2 and PKC-delta. *J Biol Chem.* 2002; 277(25):22558–22565. Epub 2002 Apr 17.

47 Russell KS, Haynes MP, Sinha D, et al. Human vascular endothelial cells contain membrane binding sites for estradiol, which mediate rapid intracellular signaling. *Proc Natl Acad Sci USA.* 2000;97(11):5930–5935.

48 Lu Q, Ebling H, Mittler J, et al. MAP kinase mediates growth factor-induced nuclear translocation of estrogen receptor alpha. *FEBS Lett.* 2002;516(1–3):1–8.

49 Li L, Haynes MP, Bender JR. Plasma membrane localization and function of the estrogen receptor alpha variant (ER46) in human endothelial cells. *Proc Natl Acad Sci USA.* 2003;100(8):4807–4812.

50 Acconcia F, Ascenzi P, Bocedi A, et al. Palmitoylation-dependent estrogen receptor alpha membrane localization: regulation by 17beta-estradiol. *Mol Biol Cell.* 2005;16(1):231–237.

51 Chambliss KL, Yuhanna IS, Mineo C, et al. Estrogen receptor alpha and endothelial nitric oxide synthase are organized into a functional signaling module in caveolae. *Circ Res.* 2000;87 (11):E44–E52.

52 Kim HP, Lee JY, Jeong JK, et al. Nongenomic stimulation of nitric oxide release by estrogen is mediated by estrogen receptor alpha localized in caveolae. *Biochem Biophys Res Commun.* 1999;263(1):257–262.

53 Gratton JP, Fontana J, O'Connor DS, et al. Reconstitution of an endothelial nitric-oxide synthase (eNOS), hsp90, and caveolin-1 complex in vitro. Evidence that hsp90 facilitates calmodulin stimulated displacement of eNOS from caveolin-1. *J Biol Chem.* 2000;275(29):22268–22272.

54 Feron O, Saldana F, Michel JB, Michel T. The endothelial nitric-oxide synthase-caveolin regulatory cycle. *J Biol Chem.* 1998;273(6):3125–3128.

55 Razandi M, Pedram A, Greene GL, Levin ER. Cell membrane and nuclear estrogen receptors (ERs) originate from a single transcript: studies of ERalpha and ERbeta expressed in Chinese hamster ovary cells. *Mol Endocrinol.* 1999;13(2):307–319.

56 Williams JK, Wagner JD, Li Z, et al. Tamoxifen inhibits arterial accumulation of LDL degradation products and progression of coronary artery atherosclerosis in monkeys. *Arterioscler Thromb Vasc Biol.* 1997;17(2):403–408.

57 Dardes RC, Schafer JM, Pearce ST, et al. Regulation of estrogen target genes and growth by selective estrogen-receptor modulators in endometrial cancer cells. *Gynecol Oncol.* 2002;85(3):498–506.

58 Hisamoto K, Ohmichi M, Kanda Y, et al. Induction of endothelial nitric-oxide synthase phosphorylation by the raloxifene analog LY117018 is differentially mediated by Akt and extracellular signal-regulated protein kinase in vascular endothelial cells. *J Biol Chem.* 2001;276(50):47642–47649.

59 Simoncini T, Genazzani AR, Liao JK. Nongenomic mechanisms of endothelial nitric oxide synthase activation by the selective estrogen receptor modulator raloxifene. *Circulation.* 2002;105(11):1368–1373.

60 Figtree GA, Lu Y, Webb CM, Collins P. Raloxifene acutely relaxes rabbit coronary arteries in vitro by an estrogen receptor-dependent and nitric oxide-dependent mechanism. *Circulation.* 1999;100(10):1095–1101.

61 Wassmann S, Laufs U, Stamenkovic D, et al. Raloxifene improves endothelial dysfunction in hypertension by reduced oxidative stress and enhanced nitric oxide production. *Circulation.* 2002;105(17):2083–2091.

62 Martel C, Provencher L, Li X, et al. Binding characteristics of novel nonsteroidal antiestrogens to the rat uterine estrogen receptors. *J Steroid Biochem Mol Biol.* 1998;64(3–4):199–205.

63 Labrie F, Labrie C, Belanger A, et al. EM-652 (SCH57068), a pure SERM having complete antiestrogenic activity in the mammary gland and endometrium. *J Steroid Biochem Mol Biol.* 2001;79(1–5):213–225.

64 Simoncini T, Varone G, Fornari L, et al. Genomic and nongenomic mechanisms of nitric oxide synthesis induction in human endothelial cells by a fourth-generation selective estrogen receptor modulator. *Endocrinology.* 2002;143(6):2052–2061.

65 Tronche F, Kellendonk C, Reichardt HM, Schutz G. Genetic dissection of glucocorticoid receptor function in mice. *Curr Opin Genet Dev.* 1998;8(5):532–538.

66 McEwen BS. Protective and damaging effects of stress mediators. *N Engl J Med.* 1998;338(3):171–179.

67 Jonat C, Rahmsdorf HJ, Park KK, et al. Antitumor promotion and antiinflammation: down-modulation of AP-1 (Fos/Jun) activity by glucocorticoid hormone. *Cell.* 1990;62(6):1189–1204.

68 Karin M. New twists in gene regulation by glucocorticoid receptor: is DNA binding dispensable? *Cell.* 1998;93(4):487–490.

69 Thiemermann C. Corticosteroids and cardioprotection. *Nat Med.* 2002;8(5):453–455.

70 Cronstein BN, Kimmel SC, Levin RI, et al. A mechanism for the antiinflammatory effects of corticosteroids: the glucocorticoid receptor regulates leukocyte adhesion to endothelial cells and expression of endothelial-leukocyte adhesion molecule 1 and intercellular adhesion molecule 1. *Proc Natl Acad Sci USA.* 1992;89(21):9991–9995.

71 Barzilai D, Plavnick J, Hazani A, et al. Use of hydrocortisone in the treatment of acute myocardial infarction. Summary of a clinical trial in 446 patients. *Chest.* 1972;61(5):488–491.

72 Libby P, Maroko PR, Bloor CM, et al. Reduction of experimental myocardial infarct size by corticosteroid administration. *J Clin Invest.* 1973;52(3):599–607.

73 Spath JA Jr, Lane DL, Lefer AM. Protective action of methylprednisolone on the myocardium during experimental myocardial ischemia in the cat. *Circ Res.* 1974;35(1):44–51.

74 Patten BM, Mendell J, Bruun B, et al. Double-blind study of the effects of dexamethasone on acute stroke. *Neurology.* 1972;22(4):377–383.

75 Bertorelli R, Adami M, Di Santo E, Ghezzi P. MK 801 and dexamethasone reduce both tumor necrosis factor levels and infarct volume after focal cerebral ischemia in the rat brain. *Neurosci Lett.* 1998;246(1):41–44.

76 de Courten-Myers GM, Kleinholz M, Wagner KR, et al. Efficacious experimental stroke treatment with high-dose methylprednisolone. *Stroke.* 1994;25(2):487–492.

77 Slivka AP, Murphy EJ. High-dose methylprednisolone treatment in experimental focal cerebral ischemia. *Exp Neurol.* 2001;167(1):166–172.

78 Braunwald E, Zipes DP, Libby P. *Heart Disease*, 6th ed. Philadelphia: W. B. Saunders, 2001.

79 Wallerath T, Witte K, Schafer SC, et al. Down-regulation of the expression of endothelial NO synthase is likely to contribute to glucocorticoid-mediated hypertension. *Proc Natl Acad Sci USA.* 1999;96(23):13357–13362.

80 Sapolsky RM, Pulsinelli WA. Glucocorticoids potentiate ischemic injury to neurons: therapeutic implications. *Science.* 1985;229(4720):1397–1400.

81 Roberts R, DeMello V, Sobel BE. Deleterious effects of methylprednisolone in patients with myocardial infarction. *Circulation.* 1976;53(Suppl 3):I204–I206.

82 Bulkley BH, Roberts WC. Steroid therapy during acute myocardial infarction. A cause of delayed healing and of ventricular aneurysm. *Am J Med.* 1974;56(2):244–250.

83 Wehling M. Specific, nongenomic actions of steroid hormones. *Annu Rev Physiol.* 1997;59:365–393.

84 Orchinik M, Murray TF, Moore FL. A corticosteroid receptor in neuronal membranes. *Science.* 1991;252(5014):1848–1851.

85 Steiner A, Vogt E, Locher R, Vetter W. Stimulation of the phosphoinositide signalling system as a possible mechanism for glucocorticoid action in blood pressure control. *J Hypertens Suppl.* 1988;6(4):S366–S368.

86 Inagaki N, Miura T, Nakajima T, et al. Studies on the anti-allergic mechanism of glucocorticoids in mice. *J Pharmacobiodyn.* 1992;15(10):581–587.

87 Smith MD, Ahern MJ, Brooks PM, et al. The clinical and immunological effects of pulse methylprednisolone therapy in

rheumatoid arthritis. III. Effects on immune and inflammatory indices in synovial fluid. *J Rheumatol.* 1988;15(2):238–241.

88 Vyden JK, Nagasawa K, Rabinowitz B, et al. Effects of methylprednisolone administration in acute myocardial infarction. *Am J Cardiol.* 1974;34(6):677–686.

89 Buttgereit F, Scheffold A. Rapid glucocorticoid effects on immune cells. *Steroids.* 2002;67(6):529–534.

90 Pitzalis C, Pipitone N, Perretti M. Regulation of leukocyte-endothelial interactions by glucocorticoids. *Ann NY Acad Sci.* 2002;966:108–118.

91 Minden A, Karin M. Regulation and function of the JNK subgroup of MAP kinases. *Biochim Biophys Acta.* 1997;1333(2): F85–F104.

92 Caelles C, Gonzalez-Sancho JM, Munoz A. Nuclear hormone receptor antagonism with AP-1 by inhibition of the JNK pathway. *Genes Dev.* 1997;11(24):3351–3364.

93 Gonzalez MV, Jimenez B, Berciano MT, et al. Glucocorticoids antagonize AP-1 by inhibiting the activation/phosphorylation of JNK without affecting its subcellular distribution. *J Cell Biol.* 2000;150(5):1199–1208.

94 Loscalzo J. Nitric oxide and vascular disease. *N Engl J Med.* 1995; 333(4):251–253.

95 De Caterina R, Gimbrone MA Jr. Leukocyte-endothelial interactions and the pathogenesis of atherosclerosis. In: Kristensen SD, Schmidt EB, De Caterina R, Endres S, eds. *n-3 Fatty Acids-Prevention and Treatment in Vascular Disease.* London: Springer Verlag; 1995:9–24.

96 Ishida A, Sasaguri T, Kosaka C, et al. Induction of the cyclin-dependent kinase inhibitor p21(Sdi1/Cip1/Waf1) by nitric oxide-generating vasodilator in vascular smooth muscle cells. *J Biol Chem.* 1997;272(15):10050–10057.

97 Dalkara T, Morikawa E, Panahian N, Moskowitz MA. Blood flow-dependent functional recovery in a rat model of focal cerebral ischemia. *Am J Physiol.* 1994;267(2 Pt 2):H678–H683.

98 Morikawa E, Moskowitz MA, Huang Z, et al. L-arginine infusion promotes nitric oxide-dependent vasodilation, increases regional cerebral blood flow, and reduces infarction volume in the rat. *Stroke.* 1994;25(2):429–435.

99 Endres M, Laufs U, Huang Z, et al. Stroke protection by 3-hydroxy-3-methylglutaryl (HMG)-CoA reductase inhibitors mediated by endothelial nitric oxide synthase. *Proc Natl Acad Sci USA.* 1998;95(15):8880–8885.

100 Kawashima S, Yamashita T, Ozaki M, et al. Endothelial NO synthase overexpression inhibits lesion formation in mouse model of vascular remodeling. *Arterioscler Thromb Vasc Biol.* 2001;21 (2):201–207.

101 Moroi M, Zhang L, Yasuda T, et al. Interaction of genetic deficiency of endothelial nitric oxide, gender, and pregnancy in vascular response to injury in mice. *J Clin Invest.* 1998;101(6):1225–1232.

102 Huang Z, Huang PL, Ma J, et al. Enlarged infarcts in endothelial nitric oxide synthase knockout mice are attenuated by nitro-L-arginine. *J Cereb Blood Flow Metab.* 1996;16(5):981–987.

103 Huang Z, Huang PL, Panahian N, et al. Effects of cerebral ischemia in mice deficient in neuronal nitric oxide synthase. *Science.* 1994;265(5180):1883–1885.

104 Langdown ML, Holness MJ, Sugden MC. Early growth retardation induced by excessive exposure to glucocorticoids in utero selectively increases cardiac GLUT1 protein expression and Akt/protein kinase B activity in adulthood. *J Endocrinol.* 2001; 169(1):11–22.

105 Datta SR, Brunet A, Greenberg ME. Cellular survival: a play in three Akts. *Genes Dev.* 1999;13(22):2905–2927.

106 Heck S, Kullmann M, Gast A, et al. A distinct modulating domain in glucocorticoid receptor monomers in the repression of activity of the transcription factor AP-1. *EMBO J.* 1994;13 (17):4087–4095.

107 Hiroi Y, Kim HH, Ying H, et al. Rapid nongenomic actions of thyroid hormone. *Proc Natl Acad Sci USA.* 2006;103:14104–14109.

181

Organic Nitrates

Exogenous Nitric Oxide Administration and Its Influence on the Vascular Endothelium

John D. Parker* and Tommaso Gori†

*University Health Network Hospitals, University of Toronto, Onatario, Canada;
†University of Siena, Italy

Nitroglycerin (GTN) and other nitrates have been used in cardiovascular medicine for more than 120 years. GTN was first synthesized in 1847 by an Italian, Ascanio Sobrero, who described a "violent headache" upon self-administration of a "minute quantity" of the drug (1). Because of this side effect, investigation of possible pharmacological applications of GTN was limited for several years, until the reports of Brunton and Murrell, who employed nitrates in the treatment of angina (2,3). Although sublingual GTN has been commonly used for more than a century to treat acute attacks of angina, the development of organic nitrates with sustained activity was limited by their poor oral bioavailability. Eventually, this difficulty was overcome and long-acting formulations of GTN and other long-acting organic nitrates, including isosorbide dinitrate, isosorbide-5-mononitrate, and pentaerythritol tetranitrate, have been developed and marketed. More details on the pharmacologic characteristics, formulations, and indications of different nitrates have been reviewed recently (4).

BIOTRANSFORMATION AND HEMODYNAMIC EFFECTS

Organic nitrates are prodrugs that release their active principle, nitric oxide (NO) or a NO-containing compound, via an intracellular enzyme-dependent denitrification (5) (Figure 181.1). The exact determination of the enzyme system involved in this bioconversion has remained elusive, despite extensive investigation. Multiple enzymatic candidates have been identified, including cytochrome P450, endothelial NO synthase, and glutathione transferase (6–10). There has been intense interest in identifying the denitrification pathway because it was felt that the development of abnormalities in this process might explain the loss of nitrate effects during sustained therapy, a phenomenon termed *nitrate tolerance* (discussed later). Spontaneous thiol-dependent denitrification of GTN also has been proposed as the mechanism of GTN biotransformation. However, today it is believed that enzyme-bound, rather than free, thiol groups play an important role in the biotransformation of GTN. Indeed, more recent and convincing studies have proposed a role for a mitochondrial aldehyde dehydrogenase that catalyzes the formation of NO or a NO-containing compound from GTN (8,11–13). This mechanism appears to depend on the oxidation of two cysteine residues in the active site of the enzyme, a finding that is compatible with the thiol-dependence of GTN bioactivation described by Needleman more than 30 years ago (14) (Figure 181.1). Evidence in support of the importance of this pathway consists of the lack of a vasodilator effect of GTN in vitro and in animals in vivo after administration of specific inhibitors for this mitochondrial enzyme, as well as significant stimulation of cyclic guanosine monophosphate (cGMP) formation by GTN in the presence of the enzyme. Furthermore, there are observations of reduced hemodynamic responses to GTN in Asian subjects who are genetically deficient in the enzyme aldehyde dehydrogenase (13,15,16).

Regardless of the mechanism, GTN biotransformation results in exogenous (nitrate-derived) NO, which like its endogenous (endothelial-derived) counterpart, then activates the soluble guanylate cyclase to increase cyclic GMP synthesis. The molecular cascade triggered by increased cGMP bioavailability can mediate vasorelaxation through the activation of protein kinases and the subsequent phosphorylation of different proteins involved in the regulation of intracellular calcium (Ca^{2+}) levels, such as the sarcoplasmic Ca^{2+}-ATPase (17). Additional proposed biochemical consequences of NO donors, which might also be involved in NO-induced vasodilation (and its other effects; see following paragraph) include thiol modification, regulation of mitochondrial respiration, modulation of potassium (K^+) channel activity (18), and

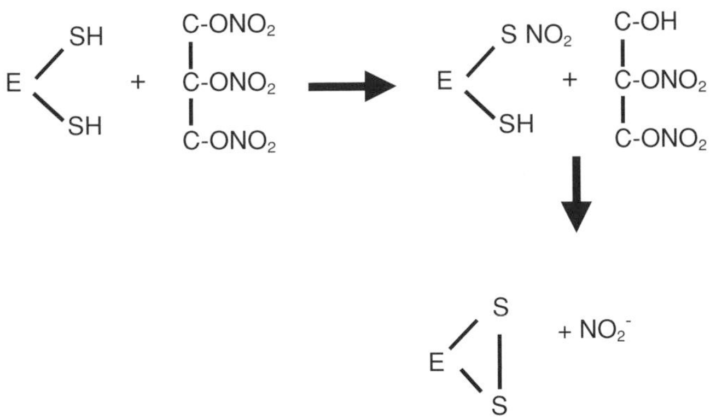

Figure 181.1. The mechanism of activation of GTN by the mitochondrial aldehyde dehydrogenase as proposed by Chen and associates (13). This process yields the dinitrate metabolites glyceryl-1, 2-dinitrate (1,2-GDN) and glyceryl-1,3-dinitrate (1,3-GDN) via a thiol-dependent denitration. Of note, several other enzymes have been identified that are capable of mediating the denitration of GTN, including the glutathione *S*-transferases and the cytochrome P450. In nitrate-tolerant tissues, the biotransformation of GTN is attenuated, an effect that might be due to oxidation of thiol groups in the active site of aldehyde dehydrogenase during GTN denitration.

protein nitration, although less evidence is available concerning the therapeutic relevance of these effects (19–21).

The potent vasodilator effects of organic nitrates are preferentially exerted at the level of capacitance veins and conductance arteries (22), resulting in a reduction in cardiac preload and oxygen consumption and, theoretically, an increase in oxygen supply to areas of myocardial ischemia (23). Nitrates have limited effects on resistance vessels (22), which may limit coronary steal phenomena.[1] Therefore, important differences exist in the tissue selectivity of the nitrates versus endothelial NO: in particular, although endothelial NO is believed to play a (relatively) more important role in resistance arteries as compared to conduit arteries and veins, possibly due to a higher metabolic activity of the arterial endothelium (flow-mediated dilation is both more potent and rapid in the peripheral circulation) (24), nitrate-derived NO is more potent at the level of conductance vessels. This site-specific differential potency of nitrates appears to result from differences in nitrate biotransformation rate, through a negative feedback inhibition of GTN biotransformation by the endogenous NO (25).

CLINICAL USE

The organic nitrates are among the most commonly used drugs in the treatment of cardiovascular disease, including congestive heart failure and coronary artery disease. In the United States, a recent survey showed that 91% of physicians treat congestive heart failure patients with drugs of this class (26), a figure that is likely to be even higher when patients with coronary artery disease are considered. The clinical applications of nitrates reflect their hemodynamic effects. Stable angina guidelines state that "in patients with exertional stable angina, nitrates improve exercise tolerance, increase the time to onset of angina, and decrease ST-segment depression during the treadmill exercise test. Combined with β-blockers or calcium antagonists, nitrates produce greater antianginal and anti-ischemic effects in patients with stable angina" (27). Short-acting nitrates such as sublingual formulations of GTN or isosorbide dinitrate are employed in the treatment and/or prevention of anginal episodes in patients with coronary artery disease (3,28), whereas long-acting nitrates serve as a chronic treatment for the prevention of these episodes. Of note, guidelines recommend the use of long-acting nitrates in the case of chronic stable angina only as second- or third-line therapy, for cases that are refractory to treatment with drugs such as calcium antagonists and/or β-blockers. In the treatment of acute coronary syndromes, nitrates are recommended by published guidelines as Class I intervention[2] for the relief of acute ischemia and associated symptoms (29), an indication that is based on "experts' consensus in the absence of randomized trials" (level of evidence C), due to the lack of any clear documentation that nitrates have a positive effect on overall clinical outcome and/or mortality in this setting (30,31). As mentioned earlier, nitrates also are commonly employed in the setting of congestive heart failure for their beneficial hemodynamic effects, mediated by a decrease in preload, which is associated with a reduction in symptoms of congestion and a decrease in the severity of mitral regurgitation (26,32,33).

Because they release NO, organic nitrates have been proposed as possible alternate NO sources in the setting of endothelial dysfunction (34,35). Some authors have postulated that the protective effects of endothelial NO might be restored by the administration of these drugs. The following paragraphs will describe the current knowledge regarding the interactions between the organic nitrates as exogenous NO donors and the endothelium and will critically analyze the hypothesis that they can act as surrogate (and replacement) NO sources.

THE "PLEIOTROPIC" EFFECTS OF ORGANIC NITRATES

Beyond their hemodynamic effects, other important beneficial effects of organic nitrates have been proposed (Table 181-1). These include antiplatelet effects and the ability to prevent

1 Blood flow steal is caused by a difference in vascular resistance between two branches of a vascular bed. Selective dilation of resistance (distal) arteries (e.g., with calcium antagonists) causes a marked decrease in vascular resistance in the vessels where no conduit artery stenosis is present. Therefore, blood flow meant for the territories distal to a conduit artery stenosis is "stolen" by the other vessels, causing ischemia.

2 Class I intervention: conditions for which there is evidence and/or general agreement that a given procedure or treatment is useful and effective. Level of evidence A: recommendation based on large, randomized trials; B: based on smaller trials or observational registries; C: based on experts' consensus.

Table 181-1: Effects of Endothelial and Nitrate-Derived NO

	Endothelial NO	*Nitrate Therapy*	
		Acute or Chronic Intermittent	*Chronic Continuous*
Platelet	Antiaggregant	Antiaggregant	Controversial: in vitro: antiaggregant; in vivo: proaggregant
Vascular tone			
Conduit arteries	Potent vasodilator	Potent vasodilator	Vasodilator
Resistance arteries	Vasodilator	Mild vasodilator	No effect (tolerance)
Veins	Weak dilator	Potent vasodilator	No effect (tolerance)
Exercise tolerance	?	Increased	No effect (tolerance)
Endothelial function	Major determinant	Potentiating	Causes dysfunction
Modulation of atherogenesis	Beneficial	Retards oxidative stress and intima-media thickening*	Possibly accelerating
Ischemic preconditioning	Acts as mediator	Protective effect†	Controversial, possibly deleterious‡

*Only animal data of intermittent, high-dose, isosorbide mononitrate eccentric administration available
†Only data of acute nitrate treatment
‡Only animal data available

atherosclerosis as well as "preconditioning-mimetic" properties.

The existence of a tonic effect of endogenous NO on platelet activity is documented by the fact that the administration of an inhibitor of NO synthase increases aggregation and the expression of membrane markers of activation. This effect is completely reversed by sublingual GTN (36). To date, both human and animal studies have documented a direct antiaggregant effect of organic nitrates (36). Although this might not be of relevance in the setting of stable angina, inhibition of platelet aggregation might be of great importance in the therapy of acute coronary syndromes. In this setting, in which endothelial NO bioavailability is reduced, the antiplatelet effect of exogenous NO could substitute for it, limiting both the thrombogenic effects and the coronary spasm induced by release of vasoactive platelet-derived mediators.

Certain organic nitrates have been demonstrated to have antioxidant and protective effects in animal models of atherogenesis. In the setting of chronic disease, it is commonly accepted that the antiproliferative, antioxidant, and anti-inflammatory effects of NO play a role in limiting the progression of vascular disease and that the loss of these effects in the presence of cardiovascular risk factors is an early occurrence in the development of atherosclerosis (37). Animal studies have reported that administration of high doses of isosorbide mononitrate and pentaerythritol tetranitrate limits the development of endothelial dysfunction, the increase in reactive oxygen species production, and plaque formation in animals genetically and environmentally prone to develop atherosclerosis (38,39). Future studies must confirm whether these findings translate into a clinical advantage in humans with coronary artery disease.

Furthermore, organic nitrates have been attributed protective effects that mimic those of ischemic preconditioning. The concept of ischemic preconditioning was introduced after the observation that brief periods of ischemia, such as in the setting of preinfarction angina, are able to reduce the extent of the necrotic area and to improve prognosis after myocardial infarction (40,41). Interestingly, the administration of several drugs, including NO donors, has been associated with the development of an analogous protective state that has been termed *pharmacologic preconditioning*. Multiple studies in animal models of peripheral and cardiac ischemic injury have reported that treatment with GTN and/or NO donors is associated with a significant reduction of infarct size, an effect that is independent of any hemodynamic effect (42–44). Similarly, a protective effect of GTN has been demonstrated in human models of both low-flow and demand ischemia (45,46). The clinical significance of these effects remains to be determined.

Collectively, these findings suggest that endothelial- and nitrate-derived NO might have synergistic effects and that, in the case of insufficient endothelial NO production, nitrate-derived NO might substitute for it. Furthermore, a human study suggested that organic nitrates might even improve the bioavailability of endothelial NO, as demonstrated by a marked potentiation of the responses to endothelium-dependent vasodilators during low-dose GTN infusion (47). From this perspective, authors have postulated that nitrate use might have clinically important implications that go beyond the treatment of symptoms of angina and/or congestive heart failure (35). These effects might be clinically relevant in the prevention of coronary and peripheral atherosclerosis. However, although such characteristics make organic nitrates unique and potentially powerful drugs in the therapy of cardiovascular disease, recent and ongoing experimental findings (discussed in the next section) suggest that this class of drugs might have unexpected negative effects on vascular and endothelial function.

NITRATE TOLERANCE AND NITRATE-INDUCED ENDOTHELIAL DYSFUNCTION

Traditionally, nitrate tolerance is defined as the loss of hemodynamic and/or symptomatic effects of organic nitrates (or the requirement of higher administered doses to maintain these effects) that appears during continuous therapy (4). Tolerance is not influenced by the dosage administered (48) and occurs with all nitrate formulations (except possibly for pentaerythritol tetranitrate) (49), as long as they are administered using dosing schedules or formulations that produce significant plasma concentrations throughout the majority of the day. Tolerance has most often been described using the pharmacodynamic effects of nitrates – for example, their effects on blood pressure, cardiac filling pressures, and other physiological variables. Tolerance also occurs to the symptomatic effects of nitrates (anecdotal evidence suggests that this is also true for the side effects of nitrates, such as headache). Thus, in patients with angina, the beneficial effects of nitrates that are observed when these agents are given acutely are rapidly lost during sustained therapy.

The mechanism of tolerance has been investigated for several decades, and multiple hypotheses have been proposed (4,50,51) (Table 181-2). A schematic classification subdivides the mechanisms of this phenomenon as either due to the progressive loss of nitrate effects during continuous use ("true" tolerance) or to the reversal of nitrate effects caused by the activation of counterregulatory mechanisms ("pseudotolerance"). These counterregulatory responses include neurohormonal activation and plasma volume expansion, which would overcome the vasodilatory effects of nitrates (51,52).

Table 181-2: Proposed Mechanisms for the Development of Nitrate Tolerance

Proposed Hypotheses for Nitrate Tolerance
True Tolerance
Reduced biotransformation of the nitrate
Reduced NO signal transduction
Increased catabolism of NO
Pseudotolerance
Systemic neurohormonal counterregulatory responses
Plasma volume expansion
Sympathetic activation
Oxidative Stress Hypothesis

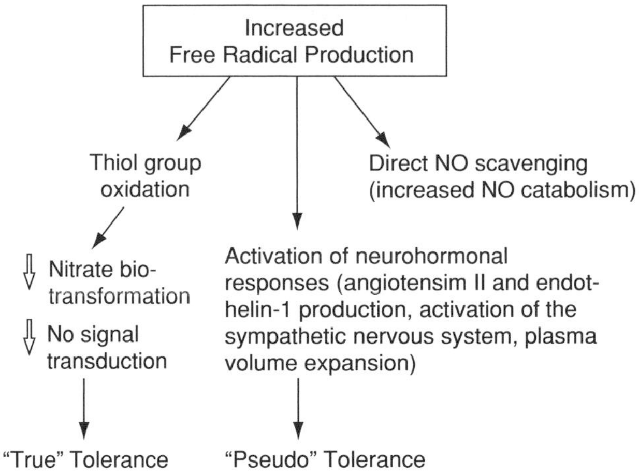

Figure 181.2. Interplay among free radical production and other proposed mechanisms for nitrate tolerance. An increased free radical production appears to be associated with many of the abnormalities observed in the setting of tolerance: these modifications include, on one side, plasma volume expansion, neurohormonal activation: these changes have the potential to reduce the ischemic threshold ("pseudotolerance"). On the other, they lead to reduced biotransformation of the nitrate to its active principle NO and to decreased signal transduction of this mediator ("true tolerance").

Recently, Munzel and colleagues have proposed the concept that sustained nitrate therapy might be associated with an increased endothelial production of reactive oxygen species, including superoxide anion and peroxynitrite (53). Since the formulation of this hypothesis, a number of enzymatic sources for this increase in oxygen free radical production have been described, including membrane-bound oxidases, the mitochondrial respiratory chain, and the endothelial NO synthase (20). Interestingly, the endothelium appears to be the source of this increased oxygen free radical production, and in vitro experiments have documented that tolerance to the vasorelaxant effects of GTN is dependent on the presence of the endothelium (53). This oxygen free radical hypothesis of nitrate tolerance provides the possibility to formulate a more complete view concerning the mechanism of nitrate tolerance (Figure 181.2). This "unifying hypothesis" was recently proposed by our group (19) and is based on the hypothesis that an increase in the bioavailability of oxidant free radical species triggered by nitrate therapy may (a) reduce the net bioavailability of NO by direct quenching, thus limiting the vasodilatory effect of the nitrate; (b) cause oxidation of important thiol groups and subsequent inactivation of several enzymes involved in the biotransformation of GTN and in NO signaling; and (c) trigger a variety of counterregulatory responses leading to pseudotolerance. The fact that nitrate therapy is associated with an increase in vascular oxygen free radical production therefore provides for the first time an explanation for other observations that have been made in the setting of tolerance, including neurohormonal activation (54), increased responsiveness and expression of vasoconstrictors such as endothelin (ET)-1 and angiotensin II (55,56), plasma volume expansion (57), and, via oxidative inactivation of enzymes, reduced GTN biotransformation (58) and NO signaling (21). At the same time, these modifications have important implications in terms of the impact of nitrate therapy on vascular function and, possibly, on patients' prognosis.

An increased oxygen free radical production in the setting of prolonged nitrate therapy might have important effects on native NO production – i.e., endothelial function. Animal studies in which GTN was administered long term in vivo have repeatedly demonstrated a marked reduction in the responses to endothelium-dependent dilators (53,59,60). This evidence was more recently confirmed in human studies documenting decreased endothelial responsiveness in healthy volunteers (61) and patients with coronary artery disease (62,63). As mentioned, studies have proven the existence of multiple sources of oxygen free radicals in the setting of sustained nitrate therapy, including membrane-bound oxidases and the mitochondrial respiratory chain. Importantly, prolonged GTN therapy also appears to lead to uncoupling of NO synthase, where the oxidation of tetrahydrobiopterin to inactive dihydrobiopterin and/or a reduced L-arginine bioavailability leads to increased superoxide anion production from NO synthase itself. Evidence for the role of NO synthase uncoupling in the development of nitrate tolerance has been provided in vitro (64,65) and in human in vivo models, where administration of L-arginine or folic acid (which can substitute for and/or regenerate tetrahydrobiopterin) was able to prevent this phenomenon (66,67). Furthermore, this impairment in NO synthase function might not be limited to the vasculature. Zanzinger and colleagues recently demonstrated in a porcine model that tolerance is associated with reduced NO production in the brain stem, leading to a loss of its restraining on sympathetic efferents (68). These data were confirmed by findings from our group, which showed increased cardiac sympathetic and decreased vagal outflow to the heart in a human model of nitrate tolerance (69).

Of interest, uncoupling of NO synthase and reduced endothelial NO production have been found in the setting of hypercholesterolemia (70), diabetes mellitus (71), and in the setting of other risk factors for atherosclerosis. Therefore, uncoupled NO might represent an important mediator of the progression of coronary artery disease in the presence of these risk factors. Furthermore, nitrate-derived free radicals have been shown to lead to other important modifications, including oxidation of the enzyme prostacyclin synthase, upregulation of ET-1 production, and, importantly, of matrix metalloproteinases, which might have important implications with respect to atherogenesis (72–74).

The modifications in vascular homeostasis that follow this nitrate-induced free radical production could have other implications. Tolerance (and tolerance-associated vascular abnormalities) has been linked to an increased platelet aggregability (75) in vivo and with a loss of the GTN preconditioning-mimetic effect associated with an impairment in the capacity to develop ischemic preconditioning in response to the exposure to brief ischemic events (76). Finally, prolonged nitrate therapy leads to the phenomenon termed *rebound*. This describes the paradoxical increase in the number of episodes of myocardial ischemia (77) and a corresponding decrease in exercise capacity (78) during periods following withdrawal of nitrate therapy. The cause of this phenomenon remains uncertain; however, it has been hypothesized that the increased free radical bioavailability during nitrate therapy would, when left unopposed following nitrate withdrawal, lead to increased vasoconstrictor responses and subsequent reduced exercise capacity (79).

KEY POINTS

- Recent data suggest that a thorough revision of the mechanism(s) of action and of the effects of organic nitrates is necessary. These data suggest that the administration of organic nitrates has potentially important beneficial effects that go beyond any change in hemodynamics. Their antiplatelet effects might have importance in the setting of acute coronary syndromes, and the antiatherosclerotic and preconditioning-mimetic effects of nitrates deserve further attention and investigation. These effects suggest that the important "pleiotropic" effects of endothelial NO can be replaced by exogenous administration of NO donors.
- On the other hand, therapy with organic nitrates for prolonged periods (or, possibly, in high doses, in a time of exposure/administered dosage interaction) leads to toxic effects that are manifested by (but not limited to) nitrate tolerance.
- Given this background, it has to be acknowledged that important pieces of information are still missing: No study has investigated whether chronic therapy with organic nitrates has an influence on cardiovascular (or overall) mortality and morbidity, and published studies are limited to short-term nitrate therapy (30,31).
- Furthermore, it is not clear whether a dose–response relationship exists in the antianginal effects of nitrates: studies in patients with coronary artery disease have demonstrated the absence of an incremental effect for increasing administered dosages of transdermal GTN (80). Because free radicals seem to be produced in a dose-dependent (possibly stoichiometric) fashion in mitochondria (16), the administration of higher nitrate dosages might be associated with a higher incidence of side effects such as the rebound phenomenon and nitrate tolerance, which

are both likely to be mediated by free radical–based mechanisms.

- In sum, although nitrates have been used in clinical practice for more than 100 years, recent findings regarding endothelial biology and the effects of nitrates have raised important questions concerning the safety of sustained nitrate therapy. Recent investigations have yielded more detailed information concerning the mechanisms of nitrate effects and the development of tolerance. These new concepts have demonstrated that nitrate tolerance does not simply represent the loss of hemodynamic effects but is accompanied by evidence of vascular toxicity mediated by increased oxygen free radical production.

REFERENCES

1 Marsh N, Marsh A. A short history of nitroglycerine and nitric oxide in pharmacology and physiology. *Clin Exp Pharmacol Physiol.* 2000;27:313–319.

2 Brunton L. On the use of nitrite of amyl in angina pectoris. *Lancet*. 1867;(ii):20–29.

3 Murrell W. Nitroglycerin as a remedy for angina pectoris. *Lancet.* 1879;(i):80–81.

4 Parker JD, Parker JO. Nitrate therapy for stable angina pectoris. *N Engl J Med.* 1998;338:520–531.

5 Ignarro LJ, Lippton H, Edwards JC, et al. Mechanism of vascular smooth muscle relaxation by organic nitrates, nitrites, nitroprusside and nitric oxide: evidence for the involvement of S-nitrosothiols as active intermediates. *J Pharmacol Exp Ther.* 1981; 218:739–749.

6 Mulsch A, Mordvintcev P, Bassenge E, et al. In vivo spin trapping of glyceryl trinitrate-derived nitric oxide in rabbit blood vessels and organs. *Circulation.* 1995;92:1876–1882.

7 Bredt DS, Hwang PM, Glatt CE, et al. Cloned and expressed nitric oxide synthase structurally resembles cytochrome P-450 reductase. *Nature.* 1991;351:714–718.

8 McDonald BJ, Bennett BM. Cytochrome P-450 mediated biotransformation of organic nitrates. *Can J Physiol Pharmacol.* 1990;68:1552–1557.

9 Ratz JD, McGuire JJ, Anderson DJ, Bennett BM. Effects of the flavoprotein inhibitor, diphenyleneiodonium sulfate, on ex vivo organic nitrate tolerance in the rat. *J Pharmacol Exp Ther.* 2000; 293:569–577.

10 Minamiyama Y, Imaoka S, Takemura S, et al. Escape from tolerance of organic nitrate by induction of cytochrome P450. *Free Radic Biol Med.* 2001;31:1498–1508.

11 Haj-Yehia AI, Benet LZ. In vivo depletion of free thiols does not account for nitroglycerin-induced tolerance: a thiol-nitrate interaction hypothesis as an alternative explanation for nitroglycerin activity and tolerance. *J Pharmacol Exp Ther.* 1996;278:1296–1305.

12 Seth P, Fung HL. Biochemical characterization of a membrane-bound enzyme responsible for generating nitric oxide from nitroglycerin in vascular smooth muscle cells. *Biochem Pharmacol.* 1993;46:1481–1486.

13 Chen Z, Zhang J, Stamler JS. Identification of the enzymatic mechanism of nitroglycerin bioactivation. *Proc Natl Acad Sci USA.* 2002;99:8306–8311.

14 Needleman P, Jakschik B, Johnson EM Jr. Sulfhydryl requirement for relaxation of vascular smooth muscle. *J Pharmacol Exp Ther.* 1973;187:324–331.

15 Sydow K, Daiber A, Oelze M, et al. Central role of mitochondrial aldehyde dehydrogenase and reactive oxygen species in nitroglycerin tolerance and cross-tolerance. *J Clin Invest.* 2004;113:482–489.

16 Daiber A, Oelze M, Coldewey M, et al. Oxidative stress and mitochondrial aldehyde dehydrogenase activity: a comparison of pentaerythritol tetranitrate with other organic nitrates. *Mol Pharmacol.* 2004;66:1372–1382.

17 Khan SA, Higdon NR, Meisheri KD. Coronary vasorelaxation by nitroglycerin: involvement of plasmalemmal calcium-activated K^+ channels and intracellular Ca^{++} stores. *J Pharmacol Exp Ther.* 1998;284:838–846.

18 Bolotina VM, Najibi S, Palacino JJ, et al. Nitric oxide directly activates calcium-dependent potassium channels in vascular smooth muscle. *Nature.* 1994;368:850–853.

19 Gori T, Parker JD. Nitrate tolerance: a unifying hypothesis. *Circulation.* 2002;106:2510–2513.

20 Gori T, Parker JD. The puzzle of nitrate tolerance: pieces smaller than we thought? *Circulation.* 2002;106:2404–2408.

21 Munzel T, Daiber A, Mulsch A. Explaining the phenomenon of nitrate tolerance. *Circ Res.* 2005;97:618–628.

22 Barba G, Mullen MJ, Donald A, MacAllister RJ. Determinants of the response of human blood vessels to nitric oxide donors in vivo. *J Pharmacol Exp Ther.* 1999;289:1662–1668.

23 Fallen EL, Nahmias C, Scheffel A, et al. Redistribution of myocardial blood flow with topical nitroglycerin in patients with coronary artery disease. *Circulation.* 1995;91:1381–1388.

24 Jones CJ, Kuo L, Davis MJ, Chilian WM. Myogenic and flow-dependent control mechanisms in the coronary microcirculation. *Basic Res Cardiol.* 1993;88:2–10.

25 Kojda G, Patzner M, Hacker A, Noack E. Nitric oxide inhibits vascular bioactivation of glyceryl trinitrate: a novel mechanism to explain preferential venodilation of organic nitrates. *Mol Pharmacol.* 1998;53:547–554.

26 Bitar F, Akhter MW, Khan S, et al. Survey of the use of organic nitrates for the treatment of chronic congestive heart failure in the United States. *Am J Cardiol.* 2004;94:1465–1468.

27 Gibbons RJ, Chatterjee K, Daley J, et al. ACC/AHA/ACP-ASIM guidelines for the management of patients with chronic stable angina: executive summary and recommendations. A Report of the American College of Cardiology/American Heart Association Task Force on Practice Guidelines (Committee on Management of Patients with Chronic Stable Angina). *Circulation.* 1999;99: 2829–2848.

28 Graboys TB, Lown B. Cardiology patient page. Nitroglycerin: the "mini" wonder drug. *Am J Cardiol.* 2003;108:E78–E79.

29 Braunwald E, Antman EM, Beasley JW, et al. ACC/AHA guidelines for the management of patients with unstable angina and non-ST-segment elevation myocardial infarction. A report of the American College of Cardiology/American Heart Association Task Force on Practice Guidelines (Committee on the Management of Patients With Unstable Angina). *J Am Coll Cardiol.* 2000;36:970–1062.

30 Gruppo Italiano per lo Studio della Sopravvivenza nell'infarto Miocardico. GISSI-3: effects of lisinopril and transdermal glyceryl trinitrate singly and together on 6-week mortality and ventricular function after acute myocardial infarction. *Lancet.* 1994;343:1115–1122.

31 ISIS-4 (Fourth International Study of Infarct Survival) Collaborative Group. ISIS-4: a randomised factorial trial assessing early oral captopril, oral mononitrate, and intravenous magnesium sulphate in 58,050 patients with suspected acute myocardial infarction. *Lancet.* 1995;345:669–685.

32 Hunt SA, Baker DW, Chin MH, et al. ACC/AHA guidelines for the evaluation and management of chronic heart failure in the adult: executive summary a report of the American College of Cardiology/American Heart Association Task Force on Practice Guidelines (Committee to Revise the 1995 Guidelines for the Evaluation and Management of Heart Failure): developed in collaboration with the International Society for Heart and Lung Transplantation; endorsed by the Heart Failure Society of America. *Am J Cardiol.* 2001;104:2996–3007.

33 Scardi S, Mazzone C. What is the role of nitrates in the treatment of heart decompensation? A rationale for their use. *G Ital Cardiol.* 1999;29:1579–1586.

34 Gewaltig MT, Kojda G. Vasoprotection by nitric oxide: mechanisms and therapeutic potential. *Cardiovasc Res.* 2002;55:250–260.

35 Abrams J. Beneficial actions of nitrates in cardiovascular disease. *Am J Cardiol.* 1996;77:C31–C37.

36 Schafer A, Wiesmann F, Neubauer S, et al. Rapid regulation of platelet activation in vivo by nitric oxide. *Am J Cardiol.* 2004;109: 1819–1822.

37 Pepine C. Endothelial dysfunction and its role in the cycle of cardiovascular disease. *Can J Cardiol.* 1998;14(Suppl D):D5–D7.

38 Muller S, Konig I, Meyer W, Kojda G. Inhibition of vascular oxidative stress in hypercholesterolemia by eccentric isosorbide mononitrate. *J Am Coll Cardiol.* 2004;44:624–631.

39 Kojda G, Hacker A, Noack E. Effects of nonintermittent treatment of rabbits with pentaerythritol tetranitrate on vascular reactivity and superoxide production. *Eur J Pharmacol.* 1998;355:23–31.

40 Ishihara M, Sato H, Tateishi H, et al. Implications of prodromal angina pectoris in anterior wall acute myocardial infarction: acute angiographic findings and long-term prognosis. *J Am Coll Cardiol.* 1997;30:970–975.

41 Murry CE, Jennings RB, Reimer KA. Preconditioning with ischemia: a delay of lethal cell injury in ischemic myocardium. *Circulation.* 1986;74:1124–1136.

42 Lochner A, Marais E, Genade S, Moolman JA. Nitric oxide: a trigger for classic preconditioning? *Am J Physiol Heart Circ Physiol.* 2000;279:H2752–H2765.

43 Heusch G. Nitroglycerin and delayed preconditioning in humans: yet another new mechanism for an old drug? *Circulation.* 2001;103:2876–2878.

44 Takano H, Tang XL, Qiu Y, et al. Nitric oxide donors induce late preconditioning against myocardial stunning and infarction in conscious rabbits via an antioxidant-sensitive mechanism. *Circ Res.* 1998;83:73–84.

45 Leesar MA, Stoddard MF, Dawn B, et al. Delayed preconditioning-mimetic action of nitroglycerin in patients undergoing coronary angioplasty. *Circulation.* 2001;103:2935–2941.

46 Penna C, Pagliaro P, Rastaldo R, et al. F0F1ATP-synthase activity is differently modulated by coronary reactive hyperemia before and after ischemic preconditioning in the goat. *Am J Physiol Heart Circ Physiol.* 2004; 287(5):H2192–200. Epub 2004 Jun 24.

47 Schwarz M, Katz SD, Demopoulos L, et al. Enhancement of endothelium-dependent vasodilation by low-dose nitroglycerin in patients with congestive heart failure. *Circulation.* 1994;89: 1609–1614.

48 Steering Committee, Transdermal Nitroglycerin Cooperative Study. Acute and chronic antianginal efficacy of continuous twenty-four-hour application of transdermal nitroglycerin. *Am J Cardiol.* 1991;68:1263–1273.

49 Jurt U, Gori T, Ravandi A, et al. Differential effects of pentaerythritol tetranitrate and nitroglycerin on the development of tolerance and evidence of lipid peroxidation: a human in vivo study. *J Am Coll Cardiol.* 2001;38:854–859.

50 Munzel T, Kurz S, Heitzer T, Harrison DG. New insights into mechanisms underlying nitrate tolerance. *Am J Cardiol.* 1996;77: 24C–30C.

51 Packer M. What causes tolerance to nitroglycerin? The 100 year old mystery continues [Editorial; comment]. *J Am Coll Cardiol.* 1990;16:932–935.

52 Armstrong PW, Moffat JA. Tolerance to organic nitrates: clinical and experimental perspectives. *Am J Med.* 1983;74:73–84.

53 Munzel T, Sayegh H, Freeman BA, et al. Evidence for enhanced vascular superoxide anion production in nitrate tolerance. A novel mechanism underlying tolerance and cross-tolerance. *J Clin Invest.* 1995;95:187–194.

54 Parker JD, Farrell B, Fenton T, et al. Counter-regulatory responses to continuous and intermittent therapy with nitroglycerin. *Circulation.* 1991;84:2336–2345.

55 Kurz S, Hink U, Nickenig G, et al. Evidence for a causal role of the renin-angiotensin system in nitrate tolerance. *Circulation.* 1999;99(24):3181–3187.

56 Munzel T, Giaid A, Kurz S, et al. Evidence for a role of endothelin 1 and protein kinase C in nitroglycerin tolerance. *Proc Natl Acad Sci USA.* 1995;92:5244–5248.

57 Parker JD, Farrell B, Fenton T, Parker JO. Effects of diuretic therapy on the development of tolerance during continuous therapy with nitroglycerin. *J Am Coll Cardiol.* 1992;20:616–622.

58 Sage PR, de la Lande I, Stafford I, et al. Nitroglycerin tolerance in human vessels: evidence for impaired nitroglycerin bioconversion. *Circulation.* 2000;102:2810–2815.

59 Molina CR, Andresen JW, Rapoport RM, et al. Effect of in vivo nitroglycerin therapy on endothelium-dependent and independent vascular relaxation and cyclic GMP accumulation in rat aorta. *J Cardiovasc Pharmacol.* 1987;10:371–378.

60 Laursen JB, Boesgaard S, Poulsen HE, Aldershvile J. Nitrate tolerance impairs nitric oxide-mediated vasodilation in vivo. *Cardiovasc Res.* 1996;31:814–819.

61 Gori T, Mak SS, Kelly S, Parker JD. Evidence supporting abnormalities in nitric oxide synthase function induced by nitroglycerin in humans. *J Am Coll Cardiol.* 2001;38:1096–1101.

62 Caramori PR, Adelman AG, Azevedo ER, et al. Therapy with nitroglycerin increases coronary vasoconstriction in response to acetylcholine. *J Am Coll Cardiol.* 1998;32:1969–1974.

63 Schulz E, Tsilimingas N, Rinze R, et al. Functional and biochemical analysis of endothelial (dys)function and NO/cGMP signaling in human blood vessels with and without nitroglycerin pretreatment. *Circulation.* 2002;105:1170–1175.

64 Munzel T, Li H, Mollnau H, et al. Effects of long-term nitroglycerin treatment on endothelial nitric oxide synthase (NOS III)

gene expression, NOS III-mediated superoxide production, and vascular NO bioavailability. *Circ Res.* 2000;86:E7–E12.

65 Abou-Mohamed G, Kaesemeyer WH, Caldwell RB, Caldwell RW. Role of L-arginine in the vascular actions and development of tolerance to nitroglycerin. *Br J Pharmacol.* 2000;130: 211–218.

66 Gori T, Burstein JM, Ahmed S, et al. Folic acid prevents nitroglycerin-induced nitric oxide synthase dysfunction and nitrate tolerance: a human in vivo study. *Circulation.* 2001;104: 1119–1123.

67 Parker JO, Parker JD, Caldwell RW, et al. The effect of supplemental L-arginine on tolerance development during continuous transdermal nitroglycerin therapy. *J Am Coll Cardiol.* 2002;39: 1199–1203.

68 Zanzinger J, Czachurski J, Seller H. Impaired modulation of sympathetic excitability by nitric oxide after long-term administration of organic nitrates in pigs. *Circulation.* 1998;97:2352–2358.

69 Gori T, Floras JS, Parker JD. Nitroglycerin blunts the gain of the baroreceptor-heart rate reflex. *Circulation.* 2001;104:II–331.

70 Stroes E, Kastelein J, Cosentino F, et al. Tetrahydrobiopterin restores endothelial function in hypercholesterolemia. *J Clin Invest.* 1997;99:41–46.

71 Hink U, Li H, Mollnau H, et al. Mechanisms underlying endothelial dysfunction in diabetes mellitus. *Circ Res.* 2001;88:E14–E22.

72 Hink U, Oelze M, Kolb P, et al. Role for peroxynitrite in the inhibition of prostacyclin synthase in nitrate tolerance. *J Am Coll Cardiol.* 2003;42:1826–1834.

73 Kahler J, Ewert A, Weckmuller J, et al. Oxidative stress increases endothelin-1 synthesis in human coronary artery smooth muscle cells. *J Cardiovasc Pharmacol.* 2001;38:49–57.

74 Kim D, Rybalkin SD, Pi X, et al. Upregulation of phosphodiesterase 1A1 expression is associated with the development of nitrate tolerance. *Circulation.* 2001;104:2338–2343.

75 Fink B, Bassenge E. Association between vascular tolerance and platelet upregulation: comparison of nonintermittent administration of pentaerithrityltetranitrate and glyceroltrinitrate. *Nitric Oxide.* 2002;40:890–897.

76 Szilvassy Z, Ferdinandy P, Bor P, et al. Loss of preconditioning in rabbits with vascular tolerance to nitroglycerin. *Br J Pharmacol.* 1994;112:999–1001.

77 Pepine CJ, Lopez LM, Bell DM, et al. Effects of intermittent transdermal nitroglycerin on occurrence of ischemia after patch removal: results of the second transdermal intermittent dosing evaluation study (TIDES-II). *J Am Coll Cardiol.* 1997;30:955–961.

78 Parker JD, Parker AB, Farrell B, Parker JO. Intermittent transdermal nitroglycerin therapy. Decreased anginal threshold during the nitrate-free interval. *Circulation.* 1995;91:973–978.

79 Heitzer T, Just H, Brockhoff C, et al. Long-term nitroglycerin treatment is associated with supersensitivity to vasoconstrictors in men with stable coronary artery disease: prevention by concomitant treatment with captopril. *J Am Coll Cardiol.* 1998;31: 83–88.

80 Parker JO, Amies MH, Hawkinson RW, et al. Intermittent transdermal nitroglycerin therapy in angina pectoris. Clinically effective without tolerance or rebound. Minitran Efficacy Study Group. *Circulation.* 1995;91:1368–1374.

182

Therapeutic Approaches to Altering Hemodynamic Forces

José A. Adams

Mount Sinai Medical Center, Miami Beach, Florida

The interaction of the endothelium with blood and its elements, circulating factors, and its active metabolic state make it an attractive target for therapeutic alterations. The location of the endothelium between blood and vascular smooth muscle is ideal for the endothelium to serve as a fine transducer of mechanical and chemical forces. Endothelial dysfunction occurs in several diseases. Many endogenous regulatory responses, such as distribution of blood flow, hemostasis, and vasomotor tone, are controlled by endothelial-derived factors. Thus, exogenous modulation of such endothelial factors is a desirable goal.

The physiological importance of the endothelium in homeostatic function has been recognized since the discovery of endothelial-derived relaxing factor (EDRF), which was later found to be endothelial-derived nitric oxide (NO). The actions of NO on adjacent endothelial cells (ECs), other circulating cells, and on adjacent myocytes make NO a valuable primary or secondary messenger of innumerable physiologic processes. Signal transduction and the importance of endothelial-derived NO and inducible NO in inflammatory processes are reviewed here. This chapter further aims to review primary methods of endothelial stimulation in vivo and therapeutic methods used in the past and currently to alter endothelial responses. Unlike pharmacologic interventions in which drug dosing, local availability, distribution, and side effects can outweigh the potential beneficial effects of the drug, stimulating the endothelium to produce beneficial factors has several advantages: (a) the endothelium is present in every organ, (b) actions are local, and (c) there is virtually no potential for overdosing or toxic effects.

HEMODYNAMIC FACTORS RESPONSIBLE FOR ENDOTHELIAL OUTPUT

The frictional force engendered on the vascular endothelium by the flowing, viscous blood is termed *hemodynamic shear stress.* Shear stress on a blood vessel can be summarized using the following equation: (Poiseuille's Law)

$$\tau_s = 4\mu Q/\pi R^3 \qquad \text{(Equation 1)}$$

Where μ is the viscosity of blood, Q is flow, and R is the radius of the vessel. Shear stress is directly proportional to blood flow viscosity and inversely proportional to R^3.

Shear stress is a critical determinant of vessel caliber and plays a role in vascular remodeling (1–4). Two main hemodynamic factors are responsible for endothelial release of mediators: (1) laminar shear stress and (2) pulsatile shear stress. Laminar shear stress is the tangential deformation produced by blood flow across the ECs. Mechanical forces of laminar shear stress cause release of endothelial NO, prostaglandins (PGs), upregulation of vascular endothelial growth factor receptor 2 (VEGFR2) and tissue-type plasminogen activator (t-PA) in cell culture studies, and vascular endothelial growth factor (VEGF) in human clinical studies (5–9). Laminar shear stress alters expression at both the protein and mRNA level (1,10–14). The mean shear stress under physiologic conditions to which the vascular endoluminal surface is exposed to is remarkably constant: These have been reported to be in the range of 10 to 15 dyne/cm^2 irrespective of the arterial network considered and size of the animal. The exception appears to be the rat and mouse, for which the values reported are closer to 30 to 35 dyn/cm^2 (15,16). Pulsatile laminar shear stress relates to the stress induced on the endothelium by the pulsation and wave reflection of the heartbeat. Amplitude and frequency of pulsations are important stresses for generating endothelial-derived NO (17,18).

Laminar shear stress is promoted by either decreasing the radius of the blood vessel or increasing fluid viscosity. Turbulent shear stress or disturbed flow may occur in atherosclerosis or hypertension, with vascular prostheses or indwelling catheters. Here, endothelial production of NO is reduced and endothelin (ET)-1 increased (19–24).

Pulsatile shear stress, which is the force on ECs generated with each contraction of the heart, has two distinct components that affect the endothelium: frequency and amplitude. The endothelial production of endothelial-derived NO appears to be frequency encoded in the range of 2 to 6 Hz. The amplitude of each pulsation or pressure pulse also has a positive or negative effect on endothelial output (17,18,25,26). A direct relationship exists between the circumferential stress to which the vessel is exposed and the structure of the vessel wall as a form of adaptive mechanism. From one animal species to another the number of lamellar units varies according to the diameter of the blood vessel so that the circumferential stress remains constant irrespective of the size of the animal (16,27,28) (see also Chapter 28).

Circulating levels of cytokines, chemokines, and reactive oxygen species suppress production of endothelial-derived NO (29,30). Thus, sepsis-induced production of cytokines and inflammatory mediators cause both local and distal endothelial dysfunction. This causes pulmonary vasoconstriction and disruption of capillary integrity with subsequent interstitial edema (29–38). Excellent reviews of endothelial mechanotransduction mechanisms can be found in the literature (1,15,16,39).

HEMODYNAMIC ALTERATIONS

Accounts of physical activity and inactivity producing diseases can be found in the medical literature (40–50). Although not typically recognized as such, physical activity to varying degrees increases shear stress to the endothelium. Increase in cardiac frequency is a powerful stimulus for preconditioning the heart from ischemia–reperfusion injury. This cardioprotection is mediated in part by NO (51–56). Using an exercise protocol in dogs, Vegh and associates showed that a single bout of exercise prior to coronary occlusion significantly decreased its consequences, presumably because exercise increases endothelial-derived NO, because administration of N(G)-nitro-L-arginine methyl ester (L-NAME) blunted the response (57). Further, in human subjects undergoing ventricular pacing, release of plasminogen activator inhibitor (PAI)-1, von Willebrand factor (vWF) antigen, t-PA, and prostacyclin have been documented in the coronary vessel bloodstream (58). In isolated perfused vessels, the frequency range that induces production of endothelial-derived NO production is encoded within a range of 2 to 6 Hz and may be developmental in origin (17,59). The endothelium also responds to amplitude changes of the pulse waveform. Increasing amplitude of pulse pressure while maintaining the flow rate unchanged in an isolated vessel augments vascular constriction. The vasodilatory frequency response is inhibited by the nonselective inhibitor of NO synthase L-NAME (60).

Green and colleagues found that in human subjects, forearm vascular response (a measure of endothelial function, associated with NO release) increased with increased amplitude of blood pressure. Further, this shear stress–mediated vascular response involved an endothelial source of NO (61). Human and animal investigations during the early research period of the National Aeronautics and Space Administration (NASA) explored the effects of whole-body periodic acceleration/sinusoidal vibration in humans and animals. Such interventions increased ventilation in animals and humans and also added pulsations to the circulation without increasing the heart rate. During whole-body x-axis sinusoidal vibration (movement of the entire body headward–footward), blood pressure decreases and blood flow increases (62–64). Although such effects were initially thought to be mediated via the autonomic nervous system, a much different interpretation of these data can be provided today with the knowledge that vascular endothelium responds to pulsatile shear stress with release of NO and subsequent vasodilatation.

In the early 1970s, Arntzenius and colleagues showed that whole-body acceleration applied synchronously with the heartbeat (BASH) had the potential to serve as a cardiac assist device. These investigators proposed a platform that moved the body synchronous with the heartbeat in one direction, either headward or footward. When applied headwardly early in systole, it lowered systolic blood pressure (decrease afterload) and increased early diastolic pressure (increased coronary filling) (65–71).

Pulsatile as contrasted to steady blood flow in extracorporeal membrane oxygenation circuits and cardiopulmonary bypass improves perfusion to vital organs, enhances oxygen utilization by the vital organs, provides better distribution of blood flow to the tissues during warming and cooling, and decreases lactic acid and systemic vascular resistance (72–74).

THERAPEUTIC EXAMPLES OF HEMODYNAMIC ALTERATIONS

Exercise

The beneficial effects of exercise on health have been appreciated for centuries (see also Chapter 56). A large body of literature cites both beneficial effects as well as deleterious effects, particularly with strenuous exercise. How and why exercise is beneficial from the perspective of endothelial function has only recently come to light. Even when normalizing exercise as a percent of maximal predicted heart rate for different exercise types, the results are complex. Bicycling increases cardiac output and heart rate, but its effects on the endothelium appear to be mediated by shear stress (primarily through changes in cardiac output) and to a lesser extent by increased pulsatility (except at high exercise loads, heart rate is insufficiently increased to play a major role). Jogging and running increase both cardiac output and heart rate, but with every step the foot strikes the ground, transmitting an additional pulse that is superimposed on the natural pulse. Therefore, jogging and running as well as jumping rope and bouncing on a trampoline have the potential of greater upregulation of endothelial NO synthase (eNOS).

Moderate exercise improves endothelial function and, through endothelial release of NO, prevents and treats atherosclerosis. The beneficial effect of NO relates, in part, to its antioxidant and and anti-inflammatory activities. In fact, regular moderate-intensity exercise can reverse endothelial dysfunction (26,75–81).

On the other hand, strenuous exercise stresses the body with subsequent activation of nuclear factor (NF)-κB, inducible NO synthase (iNOS) from leukocytes and ECs, and causes the accumulation of oxygen free radicals from leukocytes. Prolonged production of large quantities of NO from iNOS activation is deleterious to the tissues because it acts as an oxidant (82–90).

Based on the above considerations, moderate exercise may be beneficial to those diseases that have chronic inflammation as the basis for their symptoms, whereas strenuous exercise may worsen the symptoms. Another limitation of exercise relates to the patient's ability to perform. This is a problem in patients with such chronic inflammatory diseases as multiple sclerosis, Alzheimer disease, Parkinson disease, stroke, fibromyalgia, and arthritis.

Enhanced External Counterpulsation

The use of multiple bladder-like pressure cuffs around the lower extremities, which inflate in synchrony with the heartbeat in diastole and deflate during systole, dates back to the 1960s, with the work of W. C. Birtwell, who employed a catheter-based system of blood withdrawal during systole and reinjection in diastole (91). In 1968, an external counterpulsation apparatus became commercially available. In 1999, the Multicenter Study (MUST)-EECP demonstrated that enhanced external counterpulsation (EECP) was effective for the treatment of chronic angina pectoris in a randomized clinical trial (92). It was originally hypothesized that diastolic augmentation relieved angina through mechanically opening coronary blood vessels during diastole. Although this may be a valid mechanism of action, it seems more likely that pulsatile shear stress to the endothelium with increased release of NO is the major basis for EECP-mediated relief of angina.

Isolated perfused vessel experiments by Kamm and his group have shown increases in *eNOS* mRNA and NO produced during application of a modified EECP to isolated vessels (93,94). Further, the highest level of *eNOS* mRNA expression was observed in a group of blood vessels that were pulsed in addition to EECP (93–95). Additionally, human data have shown an increase in plasma nitrates after EECP, which continues to rise after 1 month of treatment (7). ET-1, an endothelial-derived counter-regulatory peptide that produces intense vasoconstriction, decreases during EECP in humans (96). In humans undergoing EECP, various angiogenic factors also have been shown to increase after treatment, such as hepatocyte growth factor, basic fibroblast growth factor (bFGF), and VEGF (7). Thus, it appears that application of EECP via changes in pulsatile shear stress enhance endothelial production of beneficial mediators in both in vivo and in vitro studies (7,93,94). EECP reduces anginal episodes, increases exercise times, and improves health-related quality of life scores (vitality, bodily pain, general health, physical functioning, social function, and others) (97). Although EECP has been limited to cardiovascular health–related pathologies, it would not be surprising that its use might extend to include other diseases with endothelial dysfunction and/or chronic inflammatory diseases processes. The effects observed with EECP are likely to be complementary to current therapy in various diseases. Investigators have reported that some of the beneficial effects of EECP in terms of mediator release can be measured for as long as 1 month after cessation of treatments (7). Adverse events reported in the MUST-EECP data set were in the order of 54.9%, with the most frequent being skin abrasion, bruise or blisters, and pain in the leg or back (92). Although the device is readily available, its cost and limited U.S. Food and Drug Administration (FDA)-approved indications have likely stifled more widespread use.

Intraaortic Balloon Pump Counterpulsation

The principal phase of coronary arterial perfusion is during cardiac diastole. Increasing diastolic aortic pressure improves coronary blood flow. In 1953, Kantrowitz and colleagues validated the concept of diastolic augmentation through the intraaortic balloon pump (IABP), an invasive means of counterpulsation (98). The IABP consists of a vascular indwelling balloon catheter within the aorta, with inflation and deflation of the balloon synchronized with diastole and systole, respectively. This device has been in clinical use since the early 1960s as a cardiac assist device in circulatory failure (99). Its primary use has been as a support mechanism. The most common indications include hemodynamic support during or after cardiac catheterization, cardiogenic shock, weaning from cardiopulmonary bypass, preoperative employment in high-risk coronary artery bypass graft (CABG) surgery, and unstable refractory angina. Although the reported benefits of this device are widely attributed to augmentation of diastolic filling and reduction of cardiac afterload, it is likely that the added pulse during diastole increases shear stress to the endothelium. This would likely tend to promote release of NO and other shear stress–dependent mediators from the endothelium. Major complications when using IABP include embolism, limb ischemia, infarction, and balloon rupture. In addition to its invasive nature, the complication rates from 2% to 11% render this method of increasing endothelial-derived mediators generally unappealing (99–107).

Whole-Body Periodic Acceleration

Motion of the body in a headward–footward direction imparted in a sinusoidal manner to a supine subject (in the z plane) can be produced by a motion platform. This motion, termed *periodic acceleration* (pGz), adds a pulse to the circulation as the body accelerates and decelerates with the frequency of the motion platform. These added pulses cause release of endothelial-derived mediators in isolated blood vessels, animal models, and humans, presumably due to increased pulsatile shear stress (10,83,108–110). The motion is carried out at frequencies of 2 to 4 Hz with displacements of 1 to 5 cm producing accelerations of ± 0.2 to 0.6 Gz. Initial work on pGz was aimed at using this device for noninvasive ventilation by causing passive diaphragmatic excursions. Our laboratory has shown that pGz can adequately ventilate paralyzed animals with normal lungs as well as diseased lungs induced by meconium aspiration (111,112).

During pGz ventilation in animals with meconium aspiration syndrome, characteristically associated with pulmonary hypertension, it was found that pulmonary artery pressure decreased after treatment (111). Further, in animals with normal vascular tone, a slight decrease exists in mean arterial pressure and pulmonary artery pressure, which could be eliminated with infusion of L-NAME, a drug that blocks NO production (108). Laminar and pulsatile flow within an isolated blood vessel increased nitrite levels in the perfusate, a metabolite of NO that reflects NO, which cannot be readily measured because it is metabolized within a few seconds. Nitrite levels are significantly greater during pulsatile than laminar flow. Addition of pGz to an isolated blood vessel during pulsatile flow increased NO production an additional threefold (108).

In whole-animal models, pGz has been shown to enhance regional blood flow to the heart, brain, kidneys, liver, and gut, but not to skeletal muscle (113). This contrasts to dynamic exercise where blood flow decreases to the kidneys, gut, and liver with modest or no increase in brain blood flow in favor of large increases of cardiac and skeletal muscle blood flow. Normal animals exposed to 1 hour of pGz produce increases in vasoactive factors (NO, prostaglandin E_2, and prostacyclin) and fibrinolytic factors (t-PA, PAI-1, and D-dimers) (10). No significant changes occur in thrombin and prothrombin times, activated plasma thromboplastin time, fibrinogen, factors VII and VIII, serum cortisol, and epinephrine/norepinephrine (10).

In a ventricular fibrillation animal model, pGz has been utilized as a sole method of cardiopulmonary resuscitation (CPR). A comparison of pGz CPR to an automated chest compression device (Thumper-CPR) indicates that animals undergoing pGz CPR show a higher rate of return to spontaneous circulation and significantly better postresuscitation cardiac function as assessed with echocardiography (114–117). The improved postresuscitation function is probably related to shear stress–induced release of NO and prostaglandin during CPR. Thus, pGz produces a conditioning effect during CPR.

In an animal model of asthma, pretreatment with pGz significantly decreases the antigen-induced immediate bronchoconstrictor response and eliminates the late response (118). Further, the elimination of the late response is associated with suppression of NF-κB in the pulmonary lavage fluid. Thus, endothelial-derived NO released with pGz probably interferes with degranulation of mast cells to account for its suppression of the immediate bronchoconstrictor response and results in NF-κB–mediated inhibition of the late inflammatory bronchoconstrictor response in asthma (118).

Can whole-body pGz be utilized for therapeutic purposes in humans? The answer is yes! pGz is analogous to what mothers do when they push a baby in a carriage back and forth or gently bounce the baby from head to foot when the baby is in a vertical posture with the head and neck supported. The rate that a mother pushes a carriage back and forth ranges from 1 to 7 Hz, with accelerations of between ± 0.2 to 0.9 Gz. The higher frequencies and Gz occur when the baby is uncontrollably crying and do not last for more than 5 seconds. Normal animals exposed to 3 hours of continuous pGz have normal behavioral scores 24 and 48 hours post-pGz application. A human motion platform (AT-101, Acceleration Therapeutics, North Bay Village, FL) has been used successfully in adults who tolerate pGz well. Serum nitrite increases during pGz in healthy subjects and patients (109). Changes of the digital pulse waveform occur during pGz, which provides a measure of the vasoactivity of pGz in an analogous way to administration of organic nitrates. Whole-body pGz shows more profound changes than nitrates and is therefore likely to be more potent in its bioactivity. In normal subjects, whole-body pGz, in terms of vasoactivity, produces an effect that falls between light and moderate exercise – for example, 56% and 67% of maximum predicted heart rate (83).

Clinical benefits of pGz in humans have also been observed in patients with chronic inflammatory diseases. After ten to fifteen 45-minute session of pGz, there was marked improvement in the Role Physical, Bodily Pain, and Vitality categories of the SF36 health-related quality-of-life questionnaire (83,110).

Thus, accumulating evidence suggests that pGz is well tolerated and safe in humans, that it causes release of NO, and that chronically beneficial clinical effects are observed in individuals with diverse inflammatory/oxidative stress diseases and endothelium dysfunction. It is unclear for how long the benefits observed with pGz persist after discontinuation of treatments.

CLINICAL IMPLICATIONS

Based on the information provided previously, and on the response of the endothelium to changes in pulsatile shear stress, significant opportunities to improve health clearly exist. Exercise has been shown to be very effective in a multitude of cardiovascular diseases, from atherosclerosis to inflammatory diseases, and its health benefits cannot be underestimated. The commonality of many of these diseases is endothelial dysfunction, thus a method that prevents, resolves, or treats endothelial dysfunction is clearly needed. For patients who cannot exercise, or cannot appropriately exercise, pGz and other methods of increasing pulsatile shear stress may offer a potential solution.

Various disease processes have a commonality: inflammation. How the inflammatory response is triggered can vary, but ultimately a common endpoint of many inflammatory processes is upregulation of the transcriptional protein NF-κB. NO can downregulate NF-κB and is the subject of therapeutic interventions and research into control of inflammation (87). The ability of pulsatile shear stress to induce NO can be utilized to achieve this beneficial effect.

Prostaglandins and endothelial-derived NO have both been shown to be important in preconditioning for an ischemic/reperfusion event (119–126). Exercise preconditioning has been shown to be effective in animal models in reducing ischemia reperfusion injury; whether this can be translated to humans is likely possible.

A general understanding that the endothelium represents a target that can be stimulated by hemodynamic alterations, that responds in a potent manner, that is localized in every single organ in the body, and in which endogenous production of any of its factors occurs at the local level, leads to a greater vision of potential targets for treatment using induced pulsatile shear stress. The ultimate challenge will be to refine these methods of "tuning" the endothelium to produce the desired effect, systemically and locally.

KEY POINTS

- The endothelium responds to laminar and pulsatile shear stress, with an output that includes among many other factors, including NO, prostaglandins, t-PA, and VEGF.
- Addition of pulsatile shear stress to isolated vessels, whole-animal models, and humans further enhances production of these factors.
- Clinical methods to increase pulsatile shear stress are feasible and safe in humans.

ACKNOWLEDGMENTS

I am grateful to Marvin Sackner, MD, for his editorial comments and guidance. I also thank the Miami Heart Research Institute for their support of this project.

REFERENCES

1 Barakat A, Lieu D. Differential responsiveness of vascular endothelial cells to different types of fluid mechanical shear stress. *Cell Biochem Biophys.* 2003;38(3):323–343.

2 Cullen JP, Sayeed S, Sawai RS, et al. Pulsatile flow-induced angiogenesis: role of G(i) subunits. *Arterioscler Thromb Vasc Biol.* 2002;22(10):1610–1616.

3 Laughlin MH, McAllister RM. Exercise training-induced coronary vascular adaptation. *J Appl Physiol.* 1992;73(6):2209–2225.

4 Prior BM, Yang HT, Terjung RL. What makes vessels grow with exercise training? *J Appl Physiol.* 2004;97(3):1119–1128.

5 Arora R, Chen HJ, Rabbani L. Effects of enhanced counterpulsation on vascular cell release of coagulation factors. *Heart Lung.* 2005;34(4):252–256.

6 Conklin BS, Zhong DS, Zhao W, et al. Shear stress regulates occludin and VEGF expression in porcine arterial endothelial cells. *J Surg Res.* 2002;102(1):13–21.

7 Feldman AM. Enhanced external counterpulsation: mechanism of action. *Clin Cardiol.* 2002:II11–II15.

8 Passerini AG, Milsted A, Rittgers SE. Shear stress magnitude and directionality modulate growth factor gene expression in preconditioned vascular endothelial cells. *J Vasc Surg.* 2003;37(1): 182–190.

9 Urbich C, Stein M, Reisinger K, et al. Fluid shear stress-induced transcriptional activation of the vascular endothelial growth factor receptor-2 gene requires Sp1-dependent DNA binding. *FEBS Lett.* 2003;535(1–3):87–93.

10 Adams JA, Bassuk J, Wu D, et al. Periodic acceleration: effects on vasoactive, fibrinolytic and coagulation factors. *J Appl Physiol.* 2004.

11 Burnstock G. Release of vasoactive substances from endothelial cells by shear stress and purinergic mechanosensory transduction. *J Anat.* 1999;194(Pt 3):335–342.

12 Chen BP, Li YS, Zhao Y, et al. DNA microarray analysis of gene expression in endothelial cells in response to 24-h shear stress. *Physiol Genomics.* 2001;7(1):55–63.

13 Ohura N, Yamamoto K, Ichioka S, et al. Global analysis of shear stress-responsive genes in vascular endothelial cells. *J Atheroscler Thromb.* 2003;10(5):304–313.

14 Osanai T, Fujita N, Fujiwara N, et al. Cross talk of shear-induced production of prostacyclin and nitric oxide in endothelial cells. *Am J Physiol Heart Circ Physiol.* 2000;278(1):H233–H238.

15 Lehoux S, Tedgui A. Signal transduction of mechanical stresses in the vascular wall. *Hypertension.* 1998;32(2):338–345.

16 Lehoux S, Tedgui A. Cellular mechanics and gene expression in blood vessels. *J Biomech.* 2003;36(5):631–643.

17 Hutcheson IR, Griffith TM. Release of endothelium-derived relaxing factor is modulated both by frequency and amplitude of pulsatile flow. *Am J Physiol.* 1991;261(1 Pt 2):H257–H262.

18 Ishida T, Takahashi M, Corson MA, Berk BC. Fluid shear stress-mediated signal transduction: how do endothelial cells transduce mechanical force into biological responses? *Ann NY Acad Sci.* 1997;811:12–23.

19 Cheng C, de Crom R, van Haperen R, et al. The role of shear stress in atherosclerosis: action through gene expression and inflammation? *Cell Biochem Biophys.* 2004;41(2):279–294.

20 Cunningham KS, Gotlieb AI. The role of shear stress in the pathogenesis of atherosclerosis. *Lab Invest.* 2005;85(1):9–23.

21 Moore JE Jr, Xu C, Glagov S, et al. Fluid wall shear stress measurements in a model of the human abdominal aorta: oscillatory behavior and relationship to atherosclerosis. *Atherosclerosis.* 1994;110(2):225–240.

22 Nazemi M, Kleinstreuer C, Archie JP, Sorrell FY. Fluid flow and plaque formation in an aortic bifurcation. *J Biomech Eng.* 1989;111(4):316–324.

23 Schaper W, Scholz D. Factors regulating arteriogenesis. *Arterioscler Thromb Vasc Biol.* 2003;23(7):1143–1151.

24 Schwartz CJ, Kelley JL, Nerem RM, et al. Pathophysiology of the atherogenic process. *Am J Cardiol.* 1989;64(13):G23–G30.

25 Hutcheson IR, Griffith TM. Mechanotransduction through the endothelial cytoskeleton: mediation of flow- but not agonist-induced EDRF release. *Br J Pharmacol.* 1996;118(3):720–726.

26 Stefano GB, Salzet M, Magazine HI. Cyclic nitric oxide release by human granulocytes, and invertebrate ganglia and immunocytes: nano-technological enhancement of amperometric nitric oxide determination. *Med Sci Monit*. 2002;1908:BR199–BR204.
27 Wolinsky H, Glagov S. Structural basis for the static mechanical properties of the aortic media. *Circ Res*. 1964;14:400–413.
28 Wolinsky H, Glagov S. Comparison of abdominal and thoracic aortic medial structure in mammals. Deviation of man from the usual pattern. *Circ Res*. 1969;25(6):677–686.
29 Henneke P, Golenbock DT. Innate immune recognition of lipopolysaccharide by endothelial cells. *Crit Care Med*. 2002;30 (Suppl 5):S207–S213.
30 Terada LS. Oxidative stress and endothelial activation. *Crit Care Med*. 2002;30(Suppl 5):S186–S191.
31 Bannerman DD, Goldblum SE. Mechanisms of bacterial lipopolysaccharide-induced endothelial apoptosis. *Am J Physiol Lung Cell Mol Physiol*. 2003;284(6):L899–L914.
32 Bradley JR, Wilks D, Rubenstein D. The vascular endothelium in septic shock. *J Infect*. 1994;28(1):1–10.
33 Cuschieri J, Gourlay D, Garcia I, et al. Modulation of endotoxin-induced endothelial function by calcium/calmodulin-dependent protein kinase. *Shock*. 2003;20(2):176–182.
34 Cuschleri J, Gourlay D, Garcia I, et al. Endotoxin-induced endothelial cell proinflammatory phenotypic differentiation requires stress fiber polymerization. *Shock*. 2003;19(5):433–439.
35 Hirano S, Rees RS, Yancy SL, et al. Endothelial barrier dysfunction caused by LPS correlates with phosphorylation of HSP27 in vivo. *Cell Biol Toxicol*. 2004;20(1):1–14.
36 Levi M, ten Cate H, van der PT. Endothelium: interface between coagulation and inflammation. *Crit Care Med*. 2002;30(Suppl 5):S220–S224.
37 Parent C, Eichacker PQ. Neutrophil and endothelial cell interactions in sepsis. The role of adhesion molecules. *Infect Dis Clin North Am*. 1999;13(2):427–447, x.
38 Peters K, Unger RE, Brunner J, Kirkpatrick CJ. Molecular basis of endothelial dysfunction in sepsis. *Cardiovasc Res*. 2003; 60(1):49–57.
39 Davies PF. Flow-mediated endothelial mechanotransduction. *Physiol Rev*. 1995;75(3):519–560.
40 Esposito K, Giugliano D. The metabolic syndrome and inflammation: association or causation? *Nutr Metab Cardiovasc Dis*. 2004;14(5):228–232.
41 Biolo G, Ciocchi B, Stulle M, et al. Metabolic consequences of physical inactivity. *J Ren Nutr*. 2005;15(1):49–53.
42 Roberts CK, Barnard RJ. Effects of exercise and diet on chronic disease. *J Appl Physiol*. 2005;98(1):3–30.
43 Garrett NA, Brasure M, Schmitz KH, et al. Physical inactivity: direct cost to a health plan. *Am J Prev Med*. 2004;27(4):304–309.
44 Suvorava T, Lauer N, Kojda G. Physical inactivity causes endothelial dysfunction in healthy young mice. *J Am Coll Cardiol*. 2004;44(6):1320–1327.
45 Physical inactivity a leading cause of disease and disability, warns WHO. *J Adv Nurs*. 2002;39(6):518.
46 Booth FW, Chakravarthy MV, Spangenburg EE. Exercise and gene expression: physiological regulation of the human genome through physical activity. *J Physiol*. 2002;543(Pt 2):399–411.
47 Kohl HW III. Physical activity and cardiovascular disease: evidence for a dose response. *Med Sci Sports Exerc*. 2001;33(Suppl 6):S472–S483.
48 Kjaer M. Physical inactivity is an underestimated risk factor for development of morbidity and mortality. *Scand J Med Sci Sports*. 2000;10(5):247–248.
49 Physical activity and cardiovascular health. NIH Consensus Development Panel on Physical Activity and Cardiovascular Health. *JAMA*. 1996;276(3):241–246.
50 Blair SN. Physical inactivity and cardiovascular disease risk in women. *Med Sci Sports Exerc*. 1996;28(1):9–10.
51 Hearse DJ, Ferrari R, Sutherland FJ. Cardioprotection: intermittent ventricular fibrillation and rapid pacing can induce preconditioning in the blood-perfused rat heart. *J Mol Cell Cardiol*. 1999;31(11):1961–1973.
52 Kis A, Vegh A, Papp JG, Parratt JR. Repeated cardiac pacing extends the time during which canine hearts are protected against ischaemia-induced arrhythmias: role of nitric oxide. *J Mol Cell Cardiol*. 1999;31(6):1229–1241.
53 Parratt JR, Vegh A. Delayed protection against ventricular arrhythmias by cardiac pacing. *Heart*. 1997;78(5):423–425.
54 Parratt JR, Vegh A, Kaszala K, Papp JG. Protection by preconditioning and cardiac pacing against ventricular arrhythmias resulting from ischemia and reperfusion. *Ann NY Acad Sci*. 1996; 793:98–107.
55 Takeda S, Satoh T, Osada M, et al. Protective effect of pacing on reperfusion-induced ventricular arrhythmias in isolated rat hearts. *Can J Cardiol*. 1995;11(7):573–579.
56 Vegh A, Szekeres L, Parratt JR. Transient ischaemia induced by rapid cardiac pacing results in myocardial preconditioning. *Cardiovasc Res*. 1991;25(12):1051–1053.
57 Hajnal A, Nagy O, Litvai A, et al. Nitric oxide involvement in the delayed antiarrhythmic effect of treadmill exercise in dogs. *Life Sci*. 2005;77(16):1960–1971.
58 Gossinger HD, Speiser W, Siostrzonek P, et al. Pacing-induced myocardial ischemia does not affect the endothelial release of coagulant and fibrinolytic factors into the coronary circulation. *Clin Cardiol*. 1991;14(3):250–256.
59 Wedgwood S, Mitchell CJ, Fineman JR, Black SM. Developmental differences in the shear stress-induced expression of endothelial NO synthase: changing role of AP-1. *Am J Physiol Lung Cell Mol Physiol*. 2003;284(4):L650–L662.
60 Shimoda LA, Norins NA, Madden JA. Responses to pulsatile flow in piglet isolated cerebral arteries. *Pediatr Res*. 1998;1943: 514–520.
61 Green D, Cheetham C, Henderson C, et al. Effect of cardiac pacing on forearm vascular responses and nitric oxide function. *Am J Physiol Heart Circ Physiol*. 2002;283(4):H1354–H1360.
62 Clark JG, Williams JD, Hood WB Jr, Murray RH. Initial cardiovascular response to low frequency whole body vibration in humans and animals. *Aerosp Med*. 1967;38(5):464–467.
63 Hood WB Jr, Higgins LS. Circulatory and respiratory effects of whole-body vibration in anesthetized dogs. *J Appl Physiol*. 1965;6(20):1157–1162.
64 Hood WB Jr, Murray RH, Urschel CW, et al. Cardiopulmonary effects of whole-body vibration in man. *J Appl Physiol*. 1966;21(6):1725–1731.
65 Arntzenius AC, Koops J, Rodrigo F, Elsbach H. Circulatory effects of body acceleration given synchronously with the heartbeat. *Br Heart J*. 1969;31(6):793.
66 Arntzenius AC, Verdouw PD. Cardiovascular responses in pigs to body acceleration applied synchronously with the heartbeat. *Bibl Cardiol*. 1971;Suppl 27:44–52.
67 Arntzenius AC, Laird JD, Huisman PH. BASH – the fundamental questions it raises. *Bibl Cardiol*. 1973;30:1–8.

68 Bhattacharya A, Knapp CF, McCutcheon EP, Evans JM. Modification of cardiac function by synchronized oscillating acceleration. *J Appl Physiol*. 1979;47(3):612–620.
69 Bhattacharya A, Knapp CF, McCutcheon EP, Evans JM. Cardiac responses of dogs to nonsynchronous and heart synchronous whole-body vibration. *J Appl Physiol*. 1979;46(3):549–555.
70 Bhattacharya A, McCutcheon EP, Shvartz E, Greenleaf JE. Body acceleration distribution and O2 uptake in humans during running and jumping. *J Appl Physiol*. 1980;49(5):881–887.
71 Tyberg JV, Parmley WW, Salzman SH, Swan HJ. Hemodynamic effects of BASH. Body acceleration synchronous with the heartbeat. *Bibl Cardiol*. 1972;29:6–13.
72 Berryessa R, Hydrick D, McCormick J, Tyndal CM. Refinements in pediatric perfusion. *J Extra Corpor Technol*. 1986;18(2):140–144.
73 Casper M. Pulsatile flow during cardiopulmonary bypass: is it beneficial? *The J Extra Corpor Technol*. 1988;20(1):25–31.
74 McCormick JS, Berryessa RG, Clark DR, et al. Lactic acid generation during pediatric cardiopulmonary bypass: a comparison of blood and crystalloids primes. *The J Extra Corpor Technol 26th Proceedings*. 1988:84–88.
75 Watts K, Beye P, Siafarikas A, et al. Exercise training normalizes vascular dysfunction and improves central adiposity in obese adolescents. *J Am Coll Cardiol*. 2004;43(10):1823–1827.
76 Walther C, Gielen S, Hambrecht R. The effect of exercise training on endothelial function in cardiovascular disease in humans. *Exerc Sport Sci Rev*. 2004;32(4):129–134.
77 Niebauer J, Clark AL, Webb-Peploe KM, et al. Home-based exercise training modulates pro-oxidant substrates in patients with chronic heart failure. *Eur J Heart Fail*. 2005;7(2):183–188.
78 Higashi Y, Yoshizumi M. Exercise and endothelial function: role of endothelium-derived nitric oxide and oxidative stress in healthy subjects and hypertensive patients. *Pharmacol Ther*. 2004;102(1):87–96.
79 Green DJ, Maiorana A, O'Driscoll G, Taylor R. Effect of exercise training on endothelium-derived nitric oxide function in humans. *J Physiol*. 2004;561(Pt 1):1–25.
80 Gill JM, Al Mamari A, Ferrell WR, et al. Effects of prior moderate exercise on postprandial metabolism and vascular function in lean and centrally obese men. *J Am Coll Cardiol*. 2004;44(12):2375–2382.
81 Gielen S, Erbs S, Schuler G, Hambrecht R. Exercise training and endothelial dysfunction in coronary artery disease and chronic heart failure. From molecular biology to clinical benefits. *Minerva Cardioangiol*. 2002;50(2):95–106.
82 Niebauer J, Cooke JP. Cardiovascular effects of exercise: role of endothelial shear stress. *J Am Coll Cardiol*. 1996;28(7):1652–1660.
83 Sackner MA, Gummels E, Adams JA. Nitric oxide is released into circulation with whole-body, periodic acceleration. *Chest*. 2005;127(1):30–39.
84 Sen CK. Antioxidants in exercise nutrition. *Sports Med*. 2001; 31(13):891–908.
85 Vider J, Lehtmaa J, Kullisaar T, et al. Acute immune response in respect to exercise-induced oxidative stress. *Pathophysiology*. 2001;7(4):263–270.
86 Vider J, Laaksonen DE, Kilk A, et al. Physical exercise induces activation of NF-kappaB in human peripheral blood lymphocytes. *Antioxid Redox Signal*. 2001;3(6):1131–1137.
87 Stefano GB, Prevot V, Cadet P, Dardik I. Vascular pulsations stimulating nitric oxide release during cyclic exercise may benefit health: a molecular approach (review). *Int J Mol Med*. 2001;1907:119–129.
88 Niess AM, Sommer M, Schneider M, et al. Physical exercise-induced expression of inducible nitric oxide synthase and heme oxygenase-1 in human leukocytes: effects of RRR-alpha-tocopherol supplementation. *Antioxid Redox Signal*. 2000;2(1): 113–126.
89 Niess AM, Sommer M, Schlotz E, et al. Expression of the inducible nitric oxide synthase (iNOS) in human leukocytes: responses to running exercise. *Med Sci Sports Exerc*. 2000;32(7): 1220–1225.
90 Yang AL, Tsai SJ, Jiang MJ, et al. Chronic exercise increases both inducible and endothelial nitric oxide synthase gene expression in endothelial cells of rat aorta. *J Biomed Sci*. 2002;9(2):149–155.
91 DeMaria AN. A historical overview of enhanced external counterpulsation. *Clin Cardiol*. 2002;25(12 Suppl 2):II3–II5.
92 Arora RR, Chou TM, Jain D, et al. The Multicenter Study of Enhanced External Counterpulsation (MUST-EECP): effect of EECP on exercise-induced myocardial ischemia and anginal episodes. *J Am Coll Cardiol*. 1999;33(7):1833–1840.
93 Dai G, Tsukurov O, Orkin RW, et al. An in vitro cell culture system to study the influence of external pneumatic compression on endothelial function. *J Vasc Surg*. 2000;32(5):977–987.
94 Dai G, Tsukurov O, Chen M, et al. Endothelial nitric oxide production during in vitro simulation of external limb compression. *Am J Physiol Heart Circ Physiol*. 2002;282(6):H2066–H2075.
95 Ozawa ET, Bottom KE, Xiao X, Kamm RD. Numerical simulation of enhanced external counterpulsation. *Ann Biomed Eng*. 2001;1929:284–297.
96 Lawson WE, Hui JC, Barsness GW, et al. Effectiveness of enhanced external counterpulsation in patients with left main disease and angina. *Clin Cardiol*. 2004;1927:459–463.
97 Beller GA. A review of enhanced external counterpulsation clinical trials. *Clin Cardiol*. 2002;25(12 Suppl 2):II6–II10.
98 Chen FY, Aklog L, Couper GS, Cohn LH. Physiology and biomechanics of intraaortic balloon pumping. *Ann Thorac Surg*. 1997;1963:294–297.
99 Litmathe J, Feindt P, Boeken U, Gams E. Mechanical heart support using intraaortic balloon counterpulsation. A retrospective view and current perspectives. *Acta Cardiol*. 2004;1959:159–164.
100 Amado LC, Kraitchman DL, Gerber BL, et al. Reduction of "no-reflow" phenomenon by intra-aortic balloon counterpulsation in a randomized magnetic resonance imaging experimental study. *J Am Coll Cardiol*. 2004;1943:1291–1298.
101 Atlee JL. Cardiac assist devices. Technology and applications. *Minerva Anestesiol*. 2004;1970:25–44.
102 Azeem T, Stephens-Lloyd A, Spyt T, et al. Intra-aortic balloon counterpulsation: variations in use and complications. *Int J Cardiol*. 2004;1994:255–259.
103 Bates ER, Stomel RJ, Hochman JS, Ohman EM. The use of intraaortic balloon counterpulsation as an adjunct to reperfusion therapy in cardiogenic shock. *Int J Cardiol*. 1998;1965:S37–S42.
104 Cohen M, Urban P, Christenson JT, et al. Intra-aortic balloon counterpulsation in US and non-US centres: results of the Benchmark Registry. *Eur Heart J*. 2003;1924:1763–1770.
105 Khan AL, Flett M, Yalamarthi S, et al. The role of the intra-aortic balloon pump counterpulsation (IABP) in emergency surgery. *Surgeon*. 2003;2001:279–282.

106 Khir AW, Price S, Henein MY, et al. Intra-aortic balloon pumping: effects on left ventricular diastolic function. *Eur J Cardiothorac Surg*. 2003;1924:277–282.
107 Nanas JN, Moulopoulos SD. Counterpulsation: historical background, technical improvements, hemodynamic and metabolic effects. *Cardiology*. 1994;1984:156–167.
108 Adams JA, Moore JE Jr, Moreno MR, et al. Effects of periodic body acceleration on the in vivo vasoactive response to N-w-nitro-L-arginine and the in vitro nitric oxide production. *Ann Biomed Eng*. 2003;31(11):1337–1346.
109 Fujita M, Tambara K, Ikemoto M, et al. Periodic acceleration enhances release of nitric oxide in healthy adults. *Int J Angiol*. 2005;14:11–14.
110 Sackner MA, Gummels E, Adams JA. Effect of moderate-intensity exercise, whole body, periodic acceleration and passive cycling on nitric oxide release into circulation. *Chest*. 2005;128(4):2794–2803.
111 Adams JA, Mangino MJ, Bassuk J, Sackner MA. Hemodynamic effects of periodic G(z) acceleration in meconium aspiration in pigs. *J Appl Physiol*. 2000;89(6):2447–2452.
112 Adams JA, Mangino MJ, Bassuk J, et al. Noninvasive motion ventilation (NIMV): a novel approach to ventilatory support. *J Appl Physiol*. 2000;89(6):2438–2446.
113 Adams JA, Mangino MJ, Bassuk J, et al. Regional blood flow during periodic acceleration. *Crit Care Med*. 2001;29(10):1983–1988.
114 Adams JA, Mangino MJ, Bassuk J, et al. Novel CPR with periodic Gz acceleration. *Resuscitation*. 2001;51(1):55–62.
115 Adams JA, Bassuk J, Wu D, Kurlansky P. Survival and normal neurological outcome after CPR with periodic Gz acceleration and vasopressin. *Resuscitation*. 2003;56(2):215–221.
116 Adams JA, Wu D, Bassuk J, Kurlansky P. Cardiopulmonary resuscitation (CPR) using periodic acceleration (pGz) in an older porcine model of ventricular fibrillation. *Resuscitation*. 2004;60(3):327–334.
117 Nava G, Adams JA, Bassuk J, et al. Echocardiographic comparison of cardiopulmonary resuscitation (CPR) using periodic acceleration (pGz) vs chest compression. *Resuscitation*. 2005;66(1):91–97.
118 Abraham WM, Ahmed A, Adams JA, Sackner MA. Periodic acceleration via nitric oxide release modifies antigen-induced airway responses in sheep. *Am J Resp Crit Care Med*. 2006;174:743–752.
119 Aitchison KA, Coker SJ. Cyclooxygenase inhibition converts the effect of nitric oxide synthase inhibition from infarct size reduction to expansion in isolated rabbit hearts. *J Mol Cell Cardiol*. 1999;31(6):1315–1324.
120 Bell RM, Yellon DM. The contribution of endothelial nitric oxide synthase to early ischaemic preconditioning: the lowering of the preconditioning threshold. An investigation in eNOS knockout mice. *Cardiovasc Res*. 2001;52(2):274–280.
121 Bolli R. Cardioprotective function of inducible nitric oxide synthase and role of nitric oxide in myocardial ischemia and preconditioning: an overview of a decade of research. *J Mol Cell Cardiol*. 2001;33(11):1897–1918.
122 Gourine AV, Bulhak AA, Gonon AT, et al. Cardioprotective effect induced by brief exposure to nitric oxide before myocardial ischemia-reperfusion in vivo. *Nitric Oxide*. 2002;7(3):210–216.
123 Hide EJ, Ney P, Piper J, et al. Reduction by prostaglandin E1 or prostaglandin E0 of myocardial infarct size in the rabbit by activation of ATP-sensitive potassium channels. *Br J Pharmacol*. 1995;116(5):2435–2440.
124 Jugdutt BI. Nitric oxide and cardioprotection during ischemia-reperfusion. *Heart Fail Rev*. 2002;7(4):391–405.
125 Jugdutt BI. Nitric oxide and cardiovascular protection. *Heart Fail Rev*. 2003;8(1):29–34.
126 Shinmura K, Tang XL, Wang Y, et al. Cyclooxygenase-2 mediates the cardioprotective effects of the late phase of ischemic preconditioning in conscious rabbits. *Proc Natl Acad Sci USA*. 2000;97(18):10197–10202.

183

Stent- and Nonstent-Based Cell Therapy for Vascular Disease

Michael R. Ward, Duncan J. Stewart, and Michael J.B. Kutryk

St. Michael's Hospital, University of Toronto, Onatario, Canada

Normally, blood vessels are quiescent structures that fulfill their crucial function delivering blood to tissues and organs throughout the body with remarkable efficiency over the course of the entire normal lifespan. This is accomplished through interactions among a wide variety of regulatory systems that control the structure and function of the vasculature. The endothelium not only makes up the innermost layer of all blood vessels and provides a nonthrombogenic surface, but it also produces a wide variety of vascular and growth regulatory signals and participates in repair, remodeling, and regeneration of the vasculature. The endothelium normally exhibits a very slow turnover, with a half-life of several years; however, in response to arterial injury and forms of pathological stress, widespread damage to the endothelium necessitates efficient mechanisms of repair and regeneration. Understanding the endogenous mechanisms that protect and repair this crucial layer may form the basis for new therapeutic strategies for the treatment of vascular diseases.

LOCAL VERSUS SYSTEMIC MECHANISMS OF VASCULAR REPAIR

Until quite recently, repair of the damaged endothelium was thought to occur largely from the migration and proliferation of nearby undamaged endothelial cells (ECs) (1), and the regeneration of new blood vessels was believed to be mainly through the sprouting of ECs from preexisting blood vessels. However, several observations challenged this concept, beginning as early as 1963, when it was found that Dacron patches that were not contiguous with the vascular endothelium could become efficiently endothelialized (2), presumably by seeding from circulating ECs or endothelial progenitor cells (EPCs). In later studies using a genetic tag, it could be shown that the colonizing cells were of bone marrow (BM) origin (3). The existence of circulating BM-derived EPCs and their role in postnatal vasculogenesis in chronic ischemia models was first demonstrated with the pioneering work of Asahara and colleagues (4,5). They reported that purified CD34$^+$ hematopoietic progenitor cells could differentiate in vitro to an endothelial phenotype and incorporate into sites of neovessel formation in ischemic models in vivo. They also were the first to coin the term *endothelial progenitor cell*. The existence of "bone marrow–derived circulating EPCs" was also reported at about the same time by Shi et al (3), who showed that a subset of CD34$^+$ cells could differentiate into an endothelial-like phenotype, expressing markers such as von Willebrand factor (vWF), and take up DiI-acetylated low-density lipoprotein (LDL). EPCs were initially defined by the expression of both CD34 and vascular endothelial growth factor receptor 2 (VEGFR2, or FLK-1) (4). CD34 is also expressed to a lesser extent by mature endothelium as well as hematopoietic stem cells (HSCs). Further studies have identified that CD133 (or prominin) is a marker of the subset of CD34$^+$ cells that appears to demonstrate the best potential for endothelial differentiation (6). CD133 expression is rapidly lost from EPCs undergoing endothelial differentiation, which makes it useful as a marker of immature EPCs versus mature ECs that may be shed from the vessel wall. Interestingly, a recent report suggested that CD14$^+$ cells (monocytes) that express low levels of CD34 (undetectable with conventional techniques) may in fact contain the "true" progenitor cell population, with the greatest regenerative capacity (7). Although a consensus has not yet been achieved as to how best to identify circulating EPCs, it is generally accepted that this population is important for vascular maintenance and repair.

THERAPEUTIC APPLICATIONS OF CELL THERAPY FOR VASCULAR REPAIR AND REGENERATION

The promotion of neovascularization represents a major focus of cell therapy for vascular disease and involves three distinct processes: angiogenesis, vasculogenesis, and arteriogenesis. In distinction to angiogenesis, vasculogenesis involves the de novo formation of blood vessels from the differentiation of precursor cells (angioblasts or EPCs) into ECs, which eventually give rise to capillaries. In contrast, arteriogenesis refers to the increasing size and complexity of preexisting arteries by intimal and medial remodeling (8,9). A second potential application of cell therapy is to accelerate the regeneration of injured or dysfunctional endothelium of native vessels. Maintenance of an intact and functional endothelium is crucial for the prevention of the progression of atherosclerosis, intimal hyperplasia, and thrombosis. The ability of EPCs to differentiate into ECs suggests that they may incorporate into sites of endothelial damage and repair the compromised endothelium (10–12). One of the challenges to the success of such cell therapies is that patients with risk factors for coronary artery disease (CAD), which are known to cause endothelial dysfunction, have reduced circulating levels and function of EPCs (13,14). This may reduce the ability of circulating EPCs to maintain endothelial integrity (15). In addition to their ability to repair preexisting endothelium, EPCs have been shown to contribute to the endothelialization of vascular implants, a phenomenon known as *fallout endothelialization* (16,17). The ability of EPCs to regenerate the endothelium has led to several applications, discussed in later sections.

CELL THERAPY WITH VASCULAR STENTS AND PROSTHESES

The mechanisms of postangioplasty restenosis have been extensively studied. The restenotic process is characterized by four distinct phases: thrombosis, inflammation, proliferation, and matrix deposition/vessel remodeling (18–20). It is known that one of the most important mechanisms that contributes to arterial restenosis is endothelial denudation at the time of intervention, and that the kinetics of reendothelialization determine the magnitude of intimal hyperplasia. Many attempts have been made to seed ECs on denuded or synthetic surfaces using a variety of devices and techniques. These have shown some positive results but also have been hampered by rapid loss of the seeded cells (21–26).

Cell Seeding of the Native Vessel Wall

In one of the first studies, Nabel and colleagues instilled ECs into balloon-injured porcine iliofemoral arteries of syngeneic animals, achieving 2% to 11% adherence after 2 to 4 weeks (23). Subsequently, it was reported that a high level of EC attachment (36%) could be obtained in damaged human saphenous veins in vitro (26). The same investigators demonstrated 17% cell retention after 100 minutes of blood flow in previously denuded external iliac arteries of rabbits (21). To improve engraftment, ECs were grown in polymer matrices, which were then implanted into a porcine model of arterial injury (27). Despite the use of nonimmune matched allogeneic (porcine) and xenogeneic (bovine) EC implants, experimental restenosis was significantly reduced 3 months after angioplasty compared with control by 56% and 31%, respectively. Conte and associates delivered genetically modified (β-galactosidase) autologous venous cells to the surface of a balloon-injured rabbit femoral artery. Four to 7 days after seeding, injured arteries displayed 40% to 90% coverage with β-galactosidase-transduced cells, even when seeded at subconfluent density, and an intact EC monolayer, as evidenced by scanning electron microscopy studies (22). This also significantly improved restenosis in a hypercholesterolemic rabbit model of iliofemoral injury (28); however, this method required surgical exposure of the vessels and complete interruption of blood flow for 30 minutes, which may not be clinically desirable.

More recent studies have used EPCs following angioplasty injury to enhance reendothelization and reduce neointimal hyperplasia (29,30). It was subsequently shown that the genetic modification of EPCs with a retroviral vector expressing endothelial nitric oxide synthase (eNOS) potentiated the therapeutic benefit of the transplanted cells in a rabbit carotid injury model (11). It also has been reported that the mobilization of EPCs can lead to accelerated re-endothelization of denuded vessels. For example, the exogenous administration of granulocyte colony stimulating factor (G-CSF) for several days to mobilize BM stem and progenitor cells before balloon injury of carotid arteries in rats resulted in enhanced reendothelization (11). Similarly, statin (12,30) and estrogen (31) therapy increased the number of peripheral blood EPCs and reduced neointimal hyperplasia in animal models of arterial injury.

Cell Seeding of Intravascular Stents

In 1988, van der Giessen and colleagues (32) seeded human umbilical vein ECs on stainless steel self-expandable stents and implanted them into pig femoral arteries. Interestingly, they reported complete reendothelialization despite the obvious immunological mismatch of this "xenograft." (32) Since this early study, there have been no further preclinical reports of cell seeding with mature ECs, other than studies demonstrating in vitro technical feasibility. There are several likely reasons for this lack of progress. Cell adherence is difficult to maintain with the restoration of blood flow, and stainless steel provides a poor platform for EC attachment. More important, stents require balloon expansion for deployment, which will damage the fragile neoendothelium. Even with the use of self-expandable stents, there is a requirement for balloon postdilatation for optimal apposition of the stent struts to the vessel wall. We have recently developed a novel technique to "autoseed" implanted vascular stents by capturing endogenous nonmobilized circulating EPCs (Figure 183.1). Stents coated with anti-CD34 antibodies efficiently capture circulating EPCs, which then mature into cells with an endothelial phenotype, completely covering the stent struts and the intrastrut denuded vessel wall (Figure 183.1). The ability of

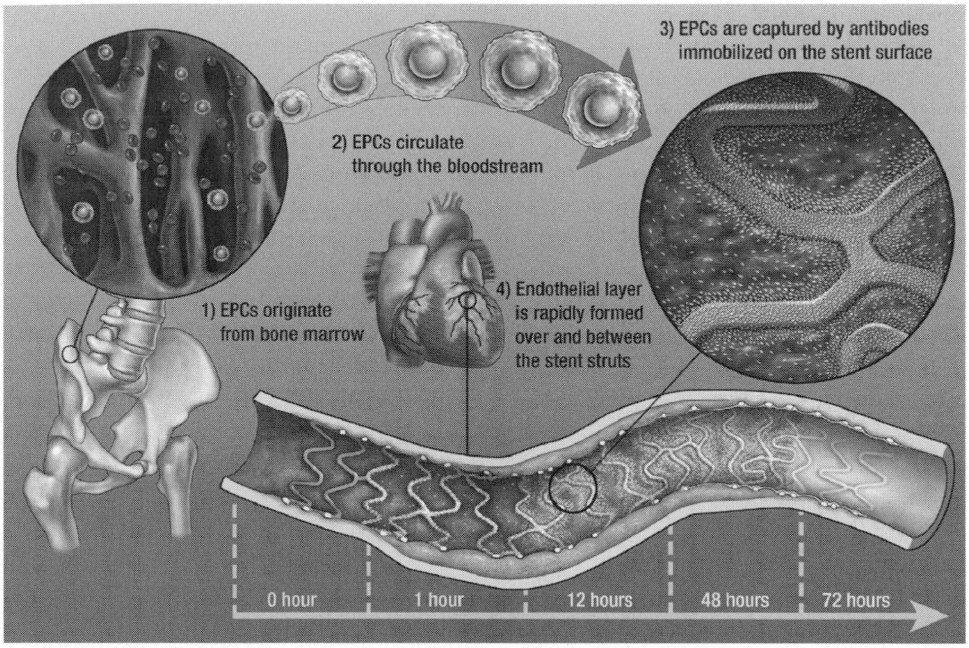

Figure 183.1. EPCs originate in the bone marrow and circulate in the blood. CD34^{+} EPCs are captured by anti-CD34 antibodies on the stent surface. The adherent cells proliferate to rapidly cover the stent struts, and subsequently differentiate to re-establish functional endothelium, reducing the chance of stentimthrombosis. Accelerated healing post-stent implantation can also result in inhibition of intimal hyperplasia and subsequent in-stent restenosis.

these devices to capture circulating EPCs and, as a consequence, to prevent restenosis and eliminate the risk of stent in situ thrombosis, has been demonstrated in a pig coronary stent model (in press). The clinical safety of these devices has been shown in the HEALING-FIM Trial (33), and the efficacy of this approach to prevent in-stent restenosis is being tested in the HEALING II clinical trial.

Cell Seeding of Prosthetic Grafts

Apart from a small territory around the anastomotic site, prosthetic grafts do not spontaneously endothelialize in humans (34). It is thought that the lack of surface ECs promotes thrombosis and ultimately leads to the failure of the prosthetic graft. Seeding of the graft lumen with ECs has been used to overcome this problem. Two strategies have been described: single-stage and two-stage seeding (35,36). The single-stage methodology uses freshly harvested ECs either of venous or microvascular origin to immediately line the graft. In two-stage seeding, ECs are extracted and undergo a prolonged period of ex vivo cell culture to increase the cell numbers before seeding.

Venous and Arterial Endothelial Cells

Mansfield and coworkers were the first to describe EC seeding in 1975 (37). They used granulation tissue from the bed of a pedicled skin flap as a cell source to seed a mixture of ECs, fibroblasts, and macrophages onto patches of polyester. In 1978, Herring and colleagues introduced the concept of single-stage seeding of prosthetic grafts using autologous ECs (35). In dogs, mechanically isolated venous ECs were mixed with blood and used to preclot polyester grafts, which were then implanted as infrarenal abdominal interposition grafts. After 2 to 4 weeks, they demonstrated that the grafts seeded with this method showed extensive lining of endothelium, with a reduction in the area affected by thrombus (38). These initial studies were followed by a series of reports examining cell seeding using both mechanically and enzymatically harvested venous ECs (39–47). It was shown that not only preclotting of polyester grafts with a mixture of ECs and blood, but also seeding of ECs onto the luminal surface of expanded polytetrafluoroethylene (ePTFE), resulted in surface endothelization in dog models. Herring and coworkers were the first to show that ECs seeded on human arterial prostheses resulted in endothelization (48). Despite encouraging preclinical results, the results of the initial clinical trials of single-cell seeding using autogenous vein as a source of ECs were variable, but overall disappointing (49–57). The failure of these single-stage seeding trials was thought to be caused by the low seeding density of the ECs. A limited number of ECs can be extracted from veins. In addition, a proportion of seeded cells are lost when exposed to pulsatile blood flow.

Compared with single-stage seeding, two-stage seeding results in superior patency rates in animals (58–62). In two-stage seeding, the ECs are extracted and undergo a period of cell culture and expansion. The superiority of this technique is thought to be due to the high density of seeding and the retention of a large number of ECs on the surface of the prosthetic material despite cell loss on exposure to pulsatile flow. Zilla and colleagues (63) published the results of the first randomized clinical trial performed with two-stage seeded and nonseeded ePTFE grafts used for femoropopliteal bypasses using ECs isolated from the external jugular vein. Seeded grafts were cultured in vitro for at least 9 days and the total period between EC isolation and graft implantation was 37 days. After 32 months, the actuarial patency rate was 85% for the seeded and 55% for the nonseeded grafts. The persistence of a confluent endothelium was shown in an explanted specimen (64). In 1999, the 9-year follow-up of the initial randomized study by Zilla and coworkers and the 5-year follow-up of the routine clinical endothelization program were reported (65). The patency rates were significantly higher for the seeded grafts (65%) than for the nonseeded control group (16%). Other clinical trials involving the two-stage endothelialization process have also shown a significantly enhanced clinical performance in small-diameter grafts (65–70). The main disadvantage of the two-stage seeding technique is the labor intensity. A 4- to 5-week delay exists between cell harvest and graft implantation, which prevents its use in emergency situations. The risk of failure of cell culture because of infection of the cells and the inability of the cells to proliferate effectively also limits its clinical applicability.

Microvascular Endothelial Cells

A shortcoming of the use of venous ECs, either as part of a single- or two-stage seeding process, is the relatively low yield, which has led to the evaluation of other EC sources. A large number of microvascular ECs can be extracted from abdominal, buttock, breast, and omental fat ($>1 \times 10^6$/g). Endothelization of prosthetic grafts utilizing such a high cell density, single-stage process, has been coined "cell sodding." Despite favorable patency rates in animal experiments (71–76), the results of clinical studies using microvascular ECs have been disappointing. In a case report, a seeded mesoatrial polyester graft explanted after 9 months showed a vascular endothelium with a substantial subendothelial layer composed of sparse cells and extracellular matrix (77). Meerbaum and associates reported the medium-term results from 34 patients after implantation of ePTFE grafts seeded with subcutaneous fat-derived microvascular ECs for peripheral vascular reconstruction (78). The cumulative patency at 36 months was 19%, similar to studies using nonseeded grafts. Williams published the results of a safety study of 11 patients with significant vascular disease who received microvascular EC-seeded ePTFE peripheral bypasses (79). Of the 11 patients, seven were available for follow-up at 4 years and the patency was 55%. The results of human studies with fat-derived microvascular cells have been less favorable than those of animal studies and of human studies with cultured vein-derived ECs. The inferiority of sodding with microvascular cells may be due to the delay in the formation of a confluent endothelium with this single-stage procedure and the presence of a high percentage of contaminating non-ECs.

Endothelial Progenitor Cells

EPCs have been shown to contribute to reendothelization of prosthetic grafts in animals (80). The ability of locally transplanted EPCs to endothelialize vascular bioprostheses has also been demonstrated (29). Using a similar approach, Kaushal and colleagues showed that seeding of EPCs into decellularized porcine iliac vessels implanted as coronary interposition grafts formed a functional endothelial layer and improved vasodilatory function and patency of the grafts (10). An advantage of using EPCs is that they can be obtained by venipuncture rather than surgically. The disadvantages of using EPCs include the risk of infection during ex vivo culture, change of phenotype, and limitations for their use in emergency situations because they require ex vivo expansion.

A novel method of "autoseeding" prosthetic vascular material, developed by our group, has been described by Rotmans and coworkers (81). Anti-CD34 antibodies immobilized on the surface of ePTFE were effective in capturing circulating EPCs and establishing an endothelial layer on the surface of the material when implanted as an arteriovenous shunt. Despite EC coverage of the graft, robust intimal hyperplasia at the outflow anastomotic site remained a problem. Mobilization of EPCs with cytokines has also been shown to be effective in promoting in vivo endothelization of prosthetic grafts. Both Bhattacharya and colleagues (82) and Shi and associates (83) reported that the mobilization of EPCs from BM by exogenous G-CSF enhances the endothelization and patency of small-caliber prosthetic grafts.

Genetically Modified Cells

Several preclinical studies have shown that it is possible to genetically modify ECs ex vivo prior to the seeding of prosthetic material to increase the efficiency of engraftment and to enhance endothelial function (84,85). However, any potential benefits from enhanced endothelial function, proliferative ability, and viability appear to be outweighed by poor retention rates compared with nongenetically modified ECs. Clinical trials using genetically modified ECs have not been initiated because of the poor results in animal studies and ethical questions concerning the role of such therapy.

CELL THERAPY AFTER ACUTE MYOCARDIAL INFARCTION

Due to advanced reperfusion management approaches, survival after myocardial infarction (MI) has improved greatly, but the benefit is critically dependent on the time it takes for patients to present for appropriate treatment. As well, even when revascularization therapy is initiated within the "window of opportunity," many patients fail to show significant recovery of cardiac function. Therefore, a large number of preclinical and clinical studies have attempted to use cell therapy to restore cardiac function post-MI. For the purpose of this chapter, we will deal only with studies related to the use of BM-derived or circulating progenitor and stem cells, and the reader is directed to recent reviews that address the use of myocyte or myoblast transplantation (86,87).

Cell Therapy in Experimental Models

In 2001, Orlic and colleagues first reported regeneration of myocardium by hematopoietic stem cells (HSCs) injected into the infarct border zone of a mouse MI model (88). However, transdifferentiation of HSCs into myocytes has been strongly refuted by several studies that show maintenance of hematopoietic characteristics of transplanted cells despite localizing to the peri-infarct regions (89,90). Despite this controversy, the ability of HSCs (as well as other cell types, such as BM or peripheral blood mononuclear cells [MNCs] and EPCs) to improve cardiac function after acute MI is widely reported (91–93), although the underlying mechanisms are still poorly understood. Recently, EPCs have also been reported to differentiate into cardiomyocytes when cocultured with myocytes (94,95), and Yeh and associates (96) showed that $CD34^{+}$ cells transdifferentiated into cardiomyocytes, ECs, and smooth muscle cells (SMCs) in a mouse model of acute MI. BM stromal cells (or mesenchymal stem cells [MSCs]) have also been shown to differentiate into cardiomyocyte-like cells in vitro and in vivo (97–99), although more typically they differentiate into osteocytes, chondrocytes, and adipocytes. It has also been suggested that MSC transplantation into the injured zone prevents remodeling of peri-infarct regions and improves regional wall motion (98,100).

An alternative strategy for "cell therapy" involves the mobilization of stem and progenitor cells from the BM, thus avoiding the need for cell isolation, processing, and delivery. In mice, G-CSF and stem cell factor (SCF) have been reported to increase bone marrow cell mobilization and stimulate myogenesis and angiogenesis with an improvement in cardiac function after acute MI (101,102). Apart from EPC recruitment, G-CSF has been shown to have direct myocardial effects contributing to improved myocardial function, including the inhibition of cardiomyocyte and endothelial apoptosis (103). It has also been shown that G-CSF can accelerate infarct healing by enhancing macrophage infiltration and matrix metalloproteinase activation (104). Despite the potential advantage of using this technique, it is possible that mobilization of BM cells could aggravate chronic inflammation of atherosclerosis and lead to plaque rupture or other complications (105–108), and thus the safety of this strategy in patients with vascular disease will have to be assessed carefully.

Clinical Trials

Despite a lack of thorough understanding of the mechanism of cardiac regeneration due to cell therapy, many safety and efficacy trials for acute MI have already been initiated based on encouraging preclinical data. Patients in many of these trials received primary angioplasty with stenting to reperfuse the ischemic myocardium, followed by infusion of cells into the coronary artery. Most studies used unselected BM cells (i.e., BM MNCs), which include stromal cells, vascular cells, MSCs, and HSCs, and only a few reports used selected (i.e., $CD34^{+}$) cell populations. Results from more than 600 patients from various trials suggest that intracoronary delivery of unselected BMCs is safe when given within several months of MI (109–114). Transplantation of Progenitor Cells and Regeneration Enhancement in Acute Myocardial Infarction (TOPCARE-AMI) also administered EPCs derived from peripheral blood with similar results (112). Cell infusions into the culprit artery after stenting did not appear to result in further damage to the myocardium, nor did this lead to a systemic inflammatory reaction based on analysis of serum troponin and C-reactive protein (CRP) levels. BMC and EPC transfer did not increase the rate of ventricular or supraventricular arrhythmias, as assessed by Holter monitoring and clinical surveillance (109,113), and in most patients, in-stent restenosis was not increased due to cell transfer (110,112,114).

Until recently, trials have not been designed to assess efficacy and, with the exception of the Bone Marrow Transfer to Enhance ST-elevation Infarct Regeneration (BOOST) trial, have not included a randomized control group (114,115). For ethical reasons, even the BOOST trial did not incorporate sham procedures, thus limiting the ability to blind patients and investigators. With these caveats in mind, most trials suggest that intracoronary injection of unselected BMCs or EPCs enhances regional wall motion within the infarct area (109,110,113, 114,116,117). In TOPCARE-AMI and BOOST, regional wall motion was associated with a significantly improved global left ventricular ejection fraction (LVEF) (109,114). In the BOOST trial, there was a 6% increase in LVEF at 6-month follow-up in patients receiving cell therapy compared with controls (114). However, after 18 months there was no significant difference in LVEF between patients receiving placebo and cell therapy, mainly due to an unexpected improvement in control patients.

In the last year, the results of three larger, randomized controlled trials have become available (118–121). The Reinfusion of Enriched Progenitor Cells and Infarct Remodeling in Acute Myocardial Infarction (REPAIR-AMI) (119) clinical trial was the first randomized double-blinded study to show efficacy of BM-MNCs in the post-MI setting (119). This 204-patient study achieved its primary endpoint of improvement in global EF at 4 months, measured by LV angiography, although the actual improvement was rather modest, with only a 2.5% difference between the cell and placebo groups (119). In contrast, two other recent cardiac cell therapy randomized, controlled trials failed to show significant improvement in LV function. The ASTAMI trial was randomized but not blinded and showed no benefit in LVEF measured by MRI (121). However, this trial used cell isolation and processing techniques that have been shown to adversely affect cell viability (122). The Leuven study (Belgium), which involved delivery of cells much earlier (within 24 hours) than the other trials, also failed to achieve its primary endpoint despite improvements in secondary endpoints (i.e., infarct size) (118). Chen and colleagues

(123,124) utilized MSCs injected into the coronary circulation in patients after acute MI and reported no adverse effects, despite evidence in dogs that this approach could accentuate ischemic damage to the myocardium (125). They also showed improved wall motion and global LVEF 6 months after MSC transfer, compared with controls (123,124). In the Magnesium in Cardiac Arrest (MAGIC) trial, peripheral blood MNCs were isolated after mobilization with G-CSF and injected into the infarct-related artery of patients undergoing percutaneous coronary intervention (126). After 6 months, LVEF, regional wall motion, and tissue perfusion were improved in patients receiving cell transplantation compared with matched controls. However, seven of ten patients followed-up at 6 months developed significant in-stent restenosis or complete reocclusion, and two developed a de novo lesion in the infarct-related artery as assessed by quantitative angiography. Whether this was related to the proinflammatory effects of BM mobilization or to an effect of the local delivery of MNCs is uncertain. Although these results raise concerns, the small scale of this trial precludes any definitive interpretations.

Controversies and Key Issues

It is certainly too early to come to any firm conclusions as to whether cell therapy, in its present form, is sufficiently effective to improve cardiac function substantially after acute MI, and if so, by what mechanisms. As we await the results of larger and more rigorously designed clinical trials, it is important to maintain realistic expectations. As well, it is unlikely that these first trials, which are using simplistic strategies, will reflect the true potential of cell therapy. Even fairly modest improvement, if truly reproducible, would be very encouraging. Moreover, it must be recognized that the use of autologous cells, which will likely be required in the foreseeable future, is associated with several significant limitations. Most patients that require cell therapy will be older, with widespread vascular disease, and multiple risk factors. All these factors will likely have a negative impact on the number and quality of autologous stem and progenitor cells – for example, by reducing their ability to proliferate, migrate, and survive. Already, some preclinical work has been performed to attempt to address some of these issues – for example, by using genetically engineered MSCs, which overexpress the survival factor Akt (98). However, whether genes that interfere with apoptosis can be used clinically is questionable in view of the potential increased risk of cell transformation and neoplastic transformation.

CELL THERAPY FOR NEOVASCULARIZATION

Preclinical Experience

A number of the preclinical studies attempting to stimulate neovascularization have been performed with BM- or peripheral blood–derived EPCs. BM transplant experiments have shown that EPCs are mobilized from the marrow in response to tissue ischemia (127,128) and home to, and incorporate into, regions of neovascularization (129). In small-animal models of limb ischemia, injection of human blood–derived EPCs improved neovascularization and subsequent limb blood flow (127,130–133). Similarly, using small-animal models of myocardial ischemia or infarction, it has been shown that systemically administered or BM-recruited EPCs stimulate angiogenesis and arteriogenesis, with subsequent improvement in myocardial perfusion (5,92). Kawamoto and colleagues (93) reported that autologous EPCs significantly increased capillary density and collateral development (by angiography) with an associated improvement in LVEF in a pig model of myocardial ischemia.

In a rat ischemic hindlimb model, intramuscular injection of BM MNCs increased the number of visible capillaries (by microangiography), limb flow, and exercise tolerance compared with controls (134). In a rabbit model of hindlimb ischemia, BM MNCs augmented small and larger vessel remodeling, with enhanced collateral development (129). Of note, MNCs cultured under hypoxic conditions had a greater effect on flow recovery than cells grown under normoxic conditions (135), suggesting that these cells are normally responsive to ischemia. In a rat model of cardiac ischemia, BMC implantation induced angiogenesis and improved the perfusion of ischemic myocardium (136). Results from large-animal studies have shown that injection of BMCs improved collateral flow (137,138), augmented capillary density, and reduced contrast echocardiography perfusion defects (91,139).

The use of different disease models, cell types, cell numbers, and species makes the comparison of results between studies very difficult. However, overall functional improvement appears similar whether cell populations are selected using CD34, CD133, or murine Sca-1 determinants or whether unselected MNCs are expanded in differential culture (6,92,127,128,130,140–143). As well, both "early-growth" and "late-growth" EPCs showed comparable in vivo neovascularization capacity despite the marked phenotypic differences between these two populations (144). These results seem to suggest that the functional benefit is relatively independent of the type of stem/progenitor cell used. However, some studies have reported that the $CD34^{-}$ fractions of BM- or blood-derived MNCs exhibited reduced neovascularization capacity (131,145). In addition, fully differentiated mature ECs have been shown to lack the ability to induce neovascularization (130,144,146). Urbich and Dimmeler (147) have suggested that chemokine- or integrin-receptor–mediated homing is needed for this effect. Certainly in arteriogenesis this may be crucial, as delivery of monocyte chemoattractant protein (MCP)-1 by itself enhanced collateral vessel formation (146), whereas depletion of monocytes has been shown to impair arteriogenesis (149,150).

The ability of EPCs to stimulate neovascularization was initially attributed to their ability to differentiate into ECs. However, studies using BM transplantation of genetically modified cells (with reporter genes such as *GFP* or *LacZ*) have shown greatly varying incorporation rates of genetically labeled, BM-derived cells coexpressing endothelial markers

(0%–90%) (151–155). Some of this variation may be attributable to different levels of tissue injury in the models of ischemia, as injury itself may promote BM mobilization and incorporation (156). The efficiency of engraftment and incorporation also may differ between the types and source of cells as well as the method of delivery. For example, intravenous infusion of ex vivo EPCs leads to a higher incorporation rate than with G-CSF mobilization alone (157,158).

Because the incorporation rate of EPCs is often quite low, it has been suggested that neovascularization may result from paracrine effects (92,159). This is supported by studies showing that these cells have the ability to secrete growth factors such as vascular endothelial growth factor (VEGF), insulin-like growth factor (IGF)-1, hepatocyte growth factor (HGF), and nitric oxide (NO) (147,160,161), which could influence the classical process of angiogenesis (162). In a model of tumor angiogenesis, it was shown that transplanted cells are not integrated in the tumor vessels per se but are found adjacent to the vessels (163). Likely, both the ability to incorporate into the vasculature and to secrete growth factors contributes to the overall neovascular response. In contrast, infusion of macrophages, which did not incorporate into vessels (164), caused only a slight increase in collateral flow after ischemia, suggesting that a combination of paracrine actions of EPCs together with their ability to physically contribute to vessel-like structures may be important for efficient vascular regeneration. Further studies must be conducted to better clarify the mechanisms by which EPCs enhance neovascularization.

Clinical Trials for Cell-Based Neovascularization

Based on encouraging preclinical results using cell therapy for neovascularization, several clinical trials have already been initiated, not only in the setting of myocardial ischemia, but also for peripheral arterial disease (PAD). In 2002, Tateishi-Yuyama and colleagues (165) published the results from the Therapeutic Angiogenesis by Cell Transplantation (TACT) study, which enrolled 25 patients with unilateral leg ischemia. Patients were treated with BM MNCs in the gastrocnemius muscle of the ischemic leg and with saline in the nonischemic leg. Following the pilot study, 22 patients with bilateral ischemia were randomized to receive BM MNCs in one leg and freshly isolated peripheral blood MNCs in the other leg (158). After 4 weeks, the authors reported a significant improvement in ankle-brachial index (ABI), transcutaneous oxygen pressure (TcO_2), rest pain, and pain-free walking time after intramuscular injection of BM MNCs compared with peripheral blood MNCs. Based on this limited-scale trial, the authors concluded that administration of BM MNCs could potentially be used to augment neovascularization of the ischemic limb of humans (165). More recently, a similar study was performed using intra-arterial administration of EPCs to rescue critical limb ischemia (CLI) (166). Seven patients were treated with an intra-arterial infusion of autologous EPCs isolated from peripheral blood following G-CSF mobilization. Twelve weeks after EPC administration, the authors reported an increase in pain-free walking distance, ABI, TcO_2, flow-dependent vasodilation, flow reserve in response to adenosine, and endothelium-dependent vasodilation. Although these early trials suggest that cell therapy may be useful for inducing neovascularization, larger trials are required to confirm the efficacy of these approaches and their applicability to the treatment of PAD.

CELL THERAPY FOR PULMONARY VASCULAR DISEASE

Pulmonary arterial hypertension (PAH) is a vascular disease of unknown cause characterized by severe endothelial dysfunction. Although an imbalance between vasodilatory and vasoconstrictive properties in the pulmonary vasculature may contribute to the initial stages in the pathogenesis of PAH, recent studies have suggested that arterial remodeling may predominate, particularly in established disease (167) (see Chapter 129). Recently, loss-of-function mutations in a member of the TGF-β family of receptors, bone morphogenetic protein receptor type II, have been found in patients with idiopathic and familial PAH (168), but how these are linked to pathogenesis of this progressive and lethal condition is uncertain. Although many patients with advanced PAH exhibit plexiform lesions characterized by EC and SMC proliferation (167), recent evidence suggests that EC apoptosis, likely as a result of environmental stress and/or genetic predisposition, may be an initiating event in PAH (167). Persistent or repeated bouts of EC apoptosis may lead primarily to microvascular rarefaction, and secondarily to the emergence of hyperproliferative and apoptosis-resistant EC "clones." Both processes may contribute to the relentless increase in pulmonary vascular resistance that culminates in severe right-heart failure and death. Although there has been considerable progress in the development of new pharmacological treatments for PAH (169), these do not appear to alter the fundamental remodeling processes that drive progression of disease.

Role of Bone Marrow–Derived Cells in Pulmonary Arterial Hypertension

Our group has conducted studies to determine whether administration of EPCs can rescue the PAH phenotype in a model induced by the injection of the plant alkaloid (170), monocrotaline (MCT) in rats, which results in reproducible and lethal PAH. We also tested the hypothesis that EPCs genetically engineered to overexpress eNOS, the gene responsible for producing NO in ECs, have a greater regenerative capacity and show greater efficacy in rescuing PAH compared to EPCs alone (170). Endothelial dysfunction is characterized by reduced NO bioavailability in ECs (171,172) and notably, NO has been shown to be reduced in severe pulmonary hypertension (173). Previously, it was shown that intrajugular delivery leads to a high percentage of cells retained within the lung

Table 183-1: Overview of Published Clinical Trials for Acute Myocardial Infarction and Peripheral Arterial Disease

Study	*Pathology*	*Design*	*N*	*Cell Type*	*Delivery*	*Outcome*
Strauer et al., 2002 (113)	Acute MI	Primarily safety	$n = 10$	BM MNCs	Intracoronary	Stroke volume↑; wall motion↑; perfusion↑; infarct size↓
TOPCARE-AMI, 2002 (109,112)	Acute MI	Primarily safety	$n = 59$	BM MNCs, PB EPCs	Intracoronary	EF↑; Infarct size↓; perfusion↑; remodeling↓
Stamm et al., 2003 (116)	Acute MI	Primarily safety	$n = 12$	BM $CD133^+$	Intramyocardial (during CABG)	Perfusion↑
Fernandez-Aviles, 2004 (110)	Acute MI	Primarily safety	$n = 20$	BM MNCs	Intracoronary	EF↑; wall motion↑; ESV↓
BOOST, 2004 (114,115)	Acute MI	Randomized, controlled	$n = 60$	BM MNCs, Saline	Intracoronary	MR: EF↑ (6 months only)
Leuven Trial (118)	Acute MI	Randomized, double-blinded	$n = 33$	BM-MNCs	Intracoronary	No increase in LV EF; ↓infract size; ↑ recovery of regional systolic function
REPAIR-AMI (119)	Acute MI	Randomized, double-blinded	$n = 102$	BM-MNCs	Intracoronary	EF;↑; greater benefit in <49% EF and >5 days after MI
ASTAMI (121)	Acute MI	Randomized, non-blinded	$n = 50$	BM-MNCs	Intracoronary	No benefit
MAGIC, 2004 (126)	Acute MI	Randomized, controlled	$n = 20$	GCSF mobilized	Intracoronary	EF↑; perfusion↑; in-stent restenosis↑
TACT, 2002 (165)	Limb ischemia	Randomized, controlled	$n = 85$	BM MNCs PB MNCs saline	Intramuscular (Gastrocnemius muscle)	ABI↑; walking dist.↑; Tc O_2 pressure↑
Lenk et al., 2005 (166)	Critical limb ischemia	Primarily safety	$n = 7$	PB MNCs (following G-CSF)	Intra-arterial	ABI↑; walking dist.↑; Tc O_2 pressure↑; vasodilation (flow- and endothelial-dependent)↑; flow reserve↑

PB EPCs, peripheral blood endothelial progenitor cells; ESV, end-systolic volume; PB MNC, peripheral blood mononuclear cells.

immediately after injection (174) and that the majority of these cells transmigrate through the endothelial layer within 24 hours to engraft into the lung (175). Fluorescent microangiography revealed that, in the MCT-injured lung, labeled EPCs engrafted into the endothelium of distal pulmonary arterioles (170), and the administration of EPCs 3 days after MCT, was able to prevent the increase in right ventricle systolic pressure (RVSP) usually seen in this model, whereas fibroblasts had no protective effect. To test the ability of this approach to reverse already established PAH, EPCs were administered 3 weeks following MCT. Compared to the administration of MCT alone, which led to widespread occlusion of precapillary arterioles, EPC administration restored capillary continuity and microvascular architecture. This also was associated with a marked improvement in survival in rats receiving EPCs, with the eNOS-transfected EPCs showing the greatest effect (170). Moreover, in rats with established PAH, eNOS-transfected cells reduced RVSP to levels that were not different from control animals that did not receive MCT, although EPC administration alone was not sufficient to fully rescue the PAH phenotype in this model (170). Similar results were reported by Nagaya and colleagues using human cord-blood-derived EPCs transfected with adrenomedullin (176).

The results of these experiments show that administration of EPCs can prevent the vascular abnormalities caused by MCT in rats and that gene delivery using EPCs may represent a feasible strategy in regenerating lung microvasculature and treating PAH. Based on this experience, an early-phase dose-ranging trial, the Pulmonary Hypertension: Assessment of Cell Therapy (PHACeT) trial, is about to begin in Toronto, which will be the world's first clinical study combining gene and cell therapy of cardiovascular disease.

THE FUTURE OF CELL THERAPY

Cell therapy for vascular diseases represents an exciting new avenue that holds tremendous promise for repair and regeneration of blood vessels and possibly even the myocardium. Three main strategies for cell-based therapies are outlined in this chapter: (a) direct delivery of selected or nonselected cells, (b) mobilization of BM stem and progenitor cells, and (c) "autoseeding" of circulating stem or progenitor cells. The suitability of these different strategies for specific applications will vary. For example, autoseeding of stents or other vascular devices using an antibody coating directed at EPCs is ideally suited for the acceleration of endothelization. In contrast, the direct delivery of regenerative cells into the tissue or into the arterial circulation may be the most efficient strategy for neovascularization of ischemic tissue. In any event, the clinical role of cell therapy for the treatment of vascular diseases has not yet been established in rigorous and adequately designed clinical trials. Although several trials are now under way, it is important to maintain realistic expectations, given our lack of understanding about which cells to use, how they best should be delivered, and perhaps, most important, how to overcome the inherent limitations of autologous cell therapy. Future research needs to be directed at these questions, in particular at ways to enhance the regenerative capacity of autologous cells harvested from generally older individuals with systemic disorders (e.g., hyperlipidemia, diabetes) or genetic factors (e.g., BMPR2 mutations) that may reduce the ability of their cells to repair or regenerate the cardiovascular system.

KEY POINTS

- The use of stem and progenitor cells for the repair or regeneration of the endothelium is a promising strategy for enhancement of the clinical benefits of stents and other vascular prostheses as well as improvement of post-MI myocardial regeneration and neovascularization in ischemic tissues.
- Strategies to promote adhesion of circulating EPCs to stents or other prosthetic vascular devices may enhance the endothelialization of the luminal surface and reduce complications (restenosis and thrombosis).
- The use of various stem and progenitor cells to improve function of damaged myocardium has shown great promise in preclinical studies and has been shown to be safe in several small-scale clinical trials. Currently, double-blinded and randomized trials are being conducted to assess the efficacy of this approach.
- In vascular disorders such as PAH or peripheral arterial disease, the delivery of bone marrow– or peripheral blood–derived endothelial progenitor cells has been shown, through various preclinical studies and limited clinical trials, to improve endothelial dysfunction and promote neovascularization of ischemic tissues.

Future Goals

- To increase the efficacy of autologous cell therapy and make it a successful clinical therapy, it will be important to reverse the reduced regenerative properties of EPCs derived from subjects with extensive cardiovascular risk factors; this may require genetic or pharmacological enhancement during the ex vivo culture.

REFERENCES

1 Ross R, Bowen-Pope D, Raines EW, Faggiotto A. Endothelial injury: blood-vessel wall interactions. *Ann NY Acad Sci.* 1982:260–264.

2 Stump MM, Jordan GL Jr, Debakey ME, Halpert B. Endothelium grown from circulating blood on isolated intravascular Dacron Hub. *Am J Pathol.* 1963:361–367.

3 Shi Q, Rafii S, Wu MH, et al. Evidence for circulating bone marrow-derived endothelial cells. *Blood.* 1998;2:362–367.

4 Asahara T, Murohara T, Sullivan A, et al. Isolation of putative progenitor endothelial cells for angiogenesis. *Science.* 1997;5302:964–967.

5 Asahara T, Masuda H, Takahashi T, et al. Bone marrow origin of endothelial progenitor cells responsible for postnatal vasculogenesis in physiological and pathological neovascularization. *Circ Res.* 1999;3:221–228.

6 Gehling UM, Ergun S, Schumacher U, et al. In vitro differentiation of endothelial cells from AC133-positive progenitor cells. *Blood.* 2000;10:3106–3112.

7 Romagnani P, Annunziato F, Liotta F, et al. CD14+CD34low cells with stem cell phenotypic and functional features are the major source of circulating endothelial progenitors. *Circ Res.* 2005;4:314–322.

8 Carmeliet P. Mechanisms of angiogenesis and arteriogenesis. *Nat Med.* 2000;4:389–395.

9 Prior BM, Lloyd PG, Ren J, et al. Arteriogenesis: role of nitric oxide. *Endothelium.* 2003;4–5:207–216.

10 Kaushal S, Amiel GE, Guleserian KJ, et al. Functional small-diameter neovessels created using endothelial progenitor cells expanded ex vivo. *Nat Med.* 2001;9:1035–1040.

11 Kong D, Melo LG, Mangi AA, et al. Enhanced inhibition of neointimal hyperplasia by genetically engineered endothelial progenitor cells. *Circulation.* 2004;109(14):1769–1775.

12 Walter DH, Rittig K, Bahlmann FH, et al. Statin therapy accelerates reendothelialization: a novel effect involving mobilization and incorporation of bone marrow-derived endothelial progenitor cells. *Circulation.* 2002;25:3017–3024.

13 Hill JM, Zalos G, Halcox JP, et al. Circulating endothelial progenitor cells, vascular function, and cardiovascular risk. *N Engl J Med.* 2003;7:593–600.
14 Hill JM, Finkel T, Quyyumi AA. Endothelial progenitor cells and endothelial dysfunction. *Vox Sang.* 2004:31–37.
15 Quyyumi AA, Hill JM. Circulating endothelial progenitor cells as novel biological determinants of vascular function and risk. *Can J Cardiol.* 2004:B44–B48.
16 Shi Q, Wu MH, Onuki Y, et al. The effect of flow shear stress on endothelialization of impervious Dacron grafts from circulating cells in the arterial and venous systems of the same dog. *Ann Vasc Surg.* 1998;4:341–348.
17 Shi Q, Wu MH, Hayashida N, et al. Proof of fallout endothelialization of impervious Dacron grafts in the aorta and inferior vena cava of the dog. *J Vasc Surg.* 1994;4:546–556.
18 Edelman ER, Rogers C. Pathobiologic responses to stenting. *Am J Cardiol.* 1998;7A:E4–E6.
19 Komatsu R, Ueda M, Naruko T, et al. Neointimal tissue response at sites of coronary stenting in humans: macroscopic, histological, and immunohistochemical analyses. *Circulation.* 1998;3:224–233.
20 Kornowski R, Hong MK, Tio FO, et al. In-stent restenosis: contributions of inflammatory responses and arterial injury to neointimal hyperplasia. *J Am Coll Cardiol.* 1998;1:224–230.
21 Thompson MM, Budd JS, Eady SL, et al. A method to transluminally seed angioplasty sites with endothelial cells using a double balloon catheter. *Eur J Vasc Surg.* 1993;2:113–121.
22 Conte MS, Birinyi LK, Miyata T, et al. Efficient repopulation of denuded rabbit arteries with autologous genetically modified endothelial cells. *Circulation.* 1994;5:2161–2169.
23 Nabel EG, Plautz G, Boyce FM, et al. Recombinant gene expression in vivo within endothelial cells of the arterial wall. *Science.* 1989;4910:1342–1344.
24 Schneider PA, Hanson SR, Price TM, Harker LA. Confluent durable endothelialization of endarterectomized baboon aorta by early attachment of cultured endothelial cells. *J Vasc Surg.* 1990;3:365–372.
25 Sterpetti AV, Schultz RD, Bailey RT. Endothelial cell seeding after carotid endarterectomy in a canine model reduces platelet uptake. *Eur J Vasc Surg.* 1992;4:390–394.
26 Thompson MM, Budd JS, Eady SL, et al. Endothelial cell seeding of damaged native vascular surfaces: prostacyclin production. *Eur J Vasc Surg.* 1992;5:487–493.
27 Nugent HM, Edelman ER. Endothelial implants provide long-term control of vascular repair in a porcine model of arterial injury. *J Surg Res.* 2001;2:228–234.
28 Conte MS, VanMeter GA, Akst LM, et al. Endothelial cell seeding influences lesion development following arterial injury in the cholesterol-fed rabbit. *Cardiovasc Res.* 2002;2:502–511.
29 Griese DP, Ehsan A, Melo LG, et al. Isolation and transplantation of autologous circulating endothelial cells into denuded vessels and prosthetic grafts: implications for cell-based vascular therapy. *Circulation.* 2003;21:2710–2715.
30 Werner N, Junk S, Laufs U, et al. Intravenous transfusion of endothelial progenitor cells reduces neointima formation after vascular injury. *Circ Res.* 2003;2:E17–E24.
31 Strehlow K, Werner N, Berweiler J, et al. Estrogen increases bone marrow-derived endothelial progenitor cell production and diminishes neointima formation. *Circulation.* 2003;24:3059–3065.
32 van der Giessen WJ, Serruys PW, Visser WJ. Endothelialization of intravascular stents. *J Interv Cardiol.* 1988;1:109–120.
33 Aoki J, Serruys PW, van Beusekom H, et al. Endothelial progenitor cell capture by stents coated with antibody against CD34: the HEALING-FIM (healthy endothelial accelerated lining inhibits neointimal growth-first in man) Registry. *J Am Coll Cardiol.* 2005;10:1574–1579.
34 Sauvage LR, Berger K, Beilin LB, et al. Presence of endothelium in an axillary-femoral graft of knitted Dacron with an external velour surface. *Ann Surg.* 1975;6:749–753.
35 Herring M, Gardner A, Glover J. A single-staged technique for seeding vascular grafts with autogenous endothelium. *Surgery.* 1978;4:498–504.
36 Zilla P, Fasol R, Dudeck U, et al. In situ cannulation, microgrid follow-up and low-density plating provide first passage endothelial cell mass cultures for in vitro lining. *J Vasc Surg.* 1990;2:180–189.
37 Mansfield PB, Wechezak AR, Sauvage LR. preventing thrombus on artificial vascular surfaces: true endothelial cell linings. *Trans Am Soc Artif Intern Organs.* 1975;264–272.
38 Herring M, Gardner A, Glover J. Seeding endothelium onto canine arterial prostheses. The effects of graft design. *Arch Surg.* 1979;6:679–682.
39 Allen BT, Long JA, Clark RE, et al. Influence of endothelial cell seeding on platelet deposition and patency in small-diameter Dacron arterial grafts. *J Vasc Surg.* 1984;1:224–233.
40 Belden TA, Schmidt SP, Falkow LJ, Sharp WV. Endothelial cell seeding of small-diameter vascular grafts. *Trans Am Soc Artif Intern Organs.* 1982:173–177.
41 Burkel WE, Vinter DW, Ford JW, et al. Sequential studies of healing in endothelial seeded vascular prostheses: histologic and ultrastructure characteristics of graft incorporation. *J Surg Res.* 1981;4:305–324.
42 Graham LM, Burkel WE, Ford JW, et al. Immediate seeding of enzymatically derived endothelium in Dacron vascular grafts. early experimental studies with autologous canine cells. *Arch Surg.* 1980;11:1289–1294.
43 Herring M, Baughman S, Glover J, et al. Endothelial seeding of Dacron and polytetrafluoroethylene grafts: the cellular events of healing. *Surgery.* 1984;4:745–755.
44 Kempczinski RF, Rosenman JE, Pearce WH, et al. Endothelial cell seeding of a new PTFE vascular prosthesis. *J Vasc Surg.* 1985;3:424–429.
45 Schmidt SP, Hunter TJ, Falkow LJ, et al. Effects of antiplatelet agents in combination with endothelial cell seeding on small-diameter Dacron vascular graft performance in the canine carotid artery model. *J Vasc Surg.* 1985;6:898–906.
46 Schmidt SP, Hunter TJ, Hirko M, et al. Small-diameter vascular prostheses: two designs of PTFE and endothelial cell-seeded and nonseeded Dacron. *J Vasc Surg.* 1985;2:292–297.
47 Stanley JC, Burkel WE, Ford JW, et al. Enhanced patency of small-diameter, externally supported Dacron iliofemoral grafts seeded with endothelial cells. *Surgery.* 1982;6:994–1005.
48 Herring M, Baughman S, Glover J. Endothelium develops on seeded human arterial prosthesis: a brief clinical note. *J Vasc Surg.* 1985;5:727–730.
49 Herring M, Gardner A, Glover J. Seeding human arterial prostheses with mechanically derived endothelium. The detrimental effect of smoking. *J Vasc Surg.* 1984;2:279–289.
50 Herring M, Smith J, Dalsing M, et al. Endothelial seeding of polytetrafluoroethylene femoral popliteal bypasses: the failure

of low-density seeding to improve patency. *J Vasc Surg*. 1994; 4:650–655.
51 Herring MB, Compton RS, LeGrand DR, et al. Endothelial seeding of polytetrafluoroethylene popliteal bypasses. A preliminary report. *J Vasc Surg*. 1987;2:114–118.
52 Jensen N, Lindblad B, Bergqvist D. Endothelial cell seeded Dacron aortobifurcated grafts: platelet deposition and long-term follow-up. *J Cardiovasc Surg (Torino)*. 1994;5:425–429.
53 Ortenwall P, Wadenvik H, Kutti J, Risberg B. Reduction in deposition of indium 111-labeled platelets after autologous endothelial cell seeding of Dacron aortic bifurcation grafts in humans: a preliminary report. *J Vasc Surg*. 1987;1:17–25.
54 Ortenwall P, Wadenvik H, Risberg B. Reduced platelet deposition on seeded versus unseeded segments of expanded polytetrafluoroethylene grafts: clinical observations after a 6-month follow-up. *J Vasc Surg*. 1989;4:374–380.
55 Ortenwall P, Wadenvik H, Kutti J, Risberg B. Endothelial cell seeding reduces thrombogenicity of Dacron grafts in humans. *J Vasc Surg*. 1990;3:403–410.
56 Smyth JV, Welch M, Carr HM, et al. Fibrinolysis profiles and platelet activation after endothelial cell seeding of prosthetic vascular grafts. *Ann Vasc Surg*. 1995;6:542–546.
57 Zilla P, Fasol R, Deutsch M, et al. Endothelial cell seeding of polytetrafluoroethylene vascular grafts in humans: a preliminary report. *J Vasc Surg*. 1987;6:535–541.
58 Hess F, Steeghs S, Jerusalem R, et al. Patency and morphology of fibrous polyurethane vascular prostheses implanted in the femoral artery of dogs after seeding with subcultivated endothelial cells. *Eur J Vasc Surg*. 1993;4:402–408.
59 Koveker GB, Graham LM, Burkel WE, et al. Extracellular matrix preparation of expanded polytetrafluoroethylene grafts seeded with endothelial cells: influence on early platelet deposition, cellular growth, and luminal prostacyclin release. *Surgery*. 1991;3(Pt 1):313–319.
60 Seeger JM, Klingman N. Improved endothelial cell seeding with cultured cells and fibronectin-coated grafts. *J Surg Res*. 1985;6:641–647.
61 Seeger JM, Klingman N. Improved in vivo endothelialization of prosthetic grafts by surface modification with fibronectin. *J Vasc Surg*. 1988;4:476–482.
62 Zilla P, Preiss P, Groscurth P, et al. In vitro-lined endothelium: initial integrity and ultrastructural events. *Surgery*. 1994;3:524–534.
63 Zilla P, Deutsch M, Meinhart J, et al. Clinical in vitro endothelialization of femoropopliteal bypass grafts: an actuarial follow-up over three years. *J Vasc Surg*. 1994;3:540–548.
64 Deutsch M, Meinhart J, Vesely M, et al. In vitro endothelialization of expanded polytetrafluoroethylene grafts: a clinical case report after 41 months of implantation. *J Vasc Surg*. 1997;4:757–763.
65 Deutsch M, Meinhart J, Fischlein T, et al. Clinical autologous in vitro endothelialization of infrainguinal EPTFE grafts in 100 patients: a 9-year experience. *Surgery*. 1999;5:847–855.
66 Hsu S, Tseng H, Wu M. Comparative in vitro evaluation of two different preparations of small diameter polyurethane vascular grafts. *Artif Organs*. 2000;2:119–128.
67 Laube HR, Duwe J, Rutsch W, Konertz W. Clinical experience with autologous endothelial cell-seeded polytetrafluoroethylene coronary artery bypass grafts. *J Thorac Cardiovasc Surg*. 2000;1:134–141.
68 Magometschnigg H, Kadletz M, Vodrazka M, et al. Prospective clinical study with in vitro endothelial cell lining of expanded polytetrafluoroethylene grafts in crural repeat reconstruction. *J Vasc Surg*. 1992;3:527–535.
69 Meinhart JG, Deutsch M, Fischlein T, et al. Clinical autologous in vitro endothelialization of 153 infrainguinal EPTFE grafts. *Ann Thorac Surg*. 2001;(Suppl 5):S327–S331.
70 Swedenborg J, Bengtsson L, Clyne N, et al. In vitro endothelialisation of arteriovenous loop grafts for haemodialysis. *Eur J Vasc Endovasc Surg*. 1997;3:272–277.
71 Williams SK, Kleinert LB, Rose D, McKenney S. Origin of endothelial cells that line expanded polytetrafluorethylene vascular grafts sodded with cells from microvascularized fat. *J Vasc Surg* 1994;4:594–604.
72 Williams SK, Rose DG, Jarrell BE. Microvascular endothelial cell sodding of EPTFE vascular grafts: improved patency and stability of the cellular lining. *J Biomed Mater Res*. 1994;2:203–212.
73 Ahlswede KM, Williams SK. Microvascular endothelial cell sodding of 1-mm expanded polytetrafluoroethylene vascular grafts. *Arterioscler Thromb*. 1994;1:25–31.
74 Wang ZG, Li G, Wu J, et al. Enhanced patency of venous Dacron grafts by endothelial cell sodding. *Ann Vasc Surg*. 1993;5:429–436.
75 Williams SK, Carter T, Park PK, et al. Formation of a multilayer cellular lining on a polyurethane vascular graft following endothelial cell sodding. *J Biomed Mater Res*. 1992;1:103–117.
76 Williams SK, Schneider T, Kapelan B, Jarrell BE. Formation of a functional endothelium on vascular grafts. *J Electron Microsc Tech*. 1991;4:439–451.
77 Park PK, Jarrell BE, Williams SK, et al. Thrombus-free, human endothelial surface in the midregion of a Dacron vascular graft in the splanchnic venous circuit – observations after nine months of implantation. *J Vasc Surg*. 1990;3:468–475.
78 Meerbaum SO, Sharp WV, Schmidt SP. Lower extremity revascularization with polytetrafluoroethylene grafts seeded with microvascular endothelial cells. In: Zilla P, Fasol R, Callow A, eds. *Applied Cardiovascular Biology*. Basel: Karger; 1990:107–119.
79 Williams SK. Human clinical trials of microvascular endothelial cell sodding. In: Zilla P, Griesler HP, eds. *Tissue Engineering of Prosthetic Vascular Grafts*. Austin, TX: RG Landes Company; 1999:143–7.
80 Shi Q, Wu MH, Fujita Y, et al. Genetic tracing of arterial graft flow surface endothelialization in allogeneic marrow transplanted dogs. *Cardiovasc Surg*. 1999;1:98–105.
81 Rotmans JI, Heyligers JM, Verhagen HJ, et al. In vivo cell seeding with anti-cd34 antibodies successfully accelerates endothelialization but stimulates intimal hyperplasia in porcine arteriovenous expanded polytetrafluoroethylene grafts. *Circulation*. 2005;1:12–18.
82 Bhattacharya V, McSweeney PA, Shi Q, et al. Enhanced endothelialization and microvessel formation in polyester grafts seeded with CD34(+) bone marrow cells. *Blood*. 2000;2:581–585.
83 Shi Q, Bhattacharya V, Hong-De Wu M, Sauvage LR. Utilizing granulocyte colony-stimulating factor to enhance vascular graft endothelialization from circulating blood cells. *Ann Vasc Surg*. 2002;3:314–320.
84 Dichek DA, Anderson J, Kelly AB, et al. Enhanced in vivo antithrombotic effects of endothelial cells expressing recombinant plasminogen activators transduced with retroviral vectors. *Circulation*. 1996;2:301–309.

85 Dunn PF, Newman KD, Jones M, et al. Seeding of vascular grafts with genetically modified endothelial cells. Secretion of recombinant TPA results in decreased seeded cell retention in vitro and in vivo. *Circulation.* 1996;7:1439–1446.

86 Fazel S, Angoulvant D, Desai N, et al. Cardiac restoration: frontier or fantasy? *Can J Cardiol.* 2005;4:355–359.

87 Menasche P. Skeletal myoblast for cell therapy. *Coron Artery Dis.* 2005;2:105–110.

88 Orlic D, Kajstura J, Chimenti S, et al. Transplanted adult bone marrow cells repair myocardial infarcts in mice. *Ann NY Acad Sci.* 2001;221–229.

89 Balsam LB, Wagers AJ, Christensen JL, et al. Haematopoietic stem cells adopt mature haematopoietic fates in ischaemic myocardium. *Nature.* 2004;6983:668–673.

90 Murry CE, Soonpaa MH, Reinecke H, et al. Haematopoietic stem cells do not transdifferentiate into cardiac myocytes in myocardial infarcts. *Nature.* 2004;6983:664–668.

91 Kamihata H, Matsubara H, Nishiue T, et al. Implantation of bone marrow mononuclear cells into ischemic myocardium enhances collateral perfusion and regional function via side supply of angioblasts, angiogenic ligands, and cytokines. *Circulation.* 2001;9:1046–1052.

92 Kawamoto A, Gwon HC, Iwaguro H, et al. Therapeutic potential of ex vivo expanded endothelial progenitor cells for myocardial ischemia. *Circulation.* 2001;5:634–637.

93 Kawamoto A, Tkebuchava T, Yamaguchi J, et al. Intramyocardial transplantation of autologous endothelial progenitor cells for therapeutic neovascularization of myocardial ischemia. *Circulation.* 2003;3:461–468.

94 Badorff C, Brandes RP, Popp R, et al. Transdifferentiation of blood-derived human adult endothelial progenitor cells into functionally active cardiomyocytes. *Circulation.* 2003;7:1024–1032.

95 Rupp S, Badorff C, Koyanagi M, et al. Statin therapy in patients with coronary artery disease improves the impaired endothelial progenitor cell differentiation into cardiomyogenic cells. *Basic Res Cardiol.* 2004;1:61–68.

96 Yeh ET, Zhang S, Wu HD, et al. Transdifferentiation of human peripheral blood CD34+-enriched cell population into cardiomyocytes, endothelial cells, and smooth muscle cells in vivo. *Circulation.* 2003;17:2070–2073.

97 Makino S, Fukuda K, Miyoshi S, et al. Cardiomyocytes can be generated from marrow stromal cells in vitro. *J Clin Invest.* 1999;5:697–705.

98 Mangi AA, Noiseux N, Kong D, et al. Mesenchymal stem cells modified with Akt prevent remodeling and restore performance of infarcted hearts. *Nat Med.* 2003;9:1195–1201.

99 Toma C, Pittenger MF, Cahill KS, et al. Human mesenchymal stem cells differentiate to a cardiomyocyte phenotype in the adult murine heart. *Circulation.* 2002;1:93–98.

100 Shake JG, Gruber PJ, Baumgartner WA, et al. Mesenchymal stem cell implantation in a swine myocardial infarct model: engraftment and functional effects. *Ann Thorac Surg.* 2002;6:1919–1925.

101 Ohtsuka M, Takano H, Zou Y, et al. Cytokine therapy prevents left ventricular remodeling and dysfunction after myocardial infarction through neovascularization. *FASEB J.* 2004;7:851–853.

102 Orlic D, Kajstura J, Chimenti S, et al. Mobilized bone marrow cells repair the infarcted heart, improving function and survival. *Proc Natl Acad Sci USA.* 2001;18:10344–10349.

103 Harada M, Qin Y, Takano H, et al. G-CSF prevents cardiac remodeling after myocardial infarction by activating the jak-stat pathway in cardiomyocytes. *Nat Med.* 2005;3:305–311.

104 Minatoguchi S, Takemura G, Chen XH, et al. Acceleration of the healing process and myocardial regeneration may be important as a mechanism of improvement of cardiac function and remodeling by postinfarction granulocyte colony-stimulating factor treatment. *Circulation.* 2004;21:2572–2580.

105 Conti JA, Scher HI. Acute arterial thrombosis after escalated-dose methotrexate, vinblastine, doxorubicin, and cisplatin chemotherapy with recombinant granulocyte colony-stimulating factor. A possible new recombinant granulocyte colony-stimulating factor toxicity. *Cancer.* 1992;11:2699–2702.

106 Fukumoto Y, Miyamoto T, Okamura T, et al. Angina pectoris occurring during granulocyte colony-stimulating factor-combined preparatory regimen for autologous peripheral blood stem cell transplantation in a patient with acute myelogenous leukaemia. *Br J Haematol.* 1997;3:666–668.

107 Hill JM, Syed MA, Arai AE, et al. Outcomes and risks of granulocyte colony-stimulating factor in patients with coronary artery disease. *J Am Coll Cardiol.* 2005;9:1643–1648.

108 Vij R, Adkins DR, Brown RA, et al. Unstable angina in a peripheral blood stem and progenitor cell donor given granulocyte-colony-stimulating factor. *Transfusion.* 1999;5:542–543.

109 Assmus B, Schachinger V, Teupe C, et al. Transplantation of progenitor cells and regeneration enhancement in acute myocardial infarction (TOPCARE-AMI). *Circulation.* 2002;24:3009–3017.

110 Fernandez-Aviles F, San Roman JA, Garcia-Frade J, et al. Experimental and clinical regenerative capability of human bone marrow cells after myocardial infarction. *Circ Res.* 2004;7:742–748.

111 Kuethe F, Richartz BM, Sayer HG, et al. Lack of regeneration of myocardium by autologous intracoronary mononuclear bone marrow cell transplantation in humans with large anterior myocardial infarctions. *Int J Cardiol.* 2004;1:123–127.

112 Schachinger V, Assmus B, Britten MB, et al. Transplantation of progenitor cells and regeneration enhancement in acute myocardial infarction: final one-year results of the TOPCARE-AMI trial. *J Am Coll Cardiol.* 2004;8:1690–1699.

113 Strauer BE, Brehm M, Zeus T, et al. Repair of infarcted myocardium by autologous intracoronary mononuclear bone marrow cell transplantation in humans. *Circulation.* 2002;15:1913–1918.

114 Wollert KC, Meyer GP, Lotz J, et al. Intracoronary autologous bone-marrow cell transfer after myocardial infarction: the BOOST randomised controlled clinical trial. *Lancet.* 2004;9429:141–148.

115 Meyer GP, Wollert KC, Lotz J, et al. Intracoronary bone marrow cell transfer after myocardial infarction; eighteen months' follow-up data from the randomized controlled BOOST (Bone Marrow Transfer to Enhance ST-Elevation Infarct Regeneration) trial. *Circulation.* 2006;113:1287–1294.

116 Stamm C, Westphal B, Kleine HD, et al. Autologous bone-marrow stem-cell transplantation for myocardial regeneration. *Lancet.* 2003;9351:45–46.

117 Stamm C, Kleine HD, Westphal B, et al. CABG and bone marrow stem cell transplantation after myocardial infarction. *Thorac Cardiovasc Surg.* 2004;3:152–158.

118 Janssens S, Dubois C, Bogaert J, et al. Autologous bone marrow-derived stem-cell transfer in patients with ST-segment elevation myocardial infarction: double-blind, randomised controlled trial. *Lancet.* 2006;367:113–121.

119 Schachinger V, Erbs S, Elsasser A, et al. Improved clinical outcome after intracoronary administration of bone-marrow-derived progenitor cells in acute myocardial infarction: final 1-year results of the REPAIR-AMI trial. *Eur Heart J*. 2006;27:2775–2783.

120 Lunde K, Solheim S, Aakhus S, et al. Autologous stem cell transplantation in acute myocardial infarction: the ASTAMI randomized controlled trial. Intracoronary transplantation of autologous mononuclear bone marrow cells, study design and safety aspects. *Scand Cardiovasc J*. 2005;39:150–158.

121 Lunde K, Solheim S, Aakhus S, et al. Intracoronary injection of mononuclear bone marrow cells in acute myocardial infarction. *N Engl J Med*. 2006;355:1199–1209.

122 Seeger F, Tonn T, Krzossok N, et al. Cell isolation procedures matter: a comparison of different isolation protocols of bone marrow mononuclear cells used for cell therapy in patients with acute myocardial infarction. *Circulation*. 2006.

123 Chen SL, Fang WW, Qian J, et al. Improvement of cardiac function after transplantation of autologous bone marrow mesenchymal stem cells in patients with acute myocardial infarction. *Chin Med J (Engl)*. 2004;10:1443–1448.

124 Chen SL, Fang WW, Ye F, et al. Effect on left ventricular function of intracoronary transplantation of autologous bone marrow mesenchymal stem cell in patients with acute myocardial infarction. *Am J Cardiol*. 2004;1:92–95.

125 Vulliet PR, Greeley M, Halloran SM, et al. Intra-coronary arterial injection of mesenchymal stromal cells and microinfarction in dogs. *Lancet*. 2004;9411:783–784.

126 Kang HJ, Kim HS, Zhang SY, et al. Effects of intracoronary infusion of peripheral blood stem-cells mobilised with granulocyte-colony stimulating factor on left ventricular systolic function and restenosis after coronary stenting in myocardial infarction: the MAGIC cell randomised clinical trial. *Lancet*. 2004;9411:751–756.

127 Murohara T, Asahara T, Silver M, et al. Nitric oxide synthase modulates angiogenesis in response to tissue ischemia. *J Clin Invest*. 1998;11:2567–2578.

128 Takahashi T, Kalka C, Masuda H, et al. Ischemia- and cytokine-induced mobilization of bone marrow-derived endothelial progenitor cells for neovascularization. *Nat Med*. 1999;4:434–438.

129 Shintani S, Murohara T, Ikeda H, et al. Augmentation of postnatal neovascularization with autologous bone marrow transplantation. *Circulation*. 2001;6:897–903.

130 Kalka C, Masuda H, Takahashi T, et al. Transplantation of ex vivo expanded endothelial progenitor cells for therapeutic neovascularization. *Proc Natl Acad Sci USA*. 2000;7:3422–3427.

131 Schatteman GC, Hanlon HD, Jiao C, et al. Blood-derived angioblasts accelerate blood-flow restoration in diabetic mice. *J Clin Invest*. 2000;4:571–578.

132 Kocher AA, Schuster MD, Szabolcs MJ, et al. Neovascularization of ischemic myocardium by human bone-marrow-derived angioblasts prevents cardiomyocyte apoptosis, reduces remodeling and improves cardiac function. *Nat Med*. 2001;4:430–436.

133 Yang C, Zhang ZH, Lu SH, et al. Transplantation of cord blood endothelial progenitor cells ameliorates limb ischemia. *Zhonghua Yi Xue Za Zhi*. 2003;16:1437–1441.

134 Ikenaga S, Hamano K, Nishida M, et al. Autologous bone marrow implantation induced angiogenesis and improved deteriorated exercise capacity in a rat ischemic hindlimb model. *J Surg Res*. 2001;2:277–283.

135 Li TS, Hamano K, Suzuki K, et al. Improved angiogenic potency by implantation of ex vivo hypoxia prestimulated bone marrow cells in rats. *Am J Physiol Heart Circ Physiol*. 2002;2:H468–H473.

136 Nishida M, Li TS, Hirata K, et al. Improvement of cardiac function by bone marrow cell implantation in a rat hypoperfusion heart model. *Ann Thorac Surg*. 2003;3:768–773.

137 Fuchs S, Baffour R, Zhou YF, et al. Transendocardial delivery of autologous bone marrow enhances collateral perfusion and regional function in pigs with chronic experimental myocardial ischemia. *J Am Coll Cardiol*. 2001;6:1726–1732.

138 Vicario J, Piva J, Pierini A, et al. Transcoronary sinus delivery of autologous bone marrow and angiogenesis in pig models with myocardial injury. *Cardiovasc Radiat Med*. 2002;2:91–94.

139 Hamano K, Li TS, Kobayashi T, et al. Therapeutic angiogenesis induced by local autologous bone marrow cell implantation. *Ann Thorac Surg*. 2002;4:1210–1215.

140 Grant MB, May WS, Caballero S, et al. Adult hematopoietic stem cells provide functional hemangioblast activity during retinal neovascularization. *Nat Med*. 2002;6:607–612.

141 Reyes M, Dudek A, Jahagirdar B, et al. Origin of endothelial progenitors in human postnatal bone marrow. *J Clin Invest*. 2002;3:337–346.

142 Schatteman GC, Awad O. In vivo and in vitro properties of $CD34^+$ and $CD14^+$ endothelial cell precursors. *Adv Exp Med Biol*. 2003:9–16.

143 Yamaguchi J, Kusano KF, Masuo O, et al. Stromal cell-derived factor-1 effects on ex vivo expanded endothelial progenitor cell recruitment for ischemic neovascularization. *Circulation*. 2003;9:1322–1328.

144 Hur J, Yoon CH, Kim HS, et al. Characterization of two types of endothelial progenitor cells and their different contributions to neovasculogenesis. *Arterioscler Thromb Vasc Biol*. 2004;2:288–293.

145 Kocher AA, Schuster MD, Szabolcs MJ, et al. Neovascularization of ischemic myocardium by human bone-marrow-derived angioblasts prevents cardiomyocyte apoptosis, reduces remodeling and improves cardiac function. *Nat Med*. 2001;4:430–436.

146 Kocher AA, Schuster MD, Szabolcs MJ, et al. Neovascularization of ischemic myocardium by human bone-marrow-derived angioblasts prevents cardiomyocyte apoptosis, reduces remodeling and improves cardiac function. *Nat Med*. 2001;4:430–436.

147 Urbich C, Dimmeler S. Endothelial progenitor cells: characterization and role in vascular biology. *Circ Res*. 2004;4:343–353.

148 Van Royen N, Hoefer I, Bottinger M, et al. Local monocyte chemoattractant protein-1 therapy increases collateral artery formation in apolipoprotein e-deficient mice but induces systemic monocytic CD11b expression, neointimal formation, and plaque progression. *Circ Res*. 2003;2:218–225.

149 Heil M, Ziegelhoeffer T, Pipp F, et al. Blood monocyte concentration is critical for enhancement of collateral artery growth. *Am J Physiol Heart Circ Physiol*. 2002;6:H2411–H2419.

150 Pipp F, Heil M, Issbrucker K, et al. VEGFR-1-selective VEGF homologue PlGF is arteriogenic: evidence for a monocyte-mediated mechanism. *Circ Res*. 2003;4:378–385.

151 Murayama T, Tepper OM, Silver M, et al. Determination of bone marrow-derived endothelial progenitor cell significance in angiogenic growth factor-induced neovascularization in vivo. *Exp Hematol*. 2002;8:967–972.

152 Crosby JR, Kaminski WE, Schatteman G, et al. Endothelial cells of hematopoietic origin make a significant contribution to adult blood vessel formation. *Circ Res*. 2000;9:728–730.

153 Jackson KA, Majka SM, Wang H, et al. Regeneration of ischemic cardiac muscle and vascular endothelium by adult stem cells. *J Clin Invest*. 2001;11:1395–1402.

154 Llevadot J, Murasawa S, Kureishi Y, et al. HMG-CoA reductase inhibitor mobilizes bone marrow – derived endothelial progenitor cells. *J Clin Invest*. 2001;3:399–405.

155 Lyden D, Hattori K, Dias S, et al. Impaired recruitment of bone-marrow-derived endothelial and hematopoietic precursor cells blocks tumor angiogenesis and growth. *Nat Med*. 2001;11:1194–1201.

156 Gill M, Dias S, Hattori K, et al. Vascular trauma induces rapid but transient mobilization of VEGFR2(+)AC133(+) endothelial precursor cells. *Circ Res*. 2001;2:167–174.

157 Aicher A, Heeschen C, Mildner-Rihm C, et al. Essential role of endothelial nitric oxide synthase for mobilization of stem and progenitor cells. *Nat Med*. 2003;11:1370–1376.

158 Urbich C, Heeschen C, Aicher A, et al. Relevance of monocytic features for neovascularization capacity of circulating endothelial progenitor cells. *Circulation*. 2003;20:2511–2516.

159 Kocher AA, Schuster MD, Szabolcs MJ, et al. Neovascularization of ischemic myocardium by human bone-marrow-derived angioblasts prevents cardiomyocyte apoptosis, reduces remodeling and improves cardiac function. *Nat Med*. 2001;4: 430–436.

160 Ii M, Nishimura H, Iwakura A, et al. Endothelial progenitor cells are rapidly recruited to myocardium and mediate protective effect of ischemic preconditioning via "imported" nitric oxide synthase activity. *Circulation*. 2005;111(9):1114–1120.

161 Rehman J, Li J, Orschell CM, March KL. Peripheral blood "endothelial progenitor cells" are derived from monocyte/macrophages and secrete angiogenic growth factors. *Circulation*. 2003;8:1164–1169.

162 Folkman J. Angiogenesis in cancer, vascular, rheumatoid and other disease. *Nat Med*. 1995;1:27–31.

163 De Palma M, Venneri MA, Roca C, Naldini L. Targeting exogenous genes to tumor angiogenesis by transplantation of genetically modified hematopoietic stem cells. *Nat Med*. 2003;6:789–795.

164 Polverini PJ, Cotran PS, Gimbrone MA Jr, Unanue ER. Activated macrophages induce vascular proliferation. *Nature*. 1977; 5631:804–806.

165 Tateishi-Yuyama E, Matsubara H, Murohara T, et al. Therapeutic angiogenesis for patients with limb ischaemia by autologous transplantation of bone-marrow cells: a pilot study and a randomised controlled trial. *Lancet*. 2002;9331:427–435.

166 Lenk K, Adams V, Lurz P, et al. Therapeutical potential of blood-derived progenitor cells in patients with peripheral arterial occlusive disease and critical limb ischaemia. *Eur Heart J*. 2005;26(18):1903–1909.

167 Voelkel NF, Cool C. Pathology of pulmonary hypertension. *Cardiol Clin*. 2004;3:343–351, v.

168 Lane KB, Machado RD, Pauciulo MW, et al. Heterozygous germline mutations in BMPR2, encoding a TGF-beta receptor, cause familial primary pulmonary hypertension. The International PPH Consortium. *Nat Genet*. 2000;1:81–84.

169 Sitbon O, Humbert M, Simonneau G. Primary pulmonary hypertension: current therapy. *Prog Cardiovasc Dis*. 2002;2:115–128.

170 Zhao YD, Courtman DW, Deng Y, et al. Rescue of monocrotaline-induced pulmonary arterial hypertension using bone marrow-derived endothelial-like progenitor cells: efficacy of combined cell and ENOS gene therapy in established disease. *Circ Res*. 2005;4:442–450.

171 Duvall WL. Endothelial dysfunction and antioxidants. *Mt Sinai J Med*. 2005;2:71–80.

172 Naseem KM. The role of nitric oxide in cardiovascular diseases. *Mol Aspects Med*. 2005;1–2:33–65.

173 Giaid A, Saleh D. Reduced expression of endothelial nitric oxide synthase in the lungs of patients with pulmonary hypertension. *N Engl J Med*. 1995;4:214–221.

174 Campbell AI, Kuliszewski MA, Stewart DJ. Cell-based gene transfer to the pulmonary vasculature: endothelial nitric oxide synthase overexpression inhibits monocrotaline-induced pulmonary hypertension. *Am J Respir Cell Mol Biol*. 1999;5:567–575.

175 Campbell AI, Zhao Y, Sandhu R, Stewart DJ. Cell-based gene transfer of vascular endothelial growth factor attenuates monocrotaline-induced pulmonary hypertension. *Circulation*. 2001;18:2242–2248.

176 Nagaya N, Kangawa K, Kanda M, et al. Hybrid cell-gene therapy for pulmonary hypertension based on phagocytosing action of endothelial progenitor Cells. *Circulation*. 2003;7:889–895.

184

Building Blood Vessels

James B. Hoying and Stuart K. Williams

BIO5 Collaborative Research Institute, University of Arizona, Tucson

BLOOD VESSEL STRUCTURE AND FUNCTION

Diffusion, Convection, and the Circulation

Vertebrate cells depend on the molecular diffusion of nutrients for metabolism and function. Single-celled organisms such as bacteria or protists obtain nutrients directly from their respective environments. Here, distances between the medium and cell interiors are small and well within diffusion distance limitations (assuming sufficient local concentrations exist). In contrast, multicellular organisms have cells that are internal to the organism structure, making access to nutrients by simple diffusion from the external environment difficult or improbable. For example, the maximum distance oxygen (O_2) can diffuse in mammalian tissues is 20 to 100 μm (due primarily to a relatively low solubility in aqueous environments) (1,2). This limitation is exacerbated for larger nutrient molecules with smaller diffusion coefficients.

Many multicellular organisms overcome this diffusion limitation through two general mechanisms. First, a transport medium is used to contain and even concentrate nutrients. Second, this medium is delivered to the internal cells through some convective means such as active "pumping" of the medium or current-driven permeation through extracellular spaces. In vertebrates, the transport medium is blood, which contains, in addition to a vast array of molecular and cellular components, hemoglobin-carrying red blood cells. Hemoglobin in the blood overcomes the relatively low solubility of O_2 in aqueous environments and creates an "O_2 reservoir" that optimizes O_2 diffusion gradients from blood to tissue cells. In vertebrates, the cardiovascular system serves to move the blood throughout the body, bringing nutrients into close association with tissue cells, where diffusion can then be effective.

The cardiovascular system is a closed-loop distribution system comprising a series of interconnected conduits – blood vessels – that facilitate convective flow initiated by the heart. Although the blood vessel performs a number of critical functions in the body, its primary function is to serve as the conduit for blood flow and nutrient delivery. Consequently, the form and function of an individual blood vessel are well suited to enable effective transport of a fluid medium (i.e., blood). Considerable effort is being directed toward building vessels (either artificially or naturally) to serve as replacements for a diseased vessel (e.g., sclerotic or aneurysmal arteries) or to provide new avenues for supplemental blood delivery to a tissue (e.g., tissue revascularization). In addition, the ability to build vessels as vascular models enables further study of vascular biology, disease mechanisms, and therapies. In building blood vessels, it is important to consider those critical morphological and functional aspects of a blood vessel in the design and fabrication of the new vessel.

The Tube as a Transport Structure

One common feature of convective conduits, whether in the marine sponge wall or in the mammalian circulatory system, is that they are circular in cross-section. In round tubes, radial stresses are evenly applied to all points in the tube wall and surface drag is uniform along the tube's interior surfaces (3–6). Furthermore, a tubular structure maintains uniform flow patterns and minimizes turbulence, unlike other possible designs such as square or oval (in cross-section) conduits, which would yield differential flow gradients within the conduit. Overall, the tube, with its circular cross-section, makes for a stable and efficient distribution conduit and is the hallmark characteristic of blood vessels.

Macrovessels versus Capillaries

To move fluids over greater distances, as must occur in multicellular organisms, conduits must be interconnected to form a distribution system. The wall structure of the sponge is such that tunnels of various calibers are distributed throughout the sponge wall with larger-diameter tunnels branching into smaller tunnels. A similar arrangement occurs in the vertebrate circulatory system. However, because it is a closed loop, the circulatory system is generally arranged as a series of larger caliber vessels branching into smaller caliber vessels which progress back into larger caliber vessels. This "duality" in the network (small- and large-caliber vessels) resolves two general challenges related to convective transport in large multicellular organisms: (a) bulk flow distribution of blood over relatively long distances to all regions of the body, including the most distal tissues, from a centralized pump (via large-caliber vessels); and (b) diffusion of O_2 and other nutrients into the tissues (via small-caliber vessels). The large-caliber vessels (also termed *conduit vessels* or *macrovessels*) are represented by arteries and veins. The smallest caliber vessels (also termed *exchange vessels*) are represented by capillaries. Large-caliber conduit vessels feed small-caliber capillaries via intermediate-caliber arterioles. Similarly, capillary blood is collected into veins via intermediate-caliber venules. In addition to mediating flow to and from the capillaries, these intermediate vessels serve additional roles. The precapillary arterioles (also termed *resistance vessels*) function to tightly regulate blood flow to the capillaries. The postcapillary venules mediate leukocyte trafficking and permeability changes. Arterioles, capillaries, and venules are collectively referred to as *microvessels.*

Blood moves through the vascular system down a pressure gradient (established by the heart), with the highest pressures occurring in the upstream portions of the vascular network, the arteries, and the lowest pressures in the most downstream portions, the veins. Because the conduit vessels must individually transport the bulk of the blood (e.g., the most upstream vessel, the aorta, must singularly transport all of the blood moving into the systemic circulation), they have large diameters to accommodate larger blood volume flows (flow is proportional to the cross-sectional area). However, the blood pressure will act to push the wall of the vessel outward, resulting in a circumferential stress within the vessel wall. According to Laplace's Law, higher vessel diameters and internal pressures lead to larger circumferential wall stresses (7). Arteries have thick, muscular walls to normalize circumferential wall stress resulting from the higher intravascular pressures associated with blood pressurization and larger diameters. Veins, on the other hand, have thinner, more compliant walls and are capable of significant stretch and dilation. Even though veins have larger diameters than the corresponding arteries in the vascular tree (e.g., the vena cava has a larger diameter than the aorta), the reduced pressures help to reduce the increased circumferential stresses associated with the relatively larger diameters (7).

In contrast to the macrovessel, which serves primarily as a transport conduit, the capillary provides the necessary interface for diffusion of nutrients from the blood to the tissue cells. For effective exchange, (a) blood flow through the vessel must slow to provide enough time for diffusion from the vascular space to take place, (b) the vessel wall must be thin enough to not be a diffusion barrier, and (c) there must be a sufficiently large surface area for exchange.

The capillaries (as with other microvessels) are uniquely designed to meet these three challenges. Capillary diameters are in the range of approximately 5 to 10 μm. Arterioles, which are immediately upstream of capillaries in the vascular network, range in size from approximately 10 to 100 μm in diameter. The small caliber of capillaries and more significantly, the arterioles, help to slow the flow of blood due to the increases in resistance associated with the smaller diameters (resistance to flow is inversely proportional to the fourth power of the radius) (8). The smaller diameter of capillaries also reduces circumferential wall stress, preventing capillary rupture. The wall structure of the capillary comprises the endothelial cell (EC) and the peri-EC (or perivascular cell). The number of perivascular cell layers is few, with the most distal vessel elements containing only one or a partial outer cell layer. The organization of ECs and perivascular cells is tightly structured, with only a basement membrane separating the two layers in the smaller vessel segments (Figure 184.1). In healthy vasculature, the capillary EC body is attenuated and may range in thickness from a few microns (across the nucleus) to submicron dimensions in the thinnest areas. Consequently, vessel diameter (nutrients must diffuse through the blood to reach the capillary wall) and the distance from the capillary wall to the target tissue cells comprise the greatest distance-related

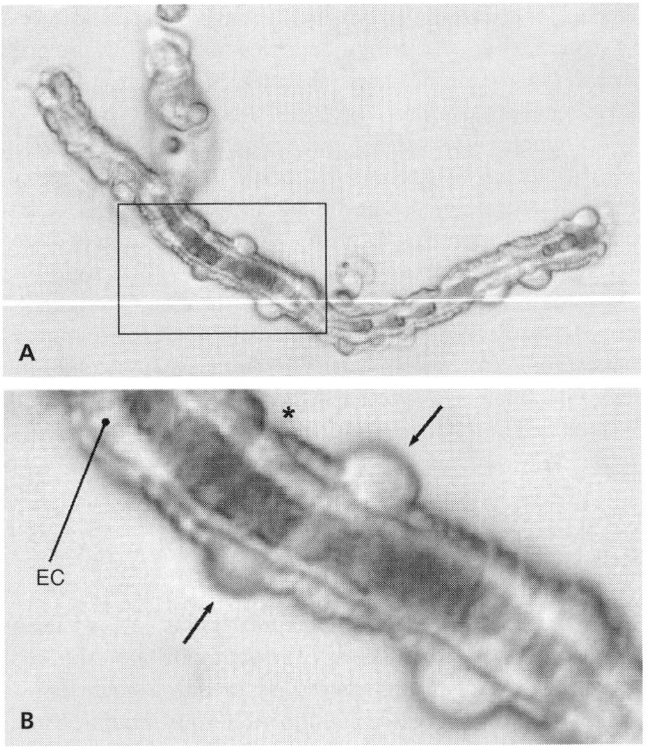

Figure 184.1. Phase micrograph of a microvessel fragment. **A** panel is a low magnification image showing a longitudinal cross-section of a 7–8 micron internal diameter microvessel comprising two cell layers. The dark cells in the lumen are red blood cells packed together. **B** is a higher magnification of the box in the upper panel showing the compact and intimate organization of perivascular cells, with prominent nuclei (*arrows*), and the EC layer. Separating these two layers is only a thin basement membrane. The perivascular layer covers most of the EC tube with only occasional gaps in the coverage (*). Any design strategy for building a microvessel must address this unique architecture.

barriers for diffusion (9). Finally, although not specific to vessel structure per se, the high densities of capillaries provides for a very large surface area (as much as 1,200 m^2) (10) for diffusion and, therefore, effective blood–tissue exchange.

Although vessel caliber and wall thickness are certainly hallmark features of specific vessel types (e.g., arteries are vessels with large diameters and thick, muscular walls), this is a relative description. Vessels of approximately 100 μm in diameter in larger animals are routinely used in studies examining arteriole physiology (11–14). However, the femoral artery of a mouse, clearly a conduit vessel in function, is also approximately 100 μm in diameter. Therefore, when establishing design parameters for a constructed vessel, it is important that the intended function and position of the vessel in the vascular tree are known.

BUILDING A BLOOD VESSEL

As discussed, macrovessels and capillaries display significant differences in vessel wall structure and composition. Although both macrovessels and capillaries are lined by a single monolayer of ECs, conduit vessels contain numerous circumferential layers of smooth muscle cells and extracellular matrix, whereas capillaries contain considerably less cell mass and extracellular matrix in their walls. Despite these significant differences, both vessel types conform to the shape of a tube. Therefore, in building a blood vessel, the vessel construct must be (a) configured as a tube, and (b) customized to serve the specific functional role of the vessel in the vasculature.

Building Macrovessels (Conduit Vessels)

The motivation for building large conduit vessels relates to the clinical need to bypass/replace diseased conduit vessels. Arterial occlusive diseases are numerous in Western countries and have a serious impact on human health. The occlusion effectively disrupts the convective delivery of blood to downstream tissues, resulting in tissue ischemia, dysfunction, and/or death. The obvious solution is to re-establish conduit flow either through reopening the occluded vessel or replacing the diseased vessel segment with a new, patent vessel substitute.

Autologous Conduit Vessels

The development of larger diameter blood vessels for bypassing or replacing diseased arterial segments was based initially on the use of autologous tissue such as a patient's own segment of vein. The redundancy of the venous circulation made it possible to utilize a vein from one region of the body to provide arterial bypass grafting. These procedures showed initial promise and success, and the use of veins for arterial bypass remains a common surgical procedure today. However, it was quickly realized that autologous vein grafting did not permit replacement of large blood vessels such as the aorta. It also has been recognized that veins must adapt to the arterial circulation, and this adaptation often results in failure of the veins once placed in the arterial position (15–18).

Synthetic Conduit Vessels

The first use of synthetic materials to build a blood vessel has been credited to Vorhees, Jaretzki, and Blakemore in 1952, with their pioneering work using Vinyon-N fabric to construct a replacement aorta for patients (19). This material was later replaced by Dacron and expanded polytetrafluoroethylene (ePTFE) for use in a wide variety of arterial bypass procedures. Of interest, these earliest studies using Vinyon-N resulted in the spontaneous formation of an EC lining in the midportion of some areas of these grafts. Identifying the source of these luminal ECs has remained a major goal since this early report and has been expanded recently to include the evaluation of circulating ECs and endothelial progenitor cells. Pioneering work by Sauvage and Wesolowski established that tissue ingrowth, including ingrowth of new blood vessels, into these materials following their implantation is critical to their extended patency (20–22). However, it was soon determined that the success of these synthetic arterial conduits was highly dependent on their internal diameter with a critical diameter cutoff of 6 mm. For the last 25 years it has been widely held that synthetic grafts with internal diameters less than 6 mm exhibit clinically unacceptably low patency rates (23). This means that patients needing arterial bypass to smaller vessels – for example, in the lower leg or the coronary circulation – must have a native vessel conduit to assure acceptable patency.

The patient needing a bypass vessel of diameter less than 6 mm who has no acceptable native vessel conduit often faces amputation of a limb or, when the target vessel is a coronary artery, significant morbidity or mortality. Based on extensive research, it is believed that the failure of synthetic grafts in small-diameter positions is related to the lack of a stable, antithrombogenic EC lining on the luminal surface. Implanted synthetic grafts form a pseudointima on the luminal surface, which is composed primarily of fibrin and platelets. These synthetic grafts, regardless of their inner diameter, never spontaneously endothelialize, and grafts explanted from patients after more than 10 years of implantation continue to exhibit a thrombogenic luminal surface. A simple solution to this thrombogenicity, hypothesized over 25 years ago, has been the accelerated formation of a stable EC lining on the blood-contacting surface of synthetic grafts. This simple premise has led to widespread efforts to create a living blood vessel from synthetic and biological components. These studies have progressed through animal studies to multiple human trials.

Biohybrid Conduit Vessels

Malcolm Herring provided the first experience with methods to isolate endothelium from a patient and prepare a synthetic graft with the patient's own cells to accelerate endothelialization (24,25). The term *seeding* was adopted to describe his method of mixing autologous ECs, isolated from the jugular vein by a simple scraping method, with whole blood, and using this EC-containing blood to preclot the graft (26). The cells and blood were "milked" into the porous structure of the graft wall, followed by graft implantation. These early studies established that the ECs could survive this transplantation, but the efficiency of the cell coverage was poor and no benefit to patency was realized in animal or human studies. Two critical lessons were learned from these studies: (a) the number of cells needed for coverage of the graft was far beyond the number of cells available from a segment of autologous vein (27), and (b) the attachment of the cells to the surface of these grafts was essential to accelerate endothelization either prior to or following graft implantation (28–33).

Methods to increase the number of ECs available for seeding focused on culture conditions and the use of growth factors to stimulate EC proliferation in vitro. The ability to expand human adult ECs in culture prior to placement on a graft was finally achieved using heparin as a co-mitogen in the presence of a crude mixture of EC growth factors (34). With the ability to expand the number of human ECs isolated from a small segment of vein, the possibility to seed large numbers of ECs onto the luminal surface of synthetic grafts was realized. This method has been championed by Peter Zilla and has established the efficacy of using large numbers of autologous ECs to establish monolayers of ECs on the luminal surface of synthetic vascular grafts prior to implantation into small diameter positions (35–39). Nonhuman animal studies established the durability and persistence of these linings and also established the increased patency of synthetic grafts prepared using this culturing system (40). The method involves a significant period of in vitro graft development, to allow expansion of ECs and to allow maturation of the EC lining. Nevertheless, these studies established the ability of cell transplantation to improve the patency of synthetic grafts in less than 6-mm applications, attaining the patency observed for autologous saphenous vein (41–43).

The function of these hybrid (cells and synthetic/nondegradable scaffold) tissue–engineered grafts has been assessed extensively in both animal trials and more recently in human trials (44–59). The results of these studies are encouraging and reflect the consistent improvement in techniques to isolate and deposit cells on the surface of grafts. Our expanded understanding of the function of endothelium in vessel homeostasis has resulted in devices that exhibit patency rates approaching those of the saphenous vein. The final proof-of-efficacy for tissue-engineered vascular grafts is the patency of an autologous artery (e.g., internal mammary artery for coronary bypass) when used as a bypass graft, a goal not yet attained but in sight. Initial animal studies established the ability to create EC linings on synthetic grafts and determined that the cells lining these grafts represented the cells originally used to treat the graft at the time of surgical implantation (60,61). This is critical, as it establishes the persistence of the cells transplanted and indicates re-endothelization is not the result of spontaneous endothelization through deposition of circulating endothelial progenitor cells. Prior to commencing human clinical trials, it was critical to establish a statistically significant improvement in the patency of endothelialized as compared to synthetic grafts alone. Of interest, a large number of studies evaluate tissue-engineered neovessel function using grafts of lengths (graft length <10 times the inner graft diameter) (62,63) that are not a critical test of patency. Using grafts of appropriate length and studies with sufficient subjects to provide a statistical assessment, the positive effect of EC treatment on patency has been established in animal studies (64).

The Endothelial Cell in Building Macrovessels

Several factors led investigators to search for alternative sources of endothelium for cell transplantation. The expansion of ECs using growth factors was reported to cause changes in the karyotype of these cells, with unknown effects on cell function following transplantation (65). The in vitro culturing and maturation process involves periods that can extend up to 8 weeks. This timeframe precludes the use of this method for patients needing immediate bypass, and the long culture times increase the cost of the procedure and the susceptibility to infection. An alternative source of cells for graft seeding was identified in human adipose tissue (66–68). This tissue is well vascularized, and methods for the isolation of pure populations of ECs had been previously reported (69). The pellet of cells isolated from adipose tissue following enzymatic dissociation and centrifugal sedimentation was found to be highly enriched in cells that can form a stable EC lining on synthetic grafts (70,71). It was also critical that this method could be accomplished in the operating room at the time of primary bypass (either coronary or peripheral) surgery. A new term, *cell sodding*, was developed to describe the method of EC isolation and immediate cell deposition on a graft to form a monolayer (72). The grafts were implanted immediately into patients. Cells isolated from autologous fat have been used to construct neovessels for use as venous replacements, peripheral arterial grafts, arteriovenous fistula grafts, and coronary artery bypass grafts.

Translation

The translation of this technology for building a blood vessel from cellular components has progressed through careful planning of studies to first assess safety and more recently efficacy. Some investigators have suggested that early successful results reported in animals using EC transplantation, including methods defined as both seeding and sodding, have not been realized in humans. The clinical results are still emerging, but results to date support a more positive outcome and have established the ability to accelerate formation of stable antithrombogenic EC linings on synthetic grafts in humans using autologous cell transplantation. New sources of cells for creating these linings continue to be evaluated, including cells derived from bone marrow and cells circulating in the blood (73–75). The use of degradable scaffolds is also under active evaluation, although serious concerns must be overcome regarding the susceptibility of these grafts to aneurysm formation (76–78).

Recently, the concept of "fallout" endothelialization has reemerged as a possible way to stimulate the formation of an antithrombogenic EC lining on implanted devices. The concept was first suggested in the 1960s, as a way to explain the difference in healing patterns and spontaneous endothelialization observed following implantation of vascular grafts into different species, including man (79). Significant work was performed modifying the internal, luminal surface of synthetic grafts to support spontaneous endothelialization by stimulating the adherence, or "fallout," of circulating cells that could become endothelium. This concept has been studied most recently using stents modified with proteins or antibodies that have affinity for ECs and/or their progenitor cell types. The results of these studies are highly variable based on the choice of animal models (80), and translation to humans is ongoing. Because these therapies will be most commonly used in the aging population, it will be most interesting to see whether the population of circulating stem cells is present in sufficient quantities in older individuals to support spontaneous endothelialization.

Building Microvessels

A clinical motivation for building microvessels arises from the need to reestablish or improve blood flow to a compromised or damaged tissue. Often, the introduction of a new conduit vessel, natural or artificial, upstream to an affected tissue reestablishes blood flow and relieves symptoms. However, in those situations where the replacement of an upstream diseased conduit vessel does not resolve the disease condition, it is necessary to regenerate a vascular supply, so-called *therapeutic angiogenesis* (81–83). In therapeutic angiogenesis, the goal is to expand an existing vascular bed, especially the downstream microvasculature, as a means to provide alternate and/or additional avenues for tissue perfusion. Additionally, new advances in tissue engineering are promising to repair or replace the dysfunctional tissues (84–87). Critical to any successful tissue engineering effort is the ability to quickly perfuse the tissue construct when implanted. Any tissue implant greater in dimension than a few millimeters is too big for nutrients to efficiently diffuse to the construct cells. Therefore, the ability to build or grow microvessels is critically important for both in vivo vascular regeneration (i.e., therapeutic angiogenesis) and in vitro tissue construction.

The Endothelial Cell in Building Microvessels

Clearly, the EC is the central player in establishing, maintaining, and regulating microvessel structure and function. At the core of the microvessel is a tube, constructed from ECs. In the case of the smallest microvessel, the capillary, a single EC wraps around, like a ring, to form a lumen. These "rings" of ECs are connected end-to-end to form the actual microvessel. During development, the first vascular cells to appear and assemble the vessel structure are ECs (or EC precursors) (88). These cells coalesce and align to form tubes organized into a primitive plexus that eventually progresses to establish the foundation for the subsequent adult vasculature (89). In establishing additional microvessel segments in the adult, again it is the EC, this time already present in the existing microvasculature, that orchestrates the new vessel formation (90).

It is unclear what factors and/or forces are critical in forming a tube from ECs (91). It may be that this ability to form a ring is inherent in the EC. There are many examples in which cultured ECs placed in three-dimensional (3D) matrix environments, even without additional growth factors, spontaneously organize and align into tube or tubelike structures (92). It is unclear what the specific induction cues are in these examples. However, it seems that biophysical forces such as stress and strain within (e.g., along the cytoskeleton) and across ECs (e.g., through cell–matrix interactions), as opposed to solely biochemical signals arising from specific extracellular matrix molecules, may be sufficient given that ECs can form tubes in a variety of 3D matrices regardless of the specific biochemical aspects of the matrices. However, the EC cytoskeleton, in enabling the EC to curve its structure, and cell–cell adhesion molecules for "latching" the EC in place, must be playing significant roles. Clearly, though, the presence of other cues, such as angiogenic factors and cytokines (93–97) or nonvascular tissue cells (89,98), can regulate EC tube formation. In the end, it is likely that these different stimuli and processes work in concert to give rise to an EC tube. In this regard, ECs might spontaneously form a tube in a 3D environment but do so at different rates or lead to varying tube structures, depending on the type of matrix present and the additional varied biochemical signals.

Perivascular Cells

Recent studies highlight the critical interactions between the ECs of microvessels and the associated perivascular cells (i.e., pericytes, smooth muscle cells) in establishing and stabilizing microvessel structure and function (99). For example, mice lacking platelet-derived growth factor (PDGF)-β develop fragile capillaries, due to a failure in recruiting perivascular cells to vessels during development (100,101). Other experiments are revealing that the release of pericytes from a capillary in response to an angiogenic factor is a necessary event for angiogenic sprouting, the natural means by which new microvessels form (102). Related to this, it appears that pericytes act to mature a neovessel and attenuate vascular proliferation. The proliferation of retinal capillaries leading to blindness is associated with the depletion of capillary pericytes in the retina (103). Finally, it is well known that the perivascular cells are an important component in the control of blood flow through the microcirculation, particularly at the precapillary level. This control involves not only indirect endothelial-to-perivascular cell communication (e.g., through soluble mediators such as nitric oxide) but also through direct cell–cell contact between the two cell layers, possibly involving gap junctions (104–109). It is becoming clear that a healthy, functional microvessel must have the appropriate perivascular cells incorporated into the vessel wall

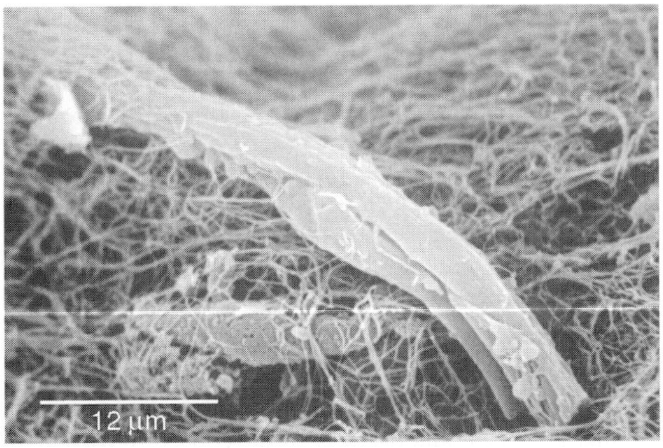

Figure 184.2. Scanning electron micrograph of a sprouting microvessel in a 3D collagen gel matrix. In the preparation, the gels containing growing microvessels were fixed and then peeled apart in layers to expose growing microvessels. Collagen gels were prepared at 3 mg/mL collagen per gel volume. At this concentration, microvessels retain normal architecture but are still able to displace collagen fibrils and grow. Used with permission from Dr. Constance Temm, Indiana University School of Medicine.

with the necessary functional connections between the different vascular cell types.

Tissue Environment in Microvessel Structure

The nature of conduit macrovessels is such that they can maintain vessel structure (i.e., remain a tube) when removed from the body. This is particularly the case with the large elastic or muscular arteries. In this respect, the extravascular environment, although important in regulating conduit vessel function, is not critical for these vessels to establish and maintain form. In contrast, removal of microvessels from their 3D environments and placing them on two-dimensional (2D) surfaces causes the collapse of the tube architecture and dispersal of the microvessels' cells. Clearly the extravascular environment plays an important role in microvessel assembly and form. Consequently, choices of environments that support microvascular organization are critical to any successful attempt at building microvessels.

One stereotypical environment that is often used to support the building of microvessels is the type I collagen gel matrix. Comprising fibrillar collagen, and representative of a simple interstitial environment, collagen gel matrices provide 3D biophysical constraints and scaffolding for microvascular organization (110,111). These gels are isotropic meshes of randomly arranged collagen fibrils (Figure 184.2) dense enough to provide sufficient structure to constrain vascular cell organization but loose enough to enable rapid diffusion and vessel dynamics. Collagen gels exhibit stereotypical viscoelastic material properties (111,112). When deformed, these gels experience initial relatively high stresses, which then relax over time. The high amount of fluid contained within the gels generates a viscous drag between the gel constituents during loading (113). Whether viscoelastic behavior is critical for EC tube formation and maintenance of microvessel structure is not clear. Given that other, manmade, scaffolds such as hydrogels share similar biomechanical properties as collagen gels, viscoelastic behavior seems likely to play some role.

Artificial scaffolds are being explored as alternatives to native matrices for use in the engineering of vascular and nonvascular tissues alike (114–119). Made from artificial materials, these scaffolds offer greater flexibility in determining the mechanical strength, compliance, porosity, and ability to degrade of environments for building microvessels (120). A polyglycolic acid-poly-L-lactic acid (PGA/PLLA) composite is one such material used to create artificial scaffolds often employed in assembling microvessels (121). A challenge in using artificial scaffolds concerns the biocompatibility of the materials. Vascular cells and other tissue cells need to interface with these scaffolds in a way that promotes normal cell behavior (e.g., tube formation by EC). One approach to improve biocompatibility is to attach biomolecules, such as matrix molecules, to scaffold materials to create a more biofunctional surface (120,122). The development of new materials and synthesis approaches for making more biocompatible scaffolds that retain design flexibility will greatly expand the ability to build microvascular systems.

Manmade Microvessels

Common strategies employed to build microvessels in vitro involve a de novo approach in which freshly isolated or cultured vascular cells are placed within a scaffolding environment (native or artificial) to form a construct containing microvascular tubes. The simple vascular tubes that subsequently form in these microvascular constructs will then progress to form true microvessels and microvascular networks, but only when implanted into a living host. The most successful approaches in forming in vitro vascular tubes have involved the use of ECs in combination with perivascular cells such as smooth muscle cells, mesenchymal smooth muscle precursors (e.g., 10T1/2 cells), and/or tissue stromal cells (123–125), often cultured separately and combined later with the scaffold. Growth factors/morphogens (e.g., fibroblast growth factor 2 [FGF2]) are often included to induce or accelerate vascular cell organization into a microvascular network (126,127). Interestingly, ECs, when used alone, will form tubes but will not effectively progress into a stable microvasculature when implanted without additional manipulation. For example, microvessel formation and survival in constructs prepared from human ECs in collagen gels or polylactic acid sponges were considerably enhanced when the ECs were engineered to overexpress the antiapoptotic protein Bcl-2 (128,129). An alternate strategy uses isolated, intact microvessel segments instead of cultured vascular cells (110). The microvessel segments retain the intimate microvessel structure, including endothelial and perivascular cells. When embedded in a 3D collagen matrix, the isolated microvessel fragments undergo angiogenesis and are capable of progressing into a mature, functional microcirculation when implanted in vivo (130). In this approach, no further manipulation or addition of factors is necessary to produce the new microvasculature.

The findings in the above example studies point to two key concepts in the establishment of a healthy microvessel structure: (a) perivascular cells are important in microvessel formation and structure, and (b) the in vivo environment provides a key stimulus in building a mature microvasculature. When cultured perivascular or perivascular precursor cells were included with pure EC cultures or as part of the vascular structure to begin with (as in the isolated microvessel fragments), no special manipulation (e.g., genetic engineering) of the system was required (125,130). Interestingly, in those examples in which Bcl-2 overexpression in cultured ECs was used, increased perivascular cell recruitment from host tissues correlated with the increased microvascular stability (129). The findings that mature, specified (i.e., arteriole, capillary, venule specification) microvessels formed only when microvascular constructs were implanted indicate that some aspect(s) of the in vivo environment, absent in vitro, is critical to the ultimate formation of a mature microvessel. Although there may be a variety of in vivo–related stimuli involved (such as blood-borne or tissue-derived molecular factors), it seems likely that blood flow, and the associated pressures and shear stresses, is the primary stimulus, in vivo, leading to microvessel maturation. It is well established that hemodynamics contribute greatly to the maturation and remodeling of preexisting macro- and microvessels (7,131,132). Presumably, similar mechanisms are at play concerning the maturation of EC tubes.

A growing effort to build microvessels has involved traditional microchip fabrication techniques. Technologies such as soft lithography (133) or micro-electricomechanical system (MEMS) (134) fabrication methods are being used to create channels of microvascular dimensions to serve as a template for microvessel construction. Typically, grooves or channels are fabricated on a silicon wafer surface with appropriate dimensions. These channels are then coated with extracellular matrix molecules to facilitate EC adhesion to the channel walls to create vascular cell–lined channels (Figure 184.3) (135,136).

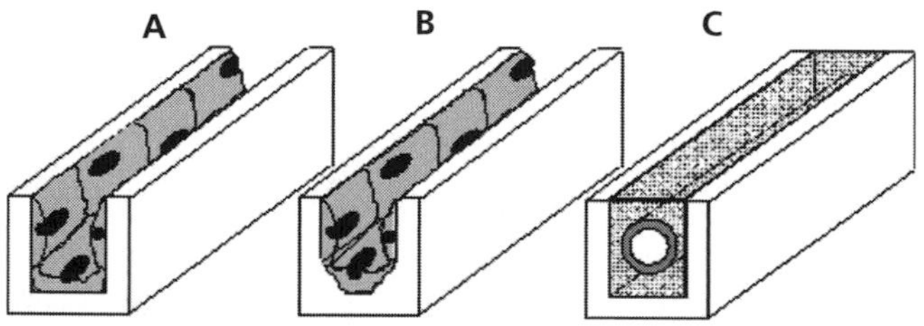

Figure 184.3. Schematics of three different soft lithography-based approaches for microvessel fabrication. **A** and **B** involve fabricating channels using dimensions that reflect the microvessel structure. The walls of the channels serve as the walls of the microvessel. **C:** The channel serves only to contain a 3D matrix within which a microvessel is allowed to form. Channel dimensions in **A** and **B** may be on the order of 8 to 20 microns, whereas the channel shown in **C** may be as large as 100 microns.

An alternative strategy would involve creating microvessels within a scaffold matrix placed within the microfabricated channels (Figure 184.3). In this approach, microvessels would form as they do in 3D matrices and are therefore not limited by channel dimensions. The channels, in this case, serve only to constrain the pattern of microvessels formed into a desired network topology (i.e., the channels do not directly determine microvessel structure).

Naturally Made Microvessels

ANGIOGENESIS. In the adult, new microvessels are built from preexisting vessels via angiogenesis. Although angiogenesis will not be discussed in detail here, it is useful to consider aspects of angiogenesis in the context of forming a new microvessel element and structure. The classic conceptual model of angiogenesis involves the signaled focal relaxation of the parent microvessel structure followed by the sprouting of an EC at the relaxed site (137). This sprout appears as a spear-like cell projection from the microvessel wall that extends out into the extravascular space (110,138). Interestingly, it does not appear that formation of the new microvessel segment involves complete dissolution of either the parent microvessel or the growing neovessel (139). In studies involving intact microvessel elements, elongated neovessels comprised ECs contiguous with the tip EC of the sprout (110,130). There was no clear evidence from these studies that ECs leave the vessel and then reassociate during neovessel growth. In this regard, the formation of a new vessel element by angiogenesis differs from the de novo approaches used in vitro to fabricate microvessels from cultured cells. In vitro, individual cells coalesce to form the prevessel tube. In sprouting angiogenesis in vivo, the "pre-vessel" is an extension of a preexisting vessel structure, which continues to elongate and finally mature. However, growing evidence supporting a role for circulating endothelial progenitor cells in building neovessels during angiogenesis suggests that individual vascular cells separate from the vessel structure do add to the neovessel wall (140,141). Perhaps the in vitro approaches model this process.

An interesting question concerning the mechanism of vessel formation during angiogenesis is whether a lumen is maintained in the sprouting neovessel or is instead reestablished as the angiogenesis event is resolved. It appears that the lumen of the neovessel forms in the elongating region behind the tip cell of the sprout (142). This suggests that the EC is capable of forming a ring soon after or even during a proliferative event. An understanding of how the EC is organized within the sprouting segment and how the sprout is integrated with the parent vessel might provide insight leading toward more effective strategies for assembling microvessels in vitro.

During sprouting, it is thought that the perivascular cells are not associated with the growing neovessel. This was directly demonstrated in a model of vascularization (130) in which α-actin–positive cells were seen to leave the angiogenic vasculature and not associate with the new sprouting neovessels. However, once angiogenesis stopped and vessel maturation began, α-actin–positive cells reassociated with the microvasculature, which correlated with reestablishment of blood flow within the newly formed network (130). These observations are consistent with findings that perivascular cells are important in establishing mature microvessels (143).

Clinically, angiogenesis can be used therapeutically to create new vasculature within a tissue. The general approach has been to deliver angiogenic factors, either as a native molecule (e.g., a peptide) or as a transgene encoding the factor (83,144). Vascular endothelial growth factor (VEGF) and FGF2 are two factors that have been actively explored as agents for therapeutic angiogenesis (145–148). Clinical trials with these agents have produced mixed results (144). Although perfusion-related evidence indicates an increase in vasculature within treated tissues (typically heart or skeletal muscle) relatively early after treatment, long-term benefits of the therapy were not realized. These findings suggest that early vessel growth occurred, but long-term stability of the new vessels (and resulting new vasculature) was compromised; either the new vasculature regressed, or it is not functioning normally with respect to perfusion capacity. Consistent with this interpretation is that, in the context of tissue vascularization (such as in therapeutic angiogenesis), simply providing new microvessel segments (i.e., angiogenesis) is ineffective at expanding the perfusion capacity of the microvasculature in the absence of the necessary postangiogenic events. These events include vessel redifferentiation, artery side–vein side specification, structural adaptation, and network remodeling. More information is needed concerning the biology driving postangiogenesis events before an effective angiogenesis-dependent therapy can be developed.

ARTERIOGENESIS. One vascular process related to postangiogenesis maturation that is being actively investigated involves the development of feed vessels from microvessels. Called *arteriogenesis*, this process results in the increase in diameter and vessel wall mass of an existing capillary or arterioles (149–153). Typically the increase in mass involves the addition of both cells (often perivascular cells) and matrix to the vessel wall. Presumably there is an analogous process for forming venules/veins (*venogenesis*) and capillaries (*capillarygenesis*) as neovessels formed by angiogenesis are integrated into the perfusion network. The term *arteriogenesis* generally is used when describing the formation of a new feed artery to a local vascular network. This is often visualized clinically by angiography as an appearance of collateral vessels in a tissue. Active research into the formation of collateral arteries has identified numerous possible biochemical regulators of arteriogenesis, including growth factors, inflammatory cytokines, and vasoactive mediators (154–157). Furthermore, changes in endothelial shear stress and circumferential stress can lead to compensatory changes in wall characteristics leading to arteriogenesis (158–160). In the case of arteriogenesis, an increase in shear (due to an increase in flow) will result in dilation of the vessel segment. As the vessel dilates, the wall will experience greater circumferential stress, leading to a compensatory thickening of the vessel wall to normalize this stress. Regardless of the mechanism, arteriogenesis is thought to occur in response to a chronic need to provide additional blood flow to the downstream vasculature such as would happen following the formation of new vessel elements via angiogenesis.

The Microvascular Network

Whether microvessels are built naturally or in the laboratory, the final intent is actually the formation of a microvascular network. Rarely, if at all, is the outcome the fabrication of a single, stand-alone microvessel. This may reflect, perhaps, significant technical challenges related to fabricating a single, small-caliber vessel that can be practically used. However, a single capillary brings blood into close proximity to an area that is a few cell layers thick. Only through a network of capillaries and other microvessels will there be sufficient exchange to support an entire tissue bed (161). Therefore, unlike with conduit vessels where a single vessel meets the primary purpose (i.e., bulk transport of blood), it is the microvessel network that fully meets the tissue needs. In addition, it is likely that proper neovessel maturation and adaptation depends on network hemodynamics and communication (131). Consequently, one might argue that successful strategies to build microvessels should include network-related considerations.

The Capillary as a Specialized Tissue–Blood Interface

As discussed earlier, the primary role of the capillaries within the cardiovascular system is to provide an effective means by which nutrients are delivered to tissue cells and serve as specialized interfaces between tissues and the blood. Consequently, the capillary is structured such that diffusive exchange between the blood and tissue is optimized. Furthermore, the endothelium of capillaries exhibit a variety of specialized structures and functions associated with specific tissue activities. For example, in tissues requiring relatively large exchange capacities, such as the kidney or an endocrine organ, the endothelium of the capillaries is fenestrated. Currently, strategies/approaches to build capillaries are focused on creating small-caliber perfusion elements and not necessarily attempting to impart additional specialization to the fabricated vessels. Certainly, most, if not all, of the studies involving the fabrication of microvessels have not evaluated the structural and functional aspects of the microvessels with respect to tissue-specific functions; emphasis is placed on the ability of the new microvasculature to carry blood. There is growing evidence that tissue cells contribute to the specialization of endothelium within the microvasculature of that tissue. For example, the presence of astrocytes can lead to the formation of tighter cell–cell junctions in EC monolayers (162,163), cardiac cells can regulate cardiac-specific gene expression in EC (164), and kidney podocytes can stimulate the formation of fenestrae in cultured ECs (165,166). It is likely that as methodologies to build microvessels improve, strategies to impart microvessel specificity in these methodologies will be explored.

KEY POINTS

- The primary function of blood vessels is to act as conduits for the transport of blood to within close proximity of tissues, thereby enabling effective nutrient delivery to tissue cells.
- The tubular nature of vessels is optimal for efficient fluid transport and vessel stability and is the hallmark structural feature of all vessels.
- The functional requirements of the vessel, be it bulk transport of blood or efficient exchange of nutrients, determine key structural features of vessels.
- The wall characteristics must be customized to serve the specific functional role of that vessel.

Future Goals

- To define the molecular, cellular, and biophysical rules that govern tube and lumen formation by the EC
- To determine how the cells of the vessel wall integrate to establish wall structure and vessel architecture. This pertains to both native vessels and biohybrid designs incorporating artificial biomaterials
- To define the relationships between the microvessel and the extravascular environment in determining microvessel structure and function

ACKNOWLEDGMENTS

We thank the members of the University of Arizona Vascular Research Group for their useful discussions and input. This work was supported in part by National Institutes of Health grants HL63732, HL63873, HL67067, HL077683, and HL60450.

REFERENCES

1 Bentley TB, Meng H, Pittman RN. Temperature dependence of oxygen diffusion and consumption in mammalian striated muscle. *Am J Physiol.* 1993;264(6 Pt 2):H1825–H1830.
2 Secomb TW, Hsu R, Park EY, Dewhirst MW. Green's function methods for analysis of oxygen delivery to tissue by microvascular networks. *Ann Biomed Eng.* 2004;32(11):1519–1529.
3 McDonald DA. Hemodynamics. *Annu Rev Physiol.* 1968;30: 525–556.
4 Taylor MG. Hemodynamics. *Annu Rev Physiol.* 1973;35:87–116.
5 Dobrin PB. Mechanical properties of arteries. *Physiol Rev.* 1978; 58(2):397–460.
6 Roach MR. Biophysical analyses of blood vessel walls and blood flow. *Annu Rev Physiol.* 1977;39:51–71.
7 Pries AR, Reglin B, Secomb TW. Remodeling of blood vessels: responses of diameter and wall thickness to hemodynamic and metabolic stimuli. *Hypertension.* 2005;46(4):725–731.
8 Pries AR, Secomb TW, Gessner T, et al. Resistance to blood flow in microvessels in vivo. *Circ Res.* 1994;75(5):904–915.
9 McGuire BJ, Secomb TW. Theoretical predictions of maximal oxygen consumption in hypoxia: effects of transport limitations. *Respir Physiol Neurobiol.* 2004;143(1):87–97.
10 Tsai AG, Johnson PC, Intaglietta M. Oxygen gradients in the microcirculation. *Physiol Rev.* 2003;83(3):933–963.
11 Nakamura K, Ikomi F, Ohhashi T. Cilostazol, an inhibitor of type 3 phosphodiesterase, produces endothelium-independent vasodilation in pressurized rabbit cerebral penetrating arterioles. *J Vasc Res.* 2006;43(1):86–94.
12 Lizanecz E, Bagi Z, Pasztor ET, et al. Phosphorylation-dependent desensitization of vanilloid receptor-1 (TRPV1) function in rat skeletal muscle arterioles and in CHO-TRPV1 cells by anandamide. *Mol Pharmacol.* 2005;69(3):1015–1023.
13 Heaps CL, Mattox ML, Kelly KA, et al. Exercise training increases basal tone in arterioles distal to chronic coronary occlusion. *Am J Physiol Heart Circ Physiol.* 2005;290(3):H1128–H1135.
14 McGahon MK, Dawicki JM, Scholfield CN, et al. A-type potassium current in retinal arteriolar smooth muscle cells. *Invest Ophthalmol Vis Sci.* 2005;46(9):3281–3287.
15 Jiang Z, Wu L, Miller BL, et al. A novel vein graft model: adaptation to differential flow environments. *Am J Physiol Heart Circ Physiol.* 2004;286(1):H240–H245.
16 Westerband A, Crouse D, Richter LC, et al. Vein adaptation to arterialization in an experimental model. *J Vasc Surg.* 2001; 33(3):561–569.
17 Fortunato JE, Glagov S, Bassiouny HS. Biomechanical factors as regulators of biological responses to vascular grafts. *Semin Vasc Surg.* 1999;12(1):27–37.
18 Fillinger MF, Cronenwett JL, Besso S, et al. Vein adaptation to the hemodynamic environment of infrainguinal grafts. *J Vasc Surg.* 1994;19(6):970–978.
19 Voorhees AB Jr, Jaretzki A III, Blakemore AH. The use of tubes constructed from vinyon "N" cloth in bridging arterial defects. *Ann Surg.* 1952;135(3):332–336.
20 Sauvage LR, Harkins HN. An experimental study of fresh arterial and venous autografts and homografts in the growing pig. *Surg Forum.* 1951;94:247–255.
21 Sauvage LR, Harkins HN. Some considerations regarding the use of vascular grafts. *West J Surg Obstet Gynecol.* 1952;60(3): 113–116.
22 Wesolowski SA, Sauvage LR. Comparison of the fates of Orion mesh prosthetic replacement of the thoracic aorta and aortic bifurcation. *Ann Surg.* 1956;143(1):65–72.
23 Sauvage LR, Berger KE, Wood SJ, et al. Interspecies healing of porous arterial prostheses: observations, 1960 to 1974. *Arch Surg.* 1974;109(5):698–705.
24 Herring MB, Dilley R, Jersild RA Jr, et al. Seeding arterial prostheses with vascular endothelium. The nature of the lining. *Ann Surg.* 1979;190(1):84–90.
25 Herring M, Gardner A, Glover J. Seeding endothelium onto canine arterial prostheses. The effects of graft design. *Arch Surg.* 1979;114(6):679–682.
26 Herring MB, Dilley R, Jersild RA Jr, et al. Seeding arterial prostheses with vascular endothelium. The nature of the lining. *Ann Surg.* 1979;190(1):84–90.
27 Sharefkin JB, Latker C, Smith M, et al. Early normalization of platelet survival by endothelial seeding of Dacron arterial prostheses in dogs. *Surgery.* 1982;92(2):385–393.
28 Callow AD. Endothelial cell seeding: problems and expectations. *J Vasc Surg.* 1987;6(3):318–319.

29 Tannenbaum G, Ahlborn T, Benvenisty A, et al. High-density seeding of cultured endothelial cells leads to rapid coverage of polytetrafluoroethylene grafts. *Curr Surg*. 1987;44(4):318–321.
30 Sterpetti AV, Hunter WJ, Schultz RD, et al. Seeding with endothelial cells derived from the microvessels of the omentum and from the jugular vein: a comparative study. *J Vasc Surg*. 1988;7(5):677–684.
31 Newman KD, Nguyen N, Dichek DA. Quantification of vascular graft seeding by use of computer-assisted image analysis and genetically modified endothelial cells. *J Vasc Surg*. 1991;14(2):140–146.
32 Baker KS, Williams SK, Jarrell BE, et al. Endothelialization of human collagen surfaces with human adult endothelial cells. *Am J Surg*. 1985;150(2):197–200.
33 Williams SK, Jarrell BE, Friend L, et al. Adult human endothelial cell compatibility with prosthetic graft material. *J Surg Res*. 1985;38(6):618–629.
34 Jarrell B, Levine E, Shapiro S et al. Human adult endothelial cell growth in culture. *J Vasc Surg*. 1984;1(6):757–764.
35 Zilla P, Fasol R, Deutsch M, et al. Endothelial cell seeding of polytetrafluoroethylene vascular grafts in humans: a preliminary report. *J Vasc Surg*. 1987;6(6):535–541.
36 Zilla P, Siedler S, Fasol R, Sharefkin JB. Reduced reproductive capacity of freshly harvested endothelial cells in smokers: a possible shortcoming in the success of seeding? *J Vasc Surg*. 1989;10(2):143–148.
37 Zilla P, Fasol R, Preiss P, et al. Use of fibrin glue as a substrate for in vitro endothelialization of PTFE vascular grafts. *Surgery*. 1989;105(4):515–522.
38 Zilla P. Endothelialization of vascular grafts. *Curr Opin Cardiol*. 1991;6(6):877–886.
39 Zilla P, Preiss P, Groscurth P, et al. In vitro-lined endothelium: initial integrity and ultrastructural events. *Surgery*. 1994;116(3):524–534.
40 Zilla P. Endothelialization of vascular grafts. *Curr Opin Cardiol*. 1991;6(6):877–886.
41 Zilla P, Deutsch M, Meinhart J. Endothelial cell transplantation. *Semin Vasc Surg*. 1999;12(1):52–63.
42 Deutsch M, Meinhart J, Fischlein T, et al. Clinical autologous in vitro endothelialization of infrainguinal ePTFE grafts in 100 patients: a 9-year experience. *Surgery*. 1999;126(5):847–855.
43 Meinhart JG, Deutsch M, Fischlein T, et al. Clinical autologous in vitro endothelialization of 153 infrainguinal ePTFE grafts. *Ann Thorac Surg*. 2001;71(Suppl 5):S327–S331.
44 Schmidt SP, Hunter TJ, Falkow LJ, et al. Effects of antiplatelet agents in combination with endothelial cell seeding on small-diameter Dacron vascular graft performance in the canine carotid artery model. *J Vasc Surg*. 1985;2(6):898–906.
45 Herring MB, Dilley R, Jersild RA Jr, et al. Seeding arterial prostheses with vascular endothelium. The nature of the lining. *Ann Surg*. 1979;190(1):84–90.
46 Graham LM, Vinter DW, Ford JW, et al. Endothelial cell seeding of prosthetic vascular grafts: early experimental studies with cultured autologous canine endothelium. *Arch Surg*. 1980;115(8):929–933.
47 Sharefkin JB, Latker C, Smith M, et al. Early normalization of platelet survival by endothelial seeding of Dacron arterial prostheses in dogs. *Surgery*. 1982;92(2):385–393.
48 Allen BT, Long JA, Clark RE, et al. Influence of endothelial cell seeding on platelet deposition and patency in small-diameter Dacron arterial grafts. *J Vasc Surg*. 1984;1(1):224–233.
49 Schmidt SP, Hunter TJ, Falkow LJ, et al. Effects of antiplatelet agents in combination with endothelial cell seeding on small-diameter Dacron vascular graft performance in the canine carotid artery model. *J Vasc Surg*. 1985;2(6):898–906.
50 Kempczinski RF, Rosenman JE, Pearce WH, et al. Endothelial cell seeding of a new PTFE vascular prosthesis. *J Vasc Surg*. 1985; 2(3):424–429.
51 Stanley JC, Burkel WE, Graham LM, Lindblad B. Endothelial cell seeding of synthetic vascular prostheses. *Acta Chir Scand Suppl*. 1985;529:17–27.
52 Zilla P, Fasol R, Deutsch M, et al. Endothelial cell seeding of polytetrafluoroethylene vascular grafts in humans: a preliminary report. *J Vasc Surg*. 1987;6(6):535–541.
53 Callow AD. Endothelial cell seeding: problems and expectations. *J Vasc Surg*. 1987;6(3):318–319.
54 Tannenbaum G, Ahlborn T, Benvenisty A, et al. High-density seeding of cultured endothelial cells leads to rapid coverage of polytetrafluoroethylene grafts. *Curr Surg*. 1987;44(4):318–321.
55 Ortenwall P, Wadenvik H, Kutti J, Risberg B. Reduction in deposition of indium 111-labeled platelets after autologous endothelial cell seeding of Dacron aortic bifurcation grafts in humans: a preliminary report. *J Vasc Surg*. 1987;6(1):17–25.
56 Douville EC, Kempczinski RF, Birinyi LK, Ramalanjaona GR. Impact of endothelial cell seeding on long-term patency and subendothelial proliferation in a small-caliber highly porous polytetrafluoroethylene graft. *J Vasc Surg*. 1987;5(4):544–550.
57 Pearce WH, Rutherford RB, Whitehill TA, et al. Successful endothelial seeding with omentally derived microvascular endothelial cells. *J Vasc Surg*. 1987;5(1):203–206.
58 Campbell JB, Glover JL, Herring B. The influence of endothelial seeding and platelet inhibition on the patency of ePTFE grafts used to replace small arteries – an experimental study. *Eur J Vasc Surg*. 1988;2(6):365–370.
59 Sterpetti AV, Hunter WJ, Schultz RD, et al. Seeding with endothelial cells derived from the microvessels of the omentum and from the jugular vein: a comparative study. *J Vasc Surg*. 1988;7(5):677–684.
60 Williams SK, Kleinert LB, Rose D, McKenney S. Origin of endothelial cells that line expanded polytetrafluorethylene vascular grafts sodded with cells from microvascularized fat. *J Vasc Surg*. 1994;19(4):594–604.
61 Williams SK, Rose DG, Jarrell BE. Microvascular endothelial cell sodding of ePTFE vascular grafts: improved patency and stability of the cellular lining. *J Biomed Mater Res*. 1994;28(2):203–212.
62 Herring M, Gardner A, Glover J. Seeding endothelium onto canine arterial prostheses. The effects of graft design. *Arch Surg*. 1979;114(6):679–682.
63 Herring M, Gardner A, Peigh P, et al. Patency in canine inferior vena cava grafting: effects of graft material, size, and endothelial seeding. *J Vasc Surg*. 1984;1(6):877–887.
64 Herring M, Gardner A, Peigh P, et al. Patency in canine inferior vena cava grafting: effects of graft material, size, and endothelial seeding. *J Vasc Surg*. 1984;1(6):877–887.
65 Nichols WW, Buynak EB, Bradt C, et al. Cytogenetic evaluation of human endothelial cell cultures. *J Cell Physiol*. 1987;132(3):453–462.
66 Jarrell BE, Williams SK, Stokes G, et al. Use of freshly isolated capillary endothelial cells for the immediate establishment of a monolayer on a vascular graft at surgery. *Surgery*. 1986;100(2):392–399.

67 Jarrell BE, Williams SK, Solomon L, et al. Use of an endothelial monolayer on a vascular graft prior to implantation. Temporal dynamics and compatibility with the operating room. *Ann Surg*. 1986;203(6):671–678.
68 Williams SK, Jarrell BE, Rose DG, et al. Human microvessel endothelial cell isolation and vascular graft sodding in the operating room. *Ann Vasc Surg*. 1989;3(2):146–152.
69 Williams SK, Jarrell BE, Rose DG, et al. Human microvessel endothelial cell isolation and vascular graft sodding in the operating room. *Ann Vasc Surg*. 1989;3(2):146–152.
70 Young C, Jarrell BE, Hoying JB, Williams SK. A porcine model for adipose tissue-derived EC transplantation. *Cell Transplant*. 1992;1(4):293–298.
71 Williams SK, Wang TF, Castrillo R, Jarrell BE. Liposuction-derived human fat used for vascular graft sodding contains endothelial cells and not mesothelial cells as the major cell type. *J Vasc Surg*. 1994;19(5):916–923.
72 Rupnick MA, Hubbard FA, Pratt K, et al. Endothelialization of vascular prosthetic surfaces after seeding or sodding with human microvascular endothelial cells. *J Vasc Surg*. 1989;9(6):788–795.
73 Hernandez DA, Townsend LE, Uzieblo MR, et al. Human endothelial cell cultures from progenitor cells obtained by leukapheresis. *Am Surg*. 2000;66(4):355–358.
74 Boyer M, Townsend LE, Vogel LM, et al. Isolation of endothelial cells and their progenitor cells from human peripheral blood. *J Vasc Surg*. 2000;31(1 Pt 1):181–189.
75 He H, Shirota T, Yasui H, Matsuda T. Canine endothelial progenitor cell-lined hybrid vascular graft with nonthrombogenic potential. *J Thorac Cardiovasc Surg*. 2003;126(2): 455–464.
76 Cho SW, Lim SH, Kim IK, et al. Small-diameter blood vessels engineered with bone marrow-derived cells. *Ann Surg*. 2005;241 (3):506–515.
77 Daly CD, Campbell GR, Walker PJ, Campbell JH. In vivo engineering of blood vessels. *Front Biosci*. 2004;9:1915–1924.
78 Bujan J, Garcia-Honduvilla N, Bellon JM. Engineering conduits to resemble natural vascular tissue. *Biotechnol Appl Biochem*. 2004;39(Pt 1):17–27.
79 Sauvage LR, Berger KE, Wood SJ, et al. Interspecies healing of porous arterial prostheses: observations, 1960 to 1974. *Arch Surg*. 1974;109(5):698–705.
80 Sauvage LR, Berger KE, Wood SJ, et al. Interspecies healing of porous arterial prostheses: observations, 1960 to 1974. *Arch Surg*. 1974;109(5):698–705.
81 Tirziu D, Simons M. Angiogenesis in the human heart: gene and cell therapy. *Angiogenesis*. 2005;8(3):241–251.
82 Bauer SM, Bauer RJ, Velazquez OC. Angiogenesis, vasculogenesis, and induction of healing in chronic wounds. *Vasc Endovascular Surg*. 2005;39(4):293–306.
83 Hughes GC, Annex BH. Angiogenic therapy for coronary artery and peripheral arterial disease. *Expert Rev Cardiovasc Ther*. 2005;3(3):521–535.
84 Kuo CK, Li WJ, Mauck RL, Tuan RS. Cartilage tissue engineering: its potential and uses. *Curr Opin Rheumatol*. 2006;18(1): 64–73.
85 Eschenhagen T, Zimmermann WH. Engineering myocardial tissue. *Circ Res*. 2005;97(12):1220–1231.
86 Yacoub MH, Takkenberg JJ. Will heart valve tissue engineering change the world? *Nat Clin Pract Cardiovasc Med*. 2005;2(2): 60–61.
87 Tsang VL, Bhatia SN. Three-dimensional tissue fabrication. *Adv Drug Deliv Rev*. 2004;56(11):1635–1647.
88 Drake CJ. Embryonic and adult vasculogenesis. *Birth Defects Res C Embryo Today*. 2003;69(1):73–82.
89 Jin SW, Beis D, Mitchell T, et al. Cellular and molecular analyses of vascular tube and lumen formation in zebrafish. *Development*. 2005;132(23):5199–5209.
90 Sholley MM, Ferguson GP, Seibel HR, et al. Mechanisms of neovascularization: vascular sprouting can occur without proliferation of endothelial cells. *Lab Invest*. 1984;51(6):624–634.
91 Egginton S, Gerritsen M. Lumen formation: in vivo versus in vitro observations. *Microcirculation*. 2003;10(1):45–61.
92 Vailhe B, Vittet D, Feige JJ. In vitro models of vasculogenesis and angiogenesis. *Lab Invest*. 2001;81(4):439–452.
93 Carter WB, Uy K, Ward MD, Hoying JB. Parathyroid-induced angiogenesis is VEGF-dependent. *Surgery*. 2000;128(3):458–464.
94 Chow J, Ogunshola O, Fan SY, et al. Astrocyte-derived VEGF mediates survival and tube stabilization of hypoxic brain microvascular endothelial cells in vitro. *Brain Res Dev Brain Res*. 2001;130(1):123–132.
95 Yan Q, Li Y, Hendrickson A, Sage EH. Regulation of retinal capillary cells by basic fibroblast growth factor, vascular endothelial growth factor, and hypoxia. *In Vitro Cell Dev Biol Anim*. 2001;37(1):45–49.
96 Haspel HC, Scicli GM, McMahon G, Scicli AG. Inhibition of vascular endothelial growth factor-associated tyrosine kinase activity with SU5416 blocks sprouting in the microvascular endothelial cell spheroid model of angiogenesis. *Microvasc Res*. 2002;63(3):304–315.
97 Parker LH, Schmidt M, Jin SW, et al. The endothelial-cell-derived secreted factor Egfl7 regulates vascular tube formation. *Nature*. 2004;428(6984):754–758.
98 Nakatsu MN, Sainson RC, Aoto JN, et al. Angiogenic sprouting and capillary lumen formation modeled by human umbilical vein endothelial cells (HUVEC) in fibrin gels: the role of fibroblasts and Angiopoietin-1. *Microvasc Res*. 2003;66(2):102–112.
99 Betsholtz C, Lindblom P, Gerhardt H. Role of pericytes in vascular morphogenesis. *EXS*. 2005;(94):115–125.
100 Lindahl P, Johansson BR, Leveen P, Betsholtz C. Pericyte loss and microaneurysm formation in PDGF-B-deficient mice. *Science*. 1997;277(5323):242–245.
101 Hellstrom M, Kalen M, Lindahl P, et al. Role of PDGF-B and PDGFR-beta in recruitment of vascular smooth muscle cells and pericytes during embryonic blood vessel formation in the mouse. *Development*. 1999;126(14):3047–3055.
102 Maisonpierre PC, Suri C, Jones PF, et al. Angiopoietin-2, a natural antagonist for Tie2 that disrupts in vivo angiogenesis (see comments). *Science*. 1997;277(5322):55–60.
103 D'Amore PA. Mechanisms of retinal and choroidal neovascularization. *Invest Ophthalmol Vis Sci*. 1994;35(12):3974–3979.
104 Rudic RD, Bucci M, Fulton D, et al. Temporal events underlying arterial remodeling after chronic flow reduction in mice: correlation of structural changes with a deficit in basal nitric oxide synthesis. *Circ Res*. 2000;86(11):1160–1166.
105 Rudic RD, Shesely EG, Maeda N, et al. Direct evidence for the importance of endothelium-derived nitric oxide in vascular remodeling. *J Clin Invest*. 1998;101(4):731–736.
106 Segal SS, Damon DN, Duling BR. Propagation of vasomotor responses coordinates arteriolar resistances. *Am J Physiol*. 1989; 256(3 Pt 2):H832–H837.

107 Isakson BE, Duling BR. Heterocellular contact at the myoendothelial junction influences gap junction organization. *Circ Res*. 2005;97(1):44–51.
108 Yashiro Y, Duling BR. Integrated Ca(2+) signaling between smooth muscle and endothelium of resistance vessels. *Circ Res*. 2000;87(11):1048–1054.
109 Dora KA, Doyle MP, Duling BR. Elevation of intracellular calcium in smooth muscle causes endothelial cell generation of NO in arterioles. *Proc Natl Acad Sci USA*. 1997;94(12):6529–6534.
110 Hoying JB, Boswell CA, Williams SK. Angiogenic potential of microvessel fragments established in three-dimensional collagen gels. *In Vitro Cell Dev Biol Anim*. 1996;32(7):409–419.
111 Krishnan L, Weiss JA, Wessman MD, Hoying JB. Design and application of a test system for viscoelastic characterization of collagen gels. *Tissue Eng*. 2004;10(1–2):241–252.
112 Hsu S, Jamieson AM, Blackwell J. Viscoelastic studies of extracellular matrix interactions in a model native collagen gel system. *Biorheology*. 1994;31(1):21–36.
113 Huang D, Chang TR, Aggarwal A, et al. Mechanisms and dynamics of mechanical strengthening in ligament-equivalent fibroblast-populated collagen matrices. *Ann Biomed Eng*. 1993; 21(3):289–305.
114 Tsang VL, Bhatia SN. Three-dimensional tissue fabrication. *Adv Drug Deliv Rev*. 2004;56(11):1635–1647.
115 Hammond JS, Beckingham IJ, Shakesheff KM. Scaffolds for liver tissue engineering. *Expert Rev Med Devices*. 2006;3(1):21–27.
116 Boccaccini AR, Blaker JJ. Bioactive composite materials for tissue engineering scaffolds. *Expert Rev Med Devices*. 2005;2(3): 303–317.
117 Levenberg S, Huang NF, Lavik E, et al. Differentiation of human embryonic stem cells on three-dimensional polymer scaffolds. *Proc Natl Acad Sci USA*. 2003;100(22):12741–12746.
118 Causton AS, Sherman JC. Design of proteins using rigid organic macrocycles as scaffolds. *Bioorg Med Chem*. 1999;7(1):23–27.
119 Madihally SV, Matthew HW. Porous chitosan scaffolds for tissue engineering (in process citation). *Biomaterials*. 1999;20(12): 1133–1142.
120 Seliktar D. Extracellular stimulation in tissue engineering. *Ann NY Acad Sci*. 2005;1047:386–394.
121 Nor JE, Peters MC, Christensen JB, et al. Engineering and characterization of functional human microvessels in immunodeficient mice. *Lab Invest*. 2001;81(4):453–463.
122 Kidd KR, Nagle RB, Williams SK. Angiogenesis and neovascularization associated with extracellular matrix-modified porous implants. *J Biomed Mater Res*. 2002;59(2):366–377.
123 Nakatsu MN, Sainson RC, Aoto JN, et al. Angiogenic sprouting and capillary lumen formation modeled by human umbilical vein endothelial cells (HUVEC) in fibrin gels: the role of fibroblasts and Angiopoietin-1. *Microvasc Res*. 2003;66(2):102–112.
124 Darland DC, D'Amore PA. TGF beta is required for the formation of capillary-like structures in three-dimensional cocultures of 10T1/2 and endothelial cells. *Angiogenesis*. 2001;4(1):11–20.
125 Levenberg S, Rouwkema J, Macdonald M, et al. Engineering vascularized skeletal muscle tissue. *Nat Biotechnol*. 2005;23 (7):879–884.
126 Hirschi KK, Rohovsky SA, Beck LH, et al. Endothelial cells modulate the proliferation of mural cell precursors via platelet-derived growth factor-BB and heterotypic cell contact. *Circ Res*. 1999;84(3):298–305.
127 Grosskreutz CL, nand-Apte B, Duplaa C, et al. Vascular endothelial growth factor-induced migration of vascular smooth muscle cells in vitro. *Microvasc Res*. 1999;58(2):128–136.
128 Nor JE, Christensen J, Mooney DJ, Polverini PJ. Vascular endothelial growth factor (VEGF)-mediated angiogenesis is associated with enhanced endothelial cell survival and induction of Bcl-2 expression. *Am J Pathol*. 1999;154(2):375–384.
129 Schechner JS, Nath AK, Zheng L, et al. In vivo formation of complex microvessels lined by human endothelial cells in an immunodeficient mouse. *Proc Natl Acad Sci USA*. 2000;97(16):9191–9196.
130 Shepherd BR, Chen HY, Smith CM, et al. Rapid perfusion and network remodeling in a microvascular construct after implantation. *Arterioscler Thromb Vasc Biol*. 2004;24(5):898–904.
131 Pries AR, Secomb TW. Control of blood vessel structure: insights from theoretical models. *Am J Physiol Heart Circ Physiol*. 2005;288(3):H1010–H1015.
132 Pries AR, Secomb TW. Microvascular adaptation – regulation, coordination and function. *Z Kardiol*. 2000;89(Suppl 9): IX/117–IX/120.
133 Kane RS, Takayama S, Ostuni E, et al. Patterning proteins and cells using soft lithography. *Biomaterials*. 1999;20(23–24):2363–2376.
134 Brunette DM, Chehroudi B. The effects of the surface topography of micromachined titanium substrata on cell behavior in vitro and in vivo. *J Biomech Eng*. 1999;121(1):49–57.
135 Kane RS, Takayama S, Ostuni E, et al. Patterning proteins and cells using soft lithography. *Biomaterials*. 1999;20(23–24):2363–2376.
136 Thakar RG, Ho F, Huang NF, et al. Regulation of vascular smooth muscle cells by micropatterning. *Biochem Biophys Res Commun*. 2003;307(4):883–890.
137 Risau W. Mechanisms of angiogenesis. *Nature*. 1997;386(6626): 671–674.
138 Gerhardt H, Golding M, Fruttiger M, et al. VEGF guides angiogenic sprouting utilizing endothelial tip cell filopodia. *J Cell Biol*. 2003;161(6):1163–1177.
139 Gerhardt H, Golding M, Fruttiger M, et al. VEGF guides angiogenic sprouting utilizing endothelial tip cell filopodia. *J Cell Biol*. 2003;161(6):1163–1177.
140 Galasso G, Schiekofer S, Sato K, et al. Impaired angiogenesis in glutathione peroxidase-1-deficient mice is associated with endothelial progenitor cell dysfunction. *Circ Res*. 2005;98(2): 254–261.
141 Roberts N, Jahangiri M, Xu Q. Progenitor cells in vascular disease. *J Cell Mol Med*. 2005;9(3):583–591.
142 Gerhardt H, Betsholtz C. How do endothelial cells orientate? *EXS*. 2005;(94):3–15.
143 Hirschi KK, D'Amore PA. Pericytes in the microvasculature. *Cardiovasc Res*. 1996;32(4):687–698.
144 Staudacher DL, Preis M, Lewis BS, et al. Cellular and molecular therapeutic modalities for arterial obstructive syndromes. *Pharmacol Ther*. 2006;109(1–2):263–273.
145 Masaki I, Yonemitsu Y, Yamashita A, et al. Angiogenic gene therapy for experimental critical limb ischemia: acceleration of limb loss by overexpression of vascular endothelial growth factor 165 but not of fibroblast growth factor-2. *Circ Res*. 2002;90(9):966–973.
146 Chiu RC. Therapeutic cardiac angiogenesis and myogenesis: the promises and challenges on a new frontier. *J Thorac Cardiovasc Surg*. 2001;122(5):851–852.

147 Schaper W. Therapeutic arteriogenesis has arrived. *Circulation.* 2001;104(17):1994–1995.

148 Bouis D, Kusumanto Y, Meijer C, et al. A review on pro- and anti-angiogenic factors as targets of clinical intervention. *Pharmacol Res.* 2006;53(2):89–103.

149 Peirce S, Skalak TC. Microvascular remodeling: a complex continuum spanning angiogenesis to arteriogenesis. *Microcirculation.* 2003;10(1):99–111.

150 Scholz D, Cai WJ, Schaper W. Arteriogenesis, a new concept of vascular adaptation in occlusive disease. *Angiogenesis.* 2001; 4(4):247–257.

151 van Royen N, Piek JJ, Buschmann I, et al. Stimulation of arteriogenesis; a new concept for the treatment of arterial occlusive disease. *Cardiovasc Res.* 2001;49(3):543–553.

152 Buschmann I, Schaper W. Arteriogenesis versus angiogenesis: two mechanisms of vessel growth. *News Physiol Sci.* 1999;14: 121–125.

153 Schaper W, Buschmann I. Arteriogenesis, the good and bad of it. *Cardiovasc Res.* 1999;43(4):835–837.

154 Hoefer IE, van Royen N, Rectenwald JE, et al. Direct evidence for tumor necrosis factor-alpha signaling in arteriogenesis. *Circulation.* 2002;105(14):1639–1641.

155 van Royen N, Hoefer I, Buschmann I, et al. Exogenous application of transforming growth factor beta 1 stimulates arteriogenesis in the peripheral circulation. *FASEB J.* 2002;16(3):432–434.

156 Deindl E, Buschmann I, Hoefer IE, et al. Role of ischemia and of hypoxia-inducible genes in arteriogenesis after femoral artery occlusion in the rabbit. *Circ Res.* 2001;89(9):779–786.

157 Carmeliet P. Mechanisms of angiogenesis and arteriogenesis. *Nat Med.* 2000;6(4):389–395.

158 Price RJ, Less JR, Van Gieson EJ, Skalak TC. Hemodynamic stresses and structural remodeling of anastomosing arteriolar networks: design principles of collateral arterioles. *Microcirculation.* 2002;9(2):111–124.

159 Skalak TC, Price RJ, Zeller PJ. Where do new arterioles come from? Mechanical forces and microvessel adaptation. *Microcirculation.* 1998;5(2–3):91–94.

160 Skalak TC, Price RJ. The role of mechanical stresses in microvascular remodeling. *Microcirculation.* 1996;3(2):143–165.

161 Pittman RN. Oxygen transport and exchange in the microcirculation. *Microcirculation.* 2005;12(1):59–70.

162 Chow J, Ogunshola O, Fan SY, et al. Astrocyte-derived VEGF mediates survival and tube stabilization of hypoxic brain microvascular endothelial cells in vitro. *Brain Res Dev Brain Res.* 2001;130(1):123–132.

163 Risau W, Esser S, Engelhardt B. Differentiation of blood-brain barrier endothelial cells. *Pathol Biol (Paris).* 1998;46(3):171–175.

164 Aird WC, Edelberg JM, Weiler-Guettler H, et al. Vascular bed-specific expression of an endothelial cell gene is programmed by the tissue microenvironment. *J Cell Biol.* 1997;138(5):1117–1124.

165 Rohr C, Prestel J, Heidet L, et al. The LIM-homeodomain transcription factor Lmx1b plays a crucial role in podocytes. *J Clin Invest.* 2002;109(8):1073–1082.

166 Eremina V, Sood M, Haigh J, et al. Glomerular-specific alterations of VEGF-A expression lead to distinct congenital and acquired renal diseases. *J Clin Invest.* 2003;111(5):707–716.

185

Gene Transfer and Expression in the Vascular Endothelium

Michael J. Passineau and David T. Curiel

The Gene Therapy Center, University of Alabama at Birmingham

Gene transfer to the endothelium represents a potentially powerful approach to research and therapy. Interest in gene transfer and expression has grown apace with the evolving understanding of the importance of the endothelium in many biological and pathological processes, as detailed elsewhere in this volume. The technological advances that have made efficient gene transfer to living cells possible are relatively recent, and therefore gene transfer to the endothelium is still a somewhat immature field with substantial technical challenges still unresolved. This is particularly true with regard to in vivo gene transfer to the vascular endothelium of an intact organism.

Transfer and expression of foreign DNA to a living cell requires two components: (a) a vector to carry, protect, and traffic the foreign DNA, and (b) a DNA sequence which, once inside the host cell, is capable of interacting with the host cell's translational machinery. In nature, this process can occur in single-cell organisms by means of transformation, but it occurs in mammalian cells efficiently only by viruses. Therefore, the advent of recombinant viruses, produced by means of molecular engineering, brought gene transfer into the scientific mainstream.

As an intact organ, the endothelium has greater access to the vascular compartment than any other cell or tissue type. Therefore, the endothelium is the first point of contact between a gene-delivery vector administered to the vascular compartment and the host. For this reason, it presents an accessible target of opportunity for systemically administered gene-delivery vectors. Conversely, the intimate relationship between the endothelium and the intravascular compartment mandates careful consideration of unintended consequences of the interaction between the endothelium and a systemically administered gene-delivery vector.

The history of attempts to transfer genetic payloads to the endothelium has clearly been technology-limited. In this regard, a troublesome disconnect persists between the considerable progress that has been made in unraveling endothelial physiology and the gene-transfer technology needed to translate these insights into robust in vivo transgene expression in the intact endothelium of a living mammal. Gene transfer to endothelial cells (ECs) ex vivo is an established and powerful research tool that has contributed to many advances in endothelial biology. Nevertheless, gene transfer to isolated, ex vivo systems has limited value as it does not take into account the organ complexity of the intact endothelium. In addition, therapeutic modification of gene expression in disease states of the endothelium will require a vector capable of effecting gene transfer following systemic infusion into the vascular space.

Meeting the technical challenge of bridging the gap between ex vivo and practicable in vivo gene transfer to the endothelium represents a key step in advancing the field of endothelial biomedicine. The following brief overview of gene transfer to the endothelium will illuminate the technical hurdles restricting progress and how recent technological developments suggest that these restrictions might soon be obviated in the research context. Finally, an experimental model will be described that makes efficient vector-mediated endothelial gene transfer possible in an intact animal.

HISTORY OF VASCULAR ENDOTHELIAL GENE EXPRESSION AND VECTOR DEVELOPMENT

The history of gene transfer to the endothelium has evolved concurrently with the development of both gene transfer and vectorology as scientific disciplines. Although gene transfer technology has generally been developed along strict demarcations between ex vivo and in vivo potential, the branching nature of the vascular tree makes possible the anatomical sequestration of a vector infused in vivo with simple surgical techniques such as clamping of vascular segments. Owing to this unique capacity for vector compartmentalization, gene transfer to the endothelium has occupied a middle ground between the in vivo and ex vivo experimental paradigms.

The key factor in the early development of gene transfer in ECs was the emergence of recombinant retroviruses as gene-transfer vectors and the optimization of such systems for endothelial expression (1). Gene transfer to ECs was first shown to be feasible in a practicable way in 1989 by Zwiebel and colleagues at the National Institutes of Health (NIH), who transduced rabbit ECs with retroviruses (2). This same group, a short time later, extended these findings to human ECs (3), and later replaced the reporter gene with a relevant "therapeutic" gene, *tissue-type plasminogen activator* (*t-PA*) (4,5).

Several laboratories reported direct gene transfer to endothelium by infusion of retroviral vectors into a surgically isolated vascular space (6,7). However, owing to the limitations of the available retroviral systems, most notably their relatively inefficient transduction rates, the in vivo potential of these systems has not yet been demonstrated. At present, retroviruses are mainly conceptualized as an ex vivo gene-transfer system (8,9) whereby target cells are extracted and purified prior to gene transfer and then reimplanted into the host. Notably, this ex vivo approach has been successfully implemented, recently forming the treatment basis of the first gene therapy "cure" for a genetic disorder, severe combined immune deficiency (SCID) (10). This clinical trial in SCID has highlighted an additional safety concern with retroviral vectors – namely, the oncogenic potential of these vectors due to the potential for insertional mutagenesis (11).

In 1995, von der Leyen and colleagues demonstrated the potential to apply previously established in vivo endothelial gene delivery to treat various diseases (12). However, the limitations imposed by poor in vivo expression with retroviral or liposome systems delayed the clinical implementation of these gene-therapy strategies. Attempts to improve the in vivo expression efficiency of these vector systems have been largely unsuccessful thus far, confining their present usefulness to the aforementioned ex vivo schema. These considerations limit the usefulness of retroviral vectors for applications in endothelial gene delivery.

Recently, lentiviral vector systems have received a great deal of attention as gene-transfer vectors. In contrast to many other Retroviridae, lentiviruses are capable of gene transfer to quiescent or slowly dividing cells. This property is particularly relevant in the central nervous system, including brain (13) and retina (14). Lentiviral vectors administered via a systemic vascular route show low but measurable gene transfer to the endothelium (15). Various strategies are being tested to improve the production efficiency, in vivo expression and targeting of lentiviral vectors (16).

As an alternative to retroviral systems, adenovirus (Ad) was proposed (17), with the expectation that the robust in vivo expression provided by this vector system would allow for direct in vivo expression (18) of therapeutic transgenes, in this case α-1 antitrypsin, and obviate the need for ex vivo systems. These studies by Lemarchand and coworkers demonstrated that Ad vectors can achieve high levels of gene expression, consistent with replacement levels of α-1 antitrypsin levels, in the intact vascular endothelium.

Lemarchand's work also highlighted the two of the main shortfalls of Ad vectors for in vivo gene transfer: (a) expression was variable and transient, peaking at 7 days and becoming extinct by 28 days, and (b) anatomical sequestration of the vector by occlusion of the vascular segment for more than 15 minutes was necessary for the achievement of significant endothelial expression. Although modern surgical techniques make this type of transient anatomical sequestration possible, many endothelial populations of clinical interest exist in vascular segments that cannot be occluded without significant risk (e.g., pulmonary and cardiac endothelium). Despite the drawbacks of the Ad system, these studies powerfully demonstrated the utility of recombinant Ad systems as a tool to achieve gene transfer to the endothelium. Based on these findings, subsequent studies examined different methodologies for selective delivery of Ad vectors to the vessel wall (19).

Additional studies have confirmed the finding that robust in vivo expression can be achieved following systemic injection of an Ad vector, but that this expression is low in endothelium. The innate hepatotropism of this vector system effectively sequesters systemically administered Ad to the liver, with resultant toxicity to the host. The relatively low affinity of Ad vectors for ECs, particularly when compared to the strong affinity of Ad vectors for the liver, constitutes the primary limitation to achieving Ad-mediated gene transfer to the vascular endothelium. On this basis, it has become clear that in order to achieve meaningful in vivo gene transfer to the vascular endothelium, retargeting of Ad vectors is mandated.

Finally, the use of adeno-associated viruses (AAVs) as gene-transfer vector systems for endothelium has recently gained widespread attention, beginning with a series of studies in 1997 (20–23) using AAV serotype 2. Potential advantages of AAVs over Ad include increased duration of transgene expression (months versus days in many models), relatively less immunogenicity in the host, and substantially diminished hepatotropism and toxicity. Difficulties with producing high-titer preparations of AAVs, combined with low levels of in vivo expression (relative to Ad), have thus far limited the clinical utility of AAVs. In addition, the relatively small genome of AAVs creates a technological limitation when considering various endothelial-specific promoters or large transgenes (as described in more detail in this chapter). Improvements in AAV technology aimed at overcoming these limitations are ongoing (Table 185-1).

It should be noted that the endothelium forms a barrier that is essentially impenetrable to viral vectors, with the exception of fenestrated endothelium in such sites as the liver and spleen. Indeed, Cefai and coworkers (15) have directly compared lentiviral vectors to Ad vectors and shown that these vectors do not penetrate beyond the endothelial layer of the arterial wall in vivo, although an inflammatory response to Ad was observed in the smooth muscle cell layer. In contrast, one of the more intriguing utilities of AAVs is an apparent ability of some serotypes to traverse the EC layer (24).

Table 185-1: Vector Systems for Endothelial Gene Transfer – Advantages and Disadvantages

Vector System	*Advantages*	*Disadvantages*	*Applications*
Retrovirus	Genome integration, long-term expression	Relatively weak in vivo expression	Ex vivo gene transfer for transplantation/ reimplantation
Liposomes	Ease of manufacture, low immunogenicity, large capacity	Extremely low expression of genetic payload	Research
Adenovirus (Ad)	Strong expression in vivo	Strongly immunogenic, transient expression	Gene transfer in many contexts, including in vivo
Adeno-associated virus (AAV)	Some genome integration, long-term expression, low immunogenicity	Small genome size, difficult to produce in high titer	Ex vivo gene transfer

ADENOVIRUS RETARGETING TO ACHIEVE IN VIVO ENDOTHELIAL GENE TRANSFER

In the in vivo context, restricting Ad vector–mediated gene expression to a particular cell type (endothelium) requires both increasing the affinity of the vector for the target cell (targeting) and simultaneously reducing ectopic expression (untargeting), particularly in the liver. With the notable exception of the liver, kidney, and spleen, the endothelial barrier tightly regulates transfer of substances from blood to extravascular compartments and thus restricts access of a systemically administered Ad vector to non-ECs. Successful endothelial "retargeting" is therefore achieved once the threshold is reached at which the affinity of the Ad gene transfer vector for the target EC exceeds its affinity for the liver (and to a lesser extent spleen and kidney). The degree to which this threshold can be exceeded by the retargeting strategy dictates the therapeutic relevance of the retargeting strategy.

To date, two general approaches to retargeting Ad have been developed and successfully validated. First, vectors may be retargeted on the *transductional* level by means of modifying the viral capsid in a manner which alters its ability to bind to cellular surfaces. Second and independent from transductional retargeting, the vector's expression cassette can be modified in such a way as to embody target cell–specific expression, termed *transcriptional* targeting. The combination of these two retargeting strategies in a single Ad vector system has been shown to be synergistic (25). This concept of transcriptional targeting, and its specific relevance to endothelial gene transfer, will be detailed later in the chapter.

Transductional Targeting

Transductional targeting has been accomplished by modifying the viral capsid's cell surface tropism in two ways: either with bispecific "adapter molecules" or by means of genetic capsid modification (Figure 185.1). The first instance of transductional retargeting to ECs (26,27) was accomplished using bispecific antibodies. In this schema, antibodies to a surface epitope on the EC target (such as αv integrins) are chemically conjugated to antibodies specific for an epitope on the viral capsid. Later studies embodying the same technique were shown to accomplish EC transductional retargeting in vivo (28). Of note, the adapter molecule approach to transductional retargeting can achieve, to some extent, the objective of liver untargeting when the portion of the bispecific adapter molecule that recognizes the viral capsid thereby blocks a liver-binding epitope normally displayed on the virus capsid (most notably the fiber knob that recognizes the coxsackie–adenovirus receptor, CAR). This general adapter molecule strategy has since been slightly modified to create single-molecule bispecific fusion proteins (29) (fusions of single chain antibodies, soluble CAR receptors, CD40 ligand) rather than chemical conjugates of two antibodies (Table 185-2).

Direct genetic modification of the viral capsid proteins responsible for cell tropism (namely those in the fiber and penton base) has been proposed as an alternative strategy to adapter molecules. One of the primary advantages of this approach with respect to potential for clinical translation is the "single agent" nature of a genetically modified vector, as opposed to a vector/adapter molecule system. The genetic retargeting strategy may be as simple as introducing an alternate Ad serotype (Ad3, Ad35, etc., as opposed to the most commonly used Ad5). Furthermore, the capsid protein genes can be selectively modified on the genomic level, yielding capsid proteins that, when assembled into virions, confer an alternate tropism (30). Such an approach can modify or ablate native tropism, generating a genetically distinct vector species with the capacity to be propagated perpetually in permissive cell lines (31).

To date, the primary technological limitation hindering the usefulness of the genetic capsid modification strategy is the relative intolerance of the fiber and penton base proteins to large additions, deletions, or modifications. Attempts to insert targeting ligands into the fiber protein have resulted in vector species that are nonviable. Similarly, penton base

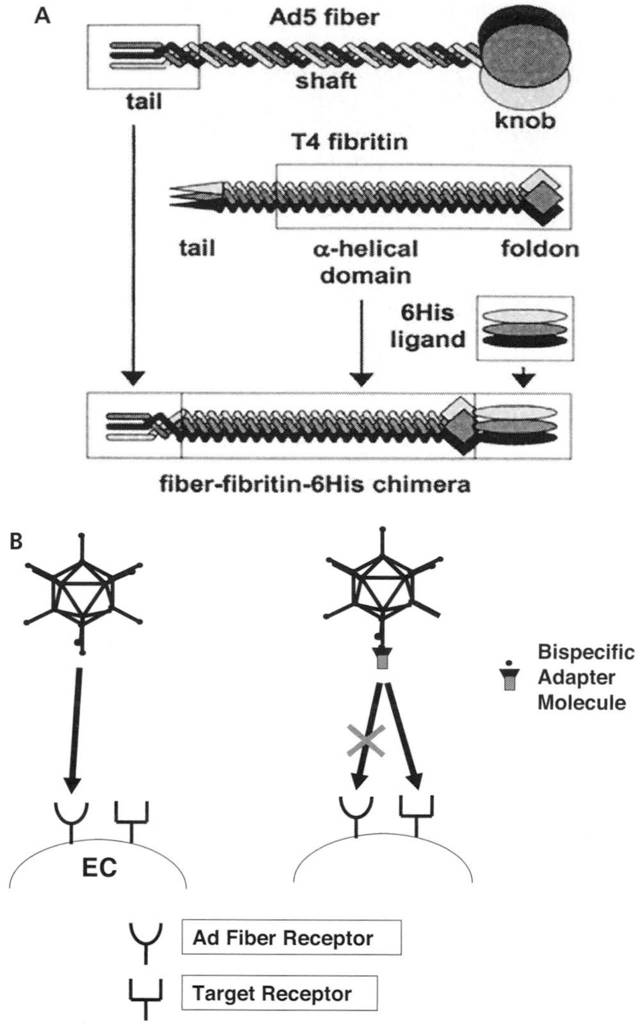

Figure 185.1. Transductional targeting approaches. (**A**) Fiber replacement strategy. The Ad5 fiber shaft sequence is replaced with the sequence for phage T4 fibritin. Fibritin forms trimers similar to those of the Ad5 shaft but is more permissive of C-terminal ligand incorporation. This example illustrates the replacement of Ad5 knob with a c-terminal 6His tag in the fiber–fibritin chimera. (**B**) Adapter molecule strategy. An adapter molecule, with motifs that recognize and bind the Ad knob and a target receptor (such as ACE) partially ablates the vector's native tropism while allowing the vector to recognize and bind an alternate receptor.

modification has been primarily restricted to deletion of the RGD motif exposed on the surface of the mature protein.

Insertion of small peptide sequences into the knob region of Ad fiber is possible, and modest retargeting has been accomplished using this approach in vitro (32–35) and in vivo (36). More recently, Nicklin and coworkers (37) have applied this principle to achieve EC transduction. However, this approach is severely limited with regard to both the size of the ligand (published upper limit of 30 amino acids) (38,39) and the potential for post-translational processing needed to produce viable ligand function (e.g., S–S bonds). Finally, incorporation of small peptide sequences into the fiber protein has

Table 185-2: Illustrative Studies Documenting Transductional Targeting of Vectors to Endothelium In Vivo

Vector	*Technique*	*References*
Retrovirus	Nonspecific uptake in anatomically isolated vascular space	6, 7
Adenovirus	Adapter molecule, hCAR model	28, 45
AAV	Small peptide insertion in viral capsid	47

not been shown to effectively ablate the native tropism of the vector.

In light of these considerations, replacement of the entire fiber protein has been proposed (40). Using the phage T4 fibritin protein as a substitute for Ad5 fiber, Belousova and coworkers (41) successfully incorporated a physiological targeting ligand, CD40 ligand (CD154), into an artificial fiber and demonstrated binding to the cognate receptor, CD40. To date, this approach represents the only viable approach to genetic ligand incorporation into the Ad fiber.

One difficulty associated with this retargeting strategy is that the genetic incorporation of targeting ligands is limited by the failure of Ad fiber proteins to pass through the Golgi apparatus prior to assembly, thus limiting the post-translational modification of the proteins. The fibritin-CD40L virus constructed by Belousova and coworkers exploited the lack of disulfide bonds in the mature CD40L protein, thus avoiding the post-translational limitations imposed by the Ad lifecycle (41). However, nearly all targeting ligands of medical relevance are not amenable to such an approach due to the requirement for disulfide bond formation for proper ligand function.

Both of the retargeting approaches described – adapter molecule retargeting and genetic incorporation of targeting ligands – are constrained by the availability of relevant genetically encoded targeting ligands. In the field of endothelial biomedicine, target definition has made considerable progress, and more progress is expected, particularly with regard to vascular bed–specific expression of endothelial surface markers. Nevertheless, it must be considered that only a subset of these markers will interact with a known genetically encoded ligand. In this regard, single-chain antibodies (scFv's) have been proposed as an alternative to the more stringent receptor–ligand schema. An additional advantage of scFv's is the lack of a requirement for post-translational modification to form the mature protein. Although promising, scFv's have yet to prove their full potential in vivo.

Transcriptional Targeting

As scientists continue to unravel the complex physiology and function of the endothelium, it has become clear that the endothelium is not homogeneous throughout the vascular tree (42). Rather, ECs differ in their specific cellular programs from

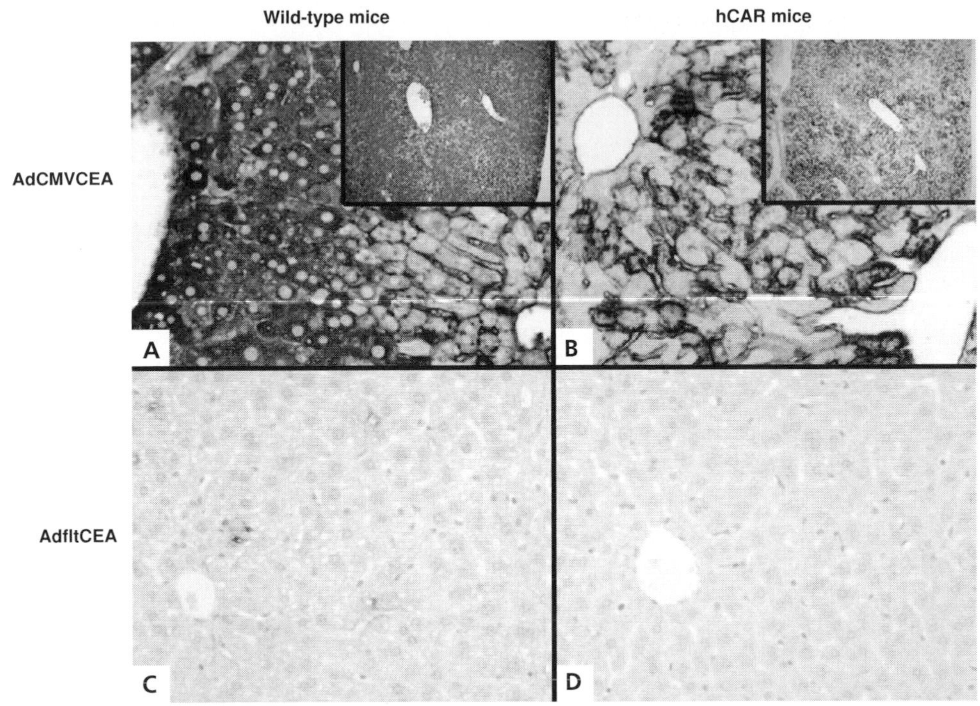

Figure 185.2. Transcriptional targeting in vivo. Transcriptional control of Ad-mediated gene transfer to the liver. Immunohistochemistry of liver from wild-type or hCAR mice 24 hours after systemic administration of a vector encoding CEA (as a reporter gene) under the control of either *CMV* or *FLT-1* promoters. The *FLT-1* promoter leads to dramatic reduction of expression in the liver. For color reproduction, see Color Plate 185.2.

one vascular bed to another. The molecular characterization of this heterogeneity is far from complete, but the implications for vascular bed–specific control of gene transfer are already significant.

Gene transfer to the endothelium, or to any cell type, can be accomplished by any number of vectors, viral or nonviral, as previously discussed. Once the transgene has been delivered, however, expression is uniformly accomplished by interaction of the target cell's transcriptional machinery with a promoter sequence. Thus, in their most minimal form, transgenes are composed of a promoter region and a coding region.

Historically, gene transfer has been undertaken using promoters with a uniformly strong activity profile in multiple cell types. Most famous among these promoters is the *cytomegalovirus* (*CMV*) promoter, which has been used in literally thousands of viral and nonviral constructs. More recently, alternative promoters have been explored that show a more tissue/organ or cell-specific activity profile. Control of transgene expression with promoters whose activity profiles differ from one vascular bed to another – e.g., the *von Willebrand Factor* (*vWF*) or *FLT-1* promoters (43) – is a reasonable strategy for achieving targeted gene expression (Figure 185.2; for color reproduction, see Color Plate 185.2).

It must be noted that any transcriptional targeting strategy, although a powerful research tool, ignores the major technological obstacles that must be overcome before endothelial gene transfer can expand beyond laboratory research to become a practicable clinical intervention. Specifically, transcriptional targeting controls only the expression profile of the vector, without affecting its biodistribution. In other words, toxicities and other complications known to result from viral vectors are not primarily dependent on the expression profile of the transgenic construct, but rather the host immune response, both preexisting and adaptive. Toxicities could therefore still result in tissues where little or no transgene expression occurs.

A Transgenic "Shortcut"

On the basis of the aforementioned considerations, it is clear that vectors that achieve robust endothelial gene expression while minimizing ectopic expression in the systemic context are not yet available. Although this technological limitation severely limits the clinical implications of gene transfer to the vascular endothelium, the research community has a broader array of options. The development of artificial systems capable of directing gene expression to the vascular endothelium in an intact animal model represents a significant research priority in the field of endothelial biomedicine.

Tallone and coworkers (44) have developed a simple transgenic mouse model which effectively enables robust endothelial gene transfer with Ad vectors. This transgenic animal expresses a truncated, nonsignaling version of human CAR (hCAR), the native receptor for Ad5, ubiquitously throughout

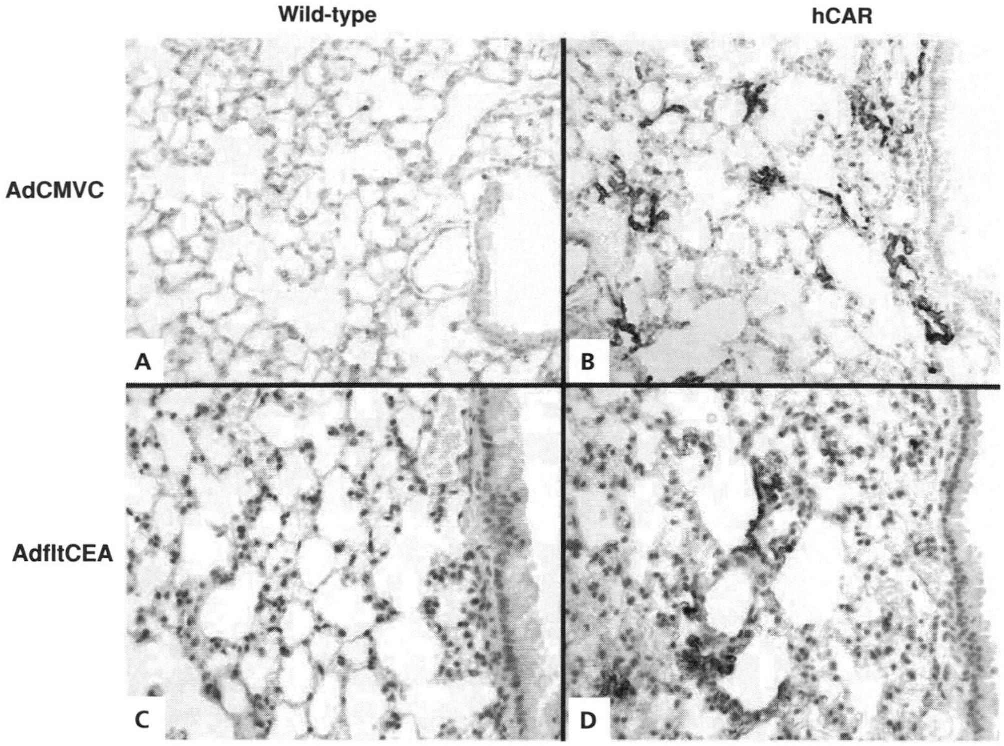

Figure 185.3. Transductional retargeting in the hCAR model. Ad-mediated gene transfer to ECs in the hCAR mouse. Strong endothelial expression of the reporter gene (CEA) is observed in lung sections from hCAR mice. Lung expression is essentially absent in the wild-type mouse. For color reproduction, see Color Plate 185.3.

all tissues. In this system, all cell and tissue types have a similar and high susceptibility to Ad5 infection. Hence, the tissue distribution of an administered Ad vector becomes dependent on tissue access rather than tissue tropism.

An Ad vector administered systemically to this "hCAR" mouse will be predicted to achieve robust transduction of the vascular endothelium. Further, the impermeability of the vascular endothelium to viral vectors restricts a systemically administered vector's access to cells that have direct contact with the vascular compartment, limiting spread of the vector beyond the endothelium. In this schema, efficient endothelial gene transfer in the intact animal should be achieved (Figure 185.3; for color reproduction, see Color Plate 185.3).

The concept that transcriptional targeting using the *FLT-1* promoter can achieve selective gene expression in vivo has been previously established (23). The availability of the hCAR transgenic mouse gives investigators the opportunity to study gains accruing to transcriptional targeting strategies exclusive of transductional considerations. In order to establish the validity of this system, Everts and coworkers (45) tested the FLT-1 Ad vector constructs used previously in Reynolds's (28) study in the hCAR transgenic model, finding that lung-specific expression could be achieved in the hCAR mouse model by replacing the *CMV* promoter with the *FLT-1* promoter. Further, this expression was found to be localized exclusively to ECs.

The results of this study validated two concepts: (a) the hCAR mouse model allows robust transduction of ECs with Ad vectors, and (b) transcriptional control alone, in the absence of tropism modification, achieves selective gene transfer to specific vascular beds. This system, therefore, represents a high-throughput experimental model in which to study gene transfer to the endothelium. Of note, various studies have illuminated the activity of human promoters in mouse endothelium, establishing the principle that the mouse system is a useful in vivo model for studying the activity of human promoters in the endothelium of different vascular beds. This cross-species relevance of the hCAR system greatly expedites efforts to screen the ever-increasing array of promoters relevant to the endothelium. The Ad vector technology allowing facile and efficient swapping of promoters in Ad constructs is currently available, establishing the hCAR system as a powerful advancement in the study of gene transfer to the endothelium.

The relevance of the hCAR system to promoter screening is apparent. However, applications of this system that go beyond simply reporter gene transfer to the endothelium must be interpreted with caution. For example Ad vector–mediated transfer of a gene of interest to the vascular endothelium with the intent of studying the resulting biological effects of the expression of such a gene must take into account a number of complicating factors. First, Ad vector expression, although robust, is transient and varies over days and weeks. In general, the time course and peak of gene expression in this system is known. However, great care must be taken to normalize doses and experimental time points before any conclusions are drawn. Further, Ad vectors are extremely immunogenic, particularly in the dose levels needed to achieve robust in vivo gene transfer and expression. This consideration is emphasized in an experimental schema that utilizes promoters other than *CMV*, particularly vascular-bed-specific promoters. The low native activities of many promoters of particular relevance to the vascular endothelium require that larger doses of vector be administered in order to reach detectable expression levels.

As previously noted, transcriptional targeting strategies that minimize ectopic expression by gene transfer vectors in nontarget tissues do not thereby minimize ectopic toxicities in those same nontarget tissues. In the case of the hCAR mouse system, it is clear that substitution of the conventional *CMV* promoter with the partially lung-specific *FLT-1* promoter does effectively redirect transgene expression to the pulmonary endothelium, at the same time dramatically reducing expression in the liver. However, the sequestration and accumulation of vector in the liver is only slightly reduced in the hCAR system relative to the wild-type mouse.

CONCLUSION

Gene transfer to EC culture or explants has been effectively employed within the research community for over a decade, and studies employing this technique have contributed to the evolving understanding of the endothelium as a dynamic organ. With the development of small interfering RNA (siRNA) technology, and its demonstrated utility in viral vectors, investigators have the capability to rapidly overexpress or eliminate expression of a gene of interest to study the biological sequelae in ECs. Combined with microarray technology, gene transfer provides a powerful scientific tool.

Ex vivo gene transfer to the endothelium has been well established and is less technology limited than in vivo gene transfer. Applications of ex vivo gene transfer include not only myriad research strategies but also select clinical interventions. For example, autologous ECs modified to overexpress cytokines or growth factors, then seeded into vascular stents, have been proposed as a strategy to reduce inflammation and rejection. At present, the clinical applications of ex vivo gene transfer to ECs is principally limited not by the available vector technology but rather by the difficulty of obtaining and expanding ECs for autologous transplantation.

With regard to in vivo gene transfer, it is clear that research tools are now emerging that will allow scientists to solve one half of the endothelial gene expression puzzle, namely the transcriptional targeting approach for targeted gene expression. However, it is equally clear that transductional targeting approaches, by far the more important consideration in contemplating clinical implications, are still lacking. The hCAR mouse model explained herein obviates the transductional issues as they relate to the endothelium, making efficient endothelial gene transfer in vivo a practicable reality. This transgenic "shortcut" offers a powerful new research tool. Nevertheless, transductional targeting issues must be addressed if in vivo gene transfer is ever to transition from bench to bedside.

It is important to note that the biological effects of viral infection in ECs, irrespective of the particular genetic payload, are mostly unknown. Various microarray studies have indicated that Ad vector infection provokes dramatic changes in the cellular expression of thousands of genes (46,47). In this milieu of cellular responses, deciphering the physiological changes due solely to the expression of a transgene of interest seems a daunting task. These issues become even more complex in conceptualizing medical applications of endothelial gene transfer.

At present, Ad vectors have yet to be usurped as the vector of choice for in vivo gene transfer. However, the disadvantages of Ad vectors including transient and variable gene expression, immunogenicity, and toxicity make Ad an imperfect vector upon which to advance endothelial gene transfer as a therapeutic strategy. Until a substitute for Ad vectors is devised that accomplishes the level of expression needed for research and medical applications, gene transfer to the endothelium will continue to be significantly complicated by ectopic gene expression and the complex postinfection physiology of the Ad vector itself. It is expected that improvements in alternative vector systems, as well as refinement of retargeted Ad vectors, will continue to emerge.

KEY POINTS

- To date, gene transfer to the endothelium has been largely confined to in vitro or ex vivo systems.
- Available vector systems, with the exception of Adenovirus, fail to achieve in vivo expression levels desired for research and clinical applications.
- Improvements in Adenoviral targeting along with improvement in other viral systems, particularly AAV and lentiviruses, are progressing.
- An artificial animal model (hCAR mouse) has been designed which facilitates robust in vivo gene transfer to the endothelium.

ACKNOWLEDGMENT

This work was supported by 5POIHL076540 and 5T32CA75930 from the National Institutes of Health.

REFERENCES

1 Kahn ML, Lee SW, Dichek DA. Optimization of retroviral vector-mediated gene transfer into endothelial cells in vitro. *Circ Res.* 1992;71(6):1508–1517.

2 Zwiebel JA, Freeman SM, Kantoff PW, et al. High-level recombinant gene expression in rabbit endothelial cells transduced by retroviral vectors. *Science.* 1989;243(4888):220–222.
3 Zwiebel JA, Freeman SM, Cornetta K, et al. Recombinant gene expression in human umbilical vein endothelial cells transduced by retroviral vectors. *Biochem Biophys Res Commun.* 1990;170(1): 209–213.
4 Dichek DA, Nussbaum O, Degen SJ, Anderson WF. Enhancement of the fibrinolytic activity of sheep endothelial cells by retroviral vector-mediated gene transfer. *Blood.* 1991;77(3):533–541.
5 Dichek DA, Lee SW, Nguyen NH. Characterization of recombinant plasminogen activator production by primate endothelial cells transduced with retroviral vectors. *Blood.* 1994;84(2):504–516.
6 Nabel EG, Plautz G, Nabel GJ. Transduction of a foreign histocompatibility gene into the arterial wall induces vasculitis. *Proc Natl Acad Sci USA.* 1992;89(11):5157–5161.
7 Lim CS, Chapman GD, Gammon RS, et al. Direct in vivo gene transfer into the coronary and peripheral vasculatures of the intact dog. *Circulation.* 1991;83(6):2007–2011.
8 Messina LM, Podrazik RM, Whitehill TA, et al. Adhesion and incorporation of lacZ-transduced endothelial cells into the intact capillary wall in the rat. *Proc Natl Acad Sci USA.* 1992;89(24): 12018–12022.
9 Nabel EG, Plautz G, Boyce FM, et al. Recombinant gene expression *in vivo* within endothelial cells of the arterial wall. *Science.* 1989;244(4910):1342–1344.
10 Cavazzana-Calvo M, Hacein-Bey S, de Saint Basile G, et al. Gene therapy of human severe combined immunodeficiency (SCID)-X1 disease. *Science.* 2000;288(5466):669–672.
11 Kaiser J. Gene therapy. Panel urges limits on X-SCID trials. *Science.* 2005;307(5715):1544–1545.
12 von der Leyen HE, Gibbons GH, Morishita R, et al. Gene therapy inhibiting neointimal vascular lesion: *in vivo* transfer of endothelial cell nitric oxide synthase gene. *Proc Natl Acad Sci USA.* 1995; 92(4):1137–1141.
13 Naldini L, Blomer U, Gallay P, et al. In vivo gene delivery and stable transduction of nondividing cells by a lentiviral vector. *Science.* 1996;272(5259):263–267.
14 Miyoshi H, Takahashi M, Gage FH, Verma IM. Stable and efficient gene transfer into the retina using an HIV-based lentiviral vector. *Proc Natl Acad Sci USA.* 1997;94(19):10319–10323.
15 Cefai D, Simeoni E, Ludunge KM, et al. Multiply attenuated, self-inactivating lentiviral vectors efficiently transduce human coronary artery cells in vitro and rat arteries in vivo. *J Mol Cell Cardiol.* 2005;38(2):333–344.
16 Bartosch B, Cosset FL. Strategies for retargeted gene delivery using vectors derived from lentiviruses. *Curr Gene Ther.* 2004; 4(4):427–443.
17 Lemarchand P, Jaffe HA, Danel C, et al. Adenovirus-mediated transfer of a recombinant human alpha 1-antitrypsin cDNA to human endothelial cells. *Proc Natl Acad Sci USA.* 1992;89(14): 6482–6486.
18 Lemarchand P, Jones M, Yamada I, Crystal RG. *In vivo* gene transfer and expression in normal uninjured blood vessels using replication-deficient recombinant adenovirus vectors. *Circ Res.* 1993;72(5):1132–1138.
19 Willard JE, Landau C, Glamann DB, et al. Genetic modification of the vessel wall. Comparison of surgical and catheter-based techniques for delivery of recombinant adenovirus. *Circulation.* 1994;89(5):2190–2197.
20 Maeda Y, Ikeda U, Ogasawara Y, et al. Gene transfer into vascular cells using adeno-associated virus (AAV) vectors. *Cardiovasc Res.* 1997;35(3):514–521.
21 Rolling F, Nong Z, Pisvin S, Collen D. Adeno-associated virus-mediated gene transfer into rat carotid arteries. *Gene Ther.* 1997; 4(8):757–761.
22 Lynch CM, Hara PS, Leonard JC, et al. Adeno-associated virus vectors for vascular gene delivery. *Circ Res.* 1997;80(4):497–505.
23 Gnatenko D, Arnold TE, Zolotukhin S, et al. Characterization of recombinant adeno-associated virus-2 as a vehicle for gene delivery and expression into vascular cells. *J Investig Med.* 1997; 45(2):87–98.
24 Wang Z, Zhu T, Qiao C, et al. Adeno-associated virus serotype 8 efficiently delivers genes to muscle and heart. *Nat Biotechnol.* 2005;23(3):321–328.
25 Reynolds PN, Nicklin SA, Kaliberova L, et al. Combined transductional and transcriptional targeting improves the specificity of transgene expression *in vivo. Nat Biotechnol.* 2001;19(9):838–842.
26 Wickham TJ, Segal DM, Roelvink PW, et al. Targeted adenovirus gene transfer to endothelial and smooth muscle cells by using bispecific antibodies. *J Virol.* 1996;70(10):6831–6838.
27 Rogers BE, Douglas JT, Ahlem C, et al. Use of a novel cross-linking method to modify adenovirus tropism. *Gene Ther.* 1997; 4(12):1387–1392.
28 Reynolds PN, Zinn KR, Gavrilyuk VD, et al. A targetable, injectable adenoviral vector for selective gene delivery to pulmonary endothelium in vivo. *Mol Ther.* 2000;2(6):562–578.
29 Kashentseva EA, Seki T, Curiel DT, Dmitriev IP. Adenovirus targeting to c-erbB-2 oncoprotein by single-chain antibody fused to trimeric form of adenovirus receptor ectodomain. *Cancer Res.* 2002;62(2):609–616.
30 Krasnykh VN, Mikheeva GV, Douglas JT, Curiel DT. Generation of recombinant adenovirus vectors with modified fibers for altering viral tropism. *J Virol.* 1996;70(10):6839–6846.
31 Douglas JT, Miller CR, Kim M, et al. A system for the propagation of adenoviral vectors with genetically modified receptor specificities. *Nat Biotechnol.* 1999;17(5):470–475.
32 Michael SI, Hong JS, Curiel DT, Engler JA. Addition of a short peptide ligand to the adenovirus fiber protein. *Gene Ther.* 1995; 2(9):660–668.
33 Dmitriev I, Krasnykh V, Miller CR, et al. An adenovirus vector with genetically modified fibers demonstrates expanded tropism via utilization of a coxsackievirus and adenovirus receptor-independent cell entry mechanism. *J Virol.* 1998;72(12):9706–9713.
34 Bouri K, Feero WG, Myerburg MM, et al. Polylysine modification of adenoviral fiber protein enhances muscle cell transduction. *Hum Gene Ther.* 1999;10(10):1633–1640.
35 Einfeld DA, Brough DE, Roelvink PW, et al. Construction of a pseudoreceptor that mediates transduction by adenoviruses expressing a ligand in fiber or penton base. *J Virol.* 1999;73(11): 9130–9136.
36 Reynolds P, Dmitriev I, Curiel D. Insertion of an RGD motif into the HI loop of adenovirus fiber protein alters the distribution of transgene expression of the systemically administered vector. *Gene Ther.* 1999;6(7):1336–1339.
37 Nicklin SA, Von Seggern DJ, Work LM, et al. Ablating adenovirus type 5 fiber-CAR binding and HI loop insertion of the SIGYPLP

peptide generate an endothelial cell-selective adenovirus. *Mol Ther*. 2001;4(6):534–542.

38 Hong JS, Engler JA. Domains required for assembly of adenovirus type 2 fiber trimers. *J Virol*. 1996;70(10):7071–7078.

39 Wickham TJ, Tzeng E, Shears LL2nd, et al. Increased *in vitro* and *in vivo* gene transfer by adenovirus vectors containing chimeric fiber proteins. *J Virol*. 1997;71(11):8221–8229.

40 Krasnykh V, Belousova N, Korokhov N, et al. Genetic targeting of an adenovirus vector via replacement of the fiber protein with the phage T4 fibritin. *J Virol*. 2001;75(9):4176–4183.

41 Belousova N, Korokhov N, Krendelshchikova V, et al. Genetically targeted adenovirus vector directed to CD40-expressing cells. *J Virol*. 2003;77(21):11367–11377.

42 Aird WC, Edelberg JM, Weiler-Guettler H, et al. Vascular bed-specific expression of an endothelial cell gene is programmed by the tissue microenvironment. *J Cell Biol*. 1997;138(5):1117–1124.

43 Minami T, Donovan DJ, Tsai JC, et al. Differential regulation of the von Willebrand factor and Flt-1 promoters in the endothelium of hypoxanthine phosphoribosyltransferase-targeted mice. *Blood*. 2002;100(12):4019–4025.

44 Tallone T, Malin S, Samuelsson A, et al. A mouse model for adenovirus gene delivery. *Proc Natl Acad Sci USA*. 2001;98(14):7910–7915.

45 Everts ME, Kim-Park SA, Preuss MA, et al. Inducing tumor associated antigens in pulmonary vasculature for analysis of targeting strategies for cancer therapy. *Gene Therapy*. 2004.

46 Volk AL, Rivera AA, Page GP, et al. Employment of microarray analysis to characterize biologic differences associated with tropism-modified adenoviral vectors: utilization of non-native cellular entry pathways. *Cancer Gene Ther*. 2005;12(2):162–174.

47 White SJ, Nicklin SA, Buning H, et al. Targeted gene delivery to vascular tissue in vivo by tropism-modified adeno-associated virus vectors. *Circulation*. 2004;109(4):513–519.

186

Drug Targeting to Endothelium

Vladimir R. Muzykantov

University of Pennsylvania School of Medicine, Philadelphia

Endothelial tissue represents both a key target for therapeutic intervention and a barrier to successful extravascular drug delivery. Alas, most drugs have no natural affinity to any therapeutic sites, including endothelial cells (ECs). Thus, only a minute fraction of injected therapeutics actually binds to the endothelium, despite its immediate accessibility to circulation. The goal of endothelial targeting is to promote the specific and safe delivery of a drug to, into, or across ECs, to localize its effects in the lumen, desired intracellular endothelial compartment, or subendothelial space. Ideally, targeting would direct drug accumulation to a desired vascular domain.

The successful pursuit of this goal depends on several parallel lines of research, including the design of drug carriers and the identification of target endothelial determinants. To date, the majority of studies have focused on cell culture systems and animal models, with few published reports in human subjects. In this chapter, we briefly review research efforts in this field and describe basic principles that are important for endothelial targeting in specific disease states. More details on this topic can be found in recent reviews (1–10) as well as in other chapters in this volume.

The pharmacokinetics of most drugs do not favor their delivery to or/and across endothelium due to their: (a) inactivation in vivo and elimination from blood by the reticuloendothelial system (RES), a problem hampering delivery of labile, large (>50 kD) biotherapeutics including proteins and nucleic acids; (b) renal excretion, a problem especially acute for small (<10 kD) drugs; and (c) lack of affinity to ECs, which is characteristic of both classes of drugs.

Based on these considerations, several goals exist for optimizing the pharmacokinetics and thus delivery of drugs: (a) protection of drugs from inactivation en route; (b) provision of sufficient circulation time to permit accumulation in the target; and (c) reduction of interaction with nontarget cells and general toxicity (11–13). These aims can be achieved by using natural (e.g., cells or proteins) or artificial (e.g., liposomes or polymer particles) drug carriers conjugated with targeting moieties that selectively bind to determinants on the cells of interest (Table 186-1).

CARRIERS FOR DRUG DELIVERY TO ENDOTHELIAL CELLS

Liposomes are bilayer submicron vesicles (~50–200 nm is the most popular size range) composed of either natural or synthetic phospholipids. They provide versatile carriers for either hydrophobic drugs that intercalate into liposomal membranes or hydrophilic drugs that can be loaded into an aqueous inner space (14–16). Cationic liposomes condense DNA, protecting it from inactivation and facilitating intracellular delivery via endocytic mechanisms and lipid fusion with the plasma membrane. Liposomes have been intensively studied for the targeting of drugs and genetic material into ECs in cell cultures and animal studies (17–23).

Most drugs and carriers, including liposomes, are rapidly cleared from the blood by renal filtration and by macrophages in the RES, which eliminate foreign, denatured, and senescent proteins and cells from circulation. Macrophages in the hepatic sinuses (termed *Kupffer cells*) have unlimited access to circulating drugs and carriers. Fixation of antibodies or complement (opsonization) by carriers greatly enhances their RES uptake. Some drugs, carriers, and targeting moieties (e.g., immunoglobulin [Ig]G antibodies) may cause adverse effects by activating host defenses, including blood coagulation, complement, cytokines, and kinins. Small antigen-binding domains of targeting antibodies lacking Fc-fragment (Fab and scFv fragments) do not activate complement, vascular cells, or phagocytes and therefore are safer for intravascular administration.

Conjugation of polyethylene glycol (PEG, a hydrophilic linear polymer) to proteins, liposomes, polymer carriers, and viral particles creates a "PEG brush" on their surface that forms a water shell that prevents their aggregation and interactions with antibodies, complement, phagocytes, and vascular cells. This "stealth technology" prolongs the circulation of drugs and drug-carrier complexes and minimizes immune reactions and side effects associated with recognition by host defense systems (16,24). Several types of PEG proteins and PEG liposomes are approved for clinical use.

Table 186-1: Advanced Drug Delivery Means

Cargoes	*Carriers*	*Masking Moiety*	*Affinity Moieties*
Drugs	Liposomes	PEG	IgG
Enzymes	Polymer particles	Glycolipids	Multivalent IgG
Toxins	Polymer chains	SACCH	Fab fragments
Isotopes	Cells		scFv:
MRI contrasts	Viruses		– Monovalent
Imaging probes	Phages		– Multimeric
Genes	Proteins		Carbohydrates
Oligonucleotides	Antibodies		Lectins
	Carbohydrates		Hormones
			Nutrients (folate)

Left column provides examples of therapeutic and imaging agents that can be loaded to, encapsulated by, intercalated into, or conjugated with carriers shown in the second column. In addition to polyethylene glycol (PEG), natural molecules including glycolipids (e.g., some gangliosides) and sialic acid–containing carbohydrates (SACCH) have been tested as potential masking moieties for coating liposomes and other carriers to achieve "stealth" features. Based on the bioconjugation format, antibodies (IgG) may serve as both carriers and affinity moieties (direct conjugation of cargoes to IgG) or as affinity moieties (e.g., IgG coupling to drug-loaded liposome). IgG, Fab, or scFv fragments can be used in either monomolecular or multivalent format, attained by either chemical conjugation or using recombinant techniques (scFv). Coupling multiple copies of affinity moieties to a carrier also converts them to multivalent format.

This stealth effect increases proportionally with the number of PEG molecules attached to a protein, DNA, or liposome; the denser the PEG brush, the more profound the stealth effect. However, conjugation of multiple PEG chains inactivates sensitive proteins and DNA, and PEG grafting of more than 5% to 10% of phospholipids destroys liposomes. To solve the problem, stable carriers consisting of biocompatible, biodegradable synthetic polymers (e.g., polylactic acid or polyglycolic acid) chemically conjugated with PEG have been designed for drug delivery (25–28). Such polymer carriers are considerably more stable and circulate for a longer time compared with PEG-coated liposomes (29). Drugs can be coupled to polymer chains (either natural, such as sugars, or synthetic), intercalated into polymer or encapsulated into the inner space. Encapsulation into a polymer shell protects drugs from inactivation and enables controlled drug release regulated by environmental factors, for example low pH in lysosomes (30).

Polymersomes synthesized from PEG-conjugated biocompatible copolymers, such as polylactic-co-glycolic acid (PLGA), are polymer analogs of liposomes with a large internal aqueous domain that can be loaded with drugs (31). The membrane bilayer of polymersomes is much thicker than that of liposomes (~8 nm compared with ~3 nm), resulting in a more durable carrier with a greatly enhanced circulation half-life. Diverse hydrophilic and hydrophobic drugs can be encapsulated into polymersomes. The efficiency of encapsulation generally decreases with the increasing molecular size of a drug. Loading of active proteins is challenging due to harsh encapsulation conditions.

In summary, drug delivery systems consist of: (a) a drug encapsulated into the inner volume of a carrier, intercalated into or coupled to the shell; (b) the carrier proper (e.g., a liposome); and, (c) masking moieties, such as PEG conjugated to the carrier surface. Some delivery means, such as protein conjugates and recombinant fusion protein constructs, do not utilize drug carriers, but instead combine drugs directly with targeting moieties (32). Ideally, a drug delivery system also includes a targeting moiety coupled to the carrier surface (preferably to ends of the masking PEG chains), which provides specific binding to endothelial surface determinants, a topic discussed in the next section.

DRUG TARGETING TO ENDOTHELIAL BINDING SITES

Diverse ligands that bind to selected determinants, including hormones, sugars, nutrients, lectin, and peptides can serve as targeting moieties, directing the binding of drugs and drug carriers to selected cells (see Table 186-1). Monoclonal antibodies (mAb) and their antigen-binding fragments represent the most popular modular class of such ligands. Humanization, the recombinant grafting of antigen-binding sites of a mouse IgG into human IgG scaffold, minimizes immune reactions. Recombinant methods also permit deimmunization (replacement of immunogenic amino acids), maturation (enhancement of binding affinity by point-directed mutagenesis of antigen recognizing sites), and production of diverse forms of antigen-binding fragments including single-chain variable fragments (scFv), which are the minimal full-affinity antigen-binding unit of IgG that combine the first hypervariable domains of light and heavy chains of IgG, connected by a peptide linker (33,34).

Bioconjugation means for coupling drugs or drug carriers with antibodies include chemical bifunctional reagents, streptavidin-biotin pairs, and construction of recombinant fusion proteins, for example, by combining an active part of a therapeutic enzyme with scFv via a flexible peptide linker. For example, biotin-streptavidin cross-linking pairs can be used to couple antibodies (or drugs) to carriers or synthesize protein conjugates characterized by high drug incorporation efficiency; a wide range of sizes and valency governed by the degree of biotinylation and molar ratios of the components; and a relatively rigid and biodegradable structure (32,35). When compared to chemical conjugation, genetically fused recombinant constructs afford even higher degrees of purity, homogeneity, and reproducibility (33,36).

For targeting to reach its full potential, endothelial determinants must meet several criteria. Accessibility, specificity, and safety are obligatory criteria for all targeting means, whereas selectivity and precision are highly desirable and, in fact, obligatory for the targeting of many therapeutic and diagnostic agents.

Accessibility

Surface density and the accessibility of target sites must suffice to accommodate therapeutic amounts of carriers ranging in size from small fusion proteins to submicron particles. Pathological factors may cause shedding, masking, or suppression of expression of some target molecules (37). On the other hand, altered ECs expose abnormal surface markers and cellular components. Such pathological endothelial determinants represent very attractive targets.

Specificity

Target counterparts in blood or non-ECs accessible to blood should be below levels at which they compromise delivery or induce side effects. For example, P-selectin, a surface marker of pathologically activated ECs also is expressed on activated platelets; hence, targeting P-selectin may direct drug delivery to platelets. Yet, absolute specificity is not necessary if competing determinants are inaccessible to blood.

Safety

Targeting should have minimal, if any, adverse side effects in endothelium. Engagement and cross-linking by carriers may inhibit or activate target molecules, induce signaling, endothelial activation, shedding, and/or endocytosis. Some of these potential side effects may be detrimental and must be rigorously tested for. Ideally, binding to a target molecule should cause therapeutically beneficial secondary effects, such as the inhibition of the target molecule's function.

Selectivity

Even indiscriminate delivery to ECs in the vasculature will improve the therapeutic effects of many drugs. Yet, ideally, targeting should direct drugs to selected vascular areas (e.g., pulmonary, cerebral, or tumor), blood vessel types (e.g., arterioles, capillaries, or veins), or domains within the EC body (e.g., caveolae, fenestrae, or intercellular junctions) to enhance specificity and effectiveness of action. Site-selective delivery is necessary for many diagnostic strategies and for the delivery of harmful therapeutics, such as antitumor agents.

Precision

Docking to a target should direct a drug to the desired subcellular site, such as the luminal surface proper, or the intracellular or subcellular compartments. In some cases, the constitutive turnover of a target molecule guides the subcellular trafficking of drugs. For example, noninternalizable adhesion molecules are ideal for anchoring drugs that are designed to exert their effect in the lumen (38), whereas caveolar antigens facilitate intracellular delivery (39,40). Valency and carrier size control the rates of their internalization and intracellular traffic. For example, the multivalent engagement of certain noninternalizable target determinants induces endothelial endocytosis and intracellular drug delivery.

QUEST FOR TARGETS: PAN-ENDOTHELIAL, SITE-SPECIFIC, CONSTITUTIVE, AND INDUCIBLE ENDOTHELIAL SURFACE DETERMINANTS

Most drugs and carriers have no specific affinity to endothelium. Thus, an important goal is to achieve EC-specific binding. Ideally, such binding should be localized to the desired vascular areas. However, site specificity may not be necessary for those drugs providing protective effects (e.g., antioxidant, antithrombotic, anti-inflammatory agents), as well as for certain systemic interventions (e.g., enzyme replacement and gene therapies). These drug delivery systems can be directed to pan-endothelial determinants indiscriminately distributed throughout the vasculature.

Pan-endothelial ligands can be directed to specific vascular beds using vascular catheters inserted into conduit arteries (18,41). Further "organ-specific" endothelial targeting using pan-endothelial determinants can be achieved in transplantation settings, in which a graft is perfused with the targeted drugs prior to or after the harvest, thus precluding recipient exposure to the agent (42).

The pulmonary vasculature is a privileged target for pan-endothelial delivery systems. It represents approximately 30% of the total endothelial surface in the body and receives the entire cardiac output of mixed venous blood; hence, carriers targeted to pan-endothelial determinants accumulate in the lungs after intravascular injections (8,10,43,44).

Although pan-endothelial drug delivery may provide therapeutic gain, targeting to specific populations of ECs in diverse types of blood vessels, organs, and tissues, either normal or pathologically altered, is considered the "Holy Grail" of vascular targeting. Site-selective targeting also may facilitate drug delivery to extravascular compartments in desired organs or tissues.

Genetic and local environmental factors dictate specific EC phenotypes, including surface expression of unique marker molecules (45–47). ECs located in different organs, blood vessel types, and regions of the same vessel (e.g., bifurcations versus straight segments of arteries) may express stable or inducible markers potentially useful for targeting these specific vascular domains. The distribution of these markers may differ in health and disease. Constitutively expressed determinants can be used for either prophylactic or therapeutic drug delivery. Inducible determinants are ideal for diagnostics

Table 186-2: Tentative Endothelial Determinants for Drug Targeting

	Pan-EC	*Relative Vascular Specificity*	*Function*	*Constitutive or Inducible*	*Internalization*
PECAM-1	Yes	No	WBC transmigration	C	Only induced by anti-CAM multimers (CAM-mediated endocytosis)
TM	+/–	No	Anti-thrombotic	C	Constitutive, stimulated by monomolecular ligands
ACE	+/–	Lung	Ang-II production, Bradykinin decay	C	Constitutive
AP	No	Lung	Bradykinin decay	C	Not known
ICAM-1	+/–	No	WBC adhesion	C/I	CAM-mediated endocytosis (multimers)
P-selectin	+/–	Postcapillary venules	WBC rolling	I	Constitutive via clathrin-coated pits
E-selectin	+/–	Postcapillary venules	WBC rolling	I	Constitutive via clathrin-coated pits
VCAM-1	–/+	Postcapillary venules	WBC adhesion	I	Constitutive via clathrin-coated pits
GP-60	?	Unknown	Albumin uptake	C	Constitutive via caveolae
GP-90	?	Lung	Unknown	C	Constitutive via caveolae
GP-85	No	Alveolar capillaries (AVZ) and heart	Unknown	C	No
Tf-R	+/–	BBB, liver	Iron transport		Constitutive via clathrin-coated pits
AN	No	Tumors	Peptides decay	Unknown	Unknown

TM, thrombomodulin; AP, aminopeptidase-P; AN, aminopeptidase-N; Tf-R, transferrin receptor; GP, glycoproteins of indicated molecular weight; AVZ, avesicular zone; WBC, white blood cell.

and therapeutics, but not for prophylactic drug delivery. Table 186-2 provides several examples of tentative endothelial determinants that can be used for drug targeting; some of these determinants will be discussed in detail in the next sections.

Even within one cell, the endothelial plasma membrane contains diverse domains, including those that are relatively static (e.g., intercellular junctions) and those that are dynamic (e.g., caveolae and intracellular fenestrae). Recent studies identified several EC markers enriched in these specific domains. For example, glycoprotein GP85 is localized in a thin part of EC body that lacks main organelles and separates alveolar and vascular space in lung capillaries ("avesicular zone") (48,49). GP85 antibodies accumulate in rat pulmonary vasculature without internalization and deliver conjugated cargoes into the pulmonary vasculature (50). In theory, such an antigen could be an interesting candidate for drug delivery to the surface of alveolar capillaries.

Targeting caveolae, which are cholesterol-rich invaginations in endothelial plasmalemma (6,51,52), may afford intracellular and, perhaps, transcellular drug delivery in the vasculature (2,40) (see Chapter 73). Caveolar-mediated endocytosis and transcytosis play an important role in endothelial transport functions (6,53,54). Figure 186.1 provides an abridged overview of principal endocytic and intracellular trafficking pathways in ECs. Ligands, including antibodies, may promote clustering of receptors within caveolae, leading to their endocytosis and transcytosis (54–56). In rats, the intravenous injection of ligands with an affinity for caveolar determinants (e.g., glycoprotein GP90) results in their accumulation in the pulmonary vasculature, followed by endothelial endocytosis and transcytosis (39,40). The functions of GP90 and effects of its alteration by targeting remain to be defined, yet this and other caveolar determinants are candidates for drug delivery to and across the endothelium.

QUEST FOR ENDOTHELIAL TARGETS: STRATEGIES FOR IDENTIFICATION OF MARKER DETERMINANTS

Staining of cells and tissues by antibodies, followed by studies of the biodistribution of these injected labeled antibodies, helps to identify candidate determinants. This relatively low-throughput hypothesis-driven strategy sets a framework for targeting in a given biomedical context, taking into account the distribution, functions, and regulation of the identified target molecules, but it is not designed to discover unknown and/or rare markers. New high-throughput discovery-driven strategies such as functional genomics, proteomics of the endothelial plasma membrane, and phage-display library selection provide a basis for vascular mapping and identification of determinants specific for defined areas of vasculature or endothelial domains.

Functional genomics of ECs obtained from normal or pathological vascular areas characterizes the relative abundance of messages for practically unlimited numbers of proteins (57). Analysis of proteins and peptide digests using

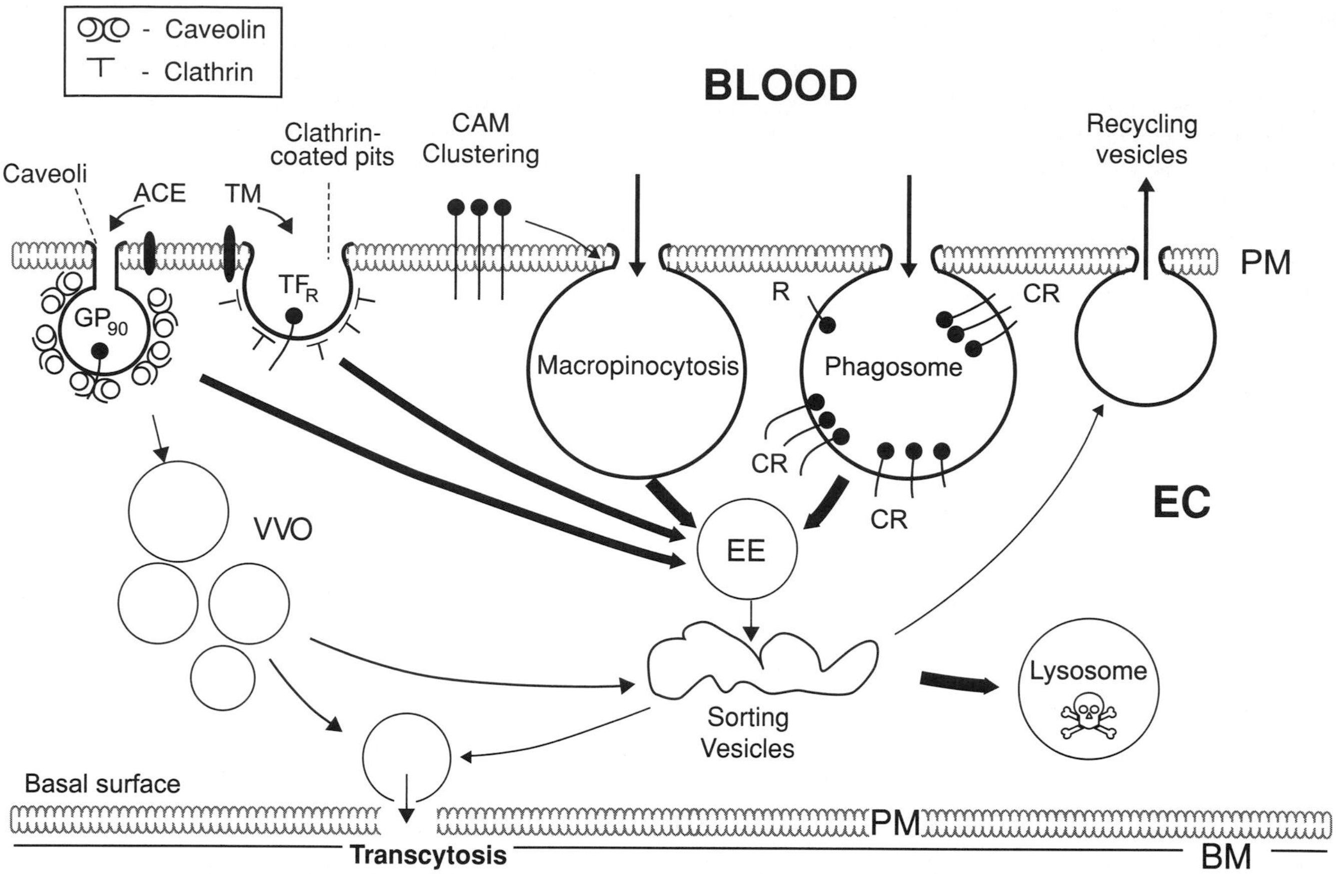

Figure 186.1. Physiological pathways for intracellular and transcellular drug delivery in ECs. Circulating cargoes can access the EC interior from blood lumen via fluid-phase uptake mechanism (pinocytosis and macropinocytosis), or via receptor-mediated endocytic (*left*) and phagocytic (*right*) pathways. Clustering of target determinants such as receptors (*R*) into cross-linked clusters (*CR*) by affinity carriers usually stimulates these pathways. Caveolae and clathrin-coated pits serve for endocytosis via specific receptors enriched in these plasma membrane (*PM*) domains, such as GP90 and transferring receptor (*TFr*). Constitutively internalized endothelial determinants such as ACE and thrombomodulin (TM) undergo endocytosis via either caveolar or clathrin-dependent pathways. Clustering of PECAM-1 and ICAM-1 (i.e., CAMs that are not constitutively internalized), induces endocytic pathway similar to, yet distinct from macropinocytosis. The formation of endocytic vesicles brings cargoes into the cell traffic in the cytosol and merges with vesicular organelles including early endosomes (EE), sorting vesicles, and the vesiculo-vacuolar organelle (VVO) compartment (the latter trafficking direction is more typical for caveolar uptake). From the sorting compartment, cargoes can traffic to lysosomes (a dominant pathway), where acidic hydrolases destroy them, or recycle to the plasma membrane via recycling vesicles. The caveolar pathway favors the transfer of some cargoes through the EC (transcytosis), although most likely other endocytic pathways also can provide transcytosis. Bold arrows show the typical destination of materials entering ECs via indicated endocytic pathways.

mass-spectrography and two-dimensional electrophoresis provides information about relative levels of proteins in the ECs of these tissues (58,59). The laser capture microdissection and analysis of ultra-small samples permit a delineation of differential gene expression in ECs from adjacent vascular areas that are subject to distinct environmental factors (60).

Further, methods used to isolate endothelial plasma membrane from different vascular beds and from specific microdomains (including caveolae [52,61]) provide a means for characterizing the surface topography of the endothelium. Such strategies yield novel site-specific determinants that may be leveraged for vascular bed-selective targeting (62,63).

Screening of phage display libraries in cell cultures helps to identify diverse epitopes potentially useful for endothelial delivery (64,65). The testing of phage display in vivo is a high-throughput method to find site-specific binding sites in selected normal or pathological vascular areas (66,67). This method is based on repetitive cycles of intravenous injections of phages encoding stochastic peptides, which may bind to sites accessible to circulation, followed by the identification of specific phages homing into selected tissues, and the eventual isolation of the homing determinants (68–70). The restriction to accessible targets – a built-in feature of this selection strategy – is an advantage in terms of drug delivery.

TARGETING CONSTITUTIVE PAN–ENDOTHELIAL CELL ADHESION MOLECULES

The endothelial adhesion molecules platelet-endothelial cell adhesion molecule (PECAM)-1/CD31 and intercellular adhesion molecule (ICAM)-1/CD54, which are transmembrane glycoproteins of the Ig superfamily, are relatively evenly distributed in ECs throughout the vasculature (71,72). PECAM-1, predominantly localized in intercellular borders, is stably and abundantly (millions of copies/cell) expressed in ECs, permitting robust targeting to normal or pathologically altered vasculature in prophylactic or therapeutic settings (10,43). ICAM-1 is expressed normally on the apical endothelial surface at levels of 2×10^4 to 2×10^5 surface copies per cell (73). Other cell types also express ICAM-1, yet the blood-accessible endothelial ICAM-1 binds most of anti–ICAM-1 after intravascular injection.

Reporter compounds, therapeutic enzymes, genetic materials, and drug carriers conjugated with anti–PECAM-1 or anti–ICAM-1 accumulate and display their functional activity in the pulmonary endothelium as soon as 10 minutes after intravenous injection in mice, rats, and pigs (21,35,38,41,74–76). Protein conjugates (32,43), liposomes (19), or polymer carriers (76) carrying multivalent anti–PECAM-1 and anti–ICAM-1 display high endothelial affinity and targeting.

Pathological stimuli including oxidants, cytokines, thrombin, and abnormal shear stress enhance the surface density of ICAM-1 on ECs (77,78); hence, inflammation and oxidative stress facilitate ICAM-1 targeting (38,74,79). Therefore, ICAM-1 is an attractive target for the endothelial delivery of diagnostic (80,81), prophylactic, and therapeutic agents (10,38).

ECs bind anti–PECAM-1 and anti–ICAM-1 without internalization, but internalize multimeric 0.1- to 0.5-micron anti-cell adhesion molecule (CAM) complexes (9,35,43,82,83). This feature permits the use of anti-CAM as affinity moieties to target drugs either to the endothelial surface (e.g., utilizing monomolecular conjugates and scFv constructs) (38) or intracellularly (e.g., utilizing multimeric anti-CAM complexes) (10,76,84). It is tempting to speculate that anti-CAM targeting can also facilitate drug delivery across endothelium.

ECs internalize anti–ICAM-1 and anti–PECAM conjugates via a novel endocytic pathway (see Figure 186.1), namely CAM-mediated endocytosis (9,82,84). ICAM-1 dissociates from anti–ICAM-1 conjugates early after internalization, recycles to the plasma membrane, and participates in multiple cycles of intracellular delivery (76). The pace of PECAM-1– and ICAM-1–mediated trafficking and the destination of its cargo may be modulated by auxiliary drugs. For example, chloroquine (a weak base) decelerates the degradation of proteolysis-susceptible cargoes, while disruption of microtubules blocks lysosomal traffic and dramatically prolongs the duration of therapeutic activity of internalized drugs (84).

Potential therapeutic applications of PECAM-1 or ICAM-1 targeting include local and systemic delivery of protective drugs. For example, targeting antioxidants conjugated with anti–PECAM-1 and anti–ICAM-1 protects ECs against oxidative stress in cell cultures (32,43,84,85), perfused organs (42,86), and in mice and rats (42,87). Targeting plasminogen activators to the endothelial surface using monomolecular, non–cross-linking anti-CAM (which minimizes internalization) facilitates fibrinolysis in the pulmonary vasculature (34,38).

ICAM-1 supports the firm adhesion of activated leukocytes to ECs, whereas PECAM-1 supports cellular recognition, adhesion, and transendothelial migration of leukocytes (71,72,88,89). Blocking these molecules attenuates inflammation and injury in animal models and clinical pathological settings associated with acute inflammation, ischemia–reperfusion, and oxidative stress (90–94). This aspect of endothelial drug delivery has been discussed in a more general context in the earlier Safety section. Targeting CAMs may offer secondary benefits in conditions associated with inflammation, oxidative stress, and thrombosis.

In summary, antibodies to PECAM-1 and ICAM-1 represent potentially practical and useful means for drug targeting to and into the ECs.

TARGETING SELECTED VASCULAR AREAS

Targeting pan-endothelial determinants will optimize the effects of many drugs by enhancing their effectiveness (dose percentage that actually acts in target cells may be increased), safety (general dose may be decreased), and specificity (the drug may be delivered to an appropriate subcellular compartment). It could be especially useful in those settings in which systemic endothelial effects are welcome or local administration of targeting is feasible (e.g., infusions via vascular catheters or application in organ transplantation). Yet, specific accumulation of drugs in selected vascular areas after systemic administration promises even greater advantages. Vascular selectivity provides a critical means for reducing the side effects of highly toxic drugs. This section provides selected examples of constitutive and inducible endothelial determinants potentially useful for site-selective drug delivery to and across ECs in several specific vascular areas.

Pulmonary Vasculature

The pulmonary compartment is a preferable site for the action of wide classes of noninjurious agents including antithrombotic, anti-inflammatory, and antioxidant drugs, as well as for enzyme replacement therapies, such as those used to alleviate genetic defects. As discussed earlier, the pulmonary vasculature represents a preferential accumulation site of drug delivery systems targeted to pan-endothelial determinants.

Endothelial determinants which are enriched in the pulmonary vasculature and potentially useful for lung-specific drug delivery, have been identified. This strategy may be useful for the treatment of pulmonary diseases, including pulmonary hypertension, edema, hyperplasia, tumor growth, and other maladies involving this predominantly vascular area.

One example of such a determinant is angiotensin-converting enzyme (ACE), a luminal endothelial transmembrane glycoprotein that converts angiotensin (Ang)-I into the vasoconstricting and proinflammatory peptide, Ang-II (95). Some ACE antibodies block its active site and facilitate ACE shedding from the endothelium (96,97), which may be beneficial in the management of hypertension and inflammation. Pulmonary vasculature is rich in ACE: approximately 100% of alveolar capillaries are ACE-positive versus less than 15% of extrapulmonary capillaries (44).

ACE is a good candidate for pulmonary endothelium targeting for diagnostic, prophylactic and, perhaps, therapeutic means. Reporter probes and drugs conjugated with anti-ACE accumulate in the lungs after intravenous injection in animals and humans (1,44,98,99). Because ECs constitutively internalize monomolecular anti-ACE (100), intracellular drug delivery is possible (albeit the endocytic pathway of anti-ACE and its regulation are not yet characterized). Antioxidant enzymes conjugated with anti-ACE accumulate in rat lungs in vivo and protect perfused rat lungs against hydrogen peroxide (H_2O_2) damage (86,101). Anti-ACE does not cause acute harmful reactions in animals and humans (1,96). Recently, anti-ACE has been used successfully for the retargeting of viruses to pulmonary endothelium in rats (102,103). Targeting of genetic materials into ECs is a rapidly developing field (70,104).

Using endothelial plasma membrane proteomics and in vivo phage display, new markers in the pulmonary vasculature have been identified in animals. For example, the injection of radiolabeled antibody to aminopeptidase-P, identified in the pulmonary endothelium using subtractive proteomics, allows the γ-camera visualization of lung blood vessels in rats and the detection of vascular abnormalities associated with pulmonary pathologies (62,63). These data, as well as similar results obtained with radiolabeled anti-ACE in animals and humans (1), imply that targeting of constitutive endothelial markers may have a diagnostic utility.

Interestingly, many pulmonary-specific endothelial markers are ecto-peptidases. This fact presumably reflects a key role of the pulmonary vasculature in metabolizing vasoactive peptides, and emphasizes the importance of understanding the potential consequences of inhibiting target molecule function. For example, ACE and aminopeptidase-P play a part in important physiological roles, such as bradykinin degradation (105); hence, the unintended inhibition of their functions by blocking antibodies may cause potentially harmful effects, such as vascular edema.

The issue of adverse side effects inflicted by inhibition of target molecules is nicely illustrated by studies that target thrombomodulin, an endothelial surface glycoprotein that converts thrombin into an antithrombotic enzyme that inactivates coagulation factors (106). This affinity moiety has been extensively used in the model targeting of liposomes, genetic materials, and injurious agents to the pulmonary endothelium in laboratory animals (17,107,108). However, thrombomodulin inhibition may increase the risk for thrombosis, which is illustrated in animal studies using antithrombomodulin conjugates (108–110). One potential application of thrombomodulin targeting is in animal models of human vascular pathologies localized in the pulmonary vasculature (87,108).

Sites of Inflammation

Diagnostic visualization and drug targeting to sites of vascular inflammation represent an enormously important biomedical goal (81). Some basic principles of inflammatory site targeting, including the secondary benefit of inhibiting the target per se, have been discussed in the context of targeting pan-endothelial determinants ICAM-1 and PECAM-1. In fact, these determinants appear to be excellent candidates for the delivery of anti-inflammatory therapies. Moreover, because inflammatory mediators result in the local induction of ICAM-1 expression on ECs, targeting this molecule will facilitate site-specific delivery of drugs to foci of inflammation.

At sites of inflammation, the endothelium displays many other phenotypic changes (111). Inflammatory mediators (e.g., cytokines, activated complement), oxidants, and proteases cause endothelial exposure of P-selectin, normally stored intracellularly and mobilized rapidly to the surface, and E-selectin, which is newly synthesized by activated endothelium (77,88). An Ig-superfamily CAM, vascular cell adhesion molecule (VCAM)-1, also is exposed on the surface of activated ECs (72). Selectins and VCAM-1 facilitate adhesive interaction of activated white blood cells with ECs (71). In theory, selectins and VCAM-1, transiently exposed on the surface of activated ECs, are attractive determinants for therapeutic (but not prophylactic) drug delivery to pathologically activated endothelium, with a potential secondary benefit of leukocyte adhesion attenuation (10,93).

Diverse carriers include polymer particles ranging from 0.1 to 0.5 microns which carry antibodies against EC adhesion molecules (e.g., P- and E-selectin, VCAM-1, and ICAM-1) adhesion molecules. These bind to cytokine-activated ECs in cultures (76,112,113). Experiments in cell cultures (19,20,114,115) and animal studies (36,115–117) showed that antibodies to selectins may permit the targeting of liposomes and drugs to cytokine-activated endothelium. It should be noted, however, that affinity carriers directed to P-selectin can target not only ECs but also – and, in many cases of vascular pathologies, predominantly – platelets (36,81).

ECs constitutively internalize E- and P-selectins and VCAM-1 (in contrast to ICAM-1 and PECAM-1) via clathrin-coated pits (118,119). This enables the intracellular delivery of liposomes (22,120), anti-inflammatory drugs (121), and genetic materials (115) targeted to these determinants. However, the rapid uptake of cell-bound drugs targeted to selectins and VCAM-1, which restricts their effects in the lumen, makes these targets unsuitable for delivery of certain drugs such as antithrombotic agents.

Targeting VCAM-1, a marker that is expressed in ECs early in the course of atherosclerosis, may eventually provide diagnostic and drug-delivery tools for site-specific containment of atherosclerosis (122,123). High-affinity antibody fragments selected by phage-display library techniques and identified as anti–VCAM-1 peptides currently are being tested in cell cultures and in animal models of atherosclerosis (65,124). ICAM-1, a hallmark of atherosclerotic lesions, is also a plausible target for the delivery of anti-inflammatory drugs in this disease condition (125,126).

The kinetics and persistence of selectin exposure on endothelial activation are difficult to follow even in animal models in which conditions are tightly controlled. Furthermore, even at their activation peak, selectins are exposed at relatively low surface densities; hence, targeting robustness may be suboptimal for therapeutic interventions requiring the delivery of large drug doses. Yet, targeting selectins is an attractive avenue for the diagnostic visualization of activated endothelium in inflammation foci by delivery of conjugated isotopes (116) or ultrasound contrasts (81).

Cerebral Vasculature

The blood–brain barrier (BBB) represents a specific, challenging, and extremely important therapeutic target. Cerebral endothelium is enriched in surface markers, including receptors for transferrin, insulin, putrescin, and some growth factors. Animal studies have demonstrated the delivery of reporter compounds and genes into the brain using antibodies and peptides binding to these determinants (127–129).

Importantly, some of these endothelial receptors apparently permit transcellular drug delivery across the BBB into the brain cells (129,130). The transferrin receptor represents one of the most extensively studied (131). Phage-display promises to guide the selection of new targeting peptides that bind to and transcytose across cerebral ECs (132).

The expression of CAMs, including ICAM-1, is elevated in cerebral endothelium in ischemia and inflammation (133). It is tempting to speculate that targeting therapeutic cargoes to ICAM-1 can improve treatment of stroke and cerebral infections, via both the delivery of cargoes and blocking ICAM-1. Moreover, blocking blood cell adhesion to ECs may provide therapeutic benefit in stroke, brain trauma and, likely, neurodegenerative diseases. More details on targeting of cerebral endothelium and means to permeate the BBB can be found in reviews (134–136).

Tumor Vasculature

ECs in solid tumors represent a key target for the delivery of agents designed to visualize tumors, inhibit angiogenesis, and/or kill malignant cells (4,137,138). Tumor vasculature is characterized by numerous functional (e.g., acidosis and hypoxia) and morphological (e.g., irregular shape and orientation of blood vessel) abnormalities (139,140). It has been suspected for decades that the ECs in tumor vasculature express abnormal determinants. Indeed, functional genomic and proteomic approaches have identified a spectrum of genes and proteins characteristic of tumor endothelium (57). Screening techniques utilized since the early 1990s, including plasma membrane proteomics and phage-display libraries in vivo, have identified several promising candidate determinants. For example, ECs in tumor vessels expose abnormal determinants including selectins, specific integrins, markers of apoptosis, and receptors for growth factors; targeting these determinants augments the accumulation of drugs in solid tumors in animal models (7,141–144). High-throughput plasma membrane proteomics and phage-display libraries in vivo have identified several promising candidate determinants characteristic of certain solid tumors, including aminopeptidase-N (143,145) and annexin-A1 (63).

Targeting endothelial determinants in tumors helps to accomplish many goals: (a) tumor localization; (b) thrombosis of tumor vessels with infarction and starvation of malignant cells; and, (c) delivery of antitumor agents to the target malignant cells (62,63,143,146–148). The latter approach involves permeation of the endothelial barrier, mostly due to enhanced vascular permeability and decelerated lymphatic drainage in solid tumors (enhanced permeability and retention, EPR) (149). However, targeting those integrins overexpressed on tumor endothelium and caveolar antigens can provide endothelial transcytosis (150). Targeting tumor endothelium is an area of very active research; for recent reviews on the identification of surface determinants characteristic of tumor vasculature see Refs. 4, 7, and 151.

KEY POINTS

- Diverse drug delivery systems such as stealth liposomes, polymers, and polypeptide carriers designed for a wide range of drug delivery applications can be utilized to target drugs and diagnostic probes to the endothelial tissue that represents both a prime therapeutic target and a barrier for drug delivery.
- The targeting is achieved by coupling drugs or drug carriers to affinity ligands of endothelial surface determinants. These agents must meet several criteria, including accessibility to circulating agents, endothelial specificity, safety of targeting effects, selectivity of addressing within the vasculature, and precision of addressing to proper cellular compartments.
- Pan-endothelial targeting, for example, directed to PECAM-1 and ICAM-1, optimizes the effectiveness, safety, and specificity of many therapeutic means aimed at the protection or correction of endothelial function and interventions within the circulation. Surface determinants expressed predominantly

in certain types of endothelia, such as in specific vascular areas or pathologies, offer site-selective delivery and greater specificity of interventions.

- Constitutive endothelial targets, such as PECAM-1 and ICAM-1, as well as ecto-peptidases (e.g., ACE, aminopeptidase-P) can be utilized for both prophylactic and therapeutic drug delivery. They may also be used as diagnostics based on the suppression of their endothelial expression in pathologies. Inducible determinants in pathologically altered endothelium (fairly common for inflammation, tumor growth, and ischemia) are ideal for diagnostics and, if the time window and surface density of their exposure suffice, therapeutic targeting.
- Hypothesis-driven strategies for the selection of endothelial targets allow a target to be tested in terms of optimal subcellular addressing and secondary effects beneficial in the context a particular pathological process. However, these strategies are low-throughput and not inclusive. High-throughput, discovery-driven strategies offer a wealth of knowledge on endothelial surface topography, yet offer rather limited, if any, insight into the functions and metabolism of newly identified determinants. A combination of these complementary strategies will guide the design of safe and effective systems for drug delivery to, into, and across selected populations of ECs.

ACKNOWLEDGMENTS

This work was supported by NHLBI RO1 grants HL71175, HL078785, and HL73940; Department of Defense Grant (PR 012262); ITMAT award; and Pennsylvania Nanotechnology Institute core project. I express my gratitude and appreciation to Drs. S. Muro and T. Dziubla (University of Pennsylvania) for productive collaboration, exciting discussions, and invaluable help in preparing this chapter.

REFERENCES

1 Muzykantov VR, Danilov SM. Targeting of radiolabeled monoclonal antibody against ACE to the pulmonary endothelium. In: Torchilin V, ed. *Handbook of Targeted Delivery of Imaging Agents.* Boca Raton, FL: CRC Press; 1995;465–485.

2 Jacobson BS, Stolz DB, Schnitzer JE. Identification of endothelial cell-surface proteins as targets for diagnosis and treatment of disease. *Nat Med.* 1996;2(4):482–484.

3 Muzykantov VR. Immunotargeting of drugs to the pulmonary vascular endothelium as a therapeutic strategy. *Pathophysiology.* 1998;5(1):15–33.

4 Schnitzer JE. Vascular targeting as a strategy for cancer therapy. *N Engl J Med.* 1998;339(7):472–474.

5 Muzykantov VR. Targeting of superoxide dismutase and catalase to vascular endothelium. *J Control Release.* 2001;71(1):1–21.

6 Schnitzer JE. Caveolae: from basic trafficking mechanisms to targeting transcytosis for tissue-specific drug and gene delivery in vivo. *Adv Drug Deliv Rev.* 2001;49(3):265–280.

7 Molema G. Tumor vasculature directed drug targeting: applying new technologies and knowledge to the development of clinically relevant therapies. *Pharm Res.* 2002;19(9):1251–1258.

8 Muzykantov VR. Targeting pulmonary endothelium. In: Muzykantov VR, Torchilin VP, eds. *Biomedical Aspects of Drug Targeting.* Boston: Kluwer Academic Publishers; 2002:129–148.

9 Muro S, Koval M, Muzykantov V. Endothelial endocytic pathways: gates for vascular drug delivery. *Curr Vasc Pharmacol.* 2004;2(3):281–299.

10 Muro S, Muzykantov VR. Targeting of antioxidant and antithrombotic drugs to endothelial cell adhesion molecules. *Curr Pharmacologic Design.* 2005;11(18):2383–2401.

11 Langer R. Drug delivery and targeting. *Nature.* 1998;392(Suppl 6679):5–10.

12 Bartus RT, Tracy MA, Emerich DF, Zale SE. Sustained delivery of proteins for novel therapeutic agents. *Science.* 1998;281 (5380):1161–1162.

13 Kopecek J, Kopeckova P, Minko T, et al. Water soluble polymers in tumor targeted delivery. *J Control Release.* 2001;74(1–3):147–158.

14 Lasic DD. Novel applications of liposomes. *Trends Biotechnol.* 1998;16(7):307–321.

15 Szoka FC Jr. The future of liposomal drug delivery. *Biotechnol Appl Biochem.* 1990;12(5):496–500.

16 Torchilin VP. Strategies and means for drug targeting: An overview. In: Muzykantov VR, Torchilin VP, eds. *Biomedical Aspects of Drug Targeting.* Boston: Kluwer Academic Pub.; 2002: 3–26.

17 Maruyama K, Kennel SJ, Huang L. Lipid composition is important for highly efficient target binding and retention of immunoliposomes. *Proc Natl Acad Sci USA.* 1990;87(15):5744–5748.

18 Muller DW, Gordon D, San H, et al. Catheter-mediated pulmonary vascular gene transfer and expression. *Circ Res.* 1994; 75(6):1039–1049.

19 Bloemen PG, Henricks PA, van Bloois L, et al. Adhesion molecules: a new target for immunoliposome-mediated drug delivery. *FEBS Lett.* 1995;357(2):140–144.

20 Spragg DD, Alford DR, Greferath R, et al. Immunotargeting of liposomes to activated vascular endothelial cells: a strategy for site-selective delivery in the cardiovascular system. *Proc Natl Acad Sci USA.* 1997;94(16):8795–8800.

21 Li S, Tan Y, Viroonchatapan E, et al. Targeted gene delivery to pulmonary endothelium by anti-PECAM antibody. *Am J Physiol Lung Cell Mol Physiol.* 2000;278(3):L504–L511.

22 Kessner S, Krause A, Rothe U, Bendas G. Investigation of the cellular uptake of E-Selectin-targeted immunoliposomes by activated human endothelial cells. *Biochim Biophys Acta.* 2001; 1514(2):177–190.

23 Benzinger P, Martiny-Baron G, Reusch P, et al. Targeting of endothelial KDR receptors with 3G2 immunoliposomes in vitro. *Biochim Biophys Acta.* 2000;1466(1–2):71–78.

24 Abuchowski A, McCoy JR, Palczuk NC, et al. Effect of covalent attachment of polyethylene glycol on immunogenicity and circulating life of bovine liver catalase. *J Biol Chem.* 1977;252 (11):3582–3586.

25 Discher DE, Eisenberg A. Polymer vesicles. *Science.* 2002;297 (5583):967–973.

26 Kabanov AV, Batrakova EV, Alakhov VY. Pluronic block copolymers as novel polymer therapeutics for drug and gene delivery. *J Control Release.* 2002;82(2–3):189–212.

27 Panyam J, Labhasetwar V. Biodegradable nanoparticles for drug and gene delivery to cells and tissue. *Adv Drug Deliv Rev.* 2003; 55(3):329–347.

28 Dziubla TD, Muzykantov V. Synthetic carriers for the delivery of protein therapeutics. *Biotechnol Genet Eng Rev.* 2005:22.

29 Jeong B, Bae YH, Lee DS, Kim SW. Biodegradable block copolymers as injectable drug-delivery systems. *Nature.* 1997;388 (6645):860–862.

30 Dziubla TD, Karim A, Muzykantov VR. Polymer nanocarriers protecting active enzyme cargo against proteolysis. *J Control Release.* 2005;102(2):427–439.

31 Discher BM, Won YY, Ege DS, et al. Polymersomes: tough vesicles made from diblock copolymers. *Science.* 1999;284(5417): 1143–1146.

32 Shuvaev VV, Dziubla T, Wiewrodt R, Muzykantov VR. Streptavidin-biotin crosslinking of therapeutic enzymes with carrier antibodies: nanoconjugates for protection against endothelial oxidative stress. *Methods Mol Biol.* 2004;283:3–19.

33 Caron PC, Laird W, Co MS, et al. Engineered humanized dimeric forms of IgG are more effective antibodies. *J Exp Med.* 1992;176(4):1191–1195.

34 Ding BS, Gottstein C, Grunow A, et al. Endothelial targeting of a recombinant construct fusing a PECAM-1 single-chain variable antibody fragment (scFv) with prourokinase facilitates prophylactic thrombolysis in the pulmonary vasculature. *Blood.* 2005; 106(13):4191–4198.

35 Scherpereel A, Wiewrodt R, Christofidou-Solomidou M, et al. Cell-selective intracellular delivery of a foreign enzyme to endothelium in vivo using vascular immunotargeting. *FASEB J.* 2001;15(2):416–426.

36 Fujise K, Revelle BM, Stacy L, et al. A tissue plasminogen activator/P-selectin fusion protein is an effective thrombolytic agent. *Circulation.* 1997;95(3):715–722.

37 Muzykantov VR, Puchnina EA, Atochina EN, et al. Endotoxin reduces specific pulmonary uptake of radiolabeled monoclonal antibody to angiotensin-converting enzyme. *J Nucl Med.* 1991;32(3):453–460.

38 Murciano JC, Muro S, Koniaris L, et al. ICAM-directed vascular immunotargeting of antithrombotic agents to the endothelial luminal surface. *Blood.* 2003;101(10):3977–3984.

39 McIntosh DP, Tan XY, Oh P, Schnitzer JE. Targeting endothelium and its dynamic caveolae for tissue-specific transcytosis in vivo: a pathway to overcome cell barriers to drug and gene delivery. *Proc Natl Acad Sci USA.* 2002;99(4):1996–2001.

40 JE CLaS. Tissue-specific pharmacodelivery and overcoming key cellular barriers in vivo: Vascular targeting of caveoli. In: Muzykantov VR, Torchilin V, eds. *Biomedical Aspects of Drug Targeting.* London: Kluwer Academic Publishers; 2002:129–148.

41 Scherpereel A, Rome JJ, Wiewrodt R, et al. Platelet-endothelial cell adhesion molecule-1-directed immunotargeting to cardiopulmonary vasculature. *J Pharmacol Exp Ther.* 2002;300(3): 777–786.

42 Kozower BD, Christofidou-Solomidou M, Sweitzer TD, et al. Immunotargeting of catalase to the pulmonary endothelium alleviates oxidative stress and reduces acute lung transplantation injury. *Nat Biotechnol.* 2003;21(4):392–398.

43 Muzykantov VR, Christofidou-Solomidou M, Balyasnikova I, et al. Streptavidin facilitates internalization and pulmonary targeting of an anti-endothelial cell antibody (platelet-endothelial cell adhesion molecule 1): a strategy for vascular immunotargeting of drugs. *Proc Natl Acad Sci USA.* 1999;96(5):2379–2384.

44 Danilov SM, Gavrilyuk VD, Franke FE, et al. Lung uptake of antibodies to endothelial antigens: key determinants of vascular immunotargeting. *Am J Physiol Lung Cell Mol Physiol.* 2001;280 (6):L1335–L1347.

45 Davies PF. Mechanisms involved in endothelial responses to hemodynamic forces. *Atherosclerosis.* 1997;(Suppl 131):S15–S17.

46 Gimbrone MA Jr. Vascular endothelium, hemodynamic forces, and atherogenesis. *Am J Pathol.* 1999;155(1):1–5.

47 Aird WC. Endothelial cell heterogeneity. *Crit Care Med.* 2003; 31(Suppl 4):S221–S230.

48 Ghitescu L, Jacobson BS, Crine P. A novel, 85 kDa endothelial antigen differentiates plasma membrane macrodomains in lung alveolar capillaries. *Endothelium.* 1999;6(3):241–250.

49 Ghitescu LD, Crine P, Jacobson BS. Antibodies specific to the plasma membrane of rat lung microvascular endothelium. *Exp Cell Res.* 1997;232(1):47–55.

50 Murciano JC, Harshaw DW, Ghitescu L, et al. Vascular immunotargeting to endothelial surface in a specific macrodomain in alveolar capillaries. *Am J Respir Crit Care Med.* 2001;164(7): 1295–1302.

51 Stan RV. Structure and function of endothelial caveolae. *Microsc Res Tech.* 2002;57(5):350–364.

52 Schnitzer JE, McIntosh DP, Dvorak AM, et al. Separation of caveolae from associated microdomains of GPI-anchored proteins. *Science.* 1995;269(5229):1435–1439.

53 Minshall RD, Tiruppathi C, Vogel SM, Malik AB. Vesicle formation and trafficking in endothelial cells and regulation of endothelial barrier function. *Histochem Cell Biol.* 2002;117(2): 105–112.

54 Predescu D, Predescu S, Malik AB. Transport of nitrated albumin across continuous vascular endothelium. *Proc Natl Acad Sci USA.* 2002;99(21):13932–13937.

55 John TA, Vogel SM, Tiruppathi C, et al. Quantitative analysis of albumin uptake and transport in the rat microvessel endothelial monolayer. *Am J Physiol Lung Cell Mol Physiol.* 2003;284 (1):L187–L196.

56 Tiruppathi C, Song W, Bergenfeldt M, et al. GP activation mediates albumin transcytosis in endothelial cells by tyrosine kinase-dependent pathway. *J Biol Chem.* 1997;272(41):25968–25975.

57 St Croix B, Rago C, Velculescu V, et al. Genes expressed in human tumor endothelium. *Science.* 2000;289(5482):1197–1202.

58 Bruneel A, Labas V, Mailloux A, et al. Proteomic study of human umbilical vein endothelial cells in culture. *Proteomics.* 2003;3(5):714–723.

59 Franzen B, Duvefelt K, Jonsson C, et al. Gene and protein expression profiling of human cerebral endothelial cells activated with tumor necrosis factor-alpha. *Brain Res Mol Brain Res.* 2003;115(2):130–146.

60 Passerini AG, Polacek DC, Shi C, et al. Coexisting proinflammatory and antioxidative endothelial transcription profiles in a disturbed flow region of the adult porcine aorta. *Proc Natl Acad Sci USA.* 2004;101(8):2482–2487.

61 Jacobson BS, Schnitzer JE, McCaffery M, Palade GE. Isolation and partial characterization of the luminal plasmalemma of

microvascular endothelium from rat lungs. *Eur J Cell Biol.* 1992; 58(2):296–306.

62 Durr E, Yu J, Krasinska KM, et al. Direct proteomic mapping of the lung microvascular endothelial cell surface in vivo and in cell culture. *Nat Biotechnol.* 2004;22(8):985–992.

63 Oh P, Li Y, Yu J, et al. Subtractive proteomic mapping of the endothelial surface in lung and solid tumours for tissue-specific therapy. *Nature.* 2004;429(6992):629–635.

64 Kennel SJ, Lankford T, Foote L, et al. Phage display selection of scFv to murine endothelial cell membranes. *Hybrid Hybridomics.* 2004;23(4):205–211.

65 Tsourkas A, Shinde-Patil VR, Kelly KA, et al. In vivo imaging of activated endothelium using an anti-VCAM-1 magnetooptical probe. *Bioconjug Chem.* 2005;16(3):576–581.

66 Pasqualini R, McDonald DM, Arap W. Vascular targeting and antigen presentation. *Nat Immunol.* 2001;2(7):567–568.

67 Pasqualini R, Arap W, McDonald DM. Probing the structural and molecular diversity of tumor vasculature. *Trends Mol Med.* 2002;8(12):563–571.

68 Rajotte D, Arap W, Hagedorn M, et al. Molecular heterogeneity of the vascular endothelium revealed by in vivo phage display. *J Clin Invest.* 1998;102(2):430–437.

69 Nicklin SA, White SJ, Watkins SJ, et al. Selective targeting of gene transfer to vascular endothelial cells by use of peptides isolated by phage display. *Circulation.* 2000;102(2):231–237.

70 Work LM, Nicklin SA, White SJ, Baker AH. Use of phage display to identify novel peptides for targeted gene therapy. *Methods Enzymol.* 2002;346:157–176.

71 Springer TA. Adhesion receptors of the immune system. *Nature.* 1990;346(6283):425–434.

72 Albelda SM. Endothelial and epithelial cell adhesion molecules. *Am J Respir Cell Mol Biol.* 1991;4(3):195–203.

73 Almenar-Queralt A, Duperray A, Miles LA, et al. Apical topography and modulation of ICAM-1 expression on activated endothelium. *Am J Pathol.* 1995;147(5):1278–1288.

74 Sasso DE, Gionfriddo MA, Thrall RS, et al. Biodistribution of indium-111-labeled antibody directed against intercellular adhesion molecule-1. *J Nucl Med.* 1996;37(4):656–661.

75 Christofidou-Solomidou M, Pietra GG, Solomides CC, et al. Immunotargeting of glucose oxidase to endothelium in vivo causes oxidative vascular injury in the lungs. *Am J Physiol Lung Cell Mol Physiol.* 2000;278(4):L794–L805.

76 Muro S, Gajewski C, Koval M, Muzykantov VR. ICAM-1 recycling in endothelial cells: a novel pathway for sustained intracellular delivery and prolonged effects of drugs. *Blood.* 2005; 105(2):650–658.

77 Doerschuk CM, Quinlan WM, Doyle NA, et al. The role of P-selectin and ICAM-1 in acute lung injury as determined using blocking antibodies and mutant mice. *J Immunol.* 1996;157(10): 4609–4614.

78 Beck-Schimmer B, Schimmer RC, Warner RL, et al. Expression of lung vascular and airway ICAM-1 after exposure to bacterial lipopolysaccharide. *Am J Respir Cell Mol Biol.* 1997;17(3):344–352.

79 Mulligan MS, Miyasaka M, Tamatani T, et al. Requirements for L-selectin in neutrophil-mediated lung injury in rats. *J Immunol.* 1994;152(2):832–840.

80 Villanueva FS, Jankowski RJ, Klibanov S, et al. Microbubbles targeted to intercellular adhesion molecule-1 bind to activated coronary artery endothelial cells. *Circulation.* 1998;98(1):1–5.

81 Lindner JR, Klibanov AL, Ley K. Targeting inflammation. In: Muzykantov VR, Torchilin VP, eds. *Biomedical Aspects of Drug Targeting.* Boston: Kluwer Academic Pub.; 2002:149–172.

82 Muro S, Wiewrodt R, Thomas A, et al. A novel endocytic pathway induced by clustering endothelial ICAM-1 or PECAM-1. *J Cell Sci.* 2003;116(Pt 8):1599–1609.

83 Wiewrodt R, Thomas AP, Cipelletti L, et al. Size-dependent intracellular immunotargeting of therapeutic cargoes into endothelial cells. *Blood.* 2002;99(3):912–922.

84 Muro S, Cui X, Gajewski C, et al. Slow intracellular trafficking of catalase nanoparticles targeted to ICAM-1 protects endothelial cells from oxidative stress. *Am J Physiol Cell Physiol.* 2003;285(5):C1339–C1347.

85 Sweitzer TD, Thomas AP, Wiewrodt R, et al. PECAM-directed immunotargeting of catalase: specific, rapid and transient protection against hydrogen peroxide. *Free Radic Biol Med.* 2003;34 (8):1035–1046.

86 Atochina EN, Balyasnikova IV, Danilov SM, et al. Immunotargeting of catalase to ACE or ICAM-1 protects perfused rat lungs against oxidative stress. *Am J Physiol.* 1998;275(4 Pt 1):L806–L817.

87 Christofidou-Solomidou M, Scherpereel A, Wiewrodt R, et al. PECAM-directed delivery of catalase to endothelium protects against pulmonary vascular oxidative stress. *Am J Physiol Lung Cell Mol Physiol.* 2003;285(2):L283–L292.

88 Kishimoto TK, Rothlein R. Integrins, ICAMs, and selectins: role and regulation of adhesion molecules in neutrophil recruitment to inflammatory sites. *Adv Pharmacol.* 1994;25:117–169.

89 Newman PJ. The biology of PECAM-1. *J Clin Invest.* 1997; 99(1):3–8.

90 Rothlein R, Mainolfi EA, Kishimoto TK. Treatment of inflammation with anti-ICAM-1. *Res Immunol.* 1993;144(9):735–739; Discussion 54–62.

91 Kavanaugh AF, Davis LS, Jain RI, et al. A phase I/II open label study of the safety and efficacy of an anti-ICAM-1 (intercellular adhesion molecule-1; CD54) monoclonal antibody in early rheumatoid arthritis. *J Rheumatol.* 1996;23(8):1338–1344.

92 Bogen S, Pak J, Garifallou M, et al. Monoclonal antibody to murine PECAM-1 (CD31) blocks acute inflammation in vivo. *J Exp Med.* 1994;179(3):1059–1064.

93 Lefer DJ, Flynn DM, Anderson DC, Buda AJ. Combined inhibition of P-selectin and ICAM-1 reduces myocardial injury following ischemia and reperfusion. *Am J Physiol.* 1996;271(6 Pt 2): H2421–H2429.

94 Vaporciyan AA, DeLisser HM, Yan HC, et al. Involvement of platelet-endothelial cell adhesion molecule-1 in neutrophil recruitment in vivo. *Science.* 1993;262(5139):1580–1582.

95 Erdos EG. Angiotensin I converting enzyme and the changes in our concepts through the years. Lewis K. Dahl memorial lecture. *Hypertension.* 1990;16(4):363–370.

96 Danilov S, Atochina E, Hiemisch H, et al. Interaction of mAb to angiotensin-converting enzyme (ACE) with antigen in vitro and in vivo: antibody targeting to the lung induces ACE antigenic modulation. *Int Immunol.* 1994;6(8):1153–1160.

97 Balyasnikova IV, Karran EH, Albrecht RF, 2nd, Danilov SM. Epitope-specific antibody-induced cleavage of angiotensin-converting enzyme from the cell surface. *Biochem J* 2002;362 (Pt 3):585–95.

98 Danilov SM, Muzykantov VR, Martynov AV, et al. Lung is the target organ for a monoclonal antibody to angiotensin-converting enzyme. *Lab Invest.* 1991;64(1):118–124.

99 Muzykantov VR, Martynov AV, Puchnina EA, Danilov SM. In vivo administration of glucose oxidase conjugated with monoclonal antibodies to angiotensin-converting enzyme. The tissue distribution, blood clearance, and targeting into rat lungs. *Am Rev Respir Dis.* 1989;139(6):1464–1473.

100 Muzykantov VR, Atochina EN, Kuo A, et al. Endothelial cells internalize monoclonal antibody to angiotensin-converting enzyme. *Am J Physiol.* 1996;270(5 Pt 1):L704–L713.

101 Muzykantov VR, Atochina EN, Ischiropoulos H, et al. Immunotargeting of antioxidant enzyme to the pulmonary endothelium. *Proc Natl Acad Sci USA.* 1996;93(11):5213–5218.

102 Reynolds PN, Zinn KR, Gavrilyuk VD, et al. A targetable, injectable adenoviral vector for selective gene delivery to pulmonary endothelium in vivo. *Mol Ther.* 2000;2(6):562–578.

103 Reynolds PN, Nicklin SA, Kaliberova L, et al. Combined transductional and transcriptional targeting improves the specificity of transgene expression in vivo. *Nat Biotechnol.* 2001;19(9):838–842.

104 White SJ, Nicklin SA, Buning H, et al. Targeted gene delivery to vascular tissue in vivo by tropism-modified adeno-associated virus vectors. *Circulation.* 2004;109(4):513–519.

105 Skidgel RA. Bradykinin-degrading enzymes: structure, function, distribution, and potential roles in cardiovascular pharmacology. *J Cardiovasc Pharmacol.* 1992;20(Suppl 9):S4–S9.

106 Esmon CT. Thrombomodulin as a model of molecular mechanisms that modulate protease specificity and function at the vessel surface. *FASEB J.* 1995;9(10):946–955.

107 Kennel SJ, Lee R, Bultman S, Kabalka G. Rat monoclonal antibody distribution in mice: an epitope inside the lung vascular space mediates very efficient localization. *Int J Rad Appl Instrum B.* 1990;17(2):193–200.

108 Christofidou-Solomidou M, Kennel S, Scherpereel A, et al. Vascular immunotargeting of glucose oxidase to the endothelial antigens induces distinct forms of oxidant acute lung injury: targeting to thrombomodulin, but not to PECAM-1, causes pulmonary thrombosis and neutrophil transmigration. *Am J Pathol.* 2002;160(3):1155–1169.

109 Brisson C, Archipoff G, Hartmann ML, et al. Antibodies to thrombomodulin induce receptor-mediated endocytosis in human saphenous vein endothelial cells. *Thromb Haemost.* 1992;68(6):737–743.

110 Guermazi S, Mellouli F, Trabelsi S, et al. Anti-thrombomodulin antibodies and venous thrombosis. *Blood Coagul Fibrinolysis.* 2004;15(7):553–558.

111 McDonald DM, Thurston G, Baluk P. Endothelial gaps as sites for plasma leakage in inflammation. *Microcirculation.* 1999; 6(1):7–22.

112 Ranney DF. Biomimetic transport and rational drug delivery. *Biochem Pharmacol.* 2000;59(2):105–114.

113 Sakhalkar HS, Dalal MK, Salem AK, et al. Leukocyte-inspired biodegradable particles that selectively and avidly adhere to inflamed endothelium in vitro and in vivo. *Proc Natl Acad Sci USA.* 2003;100(26):15895–15900.

114 Kiely JM, Cybulsky MI, Luscinskas FW, Gimbrone MA Jr. Immunoselective targeting of an anti-thrombin agent to the surface of cytokine-activated vascular endothelial cells. *Arterioscler Thromb Vasc Biol.* 1995;15(8):1211–1218.

115 Harari OA, Wickham TJ, Stocker CJ, et al. Targeting an adenoviral gene vector to cytokine-activated vascular endothelium via E-selectin. *Gene Ther.* 1999;6(5):801–807.

116 Keelan ET, Harrison AA, Chapman PT, et al. Imaging vascular endothelial activation: an approach using radiolabeled monoclonal antibodies against the endothelial cell adhesion molecule E-selectin. *J Nucl Med.* 1994;35(2):276–281.

117 Lindner JR, Song J, Christiansen J, et al. Ultrasound assessment of inflammation and renal tissue injury with microbubbles targeted to P-selectin. *Circulation.* 2001;104(17):2107–2112.

118 von Asmuth EJ, Smeets EF, Ginsel LA, et al. Evidence for endocytosis of E-selectin in human endothelial cells. *Eur J Immunol.* 1992;22(10):2519–2526.

119 Kuijpers TW, Raleigh M, Kavanagh T, et al. Cytokine-activated endothelial cells internalize E-selectin into a lysosomal compartment of vesiculotubular shape. A tubulin-driven process. *J Immunol.* 1994;152(10):5060–5069.

120 Kok RJ, Everts M, Asgeirsdottir SA, et al. Cellular handling of a dexamethasone-anti-E-selectin immunoconjugate by activated endothelial cells: comparison with free dexamethasone. *Pharm Res.* 2002;19(11):1730–1735.

121 Everts M, Kok RJ, Asgeirsdottir SA, et al. Selective intracellular delivery of dexamethasone into activated endothelial cells using an E-selectin-directed immunoconjugate. *J Immunol.* 2002;168 (2):883–889.

122 O'Brien KD, Allen MD, McDonald TO, et al. Vascular cell adhesion molecule-1 is expressed in human coronary atherosclerotic plaques. Implications for the mode of progression of advanced coronary atherosclerosis. *J Clin Invest.* 1993;92(2):945–951.

123 Iiyama K, Hajra L, Iiyama M, et al. Patterns of vascular cell adhesion molecule-1 and intercellular adhesion molecule-1 expression in rabbit and mouse atherosclerotic lesions and at sites predisposed to lesion formation. *Circ Res.* 1999;85(2):199–207.

124 Kelly KA, Allport JR, Tsourkas A, et al. Detection of vascular adhesion molecule-1 expression using a novel multimodal nanoparticle. *Circ Res.* 2005;96(3):327–336.

125 Houston P, Goodman J, Lewis A, et al. Homing markers for atherosclerosis: applications for drug delivery, gene delivery and vascular imaging. *FEBS Lett.* 2001;492(1–2):73–77.

126 Poston RN, Haskard DO, Coucher JR, et al. Expression of intercellular adhesion molecule-1 in atherosclerotic plaques. *Am J Pathol.* 1992;140(3):665–673.

127 Lee HJ, Engelhardt B, Lesley J, et al. Targeting rat anti-mouse transferrin receptor monoclonal antibodies through blood-brain barrier in mouse. *J Pharmacol Exp Ther.* 2000;292(3): 1048–1052.

128 Wengenack TM, Curran GL, Olson EE, Poduslo JF. Putrescine-modified catalase with preserved enzymatic activity exhibits increased permeability at the blood-nerve and blood-brain barriers. *Brain Res.* 1997;767(1):128–135.

129 Zhang Y, Schlachetzki F, Pardridge WM. Global non-viral gene transfer to the primate brain following intravenous administration. *Mol Ther.* 2003;7(1):11–18.

130 Song BW, Vinters HV, Wu D, Pardridge WM. Enhanced neuroprotective effects of basic fibroblast growth factor in regional brain ischemia after conjugation to a blood-brain barrier delivery vector. *J Pharmacol Exp Ther.* 2002;301(2):605–610.

131 Roberts RL, Fine RE, Sandra A. Receptor-mediated endocytosis of transferrin at the blood-brain barrier. *J Cell Sci.* 1993;104 (Pt 2):521–532.

132 Muruganandam A, Tanha J, Narang S, Stanimirovic D. Selection of phage-displayed llama single-domain antibodies that transmigrate across human blood-brain barrier endothelium. *FASEB J.* 2002;16(2):240–242.

133 Perry VH, Anthony DC, Bolton SJ, Brown HC. The blood-brain barrier and the inflammatory response. *Mol Med Today*. 1997;3(8):335–341.
134 Terasaki T, Pardridge WM. Targeted drug delivery to the brain; (blood-brain barrier, efflux, endothelium, biological transport). *J Drug Target*. 2000;8(6):353–355.
135 Prokai L. Targeting drugs into the central nervous system. In: Muzykantov V, Torchilin VP, eds. *Biomedical Aspects of Drug Targeting*. Boston: Kluwer Academic Pub.; 2002:359–377.
136 Pardridge WM. Molecular biology of the blood-brain barrier. *Mol Biotechnol*. 2005;30(1):57–70.
137 McDevitt MR, Ma D, Lai LT, et al. Tumor therapy with targeted atomic nanogenerators. *Science*. 2001;294(5546):1537–1540.
138 Jain RK. Delivery of molecular and cellular medicine to solid tumors. *Adv Drug Deliv Rev*. 2001;46(1–3):149–168.
139 Hashizume H, Baluk P, Morikawa S, et al. Openings between defective endothelial cells explain tumor vessel leakiness. *Am J Pathol*. 2000;156(4):1363–1380.
140 Dvorak HF, Nagy JA, Dvorak AM. Structure of solid tumors and their vasculature: implications for therapy with monoclonal antibodies. *Cancer Cells*. 1991;3(3):77–85.
141 Brooks PC, Clark RA, Cheresh DA. Requirement of vascular integrin alpha v beta 3 for angiogenesis. *Science*. 1994;264(5158):569–571.
142 Burrows FJ, Derbyshire EJ, Tazzari PL, et al. Up-regulation of endoglin on vascular endothelial cells in human solid tumors: implications for diagnosis and therapy. *Clin Cancer Res*. 1995;1(12):1623–1634.
143 Arap W, Pasqualini R, Ruoslahti E. Cancer treatment by targeted drug delivery to tumor vasculature in a mouse model. *Science*. 1998;279(5349):377–380.
144 Schraa AJ, Kok RJ, Moorlag HE, et al. Targeting of RGD-modified proteins to tumor vasculature: a pharmacokinetic and cellular distribution study. *Int J Cancer*. 2002;102(5):469–475.
145 Curnis F, Sacchi A, Borgna L, et al. Enhancement of tumor necrosis factor alpha antitumor immunotherapeutic properties by targeted delivery to aminopeptidase N (CD13). *Nat Biotechnol*. 2000;18(11):1185–1190.
146 Ran S, Gao B, Duffy S, et al. Infarction of solid Hodgkin's tumors in mice by antibody-directed targeting of tissue factor to tumor vasculature. *Cancer Res*. 1998;58(20):4646–4653.
147 Huang X, Molema G, King S, et al. Tumor infarction in mice by antibody-directed targeting of tissue factor to tumor vasculature. *Science*. 1997;275(5299):547–550.
148 Sugano M, Egilmez NK, Yokota SJ, et al. Antibody targeting of doxorubicin-loaded liposomes suppresses the growth and metastatic spread of established human lung tumor xenografts in severe combined immunodeficient mice. *Cancer Res*. 2000;60(24):6942–6949.
149 Maeda H, Wu J, Sawa T, et al. Tumor vascular permeability and the EPR effect in macromolecular therapeutics: a review. *J Control Release*. 2000;65(1–2):271–284.
150 Schraa AJ, Kok RJ, Berendsen AD, et al. Endothelial cells internalize and degrade RGD-modified proteins developed for tumor vasculature targeting. *J Control Release*. 2002;83(2):241–251.
151 Thorpe PE. Vascular targeting agents as cancer therapeutics. *Clin Cancer Res*. 2004;10(2):415–427.

PART V

CHALLENGES AND OPPORTUNITIES

Complexity

187

Introductory Essay
Complexity and the Endothelium

Ary L. Goldberger

Beth Israel Deaconess Medical Center, Harvard Medical School, Boston, Massachusetts

This medical textbook is unusual in that it includes a section devoted to the endothelium and "complex systems." What may at first appear an editorial eccentricity will, upon further inspection, hopefully be seen as a useful addition and perhaps even as a template for future texts in other clinical fields. Indeed, the endothelium presents a compelling case study of the wide and deep potential interconnections between biomedicine and the contemporary study of complexity (1).

What makes a system complex, and not just complicated? The endothelium exemplifies two key features of complex systems (2). First, such systems function over a broad range of time and space scales (3,4) ranging from picoseconds to minutes, months, and longer, and from the quantum to the cellular to the organismic and even up to the level of social networks. Second, such multiscale systems display dynamics that are *nonstationary* and *nonlinear* and, therefore, defy analysis using traditional tools used by biostatisticians (1). Nonstationarity refers to the finding that the statistical properties (e.g., the mean and the variance) of fluctuations generated by a system change over time. Nonlinearity means that the components interact in nonadditive ways, so that the output will not be consistently proportional to the input. Indeed, in nonlinear systems, small changes can have huge or anomalous consequences, a phenomenon sometimes referred to as the *butterfly effect.*

Nonlinear systems cannot be understood by the traditional reductionist strategy of dissecting out their components, studying them in isolation, and then "recompiling" the system. Instead, the nonlinear interactions can lead to qualitatively novel structures and dynamics, so-called *emergent properties.* Thus, nonlinear systems cannot be characterized using a "modular" type of approach. Not only are the components at different scales interdependent, but global behaviors may arise as emergent properties of more local interactions. The failure of reductionist models to fully understand a system is a useful test of whether one is dealing with a truly complex system (e.g., an organ system) versus one that is merely complicated (e.g., most man-made machines). From a practical point of view, the presence of nonlinear interactions may foil attempts to engineer safe and effective pharmacologic interventions based on linear ("domino-like") cause–effect mechanisms with fixed targets. In a nonlinear system, the targets may not only be "moving" but also "morphing," a consideration that accounts for some of the unintended consequences induced by well-intended but deadly pharmacologic interventions.

The implications of complex systems for physiology and clinical medicine have only begun to be explored in the last decade or so (2–6). The rigorous underpinnings for these analyses have come primarily from the interrelated fields of nonlinear dynamics, applied mathematics, statistical physics, and chaos theory, including the study of fractals (scale-free objects and processes) (7,8). At the same time, the challenges posed by the study of living systems is stretching the envelope of concepts and models developed to study the most complex processes known in the physical world, such as turbulent flows. The preeminent cosmologist, Stephen Hawking, has predicted that this era will be the "century of complexity," an apt moniker likely to be borne out by explorations at the emerging and unexpected interfaces of biology, engineering, mathematics, and physics.

One of the most exciting discoveries in complexity science that will help propel such interdisciplinary studies is the principle of *universality* (5). Surprisingly, nonlinear systems that appear to be very different in their specific details may exhibit certain common patterns of response or architecture. For example, nonlinear systems may change in a sudden, discontinuous fashion. One important and universal class of abrupt, nonlinear transitions is called a *bifurcation.* This term describes situations in which a very small increase or decrease in the value of some parameter controlling the system causes it to change abruptly from one type of behavior to another. The output of the same system may suddenly go from being very irregular to a highly periodic, or vice versa. Another example of the principle of universality is related to the concept of *fractals* (7,8). Classical geometric forms are smooth and regular and have integer dimensions. In contrast, fractals are highly irregular objects that have noninteger, or fractional, dimensions. Fractals are composed of subunits and these, in turn, of sub-subunits, and so on, that resemble the larger scale structure. A wide variety of natural shapes share this self-similarity property, including branching vascular trees and coral formations, crenellated coastlines, and ragged mountain ranges. The concept of self-similarity (scale-invariance) also can be applied to complex processes that lack a characteristic scale and to certain types of scale-free networks described in this section. Loss of complexity, including the breakdown of fractal organization, appears to be a widespread, if not universal, theme in pathophysiology and aging (3,4,8).

Of interest within the community of complexity scientists already investigating biologic process is a useful tension between those who embrace stochastic mechanisms related to purely random effects and those emphasizing deterministic (rule-based) mechanisms. Common sense and analysis of actual data suggest a necessary compromise based on mixed models that include both stochastic and deterministic elements (9). In some cases, the same "phenotype," for example a fractal-type tree, can be generated either by stochastic models or by deterministic ones. It is possible, from an evolutionary perspective, that complex novel behaviors or structures initially arising by purely stochastic means but conveying some adaptive advantage could become encoded as part of the subsequent deterministic blueprint of the organism. (Nonlinear mechanisms such as emergence and bifurcations appear to toss a monkey wrench into the machinery of those challenging Darwinian theory, because these well-established phenomena already provide a rigorous foundation for the discontinuous appearance of novelty and complexity in the absence of an intelligent designer. Further, extravagantly complex structures and dynamics can arise from even simple-appearing nonlinear systems [10,11]. For example, the Mandelbrot set [7], among the most complex objects in mathematics, is generated by iteration of one of the simplest-in-form nonlinear equations.)

The chapters in this section present a wide-ranging, but by no means exhaustive, sampling of state-of-the art work linking complex systems with the study of endothelium.

Classical mathematical modeling attempts to describe the rate of changes in a system. The most powerful models not only have explanatory capacity – the ability to account for observed behaviors – but also are predictive of experimentally testable changes. Clermont, Vodovotz, and Rubin concisely overview the use of differential equations (ordinary and partial) to model such changes, the classical approach to understanding the temporal and spatial properties of dynamical systems (Chapter 191). The extraordinary challenges posed by the complexity of multiscale biologic systems such as the endothelium also have generated complementary approaches, including so-called *agent (individual or entity-based) based-models* introduced in the chapter by An (Chapter 188). Such simulations attempt to capture features of the global behavior of a system based on the rules governing the interactions of its individual components. As an example, Athale and Deisboeck present an agent-based computational approach to modeling the effects of angiogenesis on multiscale tumor growth dynamics (Chapter 193). Huang considers endothelial cell fates (different phenotypes) in terms of the nonlinear conceptual framework of *attractors* and *multistability* (Chapter 190). The basic idea is that the genome-wide gene regulatory network that controls cell behavior imposes constraints on the global dynamics such that only relatively few stable network states can arise. High-dimensional "attractor states" in the gene expression state space represent these discrete phenotypic cell fates. Almaas and Barabási describe new approaches to modeling complex networks that lack a characteristic scale (Chapter 189). Such scale-free (fractal-like) networks, which are intimately associated with ubiquitous power-law distributions, have universal properties that promise to link the study of endothelium with a wide range of other networks in nature. Finally, Ingber explores the delicate and essential interplay of mechanical forces and biochemical changes in cellular biology (Chapter 192). This chapter describes how a particular form of architecture known as *tensegrity* stabilizes the shape of the constituent cells, the extracellular matrix, and the intracellular cytoskeleton. Among its far-reaching implications, this work emphasizes the physical context of multicomponent networks and underscores the inherent limitations of contemporary analytic approaches focused primarily at the gene-molecule level.

These discussions of complementary approaches to modeling will hopefully provoke new "out-of-the-box" ways of thinking and rethinking endothelial structure–function relationships and stimulate previously unplanned basic and clinical experiments and therapeutic approaches (1). The ways in which models fail to account for the complexity of real-world dynamics is as important as their successes and an essential guide in the dynamic and evolving interplay of theory and experiment. To paraphrase a remark attributed to the English astronomer, Sir Arthur Eddington: The physiologic universe is likely to be not only stranger than we imagine, but stranger than we can imagine.

ACKNOWLEDGMENTS

The author gratefully acknowledges support from the National Institutes of Health/National Center for Research Resources, the James S. McDonnell Foundation, the Defense Advanced Research Projects Agency (DARPA) under grant HR0011-05-1-0057, the Ellison Medical Foundation (Senior Scholar Award in Aging Program), and the G. Harold and Leila Y. Mathers Charitable Foundation.

REFERENCES

1 Aird WC. Endothelial cell dynamics and complexity theory. *Crit Care Med.* 2002:30:(Suppl): S180–S185.
2 Goldberger AL. Giles F. Filley lecture: complex systems. *Proc Am Thorac Soc.* 2006; 3:467–472.
3 Costa M, Goldberger AL, Peng CK. Multiscale entropy analysis of biological signals. *Phys Rev E.* 2005;71:021906.

4 Costa M, Goldberger AL, Peng CK. Broken asymmetry of the human heartbeat: loss of time irreversibility in aging and disease. *Phys Rev Lett.* 2005;95:198102.
5 Glass L, Mackey MC. *From Clocks to Chaos: The Rhythms of Life.* Princeton, NJ: Princeton University Press, 1988.
6 Beuter A, Glass L, Mackey M, Titcombe MS. *Nonlinear Dynamics in Physiology and Medicine.* New York, NY: Springer-Verlag, 2003.
7 Mandelbrot BB. *The Fractal Geometry of Nature.* New York, NY: WH Freeman, 1982.
8 Goldberger AL, Amaral LAN, Hausdorff JM, et al. Fractal dynamics in physiology: alterations with disease and aging. *Proc Natl Acad Sci USA.* 2002; 99(Suppl 1):2466–2472.
9 Schulte-Frohlinde V, Ashkenazy Y, Goldberger AL, et al. Complex patterns of abnormal heartbeats. *Phys Rev E.* 2002;66: 031901–1–12.
10 May RM. Simple mathematical models with very complicated dynamics. *Nature.* 1976; 261:459–467.
11 Amaral LAN, Diaz-Guilera A, Moreira A, et al. Emergence of complex dynamics in a simple model of signaling networks. *Proc Natl Acad Sci USA.* 2004; 101:15551–15555.

188

Agent-Based Modeling and Applications to Endothelial Biomedicine

Gary An

Northwestern University, Feinberg School of Medicine, Chicago, Illinois

The chapters in this volume have thus far primarily described the properties of individual endothelial cells (ECs) as well as the behavior of the endothelium as a whole organ. But how is that transition from individual cellular function to organ-level behavior made? Merely extrapolating the behavior of individual ECs is insufficient; the internal heterogeneity of the endothelial organ precludes linear summation of individual EC function. Rather, it is necessary to place the behavior of the ECs in the context of their local environment, be that in various tissue beds in a baseline state of health or in pathological regional disruptions associated with injury or infection. What is required is a means of formalizing the process by which the laboratory-derived data about individual ECs can be translated into the richness of behavior that is seen at the level of an organism. Systems biology and mathematical modeling can provide a mechanism for this translation, and this chapter will introduce one of these techniques, agent-based modeling (ABM), and give an example of an ABM that includes endothelial function.

WHAT IS AGENT-BASED MODELING?

ABM is a type of mathematical modeling that is dynamic (evolving over time), discrete-event (stepwise progression of time and action), and mechanistic (dependent upon rules). ABM is completely deterministic, meaning that, for a specific set of initial conditions, the model will run exactly the same for each simulation run. However, stochasticity built into the model at multiple levels (through the use of random-number generators) allows for variation in the models' dynamics. The emphasis with ABM is on the individual components of a dynamic system and the rules that govern them. A dynamic system is viewed as an aggregate of the behaviors of its constituent components, with the recognition that the components function based on an internal rule system. The components of the system are grouped into classes of "agents" by virtue of shared mechanisms, which are derived from experimental and observational data. The mechanisms can be expressed as a series of conditional ("if-then") statements, and computer programs are written to describe the rules of behavior. Agents are treated as finite-state machines that respond to inputs and outputs based on their class rules. A specific series of rules defines a particular class of agent. Another important aspect of ABM is that agent inputs are locally constrained. Thus, an agent acts based on limited knowledge of what is going on in the system as a whole. The boundary of agent knowledge is determined by the structure of the model. For example, the knowledge boundary can be spatial (as in grid-based ABMs) or a restricted group of links (as in a network ABM).

In addition to creating populations of agents, ABM defines a particular "virtual world" in which the agents carry out their functions. The structure and parameters of the virtual world depend on the system being modeled and the type of ABM being created. For instance, in spatial ABM, the virtual world is often represented as a grid. Variables representing the characteristics of the reference system are placed in the grid, and the agents move and interact over the grid. Heterogeneous distribution of "environmental" parameters/variables in the virtual world allows for divergent behavior of agents from within the same class (i.e., those agents having the same rules). As the agents execute their internal algorithms in a stepwise, parallel fashion (to simulate simultaneous behavior), they respond to these heterogeneously distributed local inputs, resulting in multiple paths/trajectories of individual agent behavior within the population of an agent class. The dynamics of the system emerges from the multiple differential interactions between the agents over time and the resulting aggregate distribution of characteristics of their populations.

Agents ideally should have well-identified, simple rules. A wide range of abstraction is used in developing the parameters of agents and their world, extending from simple two-dimensional (2D) cellular automata, to weighted nodes in a neural network, to more sophisticated nodes in a small-worlds network, all the way to extended algorithms of conditional statements (such as experimentally derived cellular/molecular mechanisms) in a multidimensional, mobile cellular automata. This last type of ABM is possible because, although the agents themselves may be complex constructs, their complexity can be managed by drawing a boundary at which well-documented responses occur to external inputs and outputs. Applying stochasticity to their algorithms through the use of random-number generators introduces the "uncertainty" that can accompany the behavior of complex systems. It should be noted, however, that this places a limit on how finely grained the analysis of model behavior can be; the probabilistic components in the model therefore represent "black boxes" in terms of analysis and intervention.

ABM can be thought of as essentially a formalized means of concurrently expressing multiple hypotheses of multiple mechanisms. By focusing on a system's components and their behavior, ABM is often fairly intuitive to nonmathematicians who may find it easier to think in terms of conditional statements than equations. As a result, ABM is well suited to translating multiple experimental conclusions into a dynamic, synthetic, analytical framework.

APPLICATIONS OF AGENT-BASED MODELS

Complex systems are often defined as those systems in which the overall behavior of the system cannot be directly inferred or predicted by knowing the characteristics of the individual components of the system (1). Implicit in this description is a nonlinear and noncontinuous relationship between component behavior and overall system behavior. This particular characterization of complex systems is well suited to the capabilities of ABM. Therefore, it is not surprising that ABM has found most of its applications in complex systems described in this way. The classic examples are that of bird flocking, fish schooling, and insect colony behavior (2). Other systems and processes that fall into this category and that have been investigated using ABM include economics (3), political science (4), social science (5), ecology (6), and evolution (7). Human physiology and pathophysiology are generally recognized as behaving as a complex system, and relatively recently there has been interest in the use of ABM to study these processes (8–11). One such ABM (9), which specifically includes EC function, is described briefly later. This model will be used as an example of the considerations that go into the process of developing ABM, will focus discussion of validation strategies for ABM, and will give examples of how such ABM can be used. Another example of ABM in the medical field is the tumor angiogenesis model of Deisboeck found in Chapter 193.

AGENT-BASED MODELING OF THE ACUTE INFLAMMATORY RESPONSE

The ABM described herein uses cells as the primary agent level and focuses on the interactions that occur at the interface between ECs and blood-borne inflammatory cells. (A more extensive description of this ABM, as well as the web site at which the model is posted, is available [9]). Cells were selected as the agent level because they can be readily subgrouped into classes based on common behavioral rules, and the responses of single types of cells to various mediators has been extensively characterized in the basic science literature. The focus is on the endothelial–blood interface because I believe that activation of endothelial surfaces (be it from injury or infection) is the initiating point at which inflammation has consequence to tissue beds, and that the interaction between the endothelium and blood-borne inflammatory cells is responsible for the propagation and regulation of the inflammatory process. The endothelial surface is therefore treated as a physiological membrane that has barrier, transport, and communication functions.

The ABM reflects this particular concept by its basic configuration as a 2D grid (Fig. 188.1A; for color reproduction, see Color Plate 188.1). The edges of the grip "wrap," meaning that agents moving off the right edge of the screen appear on the left edge, and vice versa. This is also true for the upper and lower edges of the screen, with an end result that the topological surface of the ABM is a torus. This configuration is also topologically equivalent to an aggregate of all the capillary beds in the body. On each "square" of the grid resides an agent of the EC class. As mentioned earlier, all these agents have the same rules and algorithms and, although ECs do not move, their discreteness in space allows them to manifest different trajectories of behavior. The knowledge boundary for each agent is the immediately surrounding eight grid spaces (called a *Moore neighborhood* in the terminology of cellular automata; for more information about cellular automata see Ref. 12). This neighborhood corresponds to the immediate extracellular space in which the cell interacts via its membrane. The ECs change their state depending on the variables within this space, with the end result being that the EC component of the ABM functions as a cellular automata.

A mobile component to the ABM is also present, made up of the inflammatory cell agents (ICAs). The classes of ICAs include neutrophils (polymorphonuclear cells, PMNs), circulating mononuclear cells, and T-cell subspecies. As opposed to the EC agents, which are imbedded in the grid, ICAs move freely over the surface of the EC matrix. Although no directional flow is present, the baseline random movement of these agents reflects the fact that there is localization and limitation with respect to how far a particular cell can move between capillary beds. Even though they are mobile, ICAs have the same knowledge boundary as ECs (a Moore neighborhood) that moves with them as they move. Variables that correspond to chemotactic mediators will affect the pattern of ICA movement, variables that correspond to adhesion factors will affect the rate of ICA movement, and other variables that correspond to pro- and anti-inflammatory cytokines will initiate ICA algorithms that correspond to cytokine production/secretion, respiratory burst, and apoptosis, among other cellular functions. Again, it should be noted that each individual agent runs its algorithms based on the input from its Moore neighborhood,

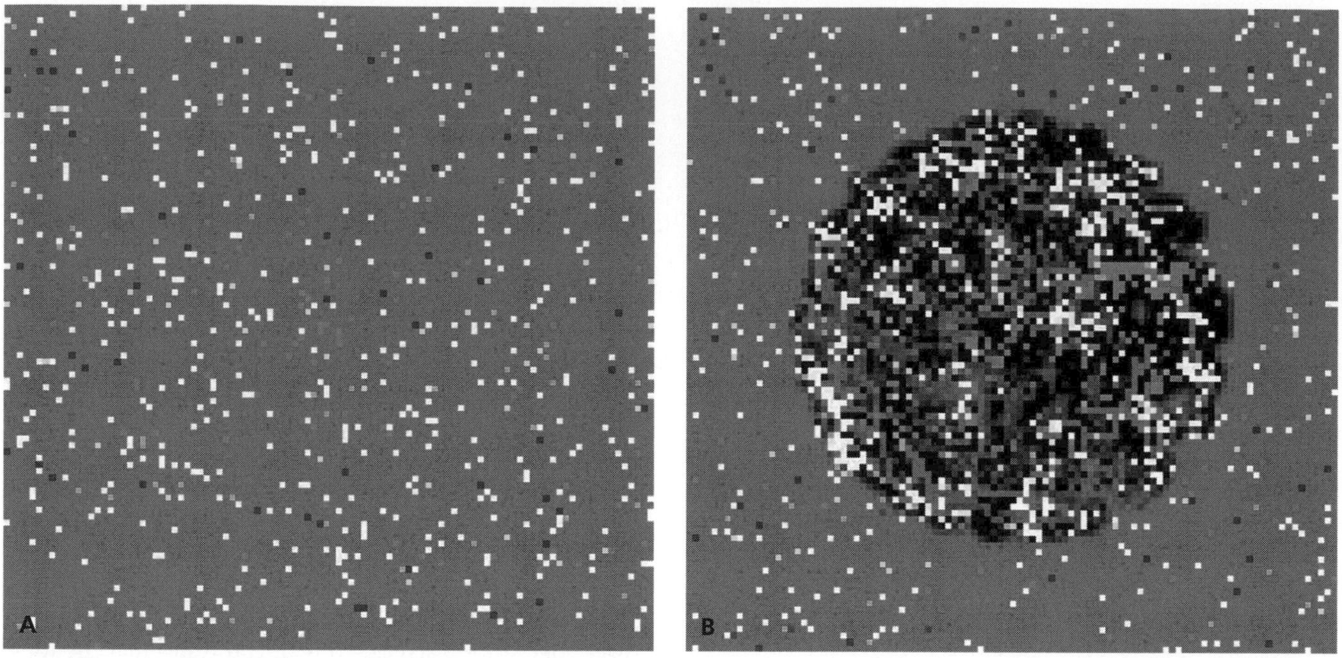

Figure 188.1. These panels are screen shots of the ABM (8) described in the text. The space is made up of 101 × 101 cells, in each of which is an EC agent, shown as gray background. The other colored squares represent ICA agents: PMNs are white, monocytes are light gray, T-cell species are black. (**A**) A nonperturbed system with only the EC/ICA agents present. (**B**) A system perturbed with a simulated infection. The initial insult is given in a roughly circular distribution of simulated infectious vectors (representing gram-negative bacteria) represented by gray squares. The effect of the simulated infection is to damage and kill EC agents, represented as patches of black and shades of black. The lighter patches in **B** represent activated PMNs that have responded to the insult and are exercising their algorithmic rules systems. The screen shot in **B** was taken at approximately 48 hours simulated time. Note the relative paucity of infectious vectors compared to the size of the area of damaged ECs. This denotes the efficacy of bacterial killing by the ICAs, but also the degree of "collateral damage" generated by the proinflammatory component of the inflammatory response. For color reproduction, see Color Plate 188.1.

and that the agents execute their rules in an emulated parallel fashion to simulate simultaneous behavior. In the course of a simulation, all the agents execute their algorithms in a stepwise fashion that is calibrated to the relative amount of "real time" it takes for the process to occur. In a nonperturbed situation, the behavior of the model is "stable," meaning that the general overall parameters of the model remain relatively constant. The model can persist in this state indefinitely. However, interesting dynamics occur when the system is perturbed. This is accomplished with a spatial distribution of damage to the underlying EC matrix. This damage can take the form of simply "killing" a set number and pattern of EC agents, or introducing simulated infectious vectors that will not only damage ECs but also replicate and spread. In either case, the introduction of the perturbation changes the state conditions of the affected agents, with subsequent consequences on the other agents in the area. The spatial distribution of the initial insult is variable even given a set value/level of the initial insult. This heterogeneity of initial conditions plays a role in the subsequent divergent behavior of the system as a whole. The dynamics of the ABM now can follow different trajectories reflecting different behaviors and outcomes that can be evaluated against what we see in the real world. An example of the ABM in the process of dealing with a simulated infection can be seen in Figure 188.1B.

Therefore, the end result is the creation of a "virtual organism" that is essentially a giant EC culture that includes active inflammatory cells. It combines the knowledge gleaned from individual EC function and individual ICA function into an interactive framework that would be difficult, if not impossible, to replicate in a basic science lab without significant artifacts. This "virtual organism" can now be subjected to in silico experiments of a type that would parallel experiments carried out in a wet lab. Because the ABM is based on mechanistic rules, any rule programmed into the ABM can be manipulated in a potential experiment. Before "new" experiments can be considered, however, it is necessary to validate the ABM.

VALIDATION AND IN-SILICO EXPERIMENTS

The validity of any model is a measure of how well the model matches the behavior of the reference system it is intended to emulate. It determines the applicability and usefulness of the model. Validation is therefore a critical step in the development of any model, be it in silico, in vitro, or in vivo. With respect to ABM, validation provides both challenges and opportunities. Validation strategies can be divided into two general categories: validation of assumptions and validation of behavior.

The validity of the assumptions of a model is a perpetual sticking point with respect to all mathematical models. This is popularly expressed with the acronym "GIGO," or "Garbage In, Garbage Out." This is especially true with mathematical models, in which the lack of intrinsic constraints (such as the laws of physics or chemistry) can lead to nonsensical results. However, there is a readily available solution: transparency. Transparency means making explicit the process by which the architecture and rules of the model are chosen and implemented. This is important, because all models represent some degree of abstraction that is at the discretion of the modeler. Making the initial modeling assumptions (as expressed in the architecture and rules of the model) explicit allows anyone examining the model and its behavior to determine for themselves to what degree they agree with the structure of the model and the results gathered from it. The potential limitations of the model are therefore apparent from the outset, and the expectations of what can be gleaned from the model are placed in an appropriate context. This is consistent with the concept of modeling as an iterative process of hypothesis generation and testing, with each model sufficient to an acknowledged point of failure, but through its failure constantly advancing the boundary of understanding of the system being examined.

The other general category of validation is that of the behavior of the model. This can be approached through a series of strategies. All of them, to some degree, consist of comparing the behavior of the model to some real-world data set of expected behavior. When the behavior of the model matches the real world data set, then the model is deemed valid for that particular test. If the model does not match, then the model is reexamined and either the basic structure of the model is reconfigured, or specific variables and relationships are fine-tuned. This latter process is called *calibration.* It should be evident by this last statement that one potential significant pitfall exists – namely, When is the problem one of calibration, and when is it one of basic structure? There is no easy answer to this, other than to make the basic structure transparent, and rely on feedback from the community at large.

Validation of ABM behavior can be accomplished at three levels: at the level of individual response (*individual dynamic*), at the level of the behavior of a population with respect to intrinsic variables (*population dynamic*), and finally with the behavior of the population with respect to an intervention (*population response*). For example, in the EC/ICA ABM described earlier, the individual dynamic behavior is the reproduction of the four possible dynamics associated with injury/infection: containment and healing, a hyperinflammatory forward feedback response, an anti-inflammatory negative feedback response, and overwhelming initial insult (9). The population dynamic is reproduced by mean patterns of pro- and anti-inflammatory cytokines generated in simulated populations that match those seen in equivalent clinical populations (9). Finally, the population response is validated by simulating existing mediator-directed therapies and performing in silico trials in simulated populations matched to those reported in the clinical trials. Up to this point, all the validation strategies are dealing with a model that reflects the baseline, unmodified functions of the reference system. No manipulation of the baseline agent rules or state variables has been done; the initial perturbation is the only intervention to the steady-state model. However, with the third level of validation, the population response, we move into the realm of in silico experiments. As mentioned before, any rule or variable in the model can be targeted for intervention, and populations of simulated runs can be generated for both the baseline ABM (the "control group") and for the modified ABM (the "intervention group"). Again, the structure of an ABM and its reliance on mechanistic rules makes this transition very intuitive, and it is not hard to imagine treating the ABMs just like any other wet lab/clinical population. In the particular case of the EC/ICA ABM, the reference system interventions are previously published clinical trials of mediator-directed therapies against sepsis, and the resulting ABM experiments are in silico simulations of those trials. In these situations, the addition of anti-tumor necrosis factor (TNF), anti-interleukin (IL)-1, and granulocyte colony stimulating factor (G-CSF) all failed to produce a statistically significant improvement in mortality, results that were seen in the clinical trials (9). Unfortunately, at this time, the single successful trial of mediator-directed therapy, activated protein C, has not been simulated, and the strength of the validation would be greatly enhanced by replication of a positive result (this is currently in development). However, despite the reliance on validation by negative outcome, the behavior of the model is significant in that the model reflects the summation of beliefs and hypotheses that existed and fomented the development of those failed trials, and is a further example of the utility of ABM as a means of hypothesis testing.

This can be carried to the next logical step by simulating hypothetical, as yet untried interventions. In the case of the EC/ICA ABM, this was done by trying two different treatment strategies. The first was a modification of the dose and duration of anti-TNF and anti–IL-1, with the idea of using lower doses and longer durations to "dampen out" the inflammatory response (this hypothesis has its basis in the trend of "supplementation therapies," such as low-dose "replacement" corticosteroids and vasopressin). The second was to use multiple drugs to attack the redundancies in the inflammatory response, specifically anti-TNF with anti–IL-1. Neither of these strategies has been implemented (yet), so these in silico experiments were truly prospective. Unfortunately, both strategies proved ineffective in the in silico experiments (9). It must be noted that, because the ABM in which these experiments were carried out is a very abstract model, one should not come away with the idea that these approaches are doomed to failure. However, one should also note that the philosophies underpinning both these strategies are currently prevalent in the critical care community. Perhaps the finding that these strategies do not work when they are implemented in a synthetic, analytical framework should be cause for some reexamination of the underlying hypotheses.

LIMITATIONS

ABM is a modeling technique that is just beginning to progress out of its infancy. In the scope of more traditional, equation-based modeling, ABM can seem imprecise. Conditional statements that constitute most agent rules do not have the same degree of mathematical precision as equations. ABM cannot be "solved" in a formal sense and, for them to be mathematically analyzed, they must be transformed into equations (if possible). The calibration of ABM can be very difficult. Because the agents are treated as "black box" input-output devices, generating their rule systems requires very extensive reductionist mechanistic data from the wet lab. Since the dynamics of the model arise through multiple agent interactions with a stochastic component, it can be very difficult to determine the effect of a single rule change (in fact, this property is precisely why certain systems are modeled using ABM). This precludes the curve fitting that can be done with coefficient modification, as in equation-based models. These characteristics of ABM make it suited primarily as a research tool, not a clinical/bedside application.

There is a paradox, however (as is the nature of complex systems!) with the limitations of ABM. The paradox is that the limitations listed here derive from the characteristics that make ABM a desirable and beneficial modeling tool. Agent rules expressed as algorithms of conditional statements are often more intuitive than deriving ordinary or partial differential equations. Furthermore, experimental conclusions are usually expressed as conditional statements, making ABM well suited as a translational, synthetic tool. The need for extensive mechanistic wet lab data integrates ABM development with ongoing scientific investigation, potentially emphasizing gaps in knowledge and directing future experiments. Often the systems under study are too complex for intuitive determination of cause-and-effect of a single rule change; this is why mathematical modeling, and ABM in particular, is being applied to the particular problem – if people could have figured it out without having to resort to computational methods, they quite likely would have done this already.

This carries us back to a point that has been revisited multiple times in this chapter: that mathematical modeling such as ABM represents a formalized way of expressing and testing hypotheses. The process that goes into developing ABM is an integral component of what every good scientist already does intuitively. However, we recognized that in some systems and problems, intuition breaks down; in those circumstances, tools like ABM have their place. As with any tool, its useful application is tied to understanding and working within its known limitations.

FUTURE DIRECTIONS FOR AGENT-BASED MODELING AND ENDOTHELIAL BIOMEDICINE

With this in mind, it is possible to place ABM in its appropriate context with respect to its applications to the emerging field of endothelial biomedicine. As with nearly all biological systems, system-level endothelial function is a complex system (13). As with all complex systems, there is a nonlinear, nonintuitive relationship between the underlying mechanisms of the system's components and the system's behavior. This is the point of breakdown with the traditional analytic-synthetic paradigm of Newtonian science. We have the analytical capabilities, the product of 500 years of the Scientific Method; what is lacking are the appropriate synthetic tools for integrating the information we obtain through analysis and placing it into a context that reflects the phenomena we observe in the real world. The mechanisms and tools of systems science, and ABM in particular, can provide the appropriate synthetic framework to form the necessary adjunct to the study of a complex system like the endothelium. More specifically, the intrinsic spatial characteristics of ABM make it well suited to handle the heterogeneity of endothelial behavior, and the flexible, modular nature of ABM can foster the collaborative integration of disparate lines of study. Toward this end, work is being done on creating a universal modeling "grammar" for the simulation of acute inflammation (14). Given the use of classes and class rules in ABM, the goal is to create a robust and flexible grammar through which the cellular and molecular data generated from basic science laboratories can be translated into a community-wide, open source ABM. It is hoped that this platform will provide the requisite transparency needed for evaluating mathematical models, and also a potential common frame of reference through which controversies can be discussed and compared. This goal will hopefully parallel the evolution of endotheliology as a discipline, because many of the same issues of integration and synthesis lie at the heart of both endeavors.

KEY POINTS

- ABM, a form of mathematical modeling, can be a useful tool in the study of complex systems, such as the endothelium.
- ABM is essentially a formalized means of hypothesis expression and testing.
- By its nature as a mechanistic, rules-driven type of modeling, ABM is well suited as a translational, synthetic adjunct to traditional research paradigms.
- ABM may be able to provide a community-wide framework for integrating collaborative approaches to studying a system as wide-ranging and extensively distributed as the endothelium.

REFERENCES

1 Auyang S. *Foundations of Complex-System Theories in Economics, Evolutionary Biology and Statistical Physics.* Cambridge, United Kingdom: Cambridge University Press, 1998.

2 Resnick M. *Turtles, Termites and Traffic Jams: Explorations in Massively Parallel Microworlds.* Cambridge, MA and London, England: Bradford Press, 1994.
3 Luna F, Perrone A. *Agent-Based Methods in Economics and Finance: Simulations in Swarm.* Amsterdam: Kluwer Academic Publishers, 2001.
4 Johnson P. Simulation modeling in political science. *Am Behav Sci.* 1999, 42:1509–1530.
5 Cohen MD, Riolo RL, Axelrod R. The role of social structure in the maintenance of cooperative regimes. *Rationality and Society.* 2001;18:5–32.
6 Railsback SF, Harvey BC. Analysis of habitat selection rules using an individual-based model. *Ecology.* 2002;83:1817–1830.
7 Marshall JAR, Rowe JE. Kin selection may inhibit the evolution of reciprocation. *J Theor Biol.* 2003;222:331–335.
8 An G. Agent-based computer simulation and SIRS: building a bridge between basic science and clinical trials. *Shock.* 2001;16(4): 266–273.
9 An G. In-silico experiments of existing and hypothetical cytokine-directed clinical trials using agent based modeling. *Crit Care Med.* 2004;32(10):2050–2060.
10 Ropella GEP, Hunt CA. Prerequisites for effective experimentation in computational biology. *Conf Proc IEEE Eng Med Biol Soc.* 2003:1272–1275.
11 Yan L, Hunt CA, Ropella GEP, Roberts MS. In silico representation of the liver – connecting function to anatomy, physiology and heterogeneous micro environments. *Conf Proc IEEE Eng Med Biol Soc.* 2004:853–856.
12 Casti J. *Reality Rules I: Picturing the World in Mathematics – The Fundamentals.* New York: John Wiley & Sons, Inc., 1992.
13 Aird WC. Endothelial cell dynamics and complexity theory. *Crit Care Med.* 2002;30(Suppl 5):S180–S185.
14 An G. Concepts for developing a collaborative in-silico model of the acute inflammatory response using agent based modeling. *J Crit Care.* 2006;21(1):105–110; Discussion 110–111.

LIMITED GLOSSARY OF MODELING TERMS

Cellular Automata (CA): Initially described by Stanislav Ulam and John Von Neumann, cellular automata are discrete dynamic systems. Each point in a regular spatial lattice, called a *cell*, can have any one of a finite number of states. The states of the cells in the lattice are updated according to a local rule, relying only on its own state one time step previously, and the states of its nearby neighbors at the previous time step. All cells on the lattice are updated synchronously. Thus, the state of the entire lattice advances in discrete time steps.

Class Rules: A set of rules for behavior, usually expressed as conditional ("if/then") statements, that is common to a particular group of agents (an "agent class").

Finite State Machine: An object in which the number of its conditions or "states" can be expressed using a finite number. This implies that each of those states can be linked to a distinct time step.

Population Dynamic: The dynamics of the behavior of a population as a whole.

Population Response: The subsequent population dynamic following a perturbation to a population.

Random-Number Generator (RNG): A mathematical algorithm that, when implemented in a computer, produces a string of "random" numbers. In truth, since they are generated via an algorithm, the numbers are "pseudorandom." However, with most available RNGs, the cycle of the algorithm is so great as to effectively be "random."

Reference System: The real world system that is the subject of modeling/simulation. Example: The reference system for a model of human acute inflammation is the human acute inflammatory system.

Simulation: A real, implemented, or instantiated model, or a situated or embedded model, or a set of actions such that, when executed, affect the environment. The use of the term *simulation* carries implications about a similarity relationship between the simulator and the simulation. There is an implication that the similarity is somewhat analytic, meaning that, if the reference system is decomposable, the simulation is decomposable, and the similarity extends down to the components. However, simulations are all abstractions, and this similarity is always incomplete.

Virtual World: An environment that is an informational construct and not physically manifest in the real world. The most common popular description is "cyberspace."

189

Scale-Free Networks in Cell Biology

Eivind Almaas and Albert-László Barabási

University of Notre Dame, Indiana

The last century brought with it unprecedented technological and scientific progress, rooted in the success of the reductionist approach. For many current scientific problems, however, it is not possible to predict the behavior of a system from an understanding of its (often identical) elementary constituents and their individual interactions. For these systems, we need to develop new methods to gain insight into their properties and dynamics. During the last few years, network approaches have shown great promise in this direction, offering new tools to analyze and understand a host of complex systems (1–4). A much studied example of the network approach concerns communication systems like the Internet and the World Wide Web, which are modeled as networks with nodes being routers (5) or web pages (6) and the links being the physical wires or URLs. The network approach also lends itself to the analysis of societies, with people as nodes and the connections between the nodes representing friendships (7), collaborations (8,9), sexual contacts (10), or coauthorship of scientific papers (11), to name a few possibilities. It seems that the more we scrutinize the world surrounding us, the more we realize that we are inextricably entangled in myriad interacting webs; to describe them we need to understand the architecture of these various networks that nature and technology offers to us.

Biological systems ranging from food webs in ecology to biochemical interactions in molecular biology can benefit greatly from being analyzed as networks. In particular, in the cell, the variety of interactions between genes, proteins, and metabolites are well captured by network representations, especially with the availability of veritable mountains of interaction data from genomics approaches. In this chapter, we discuss recent results and developments in the study and characterization of the structure and utilization of biological networks, and their relevance for the endothelium.

CHARACTERIZING NETWORK TOPOLOGY

Many tools are available to study the structure of complex networks. Here, we discuss two methods that have received the most attention: degree distribution and node clustering. Additionally, it is customary to investigate the *betweenness-centrality* (BC) of both nodes and links, and the network *assortativity*. The BC is related to the number of shortest paths going through either a node or a link; hence, a large BC value indicates that the node or link acts as a bridge by connecting different parts of the network. The assortativity describes the propensity of a node to be directly connected to other nodes of similar degree.

Degree Distribution

The representation of various complex systems as networks has revealed surprising similarities, many of which are intimately tied to power laws. The simplest network measure is the average number of nearest neighbors of a node, or the average degree $\langle k \rangle$. However, this is a rather crude property, and to gain further insight into the topological organization of real networks, we need to determine the variation in the nearest neighbors, given by the degree distribution $P(k)$. For a surprisingly large number of networks, this degree distribution is best characterized by the power law functional form (12) (Fig. 189.1A),

$$P(k) \sim k^{-\alpha} \qquad \text{(Equation 1)}$$

Important examples include the metabolic network of 43 organisms (13), the protein interaction network of *Saccharomyces cerevisiae* (14), and various food webs (15). If the degree distribution instead was single-peaked (e.g., Poisson or Gaussian), as in Figure 189.1B, the majority of the nodes would be well described by the average degree, and we could with reason talk about a "typical" node of the network. This is very different for networks with a power-law degree distribution; the majority of the nodes only have a few neighbors, while many nodes have hundreds and some even thousands of neighbors. Although average node degree values can be calculated for these networks because their size is finite, these values are not representative of a typical node. For this reason, these networks are often referred to as "scale-free."

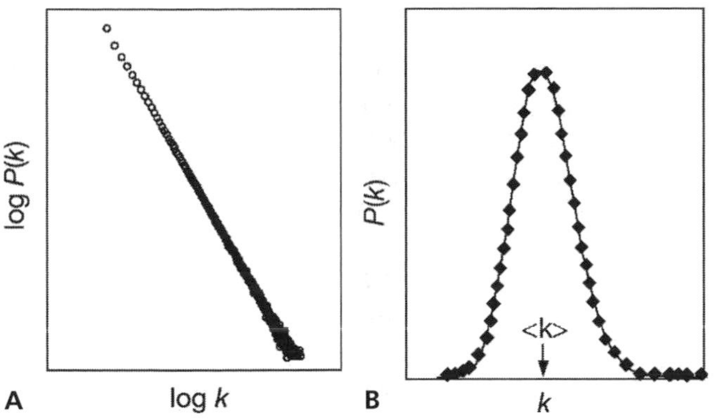

Figure 189.1. Characterizing degree distributions. For the power-law degree distribution (**A**), there exists no typical node, whereas for single-peaked distributions (**B**), most nodes are well represented by the average (typical) node with degree $\langle k \rangle$.

Clustering Coefficient

A measure that gives insight into the local structure of a network is the so-called *clustering of a node:* the degree to which the neighborhood of a node resembles a complete subgraph (16). For a node i with degree k_i the clustering is defined as

$$C_i = \frac{2n_i}{k_i(k_i - 1)} \qquad \text{(Equation 2)}$$

representing the ratio of the number of actual connections between the neighbors of node i to the number of possible connections. For a node that is part of a fully interlinked cluster $C_i = 1$, while $C_i = 0$ for a node where none of its neighbors are interconnected. Accordingly, the overall clustering coefficient of a network with N nodes is given by $\langle C \rangle = \sum C_i/N$, quantifying a network's potential modularity. By studying the average clustering of nodes with a given degree k, information about the actual modular organization of a network can be extracted (17–20): For all metabolic networks available, the average clustering follows a power-law form as

$$C(k) \sim k^{-\delta} \qquad \text{(Equation 3)}$$

suggesting the existence of a hierarchy of nodes with different degrees of modularity (as measured by the clustering coefficient) overlapping in an iterative manner (17). In summary, we have seen strong evidence that biological networks are both scale-free (13,14) and hierarchical (17).

NETWORK MODELS

As we have just seen, many biological networks are dominated by a scale-free distribution of nearest neighbors. Why is this power-law behavior so pervasive? To understand the cause of the scale-free degree distribution and the hierarchical network structure, we describe three models that serve as network paradigms. These models build on very different principles and, to varying degrees, are able to explain the observed network features.

Random Network Model

In discussing the origin of the observed power-law behavior, we must first understand the properties of the simplest available network model. Although graph theory initially focused on regular graphs, since the 1950s, large networks with no apparent design principles were described as *random graphs* (21), proposed as the simplest and most straightforward realization of a complex network. According to this Erdos-Renyi (ER) model of random networks (22), we start with N nodes and connect every pair of nodes with probability p. This creates a graph with approximately $pN(N-1)/2$ randomly distributed edges (Fig. 189.2A,D; for color reproduction, see Color Plate 189.2). The distribution of nearest neighbors follows a Poisson distribution (Fig. 189.3A; for color reproduction, see Color Plate 189.3) and, consequently, the average degree $\langle k \rangle$ of the network describes the properties of a typical node. Furthermore, for this "democratic" network model, the clustering is independent of the node degree k (Fig. 189.3D). The ER model, although simple and appealing, captures neither the properties of the degree distribution nor the clustering coefficient observed in biological networks.

Scale-Free Network Model

In the network model of Barabási and Albert (BA), two key mechanisms, which both are absent from the classical random network model, are responsible for the emergence of a power-law degree distribution (12). First, networks grow through the addition of new nodes linking to nodes already present in the system. Second, a higher probability exists to link to a node with a large number of connections, a property called *preferential attachment.* These two principles are implemented as follows: Starting from a small core graph consisting of m_0 nodes, a new node with m links is added at each time step and connected to the already existing nodes (Fig. 189.2B,E). Each of the m new links are then preferentially attached to a node i (with k_i neighbors), which is chosen according to the probability

$$\Pi_i = k_i \Big/ \sum_j k_j \qquad \text{(Equation 4)}$$

The simultaneous combination of these two network growth rules gives rise to the observed power-law degree distribution (Fig. 189.3B). In Figure 189.2B, we illustrate the growth process of the scale-free model by displaying a network at time t (green links) and then at time $(t+1)$, when we have added a new node (red links) using the preferential attachment probability. Compared to random networks, the probability that a node is highly connected is statistically significant in scale-free networks. Consequently, many network properties are determined by a relatively small number of highly connected

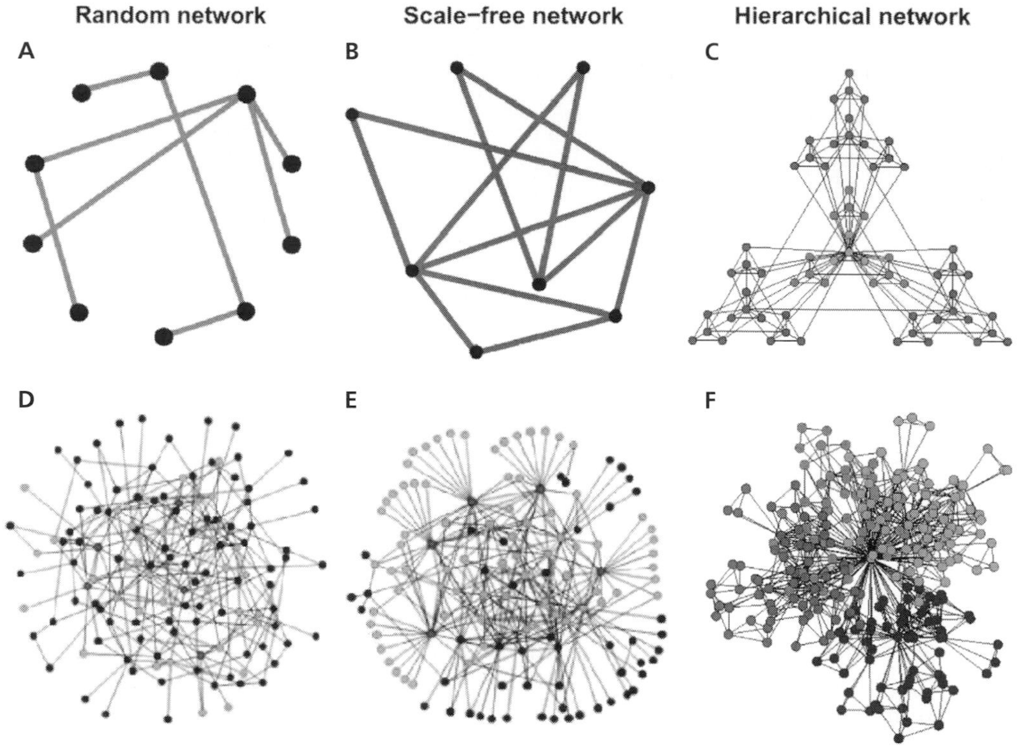

Figure 189.2. Graphical representation of three network models: (**A, D**) The ER (random) model; (**B, E**) the BA (scale-free) model; and (**C, F**) the hierarchical model. The random network model is constructed by starting from N nodes before the possible node-pairs are connected with probability p. Panel **A** shows a particular realization of the ER model with 10 nodes and connection probability $p = 0.2$. In panel **B**, we show the scale-free model at time t and at time $(t + 1)$ when we have added a new node using the preferential attachment probability (see Equation 4). Panel **C** demonstrates the iterative construction of a hierarchical network, starting from a fully connected cluster of four nodes. This cluster is then copied three times while connecting the peripheral nodes of the replicas to the central node of the starting cluster. By once more repeating this replication and connection process, we end up with a 64-node scale-free hierarchical network. In panel **D**, we display a larger version of the random network, and it is evident that most nodes have approximately the same number of links. For the scale-free model (**E**), the network is clearly inhomogeneous: Although the majority of nodes have one or two links, a few nodes have a large number of links. Whereas in the random network only 27% of the nodes are reached by the five most connected nodes, we reach more than 60% of the nodes in the scale-free network, demonstrating the key role played by the hubs. Note that the networks in **D** and **E** consist of the same number of nodes and links. Panel **F** demonstrates that the standard clustering algorithms are not that successful in uncovering the modular structure of a scale-free hierarchical network. For color reproduction, see Color Plate 189.2.

nodes, often called *hubs*. To make the effect of the hubs on the network structure visible, we have colored the five nodes with largest degrees red in Fig. 189.2D and 189.2E and their nearest neighbors green. While in the ER network only 27% of the nodes are reached by the five most connected ones, we reach more than 60% of the nodes in the scale-free network, demonstrating the key role played by the hubs. Another consequence of the hub's dominance of the network topology is that scale-free networks are highly tolerant of random failures (perturbations) while being extremely sensitive to targeted attacks (23). Comparing the properties of the BA network model with those of the ER model, we note that the clustering of the BA network is larger; however, $C(k)$ is approximately constant (Fig. 189.3E), indicating the absence of a hierarchical structure.

Hierarchical Network Model

Many real networks are expected to be fundamentally modular, meaning that the network can be seamlessly partitioned into a collection of modules in which each module performs an identifiable task, separable from the function(s) of other modules (24–27). In biology, most of these networks have a scale-free degree distribution, which initially seems at odds with the notion of modularity, given that the hubs with their many links connect to a significant fraction of nodes, making it difficult to

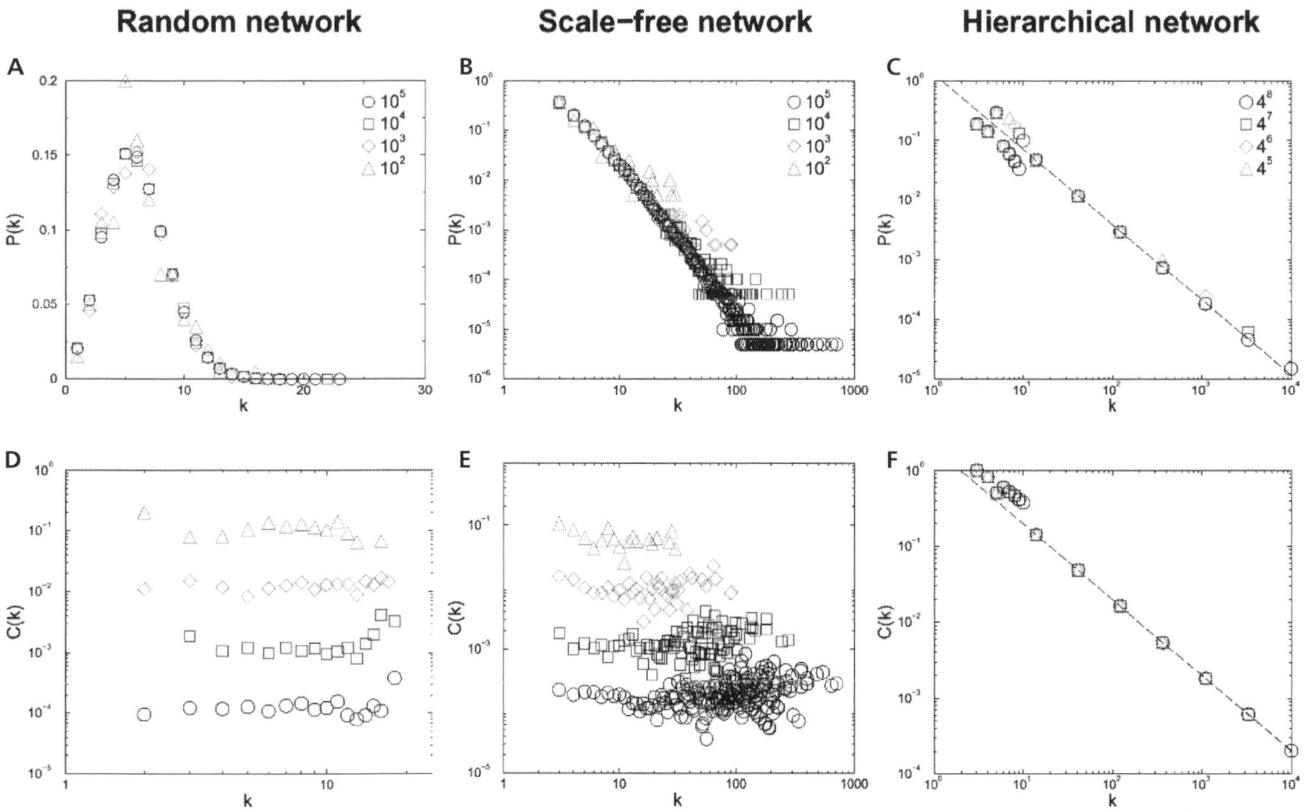

Figure 189.3. Properties of the three network models. (**A**) The ER model gives rise to a Poisson degree distribution $P(k)$ (the probability that a randomly selected node has exactly k links), which is strongly peaked at the average degree $\langle k \rangle$. The degree distributions for the scale-free (**B**) and the hierarchical (**C**) network models do not have a peak; they instead decay according to $P(k) \sim k^{-\gamma}$. The average clustering coefficient for nodes with exactly k neighbors, $C(k)$, is independent of k for both the ER (**D**) and the scale-free (**E**) network model. (**F**) In contrast, $C(k) \sim k^{-1}$ for the hierarchical network model. For color reproduction, see Color Plate 189.3.

have relatively separated modules. Therefore, we must reconcile the scale-free property with that of potential modularity. To account for the modularity as reflected in the power-law behavior of $C(k)$ and a simultaneous scale-free degree distribution, we have to assume that clusters combine in an iterative manner, generating a hierarchical network (17,20). Such a network emerges from a repeated duplication and integration process of clustered nodes (17), which in principle can be repeated indefinitely. This process is depicted in Figure 189.2C, in which we start from a small cluster of four densely linked nodes (blue). We next generate three replicas of this hypothetical initial module (green) and connect the three external nodes of the replicated clusters to the central node of the old cluster, thus obtaining a large 16-node module. Subsequently, we again generate three replicas of this 16-node module (red), and connect the 16 peripheral nodes to the central node of the old module, obtaining a new module of 64 nodes. This hierarchical network model seamlessly integrates a scale-free topology with an inherent modular structure by generating a network that has a power-law degree distribution (Fig. 189.3C) with degree exponent $\gamma = 1 + \ln 4/\ln 3 \approx 2.26$ and a clustering coefficient $C(k)$, which proves to be dependent on k^{-1} (Fig. 189.3F). However, note that modularity does not imply clear-cut subnetworks linked in well-defined ways (17,28). In fact, the boundaries of modules are often blurred considerably (see, for example, Fig. 189.2F). Finally, note that we know examples of scale-free networks, such as the router-level Internet map or the Barabási-Albert model, that are scale-free but not hierarchical. Thus, hierarchy is an additional important property of real networks, characterizing modularity, and it is not a simple consequence of a network's scale-free nature.

NETWORK UTILIZATION

Despite their impressive successes, purely topological approaches have important intrinsic limitations. For example, the activity of the various metabolic reactions or regulatory interactions differs widely, some being highly active under most growth conditions, whereas others are switched on only for some rare environmental circumstances. Therefore, an ultimate description of cellular networks requires us to consider the intensity (i.e., strength), the direction (when applicable), and the temporal aspects of the interactions. Although, so

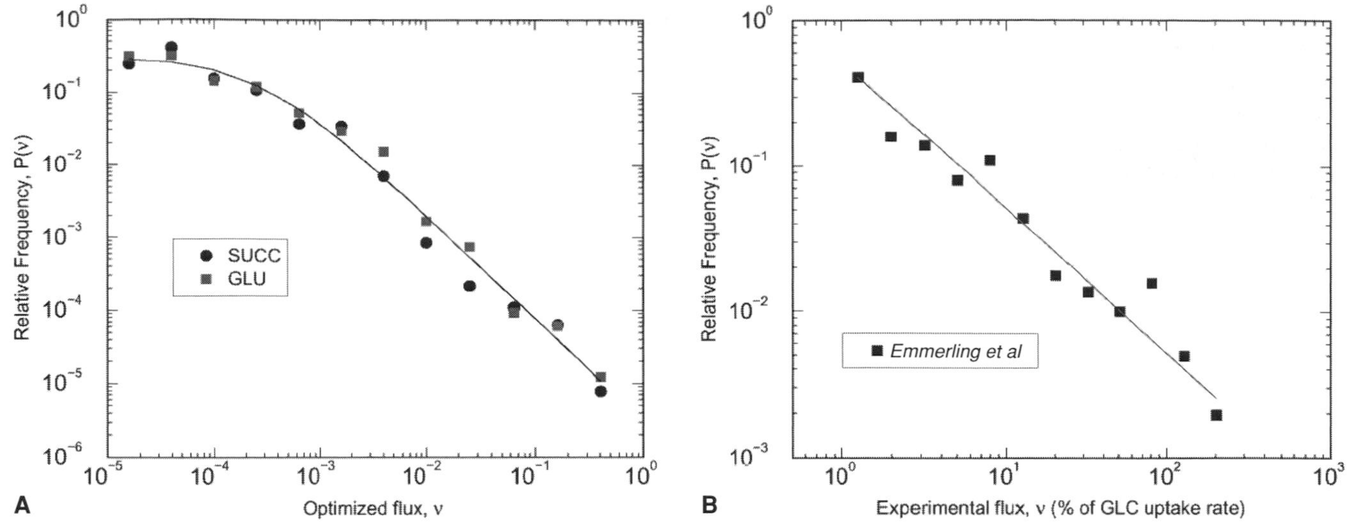

Figure 189.4. Flux distribution for the metabolism of *E. coli*. (**A**) Flux distribution for optimized biomass production on succinate- (black) and glutamate- (gray) rich uptake substrates. The solid line corresponds to the power law fit $P(\nu) \sim (\nu + \nu_0)^{-\alpha}$ with $\nu_0 = 0.0003$ and $\alpha = 1.5$. (**B**) The distribution of experimentally determined fluxes (36) from the central metabolism of *E. coli* also displays power-law behavior with a best fit to $P(\nu) \sim \nu^{-\alpha}$ with $\alpha = 1$.

far, we know little about the temporal aspects of the various cellular interactions, recent results have shed light on how the strength of the interactions is organized in metabolic and genetic-regulatory networks (29–31) and how the local network structure is correlated with these link strengths.

Flux Utilization

In metabolic networks, the flux of a given metabolic reaction, representing the amount of substrate being converted to a product within unit time, offers the best measure of interaction strength. Recent advances in metabolic flux-balance approaches (FBA) (32–35) allow us to calculate the flux for each reaction, and they have significantly improved our ability to generate quantitative predictions on the relative importance of the various reactions, thus leading to experimentally testable hypotheses. The FBA approaches can be described as follows: Starting from a stoichiometric matrix model of an organism, for example, one for *Escherichia coli* containing 537 metabolites and 739 reactions (32–34), the steady-state concentrations of all metabolites must satisfy

$$\frac{d}{dt}[A_i] = \sum_j S_{ij}\nu_j = 0 \qquad \text{(Equation 5)}$$

where S_{ij} is the stoichiometric coefficient of metabolite A_i in reaction *j*, and ν_j is the flux of reaction *j*. We use the convention that if metabolite A_i is a substrate (product) in reaction *j*, $S_{ij} < 0$ ($S_{ij} > 0$), and we constrain all fluxes to be positive by dividing each reversible reaction into two "forward" reactions with positive fluxes. Any vector of positive fluxes $\{\nu_j\}$ that satisfies Equation 5 corresponds to a state of the metabolic network, and hence, a potential state of operation of the cell.

Assuming that the cellular metabolism is in a steady state and optimized for the maximal growth rate (33,34), FBA allows us to calculate the flux for each reaction using linear optimization, providing a measure of each reaction's relative activity (29). A striking feature of the resulting flux distribution from such modeling of *Helicobacter pylori, E. coli*, and *S. cerevisiae* is its overall inhomogeneity: Reactions with fluxes spanning several orders of magnitude coexist under the same conditions (Fig. 189.4A). This is captured by the flux distribution for *E. coli*, which follows a power law in which the probability that a reaction has flux ν is given by $P(\nu) \sim (\nu + \nu_0)^{-\alpha}$. This flux exponent is predicted to be $\alpha = 1.5$ by FBA methods (29). In a recent experiment (36), the strength of the various fluxes of the *E. coli* central metabolism was measured, revealing (29) the power-law flux dependence $P(\nu) \sim \nu^{-\alpha}$ with $\alpha \cong 1$ (Fig. 189.4B). This power-law behavior indicates that the vast majority of reactions have quite small fluxes, while coexisting with a few reactions with extremely large flux values.

The observed flux distribution is compatible with two quite different potential *local* flux structures (29). A homogeneous local organization would imply that all reactions producing (consuming) a given metabolite have comparable fluxes. On the other hand, a more delocalized "hot backbone" is expected if the local flux organization is heterogeneous, such that each metabolite has a dominant source (consuming) reaction. To distinguish between these two scenarios for each metabolite *i* produced (consumed) by *k* reactions, we define the measure (37,38)

$$Y(k, i) = \sum_{j=1}^{k} \left(\frac{\hat{\nu}_{ij}}{\sum_{l=1}^{k} \hat{\nu}_{ij}} \right)^2 \qquad \text{(Equation 6)}$$

where $\hat{v}_{ij}$ is the mass carried by reaction j, which produces (consumes) metabolite i. If all reactions producing (consuming) metabolite i have comparable $\hat{v}_{ij}$ values, $Y(k, i)$ scales as $1/k$. If, however, a single reaction's activity dominates Equation 6, we expect $Y(k, i) \sim 1$; that is, $Y(k, i)$ is independent of k. For the *E. coli* metabolism optimized for succinate and glucose uptake, we find that both the *in* and *out* degrees follow the power law $Y(k, i) \sim k^{-0.27}$, representing an intermediate behavior between the two extreme cases (29). This suggests that the large-scale inhomogeneity observed in the overall flux distribution is increasingly valid at the level of the individual metabolites as well: For most metabolites, a single reaction carries the majority of the flux. Hence, the majority of the metabolic flux is carried along linear pathways instead of branching pathways. We call the set of such linear pathways the metabolic high flux backbone (HFB) (29).

Gene Interactions

One can also investigate the strength of the various genetic regulatory interactions provided by microarray data sets. Assigning each pair of genes a correlation coefficient that captures the degree to which they are coexpressed, one finds that the distribution of these pair-wise correlation coefficients follows a power law (30,31). That is, while the majority of gene pairs have only weak correlations, a few gene pairs display a significant correlation coefficient. These highly correlated pairs likely correspond to direct regulatory and protein interactions. This hypothesis is supported by the finding that the correlations are larger along the links of the protein interaction network and between proteins occurring in the same complex, than they are for pairs of proteins that are not known to interact directly (39–42).

Taken together, these results indicate that the biochemical activity in both the metabolic and genetic networks is dominated by several "hot links" that represent a few high-activity interactions embedded into a web of less active interactions. This attribute does not seem to be a unique feature of biological systems: Hot links appear in a wide range of nonbiological networks in which the activity of the links follows a wide distribution (43,44). The origin of this seemingly universal property is, again, likely rooted in the network topology. Indeed, it seems that the metabolic fluxes and the weights of the links in some nonbiological system (43,44) are uniquely determined by the scale-free nature of the network. A more general principle that could explain the correlation distribution data is currently lacking.

CONCLUSION

Power laws are abundant in nature, affecting both the construction and utilization of real networks. The power-law degree distribution has become the trademark of scale-free networks and can be explained by invoking the principles of network growth and preferential attachment. However, many biological networks are inherently modular, a fact which at first seems to be at odds with the properties of scale-free networks. However, these two concepts can coexist in hierarchical scale-free networks. In the utilization of complex networks, most links represent disparate connection strengths or transportation thresholds. For the metabolic network of *E. coli*, we can implement a flux-balance approach and calculate the distribution of link weights (fluxes), which (reflecting the scale-free network topology) displays a robust power law, independent of exocellular perturbations. Furthermore, this global inhomogeneity in link strengths is also present at the local level, resulting in a connected "hot-spot" backbone of the metabolism. Similar features also are observed in the strength of various genetic regulatory interactions.

The endothelium, consisting of cells interacting through a network with a highly complex topology, readily lends itself to a network description on multiple levels. On the level of the individual cell, the utilization and structure of the regulatory and metabolic networks is dynamically adjusting to changes in the local environment. These networks represent the topic of this chapter. Yet, the cells do not exist in isolation, but constantly integrate cellular signals from their neighborhoods to produce coherent responses at the level of multiple cells, thus creating intercellular communication networks. At this point, much less is known about the structure of these networks. We can continue this network description all the way up to the organism level. Because the endothelium is part of a fractal-like vascular architecture, we must take into account the constraints imposed by this physical layout and develop new models that properly incorporate the three-dimensional embedding. To develop realistic network models of the endothelium, it is necessary to further develop our understanding of how different networks dynamically interact and integrate information. Despite the significant advances witnessed during the last few years, network biology is still in its infancy, with future advances most notably expected from the development of theoretical tools, development of new interactive databases, and increased insights into the interplay between biological function and topology.

KEY POINTS

- Important insights into the structure and function of complex systems can be gained by using a network description.
- The degree distribution of many biological networks, most notably the protein interactions and metabolism, is scale-free.
- The structure of biological networks strongly influences how they are being used: The metabolic fluxes follow a power law, and the mass predominantly flows along linear pathways.

REFERENCES

1 Strogatz SH. Exploring complex networks. *Nature*. 2001;410: 268–276.
2 Albert R, Barabási AL. Statistical mechanics of complex networks. *Rev Mod Phys*. 2002;74:47–97.
3 Dorogovtsev SN, Mendes JFF. *Evolution of Networks: From Biological Nets to the Internet and WWW*. Oxford, England: Oxford University Press, 2003.
4 Bornholdt S, Schuster HG. *Handbook of Graphs and Networks: From the Genome to the Internet*. Berlin, Germany: Wiley-VCH, 2003.
5 Faloutsos M, Faloutsos P, Faloutsos C. On power-law relationships of the internet topology. *Comput Commun Rev*. 1999;29: 251–262.
6 Albert R, Jeong H, Barabási AL. Diameter of the World-Wide Web. *Nature*. 1999;401:130–131.
7 Milgram S. The small-world problem. *Psychol Today*. 1967;2: 60–67.
8 Kochen M, ed. *The Small-World*. Norwood, New Jersey: Ablex, 1989.
9 Wasserman S, Faust K. *Social Network Analysis: Methods and Applications*. Cambridge, New York: Cambridge University Press, 1994.
10 Liljeros F, Edling CR, Amaral LAN, et al. The web of human sexual contacts. *Nature*. 2001;411:907–908.
11 Newman MEJ. The structure of scientific collaboration networks. *Proc Natl Acad Sci USA*. 2001;98:404–409.
12 Barabási AL, Albert R. Emergence of scaling in random networks. *Science*. 1999;286:509–512.
13 Jeong H, Tombor B, Albert R, et al. The large-scale organization of metabolic networks. *Nature*. 2000;407:651–654.
14 Jeong H, Mason SP, Barabási AL, Oltvai ZN. Lethality and centrality in protein networks. *Nature*. 2001;411:41–42.
15 Montoya JM, Sole RV. Small-world patterns in food webs. *J Theor Biol*. 2002;214:405–412.
16 Watts DJ, Strogatz SH. Collective dynamics of small-world networks. *Nature*. 1998;393:440–442.
17 Ravasz E, Somera AL, Mongru DA, et al. Hierarchical organization of modularity in metabolic networks. *Science*. 2002;297: 1551–1555.
18 Ravasz E, Barabási AL. Hierarchical organization in complex networks. *Phys Rev E Stat Nonlin Soft Matter Phys*. 2003;67:026112.
19 Dorogovtsev SN, Goltsev AV, Mendes JFF. Pseudofractal scale-free web. *Phys Rev E Stat Nonlin Soft Matter Phys*. 2002;65:066122.
20 Vázquez A, Pastor-Satorras R, Vespignani A. Large-scale topological and dynamical properties of the Internet. *Phys Rev E Stat Nonlin Soft Matter Phys*. 2002;65:066130.
21 Bollobas B. *Random Graphs*. London, England: Academic Press, 1985.
22 Erdos P, Renyi A. On the evolution of random graphs. *Publ Math Inst Hung Acad Sci*. 1960;5:17–61.
23 Albert R, Jeong H, Barabási AL. Attack and error tolerance of complex networks. *Nature*. 2000;406:378–382.
24 Hartwell LH, Hopfield JJ, Leibler S, Murray AW. From molecular to modular cell biology. *Nature*. 1999;402:C47–C52.
25 Rao CV, Arkin AP. Control motifs for intracellular regulatory networks. *Annu Rev Biomed Eng*. 2001;3:391.
26 Hasty J, McMillen D, Isaacs F, Collins JJ. Computational studies of gene regulatory networks: In numero molecular biology. *Nature Rev Genet*. 2001;2:268–279.
27 Shen-Orr SS, Milo R, Mangan S, Alon U. Network motifs in the transcriptional regulation network of *Escherichia coli. Nature Genet*. 2001;31:64–68.
28 Holme P, Huss M, Jeong H. Sub-network hierarchies of biochemical pathways. *Bioinformatics*. 2003;19:532–539.
29 Almaas E, Kovacs B, Vicsek T, et al. Global organization of metabolic fluxes in the bacterium *Escherichia coli. Nature*. 2004; 427:839–843.
30 Kutznetsov VA, Knott GD, Bonner RF. General statistics of stochastic processes of gene expression in eukaryotic cells. *Genetics*. 2002;161:1321–1332.
31 Farkas IJ, Jeong H, Vicsek T, et al. The topology of the transcription regulatory network in the yeast, *Saccharomyces cerevisiae. Physica A*. 2003;318:601–612.
32 Edwards JS, Palsson BO. The *Escherichia coli* MG1655 in silico metabolic genotype: its definition, characteristics, and capabilities. *Proc Natl Acad Sci USA*. 2000;97:5528–5533.
33 Edwards JS, Ibarra RU, Palsson BO. In silico predictions of *Escherichia coli* metabolic capabilities are consistent with experimental data. *Nat Biotechnol*. 2001;19:125–130.
34 Ibarra RU, Edwards JS, Palsson BO. *Escherichia coli* K-12 undergoes adaptive evolution to achieve in silico predicted optimal growth. *Nature*. 2002;420:186–189.
35 Segre D, Vitkup D, Church GM. Analysis of optimality in natural and perturbed metabolic networks. *Proc Natl Acad Sci USA*. 2002;99:15112–15117.
36 Emmerling M, Dauner M, Ponti A, et al. Metabolic flux responses to pyruvate kinase knockout in *Escherichia coli. J Bacteriol*. 2002; 184:152–164.
37 Barthelemy M, Gondran B, Guichard E. Spatial structure of the Internet traffic. *Physica A*. 2003;319:633–642.
38 Derrida B, Flyvbjerg H. Statistical properties of randomly broken objects and of multivalley structures in disordered-systems. *J Phys A Math Gen*. 1987;20:5273–5288.
39 Dezso Z, Oltvai ZN, Barabási AL. Bioinformatics analysis of experimentally determined protein complexes in the yeast, *Saccharomyces cerevisiae. Genome Res*. 2003;13:2450–2454.
40 Grogoriev A. A relationship between gene expression and protein interactions on the proteome scale: analysis of the bacteriophage T7 and yeast *Saccharomyces cerevisiae. Nucleic Acids Res*. 2001;29:3513–3519.
41 Jansen R, Greenbaum D, Gerstein M. Relating whole-genome expression data with protein-protein interactions. *Genome Res*. 2002;12:37–46.
42 Ge H, Liu Z, Church GM, Vidal M. Correlation between transcriptome and interactome mapping data from *Saccharomyces cerevisiae. Nat Genet*. 2001;29:482–486.
43 Goh KI, Kahng B, Kim D. Fluctuation-driven dynamics of the Internet topology. *Phys Rev Lett*. 2002;88:108701.
44 deMenezes MA, Barabási AL. Fluctuations in network dynamics. *Phys Rev Lett*. 2004;92:028701.

190

Cell Fates as Attractors

Stability and Flexibility of Cellular Phenotypes

Sui Huang

Children's Hospital, Harvard Medical School, Boston, Massachusetts

The vascular endothelial cell (EC) is one of perhaps thousands of "cell types" in the body of higher vertebrates for which a definition as a distinct type is fairly straightforward. A vascular EC is naturally defined by its distinct anatomical and physiological attributes as the cell lining the blood vessels and in contact with circulating blood. In general, we do not yet have a handle for a precise, systematic, and comprehensive taxonomy for all the cell types in the body. Nevertheless, in molecular terms, it appears that a cell type can be characterized by its distinct transcriptome and proteome – that is, by a genome–scale combinatorial code of the expression status of tens of thousands of genes, akin to a very long bar code. Currently, DNA microarray-based gene expression profiling at the level of mRNA abundance and subsequent cluster analysis of the high-dimensional set of molecular attributes reveal highly robust expression profiles that can readily be grouped into similarity "clusters" that correspond to the individual tissues or cell types as defined by physiology and histology (1,2).

Morphologically and at the molecular level, the EC cell type consists of an entire class of distinct subtypes defined by their location in the hierarchy of the vasculature (arteries, arterioles, capillaries, venules, and veins) as well as by tissue bed (e.g., brain versus dermal capillary ECs). In addition to this apparently static variation that is manifest as the stable subtypes, gene expression patterns in an EC also vary with the dynamics of functional states, such as the switching from a quiescent to "activated" state during inflammation, or to the proliferating and migrating state during angiogenesis. Importantly, the heterogeneity of the EC phenotype also is reflected in characteristic similarity clusters of gene expression profiles, suggesting the existence of distinct subtypes that represent phenotypic entities (3–5).

Of the 25,000 or so genes in the human genome, perhaps 10,000 are expressed in a given cell type. Only a fraction of them are unique – that is, specific for a given cell type. In the case of ECs, roughly 100 to 400 genes can be called "EC-specific" based on expression at the mRNA level (5). (These numbers are to be taken *cum grano salis*, since they depend on measurement sensitivity and threshold definition.) The subtype identity is in part reflected in the relative expression of these EC specific genes – but also in the expression patterns of nonendothelial genes (4,5). Despite the considerable overlap of gene expression profiles *between* and the spatiotemporal heterogeneity *within* one cell type, each phenotypically identifiable cell type nevertheless forms a recognizable cluster of similarity with respect to the high-dimensional set of variables provided by gene expression profiles (1,2). Hence, the heterogeneity of gene expression within a cell type is due to variability at a hierarchically finer scale, so that one could postulate a taxonomical hierarchy that extends over multiple ontological levels.

How can we describe this intra–cell type heterogeneity within the framework of stable cell types? The static variability of a cellular phenotype may correspond to the "nature" aspect of a cell phenotype, to use the terminology proposed by William Aird (6), and to its associated gene expression profile, since it is "epigenetically fixed." In contrast, the dynamic variability may reflect the "nurture" aspect, since changes in the microenvironment in part account for additional variation in the transcriptome. However, it is important to note that this terminological distinction is not a black-and-white scheme, but depends on the height of the "epigenetic barriers": To what extent can an epigenetically fixed phenotype change? The most extreme case of "jumping" epigenetic barriers is the rare event of transdifferentiation between quite unrelated types, best epitomized in the epithelial–mesenchymal transition under highly specific conditions.

This question leads us back to the more fundamental question: What is a cell type? Because it is ultimately defined by a particular, characteristic gene expression profile, and because (virtually) all cells in the body harbor the same set of genes, why then don't we see a continuum of gene expression profiles in the vast "universe" of all combinatorially possible expression profiles, much as a bar code can encode any item of a given inventory? In other words, why can cells not smoothly "morph" into any phenotype by gradually changing the expression of individual genes so as to match any desired expression pattern? Instead, we find discrete groups of cell types that form clearly separable entities. Shifts of expression profiles do in fact happen – but mostly within some epigenetic barriers that ensure that the cell maintains its type identity. This confined variability of gene expression profiles produces the *within* cell type–heterogeneity. Concretely, despite the heterogeneity of ECs reflected in the variations of gene expression profiles caused by environmental influences ("nurture"), the EC's very "nature" is usually not challenged by these external perturbations: An EC remains recognizable as such under a wide variety of conditions, including cell culture.

Thus, an operationally appropriate question is: What is the very nature of the "epigenetic barriers" that separate the combinatorially possible expression profiles into classes of distinct, discrete, and robust cell phenotypes, and hence, define the "nature" as opposed to "nurture" in controlling the gene expression patterns? The purpose of this chapter is to help formalize this question in more precise terms by presenting a conceptual framework in which cell types represent "attractors" that can be imagined as valleys in the landscape of all possible gene expression profiles. Specifically, we will discuss concepts derived from system dynamics that explain how the gene regulatory network imposes constraints on the dynamics of gene expression profiles as a whole, so as to govern whole-cell behavior and produce the stable, yet flexible molecular definitions of cell types. But first, in the next section, we will discuss three philosophical problems of biological explanation that rely on the conventional paradigm of molecular pathways.

CELL FATE DYNAMICS AND THE PROBLEM OF MOLECULAR EXPLANATION

Molecular Chains of Causality: Who Controls the Controller?

Over the past decades, the focus on individual molecular pathways as a scheme to explain a particular cellular process has shifted the attention away from the actual system-level dynamics of cell-fate regulation. At this level of whole-cell behavior, a given cell in the body is limited to occupy a finite and relatively small number of accessible cell fates, such as proliferation, apoptosis, senescence, quiescence, or differentiation into a particular cell type (Figure 190.1A). The latter, of course, is the cell-fate decision faced by multipotential progenitor or stem cells. The embryologist Conrad H. Waddington recognized this particular dynamic in the 1940s, and proposed the "epigenetic landscape" (Figure 190.1B) as an intuitive metaphor to explain the natural discreteness of cell phenotypes, the switch-like fate transitions and the instability of "intermediate types" (for other modern metaphors of the endothelium, see Chapters 24–26). In this metaphor, the valleys represent stable cell fates. A prosaic manifestation of discreteness that is often taken for granted is the ease with which we detect distinct subpopulations of cells when monitoring the expression of a cell-fate marker, for example, a protein X, in a cell population using flow

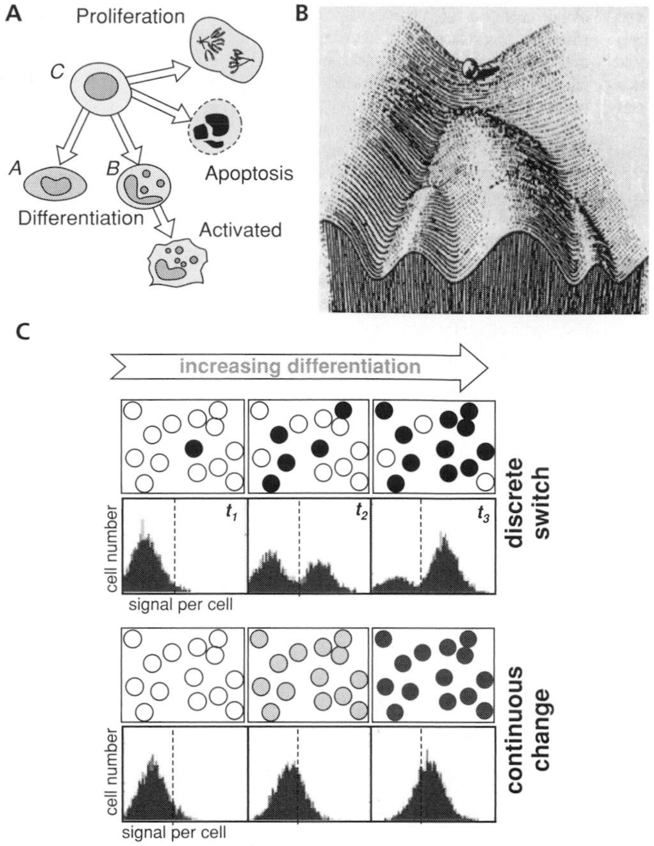

Figure 190.1. Dynamics of cell fate behavior. (**A**) Schematic example of cell-fate options and transitions available to a cell. Note the "discreteness" of alternative fates. (**B**) Waddington's "epigenetic landscape," a metaphor proposed to capture the discreteness of cell-fate regulation. (Reprinted with permission from Waddington CF. *The Strategy of Genes.* New York: Allen & Unwin, 1957.) (**C**) Schematic illustration for two fundamentally different ways of how a cell population apparently gradually increases the expression of a differentiation marker over time: discrete switching or gradual expression at single cell level. Circles represent cells, color of cell represent expression level of the differentiation marker. Computer-simulated flow cytometry histograms are shown. Note the bimodal distribution in the case of discrete switching – this is more frequently observed.

cytometry (Figure 190.1C). As cells differentiate, a *bimodal* distribution of expression of the differentiation marker X indicates that a differentiating cell population consists of two distinct subpopulations, X^{low} and X^{high} cells, representing undifferentiated and differentiated cells. The general experience of monitoring differentiation markers in flow cytometry is that individual cells "jump" from one subpopulation to another when cells differentiate, as manifested by the shift in the size of the two subpopulations or peaks in the bimodal distribution of the flow cytometry histogram (Figure 190.1C). The alternative mode of change would be a gradual increase of expression level of X, which would be presented as a shift of one single (*unimodal*) population from one position to another (bottom of Figure 190.1C). However, such gradual change is almost never observed. (Note that biochemical methods for measuring the expression of *X*, such as immunoblots, measure population averages, and will suggest a gradual response due to noise). Later in this chapter we will explain the discrete behavior of cell-fate switching based on the dynamics of gene networks.

Decades of a pathway-centered view in molecular biology have produced a particular scheme of explanation, which is based on cascades of molecular events and to which we have become so habituated that it remains unquestioned. Thus, a liver cell–specific gene is thought to be expressed in the liver because it is regulated by organ-specific transcription factors that are present in the liver, such as hepatocyte nuclear factors (HNFs) (7). Such "master genes" control the expression of genes that are considered to be part of the liver-specific "genetic program." But who controls the controller? Why are the HNFs expressed in the liver? Obviously, because of another transcription factor acting further upstream of it. This scheme of so-called *proximal explanation* only shifts the true *causa rerum* out of sight, out of mind. Well-studied examples of such "key regulators" include MyoD for muscle cells, peroxisome proliferator-activated receptor (PPAR)γ for adipocytes, GATA-1 for the erythroid cells, or PU.1 for myeloid cells. These gene products appear to play a role both in the actual dynamics of cell-fate switching as well as in controlling the activation of downstream effector genes that execute the functions of the differentiated state, for example, hemoglobin gene expression driven by the transcription factor GATA-1. In the case of ECs, no such master gene has been unambiguously identified to date. Perhaps the difficulty in identifying specific master control switches in certain cell types is due to "distributed control" of gene expression, in that a network of multiple genes jointly control a particular gene expression pattern, as will be discussed below using a two-gene network.

A similarly unsatisfactory explanation is that the stability of gene expression programs associated with a cell type – the "memory" of lineage identity – is maintained by covalent chromatin modification, such as acetylation or methylation of histones and by direct methylation of DNA (7–9). These modifications regulate chromatin remodeling and hence accessibility of genomic DNA for the transcriptional machinery, thus controlling gene expression. Although the general notion is that these covalent modifications, commonly referred to by molecular biologists as *epigenetic mechanisms*, are capable of maintaining cell type–specific gene expression programs, this explanation does not stand rigorous epistemological scrutiny: There is nothing particularly enduring about covalent bonds in biology because they are subjected to both enzymatic formation and breakage, which makes the thermodynamic stability of covalent bonds irrelevant. Indeed, the histone-modifying enzymes are part of a highly dynamic system consisting of a sophisticated balance between reactions in both directions: histone acetyltransferases versus deacetylases, or methyltransferases versus dimethylases, and so on (10). Moreover, these enzymes act in tight association with the locus-specific transcription factors to control differentiation (11,12). So, if all is dynamic, where does the stability of cell type–specific programs of gene expression come from? To theoretical biologists, the term *epigenetic* has a much broader meaning than merely covalent modification of chromatin or DNA: It encompasses all the dynamic principles that can store information based on gene interactions, as we will see in the next section.

Taken together, what is needed is a closed loop, self-consistent explanation that involves the genome-wide, complex networks of genes that regulate each other. Such an integrating concept should help us understand the governing principles underlying the global dynamics of gene expression profiles and would explain why cells get "stuck" in stable profiles that define the phenotypic cell types.

How Do Signal Transduction Pathways Coordinate Global Expression Changes?

A second problem of explanation is that our knowledge of the very fundamental organizing principles that govern the transition between cell fates, such as a switch from a proliferative state (e.g., progenitor cell) to the terminally differentiated state, is fairly limited. The current pathway-centered approach focuses on specific factors as inputs that trigger such transitions. These inputs may be such agents as growth factors or cytokines, which bind to their cognate cellular receptors and elicit signaling cascades that end in the nucleus to regulate the expression of genes to implement the output. Although this explanatory scheme has promoted the elucidation of many useful mechanistic details of signal transduction cascades, it faces multiple challenges when placed in the broader context of complex molecular networks: It does not explain how cells seamlessly compute all the pleiotropic changes triggered by the fanning-out signaling cascades, yet still maintain signal specificity that allows the cell to switch to a new cell fate by inducing the concerted changes in the activity of thousands of genes, precisely so as to produce the new gene expression profile associated with the new fate. In other words – to use the metaphor discussed in Chapter 24 – we still don't know how the input is mapped to the output of coherent cell behavior.

The challenge is even greater if we consider the larger picture of cell-fate dynamics: Despite the astronomical number of theoretically possible gene activation configurations in the universe of expression profiles, cells are able to robustly integrate multiple simultaneous, noisy, and often conflicting signals to which they are exposed and which affect scores of genes across the genome. Their response is then the unambiguous adoption of one of just a few alternative cell fates. As a consequence of this limited number of fate choices, the very same cell fate can be triggered by a broad variety of unrelated, often noisy signals, including those lacking molecular specificity, such as mechanical forces, as discussed later (13,14). Conversely, the very same signal can induce different, even opposite cell fates (such as proliferation versus apoptosis), depending on the biochemical or physical state of the cell (15). The mapping of an input to a coherent, whole-cell behavior as the output will require the coordination of thousands of parallel molecular events, in a kind of "distributed information processing." This processing is not easily accommodated by the traditional picture of individual signaling pathways, which are thought to embody a point-to-point chain of causation.

The distributed cellular information processing is well illustrated in the finding that in ECs (and probably other cells) the shape of the cell, a physical signal devoid of molecular specificity, acts as a particular input. Specifically, the state of cell spreading is a critical regulator of cell fates, in addition to the classical inputs, such as soluble growth factors and the extracellular matrix (ECM) (16). It makes sense that cell spreading is a necessary condition for entry into the cell division cycle: During angiogenesis, a cell at the sprouting tip of capillaries spreads on newly formed ECM. This cell shape change, which occurs only in cells that can extend to the new matrix, stimulates cell proliferation, whereas immediately adjacent cells in the densely packed monolayer of the vessel wall will remain quiescent despite receiving the same inputs in terms of soluble angiogenic factors. On the other hand, if a cell is spatially compressed to a small area, then, despite the presence of growth factors, it will undergo apoptosis (16,17). This regulatory mechanism that permits an individual cell to survive and divide only if it can stretch out implements an organizing principle that is known from automaton models: The behavior of a single cell as an "agent" is determined only by its local neighborhood and some rules. But when implemented by every cell of a population, a collective behavior arises, such as tissue homeostasis, which, in this case, is the maintenance of an intact endothelial layer.

Molecularly, it is assumed that, during cell spreading, the distortion of the actin cytoskeleton, which forms a mechanically continuous, force-bearing system between cell and ECM through focal adhesions, will cause rearrangements in these adhesions. Because the latter provide the scaffold for orienting arrays of signaling complexes, they could simultaneously modulate the signal transduction of multiple pathways. Again, somehow, the distributed signals are coordinated such as to produce a coherent, mitogenic stimulus to the cell. In other words, the complex apparatus of cell-fate regulation has evolved to harness the forces exerted by a physical input that lacks molecular specificity and to use it as a signal for switching between cell fates. The concept of attractors (presented later in this chapter) provides an explanation for how distributed signals are spontaneously channeled through a complex molecular machinery to produce a specific cell-fate program.

The Numerical Problem: A Huge Universe of Expression Profiles

The third problem of explanation is more abstract and fundamental. To demonstrate and understand elementary principles, it is often necessary to make (temporarily) simplifying assumptions. Hence, let us assume here that each of the biologically observable "macroscopic" cell fates is characterized by a distinct profile of gene expression, and that each gene locus has a "micro-state" that can either be active – that is, the gene is expressed (status ON $= 1$) or silenced (status OFF $= 0$). Also, let us for now assume a genome consisting of only of $N = 2$ genes, labeled X and Y. Hence, we would have $2^2 = 4$ gene activation configurations, or "macro-states" $S = (X,Y)$, each defined by the combination of micro-states of X and Y: (0,0), (1,0), (0,1), and (1,1). All four gene expression profile $S = (X,Y)$ could be mapped as one point into a two-dimensional (2D) *state space*, spanned by the axes X and Y. The state space hence represents the aforementioned "universe" of all thinkable states of the cell as defined by gene expression profiles. In this case, it contains only four states or gene expression profiles. In reality, the human genome has $N = 25{,}000$ genes, according to the last count (18). Hence, assuming again ON–OFF genes, we would have $2^{25000} = 10^{7520}$ possible genome-wide configurations of gene activation (macro-states) in a state space that now has N dimensions – one axis for each gene. Despite our simplification, which omits gradual activation levels, post-transcriptional, post-translational modifications, and the like, the number of 10^{7520} possible gene expression profiles is hyperastronomic: There are approximately 10^{80} protons in the universe, and there are approximately 10^{17} cells in our body. Assuming that each expression profile encodes a particular phenotypic cell fate, we would have an almost endless, quasicontinuous universe of cell phenotypes if all these gene activation configurations would be realizable. As discussed in the introduction, the cell could move in the N dimensional state space and *gradually* morph from one phenotype to another. Many cell types would be very similar to each other; for example, two expression profiles that differ by only the gene expression micro-state of one of N genes. The biological reality is that we have *discrete* cell phenotypes, with almost discontinuous switches between them (Figure 190.1B–C). In other words, most of the state space of expression profiles goes unused.

The obvious explanation is that not all combinatorial gene activation configurations of the genome are logically possible because of gene regulatory interactions: Genes do not change their activation status independently. Take the above two-gene example: If gene X inhibits gene Y, then the state $(X,Y) = (1,1)$ would not be "logically" allowed. It is physically unstable and will transition into the state $(X,Y) = (1,0)$ that complies with the regulatory interactions. The cell is said to "move" in state space along a *trajectory* from state (1,1) to state (1,0). This is still a hand-waving explanation, but a step toward a formal concept: A network of gene regulatory interactions enforces some dynamics upon the global gene expression profile, so that the cell (gene network) cannot occupy any point in the state space, but rather is forced to move along trajectories to resolve the conflicts arising from the regulatory interactions.

In this section we have discussed three problems that challenge current molecular biology thinking. In the next section we will show, using a two-gene circuitry, that abstraction and conceptualization are helpful and could explain in a more formal way how one system can produce multiple, in this case two or three, stable network states that correspond to discrete cell fates.

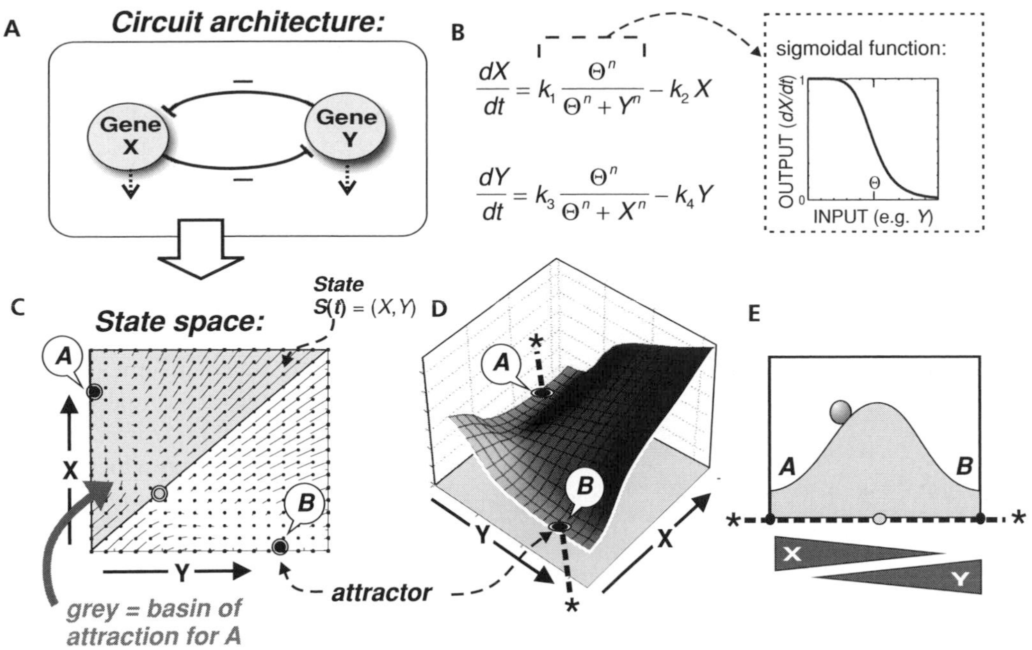

$$\frac{dX}{dt} = k_1 \frac{\Theta^n}{\Theta^n + Y^n} - k_2 X$$

$$\frac{dY}{dt} = k_3 \frac{\Theta^n}{\Theta^n + X^n} - k_4 Y$$

Figure 190.2. Bistability in two-gene circuit of mutual inhibition. (**A**) The circuit architecture of two-gene circuit as a didactical example, with mutual inhibition of the two genes/proteins, *X* and *Y*; both are also subjected to first-order inactivation (*dashed arrow*). (**B**) Set of ordinary differential equations for the activity level of the variables *X* and *Y*. *Inset*: Inverse sigmoidal curve illustrating the "INPUT–OUTPUT" relationship, as expressed in the first term of the equation, determining how the input *Y* affects the output *dX/dt*: The higher the activity *Y*, the smaller is the synthesis rate of *X*, *dX/dt*. (**C**) Two-dimensional state space (phase plane) for *X*, *Y*. Each point in the *XY* plane is a state $S(t) = [X(t), Y(t)]$. The "ticks" emanating from each dot (representing initial states) illustrate a fragment of the trajectory that the respective state would travel during Δt, collectively giving rise to a "vector field." The two large black dots, *A* and *B*, are the attractor states. The gray dot represents an unstable stationary state. The *basin of attraction* for attractor state *A* is shown as the gray triangular region of the phase plane. (**D**) Schematic representation of the state space as a landscape. The vertical axes represent some "potential function" reflecting the "flow" in the vector field (see text for details). (**E**) Another schematic view of the state space, projected as a simple "hill" corresponding to the cross-section of the state space in **D** through the dashed line. The horizontal axis reflects the relative activity of *X* and *Y*.

MULTISTABILITY AND ATTRACTORS IN GENE REGULATORY NETWORKS

A Two-Gene Circuit as Toy Example

The goal of this section is to explain how a simple system can capture the essential macroscopic dynamics of cell-fate transitions; specifically, how a progenitor cell *C* can differentiate into either one of two distinct cell types, *A* and *B*. This behavior captures the general features of cell-fate dynamics with discrete, mutually exclusive fate decisions and all-or-none transitions between these discrete cell phenotypes. Figure 190.2A shows a circuitry consisting of two genes, *X* and *Y*, a mini-network so to speak. The circuitry consists of the following interactions: The two genes inhibit each other's basal activity (e.g., gene expression or protein synthesis). Moreover, each gene spontaneously loses activity (e.g., due to decay of the protein). Unlike the example discussed earlier, the genes now can continuously change their values of activation (e.g., expression level) over time *t*. We define again the *state variables* $X(t)$ and $Y(t)$. Then, as used earlier, the dynamics of the system can be mathematically described by the *state vector* $S(t)$, which defines a position in the *state space* spanned by the two axes *X* and *Y* for each time point: $S(t) = [X(t), Y(t)]$, and hence represents a *system state S* at time *t*. To mathematically describe the continuous movement of *S* in this state space driven by the gene regulatory interactions, one describes the *change* of *X* and *Y* over time; that is, the rates *dX/dt* and *dY/dt*. The *differential dX* is the infinitesimally small difference in *X* by which *X* changes in the infinitesimally small time window *dt*.

Mathematical modeling here is all about how to describe the rate of changes in the state variables *X* and *Y* – that is, *dX/dt* and *dY/dt* – by an expressions that involves $X(t)$ and $Y(t)$ themselves, since obviously, their values influence the changes as laid out in the circuit diagram. The resulting system equations of our two-gene system represent a set of so-called *nonlinear differentiation equations*, as shown in Figure 190.2B. Such modeling, based on differential equations describing the continuous changes of variables, is discussed in detail in

Chapter 191. The goal of solving these differential equations is to find out the behavior of *S in the state space* – how X(t) and Y(t) change over time, since $S(t) = [X(t), Y(t)]$, and to evaluate the long-term behavior.

So, how can we model the regulatory influence exerted by the current level of gene expression, $X(t)$ or $Y(t)$, on the rates dY/dt or dX/dt, respectively? An essential point in establishing the differential equations for such gene regulatory systems is to model the regulatory influences as a *sigmoidal* input/output (stimulus/response) relationship (see inset in Figure 190.2B, *right*), which for example, can be written as: $dX/dt = 1/(\Theta^n + Y^n)$ for the influence of Y on the change of X. This equation captures the inverse sigmoidal relationship, in that increasing levels of Y (the input) would decrease the activation rate of X, dX/dt (the output). Θ and n are constants – Θ represents the threshold Y value in the sigmoidal curve, and n its steepness (Hill coefficient). Such sigmoidal "transfer functions" reflect *sensitivity amplification* (19), which is warranted, even in the absence of cooperativity (the best known cause of such sigmoidality), on grounds of the particular physicochemical conditions of the intracellular milieu that departs from that of ideal, well-stirred macroscopic solutions (20–22). The sigmoidality of this regulatory influence introduces the "nonlinearity" aspect and is crucial for the generation of discrete behavior.

For the spontaneous decay term, each gene (or its encoded protein) is modeled as being subjected to a *first-order* inactivation (degradation), in that the inactivation rate is proportional (linear) to the current amount of $X(t)$ or $Y(t)$, respectively, that is present. Thus, the term $-kX$ in the equation describes the inactivation of X at a rate proportional to the level of $X(t)$.

The circuit architecture of the type shown in Figure 190.2A with mutual, nonlinear regulation, either reciprocal inhibition or activation, is in fact very common among transcription factors critically involved in cell-fate decisions (23–28). In ECs, Id and Egr-1 appear to form such a feedback loop of mutual activation (29). We now discuss the solution of the equations – the dynamic behavior of this system, $S(t) = [X(t), Y(t)]$ – without going into the detailed mathematics.

The Dynamic Behavior: Multistability and Attractors

In our example, the two genes, X and Y, inhibit each other (Figure 190.2A). Thus, with the experience of our ON–OFF system discussed earlier, one would intuitively expect two possible stable *steady-states*, in which either X dominates (state A with $X >> Y$) or Y dominates (state B, with $Y >> X$). Because of the interaction in our example, not every state S, defined by a pair of values for X and Y, $S = (X,Y)$, is stationary and stable: Due to mutual (symmetrical) inhibition, if there is just slightly more Y than X, then Y would suppress X (more than X suppresses Y) and Y will increase until it reaches an equilibrium with its own decay that scales with the higher values of Y because of the first-order kinetics of decay. X similarly would decrease to an equilibrium between its activation and deactivation rate. Mathematically, to find the steady states of the system – the states in which X and Y do not change – one sets dX/dt = rate of change = 0 and dY/dt = 0 and solves the equations algebraically for X and Y. Doing so will find three *stationary* states, $A = (X_A, Y_A)$, $B = (X_B, Y_B)$, and $C = (X_C, Y_C)$.

The stationary states A, B, and C are shown as circles in the 2D state space spanned by X and Y in Figure 190.2C, which represents the 2D state space, or *XY-phase plane*, in which each point represent a state $S(t) = [X(t),Y(t)]$. We can now pick any state S around the steady-states and, using the differential equations, calculate how these arbitrary states will "move" by a small step ΔX and ΔY in both state space dimensions during a short time period Δt along a trajectory. This movement of S is driven by the interactions of the circuitry captured in the differential equations. Similar to a night photograph snapshot of cars moving on a road, in which the traces of the headlights represent the motion of cars during the exposure time Δt, we can draw trajectory fragments in the state space for various points S to obtain a "vector field" indicating where the states would move in Δt. As shown in Figure 190.2C, we see that the two points A and B in fact "attract" trajectories from starting points in a certain region in their neighborhood, confirming that these steady states are stable. Stability of a state thus means that, upon small perturbations that displace that state to a spot within its neighborhood in state space, the displaced state will fall back to that original state. In our case, the system has two stable states, A and B, and is said to exhibit *bistability* or, in general terms, *multistability*. In contrast, the steady state C is not stable: It is a "mountain pass" between A and B that attracts some trajectories along one direction (diagonal), but repels them toward the "valleys" A or B along the other direction. Thus, the two states $A = (X_A, Y_A)$ and $B = (X_B, Y_B)$ are the "attractors" of the system. The region in the state-space neighborhood around the attractor A in which all the starting states $S(t = 0)$ (so-called *initial states*) would "flow" to the attractor A represents the *basin of attraction* of attractor A (gray zone in the state space in Figure 190.2C).

The concepts of attractors and basins of attraction are fundamental ideas in the field of system dynamics and are central to the understanding of the robustness of cell-fate regulation. The vector field in Figure 190.2C suggests some natural movement like that of a marble rolling down to the attractors; hence, it can be regarded as a topographic *landscape*, as shown in Figure 190.2D. The tendency with which a given state $S(t)$ moves to the attractor states is captured in the vertical dimension to emphasize the *potential character* of the attractors. (This is not a true potential because the functions on the right-hand side of the equations in Figure 190.2B are nonintegrable.) Figure 190.2E is a simplified, schematic cross-section of the state space along the dashed line through the attractor states A and B. In this intuitive picture, the valleys represent the attractors, separated by a hill, which can be viewed as the epigenetic barrier that we alluded to at the beginning of the chapter. This attractor landscape may be what Waddington had in mind in the 1940s, when he proposed the "epigenetic landscape"

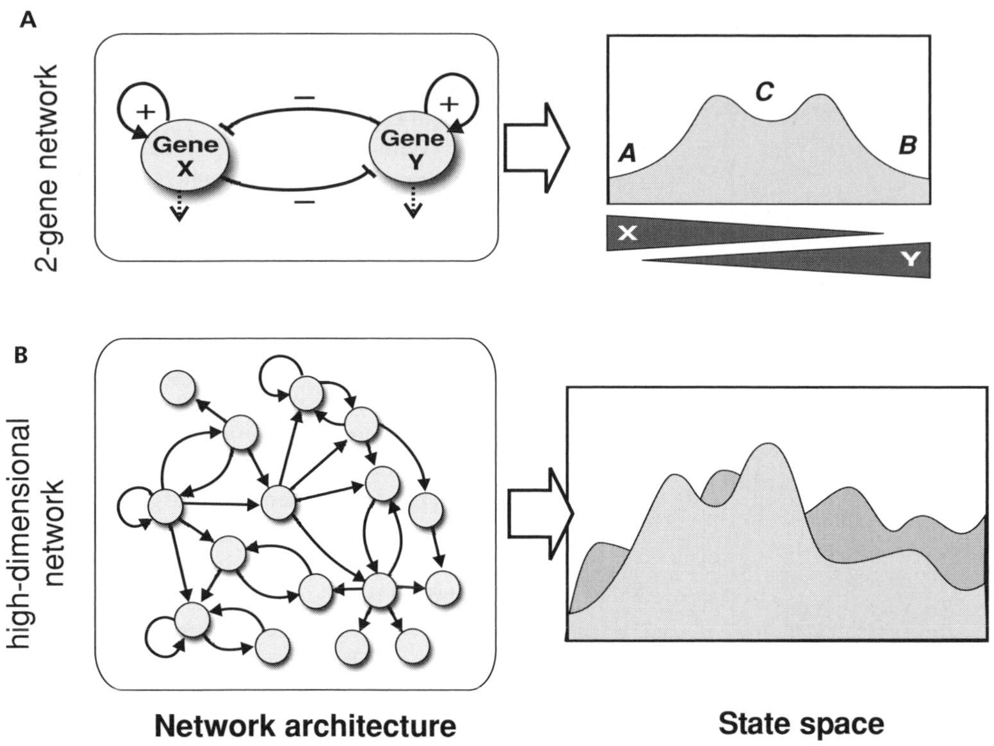

Figure 190.3. State space for more complex networks. (**A**) Two-gene network as in Figure 190.2, but each gene is now autoregulatory, promoting its own expression. Note that this leads to converting the unstable stationary state of Figure 190.2 into a local attractor *C*, corresponding to a progenitor cell that expresses both *X* and *Y* ("promiscuous expression") and can be tipped into either cell fate (attractor) *A*, expressing *X*, or *B*, expressing *Y*. (**B**) A large ("complex") network. Schematic cartoon of a high-dimensional state space with attractors, projected into an intuitive landscape image.

(Figure 190.1B) – without the notion of genes and regulatory interactions!

Although we have explained the case of mutual inhibition, it can be easily shown that a similar circuit with mutual *activation* will result also in two stable states A ($X_A = Y_B = 0$) and B ($X_A = Y_B =$ high). Now let us make the circuitry of Figure 190.2A slightly more complicated: Assume now that the genes *X* and *Y* also activate themselves via an autoregulatory loop (Figure 190.3A). Such a circuit structure, in which a transcription factor stimulates its own expression, is very often encountered in key factors that regulate cell-fate transitions: For example, GATA-1, GATA-2, MyoD, and E2F-1 are all autoregulatory factors. If we build such autoregulatory feedback loops into our two-gene system, the dynamics will be changed qualitatively, as can easily be shown by numerical solution of the differential equations. The stationary but unstable "hill top" state $C = (X_C, Y_C)$ becomes locally stable (compare Figures 190.2E and A): It has a basin of attraction between that of *A* and *B*, and is defined by intermediate-low, equal levels of *X* and *Y*. We will return to this later in this chapter when we discuss multipotential progenitor cells.

In summary, the mathematical treatment of the simple mutual regulatory circuit shows that one system (our two-gene network) can produce a system behavior that encompasses multiple stable states (attractors) given a particular "wiring" of the interactions of its parts (the genes). It can be shown that the discussed behavior exists for a large range of parameter values (k, Θ, n, where $n > 2$) that shall not be further discussed here. Which attractor the network (the cell) "occupies" at equilibrium depends on where it started from, the so-called "initial states" or the position of $S(t = 0)$ in the state space.

CELL FATES AS ATTRACTORS

It is now natural to view an attractor as a state representing a cell fate: Both are discrete, mutually exclusive entities defined by a stable gene activation configuration that arises from the dynamics of the gene regulatory network. In other words, the state space is the formal framework that helps us translate the static gene regulatory network architecture into the dynamics of whole-cell behavior. Even if the gene expression profiles overlap (e.g., state *C* in Figure 190.3A shares genes *X* and *Y* with states *A or B*, respectively), a given system (cell) can only occupy one of a set of alternative states at a time – the attractors are mutually exclusive, and the cell appears to make discrete "decisions" between distinct, genome-wide (typically overlapping) gene activation patterns. The attractors are stable to small perturbations. Larger changes in *X* or *Y*, however, can cause a system, which for instance is in state *A*, to move to state *B*, if the perturbation changed the *X* and *Y* values so much as to place the system state $S = (X, Y)$ into the basin of attraction of the attractor *B*. Such an attractor transition would correspond to a cell-fate switch in response to external signals. Importantly, the attractor model suggests that a cell-fate switch can take multiple trajectories; that is, multiple differentiation paths exist leading from *A* to *B*, provided the cell reaches the basin of attraction of *B*.

If cell fates correspond to attractor states, then the developmental potential of a cell is represented by the neighboring attractors that a cell can access. Of special biological interest in our more complicated example (Figure 190.3A) is the attractor state *C* which – because of its position between the other two – can be "kicked" out into either attractor *A* or *B*. Thus, *C* represents a multipotential precursor cell poised to (almost irreversibly) commit to the *A* or *B* cell fate by promoting expression of either *X* over Y, or vice versa. Interestingly, this multipotent cell *C* is characterized by a gene expression configuration in which both *X* and *Y* (the cell type–specific markers for *A* and *B*, respectively) are active at low levels. This property of promiscuous gene expression of multipotent progenitor cells in fact recapitulates experimental observations, notably in the hematopoietic system, in which progenitor cells have been shown to exhibit a promiscuous gene expression pattern that contains concomitantly the genes that are expressed uniquely in either of the successor cell lineages (30–33). From this example we also learn that a biological quality, such as multipotency or "stemness" need not be reduced to a particular molecule (as manifest in the current quest for the "stemness genes") (34), but can be simply and naturally explained by the dynamics of the underlying circuitry.

Multistability explains the existence of multiple alternative, nongenetic traits and their enduring nature (35) within one single genetic system. This property is at the very core of the emergence of differentiation into multiple cell types in multicellular organisms, because all cells in the body have the same genome. Because an attractor state is robust to perturbations, after cell division and random partitioning of cellular proteins into the daughter cells, the gene activation pattern of the attractor reestablishes itself from a slightly perturbed state, thereby providing nongenetic inheritance of a particular cellular phenotype. A material basis, such as a covalent modification of chromatin, is in principle not necessary to explain the "memory" of stable gene expression configurations. In 1948, Max Delbrück proposed bistability as a general principle to explain how discontinuous transitions between two stable states arise in biochemical reaction systems and how they could explain differentiation. Since then, this central theme with variations has been discussed by many investigators, mostly applied to single-cell organisms (36). The architectural requirement in a system for the emergence of multiple, distinct, stable states is that the circuit contains feedback loops (37,38). Low-dimensional systems consisting of one or two genes that exhibit multistability and can be described by a set of differential equations that assume sigmoidal regulation characteristics have recently seen a renaissance in the spirit of "systems biology," both in theory (39,40) and experiments (41–44).

With the concepts of "attractor" or "epigenetic landscape" (Figure 190.1B, Figure 190.2D,E; Figure 190.3), we also naturally introduce the aforementioned notion of the epigenetic barriers, an almost tangible metaphor but defined purely by abstract dynamics. The landscapes can become structurally quite sophisticated for systems with more than just two genes, exhibiting multiple pits, valley, planes, notches, and hills. The essence is that the epigenetic barriers separate the valleys (basins of attraction) and thus, the individual cell fates and associated "genetic programs." The epigenetic barriers account for the observation that many different input signals can produce the same output – if they displace the cell state to different positions in the state space, but within the same basin of attraction. Conversely, multistability implies that, given the same genome, a cell can occupy different stable states, and hence respond differently to the same input because of different accessibility to neighboring basins. Thus, the epigenetic barriers can be related to the "set-points" proposed by Aird (6) and would account for the "nature" aspect of EC heterogeneity.

GENERALIZATION TO LARGER NETWORKS: HIGH-DIMENSIONAL ATTRACTORS

The Concept of Contraction of State Space

We have so far only discussed how attractor states, which explain much of the cell-fate dynamics, arise in a two-gene circuit. The reality is that gene regulation is not organized in the isolated modules of few genes, but is embedded in an almost genome-wide network. In fact, emerging functional genomics data for a bacterium (*Escherichia coli*), a yeast (*Saccharomyces cerevisiae*), a nematode (*Caenorhabditis elegans*), and a fruit fly (*Drosophila melanogaster*) show that certainly protein–protein interaction networks (and hence, probably also gene regulatory networks) form genome-wide connected networks that encompass almost the entire genome (45). Graph theoretical principles of how networks evolved by the addition of new genes that then connected to the rest of the network also suggest that most probably the members of the network will form a large, connected component rather than consist of fully isolated, small subnetworks (46). So, what are the dynamics of such huge complex networks? Do they also produce "ordered" behavior with attractors?

We are now dealing with networks of $N =$ tens of thousands of elements, not just two, and the defining requirement for an attractor state is much higher: An attractor state must be stable (attracting) with respect to changes not in two but in thousands of genes. Thus, a state that we described earlier as $S(t) = [X(t), Y(t)]$ is now defined as $S(t) = [x_1(t), x_2(t), .. x_N(t)]$, where $x_i(t)$ is the expression level of gene *i*, with $i = 1, 2, .., N$. With *N* genes, the state space now has *N* dimensions and – as discussed earlier, an unfathomable number of possible states or gene expression profiles. We do not know the architecture ("wiring diagram") of the gene interactions in the genome-wide regulatory network. However, studies during the 1960s by Stuart Kauffman, in which the dynamics of statistical ensembles of randomly wired, simplified model networks consisting of several thousands of binary genes were simulated (47,48), showed that in fact, a complex, apparently irregularly wired network of thousands of genes can, if the "wiring" of the interactions exhibits certain architectural properties, give rise to stable attractors rather than behave "chaotically." (A behavior that is "chaotic," in conjunction with high-dimensional systems, would mean that the network state $S(t)$ would wander aimlessly in state space and eventually visit the entire universe of possible states.) Based on the intriguing finding that complex networks can exhibit stable attractor behavior, Kaufmann proposed that cell types in the body are attractor states (48). For high-dimensional systems, we can define an attractor in a more general way: It is the property that a N-dimensional volume of initial states S_i in the N-dimensional state space (as shown by the triangle in Figure 190.2C for the 2D case) will "contract," as the trajectories of their dynamics are followed for some time, to a state space volume that is much smaller than the initial volume. (Technically, the attractor can be a point attractor, consisting of one stable, stationary state, or $S(t)$ can oscillate periodically or chaotically, but in doing so, cover a small volume that is tiny compared to the volume of the initial states that represent its basin of attraction. Thus, in high-dimensional systems the relevant scale of observation emphasizes the contraction of state space volume rather than the actual shape of the attractor.)

Systems in which a volume of initial states collapses to a smaller volume and stays there are called *dissipative systems*. The existence of multiple such *attracting regions* would compartmentalize the huge universe of combinatorially possible expression profiles, the N-dimensional state space, into subregions that represent basins of attraction of a small number (relative to 2^N) of attractor states. These in turn would map into the phenotypic cell fates and their associated expression profiles. Thus, because of gene regulatory interactions, the hyperastronomical number of possible gene activation configurations collapses to a countable number (thousands?) of stable attractor regions. Relatively nonspecific perturbations that place a cell from state A to somewhere in the basin of attraction of another new attractor A' would result in the cell "automatically" assuming the highly specific expression profile of the new attractor states. Thus, even sloppy signals, if they perturb a certain set of genes, can induce cell phenotypes that are defined by distinct expression profiles. This is the basis for robustness, yet flexibility, of cell fates in multicellular organisms.[1]

1 Note that *robustness*, in terms of dynamics in state space, as used here, is conceptually distinct from robustness in terms of connectivity (communication between the nodes = genes) of a network that depends on the particular topology.

Experimental Data: Existence of High-Dimensional Attractors

The larger view of cell regulation with its system-level dynamic features, including the mutual exclusivity and discreteness of stable cell fates and the ability of nonspecific stimuli to trigger a switch between distinct programs, already lends support to the idea of attractor states in the complex gene regulatory network of real cells. As mentioned earlier, the ability of physical perturbations, such as the distortion of a cell, to regulate the switch of cell fates in ECs (14,16) also points to the existence of attractors that would relax the biochemical constraints for an input signal to induce a particular program. The ability of nonselective stimuli that affect the expression level of hundreds to thousands of signaling proteins and genes, such as kinase inhibitors, solvents, and modifiers of chromatin structure and methylation, to trigger a distinct cell-fate transition also supports the idea that cell fates are high-dimensional attractor states of the molecular regulatory network (22).

But are real genomic networks, with their complex network architecture, also wired in a way that such compartmentalization into alternative, globally defined stable states really occurs? Most randomly wired, complex networks would behave chaotically, without compact attracting regions. Finding the relevant architectural features of the network that can produce ordered behavior with compact attractors has been the objective of intense research activities by computational biologists. Properties that have been shown so far to increase the likelihood for generating ordered rather than chaotic behavior with few compact attractors include: (a) sparseness of interactions in the network, in that a gene on average interacts with a relatively low number k of other genes compared to N, that is, $k << N$ (48); (b) preferential use of a certain subset of logical functions that describe the "promoter logics" through which gene transcription is regulated by multiple upstream factors (48,49); and (c) a "scale-free" topology, as discussed in Chapter 189 (50,51). These architectural properties have indeed been found in real protein-interaction networks (45). In addition, a bias in the occurrence of regulatory functions also has been observed for a set of selected, well-characterized promoters (52).

Direct evidence at the molecular level for the existence of attractor states in the mammalian genome and the notion that they represent cell fates has recently been obtained using DNA microarray-based dynamic gene expression profiling of in vitro differentiating neutrophils in a promyeloid cell lines (53,54). By monitoring two different state space trajectories that were defined by $N = 2{,}800$ genes and represented two differentiation processes triggered by two different agents, it was shown that the state S of the undifferentiated cells could be displaced to two different regions of the state space. Then, as cells differentiate into neutrophils, the two trajectories would converge from different directions of the state space to a very similar expression profile, corresponding to that of differentiated neutrophils. Although only two trajectories were monitored, their convergence with respect to thousands of state-space dimensions is in agreement with the contraction of a state-space volume. Concretely, at the level of individual genes, it was found that some genes that have a characteristic expression level in neutrophils arrived at that level through initially different if not disparate time courses (e.g., temporary down- versus upregulation in response to the two triggering agents) before they were "attracted" to the final expression level. Such behavior, observed for thousands of genes, is in line with the existence of alternative, converging high-dimensional trajectories. This behavior of genes also defies the signal transduction paradigm of a unique and specific molecular pathway that is dedicated to mediating differentiation.

ATTRACTORS AND HETEROGENEITY WITHIN A CELL TYPE

If a particular cell fate, or a cell type, such as a differentiated EC, is an attractor state, then how do we explain the heterogeneity of cell phenotypes within the "organ" endothelium? First, as discussed in the beginning of this chapter, the very heterogeneity itself comes in different ontological flavors: We can operationally distinguish at least between *stable* (enduring) and *dynamic* (temporary) heterogeneity. The former is realized by the variety of subtypes of cells, such as the location-dependent diversification of ECs into morphological and biochemical subtypes. Temporary heterogeneity can act on top of static heterogeneity and modify the gene expression profile to accommodate the various functional states of the endothelium induced by external signals.

How does this relate to the concept of gene expression profile dynamics in the high-dimensional state space with attractors? As discussed in another section, an attractor is unlikely to be a simple point attractor, as we have seen for the 2D pedagogical case in Figure 190.2. In contrast, the N-dimensional attractor landscape has a complex structure, with valleys representing the attractor regions. Importantly, such attractors are not dimensionless points in state space, but have some dimensionality that can still grant the system some *degrees of freedom* with respect to selected dimensions. These state-space dimensions may represent "marginal" genes that are loosely connected to the core regulatory network of feedback interactions that establishes the attracting property. Thus, such attracting *regions* can be metaphorically thought of as a plane (in a subset of directions) at the bottom of a valley in a high-dimensional space.

The notion of degrees of freedom is now central for understanding how microenvironmental inputs ("nurture") can cause a heterogeneity of EC states. The degree of freedom with respect to some genes permits temporary changes of their expression level in response to short-term environmental signals that will shift the state S within a constrained volume of the attractor region (since affecting only a tiny fraction of the N dimensions), without overcoming the epigenetic barriers of the attractor that define the (sub)type. Therefore, despite these adaptive responses to temporary environmental inputs, the cell will stay in the attractor and not switch its type identity or "nature." Such free movements in dimensions that are virtually independent of the attracting directions of the state space, and hence, would not violate gene interaction rules, are the basis for *local* epigenetic *plasticity*. For example, in an inflammatory response, skin capillary ECs may be brought to various "activated states" in response to various input signals that target such genes in the nonattracting dimensions. These states are thus located within the flat bottom of the attractor region. Alternatively, temporary "nurture" signals may also be viewed as perturbations that "kick" a state up the steep walls of the basins of attraction (but without reaching the top of the epigenetic barriers) because of conflicting gene expression changes. This would correspond to activated states that will slowly relax back to the basal attractor states once the perturbation disappears – a central mechanism of homeostasis.[2]

A useful metaphor for the dynamic heterogeneity of a cell population of ECs is the following: If the expression profile of each individual cell were to be projected to a point S in the state space, the population would appear like a swarm of flies, each point (cell) moving individually, yet staying together as a whole, forming a cloud that covers a compact subvolume in state space, and thus, collectively representing a relatively stable EC subtype.

On the other hand, the *stable* heterogeneity of subtypes, such as brain versus skin capillary ECs, may be deterministically defined by the gene network and manifest in the distinct "subattractors" (again, with respect to a subset of state-space dimensions) that exist within the larger "EC attractor." The attractor property and associated epigenetic barriers would ensure that, despite changes in the gene expression profile caused by alterations in the expression of genes in response to changes in "nurture," the "nature" of the cell is maintained: For example, in the larger picture, a skin capillary EC remains a skin capillary EC even in an injured vasculature. In fact, the EC-specific transcriptome of scleroderma tissue still clusters very closely with that of the endothelium of normal skin (4).

As cautioned earlier, a gray zone exists between nature and nurture, because ontologically an entire hierarchy of heterogeneity can exist. Thus, although some subtype-specific features, such as genes expressed in high endothelial venules, are epigenetically maintained – that is, they belong to the attracting dimensions of the state space – others require the continuous input of environmental factors (6). Attractors and, accordingly, the expression of associated cellular phenotypes whose persistence requires the chronic contribution of an external input to maintain epigenetic barriers, have previously been termed *pseudoattractors* (53). On the other hand, a truly

2 Technically, the *dynamic* component of heterogeneity in expression profiles may also be due to oscillatory or "strange" attractors (chaos in the classical sense), where $S(t)$ is not stationary but deterministically "dances" in a small state-space volume that in high-dimensional systems may be tiny compared to the scale of the basin of attraction traversed by transients in biologically relevant time scales. Furthermore, nonsynchronized, random fluctuations in gene expression due to molecular noise also may become manifest as cell population heterogeneity and contribute to the dispersion of occupied states within the attractor region.

epigenetically "fixed" phenotype also can be lost: Strong and persistent perturbations can cause major displacements of the state *S* to jump beyond epigenetic barriers that define EC subtypes and bring the cell into a new, stable, and persisting phenotype. A classical example for such a local transdifferentiation between subtypes is the observation that venous ECs, when exposed to the pulsatile mechanical inputs normally encountered in arteries, will adopt an arterial cell phenotype. The irreversible loss of an EC's natural features which, when isolated and kept in cell culture, subsequently converge to a typical, stable culture phenotype may be due to similar processes.

In summary, the structure of attractors in a high-dimensional system can be complex enough to permit local variations of the expression profile within a small, attracting region in state space without having global effects on the cell type–defining aspects of (attracting) gene expression profiles. Such local variations represent the observed heterogeneity of gene expression profile *within* a cell type. Because the height of epigenetic barriers may vary over a broad range, the relative contribution of nature and nurture also varies. This results in a broad hierarchy of phenotypic heterogeneity that may blur the ontological notions of "cell type" and "subtype."

CONCLUSION

In this chapter, we have presented a conceptual framework for understanding cell-fate regulation based on the treatment of complex networks as high-dimensional dynamical systems. Relying largely on qualitative reasoning, we have explained the emergence of high-dimensional attractors and presented the idea that stable cell phenotypes, or cell fates, including cell types and subtypes, represent attractors in the high-dimensional gene expression state space. Observed cell-fate behavior, the architecture of gene regulatory networks found so far, and genome-scale dynamic gene expression analysis supports this idea.

The gene expression state space, compartmentalized into potential-like attractors and their basins within a complex landscape, establishes a formal basis for Waddington's intuitive metaphor of an "epigenetic landscape." The attractor landscape provides a stage on which cell behavior takes place, governed by the constraints of the regulatory network (nature) that is embodied by the deterministic landscape substructure, but susceptible to influence from environmental inputs and stochastic variability (nurture). With the reduction of the landscape metaphor to mathematical principles, we can now more comfortably use it to articulate subtle questions, such as those about phenotypic heterogeneity of cell types. Similarly, old concepts expressed in intuitive metaphors, such as epigenetic barriers, nature versus nurture, input/output, plasticity, and the like all can be formally addressed using the principles of high-dimensional state space and attractor.

Although current efforts in biology are focused around the elucidation of the details of the wiring diagram of the gene regulatory network, the ultimate goal for understanding cell behavior as an emerging property is to map out the actual structure of the "epigenetic landscape" and to correlate it with observed cell behavior. This not only requires knowledge of the wiring diagram, but must involve systematic perturbations and dynamic gene expression profiling to measure the constrained temporal evolution of transcriptomes and proteomes, coupled with phenotypic observation. Ideally, this should be achieved in single cells within the tissue context to capture the microenvironmental inputs and to identify stochastic variability as one source of heterogeneity. With a clear and concrete conceptual framework in mind, as the one presented here, there should be enough motivation to drive our multidisciplinary effort to surmount these formidable technical challenges in the near future.

KEY POINTS

- Cell fates, including cell types and stable subtypes, represent discrete cell behavior programs that correspond to high-dimensional attractor states in the dynamics of the underlying complex gene regulatory networks.
- The "attractor landscape" is a mathematical formulation of Waddington's "epigenetic landscape," and explains why cells are forced to undergo almost all-or-none transitions between a small number of discrete cell fates.
- The model naturally explains many observed phenomena that remain unexplained by the traditional pathway paradigm, such as discreteness of cell fate; limited accessibility to neighboring phenotypes (including multipotential capability and "stemness" of progenitor cells); epigenetic inheritance of phenotype; regulation by nonspecific, noninstructive signals; and the coexistence of robustness and flexibility.

Future Goals

- To begin, in addition to characterizing in detail the wiring diagram (regulatory pathways) of the genome, to map out the structure of the "epigenetic landscape" (state space with attractors) through systematic measurements of the evolution over time of genome-wide profiles of the activity of cellular molecules
- To distinguish between epigenetically fixed, environmentally induced, and stochastically varying phenotypes, using single-cell analysis in the context of the tissue

ACKNOWLEDGMENTS

This work was supported by the Air Force Office of Scientific Research.

REFERENCES

1 Hsiao LL, Dangond F, Yoshida T, et al. A compendium of gene expression in normal human tissues. *Physiol Genomics.* 2001;7(2):97–104. Epub 2001 Oct 02.

2 Ross DT, Scherf U, Eisen MB, et al. Systematic variation in gene expression patterns in human cancer cell lines. *Nat Genet.* 2000;24(3):227–235.

3 Chi JT, Chang HY, Haraldsen G, et al. Endothelial cell diversity revealed by global expression profiling. *Proc Natl Acad Sci USA.* 2003;100(19):10623–10628.

4 Barnes CM, Huang S, Kaipainen A, et al. Evidence by molecular profiling for a placental origin of hemangioma. Submitted.

5 Ho M, Yang E, Matcuk G, et al. Identification of endothelial cell genes by combined database mining and microarray analysis. *Physiol Genomics.* 2003;13(3):249–262.

6 Aird WC. Endothelial cell heterogeneity: a case for nature and nurture. *Blood.* 2004;103(11):3994.

7 Odom DT, Zizlsperger N, Gordon DB, et al. Control of pancreas and liver gene expression by HNF transcription factors. *Science.* 2004;303(5662):1378–1381.

8 Khorasanizadeh S. The nucleosome: from genomic organization to genomic regulation. *Cell.* 2004;116(2):259–272.

9 Arney KL, Fisher AG. Epigenetic aspects of differentiation. *J Cell Sci.* 2004;117(Pt 19):4355–4363.

10 Kubicek S, Jenuwein T. A crack in histone lysine methylation. *Cell.* 2004;119(7):903–906.

11 Lomvardas S, Thanos D. Opening chromatin. *Mol Cell.* 2002;9(2):209–211.

12 Georgopoulos K. Haematopoietic cell-fate decisions, chromatin regulation and ikaros. *Nat Rev Immunol.* 2002;2(3):162–174.

13 Huang S. Regulation of cellular states in mammalian cells from a genome-wide view. In: *Gene Regulation and Metabolism: Post-Genomic Computational Approach.* Cambridge, MA: MIT Press; 2002:181–220.

14 Huang S, Ingber DE. Shape-dependent control of cell growth, differentiation, and apoptosis: switching between attractors in cell regulatory networks. *Exp Cell Res.* 2000;261(1):91–103.

15 Evan GI, Brown L, Whyte M, Harrington E. Apoptosis and the cell cycle. *Curr Opin Cell Biol.* 1995;7(6):825–834.

16 Huang S, Ingber DE. The structural and mechanical complexity of cell-growth control. *Nat Cell Biol.* 1999;1(5):E131–E138.

17 Chen CS, Mrksich M, Huang S, et al. Geometric control of cell life and death. *Science.* 1997;276(5317):1425–1428.

18 Southan C. Has the yo-yo stopped? An assessment of human protein-coding gene number. *Proteomics.* 2004;4(6):1712–1726.

19 Koshland DE Jr, Goldbeter A, Stock JB. Amplification and adaptation in regulatory and sensory systems. *Science.* 1982;217(4556):220–225.

20 Paulsson J, Berg OG, Ehrenberg M. Stochastic focusing: fluctuation-enhanced sensitivity of intracellular regulation. *Proc Natl Acad Sci USA.* 2000;97(13):7148–7153.

21 Savageau MA. Michaelis-Menten mechanism reconsidered: implications of fractal kinetics. *J Theor Biol.* 1995;176(1):115–124.

22 Huang S. Genomics, complexity and drug discovery: insights from Boolean network models of cellular regulation. *Pharmacogenomics.* 2001;2(3):203–222.

23 Zingg JM, Pedraza-Alva G, Jost JP. MyoD1 promoter autoregulation is mediated by two proximal E-boxes. *Nucleic Acids Res.* 1994;22(12):2234–2241.

24 Chen H, Ray-Gallet D, Zhang P, et al. PU.1 (Spi-1) autoregulates its expression in myeloid cells. *Oncogene.* 1995;11(8):1549–1560.

25 Ohneda K, Yamamoto M. Roles of hematopoietic transcription factors GATA-1 and GATA-2 in the development of red blood cell lineage. *Acta Haematol.* 2002;108(4):237–245.

26 Grass JA, Boyer ME, Pal S, et al. GATA-1-dependent transcriptional repression of GATA-2 via disruption of positive autoregulation and domain-wide chromatin remodeling. *Proc Natl Acad Sci USA.* 2003;100(15):8811–8816.

27 Nerlov C, Querfurth E, Kulessa H, Graf T. GATA-1 interacts with the myeloid PU.1 transcription factor and represses PU.1-dependent transcription. *Blood.* 2000;95(8):2543–2551.

28 Zhang P, Zhang X, Iwama A, et al. PU.1 inhibits GATA-1 function and erythroid differentiation by blocking GATA-1 DNA binding. *Blood.* 2000;96(8):2641–2648.

29 Ruzinova MB, Benezra R. Id proteins in development, cell cycle and cancer. *Trends Cell Biol.* 2003;13(8):410–418.

30 Hu M, Krause D, Greaves M, et al. Multilineage gene expression precedes commitment in the hemopoietic system. *Genes Dev.* 1997;11(6):774–785.

31 Enver T, Greaves M. Loops, lineage, and leukemia. *Cell.* 1998;94(1):9–12.

32 Akashi K, He X, Chen J, et al. Transcriptional accessibility for genes of multiple tissues and hematopoietic lineages is hierarchically controlled during early hematopoiesis. *Blood.* 2003;101(2):383–389.

33 Bruno L, Hoffmann R, McBlane F, et al. Molecular signatures of self-renewal, differentiation, and lineage choice in multipotential hemopoietic progenitor cells in vitro. *Mol Cell Biol.* 2004;24(2):741–756.

34 Vogel G. Stem cells. "Stemness" genes still elusive. *Science.* 2003;302(5644):371.

35 Rubin H. On the nature of enduring modifications induced in cells and organisms. *Am J Physiol.* 1990;258(2 Pt 1):L19–L24.

36 Monod J, Jacob F. Teleonomic mechanisms in cellular metabolism, growth, and differentiation. *Cold Spring Harb Symp Quant Biol.* 1961;26:389–401.

37 Thomas R. Logical analysis of systems comprising feedback loops. *J Theor Biol.* 1978;73(4):631–656.

38 Cinquin O, Demongeot J. Positive and negative feedback: striking a balance between necessary antagonists. *J Theor Biol.* 2002;216(2):229–241.

39 Cherry JL, Adler FR. How to make a biological switch. *J Theor Biol.* 2000;203(2):117–133.

40 Angeli D, Ferrell JE Jr, Sontag ED. Detection of multistability, bifurcations, and hysteresis in a large class of biological positive-feedback systems. *Proc Natl Acad Sci USA.* 2004;101(7):1822–1827.

41 Gardner TS, Cantor CR, Collins JJ. Construction of a genetic toggle switch in Escherichia coli. *Nature.* 2000;403(6767):339–342.

42 Bhalla US, Ram PT, Iyengar R. MAP kinase phosphatase as a locus of flexibility in a mitogen-activated protein kinase signaling network. *Science.* 2002;297(5583):1018–1023.

43 Ozbudak EM, Thattai M, Lim HN, et al. Multistability in the lactose utilization network of Escherichia coli. *Nature.* 2004;427(6976):737–740.

44 Xiong W, Ferrell JE, Jr. A positive-feedback-based bistable "memory module" that governs a cell fate decision. *Nature.* 2003;426(6965):460–465.

45 Huang S. Back to the biology in systems biology: what can we learn from biomolecular networks. *Brief Funct Genomic Proteomic.* 2004;2(4):279–297.
46 Callaway DS, Hopcroft JE, Kleinberg JM, et al. Are randomly grown graphs really random? *Phys Rev E Stat Nonlin Soft Matter Phys.* 2001;64(4 Pt 1):041902.
47 Kauffman SA. Metabolic stability and epigenesis in randomly constructed genetic nets. *J Theor Biol.* 1969;22(3):437–467.
48 Kauffman SA. *The Origins of Order*. New York: Oxford University Press, 1993.
49 Shmulevich I, Lahdesmaki H, Dougherty ER, et al. The role of certain post classes in Boolean network models of genetic networks. *Proc Natl Acad Sci USA.* 2003;100(19):10734–10739.
50 Fox JJ, Hill CC. From topology to dynamics in biochemical networks. *Chaos.* 2001;11(4):809–815.
51 Aldana M, Cluzel P. A natural class of robust networks. *Proc Natl Acad Sci USA.* 2003;100(15):8710–8714.
52 Harris SE, Sawhill BK, Wuensche A, Kauffman SA. A model of transcriptional regulatory networks based on biases in the observed regulation rules. *Complexity.* 2002;7(4):23–40.
53 Huang S, Eichler GS, Bar-Yam Y, Ingber DE. Cell fate as high-dimensional attractor of a complex gene regulatory network. *Phys Rev Lett.* 2005;94(12):12871.
54 Huang S. Gene expression profiling, genetic networks, and cellular states: an integrating concept for tumorigenesis and drug discovery. *J Mol Med.* 1999;77(6):469–480.

191

Equation-Based Models of Dynamic Biological Systems

Gilles Clermont, Yoram Vodovotz, and Jonathan Rubin

University of Pittsburgh, Pennsylvania

The endothelium serves barrier, synthetic and catalytic functions and is a site of complex interacting processes involving a large number of biological components. Mathematical modeling might provide valuable insight into the global integration of those interactions into tissue function. The purpose of this chapter is to provide a nontechnical review of a well-established modeling platform, namely differential equations, that harnesses the powerful tools of calculus to analyze the time-dependent behavior of dynamical systems. Differential equations have been abundantly used by modelers. Yet, this framework is largely unknown to basic and clinical scientists. We will briefly describe this framework, provide examples that relate to endothelium modeling, and discuss its strengths and weaknesses (Figure 191.1).

DYNAMICAL SYSTEMS

A dynamical system is an amalgam of interacting components together with a set of rules for how the states of the components evolve in time, and so the notion of time evolution is key when thinking about such a system. Many primary or calculated useful physiological quantities, such as cardiac output and vascular resistance, are related in a static fashion. In other words, one can relate these quantities by means of algebraic equations of varying complexity. The equations resulting from drawing an analogy between electrical circuits and the circulation have led to additional appealing concepts, such as peripheral vascular resistance and vascular capacitance. However, the clinician is clearly aware that these quantities change over time as the "system" adapts to changing external and internal conditions such as fluid administration, local concentration of effectors, or drug dose. The previous chapters have described the concept of predicting the time evolution of complex systems and the implementation of rules that will dictate, given the state of a system now, the state of the system at some future time. These rules can be represented through several mathematical formalisms. This chapter will concentrate on differential equations, a powerful mathematical framework with a long and fruitful track record in helping physicists and biologists describe and predict the behavior of simple and complex systems.

DIFFERENCE AND DIFFERENTIAL EQUATIONS

Ordinary Differential Equations

Given a system described by variables, taking values $x(t) = \{x_1(t), x_2(t), \ldots, x_n(t)\}$ at time t, a difference equation is a rule specifying how to use $x(t)$ to calculate $x(t+1)$, the value of the same variable at the next time step. The "biology" of the system is embodied in the difference equation. For example, $x(t+1) = 2gx(t)$ represents the geometrical growth of dividing cells or bacteria, where the time step is taken to be 1 in some unit of time. This approach is similar to the cellular automata or agent-based models described previously, in which the rules of the biology are included in the mapping (1). There is no obstacle to including several interacting variables, or even values from multiple different time steps, within a difference equation. Although the principle is simple, the resulting simulations are often not intuitive, giving rise to interesting emergent behaviors. The natural extension of difference equations is to consider the case where the time steps become infinitely small. In this limit, the left hand side of the prediction equation consists of not the value of a variable at the next time step, but rather the velocity with which the variable is evolving. This formulation is a differential equation.

Many of the best known computational models for biological processes that evolve continuously in time are expressed as ordinary differential equations (ODEs). ODE models consist of sets of coupled differential equations of the form $dx_i/dt = h(x_1, \ldots, x_n)$, where h is a function of some subset of the variables in the model. The function h encodes some simplified version of the biology of the system being modeled.

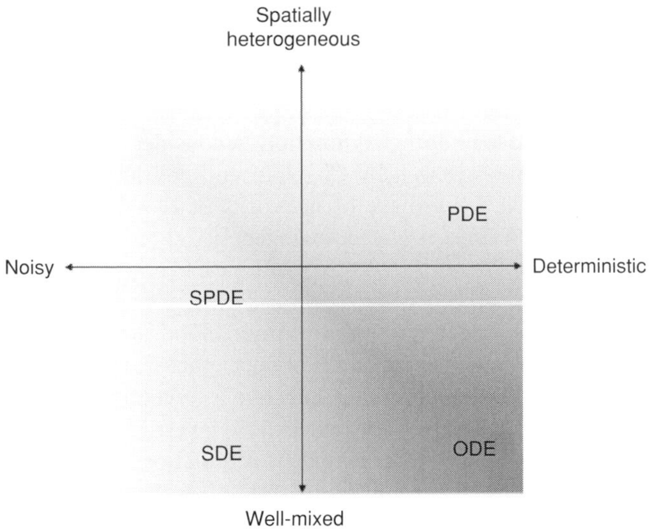

Figure 191.1. Equation-based models are particularly applicable to systems that contain large numbers of components, that evolve continuously in time, and that are not too spatially heterogeneous. The choice between ordinary differential equations (ODEs) or partial differential equations (PDEs) depends on whether precise spatial information is essential to describing the behavior of the biological system. Incorporating uncertainty generated by noise and incomplete information into differential equations will lead to stochastic ordinary differential equations (SODEs), or partial differential equations (SPDEs). The darker the shading on the figure, the more natural is the fit between the biological system and a description using differential equation-based models.

For example, this typically includes information with regard to the presence of specific interactions between components of the system and the relative strength of those interactions, quantified by model parameters. These models are based on the premise that the system involved is "well-mixed" and that there are sufficiently many components (e.g., proteins, cells, small molecules) that their numbers can be regarded as continuous quantities. Although many of these assumptions are not strictly true within an individual cell, ODE models have been found to provide a great deal of insight into the behavior of complex interacting systems. Numerical methods for the solutions to ODEs are mature, very reliable, freely available, and easily implemented on personal computers. Furthermore, ODE models are amenable to formal mathematical analysis. Although it may not be possible to write down formulas expressing the solutions to differential equations as explicit functions of time, this mathematical framework allows one to determine the possible qualitative behaviors that the system can support, depending on particular choices of initial conditions and model parameter values.

Qualitative behaviors of ODE systems may include convergence to equilibrium points, which are states at which all variables remain constant, and periodic oscillations, as well as many more complicated time courses. For a fixed parameter set, a system of ODEs may have more than one possible equilibria, each of which may be stable (in a variety of precise ways) or unstable. To every stable equilibrium, there corresponds a basin of attraction, which refers to the set of initial conditions that will ultimately evolve to this equilibrium. These basins can change with changes in parameters; indeed, equilibria and other interesting solutions can be born or cease to exist as parameters vary. Correspondingly, one can segment parameter space a priori into regions with qualitatively different equilibrium structures.

When spatial structure plays an important role in the behavior of a system, ODEs may still provide a useful modeling tool, if the system can be considered as a collection of well-mixed compartments. In this case, a key aspect of the modeling process is the characterization of the transfer of information from one compartment to the next, such as occurs via molecular diffusion or cellular trafficking. As long as inter-compartmental exchange can be reasonably modeled as being driven by differences in relative abundances of a quantity, or relative strengths of a physical force such as pressure, across adjacent compartments, ODEs remain a viable, relatively simple option for characterizing spatially structured systems.

Partial Differential Equations

There are situations where the particular geometry of a system is crucial in dictating behavior. Whether this is true or not may depend on the type of information one seeks through modeling. For example, the geometry of the axon need not be considered for an approximate analysis of action potential propagation, but it is key in determining the precise influence of gradual changes in axon width on propagation characteristics. Similarly, cellular motion along chemoattracting gradients, the physical plugging of an endothelial tear by aggregating platelets, and the thickening of lung tissue with the accumulation of interstitial edema are also exquisitely dependent upon the spatial characteristics of the biological system. When continuous aspects of geometry are important or the "well-mixed" assumption is obviously at odds with the nature of the system at hand, ODEs are insufficient to capture the underlying physical processes, and spatial dependence must be explicitly introduced into model equations. The resulting formalism is that of partial differential equations (PDEs). PDEs and ODEs are often complementary modeling tools and may coexist in analyses of the same system, especially when one is attempting to include several scales of description. PDEs are considerably more demanding computationally than ODEs, often requiring a larger computing platform than personal computers. As with ODEs, formal mathematical theory has been developed for many types of PDEs. Unfortunately, this theory is more complicated, and often further removed from insights about the dynamic structure of the underlying system, than is the theory for ODEs, and many of the PDEs relevant to detailed biological modeling are too complex for analytical treatment. However, mathematical theory may be useful in selecting appropriate numerical simulation methods to apply, even when it cannot give any direct information about possible model behavior.

In some cases, biological systems may be so fragmented that PDEs may be impractical. For example, the dissemination of an infectious disease in a city, where health care service points, schools, restaurants, and churches all represent privileged areas of disease transmission, would be difficult to simulate with equation-based models. Realistic simulations must take into account individual behavior and the precise geographic location of each of those areas. In such situations, agent-based models certainly represent a more practical simulation platform than PDEs.

Noise and Differential Equations

Traditional ODEs and PDEs are deterministic models. A given set of initial and/or boundary conditions will always lead to the same outcome in these models. However, there is some degree of randomness, or stochasticity, due to temperature fluctuation, variability in molecular interactions, and a wide variety of other factors, in all physical systems. Although some dynamic phenomena, such as convergence to globally attracting equilibria, are highly robust to noise, others are not. In addition to the types of dynamics generated by a system, the number of cells or molecules involved in the system also contributes to the importance of stochastic effects in sculpting its behaviors. The smaller the scale of description, the greater the potential impact of noise. In situations where the inclusion of noise in a model is appropriate, and where the nature of this noise is reasonably well understood, stochastic differential equations (SDEs) provide a mathematical framework for doing so. Due to the importance of stochasticity in the outcomes of many cellular processes, there has been a great deal of interest in applying stochastic simulation methods in this setting (2–4). Most of these methods rely on the Gillespie algorithm (5), which yields an exact simulation of a stochastic reaction process. There are extensions of the original algorithm, which provide the core for a number of recent simulation tools (6). Wolkenhauer and colleagues review the advantages and disadvantages of the Gillespie method of stochastic simulation relative to the standard ODE approach outlined above (7).

The difficulty with the Gillespie method occurs when a model includes many reactions and interactions and a large number of molecules and biological species. In this case the net reaction rate becomes very big, so that the time between events is exceedingly small and millions of steps must be taken to advance a significant amount in time. One way around this difficulty is to derive an approximation called the chemical Langevin equation, consisting of a classical SDE, to the overall system of reactions (8). Depending on whether spatial elements are included or upon the complexity of the system, solving such a stochastic equation may be computationally challenging. The practical tools available for analyzing the characteristics of such systems also may be quite limited.

EXAMPLE OF EQUATION-BASED MODELS

The biomedical literature contains a rich repertoire of differential equation-based models. Yet relatively few of those models relate to endothelial function. A considerable body of work, recently reviewed by Chaplain and colleagues (9,10), has addressed the growth of networks of blood vessels in the context of cancer-related angiogenesis. These models are both of the ODE and PDE types and have been used to predict the geometry of neovascular growth, the relationship between neovascularization and tumor blood flow (11), the onset and sustained stimulation of neocapillarization (12,13) and the prediction of possibly effective interventions in curtailing angiogenesis. Other models have investigated the role of the local cellular environment in stimulating blood vessel growth (14).

Other investigators used PDEs to examine the interdependence of nitric oxide (NO) and of oxygen in the microcirculation. All NO synthase isoforms (endothelial [eNOS], neuronal [nNOS], and inducible [iNOS]) require oxygen to produce NO, yet tissue oxygen consumption is reversibly inhibited by NO (15). Interestingly, the blood flow "steal" phenomenon, in which decreased vascular resistance in specific regions of the circulatory bed results in excessive flow (compared to metabolic demand) to those regions while other regions become underperfused, is an emergent behavior of these simulations.

There exist several models of intercellular communication through tight junctions, although most of these address renal or gastrointestinal epithelial tight junctions. To our knowledge, there are only two published mathematical models for endothelial permeability (16,17). Several authors have examined the dynamics of coagulation and the role of the endothelium in clot growth and propagation (18–23).

Of particular significance is the absence of published equation-based models, or any other type of simulations, of the interactions among inflammation, coagulation, and the endothelium, a significant knowledge gap because these processes are key to a number of high impact acute and chronic illnesses. Our group is currently involved in such an effort, in which we have created a mathematical model that describes the dynamics of acute inflammation in the settings of sepsis and trauma (24–27). The inflammation model was calibrated using data from mice that were subjected to endotoxemia at various doses, surgical trauma, and surgical trauma combined with endotoxemia. We have initiated a preliminary calibration of the coagulation equations, which resulted in defining initial values of all rate constants in those equations, in the setting of endotoxemia. This calibration involved an initial description of the changes over time in the concentrations of various components of coagulation (tissue factor; prothrombin; thrombin; protein C; activated protein C; protein S; factors VIIa, IX, IXa, X and Xa; and plasminogen activator inhibitor [PAI]-1), specifically as they relate to the changes in other components of the innate response previously calibrated in animal experiments. Of relevance to endothelial biology, our model contains interactions in which proinflammatory stimuli first suppress eNOS expression and later induce iNOS expression, in line both with our data and with the literature (28–31).

STRENGTHS AND PITFALLS OF EQUATION-BASED MODELS

Differential equations, as a modeling approach, have enormous appeal. They (a) provide an intuitive means to translate mechanistic concepts into a mathematical framework, (b) can be analyzed using a large body of existing techniques, (c) can be numerically simulated easily and inexpensively on a desktop computer, (d) provide *both* qualitative and quantitative predictions, and (e) allow for the systematic incorporation of higher levels of complexity. Critics are entirely justified in pointing out that equation-based models usually depend on a large number of parameters that quantify biological interaction, and that specifying these parameters is a difficult task. However, unlike in other approaches, these parameters are explicit in differential equation models. Therefore, knowledge gaps are readily identified. Further, the speed of computation with differential equation-based models allows for massive experimentation with parameters that may not be cleanly manipulable experimentally, leading to the development of hypotheses on the roles of individual parameters that can drive subsequent experimental investigations.

VALIDATION OF EQUATION-BASED MODELS

Several issues relate to the validation of complex models. At the simplest level, a model is deemed valid if the predictions furnished correspond to empirical data. The problem of assessing the validity of an equation-based model is somewhat more involved. Such a model must first be structurally valid. In other words, the biology embodied in the equations of the model must be rooted in empirical evidence. This entails the inclusion in the model of specific biological processes and their quantification. The level of detail included in such models depends entirely on the scale of description for which a particular model is intended. Interestingly enough, less detailed models might be more difficult to specify, because individual molecular or cellular species may be replaced by more abstract physiological functions, which may be difficult to associate with quantitative empirical evidence. Quantifying biological processes and interactions is clearly a formidable challenge to modelers, even when there exists good empirical evidence. For example, models commonly necessitate the quantification of molecular or cellular half-lives. Empirical data describing an observed half-life in the circulation may be inaccurate in describing the duration of the biological effect of a particular component. A model describing the proinflammatory effect of lipopolysaccharide on vascular endothelium should use a serum half-life, provided that provision is made for recirculation and the processes triggered by endotoxin binding are themselves modeled. However, if the downstream effects of the interaction of a molecule with a target cell receptor are not specifically included in the model, then the modeler must include an appropriate estimate of the biological half-life of this molecule.

This simple example should illustrate the great care that must be taken in ascribing values to parameters, irrespective of the modeling framework adopted. At best, these parameter values should serve as a starting point to further fitting of the model to empirical data, or to known qualitative behaviors in the absence of empirical data. A number of very sophisticated algorithms have been proposed and are commonly used to adjust parameters to produce optimal fits to empirical data. Unfortunately, the process differs fundamentally from statistical fits, where there are clear prescriptions as to how to achieve model fits and unique solutions typically exist. Because mechanistic models often include many parameters with unclear biological values, the problem of fitting models to empirical data is high dimensional, computationally intensive, and may produce nonunique solutions. Therefore, modelers have typically not produced quantitative model validation, except for restricted, highly simplified systems. Those models are therefore of limited usefulness. A clear research objective resides in the production of new theoretical and computational methods to conduct quantitative fits, together with the congruent development of experimental paradigms that will provide optimal data streams to accomplish this task. Indeed, the range of predictions of a mechanistic model extends well beyond highly controlled sets of initial conditions (experimental or clinical conditions under which empirical data was obtained). This situation is in sharp contrast to statistical models, which cannot offer predictions beyond the observed set of associations between initial conditions or lead to conclusions about temporal dynamics.

MULTISCALE MODELS

Multiscale models combine components relating to different scales of description of a biological system, such as molecules and tissues, tissues and organs, and the like (32). It appears inevitable that serious models of endothelial biology, for example, will have to relate NO to vascular resistance and blood pressure on one hand, requiring modeling at the level of global cardiovascular dynamics, and capillary permeability on the other hand, necessitating a link to local cellular and gaseous diffusion and blood flow and in turn affecting local and systemic energy supply. Differential equations can be used to describe several aspects of this system or subsystems, and could be combined with other modeling platforms if these were determined to be more suitable for certain aspects within particular multiscale models. Considerable work has been published on theoretical and practical aspects of merging modeling platforms describing different aspects of a multiscale model (33). Such expertise will be crucial to large-scale simulations of endothelial processes and their systemic repercussions.

CONCLUSION

Early in the twenty-first century, the scientific community is marshalling an unprecedented effort to encourage, support, fund, and implement interdisciplinary, systems-based science. Attempts at this dialogue are not new, but never has the need been so great. There is a remarkable divide in the peer-reviewed literature between the modeling community and the clinical and basic sciences communities. This divide originates from a critical lack of collaborative efforts, a shortfall within past modeling efforts in the creation of standards that would appeal to other communities, a lack of understanding by experimentalists of the type of data required for modeling, and a lack of demonstrable incremental utility of the modeling exercise in general. Recent initiatives by funding and regulatory agencies such as the National Institutes of Health (NIH Roadmap Initiative) (34) and the Food and Drug Administration (35) to bridge this gap and reward serious interdisciplinary work are both necessary and commendable. A similar step from industry is long overdue. The repeated failures in registering promising inflammation modulation drugs in late-phase clinical trials and postmarketing discoveries of disastrous side-effects may promote the emergence of modeling as a useful discovery and design tool.

Any effort at producing insightful models depends exquisitely on the availability of high quality empirical data. Therefore, hypothesis-driven, reductionist investigations that examine and quantify biological processes in highly controlled environments, remain the key foundation of any effort at gaining insight into complex biology through modeling. Computational neuroscience has a longer tradition of association between modeling, physiology and clinical medicine than other theaters where computational tools are currently applied, and therefore valuable lessons can be drawn from this literature (36). Yet, the full potential of reductionist science remains unexplored from lack of interdisciplinary effort. As a result, in the absence of such efforts, more data may fail to translate to new knowledge. Conversely, modelers should be exquisitely aware of the importance of producing relevant results and should interact with basic researchers to help identify critical data gaps and to structure experimental work to fill those gaps.

Equation-based modeling offers distinct practical advantages as a modeling platform, including its potential for inclusion of biological detail, suitability for simulation, and ease of parameter manipulation for predictive purposes, but it is not the preferred approach for simulating the stochastic interactions of small numbers of agents or the dynamics of highly fragmented systems. This modeling platform is one of several tools that will be solicited by interdisciplinary teams of expert modelers, basic scientists, and clinicians, with sufficient funding and fair access to peer-reviewed literature. The expected output of this work will be an improved translation of basic knowledge into clinical care and prevention approaches, as well as the design of new therapies. Modeling efforts that relate to an integrated understanding of endothelial function have been sparse, a frustrating state of fact for an emerging science where integrated knowledge appears to be so central.

KEY POINTS

- Equation-based models remain widely used in the mathematical description of dynamical biological systems.
- Equation-based modeling is a powerful, mature mathematical discipline well-suited to a vast array of biological problems that can shed light into mechanism, provide predictions, and lead to new hypotheses.
- The usefulness of equation-based models for specific applications is best determined by a multidisciplinary team.

Future Goals

- To develop a close collaboration between bioscientists and modelers in the use of models that translate empirical data into new knowledge, with the hypotheses to be examined and the specific data to be gathered rooted in this collaboration

REFERENCES

1 Ermentrout GB, Edelstein-Keshet L. Cellular automata approaches to biological modeling. *J Theor Biol.* 1993;160:97–133.

2 Arkin A, Ross J, McAdams HH. Stochastic kinetic analysis of developmental pathway bifurcation in phage lambda-infected Escherichia coli cells. *Genetics.* 1998;149:1633–1648.

3 Barkai N, Leibler S. Robustness in simple biochemical networks. *Nature.* 1997;387:913–917.

4 Firth CAJM, Bray D. Stochastic simulation of cell signaling pathways. In: Bower JM, Bolouri H, eds. *Computational Modeling of Genetic Biochemical Networks.* Cambridge, MA: MIT Press, 2005.

5 Gillespie DT. Exact stochastic simulation of coupled chemical reactions. *J Phys Chem.* 1997;8:2340–2361.

6 Adalsteinsson D, McMillen D, Elston TC. Biochemical Network Stochastic Simulator (BioNetS): software for stochastic modeling of biochemical networks. *BMC Bioinformatics.* 2004;5:24.

7 Wolkenhauer O, Ullah M, Kolch W, Cho KH. Modeling and simulation of intracellular dynamics: choosing an appropriate framework. *IEEE Trans Nanobioscience.* 2004;3:200–207.

8 Gillespie DT. The chemical Langevin equation. *J Chem Phys.* 2000;113:297–306.

9 Chaplain M, Anderson A. Mathematical modelling of tumour-induced angiogenesis: network growth and structure. *Cancer Treat Res.* 2004;117:51–75.

10 Chaplain MA. Mathematical modelling of angiogenesis. *J Neurooncol.* 2000;50:37–51.

11 McDougall SR, Anderson AR, Chaplain MA, Sherratt JA. Mathematical modelling of flow through vascular networks: implications for tumour-induced angiogenesis and chemotherapy strategies. *Bull Math Biol.* 2002;64:673–702.
12 Levine HA, Tucker AL, Nilsen-Hamilton M. A mathematical model for the role of cell signal transduction in the initiation and inhibition of angiogenesis. *Growth Factors.* 2002;20:155–175.
13 Levine HA, Sleeman BD, Nilsen-Hamilton M. A mathematical model for the roles of pericytes and macrophages in the initiation of angiogenesis. I. The role of protease inhibitors in preventing angiogenesis. *Math Biosci.* 2000;168:77–115.
14 Op den Buijs, Musters M, Verrips T et al. Mathematical modeling of vascular endothelial layer maintenance: the role of endothelial cell division, progenitor cell homing, and telomere shortening. *Am J Physiol Heart Circ Physiol.* 2004;287:H2651–H2658.
15 Lamkin-Kennard KA, Buerk DG, Jaron D. Interactions between NO and O_2 in the microcirculation: a mathematical analysis. *Microvasc Res.* 2004;68:38–50.
16 Carter EP, Olveczky BP, Matthay MA, Verkman AS. High microvascular endothelial water permeability in mouse lung measured by a pleural surface fluorescence method. *Biophys J.* 1998;74:2121–2128.
17 Phillips CG, Parker KH, Wang W. A model for flow through discontinuities in the tight junction of the endothelial intercellular cleft. *Bull Math Biol.* 1994;56:723–741.
18 Bungay SD, Gentry PA, Gentry RD. A mathematical model of lipid-mediated thrombin generation. *Math Med Biol.* 2003;20:105–129.
19 Xu CQ, Zeng YJ, Gregersen H. Dynamic model of the role of platelets in the blood coagulation system. *Med Eng Phys.* 2002;24:587–593.
20 Zarnitsina VI, Pokhilko AV, Ataullakhanov FI. A mathematical model for the spatio-temporal dynamics of intrinsic pathway of blood coagulation. II. results. *Thromb Res.* 1996;84:333–344.
21 Kuharsky AL, Fogelson AL. Surface-mediated control of blood coagulation: the role of binding site densities and platelet deposition. *Biophys J.* 2001;80:1050–1074.
22 Pokhilko AV, Ataullakhanov FI. Contact activation of blood coagulation: trigger properties and hysteresis hypothesis: kinetic recognition of foreign surfaces upon contact activation of blood coagulation: a hypothesis. *J Theor Biol.* 1998;191:213–219.
23 Butenas S, van 't Veer C, Cawthern K, et al. Models of blood coagulation. *Blood Coagul Fibrinolysis.* 2000;11(Suppl 1):S9–S13.
24 Clermont G, Chow CC, Constantine GM, et al. Mathematical and statistical modeling of acute inflammation. *Proc Internatl Fed Classific Soc.* 2004.
25 Lagoa CE, et al. Prediction of the acute inflammatory response from a mathematical model. *Chest.* 2003;124:S121.
26 Vodovotz Y, et al. Simulating the roles of NO in septic shock. *Nitric Oxide.* 2002;6:484.
27 Vodovotz Y, Clermont G, Chow C, An G. Mathematical models of the acute inflammatory response. *Curr Opin Crit Care.* 2004;10:383–390.
28 Parker JL, Adams HR. Selective inhibition of endothelium-dependent vasodilator capacity by Escherichia coli endotoxemia. *Circ Res.* 1993;72:539–551.
29 Myers PR, Parker JL, Tanner MA, Adams HR. Effects of cytokines tumor necrosis factor alpha and interleukin 1 beta on endotoxin-mediated inhibition of endothelium-derived relaxing factor bioactivity and nitric oxide production in vascular endothelium. *Shock.* 1994;1:73–78.
30 Parker JL, Myers PR, Zhong Q, et al. Inhibition of endothelium-dependent vasodilation by Escherichia coli endotoxemia. *Shock.* 1994;2:451–458.
31 Morrison AM, Wang P, Chaudry IH. A novel nonanticoagulant heparin prevents vascular endothelial cell dysfunction during hyperdynamic sepsis. *Shock.* 1996;6:46–51.
32 Crampin EJ, Smith NP, Hunter PJ. Multi-scale modelling and the IUPS physiome project. *J Mol Histol.* 2004;35:707–714.
33 Ye X, Chu J, Zhuang Y, Zhang S. Multi-scale methodology: a key to deciphering systems biology. *Front Biosci.* 2005;10:961–965.
34 Miller HG, Li RM. Measuring hot flashes: summary of a National Institutes of Health workshop. *Mayo Clin Proc.* 2004;79:777–781.
35 Food and Drug Administration. *Innovation or Stagnation: Challenge and Opportunity on the Critical Path to New Medical Products.* 2004:1–38.
36 Dayan P, Abbott LF. *Theoretical Neuroscience: Computational and Mathematical Modeling of Neural Systems.* Cambridge: The MIT Press, 2001.

192

Vascular Control through Tensegrity-Based Integration of Mechanics and Chemistry

Donald E. Ingber

Children's Hospital, Harvard Medical School, Boston, Massachusetts

The vascular endothelium paves the entire inner surface of the blood vessels that permeate every organ of our bodies. It is essentially the "Grand Canal" through which oxygen, nutrients, and other chemical stimuli travel to reach the cells that comprise our tissues. Although originally viewed as a passive boundary layer that interfaces with blood, the endothelium is now known to be a key integrator in human physiology, and to play a central role in control of thrombosis and the inflammatory response. The growth of new blood vessels by microvascular endothelium through angiogenesis and vasculogenesis also is critical for the development of all normal organs, in addition to contributing to the etiology of many diseases, including cancer, atherosclerosis, arthritis, and macular degeneration. Thus, understanding of vascular control can have a fundamental impact on biology and medicine.

Because biology has been dominated by a focus on genomics and molecular biochemistry over the past 50 years, there has been an enormous increase in our understanding of the molecules and genes that contribute to control of endothelial cell (EC) function. Cogent descriptions of these molecules and the important roles they play can be found in other chapters in this volume. However, in addition to being a conduit for transfer of chemical factors, the endothelium is also constantly exposed to mechanical stresses that influence its form and function. In particular, physical forces associated with changes in blood pressure, wall tension, and fluid shear stresses have been found to regulate various signal transduction pathways, change cytoskeletal structure, and alter gene expression in ECs, as well as modulate the angiogenic response (1–3). Moreover, these stress-dependent changes in cell behavior occur even in the presence of constant levels of soluble regulatory factors. Pathological changes of the vasculature that characterize some of the biggest killers of man, including atherosclerosis and hypertension, also are associated with local changes in cell or tissue mechanics (3). The challenge is therefore to understand how physical forces interplay with soluble factors and adhesive cues to regulate endothelial form and function.

The question of how living cells sense multiple simultaneous chemical and mechanical inputs, integrate this information, and produce a single concerted response (e.g., growth versus differentiation or apoptosis) cannot be addressed by classic reductionist approaches in which the goal is to identify a single critical gene or signaling molecule. To effectively attack this problem, it is necessary to understand how the whole cell and tissue process these different types of information and orchestrate complex cellular responses through collective interactions among thousands of different molecular components. But understanding cell regulation by mechanical forces involves more than information flow; we also must incorporate "physicality" into the cellular regulatory network.

This brief chapter reviews work from our laboratory carried out over the past two decades that has focused on this question of how physics and biochemistry are integrated in living tissues, and particularly in vascular ECs.

These studies have revealed that the ability of cells to coordinate their form and function, and to integrate mechanics and chemistry, is based on how they are structured architecturally at the molecular level. Specifically, we have found that living tissues use a particular form of architecture known as *tensegrity* to stabilize the shape of their constituent cells and extracellular matrix (ECM), as well as the form of the intracellular cytoskeleton (4–6). This latter observation is critical because many of the molecules and enzymes that mediate cell metabolism and signal transduction normally function when physically immobilized on insoluble cytoskeletal scaffolds (7). Because of this architectural arrangement, mechanical stresses transmitted across the surface membrane via adhesion receptors that connect the ECM to the cytoskeleton, such as integrins, are channeled through the cytoskeleton, and focused on multiple cytoskeletal-associated sensory molecules located throughout the cell and nucleus. Mechanical forces may thereby influence cellular biochemistry and signal transduction as a result of associated stress-dependent changes in molecular shape that alter chemical potentials, and hence change biochemical activities (1).

Thus, the challenge for the future is to understand molecular biochemistry in the physical context of living cells and tissues, and to develop quantitative descriptions of these processes. Only this way, will we be able to develop computational models, nanotechnologies, and engineered tissue replacements that will revolutionize how endothelial biology and vascular medicine are carried out in the future.

CONTROL OF VASCULAR MORPHOGENESIS: SOMETIMES FUNCTION FOLLOWS FORM

The old adage that "form follows function" is based on the idea that if an object performs a particular task, it must be designed to best support that function. Although this concept comes from the world of design, it also is often assumed to govern the natural world given the exquisite coupling between form and function observed in all living systems. However, in part because of this simple phrase, most people assume that the shape of a tissue is governed by the functions it must carry out. In fact, this is true in the case of the vasculature. Increases in blood pressure or hemodynamic shear stresses result in remodeling of the vascular wall, including changes in wall thickness and vessel diameter (2). But what is less well recognized is that the opposite is true as well: The form of a tissue also may govern its function.

The three dimensional (3D) shape of any structure – whether a building or a blood vessel – is determined through a dynamic balance of physical forces, and thus, every architectural form will be characterized by its own distinct pattern of internal forces. The stability of the vascular endothelium, for example, requires that tensional stresses generated in the cytoskeleton of each cell be balanced by its points of anchorage to the underlying ECM and to neighboring cells. Every position in a physical structure will therefore experience distinct architectural stresses – tension, compression, shear – which are characteristic of that point in the structural assembly. Hence, over 20 years ago, we raised the possibility that tissue shape may itself provide a physical form of positional information that could be equally important for control of tissue morphogenesis as gradients of soluble mitogens or growth factors (4,5,8).

Macroscale mechanical forces have been long known to regulate tissue form and function; the effects of compression on bone, tension on muscle, and shear on blood vessels are just a few examples. Thus, the novel point here is that *microscale* forces internal to tissues, and characteristic of their particular forms, may feed back to control cell growth and function. In particular, we suggested that because the ECM (e.g., basement membrane) is stiffer than the cell, then local changes in ECM thickness or mechanical compliance (flexibility) may feed back to alter cell shape. This is because as cells exert cytoskeletal traction forces on their ECM anchors and cell–cell adhesions, they will spontaneously remodel their internal cytoskeletal structure to minimize local stresses and strains, and this response will differ depending on the cell's position.

During angiogenesis, the endothelial basement membrane thins locally as a result of increased degradative activity at sites where new capillary sprouts will subsequently form (9). Vascular endothelial tissues are "prestressed" (i.e., exist in a state of isometric tension) as a result of the action of different cell types (e.g., adventitial fibroblasts, ECs) generating opposing forces and exerting them on a common basement membrane. Thus, we proposed that a local weak spot in the basement membrane might stretch out more than surrounding regions, and hence physically distort adherent ECs in these particular areas. EC shape and cytoskeletal structure are tightly coupled to cell growth, with more highly spread cells exhibiting enhanced growth in response to soluble mitogens (10). Changes in the mechanical force balance within small regions of the vascular tissue may therefore feed back to spatially constrain cell spreading and proliferation, and thereby drive branching morphogenesis during angiogenesis (4,5,8). These changes in physical forces might alter cellular biochemistry as a result of their transmission over discrete cytoskeletal filaments that orient much of the cellular metabolic machinery (7). Hence, to explain how these architectural forces could influence cell and tissue morphogenesis, it is first necessary to understand how cells and tissues structure themselves at the nanometer scale.

TENSEGRITY IN CELLS AND TISSUES

We based our vision of tissue form governing tissue function on a theoretical model of cell and tissue structure in which cells use a form of tensile architecture known as tensegrity (short for "tensional integrity") to maintain their shape stability (1,4–6). Tensegrity networks are composed of opposing tension and compression elements that balance each other and thereby create an internal prestress (resting isometric tension) that stabilizes the entire structure (Figure 192.1A). Tensegrity can be seen in the way pulling forces generated in muscles and resisted by bones stabilize the shape of our bodies. In tensegrities, a local stress is borne by multiple prestressed elements that rearrange at multiple size scales, rather than deform and yield locally, when a mechanical load is applied (Figure 192.1B). These rearrangements occur without compromising structural integrity, and they cause the entire network to physically strengthen when mechanically stressed (11). Because tensegrities are composed of discrete support elements, they also provide a way to transmit mechanical forces along specific paths, and to focus or concentrate stresses on distant sites and at different size scales.

Analysis of living organisms reveals the use of structural hierarchies (systems within systems) that span multiple size scales and are composed of tensed networks of muscles, bones, ECMs, cells, and cytoskeletal filaments (1,6). Moreover, the level of prestress (isometric tension) within this network tunes the whole system and governs how it will respond to external mechanical loads. This use of structural hierarchies and tensile prestress for mechanical stability and integrated

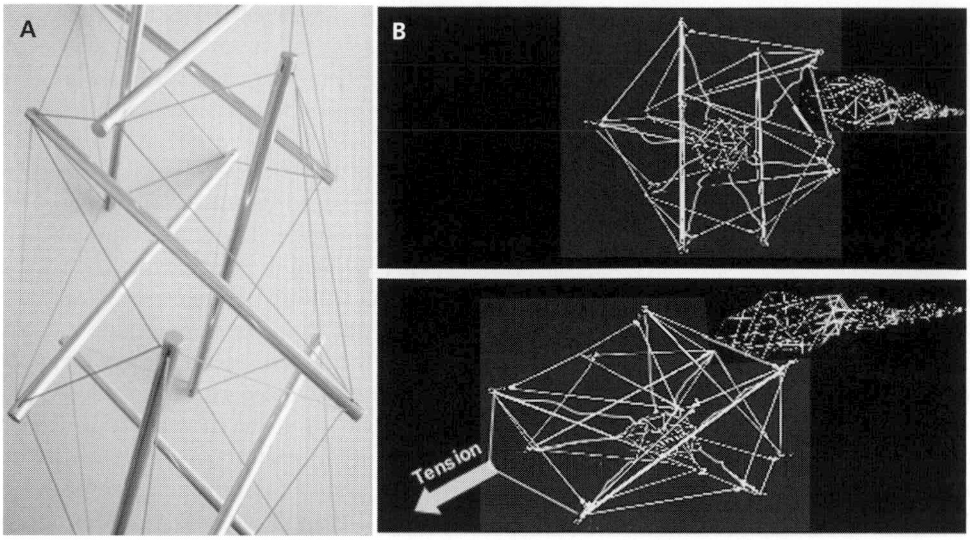

Figure 192.1. Tensegrity architecture. (**A**) A prestressed tensegrity framework composed of steel bars and tension cables. Note that the compression struts resist the inward pull of the surrounding tensed cables and thereby maintain a stabilizing prestress throughout the entire structure. (**B**) A hierarchical tensegrity structure that contains both a smaller spherical structure at its center (analogous to a nucleus in a cell) as well as elements that are themselves composed of tensegrity networks at smaller and smaller size scales. Note, as shown at the bottom, that when external tension is applied at one point on the surface of this network, all of the interconnected elements reorient to bear that load.

system-wide control is a hallmark of tensegrity architecture. In fact, blood vessels are known to be prestressed under physiological conditions (e.g., exhibit "residual strain"), and they increase their stiffness in direct proportion to the applied mechanical stress, just like living cells and tensegrity structures (11,12). As described earlier, this prestress results from a balance between contractile forces generated within the cytoskeletons of ECs and resisting forces exerted by other contractile cells (e.g., adventitial cells), by stiffened ECMs (e.g., tensed basement membranes, cross-linked collagen bundles), and by macroscale forces acting in the tissue microenvironment (e.g., hemodynamic forces in the cardiovascular system).

When a whole blood vessel is physically stressed, the ECMs (e.g., basement membranes, interstitial matrix) that hold the cells together become mechanically strained. These tensed ECMs adapt to the stress by rearranging on multiple size scales like a tensegrity structure (Figure 192.1B) and becoming stiffer. Cells sense distortion of the ECM or an increase in its rigidity as a tug on integrins or other transmembrane receptors that link the ECM to the contractile cytoskeleton (11). Integrins mechanically couple to the cytoskeleton by clustering together and recruiting a multitude of proteins (e.g., talin, vinculin, paxillin, zyxin, α-actinin) that self-assemble within focal adhesion anchoring structures to form a molecular bridge to the actin cytoskeleton. Similar force transfer occurs at cell–cell junctions through other transmembrane adhesion proteins, such as cadherins (13).

The cytoskeleton responds mechanically to forces transferred over the ECM and channeled through integrins by rearranging interlinked cytoskeletal filaments on an even smaller scale, thereby strengthening the cell against the potential injurious effects of mechanical distortion. Studies using confocal microscopy, birefringence microscopy or intracellular stress tomography of living cells expressing green fluorescent protein (GFP)-labeled molecules confirm that forces applied to the EC surface and transmitted to cytoskeletal scaffolds (i.e., microfilaments, intermediate filaments, and microtubules) result in localized stress concentrations deep in the cytoplasm and nucleus, as well as within basal focal adhesions where cells are physically anchored to the ECM (14–16). These forces also can produce stress-dependent changes in intracellular signal transduction and gene transcription (17). In contrast, these effects are not observed when similar forces are applied to other transmembrane molecules on the cell surface (e.g., metabolic receptors, growth factor receptors). Moreover, this long distance force transfer can be prevented by dissipating cytoskeletal prestress (15), the key stabilizing feature of tensegrity architecture. This "action at a distance" may explain how mitochondria that associate with microtubules may sense mechanical strain in ECs and release reactive oxygen species that are required for activation of nuclear factor (NF)-κB and induction of vascular cell adhesion molecule (VCAM)-1 gene expression (18). Conversely, increases in cytoskeletal tension inside the cell may feed back to promote structural changes in the surrounding ECM, such as unfolding of fibronectin molecules that promote ECM fibril assembly (19), and thereby add strength to the tissue at a higher systems level.

In tensegrity structures, tensional forces in the cables are balanced by compression struts and by external tethers to stiffened components that are part of larger tensegrity systems (see Figure 192.1). In cells, tensional forces that are generated within contractile microfilaments and transmitted over intermediate filaments and membranes are balanced by internal microtubules and cross-linked cytoskeletal filament bundles (e.g., microfilaments within filopodia and stereocilia, microtubules in the primary cilium), as well as by extracellular adhesions to the ECM and to other cells at the tissue level (1,6). This allows cells to shift compressive forces back and forth between microtubules and ECM adhesions, such that microtubules bear the majority of the prestress in poorly adherent cells, whereas the ECM bears most of the load in highly spread cells (20). The existence of this complementary force balance explains how forces applied at the cell surface can change the critical concentration of tubulin and control microtubule polymerization in various cells, as well as why cytoskeletal tension and ECM adhesions contribute to this response (6). Tensegrity-based mechanical interactions between microfilaments and microtubules also control the shape and mechanical stiffness of cells, as well as spatial regulation of microtubule turnover during directional cell migration (6). In addition, shifts in this cytoskeletal force balance may control activation of Rho as well as translocation of Rho and Rac to the membrane where these small guanine triphosphate (GTP)ases can influence focal adhesion assembly (21,22). In fact, both disruption of microtubules, which shifts mechanical loads onto ECM adhesions via a tensegrity force balance, and direct physical distortion of these adhesions activate Rho and stimulate focal adhesion formation.

This ability to channel mechanical forces over discrete molecular paths to sites deep inside hierarchical tensegrity structures can explain how cell distortion or mechanical stress application to cell surface integrins produces changes in nuclear shape and induces molecular realignment within nucleoli on progressively smaller size scales (14). These long-range effects are mediated largely by intermediate filaments, and to a lesser degree by the actin cytoskeleton (14–16). Importantly, focused stress transfer through the cytoskeleton and to the nucleus can activate stress-sensitive ion channels on the nuclear membrane and thereby impact gene transcription (23); it also could potentially alter chromatin folding, torsional strain within DNA, or access of key proteins (e.g., transcription factors, steroid hormones) to gene regulatory sites (1). The importance of this mechanical connectedness between the surface, cytoskeleton, and nucleus may explain why disruption, mutations, or deletions of intermediate filament proteins or nuclear lamins results in decreased mechanical stiffness of the whole cell and cell injury in response to mechanical stress, as well as reduced mechanical activation of gene transcription in the cardiovascular system (24).

In summary, if cells are hierarchical tensegrity systems, then mechanochemical conversion does not have to take place only at the cell membrane because forces applied at the surface can be transmitted to structural elements deep inside these prestressed networks. Perhaps even more importantly, stresses applied at the macroscale that mechanically strain ECMs and deform cells and their internal cytoskeleton may be able to trickle down to smaller size scales, be focused on specific molecular components, and produce structural rearrangements within these molecules that change their biochemical activities. In this context, it is interesting to note that individual proteins may be viewed and modeled computationally as tensegrity structures that globally reorient their local stiffened domains (e.g., α-chains, β-strands) when mechanically stressed (6). The tension exerted on these molecules through their prestressed cytoskeletal linkages may allow cells to sense signals in the cytoplasm in the presence of background thermal noise and to maintain optimum mechanosensitivity, not unlike the amplification scheme used by the tensed hair cells of the inner ear. Thus, the process of cellular mechanotransduction may be a manifestation of this fundamental tensegrity design principle that governs biological architecture, rather than being due to the action of any single molecular component (1).

Importantly, a mathematical formulation of the tensegrity model predicts both qualitative and quantitative results from experimental studies with various cultured cells (6,12). Recent studies also confirm that cytoskeletal prestress is critical for cell shape stability, and that microtubules bear compression in the cytoskeleton of living cells – the key hallmarks of the cellular tensegrity theory (6,25,26). Experiments using microfabricated substrates to control the shape of EC monolayers have directly confirmed that the architectural form of a tissue can feed back to control cell growth patterns as a result of local variations of internal mechanical stresses that are distributed through the cytoskeleton and resisted by cell–cell and cell–ECM adhesions (27). Cell distortion–dependent changes in cytoskeletal structure that feed back to regulate the Rho signaling pathway are key to this architectural form of growth control (28). Endothelial cell–cell adhesion molecules (e.g., vascular endothelial [VE]-cadherin) also influence cytoskeletal organization and cell proliferation by feeding into this same signaling pathway (29). Thus, tissue shape may feed back to control cell proliferation by concentrating stresses at particular sites and physically restructuring the cytoskeleton inside cells within these regions, thereby changing Rho signaling. Importantly, we recently showed that this mechanical control mechanism also operates in vivo: The local gradients of cell proliferation that drive epithelial budding during embryonic lung morphogenesis as well as endothelial branching during angiogenesis can be enhanced or inhibited by altering cytoskeletal tension using Rho signaling modulators (30). Thus, tissue development appears to be controlled through a mechanochemical mechanism, as we previously proposed (4–8).

CELLULAR COMPLEXITY: INTEGRATION BETWEEN STRUCTURAL AND INFORMATIONAL NETWORKS

The vascular endothelium is an outstanding example of a complex adaptive system in which system-level behaviors emerge through collective interactions among numerous components. The field of "Complexity" attempts to explain the behavior of systems that contain multiple interdependent subcomponents by focusing on information flow within dynamic regulatory networks. For example, it has been proposed based on study of theoretical networks that different cell types (e.g., lung versus liver) represent different "attractor" states, or "default" states, in the state space of the genome-wide gene regulatory network (31). In this model, stable high-dimensional attractor states arise at the *whole system level* as a consequence of particular regulatory interactions between the network components that impose constraints on the global dynamics of the network. As a result of these regulatory constraints, the cell cannot occupy any arbitrary network state, and tends to be attracted to only a handful of stable attractor states. This hypothesis is consistent with the observation that cells undergo all-or-none, phase transition-like switches between different fates during embryological development.

We have shown that ECs similarly switch between growth, differentiation, and apoptosis in an all-or-none manner when their physical shape is varied by being spatially constrained on microfabricated ECM islands (10,32). This finding that the cell-fate transitions normally induced by specific high affinity growth factors also can be triggered by unrelated signals that apparently lack molecular specificity (e.g., cell shape distortion) is consistent with the concept that these stable cell phenotypes represent high dimensional attractors (33,34). Importantly, we recently experimentally confirmed that cell fates represent high dimensional attractor states that self-organize within the genome-wide gene regulatory network of human cells (35; see Chapter 190 for more details).

The existence of attractors in cell regulatory networks that confer stability with respect to thousands of dimensions (e.g., gene expression levels) is important because it explains how cells can simultaneously sense multiple chemical, adhesive, and mechanical inputs and yet only switch on one of a limited number of specific, reproducible behavioral responses. But because it lacks physicality, this information flow-based theory of cell regulation cannot explain how specific 3D structures or tissue patterns come into existence. Interestingly, a key feature of the attractor model is that multiple regulatory elements (e.g., ensembles of genes and signaling proteins) must alter their values at the same time to produce an attractor switch. Given that mechanical forces and cell shape distortion simultaneously impact many cytoskeletal-associated signaling molecules, this may in part explain how mechanical forces and associated stress-dependent changes in cell shape are able to control cell-fate switching.

CONCLUSION

My laboratory has pursued both conventional and nonconventional areas of research in our pursuit of how blood vessels form and function. We have published many reports describing particular molecules and signaling components that mediate control of EC growth, differentiation, apoptosis, and directional motility, as well as contractility of vascular smooth muscle cells. But the most important insights have come from our analysis of how cells sense mechanical cues in their local microenvironment, and integrate them with chemical signals to control cell function and tissue development, as summarized above.

In simplest terms, there are two key fundamental insights relating to biological control that have emerged from our work. The first is the discovery of the critical need to bring pushes and pulls into balance to create the stability necessary for meaningful growth and development in living systems. This is true whether we are considering a single EC, a capillary network, a large blood vessel, or the entire cardiovascular system. Most investigators are aware of the importance of hemodynamic forces and other mechanical stresses for development of the vasculature. The surprising insight here is the key role that isometric tension and hidden structural forces play in vascular control.

The second theme is the importance of context: whether genes, molecules, cells, tissues or organs, all living things are context-dependent. This perspective emphasizes the importance of structural context when considering the function of multicomponent *networks*, such as where a cell is located within a tissue, where a particular molecule is positioned inside a cell, or the physical stresses these components experience at these sites. But the implications are much larger because once one understands the importance of physical context, one then realizes the innate limitations of current gene- and molecule-focused analytical approaches.

The use of tensegrity and its link to dynamic information processing networks offers potential new avenues of exploration in physiology, pharmacology, and bioengineering of the vasculature. For example, tensegrity can explain how molecular filaments (e.g., microtubules, intermediate filaments) and subcellular components (e.g., integrins, nuclear lamins) that are not contractile can contribute to the impaired whole cell contractile response that characterizes many vascular and cardiovascular diseases. Because changes in the mechanical properties of the whole cell can contribute greatly to development of vascular diseases (e.g., hypertension), it may be possible to screen for drugs that alter global changes in the integrated mechanical behavior of the cytoskeleton that will prove effective regardless of the particular molecular defect or defects that lead to development of the disease. Analyzing their effects on whole-cell mechanics also may minimize the unwanted side effects of drugs that are designed to target single molecules. Finally, because this work has uncovered fundamental design criteria that govern how nature builds and controls on the nanometer scale, this knowledge may be harnessed to create entirely new types of nanotechnology-based materials, microdevices, and control technologies for medicine in the future. Examples include self-assembling mechanochemical scaffolds for tissue engineering that respond to local mechanical cues by elaborating growth factors and altering vascular tone, or therapeutic control technologies that use dynamical inputs (chemical, mechanical, or electrical) to switch capillary cells from growth to apoptosis, and thereby shrink tumors, as well as atherosclerotic plaques.

Thus, the challenge for the future is to develop quantitative descriptions of both physical and information processing networks, and eventually to analyze how they are seamlessly integrated so as to produce vascular tissues in which structure and function can never be dissociated. With this understanding in hand, it will be possible to engineer artificial blood vessels, and to build biologically inspired materials and cellular reprogramming technologies for clinical care that meld naturally with our own bodies and provide a new handle on physiological control.

KEY POINTS

- ECs use tensegrity architecture to channel mechanical forces through the cytoskeleton and to influence cellular biochemistry so that their growth and function are optimally adapted to their physical and chemical microenvironment.
- Mechanical forces are critical regulators of cellular biochemistry and gene expression as well as tissue development.
- The cytoskeletal scaffolds that physically interconnect with extracellular matrices through cell surface integrin receptors provide a mechanism for ECs to integrate mechanical and chemical signals.

Future Goals

- To develop quantitative descriptions and computational models of how physical networks and information processing networks are integrated within living cells
- To use these models of cellular information integration to develop new therapeutic modalities and biologically inspired control technologies for clinical care

ACKNOWLEDGMENTS

This work was supported by grants from NIH, NASA, DOD, DARPA, ARO, NSF, and the Keck Foundation.

REFERENCES

1 Ingber DE. Cellular mechanotransduction: putting all the pieces together again. *FASEB J*. 2006;20:811–827.

2 Davies PF. Flow-mediated endothelial mechanotransduction. *Physiol Rev*. 1995;75:519–560.

3 Ingber DE. Mechanobiology and diseases of mechanotransduction. *Ann Med*. 2003;35:564–577.

4 Ingber DE, Madri JA, Jamieson JD. Role of basal lamina in the neoplastic disorganization of tissue architecture. *Proc Natl Acad Sci USA*. 1981;78:3901–3905.

5 Ingber DE, Jamieson JD. Cells as tensegrity structures: architectural regulation of histodifferentiation by physical forces transduced over basement membrane. In: Andersson LC, Gahmberg CG, Ekblom P, eds. *Gene Expression During Normal and Malignant Differentiation*. Orlando: Academic Press; 1985:13–32.

6 Ingber DE. Cellular tensegrity revisited I. Cell structure and hierarchical systems biology. *J Cell Sci*. 2003;116:1157–1173.

7 Ingber DE. The riddle of morphogenesis: a question of solution chemistry or molecular cell engineering? *Cell*. 1993;75:1249–1252.

8 Huang S, Ingber DE. The structural and mechanical complexity of cell growth control. *Nat Cell Biol*. 1999;1:E131–E138.

9 Ausprunk DH, Folkman J. Migration and proliferation of endothelial cells in preformed and newly formed blood vessels during tumor angiogenesis. *Microvasc Res*. 1997;14:53–65.

10 Chen CS, Mrksich M, Huang S, et al. Geometric control of cell life and death. *Science*. 1997;276:425–1428.

11 Wang N, Butler JP, Ingber DE. Mechanotransduction across the cell surface and through the cytoskeleton. *Science*. 1993;260: 1124–1127.

12 Stamenovic D, Ingber DE. Models of cytoskeletal mechanics of adherent cells. *Biomech Model Mechanobiol*. 2002;1:95–108.

13 Potard US, Butler JP, Wang N. Cytoskeletal mechanics in confluent epithelial cells probed through integrins and E-cadherins. *Am J Physiol*. 1997;272:C1654–C1663.

14 Maniotis AJ, Bojanowski K, Ingber DE. Mechanical continuity and reversible chromosome disassembly within intact genomes removed from living cells. *J Cell Biochem*. 1997;65:114–130.

15 Hu S, Chen J, Fabry B, et al. Intracellular stress tomography reveals stress focusing and structural anisotropy in cytoskeleton of living cells. *Am J Physiol Cell Physiol*. 2003;285:C1082–C1090.

16 Helmke BP, Rosen AB, Davies PF. Mapping mechanical strain of an endogenous cytoskeletal network in living endothelial cells. *Biophys J*. 2003;84:2691–2699.

17 Meyer CJ, Alenghat FJ, Rim P, et al. Mechanical control of cyclic AMP signaling and gene transcription through integrins. *Nat Cell Biol*. 2000;2:666–668.

18 Ali MH, Pearlstein DP, Mathieu CE, Schumacker PT. Mitochondrial requirement for endothelial responses to cyclic strain: implications for mechanotransduction. *Am J Physiol Lung Cell Mol Physiol*. 2004;287:L486–L496.

19 Baneyx G, Baugh L, Vogel V. Fibronectin extension and unfolding within cell matrix fibrils controlled by cytoskeletal tension. *Proc Natl Acad Sci USA*. 2002;99:5139–5143.

20 Hu S, Chen J, Wang, N. Cell spreading controls balance of prestress by microtubules and extracellular matrix. *Front Biosci*. 2004;9:2177–2182.

21 Putnam AJ, Cunningham JJ, Pillemer BB, Mooney DJ. External mechanical strain regulates membrane targeting of Rho GTPases by controlling microtubule assembly. *Am J Physiol Cell Physiol*. 2003;284:C627–C639.

22 Liu BP, Chrzanowska-Wodnicka M, Burridge K. Microtubule depolymerization induces stress fibers, focal adhesions, and DNA synthesis via the GTP-binding protein Rho. *Cell Adhes Commun*. 1998;5:249–255.

23 Itano N, Okamoto S, Zhang D, et al. Cell spreading controls endoplasmic and nuclear calcium: a physical gene regulation pathway

from the cell surface to the nucleus. *Proc Natl Acad Sci USA.* 2003;100:5181–5186.

24 Lammerding J, Schulze PC, Takahasi T, et al. Lamin A/C deficiency causes defective nuclear mechanics and mechanotransduction. *J Clin Invest*. 2004;113:370–378.

25 Kumar S, Maxwell IZ, Heisterkamp A, et al. Viscoelastic retraction of single living stress fibers and its impact on cell shape, cytoskeletal organization and extracellular matrix mechanics. *Biophys J.* 2006;90:1–12.

26 Brangwynne C, Macintosh FC, Kumar S, et al. Microtubules can bear enhanced compressive loads in living cells due to lateral reinforcement . *J Cell Biol.* 2006;173:1175–1183.

27 Nelson CM, Jean RP, Tan JL, et al. Emergent patterns of growth controlled by multicellular form and mechanics. *Proc Natl Acad Sci USA.* 2005;102:11594–11599.

28 Mammoto A, Huang S, Moore K, et al. Role of RhoA, mDia, and ROCK in cell shape-dependent control of the Skp2-p27kip1 pathway and the G1/S transition. *J Biol Chem.* 2004;279:26323–26330.

29 Nelson CM, Pirone DM, Tan JL, Chen CS. Vascular endothelial-cadherin regulates cytoskeletal tension, cell spreading, and focal adhesions by stimulating RhoA. *Mol Biol Cell.* 2004;15:2943–2953.

30 Moore KA, Polte T, Huang S, et al. Control of basement membrane remodeling and epithelial branching morphogenesis inembryonic lung by Rho and cytoskeletal tension. *Dev Dyn.* 2005;232:268–281.

31 Kauffman SA. *The Origins of Order.* New York: Oxford University Press, 1993.

32 Dike L, Chen CS, Mrkisch M, et al. Geometric control of switching between growth, apoptosis, and differentiation during angiogenesis using micropatterned substrates. *In Vitro Cell Dev Biol.* 1999;35:441–448.

33 Huang S, Ingber DE. Shape-dependent control of cell growth, differentiation, and apoptosis: switching between attractors in cell regulatory networks. *Exp Cell Res.* 2000;261:91–103.

34 Ingber DE. Tensegrity II. How structural networks influence cellular information processing networks. *J Cell Sci.* 2003;116:1397–1408.

35 Huang S, Eichler G, Bar-Yam Y, Ingber DE. Cell fates as attractors in gene expression state space. *Phys Rev Lett.* 2005;94:128701.

193

Simulating the Impact of Angiogenesis on Multiscale Tumor Growth Dynamics Using an Agent-Based Model

Chaitanya A. Athale and Thomas S. Deisboeck

Complex Biosystems Modeling Laboratory, Harvard-MIT (HST) Athinoula A. Martinos Center for Biomedical Imaging, Massachusetts General Hospital, Charlestown, Massachusetts

The growth of solid tumors and metastases beyond a certain size limit of 1.4 to 2.5 mm in diameter (1) is enabled by the tumor-induced sprouting of blood vessels, a process termed *angiogenesis.* Folkman and colleagues (2) first identified a soluble growth-factor referred to as tumor-angiogenesis factor (TAF) produced by tumors that could induce the growth of such new capillaries. This in turn led to the concept of antiangiogenesis as tumor therapy (3). Experimental verification of this mechanism in a rabbit model (4), research discovering details of the cellular processes of tumor angiogenesis (5), and in vivo inhibition of tumor-induced angiogenesis (6) form the basis of most of the work that followed.

In parallel to the experimental work, attempts were being made early on to mathematically model the endothelial cell (EC) sprouting, the interconnection and looping of blood vessels, and their growth toward the tumor induced by TAFs (7). Meanwhile, a model of the extravascular transport of reporter molecules in tumor and normal tissue showed a good fit to one-dimensional diffusion. Furthermore, real tumors were found to provide fewer hindrances to macromolecular transport by diffusion than were surrounding tissue (8). Modeling work that compared experimentally measured TAF profiles predicted that this factor could be secreted outside the neoplasm only after the tumor exceeds a threshold size (9). Patterns of angiogenesis have been examined subsequently using reaction diffusion simulations that focused (for example) on TAF diffusion (10). Furthermore, work by Byrne and Chaplain (11) examined the role of the migration rate of newly sprouted capillaries and its consequences on brush borders and the parameters contributing to successful vascularization of tumors. Another study concentrated on the role of EC proliferation and migration and showed good comparison to experimental data (12). In addition, detailed biochemistry, molecular transport models, and EC chemotaxis has been used to produce good fits with experimental data on the onset of vascularization and the rate of migration of the growing capillary tip (13). Recent work has attempted to use numerical simulations that explicitly include molecular interactions as a way to test various angiogenesis-directed chemotherapeutic approaches (14).

Work in our laboratory has examined the emergent properties of brain tumor growth using, for example, quantitative experimental measurements (15), agent-based simulations (16), and a theoretical chemotaxis model that demonstrates homotype attraction as critical for the branch formation of invasive cells (17). Further work using our agent-based model showed a correlation between the tumor's surface fracticality (the scale-invariant "self-similar" nature of the tumor surface indicated by a noninteger value of the dimension of the surface) and its expansion rate (18), and has been augmented to include a two-gene network that tests the predictive value of the gene expression values (19). Experimental observations of gliomas have shown evidence that migrating cells do not proliferate and, conversely, proliferating ones do not migrate, which in turn led Giese and colleagues (20) to propose the concept of "dichotomy" in glioma. Because both in vitro and in vivo data have implicated the epidermal growth factor receptor (EGFR) pathway in proliferative as well as migratory responses (21) and phospholipase C γ (PLCγ) in the "dichotomy" (22,23), we have recently adapted our agent-based simulation to model an EGFR gene–protein interaction-based molecular "decision" network that predicts subcellular, single cell, and multicellular patterns within the growing tumor (24). Here we continue that line of work by integrating an angiogenesis response to a TAF into our multiscale *agent-based* brain tumor modeling platform. Each cell in this particular modeling framework (see Chapter 188) is simulated as an autonomous "agent" that uses local information to make decisions about its "fate"; that is, depending on its internal gene–protein interaction network state, each tumor cell proliferates,

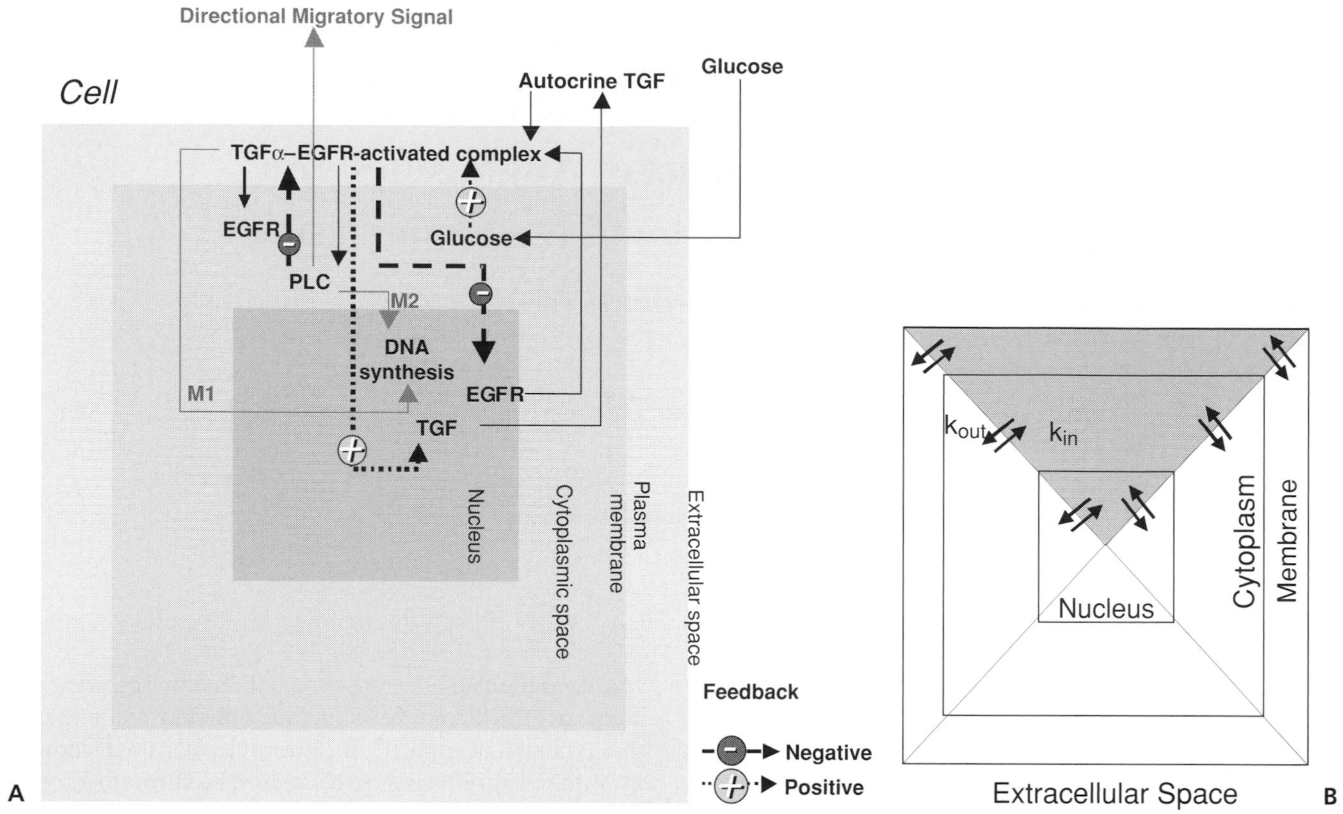

Figure 193.1. (**A**) Shown is the interaction network involving genes and proteins of the simplified EGFR pathway. The arrows indicate interactions and those with a plus or minus sign represent positive or negative feedbacks, respectively. The components of the network are restricted to specific subcellular compartments. (**B**) The directional geometry of each cellular subcompartment is displayed and mass exchange between spatially oriented subcompartments due to diffusion is demonstrated for one region. The solid arrows represent the in (k_{in}) and out (k_{out}) rates of mass exchange.

migrates, or remains stationary at a particular point in time. The dynamics of the multicellular tumor system thus emerge from the collective behavior of these autonomous cells.

METHODS

Our computational model consists of the following three *modules*: (a) the tumor growth module, (b) the extracellular grid module, and (c) the angiogenesis module.

Tumor Module

We model the tumor as a collection of cells in which the smallest unit is a molecular species. The model focuses on a simplified EGFR *gene–protein interaction network* involved in phenotypic – migration–proliferation – decision making. The molecules are localized in one of three compartments, either the cell membrane, cytoplasm, or nucleus (Figure 193.1A).

Specifically, the cell-surface EGFR protein can reversibly bind to extracellular transforming growth factor (TGF)-α. This complex dimerizes and can undergo autophosphorylation. In turn, this phosphorylated complex is capable of signaling to the mitotic pathway and phosphorylating PLCγ. PLCγ is then bound to the membrane-anchored TGF-α/EGFR–activated complex and downregulates the complex (negative feedback). The ligand–receptor (LR) active complex upregulates the gene expression of TGF-α (positive feedback) and downregulates that of EGFR (negative feedback), both in the nucleus. The RNA transcribed from the two genes, *TGF-α* and *EGFR*, is transported to the cytoplasm and translated into protein at a constant rate. In addition to synthesis, the cytoplasmic pool of EGFR and TGF-α is altered by internalization and dissociation of the LR complex. Due to receptor recycling, EGFR enters the membrane from the cytoplasmic pool at a constant rate and internalizes in the reverse direction. Similarly, TGF-α protein is secreted from the cytoplasm into the extracellular space. It is this interaction network, modeled as a set of coupled mass balance equations, that determines the exchange of material between compartments and within them. The mass balance scheme is of the generic form,

$$\frac{dX_i}{dt} = s \cdot X_i - r \cdot X_i \qquad \text{(Equation 1)}$$

where X_i is the concentration of the molecule, s is the rate at which this molecule is synthesized, and r is its rate of degradation. Coupling with other molecules occurs when the s or *r* values change due to other molecular components of the network.

The subcellular compartments of the cell (plasma membrane, cytoplasm, and nucleus) are all divided into four spatial regions in the four cardinal directions North (N), South (S), East (E), and West (W), and each subcompartment is connected to two other spatially adjacent ones (Figure 193.1B). The flux of molecular concentrations between different regions (N, S, E, and W) also is governed by a mass balance kinetic scheme, based on the values of diffusion rates of each molecular species. Thus, subcellular localization is determined by the following equation:

$$\frac{dX_{i,j}}{dt} = k_{in} \cdot \left[X_{i,j-1} + X_{i,j+1}\right] - 2 \cdot k_{out} \cdot X_{i,j} \qquad \text{(Equation 2)}$$

where $X_{i,j}$ is the concentration of the molecule X_i in a given compartment *j*, while $X_{i,j-1}$ refers to the neighboring compartment before $X_{i,j}$ and $X_{i,j+1}$ to the compartment after $X_{i,j}$. Here, *j* (1 to 4) is the compartment number and k_{in} and k_{out} are the flux rate constants into and out of the compartment, respectively.

Behavior of single cells in our model is based on converting the status of the interaction network into a particular cellular *phenotype*. Presently, we consider the following phenotypic states: cell quiescence, proliferation, or migration. Dittmar and colleagues (25) could show in breast cancer cells that PLCγ is activated transiently and to a greater extent during migration and more gradually in the proliferative "mode." Hypothesizing that such a mechanism also operates in gliomas, we adopt a simple threshold, σ_{PLC}, to decide whether the cell should undergo migration or not. Thus, the *potential to migrate* is assessed by a cell evaluating the following function:

$$M_n\left[(PLC\gamma)\right] = \left[\frac{d(PLC\gamma)}{dt}\right]_n \qquad \text{(Equation 3)}$$

where $d(PLC\gamma)/dt$ is the change in concentration of PLCγ over time (*t*) and *n* is the cell number. If $M_n > \sigma_{PLC}$, the cell becomes eligible to migrate, otherwise it could proliferate or become quiescent (i.e., it neither proliferates nor migrates, yet remains viable). If the cell decides to migrate, the direction of migration is expressed in terms of a local valuation function for each neighboring lattice point *j* in the *von Neumann* neighborhood as follows:

$$L_j\lfloor(PLC\gamma), n_j\rfloor = \left(1 - n_j\right) \cdot \left[(PLC\gamma)\right]_j - \arg\max_{j \le m}\left[(PLC\gamma)\right]_j \cdot \Psi_{PLC} \qquad \text{(Equation 4)}$$

where L_j is the value of a grid point in the neighborhood of the cell; the neighboring grid locations and compartments are numbered as *j* (1 to 4). In a given subcellular compartment *j*, $(PLC\gamma)$ is the concentration of activated PLCγ, *n* is the number of cells at that location in the neighborhood, and *m* is the total number of compartments (here: $m = 4$). $\Psi_{PLC} \in [0, 1]$ denotes the so-called *search-precision* parameter, representing the biological equivalent of a receptor-driven, nonerroneous evaluation of the permissibility of the microenvironment, as discussed in detail in our previous work (24,26,27). For example, when the search precision parameter $\Psi_{PLC} = 0$, the cell performs a pure random walk, and when $\Psi_{PLC} = 1$ the cell never commits chemotactic "mistakes" and always migrates fully biased to the most permissive location with the highest level of PLCγ (based on Equation 4).

If the change in concentration of active PLCγ is below the migration-threshold, σ_{PLC}, yet above a set noise threshold, σ_n, then the cell evaluates the *proliferative potential* $P_{prolif} \geq 0$. This potential is calculated using Equation 5,

$$P_{prolif}\left[(LR)\right] = (LR) - \sigma_{EGFR} \qquad \text{(Equation 5)}$$

where (LR) is the concentration of ligand-bound phosphorylated TGFα–EGFR complex in a cell and σ_{EGFR} is the LR concentration threshold (24). This function is derived from experimental observations citing cell proliferation in relation to an EGF-receptor threshold (28) and an experimentally validated model that relates receptor occupancy to percent maximal proliferation (29). Again, if neither the conditions for migration nor proliferation are met, the cell turns quiescent. Cell death is currently not considered in our model here.

Extracellular Module

The simulation is run in two dimensions (2D) and initialized with 100 tumor cells in the center of the grid (at [X,Y = 100,100]), in itself a rectangular section of a toroidal lattice with one side (R) of length R = 200. The grid represents the extracellular space, and the edges of the grid behave as a perfect sink. In the North East (NE) quadrant, we model a circular source of nutrients, which is representative of a cross-section of a blood vessel (as discussed later) from which the "example" nutrient (i.e., glucose) constantly diffuses down a concentration gradient. The boundary of the grid functions as a sink, and the blood vessel as a replenished source. The concentration of glucose at any point *s* on the grid changes based on,

$$J_s = -D_g \cdot \nabla C_g \qquad \text{(Equation 6)}$$

where D_g stands for the extracellular glucose diffusion constant, taken to be 6.7×10^{-7} cm^2/s from previous work (30), and ∇C_g is the glucose concentration gradient in one dimension (1D). Glucose enters the lattice or virtual tissue system through the blood vessel walls and this flux is represented by,

$$J_w = q_g\left(C_s - C_g\big|_{wall}\right) \qquad \text{(Equation 7)}$$

where vessel permeability to glucose q_g is taken to be 3×10^{-5} cm/s from Gatenby (31) and the serum concentration of glucose $C_s \approx 5$ mM. The diffusion of glucose from the vessel edge occurs in an orthogonal manner. At the same time, each

Table 193-1: Values of Parameters and Their Definitions as Used in the Angiogenesis Part of the Simulation, Based on Literature Sources Where Available

Parameter	*Value*	*Description*	*References*
D_g	6.7×10^{-7} cm^2/s	Diffusion coefficient of glucose	30
D_c	10^{-8} cm^2/s	Diffusion coefficient of TAF	30
q_g	3×10^{-5} cm/s	Blood vessel permeability to glucose	30
C_s	5 mM	Serum concentration of glucose	30
λ_c	0.1 min^{-1}	Decay rate of TAF	11
α_c	0.1 mM/min	Production rate of TAF	11
μ	2×10^{-9} to 2×10^{-8} cm^2/s · M	Coefficient of random motility of expanding blood vessel	11
χ	2.6×10^3 cm^2/s · M	Coefficient of chemotaxis of expanding blood vessel	11

viable tumor cell (regardless of its phenotype) produces TAF molecules at a constant rate, and they also diffuse through the extracellular matrix down a concentration gradient, as in Equation 6, with the grid edges serving as a sink and the individual viable tumor cells as a source. The TAF gradient formation is based on the Byrne and Chaplain model (11) as follows,

$$\frac{\partial c}{\partial t} = D_c \cdot \frac{\partial^2 c}{\partial x^2} - \lambda_c \cdot c + \alpha_c \cdot m \qquad \text{(Equation 8)}$$

where c is the concentration of TAF, D_c its diffusion constant (10^{-8} cm^2/s estimated by Byrne and Chaplain [11]), λ_c its natural decay rate, and α_c the rate of production. The number of cells is denoted by m and indicates that, in the absence of tumor cells, no TAF production occurs.

Angiogenesis Module

The blood vessel in the NE quadrant (at [X,Y = 150,50]) is modeled as a 2D cross-section with an expanding radius, accounting for the effect of angiogenesis. Specifically, the vessel's advancing tips are implicit in its increasing radius; that is, the radius of the cross-sectional blood vessel is a continuum with many small vessels that sprout and grow. Modifying the 1D model of Bryne and Chaplain (11) we model this change in vessel radius (ρ) as,

$$\frac{\partial \rho}{\partial t} = v - d \qquad \text{(Equation 9)}$$

where v is the speed of expansion due to blood vessel tips and d is the death rate of the capillaries in units of capillary length per unit area per unit time. This is further resolved by,

$$d = \gamma \cdot \rho \qquad \text{(Equation 10)}$$

where γ is the rate of length reduction in cm/min. Note that, in the absence of experimentally measured values, we set $\gamma = 0$.

The velocity of blood vessel sprouts in angiogenesis ($v(x,t)$) is a combination of random and chemotactic tip motion. The constants that determine this are referred to as *coefficients of random motility* (μ) and *chemotaxis* (χ) and the tip extension rate is,

$$v = \mu \cdot \frac{\partial n}{\partial x} + \chi \cdot \frac{\partial c}{\partial x} \qquad \text{(Equation 11)}$$

In vitro experiments measuring motility response of ECs to fibroblast growth factor (FGF) gradients (in the order of magnitude of 10^{-10} M) provide values of $\chi = 2.6 \times 10^3$ cm^2 s^{-1} M^{-1} from previous works (32–34). The TAF concentration is represented by c and evolves according to Equation 8. The variable n, which is the tip density of capillaries, has been modeled in previous work (11) to change over space, yet as a first approximation here, we consider it to be constant within the region of the blood vessel: $\partial_n/\partial_x = 0$.

PRELIMINARY RESULTS

Employing the aforementioned algorithm, implemented in Java and using the agent-based modeling toolkit Repast (http://repast.sourceforge.net/), the model is integrated using finite time stepping. The simulation is terminated when the first cancer cell enters the edge of the expanding blood vessel. A single run with the parameters listed in Table 193-1 here (revised from our earlier work [24]) took 2 hours with visualization on a single 2.40-GHz Xeon processor workstation running WinXP. The human U87 glioma cell line served as the "virtual" cell line with 2.1×10^5 EGFR molecules per cell (35).

2D snapshots of the expanding tumor system at consecutive time points (Figure 193.2; for color reproduction, see Color Plate 193.2) confirm that both the tumor system and blood vessel grow *toward each other* as a direct consequence of the dynamic feedback between tumor cells and ECs, mediated by TAF.

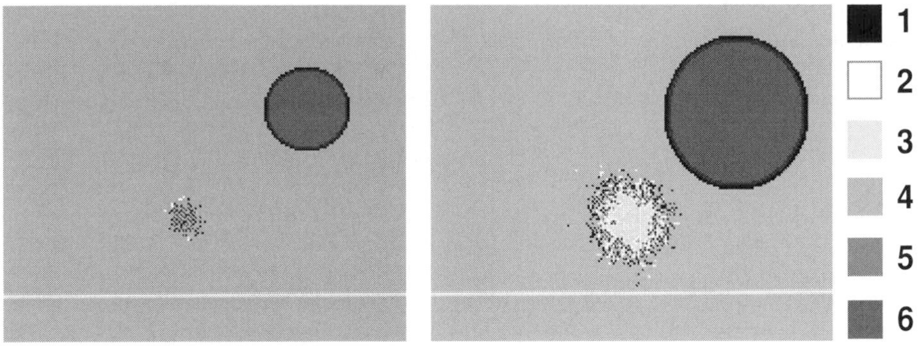

Figure 193.2. Depicted is a single-run, 2D visualization of the growing tumor at consecutive (scaled) time points $t = 50$ and 200 min, respectively. The key to the color coding (*right*) is as follows: (1) migrating, (2) proliferating and (3) quiescent tumor cells, (4) the extracellular matrix, (5) blood vessel outline and (6) glucose. Note the spatio-temporal expansion of the tumor toward the nutrient that diffuses from an expanding blood vessel located in the NE quadrant. For color reproduction, see Color Plate 193.2.

We also have studied the subcellular localization of the PLCγ protein in several tumor cells at the end-point (data not shown). These preliminary results indicate that PLCγ is *heterogeneously* distributed in its cytosolic localization.

CONCLUSION

As described in a recent review by Mantzaris and colleagues (36), tumor nourishment is generally believed to consist of two components, the initial diffusion of nutrients, typically termed as *avascular tumor growth state,* and the *vascular growth phase,* which involves the formation of blood vessels or angiogenesis. We have added such a glucose diffusion–yielding neovascularization module to our multiscale agent-based brain tumor model. This model allows us to study the microenvironmental impact on subcellular, microscopic, and macroscopic tumor growth patterns. Building on our previous gene–protein interaction network model (24), the present simulation combines in a sense both phases of cancer growth, avascular and vascular. We found that:

- Initially, the virtual tumor grows primarily based on autocrine and paracrine stimuli, here represented solely through TGF-α.
- However, once extracellular glucose reaches the tumor, diffusing from the distant blood vessel, the effects of tumor cellular uptake, changes in intracellular phosphorylation state and phenotypic alterations predominate. Typically, at this point cells at the periphery of the tumor then tend to regain their migratory phenotype in response to the strong chemotactic stimulus (Figure 193.2, left panel).
- The diffusion of TAF, which is typically 10 times slower than glucose (see diffusion constant values of glucose (D_g) and TAF (D_c) in Table 193-1), eventually reaches the edge of the blood vessel, which in turn leads the vessel diameter to expand (Figure 193.2, right panel).

We then conclude for the pilot simulation experiment that:

- Driven by the *feedback* mechanism between tumor-secreted TAF and capillary growth, angiogenesis appears to affect tumor growth *across* various scales of interest.
- In particular, the subcellular localization patterns of the PLCγ protein concentration should have an impact on the cells' phenotypic decision between proliferation and migration, and therefore ultimately alter the tumor's overall spatio-temporal growth dynamics.

Future work will have to include not only more simulation runs and detailed numerical analyses, but will also need to investigate, for example, the level of robustness of the tumor growth patterns with regards to dynamical changes of nutrient supply. Relating the tumor cells' distinct PLCγ polarization patterns and resulting phenotypic behavior to the vessel's growth dynamics is another area of interest. This may also spur in vitro and in vivo experiments aimed at comparing the dynamics of molecular profiles of proteins such as phospho-PLCγ and TGF-α in both the avascular and vascular tumor growth stages.

Necessarily, in this first approximation, the intricate processes involved in tumor angiogenesis had to be simplified to make the results tractable. However, this approach will allow us to add further complexity in future iterations, geared toward investigating the inherent nonlinearity of these biological processes in more detail. Therefore, although still at its beginning, we argue that multiscale cancer models such as this one represent a promising step toward a new set of research tools to simulate the impact of the microenvironment on subcellular signaling pathways, its phenotypic consequences, and the multicellular patterns this may trigger. This line of work generates testable hypotheses and thus contributes to our understanding about cancer biology.

ACKNOWLEDGMENTS

This work has been supported in part by NIH grants CA 085139 and CA 113004 and by the Harvard-MIT (HST) Athinoula A. Martinos Center for Biomedical Imaging and the Department of Radiology at

Massachusetts General Hospital. We thank Dr. Le Zhang (Complex Biosystems Modeling Laboratory, MGH) for critical review of the manuscript.

REFERENCES

1 Knighton D, Ausprunk D, Tapper D, Folkman J. Avascular and vascular phases of tumour growth in the chick embryo. *Br J Cancer*. 1977;35:347–356.

2 Folkman J, Merler E, Abernathy C, Williams G. Isolation of a tumor factor responsible or angiogenesis. *J Exp Med*. 1971;133: 275–288.

3 Folkman J. Anti-angiogenesis: new concept for therapy of solid tumors. *Ann Surg*. 1972;175:409–416.

4 Gimbrone MA Jr, Cotran RS, Leapman SB, Folkman J. Tumor growth and neovascularization: an experimental model using the rabbit cornea. *J Natl Cancer Inst*. 1974;52:413–427.

5 Ausprunk DH, Folkman J. Migration and proliferation of endothelial cells in preformed and newly formed blood vessels during tumor angiogenesis. *Microvasc Res*. 1977;14:53–65.

6 Preis I, Langer R, Brem H, Folkman J. Inhibition of neovascularization by an extract derived from vitreous. *Am J Ophthalmol*. 1977;84:323–328.

7 Deakin AS. Model for initial vascular patterns in melanoma transplants. *Growth*. 1976;40:191–201.

8 Nugent LJ, Jain RK. Extravascular diffusion in normal and neoplastic tissues. *Cancer Res*. 1984;44:238–244.

9 Chaplain MA, Sleeman BD. A mathematical model for the production and secretion of tumour angiogenesis factor in tumours. *IMA J Math Appl Med Biol*. 1990;7:93–108.

10 Chaplain MA. The mathematical modelling of tumour angiogenesis and invasion. *Acta Biotheor*. 1995;43:387–402.

11 Byrne HM, Chaplain MA. Mathematical models for tumour angiogenesis: numerical simulations and nonlinear wave solutions. *Bull Math Biol*. 1995;57:461–486.

12 Chaplain MA, Giles SM, Sleeman BD, Jarvis RJ. A mathematical analysis of a model for tumour angiogenesis. *J Math Biol*. 1995;33:744–770.

13 Levine HA, Pamuk S, Sleeman BD, Nilsen-Hamilton M. Mathematical modeling of capillary formation and development in tumor angiogenesis: penetration into the stroma. *Bull Math Biol*. 2001;63:801–863.

14 Chaplain M, Anderson A. Mathematical modelling of tumour-induced angiogenesis: network growth and structure. *Cancer Treat Res*. 2004;117:51–75.

15 Deisboeck TS, Berens ME, Kansal AR, et al. Pattern of self-organization in tumour systems: complex growth dynamics in a novel brain tumour spheroid model. *Cell Prolif*. 2001;34:115–134.

16 Mansury Y, Kimura M, Lobo J, Deisboeck T. Emerging patterns in tumor systems: simulating the dynamics of multicellular clusters with an agent-based spatial agglomeration model. *J Theor Biol*. 2002;219:343–370.

17 Sander LM, Deisboeck TS. Growth patterns of microscopic brain tumors. *Phys Rev E Stat Nonlin Soft Matter Phys*. 2002;66:051901. Epub 2002 Nov 06.

18 Mansury Y, Deisboeck T. Simulating "structure-function" patterns of malignant brain tumors. *Physica A*. 2004;331:219–232.

19 Mansury Y, Deisboeck T. Simulating the time series of a selected gene expression profile in an agent-based tumor model. *Physica D*. 2004;196:193–204.

20 Giese A, Loo MA, Tran N, et al. Dichotomy of astrocytoma migration and proliferation. *Int J Cancer*. 1996;67:275–282.

21 Prenzel N, Zwick E, Leserer M, Ullrich A. Tyrosine kinase signalling in breast cancer. Epidermal growth factor receptor: convergence point for signal integration and diversification. *Breast Cancer Res*. 2000;2:184–190. Epub 2000 Mar 25.

22 Piccolo E, Innominato PF, Mariggio MA, et al. The mechanism involved in the regulation of phospholipase Cgamma1 activity in cell migration. *Oncogene*. 2002;21:6520–6529.

23 Chen P, Xie H, Sekar MC, et al. Epidermal growth factor receptor-mediated cell motility: phospholipase C activity is required, but mitogen-activated protein kinase activity is not sufficient for induced cell movement. *J Cell Biol*. 1994;127:847–857.

24 Athale CA, Mansury Y, Deisboeck TS. Simulating the impact of a molecular "decision-process" on single cell phenotype and multicellular patterns in brain tumors. *J Theor Biol*. 2005;233:469–481.

25 Dittmar T, Husemann A, Schewe Y, et al. Induction of cancer cell migration by epidermal growth factor is initiated by specific phosphorylation of tyrosine 1248 of c-erbB-2 receptor via EGFR. *FASEB J*. 2002;16:1823–1825. Epub 2002 Sep 19.

26 Athale CA, Deisboeck TS. The effects of EGF-receptor density on multiscale tumor growth patterns. *J Theor Biol*. 2006;238(4):771–779. Epub 2005 Aug 26.

27 Mansury Y, Deisboeck T. The impact of "search precision" in an agent-based tumor model. *J Theor Biol*. 2003;224:325–337.

28 Knauer DJ, Wiley HS, Cunningham DD. Relationship between epidermal growth factor receptor occupancy and mitogenic response. Quantitative analysis using a steady state model system. *J Biol Chem*. 1984;259:5623–5631.

29 Lauffenburger DA, Linderman J. *Receptors: Models for Binding, Trafficking, and Signaling*. Oxford, England: Oxford University Press, 1996.

30 Jain RK. Transport of molecules across tumor vasculature. *Cancer Metastasis Rev*. 1987;6:559–593.

31 Gatenby RA, Gawlinski ET. The glycolytic phenotype in carcinogenesis and tumor invasion: insights through mathematical models. *Cancer Res*. 2003;63:3847–3854.

32 Stokes CL, Lauffenburger DA. Analysis of the roles of microvessel endothelial cell random motility and chemotaxis in angiogenesis. *J Theor Biol*. 1991;152:377–403.

33 Stokes CL, Lauffenburger DA, Williams SK. Migration of individual microvessel endothelial cells: stochastic model and parameter measurement. *J Cell Sci*. 1991;99:419–430.

34 Rupnick MA, Stokes CL, Williams SK, Lauffenburger DA. Quantitative analysis of random motility of human microvessel endothelial cells using a linear under-agarose assay. *Lab Invest*. 1988; 59:363–372.

35 Nishikawa R, Ji XD, Harmon RC, et al. A mutant epidermal growth factor receptor common in human glioma confers enhanced tumorigenicity. *Proc Natl Acad Sci USA*. 1994;91:7727–7731.

36 Mantzaris NV, Webb S, Othmer HG. Mathematical modeling of tumor-induced angiogenesis. *J Math Biol*. 2004;49:111–187. Epub 2004 Feb 06.

Future

194

New Educational Tools for Understanding Complexity in Medical Science

Grace Huang*, Michael J. Parker*, and James Gordon†

*Beth Israel Deaconess Medical Center, Harvard Medical School, Boston, Massachusetts;
†Harvard Medical School and Massachusetts General Hospital, Boston, Massachusetts

The factors that have caused the endothelium to be an underrecognized and underemphasized organ system are the same reasons that make it a challenging topic for students and teachers alike. The endothelial cell (EC) is hidden from view, seen only by microscopy rather than at the gross anatomical level. Because it participates in all vascular structures, it is often regarded as part of other organs, rather than as a distinct entity. Its distribution throughout the body results in interactions that are multiple and complex, which means that it cannot be discussed without a prior understanding of other organ systems. The endothelium is not readily evaluated by specific diagnostic tests, and at the cognitive level, the endothelium may not evoke a specific mental image as would the heart, for instance. To fully understand the complexity of endothelial biomedicine, new educational tools may be useful in complementing traditional teaching approaches.

Technological advances in the past few decades have had dramatic consequences on how the biomedical sciences are being taught. Educators can now harness the power of technology to manage enormous amounts of information and to deliver content in a manner that promotes high-level, experiential learning. Medical simulation is one area that is revolutionizing medical education. Broadly defined, medical simulation refers to the controlled reproduction of physiologic or clinical environments. For our purposes, we will consider it to be "a model of some phenomenon or activity that users learn about through interaction with the simulation" (1). In the most effective educational simulations, learner-driven exploration and experience lead to enhanced understanding of complex phenomena. Although "standardized patients" played by patient-actors can provide such a platform, in this chapter we will instead concentrate on three types of technology-enhanced medical simulation:

- Concept simulation: Computer-based animations that model scientific phenomena for the purpose of understanding complex physiology or pathophysiology
- Virtual patient simulation: Interactive computer-based programs that mimic real-life clinical scenarios to allow the user to interview, examine, diagnose, and treat a patient
- High-fidelity patient simulation: The use of realistic robot-mannequins to replicate the clinical encounter in its entirety

Used together, these forms of medical simulation can animate the entire range of biomedical processes, from molecular mechanisms to the actual delivery of health care (Figure 194.1). In a learning environment that employs all three modalities, for instance, a student can see not only how derangement at the level of the EC leads to tissue abnormalities, organ dysfunction, and systemic illness, but also how these concepts affect the patient's overall experience and course of care. Table 194-1 summarizes the advantages and relative disadvantages of the different types of medical simulation.

CONCEPT SIMULATION

One of the ways adults learn is through the development of mental models – imagined representations that can be "run" in working memory to understand, solve problems, and predict outcomes (2). Such models are critical to understanding complex phenomena or systems (3). Because the structure and function of the EC are complex, particularly when considered at multiple levels of organization (cell, vessel, organ, whole body), there is a tendency for learners to oversimplify concepts. One method of helping learners develop more complete mental models is to provide them with "conceptual" models.

Whereas a mental model is a construct in a learner's mind, a conceptual model is an instructional tool that facilitates the process of creating the mental model. Conceptual models may come in different forms, including images, animations, videos, computer diagrams, or other simulations. For example, a computer animation that illustrates how blood flows through

Table 194-1: Advantages and Disadvantages of the Different Types of Medical Simulation

Type	*Advantages*	*Disadvantages*
Concept simulation	Useful for visualization and understanding of pathology and pathophysiology Capable of allowing independent exploration of complex phenomena Ease of access Relatively low-cost for production	Limited physical interactivity Limited fidelity
Virtual patient simulation	Able to encompass multiple aspects of clinical encounter Useful for incorporating lessons about longitudinal and multidisciplinary care Ease of access Readily customizable	Limited physical interactivity Limited fidelity
High-fidelity patient simulation	Immersive, active experience Takes advantage of emotional and sensory components of learning Fosters critical thought and communication Animates basic science within the context of clinical reality	Cost and space requirements Access inherently limited by the number of simulators and staff Reliance on human teachers/operators Engineering limitations

Concept simulation, virtual patient simulation, and high-fidelity patient simulation have distinct advantages and disadvantages compared to one another, which are important considerations in curriculum development.

the heart can vividly portray aspects of anatomy and physiology that are typically described only through words and static images. Multimedia technology and simulations are well suited to help learners construct useful mental models of complex phenomena.

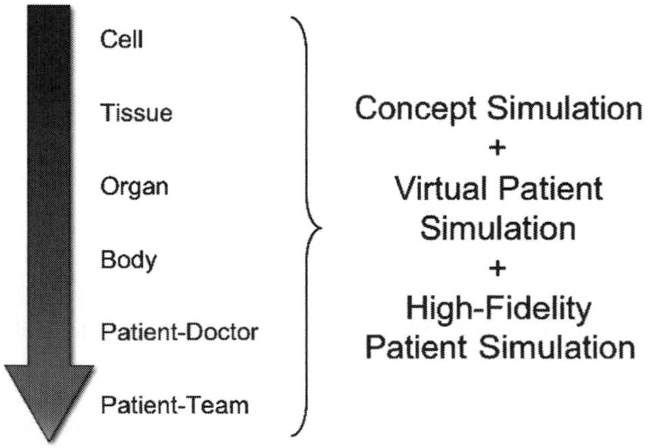

Figure 194.1. The spectrum of medical simulation. The spectrum of biomedical science from molecular processes to human-system interactions can be depicted using the various forms of medical simulation in concert.

Concept simulation refers to computer-based interactive diagrams that model phenomena for the purpose of understanding the underlying physiology or pathophysiology. "Interactivity" can range from animations in which the learner controls the pace to dynamic simulations whereby the learner influences the output by adjusting parameters. The strengths of concept simulation include its abilities to illustrate processes that are difficult to visualize, help form correlations and connections between ideas, make complex topics easier to conceptualize, and allow interactive exploration of topics. Concept simulation promotes integration of basic with clinical science by bringing molecular processes within easy reach of learners, simulating empiric methods of research.

How can we take advantage of these strengths in teaching about the endothelium? The EC can be difficult to visualize in terms of structure and function. Useful concept simulations include three-dimensional modeling of complex structures and animations of time-dependent phenomena. Three-dimensional graphic models can be developed of the endothelium at the level of the cell, vessel, organ, and whole body. A concept simulation could tie these together, allowing the learner to shift between levels, first examining the structure at the whole body level, then zooming in to individual organs, vessels, and cells. Such a construct would allow for comparisons between organs, individual vessels, or even cells, thereby emphasizing the heterogeneity of the endothelial processes.

The three-dimensional model discussed earlier need not be static. Animation at each level of organization can illustrate important concepts. Some have proposed the metaphor of endothelium as a circuit board, with ECs as input-output devices. This can be visually reinforced and clarified by animations that show how various cells are switched on or off or variably activated (e.g., the spectrum of activation states could be shown as changing shades on a color gradient). At the level of individual cells, an array of inputs and outputs can be animated along with the known signal transduction mechanisms.

Concept simulations need not be restricted to animation alone; graphical user interfaces combined with sophisticated underlying models allow a high degree of interactivity. Imagine a situation where a learner could interact with a computer-based simulation by "trying out" various hypothetical inputs to a cell; resulting signal transduction and output could be instantaneously displayed and evaluated. Such a model could be further used to illustrate how the same inputs may produce different outputs for individual ECs and even for the same EC at different times. This type of interactive exploration engages the learner and encourages active construction of knowledge (4).

Similarly, interactive concept simulations can help learners understand the idea of the endothelium as part of a network, with linkages and communication between nodes that include ECs, circulating cells, and soluble mediators. A computer model may show the learner a dynamic representation of this network and allow pathophysiology to be introduced, with resulting changes in both number and nature of nodes and links.

The possibilities for potential concept simulations are numerous. How would these be used in the spectrum of medical simulation to teach about the EC? We offer a hypothetical teaching scenario using severe sepsis as the endothelium-related topic. In this scenario, a full body simulator or virtual patient simulation could be used to introduce learners to a patient with an infection that leads to a systemic inflammatory response. In the course of interacting with the patient, students would observe signs and symptoms of this illness and other evidence of altered perfusion and organ system dysfunction. They would examine the patient for possible sites of infection, order lab tests, and administer therapeutic agents as appropriate.

Both during and after the case, the role of the endothelium could be highlighted through the use of explanatory concept simulations. As the patient's blood pressure drops, learners can view a conceptual model of the mediators involved, their signaling mechanisms, and their effects on the endothelium. As the team considers using novel therapeutic agents, they can use the computer to run a simulation illustrating how the drug affects targets in the endothelial network. The effects of changes in vascular tone and permeability on cardiac output and differential distribution of blood flow could also be shown in the same way. Other parameters, such as body temperature and heart rate, can be incorporated into a concept simulation depending on the desired level of realism.

As we discuss simulations, it is important to note that the level of complexity of a model or scenario should be appropriately adjusted to the level of the learner and to the desired learning objectives – a senior medical student or resident might be able to attend to and understand a greater level of detail than a first-year medical student. Providing too complex a model or scenario for a learner's level may result in decreased initial learning and transfer of knowledge (with transfer being defined as ability to apply the knowledge in related situations) (5,6).

VIRTUAL PATIENT SIMULATION

A medical student taking basic science classes may understand the structure and function of the endothelial cell. She may be able to identify it against the backdrop of other cell types. She might also recognize the specific cytokines and cell–cell interactions that result in striking changes in vascular permeability. In essence, the traditional student learning from lectures is equipped with skills at the basest level of clinical competence as depicted by Miller's triangle (i.e., "knows") (7) (Figure 194.2; for color reproduction, see Color Plate 194.2.). The next logical step is to assess how our hypothetical student applies her medical knowledge in the cognitive domain (i.e., "knows how"). Virtual patient simulation can be helpful in tracking such development.

Virtual patients are defined as "interactive computer programs that simulate real-life clinical scenarios in which the learner acts as a health care professional obtaining a history and physical examination and making diagnostic and therapeutic decisions" (8). The original impetus for the

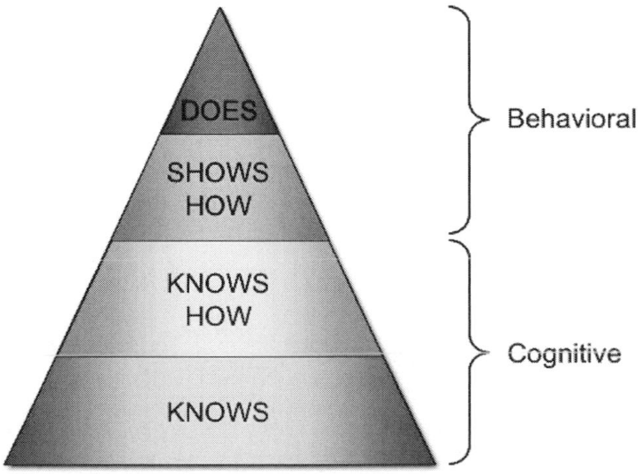

Figure 194.2. Miller's Pyramid of Clinical Competence. This framework for clinical competence depicts a hierarchy for assessing skills in order of increasing relevance. For color reproduction, see Color Plate 194.2.

development of virtual patients was the failure of the traditional undergraduate medical school curriculum to keep pace with changes in health care delivery. Economic forces have caused medical management to shift into the outpatient setting. As a consequence, medical students have diminished contact with patients in the hospital, where the bulk of clinical training still takes place. The cases encountered by students in the hospital are increasingly severe and complex and therefore less representative of the disease spectrum seen by many physicians in typical practice.

Our index disease, sepsis, would obviously be managed in the hospital, but many other disorders of the vascular endothelium, such as atherosclerosis, thrombosis, and hypertension, are chronic and therefore handled in the ambulatory arena. In addition, faculty members face constant pressure to increase clinical volume and to be productive researchers. The cumulative result is that students have less exposure to the two core elements of their training: they lose the opportunity to build relationships with patients and to learn about disease in a psychosocial context, and they lack physician role models and mentors.

Virtual patient simulation addresses these deficiencies by giving students access to both patients and faculty. A student using the virtual patient gets the opportunity to manage a multitude of diseases and to practice the skills of clinical decision-making. The patient encounters can span many "virtual years," allowing the student to gain the experience of longitudinal care in a matter of hours. He may also see events and procedures he would not ordinarily witness while on a short clinical rotation, which permits knowledge acquisition at an accelerated pace. Furthermore, virtual patient cases give specific and ongoing feedback, as would a faculty member, and they may feature videos that model exemplary behavior in physician–patient interactions.

Consider the following hypothetical virtual patient case developed to teach about the endothelium. A middle-aged male patient presents to a virtual emergency room complaining of malaise and cough. By typing questions into the computer program, the student obtains relevant data through a focused interview; she elicits symptoms of fevers, chills and dizziness and a distant history of splenectomy. A virtual faculty member reminds her to ask about recent travel or sick contacts, to elucidate the spectrum of possible organisms causing the fever. A simulated examination (Figure 194.3) combines user input with audiovisual elements and allows the student to check vital signs, to listen to the heart and lungs, and to form a general impression of the disease severity. Here, the student hears abnormal lung sounds, measures a low blood pressure, and recognizes that the patient is acutely ill. The program confirms the accuracy of her assessment.

As the clinical encounter progresses, the computer program asks her to draw up a list of possible diagnoses, and she

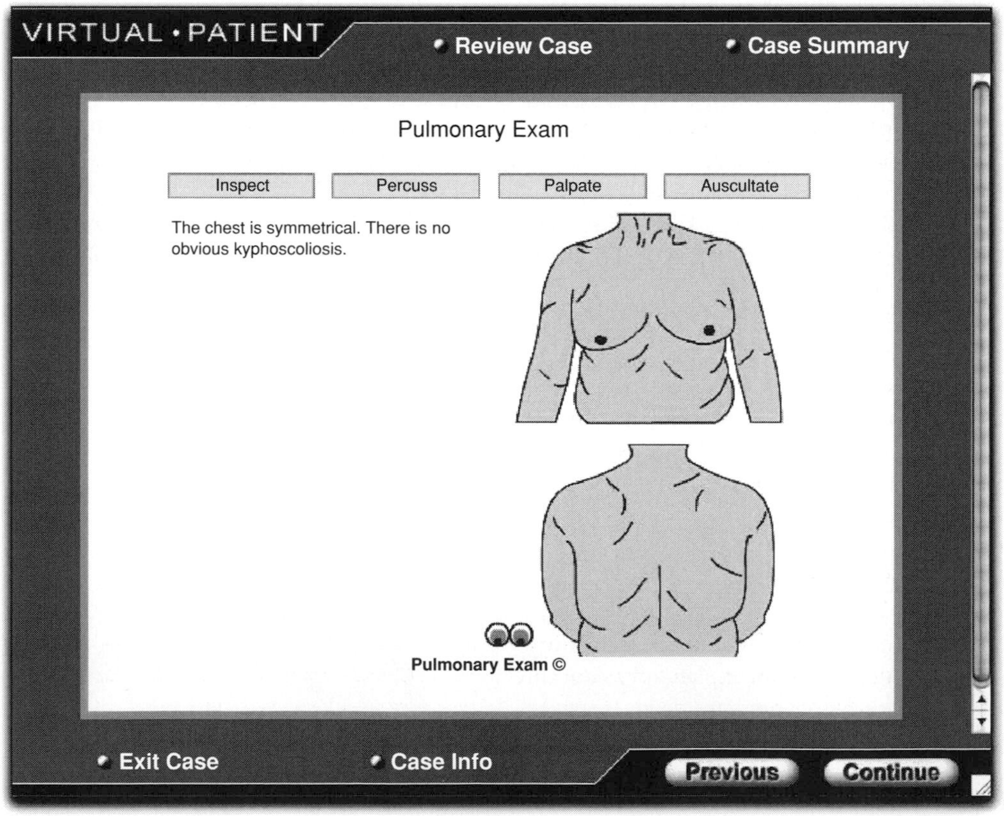

Figure 194.3. Physical examination simulation. This figure is an example of screen-based simulation of the physical examination. (Reproduced with permission from the Carl J. Shapiro Institute for Education and Research at Harvard Medical School and Beth Israel Deaconess Medical Center.)

includes bacterial pneumonia as the most likely cause of the patient's symptoms. This clinical impression will inform her choice of diagnostic tests, which includes a complete blood count and an urgent chest radiograph. Due to the acuity of the clinical presentation, the question of "where" to treat the patient is as equally pressing as "how" to treat the patient. The student astutely decides to hospitalize the patient in the intensive care unit (ICU) for presumed septic shock, rather than putting him in a hospital ward, where monitoring would occur less frequently. In the ICU, the student needs to demonstrate her understanding of the endothelium through her actions. She therefore treats sepsis by addressing the endothelial defect; she counteracts abnormal systemic vascular permeability and loss of arterial perfusion through the use of isotonic intravenous fluid and vasopressors. She must also manage the underlying etiology that is responsible; she decides to broaden the antibiotic coverage. Every action on the part of the student leads to a clinical event, whether it is an arrhythmia, a lab abnormality, an adverse drug reaction, or multi-organ failure. The final outcome, that is, survival, depends on the user's choices, including the timing and sequence of those decisions.

On the other hand, imagine that the student did not realize the severity of the patient's presentation from the beginning, and she discharged him with oral antibiotics. The virtual patient simulation might have resulted in a respiratory arrest in the patient's home. Herein lies the critical benefit of simulated learning environments: virtual patient simulation permits discovery and enforces the consequences of poor decision-making, without harm to the live patient. The richness of the experience derives from witnessing a multitude of cause-effect relationships, with the ability to rehearse the same actions or, alternatively, to experiment with another path. The complexity of clinical outcomes for virtual patients can mirror the intricacies of real patient care.

In addition, virtual patient simulation addresses the needs of adult learners. Teaching students how to take care of a virtual patient with sepsis provides both context and relevance. Adult learners are goal-oriented and therefore benefit from problem-centered education. They are also motivated by their own objectives, and in the case of medical students, their professional goals likely include the ability to manage an illness that is the common endpoint of severe infections.

Virtual patient simulation takes the benefits of case-based learning and delivers the content using a technology that is ubiquitous, interactive, and multimedia-enriched. The end result is a tool that gives a student insight into the actual management of sepsis, tailored to the individual's actions. Ultimately, virtual patient simulation provides a medium to practice critical decision-making in a learning environment that is not only safe to students but also safe to patients.

HIGH-FIDELITY PATIENT SIMULATION

High-fidelity patient simulation refers to the use of sophisticated human models, and encompasses both full-body simulation (robot-mannequins) and procedural simulation (task-specific trainers for invasive procedures). Full-body patient simulators, first popularized for anesthesia training in the early 1990s (9), can breathe, blink, talk, move, and respond like actual patients, allowing doctors to experience realistic patient care in a protected environment (10). At face value, they have remarkable appeal for health care providers, who can now practice with patient simulators like pilots can practice on flight simulators. In the case of high-fidelity patient simulation, a single immersive scenario can be as useful for the novice as for the expert when the learning objectives are tailored to the individual (i.e., the "good teaching case" can be instructive at multiple levels).

Consider the power of experience for a new intern who "floats" her first pulmonary artery catheter on a hypotensive patient in the ICU; based on the readings, the preliminary diagnosis of cardiogenic shock is changed to septic shock, completely redirecting the therapeutic rationale and plan. This level of intimate participation in real clinical care – the actual experience of doctoring – fosters a level of cognitive and emotional integration that is critical to the development of expertise (11). Through experience, the novice will learn more about the differential presentation of septic shock than ever possible through books or lectures. The question is, can high-fidelity patient simulation replicate the power of experiential learning in an artificial environment, thus enhancing the acquisition of expertise without risk? Can realistic simulation promote the progression from "knows" and "knows how" all the way up to "shows how" and "does" on the pyramid of clinical competence?

Students who participate in such scenarios think so (12,13), but research in this area has just begun. While there is an entire field of procedural simulation (and a corresponding literature of outcomes research) (14), full-body medical simulation typically concentrates on establishing a level of emotional involvement and experience that fosters complex thought (15). Students who "float" a pulmonary artery catheter in the immersive simulator lab, for example, not only practice the procedure, but also comment on the level of critical thought and communication required for optimal patient care. In most scenarios, the radio-transmitted "voice" of the simulator is critical to establishing the proper learning environment, transforming the experience from a passive or technical algorithm to an emotionally charged clinical encounter.

One of the true benefits of high-fidelity patient simulation lies in its capacity to unify multiple teaching approaches in a self-contained learning space. A classroom that contains a patient simulator, a conference table, and a large-screen computer display, for example, could allow students to seamlessly transition from a high-tech tutorial discussion to realistic bedside teaching (16). Students studying septic shock could use the display screen to work through interactive computer diagrams (i.e., concept simulation of endothelial effects), and to explore virtual patient resources, including virtual microscopy of endothelial structures. Then they could turn to the patient simulator as a force for contextualizing core concepts, using the emotionality of a simulated patient encounter as a catalyst for advanced understanding of complex material (17).

KEY POINTS

- Medical simulation is a novel approach that promises to harness the power of experiential learning as a unifying intellectual force.
- Combined effectively, the types of medical simulation can help fundamentally integrate basic and clinical science throughout the learning process.
- Medical simulation is an innovative interdisciplinary educational method that promotes contextual learning, allows practice without risk to real patients, and may accelerate development of expertise.
- Concept simulation refers to computer-based animations and models of physiologic processes.
- Virtual patient simulation refers to computer-based programs that allow the learner to play the role of a health care provider in the care of a patient.
- High-fidelity patient simulation refers to the use of realistic robot-mannequins to replicate clinical scenarios.
- Used together, these three types of medical simulation permit robust integration of basic science and clinical science as a unified whole.

Future Goals

- To better understand how medical simulation can foster the development of researchers and clinicians who are more flexible in incorporating new information, more adaptable in applying knowledge outside the classic paradigm, and more likely to conceive of new diagnostics or therapeutics based on a deeper understanding of biomedical science

REFERENCES

1 Frederikson JR, White BY, Gutwill J. Dynamic mental models in learning science: the importance of constructing derivational linkages among models. *Journal of Research in Science Teaching*. 1999;36:806–836.

2 Alessi AM, Trollip SR. *Multimedia for Learning: Methods and Development*, 3rd ed. Boston: Allyn & Bacon;2001: 213.

3 Alessi AM, Trollip SR. *Multimedia for Learning: Methods and Development*, 3rd ed. Boston: Allyn & Bacon;2001:28.

4 Duffy TM, Lowyck J, Jonassen DH, eds. *Designing Environments for Constructivist Learning*. Berlin: Springer-Verlag, 1993.

5 Goodyear P, Njoo M, Hijne H, van Berkum JJA. Learning processes, learner attributes and simulations. *Education & Computing*. 1991;6:263–304.

6 Swaak J, van Joolingen WR, de Jong T. Supporting simulation-based learning: the effects of model progression and assignments on definitional and intuitive knowledge. *Learning and Instruction*. 1998;8:235–252.

7 Miller GE. The assessment of clinical skills/competence/performance. *Acad Med*. 1990;65:S63–S67.

8 Huang, G. Association of American Medical Colleges. At http://www.aamc.org/meded/vp/.

9 Gaba D, Howard S, Fish K, et al. Simulation in anesthesia crisis management: a decade of experience. *Simulation and Gaming*. 2001;32(2):175–193.

10 Freidrich MJ. Practice makes perfect: risk-free medical training with patient simulators. *JAMA*. 2002;288:2808–2812.

11 Dewey J. Thinking in education. In: Barnes LB, Christensen CR, Hansen AJ, eds. *Teaching and the Case Method*, 3rd ed. Boston, MA: Harvard Business School Press, 1994.

12 Gordon JA, Wilkerson WM, Shaffer DW, Armstrong EG. "Practicing" medicine without risk: students' and educators' responses to high-fidelity patient simulation. *Acad Med*. 2001;76:469–472.

13 Center for Medical Simulation. Simulator: Introduction to the Use of High-fidelity Patient Simulation among Harvard Medical Students in a Clinical Clerkship [demonstration video]. Boston: Center for Medical Simulation, 1999. www.harvardmedsim.org <go to: student video>. Accessed Mar 1 2005.

14 Gallagher AG, Cates CU. Approval of virtual reality training for carotid stenting: what this means for procedural-based medicine. *JAMA*. 2004;292(24):3024–3026.

15 Gordon JA. High-fidelity patient simulation: a revolution in medical education. In: Dunn WF, ed. *Simulators in Critical Care and Beyond*. Des Plains, IL: Society for Critical Care Medicine Press, 2004.

16 Gourley L. Patient simulator goes to the head of the class. In Focus: News from Harvard Medical, Dental, and Public Health Schools, Dec. 3, 2004 [http://focus.hms.harvard.edu/2004/Dec3/_2004/forum.html]. Accessed Mar 1 2005.

17 Gordon JA, Oriol NE, Cooper JB. Bringing good teaching cases "to life": a simulator-based medical education service. *Acad Med*. 2004;79(1):23–27.

195

Endothelial Biomedicine

The Public Health Challenges and Opportunities

George A. Mensah

National Center for Chronic Disease Prevention and Health Promotion, Centers for Disease Control and Prevention, Atlanta, Georgia

Recent advances in endothelial biomedicine and vascular biology have provided renewed hope that strategies for health promotion and the prevention and control of diseases can be improved substantially. These advances hold promise for the detection, evaluation, prevention, treatment, and control of chronic diseases in general and cardiovascular diseases (CVDs) in particular. However, translating these scientific advances into effective clinical practices poses unique challenges (1–3), and translating them into effective public health practices at the community or population level is even more daunting.

This final translational step is crucial, especially for scientific advances related to chronic noncommunicable diseases and their risk factors, which constitute some of the major public health challenges of the twenty-first century. The World Health Organization has estimated that 35 million people died from chronic diseases in 2005 alone and that heart diseases, stroke, cancer, chronic respiratory diseases, and diabetes accounted for the bulk of these deaths (4). These chronic diseases and their risk factors are major epidemics that are not likely to be conquered at the bedside by using individual patient strategies. However, a combination of individual and population-based strategies built on a sound foundation of biomedical science holds tremendous promise for success.

This chapter presents a review of the major advances in vascular biology likely to provide the bases for successful public health policy, practice, and research initiatives in chronic disease prevention and control. It describes clinicians' frequent failure to recognize the endothelium as a spatially distributed organ system, addresses the relevance for continuing medical education and translational research, and discusses the practical implications for health promotion and the prevention and control of chronic diseases. It also presents specific examples of steps that can be taken immediately to apply endothelial research findings in efforts to improve population health and proposes that members of the public health workforce receive appropriate training about the importance of the endothelium in efforts to prevent or control chronic diseases. Most importantly, this chapter emphasizes the need for increased collaboration among vascular biology centers, basic scientists, clinician-investigators, and public health practitioners.

Disclaimer: The findings and conclusions this chapter are those of the author and do not necessarily represent the views of the Centers for Disease Control and Prevention.

PUBLIC HEALTH IN THE TWENTY-FIRST CENTURY

As pointed out by the Institute of Medicine, "*public health is what we, as a society, do collectively to assure the conditions for people to be healthy*" (5). Being healthy in this regard means more than just being free from disease but also having the physical, mental, social, and psychological wellbeing necessary to lead long, satisfying lives. The primary challenge to public health practitioners attempting to promote this broad concept of health is to address the leading causes of morbidity, death, and poor quality of life at the population level and to reduce or eliminate health disparities between subgroups of the population.

At the turn of the twentieth century, communicable diseases and perinatal conditions accounted for most of the deaths and cases of disability in the United States (6,7). In fact, in 1900, the three leading causes of death were pneumonia, tuberculosis, and diarrhea and enteritis (7). Together with diphtheria, these infectious diseases caused one-third of all deaths (7). Today, however, chronic noncommunicable diseases constitute the leading causes of death and disability in the United States and worldwide (8–12).

In 2002, the World Health Organization estimated that chronic diseases were responsible for 59% of global deaths (33.1 million)(11) and 47% of the global burden of disease, and projected that they would account for 73% and 60%, respectively, by 2020. In addition to their toll in human suffering, these diseases are also financially costly. In the United States, nearly 75% of the estimated $1.9 trillion in annual health care spending goes to chronic diseases (13,14). Cardiovascular diseases, including stroke and all diseases of the heart, cause almost one-third of global deaths (11). The four leading chronic diseases (cardiovascular diseases, cancer, chronic lung disease, and diabetes) caused 29 million deaths in 2002 (10). Seven of the ten leading causes of death in the United States are chronic diseases, which together accounted for nearly 70% of all U.S. deaths in 2003 (15). Furthermore, the number of diabetes cases worldwide is projected to double in the next 25 years (10).

These chronic diseases are also leading causes of disability, poor quality of life, and health disparities (16–20) and constitute some of the most serious public health challenges of the twenty-first century. Addressing these challenges will require appropriate recognition of the role of endothelial biomedicine in heath promotion and disease prevention programs; its successful integration into basic, clinical, and population science research; and the translation of basic endothelial biomedical science findings into clinical and public health practices.

CHALLENGES AND OPPORTUNITIES

In many ways, the endothelium is to individual health what public health activities are to the health of society. Like public health, the endothelium is hidden and very much underappreciated. When the endothelium functions normally and organ function is maintained and disease is prevented, society rarely takes notice. The endothelium is also like public health in that relatively inexpensive strategies that preserve function (such as increasing one's physical activity, avoiding tobacco smoke, getting proper nutrition, and obtaining essential preventive services) are often less supported than more expensive interventions such as drug therapy after function has already been compromised. These are important programmatic and practical challenges faced by those attempting to translate endothelial biomedical advances into public health practice.

Several factors further compound these challenges. First, basic scientists and clinician-investigators who work in the field of endothelial biomedicine focus on endothelial cells (ECs), and discrete organ systems, while public health practitioners in chronic disease prevention focus mostly on individual persons, families, and populations, rarely on organ systems, and almost never on cells. Thus interactions between members of these two groups are rare. Second, most clinician-scientists who work in endothelial biomedicine view endothelial function in the limited context of large artery vasomotion and its implications for ischemic cardiovascular disease. Third, clinician-scientists seldom fully appreciate the vast spatial distribution, complexity, and heterogeneity of ECs. Finally, people in traditional public health practice and research rarely consider the diagnostic and therapeutic potential of endothelial biomedicine for health promotion and disease prevention. However, before addressing the challenges of translating endothelial research findings into effective public health practices, we must begin with a review of recent advances in endothelial biomedicine.

ADVANCES IN ENDOTHELIAL BIOMEDICINE: PUBLIC HEALTH RELEVANCE

Enormous advances have been made in our understanding of the structure and function of the endothelium (21–23). The most important advances regarding the structure of the endothelium include insights into its vast spatial distribution, its heterogeneity in the different vascular beds, its complexity, and its remarkable ability to adapt to a wide variety of conditions. Similarly important advances in how we understand the function of the endothelium have shown that the endothelium does more than provide a barrier between the luminal contents of the blood vessel and the underlying vascular wall structures. The endothelium also plays a key role in cell growth, cell migration, cell survival, the release of endocrine and paracrine hormone factors, the control of vasomotor tone, leukocyte trafficking, inflammation, and hemostasis (21,24). These functions have tremendous implications for several major chronic diseases, including heart disease, stroke, cancer, diabetes, and chronic lung disease.

For example, in a recent review, Vogel (25) described four key ways in which endothelial research findings may be relevant to the prevention of coronary heart disease. First, endothelial vasodilator dysfunction has been associated with all the established coronary risk factors both among asymptomatic subjects in the general population and among patients with coronary heart disease. Second, the degree of vasodilator dysfunction has been shown to correlate with the number and severity of coronary risk factors. Third, this vasodilator dysfunction has been shown to precede the development of even the earliest anatomic atherosclerotic lesions (26). And fourth, effective treatment of controllable risk factors has been associated with a demonstrable reversal of endothelial vasodilator dysfunction (25). Taken together, these findings indicate that endothelial dysfunction is a reliable marker of atherogenic risk (as defined by conventionally determined coronary risk factor status) and that reversal of this dysfunction during the treatment of risk factors is desirable.

Another form of endothelial dysfunction, caused by oxidative damage and the presence of oxidized low-density lipoprotein (LDL)-cholesterol, results in destruction of nitric oxide, increased platelet aggregation, thrombosis, inflammation, plaque formation, proteolysis, plaque fissure, and rupture (22,27). Specific evidence that endothelial dysfunction is associated with major coronary risk factors such as age, male sex (22,28–30), hypercholesterolemia (31), cigarette smoking (32,33), diabetes (34–39), physical inactivity (40,41), and high

Table 195-1: The Relative Extent to Which Researchers Recognize the Endothelium's Role in Various Endothelial-Based Disease, and the Estimated Direct and Indirect Costs of Each Disease in the United States

Endothelial-Based Disease or Condition	*Relative Recognition of Endothelium's Role*			*Direct and Indirect Costs in Billions of Dollars, USA (Year) (Refs.)*
	Basic Science	*Clinical Science*	*Population Science*	
Total Cardiovascular Diseases	++++	++++	+/−	393.5 (2005)(75)
Coronary Artery Disease	++++	++++	+/−	142.1 (2005)(75)
Hypertensive Disease	++++	++++	+/−	59.7 (2005)(75)
Cerebrovascular Disease or Stroke	++++	++++	+/−	56.8 (2005)(75)
Peripheral Vascular Disease	++++	++++	+/−	*
Diabetes	+++	+	+/−	132 (2002)(76)
Sepsis	+++	+	None	*
Arthritis & Other Rheumatic Conditions	+	+	None	269.3 (1997)(77)
Chronic Obstructive Pulmonary Diseases	+	None	None	37.2 (2004)(78)
Asthma	+	None	None	16.1 (2004)(78)
Pulmonary Hypertension	+++	++	None	*
Sickle Cell Disease	++++	+++	+/−	*
Preeclampsia	+++	+/−	None	*
Cancer	+++	++	+/−	189.8 (2004)(79)

Key: Subjective relative recognition is designated on a continuum where "None" is relatively no recognition, "+/−" is very little recognition, and "++++" is very substantial recognition; *, data not available.

saturated fat intake all add to the public health relevance of the endothelium.

As a result of the endothelium's role as both a source and target of chronic disease mediators and because of its vast spatial distribution throughout most organ systems, studies of the endothelium and the scientific underpinnings of its activation and dysfunction could play a major role in public health research in chronic diseases and also provide novel avenues for public health programs and policies. However, as shown in Table 195-1, the endothelium's role in most chronic diseases is relatively unrecognized by clinicians and public health practitioners at the clinical and population/community levels.

ENDOTHELIAL FUNCTION AND DYSFUNCTION IN CHRONIC DISEASES

Although the potential effects of EC dysfunction are manifold, most clinicians and public health practitioners largely view endothelial dysfunction within the narrow confines of its effect on large artery vasomotor tone. They also generally fail to appreciate the heterogeneity and phenotypic variation of ECs and the implications of these variations in the role that endothelial dysfunction plays in chronic diseases (24).

Why is the endothelium so important in chronic diseases? First, because the endothelium is both victim and perpetrator (both a source and target of the mediators of chronic diseases and their risk factors), it plays an important role in the initiation, progression, and development of chronic disease complications. Second, because the endothelium is spatially distributed throughout the body, it establishes a dialogue with every tissue cell in the body. Third, because the endothelium is highly plastic, it is affected by many disease processes and their risk factors (endothelium as victim). Fourth, because the endothelium is highly active, it contributes to many disease processes (endothelium as perpetrator). The plasticity and accessibility of the endothelium make it a potentially powerful platform for diagnostic as well as therapeutic interventions (24).

ENDOTHELIUM IN SCREENING AND DIAGNOSTIC EVALUATION

Screening and early detection of disease is a crucial component of the public health strategy for disease prevention and control. The strategic position of the endothelium within the vascular tree makes it ideal for the early detection of disease before clinical manifestations appear and also for the diagnostic evaluation of established diseases. Until recently, however, the emphasis in most clinical cardiovascular research has predominantly been on the noninvasive assessment of flow-mediated dilation of large arteries as a marker of atherosclerosis in children and young adults at risk for coronary artery disease (42,43) and in adult patients with established atherosclerosis (44,45). From a public health perspective, however, the endothelium holds promise as a marker for a lot more.

For example, circulating endothelial growth factors could prove useful as early markers for cancer (46,47), just as circulating endothelium-related biomarkers may serve as novel risk markers for atherogenesis and for plaque rupture leading to acute unstable syndromes and sudden cardiac death. In fact, circulating ECs and endothelial progenitor cells have shown potential value as markers for the diagnostic evaluation of a wide variety of chronic diseases, including erectile dysfunction, rheumatoid arthritis, diabetes, stroke, cancer, systemic sclerosis, pulmonary hypertension, and end-stage kidney disease (48–53). Werner and colleagues (54) recently showed that increased levels of bone marrow–derived endothelial progenitor cells were associated with a reduced risk of death from cardiovascular causes, a first major cardiovascular event, coronary revascularization, and hospitalization even after adjustment for age and other major cardiovascular risk factors. Thus, endothelial progenitor cell levels may also have prognostic value.

ENDOTHELIUM IN DISEASE PREVENTION AND TREATMENT

Health promotion strategies and the entire spectrum of disease prevention approaches (55) – primordial, primary, and secondary prevention – all stand to gain from the appropriate translation of endothelial biomedical findings into public health practice. Health-promoting strategies such as increasing one's level of physical activity, eating a diet rich in fruits and vegetables and low in saturated fats and calorie-dense starches, not smoking, and avoiding environmental tobacco smoke have also been shown to promote endothelial health and reverse endothelial dysfunction (56–64), whereas unhealthful behavioral and lifestyle practices have been shown to impair endothelial function (65–68). These study results suggest that the scientific underpinnings of endothelial function, activation, and dysfunction could play a major role in public health research into chronic diseases and also provide novel avenues for public health programs and policies.

STRATEGIC INITIATIVES FOR ACTION

The advances made in the basic science of endothelial biomedicine need to be translated into clinical practice and into public health strategies for improving population health, including more innovative public health research. Increased collaborations among vascular biology centers, basic and clinical researchers, and the public health community can help facilitate this translation. In addition, because clinical and public health practitioners should recognize that the endothelium is a spatially distributed organ system with tremendous potential to influence the function of all organs, endothelial biomedicine should be part of continuing medical education for all practitioners.

The development of a common lexicon to describe endothelial function, activation, and dysfunction in terms appropriate for basic, clinical, and population science research and for clinical and public health practice would be an important first step towards increased research collaboration. Endothelial biomedical scientists, clinician-investigators, practicing clinicians, and public health professionals also need to integrate their scientific activities and promote an increased awareness of the diagnostic and therapeutic potential of the endothelium. Most importantly, patients, physicians, and the general public need to be educated about the importance of endothelial biomedicine and the promise it holds for health promotion and disease prevention and encouraged to support endothelial-related public health research and the translation of basic endothelial research findings into effective clinical and public health practices.

PRACTICAL STEPS FOR IMPROVING INDIVIDUAL AND POPULATION HEALTH

As a first practical step, all lifestyle interventions that enhance normal endothelial function should be encouraged as a means of overall health promotion and as a means of addressing risk factors for chronic disease. Such interventions include the adoption of diets low in saturated fats and total cholesterol but rich in antioxidants, smoking cessation or avoidance, increased regular physical activity, and maintenance of ideal body weight. People should also be encouraged to adopt healthful habits beginning in youth and adolescence so that they do not develop major risk factors for chronic disease such as dyslipidemia, obesity, and hypertension. Appropriate intervention strategies should also be designed for people who already have one or more risk factors for CVD (primary prevention) and for those with existing CVD (secondary prevention).

Pharmacologic interventions are also important means of preserving or improving endothelial function. Angiotensin-converting enzyme (ACE) inhibitors and statins have been shown to help reverse endothelial vasodilator dysfunction in patients with a variety of chronic diseases (65–68) through their effects on the renin-angiotensin-aldosterone system, amelioration of dyslipidemia, and a variety of pleiotropic effects. These two drug classes are also important in the primary and secondary prevention of chronic diseases, especially heart disease and stroke. The effectiveness of other pharmacologic interventions with agents such as aspirin, supplemental antioxidants, vitamins, L-arginine, $\omega 3$ fatty acids, and glycemic control interventions have been reviewed elsewhere (69), but in general they have been examined mostly in the context of reversing endothelial vasomotor dysfunction. Future therapeutic strategies need to address other realms of endothelial dysfunction in their appropriate spatial and temporal context (24).

UNANSWERED QUESTIONS AND FUTURE IMPLICATIONS

Much work needs to be done and many questions have to be answered for the full potential benefits of endothelial biomedicine to be tapped. For example, how can the diversity of endothelial function and dysfunction be reliably assessed and used for population research and program development? To what extent does the reversibility of endothelial dysfunction in patients with chronic diseases translate into lives saved or quality-of-life improvements? Are there endothelial biomarkers that characterize the transition of patients from preclinical states to manifest disease states or the conversion of tumors from benign, to premalignant, to malignant states? What insights from endothelial research can be applied to the development of new preventative strategies? What does biomedical research concerning endothelial function in the pulmonary vascular bed teach us about the huge burden of morbidity and death associated with chronic lung diseases and what, beyond smoking cessation, would work to reduce this burden? Can changes in endothelial factors and function help identify women at risk for eclampsia months before the clinical disease (70,71)? What role does the endothelium play in the interaction among monocytes, fat cells, and macrophages that leads to the increased secretion of proinflammatory cytokines/chemokines, adipokines, and angiogenic factors and thus to obesity and its complications (72)? Does the endothelium provide clues for the development of interventions to combat the burgeoning obesity epidemic? How can endothelial biomedicine facilitate investigation of the type, dose, and frequency of lifestyle-related interventions that promote a healthy endothelium, prevent disease, and permit people to lead long, satisfying, healthy lives? What are the appropriate genomic and proteomic applications of endothelial biomedical research (73,74) for disease prevention, health promotion, and healthy aging?

CONCLUSION

Despite the many unanswered questions about the potential benefits of endothelial biomedical research in the prevention and control of chronic diseases, what has thus far been discovered is already cause for at least cautious optimism and perhaps even excitement. Nevertheless, it must be emphasized that a major problem with chronic disease prevention efforts today is the underutilization of interventions known to be effective, principally smoking cessation, proper nutrition, moderate levels of physical activity on most days of the week, and adherence to clinical and community-level guidelines for the provision of essential preventive services. So even if we do translate the findings of basic endothelial research into effective clinical practices and public health interventions, we then must ensure that these practices and interventions are widely adopted.

Endothelial biomedicine may lead to innovative new methods of addressing many unanswered questions about chronic disease. However, the successful implementation of these methods will depend in part on clinicians and public health practitioners recognizing the unique attributes of the endothelium and the implication of endothelium dysfunction in chronic diseases. To ensure that findings from basic endothelial research are translated into effective clinical treatments and public health interventions, we must have greater collaboration among vascular biology centers, basic scientists, clinician-investigators, and public health practitioners; more continuing education about the role of endothelial biomedicine in clinical treatment and public health; and a genuine commitment to close the gap between basic endothelial research and clinical and public health practice.

KEY POINTS

- Advances in endothelial biomedicine hold great promise for chronic disease prevention, detection, treatment, and control; however, translation of the science into clinical and public health practice remains a daunting challenge.
- Traditional public health practitioners rarely consider the diagnostic and therapeutic potential of ECs for health promotion and disease prevention.
- The role of the endothelium in most chronic diseases is relatively unrecognized by clinicians and public health practitioners at the clinical and population/community levels.
- As both a source and target of the mediators of chronic diseases and their risk factors, the endothelium plays an important role in the initiation, progression, and development of chronic disease complications.
- The plasticity, accessibility, and vast spatial distribution of the endothelium make it a potentially powerful platform for diagnostic as well as therapeutic interventions.

- Evidence of this potential already exist for many chronic diseases such as coronary heart disease, cancer, high blood pressure, stroke, diabetes, and pulmonary hypertension.

Future Goals

- To seek greater collaboration among vascular biology centers, basic scientists, clinician-investigators, and public health practitioners
- To explore the diagnostic potential of EC dysfunction beyond the narrow confines of large artery vasomotor tone in the evaluation of chronic diseases
- To emphasize the heterogeneity, complexity, and phenotypic variation of ECs and their implications for chronic disease prevention and control
- To support the development of a common vocabulary to describe endothelial function, activation, and dysfunction in terms appropriate for basic, clinical, and population science research and for clinical and public health practice

REFERENCES

1 Hwa C, Sebastian A, Aird WC. Endothelial biomedicine: its status as an interdisciplinary field, its progress as a basic science, and its translational bench-to-bedside gap. *Endothelium.* 2005;12(3):139–151.

2 Lenfant C. Shattuck lecture: clinical research to clinical practice – lost in translation? *N Engl J Med.* 2003;349:868–874.

3 Sung NS, Crowley WF Jr, Genel M, et al. Central challenges facing the national clinical research enterprise. *JAMA.* 2003;289(10):1278–1287.

4 World Health Organization. *Preventing Chronic Diseases: A Vital Investment.* Geneva: World Health Organization, 2005.

5 Institute of Medicine, Committee for the Study of the Future of Public Health, Division of Health Care Services. *The Future of Public Health.* Washington, DC: National Academies Press, 1988.

6 Centers for Disease Control and Prevention (CDC). Ten great public health achievements – United States, 1900–1999. *MMWR Morb Mortal Wkly Rep.* 1999;48(12):241–243.

7 Centers for Disease Control and Prevention. Achievements in public health, 1900–1999: control of infectious diseases. *MMWR Morb Mortal Wkly Rep.* 1999;48(29):621–629.

8 Yach D, Hawkes C, Gould CL, Hofman KJ. The global burden of chronic diseases: overcoming impediments to prevention and control. *JAMA.* 2004;291(21):2616–2622.

9 Centers for Disease Control and Prevention. *The Burden of Chronic Diseases and Their Risk Factors: National and State Perspectives.* Washington, DC: Department of Health and Human Services, 2004.

10 World Health Organization. *The World Health Report 2003: Shaping the Future.* Geneva, Switzerland: WHO, 2003.

11 World Health Organization. *The World Health Report 2002: Reducing Risks, Promoting Healthy Life.* Geneva, Switzerland: WHO, 2002.

12 Murray CJ, Lopez AD. Mortality by cause for eight regions of the world: Global Burden of Disease Study. *Lancet.* 1997;349(9061):1269–1276.

13 Centers for Disease Control and Prevention. *The Power of Prevention. Reducing the Health and Economic Burden of Chronic Disease.* Atlanta: Department of Health and Human Services, Centers for Disease Control and Prevention, 2003.

14 Heffler S, Smith S, Keehan S, et al. Trends: U.S. health spending projections for 2004–2014. *Health Aff (Millwood).* 2005;Web Exclusive(Suppl):W5–74-W5–85.

15 Hoyert DL, Kung HC, Smith BL. Deaths: preliminary data for 2003. *Natl Vital Stat Rep.* 2005;53(15):1–48.

16 Lillie-Blanton M, Maddox TM, Rushing O, Mensah GA. Disparities in cardiac care: rising to the challenge of Healthy People 2010. *J Am Coll Cardiol.* 2004;44(3):503–508.

17 U.S. Commission on Civil Rights. *The Health Care Challenge: Acknowledging Disparity, Confronting Discrimination, and Ensuring Equality. Volume I: The Role of Governmental and Private Health Care Programs and Initiatives.* Washington, DC: U.S. Commission on Civil Rights, 1999.

18 U.S. Commission on Civil Rights. *The Health Care Challenge: Acknowledging Disparity, Confronting Discrimination, and Ensuring Equality. Volume II: The Role of Federal Civil Rights Enforcement Efforts.* Washington, DC: U.S. Commission on Civil Rights, 1999.

19 Liao Y, Tucker P, Okoro CA, et al. REACH 2010 Surveillance for Health Status in Minority Communities – United States, 2001 – 2002. *MMWR Surveill Summ.* 2004;53(6):1–36.

20 National Center for Health Statistics. *Health, United States, 2004. With Chartbook on Trends in the Health of Americans.* Hyattsville, MD: National Center for Health Statistics, 2004.

21 Cines DB, Pollak ES, Buck CA, et al. Endothelial cells in physiology and in the pathophysiology of vascular disorders. *Blood.* 1998;91:3527–3561.

22 Alexander RW, Dzau VJ. Vascular biology: the past 50 years. *Circulation.* 2000;102(20):IV112–IV116.

23 Luscher TF, Barton M. Biology of the endothelium. *Clin Cardiol.* 1997;20(11 Suppl 2):II10.

24 Aird WC. Spatial and temporal dynamics of the endothelium. *J Thromb Haemost.* 2005;3(7):1392–1406.

25 Vogel RA. Coronary risk factors, endothelial function, and atherosclerosis: a review. *Clin Cardiol.* 1997;20(5):426–432.

26 Reddy KG, Nair RN, Sheehan HM, Hodgson JM. Evidence that selective endothelial dysfunction may occur in the absence of angiographic or ultrasound atherosclerosis in patients with risk factors for atherosclerosis. *J Am Coll Cardiol.* 1994;23:833–843.

27 McGorisk GM, Treasure CB. Endothelial dysfunction in coronary heart disease. *Curr Opin Cardiol.* 1996;11(4):341–350.

28 Celermajer DS, Sorensen KE, Spiegelhalter DJ, et al. Aging is associated with endothelial dysfunction in healthy men years before the age-related decline in women. *J Am Coll Cardiol.* 1994;24(2):471–476.

29 Chauhan A, More RS, Mullins PA, et al. Aging-associated endothelial dysfunction in humans is reversed by L-arginine. *J Am Coll Cardiol.* 1996;28(7):1796–1804.

30 Husken BC, Hendriks MG, Pfaffendorf M, van Zwieten PA. Effects of aging and hypertension on the reactivity of isolated conduit and resistance vessels. *Microvasc Res.* 1994;48(3):303–315.

31 Mosca L, Rubenfire M, Tarshis T, et al. Clinical predictors of oxidized low-density lipoprotein in patients with coronary artery disease. *Am J Cardiol.* 1997;80(7):825–830.

32 Heitzer T, Yla-Herttuala S, Luoma J, et al. Cigarette smoking potentiates endothelial dysfunction of forearm resistance vessels in patients with hypercholesterolemia. Role of oxidized LDL. *Circulation.* 1996;93(7):1346–1353.

33 Higman DJ, Strachan AM, Powell JT. Reversibility of smoking-induced endothelial dysfunction. *Br J Surg.* 1994;81(7):977–978.

34 Calles-Escandon J, Garcia-Rubi E, Mirza S, Mortensen A. Type 2 diabetes: one disease, multiple cardiovascular risk factors. *Coron Artery Dis.* 1999;10(1):23–30.

35 Yudkin JS. The Deidesheimer meeting: significance of classical and new risk factors in non-insulin-dependent diabetes mellitus. *J Diabetes Complicat.* 1997;11(2):100–103.

36 Betteridge DJ. Lipids and atherogenesis in diabetes mellitus. *Atherosclerosis.* 1996;124:S43–S47.

37 Norgaard K. Hypertension in insulin-dependent diabetes. *Dan Med Bull.* 1996;43(1):21–38.

38 Parving HH. Microalbuminuria in essential hypertension and diabetes mellitus. *J Hypertens Suppl.* 1996;14(2):S89–S93.

39 Vallance P, Calver A, Collier J. The vascular endothelium in diabetes and hypertension. *J Hypertens Suppl.* 1992;10(1):S25–S29.

40 Charo S, Gokce N, Vita JA. Endothelial dysfunction and coronary risk reduction. *J Cardiopulm Rehabil.* 1998;18(1):60–67.

41 Gabriel HH, Heine G, Kroger K, et al. Exercise and atherogenesis: where is the missing link? *Exerc Immunol Rev.* 1999;5:96–102.

42 Celermajer DS, Sorensen KE, Gooch VM. Non-invasive detection of endothelial dysfunction in children and adults at risk of atherosclerosis. *Lancet.* 1992;340:1111–1115.

43 Raitakari OT, Adams MR, McCredie RJ, et al. Arterial endothelial dysfunction related to passive smoking is potentially reversible in healthy young adults. *Ann Intern Med.* 1999;130(7):578–581.

44 Schroeder S, Enderle MD, Ossen R, et al. Noninvasive determination of endothelium-mediated vasodilation as a screening test for coronary artery disease: pilot study to assess the predictive value in comparison with angina pectoris, exercise electrocardiography, and myocardial perfusion imaging. *Am Heart J.* 1999;138 (4 Pt 1):731–739.

45 Yataco AR, Corretti MC, Gardner AW, et al. Endothelial reactivity and cardiac risk factors in older patients with peripheral arterial disease. *Am J Cardiol.* 1999;83(5):754–758.

46 Tamura M, Ohta Y, Nakamura H, et al. Diagnostic value of plasma vascular endothelial growth factor as a tumor marker in patients with non-small cell lung cancer. *Int J Biol Markers.* 2002; 17(4):275–279.

47 Ohta Y, Ohta N, Tamura M, et al. Vascular endothelial growth factor expression in airways of patients with lung cancer: a possible diagnostic tool of responsive angiogenic status on the host side. *Chest.* 2002;121(5):1624–1627.

48 Foresta C, Caretta N, Lana A, et al. Circulating endothelial progenitor cells in subjects with erectile dysfunction. *Int J Impot Res.* 2005;17(3):288–290.

49 Herbrig K, Haensel S, Oelschlaegel U, et al. Endothelial dysfunction in patients with rheumatoid arthritis is associated with a reduced number and impaired function of endothelial progenitor cells. *Ann Rheum Dis.* 2006;65(2):157–163.

50 McClung JA, Naseer N, Saleem M, et al. Circulating endothelial cells are elevated in patients with type 2 diabetes mellitus independently of HbA(1)c. *Diabetologia.* 2005;48(2):345–350.

51 Nadar SK, Lip GY, Lee KW, Blann AD. Circulating endothelial cells in acute ischaemic stroke. *Thromb Haemost.* 2005;94(4): 707–712.

52 Bull TM, Golpon H, Hebbel RP, et al. Circulating endothelial cells in pulmonary hypertension. *Thromb Haemost.* 2003;90(4):698–703.

53 Mancuso P, Burlini A, Pruneri G, et al. Resting and activated endothelial cells are increased in the peripheral blood of cancer patients. *Blood.* 2001;97(11):3658–3661.

54 Werner N, Kosiol S, Schiegl T, et al. Circulating endothelial progenitor cells and cardiovascular outcomes. *N Engl J Med.* 2005; 353(10):999–1007.

55 Mensah GA, Dietz WH, Harris VB, et al. Prevention and control of coronary heart disease and stroke-nomenclature for prevention approaches in public health a statement for public health practice from the Centers for Disease Control and Prevention. *Am J Prev Med.* 2005;29(5 Suppl 1):152–157.

56 Lopez-Garcia E, Hu FB. Nutrition and the endothelium. *Curr Diab Rep.* 2004;4(4):253–259.

57 Brown AA, Hu FB. Dietary modulation of endothelial function: implications for cardiovascular disease. *Am J Clin Nutr.* 2001; 73(4):673–686.

58 Hennig B, Toborek M, McClain CJ. High-energy diets, fatty acids and endothelial cell function: implications for atherosclerosis. *J Am Coll Nutr.* 2001;20(Suppl 2):97–105.

59 Franco OH, Bonneux L, de Laet C, Peeters A, et al. The Polymeal: a more natural, safer, and probably tastier (than the Polypill) strategy to reduce cardiovascular disease by more than 75%. *BMJ.* 2004;329(7480):1447–1450.

60 Franzoni F, Ghiadoni L, Galetta F, et al. Physical activity, plasma antioxidant capacity, and endothelium-dependent vasodilation in young and older men. *Am J Hypertens.* 2005;18(4 Pt 1):510–516.

61 McCarty MF. Optimizing endothelial nitric oxide activity may slow endothelial aging. *Med Hypotheses.* 2004;63(4):719–723.

62 Wildman RP, Schott LL, Brockwell S, et al. A dietary and exercise intervention slows menopause-associated progression of subclinical atherosclerosis as measured by intima-media thickness of the carotid arteries. *J Am Coll Cardiol.* 2004;44(3):579–585.

63 Hamdy O, Ledbury S, Mullooly C, et al. Lifestyle modification improves endothelial function in obese subjects with the insulin resistance syndrome. *Diabetes Care.* 2003;26(7):2119–2125.

64 Fuchsjager-Mayrl G, Pleiner J, Wiesinger GF, et al. Exercise training improves vascular endothelial function in patients with type 1 diabetes. *Diabetes Care.* 2002;25(10):1795–1801.

65 Katoh A, Ikeda H, Takajo Y, et al. Coexistence of impairment of endothelium-derived nitric oxide and platelet-derived nitric oxide in patients with coronary risk factors. *Circ J.* 2002;66(9): 837–840.

66 Hashimoto M, Kozaki K, Eto M, et al. Association of coronary risk factors and endothelium-dependent flow-mediated dilatation of the brachial artery. *Hypertens Res.* 2000;23(3):233–238.

67 Celermajer DS, Sorensen KE, Bull C, et al. Endothelium-dependent dilation in the systemic arteries of asymptomatic subjects relates to coronary risk factors and their interaction. *J Am Coll Cardiol.* 1994;24(6):1468–1474.

68 Vincent HK, Taylor AG. Biomarkers and potential mechanisms of obesity-induced oxidant stress in humans. *Int J Obes (Lond).* 2006;30(3):400–418.

69 Husain S, Andrews NP, Mulcahy D, et al. Aspirin improves endothelial dysfunction in atherosclerosis. *Circulation.* 1998;97 (8):716–720.

70 Khan F, Belch JJ, Macleod M, Mires G. Changes in endothelial function precede the clinical disease in women in whom preeclampsia develops. *Hypertension.* 2005;46(5):1123–1128.

71 Myers J, Mires G, Macleod M, Baker P. In preeclampsia, the circulating factors capable of altering in vitro endothelial function precede clinical disease. *Hypertension.* 2005;45(2):258–263.

72 Neels JG, Olefsky JM. Inflamed fat: what starts the fire? *J Clin Invest*. 2006;116(1):33–35.

73 Pardridge WM. Molecular biology of the blood-brain barrier. *Mol Biotechnol.* 2005;30(1):57–70.

74 Shusta EV, Boado RJ, Mathern GW, Pardridge WM. Vascular genomics of the human brain. *J Cereb Blood Flow Metab.* 2002; 22(3):245–252.

75 American Heart Association. *Heart Disease and Stroke Statistics – 2005 Update.* Dallas, TX: American Heart Association, 2005.

76 American Diabetes Association. Cost of diabetes in the United States, 2002. 2005; Available at http://diabetes.niddk.nih.gov/dm/pubs/statistics/#14. Accessed Dec 23, 2005.

77 Yelin E, Cisternas MG, Pasta DJ, et al. Medical care expenditures and earnings losses of persons with arthritis and other rheumatic conditions in the United States in 1997: total and incremental estimates. *Arthritis Rheum.* 2004;50(7):2317–2326.

78 National Heart, Lung, and Blood Institute. *Morbidity and Mortality: 2004 Chartbook on Cardiovascular, Lung, and Blood Diseases.* Bethesda, MD: National Heart, Lung, and Blood Institute, 2004.

79 American Cancer Society. *Cancer Facts and Figures 2005.* Atlanta: American Cancer Society, 2005.

196

Conclusion

Jane Maienschein*, Manfred D. Laubichler*, and William C. Aird[†]

*Center for Biology and Society, Arizona State University, Tempe;
[†]Beth Israel Deaconess Medical Center, Harvard Medical School, Boston, Massachusetts

Here, by Chapter 196 of a volume exploring the endothelium in medicine, it has become clear that this is an astonishingly diverse and far-reaching subject. Identified for the first time only at the end of the nineteenth century, the once seemingly simple and passive cell layer is now understood to represent a dynamic system that reflects and determines much about health and function for humans and other vertebrate species. This collection of chapters shows how far we have come in our understanding, and also how rich the opportunities are for further study and more effective translation of our understanding into clinical practice.

There is great virtue in bringing together all these topics that would normally reside in the archives of separate disciplines. We have gained many different perspectives of what the endothelium is and how endothelial cells (ECs) come together as a functioning system. By reflecting on the way the endothelium develops and by comparing phenotypes across different species, we gain insights into the alternatives available and what any particular developmental pathway means for the organism. This helps us understand the endothelium as a highly evolved, integrated system. The system is by no means perfect. Rather it is an interlocking "bundle of adaptations," some of which were born of trade-offs that are no longer applicable. Indeed, the extreme path-dependence of evolutionary change has undoubtedly engendered design flaws that render the modern-day human endothelium highly vulnerable to disease.

The use of metaphors to conceptualize the endothelium is a powerful tool for understanding an otherwise hidden and under-appreciated entity. We hope that the reader will agree that the analogies between endothelium and weather, landscape, and urban design are at the same time provocative and illuminating. While the metaphor of the EC as an input-output device has obvious limitations, not least of which is the risk of overlooking what are often considered to be "emergent properties," it has nonetheless provided an important organizing theme for the book. Moreover, the use of the input-output device metaphor illuminates how our biological knowledge might be carried into clinical practice. For example, if we consider the remarkable capacity of the endothelium to sense and respond to its extracellular environment it is not a leap to assume that *every* drug we administer systemically to patients will alter EC phenotypes, for better or for worse. On one hand, such capacity to be modulated may be leveraged to develop novel agents targeted against the endothelium. On the other hand, the exquisite sensitivity of ECs to the extracellular environment may contribute to the adverse effects of many drugs.

When considered collectively, the chapters on endothelium in health and disease provide a clear message: as diverse as the endothelium is in structure and function, it is nevertheless a highly integrated system. The more we learn about the endothelium *across* the vasculature, the more successful we will be in identifying common or core properties, and the better poised we will be to understand the wider consequences of endothelial-based disease.

The collection of perspectives brings many, many new questions and new directions for research. That is clear. New knowledge together with new ways of knowing will bring richer research results, and we have opportunities to translate those results into clinical practice. But we will not do so without changes in the medical curriculum. In that spirit, it is incumbent upon us to train our physicians, starting in undergraduate programs and medical schools, to think of the endothelium not simply as a collection of cells that passively separate circulating blood from underlying tissue, but rather as an active, interactive, dynamic and directive organ system.

Based on the collective writings in this volume, we conclude that the endothelium is a powerful organizing principle in health and disease. Indeed, endothelial biomedicine is a field that warrants recognition in its own right – at all levels of training and research. In addition, the current book calls for reconceptualizing biological structure and function more generally and extending the same questions and approaches to other biological systems. And it calls for new ways of working, not in isolated and separated specialized disciplinary laboratories, but interactively and collaboratively. Perhaps through our own efforts, we can set an example on how to apply integrated research strategies to complex systems. Systems biology continues to be inspired for the most part by high-throughput genomic and proteomic approaches. However, as emphasized throughout this volume, complexity is "encoded" not so much in the component parts, but rather in the links or interactions between genes, proteins and cells. Moreover, systems biology is heavily rooted in evolution and development. Hence, an understanding of the phylogeny and ontogeny of complex systems, such as the endothelium, provides invaluable insights into the role of these systems in health and disease.

Finally, we hope to have made a case for the importance of conceptual innovation and clarification for structuring available information in such a way that new perspectives can emerge. History teaches us how the endothelium was conceptualized within certain domineering paradigms and how our ideas about the endothelium have changed in light of new developments. Today, however, this relationship might well be reversed and the focus on endothelium as an integrated complex system might lead to a reconceptualization of other areas of biomedicine.

Index

Note: entries followed by a lower-case *f* or *t* represent information to be found in figures or tables respectively.